DICTIONARY OF Chemical Names and Synonyms

Philip H. Howard
Michael Neal

Library of Congress Cataloging-in-Publication Data

Howard, Philip H. (Philip Hall), 1943-
Dictionary of chemical names and synonyms / Philip H. Howard, Michael Neal
p. cm.
Includes indexes.
1. Chemicals--Dictionaries. I. Neal, Michael, II. Title.
TP9.H65 1992
660'.03--dc20 92-9160
ISBN 0-87371-396-6

LEWIS PUBLISHERS
121 South Main Street, Chelsea, Michigan 48118

PRINTED IN THE UNITED STATES OF AMERICA 34567890
Printed on acid-free paper

Philip H. Howard joined Syracuse Research Corporation in 1970 and has served as project director for numerous environmental fate and effects projects for federal agencies and industry. Dr. Howard's current research projects include development of structure/biodegradability correlations, development of estimation techniques for environmental fate physical properties and rate constants, and databases of information to support these efforts. He received a B.S. degree in chemistry from Norwich University in 1965 and a Ph.D. in organic chemistry from Syracuse University in 1970.

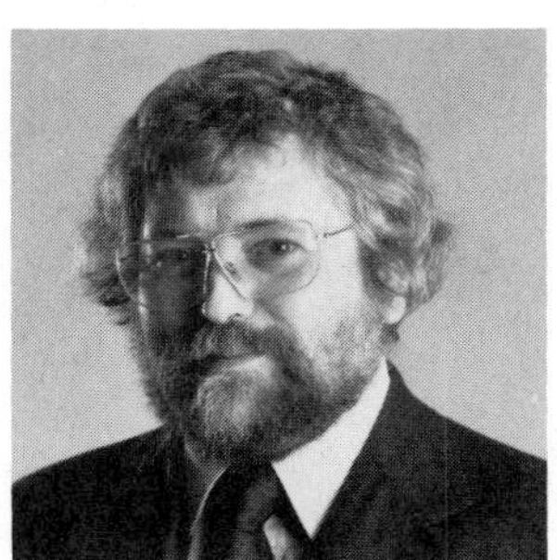

Michael Neal joined Syracuse Research Corporation in the late 1970s and has directed projects in toxicology and human risk assessment for the U. S. Environmental Protection Agency and the Agency for Toxic Substances and Disease Registry. Dr. Neal is currently involved in projects to identify areas of toxicologic data for chemicals in commerce where there is insufficient health and safety information, and provide recommendations for needed additional testing. He also manages the Toxic Substance Control Act Test Submissions (TSCATS) data base which disseminates to the public unpublished information received under the Toxic Substances Control Act. Dr. Neal received a B.A. degree in biology from Hartwick College in 1969 and a Ph.D. in biochemistry from Syracuse University in 1975.

Preface

Many scientists and engineers who work with chemicals spend an inordinate amount of time using a chemical name to identify a more appropriate chemical name, the Chemical Abstract Service (CAS) Registry Number, and/or the chemical structure. Often an individual will start out with a trade name that provides no insight into the chemical identity. One of our colleagues once told us that identifying a chemical often took him four to eight hours. The goal of this book is to make these lookup tasks cost nothing (other than the cost of this book) and take a reasonable amount of time for 95% of the chemicals that most individuals will need. For the remaining chemicals, investigators will have to use the traditional on-line sources, such as CAS On-line's Registry File [1] (>10,000,000 chemicals), the Chemical Information System's SANSS file [2] (350,000 chemicals), or the National Library of Medicine's (NLM) CHEMLINE (over 1,000,000 chemicals) or ChemID files (over 180,000 chemicals) [3], and pay the on-line fees.

In order to achieve this goal, a limited number of significant chemicals (approximately 20,000 chemicals) were selected which would allow this book to be a reasonable size, but cover most chemicals that would be of interest. Critical to this goal was the selection of the lists of chemicals. We were particularly fortunate in that we are project directors for two databases developed for the U.S. EPA, the Toxic Substances Control Act Test Submissions (TSCATS) database [4] and the Environmental Fate Data Base (EFDB) [5], and have developed criteria and support documents for thousands of chemicals for industry and federal agency sponsors. TSCATS contains approximately 3,500 chemicals for which U.S. industry has submitted test data to EPA under the Toxic Substances Control Act (TSCA). EFDB contains approximately 16,000 chemicals that have environmental degradation and transport data and/or have been monitored in food, the workplace, effluents, or the environment. These two files served as the base list of chemicals and the following lists were added to assure that all significant chemicals would be included:

• A list of approximately 8,600 chemicals in the Chemical Update System for production year 1985. This contains organic chemicals whose site production exceeds 10,000 lbs/year.

• A list of approximately 5,000 chemicals from the FATE/EXPOS file [5]. This file contains chemicals that have been monitored in the environment and/or are produced at over 100,000 lbs/year in 1977, the initial year of the TSCA Inventory.

• A list of approximately 2,300 chemicals that have been tested or considered by the National Toxicology Program.

• The SUPERLIST from the ChemID file of the NLM which contains 5,600 chemicals on "16 lists relevant to SARA Title III, FIFRA, Department of Transportation, Interagency Agency for Research on Cancer (IARC), National Toxicology Program, OSHA, and some of states" [6]. This includes all the chemicals that are in the U.S. EPA's "List of Lists."

• A list of approximately 4,200 chemicals in the Hazardous Substances Data Bank from the NLM.

This book is divided into three sections: (1) the basic individual chemical records that are ordered by CAS Registry Number and contain the chemical formula, molecular weight, the SMILES (Simplified Molecular Input Line Entry System) chemical structure notation [7] when available, a preferred name (usually the 9th Collective Index [9CI] or common name), and a list of chemical synonyms in alphabetical order; (2) an index of over 130,000 chemical synonyms arranged in alphabetical order that indicate the corresponding CAS Registry Number; and (3) an index of chemical formulas ordered by the Hill system (number of carbons, number of hydrogens and then alphabetical by element) that also indicate the corresponding CAS Registry Numbers. The basic individual chemical records and the synonyms index contains other useful information such as the language of the chemical name, the name used in the Chemical Abstracts Collective Index (e.g., 8CI, 9CI means that the name was used in both the 8th and 9th Collective Indices), the Department of Transportation name (DOT) or the United Nations (UN) definition of the chemical class, and an indication of which name is used by the American Conference of Governmental Industrial Hygienists (ACGIH), the Occupational Safety and Health Administration (OSHA), and the National Institute for Occupational Safety and Health (NIOSH). The SMILES notations were included in the base record when available because many programs are using these structural notations as input for predictions of physical properties, environmental fate, or toxicological effects [8-13].

The following briefly provides the procedure for using the book when you have different types of information;

Have CAS Registry Number and Want to Determine Chemical Name, Chemical Synonym, Molecular Weight, or SMILES Notation

Go to Section I which is ordered by CAS Registry Number and identify the page where the chemical record is located by using the headers on top of the page. The chemical you are interested in should look similar to the following example:

Field	Sample
CAS Registry Number	**90-96-0**
Chemical Formula	$C_{15}H_{14}O_3$
Molecular Weight	242.27
SMILES Notation	COc1ccc(cc1)C(=O)c2ccc(OC)cc2
Preferred Name	**Methanone, bis(4-methoxy-phenyl)- (9CI)**
Synonyms	Benzophenone, 4,4'-dimethoxy- (8CI) NSC-4191

The CAS Registry Number and preferred name are bolded in the record for emphasis and the SMILES notation is only included when appropriate and available.

Have Chemical Name and Want CAS Registry Number and Other Chemical Information

Go to Section II which contains all the chemical names and synonyms in alphabetical order. The chemical synonym index is alphabetically ordered by capital letters first and then by lower case letters (see following example). Prefixes commonly used in organic chemistry which are not normally considered part of the name for alphabetizing purposes, such as o-, ortho-, m-, meta-, p-, para-, α-, β-, γ-, Δ-, n-, sec-, tert-, cis-, trans-, N-, as well as all numbers, have not been considered for alphabetical order. Other prefixes which normally are considered part of the name, such as iso-, di-, tri-, tetra-, and cyclo-, are used for alphabetical positioning. For example, 2,4-Dinitrotoluene is under D and tert-Butyl alcohol is under B. Once you have found the chemical name you need to record the indicated CAS Registry Number and then go to Section I for the full chemical record.

Sample from Section II - Alphabetical Index

Aniline Green **(633-03-4)**
Aniline, Polymer with formaldehyde (8CI) **(25214-70-4)**
Aniline Red **(632-99-5)**
Aniline Violet **(548-62-9)**
Aniline Violet Pyoktanine **(548-62-9)**
Aniline Yellow **(60-09-3)**
Aniline, N-acetoxyethyl-N-cyanoethyl- **(22031-33-0)**
Aniline, N-acetyl- **(103-84-4)**
Aniline-ω-acid **(103-06-0)**
p-Anilinearsonic acid **(98-50-0)**
Aniline, 4,4'-azodi **(538-41-0)**
Aniline, N-benzyl-N-dimethylaminoethyl-, hydrochloride **(2045-52-5)**
Aniline, N-benzylidene- (8CI) **(538-51-2)**
Aniline, N,N-bis(2-chloroethyl) **(553-27-5)**
Aniline, 3,5-bis(trifluoromethyl)- **(328-74-5)**
Aniline, ar-bromo- **(55777-84-9)**
Aniline, p-bromo **(106-40-1)**
Aniline, m-bromo- (8CI) **(591-19-5)**
Aniline, o-bromo- (8CI) **(615-36-1)**
Aniline, p-bromo-N,N-diethyl- (8CI) **(2052-06-4)**
Aniline, 2-bromo-6-chloro-4-nitro- (8CI) **(99-29-6)**
Aniline, 2-bromo-4,6-dichloro- (8CI) **(697-86-9)**
Aniline, 4-bromo-3,5-dichloro- (8CI) **(1940-29-0)**
Aniline, 2-bromo-4,6-dinitro **(1817-73-8)**
Aniline, 4-butyl **(104-13-2)**
Aniline, n-butyl **(1126-78-9)**
Aniline, N-sec-butyl-4-tert-butyl-2,6-dinitro **(33629-47-9)**
Aniline, p-butyl-N-(p-methoxybenzylidene) **(26227-73-6)**
Aniline, N-tert-butyl-p-nitro- (7CI,8CI) **(4138-38-9)**
Aniline-3-carboxylic acid **(99-05-8)**
Aniline chloride **(142-04-1)**
Aniline, 4-chloro- **(106-47-8)**
Aniline, m-chloro **(108-42-9)**
Aniline, o-chloro **(95-51-2)**
Aniline, p-chloro **(106-47-8)**
Aniline, 4-chloro-2,5-dimethoxy **(6358-64-1)**
Aniline, o-chloro-N,N-dimethyl- **(698-01-1)**
Aniline, 6-chloro-2,4-dinitro **(3531-19-9)**
Aniline, 4-chloro-2,6-dinitro- (8CI) **(5388-62-5)**
Aniline, 3-chloro-2,6-dinitro- (7CI,8CI) **(10250-71-2)**
Aniline, N-(2-chloroethyl)-N-ethyl **(92-49-9)**
Aniline, 3-chloro-4-fluoro- **(367-21-5)**

Have Chemical Formula and Want Chemical Name, CAS Registry Number and Other Chemical Information

Go to Section III which contains all the chemical formulas ordered by the Hill system (number of carbons, number of hydrogens and then alphabetical by element). For example, C_2H_5NO would precede C_2H_5O and C_2H_5O would precede C_2H_6. The salt portion of the chemical formula is placed at the end of the chemical formula, preceded by a period, and ordered by the Hill system also (see formulas marked with * in the sample below). Once you have found the chemical formula you need to record the indicated CAS Registry Numbers and then go to Section I for the full chemical records. Note that often there will be more than one CAS Registry Number for each chemical formula, because many chemicals can have the same chemical formula (e.g., o-, m-, and p-chloronitrobenzene all have a chemical formula of $C_6H_4ClNO_2$).

Sample from Section III - Chemical Formula Index

$C_2H_4O_2.H_4N_2$*
(7335-65-1)
$C_2H_4O_2.Li$*
(546-89-4)
$C_2H_4O_2.xPb$*
(15347-57-6)
$C_2H_4O_2.1/2Na$*
(126-96-5)
$C_2H_4O_2S$
(68-11-1)
$C_2H_4O_2S.1/2Ca$*
(814-71-1)
$C_2H_4O_3$
(79-14-1)
(79-21-0)
$C_2H_4O_3.K$*
(1932-50-9)
$C_2H_4O_3.xK$*
(25904-89-6)
$C_2H_4O_3S.Na$*
(3039-83-6)
$C_2H_4O_5S$
(123-43-3)
$C_2H_4O_7P_2.4Na$*
(3794-83-0)
C_2H_4S
(420-12-2)
$C_2H_4S_3$
(289-16-7)
$C_2H_5AlCl_2$
(563-43-9)
$C_2H_5AsCl_2$
(598-14-1)
C_2H_5Br
(74-96-4)
C_2H_5BrO
(540-51-2)
$C_2H_5BrO_3S$
(26978-65-4)
C_2H_5Cl
(75-00-3)
C_2H_5ClHg
(107-27-7)
C_2H_5ClMg
(2386-64-3)
C_2H_5ClO
(107-07-3)
(107-30-2)
$C_2H_5ClO_3S$
(18024-00-5)
$C_2H_5ClO_4$
(22750-93-2)
$C_2H_5Cl_2O_2P$
(1498-51-7)
$C_2H_5Cl_2PS$
(993-43-1)
$C_2H_5Cl_3Si$
(115-21-9)
(1558-33-4)
C_2H_5F
(353-36-6)
C_2H_5FO
(371-62-0)
C_2H_5I
(75-03-6)
C_2H_5IO
(624-76-0)
C_2H_5N
(151-56-4)
C_2H_5NO
(60-35-5)
(107-29-9)
(123-39-7)
$C_2H_5NO_2$
(56-40-6)
(79-24-3)
(109-95-5)
(546-88-3)

References

1. STN Database Discriptor. CAS On-line, REGISTRY file. Chemical Abstract Services. September 1990.

2. Chemical Information Systems, Inc. USER'S GUIDE TO THE CIS, SANSS file. CIS Inc., Baltimore, MD. 1986.

3. National Library of Medicine. ELHILL Files Quick Reference Guide. CHEMLINE file, ChemID file. November 1990.

4. Santodonato, J., C. Bush, P. Howard, K. Howard, S. DelFavero, P.C. Miles, E.T. Merrick, L.K. Smith, and L.A. Travers. TSCATS: A database for chemical and subject indexing of health and environmental studies submitted under the Toxic Substances Control Act. Environ Toxicol Chem 6:921-927 (1987).

5. Howard, P.H., A.E. Hueber, B.C. Mulesky, J.S. Crisman, W. Meylan, E. Crosbie, D.A. Gray, G.W. Sage, K.P. Howard, A. LaMacchia, R. Boethling, and R. Troast. BIOLOG, BIODEG, and FATE/EXPOS: New files on microbial degradation and toxicity as well as environmental fate/exposure of chemicals. Environ Toxicol Chem 5:977-988 (1986).

6. National Library of Medicine. SUPERLIST, ChemID file. Bethesda, MD, August 1990.

7. Weininger, D. SMILES, a chemical language and information system. I. Introduction to methodology and encoding rules. J Chem Inf Comput Sci 28:31-36 (1988).

8. Meylan, W. and P.H. Howard. Bond contribution method for estimating Henry's law constants. Environ Toxicol Chem 10:1283-1293 (1991).

9. Howard, P.H., R.S. Boethling, W.M. Stiteler, W.M. Meylan, A.E. Hueber, J.A. Beauman, and M.E. Larosche. Predictive model for aerobic biodegradability developed from a file of evaluated biodegradation data. Environ Toxicol Chem 11:593-603 (1992).

10. Syracuse Research Corporation. PC Estimation Programs; Atmospheric Oxidation Program, Henry's Law Constant Program, Biodegradation Probability Program, Hydrolysis Rate Constant Program, Soil Adsorption Coefficient (Koc) Program. 1992.

11. Medicinal Chemistry Project. CLOGP Program. Pomona College, Pomona, CA (1986).

12. Health Designs, Inc. TOPKAT Program, Toxicology Newsletter, HDi, Rochester, NY. November 1991.

13. Hunter, R.S. et al. QSAR System User Manual. Joint Project of U.S.EPA and Montana State University. April 1988.

Acknowledgments

We would like to express our appreciation to Dr. D. Anthony Gray and Dr. Jay Tunkel for their assistance in preparing this book. These colleagues of ours in the Chemical Hazard Assessment Division at Syracuse Research Corporation provided assistance in computer programming and data base manipulation without which many tasks associated with this endeavor would have been much more difficult. We would also like to express our appreciation to Ms. Carol Middleton for her assistance with editing and formatting.

Table of Contents

DICTIONARY OF Chemical Names and Synonyms

50-00-0
CH_2O
30.03
O=C
Formaldehyde (ACGIH, OSHA)
Aldehyde formique (French)
Aldehyd mravenci (Czech)
Aldeide formica (Italian)
BFV
FA
Fannoform
Formaldehyd (Czech, Polish)
Formaldehyde, Gas
Formaldehyde, Solutions, Flammable [UN 1198]
Formaldehyde solution [UN 2209]
Formalin [UN 2209]
Formalin 40
Formalina (Italian)
Formaline (German)
Formalin-loesungen (German)
Formalith
Formic aldehyde
Formol
Fyde
Hoch
Ivalon
Karsan
Lysoform
Methaldehyde
Methanal
Methyl aldehyde
Methylene glycol
Methylene oxide
Morbicid
NCI-C02799
Oplossingen (Dutch)
Oxomethane
Oxymethylene
Paraform
RCRA waste number U122
Superlysoform
UN 1198 [Formaldehyde, solutions, flammable]
UN 2209 [Formaldehyde solution]

50-01-1
$CH_5N_3.ClH$
95.55
Guanidine, monohydrochloride
Guanidine chloride
Guanidine hydrochloride
Guanidinium chloride
Guanidinium hydrochloride
USAF EK-749

50-02-2
$C_{22}H_{29}FO_5$
392.51
O=C(C(O)(C(C(C(C(F)(C(C(=CC(=O)C=1)C2)(C1)C)C3O)C2)C4)(C3)C)C4C)CO
Pregna-1,4-diene-3,20-dione, 9-fluoro-11-β,17,21-trihydroxy-16-α-methyl
Aeroseb-dex
Aphtasolon
Azium
Calonat
Corsone
Decaderm
Decadron
Decasone
Decaspray
Dectancyl
1-Dehydro-16-α-methyl-9-α-fluorohydrocortisone
Dekacort
Deltafluorene
Dergramin
Deronil
Desadrene
Desametasone
Desamethasone
Desameton
Dexa
Dexacort
Dexa-Cortidelt
Dexa-Cortisyl
Dexadeltone
Dexalona
Dexameth
Dexamethasone
Dexamethasone alcohol
Dexamethazone
Dexapolcort
Dexaprol
Dexa-Scheroson
Dexason
Dexone
Dextelan
Dezone
DXMS
Fluorocort
δ[1]-9-α-Fluoro-16-α-methylcortisol
9-α-Fluoro-16-α-methylprednisolone
9-α-Fluoro-16-α-methyl-1,4-pregnadiene-11-β,17-α,21-triol-3,20-dione
4-α-Fluoro-16-α-methyl-11-β,17,21-trihydroxypregna-1,4-diene-3,20-dione
9-Fluoro-11-β,17,21-trihydroxy-16-α-methylpregna-1,4-diene-3,20-dione
9-α-Fluoro-11-β,17-α,21-trihydroxy-16-α-methylpregna-1,4-diene-3,20-dione
Fortecortin
Gammacorten
Hexadecadrol
Hexadrol
HL-Dex
Maxidex
16-α-Methyl-9-α-fluoro-1-dehydrocortisol
16-α-Methyl-9-α-fluoro-δ[1]-hydrocortisone
16-α-Methyl-9-α-fluoroprednisolone
16-α-Methyl-9-α-fluoro-1,4-pregnadiene-11-β,17-α,21-triol-3,20-dione
16-α-Methyl-9-α-fluoro-11-β,17-α,21-trihydroxypregna-1,4-diene-3,20-dione
Mexidex
Millicorten
MK 125
Oradexon
Prednisolone F
Prednisolon F
SK-Dexamethasone
Superprednol
Visumetazone

50-04-4
$C_{23}H_{30}O_6$
402.53
CC(=O)OCC(=O)C3(O)CCC4C2CCC1=CC(=O)CCC1(C)C2C(=O)CC34C
Cortisone 21-acetate
Acetate cortisone
21-Acetoxy-17,α-hydroxypregn-4-ene-3,11,20-trione
21-Acetoxy-17,α-hydroxy-3,11,20-triketopregnene-4
Adreson
Artriona
Biocort acetate
Compound E acetate
Cortadren
Cortelan
Cortisal
Cortisate
Cortisone acetate
Cortisone monoacetate
Cortistab
Cortisyl
Cortivite
Cortogen
Cortogen acetate
Cortone
Cortone acetate
11-Dehydro-17-hydroxycorticosterone acetate
11-Dehydro-17-hydroxycorticosterone-21-acetate
17,21-Dihydroxypregn-4-ene-3,11,20-trione acetate
Incortin
Irisone acetate
4-Pregnene-17,α,21-diol-3,11,20-trione 21-acetate
Pregn-4-ene-3,11,20-trione, 21-(acetyloxy)-17-hydroxy-(9CI)
Pregn-4-ene-3,11,20-trione, 17,21-dihydroxy-, 21-acetate
Ricortex
Scheroson

50-06-6
$C_{12}H_{12}N_2O_3$
232.26
CCC1(C(=O)NC(=O)NC1=O)c2ccccc2
Barbituric acid, 5-ethyl-5-phenyl
Acido 5-fenil-5-etilbarbiturico (Italian)
Adonal
Aephenal
Agrypnal
Amylofene

Aphenylbarbit
Aphenyletten
Austrominal
Barbapil
Barbellen
Barbellon
Barbenyl
Barbilehae (Barbilettae)
Barbinal
Barbiphen
Barbiphenyl
Barbipil
Barbita
Barbivis
Barbonal
Barbophen
Bardorm
Bartol
Bialminal
Blu-Phen
Cabronal
Calmetten
Calminal
Cardenal
Codibarbita
Coronaletta
Cratecil
Damoral
Dezibarbitur
Dormina
Dormiral
Doscalun
Duneryl
Ensobarb
Ensodorm
Epanal
Epidorm
Epilol
Episedal
Epsylone
Eskabarb
5-Ethyl-5-phenylbarbituric acid
5-Ethyl-5-phenyl-2,4,6-(1H,3H,
5H)pyrimidinetrione
Etilfen
Euneryl
Fenbital
Fenemal
Fenobarbital
Fenosed
Fenylettae
Gardenal
Gardepanyl
Glysoletten
Haplopan
Haplos
Helional
Hennoletten
Hypnaletten
Hypnogen
Hypnolone
Hypnoltol
Hypno-tablinetten
Hysteps
Lefebar
Leonal
Lephebar
Lepinal
Lepinaletten
Linasen
Liquital
Lixophen
Lubergal
Lubrokal
Lumen
Lumesettes
Lumesyn
Luminal
Lumofridetten
Luphenil
Luramin
Molinal
Neurobarb
Nirvonal
Noptil
Nova-Pheno
Nunol
Parkotal
Pharmetten
Phenaemal
Phen-Bar
Phenemal
Phenobal
Phenobarbital
Phenobarbitone
Phenobarbituric acid
Phenobarbyl
Phenoluric
Phenolurio
Phenomet
Phenonyl
Phenoturic
Phenylethylbarbiturate
Phenyl-ethyl-barbituric acid
5-Phenyl-5-ethylbarbituric acid
Phenylethylmalonylurea
Phenyletten
Phenyral
Phob
Polcominal
Promptonal
2,4,6(1H,3H,5H)-Pyrimidine-
trione, 5-ethyl-5-phenyl-
Sedabar
Seda-Tablinen
Sedicat
Sedizorin
Sedlyn
Sedofen
Sedonal
Sedonettes
Sedophen
Sevenal
SK-Phenobarbital
Solfoton
Sombutol
Somnolens
Somnoletten
Somnosan
Somonal
Spasepilin
Starifen
Starilettae
Stental
Stental extentabs
Talpheno
Teolaxin
Teoloxin
Thenobarbital
Theoloxin
Theominal
Triabarb
Tridezibarbitur
Triphenatol
Versomnal
Zadoletten
Zadonal

50-07-7
$C_{15}H_{18}N_4O_5$
334.37
O=C(OCC(C(=C(N1CC(N2)C23)
C(=O)C(=C4N)C)C4=O)
C13OC)N
**Azirino(2',3':3,4)pyrrolo-
(1,2-a)indole-4,7-dione,
6-amino-1,1a,2,8,8a,8b-hexa-
hydro-8-(hydroxy methyl)-
8a-methoxy-5-methyl-, car-
bamate (ester)**
Ametycin
Ametycine
7-Amino-9-α-methoxymitosane
Mit-C
Mito-C
Mitocin-C
Mitomycin
Mitomycin-C
Mitomycinum
MMC
Mutamycin
Mutamycin(mitomycin for
injection)
Mytomycin
Mitomycyna C (Polish)
NCI-C04706
NSC-26980
RCRA waste number U010

50-11-3
$C_9H_{14}N_2O_3$
198.25
CCC1(CC)C(=O)NC(=O)N(C)
C1=O
**Barbituric acid, 5,5-diethyl-
1-methyl**
AN 23
5,5-Diethyl-1-methylbarbituric
acid
Endiemalum
Gemonal
Gemonil
Gemonit
Metabarbital
Metharbital
Metharbitone
Metharbutal
Methylbarbital
N-Methylbarbital
1-Methylbarbital
2,4,6(1H,3H,5H)-Pyrimidine-
trione, 5,5-diethyl-1-methyl-
(9CI)
SCH 412

50-12-4
$C_{12}H_{14}N_2O_2$
218.28
CCC1(NC(=O)N(C)C1=O)c2cc
ccc2
**Hydantoin, 5-ethyl-3-methyl-
5-phenyl**
Epilan

5-Ethyl-5-fenyl-3-methyl-hydantoin (Czech)
5-Ethyl-3-methyl-5-phenyl-hydantoin
5-Ethyl-3-methyl-5-phenyl-2,4-imidazolidinedione
5-Ethyl-3-methyl-5-phenyl-2,4(3H,5H)-imidazoledione
5-Ethyl-3-methyl-5-phenylimidazolidin-2,4-dione
3-Ethylnirvanol
Gerot-epilan
2,4-Imidazolidinedione, 5-ethyl-3-methyl-5-phenyl-
Insulton
Mephentoin
Mephenytoin
Mesantoin
Mesontoin
Methoin
3-Methyl-5-ethyl-5-phenyl-hydantoin
3-Methyl-5,5-ethylphenyl-hydantoin
Methyl hydantoin
3-Methyl-5,5-phenylethyl-hydantoin
NSC-34652
Phenantoin
Phenylethylmethylhydantoin
Sacerno
Sedantoinal
Triantoin

50-13-5
$C_{15}H_{21}NO_2.ClH$
283.83
Cl.CCOC(=O)C1(CCN(C)CC1)c2ccccc2
Isonipecotic acid, 1-methyl-4-phenyl-, ethyl ester, hydrochloride
Algil
Alodan (Gerot)
Antidurol
Centralgin
Chlorbicyclene (French)
Chlorbycyclen
Demerol
Demerol hydrochloride
Dispadol
Dolantal
Dolantin
Dolantin hydrochloride
Dolantol
Dolaren
Dolargan
Dolcontral
Dolenal
Dolenol
Dolestine
Dologal
Dolopethin
Dolosal
Dolin
Doloneurine
Dolvanol
Endolat
Ethyl-1-methyl-4-phenylisonipecotate hydrochloride
Ethyl 1-methyl-4-phenylpiperidine-4-carboxylate hydrochloride
Ethyl 1-methyl-4-phenylpiperidyl-4-carboxylate hydrochloride
Isonipecaine hydrochloride
Lidol
Lydol
Mefedina
Mepadin
Meperidine hydrochloride
Mephedine
1-Methyl-4-carbethoxy-4-phenylpiperidine hydrochloride
N-Methyl-4-phenyl-4-carbethoxypiperidine hydrochloride
1-Methyl-4-phenyl-4-carboethoxypiperidine hydrochloride
1-Methyl-4-phenylisonipecotic acid ethyl ester hydrochloride
Operidine
Pantalgine
Pentantin
Petantin hydrochloride
Pethidine chloride
Pethidine, hydrochloride
Petidin
4-Piperidinecarboxylic acid, 1-methyl-4-phenyl-, ethyl ester, hydrochloride
Piridosal
S 140
Sauteralgyl
Spasmedal
Spasmodolin
Synelaudine
WY 554

50-14-6
$C_{28}H_{44}O$
396.72
CC(C)C(C)C=CC(C)C1CCC2C(CCCC12C)=CC=C3CC(O)CCC3=C
Ergocalciferol
D-Arthin
Calciferol
Calciferon 2
Condacaps
Condocaps
Condol
Crtron
Crystallina
Daral
Davitamon D
Davitin
Decaps
Dee-osterol
Dee-Ron
Dee-Ronal
Dee-Roual
Deltalin
Deratol
Detalup
Diactol
Divit Urto
Doral
Drisdol
Ergorone
Ergosterol activated
Ergosterol, irradiated
Ertron
Fortodyl
Geltabs
Hi-Deratol
Infron
Irradiated ergosta-5,7,22-trien-3-β-ol
Metadee
Mulsiferol
Mykostin
Oleovitamin D
Ostelin
Radiostol
Radsterin
9,10,Secoergosta-5,7,10(19),22-tetraen 3-β-ol
Shock-Ferol
Sterogyl
Vigantol
Viosterol
Vitamin D2
Vitavel-D

50-18-0
$C_7H_{15}Cl_2N_2O_2P$
261.11
ClCCN(CCCl)P1(=O)NCCCO1
2H-1,3,2-Oxazaphosphorine, 2-(bis(2-chloroethyl)amino)-tetrahydro-, 2-oxide
ASTA
ASTA B518
B 518
N,N-Bis-(β-chloraethyl)-N',O-propylen-phosphorsaeure-ester-diamid (German)
2-(Bis(2-chloroethyl)amino)-2H-1,3,2-oxazaphosphorine 2-oxide
Bis(2-chloroethyl)phosphoramide-cyclic propanolamide ester
N,N-Bis(2-chloroethyl)-N',O-propylenephosphoric acid ester diamide
N,N-Bis(2-chloroethyl)tetrahydro-2H-1,3,2-oxazaphosphorin-2-amine 2-oxide
N,N-Bis(β-chloroethyl)-N',O-trimethylenephosphoric acid ester diamide
CB 4564
Clafen
Claphene
CP
CPA
CTX
CY
Cyclophosphamid
Cyclophosphamide
(-)-Cyclophosphamide
Cyclophosphamidum
Cyclophosphan
Cyclophosphane
Cyclophosphoramide
Cyclostin
Cyklofosfamid (Czech)
Cytophosphan
Cytoxan
N,N-Di(2-chloroethyl)-N,O-pro-

pylene-phosphoric acid ester diamide
Endoxan
Endoxana
Endoxan-asta
Endoxane
Endoxan R
Enduxan
Endoxanal
Genoxal
Hexadrin
Mitoxan
NCI-C04900
Neosar
NSC-26271
2H-1,3,2-Oxazaphosphorin-2-amine, N,N-bis(2-chloroethyl)tetrahydro-, 2-oxide (9CI)
2-H-1,3,2-Oxazaphosphorinane
Phosphorodiamidic acid, N,N-bis(2-chloroethyl)-N'-(3-hydroxypropyl)-, intramol. ester
Procytox
RCRA waste number U058
Semdoxan
Sendoxan
Senduxan
SK 20501
Zyklophosphamid (German)

50-21-5
$C_3H_6O_3$
90.09
O=C(O)C(O)C
Lactic-acid
Acetonic acid
Ethylidenelactic acid
1-Hydroxyethanecarboxylic acid
2-Hydroxypropanoic acid
2-Hydroxypropionic acid
α-Hydroxypropionic acid
Kyselina 2-hydroxypropanova (Czech)
Kyselina mlecna (Czech)
DL-Lactic acid
Milchsaure (German)
Milk Acid
Ordinary lactic acid
Propanoic acid, 2-hydroxy-
Propionic acid, 2-hydroxy-
Racemic lactic acid

50-22-6
$C_{21}H_{30}O_4$
346.51
CC34CC(O)C1C(CCC2=CC(=O)CCC12C)C3CCC4C(=O)CO
Corticosterone
Compound B
Corticosteron
17-Deoxycortisol
11-β,21-Dihydroxypregn-3,20-dione
11,12-Dihydroxyprogesterone
11-β,21-Dihydroxyprogesterone
11-Hydroxycorticoaldosterone
Kendall's Compound B
4-Pregnene-11-β,21-diol-3,20-dione
Pregn-4-ene-3,20-dione, 11,21-dihydroxy-, (11-β)-(9CI)
Pregn-4-ene-3,20-dione, 11-β,21-dihydroxy-
Reichstein's Substance H

50-23-7
$C_{21}H_{30}N_{40}O_5S$
954.97
O=C(C(O)(C(C(C(C(C(C(=CC(=O)C1)C2)(C1)C)C3O)C2)C4)(C3)C)C4)CO
Cortisol
Aeroseb-HC
Anti-Inflammatory Hormone
Barseb HC
Cetacort
Cobadex
Compound F
Compound F (Kendall)
Cort-Dome
Cortef
Cortenema
Cortifan
Cortisol alcohol
Cortispray
Cortonema
Cortril
Dermacort
Efcorbin
EF Corlin
Efcortelin
Eldecort
Ficortril
Fiocortril
Genacort
HC
Heb-Cort
Hidro-Colisona
Hycort
Hycortol
Hycortole
Hydro-Adreson
Hydro-Colisona
Hydrocortisone
11-β-Hydrocortisone
Hydrocortisone free alcohol
Hydrocortistab
Hydrocortisyl
Hydrocortone
17-Hydroxycorticosterone
Hydroxycortisone
11-β-Hydroxycortisone
Hytone
Hytone Lotion
Incortin-H
Kendall's Compound F
NSC-10483
Optef
Otosone-F
Permicort
Pregn-4-ene-3,20-dione, 11,17,21-trihydroxy-, (11-β)-
4-Pregnene-11-β,17-α,21-triol 3,20-dione
Proctocort
Rectoid
Reichstein's Substance M
Scheroson F
Tarcortin
Texacort Lotion 25
Topicort
Traumaide
11-β,17,21-Trihydroxypregn-4-ene-3,20-dione
11-β,17-α,21-Trihydroxypregn-4-ene-3,20-dione
11-β,17-α,21-Trihydroxy-4-pregnene-3,20-dione

50-24-8
$C_{21}H_{28}O_5$
360.49
O=C(C(O)(C(C(C(C(C(C(=CC(=O)C=1)C2)(C1)C)C3O)C2)C4)(C3)C)C4)CO
Pregna-1,4-diene-3,20-dione, 11-β,17,21-trihydroxy
Codelcortone
Co-Hydeltra
δ[1]-Cortisol
Decortin H
δ[1]-Dehydrocortisol
δ[1]-Dehydrohydrocortisone
1-Dehydrohydrocortisone
Delcortol
δ F
δ-Cortef
Deltacortenol
Deltacortril
δ-Stab
Dexa-Cortidelt Hostacortin H
Di-Adreson F
Dicortol
Dydeltrone
Fernisolone
Hostacortin
Hydeltra
Hydeltrone
δ[1]-Hydrocortisone
Hydrodeltalone
Hydrodeltisone
Hydroretrocortin
Metacortandralone
Meticortelone
Meti-Derm
Paracortol
Paracotol
Precortancyl
Precortisyl
Predne-Dome
Prednelan
Prednis
Prednisolone
Predonin
Predonine
1,4-Pregnadiene-3,20-dione-11-β,17-α,21-triol
1,4-Pregnadiene-11-β,17-α,21-triol-3,20-dione
1,4-Pregnadien-11-β,17-α,21-triol-3,20-dione
Prenolone
Scherisolon
Sterane
Sterolone
11-β,17,21-Trihydroxypregna-1,4-diene-3,20-dione
11-β,17-α,21-Trihydroxypregna-1,4-diene-3,20-dione
11-β,17-α,21-Trihydroxy-

1,4-pregnadiene-3,20-dione
Ulacort
Ultracortene-H

50-27-1
$C_{18}H_{24}O_3$
288.42
CC34CCC1C(CCc2cc(O)ccc12)C3CC(O)C4O
Estriol
Aacifemine
Colpovister
Destriol
Deuslon-A
Estra-1,3,5(10)-triene-3,16-α, 17-β-triol
1,3,5-Estratriene-3-β,16-α,17-β-triol
(16-α,17-β)-Estra-1,3,5(10)-triene-3,16,17-triol
Estratriol
3,16-α,17-β-Estriol
16-α,17-β-Estriol
Estriolo (Italian)
Follicular hormone hydrate
Gynaesan
Hemostyptanon
Holin
Hormomed
Hormonin
16-α-Hydroxyestradiol
16-α-Hydroxyoestradiol
Klimoral
NSC-12169
OE3
Oestra-1,3,5(10)-triene-3,16-α, 17-β-triol
1,3,5-Oestratriene-3-β,16-α, 17-β-triol
(16-α,17-β)-Oestra-1,3,5(10)-triene-3,16,17-triol
Oestratriol
Oestriol
3,16-α,17-β-Oestriol
16-α,17-β-Oestriol
Orgastyptin
Ovesterin
Ovestinon
Ovestrion
Ovestin
Stiptanon
Synapause
Theelol
Thulol
Tridestrin
3,16-α,17-β-Trihydroxy-δ-1,3, 5-estratriene
3,16-α,17-β-Trihydroxy-δ-1,3, 5-oestratriene
3,16-α,17-β-Trihydroxyestra-1,3,5(10)-triene
Trihydroxyestrin
3,16-α,17-β-Trihydroxyoestra-1,3,5(10)-triene
Trihydroxyoestrin
Triodurin
Triovex

50-28-2
$C_{18}H_{24}O_2$
272.42
CC34CCC1C(CCc2cc(O)ccc12)C3CCC4O
Estradiol
Altrad
Bardiol
Dihydrofollicular hormone
Dihydrofolliculin
Dihydromenformon
Dihydrotheelin
3,17-β-Dihydroxyestra-1,3,5(10)-triene
3,17-β-Dihydroxy-1,3, 5(10)-estratriene
Dihydroxyestrin
3,17-β-Dihydroxyoestra-1,3, 5-triene
3,17-β-Dihydroxy-1,3, 5(10)-oestratriene
Dihydroxyoestrin
Dimenformon
Dimenformon prolongatum
Diogyn
Diogynets
E_2
3,17-Epidihydroxyestratriene
3,17-Epidihydroxyoestratriene
Estradiol-17-β
α-Estradiol
β-Estradiol
3,17-β-Estradiol
17-β-Estradiol
cis-Estradiol
d-Estradiol
D-3,17-β-Estradiol
Estraldine
Estra-1,3,5(10)-triene-3,17-β-diol
17-β-Estra-1,3,5(10)-triene-3,17-diol
1,3,5-Estratriene-3,17-β-diol
Estrovite
Femestral
Femogen
Gynergon
Gynestrel
Gynoestryl
Lamdiol
Macrodiol
Macrol
Microdiol
Nordicol
NSC-9895
Oestergon
Oestradiol
α-Oestradiol
β-Oestradiol
3,17-β-Oestradiol
cis-Oestradiol
d-Oestradiol
D-3,17-β-Oestradiol
Oestradiol R
Oestradiol-17-β
Oestra-1,3,5(10)-triene-3,17-β-diol
17-β-Oestra-1,3,5(10)-triene-3,17-diol
Oestroglandol
Oestrogynal
17-β-OH-Estradiol
17-β-OH-Oestradiol
Ovahormon
Ovasterol
Ovastevol
Ovociclina
Ovocyclin
Ovocycline
Ovocylin
Primofol
Profoliol
Progynon
Progynon-DH
Syndiol
Theelin, dihydro-

50-29-3
$C_{14}H_9Cl_5$
354.48
c(ccc(c1)Cl)(c1)C(c(ccc(c2)Cl)c2)C(Cl)(Cl)Cl
Ethane, 1,1,1-trichloro-2,2-bis(p-chlorophenyl)
Agritan
Anofex
Arkotine
Azotox
Benzene, 1,1'-(2,2,2-trichloroethylidene)bis(4-chloro-
α,α-Bis(p-chlorophenyl)-β,β,β-trichlorethane
1,1-Bis-(p-chlorophenyl)-2,2,2-trichloroethane
2,2-Bis(p-chlorophenyl)-1,1,1-trichloroethane
Bosan Supra
Bovidermol
Chlorophenothan
Chlorophenothane
Chlorophenotoxum
Citox
Clofenotane
p,p'-DDT
DDT (ACGIH,OSHA) [UN 2761]
Dedelo
Deoval
Detox
Detoxan
Dibovan
p,p'-Dichlorodiphenyltrichloroethane
4,4'-Dichlorodiphenyltrichloroethane
Dichlorodiphenyltrichloroethane (OSHA) [UN 2761]
Dicophane
Didigam
Didimac
Diphenyltrichloroethane
Dodat
Dykol
ENT 1,506
Estonate
Genitox
Gesafid
Gesapon
Gesarex
Gesarol
Guesapon
Guesarol
Gyron
Havero-extra
Hildit

Ivoran
Ixodex
Kopsol
Micro DDT 75
Mutoxin
NCI-C00464
Neocid
OMS 16
Parachlorocidum
PEB1
Pentachlorin
Pentech
pPzeidan
R50
RCRA waste number U061
Rukseam
Santobane
Tech DDT
1,1,1-Trichloor-2,2-bis(4-chloor fenyl)-ethaan (Dutch)
1,1,1-Trichlor-2,2-bis(4-chlor-phenyl)-aethan (German)
Trichlorobis(4-chlorophenyl)-ethane
1,1,1-Trichloro-2,2-bis(p-chloro-phenyl)ethane
1,1,1-Trichloro-2,2-di(4-chloro-phenyl)-ethane
1,1,1-Tricloro-2,2-bis(4-cloro-fenil)-etano (Italian)
UN 2761 [Organochlorine pesticides, solid toxic N.O.S.]
Zeidane
Zerdane

50-30-6
$C_7H_4Cl_2O_2$
191.01
Benzoic acid, 2,6-dichloro
2,6-Dichlorobenzoic acid

50-31-7
$C_7H_3Cl_3O_2$
225.45
OC(=O)c1c(Cl)ccc(Cl)c1Cl
Benzoic acid, 2,3,6-trichloro
Acide trichlorobenzoique (French)
Benzabar
Benzac
Benzac-1281
Benzak (Czech)
Fen-All
HC 1281
Kyselina 2,3,6-trichlorbenzoova (Czech)
NCI-C60242
T-2
TBA
2,3,6-TBA
2,3,6-TBA(The herbicide)
TCB
2,3,6-TCB
TCBA
2,3,6-TCBA
Tribac
2,3,6-Trichlorbenzoesaeure (German)
Trichlorobenzoic acid
2,3,6-Trichlorobenzoic acid
Trisben (Czech)
Tryben
Trysben
Trysben 200
Zobar

50-32-8
$C_{20}H_{12}$
252.32
c(c(c(cc1)ccc2)c2cc3)(c3cc(c4ccc5)c5)c14
Benzo(a)pyrene
Benzo(d,e,f)chrysene
3,4-Benzopirene (Italian)
3,4-Benzopyrene
6,7-Benzopyrene
Benzo(a)pyrene (OSHA)
3,4-Benzpyren (German)
Benz(a)pyrene
3,4-Benzpyrene
3,4-Benz(a)pyrene
3,4-Benzypyrene
BP
B(a)P
3,4-BP
RCRA waste number U022

50-33-9
$C_{19}H_{20}N_2O_2$
308.41
O=C(N(N(C1=O)c(cccc2)c2)c(cccc3)c3)C1CCCC
3,5-Pyrazolidinedione, 4-butyl-1,2-diphenyl
Alindor
Alkabutazona
Alqoverin
Anerval
Anpuzone
Antadol
Anuspiramin
Artrizin
Artrizone
Artropan
Azdid
Azobutyl
Azolid
Benzone
Betazed
Bizolin 200
B.T.Z.
Busone
Butacote
Butacompren
Butadion
Butadiona
Butadione
Butagesic
Butalgina
Butalan
Butalidon
Butaluy
Butaphen
Butapirazol
Butapyrazole
Butarecbon
Butartril
Butartrina
Butazina
Butazolidin
Butazolidine
Butazona
Butazone
Bute
Butidiona
Butiwas-simple
Butone
Butoz
4-Butyl-1,2-diphenyl-3,5-dioxo-pyrazolidine
4-Butyl-1,2-diphenylpyrazolidine-3,5-dione
4-Butyl-1,2-diphenyl-3,5-pyrazolidinedione
Butylpyrin
Buvetzone
Buzon
Chembutazone
DA-192
Digibutina
Diossidone
3,5-Dioxe-4 buty-1, diphenyl-pyrazolidine
3,5-Dioxo-1,2-diphenyl-4-n-butylpyrazolidene
Diozol
3,5-Dioxo-1,2-diphenyl-4-n-butyl-pyrazolidin
3,5-Dioxo-1,2-diphenyl-4-n-butylpyrazolidine
Diphebuzol
Diphenylbutazone
1,2-Diphenyl-4-butyl-3,5-dioxo-pyrazolidine
1,2-Diphenyl-4-butyl-3,5-pyrazolidinedione
1,2-Diphenyl-3,5-dioxo-4-butyl-pyrazolidine
1,2-Diphenyl-2,3-dioxo-4-N-butylpyrazoline
Ecobutazone
Elmedal
Equi Bute
Eributazone
(Esteve)
FBZ
Febuzina
Fenartil
Fenibutasan
Fenibutazona
Fenilbutine
Fenibutol
Fenilbutazona
Fenilidina
Fenotone
Fenylbutazon
Flexazone
G 13,871
IA-But
Intalbut
Intrabutazone
Ipsoflame
Kadol
Lingel
Malgesic
Mephabutazone
Merizone
Nadazone
Nadozone
NCI-C56531
Neo-Zoline
Novophenyl

PBZ
Phebuzin
Phebuzine
Phen-Buta-Vet
Phenbutazol
Phenopyrine
Phenylbutaz
Phenylbutazon (German)
Phenylbutazone
Phenylbutazonum
Phenyl-mobuzon
Pirarreumol "B"
Praecirheumin
Pyrabutol
Pyrazolidin
Rectofasa
Reudo
Reudox
Reumasyl
Reumazin
Reumazol
Reumune
Reumuzol
Reupolar
Robizon-V
Rubatone
R-3-Zon
Scanbutazone
Schemergin
Shigrodin
Tazone
Tetnor
Tevcodyne
Therazone
Ticinil
Todalgil
USAF GE-15
Uzone
Vac-10
Wescozone
Zolaphen
Zolidinum
Zorane

50-34-0
$C_{23}H_{29}NO_3.Br$
447.44
[OH].[Br-].CC(C)[N+](C)(CCOC(=O)C2c1ccccc1Oc3ccccc23)C(C)C
Ammonium, diisopropyl(2-hydroxyethyl)methyl-, bromide, xanthene-9-carboxylate
Diisopropyl(2-hydroxyethyl)-methylammonium bromide xanthene-9-carboxylate
Ercorax
Ercotina
Ketaman
Kivatin
NCI-C56257
Neometantyl
Neopepulsan
Pantas
Pantheline
Pervagal
Pro-banthine
Prodixamon
Pro-gastron
Propantel
Propantheline bromide
SC-3171
Xanthene-9-carboxylic acid, ester with (2-hydroxyethyl)diisopropylmethylammonium bromide

50-35-1
$C_{13}H_{10}N_2O_4$
258.22
Thalidomide
AI3-50606
Algosediv
Asidon 3
Asmadion
Asmaval
Bonbrain
Calmore
Calmorex
Contergan
Corronarobetin
Distaval
Distaxal
Distoval
2,6-Dioxo-3-phthalimido-piperidine
N-(2,6-Dioxo-3-piperidyl)-phthalimide
E-217
Ectiluran
Enterosediv
Gastrinide
Glupan
Glutanon
Glutarimide, 2-phthalimido-
Grippex
Hippuzon
Imida-Lab
Imidan (Peyta)
Imidene
1H-Isoindole-1,3(2H)-dione, 2-(2,6-dioxo-3-piperidinyl)-(9CI)
Isomin
K-17
Kedavon
Kevadon
Lulamin
Neaufatin
Neo
Neosedyn
Neosydyn
Nerosedyn
Neufatin
Neurodyn
Neurosedin
Neurosedym
Nevrodyn
Nibrol
Noctosediv
Noxodyn
NSC-527179
NSC-66847
Pangul
Pantosediv
Phthalimide, N-(2,6-dioxo-3-piperidyl)- (8CI)
α-Phthalimidoglutarimide
α-(N-Phthalimido)glutarimide
3-Phthalimidoglutarimide
N-Phthaloylglutamimide
N-Phthalylglutamic acid imide
N-Phthalyl-glutaminsaeure-imid (German)
α-N-Phthalylglutaramide
Poly-Giron
Polygripan
Predni-Sediv
Pro-Ban M
Profarmil
Psycholiquid
Psychotablets
Quetimid
Quietoplex
Sandormin
Sedalis Sedi-Lab
Sedimide
Sedin
Sedisperil
Sedoval
Shin-Naito S
Shinnibrol
Sleepan
Slipro
Softenil
Softenon
Talargan
Talidomida (Spanish)
Talidomide
Talimol
Talismol
Telagan
Telargan
Telargean
Tensival
Thalidomidum (Latin)
Thalin
Thalinette
Theophilcholine
Ulcerfen
Valgis
Valgraine
Yodomin

50-36-2
$C_{17}H_{21}NO_4$
303.39
COC(=O)C1C(CC2CCC1N2C)OC(=O)c3ccccc3
1-α-H,5-α-H-Tropane-2-β-carboxylic acid, 3-β-hydroxy-, methyl ester, benzoate
8-Azabicyclo(3.2.1)octane-2-carboxylic acid, 3-(benzoyloxy)-8-methyl-, methyl ester, (1R-(exo,exo))- (9CI)
Benzoylmethylecgonine
Bernice
Bernies
Burese
2-β-Carbomethoxy-3-β-benzoxytropane
"C" Carrie
Cecil
Cholly
Cocaine
(-)-Cocaine
β-Cocaine
1-Cocaine
Coke
Corine
DEA No. 9041

Ecgonine, methyl ester, benzoate (ester)
Eritroxilina
Erytroxylin
Girl
Gold dust
Happy dust
Kokain
Kokan
Kokayeen
Methyl 3-β-hydroxy-1-α-H,5-α-H-tropane-2-β-carboxylate benzoate (ester)
Neurocaine
Star dust
1αH,5αH-Tropane-2β-carboxylic acid, 3β-hydroxy-, methyl ester, benzoate (ester) (8CI)
2-β-Tropanecarboxylic acid, 3-β-hydroxy-, methyl ester, benzoate (ester)
3-Tropanylbenzoate-2-carboxylic acid methyl ester

50-37-3
$C_{20}H_{25}N_3O$
323.48
CCN(CC)C(=O)C2CN(C)C1Cc3c[nH]c4cccc(C1=C2)c34
Ergoline-8-β-carboxamide, 9,10-didehydro-N,N-diethyl-6-methyl
Acid
Cubes
Delysid
9,10-Didehydro-N,N-diethyl-6-methyl-ergoline-8-β-carboxamide
Diethylamid kyseliny lysergove (Czech)
N,N-Diethyllysergamide
D-LSD
Heavenly blue
LSD
LSD-25
Lysergamid
Lysergamide, N,N-diethyl-
Lysergaure diethylamid
D-Lysergic acid diethylamide
Lysergic acid diethylamide-25
Lysergide
Lysergsaeurediaethylamid
Pearly Gates
Royal Blue
Wedding Bells

50-41-9
$C_{26}H_{28}ClNO.C_6H_8O_7$
598.14
Triethylamine, 2-(p-(2-chloro-1,2-diphenylvinyl)phenoxy)-, citrate (1:1)
Chloramiphene
Chloramiphene citrate
2-Chloro-1-(p-(β-diethylaminoethoxy)phenyl)-1,2-diphenylethylene
2-(p-(2-Chloro-1,2-diphenylvinyl)phenoxy)triethylamine citrate
2-(p-(2-Chloro-1,2-diphenylvinyl)phenoxy)triethylamine citrate (1:1)
Clomid
Clomifen citrate
Clomifeno
Clomiphene citrate
Clomiphene dihydrogen citrate
Clomiphene-r
Clomiphine
Clomivid
Clomphid
1-(p-(β-Diethylaminoethoxy)phenyl)-1,2-diphenyl-2-chloroethylene citrate
Dyneric
Genozym
Ikaclomin
MER-41
MRL 41
NSC-35770
Omifin
Racemic clomiphene citrate

50-43-1
$C_7H_3Cl_3O_2$
225.45
Benzoic acid, 2,4,6-trichloro
2,4,6-Trichlorobenzoic acid

50-44-2
$C_5H_4N_4S$
152.19
Sc1ncnc2nc[nH]c12
Purine-6-thiol
Hypoxanthine, thio-
Ismipur
Leukeran
Leukerin
Leupurin
Mercaleukin
Mercaptopurin (German)
6-Mercaptopurin
Mercaptopurine
6-Mercaptopurine
7-Mercapto-1,3,4,6-tetrazaindene
Mercapurin
6-Merkaptopurin (Czech)
Merkaptopuryna (Polish)
Mern
MP
6 MP
NCI-C04886
NSC-755
Purimethol
Purine, 6-mercapto-
Purinethiol
3H-Purine-6-thiol
6-Purinethiol
6H-Purine-6-thione, 1,7-dihydro- (9CI)
Purinethol
6-Thiopurine
6-Thioxopurine
U-4748

50-45-3
$C_7H_4Cl_2O_2$
191.01
Benzoic acid, 2,3-dichloro
2,3-Dichlorobenzoic acid

50-47-5
$C_{18}H_{22}N_2$
266.42
CNCCCN2c1ccccc1CCc3ccccc23
5H-Dibenz(b,f)azepine-5-propanamine, 10,11-dihydro-N-methyl- (9CI)
Demethylimipramine
Desimipramine
Desimpramine
Desipramin
Desipramina (Spanish)
Desipramine
Desipraminum (Latin)
Desmethylimipramine
5H-Dibenz(b,f)azepine, 10,11-dihydro-5-(3-(methylamino)propyl) (8CI)
10,11-Dihydro-5-(3-methylaminopropyl)-5H-dibenz(b,f)azepine
Dimethylimipramine
DMI
DMI 50475
Imipramine, demethyl-
Methylaminopropyliminodibenzyl
Monodemethylimipramine
Norimipramine
Pentofran
Pertofran
Pertrofane
Sertofran

50-48-6
$C_{20}H_{23}N$
277.44
CN(C)CCC=C2c1ccccc1CCc3ccccc23
5H-Dibenzo(a,d)cycloheptene-δ^5,γ-propylamine, 10,11-dihydro-N,N-dimethyl
Amitriptylin (German)
Amitriptyline
Amitryptyline
Amytriptiline
Damilan
3-(10,11-Dihydro-5H-dibenzo-(a,d)cyclohepten-5-ylidene)-N,N-dimethyl-1-propanamine
10,11-Dihydro-5-(γ-dimethylaminopropylidene)-5H-dibenzo(a,d)cycloheptene
10,11-Dihydro-N,N-dimethyl-5H-dibenzo(a,d)heptalene-δ^5,γ-propylamine
5-(3-Dimethylaminopropylidene)-10,11-dihydro-5H-dibenzo(a,d)cycloheptatriene
5-(3'-Dimethylaminopropylidene)-dibenzo-(a,d)(1,4)-cycloheptadiene
5-(3-Dimethylaminopropylidene)-10,11-dihydro-5H-dibenzo(a,d)cycloheptene
5-(γ-Dimethylaminopropyli-

dene)-5H-dibenzo(a,d)-10,11-dihydrocycloheptene
5-(γ-Dimethylaminopropylidene)-10,11-dihydro-5H-dibenzo(a,d)cycloheptene
5-(3-Dimethylpropylidene)dibenzo(a,d)(1,4)cycloheptadiene
Elanil
Elavil
Laroxil
Laroxyl
Proheptadiene
Tryptizol

50-49-7
$C_{19}H_{24}N_2$
280.45
CN(C)CCCN2c1ccccc1CCc3ccccc23
5H-Dibenz(b,f)azepine, 5-(3-(dimethylamino)propyl)-10,11-dihydro
Antideprin
Berkomine
Censtim
Censtin
5H-Dibenz(b,f)azepine, 10,11-dihydro-5-(3-(dimethylamino)propyl)-
5,6-Dihydro-N-(3-(dimethylamino)propyl)-11H-dibenz-(b,e)azepine
1-(3-Dimethylaminopropyl)-4,5-dihydro-2,3,6,7-dibenzazepine
5-(3-(Dimethylamino)propyl)-10,11-dihydro-5H-dibenz-(b,f)azepine
5-(3-Dimethylaminopropyl)-10,11-dihydro-5H-dibenzo-(b,f)azepine
2,2'-(3-Dimethylaminopropylimino)dibenzyl
N-(γ-Dimethylaminopropyl)-iminodibenzyl
2,2'-(3-Dimethylaminopropylimino)bibenzyl
Dimipressin
DPID
Dynaprin
Dyna-Zina
Eupramin
G-22355
IM
Imidobenzyle
Imipramina (Italian)
Imipramine
Imiprin
Imizin
Imizine
Imizinum
Impramine
Intalpram
Iramil
Irmin
Melipramin
Melipramine
Nelipramin
Prazepine
Promiben
Surplix
Timolet
Tofranil

50-50-0
$C_{25}H_{28}O_3$
376.53
CC45CCC2C(CCc3cc(OC(=O)c1ccccc1)ccc23)C4CCC5O
Estradiol, 3-benzoate
Benovocylin
Benzhormovarine
Benzoate d'oestradiol (French)
Benzoestrofol
Benzofoline
Benzo-gynoestryl
Benzoic acid estradiol
De graafina
Diffollisterol
Difolliculine
Dihydroestrin benzoate
Dihydrofolliculin benzoate
Dimenformon benzoate
Dimenformone
Diogyn B
EBZ
Eston-B
Estradiol benzoate
Estradiol-17-β-benzoate
Estradiol-17-β-3-benzoate
β-Estradiol benzoate
β-Estradiol 3-benzoate
17-β-Estradiol benzoate
17-β-Estradiol 3-benzoate
Estradiol monobenzoate
17-β-Estradiol monobenzoate
Estra-1,3,5(10)-triene-3,17-diol (17-β)-3-benzoate
Estra-1,3,5(10)-triene-3,17-β-diol, 3-benzoate
1,3,5(10)-Estratriene-3,17-β-diol 3-benzoate
Femestrone
Follicormon
Follidrin
Graafina
Gynecormone
Gynformone
Hidroestron
Hormogynon
Hydroxyestrin benzoate
MEE
ODB
Oestradiol benzoate
Oestradiol 3-benzoate
β-Oestradiol benzoate
β-Oestradiol 3-benzoate
17-β-Oestradiol 3-benzoate
Oestradiol monobenzoate
1,3,5(10)-Oestratriene-3,17-β-diol 3-benzoate
Oestraform (BDH)
Ovahormon benzoate
Ovasterol-B
Ovex
Ovocyclin benzoate
Ovocyclin M
Ovocyclin-MB
Primogyn B
Primogyn Boleosum
Primogyn I
Progynon B
Progynon Benzoate
Recthormone oestradiol
Solestro
Unistradiol

50-52-2
$C_{21}H_{26}N_2S_2$
370.61
CSc4ccc3Sc1ccccc1N(CCC2CCCCN2C)c3c4
Phenothiazine, 10-((1-methyl-2-piperidyl)ethyl)-2-(methylthio)
Mallorol
Meleril
Mellaril
Mellarit
Mellerette
Melleretten
Melleril
2-Methylmercapto-10-(2-(n-methyl-2-piperidyl)ethyl)-phenothiazine
10-(2-(1-Methyl-2-piperidyl)-ethyl)-2-(methylthio)phenothiazine
10H-Phenothiazine, 10-(2-(1-methyl-2-piperidinyl)ethyl)-2-(methylthio)-
Sonapax
Thioridazin
Thioridazine
TP-21

50-53-3
$C_{17}H_{19}ClN_2S$
318.89
CN(C)CCCN2c1ccccc1Sc3ccc(Cl)cc23
Phenothiazine, 2-chloro-10-(3-(dimethylamino)propyl)
Aminasine
Aminazin
Aminazine
Ampliactil
Amplicitil
Chloro-3 (dimethylamino-3 propyl)-10 phenothiazine (French)
2-Chloro-10-(3-(dimethylamino)propyl)phenothiazine
Chloropromazine
Chlorpromazin
Chlorpromazine
ChloR-PZ
2-Cloro-10 (3-dimetilamino-propil)fenotiazina (Italian)
Clorpromazina (Italian)
CPZ
Fenactil
Fenaktyl
Fraction AB
Hibanil
Hibernal
Largactil
Largactilothiazine
Largactyl
Megaphen
Novomazina

Phenactyl
Phenathyl
10H-Phenothiazine-10-propan-amine, 2-chloro-N,N-dimethyl-
Plegomazin
Prazil
Promactil
Promazil
Propaphenin
Prozil
Prozin
4560 R.P.
SKF-2601
Thorazine
Torazina
Wintermin

50-54-4
$C_{40}H_{48}N_4O_4.H_2O_4S$
747.00
Quinidine sulfate (2:1) (Salt)
Chinidine sulfate
Cinchonan-9-ol, 6'-methoxy-, (9S)-, sulfate (2:1) (Salt) (9CI)
Cin-quin
(9S)-6'-Methoxycinchonan-9-ol sulfate (2:1) (Salt)
Quinicardine
Quinidate
Quinidex
Quinidine monosulfate
Quinidine sulfate
Quinidine sulphate
Quinitex
Quinora
Systodin

50-55-5
$C_{33}H_{40}N_2O_9$
608.75
O=C(OC(C(OC)C(C(C1CN(C2C(Nc(c3ccc4OC)c4)=C3C5)C5)C2)C(=O)OC)C1)c(cc(OC)c(OC)c6OC)c6
3-β,20-α-Yohimban-16-β-carboxylic acid, 18-β-hydroxy-11,17-α-dimethoxy-, methyl ester, 3,4,5-trimethoxybenzoate (ester)
Abesta
Abicol
Adelfan
Adelphane
Adelphin
Adelphin-esidrex-K
Alkarau
Alkaserp
Alserin
Anquil
Apoplon
Apsical
Arcum R-S
Ascoserp
Ascoserpina
Austrapine
Banasil
Banisil
Benazyl
Bendigon
Bioserpine
Brinderdin
Briserine
Broserpine
Butiserpazide-25
Butiserpazide-50
Butiserpine
Carditivo
Cardioserpin
Carrserp
Crystoserpine
Darebon
Deserpine
Diupres
Diutensen-R
Drenusil-R
Dypertane Compound
Eberpine
Eberspine
Ebserpine
Elfanex
Elserpine
Enipresser
ENT 50,146
Escaspere
Eserpine
Eskaserp
Gammaserpine
Gilucard
Gamaserpin
H 520
Helfoserpin
Hexaplin
Hiposerpil
Hiserpia
Hydromox R
Hydropres
Hydropres KA
Hygroton-reserpine
Hypercal B
Hypertane Forte
Hypertensan
Idoserp
Idsoserp
Interpina
Key-Serpine
Kitine
Klimanosid
"L," Carpserp
Lemiserp
Loweserp
Marnitension simple
Maviserpin
Mayserpine
Mephaserpin
Metatensin
Methylreserpate 3,4,5-trimethoxybenzoic acid
Methyl reserpate 3,4,5-trimethoxybenzoic acid ester
Mio-pressin
Modenol
Naquival
NCI-C50157
Nembu-serpin
Neo-Antitensol
Neo-Antitersol
Neoserfin
Neo-Serp
Neoslowten
Orthoserpina
Perskleran
Pressimedin
Purserpine
Quiescin
Raucap
Raudiford
Raudixin
Raudixoid
Raugal
Raulen
Rauloycin
Rauloydin
Raumorine
Raunervil
Raunorine
Raunormin (Orzan)
Raunova
Raupasil
Raupoid
Raurine
Rausan
Rau-sed
Rausedan
Rausedil
Rausedyl
Rauserpen-alk
Rauserpin
Rauserpin-alk
Rauserpine
Rauserpol
Rausingle
Rautrin
Rauvilid
Rauvlid
Rauwasedin
Rauwilid
Rauwiloid
Rauwiloid+
Rauwipur
Rauwoleaf
Rauwopur (BYK)
Rawilid
RCRA waste number U200
Recipin
Regroton
Renese R
R-E-S
Resaltex
Resedin
Resedril
Resedrex
Rese-LAR
Reser-AR
Reserbal
Resercaps
Resercen
Resercrine
Reserfia
Reserjen
Reserlor
Reserp
Reserpal
Reserpamed
Reserpanca
Reserpene
Reserpex
Reserpidefe
Reserpil
Reserpin
Reserpina
Reserpine
Reserpinum

Reserpka
Reserpoid
Reserpur
Reserp (Wander)
Resersana
Reserutin
Resiatric
Residine
Resine
Resocalm
Resomine
Resperin
Resperine
Respital
Restran
Rezerpin
Riserpa
Rivased
Rivasin
Rolserp
Roxel
Roxinoid
Roxynoid
Ryser
Salupres
Salutensin
Sandril
Sandron
Sarpagan
Sarpagen
Sedaraupin
Sedaraupina
Seda-recipin
Seda-salurepin
Sederaupin
Sedserp
Seominal
Serfin
Serfolia
Serolfia
Serp
Serp-AFD
Serpalan
Serpaloid
Serpaneurona
Serpanray
Serpasil
Serpasil Apresoline
Serpasil-Esidrex
Serpasil-Esidrex No. 1
Serpasil-Esidrex No. 2
Serpasil-Esidrex K
Serpasil Premix
Serpasol
Serpate
Serpatone
Serpax
Serpazil
Serpazol
Serpedin
Serpen
Serpena
Serpentil
Serpentin
Serpentina
Serpentine (pharbil)
Serpicon
Serpil
Serpiloid
Serpilum
Serpine
Serpipur
Serpine (Pharmaceutical)
Serpivite
Serplex K
Serpogen
Serpoid
Serpone
Serpresan
Serpyrit
Sertabs
Sertens
Sertensin
Sertina
Sinesalin composition
Solfo Serpine
Supergan
Tefaserpina
Tempo-Reserpina
Temposerpine
Tendoscen-compr.
Tensanyl
Tenserlix
Triserpin
Tenserpine (ASSIA)
Tenserpinie
Tensional
Tensionorme
Tepserpine
Terbolan
Transerpin
3,4,5-Trimethoxybenzoyl methyl reserpate
T-Serp
Tylandril
Unilord
Unitensen
SK-Reserpine
USAF CB-27
Veriloid
Vio-Serpine
V-Serp
Yohimban-16-carboxylic acid derivative of benz(g)indolo-(2,3-a)quinolizine

50-56-6
$C_{43}H_{66}N_{12}O_{12}S_2$
1007.33
Oxytocin
Atonin O
Di-Sipidin
Endopituitrina
α-Hypophamine
Nobitocin S
Orasthin
Oxystin
Partocon
Pitocin
Piton S
Posterior pituitary extract
Presoxin
Synpitan
Synthetic oxytocin
Syntocin
Syntocinon
Syntocinone
Utedrin
Uteracon

50-59-9
$C_{19}H_{17}N_3O_4S_2$
415.51
[O-]C(=O)C2=C(CSC3C(NC(=O)Cc1cccs1)C(=O)N23)C[n+]4ccccc4
Pyridinium, 1-((2-carboxy-8-oxo-7-(2-(2-thienyl)acetamido)-5-thia-1-azabicyclo-(4.2.0)oct- 2-en-3-yl)methyl)-, hydroxide, inner salt
Aliporina
Ampligram
Betaine cephaloridine
Cefaloridin
Cefaloridine
Ceflorin
Cepaloridin
Cepalorin
Ceph 87/4
Cephaloridin
Cephaloridine
Cefalorizin
Ceporan
Ceporin
Ceporine
CER
Cilifor
Deflorin
Faredina
Floridin
Glaxoridin
Intrasporin
Keflodin
Keflordin
Kefspor
Lilly 40602
Lloncefal
Loridine
SCH 11527
Sefacin
7-((2-Thienyl)acetamido)-3-(1-pyridylmethyl)cephalosporanic acid
7-(α-(2-Thienyl)acetamido)-3-(1-pyridylmethyl)-3-cephem-4-carboxylic acid betaine
7-(Thiophene-2-acetamido)-3-(1-pyridylmethyl)-3-cephem-4-carboxylic acid betaine

50-60-2
$C_{17}H_{19}N_3O$
281.39
Cc1ccc(cc1)N(CC2=NCCN2)c3ccc(O)c3
Phenol, 3-(((4,5-dihydro-1H-imidazol-2-yl)methyl)-(4-methylphenyl)amino)
C 7337 Ciba
2-((N-(m-Hydroxyphenyl)-p-toluidino)methyl)-2-imidazoline
2-(m-Hydroxy-N-p-tolylanilinomethyl)-2-imidazoline
2-Imidazoline, 2-((N-(m-hydroxyphenyl)-p-toluidino)-methyl)-
Phenol, m-(N-(2-imidazolin-2-ylmethyl)-p-toluidino)-
Phenotolamine
Phentalamine
Phentolamine

Regitin
Regitine
Regitipe
Rogitine
2-(N'-p-Tolyl-N'-m-hydroxy-phenylaminomethyl)-2-imidazoline

50-65-7
$C_{13}H_8Cl_2N_2O_4$
327.13
Salicylanilide, 2',5-dichloro-4'-nitro
Bay 2353
Bayer 73
Bayer 2353
Bayluscid
Benzamide, 5-chloro-N-(2-chloro-4-nitrophenyl)-2-hydroxy-
Chemagro 2353
N-(2'-Chlor-4'-nitrophenyl)-5-chlorsalicylamid (German)
5-Chloro-2'-chloro-4'-nitro-salicylanilide
2-Chloro-4-nitrophenylamide-6-chlorosalicylic acid
N-(2-Chloro-4-nitrophenyl)-5-chlorosalicylamide
Clonitralid
2',5-Dichloro-4'-nitrosalicyl-anilide
2',5-Dichlor-4'-nitro-salizyl-saeureanilid (German)
Dichlosale
ENT 25,823
Fenasal
HL 2447
2-Hydroxy-5-chloro-N-(2-chloro-4-nitrophenyl)benzamide
Iomesan
Iomezan
Niclosamide
Phenasal
Vermitin
Yomesan

50-67-9
$C_{10}H_{12}N_2O$
176.24
NCCc1c[nH]c2ccc(O)cc12
Indol-5-ol, 3-(2-aminoethyl)
3-(β-Aminoethyl)-5-hydroxy-indole
3-(2-Aminoethyl)indol-5-ol
Antemoqua
Antemovis
DS Substance
Enteramine
Hippophain
5-HT
5-HTA
5-Hydroxy-3-(β-aminoethyl)-indole
5-Hydroxytryptamine
Serotonin
Serotonine
Substance DS
Substanz DS
Thrombocytin
Thrombotonin
Tryptamine, 5-hydroxy-

50-69-1
$C_5H_{10}O_5$
150.13
O=CC(O)C(O)C(O)CO
D-Ribose
AI3-52667

50-70-4
$C_6H_{14}O_6$
182.20
OCC(O)C(O)C(O)C(O)CO
Glucitol
AI3-19424
Cholaxine
Diakarmon
Esasorb
Glucitol, D- (8CI)
D-Glucitol (9CI)
Gulitol
L-Gulitol
Hefti Sorbex-R
Hefti Sorbex-RP
Karion
Neosorb
Neosorb 70/70
Nivitin
Sionit
Sionite
Sionon
Siosan
Sorbitol
Sorbex M
Sorbex R
Sorbex RP
Sorbex S
Sorbex X
Sorbicolan
Sorbilande
Sorbit
Sorbite
Sorbitol
D-Sorbitol
D-(-)-Sorbitol
Sorbitol syrup C
Sorbo
Sorbol
Sorbostyl
Sorvilande
Unisweet CONC
Unisweet 70

50-71-5
$C_4H_2N_2O_4$
142.08
O=C(NC(=O)C(=O)C1=O)N1
Alloxan
Alloxane
Barbituric acid, 5-oxo-
Mesoxalylcarbamide
Mesoxalylurea
2,4,5,6(1H,3H)-Pyrimidinetetrone
2,4,5,6-Pyrimidintetron (Czech)
2,4,5,6-Tetraoxohexahydropyrimidine
Urea, mesoxalyl-

50-73-7
$C_7H_3Cl_3O_2$
225.45
Benzoic acid, 2,3,5-trichloro
2,3,5-Trichlorobenzoic acid

50-74-8
$C_7H_2Cl_4O_2$
259.89
Benzoic acid, 2,3,4,5-tetrachloro
2,3,4,5-Tetrachlorobenzoic acid

50-76-0
$C_{62}H_{86}N_{12}O_{16}$
1255.60
C.C.C.C.C.CC.CC
Actinomycin-D
Act
Act D
Actinomycin A IV
Actinomycin 7
Actinomycin I_1
Actinomycin I
Actinomycin IV
Actinomycin C1
Actinomycin 11 cosmegen
Actinomycindioic D acid, dilactone
Actinomycin X 1
Acto-D
AD
C1
Cosmegen
Dactinomycin
Dactinomycin D
Dilactone actinomycin D acid
Dilactone actinomycindioic D acid
HBF 386
Lyovac cosmegen
Meractinomycin
NCI-C04682
NSC-3053
Oncostatin K
X 97

50-78-2
$C_9H_8O_4$
180.17
O=C(Oc(c(ccc1)C(=O)O)c1)C
Salicylic acid, acetate
Acenterine
Acesal
Acetal
Aceticyl
Acetilsalicilico
Acetilum acidulatum
Acetisal
Acetol
Acetonyl
Acetophen
Acetosal
Acetosalic acid
Acetosalin
o-Acetoxybenzoic acid

2-Acetoxybenzoic acid
Acetylin
2-(Acetyloxy)benzoic acid
Acetylsal
Acetylsalicylic acid (ACGIH, OSHA)
Acido acetilsalicilico (Italian)
Acimetten
Acetylsalicylsaure (German)
Acide acetylsalicylique (French)
Acido o-acetil-benzoico (Italian)
Acidum acetylsalicylicum
Acisal
Acylpyrin
ASA
A.S.A.
A.S.A. Empirin
Asagran
Asatard
Aspalon
Aspergum
Aspirdrops
Aspirin
Aspirine
Aspirin (OSHA)
Aspro
Asteric
AC 5230
Benaspir
Entericin
Extren
Bialpirinia
Caprin
o-Carboxyphenyl acetate
Colfarit
Contrheuma retard
Crystar
Delgesic
Dolean pH 8
Duramax
ECM
Ecotrin
Empirin
Endydol
Enterophen
Enterosarine
Entrophen
Globoid
Helicon
Idragin
Kyselina 2-acetoxybenzoova (Czech)
Kyselina acetylsalicylova (Czech)
Measurin
Neuronika
Novid
Polopiryna
Rheumin tabletten
Rhodine
Salacetin
Salcetogen
Saletin
Solpyron
Xaxa

50-79-3
$C_7H_4Cl_2O_2$
191.01
O=C(O)c(c(ccc1Cl)Cl)c1
Benzoic acid, 2,5-dichloro
2,5-Dichlorobenzoic acid

50-81-7
$C_6H_8O_6$
176.14
O=C(OC(C=1O)C(O)CO)C1O
L-Ascorbic acid (8CI,9CI)
AA
Adenex
Acide ascorbique (French)
Acido ascorbico (Spanish)
Acidum arcorbicum (Latin)
Acidum ascorbinicum
Allercorb
Antiscorbic vitamin
Antiscorbutic vitamin
Arco-cee
Ascoltin
Ascorb
Ascorbajen
Ascorbate
Ascorbicab
Ascorbic acid
L(+)-Ascorbic acid
Ascorbicap
Ascor-B.I.D.
Ascorbin
Ascorbutina
Ascorin
Ascorteal
Ascorvit
Cantan
Cantaxin
Caswell No. 061B
Catavin C
Cebicure
Cebid
Cebion
Cebione
Cecon
Cee-Caps TD
Cee-Vite
Cegiolan
Ceglion
Celaskon
Ce Lent
Celin
Cemagyl
Ce-Mi-Lin
Cemill
Cenetone
Cenolate
Cereon
Cergona
Cescorbat
Cetamid
Cetane
Cetane-Caps TC
Cetane-Caps TD
Cetemican
Cevalin
Cevatine
Cevex
Cevi-Bid
Cevimin
Ce-Vi-Sol
Cevital
Cevitamic acid
Cevitamin
Cevitan
Cevitex
Cewin
Ciamin
Cipca
Citriscorb
C-Level
C-Long
Colascor
Concemin
C-Quin
C-Span
C-Vimin
Dora-C-500
Davitamon C
Duoscorb
L-Threo-hex-2-enonic acid, γ-lactone
Hicee
Hybrin
Ido-C
3-Keto-L-gulofuranolactone
L-3-Ketothreohexuronic acid lactone
Kyselina askorbova (Czech)
Laroscorbine
Lemascorb
Liqui-Cee
L-Lyxoascorbic acid
Meri-C
Natrascorb
Natrascorb injectable
NCI-C54808
Neo-Valdrin
NSC-33832
3-Oxo-L-gulofuranolactone
3-Oxo-L-gulofuranolactone (enol form)
Planavit C
Proscorbin
Redoxon
Ribena
Roscorbic
Scorbacid
Scorbu-C
Secorbate
Testascorbic
Uantox ASCA
Vicelat
Vicin
Vicomin C
Viforcit
Viscorin
Vitace
Vitacee
Vitacimin
Vitacin
Vitamin C
Vitamisin
Vitascorbol
Xitix
Xyloascorbic acid, L-
L-Xyloascorbic acid

50-82-8
$C_7H_3Cl_3O_2$
225.45
Benzoic acid, 2,4,5-trichloro
2,4,5-Trichlorobenzoic acid

50-84-0
$C_7H_4Cl_2O_2$

191.01
O=C(O)c(c(cc(c1)Cl)Cl)c1
Benzoic acid, 2,4-dichloro
2,4-Dichlorobenzoic acid

50-85-1
$C_8H_8O_3$
152.16
Cc1ccc(C(O)=O)c(O)c1
2,4-Cresotic acid
Benzoic acid, 2-hydroxy-4-methyl- (9CI)
m-Cresotic acid
γ-Cresotic acid
m-Cresotinic acid
m-Homosalicylic acid
4-Methylsalicylic acid

50-89-5
$C_{10}H_{14}N_2O_5$
242.26
O=C(N(C=C(C1=O)C)C(OC(C2O)CO)C2)N1
Thymidine
Deoxythymidine
2'-Deoxythymidine
DT
DTHYD
5-Methyldeoxyurindine
Thymidin
Thyminedeoxyriboside
Thymine-2-deoxyriboside

50-90-8
$C_9H_{11}ClN_2O_5$
262.67
OCC1OC(CC1O)n2cc(Cl)c(=O)[nH]c2=O
Uridine, 5-chloro-2'-deoxy
5-Chlorodeoxyuridine
5-Chloro-2'-deoxyuridine
ClUDR

50-91-9
$C_9H_{11}FN_2O_5$
246.22
OCC1OC(CC1O)n2cc(F)c(=O)[nH]c2=O
Uridine, 2'-deoxy-5-fluoro (8CI,9CI)
Deoxyfluorouridine
2'-Deoxy-5-fluorouridine
1-β-D-2'-Deoxyribofuranosyl-5-flurouracil
FDUR
Floxuridin
Floxiridina (Spanish)
Floxuridine
Floxuridinum (Latin)
5-Fluor-1-(β-2'-deoxyribofuranosyl)pyrimidin-2,4(1H,3H)-dion (Czech)
Fluorodeoxyuridine
β-5-Fluoro-2'-deoxyuridine
5-Fluorodeoxyuridine
5-Fluoro-2-deoxyuridine
5-Fluoro-2'-deoxyuridine
5-Fluorouracil deoxyriboside
5-Fluorouracil 2'-deoxyriboside
Fluoruridine deoxyribose
FUDR
5-FUDR
NSC-27640
Ro 5-0360

50-99-7
$C_6H_{12}O_6$
180.18
O=CC(O)C(O)C(O)C(O)CO
D-Glucose
Anhydrous dextrose
Cartose
Cerelose
Corn sugar
Dextropur
Dextrose
Dextrose, Anhydrous
Dextrosol
Glucolin
Glucose
Glucose, Anhydrous
D-Glucose, Anhydrous
Glucose Liquid
Grape sugar
Sirup
Sugar, Grape

51-02-5
$C_{15}H_{19}NO.ClH$
265.81
Cl.CC(C)NCC(O)c2ccc1ccccc1c2
Naphthalenemethanol, α-((isopropylamino)methyl)-, hydrochloride
Alderlin hydrochloride
ICI 38174
I.C.I. Hydrochloride
Inetol
α-((Isopropylamino)methyl)-2-naphthalenemethanol hydrochloride
2-Isopropylamino-1-(2-naphthyl)ethanol hydrochloride
α-(((1-Methylethyl)amino)-methyl)-2-naphthalene-methanol, hydrochloride
Naphthylisoproterenol hydrochloride
Nethalide hydrochloride
Pronetalol hydrochloride
Pronethalol hydrochloride

51-03-6
$C_{19}H_{30}O_5$
338.49
O(c(c(O1)cc(c2CCC)COCCOCCOCCCC)c2)C1
Toluene, α-(2-(2-butoxyethoxy)ethoxy)-4,5-(methylenedioxy)-2-propyl
1,3-Benzodioxole, 5-((2-(2-butoxyethoxy)ethoxy)methyl)-6-propyl-
Butacide
Butocide
Butoxide
α-(2-(2-Butoxyethoxy)ethoxy)-4,5-methylenedioxy-2-propyltoluene
α-(2-(2-n-Butoxyethoxy)-ethoxy)-4,5-methylenedioxy-2-propyltoluene
5-((2-(2-Butoxyethoxy)ethoxy)-methyl)-6-propyl-1,3-benzodioxole
Butyl carbitol 6-propylpiperonyl ether
Butyl-carbityl (6-propylpiperonyl) ether
ENT 14,250
FAC 5273
FMC 5273
3,4-Methylendioxy-6-propylbenzyl-n-butyl-diaethylenglykolaether (German)
(3,4-Methylenedioxy-6-propylbenzyl) (butyl) diethylene glicol ether
3,4-Methylenedioxy-6-propylbenzyl n-butyl diethyleneglycol ether
NCI-C02813
NIA 5273
Nusyn-Noxfish
PB
Piperonyl butoxide
Prentox
6-(Propylpiperonyl)-butyl carbityl ether
6-Propylpiperonyl butyl diethylene glycol ether
5-Propyl-4-(2,5,8-trioxa-dodecyl)-1,3-benzodioxol (German)
Pybuthrin
Pyrenon
Pyrenone 606
Synpren-Fish

51-05-8
$C_{13}H_{20}N_2O_2.ClH$
272.81
Cl.CCN(CC)CCOC(=O)c1ccc(N)cc1
Benzoic acid, p-amino-, 2-(diethylamino)ethyl ester, monohydrochloride
Allocaine
Alocaine
p-Aminobenzoic acid 2-diethylaminoethyl ester hydrochloride
4-Aminobenzoic acid 2-(diethylamino)ethyl ester hydrochloride
p-Aminobenzoyldiethylaminoethanol hydrochloride
Aminocaine
Anadolor
Anesthesol
Anestil
Atoxicocaine
Benzoic acid, 4-amino-, 2-(diethylamino)ethyl ester, monohydrochloride
Bernocaine
Cetain
Chlorocaine
Diethylaminoethanol 4-amino-

benzoate hydrochloride
2-Diethylaminoethyl p-amino-benzoate hydrochloride
Ethocaine
Eugerase
Irocaine
Isocaine-Asid
Isocaine-Heisler
Jenacain
Juvocaine
Kerocaine
Lactocaine
Naucaine
Neocaine
Novocain-chlorhydrat (German)
Novocaine hydrochloride
Novocain hydrochlorid (German)
Paracain
Planocaine
Procaine hydrochloride
Scurocaine
Sevicaine
Syncaine
Topokain
Westocaine

51-06-9
$C_{13}H_{21}N_3O$
235.37
CCN(CC)CCNC(=O)c1ccc(N)cc1
Benzamide, p-amino-N-(2-(diethylamino)ethyl)
p-Aminobenzoic diethyl-aminoethylamide
p-Amino-N-(2-diethylamino-ethyl)benzamide
Benzamide, 4-amino-N-(2-(diethylamino)ethyl)- (9CI)
2-Diethylaminoethylamid kyseliny p-aminobenzoove (Czech)
Novocainamid
Novocainamide
Novocaine amide
Novocamid
Procainamide
Procaine amide
Procamide
Pronestyl

51-12-7
$C_{16}H_{18}N_4O_2$
298.38
O=C(CCNNC(=O)c1ccncc1)NCc2ccccc2
Isonicotinic acid, 2-(2-(benzylcarbamoyl)ethyl)hydrazide
Beih
N[1]-β-Benzylcarbamoylethyl-N[2]-isonicotinoylhydrazine
N-(2-(Benzylcarbamyl)ethyl-amino)isonicotinamide
2-(2-(Benzylcarbamyl)ethyl)-hydrazide isonicotinic acid
1-(2-(Benzylcarbamoyl)ethyl)-2-isonicotinoylhydrazine
N-Benzyl-β-(isonicotinoyl-hydrazine)propionamide
N-Benzyl-β-(isonicotinylhydrazino)propionamide
Delmoneurina
Espril
Isalizina
Isonicotinic acid, 2-(2-(benzylcarbamoyl)ethyl)hydrazide (8CI)
N-Isonicotinoyl-N'(β-N-benzyl-carboxamidoethyl)hydrazine
Mygal
Nialamida (Spanish)
Nialamide
Nialamidum (Latin)
Niamid
Niamidal
Niamide
Niaquitil
NSC-124514
Nuredal
Nyazin
P 1133
Psicodisten
4-Pyridinecarboxylic acid 2-(3-oxo-3-((phenylmethyl)-amino)propyl)hydrazide (9CI)
Surgex

51-14-9
$C_{15}H_{22}O_6$
298.37
CCOCCOCCOC(C)Oc2ccc1OCOc1c2
Acetaldehyde, 2-(2-ethoxyethoxy)ethyl 3,4-(methylenedioxy)phenyl acetal
2-(2-Aethoxy-aethoxy)-aethy-3,6,9-trioxa-undecan (German)
AI3-20871
1,3-Benzodioxole, 5-(1-(2-(2-ethoxyethoxy)ethoxy)-ethoxy)-
ENT 20,871
5-(1-(2-(2-Ethoxyethoxy)-ethoxy)ethoxy)-1,3-benzodioxole
2-(2-Ethoxyethoxy)ethyl-3,4-(methylenedioxy)phenyl acetal of acetaldehyde
2-(3,4-Methylenedioxyphen-oxy)-3,6,9-trioxoundecane
Sesamex
Sesoxane
3,6,9-Trioxaundecane, 2-(3,4-(methylenedioxy)phen-oxy)-

51-17-2
$C_7H_6N_2$
118.15
N(c(c(N1)ccc2)c2)=C1
Benzimidazole
3-Azaindole
Azindole
1H-Benzimidazole (9CI)
o-Benzimidazole
Benziminazole
1,3-Benzodiazole
Benzoglyoxaline
Benzoimidazole
BZI
1,3-Diazaindene
N,N'-Methenyl-o-phenylene-diamine
NSC-759

51-18-3
$C_9H_{12}N_6$
204.27
C1CN1c2nc(nc(n2)N3CC3)N4CC4
s-Triazine, 2,4,6-tris(1-aziridinyl)
Aziridine, 1,1',1''-s-triazine-2,4,6-triyltris-
DRP 859025
ENT 25,296
M-9500
Melamine, triethylene-
NSC-9706
Persistol
Persistol HO 1/193
Persistol HOE 1/193
R-246
SK1133
TAT
TEM
TEM (Cytostatic)
TEM-Simes
TET
Tretamin
Tretamine
Triaethylenmelamin (German)
Triamelin
1,3,5-Triazine, 2,4,6-tris-(1-aziridinyl)-
1,1',1''-s-Triazine-2,4,6-triyl-trisaziridine
Triaziridinyl triazine
Triethanomelamine
2,4,6-Tri(ethyleneimino)-1,3,5-triazine
2,4,6-Triethyleneimino-s-triazine
Triethylenemelamine
2,4,6-Triethylenimino-s-triazine
2,4,6-Triethylenimino-1,3,5-triazine
Trisaziridinyltriazine
2,4,6-Tris(1-aziridinyl)-s-triazine
2,4,6-Tris(1'-aziridinyl)-1,3,5-triazine
2,4,6-Tris(ethyleneimino)-s-triazine
Tris(ethyleneimino)triazine
2,4,6-Tris(ethylenimino)-s-triazine
Trisethyleneimino-1,3,5-triazine

51-19-4
$C_6H_7NO.ClH$
145.60
Phenol, o-amino-, hydrochloride
o-Aminophenol hydrochloride
2-Aminophenol hydrochloride

51-20-7
$C_4H_3BrN_2O_2$
191.00

n(cc(c(n1)O)Br)c1O
Uracil, 5-bromo
5-Bromo-uracil
2,4(1H,3H)-Pyrimidinedione, 5-bromo- (9CI)

51-21-8
$C_4H_3FN_2O_2$
130.09
n(cc(F)c(n1)O)c1O
Uracil, 5-fluoro
Adrucil
Arumel
Carzonal
Effluderm (Free base)
Efudex
Efudix
Efurix
5-Fluoracil (German)
5-Fluor-2,4-dihydroxypyrimidin (Czech)
Fluoroblastin
Fluoroplex
5-Fluoropyrimidine-2,4-dione
5-Fluoro-2,4-pyrimidinedione
5-Fluoro-2,4(1H,3H)-pyrimidinedione
5-Fluor-2,4-pyrimidindiol (Czech)
5-Fluor-2,4(1H,3H)-pyrimidindion (Czech)
Fluorouracil
5-Fluorouracil
5-Fluoruracil (German)
Fluracil
Fluracilum
Fluri
Fluril
FT-207
Ftoruracil
FU
5-FU
NSC-19893
2,4(1H,3H)-Pyrimidinedione, 5-fluoro-
Queroplex
Ro 2-9757
Timazin
U-8953
ULUP

51-28-5
$C_6H_4N_2O_5$
184.12
O=N(=O)c(ccc(O)c1N(=O)=O)c1
Phenol, 2,4-dinitro
Aldifen
Chemox PE
2,4-Dinitrofenol (Dutch)
Dinitrofenolo (Italian)
α-Dinitrophenol
2,4-Dinitrophenol
Dinofan
2,4-DNP
Fenoxyl Carbon N
1-Hydroxy-2,4-dinitrobenzene
Maroxol-50
Nitro Kleenup
NSC-1532
Phenol, α-dinitro-
RCRA waste number P048
Solfo Black B
Solfo Black BB
Solfo Black 2B Supra
Solfo Black G
Solfo Black SB
Tertrosulphur Black PB
Tertrosulphur PBR

51-30-9
$C_{11}H_{17}NO_3.ClH$
247.75
Cl.CC(C)NCC(O)c1ccc(O)c(O)c1
Benzyl alcohol, 3,4-dihydroxy-α-((isopropylamino) methyl)-, hydrochloride
1,2-Benzenediol, 4-(1-hydroxy-2-((1-methylethyl)amino)-ethyl)-, hydrochloride
3,4-Dihydroxy-α-((isopropyl-amino)methyl)benzyl alcohol hydrochloride
Euspiran
Isadrine
Isadrine-hydrochloride
Isoprenaline chloride
Isoprenaline hydrochloride
α-(Isopropylaminomethyl)-3,4-dihydroxybenzyl alcohol hydrochloride
Isopropylarterenol hydrochloride
Isopropylnorepinephrine-hydrochloride
Isoproterenol hydrochloride
Isoproterenol monohydrochloride
Isuprel
Isuprel hydrochloride
Izadrin
NCI-C55630
Norisodrine hydrochloride
Saventrine
Vapo-Iso

51-34-3
$C_{17}H_{21}NO_4$
303.39
CN1C2CC(CC1C3OC23)OC(=O)C(CO)c4ccccc4
Scopolamine
Atrochin
Atroquin
6-β,7-β-Epoxy-3-α-tropanyl S-(-)-tropate
Epoxytropine tropate
Hyosceine
Hyoscine
(-)-Hyoscine
Hyosol
Isopto Hyoscine
Oscine
3-Oxa-9-azatricyclo-(3.3.1.$O^{2,4}$)nonan-7-ol, 9-methyl-, tropate (ester)
Scopine tropate
(-)-Scopolamine
Tropic acid, ester with scopine
Tropic acid, 9-methyl-3-oxa-9-azatricyclo(3.3.1.$O^{2,4}$)non-7-yl ester

51-35-4
$C_5H_9NO_3$
131.13
O=C(O)C(NCC1O)C1
Hydroxyproline
Hydroxyproline, (L)
δ-Hydroxyproline
Hydroxy-L-proline
L-Hydroxyproline
L-4-Hydroxyproline
trans-Hydroxyproline
trans-L-Hydroxyproline
trans-4-Hydroxyproline
trans-4-Hydroxy-L-proline
4-Hydroxy-L-proline
4-Hydroxy-2-pyrrolidine-carboxylic acid
4-L-Hydroxyproline
Hypro
NSC-46704
Proline, 4-hydroxy- (VAN)
Proline, 4-hydroxy-, L- (8CI)
L-Proline, 4-hydroxy-
L-Proline, 4-hydroxy-, trans- (9CI)

51-36-5
$C_7H_4Cl_2O_2$
191.01
OC(=O)c1cc(Cl)cc(Cl)c1
Benzoic acid, 3,5-dichloro
3,5-Dichlorobenzoic acid

51-41-2
$C_8H_{11}NO_3$
169.20
NCC(O)c1ccc(O)c(O)c1
Benzyl alcohol, α-(amino-methyl)-3,4-dihydroxy-, (-)
Adrenor
Aktamin
l-2-Amino-1-(3,4-dihydroxy-phenyl)ethanol
l-α-(Aminomethyl)-3,4-dihydroxybenzyl alcohol
(-)-α-(Aminomethyl)protocatechuyl alcohol
Arterenol
l-Arterenol
1,2-Benzenediol, 4-(2-amino-1-hydroxyethyl)-, (R)- (9CI)
l-1-(3,4-Dihydroxyphenyl)-2-aminoethanol
l-3,4-Dihydroxyphenylethanol-amine
Levarterenol
Levoarterenol
Levonoradrenaline
Levonorepinephrine
Levophed
(-)-Noradrec
Noradrenalin
Noradrenalina (Italian)
Noradrenaline
(-)-Noradrenaline
D-(-)-Noradrenaline
l-Noradrenaline

Noradrenline
Norartrinal
Norepinephrine
(-)-Norepinephrine
l-Norepinephrine
l-Norepinephrine
Norepirenamine
Sympathin E

51-42-3
$C_9H_{13}NO_3.C_4H_6O_6$
333.33
Benzyl alcohol, 3,4-dihydroxy-α-((methylamino)methyl)-, (-)-, tartrate (1:1), (+)
Adrenalin bitartrate
Adrenaline acid tartrate
(-)-Adrenaline acid tartrate
Adrenaline bitartrate
l-Adrenaline bitartrate
(-)-Adrenaline bitartrate
l-Adrenaline d-bitartrate
Adrenaline hydrogen tartrate
(-)-Adrenaline hydrogen tartrate
l-Adrenaline hydrogen tartrate
Adrenaline tartrate
(-)-Adrenaline tartrate
l-Adrenaline tartrate
l-Adrenaline tartrate
Adrenatrate
Asmatane Mist
(-)-3,4-Dihydroxy-α-((methylamino)methyl)benzyl alcohol (+)-tartrate (1:1) salt
Epinephrine bitartrate
Epinephrine d-bitartrate
(-)-Epinephrine bitartrate
l-Epinephrine bitartrate
l-Epinephrine bitartrate
l-Epinephrine d-bitartrate
Epinephrine hydrogen tartrate
Epinephrine tartrate
l-Epinephrine tartrate
IOP
Lyophrin
Medihaler-EPI
Suprarenin

51-43-4
$C_9H_{13}NO_3$
183.23
OC(c(ccc(O)c1O)c1)CNC
Benzyl alcohol, 3,4-dihydroxy-α-((methylamino)methyl)-, (-)
Adnephrine
Adrenal
Adrenalin
l-Adrenalin
Adrenaline
(-)-Adrenaline
l-Adrenaline
Adrenalin in oil
Adrenalin-medihaler
Adrenamine
Adrenan
Adrenapax
Adrenasol
Adrenatrate
Adrenine
Adrenodis
Adrenohorma
Adrenosan
Adrenutol
Adrin
Adrine
Antiasthmatique
Asmatane mist
Asthma Meter Mist
Asthma-Nefrin
Astmahalin
Astminhal
Balmadren
1,2-Benzenediol, 4-(1-hydroxy-2-(methylamino)ethyl)-, (R)-(9CI)
Bernarenin
Biorenine
Bosmin
Brevirenin
Bronkaid mist
Chelafrin
Corisol
3,4-Dihydroxy-α-((methylamino)methyl)benzyl alcohol
l-1-(3,4-Dihydroxyphenyl)-2-methylaminoethanol
Drenamist
Dylephrin
Dyspne-inhal
Epifrin
Epinefrin (Czech)
Epinefrina
Epinephran
Epinephrine
(-)-Epinephrine
l-Epinephrine
(R)-Epinephrine
l-Epinephrine (synthetic)
Epirenamine
l-Epirenamine
Epirenan
Epirenin
Epitrate
Esphygmogenina
Exadrin
Glaucosan
Glycirenan
Haemostasin
Haemostatin
Hektalin
Hemisine
Hemostasin
Hemostatin
Hypernephrin
Hyporenin
Intranefrin
Kidoline
Levorenin
Levorenine
Lyophrin
Medihaler-EPI
Metanephrin
Methylarterenol
Mucidrina
Myosthenine
Mytrate
Nephridine
Nieraline
Paranephrin
Primatene mist
RCRA waste number P042
Renagladin
Renaglandin
Renaglandulin
Renaleptine
Renalina
Renoform
Renostypricin
Renostypticin
Renostyptin
Scurenaline
Sindrenina
Soladren
Sphygmogenin
Stryptirenal
Styptirenal
Supracapsulin
Supradin
Supranefran
Supranephrane
Supranephrine
Supranol
Suprarenaline
Suprarenin
Suprel
Surenine
Surrenine
Susphrine
Sympathin I
Takamina
Takamine
Tokamina
Tonogen
Vaponefrin
Vasoconstrictine
Vasoconstrictor
Vasodrine
Vasoton
Vasotonin

51-44-5
$C_7H_4Cl_2O_2$
191.01
O=C(O)c(ccc(c1Cl)Cl)c1
Benzoic acid, 3,4-dichloro
3,4-Dichlorobenzoic acid
Synstigmine
Syntostigmin
Syntostigmin (injection)
Vagostigmin

51-45-6
$C_5H_9N_3$
111.17
N(C(=CN1)CCN)=C1
Histamine
β-Aminoethylglyoxaline
β-Aminoethylimidazole
Eramin
Ergamine
Ergotidine
Ethylamine, 2-imidazol-4-yl-
Free Histamine
Imidazole, 4-(2-aminoethyl)-
1H-Imidazole-4-ethanamine
Imidazole-4-ethylamine
4-Imidazoleethylamine
5-Imidazoleethylamine
β-Imidazolyl-4-ethylamine
2-(4-Imidazolyl)ethylamine
Theramine

51-48-9
$C_{15}H_{11}I_4NO_4$
776.87
NC(Cc2cc(I)c(Oc1cc(I)c(O)c(I)c1)c(I)c2)C(O)=O
L-Tyrosine, o-(4-hydroxy-3,5-diiodophenyl)-3,5-diiodo
Levothyroxine
L-T4
T4
T4 (Hormone)
Tetraiodothyronine
THX
Thyreoideum
Thyroxin
L-Thyroxin
Thyroxine
(-)-Thyroxine
L-Thyroxine
Thyroxine, L-

51-50-3
$C_{16}H_{18}ClN$
259.80
ClCCN(Cc1ccccc1)Cc2ccccc2
Dibenzylamine, N-(2-chloroethyl)
N-(2-Chloroethyl)dibenzylamine
Dibenamine
Dibenzyl chlorethylamine
N,N-Dibenzyl-β-chloroethylamine
Sympatholytin

51-52-5
$C_7H_{10}N_2OS$
170.25
n(c(O)cc(n1)CCC)c1S
Uracil, 6-propyl-2-thio
2,3-Dihydro-6-propyl-2-thioxo-4(1H)-pyrimidinone
2-Mercapto-4-hydroxy-6-n-propylpyrimidine
2-Mercapto-6-propyl-4-pyrimidone
2-Mercapto-6-propylpyrimid-4-one
Procasil
Propacil
Propilthiouracil
6-Propil-tiouracile (Italian)
Propycil
6-Propyl-2-thio-2,4(1H,3H)pyrimidinedione
Propyl-thiorist
Propyl-thiorit
Propylthiouracil
4-Propyl-2-thiouracil
6-Propylthiouracil
6-Propyl-2-thiouracil
6-n-Propylthiouracil
6-n-Propyl-2-thiouracil
Prothycil
Propyl-thyracil
Propythiouracil
Prothiucil
Prothiurone
Prothyran
Protiural
PTU
PTU (Thyreostatic)
4(1H)-Pyrimidinone, 2,3-dihydro-6-propyl-2-thioxo-
Tegretol
2-Thio-4-oxo-6-propyl-1,3-pyrimidine
2-Thio-6-propyl-1,3-pyrimidin-4-one
6-Thio-4-propyluracil
Thiuragyl
Thyreostat II
T 72

51-55-8
$C_{17}H_{23}NO_3$
289.41
O=C(OC(CC(N(C1C2)C)C2)C1)C(c(cccc3)c3)CO
Atropine
Atropin (German)
Atropina (Italian)
Atropin-flexiolen
Atropinol
Atropisol
Eyesules
α-(Hydroxymethyl)benzeneacetic acid 8-methyl-8-azabicyclo(3.2.1)oct-3-yl ester
dl-Hyoscyamine
Isopto-Atropine
2-Phenylhydracrylic acid 3-α-tropanyl ester
β-Phenyl-γ-oxypropionsaeure-tropyl-ester (German)
1-α-H,5-α-H-Tropan-3-α-ol (+-)-tropate (ester) (8CI)
DL-Tropanyl 2-hydroxy-1-phenylpropionate
Tropic acid, ester with tropine
Tropic acid, 3-α-tropanyl ester
Tropine tropate
Tropine, tropate (ester)
(+,-)-Tropyl tropate
dl-Tropyltropate

51-56-9
$C_{16}H_{21}NO_3.BrH$
356.30
1-α-H,5-α-H-Tropan-3-α-ol, mandelate (ester), hydrobromide
(+-)-Homatropine bromide
Homatropine hydrobromide

51-61-6
$C_8H_{11}NO_2$
153.20
NCCc1ccc(O)c(O)c1
Pyrocatechol, 4-(2-aminoethyl)
4-(2-Aminoethyl)pyrocatechol
Dopamine
3-Hydroxytyramine

51-63-8
$C_{18}H_{26}N_2.H_2O_4S$
368.54
CC(N)Cc1ccccc1.OS(O)(=O)=O
Phenethylamine, α-methyl-, sulfate (2:1), (+)
Acedron
Adjudets
Adrizine
Afatin
Albemap
Algo-Dex
Amdex
d-Amfetasul
Amitrene
Amphaetex
Amphedrine
Ampherex
(+)-Amphetamine sulfate
d-Amphetamine sulfate
Amphetasul
Amphex
Amptrerex
Amsustain
Apetain
Ardex
d-Ate phenyl 747
d-Benzedrine sulfate
Benzeneethanamine, α-methyl-, (S)-, sulfate (2:1)
Betafedrina
Betafedrine
d-Betaphedrine
Carrtime
Cradex
Dadex
Dadox D-Citramine
DAMS
DAS
Dellipsoids
Dephadren
Desoxyn
Dexaime
Dexaline
Dexalme
Dexalone
Dexamed
Dexamine
Dexamphamine
Dexamphetamine
Dexamphetamine sulfate
Dexamyl
Dexedrina
Dexedrine
Dexedrine sulfate
Dexies
Dex OB
Dexoval
Dex-Sule
Dexten
Dextenal
Dextroamphetamine
Dextroamphetamine sulfate
Dextroanfetamina
Dextro-profetamine
Dextrosule
Diocurb
Diphylets
Ditab
Domafate
Dura Dex
Dynaphenyl
Elastonon
Ephadren
Evrodex

Fastballs
Hearts
Hetamine
Lentanet
Lowedex
Maxiton
Maxiton sulfate
Medex
Metaphyllin
α-Methylbenzeneethanamine sulfate
(+)-α-Methylphenethylamine sulfate (2:1)
α-Methylphenethylamine sulfate, d-form
d-α-Methylphenethylamine sulfate
dextro-α-Methylphenethyl-amine sulfate
Nilox
Obesedrin
Obesonil
Oranges
Pellcafs
Pellcap
Pellcaps
Perke
Phenopromin
Phenethylamine,α-methyl-, sulfate, (+)
d-1-Phenyl-2-aminopropane sulfate
dextro-1-Phenyl-2-amino-propane sulfate
d-β-Phenylisopropylamine sulfate
dextro-β-Phenylisopropylamine sulfate
Phetadex
Pomadex
Pro-Dexter
Psychodrine
Recordati
Revidex
Simpamina-D
Sympamin
Sympamina-D
Tempodex
Tuphetamine
Tydex
Zamine

51-64-9
$C_9H_{13}N$
135.23
NC(C)Cc(cccc1)c1
Phenethylamine, α-methyl-, (+)
d-AM
d-2-Amino-1-phenylpropane
(+)-Amphetamine
d-Amphetamine
Amsustain
Dephadren
Dexamphetamine
Dexedrine
α-Methylphenethylamine, d-Form
Phenethylamine, α-methyl-, d-
Phenethylamine, α-methyl-, D-
d-1-Phenyl-2-aminopropan (German)
d-1-Phenyl-2-aminopropane

51-65-0
$C_9H_{10}FNO_2$
183.20
O=C(O)C(N)Cc(ccc(F)c1)c1
Alanine, 3-(p-fluorophenyl)-, DL
Alnasid
Fluorophenylalanine
D,L-Fluorophenylalanine
D,L-p-Fluorophenylalanine
DL-4-Fluorophenylalanine
p-Fluorophenylalanine
FPA
DL-Phenylalanine, 4-fluoro- (9CI)

51-66-1
$C_9H_{11}NO_2$
165.21
O=C(Nc(ccc(OC)c1)c1)C
p-Acetanisidide
Acetamide, N-(4-methoxy-phenyl)- (9CI)
Acetanilide, 4'-methoxy-
p-Acetanisidine
Aceto-p-anisidide
Acetyl-p-anisidine
Metacetin
Methacetin
p-Methoxyacetanilide
4-Methoxyacetanilide
4'-Methoxyacetanilide

51-67-2
$C_8H_{11}NO$
137.20
NCCc1ccc(O)cc1
Phenol, p-(2-aminoethyl)
p-β-Aminoethylphenol
p-Hydroxyphenethylamine
p-Hydroxy-β-phenethylamine
4-Hydroxyphenethylamine
α-(4-Hydroxyphenyl)-β-amino-ethane
β-Hydroxyphenylethylamine
p-Hydroxyphenylethylamine
2-(p-Hydroxyphenyl)ethylamine
4-Hydroxyphenylethylamine
Phenethylamine, p-hydroxy-
Systogene
Tenosin-wirkstoff
Tocosine
Tyramine
p-Tyramine
Tyrosamine
Uteramine

51-71-8
$C_8H_{12}N_2$
136.22
NNCCc1ccccc1
Hydrazine, phenethyl
Fenelzyna (Polish)
Fenelzyne (Polish)
Hydrazine, (2-phenylethyl)-
1-Hydrazino-2-phenylethane
Nardil
Phenelzine
Phenethylhydrazine
β-Phenylethylhydrazine
2-Phenylethylhydrazine
Stinerval
W1544
W 1544-A

51-75-2
$C_5H_{11}Cl_2N$
156.07
CN(CCCl)CCCl
Diethylamine, 2,2'-dichloro-N-methyl
Bis(β-chloroethyl)methylamine
Bis(2-chloroethyl)methylamine
N,N-Bis(2-chloroethyl)methyl-amine
Caryolysin
Chlorethazine
Chlormethine
2-Chloro-N-(2-chloroethyl)-N-methylethanamine
Cloramin
Dichlor amine
Dichloren (German)
β,β'-Dichlorodiethyl-N-methyl-amine
2,2'-Dichlorodiethyl-methyl-amine
Di(2-chloroethyl)methylamine
N,N-Di(chloroethyl)methyl-amine
2,2'-Dichloro-N-methyldiethyl-amine
Embichin
ENT 25,294
HN2
MBA
Mechlorethamine
Mecloretamina (Italian)
Methylbis(β-chloroethyl)amine
Methylbis(2-chloroethyl)amine
N-Methyl-bis-chloraethylamin (German)
N-Methyl-bis(β-chloroethyl)-amine
N-Methyl-bis(2-chloroethyl)-amine
N-Methyl-2,2'-dichlorodiethyl-amine
Methyldi(2-chloroethyl)amine
N-Methyl-lost (German)
Mustargen
Mustine
Mutagen
Nitrogen mustard
NSC-762
T-1024
TL 146

51-78-5
$C_6H_7NO.ClH$
145.60
Phenol, p-amino-, hydrochlor-ide
p-Aminophenol hydrochloride
4-Aminophenol hydrochloride

51-79-6
$C_3H_7NO_2$
89.11
O=C(OCC)N
Carbamic acid, ethyl ester
A 11032
Aethylcarbamat (German)
Aethylurethan (German)
Carbamidsaeure-aethylester (German)
Ethyl carbamate
Ethylester kyseliny karbaminove (Czech)
Ethylurethan
Ethyl urethane
o-Ethylurethane
Leucethane
Leucothane
NSC-746
Pracarbamin
Pracarbamine
RCRA waste number U238
U-Compound
Uretan
Uretan etylowy (Polish)
Urethan
Urethane
X 41

51-80-9
$C_5H_{14}N_2$
102.17
N(CN(C)C)(C)C
Methanediamine, N,N,N',N'-tetramethyl- (8CI,9CI)
AI3-26640
Bis(dimethylamino)methane
Dimethyl((dimethylamino)-methyl)amine
Methylenebis(dimethylamine)
Methylenediamine, N,N,N',N'-tetramethyl-
NA 9069
NSC-166169
Tetramethyl methylene diamine
Tetramethylmethylenediamine
N,N,N',N'-Tetramethyldiaminomethan (German)
N,N,N',N'-Tetramethyldiaminomethane
N,N,N',N'-Tetramethylmethanediamine
N,N,N',N'-Tetramethylmethylenediamine

51-82-1
$C_6H_{14}N_2S_2$
178.34
N(CSC(N(C)C)=S)(C)C
Carbamic acid, dithio-, N,N-dimethyl-, dimethylaminomethyl ester
N,N-Dimethyldithiocarbamic acid dimethylaminomethyl ester
N,N-Dimethyl-dithiocarbaminsaeure-dimethylamino-methyl-ester (German)

51-83-2
$C_6H_{15}ClN_2O_2$
182.64
Carbachol
2-((Aminocarbonyl)oxy)-N,N,N-trimethylethanaminium chloride
Carbachol chloride
Carbachol hydrochloride
Carbacholin
Carbacholine
Carbacholine chloride
Carbacholini chloridum
Carbacholinium chloratum
Carbacholinum
Carbacholum (Latin)
Carbacholum chloratum
Carbacol (Spanish)
Carbacolina
Carbacolo
Carbamic acid, ester with choline chloride
Carbaminocholine chloride
Carbaminoylcholine chloride
Carbamiotin
γ-Carbamoyl choline chloride
Carbamoylcholine chloride
Carbamoylcholine-hydrochloride
(2-Carbamoyloxyethyl)trimethylammonium chloride
Carbamylcholine chloride
Carbochol
Carbocholin
Carbocholine
Carbyl
Carcholin
CB
Choline carbamate chloride
Choline, chloride, carbamate (8CI)
Choline, chloride carbamate (Ester)
Choline, chloride, carbamate, hydrochloride
Choline chloride, carbamoyl-
Coletyl
Doryl (VAN)
Doryl (Pharmaceutical)
Ethanaminium, 2-(aminocarbonyl)oxy-N,N,N-trimethyl-, chloride
Ethanaminium, 2-((aminocarbonyl)oxy)-N,N,N-trimethyl-, chloride (9CI)
(2-Hydroxyethyl)trimethyl ammonium chloride carbamate
Isopto Carbachol
Jestryl
Karbachol (Czech)
Karbamoylcholin chlorid (Czech)
Lentin
Lentine (French)
Miostat
Mistura C
Moryl
NSC-32865
P.V. Carbachol
Rilentol
TL 457
Vasoperif

51-98-9
$C_{22}H_{28}O_3$
340.50
CC(=O)OC3(CCC4C2CCC1=CC(=O)CCC1C2CCC34C)C#C
19-Nor-17-α-pregn-4-en-20-yn-3-one, 17-acetoxy
17-Acetoxy-19-nor-17-α-pregn-4-en-20-yn-3-one
17-β-Acetoxy-19-nor-17-α-pregn-4-en-20-yn-3-one
(17-α)-17-(Acetyloxy)-19-norpregn-4-en-20-yn-3-one
17-Acetyloxy(17-α)-19-norpregn-4-estren-17-β-ol-acetate-3-one
17-ENT
ENTA
17-α-Ethinyl-19-nortestosterone acetate
17-α-Ethinyl-19-nortestosterone-17-β-acetate
17-α-Ethynyl-17-β-acetoxy-19-norandrost-4-en-3-one
17-α-Ethynyl-17-hydroxyestr-4-en-3-one acetate
17-α-Ethynyl-19-nortestosterone acetate
17-Hydroxy-19-nor-17-α-pregn-4-en-20-yn-3-one acetate
17-β-Hydroxy-19-nor-17-α-pregn-4-en-20-yn-3-one acetate
Norethindrone acetate
Norethindrone 17-acetate
Norethisteron acetate
Norethisterone acetate
19-Norethisterone acetate
Norethynyltestosterone acetate
19-Norethynyltestosterone acetate
Norethysterone acetate
Norlutate
Norlutine acetate
Orlutate

52-01-7
$C_{24}H_{32}O_4S$
416.62
O=C(OC(C(C(C(C(C(C(=CC(=O)C1)C2)(C1)C)C3)C2SC(=O)C)C4)(C3)C)(C4)C5)C5
17-α-Pregn-4-ene-21-carboxylic acid, 17-hydroxy-7-α-mercapto-3-oxo-, γ- lactone acetate
7-α-Acetylthio-3-oxo-17-α-pregn-4-ene-21,17-β-carbolactone
7-α-Acetylthio-3-oxo-17-β-pregn-4-ene-21,17-β-carbolactone
Aldactazide
Aldactide
Aldactone
Aldactone A

3-(3-Keto-7-α-acetylthio-17-β-hydroxy-4-androsten-17-α-yl)propionic acid lactone
Osiren
Osyrol
3'-(3-Oxo-7-α-acetylthio-17-β-hydroxyandrost-4-en-17-β-yl)propionic acid lactone
17-α-Pregn-4-ene-21-carboxylic acid, 1-hydroxy-7-α-mercapto-3-oxo-α-lactone
SC 9420
SC 15983
Spiresis
Spiridon
Spiroctanie
Spiro(17H-cyclopenta(a)phenanthrene-17,2'(5'H)furan), pregn-4-ene-21-carboxylic acid deriv.
Spiro(17H-cyclopenta(a)phenanthrene-17,2'-(3'H)-furan)
Spirolactone
Spirolakton
Spirolang
Spirone
Spironolactone
Spironolactone A
Uractone
Verospiron
Verospirone

52-24-4
$C_6H_{12}N_3PS$
189.24
S=P(N1CC1)(N2CC2)N3CC3
Phosphine sulfide, tris(1-aziridinyl)
Aziridine, 1,1,1''-phosphinothioylidynetris-
CBC 806495
Girostan
NCI-C01649
NSC-6396
Oncotepa
Oncothio-tepa
Oncotiotepa
1,1',1''-Phosphinothioylidynetrisaziridine
Phosphorothioic acid triethylenetriamide
Phosphorothioic triamide, N,N',N''-tri-1,2-ethanediyl-
Phosphorothioic triamide, N,N',N''-triethylene-
SK 6882
Stepa
Tespa
Tespamin
Tespamine
Thiofozil
Thiophosphamide
Thiophosphoramide, N,N',N''-tri-1,2-ethanediyl-
Thiophosphoramide, N,N',N''-triethylene-
Thiotef
Thio-TEP
Thio-TEPA
Thio-TEPA S
Thiotriethylenephosphoramide
Tifosyl
Tiofosfamid
Tiofosyl
Tiofozil
Tio TEF
Triaziridinylphosphine sulfide
N,N',N''-Tri-1,2-ethanediylphosphorothioic triamide
N,N',N''-Tri-1,2-ethanediylthiophosphoramide
Tri(ethyleneimino)thiophosphoramide
N,N',N''-Triethylenephosphorothioic triamide
Triethylene thiophosphoramide
N,N',N''-Triethylenethiophosphamide
Triethylenethiophosphoramide
N,N',N''-Triethylenethiophosphoramide
Triethylenethiophosphorotriamide
Tris(1-aziridinyl)phosphine sulfide
Tris(1-aziridinyl)phosphine sulphide
Tris(ethylenimino)thiophosphate
TSPA

52-28-8
$C_{18}H_{21}NO_3.H_3O_4P$
397.40
Morphinan-6-α-ol, 7,8-didehydro-4,5-α-epoxy-3-methoxy-17-methyl-, phosphate (1:1)
Codeine phosphate

52-31-3
$C_{12}H_{16}N_2O_3$
236.30
CCC1(C(=O)NC(=O)NC1=O)C2=CCCCC2
Barbituric acid, 5-(1-cyclohexen-1-yl)-5-ethyl
Adorm
Amnosed
Cavonyl
Cyclobarbital
Cyclobarbitol
Cyclobarbiton
Cyclobarbitone
Cyclodorm
Cyclohexenyl-ethyl barbituric acid
5-(1-Cyclohexenyl)-5-ethylbarbituric acid
5-(1-Cyclohexen-1-yl)-5-ethylbarbituric acid
5-(1-Cyclohexen-1-yl)-5-ethyl-2,4,6(1H,3H,5H)-pyrimidinetrione
Cyklodorm
5-Ethyl-5-cyclohexenylbarbituric acid
Ethylhexabital
Fanodormo
Hexemal
Irifan
Namuron
Palinum
Phanodorm
Phanodorn
Philodorm
Pralumin
Pro-Sonil
Sonaform
Tetrahydrophenobarbital

52-43-7
$C_{10}H_{12}N_2O_3$
208.24
C=CCC1(CC=C)C(=O)NC(=O)NC1=O
Barbituric acid, 5,5-diallyl
Allbarbital
Allobarbital
Allobarbitone
Allybarbitural
Alnox
Alobarbital
Barballyl
Barbidal
Curral
Díadol
Dial
Dial (barbiturate)
Diallylbarbital
Diallylbarbituric acid
5,5-Diallylbarbituric acid
Diallymal
Dorm
Dormallyl
Malil
Malilum
Novallyl
NSC-9324
5,5-Di-2-propenyl-2,4,6(1H,3H,5H)-pyrimidinetrione
2,4,6(1H,3H,5H)-Pyrimidinetrione, 5,5-di-2-propenyl-(9CI)

52-46-0
$C_{12}H_{24}N_9P_3$
387.36
C1CN1P2(=NP(=NP(=N2)(N3CC3)N4CC4)(N5CC5)N6CC6)N7CC7
1,3,5,2,4,6-Triazatriphosphorine, 2,2,4,4,6,6-hexakis-(1-aziridinyl)-2,2,4,4,6,6-hexahydro
Afolat (Czech)
Apholate
APN
Aziridine, 1,3,5,2,4,6-triazatriphosphorine derivative
1-Aziridinylphosphonitrile trimer
ENT 26,316
Hexa(1-aziridinyl)triphosphotriazine
2,2,4,4,6,6-Hexahydro-2,2,4,4,-6,6-hexakis(1-aziridinyl)-1,3,5,2,4,6-triazatriphosphorine
2,2,4,4,6,6-Hexakis(1-aziridinyl)cyclotriphosphaza-1,3,5-triene
2,2,4,4,6,6-Hexakis(1-aziri-

dinyl)-2,2,4,4,6,6-hexahydro-1,3,5,2,4,6-triazatri-phosphorine
Hexakis-(1-aziridinyl)phospho-nitrile
Hexakis(aziridinyl)phospho-triazine
NSC-26812
Olin MO. 2174
Pholate
PN6
SQ 8388

52-51-7
$C_3H_6BrNO_4$
199.98
O=N(=O)C(Br)(CO)CO
Bronopol
AI3-61639
2-Bromo-2-nitropropan-1,3-diol
2-Bromo-2-nitro-1,3-propane-diol
2-Bromo-2-nitropropane-1,3-diol
β-Bromo-β-nitrotrimethylene-glycol
Bronocot
Bronopolu (Polish)
Bronopolum (Latin)
Bronosol
Caswell No. 116A
EPA Pesticide Chemical Code 216400
Lexgard bronopol
2-Nitro-2-bromo-1,3-propane-diol
NSC-141021
Onyxide 500
1,3-Propanediol, 2-bromo-2-nitro- (8CI,9CI)

52-52-8
$C_6H_{11}NO_2$
129.18
O=C(O)C(N)(CCC1)C1
Cyclopentanecarboxylic acid, 1-amino
ACPC
1-Aminocyclopentane-1-car-boxylic acid
1-Amino-1-cyclopentanecarbox-ylic acid
CB 1639
Cycloleucin
Cycloleucine
Cyclopentanecarboxylic acid, 1-amino-, 1-
NSC-1026
WR 14,997
X 201

52-53-9
$C_{27}H_{38}N_2O_4$
454.67
COc2ccc(CCN(C)CCCC(C#N)(C(C)C)c1ccc(OC)c(OC)c1)cc2OC
Valeronitrile, 5-((3,4-di-methoxyphenethyl)methyl-amino)-2-(3,4-dimethoxy-phenyl)-2-isopropyl
Cordilox
CP-16533-1
D-365
Dilacoran
5-((3,4-Dimethoxyphenethyl)-methylamino)-2-(3,4-dimeth-oxyphenyl)-2-isopropylvalero-nitrile
Iproveratril
Isoptin
α-((N-Methyl-N-homoveratryl)-γ-aminopropyl)-3,4-dimeth-oxyphenylacetonitrile
Vasolan
Verapamil

52-67-5
$C_5H_{11}NO_2S$
149.23
CC(C)(S)C(N)C(O)=O
Valine, 3-mercapto-, D
Cuprenil
Cuprimine
Depen
Dimethylcysteine
β,β-Dimethylcysteine
D-β,β-Dimethylcysteine
D-Mercaptovaline
D-3-Mercaptovaline
Metalcaptase
PCA
D-Penamine
Penicillamin
Penicillamine
D-Penicillamine
(S)-Penicillamin
Reduced penicillamine
Reduced D-penicillamine
D-β-Thiovaline
Trolovol
D-Valine, 3-mercapto-

52-68-6
$C_4H_8Cl_3O_4P$
257.44
COP(=O)(OC)C(O)C(Cl)(Cl)Cl
Phosphonic acid, (2,2,2-tri-chloro-1-hydroxyethyl)-, dimethyl ester
Aerol 1 (Pesticide)
Aerol 1
Agroforotox
Anthon
Bay 15922
Bayer 15922
Bayer L 13/59
Bilarcil
Bovinox
Britten
Briton
Cekufon
Chlorfos
Chlorak
Chlorofos
Chloroftalm
Chlorophos
Chlorophthalm
Chloroxyphos
Ciclosom
Clorofos (Russian)
Chlorophose
Combot
Combot Equine
Danex
DEP
DEP (pesticide)
Depthon
DETF
Dimethoxy-2,2,2-trichloro-1-hydroxy-ethyl-phosphine oxide
O,O-Dimethyl-(1-hydroxy-2,2,2-trichloraethyl)phosphon-saeure ester (German)
O,O-Dimethyl-(1-hydroxy-2,2,2-trichlorathyl)-phosphat (German)
O,O-Dimethyl-(1-hydroxy-2,2,2-trichloro)ethyl phosphate
Dimethyl 1-hydroxy-2,2,2-trichloroethyl phosphon-ate
O,O-Dimethyl (1-hydroxy-2,2,2-trichloroethyl)phosphon-ate
O,O-Dimethyl-1-oxy-2,2,2-tri-chloroethyl phosphonate
O,O-Dimethyl-(2,2,2-trichloor-1-hydroxy-ethyl)-fosfonaat (Dutch)
O,O-Dimethyl-(2,2,2-trichlor-1-hydroxy-aethyl)phosphonat (German)
Dimethyltrichlorohydroxyethyl phosphonate
Dimethyl 2,2,2-trichloro-1-hydroxyethylphosphonate
O,O-Dimethyl-2,2,2-trichloro-1-hydroxyethyl phosphonate
O,O Dimetil 2,2,2-trichloro 1 hidroxietil fosfonato (Portugese)
O,O-Dimetil-(2,2,2-tricloro-1-idrossi-etil)-fosfonato (Italian)
Dimetox
Dipterax
Dipterex
Dipterex 50
Diptevur
Ditrifon
Dylox
Dylox-Metasystox-R
Dyrex
Dyvon
ENT 19,763
Equino-Acid
Equino-Aid
Flibol E
Fliegenteller
Forotox
Foschlor
Foschlor 25
Foschlorem (Polish)
Foschlor R
Foschlor R-50
1-Hydroxy-2,2,2-trichloro-ethyle phosphonate de di-methyle (French)
1-Hydroxy-2,2,2-trichloroethyl-

phosphonic acid dimethyl ester
Hypodermacid
Leivasom
Loisol
Masoten
Mazoten
Methyl chlorophos
Metifonate
Metrifonate
Metriphonate
NCI-C54831
Neguvon
Neguvon A
Phoschlor
Phoschlor R50
Phosphonic acid, (1-hydroxy-2,2,2-trichloroethyl)-, dimethyl ester
Polfoschlor
Proxol
Ricifon
Ritsifon
Satox 20WSC
Soldep
Sotipox
TCF
Trichloorfon (Dutch)
Trichlorfon (German)
Trichlorfon [UN 2783]
2,2,2-Trichloroethyl dimethyl phosphate
Trichlorofon
((2,2,2-Trichloro-1-hydroxyethyl) dimethylphosphonate)
2,2,2-Trichloro-1-hydroxyethylphosphonate, dimethyl ester
(2,2,2-Trichloro-1-hydroxyethyl)phosphonic acid dimethyl ester
Trichloropho
Trichlorphene
Trichlorphon
Trichlorphon FN
Trinex
Tugon
Tugon Fly Bait
Tugon Stable Spray
Vermicide Bayer 2349
UN 2783 [Organophosphorus pesticides, solid, toxic, N.O.S.]
Volfartol
Votexit
WEC 50
Wotexit

52-76-6
$C_{20}H_{28}O$
284.48
CC34CCC1C(CCC2=CCCCC12)C3CCC4(O)C#C
19-Nor-17-α-pregn-4-en-20-yn-17-ol
3-Desoxynorlutin
Ethinylestrenol
δ^4-17-α-Ethinylestren-17-β-ol
17-α-Ethinyl-17-β-hydroxyestr-4-ene
17-α-Ethinyl-17-β-hydroxyo-estr-4-ene
Ethinyloestranol
Ethinyl oestrenol
δ^4-17-α-Ethinyloestren-17-β-ol
17-α-Ethynil-δ-4-estrene-17-β-ol
Ethynylestrenol
17-α-Ethynylestrenol
17-α-Ethynylestr-4-en-17-β-ol
Ethynloestrenol
17-α-Ethynyloestrenol
17-α-Ethynyloestr-4-en-17-β-ol
Exluten
Exlution
Exluton
Exlutona
Linestrenol
Lynenol
Lynestrenol
Lynoestrenol
(17-α)-19-Norpregn-4-en-20-yn-17-ol
NSC-37725
ORG 485-50
Orgametil
Orgametril
Orgametrol

52-85-7
$C_{10}H_{16}NO_5PS_2$
325.36
O=S(=O)(N(C)C)c(ccc(OP(OC)(OC)=S)c1)c1
Phosphorothioic acid, O,O-dimethyl ester, o-ester with p-hydroxy-N,N-dimethylbenzene- sulfonamide
AC 38023
American Cyanamid-38023
American Cyanamid CL-38,023
Bo-Ana
CL-38023
Cyflee
O-(4-((Dimethylamino)sulfonyl)phenyl) O,O-dimethyl phosphorothioate
O,O-Dimethyl O-(p-(N,N-dimethylsulfamoyl)phenyl) phosphorothioate
O-4-Dimethylsulfamoylphenyl O,O-dimethyl phosphorothioate
Dovip
ENT 25,644
Famfos
Famofos
Famophos
Famophos warbex
Famphos
Famphur
Fanfos
RCRA waste number P097
Warbex
Warbexol

52-86-8
$C_{21}H_{23}ClFNO_2$
375.90
OC2(CCN(CCCC(=O)c1ccc(F)cc1)CC2)c3ccc(Cl)cc3
Butyrophenone, 4-(4-(p-chlorophenyl)-4-hydroxypiperidino)-4'-fluoro
Aldo
Aloperidin
Aloperidolo
Aloperidon
Brotopon
1-Butanone, 4-(4-(4-chlorophenyl)-4-hydroxy-1-piperidinyl)-1-(4-fluorophenyl)-
Butyrophenone, 4'-fluoro-4-(4-(p-chlorophenyl)-4-hydroxypiperidino)-
γ-(4-(p-Chlorphenyl)-4-hydroxpiperidino)-p-fluorbutyrophenone
4-(4-(4-Chlorophenyl)-4-hydroxy-1-piperidinyl)-1-(4-fluorophenyl)-1-butanone
Einalon S
Eukystol
1-(3-p-Fluorobenzoylpropyl)-4-p-chlorophenyl-4-hydroxypiperidine
4'-Fluoro-4-(4-(p-chlorophenyl)-4-hydroxypiperidinyl)butyrophenone
4'-Fluoro-4-(4-hydroxy-4-(4'-chlorophenyl)piperidino)-butyrophenone
Galoperidol
Haldol
Halidol
Halol
Halopal
Haloperidol
Halopidol
Halopoidol
Halosten
4-(4-Hydroxy-4'-chloro-4-phenylpiperidino)-4'-fluorobutyrophenone
Keselan
Lealgin compositum
Linton
McN-JR-1625
Peluces
Pernox
R 1625
Serenace
Serenase
Serenelfi
Sernas
Sernel
Uliolind
Ulcolind
Vesalium

52-89-1
$C_3H_7NO_2S.ClH$
157.63
Cysteine, hydrochloride, L
Cysteine chlorhydrate
Cysteine hydrochloride
l-Cysteine hydrochloride
l-Cystein-hydrochloride
l-Cysteine monohydrochloride

52-90-4
$C_3H_7NO_2S$

121.17
O=C(O)C(N)CS
Cysteine, L
Cystein
Cysteine
L-Cysteine (9CI)
L-(+)-Cysteine
Half-Cysteine
Half-Cystine
β-Mercaptoalanine
Thioserine

53-03-2
$C_{21}H_{26}O_5$
358.47
O=C(C(O)(C(C(C(C(C1=O)C(C(=CC(=O)C=2)C3)(C2)C)C3)C4)(C1)C)C4)CO
Pregna-1,4-diene-3,11,20-trione, 17,21-hydroxy
Ancortone
Bicortone
Colisone
Cortan
Cortancyl
δ-Cortelan
Cortidelt
δ-Cortisone
δ-1-Cortisone
δ^1-Cortisone
δ-Cortone
Cotone
Dacortin
Decortancyl
Decortin
Decortisyl
δ-1-Dehydrocortisone
1-Dehydrocortisone
1,2-Dehydrocortisone
Dekortin
δ Cortelan
Deltacortisone
Deltacortone
Delta-Dome
δ E
Deltasone
Deltison
Deltisone
Deltra
Di-Adreson
17,21-Dihydroxypregna-1,4-diene-3,11,20-trione
Encorton
Encortone
Enkorton
Hostacortin
In-Sone
Juvason
Lisacort
Metacortandracin
Meticorten
NCI-C04897
NSC-10023
Orasone
Paracort
Precort
Prednicen-M
Prednilonga
Prednison
Prednisone
Prednizon
1,4-Pregnadiene-17-α,21-diol-3,11,20-trione
Rectodelt
Servisone
SK-Prednisone
δ-Sone
Supercortil
U 6020
Ultracorten
Ultracortene
Wojtab
Zenadrid
Zenadrid (Veterinary)

53-06-5
$C_{21}H_{28}O_5$
360.49
CC13CCC(=O)C=C1CCC4C2CCC(O)(C(=O)CO)C2(C)CC(=O)C34
Cortisone
Adrenalex
Compound E
Corlin
Cortadren
Cortisal
Cortisate
Cortistal
Cortivite
Cortogen
Cortone
11-Dehydro-17-hydroxycorticosterone
17α,21-Dihydroxypregn-4-ene-3,11,20-trione
17α,21-Dihydroxy-4-pregnene-3,11,20-trione
17-Hydroxy-11-dehydrocorticosterone
17-α-Hydroxy-11-dehydrocorticosterone
Incortin
KE
Kendall's Compound E
Pregn-4-en-17α,21-diol-3,11,20-trione
δ^4-Pregnene-17α,21-diol-3,11,20-trione
4-Pregnene-17α,21-diol-3,11,20-trione
Pregn-4-ene-3,11,20-trione, 17,21-dihydroxy-
Reichstein's Substance FA
Scheroson
Wintersteiner's Compound F

53-16-7
$C_{18}H_{22}O_2$
270.40
O=C(C(C(C(C(c(c(cc(O)c1)C2)c1)C3)C2)C4)(C3)C)C4
Estrone
Aquacrine
Crinovaryl
Cristallovar
Crystogen
δ-1,3,5-Estratrien-3-β-ol-17-one
δ-1,3,5-Oestratrien-3-β-ol-17-one
Destrone
Disynformon
E_1
Endofolliculina
Esterone
1,3,5-Estratrien-3-ol-17-one
1,3,5(10)-Estratrien-3-ol-17-one
Estra-1,3,5(10)-trien-17-one, 3-hydroxy-
Estrin
Estrol
Estron
Estrona (Spanish)
Estrone-A
Estrugenone
Estrusol
Femestrone injection
Femidyn
Folikrin
Folipex
Folisan
Follestrine
Follicular hormone
Folliculin
Folliculine
Folliculine benzoate
Follicunodis
Follidrin
Glandubolin
Hiestrone
Hormofollin
Hormovarine
3-Hydroxyestra-1,3,5(10)-trien-17-one
3-Hydroxy-17-keto-estra-1,3,5-triene
3-Hydroxy-17-keto-oestra-1,3,5-triene
3-Hydroxy-oestra-1,3,5(10)-trien-17-one
3-Hydroxy-1,3,5(10)-oestratrien-17-one
Kestrone
Ketodestrin
Ketohydroxy-estratriene
Ketohydroxyestrin
Ketohydroxyoestrin
Kolpon
Menagen
Menformon
1,3,5-Oestratrien-3-ol-17-one
1,3,5(10)-Oestratrien-3-ol-17-one
Oestrin
Oestroform
Oestrone
Oestroperos
Ovex
Ovifollin
Perlatan
Solliculin
Theelin
Thelestrin
Thelykinin
Thynestron
Tokokin
Unden
Wynestron

53-19-0
$C_{14}H_{10}Cl_4$
320.04

ClC(Cl)C(c1ccc(Cl)cc1)c2ccc
cc2Cl
Ethane, 2-(o-chlorophenyl)-2-(p-chlorophenyl)-1,1-dichloro
Benzene, 1-chloro-2-(2,2-dichloro-1-(4-chlorophenyl)-ethyl)-
Chloditan
Chlodithane
2-(o-Chlorophenyl)-2-(p-chlorophenyl)-1,1-dichloroethane
o,p'-DDD
2,4'-DDD
1,1-Dichloro-2,2-bis(2,4'-dichlorophenyl)ethane
1,1-Dichloro-2-(o-chlorophenyl)-2-(p-chlorophenyl)-ethane
1,1-Dichloro-2-(p-chlorophenyl)-2-(o-chlorophenyl)-ethane
o,p'-Dichlorodiphenyldichloroethane
2,4'-Dichlorophenyldichlorethane
Mitotane
NCI-C04933
NSC-38721
o,p-TDE
o,p'-TDE

53-21-4
$C_{17}H_{21}NO_4.ClH$
339.85
1-α-H,5-α-H-Tropane-2-β-carboxylic acid, 3-β-hydroxy-, methyl ester, benzoate (ester), hydrochloride
Cocain-chlorhydrat (German)
Cocaine chloride
Cocaine hydrochloride
(-)-Cocaine hydrochloride
l-Cocaine hydrochloride
Cocaine muriate
Methyl 3-β-hydroxy-1-α-H,5-α-H-tropan-2-β-carboxylate, benzoate (ester) HCl
Sal de Merck

53-33-8
$C_{22}H_{29}FO_5$
392.47
Paramethasone
Alondra
Cassenne
Cortiden
CS 1483
Dilar
Dillar
Flumethone
6α-Fluoro-16α-methylprednisolone
Haldrate
16α-Methyl-6α-fluoroprednisolone
Metilar
Monocortin
Parametasona (Spanish)
Parametasone
Paramethasonum (Latin)
Paramezone
Pregna-1,4-diene-3,20-dione, 6-fluoro-11,17,21-trihydroxy-16-methyl-, (6α,11β,16α)-
Pregna-1,4-diene-3,20-dione, 6α-fluoro-11β,17,21-trihydroxy-16α-methyl-

53-34-9
$C_{21}H_{27}FO_5$
378.44
Fluprednisolone
Alphadrol
B 673
Corticosterone, 1-dehydro-6α-fluoro-
Etadrol
F. I. 6150
Flucort
6α-Fluoro-1-dehydrohydrocortisone
6α-Fluoro-1,4-pregnadiene-11β,17α,21-triol-3,20-dione
6α-Fluoro-11β,17,21-trihydroxypregna-1,4-diene-3,20-dione
6α-Fluoro-11β,17α,21-trihydroxypregna-1,4-diene-3,20-dione
6α-Fluoroprednisolone
Fluprednisolonum (Latin)
Glucocorticoid
Isopredon
NSC-47439
Prednisolone, 6α-fluoro-
Pregna-1,4-diene-3,20-dione, 6-fluoro-11,17,21-trihydroxy-, (6α,11β)- (9CI)
Pregna-1,4-diene-3,20-dione, 6α-fluoro-11β,17,21-trihydroxy- (8CI)
U 7800
Vladicort

53-36-1
$C_{24}H_{32}O_6$
416.56
Pregna-1,4-diene-3,20-dione, 11-β,17,21-trihydroxy-6-α-methyl-, 21-acetate
Depo-Medrate
Depo-Medrol
Depo-Medrone
Depo-Methylprednisolone
Depo-Methylprednisolone acetate
Depot-Medrol
Medrol acetate
Methylprednisolone acetate
Methylprednisolone 21-acetate
6-Methylprednisolone acetate
6-α-Methylprednisolone acetate
MPA
Pregna-1,4-diene-3,20-dione, 21-(acetyloxy)-11,17-dihydroxy-6-methyl-, (6-α, 11-β)-
U 8210
Urbason crystal suspension

53-39-4
$C_{19}H_{30}O_3$
306.49
CC3(O)CCC4C2CCC1CC(=O)O
CC1(C)C2CCC34C
2-Oxa-5-α-androstan-3-one, 17-β-hydroxy-17-methyl
Anavar
Oxandrolone
8075 C.B.
2-Oxaandrostan-3-one, 17-hydroxy-17-methyl-, (5-α, 17-β)- (9CI)
Protivar
Provitar
SC 11585
Vasorome

53-46-3
$C_{21}H_{26}NO_3.Br$
420.39
[OH].[Br-].CC[N+](C)(CC)CCOC
(=O)C2c1ccccc1Oc3ccccc23
Ammonium, diethyl(2-hydroxyethyl)methyl-, bromide, xanthene-9-carboxylate
Asabaine
Avagal
Banthin
Banthine
Banthine bromide
β-Diethylaminoethyl xanthene-9-carboxylate methobromide
β-Diethylaminoethyl 9-xanthenecarboxylate methobromide
Diethyl(2-hydroxyethyl)methylammoniumbromide xanthene-9-carboxylate
Doladene
Ethanaminium, N,N-diethyl-N-methyl-2-((9H-xanthen-9-ylcarbonyl)oxy)-, bromide (9CI)
Frenogastrico
Gastron
Gastrosedan
Mantheline
Metantyl
Metaxan
Methanide
Methantheline bromide
Methanthelinium bromide
Methanthine bromide
Methelina
MTB 51
Resobantin
SC 2910
Ulcine
Ulcudexter
Vagamin
Vagantin
Xanteline
Xanthene-9-carboxylic acid, ester with diethyl(2-hyhroxyethyl)methylammoniumbromide

53-57-6
$C_{21}H_{30}N_7O_{17}P_3$
745.38
O=P(OP(=O)(OCC(OC(N(c(ncnc1N)c1N=2)C2)C3OP(=O)(O)O)C3O)O)(OCC(OC(N(C=CCC=4C(=O)N)C4)C5O)C5O)O
Adenosine 5'-(trihydrogen diphosphate), 2'-(dihydrogen phosphate), 5'-5'-ester with 1,4-dihydro-1-β-D-ribofuranosyl-3-pyridinecarboxamide (9CI)
Nicotinamide adenine dinucleotide phosphate

53-60-1
$C_{17}H_{20}N_2S.ClH$
320.91
CN(C)CCCN2c1ccccc1Sc3ccccc23
Phenothiazine, 10-(3-(dimethylamino)propyl)-, monohydrochloride
10-(γ-Dimethylamino-n-propyl)-phenothiazine hydrochloride
10-(3-(Dimethylamino)propyl)-phenothiazine hydrochloride
10-(3-(Dimethylamino)propyl)-phenothiazine monohydrochloride
Esparinal
Megaphen
Phenothiazine, 10-(3-(dimethylamino)propyl)-, hydrochloride
Promazine hydrochloride
RP 4560
10H-Phenothiazine-10-propanamine, N,N-dimethyl-, monohydrochloride
Sparine hydrochloride

53-69-0
$C_{19}H_{15}N$
257.35
Benz(a)acridine, 8,10-dimethyl
5,7-Dimethyl-1,2-benzacridine

53-70-3
$C_{22}H_{14}$
278.36
c(c(c(c(c1)ccc2)c2)cc(c3c(c(c4)ccc5)c5)c4)(c1)c3
Dibenz(a,h)anthracene
1,2:5,6-Benzanthracene
DBA
DB(a,h)A
1,2,5,6-DBA
1,2,5,6-Dibenzanthraceen (Dutch)
1,2:5,6-Dibenzanthracene
1,2:5,6-Dibenz(a)anthracene
Dibenzo(a,h)anthracene
1,2:5,6-Dibenzoanthracene
RCRA waste number U063

53-86-1
$C_{19}H_{16}ClNO_4$
357.81
O=C(N(c(c(C=1CC(=O)O)cc(OC)c2)c2)C1C)c(ccc(c3)Cl)c3
Indole-3-acetic acid, 1-(p-chlorobenzoyl)-5-methoxy-2-methyl
Amuno
Artracin
Artrinovo
Artrivia
N-p-Chlorbenzoyl-5-methoxy-2-methylindole-3-acetic acid
1-(p-Chlorobenzoyl)-5-methoxy-2-methylindole-3-acetic acid
(1-p-Chlorobenzoyl-5-methoxy-2-methylindol-3-yl)acetic acid
1-(p-Chlorobenzoyl)-2-methyl-5-methoxyindole-3-acetic acid
1-(p-Chlorobenzoyl)-2-methyl-5-methoxy-3-indole-acetic acid
α-(1-(p-Chlorobenzoyl)-2-methyl-5-methoxy-3-indolyl)-acetic acid
1-p-Cloro-benzoil-5-metoxi-2-metilindol-3-acido acetico (Spanish)
Confortid
Dolovin
Idomethine
Imbrilon
Inacid
Indacin
Indocid
Indocin
Indomecol
Indomed
Indomee
Indometacin
Indometacine
Indometacyna (Polish)
Indomethacin
Indomethacine
Indomethazine
Indometicina (Spanish)
Indoptic
Indo-Rectolmin
Indo-Tablinen
Inflazon
Infrocin
Inteban SP
Kwas 1-(p-chlorobenzoilo)-2-metylo-5-metoksy-3-indolilooctowy (Polish)
Lausit
Metacen
Metartril
Methazine
Metindol
Mezolin
Mikametan
Mobilan
NCI-C56144
Reumacide
Sadoreum
Tannex

53-94-1
$C_{13}H_{11}NO$
197.25
ONc3ccc1c(Cc2ccccc12)c3
Hydroxylamine, N-fluoren-2-yl
N-Hydroxy-2-aminofluorene
2-Fluorenylhydroxylamine

53-95-2
$C_{15}H_{13}NO_2$
239.29
CC(=O)N(O)c3ccc1c(Cc2ccccc12)c3
Acetohydroxamic acid, N-fluoren-2-yl
Acetamide, N-9H-fluoren-2-yl-N-hydroxy- (9CI)
Acetamide, N-hydroxy-N-(2-fluorenyl)-
N-Acetyl-N-2-fluorenyl hydroxylamine
Fluorenyl-2-acethydroxamic acid
N-Fluoren-2-ylacetohydroxamic acid
N-2-Fluorenylacetohydroxamic acid
N-Hydroxy-AAF
N-Hydroxy-2-acetamidofluorene
2-(N-Hydroxyacetamido)-fluorene
N-Hydroxy-N-acetyl-2-aminofluorene
N-Hydroxy-2-acetylaminofluorene
N-Hydroxy-2-FAA
N-Hydroxy-N-2-fluorenylacetamide
NOHFAA

53-96-3
$C_{15}H_{13}NO$
223.29
O=C(Nc(ccc(c1Cc2cccc3)c23)c1)C
Acetamide, N-fluoren-2-yl
AAF
2-AAF
Acetamide, N-9H-fluoren-2-yl-(9CI)
2-Acetamidofluorene
2-Acetaminofluorene
Acetoaminofluorene
2-Acetylamino-fluoren (German)
N-Acetyl-2-aminofluorene
2-(Acetylamino)fluorene
2-Acetylaminofluorine (OSHA)
Azetylaminofluoren (German)
FAA
2-FAA
2-Fluorenylacetamide
N-2-Fluorenylacetamide
N-Fluoren-2-ylacetamide
RCRA waste number U005

54-04-6
$C_{11}H_{17}NO_3$
211.29
COc1cc(CCN)cc(OC)c1OC
Phenethylamine, 3,4,5-tri-

methoxy
Benzeneethanamine, 3,4,5-trimethoxy-
Ethane, 1-amino-2-(3,4,5-trimethoxyphenyl)-
Mescaline
Mescalin (German)
Mescline
Mezcaline
Mezcline
TMPEA
3,4,5-Trimethoxybenzeneethanamine
3,4,5-Trimethoxyphenethylamine

54-05-7
$C_{18}H_{26}ClN_3$
319.92
CCN(CC)CCCC(C)Nc1ccnc2cc(Cl)ccc12
Quinoline, 7-chloro-4-((4-(diethylamino)-1-methylbutyl)amino)
Amokin
Aralen
Arthrochin
Artrichin
Avlochlor
Avloclor
Bemaco
Bemaphate
Bemasulph
Benaquin
Bipiquin
Chemochin
Chingamin
Chloraquine
Chlorochin
7-Chloro-4-(4-diethylamino-1-methylbutylamino)quinoline
Chlorquin
Chloroquine
Chloroquinium
N[4]-(7-Chloro-4-quinolinyl)-N[1],N[1]-diethyl-1,4-pentanediamine
Cidanchin
Clorochina
Chloroquina
Cocartrit
Delagil
Dichinalex
Elestol
Gontochin
Heliopar
Imagon
Iroquine
Klorokin
Lapaquin
Malaquin
Malaren
Malarex
Mesylith
Neochin
Nivachine
Nivaquine
Nivaquine B
1,4-Pentanediamine, N[4]-(7-chloro-4-quinolinyl)-N[1],N[1]-diethyl-
Quinachlor
Quinercyl
Quinagamin
Quinagamine
Quinilon
Quinoscan
Resochen
Resochin
Resoquina
Resoquine
Reumachlor
Reumaquin
Roquine
RP 3377
Sanoquin
Silbesan
Siragan
SN 6718
SN 7618
Solprina
Sopaquin
Tanakan
Tresochin
Trochin
W 7618
WIN 244

54-11-5
$C_{10}H_{14}N_2$
162.26
n(cccc1C(N(CC2)C)C2)c1
Nicotine
Black Leaf
Black Leaf 40
Destruxol orchid spray
Emo-Nik
ENT 3,424
Flux Maag
Fumetobac
Mach-Nic
1-Methyl-2-(3-pyridyl)pyrrolidine
3-(N-Methylpyrrolidino)pyridine
l-3-(1-Methyl-2-pyrrolidyl)-pyridine
(-)-3-(1-Methyl-2-pyrrolidyl)-pyridine
Niagara P.A. Dust
Nicocide
Nico-Dust
Nico-Fume
Nicotina (Italian)
(-)-Nicotine
l-Nicotine
l-Nicotine
(S)-Nicotine
Nicotine alkaloid
Nicotine, Liquid [UN 3144]
Nicotine, Solid [UN 1655]
Nicotine (ACGIH,OSHA) [UN 1654]
Nikotin (German)
Nikotyna (Polish)
Ortho N-4 Dust
Ortho N-5 Dust
Ortho N-4 and N-5 Dusts
Pyridine, 3-(1-methyl-2-pyrrolidinyl)-
Pyridine, 3-(1-methyl-2-pyrrolidinyl)-, (S)- (9CI)
Pyridine, 3-(tetrahydro-1-methylpyrrol-2-yl)
β-Pyridyl-α-N-methylpyrrolidine
Pyrrolidine, 1-methyl-2-(3-pyridal)-
RCRA waste number P075
Tendust
Tetrahydronicotyrine, dl-
UN 1654 [Nicotine]
UN 3144 [Nicotine compounds, Liquid, N.O.S.]
UN 1655 [Nicotine compounds, solid, N.O.S.]
XL all insecticide

54-12-6
$C_{11}H_{12}N_2O_2$
204.25
O=C(O)C(N)CC(c(c(N1)ccc2)c2)=C1
Tryptophan, DL
DL-Tryptophan

54-21-7
$C_7H_5O_3.Na$
160.11
[Na+].Oc1ccccc1C([O-])=O
Salicylic acid, monosodium salt
Alysine
Ardall
Aroall
Benzoic acid, 2-hydroxy-, monosodium salt (9CI)
Clin
Diuretin
Enterosalicyl
Enterosalil
2-Hydroxybenzoic acid monosodium salt
o-Hydroxybenzoic sodium salt
Idocyl novum
Kerasalicyl
Kerosal
Magsalyl
Nadisal
Natrium salicylat (German)
Neo-Salicyl
Parbocyl-rev
Salicylic acid, sodium salt
Salisod
Salsonin
Sodium o-hydroxybenzoate
Sodium salicylate
Sodium salicylic acid

54-25-1
$C_8H_{11}N_3O_6$
245.22
OCC1OC(C(O)C1O)n2ncc(=O)[nH]c2=O
as-Triazine-3,5(2H,4H)-dione, 2-β-D-ribofuranosyl
6-Azauracilribosid (Czech)
6-Azauracilriboside
6-Azauracil-β-D-riboside
Azauridine
6-Azauridine
Azur

6-Azur
6-Azuridine
NSC-32074
Ribo-azauracil
Ribo-azuracil
3,5-Dioxo-2,3,4,5-tetrahydro-1,2,4-triazine riboside
2-β-D-Ribofuranosyl-as-triazine-3,5(2H,4H)-dione
2-β-D-Ribofuranosyl-1,2,4-triazin-3,5(2H,4H)-dion (Czech)
2-β-D-Ribofuranosyl-1,2,4-triazine-3,5(2H,4H)-dione

54-31-9
$C_{12}H_{11}ClN_2O_5S$
330.76
NS(=O)(=O)c2cc(C(O)=O)c(NCc1ccco1)cc2Cl
Anthranilic acid, 4-chloro-N-furfuryl-5-sulfamoyl
Aisemide
Aluzine
5-(Aminosulfonyl)-4-chloro-2-((2-furanylmethyl)amino)-benzoic acid
Benzoic acid, 5-(aminosulf onyl)-4-chloro-2-((2-furanyl-methyl)amino)- (9CI)
Beronald
Chlor-N-(2-furylmethyl)-5-sulfamylanthranilsaeure (German)
4-Chloro-N-furfuryl-5-sulfamoylanthranilic acid
4-Chloro-N-(2-furylmethyl)-5-sulfamoylanthranilic acid
Desdemin
Diural
Dryptal
Errolon
Eutensin
Frusemide
Frusemin
Frusid
Fulsix
Fuluvamide
Furanthril
Furanthryl
Furantril
Furesis
Furosedon
Furosemid
Furosemide
Furosemide "Mita"
Furosemidu (Polish)
Fursemid
Fursemide
Fusid
Hydro-Rapid
Katlex
Lasex
Lasilix
Lasix
LB 502
Lowpstron
Macasirool
Nicorol
NCI-C55936
Prefemin
Profemin
Radonna
Rosemide
Salix
Seguril
Transit
Trofurit
Urex
Urosemide

54-35-3
$C_{16}H_{18}N_2O_4S.C_{13}H_{20}N_2O_2$
570.77
4-Thia-1-azabicyclo(3.2.0)heptane-2-carboxylic acid, 3,3-dimethyl-7-oxo-6-(2- phenylacetamido)-, Compd. with 2-(diethyl-amino)ethyl p-aminobenz-oate (1:1)
Benzylpenicillin novocaine salt
Benzylpenicillin procaine
Depocillin
Duphapen
Hostacillin
Hydracillin
Jenacillin O
Micro-Pen
Nopcaine
Penicillin G procaine
Penzal N 300
Procaine penicillin G
Procaine benzylpenicillinate
Retardillin
Vetspen
Vitablend

54-36-4
$C_{14}H_{14}N_2O$
226.30
CC(C)(C(=O)c1cccnc1)c2cccnc2
1-Propanone, 2-methyl-1,2-di-3-pyridyl
1,2-Di-3-pyridyl-2-methyl-1-propanone
Mepyrapone
Methapyrapone
Methopirapone
Methopyrapone
Methopyrinine
Methopyrone
Metopiron
Metopirone
Metopyrone
Metyrapon
Metyrapone
1-Propanone, 1,2-di-3-pyridyl-2-methyl-
1-Propanone, 2-methyl-1,2-di-3-pyridinyl- (9CI)
SU-4885

54-47-7
$C_8H_{10}NO_6P$
247.16
O=P(OCc(c(c(O)c(n1)C)C=O)c1)(O)O
Pyridoxal, 5-(dihydrogen phosphate)
Apolon B6
Apolon B_6
Biosechs
Codecarboxylase
Hairoxal
Hexermin P
Hiadelon
Hi-Pyridoxin
Pal-P
Phosphopyridoxal
Phosphoridoxal coenzyme
Piodel
PLP
Pydoxal
4-Pyridinecarboxaldehyde, 3-hydroxy-2-methyl-5-((phosphonooxy)methyl)-
Pyridoxaldehyde phosphate
Pyridoxal monophosphate
Pyridoxal P
Pyridoxal phosphate
Pyridoxal 5-phosphate
Pyridoxal-5'-phosphate
Pyridoxyl phosphate
Pyromijin
Sechvitan
Vitahexin P
Vitazechs

54-49-9
$C_9H_{13}NO_2$
167.23
CC(N)C(O)c1cccc(O)c1
Benzyl alcohol, α-(1-amino-ethyl)-m-hydroxy-, (-)
1-α-(1-Aminoethyl)-m-hydroxy-benzyl alcohol
Aramine
Benzenemethanol, α-(1-amino-ethyl)-3-hydroxy-, (R-(R*,S*))- (9CI)
Hydroxynorephedrine
m-Hydroxy norephedrine
m-Hydroxypropadrine
Icoral B
Metaradrine
Metaraminol
(-)-Metaraminol
1-Metaraminol
Pressonex

54-62-6
$C_{19}H_{20}N_8O_5$
440.47
Nc3nc(N)c2nc(CNc1ccc(cc1)C(=O)NC(CCC(O)=O)C(O)=O)cnc2n3
Glutamic acid, N-(p-(((2,4-di-amino-6-pteridinyl)methyl)-amino)benzoyl)-, L
4-Amino-4-deoxypteroyl-glutamate
4-Aminofolic acid
4-Amino-PGA
Aminopteridine
Aminopterin
4-Aminopteroylglutamic acid
A-Ninopterin
APGA
N-(4-(((2,4-Diamino-6-pteridin-yl)methyl)amino)benzoyl)-l-glutamic acid
ENT 26,079

Folic acid, 4-amino-
Kyselina 4-aminolistova (Czech)
Kyselina 4-aminopteroyl-glutamova (Czech)
Kyselina N-(p-((2,4-diamino-6-pteridinylmethyl)amino)-benzoyl)-l(+)-glutamova (Czech)
NSC-739
Pteramina (Czech)

54-64-8
$C_9H_9HgO_2S.Na$
404.82
[Na+].CC[Hg]Sc1ccccc1C([O-])=O
Mercury, ((o-carboxyphenyl)-thio)ethyl-, sodium salt
((o-Carboxyphenyl)thio)ethyl-mercury sodium salt
Elcide 75
Elicide
Ethyl(2-mercaptobenzoato-s)-mercury sodium salt
o-(Ethylmercurithio)benzoic acid sodium salt
Ethylmercurithiosalicylic acid sodium salt
Ethylmerkurithiosalicilan sodny (Czech)
Ethyl (sodium o-mercapto-benzoato)mercury
Mercurothiolate
Mercury, ethyl(2-mercapto-benzoate-s)-, sodium salt
Merfamin
Merthiolate
Merthiolate salt
Merthiolate sodium
Mertorgan
Merzonin
Merzonin sodium
Merzonin, sodium salt
SET
Sodium ethylmercuric thio-salicylate
Sodium o-(ethylmercurithio)-benzoate
Sodium ethylmercurithio-salicylate
Sodium merthiolate
Thimerosal
Thimerosalate
Thimerosol
Thimersalate
Thiomerosal
Thiomersal
Thiomersalate

54-71-7
$C_{11}H_{16}N_2O_2.ClH$
244.75
Cl.CCC1C(COC1=O)Cc2cncn2C
Pilocarpine, monohydro-chloride
Almocarpine
Isopto Carpine
Ami-Pilo
Mi-Pilo Ophth Sol
Amistura P
Pilocarpine, hydrochloride
Pilocarpine muriate
Pilocar
Pilocel
Pilomiotin
Pilovisc

54-85-3
$C_6H_7N_3O$
137.16
NNC(=O)c1ccncc1
Isonicotinic-acid-hydrazide
Amidon
Andrazide
Antimicina
Antituberkulosum
Armacide
Armazide
Atcotibine
Azuren
Bacillin
BP 5015
Cedin
Cemidon
Chemiazid
Chemidon
Cotinazin
Cortinazine
Cotinizin
Defonin
Dibutin
Diforin
Dinacrin
Ditubin
Ebidene
Eralon
Ertuban
Eutizon
Evalon
Fimalene
FSR 3
GINK
HIA
Hidranizil
Hidrasonil
Hidrulta
Hidrun
Hycozid
Hydrazid
Hydrazide
Hyozid
Hyzyd
I.A.I.
Ido-Tebin
Idrazide dell'acido isonicotinico (Italian)
Idrazil
IN-73
INAH
INH
Inizid
Iscotin
Isidrina
Ismazide
Isobicina
Isocid
Isocidene
Isocotin
Isolyn
Isonerit
Isonex
Isoniacid
Isoniazid
Isoniazide
Isonicazide
Isonicid
Isonico
Isonicotan
Isonicotil
Isonicotinhydrazid
Isonicotinic acid hydrazide
Isonicotinoyl hydrazide
Isonicotinoylhydrazine
Isonicotinsaeurehydrazid (German)
Isonicotinyl hydrazide
Isonide
Isonidrin
Isonikazid
Isonilex
Isonin
Isonindon
Isonirit
Isoniton
Isonizide
Isotebe
Isotebezid
Isotinyl
Isozide
Isozyd
L 1945
Laniazid
Laniozid
Mybasan
Neoteben
Neo-Tizide
Neoxin
Neumandin
Nevin
Niadrin
Nicazide
Nicetal
Nicizina
Niconyl
Nicotibina
Nicotibine
Nicozide
Nidaton
Nidrazid
Nikozid
Niplen
Nitadon
Niteban
NSC-9659
Nydrazid
Nyscozid
Pelazid
Percin
Phthisen
Preparation 6424
Pycazide
Pyreazid
Pyricidin
Pyridicin
4-Pyridinecarboxylic acid, hydrazide
Pyrizidin
Raumanon
Razide
Retozide
Rifamate
Rimicid

Rimifon
Rimitsid
Robiselin
Robisellin
Roxifen
5015 RP
RU-EF-TB
Sanohidrazina
Sauterazid
Sauterzid
Stanozide
TB-Phlogin
TB-Razide
TB-Vis
Tebecid
Tebenic
Tebexin
Tebilon
Tebos
Teebaconin
Tekazin
Tibazide
Tibemid
Tibinide
Tibison
Tibivis
Tibizide
Tibusan
Tisin
Tisiodrazida
Tizide
Tubazid
Tubazide
Tubeco
Tubercid
Tuberian
Tubicon
Tubomel
Tyvid
Unicocyde
Unicozyde
USAF CB-2
Vazadrine
Vederon
Zinadon
Zonazide

54-86-4
$C_6H_4O_2.Na$
131.09
Nicotinic acid, sodium salt
Natriumnicotinat (German)
Sodium nicotinate

54-91-1
$C_{10}H_{16}Br_2N_2O_2$
356.10
BrCCC(=O)N1CCN(CC1)C(=O)CCBr
Piperazine, 1,4-bis(3-bromopropionyl)
A 1803
A-8103
Amedel
N,N-Bis-(3-bromopropionyl)-piperazine
1,4-Bis(3-bromopropionyl)-piperazine
NSC-25154
Piperazine, 1,4-bis(3-bromo-1-oxopropyl)-
Pipobroman
Vercyte

54-95-5
$C_6H_{10}N_4$
138.20
N(N=NN1CCCCC2)=C12
5H-Tetrazoloazepine, 6,7,8,9-tetrahydro
Angiazol
Angioton
Angiotonin
Cardiazol
Cardiazole
Cardifortan
Cardiol
Cardiotonicum
Cardosal
Cardosan
Cenalene-M
Cenazol
Centrazole
Cerebro-Nicin
Coranormal
Coranormol
Corasol
Coratoline
Corazol
Corazole
Corazole (analeptic)
Corisan
Corsedrol
Cortis
Corvasol
Corvis
Coryvet
α,β-Cyclopentamethylenetetrazole
Deamocard
Delzol-W
Diovascole
Deumacard
Gewazol
Kardiazol
Korazol
Korazole
Lepazol
Leptazol
Leptazole
Metrazol
Metrazole
Nauranzol
Naurazol
Nedcardol
Neocardol
Neurazol
Novo cora-vinco
Opticor
Pemetesan
Penetrasol
Penetratsol
Penetiazol
Pentacard
Pentacor
Pentamethazol
Pentamethazolum
Pentamethylenetetrazal
Pentamethylenetetrazol
Pentamethylenetetrazole
Pentamethylene-1,5-tetrazole
1,5-Pentamethylenetetrazole
Pentametilentetrazolo (Italian)
Pentazol
Pentazolum
Pentemesan
Pentetrazol
Pentetrazole
Pentrazol
Pentrolone
Pentrozol
Pentylenetetrazol
Pentylenetetrazole
Petazol
Petezol
Petrazole
Phrenazol
Phrenazone
PMT
PTZ
Stellacardiol
Stillcardiol
Tetracor
6,7,8,9-Tetrahydro-5-azepotetrazole
6,7,8,9-Tetrahydro-5H-tetrazoloazepine
7,8,9,10-Tetrazabicyclo(5.3.0)-8,10-decadiene
1,2,3,3a-Tetrazacyclohepta-8a,2-cyclopentadiene
Tetrasol
Tetrazol
Tetrazole, pentamethylene-
5H-Tetrazolo(1,5-a)azepine, 6,7,8,9-tetrahydro- (8CI,9CI)
TT87
Vasazol
Vasorex
Ventrazol
Yetrazol

54-96-6
$C_5H_7N_3$
109.15
Nc1ccncc1N
Pyridine, 3,4-diamino
3,4-Diaminopyridine
Diamino-3,4 pyridine
SC10

55-10-7
$C_9H_{10}O_5$
198.18
O=C(O)C(O)c(ccc(O)c1OC)c1
Vanilmandelic acid
Benzeneacetic acid, α,4-dihydroxy-3-methoxy- (9CI)
α,4-Dihydroxy-3-methoxybenzeneacetic acid

55-18-5
$C_4H_{10}N_2O$
102.16
O=NN(CC)CC
Diethylamine, N-nitroso
Dana
DEN
DENA
Diaethylnitrosamin (German)
Diethylnitrosamide

Diethylnitrosamine
Diethylnitrosoamine
N,N-Diethylnitrosamine
Ethylamine, N-nitrosodi-
N-Ethyl-N-nitroso-ethanamine
NDEA
Nitrosodiethylamine
N-Nitroso-diaethylamine (German)
N-Nitrosodiethylamine
N-Nitroso-N,N-diethylamine
RCRA waste number U174

55-21-0
C_7H_7NO
121.15
O=C(N)c(cccc1)c1
Benzamide
Amid kyseliny benzoove (Czech)
Benzoic acid amide
Benzoylamide
Phenylcarboxyamide

55-22-1
$C_6H_5NO_2$
123.12
O=C(O)c(ccnc1)c1
Isonicotinic-acid
Acide iso-nicotinique (French)
4-Carboxypyridine
α-Picolinic acid
4-Pyridinecarboxylic acid

55-27-6
$C_8H_{11}NO_3.ClH$
205.66
Benzyl alcohol, α-(aminomethyl)-3,4-dihydroxy-, hydrochloride-, (+-)
dl-Arterenol hydrochloride
1,2-Benzenediol, 4-(2-amino-1-hydroxyethyl)-, hydrochloride, (+-)- (9CI)
(+-)-Noradrenaline hydrochloride
dl-Noradrenaline hydrochloride
(+-)-Norepinephrine hydrochloride
dl-Norepinephrine hydrochloride
dl-Norepinephrine hydrochloride

55-31-2
$C_9H_{13}NO_3.ClH$
219.69
Cl.CNCC(O)c1ccc(O)c(O)c1
Benzyl alcohol, 3,4-dihydroxy α-((methylamino)methyl)-, hydrochloride, (-)
Adrenalin chloride
Adrenaline chloride
l-Adrenaline chloride
Adrenaline hydrochloride
Adrenaline hydrochloride, (-)-
l-Adrenaline hydrochloride
(-)-Adrenaline hydrochloride
Adrenalin hydrochloride
1,2-Benzenediol, 4-(1-hydroxy-2-(methylamino)ethyl)-, hydrochloride, (R)- (9CI)
l-1-(3,4-Dihydroxyphenyl)-2-methylamino-1-ethanol hydrochloride
Epinephrine chloride
l-Epinephrine chloride
Epinephrine, hydrochloride
(-)-Epinephrine hydrochloride
l-Epinephrine hydrochloride
Gelatin-epinephrine
l-Methylaminoethanolcathechol hydrochloride
NCI-C55663
Suprarenin hydrochloride
Supranephrin solution

55-38-9
$C_{10}H_{15}O_3PS_2$
278.34
COP(=S)(OC)Oc1ccc(SC)c(C)c1
Phosphorothioic acid, O,O-dimethyl O-(4-(methylthio)-m-tolyl) ester
B 29493
Bay 29493
Baycid
Bayer 9007
Bayer 29493
Bayer S-1752
Baytex
m-Cresol, 4-(methylthio)-, O-ester with O,O-dimethyl phosphorothioate
O,O-Dimethyl O-4-(methylmercapto)-3-methylphenyl phosphorothioate
O,O-Dimethyl-O-4-(methylmercapto)-3-methylphenyl thiophosphate
O,O-Dimethyl O-(3-methyl-4-methylmercaptophenyl)phosphorothioate
O,O-Dimethyl-O-(3-methyl-4-methylthio-fenyl)-monothiofosfaat (Dutch)
O,O-Dimethyl-O-(3-methyl-4-methylthiophenyl)-monothiophosphat (German)
O,O-Dimethyl O-3-methyl-4-methylthiophenyl phosphorothioate
O,O-Dimethyl-O-(3-methyl-4-methylthio-phenyl)-thionophosphat (German)
O,O-Dimethyl O-(4-methylthio 3-methylphenyl) phosphorothioate
O,O-Dimethyl O-(4-(methylthio)-m-tolyl) phosphorothioate
O,O-Dimetil-O-(3-metil-4-metiltio-fenil)-monotiofosfato (Italian)
DMTP
ENT 25,540
Entex
Fenthion (ACGIH,OSHA)
Fenthion 4E
Lebaycid
Mercaptophos
4-Methylmercapto-3-methylphenyl dimethyl thiophosphate
MPP
NCI-C08651
OMS 2
Phosphorothioic acid, O,O-dimethyl O-(3-methyl-4-(methylthio)phenyl) ester
Prentox
Queletox
S 1752
Spotton
Talodex
Thiophosphate de O,O-dimethyle et de O-(3-methyl-4-methylthiophenyle) (French)
Tiguvon

55-48-1
$C_{34}H_{46}N_2O_6.H_2O_4S$
676.90
Atropine, sulfate (2:1)
Atropette
Atropine sulfate
Atropine sulphate
Atropinium sulfate
Atropinsal
Atropin siran (Czech)
Atropinsulfat (German)
Atropiny siarczan (Polish)
Atropisal
Atropisol
Corbella
Davurtrop
Eyesule
Ichtho-bellol
Lio-Atropin
MBK
Sulfate d'atropine (French)
Sulfatropinol
1-α-H,5-α-H-Tropan-3-α-ol (+-)-tropate (ester), sulfate (2:1) salt
DL-Tropanyl 2-hydroxy-1-phenylpropionate sulfate
Tropintran

55-55-0
$C_7H_9NO.O_4S$
219.23
Phenol, p-methylamino-, sulfate (Salt)
Armol
Elon
Elon (Developer)
Genol
Graphol
Metatyl
Methyl-p-aminophenol sulfate
p-Methylaminophenolsulfate
Metol
Phenol, p-(methylamino)-, sulfate (2:1) (Salt)
Photol
Pictol
Planetol
Rhodol
Verol

55-56-1
$C_{22}H_{30}Cl_2N_{10}$
505.52
N=C(NCCCCCCNC(=N)NC(=N)Nc(ccc(c1)Cl)c1)NC(=N)Nc(ccc(c2)Cl)c2
Biguanide, 1,1'-hexamethylenebis(5-(p-chlorophenyl)
1,6-Bis(5-(p-chlorophenyl)-biguandino)hexane
1,6-Bis(p-chlorophenyl-diguanido)hexane
Chlorhexidin (Czech)
Chlorhexidine
1,6-Di(4'-chlorophenyldi-guanido)hexane
1,1'-Hexamethylenebis(5-(p-chlorophenyl)biguanide)
Hibitane
Nolvasan
Rotersept
Sterido

55-63-0
$C_3H_5N_3O_9$
227.11
O=N(=O)OCC(ON(=O)=O)CON(=O)=O
Nitroglycerin (ACGIH,OSHA)
Angibid
Anginine
Angiolingual
Angorin
Blasting Gelatin (DOT)
Blasting Oil
Cardamist
Gilucor nitro
Glonoin
Glycerintrinitrate (Czech)
Glycerol, nitric acid triester
Glyceroltrinitraat (Dutch)
Glycerol trinitrate
Glycerol(trinitrate de) (French)
Glyceryl nitrate
Glyceryl trinitrate
GTN
Klavi Kordal
Lenitral
Myocon
Myoglycerin
NA 1204 (DOT)
NG
Niglin
Niglycon
Niong
Nitora
Nitric acid triester of glycerol
Nitrin
Nitrine
Nitrine-TDC
Nitro-Dur
Nitroglicerina (Italian)
Nitrogliceryna (Polish)
Nitroglycerin, Liquid, Desensitized [UN 0143]
Nitroglycerin, Liquid, Not desensitized [UN 0143]
Nitroglycerin, Solution with 5% but not more than 10% nitroglycerin [UN 144]
Nitroglycerine
Nitroglycerine, Spirit of (1% to 5%) [UN 3064]
Nitroglycerol
Nitroglyn
Nitrol
Nitrol (Pharmaceutical)
Nitrolan
Nitro-lent
Nitroletten
Nitrolingual
Nitrolowe
Nitromel
Nitronet
Nitrong
Nitrorectal
Nitro-span
Nitrostabilin
Nitrostat
Nitrozell retard
NK-843
NTG
Nysconitrine
Perglottal
Propanetriol trinitrate
1,2,3-Propanetriol, trinitrate
1,2,3-Propanetriyl nitrate
RCRA waste number P081
SK-106N
S.N.G
Soup
Temponitrin
TNG
Trinalgon
Trinitrin
Trinitroglycerin
Trinitroglycerol
Trinitrol
UN 0143 [Nitroglycerin, desensitized with not less than 40 per cent non-volatile water insoluble phlegmatizer, by mass]
UN 0144 [Nitroglycerin, solution in alcohol, with more than 5% but not more than 10% nitroglycerin]
UN 1204 [Nitroglycerin solution in alcohol with not more than 1 per cent nitroglycerin]
UN 3064 [Nitroglycerin, solution in alcohol, with more than 1% but not more than 5% nitroglycerin]
Vasoglyn

55-65-2
$C_{10}H_{22}N_4$
198.36
NC(=N)NCCN1CCCCCCC1
Guanidine, (2-(hexahydro-1(2H)-azocinyl)ethyl)
Azocine, 1-(2-guanidinoethyl)-octahydro-
Guanethidine
Heptamethylenimine, 1-(2-guanidinoethyl)-
Ismelin
Octatensin
Octatenzine
SU 5864

55-68-5
$C_6H_5HgNO_3$
339.71
[O-]N(=O)=O.[Hg+]c1ccccc1
Mercury, nitratophenyl
Fenylmerkurinitrat (Czech)
Mercuriphenyl nitrate
Merphenyl nitrate
Mersolite 7
Nitric acid, phenylmercury salt
Phe-Mer-Nite
Phenalco
Phenitol
Phenmerzyl nitrate
Phenylmercuric nitrate [UN 1895]
Phenylmercury nitrate
Phermernite
UN 1895 [Phenylmercuric nitrate]

55-73-2
$C_{10}H_{15}N_3$
177.28
CNC(NCc1ccccc1)=NC
Guanidine, 2-benzyl-1,3-dimethyl
2-Benzyl-1,3-dimethylguanidine
Guanidine, N,N'-dimethyl-N''-(phenylmethyl)-

55-80-1
$C_{15}H_{17}N_3$
239.35
N(=Nc(cccc1C)c1)c(ccc(N(C)C)c2)c2
Aniline, N,N-dimethyl-p-(m-tolylazo)
Aniline, N,N-dimethyl-p-(3'-methylphenylazo)-
Benzenamine, N,N-dimethyl-4-((3-methylphenyl)azo)-(9CI)
4-(N,N-Dimethylamino)-3'-methylazobenzene
N,N-Dimethyl-p-(m-tolylazo)-aniline
MDAB
3'-MDAB
3'-Me-DAB
3'-Methylbuttergelb (German)
3'-Methyl-DAB
3'-Methyl-4-dimethylaminoazobenzen (Czech)
m'-Methyl-p-dimethylaminoazobenzene
3'-Methyl-4-dimethylaminoazobenzene
3'-Methyl-N,N-dimethyl-4-aminoazobenzene
3'-Methyl-4-(N,N-dimethylamino)azobenzene
3'-Methyldimethylaminoazobenzol (German)

55-86-7

$C_5H_{11}Cl_2N.ClH$
192.53
Cl.CN(CCCl)CCCl
Diethylamine, 2,2'-dichloro-N-methyl, hydrochloride
Antimit
Azotoyperite
N,N-Bis(2-chloraethyl)-methylamin-hydrochlorid (German)
Bis(2-chloroethyl)methylamine hydrochloride
N,N-Bis(2-chloroethyl)-methylamine hydrochloride
C 6866
Carolysine
Caryolysine
Caryolysine hydrochloride
Chloramin
Chloramine
Chloramin hydrochloride
Chlorethamine
Chlorethazine
Chlormethine hydrochloride
Chlormethinum
2-Chloro-N-(2-chloroethyl)-N-methylethanamine hydrochloride
DEMA
Dichloren
Dichloren hydrochloride
β,β'-Dichlorodiethyl-N-methylamine hydrochloride
Di(2-Chloroethyl)methylamine hydrochloride
1,5-Dichloro-3-methyl-3-azapentane hydrochloride
2,2'-Dichloro-N-methyldiethylamine hydrochloride
Dimitan
Embechine
Embichin
Embichin hydrochloride
Embikhine
Erasol
Erasol hydrochloride
Erasol-Ido
Ethanamine, 2-chloro-N-(2-chloroethyl)-N-methyl-, hydrochloride
HN2.HCl
HN2 Hydrochloride
Kloramin
N-Lost (German)
MBA hydrochloride
Mebichloramine
Mechlorethamine hydrochloride
Merchlorethamine
Methylbis(β-chloroethyl)amine hydrochloride
N-Methyl-bis-β-chlorethylamine hydrochloride
Methylbis(2-chloroethyl)amine hydrochloride
N-Methylbis(2-chloroethyl)-amine hydrochloride
N-Methyl-2,2'-dichlorodiethylamine hydrochloride
N-Methyl-di-2-chloroethylamine hydrochloride
Methyldi(β-chloroethyl)amine hydrochloride
Methyldi(2-chloroethyl)amine hydrochloride
Mitoxine
Mustargen
Mustargen hydrochloride
Mustine hydrochlor
Mustine hydrochloride
NCI-C56382
Nitrogen mustard hydrochloride
Nitol
Nitol "Takeda"
Nitrogranulogen
Nitrogranulogen hydrochloride
N-Mustard (German)
NSC-762
NSC-762 hydrochloride
Pliva
Stickstofflost
Zagreb

55-91-4
$C_6H_{14}FO_3P$
184.17
O=P(F)(OC(C)C)OC(C)C
Phosphorofluoridic acid, bis-(1-methylethyl) ester
DFP
Diflurophate
Diflupyl
Diisopropoxyphosphoryl fluoride
Diisopropylfluorfosfat (Czech)
Diisopropyl fluorophosphate
O,O-Diisopropyl fluorophosphate
Diisopropyl fluorophosphonate
Diisopropylfluorophosphoric acid ester
Diisopropylfluorphosphorsaeureester (German)
Diisopropyl phosphofluoridate
Diisopropyl phosphorofluoridate
O,O'-Diisopropyl phosphoryl fluoride
Dyflos
EA 1152
Floropryl
Fluophosphoric acid, diisopropyl ester
Fluorodiisopropyl phosphate
Fluostigmine
Fluoropryl
Isofluorophate
Isoflurophate
Isopropyl fluophosphate
Isopropyl phosphorofluoridate
Neoglaucit
PF-3
Phosphorofluoridic acid, diisopropyl ester
RCRA waste number P043
T-1703
TL 466

55-98-1
$C_6H_{14}O_6S_2$
246.32
CS(=O)(=O)OCCCCOS(C)(=O)=O
1,4-Butanediol, dimethanesulfonate
AN 33501
1,4-Bis(methanesulfonoxy)-butane
(1,4-Bis(methanesulfonyloxy)-butane)
Busulfan
Busulphan
Busulphane
1,4-Butanediol dimethanesulfonate
1,4-Butanediol dimethanesulphonate
Buzulfan
C.B. 2041
2041 C.B.
Citosulfan
1,4-Dimesyloxybutane
1,4-Dimethanesulfonoxybutane
1,4-Dimethanesulfonoxylbutane
1,4-Dimethanesulfonyloxybutane
1,4-Dimethanesulphonyloxybutane
1,4-Dimethylsulfonoxybutane
1,4-Dimethylsulfonyloxybutane
GT41
GT 2041
Leucosulfan
Mablin
Methanesulfonic acid, tetramethylene ester
Mielevcin
Mielosan
Mielucin
Milecitan
Mileran
Misulban
Mitosan
Mitostan
Myeleukon
Myeloleukon
Myelosan
Mylecytan
Myleran
NCI-C01592
NSC-750
NSC-750 sulphabutin
Sulfabutin
Sulphabutin
Tetramethylene bis(methanesulfonate)
Tetramethylene dimethane sulfonate
Tetramethylenester kyseliny methansulfonove (Czech)
X 149

56-03-1
$C_2H_7N_5$
101.08
Imidodicarbonimidic diamide
AI3-52571

56-04-2
$C_5H_6N_2OS$
142.19
n(c(O)cc(n1)C)c1S
Uracil, 6-methyl-2-thio
Alkiron

Antibason
Basecil
Basethyrin
2,3-Dihydro-6-methyl-2-thioxo-4(1H)-pyrimidinone
2-Mercapto-6-methylpyrimid-4-one
2-Mercapto-4-hydroxy-6-methylpyrimidine
2-Mercapto-6-methyl-4-pyrimidone
Metacil
Methacil
Methiacil
Methicil
Methiocil
6-Methyl-2-thio-2,4-(1H3H)-pyrimidinedione
Methylthiouracil
6-Methylthiouracil
4-Methyl-2-thiouracil
6-Methyl-2-thiouracil
6-Methyl-2-thiouracyl (Czech)
4-Methyluracil
6-Metil-tiouracile (Italian)
MTU
Muracil
Orcanon
4(1H)-Pyrimidinone, 2,3-dihydro-6-methyl-2-thioxo-
Prostrumyl
RCRA waste number U164
Strumacil
Thimecil
Thiomecil
2-Thio-6-methyl-1,3-pyrimidin-4-one
6-Thio-4-methyluracil
Thiomidil
2-Thio-4-oxo-6-methyl-1,3-pyrimidine
Thioryl
Thiothymin
Thiothyron
Thiuryl
Thyreonorm
Thyreostat
Thyreostat I
Tiomeracil
Tiorale M
Tiotiron
Thyril
USAF EK-6454

56-05-3
$C_4H_3Cl_2N_3$
163.98
n(c(cc(n1)Cl)Cl)c1N
2-Pyrimidinamine, 4,6-dichloro- (9CI)
AI3-52142
2-Amino-4,6-dichloropyrimidine
4,6-Dichloro-2-pyrimidinamine
NSC-18698
Py 11
Pyrimidine, 2-amino-4,6-dichloro- (8CI)

56-09-7
$C_4H_5N_3O_2$
127.08
n(c(O)cc(n1)O)c1N
4(1H)-Pyrimidinone, 2-amino-6-hydroxy- (9CI)
2-Amino-4,6-dihydroxypyrimidine
2-Amino-4,6-dioxypyrimidine
2-Amino-6-hydroxy-4(1H)-pyrimidinone
2-Amino-4,6-pyrimidinedione
4,6-Dihydroxy-2-aminopyrimidine
NSC-15920

56-12-2
$C_4H_9NO_2$
103.14
O=C(O)CCCN
Butyric acid, 4-amino
4-Aminobutanoic acid
γ-Aminobuttersaeure (German)
γ-Aminobutyric acid
γ-Amino-n-butyric acid
4-Aminobutyric acid
DF 468
Gaba
Gamarex
Gammalon
Piperidic acid
Piperidinic acid

56-18-8
$C_6H_{17}N_3$
131.26
N(CCCN)CCCN
Dipropylamine, 3,3'-diamino
Aminobis(propylamine)
Bis-(3-aminopropyl)amine
3,3-Diaminodipropylamine
3,3'-Diaminodipropylamine
Dipropylenetriamine
Dipropylentriamin (German)
Iminobis(propylamine)
3,3'-Iminobis(propylamine)
Iminobispropylamine
3,3'-Iminobispropylamine [UN 2269]
1,3-Propanediamine, N-(3-aminopropyl)-
Propylamine, 3,3'-iminobis-
UN 2269 [3,3'-Iminodipropylamine]

56-23-5
CCl_4
153.81
C(Cl)(Cl)(Cl)Cl
Carbon-tetrachloride
Benzinoform
Carbona
Carbon chloride (CCl_4)
Carbon Tet
Carbon tetrachloride (ACGIH, OSHA) [UN 1846]
Chlorid uhlicity (Czech)
Czterochlorek wegla (Polish)
ENT 4,705
ENT 27,164
Fasciolin
Flukoids
Methane tetrachloride
Methane, tetrachloro-
Necatorina
Necatorine
Perchloromethane
R 10
RCRA waste number U211
R 10 (Refrigerant)
Tetrachloorkoolstof (Dutch)
Tetrachloormetaan
Tetrachlorkohlenstoff, tetra (German)
Tetrachlormethan (German)
Tetrachlorocarbon
Tetrachloromethane
Tetrachloromethane (OSHA)
Tetrachlorure de carbone (French)
Tetraclorometano (Italian)
Tetracloruro di carbonio (Italian)
Tetrafinol
Tetraform
UN 1846 [Carbon tetrachloride]
Tetrasol
Univerm
Vermoestricid

56-24-6
$C_3H_{10}OSn$
180.82
Stannane, hydroxytrimethyl
Hydroxytrimethylstannane
Hydroxytrimethyltin
Tin, trimethyl-, hydroxide
Trimethylstannanol
Trimethyl tin hydroxide

56-25-7
$C_{10}H_{12}O_4$
196.22
O=C(OC(=O)C1(C(OC2C3)C3)C)C12C
7-Oxabicyclo(2.2.1)heptane-2,3-dicarboxylic anhydride, 2,3-dimethyl
Can
Cantharides camphor
Cantharidin
Cantharidine
Cantharone
exo-1,2-cis-Dimethyl-3,6-epoxyhexahydrophthalic anhydride
2,3-Dimethyl-7-oxabicyclo-(2.2.1)heptane-2,3-dicarboxylic anhydride
Hexahydro-3a,7a-dimethyl-4,7-epoxyisobenzofuran-1,3-dione
Kantharidin (German)

56-29-1
$C_{12}H_{16}N_2O_3$
236.30
CN2C(=O)NC(=O)C(C)(C1=CCCCC1)C2=O
Barbituric acid, 5-(1-cyclohexen-1-yl)-1,5-dimethyl
Barbidorm

Citodon
Citopan
5-(1-Cyclohexen-1-yl)-1,5-dimethylbarbituric acid
5-(1-Cyclohexen-1-yl)-1,5-dimethyl-2,4,6(1H,3H,5H)-pyrimidinetrione
5-(1-Cyclohexenyl-1)-1-methyl-5-methylbarbituric acid
5-(δ-1,2-Cyclohexenyl)-5-methyl-N-methyl-barbitursaeure (German)
Cyclonal
Cyclopan
1,5-Dimethyl-5-(1-cyclohexenyl)barbituric acid
Dorico
Enhexymal
Esobarbitale (Italian)
Evipal
Evipan
Hexabarbital
Hexanastab oral
Hexenal
Hexenal (barbiturate)
Hexobarbital
Hexobarbitone
Methexenyl
N-Methyl-5-cyclohexenyl-5-methylbarbituric acid
Methylhexabarbital
Methylhexabital
Narcosan
Noctivane
2,4,6(1H,3H,5H)-Pyrimidinetrione, 5-(1-cyclohexen-1-yl)-1,5-dimethyl-
Sombucaps
Sombulex
Somnalert

56-33-7
$C_{16}H_{22}OSi_2$
286.56
Disiloxane, 1,3-diphenyl-1,1,3,3-tetramethyl
1,3-Diphenyl-1,1,3,3-tetramethyldisiloxane

56-34-8
$C_8H_{20}N.Cl$
165.74
[Cl-].CC[N+](CC)(CC)CC
Ammonium, tetraethyl-, chloride
Etamon chloride
Ethanaminium, N,N,N-triethyl-, chloride (9CI)
TEAC
TEA chloride
Tetraethylammonium chloride

56-35-9
$C_{24}H_{54}OSn_2$
596.16
CCCC[Sn](CCCC)(CCCC)O[Sn](CCCC)(CCCC)CCCC
Distannoxane, hexabutyl
Biomet TBTO
Bis-(tri-n-butylcin)oxid (Czech)
Bis(tributyloxide) of tin
Bis(tributylstannium) oxide
Bis(tributylstannyl)oxide
Bis(tributyltin)oxide
Bis(tri-n-butyltin)oxide
Bis(tri-n-butylzinn)-oxyd (German)
BTO
Butinox
C-Sn-9
ENT 24,979
Hexabutyldistannioxan (Czech)
Hexabutyldistannoxane
Hexabutylditin
Kyslicnik tri-n-butylcinicity (Czech)
L.S. 3394
OTBE (French)
Oxybis(tributyltin)
Oxyde de tributyletain
Stannane, tri-n-butyl-, oxide
TBOT
TBTO
Tin, bis(tributyl)-, oxide
Tin, oxybis(tributyl-

56-36-0
$C_{14}H_{30}O_2Sn$
349.13
CCCC[Sn](CCCC)(CCCC)OC(C)=O
Stannane, acetoxytributyl
Tin, tributyl-, acetate
Tributylstannium acetate
Tributyltin acetate
Tri-n-butyl-zinn-acetat (German)

56-38-2
$C_{10}H_{14}NO_5PS$
291.28
CCOP(=S)(OCC)Oc1ccc(cc1)N(=O)=O
Phosphorothioic acid, O,O-diethyl O-(p-nitrophenyl) ester
AAT
AATP
AC 3422
ACC 3422
Alkron
Alleron
American Cyanamid 3422
Aphamite
Aralo
B 404
Bay E-605
Bayer E-605
Bladan
Bladan F
Compound 3422
Corothion
Corthion
Corthione
Danthion
O,O-Diaethyl-O-(4-nitro-phenyl)-monothiophosphat (German)
O,O-Diethyl-O-(4-nitro-fenil)-monothiofosfaat (Dutch)
O,O-Diethyl-O-p-nitrofenylester kyseliny thiofosforecne (Czech)
Diethyl para-nitrophenol thiophosphate
O,O-Diethyl O-(p-nitrophenyl) phosphorothioate
O,O-Diethyl-O-(4-nitrophenyl) phosphorothioate
Diethyl p-nitrophenyl phosphorothionate
Diethyl 4-nitrophenyl phosphorothionate
Diethyl p-nitrophenyl thiophosphate
O,O-Diethyl O-p-nitrophenyl thiophosphate
Diethyl p-nitrophenyl thionophosphate
O,O-Diethyl-O-(p-nitrophenyl)-thionophosphate
O,O-Diethyl O-4-nitrophenyl thiophosphate
Diethylparathion
O,O-Dietil-O-(4-nitro-fenil)-monotiofosfato (Italian)
Dietil tiofosfato de p-nitrofenila (Portuguese)
O,O-Dietyl-O-p-nitrofenyltiofosfat (Czech)
DNTP
DPP
Drexel Parathion 8E
E 605
E 605 f
Ecatox
E 605 forte
Ekatin WF & WF ULV
Ekatox
ENT 15,108
Ethlon
Ethyl parathion
Etilon
Etylparation (Czech)
Folidol
Folidol E
Folidol E605
Folidol E & E 605
Folidol Oil
Fosfex
Fosfermo
Fosferno
Fosfive
Fosova
Fostern
Fostox
Gearphos
Genithion
Kolphos
Kypthion
Lethalaire G-54
Lirothion
Murfos
NA 1967 [Parathion and compressed gas mixture]
NA 2783 [Parathion]
NCI-C00226
Niran
Niran E-4
Nitrostigmin (German)
Nitrostigmine
Nitrostygmine

Niuif-100
Nourithion
Oleofos 20
Oleoparaphene
Oleoparathion
OMS 19
Orthophos
PAC
Pacol
Panthion
Paradust
Paramar
Paramar 50
Paraphos
Parathene
Parathion (ACGIH,OSHA) [NA 2783]
Parathion and compressed gas mixture [NA 1967]
Parathion, Liquid (DOT)
Parathion mixture, Dry
Parathion mixture, Liquid (DOT)
Parathion-aethyl (German)
Parathion-ethyl
Parawet
Penncap E
Pestox Plus
Pethion
Phenol, p-nitro-, O-ester with O,O-diethylphosphorothioate
Phoskil
Phosphemol
Phosphenol
Phosphorothioic acid, O,O-diethyl O-(4-nitrophenyl) ester
Phosphostigmine
RB
RCRA waste number P089
Rhodiasol
Rhodiatox
Rhodiatrox
Selephos
Sixty-three Special E.C. Insecticide
SNP
Soprathion
Stabilized ethyl parathion
Stathion
Strathion
Sulphos
Super rodiatox
T-47
Thiofos
Thiomex
Thiophos
Thiophos 3422
Thiophosphate de O,O-diethyle et de O-(4-nitrophenyle) (French)
Tiofos
Tox 47
Vapophos
Vitrex

56-40-6
$C_2H_5NO_2$
75.08
O=C(O)CN
Glycine
Aminoacetic acid
Glycolixir
Hampshire glycine

56-41-7
$C_3H_7NO_2$
89.09
O=C(O)C(N)C
(L)-Alanine (9CI)
Alanine
Alanine, L- (8CI)
α-Alanine
(S)-Alanine
L-Alanine
L-α-Alanine
L(+)-Alanine
L-(+)-Alanine
α-Aminopropionic acid
L-α-Aminopropionic acid
L-S-Aminopropionic acid
L-2-Aminopropionic acid
2-Aminopropionic acid
NSC-206315
Propanoic acid, 2-amino-
Propanoic acid, 2-amino-, (S)-

56-45-1
$C_3H_7NO_3$
105.09
O=C(O)C(N)CO
L-Serine (9CI)
α-Amino-β-hydroxypropionic acid
2-Amino-3-hydroxypropanoic acid
(S)-2-Amino-3-hydroxypropanoic acid
β-Hydroxyalanine
NSC-118365
Propanoic acid, 2-amino-3-hydroxy-, (S)-
SER (IUPAC Abbrev)
Serine
Serine, L- (8CI)
L-(-)-Serine

56-45-1
$C_3H_7NO_3$
105.09
O=C(O)C(N)CO
L-Serine (9CI)
α-Amino-β-hydroxypropionic acid
2-Amino-3-hydroxypropanoic acid
(S)-2-Amino-3-hydroxypropanoic acid
β-Hydroxyalanine
NSC-118365
Propanoic acid, 2-amino-3-hydroxy-, (S)-
Serina (Spanish)
Serine
Serine, L- (8CI)
L-(-)-Serine
Serinum (Latin)

56-49-5
$C_{21}H_{16}$
268.37
c(c(ccc1C)cc(c2ccc3cccc4)c34)(c1CC5)c25
Cholanthrene, 3-methyl
Benz(j)aceanthrylene, 1,2-dihydro-3-methyl-
MC
3-MC
20-MC
MCA
3-MCA
Methylcholanthrene
3-Methylcholanthrene
20-Methylcholanthrene
RCRA waste number U157

56-53-1
$C_{18}H_{20}O_2$
268.38
Oc(ccc(C(=C(c(ccc(O)c1)c1)CC)CC)c2)c2
4,4'-Stilbenediol, α,α'-diethyl
Acnestrol
Agostilben
Antigestil
Bio-DES
3,4-Bis(p-hydroxyphenyl)-3-hexene
Bufon
Climaterine
Comestrol
Comestrol estrobene
Cyren
Cyren A
Dawe's Destrol
DEB
DES
DESMA
DES (Synthetic estrogen)
Destrol
Diastyl
Dibestrol
Dibestrol (2) premix
Dicorvin
Di-Estryl
trans-4,4'-(1,2-Diethyl-1,2-ethenediyl)bisphenol
4,4'-(1,2-Diethyl-1,2-ethenediyl)bis-phenol
α,α'-Diethylstilbenediol
α,α'-Diethyl-(E)-4,4'-stilbenediol
α,α'-Diethyl-4,4'-stilbenediol
trans-α,α'-Diethyl-4,4'-stilbenediol
Diethylstilbesterol
trans-Diethylstilbesterol
Diethylstilbestrol
trans-Diethylstilbestrol
Diethylstilboesterol
trans-Diethylstilboesterol
Dietilestilbestrol (Spanish)
4,4'-Dihydroxydiethylstilbene
4,4'-Dihydroxy-α,β-diethylstilbene
3,4'(4,4'-Dihydroxyphenyl)hex-3-ene
Distilbene
Domestrol
Dyestrol
Estilben
Estilbin "MCO"

Estril
Estrobene
Estrogen
Estromenin
Estrosyn
Follidiene
Fonatol
Grafestrol
Gynopharm
3-Hexene,3,4-bis(p-hydroxy-phenyl)-
Hibestrol
Idroestril
Iscovesco
Makarol
Menostilbeen
Micrest
Microest
Milestrol
Neo-oestranol 1
NSC-3070
Oekolp
Oestrogenine
Oestrol Vetag
Oestromenin
Oestromensil
Oestromensyl
Oestromienin
Oestromon
Pabestrol
Palestrol
Percutatrine oestrogenique iscovesco
Phenol 4,4'-(1,2-diethyl-1,2-ethenediyl)bis-, (E)-
Protectona
RCRA waste number U089
Rumestrol 1
Rumestrol 2
Sedestran
Serral
Sexocretin
Sibol
Sintestrol
Stibilium
Stil
4,4'-Stilbenediol, α,α'-diethyl-, (E)-
4,4'-Stilbenediol,2,2'-diethyl-
Stilbestrol
Stilbestrol, diethyl-
Stilbestrone
Stilbetin
Stilboefral
Stilboestroform
Stilboestrol
Stilbofolin
Stilbofollin
Stilbol
Stilkap
Stil-Rol
Synestrin
Synthoestrin
Synthofolin
Syntofolin
Tampovagan stilboestrol
Tylosterone
Vagestrol

56-54-2
$C_{20}H_{24}N_2O_2$
324.46
O(c(ccc(ncccl C(O)C(N(CCC2C 3C=C)C3)C2)c14)c4)C
Quinidine
Chinidin (German)
Cinchonan-9-ol, 6'-methoxy-, (9s)-
Conchinin
Conquinine
6'-Methoxycinchonan-9-ol
α-(6-Methoxy-4-quinolyl)-5-vinyl-2-quinuclidine-methanol
6-Methoxy-α-(5-vinyl-2-quinuclidinyl)-4-quinolinemethanol
NCI-C56246
Pitayine
Quinicardine
Cin-Quin
Quinidex
(+)-Quinidine
β-Quinine
2-Quinuclidinemethanol, α-(6-methoxy-4-quinolyl)-5-vinyl-

56-55-3
$C_{18}H_{12}$
228.30
c(c(c(c(c1)ccc2)c2)cc(c3ccc4)c4)(c1)c3
Benz(a)anthracene
BA
Benzanthracene
1,2-Benzanthracene
1,2-Benz(a)anthracene
1,2-Benzanthrazen (German)
Benzanthrene
1,2-Benzanthrene
Benzoanthracene
Benzo(a)anthracene
1,2-Benzoanthracene
Benzo(a)phenanthrene
Benzo(b)phenanthrene
2,3-Benzophenanthrene
2,3-Benzphenanthrene
Naphthanthracene
Tetraphene
RCRA waste number U018

56-57-5
$C_9H_6N_2O_3$
190.17
O=N(=O)c1ccn(=O)c2ccccc12
Quinoline, 4-nitro-, 1-oxide
4-Nitrochinolin N-oxid (Swedish)
4-Nitroquinoline-N-oxide
4-Nitroquinoline-1-oxide
4-NQO

56-65-5
$C_{10}H_{16}N_5O_{13}P_3$
507.22
O=P(OP(=O)(O)O)(OP(=O)(OCC (OC(N(c(ncnc1N)c1N=2)C2) C3O)C3O)O)O
Adenosine 5'-(tetrahydrogen triphosphate)
Adenosine triphosphate
Adenosine 5'-triphosphate
Adenosine 5'-triphosphoric acid
Adenylpyrophosphoric acid
Adephos
Adetol
Adynol
Ara-ATP
9-β-D-Arabinofuranosyladenine 5'-triphosphate
ATIPI
ATP
5'-ATP
ATP (Nucleotide)
Atriphos
Glucobasin
Myotriphos
Striadyne
Triadenyl
Triphosaden
Triphosphaden
Triphosphoric acid adenosine ester

56-69-9
$C_{11}H_{12}N_2O_3$
220.25
O=C(O)C(N)CC(c(c(N1)ccc2O) c2)=C1
Tryptophan, 5-hydroxy
5-HTP
Hydroxytryptophan
5-Hydroxytryptophan
5-Hydroxytryptophane
5-Hydroxytrytophan
NCI-C56644
USAF CB-96

56-72-4
$C_{14}H_{16}ClO_5PS$
362.78
CCOP(=S)(OCC)Oc2ccc1c(C)c (Cl)c(=O)oc1c2
Coumarin, 3-chloro-7-hydroxy-4-methyl-, O-ester with O,O-diethyl phosphorothioate
Agridip
Asunthol
Asuntol
Azunthol
Bay 21/199
Bayer 21/199
Baymix
Baymix 50
3-Chloro-7-hydroxy-4-methyl-coumarin O,O-diethyl phosphorothioate
3-Chloro-7-hydroxy-4-methyl-coumarin O-ester with O,O-diethyl phosphorothioate
3-Chloro-4-methyl-7-coumarinyl diethyl phosphorothioate
O-3-Chloro-4-methyl-7-coumarinyl O,O-diethyl phosphorothioate
3-Chloro-4-methyl-7-hydroxy-coumarin diethyl thiophosphoric acid ester
3-Chloro-4-methylumbelli-

ferone O-ester with O,O-diethyl phosphorothioate
Co-Ral
Coumafos
Coumaphos
Coumaphos [UN 2783]
Coumaphos mixture, Liquid [UN 2783]
Cumafos (Dutch)
O,O-Diaethyl-O-(3-chlor-4-methyl-cumarin-7-yl)-monothiophosphat (German)
O,O-Diethyl-O-(3-chloor-4-methyl-cumarin-7-yl)monothiofosfaat (Dutch)
O,O-Diethyl O-(3-chloro-4-methyl-7-coumarinyl)phosphorothioate
O,O-Diethyl O-(3-chloro-4-methylcoumarinyl-7) thiophosphate
O,O-Diethyl O-(3-chloro-4-methyl-2-oxo-2H-benzopyran-7-yl)phosphorothioate
O,O-Diethyl 3-chloro-4-methyl-7-umbelliferone thiophosphate
O,O-Diethyl O-(3-chloro-4-methylumbelliferyl)phosphorothioate
Diethyl 3-chloro-4-methylumbelliferyl thionophosphate
Diethyl thiophosphoric acid ester of 3-chloro-4-methyl-7-hydroxycoumarin
O,O-Dietil-O-(3-cloro-4-metil-cumarin-7-il-monotiofosfato) (Italian)
Diolice
ENT 17,957
Meldane
Meldone
Muscatox
NCI-C08662
Negashunt
Phosphorothioic acid, O,O-diethyl ester, O-ester with 3-chloro-7-hydroxy-4-methylcoumarin
Resistox
Resitox
Suntol
Thiophosphate de O,O-diethyle et de O-(3-chloro-4-methyl-7-coumarinyle) (French)
Umbethion
UN 2783 [Organophosphorus pesticides, solid, toxic, N.O.S.]

56-73-5
$C_6H_{13}O_9P$
260.14
O=P(OCC(O)C(O)C(O)C(O)C=O)(O)O
Glucose-6-phosphate
D-Glucose, 6-(dihydrogen phosphate) (9CI)

56-75-7
$C_{11}H_{12}Cl_2N_2O_5$
323.15
O=C(NC(C(O)c(ccc(N(=O)=O)c1)c1)CO)C(Cl)Cl
Acetamide, 2,2-dichloro-N-(β-hydroxy-α-(hydroxymethyl)-p-nitrophenethyl)-, D-(-)-threo
Acetamide, 2,2-dichloro-N-(β-hydroxy-α-(hydroxymethyl)-p-nitrophenethyl)-
Acetamide, 2,2-dichloro-N-(2-hydroxy-1-(hydroxymethyl)-2-(4-nitrophenyl)ethyl)-, (R-(R*,R*))-
Alficetyn
Ambofen
Amphenicol
Amphicol
Amseclor
Anacetin
Aquamycetin
Austracil
Austracol
Biocetin
Biophenicol
CAF
CAM
CAP
Catilan
Chemicetin
Chemicetina
Chlomin
Chlomycol
Chloramex
Chloramfenikol (Czech)
Chloramficin
Chloramfilin
Chloramphenicol
D-Chloramphenicol
D-threo-Chloramphenicol
D-(-)-threo-Chloramphenicol
Chloramsaar
Chlorasol
Chlora-tabs
Chloricol
Chlornitromycin
Chloroamphenicol
Chlorocaps
Chlorocid
Chlorocide
Chlorocidin C
Chlorocidin c tetran
Chlorocid S
Chlorocol
Chloroject L
Chloromax
Chloromycetin
Chloromycetny (Polish)
Chloronitrin
Chloroptic
Chloro-25 vetag
Chlorovules
Cidocetine
Ciplamycetin
Cloramficin
Cloramicol
Cloramidina
Cloroamfenicolo (Italian)
Clorocyn
Cloromisan
Clorosintex
Comycetin
CPH
Cylphenicol
Desphen
Detreomycine
Dextromycetin
D-(-)-threo-2-Dichloroacetamido-1-p-nitrophenyl-1,3-propanediol
D-threo-N-Dichloroacetyl-1-p-nitrophenyl-2-amino-1,3-propanediol
D-(-)-threo-2,2-Dichloro-N-(β-hydroxy-α-(hydroxymethyl))-p-nitrophenethylacetamide
D-(-)-2,2-Dichloro-N-(β-hydroxy-α-(hydroxymethyl)-p-nitrophenylethyl)acetamide
D-threo-N-(1,1'-Dihydroxy-1-p-nitrophenylisopropyl)dichloroacetamide
Doctamicina
Econochlor
Embacetin
Emetren
Enicol
Enteromycetin
Erbaplast
Ertilen
Farmicetina
Fenicol
Globenicol
Glorous
Halomycetin
Hortfenicol
I 337A
Intramycetin
Isicetin
Ismicetina
Isophenicol
Isopto Fenicol
Juvamycetin
Kamaver
Kemicetina
Kemicetine
Klorita
Klorocid S
Leukomyan
Leukomycin
Levomicetina
Levomycetin
Loromisan
Loromisin
Mastiphen
Mediamycetine
Micloretin
Micochlorine
Micoclorina
Microcetina
Mychel
Mycinol
NCI-C55709
D-(-)-threo-1-p-Nitrophenyl-2-dichloracetamido-1,3-propanediol
D-threo-1-(p-Nitrophenyl)-2-(dichloroacetylamino)-1,3-propanediol
Normimycin V
Novochlorocap
Novomycetin

Novophenicol
NSC-3069
Oftalent
Oleomycetin
Opclor
Opelor
Ophthochlor
Ophtochlor
Otachron
Otophen
Pantovernil
Paraxin
Pentamycetin
Quemicetina
Rivomycin
Romphenil
Septicol
Sificetina
Sintomicetina
Sintomicetine R
Stanomycetin
Synthomycetin
Synthomycetine
Synthomycine
Tevcocin
Tevcosin
Tifomycin
Tifomycine
Treomicetina
U-6062
Unimycetin
Veticol

56-81-5
$C_3H_8O_3$
92.11
OCC(O)CO
Glycerol
Glycerin (ACGIH,OSHA)
Glycerin, Anhydrous
Glycerine
Glycerin, Synthetic
Glyceritol
Glycyl alcohol
Clyzerin, Wasserfrei (German)
Grocolene
Moon
1,2,3-Propanetriol
Star
Superol
Synthetic glycerin
90 Technical glycerine
Trihydroxypropane
1,2,3-Trihydroxypropane

56-82-6
$C_3H_6O_3$
90.09
Glyceraldehyde, DL
Glyceraldehyde, (+-)-
DL-Glyceric aldehyde

56-84-8
$C_4H_7NO_4$
133.12
O=C(O)C(N)CC(=O)O
Aspartic acid, L
(l)-Aspartic acid

56-85-9
$C_5H_{10}N_2O_3$
146.17
O=C(O)C(N)CCC(=O)N
Glutamine, L
2-Aminoglutaramic acid
l-2-Aminoglutaramidic acid
Cebrogen
Glumin
Glutamic acid amide
Glutamic acid 5-amide
Glutamine
γ-Glutamine
l-Glutamine (9CI)
Levoglutamid
Levoglutamide
Stimulina

56-86-0
$C_5H_9NO_4$
147.15
O=C(O)C(N)CCC(=O)O
Glutamic acid, L
α-Aminoglutaric acid
l-2-Aminoglutaric acid
2-Aminopentanedioic acid
1-Aminopropane-1,3-dicarboxylic acid
Glusate
Glutacid
Glutamic acid
l-Glutamic acid
α-Glutamic acid
Glutaminic acid
D-Glutamiensuur
l-Glutaminic acid
Glutaminol
Glutaton

56-87-1
$C_6H_{14}N_2O_2$
146.22
O=C(O)C(N)CCCCN
Lysine, L
Aminutrin
α,ε-Diaminocaproic acid
2,6-Diaminohexanoic acid
Lysine
l-Lysine (9CI)
l-(+)-Lysine
Lysine acid

56-89-3
$C_6H_{12}N_2O_4S_2$
240.32
O=C(O)C(N)CSSCC(N)C(=O)O
Cystine, L
Cysteine disulfide
Cystin
Cystine
(-)-Cystine
l-Cystine (9CI)
Cystine acid
Dicysteine
β,β'-Dithiodialanine
Oxidized l-cysteine

56-92-8
$C_5H_9N_3.2ClH$
184.09
[Cl-].[Cl-].NCCc1c[nH]cn1
Histamine, dihydrochloride
Bichlorhydrate d'histamine (French)
Histamine dichloride

56-93-9
$C_{10}H_{16}N.Cl$
185.72
[Cl-].C[N+](C)(C)Cc1ccccc1
Ammonium, benzyltrimethyl-, chloride
Benzenemethanaminium, N,N,N-trimethyl-, chloride (9CI)
Benzyltrimethylammonium chloride
BTM
TMBAC
Trimethylbenzylammonium chloride

56-94-0
$C_{32}H_{52}N_4O_4.2Br$
716.70
Ammonium, (m-hydroxyphenyl)trimethyl-, bromide, decamethylenebis(methylcarbamate)
Demecarium bromide
Humorsol

56-95-1
$C_{22}H_{30}Cl_2N_{10}.2C_2H_4O_2$
625.50
Chlorhexidine diacetate
Arlacide A
Bactigras
Biguanide, 1,1'-hexamethylenebis(5-(p-chlorophenyl)-, diacetate (8CI)
1,6-Bis(p-chlorophenylbiguanido)hexane diacetate
1,6-Bis(5-(p-chlorophenyl)biguandino)hexane diacetate
Bis(p-chlorophenyldiguanidohexane) diacetate
N,N'-Bis(4-chlorophenyl)-3,12-diimino-2,4,11,13-tetraazatetradecanediimidamide, diacetate
Caswell No. 481E
Chlorhexidine acetate (VAN)
Chlorhexidine diacetate
Chlorohexidine diacetate
10,040 Diacetate
1,6-Di(4'-chlorophenyldiguanidino)hexane diacetate
EPA Pesticide Chemical Code 045502
1,1'-Hexamethylene bis(5-(p-chlorophenyl)biguanide) diacetate
1,1'-Hexamethylenebis(5-(p-chlorophenyl)biguanide) diacetate

1,1'-Hexamethylenebis(5-(p-chlorophenyl)biguanide)diacetate
Hibitane diacetate
Nolvasan
NSC-526936
2,4,11,13-Tetraazatetradecanediimidamide, N,N''-bis-(4-chlorophenyl)-3,12-diimino-, diacetate (9CI)
2,4,11,13-Tetraazatetradecanediimidamide, N,N'-bis(4-chlorophenyl)-3,12-diimino-, diacetate

57-00-1
$C_4H_9N_3O_2$
131.12
O=C(O)CN(C(=N)N)C
Creatine (8CI)
AI3-15320
N-(Aminoiminomethyl)-N-methylglycine
Creatin
Creatine, hydrate
Glycine, N-(aminoiminomethyl)-N-methyl- (9CI)
Kreatin
Krebiozon
(α-Methylguanido)acetic acid
N-Methyl-N-guanylglycine
NSC-8752

57-03-4
$C_3H_9O_6P$
172.07
O=P(OCC(O)CO)(O)O
α-Glycerophosphoric acid
Glycerol, 1-(dihydrogen phosphate) (8CI)
Glycerol α-phosphate
Glycerophosphate
1-Glycerophosphate
3-Glycerophosphate
Glycerophosphoric acid
1-Glycerophosphoric acid
NSC-9231
α-Phosphoglycerol
1,2,3-Propanetriol, 1-(dihydrogen phosphate) (9CI)

57-06-7
C_4H_5NS
99.16
N(=C=S)CC=C
Isothiocyanic acid, allyl ester
AITC
Aitk
Allyl isorhodanide
Allyl isosulfocyanate
Allyl isosulphocyanate
Allyl isothiocyanate
Allyl isothiocyanate, Stabilized [UN 1545]
Allylisothiokyanat (Czech)
Allyl mustard oil
Allylsenevol
Allylsenfoel (German)
Allyl sevenolum
Allyl thiocarbonimide
Artificial mustard oil
Artificial oil of mustard
Carbospol
Fema No. 2034
Isothiocyanate d'allyle (French)
3-Isothiocyanato-1-propene
Mustard oil
NCI-C50464
Oil of Mustard, Artificial
Oil of Mustard BPC 1949
Oleum Sinapis Volatile
Propene, 3-isothiocyanato-
2-Propenyl isothiocyanate
Redskin
Senf Oel (German)
Synthetic Mustard Oil
UN 1545 [Allyl isothiocyanate, stabilized]
Volatile Oil of Mustard

57-09-0
$C_{19}H_{42}N.Br$
364.53
[Br-].CCCCCCCCCCCCCCCC[N+](C)(C)C
Ammonium, hexadecyltrimethyl-, bromide
Acetoquat CTAB
Bromat
Cee Dee
Centimide
Cetab
Cetarol
Cetavlon
Cetrimide
Cetrimide BP
Cetrimonium bromide
Cetylamine
Cetyltrimethylammonium bromide
N-Cetyltrimethylammonium bromide
Cirrasol-OD
CTAB
CTMAB
Cycloton V
1-Hexadecanaminium, N,N,N-trimethyl-, bromide
Hexadecyltrimethylammonium bromide
n-Hexadecyltrimethylammonium bromide
n-Hexadecyl-N,N,N-trimethylammonium bromide
(1-Hexadecyl)trimethylammonium bromide
Lissolamine
Lissolamine A
Lissolamine V
Lissolamin V
Micol
Pollacid
Quamonium
Suticide
Trimethylcetylammonium bromide
N,N,N-Trimethyl-1-hexadecanaminium bromide
Trimethylhexadecylammonium bromide

57-10-3
$C_{16}H_{32}O_2$
256.48
O=C(O)CCCCCCCCCCCCCCC
Palmitic-acid
Cetylic acid
Emersol 140
Emersol 143
Hexadecanoic acid
n-Hexadecanoic acid
n-Hexadecoic acid
Hexadecylic acid
Hydrofol
Hystrene 8016
Hystrene 9016
Industrene 4516
1-Pentadecanecarboxylic acid

57-11-4
$C_{18}H_{36}O_2$
284.54
O=C(O)CCCCCCCCCCCCCCCCC
Stearic-acid
Century 1240
Cetylacetic acid
Dar-Chem 14
Emersol 120
Emersol 132
Emersol 150
Formula 300
Glycon DP
Glycon S-70
Glycon S-80
Glycon S-90
Glycon TP
Groco 54
Groco 55
Groco 58
Groco 59
Groco 551
Hydrofol Acid 1655
Hydrofol Acid 1855
Hydrofol 1895
1-Heptadecanecarboxylic acid
Hy-Phi 1199
Hy-Phi 1205
Hy-Phi 1303
Hy-Phi 1401
Hystrene 80
Hystrene 4516
Hystrene 5016
Hystrene 7018
Hystrene 9718
Industrene 5016
Industrene 8718
Industrene 9018
Kam 1000
Kam 2000
Kam 3000
Neo-Fat 18
Neo-Fat 18-S
Neo-Fat 18-53
Neo-Fat 18-54
Neo-Fat 18-55
Neo-Fat 18-59
Neo-Fat 18-61
Octadecanoic acid
n-Octadecanoic acid

Pearl stearic
Stearex beads
Stearophanic acid
Tegostearic 254
Tegostearic 255
Tegostearic 272

57-12-5
CN
26.02
N#C
Cyanide
Carbon nitride ion (CN^{1-})
Cyanide(1-)
Cyanide anion
Cyanide (CN^{1-})
Cyanide ion
Cyanide(1-) ion
Cyanure (French)
Hydrocyanic acid, ion(1-)
Isocyanide
RCRA waste number P030

57-13-6
CH_4N_2O
60.07
O=C(N)N
Urea
B-I-K
Carbamide
Carbamide resin
Carbamimidic acid
Carbonyl diamide
Carbonyldiamine
Isourea
Mocovina (Czech)
NCI-C02119
Prespersion, 75 urea
Pseudourea
Supercel 3000
Ureaphil
Ureophil
Urevert
Varioform II

57-14-7
$C_2H_8N_2$
60.12
N(N)(C)C
Hydrazine, 1,1-dimethyl
Dimazin
Dimazine
1,1-Dimethylhydrazin (German)
Dimethylhydrazine
asymmetric Dimethylhydrazine
N,N-Dimethylhydrazine
as-Dimethyl hydrazine
u-Dimethylhydrazine
uns-Dimethylhydrazine
unsym-Dimethylhydrazine
1,1-Dimethylhydrazine (ACGIH,OSHA)
Dimethylhydrazine, unsymmetrical [UN 1163]
DMH
Niesymetryczna dwu metylohydrazyna (Polish)
RCRA waste number U098
UDMH (DOT)
UN 1163 [Dimethylhydrazine, unsymmetrical]
Unsymmetrical dimethylhydrazine

57-15-8
$C_4H_7Cl_3O$
177.46
OC(C(Cl)(Cl)Cl)(C)C
2-Propanol, 1,1,1-trichloro-2-methyl
Acetone chloroform
Anhydrous chlorobutanol
Chlorbutanol
Chlorbutol
Chloretone
Chlorobutanol
Chlorobutanol, Anhydrous
Clortran
Methaform
2-Propanol, 2-methyl-1,1,1-trichloro-
Sedaform
Trichloro-tert-butanol
Trichloro-t-butyl alcohol
β,β,β-Trichloro-tert-butyl alcohol
t-Trichlorobutyl alcohol
HCP
1,1,1-Trichloro-2-methyl-2-propanol

57-22-7
$C_{46}H_{56}N_4O_{10}$
825.06
CCC9(O)CC3CN(CCc1c([nH]c2ccccc12)C(C3)(C(=O)OC)c4cc5c(cc4OC)N(C=O)C6C(O)(C(OC(C)=O)C7(CC)C=CCN8CCC56C78)C(=O)OC)C9
Leurocristine
LCR
NCI-C04864
NSC-67574
22-Oxovincaleukoblastine
Oncovin
VCR
Vincaleukoblastine, 22-oxo-
Vincrystine
Vinkristin
Vincristine

57-24-9
$C_{21}H_{22}N_2O_2$
334.45
O=C(N(c(c(C1(C(N(C2)CC(C3C4C5OC6)=C6)C3)C2)ccc7)c7)C14)C5
Strychnine (ACGIH,OSHA) [UN 1692]
Certox
Dolco mouse cereal
Kwik-Kil
Mole Death
Mouse-Nots
Mouse-Rid
Mouse-Tox
Pied Piper Mouse Seed
RCRA waste number P108
Ro-Dex
Sanaseed
Stricnina (Italian)
Strychnidin-10-one
Strychnin (German)
Strychnine, Liquid [UN 1692]
Strychnine, Solid [UN 1692]
Strychnos
UN 1692 [Strychnine or Strychnine salts]

57-27-2
$C_{17}H_{19}NO_3$
285.37
O(c(c(c(cc1)CC(N(CC2)C)C3C=CC4O)C235)c1O)C45
Morphinan-3,6-α-diol, 7,8-didehydro-4,5-α-epoxy-17-methyl
9H-9,9c-Iminoethanophenanthro(4,5-bcd)furan-3,5-diol, 4a,5,7a,8-tetrahydro-12-methyl-
Morfina (Italian)
Morphia
Morphin (German)
Morphina
Morphine
(-)-Morphine
Morphinism
Morphinum
Morphium

57-30-7
$C_{12}H_{12}N_2O_3$.Na
255.25
[Na+].CCC1(C(=O)NC(=O)[N-]C1=O)c2ccccc2
Barbituric acid, 5-ethyl-5-phenyl-, sodium salt
5-Ethyl-5-phenylbarbituric acid sodium
5-Ethyl-5-phenylbarbituric acid sodium salt
5-Ethyl-5-phenyl-2,4,6-(1H,3H,5H)pyrimidinetrione monosodium salt
Gardenal sodium
Luminal sodium
PBS
Phenemalum
Phenobal sodium
Phenobarbital elixir
Phenobarbital na
Phenobarbital sodium
Phenobarbital sodium salt
Phenobarbitone sodium
Phenobarbitone sodium salt
Phenyl-aethyl-barbitursaeure natrium (German)
Phenylethylbarbituric acid, sodium salt
2,4,6(1H,3H,5H)-Pyrimidinetrione, 5-ethyl-5-phenyl-, monosodium salt (9CI)
Sodium 5-ethyl-5-phenylbarbiturate
Sodium luminal
Sodium phenobarbital

Sodium phenobarbitone
Sodium phenylethylbarbiturate
Sodium phenylethylmalonylurea
Sol Phenobarbital
Sol Phenobarbitone
Soluble Phenobarbital
Soluble Phenobarbitone

57-33-0
$C_{11}H_{18}N_2O_3.Na$
249.30
[Na+].CCCC(C)C1(CC)C(=O)NC(=O)[N-]C1=O
Barbituric acid, 5-ethyl-5-(1-methylbutyl)-, sodium salt
Auropan
Barpental
Biosedan
Butylmethyl ethyl malonyl urea sodium
Butylone
Carbrital
Continal
Diabutal
Embutal
Etaminal sodium
Ethaminal
Ethaminal sodium
5-Ethyl-5-(1-methylbutyl)-barbituric acid sodium salt
5-Ethyl-5-(1-methylbutyl)-2,4,6(1H,3H,5H)-pyrimidinetrione monosodium salt
Euthatal
Ipral sodium
Isobarb
Mebubarbital
Mebubarbital sodium
Mebumal natrium
Mebumal sodium
Mintal
Napental
Nembutal
Nembutal sodium
Pacifan
Palapent
Penbar
Pentabarbital sodium
Pental
Pentobarbital sodium
Pentobarbitone sodium
Pentonal
Pentone
Pentyl
Propylmethylcarbinylethyl barbituric acid sodium salt
2,4,6(1H,3H,5H)-Pyrimidinetrione, 5-ethyl-5-(1-methylbutyl)-, monosodium salt
Rivadorn
Sagatal
Sodital
Sodium ethaminal
Sodium 5-ethyl-5-(1-methylbutyl)barbiturate
Sodium nembutal
Sodium-pent
Sodium pentabarbital
Sodium pentabarbitone
Sodium pentobarbital
Sodium pentobarbitone
Sodium pentobarbiturate
Soluble Pentobarbital
Somnopentyl
Sonistan
Sontobarbital nabitone
Sopental
Sotyl
Vetbutal

57-39-6
$C_9H_{18}N_3OP$
215.27
O=P(N(C1C)C1)(N(C2C)C2)N(C3C)C3
Phosphine oxide, tris(2-methyl-1-aziridinyl)
C 3172
ENT 50,003
MAPO
Metapoxide
Metepa
Methaphoxide
Methyl aphoxide
1,1',1''-Phosphinylidynetris-(2-methyl)azridine
Tris(2-methyl-1-aziridinyl)phosphine oxide
Tris(2-methylaziridin-1-yl)phosphine oxide
N,N',N''-Tris(1-methylethylene)phosphoramide
Tris(1-methylethylene)phosphoric triamide

57-41-0
$C_{15}H_{12}N_2O_2$
252.29
O=C1NC(=O)C(N1)(c2ccccc2)c3ccccc3
Hydantoin, 5,5-diphenyl
Aleviatin
Antisacer
Auranile
Causoin
Citrullamon
Citrulliamon
Comital
Comitoina
Convul
Danten
Dantinal
Dantoinal
Dantoinal klinos
Dantoine
Denyl
Didan-TDC-250
Difenin
Difetoin
Difhydan
Difenilhidantoina (Spanish)
Dihycon
Di-Hydan
Dihydantoin
Dilabid
Di-Lan
Dilantine
Dilantin
Dillantin
Dintoin
Dintoina
Diphantoin
Diphedal
Diphedan
Diphenin
Diphenine
Diphentoin
Diphentyn
Diphenylan
Diphenylhydantoin
5,5-Diphenylhydantoin
Diphenylhydantoine (French)
5,5-Diphenylimidazolidin-2,4-dione
5,5-Diphenyl-2,4-imidazolidinedione
Di-Phetine
Ditoinate
DPH
5,5-Dwufenylohydantoina (Polish)
EKKO capsules
Elepsindon
Enkelfel
Epamin
Epanutin
Epasmir "5"
Epdantoine simple
Epelin
Epifenyl
Epihydan
Epilan
Epilan-d
Epilantin
Epinat
Epised
Eptal
Eptoin
Fenantoin
Fenidantoin "S"
Fenitoina
Fentoin
Fenylepsin
Fenytoine
Gerot-epilan-D
Hidan
Hidantal
Hidantilo
Hidantina
Hidantina senosian
Hidantina vitoria
Hidantomin
Hindatal
Hydantal
Hydantin
Hydantoin
Hydantoinal
Ictalis simple
Idantoil
Idantoin
Kessodanten
Labopal
Lehydan
Lepitoin
Lepsin
Minetoin
NCI-C55765
Neos-Hidantoina
Neosidantoina
Novantoina
OM Hidantoina simple
OM-Hydantoine
Oxylan

Phanantin
Phanatine
Phenatine
Phenatoine
Phenitoin
Phentoin
Phenytoin
Ritmenal
Saceril
Sanepil
Silantin
Sodantoin
Sodanton
Solantin
Solantoin
Solantyl
Sylantoic
Tacosal
Thilophenyl
Toin
Toin unicelles
Zentronal
Zentropil

57-42-1
$C_{15}H_{21}NO_2$
247.37
CCOC(=O)C1(CCN(C)CC1)c2ccccc2
Isonipecotic acid, 1-methyl-4-phenyl-, ethyl ester
Demarol
Demerol
Dolantin
Dolcontral
Dolosal
Dolsin
Ethyl 1-methyl-4-phenyliso-nipecotate
Ethyl 1-methyl-4-phenylpiperi-dine-4-carboxylate
Isonipecaine
Lidol
Meperidine
N-Methyl-4-phenyl-4-carbe-thoxypiperidine
1-Methyl-4-phenylisonipecotic acid, ethyl ester
1-Methyl-4-phenyl-piperidin-4-carbon-saeure-aethylester (German)
Methyl phenylpiperidine car-bonic acid ethyl ester
1-Methyl-4-phenylpiperidine-4-carboxylic acid ethyl ester
Nemerol
Pethidine
Pethidineter
Petydyna (Polish)
Phetidine
Pipersal
Piridosal

57-43-2
$C_{11}H_{18}N_2O_3$
226.31
CCC1(CCC(C)C)C(=O)NC(=O)NC1=O
Barbituric acid, 5-ethyl-5-iso-pentyl
Amal
Amasust
Amital
Amobarbital
Amobarbitone
Amospan
Amybal
Amylbarbitone
Amylobarbital
Amylobarbitone
Amytal
Barbamil
Barbamyl
Barbamyl acid
Binoctal
Dorlotyn
Dormytal
5-Ethyl-5-isoamylbarbituric acid
5-Ethyl-5-isoamylmalonyl urea
Ethylisopentylbarbituric acid
5-Ethyl-5-isopentylbarbituric acid
5-Ethyl-5-(3-methylbutyl)-barbituric acid
Eunoctal
Isoamylethylbarbituric acid
5-Isoamyl-5-ethylbarbituric acid
Isomyl
Isomytal
Mylodorm
NSC-10815
Pentymal
Pentymalum
2,4,6(1H,3H,5H)-Pyrimidine-trione, 5-ethyl-5-(3-methyl-butyl)- (9CI)
Robarb
Schiwanox
Sednotic
Somnal
Stadadorm
Sumital
Talamo

57-44-3
$C_8H_{12}N_2O_3$
184.22
CCC1(CC)C(=O)NC(=O)NC1=O
Barbituric acid, 5,5-diethyl
Barbital
Barbitone
DEBA
Diemal
Diethylbarbitone
Diethyl-barbituric acid
5,5-Diethylbarbituric acid
Diethylmalonylurea
Dormonal
Ethylbarbital
Hypnogene
Kyselina 5,5-diethylbarbiturova (Czech)
Malonal
2,4,6(1H,3H,5H)-Pyrimidine-trione, 5,5-diethyl-
Sedeval
Uronal
Veroletten
Verolettin
Veronal
Vesperal

57-47-6
$C_{15}H_{21}N_3O_2$
275.39
O=C(Oc(ccc(N(C(N(CC1)C)C12C)C)c23)c3)NC
Physostigmine
Carbamic acid, methyl-, ester with eseroline
CS 58525
Erserine
Eserine
Eserolein, methylcarbamate (ester)
Fysostigmin (Czech)
Physostol

57-48-7
$C_6H_{12}O_6$
180.16
O=C(C(O)C(O)C(O)CO)CO
Fructose
AI3-23514
D-Fructose
Fructose solution
Levulose

57-50-1
$C_{12}H_{22}O_{11}$
342.34
O(C(C(O)C(O)C1O)CO)C1OC(OC(C2O)CO)(C2O)CO
Sucrose
Beet sugar
Cane sugar
Confectioner's sugar
Fructofuranoside, α-D-gluco-pyranosyl, β-d
β-D-Fructofuranoside, α-D-glucopyranosyl
Glucopyranoside, β-D-fructo-furanosyl, α-d
α-D-Glucopyranosyl β-D-fructofuranoside
(α-D-Glucosido)-β-D-fruc-tofuranoside
Granulated sugar
NCI-C56597
Rock Candy
Saccharose
Saccharum
Sucrose (OSHA)
Sugar

57-53-4
$C_9H_{18}N_2O_4$
218.29
O=C(OCC(CCC)(C)COC(=O)N)N
1,3-Propanediol, 2-methyl-2-propyl-, dicarbamate
Amepromat
Amosene
Anastress
Anathylmon
Andaksin
Andaxin
Aneural
Aneurol

57-55-6

Aneusral
Aneuxal
Aneuxral
Ansiatan
Ansietan
Ansil
Ansiowas
Anural
Anxietil
Apascil
Apasil
Appetrol-sr
Arcoban
Arpon
Artolon
Ataraxine
Atraxin
Auxietil
Ayermate
Bamd 400
Bamo 400
Biobamat
Brobamate
Calmadin
Calmax
Calmiren
Canquil-400
Cap-O-Tran
Carbamic acid, 2-methyl-2-propyltrimethylene ester
Cirpon
Cirponyl
Crestanil
Cypron
Cyrpon
Dapaz
Dicandiol
2,2-Di(carbamoyloxymethyl)-pentane
Diurnal
Diveron
Dormabrol
Ecuanil
Edenal
Enorden
Epicur
Epikur
Equanil
Equanil suspension
Equatrate
Equilium
Equinil
Equitar
Erina
Estasil
Fas-Cile
Gadexyl
Gagexyl
Harmonin
Hartol
Holbamate
Ipsotian
Kessobamate
Klort
Larten
Lepenil
Lepetown
Letyl
Libiolan
Madiol
Mar Bate
Margonil
Mendel
Mepamtin
Mepantin
Mepavlon
Mepiosine
Meposed
Mepranil
Meprin
Meprindon
Meprobam
Meprobamat (German)
Meprobamate
Meprobamato (Italian)
Meproban
Meprocompren
Meprocon CMC
Meprodil
Meprodiol
Meprol
Meproleaf
Mepron
Meprosa
Meprosan
Meprosin
Meprospan
Meprotabs
Meprotan
Meprovan
Meprozine
Meptran
Mesmar
2-Methyl-2-N-propyl-1,3-propanediol dicarbamate
2-Methyl-2-propyltrimethylene carbamate
Metractyl
Milprem
Miltamato
Miltann
Miltaun
Miltown
Miltuan
Miltwon
Morbam
Multaun
Neo-tran
Nephentine
Nervonus
Oasil
Optarket
Orlevol
Orolevol
Pancalma
Panediol
Pankalma
Pan-Tranquil
Paxin
3P Bamate
Pensive
Perequietil
Perequil
Perquietil
Pertranquil
Pimal
Placidon
Placitate
Prequil
Probamato
Probamyl
Procalmadiol
Procalmadol
Procalmidol
Procarbamide
Promate
Promato
Proquanil
Protran
Quaname
Quanane
Quanil
Quietidon
Quivet
Reostral
Restenil
Restenyl
Restinal
Restinil
Robamate
Sadanyl
Scolazil
Sedabamate
Sedanyl
Sedazil
Sedoquil
Sedoselecta
Selene
Seril
Setran
Shalvaton
Sk-Bamate
Solevione anastress
Sowell
Spantran
Stensolo
Tamate
Tensol
Tensonal
Trankvilan
Tranlisant
Tranmep
Tranquil
Tranquilan
Tranquilate
Tranquilax
Tranquiline
Tranquilsan
Tranquinol
Tranquisan
Trelmar
Urbil
Urbilat
Vio-Bamate
Vistabamate
Wardamate
Wyseals
Zirpon

57-55-6
$C_3H_8O_2$
76.11
OCC(O)C

1,2-Propanediol
1,2-Dihydroxypropane
Dowfrost
Methylethylene glycol
Methyl glycol
Monopropylene glycol
PG 12
Propane-1,2-diol
Propylene glycol
Propylene glycol USP
α-Propyleneglycol
1,2-Propylene glycol

1,2-Propylenglykol (German)
Sirlene
Solar Winter Ban
Trimethyl glycol

57-56-7
CH_5N_3O
75.09
O=C(NN)N
Semicarbazide
Aminomocovina (Czech)
Aminourea
Carbamic acid, hydrazide
Carbamylhydrazine
Carbazamide
Hydrazine, carbamoyl-
Hydrazinecarboxamide
Semikarbazid (Czech)
Urea, amino-

57-57-8
$C_3H_4O_2$
72.07
O=C(OC1)C1
2-Oxetanone
Betaprone
BPL
Hydracrylic acid β-lactone
3-Hydroxypropionic acid lactone
Propanolide
3-Propanolide
Propiolactone
β-Propiolactone
1,3-Propiolactone
3-Propiolactone
β-Propriolactone (ACGIH, OSHA)
β-Propiolakton (Czech)
Propionic acid, 3-hydroxy-, β-lactone
β-Propionolactone
β-Proprolactone

57-62-5
$C_{22}H_{23}ClN_2O_8$
478.92
O=C(N)C(=C(O)C(N(C)C)C(C1(O)C(O)=C(C2C(O)(c(c3c(O)cc4)c4Cl)C)C3=O)C2)C1=O
2-Naphthacenecarboxamide, 7-chloro-4-(dimethylamino)-1,4,4a,5,5a,6,11,12a-octahydro- 3,6,10,12,12a-pentahydroxy-6-methyl-1,11-dioxo
Acronize
Aureocina
Aureomycin
Aureomycin A-377
Aureomykoin
Biomitsin
Biomycin
7-Chlorotetracycline
Chlortetracycline
Chrysomykine
CTC
Duomycin
Flamycin

57-63-6
$C_{20}H_{24}O_2$
296.44
OC(C#C)(C(C(C(C(c(c(cc(O)c1)C2)c1)C3)C2)C4)(C3)C)C4
19-Nor-17-α-pregna-1,3,5(10)-trien-20-yne-3,17-diol
Amenoron
Amenorone
Anovlar
Aethinyoestradiol (German)
Chee-o-gen
Chee-o-genf
3,17-β-Dihydroxy-17-α-ethynyl-1,3,5(10)-oestratriene
3,17-β-Dihydroxy-17-α-ethynyl-1,3,5(10)-estratriene
Diognat-E
Diogyn-E
Diprol
Dyloform
EE
EE_2
EED
EO
Esteed
Estigyn
Estinyl
Eston-E
Estoral
Estoral (orion)
Estorals
Estradiol, 17-ethynyl-
Estra-1,3,5(10)-triene-3,17-β-diol, 17-α-ethynyl-
Estrogen
Ethidol
Ethinoral
17-α-Ethinyl-3,17-dihydroxy-$\delta^{1,3,5}$-estratriene
17-α-Ethinyl-3,17-dihydroxy-$\delta^{1,3,5}$Oestratriene
Ethinyl estradiol
17-Ethinylestradiol
17-Ethinyl-3,17-estradiol
17-α-Ethinylestradiol
17-α-Ethinyl-17-β-estradiol
17-α-Ethinylestra-1,3,5(10)-triene-3,17-β-diol
Ethinylestriol
Ethinyloestradiol
17-Ethinyl-3,17-oestradiol
Ethinyl-oestranol
17-α-Ethinyloestra-1,3,5(10)-triene-3,17-β-diol
17-α-Ethinyl-$\delta^{1,3,5(10)}$oestratriene-3,17-β-diol
Ethinyloestriol
17-Ethynyl-3,17-dihydroxy-1,3,5-oestratriene
Ethynylestradiol
17-α-Ethynylestradiol
17-α-Ethynylestradiol-17-β
17-α-Ethynyl-1,3,5(10)-estratriene-3,17-β-diol
17-α-Ethynylestra-1,3,5(10)-triene-3,17-β-diol
Ethynyloestradiol
17-Ethynyloestradiol
17-α-Ethynyloestradiol
17-α-Ethynyl-17-β-oestradiol
17-α-Ethynyloestradiol-17-β
17-Ethynyloestra-1,3,5(10)-triene-3,17-β-diol
17-α-Ethynyl-1,3,5-oestratriene-3,17-β-diol
17-α-Ethynyl-1,3,5(10)-oestratriene-3,17-β-diol
17-α-Ethynyloestra-1,3,5(10)-triene-3,17-β-diol
Eticyclin
Eticyclol
Etinestrol
Etinestryl
Etinoestryl
Etistradiol
Feminone
Ginestrene
Inestra
Linoral
Lynoral
Menolyn
Neo-estrone
(17-α)-19-Norpregna-1,3,5(10)-trien-20-yne-3,17,diol
19-Nor-17-α-pregna-1,3,5(10)-triene-20-yne-3,17-diol
Novestrol
NSC-10973
Oradiol
Orestralyn
Orestrayln
Palonyl
Perovex
Primogyn
Primogyn C
Primogyn M
Progynon C
Spanestrin
Ylestol

57-64-7
$C_{15}H_{21}N_3O_2.C_7H_6O_3$
413.52
Physostigmine, salicylate (1:1)
Antilirium
AR-44
Eserine salicylate
Physostol salicylate
Salicylic acid, Compd. with physostigmine (1:1)
Tl-1380

57-66-9
$C_{13}H_{19}NO_4S$
285.39
O=C(O)c(ccc(S(=O)(=O)N(CCC)CCC)c1)c1
Benzoic acid, p-(dipropylsulfamoyl)
Apurina
Benecid
Benemid
Benuryl
Benzoic acid, 4-((dipropylamino)sulfonyl)- (9CI)
4-((Dipropylamino)sulfonyl)-benzoic acid
p-(Dipropylsulfamoyl)benzoic acid
4-(Dipropylsulfamoyl)benzoic

acid
p-(Dipropylsulfamyl)benzoic acid
Ethamide
NCI-C56097
Probecid
Proben
Probenecid
Probenecid acid
Probenemid
Prolongine
Synergid R
Tubophan
Uricosid

57-67-0
$C_7H_{10}N_4O_2S$
214.27
O=S(=O)(NC(=N)N)c(ccc(N)c1)c1
Sulfanilamide, N[1]-amidino
Abiguanil
N[1]-Amidinosulfanilamide
4-Amino-N-(aminoimino-methyl)benzenesulfonamide
p-Aminobenzenesulfonyl-guanidine
N-p-Aminobenzenesulphonyl-guanidine monohydrate
4-Amino-N-(diaminomethyl-ene)benzenesulfonamide
Benzenesulfonamide, 4-amino-N-(diaminomethylene)-
Aterian
N[1]-(Diaminomethylene)sulfanil-amide
Ganidan
Guamide
Guanicil
Guanidan
Guanidine, sulfanilyl-
N[1]-Guanylsulfanilamide
Resulfon
RP 2275
Ruocid
Shigatox
Suganyl
Sulfaguanidine
Sulfaguine
Sulfanilguanidine
Sulfanilylguanidine
Sulfoguanidine
Sulfoguenil
Sulfoquanidine
Sulgin
Sulphaguanidine

57-68-1
$C_{12}H_{14}N_4O_2S$
278.36
O=S(=O)(Nc(nc(cc1C)C)n1)c(ccc(N)c2)c2
Sulfanilamide, N[1]-(4,6-di-methyl-2-pyrimidinyl)
A-502
2-(p-Aminobenzenesulfon-amido)-4,6-dimethylpyrim-idine
6-(4'-Aminobenzol-sulfonam-ido)-2,4-dimethylpyrimidin (German)
(p-Aminobenzolsulfonyl)-2-amino-4,6-dimethylpyrim-idin (German)
Azolmetazin
Benzenesulfonamide, 4-amino-N-(4,6-dimethyl-2-pyrimidin-yl)-
Cremomethazine
Diazil
Diazyl
N[1]-(4,6-Dimethyl-2-pyrim-idinyl)sulfanilamide
N-(4,6-Dimethyl-2-pyrim-idyl)sulfanilamide
N[1]-(4,6-Dimethyl-2-pyrim-idyl)sulfanilamide
4,6-Dimethyl-2-sulfanil-amidopyrimidine
Dimezathine
Mermeth
Metazin
NCI-C56600
Neasina
Neazina
Pirmazin
Primazin
SA III
Spanbolet
Sulfadimerazine
Sulfadimesine
Sulfadimethyldiazine
Sulfadimethylpyrimidine
Sulfadimetine
Sulfadimezine
Sulfadimidine
Sulfadine
Sulfaisodimidine
Sulfa-isodimerazine
Sulfametazyny (Polish)
Sulfamethazine
Sulfamethiazine
Sulfamethin
Sulfamezathine
2-Sulfanilamido-4,6-di-methylpyrimidine
Sulfisomidin
Sulfisomidine
Sulfodimesin
Sulfodimezine
Sulmet
Sulphadimethylpyrimidine
Sulphadimidine
Sulphamethazine
Superseptil
Vertolan

57-71-6
$C_4H_7NO_2$
101.12
O=C(C(=NO)C)C
2,3-Butanedione, monooxime
Biacetylmonoxime
2,3-Butanedione 2-oxime
DAM
Diacetylmonooxime
Diacetylmonoxime
2-Oximino-3-butanone

57-74-9
$C_{10}H_6Cl_8$
409.76
ClC1CC2C(C1Cl)C3(Cl)C(=C(Cl)C2(Cl)C3(Cl)Cl)Cl
4,7-Methanoindan, 1,2,4,5,6,7,8,8-octachloro-3a,4,7,7a-tetrahydro
Aspon-chlordane
Belt
CD 68
Chloordaan (Dutch)
Chlordan
γ-Chlordan
Chlordane (ACGIH,OSHA)
Chlordane, Liquid (DOT)
Chlorindan
Chlor Kil
Chlorodane
Chlortox
Clordan (Italian)
Clordano
Corodane
Cortilan-Neu
Dichlorochlordene
Dowchlor
ENT 9,932
ENT 25,552-X
HCS 3260
Kilex Lindane
Kypchlor
Latka 1068 (Czech)
M 140
M 410
4,7-Methano-1H-indene, 1,2,4,5,6,7,8,8-octachloro-2,3,3a,4,7,7a-hexahydro-
NA 2762
NCI-C00099
Niran
Octachlor
Octachlorodihydrodicyclo-pentadiene
1,2,4,5,6,7,8,8-Octahydro-2,3,3a,4,7,7a-hexahydro-4,7-methanoindan
1,2,4,5,6,7,8,8-Octachlor-2,3,3a,4,7,7a-hexahydro-4,7-methanoindane
1,2,4,5,6,7,8,8-Octachloro-2,3,3a,4,7,7a-hexahydro-4,7-methanoindene
1,2,4,5,6,7,8,8-Octachloro-2,3,3a,4,7,7a-hexahydro-4,7-methano-1H-indene
1,2,4,5,6,7,8,8-Octachloro-3a,4,7,7a-hexahydro-4,7-methylene indane
Octachloro-4,7-methano hydroindane
Octachloro-4,7-methanotetra-hydroindane
1,2,4,5,6,7,8,8-Octachloro-4,7-methano-3a,4,7,7a-tetrahydroindane
1,2,4,5,6,7,8,8-Octachloor-3a,4,7,7a-tetrahydro-4,7-endo-methano-indaan (Dutch)
1,2,4,5,6,7,8,8-Octachloro-3a,4,7,7a-tetrahydro-4,7-methanoindan
1,2,4,5,6,7,8,8-Octachloro-3a,4,7,7a-tetrahydro-4,7-meth-

anoindane
1,2,4,5,6,7,10,10-Octachloro-4,7,8,9-tetrahydro-4,7-methyleneindane
1,2,4,5,6,7,8,8-Octachlor-3a,4,7,7a-tetrahydro-4,7-endo-methano-indan (German)
Octa-Klor
Oktaterr
1,2,4,5,6,7,8,8-Ottochloro-3a,4,7,7a-tetraidro-4,7-endo-metano-indano (Italian)
OMS 1437
Ortho-Klor
RCRA waste number U036
SD 5532
Shell SD-5532
Starchlor
Synklor
Unexan-Koeder
Tat Chlor 4
Termi-Ded
Topichlor 20
Topiclor
Topiclor 20
Toxichlor
Velsicol 1068

57-83-0
$C_{21}H_{30}O_2$
314.51
O=C(C(C(C(C(C(C(C(=CC(=O)C1)C2)(C1)C)C3)C2)C4)(C3)C)C4)C
Progesterone
Corlutin
Corlutina
Corluvite
Corporin
Corpus luteum hormone
Cyclogest
δ[4]-Pregnene-3,20-dione
Depo-Provera
Flavolutan
Fologenon
Gesterol 100
Gestone
Gestormone
Glanducorpin
Hormoflaveine
Hormoluton
Lingusorbs
Lipo-Lutin
Lucorteum Sol
Luteal Hormone
Luteine
Luteinique
Luteodyn
Luteogan
Luteohormone
Luteol
Luteosan
Luteostab
Luteovis
Lutex
Lutidon
Lutocyclin
Lutocyclin M
Lutocylin
Lutocylol
Lutoform
Lutogyl
Lutren
Lutromone
Nalutron
NSC-9704
Percutacrine
Piaponon
3,20-Pregnene-4
Pregnenedione
Pregnene-3,20-dione
Pregn-4-ene-3,20-dione
4-Pregnene-3,20-dione
Primolut
Progekan
Progestasert
Progesterol
β-Progesterone
Progesteronum
Progestin
Progestone
Prolidon
Prolutone
Syngesterone
Syngestrets
Synovex S
Syntolutan

57-85-2
$C_{22}H_{32}O_3$
344.54
O=C(OC(C(C(C(C(C(C(C(=CC(=O)C1)C2)(C1)C)C3)C2)C4)(C3)C)C4)CC
Testosterone, propionate
Agovirin
Androgen
Androsan
δ[4]-Androstene-17-β-propionate-3-one
Androst-4-en-3-one, 17-(1-oxopropoxy)-(17-β)-
Andrusol-P
Androteston
Androtest P
Anertan
Aquaviron
Bio-Testiculina
Enarmon
Homandren (amps)
Hormoteston
Masenate
Nasdol
Neo-Hombreol
NSC-9166
Okasa-Mascul
Orchiol
Orchistin
Oreton
Oreton propionate
Pantestin
Perandren
Propiokan
Recthormone testosterone
Synandrol
Sterandryl
Synerone
Telipex
Testaform
Testex
Testodet
Vulvan
Testodrin
Testogen
Testonique
Testormol
Testosteron propionate
Testosterone-17-propionate
Testosterone-17-β-propionate
Testoviron
Testoxyl
Testrex
Tostrin
TP
Uniteston

57-87-4
$C_{28}H_{44}O$
396.66
Ergosterol
AI3-18876
δ-5,7,22-Ergostatrien-3β-ol
Ergosta-5,7,22-trien-3-ol, (3β,22E)-
Ergosta-5:6,7:8,22:23-trien-3-ol
Ergosta-5,7,22-trien-3β-ol
Ergosterin
Provitamin D
Provitamin D2

57-88-5
$C_{27}H_{46}O$
386.73
OC(CCC(C1=CCC2C(C(C(C3)C(CCCC(C)C)C)(CC4)C)C3)(C24)C)C1
Cholesterol
Cholest-5-en-3-ol (3-β)- (9CI)
Cholest-5-en-3-β-ol
δ[5]-Cholesten-3-β-ol
5-Cholesten-3-β-ol
5:6-Cholesten-3-ol
5:6-Cholesten-3-β-ol
Cholesterin
Cholesterine
Cholesterol base H
Cholestrin
Cholestrol
(-)-Cholesterol
Cholesteryl alcohol
Cordulan
Dusoline
Dusoran
Dythol
Hydrocerin
3-β-Hydroxycholest-5-ene
Kathro
Lanol
Nimco Cholesterol Base H
Nimco Cholesterol Base No. 712
Provitamin D
Super Hartolan
Tegolan

57-91-0
$C_{18}H_{24}O_2$
272.42
17-α-Estradiol
3,17-Dihydroxyestratriene
3,17-α-Dihydroxyoestra-

1,3,5(10)-triene
Estra-1,3,5(10)-triene-3,17-α-diol
1,3,5-Estratriene-3,17-α-diol
Oestradiol-17-α

57-92-1
$C_{21}H_{39}N_7O_{12}$
581.67
O=CC(O)(C(OC(OC(C(O)C1O)C O)C1NC)C(O2)OC(C(O)C (O)C(NC(=N)N)C3O)C3NC (=N)N)C2C
Streptomycin
Agrimycin 17
Chemform
Gerox
Hokko-Mycin
NSC-14083
Strepcen
Streptomicina (Italian)
Streptomycin A
Streptomycine
Streptomycinum
Streptomyzin (German)

57-95-4
$C_{38}H_{44}N_2O_6$
624.84
[Cl-].[Cl-].COc1cc2CC[NH+](C) C3Cc7ccc(Oc5c(O)c(OC) cc6CC[N+](C)(C)C(Cc4ccc (O)c(Oc1cc23)c4)c56)cc7
Tubocurarine, (+)
Tubocurarin
Tubocurarine
(+)-Tubocurarine
d-Tubocurarine
d-Tubocurarine

57-96-5
$C_{23}H_{20}N_2O_3S$
404.51
O=C2C(CCS(=O)c1ccccc1)C(=O) N(N2c3ccccc3)c4ccccc4
3,5-Pyrazolidinedione, 1,2-diphenyl-4-(2-(phenylsulfinyl)-ethyl)
Anturan
Anturane
1,2-Diphenyl-3,5-dioxo-4-(2'-phenyl-sulfinyl-aethyl)-pyrazolidin (German)
1,2-Diphenyl-4-(2'-phenylsulfinethyl)-3,5-pyrazolidinedione
Diphenylpyrazone
Enturen
G 28315
4-(Phenylsulfoxyethyl)-1,2-diphenyl-3,5-pyrazolidinedione
Sulfinpyrazine
Sulfinpyrazone
Sulfoxyphenylpyrazolidine
Sulphinpyrazone
USAF GE-13

57-97-6
$C_{20}H_{16}$
256.36
c(c(c(c(c1)ccc2)c2)c(c(c3ccc4) c4)C)(c3C)c1
Benz(a)anthracene, 7,12-dimethyl
DBA
Dimethylbenzanthracene
Dimethylbenz(a)anthracene
7,12-Dimethylbenzanthrancene
7,12-Dimethylbenz(a)anthracene
9,10-Dimethyl-benzanthracene
9,10-Dimethylbenz(a)anthracene
9,10-Dimethyl-1,2-benzanthracene
9,10-Dimethyl-1,2-benzanthrazen (German)
Dimethylbenzanthrene
7,12-Dimethylbenzo(a)anthracene
1,4-Dimethyl-2,3-benzphenanthrene
DMBA
7,12-DMBA
NCI-C03918
RCRA waste number U094

58-00-4
$C_{17}H_{17}NO_2$
267.35
CN2CCc1cccc3c1C2Cc4ccc(O)c (O)c34
6a-β-Aporphine-10,11-diol
Apomorfin
Apomorphine
Apormorphine
4H-Dibenzo(de,g)quinoline-10,11-diol, 5,6,6a,7-tetrahydro-6-methyl-

58-08-2
$C_8H_{10}N_4O_2$
194.22
O=C(N(C(=O)C(N(C=N1) C)=C12)C)N2C
Caffeine
Caffein
Coffein (German)
Coffeine
3,7-Dihydro-1,3,7-trimethyl-1H-purine-2,6-dione
Eldiatric C
Guaranine
Kofein (Czech)
Koffein (German)
Methyltheobromide
NCI-C02733
No-Doz
Organex
1H-Purine-2,6-dione, 3,7-dihydro-1,3,7-trimethyl-
Thein
Theine
Theobromine, 1-methyl-
Theophylline, 7-methyl
1,3,7-Trimethyl-2,6-dioxopurine
1,3,7-Trimethylxanthine
Xanthine, 1,3,7-trimethyl

58-14-0
$C_{12}H_{13}ClN_4$
248.74
CCc1nc(N)nc(N)c1c2ccc(Cl)cc2
2,4-Pyrimidinediamine, 5-(p-chlorophenyl)-6-ethyl
BW 50-63
CD
Chloridin
Chloridine
Chloridyn
5-(4'-Chlorophenyl)-2,4-diamino-6-ethylpyrimidine
5-(4-Chlorophenyl)-6-ethyl-2,4-pyrimidinediamine
Daraclor
Darapram
Daraprim
Daraprime
2,4-Diamino-5-p-chlorophenyl-6-ethylpyrimidine
2,4-Diamino-5-(4-chlorophenyl)-6-ethylpyrimidine
Diaminopyritamin
Erbaprelina
Khloridin
Malacid
Malocid
Malocide
Maloprim
NCI-C01683
NSC-3061
Pirimecidan
Pirimetamina (Spanish)
Pyremethamine
Pyrimethamine
4753 R.P.
Tindurin
WR 2978

58-15-1
$C_{13}H_{17}N_3O$
231.33
O=C(N(N(C=1C)C)c(cccc2)c2) C1N(C)C
Antipyrine, 4-(dimethylamino)
Amidazophen
Amidazophene
Amidofebrin
Amidofen
Amidophen
Amidophenazone
Amidopyrazoline
Amidopyrin
Amidopyrine
Aminofenazone (Italian)
Aminophenazone
Aminopyrine
Anafebrina
Brufaneuxol
DAP
Dereuma
Dimapyrin
Dimethylamino-analgesine
Dimethylaminoantipyrine
4-(Dimethylamino)antipyrine
Dimethylaminoazophene
4-Dimethylamino-2,3-dimethyl-1-phenyl-3-pyrazolin-5-one
4-Dimethylamino-2,3-dimethyl-1-phenyl-5-pyrazolone
Dimethylaminophenazon

(German)
Dimethylaminophenazone
4-Dimethylaminophenazone
Dimethylaminophenyldimethyl-pyrazolin
4-Dimethylamino-1-phenyl-2,3-dimethylpyrazolone
1,5-Dimethyl-4-dimethylamino-2-phenyl-3-pyrazolone
2,3-Dimethyl-4-dimethylamino-1-phenyl-5-pyrazolone
Dipirin
Dipyrin
Dipyrine
Febrinina
Febron
Itamidone
3-Keto-1,5-dimethyl-4-dimethyl-amino-2-phenyl-2,3-dihydro-pyrazole
Mamallet-A
Netsusarin
Novamidon
1-Phenyl-2,3-dimethyl-4-di-methylaminopyrazol-5-one
1-Phenyl-2,3-dimethyl-4-di-methylaminopyrazolone-5
Piramidon
Piridol
Piromidina
Polinalin
Pyradone
Pyramidon
Pyramidone
3H-Pyrazol-3-one, 4-(dimethyl-amino)-1,2-dihydro-1,5-di-methyl-2-phenyl- (9CI)
3-Pyrazolin-5-one, 4-(dimethyl-amino)-2,3-dimethyl-1-phenyl-

58-18-4
$C_{20}H_{30}O_2$
302.50
O=C(C=C(C(C(C(C(C(C(O)(C1)C)(C2)C)C1)C3)C2)(C4)C)C3)C4
Androst-4-en-3-one, 17-β-hydroxy-17-methyl
Andrometh
Androsan
Androsan (tablets)
4-Androstene-17-α-methyl-17-β-ol-3-one
Androst-4-en-3-one, 17-hydroxy-17-methyl-, (17-β)-(9CI)
Androsten
Anertan
Anertan (tablets)
Delatestryl
Dianabol
Dumogran
Glosso sterandryl
Homandren
Hormale
17-β-Hydroxy-17-methyl-androst-4-en-3-one
Malestrone
Malogen
Masenone
Mastestona
Mesterone
Metandren
17-Methyltestosteron
Methyltestosterone
17-Methyltestosterone
17-α-Methyltestosterone
Metrone
M.T.Mucorettes
Nabolin
Neo-Hombreol-M
NSC-9701
Nu Man
Oraviron
Oreton-M
Oreton methyl
Stenolon
Steronyl
Synandrets
Synandrotabs
Testhormone
Testora
Testoviron
Testred

58-19-5
$C_{20}H_{32}O_2$
304.52
5-α-Androstan-3-one, 17-β-hydroxy-2-α-methyl
Androstan-3-one, 17-hydroxy-2-methyl-, (2-α,5-α,17-β)-
Dihydro-2-α-methyltestosterone
Dromostanolone
Drostanolone
17-β-Hydroxy-2-α-methyl-5-α-androstan-3-one
Medrosteron
Metholone
2-α-Methyldihydrotestosterone
2-α-Methyl-17-β-hydroxy-5-α-androstan-3-one

58-22-0
$C_{19}H_{28}O_2$
288.47
O=C(C=C(C(C(C(C(C(C(O)C1)(C2)C)C1)C3)C2)(C4)C)C3)C4
Testosterone
Androlin
Androst-4-en-3-one, 17-hydroxy-, (17-β)-
Andronaq
Androst-4-en-17β-ol-3-one
δ^4-Androsten-17(β)-ol-3-one
Androst-4-en-3-one, 17-β-hydroxy-
Andrusol
Cristerone T
Geno-Cristaux Gremy
Homosteron
Homosterone
17-β-Hydroxy-δ^4-androsten-3-one
17-β-Hydroxyandrost-4-en-3-one
17-β-Hydroxy-4-androsten-3-one
7-β-Hydroxyandrost-4-en-3-one
Malestrone (AMPS)
Mertestate
Neo-Testis
Oreton
Oreton-F
Orquisteron
Perandren
Percutacrine androgenique
Primotest
Primoteston
Sustanone
Synandrol F
Teslen
Testandrone
Testiculosterone
Testobase
Testopropon
Testosteroid
Testosteron
trans-Testosterone
Testosterone hydrate
Testostosterone
Testoviron schering
Testoviron T
Testrone
Testryl
Virormone
Virosterone

58-25-3
$C_{16}H_{14}ClN_3O$
299.78
CNC2=Nc1ccc(Cl)cc1C(=N(=O)C2)c3ccccc3
3H-1,4-Benzodiazepine, 7-chloro-2-(methylamino)-5-phenyl-, 4-oxide
3H-1,4-Benzodiazepin-2-amine, 7-chloro-N-methyl-5-phenyl-, 4-oxide (9CI)
CD 2
CDO
CDP
Chlordiazepoxide
Chloridiazepide
Chloridiazepoxide
Chlorodiazepoxide
7-Chloro-2-methylamino-5-phenyl-3H-1,4-benzo-diazepine 4-oxide
7-Chloro-2-methylamino-5-phenyl-3H-1,4-benzo-diazepin-4-oxide
Clopoxide
Clordiazepossido (Italian)
7-Cloro-2-metilamino-5-fenil-3H-1,4-benzodiazepina 4-ossido (Italian)
Decacil
Eden
Elenium
Ifibrium
Kalmocaps
Librax
Librinin
Libritabs
Librium
Mesural
Methaminodiazepoxide
Mildmen
Napoton
Napton

Psicosan
Radepur
Viopsicol

58-27-5
$C_{11}H_8O_2$
172.19
O=C(c(c(C(=O)C=1)ccc2)c2)C1C
1,4-Naphthoquinone, 2-methyl
Aquakay
Aquinone
Hemodal
Juva-K
Kaergona
Kanone
Kappaxan
Kappaxin
Karcon
Kareon
Kativ-G
Kayklot
Kaykot
Kayquinone
Kipca
Kipca, Oil soluble
Klottone
Koaxin
Kolklot
K-Thrombyl
K-Vitan
Menadion
Menadione
Menaphthon
Menaphthone
Menaquinone
Menaquinone 0
Menaphtone
2-Methyl-1,4-naftochinon (Czech)
2-Methyl-1,4-naphthalendione
2-Methyl-1,4-naphthochinon (German)
2-Methyl-1,4-naphthoquinone
3-Methyl-1,4-naphthoquinone
Mitenon
Mitenone
MNQ
1,4-Naphthalenedione, 2-methyl- (9CI)
NSC-4170
Panosine
Prokayvit
Synkay
Thyloquinone
USAF EK-5185
Vitamin K2(0)
Vitamin K3

58-32-2
$C_{24}H_{40}N_8O_4$
504.72
OCCN(CCO)c4nc(N1CCCCC1)c3nc(nc(N2CCCCC2)c3n4)N(CCO)CCO
Ethanol, 2,2',2'',2'''-((4,8-dipiperidinopyrimido(5,4-d)pyrimidine-2,6-diyl)dinitrilo)tetra
Anginal
2,6-Bis(di-ethanolamino)-4,8-dipiperidinopyrimido-(5,4-d)pyrimidine
Cardoxin
Cleridium 150
Coronarine
Curantyl
2,2',2'',2'''-(4,8-Dipiperidinopyrimido(5,4-d)pyrimidine-2,6-diyldinitrilo)tetraethanol
Dipyridamine
Dipyridamol
Dipyridamole
Dipyridan
Dipyudamine
Gulliostin
Natyl
Peridamol
Persantin
Persantine
Piroan
Prandiol
RA 8
Pyrimido(5,4-d)pyrimidine, 2,6-bis(bis(2-hydroxyethyl)-amino)-4,8-dipiperidino-
USAF GE-12

58-33-3
$C_{17}H_{20}N_2S.ClH$
320.91
Cl.CC(CN2c1ccccc1Sc3ccccc23)N(C)C
Phenothiazine, 10-(2-(dimethylamino)propyl)-, monohydrochloride
Atosil
10-(3-Dimethylaminoisopropyl)phenothiazine hydrochloride
10-(2-(Dimethylamino)propyl)-phenothiazine monohydrochloride
N-(2-Dimethylaminopropyl-1)-phenothiazine hydrochloride
10-(2-Dimethylaminopropyl)-phenothiazine hydrochloride
10-(2-Dimethylamino-1-propyl)-phenothiazine hydrochloride
N-(2'-Dimethylamino-2'-methyl)ethylphenothiazine hydrochloride
Diphergan
Dorme
Fargan
Fellozine
Fenazil
Fenergan
Ganphen
HL 8700
Lergigan
Phencen
10H-Phenothiazine-10-ethanamine, N,N,α-trimethyl-, monohydrochloride
Phenergan hydrochloride
Plletia
Primine
Promantine
Promethiazin (German)
Promethazine hydrochloride
Promethazine N-(2'-dimethylamino-2'-methylethyl)phenothiazine hydrochloride
Prorex
Protazine
Remsed
3277 R.P.

58-34-4
$C_{18}H_{23}N_2S.CH_3O_4S$
410.59
10H-Phenothiazine-10-ethanaminium, N,N,N,α-tetramethyl-, methyl sulfate
Ammonium, trimethyl(1-methyl-2-phenothiazin-10-ylethyl)-, methyl sulfate
Methylphenazonium methosulfate
Multergan
Multergan methyl sulfate
Multezin
Padisal
N-(β-(10-Phenothiazinyl)-propyl)trimethylammonium methyl sulfate
PMS
Prothazin methosulfate
3354 R.P.
RP 3554
N,N,N,α-Tetramethyl-10H-phenothiazine-10-ethanaminium methyl sulfate
Thiazinamium methyl sulfate
Trimethyl (1-methyl-2-phenothiazin-10-ylethyl)ammonium methyl sulfate
Trimethyl(1-methyl-2-(10-phenothiazinyl)ethyl)ammonium methyl sulfate
Valan

58-36-6
$C_{24}H_{16}As_2O_3$
502.23
Phenoxarsine oxide
Bis(phenoxarsin-10-yl) ether
Bis(10-phenoxarsyl) oxide
Bis(10-phenoxyarsinyl) oxide
10,10'-Bis(phenoxyarsinyl) oxide
DID 47
OBPA
10,10'-Oxidiphenoxarsine
10,10'-Oxybisphenoxarsine
10-10' Oxybisphenoxyarsine
10,10'-Oxybis-10H-phenoxarsine
Phenoxaksine oxide
Phenoxarsine, 10,10'-oxydi-
Phenoxyarsine, 10,10'-oxybis-
10H-Phenoxarsine, 10,10'-oxybis- (9CI)
PXO
SA 546
Vinadine
Vinyzene (Pesticide)
Vinyzene BP 5
Vinyzene BP 5-2
Vinyzene SB 1

58-38-8
$C_{20}H_{24}ClN_3S$
373.98
CN4CCN(CCCN2c1ccccc1Sc3ccc(Cl)cc23)CC4
Phenothiazine, 2-chloro-10-(3-(4-methyl-1-piperazinyl)propyl)
Chloro-3 (N-methylpiperazinyl-3 propyl)-10 phenothiazine (French)
2-Chloro-10-(3-(1-methyl-4-piperazinyl)-propyl)-phenothiazine
2-Chloro-10-(3-(4-methyl-1-piperazinyl)propyl)phenothiazine
3-Chloro-10-(3-(1-methyl-4-piperazinyl)propyl)phenothiazine
Chlorperazine
Compazine
Meterazine
N-(γ-(4'-Methylpiperazinyl-1')propyl)-3-chlorophenothiazine
Nipodal
Novamin
Prochloroperazine
Prochlorpemazine
Prochlorperazine
Prochlorpromazine
10H-Phenothiazine, 2-chloro-10-(3-(4-methyl-1-piperazinyl)propyl)-
6140 RP
Stemetil
Tementil

58-39-9
$C_{21}H_{26}ClN_3OS$
404.01
OCCN4CCN(CCCN2c1ccccc1Sc3ccc(Cl)cc23)CC4
1-Piperazineethanol, 4-(3-(2-chlorophenothiazin-10-yl)propyl)
2-Chloro-10-3-(1-(2-hydroxyethyl)-4-piperazinyl)propyl phenothiazine
4-(3-(2-Chlorophenothiazin-10-yl)propyl)-1-piperazineethanol
Decentan
Etaperazin
Etaperazine
Ethaperazine
Fentazin
1-(2-Hydroxyethyl)-4-(3-(2-chloro-10-phenothiazinyl)-propyl)piperazine
γ-(4-(β-Hydroxyethyl)piperazin-1-yl)propyl-2-chlorophenothiazine
1',1-(2-Idrossietil)-4,3-(2-cloro-10-fenotiazil)propilpiperazina (Italian)
Perfenazina (Italian)
Perphenazin
Perphenazine
Trifaron
Trilafon

58-40-2
$C_{17}H_{20}N_2S$
284.45
Phenothiazine, 10-(3-(dimethylamino)propyl)
A 145
Ampazine
10-(3-(Dimethylamino)propyl)-phenothiazine
Esparin
Liranol
Neo-Hibernex
10H-Phenothiazine-10-propanamine, N,N-dimethyl-
Prazin
Prazine
Promazin
Promazina (Italian)
Promazine
Protactyl
RP 3276
Sparine
Verophen
WY 1094

58-54-8
$C_{13}H_{12}Cl_2O_4$
303.15
CCC(=C)C(=O)c1ccc(OCC(O)=O)c(Cl)c1Cl
Acetic acid, (2,3-dichloro-4-(2-methylenebutyryl)phenoxy)
Acetic acid, (2,3-dichloro-4-(2-methylene-1-oxobutyl)-phenoxy)- (9CI)
Crinuryl
2,3-Dichloro-4-(2-methylene-butyryl)phenoxy acetic acid
(2,3-Dichloro-4-(2-methylene-butyryl)phenoxy)acetic acid
(2,3-Dichloro-4-(2-methylene-1-oxobutyl)phenoxy)acetic acid
Edecril
Edecrin
Edecrina
Endecril
Etacrinic acid
Etakrinic acid
Ethacrynic acid
Hidromedin
Hydromedin
Kyselina 4-(2-(1-butenyl)-karbonyl)-2,3-dichlor-fenoxyoctova (Czech)
Kyselina ethakrynova (Czech)
(4-(2-Methylenebutyryl)-2,3-dichlorophenoxy)acetic acid
Methylenebutyryl phenoxyacetic acid
Mingit
MK-595
Otacril
Reomax
Taladren
Uregit

58-55-9
$C_7H_8N_4O_2$
180.19
O=C(N(C(=O)C(N=CN1)=C12)C)N2C
Theophylline
Acet-Theocin
3,7-Dihydro-1,3-dimethyl-1H-purine-2,6-dione
1,3-Dimethylxanthine
Elixicon
Elixophyllin
Elixophylline
Euphylline
Lanophyllin
Liquophylline
NSC-2066
Optiphyllin
Parkophyllin
Pseudotheophylline
1H-Purine-2,6-dione, 3,7-dihydro-1,3-dimethyl- (9CI)
Slo-Phyllin
Solosin
Tefamin
Teofilina (Polish)
Teofyllamin
Theal Tabl.
Theocin
Theo-Dur
Theofol
Theograd
Theolair
Theolix
Theophyl-225
Theophyllin
Theophyline
Theophylline, Anhydrous
Xanthine, 1,3-dimethyl-

58-56-0
$C_8H_{11}NO_3.ClH$
205.66
[Cl-].Cc1ncc(CO)c(CO)c1O
Pyridoxol-hydrochloride
Adermine hydrochloride
Becilan
Benadon
Campoviton 6
Hexabetalin
Hexabione hydrochloride
Hexavibex
Hexermin
Hexobion
3-Hydroxy-4,5-dimethylol-α-picoline hydrochloride
5-Hydroxy-6-methyl-3,4-pyridinedicarbinol hydrochloride
5-Hydroxy-6-methyl-3,4-pyridinedimethanol hydrochloride
2-Methyl-3-hydroxy-4,5-bis-(hydroxymethyl)pyridine hydrochloride
3,4-Pyridinedimethanol, 5-hydroxy-6-methyl-, hydrochloride
Pyridipca
Pyridoxine, hydrochloride

Pyridoxinium chloride
Pyridoxinum hydrochloricum (Hungarian)
Vitamin B6-hydrochloride

58-61-7
$C_{10}H_{13}N_5O_4$
267.28
O(C(C(O)C1O)CO)C1N(c(ncnc2N)c2N=3)C3
Adenosine
Adenine riboside
Adenosin (German)
β-Adenosine
β-D-Adenosine
6-Amino-9-β-D-ribofuranosyl-9H-purine
Boniton
Myocol
Nucleocardyl
9-β-D-Ribofuranosidoadenine
9-β-D-Ribofuranosyl-9H-purin-6-amine
Sandesin
USAF CB-10

58-63-9
$C_{10}H_{12}N_4O_5$
268.26
O(C(C(O)C1O)CO)C1N(c(ncnc2O)c2N=3)C3
Inosine
Atorel
HXR
Hypoxanthine nucleoside
Hypoxanthine ribonucleoside
Hypoxanthine riboside
Hypoxanthine D-riboside
Hypoxanthosine
Ino
Inosie
β-Inosine
Oxiamin
Pantholic-l
Ribonosine
Selfer
Trophicardyl

58-72-0
$C_{20}H_{16}$
256.36
c(cccc1)(c1)C=C(c(cccc2)c2)c(cccc3)c3
Ethylene, triphenyl
Triphenylethylene
1,1,2-Triphenylethylene

58-73-1
$C_{17}H_{21}NO$
255.39
O(C(c(cccc1)c1)c(cccc2)c2)CCN(C)C
Ethylamine, 2-(diphenylmethoxy)-N,N-dimethyl
Aleryl
Alledryl
Allergan B
Allergeval
Allergical
Allergin
Allergina
Allergival
Amidryl
Antistominum
Antomin
Automin
Bagodryl
Baramine
Bena
Benachlor
Benadon
Benadrin
Benadryl
Ben-Allergin
Benapon
Benodin
Benodine
Benylan
Benzantine
Benzhydramine
Benzhydraminum
Benzhydril
Benzhydroamina
Benzhydryl
o-Benzhydryldimethylaminoethanol
N-(Benzhydryloksy-etylo)-dwumetyloamina (Polish)
2-(Benzhydryloxy)-N,N-dimethylethylamine
2-(Benzohydryloxy)-N,N-dimethylethylamine
Betramin
Dabylen
Debendrin
Dermistina
Dermodrin
Desentol
Diabenyl
Diabylen
Dibendrin
Dibondrin
Difedryl
Difenhydramin
Difenidramina (Italian)
Dihidral
Dimedrol
Dimedryl
β-Dimethylamino-aethyl-benzhydryl-aether (German)
β-Dimethylaminoethanol diphenylmethyl ether
α-(2-Dimethylaminoethoxy)-diphenylmethane
β-Dimethylaminoethylbenzhydryl ether
Diphantine
Diphenhydramine
Diphenylhydramine
2-(Diphenylmethoxy)-N,N-dimethylethylamine
Dryistan
Drylistan
Dylamon
Etanautine
Ethylamine, N,N-dimethyl-2-(diphenylmethoxy)-
Histaxin
Hyadrine
Ibiodral
Medidryl
Mephadryl
Nausen
Novamina
PM 255
Probedryl
Restamin
Restamine
Rigidil
Rigidyl
S51
Syntedril
Syntodril
Vena

58-74-2
$C_{20}H_{21}NO_4$
339.42
COc3ccc(Cc1nccc2cc(OC)c(OC)cc12)cc3OC
Isoquinoline, 1-((3,4-dimethoxyphenyl)methyl)-6,7-dimethoxy
1-((3,4-Dimethoxyphenyl)-methyl)-6,7-dimethoxyisoquinoline
6,7-Dimethoxy-1-veratrylisoquinoline
Isoquinoline, 6,7-dimethoxy-1-veratryl-
Papanerine
Papaverina (Italian)
Papaverine

58-85-5
$C_{10}H_{16}N_2O_3S$
244.34
O=C(NC(C1CS2)C2CCCCC(=O)O)N1
1H-Thieno(3,4-d)imidazole-4-pentanoic acid, hexahydro-2-oxo-, (3as-(3a-α,4-β, 6a-α))
Bioepiderm
Bios II
Biotin
(+)-Biotin
d-Biotin
d-Biotin
d-(+)-Biotin
Coenzyme R
Factor S
Factor S (vitamin)
Vitamin B7
Vitamin H

58-86-6
$C_5H_{10}O_5$
150.15
O=CC(O)C(O)C(O)CO
Xylose, D
(D)-Xylose

58-89-9
$C_6H_6Cl_6$
290.82
C(C(C(C(C(C1Cl)Cl)Cl)Cl)(C1Cl)Cl
Cyclohexane, 1,2,3,4,5,6-hexa-

chloro-, γ-isomer
Aalindan
Aficide
Agrisol G-20
Agrocide
Agrocide 2
Agrocide 7
Agrocide 6G
Agrocide III
Agrocide WP
Agronexit
Ameisenatod
Ameisenmittel Merck
Aparasin
Aphtiria
Aplidal
Arbitex
BBH
Benhexol
Ben-Hex
Bentox 10
Benzene hexachloride
γ-Benzene hexachloride
γ-Benzohexachloride
Bexol
BHC
γ-BHC
Celanex
Chloresene
Codechine
DBH
Detmol-Extrakt
Detox 25
Devoran
Dol granule
Drill tox-spezial aglukon
ENT 7,796
Entomoxan
Entomoxan
Exagama
Forlin
Forst-nexen
Gallogama
Gamacarbatox
Gamacid
Gamaphex
Gamene
Gamiso
γ-Col
γ-HCH
Gammahexa
Gammahexane
Gammalin
Gammalin 20
Gammaterr
Gammex
Gammexane
Gammopaz
Geobilan
Geolin G 3
Gexane
HCC
HCCH
HCH
γ-HCH
Heclotox
Hexa
γ-Hexachlor
Hexachloran
γ-Hexachloran
Hexachlorane
γ-Hexachlorane
γ-Hexachlorobenzene
Hexachlorocyclohexane
Hexachlorocyclohexane, γ-isomer
γ-Hexachlorocyclohexane
γ-1,2,3,4,5,6-Hexachlorocyclohexane
1,2,3,4,5,6-Hexachlorocyclohexane
1,2,3,4,5,6-Hexachlorocyclohexane, γ-isomer
1-α,2-α,3-β,4-α,5-α,6-β-Hexachlorocyclohexane
Hexatox
Hexaverm
Hexicide
Hexyclan
HGI
Hilbeech
Hortex
Hungaria L7
Inexit
Isotox
Jacutin
Kokotine
Lasochron
Kwell
Lendine
Lentox
Lidenal
Lindafor
Lindagam
Lindagrain
Lindagranox
Lindane (ACGIH,DOT,OSHA)
γ-Lindane
Lindapoudre
Lindatox
Lindex
Lindosep
Lintox
Linvur
Lorexane
Mglawik L
Milbol 49
Mszycol
NA 2761
NCI-C00204
Neo-Scabicidol
Nexen FB
Nexit
Nexit-Stark
Nexol-E
Nicochloran
Novigam
Omnitox
Ovadziak
Owadziak
Pedraczak
Pflanzol
PLK
Quellada
RCRA waste number U129
Sang γ
Silvanol
Spritzlindane
Spritz-Rapidin
Spruehpflanzol
Streunex
TAP 85
TRI-6
Verindal Ultra
Viton

58-90-2
$C_6H_2Cl_4O$
231.89
Oc(c(cc(c1Cl)Cl)Cl)c1Cl
Phenol, 2,3,4,6-tetrachloro- (8CI,9CI)
Dowicide 6
NSC-2428
RCRA waste number U212
TCP
2,3,4,6-Tetrachlorophenol

58-93-5
$C_7H_8ClN_3O_4S_2$
297.75
O=S(=O)(N)c(c(cc(NCNS1(=O)=O)c12)Cl)c2
2H-1,2,4-Benzothiadiazine-7-sulfonamide, 6-chloro-3,4-dihydro-, 1,1-dioxide
Aquarills
Aquarius
Bremil
6-Chloro-3,4-dihydro-2H-1,2,4-benzothiadiazine-7-sulfonamide 1,1-dioxide
6-Chloro-3,4-dihydro-7-sulfamoyl-2H-1,2,4-benzothiadiazine 1,1-dioxide
6-Chloro-7-sulfamoyl-3,4-dihydro-2H-1,2,4-benzothiadiazine 1,1-dioxide
Chlorosulthiadil
Chlorsulfonamidodihydrobenzothiadiazine dioxide
Chlorzide
Cidrex
Dichlorosal
Dichlotiazid
Dichlotride
Diclotride
3,4-Dihydro-6-chloro-7-sulfamyl-1,2,4-benzothiadiazine-1,1-dioxide
Dihydrochlorothiazid
Dihydrochlorothiazide
3,4-Dihydrochlorothiazide
Dihydroxychlorothiazidum
Direma
Disalunil
Drenol
Dyazide
Esidrex
Esidrix
Fluvin
HCTZ
HCZ
Hidril
Hidrochlortiazid
Hidroronol
Hidrotiazida
Hydril
Hydro-Aquil
Hydrochlorothiazid
Hydrochlorothiazide
Hydrochlorthiazide
Hydrodiuretic
Hydro-diuril

Hydrosaluric
Hypothiazid
Hypothiazide
Hydrothide
Idrotiazide
Ivaugan
Jen-Diril
Maschitt
Megadiuril
NCI-C55925
Nefrix
Neo-Codema
Neoflumen
Oretic
Panurin
Ro-Hydrazide
SU 5879
Thiaretic
Thiuretic
Thlaretic
Urodiazin
Vetidrex
Zide

58-94-6
$C_7H_6ClN_3O_4S_2$
295.73
O=s(ncnc1cc(c(S(=O)(=O)N)c2)Cl)(O)c12
2H-1,2,4-Benzothiadiazine-7-sulfonamide, 6-chloro-, 1,1-dioxide
Alurene
CT
Chloriazid
6-Chloro-2H-1,2,4-benzothiadiazine-7-sulfonamide 1,1-dioxide
Chlrosal
6-Chloro-7-sulfamoyl-2H-1,2,4-benzothiadiazine 1,1-dioxide
Chlorothiazid
Chlorothiazide
Chlorthiazide
Chlorurit
Chlotride
Clotride
Diuresal
Diuril
Diurilix
Diurite
Diutrid
Flumen
Minzil
Neo-Dema
Salisan
Salunil
Saluretil
Saluric
SK-Chlorothiazide
Thiazide
Urinex
Warduzide
Yadalan

58-95-7
$C_{31}H_{52}O_3$
472.75
O=C(Oc(c(c(c(OC(CC1)(CCCC(CCCC(CCCC(C)C)C)C)C)c12)C)C)c2C)C
2H-1-Benzopyran-6-ol, 3,4-dihydro-2,5,7,8-tetramethyl-2-(4,8,12-trimethyltridecyl)-, acetate, (2R-(2R*(4R*, 8R*)))- (9CI)

58-96-8
$C_9H_{12}N_2O_6$
244.23
O=C(N(C=CC1=O)C(OC(C2O)CO)C2O)N1
Uracil, 1-β-D-ribofuranosyl
1-β-D-Ribofuranosyluracil
Uracil riboside
Uridin
Uridine
β-Uridine

59-01-8
$C_{18}H_{36}N_4O_{11}$
484.58
O(C(C(O)C(O)C1O)CN)C1OC(C(O)C(OC(OC(C(O)C2N)CO)C2O)C(N)C3)C3N
D-Streptamine, O-3-amino-3-deoxy-α-D-glucopyranosyl-(1-6)-O-(6-amino-6-deoxy-α- D-glucopyranosyl-(1-4))-2-deoxy
Cantrex
4,6-Diamino-2-hydroxy-1,3-cyclohexane 3,6'diamino-3,6'-dideoxydi-α-D-glucoside
Glucopyranoside, 4,6-diamino-2-hydroxy-1,3-cyclohexylene 3,6'-diamino-3,6'-dideoxydi-, D-
Kanamicina (Italian)
Kanamycin
Kanamycin A
Kanamytrex
Kantrex
KM
KM (the antibiotic)

59-02-9
$C_{29}H_{50}O_2$
430.79
O(c(c(c(c(O)c1C)C)CC2)c1C)C2(CCCC(CCCC(CCCC(C)C)C)C)C
2H-1-Benzopyran-6-ol, 3,4-dihydro-2,5,7,8-tetramethyl-2-(4,8,12-trimethyltridecyl)-, (2R-(2R*(4R*,8R*)))
Almefrol
Antisterility vitamin
Covi-Ox
Denamone
Emipherol
Endo E
Ephynal
Eprolin
Epsilan
Esorb
Etamican
Etavit
Evion
Evitaminum
Ilitia
Phytogermine
Profecundin
Spavit
Syntopherol
d-α-Tocopherol
(R,R,R)-α-Tocopherol
α-Tocopherol
(2R,4'R,8'R)-α-Tocopherol
Tokopharm
5,7,8-Trimethyltocol
Vascuals
Verrol
Vitamin E
Vitaplex E
Vitayonon
Viteolin

59-05-2
$C_{20}H_{22}N_8O_5$
454.50
O=C(NC(C(=O)O)CCC(=O)O)c(ccc(N(Cc(nc(c(nc(n1)N)n2)c1N)c2)C)c3)c3
Glutamic acid, N-(p-(((2,4-diamino-6-pteridinyl)methyl)-methylamino)benzoyl)-, L
Amethopterin
Amethopterine
4-Amino-4-deoxy-N^{10}-methylpteroylglutamate
4-Amino-4-deoxy-N^{10}-methylpteroylglutamic acid
4-Amino-10-methylfolic acid
4-Amino-N^{10}-methylpteroylglutamic acid
Antifolan
N-Bismethylpteroylglutamic acid
CL-14377
l-(+)-N-(p-(((2,4-Diamino-6-pteridinyl)methyl)methylamino)benzoyl)glutamic acid
N-(p-(((2,4-Diamino-6-pteridinyl)methyl)methylamino)benzoyl)-l-(+)-glutamic acid
N-(p-(((2,4-Diamino-6-pteridyl)-methyl)methylamino)benzoyl)glutamic acid
EMT 25,299
Emtexate
Glutamic acid, N-(p-(((2,4-diamino-6-pteridinyl)methyl)-methylamino)benzoyl)-, l-(+)-(8CI)
l-Glutamic acid, N-(4-(((2,4-diamino-6-pteridinyl)methyl)-methylamino)benzoyl)- (9CI)
HDMTX
Kyselina 4-amino-N^{10}-methylpteroylglutamova (Czech)
Kyselina 4-desoxy-4-amino-N^{10}-methyllistova (Czech)
Kyselina N-(p-((2,4-diamino-6-pteridinylmethyl)methylamino)benzoyl)-l-glutamova (Czech)
Ledertrexate
Metatrexan

Methopterin
Methotextrate
Methotrexat
Methotrexate
Methylaminopterin
Methylaminopterinum
MTX
NCI-C04671
NSC-740
R 9985

59-23-4
$C_6H_{12}O_6$
180.18
O=CC(O)C(O)C(O)C(O)CO
Galactose, D
Galactose

59-26-7
$C_{10}H_{14}N_2O$
178.26
CCN(CC)C(=O)c1cccnc1
Nicotinamide, N,N-diethyl
Anacardone
Anacordone
Astrocar
Betapyrimidum
Camphozone
Carbamidal
Cardamine
Cardiagen
Cardiamid
Cardiamina
Cardiamine
Cardimon
Citocor
Coracon
Coraethamide
Coraethamidum
Coralept
Coramine
Coravita
Corazone
Cordiamid
Cordiamin
Cordiamine
Corditon
Cordynil
Corediol
Corespin
Corethamide
Coretone
Cormed
Cormid
Cormotyl
Cornotone
Corotonin
Corovit
Corvitan
Corvitol
Corvotone
Corywas
Danamine
Diaethyl-nicotinamid (German)
Diethylamid kyseliny nikotinove (Czech)
Diethyl-nicotamide
N,N-Diethylnicotinamide
N,N-Diethyl-3-pyridinecarboxamide
Dietilamide-carbopiridina
Dinacoryl
Dynacoryl
Dynamicarde
Elitone
Eucoran
Hansacor
Inicardio
Kardiamid
Kardonyl
Kordiamin
Leptamin
Mediamid
Niamine
Nicamide
Nicetamide
Nicethamide
Nicor
Nicordamin
Nicorine
Nicoryl
Nicotinic acid diethylamide
Nikardin
Niketamid
Nikethamide
Niketharol
Nikethyl
Niketilamid
Nikorin
Nikotin
Niquetamida
Nisetamide
Percoral
Procardine
Procorman
Pyricardyl
Pyridine-3-carboxydiethylamide
Pyridine-3-carboxylic acid diethylamide
Reformin
Rehormin
Salvacard
Solyacord
Salvacorin
Stellamine
Stiminol
Stimulin
Tonocard
Tonocor
Vasazol
Ventramine

59-30-3
$C_{19}H_{19}N_7O_6$
441.45
O=C(NC(C(=O)O)CCC(=O)O)c(ccc(NCc(nc(c(nc(n1)N)n2)c1O)c2)c3)c3
Folic-acid
Folacin
Folate
Folcysteine
Glutamic acid, N-(p-(((2-amino-4-hydroxy-6-pteridinyl)-methyl)amino)benzoyl)-, l-
Glutamic acid, pteroyl-, l-
Kyselina listova (Czech)
NSC-3073
Pteglu
4-Pteridinol, 2-amino-6-((p-((1,3-dicarboxypropyl)carbamoyl)anilino)methyl)-
Pteroylglutamic acid
Pteroyl-l-glutamic acid
Pteroylmonoglutamic acid
Pteroyl-l-monoglutamic acid
USAF CB-13
Vitamin B11
Vitamin BC
Vitamin M

59-31-4
C_9H_7NO
145.17
Oc(nc(c(ccc1)c2)c1)c2
Carbostyril
o-Aminocinnamic acid lactam
α-Hydroxyquinoline
2-Hydroxyquinoline
2-Quinolinol
2-Quinolinone
2(1H)-Quinolinone (9CI)
α-Quinolone
2-Quinolone
2(1H)-Quinolone

59-33-6
$C_{17}H_{23}N_3O.C_4H_4O_4$
401.51
COc2ccc(CN(CCN(C)C)c1ccccn1)cc2.OC(=O)C=CC(O)=O
Pyridine, 2-((2-(dimethylamino)ethyl)(p-methoxybenzyl)amino)-, maleate (1:1)
AH
Anisopyradamine
Anthisan maleate
Antihist
Diaminide maleate
N-Dimethylaminoethyl-N-p-methoxy-α-aminopyridine maleate
2-((2-(Dimethylamino)ethyl)-(p-methoxybenzyl)amino)-pyridine bimaleate
2-((2-(Dimethylamino)ethyl)-(p-methoxybenzyl)amino)-pyridine maleate
N,N-Dimethyl-N'-(4-methoxybenzyl)-N'-(2-pyridyl)-ethylenediamine maleate
Histatex
Mepyramine maleate
N-p-Methoxybenzyl-N'-N'-dimethyl-N-α-pyridylethylenediamine maleate
Minihist
Neoantergan maleate
Paramal
Paraminyl maleate
Pymafed
Pyra maleate
Pyranilamine maleate
Pyraninyl
Pyranisamine maleate
Pyrilamine maleate
Renstamin
2786 R.P. Maleate
Stangen maleate
Statomin maleate

Thylogen maleate

59-40-5
$C_{14}H_{12}N_4O_2S$
300.36
O=S(=O)(Nc(nc(c(n1)ccc2)c2)c1)c(ccc(N)c3)c3
Sulfanilamide, N[1]-2-quinoxalinyl
2-p-Aminobenzenesulfonamid-o quinoxaline
2-p-Aminobenzenesulphonamidoquinoxaline
Compound 3-120
N-(2-Quinoxalinyl)sulfanilamide
N'-2-Quinoxalylsulfanilamide
Sulfabenzpyrazine
Sulfacox
Sulfaline
2-Sulfanilamidoquinoxaline
Sulfaquinoxaline
Sulquin

59-42-7
$C_9H_{13}NO_2$
167.23
CNCC(O)c1cccc(O)c1
Benzyl alcohol, m-hydroxy-α-((methylamino)methyl)-, (-)
Benzenemethanol, 3-hydroxy-α-((methylamino)methyl)-, (R)- (9CI)
l-α-Hydroxy-β-methylamino-3-hydroxy-l-ethylbenzene
l-m-Hydroxy-α-((methylamino)-methyl)benzyl alcohol
(-)-m-Hydroxy-α-(methylamino-methyl)benzyl alcohol
l-1-(m-Hydroxyphenyl)-2-methylaminoethanol
l-(3-Hydroxyphenyl)-N-methyl-ethanolamine
Isophrin
Mesaton
Mesatone
Metaoxedrin
Metaoxedrine
Metasympatol
Metasynephrine
m-Methylaminoethanolphenol
Mezaton
R(-)-Mezaton
Neosynephrine
m-Oxedrine
(-)-m-Oxedrine
Phenylephrine
(-)-Phenylephrine
R(-)-Phenylephrine
m-Sympathol
m-Sympatol
m-Synephrine
Visadron

59-43-8
$C_{12}H_{17}N_4OS.Cl$
300.84
[Cl-].Cc2ncc(C[n+]1csc(CCO)c1C)c(N)n2
Thiamine, chloride
b-Amin
3-((4-Amino-2-methyl-5-pyrimidinyl)methyl)-5-(2-hydroxyethyl)-4-methylthiazolium chloride
Aneurine
Apatate drape
Beivon
Betabion
Betalin S
Bethiamin
Betaxin
Bewon
Oryzanin
Oryzanine
Thiamin
Thiamine
Thiamine monochloride
Vinothiam
Vitamin B1
Vitaneuron

59-46-1
$C_{13}H_{20}N_2O_2$
236.35
O=C(OCCN(CC)CC)c(ccc(N)c1)c1
Benzoic acid, p-amino-, 2-(diethylamino)ethyl ester
Allocaine
p-Aminobenzoic acid 2-diethylaminoethyl ester
4-Aminobenzoic acid diethylaminoethyl ester
p-Aminobenzoyldiethylaminoethanol
Diethylaminoethyl p-aminobenzoate
β-Diethylaminoethyl 4-aminobenzoate
2-Diethylaminoethyl p-aminobenzoate
2-Diethylaminoethylester kyseliny p-aminobenzoove (Czech)
Gerovital
Jenacain
Jenacaine
Neocaine
Nissocaine
Norocaine
Novocain
Novocaine
Procain
Procaine
Procaine, base
Scurocaine
Spinocaine

59-49-4
$C_7H_5NO_2$
135.13
O=C(Oc(c1ccc2)c2)N1
2-Benzoxazolinone
Benzoxazole, 2-hydroxy-
Benzoxazolinone
2-Benzoxaxolol
Benzoxazolone
2(3H)-Benzoxazolone
2-Hydroxybenzoxazole
USAF EK-5429

59-50-7
C_7H_7ClO
142.59
Oc(ccc(c1C)Cl)c1
m-Cresol, 4-chloro
Aptal
Baktol
Baktolan
Candaseptic
p-Chlor-m-cresol
Chlorocresol
p-Chlorocresol
p-Chloro-m-cresol
4-Chloro-m-cresol
6-Chloro-m-cresol
2-Chloro-hydroxytoluene
6-Chloro-3-hydroxytoluene
2-Chloro-5-methylphenol
4-Chloro-3-methylphenol
3-Methyl-4-chlorophenol
Ottafact
Parmetol
Parol
PCMC
Peritonan
Phenol, 4-chloro-3-methyl-(9CI)
Preventol CMK
Raschit
Raschit K
Rasen-Anicon
RCRA waste number U039

59-51-8
$C_5H_{11}NO_2S$
149.23
O=C(O)C(N)CCSC
Methionine, DL
Acimetion
Banthionine
Cynaron
Dyprin
Lobamine
Meonine
Methilanin
(+-)-Methionine
DL-Methionine (9CI)
Metione
Mertionin
Neston

59-52-9
$C_3H_8OS_2$
124.23
OCC(S)CS
1-Propanol, 2,3-dimercapto
Bal
British Antilewisite
Dicaptol
Dimercaprol
Dimercaprol propanol
Dimercaptol
Dimercaptopropanol
2,3-Dimercaptopropanol
2,3-Dimercaptopropan-1-ol
2,3-Dimercaptol-1-propanol

Dimerkaprol (Czech)
Dithioglycerol
1,2-Dithioglycerol
2,3-Dithiopropanol
Glycerol, 1,2-dithio-
Sulfactin
USAF ME-1

59-56-3
$C_6H_{13}O_9P$
260.14
O=P(OC(OC(C(O)C1O)CO)C1O)(O)O
Glucose-1-phosphate
α-D-Glucopyranose, 1-(dihydrogen phosphate) (9CI)
α-Glucose-1-phosphate
β-Glucose-1-phosphate

59-66-5
$C_4H_6N_4O_3S_2$
222.26
O=C(N=C(SC(=N1)S(=O)(=O)N)N1)C
Acetamide, N-(5-sulfamoyl-1,3,4-thiadiazol-2-yl)
Acetamide, N-(5-(aminosulfonyl)-1,3,4-thiadiazol-2-yl)-
5-Acetamide-1,3,4-thiadiazole-2-sulfonamide
2-Acetamido-5-sulfonamido-1,3,4-thiadiazole
Acetamidothiadiazolesulfonamide
Acetamox
Acetazolamid
Acetazolamide
Acetazoleamide
Acetozalamide
2-Acetylamino-1,3,4-thiadiazole-5-sulfonamide
N-(5-(Aminosulfonyl)-1,3,4-thiadiazol-2-yl)acetamide
Carbonic anhydrase inhibitor No. 6063
Cidamex
Defiltran
Dehydratin
Diacarb
Diakarb
Diamox
4-Diamox
Didoc
Diluran
Diuramid
Diureticum-holzinger
Diuriwas
Diutazol
Donmox
Duiramid
Edemox
Eumicton
Fonurit
Glaupax
Glupax
Natrionex
Nephramid
Nephramide
Phonurit
N-(5-Sulfamoyl-1,3,4-thiadiazol-2-yl)acetamide
1,3,4-Thiadiazole-2-sulfonamide, 5-acetamido-
Vetamox

59-67-6
$C_6H_5NO_2$
123.12
O=C(O)c(cccn1)c1
Nicotinic-acid
Acide nicotinique (French)
Acidum nicotinicum
Akotin
Anti-Pellagra Vitamin
Apelagrin
Bionic
3-Carboxypyridine
Daskil
Davitamon PP
Direktan
Efacin
Kyselina nikotinova (Czech)
NAH
Naotin
Niacin
Nicacid
Nicamin
Nicangin
Nico
Nico-400
Nicobid
Nicocap
Nicocidin
Nicocrisina
Nicodan
Nicodelmine
Nicolar
Niconacid
Niconat
Niconazid
Nicorol
Nicoside
Nico-Span
Nicosyl
Nicotamin
Nicotene
Nicotil
Nicotine acid
Nicotinipca
Nicotinoylhydrazine
Nicotinsaure (German)
Nicovasan
Nicovasen
Nicovel
Nicyl
Nipellen
Pellagramin
Pellagra preventive factor
Pellagrin
Pelonin
Peviton
PP Factor
P.P. Factor-pellagra preventive factor
Pyridine-3-carbonic acid
Pyridine-β-carboxylic acid
Pyridine-3-carboxylic acid
3-Pyridinecarboxylic acid
Pyridine-carboxylique-3 (French)
S115
SK-Niacin
Tinic
Vitaplex N
Wampocap

59-85-8
$C_7H_5ClHgO_2$
357.16
Mercury, (p-carboxyphenyl)-chloro
p-Chloro-mercuric benzoic acid
Benzoic acid, p-(chloromercuri)-
USAF D-3

59-87-0
$C_6H_6N_4O_4$
198.16
O=C(N)NN=CC(OC(N(=O)=O)=C1)=C1
2-Furaldehyde, 5-nitro-, semicarbazone
Aldomycin
Alfucin
Amifur
Babrocid
Becafurazone
Biofuracina
Biofurea
Chemofuran
Chixin
Cocafurin
Coxistat
Dermofural
Dynazone
Eldezol
Eldezol F-6
Fedacin
Flavazone
Fracine
Furacilin
Furacillin
Furacin
Furacin-E
Furacine
Furacinetten
Furacin-HC
Furacoccid
Furacort
Furacycline
Furaldon
Furalone
Furametral
Furan-ofteno
Furaplast
Furaseptyl
Furaskin
Furaziline
Furazin
Furazina
Furazol W
Furazone
Furesol
Furfurin
Furosem
Fuvacillin
Hemofuran
Hydrazinecarboxamide, 2-((5-nitro-2-furanyl)methylene)-
Ibiofural

Mammex
Mastofuran
Monofuracin
NCI-C56064
Nefco
NF
NF-7
NFS
NFZ
Nifucin
Nifurid
Nifuzon
Nitrofural
Nitrofuraldehyde semicarbazone
5-Nitrofuraldehyde semicarbazide
6-Nitrofuraldehyde semicarbazide
5-Nitro-2-furaldehyde semicarbazone
5-Nitrofuran-2-aldehyde semicarbazone
5-Nitro-2-furancarboxaldehyde semicarbazone
2((5-Nitro-2-furanyl)methylene)hydrazinecarboxamide
Nitrofurazan
Nitrofurazone
5-Nitro-2-furfuraldehyde semicarbazone
5-Nitrofurfural semicarbazone
5-Nitro-2-furfural semicarbazone
(5-Nitro-2-furfurylideneamino)urea
Nitrozone
NSC-2100
Otofuran
Sanfuran
Semikarbazon 5-nitrofurfuralu (Polish)
Spray-Dermis
Spray-Foral
U-6421
USAF EA-4
Vabrocid
Vadrocid
Veterinary Nitrofurazone
Yatrocin

59-88-1
$C_6H_8N_2.ClH$
144.59
Phenylhydrazine hydrochloride
AI3-17266
Hydrazine, phenyl-, hydrochloride
Hydrazine, phenyl-, monohydrochloride (8CI,9CI)
NSC-5710
Phenylhydrazine-HCl
Phenylhydrazin hydrochlorid (German)
Phenylhydrazine hydrochloride (VAN)
Phenylhydrazine monohydrochloride
Phenylhydrazinium chloride

59-89-2
$C_4H_8N_2O_2$
116.14
O=NN1CCOCC1
Morpholine, N-nitroso
Morpholine, 4-nitroso-
N-Nitrosomorfolin (Czech)
N-Nitrosomorpholin (German)
Nitrosomorpholine
N-Nitrosomorpholine
4-Nitrosomorpholine
NMOR

59-92-7
$C_9H_{11}NO_4$
197.21
O=C(O)C(N)Cc(ccc(O)c1O)c1
Alanine, 3-(3,4-dihydroxyphenyl)-, L
Alanine, 3-(3,4-dihydroxyphenyl)-, (-)-
2-Amino-3-(3,4-dihydroxyphenyl)propanoic acid
Bendopa
Biodopa
Brocadopa
Cerepap
Cidandopa
DA
Deadopa
Dihydroxy-l-phenylalanine
l-Dihydroxyphenylalanine
(-)-3-(3,4-Dihydroxyphenyl)-l-alanine
β-(3,4-Dihydroxyphenyl)-α-alanine
l-α-Dihydroxyphenylalanine
l-β-(3,4-Dihydroxyphenyl)-alanine
l-3,4-Dihydroxyphenylalanine
l-3,4-Dihydroxyphenyl-α-alanine
β-(3,4-Dihydroxyphenyl)-l-alanine
3-(3,4-Dihydroxyphenyl)-l-alanine
3,4-Dihydroxyphenylalanine
(-)-3,4-Dihydroxyphenylalanine
3,4-Dihydroxyphenyl-l-alanine
3,4-Dihydroxy-l-phenylalanine
(-)-Dopa
Dopa
l-Dopa
Dopaflex
Dopal
Dopaidan
Dopal-Fher
Dopalina
Dopar
Doparkine
Doparl
Dopasol
Dopaston
Dopastral
Doprin
Eldopal
Eldopar
Eldopatec
Eurodopa
Helfo Dopa
l-o-Hydroxytyrosine
3-Hydroxy-l-tyrosine
Insulamina
Laradopa
Larodopa
Ledopa
Levodopa
Levopa
Maipedopa
Parda
Pardopa
Ro 4-6316
Sobiodopa
l-Tyrosine, 3-hydroxy- (9CI)
Veldopa

59-96-1
$C_{18}H_{22}ClNO$
303.86
CC(COc1ccccc1)N(CCCl)Cc2ccccc2
Benzylamine, N-(2-chloroethyl)-N-(1-methyl-2-phenoxyethyl)
A 688
Bensylyte
Benzenemethanamine, N-(2-chloroethyl)-N-(1-methyl-2-phenoxyethyl)-
2-(N-Benzyl-2-chloroethylamino)-1-phenoxypropane
Benzyl(2-chloroethyl)-(1-methyl-2-phenoxyethyl)-amine
Benzylyt
N-(2-Chloroethyl)-N-(1-methyl-2-phenoxyethyl)benzenemethanamine
N-(2-Chloroethyl)-N-(1-methyl-2-phenoxyethyl)benzylamine
Dibenylin
Dibenyline
Dibenzyline
NSC-37448
Phenoxybenzamine
N-Phenoxyisopropyl-N-benzyl-β-chloroethylamine

59-99-4
$C_{12}H_{19}N_2O_2$
223.33
CN(C)C(=O)Oc1cccc(c1)[N+](C)(C)C
Ammonium, (m-hydroxyphenyl)trimethyl-, dimethylcarbamate (Ester)
Benzenaminium, 3-(((dimethylamino)carbonyl)oxy)-N,N,N-trimethyl- (9CI)
Eustigmin
Eustigmine
(m-Hydroxyphenyl)trimethylammonium dimethylcarbamate (Ester)
Neostigmine
Prostigmin
Prostigmine
Vagostigmine

60-00-4

$C_{10}H_{16}N_2O_8$
292.28
O=C(O)CN(CCN(CC(=O)O)CC
(=O)O)CC(=O)O
**Acetic acid, (ethylenedi-
nitrilo)tetra**
Acide ethylenediaminetetra-
cetique (French)
Celon A
Celon ATH
Cheelox
Cheelox BF acid
Chemcolox 340
Complexon II
3,6-Diazaoctanedioic acid,
3,6-bis(carboxymethyl)-
Edathamil
Edetic
Edetic acid
EDTA
EDTA (chelating agent)
EDTA acid
Endrate
Ethylenediaminetetraacetate
Ethylenediaminetetraacetic acid
Ethylenediamine-N,N,N',N'-te-
traacetic acid
Ethylenedinitrilotetraacetic acid
Glycine, N,N'-1,2-ethanediyl-
bis(N-(carboxymethyl)- (9CI)
Hamp-Ene acid
Havidote
Komplexon II
Kyselina ethylendiamintetra-
octova (Czech)
Metaquest A
Nervanaid B acid
Nullapon B acid
Nullapon BF acid
Perma Kleer 50 acid
Questex 4H
Seq 100
Sequestrene AA
Sequestric acid
Sequestrol
Tetrine acid
Titriplex
Tricon BW
Trilon B
Trilon BW
Versene
Versene Acid
Vinkeil 100
Warkeelate acid

60-01-5
$C_{15}H_{26}O_6$
302.41
O=C(OCC(OC(=O)CCC)COC
(=O)CCC)CCC
Butyrin, tri
Butanoic acid, 1,2,3-propane-
triyl ester
Butyric acid triester with
glycerin
Butyryl triglyceride
Glycerol tributyrate
Kodaflex
Tributyrin
Tributyroin

60-09-3
$C_{12}H_{11}N_3$
197.26
N(=Nc(cccc1)c1)c(ccc(N)c2)c2
Aniline, p-(phenylazo)
AAB
Aminoazobenzene
p-Aminoazobenzene
4-Aminoazobenzene
4-Amino-1,1'-azobenzene
Aminoazobenzene (indicator)
p-Aminoazobenzol
4-Aminoazobenzol
p-Aminodiphenylimide
Aniline Yellow
Azobenzene, 4-amino-
Benzenamine, 4-(phenylazo)-
(9CI)
4-Benzeneazoaniline
Brasilazina Oil Yellow G
Cellitazol R
Ceres Yellow R
C.I. 11000
C.I. Solvent Blue 7
C.I. Solvent Yellow 1
Fast Spirit Yellow
Fast Spirit Yellow AAB
Fat Yellow AAB
Induline R
Oil-Sol. Aniline Yellow
Oil Soluble Aniline Yellow
Oil Yellow AAB
Oil Yellow AB
Oil Yellow AN
Oil Yellow B
Oil Yellow 2G
Oil Yellow R
Organol Yellow
Organol Yellow 2A
Paraphenolazo aniline
4-(Phenylazo)aniline
p-(Phenylazo)aniline
4-(Phenylazo)benzenamine
p-Phenylazophenylamine
Solvent Yellow 1
Somalia Yellow 2G
Stearix Brown 4R
Sudan Yellow R
Sudan Yellow RA
USAF EK-1375
Zlut Anilinova (Czech)
Zlut Rozpoustedlova 1 (Czech)

60-11-7
$C_{14}H_{15}N_3$
225.32
N(=Nc(cccc1)c1)c(ccc(N(C)C)
c2)c2
**Aniline, N,N-dimethyl-p-
phenylazo**
Atul Fast Yellow R
Azobenzene, p-dimethylamino-
Benzenamine, N,N-dimethyl-
4-(phenylazo)- (9CI)
Benzeneazodimethylaniline
Brilliant Fast Oil Yellow
Brilliant Fast Spirit Yellow
Brilliant Fast Yellow
Brilliant Oil Yellow
Butter Yellow
Cerasine Yellow GG
C.I. 11020
C.I. Solvent Yellow 2
DAB
DAB (Carcinogen)
p-Dimethylaminoazobenzen
(Czech)
Dimethylaminoazobenzene
N,N-Dimethyl-4-aminoazo-
benzene
N,N-Dimethyl-p-aminoazo-
benzene
p-Dimethylaminoazobenzene
4-Dimethylaminoazobenzene
4-(N,N-Dimethylamino)azo-
benzene
4-Dimethylaminoazobenzene
(OSHA)
Dimethylaminoazobenzol
p-Dimethylamino-azobenzol
(German)
4-Dimethylaminoazobenzol
4-Dimethylaminophenylazo-
benzene
N,N-Dimethyl-p-azoaniline
N,N-Dimethyl-p-phenylazo-
aniline
N,N-Dimethyl-4-(phenylazo)-
benzamine
N,N-Dimethyl-4-(phenylazo)-
benzenamine
Dimethyl Yellow
Dimethyl Yellow Analar
Dimethyl Yellow N,N-dimethyl-
aniline
DMAB
Enial Yellow 2G
Fast Oil Yellow B
Fast Yellow
Fat Yellow
Fat Yellow A
Fat Yellow AD OO
Fat Yellow ES
Fat Yellow ES Extra
Fat Yellow Extra Conc.
Fat Yellow R
Fat Yellow R (8186)
Grasal Brilliant Yellow
Iketon Yellow Extra
Jaune de Beurre (French)
Methyl Yellow
Oil Yellow
Oil Yellow II
Oil Yellow 20
Oil Yellow 2625
Oil Yellow 7463
Oil Yellow BB
Oil Yellow D
Oil Yellow DN
Oil Yellow FF
Oil Yellow FN
Oil Yellow G
Oil Yellow 2G
Oil Yellow G-2
Oil Yellow GG
Oil Yellow GR
Oil Yellow N
Oil Yellow Pel
Oil Yellow S
Oleal Yellow 2G
Organol Yellow ADM
Orient Oil Yellow GG
P.D.A.B.

Petrol Yellow WT
4-(Phenylazo)-N,N-dimethyl-aniline
RCRA waste number U093
Resinol Yellow GR
Resoform Yellow GGA
Silotras Yellow T2G
Somalia Yellow A
Stear Yellow JB
Sudan GG
Sudan Yellow
Sudan Yellow GG
Sudan Yellow GGA
Toyo Oil Yellow G
USAF EK-338
Waxoline Yellow AD
Waxoline Yellow ADS
Yellow G Soluble in grease
Zlut Maselna (Czech)
Zlut Rozpoustedlova 2 (Czech)

60-12-8
$C_8H_{10}O$
122.18
OCCc(cccc1)c1
Phenethyl-alcohol
Benzyl carbinol
Ethanol, 2-phenyl-
β-Fenethylalkohol (Czech)
β-Fenylethanol (Czech)
β-Hydroxyethylbenzene
Methanol, benzyl-
Orange oil
Pea
β-Pea
Phenethanol
β-Phenethyl alcohol
2-Phenethyl alcohol
β-Phenylethanol
2-Phenylethanol
Phenylethyl alcohol
β-Phenylethyl alcohol
2-Phenylethyl alcohol
Rose Oil

60-13-9
$C_{18}H_{26}N_2.H_2O_4S$
368.54
Phenethylamine, α-methyl-, sulfate (2:1), (+-)
Acetdron
Adipan
Adiparthrol
Aketdrin
Aktedrin
Alentol
Amfetamina
Amfetamine
(+-)-2-Amino-1-phenylpropane sulfate
Amphaetamin
Amphamed
Amphamine sulfate
Amphatamin
Amphate
Amphedrine
Amphetamine sulfate
dl-Amphetamine sulfate
(+-)-Amphetamine sulfate
Amphetamine sulphate
Amphetaminum
Amphezamin
Amphoids-S
Anara
Anfetamina
Anorexine
Astedin
Bennie
Benzafinyl
Benzamphetamine
Benzebar
Benzedrina
Benzedrine
Benzedryna
Benzies
Benzolone
Benzpropamine
Betafen
Betaphen
Bluzedrin
Cartwheels
Centramina
Deoxynorephedrine
Desoxynorephedrine
Didrex
Dietamine
Durophet
Elastonin
Elastonon
Euphobine
Euphodine
Euphodyn
Fabedrine
Fenamin
Fenara
Fenedrin
Fenopromin
Halloo-Wach
Hearts
Ibiozedrine
Isamin
Isoamin
Isoamyne
Isomyn
Leodrin
Levonor
Linampheta
Mecodrin
α-Methylphenethylamine sulfate, (+-)-
Mimetina
Monetamine
NCI-C55710
Noclon
Norephedrane
Norphedrane
Novydrine
Oktedrin
Oraldrina
Ortenal
Orthedrin
Peaches
Percomon
Pharmamedrine
Pharmedrine
Phenamine
Phenedrine
Phenisopropylamine sulfate
(+-)-Phenisopropylamine sulfate
Phenopromin
Phenpromin
DL-1-Phenyl-2-aminopropane sulfate
β-Phenyl isopropylamine sulfate
Profamina
Profetamine
Propenyl
Propisamine
Psychedrine
Psychedrinum
Psychedryna
Psychoton
Racephen
Rhinalator
Roses
Sedolin
Simpamina
Simpamine
Simpatedrin
Stimulan
Sympametin
Sympamine
Sympatedrine
Synsatedrine
Theptine
Vapedrine
Weckamine
Zedrine

60-15-1
$C_9H_{13}N$
135.23
Phenethylamine, α-methyl
Amfetamina (Italian)
2-Amino-1-phenylpropane
β-Aminopropylbenzene
Amphetamine
Benzedrine
Desoxynorephedrine
Elastonon
Fenopromin
α-Methylphenethylamine
Mydrial
1-Phenyl-2-amino-propan (German)
1-Phenyl-2-aminopropane
β-Phenylisopropylamin (German)
(Phenylisopropyl)amine
β-Phenylisopropylamine
Protioamphetamine
Psychoton

60-17-3
$C_9H_{10}FNO_2$
183.20
NC(Cc1ccc(F)cc1)C(O)=O
Alanine, 3-(p-fluorophenyl)
p-Fluorophenylalanine
para-Fluorphenylalanine
4-Fluorophenylalanine
p-FPHE
PFPA
Phenylalanine, 4-fluoro- (9CI)

60-18-4
$C_9H_{11}NO_3$
181.21
O=C(O)C(N)Cc(ccc(O)c1)c1
Tyrosine, L
Tyrosine

L-Tyrosine (9CI)
L-p-Tyrosine
p-Tyrosine

60-23-1
C_2H_7NS
77.16
NCCS
Ethanethiol, 2-amino
2-Aminoethanethiol
2-Aminoethyl mercaptan
Becaptan
Cisteamina (Italian)
Cysteamide
Cysteamin
Cysteamine
Cysteinamine
Decarboxycysteine
Lambraten
MEA
Mecramine
Mercamine
Mercaptamine
β-Mercaptoethylamine
(2-Mercaptoethyl)amine
Thioethanolamine

60-24-2
C_2H_6OS
78.14
OCCS
Ethanol, 2-mercapto
Emery 5791
1-Ethanol-2-thiol
Ethylene glycol, monothio-
2-Hydroxy-1-ethanethiol
2-Hydroxyethyl mercaptan
2-ME
Mercaptoethanol
β-Mercaptoethanol
2-Mercaptoethanol
Monothioethyleneglycol
2-Thioethanol
Thioglycol [UN 2966]
Thiomonoglycol
USAF EK-4196
UN 2966 [Thioglycol]

60-27-5
$C_4H_7N_3O$
113.10
O=C(NC(=N)N1C)C1
Creatinine
AI3-15321
2-Amino-1,5-dihydro-1-methyl-4H-imidazol-4-one
Creatinine (VAN) (8CI)
4H-Imidazol-4-one, 2-amino-1,5-dihydro-1-methyl- (9CI)
1-Methylglycocyamidine
1-Methylhydantoin-2-imide
NSC-13123

60-29-7
$C_4H_{10}O$
74.14
O(CC)CC
Ethane, 1,1'-oxybis
Aether
Anaesthetic ether
Anesthesia ether
Anesthetic ether
Diaethylaether (German)
Diethyl ether
Diethyl ether (OSHA) [UN 1155]
Diethyl oxide
Dwuetylowy eter (Polish)
Etere etilico (Italian)
Ether
Ether ethylique (French)
Ethoxyethane
Ethyl ether (ACGIH,OSHA) [UN 1155]
Oxyde d'ethyle (French)
RCRA waste number U117
Solvent ether
UN 1155 [Diethyl ether; (ethyl ether)]

60-32-2
$C_6H_{13}NO_2$
131.20
O=C(O)CCCCCN
Hexanoic acid, 6-amino
Acepramin
Acepramine
ACS
Afibrin
Amicar
Amikar
Aminocaproic acid
ε-Aminocaproic acid
ω-Aminocaproic acid
6-Aminocaproic acid
ω-Aminohexanoic acid
ε-Aminohexanoic acid
6-Aminohexanoic acid
Aminokapron
Capramol
Capralense
Caprocid
Caprolisin
CL 10304
CY 116
Eaca
Eaca Kabi
Eacs
Epsamon
Epsicapron
Epsikapron
Epsilcapramin
Epsilon S
Hemocaprol
Hemopar
Hepin
Ipsilon
177 J.D.
Kyselina ω-aminokapronova (Czech)
ε-Leucine
ε-Norleucine
NSC-26154
Respramin

60-33-3
$C_{18}H_{32}O_2$
280.50
O=C(O)CCCCCCCC=CCC=CCCCCC
9,12-Octadecadienoic acid, (Z)
Emersol 310
Emersol 315
Leinoleic acid
Linoleic acid
9,12-Linoleic acid
cis,cis-9,12-Octadecadienoic acid
cis-9,cis-12-Octadecadienoic acid
9,12-Octadecadienoic acid
Polylin No. 515
Telfairic acid

60-34-4
CH_6N_2
46.09
N(N)C
Hydrazine, methyl
Hydrazomethane
Methylhydrazine (ACGIH, OSHA) [UN 1244]
1-Methylhydrazine
Metylohydrazyna (Polish)
MMH
Monomethylhydrazine
Monomethylhydrazine (OSHA)
RCRA waste number P068
UN 1244 [Methylhydrazine]

60-35-5
C_2H_5NO
59.08
O=C(N)C
Acetamide
Acetic acid amide
Acetimidic acid
Amid kyseliny octove (Czech)
Ethanamide
Methanecarboxamide
NCI-C02108

60-41-3
$C_{21}H_{22}N_2O_2.1/2H_2O_4S$
383.49
Strychnine, sulfate (2:1)
Strychinine sulfate
Strychnidin-10-one, sulfate (2:1)
Strychnine sulfate

60-51-5
$C_5H_{12}NO_3PS_2$
229.27
O=C(NC)CSP(OC)(OC)=S
Phosphorodithioic acid, O,O-dimethyl ester, S-ester with 2-mercapto-N-methyl-acetamide
AC-12880
AC-18682
Acetic acid, O,O-dimethyl-dithiophosphoryl-, N-mono-methylamide salt
American Cyanamid 12880
BI-58
BI 58 EC

8014 BIS HC
Cekuthoate
CL 12880
Cygon
Cygon 2-E
Cygon 4E
Cygon Insecticide
Daphene
De-Fend
Demos-l40
Devigon
Dimate 267
Dimet
Dimetate
Dimethoaat (Dutch)
Dimethoat (German)
Dimethoate
Dimethoate-267
Dimethoat technisch 95%
Dimethogen
O,O-Dimethyldithiophosphoryl-acetic acid, N-monomethyl-amide salt
O,O-Dimethyl-dithiophosphoryl-essigsaeure monomethylamid (German)
O,O-Dimethyl S-(2-(methyl-amino)-2-oxoethyl) phos-phorodithioate
O,O-Dimethyl-S-(N-methyl-carbamoyl)-methyl-dithiofos-faat (Dutch)
(O,O-Dimethyl-S-(N-methyl-carbamoyl-methyl)-dithio-phosphat) (German)
O,O-Dimethyl methylcarbamo-ylmethyl phosphorodithioate
O,O-Dimethyl S-(N-methylcar-bamoylmethyl) dithiophos-phate
O,O-Dimethyl S-(N-methylcar-bamoylmethyl) phosphorodi-thioate
O,O-Dimethyl S-(N-methylcar-bamylmethyl) thiothiono-phosphate
O,O-Dimethyl-S-(N-mono-methyl)-carbamyl methyl dithiophosphate
O,O-Dimethyl-S-(2-oxo-3-aza-butyl)-dithiophosphat (German)
O,O-Dimetil-S-(N-metil-car-bamoil-metil)-ditiofosfato (Italian)
Dimeton
Dimevur
Dithiophosphate de O,O-di-methyle et de S(-N-methyl-carbamoyl-methyle) (French)
EI-12880
ENT 24,650
Experimental insecticide 12,880
Ferkethion
FIP
Fortion NM
Fosfamid
Fosfotox
Fosfotox R
Fosfotox R 35
Fostion MM
L-395
Lurgo
S-Methylcarbamoylmethyl O,O-dimethyl phosphorodi-thioate
N-Monomethylamide of O,O-di-methyldithiophosphorylacetic acid
NC 262
NCI-C00135
OMS 94
OMS 111
PEI 75
Perfecthion
Perfekthion
Perfektion
Phosphamid
Phosphamide
Phosphorodithioic acid, O,O-dimethyl-S-(2-(methyl-amino)-2-oxoethyl) ester (9CI)
Racusan
RCRA waste number P044
Rebelate
Rogodial
Rogor
Rogor 40
Rogor L
Rogor 20L
Rogor P
Roxion
Roxion U.A.
Sinoratox
Trimetion

60-54-8

$C_{22}H_{24}N_2O_8$
444.48
CN(C)C2C1CC4C(=C(O)C1(O)C(=O)C(=C2O)C(N)=O)C(=O)c3c(O)cccc3C4(C)O
2-Naphthacenecarboxamide, 4-(dimethylamino)-1,4,4a,5,5a,6,11,12a-octahydro-3,6,10,12,12a- penta-hydroxy-6-methyl-1,11-dioxo
Abramycin
Abricycline
Achromycin
Agromicina
Ambramicina
Ambramycin
Amycin
Bio-Tetra
Bristaciclin α
Bristaciclina
Bristacycline
Cefracycline suspension
Criseociclina
Cyclomycin
Democracin
Deschlorobiomycin
Hostacyclin
Liquamycin
6-Methyl-1,11-dioxy-2-naphtha-cenecarboxamide
Neocycline
Oletetrin
Omegamycin
Orlycycline
Panmycin
Polycycline
Purocyclina
Robitet
Sanclomycine
Sigmamycin
Steclin
Tetrabon
Tetracycline
Tetracycline I
Tetracycline II
SK-Tetracycline
Tetracyn
Tetradecin
Tetraverine
Tsiklomitsin
T-125

60-56-0

$C_4H_6N_2S$
114.18
N(C=CN1C)=C1S
Imidazole-2-thiol, 1-methyl
Basolan
Danantizol
Favistan
Frentirox
Imidazole, 1-methyl-2-mer-capto-
Imidazole-2-thio, 1-methyl-
Mercaptazole
2-Mercapto-1-methylimidazole
Mercasolyl
Mercazole
Merkazolil
Mercazolyl
Metazolo
Methamazole
Methiamazole
Methimazol
Methimazole
1-Methylimidazole-2-thiol
Methylmercaptoimidazole
1-Methyl-2-mercaptoimidazole
Metizol
Metothyrine
1-Metylo 2 merkaptoimi-dazolem (Polish)
Strumazole
Tapazole
Tapuzole
Thacapzol
Thiamazole
Thimazole
Thycapsol
Thycapzol
USAF EL-30

60-57-1

$C_{12}H_8Cl_6O$
380.90
ClC4=C(Cl)C5(Cl)C3C1CC(C2OC12)C3C4(Cl)C5(Cl)Cl
1,4:5,8-Dimethanonaphthal-ene, 1,2,3,4,10,10-hexa-chloro-6,7-epoxy-1,4,4a,5,6,7,8,8a- octahydro, endo,exo
Alvit
Compound 497
Dieldrex
Dieldrin (ACGIH,OSHA)

[NA 2761]
Dieldrine (French)
Dieldrite
ENT 16,225
Heod
Hexachloroepoxyoctahydro-endo,exo-dimethanonaphthalene
3,4,5,6,9,9-Hexachloro-1a,2,2a,3,6,6a,7,7a-octahydro-2,7:3,6-dimethanonaphth-(2,3-b)oxirene
Illoxol
Insecticide No. 497
Latka 497 (Czech)
NA 2761 [Aldrin, solid or dieldrin]
NCI-C00124
Octalox
Panoram D-31
Quintox
RCRA waste number P037

60-70-8
$C_{27}H_{39}NO_2$
409.67
CC1CNC(C(O)C1)C(C)c5ccc4C3CC=C2CC(O)CCC2(C)C3Cc4c5C
Veratramine

60-79-7
$C_{19}H_{23}N_3O_2$
325.45
CC(CO)NC(=O)C2CN(C)C1Cc3c[nH]c4cccc(C1=C2)c34
Ergoline-8-β-carboxamide, 9,10-didehydro-N-((s)-2-hydroxy-1-methylethyl)-6-methyl
Basergin
Cornocentin
9,10-Didehydro-N-(α-(hydroxymethyl)ethyl)-6-methylergoline-8-β-carboxamide
Ergoatetrine
Ergobasine
Ergoklinine
Ergotocine
Ergometrine
Ergonovine
Ergotrate
Ermetrine
N-(α-(Hydroxymethyl)ethyl)-D-lysergomide
N-(1-(Hydroxymethyl)ethyl)-D-lysergomide
D-Lysergic acid 1-hydroxymethylethylamide
Lysergic acid propanolamide
D-Lysergic acid-l,2-propanolamide
Margonovine
Neofemergen
Secacornin
Secometrin
Syntometrine

60-80-0
$C_{11}H_{12}N_2O$
188.25
O=C(N(N(C=1C)C)c(cccc2)c2)C1
Antipyrine
Analgesine
Anodynin
Anodynine
Antipirin
Antipyrin
Apirelina
Azophen
Azophene
1,2-Dihydro-1,5-dimethyl-2-phenyl-3H-pyrazol-3-one
Dimethyloxychinizin
Dimethyloxyquinazine
2,3-Dimethyl-1-phenyl-3-pyrazolin-5-one
2,3-Dimethyl-1-phenyl-5-pyrazolone
Fenazon (Czech)
Fenazone
Methozin
Oxydimethylquinazine
Parodyne
Phenazon
Phenazone
Phenazone (Pharmaceutical)
Phenazonum
Phenozone
1-Phenyl-2,3-dimethylpyrazole-5-one
1-Phenyl-2,3-dimethyl-5-pyrazolone
Phenylon
Phenylone
Pyrazophyl
3H-Pyrazol-3-one, 1,2-dihydro-1,5-dimethyl-2-phenyl- (9CI)
Sedatin
Sedatine

60-81-1
$C_{21}H_{24}O_{10}$
436.45
1-Propanone, 1-(2-(β-D-glucopyranosyloxy)-4,6-dihydroxyphenyl)-3-(4-hydroxyphenyl)
Phlorhizin
Phloridzin
Phloridzine
Phlorizin
Phlorizine
Phlorizoside
Phlorrhizin

60-82-2
$C_{15}H_{14}O_5$
274.27
Phloretin
Dihydronaringenin
β-(p-Hydroxyphenyl)phloropropiophenone
β-(p-Hydroxyphenyl)-2,4,6-trihydroxypropiophenone
NSC-407292
Phloretol
1-Propanone, 3-(4-hydroxyphenyl)-1-(2,4,6-trihydroxyphenyl)- (9CI)
Propiophenone, 2',4',6'-trihydroxy-3-(p-hydroxyphenyl)-(8CI)
2',4',6'-Trihydroxy-3-(p-hydroxyphenyl)propiophenone

60-87-7
$C_{17}H_{20}N_2S$
284.45
CC(CN2c1ccccc1Sc3ccccc23)N(C)C
Phenothiazine, 10-(2-dimethylaminopropyl)
Dimethylamino-isopropyl-phenthiazin (German)
A-91033
Aprobit
Atosil
Avomine
Dimapp
(2-Dimethylamino-2-methyl)-ethyl-N-dibenzoparathiazine
N-(2'-Dimethylamino-2'-methyl)ethylphenothiazine
10-(2-(Dimethylamino)-2-methylethyl)phenothiazine
N-Dimethylamino-2-methylethyl thiodiphenylamine
10-(2-(Dimethylamino)propyl)phenothiazine
Diprazine
Diprozin
Fargan
Fenazil
Fenergan
Fenetazina
Hiberna
Histargan
Iergigan
Isophenergan
Isopromethazine
Lercigan
Lergigan
Lilly 1516
Lilly 01516
Phargan
NCI-C60673
Phenergan
Phensedyl
Pilpophen
Pipolphen
Proazaimine
Proazamine
Procit
Promazinamide
Prometasin
Prometazin
Promethazine
Promethiazine
Promezathine
Prorex
Protazine
Prothazin
Provigan
Pyrethia
Pyrethiazine
Romergan
3277 RP
3389 R.P.
4182 R.P.

SKF 1498
Synalgos
Tanidil
Thiergan
Vallergine
WY 509

60-92-4
$C_{10}H_{12}N_5O_6P$
329.24
O=P(OCC(OC(N(c(ncnc1N)c1N=2)C2)C3O)C34)(O4)O
Adenosine, cyclic 3',5'-(hydrogenphosphate)
Adenosine cyclic monophosphate
Adenosine 3',5'-cyclic monophosphate
Adenosine cyclic 3',5'-phosphate
Adenosine 3',5'-cyclophosphate
Adenosine 3',5'-monophosphate
Adenosine 3',5'-phosphate
3',5'-AMP
CAMP
Cyclic adenosine 3',5'-phosphate
Cyclic 3',5'-adenylic acid
Cyclic AMP
Cyclic 3',5'-AMP

61-16-5
$C_{11}H_{17}NO_3.ClH$
247.75
COc1ccc(OC)c(c1)C(O)C(C)N
Benzyl alcohol, α-(1-aminoethyl)-2,5-dimethoxy-, hydrochloride
2-Amino-1-(2,5-dimethoxyphenyl)-1-propanol hydrochloride
α-(1-Aminoethyl)-2,5-dimethoxybenzyl alcohol hydrochloride
1-(2,5-Dimethoxyphenyl)-2-aminopropanol
β-(2,5-Dimethoxyphenyl)-β-hydroxyisopropylamine hydrochloride
β-Hydroxy-β-(2,5-dimethoxyphenyl)-isopropylamine hydrochloride
Methoxamine hydrochloride
Pressomin hydrochloride
Vasoxine
Vasoxine hydrochloride
Vasoxyl hydrochloride

61-19-8
$C_{10}H_{14}N_5O_7P$
347.26
O=P(OCC(OC(N(c(ncnc1N)c1N=2)C2)C3O)C3O)(O)O
5'-Adenylic acid
Adenosine 5'-monophosphate
Adenosine-5-monophosphoric acid
Adenosine-5'-monophosphoric acid
Adenosine phosphate
Adenosine 5'-phosphate
Adenosine 5'-phosphoric acid
Adenovite
Adenyl
Adenylic acid
tert-Adenylic acid
AMP
A5MP
5-AMP
5'-AMP
AMP (nucleotide)
Cardiomone
Ergadenylic acid
Lycedan
Muscle adenylic acid
My-B-Den
Myoston
NSC-20264
Phosaden
Phosphaden
Phosphentaside

61-25-6
$C_{20}H_{21}NO_4.ClH$
375.88
[Cl-].COc3ccc(Cc1nccc2cc(OC)c(OC)cc12)cc3OC
Isoquinoline, 1-((3,4-dimethoxyphenyl)methyl)-6,7-dimethoxy-, hydrochloride
Cardoverina
Cerespan
Chlorhydrate de papaverine (French)
Dispamil
Isoquinoline, 6,7-dimethoxy-1-veratryl-, hydrochloride
NCI-C56359
Pavabid
Papavarine chlorhydrate
Papaverine chlorohydrate
Papanerin-HCL (German)
Papaverine hydrochloride
Papaverine monohydrochloride
Pap H
6,7,3',4'-Tetramethoxy-1-benzylisoquinoline hydrochloride
Therapav
Vasal

61-31-4
$C_{12}H_{10}O_2.Na$
209.20
Sodium 1-naphthaleneacetate
Caswell No. 589D
EPA Pesticide Chemical Code 056007
1-Naphthaleneacetic acid, sodium salt (8CI,9CI)

61-32-5
$C_{17}H_{20}N_2O_6S$
380.45
COc1cccc(OC)c1C(=O)NC3C2SC(C)(C)C(N2C3=O)C(O)=O
4-Thia-1-azabicyclo-(3.2.0)heptane-2-carboxylic acid, 6-(2,6-dimethoxybenzamido)-3,3,- dimethyl-7-oxo
Dimethoxyphenylpenicillin
Methicillin

61-33-6
$C_{16}H_{18}N_2O_4S$
334.42
O=C(NC(C(=O)N1C(C(=O)O)C(S2)(C)C)C12)Cc(cccc3)c3
4-Thia-1-azabicyclo-(3.2.0)heptane-2-carboxylic acid, 3,3-dimethyl-7-oxo-6- (2-phenylacetamido)
Abbocillin
Benzopenicillin
Benzyl-6-aminopenicillinic acid
Benzylpenicillin
(5R,6R)-Benzylpenicillin
Benzylpenicillin G
Benzylpenicillinic acid
Cilloral
Cilopen
Compocillin G
Cosmopen
Dropcillin
Free Benzylpenicillin
Free Penicillin G
Free Penicillin II
Galofak
Gelacillin
Liquacillin
Pencillin G
Penicillin G
Penicillinic acid, benzyl-
Pharmacillin
Phenylacetamidopenicillanic acid
(Phenylmethyl)penicillin
(Phenylmethyl)penicillinic acid
Pradupen
Specilline G

61-50-7
$C_{12}H_{16}N_2$
188.30
Indole, 3-(2-(dimethylamino)ethyl)
N,N-Dimethyltryptamine
DMT

61-51-8
$C_{14}H_{20}N_2$
216.31
N,N-Diethyltryptamine
DEA No. 7434
DET
D.E.T.
Diethyltryptamine
N,N-Di-ethyltryptamine
Indole, 3-(2-(diethylamino)-ethyl)-

61-54-1
$C_{10}H_{12}N_2$
160.24
NCCc1c[nH]c2ccccc12
Indole, 3-(2-aminoethyl)

(Amino-2 ethyl)-3 indole (French)
3-(2-Aminoethyl)indole
1H-Indole-3-ethanamine
Indol-3-ethylamine
2-(3-Indolyl)ethylamine
Tryptamine

61-57-4
$C_6H_6N_4O_3S$
214.22
O=C1NCCN1c2ncc(s2)N(=O)=O
2-Imidazolidinone, 1-(5-nitro-2-thiazolyl)
Ambilhar
BA 32644
BA 32644 Ciba
Ciba 32644
Ciba 32644-BA
Niridazole
Nitridazole
Nitrothiamidazol
Nitrothiamidazole
Nitrothiazole
1-(5-Nitro-2-thiazolyl)imidazolidin-2-one
1-(5-Nitro-2-thiazolyl)-2-imida zolidinone
1-(5-Nitro-2-thiazolyl)-2-imidazolinone
1-(5-Nitro-2-thiazolyl)-2-oxotetrahydroimidazol
1-(5-Nitro-2-thiazolyl)-2-oxotetrahydroimidazole
NTOI

61-68-7
$C_{15}H_{15}NO_2$
241.31
Cc2cccc(Nc1ccccc1C(O)=O)c2C
Anthranilic acid, N-(2,3-xylyl)
Acide mefenamique (French)
Bafhameritin-M
Benzoic acid, 2-((2,3-dimethylphenyl)amino)- (9CI)
Bonabol
C.I. 473
CN-35355
Coslan
2-((2,3-Dimethylphenyl)amino)-benzoic acid
N-(2,3-Dimethylphenyl)anthranilic acid
2-Diphenylaminecarboxylic acid, 2',3'-dimethyl-
INF 3355
Lysalgo
Mefenamic acid
Mefenaminsaeure (German)
Mephenamic acid
Mephenaminic acid
Methenamic acid
Namphen
Parkemed
Ponalar
Ponstan
Ponstan forte
Ponstel
Ponstil
Ponstyl
Pontal
Tanston
Vialidon
N-(2,3-Xylyl)-2-aminobenzoic acid
N-(2,3-Xylyl)anthranilic acid

61-72-3
$C_{19}H_{18}ClN_3O_5S$
435.91
Cc2onc(c1ccccc1Cl)c2C(=O)NC4C3SC(C)(C)C(N3C4=O)C(O)=O
4-Thia-1-azabicyclo-(3.2.0)heptane-2-carboxylic acid, 6-(3-(o-chlorophenyl)-s-methyl-4-isoxazolecarboxamido)-3,3-dimethyl-7-oxo
BRL 1621
Cloxacillin
Methocillin S
Syntarpen

61-73-4
$C_{16}H_{18}N_3S.Cl$
319.88
[Cl-].CN(C)c3ccc2nc1ccc(cc1[s+]c2c3)N(C)C
3H-Phenothiazine, 7-(dimethylamino)-3-(methylimino)-, 3-methochloride
Aizen Methylene Blue BH
Aizen Methylene Blue FZ
Basic Blue 9
3,7-Bis (dimethylamino) phenazathionium chloride
3,7-Bis(dimethylamino)phenothiazin-5-ium chloride
Calcozine Blue ZF
Chromosmon
C.I. 52015 (Czech)
C.I. Basic Blue 9
D & C Blue Number 1
Ext. D & C Blue No. 1
External Blue 1
Hidaco Methylene Blue Salt Free
Leather Pure Blue HB
Methylenblau (German)
Methylene Blue
Methylene Blue A
Methylene Blue B
Methylene Blue 2B
Methylene Blue BB
Methylene Blue BBA
Methylene Blue BB (Zinc free)
Methylene Blue BD
Methylene Blue 2BF
Methylene Blue 2BN
Methylene Blue BP
Methylene Blue 2BP
Methylene Blue BPC
Methylene Blue BX
Methylene Blue BZ
Methylene Blue Chloride
Methylene Blue Chloride (Biological stain)
Methylene Blue D
Methylene Blue FZ
Methylene Blue G
Methylene Blue GZ
Methylene Blue HGG
Methylene Blue IAD
Methylene Blue I (Medicinal)
Methylene Blue JFA
Methylene Blue (Medicinal)
Methylene Blue N
Methylene Blue NF (Medicinal)
Methylene Blue NZ
Methylene Blue Polychrome
Methylene Blue SG
Methylene Blue SP
Methylene Blue USP (Medicinal)
Methylene Blue USP XII (Medicinal)
Methylene Blue ZF
Methylene Blue Zinc Free
Methylene Blue ZX
Methylenium ceruleum
Methylthionine
Methylthionine chloride
Methylthionium chloride
Mitsui Methylene Blue
Modr Methylenova (Czech)
Modr Rozpoustedlova 8 (Czech)
Modr Zasadita 9 (Czech)
Phenothiazin-5-ium, 3,7-bis-(dimethylamino)-, chloride
Sandocryl Blue BRL
Schultz No. 1038
Swiss Blue
Tetramethylene Blue
Tetramethylthionine chloride
Yamamoto Methylene Blue B
Yamamoto Methylene Blue ZF

61-76-7
$C_9H_{13}NO_2.ClH$
203.69
Cl.CNCC(O)c1cccc(O)c1
Benzyl alcohol, m-hydroxy-α-((methylamino)methyl)-, hydrochloride, (-)
Adrianol
Alcon-Efrin
Almefrin
Benzenemethanol, 3-hydroxy-α-((methylamino)methyl)-, hydrochloride, (-)-
Biomydrin
Consdrin
Consdrin hydrochloride
Derizene
Efricel
Emagrin
Fenox
Furacin
Histabid
Fenilfar
(-)-α-Hydroxy-β-(methylamino)ethyl-α-(3-hydroxybenzene) hydrochloride
(R)-3-Hydroxy-α-((methylamino)methyl)benzenemethanol hydrochloride
1-m-Hydroxy-α-(methylaminomethyl)benzyl alcohol hydrochloride
l-1-(m-Hydroxyphenyl)-2-

methylaminoethanol hydro-
chloride
Idrianol
Isophrine
Isophrin hydrochloride
Lexatol
Metaoksedrin
Metaoxedrin
Metaoxedrinum
meta-Sympathol
meta-Sympatol
meta-Synephrine
meta-Synephrine hydrochloride
m-Methylaminoethanolphenol
hydrochloride
Metroxedrine
Mezaton
Mydfrin
NCI-C55641
Neooxedrine
Neophryn
Neo-sinefrina
Neosympatol
Neosynephrine
Neosynephrine hydrochloride
Neosynesine
Newphrine
Ocusol
Oftalfrine
Op-Isophrin
m-Oxedrine
Phenistan
Phenyl-drane
Phenylephrine hydrochloride
D-(-)-Phenylephrine hydro-
chloride
Prefrin
Pyracort D
Pyristan
Rhinall
Stanephrin
Sucraphen
m-Sympathol
Synasal
Synethenate
Uri
Visadron

61-78-9
$C_9H_{10}N_2O_3$
194.18
O=C(NCC(=O)O)c(ccc(N)c1)c1
Aminohippuric Acid
AI3-52275
N-(p-Aminobenzoyl)amino-
acetic acid
N-(p-Aminobenzoyl)glycine
N-(para-Aminobenzoyl)glycine
N-(4-Aminobenzoyl)glycine
Aminohippuric acid
p-Aminohippuric acid
para-Aminohippuric acid
4-Aminohippuric acid
Glycine, N-(4-aminobenzoyl)-
(9CI)
Hippuric acid, p-amino- (8CI)
Nefrotest
NSC-13064
PAH (VAN)
PAH (Amino acid)
PAHA

61-82-5
$C_2H_4N_4$
84.10
N(C(=N)NN1)=C1
s-Triazole, 3-amino
Amerol
Aminotriazole
2-Aminotriazole
3-Aminotriazole
3-Amino-s-triazole
3-Amino-1,2,4-triazole
(ACGIH)
2-Amino-1,3,4-triazole
3-Amino-1H-1,2,4-triazole
Aminotriazole (Plant regulator)
Amino Triazole Weedkiller 90
Aminotriazol-Spritzpulver
Amitol
Amitril
Amitril T.L.
Amitrol
Amitrol 90
Amitrole (ACGIH,OSHA)
Amitrol-T
Amizol
Amizol D
Amizol dp nau
Amizol F
AT
ATA
3,A-T
AT-90
AT Liquid
Azaplant
Azaplant Kombi
Azolan
Azole
Campaprim A 1544
Cytrol
Cytrol Amitrole-T
Cytrole
Diurol
Diurol 5030
Domatol
Domatol 88
Elmasil
Emisol
Emisol 50
Emisol F
ENT 25,445
Fenamine
Fenavar
Herbidal Total
Herbizole
Kleer-Lot
Orga-414
Radoxone TL
Ramizol
RCRA waste number U011
Simazol
Solution concentree T271
Triazolamine
1H-1,2,4-Triazol-3-amine
USAF XR-22
Vorox
Vorox AA
Vorox AS
Weedar ADS
Weedar AT
Weedazin
Weedazin arginit
Weedazol
Weedazol GP2
Weedazol super
Weedazol T
Weedazol TL
Weedex Granulat
Weedoclor
X-All Liquid

61-90-5
$C_6H_{13}NO_2$
131.20
O=C(O)C(N)CC(C)C
Leucine, L
α-Aminoisocaproic acid
2-Amino-4-methylpentanoic
acid
α-Amino-γ-methylvaleric acid
Leucin (German)
Leucine
L-Leucine
Norvaline, 4-methyl-
Valeric acid, 2-amino-4-methyl-

62-23-7
$C_7H_5NO_4$
167.13
O=C(O)c(ccc(N(=O)=O)c1)c1
Benzoic acid, p-nitro
1-Carboxy-4-nitrobenzene
Kyselina p-nitrobenzoova
(Czech)
p-Nitrobenzenecarboxylic acid
p-Nitrobenzoic acid
4-Nitrobenzoic acid
4-Nitrodracylic acid

62-33-9
$C_{10}H_{12}CaN_2O_8.2Na$
374.30
[Na+].[Na+].[O-]C(=O)CN1CCN
(CC([O-])=O)CC(=O)O[Ca]
OC(=O)C1
**Calciate(2-), ((ethylenedi-
nitrilo)tetraacetato)-, di-
sodium**
Acetic acid, (ethylenedinitrilo)-
tetra-, calcium disodium salt
Adsorbonac
Antalin
Calcitetracemate disodium
Calcium disodium edathamil
Calcium disodium edetate
Calcium disodium EDTA
Calcium disodium ethylenedi-
aminetetraacetate
Calcium disodium (ethylenedi-
nitrilo)tetraacetate
Calcium disodium versenate
Calcium titriplex
Disodium calcium ethylenedi-
aminetetraacetate
Edathamil calcium disodium
Edetamin
Edetamine
Edetate calcium
Edetic acid calcium disodium
salt

Edtacal
EDTA calcium disodium salt
Ethylenediaminetetraacetic acid, calcium disodium chelate
Mosatil
Rikelate calcium
Sormetal
Tetacin
Tetacin-calcium
Tetazine

62-37-3
$C_5H_{11}ClHgN_2O_2$
367.22
COC(CNC(N)=O)C[Hg]Cl
Mercury, chloro(2-methoxy-3-ureidopropyl)
(3-((Aminocarbonyl)amino)-2-methoxypropyl)chloromercury
Chlormerodrin
Chlormerodrine
Chlormeroprin
1-(3-(Chloromercuri)-2-methoxypropyl)urea
Chloromeridin
Chloromerodrin
Diurone
HG-203 chlormerodrin
Katonil
Mercloran
Mercoral
Merilid
Neohydrin
Oricur
Percapyl
Promeran
Urea, (3-(chloromercuri)-2-methoxypropyl)-
Urea, (3-(chloromercurio)-2-methoxypropyl)-
Urea, (2-methoxypropyl)-, mercury complex

62-38-4
$C_8H_8HgO_2$
336.75
CC(=O)O[Hg]c1ccccc1
Mercury, (acetato)phenyl
Acetate phenylmercurique (French)
(Acetato)phenylmercury
Acetic acid, phenylmercury deriv.
(Acetoxymercuri)benzene
Acetoxyphenylmercury
Agrosan
Agrosand
Agrosan GN 5
Algimycin
Antimucin WDR
Benzene, (acetoxymercuri)-
Benzene, (acetoxymercurio)-
Bufen
Cekusil
Celmer
Ceresan
Ceresan Universal
Ceresol
Contra Creme
Dyanacide
Femma
Fenylmerkuriacetat (Czech)
FMA
Fungitox OR
Gallotox
HL-331
Hong Nien
Hostaquick
Hostaquik
Kwiksan
Leytosan
Liquiphene
Mercuriphenyl acetate
Mercury (II) acetate, phenyl-
Mercury, acetoxyphenyl-
Mersolite
Mersolite 8
Metasol 30
Norforms
Nylmerate
Octan fenylrtutnaty (Czech)
Pamisan
Phenmad
Phenomercuric acetate
Phenylmercuriacetate
Phenyl mercuric acetate
Phenylmercuric acetate [UN 1674]
Phenylmercury acetate
Phenylquecksilberacetat (German)
Phix
PMA
PMAC
Pmacetate
PMAL
PMAS
Purasan-SC-10
Puraturf 10
Quicksan
Quicksan 20
RCRA waste number P092
Sanitized SPG
SC-110
Scutl
Seedtox
Shimmerex
Spor-Kil
TAG
TAG 331
TAG Fungicide
TAG HL 331
Trigosan
UN 1674 [Phenylmercuric acetate]
Ziarnik

62-44-2
$C_{10}H_{13}NO_2$
179.24
O=C(Nc(ccc(OCC)c1)c1)C
p-Acetophenetidide
Acetamide, N-(4-ethoxyphenyl)-(9CI)
1-Acetamido-4-ethoxybenzene
Acetanilide, 4'-ethoxy-
Aceto-para-phenalide
p-Acetophenetide
Aceto-para-phenetidide
para-Acetophenetidide
Acetophenetidin
Acetophenetidine
p-Acetophenetidine
Aceto-4-phenetidine
Acetophenetin
Acet-p-phenalide
Acetphenetidin
p-Acetphenetidin
Acet-p-phenetidin
Acetylphenetidin
N-Acetyl-p-phenetidine
Achrocidin
Anapac
APC
ASA Compound
Bromo seltzer
Buff-A-Comp
Citra-fort
Clistanol
Codempiral
Commotional
Contradol
Contradouleur
Coricidin
Coriforte
Coryban-D
Daprisal
Darvon Compound
Dasikon
Dasin
Dasin CH
Dolostop
Dolviran
Edrisal
Empiral
Empirin Compound
Emprazil
Emprazil-C
Epragen
p-Ethoxyacetanilide
4-Ethoxyacetanilide
4'-Ethoxyacetanilide
p-Ethoxyanilid kyseliny octove (Czech)
N-para-Ethoxyphenylacetamide
N-(4-Ethoxyphenyl)acetamide
Fenacetin (Czech)
Fenacetina
Fenidina
Fenia
Fenina
Fiorinal
Fortacyl
Gelonida
Gewodin
Helvagit
Hjorton's powder
Hocophen
Kafa
Kalmin
Malex
Melabon
Melaforte
Norgesic
Pamprin
Paracetophenetidin
Paramette
Paratodol
Percobarb
Pertonal
Phenacet
Phenacetin

para-Phenacetin
Phenacetine
Phenacetinum
Phenacitin
Phenacon
Phenaphen
Phenaphen plus
Phenazetin
Phenazetina
Phenedina
p-Phenetidine, N-acetyl-
Phenidin
Phenin
Phenodyne
Pyraphen
Pyrroxate
Quadronal
RCRA waste number U187
Reformin
Robaxisal-PH
Salgydal
Sanalgine
Saridon
Seranex
Sinedal
Sinubid
Sinutab
Sinutab II
Soma
Stellacyl
Super Anahist
Supralgin
Synalgos-DC
Synalogos
Tacol
Terracydin
Tetracydin
Thephorin A-C
Treupel
Veganine
Viden
Wigraine
Xaril
Zactirin Compound

62-46-4
$C_8H_{14}O_2S_2$
206.34
1,2-Dithiolane-3-valeric acid
Acetate-replacing factor
Biletan
5-(1,2-Dithiolan-3-yl)valeric acid
6,8-Dithiooctanoic acid
Heparlipon
Lipoic acid
α-Lipoic acid
α-Liponic acid
α-Liponsaeure (German)
Liposan
Lipothion
Thioctacid
Thioctic acid
6-Thioctic acid
6,8-Thioctic acid
Thioctidase
Thioctsan
Thioktsaeure (German)
Thiooctanoic acid
6-Thiotic acid
6,8-Thiotic acid
Tioctacid
Tioctidasi
Tioctidasi acetate replacing factor

62-49-7
$C_5H_{14}NO$
104.20
OCCN(C)(C)C
Choline
Bilineurine
Choline ion
Ethanaminium, 2-hydroxy-N,N,N-trimethyl- (9CI)
(2-Hydroxyethyl)trimethylammonium

62-50-0
$C_3H_8O_3S$
124.17
O=S(=O)(OCC)C
Methanesulfonic acid, ethyl ester (8CI,9CI)
EMS
ENT 26,396
Ethylester kyseliny methansulfonove (Czech)
Ethyl ester of methanesulfonic acid
Ethyl ester of methanesulphonic acid
Ethyl ester of methylsulfonic acid
Ethyl ester of methylsulphonic acid
Ethyl methanesulfonate
Ethyl methanesulphonate
Ethyl methansulfonate
Ethyl methansulphonate
Half-Myleran
Methanesulphonic acid ethyl ester
Methylsulfonic acid, ethyl ester
NSC-26805
RCRA waste number U119

62-51-1
$C_8H_{18}NO_2.Cl$
195.72
[OH].[Cl-].CC(C[N+](C)(C)C)OC(C)=O
Ammonium, (2-hydroxypropyl)trimethyl-, chloride, acetate
o-Acetyl-β-methylcholine chloride
Amechol
(2-Hydroxypropyl)trimethylammonium chloride acetate
Mecholyl
Mecholyl chloride
Metacholine chloride
Methacholine chloride
Methacholinium chloride
Methylacetyl choline
β-Methylacetylcholine chloride
1-Propanaminium, 2-(acetyloxy)-N,N,N-trimethyl-, chloride (9CI)
Trimethyl-β-acetoxypropylammonium chloride

62-53-3
C_6H_7N
93.14
Nc(cccc1)c1
Aniline (ACGIH,OSHA) [UN 1547]
Aminobenzene
Aminophen
Anilin (Czech)
Anilina (Italian, Polish)
Aniline oil
Aniline oil, Liquid [UN 1547]
Anyvim
Benzenamine
Benzene, amino
Benzidam
Blue oil
C.I. 76000
C.I. Oxidation base 1
Cyanol
Huile d'aniline (French)
Krystallin
Kyanol
NCI-C03736
Phenylamine
RCRA waste number U012
UN 1547 [Aniline]

62-54-4
$C_4H_6O_4.Ca$
158.18
Acetic acid, calcium salt
Brown acetate
Calcium acetate
Calcium diacetate
Gray acetate
Lime acetate
Lime pyrolignite
Sorbo-Calcian
Sorbo-Calcion
Teltozan
Vinegar salts

62-55-5
C_2H_5NS
75.14
N=C(S)C
Acetamide, thio
Acetothioamide
Ethanethioamide
RCRA waste number U218
TAA
Thiacetamide
Thioacetamide
Thioacetimidic acid
USAF CB-21
USAF EK-1719

62-56-6
CH_4N_2S
76.13
N=C(S)N
Urea, 2-thio
Isothiourea
Pseudothiourea

RCRA waste number U219
Sulourea
Thiocarbamide
Thiocarbonic acid diamide
Thiomocovina (Czech)
β-Thiopseudourea
2-Thiopseudourea
Thiourea (DOT)
2-Thiourea
THU
TSIZP 34
UN 2877 (DOT)
Urea, thio- (8CI)
USAF EK-497

62-59-9
$C_{32}H_{49}NO_9$
591.82
CC=C(C)C(=O)OC6CCC4(C)C7CCC5C3(O)CC(O)C2(O)C(CN1CC(C)CCC1C2(C)O)C3(O)CC45OC67O
Cevane-3-β,4-β,12,14,16-β,17,-20-heptol, 4,9-epoxy-, 3-((Z)-2-methylcrotonate), (Z)
Cevadene
Cevadin
Cevadine
(Z)-4-α,9-Epoxycevane-3-β,4,12,14,16-β,17,20-heptol 3-(2-methyl-2-butenoate)
Veratrine
Veratrine (Crystallized)

62-67-9
$C_{19}H_{21}NO_3$
311.41
OC1C=CC2C3Cc4ccc(O)c5OC1C2(CCN3CC=C)c45
Morphinan-3,6-α-diol, 17-allyl-7,8-didehydro-4,5-α-epoxy
Acetorfin (Czech)
Acetorphine
Allorphine
N-Allyl-7,8-dehydro-4,5-epoxy-3,6-dihydroxymorphinan
N-Allyl-N-desmethylmorphine
N-Allylnormorphine
Anarcon
Antofin
Antorfin
Antorphine
Lethidrome
Lethidrone
Letidron
Lithidrone
Nalline
Nalorfina
Nalorphine
Nalorphinium
Nanm
Normorphine, N-allyl-

62-73-7
$C_4H_7Cl_2O_4P$
220.98
O=P(OC)(OC)OC=C(Cl)Cl
Phosphoric acid, 2,2-dichlorovinyl dimethyl ester
Apavap
Astrobot
Atgard
Atgard C
Atgard V
Bay-19149
Bayer 19149
Benfos
Bibesol
Brevinyl
Brevinyl E50
Canogard
Cekusan
Chlorvinphos
Ciovap
Cypona
DDVF
DDVP
DDVP (Insecticide)
Dedevap
Deriban
Derribante
Devikol
(2,2-Dichloor-vinyl)-dimethyl-fosfaat (Dutch)
Dichloorvo (Dutch)
Dichlorfos (Polish)
Dichlorman
2,2-Dichloroethenyl dimethyl phosphate
2,2-Dichloroethenyl phosphoric acid dimethyl ester
Dichlorophos
Dichlorovas
(2,2-Dichloro-vinil)dimetil-fosfato (Italian)
2,2-Dichlorovinyl dimethyl phosphate
2,2-Dichlorovinyl dimethyl phosphoric acid ester
Dichlorovos
Dichlorovos mixture, Dry (DOT)
Dichlorphos
(2,2-Dichlor-vinyl)-dimethyl-phosphat (German)
O-(2,2-Dichlorvinyl)-O,O-dimethylphosphat (German)
Dichlorvos (ACGIH,DOT, OSHA)
Dimethyl 2,2-dichloroethenyl phosphate
Dimethyl dichlorovinyl phosphate
Dimethyl 2,2-dichlorovinyl phosphate
O,O-Dimethyl dichlorovinyl phosphate
O,O-Dimethyl 2,2-dichlorovinyl phosphate
O,O-Dimethyl O-2,2-dichlorovinyl phosphate
O,O-Dimethyl-O-(2,2-dichlor-vinyl)-phosphat (German)
Divipan
Duo-Kill
Duravos
O,O-Dwumetylo-O-dwuchloro-winylofosforan (Polish)
ENT 20,738
Equigard
Equigel
Estrosel
Estrosol
Ethenol, 2,2-dichloro-, dimethyl phosphate
Fecama
Fekama
Fly-Die
Fly Fighter
Herkal
Herkol
Insectigas D
Krecalvin
Lindan
Mafu
Mafu Strip
Marvex
Mopari
NA 2783
NCI-C00113
Nefrafos
Nerkol
Nogos
Nogos 50
Nogos 50 EC
Nogos G
No-Pest
No-Pest Strip
Novotox
NSC-6738
Nuva
Nuvan
Nuvan 7
Nuvan 100EC
OKO
OMS 14
Panaplate
Phosphate de dimethyle et de 2,2-dichlorovinyle (French)
Phosphoric acid, 2,2-dichloro-ethenyl dimethyl ester
Phosvit
Prentox
Ravap
SD-1750
Szklarniak
TAP 9VP
Task
Tenac
Task Tabs
Tetravos
UDVF
Unifos
Unifos 50 EC
Unifos (Pesticide)
Unitox
Vapona
Vapona insecticide
Vaponite
Vapora II
Verdican
Verdipor
Vinyl alcohol, 2,2-dichloro-, dimethyl phosphate
Vinylofos
Vinylophos
Winylophos

62-74-8
$C_2H_2FO_2.Na$
100.03

[Na+].[O-]C(=O)CF
Acetic acid, fluoro-, sodium salt
1080
Compound No. 1080
Fluoressigaeure (German)
Fluoroacetic acid, sodium salt
Fluoroctan sodny (Czech)
Fratol
Furatol
Latka 1080 (Czech)
Monofluoressigsaures natrium (German)
Natriumfluoracetaat (Dutch)
Natriumfluoracetat (German)
Ratbane 1080
RCRA waste number P058
Sodio, fluoracetato di (Italian)
Sodium fluoacetate
Sodium fluoacetic acid
Sodium fluoracetate
Sodium fluoroacetate (ACGIH, OSHA) [UN 2629]
Sodium fluoroacetate de (French)
Sodium monofluoroacetate
TL 869
UN 2629 [Sodium fluoroacetate]
Yasoknock

62-75-9
$C_2H_6N_2O$
74.10
O=NN(C)C
Dimethylamine, N-nitroso
Dimethylnitrosamin (German)
Dimethylnitrosamine
N,N-Dimethylnitrosamine
Dimethylnitrosoamine
DMN
DMNA
N-Methyl-N-nitrosomethanamine
NDMA
Nitrosodimethylamine
N-Nitrosodimethylamine (ACGIH,OSHA)
N-Nitroso-N,N-dimethylamine
RCRA waste number P082

62-76-0
C_2O_4.2Na
134.00
Ethanedioic acid, disodium salt
Natriumoxalat (German)
Oxalic acid, disodium salt
Sodium oxalate
Stavelan sodny (Czech)

63-12-7
$C_{22}H_{32}N_2O_5$
404.49
CCN(CC)C(=O)C3CN2CCc1cc(OC)c(OC)cc1C2CC3OC(C)=O
Benzquinamide
2-Acetoxy-3-diethylcarbamyl-9,10-dimethoxy-1,2,3,4,6,7-hexahydro-11b-benzo(a)-quinolizine
2-Acetoxy-3-(N,N-diethyl-carboxamido)-9,10-dimethoxy-1,2 ,3,4,6,7-hexahydro-11bh-benzopyridocoline
2-(Acetyloxy)-N,N-diethyl-1,3,4,6,7,11b-hexahydro-9,10-dimethoxy-2H-benzo(a)-quinolizine-3-carboxamide
2H-Benzo(a)quinolizine-3-carboxamide, 2-(acetyloxy)-N,N-diethyl-1,3,4,6,7,11b-hexahydro-9,10-dimethoxy-
2H-Benzo(a)quinolizine-3-carboxamide, N,N-diethyl-1,3,4,6,7,11b-hexahydro-2-hydroxy-9,10-dimethoxy-, acetate (ester)
Benzchinamide
Benzchinamidum
Benzochinamide
Benzoquinamida (Spanish)
Benzoquinamide
Benzquinamidum (Latin)
BZQ
N,N-Diethyl-1,3,4,6,7,11b-hexahydro-2-hydroxy-9,10-dimethoxy-2H-benzo(a)quinolizine-3-carboxamide acetate
N,N-Diethyl-1,3,4,6,7,11b-hexahydro-2-hydroxy-9,10-dimethoxy-2H-benzo(a)quinolizine-3-carboxamide acetate (ester)
Emete-Con
Emeticon
2-Hydroxy-3-diethylcarbamyl-9,10-dimethoxy-1,2,3,4,6,7-hexahydro-11bh-benzoquinolizine acetate
NSC-64375
P-2647
Quantril
Quantryl

63-25-2
$C_{12}H_{11}NO_2$
201.24
O=C(Oc(c(c(ccc1)cc2)c1)c2)NC
Carbamic acid, methyl-, 1-naphthyl ester
Arilat
Arilate
Arylam
Atoxan
Bercema NMC50
Bug Master
Caprolin
Carbamine
Carbaril (Italian)
Carbaryl (ACGIH,DOT,OSHA)
Carbatox
Carbatox-60
Carbatox-75
Carbavur
Carbomate
Carpolin
Carylderm
Cekubaryl
Compound 7744
Crag Sevin
Crunch
Denapon
Devicarb
Dicarbam
ENT 23,969
Experimental insecticide 7744
Gamonil
Germain's
Hexavin
Karbaryl (Polish)
Karbaspray
Karbatox
Karbatox 75
Karbosep
Latka 7744 (Czech)
N-Methylcarbamate de 1-naphtyle (French)
Menaphtam
Methylcarbamate 1-naphthalenol
Methylcarbamate 1-naphthol
Methylcarbamic acid, 1-naphthyl ester
N-Methyl-1-naftyl-carbamaat (Dutch)
N-Methyl-1-naphthyl-carbamat (German)
N-Methyl-α-naphthylcarbamate
N-Methyl-1-naphthyl carbamate
N-Methyl-α-naphthylurethan
N-Metil-1-naftil-carbammato (Italian)
Monsur
Mugan
Murvin
NAC
NA 2757 (DOT)
1-Naftylester kyseliny methyl-karbaminove (Czech)
α-Naftyl-N-methylkarbamat (Czech)
α-Naphthalenylmethylcarbamate
1-Naphthalenyl methylcarbamate
1-Naphthol N-methylcarbamate
α-Naphthyl methylcarbamate
α-Naphthyl N-methylcarbamate
1-Naphthyl methylcarbamate
1-Naphthyl N-methylcarbamate
1-Naphthyl-N-methyl-karbamat (German)
NMC 50
Oltitox
OMS-29
Panam
Pomex
Prosevor 85
Ravyon
Rylam
Seffein
Septene
Sevimol
Sevin
Sevin 4
Sevin (OSHA)
Sewin
Sok
Tercyl
Toxan
Tricarnam
UC 7744

Union Carbide 7,744
Vioxan

63-36-5
$C_7H_5O_3$
137.12
O=C(O)c(c(O)ccc1)c1
Salicylic acid, ion(1-)
Benzoic acid, 2-hydroxy-, ion-(1-) (9CI)
o-Hydroxybenzoate
2-Hydroxybenzoate
o-Hydroxybenzoate anion
Salicylate
Salicylate anion
Salicylate ion

63-42-3
$C_{12}H_{22}O_{11}$
342.34
O=CC(O)C(O)C(OC(OC(C(O)C1O)CO)C1O)C(O)CO
Lactose
4-(β-D-Galactosido)-D-glucose
D-Glucose, 4-o-β-D-galacto-pyranosyl-
D-Lactose
Lactin
Lactobiose
Milk Sugar
Saccharum Lactin

63-56-9
$C_{16}H_{22}N_4O.ClH$
322.88
Pyrimidine, 2-((2-(dimethyl-amino)ethyl)(p-methoxy-benzyl)amino)-, hydro-chloride
Anahist
Andhist
Anshist
2-((2-(Dimethylamino)ethyl)(p-methoxy-benzyl)amino)-pyrimidine hydrochloride
N,N-Dimethyl-N'-(4-methoxy-benzyl)-N'-(2-pyrimidyl)-ethylenediamine hydrochloride
N-p-Methoxybenzyl-N',N'-di-methyl-N-2-pyrimidinyl-ethylene diamine hydro-chloride
Neohetramine hydrochloride
NH 188
Novohetramin
Resistab
Thonzylamine hydrochloride
Thonzylaminium chloride

63-68-3
$C_5H_{11}NO_2S$
149.23
O=C(O)C(N)CCSC
Methionine, L
l-α-Amino-γ-methylmercapto-butyric acid
l(-)-Amino-γ-methylthiobutyric acid
Butyric acid, 2-amino-4-(methylthio)-
Cymethion
Liquimeth
Methionine
l-Methionine
l-(-)-Methionine
l-γ-Methylthio-α-aminobutyric acid

63-74-1
$C_6H_8N_2O_2S$
172.22
O=S(=O)(N)c(ccc(N)c1)c1
Sulfanilamide
Albexan
Albosal
Ambeside
p-Aminobenzenesulfamide
p-Aminobenzenesulfonamide
4-Aminobenzenesulfonamide
p-Aminophenylsulfonamide
4-Aminophenylsulfonamide
p-Anilinesulfonamide
Aniline-p-sulfonic amide
Antistrept
Astreptine
Astrocid
Bacteramid
Bactesid
Benzenesulfonamide, p-amino-
Collomide
Colsulanyde
Copticide
Deseptyl
Dipron
Ergaseptine
Erysipan
Estreptocida
F 1162
1162 F
Fourneau 1162
Gerison
Gombardol
Lusil
Lysococcine
Neococcyl
Orgaseptine
Pabs
Prontalbin
Prontosil album
Prontosil I
Prontosil White
Prontylin
Pronzin Album
Proseptal
Proseptine
Proseptol
Pysococcine
Rubiazol A
Sanamid
Septamide Album
Septanilam
Septinal
Septolix
Septoplex
Septoplix
Stopton Album
Stramid
Strepamide
Strepsan
Streptagol
Streptamid
Streptasol
Streptocid Album
Streptocide White
Streptoclase
Streptocom
Streptol
p-Sulfamidoaniline
Strepton
Streptosil
Streptozol
Streptozone
Streptrocide
Sulfamidyl
Sulfamine
Sulfana
Sulfanalone
Sulfanil
Sulfocidin
Sulfocidine
Sulfonamide
Sulfonamide P
Sulphanilamide
Therapol
White Streptocide

63-91-2
$C_9H_{11}NO_2$
165.21
O=C(O)C(N)Cc(cccc1)c1
Phenylalanine (9CI)
Alanine, phenyl- (8CI)
Alanine, phenyl-, L
Alanine, 3-phenyl-
L-Alanine, phenyl-
(S)-α-Aminobenzenepropanoic acid
α-Aminohydrocinnamic acid
α-Amino-β-phenylpropionic acid
Antibiotic FN 1636
Benzenepropanoic acid, α-amino-, (S)-
Fenilalanina (Spanish)
Hydrocinnamic acid, α-amino-
NSC-79477
PAL
L-Phenylalanine
L-β-Phenylalanine
Phenyl-α-alanine
β-Phenylalanine
β-Phenyl-α-alanine
β-Phenyl-α-alanine, L-
β-Phenyl-L-alanine
(-)-β-Phenylalanine
(L)-Phenylalanine
(S)-Phenylalanine
3-Phenylalanine
3-Phenyl-L-alanine
Phenylalaninum (Latin)

63-92-3
$C_{18}H_{22}ClNO.ClH$
340.32
[Cl-].CC(COc1ccccc1)N(CCCl)Cc2ccccc2
Benzylamine, N-(2-chloro-ethyl)-N-(1-methyl-2-phen-oxyethyl)-, hydrochloride

688A
Bensylyt NEN
Benzenemethanamine, N-(2-chloroethyl)-N-(1-methyl-2-phenoxyethyl)-, hydrochloride
2-(N-Benzyl-2-chloroethylamino)-1-phenoxypropane hydrochloride
Benzyl(2-chloroethyl)(1-methyl-2-phenoxyethyl)amine hydrochloride
N-Benzyl-N-phenoxyisopropyl-β-chlorethylamine hydrochloride
Benzylyt
Blocadren
N-(2-Chloroethyl)-N-(1-methyl-2-phenoxyethyl)benzenemethanamine hydrochloride
N-(2-Chloroethyl)-N-(1-methyl-2-phenoxyethyl)benzylamine, hydrochloride
Dibenylin
Dibenyline
Dibenzylene
Dibenzyline
Dibenzyline hydrochloride
Dibenzyran
Fenoxybenzamin
NCI-C01661
Phenoxybenzamide hydrochloride
Phenoxybenzamine hydrochloride
N-Phenoxyisopropyl-N-benzyl-β-chloroethylamine hydrochloride
N-2-Phenoxyisopropyl-N-benzyl-chloroethylamine hydrochloride
SKF 688A

63-98-9
$C_9H_{10}N_2O_2$
178.21
NC(=O)NC(=O)Cc1ccccc1
Urea, (phenylacetyl) (8CI)
A-1348
Acetylureum
N-(Aminocarbonyl)benzeneacetamide
Benzeneacetamide, N-(aminocarbonyl)- (9CI)
Carbamide phenylacetate
Carbanmide
Cetylureum
Comitiadone
Eferon
Epheron
Epiclase
Felurea
Fenacemid
Fenacemida (Spanish)
Fenacemide
Fenacetamide
Fenacetil-Karbamide
Fenilep
Fenised
Fenostenyl
Fenural
Fenurea
Fenurone
Fenylacetylmocovina (Czech)
Fenytan
Neophedan
Neophenal
Phacetur
Phenacalum
Phenacemide
Phenacemidum (Latin)
Phenacereum
Phenacetur
Phenacetylcarbamide
Phenacetylurea
Phenarone
Phenicarb
Phenuron
Phenurone
Phenutal
(Phenylacetyl)urea
α-Phenylacetylurea
Phenylacetyluree (French)
Phenyrit
Phetylureum
Trioxanona

63-99-0
$C_8H_{10}N_2O$
150.20
Cc1cccc(NC(N)=O)c1
Urea, m-tolyl
Metatolylcarbamide
3-Methylphenylurea
MTC
m-Tolylcarbamide
m-Tolylurea
3-Tolylurea

64-00-6
$C_{11}H_{15}NO_2$
193.27
CNC(=O)Oc1cccc(c1)C(C)C
Carbamic acid, methyl-, m-cumenyl ester (8CI)
AC 5727
Carbamic acid, N-methyl-, 3-isopropylphenyl ester
Carbamic acid, methyl-, 3-(1-methylethyl)phenyl ester
Caswell No. 512A
Compound 10854
m-Cumenol methylcarbamate
m-Cumenyl methylcarbamate
ENT 25,500
ENT 25,543
EPA Pesticide Chemical Code 047801
H 5727
H 8757
HER. 5727
Hercules 5727
Hercules AC 5727
HIP
m-Isopropylphenol N-methylcarbamate
m-Isopropylphenyl methylcarbamate
m-Isopropylphenyl N-methylcarbamate
3-Isopropylphenyl methylcarbamate
m-Kumenylester kyseliny methylkarbaminove (Czech)
Methylcarbamic acid m-cumenyl ester
N-Methyl m-isopropylphenyl carbamate
N-Methyl 3-isopropylphenyl carbamate
OMS 15
OMS 162
OMS-15
1PC
Phenol, m-isopropyl-, methylcarbamate
Phenol, 3-(1-methylethyl)-, methylcarbamate (9CI)
UC 10854
Union Carbide 10854
Union Carbide UC-10,854

64-02-8
$C_{10}H_{12}N_2O_8.4Na$
380.20
[Na].[Na].[Na].[Na].[O-]C(=O)CN(CCN(CC([O-])=O)CC([O-])=O)CC([O-])=O
Acetic acid, (ethylenedinitrilo)tetra-, tetrasodium salt
Aquamoline BC
Aquamollin
Calsol
Celon E
Celon H
Celon IS
Cheelox BF
Cheelox BF-12
Cheelox BF-13
Cheelox BF-78
Cheelox BR-33
Chelon 100
Chemcolox 200
Chemcolox 240 Powder
Complexone
Conigon BC
Distol
Distol 8
Edathanil tetrasodium
Edetate sodium
Edetic acid tetrasodium salt
EDTA, sodium salt
EDTA tetrasodium salt
Endrate tetrasodium
Ergon
Ethylenebis(iminodiacetic acid) tetrasodium salt
N,N'-Ethylenediaminediacetic acid tetrasodium salt
Ethylenediaminetetraacetic acid, tetrasodium salt
Glycine, N,N'-1,2-ethanediylbis(N-(carboxymethyl)-, tetrasodium salt (9CI)
Hamp-ene 100
Hamp-ene 215
Hamp-ene 220
Hamp-ene Na4
Irgalon
Kalex
Kemplex 100

Komplexon
Metaquest C
Nervanaid B
Nervanaid B Liquid
Nervanid B
Nullapon
Nullapon B
Nullapon BF-12
Nullapon BF-78
Nullapon BFC
Nullapon BFC Conc
Nullapon BFC Conc Beads
Nullapon BFC Liquid
Perma Kleer 100
Perma Kleer 50 Crystals
Perma Kleer Tetra CP
Questex
Questex 4
Sequestrene
Sequestrene 30A
Sequestrene Na 4
Sequestrene ST
Sodium Edetate
Sodium EDTA
Sodium ethylenediaminetetra-acetate
Sodium ethylenediaminetetra-acetic acid
Sodium salt of ethylenediamine-tetraacetic acid
Syntes 12A
Syntron B
Tetracemate tetrasodium
Tetracemin
Tetranatrium ethylendiamin-tetraacetat (Czech)
Tetrasodium edetate
Tetrasodium EDTA
Tetrasodium ethylenediamine-tetraacetate
Tetrasodium ethylenediamine-tetracetate
Tetrasodium (ethylenedinitrilo)-tetraacetate
Tetrasodium salt EDTA
Tetrasodium salt of EDTA
Tetrasodium salt of ethylene-diaminetetracetic acid
Tetrine
Trilon B
TST
Tyclarosol
Versene
Versene 67
Versene 100
Versene 220
Versene Beads
Versene FE 3
Versene Flake
Versene Powder
Versene Powder tetra sodium
Warkeelate PS-42
Warkeelate PS-43
Warkeelate PS-47
Warkeelate S-42

64-04-0
$C_8H_{11}N$
121.20
NCCc(cccc1)c1
Phenethylamine
β-Aminoethylbenzene
2-Amino-fenylethan (Czech)
1-Amino-2-phenylethane
2-Amino-1-phenylethane
Benzeneethanamine
Ethylamine, 2-phenyl-
2-Fenylethylamin (Czech)
β-Phenethylamine
β-Phenylaethylamin (German)
1-Phenyl-2-amino-athan (German)
1-Phenyl-2-aminoethane
Phenylethylamine
β-Phenylethylamine
ω-Phenylethylamine
2-Phenylethylamine

64-10-8
$C_7H_8N_2O$
136.17
O=C(Nc(cccc1)c1)N
Urea, phenyl
Monophenylurea
PC
Phenylcarbamide
Phenylurea
N-Phenylurea
1-Phenylurea
Stabilisator VH
Stabilizer VH
VH

64-17-5
C_2H_6O
46.08
OCC
Ethanol
Absolute ethanol
Aethanol (German)
Aethylalkohol (German)
Alcohol
Alcohol, Anhydrous
Alcohol, Dehydrated
Alcool ethylique (French)
Alcool etilico (Italian)
Algrain
Alkohol (German)
Alkoholu etylowego (Polish)
Anhydrol
Cologne Spirit
Etanolo (Italian)
Ethanol (OSHA) [UN 1170]
Ethanol, Solution [UN 1170]
Ethanol 200 proof
Ethanol, Solution [UN 1170]
Ethyl alcohol (ACGIH,OSHA) [UN 1170]
Ethylalcohol (Dutch)
Ethyl alcohol anhydrous
Ethyl Hydrate
Ethyl hydroxide
Etylowy alkohol (Polish)
Fermentation Alcohol
Grain Alcohol
Jaysol
Jaysol S
Methylcarbinol
Molasses Alcohol
NCI-C03134
Potato Alcohol
SD Alcohol 23-hydrogen
Spirits of Wine
Spirt
Tecsol
UN 1170 [Ethanol; (ethyl alcohol) or Ethanol Solutions; (ethyl alcohol solutions)]

64-18-6
CH_2O_2
46.03
O=CO
Formic-acid
Acide formique (French)
Acido formico (Italian)
Ameisensaeure (German)
Aminic acid
Formic acid (ACGIH,OSHA) [UN 1779]
Formic acid, Solution [UN 1779]
Formylic acid
Hydrogen carboxylic acid
Kwas metaniowy (Polish)
Kyselina mravenci (Czech)
Methanoic acid
Mierenzuur (Dutch)
RCRA waste number U123
UN 1779 [Formic acid]

64-19-7
$C_2H_4O_2$
60.06
O=C(O)C
Acetic-acid
Acetic acid (ACGIH,OSHA)
Acetic acid (Aqueous solution) (DOT)
Acetic acid, glacial
Acetic acid, glacial (DOT)
Acetic acid, glacial, More than 80% acid, by weight [UN 2789]
Acetic acid solution, More than 80% acid, by weight [UN 2789]
Acetic acid solution, More than 25% but not more than 80% acid, by weight [UN 2790]
Acide acetique (French)
Acido acetico (Italian)
Azijnzuur (Dutch)
Essigsaeure (German)
Ethanoic acid
Ethylic acid
Glacial acetic acid
Kyselina octova (Czech)
Methanecarboxylic acid
Octowy kwas (Polish)
UN 2789 [Acetic acid, glacial or Acetic acid solution, more than 80 per cent acid, by mass]
UN 2790 [Acetic acid solution, more than 10 per cent but not more than 80 per cent acid, by mass]
Vinegar acid

64-20-0
$C_4H_{12}N.Br$
154.08
Ammonium, tetramethyl-, bromide
Methanaminium, N,N,N-trimethyl-, bromide (9CI)
Tetramethylammonium bromide

64-31-3
$C_{34}H_{38}N_2O_6.H_2O_4S$
668.82
Morphinan-3,6-α-diol, 7,8-didehydro-4, 5-α-epoxy-17-methyl-, sulfate
Morphine sulfate
Morphine sulphate

64-39-1
$C_{17}H_{25}NO_2$
275.38
Promedol
DEA No. 9646
Isopromedol
4-Piperidinol, 1,2,5-trimethyl-4-phenyl-, propionate (ester)
Trimeperidina (Spanish)
Trimeperidine
Trimeperidinum (Latin)
1,2,5-Trimethyl-4-phenyl-4-piperidinol, propionate (ester)
1,2,5-Trimethyl-4-phenyl-4-propionoxypiperidine

64-47-1
$C_{30}H_{42}N_6O_4.O_4S$
646.84
Pyrrolo(2,3-b)indole, 1,2,3,3a, 8,8a-hexahydro-5-hydroxy-1,3a,8-trimethyl-, methylcarbamate (ester), (3as-cis)-, sulfate (2:1)
Eserine sulfate
Eserine sulphate
Physostigmine SO_4
Physostigmine sulfate
Physostigmine sulphate

64-52-8
$C_{21}H_{25}NO_2$
323.43
DEA No. 9663
PEPAP
1-(2-Phenethyl)-4-phenyl-4-acetoxypiperidine
1-Phenethyl-4-phenyl-4-piperidinol acetate (ester)
1-(2-Phenylethyl)-4-phenyl-4-acetoxypiperidine
4-Piperidinol, 1-phenethyl-4-phenyl-, acetate (ester)
4-Piperidinol, 4-phenyl-1-(2-phenylethyl)-, acetate (ester)

64-55-1
$C_{10}H_{20}N_2O_4$
232.27
Mebutamate
Axiten
Butatensin
2-sec-Butyl-2-methyl-1,3-propanediol dicarbamate
2-sec-Butyl-2-methyltrimethylene dicarbamate
Capla
Carbamic acid, 2-sec-butyl-2-methyltrimethylene ester
Carbamic acid, 2-sec-butyl-, 2-methyltrimethylene ester
Carbuten
DEA No. 2800
Dicamoylmethtane
2,2-Dicarbamyloxymethyl-3-methylpentane
Dormate
Encapla
Ipotensivo
Mebutamat
Mebutamato (Spanish)
Mebutamatum (Latin)
Mebutina
MEGA
2-Methyl-2-sec-butyl-1,3-propanediol dicarbamate
2-Methyl-2-(1-methylpropyl)-1,3-propanediol dicarbamate
Mioartrina
No-Press
NSC-163921
Prean
Preminex
1,3-Propanediol, 2-sec-butyl-2-methyl-, dicarbamate (8CI)
1,3-Propanediol, 2-methyl-2-(1-methylpropyl)-, dicarbamate (9CI)
Sigmafon
Vallene
W-583

64-67-5
$C_4H_{10}O_4S$
154.20
O=S(=O)(OCC)OCC
Sulfuric acid, diethyl ester
Diaethylsulfat (German)
Diethylester kyseliny sirove (Czech)
Diethyl sulfate
Diethyl sulphate [UN 1594]
DS
Ethyl sulfate
UN 1594 [Diethyl sulfate]

64-69-7
$C_2H_3IO_2$
185.95
O=C(O)CI
Acetic acid, iodo
IA
Iodoacetate
Iodoacetic acid
Kyselina jodoctova (Czech)
MIA
Monoiodoacetate
Monoiodoacetic acid

64-75-5
$C_{22}H_{24}N_2O_8.ClH$
480.94
Cl.CN(C)C1C(O)C(C(N)=O)C(=O)C2(O)C1CC4C(=C2O)C(=O)c3c(O)cccc3C4(C)O
2-Naphthacenecarboxamide, 4-(dimethylamino)-1,4,4a, 5,5a,6,11,12a-octahydro-3,6,10,12,12a- pentahydroxy-6-methyl-1,11-dioxo-, monohydrochloride
Achro
Achromycin
Achromycin hydrochloride
Achromycin V
Ala Tet
Ambracyn
Amycin, hydrochloride
Artomycin
Bristacycline
Cefracycline tablets
Chlorhydrate de tetracycline (French)
Cyclopar
Diacycine
Dumocycin
Medamycin
Mephacyclin
NCI-C55561
Paltet
Panmycin hydrochloride
Partrex
Piracaps
Polycycline hydrochloride
Qidtet
Quadracycline
Quatrex
Remicyclin
Ricycline
Ro-Cycline
SK-Tetracycline
Steclin
Steclin hydrochloride
Stilciclina
Subamycin
Sumycin
Supramycin
Sustamycin
T-250 Capsules
TC hydrochloride
Tefilin
Teline
Telotrex
Tet-Cy
Tetrabakat
Tetrabid
Tetrablet
Tetracaps
Tetrachel
Tetraciclina cloridrato (Italian)
Tetracompren
Tetracycline chloride
Tetracycline hydrochloride
Tetracyn
Tetra-D
Tetrakap
Tetralution
Tetramavan
Tetramycin

Tetra-wedel
Tetrosol
Topicycline
Totomycin
Triphacyclin
U-5965
Unicin
Unimycin
Vetquamycin-324

64-77-7
$C_{12}H_{18}N_2O_3S$
270.38
O=C(NCCCC)NS(=O)(=O)c(ccc(c1)C)c1
Urea, 1-butyl-3-(p-tolylsulfonyl)
Aglicid
Arkozal
Artosin
Artozin
Butamid
1-Butyl-3-(p-methylphenylsulfonyl)urea
Benzenesulfonamide, N-((butylamino)carbonyl)-4-methyl-
N-Butyl-N'-p-toluenesulfonylurea
N-Butyl-N'-toluene-p-sulfonylurea
1-Butyl-3-(p-tolylsulfonyl)urea
N-n-Butyl-N'-tosylurea
1-Butyl-3-tosylurea
BZ 55
D 860
Diaben
Diabetamid
Diabetol
Diabuton
Dolipol
Drabet
HLS 831
Ipoglicone
N-4-Methylbenzolsulfonyl-N-butylurea
Mobenol
NCI-C01763
Orabet
Oralin
Orezan
Orinase
Orinaz
Oterben
Rastinon
N-(Sulfonyl-p-methylbenzene)-N'-n-butylurea
Tolbusal
Tolbutamid
Tolbutamide
1-p-Toluenesulfonyl-3-butylurea
Toluina
Tolumid
Toluvan
N-(p-Tolylsulfonyl)-N'-butylcarbamide
Tolylsulfonylbutylurea
3-(p-Tolyl-4-sulfonyl)-1-butylurea
Willbutamide
SK-Tolbutamide

64-86-8
$C_{22}H_{25}NO_6$
399.48
O=C(NC(C(C(c(c(cc(OC)c1OC)C2)c1OC)=CC=C(OC)C3=O)=C3)C2)C
Colchicine
Acetamide, N-(5,6,7,9-tetrahydro-1,2,3,10-tetramethoxy-9-oxobenzo(α)heptalen-7-yl)-
N-Acetyl trimethylcolchicinic acid methylether
Benzo(a)heptalen-9(5H)-one, 7-acetamido-6,7-dihydro-1,2,3,10-tetramethoxy-
Colchicin (German)
Colchicina (Italian)
7-α-H-Colchicine
Colchineos
Colchisol
Colcin
Colsaloid
Condylon
NSC-757

64-95-9
$C_{20}H_{25}NO_2$
311.46
CCN(CC)CCOC(=O)C(c1ccccc1)c2ccccc2
Acetic acid, diphenyl-, 2-(diethylamino)ethyl ester
Adiphenin
Adiphenine
Benzeneacetic acid, α-phenyl-, 2-(diethylamino)ethyl ester, (9CI)
2-Diethylaminoethyl diphenylacetate
2-Diethylaminoethylester kyseliny difenyloctove (Czech)
Difacil
Diphacil
Diphacyl
Diphenylacetic acid diethylaminoethyl ester
Diphenylacetic acid, 2-(diethylamino)ethyl ester
Diphenylacetyldiethylaminoethanol
Ester dwuetyloaminoetylowy kwasu dwufenylooctowego (Polish)
Patrovine
Spasmolytin
Transentine
Tranzetil
Trasentin
Trasentine
Trazentyna (Polish)
Vegantine
Wegantyna (Polish)

65-19-0
$C_{21}H_{26}N_2O_3.ClH$
390.95
Yohimban-16-α-carboxylic acid, 17-α-hydroxy-, methyl ester, hydrochloride
Aphrodine hydrochloride
Yohimbine hydrochloride
Yohimbine monohydrochloride
Yohimbin hydrochloride

65-22-5
$C_8H_9NO_3.ClH$
203.64
Pyridoxal, hydrochloride
3-Hydroxy-5-(hydroxymethyl)-2-methylisonicotinaldehyde, hydrochloride
2-Methyl-3-hydroxy-4-formyl-5-hydroxymethylpyridine hydrochloride
Vitamin b6 hydrochloride

65-23-6
$C_8H_{11}NO_3$
169.20
n(c(c(O)c(c1CO)CO)C)c1
Pyridoxol
Beesix
Adermine
Gravidox
Hydoxin
3-Hydroxy-4,5-dimethylol-α-picoline
5-Hydroxy-6-methyl-3,4-pyridinedimethanol
2-Methyl-4,5-bis(hydroxymethyl)-3-hydroxypyridine
2-Methyl-3-hydroxy-4,5-bis-(hydroxymethyl)pyridine
2-Methyl-3-hydroxy-4,5-dihydroxymethyl-pyridin (German)
2-Methyl-3-hydroxy-4,5-di-(hydroxymethyl)pyridine
2-Picoline-4,5-dimethanol, 3-hydroxy-
Pridoxine
3,4-Pyridinedimethanol, 5-hydroxy-6-methyl-
Pyridoxin
Pyridoxine
Vitamin B6

65-29-2
$C_{30}H_{60}N_3O_3.3I$
891.63
[I-].[I-].[I-].CC[N+](CC)(CC)CCOc1cccc(OCC[N+](CC)(CC)CC)c1OCC[N+](CC)(CC)CC
Ammonium, (v-phenenyltris-(oxyethylene)tris(triethyl-, triiodide
Benzcurine iodide
F 2559
Flacedil
Flaxedil
Fourneau 2559
Gallamin
Gallamine
Gallamine iodide
Gallamine-3ETI
Gallamine triethiodide
Gallamine triiodoethylate
Gallamin triethiodide
(v-Phenenyltris(oxyethylene))-

tris(triethylammonium)iodide
Relaxan
Remyolan
Retensin
RP 3697
3.697 R.P.
Sincurarine
Syncurarine
Tricuran
Tri(β-diethylaminoethoxy)-1,2,3-benzenetri-iodoethylate
1,2,3-Tri(β-diethylamino-ethoxy)benzene triethiodide
Triiodoethylate de gallamine (French)
Triiodoethylate of tri(diethyl-aminoethyloxy)-1,2,3-benzene
Tri(iodoethylate) de tri (β di-ethylaminoethoxy)-1,2,3 benzene (French)
Triiodure de tri(β-triethyl-ammoniumethoxy)-1,2,3 benz-ene (French)
1,2,3-Tris(2-diethylamino-ethoxy)benzene triethiodide
1,2,3-Tris(2-diethylamino-ethoxy)benzene tris(ethyl-iodide)
1,2,3-Tris(2-triethylammonium ethoxy)benzene triiodide

65-30-5
$C_{20}H_{26}N_4.O_4S$
418.56
Nicotine, sulfate (2:1)
ENT 2,435
l-1-Methyl-2-(3-pyridyl)-pyr-rolidine sulfate
(S)-3-(1-Methyl-2-pyrrolidinyl)-pyridine sulfate (2:1)
l-3-(1-Methyl-2-pyrrolidyl)-pyridine sulfate
Nicotine sulfate
Nicotine sulfate, Solution [UN 1658]
Nicotine sulfate, Solid [UN 1658]
Nicotine sulphate
Nikotinsulfat (German)
Pyridine, 3-(1-methyl-2-pyrroli-dinyl)-, (S)-, sulfate (2:1)
Pyrrolidine, 1-methyl-2-(3-pyri-dyl)-, sulfate
Sulfate de nicotine (French)
UN 1658 [Nicotine sulfate, solid or solution]

65-31-6
$C_{10}H_{14}N_2.2C_4H_6O_6$
462.46
Nicotine, tartrate (1:2)
Nicotine acid tartrate
Nicotine bitartrate
Nicotine hydrogen tartrate
(-)-Nicotine hydrogen tartrate
Nicotine tartrate [UN 1659]
Nikotinbitartrat (German)
Pyridine, 3-(1-methyl-2-pyr-rolidinyl)-, (S)-, (R-(R*,R*))-2,3-dihydroxybutanedioate (1:2)
Tartrate de nicotine (French)
UN 1659 [Nicotine tartrate]

65-45-2
$C_7H_7NO_2$
137.15
O=C(N)c(c(O)ccc1)c1
Salicylamide
Acket
Afko-sal
Algamon
Algiamida
Amid kyseliny salicylove (Czech)
Amidosal
Amid-sal
Anamid
Benesal
Benzamide, o-hydroxy-
Benzamide, 2-hydroxy-
Cidal
Dolomide
Dropsprin
H.P. 34
o-Hydroxybenzamide
2-Hydroxybenzamide
Liquiprin
Novecyl
OHB
Oramid
Panithal
Raspberin
Salamid
Salamide
Saliamin
Salicilamide (Italian)
Salicylamid
Salicim
Salipur
Salizell
Salrin
Salymid
Sam
Samid
Urtosal

65-46-3
$C_9H_{13}N_3O_5$
243.25
O=C(N(C=CC1=N)C(OC(C2O)CO)C2O)N1
2(1H)-Pyrimidinone, 4-amino-1-β-D-ribofuranosyl
4-Amino-1-β-D-ribofuranosyl-2(1H)-pyrimidinone
Cytidine
Cytosine riboside
1-β-Ribofuranosylcytosine

65-49-6
$C_7H_7NO_3$
153.15
O=C(O)c(c(O)cc(N)c1)c1
Salicylic acid, 4-amino
4-Amino-2-hydroxybenzoic acid
Aminopar
Aminosalicylic acid
para-Amino salicylic acid
p-Aminosalicylic acid
4-Aminosalicylic acid
Aminox
Apacil
Apas
Benzoic acid, 4-amino-2-hydroxy-
Deapasil
Entepas
Gabbropas
Hellipidyl
2-Hydroxy-4-aminobenzoic acid
3-Hydroxy-4-carboxyaniline
Kyselina p-aminosalicylova (Czech)
NSC-2083
Osacyl
Pamacyl
Pamisyl
Paramycin
Para-Pas
Parasal
Parasalicil
Parasalindon
Pas
Pasa
Pasalon
Pasara
Pas-C
Pascorbic
Pasem
Pask
Pasmed
Pasnodia
Pasolac
Propasa
Rezipas
Sanipriol-4

65-61-2
$C_{17}H_{19}N_3.ClH$
301.85
Cl.CN(C)c3ccc2cc1ccc(cc1nc2c3)N(C)C
Acridine, 3,6-bis(dimethyl-amino)-, monohydrochloride
3,6-Acridinediamine, N,N,N',N'-tetramethyl-, monohydrochloride
Acridine Orange
Acridine Orange No
Acridine Orange R
Basic Orange 3RN
C.I. 46005
C.I. Basic Orange 14
Rhoduline Orange NO

65-71-4
$C_5H_6N_2O_2$
126.13
n(cc(c(n1)O)C)c1O
Thymine
5-Methyluracil
Thymin (purine base)

65-85-0
$C_7H_6O_2$
122.13
O=C(O)c(cccc1)c1

Benzoic-acid
Acide benzoique (French)
Benzenecarboxylic acid
Benzeneformic acid
Benzenemethanoic acid
Benzoate
Benzoesaeure (German)
Carboxybenzene
Dracylic acid
Kyselina benzoova (Czech)
Phenyl carboxylic acid
Phenylformic acid
Retarder BA
Retardex
Tenn-Plas
Salvo Liquid
Salvo Powder

65-86-1
$C_5H_4N_2O_4$
156.11
O=C(O)c(nc(nc1O)O)c1
Orotic-acid
Animal galactose factor
6-Carboxyuracil
Orodin
Orotonin
Orotsaure (German)
Oroturic
Orotyl
4-Pyrimidinecarboxylic acid, 1,2,3,6-tetrahydro-2,6-dioxo- (9CI)
6-Uracilcarboxylic acid
Whey factor

66-02-4
$C_9H_9I_2NO_3$
432.98
Diiodotyrosine
Agontan
Apothyrin
Cemiod
Diiodo-l-tryosine
3,5-Diiodotyrosine (VAN)
DIT (VAN)
Dityrin
Flaianina
Gorgoic acid, diiodo-
Gorgonic acid, diiodo-
β-(4-Hydroxy-3,5-diiodo-phenyl)alanine
Iodogorgonic acid
Itir
Jodgorgon
Jodgorgosaeure
NSC-208959
Tyrosine, 3,5-diiodo- (9CI)

66-22-8
$C_4H_4N_2O_2$
112.10
n(ccc(n1)O)c1O
Uracil
2,4-Dihydroxypyrimidine
2,4-Dioxopyrimidine
Hybar X
Pirod
2,4-Pyrimidinediol
2,4-Pyrimidinedione
2,4(1H,3H)-Pyrimidinedione (9CI)
Pyrod

66-25-1
$C_6H_{12}O$
100.18
O=CCCCCC
Hexanal
Aldehyde C-6
Caproaldehyde
Caproic aldehyde
Capronaldehyde
n-Caproylaldehyde
Hexaldehyde [UN 1207]
1-Hexanal
Kapronaldehyd (Czech)
UN 1207 [Hexaldehyde]

66-27-3
$C_2H_6O_3S$
110.14
O=S(=O)(OC)C
Methanesulfonic acid, methyl ester
as-Dimethyl sulphite
Methanesulphonic acid methyl ester
Methylester kyseliny methan-sulfonove (Czech)
Methyl ester of methanesulfonic acid
Methyl ester of methanesulph-onic acid
Methyl mesylate
Methylmethansulfonat (German)
Methyl methanesulfonate
Methyl methanesulphonate
Methyl methansulfonate
Methyl methansulphonate
MMS
NSC-50256

66-32-0
$C_{21}H_{22}N_2O_2.HNO_3$
397.47
Strychnine, mononitrate
Strychnine nitrate

66-56-8
$C_6H_4N_2O_5$
184.12
Phenol, 2,3-dinitro
2,3-Dinitrofenol (Czech)
2,3-Dinitrophenol

66-71-7
$C_{12}H_8N_2$
180.22
n(c(c(ccc1cccn2)cc3)c12)c3
1,10-Phenanthroline
4,5-Diazaphenanthrene
1,10-Fenanthrolin (Czech)
Orthophenanthroline
β-Phenanthroline
o-Phenanthroline
1,10-o-Phenanthroline

66-72-8
$C_8H_9NO_3$
167.18
Cc1ncc(CO)c(C=O)c1O
Pyridoxal
4-Pyridinecarboxaldehyde, 3-hydroxy-5-(hydroxymethyl)-2-methyl-
Pyridoxaldehyde

66-75-1
$C_8H_{11}Cl_2N_3O_2$
252.12
ClCCN(CCCl)c1c[nH]c(=O)[nH]c1=O
Uracil, 5-(bis(2-chloroethyl)-amino)
Aminouracil mustard
5-(Bis(2-chloroethyl)amino)-2,4(1H,3H)pyrimidinedione
5-(Bis(2-chloroethyl)amino)-uracil
5-N,N-Bis(2-chloroethyl)amino-uracil
CB-4835
Chlorethaminacil
Demethyldopan
Desmethyldopan
5-(Di-(β-chloroethyl)amino)-uracil
5-(Di-2-chloroethyl)aminouracil
2,6-Dihydroxy-5-bis(2-chloro-ethyl)aminopyrimidine
ENT 50,439
NCI-C04820
NSC-34462
Nordopan
2,4(1H,3H)-Pyrimidinedione, 5-(bis(2-chloroethyl)amino)-
RCRA waste number U237
SK-19849
U-8344
Uracil lost (German)
Uracilmostaza
Uracil mustard
Uracil nitrogen mustard
Uramustin
Uramustine

66-76-2
$C_{19}H_{12}O_6$
336.31
O=C(Oc(c(C=1O)ccc2)c2)C1CC(=C(O)c(c(O3)ccc4)c4)C3=O
Coumarin, 3,3'-methylenebis-(4-hydroxy
Acadyl
Acavyl
Antitrombosin
Baracoumin
2H-1-Benzopyran-2-one, 3,3'-methylenebis(4-hydroxy- (9CI)
BHC
Bishydroxycoumarin
Bis(4-hydroxycoumarin-3-yl)methane

Cuma
Cumid
Dicoumarin
Dicoumarol
Dicuman
Dicumaol R
Dicumarine
Dicumarol
Dicumol
Di-(4-hydroxy-3-coumarinyl)-methane
Di-4-hydroxy-3,3'-methylenedi-coumarin
Dikumarol
Dufalone
Kumoran
Melitoxin
3,3'-Methyleen-bis(4-hydroxy-cumarine) (Dutch)
3,3'-Methylen-bis(4-hydroxy-cumarin) (German)
3,3'-Methylenebis(4-hydroxy-2H-1-benzopyran-2-one)
3,3'-Methylenebis(4-hydroxy-1,2-benzopyrone)
3,3'-Methylenebis(4-hydroxy-coumarin)
3,3'-Methylene-bis(4-hydroxy-coumarine) (French)
3,3'-Metilen-bis(4-idrossi-cumarina) (Italian)
Temparin
Trombosan

66-77-3
$C_{11}H_8O$
156.19
O=Cc(c(c(ccc1)cc2)c1)c2
1-Naphthaldehyde
1-Formylnaphthalene
α-Naphthal
α-Naphthaldehyde
1-Naphthalenecarboxaldehyde
α-Naphthylaldehyde
1-Naphthylaldehyde
α-Naphthylcarboxaldehyde

66-81-9
$C_{15}H_{23}NO_4$
281.39
CC1CC(C)C(=O)C(C1)C(O)CC2CC(=O)NC(=O)C2
Glutarimide, 3-(2-(3,5-dimethyl-2-oxocyclohexyl)-2-hydroxyethyl)
Acti-Aid
Actidion
Actidione
Actidione PM
Actidione TGF
Actidone
Actispray
Aktidion (Czech)
Cycloheximide
β-(2-(3,5-Dimethyl-2-oxocyclohexyl)-2-hydroxyethyl)-glutarimide
3-(2-(3,5-Dimethyl-2-oxocyclohexyl)-2-hydroxyethyl)-glutarimide
Hizarocin
Kaken
Naramycin
Naramycin A
Neocycloheximide
NSC-185
U-4527

66-97-7
$C_{11}H_6O_3$
186.17
O=c3ccc2cc1ccoc1cc2o3
7H-Furo(3,2-g)(1)benzopyran-7-one
Ficusin
Furocoumarin
Furo(2',3',7,6)coumarin
Furo(4',5',6,7)coumarin
Psoralen
Psoralene

66-99-9
$C_{11}H_8O$
156.19
2-Naphthaldehyde
β-Formylnaphthalene
2-Formylnaphthalene
β-Naphthaldehyde
2-Naphthalenecarboxaldehyde
β-Naphthylaldehyde
β-Naphthylcarboxaldehyde

67-03-8
$C_{12}H_{17}N_4OS.ClH.Cl$
337.30
Thiamine, chloride, hydrochloride
Aneurine hydrochloride
Apate Drops
Beatine
Bedome
Begiolan
Benerva
Bequin
Berin
Betabion hydrochloride
Betalin S
Betaxin
Bethiazine
Beuion
Bevitex
Bevitine
Bewon
Biuno
Bivatin
Bivita
Clotiamina
Eskapen
Eskaphen
Lixa-Beta
Metabolin
Slowten
THD
Thiadoxine
Thiaminal
Thiamin chloride
Thiamin dichloride
Thiamine chloride
Thiamine dichloride
Thiamine hydrochloride
Thiamin hydrochloride
Thiaminium chloride hydrochloride
Thiamol
Thiavit
Tiamidon
Tiaminal
Trophite
USAF CB-20
Vetalin S
Vinothiam
Vitamin B hydrochloride
Vitamin B1 hydrochloride
Vitamin B_1 hydrochloride
Vitaneuron

67-20-9
$C_8H_6N_4O_5$
238.18
O=C(N(N=CC(OC(N(=O)=O)=C1)=C1)CC2=O)N2
Hydantoin, 1-((5-nitrofurfurylidene)amino)
Benkfuran
Berkfurin
Chemiofuran
Cyantin
Dantafur
Furadantin
Furadantine
Furadantoin
Furadonin
Furadonine
Furantoin
Furatoin
Furobactina
2,4-Imidazolidinedione, 1-(((5-nitro-2-furanyl)-methylene)amino)-
Ituran
Macrodantin
NCI-C55196
Nifurantin
Nitrofurantoin
N-(5-Nitrofurfurylidene)-1-aminohydantoin
N-(5-Nitro-2-furfurylidene)-1-aminohydantoin
1-((5-Nitrofurfurylidene)amino)-hydantoin
N-(5-Nitro-2-furfurylideno)-1-aminohydantoina (Polish)
NSC-2107
N-Toin
Orafuran
Parfuran
Urizept
USAF EA-2
Welfurin
Zoofurin

67-21-0
$C_6H_{13}NO_2S$
163.26
O=C(O)C(N)CCSCC
Butyric acid, 2-amino-4-(ethylthio)-, DL
Aethionin
2-Amino-4-(ethylthio)butyric

acid
DL-2-Amino-4-(ethylthio)-butyric acid
Butyric acid, DL-2-amino-4-(ethylthio)-
CN 8676
ETH
Ethionin
Ethionine
(+-)-Ethionine
DL-Ethionine
Ethionine, DL-
Homocysteine, S-ethyl-
DL-Homocysteine, S-ethyl-
NSC-751
U-1434

67-43-6
$C_{14}H_{23}N_3O_{10}$
393.40
O=C(O)CN(CCN(CC(=O)O)CC(=O)O)CCN(CC(=O)O)CC(=O)O
Glycine, N,N-bis(2-(bis(carboxymethyl)amino)ethyl)
Acetic acid, ((carboxymethylimino)bis(ethylenenitrilo))-tetra-
CHEL 330
CHEL 330 acid
CHEL DTPA
Dabeersen 503
Detapac
Detarex
Diethylenetriaminepentaacetic acid
1,1,4,7,7-Diethylenetriamine-pentaacetic acid
(Diethylenetrinitrilo)pentaacetic acid
DTPA
Hamp-Ex Acid
Monaquest
Pentetic acid
Penthamil
Perma Kleer
3,6,9-Triazaundecanedioic acid, 3,6,9-tris(carboxymethyl)-

67-45-8
$C_8H_7N_3O_5$
225.18
O=C(OCC1)N1N=CC(OC(N(=O)=O)=C2)=C2
2-Oxazolidinone, 3-(5-nitro-furfurylidine-amino)
Bifuron
Corizium
Diafuron
Enterotoxon
Furaxon
Furaxone
Furazol
Furazolidon
Furazolidone
Furazon
Furidon
Furovag
Furox
Furoxal
Furoxane
Furoxone
Furoxone Swine Mix
Furozolidine
Giardil
Giarlam
Medaron
Neftin
NF-180
NF 180 custom mix ten
Nicolen
Nifulidone
Nifuran
3-(((5-Nitro-2-furanyl)-methylene)amino)-2-oxazol-idinone
Nitrofurazolidone
Nitrofurazolidonum
3-(5'-Nitrofurfuralamino)-2-oxa-zolidone
N-(5-Nitro-2-furfurylidene)-3-aminooxazolidine-2-one
3-((5-Nitrofurfurylidene)amino)-2-oxazolidone
N-(5-Nitro-2-furfurylidene)-3-amino-2-oxazolidone
Nitrofuroxon
3-((5-Nitrofurylidene)amino)-2-oxazolidone
5-Nitro-N-(2-oxo-3-oxazolidin-yl)-2-furanmethanimine
Optazol
Ortazol
2-Oxazolidinone, 3-(((5-nitro-2-furanyl)methylene)amino)-
Puradin
Roptazol
Sclaventerol
Tikofuran
Topazone
Trichofuron
Tricofuron
Trifurox
USAF EA-1
Viofuragyn

67-47-0
$C_6H_6O_3$
126.12
OCc1ccc(C=O)o1
2-Furaldehyde, 5-(hydroxy-methyl)
2-Furancarboxaldehyde, 5-(hydroxymethyl)- (9CI)
HMF
5-Hydroxymethylfuraldehyde
5-(Hydroxymethyl)furfural
Hydroxymethylfurfurole
5-Oxymethylfurfurole

67-48-1
$C_5H_{14}NO.Cl$
139.65
[Cl-].C[N+](C)(C)CCO
Ethanaminium, 2-hydroxy-N,N,N-trimethyl-, chloride
Ammonium, (2-hydroxyethyl)-trimethyl-, chloride
Biocolina
Choline chloride
Choline chlorhydrate
Choline hydrochloride
Cholinium chloride
Chloride de choline (French)
Hepacholine
(2-Hydroxyethyl)trimethyl-ammonium chloride
Lipotril

67-52-7
$C_4H_4N_2O_3$
128.10
O=C(NC(=O)CC1=O)N1
Barbituric-acid
Hydrouracil, 6-hydroxy-
6-Hydroxyuracil
Malonylurea
Pyrimidinetriol
2,4,6-Pyrimidinetriol
2,4,6-Pyrimidinetrione
2,4,6(1H,3H,5H)-Pyrimidine-trione
2,4,6-Trihydroxypyrimidine
2,4,6-Trioxohexahydropyr-imidine

67-56-1
CH_4O
32.05
OC
Methanol
Alcool methylique (French)
Alcool metilico (Italian)
Carbinol
Colonial Spirit
Columbian spirit
Columbian spirits (DOT)
Methanol (DOT)
Metanolo (Italian)
Methyl alcohol (ACGIH,OSHA) [UN 1230]
Methylol
Methylalkohol (German)
Methyl hydrate
Methyl hydroxide
Metylowy alkohol (Polish)
Monohydroxymethane
Pyroxylic spirit
RCRA waste number U154
UN 1230 [Methanol, or methyl alcohol]
Wood Alcohol (DOT)
Wood Naphtha
Wood Spirit

67-62-9
CH_5NO
47.07
CON
Hydroxylamine, o-methyl
Methoxyamine
o-Methylhydroxylamine
NCI-C60060

67-63-0
C_3H_8O
60.11
OC(C)C

Isopropanol [UN 1219]
Alcool isopropilico (Italian)
Alcool isopropylique (French)
Alcojel
Alcosolve
Avantin
Avantine
Chromar
Combi-schutz
Dimethylcarbinol
Hartosol
2-Hydroxypropane
Imsol A
Isohol
Isopropyl alcohol (ACGIH, OSHA) [UN 1219]
Iso-propylalkohol (German)
Lutosol
Petrohol
Pro
Propan-2-ol
2-Propanol
i-Propanol (German)
n-Propan-2-ol
Propol
sec-Propyl alcohol
2-Propyl alcohol
sec-Propyl alcohol
i-Propylalkohol (German)
Spectrar
Sterisol hand disinfectant
Takineocol
UN 1219 [Isopropanol or isopropyl alcohol]

67-64-1
C_3H_6O
58.09
O=C(C)C
Acetone (ACGIH,OSHA) [UN 1090]
Aceton (German, Dutch, Polish)
Chevron acetone
Dimethylformaldehyde
Dimethylketal
Dimethyl ketone
Ketone, dimethyl
Ketone propane
β-Ketopropane
Methyl ketone
Propanone
2-Propanone
Pyroacetic acid
Pyroacetic ether
RCRA waste number U002
UN 1090 [Acetone]

67-66-3
$CHCl_3$
119.37
C(Cl)(Cl)Cl
Chloroform (ACGIH,OSHA) [UN 1888]
Chloroforme (French)
Cloroformio (Italian)
Formyl trichloride
Methane trichloride
Methane, trichloro-
Methenyl trichloride
Methyl trichloride
NCI-C02686
R 20
R 20 (Refrigerant)
RCRA waste number U044
TCM
Trichloormethaan (Dutch)
Trichlormethan (Czech)
Trichloroform
Trichloromethane
Trichloromethane (OSHA)
Triclorometano (Italian)
UN 1888 [Chloroform]

67-68-5
C_2H_6OS
78.14
O=S(C)C
Methyl-sulfoxide
A 10846
Deltan
Demeso
Demasorb
Demavet
Demsodrox
Dermasorb
Dimethyl sulfoxide
Dimethyl sulphoxide
Dimexide
Dipirartril-Tropico
DMS-70
DMS-90
DMSO
Dolicur
Doligur
Domoso
Dromisol
Durasorb
Gamasol 90
Hyadur
Infiltrina
M 176
Methane, sulfinylbis-
Methylsulfinylmethane
NSC-763
Rimso-50
Somipront
SQ 9453
Sulfinylbis(methane)
Syntexan
Topsym

67-71-0
$C_2H_6O_2S$
94.13
O=S(=O)(C)C
Dimethyl sulfone
AI3-25306
Methane, sulfonylbis- (9CI)
Methyl sulfone (8CI)
Methylsulfonyl methane
NSC-63345
Sulfonylbismethane

67-72-1
C_2Cl_6
236.72
C(C(Cl)(Cl)Cl)(Cl)(Cl)Cl
Ethane, hexachloro
Avlothane
Carbon hexachloride
Distokal
Distopan
Distopin
Egitol
Ethane hexachloride
Ethylene hexachloride
Falkitol
Fasciolin
Hexachlor-aethan (German)
Hexachlorethane
Hexachloroethane (ACGIH, DOT,OSHA)
1,1,1,2,2,2-Hexachloroethane
Hexachloroethylene
Mottenhexe
NA 9037 (DOT)
NCI-C04604
Perchloroethane
Phenohep
RCRA waste number U131

67-73-2
$C_{24}H_{30}F_2O_6$
452.54
Pregna-1,4-diene-3,20-dione, 6,9-difluoro-11,12-dihydroxy-16,17-((1-methylethylidene) bis(oxy))-, (6-α,11-β,16-α)
Dermalar
6-α,9-α-Difluoro-16-α-hydroxyprednisolone 16,17-acetonide
Flucinar
Flucort
Fluocinolone acetonide
Fluocinolone 16,17-acetonide
Fluovitif
Jellin
Localyn
Percutina
Radiocin
RS-1401 AT
Sinalar
Synamol
Synandone
Synalar
Synandrone
Synsac

67-96-9
$C_{28}H_{46}O$
398.74
Tachysterol, dihydro
Antitanil
Anti-tetany substance 10
A.T. 10
Calcamine
DHT2
Dichystrolum
Dihydrotachysterol
Dihydrotachysterol2
Dygratyl
Hytakerol
Parterol
9,10-Secoergosta-5,7,22-trien-3-ol, (3-β,5E,7E,10-α,22E)-(9CI)
Tachysterol, dihydro- (8CI)

Tachysterol2, dihydro-
Tachystin
Vitamine D1

67-97-0
$C_{27}H_{44}O$
384.71
OC(CCC(C1=CC=C(C(C(C(C2)C(CCCC(C)C)C)(CC3)C)C2)C3)=C)C1
9,10-Secocholesta-5,7,10(19)-trien-3-β-ol
Cholecalciferol
Cholecalciferol, D3
Colecalciferol
7-Dehydrocholestrol, activated
Delsterol
Deparal
D3-Vigantol
Oleovitamin D3
Ricketon
Vigorsan
Trivitan
Vitamin D3
Vitinc Dan-Dee-3

67-99-2
$C_{13}H_{14}N_2O_4S_2$
326.41
10H-3,10a-Epidithiopyrazino-(1,2-a)indole-1,4-dione, 2,3,5a,6-tetrahydro-6-hydroxy-3- (hydroxymethyl)-2-methyl
Aspergillin
Gliotoxin
S-82
S.N. 12870

68-04-2
$C_6H_5O_7$.3Na
258.08
Citric acid, trisodium salt
Citnatin
Citreme
Citrosodine
Citrosodna
Natrocitral
1,2,3-Propanetricarboxylic acid, 2-hydroxy-, trisodium salt (9CI)
Sodium citrate
Sodium citrate anhydrous
Trisodium citrate

68-05-3
$C_8H_{20}N$.I
257.19
[I-].CC[N+](CC)(CC)CC
Ammonium, tetraethyl-, iodide
Ethanaminium, N,N,N-triethyl-, iodide (9CI)
Tetamon iodide
Tetraethylammonium iodide
Tetramon J

68-11-1
$C_2H_4O_2S$
92.12
O=C(O)CS
Acetic acid, mercapto
Acide thioglycolique (French)
Glycolic acid, thio-
Glycolic acid, 2-thio-
Kyselina merkaptooctova (Czech)
Kyselina thioglykolova (Czech)
Mercaptoacetate
Mercaptoacetic acid
α-Mercaptoacetic acid
2-Mercaptoacetic acid
Thioglycolic acid (ACGIH, OSHA) [UN 1940]
2-Thioglycolic acid
Thioglycollic acid
Thiovanic acid
UN 1940 [Thioglycolic acid]
USAF CB-35

68-12-2
C_3H_7NO
73.11
O=CN(C)C
Formamide, N,N-dimethyl
Dimethylamid kyseliny mravenci (Czech)
Dimethylformamid (German)
Dimethyl formamide
N,N-Dimethyl formamide
Dimethylformamide (ACGIH, OSHA)
N,N-Dimethylformamide [UN 2265]
Dimetilformamide (Italian)
Dimetylformamidu (Czech)
DMF
DMFA
Dwumetyloformamid (Polish)
N-Formyldimethylamine
NCI-C60913
NSC-5356
U-4224
UN 2265 [N,N-Dimethyl-formamide]

68-19-9
$C_{63}H_{88}CoN_{14}O_{14}P$
1355.55
Cobinamide, cyanide phosphate 3'-ester with 5,6-dimethyl-1-α-D- ribofuranosyl-benzimidazole, inner salt
Anacobin
B-12
Berubigen
Betalin-12
Betalin 12 crystalline
Betaline-12
Bevatine-12
Bevidox
B-Twelve
B-Twelve ORA
Byladoce
Cabadon M
$Cn-B_{12}$
Cobadoce forte
Cobalin
Cobamin
Cobione
Cotel
Covit
Crystamin
Crystwel
Cyano-B12
Cyanocobalamin
Cyanocobalamine
Cycolamin
Cykobeminet
Cyredin
Cytacon
Cytamen
Cytobion
Depinar
5,6-Dimethylbenzimidazolyl cobamide cyanide
Dimethylbenzimidazoyl-cobamide
Distivit (B12 peptide)
Dobetin
Docemine
Docibin
Docigram
Dodecabee
Dodecavite
Dodex
Ducobee
Duodecibin
Embiol
Emociclina
Eritrone
Erycytol
Erythrotin
Euhaemon
Extrinsic factor
Factor II
Factor II (vitamin)
Fresmin
Hemo-B-Doze
Hemomin
Hepagon
Hepavis
Hepcovite
Lactobacillus lactis dorner factor
Lld Factor
Macrabin
Megabion
Megabion
Megalovel
Milbedoce
Nagravon
Normocytin
Pernaemon
Pernaevit
Pernipuron
Plecyamin
Poyamin
Rebramin
Redamina
Redisol
Rhodacryst
Rubesol
Rubramin
Rubramin PC
Rubripca
Rubrocitol
Sytobex
Vibalt

Vibisone
Virubra
Vitamin B12
Vitamin B12B
Vitamin B12 complex
Vitamin B_{12} preparation
Vitarubin
Vita-Rubra
Vitral
Vi-Twel

68-22-4
$C_{20}H_{26}O_2$
298.46
CC34CCC1C(CCC2=CC(=O)CC C12)C3CCC4(O)C#C
19-Nor-17-α-pregn-4-en-20-yn-3-one, 17-hydroxy
Anhydrohydroxynorprogesterone
Ent
17-α-Ethinylestra-4-en-17-β-ol-3-one
17-α-Ethinyl-17-β-hydroxy-δ⁴-estren-3-one
Ethinylnortestosterone
Ethinyl-19-nortestosterone
17-Ethinyl-19-nortestosterone
17-α-Ethinyl-19-nortestosterone
17-α-Ethynyl-4-estren-17-ol-3-one
17-α-Ethynyl-17-hydroxy-4-estren-3-one
17-α-Ethynyl-17-β-hydroxyestr-4-en-3-one
17-α-Ethynyl-17-β-hydroxy-4-estren-3-one
17-α-Ethynyl-17-β-hydroxy-19-norandrost-4-en-3-one
17-α-Ethynyl-19-norandrost-4-en-17-β-ol-3-one
17-α-Ethynyl-19-nor-4-androsten-17-β-ol-3-one
17-α-Ethynyl-19-nortestosterone
(17-α)-17-Hydroxy-19-norpregn-4-en-20-yn-3-one
17-Hydroxy-17-α-19-norpregn-4-en-20-yn-3-one
17-Hydroxy-19-nor-17-α-pregn-4-en-20-yn-3-one
17-β-Hydroxy-19-norpregn-4-en-20-yn-3-one
Micronor
Net
Norethindrone
19-Nor-ethindrone
19-Norethinyltestosterone
19-Nor-ethinyl-4,5-testosterone
Norethisteron
Norethisterone
19-Norethisterone
Norethyndron
Norethynodrone
19-Nor-17-α-ethynylandrosten-17-β-ol-3-one
19-Nor-17-α-ethynyl-17-β-hydroxy-4-androsten-3-one
19-Nor-17-ethinyltestosterone
19-Nor-17-α-ethynyltestosterone
Norlutin
Norpregneninlone
17-α-19-Norpregn-4-en-20-yn-3-one, 17-hydroxy-
Nor-Q.D.

68-23-5
$C_{20}H_{26}O_2$
298.46
CC34CCC1C(CCC2=C1CCC (=O)C2)C3CCC4(O)C#C
19-Nor-17-α-pregn-5(10)-en-20-yn-3-one, 17-hydroxy
17-α-Ethinyl-δ⁵,10-19-nortestosterone
17-α-Ethinyl-5,10-estrenolone
17-Ethinyl-5(10)-estraeneolone
17-α-Ethinyl-estra(5,10)eneolone
17-α-Ethynyl-5(10)-estren-17-ol-3-one
17-α-Ethynylestr-5(10)-en-17-β-ol-3-one
17-α-Ethynyl-estr-5(10)-en-3-on-17-β-ol
17-α-Ethynyl-17-hydroxyestr-5(10)-en-3-one
17-α-Ethynyl-17-hydroxy-5(10)-estren-3-one
17-α-Ethynyl-17-β-hydroxy-5(10)-estren-3-one
17-α-Ethynyl-17-β-hydroxyestr-5(10)-en-3-one
17-α-Ethinyl-17-β-hydroxy-δ⁵⁽¹⁰⁾-estren-3-one
17-α-Ethynyl-17-β-hydroxy-3-oxo-δ⁵⁽¹⁰⁾-estrene
17-α-Ethynyl-19-nor-5(10)-androsten-17-β-ol-3-one
17-Hydroxy-19-nor-17-α-pregn-5(10)-en-20-yn-3-one
(17-α)-17-Hydroxy-19-norpregn-5(10)-en-20-yn-3-one
17-Hydroxy(17-α)-19-norpregn-5(10)-en-20-yn-3-one
17-β-Hydroxy-17-α-ethinyl-5(10)-estren-3-one
Lynestrol
Norethinodrel
19-Nor-ethinyl-5,10-testosterone
Norethinynodrel
Norethynodral
Norethynodrel
19-Norethynodrel
NSC-15432
SC-4642

68-26-8
$C_{20}H_{30}O$
286.50
OCC=C(C=CC=C(C=CC(=C(CC C1)C)C1(C)C)C)C
Retinol, all trans
Acon
Afaxin
Agiolan
Alphalin
Alphasterol
Anatola
Anatola A
Anti-Infective Vitamin
Antixerophthalmic Vitamin
Aoral
Apexol
Aquasynth
Atars
Atav
Avibon
Avita
Avitol
Axerophthol
Biosterol
Chocola A
3,7-Dimethyl-9-(2,6,6-trimethyl-1-cyclohexen-1-yl)-2,4,6,8-nonatetraen-1-ol
Disatabs Tabs
Dofsol
Epiteliol
Hi-A-vita
Lard Factor
Myvpack
2,4,6,8-Nonatetraen-1-ol, 3,7-dimethyl-9-(2,6,6-trimethyl-1-cyclohexen-1-yl)-, (all-E)-
Oleovitamin A
Ophthalamin
Prepalin
Retinol
all-trans Retinol
Retrovitamin A
Testavol
Vaflol
Vafol
Vi-Alpha
Vitamin A
Vitamin A1
Vitamin A Alcohol
Vitamin A1 Alcohol
all-trans-Vitamin A Alcohol
Vitavel-A
Vitpex
Vogan
Vogan-Neu

68-35-9
$C_{10}H_{10}N_4O_2S$
250.30
O=S(=O)(Nc(nccc1)n1)c(ccc(N) c2)c2
Sulfanilamide, N¹-2-pyrimidinyl
Adiazine
Benzenesulfonamide, 4-amino-N-2-pyrimidinyl-
Coco-Diazine
Codiazine
Cremodiazine
Debenal
Deltazina
Diazolone
Diazyl
Eskadiazine
Honey diazine
Lipo-Diazine
Lipo-Levazine
Liquadiazine
Microsulfon
Neazine
Pecta-diazine, Suspension
Piridisir
Pyrimal

Pyrimidine, 2-sulfanilamido-
N[1]-2-Pyrimidinyl-sulfanilamide
RP 2616
Sanodiazine
SDA
Silvadene
S.N. 112
Spofadrizine
Sterazine
Sulfadiazene
Sulfadiazine
2-Sulfanilamidopyrimidin (German)
Sulfanilamidopyrimidine
2-Sulfanilylaminopyrimidine
Sulfapyrimidin (German)
Sulfapyrimidine
2-Sulfapyrimidine
Sulfazine
Sulphadiazine
Theradiazine

68-36-0
$C_8H_4Cl_6$
312.82
c(ccc(c1)C(Cl)(Cl)Cl)(c1)C(Cl)(Cl)Cl
p-Xylene, α,α,α,α',α',α'-hexachloro
1,4-Bis-trichloromethyl benzene
α,α'-Hexachloroxylene
α,α,α,α',α',α'-Hexachloro-p-xylene

68-41-7
$C_3H_6N_2O_2$
102.11
NC1CONC1=O
3-Isoxazolidinone, 4-amino-, D
D-4-Amino-3-isossazolidone (Italian)
D-4-Amino-3-isoxazolidinone
D-4-Amino-3-isoxazolidone
Cyclomycin
Cicloserina (Italian)
Cycloserine
(+)-Cycloserine
Cyclo-D-serine
D-Cycloserine
E-733-A
Farmiserine
I-1431
3-Isoxazolidine, 4-amino-, (R)-(9CI)
K-300
Miroserina
NJ-21
Novoserin
Orientomycin
Oxamicina (Italian)
Oxamycin
D-Oxamicina (Italian)
D-Oxamycin
Oxymycin
PA 94
Ro-1-9213
Seromycin
Tisomycin
Wasserina

68-76-8
$C_{12}H_{13}N_3O_2$
231.28
O=C2C=C(N1CC1)C(=O)C(=C2N3CC3)N4CC4
p-Benzoquinone, 2,3,5-tris-(1-aziridinyl)
Aziridine, 1,1',1''-(3,6-dioxo-1,4-cyclohexadiene-1,2,4-triyl)tris-
Bay 3231
Bayer 3231
p-Benzoquinone, tris(1-aziridinyl)-
1,1',1''-(3,6-Dioxo-1,4-cyclohexadiene-1,2,4-triyl)trisaziridine
2,3,5-Ethylenimine-1,4-benzoquinone
NSC-29215
Oncoredox
Oncovedex
Prenimon
Riker 601
10257 R.P.
Teib
Trenimon
Treninon
Triazichon (German)
Triaziquinone
Triaziquinonum
Triaziquon (German)
Triaziquon
Triaziquone
2,3,5-Tri-(1-aziridinyl)-p-benzoquinone
Triethyleneiminobenzoquinone
2,3,5-Triethyleneimino-p-benzoquinone
2,3,5-Triethyleneimino-1,4-benzoquinone
Triethyleniminobenzoquinone
Trisaethyleniminobenzochinon (German)
2,3,5-Tris(aziridino)-1,4-benzoquinone
2,3,5-Tris(1-aziridino)-p-benzoquinone
Tris(aziridinyl)-p-benzoquinone
Tris(1-aziridinyl)-p-benzoquinone
2,3,5-Tris(1-aziridinyl)-p-benzoquinone
2,3,5-Tris(aziridinyl)-1,4-benzoquinone
2,3,5-Tris(1-aziridinyl)-2,5-cylohexadiene-1,4-dione
Tris(ethyleneimino)benzoquinone
2,3,5-Trisethyleneiminobenzoquinone
Trisethyleneiminoquinone
2,3,5-Tris(ethyleneimino)benzoquinone
2,3,5-Tris(ethyleneimino)-1,4-benzoquinone
2,3,5-Tris(ethylenimino)-p-benzoquinone

68-88-2
$C_{21}H_{27}ClN_2O_2$
374.95
OCCOCCN1CCN(CC1)C(c2ccccc2)c3ccc(Cl)cc3
Ethanol, 2-(2-(4-(p-chloro-α-phenylbenzyl)-1-piperazinyl)ethoxy)
Atara
Atarax
Ataraxoid
Atarazoid
Atazina
Aterax
1-(p-Chlorobenzhydryl)-4-(2-(2-hydroxyethoxy)ethyl)diethylenediamine
1-(p-Chlorobenzhydryl)-4-(2-(2-hydroxyethoxy)ethyl)piperazine
N-(4-Chlorobenzhydryl)-N'-(hydroxyethoxyethyl)piperazine
1-(p-Chlorodiphenylmethyl)-4-(2-(2-hydroxyethoxy)ethyl)-piperazine
1-(p-Chloro-α-phenylbenzyl)-4-(2-(2-hydroxyethoxy)ethyl)-piperazine
2-(2-(4-(p-Chloro-α-phenylbenzyl)-1-piperazinyl)ethoxy)-ethanol
Deinait
Equipoise
Fenarol
Hychotine
Hydroksyzyny (Polish)
Hydroxine
Hydroxycine
Hydroxyzine
Idrossizina
Neo-Calma
Neurozina
NP 212
Pamazone
Parenteral
Paxistil
Placidol
Plaxidol
Tran-Q
Traquizine
UCB 492
U.CB 4492
Vesparaz-wirkstoff

68-90-6
$C_{17}H_{12}I_2O_3$
518.09
Ketone, 3,5-diiodo-4-hydroxyphenyl 2-ethyl-3-benzofuranyl
Aethyl-2-(3',5'-dijoD-4'-oxybenzoyl)-3 cumaron (German)
Algocor
Amplivix
Benziodaron
Benziodarone
Benzofuran, 3-(3,5-diiodo-4-hydroxybenzoyl)-2-ethyl-
Cardivix
Carofam
Coronal-Crinos

3,5-Diiodo-4-hydroxyphenyl 2-ethyl-3-benzofuranyl ketone
Dilafurane
Dila-vasal
2-Ethyl-3-(3',5'-diiodo-4'-hydroxybenzoyl)-cumarone
L 2329
2329 LABAZ
Retrangor

68-94-0
$C_5H_4N_4O$
136.13
n(c(O)c(N=CN1)c1n2)c2
6H-Purin-6-one, 1,7-dihydro
1,7-Dihydro-6H-purin-6-one
HX
6-Hydroxypurine
Hypoxanthine
Hypoxanthine enol
6-Oxopurine
Purin-6-ol
9H-Purin-6-ol
Purin-6(3H)-one
6(1H)-Purinone
Sarcine
Sarkin
Sarkine

68-96-2
$C_{21}H_{30}O_3$
330.51
CC(=O)C3(O)CCC4C2CCC1=CC(=O)CCC1(C)C2CCC34C
Pregn-4-ene-3,20-dione, 17-hydroxy
Gestageno gador
17-Hydroxypregn-4-ene-3,20-dione
17-Hydroxyprogesterone
17-α-Hydroxyprogesterone
Prodix
Prodox

69-05-6
$C_{23}H_{30}ClN_3O.2ClH$
472.93
[Cl-].[Cl-].CCN(CC)CCCC(C)Nc2c1ccc(Cl)cc1nc3ccc(OC)cc23
Acridine, 6-chloro-9-((4-(diethylamino)-1-methylbutyl)amino)-2-methoxy-, dihydrochloride
Acrichine
Acriquine
Akrichin (Czech)
Arichin
Atabrine
Atabrine dihydrochloride
Atabrine hydrochloride
Atebrin
Atebrine
Atebrin hydrochloride
Chemiochin
Chinacrin
Chinacrin hydrochloride
2-Chloro-5-(ω-diethylamino-α-methylbutylamino)-7-methoxyacridine dihydrochloride
3-Chloro-9-(4'-diethylamino-1'-methylbutylamino)-7-methoxyacridine dihydrochloride
6-Chloro-9-((4-(diethylamino)-1-methylbutyl)amino)-2-methoxyacridine dihydrochloride
3-Chloro-7-methoxy-9-(1-methyl-4-diethylaminobutylamino)acridine dihydrochloride
Crinodora
Dial
Erion
Italchin
Malaricida
Mecryl
Mepacrine dihydrochloride
Mepacrine hydrochloride
Methoquine
2-Methoxy-6-chloro-9-(4-diethylamino-1-methylbutylamino)acridine dihydrochloride
Metochin
Metoquin
Metoquine
Palacrin
Palusan
Pentilen
Quinacrine dihydrochloride
Quinacrine hydrochloride
866 R.P.
SN 390
S. N. 390

69-09-0
$C_{17}H_{19}ClN_2S.ClH$
355.35
Phenothiazine, 2-chloro-10-(3-(dimethylamino)propyl)-, monohydrochloride
Aminazin monohydrochloride
Ampliactil monohydrochloride
Chloractil
Chlorazin
2-Chloro-10-(3-dimethylaminopropyl)phenothiazine monohydrochloride
Chloropromazine hydrochloride
Chloropromazine monohydrochloride
Chlorpromazine chloride
Chlorpromazine hydrochloride
Chlorpromazine monohydrochloride
Chlorpromazinium chloride
Contomin
CPZ
10-(3-Dimethylaminopropyl)-2-chlorophenothiazine monohydrochloride
Hebanil
Hibanil
Hibernal
Hybernal
Klorproman
Klorpromex
Largactil monohydrochloride
Largaktyl
Megaphen
NCI-C05210
Neurazine
Norcozine
Phenothiazine hydrochloride
10H-Phenothiazine-10-propanamine, 2-chloro-N,N-dimethyl-, monohydrochloride
Plegomazin
Promacid
Promapar
Propaphen
Propaphenin hydrochloride
Psychozine
4560 R.P. Hydrochloride
Sonazine
Taroctyl
Thorazine
Thorazine hydrochloride
Torazina
Tranzine
Unitensen

69-23-8
$C_{22}H_{26}F_3N_3OS$
437.57
OCCN4CCN(CCCN2c1ccccc1Sc3ccc(cc23)C(F)(F)F)CC4
1-Piperazineethanol, 4-(3-(2-(trifluoromethyl)phenothiazin-10-yl)propyl)
Anatensol
Fluphenazine
10-(3-(2-Hydroxyethyl)piperazinopropyl)-2-(trifluoromethyl)phenothiazine
Moditen
OMCA
Pacinol
Permitil
Phenothiazine, 10-(3-(4-(2-hydroxyethyl)-1-piperazinyl)-propyl)-2-(trifluoromethyl)-
Prolixin
Prolixine
Sevinol
Triflumethazine
4-(3-(2-Trifluoromethyl-10-phenothiazyl)-propyl)-1-piperazineethanol
Yespazine

69-53-4
$C_{16}H_{19}N_3O_4S$
349.44
CC3(C)SC2C(NC(=O)C(N)c1ccccc1)C(=O)N2C3C(O)=O
4-Thia-1-azabicyclo(3.2.0)heptane-2-carboxylic acid, 6-(2-amino-2-phenylacetamido)-3,3- dimethyl-7-oxo-, D-(-)
AB-PC
Acillin
Adobacillin
Alpen
Amblosin
Amcill
Amfipen
Aminobenzylpenicillin
D-(-)-α-Aminobenzylpenicillin
D-(-)-α-Aminopenicillin

6-(D(-)-α-Aminophenylacetamido)penicillanic acid
Amipenix S
Amperil
Ampi-Bol
Ampicillin
D-Ampicillin
D-(-)-Ampicillin
Ampicillin A
Ampicillin acid
Ampicillin anhydrate
Ampicin
Ampikel
Ampimed
Ampipenin
Amplisom
Amplital
Ampy-penyl
Austrapen
AY-6108
Binotal
Bonapicillin
Britacil
BRL
BRL 1341
Copharcilin
Cymbi
Divercillin
Doktacillin
Grampenil
Guicitrina
Guicitrine
Lifeampil
Marisilan
NSC-528986
Nuvapen
Omnipen
P-50
Penbristol
Penbritin
Penbritin paediatric
Penbritin syrup
Penbrock
Penicillin, (aminophenylmethyl)-
Penicline
Pentrex
Pentrexl
Pfizerpen A
Polycillin
Ponecil
Principen
Qidamp
Ro-Ampen
Semicillin
SK-Ampicillin
Synpenin
Tokiocillin
Tolomol
Totacillin
Totalciclina
Totapen
Ultrabion
Ultrabron
Viccillin
Viccillin S
Vicillin
WY-5103

69-57-8
$C_{16}H_{17}N_2O_4S.Na$
356.40
4-Thia-1-azabicyclo(3.2.0)heptane-2-carboxylic acid, 3,3-dimethyl-7-oxo-6-(2-phenyl acetamido)-, monosodium salt
American Penicillin
Benzylpenicillinic acid sodium salt
Benzylpenicillin sodium
Benzylpenicillin sodium salt
Crystapen
Mycofarm
Novocillin
Pen-a-brasive
Penicillin-G, monosodium salt
Penicillin G, sodium
Penicillin G, sodium salt
Penilaryn
Sodium benzylpenicillin
Sodium benzylpenicillin G
Sodium benzylpenicillinate
Sodium penicillin
Sodium penicillin G
Sodium penicillin II
Veticillin

69-65-8
$C_6H_{14}O_6$
182.20
OCC(O)C(O)C(O)C(O)CO
Mannitol, D
1,2,3,4,5,6-Hexanehexol
Manna Sugar
Mannite
D-Mannitol
NCI-C50362
Osmitrol

69-72-7
$C_7H_6O_3$
138.13
O=C(O)c(c(O)ccc1)c1
Salicylic-acid
Acido salicilico (Italian)
Benzoic acid, 2-hydroxy-
o-Hydroxybenzoic acid
2-Hydroxybenzoic acid
Keralyt
Kyselina 2-hydroxybenzoova (Czech)
Kyselina salicylova (Czech)
Orthohydroxybenzoic acid
Retarder W
SA
SAX

69-74-9
$C_9H_{13}N_3O_5.ClH$
279.71
Cytosine, 1-β-D-arabinofuranosyl-, monohydrochloride
AC 1075
Arabinofuranosylcytosine hydrochloride
1-β-D-Arabinofuranosylcytosine hydrochloride
1-β-D-Arabinofuranosylcytosine monohydrochloride
Arabinosylcytosine hydrochloride
Arabitin hydrochloride
Ara-C
Aracytidine hydrochloride
Aracytin hydrochloride
CA
Cylocide
Cytarabine hydrochloride
Cytosar hydrochloride
Cytosine, 1-β-D-arabinofuranosyl-, hydrochloride
Cytosine arabinoside hydrochloride
Iretin
NSC-63878
2(1H)-Pyrimidinone, 4-amino-1-β-D-arabinofuranosyl-, monohydrochloride
Spongocytidine hydrochloride
U 19920
U-19920a

69-79-4
$C_{12}H_{22}O_{11}$
342.34
O=CC(O)C(O)C(OC(OC(C(O)C1O)CO)C1O)C(O)CO
Maltose
Cextromaltose
4-(α-D-Glucopyranosido)-α-glucopyranose
D-Glucose, 4-o-α-D-glucopyranosyl-
4-(α-D-Glucosido)-D-glucose
Maltobiose
D-Maltose
Malt Sugar
α-Malt Sugar

69-89-6
$C_5H_4N_4O_2$
152.13
n(c(O)c(N=CN1)c1n2)c2O
Xanthine
3,7-Dihydro-1H-purine-2,6-dione
2,6-Dioxopurine
Isoxanthine
Pseudoxanthine
Purine-2,6-diol
9H-Purine-2,6-diol
2,6(1,3)-Purinedion
Purine-2,6-(1H,3H)-dione
9H-Purine-2,6-(1H,3H)-dione
USAF CB-17
Xan
Xanthic oxide
Xanthin

69-93-2
$C_5H_4N_4O_3$
168.13
O=C(Nc(nc(nc1O)O)c12)N2
Uric-acid
Lithic acid
1H-Purine-2,6,8(3H)-trione, 7,9-dihydro- (9CI)
2,6,8-Trihydroxypurine

2,6,8-Trioxopurine
2,6,8-Trioxypurine

70-00-8
$C_{10}H_{11}F_3N_2O_5$
296.23
OCC1OC(CC1O)n2cc(c(=O)[nH]c2=O)C(F)(F)F
Thymidine, α,α,α-trifluoro
2'-Deoxy-5-(trifluoromethyl)-uridine
F3DThd
F3T
F3TDR
NSC-75520
2,4(1H,3H)-Pyrimidinedione, 1-(2-deoxy-β-D-ribofuranosyl)-5-(trifluoromethyl)-
TFDU
5-Trifluoro-2'-deoxythymidine
Trifluoromethyldeoxyuridine
5-(Trifluoromethyl)deoxyuridine
5-Trifluoromethyl-2-deoxy-uridine
5-(Trifluoromethyl)-2'-deoxy-uridine
Trifluorothymidine
α,α,α-Trifluorothymidine
Trifluridine
Uridine, 2'-deoxy-5-(trifluoro-methyl)-
Viroptic

70-11-1
C_8H_7BrO
199.06
O=C(c(cccc1)c1)CBr
Acetophenone, 2-bromo
α-Bromoacetophenone
ω-Bromoacetophenone
2-Bromoacetophenone
Bromomethyl phenyl ketone
Ethanone, 2-bromo-1-phenyl-(9CI)
Phenacyl bromide [UN 2645]
Stauffer 4644
UN 2645 [Phenacyl bromide]

70-18-8
$C_{10}H_{17}N_3O_6S$
307.36
O=C(NC(C(=O)NCC(=O)O)CS)CCC(N)C(=O)O
Glycine, N-(N-l-γ-glutamyl-l-cysteinyl)
Copren
Deltathione
Glutathion
Glutathione
Glutathione (reduced)
Glutathione SH
Glutatiol
Glutatione
l-Glutatione
Glutide
Glutinal
GSH
Isethion
Neuthion
Panaron
Reduced glutathione
Tathion
Tathione
Triptide

70-25-7
$C_2H_5N_5O_3$
147.12
O=N(=O)NC(=N)N(N=O)C
Guanidine, 1-methyl-3-nitro-1-nitroso
Guanidine, N-methyl-N'-nitro-N-nitroso- (9CI)
Methylnitronitrosoguanidine
N-Methyl-N'-nitro-N-nitroso-guanidine
1-Methyl-1-nitroso-3-nitro-guanidine
1-Methyl-3-nitro-1-nitroso-guanidine
N-Methyl-N'-nitro-N-nitroso-guanidine, Not exceeding 25 grams in one outside packaging [NA 1325]
N-Methyl-N-nitrosonitro-guanidin (German)
N-Methyl-N-nitroso-N'-nitro-guanidine
N-Metylo-N'-nitro-N-nitrozo-guanidyny (Polish)
MNG
MNNG
NA 1325 [Smokeless powder for small arms (100 pounds or less)]
N-Nitroso-N-methylnitro-guanidine
N'-Nitro-N-nitroso-N-methyl-guanidine
NSC-9369
RCRA waste number U163

70-26-8
$C_5H_{12}N_2O_2$
132.19
O=C(O)C(N)CCCN
Ornithine, L
(S)-2,5-Diaminopentanoic acid
(S)-α,δ-Diaminovaleric acid
Ornithine
l-(-)-Ornithine
(S)-Ornithine
Pentanoic acid, 2,5-diamino-, (S)-

70-30-4
$C_{13}H_6Cl_6O_2$
406.89
Oc(c(c(c(c1Cl)Cl)Cl)Cc(c(c(cc2Cl)Cl)Cl)c2O)c1Cl
Phenol, 2,2'-methylenebis-(3,4,6-trichloro
Acigena
Almederm
AT 7
AT-17
B32
Bilevon
Bis(2-hydroxy-3,5,6-tri-chlorophenyl)methane
Bis-2,3,5-trichlor-6-hydroxy-fenylmethan (Czech)
Bis(3,5,6-trichloro-2-hydroxy-phenyl)methane
Compound G-11
Cotofilm
Dermadex
2,2'-Dihydroxy-3,3',5,5',6,6'-hexachlorodiphenylmethane
2,2'-Dihydroxy-3,5,6,3',5',6'-hexachlorodiphenylmethane
Exofene
Fomac
Fostril
G-11
G-II
Gamophen
Gamophene
G-Eleven
Germa-Medica
HCP
Hexabalm
2,2',3,3',5,5'-Hexachloro-6,6'-dihydroxydiphenyl methane
Hexachlorofen (Czech)
Hexachlorophane
Hexachlorophen
Hexachlorophene [UN 2875]
Hexafen
Hexide
Hexophene
Hexosan
Isobac
Isobac 20
Methane, bis(2,3,5-trichloro-6-hydroxyphenyl)
2,2'-Methylenebis(3,4,6-tri-chlorophenol)
Nabac
Nabac 25 EC
NCI-C02653
Neosept V
Phisodan
Phisohex
RCRA waste number U132
Ritosept
Septisol
Septofen
Steral
Steraskin
Surgi-cen
Surgi-cin
Surofene
Trichlorophene
Tersaseptic
Turgex
UN 2875 [Hexachlorophene]

70-34-8
$C_6H_3FN_2O_4$
186.11
O=N(=O)c(ccc(F)c1N(=O)=O)c1
Benzene, 2,4-dinitro-1-fluoro
2,4-Dinitrofluorobenzene
2,4-Dinitro-1-fluorobenzene
2,4-DNFB
1,2,4-Fluorodinitrobenzene
1-Fluoro-2,4-dinitrobenzene

70-38-2
$C_{19}H_{26}O_2$
286.45
CC(C)=CC2C(C(=O)OCc1ccc(C)cc1C)C2(C)C
Cyclopropanecarboxylic acid, 2,2-dimethyl-3-(2-methylpropenyl)-, 2,4-dimethylbenzyl ester
Dimethrin
2,4-Dimethylbenzylchrysanthemumate
2,4-Dimethylbenzyl-(i)-cis-trans-chrysanthemumate
2,4-Dimethylbenzyl 2,2-dimethyl-3-(2-methylpropenyl)-cyclopropanecarboxylate
2,4-Dimethylbenzyl ester of cis, trans-chrysanthemumic acid
2,4-Dimethylbenzylester kyseliny chrysanthemove (Czech)
(2,4-Dimethylphenyl)methyl 2,2-dimethyl-3-(2-methyl-1-propenyl)cyclopropanecarboxylate
ENT 21,170

70-43-9
$C_{18}H_{21}ClO_4$
336.84
CC(C)=CC3C(C(=O)OCc1cc(Cl)cc2OCOc12)C3(C)C
Cyclopropanecarboxylic acid, 2,2-dimethyl-3-(2-methylpropenyl)-, 6-chloropiperonyl ester, (+-)-cis, trans
Barthrin
6-Chloropiperonyl chrysanthemumate
6-Chloropiperonyl (+-)-cis/trans-chrysanthemumate
6-Chloropiperonyl 2,2-dimethyl-3-(2-methylpropenyl)cyclopropanecarboxylate
6-Chloropiperonyl ester of chrysanthemummonocarboxylic acid
Chrysanthemumic acid 6-chloropiperonyl ester
Chrysanthemummonocarboxylic acid 6-chloropiperonyl ester
ENT 21,557

70-47-3
$C_4H_8N_2O_3$
132.11
O=C(O)C(N)CC(=O)N
Asparagine, L- (8CI)
Agedoite
Altheine
Asparagine (VAN)
L-Asparagine (9CI)
L-β-Asparagine (VAN)
L-(+)-Asparagine
(-)-Asparagine
Butanoic acid, 2,4-diamino-4-oxo-, (S)-
NSC-82391

70-51-9
$C_{25}H_{48}N_6O_8$
560.79
CC(=O)N(O)CCCCCNC(=O)CCC(=O)N(O)CCCCCNC(=O)CCC(=O)N(O)CCCCCN
Propionohydroxamic acid, N-(5-(3-((5-aminopentyl)-hydroxycarbamoyl)propionamido)pentyl)- 3-((5-(N-hydroxyacetamido)pentyl)-carbamoyl)
30-Amino-3,14,25-trihydroxy-3,9,14,20,25-pentaazatriacontane-2,10,13,21,24-pentaone
N-Benzoylferrioxamine B
Deferoxamine
Deferoxaminum
Deferrioxamine
Deferrioxamine B
Desferal
Desferral
Desferrin
Desferrioxamine
Desferrioxamine B
DF B
DFO
DFOA
DFOM
Ferrioxamine B, N-benzoyl-
NSC-527604
3,9,14,20,25-Pentaazatriacontane-2,10,13,21,24-pentone, 30-amino-3,14,25-trihydroxy-

70-54-2
$C_6H_{14}N_2O_2$
146.18
O=C(O)C(N)CCCCN
DL-Lysine (9CI)
Lysine, DL- (8CI)

70-55-3
$C_7H_9NO_2S$
171.23
O=S(=O)(N)c(ccc(c1)C)c1
p-Toluenesulfonamide
p-Methylbenzenesulfonamide
4-Methylbenzenesulfonamide
p-Toluenesulfonylamide
p-Tosylamide
p-Toluenesulfanamide
4-Toluenesulfanamide
Toluene-4-sulfonamide
Toluene-p-sulphonamide
Tolylsulfonamide
p-Tolylsulfonamide
Tosylamide

70-69-9
$C_9H_{11}NO$
149.19
O=C(c(ccc(N)c1)c1)CC
4-Aminopropiophenone
AI3-14675
1-(4-Aminophenyl)-1-propanone
p-Aminopropiophenone
4'-Aminopropiophenone
Ethyl p-aminophenyl ketone
NSC-3187
PAPP
Paraminopropiophenone
1-Propanone, 1-(4-amino-phenyl)- (9CI)
Propiophenone, 4'-amino- (8CI)
Propiophenone, 4-amino-
USAF UCTL-1856

70-70-2
$C_9H_{10}O_2$
150.19
O=C(c(ccc(O)c1)c1)CC
Propiophenone, 4'-hydroxy
B 360
Ethyl p-hydroxyphenyl ketone
Frenantol
Frenohypon
H-365
p-Hydroxyphenyl-1-propanone
1-(4-Hydroxyphenyl)-1-propanone
Hydroxypropiophenone
p-Hydroxypropiophenone
4-Hydroxypropiophenone
Hypophenon
p-Oxypropiophenone
Paroxon
Paroxypropione
PHP
POP
Profenone
p-Propionylphenol
USAF EK-3302

71-00-1
$C_6H_9N_3O_2$
155.18
O=C(O)C(N)CC(N=CN1)=C1
Histidine, l
Glyoxaline-5-alanine
Histidine
L-Histidine

71-23-8
C_3H_8O
60.11
OCCC
n-Propyl alcohol (ACGIH, OSHA) [UN 1274]
Alcool propilico (Italian)
Alcool propylique (French)
Ethyl carbinol
1-Hydroxypropane
Optal
Osmosol extra
Propanol
n-Propanol
Propanol-1
1-Propanol
Propanol [UN 1274]
Propanole (German)
Propanolen (Dutch)
Propanoli (Italian)
n-Propyl alcohol

1-Propyl alcohol
Propyl alcohol [UN 1274]
n-Propyl alkohol (German)
Propylic alcohol
Propylowy alkohol (Polish)
UN 1274 [n-Propanol or propyl alcohol normal]

71-30-7
$C_4H_5N_3O$
111.12
n(ccc(n1)N)c1O
2(1H)-Pyrimidinone, 4-amino
4-Amino-2-hydroxypyrimidine
4-Amino-2(1H)-pyrimidinone
Cytosine (8CI)
Cytosinimine

71-33-0
$C_3H_3N_3O_2$
113.06
O=c1nc[nH]c(=O)[nH]1
5-Azauracil
Allantoxaidin
Allantoxaidine
NSC-56901
Oxaidin
s-Triazine-2,4(1H,3H)-dione (8CI)
1,3,5-Triazine-2,4(1H,3H)-dione (9CI)

71-36-3
$C_4H_{10}O$
74.14
OCCCC
n-Butanol
Alcool butylique (French)
Butanol
Butanol (French)
1-Butanol
Butan-1-ol
n-Butan-1-ol
Butanol [UN 1120]
Butanolen (Dutch)
Butanolo (Italian)
n-Butyl alcohol (ACGIH, OSHA)
1-Butyl alcohol
Butyl alcohol [UN 1120]
Butyl hydroxide
Butylowy alkohol (Polish)
Butyric or normal primary butyl alcohol
CCS 203
Hemostyp
1-Hydroxybutane
Methylolpropane
NA 1120 (DOT)
Propylcarbinol
Propylmethanol
RCRA waste number U031
UN 1120 [Butanols]

71-41-0
$C_5H_{12}O$
88.17
OCCCCC
Pentyl-alcohol
Alcool amylique (French)
Amyl alcohol
n-Amyl alcohol [UN 1105]
Amyl alcohol, normal
Amylol
n-Butylcarbinol
n-Amylalkohol (Czech)
Pentanol
Pentanol-1
n-Pentanol
Pentan-1-ol
1-Pentanol
Pentasol
Primary amyl alcohol
UN 1105 [Amyl alcohols]

71-43-2
C_6H_6
78.12
c(cccc1)c1
Benzene (ACGIH,OSHA) [UN 1114]
(6)Annulene
Benzeen (Dutch)
Benzen (Polish)
Benzin (Obs.)
Benzine (Obs.)
Benzol [UN 1114]
Benzole
Benzolene
Benzolo (Italian)
Bicarburet of hydrogen
Carbon Oil
Coal naphtha
Cyclohexatriene
Fenzen (Czech)
Mineral naphtha
Motor Benzol
NCI-C55276
Nitration benzene
Phene
Phenyl hydride
Pyrobenzol
Pyrobenzole
RCRA waste number U019
UN 1114 [Benzene]

71-44-3
$C_{10}H_{26}N_4$
202.40
N(CCCCNCCCN)CCCN
1,4-Butanediamine, N,N'-bis-(3-aminopropyl)
N,N'-Bis(3-aminopropyl)-1,4-butanediamine
1,4-Bis(aminopropyl) butane-diamine
1,4-Diaminobutane, N,N'-bis-(3-aminopropyl)-
Diaminopropyltetramethylene-diamine
4,9-Diaza-1,12-dodecane-diamine
Gerontine
Musculamine
Neuridine
Spermine
Spermine, puriss

71-47-6
CHO_2
45.02
O=CO
Formic acid, ion(1-) (8CI,9CI) (VAN)

71-48-7
$C_4H_6O_4$.Co
177.03
[Co].[O-]C(=O)CCC([O-])=O
Acetic acid, cobalt(2+) salt
Cobalt acetate
Cobalt(2+) acetate
Cobalt acetate (co(oac)2)
Cobalt(II) acetate
Cobalt diacetate
Cobaltous acetate
Cobaltous diacetate

71-50-1
$C_2H_3O_2$
59.04
O=C(O)C
Acetic acid, ion(1-) (8CI,9CI) (VAN)
Acetate (VAN)

71-55-6
$C_2H_3Cl_3$
133.40
C(Cl)(Cl)(Cl)C
Ethane, 1,1,1-trichloro
Aerothene tt
CF 2
Chloroetene
Chloroethene
Chloroethene NU
Chloroform, methyl-
Chlorothane NU
Chlorothene
Chlorothene(inhibited)
Chlorothene NU
Chlorothene SM
Chlorothene VG
Chlorten
Ethana NU
ICI-CF 2
Inhibisol
Methylchloroform
Methyl chloroform (ACGIH, DOT,OSHA)
Methyltrichloromethane
NCI-C04626
RCRA waste number U226
Solvent 111
Strobane
α-T
Tafclean
1,1,1-TCE
1,1,1-Trichloorethaan (Dutch)
1,1,1-Trichloraethan (German)
Trichloroethane
Trichloro-1,1,1-ethane (French)
α-Trichloroethane
1,1,1-Trichloroethane
1,1,1-Trichloroethane (OSHA) [UN 2831]

1,1,1-Tricloroetano (Italian)
Trichloromethylmethane
Tri-ethane
UN 2831 [1,1,1-Trichloroethane]

71-58-9
$C_{24}H_{34}O_4$
386.58
O=C(OC(C(C(C(C(C(C(=CC(=O)C1)C2C)(C1)C)C3)C2)C4)(C3)C)(C(=O)C)C4)C
(6-α)-Pregn-4-ene-3,20-dione, 17-(acetyloxy)-6-methyl
17-α-Acetoxy-6-α-methylpregn-4-ene-3,20-dione
17-Acetoxy-6-α-methylprogesterone
17-α-Acetoxy-6-α-methylprogesterone
(6-α)-17-(Acetyloxy)-6-methyl-preg-4-ene-3,20-dione
Depo-medroxyprogesterone acetate
Depo-MPA
Depo-Provera
DMPA
Farlutin
17-α-Hydroxy-6-α-methyl-pregn-4-ene-3,20-dione acetate
17-Hydroxy-6-α-methylpregn-4-ene-3,20-dione acetate
17-α-Hydroxy-6-α-methylprogesterone acetate
MAP
Medroxyacetate progesterone
Medroxyprogesterone acetate
6-α-Methyl-17-α-acetoxypregn-4-ene-3,20-dione
6-α-Methyl-17-acetoxy progesterone
6-α-Methyl-17-α-acetoxyprogesterone
6-α-Methyl-17-α-hydroxy-progesterone acetate
6-α-Methyl-4-pregnene-3,20-dion-17-α-ol acetate
Metipregnone
MPA
Nogest
NSC-26386
Oragest
Perlutex
Progesterone, 17-α-hydroxy-6-α-methyl-, acetate
Provera
Provera dosepak
Repromix

71-62-5
$C_{36}H_{51}NO_{11}$
673.88
Veratridine
Cevane-3-β,4-α,12,14,16-β,17,20-heptol, 4,9-epoxy-, 3-(3,4-dimethoxybenzoate)
4,9-Epoxycevane-3,4,12,14,16,17,20-heptol 3-(3,4-dimethoxybenzoate)
Veratrine (Amorphous)
3-Veratroylveracevine

71-63-6
$C_{41}H_{64}O_{13}$
765.05
CC1OC(CC(O)C1O)OC2C(O)CC(OC2C)OC8C(O)CC(OC7CC3(C)C(CCC4C3CCC5(C)C(CCC45O)C6=CC
Digitoxin
Acedoxin
Asthenthilo
Cardidigin
Cardigin
Carditalin
Carditoxin
Cristapurat
Crystalline
Crystalline digitalin
Crystodigin
Digilong
Digimed
Digimerck
Digisidin
Digitalin
Digitaline (French)
Digitaline cristallisee
Digitaline nativelle
Digitalinum verum
Digitophyllin
Digitoxigenin-tridigitoxosid (German)
Digitoxigenin tridigitoxoside
Digitoxinum
Digitoxoside
Digitrin
Ditaven
Glucodigin
Lanatoxin
Mono-digitoxid (German)
Monodigitoxoside
Mono-glycocard
Myodigin
Purodigin
Purpurid
Tardigal
Tri-digitoxoside (German)
Unidigin

71-67-0
$C_{20}H_8Br_4O_{10}S_2.2Na$
838.02
Phenolphthalein, 4,5,6,7-tetrabromo-3',3''-disulfo-, disodium salt
Bromosulfalein
Bromosulfophthalein
Bromosulphalein
Bromosulphthalein
Bromotaleina
Bromsulfalein
Bromsulfan
Bromsulfophthalein
Bromsulfthalein
Bromsulphalein
Bromsulphthalein
Brom-tetragnost
Bromthalein
BSF
BSF simes
BSP
BSP sodium
CBSP
Disodium bromosulfophthalein
Hepartest
Hepartestabrome
Hepatosulfalein
Phenoltetrabromophthaleinsulfonate
Sodium bromosulfalein
Sodium bromosulfophthalein
Sodium bromsulphalein
Sodium bromsulphthalein
Sodium phenol tetrabromophthalein
Sodium sulfobromophthalein
Sodium sulphobromophthalein
Sulfobromophthalein
Sulfobromophthalein sodium
Sulfobromphthalein
Sulphobromophthalein
Sulphobromophthalein sodium
Tetrabromophenolsulfophthalein
Tetrabromosulfophthalein
Tetrabromsulfthalein

71-91-0
$C_8H_{20}N.Br$
210.20
[Br-].CC[N+](CC)(CC)CC
Ammonium, tetraethyl-, bromide
Beparon
Bromure de tetraethylammonium (French)
Etambro
Etylon
Sympatektoman
TEA
TEAB
Tetraethyl ammonium bromide
Tetrylammonium bromide
USAF DO-32

72-00-4
$C_6H_{11}Cl_2O_4P$
249.04
Phosphoric acid, 2,2-dichlorovinyl diethyl ester
Diaethyl-dichlorvinyl-phosphat (German)
2,2-Dichloroethenyl diethyl phosphate
2,2-Dichlorovinyl diethyl phosphate
O-(2,2-Dichlorvinyl)-O,O-diethylphosphat (German)
Dichlorvos-ethyl

72-14-0
$C_9H_9N_3O_2S_2$
255.33
O=S(=O)(N=C(NC=C1)S1)c(ccc(N)c2)c2
Sulfanilamide, N[1]-2-thiazolyl
2-(p-Aminobenzenesulfonamido)thiazole
2-(p-Aminobenzenesulphonamido)thiazole

4-Amino-N-2-thiazolylbenzene-sulfonamide
Azoseptale
Cerazol (Suspension)
Chemosept
Cibazol
Duatok
Eleudron
Formosulfathiazole
M+B 760
Neostrepsan
Norsulfasol
Norsulfazole
Planomide
Poliseptil
RP 2090
Streptosilthiazole
Sulfamul
2-Sulfanilamidothiazol (German)
2-Sulfanilamidothiazole
2-(Sulfanilylamino)thiazole
Sulfathiazol
Sulfathiazole
2-Sulfonamidothiazole
Sulphathiazole
Sulzol
Thiacoccine
Thiazamide
N[1]-2-Thiazolylsulfanilamide
Thiozamide
USAF SN-9

72-17-3
$C_3H_5O_3.Na$
112.07
Lactic acid, monosodium salt
Lacolin
Lactic acid sodium salt
per-Glycerin
Propanoic acid, 2-hydroxy-, monosodium salt
Sodium lactate

72-18-4
$C_5H_{11}NO_2$
117.17
O=C(O)C(N)C(C)C
Valine, L
L(+)-α-Aminoisovaleric acid
L-Valine

72-19-5
$C_4H_9NO_3$
119.14
O=C(O)C(N)C(O)C
Threonine, L
Threonin
Threonine
L-Threonine

72-20-8
$C_{12}H_8Cl_6O$
380.90
ClC4=C(Cl)C5(Cl)C3C1CC(C2OC12)C3C4(Cl)C5(Cl)Cl
1,4:5,8-Dimethanonaphthalene, 1,2,3,4,10,10-hexachloro-6,7-epoxy-1,4,4a,5,6,7,8,8a- octahydro-, endo, endo
Compound 269
Endrex
Endrin (ACGIH,DOT,OSHA)
Endrin mixture (DOT)
Endrine (French)
ENT 17,251
Experimental Insecticide 269
Hexachloroepoxyoctahydro-endo,endo-dimethanonaphthalene
3,4,5,6,9,9-Hexachloro-1a,2,2a,3,6,6a,7,7a-octahydro-2,7:3,6-dimethanonaphth(2,3-b)-oxirene
Hexadrin
Latka 269 (Czech)
Mendrin
NA 2761
NCI-C00157
Nendrin
OMS 197
RCRA waste number P051

72-33-3
$C_{21}H_{26}O_2$
310.47
COc4ccc3C2CCC1(C)C(CCC1(O)C#C)C2CCc3c4
17-α-19-Norpregna-1,3,5(10)-trien-20-yn-17-ol, 3-methoxy
Compound 33355
δ-MVE
3,17-β-Dihydroxy-17-α-ethynyl-1,3,5(10)-estratriene-3-methyl ether
EE3ME
Estra-1,3,5(10)-trien-17-β-ol, 17-α-ethynyl-3-methoxy-
17-α-Ethinyl estradiol 3-methyl ether
Ethinylestradiol 3-methyl ether
17-α-Ethinyl oestradiol 3-methyl ether
Ethinyloestradiol 3-methyl ether
Ethynylestradiol 3-methyl ether
17-Ethynylestradiol 3-methyl ether
17-α-Ethynylestradiol 3-methyl ether
17-α-Ethynyl-3-methoxy-1,3,5(10)-estratrien-17-β-ol
(+)-17-α-Ethynyl-17-β-hydroxy-3-methoxy-1,3,5(10)-estratriene
(+)-17-α-Ethynyl-17-β-hydroxy-3-methoxy-1,3,5(10)-oestratriene
17-Ethynyl-3-methoxy-1,3,5(10)-estratrien-17-β-ol
17-α-Ethynyl-3-methoxy-17-β-hydroxy-δ-1,3,5(10)-estratriene
17-α-Ethynyl-3-methoxy-17-β-hydroxy-δ-1,3,5(10)-oestratriene
17-Ethynyl-3-methoxy-1,3,5(10)-oestratien-17-β-ol
Ethynyloestradiol 3-methyl ether
Ethynyloestradiol methyl ether
17-Ethynyloestradiol 3-methyl ether
17-α-Ethynyloestradiol 3-methyl ether
17-α-Ethynyloestradiol methyl ether
Mestranol
Mestrenol
3-Methoxy-17-α-ethinylestradiol
3-Methoxy-17-α-ethinyl-oestradiol
3-Methoxyethynylestradiol
3-Methoxy-17-α-ethynyl-estradiol
3-Methoxy-17-α-ethynyl-1,3,5(10)-estratrien-17-β-ol
3-Methoxyethynyloestradiol
3-Methoxy-17-α-ethynyloestradiol
3-Methoxy-17-ethynyloestradiol-17-β
3-Methoxy-17-α-ethynyl-1,3,5(10)-oestratrien-17-β-ol
3-Methoxy-19-nor-17-α-pregna-1,3,5(10)-trien-20-yn-17-ol
3-Methoxy-17-α-19-norpregna-1,3,5(10)-trien-20-yn-17-ol
(17-α)-3-Methoxy-19-norpregna-1,3,5(10)-trien-20-yn-17-ol
3-Methylethynylestradiol
3-Methylethynyloestradiol
SC 4725

72-43-5
$C_{16}H_{15}Cl_3O_2$
345.66
COc1ccc(cc1)C(c2ccc(OC)cc2)C(Cl)(Cl)Cl
Ethane, 1,1,1-trichloro-2,2-bis(p-methoxyphenyl)
Benzene, 1,1'-(2,2,2-trichloroethylidene)bis-(4-methoxy-
2,2-Bis(p-anisyl)-1,1,1-trichloroethane
1,1-Bis(p-methoxyphenyl)-2,2,2-trichloroethane
2,2-Bis(p-methoxyphenyl)-1,1,1-trichloroethane
Chemform
Dianisyltrichlorethane
2,2-Di-p-anisyl-1,1,1-trichloroethane
Dimethoxy-DDT
p,p'-Dimethoxydiphenyltrichloroethane
Dimethoxy-DT
2,2-Di-(p-methoxyphenyl)-1,1,1-trichloroethane
Di(p-methoxyphenyl)-trichloromethyl methane
DMDT
p,p'-DMDT
Double-M EC
p,p'-Dwumetoksydwufenylotrojchloroetan (Polish)
ENT 1,716
Ethane, 2,2-bis(p-anisyl)-1,1,1-trichloro-

Flo Pro Mcseed Protectant
Maralate
Marlate
Methoxcide
Methoxo
Methoxychlor
Methoxychlore
p,p'-Methoxychlor
Methoxychlor 2 EC
Methoxy-DDT
Metoksychlor (Polish)
Metox
Mezox K
Moxie
NCI-C00497
OMS 466
RCRA waste number U247
1,1,1-Trichlor-2,2-bis(4-methoxy-phenyl)-aethan (German)
1,1,1-Trichloro-2,2-bis(p-anisyl)ethane
1,1'-(2,2,2-Trichloroethylidene)bis(4-methoxybenzene)
1,1,1-Trichloro-2,2-bis(p-methoxyphenol)ethanol
1,1,1-Trichloro-2,2-bis(p-methoxyphenyl)ethane
1,1,1-Trichloro-2,2-bis(4-methoxyphenyl)ethane
2,2,2-Trichloro-1,1-bis(4-methoxyphenyl)ethane
1,1,1-Trichloro-2,2-di(4-methoxyphenyl)ethane
4,4-(2,2,2-Trichloroethylidene)-dianisole

72-44-6
$C_{16}H_{14}N_2O$
250.32
O=C(N(c(c(ccc1)C)c1)C(=Nc2cc cc3)C)c23
4(3H)-Quinazolinone, 2-methyl-3-o-tolyl
Cateudyl
Citexal
Ci-705
CN 38703
3,4-Dihydro-2-methyl-4-oxo-3-o-tolylquinazoline
Dormigoa
Dormogen
Dormutil
Dorsedin
Fadormir
Holodorm
Hyminal
Hypcol
Hyptor
Hyptor base
Ipnofil
Maoa
Mequin
Melsedin Base
Melsomin
Metachalon (Czech)
Metaqualon
Methaqualone
Methaqualoneinone
2-Methyl-3-(2-methylphenyl)-4-quinazolinone
2-Methyl-3-(2-methylphenyl)-4(3H)-quinazolinone
2-Methyl-3-o-tolyl-4(3H)-chinazolinon (German)
2-Methyl-3-o-tolyl-4(3H)-chinazolone
2-Methyl-3-(o-tolyl)-3,4-dihydro-4-quinazolinone
2-Methyl-3-tolyl-4-oxybensdiazine
2-Methyl-3-o-tolyl-4(3H)-quinazolinone
2-Methyl-3-o-tolyl-4-quinazolone
2-Methyl-3-(2-tolyl)quinazol-4-one
Metolquizolone
Mozambin
Mollinox
Motolon
MTQ
Nethaqualone
Nobedorm
Noctilene
Normi-Nox
Omnyl
Optinoxan
Orthonal
Ortonal
Parest
Parminal
Pro-Dorm
Quaalude
4(3H)-Quinazolinone, 2-methyl-3-(2-methylphenyl)-
QZ 2
Revonal
RIC 272
Rorer 148
Roulone
Rouqualone
Sindesvel
Somberol
Somnafac
Somnomed
Sonal
Sopor
Soverin
Torinal
TR-495
Tuazole
Tuazolone

72-48-0
$C_{14}H_8O_4$
240.22
O=C(c(c(C(=O)c1c(O)c(O)cc2) ccc3)c3)c12
Anthraquinone, 1,2-dihydroxy
Alizarin
Alizarina
Alizarin B
Alizarine
Alizarine B
Alizarine 3B
Alizarine Indicator
Alizarine Lake Red 2P
Alizarine Lake Red 3P
Alizarine Lake Red IPX
Alizarine L Paste
Alizarine NAC
Alizarine Paste 20% bluish
Alizarine Red
Alizarine Red B
Alizarine Red B2
Alizarine Red IP
Alizarine Red IPP
Alizarine Red L
Alizarin Red
9,10-Anthracenedione, 1,2-dihydroxy-
1,2-Anthraquinonediol
Certiqual Alizarine
C.I. 58000
C.I. Mordant Red 11
C.I. Pigment Red 83
D & C Orange Number 15
Deep Crimson Madder 10821
1,2-Dihydroxyanthrachinon (Czech)
1,2-Dihydroxyanthraquinone
1,2-Dihydroxy-9,10-anthraquinone
Eljon Madder
Mitsui Alizarine B
Sanyo Carmine I2B
Turkey Red

72-54-8
$C_{14}H_{10}Cl_4$
320.04
ClC(Cl)C(c1ccc(Cl)cc1)c2ccc(Cl) cc2
Ethane, 1,1-dichloro-2,2-bis-(p-chlorophenyl)
1,1-Bis(p-chlorophenyl)-2,2-dichloroethane
1,1-Bis(4-chlorophenyl)-2,2-dichloroethane
2,2-Bis(p-chlorophenyl)-1,1-dichloroethane
2,2-Bis(4-chlorophenyl)-1,1-dichloroethane
DDD
p,p'-DDD
1,1-Dichloor-2,2-bis(4-chloor fenyl)-ethaan (Dutch)
1,1-Dichlor-2,2-bis(4-chlorphenyl)-aethan (German)
1,1-Dichloro-2,2-bis(4-chlorophenyl)-ethane (French)
1,1-Dichloro-2,2-bis(para-chlorophenyl)ethane
1,1-Dichloro-2,2-bis(p-chlorophenyl)ethane (DOT)
1,1-Dichloro-2,2-di(4-chlorophenyl)ethane
Dichlorodiphenyl dichloroethane
p,p'-Dichlorodiphenyldichloroethane
1,1-Dicloro-2,2-bis(4-clorofenil)-etano (Italian)
Dilene
ENT 4,225
ME-1700
NA 2761
NCI-C00475
OMS 1078
RCRA waste number U060
Rhothane
Rhothane D-3
Rothane

TDE (DOT)
p,p'-TDE
Tetrachlorodiphenylethane

72-55-9
$C_{14}H_8Cl_4$
318.02
ClC(Cl)=C(c1ccc(Cl)cc1)c2ccc(Cl)cc2
Ethylene, 1,1-dichloro-2,2-bis-(p-chlorophenyl)
2,2-Bis(4-chlorophenyl)-1,1-dichloroethene
2,2-Bis(p-chlorophenyl)-1,1-dichloroethylene
DDE
p,p'-DDE
DDT dehydrochloride
1,1-Dichloro-2,2-bis(p-chlorophenyl)ethylene
Dichlorodiphenyldichloroethylene
p,p'-Dichlorodiphenyl dichloroethylene
1,1'-(Dichloroethenylidene)-bis(4-chlorobenzene)
NCI-C00555

72-56-0
$C_{18}H_{20}Cl_2$
307.28
CCc1ccc(cc1)C(C(Cl)Cl)c2ccc(CC)cc2
Ethane, 2,2-bis(p-ethylphenyl)-1,1-dichloro
1,1-Bis(p-ethylphenyl)-2,2-dichloroethane
2,2-Bis(p-ethylphenyl)-1,1-dichloroethane
1,1-Dichloro-2,2-bis(p-ethylphenyl)ethane
1,1-Dichloro-2,2-bis(4-ethylphenyl)ethane
2,2-Dichloro-1,1-bis(p-ethylphenyl)ethane
a,a-Dichloro-2,2-bis(p-ethylphenyl)ethane
Diethyl-diphenyl dichloroethane
Di(p-ethylphenyl)dichloroethane
Ethane, 1,1-dichloro-2,2-bis-(p-ethylphenyl)-
Ethylan
p,p-Ethyl DDD
p,p'-Ethyl-DDD
NCI-C02868
Perthane
Q-137

72-57-1
$C_{34}H_{28}N_6O_{14}S_4.4Na$
964.88
[Na+].[Na+].[Na+].[Na+].Cc1cc(ccc1N=Nc3c(O)c2c(N)cc(cc2cc3S([O-])(=O)=O)S([O-])
2,7-Naphthalenedisulfonic acid, 3,3'-((3,3'-dimethyl-4,4'-biphenylylene)bis(azo))-bis(5- amino-4-hydroxy-, tetrasodium salt
Amanil Sky Blue
Amanil Sky Blue R
Amidine Blue 4B
Azidinblau 3B
Azidine Blue 3B
Azurro Diretto 3B
Bencidal Blue 3B
Benzaminblau 3B
Benzamine Blue
Benzamine Blue 3B
Benzanil Blue 3BN
Benzanil Blue R
Benzoblau 3B
Benzo Blue
Benzo Blue 3B
Benzo Blue 3Bs
Bleu Diamine
Bleu Diazole N 3B
Bleu Directe 3B
Bleue Diretto 3B
Bleu Trypane N
Blue 3B
Blue EMB
Brasilamina Blue 3B
Brasilazina Blue 3B
Centraline Blue 3B
Chloramiblau 3B
Chloramine Blue
Chloramine Blue 3B
Chlorazol Blue 3B
Chrome Leather Blue 3B
C.I. 23850
C.I. Direct Blue 14
C.I. Direct Blue 14, Tetrasodium salt
Congoblau 3B
Congo Blue
Congo Blue 3B
Cresotine Blue 3B
Diaminblau 3B
Diamineblue
Diamine Blue 3B
Dianilblau
Dianilblau H3G
Dianil Blue
Dianil Blue H3G
Diaphtamine Blue TH
Diazine Blue 3B
Diazol Blue 3B
Diphenyl Blue 3B
Directblau 3B
Direct Blue 14
Direct Blue 3B
Direct Blue 3BX
Direct Blue D3B
Direct Blue FFN
Direct Blue H3G
Direct Blue M3B
Directakol Blue 3BL
Hispamin Blue 3BX
Modr Prima 14 (Czech)
Modr Trypanova (Czech)
Naphtamine Blue 2B
Naphtamine Blue 3BX
Naphthaminblau 3BX
Naphthamine Blue 3BX
Naphthylamine Blue
NCI-C61289
Niagara Blue
Niagara Blue 3B
Orion Blue 3B
Paramine Blue 3B
Parkibleu
Parkipan
Pontamine Blue 3BX
Pyrazol Blue 3B
Pyrotropblau
RCRA waste number U236
Renolblau 3B
Sodium ditolyldisazobis-8-amino-1-naphthol-3,6-disulfonate
Sodium ditolyldisazobis-8-amino-1-naphthol-3,6-disulphonate
TB
Trianol Direct Blue 3B
Triazolblau 3BX
Tripan Blue
Trypanblau (German)
Trypan Blue
Trypan Blue BPC
Trypan Blue sodium salt
Trypane Blue

72-63-9
$C_{20}H_{28}O_2$
300.44
Androsta-1,4-dien-3-one, 17-β-hydroxy-17-α-methyl
Abirol
Anabolin
Androsta-1,4-dien-3-one, 17-hydroxy-17-methyl-, (17-β)- (9CI)
Androsta-1,4-dien-3-one, 17-β-hydroxy-17-methyl- (8CI)
Ciba 17309 BA
Compound 17309
Crein
Danabol
Dehydromethyltesterone
a1-Dehydromethyltesterone
1-Dehydro-17-α-methyltestosterone
Dianabol
Dianabole
Geabol
17-β-Hydroxy-17-α-methyl-androstra-1,4-dien-3-one
MA
Metanabol
Metandienon
Metandienone
Metandienonum
Metandrostenolon
Metandrostenolone
Metastenol
Methandienone
Methandrolone
Methandrostenolone
17-α-Methyl-17-β-hydroxy-1,4-androstadien-3-one
δ'-17-Methyltestosterone
$δ^1$-17-α-Methyltestosterone
Nerobol
Nerobolettes
NSC-42722
Protobolin
Stenolon
Stenolone

72-69-5
$C_{19}H_{21}N$
263.41
CNCCC=C2c1ccccc1CCc3cccc c23
5H-Dibenzo(a,d)cycloheptene-δ^5,γ-propylamine, 10,11-dihydro-N-methyl
Amitryptyline, demethyl-
Avantyl
Aventyl
Demethylamitriptylene
Demethylamitriptyline
Demethylamitryptyline
Desmethylamitriptyline
10,11-Dihydro-N-methyl-5H-dibenzo(a,d)cycloheptane-δ,γ-propylamine
5-(3-(Methylamino)propylidene)dibenzo(a,e)cyclohepta-(1,5)diene
5-(3-Methylaminopropylidene)-10,11-dihydro-5H-dibenzo-(a,d)cycloheptene
Noramitriptyline
Nortriptyline
Nortryptiline

73-22-3
$C_{11}H_{12}N_2O_2$
204.25
O=C(O)C(N)CC(c(c(N1)ccc2)c2)=C1
Tryptophan, L
Alanine, 3-indol-3-yl-
α'-Amino-3-indolepropionic acid
EH 121
Indole-3-alanine
Indole-3-propionic acid, α-amino-
1-β-3-Indolylalanine
NCI-C01729
Pacitron
Propionic acid, 2-amino-3-indol-3-yl-
L-Tryptofan
TRP
L-TRP
Tryptophan
(-)-Tryptophan
L-Tryptophan
Tryptophane
L-Tryptophane
L-TTP

73-24-5
$C_5H_5N_5$
135.15
n(c(c(N=CN1)c1n2)N)c2
Adenine
ADE
Adeninimine
6-Aminopurine
6-Amino-1H-purine
6-Amino-3H-purine
6-Amino-9H-purine
1,6-Dihydro-6-iminopurine
3,6-Dihydro-6-iminopurine
Leuco-4
1H-Purin-6-amine
Purine, 6-amino-
1H-Purine, 6-amino
9H-Purine, 1,6-dihydro-6-imino-
USAF CB-18
Vitamin B4

73-31-4
$C_{13}H_{16}N_2O_2$
232.31
Acetamide, N-(2-(5-methoxy-indol-3-yl)ethyl)
Acetamide, N-(2-(5-methoxy-1H-indol-3-yl)ethyl)- (9CI)
N-Acetyl-5-methoxytryptamine
Melatonin
Melatonine
5-Methoxy-N-acetyltryptamine

73-32-5
$C_6H_{13}NO_2$
131.20
O=C(O)C(N)C(CC)C
Isoleucine, L
Acetic acid, amino-sec-butyl-
2-Amino-3-methylpentanoic acid
α-Amino-β-methylvaleric acid
Isoleucine
L-Isoleucine
Norvaline, 3-methyl-
Valeric acid, 2-amino-3-methyl-

73-40-5
$C_5H_5N_5O$
151.15
n(c(O)c(N=CN1)c1n2)c2N
Guanine
2-Aminohypoxanthine
Mearlmaid

73-48-3
$C_{15}H_{14}F_3N_3O_4S_2$
421.44
NS(=O)(=O)c3cc2c(NC(Cc1ccc cc1)NS2(=O)=O)cc3C(F)(F)F
2H-1,2,4-Benzothiadiazine-7-sulfonamide, 3-benzyl-3,4-dihydro-6-(trifluoromethyl)-, 1,1-dioxide
Aprinox
Be 724-A
Bendrofluazide
Bendroflumethiazide
Bentride
Benuron
Benzylhydroflumethiazide
3-Benzyl-3,4-dihydro-6-(trifluoromethyl)-2H-1,2,4-benzothiadiazine-7-sulfonamide 1,1-dioxide
Benzydroflumethiazide
Benzylrodiuran
3-Benzyl-6-trifluoromethyl-7-sulfamoyl-3,4-dihydro-1,2,4-benzothiadiazine, 1,1-dioxide
Berkozide
BHFT
BL H368
Bristuric
Bristuron
Centyl
Flumesil
FT 8
Intolex
Nateretin
Naturetin
Naturine
Neo-Naclex
Neo-Rontyl
Niagaril
Nikion
Orsile
Pluryl
Pluryle
Plusuril
Poliuron
Relan β
Repicin
Salural
Salures
Sinesalin
Sodiuretic
Thiazidico
6-Trifluoromethyl-3-benzyl-7-sulfamyl-3,4-dihydro-1,2,4-benzothiadiazine, 1,1-dioxide
Urlea

73-49-4
$C_{10}H_{12}ClN_3O_3S$
289.72
CCC2NC(=O)c1cc(c(Cl)cc1N2) S(N)(=O)=O
Quinethazone
Aquamox
Chinetazone
Chinethazone
Chinethazonum
7-Chloro-2-ethyl-6-sulfamoyl-1,2,3,4-tetrahydro-4-quinazolinone
7-Chloro-2-ethyl-1,2,3,4-tetrahydro-4-oxo-6-quinazoline-sulfonamide
7-Chloro-2-ethyl-1,2,3,4-tetrahydro-4-oxo-6-sulfamoylquinazoline
CL 36010
Hydromox
Idrokin
6-Quinazolinesulfonamide, 7-chloro-2-ethyl-1,2,3,4-tetrahydro-4-oxo-
Quinetazona (Spanish)
Quinethazon
Quinethazonum (Latin)

74-11-3
$C_7H_5ClO_2$
156.57
O=C(O)c(ccc(c1)Cl)c1
Benzoic acid, p-chloro
Acido p-clorobenzoico (Italian)
Benzoic acid, 4-chloro- (9CI)
p-Carboxychlorobenzene

p-Chlorbenzoic acid
p-Chlorobenzoic acid
4-Chlorobenzoic acid
Chlorodracylic acid

74-31-7
$C_{18}H_{16}N_2$
260.36
N(c(ccc(Nc(cccc1)c1)c2)c2)c(ccc c3)c3
p-Phenylenediamine, N,N'-diphenyl
Agerite
Agerite DPPD
1,4-Bis(phenylamino)benzene
N,N'-Difenyl-p-fenylendiamin (Czech)
Diphenyl-p-phenylenediamine
N,N'-Diphenyl-p-phenylenediamine
DPPD
Flexamine G
JZF
Nonox DPPD
p-Phenylaminodiphenylamine
4-Phenylaminodiphenylamine
USAF GY-2

74-39-5
$C_{12}H_9N_3O_4$
259.20
O=N(=O)c(ccc(N=Nc(c(O)cc(O) c1)c1)c2)c2
Resorcinol, 4-((p-nitrophenyl)azo)- (8CI)
AI3-09041
1,3-Benzenediol, 4-((4-nitrophenyl)azo)- (9CI)
p-Diazoviolet
2,4-Dihydroxy-4'-nitroazobenzene
Magneson
Magneson I
Magnezon I
NSC-3914
p-Nitrobenzeneazoresorcinol
4-((4-Nitrophenyl)azo)-1,3-benzenediol
p-Nitrophenylazoresorcinol
4-(p-Nitrophenylazo)resorcinol

74-55-5
$C_{10}H_{24}N_2O_2$
204.36
CCC(CO)NCCNC(CC)CO
1-Butanol, 2,2'-(1,2-ethanediyldiimino)bis-, (R)
d,N,N'-Bis(1-hydroxymethylpropyl)ethylenediamine
Dadibutol
EMB
Ethambutol
(R)-2,2'-(1,2-Ethanediyldiimino)bis-1-butanol
(+)-2,2'-(Ethylenediimino)di-1-butanol
Tibutol

74-61-3
$C_3H_8O_3S_3$
188.29
1-Propanesulfonic acid, 2,3-dimercapto
2,3-Dimercaptopropanesulfonic acid
2,3-Dimercapto-1-propanesulf onic acid

74-79-3
$C_6H_{14}N_4O_2$
174.24
O=C(O)C(N)CCCNC(=N)N
Arginine, L
Arginine
(L)-Arginine

74-82-8
CH_4
16.05
C
Methane [UN 1971]
Fire Damp
Marsh Gas
Methane, Compressed [UN 1971]
Methane, Refrigerated liquid [UN 1972]
Methyl hydride
UN 1971 [Methane, Compressed or natural gas, Compressed (with high methane content)]
UN 1972 [Methane, Refrigerated liquid (cryogenic liquid) or natural gas, Refrigerated liquid (cryogenic liquid)(with high methane content)]

74-83-9
CH_3Br
94.95
BrC
Methyl bromide (ACGIH, OSHA) [UN 1062]
Bercema
Brom-methan (German)
Brom-o-Gas
Brom-o-Gaz
Bromometano (Italian)
Bromomethane
Bromure de methyle (French)
Bromuro di metile (Italian)
Broommethaan (Dutch)
Celfume
Curafume
Dawson 100
Detia Gas EX-M
Dowfume
Dowfume MC-2
Dowfume MC-33
Dowfume MC-2 soil fumigant
Edco
Embafume
Fumigant-1 (Obs.)
Halon 1001
Haltox
Iscobrome
Kayafume
MB
MBX
MeBr
Metafume
Methane, bromo
Methogas
Methylbromid (German)
Metylu bromek (Polish)
Monobromomethane
Pestmaster (Obs.)
Profume (Obs.)
R 40B1
RCRA waste number U029
Rotox
Terabol
Terr-O-Gas 67
Terr-O-Gas 100
UN 1062 [Methyl bromide]
Zytox

74-84-0
C_2H_6
30.08
CC
Ethane
Bimethyl
Dimethyl
Ethane [UN 1035]
Ethane, Compressed [UN 1035]
Ethane, Refrigerated liquid [UN 1961]
Ethyl hydride
Methylmethane
UN 1035 [Ethane, compressed]
UN 1961 [Ethane, refrigerated liquid (cryogenic liquid)]

74-85-1
C_2H_4
28.06
C=C
Ethylene
Acetene
Athylen (German)
Bicarburretted hydrogen
Elayl
Ethene
Ethylene [UN 1962]
Ethylene, Compressed [UN 1962]
Ethylene, Refrigerated liquid [UN 1038]
Liquid ethyene
Olefiant Gas
UN 1038 [Ethylene, refrigerated liquid (cryogenic liquid)]
UN 1962 [Ethylene, compressed]

74-86-2
C_2H_2
26.04
C#C
Acetylene [UN 1001]
Acetylen
Acetylene, Dissolved [UN 1001]
Acetylene, Liquid [UN 1001]
Ethine
Ethyne

Narcylen
UN 1001 [Acetylene, dissolved, Liquefied]

74-87-3
CH_3Cl
50.49
ClC
Methyl chloride (ACGIH, OSHA) [UN 1063]
Artic
Chloor-methaan (Dutch)
Chlor-methan (German)
Chloromethane
Chlorure de methyle (French)
Clorometano (Italian)
Cloruro di metile (Italian)
Methane, chloro
Methylchlorid (German)
Metylu chlorek (Polish)
Monochloromethane
R 40
RCRA waste number U045
UN 1063 [Methyl chloride]

74-88-4
CH_3I
141.94
CI
Methyl iodide (ACGIH, OSHA) [UN 2644]
Halon 10001
Iodometano (Italian)
Iodomethane
Iodure de methyle (French)
Jod-methan (German)
Joodmethaan (Dutch)
Methane, iodo
Methyljodid (German)
Methyljodide (Dutch)
Metylu jodek (Polish)
Monoioduro di metile (Italian)
RCRA waste number U138
UN 2644 [Methyl iodide]

74-89-5
CH_5N
31.07
NC
Methylamine (ACGIH,OSHA)
Aminomethane
Carbinamine
Mercurialin
Methanamine (9CI)
Methylamine, Anhydrous [UN 1061]
Methylamine, Aqueous solution UN [1235]
Methylaminen (Dutch)
Metilamine (Italian)
Metyloamina (Polish)
Monomethylamine
Monomethylamine, Anhydrous [UN 1061]
Monomethylamine, Aqueous solution [UN 1235]
UN 1061 [Methylamine, anhydrous]
UN 1235 [Methylamine, aqueous solution]

74-90-8
CHN
27.03
N#C
Hydrocyanic-acid
Acide cyanhydrique (French)
Acido cianidrico (Italian)
Aero Liquid HCN
Blausaeure (German)
Blauwzuur (Dutch)
Carbon hydride nitride (CHN)
Cyaanwaterstof (Dutch)
Cyanwasserstoff (German)
Cyclon
Cyclone B
Cyjanowodor (Polish)
Evercyn
Fluohydric acid gas
Formic anammonide
Formonitrile
HCN
Hydrocyanic acid, Aqueous solutions not more than 20% hydrocyanic acid [UN 1613]
Hydrocyanic acid, Liquefied (DOT)
Hydrocyanic acid (Prussic), Unstabilized (DOT)
Hydrogen cyanide (ACGIH, OSHA)
Hydrogen cyanide, Anhydrous, Stabilized [UN 1051]
Hydrogen cyanide, Anhydrous, Stabilized, Absorbed in a porous inert material [UN 1614]
NA 1051 (DOT)
Prussic acid
Prussic acid (DOT)
Prussic acid, Unstabilized
RCRA waste number P063
UN 1051 [Hydrogen cyanide, anhydrous, stabilized]
UN 1613 [Hydrocyanic acid, aqueous solutions not more than 20% hydrocyanic acid]
UN 1614 [Hydrogen cyanide, anhydrous, stabilized, absorbed in a porous inert material]
Zaclondiscoids

74-93-1
CH_4S
48.11
SC
Methanethiol
Mercaptan methylique (French)
Methaanthiol (Dutch)
Methanethiol (OSHA)
Methanthiol (German)
Methvtiolo (Italian)
Methylmercaptaan (Dutch)
Methyl mercaptan (ACGIH, OSHA) [UN 1064]
Metilmercaptano (Italian)
RCRA waste number U153
Thiomethanol
UN 1064 [Methyl mercaptan]

74-95-3
CH_2Br_2
173.85
BrCBr
Methane, dibromo
Dibromomethane [UN 2664]
Methylene bromide
Methylene dibromide
RCRA waste number U068
UN 2664 [Dibromomethane]

74-96-4
C_2H_5Br
108.98
BrCC
Ethyl bromide (ACGIH, OSHA) [UN 1891]
Bromoethane
Bromure d'ethyle
Ethane, bromo
Etylu bromek (Polish)
Halon 2001
Monobromoethane
NCI-C55481
UN 1891 [Ethyl bromide]

74-97-5
CH_2BrCl
129.39
BrCCl
Methane, bromochloro
Bromochloromethane [UN 1887]
Chlorobromomethane (ACGIH, OSHA)
Halon 1011
Methylene chlorobromide
MIL-B-4394-B
Mono-chloro-mono-bromo-methane
UN 1887 [Bromochloro-methane]

74-98-6
C_3H_8
44.11
C(C)C
Propane (OSHA) [UN 1978]
Dimethylmethane
Petroleum gases, Liquefied [UN 1075]
Propyl hydride
UN 1075 [Petroleum gases, Liquefied]
UN 1978 [Propane]

74-99-7
C_3H_4
40.07
C(#C)C
Propyne
Acetylene, methyl-
Allylene
Methyl acetylene (ACGIH, OSHA)
Propine

Propyne (OSHA)

75-00-3
C_2H_5Cl
64.52
ClCC
Ethyl chloride (ACGIH, OSHA) [UN 1037]
Aethylchlorid (German)
Aethylis
Aethylis chloridum
Anodynon
Chelen
Chloorethaan (Dutch)
Chlorene
Chlorethyl
Cloretilo
Chloridum
Chloroaethan (German)
Chloroethane
Chlorure d'ethyle (French)
Chloryl
Chloryl anesthetic
Cloroetano (Italian)
Cloruro di etile (Italian)
Dublofix
Ethane, chloro
Ether chloratus
Ether hydrochloric
Ether muriatic
Etylu chlorek (Polish)
Hydrochloric ether
Kelene
Monochlorethane
Monochloroethane
Muriatic ether
Narcotile
NCI-C06224
UN 1037 [Ethyl chloride]

75-01-4
C_2H_3Cl
62.50
C(=C)Cl
Vinyl chloride (DOT,OSHA)
Chlorethene
Chlorethylene
Chloroethene
Chloroethylene
Chloroethylene (OSHA)
Chlorure de vinyle (French)
Cloruro di vinile (Italian)
Ethene, chloro-
Ethylene, chloro
Ethylene monochloride
Monochloroethene
Monochloroethylene
Monochloroethylene (DOT)
RCRA waste number U043
Trovidur
UN 1086 [Vinyl chloride, inhibited]
VC
VCM
Vinile (cloruro di) (Italian)
Vinylchlorid (German)
Vinyl chloride, Inhibited [UN 1086]
Vinyl chloride monomer
Vinyle(chlorure de) (French)
Vinyl C monomer
Winylu chlorek (Polish)

75-02-5
C_2H_3F
46.05
FC=C
Vinyl fluoride, Inhibited [UN 1860]
Ethene, fluoro-
Ethylene, fluoro- (8CI)
Fluoroethene
Fluoroethylene
Monofluoroethylene
UN 1860 [Vinyl fluoride, inhibited]

75-03-6
C_2H_5I
155.97
C(C)I
Ethane, iodo
Ethyl iodide
Ethyljodid (Czech)
Hydriodic ether
Iodoethane
Jodethan (Czech)

75-04-7
C_2H_7N
45.10
NCC
Ethylamine (ACGIH,DOT, OSHA) [UN 1036]
Aethylamine (German)
Aminoethane
1-Aminoethane
Ethanamine, (Aqueous solution)
Ethylamine, Aqueous solution with not less than 50 per cent but not more than 70 per cent ethylamine [UN 2270]
Etilamina (Italian)
Etyloamina (Polish)
Monoethylamine
UN 1036 [Ethylamine]
UN 2270 [Ethylamine, aqueous solution with not less than 50 per cent but not more than 70 per cent ethylamine]

75-05-8
C_2H_3N
41.06
N#CC
Acetonitrile (ACGIH,DOT, OSHA)
Acetonitril (German, Dutch)
Cyanomethane
Cyanure de methyl (French)
Ethanenitrile
Ethyl nitrile
Methanecarbonitrile
Methane, cyano-
Methyl cyanide [UN 1648]
Methylkyanid (Czech)
NA 1648 (DOT)
NCI-C60822
RCRA waste number U003
UN 1648 [Methyl cyanide]
USAF EK-488

75-07-0
C_2H_4O
44.06
O=CC
Acetaldehyde (ACGIH,OSHA) [UN 1089]
Acetaldehyd (German)
Acetic aldehyde
Acetylaldehyde
Aldehyde acetique (French)
Aldeide acetica (Italian)
Ethanal
Ethyl aldehyde (DOT)
NCI-C56326
Octowy aldehyd (Polish)
RCRA waste number U001
UN 1089 [Acetaldehyde]

75-08-1
C_2H_6S
62.14
SCC
Ethanethiol
Aethanethiol (German)
Aethylmercaptan (German)
Etantiolo (Italian)
Ethaanthiol (Dutch)
Ethanethiol (OSHA)
Ethyl hydrosulfide
Ethyl mercaptan (ACGIH,DOT, OSHA) [UN 2363]
Ethylmercaptaan (Dutch)
Ethylmerkaptan (Czech)
Ethyl sulfhydrate
Ethyl thioalcohol
Etilmercaptano (Italian)
LPG Ethyl mercaptan 1010
Thioethanol
Thioethyl alcohol
UN 2363 [Ethyl mercaptan]

75-09-2
CH_2Cl_2
84.93
ClCCl
Methylene chloride (ACGIH, DOT,OSHA) [UN 1593]
Aerothene MM
Chlorure de methylene (French)
DCM
Dichloromethane (DOT,OSHA)
Methane dichloride
Methane, dichloro
Methylene bichloride
Methylene dichloride
Metylenu chlorek (Polish)
Narkotil
NCI-C50102
R 30
RCRA waste number U080
Solaesthin
Solmethine
UN 1593 [Dichloromethane]

75-10-5
CH_2F_2
52.02
FCF
Methane, difluoro- (9CI)
Difluoromethane

75-11-6
CH_2I_2
267.83
C(I)I
Methane, diiodo
Diiodomethane
Dijodmethan (Czech)
Methylene diiodide
Methylene iodide
Methylenjodid (Czech)
MI-Gee

75-12-7
CH_3NO
45.05
O=CN
Formamide (ACGIH,OSHA)
Amid kyseliny mravenci (Czech)
Carbamaldehyde
Methanamide

75-13-8
CHNO
43.03
Isocyanic acid

75-15-0
CS_2
76.13
C(=S)=S
Carbon sulfide (DOT)
Carbon bisulfide (DOT)
Carbon bisulphide
Carbon disulfide (ACGIH, OSHA) [UN 1131]
Carbon disulphide
Carbone (sufure de) (French)
Carbonio (solfuro di) (Italian)
Carbon sulphide (DOT)
Dithiocarbonic anhydride
Kohlendisulfid (schwefelkohlenstoff) (German)
Koolstofdisulfide (zwavelkoolstof) (Dutch)
NCI-C04591
RCRA waste number P022
Schwefelkohlenstoff (German)
Solfuro di carbonio (Italian)
Sulphocarbonic anhydride
UN 1131 [Carbon disulfide]
Weeviltox
Wegla dwusiarczek (Polish)

75-16-1
CH_3BrMg
119.26
Magnesium, methyl-, bromide (ethyl ether solution)
Methyl magnesium bromide, In ethyl ether [UN 1928]
Methyl magnesium bromide in ethyl ether, Not over 40% concentration [UN 1928]
UN 1928 [Methyl magnesium bromide, in ethyl ether]

75-18-3
C_2H_6S
62.14
S(C)C
Dimethyl sulfide [UN 1164]
Dimethylsulfid (Czech)
Dimethyl sulphide [UN 1164]
DMS
Exact-S
Methyl sulfide [UN 1164]
Methyl sulphide [UN 1164]
Methylthiomethane
Sulfure de methyle (French)
2-Thiapropane
2-Thiopropane
UN 1164 [Dimethyl sulfide]

75-19-4
C_3H_6
42.09
C(C1)C1
Cyclopropane (DOT)
Cyclopropane, Liquefied [UN 1027]
Trimethylene
UN 1027 [Cyclopropane, liquified]

75-20-7
C_2Ca
64.10
Calcium carbide [UN 1402]
UN 1402 [Calcium carbide]

75-21-8
C_2H_4O
44.06
O(C1)C1
Ethylene oxide (ACGIH, OSHA) [UN 1040]
Aethylenoxid (German)
Amprolene
Anprolene
Anproline
Dihydrooxirene
Dimethylene oxide
ENT 26,263
E.O.
1,2-Epoxyaethan (German)
Epoxyethane
1,2-Epoxyethane
Ethene oxide
Ethox
Ethyleenoxide (Dutch)
Ethylene oxide, Containing not more than 0.2% nitrogen [UN 1040]
Ethylene (oxyde d') (French)
Etilene (ossido di) (Italian)
ETO
Etylenu tlenek (Polish)
Fema No. 2433
Merpol
NCI-C50088
Oxacyclopropane
Oxane
Oxidoethane
α,β-Oxidoethane
Oxiraan (Dutch)
Oxiran
Oxirane
Oxirene, dihydro-
Oxyfume
Oxyfume 12
RCRA waste number U115
Sterilizing gas ethylene oxide 100%
T-Gas
UN 1040 [Ethylene oxide, pure or with nitrogen at 50 degrees C]

75-22-9
$C_3H_9N.BH_3$
72.97
Methanamine, N,N-dimethyl-, Compd. with borane (1:1)
Borane, Compd. with N,N-dimethylmethanamine (1:1)
Borane, Compd. with trimethylamine (1:1)
Boron, (N,N-dimethylmethanamine)trihydro-, (t-4)- (9CI)
TMAB
Trimethylamine borane
Trimethylamine, Compd. with borane (1:1)

75-23-0
$C_2H_7BF_3N$
112.91
Ethylamine Compd. with boron fluoride (1:1)
Borontrifluoride monoethylamine
BTFMEA

75-24-1
C_3H_9Al
72.10
Aluminum, trimethyl
Trimethylalane
Trimethylaluminium (DOT)
Trimethylaluminum
UN 1103 (DOT)

75-25-2
$CHBr_3$
252.75
BrC(Br)Br
Bromoform (ACGIH,DOT, OSHA) [UN 2515]
Bromoforme (French)
Bromoformio (Italian)
Methane, tribromo
Methenyl tribromide
NCI-C55130
RCRA waste number U225
Tribrommethaan (Dutch)
Tribrommethan (German)

Tribromometan (Italian)
Tribromomethane
UN 2515 [Bromoform]

75-26-3
C_3H_7Br
123.01
BrC(C)C
2-Bromopropane [UN 2344]
Isopropyl bromide
Propane, 2-bromo
UN 2344 [2-Bromopropane]

75-27-4
$CHBrCl_2$
163.83
BrC(Cl)Cl
Methane, bromodichloro
BDCM
Bromodichloromethane
Dichlorobromomethane
Dichloromonobromomethane
Monobromodichloromethane
NCI-C55243

75-28-5
C_4H_{10}
58.14
C(C)(C)C
Propane, 2-methyl
Isobutane [UN 1969]
Liquefied Petroleum Gas [UN 1075]
UN 1075 [Liquefied petroleum gas]
UN 1969 [Isobutane or isobutane mixtures]

75-29-6
C_3H_7Cl
78.55
C(Cl)(C)C
Propane, 2-chloro
2-Chloropropane [UN 2356]
Isopropylchloride
UN 2356 [2-Chloropropane]

75-30-9
C_3H_7I
170.00
C(C)(C)I
Propane, 2-iodo
2-Iodopropane
Isopropyl iodide
2-Jodpropan (Czech)
i-Propyl iodide

75-31-0
C_3H_9N
59.13
NC(C)C
Isopropylamine (ACGIH, DOT,OSHA) [UN 1221]
2-Amino-propaan (Dutch)
2-Aminopropan (German)
2-Amino-propano (Italian)
2-Aminopropane
Isopropilamina (Italian)
1-Methylethylamine
Monoisopropylamine
2-Propanamine
Propane, 2-amino-
sec-Propylamine
2-Propylamine
UN 1221 [Isopropylamine]

75-33-2
C_3H_8S
76.17
SC(C)C
2-Propanethiol [UN 2402]
Isopropanethiol
Isopropyl mercaptan [UN 2402]
Isopropylthiol
2-Mercaptopropane
1-Methylethanethiol
2-Propylmercaptan
UN 2402 [Propanethiols]

75-34-3
$C_2H_4Cl_2$
98.96
C(Cl)(Cl)C
1,1-Dichloroethane (ACGIH, DOT,OSHA) [UN 2362]
Aethylidenchlorid (German)
Chlorinated hydrochloric ether
Chlorure d'ethylidene (French)
Cloruro di etilidene (Italian)
1,1-Dichloorethaan (Dutch)
1,1-Dichloraethan (German)
1,1-Dichlorethane
1,1-Dicloroetano (Italian)
Ethane, 1,1-dichloro
Ethylidene chloride
Ethylidene chloride (OSHA)
Ethylidene dichloride
NCI-C04535
RCRA waste number U076
UN 2362 [1,1-Dichloroethane]

75-35-4
$C_2H_2Cl_2$
96.94
C(=C)(Cl)Cl
Vinylidene chloride (ACGIH, OSHA)
Chlorure de vinylidene (French)
1,1-DCE
1,1-Dichloroethene (9CI)
1,1-Dichloroethylene
Ethene, 1,1-dichloro-
Ethylene, 1,1-dichloro- (8CI)
NCI-C54262
RCRA waste number U078
UN 1303 [Vinylidene chloride, inhibited]
Sconatex
VDC
Vinylidene chloride (II)
Vinylidene chloride, Inhibited [UN 1303]
Vinylidene dichloride
Vinylidine chloride

75-36-5
C_2H_3ClO
78.50
O=C(Cl)C
Acetyl chloride [UN 1717]
Acetic acid, chloride
Acetic chloride
Ethanoyl chloride
RCRA waste number U006
UN 1717 [Acetyl chloride]

75-37-6
$C_2H_4F_2$
66.06
FC(F)C
Ethane, 1,1-difluoro
Algofrene Type 67
Difluoroethane
1,1-Difluoroethane (DOT)
Ethylene fluoride
Ethylidene difluoride
Ethylidene fluoride
FC 152a
Freon 152
Genetron 100
Genetron 152a
Halocarbon 152A
UN 1030 (DOT)

75-38-7
$C_2H_2F_2$
64.04
FC(F)=C
Ethylene, 1,1-difluoro
1,1-Difluoroethene
1,1-Difluoroethylene [UN 1959]
Ethene, 1,1-difluoro-
Halocarbon 1132A
NCI-C60208
UN 1959 [1,1-Difluoroethylene]
VDF
Vinylidene difluoride
Vinylidene fluoride

75-39-8
$C_2H_4O.H_3N$
MW:
61.08
OC(N)C
Acetaldehyde ammonia
Acetaldehidato amonico (Spanish)
Acetaldehyde, amine salt
AI3-52423
Aldehydate d'ammoniaque (French)
Aldehyde ammonia
1-Aminoethanol
1-Amino-ethanol
α-Aminoethyl alcohol
Ethanol, 1-amino-
UN 1841 [Acetaldehyde ammonia]

75-43-4
$CHCl_2F$
102.92

FC(Cl)Cl
Methane, dichlorofluoro
Algofrene Type 5
Arcton 7
Dichlorofluoromethane (ACGIH)
Dichloromonofluoromethane (OSHA) [UN 1029]
Dwuchlorofluorometan (Polish)
Fluorodichloromethane
Freon 21
Genetron 21
UN 1029 [Dichloromonofluoromethane]

75-44-5
CCl_2O
98.91
O=C(Cl)Cl
Phosgene (ACGIH,OSHA) [UN 1076]
Carbon dichloride oxide
Carbone (oxychlorure de) (French)
Carbonic chloride
Carbonio (ossicloruro di) (Italian)
Carbon oxychloride
Carbonylchlorid (German)
Carbonyl chloride (DOT, OSHA)
Carbonyl dichloride
CG
Chloroformyl chloride
Fosgeen (Dutch)
Fosgen (Polish)
Fosgene (Italian)
Koolstofoxychloride (Dutch)
NCI-C60219
Phosgen (German)
RCRA waste number P095
UN 1076 [Phosgene]

75-45-6
$CHClF_2$
86.47
FC(F)Cl
Chlorodifluoromethane (ACGIH,OSHA) [UN 1018]
Algeon 22
Algofrene 22
Algofrene Type 6
Arcton 4
Arcton 22
CFC 22
Daiflon 22
Difluorochloromethane
Difluoromonochloromethane
Dymel 22
Electro-CF 22
Eskimon 22
F 22
FC 22
Flugene 22
Fluorocarbon-22
Forane 22
Freon
Freon 22
Frigen
Frigen 22
Genetron 22
Haltron 22
Isceon 22
Isotron 22
Khaladon 22
Methane, chlorodifluoro
Monochlorodifluormethane (DOT)
Propellant 22
R 22 (DOT)
Refrigerant 22
Ucon 22
Ucon 22/halocarbon 22
UN 1018 [Chlorodifluoromethane]

75-46-7
CHF_3
70.02
FC(F)F
Trifluoromethane [UN 1984]
Arcton
Carbon trifluoride
Fluoroform
Fluoryl
Freon 23
Freon F-23
Genetron-23
Halocarbon 23
Methane, trifluoro
Methyl trifluoride
R 23
UN 1984 [Trifluoromethane]

75-47-8
CHI_3
393.72
C(I)(I)I
Methane, triiodo
Iodoform (ACGIH,OSHA)
Jodoform (Czech)
NCI-C04568
Triiodomethane
Trijodmethane (Czech)

75-50-3
C_3H_9N
59.13
N(C)(C)C
Methanamine, N,N-dimethyl
TMA
Trimethylamine (ACGIH, OSHA)
Trimethylamine, Anhydrous [UN 1083]
Trimethylamine, Aqueous solution (DOT)
Trimethylamine, Aqueous solutions containing not more than 30% of trimethylamine (DOT)
Trimethylamine, Aqueous solutions not more than 50 per cent trimethylamine by mass [UN 1297]
UN 1083 [Trimethylamine, anhydrous]
UN 1297 [Trimethylamine, aqueous solutions not more than 50 per cent trimethylamine by mass]

75-52-5
CH_3NO_2
61.05
O=N(=O)C
Nitromethane (ACGIH, OSHA) [UN 1261]
Methane, nitro
Nitrocarbol
Nitrometan (Polish)
UN 1261 [Nitromethane]

75-54-7
CH_4Cl_2Si
115.04
C[SiH](Cl)Cl
Silane, dichloromethyl
Dichloromethylsilane
Methyldichlorosilane
Methyl dichlorosilane [UN 1242]
Methyl-dichlorsilan (Czech)
UN 1242 [Methyldichlorosilane]

75-55-8
C_3H_7N
57.11
N(C1C)C1
Aziridine, 2-methyl
2-Methylazacyclopropane
2-Methylaziridine
Methylethylenimine
2-Methylethylenimine
Propylene imine (ACGIH, OSHA)
Propylene imine, Inhibited [UN 1921]
1,2-Propyleneimine
Propylenimine
1,2-Propylenimine
RCRA waste number P067
UN 1921 [Propyleneimine, inhibited]

75-56-9
C_3H_6O
58.09
O(C1C)C1
Propane, 1,2-epoxy
AD 6
AD 6 (Suspending agent)
Epoxypropane
1,2-Epoxypropane
2,3-Epoxypropane
1,2-Epoxypropane (OSHA)
Ethylene oxide, methyl-
Methyl ethylene oxide
Methyl oxirane
NCI-C50099
Oxirane, methyl-
Oxyde de propylene (French)
Propane, epoxy-
Propene oxide
Propylene epoxide
Propylene oxide (ACGIH,

OSHA) [UN 1280]
1,2-Propylene oxide
UN 1280 [Propylene oxide]

75-57-0
$C_4H_{12}N.Cl$
109.62
Ammonium, tetramethyl-, chloride
Methanaminium, N,N,N-trimethyl-, chloride (9CI)
Tetramethylammonium chloride
USAF AN-8

75-59-2
$C_4H_{12}N.HO$
91.18
[OH-].C[N+](C)(C)C
Methanaminium, N,N,N-trimethyl-, hydroxide
Ammonium, tetramethyl-, hydroxide
Hydroxyde de tetramethylammonium (French)
TM
Tetramethylammonium hydroxide [UN 1835]
Tetramethyl ammonium hydroxide, Liquid (DOT)
UN 1835 [Tetramethylammonium hydroxide]

75-60-5
$C_2H_7AsO_2$
138.01
C[As](C)(O)=O
Cacodylic acid [UN 1572]
Acide cacodylique (French)
Acide dimethylarsinique (French)
Agent Blue
Ansar
Ansar 138
Arsan
Arsine oxide, dimethylhydroxy
Arsinic acid, dimethyl- (9CI)
Bolls-Eye
Chexmate
Dilic
Dimethylarsenic acid
Dimethylarsinic acid
DMAA
Erase
Hydroxydimethylarsine oxide
Kyselina kakodylova (Czech)
Phytar
Phytar 138
Phytar 560
Rad-E-Cate 25
RCRA waste number U136
Salvo
Silvisar 510
UN 1572 [Cacodylic acid]

75-61-6
CBr_2F_2
209.83
FC(F)(Br)Br
Methane, dibromodifluoro
Dibromodifluoromethane [UN 1941]
Difluorodibromomethane (ACGIH,OSHA)
Freon 12-B2
Halon 1202
UN 1941 [Dibromodifluoromethane]

75-62-7
$CBrCl_3$
198.27
BrC(Cl)(Cl)Cl
Methane, bromotrichloro
Bromotrichloromethane
Carbon bromotrichloride
Carbon trichlorobromide
Monobromotrichloromethane
Trichlorobromomethane
Trichloromethyl bromide

75-63-8
$CBrF_3$
148.92
FC(F)(F)Br
Methane, bromotrifluoro
Bromofluoroform
Bromotrifluoromethane [UN 1009]
F-13B1
Freon 13B1
Halon 1301
Trifluorobromomethane (ACGIH,DOT,OSHA)
Trifluoromonobromomethane
UN 1009 [Bromotrifluoromethane]

75-64-9
$C_4H_{11}N$
73.16
NC(C)(C)C
tert-Butylamine
2-Aminoisobutane
2-Amino-2-methylpropane
Butylamine, tertiary
1,1-Dimethylethylamine
Trimethylaminomethane

75-65-0
$C_4H_{10}O$
74.14
OC(C)(C)C
tert-Butyl alcohol
Alcool butylique tertiaire (French)
t-Butanol
tert-Butanol [UN 1120]
Butanol tertiaire (French)
tert-Butyl alcohol (ACGIH, OSHA) [UN 1120]
t-Butyl hydroxide
1,1-Dimethylethanol
Methanol, trimethyl-
2-Methyl-2-propanol
NCI-C55367
2-Propanol, 2-methyl-
Trimethylcarbinol
UN 1120 [Butanols]

75-66-1
$C_4H_{10}S$
90.20
SC(C)(C)C
2-Propanethiol, 2-methyl
tert-Butanethiol
tert-Butyl mercaptan
2-Methyl-2-propanethiol

75-68-3
$C_2H_3ClF_2$
100.50
FC(F)(Cl)C
Ethane, 1-chloro-1,1-difluoro
CFC 142b
Chlorodifluoroethane (DOT)
1-Chloro-1,1-difluoroethane (DOT)
Chloroethylidene fluoride
α-Chloroethylidene fluoride
1,1-Difluoro-1-chloroethane
Difluoromonochloroethane (DOT)
FC142b
Fluorocarbon FC142b
Freon 142
Freon 142b
Genetron 101
Genetron 142b
Gentron 142b
Hydrochlorofluorocarbon 142b
UN 2517 (DOT)

75-69-4
CCl_3F
137.36
FC(Cl)(Cl)Cl
Methane, trichlorofluoro
Algofrene Type 1
Arcton 9
Electro-CF 11
Eskimon 11
F 11
FC 11
Fluorocarbon No. 11
Fluorotrichloromethane
Fluorotrichloromethane (OSHA)
Fluorotrojchlorometan (Polish)
Freon 11
Freon 11A
Freon 11b
Freon HE
Freon MF
Frigen 11
Genetron 11
Halocarbon 11
Isceon 131
Isotron 11
Ledon 11
Methane, fluorotrichloro-
Monofluorotrichloromethane
NCI-C04637
Trichlorofluoromethane (ACGIH,OSHA)
Trichloromonofluoromethane
Ucon Flurocarbon 11

Ucon Refrigerant 11

75-70-7
$CHCl_3S$
151.44
Trichloromethyl mercaptan
Methanethiol, trichloro-
Perchlorinemethylmercaptan
Perchloromethanethiol
Perchloromethyl mercaptan
Trichloromethanethiol

75-71-8
CCl_2F_2
120.91
FC(F)(Cl)Cl
Methane, dichlorodifluoro
Algofrene Type 2
Arcton 6
Arcton 12
Dichlorodifluoromethane (ACGIH,OSHA) [UN 1028]
Difluorodichloromethane
Dwuchlorodwufluorometan (Polish)
Electro-CF 12
F 12
FC 12
Fluorocarbon-12
Freon 12
Eskimon 12
Freon F-12
Frigen 12
Genetron 12
Halon
Isceon 122
Isotron 12
Kaiser Chemicals 12
Ledon 12
Propellant 12
RCRA waste number U075
R 12 [UN 1028]
Refrigerant 12
Ucon 12
Ucon 12/halocarbon 12
UN 1028 [Dichlorodifluoromethane]

75-72-9
$CClF_3$
104.46
FC(F)(F)Cl
Methane, chlorotrifluoro
Arcton 3
Chlorotrifluoromethane [UN 1022]
F 13
Freon 13
Genetron 13
Halocarbon 13/Ucon 13
Monochlorotrifluoromethane (DOT)
R-13 [UN 1022]
Trifluorochloromethane (DOT)
Trifluoromethyl chloride
Trifluoromonochlorocarbon
UN 1022 [Chlorotrifluoromethane]

75-73-0
CF_4
88.01
FC(F)(F)F
Carbon-tetrafluoride
Arcton 0
Carbon fluoride
F 14
FC 14
Freon 14
Halocarbon 14
Halon 14
Methane, tetrafluoro-
Perfluoromethane
R 14
R 14 (Refrigerant)
Tetrafluoromethane [UN 1982]
UN 1982 [Tetrafluoromethane]

75-74-1
$C_4H_{12}Pb$
267.35
C[Pb](C)(C)C
Plumbane, tetramethyl
Lead, tetramethyl-
Piombo tetra-metile (Italian)
Tetramethyl lead (ACGIH, OSHA)
Tetramethylolovo (Czech)
Tetramethylplumbane
TML

75-75-2
CH_4O_3S
96.11
O=S(=O)(O)C
Methanesulfonic-acid
Kyselina methansulfonova (Czech)

75-76-3
$C_4H_{12}Si$
88.25
CSi(C)(C)C
Silane, tetramethyl
Tetramethylsilane [UN 2749]
UN 2749 [Tetramethylsilane]

75-77-4
C_3H_9ClSi
108.66
Silane, chlorotrimethyl
Chlorotrimethylsilane
Silane, trimethylchloro-
Silicane, chlorotrimethyl-
TL 1163
Trimethyl chlorosilane
Trimethylchlorosilane [UN 1298]
UN 1298 [Trimethylchloro silane]

75-78-5
$C_2H_6Cl_2Si$
129.07
CSi(C)(Cl)Cl
Silane, dichlorodimethyl
Dichlorodimethylsilane [UN 1162]
Dimethyl-dichlorsilan (Czech)
UN 1162 [Dimethyldichlorosilane]

75-79-6
CH_3Cl_3Si
149.48
CSi(Cl)(Cl)Cl
Silane, methyltrichloro
Methyltrichlorosilane [UN 1250]
Methyl-trichlorsilan (Czech)
Silane, trichloromethyl-
Trichlor-methylsilan (Czech)
UN 1250 [Methyltrichlorosilane]

75-81-0
$C_2H_2Br_2Cl_2$
256.75
BrCC(Br)(Cl)Cl
1,2-Dibromo-1,1-dichloroethane
AI3-14678
1,2-Dibromo-2,2-dichloroethane
Ethane, 1,2-dibromo-1,1-dichloro- (9CI)
NSC-6199

75-83-2
C_6H_{14}
86.20
C(CC)(C)(C)C
Butane, 2,2-dimethyl
Neohexane [UN 1208]
UN 1208 [Hexanes]

75-84-3
$C_5H_{12}O$
88.15
OCC(C)(C)C
Neopentyl alcohol
AI3-20879
tert-Butyl carbinol
tert-Butylcarbinol
2,2-Dimethyl-1-propanol
2,2-Dimethylpropyl alcohol
Neoamyl alcohol
Neopentanol
1-Propanol, 2,2-dimethyl-

75-85-4
$C_5H_{12}O$
88.17
OC(CC)(C)C
tert-Pentyl alcohol
tert-Amyl alcohol [UN 1105]
Amylene hydrate
2-Butanol, 2-methyl-
Dimethylethylcarbinol
Ethyl dimethyl carbinol
2-Methyl butanol-2
2-Methyl-2-butanol
3-Methyl-butanol-(3) (German)

3-Methylbutan-3-ol
tert-Pentanol
UN 1105 [Amyl alcohols]

75-86-5
C_4H_7NO
85.12
N#CC(O)(C)C
Lactonitrile, 2-methyl
Acetoncianidrina (Italian)
Acetoncianhidrinei (Romanian)
Acetoncyaanhydrine (Dutch)
Acetoncyanhydrin (German)
Acetonecyanhydrine (French)
Acetone cyanohydrin (DOT)
Acetone cyanohydrin, Stabilized [UN 1541]
Acetonkyanhydrin (Czech)
Cyanhydrine d'acetone (French)
α-Hydroxyisobutyronitrile
2-Hydroxy-2-methylpropionitrile
2-Methyllactonitrile
Propanenitrile, 2-hydroxy-2-methyl-
RCRA waste number P069
UN 1541 [Acetone cyanohydrin, stabilized]
USAF RH-8

75-87-6
C_2HCl_3O
147.38
O=CC(Cl)(Cl)Cl
Chloral
Acetaldehyde, trichloro- (9CI)
Anhydrous chloral
Chloral, Anhydrous, Inhibited [UN 2075]
Cloralio (Italian)
Grasex
RCRA waste number U034
Trichloroacetaldehyde
2,2,2-Trichloroacetaldehyde
Trichloroethanal
UN 2075 [Chloral, anhydrous, inhibited]

75-88-7
$C_2H_2ClF_3$
118.49
FC(F)(F)CCl
Ethane, 2-chloro-1,1,1-trifluoro
CFC 133a
1-Chloro-2,2,2-trifluoroethane
2-Chloro-1,1,1-trifluoroethane
FC 133a
Freon 133a
Genetron 133a
R 133a
1,1,1-Trifluoro-2-chloroethane
2,2,2-Trifluorochloroethane
1,1,1-Trifluoroethyl chloride

75-89-8
$C_2H_3F_3O$
100.05
FC(F)(F)CO
Ethanol, 2,2,2-trifluoro
TFE
2,2,2-Trifluoroethanol

75-91-2
$C_4H_{10}O_2$
90.14
O(O)C(C)(C)C
tert-Butyl hydroperoxide
tert-Butyl hydroperoxide, Maximum concentration 72% with water (DOT)
tert-Butyl hydroperoxide, More than 72% but not more than 90% water (DOT)
tert-Butyl hydroperoxide, More than 90% water (DOT)
terc. Butylhydroperoxid (Czech)
Cadox TBH
1,1-Dimethylethyl hydroperoxide
Hydroperoxide, tert-butyl
Hydroperoxide, 1,1-dimethylethyl- (9CI)
Hydroperoxyde de butyle tertiaire (French)
Perbutyl H
TBHP-70
Trigonox A-75 (Czech)
Trigonox A-W70
UN 2093 (DOT)
UN 2094 (DOT)

75-92-3
CH_4O_4S
112.10
Formaldehyde bisulfite
Formaldehyde sodium bisulfite
Hydroxymethanesulfonate, sodium salt
Sodium hydroxymethanesulfonate

75-93-4
CH_4O_4S
112.10
O=S(=O)(OC)O
Methyl sulfate
Monomethyl sulfate

75-94-5
$C_2H_3Cl_3Si$
161.49
ClSi(Cl)(Cl)C=C
Silane, trichloroethenyl
Silane, trichlorovinyl-
Silane, vinyl trichloro A-150
Trichloro(vinyl)silane
Trichlorovinyl silicane
UN 1305 [Vinyltrichlorosilane]
Vinylsilicon trichloride
Vinyltrichlorosilane
Vinyl trichlorosilane [UN 1305]
Vinyl trichlorosilane, Inhibited (DOT)
Union Carbide A-150

75-96-7
$C_2HBr_3O_2$
296.74
O=C(O)C(Br)(Br)Br
Tribromoacetic acid
Acetic acid, tribromo- (9CI)

75-97-8
$C_6H_{12}O$
100.18
O=C(C(C)(C)C)C
2-Butanone, 3,3-dimethyl
t-Butyl methyl ketone
3,3-Dimethyl-2-butanone
Ketone, t-butyl methyl
Methyl t-butyl ketone
Methyltert-butyl ketone
Pinacolin
Pinacoline
Pinakolin
Pinacolone
Pinakolin (German)

75-98-9
$C_5H_{10}O_2$
102.15
O=C(O)C(C)(C)C
Pivalic-acid
Acetic acid, trimethyl-
2,2-Dimethylpropanoic acid
α,α-Dimethylpropionic acid
2,2-Dimethylpropionic acid
Kyselina 2,2-dimethylpropionova (Czech)
Kyselina pivalova (Czech)
Neopentanoic acid
tert-Pentanoic acid
Propanoic acid
Propionic acid, 2,2-dimethyl-
Trimethylacetic acid

75-99-0
$C_3H_4Cl_2O_2$
142.97
O=C(O)C(Cl)(Cl)C
Propionic acid, 2,2-dichloro
Basfapon
Basfapon B
Basfapon/basfapon N
BH Dalapon
Basinex
Crisapon
Dalapon
Dalapon 85
Ded-Weed
Devipon
2,2-Dichloropropionic acid (ACGIH,DOT,OSHA)
α-Dichloropropionic acid
α,α-Dichloropropionic acid
Dowpon
Dowpon M
Gramevin
Kenapon
Kyselina 2,2-dichlorpropionova (Czech)
Liropon
Proprop

Radapon
Revenge
Unipon

76-01-7
C_2HCl_5
202.28
C(C(Cl)Cl)(Cl)(Cl)Cl
Ethane, pentachloro
Ethane pentachloride
NCI-C53894
Pentachloorethaan (Dutch)
Pentachloraethan (German)
Pentachlorethane (French)
Pentachloroethane [UN 1669]
Pentacloroetano (Italian)
Pentalin
RCRA waste number U184
UN 1669 [Pentachloroethane]

76-02-8
C_2Cl_4O
181.82
O=C(C(Cl)(Cl)Cl)Cl
Acetyl chloride, trichloro
Superpalite
Trichloroacetic acid chloride
Trichloroacetochloride
Trichloroacetyl chloride [UN 2442]
UN 2442 [Trichloroacetyl chloride]

76-03-9
$C_2HCl_3O_2$
163.38
O=C(O)C(Cl)(Cl)Cl
Trichloroacetic acid (ACGIH, OSHA)
Acetic acid, trichloro
Aceto-caustin
Acide trichloracetique (French)
Acido tricloroacetico (Italian)
Amchem Grass Killer
Konesta
Kyselina trichloroctova (Czech)
TCA
Trichloorazijnzuur (Dutch)
Trichloracetic acid
Trichloressigsaeure (German)
Trichloroacetic acid, Solid [UN 1839]
Trichloroacetic acid, Solution [UN 2564]
Trichloroethanoic acid
UN 1839 [Trichloroacetic acid]
UN 2564 [Trichloroacetic acid, solution]
Varitox

76-05-1
$C_2HF_3O_2$
114.03
O=C(O)C(F)(F)F
Acetic acid, trifluoro
Kyselina trifluoroctova (Czech)
Perfluoroacetic acid
Trifluoracetic acid
Trifluoroacetic acid [UN 2699]
Trifluoroethanoic acid
UN 2699 [Trifluoroacetic acid]

76-06-2
CCl_3NO_2
164.37
O=N(=O)C(Cl)(Cl)Cl
Chloropicrin (ACGIH,OSHA) [UN 1580]
Acquinite
Aquinite
Chloorpikrine (Dutch)
Chloroform, nitro-
Chlor-o-Pic
Chloropicrin, Absorbed (DOT)
Chloropicrin, Liquid (DOT)
Chloropicrin mixture, Flammable [UN 1582]
Chloropicrin mixtures, N.O.S. [UN 1583]
Chloropicrine (French)
Chlorpikrin (German)
Cloropicrina (Italian)
Dojyopicrin
Dolochlor
G 25
Klop
Larvacide
Methane, trichloronitro-, (Flammable mixture) [UN 1583]
Microlysin
NA 2929 (DOT)
NA 1583 (DOT)
NCI-C00533
Nitrochloroform
Nitrotrichloromethane
Nitrotrichloromethane (OSHA)
Pic-Clor
Picfume
Picride
Profume A
PS
S 1
Trichloornitromethaan (Dutch)
Trichlornitromethan (German)
Trichloronitromethane
Tri-Clor
Tricloro-nitro-metano (Italian)
UN 1580 [Chloropicrin]
UN 1582 [Chloropicrin mixture, flammable]
UN 1583 [Chloropicrin mixtures, N.O.S.]

76-08-4
$C_4H_7Br_3O$
310.81
OC(C(Br)(Br)Br)(C)C
1,1,1-Tribromo-2-methyl-2-propanol
Acetonbromoform
Acetone-bromoform
AI3-14655
Brometon
Brometone
Bromobutanol
NSC-2865
2-Propanol, 1,1,1-tribromo-2-methyl- (9CI)
Tribromo-tert-butyl alcohol
2-Tribromomethyl-2-propanol

76-09-5
$C_6H_{14}O_2$
118.20
OC(C(O)(C)C)(C)C
2,3-Butanediol, 2,3-dimethyl
2,3-Dimethyl-2,3-butanediol
Pinacol
Tetramethylethylene glycol

76-11-9
$C_2Cl_4F_2$
203.82
FC(F)(C(Cl)(Cl)Cl)Cl
Ethane, 2,2-difluoro-1,1,1,2-tetrachloro
Halocarbon 112a
Refrigerant 112a
1,1,1,2-Tetrachloro-2,2-difluoroethane (ACGIH,OSHA)

76-12-0
$C_2Cl_4F_2$
203.82
FC(C(F)(Cl)Cl)(Cl)Cl
Ethane, 1,2-difluoro-1,1,2,2-tetrachloro
1,2-Difluoro-1,1,2,2-tetrachloroethane
Ethane, 1,1,2,2-tetrachloro-1,2-difluoro-
F-112
FC 112
Freon 112
Freon R 112
Genetron 112
Halocarbon 112
Refrigerant 112
1,1,2,2-Tetrachloro-1,2-difluoroethane (ACGIH,OSHA)
UCON 112

76-13-1
$C_2Cl_3F_3$
187.37
FC(F)(C(F)(Cl)Cl)Cl
Ethane, 1,1,2-trichloro-1,2,2-trifluoro
Arcton 63
Arklone P
Asahifron 113
Daiflon S 3
F 113
FC 113
Fluorocarbon 113
Forane
Freon 113
Freon F113
Freon TF
Freon 113TR-T
Frigen 113
Frigen 113a
Frigen 113TR
Frigen 113TR-N

Frigen 113 TR-T
Genetron 113
Halocarbon 113
Isceon 113
Kaiser Chemicals 11
Khladon 113
Ledon 113
R 113
Refrigerant 113
Refrigerant R 113
R 113 (Halocarbon)
Trichlorotrifluoroethane
1,1,2-Trichlorotrifluoroethane
1,2,2-Trichlorotrifluoroethane
1,1,2-Trichloro-1,2,2-trifluoroethane (ACGIH,OSHA)
1,1,2-Trifluorotrichloroethane
1,1,2-Trifluoro-1,2,2-trichloroethane
Ucon 113
Ucon Fluorocarbon 113
Ucon 113/Halocarbon 113

76-14-2
$C_2Cl_2F_4$
170.92
FC(F)(C(F)(F)Cl)Cl
Ethane, 1,2-dichloro-1,1,2,2-tetrafluoro
Arcton 33
Arcton 114
Cryofluoran
Cryofluorane
sym-Dichlorotetrafluoroethane
1,2-Dichloro-1,1,2,2-tetrafluoroethane
Dichlorotetrafluoroethane (ACGIH,OSHA)
Ethane, 1,2-dichlorotetrafluoro-
F 114
FC 114
Fluorane 114
Fluorocarbon 114
Freon 114
Frigen 114
Frigiderm
Genetron 114
Genetron 316
Halocarbon 114
Ledon 114
Propellant 114
R 114
1,1,2,2-Tetrafluoro-1,2-dichloroethane
Ucon 114

76-15-3
C_2ClF_5
154.47
FC(F)(F)C(F)(F)Cl
Ethane, chloropentafluoro
Chloropentafluoroethane (ACGIH,DOT,OSHA) [UN 1020]
F-115
Fluorocarbon-115
Freon 115
Genetron 115
Halocarbon 115
Monochloropentafluoroethane [UN 1020]
UN 1020 [Chloropentafluoroethane]

76-16-4
C_2F_6
138.02
FC(F)(F)C(F)(F)F
Ethane, hexafluoro
F-116
Freon 116
Hexafluoroethane (DOT)
Perfluoroethane
UN 2193 (DOT)

76-19-7
C_3F_8
188.03
FC(F)(F)C(F)(F)C(F)(F)F
Propane, octafluoro
Freon 218
Genetron 218
Octafluoropropane [UN 2424]
Perfluoropropane
UN 2424 [Octafluoropropane]

76-20-0
$C_8H_{18}O_4S_2$
242.38
CCC(C)(S(=O)(=O)CC)S(=O)(=O)CC
Butane, 2,2-bis(ethylsulfonyl)
2,2-Bis(ethylsulfonyl)butane
Diethylsulfonmethylethylmethane
Ethylsulfonal
Methylsulfonal
Methylsulphonal
Sulfonethylmethane
Tional
Trional

76-22-2
$C_{10}H_{16}O$
152.26
O=C(C(C(C1C2)(C)C)(C2)C)C1
Camphor
Bicyclo(2.2.1)heptan-2-one, 1,7,7-trimethyl-
Bornane, 2-oxo-
2-Bornanone
2-Camphanone
Camphor, Synthetic (ACGIH, OSHA) [UN 2717]
Camphor, Natural [UN 1130]
Camphor oil [UN 1130]
Formosa Camphor
Gum Camphor
Huile de camphre (French)
Japan Camphor
2-Kamfanon (Czech)
Kampfer (German)
2-Keto-1,7,7-trimethylnorcamphane
Laurel Camphor
Matricaria Camphor
Norcamphor, 1,7,7-trimethyl-
1,7,7-Trimethylbicyclo(2.2.1)-2-heptanone
UN 1130 [Camphor oil]
UN 2717 [Camphor, synthetic]

76-24-4
$C_8H_6N_4O_8$
286.13
O=C(NC(=O)C(O)(C1=O)C(O)(C(=O)NC(=O)N2)C2=O)N1
Alloxantin
AI3-23206
Alloxantin, dihydrate
5,5'-Bibarbituric acid, 5,5'-dihydroxy- (8CI)
(5,5'-Bipyrimidine)-2,2',4,4',6,6'(1H,1'H,3H,3'H,5H,5'H)-hexone, 5,5'-dihydroxy- (9CI)
5,5'-Dihydroxy-5,5'-bibarbituric acid
NSC-7634
Uroxin
Uroxine

76-25-5
$C_{24}H_{31}FO_6$
434.55
O=C(C(OC(O1)(C)C)(C(C(C(C(F)(C(C(=CC(=O)C=2)C3)(C2)C)C4O)C3)C5)(C4)C)C15)CO
Pregna-1,4-diene-3,20-dione, 9-fluoro-11-β,16-α,17,21-tetrahydroxy-, cyclic 16, 17-acetal with acetone
Acetospan
Aristocort acetonide
Aristoderm
Aristogel
9-α-Fluoro-11-β,21-dihydroxy-16-α-isopropylidenedioxy-1,4-pregnadiene,3,20-dione
9-α-Fluoro-16-hydroxyprednisolone acetonide
9-α-Fluoro-16-α-hydroxyprednisolone 16-α,17-α-acetonide
9-α-Fluoro-16-α-17-α-isopropyledenedioxyprednisolone
9-α-Fluoro-16-α-17-α-isopropylidenedioxy-δ-1-hydrocortisone
Flutone
Kenacort-A
Kenalog
Tramacin
Triamcincolone acetonide
Triamcinolone acetonide
Triamcinolone 16,17-acetonide
Triamsinolone acetonide
Vetalog

76-30-2
$C_4H_6O_8$
182.09
Butanedioic acid, tetra hydroxy- (9CI)
Dihydroxytartaric acid
NSC-4647
Succinic acid, tetrahydroxy- (8CI)

76-37-9
$C_3H_4F_4O$
132.07
FC(F)C(F)(F)CO
1-Propanol, 2,2,3,3-tetrafluoro
2,2,3,3-Tetrafluoropropanol
2,2,3,3-Tetrafluoro-1-propanol
Tetrafluoropropyl alcohol

76-38-0
$C_3H_4Cl_2F_2O$
164.97
COC(F)(F)C(Cl)Cl
Ether, 2,2-dichloro-1,1-difluoroethyl methyl
Analgizer
Anecotan
2,2-Dichloro-1,1-difluoroethyl methyl ether
Ethane, 2,2-dichloro-1,1-difluoro-1-methoxy-
Ingalan
Ingalan (Russian)
Inhalan
Methoflurane
Methoxane
Methoxyfluoran
Methoxyfluorane
Methoxyflurane
Metofane
Metoxfluran
Metoxifluran
MOF
NSC-110432
Penthrane
Pentran
Pentrane

76-39-1
$C_4H_9NO_3$
119.14
O=N(=O)C(CO)(C)C
1-Propanol, 2-methyl-2-nitro
2-Nitro-2-methyl-1-propanol

76-40-4
$C_4H_7Cl_3O_2$
193.46
CC(Cl)C(Cl)(Cl)C(O)O
1,1-Butanediol, 2,2,3-trichloro- (8CI,9CI)
Butylchloral hydrate

76-41-5
$C_{17}H_{19}NO_4$
301.34
Oxymorphone
DEA No. 9652
Dihydrohydroxymorphinone
Dihydro-14-hydroxymorphinone
7,8-Dihydro-14-hydroxymorphinone
3,14-Dihydroxy-4,5-α-epoxy-17-methylmorphinan-6-one
14-Hydroxydihydromorphinone
14S)-14-Hydroxydihydromorphinone
Morphinan-6-one, 3,14-dihydroxy-4,5-α-epoxy-17-methyl-
Morphinan-6-one, 4,5-epoxy-3,14-dihydroxy-17-methyl-, (5α)- (9CI)
Morphinan-6-one, 4,5α-epoxy-3,14-dihydroxy-17-methyl- (8CI)
Morphinone, dihydro-14-hydroxy-
NSC-19045
Ossimorfone
Oximorfona (Spanish)
Oximorphonum
Oxymorphine
Oxymorphonum (Latin)

76-42-6
$C_{18}H_{21}NO_4$
315.40
COc1ccc2CC4N(C)CCC35C(Oc1c23)C(=O)CCC45O
Morphinan-6-one, 4,5-α-epoxy-14-hydroxy-3-methoxy-17-methyl
Codeinone, dihydro-14-hydroxy-
Codeinone, 7,8-dihydro-14-hydroxy-
Dihydrohydroxycodeinone
Dihydro-14-hydroxycodeinone
Dihydrone
Oxycodon
14-Hydroxydihydrocodeinone
Oxycodeinone
Oxycodone
Percobarb
Percodan

76-43-7
$C_{20}H_{29}FO_3$
336.49
Androst-4-en-3-one, 9-fluoro-11-β,17-β-dihydroxy-17-methyl
Androfluorene
Androfluorone
Androsterolo
Androst-4-en-3-one, 9-fluoro-11,17-dihydroxy-17-methyl-, (11-β,17-β)- (9CI)
11-β,17-β-Dihydroxy-9-α-fluoro-17-α-methyl-4-androster-3-one
Fluoro-9-α dihydroxy-11-β, 17-β methyl-17-α androstene-4 one-3 (French)
9-Fluoro-11-β,17-β-dihydroxy-17-methylandrost-4-en-3-one
9-α-Fluoro-11-β,17-β-dihydroxy-17-α-methyl-4-androstene-3-one
9-α-Fluoro-11-β-hydroxy-17-methyltestosterone
9-α-Fluoro-17-α-methyl-11-β,17-dihydroxy-4-androsten-3-one
Fluotestin
Fluoximesterone
Fluoxymesterone
Fluoxymestrone
Flusteron
Flutestos
Halotestin
17-α-Methyl-9-α-fluoro-11-β-hydroxytesterone
Neo-Ormonal
NSC-12165
Oralsterone
Oratestin
Ora-Testryl
Testoral
U 6040
Ultandren
Ultandrene

76-44-8
$C_{10}H_5Cl_7$
373.30
ClC1C=CC2C1C3(Cl)C(=C(Cl)C2(Cl)C3(Cl)Cl)Cl
4,7-Methanoindene, 1,4,5,6,7,8,8-heptachloro-3a,4,7,7a-tetrahydro
Agroceres
3-Chlorochlordene
Dicyclopentadiene, 3,4,5,6,7,8,8a-heptachloro-
Drinox
Drinox H-34
E 3314
ENT 15,152
Eptacloro (Italian)
1,4,5,6,7,8,8-Eptacloro-3a,4,7,7a-tetraidro-4,7-endo-metano-indene (Italian)
GPKh
H
H-34
H-60
Hepta
Heptachloor (Dutch)
1,4,5,6,7,8,8-Heptachloor-3a,4,7,7a-tetrahydro-4,7-endo-methano-indeen (Dutch)
Heptachlor (ACGIH,OSHA)
Heptachlorane
Heptachlore (French)
3,4,5,6,7,8,8-Heptachloro-dicyclopentadiene
3,4,5,6,7,8,8a-Heptachloro-dicyclopentadiene
1,4,5,6,7,8,8-Heptachloro-3a,4,7,7a-tetrahydro-4,7-endomethanoindene
1,4,5,6,7,8,8a-Heptachloro-3a,4,7,7a-tetrahydro-4,7-methanoindane
1,4,5,6,7,8,8-Heptachloro-3a,4,7,7a-tetrahydro-4,7-methanoindene
1(3a),4,5,6,7,8,8-Heptachloro-3a(1),4,7,7a-tetrahydro-4,7-methanoindene
1,4,5,6,7,8,8-Heptachloro-3a,4,7,7a-tetrahydro-4,7-methanol-1H-indene
1,4,5,6,7,8,8-Heptachloro-3a,4,7,7,7a-tetrahydro-4,7-methylene indene

1,4,5,6,7,10,10-Heptachloro-4,7,8,9-tetrahydro-4,7-methyleneindene
1,4,5,6,7,10,10-Heptachloro-4,7,8,9-tetrahydro-4,7-endomethyleneindene
1,4,5,6,7,8,8-Heptachlor-3a,4,7,7,7a-tetrahydro-4,7-endo-methano-inden (German)
Heptagran
Heptamul
Heptox
Latka 104 (Czech)
NCI-C00180
RCRA waste number P059
Rhodiachlor
Velsicol 104
Velsicol Heptachlor

76-49-3
$C_{12}H_{20}O_2$
196.29
O=C(OC(C(C(C1C2)(C)C)(C2)C)C1)C
Bicyclo(2.2.1)heptan-2-ol, 1,7,7-trimethyl-, acetate, endo- (9CI)
AI3-00665
Borneol, acetate (8CI)
Bornyl acetate
Bornyl acetic ether
2-Camphanol acetate
endo-2-Camphanyl ethanoate
NSC-407158
1,7,7-Trimethylbicyclo(2.2.1)-heptan-2-ol acetate

76-57-3
$C_{18}H_{21}NO_3$
299.40
COc1ccc2CC5C3C=CC(O)C4Oc1c2C34CCN5C
Morphinan-6-α-ol, 7,8-didehydro-4,5-α-epoxy-3-methoxy-17-methyl
Codeine
Methylmorphine
Morphine monomethyl ether
Morphine-3-methyl ether
Norcodeine, N-methyl-

76-58-4
$C_{19}H_{23}NO_3$
313.43
CCOc1ccc2CC5C3C=CC(O)C4Oc1c2C34CCN5C
Morphinan-6-α-ol, 7,8-didehydro-4,5-α-epoxy-3-ethoxy-17-methyl
Codethyline
Dionine
Diohin
Ethylmorphine
3-O-Ethylmorphine
Morphinan-6-ol, 7,8-didehydro-4,5-epoxy-3-ethoxy-17-methyl-, (5-α,6-α)- (9CI)
Morphine, ethyl-

76-59-5
$C_{27}H_{28}Br_2O_5S$
624.43
O=S(=O)(OC(c1cccc2)(c(c(c(c(O)c3C(C)C)Br)C)c3)c(c(c(c(O)c4C(C)C)Br)C)c4)c12
Phenol, 4,4'-(3H-2,1-benzoxathiol-3-ylidene)bis(2-bromo-3-methyl-6-(1-methylethyl)-, S,S-dioxide
Bromthymol Blue
Bromothymol Blue
Dibromothymolsulfophthalein
3,3'-Dibromothymolsulfonphthalein
Thymol, 6,6'-(3H-2,1-benzoxathiol-3-ylidene)bis(2-bromo-, S,S-dioxide (8CI)

76-73-3
$C_{12}H_{18}N_2O_3$
238.32
CCCC(C)C1(CC=C)C(=O)NC(=O)NC1=O
Barbituric acid, 5-allyl-5-(1-methylbutyl)
5-Allyl-5-(1-methylbutyl)-barbituric acid
5-Allyl-5-(1-methylbutyl)-malonylurea
Barbosec
Evronal
Hypotrol
Imesonal
Immenoctal
Immenox
Meballymal
2,4,6(1H,3H,5H)-Pyrimidinitrione, 5-(1-methylbutyl)-5-(2-propenyl)- (9CI)
Quinalbarbital
Quinalbarbitone
Secobarbital
Secobarbitone
Seconal
Trisomnin

76-74-4
$C_{11}H_{18}N_2O_3$
226.31
CCCC(C)C1(CC)C(=O)NC(=O)NC1=O
Barbituric acid, 5-ethyl-5-(1-methylbutyl)
Dorsital
Ethaminal
5-Ethyl-5-(1-methylbutyl)-barbituric acid
5-Ethyl-5-(1-methylbutyl)-malonylurea
Ethyl-propylmethylcarbinylbarbituric acid
Mebubarbital
Nembutal
Neodorm
Neodorm (new)
Pentabarbitone
Pentobarbital
Pentobarbitone
Pentobarbiturate
Pentobarbituric acid
2,4,6(1H,3H,5H)-Pyrimidinetrione, 5-ethyl-5-(1-methylbutyl)- (9CI)
Rivadorm

76-75-5
$C_{11}H_{18}N_2O_2S$
242.33
Thiopental
Barbituric acid, 5-ethyl-5-(1-methylbutyl)-2-thio-
DEA No. 2100
5-Ethyl-5-(1-methylbutyl)-2-thiobarbituric acid
Farmotal
Intraval
Nesdonal
Penthiobarbital
Pentothal
Pentothiobarbital
4,6,(1H,5H)-Pyrimidinedione, 5-ethyldihydro-5-(1-methylbutyl)-2-thioxo- (9CI)
Thiomebumal
Thiopentobarbital
Thiopentone
Thiothal
Tiopentale (Italian)

76-83-5
$C_{19}H_{15}Cl$
278.79
c(cccc1)(c1)C(c(cccc2)c2)(c(cccc3)c3)Cl
Methane, chlorotriphenyl
Chlorotriphenylmethane
Trityl chloride

76-84-6
$C_{19}H_{16}O$
260.34
OC(c(cccc1)c1)(c(cccc2)c2)c(cccc3)c3
Benzenemethanol, α,α-diphenyl- (9CI)
AI3-08929
α,α-Diphenylbenzenemethanol
Methanol, triphenyl- (8CI)
NSC-4050
Triphenylcarbinol
Triphenylmethanol
Triphenylmethyl alcohol
Tritanol
Trityl alcohol

76-87-9
$C_{18}H_{16}OSn$
367.03
O[Sn](c1ccccc1)(c2ccccc2)c3ccccc3
Stannane, hydroxytriphenyl
DOWCO 186
Du-Ter
Du-Ter W-50
Duter
ENT 28,009

Fenolovo
Fintine hydroxyde (French)
Fintin hydroxid (German)
Fentin hydroxide
Fintin hydroxyde (Dutch)
Fintin idrossido (Italian)
Flo-Tin 4l
Haitin
Hydroxyde de triphenyl-etain (French)
Hydroxytriphenylstannane
Hydroxytriphenyltin
Idrossido di stagno trifenile (Italian)
NCI-C00260
OMS 1017
Phenostat-H
Suzu H
Tin, hydroxytriphenyl-
TPTH
TPTOH
Trifenylstanniumhydroxid (Czech)
Trifenyl-tinhydroxyde (Dutch)
Triphenylstannium hydroxide
Triphenyltin hydroxide
Triphenyltin oxide
Triphenyl-zinnhydroxid (German)
Tubotin
Vancide KS

76-93-7
$C_{14}H_{12}O_3$
228.26
O=C(O)C(O)(c(ccccl)c1)c(cc cc2)c2
Benzilic-acid
Acide diphenylhydroxyacetique (French)
Benzeneacetic acid, α-hydroxy-α-phenyl- (9CI)
Diphenylglycolic acid
α,α-Diphenylglycolic acid
Diphenylhydroxyacetic acid
Hydroxydiphenylacetic acid

76-99-3
$C_{21}H_{27}NO$
309.49
CCC(=O)C(CC(C)N(C)C)(c1cc ccc1)c2ccccc2
3-Heptanone, 6-(dimethyl-amino)-4,4-diphenyl
Adanon
Amidon
Amidone
Diaminon
Dolophin
Dolophine
Heptadone
Heptanon
Ketalgin
Mecodin
Methadon
Methadone
Phenadone
Physeptone
Polamidon
Polamidone

77-02-1
$C_{10}H_{14}N_2O_3$
210.26
CC(C)C1(CC=C)C(=O)NC(=O) NC1=O
Barbituric acid, 5-allyl-5-iso-propyl
Allional
Allonal
5-Allyl-5-isopropylbarbiturate
Allylisopropylbarbituric acid
5-Allyl-5-isopropylbarbituric acid
Allylisopropylmalonylurea
Allylpropymal
Allypropymal
Alurate
Alurate elixir verdum
Aprobarbital
Aprobarbitone
Aprozal
Isonal
Isopropylallylbarbituric acid
Numal
2,4,6(1H,3H,5H)-Pyrimidine-trione, 5-(1-methylethyl)-5-(2-propenyl)-

77-06-5
$C_{19}H_{22}O_6$
346.41
O=C(OC(C1C(C2(CC(O)(C3=C) CC4)C3)C(=O)O)(C24)C=C C5O)C15C
Gibberellic-acid
Acide gibberellique (French)
Activol
Activol GA
Berelex
Brellin
Cekugib
Floraltone
GA
GA3
Gibberellin
Gibberellin A3
Gibberellin X
Gibbrel
Gib-Sol
Gib-Tabs
Grocel
NCI-C55823
Pro-Gibb
2,4a,7-Trihydroxy-1-methyl-8-methylenegibb-3-ene-1,10-dicarboxylic acid 1,4a-lactone

77-07-6
$C_{17}H_{23}NO$
257.41
CN2CCC13CCCCC1C2Cc4ccc (O)cc34
Morphinan-3-ol, 17-methyl-, (-)
levo-Dromoran
1,3,4,9,10,10a-Hexahydro-11-methyl-2H-10,4a-iminoeth-anophenanthren-6-ol, l-
(-)-3-Hydroxy-N-methylmorph-inan
2H-10,4a-Iminoethanophen-anthren-6-ol, 1,3,4,9,10, 10a-hexahydro-11-methyl-, l-
Levorphan
Levorphanol
Morphinan, 3-hydroxy-17-methyl-

77-08-7
$C_{21}H_{24}O_4$
340.42
Oc(c(O)cc(c1C(c(c(cc(O)c2O) C3(C)C)c2)(C4)C3)C4(C)C)c1
1,1'-Spirobi(indan)-5,5',6,6'-tetrol, 3,3,3',3'-tetramethyl- (8CI)
AI3-16787
NSC-512922
1,1'-Spirobi(indane-5,6-diol), 3,3,3',3'-tetramethyl-
1,1'-Spirobi(1H-inden)-5,5', 6,6'-tetrol, 2,2',3,3'-tetra-hydro-3,3,3',3'-tetramethyl-
1,1'-Spirobi(1H-indene)-5,5',6,6'-tetrol, 2,2',3,3'-tetra-hydro-3,3,3',3'-tetramethyl- (9CI)
TTS 5

77-09-8
$C_{20}H_{14}O_4$
318.34
O=C(OC(c1cccc2)(c(ccc(O)c3) c3)c(ccc(O)c4)c4)c12
Phenolphthalein
Agoral
3,3-Bis(p-hydroxyphenyl)-phthalide
α-Di(p-Hydroxyphenyl)-phthalide
Dihydroxyphthalophenone
Euchessina
Evac-Q-Kit
Evac-Q-Kwik
Evac-U-Gen
Feen-A-Mint Gum
Fenolftalein (Czech)
1(3H)-Isobenzofuranone, 3,3-bis(4-hydroxyphenyl)-
Koprol
Laxogen
Lilo
NCI-C55798
Phthalide 3,3,-bis(p-hydroxy-phenyl)-
Phthalimetten
Prulet
Purga
Purgen
Purgophen
Spulmako-Lax
Trilax

77-10-1
$C_{17}H_{25}N$
243.43

C1CCN(CC1)C2(CCCCC2)c3ccc
cc3
Piperidine, 1-(1-phenylcyclo-hexyl)
Cl-395
Hog
PCP
PCP (Anesthetic)
Phencyclidine
1-(1-Phenylcyclohexyl)pi-peridine

77-14-5
Unknown
Unknown
DEA No. 9643
Proheptazine

77-15-6
$C_{16}H_{23}NO_2$
261.40
CCOC(=O)C1(CCCN(C)CC1)c
2ccccc2
Azepine-4-carboxylic acid, hexahydro-1-methyl-4-phenyl-, ethyl ester
Aethoheptazin
4-Carbethoxy-1-methyl-4-phenylazacycloheptane
4-Carbethoxy-1-methyl-4-phenylhexamethylenimine
Ethoheptazine
Ethyl heptazine
Ethyl hexahydro-1-methyl-4-phenyl-azepine-4-carboxylate
Hexahydro-1-methyl-4-phenyl-4-azepinecarboxylic acid ethyl ester
1-Methyl-4-carbethoxy-4-phenylhexamethyleneimine
1-Methyl-4-carbethoxy-4-phenylhexamethylenimine
WY 401
Zactane

77-17-8
$C_{14}H_{19}NO_2$
233.31
CCOC(=O)C1(CCNCC1)c2cc
ccc2
Normeperidine
DEA No. 9233
Ethyl-4-phenyl-piperidine-4-carboxylate
Norpethidine

77-19-0
$C_{19}H_{35}NO_2$
309.49
O=C(OCCN(CC)CC)C(C(CC
CC1)C1)(CCCC2)C2
Dicyclomine
Bentylol
(Bicyclohexyl)-1-carboxylic acid, 2-(diethylamino)ethyl ester
(1,1'-Bicyclohexyl)-1-carboxylic acid, 2-(diethylamino)-ethyl ester (9CI)
Diciclorerina (Spanish)
Dicycloverin
Dicycloverine
Dicycloverinum (Latin)
Diocyl
Wyovin

77-20-3
$C_{16}H_{23}NO_2$
261.40
CCC(=O)OC1(CCN(C)CC1C)c
2ccccc2
4-Piperidinol, 1,3-dimethyl-4-phenyl-, propionate
Alphaprodine
1,3-Dimethyl-4-phenyl-4-piperidinol propionate (ester)
α-1,3-Dimethyl-4-phenyl-4-propionoxypiperidine
α-1,3-Dimethyl-4-phenyl-4-piperidinyl propionate
1,3-Dimethyl-4-phenyl-4-propionoxypiperidine
Nisentil
NU-1196
Prisilidine
α-Prodine
Propionic acid, α-1,3-dimethyl-4-phenyl-4-piperidyl ester

77-21-4
$C_{13}H_{15}NO_2$
217.29
CCC1(CCC(=O)NC1=O)c2ccccc2
Glutarimide, 2-ethyl-2-phenyl
Alfimid
CC 11511
Doriden
Doriden-Sed
Elrodorm
3-Ethyl-3-phenyl-2,6-diketopiperidine
3-Ethyl-3-phenyl-2,6-dioxopiperidine
α-Ethyl-α-phenylglutarimide
2-Ethyl-2-phenylglutarimide
3-Ethyl-3-phenyl-2,6-piperidinedione
Gimid
Glimid
Glutathimid
Glutethimid
Glutethimide
Glutetimide
Glutetimidu (Polish)
Noxyron
Phenyl-aethyl-glutarsaeureimid (German)
3-Phenyl-3-ethyl-2,6-diketopiperidine
3-Phenyl-3-ethyl-2,6-dioxopiperidine
α-Phenyl-α-ethylglutaric acid imide
2-Phenyl-2-ethylglutaric acid imide
α-Phenyl-α-ethylglutarimide
Sarodormin

77-23-6
$C_{20}H_{31}NO_3$
333.52
Cyclopentanecarboxylic acid, 1-phenyl-, 2-(2-(diethylamino)ethoxy)ethyl ester
Atussil
Carbetapentane
Pentoxyverine
Tuclase
U.C.B. 2543

77-26-9
$C_{11}H_{16}N_2O_3$
224.25
Butalbital
Alisobumal
Alisobumalum
Allylbarbital
Allylbarbitone
Allylbarbituric acid
Allylisobutylbarbital
Allylisobutylbarbiturate
5-Allyl-5-isobutylbarbituric acid
5-Allyl-5-(2'-methyl-n-propyl) barbituric acid
Barbituric acid, 5-allyl-5-isobutyl-
Butalbarbital
Butalbitale
Butalbitalum (Latin)
DEA No. 2100
Iso-butylallylbarbituric acid
Isobutylallylbarturic acid
Itobarbital
Optalidon
Profundal
2,4,6(1H,3H,5H)-Pyrimidine trione, 5-(2-methylpropyl)-5-(2-propenyl)-
Sandoptal
Tetrallobarbital

77-28-1
$C_{10}H_{16}N_2O_3$
212.28
CCCCC1(CC)C(=O)NC(=O)NC1
=O
Barbituric acid, 5-butyl-5-ethyl
Budorm
Butobarbitural
Butethal
Butobarbital
Butobarbitone
5-Butyl-5-ethylbarbituric acid
5-Butyl-5-ethyl-2,4,6(1H,3H,5H)-pyrimidinetrione
5-Ethyl-5-n-butylbarbituric acid
Etoval
Hyperbutal
Longanoct
Meonal
Monodorm
Neonal
2,4,6(1H,3H,5H)-Pyrimidine-trione, 5-butyl-5-ethyl- (9CI)
Sonerile

Soneryl

77-36-1
$C_{14}H_{11}ClN_2O_4S$
338.78
NS(=O)(=O)c1cc(ccc1Cl)C2(O)NC(=O)c3ccccc23
Benzenesulfonamide, 2-chloro-5-(1-hydroxy-3-oxo-1-isoindolinyl)
Benzenesulfonamide, 2-chloro-5-(2,3-dihydro-1-hydroxy-3-oxo-1H-isoindol-1-yl)- (9CI)
Chlorothalidone
Chlorphthalidolone
Chlorphthalidone
Chlortalidone
Chlorthalidon
Chlorthalidone
G 33182
Hygroton
Igroton
Isoren
Natriuran
Oradil
Oxodolin
Phthalamodine
Phthalamudine
Renon
Saluretin
Zambesil

77-41-8
$C_{12}H_{13}NO_2$
203.26
CN1C(=O)CC(C)(C1=O)c2ccccc2
Succinimide, N,2-dimethyl-2-phenyl
Celontin
1,3-Dimethyl-3-phenyl-2,5-dioxopyrrolidine
1,3-Dimethyl-3-phenyl-pyrrolidin-2,5-dione
N,2-Dimethyl-2-phenylsuccinimide
Mesuximide
Methsuximide
N-Methyl-α-methyl-α-phenylsuccinimide
N-Methyl-α,α-methylphenylsuccinimide
α-Methylphensuximide
α-Methyl-α-phenyl N-methyl succinimide
Metsuccimide
Petinutin
PM 396

77-47-4
C_5Cl_6
272.75
C(=C(C(=C1Cl)Cl)Cl)(C1(Cl)Cl)Cl
Hexachlorocyclopentadiene (ACGIH,OSHA) [UN 2646]
C-56
1,3-Cyclopentadiene, 1,2,3,4,5,5-hexachloro
Graphlox
HCCPD
Hexachlorcyklopentadien (Czech)
1,2,3,4,5,5-Hexachloro-1,3-cyclopentadiene
Hexachloro-1,3-cyclopentadiene
HRS 1655
NCI-C55607
PCL
Perchlorocyclopentadiene
RCRA waste number U130
UN 2646 [Hexachlorocyclopentadiene]

77-48-5
$C_5H_6Br_2N_2O_2$
285.95
O=C(N(C(C1=O)(C)C)Br)N1Br
Hydantoin, 1,3-dibromo-5,5-dimethyl
Dibromantin
Dibromantine
N,N'-Dibromodimethylhydantoin
2,4-Imidazolidinedione, 1,3-dibromo-5,5-dimethyl- (9CI)

77-49-6
$C_4H_9NO_4$
135.14
O=N(=O)C(CO)(CO)C
1,3-Propanediol, 2-methyl-2-nitro
1,1-Dimethylol-1-nitroethane
2-Methyl-2-nitropropane-1,3-diol
2-Nitro-2-methyl-1,3-propanediol
NMPD

77-53-2
$C_{15}H_{26}O$
222.37
OC(C(CC(C1CC2)(C2C)C3)C1(C)C)(C3)C
1H-3a,7-Methanoazulen-6-ol, octahydro-3,6,8,8-tetramethyl-, (3R-(3α,3aβ,6α,7β,8aα))- (9CI)
AI3-02178
8βH-Cedran-8-ol (8CI)
Cedrol
α-Cedrol
Eudesmol
NSC-403883

77-54-3
$C_{17}H_{28}O_2$
264.45
O=C(OC(C(CC(C1CC2)(C2C)C3)C1(C)C)(C3)C)C
8-β-H-Cedran-8-ol, acetate
Acetic acid, cedrol ester
Cedranyl acetate
Cedryl acetate
1H-3a,7-Methanoazulen-6-ol, octahydro-3,6,8,8-tetramethyl-, acetate

77-58-7
$C_{32}H_{64}O_4Sn$
631.65
CCCCCCCCCCCC(=O)O[Sn](CCCC)(CCCC)OC(=O)CCCCCCCCCCC
Stannane, dibutylbis-(lauroyloxy)
Advastab 52
Bis(lauroyloxy)di(n-butyl)-stannane
Butynorate
DBTL
Dibutylbis(lauroyloxy)tin
Dibutylstannium dilaurate
Dibutyltin dilaurate
Dibutyltin laurate
Dibutyl-zinn-dilaurat (German)
Fomrez sul-4
Laudran di-n-butylcinicity (Czech)
Lauric acid, dibutylstannylene deriv.
Lauric acid, dibutylstannylene salt
Lauric acid, dibutyltin deriv.
Stabilizer D-22
Stannane, bis(dodecanoyloxy) di-n-butyl-
Stannane, bis(lauroyloxy)-dibutyl-
Therm chek 820
Tin, dibutylbis(lauroyloxy)-
Tin dibutyl dilaurate
Tin, di-n-butyl-, di(dodecanoate)
Tinostat

77-61-2
$C_{15}H_{22}O$
218.34
Oc(c(cc(c1)C)C)c1C(CCCC2)(C2)C
Phenol, 2,4-dimethyl-6-(1-methylcyclohexyl)- (9CI)
2,4-Dimethyl-6-(1-methylcyclohexyl)phenol

77-62-3
$C_{29}H_{40}O_2$
420.64
Oc(c(cc(c1)C)Cc(c(O)c(cc2C)C(CCCC3)(C3)C)c2)c1C(CCCC4)(C4)C
2,2'-Methylenebis(4-methyl-6-(1-methylcyclohexyl)-phenol)
Bisalkofen MTSP
p-Cresol, 2,2'-methylenebis-(6-(1-methylcyclohexyl)-
Ionox WSP
2,2'-Methylenebis(6-(1-methylcyclohexyl)-p-cresol)
Nonox WSP
Phenol, 2,2'-methylenebis-(4-methyl-6-(1-methylcyclohexyl)- (9CI)

77-63-4
$C_{28}H_{32}O_4Si_4$
544.90
Cyclotetrasiloxane, 2,4,6,8-tetramethyl-2,4,6,8-tetraphenyl- (9CI)

77-65-6
$C_7H_{13}BrN_2O_2$
237.13
O=C(NC(=O)C(Br)(CC)CC)N
Urea, (2-bromo-ethylbutyryl)
Adalin
Addisomnol
N-(Aminocarbonyl)-2-bromo-2-ethylbutanamide
Bromacetocarbamide
Bromadal
Bromadel
Bromdiethylacetylurea
2-Brom-2-ethylbutyrylmocovina (Czech)
Bromodiethylacetylcarbamide
Bromodiethylacetylurea
(α-Bromo-α-ethylbutyryl)-carbamide
(α-Bromo-α-ethylbutyryl)urea
1-Bromo-ethyl-butyryl-urea
2-Bromo-2-ethylbutyrylurea
Carbromal
Diacid
Dormiturin
Fydalin
Hoggar
Karbromal
Kartryl
NCI-C03805
Nenesin
Nyctal
Parkosed
Pelidorm
Pianadalin
Planadalin
Tildin
Uradal
Urea, (2-bromo-2-ethylbutyryl)-(8CI)

77-67-8
$C_7H_{11}NO_2$
141.19
CCC1(C)CC(=O)NC1=O
Succinimide, 2-ethyl-2-methyl
Aethosuximide (German)
Asamid
Atysmal
Capitus
C.I. 366
Emeside
Epileo petit mal
Ethosuccimide
Ethosuccinimide
Ethosuxide
Ethosuximide
α-Ethyl-α-methylsuccinimide
3-Ethyl-3-methylpyrrolidine-2,5-dione
3-Ethyl-3-methyl-2,5-pyrrolidine-dione
2-Ethyl-2-methylsuccinimide
Ethymal
Etomal
Etosuximida
H-490
H 940
Mesentol
3-Methyl-3-ethylpyrrolidine-2,5-dione
γ-Methyl-γ-ethyl-succinimide
Pemal
Pemalin
Pentinimid
Petinimid
Petnidan
PM 671
Pyknolepsinum
Ronton
Simatin(e)
Succimal
Succimitin
Suxilep
Suximal
Suxin
Suxinutin
Thetamid
Thilopemal
Zaraondan
Zarodan
Zarondan-Saft
Zarontin
Zartalin

77-68-9
$C_{12}H_{24}O_3$
216.32
O=C(OCC(C(O)C(C)C)(C)C)C(C)C
Propanoic acid, 2-methyl-, 3-hydroxy-2,2,4-trimethylpentyl ester (9CI)

77-71-4
$C_5H_8N_2O_2$
128.15
O=C(NC(C1=O)(C)C)N1
Hydantoin, 5,5-dimethyl
DMH
5,5-Dimethylhydantoin
2,4-Imidazolidinedione, 5,5-dimethyl-
T10

77-73-6
$C_{10}H_{12}$
132.22
C(C(C=CC12)C1)(C2C=C3)C3
4,7-Methanoindene,3a,4,7,7a-tetrahydro
Bicyclopentadiene
Biscyclopentadiene
1,3-Cyclopentadiene, dimer
Dicyklopentadien (Czech)
Dicyclopentadiene (ACGIH, OSHA) [UN 2048]
Dimer cyklopentadienu (Czech)
3a,4,7,7a-Tetrahydro-4,7-methanoindene
UN 2048 [Dicyclopentadiene]

77-74-7
$C_6H_{14}O$
102.20
OC(CC)(CC)C
3-Pentanol, 3-methyl
Methyldiaethylcarbinol (German)
Methyldiethylcarbinol
3-Methyl-pentanol-(3) (German)
3-Methyl-3-pentanol

77-75-8
$C_6H_{10}O$
98.16
OC(C#C)(CC)C
1-Pentyn-3-ol, 3-methyl
2-Aethinylbutanol
Allotropal
Aniphor
Anti-Stress
Apridol
Atemorin
Atempol
(BDH)
2-Butanol, 2-ethynyl-
Citodorm
Comesa
Dalgol
Dorison
Dormalest
Dormidin
Dormigen
Dormiphen
Dormison
Dormocit
Dormosan
2-Ethinylbutanol-2
Ethinylmethylethylcarbinol
3-Ethylbutinol
3-Ethylbutynol
Ethyl ethynyl methyl carbinol
2-Ethynyl-2-butanol
Formison
Hesofen
Hexofen
Imnudorm
Insomnol
Macarol
Mecarol
Melpintol
Meparfynol
Mepentamato
Mepentil
Methylethylacetylenylcarbinol
Methylethylethynylcarbinol
Methylparafynol
Methylpentinol
3-Methyl-pentin-(1)-ol-(3) (German)
3-Methylpentin-3-ol
Methylpentynol
3-Methylpent-1-yn-3-ol
3-Methyl-1-pentyn-3-ol
Methylpentynolum
Metilparafinolo
3-Metil-pentin-3-ol (Italian)
Metilpentinolo
Miramel
M-Pentynol

N-Oblivon
Noxokratin
Oblevil
Oblivon
Oblivon C
Olosot
Olvadon
Pentadorm
Pentinol
Pentydorm
Pentydrom
Pentyrest
Perlopal
Placidal
Riposon
Sedapercut
Seral
Sintyal
Somnesin
Sonnormon
Trusono
Util

77-76-9
$C_5H_{12}O_2$
104.15
O(C(OC)(C)C)C
2,2-Dimethoxypropane
Acetone, dimethyl acetal (8CI)
Acetone dimethyl ketal
AI3-26275 -
NSC-62085
Propane, 2,2-dimethoxy- (9CI)

77-77-0
$C_4H_6O_2S$
118.16
C=CS(=O)(=O)C=C
Ethene, 1,1'-sulfonylbis
Divinyl sulfone
Sulfone, divinyl-
TL 797
Vinyl sulfone

77-78-1
$C_2H_6O_4S$
126.14
O=S(=O)(OC)OC
Dimethyl sulfate (ACGIH, OSHA) [UN 1595]
Dimethylester kyseliny sirove (Czech)
Dimethyl monosulfate
Dimethylsulfaat (Dutch)
Dimethylsulfat (Czech)
Dimethyl sulphate
Dimetilsolfato (Italian)
DMS
DMS (Methyl sulfate)
Dwumetylowy siarczan (Polish)
Methyle (sulfate de) (French)
Methyl sulfate
Methyl sulfate [UN 1595]
RCRA waste number U103
Sulfate de methyle (French)
Sulfate dimethylique (French)
Sulfuric acid, dimethyl ester
UN 1595 [Dimethyl sulfate]

77-79-2
$C_4H_6O_2S$
118.16
O=S(=O)(CC=C1)C1
Thiophene, 2,5-dihydro-, 1,1-dioxide
Butadiene sulfone
2,5-Dihydrothiophene dioxide
2,5-Dihydrothiophene 1,1-dioxide
2,5-Dihydrothiophene sulfone
NCI-C04557
Sulfol-3-ene
β-Sulfolene
3-Sulfolene

77-81-6
$C_5H_{11}N_2O_2P$
162.15
CCOP(=O)(C#N)N(C)C
Phosphoramidocyanidic acid, dimethyl-, ethyl ester
Dimethylamidoethoxyphosphoryl cyanide
Dimethylaminocyanphosphorsaeureaethylester (German)
Dimethylphosphoramidocyanidic acid, ethyl ester
EA 1205
Ethyl dimethylamidocyanophosphate
Ethyl dimethylphosphoramidocyanidate
Ethyl N,N-dimethylphosphoramidocyanidate
Ethyl N,N-dimethylamino cyanophosphate
Ethylester-dimethylamid kyseliny kyanfosfonove (Czech)
GA
Gelan I
Le-100
MCE
T-2104
TL 1578
Taboon A
Tabun
Trilon 83

77-83-8
$C_{12}H_{14}O_3$
206.26
O=C(OCC)C(O1)C1(c(cccc2)c2)C
Hydrocinnamic acid, α,β-epoxy-β-methyl-, ethyl ester
Aldehyde C-16
C-16 Aldehyde
EMPG
α,β-Epoxy-β-methylhydrocinnamic acid, ethyl ester
Ethyl α,β-epoxy-β-methylhydrocinnamate
Ethyl 2,3-epoxy-3-methyl-3-phenylpropionate
Ethyl ester of 2,3-epoxy-3-phenylbutanoic acid
Ethyl methylphenylglycidate
Fraeseol
3-Methyl-3-phenylglycidic acid ethyl ester
Strawberry aldehyde

77-85-0
$C_5H_{12}O_3$
120.15
OCC(CO)(CO)C
1,1,1-Tris(hydroxymethyl)ethane
Ethane, 1,1,1-tris(hydroxymethyl)-
2-(Hydroxymethyl)-2-methyl-1,3-propanediol
Methriol
Methyltrimethanolmethane
Methyltrimethylolmethane
Metriol
NSC-65581
Pentaglycerine
Pentaglycerol
1,3-Propanediol, 2-(hydroxymethyl)-2-methyl- (9CI)
TME
Trimet
1,1,1-Trimethanolethane
Trimethylolethane
1,1,1-Trimethylolethane
Tris(hydroxymethyl)ethane
1,1,1-Tris(methylol)ethane

77-86-1
$C_4H_{11}NO_3$
121.16
OCC(N)(CO)CO
1,3-Propanediol, 2-amino-2-(hydroxymethyl)
Addex-THAM
2-Amino-2-(hydroxymethyl)-propane-1,3-diol
2-Amino-2-(hydroxymethyl)-1,3-propanediol
2-Amino-2-methylol-1,3-propanediol
Aminotrimethylolmethane
Aminotris(hydroxymethyl)-methane
Methylamine, 1,1,1-tris-(hydroxymethyl)-
Pehanorm
Talatrol
THAM
THAM-E
THAM SET
Trimethylolaminomethane
Tris
Trisamine
Tris-Amino
Trisaminol
Tris Buffer
Tris-hydroxymethyl-amino-methan (German)
Tris-hydroxymethylamino-methane
Tris(hydroxymethyl)methan-amine
Tris(hydroxymethyl)methyl-amine

Trispuffer
Tris-Steril
Trometamol
Trometamole
Tromethamine
Tromethane
Tromethanmin
Tutofusin tris

77-89-4
$C_{14}H_{22}O_8$
318.36
O=C(OC(C(=O)OCC)(CC(=O)OCC)CC(=O)OCC)C
Citric acid, triethyl ester, acetate
Acetyl triethyl citrate
ATEC
Citric acid, acetyl triethyl ester
Citroflex A 2
1,2,3-Propanetricarboxylic acid, 2-(acetyloxy)-, triethyl ester (9CI)
Tricarballylic acid, β-acetoxytributyl ester
Triethyl acetylcitrate
Triethyl citrate, acetate
Triethylester kyseliny acetylcitronove (Czech)

77-90-7
$C_{20}H_{34}O_8$
402.49
O=C(OC(C(=O)OCCCC)(CC(=O)OCCCC)CC(=O)OCCCC)C
Acetyl tributyl citrate
2-Acetoxy-1,2,3-propanetricarboxylic acid tributyl ester
Acetylcitric acid tributyl ester
2-(Acetyloxy)-1,2,3-propanetricarboxylic acid, tributyl ester
2-Acetyltributylcitrate
AI3-01999
Blo-trol
Caswell No. 005AB
Citric acid, tributyl ester, acetate (8CI)
Citroflex A
Citroflex A 4
NSC-3894
1,2,3-Propanetricarboxylic acid, 2-(acetyloxy)-, tributyl ester (9CI)
Tributyl 2-acetoxy-1,2,3-propanetricarboxylate
Tributyl acetylcitrate
Tributyl O-acetylcitrate
Tributyl 2-(acetyloxy)-1,2,3-propanetricarboxylic acid
Tributyl citrate acetate

77-92-9
$C_6H_8O_7$
192.14
O=C(O)C(O)(CC(=O)O)CC(=O)O
Citric-acid
Aciletten
Anhydrous citric acid
Citretten
Citric acid, Anhydrous
Citro
2-Hydroxy-1,2,3-propanetricarboxylic acid
β-Hydroxytricarballylic acid
Kyselina citronova (Czech)
Kyselina 2-hydroxy-1,2,3-propantrikarbonova (Czech)
1,2,3-Propanetricarboxylic acid, 2-hydroxy-

77-93-0
$C_{12}H_{20}O_7$
276.32
O=C(OCC)C(O)(CC(=O)OCC)CC(=O)OCC
Citric acid, triethyl ester
Citroflex 2
Ethyl citrate
1,2,3-Propanetricarboxylic acid, 2-hydroxy-, triethyl ester (9CI)
TEC
Triethyl citrate
Triethylester kyseliny citronove (Czech)

77-94-1
$C_{18}H_{32}O_7$
360.45
O=C(OCCCC)C(O)(CC(=O)OCCCC)CC(=O)OCCCC
Tributyl citrate
AI3-00394
Butyl citrate (VAN)
n-Butyl citrate
Citric acid, tributyl ester (8CI)
Citroflex 4
2-Hydroxy-1,2,3-propanetricarboxylic acid, tributyl ester
NSC-8491
1,2,3-Propanetricarboxylic acid, 2-hydroxy-, tributyl ester (9CI)
Tri-n-butyl citrate

77-95-2
$C_7H_{12}O_6$
192.19
OC1CC(O)(CC(O)C1O)C(O)=O
Cyclohexanecarboxylic acid, 1,3,4,5-tetrahydroxy-, D-(-)
Chinasaure
Chinic acid
Cyclohexanecarboxylic acid, 1,3,4,5-tetrahydroxy-, (-)- (8CI)
Cyclohexanecarboxylic acid, 1,3,4,5-tetrahydroxy-, (1R-(1-α,3-α,4-α,5-β))-
Kinic acid
Quinate
Quinic acid
(-)-Quinic acid
D-Quinic acid

77-99-6
$C_6H_{14}O_3$
134.20
OCC(CC)(CO)CO
1,3-Propanediol, 2-ethyl-2-(hydroxymethyl)
Ethriol
Ethyltrimethylolmethane
Etriol
Ettriol
Hexaglycerine
TMP
TMP (Alcohol)
1,1,1-Tri(hydroxymethyl)-propane
Trimethylolpropane
1,1,1-Trimethylolpropane
Tris(hydroxymethyl)propane
1,1,1-Tris(hydroxymethyl)-propane

78-00-2
$C_8H_{20}Pb$
323.47
CC[Pb](CC)(CC)CC
Plumbane, tetraethyl
Czteroetylek olowiu (Polish)
Lead, tetraethyl-
NA 1649 [Tetraethyl lead, Liquid]
NCI-C54988
Piombo tetra-etile (Italian)
RCRA waste number P110
TEL
Tetraethyl lead (ACGIH,DOT, OSHA)
Tetraethyl lead, Liquid (including flashpoint for export shipment by water) [NA 1649]
Tetraethyl lead, Motor fuel anti-knock mixtures [UN 1649]
Tetraethylolovo (Czech)
Tetraethylplumbane
Tetraethylplumbium
UN 1649 [Motor fuel anti-knock mixtures]

78-04-6
$C_{12}H_{20}O_4Sn$
347.01
1,3,2-Dioxastannepin-4,7-dione, 2,2-dibutyl
Advastab DBTM
Advastab T290
Advastab T340
BT 31
2,2-Dibutyl-1,3,2-dioxastannepin-4,7-dione
Dibutyl(maleoyldioxy)tin
Dibutylstannylene maleate
Dibutyltin maleate
Irgastab T 4
Irgastab T 150
Irgastab T 290
KS 4B
MA300A
Markure Ul2
Nuodex V 1525
Stancleret 157
Stann RC 40F

Stavinor 1300SN
Stavinor SN 1300
TN 3J
TVS-MA 300
TVS-N 2000E

78-06-8
$C_{11}H_{22}O_2SSn$
337.08
6H-1,3,2-Oxathiastannin-6-one, 2,2-dibutyldihydro
2,2-Dibutyl-1-oxa-2-stanna-3-thiacyclohexan-6-one
Dibutyltin mercaptopropionate
Dibutyltin S,O-3 mercaptopropionate
Dibutyltin S,O-β-mercaptopropionate
Dibutyltin, O,S-mercaptopropionate
Mercaptopropionic acid, dibutyltin salt
1-Oxa-2-stanna-3-thiacyclohexan-6-one, 2,2-dibutyl-
Tin, dibutyl(3-mercaptopropionato(2-))-

78-07-9
$C_8H_{20}O_3Si$
192.37
CCOSi(CC)(OCC)OCC
Silane, ethyltriethoxy
Ethyltriethoxysilane
Silane, triethoxyethyl-
Triethoxy-ethylsilane
Union Carbide A-15

78-08-0
$C_8H_{18}O_3Si$
190.35
CCOSi(OCC)(OCC)C=C
Silane, triethoxyvinyl
Triethoxyvinylsilane
Triethoxyvinylsilicane
Union Carbide A-151
Vinyltriethoxysilane

78-10-4
$C_8H_{20}O_4Si$
208.37
CCOSi(OCC)(OCC)OCC
Silicic acid, tetraethyl ester
Ethyl orthosilicate
Ethyl silicate (ACGIH,DOT, OSHA)
Etylu krzemian (Polish)
Extrema
Silane, tetraethoxy-
Silicate d'ethyle (French)
Silicate tetraethylique (French)
TEOS
Tetraethoxysilane
Tetraethyl orthosilicate [UN 1292]
Tetraethyl silicate
Tetraethylsilikat (Czech)
UN 1292 [Tetraethyl silicate]

78-11-5
$C_5H_8N_4O_{12}$
316.17
O=N(=O)OCC(CON(=O)=O)(CON(=O)=O)CON(=O)=O
Pentaerythritol, tetranitrate
Angicap
Angitet
Antora
Arcotrate
Baritrate
2,2-Bisdihydroxymethyl-1,3-propanediol tetranitrate
2,2-Bis(hydroxymethyl)-1,3-propanediol tetranitrate
Chot
Deltrate-20
1,3-Dinitrato-2,2-bis(nitratomethyl)propane
Duotrate
El Petn
Erinit
Hasethrol
Kaytrate
Lowetrate
Martrate-45
Metranil
Mikardol
Mycardol
Myotrate "10"
NCI-C55743
Neo-Corovas
Neopentanetetrayl nitrate
NA 0150 [Pentaerythrite tetranitrate wetted with not less than 25 per cent water, by mass]
Niperyt
Niperyth
Nitrinol
Nitropenta
Nitropentaerythrite
Nitropentaerythritol
Pencard
Pentaerythrite tetranitrate [NA 0150]
Pentaerythrite tetranitrate, Desensitized, Wet [NA 0150]
Pentaerythrite tetranitrate, Desensitized with not less than 15% phlegmatizer [NA 0150]
Pentaerythrite tetranitrate, With not less than 7% wax [UN 0411]
Pentaerythrite tetranitrate, Dry [NA 0150]
Pentaerythritol, tetranitrate, Containing at least 25% water/at least 15% phlegmatizer [NA 0150]
Pentaerythritol tetranitrate, Diluted
Pentafin
Pentanitrine
Pentarit
Pentestan-80
Pentetrate unicelles
Penthrit
Penthrite
Pentitrate
Pentrate
Pentriol
Pentritol
Pentryate
Pentryate 80
Pergitral
Peridex-la
Peritrate
Perityl
PET
PETN
Prevangor
1,3-Propanediol, 2,2-bis((nitrooxy)methyl)-, dinitrate (ester)
Quintrate
Rythritol
SDM No. 23
SDM No. 35
Subicard
Tanipent
Ten
Tentrate-20
Tetranitropentaerythrite
Tetrasule
Tranite D-Lay
UN 0411 [Pentaerythrite tetranitrate; Pentaerythritol tetranitrate (PETN) with not less than 7 per cent wax by mass]
UN 0150 (DOT)
Vasitol
Vasodiatol
Vaso-80 Unicelies

78-12-6
$C_{13}H_{16}Cl_{12}O_8$
725.70
Petrichloral
DEA No. 2591
Ethanol, 1,1',1'',1'''-(neopentanetetrayltetraoxy)tetrakis(2,2,2-trichloro-
1,1',1'',1'''-(Neopentanetetrayltetraoxy)tetrakis(2,2,2-trichloroethanol)
Pentaerythritol chlorol
Perichlor
Perichloral
Pericler
Periclor
Petrichloralum (Latin)
Petricloral (Spanish)

78-13-7
$C_{24}H_{52}O_4Si$
432.85
CCC(CC)COSi(OCC(CC)CC)(OCC(CC)CC)OCC(CC)CC
Silicic acid, tetra(2-ethylbutyl) ester
2-Ethyl-1-butanol, silicate
Tetra(2-ethylbutoxy) silane
Tetra (2-ethylbutyl) orthosilicate
Tetrakis(2-ethylbutoxy)silane

78-16-0
$C_{27}H_{50}O_6$
470.69
O=C(OCC(CC)(COC(=O)CCCCCC)COC(=O)CCCCCC)CC

CCCC
Heptanoic acid, 2-ethyl-2-(((1-oxoheptyl)oxy)methyl)-1,3-propanediyl ester (9CI)

78-18-2
$C_{12}H_{22}O_5$
246.34
O(O)C(OOC(O)(CCCC1)C1)(CCCC2)C2
Cyclohexanol, 1-((1-hydroperoxycyclohexyl)dioxy)
Cyclohexanone peroxide
1-Hydroperoxycyclohexyl-1-hydroxycyclohexyl peroxide
1-Hydroxy-1-hydroperoxydicyclohexyl peroxide
1-Hydroxy-1'-hydroperoxydicyclohexyl peroxide
Peroxide, 1-hydroperoxycyclohexyl 1-hydroxycyclohexyl

78-21-7
$C_{22}H_{46}NO.C_2H_5O_4S$
465.73
Cetethyl morpholinium ethosulfate
Atlas G-263
Barquat CME-A
Caswell No. 165F
Cetyl ethyl morpholinium ethosulfate
Cetylethylmorpholinium ethosulfate
N-Cetyl-N-ethylmorpholinium ethosulfate
N-Cetyl-N-ethyl morpholinium ethyl sulfate
N-Cetyl-N-ethylmorpholinium ethyl sulfate
N-Cetyl-N-ethylmorpholinium ethylsulfate
EPA Pesticide Chemical Code 069187
4-Ethyl-4-hexadecyl morpholinium ethyl sulfate
4-Ethyl-4-hexadecylmorpholinium ethyl sulfate
Ethyl-N-hexadecylmorpholinium ethosulfate
G 251 (VAN)
G 263
N-Hexadecyl-N-ethylmorpholinium ethyl sulfate
Morpholinium, 4-ethyl-4-hexadecyl-, ethyl sulfate (9CI)
NSC-215231
Quaternium-25

78-23-9
$C_{23}H_{46}O_5$
402.62
O=C(OCC(CO)(CO)CO)CCCCCCCCCCCCCCCCC
Octadecanoic acid, 3-hydroxy-2,2-bis(hydroxymethyl)-propyl ester (9CI)
AI3-14764
Gruenau S
NSC-71130
Pentaerythritol monostearate
Pentamull 6
Stearic acid, 3-hydroxy-2,2-bis-(hydroxymethyl)propyl ester (8CI)
Stearic acid, monoester with pentaerythritol
Stearic acid, pentaerythritol ester (1:1)

78-24-0
$C_{15}H_{32}O_{10}$
372.41
O(CC(CO)(CO)COCC(CO)(CO)CO)CC(CO)(CO)CO
Tripentaerythritol (8CI)
NSC-97579
1,3-Propanediol, 2,2-bis-((3-hydroxy-2,2-bis(hydroxymethyl)propoxy)methyl)-(9CI)
Tripentaerytritol
Tris(pentaerythritol)

78-30-8
$C_{21}H_{21}O_4P$
368.39
O=P(Oc(c(ccc1)C)c1)(Oc(c(ccc2)C)c2)Oc(c(ccc3)C)c3
Phosphoric acid, tri-o-tolyl ester
o-Cresyl phosphate
Phosflex 179-C
Phosphoric acid, tri-o-cresyl ester
Phosphoric acid, tris(2-methylphenyl) ester
TOCP
TOFK
o-Tolyl phosphate
TOTP
Tricresyl phosphate
Tri-o-cresyl phosphate
o-Trikesylphosphate (German)
Tri 2-methylphenyl phosphate
Triorthocresyl phosphate (ACGIH,OSHA)
Tris(o-cresyl)-phosphate
Tris(o-methylphenyl)phosphate
Tris(o-tolyl)-phosphate
Tri-o-tolyl phosphate
Tri-2-tolyl phosphate
Trojkrezylu fosforan (Polish)

78-31-9
$C_{19}H_{17}O_4P$
340.32
p-Cresyl diphenyl phosphate
Diphenyl p-tolyl phosphate
Phosphoric acid, diphenyl p-tolyl ester
Phosphoric acid, 4-methylphenyl diphenyl ester

78-32-0
$C_{21}H_{21}O_4P$
368.37
O=P(Oc(ccc(c1)C)c1)(Oc(ccc(c2)C)c2)Oc(ccc(c3)C)c3
Tri-p-cresyl phosphate
AI3-04490
NSC-2181
Phosphoric acid, tri-p-tolyl ester (8CI)
Phosphoric acid, tri(4-tolyl)ester
Phosphoric acid, tris(4-methylphenyl) ester (9CI)
TPC
TPCP
Tri-p-tolyl phosphate
Tris(4-methylphenyl) phosphate

78-33-1
$C_{30}H_{39}O_4P$
494.61
O=P(Oc(ccc(c1)C(C)(C)C)c1)(Oc(ccc(c2)C(C)(C)C)c2)Oc(ccc(c3)C(C)(C)C)c3
Tris(p-t-butylphenyl) phosphate
AI3-17846
p-tert-Butylphenol, phosphate (3:1)
4-(1,1-Dimethylethyl)phenol phosphate (3:1)
NSC-2884
Phenol, p-tert-butyl-, phosphate (3:1) (8CI)
Phenol, 4-(1,1-dimethylethyl)-, phosphate (3:1) (9CI)
Tris(p-tert-butylphenyl) phosphate
Tris(4-tert-butylphenyl) phosphate

78-34-2
$C_{12}H_{26}O_6P_2S_4$
456.56
CCOP(=S)(OCC)SC1OCCOC1SP(=S)(OCC)OCC
Phosphorodithioic acid, S,S'-p-dioxane-2,3-diyl O,O,O',O'-tetraethyl ester
AC 528
Bis(dithiophosphate de O,O-diethyle) de S,S'-(1,4-dioxanne-2,3-diyle) (French)
Delnatex
Delnav
Delnav (OSHA)
Deltic
1,4-Diossan-2,3-diyl-bis(O,O-dietil-ditiofosfato) (Italian)
2,3-p-Dioxan-S,S'-bis(O,O-diaethyldithiophosphat) (German)
1,4-Dioxaan-2,3-diyl-bis-(O,O-diethyl-dithiofosfaat) (Dutch)
1,4-Dioxan-2,3-diyl-bis-(O,O-diaethyl-dithiophosphat) (German)
Dioxane phosphate
Dioxathion
1,4-Dioxan-2,3-diyl bis(O,O-diethylphosphorothiolothionate)
1,4-Dioxan-2,3-diyl O,O,O',O'-

tetraethyl di(phosphoromithio-ate)
2,3-p-Dioxandithiol S,S-bis-(O,O-diethyl phosphorodithio-ate)
2,3-p-Dioxane S,S-bis(O,O-di-ethylphosphoroithioate)
p-Dioxane-2,3-dithiol, S,S-di-ester with O,O-diethyl phos-phorodithioate
p-Dioxane-2,3-diyl ethyl phos-phorodithioate
Dioxation
Dioxathion (ACGIH,OSHA)
Dioxothion
ENT 22,897
Hercules 528
Hercules AC528
Kavadel
Navadel
NCI-C00395
Phosphorodithioic acid, S,S'-1,4-dioxane-2,3-diyl O,O,O',O'-tetraethyl ester (9CI)
Ruphos

78-38-6
$C_6H_{15}O_3P$
166.16
O=P(OCC)(OCC)CC
Diethyl ethylphosphonate
AI3-18558
Ethanephosphonic acid, diethyl ester
NSC-2671
Phosphonic acid, ethyl-, diethyl ester (9CI)

78-39-7
$C_8H_{18}O_3$
162.23
O(C(OCC)(OCC)C)CC
Ethane, 1,1,1-triethoxy- (9CI)
AI3-23843
Ethyl orthoacetate
NSC-5596
Orthoacetic acid, triethyl ester (8CI)
1,1,1-Triethoxyethane
Triethyl orthoacetate

78-40-0
$C_6H_{15}O_4P$
182.18
O=P(OCC)(OCC)OCC
Phosphoric acid, triethyl ester
Ethyl phosphate
TEP
Triethylfosfat (Czech)
Triethyl phosphate

78-42-2
$C_{24}H_{51}O_4P$
434.72
O=P(OCC(CCCC)CC)(OCC(CCCC)CC)OCC(CCCC)CC
1-Hexanol, 2-ethyl-, phosphate
Disflamoll TOF
2-Ethyl-1-hexanol phosphate
Flexol TOF
Kronitex TOF
NCI-C54751
Phosphoric acid, tris(2-ethyl-hexyl) ester
TOF
Triethylhexyl phosphate
Tri(2-ethylhexyl)phosphate
Trioctyl phosphate
Tris-(2-ethylhexyl)fosfat (Czech)
Tris(2-ethylhexyl)phosphate

78-43-3
$C_9H_{15}Cl_6O_4P$
430.91
O=P(OCC(CCl)Cl)(OCC(CCl)Cl)OCC(CCl)Cl
Propanol, 2,3-dichloro-, phos-phate (3:1)
Celluflex FR-2
FR 2
Phosphoric acid, tris(2,3-di-chloropropyl) ester
TDCPP
Tris-dichloropropylphosphate
Tris(2,3-dichloropropyl)-phosphate

78-44-4
$C_{12}H_{24}N_2O_4$
260.38
CCCC(C)(COC(N)=O)COC(=O)NC(C)C
Carbamic acid, isopropyl-, 2-(hydroxymethyl)-2-methyl-pentyl ester carbamate (ester)
Apesan
Arusal
Brianil
Caprodat
Carbamic acid, ester with 2-(hydroxymethyl)-2-methyl-pentyl isopropylcarbamate
Carbamic acid, ester with 2-methyl-2-propyl-1,3-pro-panediol isopropylcarbamate
Carisol
Carisoma
Carisoprodate
Carisoprodatum
Carisoprodol
Carlsoma
Carlsoprol
Carsodal
Carsodol
CB 8019
Diolene
Domarax
Flexal
Flexartal
Flexartel
Isobamate
Isomeprobamate
Isopropylcarbamic acid, ester with 2-(hydroxymethyl)-2-methylpentyl carbamate
Isopropyl meprobamate
N-Isopropyl-2-methyl-2-propyl-1,3-propanediol dicarbamate
Mediquil
(1-Methylethyl)carbamic acid 2-(((aminocarbonyl)oxy)-methyl)-2-methylpentyl ester
2-Methyl-2-propyl-1,3-propane-diol carbamate isopropylcar-bamate
Mioartrina
Miolisodal
Miolisodol
Mioratrina
Mioril
Mioriodol
NCI-C56235
Nospasm
1,3-Propanediol, 2-methyl-2-propyl-, carbamate iso-propylcarbamate (ester)
Rela
Relasom
Relax
Sanoma
SCH 7307
Soma
Somadril
Somalgit
Somanil
Tonolyt isopropyl meprobamate

78-46-6
$C_{12}H_{27}O_3P$
250.36
O=P(OCCCC)(OCCCC)CCCC
Phosphonic acid, butyl-, di-butyl ester
Dibutyl butanephosphonate
Dibutyl butylphosphonate

78-48-8
$C_{12}H_{27}OPS_3$
314.54
O=P(SCCCC)(SCCCC)SCCCC
Phosphorotrithioic acid, S,S,S-tributyl ester
B-1,776
Butifos
Butiphos
Butyl phosphorotrithioate
Chemagro 1,776
Chemagro B-1776
DEF
DEF Defoliant
De-Green
E-Z-Off D
Fos-Fall "A"
Ortho phosphate defoliant
S,S,S-Tributyl phosphorotri-thioate
S,S,S-Tributyltrithiofosfat (Czech)
S,S,S-Tributyl trithiophos-phate

78-50-2
$C_{24}H_{51}OP$
386.64
O=P(CCCCCCCC)(CCCCCCCC)

CCCCCCCC
Trioctyl phosphine oxide
Phosphine oxide, trioctyl- (9CI)
Trioctylphosphine oxide

78-51-3
$C_{18}H_{39}O_7P$
398.54
O=P(OCCOCCCC)(OCCOCCC C)OCCOCCCC
Ethanol, 2-butoxy-, phosphate (3:1)
2-Butoxyethanol, phosphate
KP 140
Phosphoric acid, tris(2-butoxyethyl) ester
Kronitex KP-140
Phosflex T-Bep
TBEP
Tri(2-butoxyethanol)phosphate
Tributoxyethyl phosphate
Tri(2-butoxyethyl) phosphate
Tributyl cellosolve phosphate
Tris-(2-butoxyethyl)fosfat (Czech)
Tris(2-butoxyethyl) phosphate

78-52-4
$C_8H_{19}O_2PS_3$
274.42
Phosphorodithioic acid, O,O-diethyl S-((isopropylthio)methyl) ester
American Cyanamid 12,008
O,O-Diethyl-S-2-isopropylmercaptomethyl dithiophosphate
O,O-Diethyl S-(isopropylmercaptomethyl) phosphorodithioate
O,O-Diethyl-S-(isopropylthio)-methylester kyseliny dithiofosforecne (Czech)
O,O-Diethyl S-(isopropylthiomethyl) phosphorodithioate
ENT 22,865
Experimental insecticide 12008
Methanethiol, (isopropylthio)-, S-ester with O,O-diethyl phosphorodithioate
TM 12008

78-53-5
$C_{10}H_{24}NO_3PS$
269.38
CCOP(=O)(OCC)SCCN(CC)CC
Phosphorothioic acid, S-(2-(diethylamino)ethyl) O,O-diethyl ester
Amiton
Chipman 6200
Citram
S-(Diethylaminoethyl) O,O-diethyl phosphorothioate
S-(2-(Diethylamino)ethyl)phosphorothioic acid O,O-diethyl ester
(2-Diethylamino)ethylphosphorothioic acid O,O-diethyl ester
O,O-Diethyl-S-2-(diethylamino)ethylester kyseliny thiofosforecne (Czech)
Diethyl S-2-diethylaminoethyl phosphorothioate
O,O-Diethyl S-2-diethylaminoethyl phosphorothioate
O,O-Diethyl S-diethylaminoethyl phosphorothiolate
O,O-Diethyl S-(β-diethylamino)ethyl phosphorothiolate
O,O-Diethyl S-2-diethylaminoethyl phosphorothiolate
O,O-Diethyl S-(2-diethylaminoethyl) thiophosphate
DSDP
ENT 24,980-X
Inferno
Metramac
Metramak
R-5,158
Rhodia-6200
Tetram

78-57-9
$C_6H_{12}N_5O_2PS_2$
281.32
COP(=S)(OC)SCc1nc(N)nc(N)n1
Phosphorodithioic acid, S-((4,6-diamino-s-triazin-2-yl)methyl) O,O-dimethyl ester
Azidithion
2,4-Diamino-6-dimethoxyphosphinothionylthiomethyl-s-triazine
S-((4,6-Diamino-1,3,5-triazin-2-yl)-methyl)-O,O-dimethyl-dithiofosfaat (Dutch)
S-((4,6-Diamino-1,3,5-triazin-2-yl)-methyl)-O,O-dimethyl-dithiophosphat (German)
4,6-Diamino-1,3,5-triazin-2-yl-methyl O,O-dimethyl phosphorodithioate
S-((4,6-Diamino-s-triazin-2-yl)-methyl)-O,O-dimethyl phosphorodithioate
S-(4,6-Diamino-1,3,5-triazin-2-ylmethyl) O,O-dimethyl phosphorodithioate
S-(4,6-Diamino-1,3,5-triazin-2-ylmethyl) dimethyl phosphorothiolothionate
S-((4,6-Diamino-1,3,5-triazin-2-yl)methyl)phosphorodithioic acid O,O-dimethyl ester
2-Dimethoxyphosphinothioyl-thiomethyl-4,6-diamino-s-triazine
O,O-Dimethyl S-(4,6-diamino-1,3,5-triazinyl-2-methyl) dithiophosphate
O,O-Dimethyl S-(4,6-diamino-s-triazin-2-ylmethyl)phosphorodithioate
O,O-Dimethyl S-(4,6-diamino-1,3,5-triazin-2-yl)methyl phosphorodithioate
O,O-Dimethyl S-(4,6-diamino-1,3,5-triazin-2-yl)methyl phosphorothiolothionate
Dithiophosphate de O,O-dimethyle et de S-((4,6-diamino-1,3,5-triazine-2-yl)-methyle) (French)
ENT 25,760
Menazon
PP175
R 15,175
Saiphos
Saphicol
Saphizon
Saphizon-DP
Saphos
Sayfor
Sayfos
Sayphos
Syphos
s-Triazine-2-methanethiol, 4,6-diamino-, S-ester with O,O-dimethylphosphorodi thioate

78-59-1
$C_9H_{14}O$
138.21
O=C(C=C(CC1(C)C)C)C1
Isophorone (PEL:REL)
AI3-00046
α-Isophoron
α-Isophorone
Caswell No. 506
2-Cyclohexen-1-one, 3,5,5-trimethyl- (9CI)
EPA Pesticide Chemical Code 047401
Isoacetophorone
Isoforon
Isoforone (Italian)
Isooctopherone
Isophoron
Isophorone (ACGIH,OSHA)
Izoforon (Polish)
NCI-C55618
NSC-403657
3,5,5-Trimethyl-2-cyclohexen-one
1,1,3-Trimethyl-3-cyclohexene-5-one
3,5,5-Trimethyl-2-cyclohexene-1-one
3,5,5-Trimethyl-2-cyclohexen-1-on (German, Dutch)
3,5,5-Trimethyl-2-cyclohexen-1-one
3,3,5-Trimethyl-5-cyclohexen-1-one
3,5,5-Trimetil-2-cicloesen-1-one (Italian)

78-63-7
$C_{16}H_{34}O_4$
290.50
O(OC(C)(C)C)C(CCC(OOC(C)(C)C)(C)C)(C)C
Hexane, 2,5-dimethyl-2,5-di-(t-butylperoxy)
2,5-Dimethyl-2,5-di(t-butylperoxy)hexane
2,5-Dimethyl-2,5-di-(tert-butyl-

peroxy)hexane, Technically pure (DOT)
Hexane, 2,5-dimethyl-2,5-di-(t-butylperoxy)-, Maximum concentration 52% with inert solid
Trigonox 101-101/45
UN 2155 (DOT)
UN 2156 (DOT)
Varox

78-66-0
$C_{10}H_{18}O_2$
170.28
OC(C#CC(O)(CC)C)(CC)C
4-Octyn-3,6-diol, 3,6-dimethyl
3,6-Dimethyl-octin-4-diol-(3,6) (German)

78-67-1
$C_8H_{12}N_4$
164.24
N#CC(N=NC(C#N)(C)C)(C)C
Propionitrile, 2,2'-azobis-(2-methyl)
Aceto azib
Aibn
Azobisisobutylonitrile
Azobisisobutyronitrile
α,α'-Azobisisobutylonitrile
2,2'-Azobis(isobutyronitrile)
2,2'-Azobis(2-methylpropionitrile)
Azodiisobutyronitrile [UN 2952]
α,α'-Azodiisobutyronitrile
2,2'-Azodiisobutyronitrile
2,2'-Dicyano-2,2'-azopropane
Poly-Zole AZDN
Porofor 57
Porophor N
UN 2952 [Azodiisobutyronitrile]
Vazo 64

78-69-3
$C_{10}H_{22}O$
158.32
OC(CCCC(C)C)(CC)C
3-Octanol, 3,7-dimethyl
3,7-Dimethyloctanol-3
Linalool tetrahydride
Tetrahydrolinalool

78-70-6
$C_{10}H_{18}O$
154.28
OC(C=C)(CCC=C(C)C)C
1,6-Octadien-3-ol, 3,7-dimethyl
Allo-ocimenol
2,6-Dimethyl-2,7-octadiene-6-ol
2,6-Dimethylocta-2,7-dien-6-ol
3,7-Dimethylocta-1,6-dien-3-ol
3,7-Dimethyl-1,6-octadien-3-ol
Linalol
Linalool
Linalyl alcohol

78-71-7
$C_5H_8Cl_2O$
155.03
ClCC1(CCl)COC1
Oxetane, 3,3-bis(chloromethyl)
3,3-Bis(chloromethyl)oxetane
3,3-Dichloromethyloxycyclobutane

78-72-8
$C_8H_{16}O_2$
144.24
CCCC1OC1(CC)CO
1-Hexanol, 2,3-epoxy-2-ethyl
2,3-Epoxy-2-ethyl hexanol

78-73-9
$C_5H_{14}NO.CHO_3$
165.18
Choline, carbonate (1:1) (Salt) (8CI)
Choline bicarbonate
Choline, hydrogen carbonate
Ethanaminium, 2-hydroxy-N,N,N-trimethyl-, carbonate (1:1) (Salt) (9CI)
Hydroxyethyltrimethylammonium bicarbonate
(2-Hydroxyethyl)trimethylammonium bicarbonate
NSC-163321

78-75-1
$C_3H_6Br_2$
201.91
BrCC(Br)C
Propane, 1,2-dibromo
1,2-Dibromopropane
Propylene dibromide

78-76-2
C_4H_9Br
137.04
BrC(CC)C
Butane, 2-bromo
2-Bromobutane [UN 2339]
sec-Butyl bromide
Methylethylbromomethane
UN 2339 [2-Bromobutane]

78-77-3
C_4H_9Br
137.04
BrCC(C)C
Propane, 1-bromo-2-methyl
1-Bromo-2-methylpropane [UN 2342]
I-Butyl bromide
Isobutyl bromide
UN 2342 [Bromomethylpropanes]

78-78-4
C_5H_{12}
72.17
C(CC)(C)C
Butane, 2-methyl
Ethyldimethylmethane
Isoamylhydride
Isopentane [UN 1265]
2-Methylbutane
UN 1265 [n-Pentanes or iso-pentane]

78-79-5
C_5H_8
68.13
C(C=C)(=C)C
Isoprene (DOT)
1,3-Butadiene, 2-methyl-
Isoprene, Inhibited [UN 1218]
β-Methylbivinyl
2-Methylbutadiene
2-Methyl-1,3-butadiene (DOT)
UN 1218 [Isoprene, Inhibited]

78-80-8
C_5H_6
66.10
Isopropenyl acetylene
AI3-25132
1-Buten-3-yne, 2-methyl- (8CI, 9CI)
Isopropenylacetylene
2-Methyl-1-buten-3-yne
NSC-9296
Valylene

78-81-9
$C_4H_{11}N$
73.16
NCC(C)C
Isobutylamine
1-Amino-2-methylpropane
Isobutylamine [UN 1214]
Monoisobutylamine
UN 1214 [Isobutylamine]
Valamine

78-82-0
C_4H_7N
69.12
N#CC(C)C
Propanenitrile, 2-methyl
Isobutyronitrile [UN 2284]
Isopropyl cyanide
Isopropylkyanid (Czech)
2-Methylpropionitrile
UN 2284 [Isobutyronitrile]

78-83-1
$C_4H_{10}O$
74.14
OCC(C)C
Isobutyl-alcohol
Alcool isobutylique (French)
Fermentation butyl alcohol
1-Hydroxymethylpropane
Isobutanol [UN 1212] (78-83-1)
Isobutyl alcohol (ACGIH,DOT, OSHA)
Isobutylalkohol (Czech)

Isopropylcarbinol
2-Methyl propanol
2-Methyl-1-propanol
2-Methylpropan-1-ol
2-Methylpropyl alcohol
1-Propanol, 2-methyl-
RCRA waste number U140
UN 1212 [Isobutanol or isobutyl alcohol]

78-84-2
C_4H_8O
72.12
O=CC(C)C
Isobutyraldehyde
Isobutaldehyde
Isobutanal
Isobutyral
Isobutyraldehyd (Czech)
Isobutyraldehyde (DOT)
Isobutylaldehyde
Isobutyl aldehyde (DOT)
Isobutyric aldehyde
Isobutyryl aldehyde
Isopropyl aldehyde
Isopropyl formaldehyde
Methyl propanal
2-Methylpropanal
2-Methyl-1-propanal
α-Methylpropionaldehyde
2-Methylpropionaldehyde
NCI-C60968
Propanal, 2-methyl-
Propionaldehyde, 2-methyl-
UN 2045 (DOT)
Valine aldehyde

78-85-3
C_4H_6O
70.10
O=CC(=C)C
Methacraldehyde [UN 2396]
Acrolein, 2-methyl-
Isobutenal
Methacrolein
Methacrylic aldehyde
Methakrylaldehyd (Czech)
α-Methylacrolein
2-Methylacrolein
Methylacrylaldehyde
2-Methylpropenal
UN 2396 [Methacrylaldehyde]

78-86-4
C_4H_9Cl
92.58
C(Cl)(CC)C
Butane, 2-chloro
sec-Butyl chloride
2-Chlorobutane[UN 1127]
UN 1127 [Chlorobutanes]

78-87-5
$C_3H_6Cl_2$
112.99
ClCC(Cl)C
Propane, 1,2-dichloro
Bichlorure de propylene (French)
α,β-Dichloropropane
1,2-Dichloropropane
1,2-Dichloropropane (OSHA)
Dwuchloropropan (Polish)
ENT 15,406
NCI-C55141
Propylene chloride
Propylene dichloride (ACGIH, OSHA)
α,β-Propylene dichloride
RCRA waste number U083

78-88-6
$C_3H_4Cl_2$
110.97
C(=C)(CCl)Cl
Propene, 2,3-dichloro
1,2-Dichloro-2-propene
2,3-Dichloropropene
2,3-Dichloro-1-propene
2,3-Dichloropropylene

78-89-7
C_3H_7ClO
94.55
OCC(Cl)C
1-Propanol, 2-chloro
2-Chloropropanol
2-Chloro-1-propanol
2-Chloropropyl alcohol
Propylenechlorohydrin
Propylene chlorohydrin [UN 2611]
UN 2611 [Propylene chlorohydrin]

78-90-0
$C_3H_{10}N_2$
74.15
NCC(N)C
1,2-Propanediamine
1,2-Diaminopropane
Propylenediamine
1,2-Propylenediamine [UN 2258]
Propylene diamine
UN 2258 [1,2-Propylenediamine]

78-92-2
$C_4H_{10}O$
74.14
OC(CC)C
sec-Butyl alcohol
Alcool butylique secondaire (French)
s-Butanol
sec-Butanol
Butan-2-ol
Butanol-2
2-Butanol
sec-Butanol [UN 1120]
Butanol secondaire (French)
s-Butyl alcohol
2-Butyl alcohol
sec-Butyl alcohol (ACGIH, OSHA) [UN 1120]
Butylene hydrate
CCS 301
Ethylmethyl carbinol
2-Hydroxybutane
Methylethylcarbinol
1-Methyl-1-propanol
1-Methyl propanol
1-Methylpropyl alcohol
S.B.A.
UN 1120 [Butanols]

78-93-3
C_4H_8O
72.12
O=C(CC)C
2-Butanone
Acetone, methyl-
Aethylmethylketon (German)
Butanone
Butanone 2 (French)
2-Butanone (OSHA)
3-Butanone
Ethyl methyl cetone (French)
Ethylmethylketon (Dutch)
Ethyl methyl ketone [UN 1193]
Ketone, ethyl methyl
Meetco
MEK (OSHA)
Methyl acetone (DOT)
Methyl ethyl ketone (ACGIH, OSHA) [UN 1193]
Metiletilchetone (Italian)
Metyloetyloketon (Polish)
RCRA waste number U159
UN 1193 [Ethyl methyl ketone or methyl ethyl ketone]
UN 1232 (DOT)

78-94-4
C_4H_6O
70.10
O=C(C=C)C
3-Buten-2-one
Acetone, methylene-
Acetyl ethylene
3-Butene-2-one
Butenone
δ^3-2-Butenone
Ketone, methyl vinyl
Methylene acetone
Methyl-vinyl-cetone (French)
Methylvinylketon (German)
Methylvinyl ketone
Methyl vinyl ketone [UN 1251]
Methyl vinyl ketone, Inhibited (DOT)
γ-Oxo-α-butylene
UN 1251 [Methyl vinyl ketone]
Vinyl methyl ketone

78-95-5
C_3H_5ClO
92.53
O=C(CCl)C
2-Propanone, 1-chloro
Acetone, chloro-
Acetonyl chloride
A-Stoff

Chloracetone (French)
Chloroacetone
Chloroacetone, Stabilized [UN 1695]
Chloropropanone
1-Chloro-2-propanone
Monochloracetone
Monochloroacetone
Monochloroacetone, Inhibited [UN 1695]
Monochloroacetone, Stabilized [UN 1695]
Monochloroacetone, Unstabilized [UN 1695]
Tonite
UN 1695 [Chloroacetone, stabilized]

78-96-6
C_3H_9NO
75.13
OC(CN)C
2-Propanol, 1-amino
α-Aminoisopropyl alcohol
1-Aminopropan-2-ol
1-Amino-2-propanol
2-Hydroxypropylamine
2-Hydroxy-1-propylamine
Isopropanolamine
mono-Iso-propanolamine
1-Methyl-2-aminoethanol
Threamine

78-97-7
C_3H_5NO
71.09
N#CC(O)C
Lactonitrile
2-Hydroxypropannitril (Czech)
2-Hydroxypropionitrile
Laktonitril (Czech)
Propionitrile, 2-hydroxy-

78-98-8
$C_3H_4O_2$
72.07
O=CC(=O)C
Pyruvaldehyde
Acetylformaldehyde
Acetylformyl
Glyoxal, methyl
α-Ketopropionaldehyde
1-Ketopropionaldehyde
2-Ketopropionaldehyde
Methylglyoxal
NSC-79019
2-Oxopropanal
Propanal, 2-oxo- (9CI)
Propanedione
Propanolone
Propionaldehyde, 2-keto
Propionaldehyde, 2-oxo-
Pyroracemic aldehyde
Pyruvic aldehyde

78-99-9
$C_3H_6Cl_2$
112.99
CCC(Cl)Cl
Propane, 1,1-dichloro
1,1-Dichloropropane
Propylidene chloride

79-00-5
$C_2H_3Cl_3$
133.40
ClCC(Cl)Cl
Ethane, 1,1,2-trichloro
Ethane trichloride
NCI-C04579
RCRA waste number U227
RCRA waste number U359
β-T
β-Trichloroethane
1,1,2-Trichlorethane
1,1,2-Trichloroethane (ACGIH, OSHA)
1,2,2-Trichloroethane
Trojchloroetan(1,1,2) (Polish)
Vinyl trichloride

79-01-6
C_2HCl_3
131.38
C(=CCl)(Cl)Cl
Ethylene, trichloro
Acetylene trichloride
Algylen
Anamenth
Benzinol
Blacosolv
Blancosolv
Cecolene
Chlorilen
1-Chloro-2,2-dichloroethylene
Chlorylea
Chlorylen
Chorylen
Circosolv
Crawhaspol
Densinfluat
1,1-Dichloro-2-chloroethylene
Dow-Tri
Dukeron
Ethinyl trichloride
Ethylene trichloride
Fleck-Flip
Flock Flip
Fluate
Gemalgene
Germalgene
Lanadin
Lethurin
Narcogen
Narkogen
Narkosoid
NCI-C04546
Nialk
Perm-A-Chlor
Perm-A-Clor
Petzinol
Philex
RCRA waste number U228
TCE
Threthylen
Threthylene
Trethylene
Tri
Triad
Trial
Triasol
Trichlooretheen (Dutch)
Trichloorethyleen, tri (Dutch)
Trichloraethen (German)
Trichloraethylen, tri (German)
Trichloran
Trichloren
Trichlorethene (French)
Trichlorethylene
Trichlorethylene, tri (French)
Trichloroethene
Trichloroethylene (ACGIH, OSHA) [UN 1710]
1,1,2-Trichloroethylene
1,2,2-Trichloroethylene
Tri-Clene
Tricloretene (Italian)
Tricloroetilene (Italian)
Trielene
Trielin
Trielina (Italian)
Trieline
Triklone
Trilen
Trilene
Triline
Trimar
Triol
Tri-Plus
Tri-Plus M
UN 1710 [Trichloroethylene]
Vestrol
Vitran
Westrosol

79-02-7
$C_2H_2Cl_2O$
112.94
O=CC(Cl)Cl
Acetaldehyde, 2,2-dichloro
Acetaldehyde, dichloro-
Chloraldehyde
α,α-Dichloroacetaldehyde
2,2-Dichloroacetaldehyde

79-03-8
C_3H_5ClO
92.52
O=C(CC)Cl
Propionyl chloride (8CI)
Chlorure de propionyle (French)
Cloruro de propionilo (Spanish)
NSC-83547
Propanoyl chloride (9CI)
Propionic acid chloride
Propionic chloride
UN 1815 [Propionyl chloride]

79-04-9
$C_2H_2Cl_2O$
112.94
O=C(CCl)Cl
Acetyl chloride, chloro
Chloracetyl chloride
Chlorid kyseliny chloroctove (Czech)
Chloroacetic acid chloride

Chloroacetic chloride
Chloroacetyl chloride (ACGIH, OSHA) [UN 1752]
Chlorure de chloracetyle (French)
Monochloroacetyl chloride
UN 1752 [Chloroacetyl chloride]

79-05-0
C_3H_7NO
73.11
O=C(N)CC
Propionamide
Amid kyseliny propionove (Czech)
Propanamide
Propionic acid amide
Propionic amide

79-06-1
C_3H_5NO
71.09
O=C(N)C=C
Acrylamide (ACGIH,OSHA) [UN 2074]
Acrylic amide
Akrylamid (Czech)
Amid kyseliny akrylove (Czech)
Ethylenecarboxamide
Propenamide
2-Propenamide (9CI)
RCRA waste number U007
UN 2074 [Acrylamide]
Vinyl amide

79-07-2
C_2H_4ClNO
93.52
O=C(N)CCl
Acetamide, 2-chloro
Chloracetamid (German)
Chloracetamide
Chloroacetamide
α-Chloroacetamide
2-Chloroacetamide
2-Chloroethanamide
Mergal AF
Microcide
USAF DO-29

79-08-3
$C_2H_3BrO_2$
138.96
O=C(O)CBr
Acetic acid, bromo
Acide bromacetique (French)
Bromoacetate ion
Bromoacetic acid
α-Bromoacetic acid
Bromoacetic acid, Solid [UN 1938]
Bromoacetic acid, Solution [UN 1938]
Bromoethanoic acid
α-Bromoethanoic acid
Kyselina bromoctova (Czech)
Monobromessigsaeure (German)
Monobromoacetic acid
TO NTU
UN 1938 [Bromoacetic acid, solution]

79-09-4
$C_3H_6O_2$
74.09
O=C(O)CC
Propionic-acid
Acide propionique (French)
Carboxyethane
Ethanecarboxylic acid
Ethylformic acid
Kyselina propionova (Czech)
Luprosil
Metacetonic acid
Methyl acetic acid
Propionic acid (ACGIH,OSHA) [UN 1848]
Propionic acid, Solution (DOT)
Propionic acid, Solution containing not less than 80% acid (DOT)
Propionic Acid Grain Preserver
Prozoin
Pseudoacetic acid
Sentry Grain Preserver
Tenox P Grain Preservative
UN 1848 [Propionic acid]

79-10-7
$C_3H_4O_2$
72.07
O=C(O)C=C
Acrylic-acid
Acroleic acid
Acrylic acid (ACGIH,DOT, OSHA)
Acrylic acid, Inhibited (DOT)
Acrylic acid, glacial
Ethylenecarboxylic acid
Glacial acrylic acid
Kyselina akrylova (Czech)
Propene acid
Propenoic acid
2-Propenoic acid (9CI)
RCRA waste number U008
UN 2218 (DOT)
Vinylformic acid

79-11-8
$C_2H_3ClO_2$
94.50
O=C(O)CCl
Acetic acid, chloro
Acide chloracetique (French)
Acide monochloracetique (French)
Acidomonocloroacetico (Italian)
Chloracetic acid
Chloroacetic acid
α-Chloroacetic acid
Chloroacetic acid, Liquid [UN 1750]
Chloroacetic acid, Solid [UN 1751]
Chloroacetic acid, Solution [UN 1750]
Chloroethanoic acid
Kyselina chloroctova (Czech)
MCA
Monochloorazijnzuur (Dutch)
Monochloracetic acid
Monochloressigsaeure (German)
Monochloroacetic acid
Monochloroethanoic acid
NCI-C60231
UN 1750 [Chloroacetic acid, Liquid]
UN 1751 [Chloroacetic acid, solid]

79-14-1
$C_2H_4O_3$
76.06
O=C(O)CO
Glycolic-acid
Acetic acid, hydroxy-
Kyselina glykolova (Czech)
Kyselina hydroxyoctova (Czech)
Hydroxyacetic acid
Hydroxyethanoic acid

79-15-2
C_2H_4BrNO
137.96
O=C(NBr)C
N-Bromoacetamide
Acetamide, N-bromo- (9CI)

79-16-3
C_3H_7NO
73.11
O=C(NC)C
Acetamide, N-methyl
Methylacetamide
N-Methylacetamide
Monomethylacetamide

79-17-4
CH_6N_4
74.11
N=C(NN)N
Guanidine, amino
Aminate base
Aminoguanidine
Guanyl hydrazine
Hydrazinecarboximidamide

79-19-6
CH_5N_3S
91.15
N=C(S)NN
Semicarbazide, thio
N-Aminothiourea
Hydrazinecarbothioamide
RCRA waste number P116
Semicarbazide, 3-thio-
Thiocarbamylhydrazine
Thiosemicarbazide
TSC
USAF EK-1275

79-20-9
$C_3H_6O_2$
74.09
O=C(OC)C
Acetic acid, methyl ester
Acetate de methyle (French)
Devoton
Ethyl ester of monoacetic acid
Methylacetaat (Dutch)
Methylacetat (German)
Methyl acetate (ACGIH,OSHA) [UN 1231]
Methyle (acetate de) (French)
Methylester kiseliny octove (Czech)
Methyl ethanoate
Metile (acetato di) (Italian)
Octan metylu (Polish)
Tereton
UN 1231 [Methyl acetate]

79-21-0
$C_2H_4O_3$
76.06
O=C(OO)C
Peroxoacetic acid
Acetic peroxide
Acetyl hydroperoxide
Acide peracetique (French)
Acide peroxyacetique (French)
Acido peroxiacetico (Spanish)
Caswell No. 644
Desoxon 1
Estosteril
Ethane peroxoic acid
Ethaneperoxoic acid (9CI)
EPA Pesticide Chemical Code 063201
Hydroperoxide, acetyl
Kyselina peroxyoctova (Czech)
Monoperacetic acid
NA 2131 (DOT)
Osbon AC
Peracetic acid
Peroxyacetic acid (8CI)
Peroxyacetic-acid
Peroxyacetic acid, More than 43% with more than 6% hydrogen peroxide (DOT)
Peracetic acid, Solution not over 43% acid and not over 6% hydrogen peroxide (DOT)
Proxitane 4002
UN 2131 (DOT)

79-22-1
$C_2H_3ClO_2$
94.50
O=C(OC)Cl
Carbonochloridic acid, methyl ester
Chlorameisensaeure methylester (German)
Chlorocarbonate de methyle (French)
Chlorocarbonic acid methyl ester
Chloroformiate de methyle (French)
Chloroformic acid methyl ester
Formic acid, chloro-, methyl ester
MCF
Methoxycarbonyl chloride
Methylchloorformiaat (Dutch)
Methyl chlorocarbonate [UN 1238]
Methyl chloroformate [UN 1238]
Methylester kyseliny chlorm-ravenci (Czech)
Methylester kyseliny chloruhli-cite (Czech)
Metilcloroformiato (Italian)
RCRA waste number U156
TL 438
UN 1238 [Methyl chloroform-ate]

79-24-3
$C_2H_5NO_2$
75.08
O=N(=O)CC
Ethane, nitro
Nitroetan (Polish)
Nitroethane (ACGIH,OSHA) [UN 2842]
UN 2842 [Nitroethane]

79-27-6
$C_2H_2Br_4$
345.68
BrC(Br)C(Br)Br
Ethane, 1,1,2,2-tetrabromo
Acetylene tetrabromide (ACGIH,OSHA) [UN 2504]
Muthmann's liquid
TBE
1,1,2,2-Tetrabromaethan (German)
Tetrabromoacetylene
1,1,2,2-Tetrabromoetano (Italian)
Tetrabromoethane [UN 2504]
s-Tetrabromoethane
1,1,2,2-Tetrabromoethane
1,1,2,2-Tetrabroomethaan (Dutch)
UN 2504 [Tetrabromoethane]

79-28-7
C_2Br_4
343.64
C(=C(Br)Br)(Br)Br
Ethene, tetrabromo- (9CI)
Ethylene, tetrabromo- (8CI)
NSC-328430
Tetrabromoethene
Tetrabromoethylene

79-29-8
C_6H_{14}
86.20
C(C(C)C)(C)C
Butane, 2,3-dimethyl
2,3-Dimethylbutane [UN 2457]
UN 2457 [2,3-Dimethylbutane]

79-31-2
$C_4H_8O_2$
88.12
O=C(O)C(C)C
Isobutyric-acid
Acetic acid, dimethyl-
Dimethylacetic acid
Isobutyric acid [UN 2529]
Isopropylformic acid
Kyselina isomaselna (Czech)
α-Methylpropionic acid
2-Methylpropanoic acid
2-Methylpropionic acid
Propionic acid, 2-methyl-
UN 2529 [Isobutyric acid]

79-34-5
$C_2H_2Cl_4$
167.84
C(C(Cl)Cl)(Cl)Cl
Ethane, 1,1,2,2-tetrachloro
Acetylene tetrachloride
Bonoform
Cellon
1,1,2,2-Czterochloroetan (Polish)
1,1-Dichloro-2,2-dichloroethane
NCI-C03554
RCRA waste number U209
TCE
1,1,2,2-Tetrachloorethaan (Dutch)
1,1,2,2-Tetrachloraethan (German)
Tetrachlorethane
1,1,2,2-Tetrachlorethane (French)
Tetrachloroethane [UN 1702]
1,1,2,2-Tetrachloroethane (ACGIH,DOT,OSHA)
s-Tetrachloroethane
sym-Tetrachloroethane
1,1,2,2-Tetracloroetano (Italian)
Tetrachlorure d'acetylene (French)
UN 1702 [Tetrachloroethane]
Westron

79-35-6
$C_2Cl_2F_2$
132.92
FC(F)=C(Cl)Cl
Ethylene, 1,1-dichloro-2,2-difluoro
1,1-Dichloro-2,2-difluoro-ethylene
1,1-Difluoro-2,2-dichloro-ethylene
Ethene, 1,1-dichloro-2,2-di-fluoro-
Genetron 1112A
Genetrone 1112A

79-36-7
C_2HCl_3O
147.38
O=C(C(Cl)Cl)Cl
Acetyl chloride, dichloro

Chlorid kyseliny dichloroctove (Czech)
Chlorure de dichloracetyle (French)
Dichloracetyl chloride
Dichloroacetyl chloride [UN 1765]
α,α-Dichloroacetyl chloride
2,2-Dichloroacetyl chloride
Dichloroethanoyl chloride
UN 1765 [Dichloroacetyl chloride]

79-38-9
C_2ClF_3
116.47
FC(F)=C(F)Cl
Ethylene, chlorotrifluoro
Chlorotrifluoroethylene (DOT)
1-Chloro-1,2,2-trifluoroethylene
2-Chloro-1,1,2-trifluoroethylene
Chlortrifluoraethylen (German)
CTFE
Daiflon
Ethylene, trifluorochloro-
Fluoroplast 3
Genetron 1113
Monochlorotrifluoroethylene
Trifluorchlorethylen (Czech)
Trifluorochloroethylene (DOT)
1,1,2-Trifluoro-2-chloroethylene
Trifluorochloroethylene, Inhibited [UN 1082]
Trifluoromonochloroethylene
Trifluorovinyl chloride
Trithene
UN 1082 [Trifluorochloroethylene, inhibited]

79-39-0
C_4H_7NO
85.12
O=C(N)C(=C)C
2-Propenamide, 2-methyl
Amid kyseliny methakrylove (Czech)
Methacrylamide (8CI)
Methacrylic acid amide
Methacrylic amide
2-Methylacrylamide
α-Methyl acrylic amide
2-Methylpropenamide
Mhoromer BM801
USAF RH-1

79-40-3
$C_2H_4N_2S_2$
120.20
N=C(S)C(=N)S
Oxamide, dithio
Dithiooxamide
Dithioxamide
Ethanedithioamide
Hydrorubeanic acid
Oxaldiimidic acid, dithio-
Rubeane
Rubeanic acid
RVK
USAF EK-4394
USAF MK-6

79-41-4
$C_4H_6O_2$
86.10
O=C(O)C(=C)C
Methacrylic-acid
Acrylic acid, 2-methyl-
Kyselina methakrylova (Czech)
Methacrylic acid (ACGIH, OSHA)
Methacrylic acid, Inhibited [UN 2531]
α-Methylacrylic acid
2-Methylpropenoic acid
Propionic acid, 2-methylene-
UN 2531 [Methacrylic acid, inhibited]

79-43-6
$C_2H_2Cl_2O_2$
128.94
O=C(O)C(Cl)Cl
Acetic acid, dichloro
Bichloracetic acid
DCA
Dichloracetic acid
Dichlorethanoic acid
Dichloroacetic acid
2,2-Dichloroacetic acid
Dichloroacetic acid [UN 1764]
Dichloroethanoic acid
Kyselina dichloroctova (Czech)
UN 1764 [Dichloroacetic acid]
Urner's liquid

79-44-7
C_3H_6ClNO
107.55
O=C(N(C)C)Cl
Carbamoyl chloride, dimethyl
Carbamyl chloride, N,N-dimethyl-
Chlorid kyseliny dimethylkarbaminove (Czech)
Chloroformic acid dimethylamide
DDC
Dimethylamid kyseliny chlormravenci (Czech)
(Dimethylamino)carbonyl chloride
N,N-Dimethylaminocarbonyl chloride
Dimethylcarbamic acid chloride
N,N-Dimethylcarbamic acid chloride
Dimethylcarbamic chloride
Dimethylcarbamidoyl chloride [UN 2262]
N,N-Dimethylcarbamidoyl chloride
Dimethylcarbamoyl chloride
N,N-Dimethylcarbamoyl chloride
Dimethyl carbamoyl chloride (ACGIH)
Dimethylcarbamyl chloride
N,N-Dimethylcarbamyl chloride
Dimethylchloroformamide
Dimethylkarbamoylchlorid (Czech)
DMCC
RCRA waste number U097
TL 389
UN 2262 [Dimethylcarbamoyl chloride]

79-45-8
$C_3H_7NS_2$
121.23
N(C(S)=S)(C)C
Carbamic acid, dimethyldithio
Carbamodithioic acid, dimethyl- (9CI)
Dimethylcarbamodithioic acid
Dimethyldithiocarbamate
Dimethyldithiocarbamic acid
N,N-Dimethyldithiocarbamic acid

79-46-9
$C_3H_7NO_2$
89.11
O=N(=O)C(C)C
Propane, 2-nitro
Dimethylnitromethane
Isonitropropane
Nipar S-20
Nipar S-20 solvent
Nipar S-30 solvent
Nitroisopropane
β-Nitropropane
2-Nitropropane (ACGIH,OSHA)
2-NP
RCRA waste number U171

79-53-8
C_3ClF_5O
182.48
FC(F)(F)C(=O)C(F)(F)Cl
2-Propanone, 1-chloro-1,1,3,3,3-pentafluoro
Acetone, 1-chloro-1,1,3,3,3-pentafluoro-
Acetone, monochloropentafluoro-
Chloropentafluoroacetone
Pentafluoromonochloroacetone
2-Propanone, 1-monochloro-1,1,3,3,3-pentafluoro-

79-54-9
$C_{20}H_{30}O_2$
302.46
CC(C)C1=CCC2C(=C1)CCC3C2(C)CCCC3(C)C(O)=O
Podocarpa-8(14),12-dien-15-oic acid, 13-isopropyl- (8CI)
δ 6,8(14)-Abietadienoic acid
Levopimaric acid
NSC-4827
1-Phenanthrenecarboxylic acid, 1,2,3,4,4a,4b,5,9,10,10a-decahydro-1,4a-dimethyl-7-(1-methylethyl)-, (1R-(1α,4aβ,

4bα,10aα))- (9CI)
β-Pimaric acid
l-Pimaric acid
l-Sapietic acid

79-57-2
$C_{22}H_{24}N_2O_9$
460.48
O=C(N)C(=C(O)C(N(C)C)C(C1(O)C(O)=C(C2C(O)(c(c3c(O)cc4)c4)C)C3=O)C2O)C1=O
2-Naphthacenecarboxamide, 4-(dimethylamino)-1,4,4a,5,5a,6,11,12a-octahydro-3,5,6,10,12,12a- hexahydroxy-6-methyl-1,11-dioxo
Biostat
Biostat PA
5-Hydroxytetracycline
NCI-C56473
OTC
Oxitetracyclin
Oxymykoin
Oxyterracin
Oxyterracine
Oxyterracyne
Oxytetracycline
Oxytetracycline amphoteric
Riomitsin
Ryomycin
Taomycin
Taomyxin
Terrafungine
Terramitsin
Terramycin
Tetracycline, 5-hydroxy-
Tetran

79-64-1
$C_{23}H_{32}O_2$
340.55
CC#CC3(O)CCC4C2CC(C)C1=CC(=O)CCC1(C)C2CCC34C
Androst-4-en-3-one, 17-β-hydroxy-6-α-methyl-17-(1-propynyl)
Androst-4-en-3-one, 17-hydroxy-6-methyl-17-(1-propynyl)-, (6-α,17-β)-
Dimethesterone
Dimethisteron
Dimethisterone
6-α,21-Dimethyl-17-α-ethinyl-testosterone
6-α,21-Dimethylethisterone
6-α,21-Dimethyl-17-β-hydroxy-17-α-pregn-4-en-20-yn-3-one
17-α-Ethynyl-6-α,21-dimethyl-testosterone
17-α-Ethynyl-17-hydroxy-6-α,21-dimethylandrost-4-en-3-one
(6-α,17-β)-17-Hydroxy-6-methyl-17-(1-propynyl)-androst-4-en-3-one
17-β-Hydroxy-6-α-methyl-17-(1-propynyl)androst-4-en-3-one
Lutogan
Lutosan
6-α-Methyl-17-α-propynyltest-osterone
6-α-Methyl-17-(1-propynyl)test-osterone
P-5048
17-α-Pregn-4-en-20-yn-3-one, 6-α,21-dimethyl-17-hydroxy-
Secrosteron

79-69-6
$C_{14}H_{22}O$
206.36
O=C(C=CC(C(=CCC1C)C)C1(C)C)C
3-Buten-2-one, 4-(2,5,6,6-tetramethyl-2-cyclohexen-1-yl)-, cis
3-Buten-2-one, 4-(2,5,6,6-tetramethyl-2-cyclohexen-1-yl)-(9CI)
α-Irone
4-(2,5,6,6-Tetramethyl-2-cyclohexen-1-yl)-3-buten-2-one

79-74-3
$C_{16}H_{26}O_2$
250.42
Oc(c(cc(O)c1C(CC)(C)C)C(CC)(C)C)c1
Hydroquinone, 2,5-di-t-pentyl
2,5-Bis(1,1-dimethylpropyl)-hydroquinone
2,5-Di-tert-amylbenzene-1,4-diol
2,5-Di-t-amylhydroquinone
2,5-Di-tert-pentylbenzene-1,4-diol
2,5-Di-t-pentylhydroquinone
Hydroquinone, 2,5-di-tert-amyl-
Santouar A
Santovar A
USAF B-21

79-77-6
$C_{13}H_{20}O$
192.33
O=C(C=CC(=C(CCC1)C)C1(C)C)C
3-Buten-2-one, 4-(2,6,6-trimethyl-1-cyclohexen-1-yl)-, (e)
β-Ionone
(e)-β-Ionone
trans-β-Ionone
4-(2,6,6-Trimethyl-1-cyclohexen-1-yl)-3-buten-2-one

79-78-7
$C_{16}H_{24}O$
232.40
O=C(C=CC(C(=CCC1)C)C1(C)C)CCC=C
1,6-Heptadien-3-one, 1-(2,6,6-trimethyl-2-cyclohexen-1-yl)
Allyl α-ionone
Ionone, allyl α-
1-(2,6,6-Trimethyl-2-cyclohexen-1-yl)-1,6-heptadien-3-one

79-83-4
$C_9H_{17}NO_5$
219.27
CC(C)(CO)C(O)C(=O)NCCC(O)=O
Pantothenic acid, D
β-Alanine, N-(2,4-dihydroxy-3,3-dimethyl-1-oxobutyl)-, (R)- (9CI)
Chick Antidermatitis Factor
Kyselina pantothenova (Czech)
Pantothenic acid
(+)-Pantothenic acid
D-Pantothenic acid
Vitamin B3
Vitamin B5

79-92-5
$C_{10}H_{16}$
136.26
C(C(CC1C2)C2)(C1(C)C)=C
Camphene (DOT)
NA 9011 (DOT)

79-94-7
$C_{15}H_{12}Br_4O_2$
543.88
Oc(c(cc(c1)C(c(cc(c(O)c2Br)Br)c2)(C)C)Br)c1Br
Tetrabromobisphenol A
BA 59
2,2-Bis(3,5-dibromo-4-hydroxy-phenyl)propane
2,2-Bis(4-hydroxy-3,5-dibromo-phenyl)propane
Bromdian
FG 2000
Fire Guard 2000
Firemaster BP4A
4,4'-Isopropylidenebis(2,6-dibromophenol)
4,4'-Isopropylylidenebis(2,6-dibromophenol)
4,4'-(1-Methylethylidene)bis-(2,6-dibromophenol)
NSC-59775
Phenol, 4,4'-isopropylidenebis-(dibromo- (VAN)
Phenol, 4,4'-isopropylidenebis-(2,6-dibromo- (8CI)
Phenol, 4,4'-(1-methylethylidene)bis(2,6-dibromo- (9CI)
2,2',6,6'-Tetrabromobisphenol A
3,5,3',5'-Tetrabromobisphenol A
Tetrabromodian

79-95-8
$C_{15}H_{12}Cl_4O_2$
366.07
Oc(c(cc(c1)C(c(cc(c(O)c2Cl)Cl)c2)(C)C)Cl)c1Cl
Phenol, 4,4'-isopropylidene-

bis(2,6-dichloro
2,2-Bis(3,5-dichloro-4-hydroxyphenyl)propane
2,2-Bis(4-hydroxy-3,5-dichlorophenyl)propane
4,4'-Isopropylidenebis(2,6-dichlorophenol)
4,4'-(1-Methylethylidene)-bis(2,6-dichlorophenol)
Phenol, 4,4'-(1-methylethylidene)bis(2,6-dichloro-
Tetrachlordian (Czech)
2,2',6,6'-Tetrachlorobisphenol A

79-96-9
$C_{23}H_{32}O_2$
340.55
Oc(c(cc(c1)C(c(ccc(O)c2C(C)(C)C)c2)(C)C)C(C)(C)C)c1
Phenol, (2,2'-di-tert-butyl-4,4'-isopropylene)di
4,4'-Isopropylidene-bis-(2-t-butylphenol)

80-00-2
$C_{12}H_9ClO_2S$
252.72
O=S(=O)(c(ccc(c1)Cl)c1)c(cccc2)c2
Sulfone, p-chlorophenyl phenyl
4-Chlordifenylsulfon (Czech)
4-Chlorodiphenyl sulfone
4-Chlorodiphenyl sulphone
p-Chlorophenyl phenyl sulphone
1-Chloro-4-(phenylsulfonyl)-benzene
Compound R-242
ENT 17,941
p-Monochlorophenyl phenyl sulfone
R-242
R-242-B
Sulfenone
Sulphenone

80-04-6
$C_{15}H_{28}O_2$
240.39
OC(CCC(C(C(CCC(O)C1)C1)(C)C)C2)C2
Cyclohexanol, 4,4'-(1-methylethylidene)bis- (9CI)
AI3-25180
2,2-Bis(4-hydroxycyclohexyl)-propane
Cyclohexanol, 4,4'-isopropylidenedi- (8CI)
Dodecahydrobisphenol A
HBPA
Hydrogenated bisphenol A
1,1'-Isopropylidenebis(4-cyclohexanol)
4,4'-Isopropylidenedicyclohexanol
4,4'-(1-Methylethylidene)bis-cyclohexanol
NSC-8990

80-05-7
$C_{15}H_{16}O_2$
228.31
Oc(ccc(c1)C(c(ccc(O)c2)c2)(C)C)c1
Phenol, 4,4'-isopropylidenedi
Bisferol A (Czech)
2,2-Bis-4'-hydroxyfenylpropan (Czech)
Bis(4-hydroxyphenyl) dimethylmethane
Bis(4-hydroxyphenyl)propane
2,2-Bis(p-hydroxyphenyl)-propane
2,2-Bis(4-hydroxyphenyl)-propane
Bisphenol
Bisphenol A
4,4'-Bisphenol A
Dian
p,p'-Dihydroxydiphenyldimethylmethane
4,4'-Dihydroxydiphenyldimethylmethane
p,p'-Dihydroxydiphenylpropane
2,2-(4,4'-Dihydroxydiphenyl)-propane
4,4'-Dihydroxydiphenylpropane
4,4'-Dihydroxydiphenyl-2,2-propane
4,4'-Dihydroxy-2,2-diphenylpropane
Dimethylmethylene-p,p'-diphenol
β-Di-p-hydroxyphenylpropane
2,2-Di(4-hydroxyphenyl)-propane
Dimethyl bis(p-hydroxyphenyl)methane
Diphenylolpropane
2,2-Di(4-phenylol)propane
p,p'-Isopropylidenebisphenol
4,4'-Isopropylidenebisphenol
p,p'-Isopropylidenediphenol
NCI-C50635
Phenol, 4,4'-dimethylmethylenedi-
Propane, 2,2-bis(p-hydroxyphenyl)-

80-06-8
$C_{14}H_{12}Cl_2O$
267.16
CC(O)(c1ccc(Cl)cc1)c2ccc(Cl)cc2
Benzhydrol, 4,4'-dichloro-α-methyl
BCPE
1,1-Bis(p-chlorophenyl)ethanol
1,1-Bis(4-chlorophenyl)ethanol
Bis(p-chlorophenyl)methyl carbinol
1,1-Bis(p-chlorophenyl)methylcarbinol
1,1-Bis(4-chlorphenyl)-aethanol (German)
Chlorfenethol
DCPC
DCPE
Dichlorodiphenylethanol
p,p'-Dichlorodiphenylmethylcarbinol
4,4'-Dichloro-(methyl benzhydrol)
4,4'-Dichloro-α-methylbenzhydrol
4,4'-Dichloro-α-methylbenzohydrol
Di-(p-chlorophenyl)-ethanol
Di(p-chlorophenyl) methylcarbinol
Dimit
Dimite
DMC
ENT 9,624
Ethanol, 1,1-bis(p-chlorophenyl)-
Qikron

80-07-9
$C_{12}H_8Cl_2O_2S$
287.16
O=S(=O)(c(ccc(c1)Cl)c1)c(ccc(c2)Cl)c2
Sulfone, bis(p-chlorophenyl)

80-08-0
$C_{12}H_{12}N_2O_2S$
248.32
O=S(=O)(c(ccc(N)c1)c1)c(ccc(N)c2)c2
Aniline, 4,4'-sulfonyldi
Avlosulfon
Avlosulphone
Benzenamine, 4,4'-sulfonylbis-(9CI)
Bis(p-aminophenyl) sulfone
Bis(4-aminophenyl) sulfone
Bis(p-aminophenyl)sulphone
Bis(4-aminophenyl)sulphone
Croysulfone
Croysulphone
Dadps
Dapson
Dapsone
Dapsonum
DDS
DDS (Pharmaceutical)
DDS (Van)
4,4-Diaminodifenylsulfon (Czech)
Diaminodifenilsulfona (Spanish)
Diamino-4,4'-diphenyl sulfone
p,p'-Diaminodiphenyl sulfone
4,4'-Diaminodiphenyl sulfone
Diamino-4,4'-diphenyl sulphone
p,p-Diaminodiphenyl sulphone
4,4'-Diaminodiphenyl sulphone
Di(p-aminophenyl) sulfone
Di(4-aminophenyl)sulfone
Di(p-aminophenyl)sulphone
Di(4-aminophenyl)sulphone
Diaphenylsulfon
Diaphenylsulfone
Diaphenylsulphon
Diaphenylsulphone
Diphenasone
Diphone
Disulone

DSS
Dubronax
Dumitone
Eporal
1358F
F 1358
ICI
Maloprim
Metabolite C
NCI-C01718
Novophone
NSC-6091
NSC-6091D
Sulfona
Sulfona-MAE
Sulfone UCB
1,1'-Sulfonylbis(4-aminobenzene)
4,4'-Sulfonylbisaniline
p,p-Sulfonylbisbenzamine
4,4'-Sulfonylbisbenzamine
p,p-Sulfonylbisbenzenamine
p,p'-Sulfonyldianiline
4,4'-Sulfonyldianiline
Sulphadione
Sulphon-Mere
1,1'-Sulphonylbis(4-aminobenzene)
p,p-Sulphonylbisbenzamine
4,4'-Sulphonylbisbenzamine
p,p-Sulphonylbisbenzenamine
4,4'-Sulphonylbisbenzenamine
Sulphonyldianiline
p,p-Sulphonyldianiline
4,4'-Sulphonyldianiline
Tarimyl
Udolac
WR 448

80-09-1
$C_{12}H_{10}O_4S$
250.28
O=S(=O)(c(ccc(O)c1)c1)c(ccc(O)c2)c2
Phenol, 4,4'-sulfonyldi
4,4'-Sulfonyldiphenol

80-10-4
$C_{12}H_{10}Cl_2Si$
253.21
ClSi(Cl)(c1ccccc1)c2ccccc2
Silane, dichlorodiphenyl
Dichlor-difenylsilan (Czech)
Dichlorodiphenylsilane
Diphenyldichlorosilane [UN 1769]
Diphenyl dichlorosilane
UN 1769 [Diphenyldichlorosilane]

80-11-5
$C_8H_{10}N_2O_3S$
214.26
O=S(=O)(N(N=O)C)c(ccc(c1)C)c1
p-Toluenesulfonamide, N-methyl-N-nitroso
Diazald
Diazale
Methylnitroso-p-toluenesulfonamide
N-Methyl-N-nitroso-p-toluenesulfonamide
N-Nitroso-N-methyl-4-tolylsulfonamide
Toluene-p-sulfonylmethylnitrosamide
p-Tolylsulfonyl-methyl-nitrosamid (German)
p-Tolylsulfonylmethylnitrosamide
p-Tolylsulfonylmethylnitrosamine

80-13-7
$C_7H_5Cl_2NO_4S$
270.09
Benzoic acid, p-(dichlorosulfamoyl)
p-Carboxybenzenesulfondichloroamide
p-Dichlorosulfamoylbenzoic acid
p-(N,N-Dichlorosulfamyl)-benzoic acid
Halazone
Kyselina p-N,N-dichlorsulfamoylbenzoova (Czech)
Pantocid
Parasulfondichloramido benzoic acid
p-Sulfondichloramidobenzoic acid

80-15-9
$C_9H_{12}O_2$
152.21
O(O)C(c(cccc1)c1)(C)C
Hydroperoxide, α,α-dimethylbenzyl
Cumeenhydroperoxyde (Dutch)
Cumene hydroperoxide (DOT)
Cumene hydroperoxide, Technically pure (DOT)
Cument hydroperoxide
Cumenyl hydroperoxide
Cumolhydroperoxid (German)
Cumyl hydroperoxide
α-Cumyl hydroperoxide
Cumyl hydroperoxide, Technically pure (DOT)
α,α-Dimethylbenzyl hydroperoxide
Hydroperoxyde de cumene (French)
Hydroperoxyde de cumyle (French)
7-Hydroperoxykumen (Czech)
Idroperossido di cumene (Italian)
Idroperossido di cumolo (Italian)
Isopropylbenzene hydroperoxide
Kumenylhydroperoxid (Czech)
RCRA waste number U096
UN 2116 (DOT)

80-17-1
$C_6H_8N_2O_2S$
172.22
O=S(=O)(NN)c(cccc1)c1
Benzenesulfonic acid, hydrazide
Benzenesulfohydrazide
Benzenesulfonic hydrazide
Benzenesulfonohydrazide
Benzenesulfonyl hydrazide
Benzenesulfonyl hydrazine
Benzene sulphonohydrazide
Celogen BSH
Chkhz 9
Genitron BSH
Hydrazide BSG
Nitropore OBSH
Phenylsulfohydrazide
Phenylsulfonyl hydrazide
Phenylsulfonylhydrazine
Porofor BSH
Porofor-BSH-Pulver
Porofor ChKhZ 9

80-18-2
$C_7H_8O_3S$
172.21
O=S(=O)(OC)c(cccc1)c1
Benzenesulfonic acid, methyl ester
Methyl benzenesulfonate
20ND3-5

80-23-9
$C_7H_{10}N_2O_3S$
202.22
O=S(=O)(NC)c(ccc(O)c1N)c1
Benzenesulfonamide, 3-amino-4-hydroxy-N-methyl- (9CI)
3-Amino-4-hydroxy-N-methylbenzenesulfonamide

80-26-2
$C_{12}H_{20}O_2$
196.32
O=C(OC(C(CCC(=C1)C)C1)(C)C)C
p-Menth-1-en-8-ol, acetate
α-Terpineol, acetate
Terpinyl acetate

80-30-8
$C_{13}H_{19}NO_2S$
253.36
O=S(=O)(NC(CCCC1)C1)c(ccc(c2)C)c2
Benzenesulfonamide, N-cyclohexyl-4-methyl- (9CI)
AI3-15119
N-Cyclohexyl-4-methylbenzenesulfonamide
N-Cyclohexyl-p-toluenesulfonamide
NSC-14856
Santicizer 1H
p-Toluenesulfonamide, N-cyclohexyl- (8CI)

80-33-1

80-35-3

$C_{12}H_8Cl_2O_3S$
303.16
Clc2ccc(OS(=O)(=O)c1ccc(Cl)
cc1)cc2
Benzenesulfonic acid, 4-chloro-, 4-chlorophenyl ester
Acaricydol E 20
C-854
C 1,006
CCS
Chloorfenson (Dutch)
(4-Chloor-fenyl)-4-chloor-benzeen-sulfonaat (Dutch)
Chlorefenizon (French)
Chlorfenson
p-Chlorfenylester kyseliny p-chlorbenzensulfonove (Czech)
4-Chlorobenzenesulfonate de 4-chlorophenyle (French)
p-Chlorobenzenesulfonic acid, p-chlorophenyl ester
Chlorofenizon
Chlorofensone
p-Chlorophenyl p-chlorobenzenesulfonate
4-Chlorophenyl 4-chlorobenzenesulfonate
p-Chlorophenyl p-chlorobenzenesulphonate
4-Chlorophenyl 4-chlorobenzenesulphonate
4-Chlorphenyl-4'-chlorbenzolsulfonat (German)
(4-Chlor-phenyl)-4-chlor-benzol-sulfonate (German)
(4-Cloro-fenil)-4-cloro-venzolsolfonato (Italian)
Corotran
CPCBS
D 854
Difenson
ENT 16,358
Ephirsulphonate
Ester sulfonate
Estonmite
Ethersulfonate
Genite 883
K 6451
Lethalaire G-58
Miticide K-101
Niagaratran
Onex
Orthotran
Otracid
Ovatran
Ovatron
Ovex
Ovochlor
Ovotox
Ovotran
Parachlorophenyl-parachlorobenzene-sulfonate
PCPCBS
Sappilan
Sappiran
Trichlorfenson (Obs.)

80-35-3
$C_{11}H_{12}N_4O_3S$
280.33
COc2ccc(NS(=O)(=O)c1ccc(N)
cc1)nn2
Sulfanilamide, N[1]-(6-methoxy-3-pyridazinyl)
Altezol
4-Amino-N-(6-methoxy-3-pyridazinyl)-benzenesulfonamide (9CI)
3-(p-Aminobenzenesulfamido)-6-methoxypyridazine
3-p-Aminobenzenesulphonamido-6-methoxypyridazine
CL 13494
Davosin
Depovernil
Durox
Kineks
Kinex
Kynex
Lederkyn
Lentac
Lisulfen
Longin
Medicel
N[1]-(6-Methoxy-3-pyridazinyl)-sulfanilamide
6-Methoxy-3-sulfanilamidopyridazine
3-Methoxy-6-sulfanylamidopyridazine
Midicel
Midikel
Myasul
Mylosul
Opinsul
Paramid
Paramid Supra
Petrisul
Piridolo
Quinoseptyl
Retamid
Retasulfin
RP 7522
Slosul
SMOP
SMP
Spofadazine
Sulfalex
Sulfamethoxypyridazine
3-Sulfa-6-methoxypyridazine
Sulfametoxipiridazine
3-Sulfanilamide-6-methoxypyridazine
3-Sulfanilamido-6-methoxypyridazine
6-Sulfanilamido-3-methoxypyridazine
Sulfdurazin
Sulfapyridazine
Sulfozona
Sulphamethoxypyridazine
Sultirene
Surirene
Vinces

80-38-6
$C_{12}H_9ClO_3S$
268.72
Clc2ccc(OS(=O)(=O)c1ccccc1)
cc2
Benzenesulfonic acid, p-chlorophenyl ester
Aracid
Benzenesulfonate de 4-chlorophenyle (French)
Benzenesulfonic acid, 4-chlorophenyl ester
(4-Chloor-fenyl)-benzeen-sulfonaat (Dutch)
p-Chlorofenylester kyseliny benzensulfonove (Czech)
p-Chlorophenyl benzenesulfonate
4-Chlorophenyl benzenesulfonate
p-Chlorophenyl benzenesulphonate
4-Chlorophenyl benzenesulphonate
(4-Chlor-phenyl)-benzolsulfonat (German)
(4-Cloro-fenil)-benzol-solfonato (Italian)
CPB
CPBS
ENT 4,585
Fenizon (French)
Fenson
Fensone
GC-928
Murvesco
PCBS
PCI
PCPB
PCPBS
Trifenson

80-39-7
$C_9H_{13}NO_2S$
199.27
O=S(=O)(NCC)c(ccc(c1)C)c1
Ethyl Toluenesulfonamide
Benzenesulfonamide, N-ethyl-4-methyl- (9CI)
N-Ethyl-p-methylbenzenesulfonamide
N-Ethyl-4-methylbenzenesulfonamide
N-Ethyl-p-toluenesulfonamide
N-Ethyl-p-tolylsulfonamide
N-Tosylethylamine
NSC-68803
Santicizer 3
p-Tolueneethylsulfonamide
p-Toluenesulfonamide, N-ethyl- (8CI)
p-Toluenesulfonyl-N-ethylamide

80-40-0
$C_9H_{12}O_3S$
200.27
O=S(=O)(OCC)c(ccc(c1)C)c1
p-Toluenesulfonic acid, ethyl ester
Ethylester kyseliny p-toluensulfonove (Czech)
Ethyl p-methyl benzenesulfonate
Ethyl PTS
Ethyl-p-toluenesulfonate

Ethyl tosylate
Ethyl p-tosylate
p-Toluolsulfonsaeure aethyl ester (German)

80-42-2
$C_9H_{12}O_3S$
200.26
O=S(=O)(OCCC)c(cccc1)c1
Benzenesulfonic acid, propyl ester (9CI)
NSC-3216
Propyl benzenesulfonate

80-43-3
$C_{18}H_{22}O_2$
270.40
O(OC(c(cccc1)c1)(C)C)C(c(cccc2)c2)(C)C
Peroxide, bis(α,α-dimethylbenzyl)
Active dicumyl peroxide
Bis(α,α-dimethylbenzyl)peroxide
Cumene peroxide
Cumyl peroxide
Dicumyl peroxide
Dicumyl peroxide, Dry (DOT)
Dicumyl peroxide, Technically pure or with inert solid (DOT)
Di-α-Cumyl peroxide
Di-Cup
Di-Cup 40 KE
Di-Cupr
Diisopropylbenzene peroxide
Isopropylbenzene peroxide
Luperco
Luperox
Luperox 500R
Luperox 500T
UN 2121 (DOT)
Varox DCP-R
Varox DCP-T

80-44-4
$C_{10}H_{14}O_3S$
214.29
Benzenesulfonic acid, butyl ester (8CI,9CI)
NSC-3215

80-46-6
$C_{11}H_{16}O$
164.27
Oc(ccc(c1)C(CC)(C)C)c1
Phenol, p-(tert-pentyl)
Amilphenol
p-tert-Amylphenol
4-tert-Amylphenol
Amyl phenol 4T
p-(α,α-Dimethylpropyl)phenol
p-(1,1-Dimethylpropyl)phenol
2-Methyl-2-p-hydroxyphenylbutane
Pentaphen
p-t-Pentylphenol
PTAP
UCAR amyl phenol 4T

80-47-7
$C_{10}H_{20}O_2$
172.30
CC1CCC(CC1)C(C)(C)OO
p-Menthane-8-hydroperoxide
p-Menthane hydroperoxide

80-48-8
$C_8H_{10}O_3S$
186.24
O=S(=O)(OC)c(ccc(c1)C)c1
p-Toluenesulfonic acid, methyl ester
Methylester kyseliny p-toluensulfonove (Czech)
Methyl p-methylbenzenesulfonate
Methyl 4-methylbenzenesulfonate
Methyl-p-toluenesulfonate
Methyl toluene-4-sulfonate
Methyl tosylate
Methyl p-tosylate
p-Toluolsulfonsaeure methyl ester (German)

80-50-2
$C_{17}H_{32}NO_2.Br$
362.41
1-α-H,5-α-H-Tropanium, 3-α-hydroxy-8-methyl-, bromide, 2-propylvalerate
Anisotropine methobromide
Anisotropine methylbromide
8-Azoniabicyclo(3.2.1)octane, 8,8-dimethyl-3-((1-oxo-2-propylpentyl)oxy)-, bromide, endo-
3-α-Hydroxy-8-methyl-1-α-h, 5-α-h-tropanium bromide 2-propylvalerate
Lytispasm
8-Methyl-3-(2-propylpentanoyloxy)tropinium bromide
8-Methyltropinium bromide 2-propylvalerate
Octatropine methylbromide
2-Propylpentanoyltropinium methylbromide
Valpin
Vapin

80-51-3
$C_{12}H_{14}N_4O_5S_2$
358.42
O=S(=O)(NN)c(ccc(Oc(ccc(S(=O)(=O)NN)c1)c1)c2)c2
Benzenesulfonic acid, 4,4'-oxybis-, dihydrazide
Benzenesulfonic acid, oxybis-, dihydrazide (9CI)
Cellmic S
Celogen OT
Cenitron OB
Nitropore OBSH
OBSH
p,p'-Oxybisbenzene disulfonylhydrazide
Oxybis(benzenesulfonylhydrazide)
p,p'-Oxybis(benzenesulfonyl hydrazide)

80-52-4
$C_{10}H_{22}N_2$
170.34
NC(C(CCC(N)(C1)C)C1)(C)C
p-Menthane-1,8-diamine
4-Amino-a,a,4-trimethylcyclohexanemethamine
1,8-Diamino-p-menthane
Menthane diamine
USAF RH-4

80-53-5
$C_{10}H_{20}O_2$
172.27
OC(C(CCC(O)(C1)C)C1)(C)C
Terpin
Cyclohexanemethanol, 4-hydroxy-α,α,4-trimethyl- (9CI)
p-Menthane-1,8-diol (8CI)
NSC-403856
Terpin (VAN)
1,8-Terpin

80-54-6
$C_{14}H_{20}O$
204.34
O=CC(C)Cc(ccc(c1)C(C)(C)C)c1
Hydrocinnamaldehyde, p-tert-butyl-α-methyl
Benzenepropanal, 4-(1,1-dimethylethyl)-α-methyl-
p-tert-Butyl-α-methylhydrocinnamaldehyde
p-tert-Butyl-α-methylhydrocinnamic aldehyde
Lilial
Lilyal
α-Methyl-p-(tert-butyl)hydrocinnamaldehyde
α-Methyl, β-(p-tert-butylphenyl)propionaldehyde
Propionaldehyde, β-(4-tert-butylphenyl)-α-methyl-

80-56-8
$C_{10}H_{16}$
136.26
C(C(CC1C2)C1(C)C)(=C2)C
Bicyclo(3.1.1)hept-2-ene, 2,6,6-trimethyl
Acintene A
α-Pinene
2-Pinene
α-Pinene [UN 2368]
2,6,6-Trimethylbicyclo(3.1.1)-2-hept-2-ene
4,6,6-Trimethylbicyklo-(3,1,1)hept-3-en (Czech)
UN 2368 [α-Pinene]

80-58-0

$C_4H_7BrO_2$
167.02
O=C(O)C(Br)CC
Butyric acid, 2-bromo
2-Bromobutanoic acid
α-Bromobutyric acid
dl-2-Bromobutyric acid
2-Bromobutyric acid
Butyric acid, α-bromo-

80-59-1
$C_5H_8O_2$
100.13
O=C(O)C(=CC)C
Crotonic acid, 2-methyl-, (E)
2-Butenoic acid, 2-methyl-, (E)- (9CI)
Cevadic acid
(E)-2,3-Dimethylacrylic acid
trans-α,β-Dimethylacrylic acid
trans-2,3-Dimethylacrylic acid
trans-2-Methyl-2-butenoic acid
(E)-2-Methylcrotonic acid
trans-2-Methylcrotonic acid
Tiglic acid
Tiglinic acid

80-62-6
$C_5H_8O_2$
100.13
O=C(OC)C(=C)C
Methacrylic acid, methyl ester
Acrylic acid, 2-methyl-, methyl ester
Diakon
Metakrylan metylu (Polish)
Methacrylate de methyle (French)
Methacrylsaeuremethyl ester (German)
Methylester kyseliny methakrylove (Czech)
Methylmethacrylaat (Dutch)
Methyl-methacrylat (German)
Methyl methacrylate (ACGIH, OSHA)
Methyl methacrylate monomer
Methyl methacrylate monomer, Inhibited [UN 1247]
Methyl methacrylate monomer, Uninhibited (DOT)
Methyl methylacrylate
Methyl α-methylacrylate
Methyl 2-methyl-2-propenoate
2-Methyl-2-propenoic acid methyl ester
Metil metacrilato (Italian)
MME
"Monocite" methacrylate monomer
NA 1247 (DOT)
NCI-C50680
2-Propenoic acid, 2-methyl-, methyl ester
RCRA waste number U162
UN 1247 [Methyl methacrylate monomer, inhibited]

80-63-7
$C_4H_5ClO_2$
120.54
O=C(OC)C(=C)Cl
Acrylic acid, 2-chloro-, methyl ester
2-Chloroacrylic acid, methyl ester
Methyl-α-chloroacrylate
Methyl-2-chloroacrylate
Methylester kyseliny 2-chlorakrylove (Czech)
2-Propenoic acid, 2-chloro-, methyl ester (9CI)

80-70-6
$C_5H_{13}N_3$
115.16
N=C(N(C)C)N(C)C
Guanidine, N,N,N',N'-tetramethyl- (9CI)
AI3-51030
Guanidine, 1,1,3,3-tetramethyl- (8CI)
NSC-148309
N,N,N',N'-Tetramethylguanidine
1,1,3,3-Tetramethylguanidine

80-71-7
$C_6H_8O_2$
112.14
O=C(C(O)=C(C1)C)C1
2-Cyclopenten-1-one, 2-hydroxy-3-methyl
Corylon
Corylone
Cycloten
Cyclotene
2-Hydroxy-3-methyl-2-cyclopenten-1-one
Maple Lactone
3-Methylcyclopentane-1,2-dione
Methylcyclopentenolone

80-77-3
$C_{11}H_{12}ClNO_3S$
273.75
CN2C(c1ccc(Cl)cc1)S(=O)(=O)CCC2=O
4H-1,3-Thiazin-4-one, 2-(p-chlorophenyl)tetrahydro-3-methyl-, 1,1,-dioxide
Banabin
Banabin-Sintyal
Bisina
Chlormethazanone
Chlormethazone
Chlormezanone
2-(4-Chlorophenyl)-3-methyl-4-metathiazanone-1,1-dioxide
2-(p-Chlorophenyl)tetrahydro-3-methyl-4H-1,3-thiazin-4-one 1,1-dioxide
2-(p-Chlorphenyl)-3-methyl-1,3-perhydrothiazin-4-on-1,1-dioxide
Clorilax
Clormetazanone
Clormetazon
Dichloromethazanone
Fenarol
Lobak
Miorilax
Mio-Sed
Muskel
Muskel-Trancopal
Phenarol
Rexan
Rilansyl
Rilaquil
Rilassol
Rilax
Rillasol
Supotran
Suprotan
Tanafol
4H-1,3-Thiazin-4-one, tetrahydro-2-(p-chlorophenyl)-3-methyl-, 1,1-dioxide
Trancopal
WIN 4692

80-82-0
$C_6H_5NO_5S$
203.17
Benzenesulfonic acid, 2-nitro- (9CI)
Benzenesulfonic acid, o-nitro- (8CI)

80-97-7
$C_{27}H_{48}O$
388.68
OC(CCC(C1CCC2C(C(C(C3)C(CCC(C)C)C)(CC4)C)C3)(C24)C)C1
Dihydrocholesterol
Cholestanol (VAN)
Cholestan-3-ol
Cholestan-3-ol, (3β,5α)- (9CI)
Cholestan-3β-ol (VAN)
5α-Cholestanol (VAN)
5α-Cholestan-3β-ol (8CI)
Dihydrocholesterin
5α-Dihydrocholesterol
3β-Hydroxycholestane
3β-Hydroxy-5α-cholestane
NSC-18188
Zymostanol

81-03-8
$C_{14}H_{20}$
188.31
c(c(ccc1C)C(C2)(C)C)(c1)C2(C)C
1H-Indene, 2,3-dihydro-1,1,3,3,5-pentamethyl- (9CI)
2,3-Dihydro-1,1,3,3,5-pentamethyl-1H-indene

81-04-9
$C_{10}H_8O_6S_2$
288.30
O=S(=O)(O)c(c(c(c(S(=O)(=O)O)cc1)cc2)c1)c2
1,5-Naphthalenedisulfonic acid
Armstrong's Acid

Armstrong's S Acid
1,5-Naphthylene disulfonic acid

81-06-1
$C_{10}H_9NO_3S$
223.25
O=S(=O)(O)c(c(c(c(ccc1)c2)c1)N)c2
2-Naphthalenesulfonic acid, 1-amino- (9CI)
1-Amino-2-naphthalenesulfonic acid

81-07-2
$C_7H_5NO_3S$
183.19
O=C(NS(=O)(=O)c1cccc2)c12
1,2-Benzisothiazolin-3-one, 1,1-dioxide
Anhydro-o-sulfaminebenzoic acid
3-Benzisothiazolinone 1,1-dioxide
1,2-Benzisothiazol-3(2H)-one 1,1-dioxide
o-Benzoic sulfimide
Benzoic sulphimide
o-Benzoic sulphimide
o-Benzosulfimide
Benzosulphimide
o-Benzosulphimide
Benzo-2-sulphimide
Benzo-sulphinide
o-Benzoyl sulfimide
o-Benzoyl sulphimide
1,2-Dihydro-2-ketobenzisosulfonazole
1,2-Dihydro-2-ketobenzisosulphonazole
2,3-Dihydro-3-oxobenzisosulfonazole
2,3-Dihydro-3-oxobenzisosulphonazole
Garantose
Glucid
Gluside
Hermesetas
3-Hydroxybenzisothiazole-S,S-dioxide
Insoluble saccharin
Kandiset
Natreen
RCRA waste number U202
Sacarina
Saccharimide
Saccharin
Saccharina
Saccharin acid
Saccharine
Saccharin Insoluble
Saccharinol
Saccharinose
Saccharol
Sacharin (Czech)
Saxin
Sucre Edulcor
Sucrette
o-Sulfobenzimide
o-Sulfobenzoic acid imide
2-Sulphobenzoic imide
Sykose
Syncal
Zaharina

81-08-3
$C_7H_4O_4S$
184.17
O=C(OS(=O)(=O)c1cccc2)c12
o-Sulfobenzoic anhydride
Benzoic acid, o-sulfo-, cyclic anhydride (8CI)
Benzoic acid, 2-sulfo-, cyclic anhydride
3H-2,1-Benzoxathiol-3-one, 1,1-dioxide (9CI)
NSC-11208
o-Sulfobenzoic acid anhydride
o-Sulfobenzoic acid, cyclic anhydride
Sulfobenzoic anhydride

81-10-7
$C_{14}H_{16}N_2O_2S$
276.35
O=S(=O)(N(c(cccc1)c1)CC)c(c(N)ccc2)c2
Benzenesulfonamide, 2-amino-N-ethyl-N-phenyl- (9CI)
2-Amino-N-ethylbenzenesulfonanilide
2-Amino-N-ethyl-N-phenylbenzenesulfonamide
Benzenesulfonanilide, 2-amino-N-ethyl- (8CI)
N-Ethyl-N-phenyl-o-aminobenzenesulfonamide
NSC-81263

81-11-8
$C_{14}H_{14}N_2O_6S_2$
370.42
O=S(=O)(O)c(c(ccc1N)C=Cc(c(S(=O)(=O)O)cc(N)c2)c2)c1
2,2'-Stilbenedisulfonic acid, 4,4'-diamino
Amsonic acid
Benzenesulfonic acid, 2,2'-(1,2-ethylenediyl)bis(5-amino- (9CI)
DASD
4,4'-Diamino-2,2'-stilbenedisulfonic acid
2,2'-(1,2-Ethylenediyl)bis-(5-aminobenzenesulfonic acid)
Flavonic acid
NCI-C60162
Tinopal BHS

81-13-0
$C_9H_{19}NO_4$
205.29
O=C(NCCCO)C(O)C(CO)(C)C
Butyramide, 2,4-dihydroxy-N-(3-hydroxypropyl)-3,3-dimethyl-, D-(+)
Bepanthen
Bepanthene
Bepantol
Butanamide, 2,4-dihydroxy-N-(3-hydroxypropyl)-3,3-dimethyl-, (R)- (9CI)
Cozyme
Dexpanthenol
D-(+)-2,4-Dihydroxy-N-(3-hydroxypropyl)-3,3-dimethylbutyramide
D-P-A Injection
Ilopan
Motilyn
Panadon
Panthenol
D-Panthenol
D(+)-Panthenol
Panthoderm
Pantol
Pantothenol
d-Pantothenol
Pantothenyl alcohol
d-Pantothenyl alcohol
D(+)-Pantothenyl alcohol
Thenalton
Zentinic

81-14-1
$C_{14}H_{18}N_2O_5$
294.30
O=C(c(c(c(N(=O)=O)c(c1N(=O)=O)C(C)(C)C)C)c1C)C
Musk Ketone
Acetophenone, 4'-tert-butyl-2',6'-dimethyl-3',5'-dinitro- (8CI)
AI3-02440
4'-tert-Butyl-2',6'-dimethyl-3',5'-dinitroacetophenone
Ethanone, 1-(4-(1,1-dimethylethyl)-2,6-dimethyl-3,5-dinitrophenyl)- (9CI)
NSC-15339

81-15-2
$C_{12}H_{15}N_3O_6$
297.30
O=N(=O)c(c(c(N(=O)=O)c(c1N(=O)=O)C(C)(C)C)C)c1C
Benzene, 1-(1,1-dimethylethyl)-3,5-dimethyl-2,4,6-trinitro
Benzene, 1-tert-butyl-3,5-dimethyl-2,4,6-trinitro-
5-tert-Butyl-2,4,6-trinitroxylene
Musk Xyldl
Musk Xylene
2,4,6-Trinitro-1,3-dimethyl-5-tert-butylbenzene
2,4,6-Trinitro-3,5-dimethyl-tert-butylbenzene
m-Xylene, 5-tert-butyl-2,4,6-trinitro-
Xylene Musk

81-16-3
$C_{10}H_9NO_3S$
223.26
O=S(=O)(O)c(c(c(ccc1)cc2)c1)c2N

1-Naphthalenesulfonic acid, 2-amino
2-Amino-1-naphthalenesulfonic acid
Kyselina 2-naftylamin-1-sulfonova (Czech)
Kyselina tobiasova (Czech)
2-Naphthylamine-1-sulfonic acid
Tobias acid

81-19-6
$C_7H_4Cl_4$
229.92
c(c(c(cc1)Cl)C(Cl)Cl)(c1)Cl
Benzene, 1,3-dichloro-2-(dichloromethyl)- (9CI)
2,6-Dichlorobenzal chloride
2,6-Dichlorobenzylidene chloride
1,3-Dichloro-2-(dichloromethyl)benzene
NSC-79866
α,α2,6-Tetrachlorotoluene
Toluene, α,α2,6-tetrachloro- (8CI)

81-20-9
$C_8H_9NO_2$
149.16
O=N(=O)c(c(ccc1)C)c1C
m-Xylene, 2-nitro
2,6-Dimethylnitrobenzene
2-Nitro-m-xylene

81-21-0
$C_{10}H_{12}O_2$
164.22
O(C1C(C(C(C23)CC4O5)C45)C3)C12
4,7-Methanoindan, 3a,4,5,6,7,7a-hexahydro-1,2:5,6-diepoxy
Bicyclopentadiene dioxide
1,2:5,6-Diepoxyhexahydro-4,7-methanoindan
1,2:5,6-Diepoxy-3a,4,5,6,7,7a-hexahydro-4,7-methano indan
Dicyclopentadiene diepoxide
Dicyclopentadiene dioxide
EP 207
Epoxide 207
2,4-Methano-2H-bisoxireno-(a,f)indene, octahydro-
4,7-Methanoindan, 1,2:5,6-diepoxyhexahydro-
4,7-Methanoindan, 1,2:5,6-diepoxy-3a,4,5,6,7,7a-hexahydro-
2,4-Methano-2H-indeno-(1,2-b:5,6-b')bisoxirene, octahydro-
Unox 207
Unox Epoxide 207
Unox 207X

81-24-3
$C_{26}H_{45}NO_7S$
515.71
Taurocholic acid
Cholaic acid
Cholic acid taurine conjugate
N-Choloyltaurine
Cholyltaurine
Ethanesulfonic acid, 2-(((3α,5β,7α,12α)-3,7,12-trihydroxy-24-oxocholan-24-yl)amino)- (9CI)
NSC-25505
Taurine, N-choloyl- (8CI)
Taurocholate
3α,7α,12α-Trihydroxy-5β-cholanic acid 24-taurine
2-((3-α,7-α,12-α-Trihydroxy-24-oxo-5-be ta-cholan-24-yl)amino)ethanesulfonic acid

81-25-4
$C_{24}H_{40}O_5$
408.64
O=C(O)CCC(C(C(C(C(C(C(C(C1)CC(O)C2)(C2)C)C3)C1O)C4)(C3O)C)C4)C
Cholic-acid
Cholalin
Cholan-24-oic acid, 3,7,12-trihydroxy-, (3-α,5-β,7-α,12-α)- (9CI)
Cholsaeure (German)
Colalin
3-α,7-α,12-α-Trihydroxy-5-β-cholan-24-oic acid
3,7,12-Trihydroxy-cholan-24-oic acid (3-α,5-β,7-α,12-α)
3-α,7-α,12-α-Trihydroxycholansaeure (German)

81-30-1
$C_{14}H_4O_6$
268.18
O=C(OC(=O)c(c1c(c(cc2)C(=O)OC3=O)c3c4)c4)c12
(2)Benzopyrano(6,5,4-def)-(2)benzopyran-1,3,6,8-tetrone (9CI)
Naphthalenetetracarboxylic dianhydride
1,4,5,8-Naphthalenetetracarboxylic acid, 1,8:4,5-dianhydride (8CI)
NSC-84241

81-33-4
$C_{24}H_{10}N_2O_4$
390.34
O=C(NC(=O)c(c1c(c(c(c(c(c(C(=O)NC2=O)c3)c2cc4)c45)c3)cc6)c5c7)c7)c16
Anthra(2,1,9-def:6,5,10-d'e'f')diisoquinoline-1,3,8,10(2H,9H)-tetrone (9CI)
NSC-16842
3,4,9,10-Perylenetetracarboxylic acid diimide
3,4,9,10-Perylenetetracarboxylic 3,4:9,10-diimide (8CI)
Perylimid

81-42-5
$C_{14}H_8Cl_2N_2O_2$
307.12
O=C(c(c(C(=O)c1c(N)c(c(c2N)Cl)Cl)ccc3)c3)c12
1,4-Diamino-2,3-dichloroanthraquinone
9,10-Anthracenedione, 1,4-diamino-2,3-dichloro- (9CI)
1,4-Diamino-2,3-dichloro-9,10-anthracenedione

81-48-1
$C_{21}H_{15}NO_3$
329.37
O=C(c(c(C(=O)c1c(O)ccc2Nc(ccc(c3)C)c3)ccc4)c4)c12
Anthraquinone, 1-hydroxy-4-(p-toluidino)
Ahcoquinone Blue IR Base
Alizarine Irisol R Base
Alizarine Violet 3B Base
C.I. 60725
C.I. Solvent Violet 13
D & C Violet No. 2
Disperse Blue 72
N-(4-Hydroxy-1-anthraquinonyl)-4-methylaniline
N-(4-Hydroxy-1-anthraquinonyl)-p-toluidine
Irisol Base
Modr Disperzni 72 (Czech)
Oil Violet IRS
Oil Violet ZIRS
N-(p-Tolyl)-4-hydroxy-1-anthraquinonylamine
11092 Violet
Violet Rozpoustedlova 13 (Czech)
Waxoline Purple A

81-49-2
$C_{14}H_7Br_2NO_2$
381.04
O=C(c(c(C(=O)c1c(cc(c2N)Br)Br)ccc3)c3)c12
Anthraquinone, 1-amino-2,4-dibromo
1-Amino-2,4-dibromanthrachinon (Czech)
1-Amino-2,4-dibromoanthraquinone
9,10-Anthracenedione, 1-amino-2,4-dibromo-
2,4-Dibromo-1-anthraquinonylamine
NCI-C55458

81-54-9
$C_{14}H_8O_5$
256.22
Oc3cc(O)c2C(=O)c1ccccc1C(=O)c2c3O
Anthraquinone, 1,2,4-trihydroxy
9,10-Anthracenedione,

1,2,4-trihydroxy-
C.I. 1037
C.I. 58205
C.I. 75410
Hydroxylizaric acid
Purpurin
Purpurine
Smoke Brown G
Verantin
1,2,4-Trihydroxyanthrachinon (Czech)
1,2,4-Trihydroxyanthraquinone

81-55-0
$C_{14}H_6N_2O_8$
330.22
O=C(c(c(c(N(=O)=O)cc1)C(=O)c2c(N(=O)=O)ccc3O)c1O)c23
Anthraquinone, 1,8-dihydroxy-4,5-dinitro
9,10-Anthracenedione, 1,8-dihydroxy-4,5-dinitro- (9CI)
1,8-Dihydroxy-4,5-dinitroanthraquinone
4,5-Dinitrochrysazin
NCI-C60742

81-61-8
$C_{14}H_8O_6$
272.22
O=C(c(c(c(O)c(O)c1)C(=O)c2c(O)ccc3O)c1)c23
Anthraquinone, 1,2,5,8-tetrahydroxy
Alizarinbordeaux
Alizarine Bordeaux
Alizarine Bordeaux B
9,10-Anthracenedione, 1,2,5,8-tetrahydroxy- (9CI)
C.I. 58500
C.I. Mordant Violet 26
Khinalizarin
Quinalizarin
Quinalizarine
1,2,5,8-Tetrahydroxyanthraquinone
1,4,5,6-Tetrahydroxyanthraquinone

81-63-0
$C_{14}H_{12}N_2O_2$
240.28
O=C(c(c(C(=O)C1=C(N)CCC=2N)ccc3)c3)C12
Anthraquinone, 2,3-dihydro-1,4-diamino
9,10-Anthracenedione, 1,4-diamino-2,3-dihydro-
1,4-Diamino-2,3-dihydroanthraquinone

81-64-1
$C_{14}H_8O_4$
240.22
O=C(c(c(C(=O)c1c(O)ccc2O)cc3)c3)c12
Anthraquinone, 1,4-dihydroxy
9,10-Anthracenedione, 1,4-dihydroxy-
Chinizarin
C.I. 58050
1,4-Dihydroxyanthraquinone
1,4-Dihydroxyanthrachinon (Czech)
1,4-Dioxyanthraquinone (Russian)
1,4-Doa (Russian)
1,4-Dihydroxy-9,10-anthraquinone
Quinizarin
Quinizarine
Smoke Orange R

81-77-6
$C_{28}H_{14}N_2O_4$
442.44
O=C(c(c(c(Nc(c(N1)c(c(C(=O)c(c2ccc3)c3)c4)C2=O)c4)c1c5)C(=O)c6cccc7)c5)c67
5,9,14,18-Anthrazinetetrone, 6,15-dihydro
Anthraquinone Blue
Anthraquinone Deep Blue
Atic Vat Blue XRN
Benzadone Blue RS
Bleu Solanthrene (French)
Blue Anthraquinone Pigment
Blue O
Calcoloid Blue RS
Caledon Blue RN
Caledon Blue XRN
Caledon Brilliant Blue RN
Caledon Paper Blue RN
Caledon Printing Blue RN
Caledon Printing Blue XRN
Carbanthrene Blue 2R
Carbanthrene Blue RS
Carbanthrene Blue RSP
Celliton Blue RN
C.I. 1106
C.I. 69800
Cibanone Blue FRS
Cibanone Blue FRSN
Cibanone Blue RS
Cibanone Brilliant Blue FR
C.I. Pigment Blue 60
C.I. Vat Blue 4
Cromophtal Blue A 3R
N,N-Dihydro-1,1,1',2'-anthraquinone-azine
E 130
Fenan Blue RSN
Fenanthren Blue RS
Food Blue 4
Graphtol Blue RL
Helianthrene Blue RS
Heliogen Blue 6470
Indanthren Blue
Indanthren Blue GP
Indanthren Blue GPT
Indanthren Blue RPT
Indanthren Blue RS
Indanthren Blue RSN
Indanthren Blue RSP
Indanthren Brilliant Blue R
Indanthrene
Indanthrene Blue
Indanthrene Blue GP
Indanthrene Blue RP
Indanthrene Blue RS
Indanthrene Blue RSA
Indanthrene Blue RSN
Indanthren Printing Blue FRS
Indanthren Printing Blue KRS
Indanthrone
Lake Fast Blue BS
Lake Fast Blue GGS
Latexol Fast Blue SD
L-Blau 1 (German)
Lionogen Blue R
Lutetia Fast Blue RS
Medium Blue
Mikethrene Blue RSN
Mikethrene Brilliant Blue R
Modr Kypova 4 (Czech)
Modr Pigment 60 (Czech)
Modr Potravinarska 4 (Czech)
Monolite Fast Blue 3R
Monolite Fast Blue 3RD
Monolite Fast Blue RV
Monolite Fast Blue SRS
Navinon Blue RSN
Navinon Blue RSN Reddish Special
Nihonthrene Blue RSN
Nihonthrene Brilliant Blue RP
Ostanthren Blue RS
Ostanthren Blue RSN
Ostanthren Blue RSZ
Ostanthrene Blue RS
Palanthrene Blue GPT
Palanthrene Blue GPZ
Palanthrene Blue RPT
Palanthrene Blue RPZ
Palanthrene Blue RSN
Palanthrene Brilliant Blue R
Palanthrene Printing Blue KRS
Paradone Blue RS
Paradone Brilliant Blue R
Paradone Printing Blue FRS
Pernithrene Blue RS
Pigment Anthraquinone Deep Blue
Pigment Blue 60
Pigment Blue Anthraquinone
Pigment Blue Anthraquinone V
Pigment Deep Blue Anthraquinone
Polymon Blue 3R
Ponsol Blue GZ
Ponsol Blue RCL
Ponsol Blue RPC
Ponsol Brilliant Blue R
Ponsol RP
Romanthrene Blue FRS
Romantrene Blue FRS
Romantrene Blue GGSL
Romantrene Blue RSZ
Romantrene Brilliant Blue FR
Romantrene Brilliant Blue R
Sandothrene Blue NRSC
Sandothrene Blue NRSN
Sanyo Threne Blue IRN
Schultz No. 1228
Solanthrene Blue RS
Solanthrene Blue RSN
Solanthrene R for sugar
Symuler Fast Blue 6011
Tinon Blue RS
Tinon Blue RSN

Tyrian Blue I-RSN
Tyrian Brilliant Blue I-R
Vat Blue 4
Vat Blue O
Vat Blue OD
Vat Fast Blue R
Versal Blue GGSL
Vulcafix Fast Blue SD
Vulcafor Fast Blue 3R
Vulcanosine Fast Blue GG
Vulcol Fast Blue S
Vynamon Blue 3R

81-78-7
$C_{28}H_{18}N_2O_6$
478.45
O=C(O)c(c(Nc(c(c(C(=O)c(c1ccc2)c2Nc(c(ccc3)C(=O)O)c3)cc4)C1=O)c4)ccc5)c5
Anthranilic acid, N,N'-1,5-anthraquinonylenedi- (8CI)
N,N'-(1,5-Anthraquinonylene)-dianthranilic acid
Benzoic acid, 2,2'-((9,10-dihydro-9,10-dioxo-1,5-anthracenediyl)diimino)bis- (9CI)
1,5-Bis(2-carboxyanilino)anthraquinone
C.I. 61720
NSC-16221
Violet BN Acid Anthraquinone
Vulcan Violet BN

81-81-2
$C_{19}H_{16}O_4$
308.35
O=C(Oc(c(C=1O)ccc2)c2)C1C(c(cccc3)c3)CC(=O)C
Coumarin, 3-(α-acetonylbenzyl)-4-hydroxy
3-(Acetonylbenzyl)-4-hydroxycoumarin
3-(α-Acetonylbenzyl)-4-hydroxycoumarin
Arab Rat Deth
Athrombine-K
Athrombin-K
2H-1-Benzopyran-2-one, 4-hydroxy-3-(3-oxo-1-phenylbutyl)- (9CI)
Brumolin
Compound 42
d-Con
Co-Rax
Coumadin
Coumafen
Coumafene
Coumarins
Coumefene
Cov-R-Tox
Dethmor
Dethnel
Eastern States Duocide
Fasco Fascrat Powder
1-(4'-Hydroxy-3'-coumarinyl)-1-phenyl-3-butanone
4-Hydroxy-3-(3-oxo-1-fenylbutyl) cumarine (Dutch)
4-Hydroxy-3-(3-oxo-1-phenylbutyl)-2H-1-benzopyran-2-one
4-Hydroxy-3-(3-oxo-1-phenylbutyl)-cumarin (German)
4-Idrossi-3-(3-oxo-1-fenil-butil)-cumarine (Italian)
Kumader
Kumadu
Kypfarin
Latka 42 (Czech)
Liqua-Tox
Mar-Frin
Martin's Mar-Frin
Maveran
Mouse Pak
3-(α-Phenyl-β-acetylaethyl)-4-hydroxycumarin (German)
3-(α-Phenyl-β-acetylethyl)-4-hydroxycoumarin
3-(1'-Phenyl-2'-acetylethyl)-4-hydroxycoumarin
(Phenyl-1 acetyl-2 ethyl) 3-hydroxy-4 coumarine (French)
Prothromadin
Rat-A-Way
Rat-B-Gon
Rat-Gard
Rat-Kill
Rat & Mice Bait
Rat-Mix
Rat-O-Cide #2
Rat-Ola
Ratorex
Ratox
Ratoxin
Ratron
Ratron G
Rats-No-More
Rat-Trol
Rattunal
Rax
RCRA waste number P001
Rodafarin
Ro-Deth
Rodex
Rodex Blox
Rosex
Rough & Ready Mouse Mix
Solfarin
Spray-Trol Brand Roden-Trol
Temus W
Tox-Hid
Twin Light Rat Away
Vampirinip II
Vampirinip III
Waran
W.A.R.F. 42
Warfarat
Warfarin (ACGIH,OSHA)
Warfarine (French)
Warfarin Plus
Warfarin Q
Warf Compound 42
Warficide
Zoocoumarin (Russian)

81-82-3
$C_{19}H_{15}ClO_4$
342.79
CC(=O)CC(c1ccc(Cl)cc1)c3c(O)c2ccccc2oc3=O
Coumarin, 3-(α-acetonyl-p-chlorobenzyl)-4-hydroxy
3-(α-Acetonyl-p-chlorobenzyl)-4-hydroxycoumarin
3-(α-Acetonyl-4-chlorobenzyl)-4-hydroxycoumarin
3-(1-Acetyl-2-(p-chlorophenyl)-ethyl)-4-hydroxycoumarin
2H-1-Benzopyran-2-one, 3-(1-(4-chlorophenyl)-3-oxobutyl)-4-hydroxy- (9CI)
3-(1-(4-Chloorfenyl)-3-oxobutyl)-4-hydroxy-cumarine (Dutch)
3-(α-p-Chlorophenyl-β-acetylethyl)-4-hydroxycoumarin
3-(1-(4-Chlorophenyl)-3-oxobutyl)-4-hydroxy-coumarine (French)
3-(α-(p-Chlorphenyl)-β-acetylaethyl)-4-hydroxycumarin (German)
3-(1-(4-Chlor-phenyl)-3-oxobutyl)-4-hydroxy-cumarin (German)
3-(1-(4-Cloro-fenil)-3-oxobutil)-4-idrossicumarina (Italian)
Coumachlor
Coumachlore (French)
Cumachloor (Dutch)
Cumachlor (German)
Experimental Rodenticide 332
G-23133
Geigy Rodenticide Exp. 332
Kumachlor (Czech)
Ratilan
Tomorin

81-83-4
$C_{12}H_7NO_2$
197.19
O=C(NC(=O)c(c1c(ccc2)cc3)c3)c12
Naphthalimide (8CI)
1H-Benz(de)isoquinoline-1,3(2H)-dione (9CI)
NSC-11011
1,8-Naphthalenedicarboximide
1,8-Naphthalimide

81-84-5
$C_{12}H_6O_3$
198.18
O=C(OC(=O)c(c1c(ccc2)cc3)c3)c12
1,8-Naphthalic anhydride
Protect

81-88-9
$C_{28}H_{31}N_2O_3.Cl$
479.06
[Cl-].CCN(CC)c4ccc3c(c1ccccc1C(O)=O)c2ccc(cc2oc3c4)=[N+](CC)CC
Ammonium, (9-(o-carboxyphenyl)-6-(diethylamino)-3H-xanthen-3-ylidene)diethyl-, chloride
Acid Brilliant Pink B
ADC Rhodamine B

Akiriku Rhodamine B
Aizen Rhodamine BH
Aizen Rhodamine BHC
Basic Violet 10
Brilliant Pink B
Calcozine Red BX
Calcozine Rhodamine BL
Calcozine Rhodamine BX
Calcozine Rhodamine BXP
9-o-Carboxyphenyl-6-diethyl-amino-3-ethylimino-3-iso-xanthene, 3-ethochloride
(9-(o-Carboxyphenyl)-6-(di-ethylamino)-3H-xanthen-3-yl-idene) diethylammonium chloride
Cerise Toner X1127
Certiqual Rhodamine
C.I. 749
C.I. 45170
C.I. Basic Violet 10
C.I. Food Red 15
Cogilor Red 321.10
Cosmetic Brilliant Pink Bluish D Conc
D & C Red No. 19
Diabasic Rhodamine B
Diethyl-m-amino-phenol-phthalein hydrochloride
Edicol Supra Rose B
Edicol Suppa Rose BS
Elcozine Rhodamine B
Eriosin Rhodamine B
FD & C Red No. 19
Food Red 15
Geranium Lake N
Hexacol Rhodamine B Extra
Ikada Rhodamine B
Iragen Red L-U
Mitsui Rhodamine BX
11411 Red
Red No. 213
Rheonine B
Rhodamine
Rhodamine B
Rhodamine B 20-7470
Rhodamine BA
Rhodamine BA Export
Rhodamine B Chloride
Rhodamine B Extra
Rhodamine B Extra M 310
Rhodamine B Extra S
Rhodamine BF
Rhodamine B500
Rhodamine B500 hydrochloride
Rhodamine BL
Rhodamine, Blue Shade
Rhodamine BN
Rhodamine BS
Rhodamine BX
Rhodamine BXL
Rhodamine BXP
Rhodamine FB
Rhodamine FB CI
Rhodamine Lake Red B
Rhodamine O
Rhodamine, tetraethyl-
Rhodamine S (Russian)
Sicilian Cerise Toner A-7127
Symulex Magenta F
Symulex Pink F
Symulex Rhodamine B Toner F
Takaoka Rhodamine B
Tetraethyldiamino-o-carboxy-phenyl-xanthenyl chloride
Tetraethylrhodamine
Violet Zasadita 10 (Czech)

82-03-1
$C_{18}H_{12}O$
244.29
7H-Benz(de)anthracen-7-one, 2-methyl- (7CI, 8CI,9CI)
2-Methyl-7H-benz(de)-anthracen-7-one

82-05-3
$C_{17}H_{10}O$
230.27
O=C(c(c(c(c1c(ccc2)cc3)c3)ccc4)c4)c12
7H-Benz(de)anthracen-7-one
7H-Benz(de)anthracene-7-one
Benzanthrenone
Benzanthrone
7H-Benzo(de)anthracen-7-one
Benzoanthrone
Dye, Benzanthrone
MS-Benzanthrone
Naphthanthrone
7-Oxobenz(de)anthracene

82-17-7
$C_{26}H_{16}O_4$
392.41
O=C(c(c(c(Oc(cccc1)c1)cc2)C(=O)c3c(Oc(cccc4)c4)ccc5)c2)c35
9,10-Anthracenedione, 1,8-di-phenoxy- (9CI)
1,8-Diphenoxy-9,10-anthracene-dione

82-21-3
$C_{26}H_{16}O_4$
392.42
O=C(c(c(c(Oc(cccc1)c1)cc2)C(=O)c3cccc4Oc(cccc5)c5)c2)c34
Anthraquinone, 1,5-diphenoxy
9,10-Anthracenedione, 1,5-di-phenoxy-
1,5-Difenoxyanthrachinon (Czech)
1,5-Diphenoxyanthraquinone

82-24-6
$C_{15}H_9NO_4$
267.25
O=C(O)c(c(c(c(C(=O)c(c1ccc2)c2)c3)C1=O)N)c3
2-Anthracenecarboxylic acid, 9,10-dihydro-1-amino-9,10-dioxo
1-Amino-2-carboxylate-4-nitro-anthraquinone
1-Amino-9,10-dioxo-9,10-di-hydro-2-anthracenecarboxylic acid

82-28-0
$C_{15}H_{11}NO_2$
237.27
O=C(c(c(C(=O)c1c(N)c(cc2)C)ccc3)c3)c12
Anthraquinone, 1-amino-2-methyl
Acetate Fast Orange R
Acetoquinone Light Orange JL
1-Amino-2-methylanthraquinone
9,10-Anthracenedione, 1-amino-2-methyl- (9CI)
Artisil Orange 3RP
Celliton Orange R
C.I. 60700
C.I. Disperse Orange 11
Cilla Orange R
Disperse Orange
Duranol Orange G
2-Methyl-1-anthraquinonyl-amine
Microsetile Orange RA
NCI-C01901
Nyloquinone Orange JR
Oranz Disperzni 11 (Czech)
Perliton Orange 3R
Serisol Orange YL
Supracet Orange R

82-31-5
$C_{22}H_{18}N_2O_4S$
406.45
O=S(=O)(O)c(c(c(c(Nc(ccc(O)c1)c1)cc2)cc3)c2Nc(cccc4)c4)c3
1-Naphthalenesulfonic acid, 5-((4-hydroxyphenyl)amino)-8-(phenylamino)- (9CI)

82-33-7
$C_{14}H_9N_3O_4$
283.26
O=C(c(c(c(N(=O)=O)cc1)C(=O)c2c(N)ccc3N)c1)c23
Anthraquinone, 1,4-diamino-5-nitro
9,10-Anthracenedione, 1,4-di-amino-5-nitro- (9CI)
Celliton Fast Violet B
Celliton Fast Violet BA-CF
Celliton Violet B
Cibacet Brilliant Violet 3B
Cilla Fast Violet B
C.I. 62030
C.I. Disperse Violet 8
Diacelliton Fast Violet B
1,4-Diamino-5-nitroanthra-quinone
Dianix Fast Violet B
Disperse Violet 2S
Duranol Brilliant Blue Violet BR
Duranol Brilliant Violet BR
Fenacet Fast Violet B
Kayalon Fast Violet BR
Miketon Fast Violet B
Nitrocresolamine
Palanil Violet 3B

Perliton Violet B
Samaron Brilliant Violet B
Serisol Fast Violet B
Supracet Fast Violet B
Terasil Brilliant Violet 3B
Violet 2S
Vonteryl Violet 2B

82-34-8
$C_{14}H_7NO_4$
253.22
O=C(c(c(c(N(=O)=O)cc1)C(=O)c2cccc3)c1)c23
Anthraquinone, 1-nitro
9,10-Anthracenedione, 1-nitro-
1-Nitroanthrachinon (Czech)
α-Nitroanthraquinone
1-Nitroanthraquinone

82-38-2
$C_{15}H_{11}NO_2$
237.25
O=C(c(c(c(NC)cc1)C(=O)c2cccc3)c1)c23
Disperse Red 9
9,10-Anthracenedione, 1-(methylamino)- (9CI)
Anthraquinone, 1-(methylamino)- (8CI)
AI3-18871
C.I. Disperse Red 9
C.I. Solvent Red 111
C.I. 60505
Calco Oil Red ZMQ
Celanthrene Red Y
Celliton Pink R
Diacelliton Fast Pink R
Duranol Red GN
Macrolex Red G
Macro-lex Red G
Methane quinone
1-(Methylamino)-9,10-anthracenedione
α-Methylaminoanthraquinone
1-(Methylamino)anthraquinone
1-(Methylamino)-9,10-anthraquinone
1-(N-Methylamino)anthraquinone
1-(N-Methylamino)-9,10-anthraquinone
N-Methyl-1-anthraquinonylamine
NSC-3721
Oil Red ZMQ
Serilene Fast Pink BT
Supracet Pink R
Waxoline Red MAA
Waxoline Red MP

82-39-3
$C_{15}H_{10}O_3$
238.25
O=C(c(c(c(OC)cc1)C(=O)c2cccc3)c1)c23
Anthraquinone, 1-methoxy
9,10-Anthracenedione, 1-methoxy- (9CI)
1-Methoxyanthraquinone

82-43-9
$C_{14}H_6Cl_2O_2$
277.10
O=C(c(c(c(cc1)Cl)C(=O)c2c(ccc3)Cl)c1)c23
Anthraquinone, 1,8-dichloro
9,10-Anthracenedione, 1,8-dichloro-
1,8-Dichloranthrachinon (Czech)
1,8-Dichloroanthraquinone
1,8-Dichloro-9,10-anthraquinone

82-44-0
$C_{14}H_7ClO_2$
242.66
O=C(c(c(c(cc1)Cl)C(=O)c2cccc3)c1)c23
Anthraquinone, 1-chloro
9,10-Anthracenedione, 1-chloro-
1-Chloranthrachinon (Czech)
α-Chloroanthraquinone
1-Chloroanthraquinone
1-Chloro-9,10-anthraquinone
α-Monochloroanthraquinone

82-45-1
$C_{14}H_9NO_2$
223.24
O=C(c(c(c(N)cc1)C(=O)c2cccc3)c1)c23
Anthraquinone, 1-amino
1-Amino-9,10-anthracenedione
1-Aminoanthrachinon (Czech)
α-Aminoanthraquinone
1-Aminoanthraquinone
1-Amino-9,10-anthraquinone
9,10-Anthracenedione, 1-amino-
α-Anthraquinonylamine
C.I. 37275
Diazo Fast Red AL

82-46-2
$C_{14}H_6Cl_2O_2$
277.10
O=C(c(c(c(cc1)Cl)C(=O)c2cccc3Cl)c1)c23
Anthraquinone, 1,5-dichloro
9,10-Anthracenedione, 1,5-dichloro-
1,5-Dichloranthrachinon (Czech)
1,5-Dichloroanthraquinone
1,5-Dichloro-9,10-anthraquinone

82-47-3
$C_{10}H_9NO_7S_2$
319.31
O=S(=O)(O)c(c(c(c(c1S(=O)(=O)O)ccc2)c2O)N)c1
Chicago Acid
4-Amino-5-hydroxy-1,3-naphthalenedisulfonic acid
8-Amino-1-naphthol-5,7-disulfonic acid
1,3-Naphthalenedisulfonic acid, 4-amino-5-hydroxy- (9CI)
NSC-7111
SS Acid

82-48-4
$C_{14}H_8O_8S_2$
368.34
O=C(c(c(c(S(=O)(=O)O)cc1)C(=O)c2c(S(=O)(=O)O)ccc3)c1)c23
1,8-Anthracenedisulfonic acid, 9,10-dihydro-9,10-dioxo
1,8-Anthraquinonedisulfonic acid
1,8-Disulfoanthraquinone

82-49-5
$C_{14}H_8O_5S$
288.28
O=C(c(c(c(S(=O)(=O)O)cc1)C(=O)c2cccc3)c1)c23
1-Anthracenesulfonic acid, 9,10-dihydro-9,10-dioxo- (9CI)

82-58-6
Unknown
Unknown
Lysergic acid
Acid, lysergic
DEA No. 7300

82-62-2
$C_6H_2Cl_3NO_3$
242.44
Phenol, 2-nitro-3,4,6-trichloro
2-Nitro-3,4,6-trichlorophenol
3,4,6-Trichloro-2-nitrophenol

82-66-6
$C_{23}H_{16}O_3$
340.39
O=C(C2C(=O)c1ccccc1C2=O)C(c3ccccc3)c4ccccc4
1,3-Indandione, 2-diphenylacetyl
Didandin
Dipaxin
Diphacin
Diphacinone
Diphenacin
Diphenadione
2-Diphenylacetyl-1,3-diketohydrindene
2-Diphenyl-acetyl-indan-1,3-dion (German)
2-Diphenylacetyl-1,3-indandione
2-(Diphenylacetyl)indan-1,3-dione
2-(Diphenylacetyl)-1H-indene-1,3(2H)-dione
Kill-Ko Rat Killer
Pid
Promar
Ramik
Ratindan 1
U 1363

82-68-8
$C_6Cl_5NO_2$
295.32
O=N(=O)c(c(c(c(c1Cl)Cl)Cl)Cl)c1Cl
Benzene, pentachloronitro
Avicol
Batrilex
Benzene, nitropentachloro-
Botrilex
Brassicol
Earthcide
Fartox
Folosan
Fomac 2
Fungiclor
GC 3944-3-4
HOE 026014
Kobu
Kobutol
KP 2
NCI-C00419
Olpisan
PCNB
Pentachlornitrobenzol (German)
Pentachloronitrobenzene
Pentagen
PKhNB
Quintocene
Quintox
Quintozen
Quintozene
RCRA waste number U185
Saniclor 30
Terrachlor
Terraclor
Terrafun
Tilcarex
Tri-PCNB
Tritisan

82-71-3
$C_6H_3N_3O_8$
245.12
O=N(=O)c(c(O)c(N(=O)=O)c(O)c1N(=O)=O)c1
Resorcinol, 2,4,6-trinitro
1,3-Benzenediol, 2,4,6-trinitro- (9CI)
2,4-Dihydroxy-1,3,5-trinitrobenzene
1,3-Dihydroxy-2,4,6-trinitrobenzene
3-Hydroxy-2,4,6-trinitrophenol
Styphnic acid
2,4,6-Trinitro-1,3-benzenediol
2,4,6-Trinitroresorcinol
Trinitroresorcinol [UN 0219]
Trinitroresorcinol, Dry [UN 0219]
Trinitroresorcinol, Wetted with less than 20% water [UN 0219]
UN 0219 [Trinitroresorcinol (Styphnic acid), Dry or wetted with less than 20 per cent water, or mixture of alcohol and water, by mass]

82-75-7
$C_{10}H_9NO_3S$
223.25
O=S(=O)(O)c(c(c(ccc1)cc2)c1N)c2
1-Naphthylamine-8-sulfonic acid
1-Amino-8-naphthalene sulfonate
1-Aminonaphthalene-8-sulfonic acid
8-Amino-1-naphthalenesulfonic acid
1-Naphthalenesulfonic acid, 8-amino- (9CI)
Naphthylaminemonosulfonic acid S
α-Naphthylamine-8-sulfonic acid
NSC-7798
Peri acid
Schollkopf's acid (VAN)

82-76-8
$C_{16}H_{13}NO_3S$
299.34
O=S(=O)(O)c(c(c(cc1)ccc2)c2Nc(cccc3)c3)c1
1-Anilino-8-naphthalenesulfonate
1-Aniline-8-naphthalene sulfonate
8-Anilinonaphthalene-1-sulfonate
Anilinonaphthalenesulfonic acid
1-Anilino-8-naphthalenesulfonic acid
8-Anilino-1-naphthalenesulfonic acid
1-Anilino-8-napthalenesulfonate
ANS
1-Naphthalenesulfonic acid, 8-anilino- (8CI)
1-Naphthalenesulfonic acid, 8-(phenylamino)- (9CI)
NSC-1746
Peri acid, phenyl-
1-(Phenylamino)-8-naphthalenesulfonic acid
8-(Phenylamino)-1-naphthalenesulfonic acid
Phenylperi acid

82-86-0
$C_{12}H_6O_2$
182.18
O=C(c(c(c(cc1)ccc2)c23)c1)C3=O
Acenaphthenedione
1,2-Acenaphthylenedione

82-92-8
$C_{18}H_{22}N_2$
266.42
CN1CCN(CC1)C(c2ccccc2)c3ccccc3
Piperazine, 1-(diphenylmethyl)-4-methyl
(N-Benzhydryl)(N'-methyl)diethylenediamine
N-Benzhydryl-N-methyl piperazine
BW 47-83
Compound 47-83
Ciclizina
Cyclizine
1-Diphenylmethyl-4-methylpiperazine
Emoquil
Marazine
Marezine
Marzine
N-Methyl-N'-benzyhydrylpiperazine
Nautazine
Neo-Devomit
Valoid
Wellcome preparation 47-83

82-93-9
$C_{18}H_{21}ClN_2$
300.86
CN1CCN(CC1)C(c2ccccc2)c3ccc(Cl)cc3
Piperazine, 1-(p-chloro-α-phenylbenzyl)-4-methyl
Chlorcycline
Chlorcyclizine
1-(4-Chlorobenzhydryl)-4-methylpiperazine
Chlorocycline
Chlorocyclizine
1-(p-Chloro-α-phenylbenzyl)-4-methylpiperazine
Di-Paralen
Diparalene
Histantin
Histantine

82-94-0
$C_{26}H_{33}N_3.2Cl$
458.49
Methyl Green
Benzenaminium, 4-((4-(dimethylamino)phenyl)(4-(dimethyliminio)-2,5-cyclohexadien-1-ylidene)methyl)-N,N,N-trimethyl-, dichloride (9CI)
Methyl Green Chloride

83-05-6
$C_{14}H_{10}Cl_2O_2$
281.14
OC(C(=O)c1ccc(Cl)cc1)c2ccc(Cl)cc2
Acetic acid, bis(p-chlorophenyl)
Benzeneacetic acid, 4-chloro-α-(4-chlorophenyl)- (9CI)
Bis(p-chlorophenyl)acetic acid
Bis(4-chlorophenyl)acetic acid
2,2-Bis(p-chlorophenyl)acetic acid
Bis(p-chlorphenyl)essigsaeure (German)
DDA
p,p'-DDA
Dichlorodiphenylacetic acid
p,p'-Dichlorodiphenylacetic

acid
Di(p-chlorophenyl)acetic acid

83-08-9
$C_{18}H_{11}NO_2$
273.28
O=C(c(c(C1=O)ccc2)c2)C1c(nc(c(ccc3)c4)c3)c4
Quinophthalone (VAN) (8CI)
Erio Chinoline Yellow 4G
1H-Indene-1,3(2H)-dione, 2-(2-quinolinyl)- (9CI)
NSC-18950
Quinoline Yellow 2SF
2-(2-Quinolyl)-1,3-indandione
2-(2-Quinolyl)-1,3-indanedione
11641 Yellow

83-12-5
$C_{15}H_{10}O_2$
222.25
O=C1C(C(=O)c2ccccc12)c3ccccc3
1,3-Indandione, 2-phenyl
Athrombon
Bindan
Danilone
Diadilan
Dindevan
Dineval
Emandione
Fenhydren
Fenilin
Hedulin
Indema
1H-Indene-1,3(2H)-dione, 2-phenyl- (9CI)
Indion
Indon
Phenhydren
Phenindione
2-Phenyl-1,3-diketohydrindene
Phenylen
2-Phenylindan-1,3-dione
2-Phenyl-1,3-indandione
Phenylindione
Phenyline
Phenyllin
Pid
Pindione

83-26-1
$C_{14}H_{14}O_3$
230.28
O=C(C(C(=O)c(c1ccc2)c2)C1=O)C(C)(C)C
1,3-Indandione, 2-pivaloyl
Chemrat
2-(2,2-Dimethyl-1-oxopropyl)-1H-indene-1,3(2H)-dione
1,3-Indandione, 2-pivalyl-
Latka 333 (Czech)
Pindon (Dutch)
Pindone (ACGIH,OSHA)
Pindone, Liquid (DOT)
Pindone, Solid (DOT)
Pivacin
Pival
Pivaldion (Italian)
Pivaldione (French)
2-Pivaloyl-indaan-1,3-dion (Dutch)
2-Pivaloyl-indan-1,3-dion (German)
2-Pivaloyl-1,3-indandione
2-Pivaloylindane-1,3-dione
2-Pivalyl-1,3-indandione
2-Pivalyl-1,3-indandione (OSHA)
Pivalyl valone
Pivalyn
Tri-BAN
2-(Trimetil-acetil)-indan-1,3-dione (Italian)
UN 2472 (DOT)

83-28-3
$C_{14}H_{14}O_3$
230.28
1,3-Indandione, 2-isovaleryl
2-Isopentanoyl-1,3-indanedione
Isoval
Isovaleryl indandione
2-Isovalerylindan-1,3-dione
2-Isovaleryl-1,3-indandione
2-Isovaleryl-1,3-indanedione
Motomco Tracking Powder
PMP
Valone

83-31-8
$C_{10}H_6O_3S$
206.22
O=S(=O)(Oc(c1c(cc2)ccc3)c2)c13
Naphthosultone
Naphth(1,8-cd)-1,2-oxathiole, 2,2-dioxide (9CI)
NSC-26341

83-32-9
$C_{12}H_{10}$
154.22
c(c(ccc1)ccc2)(c1CC3)c23
Acenaphthene
Acenaphthylene, 1,2-dihydro-
1,8-Ethylenenaphthalene
Naphthyleneethylene
Periethylenenaphthalene

83-33-0
C_9H_8O
132.16
O=C(c(c(ccc1)C2)c1)C2
1-Indanone (8CI)
AI3-11798
2,3-Dihydro-1H-inden-1-one
α-Hydrindone
Indanone (VAN)
Indan-1-one
α-Indanone
1H-Inden-1-one, 2,3-dihydro-(9CI)
1-Indone
NSC-2581

83-34-1
C_9H_9N
131.19
N(c(c(C=1C)ccc2)c2)C1
Indole, 3-methyl
β-Methylindole
3-Methylindole
3-Methyl-1H-indole
3-MI
Scatole
Skatol
Skatole

83-38-5
$C_7H_4Cl_2O$
175.01
O=Cc(c(ccc1)Cl)c1Cl
Benzaldehyde, 2,6-dichloro-(9CI)
2,6-Dichlorobenzaldehyde
NSC-7193

83-40-9
$C_8H_8O_3$
152.16
O=C(O)c(c(O)c(cc1)C)c1
2,3-Cresotic acid
Acido ortocresotinico (Italian)
Acido 3-ossi-5-metil-benzoico (Italian)
Benzoic acid, 2-hydroxy-3-methyl- (9CI)
Cresotic acid
o-Cresotic acid
Cresotinic acid
β-Cresotinic acid
o-Cresotinic acid
2,3-Cresotinic acid
Homosalicylic acid
3-Methylsalycilic acid
3-MS

83-41-0
$C_8H_9NO_2$
151.16
O=N(=O)c(c(c(cc1)C)C)c1
Benzene, 1,2-dimethyl-3-nitro-(9CI)
AI3-29558
1,2-Dimethyl-3-nitrobenzene
2,3-Dimethylnitrobenzene
3-Nitro-o-xylene
NSC-5402
o-Xylene, 3-nitro-

83-42-1
$C_7H_6ClNO_2$
171.59
O=N(=O)c(c(c(cc1)Cl)C)c1
Toluene, 2-chloro-6-nitro
Benzene, 1-chloro-2-methyl-3-nitro-
2-Chloro-6-nitrotoluene
6-Chloro-2-nitrotoluene

83-43-2
$C_{22}H_{30}O_5$

374.52
O=C(C(O)(C(C(C(C(C(C(=CC(=O)C=1)C2C)(C1)C)C3O)C2)C4)(C3)C)C4)CO
Pregna-1,4-diene-3,20-dione, 6-α-methyl-11-β-17,21-trihydroxy
Medrol
Medrol ADT Pak
Medrol Dosepak
Medrone
δ[1]-6-α-Methylhydrocortisone
Methylprednisolone
6-α-Methylprednisolone
Metrisone
NSC-19987
Prednisolone, methyl-
Pregna-1,4-diene-3,20-dione, 11-β,17,21-trihydroxy-6-α-methyl-
11-β,17,21-Trihydroxy-6-α-methylpregna-1,4-diene-3,20-dione
11-β,17-α,21-Trihydroxy-6-α-methyl-1,4-pregnadiene-3,20-dione
Urbason
Urbasone
Wyacort

83-44-3
$C_{24}H_{40}O_4$
392.64
O=C(O)CCC(C(C(C(C(C(C(C(C1)CC(O)C2)(C2)C)C3)C1)C4)(C3O)C)C4)C
5-β-Cholan-24-oic acid, 3-α,12-α-dihydroxy
Cholan-24-oic acid, 3,12-dihydroxy-, (3-α,5-β,12-α)-(9CI)
Choleic acid
Cholerebic
Cholic acid, deoxy-
Cholorebic
Degalol
Deoxycholatic acid
Deoxycholic acid
7-α-Deoxycholic acid
Desoxycholic acid
Desoxycholsaeure (German)
3,12-Dihydroxycholanic acid
3-α,12-α-Dihydroxycholanic acid
3-α,12-α-Dihydroxy-5-β-cholanoic acid
3-α,12-α-Dihydroxycholansaeure (German)
Droxolan
17-β-(1-Methyl-3-carboxypropyl)-etiocholane-3-α,12-α-diol
Pyrochol
Septochol

83-46-5
$C_{29}H_{50}O$
414.79
Stigmast-5-en-3-β-ol
Angelicin
Angelicin (steroid)
Cinchol
Cupreol
α-Dihydrofucosterol
22,23-Dihydrostigmasterol
24-α-Ethylcholesterol
Quebrachol
Rhamnol
β-Sitosterin
β-Sitosterol
SKF 14463
Stigmast-5-en-3-ol, (3-β)-(9CI)

83-47-6
$C_{29}H_{50}O$
414.72
Stigmast-5-en-3-ol, (3β,24S)-(9CI)
Clionasterol
β-Dihydrofucosterol
Fucosterol, β-dihydro-
γ-Sitosterol
Stigmast-5-en-3β-ol, (24S)-(8CI)

83-48-7
$C_{29}H_{48}O$
412.70
Stigmasterol
NSC-8095
Stigmasta-5,22-dien-3-ol, (3β)-
Stigmasta-5,22-dien-3-ol, (3β,22E)- (9CI)
Stigmasta-5,22-dien-3β-ol (8CI)
(24S)-5,22-Stigmastadien-3β-ol
Stigmasterin
β-Stigmasterol

83-56-7
$C_{10}H_8O_2$
160.18
Oc(c(c(c(O)cc1)cc2)c1)c2
1,5-Naphthalenediol
C.I. 76625
1,5-Dihydroxynaphthalene
1,5-Dihydroxynapthalene
Durafur Developer E
Oxidation base

83-59-0
$C_{20}H_{26}O_6$
362.46
CCCOC(=O)C3C(C)Cc2cc1OCOc1cc2C3C(=O)OCCC
Naphthalene-1,2-dicarboxyic acid, 1,2,3,4-tetrahydro-3-methyl-6,7-methylenedioxy-, dipropyl ester
Di-n-propyl maleate
Di-n-propyl maleate-isosafrole condensate
Di-n-propyl 6,7-methylenedioxy-3-methyl-1,2,3,4-tetrahydronaphthalene
Di-n-propyl-3-methyl-6,7-methylenedioxy-1,2,3,4-tetrahydronaphthalene-1,2-dicarboxylate
Dipropyl-5,6,7,8-tetrahydro-7-methylnaphtho(2,3-d)-1,3-dioxole-5,6-dicarboxylate
ENT 15,266
Propilizon
Propyl isome
n-Propylisome
Propyl isomer
n-Propyl isomer

83-63-6
$C_{18}H_{19}N_3O_2$
309.40
CC(=O)N(C(C)=O)c2ccc(N=Nc1ccccc1C)cc2C
Acetamide, N-acetyl-N-(2-methyl-4-((2-methylphenyl)azo)phenyl)
N-Acetyl-N-(2-methyl-4-((2-methylphenyl)azo)phenyl)-acetamide
Dermagan
Dermagen
Diacetazotol
Diacetotoluide
o-Diacetotoluidide, 4''-(o-tolylazo)- (8CI)
Diacetylaminoazotoluene
N,N-Diacetyl-o-tolylazo-o-toluidine
Diamazo
Dimazon
Epidermol
Epithelone
Granulin
Pellidol
Pellidole
Peripehrmin
4-o-Tolylazo-o-diacetotoluide
4'-(o-Tolylazo)-o-diacetotoluidide

83-66-9
$C_{12}H_{16}N_2O_5$
268.30
O=N(=O)c(c(c(N(=O)=O)c(OC)c1C(C)(C)C)C)c1
Anisole, 6-t-butyl-3-methyl-2,4-dinitro
2,6-Dinitro-3-methoxy-4-tert-butyltoluene
Musk Ambrette

83-67-0
$C_7H_8N_4O_2$
180.19
O=C(N(C(N=CN1C)=C1C2=O)C)N2
Theobromine
3,7-Dihydro-3,7-dimethyl-1H-purine-2,6-dione
3,7-Dimethylxanthine
Diurobromine
Santheose
SC 15090
Teobromin
Theosalvose
Theostene

Thesal
Thesodate
Xanthine, 3,7-dimethyl-

83-72-7
$C_{10}H_6O_3$
174.16
O=C(c(c(C(=O)C=1)ccc2)c2)C1O
1,4-Naphthoquinone, 2-hydroxy
C.I. 75480
C.I. Natural Orange 6
Flower of Paradise
Hana
Henna
2-Hydroxynaphthoquinone
2-Hydroxy-1,4-naphthoquinone
Lawsone
1,4-Naphthalenedione, 2-hydroxy- (9CI)
Mehendi
Mendi

83-73-8
$C_9H_5I_2NO$
396.95
n(c(c(c(cc1I)I)cc2)c1O)c2
8-Quinolinol, 5,7-diiodo
Diiodohydroxyquin
Diiodohydroxyquinoline
5,7-Diiodo-8-hydroxyquinoline
5,7-Diiodo-oxine
Diiodoquin
5,7-Diiodo-8-quinolinol
Dinoleine
Diodohydroxyquin
Diodoquin
Diodoxylin
Di-Quinol
Direxiode
Disoquin
Dyodin
Embequin
Enterosept
Floraquin
Fluoraquin
8-Hydroxy-5,7-diiodoquinoline
Iodoquinol
Ioquin suspension
Lanodoxin
Moebiquin
Quinadome
Searlequin
Sebaquin
SS 578
Yodoxin
Zoaquin

83-74-9
$C_{20}H_{26}N_2O$
310.43
Ibogaine (8CI)
DEA No. 7260
7-Ethyl-6,6β,7,8,9,10,12,13-octahydro-2-methoxy-6,9-methano-5H-pyrido(1',2':1,2)azepino(5,4-b)indole
(-)-Ibogaine
Ibogamine, 12-methoxy- (9CI)
NSC-249764
Tabernanthe iboga

83-79-4
$C_{23}H_{22}O_6$
394.45
COc5cc4OCC3Oc2c1CC(Oc1cc c2C(=O)C3c4cc5OC)C(C)=C
(1)Benzopyrano(3,4-b)furo-(2,3-h)(1)benzopyran-6(6ah)-one, 1,2,12,12a-tetrahydro-2-α- isopropenyl-8,9-dimethoxy
Barbasco
Cenol Garden Dust
Chem Fish
Chem-Mite
Cube
Cube Extract
Cube-Pulver
Cube Root
Cubor
Curex Flea Duster
Dactinol
Deril
Derrin
Derris
Dri-Kil
ENT 133
Extrax
Fish-Tox
Green Cross Warble Powder
Haiari
Liquid Derris
Mexide
NCI-C55210
Nicouline
Noxfish
Paraderil
Powder and Root
Prentox
Pro-Nox Fish
Ro-Ko
Ronone
Rotefive
Rotefour
Rotenon
Rotenona (Spanish)
Rotenone (ACGIH,OSHA)
(-)-Rotenone
Rotessenol
Rotocide
Tubatoxin
Tubotoxin

83-86-3
$C_6H_{18}O_{24}P_6$
660.06
O=P(OC(C(OP(=O)(O)O)C(OP(=O)(O)O)C(OP(=O)(O)O)C1OP(=O)(O)O)C1OP(=O)(O)O)(O)O
Inositol, hexakis(dihydrogen phosphate), myo
Alkovert
Fytic acid
Hexakis(dihydrogen phosphate) myo-inositol (9CI)
Inosithexaphosphorsaure (German)
myo-Inosistol hexakisphosphate
Inositol hexaphosphate
myo-Inositol hexaphosphate
Phytic acid
Saure des phytins (German)

83-88-5
$C_{17}H_{20}N_4O_6$
376.41
O=C(N=C(N(c(c(N=1)cc(c2C)C)c2)CC(O)C(O)C(O)CO)C1C3=O)N3
Riboflavine
Beflavine
6,7-Dimethyl-9-D-ribitylisoalloxazine
Flavaxin
Flaxain
Hyflavin
Hyre
Isoalloxazine, 7,8-dimethyl-10-d-ribityl-
Isoalloxazine, 7,8-dimethyl-10-(d-ribo-2,3,4,5-tetrahydroxypentyl)-
Lactoflavin
Lactoflavine
Ribipca
Riboderm
Riboflavin
Riboflavinequinone
Vitamin B2
Vitamin G

83-89-6
$C_{23}H_{30}ClN_3O$
400.01
CCN(CC)CCCC(C)Nc2c1ccc(Cl)cc1nc3ccc(OC)cc23
Acridine, 6-chloro-9-((4-(diethylamino)-1-methylbutyl)-amino)-2-methoxy
Acrichine
Acrinamine
Acriquine
Akrichin
Antimalarina
Atabrine
Atebrin
Atebrine
6-Chloro-9-((4-(diethylamino)-1-methylbutyl)amino)-2-methoxyacridine
3-Chloro-7-methoxy-9-(1-methyl-4-diethylaminobutyl-amino)acridine
Erion
Haffkinine
Mepacrine
2-Methoxy-6-chloro-9-diethyl-aminopentylaminoacridine
Quinacrine
Quinactine

83-98-7
$C_{18}H_{23}NO$
269.42
CN(C)CCOC(c1ccccc1)c2cc

ccc2C
Ethylamine, N,N-dimethyl-2-((o-methyl-α-phenyl benzyl)oxy)
Brocadisipal
β-Dimethylaminoethyl-2-methylbenzhydryl ether
o-Methyldiphenhydramine
2-Methyldiphenhydramine
Orphenadine
Orphenadrin
Orphenadrine
WS 2434

84-06-0
$C_{23}H_{28}ClN_3O_2S$
446.05
CC(=O)OCCN4CCN(CCCN2c1ccccc1Sc3ccc(Cl)cc23)CC4
1-Piperazineethanol, 4-(3-(2-chlorophenothiazin-10-yl)-propyl)-, acetate (ester)
4-(3-(2-Chlorophenothiazin-10-yl)propyl)-1-piperazine-ethanol acetate
Dartal
Dartalan
Perphenazine acetate
1-Piperazineethanol, 4-(3-(2-chloro-10h-phenothiazin-10-yl)propyl)-, acetate (ester) (9CI)
Thiopropazate

84-11-7
$C_{14}H_8O_2$
208.22
O=C(c(c(c(c1ccc2)c2)ccc3)c3)C1=O
Phenanthrenequinone
9,10-Phenanthraquinone
9,10-Phenanthrenedione
Phenanthrene, 9,10-dihydro-9,10-dioxo-
9,10-Phenanthrenequinone

84-15-1
$C_{18}H_{14}$
230.32
c(c(c(cccc1)c1)ccc2)(c(cccc3)c3)c2
o-Terphenyl
1,2-Diphenylbenzene

84-16-2
$C_{18}H_{22}O_2$
270.40
Phenol, 4,4'-(1,2-diethylethylene)di-, meso
Bibenzyl, α,α'-diethyl-4,4'-dihydroxy-
meso-3,4-Bis(p-hydroxyphenyl)-n-hexane
3,4-Bis(p-hydroxyphenyl)hexane
Cycloestrol
4,4'-(1,2-Diethylethylene)diphenol
Dihydrodiethylstilbestrol
Dihydrostilbestrol
4,4'-Dihydroxy-α,β-diethyl-diphenylethane
4,4'-Dihydroxy-γ,δ-diphenyl-hexane
γ,δ-Di(p-hydroxyphenyl)-hexane
meso-3,4-Di(p-hydroxyphenyl)-n-hexane
Extra-Plex
Hexane, 3,4-bis(p-hydroxy-phenyl)-
Hexanoestrol
Hexestrol
meso-Hexestrol
Hexoestrol
Hormoestrol
Phenol, 4,4'-(1,2-diethyl-1,2-ethylenediyl)bis-, (R*,S*)-(9CI)
Sinestrol
Stilbestrol, dihydro-
Synestrol
Synthovo
Syntrogene
Vitestrol

84-17-3
$C_{18}H_{18}O_2$
266.36
CC=C(C(=CC)c1ccc(O)cc1)c2ccc(O)cc2
Phenol, 4,4'-(diethylidene-ethylene)di
Agaldog
3,4-Bis(p-hydroxyphenyl)-2,4-hexadiene
3,4-Bis(4-hydroxyphenyl)-2,4-hexadiene
Cycladiene
Dehydrostilbestrol
Dehydrostilboestrol
Dienestrol
Dienoestrol
β-Dienoestrol
Dienol
4,4'-(1,2-Diethylidene-1,2-ethanediyl)bisphenol
p,p'-(Diethylideneethylene)diphenol
4,4'-(Diethylideneethylene)diphenol
Dinovex
Di(p-oxyphenyl)-2,4-hexadiene
DV
Estragard
Estrodienol
Estroral
Follidiene
Follormon
Gynefollin
Hormofemin
4,4'-Hydroxy-γ,δ-diphenyl-β,δ-hexadiene
Isodienestrol
Oestrasid
Oestrodiene
Oestrodienol
Oestroral
Oestrovis
Para-Dien
Phenol, 4,4'-(1,2-diethylidene-1,2-ethanediyl)bis- (9CI)
Restrol
Retalon
Sexadien
Synestrol
Teserene
Willnestrol

84-23-1
$C_{10}H_6N_2O_4S$
250.22
O=S(=O)(O)c(c(c(c(N=NO1)c12)ccc3)c3)c2
Naphth(1,2-d)(1,2,3)oxadiazole-5-sulfonic acid (9CI)
NSC-65884

84-31-1
$C_{20}H_{22}N_2O_2$
322.39
Cinchonan-9-one, 6'-methoxy-, (8α)- (9CI)
Quininone (8CI)

84-51-5
$C_{16}H_{12}O_2$
236.28
O=C(c(c(C(=O)c1cccc2)ccc3CC)c3)c12
9,10-Anthracenedione, 2-ethyl
Anthraquinone, 2-ethyl-
2-Ethylanthraquinone
2-Ethyl-9,10-anthraquinone
USAF SO-1

84-54-8
$C_{15}H_{10}O_2$
222.24
O=C(c(c(C(=O)c1cccc2)ccc3C)c3)c12
Anthraquinone, 2-methyl- (8CI)
AI3-15182
9,10-Anthracenedione, 2-methyl- (9CI)
2-Methyl-9,10-anthracenedione
β-Methylanthraquinone
2-Methylanthraquinone
NSC-607
Techtoquinone
Tectochinon
Tectoquinone

84-56-0
$C_{17}H_{11}N$
229.29
Naphtho(2,3-h)quinoline
1-Azabenz(a)anthracene

84-57-1
$C_{10}H_8Cl_2N_2O_4S$
323.16
O=C(N(N=C1C)c(c(cc(S(=O)(=O)O)c2Cl)Cl)c2)C1
Benzenesulfonic acid, 2,5-dichloro-4-(3-methyl-5-oxo-2-pyrazolin-1-yl)

Benzenesulfonic acid, 2,5-dichloro-4-(4,5-dihydro-3-methyl-5-oxo-1H-pyrazol-1-yl)-
2,5-Dichloro-4-(3-methyl-5-oxo-2-pyrazolin-1-yl)benzenesulfonic acid
Dichlorsulfofenyl-methylpyrazolon (Czech)
Kyselina 2,5-dichlor-4-(3'-methyl-5'-pyrazolon-1'-yl)-benzensulfonova (Czech)

84-60-6
$C_{14}H_8O_4$
240.22
O=C(c(c(C(=O)c1cc(O)cc2)ccc3O)c3)c12
Anthraquinone, 2,6-dihydroxy
9,10-Anthracenedione, 2,6-dihydroxy-
Anthraflavic acid
Anthraflavin
2,6-Dihydroxyanthraquinone
NSC-33531

84-61-7
$C_{20}H_{26}O_4$
330.46
O=C(OC(CCCC1)C1)c(c(ccc2)C(=O)OC(CCCC3)C3)c2
Phthalic acid, dicyclohexyl ester
1,2-Benzenedicarboxylic acid, dicyclohexyl ester
Dicyclohexyl phthalate
Ergoplast.FDC
HF 191
KP 201
Unimoll 66

84-62-8
$C_{20}H_{14}O_4$
318.34
O=C(Oc(cccc1)c1)c(c(ccc2)C(=O)Oc(cccc3)c3)c2
Phthalic acid, diphenyl ester
1,2-Benzenedicarboxylic acid, diphenyl ester (9CI)
Diphenyl phthalate
Phenyl phthalate

84-64-0
$C_{18}H_{24}O_4$
304.39
O=C(OC(CCCC1)C1)c(c(ccc2)C(=O)OCCCC)c2
Butyl cyclohexyl phthalate
1,2-Benzenedicarboxylic acid, butyl cyclohexyl ester (9CI)
Cyclohexyl butyl phthalate
NSC-69994
Phthalic acid, butyl cyclohexyl ester (8CI)

84-65-1
$C_{14}H_8O_2$
208.22
O=C(c(c(C(=O)c1cccc2)ccc3)c3)c12
Anthraquinone
Anthracene, 9,10-dihydro-9,10-dioxo-
9,10-Anthracenedione
9,10-Anthrachinon (Czech)
Anthradione
9,10-Anthraquinone
Corbit
9,10-Dioxoanthracene
Hoelite
Morkit

84-66-2
$C_{12}H_{14}O_4$
222.26
O=C(OCC)c(c(ccc1)C(=O)OCC)c1
Phthalic acid, diethyl ester
Anozol
1,2-Benzenedicarboxylic acid, diethyl ester
Diethylester kyseliny ftalove (Czech)
Diethyl phthalate (ACGIH, OSHA)
Estol 1550
Ethyl phthalate
NCI-C60048
Neantine
Palatinol A
Phthalol
Phthalsaeurediaethylester (German)
Placidol E
RCRA waste number U088
Solvanol

84-68-4
$C_{12}H_{10}Cl_2N_2$
253.14
Nc(ccc(c1Cl)c(c(cc(N)c2)Cl)c2)c1
Benzidine, 2,2'-dichloro
(1,1'-Biphenyl)-4,4'-diamine, 2,2'-dichloro- (9CI)
2,2'-Dichlorobenzidine

84-69-5
$C_{16}H_{22}O_4$
278.38
O=C(OCC(C)C)c(c(ccc1)C(=O)OCC(C)C)c1
Phthalic acid, diisobutyl ester
DIBP
Diisobutylester kyseliny ftalove (Czech)
Diisobutyl phthalate
Hexaplas M/1B
Palatinol IC

84-71-9
$C_{24}H_{44}O_4$
396.61
1,2-Cyclohexanedicarboxylic acid, bis(2-ethylhexyl) ester (8CI,9CI)

84-72-0
$C_{14}H_{16}O_6$
280.30
CCOC(=O)COC(=O)c1ccccc1C(=O)OCC
Phthalic acid, ethyl ester, ester with ethyl glycolate
1,2-Benzenedicarboxylic acid, 2-ethoxy-2-oxoethyl-, ethyl ester (9CI)
Carbethoxymethyl ethyl phthalate
Diethyl o-carboxybenzoyloxyacetate
Ethyl carbethoxymethyl phthalate
Ethyl phthalyl ethyl glycolate
Santicizer E-15

84-74-2
$C_{16}H_{22}O_4$
278.38
O=C(OCCCC)c(c(ccc1)C(=O)OCCCC)c1
Phthalic acid, dibutyl ester
o-Benzenedicarboxylic acid, dibutyl ester
Benzene-o-dicarboxylic acid di-n-butyl ester
Di-n-butylester kyseliny ftalove (Czech)
Celluflex DPB
DBP
Dibutyl 1,2-benzenedicarboxylate
Dibutyl phthalate (ACGIH, OSHA)
Di-n-butyl phthalate
Elaol
Hexaplas M/B
Palatinol C
Polycizer DBP
PX 104
RCRA waste number U069
Staflex DBP
Witcizer 300

84-75-3
$C_{20}H_{30}O_4$
334.50
O=C(OCCCCCC)c(c(ccc1)C(=O)OCCCCCC)c1
Phthalic acid, dihexyl ester
1,2-Benzenedicarboxylic acid, dihexyl ester
Dihexylester kyseliny ftalove (Czech)
Dihexyl phthalate
Di-n-hexyl phthalate

84-76-4
$C_{26}H_{42}O_4$
418.68
O=C(OCCCCCCCCC)c(c(ccc1)C(=O)OCCCCCCCCC)c1
Phthalic acid, dinonyl ester
Bisoflex 91

Dinonyl 1,2-benzenedicarboxylate
Dinonyl phthalate
Di-n-nonyl phthalate

84-77-5
$C_{28}H_{46}O_4$
446.74
O=C(OCCCCCCCCCC)c(c(ccc1)C(=O)OCCCCCCCCCC)c1
Phthalic acid, didecyl ester
Didecyl phthalate
Di-n-decyl phthalate

84-78-6
$C_{20}H_{30}O_4$
334.46
O=C(OCCCCCCCC)c(c(ccc1)C(=O)OCCCC)c1
Phthalic acid, butyl octyl ester (8CI)
1,2-Benzenedicarboxylic acid, butyl octyl ester (9CI)
Butyl octyl 1,2-benzenedicarboxylate
Butyl octyl phthalate
NSC-69894
Octyl butyl phthalate
Plasticizer BOP
Plasticizer OBP
PX 914
Staflex BOP
Truflex OBP

84-79-7
$C_{15}H_{14}O_3$
242.29
1,4-Naphthoquinone, 2-hydroxy-3-(3-methyl-2-butenyl)
Bethabarra wood
C.I. 75490
C.I. Natural Yellow 16
Greenharten
2-Hydroxy-3-(3-methyl-2-butenyl)-1,4-naphthalenedione
2-Hydroxy-3-(3-methyl-2-butenyl)-1,4-naphthoquinone
Ipe-Tobacco Wood
Lapachol
Lapachic acid
Lapachol Wood
NSC-11905
Surinam Greenheart Wood
Taigu Wood
Taiguic acid
Tecomin
Zlut Prirodni 16 (Czech)

84-80-0
$C_{31}H_{46}O_2$
450.77
O=C(c(c(C(=O)C=1C)ccc2)c2)C1CC=C(CCCC(CCCC(CCCC(C)C)C)C)C
1,4-Naphthalenedione, 2-methyl-3-(3,7,11,15-tetramethyl-2-hexadecenyl)
Antihemorrhagic vitamin
Aqua Mephyton
Combinal K1
Kativ N
Kephton
Kinadion
Konakion
Mephyton
2-Methyl-3-phytyl-1,4-naphthochinon (German)
2-Methyl-3-(3,7,11,15-tetramethyl-2-hexadecenyl)-1,4-naphthalenedione
1,4-Naphthoquinone, 2-methyl-3-phytyl-
Monodion
Mono-Kay
Phyllochinon (German)
Phylloquinone
α-Phylloquinone
trans-Phylloquinone
Phytomenadione
Phytonadione
Vitamin K1
Synthex P

84-83-3
$C_{13}H_{15}NO$
201.26
O=CC=C(N(c(c1ccc2)c2)C)C1(C)C
Acetaldehyde, (1,3-dihydro-1,3,3-trimethyl-2H-indol-2-ylidene)- (9CI)
Fisher's Aldehyde
2-(Formylmethylene)-1,3,3-trimethylindoline
δ2,α-Indolineacetaldehyde, 1,3,3-trimethyl- (8CI)
NSC-68048
1,3,3-Trimethyl-2-(formylmethylene)indoline
1,3,3-Trimethyl-δ2,α-indolineacetaldehyde
(1,3,3-Trimethylindolin-2-ylidene)acetaldehyde

84-86-6
$C_{10}H_9NO_3S$
223.26
O=S(=O)(O)c(c(c(c(N)c1)ccc2)c2)c1
1-Naphthalenesulfonic acid, 4-amino
1-Aminonaphthalene-4-sulfonic acid
4-Amino-1-naphthalenesulfonic acid
1-Amino-4-sulfonaphthalene
Kyselina nafthionova (Czech)
Kyselina 1-naftylamin-4-sulfonova (Czech)
Naphthionic acid
1,4-Naphthionic acid
α-Naphthylamine-p-sulfonic acid
1-Naphthylamine-4-sulfonic acid
Piria's acid
USAF M-5

84-87-7
$C_{10}H_8O_4S$
224.24
O=S(=O)(O)c(c(c(c(O)c1)ccc2)c2)c1
1-Naphthol-4-sulfonic acid
1-Hydroxynaphthalene-4-sulfonic acid
1-Hydroxy-4-naphthalenesulfonic acid
4-Hydroxy-1-naphthalenesulfonic acid
1-Naphthalenesulfonic acid, 4-hydroxy- (9CI)
α-Naphthol-4-sulfonic acid
1-Naphtho-4-sulfonic acid
Nevile and Winther's acid
Neville-winther acid
NSC-9587
NW Acid
1,4-Oxy Acid

84-89-9
$C_{10}H_9NO_3S$
223.26
O=S(=O)(O)c(c(c(c(N)cc1)cc2)c1)c2
1-Naphthalenesulfonic acid, 5-amino
5-Amino-1-naphthalenesulfonic acid

84-91-3
$C_{10}H_5N_3O_6S$
295.21
O=S(=O)(O)c(c(c(c(N=NO1)c12)ccc3N(=O)=O)c3)c2
Naphth(1,2-d)(1,2,3)oxadiazole-5-sulfonic acid, 7-nitro- (9CI)
7-Nitronaphth(1,2-d)(1,2,3)oxadiazole-5-sulfonic acid

84-96-8
$C_{18}H_{22}N_2S$
298.48
CC(CN(C)C)CN2c1ccccc1Sc3ccccc23
Phenothiazine, 10-(3-(di methylamino)-2-methylpropyl)
Alimemazine
Alimezine
Bayer 1219
Methylpromazine
10H-Phenothiazine-10-propanamine, N,N,β-trimethyl-
Repeltin
Teralen
Trimeprazine

85-00-7
$C_{12}H_{12}N_2.2Br$
344.08
[Br-].[Br-].C2C[n+]1ccccc1c3cccc[n+]23
Dipyrido(1,2-a;2',1'-c)pyra-

zinediium, 6,7-dihydro-, dibromide
1,1'-Aethylen-2,2'-bipyridinium-dibromid (German)
Aquacide
Cleansweep
Deiquat
Dextrone
9,10-Dihydro-8a,10,-diazoniaphenanthrene dibromide
9,10-Dihydro-8a,10a-diazoniaphenanthrene(1,1'-ethylene-2,2'-bipyridylium)dibromide
5,6-Dihydro-dipyrido(1,2a;2,1c)-pyrazinium dibromide
6,7-Dihydropyrido(1,2-a;2',1'-c)pyrazinedium dibromide
Diquat (ACGIH,OSHA)
Diquat dibromide
1,1'-Ethylene-2,2'-bipyridylium dibromide
Ethylene dipyridylium dibromide
1,1-Ethylene 2,2-dipyridylium dibromide
1,1'-Ethylene-2,2'-dipyridylium dibromide
FB/2
Preeglone
Reglon
Reglone
Reglox
Weedol
Weedtrine-D

85-01-8
$C_{14}H_{10}$
178.24
c(c(c(c(c1)ccc2)c2)ccc3)(c1)c3
Phenanthrene
Coal tar pitch volatiles: phenanthrene
Phenanthren (German)
Phenantrin

85-02-9
$C_{13}H_9N$
179.23
c1ccc2c(c1)ccc3ncccc23
Benzo(f)quinoline
1-Azaphenanthrene
5,6-Benzoquinoline
β-Naphthoquinaldine

85-06-3
$C_{14}H_{11}N$
193.24
n(c(c(c(c(c1)ccc2)c2)cc3)c1)c3C
Benzo(f)quinoline, 3-methyl- (9CI)
3-Methylbenzo(f)quinoline
β-Naphthoquinaldine
NSC-43316

85-19-8
$C_{13}H_9ClO_2$
232.67
Benzophenone, 5-chloro-2-hydroxy- (8CI)
2-Benzoyl-4-chlorophenol
Chlorohydroxy benzophenone
3-Chloro-6-hydroxybenzophenone
5-Chloro-2-hydroxybenzophenone
(5-Chloro-2-hydroxyphenyl)-phenylmethanone
2-Hydroxy-5-chlorobenzophenone
DOW HCB
Methanone, (5-chloro-2-hydroxyphenyl)phenyl- (9CI)
NSC-33407
UV Absorber NL/5

85-22-3
$C_8H_5Br_5$
500.65
c(c(c(c(c1Br)Br)Br)Br)(c1Br)CC
Benzene, pentabromoethyl- (9CI)
Pentabromoethylbenzene
2,3,4,5,6-Pentabromoethylbenzene

85-29-0
$C_{13}H_8Cl_2O$
251.11
O=C(c(ccc(c1)Cl)c1)c(c(ccc2)Cl)c2
Benzophenone, 2,4'-dichloro- (8CI)
AI3-15228
(2-Chlorophenyl)(4-chlorophenyl)methanone
2,4'-Dichlorobenzophenone
Methanone, (2-chlorophenyl)-(4-chlorophenyl)- (9CI)
NSC-3221

85-32-5
$C_{10}H_{14}N_5O_8P$
363.26
O=P(OCC(OC(N(c(nc(nc1O)N)c1N=2)C2)C3O)C3O)(O)O
5'-Guanylic acid
GMP
5'-GMP
Guanidine monophosphate
Guanosine monophosphate
Guanosine 5'-monophosphate
Guanosine 5'-monophosphoric acid
Guanosine 5'-phosphate
Guanylic acid

85-34-7
$C_8H_5Cl_3O_2$
239.48
O=C(O)Cc(c(ccc1Cl)Cl)c1Cl
Acetic acid, (2,3,6-trichlorophenyl)
Benzeneacetic acid, 2,3,6-trichloro- (9CI)
Chlorfenac
Fenac
Fenae
Fenatrol
Kanepar
Kyselina 2,3,6-trichlorfenyloctova (Czech)
TCPA
2,3,6-Trichlorobenzeneacetic acid
2,3,6-Trichlorophenylacetic acid
2,3,6-Trichlorphenylessigsaeure (German)
Tri-Fen
Tri Fene

85-38-1
$C_7H_5NO_5$
183.13
Salicylic acid, 3-nitro
3-Nitrosalicylic acid

85-40-5
$C_8H_9NO_2$
151.18
O=C(NC(=O)C1CC=CC2)C12
4-Cyclohexene-1,2-dicarboximide
1H-Isoindole-1,3(2H)-dione, 3-α,4,7,7-α-tetrahydro- (9CI)
Tetrahydrophthalic acid imide
Tetrahydrophthalimide
δ^4-Tetrahydrophthalimide
1,2,3,6-Tetrahydrophthalimide

85-41-6
$C_8H_5NO_2$
147.14
O=C(NC(=O)c1cccc2)c12
Phthalimide
Ftalimmide (Italian)
Isoindole-1,3-dione
1,3-Isoindoledione
1,3-Isoindolinedione
o-Phthalic imide
Phthalimid (German)

85-42-7
$C_8H_{10}O_3$
154.17
O=C(OC(=O)C1CCCC2)C12
1,2-Cyclohexanedicarboxylic anhydride (8CI)
Araldite HT 907
1,2-Cyclohexanedicarboxylic acid anhydride
Hexahydro-1,3-isobenzofurandione
Hexahydrophthalic acid anhydride
Hexahydrophthalic anhydride
HHPA
1,3-Isobenzofurandione, hexahydro- (9CI)
Lekutherm Hardener H
NSC-8622
NT 907

85-43-8
$C_8H_8O_3$
152.16
O=C(OC(=O)C1CC=CC2)C12
4-Cyclohexene-1,2-dicarboxylic anhydride
Anhydrid kyseliny tetrahydroftalove (Czech)
1,3-Isobenzofurandione, 3a,4,7,7a-tetrahydro-
Maleic anhydride adduct of butadiene
Phthalic anhydride, 1,2,3,6-tetrahydro-
Tetrahydroftalanhydrid (Czech)
Tetrahydrophthalic acid anhydride
Tetrahydrophthalic anhydride
δ^4-Tetrahydrophthalic anhydride
1,2,3,6-Tetrahydrophthalic anhydride
Tetrahydrophthalic anhydride (DOT)
Tetrahydrophthalic anhydrides with more than 0.05 percent of maleic anhydride [UN 2698]
THPA
UN 2698 [Tetrahydrophthalic anhydrides with more than 0.05 percent of maleic anhydride]

85-44-9
$C_8H_4O_3$
148.12
O=C(OC(=O)c1cccc2)c12
Phthalic-anhydride
Anhydride phtalique (French)
Anhydrid kyseliny ftalove (Czech)
Anidride ftalica (Italian)
1,2-Benzenedicarboxylic acid anhydride
1,3-Dioxophthalan
Esen
Ftaalzuuranhydride (Dutch)
Ftalanhydrid (Czech)
Ftalowy bezwodnik (Polish)
Isobenzofuran, 1,3-dihydro-1,3-dioxo-
1,3-Isobenzofurandione
NCI-C03601
Phthalandione
1,3-Phthalandione
Phthalic acid anhydride
Phthalic anhydride (ACGIH, OSHA)
Phthalic anhydride, Solid or molten (DOT)
Phthalsaeureanhydrid (German)
RCRA waste number U190
Retarder AK
Retarder Esen
Retarder PD
UN 2214 (DOT)

85-47-2
$C_{10}H_8O_3S$
208.24
O=S(=O)(O)c(c(c(ccc1)cc2)c1)c2
1-Naphthalenesulfonic acid (9CI)

85-48-3
$C_9H_7NO_3S$
209.22
O=S(=O)(O)c(c(nccc1)c1cc2)c2
8-Quinolinesulfonic acid (9CI)
NSC-10433
Quinoline-8-sulfonic acid

85-52-9
$C_{14}H_{10}O_3$
226.24
O=C(O)c(c(ccc1)C(=O)c(cccc2)c2)c1
Benzoic acid, o-benzoyl
Benzoic acid, 2-benzoyl-
Benzophenone-2-carboxylic acid
2-Benzoylbenzoic acid

85-60-9
$C_{26}H_{38}O_2$
382.64
Oc(c(cc(c1C)C(c(c(cc(O)c2C(C)(C)C)C)c2)CCC)C(C)(C)C)c1
m-Cresol, 4,4'-butylidenebis-(6-tert-butyl
Anullex PBA 15
BBM
1,1-Bis(2-methyl-4-hydroxy-5-tert-butylphenyl)butane
4,4'-Butylidenebis(6-tert-butyl-m-cresol)
4,4'-Butylidenebis(6-tert-butyl-3-methylphenyl)
4,4'-Butylidenebis(3-methyl-6-tert-butylphenol)
Phenol, 4,4'-butylidenebis-(2-(1,1-dimethylethyl)-5-methyl- (9CI)
Santowhite
Santowhite powder
Sumilit BBM
Sumilizer BBM
SWP
SWP (Antioxidant)

85-68-7
$C_{19}H_{20}O_4$
312.39
O=C(OCc(cccc1)c1)c(c(ccc2)C(=O)OCCCC)c2
Phthalic acid, benzyl butyl ester
BBP
1,2-Benzenedicarboxylic acid, butyl phenylmethyl ester
Benzyl-butylester kyseliny ftalove (Czech)
Benzyl butyl phthalate
Benzyl n-butyl phthalate
Butyl benzyl phthalate
n-Butyl benzyl phthalate
Butyl phenylmethyl 1,2-benzenedicarboxylate
NCI-C54375
Palatinol bb
Santicizer 160
Sicol 160
Unimoll BB

85-69-8
$C_{20}H_{30}O_4$
334.46
O=C(OCC(CCCC)CC)c(c(ccc1)C(=O)OCCCC)c1
2-Ethylhexyl butyl phthalate
1,2-Benzenedicarboxylic acid, butyl 2-ethylhexyl ester (9CI)
Butyl 2-ethylhexyl phthalate
Phthalic acid, butyl 2-ethylhexyl ester

85-70-1
$C_{18}H_{24}O_6$
336.42
O=C(OCC(=O)OCCCC)c(c(ccc1)C(=O)OCCCC)c1
Phthalic acid, butyl ester, ester with butyl glycolate
Butyl carbobutoxymethyl phthalate
Butyl glycolyl butyl phthalate
Butyl phthalate butyl glycolate
Butyl phthalyl butyl glycolate
Dibutyl o-(o-carboxybenzoyl) glycolate
Dibutyl o-carboxybenzoyloxyacetate
Glycolic acid, butyl ester, butyl phthalate
Glycolic acid, phthalate, dibutyl ester
Phthalic acid, butoxycarbonylmethyl butyl ester
Phthalic acid, butyl ester, butyl glycolate
Santicizer B-16

85-71-2
$C_{13}H_{14}O_6$
266.27
CCOC(=O)COC(=O)c1ccccc1C(=O)OC
Phthalic acid, monomethyl ester, ester with ethyl glycolate
Ethoxykarbonylmethylmethylester kyseliny ftalove (Czech)
Ethyl o-(o-(methoxycarbonyl)-benzoyl)glycolate
Ethyl o-(methoxycarbonyl)benzoyloxyacetate
Glycolic acid, ethyl ester, methyl phthalate
Methyl phthalyl ethyl glycolate
Santicizer M-17

85-72-3
$C_{15}H_{13}NO_3$
255.29
Cc2cccc(NC(=O)c1ccccc1C(O)=O)c2

Phthalanilic acid, 3'-methyl
Duraset
Duraset 20W
Kyselina N-m-tolylftalamova (Czech)
N-Metatolyl phthalamic acid
2-(((3-Methylphenyl)amino)carbonyl)benzoic acid
3'-Methylphthalanilic acid
N-M-T
Phthalamate
N-m-Tolylphthalamic acid
N-meta-Tolylphthalamic acid
Tomaset

85-79-0
$C_{20}H_{29}N_3O_2$
343.52
CCCCOc2cc(C(=O)NCCN(CC)CC)c1ccccc1n2
Cinchoninamide, 2-butoxy-N-(2-(diethylamino)ethyl)
2-Butoxy-N-(β-diethylaminoethyl)cinchoninamide
2-Butoxy-N-(2-(diethylamino)-ethyl)cinchoninamide
2-Butoxyquinoline-4-carboxylic acid diethylaminoethylamide
α-Butyloxycinchoninic acid diethylethylenediamide
Cinchocaine
Dermacaine
Dibucain
Dibucaine
Dibucaine base
N-(2-(Diethylamino)ethyl)-2-butoxycinchoninamide
Nupercainal
Nupercaine
4-Quinolinecarboxamide, 2-butoxy-N-(2-(diethylamino)-ethyl)- (9CI)
Sovcaine

85-83-6
$C_{24}H_{20}N_4O$
380.48
Oc(ccc(c1ccc2)c2)c1N=Nc(c(cc(N=Nc(c(ccc3)C)c3)c4)C)c4
2-Naphthol, 1-((4-(o-tolylazo)-o-tolyl)azo)
Biebrich Scarlet BPC
Biebrich Scarlet Red
Biebrich Scarlet R Medicinal
Brasilazina Oil Red B
Calco Oil Red D
Candle Scarlet B
Candle Scarlet 2B
Candle Scarlet G
Ceres Red BB
Cerotine Ponceau 3B
Cerven Rozpoustedlova 24 (Czech)
C.I. 258
C.I. 26105
C.I. Solvent Red 24
2',3-Dimethyl-4-(2-hydroxy-naphthylazo)azobenzene
Dispersol Red PP
Enial Red IV
Fast Oil Red B
Fast Red BB
Fat Ponceau R
Fat Red B
Fat Red 2B
Fat Red BB
Fat Red BS
Fat Red TS
Grasal Brilliant Red B
Grasan Brilliant Red B
Hidaco Oil Red
Lacquer Red V
Lacquer Red VS
Lipid Crimson
1-((2-Methyl-4-((2-methyl-phenyl)azo)phenyl)azo)-2-naphthalenol
Oil Red IV
Oil Red 3
Oil Red 7
Oil Red 47
Oil Red 282
Oil Red A
Oil Red APT
Oil Red 2B
Oil Red 3B
Oil Red BB
Oil Red 4B
Oil Red BS
Oil Red D
Oil Red ED
Oil Red F
Oil Red GO
Oil Red PEL
Oil Red RC
Oil Red RR
Oil Red S
Oil Red TAX
Oil Red ZD
Oil Scarlet
Oil Scarlet 48
Oleal Red BB
Organol Red B
Orient Oil Red RR
Phenoplaste Organol Red B
Plastoresin Red F
Red 3R soluble in grease
Resinol Red 2B
Rubrum Scarlatinum
Resoform Red G
Sarlach R (Czech)
Scarlet Red
Scarlet Red, Biebrich
Scarlet R (Michaelis)
Scharlachrot
Schultz No. 541
Silotras Red T3B
Somalia Red IV
Stearix Red 4B
Stearix Red 4S
Sudan IV
Sudan P
Sudan Red IV
Sudan Red 4BA
Sudan Red BB
Sudan Red BBA
Tertrogras Red N
o-Tolueneazo-o-tolueneazo-β-naphthol
o-Tolueneazo-o-toluene-β-naphthol
o-Tolylazo-o-tolylazo-β-naphthol
o-Tolylazo-o-tolylazo-2-naphthol
1-(4-o-Tolylazo-o-tolylazo)-2-naphthol
Toyo Oil Red BB
Waxakol Red BL
Waxoline Red O
Waxoline Red OM
Waxoline Red OS

85-84-7
$C_{16}H_{13}N_3$
247.32
N(=Nc(cccc1)c1)c(c(c(ccc2)cc3)c2)c3N
2-Naphthylamine, 1-(phenylazo)
A.F. Yellow No. 2
1-Benzene-azo-β-naphthylamine
1-Benzeneazo-2-naphthylamine
Cerisol Yellow AB
C.I. 11380
C.I. Food Yellow 10
C.I. Solvent Yellow 5
Dolkwal Yellow AB
Ext. D & C Yellow No. 9
FD & C Yellow 3
FD & C Yellow No. 3
Grasal Yellow
Jaune AB
Oil Yellow A
Oil Yellow AB Pure
Oil Yellow OB
1-(Phenylazo)-2-naphthalenamine
1-Phenylazo-2-naphthylamine
Yellow AB
Yellow No. 2
Zlut Maselna AB (Czech)
Zlut Rozpoustedlova 5 (Czech)

85-85-8
$C_{15}H_{11}N_3O$
249.25
n(c(N=Nc(c(c(ccc1)cc2)c1)c2O)ccc3)c3
1-(2-Pyridylazo)-2-naphthol
2-Hydroxy-1-(2-pyridylazo)-naphthalene
2-Naphthalenol, 1-(2-pyridinyl-azo)- (9CI)
2-Naphthol, 1-(2-pyridylazo)-(8CI)
NSC-5332
PAN (VAN)
PAN (Indicator)
1-(2-Pyridinylazo)-2-naphthalenol
1-(2-Pyridylazo)-2-hydroxy-naphthalene

85-86-9
$C_{22}H_{16}N_4O$
352.42
Oc(ccc(c1ccc2)c2)c1N=Nc(ccc(N=Nc(cccc3)c3)c4)c4
2-Naphthalenol, 1-((4-(phenylazo)phenyl)azo)

Atul Oil Red G
Benzeneazobenzeneazo-β-naphthol
Brasilazina Oil Scarlet
Cerasin Red
Cerasinrot
Cerotinscharlach R
Certiqual Oil Red
Cerven rozpoustedlova 23 (Czech)
C.I. 23
C.I. 26100
C.I. Solvent Red 23
D & C Red No. 17
Fast Oil Scarlet III
Fast Red R
Fat Red (bluish)
Fat Red G
Fat Red HRR
Fat Red R
Fat Red RS
Fat Scarlet LB
Fat Soluble Red ZH
Grasal Brilliant Red G
Fettponceau G
Fettrot
Fettscharlach
Fettscharlach LB
Motirot 2R
Oil Red
Oil Red 6566
Oil Red AS
Oil Red B
Oil Red 3B
Oil Red G
Oil Red 3G
Oil Red O
Oil Scarlet
Oil Scarlet AS
Oil Scarlet G
Organol Red BS
Organol Scarlet
1-((4-(Phenylazo)phenyl)azo)-2-naphthalenol
1-((p-Phenylazo)phenyl)azo-2-naphthol
Ponceau Insoluble OLG
Pyronalrot B
111440 Red
Red ZH
Rot C
Rot G
Rouge cerasine
Scarlet B Fat Soluble
Schultz No. 31
Silotras Scarlet TB
Somalia Red III
Soudan III
Stearix Scarlet
Sudan III
Sudan G
Sudan G III
Sudan III (G)
Sudan P III
Sudan Red III
Tetrazobenzene-β-naphthol
Toney Red
Tony Red

85-87-0
$C_8H_{12}N_2O_2$
168.20
Pyridoxamine

85-91-6
$C_9H_{11}NO_2$
165.21
O=C(OC)c(c(NC)ccc1)c1
Anthranilic acid, N-methyl-, methyl ester
Dimethyl anthranilate
2-Methylamino methyl benzoate
N-Methylanthranilic acid, methyl ester
Methyl methylaminobenzoate
Methyl N-methyl anthranilate
MMA

85-97-2
$C_{12}H_9ClO$
204.66
6-Chloro-2-phenylphenol
AI3-03271
2-Biphenylol, 3-chloro- (8CI)
(1,1'-Biphenyl)-2-ol, 3-chloro- (9CI)
Caswell No. 211
3-Chloro-2-biphenylol
3-Chloro-(1,1'-biphenyl)-2-ol
2-Chloro-6-phenylphenol
Dowcide 31
Dowcide 32
Dowicide 31
Dowicide 32
EPA Pesticide Chemical Code 062210
2-Hydroxy-3-chlorobiphenyl
NSC-2600
2-Phenyl-6-chlorophenol

85-98-3
$C_{17}H_{20}N_2O$
268.39
O=C(N(c(cccc1)c1)CC)N(c(cccc2)c2)CC
Carbanilide, N,N'-diethyl
Bis-(N-ethyl-N-fenyl)mocovina (Czech)
Bis(N-ethyl-N-phenyl)urea
Carbamite
Centralite
Centralite-1
N,N-Diethylcarbanilide
1,3-Diethyl-1,3-difenyl-mocovina (Czech)
N,N'-Diethyl-N,N'-diphenylurea
sym-Diethyldiphenylurea
Ethyl centralite
Urea, N,N'-diethyl-N,N'-diphenyl- (9CI)
Urea, 1,3-diethyl-1,3-diphenyl-
USAF EK-1047

86-00-0
$C_{12}H_9NO_2$
199.22
O=N(=O)c1ccccc1c2ccccc2
Biphenyl, 2-nitro
o-Nitrobiphenyl
2-Nitrobiphenyl
o-Nitrodiphenyl
2-Nitrodiphenyl

86-13-5
$C_{21}H_{25}NO$
307.47
1-α-H,5-α-H-Tropane, 3-α-(diphenylmethoxy)
Benzotropine
Benztropine
Cogentine

86-14-6
$C_{16}H_{21}NS_2$
291.48
Diethylthiambutene
Allylamine, N,N-diethyl-3,3-di-2-thienyl-1-methyl-
3-Buten-2-amine, N,N-diethyl-4,4-di-2-thienyl- (9CI)
DEA No. 9616
Diaethylthiambutenum
3-Diethylamino-1,1-bis(2-thienyl)-1-butene
3-Diethylamino-1,1-dithienylbut-1-ene
3-Diethylamino-1,1-di(2'-thienyl)but-1-ene
3-Diethylamino-1,1-di(2'-thienyl)-1-butene
N,N-Diethyl-4,4-di-2-thienyl-3-buten-2-amine
N,N-Diethyl-1-methyl-3,3-di-2-thienylallylamine
Diethylthiambutenum (Latin)
Diethibutin
Dietiltiambutene
Dietiltiambuteno (Spanish)
NIH-4185
Themalon
Thiambutene
191C49

86-20-4
$C_{14}H_{12}N_2O_2$
240.25
O=N(=O)c(ccc(N(c(c1ccc2)c2)CC)c13)c3
Carbazole, 9-ethyl-3-nitro- (8CI)
9H-Carbazole, 9-ethyl-3-nitro- (9CI)
9-Ethyl-3-nitrocarbazole
9-Ethyl-3-nitro-9H-carbazole
NSC-17817
3-Nitro-N-ethylcarbazole
3-Nitro-9-ethylcarbazole

86-21-5
$C_{16}H_{20}N_2$
240.38
n(c(ccc1)C(c(cccc2)c2)CCN(C)C)c1
Pyridine, 2-(α-(2-(dimethylamino)ethyl)benzyl)
Avil
2-(α-(2-Dimethylaminoethyl)-

benzyl)pyridine
2-(3-Dimethylamino-1-phenylpropyl)pyridine
N,N-Dimethyl-3-phenyl-3-(2-pyridyl)propylamine
Inhiston
Metron
NCI-C60695
Pheniramine
1-Phenyl-1-(2-pyridyl)-3-dimethylaminopropane
3-Phenyl-3-(2-pyridyl)-N,N-dimethylpropylanine
PM 241
Prophenpyridamine
Trimeton
Tripoton

86-22-6
$C_{16}H_{19}BrN_2$
319.23
n(c(ccc1)C(c(ccc(c2)Br)c2)CCN(C)C)c1
Brompheniramine
Bromfeniramina (Spanish)
2-(p-Bromo-α-(2-dimethylaminoethyl)benzyl)pyridine
Bromopheniramine
γ-(4-Bromophenyl)-N,N-dimethyl-2-pyridinepropanamine
1-(p-Bromophenyl)-1-(2-pyridyl)-3-dimethylaminopropane
3-(p-Bromophenyl)-3-(2-pyridyl)-N,N-dimethylpropylamine
Brompheniraminum (Latin)
Ilvin
Parabromdylamine
Parabromodylamine
Pyridine, 2-(p-bromo-α-(2-(dimethylamino)ethyl)benzyl)-
2-Pyridinepropanamine, γ-(4-bromophenyl)-N,N-dimethyl-(9CI)

86-26-0
$C_{13}H_{12}O$
184.25
O(c(c(c(cccc1)c1)ccc2)c2)C
Anisole, o-phenyl
1,1'-Biphenyl, 2-methoxy-(9CI)
2-Methoxy biphenyl
Methyl diphenyl ether
o-Phenyl anisole
2-Phenylanisole

86-28-2
$C_{14}H_{13}N$
195.26
N(c(c(c1cccc2)ccc3)c3)(c12)CC
Carbazole, 9-ethyl- (8CI)
AI3-14686
9H-Carbazole, 9-ethyl- (9CI)
N-Ethylcarbazole
9-Ethylcarbazole
9-Ethyl-9H-carbazole
NSC-60585

86-29-3
$C_{14}H_{11}N$
193.26
N#CC(c(cccc1)c1)c(cccc2)c2
Acetonitrile, diphenyl
Benzeneacetonitrile, α-phenyl-(9CI)
Benzyhydrylcyanide
α-Cyanodiphenylmethane
Difenylacetonitril (Czech)
Dipan
Diphenatrile
Diphenylacetonitrile
Diphenyl-α-cyanomethane
Diphenylmethylcyanide
α-Phenylbenzylcyanide
α-Phenylphenylacetonitrile
USAF KF-13

86-30-6
$C_{12}H_{10}N_2O$
198.24
O=NN(c(cccc1)c1)c(cccc2)c2
Diphenylamine, N-nitroso
Benzenamine, N-nitroso-N-phenyl- (9CI)
Curetard A
Delac J
Difenylnitrosamin (Czech)
Diphenylnitrosamin (German)
Diphenylnitrosamine
Diphenyl N-nitrosoamine
N,N-Diphenylnitrosamine
Naugard TJB
NCI-C02880
NDPA
NDPhA
N-Nitrosodifenylamin (Czech)
Nitrosodiphenylamine
N-Nitrosodiphenylamine
N-Nitroso-N-phenylaniline
Nitrous diphenylamide
Redax
Retarder J
TJB
Vulcalent A
Vulcatard
Vulcatard A
Vulkalent A (Czech)
Vultrol

86-34-0
$C_{11}H_{11}NO_2$
189.23
CN1C(=O)CC(C1=O)c2ccccc2
Succinimide, N-methyl-2-phenyl
Epimid
Fenosuccimide
Lifene
Methylphenylsuccinimide
1-Methyl-3-phenylpyrrolidin-2,5-dione
N-Methyl-α-phenylsuccinimide
N-Methyl-2-phenyl-succinimide
Milontin
Mirotin
Phenosuccimide
Phensuximide
Phenylsuximide
PM 334
2,5-Pyrrolidinedione, 1-methyl-3-phenyl- (9CI)
Succitimal

86-40-8
$C_{14}H_{14}N_3.Cl$
259.76
[Cl-].C[n+]2c1cc(N)ccc1cc3ccc(N)cc23
Acridinium, 3,6-diamino-10-methyl-, chloride
Acriflavine
Acriflavine neutral
Acriflavon
AF
Avlon
Burnol
Chromoflavine
C.I. 46000
2,8-Diamino-10-methylacridinium chloride
3,6-Diamino-10-methylacridinium chloride
Euflavine
Flavin
Flavine
Flavosan
Gonacrine
Gonocrin
Neutral Acriflavine
Neutroflavine
Panflavin
Trypaflavin
Xanthacridine

86-48-6
$C_{11}H_8O_3$
188.18
O=C(O)c(c(O)c(c(ccc1)c2)c1)c2
1-Hydroxy-2-naphthoic acid
AI3-28524
2-Carboxy-1-naphthol
1-Hydroxy-2-naphthalenecarboxylic acid
2-Naphthalenecarboxylic acid, 1-hydroxy- (9CI)
2-Naphthoic acid, 1-hydroxy-(8CI)
1-Naphthol-2-carboxylic acid
NSC-3717

86-50-0
$C_{10}H_{12}N_3O_3PS_2$
317.34
COP(=S)(OC)SCn2nnc1ccccc1c2=O
Phosphorodithioic acid, O,O-dimethyl ester, S-ester with 3-(mercaptomethyl)-1,2,3-benzotriazin-4(3H)-one
Azinfos-methyl (Dutch)
Azinophos-methyl
Azinphos-metile (Italian)
Azinphos methyl
Azinphos-methyl (ACGIH, DOT,OSHA)

Azinphos methyl mixture, Liquid (DOT)
Bay 9027
Bay 17147
Bayer 9027
Bayer 17147
Benzotriazinedithiophosphoric acid dimethoxy ester
Benzotriazine derivative of a methyl dithiophosphate
1,2,3-Benzotriazin-4(3H)-one, 3-(mercaptomethyl)-, O,O-dimethyl phosphorodithioate
Carfene
Cotneon
Cotnion methyl
Crysthion 2L
Crysthyon
DBD
S-(3,4-Dihydro-4-oxo-benzo-(α)(1,2,3)triazin-3-ylmethyl) O,O-dimethyl phosphorodithioate
S-(3,4-Dihydro-4-oxo-1,2,3-benzotriazin-3-ylmethyl) O,O-dimethyl phosphorodithioate
O,O-Dimethyl-S-(benzazimino-methyl) dithiophosphate
O,O-Dimethyl-S-(1,2,3-benzotriazinyl-4-keto)methyl phosphorodithioate
O,O-Dimethyl S-(3,4-dihydro-4-keto-1,2,3-benzotriazinyl-3-methyl) dithiophosphate
Dimethyldithiophosphoric acid N-methylbenzazimide ester
O,O-Dimethyl S-(4-oxo-3H-1,2,3-benzotriazine-3-methyl)-phosphorodithioate
O,O-Dimethyl S-(4-oxobenzotriazino-3-methyl)phosphorodithioate
O,O-Dimethyl S-(4-oxo-1,2,3-benzotriazino(3)-methyl) thiothionophosphate
O,O-Dimethyl-S-(4-oxobenzotriazin-3-methyl)-dithiophosphat (German)
O,O-Dimethyl-S-((4-oxo-3H-1,2,3-benzotriazin-3-yl)-methyl)-dithiofosfaat (Dutch)
O,O-Dimethyl-S-((4-oxo-3H-1,2,3-benzotriazin-3-yl)-methyl)-dithiophosphat (German)
O,O-Dimethyl S-4-oxo-1,2,3-benzotriazin-3(4H)-ylmethyl phosphorodithioate
O,O-Dimetil-S-((4-oxo-3H-1,2,3-benzotriazin-3-il)-metil)-ditiofosfato (Italian)
ENT 23,233
Gothnion
Gusathion
Gusathion-20
Gusathion 25
Gusathion K
Gusathion M
Gusathion methyl
Guthion (DOT,OSHA)
Guthion mixture, Liquid (DOT)
3-(Mercaptomethyl)-1,2,3-benzotriazin-4(3H)-one O,O-dimethyl phosphorodithioate S-ester
Methylazinphos
N-Methylbenzazimide, dimethyldithiophosphoric acid ester
Methyl guthion
Metiltriazotion
NA 2783 (DOT)
NCI-C00066
R 1582

86-52-2
$C_{11}H_9Cl$
176.65
c(c(c(cc1)CCl)ccc2)(c2)c1
Naphthalene, 1-chloromethyl
1-(Chlormethyl)naftalen (Czech)
1-Chloromethyl naphthalene
Naphthalene, α-chloromethyl-

86-53-3
$C_{11}H_7N$
153.19
N#Cc(c(c(ccc1)cc2)c1)c2
1-Naphthonitrile
α-Cyanonaphthalene
1-Cyanonaphthalene
1-Naphthalenecarbonitrile
1-Naphthalenenitrile
α-Naphthonitrile
α-Naphthylnitrile
1-Naphthylnitrile

86-54-4
$C_8H_8N_4$
160.20
NNc1nncc2ccccc12
Phthalazine, 1-hydrazino
Apresolin
Apresoline
Apressin
Aprezolin
BA 5968
C-5068
C 5968
Ciba 5968
Hidralazin
Hidralazina (Spanish)
Hipoftalin
Hydralazin
Hydralazine
Hydrallazine
Hydrazinophthalazine
1-Hydrazinophthalazine
Hypophthalin
Idralazina (Italian)
1(2H)-Phthalazinone hydrazone

86-55-5
$C_{11}H_8O_2$
172.19
O=C(O)c(c(c(ccc1)cc2)c1)c2
1-Naphthoic acid
1-Carboxynaphthalene
Naphthalene-α-carboxylic acid
1-Naphthalenecarboxylic acid (9CI)
α-Naphthoic acid
α-Naphthylcarboxylic acid

86-56-6
$C_{12}H_{13}N$
171.26
N(c(c(c(ccc1)cc2)c1)c2)(C)C
1-Naphthylamine, N,N-dimethyl
1-Dimethylaminonaphthalene
N,N-Dimethyl-1-naftylamin (Czech)
Dimethyl-α-naphthylamine
α-Dimethylnaphthylamine
N,N-Dimethyl-α-naphthylamine
N,N-Dimethyl-1-naphthylamine

86-57-7
$C_{10}H_7NO_2$
173.18
O=N(=O)c(c(c(ccc1)cc2)c1)c2
Naphthalene, 1-nitro
NCI-C01956
Nitrol
1-Nitronaftalen (Czech)
α-Nitronaphthalene
1-Nitronaphthalene

86-65-7
$C_{10}H_9NO_6S_2$
303.32
O=S(=O)(O)c(cc(c(c1ccc2N)c2)S(=O)(=O)O)c1
1,3-Naphthalenedisulfonic acid, 7-amino
Amido-G-acid
7-Amino-1,3-naphthalenedisulfonic acid
2-Naphthylamine-6,8-disulfonic acid

86-72-6
$C_{18}H_{14}N_2O$
274.34
Oc(ccc(Nc(ccc(Nc(c1ccc2)c2)c13)c3)c4)c4
Phenol, 4-(3-carbazolylamino)
Carbazole, 3-(p-hydroxyanilino)-
4-(3-Carbazolylamino)phenol
3-(4'-Hydroxyfenyl)aminokarbazol (Czech)
R-Base (Czech)

86-73-7
$C_{13}H_{10}$
166.23
c(c(c(c1ccc2)c2)ccc3)(c3)C1
Fluorene
o-Biphenylenemethane
o-Biphenylmethane
Diphenylenemethane
9H-Fluorene
2,2'-Methylenebiphenyl

86-74-8
$C_{12}H_9N$

167.22
N(c(c(c1cccc2)ccc3)c3)c12
Carbazole
9-Azafluorene
9H-Carbazole (9CI)
Dibenzopyrrole
Dibenzo(b,d)pyrrole
Diphenyleneimine
Diphenylenimide
Diphenylenimine
USAF EK-600

86-76-0
$C_{12}H_7BrO$
247.09
Dibenzofuran, 2-bromo- (9CI)
2-Bromodibenzofuran
NSC-1735

86-81-7
$C_{10}H_{12}O_4$
196.22
O=Cc(cc(OC)c(OC)c1OC)c1
Benzaldehyde, 3,4,5-trimethoxy
3,4,5-Trimethoxybenzaldehyde

86-85-1
$C_{10}H_9HgNO$
359.79
C[Hg]Oc1cccc2cccnc12
Mercury, methyl(8-quinolinolato)
Artho LM
Liqui-San
LM Seed Protectant
Mercury, methyl(8-quinolyloxy)-
Mercury, (8-quinolinolato)-methyl-
Metasol
Metasol MMH
Metazol
8-(Methylmercurioxy)quinoline
Methylmercury 8-hydroxyquinolinate
Methylmercury β-hydroxyquinolate
Methylmercury oxinate
Methylmercury oxyquinolinate
Methylmercury quinolinolate
Methylmerkuri-8-chinolinolat (Czech)
Ortho-LM Apple Spray
Ortho LM Concentrate
Ortho LM Seed Protectant
Quinoline, 8-((methylmercuri)oxy)-
8-(Quinolinolato)methyl mercury
8-Quinolinol, mercury complex

86-86-2
$C_{12}H_{11}NO$
185.24
O=C(N)Cc(c(c(ccc1)cc2)c1)c2
1-Naphthaleneacetamide
Amid kyseliny 1-naftyloctove (Czech)
Amid-Thin
Amid-Thin W
Fruitone
NAAM
NAD
Naphthalene acetamide
α-Naphthaleneacetamide
α-Naphthylacetamide
1-Naphthylacetamide
1-Naphthylamine, N-acetyl-
Rootone
Rosetone

86-87-3
$C_{12}H_{10}O_2$
186.22
O=C(O)Cc(c(c(ccc1)cc2)c1)c2
1-Naphthaleneacetic acid
Agronaa
Alphaspra
ANA
ANU
Appl-Set
Biokor
Celmone
Fruitofix
Fruitone
Fruitone N
Kyselina 1-naftyloctova (Czech)
Klingtite
Liqui-Stik
α-NA
NAA
α-NAA
1-NAA
NAA 800
Nafusaku
Naphthaleneacetic acid
Naphthalene-1-acetic acid
α-Naphthaleneacetic acid
α-Naphthylacetic
Naphthylacetic acid
α-Naphthylacetic acid
1-Naphthylacetic acid
2-(1-Naphthyl)acetic acid
α-Naphthyleneacetic acid
α-Naphthylessigsaeure (German)
Naphyl-1-essigsaeure (German)
Niagara-Stik
Nu-Tone
Parmone
Phymone
Phyomone
Pimacol-Sol
Planofix
Planofixe
Plucker
Primacol
Rhodofix
Rootone
Stafast
Stik
Stop-Drop
Tekkam
Tip-Off
Transplantone
Tre-Hold
Vardhak

86-88-4
$C_{11}H_{10}N_2S$
202.29
N=C(S)Nc(c(c(ccc1)cc2)c1)c2
Urea, 1-(1-naphthyl)-2-thio
Alphanaphthyl thiourea
Alphanaphtyl thiouree (French)
Alrato
Antu (ACGIH,OSHA)
Anturat
Bantu
Chemical 109
Dirax
Kill kantz
Krysid
Krysid pi
1-Naftil-tiourea (Italian)
α-Naftylthiomocovina (Czech)
1-Naftylthioureum (Dutch)
1-Naphthalenylthiourea
α-Naphthalthioharnstoff (German)
Naphthothiourea (OSHA) [UN 1651]
α-Naphthothiourea
α-Naphthylthiocarbamide
1-Naphthyl-thioharnstoff (German)
1-Naphthyl thiourea
N-(1-Naphthyl)-2-thiourea
1-(1-Naphthyl)-2-thiourea
α-Naphthylthiourea (DOT, OSHA)
1-Naphthyl-thiouree (French)
Naphtox
Rattrack
Rat-Tu
RCRA waste number P072
Smeesana
Thiourea, 1-naphthalenyl-
U-5227
UN 1651 [Naphthylthiourea]
USAF EK-P-5976

86-92-0
$C_{11}H_{12}N_2O$
188.25
O=C(N(N=C1C)c(ccc(c2)C)c2)C1
2-Pyrazolin-5-one, 3-methyl-1-p-tolyl
3-Methyl-1-p-tolyl-pyrazolin-5-one
1-(p-Tolyl)-3-methylpyrazolone-5

86-93-1
$C_7H_6N_4S$
178.23
N(N=NN1c(cccc2)c2)=C1S
1H-Tetrazole-5-thiol, 1-phenyl
Mercaptophenyltetrazole
5-Mercapto-1-phenyltetrazole
Phenylmercaptotetrazole
1-Phenyl-5-mercaptotetrazole
1-Phenyl-5-mercapto-1,2,3,4-tetrazole
1-Phenyltetrazole-5-thiol

86-98-6
$C_9H_5Cl_2N$
198.05
n(c(c(c(c1)Cl)ccc2Cl)c2)c1
Quinoline, 4,7-dichloro
4,7-Dichloroquinoline
TL 1473

87-00-3
$C_{16}H_{21}NO_3$
275.38
CN1C2CCC1CC(C2)OC(=O)C(O)c3ccccc3
1-α-H,5-α-H-Tropan-3-α-ol, mandelate (ester)
Homatropin
Homatropine
Homoatropine
Homotropine
Mandelic acid, 3d-tropanyl ester
Mandelyltropeine
Mandelytropeine
Tropine, mandelate (ester)

87-01-4
$C_{12}H_{13}NO_2$
203.26
O=C(Oc(c(C=1C)ccc2N(C)C)c2)C1
Coumarin, 7-dimethylamino-4-methyl
2H-1-Benzopyran-2-one, 7-(dimethylamino)-4-methyl- (9CI)
Coumarin 311
DAMC
7-Dimethylamino-4-methylcoumarin
FBA 52

87-02-5
$C_{10}H_9NO_4S$
239.26
O=S(=O)(O)c(cc(O)c(c1cc(N)c2)c2)c1
2-Naphthalenesulfonic acid, 7-amino-4-hydroxy
7-Amino-4-hydroxy-2-naphthalenesulfonic acid
Aminonaphthol sulfonic acid J
I Acid
Isogamma Acid
J Acid
Kyselina 2-amino-5-naftol-7-sulfonova (Czech)
Kyselina 6-amino-1-naftol-3-sulfonova (Czech)
Kyselina I (Czech)

87-08-1
$C_{16}H_{18}N_2O_5S$
350.42
CC3(C)SC2C(NC(=O)COc1ccccc1)C(=O)N2C3C(O)=O
Penicillanic acid, 6-phenoxyacetamido
Acipen V
Apopen
Beromycin
Distaquaine V
Eskacillian V
Fenacilin
Fenospen
Fenoxypen
Meropenin
Oracillin
Oratren
Ospen
Penicillin phenoxymethyl
Penicillin V
Pen-Oral
Pen V
Pen-Vee
Phenopenicillin
6-Phenoxyacetamidopenicillanic acid
Phenoxymethylenepenicillinic acid
Phenoxymethylpenicillin
Stabicillin
V-Cil
V-Cillin
Vebecillin

87-10-5
$C_{13}H_8Br_3NO_2$
449.92
O=C(Nc(ccc(c1)Br)c1)c(c(O)c(cc2Br)Br)c2
Tribromsalan
Agramed
AI3-25516
ASC-4
Benzamide, 3,5-dibromo-N-(4-bromophenyl)-2-hydroxy-(9CI)
Caswell No. 863
3,5-Dibromo-N-(4-bromophenyl)-2-hydroxybenzamide
3,5-Dibromosalicylic acid p-bromoanilide
ENT 25,516
EPA Pesticide Chemical Code 077404
ET-394
NSC-20526
Polybrominated salicylanilide
Salicylanilide, 3,4',5-tribromo-(8CI)
Sherstat TBS
Stecker ASC-4
TBS
TBS 95
Temasept
Temasept II
Temasept IV
Tempasept II
Tribromosalicylanilide
3,4',5-Tribromosalicylanide and 4,5-dibromosalicylanide mixtures
3,4',5-Tribromosalicylanilide
Tribromsalanum (Latin)
Tribromsalen
Trisanil
Trisanyl
Tuasal 100
Tuasol
Tuasol 100
Vancide TBS

87-12-7
$C_{13}H_9Br_2NO_2$
371.02
O=C(Nc(ccc(c1)Br)c1)c(c(O)ccc2Br)c2
Dibromsalan
Benzamide, 5-bromo-N-(4-bromophenyl)-2-hydroxy- (9CI)
p-Bromanilid kyseliny 5-bromsalicylove (Czech)
5-Bromo-N-(4-bromophenyl)-2-hydroxybenzamide
3-Bromo-6-hydroxybenz-p-bromanilide
Bromosalicylanilide
5-Bromosalicyl-4-bromoanilide
5-Bromosalicylic acid p-bromoanilide
Bromsalicylanilide
Caswell No. 287C
4',5-Dibromosalicylanilide
Dibromsalanum (Latin)
Dibromsalen
Dibask
Dibronsalan (Spanish)
Disanyl
EPA Pesticide Chemical Code 077402
NSC-20527
Salicylanilide, 4',5-dibromo-(8CI)
Temasept

87-17-2
$C_{13}H_{11}NO_2$
213.25
O=C(Nc(cccc1)c1)c(c(O)ccc2)c2
Salicylanilide
Anilid kyseliny salicylove (Czech)
Ansadol
Shirlan Extra

87-18-3
$C_{17}H_{18}O_3$
270.35
O=C(Oc(ccc(c1)C(C)(C)C)c1)c(c(O)ccc2)c2
Salicylic acid, p-tert-butylphenyl ester
Benzoic acid, 2-hydroxy-, 4-(1,1-dimethylethyl)phenyl ester
p-terc.Butylfenylester kyseliny salicylove (Czech)
p-tert-Butylphenyl salicylate

87-19-4
$C_{11}H_{14}O_3$
194.25
O=C(OCC(C)C)c(c(O)ccc1)c1
Salicylic acid, isobutyl ester
Isobutyl o-hydroxybenzoate
Isobutyl salicylate

87-20-7
$C_{12}H_{16}O_3$
208.28
O=C(OCCC(C)C)c(c(O)ccc1)c1
Salicylic acid, isopentyl ester
Isoamylester kyseliny salicylove (Czech)
Isoamyl o-hydroxybenzoate
Isoamyl salicylate
Isopentyl-2-hydroxyphenyl methanoate
Isopentyl salicylate
3-Methylbutyl 2-hydroxybenzoate

87-25-2
$C_9H_{11}NO_2$
165.21
O=C(OCC)c(c(N)ccc1)c1
Benzoic acid, o-amino-, ethyl ester
Ethyl o-aminobenzoate
Ethyl anthranilate

87-29-6
$C_{16}H_{15}NO_2$
253.32
O=C(OCC=Cc(cccc1)c1)c(c(N)ccc2)c2
Anthranilic acid, cinnamyl ester
2-Aminobenzoic acid 3-phenyl-2-propenyl ester
Benzoic acid, 2-amino-, 3-phenyl-2-propenyl ester
Cinnamyl alcohol, anthranilate
Cinnamyl 2-aminobenzoate
Cinnamyl o-aminobenzoate
Cinnamyl anthranilate
Cinnamylester kyseliny anthranilove (Czech)
NCI-C03510
3-Phenyl-2-propenylanthranilate
3-Phenyl-2-propen-1-yl anthranilate

87-33-2
$C_6H_8N_2O_8$
236.16
O=N(=O)OC(C(OCC1ON(=O)=O)C1O2)C2
Glucitol, 1,4:3,6-dianhydro-, dinitrate, D
Cardis
Carvanil
Carvasin
Cedocard
Claodical
Corosorbide
1,4:3,6-Dianhydrosorbitol 2,5-dinitrate
Dilatrate-SR
Dinitrosorbide
Flindix
Harrical
Iso-bid
Isochron
Isoket
Isorbid
Isordil
Isordil Tembids
Isosorbide dinitrate
Isotrate
Korodil
Maycor
Nitrosorbid
Nitrosorbide
Resoidan
Rigedal
Sorate-5
Sorate-10
Sorbangil
Sorbide
Sorbide nitrate
Sorbidilat
Sorbidinitrate
Sorbitrate
Sorbonit
Sorquat
SST-101
Vascardin
Vasorbate

87-40-1
$C_7H_5Cl_3O$
211.47
Anisole, 2,4,6-trichloro- (8CI)
AI3-09173
Benzene, 1,3,5-trichloro-2-methoxy- (9CI)
Methyl 2,4,6-trichlorophenyl ether
NSC-35142
2,4,6-Trichloroanisole
1,3,5-Trichloro-2-methoxybenzene
Tyrene

87-41-2
$C_8H_6O_2$
134.13
O=C(OCc1cccc2)c12
Phthalide (8CI)
AI3-05785
2-Hydroxymethylbenzoic acid, γ-lactone
1(3H)-Isobenzofuranone (9CI)
NSC-1469
1-Phthalanone

87-44-5
$C_{15}H_{26}$
206.41
C(=CCCC(C(C(C1(C)C)C2)C1)=C)(C2)C
Bicyclo(7.2.0)undec-4-ene, 8-methylene-4,11,11-trimethyl-, (e)-(1R,9S)-(-)
Caryophyllene
β-Caryophyllene
8-Methylene-4,11,11-(tri methyl)bicyclo(7.2.0)undec-4-ene

87-47-8
$C_{13}H_{15}N_3O_2$
245.31
CN(C)C(=O)Oc1cc(C)nn1c2ccccc2
Carbamic acid, dimethyl-, 3-methyl-1-phenylpyrazol-5-yl ester
OMS 20
Dimethylcarbamic acid, 3-methyl-1-phenylpyrazol-5-yl ester
Dimethylcarbamic acid 3-methyl-1-phenyl-1H-pyrazol-5-yl
ester
Dimethyl 5-(3-methyl-1-phenylpyrazolyl) carbamate
ENT 17,588
1-Fenyl-3-methyl-5-pyrazolylester kyseliny dimethylkarbaminove (Czech)
G-22008
Geigy G-22008
3-Methyl-1-phenylpyrazol-5-yl dimethyl carbamate
3-Methyl-1-phenyl-5-pyrazolyl dimethyl carbamate
1-Phenyl-3-methyl-5-pyrazolyl N,N-dimethyl carbamate
Pyralan
Pyrazol-5-ol, 3-methyl-1-phenyl-, dimethyl carbamate (ester)
Pyrolan

87-51-4
$C_{10}H_9NO_2$
175.20
O=C(O)CC(c(c(N1)ccc2)c2)=C1
1H-Indole-3-acetic acid
Indol-3-ylacetic acid
Acetic acid, indolyl-
Heteroauxin
Hexteroauxin
IAA
3-IAA
β-Indoleacetic acid
β-Indole-3-acetic acid
3-Indoleacetic acid
α-Indol-3-yl-acetic acid
β-Indolylacetic acid
Indolyl-3-acetic acid
3-Indolylacetic acid
Kyselina 3-indolyloctova (Czech)
Rhizopin
Rhizopon A
ω-Skatole carboxylic acid

87-59-2
$C_8H_{11}N$
121.20
Nc(c(c(cc1)C)C)c1
2,3-Xylidine
Benzenamine, 2,3-dimethyl- (9CI)
2,3-Dimethylaniline
2,3-Dimethylbenzenamine
2,3-Dimethylphenylamine
UN 1711 [Xylidines, solid or solution]
o-Xylidine

o-Xylidine [UN 1711]
2,3-Xylylamine

87-60-5
C_7H_8ClN
141.61
Nc(c(c(cc1)Cl)C)c1
o-Toluidine, 3-chloro
1-Amino-2-chloro-6-methylbenzene
1-Amino-3-chloro-2-methylbenzene
2-Amino-6-chlorotoluene
Azoic Diazo Component 46
3-Chloro-2-methylaniline
3-Chlor-2-toluidin (Czech)
3-Chloro-o-toluidine
Fast Scarlet TR Base
Scarlet TR Base

87-61-6
$C_6H_3Cl_3$
181.44
c(c(c(cc1)Cl)Cl)(c1)Cl
Benzene, 1,2,3-trichloro
vic-Trichlorobenzene
1,2,3-Trichlorobenzene
1,2,3-Trichlorobenzene, Liquid [UN 2321]
1,2,6-Trichlorobenzene
UN 2321 [Trichlorobenzenes, Liquid]

87-62-7
$C_8H_{11}N$
121.20
Nc(c(ccc1)C)c1C
2,6-Xylidine
Aniline, 2,6-dimethyl-
2,6-Dimethylaniline
2,6-Dimethylbenzenamine
NCI-C56188
o-Xylidine
2,6-Xylylamine

87-63-8
C_7H_8ClN
141.61
Nc(c(ccc1)C)c1Cl
o-Toluidine, 6-chloro
2-Amino-3-chlorotoluene
3-Chloro-2-aminotoluene
6-Chloro-2-methylaniline
6-Chloro-o-toluidine
6-Chloro-2-toluidine

87-64-9
C_7H_7ClO
142.59
Cc1cccc(Cl)c1O
o-Cresol, 6-chloro
6-Chloro-o-cresol
2-Chloro-6-methylphenol

87-65-0
$C_6H_4Cl_2O$
163.00
Oc(c(ccc1)Cl)c1Cl
Phenol, 2,6-dichloro
2,6-Dichlorfenol (Czech)
2,6-Dichlorophenol
RCRA waste number U082

87-66-1
$C_6H_6O_3$
126.12
Oc(c(O)ccc1)c1O
Pyrogallol
Benzene, 1,2,3-trihydroxy-
1,2,3-Benzenetriol
C.I. 76515
C.I. Oxidation Base 32
Fouramine Brown AP
Fourrine PG
Fourrine 85
Pyrogallic acid
1,2,3-Trihydroxybenzen (Czech)
1,2,3-Trihydroxybenzene

87-67-2
$C_5H_{14}NO.C_4H_5O_6$
253.25
Choline bitartrate
Choline hydrogen tartrate
Choline tartrate
Choline, tartrate (1:1) salt
Ethanaminium, 2-hydroxy-N,N,N-trimethyl-, Salt with (R-(R+,R+))-2,3-dihydroxybutanedioic acid (1:1)
Ethanaminium, 2-hydroxy-N,N,N-trimethyl-, Salt with (R-(R*,R*))-2,3-dihydroxybutanedioic acid (1:1) (9CI)
(2-Hydroxyethyl)trimethylammonium bitartrate

87-68-3
C_4Cl_6
260.74
C(=C(C(=C(Cl)Cl)Cl)Cl)(Cl)Cl
1,3-Butadiene, hexachloro
Dolen-Pur
GP-40-66:120
HCBD
Hexachlor-1,3-butadien (Czech)
Hexachlorbutadiene
Hexachlorobutadiene (ACGIH, OSHA) [UN 2279]
1,3-Hexachlorobutadiene
1,1,2,3,4,4-Hexachloro-1,3-butadiene
Perchlorobutadiene
RCRA waste number U128
UN 2279 [Hexachlorobutadiene]

87-69-4
$C_4H_6O_6$
150.10
O=C(O)C(O)C(O)C(=O)O
Tartaric-acid
Butanedioic acid, 2,3-dihydroxy-
2,3-Dihydrosuccinic acid
2,3-Dihydroxybutanedioic acid
Kyselina 2,3-dihydroxybutandiova (Czech)
Kyselina vinna (Czech)
Malic acid, 3-hydroxy-
Succinic acid, 2,3-dihydroxy-
Threaric acid
l-(+)-Tartaric acid

87-72-9
$C_5H_{10}O_5$
150.13
O(C(O)C(O)C(O)C1O)C1
L-Arabinopyranose (9CI)

87-78-5
$C_6H_{14}O_6$
182.20
OCC(O)C(O)C(O)C(O)CO
Mannitol

87-79-6
$C_6H_{12}O_6$
180.16
O=C(C(O)C(O)C(O)CO)CO
Sorbose
AI3-19425
Esorben
NSC-97195
L-1,3,4,5,6-Pentahydroxyhexan-2-one
Sorbin
Sorbinose
L-Sorbinose
Sorbose, L- (VAN) (8CI)
L-Sorbose (9CI)
L-(-)-Sorbose
L-Xylo-2-hexulose

87-82-1
C_6Br_6
551.49
c(c(c(c(c1Br)Br)Br)Br)(c1Br)Br
Hexabromobenzene
AI3-60220
Benzene, hexabromo- (9CI)
NSC-113975

87-83-2
$C_7H_3Br_5$
486.62
c(c(c(c(c1Br)Br)Br)Br)(c1Br)C
2,3,4,5,6-Pentabromotoluene
Benzene, pentabromomethyl- (9CI)
Flammex 5BT
Pentabromomethylbenzene
Pentabromotoluene
Toluene, 2,3,4,5,6-pentabromo-

87-84-3
$C_6H_6Br_5Cl$
513.09
BrC(C(C(Br)C(Br)C1Br)Cl)C1Br
1,2,3,4,5-Pentabromo-6-

chlorocyclohexane
Cyclohexane, 1-chloro-2,3,4,5,6-pentabromo-
Cyclohexane, 1,2,3,4,5-pentabromo-6-chloro- (9CI)

87-85-4
$C_{12}H_{18}$
162.30
c(c(c(c(c1C)C)C)C)(c1C)C
Benzene, hexamethyl
Hexamethylbenzene

87-86-5
C_6HCl_5O
266.32
Oc(c(c(c(c1Cl)Cl)Cl)Cl)c1Cl
Phenol, pentachloro
Acutox
Chem-Penta
Chem-Tol
Chlorophen
Cryptogil ol
Dowcide 7
Dowicide 7
Dowicide EC-7
Dowicide G
Dow Pentachlorophenol DP-2 Antimicrobial
Durotox
EP 30
Fungifen
Glazd Penta
Grundier arbezol
Lauxtol
Lauxtol A
Liroprem
NA 2020 (DOT)
NCI-C54933
NCI-C55378
NCI-C56655
PCP
Penchlorol
Penta
Pentachloorfenol (Dutch)
Pentachlorofenol
Pentaclorofenolo (Italian)
Pentachlorophenate
Pentachlorophenol (ACGIH, OSHA)
2,3,4,5,6-Pentachlorophenol
Pentachlorophenol, Dowicide EC-7
Pentachlorophenol, DP-2
Pentachlorphenol (German)
Pentachlorophenol, Technical
Pentacon
Penta-Kil
Penta Ready
Pentasol
Penta WR
Penwar
Peratox
Permacide
Permagard
Permasan
Permatox DP-2
Permatox Penta
Permite
Prevenol
Priltox
RCRA waste number U242
Santobrite
Santophen 20
Sinituho
Term-I-Trol
Thompson's Wood Fix
Weedone
Witophen P

87-87-6
$C_6H_2Cl_4O_2$
247.88
Oc(c(c(c(O)c1Cl)Cl)Cl)c1Cl
Hydroquinone, tetrachloro
Tetrachlorohydroquinone
USAF DO-62

87-88-7
$C_6H_2Cl_2O_4$
208.98
O=C(C(=C(O)C(=O)C=1Cl)Cl)C1O
Chloranilic acid
AI3-61846
p-Benzoquinone, 2,5-dichloro-3,6-dihydroxy- (8CI)
2,5-Cyclohexadiene-1,4-dione, 2,5-dichloro-3,6-dihydroxy- (9CI)
2,5-Dihydroxy-3,6-dichlorobenzoquinone
NSC-6108

87-89-8
$C_6H_{12}O_6$
180.16
OC(C(O)C(O)C(O)C1O)C1O
Inositol, myo- (8CI)
AI3-16111
1,2,3,5/4,6-Cyclohexanehexol
cis-1,2,3,5-trans-4,6-Cyclohexanehexol
Dambose
Inosital
Inositene
Inositina
Inositol (VAN)
Inositol, meso
i-Inositol
iso-Inositol
meso-Inositol
myo-Inositol (9CI)
Insitolum
Meat Sugar
Mesoinosit
Mesoinosite
Mesoinositol
Mesol
Mesovit
MI
Myoinosite
Myoinositol
NSC-404118
Nucite
Phaseomannite
Phaseomannitol
Scyllite

87-90-1
$C_3Cl_3N_3O_3$
232.41
O=C(N(C(=O)N(C1=O)Cl)Cl)N1Cl
s-Triazine-2,4,6(1H,3H,5H)-trione, 1,3,5-trichloro
ACL 85
CBD 90
Fichlor 91
FI Clor 91
Isocyanuric chloride
Kyselina trichloisokyanurova (Czech)
NA 2468 [(Mono-(trichloro) tetra- (monopotassium dichloro)-penta-s- triazinetrione, Dry (containing over 39% available chlorine)]
NSC-405124
Symclosen
Symclosene
1,3,5-Triazine-2,4,6(1H,3H,5H)-trione, 1,3,5-trichloro-
Trichlorinated isocyanuric acid
Trichlorocyanuric acid
Trichloroisocyanic acid
Trichloroisocyanurate
Trichloroisocyanuric acid
N,N',N''-Trichloroisocyanuric acid
1,3,5-Trichloroisocyanuric acid
Trichloroisocyanuric acid, Dry [UN 2468]
Trichloro-s-triazinetrione
1,3,5-Trichloro-s-triazine-2,4,6(1H,3H,5H)-trione
Trichloro-s-triazine-2,4,6(1H,3H,5H)-trione
Trichloro-s-triazinetrione, Dry, Containing over 39% available chlorine [NA 2468]
1,3,5-Trichloro-2,4,6-trioxohexahydro-s-triazine
UN 2468 [Trichloroisocyanuric acid, dry]

87-92-3
$C_{12}H_{22}O_6$
262.30
O=C(OCCCC)C(O)C(O)C(=O)OCCCC
Butanedioic acid, 2,3-dihydroxy- (R-(R*,R*))-, dibutyl ester (9CI)
Tartaric acid, dibutyl ester (8CI)

87-99-0
$C_5H_{12}O_5$
152.17
OCC(O)C(O)C(O)CO
Xylitol
Klinit
Xylite
Xylite (sugar)

88-04-0
C_8H_9ClO

156.62
Oc(cc(c(c1C)Cl)C)c1
3,5-Xylenol, 4-chloro
Benzytol
4-Chloro-3,5-dimethylphenol
2-Chloro-5-hydroxy-1,3-di methylbenzene
4-Chloro-1-hydroxy-3,5-di methylbenzene
2-Chloro-5-hydroxy-m-xylene
Chloro-xylenol
p-Chloro-m-xylenol
2-Chloro-m-xylenol
4-Chloro-3,5-xylenol
Desson
Dettol
Espadol
Husept extra
Nipacide MX
Ottasept
Ottasept extra
PCMX
Phenol, 4-chloro-3,5-dimethyl-
RBA 777

88-05-1
$C_9H_{13}N$
135.23
Nc(c(cc(c1)C)C)c1C
Aniline, 2,4,6-trimethyl
Aminomesitylene
2-Aminomesitylene
1-Amino-2,4,6-trimethylbenzen (Czech)
2-Amino-1,3,5-trimethylbenzene
Benzenamine, 2,4,6-trimethyl- (9CI)
Mesidin (Czech)
Mesidine
Mesitylamine
Mesitylene, 2-amino-
Mezidine
2,4,6-Trimethylaniline

88-06-2
$C_6H_3Cl_3O$
197.44
Oc(c(cc(c1)Cl)Cl)c1Cl
Phenol, 2,4,6-trichloro
Dowcide 2S
Dowicide 2S
NCI-C02904
Omal
Phenachlor
RCRA waste number U231
2,4,6-Trichlorfenol (Czech)
2,4,6-Trichlorophenol

88-09-5
$C_6H_{12}O_2$
116.18
O=C(O)C(CC)CC
Butyric acid, 2-ethyl
Acetic acid, diethyl-
Diethylacetic acid
2-Ethyl butanoic acid
α-Ethylbutyric acid
2-Ethylbutyric acid
Kyselina diethyloctova (Czech)
3-Pentanecarboxylic acid

88-10-8
$C_5H_{10}ClNO$
135.61
O=C(N(CC)CC)Cl
Carbamoyl chloride, diethyl
Carbamic chloride, diethyl- (9CI)
Carbamidoyl chloride, diethyl-
Diethylamid kyseliny chlorm- ravenci (Czech)
Diethylcarbamoyl chloride
N,N-Diethylcarbamoyl chloride
Diethylcarbamyl chloride

88-12-0
C_6H_9NO
111.16
O=C(N(C=C)CC1)C1
2-Pyrrolidinone, 1-vinyl
1-Ethenyl-2-pyrrolidinone (9CI)
Vinylbutyrolactam
N-Vinylpyrrolidinone
N-Vinyl-2-pyrrolidinone
1-Vinyl-2-pyrrolidinone
Vinylpyrrolidone
N-Vinylpyrrolidone
N-Vinyl-2-pyrrolidone
1-Vinyl-2-pyrrolidone
V-Pyrol

88-13-1
$C_5H_4O_2S$
128.15
OC(=O)c1ccsc1
3-Thiophenecarboxylic acid
3-Thenoic acid
β-Thiophenecarboxylic acid
β-Thiophenic acid

88-14-2
$C_5H_4O_3$
112.09
O=C(O)C(OC=C1)=C1
2-Furoic acid
2-Carboxyfuran
α-Furancarboxylic acid
2-Furancarboxylic acid (9CI)
α-Furoic acid
Kyselina 2-furoova (Czech)
Kyselina pyroslizova (Czech)
Pyromucic acid

88-15-3
C_6H_6OS
126.18
O=C(C(SC=C1)=C1)C
Ketone, methyl 2-thienyl
2-Acetothienone
2-Acetothiophene
2-Acetylthiophene

88-16-4
$C_7H_4ClF_3$
180.56
FC(F)(F)c(c(ccc1)Cl)c1
Toluene, o-chloro-α,α,α-tri- fluoro
Benzene, 1-chloro-2-(trifluoro- methyl)- (9CI)
o-Chlorobenzotrifluoride
2-Chlorobenzotrifluoride
Chlorobenzotrifluorides [UN 2234]
2-Chloro(trifluoromethyl)- benzene
o-(Trifluoromethyl)chloro- benzene
o-Trifluoromethylphenyl chloride
UN 2234 [Chlorobenzotri- fluorides]

88-17-5
$C_7H_6F_3N$
161.14
FC(F)(F)c(c(N)ccc1)c1
o-Toluidine, α,α,α-trifluoro
o-Aminobenzotrifluoride
Benzenamine, 2-(trifluoro methyl)- (9CI)
2-Trifluoromethyl aniline [UN 2942]
UN 2942 [2-Trifluoromethyl- aniline]

88-18-6
$C_{10}H_{14}O$
150.24
Oc(c(ccc1)C(C)(C)C)c1
Phenol, o-(tert-butyl)
2-t-Butylphenol

88-19-7
$C_7H_9NO_2S$
171.23
O=S(=O)(N)c(c(ccc1)C)c1
o-Toluenesulfonamide
Benzenesulfonamide, 2-methyl-
o-Methylbenzenesulfonamide
2-Methylbenzenesulfonamide
Onco-Carbide
Oxyurea
OTS
Toluene-2-sulfonamide
ortho-Toluol-sulfonamid (German)

88-20-0
$C_7H_8O_3S$
172.20
O=S(=O)(O)c(c(ccc1)C)c1
Benzenesulfonic acid, 2-methyl- (9CI)
2-Methylbenzenesulfonic acid

88-21-1
$C_6H_7NO_3S$
173.20
O=S(=O)(O)c(c(N)ccc1)c1
Benzenesulfonic acid, o-amino
o-Aminobenzenesulfonic acid
2-Aminobenzenesulfonic acid

o-Aminophenylsulfonic acid
Anilino-o-sulfonic acid
Anilino-2-sulfonic acid
Anilino-o-sulphonic acid
Orthanilic acid
o-Sulfanilic acid

88-23-3
$C_6H_6ClNO_4S$
223.63
O=S(=O)(O)c(c(O)c(N)cc1Cl)c1
Benzenesulfonic acid, 3-amino-5-chloro-2-hydroxy- (9CI)
3-Amino-5-chloro-2-hydroxy-benzenesulfonic acid
Metanilic acid, 5-chloro-2-hydroxy- (8CI)
NSC-7539

88-24-4
$C_{25}H_{36}O_2$
368.61
Oc(c(cc(c1)CC)Cc(c(O)c(cc2CC)C(C)(C)C)c2)c1C(C)(C)C
Phenol, 2,2'-methylenebis-(6-tert-butyl-4-ethyl
Antage w 500
Antioxidant 425
Bis(2-hydroxy-3-tert-butyl-5-ethylphenyl)methane
Cyanox 425
2,2'-Methylenebis(6-tert-butyl-4-ethylphenol)
2,2'-Methylenebis(4-ethyl-6-tert-butylphenol)
Plastanox 425 Antioxidant
Nocrac NS 5
USAF CY-6

88-26-6
$C_{15}H_{24}O_2$
236.39
OCc(cc(c(O)c1C(C)(C)C)C(C)(C)C)c1
Benzyl alcohol, 3,5-di-tert-butyl-4-hydroxy
Antioxidant 754
AO 754
Benzenemethanol, 3,5-bis-(1,1-dimethylethyl)-4-hydroxy-
3,5-Di-tert-butyl-4-hydroxy-benzyl alcohol
2,6-Di-tert-butyl-4-hydroxy-methylphenol
Ionox 100
Ionox 100 Antioxidant

88-30-2
$C_7H_4F_3NO_3$
207.12
m-Cresol, α,α,α-trifluoro-4-nitro
m-Cresol, 4-nitro-α,α,α-tri-fluoro-
Dowlap F
Lamprecid
4-Nitro-3-trifluoromethylphenol
Phenol, 4-nitro-3-(trifluoro-methyl)-
Phenol, m-trifluoromethyl-
3-Trifluoromethyl-4-nitrophenol
USAF MA-6

88-32-4
$C_{11}H_{16}O_2$
180.27
O(c(c(cc(O)c1)C(C)(C)C)c1)C
Phenol, 3-tert-butyl-4-methoxy
3-tert-BHA
3-tert-Butylated hydroxyanisole
3-tert-Butyl-4-methoxyphenol

88-41-5
$C_{12}H_{22}O_2$
198.31
O=C(OC(C(C(C)(C)C)CCC1)C1)C
Cyclohexanol, 2-(1,1-dimethyl-ethyl)-, acetate (9CI)
2-(1,1-Dimethylethyl)cyclo-hexanol acetate

88-43-7
$C_6H_6ClNO_3S$
207.63
O=S(=O)(O)c(c(ccc1N)Cl)c1
6-Chlorometanilic acid
AI3-28528
5-Amino-2-chlorobenzenesulf-onic acid
Benzenesulfonic acid, 5-amino-2-chloro- (9CI)
p-Chloroaniline-m-sulfonic acid
4-Chloroaniline-3-sulfonic acid
Metanilic acid, 6-chloro- (8CI)
NSC-15346

88-44-8
$C_7H_9NO_3S$
187.23
O=S(=O)(O)c(c(N)ccc1C)c1
m-Toluenesulfonic acid, 6-amino
2-Amino-5-methylbenzene-sulfonic acid
4-Aminotoluene-3-sulfonic acid
6-Amino-m-toluenesulfonic acid
Benzenesulfonic acid, 2-amino-5-methyl-
Kyselina 4-toluidin-3-sulfonova (Czech)
PTMS
PTMSA
Red 4B Acid
p-Toluidine-m-sulfonic acid

88-50-6
$C_6H_5Cl_2NO_3S$
242.08
O=S(=O)(O)c(c(cc(N)c1Cl)Cl)c1
Benzenesulfonic acid, 4-amino-2,5-dichloro- (9CI)
4-Amino-2,5-dichlorobenzene-sulfonic acid
2,5-Dichloroaniline-4-sulfonic acid
2,5-Dichlorosulfanilic acid
NSC-1128
Sulfanilic acid, 2,5-dichloro- (8CI)

88-51-7
$C_7H_8ClNO_3S$
221.67
O=S(=O)(O)c(c(N)cc(c1C)Cl)c1
m-Toluenesulfonic acid, 6-amino-4-chloro
6-Amino-4-chloro-m-toluene-sulfonic acid
Brilliant Toning Red Amine
Kyselina 2-chlor-4-toluidin-5-sulfonova (Czech)

88-53-9
$C_7H_8ClNO_3S$
221.66
O=S(=O)(O)c(c(N)cc(c1Cl)C)c1
Lake Red C Amine
AI3-28529
2-Amino-5-chloro-4-methyl-benzenesulfonic acid
Benzenesulfonic acid, 2-amino-5-chloro-4-methyl-
Red Lake Camine
p-Toluenesulfonic acid, 2-amino-5-chloro-

88-56-2
$C_8H_{10}ClNO_3S$
235.68
O=S(=O)(O)c(c(N)cc(c1Cl)CC)c1
Benzenesulfonic acid, 2-amino-5-chloro-4-ethyl- (9CI)
2-Amino-5-chloro-4-ethylbenz-enesulfonic acid
NSC-81226

88-58-4
$C_{14}H_{22}O_2$
222.36
Oc(c(cc(O)c1C(C)(C)C)C(C)(C)C)c1
Hydroquinone, 2,5-di-tert-butyl
2,5-Di-tert-butylbenzene-1,4-diol
2,5-Di-t-butylhydroquinone

88-60-8
$C_{11}H_{16}O$
164.25
Oc(c(ccc1C)C(C)(C)C)c1
2-(1,1-Dimethylethyl)-5-methylphenol
Benzene, 1-tert-butyl-2-hydroxy-4-methyl-
6-tert-Butyl-m-cresol
2-tert-Butyl-5-methylphenol
6-tert-Butyl-3-methylphenol
m-Cresol, 6-tert-butyl- (8CI)

3-Methyl-6-tert-butylphenol
NSC-48467
Phenol, 2-(1,1-dimethylethyl)-5-methyl- (9CI)

88-61-9
$C_8H_{10}O_3S$
186.24
O=S(=O)(O)c(c(cc(c1)C)C)c1
2,4-Xylenesulfonic acid
Benzenesulfonic acid, 2,4-dimethyl-
2,4-Dimethylbenzenesulfonic acid
m-Xylenesulfonic acid
m-Xylene-4-sulfonic acid

88-63-1
$C_6H_8N_2O_3S$
188.22
O=S(=O)(O)c(c(N)cc(N)c1)c1
Benzenesulfonic acid, 2,4-diamino
o-Aminosulfanilic acid
1,3-Diaminobenzenesulfonic acid
1,3-Diaminobenzene-4-sulfonic acid
1,3-Diaminobenzene-6-sulfonic acid
2,4-Diaminobenzenesulfonic acid
Kyselina 2,4-diaminobenzensulfonova (Czech)
Kyselina 1,3-fenylendiamin-4-sulfonova (Czech)
m-Phenylenediaminesulfonic acid
m-Phenylenediamine-4-sulfonic acid
1,3-Phenylenediamine-4-sulfonic acid

88-65-3
$C_7H_5BrO_2$
201.02
O=C(O)c(c(ccc1)Br)c1
2-Bromobenzoate
AI3-03699
Benzoic acid, o-bromo- (8CI)
Benzoic acid, 2-bromo- (9CI)
o-Bromobenzoic acid
2-Bromobenzoic acid
NSC-6976

88-67-5
$C_7H_5IO_2$
248.02
O=C(O)c(c(ccc1)I)c1
Benzoic acid, o-iodo
o-Iodobenzoic acid
Kyselina o-jodbenzoova (Czech)
USAF EK-572

88-68-6
$C_7H_8N_2O$
136.17
O=C(N)c(c(N)ccc1)c1
Benzamide, o-amino
o-Aminobenzamide
2-Aminobenzamide
Anthranilamide
Anthranilimidic acid
Benzamide, 2-amino- (9CI)
2-Carbamoylaniline

88-69-7
$C_9H_{12}O$
136.21
Oc(c(ccc1)C(C)C)c1
Phenol, o-isopropyl
o-Isopropylphenol
2-Isopropylphenol
Phenol, 2-(1-methylethyl)- (9CI)
Prodox 131

88-72-2
$C_7H_7NO_2$
137.15
O=N(=O)c(c(ccc1)C)c1
Toluene, o-nitro
o-Methylnitrobenzene
2-Methylnitrobenzene
o-Nitrotoluene (ACGIH,OSHA) [UN 1664]
2-Nitrotoluene
ONT
UN 1664 [Nitrotoluenes, Liquid o-; m-; p-;]

88-73-3
$C_6H_4ClNO_2$
157.56
O=N(=O)c(c(ccc1)Cl)c1
Benzene, 1-chloro-2-nitro
Chloro-o-nitrobenzene
o-Chloronitrobenzene (DOT)
1-Chloro-2-nitrobenzene
2-Chloronitrobenzene
2-Chloro-1-nitrobenzene
Nitrochlorobenzene, ortho, Liquid [UN 1578]
o-Nitrochlorobenzene
ONCB
UN 1578 [Chloronitrobenzene, ortho, Liquid]

88-74-4
$C_6H_6N_2O_2$
138.14
O=N(=O)c(c(N)ccc1)c1
Aniline, o-nitro
1-Amino-2-nitrobenzene
Azoene Fast Orange GR Base
Azoene Fast Orange GR Salt
Azofix Orange GR Salt
Azogene Fast Orange GR
Azoic Diazo Component 6
Brentamine Fast Orange GR Base
Brentamine Fast Orange GR Salt
C.I. 37025
C.I. Azoic Diazo Component 6
Devol Orange B
Devol Orange Salt B
Diazo Fast Orange GR
Fast Orange Base GR
Fast Orange Base JR
Fast Orange GR Base
Fast Orange GR Salt
Fast Orange O Base
Fast Orange O Salt
Fast Orange Salt GR
Fast Orange Salt JR
Hiltonil Fast Orange GR Base
Hiltosal Fast Orange GR Salt
Hindasol Orange GR Salt
Natasol Fast Orange GR Salt
o-Nitraniline
o-Nitroaniline [UN 1661]
2-Nitroaniline
Orange Base Ciba II
Orange Base Irga II
Orange GRS Salt
Orange Salt Ciba II
Orange Salt Irga II
Orthonitroaniline (DOT)
UN 1661 [Nitroanilines (o-; m-; p-;)]

88-75-5
$C_6H_5NO_3$
139.12
O=N(=O)c(c(O)ccc1)c1
Phenol, o-nitro
2-Hydroxynitrobenzene
o-Nitrofenol (Czech)
o-Nitrophenol [UN 1663]
2-Nitrophenol
UN 1663 [Nitrophenols (o-; m-; p-;)]

88-82-4
$C_7H_3I_3O_2$
499.80
O=C(O)c(c(c(cc1I)I)I)c1
Benzoic acid, 2,3,5-triiodo
Floraltone
Johnkolor
Kyselina 2,3,5-trijodbenzoova (Czech)
Regim 8
Regin 8
TIB
TIBA
2,3,5-TIBA
Triiodobenzoic acid
2,3,5-Triiodobenzoic acid

88-84-6
$C_{15}H_{24}$
204.39
C(=C(C(CCC1=C(C)C)C)CC2)(C2C)C1
Guaia-1(5),7(11)-diene
Azulene, 1,2,3,4,5,6,7,8-octahydro-1,4-dimethyl-7-(1-methylethylidene)-, (1S,cis)-(9CI)
Guaiene
β-Guaiene
1,2,3,4,5,6,7,8-Octahydro-1,4-dimethyl-7-(1-methylethyl-

idene)-azulene, (1S,cis)

88-85-7
$C_{10}H_{12}N_2O_5$
240.24
O=N(=O)c(cc(N(=O)=O)c(O)c1C(CC)C)c1
Phenol, 2-sec-butyl-4,6-dinitro
Aretit
Basanite
Butaphene
BNP 30
2-sec-Butyl-4,6-dinitrophenol
Caldon
Chemox General
Chemox P.E.
Dinitro-3
Dinitro
4,6-Dinitro-2-sec.butylfenol (Czech)
Dinitrobutylphenol
2,4-Dinitro-6-sec-butylphenol
4,6-Dinitro-o-sec-butylphenol
4,6-Dinitro-2-sec-butylphenol
2,4-Dinitro-6-(1-methyl-propyl)-phenol (French)
4,6-Dinitro-2-(1-methyl-n-propyl)phenol
Dinoseb
Dinosebe (French)
DN 289
DNBP
DNOSBP
DNSBP
Dow General
Dow General Weed Killer
Dow Selective Weed Killer
Elgetol
Elgetol 318
ENT 1,122
Gebutox
Hel-Fire
Ivosit
Kiloseb
6-(1-Methyl-propyl)-2,4-dinitro-fenol (Dutch)
2-(1-Methylpropyl)-4,6-dinitro-phenol
6-(1-Metil-propil)-2,4-dinitro-fenolo (Italian)
Nitropone C
Phenotan
Premerge
Premerge 3
RCRA waste number P020
Sinox General
Sparic
Spurge
Subitex
Unicrop DNBP
Vertac Dinitro Weed Killer
Vertac General Weed Killer
Vertac Selective Weed Killer

88-86-8
$C_7H_3Cl_2NO_4$
236.01
OC(=O)c1cc(Cl)cc(N(=O)=O)c1Cl
Benzoic acid, 2,5-dichloro-3-nitro
2,5-Dichloro-3-nitrobenzoic acid
Dinoben
Kyselina 2,5-dichlor-3-nitro-benzoova (Czech)
3-Nitro-2,5-dichlorobenzoic acid

88-87-9
$C_6H_3ClN_2O_5$
218.54
O=N(=O)c(c(O)c(N(=O)=O)cc1Cl)c1
Phenol, 4-chloro-2,6-dinitro-(9CI)
AI3-08954
4-Chloro-2,6-dinitrophenol
2,6-Dinitro-4-chlorophenol
NSC-6212

88-88-0
$C_6H_2ClN_3O_6$
247.56
O=N(=O)c(cc(N(=O)=O)c(c1N(=O)=O)Cl)c1
Benzene, 2-chloro-1,3,5-tri-nitro
2-Chloro-1,3,5-trinitrobenzene
Picryl chloride
TNCB
2,4,6-Trinitrochlorobenzene

88-89-1
$C_6H_3N_3O_7$
229.12
O=N(=O)c(cc(N(=O)=O)c(O)c1N(=O)=O)c1
Picric-acid
Acide picrique (French)
Acido picrico (Italian)
AI3-15403
Carbazotic acid
C.I. 10305
2-Hydroxy-1,3,5-trinitrobenzene
Kyselina pikrova (Czech)
Melinite
Nitroxanthic acid
NA 1344 [Picric acid, Wet, with not less than 10% water]
Phenol trinitrate
Phenol, 2,4,6-trinitro-
Picral
Picric acid (ACGIH,OSHA) (8CI)
Picric acid, Dry (DOT)
Picric acid, Wet with not less than 10% water, over 25 pounds [NA 1344]
Picric acid, Wetted with at least 10% water [NA 1344]
Picric acid, Wetted with at least 30% water [UN 1344]
Picric acid, Wetted with 10% to 30% water
Picronitric acid
Pikrinezuur (Dutch)
Pikrinsaeure (German)
Pikrynowy kwas (Polish)
2,4,6-Trinitrofenol (Dutch)
2,4,6-Trinitrofenolo (Italian)
1,3,5-Trinitrophenol
2,4,6-Trinitrophenol
Trinitrophenol, Dry [UN 0154]
Trinitrophenol, Wetted with at least 10% water [NA 1344]
Trinitrophenol, Wetted with at least 30% water [UN 1344]
Trinitrophenol, Wetted with less than 30% water [UN 0154]
2,4,6-Trinitrophenyl (OSHA)
UN 0154 [Trinitrophenol (Picric acid), Dry or wetted with less than 30 per cent water, by mass]
UN 1344 [Trinitrophenol, Wetted with not less than 30 per cent water, by mass]

88-95-9
$C_8H_4Cl_2O_2$
203.02
O=C(c(c(ccc1)C(=O)Cl)c1)Cl
Phthaloyl chloride (8CI)
1,2-Benzenedicarbonyl di-chloride (9CI)
1,2-Bis(chlorocarbonyl)benzene
NSC-44611
Phthalic acid dichloride
Phthalic chloride
Phthalic dichloride
Phthaloyl dichloride
Phthalyl chloride
Phthalyl dichloride

88-96-0
$C_8H_8N_2O_2$
164.18
O=C(N)c(c(ccc1)C(=O)N)c1
1,2-Benzenedicarboxamide
NCI-C03612
P-D
Phthalamide
o-Phthalic acid diamide

88-97-1
$C_8H_7NO_3$
165.14
O=C(O)c(c(ccc1)C(=O)N)c1
Benzoic acid, 2-(aminocarbon-yl)- (9CI)
AI3-26413
2-(Aminocarbonyl)benzoic acid
Benzoic acid, o-carbamoyl-
o-Carbamoylbenzoic acid
NSC-5344
Phthalamic acid (8CI)
Phthalamide acid
Phthalamidic acid
Phthalic acid monoamide
Phthalic monoamide

88-98-2
$C_8H_{10}O_4$
170.17
O=C(O)C(C(C(=O)O)CC=C1)C1
4-Cyclohexene-1,2-dicarboxyl-ic acid (9CI)
NSC-239116

Tetrahydrophthalic acid
δ4-Tetrahydrophthalic acid
1,2,3,6-Tetrahydrophthalic acid

88-99-3
$C_8H_6O_4$
166.14
O=C(O)c(c(ccc1)C(=O)O)c1
Phthalic-acid
Acide phtalique (French)
Benzene-1,2-dicarboxylic acid
o-Benzenedicarboxylic acid
1,2-Benzenedicarboxylic acid
o-Dicarboxybenzene
Kyselina ftalova (Czech)
o-Phthalic acid

89-00-9
$C_7H_5NO_4$
167.11
O=C(O)c(cccn1)c1C(=O)O
Quinolinic acid
AI3-63017
NSC-13127
Pyridine-2,3-dicarboxylic acid
2,3-Pyridinedicarboxylic acid (9CI)

89-04-3
$C_{33}H_{54}O_6$
546.79
O=C(OCCCCCCCC)c(ccc(c1C(=O)OCCCCCCCC)C(=O)OCCCCCCCC)c1
Tri-n-octyl trimellitate
Benzene tricarboxylic acid, trioctyl ester
1,2,4-Benzenetricarboxylic acid, trioctyl ester (9CI)
PX 338
TOTM
Trimex N 08
Trioctyl trimellitate

89-05-4
$C_{10}H_6O_8$
254.16
O=C(O)c(c(cc(c1C(=O)O)C(=O)O)C(=O)O)c1
1,2,4,5-Benzenetetracarboxylic acid
Pyromellitic acid
1,2,4,5-Tetracarboxybenzene
USAF XR-20

89-08-7
$C_8H_6O_7S$
246.20
O=C(O)c(c(ccc1S(=O)(=O)O)C(=O)O)c1
4-Sulfophthalic acid
1,2-Benzenedicarboxylic acid, 4-sulfo- (9CI)
NSC-100615
Phthalic acid, 4-sulfo- (8CI)
4-Sulfo-1,2-benzenedicarboxylic acid

89-19-0
$C_{22}H_{34}O_4$
362.56
CCCCCCCCCCOC(=O)c1ccccc1C(=O)OCCCC
Phthalic acid, butyl decyl ester
Butyl decyl phthalate
Decyl butyl phthalate
Plasticizer BDP
PX 114

89-20-3
$C_8H_5ClO_4$
200.58
O=C(O)c(c(ccc1Cl)C(=O)O)c1
Phthalic acid, 4-chloro- (8CI)
1,2-Benzenedicarboxylic acid, 4-chloro- (9CI)
4-Chloro-1,2-benzenedicarboxylic acid
4-Chlorophthalic acid
NSC-57755

89-25-8
$C_{10}H_{10}N_2O$
174.22
O=C(N(N=C1C)c(cccc2)c2)C1
2-Pyrazolin-5-one, 3-methyl-1-phenyl
C.I. Developer 1
Developer Z
1-Fenyl-3-methyl-2-pyrazolin-5-on (Czech)
3-Methyl-1-phenyl-2-pyrazolin-5-one
3-Methyl-1-phenyl-5-pyrazolone
NCI-C03952
Norphenazone
1-Phenyl-3-methylpyrazolone-5
1-Phenyl-3-methyl-5-pyrazolone
5-Pyrazolone, 3-methyl-1-phenyl-

89-28-1
$C_{24}H_{39}N$
341.64
N(c(c(C(=C1)C)cc(c2)CCCCCCCCCCCC)c2)C1(C)C
Quinoline, 6-dodecyl-1,2-dihydro-2,2,4-trimethyl
6-Dodecyl-2,2,4-trimethyl-1,2-dihydroquinoline
Santoflex DD

89-29-2
$C_{10}H_{11}N_3O_3S$
253.26
O=C(N(N=C1C)c(cccc2S(=O)(=O)N)c2)C1
Benzenesulfonamide, m-(3-methyl-5-oxo-2-pyrazolin-1-yl)- (8CI)
Benzenesulfonamide, 3-(4,5-dihydro-3-methyl-5-oxo-1H-pyrazol-1-yl)- (9CI)
NSC-15355

89-32-7
$C_{10}H_2O_6$
218.12
O=C(OC(=O)c1cc(c(c2)C(=O)O3)C3=O)c12
1,2,4,5 Benzenetetracarboxylic 1,2:4,5 dianhydride
1H,3H-Benzo(1,2-c:4,5-c')difuran-1,3,5,7-tetrone
Pyromellitic acid anhydride
Pyromellitic acid dianhydride
Pyromellitic anhydride
Pyromellitic dianhydride

89-33-8
$C_{12}H_{12}N_2O_3$
232.23
O=C(OCC)C(=NN(C1=O)c(cccc2)c2)C1
1H-Pyrazole-3-carboxylic acid, 4,5-dihydro-5-oxo-1-phenyl-, ethyl ester (9CI)
3-(Ethoxycarbonyl)-1-phenyl-5-pyrazolone
NSC-49150
1-Phenyl-3-carbethoxypyrazolone
1-Phenyl-5-oxo-2-pyrazoline-3-carboxylic acid, ethyl ester
2-Pyrazoline-3-carboxylic acid, 5-oxo-1-phenyl-, ethyl ester (8CI)

89-36-1
$C_{10}H_{10}N_2O_4S$
254.25
O=C(N(N=C1C)c(ccc(S(=O)(=O)O)c2)c2)C1
Benzenesulfonic acid, p-(3-methyl-5-oxo-2-pyrazolin-1-yl)- (8CI)
AI3-08532
Benzenesulfonic acid, 4-(4,5-dihydro-3-methyl-5-oxo-1H-pyrazol-1-yl)- (9CI)
NSC-26429
Pyrazoline G

89-40-7
$C_8H_4N_2O_4$
192.14
O=C(NC(=O)c1ccc(N(=O)=O)c2)c12
Phthalimide, 4-nitro
1H-Isoindole-1,3(2H)-dione, 5-nitro-
4-Nitrophthalimide

89-52-1
$C_9H_9NO_3$
179.19
O=C(Nc(c(ccc1)C(=O)O)c1)C
Anthranilic acid, N-acetyl
o-Acetamidobenzoic acid
2-Acetamidobenzoic acid

o-Acetoaminobenzoic acid
N-Acetylaminobenzoic acid
2-(Acetylamino)benzoic acid
Acetylanthranilic acid
N-Acetylanthranilic acid
Benzoic acid, 2-(acetylamino)- (9CI)
2-Carboxyacetanilide

89-56-5
$C_8H_8O_3$
152.16
Cc1ccc(O)c(c1)C(O)=O
2,5-Cresotic acid
Benzoic acid, 2-hydroxy-5-methyl- (9CI)
p-Cresotic acid
p-Cresotinic acid
α-Cresotinic acid
p-Homosalicylic acid
5-Methylsalicylic acid

89-58-7
$C_8H_9NO_2$
151.18
O=N(=O)c(c(ccc1C)C)c1
p-Xylene, 2-nitro
Benzene, 1,4-dimethyl-2-nitro-
1,4-Dimethyl-2-nitrobenzene
Nitro-p-xylene

89-59-8
$C_7H_6ClNO_2$
171.58
O=N(=O)c(c(ccc1Cl)C)c1
Benzene, 4-chloro-1-methyl-2-nitro- (9CI)
AI3-00494
Benzene, 4-chloro-1-methyl-3-nitro-
4-Chloro-1-methyl-2-nitrobenzene
p-Chloro-o-nitrotoluene
4-Chloro-2-nitrotoluene
2-Nitro-4-chlorotoluene
NSC-5386
Toluene, 4-chloro-2-nitro- (8CI)

89-60-1
$C_7H_6ClNO_2$
171.58
O=N(=O)c(c(ccc1C)Cl)c1
Benzene, 1-chloro-4-methyl-2-nitro- (9CI)
1-Chloro-4-methyl-2-nitrobenzene
2-Chloro-5-methylnitrobenzene
4-Chloro-3-nitrotoluene
3-Nitro-4-chlorotoluene
NSC-60721
Toluene, 4-chloro-3-nitro- (8CI)

89-61-2
$C_6H_3Cl_2NO_2$
192.00
O=N(=O)c(c(ccc1Cl)Cl)c1
Benzene, 1,4-dichloro-2-nitro
2,5-Dichlornitrobenzen (Czech)
1,4-Dichloro-2-nitrobenzene
2,5-Dichloronitrobenzene
Nitro-p-dichlorobenzene

89-62-3
$C_7H_8N_2O_2$
152.14
O=N(=O)c(c(N)ccc1C)c1
2-Nitro-p-toluidine
AI3-09044
Amarthol Fast Red GL Base
Amarthol Fast Red GL Salt
1-Amino-2-nitro-4-methylbenzene
4-Amino-3-nitrotoluene
Azoamine Red A
Azobase NAT
Azoene Fast Red Red GL Salt
Azofix Red GL Salt
Azoic Diazo Component 8
Benzenamine, 4-methyl-2-nitro- (9CI)
C.I. Azoic Diazo Component 8
C.I. 37110
Devol Red G
Devol Red Salt G
Diazo Fast Red GL
Fast Red Base GL
Fast Red Base JL
Fast Red G Base
Fast Red GL
Fast Red GL Base
Fast Red MGL Base
Fast Red 3NT Base
Fast Red 3NT Salt
HD Fast Red GL Base
Hiltonil Fast Red GL Base
Hiltosal Fast Red GL Salt
Lake Red G Base
Lithosol Scarlet Base M
Lithosol Scarlet Base MB
Lithosol Scarlet Base MBW
Lithosol Scarlet Base MW
4-Methyl-2-nitroaniline
4-Methyl-2-nitrobenzenamine
Mitsui Red GL Base
MNPT
Naphthanil Red G Base
Naphtoelan Fast Red GL Base
3-Nitro-4-aminotoluene
2-Nitro-4-methylaniline
3-Nitro-4-toluidine
NSC-2759
Red Base Ciba VII
Red Base Irga VII
Red Base NGL
Red G Base
Red G Salt
Red Salt Ciba VII
Red Salt Irga VII
Sanyo Fast Red GL Base
Shinnippon Fast Red GL Base
p-Toluidine, 2-nitro- (8CI)
Toyo Fast Red GL Base
Tulabase Fast Red GL

89-63-4
$C_6H_5ClN_2O_2$
172.58
O=N(=O)c(c(N)ccc1Cl)c1
Aniline, 4-chloro-2-nitro
Azoene Fast Red 3GL Base
Azoene Fast Red 3GL Salt
Azofix Red 3GL Salt
Azoic Diazo Component 9
Benzenamine, 4-chloro-2-nitro- (9CI)
p-Chloro-o-nitroaniline
4-Chloro-2-nitroaniline
C.I. 37040
C.I. Azoic Diazo Component 9
Daito Red Base 3GL
Daito Red Salt 3GL
Devol Red F
Devol Red Salt F
Diazo Fast Red 3GL
Fast Red Base 3GL Special
Fast Red Base 3JL
Fast Red 3GL Base
Fast Red 2NC Base
Fast Red 3GL Salt
Fast Red 3GL special Base
Fast Red 3GL special Salt
Fast Red 2NC Salt
Fast Red Salt 3GL
Fast Red Salt 3JL
Hiltonil Fast Red 3GL Base
Hiltosal Fast Red 3GL Salt
Kayaku Fast Red 3GL Base
Kayaku Red Salt 3GL
Mitsui Red 3GL Base
Mitsui Red 3GL Salt
Naphthanil Red 3G Base
Naphtoelan Fast Red 3GL Base
Naphtoelan Fast Red 3GL Salt
NCI-C60355
2-Nitro-4-chloroaniline
Pcon
Pcona
Red 3G Base
Red Base Ciba VI
Red Base 3 GL
Red Base Irga VI
Red 3G Salt
Red 3GS Salt
Red Salt Ciba VI
Red Salt Irga VI
Red Salt NBGL
Sanyo Fast Red Salt 3GL
Shinnippon Fast Red 3GL Base
Symulon Red 3GL Salt

89-64-5
$C_6H_4ClNO_3$
173.55
O=N(=O)c(c(O)ccc1Cl)c1
Phenol, 4-chloro-2-nitro- (9CI)
AI3-28527
4-Chloro-2-nitrophenol
2-Nitro-4-chlorophenol
NSC-520345

89-65-6
$C_6H_8O_6$
176.14
O=C(OC(C=1O)C(O)CO)C1O
D-erythro-Hex-2-enonic acid, γ-lactone
Araboascorbic acid

D-Araboascorbic acid
Erycorbin
Erythorbic acid
D-Erythorbic acid
Glucosaccharonic acid
Isoascorbic acid
D-Isoascorbic acid
Isovitamin C
Mercate 5
Neo-Cebicure
Saccharosonic acid

89-69-0
$C_6H_2Cl_3NO_2$
226.44
O=N(=O)c(c(cc(c1Cl)Cl)Cl)c1
Benzene, 1,2,4-trichloro-5-nitro
1,2,4-Trichloro-5-nitrobenzene
2,4,5-Trichloronitrobenzene

89-72-5
$C_{10}H_{14}O$
150.24
Oc(c(ccc1)C(CC)C)c1
Phenol, o-sec-butyl
2-sec.Butylfenol (Czech)
o-sec-Butylphenol (ACGIH, OSHA)

89-74-7
$C_{10}H_{12}O$
148.20
O=C(c(c(cc(c1)C)C)c1)C
Acetophenone, 2',4'-dimethyl- (8CI)
AI3-20801
2,4-Dimethylacetophenone
2',4'-Dimethylacetophenone
1-(2,4-Dimethylphenyl)ethanone
Ethanone, 1-(2,4-dimethyl-phenyl)- (9CI)
NSC-15333

89-75-8
$C_7H_3Cl_3O$
209.46
O=C(c(c(cc(c1)Cl)Cl)c1)Cl
Benzoyl chloride, 2,4-dichloro-
AI3-14890
2,4-Dichlorobenzoyl chloride

89-78-1
$C_{10}H_{20}O$
156.30
OC(C(CCC1C)C(C)C)C1
Menthol
Cyclohexanol, 2-isopropyl-5-methyl-
Hexahydrothymol
2-Isopropyl-5-methylcyclo-hexanol
p-Menthan-3-ol
l-Menthol
5-Methyl-2-(1-methylethyl)-cyclohexanol
Peppermint Camphor

89-80-5
$C_{10}H_{18}O$
154.28
O=C(C(CCC1C)C(C)C)C1
p-Menthan-3-one, trans
Cyclohexanone, 5-methyl-2-(1-methylethyl)-, trans-
Menthone
p-Menthone
trans-Menthone

89-81-6
$C_{10}H_{16}O$
152.26
O=C(C=C(CC1)C)C1C(C)C
p-Menth-1-en-3-one
3-Carvomenthenone
2-Cyclohexen-1-one, 3-methyl-6-(1-methylethyl)-
p-Menth-1-en-3-one
1-Methyl-4-isopropyl-1-cyclo-hexen-3-one
Piperitone

89-83-8
$C_{10}H_{14}O$
150.24
Oc(c(ccc1C)C(C)C)c1
Thymol
m-Cresol, 6-isopropyl-
p-Cymene, 3-hydroxy-
p-Cymen-3-ol
3-p-Cymenol
3-Hydroxy-p-cymene
3-Hydroxy-1-methyl-4-iso-propylbenzene
Isopropyl cresol
2-Isopropyl-5-methylphenol
1-Methyl-3-hydroxy-4-iso-propylbenzene
3-Methyl-6-isopropylphenol
5-Methyl-2-isopropyl-1-phenol
5-Methyl-2-(1-methylethyl)-phenol
Phenol, 2-isopropyl-5-methyl-
Phenol, 5-methyl-2-(1-methyl-ethyl)- (9CI)
Thyme camphor
Thymic acid
m-Thymol

89-84-9
$C_8H_8O_3$
152.16
O=C(c(c(O)cc(O)c1)c1)C
Acetophenone, 2',4'-dihydroxy
4-Acetylresorcinol
2,4-Dihydroxyacetophenone
2',4'-Dihydroxyacetophenone
Ethanone, 1-(2,4-dihydroxy-phenyl)- (9CI)
Resacetophenone
β-Resacetophenone
Resoacetophenone
Resorcinol, 4-acetyl-

89-86-1
$C_7H_6O_4$
154.13
O=C(O)c(c(O)cc(O)c1)c1
β-Resorcylic acid
Benzoic acid, 2,4-dihydroxy- (9CI)
4-Carboxyresorcinol
2,4-DHBA
2,4-Dihydroxybenzoic acid
p-Hydroxysalicylic acid
4-Hydroxysalicylic acid
β-Resorcinolic acid

89-87-2
$C_8H_9NO_2$
151.16
O=N(=O)c(c(cc(c1)C)C)c1
Benzene, 2,4-dimethyl-1-nitro- (9CI)
1,3-Dimethyl-4-nitrobenzene
2,4-Dimethyl-1-nitrobenzene
1-Nitro-2,4-dimethylbenzene
4-Nitro-1,3-dimethylbenzene
4-Nitro-m-xylene
4-Nitro-1,3-xylene
NSC-50661
m-Xylene, 4-nitro- (8CI)

89-92-9
C_8H_9Br
185.06
o-Xylyl bromide
Benzene, 1-(bromomethyl)-2-methyl- (9CI)
1-(Bromomethyl)-2-methyl-benzene
2-(Bromomethyl)toluene
α-Bromo-o-xylene
α-Bromo-ortho-xylene
o-Methylbenzyl bromide
2-Methylbenzyl bromide
NSC-60145
o-Xylene, α-bromo- (8CI)
2-Xylyl bromide

89-95-2
$C_8H_{10}O$
122.17
2-Methylbenzyl alcohol
AI3-21536
Benzenemethanol, 2-methyl- (9CI)
Benzyl alcohol, o-methyl- (8CI)
o-Methylbenzyl alcohol
NSC-91
o-Tolyl alcohol

89-96-3
C_8H_9Cl
140.61
Benzene, 1-chloro-2-ethyl- (8CI,9CI)
NSC-72739

89-98-5
C_7H_5ClO

140.57
O=Cc(c(ccc1)Cl)c1
Benzaldehyde, o-chloro
Benzaldehyde, 2-chloro- (9CI)
o-Chloorbenzaldehyde (Dutch)
2-Chloorbenzaldehyde (Dutch)
2-Chlorbenzaldehyd (German)
o-Chlorobenzaldehyde
2-Chlorobenzaldehyde
o-Chlorobenzenecarboxaldehyde
2-Clorobenzaldeide (Italian)
USAF M-7

90-00-6
$C_8H_{10}O$
122.18
Oc(c(ccc1)CC)c1
Phenol, o-ethyl
o-Ethylphenol
2-Ethylphenol
Florol (Czech)
Phlorol

90-01-7
$C_7H_8O_2$
124.15
OCc(c(O)ccc1)c1
Benzyl alcohol, o-hydroxy
Benzenemethanol, 2-hydroxy- (9CI)
Diathesin
α,2-Dihydroxytoluene
2-Hydroxybenzenemethanol
o-Hydroxybenzyl alcohol
2-Hydroxybenzyl alcohol
o-(Hydroxymethyl)phenol
2-Hydroxymethylphenol
o-Methylolphenol
2-Methylolphenol
Sal
Salicyl alcohol
Saligenin
Saligenol

90-02-8
$C_7H_6O_2$
122.13
O=Cc(c(O)ccc1)c1
Salicylaldehyde
Benzaldehyde, o-hydroxy-
o-Formylphenol
2-Formylphenol
o-Hydroxybenzaldehyde
2-Hydroxybenzaldehyde
Sah
Salicyladehyde
Salicylal
Salicylic aldehyde

90-04-0
C_7H_9NO
123.17
O(c(c(N)ccc1)c1)C
o-Anisidine (ACGIH,OSHA) [UN 2431]
o-Aminoanisole
2-Aminoanisole
1-Amino-2-methoxybenzene
2-Anisidine
o-Anisylamine
Benzenamine, 2-methoxy- (9CI)
2-Methoxy-1-aminobenzene
o-Methoxyaniline
2-Methoxyaniline
2-Methoxybenzenamine
o-Methoxyphenylamine
UN 2431 [Anisidines]

90-05-1
$C_7H_8O_2$
124.15
O(c(c(O)ccc1)c1)C
Phenol, o-methoxy
Guaiacol
Guaicol
Guajakol (Czech)
o-Hydroxyanisole
2-Hydroxyanisole
1-Hydroxy-2-methoxybenzene
o-Methoxyphenol
2-Methoxyphenol
Methylcatechol
Pyroguaiac acid

90-11-9
$C_{10}H_7Br$
207.08
c(c(c(cc1)Br)ccc2)(c2)c1
Naphthalene, 1-bromo
α-Bromonaphthalene
1-Bromonaphthalene

90-12-0
$C_{11}H_{10}$
142.21
c(c(c(cc1)C)ccc2)(c2)c1
Naphthalene, 1-methyl
α-Methylnaphthalene
1-Methylnaphthalene

90-13-1
$C_{10}H_7Cl$
162.62
c(c(c(cc1)Cl)ccc2)(c2)c1
Naphthalene, 1-chloro
1-Chlornaftalen (Czech)
α-Chlornaphthalene
α-Chloronaphthalene
1-Chloronaphthalene

90-14-2
$C_{10}H_7I$
254.07
c(c(c(cc1)I)ccc2)(c2)c1
Naphthalene, 1-iodo- (9CI)
α-Iodonaphthalene
1-Iodonaphthalene
1-Naphthyl iodide
NSC-9275

90-15-3
$C_{10}H_8O$
144.18
Oc(c(c(ccc1)cc2)c1)c2
1-Naphthol
BASF Ursol Ern
C.I. 76605
C.I. Oxidation Base 33
Durafur Developer D
Fouramine ERN
Fourrine 99
Fourrine ERN
Furro ER
α-Hydroxynaphthalene
1-Hydroxynaphthalene
Nako TRB
1-Naphthalenol
α-Naphthol
Tertral ERN
Ursol ERN
Zoba ERN

90-16-4
$C_7H_5N_3O$
147.15
Oc(nnnc1cccc2)c12
1,2,3-Benzotriazin-4(1H)-one
Benzazimide
Benzazimidone
Benzoketotriazine
3H-1,2,3-Benzotriazin-4-one
4-Ketobenzotriazine
USAF MA-2

90-17-5
$C_{10}H_9Cl_3O_2$
267.54
O=C(OC(c(cccc1)c1)C(Cl)(Cl)Cl)C
Acetic acid, α-(trichloromethyl)benzyl ester
Benzenemethanol, α-(trichloromethyl)-, acetate
Benzyl alcohol, α-(trichloromethyl)-, acetate
Rosacetol
Rose Crystals
Trichloromethylphenylcarbinyl acetate

90-20-0
$C_{10}H_9NO_7S_2$
319.32
O=S(=O)(O)c(cc(c(c1cc(S(=O)(=O)O)c2)c2O)N)c1
2,7-Naphthalenedisulfonic acid, 4-amino-5-hydroxy
4-Amino-5-hydroxy-2,7-naphthalenedisulfonic acid
C.I. 35570
H Acid
Kyselina 1-amino-8-naftol-3,6-disulfonova (Czech)
Kyselina 8-amino-1-naftol-3,6-disulfonova (Czech)
Kyselina H (Czech)

90-27-7
$C_{10}H_{12}O_2$
164.22
O=C(O)C(c(cccc1)c1)CC
Butyric acid, 2-phenyl
2-Phenylbutyric acid

α-Phenyl butyric acid

90-30-2
$C_{16}H_{13}N$
219.30
N(c(c(c(ccc1)cc2)c1)c2)c(cccc3)c3
1-Naphthylamine, N-phenyl
Aceto Pan
Additin 30
1-Anilinonaphthalene
C.I. 44050
N-Fenyl-1-aminonaftalen (Czech)
Fenyl-α-naftylamin (Czech)
N-(1-Naphthyl)aniline
Neozon A
Neozone A
PAN
PANA
Phenylnaphthylamine
Phenyl-α-naphthylamine
N-Phenyl-α-naphthylamine
N-Phenyl-1-naphthylamine
α-Phenylnaphthylamine
Vulkanox Pan

90-33-5
$C_{10}H_8O_3$
176.18
O=C(Oc(c(C=1C)ccc2O)c2)C1
Coumarin, 7-hydroxy-4-methyl
2H-1-Benzopyran, 7-hydroxy-4-methyl-2-oxo-
2H-1-Benzopyren-2-one, 7-hydroxy-4-methyl-
Bilcolic
Bilicante
Cantabilin
Cantabiline
Coumarin 4
Eurogale
7-Hydroxy-4-methylcoumarin
7-Hydroxy-4-methyl-2-oxo-2H-1-benzopyran
Hymecromon
Hymecromone
Imecromone
LM 94
Medilla
Mendiaxon
4-Methyl-7-hydroxycoumarin
4-Methylumbelliferon (Czech)
β-Methylumbelliferone
4-Methylumbelliferone
Omega 127
Pilot 447

90-39-1
$C_{15}H_{26}N_2$
234.43
Sparteine, (-)
Lupinidine
7,14-Methano-2H,6H-dipyrido(1,2-a:1',2'-e)(1,5)diazocine, dodecahydro-
6-β,7-α,9-α,11-α-Pachycarpine
Sparteine
(-)-Sparteine
l-Sparteine

90-41-5
$C_{12}H_{11}N$
169.24
Nc(c(c(cccc1)c1)ccc2)c2
2-Biphenylamine
2-Aminobifenyl (Czech)
o-Aminobiphenyl
2-Aminobiphenyl
o-Aminodiphenyl
2-Aminodiphenyl
o-Biphenylamine
(1,1'-Biphenyl)-2-amine (9CI)
o-Phenylaniline
2-Phenylaniline

90-42-6
$C_{12}H_{20}O$
180.32
O=C(C(C(CCCC1)C1)CCC2)C2
(1,1'-Bicyclohexyl)-2-one
2-Cyclohexylcyclohexanone
Lavamenthe

90-43-7
$C_{12}H_{10}O$
170.22
Oc(c(c(cccc1)c1)ccc2)c2
2-Biphenylol
Biphenyl, 2-hydroxy-
(1,1'-Biphenyl)-2-ol
o-Biphenylol
o-Diphenylol
Dowcide 1
Dowicide 1
Dowicide 1 Antimicrobial
2-Fenylfenol (Czech)
2-Hydroxybifenyl (Czech)
o-Hydroxybiphenyl
2-Hydroxybiphenyl
o-Hydroxydiphenyl
2-Hydroxydiphenyl
Kiwi Lustr 277
NCI-C50351
OPP
Orthohydroxydiphenyl
Orthophenylphenol
Orthoxenol
Phenol, o-phenyl-
Phenylphenol
o-Phenylphenol
2-Phenylphenol
Preventol O Extra
Remol TRF
Tetrosin OE
Torsite
Tumescal OPE
USAF EK-2219
o-Xenol

90-44-8
$C_{14}H_{10}O$
194.24
O=C(c(c(ccc1)Cc2cccc3)c1)c23
Anthrone
9(10H)-Anthracenone
Carbothrone
9,10-Dihydro-9-oxoanthracene

90-45-9
$C_{13}H_{10}N_2$
194.25
Nc2c1ccccc1nc3ccccc23
Acridine, 9-amino
9AA
9-Acridinamine (9CI)
Aminacrin
Aminacrine
5-Aminoacridine
9-Aminoacridine
9-Aminoakridin (Czech)
Izoacridina
Monacrin

90-47-1
$C_{13}H_8O_2$
196.21
o(c(c(c(O)c1cccc2)ccc3)c3)c12
Xanthen-9-one
9-Xanthenone

90-50-6
$C_{12}H_{14}O_5$
238.26
O=C(O)C=Cc(cc(OC)c(OC)c1OC)c1
Cinnamic acid, 3,4,5-trimethoxy
o-Methylsinapic acid
2-Propenoic acid, 3-(3,4,5-trimethoxyphenyl)- (9CI)
3,4,5-Trimethoxycinnamic acid
3,4,5-Trimethoxyphenylacrylic acid

90-51-7
$C_{10}H_9NO_4S$
239.25
O=S(=O)(O)c(cc(O)c(c1ccc2N)c2)c1
2-Naphthalenesulfonic acid, 6-amino-4-hydroxy- (9CI)
AI3-19502
6-Amino-4-hydroxy-2-naphthalenesulfonic acid
6-Amino-4-hydroxy-2-naphthalenesulfonic acid (γ acid)
Aminonaphthol sulfonic acid, γ-
C.I. Developer 3
NSC-31508

90-64-2
$C_8H_8O_3$
152.16
O=C(O)C(O)c(cccc1)c1
Mandelic-acid
Amygdalic acid
Amygdalinic acid
Glycolic acid, phenyl-
α-Hydroxyphenylacetic acid
α-Hydroxy-α-toluic acid
Kyselina 2-fenyl-2-hydroxy-

ethanova (Czech)
Kyselina mandlova (Czech)
Paramandelic acid
Phenylglycolic acid
Phenylhydroxyacetic acid
Racemic mandelic acid
α-Toluic acid, α-hydroxy-
Uromaline

90-65-3
$C_8H_{10}O_4$
170.18
COC(=CC(O)=O)C(=O)C(C)=C
2,5-Hexadienoic acid, 3-methoxy-5-methyl-4-oxo
γ-Keto-β-methoxy-δ-methylene-δ$^{\alpha}$-hexenoic acid
Kyselina penicilova (Czech)
3-Methoxy-5-methyl-4-oxo-2,5-hexadienoic acid
PA
Pencillic acid
Penicillic acid

90-66-4
$C_{22}H_{30}O_2S$
358.55
Oc(c(cc(c1)C)C(C)(C)C)c1Sc(c(O)c(cc2C)C(C)(C)C)c2
2,2'-Thiobis(4-methyl-6-tert-butylphenol)
Advastab 406
AI3-63213
CaO 6
p-Cresol, 2,2'-thiobis(6-tert-butyl- (8CI)
NSC-67488
Phenol, 2,2'-thiobis(6-(1,1-dimethylethyl)-4-methyl- (9CI)
Thioalkofen BP
Thioalkophene BP

90-69-7
$C_{22}H_{27}NO_2$
337.50
CN1C(CCCC1CC(=O)c2ccccc2)CC(O)c3ccccc3
Lobeline
8,10-Diphenyllobelionol
2-(6-(β-Hydroxy-phenethyl)-1-methyl-2-piperidyl)acetophenone
2-(6-(2-Hydroxy-2-phenylethyl)-1-methyl-2-piperidinyl)-1-phenylethanone
Inflatine
Lobelin
(-)-Lobeline
α-Lobeline
Lobnico

90-72-2
$C_{15}H_{27}N_3O$
265.45
Oc(c(cc(c1)CN(C)C)CN(C)C)c1CN(C)C
Phenol, 2,4,6-tris(dimethylaminomethyl)
DMP
DMP-30
2,4,6-Tri(dimethylaminomethyl)phenol
2,4,6-Tris-N,N-dimethylaminomethylfenol (Czech)
2,4,6-Tris(dimethylaminomethyl)phenol

90-80-2
$C_6H_{10}O_6$
178.16
O=C(OC(C(O)C1O)CO)C1O
Gluconic acid, δ-lactone, D
D-Gluconic delta-lactone
Gluconolactone
δ Gluconolactone
Glucono-δ-lactone

90-82-4
$C_{10}H_{15}NO$
165.26
OC(c(cccc1)c1)C(NC)C
Pseudoephedrine, l-(+)
Besan
d-psi-Ephedrine
psi-Ephedrine
d-Isoephedrine
α-(1-(Methylamino)ethyl)-benzyl alcohol
d-psi-2-Methylamino-1-phenyl-1-propanol
Pseudoephedrine
(+)-Pseudoephedrine
l-(+)-Pseudoephedrine
Novafed
Sudafed

90-84-6
$C_{13}H_{19}NO$
205.33
CCN(CC)C(C)C(=O)c1ccccc1
Propiophenone, 2-diethylamino
Amfepramone
Amphepramone
Danylen
Derfon
α-Diethylaminopropiophenone
2-(Diethylamino)propiophenone
Diethylpropion
Diethylpropione
Dobesin
Fenyl-(1-diethylaminoethyl)-keton (Czech)
Nopropiophenone
1-Phenyl-2-diethylamino-1-propanone
Regenon
Reginon
Tepanil
Tylinal

90-93-7
$C_{21}H_{28}N_2O$
324.45
O=C(c(ccc(N(CC)CC)c1)c1)c(ccc(N(CC)CC)c2)c2
Benzophenone, 4,4'-bis(diethylamino)- (8CI)
p,p'-Bis(diethylamino)benzophenone
4,4'-Bis(N,N-diethylamino)-benzophenone
Bis(4-(diethylamino)phenyl)-methanone
Methanone, bis(4-(diethylamino)phenyl)- (9CI)
Michler's ethyl ketone
NSC-36365
p,p'-(Tetraethyldiamino)benzophenone
4,4'-(Tetraethyldiamino)benzophenone

90-94-8
$C_{17}H_{20}N_2O$
268.39
O=C(c(ccc(N(C)C)c1)c1)c(ccc(N(C)C)c2)c2
Benzophenone, 4,4'-bis(dimethylamino)
p,p'-Bis(N,N-dimethylamino)-benzophenone
4,4'-Bis(dimethylamino)-benzophenone
Bis(p-(N,N-dimethylamino)-phenyl)ketone
Bis(4-(dimethylamino)phenyl)-methanone
Michler Ketone
Michler's ketone
p,p'-Michler's Ketone
NCI-C02006
Tetramethyldiaminobenzophenone

90-96-0
$C_{15}H_{14}O_3$
242.27
COc1ccc(cc1)C(=O)c2ccc(OC)cc2
Methanone, bis(4-methoxyphenyl)- (9CI)
Benzophenone, 4,4'-dimethoxy- (8CI)
NSC-4191

90-97-1
$C_{13}H_{10}Cl_2O$
253.13
4,4'-Dichlorobenzhydrol
AI3-05090
Benzenemethanol, 4-chloro-α-(4-chlorophenyl)-
NSC-121779

90-98-2
$C_{13}H_8Cl_2O$
251.11
O=C(c(ccc(c1)Cl)c1)c(ccc(c2)Cl)c2
Benzophenone, 4,4'-dichloro
DBP
DCB
p,p'-Dichlorobenzophenone
4,4'-Dichlorobenzophenone

USAF DO-4

90-99-3
$C_{13}H_{11}Cl$
202.68
c(cccc1)(c1)C(c(cccc2)c2)Cl
Chlorodiphenylmethane
AI3-11230
Benzene, 1,1'-(chloromethylene)bis- (9CI)
Benzhydryl chloride
α-Chloroditan
1,1'-(Chloromethylene)bis-benzene
Diphenylchloromethane
Diphenylmethyl chloride
Methane, chlorodiphenyl- (8CI)
NSC-76584

91-01-0
$C_{13}H_{12}O$
184.25
OC(c(cccc1)c1)c(cccc2)c2
Benzhydrol
Benzhydryl alcohol
Benzohydrol
Diphenyl carbinol
Diphenylmethanol
Diphenylmethyl alcohol
Hydroxydiphenylmethane

91-08-7
$C_9H_6N_2O_2$
174.17
O=C=Nc(c(c(N=C=O)cc1)C)c1
Benzene, 2,6-diisocyanato-1-methyl
2,6-Diisocyanato-1-methyl-benzene
2,6-Diisocyanatotoluene
Hylene TCPA
Hylene TIC
Hylene TM
Hylene TM-65
Hylene TRF
Isocyanic acid, 2-methyl-meta-phenylene ester
2-Methyl-meta-phenylene diisocyanate
2-Methyl-meta-phenylene isocyanate
Niax TDI
Niax TDI-P
2,6-TDI
Toluene 2,6-diisocyanate
2,6-Toluene diisocyanate
Tolylene 2,6-diisocyanate
meta-Tolylene diisocyanate

91-10-1
$C_8H_{10}O_3$
154.18
O(c(c(O)c(OC)cc1)c1)C
Phenol, 2,6-dimethoxy
Aldrich
2,6-Dimethoxyphenol
1,3-Dimethyl pyrogallate
2,6-Dwumetoksyfenol (Polish)
Pyrogallol dimethylether
Pyrogallol 1,3-dimethyl ether
Syringol

91-15-6
$C_8H_4N_2$
128.14
N#Cc(c(C#N)ccc1)c1
Phthalonitrile
1,2-Benzendikarbonitril (Czech)
o-Dicyanobenzene
1,2-Dicyanobenzene
Ftalodinitril (Czech)
Ftalonitril (Czech)
o-PDN
Phthalic acid dinitrile
Phthalodinitrile
o-Phthalodinitrile
USAF ND-09

91-16-7
$C_8H_{10}O_2$
138.18
O(c(c(OC)ccc1)c1)C
Benzene, o-dimethoxy
Benzene, 1,2-dimethoxy-
o-Dimethoxybenzene
1,2-Dimethoxybenzene
Dimethylether pyrokatechinu (Czech)
Pyrocatechol dimethyl ether
Veratrol
Veratrole

91-17-8
$C_{10}H_{18}$
138.28
C(C(CCC1)CCC2)(C1)C2
Naphthalene, decahydro
Bicyclo(4.4.0)decane
Dec
Decahydronaphthalene [UN 1147]
Decalin (DOT)
Decalin Solvent
De-Kalin
Dekalina (Polish)
Naphthalane
Naphthane
Perhydronaphthalene
UN 1147 [Decahydronaphthalene]

91-19-0
$C_8H_6N_2$
130.16
n(c(c(nc1)ccc2)c2)c1
Quinoxaline
1,4-Benzodiazine
Benzoparadiazine
Benzo(a)pyrazine
1,4-Diazanaphthalene
Phenopiazine
Phenpiazine
Quinazine
USAF EK-7094

91-20-3
$C_{10}H_8$
128.18
c(c(ccc1)ccc2)(c1)c2
Naphthalene
Camphor Tar
Mighty 150
Mighty RD1
Moth Balls
Moth Flakes
Naftalen (Polish)
Naphthalene (ACGIH,OSHA) [UN 1334]
Naphthalene, Crude or refined [UN 1334]
Naphthalene, Molten [UN 2304]
Naphthalin
Naphthaline
Naphthene
NCI-C52904
RCRA waste number U165
Tar Camphor
UN 1334 [Naphthalene, crude or refined]
UN 2304 [Naphthalene, molten]
White Tar

91-22-5
C_9H_7N
129.17
n(c(c(ccc1)cc2)c1)c2
Quinoline
1-Azanaphthalene
B-500
1-Benzazine
1-Benzine
Benzo(b)pyridine
Chinoleine
Chinolin (Czech)
Chinoline
Leucol
Leucoline
Leukol
Quinoline [UN 2656]
UN 2656 [Quinoline]
USAF EK-218

91-23-6
$C_7H_7NO_3$
153.15
O=N(=O)c(c(OC)ccc1)c1
Anisole, o-nitro
Benzene, 1-methoxy-2-nitro-(9CI)
2-Methoxynitrobenzene
1-Methoxy-2-nitrobenzene
NCI-C60388
o-Nitroanisole [UN 2730]
2-Nitroanisole
o-Nitrophenyl methyl ether
UN 2730 [Nitroanisole]

91-25-8
$C_7H_6O_4S$
186.19
O=Cc(c(S(=O)(=O)O)ccc1)c1
Benzenesulfonic acid, 2-formyl- (9CI)
2-Formylbenzenesulfonic acid

91-29-2
$C_{12}H_{11}N_3O_5S$
309.28
O=S(=O)(O)c(c(Nc(ccc(N)c1)c1)ccc2N(=O)=O)c2
Benzenesulfonic acid, 2-((4-aminophenyl)amino)-5-nitro- (9CI)
2-(p-Aminoanilino)-5-nitrobenzenesulfonic acid
Aniline nitro nerol acid
Benzenesulfonic acid, 2-(p-aminoanilino)-5-nitro-(8CI)
NSC-5534
Urese

91-33-8
$C_{15}H_{14}ClN_3O_4S_3$
431.95
NS(=O)(=O)c3cc2c(N=C(CSCc1ccccc1)NS2(=O)=O)cc3Cl
2H-1,2,4-Benzothiadiazine-7-sulfonamide, 3-((benzylthio)menthyl)-6-chloro-, 1,1-dioxide
Aquatag
Benzothiazide
2H-1,2,4-Benzothiadiazine-7-sulfonamide, 6-chloro-3-(((phenylmethyl)thio)-methyl)-, dioxide,
Benzthiazide
3-((Benzylthio)methyl)-6-chloro-1,2,4-benzothiadiazine-7-sulfonamide 1,1-dioxide
3-Benzylthiomethyl-6-chloro-2H-1,2,4-benzothiadiazine-7-sulfonamide 1,1-dioxide
3-Benzylthiomethyl-6-chloro-7-sulfamoyl-1,2,4-benzothiadiazine 1,1-dioxide
3-Benzylthiomethyl-6-chloro-7-sulfamoyl-1,2,4-benzothiadiazine 1,1-dioxide
Edemex
Exna
Exosalt
Fovane
Freeuril
Naclex
P 1393
Pfizer 1393

91-34-9
$C_{26}H_{20}N_4O_8S_2$
580.57
O=S(=O)(O)c(c(ccc1N=Nc(ccc(O)c2)c2)C=Cc(c(S(=O)(=O)O)cc(N=Nc(ccc(O)c3)c3)c4)c4)c1
C.I. Direct Yellow 4
Benzenesulfonic acid, 2,2'-(1,2-ethenediyl)bis(5-((4-hydroxyphenyl)azo)- (9CI)
Brilliant Yellow
Direct Yellow 4 Dye

91-40-7
$C_{13}H_{11}NO_2$
213.25
O=C(O)c(c(Nc(cccc1)c1)ccc2)c2
Anthranilic acid, N-phenyl
o-Anilinobenzoic acid
2-Anilinobenzoic acid
Benzoic acid, 2-(phenylamino)-(9CI)
2-Carboxydiphenylamine
Diphenylamine-2-carboxylic acid
Fenamic acid
PA
2-(Phenylamino)benzoic acid
Phenylanthranilic acid
N-Phenylanthranilic acid

91-43-0
$C_{10}H_{12}ClNO_4$
245.66
O=N(=O)c(c(OCC)cc(c1OCC)Cl)c1
Benzene, 1-chloro-2,5-diethoxy-4-nitro- (9CI)
1-Chloro-2,5-diethoxy-4-nitrobenzene
5-Chloro-2-nitro-p-diethoxybenzene
2,5-Diethoxy-4-nitrochlorobenzene
NSC-60284

91-44-1
$C_{14}H_{17}NO_2$
231.32
O=C(Oc(c(C=1C)ccc2N(CC)CC)c2)C1
Coumarin, 7-diethylamino-4-methyl
2H-1-Benzopyran-2-one, 7-(diethylamino)-4-methyl- (9CI)
Coumarin 1
7-Diethylamino-4-methylcoumarin

91-49-6
$C_{12}H_{17}NO$
191.30
O=C(N(c(cccc1)c1)CCCC)C
Acetanilide, N-butyl
Acetamide, N-butyl-N-phenyl-(9CI)
BAA
Butylacetanilide
N-Butylacetanilide

91-52-1
$C_9H_{10}O_4$
182.18
O=C(O)c(c(OC)cc(OC)c1)c1
Benzoic acid, 2,4-dimethoxy-(9CI)
AI3-27498
2,4-Dimethoxybenzoic acid
NSC-6316

91-53-2
$C_{14}H_{19}NO$
217.34
O(c(ccc(NC(C=C1C)(C)C)c12)c2)CC
Quinoline, 6-ethoxy-1,2-dihydro-2,2,4-trimethyl
1,2-Dihydro-6-ethoxy-2,2,4-trimethylquinoline
1,2-Dihydro-2,2,4-trimethyl-6-ethoxyquinoline
EMQ
EQ
Ethoxychin (Czech)
6-Ethoxy-1,2-dihydro-2,2,4-trimethylquinoline
Ethoxyquin
Ethoxyquine
6-Ethoxy-2,2,4-trimethyl-1,2-dihydroquinoline
Niflex
Nix-Scald
Santoflex A
Santoflex AW
Santoquin
Santoquine
Stop-Scald
2,2,4-Trimethyl-6-ethoxy-1,2-dihydroquinoline
USAF B-24

91-56-5
$C_8H_5NO_2$
147.14
O=C(Nc(c1ccc2)c2)C1=O
Indole-2,3-dione
o-Aminobenzoylformic anhydride
2,3-Diketoindoline
2,3-Dioxoindoline
2,3-Indolinedione
Isatic acid lactam
Isatin
Isatinic acid anhydride
2,3-Ketoindoline
Pseudoisatin

91-57-6
$C_{11}H_{10}$
142.21
c(c(ccc1C)ccc2)(c2)c1
Naphthalene, 2-methyl
β-Methylnaphthalene
2-Methylnaphthalene

91-58-7
$C_{10}H_7Cl$
162.62
c(c(ccc1Cl)ccc2)(c2)c1
Naphthalene, 2-chloro
2-Chlornaftalen (Czech)
β-Chloronaphthalene
2-Chloronaphthalene
RCRA waste number U047

91-59-8
$C_{10}H_9N$

143.20
c(c(ccc1N)ccc2)(c2)c1
2-Naphthylamine
2-Aminonaftalen (Czech)
2-Aminonaphthalene
β-Naftyloamina (Polish)
C.I. 37270
Fast Scarlet Base B
β-Naftalamin (Czech)
β-Naftilamina (Italian)
2-Naftylamin (Czech)
2-Naftylamine (Dutch)
β-Naftyloamina (Polish)
2-Naphthalamine
2-Naphthalenamine
β-Naftylamin (Czech)
β-Naphthylamin (German)
2-Naphthylamin (German)
β-Naphthylamine (ACGIH, OSHA) [UN 1650]
6-Naphthylamine
2-Naphthylamine mustard
RCRA waste number U168
UN 1650 [β-Naphthylamine]
USAF CB-22

91-60-1
$C_{10}H_8S$
160.24
Sc(ccc(c1ccc2)c2)c1
2-Naphthalenethiol
β-Mercaptonaphthalene
2-Mercaptonaphthalene
β-Naphthalenethiol
Naphthalene-2-thiol
β-Naphthyl mercaptan
2-Naphthyl mercaptan
2-Naphthyl thiol
Renacit 1
RPA 2
RPA No. 2
Thionaphthol
Thio-β-naphthol
β-Thionaphthol
2-Thionaphthol
USAF CY-4
Vulcamel TBN

91-61-2
$C_{10}H_{13}N$
147.21
N(c(c(cc(c1)C)CC2)c1)C2
Quinoline, 1,2,3,4-tetrahydro-6-methyl- (9CI)
AI3-36188
Civettal
6-Methyl-1,2,3,4-tetrahydroquinoline
NSC-65606
1,2,3,4-Tetrahydro-6-methylquinoline

91-62-3
$C_{10}H_9N$
143.20
n(c(c(cc(c1)C)cc2)c1)c2
Quinoline, 6-methyl
6-Methylquinoline
p-Methylquinoline
p-Toluquinoline

91-63-4
$C_{10}H_9N$
143.20
n(c(c(ccc1)cc2)c1)c2C
Quinaldine
Chinaldine
2-Methylchinolin (Czech)
2-Methylquinoline
Quinoline, 2-methyl-

91-64-5
$C_9H_6O_2$
146.15
O=C(Oc(c(C=1)ccc2)c2)C1
Coumarin
2H-1-Benzopyran-2-one
2H-1-Benzopyran, 2-oxo-
Benzo-α-pyrone
1,2-Benzopyrone
Cinnamic acid, o-hydroxy-, δ-lactone
cis-o-Coumarinic acid lactone
Coumarinic anhydride
Cumarin
o-Hydroxycinnamic acid lactone
o-Hydroxycinnamic lactone
o-Hydroxyzimtsaure-lacton (German)
Kumarin (Czech)
NCI-C07103
2-Oxo-1,2-benzopyran
Rattex
Tonka Bean Camphor

91-65-6
$C_{10}H_{21}N$
155.28
N(C(CCCC1)C1)(CC)CC
Cyclohexanamine, N,N-diethyl- (9CI)
Cyclohexylamine, N,N-diethyl- (8CI)
Cyclohexyldiethylamine
N,N-Diethylcyclohexanamine
Diethylcyclohexylamine
N,N-Diethylcyclohexylamine
NSC-5313

91-66-7
$C_{10}H_{15}N$
149.26
N(c(cccc1)c1)(CC)CC
Aniline, N,N-diethyl
Benzenamine, N,N-diethyl- (9CI)
DEA
N,N-Diethylaminobenzene
Diaethylanilin (German)
N,N-Diethylanilin (Czech)
Diethylaniline
N,N-Diethylaniline [UN 2432]
Diethylphenylamine
UN 2432 [N,N-Diethyl aniline]

91-67-8
$C_{11}H_{17}N$
163.26
N(c(cccc1C)c1)(CC)CC
Benzenamine, N,N-diethyl-3-methyl- (9CI)
AI3-28462
1-(Diethylamino)-3-methylbenzene
3-(Diethylamino)-1-methylbenzene
3-(Diethylamino)toluene
3-(N,N-Diethylamino)toluene
N,N-Diethyl-3-methylaniline
N,N-Diethyl-3-methylbenzenamine
N,N-Diethyl-m-toluidine
N,N-Diethyl-m-toluidinium ion
m-Methyl-N,N-diethylaniline
NSC-96629
m-Toluidine, N,N-diethyl- (8CI)

91-68-9
$C_{10}H_{15}NO$
165.26
Oc(cccc1N(CC)CC)c1
Phenol, m-(diethylamino)
m-(Diethylamino)phenol
3-(Diethylamino)phenol
Phenol, 3-(diethylamino)- (9CI)

91-76-9
$C_9H_9N_5$
187.23
n(c(nc(n1)c(cccc2)c2)N)c1N
s-Triazine, 2,4-diamino-6-phenyl
Benzoguanamine
Benzoguanimine
2,4-Diamino-6-phenyl-s-triazine
4,6-Diamino-2-phenyl-s-triazine
2-Phenyl-4,6-diamino-s-triazine
2-Phenyl-4,6-diamino-1,3,5-triazine
USAF RH-5

91-78-1
$C_{21}H_{21}N_3$
315.40
N(c(cccc1)c1)(CN(c(cccc2)c2)CN3c(cccc4)c4)C3
s-Triazine, hexahydro-1,3,5-triphenyl- (8CI)
Anhydroformaldehyde aniline (VAN)
Hexahydro-1,3,5-triphenyl-1,3,5-triazine
NSC-9419
1,3,5-Triazine, hexahydro-1,3,5-triphenyl- (9CI)
1,3,5-Triphenylhexahydro-s-triazine
1,3,5-Triphenylhexahydro-1,3,5-triazine

91-79-2
$C_{14}H_{19}N_3S$
261.42

Pyridine, 2-((2-(dimethyl-amino)ethyl)-3-thenylamino)
Diethylandiamine
N-(2-Dimethylaminoethyl)-N-2-pyridyl-3-thenylamine
2-((2-Dimethylaminoethyl)-3-thenylamino)pyridine
Methapyrilene
N-(α-Pyridyl)-N-(β-thenyl)-N',N'-dimethylethylene-diamine
NCI-C60640
Tenfidil
Thefanil
Thenfadil
Thenyldiamine
WIN-2848
Thenylpyramine
Thionylan

91-80-5
$C_{14}H_{19}N_3S$
261.42
CN(C)CCN(Cc1cccs1)c2ccccn2
Pyridine, 2-((2-(dimethyl-amino)ethyl)-2-thenylamino)
A 3322
AH-42
2-((2-(Dimethylamino)ethyl)-2-thenylamino)pyridine
N,N-Dimethyl-N'-pyrid-2-yl-N'-2-thenylethylenediamine
N,N-Dimethyl-N'-2-pyridinyl-N'-(2-thienylmethyl)-1,2-ethanediamide
Dormin
Histadyl
Lulamin
Lullamin
Methapyrilene
NCI-C55550
Paradormalene
Pyrathyn
N-(α-Pyridyl)-N-(α-thenyl)-N',N'-dimethylethylene-diamine
Pyrinistab
Pyrinistol
RCRA waste number U155
Rest-On
Restryl
Semikon
Sleepwell
Tenalin
Thenylene

91-81-6
$C_{16}H_{21}N_3$
255.40
n(c(N(Cc(cccc1)c1)CCN(C)C)ccc2)c2
Pyridine, 2-(benzyl(2-(di-methylamino)ethyl)amino)
Benzoxale
2-(Benzyl(2-dimethylamino-ethyl)amino)pyridine
2-(N-Benzyl-N-(2-dimethyl-aminoethyl)amino)pyridine
N-Benzyl-N',N'-dimethyl-N-2-pyridylethylenediamine
Benzyl-(α-pyridyl)-dimethyla-ethylendiamin (German)
N-Benzyl-N-(2-pyridyl)-N',N'-dimethyl ethylenediamine
Cizaron
Dehistin
β-Dimethylaminoethyl-2-pyrid-ylaminotoluene
β-Dimethylaminoethyl-2-pyrid-ylbenzylamine
N,N-Dimethyl-N'-benzyl-N'-(α-pyridyl)ethylenediamine
N,N-Dimethyl-N'-benzyl-N'-(2-pyridyl)ethylenediamine
Ethylenediamine, N-benzyl-N',N'-dimethyl-N-(2-pyridyl)-
NCI-C60662
PBZ
Piribenzil
Pyribenzamine
Pyrinamine Base
Resistamine
2750 R.P.
Tonaril
Tripelenamine
Tripelennamina (Italian)
Tripelennamine

91-84-9
$C_{17}H_{23}N_3O$
285.43
O(c(ccc(c1)CN(c(nccc2)c2)CCN(C)C)c1)C
Pyridine, 2-((2-(dimethyl-amino)ethyl)(p-methoxy-benzyl)amino)
Afko-Hist
Anhistabs
Anhistol
Antalergan
Mepyramin (German)
Antallergan
Antamine
Anthisan
Copsamine
Coradon
2-((2-(Dimethylamino)ethyl)-(p-methoxybenzyl)amino)pyridine
Dipane
Dorantamin
Enrumay
Harvamine
Histan
Histacap
Histalon
Histapyran
Histasan
Isamin
Kriptin
Maranhist
Minihist
Mepiramine
Mepyramine
Mepyren
N-p-Methoxybenzyl-N',N'-di-methyl-N-α-pyridylethylene-diamine
N-(p-Methoxybenzyl)-N',N'-di-methyl-N-2-pyridylethylene-diamine
p-Methoxybenzyl-α-pyridyl-di-methyl-aethylendiamin (German)
NCI-C60651
Neoantergan
Neobridal
Nyscaps
Paraminyl
Parmal
Pyra
Pymafed
Pyramal
Pyrilamine
Pyranisamine
Pyrilamine
R.D. 2786
RP 2786
Stamine
Stangen
Statomin
Thylogen
Wait's Green Mountain Anti-histamine

91-85-0
$C_{16}H_{22}N_4O$
286.42
COc2ccc(CN(CCN(C)C)c1ncccn1)cc2
Pyrimidine, 2-((2-(dimethyl-amino)ethyl)(p-methoxy-benzyl)amino)
2-((2-(Dimethylamino)ethyl)-(p-methoxybenzyl)amino)-pyrimidine
N,N-Dimethyl-N'-(p-methoxy-benzyl)-N'-(2-pyrimidyl)-ethylenediamine
Ethylenediamine, N-(p-meth-oxybenzyl)-N',N'-dimethyl-N-2-pyrimidinyl-
NCI-C60708
Neohetramine
Thonzylamine
Tonzilamine

91-88-3
$C_{11}H_{17}NO$
179.29
OCCN(c(cccc1C)c1)CC
Ethanol, 2-(N-ethyl-m-tolui-dino)
Emery 5714
N-Ethyl-N-(2-hydroxyethyl)-m-toluidine
2-(N-Ethyl-m-toluidino)ethanol
N-Hydroxyethyl-N-ethyl-m-toluidine

91-92-9
$C_{36}H_{28}N_2O_6$
584.62
O=C(Nc(c(OC)cc(c(ccc(NC(=O)c(c(O)cc(c1ccc2)c2)c1)c3OC)c3)c4)c4)c(c(O)cc(c5ccc6)c6)c5
2-Naphthalenecarboxamide, N,N'-(3,3'-dimethoxy(1,1'-bi-phenyl)-4,4'-diyl)bis-

(3-hydroxy- (9CI)
Acco Naphthol AS-BR
Amanil Naphthol AS-BR
Amarthol AS-BR
Azoic Coupling Component 3
Azotol DA
4',4'''-Bi-2-naphth-o-anisidide, 3,3''-dihydroxy- (8CI)
C.I. Azoic Coupling Component 3
C.I. 37575
Hiltonaphthol AS-BR
Naphtanilide BR
Naphthol AS-BR
Naphtol AS-BR
NSC-37224
Sanatol BR

91-93-0
$C_{16}H_{12}N_2O_4$
296.30
COc1cc(ccc1N=C=O)c2ccc(N=C=O)c(OC)c2
Isocyanic acid, 3,3'-dimethoxy-4,4'-biphenylene ester
1,1'-Biphenyl, 4,4'-diisocyanato-3,3'-dimethoxy-
Dianisidine diisocyanate
4,4'-Diisocyanato-3,3'-dimethoxy-1,1'-biphenyl
3,3'-Dimethoxybenzidine-4,4'-diisocyanate
3,3'-Dimethoxy-4,4'-biphenylene diisocyanate
3,3'-Dimethoxy-4,4'-biphenylylene isocyanate
3,3'-Dimethoxy-4,4'-biphenylylene isocyanic acid ester
NCI-C02175

91-94-1
$C_{12}H_{10}Cl_2N_2$
253.14
Nc(c(cc(c(ccc(N)c1Cl)c1)c2)Cl)c2
Benzidine, 3,3'-dichloro
(1,1'-Biphenyl)-4,4'-diamine, 3,3'-dichloro-
C.I. 23060
Curithane C126
DCB
4,4'-Diamino-3,3'-dichlorobiphenyl
4,4'-Diamino-3,3'-dichlorodiphenyl
3,3'-Dichlorbenzidin (Czech)
3,3'-Dichlorobenzidina (Spanish)
Dichlorobenzidine
o,o'-Dichlorobenzidine
3,3'-Dichlorobenzidine
3',3'-Dichlorobenzidine (ACGIH)
Dichlorobenzidine Base
3,3'-Dichlorobenzidine (OSHA)
3,3'-Dichloro-4,4'-biphenyldiamine
3,3'-Dichlorobiphenyl-4,4'-diamine
3,3'-Dichloro-4,4'-diaminobiphenyl
3,3'-Dichloro-4,4'-diamino-(1,1-biphenyl)
RCRA waste number U073

91-95-2
$C_{12}H_{14}N_4$
214.30
Nc(c(N)cc(c(ccc(N)c1N)c1)c2)c2
3,3'4,4'-Biphenyltetramine
3,3'-Diaminobenzidene
3,3',4,4'-Diphenyltetramine
3,3',4,4'-Tetraaminobiphenyl

91-96-3
$C_{22}H_{24}N_2O_4$
380.43
O=C(Nc(c(cc(c(ccc(NC(=O)CC(=O)C)c1C)c1)c2)C)c2)CC(=O)C
Picramic acid
Acna Naphthol G
Amanil Naphthol AS-G
Amarthol AS-G
Arlanthol ASG
4',4'''-Bi-o-acetoacetotoluidide (8CI)
4,4'-Biphenyldiamine, N,N'-diacetoacetyl-3,3'-dimethyl-
4,4'-Bis(acetoacetamido)-3,3'-dimethylbiphenyl
p,p'-Bis(o-acetoacetotoluidide)
4,4'-Bis(o-acetoacetotoluidide)
4,4'-Bis(acetoacetylamino)-3,3'-dimethylbiphenyl
N,N'-Bis(acetoacetyl)-3,3'-dimethylbenzidine
Butanamide, N,N'-(3,3'-dimethyl(1,1'-biphenyl)-4,4'-diyl)bis(3-oxo- (9CI)
C.I. Azoic Coupling Component 5
C.I. 37610
Cibanaphthol AG
Daito Grounder G
Diacetylacetotolidide
3,3'-Dimethyl-4,4'-bis(acetoacetylamino)biphenyl
Hiltonaphthol AS-G
Kambothol ASG
Kiwa Grounder G
Mitsui Naphthozol G
Naphtanilide G
Naphtazol J
Naphthanil G
Naphthoide G
Naphthol AS-G
Naphthol AS-G Dispersible
Naphthol AS-G Supra
Naphtoelan G
Naphtol AS-G
Naphtol AS-G Supra
NSC-37212
Sanatol AS-G
Sanatol G
Solunaptol YL
Tulathol AS-G
Ultrazol G

91-97-4
$C_{16}H_{12}N_2O_2$
264.30
O=C=Nc(c(cc(c(ccc(N=C=O)c1C)c1)c2)C)c2
1,1'-Biphenyl, 4,4'-diisocyanato-3,3'-dimethyl
3,3'-Dimethyl-4,4'-biphenylene diisocyanate
Isocyanic acid, 3,3'-dimethyl-4,4'-biphenylene ester

91-99-6
$C_{11}H_{17}NO_2$
195.29
OCCN(c(cccc1C)c1)CCO
Ethanol, 2,2'-(m-tolylimino)di
N,N-Bis(β-hydroxyethyl)-3-methylaniline
N,N-Bis(2-hydroxyethyl)-3-methylaniline
N,N-Bis(2-hydroxyethyl)-m-toluidine
Diethanol-m-toluidine
N,N-Dihydroxyethyl-m-toluidine
Emery 5709
Ethanol, 2,2'-((3-methylphenyl)-imino)bis- (9CI)
m-Toluidine, N,N-bis(2-hydroxyethyl)-
m-Tolyldiethanolamine
2,2'-(m-Tolylimino)diethanol

92-00-2
$C_{10}H_{14}ClNO_2$
215.70
OCCN(c(cccc1Cl)c1)CCO
Ethanol, 2,2'-((3-chlorophenyl)imino)bis
N,N-Bis(2-hydroxyethyl)chloroanilide
Diethanolaminochlorobenzene
N,N-Diethanolanilide, 3-chloro-
Diethanolchloroanilide
N-(m-Chlorophenyl)diethanolamine
N-(3-Chlorophenyl)diethanolamine
N,N-Dihydroxyethyl-m-chloroaniline
N,N-Dihydroxyethyl-3-chloroaniline
Emery 5715
Emery 5717

92-04-6
$C_{12}H_9ClO$
204.66
Oc(c(cc(c(cccc1)c1)c2)Cl)c2
4-Biphenylol, 3-chloro
3-Chlor-4-hydroxybifenyl (Czech)
3-Chloro-4-hydroxybiphenyl
3-Chloro-4-hydroxydiphenyl
2-Chloro-4-phenylphenol
Dowicide 4
Phenol, 2-chloro-4-phenyl-
4-Phenyl-2-chlorophenol

92-05-7
$C_{12}H_{10}O_2$
186.21
Oc(c(O)cc(c(cccc1)c1)c2)c2
(1,1'-Biphenyl)-3,4-diol (9CI)
AI3-17381
1,2-Benzenediol, 4-phenyl-
3,4-Biphenyldiol (8CI)
3,4-Dihydroxybiphenyl
NSC-1267
4-Phenyl-1,2-benzenediol
4-Phenylcatechol
4-Phenylpyrocatechol

92-06-8
$C_{18}H_{14}$
230.32
c(c(cccc1)c1)(cccc2c(cccc3)c3)c2
m-Terphenyl
m-Diphenylbenzene
Isodiphenylbenzene
Santowax M
1,3-Terphenyl
m-Triphenyl

92-11-5
$C_{12}H_{19}NO_2$
209.28
OCC(O)CN(c(cccc1C)c1)CC
1,2-Propanediol, 3-(ethyl-(3-methylphenyl)amino)-(9CI)
3-(Ethyl(3-methylphenyl)-amino)-1,2-propanediol

92-13-7
$C_{11}H_{16}N_2O_2$
208.29
O=C(OCC1CC(N(C=N2)C)=C2)C1CC
Pilocarpine
Almocarpine
(3S-cis)-3-Ethyldihydro-4-((1-methyl-1H-imidazol-5-yl)-methyl)-2(3H)-furanone
Imidazole-5-butyric acid, α-ethyl-β-(hydroxymethyl)-1-methyl-, γ-lactone
Pilocarpin
Pilocarpol

92-15-9
$C_{11}H_{13}NO_3$
207.25
O=C(Nc(c(OC)ccc1)c1)CC(=O)C
o-Anisidine, acetoacetyl
o-Acetoacetanisidide (8CI)
Acetoacetic acid o-anisidide
2-Acetoacetylaminoanisole
Acetoacetyl-o-aniside
Acetoacetyl-o-anisine
Acetoacet-o-anisidin (Czech)
Butanamide, N-(2-methoxy-phenyl)-3-oxo- (9CI)
o-Methoxyacetoacetanilide
2-Methoxyacetoacetanilide
2'-Methoxyacetoacetanilide
2-Methoxyanilid kyseliny acetoctove (Czech)

92-24-0
$C_{18}H_{12}$
228.30
c(c(cc(c1ccc2)c2)cc(c3ccc4)c4)(c1)c3
Naphthacene
Benz(b)anthracene
2,3-Benzanthracene
2,3-Benzanthrene
Chrysogen
Rubene
Tetracene
Tetracene (hydrocarbon)

92-27-3
$C_{10}H_8O_5S$
240.24
O=S(=O)(O)c(ccc(c1cc(O)c2O)c2)c1
2-Naphthalenesulfonic acid, 6,7-dihydroxy- (9CI)
6,7-Dihydroxy-2-naphthalene-sulfonic acid

92-36-4
$C_{14}H_{12}N_2S$
240.34
N(c(c(S1)cc(c2)C)c2)=C1c(ccc(N)c3)c3
Benzothiazole, 2-(p-amino-phenyl)-6-methyl
2-(p-Aminophenyl)-6-methyl-benzothiazole
Benzenamine, 4-(6-methyl-2-benzothiazolyl)- (9CI)
Dehydro-p-toluidine
DHPT
p-(6-Methylbenzothiazol-2-yl)-aniline
4-(6-Methyl-2-benzothiazolyl)-benzenamine

92-40-0
$C_{10}H_8O_4S$
224.24
O=S(=O)(O)c(ccc(c1cc(O)c2)c2)c1
2-Naphthalenesulfonic acid, 7-hydroxy- (9CI)
Cassella's Acid
F Acid
2-Hydroxynaphthalene-7-sulfonic acid
7-Hydroxy-2-naphthalenesulfonic acid
Mono Acid F
Mono F Acid
Monosulfonic acid F
β-Naphthol-7-sulfonic acid
2-Naphthol, 7-sulfo-
2-Naphthol-7-sulfonic acid
NSC-1704

92-43-3
$C_9H_{10}N_2O$
162.21
O=C(NN(c(cccc1)c1)C2)C2
3-Pyrazolidinone, 1-phenyl
Phenidone
1-Phenyl-3-oxopyrazolidine
1-Phenyl-3-pyrazolidinone
1-Phenyl-3-pyrazolidone

92-44-4
$C_{10}H_8O_2$
160.18
Oc(c(O)cc(c1ccc2)c2)c1
2,3-Naphthalenediol

92-48-8
$C_{10}H_8O_2$
160.18
O=C(Oc(c(C=1)cc(c2)C)c2)C1
Coumarin, 6-methyl
2H-1-Benzopyran-2-one, 6-methyl-
6-MC
6-Methylbenzopyrone
6-Methyl-1,2-benzopyrone
6-Methylcoumarin
6-Methylcoumarinic anhydride
NCI-C55812
Toncarine

92-49-9
$C_{10}H_{14}ClN$
183.70
N(c(cccc1)c1)(CCCl)CC
Aniline, N-(2-chloroethyl)-N-ethyl
Benzenamine, N-(2-chloro-ethyl)-N-ethyl- (9CI)
N-(2-Chloroethyl)-N-ethyl-aniline
Emery 5770
Ethyl(chloroethyl)aniline

92-50-2
$C_{10}H_{15}NO$
165.23
OCCN(c(cccc1)c1)CC
N-Phenyl-N-ethylethanolamine
AI3-01463
Ethanol, 2-(N-ethylanilino)-(8CI)
Ethanol, 2-(ethylphenylamino)-(9CI)
N-Ethylanilinoethanol
2-(N-Ethylanilino)ethanol
β-(Ethylanilino)ethyl alcohol
Ethyl(β-hydroxyethyl)aniline
N-Ethyl(β-hydroxyethyl)aniline
N-Ethyl-N-(hydroxyethyl)aniline
N-Ethyl-N-(β-hydroxyethyl)-aniline
N-Ethyl-N-(2-hydroxyethyl)-aniline
N-Ethyl-N-phenylaminoethanol
2-(Ethylphenylamino)ethanol
2-(N-Ethyl-N-phenylamino)-

ethanol
Ethylphenylethanolamine
N-Ethyl-N-phenylethanolamine
Hydroxyethylethylaniline
N-(2-Hydroxyethyl)-N-ethyl-aniline
NSC-7485
Phenylethylethanolamine

92-51-3
$C_{12}H_{22}$
166.31
C(C(CCCC1)C1)(CCCC2)C2
Bicyclohexyl (8CI)
AI3-01174
Bicyclohexane
1,1'-Bicyclohexyl (9CI)
1,1'-Biphenyl, dodecahydro-
Cyclohexane, cyclohexyl-
Cyclohexylcyclohexane
Dicyclohexane
Dicyclohexyl
Dodecahydrobiphenyl
NSC-59855

92-52-4
$C_{12}H_{10}$
154.22
c(c(cccc1)c1)(cccc2)c2
Biphenyl (ACGIH,OSHA)
Bibenzene
1,1'-Biphenyl
Carolid AL
Diphenyl
Diphenyl (OSHA)
Lemonene
Phenador-X
Phenylbenzene
Phph
Tetrosin LY
Xenene

92-53-5
$C_{10}H_{13}NO$
163.21
O(CCN(c(cccc1)c1)C2)C2
4-Phenylmorpholine
AI3-01091
Morpholine, 4-phenyl- (8CI, 9CI)
NSC-2628
Phenyl morpholine
Phenylmorpholine
N-Phenylmorpholine

92-54-6
$C_{10}H_{14}N_2$
162.26
N(c(cccc1)c1)(CCNC2)C2
Piperazine, 1-phenyl
1-Fenylpiperazin (Czech)
N-Phenylpiperazine
1-Phenylpiperazine

92-59-1
$C_{15}H_{17}N$
211.33
N(c(cccc1)c1)(Cc(cccc2)c2)CC
Benzylamine, N-ethyl-N-phenyl
Benzylethylphenylamine
Ethylbenzylaniline
N-Ethyl-N-benzylaniline [UN 2274]
Phenylethylbenzylamine
UN 2274 [N-Ethyl-N-benzyl-aniline]

92-62-6
$C_{13}H_{11}N_3$
209.27
n(c(c(cccc1N)cc2ccc(N)c3)c1)c23
Acridine, 3,6-diamino
3,6-Acridinediamine (9CI)
2,8-Diaminoacridine (European)
3,6-Diaminoacridine
2,8-Diaminoacridinium
3,6-Diaminoacridinium
3,7-Diamino-5-azaanthracene
Isoflav Base
Proflavin
Proflavine
Profoliol
Profoliol-B
Proformiphen
Profundol
Profura
Progarmed
Pro-Gen
Progesic

92-64-8
$C_{11}H_{14}N_2O$
190.27
N#CCCN(c(cccc1)c1)CCO
Propionitrile, 3-(N-(2-hydroxyethyl)anilino)
Aniline, N-(β-cyanoethyl)-N-(β-hydroxyethyl)-
N-β-Cyanoethyl-N-β-hydroxy-ethylaniline
Emery 5724
N-β-Hydroxyethyl-N-β-kyan-ethylanilin (Czech)
N-2-Hydroxyethyl-N-2-kyan-ethylanilin (Czech)
Propionitrile, 3-((2-hydroxy-ethyl)phenylamino)-

92-66-0
$C_{12}H_9Br$
233.12
c(ccc(c(cccc1)c1)c2)(c2)Br
Biphenyl, 4-bromo

92-67-1
$C_{12}H_{11}N$
169.24
Nc(ccc(c(cccc1)c1)c2)c2
4-Biphenylamine
4-Aminobifenyl (Czech)
p-Aminobiphenyl
4-Aminobiphenyl
4-Aminodifenil (Spanish)
p-Aminodiphenyl
4-Aminodiphenyl (ACGIH, OSHA)
4-Bifenylamin (Czech)
Biphenylamine
(1,1'-Biphenyl)-4-amine
p-Biphenylamine
Paraaminodiphenyl
p-Phenylaniline
Xenylamin (Czech)
Xenylamine

92-69-3
$C_{12}H_{10}O$
170.22
Oc(ccc(c(cccc1)c1)c2)c2
4-Biphenylol
p-Hydroxybiphenyl
4-Hydroxybiphenyl
p-Hydroxydiphenyl
4-Hydroxydiphenyl
Paraxenol
p-Phenylphenol
4-Phenylphenol

92-70-6
$C_{11}H_8O_3$
188.19
O=C(O)c(c(O)cc(c1ccc2)c2)c1
2-Naphthoic acid, 3-hydroxy
Bon
Bona
Bon acid
C.I. Developer 8
C.I. Developer 20 (Obs.)
Developer Bon
3-Hydroxy-2-naphthoic acid
Kyselina 3-hydroxy-2-naftoova (Czech)
Miketazol Developer ONS
2-Naphthalenecarboxylic acid, 3-hydroxy- (9CI)
Naphthol B.O.N.

92-74-0
$C_{19}H_{17}NO_3$
307.34
O=C(Nc(c(OCC)ccc1)c1)c(c(O)cc(c2ccc3)c3)c2
2-Naphthalenecarboxamide, N-(2-ethoxyphenyl)-3-hydroxy- (9CI)
Acco Naf-Sol AS-phenyl
Acco Naphthol AS-phenyl
Acna Naphthol OF
Amanil Naphthol AS-phenyl
Amarthol AS-phenyl
Azotol OF
C.I. Azoic Coupling Component 14
C.I. 37558
Daito Grounder Phenyl
2'-Ethoxy-3-hydroxy-2-naphthanilide
Hiltonaphthol AS-phenyl
Kambothol ASPH
Naphtanilide Phenyl
Naphtanilide Phenyl Supra
Naphtazol OP
Naphthanil OP

Naphthol AS-OP
Naphthol AS-PH
Naphthol AS-phenyl
Naphthol AS-phenyl supra
Naphthol AS-RO
Naphtoelan Phenyl
Naphtol AS-phenyl
2-Naphtho-o-phenetidide, 3-hydroxy- (8CI)
NSC-50681
Tulathol AS-phenyl

92-77-3
$C_{17}H_{13}NO_2$
263.29
O=C(Nc(cccc1)c1)c(c(O)cc(c2ccc3)c3)c2
Naphthol AS
Acco Naphthol AS
Acna Naphthol C
Amanil Naphthol AS
Amarthol AS
Anthonaphthol AS
Azoground AS
Azonaphtol A
Azotol A
Brenthol AS
C.I. Azoic Coupling Component 2
C.I. 37505
Celcot RF
Cibanaphthol RF
Diathol AS
Diathol ASF
Dragonthol A
Hiltonaphthol AS
2-Hydroxy-3-naphthalene-carboxanilide
3-Hydroxy-2-naphthalene-carboxanilide
2-Hydroxy-3-naphthanilide
β-Hydroxynaphthoic anilide
2-Hydroxy-3-naphthoic acid anilide
2-Hydroxy-3-naphthoic anilide
3-Hydroxy-2-naphthoylanilide
3-Hydroxy-N-phenyl-2-naphthalenecarboxamide
Kambothol AS
Lake Developer A
Mitsui Naphthozol AS
Naftoelan A
Naftolo MM
Naphtanilide OL Supra
Naphtanilide RC
Naphtanilide RC Supra
Naphtazol A
2-Naphthalenecarboxamide, 3-hydroxy-N-phenyl- (9CI)
Naphthanil AS
2-Naphthanilide, 3-hydroxy- (8CI)
Naphthoide AS
Naphthol ACNA C
Naphthol AS-A
Naphthol AS Supra
Naphtholate AS
Naphtholate AS Soln
Naphtoelan A
Naphtol AS
Naphtolate AS Soln
NSC-45173
3-(N-Phenylcarbamoyl)-2-naphthol
Solunaptol A
Tulathol AS
Ultrazol I-AS

92-81-9
$C_{13}H_{11}N$
181.25
C2c1ccccc1Nc3ccccc23
Acridan
Acridane
Acridine, 9,10-dihydro- (9CI)
Carbazine

92-82-0
$C_{12}H_8N_2$
180.22
n(c(c(nc1cccc2)ccc3)c3)c12
Phenazine
Azophenylene
Dibenzoparadiazine
Dibenzopyrazine

92-83-1
$C_{13}H_{10}O$
182.23
O(c(c(ccc1)Cc2cccc3)c1)c23
Xanthene
10H-9-Oxaanthracene
9H-Xanthene (9CI)

92-84-2
$C_{12}H_9NS$
199.28
N(c(c(Sc1cccc2)ccc3)c3)c12
Phenothiazine
Afi-Tiazin
Agrazine
Antiverm
Biverm
Contaverm
Dibenzoparathiazine
Dibenzothiazine
Dibenzo-1,4-thiazine
ENT 38
Feeno
Fenothiazine (Dutch)
Fenotiazina (Italian)
Fenoverm
Fentiazin
Helmetina
Lethelmin
Nemazene
Nemazine
Orimon
Padophene
Penthazine
Phenegic
Phenosan
Phenothiazine (ACGIH,OSHA)
Phenoverm
Phenovis
Phenoxur
Phenthiazine
Reconox
Souframine
Thiodifenylamine (Dutch)
Thiodiphenylamin (German)
Thiodiphenylamine
Tiodifenilamina (Italian)
Vermitin
Wurm-Thional
XL-50

92-85-3
$C_{12}H_8S_2$
216.33
S(c(c(Sc1cccc2)ccc3)c3)c12
Thianthrene (9CI)
AI3-00638
9,10-Dithiaanthracene
NSC-439
Thiaanthrene
Thianthren

92-86-4
$C_{12}H_8Br_2$
312.00
c(ccc(c(ccc(c1)Br)c1)c2)(c2)Br
Biphenyl, 4,4'-dibromo- (8CI)
AI3-17378
1,1'-Biphenyl, 4,4'-dibromo- (9CI)
p,p'-Dibromobiphenyl
4,4'-Dibromobiphenyl
4,4'-Dibromo-1,1'-biphenyl
4,4'-Dibromodiphenyl
NSC-2098

92-87-5
$C_{12}H_{12}N_2$
184.26
Nc(ccc(c(ccc(N)c1)c1)c2)c2
Benzidine (ACGIH,OSHA) [UN 1885]
Benzidin (Czech)
Benzidina (Italian)
Benzydyna (Polish)
p,p-Bianiline
4,4'-Bianiline
(1,1'-Biphenyl)-4,4'-diamine (9CI)
4,4'-Biphenyldiamine
Biphenyl, 4,4'-diamino-
4,4'-Biphenylenediamine
C.I. 37225
C.I. Azoic Diazo Component 112
p,p'-Diaminobiphenyl
4,4'-Diaminobiphenyl
4,4'-Diamino-1,1'-biphenyl
p-Diaminodiphenyl
4,4'-Diaminodiphenyl
p,p'-Dianiline
4,4'-Diphenylenediamine
Fast Corinth Base B
NCI-C03361
RCRA waste number U021
UN 1885 [Benzidine]

92-88-6
$C_{12}H_{10}O_2$
186.22
Oc(ccc(c(ccc(O)c1)c1)c2)c2
4,4'-Biphenyldiol
p,p'-Biphenol

4,4'-Biphenol
USAF DO-30

92-91-1
$C_{14}H_{12}O$
196.25
O=C(c(ccc(c(cccc1)c1)c2)c2)C
Acetophenone, 4'-phenyl- (8CI)
p-Acetylbiphenyl
4-Acetylbiphenyl
AI3-00897
Biphenyl-4-acetophenone
1-(1,1'-Biphenyl)-4-ylethanone
4-Biphenylyl methyl ketone
Ethanone, 1-(1,1'-biphenyl)-4-yl- (9CI)
Ketone, 4-biphenylyl methyl
NSC-1875
p-Phenylacetophenone
4-Phenylacetophenone
4'-Phenylacetophenone

92-93-3
$C_{12}H_9NO_2$
199.22
O=N(=O)c(ccc(c(cccc1)c1)c2)c2
Biphenyl, 4-nitro
p-Nitrobiphenyl
4-Nitrobiphenyl
p-Nitrodiphenyl
4-Nitrodiphenyl (ACGIH, OSHA)
p-Phenyl-nitrobenzene
4-Phenyl-nitrobenzene
PNB

92-94-4
$C_{18}H_{14}$
230.32
c(c(cccc1)c1)(ccc(c(cccc2)c2)c3)c3
p-Terphenyl
Biphenyl, 4-phenyl-
p-Diphenylbenzene
1,4-Diphenylbenzene
4-Phenylbiphenyl
4-Phenyldiphenyl
Santowax P
p-Triphenyl

93-00-5
$C_{10}H_9NO_3S$
223.25
O=S(=O)(O)c(ccc(c1ccc2N)c2)c1
6-Aminonaphthalene-2-sulfonic acid
2-Aminonaphthalene-6-sulfonate
2-Aminonaphthalene-6-sulfonic acid
2-Amino-6-naphthalenesulfonic acid
6-Amino-2-naphthalenesulfonic acid
2-Amino-6-naphthylsulfonic acid
2-Amino-6-sulfonaphthalene
Broenner's acid
Bronner acid
Bronner's acid
1,6-Clev's Acid
2-Naphthalenesulfonic acid, 6-amino- (9CI)
2-Naphthylamine-6-sulfonic acid
6-Naphthylamine-2-sulfonic acid
6-Naphthylamine-2-sulphonic acid
NSC-31511
Phenamine Fast Scarlet 4BGP
Pontamine Fast Scarlet 4BA

93-01-6
$C_{10}H_8O_4S$
224.24
O=S(=O)(O)c(ccc(c1ccc2O)c2)c1
2-Naphthalenesulfonic acid, 6-hydroxy
2-Hydroxy-6-naphthalene-sulfonic acid
6-Hydroxy-2-naphthalene-sulfonic acid
Kyselina 2-naftol-6-sulfonova (Czech)
Kyselina schaferova (Czech)
β-Naphtholsulfonic acid S
β-Naphthol-6-sulfonic acid
2-Naphthol-6-sulfonic acid
2-Naphtol-6-sulfosaure (German)
Schaeffer's Acid
Schaeffer's β-Acid
Schaeffer's β-Naphtholsulfonic acid

93-02-7
$C_9H_{10}O_3$
166.18
O=Cc(c(OC)ccc1OC)c1
Benzaldehyde, 2,5-dimethoxy- (9CI)
AI3-19307
2,5-Dimethoxybenzaldehyde
NSC-6315

93-03-8
$C_9H_{12}O_3$
168.19
O(c(c(OC)cc(c1)CO)c1)C
Veratryl alcohol
AI3-24181
Benzenemethanol, 3,4-dimethoxy- (9CI)
3,4-Dimethoxybenzenemethanol
3,4-Dimethoxyphenylmethyl alcohol
NSC-6317

93-04-9
$C_{11}H_{10}O$
158.20
O(c(ccc(c1ccc2)c2)c1)C
Naphthalene, 2-methoxy- (9CI)
AI3-21213
β-Methoxynaphthalene
2-Methoxynaphthalene
Methyl β-naphthyl ether
Methyl 2-naphthyl ether
β-Naphthol methyl ether
2-Naphthol methyl ether
β-Naphthyl methyl ether
2-Naphthyl methyl ether
NSC-4171
Yara Yara
Yara-Yara
Yura Yara

93-05-0
$C_{10}H_{16}N_2$
164.28
N(c(ccc(N)c1)c1)(CC)CC
p-Phenylenediamine, N,N-diethyl
p-Aminodiethylaniline
p-(Diethylamino)aniline
4-(Diethylamino)aniline
N,N'-Diethyl-p-fenylendiamin (Czech)
N,N'-Diethyl-p-phenylene-diamine
Diethyl-para-phenylenediamine
DPD

93-07-2
$C_9H_{10}O_4$
182.19
O=C(O)c(ccc(OC)c1OC)c1
Benzoic acid, 3,4-dimethoxy
3,4-Dimethoxybenzoic acid
3,4-Dimethylprotocatechuic acid
Veratric acid

93-08-3
$C_{12}H_{10}O$
170.22
O=C(c(ccc(c1ccc2)c2)c1)C
2'-Acetonaphthone
β-Acetonaphthalene
Acetonaphthone
β-Acetonaphthone
2-Acetonaphthone
β-Acetylnaphthalene
2-Acetylnaphthalene
Ethanone, 1-(2-naphthalenyl)- (9CI)
Ketone, methyl 2-naphthyl
Methyl β-naphthyl ketone
Methyl 2-naphthyl ketone
β-Methyl naphthyl ketone
1-(2-Naphthalenyl)ethanone
β-Naphthyl methyl ketone
2-Naphthyl methyl ketone
Oranger Crystals

94-59-4
$C_{11}H_8O_2$
172.19
O=C(O)c(ccc(c1ccc2)c2)c1
2-Naphthoic acid
Isonaphthoic acid
2-Maythic acid
2-Naphthalenecarboxylic acid
Naphthalene-β-carboxylic acid

β-Naphthoic acid

93-11-8
$C_{10}H_7ClO_2S$
226.68
O=S(=O)(c(ccc(c1ccc2)c2)c1)Cl
2-Naphthalenesulfonyl chloride (9CI)
β-Naphthalenesulfochloride
Naphthalene-2-sulfonic acid chloride
Naphthalene-2-sulfonyl chloride
β-Naphthalenesulfonyl chloride
2-Naphthylsulfonyl chloride
NSC-133893

93-13-0
$C_8H_{11}NO_4S$
217.24
O=S(=O)(O)CNc(c(OC)ccc1)c1
o-Anisidyl-N-methanesulfonic acid
o-Anisidinomethanesulfonic acid
Methanesulfonic acid, o-anisidino- (8CI)
Methanesulfonic acid, ((2-methoxyphenyl)amino)- (9CI)
((2-Methoxyphenyl)amino)-methanesulfonic acid
NSC-7549

93-14-1
$C_{10}H_{14}O_4$
198.24
O(c(c(OC)ccc1)c1)CC(O)CO
1,2-Propanediol, 3-(o-methoxyphenoxy)
Aresol
Cortussin
Creson
1,2-Dihydroxy-3-(2-methoxyphenoxy)propane
Dilyn
2-G
G 87
Gaiamar
GGE
GGG
Glycerin guaiacolate
Glycerinmonoguaiacol ether
Glycero-guaiacol ether
Glycerol guaiacolate
Glycerol-α-guajakolether (Czech)
Glycerol α-(o-methoxyphenyl)ether
Glycerol-α-monoguaiacol ether
Glycerol mono(2-methoxyphenyl)ether
Glyceryl guaiacolate
α-Glyceryl guaiacolate ether
Glyceryl guaiacyl ether
Glycotuss
Gnaifenesin
Guajacol-glycerinaether (German)
Guaiacolglicerinetere
Guaiacol glyceryl ether
Guaiacurane
Guaiacyl glyceryl ether
Guaiamar
Guaianesin
Guaifenesin
Guaiphenesine
Guaiphesin
Guajacol-α-glycerinether
Guajacuran
Guajamar
Guanar
Guayanesin
Guiaphenesin
Hustodil
Hustosil
Metfenossidiolo
Methoxypropanediol
3-o-Methoxyphenoxypropane 1:2-diol
3-(o-Methoxyphenoxy)-1,2-propanediol
3-(2-Methoxyphenoxy)-1,2-propanediol
o-Methoxyphenyl glyceryl ether
Methphenoxydiol
Metossipropandiolo
Miocurin
Miorelax
Mucostop
MY 301
Myocaine
Myorelax
Myoscaine
Neurotone
Oresol
Oreson
1,2-Propanediol, 3-(2-methoxyphenoxy)-
Propanosedyl
Reduton
Relaxyl-G
Reorganin
Resil
Respenyl
Respil
Resyl
Ritussin
Robitussin
Sirotol
SL-90
Tolseron
Tolyn
Tulyl
XL-90

93-15-2
$C_{11}H_{14}O_2$
178.25
O(c(c(OC)cc(c1)CC=C)c1)C
Benzene, 4-allyl-1,2-dimethoxy
1-Allyl-3,4-dimethoxybenzene
4-Allyl-1,2-dimethoxybenzene
4-Allylveratrole
1,2-Dimethoxy-4-allylbenzene
1-(3,4-Dimethoxyphenyl)-2-propene
ENT 21,040
1,3,4-Eugenol methyl ether
Eugenyl methyl ether
Methyl eugenol
Veratrole methyl ether

93-16-3
$C_{11}H_{14}O_2$
178.25
O(c(c(OC)cc(c1)C=CC)c1)C
Benzene, 1,2-dimethoxy-4-propenyl
1,3,4-Isoeugenol methyl ether
Isoeugenyl methyl ether
Isohomogenol
Methyl isoeugenol
1-Propene, 1-(3,4-dimethoxyphenyl)-
4-Propenyl veratrole

93-18-5
$C_{12}H_{12}O$
172.24
O(c(ccc(c1ccc2)c2)c1)CC
Naphthalene, 2-ethoxy
Bromelia
2-Ethoxynaphthalene
Ethyl β-naphtholate
Ethyl β-naphthyl ether
Ethyl 2-naphthyl ether
β-Naphthol ethyl ether
2-Naphthol ethyl ether
Nerolin
Neroline
Nerolin II
Nerolin New

93-25-4
$C_9H_{10}O_3$
166.18
O=C(O)Cc(c(OC)ccc1)c1
Acetic acid, (o-methoxyphenyl)- (8CI)
Benzeneacetic acid, 2-methoxy- (9CI)
2-Methoxybenzeneacetic acid
(o-Methoxyphenyl)acetic acid
NSC-110708

93-28-7
$C_{12}H_{14}O_3$
206.26
O=C(Oc(c(OC)cc(c1)CC=C)c1)C
Phenol, 4-allyl-2-methoxy-, acetate
Aceteugenol
1-Acetoxy-2-methoxy-4-allylbenzene
Acetyleugenol
4-Allyl-2-methoxyphenol acetate
4-Allyl-2-methoxyphenyl acetate
1,3,4-Eugenol acetate
Eugenol acetate
Eugenyl acetate

93-35-6
$C_9H_6O_3$
162.15
Coumarin, 7-hydroxy
2H-1-Benzopyran-2-one,

7-hydroxy-
Hydrangin
Hydrangine
7-Hydroxycoumarin
7-Oxycoumarin
Skimmetin
Skimmetine
Umbelliferon
Umbelliferone

93-37-8
$C_{11}H_{11}N$
157.21
n(c(c(ccc1C)cc2)c1)c2C
Quinoline, 2,7-dimethyl- (9CI)
2,7-Dimethylquinoline
NSC-5240
m-Toluquinaldine

93-40-3
$C_{10}H_{12}O_4$
196.22
O=C(O)Cc(ccc(OC)c1OC)c1
Acetic acid, (3,4-dimethoxy-phenyl)
Benzeneacetic acid, 3,4-dimethoxy- (9CI)
3,4-Dimethoxybenzeneacetic acid
3,4-Dimethoxyphenyl acetic acid
Homoveratric acid

93-43-6
$C_{13}H_{12}O_3$
216.24
O=C(Oc(ccc(c1ccc2)c2)c1)C(O)C
Propanoic acid, 2-hydroxy-, 2-naphthalenyl ester (9CI)
2-Naphthalenyl 2-hydroxypropanoate

93-45-8
$C_{16}H_{13}NO$
235.30
Oc(ccc(Nc(ccc(c1ccc2)c2)c1)c3)c3
Phenol, p-(2-naphthylamino)
N-p-Hydroxyphenyl-β-naphthylamine
N-p-Hydroxyphenyl-2-naphthylamine
p-Hydroxyphenyl-β-naphthylamine
p-Hydroxyphenyl-2-naphthylamine
4-(2-Naftylamino)fenol (Czech)
p-(2-Naphthylamino)phenol
4-(2-Naphthylamino)phenol
p-Oxinozon

93-46-9
$C_{26}H_{20}N_2$
360.48
N(c(ccc(Nc(ccc(c1ccc2)c2)c1)c3)c3)c(ccc(c4ccc5)c5)c4
p-Phenylenediamine, N,N'-(di-2-naphthyl)
Aceto DIPP
Agerite White
N,N'-Bis-(2-naftyl)-p-fenylendiamin (Czech)
N,N'-Bis(2-naphthyl)-p-phenylenediamine
Di-β-naphthyl-p-phenyldiamine
Di-β-naphthyl-p-phenylenediamine
N,N'-Di-β-naphthyl-p-phenylenediamine
N,N'-Di-2-naphthyl-p-phenylenediamine
sym-Di-β-naphthyl-p-phenylenediamine
DNPD
DNPDA
Dwu-β-naftylo-p-fenylodwuamina (Polish)
2-Naphthyl-p-phenylenediamine
Nonox CL
Tisperse MB-2X

93-51-6
$C_8H_{10}O_2$
138.18
O(c(c(O)ccc1C)c1)C
p-Cresol, 2-methoxy
p-Creosol
Homoguaiacol
4-Hydroxy-3-methoxy-1-methylbenzene
4-Hydroxy-3-methoxytoluene
2-Methoxy-p-cresol
3-Methoxy-4-hydroxytoluene
2-Methoxy-4-methylphenol
p-Methylguaiacol
4-Methylguaiacol
Phenol, 2-methoxy-4-methyl-

93-52-7
$C_8H_8Br_2$
263.98
BrC(c(cccc1)c1)CBr
Benzene, (1,2-dibromoethyl)
α,β-Dibromoethylbenzene
(1,2-Dibromoethyl)benzene
1,2-Dibromo-1-phenylethane
Dowspray 9

93-53-8
$C_9H_{10}O$
134.19
O=CC(c(cccc1)c1)C
Benzeneacetaldehyde, α-methyl
Aldehyd hydratropovy (Czech)
Cumene aldehyde
2-Fenyl-1-propanal (Czech)
α-Formylethylbenzene
Hyacinthal
Hydratropa aldehyde
Hydratropaldehyde
Hydratropic aldehyde
α-Methyl phenylacetaldehyde
α-Methyl-α-toluic aldehyde
2-Phenylpropanal
2-Phenyl-1-propanal
α-Phenyl propionaldehyde
2-Phenyl propionaldehyde
Propionaldehyde, 2-phenyl-
α-Tolualdehyde, α-methyl-

93-54-9
$C_9H_{12}O$
136.21
CCC(O)c1ccccc1
Benzyl alcohol, α-ethyl
Benzenemethanol, α-ethyl- (9CI)
Ejibil
α-Ethylbenzenemethanol
α-Ethylbenzyl alcohol
Ethyl phenyl carbinol
Felicur
Felitrope
Fenicol
α-Hydroxypropylbenzene
Livonal
Phenicol
Phenychol
Phenylaethylcarbinol (German)
Phenylcholon
1-Phenylpropanol
1-Phenyl-1-propanol
1-Phenylpropyl alcohol
SH 261

93-55-0
$C_9H_{10}O$
134.19
O=C(c(cccc1)c1)CC
Propiophenone
Ethyl phenyl ketone
Ketone, ethyl phenyl
Phenyl ethyl ketone
1-Phenyl-1-propanone
1-Propanone, 1-phenyl-
Propionylbenzene
USAF EK-1235

93-56-1
$C_8H_{10}O_2$
138.18
OC(c(cccc1)c1)CO
1,2-Ethanediol, 1-phenyl
α,β-Dihydroxyethylbenzene
1,2-Dihydroxyethylbenzene
1,2-Ethanediol, phenyl-
1-Fenyl-1,2-ethandiol (Czech)
Fenylglycol (Czech)
Phenylethanediol
Phenylethylene glycol
1-Phenylethylene glycol
Phenyl glycol
Styrene glycol
Styrolyl alcohol

93-58-3
$C_8H_8O_2$
136.16
O=C(OC)c(cccc1)c1
Benzoic acid, methyl ester
Essence of Niobe
Methyl benzenecarboxylate
Methyl benzoate [UN 2938]

Methylbenzoate
Methylester kyseliny benzoove (Czech)
Niobe Oil
Oil of Niobe
Oxidate LE
UN 2938 [Methyl benzoate]

93-60-7
$C_7H_7NO_2$
137.15
O=C(OC)c(cccn1)c1
Nicotinic acid, methyl ester
Methyl nicotinate
Nicometh

93-62-9
$C_6H_{12}NO_5$
178.19
O=C(O)CN(CC(=O)O)CCO
Acetic acid, 2-(hydroxyethyl-imino)di (9CI)
Ethanolamine-N,N-diacetic acid
Glycine, N-(carboxymethyl)-N-(2-hydroxyethyl)- (8CI)
HEIDA
2-Hydroxyethylaminodiacetic acid
(2-Hydroxyethyl)iminodiacetic acid
USAF DO-37

93-65-2
$C_{10}H_{11}ClO_3$
214.66
O=C(O)C(Oc(c(cc(c1)Cl)C)c1)C
Propionic acid, 2-((4-chloro-o-tolyl)oxy)
Acide 2-(4-chloro-2-methyl-phenoxy)propionique (French)
Acido 2-(4-cloro-2-metil-fenossi)-propionico (Italian)
BH Mecoprop
2-(4-Chloor-2-methyl-fenoxy)-propionzuur (Dutch)
2-(4-Chlor-2-methyl-phenoxy)-propionsaeure (German)
2-(4-Chloro-2-methylphenoxy)-propanoic acid
4-Chloro-2-methylphenoxy-α-propionic acid
α-(4-Chloro-2-methylphenoxy)-propionic acid
(+)-α-(4-Chloro-2-methylphen-oxy) propionic acid
2-(4-Chloro-2-methylphenoxy)-propionic acid
2-(4-Chlorophenoxy-2-methyl)-propionic acid
2-(p-Chloro-o-tolyloxy)pro-pionic acid
Chipco Turf Herbicide MCPP
CMPP
Compitox
FBC CMPP
Hedonal MCPP
Iso-Cornox
Kilprop
Kwas 4-chloro-2-metyl-ofenoksypropionowy (Polish)
Kyselina 2-(4-chlor-2-methyl-fenoxy)propionova (Czech)
Liranox
2M-4CP
MCPP
2-MCPP
MCPP 2,4-D
MCPP-D-4
MCPP-K-4
Mecomec
Mecopeop
Mecoper
Mecopex
Mecoprop
Mecoturf
Mecprop
Mepro
Methoxone
α-(2-Methyl-4-chlorophenoxy)-propionic acid
2-(2-Methyl-4-chlorophenoxy)-propionic acid
2-Methyl-4-chlorophenoxy-α-propionic acid
2-(2-Methyl-4-chlorphenoxy)-propionsaeure (German)
2M 4KHP
N.B. Mecoprop
Propal
Propanoic acid, 2-(4-chloro-2-methylphenoxy)-
Propionic acid, 2-(4-chloro-2-methylphenoxy)
Propionic acid, 2-(2-methyl-4-chlorophenoxy)-
Proponex-Plus
Rankotex
Runcatex
RD 4593
U 46
U 46 KV-Ester
U 46 KV-Fluid
Vi-Par
Vi-Pex

93-68-5
$C_{11}H_{13}NO_2$
191.25
O=C(Nc(c(ccc1)C)c1)CC(=O)C
o-Acetoacetotoluidide
Acetoacet-o-toluidide
Acetoacet-ortho-toluidide
2-Acetoacetylaminotoluene
Acetoacetyl-2-methylanilide
Butanamide, N-(2-methyl-phenyl)-3-oxo- (9CI)
2'-Methylacetoacetanilide

93-70-9
$C_{10}H_{10}ClNO_2$
211.66
O=C(Nc(c(ccc1)Cl)c1)CC(=O)C
Acetoacetanilide, 2'-chloro
AAoC
Acetoacetanilide, o-chloro-
Acetoacet-o-chloranilide
Acetoacet-o-chloroanilide
o-Acetoacetochloranilide
Acetoacetyl-2-chloroanilide
Butaneamide, N-(2-chloro-phenyl)-3-oxo-
o-Chloroacetoacetanilide
2'-Chloroacetoacetanilide
N-(2-Chlorophenyl)acetoacet-amide
3-Oxo-N-(2-chlorophenylbutan-amide)

93-71-0
$C_8H_{12}ClNO$
173.66
O=C(N(CC=C)CC=C)CCl
Acetamide, 2-chloro-N,N-di-allyl
Acetamide, 2-chloro-N,N-di-2-propenyl- (9CI)
Acetamide, N,N-diallyl-2-chloro- (8CI)
Alidochlor
Alidochlore
Allidochlor
CDAA
CDAAT
α-Chloro-N,N-diallylacetamide
2-Chloro-N,N-diallylacetamide
2-Chloro-N,N-di-2-propenyl-acetamide
CP 6,343
Diallylamid kyseliny chloroc-tove (Czech)
Diallylchloroacetamide
N,N-Diallylchloroacetamide
N,N-Diallyl-α-chloroacetamide
N,N-Diallyl-2-chloroacetamide
NCI-C04035
Radox
Randox
Rantox T

93-72-1
$C_9H_7Cl_3O_3$
269.51
CC(Oc1cc(Cl)c(Cl)cc1Cl)C(O)=O
Propionic acid, 2-(2,4,5-tri-chlorophenoxy)
Acide 2-(2,4,5-trichloro-phen-oxy) propionique (French)
Acido 2-(2,4,5-tricloro-fenossi)-propionico (Italian)
Amchem 2,4,5-tp
Aqua-Vex
Color-Set
Ded-Weed
Double Strength
Fenoprop
Fenormone
Fruitone T
Herbicides, Silvex
Kuran
Kuron
Kurosal
Kurosal G
Kurosal SL
Kwas 2,4,5-trojchlorofenoksy-propionowy (Polish)
Kyselina 2-(2,4,5-trichlor-fenoxy)propionova (Czech)
Miller Nu Set
Propon

RCRA waste number U233
Silvex
Silvi-RHAP
Sta-Fast
2,4,5-TC
2,4,5-TCPPA
2,4,5-TP
2-(2,4,5-Trichloor-fenoxy)-propionzuur (Dutch)
α-(2,4,5-Trichlorophenoxy)propionic acid
2-(2,4,5-Trichlorophenoxy)propionic acid
2,4,5-Trichlorophenoxy-α-propionic acid
2-(2,4,5-Trichlor-phenoxy)-propionsaeure (German)
Weed-B-Gon

93-75-4
$C_9H_4N_2S_3$
236.33
S=c3sc2nc1ccccc1nc2s3
Carbonic acid, trithio-, cyclic ester with 2,3-quinoxalinedithiol
Bay 30686
Bayer 4935
Bayer 30686
Bayer 31686
Chinothionat
Chinoxalin-2,3-dithiol-cyclo-thio-carbonat (German)
Chinoxalin-2,3-diyl-trithio-carbonat (German)
1,3-Dithiolo(4,5-b)quinoxaline-2-thione
ENT 25,579
Eradex
Eraditon
Erazidon
Quinothionate
2,3-Quinoxalinedithiol, cyclic trithiocarbonate
Quinoxaline-2,3-diyl trithio-carbonate
Readex
SS 1451
Thiaquinox
2-Thio-1,3-dithio(4,5-b)quinoxaline
Thioquinox
Trithiocarbonic acid, cyclic ester with 2,3-quinoxalinedithiol

93-76-5
$C_8H_5Cl_3O_3$
255.48
O=C(O)COc(c(cc(c1Cl)Cl)Cl)c1
Acetic acid, (2,4,5-trichlorophenoxy)
Acide 2,4,5-trichloro phenoxy-acetique (French)
Acido (2,4,5-tricloro-fenossi)-acetico (Italian)
Amine 2,4,5-t for rice
BCF-Bushkiller
Brush-Off 445 low volatile brush killer
Brush Rhap
Brushtox
Dacamine
Debroussaillant Concentre
Debroussaillant Super Concentre
Decamine 4T
Ded-Weed Brush Killer
Ded-Weed LV-6 Brush Kil and T-5 Brush Kil
Dinoxol
Envert-T
Estercide T-2 and T-245
Esteron
Esteron 245
Esteron 245 BE
Esteron Brush Killer
Farmco Fence Rider (93-76-5)
Fence Rider
Forron
Forst U 46
Fortex
Fruitone A
Inverton 245
Kwas 2,4,5-trojchlorofenoksy-octowy (Polish)
Line Rider
NA 2765 (DOT)
Phortox
RCRA waste number U232
Reddon
Reddox
Spontox
Super D Weedone
2,4,5-T (ACGIH,DOT,OSHA)
Tippon
Tormona
Transamine
Tributon
(2,4,5-Trichloor-fenoxy)-azijnzuur (Dutch)
2,4,5-Trichlorophenoxyacetic acid (DOT)
(2,4,5-Trichlor-phenoxy)-essigsaeure (German)
Trinoxol
Trioxon
Trioxone
U 46
Veon
Veon 245
Verton 2T
Visko Rhap Low Volatile Ester
Weedar
Weedone
Weedone 2,4,5-T

93-78-7
$C_{11}H_{11}Cl_3O_3$
297.56
CC(C)OC(=O)COc1cc(Cl)c(Cl)cc1Cl
Acetic acid, (2,4,5-trichlorophenoxy)-, isopropyl ester (8CI)
Acetic acid, (2,4,5-trichlorophenoxy)-, 1-methylethyl ester (9CI)
Caswell No. 881T
EPA Pesticide Chemical Code 082066
2,4,5-T, isopropyl ester
2,4,5-Trichlorophenoxyacetic acid, isopropyl ester

93-79-8
$C_{12}H_{13}Cl_3O_3$
311.60
CCCCOC(=O)COc1cc(Cl)c(Cl)cc1Cl
Acetic acid, (2,4,5-trichlorophenoxy)-, butyl ester
Arboricid
Butylate 2,4,5-T
n-Butylester kyselini 2,4,5-trichlorfenoxyoctove (Czech)
Butyl 2,4,5-T
Butyl 2,4,5-trichlorophenoxyacetate
n-Butyl (2,4,5-trichlorophenoxy)acetate
Flomore
Kilex 3
Krzewotoks
Krzewotox
2,4,5-T butyl ester
2,4,5-T n-butyl ester
Tormona
2,4,5-Trichlorophenoxyacetic acid, butyl ester
Trioxone
U46KW

93-80-1
$C_{10}H_9Cl_3O_3$
283.54
Butyric acid, 4-(2,4,5-trichlorophenoxy)
4-(2,4,5-Trichlorophenoxy)-butyric acid

93-82-3
$C_{22}H_{45}NO_3$
371.60
O=C(N(CCO)CCO)CCCCCCCCCCCCCCCC
Stearamide DEA
N,N-Bis(2-hydroxyethyl)octadecanamide
N,N-Bis(2-hydroxyethyl)stearamide
Diethanolamine stearic acid amide
Felsamid-SPD
Octadecanamide, N,N-bis-(2-hydroxyethyl)- (9CI)
Stearic acid diethanolamide
Stearoyl diethanolamide

93-83-4
$C_{22}H_{43}NO_3$
369.58
O=C(N(CCO)CCO)CCCCCCCC=CCCCCCCCC
Oleamide DEA
N,N-Bis(2-hydroxyethyl)-9-octadecenamide
(Z)-N,N-Bis(2-hydroxyethyl)-9-octadecenamide

N,N-Bis(2-hydroxyethyl)oleamide
Diethanolamine oleic acid amide
Felsamid-OPD
Incromide OPD
Jordamide 201
9-Octadecenamide, N,N-bis-(2-hydroxyethyl)-
9-Octadecenamide, N,N-bis-(2-hydroxyethyl)-, (Z)- (9CI)
Oleic acid diethanolamide
Oleic diethanolamide
Upamide O-20
Upamide OD
Witcamide 511C
Witcamide 4120
Witcamide 5085

93-89-0
$C_9H_{10}O_2$
150.19
O=C(OCC)c(cccc1)c1
Benzoic acid, ethyl ester
Benzoic ether
Essence of Niobe
Ethyl benzoate
Ethylester kyseliny benzoove (Czech)

93-90-3
$C_9H_{13}NO$
151.20
OCCN(c(cccc1)c1)C
Phenylmethyl ethanol amine
AI3-23878
Ethanol, 2-(N-methylanilino)-(8CI)
Ethanol, 2-(methylphenyl-amino)- (9CI)
2-(N-Fenyl-N-methylamino)-ethanol (Czech)
2-(N-Methylaniline)ethanol
2-(N-Methylanilino)ethanol
N-Methyl-N-hydroxyethyl-aniline
N-Methyl-N-(hydroxyethyl)-aniline
N-Methyl-N-β-hydroxyethyl-aniline
N-Methyl-N-(2-hydroxyethyl)-aniline
2-(Methylphenylamino)ethanol
2-(N-Methyl-N-phenylamino)-ethanol
NSC-9274
Phenylmethylethanolamine

93-91-4
$C_{10}H_{12}O_2$
164.22
O=C(c(cccc1)c1)CC(=O)C
1,3-Butanedione, 1-phenyl
Acetoacetophenone
α-Acetylacetophenone
2-Acetylacetophenone
Acetylbenzoylmethane
Benzoyl-aceton (German)
Benzoylacetone

93-92-5
$C_{10}H_{12}O_2$
164.20
O=C(OC(c(cccc1)c1)C)C
Methylphenylcarbinyl acetate
AI3-18152
Benzenemethanol, α-methyl-, acetate (9CI)
Benzyl alcohol, α-methyl-, acetate (8CI)
Gardeniol II
Gardenol
α-Methylbenzenemethanol acetate
α-Methylbenzyl acetate
Methylphenylcarbinol acetate
NSC-2397
α-Phenylethyl acetate
sec-Phenylethyl acetate
1-Phenylethyl acetate
Phenylmethylcarbinyl acetate
Styralyl acetate

93-94-7
$C_{16}H_{18}O_6$
306.32
O=C(c(ccc(c1)C(=O)CC(=O)OCC)c1)CC(=O)OCC
Acetic acid, terephthaloyldi-, diethyl ester (8CI)
1,4-Benzenedipropanoic acid, β,β'-dioxo-, diethyl ester (9CI)
Diethyl terephthaloylbisacetate
NSC-50655
Terephthaloyldiacetic acid, diethyl ester

93-96-9
$C_{16}H_{18}O$
226.32
α-Methylbenzyl ether
AI3-02097
Benzene, 1,1'-(oxydiethylidene)bis- (9CI)
Bis(α-methylbenzyl) ether
Bis(α-phenylethyl) ether
1,1'-Diphenyldiethyl ether
Ether, bis(α-methylbenzyl) (8CI)
α-Methyl benzyl ether
NSC-403888

93-97-0
$C_{14}H_{10}O_3$
226.23
O=C(OC(=O)c(cccc1)c1)c(cccc2)c2
Benzoic acid, anhydride (9CI)
AI3-03698
Benzoic anhydride (8CI)
Benzoyl anhydride
NSC-37116

93-98-1
$C_{13}H_{11}NO$
197.23
O=C(Nc(cccc1)c1)c(cccc2)c2
Benzanilide
AI3-01046
Benzamide, N-phenyl- (9CI)
Benzoic acid anilide
N-Benzoylaniline
N-Phenylbenzamide
NSC-3131

93-99-2
$C_{17}H_{10}O_2$
246.27
O=C(Oc(cccc1)c1)c(cccc2)c2
Benzoic acid, phenyl ester
Phenyl benzoate

94-02-0
$C_{11}H_{12}O_3$
192.23
O=C(OCC)CC(=O)c(cccc1)c1
Acetic acid, benzoyl-, ethyl ester
Benzenepropanoic acid, β-oxo-, ethyl ester (9CI)
Benzoylacetic acid ethyl ester
Ethyl benzoyl acetate
Ethyl β-oxobenzenepropanoate

94-04-2
$C_{10}H_{18}O_2$
170.25
Vinyl 2-ethylhexoate
AI3-24890
2-Ethylhexanoic acid, vinyl ester
2-Ethylhexoic acid, vinyl ester
Hexanoic acid, 2-ethyl-, ethenyl ester (9CI)
Hexanoic acid, 2-ethyl-, vinyl ester (8CI)
NSC-5312
Vinylester kyseliny 2-ethyl-kapronove (Czech)
Vinyl 2-ethylhexanoate
Vinyl-2-ethylhexoate

94-06-4
$C_{11}H_{16}O$
164.25
Phenol, 4-(1-methylbutyl)- (9CI)
NSC-7947
Phenol, p-(1-methylbutyl)-(8CI)

94-09-7
$C_9H_{11}NO_2$
165.21
O=C(OCC)c(ccc(N)c1)c1
Benzoic acid, p-amino-, ethyl ester
Americaine
p-Aminobenzoic acid ethyl ester
4-Aminobenzoic acid, ethyl ester
Anaesthesin
Anesthesin

Anesthone
Benzocaine
Ethyl aminobenzoate
Ethyl p-aminobenzoate
Ethyl 4-aminobenzoate
Ethylester kyseliny p-aminobenzoove (Czech)
Keloform
Norcain
Orthesin
Parathesin
Topcaine

94-11-1
$C_{11}H_{12}Cl_2O_3$
263.13
CC(C)OC(=O)COc1ccc(Cl)cc1Cl
Acetic acid, (2,4-dichlorophenoxy)-, isopropyl ester
Acetic acid, (2,4-dichlorophenoxy)-, 1-methylethyl ester (9CI)
2,4-Dichlorophenoxyacetic acid, isopropyl ester
2,4-D, isopropyl ester
Esteron 44
Isopropyl 2,4-D ester
Isopropylester kyseliny 2,4-dichlorfenoxyoctove (Czech)
Weedone 128

94-12-2
$C_{10}H_{13}NO_2$
179.24
CCCOC(=O)c1ccc(N)cc1
Benzoic acid, p-amino-, propyl ester (9CI)
Benzoic acid, 4-amino-, propyl ester (7CI,8CI)
p-Aminobenzoic acid propyl ester
Keloform P
NSC-23516
Propaesin
Propazyl
Propesin (6CI)
Propesine
4-(Propoxycarbonyl)aniline
Propyl p-aminobenzoate
Propyl 4-aminobenzoate
n-Propyl p-aminobenzoate
Propylcain
Raythesin
Risocaine

94-13-3
$C_{10}H_{12}O_3$
180.22
O=C(OCCC)c(ccc(O)c1)c1
Benzoic acid, p-hydroxy-, propyl ester (6CI,8CI)
Aseptoform P
Benzoic acid, 4-hydroxy-, propyl ester (9CI)
Betacide P
Bonomold OP
Chemocide PK
p-Hydroxybenzoic acid propyl ester
4-Hydroxybenzoic acid propyl ester
p-Hydroxypropyl benzoate
Nipagin P
Nipasol
Nipasol M
Nipasol P
Nipazol
p-Oxybenzoesaurepropylester (German)
Paraben
Parasept
Paseptol
Preserval P
Propagin
Propyl aseptoform
Propyl Butex
Propyl chemosept
Propylester kyseliny p-hydroxybenzoove (Czech)
Propyl p-hydroxybenzoate
Propyl 4-hydroxybenzoate
n-Propyl p-hydroxybenzoate
Propylparaben
Propylparasept
Protaben P
Pulvis conservans
Solbrol P
Tegosept P

94-17-7
$C_{14}H_8Cl_2O_4$
311.12
O=C(OOC(=O)c(ccc(c1)Cl)c1)c(ccc(c2)Cl)c2
Peroxide, bis(p-chlorobenzoyl)
Cadox PS
Bis(p-chlorobenzoyl) peroxide
p-Chlorobenzoyl peroxide (DOT)
p-Chlorobenzoyl peroxide, Not more than 52% as a paste (DOT)
p-Chlorobenzoyl peroxide, Not more than 52% in solution (DOT)
p-Chlorobenzoyl peroxide, Not more than 75% with water (DOT)
p,p'-Dichlorobenzoyl peroxide
Di-(4-chlorobenzoyl)peroxide, Not more than 52% as a paste (DOT)
Di-(4-chlorobenzoyl)peroxide, Not more than 52% in solution (DOT)
Di-(4-chlorobenzoyl) peroxide, Not more than 75% with water (DOT)
Peroxide, bis(p-chlorobenzoyl)-, Not more than 52% as a paste or in solution
UN 2113 (DOT)
UN 2114 (DOT)
UN 2115 (DOT)

94-18-8
$C_{14}H_{12}O_3$
228.25
O=C(OCc(cccc1)c1)c(ccc(O)c2)c2
Benzylparaben
AI3-02955
Benzoic acid, p-hydroxy-, benzyl ester (8CI)
Benzoic acid, 4-hydroxy-, phenylmethyl ester (9CI)
Benzyl p-hydroxybenzoate
Benzyl 4-hydroxybenzoate
Benzyl paraben
Benzyl parahydroxybenzoate
Benzyl parasept
Benzyl-parasept
Benzyl tegosept
p-Hydroxybenzoic acid benzyl ester
Nipabenzyl
NSC-8080
Parosept
Phenylmethyl 4-hydroxybenzoate
Preserval B
Solbrol Z
Unisept BZ

94-19-9
$C_{10}H_{12}N_4O_2S_2$
284.38
CCc2nnc(NS(=O)(=O)c1ccc(N)cc1)s2
Sulfanilamide, N-(5-ethyl-1,3,4-thiadiazol-2-yl)
Aethazol
Benzenesulfonamide, 4-amino-N-(5-ethyl-1,3,4-thiadiazol-2-yl)- (9CI)
Berlophen
Etazol
Etazole
Ethazole
Ethazole (Pharmaceutical)
N-(5-Ethyl-1,3,4-thiadiazol-2-yl)sulfanilamide
Globucid
Globucin
Globuzid (German)
SETD
Sethadil
Sulfaethidiole
Sulfaethidol
Sulfaethidole
Sulfaethylthiadiazole
2-Sulfanilamidoaethylthiodiazol (German)
Sulfa-Perlongit
Sul-Spansion
Sul-Spantab
VK 55

94-20-2
$C_{10}H_{13}ClN_2O_3S$
276.76
CCCNC(=O)NS(=O)(=O)c1ccc(Cl)cc1
Urea, 1-((p-chlorophenyl)sulfonyl)-3-propyl
Adiaben
Asucrol
Benzenesulfonamide, 4-chloro-N-((propylamino)carbonyl)-

Catanil
N-(p-Chlorobenzenesulfonyl)-N'-propylurea
1-(p-Chlorobenzenesulfonyl)-3-propylurea
Chlorodiabina
Chloronase
1-p-Chlorophenyl-3-(propylsulfonyl)urea
1-(p-Chlorophenylsulfonyl)-3-propylurea
Chloropropamide
4-Chloro-4-((propylamino)carbonyl)benzenesulfonamide
Chlorpropamid
Chlorpropamide
Clorpropamide (Italian)
Diabaril
Diabechlor
Diabenal
Diabenese
Diabeneza
Diabetoral
Diabet-Pages
Diabinese
Glisema
Meldian
Melitase
Millinese
NCI-C01752
Oradian
P 607
N-Propyl-N'-(p-chlorobenzenesulfonyl)urea
1-Propyl-3-(p-chlorobenzenesulfonyl)urea
N-Propyl-N'-p-chlorphenylsulfonylcarbamide
Stabinol
U-3818

94-22-4
$C_{10}H_{11}NO_4$
209.20
Benzoic acid, 4-nitro-, propyl ester (9CI)
Benzoic acid, p-nitro-, propyl ester (8CI)
NSC-406847

94-25-7
$C_{11}H_{15}NO_2$
193.27
O=C(OCCCC)c(ccc(N)c1)c1
Benzoic acid, p-amino-, butyl ester
p-Aminobenzoic acid butyl ester
Butamben
Butesin
Butyl p-aminobenzoate
Butylester kyseliny p-aminobenzoove (Czech)

94-26-8
$C_{11}H_{14}O_3$
194.25
O=C(OCCCC)c(ccc(O)c1)c1
Benzoic acid, 4-hydroxy-, butyl ester (9CI)
Aseptoform butyl
Benzoic acid, p-hydroxy-, butyl ester (8CI)
Butoben
4-(Butoxycarbonyl)phenol
Butyl butex
Butyl chemosept
Butyl p-hydroxybenzoate
Butyl 4-hydroxybenzoate
Butyl paraben
Butylparaben
n-Butyl parahydroxybenzoate
Butyl parasept
Butyl tegosept
Caswell No. 130A
EPA Pesticide Chemical Code 061205
p-Hydroxybenzoic acid butyl ester
4-Hydroxybenzoic acid butyl ester
Nipabutyl
NSC-8475
Parasept
Preserval B
Preserval Butylique
Solbrol B
SPF
Tegosept B
Tegosept butyl
Unisept B

94-28-0
$C_{22}H_{42}O_6$
402.64
O=C(OCCOCCOCCOC(=O)C(CCCC)CC)C(CCCC)CC
Hexanoic acid, ethyl-, diester with triethylene glycol
Flexol plasticizer 3go
Triethylene glycol, bis(ethylhexanoate)
Triethylene glycol di(2-ethylhexoate)

94-34-8
$C_{10}H_{12}N_2$
160.21
N#CCCN(c(cccc1)c1)C
Propanenitrile, 3-(methylphenylamino)- (9CI)
AI3-28714
N-β-Cyanoethyl-N-methylaniline
β-N-Methylanilinopropionitrile
β-(N-Methylanilino)propionitrile
3-(N-Methylanilino)propionitrile
N-Methyl-N-(2-cyanoethyl)-aniline
3-(Methylphenylamino)propanenitrile
NSC-91616
Propionitrile, 3-(N-methylanilino)- (8CI)

94-36-0
$C_{14}H_{10}O_4$
242.24
O=C(OOC(=O)c(cccc1)c1)c(cccc2)c2
Benzoyl-peroxide
Acetoxyl
Acnegel
Aztec BPO
Benoxyl
Benzac
Benzaknew
Benzoic acid, peroxide
Benzoperoxide
Benzoyl
Benzoylperoxid (German)
Benzoyl peroxide (ACGIH, DOT,OSHA)
Benzoyl-peroxide, More than 52% with inert solid (DOT)
Benzoyl peroxide, More than 72% but less than 95% as a paste (DOT)
Benzoyl peroxide, More than 77% but less than 95% with water (DOT)
Benzoyl peroxide, Not less than 30% but not more than 52% with inert solid [UN 2089]
Benzoyl peroxide, Not more than 72% as a paste [UN 2087]
Benzoyl peroxide, Not more than 77% with water [UN 2090]
Benzoyl peroxide, Technically pure (DOT)
Benzoylperoxyde (Dutch)
Benzoyl superoxide
BZF-60
Cadet
Cadox
Cadox BS
Clearasil Benzoyl Peroxide Lotion
Clearasil BP Acne Treatment
Cuticura Acne Cream
Debroxide
Dibenzoylperoxid (German)
Dibenzoyl peroxide
Dibenzoylperoxyde (Dutch)
Diphenylglyoxal peroxide
Dry and Clear
Epi-Clear
Fostex
Garox
Incidol
Loroxide
Lucidol
Luperco
Luperox FL
NA 2085 (DOT)
Nayper B and BO
Norox BZP-250
Norox BZP-C-35
Novadelox
Oxy-5
Oxy-10
Oxylite
Oxy Wash
Panoxyl
Perossido di benzoile (Italian)
Peroxide, dibenzoyl
Peroxyde de benzoyle (French)
Persadox

Quinolor Compound
Sulfoxyl
Superox
Theraderm
Topex
UN 2085 (DOT)
UN 2086
UN 2087 [Benzoyl peroxide, Not more than 72% as a paste]
UN 2088
UN 2089 [Benzoyl peroxide, Not less than 30% but not more than 52% with inert solid]
UN 2090 [Benzoyl peroxide, Not more than 77% with water]
Vanoxide
Xerac

94-41-7
$C_{15}H_{12}O$
208.27
O=C(c(cccc1)c1)C=Cc(cccc2)c2
Chalcone
Acrylophenone, 3-phenyl-
Benzalacetophenone
2-Benzalacetophenone
1-Benzoyl-1-phenylethene
β-Benzoylstyrene
Benzylideneacetophenone
α-Benzylideneacetophenone
2-Benzylideneacetophenone
Benzylideneacetophenone
Chalkone
Cinnamophenone
1,3-Diphenylpropenone
1,3-Diphenyl-1-propen-3-one
β-Phenylacrylophenone
3-Phenylacrylophenone
1-Phenyl-2-benzoylethylene
Phenyl 2-phenylvinyl ketone
Phenyl styryl ketone
2-Propen-1-one, 1,3-diphenyl-

94-45-1
$C_9H_{10}N_2OS$
194.25
Benzothiazole, 2-amino-6-ethoxy- (8CI)
2-Amino-6-ethoxybenzothiazole
2-Benzothiazolamine, 6-ethoxy- (9CI)
6-Ethoxy-2-aminobenzothiazole
NSC-28731

94-52-0
$C_7H_5N_3O_2$
163.15
O=N(=O)c(ccc(NC=N1)c12)c2
Benzimidazole, 6-nitro
NCI-C01912
5-Nitro-1H-benzimidazole
6-Nitro-benzimidazole

94-53-1
$C_8H_6O_4$
166.13
O=C(O)c(ccc(OCO1)c12)c2
Piperonylic acid (8CI)
AI3-05972
1,3-Benzodioxole-5-carboxylic acid (9CI)
Benzoic acid, 3,4-(methylenedioxy)-
Heliotropic acid
3,4-Methylenedioxybenzoic acid
NSC-10072
Protocatechuic acid methylene ether

94-57-5
$C_8H_{11}NO_3S$
201.24
O=S(=O)(O)CNc(c(ccc1)C)c1
Methanesulfonic acid, ((2-methylphenyl)amino)- (9CI)
Methanesulfonic acid, o-toluidino- (8CI)
((2-Methylphenyl)amino)-methanesulfonic acid
NSC-11212
o-Toluidinomethanesulfonic acid

94-58-6
$C_{10}H_{12}O_2$
164.22
O(c(c(O1)cc(c2)CCC)c2)C1
Benzene, 1,2-methylenedioxy-4-propyl
Dihydrosafrole
1,2-(Methylenedioxy)-4-propylbenzene
5-Propyl-1,3-benzodioxole
4-Propyl-1,2-methylenedioxybenzene
RCRA waste number U090
Safrole, dihydro-

94-59-7
$C_{10}H_{10}O_2$
162.20
O(c(c(O1)cc(c2)CC=C)c2)C1
Benzene, 4-allyl-1,2-(methylenedioxy)
5-Allyl-1,3-benzodioxole
Allylcatechol methylene ether
Allyldioxybenzene methylene ether
1-Allyl-3,4-methylenedioxybenzene
4-Allyl-1,2-methylenedioxybenzene
m-Allylpyrocatechin methylene ether
4-Allylpyrocatechol formaldehyde acetal
Allylpyrocatechol methylene ether
Benzene, 1,2-methylenedioxy-4-allyl-
1,3-Benzodioxole, 5-allyl-
1,2-Methylenedioxy-4-allylbenzene
3,4-Methylenedioxy-allybenzene
5-(2-Propenyl)-1,3-benzodioxole
RCRA waste number U203
Rhyuno Oil
Safrol
Safrole
Safrole MF
Shikimole
Shikomol

94-60-0
$C_{10}H_{16}O_4$
200.23
O=C(OC)C(CCC(C(=O)OC)C1)C1
Dimethyl hexahydroterephthalate
AI3-28580
1,4-Cyclohexanedicarboxylic acid, dimethyl ester
1,4-Cyclohexanedicarboxylic dimethyl ester
Dimethyl 1,4-cyclohexanedicarboxylate

94-62-2
$C_{17}H_{19}NO_3$
285.37
O=C(N(CCCC1)C1)C=CC=Cc(ccc(OCO2)c23)c3
Piperidine, 1-piperoyl-, (E,E)
1,3-Benzodioxol-5-yl-1-oxo-2,4-pentadienyl-piperine
Piperidine, 1-(5-(1,3-benzodioxol-5-yl)-1-oxo-2,4-pentadienyl)-, (E,E)- (9CI)
Piperin
Piperine
1-Piperoylpiperidine

94-65-5
$C_9H_{16}O$
140.23
O=C(C(CCC1)CCC)C1
Cyclohexanone, 2-propyl- (9CI)
NSC-54015
2-Propylcyclohexanone

94-68-8
$C_9H_{13}N$
135.23
N(c(c(ccc1)C)c1)CC
o-Toluidine, N-ethyl
Benzenamine, N-ethyl-2-methyl- (9CI)
2-(Ethylamino)toluene
N-Ethyl-o-toluidine [UN 2754]
UN 2754 [N-Ethyltoluidines]

94-69-9
C_8H_9NO
135.16
O=CNc(c(ccc1)C)c1
Formamide, N-(2-methylphenyl)- (9CI)
AI3-01417

o-Formotoluidide (8CI)
o-Methylformanilide
o-Methyl-N-formylaniline
N-(2-Methylphenyl)formamide
NSC-406128
o-Tolylformamide

94-70-2
$C_8H_{11}NO$
137.20
O(c(c(N)ccc1)c1)CC
PN:o-Phenetidine [UN 2311]
2-Aminophenetole
o-Ethoxyaniline
2-Ethoxyaniline
UN 2311 [Phenetidines]

94-71-3
$C_8H_{10}O_2$
138.17
O(c(c(O)ccc1)c1)CC
Phenol, o-ethoxy- (8CI)
Catechol monoethyl ether
o-Ethoxyphenol
2-Ethoxyphenol
2-Ethyloxyphenol
Guaethol
Guaiethol
Guethol
NSC-1809
Phenol, 2-ethoxy- (9CI)

94-74-6
$C_9H_9ClO_3$
200.63
O=C(O)COc(c(cc(c1)Cl)C)c1
Acetic acid, ((4-chloro-o-tolyl)-oxy)
Acetic acid, (4-chloro-2-methyl-phenoxy)-
Acme MCPA Amine 4
Agritox
Agroxon
Agroxone
Anicon Kombi
Anicon M
BH MCPA
Bordermaster
Brominal M & Plus
B-Selektonon M
Chiptox
4-Chloro-o-cresoxyacetic acid
(4-Chloro-2-methylphenoxy)-acetic acid
4-Chloro-o-toloxyacetic acid
((4-Chloro-o-tolyl)oxy)acetic acid
Chwastox
Cornox-M
Ded-Weed
Dicopur-M
Dicotex
Dikotes
Dikotex
Dow MCP Amine Weed Killer
Emcepan
Empal
Hedapur M 52
Hedarex M
Hedonal M
Herbicide M
Hormotuho
Hornotuho
Kilsem
4K-2M
Krezone
Kyselina 4-chlor-2-methyl-fenoxyoctova (Czech)
Kwas 4-chloro-2-metylofeno-ksyoctowy (Polish)
Legumex DB
Leuna M
Leyspray
Linormone
M 40
2M-4C
2M-4CH
MCP
MCPA
2,4-MCPA
Mephanac
Metaxon
Methoxone
2-Methyl-4-chlorophenoxy-acetic acid
2-Methyl-4-chlorphenoxyessig-saeure (German)
2M-4KH
Netazol
Okultin M
Phenoxylene 50
Phenoxylene Plus
Phenoxylene Super
Raphone
Razol Dock Killer
Rhomenc
Rhomene
Rhonox
Shamrox
Seppic MMD
Soviet Technical Herbicide 2M-4C
Trasan
U 46
U 46 M-Fluid
Ustinex
Vacate
Vesakontuho MCPA
Verdone
Weedar
Weedar MCPA Concentrate
Weedone
Weedone MCPA Ester
Weed-Rhap
Zelan

94-75-7
$C_8H_6Cl_2O_3$
221.04
O=C(O)COc(c(cc(c1)Cl)Cl)c1
Acetic acid, (2,4-dichloro-phenoxy)
Acide 2,4-dichloro phenoxy-acetique (French)
Acido(2,4-dicloro-fenossi)-acetico (Italian)
Acme Amine 4
Acme butyl ester 4
Acme LV 4
Agrotect
Amidox
Amoxone
Aqua-Kleen
BH 2,4-D
Brush-Rhap
B-Selektonon
Chipco Turf Herbicide "D"
Chloroxone
Crop Rider
Crotilin
2,4-D (ACGIH,DOT,OSHA)
D 50
Dacamine
2,4-D Acid
Debroussaillant 600
Decamine
Ded-Weed
Ded-Weed LV-69
Desormone
(2,4-Dichloor-fenoxy)-azijnzuur (Dutch)
Dichlorophenoxyacetic acid
2,4-Dichlorophenoxyacetic acid (DOT)
Dichlorophenoxyacetic acid (OSHA)
2,4-Dichlorphenoxyacetic acid
(2,4-Dichlor-phenoxy)-essig-saeure (German)
Dicopur
Dicotox
Dinoxol
DMA-4
Dormone
2,4-Dwuchlorofenoksyoctowy kwas (Polish)
Emulsamine BK
Emulsamine E-3
ENT 8,538
Envert 171
Envert DT
Esteron
Esteron 99
Esteron 76 BE
Esteron Brush Killer
Esteron 99 Concentrate
Esterone Four
Esteron 44 Weed Killer
Estone
Farmco
Fernesta
Fernimine
Fernoxone
Ferxone
Foredex 75
Formula 40
Hedonal
Hedonal (The herbicide)
Herbidal
Ipaner
Kwasu 2,4-dwuchlorofenoksy-octowego (Polish)
Kwas 2,4-dwuchlorofenoksy-octowy (Polish)
Kyselina 2,4-dichlorfenoxy-octova (Czech)
Lawn-Keep
Macrondray
Miracle
Monosan
Moxone
NA 2765 (DOT)

Netagrone
Netagrone 600
NSC-423
Pennamine
Pennamine D
Phenox
Pielik
Planotox
Plantgard
RCRA waste number U240
Rhodia
Salvo
Spritz-Hormin/2,4-D
Spritz-Hormit/2,4-D
Super D Weedone
Superormone Concentre
Transamine
Tributon
Trinoxol
U 46
U-5043
U 46DP
Vergemaster
Verton
Verton D
Verton 2D
Vertron 2D
Vidon 638
Visko-Rhap
Visko-Rhap Low Drift Herbicides
Visko-Rhap Low Volatile 4L
Weed-AG-Bar
Weedar
Weedar-64
Weedatul
Weed-B-Gon
Weedez Wonder Bar
Weedone
Weedone LV4
Weed-Rhap
Weed Tox
Weedtrol

94-76-8
$C_9H_9ClO_2S$
216.69
Acetic acid, ((4-chloro-2-methyl)phenyl)thio
4-Chloro-2-methylphenylthioglycolic acid
Chlorotolylthioglycolic acid
Red 3B Acid

94-78-0
$C_{11}H_{11}N_5$
213.27
Nc2ccc(N=Nc1ccccc1)c(N)n2
Pyridine, 2,6-diamino-3-(phenylazo)
2,6-Diamino-3-phenylazopyridine
Diridone
DPP
Gastracid
Gastrotest
Mallophene
NC 150
AP
Phenazodine
3-(Phenylazo)-2,6-pyridinediamine
Phenylazo tablet
Phenazopyridine
Pirid
Pyrazofen
Pyridacil
2,6-Pyridinediamine, 3-(phenylazo)-
Pyridium
Pyripyridium
Sedural
Uridinal
Urodine
W 1655

94-80-4
$C_{12}H_{14}Cl_2O_3$
277.16
CCCCOC(=O)COc1ccc(Cl)cc1Cl
Acetic acid, (2,4-dichlorophenoxy)-, n-butyl ester (8CI,9CI)
AI3-08686
Butapon
Butyl 2,4-D
Butyl dichlorophenoxyacetate
Butyl (2,4-dichlorophenoxy)-acetate
Butyl ester 2,4-D
Butyl ester of 2,4-D
n-Butylester kyseliny 2,4-dichlorfenoxyoctove (Czech)
Caswell No. 315AL
2,4-D-butyl
2,4-D butyl ester
2,4-D n-butyl ester
2,4-DBE
2,4-Dichlorophenoxyacetic acid butyl ester
(2,4-Dichlorophenoxy)acetic acid, butyl ester
EPA Pesticide Chemical Code 030056
Esso Herbicide 10
Fernesta
Hi-Ester 2,4-D
Lironox
NSC-409767
Shell 40

94-81-5
$C_{11}H_{13}ClO_3$
228.69
Cc1cc(Cl)ccc1OCCCC(O)=O
Butyric acid, 4-((4-chloro-o-tolyl)oxy)
Bexane
Bexone
Butanoic acid, 4-(4-chloro-2-methylphenoxy)-
Can-Trol
4-(4-Chlor-2-methylphenoxy)-buttersaeure (German)
4-(4-Chlor-2-methyl-phenoxy)-buttersaeure (German)
γ-(4-Chloro-2-methylphenoxy)-butyric acid
4-(4-Chloro-2-methylphenoxy)-butyric acid
(4-Chloro-o-tolyloxy)butyric acid
4-((4-Chloro-o-tolyl)oxy)butyric acid
Kyselina 4-(4-chlor-2-methyl-fenoxy)maselna (Czech)
Legumex
MB 3046
4-(MCB)
MCPB
γ-MCPB
2,4-MCPB
4MCPB
MCP-Butyric
2-Methyl-4-chlorophenoxy-butyric acid
γ-2-Methyl-4-chlorophenoxy-butyric acid
4-(2-Methyl-4-chlorophenoxy)-butyric acid
4-(2-Methyl-4-chlorphenoxy)-buttersaeure (German)
2M 4KhM
PDQ
Thistrol
Trifolex
Tritrol
Tropotox
Trotox
U46 MCPB

94-82-6
$C_{10}H_{10}Cl_2O_3$
249.10
OC(=O)CCCOc1ccc(Cl)cc1Cl
Butyric acid, 4-(2,4-dichlorophenoxy)
Butoxon
Butoxone
Butoxone amine
Butoxone ester
Butyrac
Butyrac 118
Butyrac 200
Butyrac ester
2,4-DB
4(2,4-DB)
2,4-D butyric
γ-(2,4-Dichlorophenoxy)butyric acid
4-(2,4-Dichlorophenoxy)butyric acid
2,4-DM
Embutox
Embutox E
Embutox Klean-Up
Kyselina 4-(2,4-dichlorfenoxy)-maselna (Czech)
Legumex D

94-83-7
$C_{15}H_{12}Cl_2O_3$
311.17
Clc2ccc(OCCOC(=O)c1ccccc1)c(Cl)c2
Ethanol, 2-(2,4-dichlorophenoxy)-, benzoate
2,4-DEB
2-(2,4-Dichlorophenoxy)ethyl benzoate

94-84-8
$C_{24}H_{21}Cl_6O_6P$
649.12
Ethanol, 2-(2,4-dichlorophenoxy)-, phosphite (3:1)
2,4-DEP
Falone-44-E
Falone E 44
Phosphorous acid, tris(2-(2,4-dichlorophenoxy)ethyl ester
Tri(2,4-dichlorophenoxyethyl) phosphite
Tris(2,4-dichlorophenoxyethyl) phosphite
3Y9

94-91-7
$C_{17}H_{18}N_2O_2$
282.37
Oc(c(ccc1)C=NCC(N=Cc(c(O)ccc2)c2)C)c1
o-Cresol, α,α'-(propylenedinitrilo)di
Carlisle metal deactivator
Copper Inhibitor 50
Cuvan 80
α,α'-Dipropylenedinitrilodi-o-cresol
Disalicylalpropylenediimine
N,N'-Disalicylidene-1,2-diaminopropane
N,N'-Disalicylidene-1,2-propanediamine
DMD
Du Pont Metal Deactivator
Keromet MD
Phenol, 2,2'-((1-methyl-1,2-ethanediyl)bis(nitrilomethylidyne))bis- (9CI)
Tenamene 60
UOP Copper Deactivator

94-96-2
$C_8H_{18}O_2$
146.26
OCC(C(O)CCC)CC
1,3-Hexanediol, 2-ethyl
Carbide 6-12
Compound 6-12 insect repellent
ENT 375
Ethohexadiol
Ethyl hexanediol
2-Ethyl-1,3-hexanediol
2-Ethylhexane-1,3-diol
2-Ethylhexanediol-1,3
Ethyl hexylene glycol
2-Ethyl-3-propyl-1,3-propanediol
Latka 612 (Czech)
3-Hydroxymethyl-n-heptan-4-ol
6-12-Insect Repellent
Octylene glycol
Repellent 612
Rutgers 612

94-97-3
$C_6H_4ClN_3$
153.55
N(=NNc1ccc(c2)Cl)c12
1H-Benzotriazole, 5-chloro- (9CI)
AI3-52175
5-Chlorobenzotriazole
5-Chloro-1H-benzotriazole
6-Chlorobenzotriazole
NSC-16507

94-99-5
$C_7H_5Cl_3$
195.47
c(ccc(c1Cl)CCl)(c1)Cl
Benzene, 2,4-dichloro-1-(chloromethyl)- (9CI)
AI3-14886
2,4-Dichlorobenzyl chloride
2,4-Dichloro-1-(chloromethyl)-benzene
NSC-406892
Toluene, α2,4-trichloro- (8CI)
α,2,4-Trichlorotoluene

95-01-2
$C_7H_6O_3$
138.13
O=Cc(c(O)cc(O)c1)c1
β-Resorcylaldehyde
Benzaldehyde, 2,4-dihydroxy-
2,4-Dihydroxybenzaldehyde
2,4-Dihydroxybenzenecarbonal
4-Formylresorcinol
4-Hydroxysalicyladehyde
β-Resorcinaldehyde
β-Resorcylic aldehyde
β-Rosorcaldehyde

95-06-7
$C_8H_{14}ClNS_2$
223.80
CCN(CC)C(=S)SCC(Cl)=C
Carbamic acid, diethyldithio-, 2-chloroallyl ester
Carbamodithioic acid, diethyl-, 2-chloro-2-propenyl ester (9CI)
CDEC
Chlorallyl diethyldithiocarbamate
2-Chloroallyl diethyldithiocarbamate
2-Chloroallyl N,N-diethyldithiocarbamate
2-Chloro-2-propene-1-thiol diethyldithiocarbamate
2-Chloro-2-propenyl diethylcarbamodithioate
CP 4572
CP 4,742
Diethylcarbamodithioic acid 2-chloro-2-propenyl ester
Diethyldithiocarbamic acid 2-chloroallyl ester
NCI-C00453
2-Propene-1-thiol, 2-chloro-, diethyldithiocarbamate
Sulfallate
Thioallate
Vegadex
Vegadex Super
Vegedex

95-08-9
$C_{18}H_{34}O_6$
346.52
O=C(OCCOCCOCCOC(=O)C(CC)CC)C(CC)CC
Butyric acid, 2-ethyl-, diester with triethylene glycol
2-Ethylbutyric acid, diester with triethylene glycol
2-Ethylbutyric acid, triethylene glycol diester
2,2'-(Ethylenedioxy)di(ethyl 2-ethylbutyrate)
Flexol Plasticizer 3GH
Plasticizer 3GH
Triethylene glycol, bis(2-ethylbutyrate)
Triethyleneglycol diethyl butyrate
Triethylene glycol di(2-ethyl butyrate)
Triglycol dicaproate
Triglycol dihexoate

95-12-5
$C_8H_{12}O$
124.18
OCC(C(C=CC12)C1)C2
Bicyclo(2.2.1)hept-5-ene-2-methanol (9CI)
AI3-08981
Cyclol
2-(Hydroxymethyl)bicyclo-(2.2.1)hept-5-ene
5-Hydroxymethylbicyclo(2.2.1)-hept-2-ene
2-Hydroxymethyl-5-norbornene
5-Hydroxymethyl-2-norbornene
5-Norbornene-2-methanol (8CI)
NSC-403110

95-13-6
C_9H_8
116.17
c(c(C=C1)ccc2)(c2)C1
Indene (ACGIH,OSHA) (8CI)
Inden
1H-Indene (9CI)
Indonaphthene

95-14-7
$C_6H_5N_3$
119.11
N(=NNc1cccc2)c12
Benzotriazole
AI3-15984
1,2-Aminoazophenylene
1,2-Aminozophenylene
1,2,-Aminozophenylene
Aziminobenzene
Benzene azimide
Benzisotriazole
Benzotriazole (VAN)
1H-Benzotriazole (VAN) (9CI)
1H-1,2,3-Benzotriazole

1,2,3-Benzotriazole
Benztriazole
Cobratec 99
Cobratec #99
2,3-Diazaindole
NCI-C03521
NSC-3058
NSC-3058
1,2,3-Triazaindene
1,2,3-Triaza-1H-indene
U-6233

95-15-8
C_8H_6S
134.20
S(c(c(C=1)ccc2)c2)C1
Benzo(b)thiophene (9CI)
AI3-15523
Benzothiofuran
Benzothiophen
1-Benzothiophene
2,3-Benzothiophene
NSC-47196
1-Thiaindene
Thianaphtene
Thianaphthen
Thianaphthene
Thionaphthene

95-16-9
C_7H_5NS
135.19
N(c(c(S1)ccc2)c2)=C1
Benzothiazole
Benzosulfonazole
O-2857
1-Thia-3-azaindene
USAF EK-4812

95-19-2
$C_{22}H_{46}N_2O$
354.70
OCCN(C(=NC1)CCCCCCCCCCCCCCCCC)C1
2-Imidazoline-1-ethanol, 2-heptadecyl
Fungacide 337
2-Heptadecyl-1-hydroxyethyl-imidazoline
2-Imidazoline, 2-heptadecyl-1-hydroxyethyl-

95-20-5
C_9H_9N
131.17
N(c(c(C=1)ccc2)c2)C1C
2-Methylindole
AI3-03945
Indole, 2-methyl- (8CI)
1H-Indole, 2-methyl- (9CI)
2-Methyl-1H-indole
NSC-7514

95-29-4
$C_{13}H_{18}N_2OS_2$
282.45
N(c(c(S1)ccc2)c2)=C1SN(C(C)C)C(C)C
2-Benzothiazolesulfenamide, N,N-diisopropyl
N,N-Diisopropyl-2-benzothiazolesulfenamide
Dipac

95-31-8
$C_{11}H_{14}N_2S$
206.33
N(c(c(S1)ccc2)c2)=C1SNC(C)(C)C
2-Benzothiazolesulfenamide, N-tert-butyl
Accel BNS
N-tert-Butyl-2-benzothiazole-sulfenamide
Benzothiazolesulfenamide, N-(1,1-dimethylethyl)- (9CI)
Nocceler NS
Pennac TBBS
Santocure NS
Vanax NS
Vulkacit NZ

95-32-9
$C_{11}H_{12}N_2OS_3$
284.42
O(CCN(SSC(=Nc(c1ccc2)c2)S1)C3)C3
2-(4-Morpholinyldithio)benzothiazole
AI3-27133
Benzothiazole, 2-(morpholinodithio)- (8CI)
Benzothiazole, 2-(4-morpholinyldithio)- (9CI)
2-Benzothiazolyl morpholino disulfide
Morfax
Morpholino 2-benzothiazolyl disulfide
2-(Morpholinodithio)benzothiazole
2-(4-Morpholinodithio)benzothiazole
2-(Morpholinothio)benzothiazole
N-Morpholinyl-2-benzothiazolyl disulfide
4-Morpholinyl 2-benzothiazyl disulfide
N-Oxydiethyl-2-benzthiazolsulfenamid (Czech)
NSC-519695
Sulfenax MOB (Czech)
Vulcuren 2
Vulcuren-2

95-33-0
$C_{13}H_{16}N_2S_2$
264.43
N(c(c(S1)ccc2)c2)=C1SNC(CCCC3)C3
2-Benzothiazolesulfenamide, N-cyclohexyl
N-Cyclohexyl-2-benzothiazole-sulfenamide
Durax
Pennac CBS
Santocure
Sulfenamide TS

95-35-2
$C_{17}H_{14}N_4OS_4$
418.59
O=C(NCSC(=Nc(c1ccc2)c2)S1)NCSC(=Nc(c3ccc4)c4)S3
Urea, 1,3-bis(2-benzothiazolylthiomethyl)
N,N'-Bis(2-benzothiazolylthiomethylene)urea
1,3-Bis((2-benzothiazolylthio)-methyl)urea
El 60

95-38-5
$C_{22}H_{42}N_2O$
350.66
OCCN(C(=NC1)CCCCCCCC=CCCCCCCCC)C1
2-Imidazoline-1-ethanol, 2-(8-heptadecenyl)
Amine 220
2-(8-Heptadecenyl)-2-imidazoline-1-ethanol
1-Hydroxyethyl-2-heptadecenyl-glyoxalidine
1-(2-Hydroxyethyl)-2-heptadecenylglyoxalidine
1-(2-Hydroxyethyl)-2-n-heptadecenyl-2-imidazoline
1-(2-Hydroxyethyl)-2-(8-heptadecenyl)-2-imidazoline
1-(2-Hydroxyethyl)-2-heptadecenyl-2-imidazoline
Nalcamine G-13

95-39-6
$C_{11}H_{14}O_2$
178.25
O=C(OCC(C(C=CC12)C1)C2)C=C
Acrylic acid, 5-norbornen-2-ylmethyl ester
Acrylic acid, 5-norbornen-2-methyl ester
Bicyclo(2.2.1)hept-5-ene-2-methylol acrylate
Cyclol acrylate
2,5-endo-Methylene-δ^3-tetrahydrobenzyl acrylate
5-Norbornene-2-methanol, acrylate
5-Norbornene-2-methylol-acrylate
2-Propenoic acid, bicyclo-(2,2,1)hept-5-en-2-ylmethyl ester (9CI)

95-45-4
$C_4H_8N_2O_2$
116.14
N(O)=C(C(=NO)C)C
2,3-Butanedione, dioxime
Biacetyl, dioxime
Diacetyldioxime
2,3-Diisonitrosobutane

Dimethylglyoxime
Glyoxime, dimethyl-

95-46-5
C_7H_7Br
171.05
c(c(ccc1)Br)(c1)C
Toluene, o-bromo
Benzene, 1-bromo-2-methyl- (9CI)
o-Bromotoluene
2-Bromotoluene
2-Methylbromobenzene
o-Methylphenyl bromide
o-Tolylbromide
2-Tolyl bromide

95-47-6
C_8H_{10}
106.18
c(c(ccc1)C)(c1)C
o-Xylene (ACGIH,DOT, OSHA) [UN 1307]
o-Dimethylbenzene
1,2-Dimethylbenzene
o-Methyltoluene
UN 1307 [Xylenes]
1,2-Xylene
o-Xylol [UN 1307]

95-48-7
C_7H_8O
108.15
Oc(c(ccc1)C)c1
o-Cresol (OSHA) [UN 2076]
2-Cresol
o-Cresylic acid
1-Hydroxy-2-methylbenzene
o-Hydroxytoluene
2-Hydroxytoluene
o-Kresol (German)
o-Methylphenol
2-Methylphenol
o-Methylphenylol
Orthocresol
o-Oxytoluene
Phenol, 2-methyl- (9CI)
RCRA waste number U052
o-Toluol
UN 2076 [Cresols (o-; m-; p-)]

95-49-8
C_7H_7Cl
126.59
c(c(ccc1)Cl)(c1)C
Toluene, o-chloro
Benzene, 1-chloro-2-methyl- (9CI)
2-Chloro-1-methylbenzene
o-Chlorotoluene (ACGIH, OSHA) [UN 2238]
2-Chlorotoluene
Halso 99
1-Methyl-2-chlorobenzene
2-Methylchlorobenzene
o-Tolyl chloride
UN 2238 [Chlorotoluenes]

95-50-1
$C_6H_4Cl_2$
147.00
c(c(ccc1)Cl)(c1)Cl
Benzene, o-dichloro
Benzene, 1,2-dichloro-
Chloroben
Chloroden
Cloroben
DCB
o-Dichlorbenzene
o-Dichlor benzol
Dichlorobenzene, ortho, Liquid [UN 1591]
o-Dichlorobenzene (ACGIH, OSHA) [UN 1591]
1,2-Dichlorobenzene
Dilantin DB
Dilatin DB
Dizene
Dowtherm E
NCI-C54944
ODB
ODCB
Orthodichlorobenzene
Orthodichlorobenzol
Special Termite Fluid
Termitkil
UN 1591 [o-Dichlorobenzene]

95-51-2
C_6H_6ClN
127.58
Nc(c(ccc1)Cl)c1
Aniline, o-chloro
1-Amino-2-chlorobenzene
Benzenamine, 2-chloro- (9CI)
o-Chloraniline
o-Chloroaniline
2-Chloroaniline
o-Chloroaniline, Liquid [UN 2019]
o-Chloroaniline, Solid [UN 2018]
Fast Yellow GC Base
UN 2018 [Chloroanilines, solid]
UN 2019 [Chloroanilines, Liquid]

95-52-3
C_7H_7F
110.14
Fc(c(ccc1)C)c1
Toluene, o-fluoro
Benzene, 1-fluoro-2-methyl- (9CI)
o-Fluorotoluene [UN 2388]
2-Fluorotoluene
UN 2388 [Fluorotoluenes]

95-53-4
C_7H_9N
107.17
Nc(c(ccc1)C)c1
o-Toluidine (ACGIH,DOT, OSHA)
1-Amino-2-methylbenzene
2-Amino-1-methylbenzene
o-Aminotoluene
2-Aminotoluene
Aniline, 2-methyl-
Benzenamine, 2-methyl- (9CI)
C.I. 37077
1-Methyl-2-aminobenzene
2-Methyl-1-aminobenzene
o-Methylaniline
2-Methylaniline
o-Methylbenzenamine
2-Methylbenzenamine
RCRA waste number U328
o-Toluidin (Czech)
2-Toluidine
o-Toluidine, Solid [UN 1708]
o-Toluidyna (Polish)
o-Tolylamine
UN 1708 [Toluidines solid]

95-54-5
$C_6H_8N_2$
108.16
Nc(c(N)ccc1)c1
o-Phenylenediamine
2-Aminoaniline
o-Benzenediamine
1,2-Benzenediamine
C.I. 76010
C.I. Oxidation Base 16
o-Diaminobenzene
1,2-Diaminobenzene
EK 1700
o-Fenylendiamin (Czech)
1,2-Fenylendiamin (Czech)
NSC-5354
Orthamine
1,2-Phenylenediamine
o-Phenylenediamine [UN 1673]
SQ 15500
UN 1673 [Phenylenediamines (o-; m-; p-;)]

95-55-6
C_6H_7NO
109.14
Oc(c(N)ccc1)c1
Phenol, o-amino
2-Amino-1-hydroxybenzene
o-Aminophenol [UN 2512]
2-Aminophenol
BASF Ursol 3GA
Benzofur GG
C.I. 76520
C.I. Oxidation Base 17
Fouramine OP
o-Hydroxyaniline
2-Hydroxyanaline
Nako Yellow 3GA
Paradone Olive Green B
Pelagol 3GA
Pelagol Grey GG
Questiomycin B
UN 2512 [Aminophenols (o-; m-; p-)]
Zoba 3GA

95-56-7
C_6H_5BrO
173.02
Oc(c(ccc1)Br)c1
Phenol, o-bromo

2-Bromfenol (Czech)
o-Bromophenol
2-Bromophenol

95-57-8
C_6H_5ClO
128.56
Oc(c(ccc1)Cl)c1
Phenol, o-chloro
o-Chlorophenol
2-Chlorophenol
o-Chlorophenol, Liquid [UN 2021]
o-Chlorophenol, Solid [UN 2020]
o-Chlorphenol (German)
Phenol, 2-chloro-
RCRA waste number U048
UN 2020 [Chlorophenols, solid]
UN 2021 [Chlorophenols, Liquid]

95-59-0
$C_4H_6Cl_2O_2$
157.00
O(C(Cl)C(OC1)Cl)C1
2,3-Dichloro-1,4-dioxane
AI3-26714
p-Dioxane, 2,3-dichloro- (8CI)
1,4-Dioxane, 2,3-dichloro- (9CI)
2,3-Dichloro-p-dioxane
NSC-2293

95-63-6
C_9H_{12}
120.21
c(ccc(c1C)C)(c1)C
Benzene, 1,2,4-trimethyl
Asymmetrical trimethylbenzene
Benzene, 1,2,5-trimethyl-
psi-Cumene
Pseudocumene
Pseudocumol
as-Trimethylbenzene
1,2,4-Trimethylbenzene
Trimethyl benzene (ACGIH)

95-64-7
$C_8H_{11}N$
121.20
Nc(ccc(c1C)C)c1
3,4-Xylidine
Aniline, 3,4-dimethyl
Benzenamine, 3,4-dimethyl- (9CI)
3,4-Dimethylaminobenzene
3,4-Dimethylaniline
3,4-Dimethylphenylamine
3,4-Xylylamine

95-65-8
$C_8H_{10}O$
122.18
Oc(ccc(c1C)C)c1
3,4-Xylenol
3,4-Dimethylphenol
4,5-Dimethylphenol
3,4-DMP
1,3,4-Xylenol

95-68-1
$C_8H_{11}N$
121.20
Nc(c(cc(c1)C)C)c1
2,4-Xylidine
1-Amino-2,4-dimethylbenzene
4-Amino-1,3-dimethylbenzene
4-Amino-3-methyltoluene
4-Amino-1,3-xylene
Aniline, 2,4-dimethyl-
2,4-Dimethylaniline
2,4-Dimethylbenzenamine
2,4-Dimethylphenylamine
2-Methyl-p-toluidine
4-Methyl-o-toluidine
UN 1711 [Xylidines, solid or solution]
m-Xylidine
m-4-Xylidine
m-Xylidine [UN 1711]

95-69-2
C_7H_8ClN
141.61
Nc(c(cc(c1)Cl)C)c1
o-Toluidine, 4-chloro
Amarthol Fast Red TR Base
2-Amino-5-chlorotoluene
Azoene Fast Red TR Base
Azogene Fast Red TR
Azoic Diazo Component 11, Base
Brentamine Fast Red TR Base
5-Chloro-2-aminotoluene
4-Chloro-2-methylaniline
4-Chloro-2-methylbenzeneamine
4-Chloro-2-toluidine
4-Chloro-6-methylaniline
4-Chloro-o-toluidine
Daito Red Base TR
Deval Red K
Deval Red TR
Diazo Fast Red TRA
Fast Red Base TR
Fast Red 5CT Base
Fast Red TR
Fast Red TR11
Fast Red TR Base
Fast Red TRO Base
Kako Red TR Base
Kambamine Red TR
2-Methyl-4-chloroaniline
Mitsui Red TR Base
Red Base Ciba IX
Red Base Irga IX
Red Base NTR
Red TR Base
Sanyo Fast Red TR Base
Tulabase Fast Red TR

95-70-5
$C_7H_{10}N_2$
122.19
Nc(c(cc(N)c1)C)c1
Toluene-2,5-diamine
4-Amino-2-methylaniline
1,4-Benzenediamine, 2-methyl-
C.I. 76042
2,5-Diaminotoluene
2-Methyl-1,4-benzenediamine
2-Methyl-p-phenylenediamine
p-Toluenediamine
p-Toluylendiamine
Toluylene-2,5-diamine
p,m-Tolylenediamine

95-71-6
$C_7H_8O_2$
124.15
Oc(c(cc(O)c1)C)c1
Hydroquinone, methyl
1,4-Benzenediol, 2-methyl-
2,5-Dihydroxytoluene
2-Methyl-1,4-benzenediol
Methylhydroquinone
Methyl-p-hydroquinone
2-Methylhydroquinone
THQ
2,5-Toluenediol
p-Toluhydroquinol
Toluhydroquinone
p-Toluhydroquinone
Toluquinol
p-Toluquinol
Tolylhydroquinone

95-73-8
$C_7H_6Cl_2$
161.03
c(ccc(c1Cl)C)(c1)Cl
Toluene, 2,4-dichloro
Benzene, 2,4-dichloro-1-methyl- (9CI)
2,4-Dichlorotoluene

95-74-9
C_7H_8ClN
141.61
Nc(ccc(c1Cl)C)c1
p-Toluidine, 3-chloro
1-Amino-3-chloro-4-methyl-benzene
4-Amino-2-chlorotoluene
2-Chloro-4-aminotoluene
3-Chloro-4-methylaniline
3-Chloro-p-toluidine
CPT
DKC 1347
DRC 1339
NCI-C02040

95-75-0
$C_7H_6Cl_2$
161.03
c(ccc(c1Cl)Cl)(c1)C
Benzene, 1,2-dichloro-4-methyl- (9CI)
AI3-02581
1,2-Dichloro-4-methylbenzene
3,4-Dichlorotoluene
NSC-8765
Toluene, 3,4-dichloro- (8CI)

95-76-1
$C_6H_5Cl_2N$
162.02
Nc(ccc(c1Cl)Cl)c1
Aniline, 3,4-dichloro
1-Amino-3,4-dichlorobenzene
Benzenamine, 3,4-dichloro- (9CI)
DCA
3,4-DCA
3,4-Dichloranilin
3,4-Dichloraniline
3,4-Dichloroaniline
4,5-Dichloroaniline
3,4-Dichlorobenzenamine

95-77-2
$C_6H_4Cl_2O$
163.00
Oc1ccc(Cl)c(Cl)c1
Phenol, 3,4-dichloro
3,4-Dichlorophenol

95-78-3
$C_8H_{11}N$
121.20
Nc(c(ccc1C)C)c1
2,5-Xylidine
1-Amino-2,5-dimethylbenzene
3-Amino-1,4-dimethylbenzene
2-Amino-1,4-xylene
Aniline, 2,5-dimethyl-
2,5-Dimethylaniline
2,5-Dimethylbenzenamine
2,5-Dimethylphenylamine
5-Methyl-o-toluidine
6-Methyl-m-toluidine
UN 1711 [Xylidines, solid or solution]
p-Xylidine
p-Xylidine [UN 1711]

95-79-4
C_7H_8ClN
141.61
Nc(c(ccc1Cl)C)c1
o-Toluidine, 5-chloro
Acco Fast Red KB Base
1-Amino-3-chloro-6-methyl-benzene
2-Amino-4-chlorotoluene
Ansibase Red KB
Azoene Fast Red KB Base
Azoic Diazo Component 32
4-Chloro-2-aminotoluene
3-Chloro-6-methylaniline
5-Chloro-2-methylaniline
5-Chloro-o-toluidine
Fast Red KB Amine
Fast Red KB Base
Fast Red KB Salt
Fast Red KB Salt Supra
Fast Red KBS Salt
Genazo Red KB Soln
Hiltonil Fast Red KB Base
Lake Red BK Base
Metrogen Red former KB Soln
Naphthosol Fast Red KB Base
NCI-C02051
Pharmazoid Red KB
Red KB Base
Spectrolene Red KB
Stable Red KB Base

95-80-7
$C_7H_{10}N_2$
122.19
Nc(c(ccc1N)C)c1
Toluene-2,4-diamine
3-Amino-p-toluidine
5-Amino-o-toluidine
Azogen Developer H
Benzofur MT
C.I. 76035
C.I. Oxidation Base
C.I. Oxidation Base 20
C.I. Oxidation Base 35
C.I. Oxidation Base 200
Developer 14
Developer B
Developer DB
Developer DBJ
Developer H
Developer MC
Developer MT
Developer MT-CF
Developer MTD
Developer T
1,3-Diamino-4-methylbenzene
2,4-Diamino-1-methylbenzene
2,4-Diaminotoluen (Czech)
Diaminotoluene
2,4-Diaminotoluene
2,4-Diamino-1-toluene
2,4-Diaminotoluol
Eucanine GB
Fouramine
Fouramine J
Fourrine 94
Fourrine M
4-Methyl-1,3-benzenediamine
4-Methyl-m-phenylenediamine
MTD
Nako TMT
NCI-C02302
Pelagol Grey J
Pelagol J
Pontamine Developer TN
RCRA waste number U221
Renal MD
TDA
Tertral G
2,4-Tolamine
m-Toluenediamine
meta Toluylene diamine
2,4-Toluenediamine
m-Toluylendiamin (Czech)
m-Toluylenediamine
2,4-Toluylenediamine [UN 1709]
m-Tolyenediamine
m-Tolylenediamine
Tolylene-2,4-diamine
2,4-Tolylenediamine
4-m-Tolylenediamine
UN 1709 [2,4-Toluylene-diamine]
Zoba GKE
Zogen Developer H

95-82-9
$C_6H_5Cl_2N$
162.02
Nc(c(ccc1Cl)Cl)c1
Aniline, 2,5-dichloro
Amarthol Fast Scarlet GG Base
Amarthol Fast Scarlet GGS Base
Azobase DCA
Azoene Fast Scarlet 2G Base
Azogene Fast Scarlet GGC
Azogene Fast Scarlet GG (Free base)
Azoic Diazo Component 3
Benzenamine, 2,5-dichloro- (9CI)
C.I. 37010
C.I. Azoic Diazo Component 3
Daito Scarlet Base GG
Devol Scarlet A (Free base)
Devol Scarlet 2GS Base
Diazo Fast Scarlet GG
2,5-Dichloranilin (Czech)
2,5-Dichloroaniline
Fast Red SGG Base
Fast Scarlet Base GGT
Fast Scarlet Base 2J
Fast Scarlet Base 2JS
Fast Scarlet DS Base
Fast Scarlet 2G
Fast Scarlet 2G Base
Fast Scarlet GG Base
Fast Scarlet GGS Base
Fast Scarlet MDC Base
Hiltonil Fast Scarlet 2G Base
Hiltonil Fast Scarlet 2GS Base
Hindamine Scarlet GG
Kambamine Scarlet GG Base
Kayaku Scarlet GG Base
Lake Scarlet GG Base
Mitsui Scarlet GG Base
Naphthanil Scarlet 2G Base
Naphtoelan Fast Scarlet GG Base
Sanyo Fast Scarlet GG Base
Scarlet Base Ciba I
Scarlet Base GG
Scarlet Base Irga I (Free base)
Scarlet Base NGG
Scarlet 2G Base
Spectrolene Scarlet 2G
Stabamine Scarlet GG
Symulon Scarlet 2G Base

95-83-0
$C_6H_7ClN_2$
142.60
Nc(c(N)cc(c1)Cl)c1
o-Phenylenediamine, 4-chloro
2-Amino-4-chloroaniline
1,2-Benzenediamine, 4-chloro- (9CI)
C.I. 76015
4-Chloro-1,2-diaminobenzene
p-Chloro-o-phenylenediamine
4-Chloro-o-phenylenediamine
4-Chloro-1,2-phenylenediamine
4-Cl-o-PD
1,2-Diamino -4-chlorobenzene
3,4-Diaminochlorobenzene

3,4-Diamino-1-chlorobenzene
NCI-C03292
Ursol Olive 6G

95-84-1
C_7H_9NO
123.17
Oc(c(N)cc(c1)C)c1
Phenol, 2-amino-4-methyl

95-85-2
C_6H_6ClNO
143.58
Oc(c(N)cc(c1)Cl)c1
Phenol, 2-amino-4-chloro
2-Amino-4-chlorophenol [UN 2673]
p-Chloro-o-aminophenol
C.I. 76525
C.I. Oxidation Base 18
Fouramine PY
UN 2673 [2-Amino-4-chloro-phenol]

95-86-3
$C_6H_8N_2O$
124.16
Nc1ccc(O)c(N)c1
Phenol, 2,4-diamino
2,4-Diaminophenol

95-87-4
$C_8H_{10}O$
122.18
Oc(c(ccc1C)C)c1
2,5-Xylenol
2,5-Dimethylphenol
3,6-Dimethylphenol
2,5-DMP
6-Methyl-m-cresol
UN 2261 [Xylenols]
p-Xylenol
1,2,5-Xylenol
p-Xylenol [UN 2261]

95-88-5
$C_6H_5ClO_2$
144.56
Oc(c(ccc1O)Cl)c1
Resorcinol, 4-chloro
4-Chlororesorcinol

95-92-1
$C_6H_{10}O_4$
146.16
O=C(OCC)C(=O)OCC
Oxalic acid, diethyl ester
Diethylester kyseliny stavelove (Czech)
Diethyl ethanedioate
Diethyl oxalate
Ethyl oxalate [UN 2525]
UN 2525 [Ethyl oxalate]

95-93-2
$C_{10}H_{14}$
134.24
c(c(cc(c1C)C)C)(c1)C
Benzene, 1,2,4,5-tetramethyl
Durene
1,2,4,5-Tetramethylbenzene

95-94-3
$C_6H_2Cl_4$
215.88
c(c(cc(c1Cl)Cl)Cl)(c1)Cl
Benzene, 1,2,4,5-tetrachloro
RCRA waste number U207
1,2,4,5-Tetrachlorobenzene

95-95-4
$C_6H_3Cl_3O$
197.44
Oc(c(cc(c1Cl)Cl)Cl)c1
Phenol, 2,4,5-trichloro
Collunosol
Dowcide 2
Dowicide 2
Dowicide B
NCI-C61187
Nurelle
Preventol I
RCRA waste number U230
2,4,5-Trichlorophenol

96-05-9
$C_7H_{10}O_2$
126.17
O=C(OCC=C)C(=C)C
2-Propenoic acid, 2-methyl-, 2-propenyl ester
Ageflex AMA
Allylester kyseliny methakryl-ove (Czech)
Allyl methacrylate
Methacrylic acid, allyl ester

96-06-0
$C_8H_{16}O$
128.21
Oxirane, 3-(1,1-dimethylethyl)-2,2-dimethyl- (9CI)
Pentane, 2,3-epoxy-2,4,4-tri-methyl- (8CI)

96-08-2
$C_{10}H_{16}O_2$
168.26
O(C1(C(CC(O2)C2(C3)C)C3)C)C1
p-Menthane, 1,2:8,9-diepoxy
1,2,8,9-Diepoxylimonene
1,2:8,9-Diepoxy-p-menthane
Dipentene dioxide
Epoxide 269
4-(1,2-Epoxy-1-methylethyl)-1-methyl-7-oxabicyclo-(4.1.0)heptane
Limonene dioxide
Menthane, 1,2:8,9-diepoxy-
7-Oxabicyclo(4.1.0)heptane, 4-(1,2-epoxy-1-methylethyl)-1-methyl-
Unoxat Epoxide 269

96-09-3
C_8H_8O
120.16
O(C1c(cccc2)c2)C1
Styrene oxide
Benzene, (epoxyethyl)
Epoxyethylbenzene (8CI)
1,2-Epoxyethylbenzene
1,2-Epoxy-1-phenylethane
Epoxystyrene
α,β-Epoxystyrene
Fenyloxiran (Czech)
NCI-C54977
Oxirane, phenyl-
Phenethylene oxide
1-Phenyl-1,2-epoxyethane
Phenylethylene oxide
Phenyl oxirane
1-Phenyloxirane
2-Phenyloxirane
Styrene epoxide
Styrene-7,8-oxide
Styryl oxide

96-10-6
$C_4H_{10}AlCl$
120.57
Aluminum, chlorodiethyl
Chlorodiethylaluminum
Diethylaluminium chloride (DOT)
Diethylaluminum chloride
Diethylaluminum monochloride
Diethylchloroaluminum
UN 1101 (DOT)

96-11-7
$C_3H_5Br_3$
280.81
BrCC(Br)CBr
Propane, 1,2,3-tribromo
Glycerol tribromohydrin
Glyceryl tribromohydrin
sym-Tribromopropane
1,2,3-Tribromopropane

96-12-8
$C_3H_5Br_2Cl$
236.35
BrCC(Br)CCl
Propane, 1,2-dibromo-3-chloro
BBC 12
1-Chloro-2,3-dibromopropane
3-Chloro-1,2-dibromopropane
DBCP
Dibromochloropropane
Dibromchlorpropan (German)
1,2-Dibrom-3-chlor-propan (German)
1,2-Dibromo-3-chloropropane (DOT,OSHA)
1,2-Dibromo-3-cloro-propano (Italian)
1,2-Dibroom-3-chloorpropaan

(Dutch)
Fumagon
Fumazone
Fumazone 86
Fumazone 86E
NCI-C00500
Nemabrom
Nemafume
Nemagon
Nemagon 20
Nemagone
Nemagon 20G
Nemagon 90
Nemagon 206
Nemagon Soil Fumigant
Nemanax
Nemapaz
Nemaset
Nematocide
Nematox
Nemazon
OS 1897
Oxy DBCP
Propane, 1-chloro-2,3-dibromo-
RCRA waste number U066
SD 1897
UN 2872 [Dibromochloropropane]

96-13-9
$C_3H_6Br_2O$
217.91
OCC(Br)CBr
1-Propanol, 2,3-dibromo
2,3-Dibromopropanol
2,3-Dibromo-1-propanol
NCI-C55436
USAF DO-42

96-14-0
C_6H_{14}
86.20
C(CC)(CC)C
Pentane, 3-methyl
3-Methylpentane (DOT)
UN 1208
UN 2462 (DOT)

96-15-1
$C_5H_{13}N$
87.16
NCC(CC)C
1-Butanamine, 2-methyl- (9CI)
2-Methyl-1-butanamine

96-17-3
$C_5H_{10}O$
86.15
O=CC(CC)C
Butyraldehyde, 2-methyl
Acetaldehyde, ethylmethyl-
α-Methylbutanal
2-Methylbutanal
2-Methyl-1-butanal
α-Methylbutyraldehyde
2-Methylbutyraldehyde
α-Methylbutyric aldehyde
Methylethylacetaldehyde

96-18-4
$C_3H_5Cl_3$
147.43
ClCC(Cl)CCl
Propane, 1,2,3-trichloro
Allyl trichloride
Glycerol trichlorohydrin
Glyceryl trichlorohydrin
NCI-C60220
Trichlorohydrin
1,2,3-Trichloropropane (ACGIH,OSHA)

96-19-5
$C_3H_3Cl_3$
145.41
C(=CCl)(CCl)Cl
Propene, 1,2,3-trichloro
1,2,3-Trichloropropene

96-20-8
$C_4H_{11}NO$
89.16
OCC(N)CC
1-Butanol, 2-amino
2-Aminobutan-1-ol
2-Amino-1-butanol
2-Amino-n-butyl alcohol
Butanol-2-amine

96-21-9
$C_3H_6Br_2O$
217.91
OC(CBr)CBr
2-Propanol, 1,3-dibromo
1,3-Dibromo-2-propanol
Glycerol-α,γ-dibromohydrine

96-22-0
$C_5H_{10}O$
86.15
O=C(CC)CC
3-Pentanone
DEK
Diethylcetone (French)
Diethyl ketone (ACGIH, OSHA) [UN 1156]
Dimethylacetone
Metacetone
Methacetone
Pentanone-3
Propione
UN 1156 [Diethyl ketone]

96-23-1
$C_3H_6Cl_2O$
128.99
OC(CCl)CCl
2-Propanol, 1,3-dichloro
Dichlorohydrin
α-Dichlorohydrin
sym-Dichloroisopropyl alcohol
1,3-Dichloro-2-propanol
1,3-Dichloropropanol-2 [UN 2750]
Enodrin
Glycerol α,γ-dichlorohydrin
sym-Glycerol dichlorohydrin
UN 2750 [1,3-Dichloropropanol-2]
U 25,354

96-24-2
$C_3H_7ClO_2$
110.55
OCC(O)CCl
1,2-Propanediol, 3-chloro
Chlorhydrin
α-Chlorhydrin
Chlorodeoxyglycerol
1-Chloro-2,3-dihydroxypropane
3-Chloro-1,2-dihydroxypropane
α-Chlorohydrin
1-Chloropropane-2,3-diol
1-Chloro-2,3-propanediol
3-Chloropropane-1,2-diol
3-Chloro-1,2-propanediol
3-Chloropropylene glycol
β,β'-Dihydroxyisopropyl chloride
2,3-Dihydroxypropyl chloride
Epibloc
Glycerin α-monochlorhydrin
Glycerol chlorohydrin
Glycerol α-chlorohydrin
Glycerol-α-monochlorohydrin [UN 2689]
Glyceryl-α-chlorohydrin
Monochlorhydrin
Monochlorohydrin
α-Monochlorohydrin
U-5897
UN 2689 [Glycerol α-monochlorohydrin]

96-26-4
$C_3H_6O_3$
90.09
O=C(CO)CO
2-Propanone, 1,3-dihydroxy
Dihydroxyacetone
NSC-24343
Oxatone
Chromelin
1,3-Dihydroxyacetone
1,3-Dihydroxypropanone
Dihyxal
Otan
Oxantin
Soleal
Triulose
Viticolor

96-27-5
$C_3H_8O_2S$
108.17
OCC(O)CS
1,2-Propanediol, 3-mercapto
Glycerol, 1-thio-
1-Mercaptoglycerol
1-Mercapto-2,3-propanediol
3-Mercapto-1,2-propanediol
Monothioglycerin

Monothioglycerol
α-Monothioglycerol
Thioglycerin
Thioglycerol
α-Thioglycerol
1-Thioglycerol
Thiovanol
USAF B-40
USAF CB-37

96-29-7
C_4H_9NO
87.14
N(O)=C(CC)C
2-Butanone, oxime
Ethyl methyl ketone oxime
Ethyl-methylketonoxim (Czech)
Ethyl methyl ketoxime
MEK-Oxime
Methyl ethyl ketoxime
Skino #2
Troykyd Anti-Skin B
USAF AM-3
USAF DO-44
USAF EK-906

96-31-1
$C_3H_8N_2O$
88.13
O=C(NC)NC
Urea, 1,3-dimethyl
N,N'-Dimethylharnstoff (German)
N,N'-Dimethylurea
sym-Dimethylurea
1,3-Dimethylurea
Symmetric dimethylurea

96-32-2
$C_3H_5BrO_2$
152.99
O=C(OC)CBr
Acetic acid, bromo-, methyl ester
Bromoacetic acid methyl ester
Methyl bromoacetate
Methyl α-bromoacetate
Methyl bromoacetate [UN 2643]
Methylester kyseliny bromoctove (Czech)
Methyl monobromoacetate
UN 2643 [Methyl bromoacetate]

96-33-3
$C_4H_6O_2$
86.10
O=C(OC)C=C
Acrylic acid, methyl ester
Acrylate de methyle (French)
Acrylsaeuremethylester (German)
Curithane 103
Methoxycarbonylethylene
Methylacrylaat (Dutch)
Methyl-acrylat (German)
Methyl acrylate (ACGIH, OSHA)
Methyl acrylate, Inhibited [UN 1919]
Methylester kyseliny akrylove (Czech)
Methyl propenate
Methyl propenoate
Methyl-2-propenoate
Metilacrilato (Italian)
Propenoic acid methyl ester
2-Propenoic acid, methyl ester
UN 1919 [Methyl acrylate, inhibited]

96-34-4
$C_3H_5ClO_2$
108.53
O=C(OC)CCl
Acetic acid, chloro-, methyl ester
Chloroacetic acid methyl ester
Methyl chloroacetate [UN 2295] (96-34-4)
Methylester kyseliny chloroctove (Czech)
Methyl monochloracetate
Methyl monochloroacetate
Monochloroacetic acid methyl ester
UN 2295 [Methyl chloroacetate]

96-37-7
C_6H_{12}
84.18
C(CCC1)(C1)C
Cyclopentane, methyl [UN 2298]
Methylcyclopentane (DOT)
UN 2298 [Methyl cyclopentane]

96-40-2
C_5H_7Cl
102.56
Cyclopentene, 3-chloro- (8CI, 9CI)

96-41-3
$C_5H_{10}O$
86.15
OC(CCC1)C1
Cyclopentanol [UN 2244]
UN 2244 [Cyclopentanol]

96-43-5
C_4H_3ClS
118.59
S(C(=CC=1)Cl)C1
Thiophene, 2-chloro- (9CI)
2-Chlorothiophene
NSC-8747

96-45-7
$C_3H_6N_2S$
102.17
N(=C(S)NC1)C1
2-Imidazolidinethione
4,5-Dihydroimidazole-2(3H)-thione
Ethylene thiourea
N,N'-Ethylenethiourea
1,3-Ethylene-2-thiourea
ETU
Imidazole-2(3H)-thione, 4,5-dihydro-
Imidazoline, 2-mercapto-
L'ethylene thiouree (French)
2-Mercaptoimidazoline
2-Merkaptoimidazolin (Czech)
NA 22
NA 22-D
NCI-C03372
Pennac CRA
RCRA waste number U116
Rodanin S-62 (Czech)
Sodium-22 neoprene accelerator
2-Thiol-dihydroglyoxaline
Urea, 1,3-ethylene-2-thio-
USAF EL-62
Vulkacit NPV/C2
Warecure C

96-47-9
$C_5H_{10}O$
86.15
O(C(CC1)C)C1
Furan, 2-methyl-tetrahydro
2-Methyloxolane
Methyltetrahydrofuran
2-Methyltetrahydrofuran

96-48-0
$C_4H_6O_2$
86.10
O=C(OCC1)C1
2(3H)-Furanone, dihydro
γ-Bl
Blo
Blon
4-Butanolide
1,2-Butanolide
1,4-Butanolide
Butyric acid, 4-hydroxy-, γ-lactone
Butyric acid lactone
Butyrolactone
α-Butyrolactone
γ-Butyrolactone
4-Butyrolactone
Butyrylactone
Butyryl lactone
4-Deoxytetronic acid
Dihydro-2(3H)-furanone
Gamma-6480
4-Hydroxybutanoic acid lactone
4-Hydroxybutanoic acid, γ-lactone
γ-Hydroxybutyric acid cyclic ester
4-Hydroxybutyric acid lactone
4-Hydroxybutyric acid, γ-lactone
γ-Hydroxybutyric acid lactone
γ-Hydroxybutyrolactone
NCI-C55878
2-Oxolanone

Tetrahydro-2-furanone

96-49-1
$C_3H_4O_3$
88.07
O=C(OCC1)O1
Carbonic acid, cyclic ethylene ester
Cyclic ethylene carbonate
1,3-Dioxolan-2-one
Dioxolone-2
Ethylene carbonate
Ethylene carbonic acid
Ethylene glycol carbonate
Ethylene glycol, cyclic carbonate
Ethylenester kyseliny uhlicite (Czech)
Glycol carbonate

96-50-4
$C_3H_4N_2S$
100.15
N=C(NC=C1)S1
Thiazole, 2-amino
Abadol
Abadole
Aminothiazole
2-Aminothiazole
Basedol
2-Thiazolamine
2-Thiazolylamine
2-Thiazylamine
USAF EK-P-5501

96-54-8
C_5H_7N
81.11
N(C=CC=1)(C1)C
Methylpyrrole
N-Methylpyrrole
1-Methylpyrrole
1-Methyl-1H-pyrrole
NSC-65440
Pyrrole, 1-methyl- (8CI)
1H-Pyrrole, 1-methyl- (9CI)

96-56-0
$C_{21}H_{43}NO_2$
341.57
O=C(O)C(N(C)(C)C)CCCCCCC CCCCCCCCC
1-Heptadecanaminium, 1-carboxy-N,N,N-trimethyl-, hydroxide, inner salt (9CI)

96-64-0
$C_7H_{16}FO_2P$
182.20
CC(OP(C)(F)=O)C(C)(C)C
Phosphonofluoridic acid, methyl-, 1,2,2-trimethylpropyl ester
2-Butanol, 3,3-dimethyl-, methylphosphonofluoridate
3,3-Dimethyl-n-but-2-yl methylphosphonofluoridate
3,3-Dimethyl-2-butyl methylphosphonofluoridate
EA 1210
GD
Methylfluorphosphorsaeurepinakolylester (German)
Methylphosphonofluoridic acid, 3,3-dimethyl-2-butyl ester
Methylphosphonofluoridic acid 1,2,2-trimethylpropyl ester
Methyl pinacolyloxy phosphorylfluoride
Methyl pinacolyl phosphonofluoridate
Phosphine oxide, fluoromethyl-(1,2,2-trimethylpropoxy)-
Pinacoloxymethylphosphoryl fluoride
Pinacolyl methylfluorophosphonate
Pinacolyl methylphosphonofluoridate
Pinacolyl methylphosphonofluoride
Pinacolyloxy methylphosphoryl fluoride
Pynacolyl methylfluorophosphonate
PMFP
Soman
T.2107
1,2,2-Trimethylpropylester kyseliny methylfluorfosfonove (Czech)
1,2,2-Trimethylpropyl methylphosphonofluoridate

96-67-3
$C_6H_6N_2O_6S$
234.20
O=S(=O)(O)c(c(O)c(N)cc1N (=O)=O)c1
Metanilic acid, 2-hydroxy-5-nitro
3-Amino-2-hydroxy-5-nitrobenzenesulfonic acid
Benzenesulfonic acid, 3-amino-2-hydroxy-5-nitro-
2-Hydroxy-5-nitrometanilic acid
Kyselina 2-amino-4-nitrofenol-6-sulfonova (Czech)
Kyselina 4-nitro-2-aminofenol-6-sulfonova (Czech)

96-69-5
$C_{22}H_{30}O_2S$
358.58
Oc(c(cc(Sc(c(cc(O)c1C(C)(C)C) C)c1)c2C)C(C)(C)C)c2
m-Cresol, 4,4'-thiobis(6-tert-butyl
Bis(3-tert-butyl-4-hydroxy-6-methylphenyl) sulfide
Bis(4-hydroxy-5-tert-butyl-2-methylphenyl) sulfide
Disperse MB-61
Phenol, 4,4'-thiobis(2-(1,1-dimethylethyl)-5-methyl- (9CI)
Santonox
Santonox BM
Santonox R
Santowhite Crystals
Santox
Sumilizer WX
Sumilizer WX-R
Thioalkofen BM 4
Thioalkofen BMCh
Thioalkofen MBCh
Thioalkophene BM-4
4,4'-Thiobis(6-tert-butyl-m-cresol) (ACGIH,OSHA)
4,4'-Thiobis(2-tert-butyl-5-methylphenol)
4,4'-Thiobis(6-tert-butyl-3-methylphenol)
4,4'-Thiobis(3-methyl-6-tert-butylphenol)
1,1'-Thiobis(2-methyl-4-hydroxy-5-tert-butylbenzene)
USAF B-15
Yoshinox S
Yoshinox SR

96-70-8
$C_{12}H_{18}O$
178.27
Oc(c(cc(c1)CC)C(C)(C)C)c1
2-t-Butyl-4-ethylphenol
Benzene, 2-tert-butyl-4-ethyl-1-hydroxy-
2-tert-Butyl-4-ethylphenol
2-(1,1-Dimethylethyl)-4-ethylphenol
4-Ethyl-2-tert-butylphenol
Phenol, 2-tert-butyl-4-ethyl-
Phenol, 2-(1,1-dimethylethyl)-4-ethyl- (9CI)

96-73-1
$C_6H_4ClNO_5S$
237.61
O=S(=O)(O)c(c(ccc1N(=O)=O) Cl)c1
Benzenesulfonic acid, 2-chloro-5-nitro- (9CI)
AI3-08898
2-Chloro-5-nitrobenzenesulfonic acid
NSC-5375

96-75-3
$C_6H_6N_2O_5S$
218.20
O=S(=O)(O)c(c(N)ccc1N(=O) =O)c1
Benzenesulfonic acid, 2-amino-5-nitro
4-Nitroaniline-2-sulfonic acid

96-76-4
$C_{14}H_{22}O$
206.36
Oc(c(cc(c1)C(C)(C)C)C(C)(C) C)c1
Phenol, 2,4-di-tert-butyl
Antioxidant No. 33
2,4-Di-tert-butylphenol
Prodox 146
Prodox 146A-85X

96-80-0
$C_8H_{19}NO$
145.28
OCCN(C(C)C)C(C)C
Ethanol, 2-(diisopropylamino)
2-Diisopropylaminoethanol
Diisopropyl ethanolamine
N,N-Diisopropylethanolamine
N,N-Diisopropyl ethanolamine (DOT)
UN 2825 (DOT)

96-83-3
$C_{11}H_{12}I_3NO_2$
570.94
CCC(Cc1c(I)cc(I)c(N)c1I)C(O)=O
Hydrocinnamic acid, 3-amino-α-ethyl-2,4,6-triiodo
3-Amino-α-ethyl-2,4,6-triiodohydrocinnamic acid
2-(3-Amino-2,4,6-triiodobenzyl)butyric acid
3-(3-Amino-2,4,6-triiodophenyl)-2-ethylpropanoic acid
β-(3-Amino-2,4,6-triiodophenyl)-α-ethylpropionic acid
Choladine
Cholevid
Cistobil
Colepax
Bilijodon
Copanoic
2-Ethyl-3-(3-amino-2,4,6-triiodophenyl)propionic acid
Iodopanic acid
Iodopanoic acid
Iopanoic acid
Jopagnost
Telepaque
Teletrast

96-91-3
$C_6H_5N_3O_5$
199.14
Nc1cc(cc(N(=O)=O)c1O)N(=O)=O
Phenol, 2-amino-4,6-dinitro
Acide picramique (French)
C.I. Oxidation Base 21
4,6-Dinitro-2-aminophenol
Fourrine 93
Fourrine 4R
Furro 4R
Picramic acid
Zoba 4R

96-93-5
$C_6H_6N_2O_6S$
234.20
Nc1cc(cc(N(=O)=O)c1O)S(O)(=O)=O
Benzenesulfonic acid, 3-amino-4-hydroxy-5-nitro
3-Amino-4-hydroxy-5-nitrobenzenesulfonic acid
Kyselina 2-amino-6-nitrofenol-4-sulfonova (Czech)
Kyselina 6-nitro-2-aminofenol-4-sulfonova (Czech)

96-96-8
$C_7H_8N_2O_3$
168.17
O=N(=O)c(c(N)ccc1OC)c1
Aniline, 4-methoxy-2-nitro
Gp-Amin (Czech)
4-Methoxy-2-nitroanilin (Czech)
4-Methoxy-2-nitroaniline

96-97-9
$C_7H_5NO_5$
183.11
O=C(O)c(c(O)ccc1N(=O)=O)c1
Benzoic acid, 2-hydroxy-5-nitro- (9CI)
AI3-08840
Anilotic acid
2-Hydroxy-5-nitrobenzoic acid
5-Nitro-2-hydroxybenzoic acid
5-Nitrosalicylic acid
NSC-183
Salicylic acid, 5-nitro- (8CI)

96-98-0
$C_8H_7NO_4$
181.14
O=C(O)c(ccc(c1N(=O)=O)C)c1
Benzoic acid, 4-methyl-3-nitro- (9CI)
4-Methyl-3-nitrobenzoic acid
3-Nitro-4-methylbenzoic acid
m-Nitro-p-toluic acid
3-Nitro-para-toluic acid
NSC-50659
p-Toluic acid, 3-nitro- (8CI)

97-00-7
$C_6H_3ClN_2O_4$
202.56
O=N(=O)c(ccc(c1N(=O)=O)Cl)c1
Benzene, 1-chloro-2,4-dinitro
CDNB
1-Chloor-2,4-dinitrobenzeen (Dutch)
1-Chlor-2,4-dinitrobenzene
1-Chloro-2,4-dinitrobenzene
4-Chloro-1,3-dinitrobenzene
6-Chloro-1,3-dinitrobenzene
1-Chloro-2,4-dinitrobenzol (German)
1-Cloro-2,4-dinitrobenzene (Italian)
2,4-Dinitrochlorobenzene
1,3-Dinitro-4-chlorobenzene
2,4-Dinitro-1-chlorobenzene
Dinitrochlorobenzol
DNCB

97-02-9
$C_6H_5N_3O_4$
183.14
O=N(=O)c(ccc(N)c1N(=O)=O)c1
Aniline, 2,4-dinitro
Benzenamine, 2,4-dinitro- (9CI)
2,4-Dinitraniline
2,4-Dinitroanilin (German)
2,4-Dinitroanilina (Italian)
2,4-Dinitroaniline
DNA
NCI-C60753

97-05-2
$C_7H_6O_6S$
218.19
O=C(O)c(c(O)ccc1S(=O)(=O)O)c1
Salicylic acid, 5-sulfo
3-Carboxy-4-hydroxybenzenesulfonic acid
2-Hydroxybenzoic-5-sulfonic acid
Salicylsulfonic acid
Sulfosalicylic acid
5-Sulfosalicylic acid

97-08-5
$C_6H_3Cl_2NO_4S$
256.06
O=S(=O)(c(ccc(c1N(=O)=O)Cl)c1)Cl
Benzenesulfonyl chloride, 4-chloro-3-nitro- (9CI)
4-Chloro-3-nitrobenzenesulfonyl chloride

97-09-6
$C_6H_5ClN_2O_4S$
236.62
O=S(=O)(N)c(ccc(c1N(=O)=O)Cl)c1
4-Chloro-3-nitrobenzenesulfonamide
Benzenesulfonamide, 4-chloro-3-nitro- (9CI)
NSC-512314

97-16-5
$C_{12}H_8Cl_2O_3S$
303.16
Clc2ccc(OS(=O)(=O)c1ccccc1)c(Cl)c2
Phenol, 2,4-dichloro-, benzenesulfonate
Benzenesulfonic acid, 2,4-dichlorophenyl ester
Benzenesulphonic acid, 2,4-dichlorophenyl ester
Compound 923
2,4-Dichlorfenylester kyseliny benzensulfonove (Czech)
2,4-Dichlorophenol, benzenesulfonate
2,4-Dichlorophenyl benzenesulfonate
2,4-Dichlorophenyl benzenesulphonate
2,4-Dichlorophenyl ester of benzenesulfonic acid
DPBS
EM 923
Genite

Genite 923
Genite EM-923
Genite-R99
Genitol
Genitol 923
Latka 923 (Czech)

97-17-6
$C_{10}H_{13}Cl_2O_3PS$
315.16
CCOP(=S)(OCC)Oc1ccc(Cl)cc1Cl
Phosphorothioic acid, O-(2,4-dichlorophenyl)-O,O-diethyl ester
Bromex
O,O-Diaethyl-O-2,4-dichlorphenyl-monothiophosphat (German)
O,O-Diaethyl-O-2,4-dichlorphenyl-thionophosphat (German)
Dichlofenthion
Dichlofention
Dichlorfenthion
Dichlorofenthion
O-2,4-Dichlorophenyl O,O-diethyl phosphorothioate
2,4-Dichloro-phenyl diethyl phosphorothionate
O,O-Diethyl-O-(2,4-dichloorfenyl)-monothiofosfaat (Dutch)
O,O-Diethyl O-(2,4-dichlorophenyl) phosphorothioate
Diethyl 2,4-dichlorophenyl phosphorothionate
O,O-Diethyl O-2,4-dichloro phenyl thiophosphate
O,O-Dietil-O-(2,4-dicloro-fenil)-monotiofosfato (Italian)
ECP
ENT 17,470
Hexa-Nema
Mobilawn
Nemacide
Nemacide VC-13
Phenol, 2,4-dichloro-, O-ester with O,O-diethyl phosphorothioate
Thiophosphate de O-2,4-dichlorophenyle et de O,O-diethyle (French)
Tri-VC 13
V-C-13
V-C 1-13
VC13 Nemacide

97-18-7
$C_{12}H_6Cl_4O_2S$
356.04
Oc(c(cc(c1)Cl)Cl)c1Sc(c(O)c(cc2Cl)Cl)c2
Phenol, 2,2'-thiobis(4,6-dichloro
Actamer
Bidiphen
Bis(3,5-dichloro-2-hydroxyphenyl) sulfide
Bis(2-hydroxy-3,5-dichlorophenyl) sulfide
Bithionol
Bithionol sulfide
Bitin
CP 3438
2,2'-Dihydroxy-3,3',5,5'-tetrachlorodiphenylsulfide
2-Hydroxy-3,5-dichlorophenyl sulphide
Lorothidol
Lorothiodol
NCI-C60628
Neopellis
TBP
2,2'-Thiobis(4,6-dichlorophenol)
USAF B-22
Vancide BL
XL 7

97-23-4
$C_{13}H_{10}Cl_2O_2$
269.13
Oc(c(cc(c1)Cl)Cc(c(O)ccc2Cl)c2)c1
Phenol, 2,2'-methylenebis-(4-chloro
Anthiphen
Antiphen
Bis(5-chlor-2-hydroxyphenyl)-methan (German)
Bis(5-chloro-2-hydroxyphenyl)-methane
Bis-2-hydroxy-5-chlorfenyl-methan (Czech)
Bis(2-hydroxy-5-chlorophenyl)-methane
DDDM
DDM
Dicestal
Dichloorfeen (Dutch)
5,5'-Dichloro-2,2'-dihydroxy-diphenylmethane
Di-(5-chloro-2-hydroxyphenyl)-methane
Dichlorofen (Czech)
4,4'-Dichloro-2,2'-methylene-diphenol
Dichlorophen
Dichlorophen B
Dichlorophene
Dichlorphen
Didroxan
Didroxane
2,2'-Dihydroxy-5,5'-dichlorodiphenylmethane
Diphenthane 70
Fungicide FX
G 4
GH
Hyosan
Korium
O,O-Methyleen-bis(4-chloorfenol) (Dutch)
2,2'-Methylenebis(4-chlorophenol)
O,O-Metilen-bis(4-cloro-fenolo) (Italian)
Panacide
Parabis
Plath-Lyse
Prevental
Preventol
Preventol GD
Preventol GDC
Super Mosstox
Taeniatol
Teniathane
Teniatol
Wespuril

97-24-5
$C_{12}H_8Cl_2O_2S$
287.16
Oc1ccc(Cl)cc1Sc2cc(Cl)ccc2O
Phenol, 2,2'-thiobis(4-chloro
Bis(2-hydroxy-5-chlorophenyl)-sulfide
CR 305
D 25
2,2'-Dihydroxy-5,5'-dichlorodiphenyl sulfide
Novex
2,2'-Thiobis(4-chlorophenol)

97-30-3
$C_7H_{14}O_6$
194.18
O(C(C(O)C(O)C1O)CO)C1OC
α-Methylglucoside
Glucopyranoside, methyl, α-D-(8CI)
α-D-Glucopyranoside, methyl (9CI)
Methyl α-D-glucopyranoside
α-Methyl D-glucose ether
NSC-102101

97-32-5
$C_{14}H_{12}N_2O_4$
272.25
O=C(Nc(cccc1)c1)c(ccc(OC)c2N(=O)=O)c2
Benzamide, 4-methoxy-3-nitro-N-phenyl- (9CI)
p-Anisanilide, 3-nitro- (8CI)
4-Methoxy-3-nitro-N-phenyl-benzamide
3-Nitro-p-anisanilide
NSC-50649

97-36-9
$C_{12}H_{15}NO_2$
205.28
O=C(Nc(c(cc(c1)C)C)c1)CC(=O)C
Acetoacetanilide, 2',4'-dimethyl
Acetoacet-2,4-dimethylphenyl
2,4-Acetoacetoxylidide
2',4'-Acetoacetoxylidide
Acetoaceto-m-xylidide
Acetoacet-m-xylidide
Acetoacetyl-m-xylidide
Butanamide, N-(2,4-dimethylphenyl)-3-oxo- (9CI)
2',4'-Dimethylacetoacetanilide
3-Oxo-N-(2,4-methylphenyl)-butanamide

97-39-2
$C_{15}H_{17}N_3$
239.35
N=C(Nc(c(ccc1)C)c1)Nc(c(ccc2)C)c2
Guanidine, 1,3-di-o-tolyl
Diorthotolylguanidine
Di-o-tolylguanidine
1,3-Di-o-tolylguanidine
DOTG Accelerator
USAF A-6598
Vulkacit DOTG/C

97-41-6
$C_{12}H_{20}O_2$
196.32
O=C(OCC)C(C1(C)C)C1C=C(C)C
Cyclopropanecarboxylic acid, 2,2-dimethyl-3-(2-methyl-1-propenyl)-, ethyl ester
Cyclopropanecarboxylic acid, 2,2-dimethyl-3-(2-methylpropenyl)-, ethyl ester (8CI)
Cyclopropanecarboxylic acid, 2,2-dimethyl-3-(2-methyl-1-propenyl)-, ethyl ester (9CI)
2,2-Dimethyl-3-(2-methylpropenyl)cyclopropanecarboxylic acid ethyl ester
Ethyl chrysanthemate
Ethyl chrysanthemumate

97-42-7
$C_{12}H_{18}O_2$
194.30
O=C(OC(C(=CCC1C(=C)C)C)C1)C
p-Mentha-6,8-dien-2-ol, acetate
l-Carvyl acetate
2-Cyclohexen-1-ol, 2-methyl-5-(1-methylethenyl)-, acetate
1-p-Mentha-6(8,9)-dien-2-yl acetate

97-52-9
$C_7H_8N_2O_3$
168.17
O=N(=O)c(ccc(N)c1OC)c1
o-Anisidine, 4-nitro
Amarthol Fast Red B Base
2-Amino-5-nitroanisol (Czech)
2-Amino-5-nitroanisole
Aniline, 2-methoxy-4-nitro-
Anisole, 2-amino-5-nitro-
Azoamine Pink O
Azoene Fast Red B Base
Benzenamine, 2-methoxy-4-nitro- (9CI)
Brentamine Fast Red B Base
C.I. 37125
C.I. Azoic Diazo Component 5
Dainichi Fast Red B Base
Daito Red Base B
Devol Red E
Diabase Red B
Diazo Fast Red B
Fast Red B
Fast Red Base B
Fast Red B Base
Fast Red 5NA Base
Hiltonil Fast Red B Base
Kako Red B Base
Kayaku Red B Base
2-Methoxy-4-nitroaniline
Mitsui Red B Base
Naphthanil Red B Base
Naphthoelan Red B Base
4-Nitro-o-anisidine
PNOA
Red Base Ciba V
Red Base Irga V
Red Base NB
Red B Base
Sanyo Fast Red B Base
Shinnippon Fast Red B Base
Showa Fast Red B Base
Symulon Red B Base

97-53-0
$C_{10}H_{12}O_2$
164.22
O(c(c(O)ccc1CC=C)c1)C
Phenol, 4-allyl-2-methoxy
4-Allylcatechol-2-methyl ether
p-Allylguaiacol
4-Allylguaiacol
4-Allyl-1-hydroxy-2-methoxybenzene
4-Allyl-2-methoxyphenol
Caryophyllic acid
Eugenic acid
Eugenol
p-Eugenol
1,3,4-Eugenol
FA 100
Fema No. 2467
1-Hydroxy-2-methoxy-4-allylbenzene
4-Hydroxy-3-methoxyallylbenzene
1-Hydroxy-2-methoxy-4-prop-2-enylbenzene
2-Methoxy-4-allylphenol
2-Methoxy-1-hydroxy-4-allylbenzene
2-Methoxy-4-prop-2-enylphenol
2-Methoxy-4-(2-propenyl)-phenol
2-Metoksy-4-allilofenol (Polish)
NCI-C50453
Phenol, 2-methoxy-4-(2-propenyl)-
Synthetic Eugenol

97-54-1
$C_{10}H_{12}O_2$
164.22
O(c(c(O)ccc1C=CC)c1)C
Phenol, 2-methoxy-4-propenyl
1-Hydroxy-2-methoxy-4-propenylbenzene
4-Hydroxy-3-methoxy-1-propenylbenzene
Isoeugenol
2-Methoxy-4-propenylphenol
NCI-C60979
4-Propenylguaiacol

97-55-2
$C_5H_{13}O_{14}P_3$
390.07
5-Phosphorylribose-1-pyrophosphate
Ribofuranose, 5-(dihydrogen phosphate) 1-(trihydrogen diphosphate)
Ribofuranose, 5-(dihydrogen phosphate) 1-(trihydrogen pyrophosphate)
Ribofuranose, 5-phosphate 1-pyrophosphate
Ribose, 5-(dihydrogen phosphate) 1-(trihydrogen diphosphate)
Ribose, 5-(dihydrogen phosphate) 1-(trihydrogen pyrophosphate)

97-56-3
$C_{14}H_{15}N_3$
225.32
N(=Nc(c(ccc1)C)c1)c(ccc(N)c2C)c2
o-Toluidine, 4-(o-tolylazo)
AAT
Aminoazotoluene (indicator)
o-AAT
o-Amidoazotoluol (German)
Aminoazotoluene
o-Aminoazotoluene
4'-Amino-2,3'-azotoluene
2-Amino-5-azotoluene
o-Aminoazotolueno (Spanish)
4'-Amino-2:3'-azotoluene
o-Aminoazotoluol
4-Amino-2',3-dimethylazobenzene
4'-Amino-2,3'-dimethylazobenzene
o-AT
Brasilazina Oil Yellow R
Butter Yellow
C.I. 11160
C.I. 11160B
C.I. Solvent Yellow 3
2',3-Dimethyl-4-aminoazobenzene
Fast Garnet GBC Base
Fast Oil Yellow
Fast Yellow AT
Fat Yellow B
Hidaco Oil Yellow
2-Methyl-4-((2-methylphenyl)-azo)benzenamine
OAAT
Oil Yellow 21
Oil Yellow 2681
Oil Yellow AT
Oil Yellow
Oil Yellow A
Oil Yellow C
Oil Yellow I
Oil Yellow 2R
Oil Yellow T
Organol Yellow 2T
Somalia Yellow R

Sudan Yellow RRA
Toluazotoluidine
o-Tolueneazo-o-toluidine
o-Toluol-azo-o-toluidin (German)
5-(o-Tolylazo)-2-aminotoluene
4-(o-Tolylazo)-o-toluidine
Tulabase Fast Garnet GB
Tulabase Fast Garnet GBC
Waxakol Yellow NL
Zlut Rozpoustedlova 3 (Czech)

97-59-6
$C_4H_6N_4O_3$
158.09
O=C(NC(NC(=O)N1)C1=O)N
Allantoin (8CI)
AI3-15281
Alantan
Allantol
AVC/Dienestrol cream
Caswell No. 024
Cordianine
(2,5-Dioxo-4-imidazolidinyl)-urea
EPA Pesticide Chemical Code 085701
Glyoxyldiureid
Glyoxyldiureide
Hydantoin, 5-ureido-
NSC-7606
Sebical
Uniderm A
Urea, (2,5-dioxo-4-imidazolid-inyl)- (9CI)
5-Ureidohydantoin

97-61-0
$C_6H_{12}O_2$
116.18
O=C(O)C(CCC)C
Valeric acid, 2-methyl
Kyselina 2-methylvalerova (Czech)
2-Methylpentanoic acid
Methylpropylacetic acid
α-Methylvaleric acid
2-Methylvaleric acid
2-Pentanecarboxylic acid

97-62-1
$C_6H_{12}O_2$
116.18
O=C(OCC)C(C)C
Isobutyric acid, ethyl ester
Ethyl isobutanoate
Ethyl isobutyrate [UN 2385]
Ethylisobutyrate
Ethyl 2-methylpropanoate
Ethyl 2-methylpropionate
Propionic acid, 2-methyl-, ethyl ester
UN 2385 [Ethyl isobutyrate]

97-63-2
$C_6H_{10}O_2$
114.16
O=C(OCC)C(=C)C
Methacrylic acid, ethyl ester
Ethylester kyseliny metha-krylove (Czech)
Ethyl methacrylate [UN 2277]
Ethyl methacrylate, Inhibited (DOT)
Ethyl α-methyl acrylate
Ethyl 2-methylacrylate
Ethyl 2-methyl-2-propenoate
2-Propenoic acid, 2-methyl-, ethyl ester
RCRA waste number U118
Rhoplex AC-33 (Rohm and Haas)
UN 2277 [Ethyl methacrylate]

97-64-3
$C_5H_{10}O_3$
118.15
O=C(OCC)C(O)C
Lactic acid, ethyl ester
Actylol
Acytol
Ethylester kyseliny mlecne (Czech)
Ethyl α-hydroxypropionate
Ethyl 2-hydroxypropionate
Ethyl lactate [UN 1192]
Lactate d'ethyle (French)
Solactol
UN 1192 [Ethyl lactate]

97-65-4
$C_5H_6O_4$
130.10
O=C(O)C(=C)CC(=O)O
Itaconic acid
AI3-16901
Butanedioic acid, methylene- (9CI)
Methylenebutanedioic acid
Methylenesuccinic acid
NSC-3357
2-Propene-1,2-dicarboxylic acid
Propylenedicarboxylic acid
Succinic acid, methylene- (8CI)

97-69-8
$C_5H_9NO_3$
131.13
CC(NC(C)=O)C(O)=O
N-Acetylalanine
Acetylalanine
N-Acetyl-S-alanine
Alanine, N-acetyl-, L- (8CI)
L-Alanine, N-acetyl- (9CI)
NSC-186892

97-72-3
$C_8H_{14}O_3$
158.22
O=C(OC(=O)C(C)C)C(C)C
Isobutyric-anhydride
Isobutyric anhydride [UN 2530]
UN 2530 [Isobutyric anhydride]

97-74-5
$C_6H_{12}N_2S_3$
208.38
N(C(=S)SC(N(C)C)=S)(C)C
Sulfide, bis(dimethylthio-carbamoyl)
Aceto TMTM
Bis(dimethylthiocarbamoyl)-sulfide
Bis(dimethylthiocarbamyl) monosulfide
Carbamic acid, dimethyldithio-, anhydrosulfide
Carbamic anhydride, tetra-methyltrithio-
Formamide, 1,1'-thiobis-(N,N-dimethylthio-
Monex
Monosulfure de tetramethyl-thiurame (French)
Mono-Thiurad
Monothiuram
Pennac MS
Tetramethylthiurammonium sulfide
Tetramethylthiuram mono-sulfide
Tetramethylthiuramonosulfide
Tetramethylthiuram sulfide
Thionex
Thionex Rubber Accelerator
Thiuram monosulfide, tetra-methyl-
TMTM
TMTMS
UNADS
USAF B-32
USAF EK-P-6255
Vulkacit MS
Vulkacit thiuram MS/C

97-77-8
$C_{10}H_{20}N_2S_4$
296.56
N(C(=S)SSC(N(CC)CC)=S)(CC)CC
Disulfide, bis(diethylthio-carbamoyl)
Abstensil
Abstinil
Abstinyl
Alcophobin
Alk-aubs
Antabus
Antabuse
Antadix
Antaenyl
Antaethan
Antaethyl
Antaetil
Antalcol
Antetan
Antethyl
Antetil
Anteyl
Antiaethan
Antietanol
Anti-Ethyl
Antietil
Antikol
Antivitium
Aversan

Averzan
Bis((diethylamino)thioxo-methyl)disulphide
Bis(diethylthiocarbamoyl) di-sulfide
Bis(N,N-diethylthiocarbamoyl) disulfide
Bis(diethylthiocarbamoyl)-disulphide
Bis(N,N-diethylthiocarbamoyl)-disulphide
Bonibal
Contralin
Contrapot
Cronetal
Dicupral
Disetil
Disulfan
Disulfiram (ACGIH,OSHA)
Disulfuram
Disulphuram
1,1'-Dithiobis(N,N-diethyl-thioformamide)
Dupon 4472
Dupont Fungicide 4472
Ekagom Teds
ENT 27,340
Ephorran
Espenal
Esperal
Etabus
Ethyldithiourame
Ethyldithiurame
Ethyl thiram
Ethyl thiudad
Ethyl thiurad
Ethyl tuads
Ethyl tuex
Exhoran
Exhorran
Formamide, 1,1'-dithiobis-(N,N-diethylthio-
Hoca
Krotenal
NCI-C02959
Nocbin
Noxal
NSC-190940
Refusal
Ro-Sulfiram
Stopaethyl
Stopethyl
Stopetyl
TATD
Tenurid
Tenutex
TETD
Tetidis
Tetradin
Tetradine
Tetraethylthioperoxydicarbonic diamide
Tetraethylthiram disulphide
Tetraethylthiuram
Tetraethylthiuram disulfide
Tetraethylthiuram disulphide
N,N,N',N'-Tetraethylthiuram disulphide
Tetraetil
Teturam
Teturamin
Thiosan
Thioscabin
Thireranide
Thiuram E
Thiuram disulfide, tetraethyl-
Thiuranide
Tillram
Tiuram
TTD
TTS
TUADS, ethyl
USAF B-33

97-80-3
$C_{21}H_{41}NO_4S$
403.62
O=C(N(CCS(=O)(=O)O)C)CCC CCCCC=CCCCCCCCC
N-Methyl-N-oleoyltaurine
Arkopon
Ethanesulfonic acid, 2-(methyl-(1-oxo-9-octadecenyl)amino)-, (Z)- (9CI)
Oleylmethyltaurine
N-Oleoyl-N-methyltaurine
Sapogen T
Taurine, N-methyl-N-oleoyl-

97-84-7
$C_8H_{20}N_2$
144.30
N(CCC(N(C)C)C)(C)C
1,3-Butanediamine, N,N,N', N'-tetramethyl
1,3-Diaminobutane, N,N,N', N'-tetramethyl-
Tetramethyl butanediamine
N,N,N',N'-Tetramethyl-1,3-butanediamine

97-85-8
$C_8H_{16}O_2$
144.24
O=C(OCC(C)C)C(C)C
Isobutyric acid, isobutyl ester
Isobutylester kyseliny iso-maselne (Czech)
Isobutyl isobutyrate [UN 2528]
2-Methylpropyl isobutyrate
Propanoic acid, 2-methyl-, 2-methylpropyl ester (9CI)
UN 2528 [Isobutyl isobutyrate]

97-86-9
$C_8H_{14}O_2$
142.22
O=C(OCC(C)C)C(=C)C
Methacrylic acid, isobutyl ester
Isobutylester kyseliny metha-krylove (Czech)
Isobutyl methacrylate [UN 2283]
Isobutyl α-methacrylate
Isobutyl methacrylate, Inhibited (DOT)
2-Methylpropyl methacrylate
2-Propenoic acid, 2-methyl-, 2-methylpropyl ester
UN 2283 [Isobutyl methacryl-ate]

97-87-0
$C_8H_{16}O_2$
144.21
iso-Butyl isobutyrate
AI3-24261
Propanoic acid, 2-methyl-, butyl ester

97-88-1
$C_8H_{14}O_2$
142.22
O=C(OCCCC)C(=C)C
Methacrylic acid, butyl ester
Butil metacrilato (Italian)
Butylester kyseliny metha-krylove (Czech)
Butylmethacrylaat (Dutch)
Butyl methacrylate
Butyl 2-methacrylate
n-Butyl methacrylate [UN 2227]
Butyl 2-methyl-2-propenoate
Methacrylate de butyle (French)
Methacrylsaeurebutylester (German)
2-Methyl-butylacrylaat (Dutch)
2-Methyl-butylacrylat (German)
2-Methyl-butylacrylate
2-Propenoic acid, 2-methyl-, butyl ester
UN 2227 [n-Butyl methacryl-ate]

97-90-5
$C_{10}H_{14}O_4$
198.24
O=C(OCCOC(=O)C(=C)C)C (=C)C
Methacrylic acid, ethylene ester
Ageflex EGDM
1,2-Bis(methacryloyloxy)ethane
Diglycol dimethacrylate
Ethanediol dimethacrylate
Ethyldiol metacrylate
Ethylene glycol bis(methacryl-ate)
Ethylene glycol dimethacrylate
Ethylene methacrylate
Glycol dimethacrylate
2-Propenoic acid, 2-methyl-, 1,2-ethanediyl ester (9CI)
Sartomer SR 206
SR 206

97-91-6
$C_6H_{12}N_2S_6$
304.56
N(C(=S)SSSSC(N(C)C)=S)(C)C
Tetrasulfide, bis((dimethyl-amino)thioxomethyl) (9CI)
Bis(dimethylthiocarbamoyl) tetrasulfide
NSC-15257
Tetramethylthiuram tetrasulfide

Tetrasulfide, bis(dimethylthio-carbamoyl) (8CI)

97-93-8
$C_6H_{15}Al$
114.19
Aluminum, triethyl
TEA
Triethylaluminium (DOT)
Triethylaluminum
UN 1102 (DOT)

97-94-9
$C_6H_{15}B$
98.02
CCB(CC)CC
Borane, triethyl
Triethylborane
Triethylborine

97-95-0
$C_6H_{14}O$
102.20
OCC(CC)CC
1-Butanol, 2-ethyl
2-Ethylbutanol [UN 2275]
2-Ethylbutanol-1
2-Ethyl-1-butanol
2-Ethylbutyl alcohol
sec-Hexanol [UN 2282]
sec-Hexyl alcohol
3-Methylolpentane
sec-Pentylcarbinol
3-Pentylcarbinol
Pseudohexyl alcohol
UN 2275 [2-Ethylbutanol]
UN 2282 [Hexanols]

97-96-1
$C_6H_{12}O$
100.18
O=CC(CC)CC
Butyraldehyde, 2-ethyl
Aldehyde 2-ethylbutyrique (French)
Diethyl acetaldehyde
2-Ethylbutanal
Ethyl butyraldehyde (DOT)
α-Ethylbutyraldehyde
2-Ethylbutyraldehyde [UN 1178]
2-Ethylbutyric aldehyde
UN 1178 [2-Ethylbutyraldehyde]

97-97-2
$C_4H_9ClO_2$
124.57
O(C(OC)CCl)C
Dimethyl chloracetal
Acetaldehyde, chloro-, dimethyl acetal (8CI)
Chloroacetaldehyde, dimethyl acetal
2-Chloroacetaldehyde, dimethyl acetal
2-Chloro-1,1-dimethoxyethane
Ethane, 2-chloro-1,1-dimethoxy- (9CI)
NSC-60388

97-99-4
$C_5H_{10}O_2$
102.15
O(C(CC1)CO)C1
2-Furanmethanol, tetrahydro
Furfuryl alcohol, tetrahydro-
QO THFA
Tetrahydro-2-furancarbinol
Tetrahydro-2-furanmethanol
Tetrahydrofurfuryl alcohol
Tetrahydrofurylalkohol (Czech)
Tetrahydrofurfurylalkohol (Czech)
Tetrahydro-2-furylmethanol
THFA

98-00-0
$C_5H_6O_2$
98.11
O(C(=CC=1)CO)C1
Furfuryl-alcohol
2-Furancarbinol
2-Furanmethanol
Furfural alcohol
Furfuralcohol
Furfuryl alcohol (ACGIH, OSHA) [UN 2874]
2-Furfurylalkohol (Czech)
Furyl alcohol
α-Furylcarbinol
2-Furylcarbinol
2-Furylmethanol
2-Hydroxymethylfuran
Methanol, (2-furyl)-
NCI-C56224
UN 2874 [Furfuryl alcohol]

98-01-1
$C_5H_4O_2$
96.09
O=CC(OC=C1)=C1
2-Furaldehyde
Artificial Ant Oil
Artificial Oil of Ants
2-Formylfuran
Fural
Furale
2-Furanaldehyde
2-Furancarbonal
2-Furancarboxaldehyde
2-Furankarbaldehyd (Czech)
Furfural (ACGIH,OSHA) [UN 1199]
2-Furfural
Furfuraldehyde
Furfurale (Italian)
Furfuralu (Polish)
Furfurol
Furfurole
2-Furil-metanale (Italian)
Furole
α-Furole
2-Furylaldehyde
2-Furyl-methanal
NCI-C56177
Pyromucic aldehyde
RCRA waste number U125
UN 1199 [Furfural]

98-03-3
C_5H_4OS
112.15
O=CC(SC=C1)=C1
2-Thiophenecarboxaldehyde (9CI)
AI3-16611
α-Formylthiophene
2-Formylthiophene
NSC-2162
2-Thienylaldehyde
2-Thienylcarboxaldehyde
2-Thiophenealdehyde
α-Thiophenecarboxaldehyde

98-05-5
$C_6H_7AsO_3$
202.05
O[As](O)(=O)c1ccccc1
Benzenearsonic-acid
Kyselina benzenarsonova (Czech)
Phenyl arsenic acid
Phenylarsonic acid

98-06-6
$C_{10}H_{14}$
134.24
c(cccc1)(c1)C(C)(C)C
Benzene, tert-butyl
tert-Butylbenzene [UN 2709]
2-Methyl-2-phenylpropane
Pseudobutylbenzene
Trimethylphenylmethane
UN 2709 [Butyl benzenes]

98-07-7
$C_7H_5Cl_3$
195.47
c(cccc1)(c1)C(Cl)(Cl)Cl
Toluene, α,α,α-trichloro
Benzene, (trichloromethyl)-
Benzenyl chloride
Benzenyl trichloride
Benzoic trichloride
Benzotrichloride [UN 2226]
Benzylidyne chloride
Benzyl trichloride
Chlorure de benzenyle (French)
Phenyl chloroform
Phenyltrichloromethane
RCRA waste number U023
Toluene trichloride
Trichloormethylbenzeen (Dutch)
Trichlormethylbenzol (German)
Trichloromethylbenzene
1-(Trichloromethyl)benzene
Trichlorophenylmethane
α,α,α-Trichlorotoluene
ω,ω,ω-Trichlorotoluene
Triclorometilbenzene (Italian)
Triclorotoluene (Italian)
UN 2226 [Benzotrichloride]

98-08-8
$C_7H_5F_3$
146.12
FC(F)(F)c(cccc1)c1
Toluene, α,α,α-trifluoro
Benzenyl fluoride
Benzotrifluoride [UN 2338]
Benzylidyne fluoride
Phenylfluoroform
(Trifluoromethyl)benzene
α,α,α-Trifluorotoluene
ω-Trifluorotoluene
UN 2338 [Benzotrifluoride]
USAF MA-16

98-09-9
$C_6H_5ClO_2S$
176.62
O=S(=O)(c(cccc1)c1)Cl
Benzenesulfonyl-chloride
Benzene sulfonechloride
Benzenesulfonic chloride
Benzenesulfonic (acid) chloride
Benzenosulfochlorek (Polish)
Benzenosulphochloride
Benzene sulphonyl chloride [UN 2225]
BSC-Refine D
Phenylsulfonyl chloride
RCRA waste number U020
UN 2225 [Benzene sulfonyl chloride]

98-10-2
$C_6H_7NO_2S$
157.20
O=S(=O)(N)c(cccc1)c1
Benzenesulfonamide
Benzenesulphonamide
Benzosulfonamide
BSA

98-11-3
$C_6H_6O_3S$
158.18
O=S(=O)(O)c(cccc1)c1
Benzenesulfonic-acid
Kyselina benzensulfonova (Czech)
Phenylsulfonic acid

98-12-4
$C_6H_{11}Cl_3Si$
217.60
Cyclohexyltrichlorosilane
Ciclohexiltriclorosilano (Spanish)
Cyclohexane, 1-(trichlorosilyl)-
Cyclohexyl trichlorosilane
Silane, cyclohexyltrichloro-
Silane, trichlorocyclohexyl- (8CI,9CI)
Trichlorocyclohexylsilane
UN 1763 [Cyclohexyltrichlorosilane]

98-13-5
$C_6H_5Cl_3Si$
211.55
ClSi(Cl)(Cl)c1ccccc1
Silane, trichlorophenyl
Phenylsilicon trichloride
Phenyltrichlorosilane [UN 1804]
Phenyl trichlorosilane
Silane, phenyltrichloro-
Silicon phenyl trichloride
Trichlorophenylsilane
UN 1804 [Phenyltrichlorosilane]

98-15-7
$C_7H_4ClF_3$
180.56
FC(F)(F)c(cccc1Cl)c1
Toluene, m-chloro-α,α,α-trifluoro
Benzene, 1-chloro-3-(trifluoromethyl)- (9CI)
m-Chlorobenzotrifluoride [UN 2234]
3-Chlorobenzotrifluoride
m-Trifluoromethylphenyl chloride
UN 2234 [Chlorobenzotrifluorides]

98-16-8
$C_7H_6F_3N$
161.14
FC(F)(F)c(cccc1N)c1
m-Toluidine, α,α,α-trifluoro
m-ABTF
m-Aminobenzal fluoride
m-Aminobenzotrifluoride
3-Aminobenzotrifluoride
Toluene, 3-amino-α,α,α-trifluoro-
3-Trifluoromethyl aniline [UN 2948]
m-(Trifluoromethyl)aniline
3-(Trifluoromethyl)aniline
3-(Trifluoromethyl)benzenamine
UN 2948 [3-Trifluoromethylaniline]
USAF MA-4

98-17-9
$C_7H_5F_3O$
162.12
FC(F)(F)c(cccc1O)c1
m-Cresol, α,α,α-trifluoro
m-Hydroxybenzotrifluoride
3-Hydroxybenzotrifluoride
Phenol, 3-(trifluoromethyl)- (9CI)
m-(Trifluoromethyl)phenol
3-(Trifluoromethyl)phenol

98-18-0
$C_6H_8N_2O_2S$
172.22
O=S(=O)(N)c(cccc1N)c1
Metanilamide
m-Aminobenzenesulfonamide
3-Aminobenzenesulfonamide
m-Aminobenzenesulphonamide
Benzenesulfonamide, 3-amino-

98-19-1
$C_{12}H_{18}$
162.30
c(cc(cc1C)C(C)(C)C)(c1)C
m-Xylene, 5-tert-butyl
Benzene, 5-tert-butyl-1,3-dimethyl-
1,3-Dimethyl-5-tert-butylbenzene

98-23-7
$C_{13}H_{20}$
176.30
c(cc(c(c1C)C)C)(c1)C(C)(C)C
Benzene, 5-(1,1-dimethylethyl)-1,2,3-trimethyl- (9CI)
5-(1,1-Dimethylethyl)-1,2,3-trimethylbenzene

98-27-1
$C_{11}H_{16}O$
164.27
Oc(c(cc(c1)C(C)(C)C)C)c1
Phenol, 4-tert-butyl-2-methyl
4-tert-Butyl-2-methylphenol

98-28-2
$C_{10}H_{13}ClO$
184.67
Oc(c(cc(c1)C(C)(C)C)Cl)c1
4-tert-Butyl-2-chlorophenol
AI3-00060
2-Chloro-4-tert-butylphenol
2-Chloro-4-(1,1-dimethylethyl)-phenol
NSC-8464
Phenol, 4-tert-butyl-2-chloro- (8CI)
Phenol, 2-chloro-4-(1,1-dimethylethyl)- (9CI)

98-29-3
$C_{10}H_{14}O_2$
166.24
Oc(c(O)cc(c1)C(C)(C)C)c1
Pyrocatechol, 4-tert-butyl
1,2-Benzenediol, 4-(1,1-dimethylethyl)-
4-tert-Butyl-1,2-benzenediol
4-t-Butylcatechol
p-t-Butylpyrocatechol
4-t-Butylpyrocatechol
4-tert-Butylpyrokatechin (Czech)
Synox TBC

98-30-6
$C_7H_9NO_3S$
187.21
O=S(=O)(c(ccc(O)c1N)c1)C
Phenol, 2-amino-4-(methyl-

sulfonyl)- (9CI)
2-Amino-4-(methylsulfonyl)-phenol

98-31-7
$C_6H_3Cl_3O_2S$
245.51
O=S(=O)(c(ccc(c1Cl)Cl)c1)Cl
Benzenesulfonyl chloride, 3,4-dichloro- (9CI)
AI3-19279
3,4-Dichlorobenzenesulfonyl chloride
3,4-Dichlorophenylsulfonyl chloride
NSC-2651

98-32-8
$C_6H_8N_2O_3S$
188.20
O=S(=O)(N)c(ccc(O)c1N)c1
4-Hydroxymetanilamide
3-Amino-4-hydroxybenzene-sulfonamide
Aminophenol sulfamide
o-Aminophenol-p-sulfonamide
2-Aminophenol-4-sulfonamide
Benzenesulfonamide, 3-amino-4-hydroxy- (9CI)
Metanilamide, 4-hydroxy- (8CI)
NSC-4976

98-33-9
$C_7H_9NO_3S$
187.21
O=S(=O)(O)c(ccc(N)c1C)c1
Benzenesulfonic acid, 4-amino-3-methyl- (9CI)
AI3-16579
4-Amino-3-methylbenzene-sulfonic acid
4-Amino-m-toluenesulfonic acid
4-Amino-m-toluenesulfonic acid (SO_3H=1)
4-Amino-meta-toluenesulfonic acid (SO_3H=1)
2-Amino-5-toluenesulfonic acid
NSC-7545
m-Toluenesulfonic acid, 4 -amino- (8CI)
o-Toluidine-m-sulfonic acid

98-36-2
$C_6H_6ClNO_3S$
207.63
O=S(=O)(O)c(ccc(c1N)Cl)c1
Benzenesulfonic acid, 3-amino-4-chloro- (9CI)
3-Amino-4-chlorobenzene-sulfonic acid
4-Chlorometanilic acid
Metanilic acid, 4-chloro- (8CI)
NSC-59702

98-37-3
$C_6H_7NO_4S$
189.19
O=S(=O)(O)c(ccc(O)c1N)c1
Benzenesulfonic acid, 3-amino-4-hydroxy- (9CI)
AI3-14898
3-Amino-4-hydroxybenzene-sulfonic acid
o-Aminophenol-p-sulfonic acid
2-Aminophenol-4-sulfonic acid
2-Amino-4-sulfophenol
4-Hydroxy-3-aminobenzene-sulfonic acid
4-Hydroxymetanilic acid
2-Hydroxy-5-sulfoaniline
Metanilic acid, 4-hydroxy- (8CI)
NSC-1491

98-46-4
$C_7H_4F_3NO_2$
191.12
O=N(=O)c(cccc1C(F)(F)F)c1
Toluene, 3-nitro-α,α,α-trifluoro
m-Nitrobenzotrifluoride [UN 2306]
3-Nitrobenzotrifluoride
m-Nitrotrifluorotoluene
m-Nitrotrifluortoluol (German)
Toluene, α,α,α-trifluoro-m-nitro- (8CI)
m-(Trifluoromethyl)nitro-benzene
3-Trifluoromethylnitrobenzene
UN 2306 [Nitrobenzotri-fluorides]
USAF MA-5

98-47-5
$C_6H_5NO_5S$
203.18
O=S(=O)(O)c(cccc1N(=O)=O)c1
Benzenesulfonic acid, m-nitro
Benzenesulfonic acid, 3-nitro-
Kyselina nitrobenzen-m-sulf-onova (Czech)
Kyselina 3-nitrobenzen-sulfonova (Czech)
m-Nitrobenzenesulfonic acid
3-Nitrobenzenesulfonic acid

98-48-6
$C_6H_6O_6S_2$
238.24
O=S(=O)(O)c(cccc1S(=O)(=O)O)c1
1,3-Benzenedisulfonic acid
AI3-28531
m-Benzenedisulfonic acid

98-50-0
$C_6H_8AsNO_3$
217.07
Nc1ccc(cc1)[As](O)(O)=O
Arsanilic-acid
Acide p-arsanilique (French)
p-Aminobenzenearsonic acid
4-Aminobenzenearsonic acid
Aminophenylarsine acid
p-Aminophenylarsine acid
p-Aminophenylarsinic acid
p-Aminophenylarsonic acid
4-Aminophenylarsonic acid
p-Anilinearsonic acid
Atoxyl
Antoxylic acid
p-Arsanilic acid
4-Arsanilic acid
Arsanilic acid-100
Arsonic acid, (4-aminophenyl)-(9CI)
AS-101
Atoxylic acid
Benzenearsonic acid, p-amino-
Kyselina arsanilova (Czech)
Pro-Gen
Progen 90
Premix
Pro-Gen 227

98-51-1
$C_{11}H_{16}$
148.27
c(ccc(c1)C)(c1)C(C)(C)C
Toluene, p-tert-butyl
p-tert-Butyltoluene (ACGIH, OSHA)
p-Methyl-tert-butylbenzene
1-Methyl-4-tert-butylbenzene
TBT

98-52-2
$C_{10}H_{20}O$
156.30
OC(CCC(C(C)(C)C)C1)C1
Cyclohexanol, 4-tert-butyl
4-tert-Butylcyclohexanol
Padaryl
USAF DO-20

98-53-3
$C_{10}H_{18}O$
154.28
O=C(CCC(C(C)(C)C)C1)C1
Cyclohexanone, p-tert-butyl
p-tert-Butylcyclohexanone

98-54-4
$C_{10}H_{14}O$
150.24
Oc(ccc(c1)C(C)(C)C)c1
Phenol, p-(tert-butyl)
p-terc.Butylfenol (Czech)
Butylphen
p-tert-Butylphenol
4-t-Butylphenol
4-(1,1-Dimethylethyl)phenol
1-Hydroxy-4-tert-butylbenzene
Phenol, 4-(1,1-dimethylethyl)-(9CI)
PTBP
UCAR butylphenol 4-T Flake
UCAR butylphenol 4-T

98-55-5
$C_{10}H_{18}O$
154.28
OC(C(CCC(=C1)C)C1)(C)C
α-Terpineol

3-Cyclohexene-1-methanol, α,α,4-trimethyl- (9CI)
p-Menth-1-en-8-ol (8CI)
Terpenol
Terpineol schlechthin

98-56-6
$C_7H_4ClF_3$
180.56
FC(F)(F)c(ccc(c1)Cl)c1
Toluene, p-chloro-α,α,α-trifluoro
Benzene, 1-chloro-4-(trimethyl)- (9CI)
p-Chlorobenzotrifluoride [UN 2234]
(p-Chlorophenyl)trifluoromethane
p-Chlorotrifluoromethylbenzene
p-(Trifluoromethyl)chlorobenzene
α,α,α-Trifluoro-4-chlorotoluene
p-Trifluoromethylphenyl chloride
UN 2234 [Chlorobenzotrifluorides]

98-57-7
$C_7H_7ClO_2S$
190.65
O=S(=O)(c(ccc(c1)Cl)c1)C
Sulfone, p-chlorophenyl methyl
Benzene, 1-chloro-4-(methylsulfonyl)- (9CI)
p-Chlorophenyl methyl sulfone
4-Chlorophenyl methyl sulfone
Methyl 4-chlorophenyl sulfone

98-58-8
$C_6H_4BrClO_2S$
255.52
O=S(=O)(c(ccc(c1)Br)c1)Cl
Benzenesulfonyl chloride, p-bromo- (8CI)
Benzenesulfonyl chloride, 4-bromo- (9CI)
p-Bromobenzenesulfonyl chloride
4-Bromobenzenesulfonyl chloride
p-Bromophenylsulfonyl chloride
NSC-4506

98-59-9
$C_7H_7ClO_2S$
190.65
O=S(=O)(c(ccc(c1)C)c1)Cl
p-Toluenesulfonyl chloride
AI3-52254
Benzenesulfonyl chloride, 4-methyl- (9CI)
4-Methylbenzenesulfonyl chloride
p-Methylbenzenesulfonyl chloride
m-Methylbenzenesulfonyl chloride
NSC-175822
para-Toluenesulfochloride
p-Toluenesulfochloride
para-Toluenesulfonchloride
p-Toluenesulfonic acid chloride
4-Toluenesulfonic acid, chloride
para-Toluenesulfonyl chloride
Toluenesulfonyl chloride (VAN)
4-Toluenesulfonyl chloride
4-Toluenesulfonyl chloride
p-Toluenesulphonyl chloride
p-Tolylsulfonyl chloride
Tosyl chloride
p-Tosyl chloride

98-60-2
$C_6H_4Cl_2O_2S$
211.06
O=S(=O)(c(ccc(c1)Cl)c1)Cl
Benzenesulfonyl chloride, p-chloro
p-Chlorbenzensulfochlorid (Czech)
Chlorid kyseliny p-chlorbensulfonove (Czech)
p-Chlorobenzenesulfonyl chloride

98-61-3
$C_6H_4ClIO_2S$
302.52
O=S(=O)(c(ccc(c1)I)c1)Cl
Pipsyl chloride
Benzenesulfonyl chloride, p-iodo- (8CI)
Benzenesulfonyl chloride, 4-iodo- (9CI)
Iodobenzene-p-sulfonyl chloride
p-Iodobenzenesulfonyl chloride
4-Iodobenzenesulfonyl chloride
4-Iodobenzenesulphonyl chloride
NSC-77079

98-64-6
$C_6H_6ClNO_2S$
191.64
O=S(=O)(N)c(ccc(c1)Cl)c1
Benzenesulfonamide, p-chloro
p-Chlorobenzenesulfonamide
USAF MA-3

98-66-8
$C_6H_5ClO_3S$
192.62
O=S(=O)(O)c(ccc(c1)Cl)c1
Benzenesulfonic acid, 4-chloro-
AI3-50012
4-Chlorobenzenesulfonic acid

98-67-9
$C_6H_6O_4S$
174.18
O=S(=O)(O)c(ccc(O)c1)c1
Benzenesulfonic acid, p-hydroxy
4-Hydroxyphenylsulfonic acid

98-68-0
$C_7H_7ClO_3S$
206.65
Benzenesulfonyl chloride, 4-methoxy- (9CI)
Benzenesulfonyl chloride, p-methoxy- (8CI)
NSC-403292

98-69-1
$C_8H_{10}O_3S$
186.23
O=S(=O)(O)c(ccc(c1)CC)c1
4-Ethylbenzenesulfonic acid
Benzenesulfonic acid, p-ethyl-
Benzenesulfonic acid, 4-ethyl- (9CI)
p-Ethylbenzenesulfonic acid
Phenylethane-p-sulfonate

98-73-7
$C_{11}H_{14}O_2$
178.25
O=C(O)c(ccc(c1)C(C)(C)C)c1
Benzoic acid, p-tert-butyl
p-tert-Butyl benzoic acid
Kyselina p-terc.Butylbenzoova (Czech)
TBBA

98-74-8
$C_6H_4ClNO_4S$
221.61
O=S(=O)(c(ccc(N(=O)=O)c1)c1)Cl
Benzenesulfonyl chloride, p-nitro- (8CI)
AI3-52248
Benzenesulfonyl chloride, 4-nitro- (9CI)
4-Nitrobenzenesulfonic acid chloride
p-Nitrobenzenesulfonyl chloride
4-Nitrobenzenesulfonyl chloride
p-Nitrophenylsulfonyl chloride
4-Nitrophenylsulfonyl chloride
NSC-13065

98-77-1
$C_6H_{11}NS_2.C_5H_{11}N$
246.47
SC(=S)N1CCCCC1.C1CCNCC1
1-Piperidinecarbodithioic acid, Compd. with piperidine
Accelerator 552
Piperidinium pentamethylenedithiocarbamate
Pip-Pip
"522" Rubber Accelerator

98-79-3
$C_5H_7NO_3$
129.11

O=C(O)C(NC(=O)C1)C1
PCA
Acide pidolique (French)
Acido pidolico (Spanish)
Acidum pidolicum (Latin)
5-Oxo-L-proline
Pidolic acid
L-Proline, 5-oxo- (9CI)
L-Pyroglutamic acid
Pyrrolidonecarboxylic acid
2-Pyrrolidone-5-carboxylic acid

98-82-8
C_9H_{12}
120.21
c(cccc1)(c1)C(C)C
Cumene (ACGIH,OSHA)
Benzene, isopropyl
Benzene, (1-methylethyl)- (9CI)
Cumeen (Dutch)
Cumol
2-Fenilpropano (Italian)
2-Fenyl-propaan (Dutch)
Isopropylbenzeen (Dutch)
Isopropilbenzene (Italian)
Isopropyl benzene
Isopropylbenzene [UN 1918]
Isopropylbenzol
Isopropyl-benzol (German)
(1-Methylethyl)benzene
2-Phenylpropane
Propane, 2-phenyl
RCRA waste number U055
UN 1918 [Isopropylbenzene]

98-83-9
C_9H_{10}
118.19
c(C(=C)C)(cccc1)c1
Styrene, α-methyl
Benzene, (1-methylethenyl)-
Isopropenil-benzolo (Italian)
Isopropenylbenzene [UN 2303]
Isopropenyl-benzeen (Dutch)
Isopropenyl-benzol (German)
α-Methylstyreen (Dutch)
α-Methylstyrene
α-Methyl styrene (ACGIH, OSHA)
α-Methyl-styrol (German)
α-Metil-stirolo (Italian)
as-Methylphenylethylene
2-Phenylpropene
β-Phenylpropene
2-Phenylpropylene
β-Phenylpropylene
UN 2303 [Isopropenylbenzene]

98-84-0
$C_8H_{11}N$
121.20
CC(N)c1ccccc1
Benzylamine, α-methyl
α-Aminoethylbenzene
1-Amino-1-phenylethane
Benzenemethaneamine, α-methyl-
Ethylamine, 1-phenyl-
1-Fenylethylamin (Czech)
α-Methylbenzenemethanamine
α-Methylbenzylamine
α-Phenylethylamine
1-Phenylethylamine
Sumine 2079

98-85-1
$C_8H_{10}O$
122.18
OC(c(cccc1)c1)C
Benzyl alcohol, α-methyl
Benzenemethanol, α-methyl-
Ethanol, 1-phenyl-
1-Fenylethanol (Czech)
Fenyl-methylkarbinol (Czech)
Methanol, methylphenyl-
α-Methylbenzyl alcohol [UN 2937]
Methylphenylcarbinol
NCI-C55685
α-Phenethyl alcohol
1-Phenylethanol
Phenylmethylcarbinol
Styrallyl alcohol
Styralyl alcohol
UN 2937 [α-Methylbenzyl alcohol]

98-86-2
C_8H_8O
120.16
O=C(c(cccc1)c1)C
Acetophenone
Acetofenon (Czech)
Acetophenon
Acetylbenzene
Benzene, acetyl-
Benzoyl methide
Dymex
Ethanone, 1-phenyl- (9CI)
Hypnon
Hypnone
Ketone, methyl phenyl
Methyl phenyl ketone
1-Phenylethanone
Phenyl methyl ketone
RCRA waste number U004
USAF EK-496

98-87-3
$C_7H_6Cl_2$
161.03
c(cccc1)(c1)C(Cl)Cl
Benzene, (dichloromethyl)
Benzal chloride
Benzyl dichloride
Benzylene chloride
Benzylidene chloride [UN 1886]
Chlorobenzal
Chlorure de benzylidene (French)
(Dichloromethyl)benzene
α,α-Dichlorotoluene
RCRA waste number U017
Toluene, α,α-dichloro-
UN 1886 [Benzylidene chloride]

98-88-4
C_7H_5ClO
140.57
O=C(c(cccc1)c1)Cl
Benzoyl-chloride
Benzaldehyde, α-chloro-
Benzenecarbonyl chloride
Benzoic acid, chloride
Benzoyl chloride [UN 1736]
α-Chlorobenzaldehyde
UN 1736 [Benzoyl chloride]

98-89-5
$C_7H_{12}O_2$
128.19
O=C(O)C(CCCC1)C1
Cyclohexanecarboxylic-acid
Carboxycyclohexane
Cyclohexanoic acid
Cyclohexylcarboxylic acid
Hexahydrobenzoic acid

98-92-0
$C_6H_6N_2O$
122.14
O=C(N)c(cccn1)c1
Nicotinamide
Acid amide
Amide PP
Amid kyseliny nikotinove (Czech)
Aminicotin
Amixicotyn
Amnicotin
Austrovit PP
Benicot
Delonin Amide
Dipegyl
Dipigyl
Endobion
Factor PP
Hansamid
Inovitan PP
Nam
Niacevit
Niacinamide
Niamide
Nicamide
Nicamina
Nicamindon
Nicasir
Nicobion
Nicofort
Nicogen
Nicomidol
Nicosan 2
Nicota
Nicotamide
Nicotilamide
Nicotililamido
Nicotine acid amide
Nicotinic acid amide
Nicotinic amide
Nicotinsaureamid (German)
Nicotol
Nicotylamide
Nicovel
Nicovit
Nicovitol

Nicozymin
Niko-Tamin
Nikotinsaeureamid (German)
Niocinamide
Niozymin
Pelmin
Pelmine
Pelonin amide
PP-Faktor
Pyridine-3-carboxylic acid amide
3-Pyridinecarboxylic acid amide
Savacotyl
Nandervit-N
Vi-Nicotyl
Vi-Noctyl
Vitamin B3
Vitamin PP
Witamina PP

98-94-2
$C_8H_{17}N$
127.26
N(C(CCCC1)C1)(C)C
Cyclohexylamine, N,N-dimethyl
Cyclohexanamine, N,N-dimethyl- (9CI)
Cyclohexyldimethylamine
N-Cyclohexyldimethylamine
(Dimethylamino)cyclohexane
N,N-Dimethylaminocyclohexane
Dimethylcyclohexylamine
N,N-Dimethylcyclohexanamine
N,N-Dimethylcyclohexylamine [UN 2264]
Polycat 8
UN 2264 [Dimethylcyclohexylamine]

98-95-3
$C_6H_5NO_2$
123.12
O=N(=O)c(cccc1)c1
Benzene, nitro
Essence of Mirbane
Essence of Myrbane
Mirbane Oil
NCI-C60082
Nitrobenzeen (Dutch)
Nitrobenzen (Polish)
Nitrobenzene (ACGIH,OSHA) [UN 1662]
Nitrobenzene, Liquid (DOT)
Nitrobenzol
Nitrobenzol (DOT)
Nitrobenzol, Liquid (DOT)
Oil of Mirbane (DOT)
Oil of Myrbane
RCRA waste number U169
UN 1662 [Nitrobenzene]

98-96-4
$C_5H_5N_3O$
123.13
O=C(N)c(nccn1)c1
Pyrazinecarboxamide
Aldinamid
Aldinamide
2-Carbamyl pyrazine
D-50
Eprazin
MK 56
NCI-C01785
Pyrazinamide
Pyrazineamide
Pyrazine carboxylamide
Pyrazinoic acid amide
Tebrazid

98-98-6
$C_6H_5NO_2$
123.12
O=C(O)c(nccc1)c1
Picolinic-acid
Acide picolique (French)
2-Carboxypyridine
o-Pyridinecarboxylic acid
α-Pyridinecarboxylic acid
2-Pyridinecarboxylic acid (9CI)
Pyridine-carboxylique-2 (French)

99-03-6
C_8H_9NO
135.18
O=C(c(cccc1N)c1)C
Acetophenone, 3'-amino
Acetophenone, m-amino-
m-Acetylaniline
3-Acetylaniline
3-Aminoacetofenon (Czech)
β-Aminoacetophenone
m-Aminoacetophenone
3'-Aminoacetophenone
m-Aminoacetylbenzene
Ethanone, 1-(3-aminophenyl)- (9CI)

99-04-7
$C_8H_8O_2$
136.16
O=C(O)c(cccc1C)c1
m-Toluic acid
m-Methylbenzoic acid
3-Methylbenzoic acid
meta-Toluic acid
m-Toluylic acid

99-05-8
$C_7H_7NO_2$
137.15
O=C(O)c(cccc1N)c1
Benzoic acid, m-amino
m-Aminobenzoic acid
3-Aminobenzoic acid
Aniline-3-carboxylic acid
3-Carboxyaniline
MABA

99-06-9
$C_7H_6O_3$
138.13
O=C(O)c(cccc1O)c1
Benzoic acid, m-hydroxy
Acido m-idrossibenzoico (Italian)
Benzoic acid, 3-hydroxy- (9CI)
3-Carboxyphenol
m-HBA
m-Hydroxybenzoic acid
3-Hydroxybenzoic acid
Kyselina 3-hydroxybenzoova (Czech)
m-Salicylic acid

99-07-0
$C_8H_{11}NO$
137.20
Oc(cccc1N(C)C)c1
Phenol, m-(dimethylamino)
m-(Dimethylamino)phenol
3-(Dimethylamino)phenol
(3-Hydroxyphenyl)dimethylamine

99-08-1
$C_7H_7NO_2$
137.15
O=N(=O)c(cccc1C)c1
Toluene, m-nitro
3-Methylnitrobenzene
m-Methylnitrobenzene
MNT
m-Nitrotoluene (ACGIH,OSHA) [UN 1664]
3-Nitrotoluene
3-Nitrotoluol
UN 1664 [Nitrotoluenes, solid m-, or p-]

99-09-2
$C_6H_6N_2O_2$
138.14
O=N(=O)c(cccc1N)c1
Aniline, m-nitro
Amarthol Fast Orange R Base
m-Aminonitrobenzene
1-Amino-3-nitrobenzene
Azobase MNA
Benzenamine, 3-nitro- (9CI)
C.I. 37030
C.I. Azoic Diazo Component 7
Daito Orange Base R
Devol Orange R
Diazo Fast Orange R
Fast Orange Base R
Fast Orange M Base
Fast Orange MM Base
Fast Orange R Base
Fast Orange R Salt
Hiltonil Fast Orange R Base
MNA
Naphtoelan Orange R Base
Nitranilin
m-Nitroaminobenzene
m-Nitraniline
m-Nitroaniline [UN 1661]
3-Nitroaniline
3-Nitrobenzenamine
m-Nitrophenylamine
Orange Base Irga I
UN 1661 [Nitroanilines (o-; m-; p-;)]

99-10-5
$C_7H_6O_4$
154.13
O=C(O)c(cc(O)cc1O)c1
α-Resorcylic acid
Benzoic acid, 3,5-dihydroxy- (9CI)
5-Carboxyresorcinol
3,5-Dihydroxybenzoic acid

99-11-6
$C_6H_5NO_4$
155.12
O=C(O)c(cc(nc1O)O)c1
Isonicotinic acid, 2,6-dihydroxy
Citrazinic acid
2,6-Dihydroxy-4-carboxypyridine
2,6-Dihydroxyisonicotinic acid
Kyselina citrazinova (Czech)

99-12-7
$C_8H_9NO_2$
151.16
O=N(=O)c(cc(cc1C)C)c1
Benzene, 1,3-dimethyl-5-nitro- (9CI)
3,5-Dimethylnitrobenzene
1,3-Dimethyl-5-nitrobenzene
3,5-Dimethyl-1-nitrobenzene
5-Nitro-m-xylene
NSC-5403
m-Xylene, 5-nitro- (8CI)

99-20-7
$C_{12}H_{22}O_{11}$
342.30
O(C(C(O)C(O)C1O)CO)C1OC(OC(C(O)C2O)CO)C2O
Trehalose (8CI)
Ergot sugar
α-D-Glucopyranoside, α-D-glucopyranosyl (9CI)
Mycose
Natural Trehalose
NSC-2093
Trehalose, dihydrate
α-Trehalose
α,α-Trehalose
α-D-Trehalose
D-(+)-Trehalose

99-24-1
$C_8H_8O_5$
184.16
O=C(OC)c(cc(O)c(O)c1O)c1
Gallic acid, methyl ester
Benzoic acid, 3,4,5-trihydroxy-, methyl ester
Methyl gallate
Methyl 3,4,5-trihydroxybenzoate

99-28-5
$C_6H_3Br_2NO_3$
296.92
O=N(=O)c(cc(c(O)c1Br)Br)c1
Phenol, 2,6-dibromo-4-nitro
2,6-Dibromo-4-nitrophenol

99-29-6
$C_6H_4BrClN_2O_2$
251.45
O=N(=O)c(cc(c(N)c1Cl)Br)c1
Aniline, 2-bromo-6-chloro-4-nitro- (8CI)
Benzenamine, 2-bromo-6-chloro-4-nitro- (9CI)
2-Bromo-6-chloro-4-nitroaniline
2-Bromo-6-chloro-4-nitrobenzenamine
NSC-88985

99-30-9
$C_6H_4Cl_2N_2O_2$
207.02
O=N(=O)c(cc(c(N)c1Cl)Cl)c1
Aniline, 2,6-dichloro-4-nitro
AL-50
Allisan
Benzenamine, 2,6-dichloro-4-nitro- (9CI)
Bortran
Botran
CDNA
CNA
DCNA
DCNA (fungicide)
Dichloran
Dichloran (amine fungicide)
Dicloran
2,6-Dichlor-4-nitroanilin (Czech)
2,6-Dichloro-4-nitroaniline
2,6-Dichloro-4-nitrobenzenamine
Ditranil
4-Nitroaniline, 2,6-dichloro-
RD-6584
Resisan
U-2069

99-32-1
$C_7H_4O_6$
184.11
4H-Pyran-2,6-dicarboxylic acid, 4-oxo (8CI)
Chelidonic acid
Compound XI*
Jerva acid
Jervaic acid
Jervasic acid
4-Oxo-1,4-pyran-2,6-dicarboxylic acid
γ-Pyrone-2,6-dicarboxylic acid

99-34-3
$C_7H_4N_2O_6$
212.13
O=C(O)c(cc(N(=O)=O)cc1N(=O)=O)c1
Benzoic acid, 3,5-dinitro
3,5-Dinitrobenzoic acid
DNBA

99-35-4
$C_6H_3N_3O_6$
213.12
O=N(=O)c(cc(N(=O)=O)cc1N(=O)=O)c1
Benzene, 1,3,5-trinitro
RCRA waste number U234
TNB
Trinitrobenzeen (Dutch)
Trinitrobenzene
1,3,5-Trinitrobenzene
Trinitrobenzol (German)

99-36-5
$C_9H_{10}O_2$
150.18
O=C(OC)c(cccc1C)c1
Methyl 3-toluate
AI3-24382
Benzoic acid, 3-methyl-, methyl ester (9CI)
Methyl m-methylbenzoate
Methyl 3-methylbenzoate
Methyl m-toluate
NSC-20004
m-Toluic acid, methyl ester (8CI)
meta-Toluic acid, methyl ester

99-42-3
$C_8H_7NO_5$
197.14
Benzoic acid, 4-hydroxy-3-nitro-, methyl ester (9CI)
4-Hydroxy-3-nitrobenzoic acid methyl ester
Methyl 4-hydroxy-3-nitrobenzoate
NSC-28667

99-47-8
$C_8H_6ClNO_3$
199.60
Acetophenone, 2-chloro-m-nitro
α-Chloro-m-nitro-acetophenone
USAF MA-8

99-48-9
$C_{10}H_{16}O$
152.26
OC(C(=CCC1C(=C)C)C)C1
p-Mentha-6,8-dien-2-ol, l
l-Carveol
1-Methyl-4-isopropenyl-6-cyclohexen-2-ol

99-49-0
$C_{10}H_{14}O$
150.24
O=C(C(=CCC1C(=C)C)C)C1
p-Mentha-6,8-dien-2-one
Carvol

Carvone
6,8(9)-p-Menthadien-2-one
δ-1-Methyl-4-isopropenyl-6-cyclohexen-2-one
$\delta^{6,8}$-(9)-Terpadienone-2
NCI-C55867

99-50-3
$C_7H_6O_4$
154.13
OC(=O)c1ccc(O)c(O)c1
Protocatechuic-acid
Benzoic acid, 3,4-dihydroxy-
3,4-Dihydroxybenzoic acid

99-51-4
$C_8H_9NO_2$
151.16
O=N(=O)c(ccc(c1C)C)c1
1,2-Dimethyl-4-nitrobenzene
Benzene, 1,2-dimethyl-4-nitro- (9CI)
3,4-Dimethyl-1-nitrobenzene
4-Nitro-1,2-dimethylbenzene
p-Nitro-o-xylene
4-Nitro-o-xylene
para-Nitro-ortho-xylene
NSC-66555
o-Xylene, 4-nitro- (8CI)

99-52-5
$C_7H_8N_2O_2$
152.17
O=N(=O)c(ccc(N)c1C)c1
o-Toluidine, 4-nitro
Aniline, 2-methyl-4-nitro-
Ansibases Red RL
Azoene Fast Red GL Base
Azogene Fast Red NRL Salt
Azogene Fast Red RL
Benzenamine, 2-methyl-4-nitro-
C.I. 37100
Daito Red Base RL
Devol Red Salt E
Diabase Red RL
Diazo Fast Red RL
Fast Red Base RL
Fast Red 5NT
Fast Red 5NT Salt
Fast Red RL Base
Fast Red Salt RL
Hiltonil Fast Red RL Base
Hiltosal Fast Red RL Salt
Kako Red RL Base
Kayaku Red RL Base
Meisi Fast Red RL Base
2-Methyl-4-nitroaniline
Mitsui Red RL Base
Naphthoelan Red RL Base
Red Base Ciba X
Red Base Irga X
Red Base NRL
Red RL Base
Sanyo Fast Red RL Base
Sanyo Fast Red Salt RL
Spectrolene Red RL
Symulon Red RL Base
Tulabase Fast Red RL
Yamada Fast Red RL Base

99-53-6
$C_7H_7NO_3$
153.15
Cc1cc(ccc1O)N(=O)=O
o-Cresol, 4-nitro
2-Methyl-4-nitrophenol
4-Nitro-o-cresol

99-54-7
$C_6H_3Cl_2NO_2$
192.00
O=N(=O)c(ccc(c1Cl)Cl)c1
Benzene, 1,2-dichloro-4-nitro (8CI,9CI)
DCNB
1,2-Dichloro-4-nitrobenzene
3,4-Dichlornitrobenzen (Czech)
3,4-Dichloronitrobenzen (Czech)
3,4-Dichloronitrobenzene
1-Nitro-3,4-dichlorobenzene
NSC-6295
NSC-99806

99-55-8
$C_7H_8N_2O_2$
152.17
O=N(=O)c(ccc(c1N)C)c1
o-Toluidine, 5-nitro
Amarthol Fast Scarlet G Base
Amarthol Fast Scarlet G Salt
2-Amino-4-nitrotoluene
Azoene Fast Scarlet GC Base
Azoene Fast Scarlet GC Salt
Azofix Scarlet G Salt
Azogene Fast Scarlet G
Benzenamine, 2-methyl-5-nitro-
C.I. 37105
C.I. Azoic Diazo Component 12
Dainichi Fast Scarlet G Base
Daito Scarlet Base G
Devol Scarlet B
Devol Scarlet G Salt
Diabase Scarlet G
Diazo Fast Scarlet G
Fast Red SG Base
Fast Scarlet Base G
Fast Scarlet Base J
Fast Scarlet G
Fast Scarlet G Base
Fast Scarlet GC Base
Fast Scarlet G Salt
Fast Scarlet J Salt
Fast Scarlet M4NT Base
Fast Scarlet T Base
Hiltonil Fast Scarlet G Base
Hiltonil Fast Scarlet GC Base
Hiltonil Fast Scarlet G Salt
Kayaku Scarlet G Base
Lake Scarlet G Base
Lithosol Orange R Base
6-Methyl-3-nitroaniline
2-Methyl-5-nitro-benzeneamine
Mitsui Scarlet G Base
Naphthanil Scarlet G Base
Naphtoelan Fast Scarlet G Base
Naphtoelan Fast Scarlet G Salt
NCI-C01843
4-Nitro-2-aminotoluene
5-Nitro-o-toluidine
PNOT
RCRA waste number U181
Scarlet Base Ciba II
Scarlet Base Irga II
Scarlet Base NSP
Scarlet G Base
Sugai Fast Scarlet G Base
Symulon Scarlet G Base

99-56-9
$C_6H_7N_3O_2$
153.16
O=N(=O)c(ccc(N)c1N)c1
o-Phenylenediamine, 4-nitro
C.I. 76020
2-Amino-4-nitroaniline
1,2-Diamino-4-nitrobenzene
3,4-Diaminonitrobenzene
NCI-C03941
4NDB
4-Nitro-1,2-benzenediamine
4-Nitro-1,2-diaminobenzene
4-Nitro-1,2-fenylendiamin (Czech)
p-Nitro-o-phenylenediamine
4-Nitro-o-phenylene-diamine
4-Nitro-1,2-phenylenediamine
4-NO
4-NOP
4-NOPD
4-N-o-PDA

99-57-0
$C_6H_6N_2O_3$
154.14
O=N(=O)c(ccc(O)c1N)c1
Phenol, 2-amino-4-nitro
3-Amino-4-hydroxynitrobenzene
2-Amino-4-nitrofenol (Czech)
2-Amino-4-nitrophenol
2-Hydroxy-5-nitroaniline
NCI-C55958
p-Nitroaminofenol (Polish)
4-Nitro-2-aminofenol (Czech)
p-Nitro-o-aminophenol
4-Nitro-2-aminophenol

99-59-2
$C_7H_8N_2O_3$
168.17
O=N(=O)c(ccc(OC)c1N)c1
o-Anisidine, 5-nitro
2-Amino-1-methoxy-4-nitrobenzene
3-Amino-4-methoxynitrobenzene
2-Amino-4-nitroanisole
Aniline, 2-methoxy-5-nitro-
o-Anisidine nitrate
Azoamine Scarlet
Azoamine Scarlet K
Azogene Ecarlate R
Azoic Diazo Component 13, Base
Benzenamine, 2-methoxy-5-nitro- (9CI)
C.I. Azoic Diazo Component 13

C.I. 37130
Fast Scarlet R
2-Methoxy-5-nitroaniline
NCI-C01934
2-Methoxy-5-nitrobenzenamine
5-Nitro-o-anisidine
3-Nitro-6-methoxyaniline
5-Nitro-2-methoxyaniline

99-60-5
$C_7H_4ClNO_4$
201.57
O=C(O)c(c(cc(N(=O)=O)c1)Cl)c1
Benzoic acid, 2-chloro-4-nitro
2-Chloro-4-nitrobenzoic acid
Kyselina 2-chloro-4-nitro-benzoova (Czech)

99-61-6
$C_7H_5NO_3$
151.13
O=Cc(cccc1N(=O)=O)c1
Benzaldehyde, m-nitro
Benzaldehyde, 3-nitro- (9CI)
3-Formylnitrobenzene
m-Nitrobenzaldehyde
3-Nitrobenzaldehyde

99-62-7
$C_{12}H_{18}$
162.30
c(cccc1C(C)C)(c1)C(C)C
Benzene, m-diisopropyl
1,3-Diisopropylbenzene

99-63-8
$C_8H_4Cl_2O_2$
203.02
O=C(c(cccc1C(=O)Cl)c1)Cl
Isophthaloyl-chloride
1,3-Benzenedicarbonyl chloride
Dichlorid kyseliny isoftalove (Czech)
Isophthalic acid chloride
Isophthalic acid dichloride
Isophthaloyl dichloride
Isophthalyl chloride
Isophthalyl dichloride
m-Phthalic dichloride
m-Phthaloyl chloride

99-64-9
$C_9H_{11}NO_2$
165.19
O=C(O)c(cccc1N(C)C)c1
3-(Dimethylamino)benzoic acid
Benzoic acid, m-(dimethyl-amino)- (8CI)
Benzoic acid, 3-(dimethyl-amino)- (9CI)
m-(Dimethylamino)benzoic acid
N,N-Dimethyl-m-aminobenzoic acid
NSC-7197

99-65-0
$C_6H_4N_2O_4$
168.12
O=N(=O)c(cccc1N(=O)=O)c1
Benzene, m-dinitro
Benzene, 1,3-dinitro-
Binitrobenzene
m-Dinitrobenzene (ACGIH, DOT,OSHA)
1,3-Dinitrobenzene
2,4-Dinitrobenzene
Dinitrobenzenes, Liquid [UN 1597]
1,3-Dinitrobenzol
Dwunitrobenzen (Polish)
UN 1597 [Dinitrobenzenes, Liquid]

99-66-1
$C_8H_{16}O_2$
144.24
CCCC(CCC)C(O)=O
Valeric acid, 2-propyl
Abbott 44090
Acetic acid, dipropyl-
Depakene
Depakine
Dipropylacetic acid
Di-n-propylacetic acid
n-Dipropylacetic acid
Di-n-propylessigsaure (German)
n-DPA
Epilim
Kyselina 2-propylvalerova (Czech)
2-Propylpentanoic acid
2-Propylvaleric acid
Valproate
Valproic acid

99-71-8
$C_{10}H_{14}O$
150.24
Oc(ccc(c1)C(CC)C)c1
Phenol, p-(sec-butyl)
p-(sec-Butyl)phenol
4-sec-Butylphenol

99-73-0
$C_8H_6Br_2O$
277.96
O=C(c(ccc(c1)Br)c1)CBr
Acetophenone, 2,4'-dibromo
p-Bromophenacyl-8
p-Bromophenacyl bromide
4-Bromophenacyl bromide
α,p-Dibromoacetophenone
2,4'-Dibromoacetophenone

99-75-2
$C_9H_{10}O_2$
150.19
O=C(OC)c(ccc(c1)C)c1
p-Toluic acid, methyl ester
p-Carbomethoxytoluene
Methyl p-methylbenzoate
Methyl 4-methylbenzoate
Methyl p-toluate
Methyl 4-toluate
MPT

99-76-3
$C_8H_8O_3$
152.16
O=C(OC)c(ccc(O)c1)c1
Benzoic acid, p-hydroxy-, methyl ester
Abiol
Aseptoform
p-Hydroxybenzoic acid methyl ester
Maseptol
Methylben
Methyl chemosept
Methyl ester of p-hydroxy-benzoic acid
Methylester kyseliny p-hydroxybenzoove (Czech)
Methyl p-hydroxybenzoate
Methyl p-oxybenzoate
Methylparaben
Methyl parahydroxybenzoate
Methyl parasept
Metoxyde
Moldex
Nipagin
Nipagin M
p-Oxybenzoesauremethylester (German)
Paraben
Parasept
Paridol
Preserval M
Septos
Solbrol M
Tegosept M

99-77-4
$C_9H_9NO_4$
195.19
O=C(OCC)c(ccc(N(=O)=O)c1)c1
Benzoic acid, p-nitro-, ethyl ester
Ethyl nitrobenzoate, para ester
Ethyl p-nitrobenzoate
p-Nitrobenzoic acid, ethyl ester

99-79-6
$C_{19}H_{29}IO_2$
416.38
Undecanoic acid, 10-(p-iodo-phenyl)-, ethyl ester
Ethiodan
Ethyl 10-(p-iodophenyl)unde-canoate
Ethyl 10-(p-iodophenyl)unde-cylate
Iofendylate
Iophendylate
Mulsopaque
Myodil
Myodyl
Neurotrast
Pantopaque

99-80-9
$C_7H_7N_3O_2$

165.17
CN(N=O)c1ccc(N=O)cc1
Aniline, N,p-dinitroso-N-methyl
Aniline, N-methyl-N,p-dinitroso
Benzenamine, N-methyl-N,4-dinitroso-
N,4-Dinitroso-N-methylaniline
Elastopar
Elastopax
Heat Pre
N-Methyl-N,p-dinitrosoaniline
N-Methyl-N,4-dinitrosoaniline
N-Methyl-N,4-dinitrosobenzenamine
Methyl-(4-nitrosophenyl)nitrosamine
N-Nitroso-N-methyl-4-nitroso-aniline
Nitrozan K

99-82-1
$C_{10}H_{20}$
140.27
C(CCC(C1)C(C)C)(C1)C
1-Isopropyl-4-methylcyclohexane
AI3-24486
Cyclohexane, 1-methyl-4-(1-methylethyl)-
p-Menthane
1-Methyl-4-(1-methylethyl)-cyclohexane

99-83-2
$C_{10}H_{16}$
136.26
C(=CCC(C=1)C(C)C)(C1)C
p-Mentha-1,5-diene
1,3-Cyclohexadiene, 2-methyl-5-(1-methylethyl)-
α-Fellandrene
4-Isopropyl-1-methyl-1,5-cyclohexadiene
5-Isopropyl-2-methyl-1,3-cyclohexadiene
2-Methyl-5-isopropyl-1,3-cyclohexadiene
α-Phellandrene

99-85-4
$C_{10}H_{16}$
136.26
C(=CCC(=C1)C)(C1)C(C)C
p-Mentha-1,4-diene
1,4-Cyclohexadiene, 1-methyl-4-isopropyl-
1-Methyl-4-isopropylcyclohexadiene-1,4
γ-Terpinene

99-86-5
$C_{10}H_{16}$
136.26
C(=CC=C(C1)C)(C1)C(C)C
p-Mentha-1,3-diene
1,3-Cyclohexadiene, 1-methyl-4-isopropyl-
1-Methyl-4-isopropylcyclohexadiene-1,3
α-Terpinene

99-87-6
$C_{10}H_{14}$
134.24
c(ccc(c1)C)(c1)C(C)C
p-Cymene
Benzene, 1-isopropyl-4-methyl-
Camphogen
Cumene, p-methyl-
Cymene
p-Cymene [UN 2046]
Cymol
Dolcymene
p-Isopropylmethylbenzene
4-Isopropyl-1-methylbenzene
p-Isopropyltoluene
p-Methylisopropyl benzene
1-Methyl-4-isopropylbenzene
Paracymene
Paracymol
UN 2046 [Cymenes]

99-88-7
$C_9H_{13}N$
135.20
CC(C)c1ccc(N)cc1
Benzenamine, 4-(1-methylethyl)- (9CI)
AI3-04696
4-Aminocumene
4-Amino-1-isopropylbenzene
β-(4-Aminophenyl)propane
Aniline, p-isopropyl-
Cumene, p-amino-
Cumidine (8CI)
p-Cumidine
p-Isopropylaniline
4-Isopropylaniline
NSC-7198

99-89-8
$C_9H_{12}O$
136.21
Oc(ccc(c1)C(C)C)c1
Phenol, p-isopropyl
Australol
p-Cumenol
p-Isopropylphenol
4-Isopropylphenol
4-(1-Methylethyl)phenol
Phenol, 4-(1-methylethyl)-
Prodox 133

99-90-1
C_8H_7BrO
199.05
O=C(c(ccc(c1)Br)c1)C
4-Bromoacetophenone
Acetophenone, 4'-bromo- (8CI)
AI3-00489
4'-Bromoacetophenone
1-(4-Bromophenyl)ethanone
p-Bromophenyl methyl ketone
Ethanone, 1-(4-bromophenyl)-(9CI)
Methyl p-bromophenyl ketone
NSC-17541

99-91-2
C_8H_7ClO
154.60
O=C(c(ccc(c1)Cl)c1)C
Ethanone, 1-(4-chlorophenyl)
Acetophenone, 4'-chloro-
p-Chloroacetophenone
4-Chloroacetophenone
USAF DO-1

99-92-3
C_8H_9NO
135.18
O=C(c(ccc(N)c1)c1)C
Acetophenone, p-amino
Acetophenone, p-amino-
Acetophenone, 4'-amino- (8CI)
p-Acetylaniline
4-Acetylaniline
p-Aminoacetofenonu (Polish)
p-Aminoacetophenone
4'-Aminoacetophenone
p-Aminoacetylbenzene
Ethanone, 1-(4-aminophenyl)-(9CI)
USAF EK-631

99-93-4
$C_8H_8O_2$
136.16
O=C(c(ccc(O)c1)c1)C
Acetophenone, p-hydroxy
Acetophenone, 4'-hydroxy-
p-Acetylphenol
4-Acetylphenol
Ethanone, 1-(4-hydroxyphenyl)-(9CI)
4-Hydroksyacetofenol (Polish)
p-Hydroxyacetophenone
4'-Hydroxyacetophenone
1-(4-Hydroxyphenyl)ethanone
p-Hydroxyphenyl methyl ketone
Methyl-p-hydroxyphenyl ketone
p-Oxyacetophenone
Piceol
USAF KF-15

99-94-5
$C_8H_8O_2$
136.16
O=C(O)c(ccc(c1)C)c1
p-Toluic acid
p-Toluylic acid

99-96-7
$C_7H_6O_3$
138.13
O=C(O)c(ccc(O)c1)c1
Benzoic acid, p-hydroxy
Benzoic acid, 4-hydroxy- (9CI)
4-Carboxyphenol
p-Hydroxybenzoic acid
4-Hydroxybenzoic acid
Kyselina 4-hydroxybenzoova

(Czech)
p-Oxybenzoesaure (German)
p-Salicylic acid

99-97-8
$C_9H_{13}N$
135.23
N(c(ccc(c1)C)c1)(C)C
p-Toluidine, N,N-dimethyl
Benzenamine, N,N,4-trimethyl-
Dimetil-p-toluidina (Italian)
Dimethyl-p-toluidine
N,N,4-Trimethylaniline
p,N,N-Trimethylaniline

99-98-9
$C_8H_{12}N_2$
136.22
N(c(ccc(N)c1)c1)(C)C
p-Phenylenediamine, N,N-dimethyl
Dimethyl-paraphenylenediamine
N,N-Dimethyl-p-fenylendiamin (Czech)
N,N-Dimethyl-p-phenylene-diamine

99-99-0
$C_7H_7NO_2$
137.15
O=N(=O)c(ccc(c1)C)c1
Toluene, p-nitro
p-Methylnitrobenzene
4-Methylnitrobenzene
NCI-C60537
p-Nitrotoluene (ACGIH,OSHA) [UN 1664]
4-Nitrotoluene
4-Nitrotoluol
PNT
UN 1664 [Nitrotoluenes, solid m-, or p-]

100-00-5
$C_6H_4ClNO_2$
157.56
O=N(=O)c(ccc(c1)Cl)c1
Benzene, 1-chloro-4-nitro
1-Chloor-4-nitrobenzeen (Dutch)
1-Chlor-4-nitrobenzol (German)
p-Chloronitrobenzene (DOT)
1-Chloro-4-nitrobenzene
4-Chloronitrobenzene
4-Chloro-1-nitrobenzene
1-Cloro-4-nitrobenzene (Italian)
p-Nitrochlorobenzol (German)
p-Nitrochloorbenzeen (Dutch)
p-Nitrochlorobenzene (ACGIH, OSHA)
Nitrochlorobenzene, para, Solid [UN 1578]
p-Nitroclorobenzene (Italian)
PNCB
UN 1578 [Chloronitrobenzenes meta or para, solid]

100-01-6
$C_6H_6N_2O_2$
138.14
O=N(=O)c(ccc(N)c1)c1
Aniline, p-nitro
p-Aminonitrobenzene
1-Amino-4-nitrobenzene
Aniline, 4-nitro-
Azoamine Red zh
Azofix Red GG Salt
Azoic Diazo Component 37
Benzenamine, 4-nitro- (9CI)
C.I. 37035
C.I. Azoic Diazo Component 37
C.I. Developer 17
Developer P
Devol Red GG
Diazo Fast Red GG
Fast Red Base GG
Fast Red Base 2J
Fast Red 2G Base
Fast Red 2G Salt
Fast Red GG Base
Fast Red GG Salt
Fast Red MP Base
Fast Red P Base
Fast Red P Salt
Fast Red Salt GG
Fast Red Salt 2J
Naphtoelan Red GG Base
NCI-C60786
p-Nitraniline
4-Nitraniline
Nitrazol CF Extra
p-Nitroanilina (Polish)
p-Nitroaniline (ACGIH,OSHA) [UN 1661]
4-Nitroaniline
4-Nitrobenzenamine
p-Nitrophenylamine
Paranitroaniline, Solid (DOT)
PNA
RCRA waste number P077
Red 2G Base
Shinnippon Fast Red GG Base
UN 1661 [Nitroanilines (o-; m-; p-;)]

100-02-7
$C_6H_5NO_3$
139.12
O=N(=O)c(ccc(O)c1)c1
Phenol, p-nitro
4-Hydroxynitrobenzene
NCI-C55992
p-Nitrofenol (Czech)
4-Nitrofenol (Dutch)
p-Nitrophenol [UN 1663]
4-Nitrophenol
Paranitrofenol (Dutch)
Paranitrofenolo (Italian)
Paranitrophenol (French, German)
RCRA waste number U170
UN 1663 [Nitrophenols (o-; m-; p-;)]

100-06-1
$C_9H_{10}O_2$
150.19
O=C(c(ccc(OC)c1)c1)C
Acetophenone, 4'-methoxy
Acetanisole
p-Acetylanisole
4-Acetylanisole
Ethanone, 1-(4-methoxyphenyl)- (9CI)
Linarodin
4-Methoxyacetofenon (Czech)
p-Methoxyacetophenone
4-Methoxyacetophenone
4'-Methoxyacetophenone
p-Methoxyphenyl methyl ketone
4-Methoxyphenyl methyl ketone
Novatone
Vananote

100-07-2
$C_8H_7ClO_2$
170.60
O=C(c(ccc(OC)c1)c1)Cl
Anisoyl-chloride
Anisoyl chloride [UN 1729]
Benzoyl chloride, methoxy- (9CI)
Methoxybenzoyl chloride
UN 1729 [Anisoyl chloride]

100-09-4
$C_8H_8O_3$
152.16
O=C(O)c(ccc(OC)c1)c1
p-Anisic acid
4-Anisic acid
Draconic acid
Kyselina 4-methoxybenzoova (Czech)
p-Methoxybenzoic acid
4-Methoxybenzoic acid

100-10-7
$C_9H_{11}NO$
149.21
O=Cc(ccc(N(C)C)c1)c1
Benzaldehyde, p-(dimethyl-amino)
p-(Dimethylamino)benzaldehyde
4-(Dimethylamino)benzaldehyde
4-Dimethylaminobenzene-carbonal
Ehrlich's Reagent
p-Formyldimethylaniline
Reagens Ehrlichovo (Czech)

100-12-9
$C_8H_9NO_2$
151.16
O=N(=O)c(ccc(c1)CC)c1
Benzene, 1-ethyl-4-nitro- (9CI)
p-Ethylnitrobenzene
4-Ethylnitrobenzene
1-Ethyl-4-nitrobenzene
p-Nitroethylbenzene
p-Nitrophenylethane
NSC-858

100-14-1

$C_7H_6ClNO_2$
171.59
O=N(=O)c(ccc(c1)CCl)c1
Toluene, α-chloro-p-nitro
Benzene, 1-(chloromethyl)-4-nitro- (9CI)
p-(Chloromethyl)nitrobenzene
4-(Chloromethyl)nitrobenzene
α-Chloro-p-nitrotoluene
p-Nitrobenzyl chloride

100-15-2
$C_7H_8N_2O_2$
152.14
O=N(=O)c(ccc(NC)c1)c1
Aniline, N-methyl-p-nitro- (8CI)
AI3-08826
Benzenamine, N-methyl-4-nitro- (9CI)
p-(Methylamino)nitrobenzene
N-Methyl-p-nitraniline
N-Methyl-p-nitroaniline
N-Methyl-4-nitroaniline
N-Methyl-4-nitrobenzenamine
N-Monomethyl-p-nitroaniline
4-Nitro-N-methylaniline
NSC-5390

100-17-4
$C_7H_7NO_3$
153.15
O=N(=O)c(ccc(OC)c1)c1
Anisole, p-nitro
Benzene, 1-methoxy-4-nitro- (9CI)
p-Methoxynitrobenzene
4-Methoxynitrobenzene
p-Nitroanisol
p-Nitroanisole [UN 2730]
4-Nitroanisole
UN 2730 [Nitroanisole]

100-18-5
$C_{12}H_{18}$
162.30
c(ccc(c1)C(C)C)(c1)C(C)C
Benzene, p-diisopropyl
Benzene, 1,4-bis(1-methylethyl)- (9CI)
p-Diisopropylbenzene
1,4-Diisopropylbenzene

100-19-6
$C_8H_7NO_3$
165.16
CC(=O)c1ccc(cc1)N(=O)=O
Acetophenone, 4'-nitro
p-Acetylnitrobenzene
Ethanone, 1-(4-nitrophenyl)-
Methyl-p-nitrophenyl ketone
p-Nitroacetophenone
4'-Nitroacetophenone
p-Nitrophenyl methyl ketone
PNAP

100-20-9
$C_8H_4Cl_2O_2$
203.02
O=C(c(ccc(c1)C(=O)Cl)c1)Cl
Terephthaloyl-chloride
1,4-Benzenedicarbonyl chloride
1,4-Benzenedicarbonyl dichloride
p-Phenylenedicarbonyl dichloride
p-Phthaloyl chloride
p-Phthaloyl dichloride
Terephthalic acid chloride
Terephthalic acid dichloride
Terephthalic dichloride
Terephthaloyl dichloride

100-21-0
$C_8H_6O_4$
166.14
O=C(O)c(ccc(c1)C(=O)O)c1
Terephthalic-acid
Acide terephtalique (French)
p-Benzenedicarboxylic acid
1,4-Benzenedicarboxylic acid
Kyselina tereftalova (Czech)
TA 12
TA-33MP

100-22-1
$C_{10}H_{16}N_2$
164.28
N(c(ccc(N(C)C)c1)c1)(C)C
p-Phenylenediamine, N,N,N',N'-tetramethyl
Benzene, 1,4-bis(dimethylamino)-
1,4-Benzenediamine, N,N,N',N'-tetramethyl-
p-Bis(dimethylamino)benzene
1,4-Bis(dimethylamino)benzene
N,N,N',N'-Tetramethyl-p-fenylendiamin (Czech)
Tetramethyl-p-phenylenediamine
Tetramethyl-p-phenyldiamine
N,N,N',N'-Tetramethyl-p-phenylenediamine
TL 85
TMPD
Wurster's Blue
Wurster's Reagent

100-23-2
$C_8H_{10}N_2O_2$
166.17
O=N(=O)c(ccc(N(C)C)c1)c1
Aniline, N,N-dimethyl-p-nitro- (8CI)
AI3-08886
Benzenamine, N,N-dimethyl-4-nitro- (9CI)
p-(Dimethylamino)nitrobenzene
4-(Dimethylamino)nitrobenzene
1-(Dimethylamino)-4-nitrobenzene
N,N-Dimethyl-p-nitroaniline
N,N-Dimethyl-4-nitroaniline
N,N-Dimethyl-4-nitrobenzenamine
p-Nitrodimethylaniline
4-Nitrodimethylaniline
p-Nitro-N,N-dimethylaniline
4-Nitro-N,N-dimethylaniline
NSC-9815

100-25-4
$C_6H_4N_2O_4$
168.12
O=N(=O)c(ccc(N(=O)=O)c1)c1
Benzene, p-dinitro
p-Dinitrobenzene (ACGIH, OSHA) [UN 1597]
Dithane A-4
UN 1597 [Dinitrobenzenes, solid]

100-26-5
$C_7H_5NO_4$
167.11
O=C(O)c(ccc(n1)C(=O)O)c1
Pyridine-2,5-dicarboxylic acid
AI3-19238
Isocinchomeronic acid
NSC-177
2,5-Pyridinedicarboxylic acid (9CI)

100-27-6
$C_8H_9NO_3$
167.16
O=N(=O)c(ccc(c1)CCO)c1
Benzeneethanol, 4-nitro- (9CI)
AI3-36320
4-Nitrobenzeneethanol
p-Nitrophenethyl alcohol
4-Nitrophenethyl alcohol
2-(p-Nitrophenyl)ethanol
2-(4-Nitrophenyl)ethanol
NSC-55519
Phenethyl alcohol, p-nitro- (8CI)

100-29-8
$C_8H_9NO_3$
167.18
O=N(=O)c(ccc(OCC)c1)c1
Benzene, 1-ethoxy-4-nitro
p-Ethoxynitrobenzene
4-Ethoxynitrobenzene
Ethyl p-nitrophenyl ether
p-Nitrophenetol (German)
p-Nitrophenetole
Phenetole, p-nitro-

100-34-5
$C_6H_5N_2.Cl$
140.58
Benzenediazonium, chloride
Benzenediazonium chloride, Dry (DOT)

100-36-7
$C_6H_{16}N_2$
116.19
N(CCN)(CC)CC
N,N-Diethylethylenediamine

AI3-26638
1-Amino-2-(diethylamino)-ethane
1-Amino-2-(N,N-diethylamino)-ethane
N-(2-Aminoethyl)-N,N-diethyl-amine
N,N-(Diethylamino)ethylamine
N-(2-(Diethylamino)ethyl)amine
2-(Diethylamino)ethylamine
2-(N,N-Diethylamino)ethyl-amine
(Diethylamino)ethylamino
N,N-Diethyl-1,2-diaminoethane
N,N-Diethylethanediamine
N,N-Diethyl-1,2-ethanediamine
N,N-(Diethylethyl)diamine
N,N-Diethylethylene diamine
N,N-Diethylethylenediamine (French)
N,N-(Diethylethylene)diamine
N,N-Dietiletilendiamina (Spanish)
β-(Dimethylamino)ethylamine
1,2-Ethanediamine, N,N-di-ethyl- (9CI)
Ethylenediamine, N,N-diethyl- (8CI)
NSC-19675
UN 2685 [N,N-Diethylethylene-diamine]
USAF AM-1

100-37-8
$C_6H_{15}NO$
117.22
OCCN(CC)CC
Ethanol, 2-(diethylamino)
DEAE
Diaethylaminoaethanol (German)
Diethylaminoethanol
β-Diethylaminoethanol
N-Diethylaminoethanol
2-(Diethylamino)ethanol
2-N-Diethylaminoethanol
2-Diethylaminoethanol (ACGIH,OSHA)
Diethylaminoethanol [UN 2686]
β-Diethylaminoethyl alcohol
Diethylethanolamine
N,N-Diethylethanolamine
N,N-Diethyl-N-(β-hydroxy-ethyl)amine
2-Hydroxytriethylamine
UN 2686 [Diethylamino-ethanol]

100-39-0
C_7H_7Br
171.05
BrCc(cccc1)c1
Toluene, α-bromo
Benzene, (bromomethyl)-
Benzyl bromide [UN 1737]
(Bromomethyl)benzene
Bromophenylmethane
α-Bromotoluene
ω-Bromotoluene
Bromotoluene, α [UN 1737]
p-(Bromomethyl)nitrobenzene
UN 1737 [Benzyl bromide]

100-40-3
C_8H_{12}
108.20
C(=CCCC1C=C)C1
1-Cyclohexene, 4-vinyl
Butadiene dimer
Cyclohexene, 4-ethenyl-
Cyclohexenylethylene
4-Ethenyl-1-cyclohexene
NCI-C54999
1,2,3,4-Tetrahydrostyrene
1-Vinylcyclohexene-3
1-Vinylcyclohex-3-ene
4-Vinylcyclohexene
4-Vinylcyclohexene-1
4-Vinyl-1-cyclohexene

100-41-4
C_8H_{10}
106.18
c(cccc1)(c1)CC
Benzene, ethyl
Aethylbenzol (German)
EB
Ethylbenzeen (Dutch)
Ethyl benzene (ACGIH,OSHA) [UN 1175]
Ethylbenzol
Etilbenzene (Italian)
Etylobenzen (Polish)
NCI-C56393
Phenylethane
UN 1175 [Ethylbenzene]

100-42-5
C_8H_8
104.16
c(cccc1)(c1)C=C
Styrene
Benzene, vinyl-
Cinnamene
Cinnamenol
Cinnamol
Diarex HF 77
Ethenylbenzene
Ethylene, phenyl-
NCI-C02200
Phenethylene
Phenylethene
Phenylethylene
Phenylethylene (OSHA)
Stirolo (Italian)
Styreen (Dutch)
Styren (Czech)
Styrene, monomer (ACGIH)
Styrene monomer, Inhibited [UN 2055]
Styrene (OSHA)
Styrol (German)
Styrole
Styrolene
Styron
Styropol
Styropor
UN 2055 [Styrene monomer, inhibited]
Vinylbenzen (Czech)
Vinylbenzene
Vinyl benzene (OSHA)
Vinylbenzol

100-43-6
C_7H_7N
105.15
n(ccc(c1)C=C)c1
Pyridine, 4-vinyl
4-Ethenylpyridine
4-Vinylpyridine

100-44-7
C_7H_7Cl
126.59
c(cccc1)(c1)CCl
Toluene, α-chloro
Benzene, (chloromethyl)-
Benzile (cloruro di) (Italian)
Benzyl chloride (ACGIH,DOT, OSHA)
Benzyl chloride unstabilized [UN 1738]
Benzyle (chlorure de) (French)
Benzylchlorid (German)
Chloromethylbenzene
Chlorophenylmethane
α-Chlorotoluene
ω-Chlorotoluene
α-Chlortoluol (German)
Chlorure de benzyle (French)
NCI-C06360
RCRA waste number P028
Tolyl chloride
UN 1738 [Benzyl chloride unstabilized]

100-46-9
C_7H_9N
107.15
NCc(cccc1)c1
Benzylamine (8CI)
AI3-15299
α-Amino toluene
Aminotoluene
α-Aminotoluene
ω-Aminotoluene
Benzenemethanamine (9CI)
Monobenzylamine
Moringine
NSC-8046
(Phenylmethyl)amine

100-47-0
C_7H_5N
103.13
N#Cc(cccc1)c1
Benzonitrile [UN 2224]
Benzene, cyano-
Benzenenitrile
Benzoic acid nitrile
Cyanobenzene
Fenylkyanid (Czech)
Phenyl cyanide
UN 2224 [Benzonitrile]

100-48-1
$C_6H_4N_2$
104.10
N#Cc(ccnc1)c1
Isonicotinonitrile (8CI)
AI3-19232
4-Cyanopyridine
γ-Cyanopyridine
Isonicotinic acid nitrile
NSC-60681
4-Pyridinecarbonitrile (9CI)
4-Pyridinenitrile

100-49-2
$C_7H_{14}O$
114.21
OCC(CCCC1)C1
Cyclohexanemethanol
Benzyl alcohol, hexahydro-
Cyclohexanecarbinol
Cyclohexylcarbinol
Cyclohexylmethanol
Hexahydrobenzyl alcohol
Hydroxymethylcyclohexane
Methanol, cyclohexyl-
USAF DO-49

100-50-5
$C_7H_{10}O$
110.17
O=CC(CCC=C1)C1
4-Cyclohexene-1-carboxaldehyde
3-Cyclohexene-1-carboxaldehyde (8CI,9CI)
4-Formylcyclohexene
1,2,5,6-Tetrahydrobenzaldehyde [UN 2498]
UN 2498 [1,2,3,6-Tetrahydrobenzaldehyde]

100-51-6
C_7H_8O
108.15
OCc(cccc1)c1
Benzyl-alcohol
Benzal alcohol
Benzenecarbinol
Benzenemethanol
Benzoyl alcohol
Hydroxytoluene
α-Hydroxytoluene
Methanol, phenyl-
NCI-C06111
Phenolcarbinol
Phenylcarbinol
Phenylmethanol
Phenylmethyl alcohol
α-Toluenol

100-52-7
C_7H_6O
106.13
O=Cc(cccc1)c1
Benzaldehyde
Almond Artificial Essential Oil
Artificial Almond Oil
Artificial Essential Oil of Almond
Benzaldehyde (DOT)
Benzaldehyde FFC
Benzene carbaldehyde
Benzenecarbonal
Benzoic aldehyde
NA 1989 (DOT)
NCI-C56133
Phenylmethanal

100-53-8
C_7H_8S
124.21
SCc(cccc1)c1
α-Toluenethiol
Benzylhydrosulfide
Benzyl mercaptan
Benzylthiol
(Mercaptomethyl)benzene
α-Mercaptotoluene
Methanethiol, phenyl-
Phenylmethanethiol
Phenylmethyl mercaptan
Thiobenzyl alcohol
α-Toluolthiol
α-Tolyl mercaptan
USAF EK-1509

100-54-9
$C_6H_4N_2$
104.12
N#Cc(cccn1)c1
Nicotinonitrile
3-Azabenzonitrile
3-Cyanopyridine
3-Cyjanopirydyna (Polish)
Nicotinic acid nitrile
Nitryl kwasu nikotynowego (Polish)
3-Pyridinecarbonitrile (9CI)
3-Pyridinenitrile
3-Pyridylcarbonitrile

100-55-0
C_6H_7NO
109.14
n(cccc1CO)c1
3-Pyridinemethanol
3-(Hydroxymethyl)pyridine
Nicotinic alcohol
Nicotinyl alcohol
NU-2121
β-Picolyl alcohol
Pyridine-3-carbinol
3-Pyridylcarbinol
3-Pyridylmethanol
Ro-1-5155
Roniacol

100-56-1
C_6H_5ClHg
313.15
Cl[Hg]c1ccccc1
Mercury, chlorophenyl
Benzene, (chloromercuri)-
Chlorid fenylrtutnaty (Czech)
(Chloromercuri)benzene
Chlorophenylmercury
Fenylmercurichlorid (Czech)
Mercuriphenyl chloride
Merfazin
Mersolite 2
Phenyl chloromercury
Phenyl mercuric chloride
Phenylmercury chloride
Phenylquecksilberchlorid (German)
PMC
Stopspot

100-57-2
C_6H_6HgO
294.71
Mercury, hydroxyphenyl
Hydroxyphenylmercury
Mersolite 1
Phenyl hydroxymercury
Phenylmercuric hydroxide [UN 1894]
Phenylmercury hydroxide
UN 1894 [Phenylmercuric hydroxide]

100-59-4
C_6H_5ClMg
136.86
Phenylmagnesium chloride
Chlorophenylmagnesium
Magnesium, chlorophenyl- (9CI)

100-60-7
$C_7H_{15}N$
113.23
N(C(CCCC1)C1)C
Cyclohexylamine, N-methyl
Cyclohexanamine, N-methyl- (9CI)
Cyclohexylmethylamine
N-Methylcyclohexanamine
Methylcyclohexylamine
N-Methylcyclohexylamine

100-61-8
C_7H_9N
107.17
N(c(cccc1)c1)C
Aniline, N-methyl
Anilinomethane
Benzenamine, N-methyl- (9CI)
(Methylamino)benzene
N-Methylaminobenzene
Methylaniline
N-Methylaniline (ACGIH) [UN 2294]
N-Methylbenzenamine
Methylphenylamine
N-Methylphenylamine
Monomethylaniline
N-Monomethylaniline
Monomethyl aniline (OSHA)
N-Phenylmethylamine
UN 2294 [N-Methylaniline]

100-63-0
$C_6H_8N_2$
108.16
N(N)c(cccc1)c1
Hydrazine, phenyl
Fenilidrazina (Italian)
Fenylhydrazine (Dutch)
Hydrazine-benzene
Hydrazinobenzene
Phenylhydrazin (German)
Phenylhydrazine (ACGIH, OSHA) [UN 2572]
UN 2572 [Phenylhydrazine]

100-64-1
$C_6H_{11}NO$
113.18
N(O)=C(CCCC1)C1
Cyclohexanone, oxime
Antioxidant D
(Hydroxyimino)cyclohexane

100-65-2
C_6H_7NO
109.14
ONc(cccc1)c1
Hydroxylamine, N-phenyl
Aniline, N-hydroxy-
NCI-C60093
β-Phenylhydroxylamine
N-Phenylhydroxylamine

100-66-3
C_7H_8O
108.15
O(c(cccc1)c1)C
Anisole [UN 2222]
Benzene, methoxy
Ether, methyl phenyl
Methoxybenzene
Methyl phenyl ether
Phenyl methyl ether
UN 2222 [Anisole]

100-67-4
$C_6H_6O.K$
133.21
Phenol, potassium salt (9CI)

100-68-5
C_7H_8S
124.21
S(c(cccc1)c1)C
Benzene, (methylthio)
Methyl phenylsulfide
(Methylthio)benzene
Thioanisole

100-69-6
C_7H_7N
105.15
n(c(ccc1)C=C)c1
Pyridine, 2-vinyl
2-Vinylpyridine

100-70-9
$C_6H_4N_2$
104.10
N#Cc(nccc1)c1
2-Pyridinecarbonitrile (9CI)
2-Cyanopyridine
NSC-59697
Picolinic acid nitrile
Picolinonitrile (8CI)
2-Pyridinecarboxylic acid, nitrile
2-Pyridyl nitrile

100-71-0
C_7H_9N
107.15
n(c(ccc1)CC)c1
2-Ethylpyridine
α-Ethylpyridine
NSC-964
Pyridine, 2-ethyl- (9CI)

100-72-1
$C_6H_{12}O_2$
116.18
OCC1CCCCO1
2H-Pyran-2-methanol, tetrahydro
2-Hydroxymethyltetrahydro-pyran
2-Methanol tetrahydropyran
2-Methyloltetrahydro-1,4-pyran
Pyran-2-methanol, tetrahydro-
Tetrahydropyran-2-carbinol
Tetrahydropyran-2-methanol
Tetrahydropyranyl-2-methanol
2-Tetrahydropyranilcarbinol

100-73-2
$C_6H_8O_2$
112.13
Acrolein dimer
Acrolein dimer, Stabilized
Acroleine dimere stabilisee (French)
3,4-Dihydro-2H-pyran-2-carboxaldehyde
2,3-Dihydro-1,4-pyran-2-karboxaldehyd (Czech)
Dimero de la acroleina (Spanish)
2-Formyl-3,4-dihydro-2H-pyran
5-Hexenal, 2,6-epoxy-
NSC-95413
2-Propenal dimer
Pyran aldehyde
2H-Pyran-2-carboxaldehyde, 3,4-dihydro- (8CI,9CI)
UN 2607 [Acrolein dimer, Stabilized]

100-74-3
$C_6H_{13}NO$
115.20
O(CCN(C1)CC)C1
Morpholine, 4-ethyl
N-Ethylmorfolin (Czech)
N-Ethylmorpholine (ACGIH, OSHA)
4-Ethylmorpholine
NEM

100-75-4
$C_5H_{10}N_2O$
114.17
O=NN(CCCC1)C1
Piperidine, 1-nitroso
Nitrosopiperidin (German)
N-Nitroso-piperidin (German)
N-Nitrosopiperidine
1-Nitrosopiperidine
N-N-PIP
No-Pip
NPIP
Pyridine, hexahydro-N-nitroso-
RCRA waste number U179

100-76-5
$C_7H_{13}N$
111.18
N(CCC(C1)C2)(C1)C2
1-Azabicyclo(2.2.2)octane (9CI)

100-79-8
$C_6H_{12}O_3$
132.18
O(CC(O1)CO)C1(C)C
1,3-Dioxolane-4-methanol, 2,2-dimethyl
Acetone, cyclic (hydroxymethyl)ethylene acetal
2,2-Dimethyl-1,3-dioxolane-4-methanol
2,2-Dimethyl-5-hydroxymethyl-1,3-dioxolane
2,2-Dimethyl-4-oxymethyl-1,3-dioxolane
Dioxolan
Dioxolane [UN 1166]
Gie
Glycerolacetone
Glycerol dimethylketal
Glycerol, 1,2-O-isopropylidene
4-Hydroxymethyl-2,2-dimethyl-1,3-dioxolane
Isopropylidene glycerol
Solketal
UN 1166 [Dioxolane]

100-80-1
C_9H_{10}
118.19
c(cccc1C=C)(c1)C
Styrene, m-methyl
Benzene, 1-ethenyl-3-methyl- (9CI)
m-Methylstyrene
3-Methylstyrene
m-Vinyltoluene
3-Vinyltoluene

100-83-4
$C_7H_6O_2$
122.13

O=Cc(cccc1O)c1
Benzaldehyde, m-hydroxy
Benzaldehyde, 3-hydroxy- (9CI)
m-Formylphenol
3-Formylphenol
m-Hydroxybenzaldehyde
meta-Hydroxybenzaldehyde
3-Hydroxybenzaldehyde

100-84-5
$C_8H_{10}O$
122.17
O(c(cccc1C)c1)C
Anisole, m-methyl- (8CI)
AI3-19476
Benzene, 1-methoxy-3-methyl- (9CI)
m-Cresol methyl ether
m-Methylanisole
3-Methylanisole
Methyl m-cresyl ether
1-Methoxy-3-methylbenzene
m-Methoxytoluene
3-Methoxytoluene
3-Methylmethoxybenzene
1-Methyl-3-methoxybenzene
Methyl 3-methylphenyl ether
Methyl m-tolyl ether
NSC-6255

100-85-6
$C_{10}H_{16}N.HO$
167.28
Ammonium, benzyltrimethyl-, hydroxide
Benzenemethanaminium, N,N,N-trimethyl-, hydroxide (9CI)
Benzyltrimethylammonium hydroxide
Trimethylbenzylammonium hydroxide
Triton B

100-86-7
$C_{10}H_{14}O$
150.24
OC(Cc(cccc1)c1)(C)C
Phenethyl alcohol, α,α-dimethyl
Dimethylbenzylcarbinol
α,α-Dimethylphenethyl alcohol
1,1-Dimethyl-2-phenylethanol
DMBC

100-87-8
$C_7H_8O_3S$
172.20
O=S(=O)(O)Cc(cccc1)c1
Benzylsulfonic acid
Benzenemethanesulfonic acid (9CI)
Benzyl-α-sulfonic acid
Methanesulfonic acid, phenyl-
Phenylmethanesulfonic acid
α-Toluenesulfonic acid

100-88-9
$C_6H_{13}NO_3S$
179.26
O=S(=O)(O)NC(CCCC1)C1
Cyclohexanesulfamic-acid
Cyclamate
Cyclamic acid
Cyclohexanesulphamic acid
Cyclohexylamidosulphuric acid
Cyclohexylaminesulphonic acid
Cyclohexylsulphamic acid
N-Cyclohexylsulphamic acid
Hexamic acid
Sucaryl
Sucaryl acid
Sulfamic acid, cyclohexyl- (9CI)

100-92-5
$C_{11}H_{17}N$
163.29
CNC(C)(C)Cc1ccccc1
Phenethylamine, N,α,α-trimethyl
Mefenterdrin
Mefentermin
Mephenterdrine
Mephenterdrinum
Mephentermine
Mephetedrine
Mephine
2-Methylamino-2-methyl-1-phenylpropane
2-Methyl-2-methylamino-1-phenylpropane
N-Methyl-ω-phenyl-t-butylamine
N,α,α-Trimethylbenzeneethanamine
N,α,α-Trimethylphenethylamine
N,α,α-Trimethyl-β-phenethylamine
Vialin
WY-585
Wyamine
Wyfentermina

100-93-6
$C_{19}H_{18}N_2O_2S$
338.42
O=S(=O)(Nc(ccc(Nc(cccc1)c1)c2)c2)c(ccc(c3)C)c3
Benzenesulfonamide, 4-methyl-N-(4-(phenylamino)phenyl)- (9CI)
Aranox
NSC-41053
p-(p-Toluenesulfonamido)diphenylamine
4-(p-Toluenesulfonamido)diphenylamine
p-Toluenesulfonanilide, 4'-anilino- (8CI)
p-(p-Tolylsulfonylamino)diphenylamine

100-97-0
$C_6H_{12}N_4$
140.22
N(CN(CN1CN23)C3)(C1)C2
Hexamethylenetetramine
Aceto HMT
Aminoform
Ammoform
Ammonioformaldehyde
Cystamin
Cystogen
Esametilentetramina (Italian)
Formamine
Formin
Hexaform
Hexamethylenamine
Hexamethyleneamine
Hexamethylenetetraamine
Hexamethylentetramin (German)
Hexamethylentetramine
Hexamine [UN 1328]
Hexilmethylenamine
HMT
Methamin
Methenamine
Preparation AF
Resotropin
1,3,5,7-Tetraazaadamantane
1,3,5,7-Tetraazatricyclo-$(3.3.1.1^{37})$decane
UN 1328 [Hexamine]
Uritone
Urotropin
Urotropine

100-99-2
$C_{12}H_{27}Al$
198.37
Aluminum, triisobutyl
Aluminum, tris(2-methylpropyl)- (9CI)
Triisobutylalane
Triisobutylaluminium (DOT)
Triisobutylaluminum
UN 1930 (DOT)

101-02-0
$C_{18}H_{15}O_3P$
310.30
O(c(cccc1)c1)P(Oc(cccc2)c2)Oc(cccc3)c3
Phosphorous acid, triphenyl ester
EFED
Trifenoxyfosfin (Czech)
Trifenylfosfit (Czech)
Triphenyl phosphite

101-05-3
$C_9H_5Cl_3N_4$
275.53
Clc2nc(Cl)nc(Nc1ccccc1Cl)n2
s-Triazine, 2,4-dichloro-6-(o-chloroanilino)
Anilazin
Anilazine
B-622
Bortrysan
2-(2-Chloranilin)-4,6-dichlor-1,3,5-triazin (German)
(o-Chloroanilino)dichloro-

triazine
2,4-Dichloro-6-o-chloranilino-s-triazine
2,4-Dichloro-6-(o-chloro-anilino)-s-triazine
2,4-Dichloro-6-(2-chloro-anilino)-1,3,5-triazine
4,6-Dichloro-N-(2-chloro-phenyl)-1,3,5-triazin-2-amine
Direz
Dyrene
Dyrene 50W
ENT 26,058
Kemate
NCI-C08684
Triasyn
Triazin
Triazine
Triazine (pesticide)
Zinochlor

101-10-0
$C_9H_9ClO_3$
200.62
O=C(O)C(Oc(cccc1Cl)c1)C
Cloprop
Amchem 3-CP
Caswell No. 206
2-(3-Chlorophenoxy)propanoic acid
2-(m-Chlorophenoxy)propionic acid
EPA Pesticide Chemical Code 021201
Propanoic acid, 2-(3-chloro-phenoxy)- (9CI)

101-11-1
$C_{15}H_{17}NO_3S$
291.36
O=S(=O)(O)c(cccc1CN(c(cccc2)c2)CC)c1
Benzenesulfonic acid, 3-((ethylphenylamino)-methyl)- (9CI)

101-14-4
$C_{13}H_{12}Cl_2N_2$
267.17
Nc(c(cc(c1)Cc(ccc(N)c2Cl)c2)Cl)c1
Benzenamine, 4,4'-methylene-bis(2-chloro
Aniline, 4,4'-methylenebis-(2-chloro-
Bis Amine
Bis-Amine A
Bis(4-amino-3-chlorophenyl)-methane
Cl-MDA
Curalin M
Curene 442
Cyanaset
DACPM
Di(-4-amino-3-chlorophenyl)-methane
Di-(4-amino-3-clorofenil)-metano (Italian)
4,4'-Diamino-3,3'-dichlorodi-phenylmethane
3,3'-Dichlor-4,4'-diaminodi-phenylmethan (German)
3,3'-Dichloro-4,4'-diaminodi-phenylmethane
3,3'-Dichloro-4,4'-diaminodi-fenilmetano (Italian)
MBOCA
MBOCA (OSHA)
4,4'-Methylene(bis)-chloro-aniline
Methylene 4,4'-bis(o-chloro-aniline)
p,p'-Methylenebis(α-chloro-aniline)
4,4'-Methylenebis(o-chloro-aniline)
p,p'-Methylenebis(o-chloro-aniline)
4,4'-Methylenebis(2-chloro-aniline)
4,4'-Methylene bis(2-chloro-aniline) (ACGIH,OSHA)
4,4'-Methylenebis-2-chloro-benzenamine
Methylene-bis-orthochloro-aniline
4,4-Metilene-bis-o-cloroanilina (Italian)
MOCA
RCRA waste number U158

101-15-5
$C_{13}H_{12}N_2O$
212.24
O=C(C=CC(=Nc(ccc(N)c1C)c1)C=2)C2
2,5-Cyclohexadien-1-one, 4-((4-amino-3-methylphenyl)-imino)- (9CI)

101-16-6
$C_{13}H_{13}NO$
199.25
Benzenamine, 3-methoxy-N-phenyl- (9CI)
m-Anisidine, N-phenyl- (8CI)

101-17-7
$C_{12}H_{10}ClN$
203.66
Benzenamine, 3-chloro-N-phenyl- (9CI)
Diphenylamine, 3-chloro-(8CI)

101-18-8
$C_{12}H_{11}NO$
185.22
Oc(cccc1Nc(cccc2)c2)c1
Phenol, m-anilino- (8CI)
m-Anilinophenol
m-Hydroxydiphenylamine
3-Hydroxydiphenylamine
NSC-56930
Phenol, 3-(phenylamino)- (9CI)
3-(Phenylamino)phenol

101-20-2
$C_{13}H_9Cl_3N_2O$
315.59
O=C(Nc(ccc(c1)Cl)c1)Nc(ccc(c2Cl)Cl)c2
Carbanilide, 3,4,4'-trichloro
Cusiter
Cutisan
N-(3,4-Dichlorophenyl)-N'-(4-chlorophenyl)urea
ENT 26,925
Genoface
Procutene
NSC-72005
TCC
3,4,4'-Trichlorocarbanilide
3,4,4'-Trichlorodiphenylurea
Triclocarban
Urea, N-(4-chlorophenyl)-N'-(3,4-dichlorophenyl)- (9CI)

101-21-3
$C_{10}H_{12}ClNO_2$
213.68
CC(C)OC(=O)Nc1cccc(Cl)c1
Carbanilic acid, m-chloro-, isopropyl ester
Beet-Kleen
Bud-Nip
N-(3-Chloor-fenyl)-isopropyl carbamaat (Dutch)
Chlor-IFC
Chlor-IPC
m-Chlorocarbanilic acid, isopropyl ester
3-Chlorocarbanilic acid, isopropyl ester
Chloro-IFK
Chloro-IPC
N-(3-Chloro phenyl) carbamate d'isopropyle (French)
N-(3-Chlorophenyl)carbamic acid, isopropyl ester
(3-Chlorophenyl)carbamic acid, 1-methylethyl ester
N-3-Chlorophenylisopropyl-carbamate
Chloropropham
N-(3-Chlor-phenyl)-isopropyl-carbamat (German)
Chlorpropham
Chlorprophame (French)
CICP
CI-IPC
CIPC
N-(3-Cloro-fenil)-isopropil-carbammato (Italian)
Elbanil
ENT 18,060
Fasco Wy-HOE
Furloe
Furloe 4EC
Isopropyl meta-chlorocarbanil-ate
Isopropyl 3-chlorocarbanilate
Isopropyl 3-chlorophenylcar-bamate
Isopropyl-N-(3-chlorophenyl)-carbamate
Isopropyl-N-m-chlorophenyl-

carbamate
o-Isopropyl N-(3-chlorophenyl)-carbamate
Isopropyl-N-(3-chlorphenyl)-carbamat (German)
Jack Wilson Chloro 51 (Oil)
Liro CIPC
Metoxon
1-Methylethyl(3-chlorophenyl)-carbamate
Nexoval
Prevenol
Prevenol 56
Preventol
Preventol 56
Preweed
Sprout Nip
Sprout-Nip EC
Spud-Nic
Spud-Nie
Stopgerme-S
Taterpex
Triherbicide CIPC
Unicrop CIPC
Y 3

101-23-5
$C_{13}H_{10}F_3N$
237.24
Toluidine, N-phenyl-α,α,α-trifluoro
N-Phenyl-α,α,α-trifluoro-toluidide
3-Trifluoromethyl diphenyl-amine

101-25-7
$C_5H_{10}N_6O_2$
186.21
O=NN(CN(CN1CN2N=O)C2)C1
1,3,5,7-Tetraazabicyclo-(3.3.1)nonane, 3,7-dinitroso
Aceto DNPT 40
Aceto DNPT 80
Aceto DNPT 100
ChKhZ 18
Dinitrosopentamethylenetet-ramine
N,N-Dinitrosopentamethylenete-tramine
N[1],N[3]-Dinitrosopentamethylene-tetramine
N,N'-Dinitrosopentamethylene-tetramine
3,4-Di-N-nitrosopenta-methylenetetramine
3,7-Di-N-nitrosopentamethyl-enetetramine
3,7-Dinitroso-1,3,5,7-tetraaza-bicyclo-(3,3,1)-nonane
DNPMT
DNPT
1,5-Endomethylene-3,7-dinitro-so-1,3,5,7-tetraazacyclooctane
1,5-Methylene-3,7-dinitroso-1,3,5,7-tetraazacyclooctane
Micropor
Mikrofor N
NSC-73599
Opex
Pentamethylenetetramine, di-nitroso-
Porofor ChKhZ-18
Porophor B
Unicel-ND
Unicel NDX
Vulcacel B-40
Vulcacel BN

101-26-8
$C_9H_{13}N_2O_2.Br$
261.15
Pyridinium, 3-hydroxy-1-methyl-, bromide, di-methylcarbamate (ester)
Dimethylcarbamic acid ester of 3-hydroxy-1-methylpyridin-ium bromide
3-Hydroxy-1-methylpyridinium bromide dimethylcarbamate (ester)
Mestinon
Mestinone bromide
Pyridinium, 3-(((dimethyl-amino)carbonyl)oxy)-1-methyl-, bromide
Pyridostigmine bromide
Regonal
Ro 1-5130

101-27-9
$C_{11}H_9Cl_2NO_2$
258.11
ClCC#CCOC(=O)Nc1cccc(Cl)c1
Carbanilic acid, m-chloro-, 4-chloro-2-butynyl ester
A-980
Barbamate
Barban
Barbanate
Barbane
2-Butyn-1-ol, 4-chloro-, m-chlorocarbanilate
2-Butynyl-4-chloro-m-chloro-carbanilate
C-847
Carbin
Carbyne
Caryne
CBN
(4-Chloor-but-2-yn-yl)-N-(3-chloor-fenyl)-carbamaat (Dutch)
(4-Chlor-but-2-in-yl)-N-(3-chlor-phenyl)-carbamat (German)
Chlorinat
Chloro-2-butynyl m-chloro-carbamate
4-Chlorobut-2-ynyl-m-chloro-carbanilate
4-Chloro-2-butynyl-m-chloro-carbanilate
4-Chlorobut-2-ynyl 3-chloro-phenylcarbamate
4-Chloro-2-butynyl N-(3-chloro-phenyl)carbamate
m-Chlorocarbanilic acid, 4-chloro-2-butynyl ester
N-(3-Chloro phenyl) carbamate de 4-chloro 2-butynyle (French)
(4-Cloro-but-2-in-il)-N-(3-cloro-fenil)-carbammato (Italian)
(3-Chlorophenyl)carbamic acid 4-chloro-2-butynyl ester
CS-847
Fisons B25
Neoban
S-847

101-31-5
$C_{17}H_{23}NO_3$
289.41
O=C(OC(CC(N(C1C2)C)C2)C1)C(c(cccc3)c3)CO
Hyoscyamine, (-)
(-)-Atropine
Daturine
Duboisine
Hyoscyamine
(-)-Hyoscyamine
1-Hyoscyamine
Levsin
Tropic acid, (-)-, ester with tropine
Tropine, (-)-tropate (ester)

101-32-6
$C_{11}H_{17}NO_2$
195.26
Ethanol, 2,2'-((phenylmethyl)-imino)bis- (9CI)
Ethanol, 2,2'-(benzylimino)di-(8CI)
NSC-60297

101-37-1
$C_{12}H_{15}N_3O_3$
249.30
O(c(nc(OCC=C)nc1OCC=C)n1)CC=C
s-Triazine, 2,4,6-tris(allyloxy)
Triallyl cyanaurate
Triallyl cyanurate
Tripropargyl cyanurate
2,4,6-Triprop-2-ynyloxy-s-triazine
2,4,6-Tris(allyloxy)triazine

101-39-3
$C_{10}H_{10}O$
146.20
O=CC(=Cc(cccc1)c1)C
Cinnamaldehyde, α-methyl
α-Methylcinnamaldehyde
Methyl cinnamic aldehyde
α-Methylcinnamic aldehyde
α-Methylcinnimal
2-Methyl-3-phenyl-2-propenal
2-Propenal, 2-methyl-3-phenyl-(9CI)

101-40-6
$C_{10}H_{21}N$
155.28
Propylhexedrine

Benzedrex
CHP-Depot
Cyclohexaneethanamine, N,α-dimethyl- (9CI)
Cyclohexaneethanamine, N,α-dimethyl-, (+-)
Cyclohexaneethylamine, α,N-dimethyl-
Cyclohexaneethylamine, N-α-dimethyl- (8CI)
(Cyclohexaneethyl)amine, N-α-dimethyl-
1-Cyclohexyl-2-methylamino-propan (German)
1-Cyclohexyl-2-(methylamino)-propane
1-Cyclohexyl-N-methyl-2-propanamine
α,N-Dimethylcyclohexaneethyl amine
N,α-Dimethylcyclohexaneethyl-amine
N-α-(Dimethylcyclohexane)-ethylamine
(+-)-N,α-Dimethylcyclohexane-ethylamine
DEA No. 8161
Dristan
Dristan Inhaler
Hexahydrodesoxyephedrine
Hydromethamphetamine
Methyl-(1-methyl-2-cyclohexyl-ethyl)amine
NSC-32410
Obesin
Obesine
Propylhexadrine
Propylhexedrin

101-41-7
$C_9H_{10}O_2$
150.19
O=C(OC)Cc(cccc1)c1
Acetic acid, phenyl-, methyl ester
Benzeneacetic acid, methyl ester (9CI)
Methyl benzeneacetate
Methyl phenylacetate
Methyl α-toluate
Phenylacetic acid, methyl ester

101-42-8
$C_9H_{12}N_2O$
164.23
O=C(N(C)C)Nc(cccc1)c1
Urea, 1,1-dimethyl-3-phenyl
Beet-Kleen
Dibar
N,N-Dimethyl-N'-phenylurea
1,1-Dimethyl-3-phenylurea
Dybar
Fenidin
Fenulon
Fenuron
N-Phenyl-N',N'-dimethylurea
1-Phenyl-3,3-dimethylurea
3-Phenyl-1,1-dimethylurea
PDU
PUD (Herbicide)

101-43-9
$C_{10}H_{16}O_2$
168.24
O=C(OC(CCCC1)C1)C(=C)C
Cyclohexyl methacrylate
AI3-33324
Cyclohexyl 2-methyl-2-propenoate
Methacrylic acid, cyclohexyl ester (8CI)
NSC-20968
2-Propenoic acid, 2-methyl-, cyclohexyl ester (9CI)

101-48-4
$C_{10}H_{14}O_2$
166.24
O(C(OC)Cc(cccc1)c1)C
Acetaldehyde, phenyl-, dimethyl acetal
Benzene, (2,2-dimethoxyethyl)- (9CI)
Ethane, 1,1-dimethoxy-2-phenyl-
Hyscylene P
Phenacetaldehyde dimethyl acetal
Phenylacetaldehyde dimethyl acetal
α-Tolyl aldehyde dimethyl acetal
Viridine

101-50-8
$C_{12}H_{11}N_3O_6S_2$
357.38
O=S(=O)(O)c(ccc(N=Nc(ccc(N)c1S(=O)(=O)O)c1)c2)c2
Benzenesulfonic acid, 4-(4-amino-3-sulfophenylazo)
4-Aminoazobenzene-3,4'-disulfonic acid
4-Amino-3,4'-disulfoazobenzene
4-(4-Amino-3-sulfophenylazo)-benzenesulfonic acid
Azobenzene-3,4'-disulfonic acid, 4-amino-
Benzenesulfonic acid, 6-amino-3,4'-azodi-
Benzenesulfonic acid, 2-amino-5-((4-sulfophenyl)azo)-
Kyselina 4-aminoazobenzen-3,4'-disulfonova (Czech)

101-52-0
$C_{13}H_{12}N_4O_3$
272.24
O=N(=O)c(ccc(N=Nc(ccc(N)c1OC)c1)c2)c2
Benzenamine, 2-methoxy-4-((4-nitrophenyl)azo)- (9CI)
2-Methoxy-4-((4-nitrophenyl)-azo)benzenamine

101-53-1
$C_{13}H_{12}O$
184.24
Oc(ccc(c1)Cc(cccc2)c2)c1
4-Benzylphenol
AI3-01932
p-Cresol, α-phenyl- (8CI)
Fesia-sept
Fesiasept
p-Hydroxydiphenyl methane
4-Hydroxyditane
NSC-8078
Phenol, 4-(phenylmethyl)- (9CI)
α-Phenyl-p-cresol
4-(Phenylmethyl)phenol

101-54-2
$C_{12}H_{12}N_2$
184.26
N(c(ccc(N)c1)c1)c(cccc2)c2
p-Phenylenediamine, N-phenyl
Acna Black DF Base
p-Aminodifenylamin (Czech)
p-Aminodiphenylamine
p-Anilinoaniline
Azosalt R
1,4-Benzenediamine, N-phenyl- (9CI)
C.I. Azoic Diazo Component 22
C.I. Developer 15
C.I. Oxidation Base 2
C.I. 37240
C.I. 76085
Diphenylamine, p-amino-
Diphenylamine, 4-amino-
Diphenyl Black
Fast Blue R Salt
N-Fenyl-p-fenylendiamin (Czech)
Luxan Black R
Naphthoelan Navy Blue
NCI-C02233
Oxy Acid Black Base
Peltol BR
Peltol BR II
N-Phenyl-p-aminoaniline
N-Phenyl-p-phenylenediamine
p-Semidine
Variamine Blue Salt RT

101-55-3
$C_{12}H_9BrO$
249.11
O(c(ccc(c1)Br)c1)c(cccc2)c2
p-Bromophenyl phenyl ether
AI3-23460
Benzene, 1-bromo-4-phenoxy- (9CI)
1-Bromo-4-phenoxy benzene
1-Bromo-4-phenoxybenzene
p-Bromodiphenyl ether
4-Bromodiphenyl ether
p-Bromophenoxybenzene
4-Bromophenoxybenzene
4-Bromophenyl phenyl ether
Diphenyl ether, 4-bromo-
Ether, p-bromophenyl phenyl (8CI)
Ether, 4-bromophenyl phenyl
Ether, diphenyl, 4-bromo-
NSC-5619
p-Phenoxybromobenzene
p-(Phenoxy)bromobenzene

Phenyl ether, 4-bromo-
RCRA waste No. U030

101-61-1
$C_{17}H_{22}N_2$
254.41
N(c(ccc(c1)Cc(ccc(N(C)C)c2)c2)c1)(C)C
Aniline, 4,4'-methylene-bis(N,N-dimethyl
Baze michlerova (Czech)
Benzenamine, 4-4'-methylene-bis(N,N-dimethyl)- (9CI)
p,p'-Bis(dimethylamino)di-phenylmethane
4,4'-Bis(dimethylamino)di-phenylmethane
Bis(p-dimethylaminophenyl)-methane
Bis(4-(dimethylamino)phenyl)-methane
Bis(p-(N,N-dimethylamino)-phenyl)methane
Bis(4-(N,N-dimethylamino)-phenyl)methane
p,p'-Bis(N,N-dimethylamino-phenyl)methane
p,p-Dimethylaminodiphenyl-methane
Diphenylmethane, tetramethyl-diamino-
Methane Base
Methane, bis(p-(dimethyl-amino)phenyl)-
Methylene Base
4,4'-Methylenebis(N,N-di-methylaniline)
4,4'-Methylenebis(N,N-di-methyl)benzenamine
Michler's Base
Michler's Hydride
Michler's Methane
NCI-C01990
Reduced Michler's Ketone
Tetra-Base
Tetramethyldiaminodiphenyl-methane
p,p-Tetramethyldiaminodi-phenylmethane
N,N,N'N'-Tetramethyl-p,p'-di-aminodiphenylmethane
N,N,N'N'-Tetramethyl-4,4'-di-aminodiphenylmethane

101-63-3
$C_{12}H_8N_2O_5$
260.22
O=N(=O)c(ccc(Oc(ccc(N(=O)=O)c1)c1)c2)c2
Ether, bis(p-nitrophenyl)
Benzene, 1,1'-oxybis(4-nitro-(9CI)
Bis(p-nitrophenyl)ether
Bis(4-nitrophenyl)ether
p,p'-Dinitrodiphenyl ether
4,4'-Dinitrodiphenyl ether
4,4'-Dinitrodiphenyl oxide
Di-4-nitrophenyl ether
p-Nitrophenyl ether
Oxybis(4-nitrobenzene)

101-65-5
$C_{27}H_{22}N_2O_4$
438.47
O=C(Oc(cccc1)c1)Nc(ccc(c2)Cc(ccc(NC(=O)Oc(cccc3)c3)c4)c4)c2
Carbamic acid, (methylenedi-4,1-phenylene)bis-, diphenyl ester (9CI)
Diphenyl 4,4'-methylenebis-(phenylcarbamate)

101-67-7
$C_{28}H_{43}N$
393.65
N(c(ccc(c1)CCCCCCCC)c1)c(ccc(c2)CCCCCCCC)c2
4,4'-Dioctyldiphenylamine
AI3-17279
Benzenamine, 4-octyl-N-(4-octylphenyl)- (9CI)
Bis(p-octylphenyl)amine
Di-N-octyl diphenylamine
p,p-Dioctyldiphenylamine
p,p'-Dioctyldiphenylamine
4,4'-Dioctylphenylamine
Diphenylamine, 4,4'-dioctyl-
NSC-79268
4-Octyl-N-(4-octylphenyl)-benzenamine
Vanlube 81

101-68-8
$C_{15}H_{10}N_2O_2$
250.27
O=C=Nc(ccc(c1)Cc(ccc(N=C=O)c2)c2)c1
Isocyanic acid, methylenedi-p-phenylene ester
Benzene, 1,1'-methylenebis-(4-isocyanato- (9CI)
Bis(p-isocyanatophenyl)methane
Bis(1,4-isocyanatophenyl)-methane
Bis(4-isocyanatophenyl)methane
Caradate 30
Desmodur 44
Difenylmethaan-dissocyanaat (Dutch)
Difenil-metan-diisocianato (Italian)
4-4'-Diisocyanate de diphenyl-methane (French)
4,4'-Diisocyanatodiphenyl-methane
Diphenylmethan-4,4'-diiso-cyanat (German)
Diphenyl methane diisocyanate
p,p'-Diphenylmethane diiso-cyanate
4,4'-Diphenylmethane diiso-cyanate
Diphenylmethane 4,4'-diiso-cyanate [UN 2489]
Diphenylmethane diisocyanate (OSHA)
Hylene M50
Isonate
Isonate 125M
Isonate 125 MF
MDI
MDI (OSHA)
Methylenebis(4-isocyanato-benzene)
1,1-Methylenebis(4-isocyanato-benzene)
Methylenebis(p-phenylene iso-cyanate)
Methylenebis(4-phenylene iso-cyanate)
p,p'-Methylenebis(phenyl iso-cyanate)
Methylenebis(p-phenyl iso-cyanate)
Methylenebis(4-phenyl iso-cyanate)
4,4'-Methylenebis(phenyl iso-cyanate)
Methylene bisphenyl isocyanate (ACGIH,OSHA)
4,4'-Methylenediphenyl diiso-cyanate
Methylenedi-p-phenylene diiso-cyanate
Methylenedi-p-phenylene iso-cyanate
4,4'-Methylenediphenylene iso-cyanate
4,4'-Methylenediphenyl iso-cyanate
Methylene di(phenylene iso-cyanate) (DOT)
Nacconate 300
NCI-C50668
Rubinate 44
UN 2489 [Diphenylmeth-ane-4,4'diisocyanate]

101-70-2
$C_{14}H_{15}NO_2$
229.30
O(c(ccc(Nc(ccc(OC)c1)c1)c2)c2)C
4-Biphenylamine, 4,4'-di-methoxy
Benzenamine, 4-methoxy-N-(4-methoxyphenyl)-
Bis(p-anisylamine)
Bis(p-methoxyphenyl)amine
Bis(4-methoxyphenyl)amine
Di-p-anisylamine
p,p'-Dimethoxydiphenylamine
4,4'-Dimethoxydiphenylamine
Di-p-methoxyphenylamine
Termofleks A (Czech)

101-71-3
$C_{14}H_{13}NO_2$
227.26
Phenol, 4-(phenylmethyl)-, carbamate (9CI)
p-Cresol, α-phenyl-, carbamate (8CI)
Difenano (Spanish)
Diphenan (VAN)
Diphenane (French)
Diphenanum (Latin)
NSC-60023

101-72-4
$C_{15}H_{18}N_2$
226.35
N(c(ccc(Nc(cccc1)c1)c2)c2)C(C)C
p-Phenylenediamine, N-isopropyl-N'-phenyl
Cyzone
Elastozone 34
N-Fenyl-N'-isopropyl-p-fenylendiamin (Czech)
Flexzone 3C
4-Isopropylaminodiphenylamine
N-Isopropyl-N'-fenyl-p-fenylendiamin (Czech)
N-Isopropyl-N'-phenyl-p-phenylenediamine
4010 NA
NCI-C56304
Nonox ZA
N-Phenyl-N'-isopropyl-p-phenylenediamine
N-2-Propyl-N'-phenyl-p-phenylenediamine
Santoflex 36

101-73-5
$C_{15}H_{17}NO$
227.33
O(c(ccc(Nc(cccc1)c1)c2)c2)C(C)C
Diphenylamine, 4-isopropoxy
Agerite 150
Agerite iso
p-Hydroxydiphenylamine isopropyl ether
p-Isopropoxydiphenylamine
4-Isopropoxydiphenylamine
N-(4-Isopropoxyphenyl)aniline

101-74-6
$C_{18}H_{16}N_2O$
276.33
Oc(ccc(Nc(ccc(Nc(cccc1)c1)c2)c2)c3)c3
Phenol, 4-((4-(phenylamino)-phenyl)amino)- (9CI)
4-((4-(Phenylamino)phenyl)-amino)phenol

101-76-8
$C_{13}H_{10}Cl_2$
237.13
c(ccc(c1)Cl)(c1)Cc(ccc(c2)Cl)c2
Methane, bis(4-chlorophenyl)
Bis(p-chlorophenyl)methane
Di-(4-chlorophenyl)methane
Di-(p-chlorophenyl)methane

101-77-9
$C_{13}H_{14}N_2$
198.29
Nc(ccc(c1)Cc(ccc(N)c2)c2)c1
Aniline, 4,4'-methylenedi
4-(4-Aminobenzyl)aniline
Ancamine TL
Araldite Hardener 972
Benzenamine, 4,4'-methylenebis-
Bis-p-aminofenylmethan (Czech)
Bis(p-aminophenyl)methane
Bis(4-aminophenyl)methane
Curithane
DADPM
DAPM
DDM
p,p'-Diaminodifenylmethan (Czech)
4,4'-Diaminodiphenylmethan (German)
Diaminodiphenylmethane
p,p'-Diaminodiphenylmethane
4,4'-Diaminodiphenylmethane
4,4'-Diaminodiphenyl methane [UN 2651]
Di-(4-aminophenyl)methane
Dianilinemethane
Dianilinomethane
4,4'-Diphenylmethanediamine
Epicure DDM
Epikure DDM
HT 972
Jeffamine AP-20
MDA
Methylenebis(aniline)
4,4'-Methylenebisaniline
4,4'-Methylenebis(benzeneamine)
Methylenedianiline
p,p'-Methylenedianiline
4,4'-Methylenedianiline
4,4-Methylenedianiline (ACGIH)
Sumicure M
Tonox
UN 2651 [4,4'-Diaminodiphenyl methane]

101-80-4
$C_{12}H_{12}N_2O$
200.26
O(c(ccc(N)c1)c1)c(ccc(N)c2)c2
Aniline, 4,4'-oxydi
p-Aminophenyl ether
4-Aminophenyl ether
Benzenamine, 4,4'-oxybis-
Bis(4-aminophenyl)ether
Bis(p-aminophenyl)ether
DADPE
4,4-DADPE
4,4'-Diaminobiphenyloxide
Diaminodiphenyl ether
4,4-Diaminodiphenyl ether
p,p'-Diaminodiphenyl ether

4,4'-Diaminodiphenyl oxide
4,4'-Diaminophenyl ether
4,4'-Diaminophenyl oxide
Ether, 4,4'-diaminodiphenyl
NCI-C50146
Oxybis(4-aminobenzene)
4,4'-Oxybisaniline
p,p'-Oxybis(aniline)
4,4'-Oxybisbenzenamine
Oxydianiline
4,4-Oxydianiline
p,p'-Oxydianiline
4,4'-Oxydiphenylamine
Oxydi-p-phenylenediamine

101-81-5
$C_{13}H_{12}$
168.25
c(cccc1)(c1)Cc(cccc2)c2
Methane, diphenyl
Benzene, benzyl-
Benzylbenzene
Diphenylmethane
Ditan
Ditane
Toluene, α-phenyl-

101-82-6
$C_{12}H_{11}N$
169.24
n(c(ccc1)Cc(cccc2)c2)c1
Pyridine, 2-benzyl
2-Benzylpyridine
Pyridine, 2-(phenylmethyl)- (9CI)

101-83-7
$C_{12}H_{23}N$
181.36
N(C(CCCC1)C1)C(CCCC2)C2
Dicyclohexylamine [UN 2565]
N-Cyclohexylcyclohexanamine
DCHA
DICHA
N,N-Diclohexylamine
Dicyklohexylamin (Czech)
Dodecahydrodiphenylamine
UN 2565 [Dicyclohexylamine]

101-84-8
$C_{12}H_{10}O$
170.22
O(c(cccc1)c1)c(cccc2)c2
Ether, diphenyl
Benzene, phenoxy-
Biphenyl oxide
Diphenyl ether
Diphenyl oxide
Geranium crystals
Phenoxybenzene
Phenyl ether (ACGIH,OSHA)
Oxydiphenyl

101-86-0
$C_{15}H_{20}O$
216.35
O=CC(=Cc(cccc1)c1)CCCCCC
Cinnamaldehyde, α-hexyl (8CI)
Hexyl cinnamic aldehyde
Hexyl cinnamaldehyde
α-Hexylcinnamaldehyde
α-Hexylcinnamic aldehyde
α-n-Hexyl-β-phenylacrolein
NSC-46150
NSC-406799
Octanal, 2-(phenylmethylene)- (9CI)
2-(Phenylmethylene)octanal

101-87-1
$C_{18}H_{22}N_2$
266.42
N(c(ccc(Nc(cccc1)c1)c2)c2)C(CCCC3)C3
p-Phenylenediamine, N-phenyl-N'-cyclohexyl
N-Cyclohexyl-N'-phenyl-p-phenylendiamine
N-Cyklohexyl-N'-fenyl-p-fenylendiamin (Czech)
N-Fenyl-N'-cyklohexyl-p-fenylendiamin (Czech)
Flexzone 6H
p-Phenylenediamine, N-cyclohexyl-N'-phenyl-
Vulkacit 4010 (Czech)

101-90-6
$C_{12}H_{14}O_4$
222.26
O(C1COc(cc(OCC(O2)C2)cc3)c3)C1
Resorcinol, diglycidyl
Araldite ERE 1359
m-Bis(2,3-epoxypropoxy)-benzene
1,3-Bis(2,3-epoxypropoxy)-benzene
meta-Bis(glycidyloxy)benzene
Diglycidyl ether of resorcinol
1,3-Diglycidyloxybenzene
Diglycidyl resorcinol ether
ERE 1359
NCI-C54966
Oxirane, 2,2'-(1,3-phenylenebis-(oxymethylene))bis-
2,2'-(1,3-Phenylenebis(oxy-methylene))bisoxirane
RDGE
Resorcinol bis(2,3-epoxy-propyl)ether
Resorcinol diglycidyl ether
Resorcinyl diglycidyl ether

101-92-8
$C_{10}H_{10}ClNO_2$
211.66
O=C(Nc(ccc(c1)Cl)c1)CC(=O)C
Acetoacetanilide, 4'-chloro
Acetoacetanilide, p-chloro-
Acetoacet-p-chloroanilide
Acetoacetyl-4-chloroanilide
Butanamide, N-(4-chloro-phenyl)-3-oxo- (9CI)
p-Chloroacetoacetanilide
4'-Chloroacetoacetanilide

101-96-2
$C_{14}H_{24}N_2$
220.40
N(c(ccc(NC(CC)C)c1)c1)C(CC)C
p-Phenylenediamine, N,N'-di-sec-butyl
N,N'-Di-sek.butyl-p-fenylen-diamin (Czech)
N,N'-Di-s-butyl-p-phenylene-diamine
Tenamene 2

101-97-3
$C_{10}H_{12}O_2$
164.22
O=C(OCC)Cc(cccc1)c1
Acetic acid, phenyl-, ethyl ester
Benzeneacetic acid, ethyl ester (9CI)
Ethyl benzeneacetate
Ethyl phenacetate
Ethyl phenylacetate
Ethyl 2-phenylethanoate
Ethyl α-toluate
Phenylacetic acid, ethyl ester
α-Toluic acid, ethyl ester

101-99-5
$C_9H_{11}NO_2$
165.21
O=C(OCC)Nc(cccc1)c1
Carbanilic acid, ethyl ester
EPC (The plant regulator)
Ethyl carbanilate
Ethylester kyseliny karbanilove (Czech)
Ethyl-N-phenylcarbamate
Euphorin
Keimstop
Phenylethyl carbamate
Phenylurethan(e)
N-Phenylurethane
Urethan, phenyl-

102-01-2
$C_{10}H_{11}NO_2$
177.22
O=C(Nc(cccc1)c1)CC(=O)C
Acetoacetanilide
AAN
Acetanilide, 2-acetyl-
Acetoacetamidobenzene
Acetoacetanilid
Acetoacetic acid anilide
Acetoacetic anilide
((Acetoacetyl)amino)benzene
Acetoacetylaniline
Acetylacetanilide
α-Acetylacetanilide
N-(Acetylacetyl)aniline
Anilid kyseliny acetoctove (Czech)
Butanamide, 3-oxo-N-phenyl- (9CI)
β-Ketobutyranilide
N-Phenylacetoacetamide
USAF EK-1239

102-04-5
$C_{15}H_{14}O$
210.28
O=C(Cc(cccc1)c1)Cc(cccc2)c2
2-Propanone, 1,3-diphenyl- (9CI)
AI3-05001
Benzyl ketone
Dibenzyl ketone
α,α'-Diphenylacetone
1,3-Diphenylacetone
1,3-Diphenylpropanone
1,3-Diphenyl-2-propanone
NSC-220312

102-06-7
$C_{13}H_{13}N_3$
211.29
N=C(Nc(cccc1)c1)Nc(cccc2)c2
Guanidine, 1,3-diphenyl
1,3-Difenylguanid (Czech)
Diphenylguanidine
N,N'-Diphenylguanidine
1,3-Diphenylguanidine
DPG
DPG Accelerator
Dwufenyloguanidyna (Polish)
Melaniline
NCI-C60924
USAF B-19
USAF EK-1270
Vulcacid D
Vulkacit D
Vulkacit D/C
Vulkazit

102-07-8
$C_{13}H_{12}N_2O$
212.27
O=C(Nc(cccc1)c1)Nc(cccc2)c2
Carbanilide
Acardite
N,N'-Difenylmocovina (Czech)
N,N'-Diphenylurea
s-Diphenylurea
sym-Diphenylurea
1,3-Diphenylurea
Karbanilid (Czech)
Urea, N,N'-diphenyl-
Urea, 1,3-diphenyl-
USAF EK-534

102-08-9
$C_{13}H_{12}N_2S$
228.33
N(c(cccc1)c1)=C(S)Nc(cccc2)c2
Carbanilide, thio
A 1
DFT
1,3-Difenylthiomocovina (Czech)
N,N'-Diphenylthiocarbamide
s-Diphenylthiocarbamide
Diphenylthiourea
N,N'-Diphenylthiourea
sym-Diphenylthiourea
1,3-Diphenylthiourea
1,3-Diphenyl-2-thiourea
2-Fenylotiomocznik (Polish)
Rhenocure CA
Stabilisator C
Sulfocarbanilide
Thiocarbanilide
Thiokarbanilid (Czech)
Thiourea, N,N'-diphenyl- (9CI)
Thiourea, sym-diphenyl-
Urea, 1,3-diphenyl-2-thio-

USAF EK-245
Vulkacit CA

102-09-0
$C_{13}H_{10}O_3$
214.23
O=C(Oc(cccc1)c1)Oc(cccc2)c2
Carbonic acid, diphenyl ester
Diphenyl carbonate
Phenyl carbonate

102-13-6
$C_{12}H_{16}O_2$
192.26
O=C(OCC(C)C)Cc(cccc1)c1
Acetic acid, phenyl-, isobutyl ester (8CI)
AI3-01969
Benzeneacetic acid, 2-methylpropyl ester (9CI)
Isobutyl phenylacetate
Isobutyl phenylethanoate
Isobutyl α-toluate
2-Methylpropyl benzeneacetate
NSC-6602
Phenylacetic acid, isobutyl ester

102-20-5
$C_{16}H_{16}O_2$
240.32
O=C(OCCc(cccc1)c1)Cc(cccc2)c2
Acetic acid, phenyl-, phenethyl ester
Benzeneacetic acid, 2-phenylethyl ester (9CI)
Benzylcarbinyl α-toluate
Phenethyl phenylacetate
Phenethyl α-toluate
Phenylacetic acid, phenethyl ester
Phenylethyl phenylacetate
β-Phenylethyl phenylacetate
2-Phenylethyl phenylacetate
2-Phenylethyl α-toluate

102-25-0
$C_{12}H_{18}$
162.27
c(cc(cc1CC)CC)(c1)CC
Benzene, 1,3,5-triethyl- (9CI)
AI3-04219
NSC-406584
1,3,5-Triethylbenzene

102-27-2
$C_9H_{13}N$
135.23
N(c(cccc1C)c1)CC
m-Toluidine, N-ethyl
Benzenamine, N-ethyl-3-methyl- (9CI)
N-Ethyl-3-methylaniline
N-Ethyltoluidine
N-Ethyl-m-toluidine [UN 2754]
UN 2754 [N-Ethyltoluidines]

102-28-3
$C_8H_{10}N_2O$
150.20
O=C(Nc(cccc1N)c1)C
Acetanilide, 3'-amino
Acetamide, N-(3-aminophenyl)-(9CI)
m-Acetaminoaniline
m-(Acetylamino)aniline
3-Acetylaminoaniline
N-Acetyl-m-fenylendiamin (Czech)
N-Acetyl-m-phenylenediamine
3-Aminoacetanilid (Czech)
m-Aminoacetanilide
3'-Aminoacetanilide

102-29-4
$C_8H_8O_3$
152.16
O=C(Oc(cccc1O)c1)C
Resorcinol, monoacetate
3-Acetoxyphenol
Acetylresorcinol
1,3-Benzenediol, monoacetate
Euresol
m-Hydroxyphenyl acetate
Remonol
Resorcin acetate
Resorcin monoacetate
Resorcitate

102-30-7
$C_{21}H_{36}Cl_2N.Cl$
408.88
Dodecyl dimethyl 3,4-dichlorobenzyl ammonium chloride
Ammonium, (3,4-dichlorobenzyl)dodecyldimethyl-, chloride
Aralkonium chloride
Benzenemethanaminium, 3,4-dichloro-n-dodecyl-N,N-dimethyl-, chloride (9CI)
Dichloran
(3,4-Dichlorobenzyl)dodecyldimethylammonium chloride
3,4-Dichlorobenzyl-lauryl-dimethylammonium chloride
Dodecyldimethyl(3,4-dichlorobenzyl)ammonium chloride
Dynalione
Dynaltone
Dynium chloride
Ko 18
Lauryldimethyldichlorobenzylammonium chloride
Riseptin
Tetrasan

102-32-9
$C_8H_8O_4$
168.16
OC(=O)Cc1ccc(O)c(O)c1
Acetic acid, (3,4-dihydroxyphenyl)
BA 2773
Benzeneacetic acid, 3,4-dihydroxy- (9CI)
3,4-Dihydroxybenzeneacetic acid
Dihydroxyphenylacetic acid
3,4-Dihydroxy-phenylacetic acid
DOPAC
Dopacetic acid
Homoprotocatechuic acid

102-36-3
$C_7H_3Cl_2NO$
188.01
O=C=Nc(ccc(c1Cl)Cl)c1
Isocyanic acid, 3,4-dichlorophenyl ester
3,4-Dichlorfenylisokyanat (Czech)
3,4-Dichlorophenyl isocyanate

102-47-6
$C_7H_5Cl_3$
195.47
c(ccc(c1Cl)Cl)(c1)CCl
Benzene, 1,2-dichloro-4-(chloromethyl)- (9CI)
AI3-14887
3,4-Dichlorobenzyl chloride
1,2-Dichloro-4-(chloromethyl)-benzene
NSC-406893
Toluene, α,3,4-trichloro- (8CI)
α,3,4-Trichlorotoluene

102-50-1
$C_8H_{11}NO$
137.20
O(c(ccc(N)c1C)c1)C
p-Anisidine, 2-methyl
Benzenamine, 4-methoxy-2-methyl-
m-Cresidine
4-Methoxy-2-methylaniline
4-Methoxy-2-methylbenzenamine
2-Methyl-p-anisidine
2-Methyl-4-methoxyaniline
NCI-C02993

102-52-3
$C_7H_{16}O_4$
164.20
O(C(OC)CC(OC)OC)C
1,1,3,3-Tetramethoxypropane
AI3-28938
Malonaldehyde, bis(dimethyl acetal) (8CI)
Malonaldehyde tetramethyl acetal
NSC-27794
Propane, 1,1,3,3-tetramethoxy-(9CI)
Tetramethoxypropane

102-54-5
$C_{10}H_{10}Fe$
186.05

[Fe+2].c1cc[cH-]c1.c1cc[cH-]c1
Ferrocene
Biscyclopentadienyliron
Di-2,4-cyclopentadien-1-yliron
Dicyclopentadienyl iron (OSHA)
Iron bis(cyclopentadiene)
Iron dicyclopentadienyl

102-56-7
$C_8H_{11}NO_2$
153.20
O(c(c(N)cc(OC)c1)c1)C
Aniline, 2,5-dimethoxy
Aminohydroquinone dimethyl ether
Benzenamine, 2,5-dimethoxy- (9CI)
C.I. 35811
2,5-Dimethoxyaniline

102-60-3
$C_{14}H_{32}N_2O_4$
292.48
OC(C)CN(CCN(CC(O)C)CC(O)C)CC(O)C
2-Propanol, 1,1',1'',1'''-(ethylenedinitrilo)tetra
Entprol
1,1',1'',1'''-(Ethylenedinitrilo)-tetra-2-propanol
2-Propanol, 1',1',1'',1'''-(1,2-ethanediyldinitrilo)-tetrakis-
Quadrol
N,N,N',N'-Tetrakis(2-hydroxypropyl)ethylenediamine

102-63-6
$C_{15}H_{17}N_3$
239.30
N(=Nc(ccc(N)c1C)c1)c(c(cc(c2)C)C)c2
Benzenamine, 4-((2,4-dimethylphenyl)azo)-2-methyl- (9CI)

102-67-0
$C_9H_{21}Al$
156.25
Tripropyl aluminum
Aluminum, tripropyl- (8CI,9CI)
Tripropylaluminum
UN 2718

102-69-2
$C_9H_{21}N$
143.31
N(CCC)(CCC)CCC
1-Propanamine, N,N-dipropyl
Tripropylamine (8CI) [UN 2260]
Tri-n-propylamine
UN 2260 [Tripropylamine]

102-70-5
$C_9H_{15}N$
137.25
N(CC=C)(CC=C)CC=C
Triallylamine [UN 2610]
2-Propen-1-amine, N-N-di-2-propenyl- (9CI)
UN 2610 [Triallylamine]

102-71-6
$C_6H_{15}NO_3$
149.22
OCCN(CCO)CCO
Ethanol, 2,2',2''-nitrilotri
Daltogen
Nitrilo-2,2',2''-triethanol
2,2',2''-Nitrilotriethanol
Sterolamide
TEA
Thiofaco T-35
Triaethanolamin-NG
Triethanolamin
Triethanolamine
Triethylamine, 2,2',2''-trihydroxy-
Triethylolamine
Tri(hydroxyethyl)amine
Trihydroxytriethylamine
Tris(2-hydroxyethyl)amine
Trolamine

102-76-1
$C_9H_{14}O_6$
218.23
O=C(OCC(OC(=O)C)COC(=O)C)C
Acetin, tri
Enzactin
Fungacetin
Glycerin triacetate
Glycerol triacetate
Glyceryl triacetate
Glyped
Kesscoflex TRA
Kodaflex Triacetin
1,2,3-Propanetriol triacetate
Triacetin
Triacetine
Triacetyl glycerine
Vanay

102-77-2
$C_{11}H_{12}N_2OS_2$
252.37
O(CCN(SC(=Nc(c1ccc2)c2)S1)C3)C3
Benzothiazole, 2-(morpholinothio)
Amax
2-Benzothiazolyl N-morpholino sulfide
2-Benzothiazolylsulfenyl morpholine
4-(2-Benzothiazolylthio)-morpholine
2-(Morpholinothio)benzothiazole
Morpholinylmercaptobenzothiazole
2-(4-Morpholinylthio)benzothiazole
N-Oxydiethyl-2-benzthiazolsulfenamid (Czech)
N-(Oxydiethylene)benzothiazole-2-sulfenamide
Santocure MOR
Sulfenamide M
Sulfenax MOB (Czech)
USAF CY-7
Vulcafor BSM
Vulkacit MOZ

102-79-4
$C_8H_{19}NO_2$
161.28
OCCN(CCCC)CCO
Ethanol, 2,2'-(butylimino)di
Bide
Bis(β-hydroxyethyl)butylamine
Butylbis(2-hydroxyethyl)amine
n-Butyl-N,N-bis(hydroxyethyl)-amine
Butyldiethanolamine
n-Butyldiethanolamine
2-(n-Butyl-N-2-hydroxyethyl-amino)ethanol
n-Butyl-2,2'-iminodiethanol
2,2'-(Butylimino)diethanol
Ethanol, 2,2'-(butylimino)bis- (9CI)

102-81-8
$C_{10}H_{23}NO$
173.34
OCCN(CCCC)CCCC
Ethanol, 2-(dibutylamino)
DBAE
Dibutylaminoethanol [UN 2873]
2-Dibutylaminoethanol
2-Di-n-butylaminoethanol
2-N-Dibutylaminoethanol (ACGIH,OSHA)
N,N-Di-n-butylaminoethanol (DOT)
β-n-Dibutylaminoethyl alcohol
N,N-Dibutylethanolamine
N,N-Dibutyl-N-(2-hydroxyethyl)amine
BU2AE
UN 2873 [Dibutylaminoethanol]

102-82-9
$C_{12}H_{27}N$
185.40
N(CCCC)(CCCC)CCCC
Tributylamine [UN 2542]
Tributilamina (Romanian)
Tri-n-butylamine
Tris-n-butylamine
UN 2542 [Tributylamine]

102-85-2
$C_{12}H_{27}O_3P$
250.36
O(P(OCCCC)OCCCC)CCCC
Phosphorous acid, tributyl ester
Tributylfosfit (Czech)

Tributyl phosphite

102-86-3
$C_{18}H_{39}N$
269.51
N(CCCCCC)(CCCCCC)C
CCCCC
Trihexylamine (8CI)
AI3-16575
N,N-Dihexyl-1-hexanamine
1-Hexanamine, N,N-dihexyl- (9CI)
NSC-409786
Tri-n-hexylamine

102-87-4
$C_{36}H_{75}N$
521.99
N(CCCCCCCCCCCC)(CCCCCC
CCCCCC)CCCCCCCCCCCC
Tri-n-dodecylamine
Adogen 360
Alamine 304
N,N-Didodecyl-1-dodecanamine
1-Dodecanamine, N,N-didodecyl- (9CI)
NSC-35134
Tridodecylamine (8CI)
Trilaurylamine

102-92-1
C_9H_7ClO
166.61
O=C(C=Cc(cccc1)c1)Cl
Cinnamoyl chloride (8CI)
Cinnamic acid chloride
Cinnamic chloride
Cinnamoylchloride
NSC-4683
β-Phenylacryloyl chloride
3-Phenyl-2-propenoyl chloride
2-Propenoyl chloride, 3-phenyl- (9CI)

102-96-5
$C_8H_7NO_2$
149.16
O=N(=O)C=Cc(cccc1)c1
Styrene, β-nitro
BNS
NCI-C02211
β-Nitrostyrene
γ-Nitrostyrene

102-98-7
$C_6H_5BHgO_3.2H$
338.53
Boric acid, phenylmercury deriv.
Exomycol Gel
Famosept
Fenosept
Formasept
Mercurate(2-), (orthoborato-(3-)-o)phenyl-, dihydrogen
Mercury, (dihydrogen borato)-phenyl-
Merfen
Merphen
Metasol BT
Phenylmercury borate
Ryfen
Spidox
Spidoxol

103-00-4
$C_9H_{19}NO$
157.29
OC(C)CNC(CCCC1)C1
2-Propanol, 1-(cyclohexylamino)
1-Cyclohexylamino-2-propanol
USAF DO-19

103-03-7
$C_7H_9N_3O$
151.19
O=C(NNc(cccc1)c1)N
Carbamic acid, 2-phenylhydrazide
1-Carbamoyl-2-phenylhydrazine
1-Carbamyl-2-phenylhydrazine
CPH
Cryogenine
Fenylsemikarbazid (Czech)
Hydrazine, 1-carbamoyl-2-phenyl-
Hydrazinecarboxamide, 2-phenyl-
Kryogenin
Phenicarbazide
2-Phenyldiazenecarboxamide
2-Phenylhydrazide, carbamic acid
1-Phenylhydrazine carboxamide
2-Phenylhydrazinecarboxamide
Phenylsemicarbazide
1-Phenylsemicarbazide

103-05-9
$C_{11}H_{16}O$
164.27
OC(CCc(cccc1)c1)(C)C
2-Butanol, 2-methyl-4-phenyl
Benzyl-t-butanol
Dimethylphenylethyl carbinol
1,1-Dimethyl-3-phenylpropanol
1,1-Dimethyl-3-phenyl-1-propanol
α,α-Dimethyl-δ-phenylpropyl alcohol
2-Methyl-4-phenyl-2-butanol
Phenylethyl dimethyl carbinol

103-06-0
$C_7H_9NO_3S$
187.21
O=S(=O)(O)CNc(cccc1)c1
Anilinomethanesulfonate
Aniline-ω-acid
Aniline-ω-sulfonic acid
Anilinomethanesulfonic acid
Methanesulfonic acid, anilino- (8CI)
Methanesulfonic acid, (phenylamino)- (9CI)
NSC-7802
(Phenylamino)methanesulfonic acid
N-(Sulfomethyl)aniline

103-08-2
$C_{11}H_{24}O$
172.35
CCCCC(CC)CCC(C)O
2-Nonanol, 5-ethyl
(3-Ethyl-n-heptyl)methylcarbinol
5-Ethyl-2-nonanol

103-09-3
$C_{10}H_{20}O_2$
172.30
O=C(OCC(CCCC)CC)C
Acetic acid, 2-ethylhexyl ester
Acetic acid α-ethylhexyl ester
2-Ethylhexanyl acetate
β-Ethylhexyl acetate
2-Ethylhexyl acetate
2-Ethylhexylester kyseliny octove (Czech)
2-Ethylhexyl ethanoate
Octyl acetate

103-11-7
$C_{11}H_{20}O_2$
184.31
O=C(OCC(CCCC)CC)C=C
Acrylic acid, 2-ethylhexyl ester
2-Ethylhexyl acrylate
2-Ethylhexylester kyseliny akrylove (Czech)
2-Ethylhexyl 2-propenoate
1-Hexanol, 2-ethyl-, acrylate
Octyl acrylate
2-Propenoic acid, 2-ethylhexyl ester (9CI)

103-16-2
$C_{13}H_{12}O_2$
200.25
O(c(ccc(O)c1)c1)Cc(cccc2)c2
Phenol, p-(benzyloxy)
Agerite
Agerite Alba
Alba-Dome
Monobenzyl hydroquinone
Benoquin
Benzoquin
Benzyl hydroquinone
p-Benzyloxyphenol
4-Benzyloxyphenol
Depigman
Hydrochinon monobenzylether (Czech)
Hydroquinone benzyl ether
Hydroquinone monobenzyl ether
p-Hydroxyphenyl benzyl ether
Monobenzone
Monobenzyl ether hydroquinone
4-(Phenylmethoxy)phenol

Pigmex

103-17-3
$C_{13}H_{10}Cl_2S$
269.19
Clc2ccc(CSc1ccc(Cl)cc1)cc2
Sulfide, p-chlorobenzyl p-chlorophenyl
Chloorbenzide (Dutch)
(4-Chloor-benzyl)-(4-chloor-fenyl)-sulfide (Dutch)
Chlorbensid (German)
Chlorbenside
Chlorbenxide
Chlorbenzide
(4-Chlor-benzyl)-(4-chlor-phenyl)-sulfid (German)
p-Chlorobenzyl p-chlorophenyl sulfide
p-Chlorobenzyl p-chlorophenyl sulphide
4-Chlorobenzyl 4-chlorophenyl sulphide
1-Chloro-4-(((4-chlorophenyl)-methyl)thio)benzene
Chlorocide
Chloroparacide
4-Chlorophenyl 4'-chlorobenzyl sulfide
Chlorosulfacide
Chlorparacide
Chlorsulphacide
(4-Cloro-benzil)-(4-cloro-fenil)-solfuro (Italian)
p,p'-Dichlorodiphenyl sulfide
ENT 20,696
HRS 860
Metox
Mitox
RD 2195
Sulfure de 4-chlorobenzyle et de 4-chlorophenyle (French)

103-18-4
$C_{12}H_{11}N_3O$
213.26
Oc(ccc(N=Nc(ccc(N)c1)c1)c2)c2
Phenol, p-((p-aminophenyl)-azo)
4-Amino-4'-hydroxyazobenzene
Azobenzene, 4-amino-4'-hydroxy-

103-23-1
$C_{22}H_{42}O_4$
370.64
O=C(OCC(CCCC)CC)CCCCC(=O)OCC(CCCC)CC
Adipic acid, bis(2-ethylhexyl) ester
Adipol 2EH
BEHA
Bis(2-ethylhexyl) adipate
Bis-(2-ethylhexyl)ester kyseliny adipove (Czech)
Bisoflex DOA
DEHA
Di-2-ethylhexyl adipate
Dioctyl adipate
DOA
Effemoll DOA
Effomoll DOA
Ergoplast ADDO
Flexol A 26
Flexol Plasticizer 10-A
Flexol Plasticizer A-26
Hexanedioic acid, bis(2-ethyl-hexyl) ester (9CI)
Hexanedioic acid, dioctyl ester
Kemester 5652
Kodaflex DOA
Mollan S
Monoplex DOA
NCI-C54386
Octyl adipate
Plastomoll DOA
PX-238
Reomol DOA
Rucoflex Plasticizer DOA
Sicol 250
Staflex DOA
Truflex DOA
Uniflex DOA
Vestinol OA
Wickenol 158
Witamol 320

103-24-2
$C_{25}H_{48}O_4$
412.73
O=C(OCC(CCCC)CC)CCCCCCCC(=O)OCC(CCCC)CC
Azelaic acid, bis(2-ethylhexyl) ester
Azelaic acid, di(2-ethylhexyl)-ester
Bis(2-ethylhexyl)azelate
Bis-(2-ethylhexyl)ester kyseliny azelaove (Czech)
Dioctyl azelate
Plastolein 9058
Plastolein 9058 DOZ
Staflex DOX
Truflex DOX

103-25-3
$C_{10}H_{12}O_2$
164.20
O=C(OC)CCc(cccc1)c1
Benzenepropanoic acid, methyl ester (9CI)
AI3-02453
Hydrocinnamic acid, methyl ester (8CI)
NSC-10128
Methyl benzenepropanooate
Methyl hydrocinnamate
Methyl 3-phenylpropanoate
Methyl β-phenylpropionate
β-Phenylpropionic acid methyl ester

103-26-4
$C_{10}H_{10}O_2$
162.20
O=C(OC)C=Cc(cccc1)c1
Cinnamic acid, methyl ester
Methyl cinnamate
Methyl cinnamylate
Methyl 3-phenylpropenoate
2-Propenoic acid, 3-phenyl-, methyl ester (9CI)

103-29-7
$C_{14}H_{14}$
182.28
c(cccc1)(c1)CCc(cccc2)c2
Bibenzyl
Dibenzyl
1,2-Diphenylethane
Ethane, 1,2-diphenyl-

103-30-0
$C_{14}H_{12}$
180.26
c(cccc1)(c1)C=Cc(cccc2)c2
Stilbene, (E)
Benzene, 1,1'-(1,2-ethene-diyl)bis-, (E)- (9CI)
trans-Diphenylethene
trans-1,2-Diphenylethene
(E)-1,2-Diphenylethylene
trans-α,β-Diphenylethylene
trans-1,2-Diphenylethylene
(E)-Stilbene
trans-Stilbene

103-33-3
$C_{12}H_{10}N_2$
182.24
N(=Nc(cccc1)c1)c(cccc2)c2
Azobenzene
Azobenzeen (Dutch)
Azobenzide
Azobenzol
Azobisbenzene
Azodibenzene
Azodibenzeneazofume
Azofume
Benzeneazobenzene
Benzene, azodi
Benzofume
Diazobenzene
Diphenyldiazene
1,2-Diphenyldiazene
Diphenyldiimide
ENT 14,611
NCI-C02926
USAF EK-704

103-34-4
$C_8H_{16}N_2O_2S_2$
236.38
O(CCN(SSN(CCOC1)C1)C2)C2
Morpholine, 4,4'-dithiodi
ACCEL R
N,N'-Bismorpholine disulfide
Bismorpholino disulfide
Dimorpholine disulfide
Dimorpholino disulfide
Disulfide, dimorpholino-
Dithiobismorpholine
4,4'-Dithiobis(morpholine)
N,N'-Dithiodimorfolin (Czech)
N,N-Dithiodimorpholine
4,4'-Dithiodimorpholine
4,4'-Dithiomorpholine

Morpholine disulfide
Morpholinodisulfide
Sulfasan
Sulfasan R
Sulfasan R powder
Sulfazan R
USAF B-17
USAF EK-T-6645

103-36-6
$C_{11}H_{12}O_2$
176.23
O=C(OCC)C=Cc(cccc1)c1
Cinnamic acid, ethyl ester
Ethylcinnamate
Ethyl trans-cinnamate
Ethyl β-phenylacrylate
Ethyl 3-phenylpropenoate
2-Propenoic acid, 3-phenyl-, ethyl ester (9CI)

103-37-7
$C_{11}H_{14}O_2$
178.25
O=C(OCc(cccc1)c1)CCC
Butyric acid, benzyl ester
Benzyl n-butanoate
Benzyl butyrate
Benzyl n-butyrate
Benzylester kyseliny maselne (Czech)

103-41-3
$C_{16}H_{14}O_2$
238.30
O=C(OCc(cccc1)c1)C=Cc(cccc2)c2
Cinnamic acid, benzyl ester
Benzyl alcohol, cinnamate
Benzyl alcohol, cinnamic ester
Benzyl cinnamate
Benzylester kyseliny skoricove (Czech)
Benzyl γ-phenylacrylate
Cinnamein
trans-Cinnamic acid benzyl ester
3-Phenyl-2-propenoic acid phenylmethyl ester
2-Propenoic acid, 3-phenyl-, phenylmethyl ester (9CI)

103-43-5
$C_{18}H_{18}O_4$
298.34
Butanedioic acid, bis(phenylmethyl) ester (9CI)
NSC-4047
Succinic acid, dibenzyl ester (8CI)

103-44-6
$C_{10}H_{20}O$
156.27
O(CC(CCCC)CC)C=C
Vinyl 2-ethylhexyl ether
AI3-25059
1-Ethenoxy-2-ethylhexane
Ether, 2-ethylhexyl vinyl (8CI)
2-Ethylhexyl vinyl ether
Heptane, 3-((ethenyloxy)-methyl)- (9CI)
NSC-24170

103-45-7
$C_{10}H_{12}O_2$
164.22
O=C(OCCc(cccc1)c1)C
Acetic acid, phenethyl ester
Acetic acid, 2-phenylethyl ester
Benzylcarbinyl acetate
Ethanol, 2-phenyl-, acetate
Phenethyl acetate
β-Phenethyl acetate
2-Phenethyl acetate
β-Phenylethyl acetate
2-Phenylethyl acetate

103-49-1
$C_{14}H_{15}N$
197.27
Dibenzylamine
AI3-15327
Benzenemethanamine, N-(phenylmethyl)- (9CI)
N-Benzylbenzylamine
Dibenzylamine (8CI)
DMA (VAN)
NSC-4811
N-(Phenylmethyl)benzene-methanamine

103-50-4
$C_{14}H_{14}O$
198.28
O(Cc(cccc1)c1)Cc(cccc2)c2
Benzyl-ether
Benzyl oxide (Czech)
Dibenzyl ether

103-60-6
$C_{12}H_{16}O_3$
208.26
O=C(OCCOc(cccc1)c1)C(C)C
Isobutyric acid, 2-phenoxyethyl ester (8CI)
AI3-02711
NSC-227210
Phenoxyethyl isobutyrate
2-Phenoxyethyl 2-methylpropanoate
Propanoic acid, 2-methyl-, 2-phenoxyethyl ester (9CI)

103-62-8
$C_{10}H_{15}NO$
165.23
Phenol, p-(butylamino)- (8CI)
AI3-16924
Du Pont Gasoline Antioxidant No. 5
p-(Butylamino)phenol
N-Butyl-p-aminophenol
N-n-Butyl-p-aminophenol
N-Butylated-para-aminophenol
p-Hydroxyphenyl-n-butylamine
NSC-404034
Phenol, 4-(butylamino)- (9CI)

103-63-9
C_8H_9Br
185.06
BrCCc(cccc1)c1
(2-Bromoethyl)benzene
AI3-11264
Benzene, (2-bromoethyl)- (9CI)
β-Bromoethylbenzene
1-Bromo-2-phenylethane
NSC-33926
Phenethyl bromide
β-Phenethyl bromide
2-Phenethyl bromide
2-Phenyl-1-bromoethane
Phenylethyl bromide (VAN)
β-Phenylethyl bromide
2-Phenylethyl bromide

103-64-0
C_8H_7Br
183.06
c(cccc1)(c1)C=CBr
Styrene, β-bromo
α-Bromo-β-phenylethylene
β-Bromostyrene
ω-Bromostyrene
Bromostyrol
Bromostyrolene
β-Bromstyrol
Bromstyrole
Hyacinth Base

103-65-1
C_9H_{12}
120.21
c(cccc1)(c1)CCC
Benzene, propyl
Isocumene
1-Phenylpropane
n-Propyl benzene [UN 2364]
n-Propylbenzene
UN 2364 [n-Propyl benzene]

103-67-3
$C_8H_{11}N$
121.18
N(Cc(cccc1)c1)C
N-Methylbenzylamine
AI3-26793
Benzenemethanamine, N-methyl- (9CI)
Benzylamine, N-methyl- (8CI)
Benzylmethylamine
N-Benzylmethylamine
N-Methylbenzenemethanamine
Methylbenzylamine
N-Methyl-N-benzylamine
NSC-8059

103-69-5
$C_8H_{11}N$
121.20
N(c(cccc1)c1)CC
Aniline, N-ethyl

Aethylanilin (German)
Anilinoethane
Benzenamine, N-ethyl- (9CI)
N-Ethylaminobenzene
Ethylaniline
N-Ethylaniline [UN 2272]
N-Ethylbenzenamine
N-Ethylbenzenamino
Ethylphenylamine
UN 2272 [N-Ethylaniline]

103-70-8
C_7H_7NO
121.15
O=CNc(cccc1)c1
Aniline, N-formyl
Carbanilaldehyde
Formamide, N-phenyl- (9CI)
Formamidobenzene
Formanilide
Formylaniline
N-Formylaniline
Phenyl formamide
N-Phenylformamide

103-71-9
C_7H_5NO
119.13
O=C=Nc(cccc1)c1
Benzene, isocyanato
Carbanil
Fenylisokyanat (Czech)
Isocyanic acid, phenyl ester
Karbanil (Czech)
Mondur P
Phenylcarbimide
Phenyl carbonimide
Phenyl isocyanate [UN 2487]
UN 2487 [Phenyl isocyanate]

103-72-0
C_7H_5NS
135.19
N(c(cccc1)c1)=C=S
Isothiocyanic acid, phenyl ester
Benzene-1-isothiocyanate
Benzene, isothiocyanato-
Fenylisothiokyanat (Czech)
Phenyl isothiocyanate
Phenyl Mustard Oil
Phenylsenfoel (German)
PITC
Thiocarbanil
USAF M-4

103-73-1
$C_8H_{10}O$
122.18
O(c(cccc1)c1)CC
Phenetole
Benzene, ethoxy-
Ether, ethyl phenyl
Ethoxybenzene
Ethyl phenyl ether
Phenyl ethyl ether

103-74-2
C_7H_9NO
123.15
n(c(ccc1)CCO)c1
2-Pyridineethanol (9CI)
AI3-52671
2-(2-Hydroxyethyl)pyridine
NSC-2144
Pyridine-2-ethanol
Pyridine, 2-(2-hydroxyethyl)-
2-(2-Pyridyl)ethanol

103-75-3
$C_7H_{12}O_2$
128.19
O(C=CCC1)C1OCC
2H-Pyran, 2-ethoxy-3,4-dihydro
2-Ethoxy-2,3-dihydro-γ-pyran
2-Ethoxy-3,4-dihydro-1,2-pyran
2-Ethoxy-3,4-dihydro-2H-pyran
2-Ethoxydihydropyran, In pregnancy diagnosis

103-76-4
$C_6H_{14}N_2O$
130.22
OCCN(CCNC1)C1
1-Piperazineethanol
Ethanol, 2-(1-piperazinyl)-
N-(β-Hydroxyethyl)piperazine
1-(2-Hydroxyethyl)piperazine
USAF DO-22

103-79-7
$C_9H_{10}O$
134.19
O=C(Cc(cccc1)c1)C
2-Propanone, 1-phenyl
Benzyl methyl ketone
Methyl benzyl ketone
Phenylacetone
α-Phenylacetone
Phenylmethyl methyl ketone
1-Phenyl-2-propanone

103-80-0
C_8H_7ClO
154.60
O=C(Cc(cccc1)c1)Cl
Acetyl chloride, phenyl
Benzeneacetyl chloride (9CI)
Phenacetyl chloride
Phenylacetic acid chloride
Phenylacetyl chloride [UN 2577]
α-Phenylacetyl chloride
UN 2577 [Phenylacetyl chloride]

103-81-1
C_8H_9NO
135.18
O=C(N)Cc(cccc1)c1
Acetamide, 2-phenyl
Benzeneacetamide (9CI)
α-Phenylacetamide
2-Phenylacetamide
Phenylacetic acid amide
Phenyl-β-acetylamine
α-Toluamide
α-Toluimidic acid

103-82-2
$C_8H_8O_2$
136.16
O=C(O)Cc(cccc1)c1
Acetic acid, phenyl
Benzenacetic acid
Benzeneacetic acid
Kyselina fenyloctova (Czech)
Phenylacetic acid
ω-Phenylacetic acid
α-Toluic acid

103-83-3
$C_9H_{13}N$
135.23
N(Cc(cccc1)c1)(C)C
Benzylamine, N,N-dimethyl
Araldite Accelerator 062
BDMA
Benzenemethanamine, N,N-dimethyl- (9CI)
Benzyldimethylamine [UN 2619]
Benzyl-N,N-dimethylamine
N-Benzyldimethylamine
Benzyl dimethylamine (DOT)
N,N-Dimethylbenzenemethanamine
Dimethylbenzylamine
N,N-Dimethylbenzylamine
N-(Phenylmethyl)dimethylamine
Sumine 2015
UN 2619 [Benzyldimethylamine]

103-84-4
C_8H_9NO
135.18
O=C(Nc(cccc1)c1)C
Acetanilide
Acetamide, N-phenyl-
Acetamidobenzene
Acetanil
Acetanilid
Acetic acid anilide
Acetoanilide
Acetylaminobenzene
Acetylaniline
N-Acetylaniline
AN
Aniline, N-acetyl-
Antifebrin
Phenalgene
Phenalgin
N-Phenylacetamide
USAF EK-3

103-85-5
$C_7H_8N_2S$
152.23
N=C(S)Nc(cccc1)c1
Urea, 1-phenyl-2-thio
Fenylthiomocovina (Czech)
NCI-C02017

Phenylthiocarbamide
Phenylthiourea
α-Phenylthiourea
N-Phenylthiourea
1-Phenylthiourea
1-Phenyl-2-thiourea
PTC
PTU
RCRA waste number P093
U 6324
USAF EK-1569

103-88-8
C_8H_8BrNO
214.08
O=C(Nc(ccc(c1)Br)c1)C
Acetanilide, 4'-bromo
Acetamide, N-(4-bromophenyl)-
Acetanilide, p-bromo-
Antisepsin
Asepsin
p-Bromoacetanilide
4-Bromoacetanilide
4'-Bromoacetanilide
p-Bromo-N-acetanilide
Bromoanilide
Bromoantifebrin
USAF DO-40

103-89-9
$C_9H_{11}NO$
149.21
O=C(Nc(ccc(c1)C)c1)C
p-Acetotoluidide
Acetamide, N-(4-methylphenyl)- (9CI)
p-Acetamidotoluene
p-Acetotoluidide
4-Acetotoluidide
4-(Acetylamino)toluene
Acetyl-p-toluidine
N-Acetyl-p-toluidide
p-Methylacetanilide
4-Methylacetanilide
4'-Methylacetanilide

103-90-2
$C_8H_9NO_2$
151.18
O=C(Nc(ccc(O)c1)c1)C
Acetanilide, 4'-hydroxy
Abensanil
Acamol
Acetagesic
Acetalgin
Acetamide, N-(p-hydroxy-phenyl)-
Acetamide, N-(4-hydroxy-phenyl)-
p-Acetamidophenol
4-Acetamidophenol
Acetaminofen
Acetaminophen
p-Acetaminophenol
N-Acetyl-p-aminophenol
p-Acetylaminophenol
Algotropyl
Alpinyl
Alvedon
Amadil
Anaflon
Anelix
Anhiba
Apadon
Apamid
Apamide
APAP
Ben-U-Ron
Bickie-Mol
Calpol
Cetadol
Clixodyne
Datril
Dial-A-Gesic
Dirox
Doliprane
Dymadon
Enelfa
Eneril
Exdol
Febrilix
Febro-Gesic
Febrolin
Fendon
Finimal
G 1
Gelocatil
Hedex
Homoolan
p-Hydroxyacetanilide
4-Hydroxyacetanilide
4'-Hydroxyacetanilide
4-Hydroxyanilid kyseliny octove (Czech)
N-(4-Hydroxyphenyl)acetamide
Janupap
Korum
Lestemp
Liquagesic
Lonarid
Lyteca
Lyteca Syrup
Momentum
Multin
Napa
Napafen
Napap
Naprinol
NCI-C55801
Nobedon
Pacemo
Panadol
Panets
Panex
Panofen
Paracetamol
Paracetamole
Paracetamolo (Italian)
Paracetanol
Parapan
Paraspen
Parmol
Pedric
Phendon
Phenol, p-acetamido-
Pyrinazine
SK-Apap
Tabalgin
Tapar
Temlo
Tempanal
Tempra
Tralgon
Tussapap
Tylenol
Valadol
Valgesic

103-95-7
$C_{13}H_{18}O$
190.31
O=CC(C)Cc(ccc(c1)C(C)C)c1
Hydrocinnamaldehyde, p-iso-propyl-α-methyl
Aldehyde B
Cyclamal
Cyclamen aldehyde
p-Isopropyl-α-methylhydro-cinnamaldehyde
p-Isopropyl-α-methylhydro-cinnamic aldehyde
p-Isopropyl-α-methylphenyl-propyl aldehyde
α-Methyl-p-isopropylhydro-cinnamaldehyde
2-Methyl-3-(p-isopropylphenyl)-propionaldehyde

103-96-8
$C_{22}H_{40}N_2$
332.56
N(c(ccc(NC(CCCCCC)C)c1)c1)C(CCCCCC)C
Di-2-octyl-p-phenylenediamine
Antozite 1
1,4-Benzenediamine, N,N'-bis-(1-methylheptyl)- (9CI)
N,N'-Bis(1-methylheptyl)-1,4-benzenediamine
N,N'-Bis(1-methylheptyl)-p-phenylenediamine
N,N'-Bis(2-octyl)-p-phenylene-diamine
N,N'-Di(1-methylheptyl)-p-phenylenediamine
N,N'-Di(2-octyl)-p-phenylene-diamine
N,N'-Di(2-octyl)-para-phenyl-enediamine
Elastozone 30
NSC-56774
p-Phenylenediamine, N,N'-bis-(1-methylheptyl)- (8CI)
Santoflex 217
Tenemene 30
UOP 288

103-99-1
$C_{24}H_{41}NO_2$
375.59
O=C(Nc(ccc(O)c1)c1)CCCCCCCCCCCCCCCCC
Octadecanamide, N-(4-hydroxyphenyl)- (9CI)
N-(4-Hydroxyphenyl)octadecan-amide
NSC-166354
Octadecananilide, 4'-hydroxy- (8CI)
Stearic acid-p-hydroxyanilide

p-(Steroylamino)phenol
Stearoyl-p-aminophenol
N-Stearoyl-p-aminophenol
N-Stearoyl-4-aminophenol
Suconox 18, 4'-hydroxy-
Suconox-18

104-01-8
$C_9H_{10}O_3$
166.19
O=C(O)Cc(ccc(OC)c1)c1
Acetic acid, p-methoxyphenyl
2-(p-Anisyl)acetic acid
Anisyl formate
Benzeneacetic acid, 4-methoxy- (9CI)
Homoanisic acid
4-Methoxybenzeneacetic acid
p-Methoxybenzyl formate
p-Methoxyphenylacetic acid
4-Methoxyphenylacetic acid
MOPA

104-04-1
$C_8H_8N_2O_3$
180.18
O=C(Nc(ccc(N(=O)=O)c1)c1)C
Acetanilide, p-nitro
Acetamide, N-(4-nitrophenyl)- (9CI)
p-Acetamidonitrobenzene
p-Nitroacetanilide
4-Nitroacetanilide
4'-Nitroacetanilide
N-(4-Nitrophenyl)acetamide

104-09-6
$C_9H_{10}O$
134.18
O=CCc(ccc(c1)C)c1
Benzeneacetaldehyde, 4-methyl- (9CI)
4-Methylbenzeneacetaldehyde

104-10-9
$C_8H_{11}NO$
137.18
OCCc(ccc(N)c1)c1
Benzeneethanol, 4-amino- (9CI)
AI3-18010
4-Aminobenzeneethanol
p-Aminophenethyl alcohol
2-(p-Aminophenyl)ethanol
2-(4-Aminophenyl)ethanol
p-(2-Hydroxyethyl)aniline
4-(2-Hydroxyethyl)aniline
Phenethyl alcohol, p-amino- (8CI)
NSC-409780

104-12-1
C_7H_4ClNO
153.57
O=C=Nc(ccc(c1)Cl)c1
Isocyanic acid, p-chlorophenyl ester
p-Chlorfenylisokyanat (Czech)
p-Chlorophenyl isocyanate
PCPI

104-13-2
$C_{10}H_{15}N$
149.26
Nc(ccc(c1)CCCC)c1
Aniline, 4-butyl
p-Aminobutylbenzene
1-Amino-4-butylbenzene
Benzenamine, 4-butyl- (9CI)
p-n-Butylaniline
4-Butylbenzeneamine

104-14-3
$C_8H_{11}NO_2$
153.20
NCC(O)c1ccc(O)cc1
Benzyl alcohol, α-(aminomethyl)-p-hydroxy
α-(Aminomethyl)-p-hydroxybenzyl alcohol
1-(p-Hydroxyphenyl)-2-aminoethanol
p-Hydroxyphenylethanolamine
Norden
Norphen
Norsympathol
Norsynephrine
Octopamine
Paraoxyphenyl aminoethanol
WIN 5512

104-15-4
$C_7H_8O_3S$
172.21
O=S(=O)(O)c(ccc(c1)C)c1
p-Toluenesulfonic acid
Kyselina p-toluensulfonova (Czech)
Manro PTSA 65 E
Manro PTSA 65 H
Manro PTSA 65 LS
p-Methylbenzenesulfonic acid
4-Methylbenzenesulfonic acid
p-Methylphenylsulfonic acid
Toluenesulfonic acid
4-Toluenesulfonic acid
p-Toluenesulphonic acid
p-Tolylsulfonic acid
Tosic Acid
TSA-HP
TSA-MH

104-21-2
$C_{10}H_{12}O_3$
180.20
O=C(OCc(ccc(OC)c1)c1)C
Benzenemethanol, 4-methoxy-, acetate (9CI)
AI3-04097
Benzyl alcohol, p-methoxy-, acetate (8CI)
Cassie ketone
NSC-46102
4-Methoxybenzenemethanol acetate
p-Methoxybenzyl acetate
4-Methoxybenzyl acetate
p-Methoxybenzyl alcohol acetate

104-23-4
$C_{12}H_{11}N_3O_3S$
277.32
O=S(=O)(O)c(ccc(N=Nc(ccc(N)c1)c1)c2)c2
Benzenesulfonic acid, p-((p-aminophenyl)azo)
4'-Aminoazobenzene-4-sulfonic acid
4-Aminoazobenzene-4'-sulphonic acid
p-((p-Aminophenyl)azo)-benzenesulfonic acid
C.I. Food Yellow 6

104-28-9
$C_{14}H_{18}O_4$
250.29
O=C(OCCOCC)C=Cc(ccc(OC)c1)c1
Cinoxate
Caswell No. 427E
Cinoxato (Spanish)
Cinoxatum (Latin)
2-Ethoxyethyl p-methoxycinnamate
EPA Pesticide Chemical Code 076604
Giv-Tan F
3-(4-Methoxyphenyl)-2-propenoic acid 2-ethoxyethyl ester
Phiasol
Propenoic acid, 3-(4-methoxyphenyl)-, 2-ethoxyethyl ester
2-Propenoic acid, 3-(4-methoxyphenyl)-, 2-ethoxyethyl ester (9CI)

104-30-3
$C_{10}H_{11}NO_4$
209.20
O=C(OCCc(ccc(N(=O)=O)c1)c1)C
Benzeneethanol, 4-nitro-, acetate (ester) (9CI)
AI3-35598
4-Nitrobenzeneethanol acetate (ester)
p-Nitrophenethyl alcohol, acetate
NSC-190939
Phenethyl alcohol, p-nitro-, acetate (8CI)

104-35-8
$C_{17}H_{28}O_2$
264.41
Ethanol, 2-(4-nonylphenoxy)- (9CI)
Ethanol, 2-(p-nonylphenoxy)- (8CI)

104-38-1

$C_{10}H_{14}O_4$
198.22
O(c(ccc(OCCO)c1)c1)CCO
Ethanol, 2,2'-(p-phenylenedioxy)di- (8CI)
1,4-Bis(β-hydroxyethoxy)benzene
1,4-Bis(2-hydroxyethoxy)benzene
Bis(β-hydroxyethyl) hydroquinone ether
Ethanol, 2,2'-(1,4-phenylenebis(oxy))bis- (9CI)
Hydroquinone bis(β-hydroxyethyl) ether
Hydroquinone bis(2-hydroxyethyl) ether
Hydroquinone diethylol ether
Hydroquinone di(β-hydroxyethyl) ether
Hydroquinone, di(β-hydroxyethyl) ether
Hydroquinone di(2-hydroxyethyl) ether
NSC-1862
p-Phenylenebis(β-hydroxyethyl) ether
2,2'-(1,4-Phenylenebis(oxy))bisethanol
2,2'-(Phenylenedioxy)diethanol
2,2'-(p-Phenylenedioxy)diethanol
Vernatzer 30/10

104-40-5
$C_{15}H_{24}O$
220.39
Oc(ccc(c1)CCCCCCCCC)c1
Phenol, p-nonyl
4-Nonylphenol
para Nonyl phenol

104-41-6
$C_{19}H_{32}$
260.46
c(ccc(c1)C)(c1)CCCCCCCCCCCC
Benzene, 1-dodecyl-4-methyl- (9CI)
1-Dodecyl-4-methylbenzene

104-42-7
$C_{18}H_{31}N$
261.44
Nc(ccc(c1)CCCCCCCCCCCC)c1
Benzenamine, 4-dodecyl- (9CI)
p-Dodecylaniline
4-Dodecylbenzenamine

104-45-0
$C_{10}H_{14}O$
150.24
O(c(ccc(c1)CCC)c1)C
Anisole, p-propyl
Dihydroanethole
1-Methoxy-4-propylbenzene
p-n-Propyl anisole
4-Propylanisole
4-n-Propylanisole
p-Propylmethoxybenzene

104-46-1
$C_{10}H_{12}O$
148.22
O(c(ccc(c1)C=CC)c1)C
Anisole, p-propenyl
Acintene O
Anethol
Anethole
Anise camphor
Arizole
Isoestragole
p-Methoxy-β-methylstyrene
1-Methoxy-4-propenylbenzene
1-Methoxy-4-(1-propenyl)benzene
4-Methoxypropenylbenzene
Monasirup
Nauli "Gum"
Oil of Aniseed
Propene, 1-(p-methoxyphenyl)-
p-Propenyl anisole
p-1-Propenylanisole
4-Propenylanisole
p-Propenylmethoxybenzene
p-Propenylphenyl methyl ether

104-47-2
C_9H_9NO
147.19
N#CCc(ccc(OC)c1)c1
Acetonitrile, (p-methoxypenyl)
Anisylacetonitrile
Benzeneacetonitrile, 4-methoxy-
p-Methoxybenzeneacetonitrile
p-Methoxybenzyl cyanide
p-Methoxyphenylacetonitrile
4-Methoxyphenylacetonitrile

104-50-7
$C_8H_{14}O_2$
142.22
O=C(OC(C1)CCCC)C1
2(3H)-Furanone, dihydro-5-butyl
γ-n-Butyl-γ-butyrolactone
5-Hydroxyoctanoic acid lactone
γ-Octalactone
Octanolide-1,4
Tetrahydro-6-propyl-2H-pyran-2-one

104-51-8
$C_{10}H_{14}$
134.24
c(cccc1)(c1)CCCC
Benzene, butyl
n-Butylbenzene
n-Butylbenzene [UN 2709]
1-Phenylbutane
UN 2709 [Butyl benzenes]

104-52-9
$C_9H_{11}Cl$
154.64
ClCCCc1ccccc1
Benzene, (3-chloropropyl)- (8CI,9CI)
NSC-16939

104-54-1
$C_9H_{10}O$
134.19
OCC=Cc(cccc1)c1
Cinnamyl-alcohol
Alkohol skoricovy (Czech)
Cinnamic alcohol
3-Fenyl-2-propen-1-ol (Czech)
γ-Phenylallyl alcohol
3-Phenylallyl alcohol
3-Phenyl-2-propen-1-ol
2-Propen-1-ol, 3-phenyl-
Styrone
Styryl carbinol

104-55-2
C_9H_8O
132.17
O=CC=Cc(cccc1)c1
Cinnamaldehyde
Acrolein, 3-phenyl-
Aldehyd skoricovy (Czech)
Benzylideneacetaldehyde
Cassia aldehyde
Cinnamal
Cinnamic aldehyde
Cinnamyl aldehyde
3-Fenylpropenal (Czech)
NCI-C56111
Phenylacrolein
3-Phenylacrolein
3-Phenylpropenal
3-Phenyl-2-propenal
2-Propenal, 3-phenyl- (9CI)
Zimtaldehyde

104-57-4
$C_8H_8O_2$
136.16
O=COCc(cccc1)c1
Formic acid, benzyl ester
Benzyl alcohol, formate
Benzylester kyseliny mravenci (Czech)
Benzyl formate
Benzyl methanoate

104-60-9
$C_{24}H_{38}HgO_2$
559.15
Phenylmercuric oleate
Caswell No. 657J
EPA Pesticide Chemical Code 066022
Mercury, (9-octadecenoato-O)phenyl-, (Z)- (9CI)
Mercury, (oleato)phenyl-
(Z)-(9-Octadecenoato-O)-phenylmercury
(Z)-(9-Octadecenoato-O)phenylmercury
Phenylmercury oleate

PMO 10

104-61-0
$C_9H_{16}O_2$
156.25
O=C(OC(C1)CCCCC)C1
2(3H)-Furanone, dihydro-5-pentyl
Aldehyde C-18
γ-n-Amylbutyrolactone
Coconut aldehyde
4-Hydroxynonanoic acid, γ-lactone
γ-Nonalactone
1,4-Nonalolide
Prunolide

104-63-2
$C_9H_{13}NO$
151.20
OCCNCc(cccc1)c1
N-Benzylethanolamine
AI3-26796
Benzylaminoethanol
2-Benzylaminoethanol
2-(Benzylamino)ethanol
Benzyl ethanolamine
Benzylethanolamine
Ethanol, 2-((phenylmethyl)-amino)- (9CI)
Ethanol, 2-(benzylamino)- (8CI)
NSC-11271
2-((Phenylmethyl)amino)ethanol

104-66-5
$C_{14}H_{14}O_2$
214.26
O(c(cccc1)c1)CCOc(cccc2)c2
Benzene, 1,1'-(1,2-ethanediyl-bis(oxy))bis- (9CI)
AI3-00789
1,2-Diphenoxyethane
Ethane, 1,2-diphenoxy- (8CI)
1,1'-(1,2-Ethanediylbis(oxy))-bisbenzene
Ethylene glycol diphenyl ether
NSC-6794

104-67-6
$C_{11}H_{20}O_2$
184.31
O=C(OC(C1)CCCCCCC)C1
Undecanoic acid, 4-hydroxy-, γ-lactone
Aldehyde C-14
Aldehyde C-14 Peach
2(3H)-Furanone, 5-heptyldi-hydro-
γ-Heptylbutyrolactone
γ-n-Heptylbutyrolactone
4-Hydroxyundecanoic acid lactone
4-Hydroxyundecanoic acid, γ-lactone
Peach aldehyde
Peach lactone
Persicol
γ-Undecalactone
γ-Undecanolactone
1,4-Undecanolide
γ-Undecanolide
4-Undecanolide
γ-Undekalakton (Czech)

104-68-7
$C_{10}H_{14}O_3$
182.24
O(CCOc(cccc1)c1)CCO
Ethanol, 2-(2-phenoxyethoxy)
Diethylene glycol monophenyl ether
Diethylene glycolphenyl ether
Fenylkarbitol (Czech)
2-(2-Phenoxyethoxy)ethanol
Phenyl carbitol

104-69-8
$C_{15}H_{18}N_2$
226.31
N(c(cccc1)c1)CCCNc(cccc2)c2
1,3-Propanediamine, N,N'-di-phenyl- (9CI)
N,N'-Diphenyl-1,3-propanedi-amine

104-72-3
$C_{16}H_{26}$
218.38
c(cccc1)(c1)CCCCCCCCCC
Decylbenzene
Benzene, decyl- (9CI)
Decane, 1-phenyl- (8CI)
Decyl benzene, n-
n-Decylbenzene
NSC-74191
1-Phenyldecane

104-74-5
$C_{17}H_{30}N.Cl$
283.93
Pyridinium, 1-dodecyl-, chloride
C 2
Dehyquart C
Dodecylpyridinium chloride
N-Dodecylpyridinium chloride
1-Dodecylpyridinium chloride
DPC
Eltren
Laurylpyridinium chloride
1-Laurylpyridinium chloride
LPC
Quaternario LPC

104-75-6
$C_8H_{19}N$
129.28
NCC(CCCC)CC
Hexylamine, 2-ethyl
1-Amino-2-ethylhexan (Czech)
2-Ethyl hexylamine
2-Ethylhexylamine [UN 2276]
UN 2276 [2-Ethylhexylamine]

104-76-7
$C_8H_{18}O$
130.26
OCC(CCCC)CC
1-Hexanol, 2-ethyl
2-Aethylhexanol (German)
Ethylhexanol
2-Ethylhexanol
2-Ethyl-1-hexanol
2-Ethylhexyl alcohol

104-78-9
$C_7H_{18}N_2$
130.27
N(CCCN)(CC)CC
1,3-Propanediamine, N,N-di-ethyl
1-Amino-3-(diethylamino)-propane
N-(3-Diethylaminopropyl)amine
N,N-Diethylaminopropylamine
3-(Diethylamino)propylamine [UN 2684]
Diethylaminotrimethylenamine
N,N-Diethyl-1,3-diaminopro-pane
UN 2684 [Diethylaminopropyl-amine]

104-81-4
C_8H_9Br
185.06
BrCc(ccc(c1)C)c1
p-Xylyl bromide
Benzene, 1-(bromomethyl)-4-methyl- (9CI)
1-(Bromomethyl)-4-methyl-benzene
p-(Bromomethyl)toluene
4-(Bromomethyl)toluene
α-Bromo-p-xylene
ω-Bromo-p-xylene
α-Bromo-p-xylol
p-Methylbenzyl bromide
4-Methylbenzyl bromide
NSC-8050
p-Xylene, α-bromo- (8CI)
p-Xylyl-α-bromide

104-82-5
C_8H_9Cl
140.61
c(ccc(c1)C)(c1)CCl
α-Chloro-p-xylene
Benzene, 1-(chloromethyl)-4-methyl- (9CI)
1-(Chloromethyl)-4-methyl-benzene
(Chloromethyl)toluene (2-(chloromethyl) plus 4-(chloro-methyl))
p-Chloromethyltoluene
(p-Chloromethyl)toluene
4-(Chloromethyl)toluene
p-Methylbenzyl chloride
4-Methylbenzyl chloride
1-Methyl-4-(chloromethyl)-benzene
(4-Methylphenyl)methyl

chloride
NSC-46590
p-Xylene, α-chloro- (8CI)
p-Xylyl chloride
p-Xylyl-α-chloride

104-83-6
$C_7H_6Cl_2$
161.03
c(ccc(c1)Cl)(c1)CCl
Toluene, p,α-dichloro
p-Chlorobenzyl chloride [UN 2235]
1-Chloro-4-chloromethylbenzene
UN 2235 [Chlorobenzylchlorides]

104-85-8
C_8H_7N
117.16
N#Cc(ccc(c1)C)c1
p-Tolunitrile
p-Cyanotoluene
4-Cyanotoluene
p-Methylbenzonitrile
4-Methylbenzonitrile
4-Methylcyanobenzene
Nitril kyseliny p-toluylove (Czech)
p-Toluenenitrile
4-Toluenkarbonitril (Czech)
p-Toluic nitrile
p-Tolunitril (Czech)
4-Tolunitrile
p-Toluonitrile
p-Tolylnitrile

104-87-0
C_8H_8O
120.15
O=Cc(ccc(c1)C)c1
4-Methylbenzaldehyde
AI3-24380
Benzaldehyde, 4-methyl- (9CI)
p-Formyltoluene
p-Methylbenzaldehyde
para-Methylbenzaldehyde
NSC-2224
p-Tolualdehyde (8CI)
para-Tolualdehyde
4-Tolualdehyde
p-Toluyl aldehyde
para-Toluyl aldehyde
p-Toluylaldehyde
p-Tolylaldehyde

104-88-1
C_7H_5ClO
140.57
O=Cc(ccc(c1)Cl)c1
Benzaldehyde, p-chloro
p-Chlorobenzaldehyde
4-Chlorobenzaldehyde
p-Chlorobenzenecarboxaldehyde

104-89-2
$C_8H_{17}N$
127.26
N(C(CCC1CC)C)C1
Piperidine, 5-ethyl-2-methyl
Copellidin
3-Ethyl-6-methylpiperidine
5-Ethyl-2-methylpiperidine

104-90-5
$C_8H_{11}N$
121.20
n(c(ccc1CC)C)c1
2-Picoline, 5-ethyl
Aldehydecollidine
Aldehydine
2,5-Aldehydine
Collidine, aldehydecollidine
3-Ethyl-6-methylpyridine
5-Ethyl-2-methylpyridine
5-Ethyl-α-picoline
5-Ethyl-2-picoline
MEP
2-Methyl-5-ethylpyridine
6-Methyl-3-ethylpyridine
Methyl ethyl pyridine (DOT)
2-Methyl-5-ethylpyridine [UN 2300]
Pyridine, 5-ethyl-2-methyl-
UN 2300 [2-Methyl-5-ethylpyridine]

104-91-6
$C_6H_5NO_2$
123.12
O=Nc(ccc(O)c1)c1
Phenol, p-nitroso
p-Chinonmonoxim (Czech)
4-Nitrosofenol (Czech)
Nitrosophenol
p-Nitrosophenol
4-Nitrosophenol
Phenol, 4-nitroso- (9CI)
Quinone monoxime
Quinone oxime

104-92-7
C_7H_7BrO
187.05
O(c(ccc(c1)Br)c1)C
Anisole, p-bromo
Anisyl bromide
Benzene, 1-bromo-4-methoxy- (9CI)
p-Bromanisole
p-Bromoanisole
4-Bromoanisole
p-Bromophenyl methyl ether
p-Methoxybromobenzene
4-Methoxybromobenzene
p-Methoxyphenyl bromide
4-Methoxyphenyl bromide

104-93-8
$C_8H_{10}O$
122.18
O(c(ccc(c1)C)c1)C
Anisole, p-methyl
Benzene, 1-methoxy-4-methyl- (9CI)
p-Cresol methyl ether
p-Cresyl methyl ether
p-Methoxytoluene
4-Methoxytoluene
p-Methylanisole
4-Methyl-1-methoxybenzene
4-Methylphenol methyl ether
Methyl p-tolyl ether
p-Tolyl methyl ether

104-94-9
C_7H_9NO
123.17
O(c(ccc(N)c1)c1)C
p-Anisidine
p-Aminoanisole
4-Aminoanisole
1-Amino-4-methoxybenzene
Aniline, p-methoxy-
4-Anisidine
p-Anisidine (ACGIH,OSHA)
Anisole, p-amino-
p-Anisylamine
Benzenamine, 4-methoxy- (9CI)
4-Methoxy-1-aminobenzene
p-Methoxyaniline
4-Methoxyaniline
4-Methoxybenzenamine
4-Methoxybenzeneamine
p-Methoxyphenylamine

105-05-5
$C_{10}H_{14}$
134.22
c(ccc(c1)CC)(c1)CC
1,4-Diethylbenzene
Benzene, p-diethyl-
Benzene, 1,4-diethyl- (9CI)
p-Diethyl benzene
p-Diethylbenzene
p-Ethylethylbenzene

105-06-6
$C_{10}H_{10}$
130.19
c(ccc(c1)C=C)(c1)C=C
1,4-Diethenylbenzene
Benzene, 1,4-diethenyl- (9CI)

105-08-8
$C_8H_{16}O_2$
144.24
OCC(CCC(C1)CO)C1
1,4-Cyclohexanedimethanol
1,4-Bis(hydroxymethyl)cyclohexane
1,4-CHIDM
1,4-Cyclohexanedimethanol, hexahydro-2-oxo-

105-10-2
$C_8H_{12}N_2$
136.22
p-Phenylenediamine, N,N'-dimethyl

p-Aminodimethylaniline
1,4-Benzenediamine, N,N-dimethyl- (9CI)
C.I. 76075
p-Dimethylaminophenylamine
Dimethyl-p-phenylenediamine
N,N'-Dimethyl-p-fenylendiamin (Czech)
N,N'-Dimethyl-p-phenylenediamine
DMPD

105-11-3
$C_6H_6N_2O_2$
138.14
N(O)=C(C=CC(=NO)C=1)C1
p-Benzoquinone, dioxime
Actor Q
1,4-Benzochinondioxim (Czech)
1,4-Benzoquinone dioxime
2,5-Cyclohexadiene-1,4-dione, dioxime
Dibenzo PQD
Dioxime p-benzoquinone
Dioxime 1,4-cyclohexadienedione
Dioxime 2,5-cyclohexadiene-1,4-dione
G-M-F
NCI-C03850
PQD
QDO
Quinone dioxime
p-Quinone dioxime
para-Quinone oxime

105-12-4
$C_6H_4N_2O_2$
136.12
O=Nc(ccc(N=O)c1)c1
Benzene, p-dinitroso
Benzene, 1,4-dinitroso- (9CI)
p-Dinitrosobenzene
1,4-Dinitrosobenzene

105-13-5
$C_8H_{10}O_2$
138.18
O(c(ccc(c1)CO)c1)C
Benzyl alcohol, p-methoxy
Anise alcohol
Anisic alcohol
p-Anisol alcohol
Anisyl alcohol
4-Methoxybenzenemethanol
p-Methoxybenzyl alcohol

105-16-8
$C_{10}H_{19}NO_2$
185.30
O=C(OCCN(CC)CC)C(=C)C
Methacrylic acid, 2-(diethylamino)ethyl ester
Daktose B
2-Diethylaminoethylester kyseliny methakrylove (Czech)
Diethylaminoethyl methacrylate
β-(Diethylamino)ethyl methacrylate
2-(Diethylamino)ethyl methacrylate
2-(N,N-Diethylamino)ethyl methacrylate
2-Propenoic acid, 2-methyl-, 2-(diethylamino)ethyl ester (9CI)

105-21-5
$C_7H_{12}O_2$
128.19
O=C(OC(C1)CCC)C1
2(3H)-Furanone, dihydro-5-propyl
γ-Heptalactone
γ-Heptanolactone
Heptanolide-4,1
4-Hydroxyheptanoic acid lactone
4-Hydroxyheptanoic acid, γ-lactone
γ-Propiobutyrolactone

105-30-6
$C_6H_{14}O$
102.20
OCC(CCC)C
1-Pentanol, 2-methyl
Amyl methyl alcohol
1,3-Dimethyl butanol
Isohexyl alcohol
Isopropyl dimethyl carbinol
Methylamyl alcohol
Methyl isobutyl carbinol
2-Methylpentanol-1
2-Methyl-2-propylethanol
M.I.B.C.

105-31-7
$C_6H_{10}O$
98.16
OC(C#C)CCC
1-Hexyn-3-ol
Hexynol

105-34-0
$C_4H_5NO_2$
99.10
O=C(OC)CC#N
Acetic acid, cyano-, methyl ester
Cyanoacetic acid methyl ester
Methyl cyanoacetate
Methyl 2-cyanoacetate
Methyl cyanoethanoate
Methylester kyseliny kyanoctove (Czech)
USAF KF-22

105-36-2
$C_4H_7BrO_2$
167.02
O=C(OCC)CBr
Acetic acid, bromo-, ethyl ester
Antol
Bromoacetic acid, ethyl ester
Ethoxycarbonylmethyl bromide
Ethyl bromacetate
Ethyl bromoacetate [UN 1603]
Ethyl α-bromoacetate
Ethyl monobromoacetate
UN 1603 [Ethyl bromoacetate]

105-37-3
$C_5H_{10}O_2$
102.15
O=C(OCC)CC
Propionic acid, ethyl ester
Ethylester kyseliny propionove (Czech)
Ethyl propionate [UN 1195]
Propionate d'ethyle (French)
Propionic ether
UN 1195 [Ethyl propionate]

105-38-4
$C_5H_8O_2$
100.13
O=C(OC=C)CC
Propionic acid, vinyl ester
Propanoic acid, ethenyl ester
Vinylester kyseliny propionove (Czech)
Vinyl propionate

105-39-5
$C_4H_7ClO_2$
122.56
O=C(OCC)CCl
Acetic acid, chloro-, ethyl ester
Chloroacetic acid, ethyl ester
Ethyl chloracetate
Ethyl chloroacetate [UN 1181]
Ethyl α-chloroacetate
Ethyl chloroethanoate
Ethylester kyseliny chloroctove (Czech)
Ethyl monochloracetate
Ethyl monochloroacetate
UN 1181 [Ethyl chloroacetate]

105-40-8
$C_4H_9NO_2$
103.14
O=C(OCC)NC
Carbamic acid, methyl-, ethyl ester
Ethylester kyseliny methylkarbaminove (Czech)
Ethyl methylcarbamate
Methylcarbamic acid, ethyl ester
N-Methyl urethan

105-42-0
$C_7H_{14}O$
114.19
2-Hexanone, 4-methyl- (9CI)
4-Methyl-2-hexanone
Methyl 2-methylbutyl ketone

NSC-128218

105-43-1
$C_6H_{12}O_2$
116.16
O=C(O)CC(CC)C
3-Methylvaleric acid
3-Methylpentanoic acid
Pentanoic acid, 3-methyl- (9CI)

105-45-3
$C_5H_8O_3$
116.13
O=C(OC)CC(=O)C
Acetoacetic acid, methyl ester
Acetoacetic methyl ester
Butanoic acid, 3-oxo-, methyl ester (9CI)
Methylacetoacetate
Methyl acetylacetate
Methyl acetylacetonate
Methylester kyseliny acetoctove (Czech)
Methyl 3-oxobutyrate
3-Oxobutanoic acid methyl ester

105-46-4
$C_6H_{12}O_2$
116.18
O=C(OC(CC)C)C
Acetic acid, sec-butyl ester
Acetate de butyle secondaire (French)
Acetic acid, 2-butoxy ester
Acetic acid, 1-methylpropyl ester (9CI)
2-Butanol acetate
sec-Butyl acetate (ACGIH, DOT,OSHA) [UN 1123]
2-Butyl acetate
sec-Butyl alcohol acetate
UN 1123 [Butyl acetates]

105-48-6
$C_5H_9ClO_2$
136.59
O=C(OC(C)C)CCl
Acetic acid, chloro-, isopropyl ester
Chloroacetic acid isopropyl ester
Isopropyl chloroacetate [UN 2947]
UN 2947 [Isopropyl chloroacetate]

105-52-2
$C_{16}H_{28}O_4$
284.44
O=C(OC(CC(C)C)C)C=CC(=O)OC(CC(C)C)C
Maleic acid, bis(1,3-dimethylbutyl) ester
Bis-(1,3-dimethylbutyl)ester kyseliny maleinove (Czech)
Bis(1,3-dimethylbutyl) maleate
2-Butenedioic acid, bis(1,3-dimethylbutyl) ester
Dihexyl maleate
Di(4-methyl-2-amyl) maleate
Di(4-methyl-2-pentyl) maleate
DMAM
Maleic acid, di(1,3-dimethylbutyl) ester
Maleic acid, dihexyl ester

105-53-3
$C_7H_{12}O_4$
160.19
O=C(OCC)CC(=O)OCC
Malonic acid, diethyl ester
Carbethoxyacetic ester
Dicarbethoxymethane
Diethyl malonate
Diethyl propanedioate
Ethyl malonate
Malonic ester
Methanedicarboxylic acid, diethyl ester
Propanedioic acid, diethyl ester

105-54-4
$C_6H_{12}O_2$
116.18
O=C(OCC)CCC
Butyric acid, ethyl ester
Butanoic acid ethyl ester
Butyric ether
Ethyl butanoate
Ethyl butyrate [UN 1180]
Ethyl n-butyrate
UN 1180 [Ethyl butyrate]

105-55-5
$C_5H_{12}N_2S$
132.25
N(=C(S)NCC)CC
Urea, 1,3-diethyl-2-thio
N,N'-Diethylthiocarbamide
N,N'-Diethylthiourea
1,3-Diethylthiourea
1,3-Diethyl-2-thiourea
NCI-C03816
Pennzone E
Thiate H
Thiourea, N,N'-diethyl-
U 15030
USAF EK-1803

105-56-6
$C_5H_7NO_2$
113.13
O=C(OCC)CC#N
Acetic acid, cyano-, ethyl ester
Cyanacetate ethyle (German)
Cyanoacetic acid ethyl ester
Cyanoacetic ester
Estere cianoacetico (Italian)
Ethyl cyanoacetate [UN 2666]
Ethyl cyanoethanoate
Ethylester kyseliny kyanoctove (Czech)
Malonic acid ethyl ester nitrile
UN 2666 [Ethyl cyanoacetate]
USAF KF-25

105-57-7
$C_6H_{14}O_2$
118.20
O(C(OCC)C)CC
Acetaldehyde, diethyl acetal
Acetaal (Dutch)
Acetal [UN 1088]
Acetal diethylique (French)
Acetale (Italian)
Acetol
1,1-Diaethoxy-aethan (German)
Diaethylacetal (German)
1,1-Diethoxy-ethaan (Dutch)
1,1-Diethoxyethane
Diethyl acetal
1,1-Dietossietano (Italian)
Ethane, 1,1-diethoxy-
Ethylidene diethyl ether
UN 1088 [Acetal]
USAF DO-45

105-58-8
$C_5H_{10}O_3$
118.15
O=C(OCC)OCC
Carbonic acid, diethyl ester
DEC
Diaethylcarbonat (German)
Diatol
Diethyl carbonate [UN 2366]
Diethylester kyseliny uhlicite (Czech)
Diethylkarbonat (Czech)
Ethoxyformic anhydride
Ethyl carbonate
Eufin
NCI-C60899
UN 2366 [Diethyl carbonate]

105-59-9
$C_5H_{13}NO_2$
119.19
OCCN(CCO)C
Ethanol, 2,2'-(methylimino)di
Bis(2-hydroxyethyl)methylamine
Diethanolmethylamine
Ethanol, 2,2'-(methylimino)bis-
2-(N-2-Hydroxyethyl-N-methylamino)ethanol
MDEA
N-Methylaminodiglycol
Methylbis(2-hydroxyethyl)amine
Methyldiethanolamine
N-Methyldiethanolamine
N-Methyldiethanolimine
Methyliminodiethanol
N-Methyliminodiethanol
N-Methyl-2,2'-iminodiethanol
2,2'-(Methylimino)diethanol
USAF DO-52

105-60-2
$C_6H_{11}NO$

113.18
O=C(NCCCC1)C1
2H-Azepin-2-one, hexahydro
A1030
Akulon
Akulon M 2W
Alkamid
Amilan CM 1001
Amilan CM 1011
Amilan CM 1001C
Amilan CM 1001G
6-Aminocaproic acid lactam
Aminocaproic lactam
6-Aminohexanoic acid cyclic lactam
A1030n0
ATM 2(Nylon)
1-Aza-2-cycloheptanone
2-Azacycloheptanone
2H-Azepin-7-one, hexahydro-
Bonamid
Capran 80
Capran 77C
Caprolactam
ε-Caprolactam (ACGIH)
ω-Caprolactam
6-Caprolactam
Caprolactam monomer
Caprolactam (OSHA)
Caprolattame (French)
Caprolon B
Caprolon V
Capron
Capron 8250
Capron 8252
Capron 8253
Capron 8256
Capron 8257
Capron B
Capron GR 8256
Capron GR 8258
Capron PK4
Chemlon
CM 1001
CM 1011
CM 1031
CM 1041
Cyclohexanone iso-oxime
Danamid
Dull 704
Durethan BK
Durethan BK 30S
Durethan BKV 30H
Durethan BKV 55H
Epsylon Kaprolaktam (Polish)
Ertalon 6SA
Extrom 6N
Grilon
Hexahydro-2-azepinone
Hexahydro-2H-azepin-2-one (9CI)
Hexamethylenimine, 2-oxo-
6-Hexanelactam
Hexanoic acid, 6-amino-, cyclic lactam
Hexanoic acid, 6-amino-, lactam
Hexanolactam
Hexanone isoxime
Hexanonisoxim (German)
1,6-Hexolactam
Itamid
Itamid 250
Itamide 25
Itamide 35
Itamide 250
Itamide 350
Itamide 250G
Itamide S
e-Kaprolaktam (Czech)
Kaprolit
Kaprolit B
Kaprolon
Kaprolon B
Kapromine
Kapron
Kapron A
Kapron B
2-Ketohexamethyleneimine
2-Ketohexamethylenimine
KS 30P
Maranyl F 114
Maranyl F 124
Maranyl F 500
Metamid
Miramid H 2
Miramid WM 55
NCI-C50646
Nylon A1035SF
Nylon CM 1031
Nylon X 1051
Orgamide
Orgamid RMNOCD
2-Oxohexamethyleneimine
2-Oxohexamethylenimine
PA 6
PA 6 (Polymer)
2-Perhydroazepinone
PK 4
PKA
Plaskin 8200
Plaskon 201
Plaskon 8201
Plaskon 8201HS
Plaskon 8205
Plaskon 8207
Plaskon 8252
Plaskon 8202C
Plaskon XP 607
Polyamide PK 4
P 6 (Polyamide)
Relon P
Renyl MV
Sipas 60
Spencer 401
Spencer 601
Steelon
Stilon
Stylon
Tarlon X-A
Tarlon XB
Tarnamid T
Tarnamid T 2
Tarnamid T 27
Tnk 2G5
Torayca N 6
UBE 1022B
Ultramid B 3
Ultramid B 4
Ultramid B 5
Ultramid BMK
Vidlon
Widlon
Zytel 211

105-62-4
$C_{39}H_{72}O_4$
605.00
O=C(OCC(OC(=O)CCCCCCC C=CCCCCCCCC)C)CCCCC CCC=CCCCCCCCC
9-Octadecenoic acid (Z)-, 1-methyl-1,2-ethanediyl ester (9CI)
AI3-09503

105-64-6
$C_8H_{14}O_6$
206.22
O=C(OC(C)C)OOC(=O)OC(C)C
Peroxydicarbonic acid, diisopropyl ester
Diisopropyl perdicarbonate
Diisopropyl peroxydicarbonate
Diisopropyl peroxydicarbonate, Maximum concentration 52% in solution (DOT)
Diisopropyl peroxydicarbonate, Technically pure (DOT)
Isopropyl percarbonate
Isopropyl percarbonate, Stabilized (DOT)
Isopropyl percarbonate, Unstabilized (DOT)
Isopropyl peroxydicarbonate
Isopropyl peroxydicarbonate, Not more than 52% in solution (DOT)
Isopropyl peroxydicarbonate, Technically pure (DOT)
NA 2133 (DOT)
NA 2134 (DOT)
Peroxydicarbonate d'isopropyle (French)
Peroxydicarbonic acid, bis-(1-methylethyl) ester
Peroxydicarbonic acid, diisopropyl ester, Not more than 52% in solution
UN 2133 (DOT)
UN 2134 (DOT)

105-66-8
$C_7H_{14}O_2$
130.21
O=C(OCCC)CCC
Butyric acid, propyl ester
Butanoic acid, propyl ester (9CI)
Propyl butanoate
Propyl butyrate
Propylester kyseliny maselne (Czech)

105-67-9
$C_8H_{10}O$
122.18
Oc(c(cc(c1)C)C)c1
2,4-Xylenol
2,4-Dimethylphenol
4,6-Dimethylphenol
1-Hydroxy-2,4-dimethylbenzene

RCRA waste number U101
UN 2261 [Xylenols]
m-Xylenol [UN 2261]

105-68-0
$C_8H_{16}O_2$
144.24
O=C(OCCC(C)C)CC
Isopentyl alcohol, propionate
Isoamyl propionate
Isopentyl propionate
Propionic acid, isopentyl ester

105-73-7
Unknown
Unknown
Ethyl propionate

105-74-8
$C_{24}H_{46}O_4$
398.70
O=C(OOC(=O)CCCCCCCCC CC)CCCCCCCCCCC
Lauroyl-peroxide
Alperox C
Dilauroyl peroxide
Dilauroyl peroxide, Not more than 42% (DOT)
Dilauroyl peroxide, Technically pure (DOT)
Dilauryl peroxide
Dodecanoyl peroxide
DYP-97F
Laurox
Lauroyl peroxide (DOT)
Lauroyl peroxide, Not more than 42% (DOT)
Lauroyl peroxide, Technically pure (DOT)
Laurydol
LYP 97
LYP 97F
Peroxide, bis(1-oxododecyl)-
Peroxyde de lauroyle (French)
UN 2124 (DOT)
UN 2893 (DOT)

105-75-9
$C_{12}H_{20}O_4$
228.32
O=C(OCCCC)C=CC(=O)OCCCC
Fumaric acid, dibutyl ester
Dibutylester kyseliny fumarove (Czech)
Dibutyl fumarate

105-76-0
$C_{12}H_{20}O_4$
228.32
O=C(OCCCC)C=CC(=O)OCCCC
Maleic acid, dibutyl ester
2-Butenedioic acid, dibutyl ester
DBM
Dibutylester kyseliny maleinove (Czech)
Dibutyl maleate
RC Comonomer DBM
Staflex DBM

105-79-3
$C_{10}H_{20}O_2$
172.30
O=C(OCC(C)C)CCCCC
Hexanoic acid, isobutyl ester
Hexanoic acid, 2-methylpropyl ester
Isobutyl caproate
Isobutyl hexanoate
2-Methylpropyl hexanoate

105-80-6
$C_{17}H_{32}O_4$
300.44
O=C(OCC(C)C)CCCCCCCC(=O) OCC(C)C
Nonanedioic acid, bis-(2-methylpropyl) ester
AI3-07964
Bis(2-methylpropyl) nonanedioate

105-83-9
$C_7H_{19}N_3$
145.29
N(CCCN)(CCCN)C
Dipropylamine, 3,3'-diamino-N-methyl
Bis(γ-aminopropyl)methylamine
Bis(ω-aminopropyl)methylamine
Bis(3-aminopropyl)methylamine
N,N-Bis(γ-aminopropyl)methylamine
N,N-Bis(3-aminopropyl)methylamine
3,7'-Diamino-N-methyldipropylamine
Methylbis(3-aminopropyl)amine
Methylamine, N,N-bis(3-aminopropyl)-

105-86-2
$C_{11}H_{18}O_2$
182.29
O=COCC=C(CCC=C(C)C)C
2,6-Octadien-1-ol, 3,7-dimethyl-, formate, (E)
trans-3,7-Dimethyl-2,6-octadien-1-ol formate
trans-3,7-Dimethyl-2,6-octadien-1-yl formate
Formic acid, 3,7-dimethyl-2,6-octadienyl ester, (E)-
Formic acid, geraniol ester
Geraniol formate
Geranyl formate

105-87-3
$C_{12}H_{20}O_2$
196.32
O=C(OCC=C(CCC=C(C)C)C)C
2,6-Octadien-1-ol, 3,7-dimethyl-, acetate, (e)
Acetic acid, geraniol ester
trans-3,7-Dimethyl-2,6-octadien-1-ol, acetate
3,7-Dimethyl-2-trans, 6-octadienyl acetate
trans-3,7-Dimethyl-2,6-octadien-1-yl acetate
trans-2,6-Dimethyl-2,6-octadien-8-yl ethanoate
Geraniol acetate
Geranyl acetate
NCI-C54728
2,6-Octadien-1-ol, 3,7-dimethyl-, acetate, trans-

105-97-5
$C_{26}H_{50}O_4$
426.68
O=C(OCCCCCCCCCC)CCCCC (=O)OCCCCCCCCCC
Dicapryl adipate
Adipic acid, didecyl ester (8CI)
Didecyl adipate
Di-n-decyl adipate
Didecyl hexanedioate
Hexanedioic acid, didecyl ester (9CI)
NSC-4445
Polycizer 632

105-99-7
$C_{14}H_{26}O_4$
258.40
O=C(OCCCC)CCCCC(=O) OCCCC
Adipic acid, dibutyl ester
Butyl adipate
Dibutyl adipate
Di-n-butyl adipate
Dibutyl adipinate
Dibutylester kyseliny adipove (Czech)
Dibutyl hexanedioate
Experimental Tick Repellent 3
Experimental Tick Repellent 3PS
Hexanedioic acid, dibutyl ester
3PS

106-02-5
$C_{15}H_{28}O_2$
240.39
O=C(OCCCCCCCCCCCCCC1)C1
Exaltolide
AI3-30956
Cyclopentadecanolide
Exaltex
15-Hydroxypentadecanoic acid, lactone
15-Hydroxypentadecanoic acid-ε-lactone
Muskalactone
NSC-36763
Oxacyclohexadecan-2-one (9CI)
Pentadecalactone
2-Pentadecalone
Pentadecanoic acid, 15-hydroxy-, XI-lactone
Pentadecanolide

1,15-Pentadecanolide

106-06-9
$C_{24}H_{46}O_6$
430.63
O=C(OCCOCCOCCOC(=O)CCC
CCCCC)CCCCCCCC
Nonanoic acid, 1,2-ethanediyl-bis(oxy-2,1-ethanediyl) ester
AI3-01988

106-10-5
$C_{22}H_{42}O_6$
402.57
O=C(OCCOCCOCCOC(=O)CCC
CCCC)CCCCCCC
Octanoic acid, ethylenebis-(oxyethylene) ester (8CI)
NSC-6380
Octanoic acid, diester with triethylene glycol
Octanoic acid, 1,2-ethanediyl-bis(oxy-2,1-ethanediyl) ester (9CI)
Triethylene glycol dicaprylate
Triethylene glycol dioctanoate

106-11-6
$C_{22}H_{44}O_4$
372.66
O=C(OCCOCCO)CCCCCCCCC
CCCCCCCC
Stearic acid, 2-(2-hydroxy-ethoxy)ethyl ester
Aqua Cera
Atlas G 2146
Cerasynt
Cerasynt Special
Clindrol SDG
Diethylene glycol monostearate
Diethylene glycol stearate
Diethylene glycol, monoester with stearic acid
Diglycol monostearate
Diglycol stearate
Emcol DS-50 CAD
Emcol ETS
Glyco stearin
Nonex 411
Promul 5080
USAF KE-8

106-12-7
$C_{22}H_{42}O_4$
370.57
O=C(OCCOCCO)CCCCCCC
C=CCCCCCCCC
PEG-2 Oleate
AI3-00971
Diethylene glycol monooleate
Diglycol oleate
Hefti DMO-33
9-Octadecenoic acid, 2-(2-hydroxyethoxy)ethyl ester
9-Octadecenoic acid (Z)-, 2-(2-hydroxyethoxy)ethyl ester (9CI)
Polyethylene glycol 100 mono-oleate
Polyoxyethylene (2) monooleate

106-13-8
$C_{16}H_{32}O_3$
272.43
O=C(OCCOCC)CCCCCCC
CCCC
Dodecanoic acid, 2-ethoxy-ethyl ester (9CI)
2-Ethoxyethyl dodecanoate
Lauric acid, 2-ethoxyethyl ester (8CI)
NSC-406279

106-14-9
$C_{18}H_{36}O_3$
300.54
O=C(O)CCCCCCCCCCC(O)
CCCCCC
Stearic acid, 12-hydroxy
Barolub FTO
Cerit Fac 3
Ceroxin GL
Harwax A
Hydrofol Acid 200
12-Hydroxyoctadecanoic acid
12-Hydroxystearic acid
KOW
Loxiol G 21

106-19-4
$C_{12}H_{22}O_4$
230.34
O=C(OCCC)CCCCC(=O)OCCC
Adipic acid, dipropyl ester
Dipropyl adipate
Di-n-propyl adipate
Hexanedioic acid, dipropyl ester (9CI)

106-20-7
$C_{16}H_{35}N$
241.52
N(CC(CCCC)CC)CC(CCCC)CC
Dihexylamine, 2,2'-diethyl
Bis-2-ethylhexylamin
2,2'-Diethyldihexylamine
Di(2-ethylhexyl)amine
1-Hexanamine, 2-ethyl-N-(2-ethylhexyl)-

106-21-8
$C_{10}H_{22}O$
158.32
OCCC(CCCC(C)C)C
1-Octanol, 3,7-dimethyl
Citronellol, dihydro-
Dihydrocitronellol
Dimethyloctanol
2,6-Dimethyl-8-octanol
3,7-Dimethyl-1-octanol
Geraniol, perhydro-
Geraniol, tetrahydro-
Geraniol tetrahydride
Pelargol
Perhydrogeraniol
Tetrahydrogeraniol

106-22-9
$C_{10}H_{20}O$
156.30
OCCC(CCC=C(C)C)C
6-Octen-1-ol, 3,7-dimethyl
Cephrol
Citronellol
2,6-Dimethyl-2-octen-8-ol
3,7-Dimethyl-6-octen-1-ol
Rhodinol
Rodinol

106-23-0
$C_{10}H_{18}O$
154.25
O=CCC(CCC=C(C)C)C
Citronellal
AI3-00203
β-Citronellal
Citronellel
Citronellol,(d)
2,3-Dihydrocitral
3,7-Dimethyl-6-octenal
3,7-Dimethyl-6-octen-1-al
NSC-46106
6-Octenal, 3,7-dimethyl- (8CI, 9CI)
Rhodinal (VAN)
D-Rhodinal

106-24-1
$C_{10}H_{18}O$
154.28
OCC=C(CCC=C(C)C)C
2,6-Octadien-1-ol, 3,7-di-methyl-, (E)
2,6-Dimethyl-trans-2,6-octadien-8-ol
3,7-Dimethyl-trans-2,6-octadien-1-ol
Geraniol
Geraniol alcohol
Geraniol extra
Geranyl alcohol
Guaniol
Lemonol
2,6-Octadien-1-ol, 3,7-di-methyl-, trans-

106-25-2
$C_{10}H_{18}O$
154.28
OCC=C(CCC=C(C)C)C
2,6-Octadien-1-ol, 3,7-di-methyl-, (Z)
2-cis-3,7-Dimethyl-2,6-octadien-1-ol
Nerol

106-26-3
$C_{10}H_{16}O$
152.24
O=CC=C(CCC=C(C)C)C
2,6-Octadienal, 3,7-dimethyl-, (Z)-
AI3-28518

(Z)-3,7-Dimethyl-2,6-octadienal

106-27-4
$C_9H_{18}O_2$
158.27
O=C(OCCC(C)C)CCC
Butyric acid, isopentyl ester
Butanoic acid, 3-methylbutyl ester (9CI)
Isoamyl butanoate
Isoamyl butylate
Isoamyl butyrate (DOT)
Isoamyl-n-butyrate
Isopentyl butanoate
Isopentyl butyrate
3-Methylbutyl butyrate
UN 2620 [Amyl butyrates]

106-30-9
$C_9H_{18}O_2$
158.24
O=C(OCC)CCCCCC
Ethyl heptanoate
Aether oenanthicus
AI3-24251
Cognac Oil
Enanthylic ether
Ethyl enanthate
Ethyl n-heptanoate
Ethyl oenanthate
Ethyl oenanthylate
Grape Oil
Heptanoic acid, ethyl ester (9CI)
NSC-8891
Oenanthic ether
Oleum Vitis Viniferae
Wine Oil

106-31-0
$C_8H_{14}O_3$
158.22
O=C(OC(=O)CCC)CCC
Butyric-anhydride
Anhydrid kyseliny maselne (Czech)
Butanoic acid, anhydride (9CI)
Butanoic anhydride
Butyranhydrid (Czech)
Butyric acid anhydride
n-Butyric acid anhydride
n-Butyric anhydride
Butyric anhydride [UN 2739]
Butyryl oxide
UN 2739 [Butyric anhydride]

106-32-1
$C_{10}H_{20}O_2$
172.30
O=C(OCC)CCCCCCC
Octanoic acid, ethyl ester
Ethyl caprylate
Ethyl octanoate
Ethyl octylate

106-33-2
$C_{14}H_{28}O_2$
228.38
O=C(OCC)CCCCCCCCCCC
Ethyl laurate
AI3-00645
Dodecanoic acid, ethyl ester (9CI)
Ethyl dodecanoate
Ethyl dodecylate
Ethyl laurinate
Lauric acid, ethyl ester (8CI)
NSC-83467

106-34-3
$C_6H_6O_2.C_6H_6O_2$
220.24
Quinhydrone
p-Benzoquinone, Compd. with hydroquinone
Chinhydron (Czech)
2,5-Cyclohexadiene-1,4-dione Compd. with 1,4-benzenediol (1:1)
Green Hydroquinone
Hydroquinone, Compd. with p-benzoquinone

106-35-4
$C_7H_{14}O$
114.21
O=C(CCCC)CC
3-Heptanone
Aethylbutylketon (German)
Butyl ethyl ketone
n-Butyl ethyl ketone
Eptan-3-one (Italian)
Ethylbutylcetone (French)
Ethylbutylketon (Dutch)
Ethyl butyl ketone (ACGIH, OSHA)
Etilbutilchetone (Italian)
Heptan-3-on (Dutch, German)
Heptan-3-one
3-Heptanone (OSHA)

106-36-5
$C_6H_{12}O_2$
116.18
O=C(OCCC)CC
Propionic acid, propyl ester
Propylester kyseliny propionove (Czech)
Propyl propanoate
Propyl propionate
n-Propyl propionate

106-37-6
$C_6H_4Br_2$
235.92
c(ccc(c1)Br)(c1)Br
Benzene, p-dibromo
Benzene, 1,4-dibromo- (9CI)
p-Bromophenyl bromide
p-Dibromobenzene
1,4-Dibromobenzene

106-38-7
C_7H_7Br
171.05
c(ccc(c1)Br)(c1)C
Toluene, p-bromo
p-Bromotoluene
Parabromotoluene

106-39-8
C_6H_4BrCl
191.46
c(ccc(c1)Br)(c1)Cl
Benzene, 1-bromo-4-chloro
p-Bromochlorobenzene
4-Bromochlorobenzene
p-Bromophenyl chloride
p-Chlorobromobenzene
4-Chlorobromobenzene
1-Chloro-4-bromobenzene
p-Chlorophenyl bromide
4-Chloro-1-bromobenzene
4-Chlorophenyl bromide

106-40-1
C_6H_6BrN
172.04
Nc(ccc(c1)Br)c1
Aniline, p-bromo
Benzenamine, 4-bromo-
4-Bromanilinu (Czech)
p-Bromoaniline
4-Bromoaniline
p-Bromophenylamine

106-41-2
C_6H_5BrO
173.02
Oc(ccc(c1)Br)c1
Phenol, p-bromo
p-Bromophenol
4-Bromophenol

106-42-3
C_8H_{10}
106.18
c(ccc(c1)C)(c1)C
p-Xylene (ACGIH,OSHA) [UN 1307]
Chromar
p-Dimethylbenzene
1,4-Dimethylbenzene
p-Methyltoluene
Scintillar
UN 1307 [Xylenes]
1,4-Xylene
p-Xylol
p-Xylol (DOT)

106-43-4
C_7H_7Cl
126.59
c(ccc(c1)Cl)(c1)C
Toluene, p-chloro
Benzene, 1-chloro-4-methyl-
4-Chloro-1-methylbenzene
p-Chlorotoluene [UN 2238]
4-Chlorotoluene
p-Tolyl chloride
UN 2238 [Chlorotoluenes]

106-44-5
C_7H_8O
108.15
Oc(ccc(c1)C)c1
p-Cresol (OSHA) [UN 2076]
4-Cresol
para-Cresol
p-Cresylic acid
1-Hydroxy-4-methylbenzene
p-Hydroxytoluene
4-Hydroxytoluene
p-Kresol
p-Methylhydroxybenzene
1-Methyl-4-hydroxybenzene
p-Methylphenol
4-Methylphenol
p-Oxytoluene
Paramethyl phenol
Phenol, 4-methyl- (9CI)
RCRA waste number U052
p-Toluol
p-Tolyl alcohol
UN 2076 [Cresols (o-; m-; p-)]

106-45-6
C_7H_8S
124.21
Sc(ccc(c1)C)c1
p-Toluenethiol
p-Mercaptotoluene
p-Methylbenzenethiol
4-Methylbenzenethiol
p-Methylphenylmercaptan
4-Methylphenylmercaptan
p-Methylthiophenol
4-Methylthiophenol
p-Thiocresol
4-Thiocresol
4-Toluenethiol
p-Tolyl mercaptan
p-Tolylthiol
USAF EK-510

106-46-7
$C_6H_4Cl_2$
147.00
c(ccc(c1)Cl)(c1)Cl
Benzene, p-dichloro (8CI)
Benzene, 1,4-dichloro- (9CI)
Caswell No. 632
p-Chlorophenyl chloride
p-Dichloorbenzeen (Dutch)
1,4-Dichloorbenzeen (Dutch)
p-Dichlorbenzol (German)
1,4-Dichlor-benzol (German)
Di-Chloricide
p-Dichlorobenzene (ACGIH, OSHA) [UN 1592]
1,4-Dichlorobenzene
p-Dichlorobenzol
Dichlorobenzene, para, Solid (DOT)
p-Diclorobenceno (Spanish)
1,4-Diclorobenzene (Italian)
p-Diclorobenzene (Italian)
EPA Pesticide Chemical Code 061501
Evola
Globol
NCI-C54955
NSC-36935
Paracide
Para Crystals
Paradi
Paradichlorbenzol (German)
Paradichlorobenzene
Paradichlorobenzol
Paradow
Paramoth
Paranuggets
Parazene
PDB
PDCB
Persia-Perazol
RCRA waste number U070
RCRA waste number U071
RCRA waste number U072
Santochlor
UN 1592 [p-Dichlorobenzene]

106-47-8
C_6H_6ClN
127.58
Nc(ccc(c1)Cl)c1
Aniline, p-chloro
1-Amino-4-chlorobenzene
Aniline, 4-chloro-
Benzeneamine, 4-chloro
4-Chloranilin (Czech)
p-Chloraniline
p-Chloroaniline
4-Chloroaniline
p-Chloroaniline, Liquid [UN 2019]
4-Chlorobenzenamine
p-Chloroaniline, Solid [UN 2018]
4-Chlorophenylamine
NCI-C02039
RCRA waste number P024
UN 2018 [Chloroanilines, solid]
UN 2019 [Chloroanilines, Liquid]

106-48-9
C_6H_5ClO
128.56
Oc(ccc(c1)Cl)c1
Phenol, p-chloro
p-Chlorfenol (Czech)
p-Chlorophenol
4-Chlorophenol
p-Chlorophenol, Liquid [UN 2021]
p-Chlorophenol, Solid [UN 2020]
Parachlorophenol
Phenol, 4-chloro-
UN 2020 [Chlorophenols, solid]
UN 2021 [Chlorophenols, Liquid]

106-49-0
C_7H_9N
107.17
Nc(ccc(c1)C)c1
p-Toluidine
4-Amino-1-methylbenzene
p-Aminotoluene
4-Aminotoluene
4-Aminotoluen (Czech)
Aniline, p-methyl-
C.I. 37107
C.I. Azoic Coupling Component 107
p-Methylaniline
4-Methylaniline
p-Methylbenzenamine
4-Methylbenzenamine
Naphtol AS-KG
Naphtol AS-KGLL
RCRA waste number U353
p-Toluidin (Czech)
4-Toluidine
p-Toluidine, Liquid (ACGIH, OSHA) [UN 1708]
Tolylamine
p-Tolylamine
UN 1708 [Toluidines liquid]

106-50-3
$C_6H_8N_2$
108.16
Nc(ccc(N)c1)c1
p-Phenylenediamine
p-Aminoaniline
4-Aminoaniline
BASF Ursol D
p-Benzenediamine
1,4-Benzenediamine
Benzofur D
C.I. 76060
C.I. Developer 13
C.I. Oxidation Base 10
Developer 13
Developer PF
p-Diaminobenzene
1,4-Diaminobenzene
Durafur Black R
p-Fenylendiamin (Czech)
Fenylenodwuamina (Polish)
Fouramine D
Fourrine D
Fourrine 1
Fur Black 41867
Fur Brown 41866
Furro D
Fur Yellow
Futramine D
Nako H
Orsin
Oxidation Base 10
Para
Paraphenylen-diamine
Pelagol D
Pelagol DR
Pelagol Grey D
Peltol D
p-Phenylenediamine (ACGIH, OSHA) [UN 1673]
1,4-Phenylenediamine
Phenylenediamine, para, Solid (DOT)
PPD
Renal PF
Rodol D
Santoflex IC
Tertral D

UN 1673 [Phenylenediamines (o-; m-; p-;)]
Ursol D
USAF EK-394
Vulkanox 4020
Zoba Black D

106-51-4
$C_6H_4O_2$
108.10
O=C(C=CC(=O)C=1)C1
p-Benzoquinone
Benzo-chinon (German)
1,4-Benzoquine
Benzoquinone [UN 2587]
1,4-Benzoquinone
p-Benzoquinone (OSHA)
Chinon (Dutch, German)
p-Chinon (German)
Chinone
Cyclohexadienedione
1,4-Cyclohexadienedione
2,5-Cyclohexadiene-1,4-dione
1,4-Cyclohexadiene dioxide
1,4-Diossibenzene (Italian)
1,4-Dioxybenzene
1,4-Dioxy-benzol (German)
NCI-C55845
Quinone (ACGIH,OSHA)
p-Quinone
RCRA waste number U197
UN 2587 [Benzoquinone]
USAF P-220

106-54-7
C_6H_5ClS
144.62
Sc(ccc(c1)Cl)c1
Benzenethiol, p-chloro
4-Chlorobenzenethiol
p-Chlorothiophenol
4-Chlorothiophenol
p-Chlorthiofenol (Czech)
Phenyl mercaptan, p-chloro-

106-55-8
$C_6H_{14}N_2$
114.22
N(C(CNC1C)C)C1
Piperazine, 2,5-dimethyl
2,5-Dimethylpiperazine

106-57-0
$C_4H_6N_2O_2$
114.09
O=C(NCC(=O)N1)C1
2,5-Dioxopiperazine
Cyclic(glycylglycyl)
Cyclodiglycine
Cycloglycylglycine
Cyclo(glycylglycyl)
α,γ-Diacipiperazine
Diglycolyl diamide
Diketopiperazine
2,5-Diketopiperazine
Glycine, Bimol. cyclic peptide
Glycine, N-glycyl-, cyclic peptide
Glycylglycine lactam
NSC-26345
2,5-Piperazinedione (9CI)

106-58-1
$C_6H_{14}N_2$
114.22
N(CCN(C1)C)(C1)C
Piperazine, 1,4-dimethyl
N,N'-Dimethylpiperazine
1,4-Dimethylpiperazine
Lupetazine

106-60-5
$C_5H_9NO_3$
131.13
O=C(O)CCC(=O)CN
Aminolevulinic Acid
5-Amino-4-oxopentanoic acid
Pentanoic acid, 5-amino-4-oxo- (9CI)

106-62-7
$C_6H_{14}O_3$
134.18
O(C(CO)C)CC(O)C
1-Propanol, 2-(2-hydroxypropoxy)- (8CI,9CI)

106-63-8
$C_7H_{12}O_2$
128.19
O=C(OCC(C)C)C=C
Acrylic acid, isobutyl ester
Isobutyl acrylate [UN 2527]
Isobutyl acrylate, Inhibited (DOT)
Isobutylester kyseliny akrylove (Czech)
Isobutyl propenoate
Isobutyl 2-propenoate
z-Methylpropyl acrylate
2-Propenoic acid, 2-methylpropyl ester (9CI)
UN 2527 [Isobutyl acrylate]

106-65-0
$C_6H_{10}O_4$
146.14
O=C(OC)CCC(=O)OC
Dimethyl succinate
AI3-02480
Butanedioic acid, dimethyl ester (9CI)
Dimethyl butanedioate
Methyl succinate
NSC-52209
Succinic acid, dimethyl ester (8CI)

106-67-2
$C_8H_{18}O$
130.26
OCC(CC)CC(C)C
1-Pentanol, 2-ethyl-4-methyl
2-Ethylisohexanol
2-Ethyl-4-methylpentanol
2-Ethyl-4-methyl-1-pentanol

106-68-3
$C_8H_{16}O$
128.24
O=C(CCCCC)CC
3-Octanone
Amyl ethyl ketone
EAK
Ethyl amyl ketone (OSHA) [UN 2271]
5-Methyl-3-heptanone (OSHA)
UN 2271 [Ethyl amyl ketone]

106-69-4
$C_6H_{14}O_3$
134.20
OCC(O)CCCCO
1,2,6-Hexanetriol
Hexanetriol-1,2,6
Hexane-1,2,6-triol

106-70-7
$C_7H_{14}O_2$
130.21
O=C(OC)CCCCC
Hexanoic acid, methyl ester
Methyl caproate
Methyl capronate
Methyl hexanoate
Methyl n-hexanoate
Methyl hexoate
Methyl hexylate

106-71-8
$C_6H_7NO_2$
125.14
O=C(OCCC#N)C=C
Acrylic acid, ester with hydracrylonitrile
Acrylic acid, 2-cyanoethyl ester
Cyanoethyl acrylate
2-Cyanoethyl acrylate
2-Cyanoethyl propenoate
Hydracrylonitrile, acrylate
2-Propenoic acid, 2-cyanoethyl ester (9CI)

106-73-0
$C_8H_{16}O_2$
144.21
O=C(OC)CCCCCC
Heptanoic acid, methyl ester
AI3-33581
Methyl heptanoate

106-74-1
$C_7H_{12}O_3$
144.19
O=C(OCCOCC)C=C
Acrylic acid, 2-ethoxyethyl ester
Acrylic acid, 2-ethoxyethanol ester
Cellosolve acrylate
Ethanol, 2-ethoxy-, acrylate

Ethoxyethyl acrylate
2-Ethoxyethyl acrylate
2-Ethoxyethylester kyseliny akrylove (Czech)
2-Ethoxyethyl-2-propenoate
Ethylene glycol monoethyl ether acrylate
Ethylene glycol monoethyl ether propenoate
2-Propenoic acid, 2-ethoxyethyl ester

106-75-2
$C_6H_8Cl_2O_5$
231.04
O=C(OCCOCCOC(=O)Cl)Cl
Formic acid, chloro-, oxydiethylene ester
Carbonochloridic acid, oxydi-2,1-ethanediyl ester
Diethylene glycol, bischloroformate
Oxydiethylene bis(chloroformate)
Oxydiethylene chloroformate

106-79-6
$C_{12}H_{22}O_4$
230.30
O=C(OC)CCCCCCCCC(=O)OC
Sebacic acid dimethyl ester
AI3-00662
Decanedioic acid, dimethyl ester (9CI)
Dimethyl decanedioate
Dimethyl octane-1,8-dicarboxylate
Dimethyl sebacate
Methyl sebacate
NSC-9415
Sebacic acid, dimethyl ester (8CI)

106-81-0
$C_{57}H_{104}O_{12}$
981.45
Oxiraneoctanoic acid, 3-(2-hydroxyoctyl)-, 1,2,3-propanetriyl ester (9CI)
Estynox 330
Octadecanoic acid, 9,10-epoxy-12-hydroxy-, triester with glycerol (6CI,8CI)

106-83-2
$C_{22}H_{42}O_3$
354.64
O=C(OCCCC)CCCCCCCC(O1)C1CCCCCCCCC
Octadecanoic acid, 9,10-epoxy-, butyl ester
Butyl-9,10-epoxystearate
9,10-Epoxyoctadecanoic acid, butyl ester

106-84-3
$C_{26}H_{50}O_3$
410.68
O=C(OCCCCCCCC)CCCCCCC C(O1)C1CCCCCCCC
Oxiraneoctanoic acid, 3-octyl-, octyl ester (9CI)
Drapex 3.2
Drapex 4.4
Draplex 3.2
Lankroflex ED 3
Octadecanoic acid, 9,10-epoxy-, octyl ester
Octyl 9,10-epoxystearate
Octyl 3-octyloxiraneoctanoate
Octyl oleate epoxide

106-86-5
$C_8H_{12}O$
124.20
O(C1CC(C=C)CC2)C12
7-Oxabicyclo(4.1.0)heptane, 3-vinyl
EP-101
Epoxide 101
1,2-Epoxy-4-vinylcyclohexane
Unoxat Epoxide 101
4-Vinylcyclohexane, 1,2-epoxide
Vinylcyclohexane monoxide
4-Vinylcyclohexene-1,2-epoxide
Vinylcyclohexene monoxide
4-Vinylcyclohexene monoxide
1-Vinyl-3,4-epoxycyclohexane
3-Vinyl-7-oxabicyclo(4.1.0)heptane

106-87-6
$C_8H_{12}O_2$
140.20
O(C1C(CC(O2)C2C3)C3)C1
7-Oxabicyclo(4.1.0)heptane, 3-(epoxyethyl)
Chissonox 206
EP-206
1,2-Epoxy-4-(epoxyethyl)cyclohexane
1-Epoxyethyl-3,4-epoxycyclohexane
3-(Epoxyethyl)-7-oxabicyclo-(4.1.0)heptane
3-(1,2-Epoxyethyl)-7-oxabicyclo(4.1.0)heptane
4-(1,2-Epoxyethyl)-7-oxabicyclo(4.1.0)heptane
4-(Epoxyethyl)-7-oxabicyclo(4.1.0)heptane
ERLA-2270
ERLA-2271
1-Ethyleneoxy-3,4-epoxycyclohexane
NCI-C60139
3-Oxiranyl-7-oxabicyclo-(4.1.0)heptene
Ucet Textile Finish 11-74 (Obs.)
Unox Epoxide 206
Vinyl cyclohexene diepoxide
4-Vinylcyclohexene diepoxide
4-Vinyl-1-cyclohexene diepoxide
4-Vinyl-1,2-cyclohexene diepoxide
1-Vinyl-3-cyclohexene dioxide
4-Vinlycyclohexene dioxide
4-Vinyl-1-cyclohexene dioxide
Vinyl cyclohexene dioxide (ACGIH,OSHA)

106-88-7
C_4H_8O
72.12
O(C1CC)C1
Butane, 1,2-epoxy
1-Butene oxide
1,2-Butene oxide
Butylene oxide
1,2-Butylene oxide
Epoxybutane
1,2-Epoxybutane
Ethylene oxide, ethyl-
Ethyloxirane
Ethyl ethylene oxide
NCI-C55527

106-89-8
C_3H_5ClO
92.53
O(C1CCl)C1
Propane, 1-chloro-2,3-epoxy
1-Chloor-2,3-epoxy-propaan (Dutch)
1-Chlor-2,3-epoxy-propan (German)
1-Chloro-2,3-epoxypropane
3-Chloro-1,2-epoxypropane
1-Chloro-2,3-epoxypropane (OSHA)
epi-Chlorohydrin
(Chloromethyl)ethylene oxide
Chloromethyloxirane
2-(Chloromethyl)oxirane
Chloropropylene oxide
γ-Chloropropylene oxide
3-Chloro-1,2-propylene oxide
1-Cloro-2,3-epossipropano (Italian)
ECH
Epichloorhydrine (Dutch)
Epichlorhydrin (German)
Epichlorhydrine (French)
Epichlorohydrin (ACGIH, OSHA) [UN 2023]
α-Epichlorohydrin
(DL)-α-Epichlorohydrin
Epichlorohydryna (Polish)
Epichlorophydrin
Epicloridrina (Italian)
1,2-Epoxy-3-chloropropane
2,3-Epoxypropyl chloride
Glycerol epichlorhydrin
Oxirane, (chloromethyl)-
Oxirane, 2-(chloromethyl)
RCRA waste number U041
Skekhg
UN 2023 [Epichlorohydrin]

106-90-1
$C_6H_8O_3$

128.14
O=C(OCC(O1)C1)C=C
Acrylic acid, 2,3-epoxypropyl ester
Acrylic acid glycidyl ester
2,3-Epoxypropyl acrylate
Glycidyl acrylate
Glycidylester kyseliny akrylove (Czech)
Glycidyl propenate
1-Propanol, 2,3-epoxy-, acrylate
2-Propenoic acid, oxiranylmethyl ester (9CI)

106-91-2
$C_7H_{10}O_3$
142.17
O=C(OCC(O1)C1)C(=C)C
Methacrylic acid, 2,3-epoxypropyl ester
CP 105
2,3-Epoxypropyl methacrylate
Glycidyl methacrylate
Glycidyl α-methyl acrylate
1-Propanol, 2,3-epoxy-, methacrylate

106-92-3
$C_6H_{10}O_2$
114.16
O(C1COCC=C)C1
Oxirane, ((2-propenyloxy)-methyl)
AGE
AGE (OSHA)
Allil-glicidil-etere (Italian)
1-Allilossi-2,3 epossipropano (Italian)
Allyl 2,3-epoxypropyl ether
Allylglycidaether (German)
Allyl glycidyl ether (ACGIH, DOT,OSHA)
1-Allyloxy-2,3-epoxy-propaan (Dutch)
1-Allyloxy-2,3-epoxypropan (German)
1-(Allyloxy)-2,3-epoxypropane
Ether, allyl 2,3-epoxypropyl
NCI-C56666
Oxyde d'allyle et de glycidyle (French)
Propane, 1-(allyloxy)-2,3-epoxy-
UN 2219 (DOT)

106-93-4
$C_2H_4Br_2$
187.88
BrCCBr
Ethane, 1,2-dibromo
Aethylenbromid (German)
Bromofume
Bromuro di etile (Italian)
Celmide
DBE
1,2-Dibromaethan (German)
1,2-Dibromoetano (Italian)
Dibromoethane
α,β-Dibromoethane
sym-Dibromoethane
1,2-Dibromoethane (DOT)
Dibromure d'ethylene (French)
1,2-Dibroomethaan (Dutch)
Dowfume
Dowfume 40
Dowfume EDB
Dowfume W-8
Dowfume W-85
Dowfume W-90
Dowfume W-100
Dwubromoetan (Polish)
EDB
EDB-85
E-D-Bee
ENT 15,349
Ethylene bromide
Ethylene dibromide (ACGIH, OSHA) [UN 1605]
1,2-Ethylene dibromide
Fumo-Gas
Glycol bromide
Glycol dibromide
Iscobrome D
Kopfume
NCI-C00522
Nephis
Pestmaster
Pestmaster EDB-85
RCRA waste number U067
Soilbrom
Soilbrom-40
Soilbrom-85
Soilbrom-90
Soilbrom-100
Soilbrome-85
Soilbrom-90EC
Soilfume
UN 1605 [Ethylene dibromide]
Unifume

106-94-5
C_3H_7Br
123.01
BrCCC
Propane, 1-bromo
1-Bromopropane (DOT)
Propyl bromide
UN 2344

106-95-6
C_3H_5Br
120.99
BrCC=C
Propene, 3-bromo
Allyl bromide [UN 1099]
Bromallylene
1-Bromo-2-propene
3-Bromopropene
3-Bromopropylene
UN 1099 [Allyl bromide]

106-96-7
C_3H_3Br
118.97
C(#C)CBr
Propyne, 3-bromo
γ-Bromoallylene
1-Brom-2-propin (Czech)
1-Bromo-2-propyne
3-Bromopropyne [UN 2345]
3-Bromo-1-propyne
Propargyl bromide
UN 2345 [3-Bromopropyne]

106-97-8
C_4H_{10}
58.14
C(CC)C
Butane (ACGIH,OSHA) [UN 1011]
n-Butane
Butanen (Dutch)
Butani (Italian)
Diethyl
Liquefied Petroleum Gas [UN 1075]
Methylethylmethane
UN 1011 [Butane or Butane mixtures see also Petroleum gases, Liquified]
UN 1075 [Liquefied petroleum gas]

106-98-9
C_4H_8
56.11
C(=C)CC
1-Butene (9CI)
Butene-1
α-Butene
α-Butylene
1-Butylene
Ethylethylene

106-99-0
C_4H_6
54.10
C(C=C)=C
1,3-Butadiene (ACGIH)
Biethylene
Bivinyl
Butadieen (Dutch)
Buta-1,3-dieen (Dutch)
Butadien (Polish)
Buta-1,3-dien (German)
Butadiene
Buta-1,3-diene
α,γ-Butadiene
Butadiene (OSHA)
Divinyl
Erythrene
NCI-C50602
Pyrrolylene
Vinylethylene

107-00-6
C_4H_6
54.10
C(#C)CC
1-Butyne
Ethylacetylene
Ethyl acetylene, Inhibited [UN 2452]
Ethylethyne
UN 2452 [Ethylacetylene, inhibited]

107-01-7
C_4H_8
56.12
C(=CC)C
2-Butene
2-Butene
β-Butylene
Pseudobutylene

107-02-8
C_3H_4O
56.07
O=CC=C
Acrolein (ACGIH,OSHA)
Acquinite
Acraldehyde
Acraldehydeacroleina (Italian)
trans-Acrolein
Acroleina (Italian)
Acroleine (Dutch, French)
Acrolein, Inhibited [UN 1092]
Acrylaldehyd (German)
Acrylaldehyde
Acrylic aldehyde
Akrolein (Czech)
Akroleina (Polish)
Aldehyde acrylique (French)
Aldeide acrilica (Italian)
Allyl aldehyde
Aqualin
Aqualine
Biocide
Crolean
Ethylene aldehyde
Magnacide
Magnacide H
NSC-8819
Propenal
2-Propenal
Prop-2-en-1-al
2-Propen-1-one
Propylene aldehyde
RCRA waste number P003
Slimicide
UN 1092 [Acrolein, Inhibited]

107-03-9
C_3H_8S
76.17
SCCC
Propanethiol
3-Mercaptopropanol
Propane-1-thiol
1-Propanethiol [UN 2402]
Propyl mercaptan
n-Propyl mercaptan
Propyl mercaptan [UN 2402]
UN 2402 [Propanethiols]

107-04-0
C_2H_4BrCl
143.42
BrCCCl
Ethane, 1-bromo-2-chloro
1-Bromo-2-chloroethane
sym-Chlorobromoethane
1-Chloro-2-bromoethane
Ethylene chlorobromide

107-05-1
C_3H_5Cl
76.53
C(=C)CCl
Propene, 3-chloro
Allile (cloruro di) (Italian)
p-Aminopropiofenon (Czech)
Allylchlorid (German)
Allyl chloride (ACGIH,OSHA) [UN 1100]
Allyle (chlorure d') (French)
Chlorallylene
Chloroallylene
3-Chloroprene
1-Chloro propene-2
3-Chloropropene-1
3-Chloropropene
1-Chloro-2-propene
α-Chloropropylene
3-Chloropropylene
3-Chloro-1-propylene
3-Chlorpropen (German)
NCI-C04615
1-Propene, 3-chloro-
2-Propenyl chloride
UN 1100 [Allyl chloride]

107-06-2
$C_2H_4Cl_2$
98.96
ClCCCl
Ethane, 1,2-dichloro
Aethylenchlorid (German)
1,2-Bichloroethane
Bichlorure d'ethylene (French)
Borer Sol
Brocide
Chlorure d'ethylene (French)
Cloruro di ethene (Italian)
1,2-DCE
Destruxol Borer-Sol
1,2-Dichloorethaan (Dutch)
1,2-Dichlor-aethan (German)
Dichloremulsion
1,2-Dichlorethane
Di-Chlor-Mulsion
Dichloro-1,2-ethane (French)
α,β-Dichloroethane
sym-Dichloroethane
1,2-Dichloroethane
Dichloroethylene
1,2-Dicloroetano (Italian)
Dutch Liquid
Dutch Oil
EDC
ENT 1,656
Ethane dichloride
Ethyleendichloride (Dutch)
Ethylene chloride
Ethylene dichloride (ACGIH, OSHA) [UN 1184]
1,2-Ethylene dichloride
Glycol dichloride
NCI-C00511
RCRA waste number U077
UN 1184 [Ethylene dichloride]

107-07-3
C_2H_5ClO
80.52
OCCCl
Ethanol, 2-chloro
Aethylenechlorhydrin (German)
2-Chloorethanol (Dutch)
2-Chloraethanol (German)
2-Chlorethanol (German)
δ-Chloroethanol
2-Chloroethanol
2-Chloroethanol (OSHA)
β-Chloroethyl alcohol
2-Chloroethyl alcohol
Chloroethylowy alkohol (Polish)
2-Cloroetanolo (Italian)
Ethyleen-chloorhydrine (Dutch)
Ethylene chlorhydrin
Ethylene chlorohydrin (ACGIH, OSHA) [UN 1135]
Ethylene glycol, chlorohydrin
Glicol monocloridrina (Italian)
Glycol chlorohydrin
Glycolmonochloorhydrine (Dutch)
Glycomonochlorhydrin
Glycol monochlorohydrin
Monochlorhydrine du glycol (French)
2-Monochloroethanol
NCI-C50135
UN 1135 [Ethylene chlorohydrin]

107-08-4
C_3H_7I
170.00
C(CC)I
Propane, 1-iodo
1-Iodopropane
1-Jodpropan (Czech)
n-Propyl iodide

107-10-8
C_3H_9N
59.13
NCCC
Propylamine [UN 1277]
1-Aminopropane
Mono-n-propylamine
Monopropylamine (DOT)
Propanamine
1-Propylamine
n-Propylamine
RCRA waste number U194
UN 1277 [Propylamine]

107-11-9
C_3H_7N
57.11
NCC=C
Allylamine [UN 2334]
3-Aminopropene
3-Aminopropylene
Monoallylamine
2-Propenamine
2-Propen-1-amine (9CI)
UN 2334 [Allylamine]

107-12-0
C_3H_5N
55.09
N#CCC
Propionitrile [UN 2404]
Cyanoethane
Ether cyanatus
Ethyl cyanide
Ethylkyanid (Czech)
Hydrocyanic ether
Propanenitrile
Propannitril (Czech)
Propionic nitrile
Propiononitrile
RCRA waste number P101
UN 2404 [Propionitrile]

107-13-1
C_3H_3N
53.07
N#CC=C
Acrylonitrile (ACGIH,DOT, OSHA)
Acritet
Acrylnitril (German, Dutch)
Acrylon
Acrylonitrile, Inhibited [UN 1093]
Acrylonitrile monomer
Akrylonitril (Czech)
Akrylonitryl (Polish)
Carbacryl
Cianuro di vinile (Italian)
Cyanoethylene
Cyanure de vinyle (French)
ENT 54
Fumigrain
Miller's Fumigrain
Nitrile acrilico (Italian)
Nitrile acrylique (French)
Propenenitrile
2-Propenenitrile
RCRA waste number U009
TL 314
UN 1093 [Acrylonitrile, Inhibited]
VCN
Ventox
Vinyl cyanide
Vinylcyanide (OSHA)
Vinylkyanid (Czech)

107-14-2
C_2H_2ClN
75.50
N#CCCl
Acetonitrile, chloro
Chloracetonitrile
Chloroacetonitrile [UN 2668]
α-Chloroacetonitrile
2-Chloroacetonitrile
Chloromethyl cyanide
Monochloroacetonitrile
Monochloromethyl cyanide
UN 2668 [Chloroacetonitrile]
USAF KF-5

107-15-3
$C_2H_8N_2$
60.12
NCCN
1,2-Ethanediamine
Aethaldiamin (German)
Aethylenediamin (German)
1,2-Diaminoaethan (German)
1,2-Diamino-ethaan (Dutch)
1,2-Diaminoethane
1,2-Diaminoethane (OSHA)
1,2-Diamino-ethano (Italian)
Dimethylenediamine
Ethyleendiamine (Dutch)
Ethylendiamine
1,2-Ethylenediamine
Ethylenediamine (ACGIH, OSHA) [UN 1604]
Ethylene-diamine (French)
NCI-C60402
UN 1604 [Ethylenediamine]

107-16-4
C_2H_3NO
57.06
N#CCO
Acetonitrile, hydroxy
Cyanomethanol
Formaldehyde cyanohydrin
Glycolic nitrile
Glycolonitrile (8CI)
Glyconitrile
Glykolonitril (Czech)
Hydroxyacetonitrile
2-Hydroxyacetonitrile
Hydroxymethylkyanid (Czech)
Hydroxymethylnitrile
USAF A-8565

107-18-6
C_3H_6O
58.09
OCC=C
Allyl-alcohol
AA
Alcool allilco (Italian)
Alcool allylique (French)
Allilowy alkohol (Polish)
Allyl Al
Allyl alcohol (ACGIH,OSHA) [UN 1098]
Allylalkohol (German)
Allylic alcohol
3-Hydroxypropene
Orvinylcarbinol
2-Propene-1-ol
Propenol
1-Propenol-3
Propen-1-ol-3
1-Propen-3-ol
2-Propen-1-ol
Propenyl alcohol
2-Propenyl alcohol
RCRA waste number P005
Shell Unkrautted A
UN 1098 [Allyl alcohol]
Vinylcarbinol
Weed Drench

107-19-7
C_3H_4O
56.07
OCC#C
2-Propyn-1-ol
Ethynylcarbinol
Methanol, ethynyl-
NA 1986
Propargyl alcohol (ACGIH, DOT,OSHA)
1-Propyne-3-ol
2-Propynyl alcohol
RCRA waste number P102

107-20-0
C_2H_3ClO
78.50
O=CCCl
Acetaldehyde, chloro
Chloroacetaldehyde (ACGIH, OSHA) [UN 2232]
Chloroacetaldehyde monomer
2-Chloroacetaldehyde
2-Chloroethanal
2-Chloro-1-ethanal
Monochloroacetaldehyde
RCRA waste number P023
UN 2232 [Chloroacetaldehyde]

107-21-1
$C_2H_6O_2$
62.08
OCCO
Ethylene-glycol
Athylenglykol (German)
1,2-Dihydroxyethane
1,2-Ethandiol
1,2-Ethanediol
Ethane-1,2-diol
Ethylene alcohol
Ethylene dihydrate
Ethylene glycol (ACGIH, OSHA)
Glycol
Glycol alcohol
Lutrol-9
Macrogol 400 BPC
M.E.G.
Monoethylene glycol
NCI-C00920
Norkool
Tescol
Dowtherm SR 1
UCAR 17

107-22-2
$C_2H_2O_2$
58.04
O=CC=O
Glyoxal, 40%
Aerotex glyoxal 40
Biformal
Biformyl
Diformal
Diformyl
Ethandial
Ethanedial
1,2-Ethanedione
Glyoxal
Glyoxal, 29.2%

Glyoxylaldehyde
Oxal
Oxalaldehyde

107-25-5
C_3H_6O
58.09
O(C=C)C
Ether, methyl vinyl
Methyl vinyl ether
UN 1087 [Vinyl methyl ether, inhibited]
Vinyl methyl ether (DOT)
Vinyl methyl ether, Inhibited [UN 1087]

107-27-7
C_2H_5ClHg
265.11
CC[Hg]Cl
Mercury, chloroethyl
Ceresan
Chloroethylmercury
EMC
Ethylmercuric chloride
Ethylmercury chloride
Ethylmerkurichlorid (Czech)
Ganozan
Granosan

107-29-9
C_2H_5NO
59.08
N(O)=CC
Acetaldehyde-oxime
Acetaldehyde oxime [UN 2332]
Acetaldoxime
Aldoxime
Ethanal oxime
Ethylidenehydroxylamine
UN 2332 [Acetaldehyde oxime]
USAF AM-5

107-30-2
C_2H_5ClO
80.52
O(CCl)C
Ether, chloromethyl methyl
Chlordimethylether (Czech)
Chloromethyl methyl ether (ACGIH,OSHA)
CMME
Dimethylchloroether
Ether, dimethyl chloro
Ether methylique monochlore (French)
Methylchloromethyl ether [UN 1239]
Methyl chloromethyl ether, Anhydrous (DOT)
Monochlorodimethyl ether
RCRA waste number U046
UN 1239 [Methylchloromethyl ether]

107-31-3
$C_2H_4O_2$
60.06
O=COC
Formic acid, methyl ester
Formiate de methyle (French)
Methylester kyseliny mravenci (Czech)
Methyle (formiate de) (French)
Methyl formate (ACGIH, OSHA) [UN 1243]
Methylformiaat (Dutch)
Methylformiat (German)
Methyl methanoate
Metil (formiato di) (Italian)
Mravencan methylnaty (Czech)
UN 1243 [Methyl formate]

107-35-7
$C_2H_7NO_3S$
125.16
O=S(=O)(O)CCN
Taurine
2-Aminoethanesulfonic acid
Ethanesulfonic acid, 2-amino-
NCI-C60606

107-37-9
$C_3H_5Cl_3Si$
175.52
Silane, allyltrichloro
Allyl trichlorosilane (DOT)
Allyltrichlorosilane, Stabilized [UN 1724]
Silane, trichloroallyl-
UN 1724 [Allyltrichlorosilane, stabilized]

107-39-1
C_8H_{16}
112.24
C(=C)(CC(C)(C)C)C
1-Pentene, 2,4,4-trimethyl
2,4,4-Trimethyl-1-pentene

107-40-4
C_8H_{16}
112.22
C(=CC(C)(C)C)(C)C
2,4,4-Trimethyl-2-pentene
AI3-16047
Diisobutylene
β-Diisobutylene
2-Pentene, 2,4,4-trimethyl-
2,4,4-Trimethylpentene-2
2,2,4-Trimethyl-3-pentene

107-41-5
$C_6H_{14}O_2$
118.20
OC(CC(O)(C)C)C
2,4-Pentanediol, 2-methyl
2,4-Dihydroxy-2-methylpentane
Diolane
1,2-Hexanediol
Hexylene glycol (ACGIH, OSHA)
Isol
2-Methyl pentane-2,4-diol
2-Methyl-2,4-pentanediol
Pinakon
α,α,α'-Trimethyltrimethylene glycol

107-43-7
$C_5H_{11}NO_2$
117.17
O=C(O)CN(C)(C)C
Betaine
Abromine
(Carboxymethyl)trimethyl-ammonium hydroxide, inner salt
α-Earleine
Glycine betaine
Glycocoll betaine
Glycylbetaine
Glykokollbetain (German)
Jortaine
Loramine AMB 13
Lycine
Oxyneurine
Rubrine C
Trimethylglycine
Trimethylglycocoll

107-44-8
$C_4H_{10}FO_2P$
140.11
CC(C)OP(C)(F)=O
Phosphonofluoridic acid, methyl-, isopropyl ester
EA 1208
GB
IMPF
Isopropoxymethylphosphoryl fluoride
Isopropylester kyseliny methyl-fluorfosfonove (Czech)
Isopropyl methanefluorophos-phonate
Isopropyl methylfluorophos-phate
Isopropyl methylphosphono-fluoridate
o-Isopropyl methylphosphono-fluoridate
Isopropyl-methyl-phosphoryl fluoride
Methylfluorphosphorsaeure-isopropylester (German)
Methylphosphonofluoridic acid isopropyl ester
Methylphosphonofluoridic acid 1-methylethyl ester
MFI
Phosphine oxide, fluoroiso-propoxymethyl-
Phosphoric acid, methylfluoro-, isopropyl ester
Sarin
Sarin II
T-144
T-2106
TL 1618
Trilone 46

107-45-9

$C_8H_{19}N$
129.28
NC(CC(C)(C)C)(C)C
Butylamine, bis(1,3-dimethyl)
1,1,3,3-Tetramethylbutylamine

107-46-0
$C_6H_{18}OSi_2$
162.42
CSi(C)(C)OSi(C)(C)C
Disiloxane, hexamethyl
DC 200 Fluid
Dow Corning 200
Hexamethyldisiloxane
Oxybis(trimethylsilane)
Silane, oxybis(trimethyl-

107-47-1
$C_8H_{18}S$
146.30
S(C(C)(C)C)C(C)(C)C
Propane, 2,2'-thiobis(2-methyl- (9CI)
tert-Butyl sulfide (8CI)
Di-tert-butyl sulfide
NSC-4549
2,2,4,4-Tetramethyl-3-thiapentane
2,2'-Thiobis(2-methylpropane)

107-49-3
$C_8H_{20}O_7P_2$
290.22
CCOP(=O)(OCC)OP(=O)(OCC)OCC
Pyrophosphoric acid, tetraethyl ester
Bis-O,O-diethylphosphoric anhydride
Bladan
Diphosphoric acid tetraethyl ester
EA 1285
ENT 18,771
Ethyl pyrophosphate, tetra-
Fosvex
Grisol
Hept
Hexamite
Killax
Kilmite 40
Lethalaire G-52
Lirohex
Mortopal
NA 2783
Nifos
Nifos T
Nifost
Pyrophosphate de tetraethyle (French)
RCRA waste number P111
TEP
TEPP (ACGIH,OSHA)
O,O,O,O-Tetraethyl-difosfaat (Dutch)
Tetraethyldifosfat (Czech)
Tetraethyl diphosphate
O,O,O,O-Tetraetil-pirofosfato (Italian)
Tetraethyl pyrofosfaat (Dutch)
Tetraethylpyrofosfat (Czech)
Tetraethylpyrophosphate
Tetraethyl pyrophosphate, Liquid (DOT)
Tetraethyl pyrophosphate Mixture, Dry (DOT)
Tetraethyl pyrophosphate Mixture, Liquid (DOT)
Tetraaethylpyrophosphorsaeureester (German)
Tetrastigmine
Tetron
Tetron-100
Vapotone

107-50-6
$C_{14}H_{42}O_7Si_7$
519.08
Cycloheptasiloxane, tetradecamethyl- (8CI,9CI)

107-51-7
$C_8H_{24}O_2Si_3$
236.53
CSi(C)(C)OSi(C)(C)OSi(C)(C)C
Trisiloxane, octamethyl- (9CI)
Octamethyltrisiloxane

107-52-8
$C_{14}H_{42}O_5Si_6$
459.00
CSi(C)(C)OSi(C)(C)OSi(C)(C)OSi(C)(C)OSi(C)(C)OSi(C)(C)C
Hexasiloxane, tetradecamethyl- (9CI)
Tetradecamethylhexasiloxane

107-54-0
$C_8H_{14}O$
126.20
OC(C#C)(CC(C)C)C
1-Hexyn-3-ol, 3,5-dimethyl- (9CI)
AI3-23126
3,5-Dimethyl-1-hexyn-3-ol
NSC-978
Surfynol (VAN)
Surfynol 61

107-55-1
$C_8H_{19}O_2PS_2$
242.34
O(P(OC(CC)C)(S)=S)C(CC)C
Phosphorodithioic acid, O,O-bis(1-methylpropyl) ester (9CI)

107-56-2
$C_6H_{15}O_2PS_2$
214.29
O(P(OC(C)C)(S)=S)C(C)C
O,O-Diisopropyl dithiophosphate
O,O-Diisopropyl dithiophosphoric acid
O,O-Diisopropyl hydrogen dithiophosphate
O,O-Diisopropyl hydrogen phosphorodithioate
Diisopropyl phosphorodithioate
O,O-Diisopropyl phosphorodithioate
O,O-Diisopropylphosphorodithioic acid
Isopropyl aerofloat
NSC-15258
Phosphorodithioic acid, O,O-bis(1-methylethyl) ester (9CI)
Phosphorodithioic acid, O,O-diisopropyl ester (8CI)

107-58-4
$C_7H_{13}NO$
127.21
O=C(NC(C)(C)C)C=C
Acrylamide, N-tert-butyl
N-tert-Butylacrylamide
2-Propenamide, N-(1,1-dimethylethyl)- (9CI)

107-64-2
$C_{38}H_{80}N.Cl$
586.64
Ammonium, dimethyldioctadecyl-, chloride
Aliquat 207
Arosurf TA 100
Arquad R 40
Dimethyldioctadecylammonium chloride
Distearyl dimethylammonium chloride
Genamin DSAC
KD 83
1-Octadecanaminium, N,N-dimethyl-N-octadecyl-, chloride (9CI)
Q-D 86P
Quaternium-5
Talofloc
Varisoft 100

107-66-4
$C_8H_{19}O_4P$
210.24
O=P(OCCCC)(OCCCC)O
Phosphoric acid, dibutyl ester
Dibutyl acid phosphate
Dibutyl hydrogen phosphate
Dibutyl phosphate (ACGIH, OSHA)
Di-n-butyl phosphate

107-68-6
$C_3H_9NO_3S$
139.17
Ethanesulfonic acid, 2-(methylamino)- (9CI)
N-Methyltaurine
NSC-10479
Taurine, N-methyl- (8CI)

107-70-0
$C_7H_{14}O_2$
130.21
O=C(CC(OC)(C)C)C
2-Pentanone, 4-methoxy-4-methyl
4-Methoxy-4-methyl-2-pentanone
4-Methoxy-4-methylpentan-2-one [UN 2293]
UN 2293 [4-Methoxy-4-methylpentan-2-one]

107-71-1
$C_6H_{12}O_3$
132.18
O=C(OOC(C)(C)C)C
Peroxyacetic acid, t-butyl ester
t-Butyl peracetate
tert-Butyl peracetate
t-Butyl peroxyacetate
tert-Butyl peroxyacetate
tert-Butyl peroxyacetate, More than 76% in solution (DOT)
tert-Butyl peroxyacetate, More than 52% to a maximum concentration of 76% (UN 2095 DOT)
tert-Butyl peroxyacetate, Not more than 52% in solution (UN 2096 DOT)
1,1-Dimethylethyl ethaneperoxoate
Ethaneperoxoic acid, 1,1-dimethylethyl ester (9CI)
Lupersol 70
NSC-118417
Peroxiacetato de terc-butilo (Spanish)
Peroxyacetate de tert-butyle (French)
Peroxyacetic acid, tert-butyl ester (8CI)
Peroxyacetic acid, tert-butyl ester, More than 52% to a maximum concentration of 76%
Peroxyacetic acid, tert-butyl ester, More than 76% in solution
Peroxyacetic acid, tert-butyl ester, Not more than 52% in solution
Trigonox F-C50
UN 2095 (DOT)
UN 2096 (DOT)

107-72-2
$C_5H_{11}Cl_3Si$
205.60
CCCCCSi(Cl)(Cl)Cl
Silane, pentyltrichloro
Amyltrichlorosilane [UN 1728]
Amyl trichlorosilane
Pentyltrichlorosilane
Silane, trichloropentyl-
UN 1728 [Amyltrichlorosilane]

107-74-4
$C_{10}H_{22}O_2$
174.32
OCCC(CCCC(O)(C)C)C
1,2-Octanediol, 3,7-dimethyl
Citronellol, hydroxy-
3,7-Dimethyl-1,7-octanediol
Hydroxycitronellol
7-Hydroxy-3,7-dimethyloctan-1-ol
1-Octanol, 3,7-dimethyl-7-hydroxy-

107-75-5
$C_{10}H_{20}O_2$
172.30
O=CCC(CCCC(O)(C)C)C
1-Octanal, 3,7-dimethyl-7-hydroxy
Citronellal hydrate
Citronellal, hydroxy-
Cyclalia
Cyclosia
3,7-Dimethyl-7-hydroxyoctanal
Fixol
Hydroxycitronellal
7-Hydroxycitronellal
7-Hydroxy-3,7-dimethyloctan-1-al
Laurine
Lilyl aldehyde
Musuet synthetic
Musuettine principle
Octanal, 7-hydroxy-3,7-dimethyl-
Phixia

107-80-2
$C_4H_8Br_2$
215.92
BrCCC(Br)C
Butane, 1,3-dibromo- (9CI)
1,3-Dibromobutane

107-81-3
$C_5H_{11}Br$
151.07
BrC(CCC)C
Pentane, 2-bromo
2-Bromopentane [UN 2343]
UN 2343 [2-Bromopentane]

107-82-4
$C_5H_{11}Br$
151.07
BrCCC(C)C
Butane, 1-bromo-3-methyl
1-Bromo-3-methylbutane [UN 2341]
Isoamyl bromide
Isopentyl bromide
3-Methylbutyl bromide
UN 2341 [1-Bromo-3-methylbutane]

107-83-5
C_6H_{14}
86.20
C(CCC)(C)C
Pentane, 2-methyl
Isohexane
Isohexane [UN 1208]
2-Methylpentane (DOT)
UN 1208 [Hexanes]
UN 2462 (DOT)

107-84-6
$C_5H_{11}Cl$
106.60
C(CCCl)(C)C
Isoamyl chloride
Butane, 1-chloro-3-methyl- (9CI)
1-Chloro-3,3-dimethylpropane
1-Chloro-3-methylbutane
4-Chloro-2-methylbutane
Isopentyl chloride
3-Methylbutyl chloride
NSC-6528

107-85-7
$C_5H_{13}N$
87.16
NCCC(C)C
Isoamylamine
AI3-24040
1-Amino-3-methylbutane
1-Butanamine, 3-methyl- (9CI)
Butylamine, 3-methyl-
γ-Isoamylamine
Isopentylamine (8CI)
Leucamine
3-Methylbutanamine
3-Methyl-1-butanamine
Monoisoamylamine
NSC-7907

107-87-9
$C_5H_{10}O$
86.15
O=C(CCC)C
2-Pentanone
Ethyl acetone
Methyl-propyl-cetone (French)
Methylpropyl ketone
Methyl-n-propyl ketone
Methyl propyl ketone (ACGIH, OSHA) [UN 1249]
Metylopropyloketon (Polish)
MPK
2-Pentanone (OSHA)
UN 1249 [Methyl propyl ketone]

107-88-0
$C_4H_{10}O_2$
90.14
OCCC(O)C
1,3-Butanediol
1,3-Butandiol (German)
Butane-1,3-diol
β-Butylene glycol
1,3-Butylene glycol
1,3-Butylenglykol (German)

1,3-Dihydroxybutane
Methyltrimethylene glycol

107-89-1
$C_4H_8O_2$
88.12
O=CCC(O)C
Butyraldehyde, 3-hydroxy
Acetaldol
Aldol [UN 2839]
3-Butanolal
3-Hydroxybutanal
β-Hydroxybutyraldehyde
3-Hydroxybutyraldehyde
Oxybutanal
Oxybutyric aldehyde
UN 2839 [Aldol]

107-91-5
$C_3H_4N_2O$
84.09
O=C(N)CC#N
Acetamide, 2-cyano
Amid kyseliny kyanoctove (Czech)
CAA
Cyanacetamide
Cyanoacetamide
α-Cyanoacetamide
2-Cyanoacetamide
Cyanoiminoacetic acid
Kyanacetamid (Czech)
Malonamide nitrile
Malonamonitrile
Nitrilomalonamide
3-Nitrilo-propionamide
USAF KF-14

107-92-6
$C_4H_8O_2$
88.12
O=C(O)CCC
Butyric-acid
Butanic acid
Butanoic acid
Buttersaeure (German)
Butyric acid [UN 2820]
n-Butyric acid (DOT)
Ethylacetic acid
Kyselina maselna (Czech)
1-Propanecarboxylic acid
Propylformic acid
UN 2820 [Butyric acid]

107-93-7
$C_4H_6O_2$
86.09
O=C(O)C=CC
Crotonic acid, (E)- (8CI)
2-Butenoic acid, (E)- (9CI)
(E)-2-Butenoic acid
trans-2-Butenoic acid
Crotonic acid
trans-Crotonic acid
(E)-Crotonic acid
NSC-8751

107-94-8
$C_3H_5ClO_2$
108.53
O=C(O)CCCl
Propionic acid, 3-chloro
3-Chloropropanoic acid
3-Chloropropionic acid
β-Chloropropionic acid
β-Monochloropropionic acid

107-95-9
$C_3H_7NO_2$
89.09
O=C(O)CCN
3-Aminopropionic acid
Abufene
AI3-18470
Alanine, β
β-Alanine (9CI)
3-Aminopropanoic acid
β-Aminopropionic acid
NSC-7603
Propanoic acid, 3-amino-

107-96-0
$C_3H_6O_2S$
106.15
O=C(O)CCS
Propionic acid, 3-mercapto
3-Mercaptopropionic acid
3MPA

107-97-1
$C_3H_7NO_2$
89.11
O=C(O)CNC
Sarcosine
Acetic acid, (methylamino)-
Glycine, N-methyl-
(Methylamino)acetic acid
N-Methylaminoacetic acid
(Methylamino)ethanoic acid
Methylglycine
N-Methylglycine
Sarcosin
Sarcosinic acid

107-98-2
$C_4H_{10}O_2$
90.14
O(CC(O)C)C
2-Propanol, 1-methoxy
Dowanol 33B
Dowanol PM
Dowanol PM Glycol Ether
Dowtherm 209
Glycol Ether PM
Methoxy ether of propylene glycol
1-Methoxy-2-propanol
Poly-Solve MPM
Propasol Solvent M
Propylene glycol methyl ether
Propylene glycol monomethyl ether (ACGIH,OSHA)
α-Propylene glycol monomethyl ether
Propylenglykol-monomethylaether (German)
UCAR Solvent LM (Obs.)

108-01-0
$C_4H_{11}NO$
89.16
OCCN(C)C
Ethanol, 2-dimethylamino
Deanol
Dimethylaethanolamin (German)
Dimethylaminoaethanol (German)
Dimethylaminoethanol
β-Dimethylaminoethanol
N-Dimethylaminoethanol
N,N-Dimethylaminoethanol
2-Dimethylaminoethanol
2-(Dimethylamino)ethanol
β-Dimethylaminoethyl alcohol
Dimethylethanolamine [UN 2051]
N,N-Dimethylethanolamine
N,N-Dimethyl-N-(2-hydroxyethyl)amine
N,N-Dimethyl-2-hydroxyethylamine
DMAE
β-Hydroxyethyldimethylamine
UN 2051 [Dimethylethanolamine]

108-03-2
$C_3H_7NO_2$
89.11
O=N(=O)CCC
Propane, 1-nitro
1-Nitropropane (ACGIH,OSHA)
1-NP

108-05-4
$C_4H_6O_2$
86.10
O=C(OC=C)C
Acetic acid, vinyl ester
Acetate de vinyle (French)
Acetic acid, ethenyl ester
Acetic acid, ethylene ether
1-Acetoxyethylene
Ethanoic acid, ethenyl ester
Ethenyl acetate
Ethenyl ethanoate
Octan winylu (Polish)
UN 1301 [Vinyl acetate, inhibited]
Vac
Vinile (acetato di) (Italian)
Vinylacetaat (Dutch)
Vinylacetat (German)
Vinyl acetate (ACGIH,DOT, OSHA)
Vinyl acetate, Inhibited [UN 1301]
Vinyl acetate H.Q.
Vinyl A Monomer
Vinyle (acetate de) (French)
Vinylester kyseliny octove (Czech)
Vinyl ethanoate

VYAC
Zeset T

108-08-7
C_7H_{16}
100.20
C(CC(C)C)(C)C
2,4-Dimethylpentane
NSC-61989
Pentane, 2,4-dimethyl- (9CI)

108-09-8
$C_6H_{15}N$
101.22
NC(CC(C)C)C
Butylamine, 1,3-dimethyl
1,3-Dimethyl butylamine
1,3-Dimethylbutylamine [UN 2379]
UN 2379 [1,3-Dimethylbutylamine]

108-10-1
$C_6H_{12}O$
100.18
O=C(CC(C)C)C
2-Pentanone, 4-methyl
Hexon (Czech)
Hexone
Hexone (OSHA)
Isobutyl-methylketon (Czech)
Isobutyl methyl ketone
Isopropylacetone
Ketone, isobutyl methyl
Methyl-isobutyl-cetone (French)
Methylisobutylketon (Dutch, German)
Methyl isobutyl ketone (ACGIH,OSHA) [UN 1245]
Metyloizobutyloketon (Polish)
4-Methyl-pentan-2-on (Dutch, German)
2-Methyl-4-pentanone
4-Methyl-2-pentanon (Czech)
4-Methyl-2-pentanone
Metilisobutilchetone (Italian)
4-Metilpentan-2-one (Italian)
MIBK
MIK
RCRA waste number U161
Shell MIBK
UN 1245 [Methyl isobutyl ketone]

108-11-2
$C_6H_{14}O$
102.20
OC(CC(C)C)C
2-Pentanol, 4-methyl
Alcool methyl amylique (French)
Isobutylmethylcarbinol
Isobutylmethylmethanol
MAOH
Methyl amyl alcohol
Methyl amyl alcohol (OSHA)
Methylisobutyl carbinol
Methyl isobutyl carbinol (ACGIH,OSHA) [UN 2053]
Methyl-isobutylkarbinol (Czech)
2-Methyl-4-pentanol
4-Methylpentanol-2
4-Methyl-2-pentanol
Metilamil alcohol (Italian)
4-Metilpentan-2-olo (Italian)
MIBC
MIC
3-MIC
4-Pentanol, 2-methyl-
UN 2053 [Methyl isobutyl carbinol]

108-16-7
$C_5H_{14}N_2O$
118.21
OC(C)CN(C)C
2-Propanol, 1-(dimethylamino)
1,1-Dimethylaminopropanol-2
1,1-Dimethylaminopropan-2-ol
Dimethyl(2-hydroxypropyl)-amine
Dimethylisopropanolamine

108-18-9
$C_6H_{15}N$
101.22
N(C(C)C)C(C)C
Diisopropylamine (ACGIH, OSHA) [UN 1158]
DIPA
N-(1-Methylethyl)-2-propanamine
2-Propanamine, N-(1-methylethyl)-
UN 1158 [Diisopropylamine]

108-19-0
$C_2H_5N_3O_2$
103.06
O=C(NC(=O)N)N
Biuret (8CI)
AI3-14905
Allophanamide
Allophanic acid amide
Allophanimidic acid (VAN)
Carbamoylurea
Carbamylurea
Caswell No. 106A
Caswell No. 159A
Dicarbamylamine
EPA Pesticide Chemical Code 206200
Imidodicarbonic diamide (9CI)
Isobiuret (VAN)
NSC-8020
Urea, (aminocarbonyl)-
Ureidoformamide

108-20-3
$C_6H_{14}O$
102.20
O(C(C)C)C(C)C
Propane, 2,2'-oxybis
Diisopropyl ether
Diisopropyl ether [UN 1159]
Diisopropyl oxide
Ether, isopropyl
Ether isopropylique (French)
2-Isopropoxypropane
Isopropyl ether (ACGIH,OSHA)
Izopropylowy eter (Polish)
UN 1159 [Diisopropyl ether]

108-21-4
$C_5H_{10}O_2$
102.15
O=C(OC(C)C)C
Acetic acid, isopropyl ester
Acetate d'isopropyle (French)
Acetic acid, 1-methylethyl ester (9CI)
2-Acetoxypropane
Isopropile (acetato di) (Italian)
Isopropylacetaat (Dutch)
Isopropylacetat (German)
Isopropyl acetate (ACGIH, OSHA) [UN 1220]
Isopropyl (acetate d') (French)
Isopropylester kyseliny octove (Czech)
2-Propyl acetate
UN 1220 [Isopropyl acetate]

108-22-5
$C_5H_8O_2$
100.13
O=C(OC(=C)C)C
1-Propen-2-ol, acetate
Acetic acid, isopropenyl ester
Isopropenyl acetate [UN 2403]
Isopropenylester kyseliny octove (Czech)
Methylvinyl acetate
1-Propen-2-yl acetate
UN 2403 [Isopropenyl acetate]

108-23-6
$C_4H_7ClO_2$
122.56
O=C(OC(C)C)Cl
Formic acid, chloro-, isopropyl ester
Carbonochloride acid, 1-methylethyl ester
Chloroformic acid isopropyl ester
Isopropyl chlorocarbonate
Isopropyl chloroformate [UN 2407]
Isopropyl chloromethanoate
Isopropylester kyseliny chlormravenci (Czech)
UN 2407 [Isopropyl chloroformate]

108-24-7
$C_4H_6O_3$
102.10
O=C(OC(=O)C)C
Acetic-anhydride
Acetanhydride
Acetic acid, anhydride (9CI)
Acetic anhydride (ACGIH,-OSHA) [UN 1715]

Acetic oxide
Acetyl anhydride
Acetyl ether
Acetyl oxide
Anhydride acetique (French)
Anhydrid kyseliny octove (Czech)
Anidride acetica (Italian)
Azijnzuuranhydride (Dutch)
Essigsaeureanhydrid (German)
Ethanoic anhydrate
Octowy bezwodnik (Polish)
UN 1715 [Acetic anhydride]

108-25-8
$C_4H_8OS_2$
136.24
O-Isopropyl xanthate
Isopropylxanthic acid

108-26-9
$C_4H_6N_2O$
98.12
O=C(NN=C1C)C1
2-Pyrazolin-5-one, 3-methyl
3-Methyl-2-pyrazolin-5-one
3-Methyl-pyrazolon-(5) (German)

108-29-2
$C_5H_8O_2$
100.13
O=C(OC(C1)C)C1
2(3H)-Furanone, dihydro-5-methyl
4-Hydroxypentanoic acid lactone
4-Hydroxyvaleric acid lactone
γ-Methyl-γ-butyrolactone
4-Methyl-γ-butyrolactone
γ-Pentalactone
4-Pentanolide
γ-Valerolactone
4-Valerolactone
γ-Valerolakton (Czech)

108-30-5
$C_4H_4O_3$
100.08
O=C(OC(=O)C1)C1
Succinic-anhydride
Bernsteinsaure-anhydrid (German)
Butanedioic anhydride
Dihydro-2,5-furandione
2,5-Diketotetrahydrofuran
2,5-Furandione, dihydro-
NCI-C55696
Succinic acid anhydride
Succinyl oxide
Tetrahydro-2,5-dioxofuran

108-31-6
$C_4H_2O_3$
98.06
O=C(OC(=O)C=1)C1
Maleic-anhydride
Anhydrid kyseliny maleinove (Czech)
cis-Butenedioic anhydride
2,5-Furandione
Maleic acid anhydride
Maleic anhydride (ACGIH, DOT,OSHA)
Maleic anhydride, Solid or molten (DOT)
Maleinanhydrid (Czech)
Toxilic Anhydride
RCRA waste number U147
UN 2215 (DOT)

108-32-7
$C_4H_6O_3$
102.10
O=C(OCC1C)O1
Carbonic acid, cyclic propylene ester
Carbonic acid cyclic methylethylene ester
Cyclic methylethylene carbonate
Cyclic propylene carbonate
Cyclic 1,2-propylene carbonate
1,3-Dioxolan-2-one, 4-methyl-
Dipropylene carbonate
4-Methyldioxalone-2
4-Methyl-1,3-dioxolan-2-one
1-Methylethylene carbonate
4-Methyl-2-oxo-1,3-dioxolane
1,2-PDC
1,2-Propanediol carbonate
1,2-Propanediol cyclic carbonate
1,2-Propanediyl carbonate
Propylene carbonate
1,2-Propylene carbonate
Propylene glycol cyclic carbonate
Propylenester kyseliny uhlicite (Czech)

108-34-9
$C_8H_{15}N_2O_4P$
234.22
Phosphoric acid, diethyl 5(or 3)-methylpyrazole-3(or 5)-yl ester
O,O-Diaethyl-O-(3-methyl-1H-pyrazol-5-yl)-phosphat (German)
O,O-Diethyl-O-(3-methyl-1H-pyrazol-5-yl)-fosfaat (Dutch)
Diethyl 3-methyl-5-pyrazolyl phosphate
O,O-Diethyl O-(3-methyl-5-pyrazolyl) phosphate
O,O-Dietil-O-(3-metil-1H-pirazol-5-il)-fosfato (Italian)
ENT 24,723
G-24483
Geigy G-24483
Methylpyrazolyl diethylphosphate
3-Methylpyrazolyl-5-diethylphosphate
Phosphate de diethyle et de 3-methyl-5-pyrazolyle (French)
Phosphoric acid, diethyl-(3-methyl-5-pyrazolyl) ester
Phosphoric acid, diethyl-, 3-methylpyrazole-5-yl ester
Pirazoxon (Italian)
Pyrazol-3(or 5)-ol, 5(or 3)-methyl-, diethyl phosphate
Pyrazoxon
Pyrazoxone

108-35-0
$C_8H_{15}N_2O_3PS$
250.28
CCOP(=S)(OCC)Oc1cc(C)n[nH]1
Phosphorothioic acid, O,O-diethyl-O-(5-methyl-3-pyrazolyl) ester
ENT 25,557-X
G-24027
Geigy G-24027
O,O-Diethyl O-(3-methyl-5-pyrazolyl) phosphorothionate
Diethyl 3-methyl-5-pyrazolyl thionophosphate
Methylpyrazolyl diethylthiophosphate
Phosphorothioic acid, O,O-diethyl O-(3-methyl-1H-pyrazol-5-yl) ester
Pyrazothion

108-36-1
$C_6H_4Br_2$
235.92
c(cccc1Br)(c1)Br
Benzene, m-dibromo
Benzene, 1,3-dibromo- (9CI)
m-Dibromobenzene
1,3-Dibromobenzene

108-37-2
C_6H_4BrCl
191.46
c(cccc1Br)(c1)Cl
Benzene, 1-bromo-3-chloro
m-Bromochlorobenzene
1-Bromo-3-chlorobenzene
3-Bromochlorobenzene
m-Bromophenyl chloride
m-Chlorobromobenzene
1-Chloro-3-bromobenzene
3-Chlorobromobenzene
m-Chlorophenyl bromide
3-Chlorophenyl bromide

108-38-3
C_8H_{10}
106.18
c(cccc1C)(c1)C
m-Xylene (ACGIH,OSHA) [UN 1307]
m-Dimethylbenzene
1,3-Dimethylbenzene
UN 1307 [Xylenes]
1,3-Xylene

m-Xylol (DOT)

108-39-4
C_7H_8O
108.15
Oc(cccc1C)c1
m-Cresol (OSHA) [UN 2076]
3-Cresol
m-Cresole
m-Cresylic acid
1-Hydroxy-3-methylbenzene
m-Hydroxytoluene
3-Hydroxytoluene
m-Kresol
m-Methylphenol
3-Methylphenol
m-Oxytoluene
Phenol, 3-methyl- (9CI)
RCRA waste number U052
m-Toluol
UN 2076 [Cresols (o-; m-; p-)]

108-40-7
C_7H_8S
124.21
Sc(cccc1C)c1
m-Toluenethiol
Benzenethiol, 3-methyl- (9CI)
m-Mercaptotoluene
m-Methylbenzenethiol
3-Methylbenzenethiol
m-Methylthiophenol
3-Methylthiophenol
m-Thiocresol
3-Thiocresol
m-Tolylmercaptan
USAF EK-2680

108-41-8
C_7H_7Cl
126.59
c(cccc1Cl)(c1)C
Toluene, m-chloro
Benzene, 1-chloro-3-methyl-
m-Chlorotoluene [UN 2238]
3-Chlorotoluene
m-Tolyl chloride
UN 2238 [Chlorotoluenes]

108-42-9
C_6H_6ClN
127.58
Nc(cccc1Cl)c1
Aniline, m-chloro
m-Aminochlorobenzene
1-Amino-3-chlorobenzene
Benzenamine, 3-chloro- (9CI)
3-Chlooranilinen (Dutch)
m-Chloraniline
m-Chloroaniline
3-Chloroaniline
m-Chloroaniline, Liquid [UN 2019]
m-Chloroaniline, Solid [UN 2018]
3-Chlorobenzenamine
m-Chlorophenylamine
3-Chlorophenylamine
3-Cloroaniline (Italian)
Fast Orange GC Base
Orange GC Base
UN 2018 [Chloroanilines, solid]
UN 2019 [Chloroanilines, Liquid]

108-43-0
C_6H_5ClO
128.56
Oc(cccc1Cl)c1
Phenol, m-chloro
m-Chlorophenol
3-Chlorophenol
m-Chlorophenol, Liquid [UN 2021]
m-Chlorophenol, Solid [UN 2020]
UN 2020 [Chlorophenols, solid]
UN 2021 [Chlorophenols, Liquid]

108-44-1
C_7H_9N
107.17
Nc(cccc1C)c1
m-Toluidine (ACGIH,OSHA) [UN 1708]
3-Amino-1-methylbenzene
3-Aminophenylmethane
3-Aminotoluen (Czech)
m-Aminotoluene
3-Aminotoluene
Aniline, 3-methyl-
m-Methylaniline
3-Methylaniline
m-Methylbenzenamine
3-Methylbenzenamine
m-Toluidin (Czech)
3-Toluidine
m-Tolylamine
UN 1708 [Toluidines, liquid or solid]

108-45-2
$C_6H_8N_2$
108.16
Nc(cccc1N)c1
m-Phenylenediamine
m-Aminoaline
3-Aminoaniline
APCO 2330
m-Benzenediamine
1,3-Benzenediamine
C.I. 76025
Developer 11
Developer C
Developer H
Developer M
m-Diaminobenzene
1,3-Diaminobenzene
Direct Brown BR
Direct Brown GG
m-Fenylendiamin (Czech)
Metaphenylenediamine
1,3-Phenylenediamine
m-Phenylenediamine [UN 1673]
Phenylenediamine, meta, Solid (DOT)
UN 1673 [Phenylenediamines (o-; m-; p-;)]

108-46-3
$C_6H_6O_2$
110.12
Oc(cccc1O)c1
Resorcinol (ACGIH,OSHA) [UN 2876]
Benzene, m-dihydroxy-
m-Benzenediol
1,3-Benzenediol
C.I. 76505
C.I. Developer 4
C.I. Oxidation Base 31
Developer O
Developer R
Developer RS
m-Dihydroxybenzene
1,3-Dihydroxybenzene
m-Dioxybenzene
Durafur Developer G
Fouramine RS
Fourrine 79
Fourrine EW
m-Hydroquinone
3-Hydroxycyclohexadien-1-one
m-Hydroxyphenol
3-Hydroxyphenol
Nako TGG
NCI-C05970
Pelagol Grey RS
Pelagol RS
Phenol, m-hydroxy-
RCRA waste number U201
Resorcin
Resorcine
UN 2876 [Resorcinol]

108-47-4
C_7H_9N
107.17
n(c(cc(c1)C)C)c1
2,4-Lutidine
α,γ-Dimethylpyridine
2,4-Dimethylpyridine
Pyridine, 2,4-dimethyl- (9CI)

108-48-5
C_7H_9N
107.17
n(c(ccc1)C)c1C
2,6-Lutidine
2,6-Dimethylpyridine
α,α'-Dimethylpyridine
2,6-Dimethypyridine
α,α'-Lutidine
Pyridine, 2,6-dimethyl- (9CI)

108-50-9
$C_6H_8N_2$
108.16
n(cc(nc1C)C)c1
Pyrazine, 2,6-dimethyl
2,6-Dimethylpyrazine

108-52-1

$C_5H_7N_3$
109.15
Cc1ccnc(N)n1
Pyrimidine, 2-amino-4-methyl
2-Amino-4-methylpyrimidine

108-55-4
$C_5H_6O_3$
114.11
O=C(OC(=O)CC1)C1
Glutaric-anhydride
Anhydrid kyseliny glutarove (Czech)
2H-Pyran-2,6(3H)-dione, dihydro-

108-57-6
$C_{10}H_{10}$
130.20
c(cccc1C=C)(c1)C=C
Benzene, m-divinyl
m-Divinylbenzen (Czech)
Divinylbenzene (ACGIH, OSHA)
m-Divinylbenzene
m-Vinylstyrene

108-59-8
$C_5H_8O_4$
132.13
O=C(OC)CC(=O)OC
Malonic acid, dimethyl ester
Dimethyl malonate
Dimethyl propanedioate
Methyl malonate
Propanedioic acid, dimethyl ester (9CI)

108-60-1
$C_6H_{12}Cl_2O$
171.08
O(C(CCl)C)C(CCl)C
Ether, bis(2-chloro-1-methylethyl)
Bis(β-chloroisopropyl)ether
Bis(2-chloroisopropyl) ether
Bis(2-chloro-1-methylethyl) ether
Bis(1-chloro-2-propyl) ether
(2-Chloro-1-methylethyl) ether
DCIP
DCIP (Nematocide)
Dichlorodiisopropyl ether
β,β'-Dichlorodiisopropyl ether
Dichloroisopropyl ether
2,2'-Dichloroisopropyl ether
Dichloroisopropyl ether [UN 2490]
NCI-C50044
Nemamort
2,2'-Oxybis(1-chloropropane)
Propane, 2,2'-oxybis(1-chloro-
RCRA waste number U027
UN 2490 [Dichloroisopropyl ether]

108-62-3
$C_8H_{16}O_4$
176.21
Metaldehyde
Acetaldehyde, tetramer
AI3-15376
Antimilace
Ariotox
Caswell No. 548
Cekumeta
Corry's Slug Death
EPA Pesticide Chemical Code 053001
Halizan
META
Metacetaldehyde
Metaldehyd (German)
Metaldehyde (VAN)
Metaldeide (Italian)
Metason
Namekil
Slug-Tox
2,4,6,8-Tetramethyl-1,3,5,7-tetroxocane
r-2,c-4,c-6,c-8-Tetramethyl-1,3,5,7-tetroxocane
1,3,5,7-Tetroxocane, 2,4,6,8-tetramethyl-
UN 1332 [Metaldehyde]

108-63-4
$C_{22}H_{42}O_4$
370.57
O=C(OC(CCCCCC)C)CCCCC(=O)OC(CCCCCC)C
Adipic acid, bis(1-methylheptyl) ester (8CI)
Bis(1-methylheptyl) hexanedioate
Hexanedioic acid, bis(1-methylheptyl) ester (9CI)
NSC-6197

108-64-5
$C_7H_{14}O_2$
130.21
O=C(OCC)CC(C)C
Isovaleric acid, ethyl ester
Butanoic acid, 3-methyl-, ethyl ester
Butyric acid, 3-methyl-, ethyl ester
Ethyl isovalerate

108-65-6
$C_6H_{12}O_3$
132.18
Acetic acid, 2-methoxy-1-methylethyl ester
Propyleneglycol monomethyl ether acetate

108-67-8
C_9H_{12}
120.21
c(cc(cc1C)C)(c1)C
Mesitylene
Benzene, 1,3,5-trimethyl-
Fleet-X
TMB
sym-Trimethylbenzene
1,3,5-Trimethylbenzene [UN 2325]
Trimethyl benzene (ACGIH)
Trimethylbenzol
UN 2325 [1,3,5-Trimethylbenzene]

108-68-9
$C_8H_{10}O$
122.18
Oc(cc(cc1C)C)c1
3,5-Xylenol
3,5-Dimethylphenol
3,5-DMP
1,3,5-Xylenol

108-69-0
$C_8H_{11}N$
121.20
Nc(cc(cc1C)C)c1
3,5-Xylidine
3,5-Dimethylaniline
3,5-Dimethylbenzenamine
3,5-Dimethylphenylamine
3,5-Xylylamine

108-70-3
$C_6H_3Cl_3$
181.44
c(cc(cc1Cl)Cl)(c1)Cl
Benzene, 1,3,5-trichloro
s-Trichlorobenzene
sym-Trichlorobenzene
1,3,5-Trichlorobenzene
1,3,5-Trichlorobenzene, Liquid [UN 2321]
UN 2321 [Trichlorobenzenes, Liquid]

108-71-4
$C_7H_{10}N_2$
122.16
Nc(cc(N)cc1C)c1
Toluene-3,5-diamine
1,3-Benzenediamine, 5-methyl-(9CI)
1,3-Diamino-5-methylbenzene
3,5-Diaminotoluene
5-Methyl-1,3-benzenediamine
Toluene, 3,5-diamino-
3,5-Toluenediamine

108-73-6
$C_6H_6O_3$
126.12
Oc(cc(O)cc1O)c1
Phloroglucinol
Benzene, trihydroxy
Benzene, 1,3,5-trihydroxy-
Benzene-s-triol
Benzene-1,3,5-triol
1,3,5-Benzenetriol
3,5-Dihydroxyphenol
Dilospan S
Floroglucin (Czech)
Floroglucinol (Czech)

5-Hydroxyresorcinol
5-Oxyresorcinol
Phloroglucin
Phloroglucine
s-Trihydroxybenzene
sym-Trihydroxybenzene
1,3,5-Trihydroxybenzene
1,3,5-Trihydroxycyclohexatriene

108-74-7
$C_6H_{15}N_3$
129.24
N(CN(CN1C)C)(C1)C
s-Triazine, hexahydro-1,3,5-trimethyl
F 7771
Hexahydro-1,3,5-trimethyl-s-triazine
1,3,5-Triazine, hexahydro-1,3,5-trimethyl-
1,3,5-Trimethylhexahydro-sym-triazine
1,3,5-Trimethylhexahydro-1,3,5-triazine

108-75-8
$C_8H_{11}N$
121.20
n(c(cc(c1)C)C)c1C
Pyridine, 2,4,6-trimethyl
α,γ,α'-Collidine
γ-Collidine
s-Collidine
sym-Collidine
2,4,6-Collidine
2,4,6-Kollidin (Czech)
2,4,6-Trimethylpyridine

108-77-0
$C_3Cl_3N_3$
184.41
n(c(nc(n1)Cl)Cl)c1Cl
s-Triazine, 2,4,6-trichloro
Chlorotriazine
Cyanurchloride
Cyanuric acid chloride
Cyanuric chloride [UN 2670]
Cyanuric trichloride
Cyanuryl chloride
Kyanurchlorid (Czech)
s-Triazine trichloride
Trichlorocyanidine
Trichloro-s-triazine
sym-Trichlorotriazine
1,3,5-Trichlorotriazine
2,4,6-Trichlorotriazine
2,4,6-Trichloro-s-triazine
2,4,6-Trichloro-1,3,5-triazine
syn-Trichlotriazin (Czech)
Tricyanogen chloride
UN 2670 [Cyanuric chloride]

108-78-1
$C_3H_6N_6$
126.15
n(c(nc(n1)N)N)c1N
Melamine
Aero
Ammelide
Cyanuramide
Cyanuric triamide
Cyanurotriamide
Cyanurotriamine
Cyanurtriamide
Cymel
Hicophor PR
Isomelamine
NCI-C50715
Teoharn
Theoharn
TR
2,4,6-Triamino-s-triazine
2,4,6-Triamino-1,3,5-triazine
1,3,5-Triazine-2,4,6-triamine
s-Triazine, 2,4,6-triamino-
Virset 656-4

108-80-5
$C_3H_3N_3O_3$
129.09
n(c(nc(n1)O)O)c1O
s-triazine-2,4,6-triol
Cyanuric acid
Isocyanuric acid
Kyselina kyanurova (Czech)
Pseudocyanuric acid
sym-Triazinetriol
s-2,4,6-Triazinetriol
s-Triazine-2,4,6(1H,3H,5H)-trione
Tricyanic acid
Trihydroxycyanidine
2,4,6-Trihydroxy-1,3,5-triazine

108-82-7
$C_9H_{20}O$
144.29
OC(CC(C)C)CC(C)C
4-Heptanol, 2-6-dimethyl
Diisobutyl carbinol
2,6-Dimethyl-4-heptanol
2,6-Dimethyl heptanol-4
sec-Nonyl alcohol

108-83-8
$C_9H_{18}O$
142.27
O=C(CC(C)C)CC(C)C
4-Heptanone, 2,6-dimethyl
Diisobutilchetone (Italian)
Di-Isobutylcetone (French)
Diisobutylketon (Dutch, German)
Diisobutyl ketone (ACGIH, DOT,OSHA) [UN 1157]
s-Diisopropylacetone
sym-Diisopropylacetone
2,6-Dimethyl-heptan-4-on (Dutch, German)
2,6-Dimethylheptan-4-one
2,6-Dimethyl-4-heptanone
2,6-Dimethyl-4-heptanone (OSHA)
2,6-Dimetil-eptan-4-one (Italian)
Isobutyl ketone
Isovalerone
UN 1157 [Diisobutyl ketone]
Valerone

108-84-9
$C_8H_{16}O_2$
144.24
O=C(OC(CC(C)C)C)C
2-Pentanol, 4-methyl-, acetate
Acetic acid, 1,3-dimethylbutyl ester
1,3-Dimethylbutyl acetate
1,3-Dimethylbutylester kyseliny octove (Czech)
sec-Hexyl acetate (ACGIH, OSHA)
sek.Isohexylester kyseliny octove (Czech)
MAAC
Methyl amyl acetate
Methylamyl acetate [UN 1233]
Methylisoamyl acetate
Methylisobutylcarbinol acetate
Methylisobutylcarbinyl acetate
4-Methyl-2-pentanol, acetate
4-Methyl-2-pentyl acetate
UN 1233 [Methylamyl acetate]

108-85-0
$C_6H_{11}Br$
163.06
BrC(CCCC1)C1
Cyclohexane, bromo- (9CI)
AI3-28585
Bromocyclohexane
1-Bromocyclohexane
Cyclohexyl bromide
NSC-11207

108-86-1
C_6H_5Br
157.02
c(cccc1)(c1)Br
Benzene, bromo
Bromobenzene [UN 2514]
Monobromobenzene
NCI-C55492
Phenyl bromide
UN 2514 [Bromobenzene]

108-87-2
C_7H_{14}
98.21
C(CCCC1)(C1)C
Cyclohexane, methyl
Cyclohexylmethane
Hexahydrotoluene
Methyl cyclohexane [UN 2296]
Methylcyclohexane (ACGIH, OSHA)
Metylocykloheksan (Polish)
Sextone B
Toluene hexahydride
Toluene, hexahydro-
UN 2296 [Methyl cyclohexane]

108-88-3

C_7H_8
92.15
c(cccc1)(c1)C
Toluene (ACGIH,OSHA) [UN 1294]
Antisal 1a
Benzene, methyl-
Methacide
Methane, phenyl-
Methylbenzene
Methylbenzol
NCI-C07272
Phenylmethane
RCRA waste number U220
Tolueen (Dutch)
Toluen (Czech)
Toluol (DOT)
Toluolo (Italian)
Tolu-Sol
UN 1294 [Toluene]

108-89-4
C_6H_7N
93.14
n(ccc(c1)C)c1
Pyridine, 4-methyl
4-Methylpyridine
γ-Picoline
p-Picoline [UN 2313]
4-Picoline
UN 2313 [Picolines]

108-90-7
C_6H_5Cl
112.56
c(cccc1)(c1)Cl
Benzene, chloro
Benzene chloride
Chloorbenzeen (Dutch)
Chlorbenzene
Chlorbenzol
Chlorobenzen (Polish)
Chlorobenzene (ACGIH,OSHA) [UN 1134]
Chlorobenzenu (Czech)
Chlorobenzol (DOT)
Clorobenzene (Italian)
MCB
Monochloorbenzeen (Dutch)
Monochlorbenzene
Monochlorbenzol (German)
Monochlorobenzene
Monoclorobenzene (Italian)
NCI-C54886
Phenyl chloride
UN 1134 [Chlorobenzene]

108-91-8
$C_6H_{13}N$
99.20
NC(CCCC1)C1
Cyclohexylamine (ACGIH, OSHA) [UN 2357]
Aminocyclohexane
Aminohexahydrobenzene
Aniline, hexahydro-
CHA
Hexahydroaniline
Hexahydrobenzenamine
UN 2357 [Cyclohexylamine]

108-93-0
$C_6H_{12}O$
100.18
OC(CCCC1)C1
Cyclohexanol (ACGIH,OSHA)
Adronal
Anol
Cicloesanolo (Italian)
Cyclohexyl alcohol
Cykloheksanol (Polish)
Hexahydrophenol
Hexalin
Hydralin
Hydrophenol
Hydroxycyclohexane
Naxol
Phenol, hexahydro-

108-94-1
$C_6H_{10}O$
98.16
O=C(CCCC1)C1
Cyclohexanone (ACGIH, OSHA) [UN 1915]
Anone
Cicloesanone (Italian)
Cyclohexanon (Dutch)
Cykloheksanon (Polish)
Hexanon
Hytrol O
Ketohexamethylene
Nadone
NCI-C55005
Pimelic ketone
Pimelin ketone
RCRA waste number U057
Sextone
UN 1915 [Cyclohexanone]

108-95-2
C_6H_6O
94.12
Oc(cccc1)c1
Phenol (ACGIH,DOT,OSHA)
Acide carbolique (French)
Baker's P and S Liquid and Ointment
Benzenol
Carbolic acid (DOT)
Carbolic acid, Liquid (Liquid tar acid containing over 50% phenol) [UN 2821]
Carbolsaure (German)
Fenol (Dutch, Polish)
Fenolo (Italian)
Hydroxybenzene
Monohydroxybenzene
Monophenol
NA 2821 (DOT)
NCI-C50124
Oxybenzene
Phenic acid
Phenol alcohol
Phenol, Liquid or solution (Liquid tar acid containing over 50% phenol) [UN 2821]
Phenol, Molten [UN 2312]
Phenol, Solid [UN 1671]
Phenole (German)
Phenyl hydrate
Phenyl hydroxide
Phenylic acid
Phenylic alcohol
RCRA waste number U188
UN 1671 [Phenol, solid]
UN 2312 [Phenol, molten]
UN 2821 [Phenol solutions]

108-97-4
$C_5H_4O_2$
96.09
4H-Pyran-4-one (8CI,9CI)
4H-Pyran, 4-oxo-
4-Pyrone

108-98-5
C_6H_6S
110.18
Sc(cccc1)c1
Benzenethiol (DOT)
Phenol, thio-
Phenyl mercaptan (ACGIH, OSHA) [UN 2337]
RCRA waste number P014
Thiofenol (Czech)
Thiophenol (DOT)
UN 2337 [Phenyl mercaptan]
USAF XR-19

108-99-6
C_6H_7N
93.14
n(cccc1C)c1
3-Picoline
B-Picoline
3-Methylpyridine
m-Picoline [UN 2313]
Pyridine, 3-methyl- (9CI)
UN 2313 [Picolines]

109-00-2
C_5H_5NO
95.11
n(cccc1O)c1
3-Pyridinol
β-Hydroxypyridine
3-Hydroxypyridine
3-Pyridol

109-01-3
$C_5H_{12}N_2$
100.19
N(CCNC1)(C1)C
Piperazine, 1-methyl
N-Methylpiperazine
1-Methylpiperazine

109-02-4
$C_5H_{11}NO$
101.17
O(CCN(C1)C)C1
Morpholine, 4-methyl
4-Methylmorfolin (Czech)
N-Methylmorpholine

4-Methylmorpholine
Methylmorpholine [UN 2535]
Morpholine, N-methyl-
UN 2535 [Methylmorpholine]

109-04-6
C_5H_4BrN
158.01
n(c(Br)ccc1)c1
Pyridine, 2-bromo
2-Bromopyridine

109-05-7
$C_6H_{13}N$
99.20
N(C(CCC1)C)C1
Piperidine, 2-methyl
2-Methylpiperidine
α-Pipecolin

109-06-8
C_6H_7N
93.14
n(c(ccc1)C)c1
2-Picoline
α-Methylpyridine
2-Methylpyridine
Picoline
α-Picoline
o-Picoline [UN 2313]
Pyridine, 2-methyl-
RCRA waste number U191
UN 2313 [Picolines]

109-08-0
$C_5H_6N_2$
94.13
n(ccnc1C)c1
Pyrazine, 2-methyl
2-Methylpyrazine

109-09-1
C_5H_4ClN
113.55
n(c(ccc1)Cl)c1
Pyridine, 2-chloro
o-Chloropyridine
α-Chloropyridine
2-Chloropyridine [UN 2822]
UN 2822 [2-Chloropyridine]

109-12-6
$C_4H_5N_3$
95.12
n(cccn1)c1N
Pyrimidine, 2-amino
2-Aminopyrimidine
2-Pyrimidinamine (9CI)
2-Pyridiylamine

109-13-7
$C_8H_{16}O_3$
160.24
O=C(OOC(C)(C)C)C(C)C
Peroxyisobutyric acid, tert-butyl ester, More than 77% in solution
tert-Butyl perisobutyrate
tert-Butyl peroxyisobutyrate
tert-Butyl peroxyisobutyrate, More than 52% but not more than 77% in solution (DOT)
tert-Butyl peroxyisobutyrate, More than 77% in solution (DOT)
tert-Butyl peroxyisobutyrate, Not more than 52% in solution (DOT)
Esperox 24M
Lupersol 8
Peroxyisobutyric acid, tert-butyl ester, More than 52% but not more than 77% in solution
Peroxyisobutyric acid, tert-butyl ester, Not more than 52% in solution
Propaneperoxoic acid, 2-methyl-, 1,1-dimethylethyl ester
UN 2142 (DOT)
UN 2562 (DOT)

109-16-0
$C_{14}H_{22}O_6$
286.36
O=C(OCCOCCOCCOC(=O)C(=C)C)C(=C)C
Methacrylic acid, diester with triethylene glycol
NK Ester 3G
Polyester TGM 3
2-Propenoic acid, 2-methyl-, 1,2-ethanediylbis(oxy-2,1-ethanediyl) ester (9CI)
TEDMA
TGM 3
TGM 3PC
TGM 3S
Triethylene glycol dimethacrylate

109-17-1
$C_{16}H_{26}O_7$
330.42
O=C(OCCOCCOCCOCCOC(=O)C(=C)C)C(=C)C
Methacrylic acid, diester with tetraethylene glycol
2-Propenoic acid, 2-methyl-, oxybis(2,1-ethanediyloxy-2,1-ethanediyl) ester
SR 209
Tetraethylene glycol dimethacrylate
TGM 4

109-19-3
$C_9H_{18}O_2$
158.24
O=C(OCCCC)CC(C)C
Butyl isovalerate
AI3-33584
Butanoic acid, 3-methyl-, butyl ester (9CI)
n-Butyl isopentanoate
1-Butyl isovalerate
n-Butyl isovalerate
Butyl isovalerianate
Butyl 3-methylbutanoate
n-Butyl 3-methylbutanoate
Butyl 3-methylbutyrate
Isovaleric acid, butyl ester (8CI)
NSC-6187

109-21-7
$C_8H_{16}O_2$
144.24
O=C(OCCCC)CCC
Butyric acid, butyl ester
n-Butyl n-butanoate
Butyl butyrate
n-Butyl butyrate
n-Butyl n-butyrate

109-27-3
$C_2H_8N_{10}O$
188.20
O=NNNC(NNNNC(N)N)N
1-Tetrazene, 1-guanyl-4-nitrosaminoguanyl
4-Amidino-1-(nitrosamino-amidino)-1-tetrazene
Guanyl nitrosamino guanyl tetrazene
1-Guanyl-4-nitrosaminoguanyl-tetrazene
Guanyl nitrosamino guanyl tetrazene (DOT)
Guanyl nitrosaminoguanyl-tetrazene (Tetrazene), Wetted with not less than 30 per cent water or mixture of alcohol and water, by mass [UN 0114]
Initiating Explosive-guanyl nitrosamine guanyl tetrazene (DOT)
Initiating Explosive-tetrazene (DOT)
Tetracene Explosive
Tetracene
Tetrazene (DOT)
1-Tetrazene, 4-amidino-1-(nitrsoaminoamidino)- (8CI)
1-Tetrazene-1-carboximidic acid, 4-(aminoimino-methyl)-,2-nitrosohydrazide (9CI)
UN 0114 [Guanyl nitrosamino-guanyltetrazene (Tetrazene), Wetted with not less than 30 per cent water or mixture of alcohol and water, by mass]

109-30-8
$C_{40}H_{78}O_5$
639.06
O=C(OCCOCCOC(=O)CCCCCCCCCCCCCCCC)CCCCCCCCCCCCCCCCC
Diethylene glycol distearate
Octadecanoic acid, oxydi-

2,1-ethanediyl ester (9CI)
Oxydi-2,1-ethanediyl octadecanoate
PEG-2 Distearate
Polyethylene glycol 100 distearate
Polyoxyethylene (2) distearate

109-31-9
$C_{21}H_{40}O_4$
356.61
O=C(OCCCCCC)CCCCCCC(=O)OCCCCCC
Azelaic acid, dihexyl ester
DnHA
Di-n-hexyl azelate
Di-n-hexylester kyseliny azelaove (Czech)
Nonanedioic acid, dihexyl ester
Plastolein 9050
Plastolein 9051
Plastolein 9051 DHNZ

109-32-0
$C_{19}H_{38}O_2$
298.51
O=C(OCCCCCCCC(C)C)CCCCCCCC
Nonanoic acid, 8-methylnonyl ester (9CI)
8-Methylnonyl nonanoate

109-38-6
$C_{24}H_{48}O_3$
384.64
O=C(OCCOCCCC)CCCCCCCCCCCCCCCCC
Octadecanoic acid, 2-butoxyethyl ester
AI3-04494
2-Butoxyethyl octadecanoate

109-39-7
$C_{24}H_{46}O_3$
382.63
O=C(OCCOCCCC)CCCCCCCC=CCCCCCCCC
9-Octadecenoic acid (Z)-, 2-butoxyethyl ester (9CI)

109-43-3
$C_{18}H_{34}O_4$
314.52
O=C(OCCCC)CCCCCCCCC(=O)OCCCC
Sebacic acid, dibutyl ester
Bis(n-butyl)sebacate
Decanedioic acid, dibutyl ester
Dibutylester kyseliny sebakove (Czech)
Dibutyl sebacate
Di-n-butyl sebacate
Kodaflex DBS
Monoplex DBS
Polycizer DBS
PX 404
Staflex DBS

109-46-6
$C_9H_{20}N_2S$
188.37
N(=C(S)NCCCC)CCCC
Urea, 1,3-di-n-butyl-2-thio
N,N'-Dibutylthiourea
1,3-Dibutylthiourea
1,3-Di-n-butyl-2-thiourea
Pennzone B
Thiate U
Thiourea, N,N'-dibutyl-
Urea,1,3-dibutyl-2-thio- (8CI)
USAF EK-2138

109-49-9
$C_6H_{10}O$
98.14
O=C(CCC=C)C
5-Hexen-2-one (9CI)
AI3-21995
Allylacetone
1-Hexen-5-one
NSC-6973

109-50-2
$C_6H_{10}O$
98.14
OC(C#CCC)C
3-Hexyn-2-ol (9CI)

109-52-4
$C_5H_{10}O_2$
102.15
O=C(O)CCCC
Valeric-acid
Butanecarboxylic acid
1-Butanecarboxylic acid
Kyselina valerova (Czech)
NA 1760 [Corrosive liquids, N.O.S.]
Pentanoic acid
n-Pentanoic acid
Propylacetic acid
Valerianic acid
Valeric acid [NA 1760]
n-Valeric acid

109-53-5
$C_6H_{12}O$
100.18
O(C=C)CC(C)C
Ether, isobutyl vinyl
Isobutyl vinyl ether
IVE
UN 1304 [Vinyl isobutyl ether, inhibited]
Vinoflex MO 400*
Vinyl isobutyl ether (DOT)
Vinyl isobutyl ether, Inhibited [UN 1304]

109-55-7
$C_5H_{14}N_2$
102.21
N(CCCN)(C)C
1,3-Propanediamine, N,N-dimethyl
1-Amino-3-dimethylamino-propane
N,N-Dimethyl-N-(3-amino propyl)amine
3-(Dimethylamino)propylamine
N,N-Dimethyl-1,3-diaminopropane
N,N-Dimethyl-1,3-propanediamine
N,N-Dimethyl-1,3-propylenediamine
Propylamine, 3-(N,N-dimethylamino)-

109-56-8
$C_5H_{13}NO$
103.19
OCCNC(C)C
Ethanol, 2-(isopropylamino)
Ethanolisopropylamine
Ethanol, 2-((1-methylethyl)-amino)- (9CI)
(N-Hydroxyethyl)isopropylamine
Isopropylaminoethanol
N-Isopropylaminoethanol
2-Isopropylaminoethanol
N-Isopropylethanolamine
Monoisopropylaminoethanol

109-57-9
$C_4H_8N_2S$
116.20
N=C(S)NCC=C
Urea, 1-allyl-2-thio
Allylthiocarbamide
Allylthiomocovina (Czech)
Allylthiourea
N-Allylthiourea
1-Allylthiourea
1-Allyl-2-thiourea
Aminosin
(2-Propenyl)thiourea
Rhodallin
Rhodalline
Thiocynamine
Thiosinamin
Thiosinamine
U 19571

109-58-0
$C_3H_8N_2O_2$
104.11
O=C(O)NCCN
(2-Aminoethyl)carbamic acid
Carbamic acid, (2-aminoethyl)-(9CI)
Carbamic acid, Compd with 1,2-ethanediamine
Carbamic acid, Compd with ethylenediamine
Ethylenediamine carbamate

109-59-1
$C_5H_{12}O_2$
104.17
O(C(C)C)CCO

Ethanol, 2-isopropoxy
Dowanol Eipat
Ethylene glycol isopropyl ether
Ethylene glycol, monoisopropyl ether
β-Hydroxyethyl isopropyl ether
2-Isopropoxyethanol (ACGIH, OSHA)
Isopropyl cellosolve
Isopropyl glycol
Monoisopropyl ether of ethylene glycol

109-60-4
$C_5H_{10}O_2$
102.15
O=C(OCCC)C
Acetic acid, propyl ester
Acetate de propyle normal (French)
Acetic acid n-propyl ester
1-Acetoxypropane
Octan propylu (Polish)
Propyl acetate (DOT)
n-Propyl acetate (ACGIH, OSHA) [UN 1276]
1-Propyl acetate
Propylester kyseliny octove (Czech)
UN 1276 [n-Propyl acetate]

109-61-5
$C_4H_7ClO_2$
122.56
O=C(OCCC)Cl
Formic acid, chloro-, propyl ester
Carbonochloridic acid, propyl ester
Propyl chlorocarbonate
Propyl chloroformate
n-Propyl chloroformate [UN 2740]
UN 2740 [n-Propyl chloroformate]

109-62-6
$C_4H_8HgO_2$
288.71
Mercury, (acetato)ethyl
(Acetato-o)ethylmercury
Ethylmercuric acetate
Ethylmerkuriacetat (Czech)

109-63-7
$C_4H_{10}BF_3O$
141.95
Ethyl ether, Compd. with boron fluoride (BF_3) (1:1)
Borane, trifluoro-, Compd. with 1,1'-oxybis(ethane) (1:1)
Boron fluoride diethyl etherate
Boron fluoride etherate
Boron fluoride-ethyl etherate
Boron fluoride-ethyl ether complex
Boron fluoride monoetherate
Boron trifluoride-diethyl etherate
Boron trifluoride diethyletherate [UN 2604]
Boron trifluoride etherate
Boron trifluoride-ether complex
Boron trifluoride-ethyl ether
Boron trifluoride-ethyl etherate
Boron, trifluoro(1,1'-oxybis-(ethane))-, (t-4)- (9CI)
Trifluoroborane diethyl etherate
Trifluoroboron etherate
UN 2604 [Boron trifluoride diethyl etherate]

109-64-8
$C_3H_6Br_2$
201.91
BrCCCBr
Propane, 1,3-dibromo
α,γ-Dibromopropane
ω,ω'-Dibromopropane
1,3-Dibromopropane
Trimethylene bromide
Trimethylene dibromide

109-65-9
C_4H_9Br
137.04
BrCCCC
Butane, 1-bromo
1-Bromobutane
n-Butyl bromide [UN 1126]
Butyl bromide, normal (DOT)
UN 1126 [n-Butyl bromide]

109-66-0
C_5H_{12}
72.17
C(CCC)C
Pentane (ACGIH,OSHA) [UN 1265]
Pentan (Polish)
Pentanen (Dutch)
Pentani (Italian)
UN 1265 [n-Pentanes or isopentane]

109-67-1
C_5H_{10}
70.13
C(=C)CCC
1-Pentene (9CI)
α-Amylene
α-n-Amylene
1-Pentylene
Propylethylene

109-68-2
C_5H_{10}
70.13
C(=CC)CC
2-Pentene (9CI)
β-n-Amylene
Methylethylethylene
sym-Methylethylethylene
NSC-7894
3-Pentene

109-69-3
C_4H_9Cl
92.58
ClCCCC
Butane, 1-chloro
Butyl chloride (DOT)
n-Butyl chloride [UN 1127]
1-Chlorobutane [UN 1127]
Chlorure de butyle (French)
NCI-C06155
n-Propylcarbinyl chloride
UN 1127 [Chlorobutanes]

109-70-6
C_3H_6BrCl
157.45
BrCCCCl
Propane, 1-bromo-3-chloro
1-Bromo-3-chloropropane
3-Bromopropyl chloride
1,3-CHBP
ω-Chlorobromopropane
1-Chloro-3-bromopropane [UN 2688]
3-Chloropropyl bromide
Trimethylene bromide chloride
Trimethylene chlorobromide
UN 2688z [1-Chloro-3-bromopropane]

109-72-8
C_4H_9Li
64.06
Butyllithium
Lithium, butyl- (9CI)

109-73-9
$C_4H_{11}N$
73.16
NCCCC
Butylamine (DOT,OSHA)
1-Amino-butaan (Dutch)
1-Aminobutan (German)
1-Aminobutane
1-Butanamine
n-Butilamina (Italian)
n-Butylamin (German)
n-Butylamine (ACGIH) [UN 1125]
Monobutilamina (Romanian)
Monobutylamine
Mono-n-butylamine
Norvalamine
UN 1125 [n-Butylamine]

109-74-0
C_4H_7N
69.12
N#CCCC
Butyronitrile [UN 2411]
Butanenitrile
n-Butanenitrile
Butyric acid nitrile
n-Butyronitrile
Butyrylonitrile
1-Cyanopropane
Propyl cyanide

Propylkyanid (Czech)
UN 2411 [Butyronitrile]

109-75-1
C_4H_5N
67.10
N#CCC=C
3-Butenenitrile
Acetonitrile, vinyl-
Allyl cyanide
Allylkyanid (Czech)
Allylnitrile
1-Butene-4-nitrile
β-Butenonitrile
TL 350
Vinylacetonitrile

109-76-2
$C_3H_{10}N_2$
74.15
NCCCN
1,3-Propanediamine
1,3-Diaminopropane
1,3-Propylenediamine
Trimethylenediamine

109-77-3
$C_3H_2N_2$
66.07
N#CCC#N
Malononitrile [UN 2647]
Cyanoacetonitrile
Dicyanomethane
Malonic acid dinitrile
Malonic dinitrile
Methane, dicyano-
Methylene cyanide
Nitril kyseliny malonove (Czech)
Dwumetylosulfotlenku (Polish)
Propanedinitrite
RCRA waste number U149
UN 2647 [Malononitrile]
USAF A-4600
USAF KF-19

109-78-4
C_3H_5NO
71.09
N#CCCO
Hydracrylonitrile
2-Cyanoethanol
2-Cyanoethyl alcohol
Ethylene cyanohydrin
Glycol cyanohydrin
β-HPN
2-Hydroxyethylkyanid (Czech)
3-Hydroxypropanenitrile
β-Hydroxypropionitrile
3-Hydroxypropionitrile
Methanolacetonitrile
Propanenitrile, 3-hydroxy-
Propionitrile, 3-hydroxy-
USAF RH-7

109-79-5
$C_4H_{10}S$
90.20
SCCCC
n-Butanethiol
Butanethiol
Butanethiol (OSHA)
Butyl mercaptan (OSHA) [UN 2347]
n-Butyl mercaptan (ACGIH)
NCI-C60866
UN 2347 [Butyl mercaptans]

109-80-8
$C_3H_8S_2$
108.23
SCCCS
1,3-Propanedithiol
1,3-Dimercaptopropane
Dithiotrimethyleneglycol
NDR-132
1,3-Propanedimercaptan
Trimethylene dimercaptan
Trimethylenedithioglycol
Trimethylenedithiol

109-83-1
C_3H_9NO
75.13
OCCNC
Ethanol, 2-(methylamino)
β-(Methylamino)ethanol
N-Methylaminoethanol
2-Methylaminoethanol
Methylethanolamine
N-Methylethanolamine
Methylethylolamine
Methyl(β-hydroxyethyl)amine
Monomethyl-aminoaethanol (German)
Monomethylaminoethanol
N-Monomethylaminoethanol
USAF DO-50

109-84-2
$C_2H_8N_2O$
76.12
OCCNN
Ethanol, 2-hydrazino
BOH
Brombloom
2-Hydrazinoethanol
Hydroxyethyl hydrazine
β-Hydroxyethylhydrazine
N-(2-Hydroxyethyl)hydrazine
Omaflora

109-85-3
C_3H_9NO
75.13
O(CCN)C
Ethylamine, 2-methoxy
2-Methoxyethylamine

109-86-4
$C_3H_8O_2$
76.11
O(CCO)C
Ethanol, 2-methoxy
Aethylenglykol-monomethylaether (German)
Dowanol EM
EGM
EGME
Ether monomethylique de l' ethylene-glycol (French)
Ethylene glycol methyl ether
Ethylene glycol monomethyl ether [UN 1188]
Glycol Ether EM
Glycolmethyl ether
Glycol monomethyl ether
Jeffersol EM
MECS
2-Methoxy-aethanol (German)
2-Methoxyethanol (ACGIH, OSHA)
Methoxyhydroxyethane
Methyl cellosolve (DOT,OSHA)
Methylcelosolv (Czech)
Methyl ethoxol
Methyl glycol
Methylglykol (German)
Methyl oxitol
Metil cellosolve (Italian)
Metoksyetylowy alkohol (Polish)
2-Metossietanolo (Italian)
Monomethyl ether of ethylene glycol
Poly-Solv EM
Prist
UN 1188 [Ethylene glycol monomethyl ether]

109-87-5
$C_3H_8O_2$
76.11
O(COC)C
Methane, dimethoxy
Anesthenyl
Dimethoxymethane
Dimethoxymethane (OSHA)
Dimethyl formal
Formal
Formaldehyde dimethylacetal
Methylal (ACGIH,OSHA) [UN 1234]
Methylene dimethyl ether
Metylal (Polish)
UN 1234 [Methylal]

109-89-7
$C_4H_{11}N$
73.16
N(CC)CC
Diethylamine (ACGIH,OSHA) [UN 1154]
DEN
Diaethylamin (German)
Diethamine
N,N-Diethylamine
Dietilamina (Italian)
Dwuetyloamina (Polish)
Ethanamine, N-ethyl-
UN 1154 [Diethylamine]

109-90-0

C_3H_5NO
71.09
O=C=NCC
Isocyanic acid, ethyl ester
Ethane, isocyanato-
Ethyl isocyanate [UN 2481]
Isocyanatoethane
UN 2481 [Ethyl isocyanate]

109-92-2
C_4H_8O
72.12
O(C=C)CC
Ether, ethyl vinyl
Ethene, ethoxy
Ether, vinyl ethyl
Ethyl vinyl ether
EVE
UN 1302 [Vinyl ethyl ether, inhibited]
Vinamar
Vinyl ethyl ether
Vinyl ethyl ether, Inhibited [UN 1302]

109-93-3
C_4H_6O
70.10
O(C=C)C=C
Vinyl-ether
Divinyl ether (DOT)
Divinyl ether, Inhibited [UN 1167]
Divynyl oxide
Ethene, 1,1'-oxybis-
Ethenyloxyethene
Ether, divinyl
1,1'-Oxybisethene
UN 1167 [Divinyl ether, inhibited]
Vinesthene
Vinesthesin
Vinethen
Vinethene
Vinether
Vinidyl
Vinydan

109-94-4
$C_3H_6O_2$
74.09
O=COCC
Formic acid, ethyl ester
Aethylformiat (German)
Areginal
Ethylester kyseliny mravenci (Czech)
Ethyle (formiate d') (French)
Ethyl formate (ACGIH,OSHA) [UN 1190]
Ethylformiaat (Dutch)
Ethyl formic ester
Ethyl methanoate
Etile (formiato di) (Italian)
Formic ether
Mrowczan etylu (Polish)
UN 1190 [Ethyl formate]

109-95-5
$C_2H_5NO_2$
75.08
O=NOCC
Nitrous acid, ethyl ester
Ethylester kyseliny dusite (Czech)
Ethyl nitrite (DOT)
Ethyl nitrite, Solution [UN 1194]
Nitrosyl ethoxide
Nitrous ether (DOT)
Nitrous ethyl ether
UN 1194 [Ethyl nitrite solutions]

109-97-7
C_4H_5N
67.10
N(C=CC=1)C1
Pyrrole
1-Aza-2,4-cyclopentadiene
Azole
Divinylenimine
Imidole
Monopyrrole
Pyrrol

109-99-9
C_4H_8O
72.12
O(CCC1)C1
Furan, tetrahydro
Butane, 1,4-epoxy-
Butylene oxide
Cyclotetramethylene oxide
Diethylene oxide
Furanidine
Hydrofuran
NCI-C60560
Oxacyclopentane
Oxolane
RCRA waste number U213
Tetrahydrofuraan (Dutch)
Tetrahydrofuran (ACGIH, OSHA) [UN 2056]
Tetrahydrofuranne (French)
Tetraidrofurano (Italian)
Tetramethylene oxide
THF
UN 2056 [Tetrahydrofuran]

110-00-9
C_4H_4O
68.08
O(C=CC=1)C1
Furan [UN 2389]
Divinylene oxide
Furfuran
NCI-C56202
Oxacyclopentadiene
Oxole
RCRA waste number U124
Tetrole
UN 2389 [Furan]

110-01-0
C_4H_8S
88.18
S(CCC1)C1
Thiophene, tetrahydro
Tetrahydrothiofen (Czech)
Tetrahydrothiophene [UN 2412]
Tetramethylenesulfide
Thiacyclopentane
Thilane
Thiofan (Czech)
Thiolane
Thiophane
UN 2412 [Tetrahydrothiophene]

110-02-1
C_4H_4S
84.14
S(C=CC=1)C1
Thiophene [UN 2414]
CP 34
Divinylene sulfide
Huile H50
Huile HSO
Thiacyclopentadiene
Thiaphene
Thiofen (Czech)
Thiofuran
Thiofuram
Thiofurfuran
Thiole
Thiophen
Thiotetrole
UN 2414 [Thiophene]
USAF EK-1860

110-03-2
$C_8H_{18}O_2$
146.23
OC(CCC(O)(C)C)(C)C
2,5-Dimethyl-2,5-hexanediol
AI3-20685
Dimethylhexanediol
2,5-Dimethylhexane-2,5-diol
2,5-Hexanediol, 2,5-dimethyl- (9CI)
NSC-5595
1,1,4,4-Tetramethyl-1,4-butanediol

110-05-4
$C_8H_{18}O_2$
146.26
O(OC(C)(C)C)C(C)(C)C
tert-Butyl peroxide (DOT)
Cadox
Di-tert-butylperoxid (German)
Di-tert-butyl peroxide
Di-tert-butyl peroxide, technically pure (DOT)
Di-tert-butyl peroxyde (Dutch)
DTBP
Perossido di butile terziario (Italian)
Peroxyde de butyle tertiaire (French)
(Tributyl)peroxide
Trigonox B
UN 2102 (DOT)

110-06-5
$C_8H_{18}S_2$
178.36
S(SC(C)(C)C)C(C)(C)C
Disulfide, bis(1,1-dimethyl-ethyl)
AI3-32576

110-12-3
$C_7H_{14}O$
114.21
O=C(CCC(C)C)C
2-Hexanone, 5-methyl
Isoamyl methyl ketone
Isopentyl methyl ketone
Ketone, methyl isoamyl
2-Methyl-5-hexanone
5-Methyl-2-hexanone
5-Methylhexan-2-one [UN 2302]
Methyl isoamyl ketone (ACGIH,OSHA)
MIAK
UN 2302z [5-Methylhexan-2-one]

110-13-4
$C_6H_{10}O_2$
114.16
O=C(CCC(=O)C)C
2,5-Hexanedione

Acetone, acetonyl-
Acetonyl acetone
Diacetonyl
α,β-Diacetylethane
1,2-Diacetylethane
2,5-Diketohexane

110-14-5
$C_4H_8N_2O_2$
116.11
O=C(N)CCC(=O)N
Succinamide (8CI)
Butanediamide (9CI)
NSC-8157
Succindiamide
Succinic acid diamide
Succinic amide

110-15-6
$C_4H_6O_4$
118.10
O=C(O)CCC(=O)O
Succinic-acid
Amber acid
Asuccin
Bernsteinsaure (German)
Butanedioic acid
1,2-Ethanedicarboxylic acid
Ethylenesuccinic acid
Kyselina jantarova (Czech)
Wormwood
Wormwood Acid

110-16-7
$C_4H_4O_4$
116.08
O=C(O)C=CC(=O)O
Maleic-acid
Butenedioic acid, (Z)-
cis-Butenedioic acid
cis-1,2-Ethylenedicarboxylic acid
1,2-Ethylenedicarboxylic acid, (Z)
Kyselina maleinova (Czech)
Maleic acid [NA 2215]
Maleinic acid
Malenic acid
NA 2215 [Maleic acid]
Toxilic acid

110-17-8
$C_4H_4O_4$
116.08
O=C(O)C=CC(=O)O
Fumaric-acid
Allomaleic acid
Boletic acid
trans-Butenedioic acid
Butenedioic acid, (E)-
1,2-Ethenedicarboxylic acid, trans-
trans-1,2-Ethylenedicarboxylic acid
1,2-Ethylenedicarboxylic acid, (E)
Kyselina fumarova (Czech)
Lichenic acid
NSC-2752
U-1149
USAF EK-P-583

110-18-9
$C_6H_{16}N_2$
116.24
N(CCN(C)C)(C)C
Ethylenediamine, N,N,N',N'-tetramethyl
1,2-Bis-(dimethylamino)ethane
1,2-Di-(dimethylamino)ethane [UN 2372]
1,2-Ethanediamine, N,N,N',N'-tetramethyl- (9CI)
Propamine D
TEMED
Tetrameen
N,N,N',N'-Tetramethyl-1,2-diaminoethane
N,N,N',N'-Tetramethylethanediamine
Tetramethyl ethylene diamine
N,N,N',N'-Tetramethylethylenediamine
TMEDA
UN 2372 [1,2-Di-(dimethylamino) ethane]

110-19-0
$C_6H_{12}O_2$
116.18
O=C(OCC(C)C)C
Acetic acid, isobutyl ester
Acetate d'isobutyle (French)
Acetic acid, 2-methylpropyl ester
Isobutyl acetate (ACGIH, OSHA) [UN 1213]
Isobutylester kyseliny octove (Czech)
2-Methylpropyl acetate
2-Methyl-1-propyl acetate
β-Methylpropyl ethanoate
UN 1213 [Isobutyl acetate]

110-20-3
$C_4H_9N_3O$
115.16
O=C(NN=C(C)C)N
Acetone, semicarbazone

110-21-4
$C_2H_6N_4O_2$
118.07
O=C(NNC(=O)N)N
1,2-Hydrazinedicarboxamide (9CI)
AI3-28537
Bicarbamamide
Bicarbamimidic acid (VAN)
Biurea (8CI)
Formamide, 1,1'-hydrazobis-
Formimidic acid, 1-semicarbazido-
Hydrazine, 1,2-bis(aminocarbonyl)-
Hydrazinecarboximidic acid, 2-(aminocarbonyl)-
Hydrazine, 1,2-dicarbamoyl-
Hydrazocarbonamide
Hydrazodicarbonamide
Hydrazodicarboxamide
NSC-1897
Pseudourea, 3-ureido-
Semicarbazide, 1-carbamoyl-
Semicarbazide, 1-(1-hydroxyformimidoyl)-
Urea, ureido-
Ureidourea

110-22-5
$C_4H_6O_4$
118.10
O=C(OOC(=O)C)C
Acetyl-peroxide
Acetyl peroxide, More than 25% in solution (DOT)
Acetyl peroxide, Solid (DOT)
Acetyl peroxide (Solution)
Acetyl peroxide solution, Not over 25% peroxide (DOT)
Diacetyl peroxide
Diacetyl peroxide (Solution)
UN 2084 (DOT)

110-25-8
$C_{21}H_{39}NO_3$
353.54
O=C(N(CC(=O)O)C)CCCCCCCC=CCCCCCCCC
Oleoyl Sarcosine
Glycine, N-methyl-N-(1-oxo-9-octadecenyl)-

Glycine, N-methyl-N-(1-oxo-9-octadecenyl)-, (Z)- (9CI)
Hamposyl O
Maprosyl O
Medialanic acid (VAN)
N-Methyl-N-(1-oxo-9-octadecenyl)glycine
(Z)-N-Methyl-N-(1-oxo-9-octadecenyl)glycine
NSC-96995
Oleic sarcosine
Oleoyl N-methylaminoacetic acid
Oleoyl sarcosine
Oleoylsarcosine
N-Oleoylsarcosine
Oleyl methylaminoethanoic acid
Oleyl N-methylglycine
Oleyl sarcosine
Sarcosine, N-oleoyl- (8CI)
Sarkosyl O

110-26-9
$C_7H_{10}N_2O_2$
154.19
O=C(NCNC(=O)C=C)C=C
Acrylamide, N,N'-methylenebis
Methylenebisacrylamide
N,N'-Methylenebis(acrylamide)
N,N'-Methylenediacrylamide
N,N'-Methylidenebisacrylamide
2-Propenamide, N,N'-methylenebis-

110-27-0
$C_{17}H_{34}O_2$
270.51
O=C(OC(C)C)CCCCCCCCCCCCC
Tetradecanoic acid, isopropyl ester
Bisomel
Crodamol IPM
Deltylextra
Emcol-IM
Emerest 2314
Isomyst
Isopropyl myristate
Ja-Fa IPM
Kessco isopropyl myristate
Kesscomir
Myristic acid, isopropyl ester
Plymoutm IPM
Promyr
Starfol IPM
Stepan D-50
Tegester
Tetradecanoic acid, isopropyl
Tetradecanoic acid, 1-methylethyl ester
1-Tridecanecarboxylic acid, isopropyl ester
Unimate IPM
Wickenol 101

110-29-2
$C_{24}H_{46}O_4$
398.63
O=C(OCCCCCCCCCC)CCCCC(=O)OCCCCCCCC
n-Decyl n-octyl adipate
Adipic acid, decyl octyl ester
Adipol ody
Decyl octyl adipate
Decyl octyl hexanedioate
Hercoflex 290
Hexanedioic acid, decyl octyl ester (9CI)
Monoplex NODA
Octyl decyl adipate
n-Octyl decyl adipate
PX-202
Staflex Noda
Truflex 146

110-30-5
$C_{38}H_{76}N_2O_2$
593.02
O=C(NCCNC(=O)CCCCCCCCCCCCCCCCC)CCCCCCCCCCCCCCCCC
Ethylene distearamide
Abril wax 10DS
Acrawax CT
Acrowax C
Advawachs 280
Advawax
Advawax 275
Advawax 280
AI3-08515
Armowax EBS-P
1,2-Bis(octadecanamido)ethane
Carlisle Wax 280
Carlisle 280
Chemetron 100
N,N'-Distearoylethylenediamine
N,N'-1,2-Ethanediylbisoctadecanamide
Ethylenebisstearamide
Ethylenebis(stearamide)
N,N'-Ethylene bisstearamide
N,N'-Ethylenebisstearamide
N,N'-Ethylenebis(stearamide)
Ethylenebisstearoamide
Ethylenebis(stearylamide)
Ethylenediamine bisstearamide
Ethylenediamine steardiamide
Ethylenedistearamide
N,N'-Ethylenedistearamide
N,N'-Ethylene distearylamide
Kemamide W 40
Lubrol EA
Microtomic 280
Nopcowax 22-DS
NSC-83613
Octadecanamide, N,N'-1,2-ethanediylbis- (9CI)
Octadecanamide, N,N'-ethylenebis- (8CI)
Plastflow
Stearic acid, ethylenediamine diamide
Wax C

110-31-6
$C_{38}H_{72}N_2O_2$
588.99
O=C(NCCNC(=O)CCCCCCCC=CCCCCCCCC)CCCCCCCC=CCCCCCCCC
Ethylene dioleamide
N,N'-Dioleoylethylenediamine
N,N'-1,2-Ethanediylbis-9-octadecenamide
(Z,Z)-N,N'-1,2-Ethanediylbis-9-octadecenamide
Ethylene bis(oleamide)
N,N'-Ethylenebisoleamide
N,N'-Ethylenedioleamide
NSC-131419
Oleamide, N,N'-ethylenebis- (8CI)
Oleic acid-ethylenediamine condensate
9-Octadecenamide, N,N'-1,2-ethanediylbis-, (Z,Z)- (9CI)

110-32-7
$C_{22}H_{42}O_6$
402.64
CCCCCCOCCOC(=O)CCCCC(=O)OCCOCCCCCC
Adipic acid, bis(2-(hexyloxy)-ethyl) ester
Adipic acid, di(2-hexyloxyethyl) ester
Bis(2-(hexyloxy)ethyl)adipate
Dihexyloxyethyl adipate
Di-(2-(2-hexyloxy)ethyl)ester kyseliny adipove (Czech)
Hexanedioic acid, bis(2-(hexyloxy)ethyl)ester (9CI)

110-33-8
$C_{18}H_{34}O_4$
314.47
O=C(OCCCCCC)CCCCC(=O)OCCCCCC
Dihexyl adipate
AI3-07963
Di-n-hexyl adipate
Dihexyl hexanedioate
Hexanedioic acid, dihexyl ester

110-34-9
$C_{20}H_{40}O_2$
312.54
O=C(OCC(C)C)CCCCCCCCCCCCCCC
Isobutyl palmitate
AI3-31576
Hexadecanoic acid, 2-methylpropyl ester
2-Methylpropyl hexadecanoate
Palmitic acid, isobutyl ester

110-36-1
$C_{18}H_{36}O_2$
284.48
O=C(OCCCC)CCCCCCCCCCCCC
Butyl myristate
AI3-07958
Bumyr
Butyl tetradecanoate
Butyl n-tetradecanoate
Myristic acid, butyl ester (8CI)

NSC-4814
Tetradecanoic acid, butyl ester (9CI)

110-38-3
$C_{12}H_{24}O_2$
200.36
O=C(OCC)CCCCCCCCC
Decanoic acid, ethyl ester
Capric acid ethyl ester
Ethyl caprate
Ethyl caprinate
Ethyl decanoate
Ethyl decylate

110-40-7
$C_{14}H_{26}O_4$
258.40
O=C(OCC)CCCCCCCCC(=O)OCC
Sebacic acid, diethyl ester
Bisoflex DES
Diethyl decanedioate
Diethyl 1,10-decanedioate
Diethyl sebacate
Ethyl sebacate

110-41-8
$C_{12}H_{24}O$
184.32
O=CC(CCCCCCCCC)C
Undecanal, 2-methyl- (9CI)
AI3-03960
Aldehyde C-12
Aldehyde M.N.A.
Methylnonylacetaldehyde
Methyl-n-nonylacetaldehyde
Methylnonylacetic aldehyde
2-Methylundecanal
2-Methyl-1-undecanal
MNA
NSC-46127

110-42-9
$C_{11}H_{22}O_2$
186.29
O=C(OC)CCCCCCCCC
Methyl decanoate
AI3-26168
Capric acid methyl ester
Decanoic acid methyl ester
Decanoic acid, methyl ester (9CI)
Metholene 2095
Methyl caprate
Methyl-n-caprate
Methyl caprinate
NSC-3713
Uniphat A30

110-43-0
$C_7H_{14}O$
114.21
O=C(CCCCC)C
2-Heptanone
Amyl-methyl-cetone (French)
Amyl methyl ketone [UN 1110]
n-Amyl methyl ketone
Ketone, methyl pentyl
Methyl-amyl-cetone (French)
Methyl n-amyl ketone (ACGIH, OSHA)
Methyl amyl ketone [UN 1110]
Methyl pentyl ketone
UN 1110 [Amyl methyl ketone]

110-44-1
$C_6H_8O_2$
112.14
O=C(O)C=CC=CC
Sorbic-acid
Acetic acid, (2-butenylidene)-
Acetic acid, crotylidene-
Hexadienic acid
Hexadienoic acid
2,4-Hexadienoic acid
trans-trans-2,4-Hexadienoic acid
Kyselina 1,3-pentadien-1-karboxylova (Czech)
Kyselina sorbova (Czech)
1,3-Pentadiene-1-carboxylic acid
2-Propenylacrylic acid
Sorbistat

110-45-2
$C_6H_{12}O_2$
116.18
O=COCCC(C)C
Isopentyl alcohol, formate
Formic acid, isopentyl ester
Isoamyl formate [UN 1109]
Isoamyl methanoate
Isopentyl formate
3-Methylbutyl formate
UN 1109 [Amyl formates]

110-46-3
$C_5H_{11}NO_2$
117.17
O=NOCCC(C)C
Isopentyl alcohol, nitrite
Amilnitrit
Amyl nitrit
Amyl nitrite
Aspiral
IPN
Isoamyl nitrite
Isopentyl nitrite
3-Methylbutanol nitrite
3-Methylbutyl nitrite
Nitramyl
Nitrous acid, 3-methylbutyl ester
Vaporole

110-49-6
$C_5H_{10}O_3$
118.15
O=C(OCCOC)C
Ethanol, 2-methoxy-, acetate
Acetate de l'ether monomethylique de l'ethylene-glycol (French)
Acetate de methyle glycol (French)
Acetato di metil cellosolve (Italian)
Acetic acid 2-methoxyethyl ester
Aethylenglykolmethylaether-acetat (German)
Ethylene glycol methyl acetate (OSHA)
Ethylene glycol methyl ether acetate
Ethylene glycol monomethyl ether acetate [UN 1189]
Glycol ether em acetate
Glycol monomethyl ether acetate
MeCsAc
2-Methoxyaethylacetat (German)
2-Methoxyethanol, acetate
2-Methoxy-ethyl acetaat (Dutch)
2-Methoxyethyl acetate (ACGIH,OSHA)
2-Methoxyethyle, acetate de (French)
2-Methoxyethylester kyseliny octove (Czech)
Methyl cellosolye acetaat (Dutch)
Methyl cellosolve acetate (DOT,OSHA)
Methylcelosolvacetat (Czech)
Methyl glycol acetate
Methyl glycol monoacetate
Methylglykolacetat (German)
2-Metossietilacetato (Italian)
UN 1189 [Ethylene glycol monomethyl ether acetate]

110-50-9
$C_5H_{10}OS_2$
150.26
Butylxanthate
Butyl xanthate
Carbonodithioic acid, O-butyl ester

110-52-1
$C_4H_8Br_2$
215.94
BrCCCCBr
Butane, 1,4-dibromo
DBB
1,4-Dibrombutan (German)
1,4-Dibromobutane

110-53-2
$C_5H_{11}Br$
151.07
BrCCCCC
Pentane, 1-bromo
Amyl bromide
n-Amyl bromide
1-Bromopentane
Pentyl bromide
n-Pentyl bromide
1-Pentyl bromide

110-54-3
C_6H_{14}
86.20
C(CCCC)C
Hexane
Esani (Italian)
Gettysolve-B
Heksan (Polish)
Hexane [UN 1208]
n-Hexane (ACGIH,OSHA)
Hexanen (Dutch)
NCI-C60571
UN 1208 [Hexanes]

110-56-5
$C_4H_8Cl_2$
127.01
ClCCCCCl
1,4-Dichlorobutane
Butane, 1,4-dichloro- (9CI)
NSC-6288

110-57-6
$C_4H_6Cl_2$
125.00
C(=CCCl)CCl
2-Butene, 1,4-dichloro-, (E)
2-Butylene dichloride
1,4-Dichlorobutene-2 (trans)
1,4-Dichloro-2-butene

110-58-7
$C_5H_{13}N$
87.19
NCCCCC
Pentylamine
1-Aminopentane
Amylamine
n-Amylamine
Monoamylamine
Norleucamine
1-Pentanamine
n-Pentylamine
1-Pentylamine

110-59-8
C_5H_9N
83.15
N#CCCCC
Valeronitrile
1-Cyanobutane
Pentanenitrile (9CI)
n-Valeronitrile

110-60-1
$C_4H_{12}N_2$
88.18
NCCCCN
1,4-Butanediamine
Butylenediamine
1,4-Butylenediamine
1,4-Diaminobutane
Putrescin
Putrescine
Tetramethylenediamine
1,4-Tetramethylenediamine

110-61-2
$C_4H_4N_2$
80.10
N#CCCC#N
Succinonitrile
Butanedinitrile
Deprelin
s-Dicyanoethane
Dinile
Ethane, 1,2-dicyano-
Ethylene cyanide
Ethylene dicyanide
Succinic acid dinitrile
Succinic dinitrile
Succinodinitrile
Sukcinonitril (Czech)
Suxil
USAF A-9442

110-62-3
$C_5H_{10}O$
86.15
O=CCCCC
Valeraldehyde [UN 2058]
Amyl aldehyde
Butyl formal
Pentanal
n-Pentanal
UN 2058 [Valeraldehyde]
Valeric aldehyde
Valeral
n-Valeraldehyde (ACGIH, OSHA)
Valerianic aldehyde
Valeric acid aldehyde
n-Valeric aldehyde
Valerylaldehyde

110-63-4
$C_4H_{10}O_2$
90.14
OCCCCO
1,4-Butanediol
Butanediol
Butane-1,4-diol
1,4-Butylene glycol
1,4-Dihydroxybutane
Diol 14B
Sucol B
Tetramethylene 1,4-diol
1,4-Tetramethylene glycol

110-64-5
$C_4H_8O_2$
88.12
OCC=CCO
2-Butene, 1,4-dihydroxy
2-Butene-1,4-diol
1,4-Dihydroxy-2-butene

110-65-6
$C_4H_6O_2$
86.10
OCC#CCO
2-Butyne-1,4-diol
Bis(hydroxymethyl)acetylene
2-Butin-1,4-diol (Czech)
Butynediol
1,4-Butynediol [UN 2716]
UN 2716 [1,4-Butynediol]

110-66-7
$C_5H_{12}S$
104.23
SCCCCC
1-Pentanethiol
Amyl hydrosulfide
Amyl mercaptan [UN 1111]
n-Amyl mercaptan
Amyl sulfhydrate
Amyl thioalcohol
Mercaptan amylique (French)
Pentalarm
Pentyl mercaptan
UN 1111 [Amyl mercaptans]

110-67-8
C_4H_7NO
85.12
N#CCCOC
Propionitrile, 3-methoxy
3-Methoxypropannitril (Czech)
3-Methoxypropionitrile

110-68-9
$C_5H_{13}N$
87.19
N(CCCC)C
Butylamine, N-methyl
Methylbutylamine
N-(Methyl) butyl amine
N-Methyl-n-butylamine
N-Methylbutylamine [UN 2945]
UN 2945 [N-Methylbutylamine]

110-69-0
C_4H_9NO
87.14
N(O)=CCCC
Butyraldehyde, oxime (8CI)
Butanal oxime (9CI)
Butyraldoxime [UN 2840]
n-Butyraldoxime
Exkin 1
Exkin No. 1 Anti-skinning agent
NSC-1487
Skino #1
Troykyd Anti-Skin BTO
UN 2840 [Butyraldoxime]
USAF AM-6

110-71-4
$C_4H_{10}O_2$
90.14
O(CCOC)C
Ethane, 1,2-dimethoxy
Dimethoxyethane
α,β-Dimethoxyethane
1,2-Dimethoxyethane [UN 2252]
Dimethylcellosolve
2,5-Dioxahexane

EGDME
Ethylene dimethyl ether
Ethylene glycol dimethyl ether
Glycol dimethyl ether
Glyme
Monoethylene glycol dimethyl ether
Monoglyme
UN 2252 [1,2-Dimethoxy-ethane]

110-73-6
$C_4H_{11}NO$
89.16
OCCNCC
Ethanol, 2-(ethylamino)
2-Ethylaminoethanol
2-N-Monoethylaminoethanol

110-74-7
$C_4H_8O_2$
88.12
O=COCCC
Formic acid, propyl ester
Formiate de propyle (French)
Propylester kyseliny mravenci (Czech)
n-Propyl formate
Propyl formate [UN 1281]
Propyl methanoate
UN 1281 [Propyl formates]

110-75-8
C_4H_7ClO
106.56
O(C=C)CCCl
Ether, 2-chloroethyl vinyl
2-Chlorethyl vinyl ether
(2-Chloroethoxy)ethene
2-Chloroethyl vinyl ether
Ethene, 2-chloroethoxy-
RCRA waste number U042
Vinyl β-chloroethyl ether
Vinyl 2-chloroethyl ether

110-77-0
$C_4H_{10}OS$
106.20
OCCSCC
Ethanol, 2-(ethylthio)
Ethyl 2-hydroxyethyl sulfide
Ethyl 2-hydroxyethyl thioether
β-Ethylmerkaptoethanol (Czech)
2-(Ethylthio)ethanol
β-Hydroxydiethyl sulfide

110-78-1
C_4H_7NO
85.12
O=C=NCCC
Isocyanic acid, propyl ester
1-Isocyanatopropane
Propane, 1-isocyanato-
Propyl isocyanate
n-Propyl isocyanate [UN 2482]
1-Propyl isocyanate
UN 2482 [n-Propyl isocyanate]

110-80-5
$C_4H_{10}O_2$
90.14
O(CCO)CC
Ethanol, 2-ethoxy
Athylenglykol-monoathylather (German)
Cellosolve (DOT)
Cellosolve solvent
Celosolv (Czech)
Dowanol EE
Ektasolve EE
Ether monoethylique de l'ethyl-ene-glycol (French)
2-Ethoxyethanol (ACGIH, OSHA)
Ethyl cellosolve
Ethylene glycol ethyl ether
Ethylene glycol monoethyl ether [UN 1171]
Etoksyetylowy alkohol (Polish)
Glycol Ether EE
Glycol ethyl ether
Glycol monoethyl ether
Glycol monoethyl ether (OSHA)
Hydroxy ether
Jeffersol EE
NCI-C54853
Oxitol
Poly-Solv EE
RCRA waste number U227
RCRA waste number U359
UN 1171 [Ethylene glycol monoethyl ether]

110-81-6
$C_4H_{10}S_2$
122.26
S(SCC)CC
Disulfide, diethyl
Diethyldisulfid (Czech)
Diethyl disulfide

110-82-7
C_6H_{12}
84.18
C(CCCC1)C1
Cyclohexane (ACGIH,OSHA) [UN 1145]
Benzene, hexahydro-
Cicloesano (Italian)
Cyclohexaan (Dutch)
Cyclohexan (German)
Cykloheksan (Polish)
Hexahydrobenzene
Hexamethylene
Hexanaphthene
RCRA waste number U056
UN 1145 [Cyclohexane]

110-83-8
C_6H_{10}
82.16
C(=CCCC1)C1
Cyclohexene (ACGIH,OSHA) [UN 2256]
Benzenetetrahydride
Benzene, tetrahydro-
Cykloheksen (Polish)
Hexanaphthylene
Tetrahydrobenzene
1,2,3,4-Tetrahydrobenzene
UN 2256 [Cyclohexene]

110-85-0
$C_4H_{10}N_2$
86.16
N(CCNC1)C1
Piperazine [UN 2579]
Antiren
1,4-Diethylenediamine
Dispermine
Hexahydro-1,4-diazine
Hexahydropyrazine
Lumbrical
Piperazidine
Piperazin (German)
Piperazine, Anhydrous
Pyrazine hexahydride
Pyrazine, hexahydro-
UN 2579 [Piperazine]

110-86-1
C_5H_5N
79.11
n(cccc1)c1
Pyridine (ACGIH,OSHA) [UN 1282]
Azabenzene
Azine
NCI-C55301
Pirydyna (Polish)
Pyridin (German)
Piridina (Italian)
RCRA waste number U196
UN 1282 [Pyridine]

110-87-2
C_5H_8O
84.12
O(C=CCC1)C1
1,4-Dichlorobutane
AI3-16497
2,3-Dihidropirano (Spanish)
Dihydropyran (VAN)
δ2-Dihydropyran
2H-3,4-Dihydropyran
2,3-Dihydropyran
3,4-Dihydropyran [UN 2376]
2,3-Dihydro-4H-pyran
3,4-Dihydro-2H-pyran
5,6-Dihydro-4H-pyran
Dihydro-2,3 pyranne (French)
2H-Pyran, 3,4-dihydro- (9CI)
UN 2376 [2,3-Dihydropyran]
NSC-57860

110-88-3
$C_3H_6O_3$
90.09
O(COCO1)C1
s-Trioxane
Triformol
Trioxan

Trioxane
sym-Trioxane
1,3,5-Trioxane
Trioxymethylene

110-89-4
$C_5H_{11}N$
85.17
N(CCCC1)C1
Piperidine [UN 2401]
Azacyclohexane
Cyclopentimine
Cypentil
Hexahydropyridine
Hexazane
Pentamethyleneimine
Pentamethylenimine
Piperidin (German)
Pyridine, hexahydro-
UN 2401 [Piperidine]

110-91-8
C_4H_9NO
87.14
O(CCNC1)C1
Morpholine (ACGIH,OSHA) [UN 2054]
BASF 238
Diethyleneimide oxide
Diethylene imidoxide
Diethylene oximide
Diethylenimide oxide
Drewamine
p-Isoxazine, tetrahydro-
Morpholine, Aqueous, Mixture [NA 1760]
NA 1760 [Corrosive liquid, N.O.S.]
NA 2054 (DOT)
1-Oxa-4-azacyclohexane
2H-1,4-Oxazine, tetrahydro-
4H-1,4-Oxazine, tetrahydro-
Tetrahydro-1,4-isoxazine
Tetrahydro-1,4-oxazine
Tetrahydro-2H-1,4-oxazine
UN 2054 [Morpholine]

110-93-0
$C_8H_{14}O$
126.20
O=C(CCC=C(C)C)C
Methylheptenone
AI3-05639
5-Hepten-2-one, 6-methyl- (9CI)
Methyl heptenone
6-Methyl-5-hepten-2-one
6-Methyl-5-heptene-2-one
NSC-15294

110-94-1
$C_5H_8O_4$
132.13
O=C(O)CCCC(=O)O
Glutaric-acid
Pentandioic acid
Pentanedioic acid
1,5-Pentanedioic acid
1,3-Propanedicarboxylic acid

110-96-3
$C_8H_{19}N$
129.28
N(CC(C)C)CC(C)C
1-Propanamine, 2-methyl-N-(2-methylpropyl)
Diisobutylamine [UN 2361]
UN 2361 [Diisobutylamine]

110-97-4
$C_6H_{15}NO_2$
133.22
OC(C)CNCC(O)C
2-Propanol, 1,1'-iminodi
Bis(2-hydroxypropyl)amine
Bis(2-propanol)amine
Diisopropanolamine
DIPA
Dipropyl-2,2'-dihydroxy-amine
1,1'-Iminobis-2-propanol
1,1'-Iminodi-2-propanol
2-Propanol, 1,1'-iminobis- (9CI)

110-98-5
$C_6H_{14}O_3$
134.20
O(CC(O)C)CC(O)C
2-Propanol, 1,1'-oxydi
2,2'-Dihydroxydipropyl ether
2,2'-Dihydroxyisopropyl ether
Dipropylene glycol
Dipropylenglykol (Czech)
1,1'-Oxydi-2-propanol

110-99-6
$C_4H_6O_5$
134.10
O=C(O)COCC(=O)O
Acetic acid, oxydi
Acetic acid, 2,2'-oxybis- (9CI)
Bis(carboxymethyl)ether
Diglycolic acid (6CI)
3-Oxapentanedioic acid
Oxodiacetic acid
Oxybisacetic acid
Oxydiacetic acid
2,2'-Oxydiacetic acid
Oxydiethanolic acid

111-01-3
$C_{30}H_{62}$
422.92
C(CCCC(CCCC(CCCCC(CCCC(CCCC(C)C)C)C)C)C)(C)C
Tetracosane, 2,6,10,15,19,23-hexamethyl
Cosbiol
2,6,10,15,19,23-Hexamethyltetracosane
Squalane

111-02-4
$C_{30}H_{50}$
410.73
C(=CCCC=C(CCC=C(CCC=C(C)C)C)C)(CCC=C(CCC=C(C)C)C)C
2,6,10,14,18,22-Tetracosahexaene, 2,6,10,15,19,23-hexamethyl-, (all-E)
Spinacen
Spinacene
Squalen
Squalene
(E,E,E,E)-Squalene
trans-Squalene

111-03-5
$C_{21}H_{40}O_4$
356.55
O=C(OCC(O)CO)CCCCCCC
C=CCCCCCCCC
Glyceryl oleate
Aldo HMO
Aldo MO
Glycerin 1-monooleate
Glycerol α-monooleate
Glycerol 1-monooleate
Glycerol α-cis-9-octadecenate
Glyceryl monooleate (VAN)
1-Glyceryl oleate
Hefti GMO-33
Monoolein (VAN)
α-Monoolein
1-Monoolein
1-Monooleoylglycerol
NSC-406285
9-Octadecenoic acid, 2,3-dihydroxypropyl ester
9-Octadecenoic acid (Z)-, 2,3-dihydroxypropyl ester (9CI)
9-Octadecenoic acid, monoester with 1,2,3-propanetriol
Olein, 1-mono- (8CI)
1-Oleoylglycerol
1-Oleylglycerol
Unitina GMO

111-05-7
$C_{21}H_{41}NO_2$
339.56
O=C(NCC(O)C)CCCCCCCC=C
CCCCCCCC
Oleamide MIPA
Felsamid-OI
N-(2-Hydroxypropyl)-9-octadecenamide
(Z)-N-(2-Hydroxypropyl)-9-octadecenamide
Monoisopropanolamine oleic acid amide
9-Octadecenamide, N-(2-hydroxypropyl)-
9-Octadecenamide, N-(2-hydroxypropyl)-, (Z)- (9CI)
Oleic monoisopropanolamide
Witcamide 61

111-06-8
$C_{20}H_{40}O_2$
312.54
O=C(OCCCC)CCCCCCCCCCC

CCCC
Hexadecanoic acid, butyl ester (9CI)
AI3-07959
Butyl hexadecanoate
n-Butyl hexadecanoate
Butyl palmitate
n-Butyl palmitate
NSC-4815
Palmitic acid, butyl ester (8CI)

111-07-9
$C_{19}H_{38}O_3$
314.51
O=C(OCCOC)CCCCCCCCCCC
CCCC
Hexadecanoic acid, 2-methoxyethyl ester (9CI)
2-Methoxyethyl hexadecanoate

111-11-5
$C_9H_{18}O_2$
158.24
O=C(OC)CCCCCCC
Methyl caprylate
AI3-01979
Caprylic acid methyl ester
Methyl octanoate
Methyl n-octanoate
Methyl octylate
NSC-3710
Octanoic acid, methyl ester (9CI)
Uniphat A20

111-13-7
$C_8H_{16}O$
128.24
O=C(CCCCCC)C
2-Octanone
Methyl hexyl ketone

111-14-8
$C_7H_{14}O_2$
130.21
O=C(O)CCCCCC
Heptanoic-acid
Enanthic acid
Enanthylic acid
n-Heptoic acid
Heptylic acid
n-Heptylic acid
Hexacid C-7
1-Hexanecarboxylic acid
Oenanthic acid
Oenanthylic acid

111-15-9
$C_6H_{12}O_3$
132.18
O=C(OCCOCC)C
Ethanol, 2-ethoxy-, acetate
Acetate de cellosolve (French)
Acetate de l'ether monoethylique de l'ethylene-glycol (French)
Acetate d'ethylglycol (French)
Acetato di cellosolve (Italian)
Acetic acid, 2-ethoxyethyl ester
2-Aethoxy-aethylacetat (German)
Aethylenglykolaetheracetat (German)
Cellosolve acetate (DOT, OSHA)
Celosolvacetat (Czech)
CSAC
Ektasolve EE Acetate Solvent
Ethoxy acetate
2-Ethoxyethanol acetate
2-Ethoxyethanol, ester with acetic acid
2-Ethoxy-ethylacetaat (Dutch)
Ethoxyethyl acetate
β-Ethoxyethyl acetate
2-Ethoxyethylacetate
2-Ethoxyethyl acetate (ACGIH, OSHA)
2-Ethoxyethyle, acetate de (French)
2-Ethoxyethylester kyseliny octove (Czech)
Ethyl cellosolve acetaat (Dutch)
Ethylene glycol ethyl ether acetate
Ethylene glycol monoethyl ether acetate [UN 1172]
Ethylglycol acetate
Ethylglykolacetat (German)
2-Etossietil-acetato (Italian)
Glycol Ether EE Acetate
Glycol monoethyl ether acetate
Octan etoksyetylu (Polish)
Oxytol acetate
Poly-Solv EE Acetate
UN 1172 [Ethylene glycol monoethyl ether acetate]

111-16-0
$C_7H_{12}O_4$
160.19
O=C(O)CCCCCC(=O)O
Pimelic-acid
Heptandioic acid
Heptanedioic acid
Heptane-1,7-dioic acid
1,7-Heptanedioic acid
1,5-Pentanedicarboxylic acid
Pileric acid

111-17-1
$C_6H_{10}O_4S$
178.22
O=C(O)CCSCCC(=O)O
Propionic acid, 3,3'-thiodi
Bis(2-carboxyethyl) sulfide
Diethyl sulfide 2,2'-dicarboxylic acid
Kyselina β,β'-thiodipropionova (Czech)
Kyselina 3,3-thiodipropionova (Czech)
Sulfide, bis(2-carboxyethyl)
TDPA
4-Thiaheptanedioic acid
Thiodihydracrylic acid
Thiodipropionic acid
2-(2,3,5,6-Tetramethylphenoxy)-propionic acid
β,β'-Thiodipropionic acid
3,3'-Thiodipropionic acid
Tyox A

111-18-2
$C_{10}H_{24}N_2$
172.36
N(CCCCCCN(C)C)(C)C
1,6-Hexanediamine, N,N,N',N'-tetramethyl
N,N,N',N'-Tetramethylhexamethylene diamine
N,N,N',N'-Tetramethyl-1,6-hexanediamine

111-19-3
$C_{10}H_{16}Cl_2O_2$
239.14
O=C(CCCCCCCCC(=O)Cl)Cl
Decanedioyl dichloride (9CI)
Decanedioyl chloride
NSC-56763
Sebacic acid dichloride
Sebacoyl chloride (8CI)
Sebacoyl dichloride
Sebacyl chloride

111-20-6
$C_{10}H_{18}O_4$
202.28
O=C(O)CCCCCCCCC(=O)O
Sebacic-acid
Decanedioic acid
1,8-Octanedicarboxylic acid
USAF HC-1

111-21-7
$C_{10}H_{18}O_6$
234.28
O=C(OCCOCCOCCOC(=O)C)C
Triethylene glycol, diacetate
Acetic acid, triethylene glycol diester
Ethanol, 2,2'-ethylenedioxydi-, diacetate
2,2'-(Ethylenedioxy)di(ethyl acetate)
Triglycol, diacetate

111-22-8
$C_6H_{12}O_4.N_2O_4$
240.20
O=N(=O)OCCOCCOCCON
(=O)=O
Triethylene glycol, dinitrate
TEGDN
TEGON

111-24-0
$C_5H_{10}Br_2$
229.97
BrCCCCCBr
Pentane, 1,5-dibromo
1,5-Dibromopentane
Pentamethylene bromide

Pentamethylene dibromide

111-25-1
$C_6H_{13}Br$
165.10
BrCCCCCC
Hexane, 1-bromo
Bromohexane
1-Bromohexane
Hexyl bromide
n-Hexyl bromide
1-Hexyl bromide

111-26-2
$C_6H_{15}N$
101.22
NCCCCCC
Hexylamine
1-Aminohexane
1-Hexanamine
n-Hexylamine
Mono-n-hexylamine

111-27-3
$C_6H_{14}O$
102.20
OCCCCCC
Hexyl-alcohol
Amylcarbinol
Caproyl alcohol
Epal 6
Hexanol
n-Hexanol [UN 2282]
1-Hexanol
n-Hexyl alcohol
1-Hydroxyhexane
Pentylcarbinol
UN 2282 [Hexanols]

111-28-4
$C_6H_{10}O$
98.16
OCC=CC=CC
2,4-Hexadien-1-ol
Hexacose
Hexadenol
2,4-Hexadienol
Hexakose
Hexene-ol
Hexenol
1-Hydroxy-2,4-hexadiene
Sorbic alcohol
Sorbinic alcohol
Sorbyl alcohol

111-29-5
$C_5H_{12}O_2$
104.17
OCCCCCO
1,5-Pentanediol
1,5-Dihydroxypentane
Pentamethylene glycol
Pentane-1,5-diol
1,5-Pentylene glycol

111-30-8
$C_5H_8O_2$
100.13
O=CCCCC=O
Glutaraldehyde (ACGIH, OSHA)
Cidex
Glutaral
Glutaraldehyd (Czech)
Glutardialdehyde
Glutaric dialdehyde
NCI-C55425
1,5-Pentanedial
1,5-Pentanedione
Potentiated acid glutaraldehyde
Sonacide

111-31-9
$C_6H_{14}S$
118.26
SCCCCCC
1-Hexanethiol
Hexyl mercaptan
USAF EK-4628

111-32-0
$C_5H_{12}O_2$
104.17
1-Butanol, 4-methoxy
Butylene glycol methyl ether
Butylene glycol monomethyl ether
Dowanol BM
4-Methoxy-1-butanol

111-34-2
$C_6H_{12}O$
100.18
O(C=C)CCCC
Ether, butyl vinyl
Butane, 1-(ethenyloxy)-
Butoxyethene
Butyl vinyl ether
Butyl vinyl ether, Inhibited [UN 2352]
UN 2352 [Butyl vinyl ether, inhibited]
Vinyl butyl ether
Vinyl n-butyl ether

111-35-3
$C_5H_{12}O_2$
104.17
CCOCCCO
1-Propanol, 3-ethoxy
3-Ethoxy-1-propanol
Propylene glycol monoethyl ether, β
Propylene glycol β-monoethyl ether

111-36-4
C_5H_9NO
99.15
O=C=NCCCC
Isocyanic acid, butyl ester
BIC
n-Butyl isocyanate [UN 2485]
UN 2485 [n-Butyl isocyanate]

111-40-0
$C_4H_{13}N_3$
103.20
N(CCN)CCN
Diethylenetriamine (ACGIH, OSHA) [UN 2079]
Aminoethylethandiamine
3-Azapentane-1,5-diamine
Bis(2-aminoethyl)amine
Bis(β-aminoethyl)amine
D.E.H. 20
DETA
2,2'-Diaminodiethylamine
Diethylamine, 2,2'-diamino-
Ethylamine, 2,2'-iminobis-
Ethylenediamine, N-(2-aminoethyl)-
UN 2079 [Diethylenetriamine]

111-41-1
$C_4H_{12}N_2O$
104.18
OCCNCCN
Ethanol, 2-((2-aminoethyl)-amino)
Aminoethyl ethanolamine
N-Aminoethylethanolamine
Ethanolethylene diamine
N-Hydroxyethyl-1,2-ethanediamine
N-(β-Hydroxyethyl)ethylenediamine
N-(2-Hydroxyethyl)ethylenediamine
Monoethanolethylenediamine

111-42-2
$C_4H_{11}NO_2$
105.16
OCCNCCO
Ethanol, 2,2'-iminodi
Bis(2-hydroxyethyl)amine
DEA
Diaethanolamin (German)
Diethanolamin (Czech)
Diethanolamine (ACGIH, OSHA)
Diethylamine, 2,2'-dihydroxy-
Diethylolamine
2,2'-Dihydroxydiethylamine
Di(2-hydroxyethyl)amine
Diolamine
2-(2-Hydroxyethylamino)-ethanol
2,2'-Iminobisethanol
2,2'-Iminodiethanol
NCI-C55174

111-43-3
$C_6H_{14}O$
102.20
O(CCC)CCC
Propyl-ether
Dipropyl ether [UN 2384]
Dipropyl oxide
Ether, di-n-propyl-
Propane, 1,1'-oxybis- (9CI)

UN 2384 [Dipropyl ether]

111-44-4
$C_4H_8Cl_2O$
143.02
O(CCCl)CCCl
Ether, bis(2-chloroethyl)
Bis(β-chloroethyl) ether
Bis(2-chloroethyl) ether
Chlorex
1-Chloro-2-(β-chloroethoxy)-ethane
Chloroethyl ether
Clorex
DCEE
2,2'-Dichloorethylether (Dutch)
2,2'-Dichlor-diaethylaether (German)
2,2'-Dichlorethyl ether
2,2-Dichlorodiethyl ether [UN 1916]
β,β-Dichlorodiethyl ether
Dichloroether
Dichloroethyl ether
Di(β-chloroethyl)ether
Di(2-chloroethyl) ether
β,β'-Dichloroethyl ether
sym-Dichloroethyl ether
2,2'-Dichloroethyl ether
Dichloroethyl ether (ACGIH, DOT,OSHA)
Dichloroethyl oxide
2,2'-Dicloroetiletere (Italian)
Dwuchlorodwuetylowy eter (Polish)
ENT 4,504
Ethane, 1,1'-oxybis(2-chloro-
Ether dichlore (French)
1,1'-Oxybis(2-chloro)ethane
Oxyde de chlorethyle (French)
RCRA waste number U025
UN 1916 [2,2-Dichlorodiethyl ether]

111-46-6
$C_4H_{10}O_3$
106.14
O(CCO)CCO
Diethylene-glycol
Bis(2-hydroxyethyl) ether
Brecolane NDG
Carbitol
Deactivator E
Deactivator H
DEG
Dicol
Diethylenglykol (Czech)
Diglycol
Dihydroxydiethyl ether
β,β'-Dihydroxydiethyl ether
2,2'-Dihydroxyethyl ether
Dissolvant APV
Ethanol, 2,2'-oxydi-
Ethylene diglycol
Glycol ether
Glycol ethyl ether
3-Oxapentane-1,5-diol
3-Oxa-1,5-pentanediol
2,2'-Oxybisethanol
2,2'-Oxydiethanol
TL4N

111-47-7
$C_6H_{14}S$
118.24
S(CCC)CCC
Propane, 1,1'-thiobis- (9CI)
AI3-18787
Dipropyl sulfide
Di-n-propyl sulfide
Dipropyl thioether
NSC-78429
Propyl monosulfide
Propyl sulfide (8CI)
4-Thiaheptane
1,1'-Thiobispropane

111-48-8
$C_4H_{10}O_2S$
122.20
OCCSCCO
Ethanol, 2,2'-thiodi
Bis(β-hydroxyethyl)sulfide
Bis(2-hydroxyethyl)sulfide
β,β'-Dihydroxydiethyl sulfide
β,β'-Dihydroxyethyl sulfide
Glyecine A
β-Hydroxyethyl sulfide
Kromfax Solvent
Sulfide, bis(2-hydroxyethyl)
2,2'-Thiodiethanol
Thiodiethylene glycol
Thiodiglycol
β-Thiodiglycol

111-49-9
$C_6H_{13}N$
99.20
N(CCCCC1)C1
1H-Azepine, hexahydro
Azacycloheptane
1-Azacycloheptane
Cyclohexamethylenimine
G 0
Hexahydroazepine
Hexahydro-1H-azepine
Hexamethyleneimine [UN 2493]
Hexamethylenimine
Homopiperidine
Perhydroazepine
UN 2493 [Hexamethylene-imine]

111-50-2
$C_6H_8Cl_2O_2$
183.03
O=C(CCCCC(=O)Cl)Cl
Adipoyl chloride
AI3-52262
Hexanedioyl dichloride

111-54-6
$C_4H_8N_2S_4$
212.37
Ethylenebisdithiocarbamic acid
Carbamic acid, ethylenebis-(dithio-
Carbamodithioic acid, 1,2-ethanediylbis-
Carbamodithioic acid, 1,2-ethanediylbis-, salts and esters
EBDC
1,2-Ethanedicarbamic acid, tetrathio-
Ethylenebis(dithiocarbamic acid)
Ethylenebisdithiocarbamic acid, salts & esters
Nabam
RCRA waste No. U114

111-55-7
$C_6H_{10}O_4$
146.16
O=C(OCCOC(=O)C)C
Ethylene glycol, diacetate
1,2-Ethanediol diacetate
Ethylene acetate
Ethylene glycol acetate
Glycol diacetate

111-57-9
$C_{20}H_{41}NO_2$
327.55
O=C(NCCO)CCCCCCCCCCCCCCCCC
Stearamide MEA
Clindrol 200-MS
Comperlan HS
Cycloamide SM
Felsamid-SM
N-(2-Hydroxyethyl)octadecan-amide
N-(Hydroxyethyl)stearamide
N-(2-Hydroxyethyl)stearamide
Incromide SM
Loramine S 280
Marlamid M 18
Monoethanolamine stearic acid amide
NSC-3377
Octadecanamide, N-(2-hydroxy-ethyl)- (9CI)
Onyx Wax EL
Stearamyl
Stearic acid monoethanolamide
Stearic ethanolamide
Stearic ethylolamide
Stearic monoethanolamine
Stearoyl monoethanolamide
Stearoylethanolamine
N-Stearoylethanolamine
Stearoylmonoethanolamide
Upamide SME-M
Witcamide 70

111-59-1
$C_{21}H_{40}O_2$
324.55
O=C(OCCC)CCCCCCCC=CCCCCCCC
Oleic acid, propyl ester (8CI)
AI3-26171
Emery Oleic Acid Ester 2302

Emery 2302
NSC-50932
9-Octadecenoic acid (Z)-, propyl ester (9CI)
Propyl 9-octadecenoate
Propyl oleate

111-60-4
$C_{20}H_{40}O_3$
328.60
O=C(OCCO)CCCCCCCCCCCCCCCCC
Stearic acid, 2-hydroxyethyl ester
Clindrol Seg
Emerest 2350
Empilan 2848
Ethylene glycol, monostearate
Ethylene glycol stearate
Glycol monostearate
Glycol stearate
Ivorit
Lipo EGMS
Monthybase
Monthyle
Parastarin
Prodhybas N
Prodhybase ethyl
S 151
Sedetol
Stearic acid, monoester with ethylene glycol
Tego-Stearate
USAF KE-11

111-61-5
$C_{20}H_{40}O_2$
312.60
O=C(OCC)CCCCCCCCCCCCCCCCC
Stearic acid, ethyl ester
Ethyl octadecanoate
Ethyl n-octadecanoate
Ethyl stearate

111-62-6
$C_{20}H_{38}O_2$
310.52
O=C(OCC)CCCCCCCC=CCCCCCCCC
Ethyl oleate
AI3-00657
Ethyl cis-9-octadecenoate
NSC-229428
9-Octadecenoic acid (Z)-, ethyl ester (9CI)
Oleic acid, ethyl ester (8CI)

111-63-7
$C_{20}H_{38}O_2$
310.52
O=C(OC=C)CCCCCCCCCCCCCCCCC
Octadecanoic acid, ethenyl ester (9CI)
AI3-23120
Ethenyl octadecanoate
NSC-20891
Stearic acid, vinyl ester (8CI)
Vinyl stearate

111-64-8
$C_8H_{15}ClO$
162.66
O=C(CCCCCCC)Cl
Caprylyl chloride
Octanoyl chloride (9CI)

111-65-9
C_8H_{18}
114.26
C(CCCCCC)C
Octane (ACGIH,OSHA) [UN 1262]
Oktan (Polish)
Oktanen (Dutch)
Ottani (Italian)
UN 1262 [Octanes]

111-66-0
C_8H_{16}
112.22
C(=C)CCCCCC
1-Octene (9CI)
AI3-28403
Caprylene
1-Caprylene
NSC-8457
Octene-1
α-Octene
n-1-Octene
1-n-Octene
Octylene
α-Octylene
1-Octylene

111-67-1
C_8H_{16}
112.22
C(=CC)CCCCC
2-Octene, Mixed cis & trans
NSC-66572
Octene-2
2-Octene (9CI)
2-Octene (Mixed cis, trans isomers)

111-68-2
$C_7H_{17}N$
115.25
NCCCCCCC
Heptylamine
1-Aminoheptane
1-Heptanamine
n-Heptylamine
1-Heptylamine

111-69-3
$C_6H_8N_2$
108.16
N#CCCCCC#N
Adiponitrile
Adipic acid dinitrile
Adipic acid nitrile
Adipodinitrile
Adiponitrile (DOT)
1,4-Dicyanobutane
Hexanedinitrile
Hexanedioic acid, dinitrile
Nitrile adipico (Italian)
Tetramethylene cyanide
UN 2205 (DOT)

111-70-6
$C_7H_{16}O$
116.23
OCCCCCCC
Heptyl-alcohol
Enanthic alcohol
n-Heptanol
n-Heptanol-1 (French)
1-Heptanol
1-Hydroxyheptane
L'Alcool N-heptylique primaire (French)

111-71-7
$C_7H_{14}O$
114.21
O=CCCCCCC
Heptanal
Enanthal
Enanthaldehyde
Enanthole
Heptaldehyde
Heptyl aldehyde
Oenanthal
Oenanthaldehyde
Oenanthic aldehyde
Oenanthol

111-76-2
$C_6H_{14}O_2$
118.20
O(CCCC)CCO
Ethanol, 2-butoxy
BUCS
Butoksyetylowy alkohol (Polish)
2-Butossi-etanolo (Italian)
2-Butoxy-aethanol (German)
Butoxyethanol
n-Butoxyethanol
2-Butoxyethanol (ACGIH, OSHA)
2-Butoxy-1-ethanol
Butyl cellosolve
Butylcelosolv (Czech)
O-Butyl ethylene glycol
Butyl glycol
Butylglycol (French,German)
Butyl oxitol
Dowanol EB
Ektasolve EB
Ethylene glycol n-butyl ether
Ethylene glycol, monobutyl ether
Ethylene glycol monobutyl ether [UN 2369]
Gafcol EB
Glycol butyl ether
Glycol Ether EB
Glycol Ether EB Acetate

Glycol monobutyl ether
Jeffersol EB
Monobutyl ether of ethylene glycol
Monobutyl glycol ether
3-Oxa-1-heptanol
Poly-Solv EB
UN 2369 [Ethylene glycol monobutyl ether]

111-77-3
$C_5H_{12}O_3$
120.17
O(CCOC)CCO
Ethanol, 2-(2-methoxyethoxy)
Diethylene glycol methyl ether
Diethylene glycol monomethyl ether
Diglycol monomethyl ether
Dowanol DM
Ethylene diglycol monomethyl ether
MECB
Methoxydiglycol
2-(2-Methoxyethoxy)ethanol
β-Methoxy-β'-hydroxydiethyl ether
Methyl carbitol
Methyl karbitol (Czech)
Poly-Solv DM

111-78-4
C_8H_{12}
108.18
C(=CCCC=CC1)C1
1,5-Cyclooctadiene (9CI)
AI3-26692
Cycloocta-1,5-diene
NSC-60155

111-81-9
$C_{12}H_{22}O_2$
198.31
O=C(OC)CCCCCCCCC=C
10-Undecenoic acid, methyl ester (9CI)
AI3-00647
Methyl undecenate
Methyl 10-undecenate
Methyl undecenoate
Methyl 10-undecenoate
NSC-1273
Undecenoic acid, methyl ester
Undecylenic acid, methyl ester

111-82-0
$C_{13}H_{26}O_2$
214.35
O=C(OC)CCCCCCCCCCC
Methyl dodecanoate
AI3-00669
Dodecanoic acid, methyl ester (9CI)
Lauric acid, methyl ester (8CI)
Metholene 2296
Methyl n-dodecanoate
Methyl dodecylate
Methyl laurate
Methyl laurinate
NSC-5027
Uniphat A40

111-83-1
$C_8H_{17}Br$
193.16
BrCCCCCCCC
Octane, 1-bromo
1-Bromooctane
n-Octyl bromide

111-84-2
C_9H_{20}
128.29
C(CCCCCCC)C
Nonane (ACGIH,OSHA) [UN 1920]
n-Nonane
UN 1920 [Nonanes]
Shellsol 140

111-85-3
$C_8H_{17}Cl$
148.68
ClCCCCCCCC
1-Chlorooctane
Capryl chloride
NSC-5406
Octane, 1-chloro- (9CI)
Octyl chloride
n-Octyl chloride
1-Octyl chloride

111-86-4
$C_8H_{19}N$
129.28
NCCCCCCCC
1-Octanamine
1-Aminooctane
Armeen 8
Armeen 8D
Caprylamine
Caprylylamine
Octylamine
n-Octylamine

111-87-5
$C_8H_{18}O$
130.26
OCCCCCCCC
Octyl-alcohol
Alcohol C-8
Alfol 8
Caprylic alcohol
Dytol M-83
Epal 8
Heptyl carbinol
1-Hydroxyoctane
Lorol 20
Octanol
n-Octanol
1-Octanol
Octilin
Octyl alcohol, normal-primary
n-Octyl alcohol
Primary octyl alcohol
Sipol l8

111-88-6
$C_8H_{18}S$
146.30
SCCCCCCCC
1-Octanethiol (9CI)
AI3-06557
1-Mercaptooctane
NSC-41903
Octane-1-thiol
Octyl mercaptan
n-Octyl mercaptan
1-Octyl thiol
Octylthiol
1-Octylthiol

111-89-7
$C_7H_{16}O_2$
132.20
Pentane, 1,5-dimethoxy- (6CI,7CI,8CI,9CI)
1,5-Dimethoxypentane
1,5-Pentanediol dimethyl ether

111-90-0
$C_6H_{14}O_3$
134.20
O(CCOCC)CCO
Ethanol, 2-(2-ethoxyethoxy)
APV
Carbitol
Carbitol cellosolve
Carbitol solvent
Diethylene glycol ethyl ether
Diethylene glycol monoethyl ether
Diglycol monoethyl ether
3,6-Dioxa-1-octanol
3,6-Dioxa-1-oktanol (Czech)
Dioxitol
Dowanol
Dowanol DE
Ethoxy diglycol
2-(2-Ethoxyethoxy)ethanol
Ethyl carbitol
Ethyl diethylene glycol
Ethylene diglycol monoethyl ether
Karbitol (Czech)
Losungsmittel APV
Monoethyl ether of diethylene glycol
Poly-Solv
Solvolsol
Transcutol

111-91-1
$C_5H_{10}Cl_2O_2$
173.05
O(CCCl)COCCCl
Methane, bis(2-chloroethoxy)
Bis(2-chloroethoxy)-methane
Bis(β-chloroethyl)formal
Bis(2-chloroethyl)formal
Dichloroethyl formal
Di-2-chloroethyl formal
Ethane, 1,1'-(methylenebis-

(oxy))bis(2-chloro-
Formaldehyde bis(β-chloro-ethyl) acetal
RCRA waste number U024

111-92-2
$C_8H_{19}N$
129.28
N(CCCC)CCCC
Dibutylamine
1-Butanamine, n-butyl-
n-Butyl-1-butanamine
Dibutilamina (Romanian)
Di-n-butylamine [UN 2248]
n-Dibutylamine
Di(n-butyl)amine (DOT)
UN 2248 [Di-n-butylamine]

111-94-4
$C_6H_9N_3$
123.18
N#CCCNCCC#N
Propionitrile, 3,3'-iminodi
BBCE
Bis(β-cyanoethyl)amine
Bis-(2-cyanoethyl)amine
N,N-Bis(2-cyanoethyl)amine
Bis-(2-kyanethyl)amin (Czech)
Di(2-cianoetil)ammina (Italian)
Di-(2-cyanoethyl)amine
Diethylamine, 2,2'-dicyano-
Ethanamine, 2-cyano-N-(2-cyanoethyl)-
IDPN
Imino-β,β'-dipropionitrile
β,β-Iminodipropionitrile
β,β'-Iminodipropionitrile
3,3'-Iminodipropionitrile
2341 I.S.
Propanenitrile, 3,3'-iminobis-
USAF A-8564

111-95-5
$C_6H_{15}NO_2$
133.18
O(CCNCCOC)C
Dimethoxyethylamine
Bis(methoxyethyl)amine
Bis(2-methoxyethyl)amine
Diethylamine, 2,2'-dimethoxy-(8CI)
2,2'-Dimethoxydiethylamine
Ethanamine, 2-methoxy-N-(2-methoxyethyl)- (9CI)
2-Methoxy-N-(2-methoxyethyl)-ethanamine
NSC-78431

111-96-6
$C_6H_{14}O_3$
134.20
O(CCOC)CCOC
Ether, bis(2-methoxyethyl)
Bis(2-methoxyethyl)ether
Diethylene glycol dimethyl ether
Diethyl glycol dimethyl ether
Diglyme
Ethane, 1,1'-oxybis(2-methoxy-(9CI)

111-97-7
$C_6H_8N_2S$
140.22
N#CCCSCCC#N
Propionitrile, 3,3'-thiodi
β,β'-Dicyanodiethyl sulfide
Di(2-cyanoethyl)sulfide
Nitril kyseliny β,β'-thiodipropionove (Czech)
Sulfide, bis(2-cyanoethyl)
2,2'-Thiodiethylkyanid (Czech)
β,β'-Thiodipropionitrile
USAF HA-5

112-00-5
$C_{15}H_{34}N.Cl$
263.89
[Cl-].CCCCCCCCCCCC[N+](C)(C)C
Laurtrimonium chloride
Alicop
Aliquat 4
Ammonium, dodecyltrimethyl-, chloride (8CI)
Arquad 12
Arquad 12D
Arquad 12/50
Cation BB
Cation FB
Chemquat 12-33
Chemquat 12-50
Dehyquart LT
1-Dodecanaminium, N,N,N-trimethyl-, chloride (9CI)
Dodecyltrimethylammonium chloride
n-Dodecyltrimethylammonium chloride
Lauryl trimethyl ammonium chloride
Lauryltrimethylammonium chloride
Nissan Cation BB
NSC-6931
Rewoquat B18
N,N,N-Trimethyl-1-dodecanaminium chloride
Trimethyldodecylammonium chloride
Trimethyllaurylammonium chloride

112-02-7
$C_{19}H_{42}N.Cl$
320.00
[Cl-].CCCCCCCCCCCCCCCC[N+](C)(C)C
Cetrimonium chloride
Adogen 444
AI3-25173-X
Aliquat 6
Ammonium, hexadecyltrimethyl-, chloride
Arquad 16/28
Arquad 16-29
Arquad 16-50
Barquat CT 29
Caswell No. 167A
Cation PB 40
Cetyl trimethyl ammonium chloride
Cetyltrimethylammonium chloride
Chemquat 16-29
Chemquat 16-50
Dehyquart A
EPA Pesticide Chemical Code 069133
Genamin CTAC
1-Hexadecanaminium, N,N,N-trimethyl-, chloride (9CI)
1-Hexadecanaminium, N,N,N-trimethyl-, chloride (50% in 2-propanol)
Hexadecyl trimethyl ammonium chloride
Hexadecyltrimethylammonium chloride
n-Hexadecyltrimethylammonium chloride
HTAC
Intexan CTC 29
Intexsan CTC 29
Intexsan CTC 50
Morpan CHA
Nissan Cation PB 40
Palmityltrimethylammonium chloride
Trimethylcetylammonium chloride
N,N,N-Trimethyl-1-hexadecanaminium chloride
Trimethylhexadecylammonium chloride
Variquat E 228

112-03-8
$C_{21}H_{46}N.Cl$
348.13
[Cl-].CCCCCCCCCCCCCCCCCC[N+](C)(C)C
Ammonium, trimethyloctadecyl-, chloride
Aliquat 7
Arquad 18
Arquad 18-50
Cation AB
Nissan Cation AB
Octadecyltrimethylammonium chloride
Quaternium-10
STAC
Stearyltrimethylammonium chloride
Trimethyloctadecylammonium chloride
Trimethylstearylammonium chloride

112-04-9
$C_{18}H_{37}Cl_3Si$
387.99
Silane, octadecyltrichloro
Octadecyltrichlorosilane [UN 1800]
Silane, trichlorooctadecyl-

UN 1800 [Octadecyltrichlorosilane]

112-05-0
$C_9H_{18}O_2$
158.27
O=C(O)CCCCCCCC
Nonanoic-acid
Cirrasol 185A
Emfac 1202
Hexacid C-9
n-Nonoic acid
n-Nonylic acid
Pelargic acid
1-Octanecarboxylic acid
Pelargon (Russian)
Pelargonic acid

112-06-1
$C_9H_{18}O_2$
158.27
O=C(OCCCCCCC)C
Acetic acid, heptyl ester
Acetate C-7
Heptanyl acetate
Heptyl acetate
n-Heptyl acetate
1-Heptyl acetate

112-07-2
$C_8H_{16}O_3$
160.24
O=C(OCCOCCCC)C
Ethanol, 2-butoxy-, acetate
Acetic acid, 2-butoxyethyl ester
2-Butoxyethanol acetate
2-Butoxyethyl acetate
2-Butoxyethylester kyseliny octove (Czech)
Butyl cellosolve acetate
Butylcelosolvacetat (Czech)
Butylglycol acetate
Ektasolve EB Acetate
Ethylene glycol monobutyl ether acetate
Glycol monobutyl ether acetate

112-10-7
$C_{21}H_{42}O_2$
326.63
O=C(OC(C)C)CCCCCCCCCCCCCCCCC
Stearic acid, isopropyl ester
Isopropyl stearate
1-Methylethyl octadecanoate
Octadecanoic acid, 1-methylethyl ester (9CI)
Wickenol 127

112-11-8
$C_{21}H_{40}O_2$
324.55
O=C(OC(C)C)CCCCCCCC=CCCCCCCCC
Isopropyl oleate
AI3-32462
Isopropyl oleate
1-Methylethyl-9-octadecenoate
NSC-50952
9-Octadecenoic acid, 1-methylethyl ester
9-Octadecenoic acid (Z)-, 1-methylethyl ester (9CI)
Oleic acid, isopropyl ester (8CI)

112-12-9
$C_{11}H_{22}O$
170.33
O=C(CCCCCCCCC)C
2-Undecanone
2-Hendecanone
Methyl nonyl ketone
Methyl n-nonyl ketone
MGK Dog and Cat Repellent
Nonyl methyl ketone
Undecan-2-one

112-13-0
$C_{10}H_{19}ClO$
190.71
O=C(CCCCCCCCC)Cl
Decanoyl chloride (9CI)

112-14-1
$C_{10}H_{20}O_2$
172.30
O=C(OCCCCCCCC)C
Acetic acid, octyl ester
Acetate C-8
Caprylyl acetate
1-Octanol acetate
n-Octanyl acetate
Octyl acetate
n-Octyl acetate
1-Octyl acetate
Octyl alcohol acetate

112-15-2
$C_8H_{16}O_4$
176.24
O=C(OCCOCCOCC)C
Ethanol, 2-(2-ethoxyethoxy)-, acetate
Acetic acid 2-(2-ethoxyethoxy)-ethyl ester
Carbitol acetate
Diethylene glycol monoethyl ether acetate
Diglycol monoethyl ether acetate
Ektasolve DE Acetate
2-(2-Ethoxyethoxy)ethanol acetate
2-(2-Ethoxyethoxy)ethyl acetate
2-(2-Ethoxyethoxy)ethylester kyseliny octove (Czech)
Glycol ether de acetate
Karbitolacetat (Czech)

112-16-3
$C_{12}H_{23}ClO$
218.77
O=C(CCCCCCCCCCC)Cl
Dodecanoyl chloride
AI3-52409
Dodecanoic acid, chloride
n-Dodecanoyl chloride
Lauric acid chloride
Lauroyl chloride

112-17-4
$C_{12}H_{24}O_2$
200.32
O=C(OCCCCCCCCCC)C
Acetic acid, decyl ester (9CI)
Acetate C 10
Acetate C-10
AI3-11098
1-Decanol acetate
Decyl acetate
n-Decyl acetate
n-Decyl ethanoate
NSC-46131

112-18-5
$C_{14}H_{31}N$
213.46
N(CCCCCCCCCCCC)(C)C
Dodecylamine, N,N-dimethyl
ADMA 2
Armeen DM-12D
Barlene 125
DDA
N,N-Dimethyldodecylamine
N,N-Dimethyllaurylamine
1-Dodecanamine, N,N-dimethyl-
Dodecyldimethylamine
N-Dodecyldimethylamine
Lauryldimethylamine
N-Lauryldimethylamine
Monolauryl dimethylamine
RC 5629

112-20-9
$C_9H_{21}N$
143.27
NCCCCCCCCC
1-Nonylamine
AI3-16562
1-Nonanamine

112-24-3
$C_6H_{18}N_4$
146.28
N(CCNCCN)CCN
Triethylenetetramine [UN 2259]
Araldite Hardener HY 951
Araldite HY 951
N,N-Bis(2-aminoethyl)-1,2-diaminoethane
N,N'-Bis(2-aminoethyl)ethylenediamine
DEH 24
3,6-Diazaoctane-1,8-diamine
1,2-Ethanediamine, N,N'-bis-(2-aminoethyl)-
Ethylenediamine, N,N'-bis-(2-aminoethyl)-
HY 951
Tecza

Teta
1,4,7,10-Tetraazadecane
Trien
Trientine
UN 2259 [Triethylenetetramine]

112-25-4
$C_8H_{18}O_2$
146.26
O(CCCCCC)CCO
Ethanol, 2-(hexyloxy)
Cellosolve, n-hexyl-
Ethylene glycol n-hexyl ether
Ethylene glycol monohexyl ether
Glycol monohexyl ether
Hexyl cellosolve
n-Hexyl cellosolve
2-(Hexyloxy)ethanol

112-26-5
$C_6H_{12}Cl_2O_2$
187.08
O(CCOCCCl)CCCl
Ethane, 1,2-bis(2-chloroethoxy)
1,2-Bis(2-chloroethoxy)ethane
2-(2-Chlorethoxy)ethyl 2'-chlorethyl ether
2-(2-Chloroethoxy)ethyl 2'-chloroethyl ether
Triethylene glycol dichloride
Triglycol dichloride

112-27-6
$C_6H_{14}O_4$
150.20
O(CCOCCO)CCO
Triethylene-glycol
1,2-Bis(2-hydroxyethoxy)ethane
Di-β-Hydroxyethoxyethane
3,6-Dioxaoctane-1,8-diol
2,2'-(1,2-Ethanediylbis(oxy))-bisethanol
Ethanol, 2,2'-(ethylenedioxy)di-
2,2'-Ethylenedioxybis(ethanol)
2,2'-Ethylenedioxydiethanol
2,2'-Ethylenedioxyethanol
Ethylene glycol-bis-(2-hydroxyethyl ether)
Ethylene glycol dihydroxydiethyl ether
Glycol bis(hydroxyethyl) ether
TEG
Triethylenglykol (Czech)
Trigen
Triglycol

112-29-8
$C_{12}H_{21}Br$
245.24
BrCCCCCCCCCC
Decane, 1-bromo
1-Bromodecane
Decyl bromide
n-Decyl bromide
1-Decyl bromide

112-30-1
$C_{10}H_{22}O$
158.32
OCCCCCCCCCC
Decyl-alcohol
Agent 504
Alcohol C-10
Antak
C 10 Alcohol
Capric alcohol
Caprinic alcohol
Decanol
n-Decanol
1-Decanol
n-Decatyl alcohol
n-Decyl alcohol
Decylic alcohol
Dytol S-91
Epal 10
Lorol 22
Nonylcarbinol
Panorama
Paranol
Primary decyl alcohol
Royaltac
Royaltac-85
Royaltac M-2
Sipol 110

112-31-2
$C_{10}H_{20}O$
156.30
O=CCCCCCCCCC
1-Decanal
Aldehyde C10
C-10 Aldehyde
Capraldehyde
Capric aldehyde
Caprinaldehyde
Caprinic aldehyde
Decaldehyde
n-Decaldehyde
Decanal
n-Decanal
Decanaldehyde
Decyl aldehyde
n-Decyl aldehyde
1-Decyl aldehyde
Decylic aldehyde

112-34-5
$C_8H_{18}O_3$
162.26
O(CCOCCO)CCCC
Ethanol, 2-(2-butoxyethoxy)
BUCB
Butoxydiethylene glycol
Butoxydiglycol
2-(2-Butoxyethoxy)ethanol
Butyl carbitol
O-Butyl diethylene glycol
Butyl dioxitol
Diethylene glycol n-butyl ether
Diethylene glycol monobutyl ether
Diglycol monobutyl ether
Dowanol DB
Ektasolve DB
Glycol Ether DB
Jeffersol DB
Poly-Solv DB

112-35-6
$C_7H_{16}O_4$
164.23
O(CCOCCOC)CCO
Ethanol, 2-(2-(2-methoxyethoxy)ethoxy)
Dowanol TMAT
2-(2-(2-Methoxyethoxy)ethoxy)-ethanol
Methoxytriglycol
Poly-Solv TM
Triethylene glycolmonomethyl ether
Triglycol monomethyl ether
3,6,9-Trioxa-1-decanol

112-36-7
$C_8H_{18}O_3$
162.26
O(CCOCC)CCOCC
Ether, bis(2-ethoxyethyl)
Bis(2-ethoxyethyl) ether
Diethyl carbitol
Diethylene glycol diethyl ether
Diethylether diethylenglykolu (Czech)
1-Ethoxy-2-(β-ethoxyethoxy)-ethane
Ethyl diglyme
3,6,9-Trioxaundecane

112-37-8
$C_{11}H_{22}O_2$
186.33
O=C(O)CCCCCCCCCC
Undecanoic-acid
1-Decanecarboxylic acid
Hendecanoic acid
n-Undecanoic acid
n-Undecoic acid
Undecylic acid
n-Undecylic acid

112-38-9
$C_{11}H_{20}O_2$
184.31
O=C(O)CCCCCCCCC=C
10-Undecenoic acid
Declid
Desenex
Desenex solution
10-Hendecenoic
10-Henedecenoic acid
Kyselina 9-decen-1-karboxylova (Czech)
Kyselina undecylenova (Czech)
Renselin
Sevinon
Undecylenic acid
Undecyl-10-enic acid
9-Undecylenic acid
10-Undecylenic acid

112-39-0

$C_{17}H_{34}O_2$
270.46
O=C(OC)CCCCCCCCCCC
CCCC
Methyl palmitate
AI3-03509
Hexadecanoic acid, methyl ester (9CI)
n-Hexadecanoic acid methyl ester
Metholene 2216
Methyl hexadecanoate
Methyl n-hexadecanoate
NSC-4197
Palmitic acid, methyl ester (8CI)
Uniphat A60

112-40-3
$C_{12}H_{26}$
170.38
C(CCCCCCCCCC)C
Dodecane
Adakane 12
Bihexyl
Dihexyl
n-Dodecan (German)
Duodecane

112-41-4
$C_{12}H_{24}$
168.32
C(=C)CCCCCCCCCC
1-Dodecene (9CI)
Adacene 12
α-Dodecene
n-Dodec-1-ene
Dodecylene α-
α-Dodecylene
NSC-12016

112-42-5
$C_{11}H_{24}O$
172.35
OCCCCCCCCCCC
Undecyl-alcohol
Alcohol C-11
Hendecanoic alcohol
1-Hendecanol
Hendecyl alcohol
n-Hendecylenic alcohol
Tip-Nip
Undecanol
n-Undecanol
n-Undecyl alcohol

112-44-7
$C_{11}H_{22}O$
170.33
O=CCCCCCCCCCC
1-Undecanal
Aldehyde C-11, undecylic
C-11 Aldehyde, undecylic
1-Decyl aldehyde
Hendecanal
Hendecanaldehyde
Hendecenal
Undecanal
n-Undecanal
Undecanaldehyde
Undecyl aldehyde
n-Undecyl aldehyde
Undecylic aldehyde

112-45-8
$C_{11}H_{20}O$
168.31
O=CCCCCCCCCC=C
10-Undecen-1-al
Aldehyde C-11, undecylenic
C-11 Aldehyde, undecylenic
Hendecenal
1-Undecen-10-al
10-Undecenal (8CI,9CI)
Undecylenaldehyde
10-Undecylenealdehyde
Undecylenic aldehyde

112-48-1
$C_{10}H_{22}O_2$
174.32
CCCCOCCOCCCC
Ethane, 1,2-dibutoxy
Butane, 1,1'-(1,2-ethanediylbis(oxy))bis-
1,2-Dibutoxyethane
Dibutyl cellosolve
Dibutylether ethylenglykolu (Czech)
Ethylene glycol dibutyl ether

112-49-2
$C_8H_{18}O_4$
178.26
O(CCOCCOC)CCOC
2,5,8,11-Tetraoxadodecane
Ansul ether 161
Glyme-3
Triethylene glycol dimethyl ether
Triglyme

112-50-5
$C_8H_{18}O_4$
178.26
O(CCOCCOCC)CCO
Ethanol, 2-(2-(2-ethoxyethoxy)ethoxy)
Dowanol TE
2-(2-(2-Ethoxyethoxy)ethoxy)-ethanol
Ethoxytriethylene glycol
Ethoxytriglycol
Poly-Solv TE
Triethylene glycol ethyl ether
Triethylene glycol monoethyl ether
Triglycol monoethyl ether
3,6,9-Trioxaundecanol

112-51-6
$C_{10}H_{22}S_2$
206.42
Disulfide, dipentyl
AI3-26484

112-52-7
$C_{12}H_{25}Cl$
204.78
ClCCCCCCCCCCCC
Dodecane, 1-chloro- (9CI)
1-Chlorododecane
Dodecyl chloride
n-Dodecyl chloride
Lauryl chloride
NSC-57107

112-53-8
$C_{12}H_{26}O$
186.38
OCCCCCCCCCCCC
Dodecyl-alcohol
Alcohol C-12
Alfol 12
Cachalot L-50
Cachalot L-90
CO 12
CO-1214
CO-1214N
CO-1214S
n-Dodecanol
1-Dodecanol
n-Dodecyl alcohol
Duodecyl alcohol
Dytol J-68
Epal 12
Lauric alcohol
Laurinic alcohol
Lauryl 24
Lauryl alcohol
n-Lauryl alcohol, primary
Lorol
Lorol 5
Lorol 7
Lorol 11
MA-1214
Sipol l12

112-54-9
$C_{12}H_{24}O$
184.36
O=CCCCCCCCCCCC
1-Dodecanal
Aldehyde C-12, lauric
C-12 Aldehyde, lauric
1-Dodecyl aldehyde
Duodecylic aldehyde
Lauryl aldehyde

112-55-0
$C_{12}H_{26}S$
202.44
SCCCCCCCCCCCC
1-Dodecanethiol
Dodecyl mercaptan
m-Dodecyl mercaptan
1-Dodecyl mercaptan
Lauryl mercaptan
m-Lauryl mercaptan
1-Mercaptododecane
NCI-C60935
Pennfloat M
Pennfloat S

112-56-1
$C_9H_{17}NO_2S$
203.33
CCCCOCCOCCSC#N
Thiocyanic acid, 2-(2-butoxyethoxy)ethyl ester
2-(2-Butoxyethoxy)ethyl thiocyanate
2-(2-(Butoxy)ethoxy)ethyl thiocyanic acid ester
2-(2-Butoxyethoxy)ethylthiokyanat (Czech)
Butoxyrhodanodiethyl ether
β-Butoxy-β'-thiocyanodiethyl ether
2-Butoxy-2'-thiocyanodiethyl ether
1-Butoxy-α-(2-thiocyanoethoxy)ethane
1-Butoxy-2-(2-thiocyanoethoxy)ethane
Butyl carbitol rhodanate
Butyl carbitol thiocyanate
ENT 6
Ethane, 1-butoxy-2-(2-thiocyanatoethoxy)-
Ethanol, 2-(2-butoxyethoxy)-, thiocyanate
Lethane
Lethane 384
Lethane 384 Regular

112-57-2
$C_8H_{23}N_5$
189.36
N(CCNCCNCCN)CCN
1,2-Ethanediamine, N-(2-aminoethyl)-N'-(2-((2-aminoethyl)amino)ethyl)
D.E.H. 26
1,4,7,10,13-Pentaazatridecane
Tetraethylenepentamine [UN 2320]
UN 2320 [Tetraethylenepentamine]

112-58-3
$C_{12}H_{26}O$
186.38
O(CCCCCC)CCCCCC
Ether, dihexyl
Dihexyl ether
Di-n-hexyl ether
Hexane, 1,1'-oxybis-
Hexyl ether
n-Hexyl ether

112-59-4
$C_{10}H_{22}O_3$
190.32
O(CCOCCO)CCCCCC
Ethanol, 2-((2-hexyloxy)-ethoxy)
Diethylene glycol n-hexyl ether
Diethylene glycol monohexyl ether
3,6-Dioxadodecanol-1
3,6-Dioxa-1-dodecanol
n-Hexoxyethoxyethanol
Hexyl carbitol
n-Hexyl carbitol
Hexylkarbitol (Czech)
2-((2-Hexyloxy)ethoxy)ethanol

112-60-7
$C_8H_{18}O_5$
194.26
O(CCOCCOCCO)CCO
Tetraethylene-glycol
Ethanol, 2,2'-(oxybis(ethyleneoxy))di-
Hi-Dry

112-61-8
$C_{19}H_{38}O_2$
298.57
O=C(OC)CCCCCCCCCCCCCCCC
Stearic acid, methyl ester
Emery 2218
Kemester 9018
Metholene 2218
Methyl octadecanoate
Methyl stearate
Octadecanoic acid, methyl ester

112-62-9
$C_{19}H_{36}O_2$
296.55
O=C(OC)CCCCCCCC=CCCCCCCCC
Oleic acid, methyl ester, cis
Emerest 2301
Emerest 2801
Emery (ACGIH,OSHA)
Emery 2219
Emery 2310
Emery Oleic Acid Ester 2301
Kemester 105
Kemester 115
Kemester 205
Kemester 213
Methyl 9-octadecenoate
Methyl cis-9-octadecenoate
Methyl (Z)-9-octadecenoate
Methyl oleate
(Z)-9-Octadecenoic acid, methyl ester

112-63-0
$C_{19}H_{34}O_2$
294.48
O=C(OC)CCCCCCCC=CCC=CCCCCC
Methyl linoleate
AI3-03520
Linoleic acid, methyl ester (8CI)
Methyl octadecadienoate
Methyl cis,cis-9,12-octadecadienoate
Methyl 9-cis,12-cis-octadecadienoate
NSC-93981
9,12-Octadecadienoic acid, methyl ester
9,12-Octadecadienoic acid, methyl ester, (Z,Z)-
9,12-Octadecadienoic acid (Z,Z)-, methyl ester (9CI)

112-65-2
$C_{13}H_{29}N_3$
227.40
N=C(NCCCCCCCCCCCC)N
Dodine
Carpen
Dodecylguanidine acetate
N-Dodecylguanidine acetate

112-66-3
$C_{14}H_{28}O_2$
228.42
O=C(OCCCCCCCCCCCC)C
Acetic acid, dodecyl ester
Acetate C-12
Dodecanol acetate
1-Dodecanol acetate
Dodecan-1-yl acetate
Dodecyl acetate
n-Dodecyl acetate
Dodecyl alcohol acetate
Lauryl acetate

112-67-4
$C_{16}H_{31}ClO$
274.87
O=C(CCCCCCCCCCCCCCC)Cl
Hexadecanoyl chloride (9CI)
AI3-52614
Hexadecanoic acid, chloride
NSC-9854
Palmitic acid chloride
Palmitoyl chloride (8CI)
Palmityl chloride

112-69-6
$C_{18}H_{39}N$
269.51
N(CCCCCCCCCCCCCCCC)(C)C
Dimethyl palmitamine
AI3-16727
Armeen DM 16D
Cetyldimethylamine
DCA
Dimethylcetylamine
N,N-Dimethylcetylamine
N,N-Dimethyl-1-hexadecanamine
Dimethylhexadecylamine
Dimethyl-n-hexadecylamine
N,N-Dimethylhexadecylamine
N,N-Dimethyl-n-hexadecylamine
Dimethyl palmitylamine
1-Hexadecanamine, N,N-dimethyl- (9CI)
Hexadecylamine, N,N-dimethyl- (8CI)
Hexadecyldimethylamine
NSC-404177
Palmityl dimethyl amine

112-70-9
$C_{13}H_{28}O$
200.41
OCCCCCCCCCCCCC
1-Tridecanol
Tridecanol
n-Tridecanol
Tridecyl alcohol
n-Tridecyl alcohol

112-71-0
$C_{14}H_{29}Br$
277.29
BrCCCCCCCCCCCCCC
Tetradecane, 1-bromo- (9CI)
1-Bromotetradecane
Myristyl bromide
NSC-83468
Tetradecyl bromide
n-Tetradecyl bromide
1-Tetradecyl bromide

112-72-1
$C_{14}H_{30}O$
214.44
OCCCCCCCCCCCCCC
1-Tetradecanol
Dytol R-52
Lanette Wax KS
Loxanol V
Myristic alcohol
Myristyl alcohol
n-Tetradecanol-1
Tetradecyl alcohol
n-Tetradecyl alcohol

112-73-2
$C_{12}H_{26}O_3$
218.38
O(CCOCCOCCCC)CCCC
Ether, bis(2-butoxyethyl)
Bis(2-butoxyethyl) ether
Butane, 1,1'-(oxybis(2,1-ethanediyloxy))bis-
Butyl diglyme
2,2'-Dibutoxyethyl ether
Dibutyl carbitol
Dibutylether diethylenglykolu (Czech)
Diethyleneglycoldibutyl ether
Diethylene glycol di-n-butyl ether
Ether, bis(butoxyethyl)
5,8,11-Trioxapentadecane

112-75-4
$C_{16}H_{35}N$
241.45
N(CCCCCCCCCCCCCC)(C)C
Dimethyl myristamine
Armeen DM 14D
Armine DM14D
Dimethyl myristylamine
Dimethylmyristamine
Dimethylmyristylamine
N,N-Dimethylmyristylamine
N,N-Dimethyltetradecanamine
N,N-Dimethyl-1-tetradecanamine
Dimethyltetradecylamine
Dimethyl-n-tetradecylamine
N,N-Dimethyltetradecylamine
N,N-Dimethyl-N-tetradecylamine
Myristyl dimethyl amine
Myristyldimethylamine
NSC-78319
1-Tetradecanamine, N,N-dimethyl- (9CI)
Tetradecylamine, N,N-dimethyl- (8CI)
Tetradecyldimethylamine

112-76-5
$C_{18}H_{35}ClO$
302.93
O=C(CCCCCCCCCCCCCCCCC)Cl
Stearoyl chloride
Octadecanoic acid, chloride
Octadecanoyl chloride (9CI)
n-Octadecanoyl chloride
Stearic acid chloride
Stearic chloride
Stearyl chloride

112-77-6
$C_{18}H_{33}ClO$
300.91
O=C(CCCCCCCC=CCCCCCCCC)Cl
Oleoyl chloride (8CI)
NSC-97299
9-Octadecenoyl chloride, (Z)- (9CI)
(Z)-9-Octadecenoyl chloride
Oleic acid chloride

112-79-8
$C_{18}H_{34}O_2$
282.52
Elaidic-acid
trans-δ^9-Octadecenoic acid
trans-Octadec-9-enoic acid
trans-9-Octadecenoic acid
9-Octadecenoic acid, (E)-

112-80-1
$C_{18}H_{34}O_2$
282.52
O=C(O)CCCCCCC=CCCCCCCC
9-Octadecenoic acid, (Z)
L'acide oleique (French)
Century CD fatty acid
Elaic acid
Emersol 210
Emersol 213
Emersol 6321
Emersol 233ll
Emersol 221 Low titer white oleic acid
Emersol 220 White oleic acid
Glycon RO
Glycon WO
Groco 2
Groco 4
Groco 6
Groco 5l
Hy-Phi 1055
Hy-Phi 1088
Hy-Phi 2066
Hy-Phi 2088
Hy-Phi 2102
Industrene 105
Industrene 205
Industrene 206
K 52
Metaupon
Neo-Fat 90-04
Neo-Fat 92-04
cis-δ^9-Octadecenoic acid
cis-Octadec-9-enoic acid
cis-9-Octadecenoic acid
9-Octadecenoic acid, cis-
9,10-Octadecenoic acid
Oleic acid
Oleinic acid
Pamolyn
Red Oil
Tego-Oleic 130
Vopcolene 27
Wecoline OO
Wochem No. 320

112-82-3
$C_{16}H_{33}Br$
305.34
BrCCCCCCCCCCCCCCCC
Hexadecane, 1-bromo- (9CI)
AI3-11181
1-Bromohexadecane
Cetyl bromide
Hexadecyl bromide
n-Hexadecyl bromide
n-Hexadecyl-1-bromide
1-Hexadecyl bromide
NSC-4193

112-84-5
$C_{22}H_{43}NO$
337.58
O=C(N)CCCCCCCCCCC=CCCCCCCCC
Erucamide
13-Docosenamide
13-Docosenamide, cis-
13-Docosenamide, (Z)- (9CI)
(Z)-13-Docosenamide
Erucic acid amide
Erucyl amide
Kenamide E

112-85-6
$C_{22}H_{44}O_2$
340.59
O=C(O)CCCCCCCCCCCCCCCCCCCC
Behenic acid
AI3-52709
Docosanoic acid (9CI)
n-Docosanoic acid

1-Docosanoic acid
Docosoic acid
Glycon B-70
Hydrofol Acid 560
Hydrofol 2022-55
NSC-32364
1,2,3-Propanetriol tri(12-hydroxystearate)

112-86-7
$C_{22}H_{42}O_2$
338.57
O=C(O)CCCCCCCCCCCC=CCCCCCCCC
Erucic acid
AI3-18180
13-Docosenoic acid (cis)
13-Docosenoic acid, (Z)- (9CI)
cis-13-Docosenoic acid
(Z)-13-Docosenoic acid
δ 13:14-Docosenoic acid
δ(13)-cis-Docosenoic acid
NSC-6814

112-88-9
$C_{18}H_{36}$
252.48
C(=C)CCCCCCCCCCCCCCCC
1-Octadecene (9CI)
AI3-06521
NSC-66460
α-Octadecene
Octadecylene α-

112-89-0
$C_{18}H_{37}Br$
333.40
BrCCCCCCCCCCCCCCCCCC
Octadecane, 1-bromo- (9CI)
AI3-00994
1-Bromooctadecane
NSC-5542
Octadecyl bromide
n-Octadecyl bromide
Stearyl bromide

112-90-3
$C_{18}H_{37}N$
267.56
NCCCCCCCCC=CCCCCCCCC
9-Octadecenylamine, (Z)
Alamine 11
Armeen O
Kemamine P 989
Noram O
9-Octadecen-1-amine, (Z)- (9CI)
cis-9-Octadecenylamine
Oleamine
Oleinamine
Oleyl amine
Oleylamin (German)

112-91-4
$C_{18}H_{33}N$
263.46
N#CCCCCCCCC=CCCCCCCCC
(Z)-9-Octadecenenitrile
9-Octadecenenitrile, (Z)- (9CI)
9-Octadecenoic acid(cis), nitrile(cis)
Oleic acid nitrile
Oleonitrile
Oleoylnitrile
Oleylonitrile

112-92-5
$C_{18}H_{38}O$
270.56
OCCCCCCCCCCCCCCCCCC
1-Octadecanol
Adol
Adol 68
Atalco S
CO-1895
CO-1897
Crodacol-S
Decyl octyl alcohol
Dytol E-46
Lorol 28
Octadecanol
n-Octadecanol
Octa decyl alcohol
n-Octadecyl alcohol
Polaax
Sipol S
Siponol S
Stearol
Stearyl alcohol
Steraffine
USP XIII Stearyl alcohol

112-95-8
$C_{20}H_{42}$
282.55
C(CCCCCCCCCCCCCCCCCC)C
Eicosane (9CI)
AI3-28404
n-Eicosane
NSC-62789

112-96-9
$C_{19}H_{37}NO$
295.57
O=C=NCCCCCCCCCCCCCCCCCC
Isocyanic acid, octadecyl ester
Octadecyl isocyanate
Tonco-70

112-98-1
$C_{16}H_{34}O_5$
306.50
CCCCOCCOCCOCCOCCOCCCC
5,8,11,14,17-Pentaoxa-heneicosane
Dibutylether tetraethylenglykolu (Czech)
Tetraethylene glycol, dibutyl ether

113-00-8
CH_5N_3
59.09
N=C(N)N
Guanidine
Aminoformamidine
Aminomethanamidine
Carbamamidine
Carbamidine
Iminourea

113-15-5
$C_{33}H_{35}N_5O_5$
581.73
O=C(N(C(O1)(O)C(N(C2=O)CC3)C3)C2Cc(cccc4)c4)C1(NC(=O)C(C=C(c(c(C(=CN5)C6)c5cc7)c7)C6N8C)C8)C
Ergotamine

113-18-8
C_7H_9ClO
144.61
CCC(O)(C=CCl)C#C
1-Penten-4-yn-3-ol, 1-chloro-3-ethyl
A 71
Aethyl-chlorvynol
Alvinol
Arvynol
1-Chloro-3-ethyl-1-penten-4-yn-3-ol
β-Chlorovinyl ethyl ethynyl carbinol
3-(β-Chlorovinyl)-1-pentyn-3-ol
Etchlorvinolo
Ethchlorvinyl
Ethclorvynol
Ethchlorovynol
Ethochlorvynol
Ethychlorvynol
Ethyl β-chlorovinyl ethynyl carbinol
Normonson
Normosan
Normoson
Nostel
Placidil
Placidyl
Roeridorm
Serenil
Serensil

113-24-6
$C_3H_4O_3.Na$
111.05
[Na+].CC(=O)C([O-])=O
Propanoic acid, 2-oxo-, sodium salt (9CI)
Pyruvic acid, sodium salt
Sodium 2-oxopropanoate

113-38-2
$C_{24}H_{32}O_4$
384.56
CCC(=O)OC3CCC4C2CCc1cc(OC(=O)CC)ccc1C2CCC34C
Estradiol, dipropionate
Agofollin
Dimenformon dipropionate
Diovocyclin
Diovocylin

Dipropionate d'oestradiol (French)
Diprostron
Endofollicolina D.P.
Estradiol 3,17-dipropionate
β-Estradiol dipropionate
β-Estradiol 3,17-dipropionate
3,17-β-Estradiol dipropionate
17-β-Estradiol dipropionate
Estra-1,3,5(10)-triene-3,17-diol (17-β)-dipropionate
1,3,5(10)-Estratriene-3,17β-diol dipropionate
Estroici
Estronex
Follicyclin P
Nacyclyl
Oestradiol dipropionate
Oestradiol-3,17-dipropionate
β-Oestradiol dipropionate
3,17-β-Oestradiol dipropionate
17-β-Oestradiol dipropionate
Ovocyclin dipropionate
Ovocyclin-P
Progynon-DP

113-42-8
$C_{20}H_{25}N_3O_2$
339.48
CCC(CO)NC(=O)C2CN(C)C1Cc3c[nH]c4cccc(C1=C2)c34
Ergoline-8-β-carboxamide, 9,10-didehydro-N-((S)-1-(hydroxymethyl)propyl)-6-methyl
Basofortina
9,10-Didehydro-N-(α-(hydroxymethyl)propyl)-6-methyl-ergoline-8-β-carboxamide
ME 277
Methergine
Methylergobasine
Methylergobrevin
Methylergometrin
Methylergometrine
Methylergonovin
Methylergonovine
Partergin

113-45-1
$C_{14}H_{19}NO_2$
233.34
COC(=O)C(C1CCCCN1)c2ccccc2
2-Piperidineacetic acid, α-phenyl-, methyl ester
4311/B Ciba
Calocain
Centedein
Centredin
Meridil
Methylphenidan
Methyl phenidate
Methyl phenidyl acetate
Methyl α-phenyl-α-(2-piperidyl)acetate
NCI-C56280
Phenidylate
α-Phenyl-2-piperidineacetic acid methyl ester
Plimasine
Ritalin
Ritaline
Ritcher Works

113-48-4
$C_{17}H_{25}NO_2$
275.43
CCCCC(CC)CN3C(=O)C2C1CC(C=C1)C2C3=O
5-Norbornene-2,3-dicarboximide, N-(2-ethylhexyl)
Bicyclo(2.2.1)heptene-2-dicarboxylic acid, 2-ethylhexylimide
Endomethylenetetrahydrophthalic acid, N-2-ethylhexyl imide
ENT 8,184
N-(2-Ethylhexyl)bicyclo-(2,2,1)-hept-5-ene-2,3-dicarboximide
N-2-Ethylhexylimid kyseliny bicyklo-(2,2,1)-5-hepten-2,3-dikarboxylove (Czech)
N-2-Ethylhexylimide endomethylenetetrahydrophthalic acid
N-(2-Ethylhexyl)-5-norbornene-2,3-dicarboximide
2-(2-Ethylhexyl)-3a,4,7,7a-tetrahydro-4,7-methano-1H-isoindole-1,3(2H)-dione
MGK-264
Octacide 264
n-Octyl bicycloheptene dicarboximide
n-Octylbicyclo-(2.2.1)-5-heptene-2,3-dicarboximide
Pyrodone
Sinepyrin 222
Synepirin 222
Synergist 264
Van Dyk 264

113-53-1
$C_{19}H_{21}NS$
295.47
CN(C)CCC=C2c1ccccc1CSc3ccccc23
Propylamine, N,N-dimethyl-3-(dibenzo(b,e)thiepin-δ-sup(11(6H),γ))
3-Dibenzo(b,e)thiepin-11(6H)-ylidene-N,N-dimethyl-1-propamine
11-(3-Dimethylaminopropylidene)-6,11-dihydrodibenzo-(b,e)thiepin
N,N-Dimethyldibenzo(b,e)-thiepin-δ-sup(11(6H),γ)propylamine
Dosulepin
Dothiepin
IZ 914
Prothiaden
Prothiaden Spofa

113-59-7
$C_{18}H_{18}ClNS$
315.88
CN(C)CCC=C2c1ccccc1Sc3ccc(Cl)cc23
Thioxanthene-δ⁹, γ-propylamine, 2-chloro-N,N-dimethyl
(α-2-Chloro-9-ω-dimethylamino-propylamine)thioxanthene
2-Chloro-9-(ω-di-methylamino-propylidene)thioxanthene
2-Chloro-9-(3-(dimethylamino)-propylidene)-thioxanthene
cis-2-Chloro-9-(3-dimethylaminopropylidene)thioxanthene
2-Chloro-N,N-dimethylthioxanthene-δ⁹, γ-propylamine
Chloroprothixene
Chlorprothixen
Chlorprothixene
α-Chlorprothixene
cis-Chlorprothixene
Chlorprotixen
Chlorprotixene
Chlorprotixine
Chlothixen
CPT
CPX
Iaractan
MK 184
N 714
N 714C
Paxyl
1-Propanamine, 3-(2-chloro-9H-thioxanthen-9-ylidene)-N,N-dimethyl-, (Z)-
Rentovet
Ro 4-0403
Tactaran
Taractan
Tarasan
Tranquilan
Traquilan
Trictal
Truxal
Truxaletten
Truxil
Vetacalm

113-92-8
$C_{16}H_{19}ClN_2.C_4H_4O_4$
390.90
CN(C)CCC(c1ccc(Cl)cc1)c2ccccn2.OC(=O)C=CC(O)=O
Pyridine, 2-(p-chloro-α-(2-(dimethylamino)ethyl)benzyl)-, maleate (1:1)
Allerclor
Allergin
Allergisan
Alunex
Antagonate
Carbinoxamide maleate
Chlormene
dl-2(-p-Chloro-α-2-(dimethylamino)ethylbenzyl)pyridine bimaleate
Chloropiril
Chloroprophenpyridamine

maleate
Chlorpheniramine maleate
1-p-Chlorophenyl-1-(2-pyridyl)-3-dimethylaminopropane maleate
1-(p-Chlorphenyl)-1-(2-pyridyl)-3-dimethylaminopropan maleat (German)
Chlorprophenpyridamine maleate
Chlor-Trimeton
Chlor-Trimeton Maleate
Chlor-Tripolon
Cloropiril
C-Meton
1-(N,N-Dimethylamino)-3-(p-chlorophenyl-3-α-pyridyl)-propane maleate
Histadur
Histadur Dura-Tabs
Histalen
Histapan
Histaspan
Ibioton
Lorphen
M.P. Chlorcaps T.D.
NCI-C55265
Neorestamin
Piriex
Piriton
Polaronil (German)
Pyridamal-100
2-Pyridinepropanamine, γ-(4-chlorophenyl)-N,N-dimethyl-, (Z)-2-butenedioate (1:1) (9CI)
Synistamin
Teldrin

113-98-4
$C_{16}H_{17}N_2O_4S.K$
372.51
[K+].CC3(C)SC2C(NC(=O)Cc1ccccc1)C(=O)N2C3C([O-])=O
4-Thia-1-azabicyclo(3.2.0)heptane-2-carboxylic acid, 3,3-dimethyl-7-oxo-6-(2-phenylacetamido)-, monopotassium salt
Benzylpenicillinic acid potassium salt
Benzylpenicillin potassium
Benzylpenicillin potassium salt
Cilloral
Cosmopen
Cristapen
Crystapen
Eskacillin
Falapen
Forpen
Hipercilina
Hyasorb
Hylenta
Megacillin Tablets
Monopen
Notaral
Penalev
Penicillin G Potassium
Penicillin G Potassium salt
Penisem
Pentid
Pentids
Pfizerpen
Potassium benzylpenicillin
Potassium benzylpenicillinate
Potassium benzylpenicillin G
Potassium Penicillin G
Potassium salt of benzylpenicillin
Qidpen G
Scotcil
Sk-Penicillin G
Sugracillin
Tabilin
Tu Cillin

114-07-8
$C_{37}H_{67}NO_{13}$
734.05
CCC3OC(=O)C(C)C(OC1CC(C)(OC)C(O)C(C)O1)C(C)C(OC2OC(C)CC(C2O)N(C)C)C(C)(O)CC(C)C(=O)C(C)C(O)C3(C)O
Erythromycin
Dotycin
EM
E-Mycin
Erycin
Erythrocin
Erythrogran
Erythroguent
Erythromycin A
Ilotycin
Pantomicina
Propiocine
Robimycin

114-26-1
$C_{11}H_{15}NO_3$
209.27
CNC(=O)Oc1ccccc1OC(C)C
Carbamic acid, methyl-, o-isopropoxyphenyl ester
Aprocarb
Arprocarb
Bay 9010
Bay 39007
Bayer 39007
Baygon (OSHA)
Bifex
Blattanex
Boygon
Brygou
Carbamic acid, methyl-, 2-(1-methylethoxy)phenyl ester
Chemagro 9010
ENT 25,671
o-IMPC
Invisi-Gard
Isocarb
2-Isopropoxyphenyl-N-methylcarbamat (German)
o-Isopropoxyphenyl methylcarbamate
o-Isopropoxyphenyl N-methylcarbamate
2-Isopropoxyphenyl methylcarbamate
2-Isopropoxyphenyl N-methylcarbamate
2-(1-Methylethoxy)phenol methylcarbamate
n-2-(1-Methylethoxy)phenyl methylcarbamate
N-Methyl-2-isopropoxyphenylcarbamate
OMS-33
PHC
Phenol, o-isopropoxy-, methylcarbamate
Propogon
Propoksuru (Polish)
Propotox M
Propoxur (ACGIH,OSHA)
Propoxure
Propyon
Sendran
Suncide
Tugon fliegenkugel
Unden

114-38-5
$C_7H_7ClN_2O$
170.59
NC(=O)Nc1ccccc1Cl
2-Chlorophenylurea
AI3-20196
(o-Chlorophenyl)urea
NSC-42111
Urea, (o-chlorophenyl)- (8CI)
Urea, (2-chlorophenyl)- (9CI)

114-49-8
$C_{17}H_{21}NO_4.BrH$
384.31
1-α-H,5-α-H-Tropan-3-α-ol, 6-β,7-β-epoxy-, (-)-tropate (ester), hydrobromide
Beldavrin
Euscopol
Hydroscine hydrobromide
Hyoscine bromide
Hyoscine hydrobromide
(-)-Hyoscine hydrobromide
l-Hyoscine hydrobromide
Hyocine F hydrobromide
Hyoscyine hydrobromide
Hysco
Isoscopil
Kwells
Scopamin
Scopolamine bromide
(-)-Scopolamine bromide
Scopolamine hydrobromide
(-)-Scopolamine hydrobromide
Scopolaminium bromide
Scopolammonium bromide
Scopos
Sereen
Triptone

114-70-5
$C_8H_7O_2.Na$
158.14
Acetic acid, phenyl-, sodium salt
Benzeneacetic acid, sodium salt (9CI)
Phenylacetate sodium salt

Phenylacetic acid sodium salt
Phenylessigsaure natrium-salz (German)
Sodium benzeneacetate
Sodium phenylacetate

114-80-7
$C_{12}H_{19}N_2O_2.Br$
303.24
[Br-].CN(C)C(=O)Oc1cccc(c1)[N+](C)(C)C
Ammonium, (m-hydroxyphenyl)trimethyl-, bromide, dimethylcarbamate
Benzenaminium, 3-(((dimethylamino)carbonyl)oxy)-N,N,N-trimethyl-, bromide (9CI)
Carbamic acid, dimethyl-, ester with (m-hydroxyphenyl)trimethylammonium bromide
3-Dimethylcarbamoxyphenyl trimethyl ammonium bromide
Proserine
Eustigmin bromide
(m-Hydroxyphenyl)trimethylammonium bromide dimethylcarbamate
3-Hydroxyphenyltrimethylammonium bromide dimethylcarbamic ester
Kirkstigmine bromide
Syntostigmin (tablet)
Leostigmine bromide
Neoeserine bromide
Neoserine bromide
Neostigmine bromide
Neostigmine methyl bromide
Philostigmin bromide
Proserine bromide
Prostigmin bromide
Prostigmine bromide
RCRA waste number U053
Stigmanol bromide
Stigmosan bromide
Synstigmin bromide
Synthostigmine bromide
Syntostigmin bromide
Syntostigmine bromide
Vagostigmine bromide

114-83-0
$C_8H_{10}N_2O$
150.20
O=C(NNc(cccc1)c1)C
Acetic acid, 2-phenylhydrazide
Acetic acid phenylhydrazone
Acetylphenylhydrazine
β-Acetylphenylhydrazine
N-Acetyl-N'-phenylhydrazine
1-Acetyl-2-phenylhydrazine
APH
Fenylhydrazid kyseliny octove (Czech)
Hydracetin
N'-Phenylacethydrazide
Pyrodin
Pyrodine

114-86-3
$C_{10}H_{15}N_5$
205.30
NC(=N)NC(=N)NCCc1ccccc1
Biguanide, 1-phenethyl
DBI
Fenformina
N'-β-Fenetilformamidinilimino-urea (Italian)
NCI-C01741
PEDG
β-Phenethybiguanide
β-Phenethylbiguanide
1-Phenethylbiguanide
Phenethyldiguanide
N'-β-Phenethylformamidinyliminourea
Phenformin
Phenformine
Phenformix
Phenylethylbiguanide
W 32

115-02-6
$C_5H_7N_3O_4$
173.15
NC(COC(=O)C=N#N)C(O)=O
Serine, diazoacetate (ester)
Acetic acid, diazo-, ester with serine
Azaserin
Azaserine
l-Azaserine
AZS
C.I. 337
CN-15,757
Diazoacetate (ester) l-serine
l-Diazoacetate (ester) serine
o-Diazoacetyl-l-serine
NSC-742
P-165
l-Serine, diazoacetate
Serine, diazoacetate (ester), l-
l-Serine diazoacetate (ester)
RCRA waste number U015

115-07-1
C_3H_6
42.09
C(=C)C
Propene
Methylethene
Methylethylene
NCI-C50077
1-Propene (9CI)
Propylene [UN 1077]
1-Propylene
UN 1077 [Propylene see also Petroleum gases, Liquefied]

115-09-3
CH_3ClHg
251.08
C[Hg]Cl
Mercury, chloromethyl
Caspan
Chloromethylmercury
Mercurymethylchloride
Methylmercuric chloride
Methylmercury chloride
Methylmerkurichlorid (Czech)
MMC
Monomethyl mercury chloride

115-10-6
C_2H_6O
46.08
O(C)C
Methyl-ether
Dimethyl ether [UN 1033]
Ether, dimethyl
Ether, methyl
UN 1033 [Dimethyl ether]
Wood ether

115-11-7
C_4H_8
56.12
C(=C)(C)C
Propene, 2-methyl
γ-Butylene
Isobutene
Isobutylene [UN 1055]
Liquefied Petroleum Gas [UN 1075]
UN 1055 [Isobutylene]
UN 1075 [Petroleum gases, liquefied see also Liquefied petroleum gas]

115-17-3
C_2HBr_3O
280.74
O=CC(Br)(Br)Br
Tribromoacetaldehyde
Acetaldehyde, tribromo- (9CI)
Bromal
NSC-66406
2,2,2-Tribromoacetaldehyde

115-18-4
$C_5H_{10}O$
86.15
OC(C=C)(C)C
1-Buten-3-ol, 3-methyl
Methylbutenol
2-Methyl-3-buten-2-ol
3-Methyl-buten-(1)-ol-(3) (German)
3-Methyl-1-buten-3-ol

115-19-5
C_5H_8O
84.13
OC(C#C)(C)C
3-Butyn-2-ol, 2-methyl
1-Butyn-3-ol, 3-methyl-
Dimethylacetylenecarbinol
Dimethylacetylenylcarbinol
Dimethylethynylcarbinol
Dimethylethynylmethanol
α,α-Dimethylpropargyl alcohol
1,1-Dimethylpropargyl alcohol
1,1-Dimethylpropynol
Ethynyldimethylcarbinol
2-Hydroxy-2-methyl-3-butyne

MBY
3-Methyl-butin-(1)-ol-(3) (German)
2-Methylbutyn-3-ol-2
2-Methyl-3-butyn-2-ol
3-Methyl-1-butyn-3-ol

115-20-8
$C_2H_3Cl_3O$
149.40
OCC(Cl)(Cl)Cl
Ethanol, 2,2,2-trichloro
Trichlorethanol
Trichloroethanol
2,2,2-Trichloroethanol
Trichloroethyl alcohol
2,2,2-Trichloroethyl alcohol

115-21-9
$C_2H_5Cl_3Si$
163.51
CCSi(Cl)(Cl)Cl
Silane, ethyltrichloro
Ethyl silicon trichloride
Ethyl trichlorosilane [UN 1196]
Silane, trichloroethyl-
Silicane, trichloroethyl-
Trichloroethylsilane
Trichloroethylsilicane
UN 1196 [Ethyltrichlorosilane]

115-22-0
$C_5H_{10}O_2$
102.13
O=C(C(O)(C)C)C
2-Butanone, 3-hydroxy-3-methyl- (9CI)
AI3-23404
Dimethylacetylcarbinol
2-Hydroxy-2-methyl-3-butanone
3-Hydroxy-3-methyl-2-butanone
1-Hydroxy-1-methylethyl methyl ketone
Methylacetoin
3-Methylacetoin
2-Methyl-2-hydroxybutan-3-one
NSC-5576

115-24-2
$C_7H_{16}O_4S_2$
228.35
CCS(=O)(=O)C(C)(C)S(=O)(=O)CC
Propane, 2,2-bis(ethylsulfonyl)
Acetone, bis(ethyl sulfone)
Acetone diethylsulfone
2,2-Bis(ethylsulfonyl)propane
Diethylsulfondimethylmethane
Propane-diethyl sulfone
Sulfonal
Sulfonmethane

115-25-3
C_4F_8
200.04
FC(F)(C(F)(F)C1(F)F)C1(F)F
Cyclobutane, octafluoro
Cyclooctafluorobutane
FC-C 318
Freon C-318
Halocarbon C-138
Octafluorocyclobutane [UN 1976]
Perfluorocyclobutane
Propellant C318
R-C 318
UN 1976 [Octafluorocyclobutane]

115-26-4
$C_4H_{12}FN_2OP$
154.15
CN(C)P(F)(=O)N(C)C
Phosphorodiamidic fluoride, tetramethyl
BFP
BFPO
Bis(dimethylamido)fluorophosphate
Bis(dimethylamido)-phosphoryl fluoride
Bis(dimethylamino)fluorophosphate
Bisdimethylaminofluorophosphine oxide
CR 409
Difo
Dimefox
DMF
ENT 19,109
Fluophosphoric acid di(dimethylamide)
Fluorure de N,N,N',N'-tetramethyle phosphoro-diamide (French)
Hanane
Pestox 14
Pestox IV
Pestox XIV
Phosphine oxide, bis(dimethylamino)fluoro-
S-14
T-2002
Terra-Systam
Terra-Sytam
Terrasytum
N,N,N',N'-Tetramethyl-diamido-fluor-phosphin-oxid (German)
N,N,N',N'-Tetramethyl-diamido-fosforzuur-fluoride (Dutch)
Tetramethyldiamidophosphoric fluoride
N,N,N',N'-Tetramethyl-diamido-phosphorsaeure-fluorid (German)
Tetramethylphosphorodiamidic fluoride
N,N,N,N-Tetramethylphosphorodiamidic fluoride
N,N,N',N'-Tetrametil-fosforo-diammido-fluoruro (Italian)
Tetra Sytam
TL 792
Wacker S 14/10

115-27-5
$C_9H_2Cl_6O_3$
370.83
O=C(OC(=O)C1C(C(=C(C23Cl)Cl)Cl)(C3(Cl)Cl)Cl)C12
4,7-Methanoisobenzofuran-1,3-dione, 4,5,6,7,8,8-hexachloro-3a,4,7,7a-tetrahydro-(9CI)
Chloran 542
Chlorendic anhydride
HET anhydride
1,4,5,6,7,7-Hexachlorobicyclo(2.2.1)-5-heptene-2,3-dicarboxylic acid anhydride
Hexachloroendomethylene tetrahydrophthalic anhydride
Hexachloro-5-norbornene-2,3-dicarboxylic anhydride
5-Norbornene-2,3-dicarboxylic anhydride, 1,4,5,6,7,7-hexachloro- (8CI)
NSC-22229

115-28-6
$C_9H_4Cl_6O_4$
388.83
O=C(O)C(C(C(=C(C12Cl)Cl)Cl)(C1(Cl)Cl)Cl)C2C(=O)O
5-Norbornene-2,3-dicarboxylic acid, 1,4,5,6,7,7-hexachloro
Bicyclo(2.2.1)hept-5-ene-2,3-dicarboxylic acid, 1,4,5,6,7,7-hexachloro- (9CI)
Chlorendic acid
Het Acid
Kyselina 3,6-endomethylen-3,4,5,6,7,7-hexachlor-δ^4-tetrahydroftalova (Czech)
Kyselina 1,2,3,4,7,7-hexachlorbicyklo(2,2,1)hept-2-en-5,6-dikarboxylova (Czech)
NCI-C55072

115-29-7
$C_9H_6Cl_6O_3S$
406.91
ClC2=C(Cl)C3(Cl)C1COS(=O)OCC1C2(Cl)C3(Cl)Cl
5-Norbornene-2,3-dimethanol, 1,4,5,6,7,7-hexachloro-, cyclic sulfite
Benzoepin
Beosit
Bio 5,462
Chlorthiepin
Crisulfan
Cyclodan
Devisulphan
Endocel
Endocide
Endosol
Endosulfan (ACGIH,DOT, OSHA)
Endosulfan mixture, Liquid (DOT)
Endosulphan
Ensure
ENT 23,979

FMC 5462
Goldenleaf tobacco spray
α,β-1,2,3,4,7,7-Hexachloro-bicyclo(2.2.1)-2-heptene-5,6-bisoxymethylene sulfite
1,2,3,4,7,7-Hexachlorobicyclo-(2.2.1)hepten-5,6-bioxymethyl-enesulfite
Hexachlorohexahydromethano 2,4,3-benzodioxathiepin-3-oxide-
6,7,8,9,10,10-Hexachloro-1,5,5a,6,9,9a-hexahydro-6,9-methano-2,4,3-benzodioxa-thiepin-3-oxide
1,4,5,6,7,7-Hexachloro-5-nor-bornene-2,3-dimethanol cyclic sulfite
Hildan
HOE 2,671
Insectophene
Kop-Thiodan
Malix
NA 2761
NCI-C00566
Nia 5462
Niagara 5,462
OMS 570
Rasayansulfan
RCRA waste number P050
SD-4314
Sulfurous acid, cyclic ester with 1,4,5,6,7,7-hexachloro-5-nor-bornene-2,3-dimethanol
Thifor
Thimul
Thiodan
Thiodan 35
Thiofor
Thiomul
Thionate
Thionex
Thiosulfan
Thiosulfan tionel
Tiovel

115-31-1
$C_{13}H_{19}NO_2S$
253.39
O=C(OC(C(C(C1C2)(C)C)(C2)C)C1)CSC#N
Acetic acid, thiocyanato-, isobornyl ester
Acetic acid, thiocyanato-, 1,7,7-trimethylbicyclo-(2,2,1)hept-2-yl ester, exo-(9CI)
Bornate
Cidalon
ENT 92
Isoborneol, thiocyanatoacetate
Isobornylester kyseliny thio-kyanatooctove (Czech)
Isobornyl thiocyanatoacetate
Isobornyl thiocyanoacetate
Terpinyl thiocyanoacetate
Thanisol
Thanite
Thiocyanatoacetic acid iso-bornyl ester

115-32-2
$C_{14}H_9Cl_5O$
370.48
OC(c1ccc(Cl)cc1)(c2ccc(Cl)cc2)C(Cl)(Cl)Cl
Benzhydrol, 4,4'-dichloro-α-(trichloromethyl)
Acarin
Benzenemethanol, 4-chloro-α-(4-chlorophenyl)-α-(tri-chloromethyl)-
1,1-Bis(chlorophenyl)-2,2,2-tri-chloroethanol
1,1-Bis(p-chlorophenyl)-2,2,2-trichloroethanol
1,1-Bis(4-chlorophenyl)-2,2,2-trichloroethanol
Carbax
Cekudifol
4-Chloro-α-(4-chlorophenyl)-α-(trichloromethyl)benzene-methanol
CPCA
Dichlorokelthane
Decofol
Di-(p-chlorophenyl)trichloro-methylcarbinol
4,4'-Dichloro-α-(trichloro-methyl)benzhydrol
Dicofol
DTMC
ENT 23,648
Ethanol, 2,2,2-trichloro-1,1-bis-(p-chlorophenyl)-
FW 293
Hifol
Hilfol 18.5 EC
Keltane
Kelthane
para,para'-Kelthane
Kelthane A
Kelthane Dust Base
Kelthanethanol
Milbol
Mitigan
NCI-C00486
2,2,2-Trichloor-1,1-bis(4-chloor fenyl)-ethanol (Dutch)
1,1,1-Trichlor-2,2-bis(4-chlor-phenyl)-aethanol (German)
2,2,2-Trichlor-1,1-bis(4-chlor-phenyl)-aethanol (German)
2,2,2-Trichloro-1,1-bis(4-chloro-phenyl)-ethanol (French)
2,2,2-Trichloro-1,1-bis(4-cloro-fenil)-etanolo (Italian)
2,2,2-Trichloro-1,1-di-(4-chloro-phenyl)ethanol

115-37-7
$C_{19}H_{21}NO_3$
311.41
COC1=CC=C2C3Cc4ccc(OC)c5OC1C2(CCN3C)c45
Morphinan, 6,7,8,14-tetra-dehydro-4,5-α-epoxy-3,6-di-methoxy-17-methyl
Thebaine
Paramorphine

115-38-8
$C_{13}H_{14}N_2O_3$
246.29
CCC1(C(=O)NC(=O)N(C)C1=O)c2ccccc2
Barbituric acid, 5-ethyl-1-methyl-5-phenyl
Enfenemal
Enphenemal
N-Ethylmethylphenylbarbituric acid
5-Ethyl-N-methyl-5-phenylbar-bituric acid
5-Ethyl-1-methyl-5-phenylbar-bituric acid
5-Ethyl-5-phenyl-N-methyl-bar-tituric acid
Isonal
Isonal (Roussel)
Mebaral
Meberal
Menta-Bal
Mephobarbital
Mephobarbitone
Mephytal
Methyl-calminal
1-Methyl-5-ethyl-5-phenylbar-bituric acid
Methylphenobarbital
N-Methylphenobarbital
1-Methylphenobarbital
Methylphenobarbitone
N-Methylphenolbarbitol
Methylphenylbarbituric acid
N-Methyl-5-phenyl-5-ethyl-barbital
1-Methyl-5-phenyl-5-ethylbar-bituric acid
Metylfenemal
Metyna
Morbusan
Phemetone
Phemiton
Phemitone
5-Phenyl-5-ethyl-3-methylbar-bituric acid
Prominal
2,4,6(1H,3H,5H)-Pyrimidine-trione, 5-ethyl-1-methyl-5-phenyl- (9CI)

115-39-9
$C_{19}H_{10}Br_4O_5S$
669.99
O=S(=O)(OC(c1cccc2)(c(cc(c(O)c3Br)Br)c3)c(cc(c(O)c4Br)Br)c4)c12
Phenol, 4,4'-(3H-2,1-benzoxa-thiol-3-ylidene)bis(2,6-di-bromo-, S,S-dioxide
Albutest
Bromophenol Blue
Bromphenol Blue
Tetrabromophenolsulfophthalein
3',3'',5',5''-Tetrabromophenol-sulfophthalein

115-44-6
$C_{11}H_{16}N_2O_3$

224.29
CCC(C)C1(CC=C)C(=O)NC(=O)NC1=O
Barbituric acid, 5-allyl-5-sec-butyl
5-Allyl-5-sec-butylbarbituric acid
5-Allyl-5-(1-methylpropyl) barbituric acid
Butalbital
sec-Butyl allyl barbituric acid
Latusate
Lotusate
5-(1-Methylpropyl)-5-(2-propenyl)-2,4,6(1H,3H,5H)-pyrimidinetrione
Profundol
2,4,6(1H,3H,5H)-Pyrimidinetrione, 5-(1-methylpropyl)-5-(2-propenyl)- (9CI)
Talbutal
WIN 5095

115-58-2
$C_{11}H_{18}N_2O_3$
226.26
DEA No. 2271
5-Aethyl-5-pentyl-(2')-barbitursaeure (German)
Barbituric acid, 5-ethyl-5-pentyl- (8CI)
DEA No. 2316
5-Ethyl-5-pentylbarbituric acid
NSC-32305
Pentobarbital
2,4,6(1H,3H,5H)-Pyrimidinetrione, 5-ethyl-5-pentyl- (9CI)

115-67-3
$C_7H_{11}NO_3$
157.19
CCC1(C)OC(=O)N(C)C1=O
2,4-Oxazolidinedione, 5-ethyl-3,5-dimethyl
A 348
3,5-Dimethyl-5-ethyloxazolidine-2,4-dione
5-Ethyl-3,5-dimethyloxazolidine-2,4-dione
Isoethadione
Paradione
Parametadione
Paramethadione

115-69-5
$C_4H_{11}NO_2$
105.16
OCC(N)(CO)C
1,3-Propanediol, 2-amino-2-methyl
Aminoglycol
2-Amino-2-methyl-1,3-propanediol
AMPD
Gentimon
Isobutandiol-2-amine
Pentaerythritol dichlorohydrin

115-70-8
$C_5H_{13}NO_2$
119.16
OCC(N)(CC)CO
1,3-Propanediol, 2-amino-2-ethyl- (9CI)
AEPD
AI3-03358
2-Amino-2-ethyl-1,3-propanediol
2-Ethyl-2-aminopropanediol
NSC-8803

115-71-9
$C_{15}H_{24}O$
220.39
OCC(=CCCC(C(C1C2)(C1C3)C)(C23)C)C
2-Penten-1-ol, 5-(2,3-dimethyltricyclo(2.2.1.0^{2,6})hept-3-yl)-2-methyl-, (R(Z))
Sandal
Santalol A
(+)-α-Santalol
cis-α-Santalol
d-α-Santalol

115-76-4
$C_7H_{16}O_2$
132.23
CCC(CC)(CO)CO
1,3-Propanediol, 2,2-diethyl
3,3-Bis(hydroxymethyl)pentane
DEP
Di-Aethyl-propanediol (German)
2,2-Diethylpropanediol-1,3
2,2-Diethylpropane-1,3-diol
2,2-Diethyl-1,3-propanediol
Dietilpropandiolo
MC 1415
Penderol
Prenderol
Prendiol

115-77-5
$C_5H_{12}O_4$
136.17
OCC(CO)(CO)CO
Pentaerythritol (ACGIH, OSHA)
Auxinutril
2,2-Bis(hydroxymethyl)-1,3-propanediol
Hercules P6
Maxinutril
Methane tetramethylol
Monopentek
PE
Pentaerythrite
Pentek
1,3-Propanediol, 2,2-bis (hydroxymethyl)-
Tetrahydroxymethylmetane
Tetrakis(hydroxymethyl) methane
Tetramethylolmethane

115-78-6
$C_{19}H_{32}Cl_2P.Cl$
397.83
Phosphonium, tributyl(2,4-dichlorobenzyl)-, chloride
CBBP
Chlorfonium
Chlorphonium
Chlorphonium chloride
2,4-Dichlorobenzyltributylphosphonium chloride
Fosfon D
Phosfleur
Phosfon
Phosfon D
Phosphon
Phosphon D
Phosphone D
Tributyl(2,4-dichlorobenzyl)-phosphonium chloride

115-80-0
$C_9H_{20}O_3$
176.26
O(C(OCC)(OCC)CC)CC
Orthopropionic acid, triethyl ester (8CI)
AI3-23844
Ethyl orthopropionate
NSC-5604
Orthopropionic acid ethyl ester
Propane, 1,1,1-triethoxy- (9CI)
Triethyl orthopropionate
Triethyl o-propionate
1,1,1-Triethoxypropane

115-83-3
$C_{77}H_{148}O_8$
1202.02
Pentaerythrityl tetrastearate
Octadecanoic acid, 2,2-bis-(((1-oxooctadecyl)oxy)-methyl)-1,3-propanediyl ester (9CI)
Pentaerythritol Tetrastearate

115-84-4
$C_9H_{20}O_2$
160.26
OCC(CCCC)(CC)CO
2-Butyl-2-ethyl-1,3-propanediol
AI3-03775
BEP
3,3-Bis(hydroxymethyl)heptane
Caswell No. 129
EPA Pesticide Chemical Code 041003
2-Ethyl-2-butylpropanediol-1,3
2-Ethyl-2-butyl-1,3-propanediol
NSC-406603
1,3-Propanediol, 2-butyl-2-ethyl- (8CI,9CI)

115-86-6
$C_{18}H_{15}O_4P$
326.30

O=P(Oc(cccc1)c1)(Oc(cccc2)c2)Oc(cccc3)c3
Phosphoric acid, triphenyl ester
Celluflex tpp
TPP
Trifenylfosfat (Czech)
Triphenyl phosphate (ACGIH, OSHA)

115-87-7
$C_{26}H_{31}O_4P$
438.50
Phosphoric acid, bis(p-tert-butylphenyl) phenyl ester (8CI)
AI3-18185
Bis(p-tert-butylphenyl) phenyl phosphate
NSC-44042
Phosphoric acid, bis(4-(1,1-dimethylethyl)phenyl) phenyl ester (9CI)

115-88-8
$C_{20}H_{27}O_4P$
362.41
Diphenyl octyl phosphate
Disflamoll DPO
Octyl diphenyl phosphate
Phosphoric acid, octyl diphenyl ester

115-89-9
$C_{13}H_{13}O_4P$
264.22
O=P(OC)(Oc(cccc1)c1)Oc(cccc2)c2
Diphenyl methyl phosphate
AI3-07854
Methyl diphenyl phosphate
Methyl phenyl phosphate
NSC-96635
Phosphoric acid, methyl diphenyl ester (9CI)

115-90-2
$C_{11}H_{17}O_4PS_2$
308.37
CCOP(=S)(OCC)Oc1ccc(cc1)S(C)=O
Phosphorothioic acid, O,O-diethyl O-(p-(methylsulfinyl)phenyl) ester
Bay 25141
Bayer 25141
Bayer S767
Chemagro 25141
Dasanit (OSHA)
O,O-Diaethyl-O-4-methyl-sulfinyl-phenyl-monothiophosphat (German)
O,O-Diaethyl-O-(4-methyl-sulfinyl-phenyl)-thionophosphat (German)
O,O-Diethyl O-(p-(methylsulfinyl)phenyl) phosphorothioate
O,O-Diethyl O-p-(methylsulfinyl)phenyl thiophosphate
DMSP
ENT 24,945
Fensulfothion (ACGIH,OSHA)
OMS 37
Phenol, p-(methylsulfinyl)-, O-ester with O,O-diethyl phosphorothioate
S 767
Terracur p

115-93-5
$C_8H_{12}NO_5PS_2$
297.30
O=S(=O)(N)c(ccc(OP(OC)(OC)=S)c1)c1
Phosphorothioic acid, O,O-dimethyl ester, O-ester with p-hydroxybenzenesulfonamide
AC 26,691
American CL-26691
American Cyanamid CL-26,691
O-(4-(Aminosulfonyl)phenyl) O,O-dimethyl phosphorothioate
Benzenesulfonamide, p-hydroxy-, O-ester with O,O-dimethyl phosphorothioate
Cl 26691
Cyflee
Cythioate
O,O-Dimethyl O-p-sulfamoylphenyl phosphorothioate
ENT 25,640
Phosphorothioic acid, O-(4-(aminosulfonyl)phenyl) O,O-dimethyl ester (9CI)
Proban

115-95-7
$C_{12}H_{20}O_2$
196.32
O=C(OC(C=C)(CCC=C(C)C)C)C
1,6-Octadien-3-ol, 3,7-dimethyl-, acetate
Acetic acid linalool ester
Bergamiol
3,7-Dimethyl-1,6-octadien-3-ol acetate
3,7-Dimethyl-1,6-octadien-3-yl acetate
Licareol acetate
Linalol acetate
Linalool acetate
Linalyl acetate

115-96-8
$C_6H_{12}Cl_3O_4P$
285.50
O=P(OCCCl)(OCCCl)OCCCl
Ethanol, 2-chloro-, phosphate (3:1)
Celluflex
2-Chloroethanol phosphate
Fyrol CEF
NCI-C60128
Niax Flame Retardant 3 CF
Phosphoric acid, tris(2-chloroethyl)ester
Trichlorethyl phosphate
Tri-β-chloroethyl phosphate
Tri(2-chloroethyl)phosphate
Tris-(2-chlorethyl)fosfat (Czech)
Tris(β-chloroethyl) phosphate
Tris(2-chloroethyl) phosphate

116-01-8
$C_6H_{14}NO_3PS_2$
243.30
CCNC(=O)CSP(=S)(OC)OC
Phosphorodithioic acid, O,O-dimethyl ester, S-ester with N-ethyl-2-mercaptoacetamide
AC 18706
American Cyanamid 18706
B/77
Cl 18706
Dimethoate-ethyl
O,O-Dimethyl S-(N-ethylcarbamoylmethyl) dithiophosphate
O,O-Dimethyl S-(N-ethylcarbamoylmethyl) phosphorodithioate
Dwutiofosforan S-N-etylokarbamylometylo-O,O-dwumetylowy (Polish)
EI-18706
ENT 25,506
Ethoate methyl
S-(2-(Eethylamino)-2-oxoethyl) O,O-dimethyl phosphorodithioate
S-(N-Eethylcarbamoylmethyl) dimethyl phosphorodithioate
Etoat metylowy (Polish)
Fitios
Fitios B/77
N-Monoethylamide of O,O-dimethyldithiophosphorylacetic acid
OMS 252
Phoshorothioic acid, S-(2-(ethylamino)-2-oxoethyl) O,O-dimethyl ester
Vel 88

116-02-9
$C_9H_{18}O$
142.27
OC(CC(CC1(C)C)C)C1
Cyclohexanol, 3,3,5-trimethyl
Cyclonol
Homomenthol
Isophorol, dihydro-
3,3,5-Trimethylcyclohexanol
3,3,5-Trimethyl-1-cyclohexanol
3,5,5-Trimethylcyclohexanol

116-06-3
$C_7H_{14}N_2O_2S$
190.29
O=C(ON=CC(SC)(C)C)NC
Propionaldehyde, 2-methyl-2-(methylthio)-, O-(methyl

carbamoyl)oxime
Aldecarb
Aldicarb
Aldicarbe (French)
Ambush
Carbamic acid, methyl-, O-((2-methyl-2-(methylthio)-propylidene)amino) deriv.
ENT 27,093
2-Methyl-2-(methylthio)propanal, O-((methylamino)carbonyl)oxime
2-Methyl-2-(methylthio)propionaldehyde O-(methylcarbamoyl)oxime
2-Methyl-2-methylthio-propionaldehyd-O-(N-methyl-carbamoyl)-oxim (German)
2-Metil-2-tiometil-propionaldeid-O-(N-metila-carbamoil)-ossima (Italian)
NCI-C08640
OMS-771
Propanal, 2-methyl-2-(methylthio)-, O-((methylamino)carbonyl)oxime
RCRA waste number P070
Temic
Temik
Temik G10
Temik 10 G
UC-21149
Union Carbide 21149
Union Carbide UC-21149

116-09-6
$C_3H_6O_2$
74.09
O=C(C)CO
2-Propanone, 1-hydroxy
Acetol
Hydroxyacetone

116-11-0
C_4H_8O
72.12
O(C(=C)C)C
1-Propene, 2-methoxy
2-Methoxypropene

116-14-3
C_2F_4
100.02
FC(F)=C(F)F
Ethylene, tetrafluoro
Fluoroplast 4
Perfluoroethene
Perfluoroethylene
Tetrafluorethylene
Tetrafluoroethene (9CI)
Tetrafluoroethylene
Tetrafluoroethylene, Inhibited [UN 1081]
UN 1081 [Tetrafluoroethylene, inhibited]

116-15-4
C_3F_6
150.03
FC(F)(F)C(F)=C(F)F
Propene, hexafluoro
Hexafluoropropene
Hexafluoropropylene [UN 1858]
Perfluoropropene
Perfluoropropylene
Propylene, hexafluoro-
UN 1858 [Hexafluoropropylene]

116-16-5
C_3Cl_6O
264.73
O=C(C(Cl)(Cl)Cl)C(Cl)(Cl)Cl
2-Propanone, 1,1,1,3,3,3-hexachloro
Acetone, hexachloro-
GC-1106
HCA
HCA Weedkiller
Hexachloroacetone [UN 2661]
Hexachloro-2-propanone
1,1,1,3,3,3-Hexachloro-2-propanone
2-Propanone, hexachloro-
UN 2661 [Hexachloroacetone]

116-17-6
$C_9H_{21}O_3P$
208.27
O(P(OC(C)C)OC(C)C)C(C)C
Phosphorous acid, tris-(1-methylethyl) ester
Isopropyl phosphite, tri-
Phosphorous acid, triisopropyl ester
Triisopropyl phosphite

116-25-6
$C_6H_{10}N_2O_3$
158.14
O=C(N(C(C1=O)(C)C)CO)N1
MDM hydantoin
Hydantoin, 1-(hydroxymethyl)-5,5-dimethyl- (8CI)
1-(Hydroxymethyl)-5,5-dimethyl hydantoin
1-(Hydroxymethyl)-5,5-dimethylhydantoin
1-(Hydroxymethyl)-5,5-dimethyl-2,4-imidazolidinedione
2,4-Imidazolidinedione, 1-(hydroxymethyl)-5,5-dimethyl- (9CI)
MDMH
Monomethylol dimethyl hydantoin
1-Monomethylol-5,5-dimethylhydantoin
NSC-9185

116-29-0
$C_{12}H_6Cl_4O_2S$
356.04
Clc1ccc(cc1)S(=O)(=O)c2cc(Cl)c(Cl)cc2Cl
Sulfone, p-chlorophenyl 2,4,5-trichlorophenyl
Akaritox
Aredion
Benzene, 1,2,4-trichloro-5-((4-chlorophenyl)sulfonyl)-
p-Chlorophenyl 2,4,5-trichlorophenyl sulfone
4-Chlorophenyl 2,4,5-trichlorophenyl sulfone
p-Chlorophenyl 2,4,5-trichlorophenyl sulphone
Duphar
ENT 23,737
Fmc 5488
Mition
NIA 5488
Polacaritox
Roztozol
Sulfone, 2,4,4',5-tetrachlorodiphenyl
Tedion
Tedion V-18
2,4,4',5-Tetrachloor-difenylsulfon (Dutch)
3,4,6,4'-Tetrachlor-diphenylsulfon (German)
2,4,4',5-Tetrachlor-diphenylsulfon (German)
2,4,4',5-Tetrachlorodiphenyl sulfone
2,4,5,4'-Tetrachlorodiphenyl sulfone
2,4,5,4'-Tetrachlorodiphenylsulphone
2,4,4',5-Tetracloro-difenil-solfone (Italian)
Tetradichlone
Tetradifon
Tetradiphon
Tetrafidon
V-18

116-37-0
$C_{21}H_{28}O_4$
344.45
O(c(ccc(c1)C(c(ccc(OCC(O)C)c2)c2)(C)C)c1)CC(O)C
Bisphenol A bis(2-hydroxypropyl) ether
AI3-15588
2,2-Bis(p-(2-hydroxy-2-methylethoxy)phenyl)propane
2,2-Bis(p-(β-hydroxypropoxy)-phenyl)propane
2,2-Bis(p-(2-hydroxypropoxy)-phenyl)propane
2,2-Bis(4-(β-hydroxypropoxy)-phenyl)propane
2,2-Bis(4-(2-hydroxypropoxy)-phenyl)propane
Bisphenol A bis(β-hydroxypropyl) ether
Bisphenol A-propylene oxide adduct (1:2)
Dianol 33
Dow Resin 565
Hydroxypropylated diphenylolpropane
1,1'-Isopropylidenebis(p-phenyleneoxy)di-2-propanol

1,1'-(Isopropylidenebis(p-phenyleneoxy))di-2-propanol
Isopropylidenediphenoxypropanol
NSC-408494
Oxypropyldiphenylolpropane
2-Propanol, 1,1'-(isopropylidenebis(p-phenyleneoxy))di- (8CI)
2-Propanol, 1,1'-((1-methylethylidene)bis(4,1-phenyleneoxy))bis- (9CI)

116-44-9
$C_{10}H_{10}N_4O_2S$
250.26
Benzenesulfonamide, 4-amino-N-pyrazinyl- (9CI)
NSC-25872
Pyrazine, sulfanilamido-
N1-2-Pyrazinylsulfanilamide
Sulfanilamide, N1-(pyrazinyl)- (8CI)
2-Sulfanilamidopyrazine
Sulfapyrazine

116-52-9
$C_5H_6Cl_6N_2O_3$
354.83
OC(NC(=O)NC(O)C(Cl)(Cl)Cl)C(Cl)(Cl)Cl
Urea, 1,3-bis(2,2,2-trichloro-1-hydroxyethyl)
1,3-Bis(1-hydroxy-2,2,2-trichloroethyl)urea
1,3-Bis(2,2,2-trichloro-1-hydroxyethyl)urea
Crag 2
Crag DCU-73W
Crag Experimental Herbicide 2
Crag Herbicide 2
DCM
DCU
Dichloral urea
Dicloralurea
Dichloraluree
DKhM
EH2
Experimental Herbicide 2

116-53-0
$C_5H_{10}O_2$
102.13
O=C(O)C(CC)C
2-Methylbutyrate
Active valeric acid
AI3-24202
Butanoic acid, 2-methyl- (9CI)
Butyric acid, 2-methyl- (8CI)
Ethylmethylacetic acid
2-Methylbutanoic acid
α-Methylbutyric acid
2-Methylbutyric acid
Methylethylacetic acid
NSC-7304

116-54-1
$C_3H_4Cl_2O_2$
142.97
O=C(OC)C(Cl)Cl
Acetic acid, dichloro-, methyl ester
Dichloroacetic acid methyl ester
Methyl dichloroacetate [UN 2299]
UN 2299 [Methyl dichloroacetate]

116-63-2
$C_{10}H_9NO_4S$
239.26
O=S(=O)(O)c(c(c(c(N)c1O)ccc2)c2)c1
1-Naphthalenesulfonic acid, 4-amino-3-hydroxy
4-Amino-3-hydroxy-1-naphthalenesulfonic acid
1-Amino-4-sulfo-2-naphthol
1-Amino-2-naphthol-4-sulfonic acid
2-Hydroxy-4-sulfo-1-naphthylamine

116-66-5
$C_{14}H_{18}N_2O_4$
278.34
O=N(=O)c(c(c(N(=O)=O)c(c1C(C2)(C)C)C2(C)C)C)c1
Indan, 4,6-dinitro-1,1,3,3,5-pentamethyl
Moskene
1,1,3,3,5-Pentamethyl-4,6-dinitroindane

116-71-2
$C_{34}H_{16}O_2$
456.50
O=C(c(c(c(c1c(c(c(c(c(c(c(c(C2=O)ccc3)c3)c4)c2cc5)c56)c4)c7)c6cc8)c7)ccc9)c9)c18
Dinaphtho(1,2,3-cd:3',2',1'-lm)perylene-5,10-dione
Dibenzanthrone
Violanthrone

116-75-6
$C_{32}H_{30}N_2O_2$
474.59
O=C(c(c(C(=O)c1c(Nc(c(cc(c2)C)C)c2C)ccc3Nc(c(cc(c4)C)C)c4C)ccc5)c5)c13
Anthraquinone, 1,4-bis(2,4,6-trimethylanilino)- (8CI)
9,10-Anthracenedione, 1,4-bis((2,4,6-trimethylphenyl)amino)- (9CI)
1,4-Bis(2,4,6-trimethylanilino)anthraquinone
1,4-Dimesidinoanthraquinone
NSC-135500

116-78-9
$C_{26}H_{10}N_6O_{16}$
662.36
O=C(c(c(c(N(=O)=O)cc1)C(=O)c2c(Oc(c(N(=O)=O)cc(N(=O)=O)c3)c3)ccc4N(=O)=O)c1Oc(c(N(=O)=O)cc(N(=O)=O)c5)c5)c
9,10-Anthracenedione, 1,5-bis(2,4-dinitrophenoxy)-4,8-dinitro- (9CI)

116-81-4
$C_{14}H_8BrNO_5S$
382.19
O=C(c(c(C(=O)c1c(cc(S(=O)(=O)O)c2N)Br)ccc3)c3)c12
2-Anthracenesulfonic acid, 1-amino-4-bromo-9,10-dihydro-9,10-dioxo- (9CI)
Alizarine Cyanol Grey G (VAN)
1-Amino-4-bromo-2-anthraquinonesulfonic acid, sodium salt acid
Bromamine acid
Bromaminic acid
NSC-7574

116-82-5
$C_{14}H_8BrNO_3$
318.14
O=C(c(c(C(=O)c1c(O)cc(c2N)Br)ccc3)c3)c12
Anthraquinone, 1-amino-2-bromo-4-hydroxy
1-Amino-2-brom-4-hydroxyanthrachinon (Czech)
1-Amino-2-bromo-4-hydroxyanthraquinone
9,10-Anthracenedione, 1-amino-2-bromo-4-hydroxy-

116-85-8
$C_{14}H_9NO_3$
239.24
O=C(c(c(C(=O)c1c(N)ccc2O)ccc3)c3)c12
Anthraquinone, 1-amino-4-hydroxy
1A-4OA (Russian)
Acetate Fast Red 2B
Acetoquinone Light Gooseberry RL
Acetylon Fast Pink B
Amacel Pink B
1-Amino-4-hydroxyanthraquinone
1-Amino-4-oxyanthraquinone (Russian)
9,10-Anthracenedione, 1-amino-4-hydroxy- (9CI)
Anthraquinone, 1-amino-4-hydroxy-
Artisil Direct Red 3BP
Artisil Red 3BP
Calcosyn Pink B
Celanthrene Red 3BN
Celliton Fast Pink BA-CF
Celliton Fast Pink BN
Celutate Pink B
Celutate Pink BN

Celutate Pink BY
Cerven disperzni 15 (Czech)
Cibacete Red 3B
Cibacet Red 3B
Cibacet Red E3B
Cilla Fast Pink BN
C.I. 60710
C.I. Disperse Red 15
C.I. Solvent Red 53
Diacelliton Fast Pink B
Disperse Fast Pink B
Disperse Red 15
Disperse Red 25
Dispersol Orange D-G
Duranol Red 2B
Fenacet Fast Pink B
1-Hydroxy-4-aminoanthraquinone
4-Hydroxy-1-anthraquinonylamine
Interchem Acetate Pink BLF
Interchem Hisperse Pink BH
Microsetile Pink BN
Nacelan Pink B
Neosetile Pink BN
Oracet Red 3B
Para M
Perliton Pink 3B
Serisol Fast Red 2B
Setacyl Pink 3B
Supracet Brilliant Red 2B

117-08-8
$C_8Cl_4O_3$
285.88
O=C(OC(=O)c1c(c(c(c2Cl)Cl)Cl)Cl)c12
Phthalic anhydride, tetrachloro
1,3-Isobenzofurandione, 4,5,6,7-tetrachloro- (9CI)
NCI-C61585
Niagathal
Tetrachlorophthalic anhydride

117-10-2
$C_{14}H_8O_4$
240.22
O=C(c(c(c(O)cc1)C(=O)c2c(O)ccc3)c1)c23
Anthraquinone, 1,8-dihydroxy
Altan
9,10-Anthracenedione, 1,8-dihydroxy-
Antrapurol
Chrysazin
Criasazin
Danthron
Dantron
Diaquone
1,8-Dihydroxy-9,10-anthracenedione
1,8-Dihydroxyanthrachinon (Czech)
1,8-Dihydroxyanthraquinone
Dionone
Dorbane
Dorbanex
Duolax
Istin
Istizin
Laxanorm
Laxanthreen
Laxipur
Laxipurin
Ltan
Modane
1,4,5,8-Tetroxyantraquinone
USAF ND-59
Zwitsalax

117-11-3
$C_{14}H_8ClNO_2$
257.68
O=C(c(c(c(cc1)Cl)C(=O)c2cccc3N)c1)c23
Anthraquinone, 1-amino-5-chloro
1-Amino-5-chloroanthraquinone
9,10-Anthracenedione, 1-amino-5-chloro-
1-Chlor-5-aminoanthrachinon (Czech)
5-Chloro-1-aminoanthraquinone

117-12-4
$C_{14}H_8O_4$
240.22
O=C(c(c(c(O)cc1)C(=O)c2cccc3O)c1)c23
Anthraquinone, 1,5-dihydroxy
9,10-Anthracenedione, 1,5-dihydroxy-
Anthrarufin
1,5-Dihydroxyanthrachinon (Czech)
1,5-Dihydroxyanthraquinone
1,5-Dihydroxy-9,10-anthraquinone

117-14-6
$C_{14}H_8O_8S_2$
368.34
O=C(c(c(c(S(=O)(=O)O)cc1)C(=O)c2cccc3S(=O)(=O)O)c1)c23
1,5-Anthracenedisulfonic acid, 9,10-dihydro-9,10-dioxo
1,5-Anthraquinonedisulfonic acid
1,5-Disulfoanthraquinone

117-18-0
$C_6HCl_4NO_2$
260.88
Clc1cc(Cl)c(Cl)c(N(=O)=O)c1Cl
Benzene, 1,2,4,5-tetrachloro-3-nitro
Benzene, 3-nitro-1,2,4,5-tetrachloro-
Chipman 3,142
DB 905
Folosan
Fusarex
TCNB
Tecnazen (German)
Tecnazene
Teknazen (Czech)
2,3,5,6-Tetrachlor-3-nitrobenzol (German)
1,2,4,5-Tetrachloro-3-nitrobenzene
2,3,5,6-Tetrachloronitrobenzene

117-26-0
$C_{16}H_{15}Cl_2NO_2$
324.22
CCC(C(c1ccc(Cl)cc1)c2ccc(Cl)cc2)N(=O)=O
Butane, 1,1-bis(p-chlorophenyl)-2-nitro
1,1-Bis(p-chlorophenyl)-2-nitrobutane
1,1-Bis(4-chlorophenyl)-2-nitrobutane
BNB
Bulan
CS 674A
DNB
ENT 18,065
2-Nitro-1,1-bis(p-chlorophenyl)-butane
1,1'-(2-Nitrobutylidene)bis-(4-chlorobenzene)

117-27-1
$C_{15}H_{13}Cl_2NO_2$
310.19
CC(C(c1ccc(Cl)cc1)c2ccc(Cl)cc2)N(=O)=O
Propane, 1,1-bis(p-chlorophenyl)-2-nitro
1,1-Bis(p-chlorophenyl)-2-nitropropane
1,1-Bis(4-chlorophenyl)-2-nitropropane
BNP
C.I. Azoic Diazo Component 37
CS 645A
DNP
ENT 22,784
2-Nitro-1,1-bis(p-chlorophenyl)-propane
1,1'-(2-Nitropropylidene)bis-(4-chlorobenzene)
Prolan
Prolan (CSC)

117-34-0
$C_{14}H_{12}O_2$
212.26
O=C(O)C(c(cccc1)c1)c(cccc2)c2
Acetic acid, diphenyl
Benzeneacetic acid, α-phenyl-
Diphenylacetic acid
α,α-Diphenylacetic acid
DPA

117-37-3
$C_{16}H_{12}O_3$
252.28
COc1ccc(cc1)C3C(=O)c2ccccc2C3=O
1,3-Indandione, 2-(p-methoxy-

phenyl)
Anisindione
Anisin indandione
2-p-Anisyl-1,3-indandione
2-(p-Methoxyphenyl)-1,3-indandione
2-(4-Methoxyphenyl)-1H-indene-1,3(2H)-dione
2-(p-Methoxyphenyl)indane-1,3-dione
Miradon
SPE 2792
Unidone

117-39-5
$C_{15}H_{10}O_7$
302.25
o(c(c(c(O)cc1O)c(O)c2O)c1)c2cz(ccc(O)c3O)c3
Flavone, 3,3',4',5,7-pentahydroxy
4H-1-Benzopyran-4-one, 2-(3,4-dihydroxyphenyl)-3,5,7-trihydroxy- (9CI)
C.I. 75670
C.I. Natural Red 1
C.I. Natural Yellow 10
C.I. Natural Yellow 10 & 13
Cyanidelonon 1522
Kvercetin (Czech)
Meletin
NCI-C60106
3,5,7,3',4'-Pentahydroxyflavon
3,5,7,3',4'-Pentahydroxyflavone
Quercetin
Quercetine
Quercetol
Quercitin
Quertine
Sophoretin
3',4',5,7-Tetrahydroxyflavan-3-ol
T-Gelb BZW. Grun 1
Xanthaurine

117-51-1
$C_{22}H_{32}O_2$
328.49
Synhexyl
DEA No. 7374
6H-Dibenzo(b,d)pyran-1-ol, 3-hexyl-7,8,9,10-tetrahydro-6,6,9-trimethyl-
3-Hexyl-1-hydroxy-7,8,9,10-tetrahydro-6,6,9-trimethyl-6H-dibenzo(b,d)pyran
3-Hexyl-7,8,9,10-tetrahydro-6,6,9-trimethyl-6H-dibenzo-(b,d)pyran-1-ol
1-Hydroxy-3-n-hexyl-6,6,9-trimethyl-7,8,9,10-tetrahydro-6-dibenzopyran
3-Homotetrahydrocannibinol
Parahexyl

117-52-2
$C_{17}H_{14}O_5$
298.31
CC(=O)CC(c1ccco1)c3c(O)c2ccccc2oc3=O
Coumarin, 3-(α-acetonylfurfuryl)-4-hydroxy
3-(α-Acetonylfurfuryl)-4-hydroxycoumarin
2H-1-Benzopyran-2-one, 3-(1-(2-furanyl)-3-oxobutyl)-4-hydroxy- (9CI)
Coumafuryl
Cumafuryl (German)
Foumarin
Fumarin
Fumasol
Furmarin
3-(α-Furyl-β-acetylaethyl)-4-hydroxycumarin (German)
3-(1-Furyl-3-acetylethyl)-4-hydroxycoumarin
4-(2-Furyl)-4-(4-hydroxy-3-coumarinyl)-2-butanone
4-(2-Furyl)-4-(4-hydroxy-3-kumarinyl)-2-butanon (Czech)
Kill-Ko Rat
Krumkil
Kumatox
Lurat
Mouse Blues
Ratafin
Rat-A-Way
Tomarin

117-57-7
$C_{11}H_9NO_3$
203.19
O=C(O)c(c(c(c(nc1C)ccc2)c2)c1O
4-Quinolinecarboxylic acid, 3-hydroxy-2-methyl- (9CI)
3-Hydroxy-2-methyl-4-quinolinecarboxylic acid

117-61-3
$C_{12}H_{12}N_2O_6S_2$
344.38
O=S(=O)(O)c(c(c(c(S(=O)(=O)O)cc(N)c1)c1)ccc2N)c2
2,2'-Biphenyldisulfonic acid, 4,4'-diamino
Benzidine, 2,2'-disulfo-
2,2'-Benzidinedisulfonic acid
6,6'-Bimetanilic acid
(1,1'-Biphenyl)-2,2'-disulfonic acid, 4,4'-diamino-
4,4'-Biphenyldisulfonic acid, 2,2'-diamino-
4,4'-Diaminobiphenyl-2,2'-disulfonic acid
4,4'-Diamino-2,2'-biphenyldisulfonic acid
4,4'-Diaminodiphenyl-2,2'-disulfonic acid
2,2'-Disulfobenzidine
Kyselina benzidin-2,2'-disulfonova (Czech)

117-62-4
$C_{10}H_9NO_6S_2$
303.32
O=S(=O)(O)c(c(c(c(S(=O)(=O)O)c(N)c1)cc2)c1)c2
1,5-Naphthalenedisulfonic acid, 2-amino
2-Amino-1,5-naphthalenedisulfonic acid
Kyselina 2-naftylamin-1,5-disulfonova (Czech)
Kyselina sulfo-tobiasova (Czech)
Tobias acid, 5-sulfo-

117-79-3
$C_{14}H_9NO_2$
223.24
O=C(c(c(c(C(=O)c1cccc2)ccc3N)c3)c12
Anthraquinone, 2-amino
AAQ
2-Amino-9,10-anthracenedione
Aminoanthraquinone
β-Aminoanthraquinone
2-Aminoanthraquinone
2-Amino-9,10-anthraquinone
β-Anthraquinonylamine
9,10-Anthracenedione, 2-amino-(9CI)
NCI-C01876

117-80-6
$C_{10}H_4Cl_2O_2$
227.04
O=C(c(c(C(=O)C=1Cl)ccc2)c2)C1Cl
1,4-Naphthoquinone, 2,3-dichloro
Algistat
Compound 604
Dichlone
2,3-Dichlor-1,4-naftochinon (Czech)
2,3-Dichloro-1,4-naphthalenedione
2,3-Dichloro-1,4-naphthaquinone
2,3-Dichlor-1,4-naphthochinon (German)
Dichloronaphthoquinone
2,3-Dichloronaphthoquinone
2,3-Dichloro-α-naphthoquinone
2,3-Dichloro-1,4-naphthoquinone
2,3-Dichloronaphthoquinone-1,4
ENT 3,776
Latka 604 (Czech)
Phygon
Phygon Paste
Phygon Seed Protectant
Phygon XL
Quintar
Quintar 540F
Sanquinon
U.S. Rubber 604
Uniroyal
USR 604

117-81-7
$C_{24}H_{38}O_4$
390.62
O=C(OCC(CCCC)CC)c(c(ccc1)

C(=O)OCC(CCCC)CC)c1
Phthalic acid, bis(2-ethylhexyl) ester
BEHP
1,2-Benzenedicarboxylic acid, bis(2-ethylhexyl) ester
Bis(2-ethylhexyl)-1,2-benzenedicarboxylate
Bis-(2-ethylhexyl)ester kyseliny ftalove (Czech)
Bis(2-ethylhexyl)phthalate
Bisoflex 81
Bisoflex DOP
Compound 889
DAF 68
DEHP
Di-2-ethylhexylphthalate (OSHA)
Di(2-ethylhexyl)orthophthalate
Di(2-ethylhexyl)phthalate
Dioctyl phthalate
Di-sec-octyl phthalate (ACGIH, OSHA)
DOP
Ergoplast FDO
Ethylhexyl phthalate
2-Ethylhexyl phthalate
Eviplast 80
Eviplast 81
Fleximel
Flexol DOP
Flexol Plasticizer DOP
Good-Rite GP 264
Hatcol DOP
Hercoflex 260
Kodaflex DOP
Mollan O
NCI-C52733
Nuoplaz DOP
Octoil
Octyl phthalate
Palatinol AH
Phthalic acid dioctyl ester
Pittsburgh PX-138
Platinol AH
Platinol DOP
RC Plasticizer DOP
RCRA waste number U028
Reomol DOP
Reomol D 79P
Sicol 150
Staflex DOP
Truflex DOP
Vestinol AH
Vinicizer 80
Witcizer 312

117-82-8
$C_{14}H_{18}O_6$
282.32
O=C(OCCOC)c(c(ccc1)C(=O)OCCOC)c1
Phthalic acid, di(methoxyethyl) ester
1,2-Benzenedicarboxylic acid, bi(2-methoxyethyl) ester (9CI)
Bis(methoxyethyl) phthalate
Bis(2-methoxyethyl) phthalate
Di-(2-methoxyethyl)ester kyseliny ftalove (Czech)
Dimethoxy ethyl phthalate
Di(2-methoxyethyl)phthalate
DMEP
Kesscoflex MCP
2-Methoxyethyl phthalate
Phthalic acid, bis(2-methoxyethyl) ester

117-83-9
$C_{20}H_{30}O_6$
366.50
O=C(OCCOCCCC)c(c(ccc1)C(=O)OCCOCCCC)c1
Phthalic acid, bis(2-butoxyethyl) ester
Bis(2-butoxyethyl)phthalate
β-Butoxyethyl phthalate
Butyl "cellosolve" phthalate
Butyl glycol phthalate
Di-(2-butoxyethyl)ester kyseliny ftalove (Czech)
Di(butoxyethyl)phthalate
Dibutylcellosolve ftalat (Czech)
Dibutyl cellosolve phthalate
Dibutylglycol phthalate
Ethanol, 2-butoxy-, phthalate (2:1)
Kesscoflex
Kronisol

117-84-0
$C_{24}H_{38}O_4$
390.62
O=C(OCCCCCCCC)c(c(ccc1)C(=O)OCCCCCCCC)c1
Phthalic acid, dioctyl ester
o-Benzenedicarboxylic acid, dioctyl ester
1,2-Benzenedicarboxylic acid, dioctyl ester
Celluflex DOP
Dinopol NOP
Dioctyl o-benzenedicarboxylate
Dioctyl phthalate
Di-n-octyl phthalate
Dioktylester kyseliny ftalove (Czech)
DNOP
Octyl phthalate
n-Octyl phthalate
Polycizer 162
PX-138
RCRA waste number U107
Vinicizer 85

117-89-5
$C_{21}H_{24}F_3N_3S$
407.54
CN4CCN(CCCN2c1ccccc1Sc3ccc(cc23)C(F)(F)F)CC4
Phenothiazine, 10-(3-(4-methyl-1-piperazinyl)-propyl)-2-(trifluoromethyl)
Fluoperazine
Jatroneural
10-(γ-(N'-Methylpiperazino)-propyl)-2-trifluoromethyl-phenothiozine
10-(3-(4-Methyl-1-piperazinyl)-propyl)-2-(trifluoromethyl)-phenothiazine
RP 7623
Stelazine
Stellazine
Terfluzine
Trifluoperazina (Italian)
Trifluoperazine
Trifluoromethyl-10-(3'-(1-methyl-4-piperazinyl)propyl)-phenothiazine
Trifluoromethylperazine
Trifluperazine
Triphthazine

117-96-4
$C_{11}H_9I_3N_2O_4$
613.92
CC(=O)Nc1c(I)c(NC(C)=O)c(I)c(C(O)=O)c1I
Benzoic acid, 3,5-diacetamido-2,4,6-triiodo
Amidotrizoic acid
Benzoic acid, 3,5-bis(acetylamino)-2,4,6-triiodo- (9CI)
3,5-Diacetamido-2,4,6-triiodobenzoic acid
Diat (German)
Diatrizoesaure (German)
Diatrizoic acid
Odiston
Urografin acid
Urogranoic acid
Urotrast

117-98-6
$C_{17}H_{27}O_2$
263.44
O=C(OC(C=C(C(C(C1C)CC2=C(C)C)C2)C)C1)C
6-Azulenol, 1,2,3,3a,4,5,6,8a-octahydro-2-isopropylidene-4,8-dimethyl-, acetate
Acetic acid, vetiverol ester
6-Azulenol, 1,2,3,3a,4,5,6,8a-octahydro-4,8-dimethyl-2-(1-methylethylidene)-, acetate
Vetiver acetate
Vetiverol, acetate
Vetivert acetate
Vetiveryl acetate

117-99-7
$C_{13}H_{10}O_2$
198.22
O=C(c(cccc1)c1)c(c(O)ccc2)c2
Benzophenone, 2-hydroxy- (8CI)
o-Benzoylphenol
o-Hydroxybenzophenone
2-Hydroxybenzophenone
(2-Hydroxyphenyl)phenylmethanone
Methanone, (2-hydroxyphenyl)-phenyl- (9CI)
NSC-623
Phenyl 2-hydroxyphenyl ketone

118-00-3
$C_{10}H_{13}N_5O_5$
283.28
O(C(C(O)C1O)CO)C1N(c(nc(nc2O)N)c2N=3)C3
Guanosine
GR
Guanine, 9-β-D-ribofuranosyl-
Guanine riboside
2(3H)-Imino-9-β-D-ribofuranosyl-9H-purin-6(1H)-one
Inosine, 2-amino-
Ribofuranoside, guanine-9, β-D-
9-β-D-Ribofuranosylguanine
Vernine
USAF CB-11

118-08-1
$C_{21}H_{21}NO_6$
383.43
COc2ccc1C(OC(=O)c1c2OC)C4N(C)CCc5cc3OCOc3cc45
Hydrastine
β-Hydrastine
Phthalide, 6,7-dimethoxy-3-(5,6,7,8-tetrahydro-6-methyl-1,3-dioxolo(4,5-g)isoquinolin-5-yl)-

118-10-5
$C_{19}H_{22}N_2O$
294.43
OC(c(c(c(nc1)ccc2)c2)c1)C(N(CCC3C4C=C)C4)C3
Cinchonine
Cinchonan-9-ol, (9s)- (9CI)
d-Cinchonine
2-Quinuclidinemethanol, α-4-quinolyl-5-vinyl-
2-Quinuclidinemethanol, α-(5-vinyl-2-quinolyl)-

118-12-7
$C_{12}H_{15}N$
173.28
N(c(c(ccc1)C2(C)C)c1)(C2=C)C
Indoline, 2-methylene-1,3,3-trimethyl
2-Methylene-1,3,3-trimethyl-indoline
1,3,3-Trimethyl-2-methylene-indoline

118-29-6
$C_9H_7NO_3$
177.17
O=C(N(C(=O)c1cccc2)CO)c12
Phthalimide, N-(hydroxy-methyl)
Hydroxymethylphthalimide
N-(Hydroxymethyl)phthalimide
1H-Isoindole-1,3(2H)-dione, 2-(hydroxymethyl)-
N-Methylolphthalimide
Oxymethylphthalimide
Phthalimidomethyl alcohol

118-31-0
$C_{11}H_{11}N$
157.21
c(c(c(cc1)CN)ccc2)(c2)c1
1-Naphthalenemethanamine (9CI)

118-32-1
$C_{10}H_8O_7S_2$
304.30
O=S(=O)(O)c(cc(c(c1ccc2O)c2)S(=O)(=O)O)c1
1,3-Naphthalenedisulfonic acid, 7-hydroxy- (9CI)
G Acid
2-Hydroxynaphthalene-6,8-disulfonic acid
2-Hydroxy-6,8-naphthalenedisulfonic acid
7-Hydroxy-1,3-naphthalenedisulfonic acid
β-Naphthol-γ-disulfonic acid
2-Naphthol-6,8-disulfonic acid
β-Naphtholdisulfonic acid G
NSC-2073

118-33-2
$C_{10}H_9NO_6S_2.Na$
326.31
O=S(=O)(O)c(cc(c(c1cc(N)c2)c2)S(=O)(=O)O)c1
Naphthalene-1,3-disulfonic acid, 6-amino
2-Naftylamin-5,7-disulfonan sodny (Czech)
Kyselina amino-I

118-41-2
$C_{10}H_{12}O_5$
212.20
O=C(O)c(cc(OC)c(OC)c1OC)c1
3,4,5-Trimethoxybenzoic acid
AI3-21153
Benzoic acid, 3,4,5-trimethoxy- (9CI)
Eudesmic acid
Gallic acid trimethyl ether
NSC-2525
Tri-O-methylgallic acid
Veratric acid, 5-methoxy-

118-44-5
$C_{12}H_{13}N$
171.24
N(c(c(c(ccc1)cc2)c1)c2)CC
1-Naphthalenamine, N-ethyl- (9CI)
AI3-20321
2-(Ethylamino)naphthalene
N-Ethyl-1-naphthalenamine
Ethyl-α-naphthylamine
N-Ethyl-α-naphthylamine
N-Ethyl-1-naphthylamine
1-Naphthylamine, N-ethyl- (8CI)
NSC-8636

118-48-9
$C_8H_5NO_3$
163.14
O=C(OC(=O)c(c1ccc2)c2)N1
2H-3,1-Benzoxazine-2,4(1H)-dione
IA
Isatoic acid anhydride
Isatoic anhydride

118-52-5
$C_5H_6Cl_2N_2O_2$
197.03
O=C(N(C(C1=O)(C)C)Cl)N1Cl
Hydantoin, 1,3-dichloro-5,5-dimethyl
Dactin
Daktin
DCA
Dantoin
Dichlorantin
1,3-Dichloro-5,5-dimethyl-hydantoin
1,3-Dichloro-5,5-dimethyl hydantoin (ACGIH,OSHA)
1,3-Dichloro-5,5'-methyl hydantoin
Dwuchlorantyny (Polish)
1,3-Dwuchloro-5,5-dwumetylo-hydantoina (Polish)
Halane
Hydan
Hydan (antiseptic)
Hydantoin, dichlorodimethyl-
2,4-Imidazolidinedione, 1,3-di-chloro-5,5-dimethyl- (9CI)
NCI-C03054
Omchlor

118-55-8
$C_{13}H_{10}O_3$
214.23
O=C(Oc(cccc1)c1)c(c(O)ccc2)c2
Salicylic acid, phenyl ester
Fenylester kyseliny salicylove (Czech)
Phenyl salicylate
Salol

118-56-9
$C_{16}H_{22}O_3$
262.35
O=C(OC(CC(CC1(C)C)C)C1)c(c(O)ccc2)c2
Homosalate
Benzoic acid, 2-hydroxy-, 3,3,5-trimethylcyclohexyl ester (9CI)
Caswell No. 482B
Coppertone
EPA Pesticide Chemical Code 076603
Filtersol (A)
Heliopan
Heliophan
Homomenthyl salicylate
m-Homomenthyl salicylate

Homosalato (Spanish)
Homosalatum (Latin)
Metahomomenthyl salicylate
NSC-164918
Salicylic acid, m-homomenthyl ester
Salicylic acid, 3,3,5-trimethylcyclohexyl ester (8CI)
3,3,5-Trimethylcyclohexyl 2-hydroxybenzoate
3,3,5-Trimethylcyclohexyl salicylate
Uniderm Homsal

118-58-1
$C_{14}H_{12}O_3$
228.26
O=C(OCc(cccc1)c1)c(c(O)ccc2)c2
Salicylic acid, benzyl ester
Benzyl o-hydroxybenzoate
Benzyl salicylate

118-61-6
$C_9H_{10}O_3$
166.19
O=C(OCC)c(c(O)ccc1)c1
Salicylic acid, ethyl ester
Ethyl o-hydroxybenzoate
Ethyl salicylate
Sal ethyl
Salicylic ether
Salicylic ethyl ester

118-69-4
$C_7H_6Cl_2$
161.03
c(c(c(cc1)Cl)C)(c1)Cl
2,6-Dichlorotoluene
AI3-26487
Benzene, 1,3-dichloro-2-methyl- (9CI)
1,3-Dichloro-2-methylbenzene
NSC-60722
Toluene, 2,6-dichloro- (8CI)

118-71-8
$C_6H_6O_3$
126.12
o(c(c(O)c(O)c1)C)c1
4H-Pyran-4-one, 3-hydroxy-2-methyl
Corps Praline
Larixic acid
Larixinic acid
3-Hydroxy-2-methyl-4H-pyran-4-one
3-Hydroxy-2-methyl-γ-pyrone
3-Hydroxy-2-methyl-4-pyrone
Maltol
2-Methyl-3-hydroxy-4-pyrone
2-Methyl pyromeconic acid
Palatone
Vetol
2-Methyl-3-oxy-γ-pyrone
Talmon

118-74-1
C_6Cl_6
284.76
c(c(c(c(c1Cl)Cl)Cl)Cl)(c1Cl)Cl
Benzene, hexachloro
Amatin
Anticarie
Bunt-Cure
Bunt-No-More
Ceku C.B.
Co-Op Hexa
Esaclorobenzene (Italian)
Granox NM
HCB
Hexa C.B.
Hexachlorbenzol (German)
Hexachlorobenzene [UN 2729]
Julin's Carbon Chloride
No Bunt
No Bunt 40
No Bunt 80
No Bunt Liquid
Pentachlorophenyl chloride
Perchlorobenzene
Phenyl perchloryl
Sanocide
Smut-Go
RCRA waste number U127
Saatbeizfungizid (German)
Sanocid
Snieciotox
UN 2729 [Hexachlorobenzene]

118-75-2
$C_6Cl_4O_2$
245.86
O=C(C(=C(C(=O)C=1Cl)Cl)Cl)C1Cl
p-Benzoquinone, 2,3,5,6-tetrachloro
1,4-Benzoquinone, 2,3,5,6-tetrachloro-
Chloranil
Dow Seed Disinfectant No. 5
ENT 3,797
G-25804
G-444E
Geigy-444E
Reranil
Spergon
Spergon I
Spergon Technical
Tetrachlorobenzoquinone
Tetrachloro-p-benzoquinone
Tetrachloro-1,4-benzoquinone
2,3,5,6-Tetrachloro-p-benzoquinone
2,3,5,6-Tetrachloro-1,4-benzoquinone
2,3,5,6-Tetrachloro-2,5-cyclohexadiene-1,4-dione
Tetrachloroquinone
Tetrachloro-p-quinone
Vulklor

118-79-6
$C_6H_3Br_3O$
330.82
Oc(c(cc(c1)Br)Br)c1Br
Phenol, 2,4,6-tribromo
Bromol
Tribromophenol
2,4,6-Tribromophenol

118-82-1
$C_{29}H_{44}O_2$
424.67
Oc(c(cc(c1)Cc(cc(c(O)c2C(C)(C)C)C(C)(C)C)c2)C(C)(C)C)c1C(C)(C)C
4,4'-Methylenebis(2,6-di-tert-butylphenol)
Antioxidant E 702
Bimox M
Binox M
Binox-M
Di(4-hydroxy-3,5-di-tert-butylphenyl)methane
E 702
Ethyl 702
Etil 702
Ionox 220
Ionox 220 Antioxidant
L 3MB1
LZ-MB 1
MB 1 (Antioxidant) (VAN)
NSC-30551
Phenol, 4,4'-methylenebis(2,6-bis(1,1-dimethylethyl)- (9CI)
Phenol, 4,4'-methylenebis(2,6-di-tert-butyl- (8CI)

118-90-1
$C_8H_8O_2$
136.16
O=C(O)c(c(ccc1)C)c1
o-Toluic acid
o-Methylbenzoic acid
2-Methylbenzoic acid
Orthotoluic acid
o-Toluylic acid

118-91-2
$C_7H_5ClO_2$
156.57
O=C(O)c(c(ccc1)Cl)c1
Benzoic acid, o-chloro
Benzoic acid, 2-chloro-
2-CBA
o-Chlorobenzoic acid
2-Chlorobenzoic acid
Kyselina o-chlorbenzoova (Czech)

118-92-3
$C_7H_7NO_2$
137.15
O=C(O)c(c(N)ccc1)c1
Anthranilic-acid
AA
o-Aminobenzoic acid
2-Aminobenzoic acid
1-Amino-2-carboxybenzene
o-Anthranilic acid
Benzoic acid, o-amino-
o-Carboxyaniline
2-Carboxyaniline
Carboxyaniline

Kyselina o-aminobenzoova (Czech)
Kyselina anthranilova (Czech)
NCI-C01730
ortho-Amidobenzoic acid
ortho-Aminobenzoic acid
Vitamin L

118-93-4
$C_8H_8O_2$
136.16
O=C(c(c(O)ccc1)c1)C
Acetophenone, o-hydroxy
Acetophenone, 2'-hydroxy- (8CI)
o-Acetylphenol
2-Acetylphenol
Ethanone, 1-(2-hydroxyphenyl)- (9CI)
o-Hydroxyacetophenone
2'-Hydroxyacetophenone
o-Hydroxyphenyl methyl ketone
USAF KE-20

118-96-7
$C_7H_5N_3O_6$
227.15
O=N(=O)c(cc(N(=O)=O)c(c1N(=O)=O)C)c1
Toluene, 2,4,6-trinitro- (Wet)
Benzene, 2-methyl-1,3,5-trinitro-
Entsufon
NCI-C56155
TNT (OSHA)
α-TNT
TNT-tolite (French)
Tolit
Tolite
2,4,6-Trinitrotolueen (Dutch)
Trinitrotoluene
Trinitrotoluene, Dry or containing, by weight, less than 30% water [UN 0209]
Trinitrotoluene, Wet containing at least 10% water
Trinitrotoluene, Wet containing at least 10% water, over 16 ozs. in one outside packaging
Trinitrotoluene, Wetted with not less than 30% water [UN 1356]
s-Trinitrotoluene
sym-Trinitrotoluene
2,4,6-Trinitrotoluene (ACGIH, OSHA)
s-Trinitrotoluol
sym-Trinitrotoluol
2,4,6-Trinitrotoluol (German)
Tritol
Triton
Trojnitrotoluen (Polish)
Trotyl
Trotyl Oil
UN 0209 [Trinitrotoluene (TNT), Dry or wetted with less than 30 per cent water, by mass]
UN 1356 [Trinitrotoluene, Wetted with not less than 30% water]

118-97-8
$C_7H_3ClN_2O_6$
246.55
O=C(O)c(cc(N(=O)=O)c(c1N(=O)=O)Cl)c1
4-Chloro-3,5-dinitrobenzoate
Benzoic acid, 4-chloro-3,5-dinitro- (9CI)
4-Chloro-3,5-dinitrobenzoic acid
3,5-Dinitro-4-chlorobenzoic acid
NSC-76583

119-06-2
$C_{34}H_{58}O_4$
530.92
O=C(OCCCCCCCCCCCCC)c(c(ccc1)C(=O)OCCCCCCCCCCCCC)c1
Phthalic acid, ditridecyl ester
1,2-Benzenedicarboxylic acid, ditridecyl ester
Ditridecyl phthalate
DTDP
Jayflex DTDP
Nuoplaz
Polycizer 962-BPA
Staflex DTDP
1-Tridecanol, phthalate
Truflex DTDP

119-07-3
$C_{26}H_{42}O_4$
418.68
O=C(OCCCCCCCCCC)c(c(ccc1)C(=O)OCCCCCCCC)c1
Phthalic acid, decyl octyl ester
1,2-Benzenedicarboxylic acid, decyl octyl ester
Decyl octyl phthalate
n-Decyl n-octyl phthalate
Dinopol 235
Octyl decyl phthalate
n-Octyl n-decyl phthalate
Polycizer 532
Polycizer 562
Staflex 500

119-10-8
$C_8H_9NO_3$
167.16
O=N(=O)c(c(OC)ccc1C)c1
Benzene, 1-methoxy-4-methyl-2-nitro- (9CI)
1-Methoxy-4-methyl-2-nitrobenzene

119-12-0
$C_{14}H_{17}N_2O_4PS$
340.36
CCOP(=S)(OCC)Oc1ccc(=O)n(n1)c2ccccc2
Phosphorothioic acid, O,O-diethyl ester, O-ester with 6-hydroxy-2-phenyl-3(2H)pyridazinone
American Cyanamid 12,503
Cl 12503
O,O-Diethyl O-(2,3-dihydro-3-oxo-2-phenyl-6-pyridazinyl)phosphorothioate
O,O-Diethylphosphorothioate, O-ester with 6-hydroxy-2-phenyl-3(2H)-pyridazinone
O-(1,6-Dihydro-6-oxo-1-phenylpyridazin-3-yl) O,O-diethyl phosphorothioate
ENT 23,968
Ofnack
Ofnak
Ofunack
Pyridafenthion
Pyridaphenthion
3(2H)-Pyridazinone, 6-hydroxy-2-phenyl-, O-ester with O,O-diethyl phosphorothioate

119-15-3
$C_{12}H_9N_3O_5$
275.24
O=N(=O)c(ccc(Nc(ccc(O)c1)c1)c2N(=O)=O)c2
Phenol, p-(2,4-dinitroanilino)
Acetamine Yellow 2R
Acetoquinone Light Yellow 2RZ
Amacel Yellow RR
Celliton Fast Yellow RR
Cilla Fast Yellow RR
C.I. 10345
C.I. Disperse Yellow 1
C.I. Solvent Yellow 52
2,4-Dinitro-p-hydroxydiphenylamine
Disperse Fast Yellow 2K
Disperse Yellow R
Dispersol Fast Yellow A
Dispersol Printing Yellow A
Dispersol Yellow B-A
Disperse Yellow Stable 2K
Fast Disperse Yellow 2K
Fenacet Fast Yellow 2R
Kayalon Fast Yellow RR
Microsetile Yellow 2R
Nyloquinone Yellow 2R
Perliton Yellow RR
Permanent Yellow 2K
Reliton Yellow R
Serisol Fast Yellow A
Setacyl Yellow P-BS
Sra Golden Yellow VIII
Supracet Fast Yellow 2R
Supracet Yellow RR
Synten Yellow P 2R

119-27-7
$C_7H_6N_2O_5$
198.15
O=N(=O)c(ccc(OC)c1N(=O)=O)c1
Benzene, 1-methoxy-2,4-dinitro
Anisole, 2,4-dinitro-
2,4-Dinitroanisol
Dinitroanisole
α-Dinitroanisole

2,4-Dinitroanisole
2,4-Dinitrophenylmethyl ether
1-Methoxy-2,4-dinitrobenzene

119-28-8
$C_{10}H_9NO_3S$
223.25
O=S(=O)(O)c(ccc(c1c(N)cc2)c2)c1
Cleve's Acid
1-Aminonaphthalene-7-sulfonic acid
1-Amino-7-naphthalenesulfonic acid
8-Aminonaphthalene-2-sulfonic acid
8-Amino-2-naphthalenesulfonic acid
1-Amino-7-sulfonaphthalene
C.I. 26135
1,7-Cleve's Acid
Cleve's Theta-Acid
2-Naphthalenesulfonic acid, 8-amino- (9CI)
1-Naphthylamine-7-sulfonic acid
8-Naphthylamine-2-sulfonic acid
NSC-4983

119-31-3
$C_9H_{13}NO$
151.21
Phenol, 3-(dimethylamino)-4-methyl- (9CI)
p-Cresol, 3-(dimethylamino)- (6CI,8CI)
3-(Dimethylamino)-p-cresol
2-Dimethylamino-4-hydroxytoluene
3-Dimethylamino-4-methylphenol

119-32-4
$C_7H_8N_2O_2$
152.17
O=N(=O)c(c(ccc1N)C)c1
p-Toluidine, 3-nitro
4-Amino-2-nitrotoluene
Benzenamine, 4-methyl-3-nitro-
Gl-Amin (Czech)
4-Methyl-3-nitroaniline
2-Nitro-4-aminotoluene
3-Nitro-4-methylaniline
3-Nitro-4-toluidin (Czech)
m-Nitro-p-toluidine
3-Nitro-p-toluidine
5-Nitro-4-toluidine

119-33-5
$C_7H_7NO_3$
153.15
O=N(=O)c(c(O)ccc1C)c1
p-Cresol, 2-nitro
2-Nitro-p-cresol
2-Nitro-4-cresol

119-34-6
$C_6H_6N_2O_3$
154.14
O=N(=O)c(c(O)ccc1N)c1
Phenol, 4-amino-2-nitro
4-Amino-2-nitrofenol (Czech)
4-Amino-2-nitrophenol
C.I. 76555
Fourrine 57
Fourrine Brown PR
Fourrine Brown propyl
4-Hydroxy-3-nitroaniline
NCI-C03963
o-Nitro-p-aminophenol
2-Nitro-4-aminophenol
Oxidation Base 25

119-36-8
$C_8H_8O_3$
152.16
O=C(OC)c(c(O)ccc1)c1
Salicylic acid, methyl ester
Acide anisique (French)
Acide methyl-o-benzoique (French)
o-Anisic acid
Benzoic acid, 2-methoxy-
Betula
Betula Oil
Gaultheria Oil, Artificial
Gaultheria Oil
o-Hydroxybenzoic acid, methyl ester
2-Hydroxybenzoic acid methyl ester
o-Methoxybenzoic acid
Methylester kyseliny salicylove (Czech)
Methyl o-hydroxybenzoate
Methyl salicylate
Natural Wintergreen Oil
Oil of Wintergreen
Sweet Birch Oil
Synthetic Wintergreen Oil
Teaberry Oil
Wintergreen Oil
Wintergreen Oil, Synthetic

119-38-0
$C_{10}H_{17}N_3O_2$
211.30
CC(C)n1nc(C)cc1OC(=O)N(C)C
Carbamic acid, dimethyl-, 1-isopropyl-3-methylpyrazol-5-yl ester
Dimethylcarbamic acid 3-methyl-1-(1-methylethyl)-1H-pyrazol-5-yl ester
Dimethylcarbamate d'l-isopropyl 3-methyl 5-pyrazolyle (French)
Dimethyl-5-(l-isopropyl-3-methyl-pyrazolyl)-carbamate
ENT 19,060
G 23611
Geigy G-23611
Isolan
Isolane (French)
(1-Isopropil-3-metil-1H-pirazol-5-il)-N,N-dimetil-carbammato (Italian)
(1-Isopropyl-3-methyl-1H-pyrazol-5-yl)-N,N-dimethylcarbamaat (Dutch)
(1-Isopropyl-3-methyl-1H-pyrazol-5-yl)-N,N-dimethyl-carbamat (German)
Isopropylmethylpyrazolyl dimethylcarbamate
1-Isopropyl-3-methyl-5-pyrazolyl dimethylcarbamate
1-Isopropyl-3-methylpyrazolyl-(5)-dimethylcarbamate
1-Isopropyl-3-methyl-5-pyrazolylester kyseliny dimethylkarbaminove (Czech)
5-Methyl-2-isopropyl-3-pyrazolyl dimethylcarbamate
OMS 62
Primin
Pyrazol-5-ol, 1-isopropyl-3-methyl-, dimethylcarbamate
Saolan

119-39-1
$C_8H_6N_2O$
146.16
Oc(nncc1cccc2)c12
1(2H)Phthalazinone
Phthalazinone
Phthalazone

119-40-4
$C_{16}H_{13}NO_4S$
315.34
O=S(=O)(O)c(cc(O)c(c1cc(Nc(cccc2)c2)c3)c3)c1
2-Naphthalenesulfonic acid, 4-hydroxy-7-(phenylamino)- (9CI)
7-Anilino-4-hydroxy-2-naphthalenesulfonic acid
2-Naphthalenesulfonic acid, 7-anilino-4-hydroxy- (8CI)
NSC-10451
Phenyl J Acid

119-42-6
$C_{12}H_{16}O$
176.26
Oc(c(ccc1)C(CCCC2)C2)c1
o-Cyclohexylphenol
AI3-09047
2-Cyclohexylphenol
NSC-6093
Phenol, o-cyclohexyl- (8CI)
Phenol, 2-cyclohexyl- (9CI)

119-47-1
$C_{23}H_{32}O_2$
340.55
Oc(c(cc(c1)C)Cc(c(O)c(cc2C)C(C)(C)C)c2)c1C(C)(C)C
Methane, 2,2'-bis(6-t-butyl-p-cresyl)
Advastab 405
Antage W 400
Antioxidant 1

Anti OX
Antioxidant 2246
Antioxidant BKF
Antioxidant NG-2246
AO1
AO 2246
AO 1 (Antioxidant)
Bisaklofen BP
2,2'-Bis-6-terc.butyl-p-kresyl-methan (Czech)
BKF
Cao 5
Cao 14
Calco 2246
Catolin 14
Chemanox 21
p-Cresol, 2,2'-methylenebis-(6-tert-butyl-
Lederle 2246
2,2'-Methylene-bis(6-tert-butyl-4-methylphenol)
2,2''-Methylenebis(4-methyl-6-tert-butylphenol)
NG 2246
Nocrac NS 6
Oxy Chek 114
Plastanox 2246
Synox 5LT
Vulkanox BKF

119-53-9
$C_{14}H_{12}O_2$
212.26
O=C(c(cccc1)c1)C(O)c(cccc2)c2
Benzoin
Acetophenone, 2-hydroxy-2-phenyl-
Benzoylphenylcarbinol
Bitter Almond Oil Camphor
Ethanone, 2-hydroxy-1,2-diphenyl-
Fenyl-α-hydroxybenzylketon (Czech)
α-Hydroxybenzyl phenyl ketone
α-Hydroxy-α-phenylacetophenone
2-Hydroxy-2-phenylacetophenone
Ketone, α-hydroxybenzyl phenyl
NCI-C50011

119-56-2
$C_{13}H_{11}ClO$
218.68
OC(c(cccc1)c1)c(ccc(c2)Cl)c2
Benzenemethanol, 4-chloro-α-phenyl- (9CI)
AI3-20881
Benzhydrol, p-chloro-
Benzhydrol, 4-chloro- (8CI)
Chlorobenzhydrol
p-Chlorobenzhydrol
4-Chlorobenzhydrol
4-Chloro-α-phenylbenzenemethanol
(4-Chlorophenyl)phenylmethanol
NSC-59990

119-58-4
$C_{17}H_{22}N_2O$
270.41
OC(c(ccc(N(C)C)c1)c1)c(ccc(N(C)C)c2)c2
Benzhydrol, 4,4'-bis-(dimethylamino)
4,4'-Bis(dimethylamino)benzohydrol
α,α-Bis(p-dimethylaminophenyl)methanol
Michler's Hydrol
p,p'-Michler's Hydrol
Tetramethyldiaminobenzhydrol

119-61-9
$C_{13}H_{10}O$
182.23
O=C(c(cccc1)c1)c(cccc2)c2
Benzophenone
Benzene, benzoyl-
Benzoylbenzene
Diphenyl ketone
Diphenylmethanone
Ketone, diphenyl
α-Oxodiphenylmethane
α-Oxoditane
Phenyl ketone

119-64-2
$C_{10}H_{12}$
132.22
c(c(ccc1)CCC2)(c1)C2
Naphthalene, 1,2,3,4-tetrahydro
Naphthalene 1,2,3,4-tetrahydride
$\delta^{5,7,9}$-Naphthantriene
Tetrahydronaphthalene
1,2,3,4-Tetrahydronaphthalene
Tetralin
Tetralina (Polish)
Tetraline
Tetranap

119-65-3
C_9H_7N
129.17
n(ccc(c1ccc2)c2)c1
Isoquinoline
2-Azanaphthalene
2-Benzazine
Benzo(c)pyridine
Isochinolin (Czech)
Leucoline

119-70-0
$C_{12}H_{13}N_3O_3S$
279.30
O=S(=O)(O)c(c(Nc(ccc(N)c1)c1)ccc2N)c2
Benzenesulfonic acid, 5-amino-2-((4-aminophenyl)-amino)- (9CI)
AI3-23227
5-Amino-2-(p-aminoanilino)-benzenesulfonic acid
6-(p-Aminoanilino)metanilic acid
4-(p-Aminoanilino)-3-sulfoaniline
Benzenesulfonic acid, 5-amino-2-(p-aminoanilino)- (8CI)
4,4'-Diaminodiphenylamine-2'-sulfonic acid
4,4'-Diamino-2-sulfodiphenylamine
Diphenylamine-2-sulfonic acid, 4,4'-diamino-
NSC-4706

119-72-2
$C_{14}H_{12}N_2O_8S_2$
400.40
O=S(=O)(O)c(c(ccc1N(=O)=O)C=Cc(c(S(=O)(=O)O)cc(N)c2)c2)c1
2,2'-Stilbenedisulfonic acid, 4-amino-4'-nitro
4-Amino-4'-nitro-2,2'-stilbenedisulfonic acid
Kyselina 4-amino-4'-nitrostilben-2,2'-disulfonova (Czech)
Kyselina 4-nitro-4'-aminostilben-2,2'-disulfonova (Czech)

119-75-5
$C_{12}H_{10}N_2O_2$
214.21
O=N(=O)c(c(Nc(cccc1)c1)ccc2)c2
Benzenamine, 2-nitro-N-phenyl- (9CI)
AI3-08882
C.I. 10335
Diphenylamine, 2-nitro- (8CI)
Nitrodiphenylamine
o-Nitrodiphenylamine
2-Nitrodiphenylamine
o-Nitro-N-phenylaniline
2-Nitro-N-phenylbenzenamine
NSC-105613
N-Phenyl-o-nitroaniline
Sudan Yellow 1339

119-79-9
$C_{10}H_9NO_3S$
223.26
O=S(=O)(O)c(ccc(c1ccc2)c2N)c1
2-Naphthalenesulfonic acid, 5-amino
1-Amino-6-naphthalenesulfonic acid
5-Amino-2-naphthalenesulfonic acid
1-Amino-6-sulfonaphthalene
Cleve's Acid-1,6
1,6-Cleve's Acid
Cleve's β-Acid
Kyselina cleve (Czech)
Kyselina 1-naftylamin-6-sulfonova (Czech)
1-Naphthylamine-6-sulfonic acid

5-Naphthylamine-2-sulfonic acid

119-84-6
$C_9H_8O_2$
148.17
O=C(Oc(c(ccc1)C2)c1)C2
Hydrocoumarin
1,2-Benzodihydropyrone
2-Chromanone
Chroman, 2-oxo-
Coumarin, 3,4-dihydro-
Dihydrocoumarin
3,4-Dihydrocoumarin
Hydrocinnamic acid, o-hydroxy-, δ-lactone
Melilotin
Melilotine
Melilotol
NCI-C55890
USAF DO-12

119-89-1
$C_{19}H_{24}O_3$
300.43
CCCCCCC1=CC(=O)CC(C1)c3ccc2OCOc2c3
2-Cyclohexen-1-one, 3-hexyl-5-(3,4-(methylenedioxy)-phenyl)
Ent 2,818
3-Hexyl-5-(3,4-methylenedioxy-phenyl)-2-cyclohexen-1-one
Piperonyl cyclonene
Piperonylcyklonen (Czech)

119-90-4
$C_{14}H_{16}N_2O_2$
244.32
O(c(c(N)ccc1c(ccc(N)c2OC)c2)c1)C
Benzidine, 3,3'-dimethoxy
Acetamine Diazo Black RD
Acetamine Diazo Navy RD
Amacel Developed Navy SD
Azoene Fast Blue Base
Azoene Fast Blue Salt
Azofix Blue B Salt
Azogene Fast Blue B
Azogene Fast Blue B Salt
Blue Base Irga B
Blue Base NB
Blue BN Base
Blue BN Salt
Blue Salt NB
Brentamine Fast Blue B Base
Brentamine Fast Blue B Salt
Cellitazol B
Cellitazol BN
C.I. 24110
C.I. Azoic Diazo Component 48
C.I. Azoic Diazo Component 48, Fast Blue B Salt
Cibacete Diazo Navy Blue 2B
C.I. Disperse Black 6
Diacelliton Fast Grey G
Diacel Navy DC
4,4'-Diamino-3,3'-dimethoxy-biphenyl
o-Dianisidin (Czech, German)
o-Dianisidina (Italian)
Dianisidine
o-Dianisidine
o,o'-Dianisidine
3,3'-Dianisidine
Diato Blue Base B
Diato Blue Salt B
Diazo Fast Blue B
3,3'-Dimethoxybenzidin (Czech)
3,3'-Dimethoxybenzidine
3,3'-Dimetossibenzodina (Italian)
Fast Blue B Base
Fast Blue BN Salt
Fast Blue DSC Base
Fast Blue DS Salt
Fast Blue Salt B
Fast Blue Salt BN
Hiltonil Fast Blue B Base
Hiltosal Fast Blue B Salt
Hindasol Blue B Salt
Kako Blue B Salt
Kayaku Blue B Base
Kayaku Blue B Salt
Lake Blue B Base
Meisei Teryl Diazo Blue HR
Mitsui Blue B Base
Mitsui Blue B Salt
Naphthanil Blue B Base
Natasol Blue B Salt
Neutrosel Navy BN
RCRA waste number U091
Sanyo Fast Blue Salt B
Setacyl Diazo Navy R
Spectrolene Blue B

119-91-5
$C_{18}H_{12}N_2$
256.29
n(c(c(ccc1)cc2)c1)c2c(nc(c(ccc3)c4)c3)c4
2,2'-Biquinoline (9CI)
2,2'-Biquinolyl
Cuproin
Cuproine
2,2'-Diquinolyl
NSC-1533

119-93-7
$C_{14}H_{16}N_2$
212.32
Nc(c(cc(c(ccc(N)c1C)c1)c2)C)c2
Benzidine, 3,3'-dimethyl
Bianisidine
(1,1'-Biphenyl)-4,4'-diamine-3,3'-dimethyl-
4,4'-Bi-o-toluidine
C.I. 37230
C.I. Azoic Diazo Component 113
4,4'-Diamino-3,3'-dimethylbi-phenyl
4,4'-Diamino-3,3'-dimethyldi-phenyl
Diaminoditolyl
3,3'-Dimethylbenzidin
3,3'-Dimethylbenzidine
3,3'-Dimethyl-4,4'-biphenyl-diamine
3,3'-Dimethylbiphenyl-4,4'-di-amine
3,3'-Dimethyl-4,4'-diphenyl-diamine
3,3'-Dimethyldiphenyl-4,4'-di-amine
4,4'-Di-o-toluidine
Fast Dark Blue Base R
RCRA waste number U095
o-Tolidin
2-Tolidin (German)
2-Tolidina (Italian)
Tolidine
o-Tolidine (ACGIH)
o,o'-Tolidine
2-Tolidine
3,3'-Tolidine

119-97-1
$C_{13}H_{16}N_2O$
216.27
O=Cc(c(cc(N(CCC#N)CC)c1)C)c1
Propanenitrile, 3-(ethyl-(4-formyl-3-methylphenyl)-amino)- (9CI)

120-00-3
$C_{17}H_{20}N_2O_3$
300.35
O=C(Nc(c(OCC)cc(N)c1OCC)c1)c(cccc2)c2
Benzamide, N-(4-amino-2,5-di-ethoxyphenyl)- (9CI)
Amarthol Fast Blue BB Base
2-Amino-5-benzoylaminohydro-quinone diethyl ether
N-(4-Amino-2,5-diethoxy-phenyl)benzamide
Azoene Fast Blue BB Base
Benzanilide, 4'-amino-2',5'-di-ethoxy- (8CI)
Blue Salt NBB
Blue 2B base
Brentamine Fast Blue BB Base
Brentamine Fast Blue 2B Base
C.I. Azoic Diazo Component 20
C.I. Azoic Diazo Component 20, Base
C.I. 37175
Daito Blue Base BB
Diazo Fast Blue BB
2,5-Diethoxy-4-benzamido-aniline
Fast Blue Base BB
Fast Blue BB Base
Fast Blue BBN
Fast Blue EB Base
Hiltonil Fast Blue BB Base
Naphtoelan Blue BB Base
NSC-37184
Sanyo Fast Blue BB Base
Spectrolene Blue BB
Stabamine Blue BB
Tulabase Fast Blue BB

120-07-0
$C_{10}H_{15}NO_2$
181.26

120-08-1

OCCN(c(cccc1)c1)CCO
Ethanol, 2,2'-(phenylimino)di
N,N-Bis(2-hydroxyethyl)aniline
Diethanolaminobenzene
Diethanolaniline
N,N-Diethanolaniline
Dihydroxyethylaniline
N,N-Di(β-hydroxyethyl)aniline
N,N-Di(2-hydroxyethyl)aniline
N,N-Dioxyethylaniline
Emery 5703
2,2'-(Phenylamino)diethanol
Phenyl diethanolamine
N-Phenyldiethanolamine
2,2'-(Phenylimino)diethanol

120-08-1
$C_{11}H_{10}O_4$
206.21
COc2cc1ccc(=O)oc1cc2OC
Coumarin, 6,7-dimethoxy
Aesculetin dimethyl ether
Benzopyran-2-one, 6,7-dimethoxy- (9CI)
6,7-Dimethoxycoumarin
6,7-Dimethylesculetin
Escoparone
Esculetin dimethyl ether
Scoparon
Scoparone

120-12-7
$C_{14}H_{10}$
178.24
c(c(ccc1)cc(c2ccc3)c3)(c1)c2
Anthracene
Anthracen (German)
Anthracin
Green Oil
Paranaphthalene
Tetra Olive N2G

120-14-9
$C_9H_{10}O_3$
166.19
O=Cc(ccc(OC)c1OC)c1
Veratraldehyde
Benzaldehyde, 3,4-dimethoxy-
3,4-Dimethoxybenzaldehyde
3,4-Dimethoxybenzenecarbonal
Methylvanillin
4-o-Methylvanillin
Protocatechualdehyde dimethyl ether
Protocatechuecaldehyde dimethyl ether
Protocatechuic aldehyde dimethyl ether
Vanillin methyl ether
Veratric aldehyde
p-Veratric aldehyde
Veratryl aldehyde

120-18-3
$C_{10}H_8O_3S$
208.24
O=S(=O)(O)c(ccc(c1ccc2)c2)c1
2-Naphthalenesulfonic acid
Kyselina 2-naftalensulfonova (Czech)
β-Naphthalenesulfonic acid
Naphthalene-2-sulfonic acid
β-Naphthylsulfonic acid

120-21-8
$C_{11}H_{15}NO$
177.24
O=Cc(ccc(N(CC)CC)c1)c1
4-(Diethylamino)benzaldehyde
AI3-05886
Benzaldehyde, p-(diethylamino)- (8CI)
Benzaldehyde, 4-(diethylamino)- (9CI)
p-(Diethylamino)benzaldehyde
p-Formyl-N,N-diethylaniline
NSC-8782

120-23-0
MF:
MW:
O=C(O)COc(ccc(c1ccc2)c2)c1
Acetic acid, (2-naphthyloxy)
Acide naphthyloxyacetique (French)
Betapal
Betoxon
Gerlach 1396
2-Naphthalenoxyacetic acid
(β-Naphthalenyloxy)acetic acid
β-Naphthoxyacetic acid
2-Naphthoxyacetic acid
o-(2-Naphthyl)glycolic acid
Noxa
2-Noxa
(2-Naphthyloxy)acetic acid

120-29-6
$C_8H_{15}NO$
141.14
OC(CC(N(C1C2)C)C2)C1
Tropine
AI3-52686
8-Azabicyclo(3.2.1)octanol, 8-methyl-
8-Azabicyclo(3.2.1)octan-3-ol, 8-methyl-, endo- (9CI)
2,3-Dihydro-3α-hydroxytropidine
endo-8-Methyl-8-azabicyclo-(3.2.1)octan-3-ol
NSC-43870
3α-Tropanol
1αH,5αH-Tropan-3α-ol (8CI)
Tropin

120-32-1
$C_{13}H_{11}ClO$
218.69
Oc(c(cc(c1)Cl)Cc(cccc2)c2)c1
o-Cresol, 4-chloro-α-phenyl
Benzylchlorophenol
o-Benzyl-p-chlorophenol
2-Benzyl-4-chlorophenol
Bio-Clave
5-Chloro-2-hydroxydiphenylmethane
Chlorophene
4-Chloro-α-phenyl-o-cresol
Clorofene
Clorophene
Ketolin-H
NCI-C61201
Neosabenyl
Orthobenzyl-p-chlorophenol
Orthobenzylparachlorophenol
Phenol, 4-Chloro-2-(phenylmethyl)- (9CI)
Santophen
Santophen 1
Santophen 1 Flake
Santophen 1 Solution
Santophen I
Santophen I Germicide
Septiphene

120-35-4
$C_{14}H_{14}N_2O_2$
242.30
O=C(Nc(cccc1)c1)c(ccc(OC)c2N)c2
Benzanilide, 3-amino-4-methoxy
3-Amino-4-methoxy benzanilide

120-36-5
$C_9H_8Cl_2O_3$
235.07
O=C(O)C(Oc(c(cc(c1)Cl)Cl)c1)C
Propionic acid, 2-(2,4-dichlorophenoxy)
Acide 2-(2,4-dichloro-phenoxy) propionique (French)
Acido 2-(2,4-dicloro-fenossi)-propionico (Italian)
BH 2,4-DP
Cornox RD
Cornox RK
Desormone
2-(2,4-Dichloor-fenoxy)-propionzuur (Dutch)
α-(2,4-Dichlorophenoxy) propionic acid
2-(2,4-Dichlorophenoxy) propionic acid
Dichloroprop
2-(2,4-Dichlor-phenoxy)-propionsaeure (German)
Dichlorprop
2,4-DP
2-(2,4-DP)
Graminon-Plus
Hedonal
Hedonal DP
Herbatox
Hormatox
Kildip
Kwas 2,4-dwuchlorofenoksypropionowy (Polish)
Kyselina 2-(2,4-dichlorfenoxy)-propionova (Czech)
Polyclene
Polymone
Polytox
RD 406
Seritox 50

U46
U46 DP-Fluid
Visko-Rhap
Weedone DP
Weedone 170

120-37-6
$C_9H_{13}NO$
151.23
Oc(ccc(c1NCC)C)c1
Phenol, 3-(ethylamino)-4-methyl
3-Ethylamino-4-methylphenol

120-39-8
$C_{13}H_{15}Cl_3O_3$
325.63
Acetic acid, (2,4,5-trichlorophenoxy)-, pentyl ester (8CI, 9CI)
Amyl 2,4,5-trichlorophenoxy acetate
2,4,5-T amyl ester
Tormona 80
(2,4,5-Trichlorophenoxy)acetic acid phentyl ester
Trifenox
Trifenox 80

120-40-1
$C_{16}H_{33}NO_3$
287.50
O=C(N(CCO)CCO)CCCCCC CCCCC
Dodecanamide, N,N-bis(2-hydroxyethyl)
N,N-Bis(2-hydroxyethyl)dodecanamide
Bis(2-hydroxyethyl)lauramide
N,N-Bis(hydroxyethyl)lauramide
N,N-Bis(β-hydroxyethyl)lauramide
N,N-Bis(2-hydroxyethyl)lauramide
Clindrol 101CG
Clindrol 203CG
Clindrol 210CGN
Clindrol 200L
Clindrol Superamide 100L
Coco diethanolamide
Coconut oil amide of diethanolamine
Comperlan LD
Condensate PL
Crillon I.D.E.
Diethanollauramide
N,N-Diethanollauramide
N,N-Diethanollauric acid amide
Emid 6511
Emid 6541
Ethylan MLD
Hetamide ML
Lauramide DEA
Lauric acid diethanolamide
Lauric acid diethanolamine condensate
Lauric diethanolamide
Lauroyl diethanolamide
Lauryl diethanolamide
LDA
LDE
Monamid 150-LW
NCI-C55323
Ninol 4821
Ninol AA62
Ninol AA-62 Extra
Ninol P-621
Onyxol 345
Rewomid DLMS
Rewomid DL 203/S
Richamide 6310
Rolamid CD
Standamidd LD
Steinamid DL 203 S
Super Amide L-9A
Super Amide L-9C
Synotol L-60
Unamide J-56
Varamid ML 1

120-41-2
$C_{22}H_{46}N_2O_2$
370.61
O=C(N(CCO)CCN)CCCCCCCC CCCCCCCCC
Octadecanamide, N-(2-aminoethyl)-N-(2-hydroxyethyl)-(9CI)

120-46-7
$C_{15}H_{12}O_2$
224.26
O=C(c(cccc1)c1)CC(=O)c(cc cc2)c2
1,3-Propanedione, 1,3-diphenyl- (9CI)
AI3-19022
ω-Benzoylacetophenone
2-Benzoylacetophenone
Dibenzoylmethane
Dibenzoyl-methane
1,3-Diphenyl-1,3-propanedione
NSC-6266
Phenyl phenacyl ketone

120-47-8
$C_9H_{10}O_3$
166.19
O=C(OCC)c(ccc(O)c1)c1
Benzoic acid, p-hydroxy-, ethyl ester
Aseptoform E
Bonomold OE
p-Carbethoxyphenol
Easeptol
Ester etylowykwasu p-hydroksybenzoesowego (Polish)
Ethylester kyseliny p-hydroxybenzoove (Czech)
Ethyl p-hydroxybenzoate
Ethyl p-oxybenzoate
Ethyl paraben
Ethyl parasept
p-Hydroxybenzoic acid ethyl ester
p-Hydroxybenzoic ethyl ester
Nipagin A
Nipagina A
Nipazin A
p-Oxybenzoesaeureaethylester (German)
Solbrol A
Tegosept E

120-51-4
$C_{14}H_{12}O_2$
212.26
O=C(OCc(cccc1)c1)c(cccc2)c2
Benzoic acid, benzyl ester
Ascabin
Ascabiol
Benylate
Benzoic acid, phenylmethyl ester
Benzyl alcohol benzoic ester
Benzyl benzenecarboxylate
Benzyl benzoate
Benzylester kyseliny benzoove (Czech)
Benzylets
Benzyl phenylformate
Colebenz
Novoscabin
Peruscabin
Scabanca
Vanzoate
Venzonate

120-54-7
$C_{12}H_{20}N_2S_6$
384.70
N(C(=S)SSSSC(N(CCCC1)C1)=S)(CCCC2)C2
Piperidine, 1,1'-(tetrathiodicarbonothioyl)bis
Bis(pentamethylenethiuram)-tetrasulfide
Bis(piperidinothiocarbonyl) tetrasulfide
Dipentamethylenethiuram tetrasulfide
Tetrasulfide, bis(pentamethylenethiuram)-
Tetrasulfide, bis(piperidinothiocarbonyl)
Tetrone A
Thiuram tetrasulfide, bis(piperidinothiocarbonyl)
USAF B-31

120-55-8
$C_{18}H_{18}O_5$
314.36
O=C(OCCOCCOC(=O)c(cccc1)c1)c(cccc2)c2
Diethylene glycol, dibenzoate
Benzo Flex 2-45
Benzoic acid, diester with diethylene glycol
Dibenzoyldiethyleneglycol ester

120-56-9
$C_{20}H_{22}O_6$
358.39
O=C(OCCOCCOCCOC(=O)c(ccc

c1)c1)c(cccc2)c2
Ethanol, 2,2'-(1,2-ethanediylbis(oxy))bis-, dibenzoate (9CI)
Benzoflex T 150
3,6-Dioxaoctane-1,8-diyl dibenzoate
NSC-166503
Triethylene glycol, dibenzoate (8CI)

120-57-0
$C_8H_6O_3$
150.14
O=Cc(ccc(OCO1)c12)c2
Piperonal
Benzaldehyde, 3,4-(methylenedioxy)-
1,3-Benzodioxole-5-carboxaldehyde
3,4-Dihydroxybenzaldehyde methylene ketal
Dioxymethylene-protocatechuic aldehyde
Heliotropin
Heliotropine
3,4-Methylene-dihydroxybenzaldehyde
3,4-Methylenedioxybenzaldehyde
Piperonaldehyde
Piperonyl aldehyde
Protocatechuic aldehyde methylene ether

120-58-1
$C_{10}H_{10}O_2$
162.20
O(c(c(O1)cc(c2)C=CC)c2)C1
Benzene, 1,2-(methylenedioxy)-4-propenyl
1,3-Benzodioxole, 5-(1-propenyl)-
Isosafrole
1,2-Methylenedioxy-4-propenylbenzene
3,4-Methylenedioxy-1-propenyl benzene
5-(1-Propenyl)-1,3-benzodioxole
4-Propenylcatechol methylene ether
4-Propenyl-1,2-methylenedioxybenzene
RCRA waste number U141

120-61-6
$C_{10}H_{10}O_4$
194.20
O=C(OC)c(ccc(c1)C(=O)OC)c1
Terephthalic acid, dimethyl ester
1,4-Benzenedicarboxylic acid, dimethyl ester (9CI)
Dimethyl 1,4-benzenedicarboxylate
Dimethylester kyseliny tereftalove (Czech)
Dimethyl p-phthalate
Dimethyl terephthalate
DMT
Methyl 4-carbomethoxybenzoate
NCI-C50055
Terephthalic acid methyl ester

120-62-7
$C_{18}H_{28}O_3S$
324.52
CCCCCCCCS(=O)C(C)Cc2ccc1OCOc1c2
Benzene, 1,2-(methylenedioxy)-4-(2-(octylsulfinyl)propyl)
2-(1,3-Benzodioxol-5-yl)ethyl octyl sulfoxide
ENT 16,634
Isosafrole, octyl sulfoxide
Isosafrole n-octylsulfoxide
1-(2,3-Methylendioxyfenyl)-2-(oktylsufinyl)propan (Czech)
1,2-(Methylenedioxy)-4-(2-(octylsulfinyl)propyl)benzene
1-Methyl-2-(3,4-methylenedioxyphenyl)ethyl octyl sulfoxide
NCI-C02824
n-Octylisosafrole sulfoxide
5-(2-(Octylsulfinyl)propyl)-1,3-benzodioxole
n-Octylsulfoxide of isosafrole
Piperonyl sulfoxide
Sulfocide
Sulfox-Cide
Sulfoxide
Sulfoxide, α-methyl-3,4-(methylenedioxy)phenethoctyl
Sulfoxyl
Sulphoxide

120-66-1
$C_9H_{11}NO$
149.21
O=C(Nc(c(ccc1)C)c1)C
o-Acetotoluidide
Acetamide, N-(2-methylphenyl)- (9CI)
o-Acetotoluidide
Acetyl-o-toluidine
o-Methylacetanilide
2-Methylacetanilide
2'-Methylacetanilide

120-67-2
$C_8H_8Cl_2O_2$
207.06
OCCOc1ccc(Cl)cc1Cl
Ethanol, 2-(2,4-dichlorophenoxy)
Cellosolve, 2,4-dichlorophenyl-
2-(2,4-Dichlorophenoxy)ethanol
2,4-Dichlorphenyl cellosolve

120-71-8
$C_8H_{11}NO$
137.20
O(c(c(N)cc(c1)C)c1)C
o-Anisidine, 5-methyl
m-Amino-p-cresol, methyl ester
3-Amino-p-cresol methyl ether
1-Amino-2-methoxy-5-methylbenzene
3-Amino-4-methoxytoluene
2-Amino-4-methylanisole
Azoic Red 36
Benzenamine, 2-methoxy-5-methyl- (9CI)
C.I. Azoic Red 83
Cresidine
p-Cresidine
p-Kresidin (Czech)
Krezidine
2-Methoxy-5-methylaniline
2-Methoxy-5-methylbenzenamine
4-Methoxy-m-toluidine
4-Methyl-2-aminoanisole
5-Methyl-o-anisidine
NCI-C02982

120-72-9
C_8H_7N
117.16
N(c(c(C=1)ccc2)c2)C1
Indole
1-Azaindene
1-Benzazole
Benzopyrrole
2,3-Benzopyrrole
Indol (German)
Ketole

120-73-0
$C_5H_4N_4$
120.13
c2ncc1[nH]cnc1n2
Purine
7H-Imidazo(4,5-d)pyrimidine
NSC-753
β-Purine
7H-Purine
9H-Purine
Isopurine
3,5,7-Triazaindole

120-75-2
C_8H_7NS
149.22
N(c(c(S1)ccc2)c2)=C1C
Benzothiazole, 2-methyl
2-Methylbenzothiazole
USAF EK-1853

120-78-5
$C_{14}H_8N_2S_4$
332.48
N(c(c(S1)ccc2)c2)=C1SSC(=Nc(c3ccc4)c4)S3
Benzothiazole, 2,2'-dithiobis
Altax
Benzothiazole disulfide
Benzothiazolyl disulfide
2-Benzothiazolyl disulfide
Benzothiazyl disulfide
Bis(benzothiazolyl) disulfide

Bis(2-benzothiazyl) disulfide
Di-2-benzothiazolyl disulfide
Dibenzothiazolyl disulphide
Dibenzothiazyl disulfide
2,2'-Dibenzothiazyl disulfide
Dibenzoylthiazyl disulfide
Dibenzthiazyl disulfide
2,2'-Dithiobis(benzothiazole)
Dwusiarczek dwubenzotiazylu (Polish)
MBTS
MBTS Rubber Accelerator
2-Mercaptobenzothiazole disulfide
2-Mercaptobenzothiazyl disulfide
Royal MBTS
Thiofide
USAF B-33
USAF CY-5
USAF EK-5432
Vulkacit DM
Vulkacit DM/MGC

120-80-9
$C_6H_6O_2$
110.12
Oc(c(O)ccc1)c1
Pyrocatechol (OSHA)
Benzene, o-dihydroxy-
o-Benzenediol
1,2-Benzenediol
Catechin
Catechol (ACGIH,OSHA)
C.I. 76500
C.I. Oxidation Base 26
o-Dihydroxybenzene
1,2-Dihydroxybenzene
o-Dioxybenzene
o-Diphenol
Durafur Developer C
Fouramine PCH
Fourrine 68
o-Hydroquinone
o-Hydroxyphenol
2-Hydroxyphenol
Katechol (Czech)
NCI-C55856
Oxyphenic acid
Pelagol Grey C
o-Phenylenediol
Pyrocatechin
Pyrocatechine
Pyrocatechinic acid
Pyrocatechuic acid
Pyrokatechin (Czech)
Pyrokatechol (Czech)

120-82-1
$C_6H_3Cl_3$
181.44
c(ccc(c1Cl)Cl)(c1)Cl
Benzene, 1,2,4-trichloro
unsym-Trichlorobenzene
Hostetex l-Pec
1,2,4-Trichlorobenzene (ACGIH,OSHA)
1,2,4-Trichlorobenzene, Liquid [UN 2321]
1,2,5-Trichlorobenzene
1,3,4-Trichlorobenzene
1,2,4-Trichlorobenzol
Trojchlorobenzen (Polish)
UN 2321 [Trichlorobenzenes, Liquid]

120-83-2
$C_6H_4Cl_2O$
163.00
Oc(c(cc(c1)Cl)Cl)c1
Phenol, 2,4-dichloro
DCP
2,4-DCP
2,4-Dichlorophenol
NCI-C55345
RCRA waste number U081

120-92-3
C_5H_8O
84.13
O=C(CCC1)C1
Cyclopentanone [UN 2245]
Adipic ketone
Dumasin
Ketocyclopentane
Ketopentamethylene
UN 2245 [Cyclopentanone]

120-93-4
$C_3H_6N_2O$
86.11
O=C(NCC1)N1
2-Imidazolidinone
Ethylene urea
2-Imidazolidone
Urea, 1,3-ethylene-

120-94-5
$C_5H_{11}N$
85.17
N(CCC1)(C1)C
Pyrrolidine, 1-methyl
N-Methylpyrrolidine
1-Methylpyrrolidine
N-Methyltetrahydropyrrole

120-95-6
$C_{16}H_{26}O$
234.42
Oc(c(cc(c1)C(CC)(C)C)C(CC)(C)C)c1
Phenol, 2,4-di-tert-pentyl
Di-tert-amylphenol
2,4-Di-tert-amylphenol
2,4-Di-tert-pentylphenol
Prodox 156

120-97-8
$C_6H_6Cl_2N_2O_4S_2$
305.16
NS(=O)(=O)c1cc(Cl)c(Cl)c(c1)S(N)(=O)=O
m-Benzenedisulfonamide, 4,5-dichloro
1,3-Benzenedisulfonamide, 4,5-dichloro-
CB 8000
Daranide
Dasanide
Dichlofenamide
4,5-Dichloro-m-benzenedisulfonamide
4,5-Dichloro-1,3-disulfamoylbenzene
Dichlorophenamide
3,4-Dichloro-5-sulfamylbenzenesulfonamide
Dichlorphenamide
1,3-Disulfamyl-4,5-dichlorobenzene
Oratrol

121-00-6
$C_{11}H_{16}O_2$
180.27
O(c(ccc(O)c1C(C)(C)C)c1)C
Phenol, 2-tert-butyl-4-methoxy
2-tert-BHA
3-BHA
2-tert-Butylated hydroxyanisole
3-tert-Butyl-4-hydroxyanisole
2-tert-Butyl-4-methoxyphenol
4-Methoxy-2-tert-butylphenol

121-02-8
$C_7H_6ClNO_4S$
235.65
Cc1ccc(cc1S(Cl)(=O)=O)N(=O)=O
o-Toluenesulfonyl chloride, 5-nitro
Benzenesulfonyl chloride, 2-methyl-5-nitro-
2-Methyl-5-nitrobenzenesulfonyl chloride
5-Nitro-o-toluenesulfonyl chloride
4-Nitrotoluen-2-sulfochlorid (Czech)
4-Nitrotoluen-2-sulfonylchlorid (Czech)

121-03-9
$C_7H_7NO_5S$
217.21
O=S(=O)(O)c(c(ccc1N(=O)=O)C)c1
Benzenesulfonic acid, 2-methyl-5-nitro
Kyselina 4-nitrotoluen-2-sulfonova (Czech)

121-06-2
$C_{24}H_{27}O_4P$
410.45
O=P(Oc(c(ccc1)C)c1C)(Oc(c(ccc2)C)c2C)Oc(c(ccc3)C)c3C
Tris(2,6-xylenyl)phosphate
AI3-26997
2,6-Dimethylphenol phosphate (3:1)
NSC-15282
Phenol, 2,6-dimethyl-, phosphate (3:1) (9CI)

Phosphoric acid, tris(dimethylphenyl)ester
Tri(2,6-dimethylphenyl) phosphate
Tri(2,6-xylenyl)phosphate
Tri-2,6-xylyl phosphate
2,6-Xylenol, phosphate (3:1)
(8CI)
2,6-Xylyl phosphate,
$(C_8H_9O)_3PO$

121-14-2
$C_7H_6N_2O_4$
182.15
O=N(=O)c(ccc(c1N(=O)=O)C)c1
Toluene, 2,4-dinitro
Benzene, 1-methyl-2,4-dinitro-
Dinitrotoluene (ACGIH,OSHA)
2,4-Dinitrotoluene
2,4-Dinitrotoluol
DNT
2,4-DNT
1-Methyl-2,4-dinitrobenzene
NCI-C01865
RCRA waste number U105

121-17-5
$C_7H_3ClF_3NO_2$
225.56
O=N(=O)c(c(ccc1C(F)(F)F)Cl)c1
Toluene, 4-chloro-3-nitro-α,α,α-trifluoro
Benzotrifluoride, 4-chloro-
3-nitro-
4-Chloro-3-nitrobenzotrifluoride
4-Chloro-3-nitro-α,α,α-tri-
fluorotoluene
3-Nitro-4-chlorobenzotrifluoride
[UN 2307]
3-Nitro-4-chloro-α,α,α-tri-
fluorotoluene
UN 2307 [3-Nitro-4-chloro-
benzotrifluoride]

121-19-7
$C_6H_6AsNO_6$
263.05
Oc1ccc(cc1N(=O)=O)[As](O)(O)=O
Benzenearsonic acid, 4-hydroxy-3-nitro
Aklomix-3
Arsonic acid, (4-hydroxy-
3-nitrophenyl)-
4-Hydroxy-3-nitrobenzene-
arsonic acid
4-Hydroxy-3-nitrophenylarsonic
acid
Kyselina 4-hydroxy-3-nitro-
fenylarsonova (Czech)
NCI-C56508
3N4HPA
3-Nitro-10
3-Nitro-20
3-Nitro-50
3-Nitro-80
Nitro Acid 100 Per Cent
2-Nitro-1-hydroxybenzene-
4-arsonic acid
3-Nitro-4-hydroxybenzene-
arsonic acid
3-Nitro-4-hydroxyphenylarsonic
acid
Nitrophenolarsonic acid
NSC-2101
Ren O-Sal
Ristat
Roxarsone

121-21-1
$C_{21}H_{28}O_3$
328.49
CC(C)=CC2C(C(=O)OC1CC(=O)C(=C1C)CC=CC=C)C2(C)C
Cyclopropanecarboxylic acid, 2,2-dimethyl-3-(2-methylpropenyl)-, ester with 4-hydroxy- 3-methyl-2-(2,4-pentadienyl)-2-cyclopenten-1-one
Chrysanthemum monocarboxyl-
ic acid pyrethrolone ester
Piretrina 1 (Portuguese)
Pyrethrin I
Pyrethrolone, chrysanthemum
monocarboxylic acid ester
(+)-Pyrethronyl (+)-trans-chrys-
anthemate
RCRA waste number P008

121-29-9
$C_{22}H_{28}O_5$
372.50
O=C(OC(C(=C(C1=O)CC=CC=C)C)C1)C(C2(C)C)C2C=C(C(=O)OC)C
Cyclopropaneacrylic acid, 3-carboxy-α,2,2-trimethyl-, 1-methyl ester, ester with 4- hydroxy-3-methyl-2-(2,4-pentadienyl)-2-cyclopenten-1-one
Chrysanthemumdicarboxylic
acid monomethyl ester
pyrethrolone ester
ENT 7,543
Pyrethrin
Pyrethrin II
Pyrethrolone, chrysanthemum
dicarboxlic acid methyl ester
ester
Pyrethrolone ester of chrsanthe-
mumdicarboxylic acid mono-
methyl ester
(+)-Pyrethronyl (+)-pyrethrate
Pyretrin II

121-32-4
$C_9H_{10}O_3$
166.19
O=Cc(ccc(O)c1OCC)c1
Benzaldehyde, 3-ethoxy-4-hydroxy
Bourbonal
Ethavan
Ethovan
3-Ethoxy-4-hydroxybenz-
aldehyde
Ethylprotal
Ethyl vanillin
4-Hydroxy-3-ethoxybenz-
aldehyde
Protocatechuic aldehyde ethyl
ether
Quantrovanil
Vanillal
Vanillin, ethyl-
Vanirom

121-33-5
$C_8H_8O_3$
152.16
O=Cc(ccc(O)c1OC)c1
Vanillin
m-Anisaldehyde, 4-hydroxy-
Benzaldehyde, 4-hydroxy-
3-methoxy-
p-Hydroxy-m-methoxybenz-
aldehyde
4-Hydroxy-3-methoxybenz-
aldehyde
Lioxin
3-Methoxy-4-hydroxybenz-
aldehyde
Protocatechualdehyde, methyl-
Vanilla
Vanillaldehyde
Vanillic aldehyde
p-Vanillin
Zimco

121-34-6
$C_8H_8O_4$
168.16
O=C(O)c(ccc(O)c1OC)c1
Vanillic-acid
Acide vanillique
m-Anisic acid, 4-hydroxy-
Benzoic acid, 4-hydroxy-
3-methoxy-
4-Hydroxy-3-methoxybenzoic
acid
3-Methoxy-4-hydroxybenzoic
acid
Protocatechuic acid, 3-methyl
ester
VA
p-Vanillic acid

121-39-1
$C_{11}H_{12}O_3$
192.23
O=C(OCC)C(O1)C1c(cccc2)c2
Glycidic acid, 3-phenyl-, ethyl ester
Ethyl α,β-epoxyhydrocinnamate
Ethyl α,β-epoxy-α-phenylpro-
pionate
Ethyl phenylglycidate
Ethyl 3-phenylglycidate
Oxiranecarboxylic acid, 3-
phenyl-, ethyl ester (9CI)

121-43-7
$C_3H_9BO_3$
103.91

COB(OC)OC
Trimethyl borate [UN 2416]
AI3-60245
Borate de trimethyle (French)
Borato de trimetilo (Spanish)
Borester O
Boric acid, trimethyl ester (9CI)
Methyl borate
NSC-777
Trimethoxyborane
Trimethoxyborine
Trimethoxyboron
Trimethylester kyseliny borite (Czech)
UN 2416 [Trimethyl borate]

121-44-8
$C_6H_{15}N$
101.22
N(CC)(CC)CC
Triethylamine (ACGIH, OSHA) [UN 1296]
(Diethylamino)ethane
N,N-Diethylethanamine
Ethanamine, N,N-diethyl-
TEN
Triaethylamin (German)
Trietilamina (Italian)
UN 1296 [Triethylamine]

121-45-9
$C_3H_9O_3P$
124.09
O(P(OC)OC)C
Phosphorous acid, trimethyl ester
Fosforyn trojmetylowy (Czech)
Methyl phosphite
Trimethoxyfosfin (Czech)
Trimethoxyphosphine
Trimethylfosfit (Czech)
Trimethyl phosphite (ACGIH, OSHA) [UN 2329]
UN 2329 [Trimethyl phosphite]

121-46-0
C_7H_8
92.15
C(C=CC1C=2)(C2)C1
Norbornadiene
Bicyclo(2.2.1)heptadiene
2,5-Norbornadiene

121-47-1
$C_6H_7NO_3S$
173.20
O=S(=O)(O)c(cccc1N)c1
Metanilic-acid
m-Aminobenzenesulfonic acid
1-Aminobenzene-3-sulfonic acid
3-Amino-benzenesulfonic acid
m-Anilinesulfonic acid
Benzenesulfonic acid, 3-amino-
Kyselina anilin-3-sulfonova (Czech)
Kyselina metanilova (Czech)
m-Sulfanilic acid

121-51-7
$C_6H_4ClNO_4S$
221.61
O=S(=O)(c(cccc1N(=O)=O)c1)Cl
Benzenesulfonyl chloride, m-nitro- (8CI)
Benzenesulfonyl chloride, 3-nitro- (9CI)
m-Nitrobenzenesulfonyl chloride
3-Nitrobenzenesulfonyl chloride
m-Nitrophenylsulfonyl chloride
3-Nitrophenylsulfonyl chloride
NSC-9806

121-53-9
$C_7H_6O_5S$
202.19
O=C(O)c(cccc1S(=O)(=O)O)c1
Benzoic acid, m-sulfo- (8CI)
AI3-16939
Benzoic acid, 3-sulfo- (9CI)
m-Carboxybenzenesulfonic acid
NSC-2853
m-Sulfobenzoic acid
3-Sulfobenzoic acid

121-54-0
$C_{27}H_{42}NO_2.Cl$
448.15
[Cl-].CC(C)(C)CC(C)(C)c2ccc(OCCOCC[N+](C)(C)Cc1ccccc1)cc2
Ammonium, benzyldimethyl-(2-(2-(p-(1,1,3,3-tetramethylbutyl)phenoxy)ethoxy)-ethyl)-, chloride
Anti-Germ 77
Antiseptol
Benzathonium chloride
Benzethonium
Benzethonium chloride
Benzethonium chloride 1622
Benzetonium chloride
Benzyldimethyl-p-(1,1,3,3-tetramethylbutyl)phenoxyethoxyethylammonium chloride
Benzyldimethyl(2-(2-(p-(1,1,3,3-tetramethylbutyl)phenoxy)ethoxy)ethyl)ammonium chloride
BZT
DIAPP
Diisobutylphenoxyethoxyethyl dimethyl benzyl ammonium chloride
Disilyn
Hyamine
Hyamine 1622
NCI-C61494
p-tert-Octylphenoxyethoxyethyldimethylbenzylammonium chloride
Phemeride
Phemerol
Phemerol chloride
Phemersol chloride
Phemithyn
Polymine D
Quatrachlor
Solamin
Solamine

121-57-3
$C_6H_7NO_3S$
173.20
O=S(=O)(O)c(ccc(N)c1)c1
Sulfanilic-acid
p-Aminobenzenesulfonic acid
4-Aminobenzenesulfonic acid
p-Aminophenylsulfonic acid
Aniline-p-sulfonic acid
Aniline-4-sulfonic acid
Aniline-p-sulphonic acid
Benzenesulfonic acid, 4-amino- (9CI)
Kyselina sulfanilova (Czech)
Sulfanilsaeure (German)
Sulphanilic acid

121-59-5
$C_7H_9AsN_2O_4$
260.10
NC(=O)Nc1ccc(cc1)[As](O)(O)=O
Arsanilic acid, N-carbamoyl
Amabevan
Ameban
Amebarsone
Amibiarson
Aminarson
Aminarsone
Aminoarson
(4-((Aminocarbonyl)amino)-phenyl)arsonic acid
Arsambide
p-Arsonophenylurea
Benzenearsonic acid, p-ureido-
p-Carbamidobenzenearsonic acid
p-Carbamino phenyl arsonic acid
Carbaminophenyl-p-arsonic acid
N-Carbamoylarsanilic acid
4-Carbamylaminophenylarsonic acid
N-Carbamyl arsanilic acid
Carbarsone
Carbasone
Carb-O-Sep
Fenarsone
Histocarb
Kyselina N-karbamylarsanilova (Czech)
Leucarsone
p-Ureidobenzenearsonic acid
4-Ureido-1-phenylarsonic acid

121-60-8
$C_8H_8ClNO_3S$
233.67
O=C(Nc(ccc(S(=O)(=O)Cl)c1)c1)C
4-(Acetylamino)benzenesulfonyl chloride
p-Acetamidobenzenesulfonyl chloride
4-Acetamidobenzenesulfonyl

chloride
p-Acetamidophenylsulfonyl chloride
4-Acetamidophenylsulfonyl chloride
p-Acetaminobenzenesulfonyl chloride
Acetanilide-p-sulfonyl chloride
p-Acetylaminobenzenesulfo-chloride
p-Acetylaminobenzenesulfonyl chloride
Acetylsulfanilyl chloride
N-Acetylsulfanilyl chloride
N4-Acetylsulfanilyl chloride
N(4)-Acetylsulfanilyl chloride
ASC
Benzenesulfonic acid, 4-acetamido-, chloride
Benzenesulfonyl chloride, 4-(acetylamino)- (9CI)
4-Chlorosulfonylacetanilide
4'-(Chlorosulfonyl)acetanilide
Dagenan chloride
NSC-127860
Sulfanilyl chloride, N-acetyl- (8CI)

121-62-0
$C_8H_9NO_4S$
215.22
O=C(Nc(ccc(S(=O)(=O)O)c1)c1)C
Benzenesulfonic acid, 4-(acetylamino)- (9CI)
4-(Acetylamino)benzenesulfonic acid

121-63-1
$C_{12}H_8Cl_2O_5S_2$
367.23
O=S(=O)(c(ccc(Oc(ccc(S(=O)(=O)Cl)c1)c1)c2)c2)Cl
4,4'-Oxybis(benzenesulfonyl chloride)
Benzenesulfonyl chloride, 4,4'-oxybis- (9CI)
Benzenesulfonyl chloride, 4,4'-oxydi- (8CI)
Bis(4-chlorosulfonylphenyl) ether
Diphenyl ether 4,4'-disulfonyl chloride
NSC-212
4,4'-Oxybenzenesulfonyl-chloride
Oxybis(benzenesulfonyl chloride)
Oxybis(4-benzenesulfonyl chloride)
p,p'-Oxybis(benzenesulfonyl chloride)
4,4'-Oxybisbenzenesulfonyl chloride
4,4'-Oxydibenzenesulfonyl chloride
Phenoxybenzene-4,4'-disulfonyl chloride

121-65-3
$C_{18}H_{30}O_3S$
326.50
O=S(=O)(O)c(ccc(c1)CCCCCCCCCCCC)c1
Benzenesulfonic acid, 4-dodecyl- (9CI)
4-Dodecylbenzenesulfonic acid

121-66-4
$C_3H_3N_3O_2S$
145.15
O=N(=O)C(SC(=N)N1)=C1
Thiazole, 2-amino-5-nitro
Aminonitrothiazole
2-Amino-5-nitrothiazole
Aminonitrothiazolum
Aminzol soluble
Enheptin
Enheptin Premix
Enheptin-T
Entramin
NCI-C03065
Nitramin
Nitramine
Nitramin Ido
5-Nitro-2-aminothiazole
Nitromin Ido
5-Nitro-2-thiazolylamine
2-Thiazolamine, 5-nitro- (9CI)
USAF EK-6561

121-69-7
$C_8H_{11}N$
121.20
N(c(cccc1)c1)(C)C
Aniline, N,N-dimethyl
Benzenamine, N,N,-dimethyl- (9CI)
(Dimethylamino)benzene
Dimethylaniline (ACGIH, OSHA)
N,N-Dimethylaniline [UN 2253]
N-Dimethyl-aniline (OSHA)
N,N-Dimethylbenzeneamine
Dimethylphenylamine
N,N-Dimethylphenylamine
Dwumetyloanilina (Polish)
NCI-C56428
UN 2253 [N,N-Dimethylaniline]
Versneller NL 63/10

121-71-1
$C_8H_8O_2$
136.15
O=C(c(cccc1O)c1)C
Acetophenone, 3'-hydroxy- (8CI)
AI3-14650
m-Acetylphenol
3-Acetylphenol
Ethanone, 1-(3-hydroxyphenyl)- (9CI)
m-Hydroxyacetophenone
3'-Hydroxyacetophenone
1-(3-Hydroxyphenyl)ethanone
1-(3-Hydroxyphenyl)ethan-1-one
NSC-2440

121-73-3
$C_6H_4ClNO_2$
157.56
O=N(=O)c(cccc1Cl)c1
Benzene, 1-chloro-3-nitro
Chloro-m-nitrobenzene
m-Chloronitrobenzene, Solid [UN 1578]
1-Chloro-3-nitrobenzene
m-Nitrochlorobenzene
Nitrochlorobenzene, meta, Solid (DOT)
UN 1578 [Chloronitrobenzenes meta or para, solid]

121-75-5
$C_{10}H_{19}O_6PS_2$
330.38
CCOC(=O)CC(SP(=S)(OC)OC)C(=O)OCC
Succinic acid, mercapto-, diethyl ester, S-ester with O,O-dimethylphosphorodithioate
American Cyanamid 4,049
S-(1,2-Bis(aethoxy-carbonyl)-aethyl)-O,O-dimethyl-dithiophosphat (German)
S-(1,2-Bis(carbethoxy)ethyl) O,O-dimethyl dithiophosphate
S-(1,2-Bis(ethoxy-carbonyl)-ethyl)-O,O-dimethyl-dithiofosfaat (Dutch)
S-(1,2-Bis(ethoxycarbonyl)-ethyl) O,O-dimethyl phosphorodithioate
S-1,2-Bis(ethoxycarbonyl)ethyl-O,O-dimethyl thiophosphate
S-(1,2-Bis(etossi-carbonil)-etil)-O,O-dimetil-ditiofosfato (Italian)
Calmathion
Carbetox
Carbethoxy malathion
Carbetovur
Carbofos
Carbophos
Celthion
Chemathion
Cimexan
Compound 4049
Cythion
Detmol MA
Detmol MA 96%
S-(1,2-Dicarbethoxyethyl) O,O-dimethyldithiophosphate
Dicarboethoxyethyl O,O-dimethyl phosphorodithioate
1,2-Di(ethoxycarbonyl)ethyl O,O-dimethyl phosphorodithioate
S-(1,2-Di(ethoxycarbonyl)ethyl) dimethyl phosphorothiolothionate
Diethyl (dimethoxyphosphinothioylthio) butanedioate
Diethyl (dimethoxyphosphinothioylthio)succinate
Diethyl mercaptosuccinate,

O,O-dimethyl dithiophosphate, S-ester
Diethyl mercaptosuccinate, O,O-dimethyl phosphorodithioate
Diethyl mercaptosuccinate, O,O-dimethyl thiophosphate
Diethyl mercaptosuccinate S-ester with O,O-dimethylphosphorodithioate
Diethyl mercaptosuccinic acid O,O-dimethyl phosphorodithioate
((Dimethoxyphosphinothioyl)-thio)butanedioic acid diethyl ester
O,O-Dimethyl S-(1,2-bis(ethoxycarbonyl)ethyl)dithiophosphate
O,O-Dimethyl-S-1,2-(dicarbaethoxyaethyl)-dithiophosphat (German)
O,O-Dimethyl-S-(1,2-dicarbethoxyethyl) dithiophosphate
O,O-Dimethyl S-(1,2-dicarbethoxyethyl)phosphorodithioate
O,O-Dimethyl S-(1,2-dicarbethoxyethyl) thiothionophosphate
O,O-Dimethyl S-1,2-di(ethoxycarbamyl)ethyl phosphorodithioate
O,O-Dimethyl-S-1,2-dikarbetoxylethylditiofosfat (Czech)
O,O-Dimethyldithiophosphate diethylmercaptosuccinate
Dithiophosphate de O,O-dimethyle et de S-(1,2-dicarboethoxyethyle) (French)
O,O-Dwumetylo-S-1,2-bis-(karboetoksyetylo)-dwutiofosforan (Polish)
EL 4049
Emmatos
Emmatos Extra
ENT 17,034
Ethiolacar
Etiol
Experimental Insecticide 4049
Extermathion
Formal
Forthion
Fosfothion
Fosfotion
Fosfotion 550
Four Thousand Forty-nine
Fyfanon
Hilthion
Hilthion 25WDP
Insecticide No. 4049
Karbofos
Kop-Thion
Kypfos
Latka 4049 (Czech)
Malacide
Malafor
Malakill
Malagran
Malamar
Malamar 50
Malaphele
Malaphos
Malasol
Malaspray
Malathion (ACGIH,OSHA)
Malathion E50
Malathion LV Concentrate
Malathion ULV Concentrate
Malathiozoo
Malathon
Malathyl LV Concentrate & ULV Concentrate
Malation (Polish)
Malatol
Malatox
Maldison
Malmed
Malphos
Maltox
Maltox MLT
Mercaptosuccinic acid diethyl ester
Mercaptothion
Mercaptotion (Spanish)
MLT
Moscarda
NA 2783
NCI-C00215
Oleophosphothion
OMS 1
Ortho Malathion
Phosphorodithioic acid, O,O-dimethyl ester, S-ester with diethyl mercaptosuccinate
Phosphothion
Prentox Malathion 95% Spray
Prioderm
Sadofos
Sadophos
Sadofos 30
SF 60
Siptox I
Sumitox
TAK
TM-4049
Vegfru Malatox
Vetiol
Zithiol

121-79-9
$C_{10}H_{12}O_5$
212.22
O=C(OCCC)c(cc(O)c(O)c1O)c1
Gallic acid, propyl ester
Benzoic acid, 3,4,5-trihydroxy-, propyl ester
Nipa 49
Nipagallin P
Progallin P
Propylester kyseliny gallove (Czech)
n-Propyl ester of 3,4,5-trihydroxybenzoic acid
Propyl gallate
n-Propyl gallate
Propyl 3,4,5-trihydroxybenzoate
n-Propyl 3,4,5-trihydroxybenzoate
Tenox PG
3,4,5-Trihydroxybenzene-1-propylcarboxylate
3,4,5-Trihydroxybenzoic acid n-propyl ester

121-82-4
$C_3H_6N_6O_6$
222.15
O=N(=O)N(CN(N(=O)=O)CN1N(=O)=O)C1
s-Triazine, hexahydro-1,3,5-trinitro
Cyclonite (ACGIH,OSHA)
Cyclotrimethylenenitramine
Cyclotrimethylenetrinitramine
Cyclotrimethylenetrinitramine, Desensitized (DOT)
Cyclotrimethylenetrinitramine, Wetted with not less than 15 per cent water, by mass [UN 0072]
Cyklonit (Czech)
Esaidro-1,3,5-trinitro-1,3,5-triazina (Italian)
Heksogen (Polish)
Hexahydro-1,3,5-trinitro-s-triazine
Hexahydro-1,3,5-trinitro-1,3,5-triazin (German)
Hexahydro-1,3,5-trinitro-1,3,5-triazine
Hexogeen (Dutch)
Hexogen (Explosive), Wetted with not less than 15 per cent water, by mass [UN 0072]
Hexogen 5W
Hexolite, Dry or wetted with less than 15 per cent water, by mass [UN 0118]
PBX(AF) 108
PBXW 108(E)
RDX
T4
1,3,5-Triazine, hexahydro-1,3,5-trinitro- (9CI)
Trimethyleentrinitramine (Dutch)
Trimethylenetrinitramine
sym-Trimethylenetrinitramine
Trinitrocyclotrimethylene triamine
1,3,5-Trinitrohexahydro-s-triazine
1,3,5-Trinitro-1,3,5-triazacyclohexane
UN 0072 [Cyclotrimethylenetrinitramine (Cyclonite; Hexogen; RDX) wetted with not less than 15 per cent water, by mass]
UN 0118 [Hexolite, Dry or wetted with less than 15 per cent water, by mass]

121-86-8
$C_7H_6ClNO_2$
171.59
O=N(=O)c(ccc(c1Cl)C)c1
Toluene, 2-chloro-4-nitro
2-Chloro-4-nitrotoluene

121-87-9
$C_6H_5ClN_2O_2$
172.58

O=N(=O)c(ccc(N)c1Cl)c1
Aniline, 2-chloro-4-nitro
1-Amino-2-chloro-4-nitro-benzene
o-Chloro-p-nitroaniline
2-Chloro-4-nitroaniline
4-Nitro-2-chloroaniline
OCPNA

121-88-0
$C_6H_6N_2O_3$
154.14
O=N(=O)c(ccc(N)c1O)c1
Phenol, 2-amino-5-nitro
2-Amino-5-nitrophenol
C.I. 76535
NCI-C55970
Ursol Yellow Brown A

121-89-1
$C_8H_7NO_3$
165.16
O=C(c(cccc1N(=O)=O)c1)C
Acetophenone, 3'-nitro
m-Acetylnitrobenzene
Ethanone, 1-(3-nitrophenyl)- (9CI)
Methyl 3-nitrophenyl ketone
3-Nitroacetofenon (Czech)
m-Nitroacetophenone
3'-Nitroacetophenone
(3-Nitrophenyl) methyl ketone
USAF MA-1

121-90-4
$C_7H_4ClNO_3$
185.57
O=C(c(cccc1N(=O)=O)c1)Cl
Benzoyl chloride, m-nitro
Benzoyl chloride, 3-nitro- (9CI)
Chlorid kyseliny m-nitro-benzoove (Czech)
m-Nitrobenzoyl chloride
3-Nitrobenzoyl chloride

121-91-5
$C_8H_6O_4$
166.14
O=C(O)c(cccc1C(=O)O)c1
Isophthalic-acid
Acide isophtalique (French)
Benzene-1,3-dicarboxylic acid
m-Benzenedicarboxylic acid
IPA
Kyselina isoftalova (Czech)
m-Phthalic acid

121-92-6
$C_7H_5NO_4$
167.13
O=C(O)c(cccc1N(=O)=O)c1
Benzoic acid, m-nitro
Benzoic acid, 3-nitro- (9CI)
m-Nitrobenzenecarboxylic acid
m-Nitrobenzenecarboxylic acid
m-Nitrobenzoic acid
3-Nitrobenzoic acid

121-93-7
$C_7H_{17}NO_2$
147.21
OCCN(C(C)C)CCO
Ethanol, 2,2'-((1-methylethyl)-imino)bis- (9CI)
Diethanolisopropylamine
Ethanol, 2,2'-(isopropyl-imino)di- (8CI)
Isopropyldiethanolamine
2,2'-(Isopropylimino)diethanol
2,2'-((1-Methylethyl)imino)bis-ethanol
NSC-19185

121-98-2
$C_9H_{10}O_3$
166.19
O=C(OC)c(ccc(OC)c1)c1
p-Anisic acid, methyl ester
Benzoic acid, p-methoxy-, methyl ester
Methyl p-anisate
Methyl p-methoxybenzoate

122-00-9
$C_9H_{10}O$
134.19
O=C(c(ccc(c1)C)c1)C
Acetophenone, 4'-methyl
p-Acetyltoluene
Ethanone, 1-(4-methylphenyl)- (9CI)
Melilotal
p-Methylacetophenone
4'-Methylacetophenone
1-Methyl-4-acetylbenzene
Methyl p-tolyl ketone

122-01-0
$C_7H_4Cl_2O$
175.01
O=C(c(ccc(c1)Cl)c1)Cl
Benzoyl chloride, 4-chloro-
AI3-14889
4-Chlorobenzoyl chloride

122-03-2
$C_{10}H_{12}O$
148.22
O=Cc(ccc(c1)C(C)C)c1
Benzaldehyde, p-isopropyl
Benzaldehyde, 4-(1-methyl-ethyl)- (9CI)
Cumaldehyde
Cumic aldehyde
p-Cumic aldehyde
Cuminal
Cuminaldehyde
Cuminic aldehyde
Cuminyl aldehyde
p-Isopropylbenzaldehyde
4-Isopropylbenzaldehyde
p-Isopropylbenzenecarbox-aldehyde
4-(1-Methylethyl)benzaldehyde

122-04-3
$C_7H_4ClNO_3$
185.57
O=C(c(ccc(N(=O)=O)c1)c1)Cl
Benzoyl chloride, p-nitro
p-Nitrobenzoic acid chloride
4-Nitrobenzoic acid chloride
p-Nitrobenzoyl chloride
4-Nitrobenzoyl chloride

122-07-6
$C_5H_{13}NO_2$
119.16
O(C(OC)CNC)C
2,2-Dimethoxyethyl(methyl)-amine
Acetaldehyde, (methylamino)-, dimethyl acetal (8CI)
2,2-Dimethoxy-N-methylethan-amine
Ethanamine, 2,2-dimethoxy-N-methyl- (9CI)
Methylaminoacetaldehyde di-methyl acetal
N-Methylaminoacetaldehyde di-methyl acetal
2-(Methylamino)acetaldehyde dimethyl acetal
NSC-66270

122-09-8
$C_{10}H_{15}N$
149.26
NC(Cc(cccc1)c1)(C)C
Phenethylamine, α,α-dimethyl
Benzeethanamine, α,α-di-methyl (9CI)
α,α-Dimethylphenethylamine
α,α-Dimethyl-β-phenylethyl-amine
Duromine
Ethanamine, 1,1-dimethyl-2-phenyl-
Lipopill
Lonamin
MG 18370
MG 18570
Mirapront
Phentermine
2-Phenyl-tert-butylamine
RCRA waste number P046
Wilpo

122-10-1
$C_9H_{15}O_8P$
282.21
COC(=O)CC(OP(=O)(OC)OC)=CC(=O)OC
Glutaconic acid, 3-hydroxy-, dimethyl ester, dimethyl phosphate
1,3-Bis-(carboxymethoxy)-1-propen-2-yl-dimethylfosfat (Czech)
Bomyl
Dimethyl 1,3-bis(carbometh-oxy)-1-propen-2-yl phosphate

Dimethyl-1,3-di(carbomethoxy)-1-propen-2-yl phosphate
Dimethyl 3-(dimethoxyphosphinyloxy)glutaconate
Dimethyl 3-hydroxyglutaconate dimethyl phosphate
ENT 24,833
Fly Bait Grits
GC 3707
General Chemicals 3707
3-Hydroxyglutaconic acid, dimethyl ester, dimethyl phosphate
3-Hydroxy-2-pentanedioic acid, dimethyl ester, dimethyl phosphate
Phosphoric acid, dimethyl ester, ester with dimethyl 3-hydroxyglutaconate
Swat

122-14-5
$C_9H_{12}NO_5PS$
277.25
COP(=S)(OC)Oc1ccc(N(=O)=O)c(C)c1
Phosphorothioic acid, O,O-dimethyl O-(4-nitro-m-tolyl) ester
AC-47300
Accothion
Aceothion
Agria 1050
Agriya 1050
Agrothion
American Cyanamid CL-47,300
Arbogal
Bay 41831
Bayer 41831
Bayer S 5660
Bay S 5660
Cekutrothion
CL 47300
CP 47114
m-Cresol, 4-nitro-, O-ester with O,O-dimethyl phosphorothioate
Cyfen
Cytel
Cyten
O,O-Dimethyl-O-(3-methyl-4-nitrofenyl)-monothiofosfaat (Dutch)
O,O-Dimethyl-O-(3-methyl-4-nitro-phenyl)-monothiophosphat (German)
Dimethyl 3-methyl-4-nitrophenyl phosphorothionate
O,O-Dimethyl O-(3-methyl-4-nitrophenyl) phosphorothioate
O,O-Dimethyl O-(3-methyl-4-nitrophenyl) thiophosphate
O,O-Dimethyl O-(3-methyl) phosphorothioate
O,O-Dimethyl-O-(4-nitro-5-methylphenyl)-thionophosphat (German)
O,O-Dimethyl O-(4-nitro-3-methylphenyl)thiophosphate
O,O-Dimethyl O-4-nitro-m-tolyl phosphorothioate
Dimethyl 4-nitro-m-tolyl phosphorothionate
O,O-Dimetil-O-(3-metil-4-nitro-fenil) fosforotioato (Portugese)
O,O-Dimetil-O-(3-metil-4-nitro-fenil)-monotiofosfato (Italian)
Dybar
EI 47300
ENT 25,715
Falithion
Fenitox
Fenitrothion
Fenitrotion (Hungarian)
Folithion
Folithion EC 50
H-35-F 87 (BVM)
8057HC
Kotion
Macbar
MEP
MEP (Pesticide)
Metathio E-50
Metathion
Metathione
Metathion E 50
Metathionine
Metathionine E50
Metation
Metation E50
Methylnitrophos
Mglawik F
Monsanto CP 47114
Nitrophos
Novathion
Nuvanol
Ovadofos
Owadofos
Oleosumifene
OMS 43
Pennwalt C-4852
Phenitrothion
Phosphorothioic acid, O,O-dimethyl O-(3-methyl-4-nitrophenyl) ester
S 5660
S 112A
S-1102A
Sumithian
Sumithion
Sumitomo S-1102A
Thiophosphate de O,O-dimethyle et de O-(3-methyl-4-nitrophenyle) (French)
Verthion

122-15-6
$C_{11}H_{17}NO_3$
211.29
CN(C)C(=O)OC1=CC(=O)CC(C)(C)C1
Carbamic acid, dimethyl-, ester with 3-hydroxy-5,5-dimethyl-2-cyclohexen-1-one
2-Cyclohexen-1-one, 3-hydroxy-5,5-dimethyl-, dimethylcarbamate
Dimetan
Dimethylcarbamate de 5,5-dimethyl dihydroresorcinol (French)
5,5-Dimethyl-dihydroresorcinol-N,N-dimethylcarbamat (German)
5,5-Dimethyldihydroresorcinol dimethylcarbamate
5,5-Dimethyl-4,5-dihydro-3-resorcyl-dimethyl-carbamat (German)
(5,5-Dimethyl-3-oxo-cyclohex-1-en-yl)-N,N-dimethyl-carbamaat (Dutch)
(5,5-Dimethyl-3-oxo-cyclohex-1-en-yl)-N,N-dimethyl-carbamat (German)
5,5-Dimethyl-3-oxo-1-cyclohexen-1-yl dimethylcarbamate
5,5-Dimethyl-3-oxocyclohex-1-enyl dimethylcarbamate
5,5-Dimethyl-3-oxo-1-cyklohexenylester kyseliny dimethylkarbaminove (Czech)
(5,5-Dimetil-3-oxo-cicloes-1-enil)-N,N-dimetil-carbammato (Italian)
ENT 24,738
G 19258
Geigy 19258

122-18-9
$C_{25}H_{46}N.Cl$
396.17
[Cl-].CCCCCCCCCCCCCCCC[N+](C)(C)Cc1ccccc1
Ammonium, benzyldimethylhexadecyl-, chloride
Benzenemethanaminium, N-hexadecyl-N,N-dimethyl-, chloride
Benzyldimethylcetylammonium chloride
Benzyldimethylhexadecylammonium chloride
Winzer Solution

122-19-0
$C_{27}H_{50}N.Cl$
424.23
[Cl-].CCCCCCCCCCCCCCCCCC[N+](C)(C)Cc1ccccc1
Ammonium, benzyldimethyloctadecyl-, chloride
Ammonyx 4
Ammonyx 485
Ammonyx 490
Ammonyx 4002
Ammonyx CA Special
Arquad DM18B-90
2B
Barquat SB-25
Benzenemethanaminium, N,N-dimethyl-N-octadecyl-, chloride (9CI)
Benzyldimethylstearylammonium chloride
Benzylstearyldimethylammonium chloride
Carsoquat SDQ-25
Carsoquat SDQ-85

Dehyquart STC-25
Dimethylbenzyloctadecyl-ammonium chloride
Dimethyloctadecylbenzyl-ammonium chloride
Intexan SB-85
Intexsan SB-85
J Soft C 4
Katamine AB
Nissan Cation S2-100
n-Octadecyl-N-benzyl-N,N-dimethylammonium chloride
Octadecyldimethylbenzyl-ammonium chloride
Orthosan MB
Quaternol 1
Stearalkonium chloride
Stearyldimethylbenzyl-ammonium chloride
Stebac
Stedbac
Tallow benzyl dimethyl ammonium chloride
Triton X-40
Triton X-400
Varisoft SDC

122-20-3
$C_9H_{21}NO_3$
191.31
OC(C)CN(CC(O)C)CC(O)C
2-Propanol, 1,1',1''-nitrilotri
1,1',1''-Nitrilotri-2-propanol
Triisopropanolamine
Tris(2-hydroxypropyl)amine
Tris(2-hydroxy-1-propyl)amine

122-28-1
$C_8H_8N_2O_3$
180.15
O=C(Nc(cccc1N(=O)=O)c1)C
Acetamide, N-(3-nitrophenyl)- (9CI)
Acetanilide, 3'-nitro- (8CI)
N-Acetyl-m-nitroaniline
AI3-08832
m-Nitroacetanilide
3'-Nitroacetanilide
3-Nitro-N-acetylaniline
N-(3-Nitrophenyl)acetamide
NSC-1314

122-32-7
$C_{57}H_{104}O_6$
885.45
Glyceryl trioleate
Aldo To
Emery Oleic Acid Ester 2230
Emery 2423
Glycerin trioleate
Glycerol trioleate
Glycerol triolein
Glycerol, tri(cis-9-octadecenoate)
Glyceryl-1,2,3-trioleate
9-Octadecenoic acid, 1,2,3-propanetriyl ester
9-Octadecenoic acid (Z)-, 1,2,3-propanetriyl ester (9CI)
Oleic acid triglyceride
Oleic triglyceride
Olein
Olein, tri-
Oleyl triglyceride
Raoline
Triolein
Trioleoylglyceride
Trioleoylglycerol

122-34-9
$C_7H_{12}ClN_5$
201.69
n(c(nc(n1)NCC)NCC)c1Cl
s-Triazine, 2-chloro-4,6-bis-(ethylamino)
A 2079
Aktinit S
Aquazine
Batazina
2,4-Bis(aethylamino)-6-chlor-1,3,5-triazin (German)
2,4-Bis(ethylamino)-6-chloro-s-triazine
Bitemol
Bitemol S 50
Cat
Cat (Herbicide)
CDT
Cekusan
Cekuzina-S
CET
1-Chloro, 3,5-bisethylamino-2,4,6-triazine
2-Chloro-4,6-bis(ethylamino)-s-triazine
2-Chloro-4,6-bis(ethylamino)-1,3,5-triazine
6-Chloro-N,N'-diethyl-1,3,5-triazine-2,4-diyldiamine
Framed
G 27692
Geigy 27,692
Gesaran
Gesatop
Gesatop 50
H 1803
Herbazin
Herbazin 50
Herbex
Herboxy
Hungazin DT
Premazine
Primatol S
Princep
Printop
Radocon
Radokor
Simadex
Simanex
Simazin
Simazine
Simazine 80W
Symazine
Tafazine
Tafazine 50-W
Taphazine
Triazine A 384
W 6658
Weedex
Zeapur
Yrodazin

122-37-2
$C_{12}H_{11}NO$
185.24
Oc(ccc(Nc(cccc1)c1)c2)c2
Phenol, p-anilino
p-Anilinophenol
4-Anilinophenol
Diphenylamine, 4-hydroxy-
p-Hydroxydifenylamin (Czech)
para-Hydroxydifenylamin (Czech)
p-Hydroxydiphenylamine
4-Hydroxydiphenylamine
p-Oxydiphenylamine
Phenol, 4-(phenylamino)-
Phenyl-p-aminophenol
N-Phenyl-p-aminophenol
4-Phenylaminophenol
VTI 1

122-39-4
$C_{12}H_{11}N$
169.24
N(c(cccc1)c1)c(cccc2)c2
Diphenylamine
Aniline, N-phenyl-
Anilinobenzene
Benzenamine, N-phenyl- (9CI)
Benzene, anilino-
Benzene, (phenylamino)-
Big Dipper
C.I. 10355
Deccoscald 282
DFA
Difenylamin (Czech)
N,N-Diphenylamine
Diphenylamine (ACGIH,OSHA)
DPA
N-Fenylanilin (Czech)
No Scald
No Scald DPA 283
N-Phenylaniline
N-Phenylbenzenamine
Scaldip

122-40-7
$C_{14}H_{18}O$
202.32
O=CC(=Cc(cccc1)c1)CCCCC
Heptanal, 2-(phenyl-methylene)- (9CI)
Amyl cinnamic aldehyde
α-Amyl cinnamaldehyde
α-Amyl cinnamic aldehyde
Amylcinnamaldehyde
α-Amylcinnamaldehyde
Amylcinnamic acid aldehyde
Amylcinnamic aldehyde
α-Amyl-β-phenylacrolein
2-Benzylideneheptanal
Cinnamaldehyde, α-pentyl- (6CI,7CI,8CI)
Flomine
Heptanal, 2-benzylidene-
Jasminaldehyde
NSC-6649
Pentylcinnamaldehyde
α-Pentylcinnamaldehyde

2-(Phenylmethylene)heptanal
3-Phenyl-2-propenal mono-pentyl deriv.
2-Propenal, 3-phenyl-, mono-pentyl deriv.

122-42-9
$C_{10}H_{13}NO_2$
179.24
O=C(OC(C)C)Nc(cccc1)c1
Carbanilic acid, isopropyl ester
Ban-HOE
Beet-Kleen
Chem-HOE
IFC
IFK
INPC
IPC
IPPC
Iso.PPC.
Isopropil-N-fenil-carbammato (Italian)
Isopropyl carbanilate
Isopropyl carbanilic acid ester
Isopropylester kyseliny karbanilove (Czech)
Isopropyl-N-fenyl-carbamaat (Dutch)
Isopropyl-N-phenyl-carbamat (German)
Isopropyl phenylcarbamate
Isopropyl-N-phenylcarbamate
o-Isopropyl N-phenyl carbamate
Isopropyl-N-phenylurethan (German)
Ortho Grass Killer
N-Phenylcarbamate d'isopropyle (French)
N-Phenylcarbamic acid, isopropyl ester
Phenylcarbamic acid, 1-methylethyl ester
N-Phenyl isopropyl carbamate
Premalox
Profam
Propham
Prophame
Triherbide
Triherbide-IPC
Tuberit
Tuberite
USAF D-9
Y 2

122-48-5
$C_{11}H_{14}O_3$
194.25
O=C(CCc(ccc(O)c1OC)c1)C
2-Butanone, 4-(4-hydroxy-3-methoxyphenyl)
Gingerone
4-(4-Hydroxy-3-methoxyphenyl)-2-butanone
(4-Hydroxy-3-methoxyphenyl)-ethyl methyl ketone
3-Methoxy-4-hydroxy-benzylacetone
(0)-Paradol
Vanillyl acetone
Zingerone
Zingiberone

122-51-0
$C_7H_{16}O_3$
148.23
O(C(OCC)OCC)CC
Orthoformic acid, triethyl ester
Aethon
Ethone
Ethylester kyseliny orthomravenci (Czech)
Ethyl orthoformate [UN 2524]
Methane, triethoxy-
1,1',1'-(Methylidynetris(oxy))-tris(ethane)
Orthoformic acid, ethyl ester
Orthomravencan ethylnaty (Czech)
Triethoxymethane
Triethylester kyseliny orthomravenci (Czech)
Triethyl orthoformate
UN 2524 [Ethyl orthoformate]

122-52-1
$C_6H_{15}O_3P$
166.18
O(P(OCC)OCC)CC
Phosphorous acid, triethyl ester
Fosforyn trojetylowy (Czech)
Triethyl phosphite [UN 2323]
UN 2323 [Triethyl phosphite]

122-57-6
$C_{10}H_{10}O$
146.20
O=C(C=Cc(cccc1)c1)C
3-Buten-2-one, 4-phenyl
Benzalaceton (German)
Benzalacetone
Benzylidene acetone
4-Phenyl-3-buten-2-one
Styryl methyl ketone

122-59-8
$C_8H_8O_3$
152.16
O=C(O)COc(cccc1)c1
Acetic acid, phenoxy
Acide phenoxyacetique (French)
Glycolic acid phenyl ether
Glycollic acid phenyl ether
Phenoxyacetic acid
Phenoxyethanoic acid
o-Phenylglycolic acid

122-60-1
$C_9H_{10}O_2$
150.19
O(C1COc(cccc2)c2)C1
Propane, 1,2-epoxy-3-phenoxy
Ageflex PGE
Benzene, (2,3-epoxypropoxy)-
1,2-Epoxy-3-phenoxypropane
2,3-Epoxypropylphenyl ether
Ether, 2,3-epoxypropyl phenyl
Ether, phenylglycidyl
Fenyl-glycidylether (Czech)
Glycidyl phenyl ether
Oxirane, (phenoxymethyl)-
PGE (OSHA)
Phenol-glycidaether (German)
Phenol glycidyl ether
3-Phenoxy-1,2-epoxypropane
Phenoxypropene oxide
Phenoxypropylene oxide
Phenyl 2,3-epoxypropyl ether
Phenyl glycidyl ether (ACGIH, OSHA)
Phenylglycydyl ether

122-62-3
$C_{26}H_{50}O_4$
426.76
O=C(OCC(CCCC)CC)CCCCCCCCC(=O)OCC(CCCC)CC
Sebacic acid, bis(2-ethylhexyl)ester
Bis-(2-ethylhexyl)ester kyseliny sebakove (Czech)
Bis(2-ethylhexyl)sebacate
Bisoflex DOS
Decanedioic acid, bis(2-ethylhexyl) ester
Di(2-ethylhexyl)sebacate
Dioctyl sebacate
DOS
2-Ethylhexyl sebacate
1-Hexanol, 2-ethyl-, sebacate
Monoplex DOS
Octoil S
Octyl sebacate
PX 438
Staflex DOS
Uniflex DOS

122-63-4
$C_{10}H_{12}O_2$
164.20
O=C(OCc(cccc1)c1)CC
Propanoic acid, phenylmethyl ester (9CI)
AI3-02952
Benzyl propionate
NSC-46100
Phenylmethyl propanoate
Propionic acid, benzyl ester (8CI)

122-64-5
$C_9H_{10}HgO_3$
366.77
Phenylmercuric lactate
Caswell No. 657E
EPA Pesticide Chemical Code 066012
Mercury, (2-hydroxypropanoato)phenyl- (9CI)
Mercury, (lactato)phenyl- (8CI)
Mercury, (lactoyloxy)phenyl-
NSC-48909
Phenylmercury lactate

122-66-7
$C_{12}H_{12}N_2$
184.26
N(Nc(cccc1)c1)c(cccc2)c2
Hydrazobenzene
Benzene, hydrazodi-
N,N'-Bianiline
N,N'-Diphenylhydrazine
(sym)-Diphenylhydrazine
1,2-Diphenylhydrazine (9CI)
Hydrazine, 1,2-diphenyl-
Hydrazobenzen (Czech)
NCI-C01854
RCRA waste number U109

122-68-9
$C_{18}H_{18}O_2$
266.34
O=C(OCCCc(cccc1)c1)C=Cc(ccc c2)c2
2-Propenoic acid, 3-phenyl-, 3-phenylpropyl ester (9CI)
3-Phenylpropyl 3-phenyl-2-propenoate

122-70-3
$C_{11}H_{14}O_2$
178.23
O=C(OCCc(cccc1)c1)CC
2-Phenylethyl propionate
AI3-18544
Benzylcarbinyl propionate
Caswell No. 655D
ENT 18,544
EPA Pesticide Chemical Code 102601
NSC-404457
Phenethyl alcohol, propionate
Phenethyl propionate
2-Phenethyl propionate
Phenylethyl propionate
2-Phenylethyl propanoate
Propanoic acid, 2-phenylethyl ester (9CI)
Propionic acid, phenethyl ester (8CI)
Propionic acid, 2-phenylethyl ester

122-73-6
$C_{12}H_{18}O$
178.30
O(Cc(cccc1)c1)CCC(C)C
Ether, benzyl isopentyl
Benzyl isoamyl ether
Benzyl isopentyl ether
Isoamyl benzyl ether

122-75-8
$C_{16}H_{20}N_2.2C_2H_4O_2$
360.50
Ethylenediamine, N,N'-dibenzyl-, diacetate
DBED diacetate
N,N'-Dibenzylethylenediamine diacetate

122-78-1
C_8H_8O
120.16
O=CCc(cccc1)c1
Benzeneacetaldehyde
Acetaldehyde, phenyl-
Hyacinthin
PAA
Phenylacetaldehyde
Phenylacetic aldehyde
Phenylethanal
α-Tolualdehyde
α-Toluic aldehyde

122-79-2
$C_8H_8O_2$
136.16
O=C(Oc(cccc1)c1)C
Acetic acid, phenyl ester
Acetyl phenol
Fenylester kyseliny octove (Czech)
Phenol acetate
Phenyl acetate

122-80-5
$C_8H_{10}N_2O$
150.20
O=C(Nc(ccc(N)c1)c1)C
Acetanilide, 4'-amino
Acetamide, N-(4-aminophenyl)- (9CI)
p-Acetamidoaniline
4-Acetamidoaniline
p-Acetoaminoaniline
Acetparamin
p-(Acetylamino)aniline
4-(Acetylamino)aniline
N-Acetyl-p-fenylendiamin (Czech)
Acetyl-p-phenylenediamine
4'-Aminoacetanilid (Czech)
p-Aminoacetanilide
4-Aminoacetanilide
4'-Aminoacetanilide
N-(p-Aminophenyl)acetamide
N-(4-Aminophenyl)acetamide
C.I. 76005
C.I. Oxidation Base 19
Fourrine 88
Fourrine A

122-82-7
$C_{12}H_{15}NO_3$
221.28
O=C(Nc(ccc(OCC)c1)c1)CC (=O)C
Acetoacetic acid, p-phenetidide
Acetoacetanilide, 4'-ethoxy-
p-Acetoacetophenetidide
Acetoacet-p-phenetidide
Butanamide, N-(4-ethoxyphenyl)-3-oxo- (9CI)
4-Ethoxyacetoacetanilide
4'-Ethoxyacetoacetanilide

122-88-3
$C_8H_7ClO_3$
186.60
O=C(O)COc(ccc(c1)Cl)c1
Acetic acid, (p-chlorophenoxy)
Acetic acid, (4-chlorophenoxy)- (9CI)
p-Chlorophenoxyacetic acid
(4-Chlorophenoxy)acetic acid
4-CP
CPA
4-CPA
Kyselina 4-chlorfenoxyoctova (Czech)
Marks 4-CPA
Parachlorophenoxyacetic acid
PCPA
Sure-Set
Tomato Hold
Tomato Fix
Tomato Fix Concentrate
Tomatotone

122-94-1
$C_{10}H_{14}O_2$
166.22
O(c(ccc(O)c1)c1)CCCC
Phenol, p-butoxy- (8CI)
p-Butoxyphenol
4-Butoxyphenol
NSC-60292
Phenol, 4-butoxy- (9CI)

122-96-3
$C_8H_{18}N_2O_2$
174.28
OCCN(CCN(C1)CCO)C1
1,4-Piperazinediethanol
N,N'-Bis(β-hydroxyethyl)-piperazine
1,4-Bis(2-hydroxyethyl)-piperazine
N,N'-Di(2-hydroxyethyl)-piperazine
1,4-Di(2-hydroxyethyl)-piperazine
Piperazine, N,N'-bis(2-hydroxyethyl)-

122-97-4
$C_9H_{12}O$
136.21
OCCCc(cccc1)c1
1-Propanol, 3-phenyl
3-Benzenepropanol
Hydrocinnamic alcohol
Hydrocinnamyl alcohol
(3-Hydroxypropyl)benzene
γ-Phenylpropanol
3-Phenylpropanol
3-Phenyl-1-propanol
Phenylpropyl alcohol
γ-Phenylpropyl alcohol
3-Phenylpropyl alcohol

122-98-5
$C_8H_{11}NO$

137.20
OCCNc(cccc1)c1
Ethanol, 2-anilino
Aniline, N-(β-hydroxyethyl)-
Aniline, N-(2-hydroxyethyl)-
2-Anilinoethanol
Emery 5700
Ethanol, 2-(phenylamino)-
N-(2-Hydroxyethyl)phenylamine
2-(Phenylamino)ethanol
Phenyl ethanolamine
N-Phenylethanolamine

122-99-6
$C_8H_{10}O_2$
138.18
O(c(cccc1)c1)CCO
Ethanol, 2-phenoxy
Arosol
Dowanol EP
Dowanol EPH
Emeressence 1160
Emery 6705
Ethylene glycol monophenyl ether
Ethylene glycol phenyl ether
2-Fenoxyethanol (Czech)
Fenyl-cellosolve (Czech)
Fenylcelosolv (Czech)
Glycol monophenyl ether
β-Hydroxyethyl phenyl ether
1-Hydroxy-2-phenoxyethane
Phenoxethol
Phenoxetol
Phenoxyethanol
2-Phenoxyethanol
Phenoxyethyl alcohol
Phenoxytol
Phenyl cellosolve
Phenylmonoglycol ether
Rose Ether

123-00-2
$C_7H_{16}N_2O$
144.25
O(CCN(C1)CCCN)C1
Morpholine, 4-aminopropyl
N-(3-Aminopropyl)morfolin (Czech)
N-(3-Aminopropyl)morpholine
4-Aminopropylmorpholine
N-Aminopropylmorpholine
(DOT)
Morpholine, N-aminopropyl-

123-01-3
$C_{18}H_{30}$
246.48
c(cccc1)(c1)CCCCCCCCCCCC
Benzene, dodecyl
Detergent Alkylate
Dodecylbenzene
Phenyldodecan (German)
1-Phenyldodecane

123-02-4
$C_{19}H_{32}$
260.46
c(cccc1)(c1)CCCCCCCCCCCCC
Benzene, tridecyl- (9CI)
Tridecane, 1-phenyl- (8CI)
Tridecylbenzene

123-03-5
$C_{21}H_{38}N.Cl$
340.05
Pyridinium, 1-hexadecyl-, chloride
Acetoquat CPC
Aktivex
Ammonyx CPC
Biosept
Bis(3-hydroxy-4-hydroxy-methyl-2-methylpyridyl-(5)-methyl)disulfide dihydro-chloride monohydrate
Ceeprin chloride
Ceepryn
Ceepryn chloride
Cepacol chloride
Ceprim
Cetamium
Cetylpyridinium chloride
N-Cetylpyridinium chloride
1-Cetylpyridinium chloride
Dobendan
Germidine
Hexadecylpyridinium chloride
n-Hexadecylpyridinium chloride
Encephabol
1-Hexadecylpyridinium chloride
Intexsan CPC
Pristacin
Pyrisept
Quaternario CPC

123-04-6
$C_8H_{17}Cl$
148.70
ClCC(CCCC)CC
Heptane, 3-chloromethyl
1-Chloro-2-ethylhexane
3-Chloromethylheptane
2-Ethylhexyl chloride

123-05-7
$C_8H_{16}O$
128.24
O=CC(CCCC)CC
Hexanal, 2-ethyl
Butyl ethyl acetaldehyde
Ethylbutylacetaldehyde
α-Ethylcaproaldehyde
Ethylhexaldehyde [UN 1191]
2-Ethylhexaldehyde
2-Ethylhexanal
β-Propyl-α-ethylacrolein
UN 1191 [Octyl aldehydes, flammable]

123-07-9
$C_8H_{10}O$
122.18
Oc(ccc(c1)CC)c1
Phenol, p-ethyl
4-Ethylphenol

123-08-0
$C_7H_6O_2$
122.13
O=Cc(ccc(O)c1)c1
Benzaldehyde, p-hydroxy
Benzaldehyde, 4-hydroxy- (9CI)
p-Formylphenol
4-Formylphenol
p-Hydroxybenzaldehyde
4-Hydroxybenzaldehyde
p-Oxybenzaldehyde
Parahydroxybenzaldehyde
USAF M-6

123-09-1
C_7H_7ClS
158.65
S(c(ccc(c1)Cl)c1)C
Sulfide, p-chlorophenyl methyl
Benzene, 1-chloro-4-(methyl-thio)- (9CI)
p-Chlorophenyl methyl sulfide
p-Chlorothioanisole
4-Chlorothioanisole
Methyl p-chlorophenyl sulfide
Methyl 4-chlorophenyl sulfide

123-11-5
$C_8H_8O_2$
136.16
O=Cc(ccc(OC)c1)c1
p-Anisaldehyde
Anisic aldehyde
Aubepine
Crategine
p-Methoxybenzaldehyde
4-Methoxybenzaldehyde

123-15-9
$C_6H_{12}O$
100.18
O=CC(CCC)C
Valeraldehyde, 2-methyl
2-Methylpentaldehyde
α-Methylvaleraldehyde [UN 2367]
UN 2367 [α-Methylvaler-aldehyde]

123-17-1
$C_{12}H_{26}O$
186.34
OC(CC(CC(C)C)C)CC(C)C
2,6,8-Trimethyl-4-nonanol
4-Nonanol, 2,6,8-trimethyl- (8CI,9CI)
NSC-60574
2,6,8-Trimethylnonanol-4
2,4,8-Trimethyl-6-nonanol

123-18-2
$C_{12}H_{24}O$
184.32

Isobutyl heptyl ketone
4-Nonanone, 2,6,8-trimethyl-(9CI)
NSC-66186
2,6,8-Trimethyl-4-nonanone

123-19-3
$C_7H_{14}O$
114.21
O=C(CCC)CCC
4-Heptanone
Butyrone
Dipropylketone (ACGIH, OSHA) [UN 2710]
GBL
Heptan-4-one
Propyl ketone
UN 2710 [Dipropylketone]

123-20-6
$C_6H_{10}O_2$
114.16
CCCC(=O)OC=C
Butyric acid, vinyl ester
UN 2838 [Vinyl butyrate, inhibited]
Vinyl butyrate
Vinyl butyrate, Inhibited [UN 2838]
Vinylester kyseliny maselne (Czech)

123-23-9
$C_8H_{10}O_8$
234.18
O=C(OOC(=O)CCC(=O)O)CCC(=O)O
Propionic acid, 3,3'-(dioxydicarbonyl)di-, Maximum concentration 72%
Alphozone
Bis(3-carboxypropionyl) peroxide
Disuccinic acid peroxide, Maximum concentration 72%
Disuccinic acid peroxide, Technically pure (DOT)
Peroxide, bis(3-carboxypropionyl)
Succinic acid peroxide (DOT)
Succinic acid peroxide, Technically pure (DOT)
Succinic peroxide
Succinyl peroxide
UN 2135 (DOT)
UN 2962 (DOT)

123-25-1
$C_8H_{14}O_4$
174.22
O=C(OCC)CCC(=O)OCC
Succinic acid, diethyl ester
Butanedioic acid, diethyl ester
Diethylester kyseliny jantarove (Czech)
Diethyl succinate
Ethyl succinate

123-28-4
$C_{30}H_{58}O_4S$
514.94
O=C(OCCCCCCCCCCCC)CCSCCC(=O)OCCCCCCCCCCCC
Propionic acid, 3,3'-thiodi-, didodecyl ester
Advastab 800
Antioxidant AS
Antioxidant LTDP
Bis(dodecyloxycarbonylethyl) sulfide
Carstab DLTDP
Cyanox LTDP
Didodecyl 3,3'-thiodipropionate
Dilaurylester kyseliny β',β'-thiodipropionove (Czech)
Dilauryl thiodipropionate
Dilauryl β-thiodipropionate
Dilauryl β',β'-thiodipropionate
Dilauryl 3,3'-thiodipropionate
DLT
DLTDP
DLTP
DMPTP
Ipognox 89
Irganox PS 800
Lauryl 3,3'-thiodipropionate
Lusmit
Milban F
Neganox DLTP
Plastanox LTDP
Plastanox LTDP Antioxidant
Propanoic acid, 3,3'-thiobis-, didodecyl ester
Stabilizer DLT
Thiobis(dodecyl propionate)
Tyox B

123-29-5
$C_{11}H_{22}O_2$
186.33
O=C(OCC)CCCCCCCC
Nonanoic acid, ethyl ester
Ethyl nonanoate
Ethyl nonylate
Ethyl pelargonate
Wine Ether

123-30-8
C_6H_7NO
109.14
Oc(ccc(N)c1)c1
Phenol, p-amino
Activol
p-Aminofenol (Czech)
4-Amino-1-hydroxybenzene
p-Aminophenol [UN 2512]
4-Aminophenol
Azol
BASF Ursol P Base
Benzofur P
Certinal
C.I. Oxidation Base 6A
Citol
Durafur Brown RB
Fouramine P
Fourrine 84
Fourrine P Base
Furro P Base
p-Hydroxyaniline
4-Hydroxyaniline
Nako Brown R
PAP
Paranol
Pelagol Grey P Base
Pelagol P Base
Renal AC
Rodinal
Tertral P Base
Unal
UN 2512 [Aminophenols (o-; m-; p-)]
Ursol P
Ursol P Base
Zoba Brown P Base

123-31-9
$C_6H_6O_2$
110.12
Oc(ccc(O)c1)c1
Hydroquinone, Solid or liquid (ACGIH,OSHA) [UN 2662]
Arctuvin
Benzene, p-dihydroxy-
p-Benzenediol
1,4-Benzenediol
Benzohydroquinone
Benzoquinol
Black and White Bleaching Cream
1,4-Dihydroxy-benzeen (Dutch)
1,4-Dihydroxybenzen (Czech)
Dihydroxybenzene (OSHA)
p-Dihydroxybenzene
1,4-Dihydroxybenzene
1,4-Dihydroxy-benzol (German)
1,4-Diidrobenzene (Italian)
p-Dioxobenzene
Eldopaque
Eldoquin
Hydrochinon (Czech, Polish)
Hydroquinol
Hydroquinole
α-Hydroquinone
p-Hydroquinone
p-Hydroxyphenol
Idrochinone (Italian)
NCI-C55834
Quinol
β-Quinol
Tecquinol
Tenox HQ
Tequinol
UN 2662 [Hydroquinone, solid or liquid]
USAF EK-356

123-32-0
$C_6H_8N_2$
108.16
n(c(cnc1C)C)c1
Pyrazine, 2,5-dimethyl
2,5-Dimethylpyrazine

123-33-1
$C_4H_4N_2O_2$
112.10

Oc(nnc(O)c1)c1
3,6-Pyridazinedione, 1,2-dihydro
Burtolin
Chemform
De-Cut
De-Sprout
1,2-Dihydropyridazine-3,6-dione
1,2-Dihydro-3,6-pyradizinedione
1,2-Dihydro-3,6-pyridazinedione
Drexel-Super P
ENT 18,870
Fair 30
Fair PS
Hydrazid kyseliny maleinove (Czech)
6-Hydroxy-3(2H)-pyridazinone
KMH
MAH
Maintain 3
Malazide
Maleic acid hydrazide
Maleic hydrazide
Maleic hydrazide 30%
Maleic hydrazine
Malein 30
Maleinsaeurehydrazid (German)
N,N-Maleoylhydrazine
Malzid
MH
MH 30
MH-40
MH 36 Bayer
RCRA waste number U148
Regulox
Regulox W
Regulox 50 W
Retard
Royal MH-30
Royal Slo-Gro
Slo-Gro
Sprout/Off
Sprout-Stop
Stuntman
Sucker-Stuff
Super-De-Sprout
Super Sprout Stop
Super Sucker-Stuff
Super Sucker-Stuff HC
1,2,3,6-Tetrahydro-3,6-dioxopyridazine
Vondalhyde
Vondrax

123-34-2
$C_6H_{12}O_3$
132.18
O(CC(O)CO)CC=C
1,2-Propanediol, 3-allyloxy
α-Allyl glycerol ether
1-Allyloxy-2,3-propanediol
3-Allyloxy-1,2-propanediol
Ether, allyl glyceryl
Glycerol α-allyl ether

123-35-3
$C_{10}H_{16}$
136.26
C(C=C)(=C)CCC=C(C)C
1,6-Octadiene, 7-methyl-3-methylene
Myrcene
3-Methylene-7-methyl-1,6-octadiene
7-Methyl-3-methylene-1,6-octadiene

123-36-4
$C_{21}H_{38}O_3$
338.59
CCCCCCCCC1OC1CCCCCCCC(=O)OCC=C
Stearic acid, 9,10-epoxy-, allyl ester
Allyl-9,10-epoxystearate
EP-145
9,10-Epoxystearic acid, allyl ester

123-38-6
C_3H_6O
58.09
O=CCC
Propionaldehyde
Aldehyde propionique (French)
Methylacetaldehyde
NCI-C61029
Propaldehyde
Propanal
Propionaldehyde [UN 1275]
Propionic aldehyde
Propyl aldehyde
Propylic aldehyde
UN 1275 [Propionaldehyde]

123-39-7
C_2H_5NO
59.08
O=CNC
Formamide, N-methyl
EK 7011
Methylformamide
N-Methylformamide
Monomethylformamide
NSC-3051
X 188

123-41-1
$C_5H_{14}NO.HO$
121.21
Choline-hydroxide
Bursine
Fagine
Gossypine
Luridine
Sincaline
Sinkalin
Sinkaline
Vidine

123-42-2
$C_6H_{12}O_2$
116.18
O=C(CC(O)(C)C)C
2-Pentanone, 4-hydroxy-4-methyl
Diacetonalcohol (Dutch)
Diacetonalcool (Italian)
Diacetonalkohol (German)
Diacetone alcohol (ACGIH, OSHA) [UN 1148]
Diacetone-alcool (French)
Diketone alcohol
4-Hydroxy-2-keto-4-methylpentane
4-Hydroxy-4-methyl-pentan-2-on (German, Dutch)
4-Hydroxy-4-methylpentanone-2
4-Hydroxy-4-methyl-2-pentanone (OSHA)
4-Hydroxy-4-methyl pentan-2-one
4-Idrossi-4-metil-pentan-2-one (Italian)
2-Methyl-2-pentanol-4-one
Pyranton
Tyranton
UN 1148 [Diacetone alcohol]

123-43-3
$C_2H_4O_5S$
140.12
O=C(O)CS(=O)(=O)O
Acetic acid, sulfo
Kyselina sulfooctova (Czech)
Sulfoacetic acid
Sulfoethanoic acid

123-44-4
$C_8H_{18}O$
130.26
1-Pentanol, 2,2,4-trimethyl
2,2,4-Trimethylpentanol

123-48-8
$C_{12}H_{24}$
168.32
C(=CC(C)(C)C)(CC(C)(C)C)C
2,2,4,6,6-Pentamethyl-3-heptene
3-Heptene, 2,2,4,6,6-pentamethyl- (9CI)

123-51-3
$C_5H_{12}O$
88.17
OCCC(C)C
1-Butanol, 3-methyl
Alcool amilico (Italian)
Alcool isoamylique (French)
Amylowy alkohol (Polish)
Fermentation amyl alcohol
Isoamyl alcohol (ACGIH) [UN 1105]
Isoamyl alcohol, primary (OSHA)
Isoamyl alkohol (Czech)
Iso-amylalkohol (German)
Isoamylol
Isobutylcarbinol
Isopentanol
Isopentyl alcohol
2-Methyl-4-butanol
3-Methyl butanol
3-Methylbutan-1-ol
3-Methyl-1-butanol
3-Metil-butanolo (Italian)

UN 1105 [Amyl alcohols]

123-54-6
$C_5H_8O_2$
100.13
O=C(CC(=O)C)C
2,4-Pentanedione
Acetoacetone
Acetone, acetyl-
Acetylacetone
Acetyl 2-propanone
Diacetylmethane
Pentanedione
Pentanedione-2,4
Pentan-2,4-dione [UN 2310]
UN 2310 [Pentan-2,4-dione]

123-56-8
$C_4H_5NO_2$
99.10
O=C(NC(=O)C1)C1
Succinimide
Butanimide
3,4-Dihydropyrrole-2,5-dione
3,4-Dihydropyrrolidine
Dihydro-3-pyrroline-2,5-dione
2,5-Diketopyrrolidine
2,5-Dioxopyrrolidine
2,5-Pyrrolidinedione
Succinic imide
Succinimide-Sauba

123-62-6
$C_6H_{10}O_3$
130.16
O=C(OC(=O)CC)CC
Propionic-anhydride
Anhydrid kyseliny propionove (Czech)
Methylacetic anhydride
Propionic anhydride [UN 2496]
Propionic acid anhydride
Propionyl oxide
UN 2496 [Propionic anhydride]

123-63-7
$C_6H_{12}O_3$
132.18
O(C(OC(O1)C)C)C1C
s-Trioxane, 2,4,6-trimethyl
Acetaldehyde, trimer
Elaldehyde
Paraacetaldehyde
Paracetaldehyde
Paral
Paraldehyd (German)
Paraldehyde [UN 1264]
Paraldeide (Italian)
PCHO
RCRA waste number U182
Triacetaldehyde (French)
2,4,6-Trimethyl-1,3,5-trioxaan (Dutch)
2,4,6-Trimethyl-s-trioxane
2,4,6-Trimethyl-1,3,5-trioxane
s-Trimethyltrioxymethylene
2,4,6-Trimetil-1,3,5-triossano (Italian)
UN 1264 [Paraldehyde]

123-66-0
$C_8H_{16}O_2$
144.24
O=C(OCC)CCCCC
Hexanoic acid, ethyl ester
Ethylbutyl acetate [UN 1177]
Ethyl caproate
Ethyl hexanoate
UN 1177 [2-Ethylbutyl acetate]

123-68-2
$C_9H_{16}O_2$
156.25
O=C(OCC=C)CCCCC
Hexanoic acid, allyl ester
Allyl caproate
Allylester kyseliny kapronove (Czech)
Allyl hexanoate
2-Propenyl n-hexanoate

123-72-8
C_4H_8O
72.12
O=CCCC
Butyraldehyde
Aldehyde butyrique (French)
Aldeide butirrica (Italian)
Butal
Butaldehyde
Butalyde
Butanal
n-Butanal (Czech)
Butyl aldehyde
n-Butyl aldehyde
Butyral
Butyraldehyd (German)
Butyraldehyde [UN 1129]
n-Butyraldehyde
Butyric aldehyde
NCI-C56291
UN 1129 [Butyraldehyde]

123-73-9
C_4H_6O
70.10
O=CC=CC
Crotonaldehyde, (E)
2-Butenal, (E)-
trans-2-Butenal
Crotenaldehyde
Crotonal
Crotonaldehyde (ACGIH, OSHA)
trans-Crotonaldehyde
Crotonic aldehyde
1,2-Ethanediol, dipropanoate (9CI)
β-Methyl acrolein
NCI-C56279
Propylene aldehyde
RCRA waste number U053
Topanel

123-75-1
C_4H_9N
71.14
N(CCC1)C1
Pyrrolidine [UN 1922]
Azacyclopentane
Azolidine
Prolamine
Pyrrole, tetrahydro-
Tetrahydropyrrole
Tetramethylenimine
UN 1922 [Pyrrolidine]

123-76-2
$C_5H_8O_3$
116.13
O=C(O)CCC(=O)C
Levulinic-acid
Acetopropionic acid
β-Acetylpropionic acid
γ-Ketovaleric acid
Laevulinic acid
Levulic acid
4-Oxopentanoic acid
4-Ketovaleric acid
Laevulic acid
Propionic acid, 3-acetyl-
4-Oxovaleric acid
USAF CZ-1
Valeric acid, 4-oxo-

123-77-3
$C_2H_4N_4O_2$
116.10
O=C(N=NC(=O)N)N
Formamide, 1,1'-azobis
1,1'-Azobiscarbamide
Azobiscarbonamide
Azobiscarboxamide
1,1'-Azobis(formamide)
Azodicarbamide
Azodicarboamide
Azodicarbonamide
Azodicarboxamide
Azodicarboxylic acid diamide
δ(1,1')-Biurea
Celosen AZ
ChKhZ 21
ChKhZ 21R
Diazenedicarboxamide
Genitron AC
Genitron AC 2
Genitron AC 4
Kempore
Kempore 125
Kempore R 125
Lucel ADA
NCI-C55981
Nitropore
Pinhole AK 2
Porofor 505
Porofor ADC/R
Porofor ChKhZ 21
Porofor ChKhZ 21R
Unifoam AZ
Uniform AZ
Yunihomu AZ

123-79-5
$C_{22}H_{42}O_4$

370.57
O=C(OCCCCCCCC)CCCCC(=O)OCCCCCCCC
Dioctyl adipate
Adimoll DO
Adipic acid, dioctyl ester (8CI)
AI3-17824
Bis(2-ethylhexyl) adipate
Dicaprylyl adipate
Di(2-ethylhexyl) adipate
Di-n-octyl adipate
Dioctyl ester hexanedioic acid
Dioctyl hexanedioate
Hexanedioic acid, dioctyl ester (9CI)
NSC-16201
Octyl adipate (VAN)

123-81-9
$C_6H_{10}O_4S_2$
210.28
O=C(OCCOC(=O)CS)CS
Glycol dimercaptoacetate
Acetic acid, mercapto-, 1,2-ethanediyl ester (9CI)
Acetic acid, mercapto-, ethylene ester (8CI)
AI3-26087
1,2-Ethanediyl mercaptoacetate
Ethylene bis(mercaptoacetate)
Ethylene bis(thioglycolate)
Ethylenebis(thioglycolate)
Ethylene glycol bis(mercaptoacetate)
Ethylene glycol bis(thioglycolate)
Ethylene glycol bis(thioglycolic ester)
Ethylene mercaptoacetate
GDMA
Glycol bis(mercaptoacetate)
NSC-30032

123-82-0
$C_7H_{17}N$
115.25
NC(CCCCC)C
Hexylamine, 1-methyl
dl-2-Aminoheptane
Armeen l-7
Heptamine
2-Heptanamine
Heptedrine
2-Heptylamine
1-Methylhexylamine
Rineptil
Tuamine
Tuaminoheptane

123-86-4
$C_6H_{12}O_2$
116.18
O=C(OCCCC)C
Acetic acid, butyl ester
Acetate de butyle (French)
Acetic acid n-butyl ester
Butile (acetati di) (Italian)
Butylacetat (German)
Butyl acetate [UN 1123]
n-Butyl acetate (ACGIH,OSHA)
1-Butyl acetate
Butylacetaten (Dutch)
Butyle (acetate de) (French)
Butylester kyseliny octove (Czech)
Butyl ethanoate
Octan n-butylu (Polish)
UN 1123 [Butyl acetates]

123-88-6
C_3H_7ClHgO
295.14
COCC[Hg]Cl
Mercury, chloro(2-methoxyethyl)
Agallol
Agallolat
Agalol
Aratan
Aretan
Aretan 6
Atiran
Baytan
Cekusil Universal C
Celmer
Ceresan-Universal Nassbeize
Ceresan Universal Nazbeize
Chloro(2-methoxyethyl)mercury
Curesan
Emisan 6
Falisan
Gramisan
Higosan
MEMC
Merchlorate
Methoxyaethylquecksilberchlorid (German)
(β-Methoxyethyl)mercuric chloride
Methoxyethyl mercuric chloride
2-Methoxyethylmercuric chloride
β-Methoxyethylmercury chloride
Methoxyethylmercury chloride
2-Methoxyethylmercury chloride
2-Methoxyethylmerkurichlorid (Czech)
Sedresan
Tafasan
Tafasan 6W
Tayssato
Triadimenol

123-91-1
$C_4H_8O_2$
88.12
O(CCOC1)C1
p-Dioxane
Diethylene dioxide (OSHA)
1,4-Diethylene dioxide
Diethylene ether
Di(ethylene oxide)
Diokan
Dioksan (Polish)
Diossano-1,4 (Italian)
Dioxaan-1,4 (Dutch)
1,4-Dioxacyclohexane
Dioxan
Dioxan-1,4 (German)
p-Dioxan (Czech)
Dioxane (ACGIH,OSHA) [UN 1165]
Dioxane-1,4
1,4-Dioxane
Dioxanne (French)
p-Dioxin, tetrahydro-
Dioxyethylene ether
Glycol ethylene ether
NCI-C03689
RCRA waste number U108
Tetrahydro-p-dioxin
Tetrahydro-1,4-dioxin
UN 1165 [Dioxane]

123-92-2
$C_7H_{14}O_2$
130.21
O=C(OCCC(C)C)C
Isopentyl alcohol, acetate
Acetic acid, isopentyl ester
Banana Oil
Isoamyl acetate (ACGIH, OSHA)
Isoamylester kyseliny octove (Czech)
Isoamyl ethanoate
Isopentyl acetate
3-Methylbutyl acetate
3-Methyl-1-butyl acetate
3-Methylbutyl ethanoate
Pear Oil

123-93-3
$C_4H_6O_4S$
150.16
O=C(O)CSCC(=O)O
Acetic acid, thiodi
Acetic acid, 2,2'-thiobis- (9CI)
(Carboxymethylthio)acetic acid
Dimethylsulfide-α-α'-dicarboxylic acid
Mercaptodiacetic acid
Thiodiglycolic acid
β,β'-Thiodiglycolic acid
2,2'-Thiodiglycolic acid
Thiodiglycollic acid
USAF CB-36
USAF E-2

123-94-4
$C_{21}H_{42}O_4$
358.56
O=C(OCC(O)CO)CCCCCCCCCCCCCCCCC
Glyceryl stearate
Aldo MSD
Aldo MSLG
Aldo 33
Aldo 75
Arlacel 165
2,3-Dihydroxypropyl octadecanoate
Emerest 2407
Glycerin 1-monostearate
Glycerin 1-stearate
Glycerol α-monostearate

Glycerol 1-monostearate
Glycerol 1-stearate
1-Glyceryl stearate
Glyceryl monostearate
Glyceryl 1-monostearate
Hefti GMS-33
Hefti GMS-66
Hefti GMS-99
Monostearin
Monostearin (L)
α-Monostearin
1-Monostearin
1-Monostearoylglycerol
NSC-3875
Octadecanoic acid, 2,3-dihydroxypropyl ester (9CI)
Octadecanoic acid, Monoester with 1,2,3-propanetriol
Sandin EU
Stearic acid α-monoglyceride
Stearic acid 1-monoglyceride
Stearin, 1-mono- (8CI)
3-Stearoyloxy-1,2-propanediol
Tegin
Tegin 55G
Tegin 515
Unitina MD
Unitina MD-A
Unitolate GS
Unitolate 165-C
Witconol MS
Witconol MST

123-95-5
$C_{22}H_{44}O_2$
340.66
O=C(OCCCC)CCCCCCCCCCCCCCCCC
Stearic acid, butyl ester
Apex 4
BS
Butyl octadecanoate
n-Butyl octadecanoate
Butyl stearate
n-Butyl stearate
Emerest 2325
Groco 5810
Kessco BSC
Kesscoflex BS
Polycizer 332
Octadecanoic acid, butyl ester (9CI)
RC Plasticizer B-17
Starfol BS-100
Tegester butyl stearate
Uniflex BYS
Wickenol 122
Witcizer 200
Witcizer 201

123-96-6
$C_8H_{18}O$
130.26
OC(CCCCCC)C
2-Octanol
Capryl alcohol

123-99-9
$C_9H_{16}O_4$
188.25
O=C(O)CCCCCCCC(=O)O
Azelaic-acid
Anchoic acid
Azelaic acid, Technical grade
Emerox 1110
Emerox 1144
Heptanedicarboxylic acid
1,7-Heptanedicarboxylic acid
Lepargylic acid
Nonanedioic acid

124-02-7
$C_6H_{11}N$
97.18
N(CC=C)CC=C
2-Propen-1-amine, N-2-propenyl
Diallylamine [UN 2359]
Di-2-propenylamine
UN 2359 [Diallylamine]

124-03-8
$C_{20}H_{44}N.Br$
378.56
[OH].[Br-].CCCCCCCCCCCCCCCC[N+](C)(C)CC
Ammonium, ethylhexadecyldimethyl-, bromide
Ammonyx DME
Bretol
CDA
Cetylcide
Cetyldimethylethylammonium bromide
Cetylethyldimethylammonium bromide
Dimethylethylhexadecylammonium bromide
Ethyl Cetab
N-Ethyl-N,N-dimethyl-1-hexadecanaminium bromide
Ethylhexadecyldimethylammonium bromide
Radiol Germicidal Solution

124-04-9
$C_6H_{10}O_4$
146.16
O=C(O)CCCCC(=O)O
Adipic-acid
Acifloctin
Acinetten
Adilactetten
Adipinic acid
1,4-Butanedicarboxylic acid
Hexanedioic acid
1,6-Hexanedioic acid
Kyselina adipova (Czech)
Molten adipic acid

124-06-1
$C_{16}H_{32}O_2$
256.43
O=C(OCC)CCCCCCCCCCCCC
Ethyl myristate
AI3-01024
Ethyl tetradecanoate
Myristic acid, ethyl ester (8CI)
NSC-8917
Tetradecanoic acid, ethyl ester (9CI)

124-07-2
$C_8H_{16}O_2$
144.24
O=C(O)CCCCCCC
Octanoic-acid
Hexacid 898
C-8 Acid
Caprylic acid
n-Caprylic acid
neo-Fat 8
1-Heptanecarboxylic acid
Kyselina kaprylova (Czech)
Octic acid
n-Octoic acid
n-Octylic acid

124-09-4
$C_6H_{16}N_2$
116.24
NCCCCCCN
1,6-Hexanediamine
1,6-Diaminohexane
Hexamethylenediamine
1,6-Hexamethylenediamine
Hexamethylenediamine, Solid [UN 2280]
Hexamethylenediamine, Solution [UN 1783]
HMDA
NCI-C61405
UN 1783 [Hexamethylenediamine solution]
UN 2280 [Hexamethylenediamine, solid]

124-10-7
$C_{15}H_{30}O_2$
242.40
O=C(OC)CCCCCCCCCCCCC
Methyl myristate
AI3-01980
Metholeneat 2495
Methyl tetradecanoate
Methyl n-tetradecanoate
Myristic acid, methyl ester (8CI)
NSC-5029
Tetradecanoic acid, methyl ester (9CI)
Uniphat A50

124-11-8
C_9H_{18}
126.24
C(=C)CCCCCCC
1-Nonene (9CI)
α-Nonene
n-Non-1-ene
Nonylene
NSC-73961
Propylene trimer

124-12-9
$C_8H_{15}N$
125.24
N#CCCCCCCC
Octanenitrile
Arneel 8
Caprylnitrile
Caprylonitrile
Octanonitrile

124-13-0
$C_8H_{16}O$
128.24
O=CCCCCCCC
1-Octanal
Aldehyde C-8
C-8 aldehyde
Caprylic aldehyde
Octanal
Octanaldehyde
n-Octyl aldehyde

124-16-3
$C_9H_{20}O_3$
176.29
O(CCOCC(O)C)CCCC
2-Propanol, 1-(2-butoxy-ethoxy)
1-Butoxyethoxy-2-propanol
1-(2-Butoxyethoxy)-2-propanol

124-17-4
$C_{10}H_{20}O_4$
204.30
O=C(OCCOCCOCCCC)C
Ethanol, 2-(2-butoxyethoxy)-, acetate
Acetic acid 2-(2-butoxyethoxy)-ethyl ester
2-(2-Butoxyethoxy)ethanol acetate
2-(2-Butoxyethoxy)ethyl acetate
2-(2-Butoxyethoxy)ethylester kyseliny octove (Czech)
Butyl carbitol acetate
Butylkarbitolacetat (Czech)
Diethylene glycol butyl ether acetate
Diethylene glycol, monobutyl ether, acetate
Diglycol monobutyl ether acetate
Ektasolve DB Acetate
Glycol Ether DB Acetate

124-18-5
$C_{10}H_{22}$
142.32
C(CCCCCCCC)C
Decane
n-Decane [UN 2247]
UN 2247 [n-Decane]

124-19-6
$C_9H_{18}O$
142.27
O=CCCCCCCCC
1-Nonanal
Aldehyde C-9
C-9 Aldehyde
NCI-C61018
1-Nonaldehyde
Nonanal
1-Nonyl aldehyde
Pelargonic aldehyde

124-20-9
$C_7H_{19}N_3$
145.29
N(CCCCN)CCCN
1,4-Butanediamine, N-(3-aminopropyl)
1,4-Diaminobutane, N-(3-aminopropyl)-
Spermidine
1,5,10-Triazadecane

124-22-1
$C_{12}H_{27}N$
185.40
NCCCCCCCCCCCC
Dodecylamine
n-Dodecylamine
Laurylamine

124-25-4
$C_{14}H_{28}O$
212.42
O=CCCCCCCCCCCCCC
1-Tetradecanal
Aldehyde C-14, Myristic
C-14 Aldehyde, Myristic
Myristaldehyde
Myristic aldehyde
Tetradecanal
1-Tetradecyl aldehyde

124-26-5
$C_{18}H_{37}NO$
283.56
O=C(N)CCCCCCCCCCCCCCCCC
Octadecanamide
Kemamide S
Stearamide

124-28-7
$C_{20}H_{43}N$
297.64
N(CCCCCCCCCCCCCCCCCC)(C)C
Octadecylamine, N,N-dimethyl
N,N-Dimethyloctadecylamine
N,N-Dimethyloktadecylamin (Czech)

124-30-1
$C_{18}H_{39}N$
269.58
NCCCCCCCCCCCCCCCCCC
Octadecylamine
Adogenen 142
Alamine 7
Armeen 118D
n-Octadecylamine
Oktadecylamin (Czech)
Stearylamine

124-38-9
CO_2
44.01
O=C=O
Carbon-dioxide
Anhydride carbonique (French)
Carbon dioxide (ACGIH, OSHA) [UN 1013]
Carbon dioxide, Refrigerated liquid [UN 2187]
Carbon dioxide, Solid [UN 1845]
Carbonic acid gas
Carbonic anhydride
Carbonice (DOT)
Dry Ice [UN 1845]
Kohlendioxyd (German)
Kohlensaure (German)
UN 1013 [Carbon dioxide]
UN 1845 [Carbon dioxide, solid; (Dry ice)]
UN 2187 [Carbon dioxide, Refrigerated liquid (cryogenic liquid)]

124-40-3
C_2H_7N
45.10
N(C)C
Dimethylamine (ACGIH, OSHA)
Dimethylamine, Anhydrous [UN 1032]
Dimethylamine, Solution [UN 1160]
DMA
Methanamine, N-methyl- (9CI)
N-Methylmethanamine
RCRA waste number U092
UN 1032 [Dimethylamine, anhydrous]
UN 1160 [Dimethylamine solution]

124-41-4
$CH_3O.Na$
54.03
Methanol, sodium salt
Sodium methoxide
Sodium methylate [UN 1431]
Sodium methylate, Dry (DOT)
UN 1431 [Sodium methylate]

124-43-6
$CH_4N_2O.H_2O_2$
94.06
Carbamide peroxide
Carbamide peroxide, Solution
Gly-oxide
Glyoxide
Hydrogen peroxide, Compd. with urea (1:1)
Hydrogen peroxide carbamide

Hydroperit
Hydroperite
Hyperol
Murine Ear Drops
NA 1511 [Urea hydrogen peroxide]
NSC-24852
Ortizon
Percarbamid
Percarbamide
Perhydrit
Perhydrol-Urea
Proxigel
Thenardol
UN 1511 [Urea hydrogen peroxide]
Urea-agua oxigenada (Spanish)
Urea, Compd. with hydrogen peroxide (H_2O_2) (1:1) (9CI)
Urea, Compd. with hydrogen peroxide (1:1) (8CI)
Urea dioxide
Urea hydrogen peroxide
Urea hydrogen peroxide salt
Urea hydroperoxide
Urea peroxide
Uree-eau oxygenee (French)

124-47-0
$CH_5N_3O_4$
123.09
Urea nitrate (Wet)
Acidogen nitrate
UN 0220 [Urea nitrate, Dry or wetted with less than 20 per cent water, by mass]
UN 1357 [Urea nitrate, Wetted with not less than 20 per cent water, by mass]
Urea, mononitrate (8CI,9CI)
Urea nitrate
Urea nitrate, Dry or containing less than 20% water [UN 0220]
Urea nitrate, Wet with 10% or more water
Urea nitrate, Wet with 10% or more water, over 25 lbs. in one outside packaging
Urea nitrate, Wetted with not less than 20% water [UN 1357]

124-48-1
$CHBr_2Cl$
208.29
BrC(Br)Cl
Methane, chlorodibromo
CDBM
Chlorodibromomethane
Dibromochloromethane
NCI-C55254

124-58-3
CH_5AsO_3
139.98
C[As](O)(O)=O
Methanearsonic-acid
Kyselina methylarsonova (Czech)
MAA
Methylarsenic acid
Methylarsinic acid
Methylarsonic acid
Monomethylarsinic acid

124-63-0
CH_3ClO_2S
114.55
O=S(=O)(C)Cl
Methanesulfonyl chloride (9CI)
AI3-52234
Chloro methyl sulfone
Mesyl chloride
Methanesulfonic acid chloride
Methanesulfuryl chloride
Methanesulphonyl chloride
Methyl sulfochloride
Methyl sulfonyl chloride
Methylsulfonyl chloride
NSC-15039

124-64-1
$C_4H_{12}O_4P.Cl$
190.58
[Cl-].OC[P+](CO)(CO)CO
Phosphonium, tetrakis-(hydroxymethyl)-, chloride
NCI-C55061
Tetrakis-(hydroxymethyl)fosfoniumchlorid (Czech)
Tetrakis(hydroxymethyl)-phosphonium chloride
THPC

124-65-2
$C_2H_6AsO_2.Na$
159.99
[Na+].C[As](C)(O)=O
Arsine oxide, dimethyl-hydroxy-, sodium salt
Alkarsodyl
Ansar 160
Ansar 560
Arsecodile
Arsicodile
Arsinic acid, dimethyl-, sodium salt (9CI)
Arsycodile
Bolls-Eye
Cacodylate de sodium (French)
Cacodylic acid sodium salt
Chemaid
Dimethylarsinat sodny (Czech)
((Dimethylarsino)oxy)sodium-as-oxide
Dutch-Treat
Hydroxydimethylarsine oxide, sodium salt
Kakodylan dodny (Czech)
Phytar 560
Rad-E-Cate
Rad-E-Cate 16
Rad-E-Cate 25
Rad-E-Cate 35
Silvisar
Sodium cacodylate [UN 1688]
Sodium dimethylarsinate
Sodium dimethylarsonate
Sodium salt of cacodylic acid
UN 1688 [Sodium cacodylate]

124-68-5
$C_4H_{11}NO$
89.16
OCC(N)(C)C
1-Propanol, 2-amino-2-methyl
2-Aminodimethylethanol
β-Aminoisobutanol
2-Amino-2-methylpropanol
2-Amino-2-methylpropan-1-ol
2-Amino-2-methyl-1-propanol
AMP
AMP-95
Isobutanolamine
Isobutanol-2-amine

124-70-9
$C_3H_6Cl_2Si$
141.08
Silane, dichloromethylvinyl
Dichloromethylvinylsilane

124-73-2
$C_2Br_2F_4$
259.84
FC(F)(Br)C(F)(F)Br
Ethane, 1,2-dibromotetra-fluoro
1,2-Dibromoperfluoroethane
sym-Dibromotetrafluoroethane
1,2-Dibromotetrafluoroethane
Ethane, 1,2-dibromo-1,1,2,2-tetrafluoro- (9CI)
F-114B2
FC 114B2
Fluobrene
Freon 114B2
Halon 2402
Khladon 114B2
R 114B2

124-76-5
$C_{10}H_{18}O$
154.28
OC(C(C(C1C2)(C)C)(C2)C)C1
Isoborneol, DL
Bicyclo(2.2.1)heptan-2-ol, 1,7,7-trimethyl-, exo-
Isoborneol
DL-Isoborneol
Isobornyl alcohol
Isocamphol

124-83-4
$C_{10}H_{16}O_4$
200.23
O=C(O)C(C(C(C(=O)O)C1)(C)C)(C1)C
1,3-Cyclopentanedicarboxylic acid, 1,2,2-trimethyl-, (1R-cis)-
AI3-18160
1,2,2-Trimethyl-1,3-cyclopentanedicarboxylic acid, (1R-cis)-

124-87-8
$C_{13}H_{18}O_7.C_{15}H_{16}O_6$
578.62
CC(=C)C1C3OC(=O)C1C4(O)CC2OC25C(=O)OC3C45C
Picrotoxin
Cocculin
Cocculus [UN 1584]
Cocculus, Solid (DOT)
Cocculus, Solid (Fishberry) (DOT)
Coques du levant (French)
Fish Berry
Indian Berry
Oriental Berry
Picrotin, Compd. with picrotoxinin (1:1)
Picrotoxine
Picrotoxinin, Compd. with picrotin (1:1)
UN 1584 [Cocculus]

124-94-7
$C_{21}H_{27}FO_6$
394.48
CC34CC(O)C1(F)C(CCC2=CC(=O)C=CC12C)C3CC(O)C4(O)C(=O)CO
Pregna-1,4-diene-3,20-dione, 9-fluoro-11-β,16-α,17,21-tetrahydroxy
Aristocort
9-α-Fluoro-16-α-hydroxyprednisolone
9-α-Fluoro-11-β,16-α,17,21-tetrahydroxypregna-1,4-diene-3,20-dione
9-α-Fluoro-11-β,16-α,17,21-tetrahydroxy-1,4-pregnadiene-3,20-dione
9-α-Fluoro-11-β,16-α,17-α,21-tetrahydroxypregna-1,4-diene-3,20-dione
9-Fluoro-11-β,16-α,17,21-tetrahydroxypregna-1,4-diene-3,20-dione
Fluoxyprednisolone
Kenacort
Pregna-1,4-diene-3,20-dione, 9-fluoro-11,16,17,21-tetrahydroxy-, (11-β,16-α)-
SK-Triamcinolone
Rodinolone
11-β,16-α,17-α,21-Tetrahydroxy-9-α-fluoro-1,4-pregnadiene-3,20-dione
Triamcinolone

124-96-9
$C_{12}H_8Cl_6$
364.91
1,4:5,8-Dimethanonaphthalene, 1,2,3,4,10,10-hexachloro-1,4,4a,5,8,8a-hexahydro-(8CI,9CI)

125-12-2
$C_{12}H_{20}O_2$
196.29
O=C(OC(C(C(C1C2)(C)C)(C2)C)C1)C
Pichtosin
Acetic acid, isobornyl ester
AI3-02940
Bicyclo(2.2.1)heptan-2-ol, 1,7,7-trimethyl-, acetate, exo- (9CI)
Isoborneol, acetate (8CI)
Isobornyl acetate
NSC-62486
Pichtosine

125-23-5
$C_{18}H_{22}NO_3.Br$
380.28
DEA No. 9305
7,8-Didehydro-4,5-α-epoxy-3,6-α-dihydroxy-17,17-dimethylmorphinanium bromide
Morphinanium, 7,8-didehydro-3,6-α-dihydroxy-17,17-dimethyl-4,5-α-epoxy-, bromide
Morphine bromomethylate
Morphine methylbromide
Morphosan

125-27-9
$C_{19}H_{24}NO_3.Br$
394.30
Codeine methylbromide
Codeine bromomethylate
DEA No. 9070
Eucodin
Morphinanium, 7,8-didehydro-17,17-dimethyl-4,5-α-epoxy-6-α-hydroxy-3-methoxy-, bromide

125-28-0
$C_{18}H_{23}NO_3$
301.42
COc1ccc2CC5C3CCC(O)C4Oc1c2C34CCN5C
Morphinan-6-α-ol, 4,5-α-epoxy-3-methoxy-17-methyl
Codeine, dihydro
Codhydrine
Cohydrin
DF 118
Dehacodin
Didrate
Dihydrin
Dihydrocodeine
7,8-Dihydrocodeine
Dihydrokodein (Czech)
Dihydroneopine
Drocode
Hydrocodin
6-Hydroxy-3-methoxy-N-methyl-4,5-epoxymorphinan
Nadeine
Novicodin
Paracodin
Paracodine
Parzone
Rapacodin

125-29-1
$C_{18}H_{21}NO_3$
299.40
COc1ccc2CC5C3CCC(=O)C4Oc1c2C34CCN5C
Morphinan-6-one, 4,5-α-epoxy-3-methoxy-17-methyl
Bekadid
Dico
Dicodid
Dihydrocodeinone
4,5-α-Epoxy-3-methoxy-17-methylmorphinan-6-one
Hydrocodone
Multacodin
6-Oxo-3-methoxy-N-methyl-4,5-epoxymorphinan

125-30-4
$C_{19}H_{23}NO_3.ClH$
349.89
Morphinan-6-α-ol, 7,8-didehydro-4,5-α-epoxy-3-ethoxy-17-methyl-, hydrochloride
Codethyline hydrochloride
Dionine hydrochloride
Dionin hydrochloride
Ethylmorphine hydrochloride
o-Ethylmorphine hydrochloride

125-33-7
$C_{12}H_{14}N_2O_2$
218.28
CCC1(C(=O)NCNC1=O)c2ccccc2
4,6(1H,5H)-Pyrimidinedione, 5-ethyldihydro-5-phenyl
5-Aethyl-5-phenyl-hexahydropyrimidin-4,6-dion (German)
Cyral
2-Deoxyphenobarbital
2-Desoxyphenobarbital
Desoxyphenobarbitone
5-Ethyldihydro-5-phenyl-4,6(1H,5H)-pyrimidinedione
5-Ethylhexahydro-4,6-dioxo-5-phenylphrimidine
5-Ethylhexahydro-5-phenylpyrimidine-4,6-dione
5-Ethyl-5-phenylhexahydropyrimidine-4,6-dione
Hexadiona
Hexamidine
Hexamidine (The antispasmodic)
Lepimidin
Lepsiral
Majsolin
Midone
Milepsin
Misodine
Misolyne
Mizodin
Mizolin
Mylepsin
Mylepsinum
Mysedon
Mysoline

NCI-C56360
5-Phenyl-5-aethylhexahydropyr-imidindion-(4,6) (German)
5-Phenyl-5-ethyl-hexahydropyr-imidine-4,6-dione
Prilepsin
Primacione
Primaclone
Primacone
Primakton
Primidon
Primidone
Prysoline
Pyrimidone Medi-Pets
ROE 101
Sertan

125-40-6
$C_{10}H_{16}N_2O_3$
212.28
CCC(C)C1(CC)C(=O)NC(=O)NC1=O
Barbituric acid, 5-sec-butyl-5-ethyl
Butabarb
Butabarbital
Butabarbitone
Butatab
Butatal
Buticaps
Butisol
Butrate
5-sec-Butyl-5-ethylbarbituric acid
5-sec-Butyl-5-ethylmalonyl urea
5-Ethyl-5-(1-methylpropyl)barbiturate
5-Ethyl-5-(1-methylpropyl)barbituric acid
Medarsed
Nilox
2,4,6(1H,3H,5H)-Pyrimidinetrione, 5-ethyl-5-(1-methylpropyl)- (9CI)
Secbubarbital
Secbutabarbital
Secbutobarbitone
Unicelles

125-42-8
$C_{11}H_{16}N_2O_3$
224.25
Vinbarbital
Barbituric acid, 5-ethyl-5-(1-methyl-1-butenyl)- (8CI)
Butenemal
Butenemalum
Delvinal
Devinal
5-Ethyl-5-(1-methyl-1-butenyl)-barbiturate
5-Ethyl-5-(1-methyl-1-butenyl)-barbituric acid
NSC-117442
2,4,6(1H,3H,5H)-Pyrimidinetrione, 5-ethyl-5-(1-methyl-1-butenyl)- (9CI)
Suppoptanox
Vinbarbital (VAN)
Vinbarbitalum (Latin)
Vinbarbitone

125-46-2
$C_{18}H_{16}O_7$
344.34
O=C(c(c(OC(C1(C(=O)C(C2=O)C(=O)C)C)=C2)c1c(O)c3C)c3O)C
1,3(2H,9bH)-Dibenzofurandione, 2,6-diacetyl-7,9-dihydroxy-8,9b-dimethyl
2,6-Diacetyl-7,9-dihydroxy-8,9b-dimethyl-1,3(2H,9bh)-dibenzofurandione
Usnein
Usniacin
Usnic acid
Usninic acid
Usninsaeure (German)

125-64-4
$C_{10}H_{17}NO_2$
183.28
CCC1(CC)C(=O)NCC(C)C1=O
2,4-Piperidinedione, 3,3-diethyl-5-methyl
3,3-Diethyl-2,4-dioxo-5-methyl-piperidine
3,3-Diethyl-5-methyl-2,4-piperidinedione
3,3-Diethyl-5-methylpiperidine-2,4-dione
Dimerin
2,4-Dioxy-3,3-diethyl-5-methyl-piperidine
Methyprolon
Metiprilone
Methyprylon
Noctan
Noludar
Ro 1-6463

125-67-7
$C_{19}H_{22}O_6.K$
385.48
Potassium gibberellate
Caswell No. 692A
EPA Pesticide Chemical Code 043802
Gibberellic acid, monopotassium salt
Gibberellic acid potassium salt
Gibberellin A3 potassium salt
Gibrel
Gibrofit
K-GA

125-70-2
$C_{18}H_{25}NO$
271.40
Levomethorphan
DEA No. 9210
Levomethorphane (French)
Levomethorphanum (Latin)
Levometorfano (Spanish)
Methorphan
l-Methorphan
Morphinan, 3-methoxy-17-methyl-, l-

125-71-3
$C_{18}H_{25}NO$
271.44
9-α,13-α,14-α-Morphinan, 3-methoxy-17-methyl
BA 2666
Dextromethorfan (Czech)
Dextromethorphan
d-Methorphan
δ-Methorphan
Morphinan, 3-methoxy-17-methyl-, (9-α,13-α,14-α)-(9CI)
Romilar

125-93-9
$C_{63}H_{98}O_6$
951.47
Abietic acid, dihydro-, triester with glycerol
1-Phenanthrenecarboxylic acid, 1,2,3,4,4a,4b,5,6,7,9,10,10a-dodecahydro-1,4a-dimethyl-7-(1-methylethyl)-, 1,2,3-propanetriyl ester, (1R-(1α,4aβ,4bα,10aα))- (9CI)

126-00-1
$C_{17}H_{18}O_4$
286.33
O=C(O)CCC(c(ccc(O)c1)c1)(c(ccc(O)c2)c2)C
Diphenolic acid
Benzenebutanoic acid, 4-hydroxy-γ-(4-hydroxyphenyl)-γ-methyl- (9CI)
4,4-Bis(p-hydroxyphenyl)pentanoic acid
4,4-Bis(4-hydroxyphenyl)pentanoic acid
γ,γ-Bis(p-hydroxyphenyl)valeric acid
4,4-Bis(p-hydroxyphenyl-)valeric acid
4,4-Bis(4-hydroxyphenyl)-valeric acid
DPA (VAN)
4-Hydroxy-γ-(4-hydroxyphenyl)-γ-methylbenzenebutanoic acid
NSC-3371
Valeric acid, 4,4-bis(p-hydroxyphenyl)- (8CI)

126-06-7
$C_5H_6BrClN_2O_2$
241.49
O=C(N(C(C1=O)(C)C)Cl)N1Br
Hydantoin, 3-bromo-1-chloro-5,5-dimethyl
3-Bromo-1-chloro-5,5-dimethylhydantoin
2,4-Imidazolidinedione, 3-bromo-1-chloro-5,5-dimethyl-

126-07-8
$C_{17}H_{17}ClO_6$
352.79
COC1=CC(=O)CC(C)C13Oc2c(Cl)c(OC)cc(OC)c2C3=O
Spiro(benzofuran-2(3H),1'-(2)cyclohexene)-3,4'-dione, 7-chloro-2'4,6-trimethoxy-6'-β- methyl
Amudane
Biogrisin-FP
7-Chloro-4,6,2'-trimethoxy-6'-methylgris-2'-en-3,4'-dione
Curling Factor
Delmofulvina
Fulcin
Fulcine
Fulvican Grisactin
Fulvicin
Fulvicin-P/G
Fulvicin-U/F
Fulvina
Fulvistatin
Fungivin
Greosin
Gresfeed
Gricin
Grifulvin
Grifulvin V
Grisactin
Griscofulvin
Grisefuline
Griseo
Griseofulvin
(+)-Griseofulvin
Griseofulvin-Forte
Griseofulvinum
Grisetin
Grisofulvin
Grisovin
Gris-Peg
Grysio
Guservin
Lamoryl
Likuden
Murfulvin
Neo-Fulcin
NSC-34533
Poncyl
Spirofulvin
Sporostatin
USAF SC-2

126-11-4
$C_4H_9NO_5$
151.14
O=N(=O)C(CO)(CO)CO
1,3-Propanediol, 2-(hydroxymethyl)-2-nitro
Cimcool Wafers
2-Hydroxymethyl-2-nitropropane-1,3-diol
2-(Hydroxymethyl)-2-nitro-1,3-propanediol
Isobutylglycerol, nitro-
Methane, trimethylolnitro-
2-Nitro-2-(hydroxymethyl)-1,3-propanediol
Trihydroxymethylnitromethane
Trimethylolnitromethane
Tris(hydroxymethyl)nitromethane
Tris Nitro

126-13-6
$C_{40}H_{62}O_{19}$
847.02
Sucrose, diacetate hexaisobutyrate
Saccharose acetate isobutyrate
SAIB
Sucrose acetate isobutyrate
Sucrose acetoisobutyrate

126-14-7
$C_{28}H_{38}O_{19}$
678.66
O=C(OCC(OC(OC(OC(C(OC(=O)C)C1OC(=O)C)COC(=O)C)C1OC(=O)C)(C2OC(=O)C)COC(=O)C)C2OC(=O)C)C
Sucrose, octaacetate
α-D-Glucopyranoside, 1,3,4,6-tetra-O-acetyl-β-D-fructofuranosyl-, tetraacetate (9CI)
Octaacetylsucrose

126-15-8
$C_{13}H_{16}O_2$
204.29
O=CC(OC(C1CC=C2)C2)(C1CC=C3)C3
4a(4H)-Dibenzofurancarboxaldehyde, 1,5a,6,9,9a,9b-hexahydro
AC-R-11
Bisbutenylenetetrahydrofurfural
2,3,4,5-Bis(δ²-butenylene)-tetrahydrofurfural
2,3,4,5-Bis(2-butenylene)tetrahydrofurfural
2,3,4,5-Bis(2-butylene)tetrahydro-2-furaldehyde
Bis-δ²-butylenetetrahydrofurfural
2,3,4,5-Bis(δ²-butylene)tetrahydrofurfural
2,3:4,5-Bis(2-butylene)tetrahydro-2-furfural
Butadien-Furfural Copolymer
2,3:4,5-Di(2-butenyl)tetrahydrofurfural
ENT 17,596
4a-Formyl-1,4,4a,5a,6,9,9a,9b-octahydrodibenzofuran
2-Furaldehyde, 2,3:4,5-bis-(2-butenylene)tetrahydro-
1,5a,6,9,9a,9b-Hexahydro-4a(4H)-dibenzofurancarboxaldehyde
MGK 11
MGK Repellent 11
MGK Repellent II
Phillips R-11
Phillips Repellent 11
R-11

126-17-0
$C_{27}H_{43}NO_2$
413.71
Solasod-5-en-3-β-ol
Purapuridine
Solancarpidine
Solanidine-S
Solasodine

126-22-7
$C_8H_{14}Cl_3O_5P$
327.54
CCCC(=O)OC(C(Cl)(Cl)Cl)P(=O)(OC)OC
Butyric acid, ester with dimethyl (2,2,2-trichloro-1-hydroxyethyl)phosphonate
Butanoic acid, 2,2,2-trichloro-1-(dimethoxyphosphinyl)ethyl este R
Butilchlorofos
Butonat (German)
Butonate
Butyryltrichlorfon
Dimethoxy-2,2,2-trichloro-l-n-butyryloxy-ethylphosphine oxide
O,O-Dimethyl-(1-n-butyryloxy-2,2,2-trichloraethyl)-phosphonsaeureester
O,O-Dimethyl-(1-butyryloxy-2,2,2-trichloroethyl) phos phonate
O,O-Dimethyl 2,2,2-trichloro-1-(n-butyryloxy)ethylphosphonate
ENT 20,852
F-139
Phosphonic acid, (2,2,2-trichloro-1-hydroxyethyl)-, dimethyl ester, butyrate
T-113
Tribufon

126-27-2
$C_{28}H_{41}N_3O_3$
467.72
CN(C(=O)CN(CCO)CC(=O)N(C)C(C)(C)Cc1ccccc1)C(C)(C)Cc2ccccc2
Acetamide, 2,2'-((2-hydroxyethyl)imino)bis(N-(α,α-dimethylphenethyl)-N-methyl
Acetamide, 2,2'-(2-hydroxyethyl)imino)bis(N-(1,1-dimethylphenylethyl)-N-methyl-(9CI)
Betalgil
N,N-Bis(N-methyl-N-phenyl-tert-butylacetamido)-β-hydroxyethylamine
2-Di(N-methyl-N-phenyl-tert-butyl-carbamoylmethyl)amino-ethanol
Emoren
FH 099
H4 099
2,2'-((2-Hydroxyethyl)imino)-bis(N-(α,α-dimethylphenethyl)-N-methylacetamide)
2,2'-((2-Hydroxyethyl)imino)-

bis(N-(1,1-dimethyl-2-phenyl-ethyl)-N-methylacetamide)
Mucaine
Mucoxin
Mutesa
Muthesa
Oxaine
Oxetacaine
Oxethacaina (Italian)
Oxethacaine
Oxethazaine
Oxethazine
Stomacain
Tepilta
Topicain
WY 806

126-30-7
$C_5H_{12}O_2$
104.17
OCC(CO)(C)C
1,3-Propanediol, 2,2-dimethyl
Dimethylolpropane
2,2-Dimethyl-1,3-propanediol
Dimethyltrimethylene glycol
Neol
Neopentyl glycol
Neopentylene glycol
NPG

126-31-8
$CH_3IO_3S.Na$
244.99
Methanesulfonic acid, iodo-, sodium salt
Abroden
Abrodil
Conturex
Diagnorenol
Iodomethanesulfonic acid sodium salt
Kontrast-U
Methiodal sodium
Methoidal sodium
Monoiodomethanesulfonic acid, sodium salt
Myelotrast
NCI-C03849
Radiographol
Sergosin
Skiodan
Sodium iodomethanesulfonate
Sodium methiodal
Sodium monoiodomethane-sulfonate

126-33-0
$C_4H_8O_2S$
120.18
O=S(=O)(CCC1)C1
Thiophene, tetrahydro-, 1,1-dioxide
Bondelane A
Bondolane A
Cyclic tetramethylene sulfone
Cyclotetramethylene sulfone
Dihydrobutadiene sulphone
1,1-Dioxidetetrahydrothiofuran
1,1-Dioxidetetrahydrothiophene
1,1-Dioxothiolan
Dioxothiolan
Sulfalone
Sulfolan
Sulfolane
Sulpholane
Sulphoxaline
Tetrahydrothiofen-1,1-dioxid (Czech)
Tetrahydrothiophene dioxide
Tetrahydrothiophene 1,1-dioxide
2,3,4,5-Tetrahydrothiophene-1,1-dioxide
Tetramethylene sulfone
Thiacyclopentane dioxide
Thiocyclopentane-1,1-dioxide
Thiolane-1,1-dioxide
Thiophane dioxide
Thiophan sulfone

126-39-6
$C_6H_{12}O_2$
116.16
O(CCO1)C1(CC)C
2-Methyl-2-ethyl-1,3-dioxolane
AI3-25310
2-Butanone, cyclic 1,2-ethane-diyl acetal
2-Butanone, cyclic ethylene acetal
1,3-Dioxolane, 2-ethyl-2-methyl- (8CI,9CI)
Ethyleneacetic acid
2-Ethyl-2-methyldioxolane
2-Ethyl-2-methyl-1,3-dioxolane
2-Methyl-2-ethyldioxolane
NSC-829

126-44-3
$C_6H_5O_7$
189.10
O=C(O)C(O)(CC(=O)O)CC(=O)O
1,2,3-Propanetricarboxylic acid, 2-hydroxy-, ion(3-) (9CI)
Citric acid, ion(3-) (8CI)

126-52-3
$C_9H_{13}NO_2$
167.23
O=C(OC(C#C)(CCCC1)C1)N
Cyclohexanol, 1-ethynyl-, carbamate
Aethinyl-cyclohexyl-carbamat (German)
Carbamate de l'ethinylcyclo-hexanol (French)
Carbamic acid, 1-ethynylcyclo-hexyl ester
Ethinamate
1-Ethinylcyclohexyl carbamate
1-Ethynylcyclohexanol carbam-ate
1-Ethynylcyclohexyl carbamate
1-Ethinylcyclohexyl carbonate
Etinamate
USAF EL-42
Valamin
Valamina
Valaminettae
Valaminetten
Valmid
Valmidate
Volamin

126-57-8
$C_{33}H_{62}O_6$
554.85
O=C(OCC(CC)(COC(=O)CCCC CCCC)COC(=O)CCCCCCC C)CCCCCCCC
Nonanoic acid, 2-ethyl-2-(((1-oxononyl)oxy)methyl)-1,3-propanediyl ester (9CI)
Trimethylolpropane tripelargon-ate

126-58-9
$C_{10}H_{22}O_7$
254.28
O(CC(CO)(CO)CO)CC(CO)(CO)CO
Dipentaerythritol (8CI)
Bis(pentaerythritol)
Dipentek
NSC-65881
1,3-Propanediol, 2,2'-(oxybis-(methylene))bis(2-(hydroxy-methyl)- (9CI)

126-68-1
$C_6H_{15}O_3PS$
198.24
Phosphorothioic acid, O,O,O-triethyl ester
O,O,O-Triethylester kyseliny thiofosforecne (Czech)
Triethyl phosphorothioate
O,O,O-Triethyl phosphoro-thioate
Triethylthiofosfat (Czech)
O,O,O-Triethylthiofosfat (Czech)

126-71-6
$C_{12}H_{27}O_4P$
266.32
O=P(OCC(C)C)(OCC(C)C)OCC(C)C
Phosphoric acid, triisobutyl ester (8CI)
AI3-07850
Isobutyl phosphate
NSC-62222
Phosphoric acid, tris(2-methyl-propyl) ester (9CI)
Triisobutyl phosphate
Tris(2-methylpropyl) phosphate

126-72-7
$C_9H_{15}Br_6O_4P$
697.67
O=P(OCC(Br)CBr)(OCC(Br)CBr)OCC(Br)CBr
1-Propanol, 2,3-dibromo-, phosphate (3:1)
Anfram 3PB
Apex 462-5

Bromkal P 67-6HP
2,3-Dibromo-1-propanol phosphate
(2,3-Dibromopropyl) phosphate
ES685
Firemaster LV-T 23P
Firemaster T23P
Firemaster T23P-LV
Flacavon R
Flamex T 23P
Flammex AP
Flammex LV-T 23P
Flammex T 23P
Fyrol HB32
NCI-C03270
Phosphoric acid, tris(2,3-dibromopropyl) ester
RCRA waste number U235
TDBP (Czech)
TDBPP
T 23P
Tris
Tris-BP
Tris(dibromopropyl)phosphate
Tris(2,3-dibromopropyl) phosphate
Tris(2,3-dibromopropyl) phosphoric acid ester
Tris-2,3-dibrompropyl ester kyseliny fosforecne (Czech)
Tris-(2,3-dibrompropyl)fosfat (Czech)
Tris (Flame retardant)
USAF DO-41
Zetifex ZN

126-73-8
$C_{12}H_{27}O_4P$
266.36
O=P(OCCCC)(OCCCC)OCCCC
Phosphoric acid, tributyl ester
Butyl phosphate, tri-
Celluphos 4
TBP
Tributilfosfato (Italian)
Tributyle (phosphate de) (French)
Tributylfosfaat (Dutch)
Tributylfosfat (Czech)
Tributylphosphat (German)
Tributyl phosphate (ACGIH, OSHA)
Tri-n-butyl phosphate

126-75-0
$C_8H_{19}O_3PS_2$
258.36
Phosphorothioic acid, O,O-diethyl S-(2-(ethylthio)ethyl) ester
Demeton-S
O,O-Diaethyl-S(2-aethylthio-aethyl)-monothiophosphat (German)
Diaethylthiophosphorsaeureester des aethylthioglykol (German)
Diethyl S-(2-ethioethyl)thiophosphate
O,O-Diethyl S-(2-eththioethyl)-phosphorothioate
O,O-Diethyl S-ethyl-2-ethylmercaptophosphorothiolate
O,O-Diethyl S-2-(ethylthio)-ethyl phosphorothioate
O,O-Diethyl S-(2-(ethylthio)-ethyl) phosphorothiolate
O,O-Diethyl-S-(2-ethylthio-ethyl)-monothiofosfaat (Dutch)
O,O-Dietil-S-(2-etiltio-etil)-monotiofosfato (Italian)
O,O-Dietyl-S-2-etylmerkapto-etyltiofosfat (Czech)
Ethanethiol, 2-(ethylthio)-, S-ester with O,O-diethyl phosphorothioate
Isodemeton
Isosystox
Izosystox (Czech)
Po-Systox
Thioldemeton
Thiol Systox
Thiophosphate de O,O-diethyle et de S-(2-ethylthio-ethyle) (French)

126-80-7
$C_{16}H_{34}O_5Si_2$
362.61
Disiloxane, 1,1,3,3-tetramethyl-1,3-bis(3-(oxiranylmethoxy)propyl)- (9CI)
1,3-Bis(3-(2,3-epoxypropoxy)-propyl)tetramethyldisiloxane
Bis(3-glycidoxypropyl)tetramethyldisiloxane
1,3-Bis(3-(glycidyloxy)propyl)-1,1,3,3-tetramethyldisiloxane
Disiloxane, 1,3-bis(3-(2,3-epoxypropoxy)propyl)-1,1,3,3-tetramethyl- (8CI)
NSC-93976

126-81-8
$C_8H_{12}O_2$
140.18
O=C(CC(CC1=O)(C)C)C1
Dimedone
AI3-19939
1,3-Cyclohexanedione, 5,5-dimethyl- (9CI)
Cyclomethone
Dimedon
5,5-Dimethylcyclohexane-1,3-dione
1,1-Dimethyl-3,5-cyclohexanedione
5,5-Dimethyl-1,3-cyclohexanedione
5,5-Dimethyldihydroresorcinol
1,1-Dimethyl-3,5-diketocyclohexane
5,5-Dimethylhydroresorcinol
Lu 274
Medon
Methon
Methone
NSC-14984

126-83-0
$C_3H_7ClO_4S.Na$
197.59
Sodium 3-chloro-2-hydroxypropylsulfonate
3-Chloro-2-hydroxy-1-propanesulfonic acid, sodium salt
NSC-52602
1-Propanesulfonic acid, 3-chloro-2-hydroxy-, monosodium salt (9CI)
Sodium-3-chloro-2-hydroxypropane sulfonate
Sodium 3-chloro-2-hydroxypropanesulfonate
Sodium 3-chloro-2-hydroxy-1-propanesulfonate
Sodiumchlorooxypropanesulfonate
Sodium epichlorohydrinsulfonate
Sodium 2-hydroxy-3-chloropropanesulfonate

126-84-1
$C_7H_{16}O_2$
132.23
O(C(OCC)(C)C)CC
Acetone, diethyl acetal
Acetone diethyl ketal
2,2-Diethoxypropane
Propane, 2,2-diethoxy-
USAF DO-44

126-85-2
$C_5H_{11}Cl_2NO$
172.07
CN(=O)(CCCl)CCCCl
Diethylamine, 2,2'-dichloro-N-methyl-, N-oxide
2-Chloro-N-(2-chloroethyl)-N-methylethanamine-N-oxide
2,2'-Dichloro-N-methyldiethylamine-N-oxide
Diethylamine, 2,2'-dichloro-N-methyl-, oxide
HN2 Amine Oxide
HN2 Oxide Mustard
MBAO
Mechlorethamine oxide
Methyl-bis(β-chloroethyl)amine oxide
Methylbis(β-chloroethyl)amine N-oxide
N-Methyl-di-2-chloroethyl-amine-N-oxide
Mitomen
Mitomin
Nitrogen Mustard Amine Oxide
Nitrogen Mustard Oxide
Nitrogen Mustard N-Oxide
Nitromin
NMO
N-Oxyd-Lost (German)
N-Oxyd-Mustard (German)
NSC-10107
Oxy-NH2

126-86-3

126-92-1

$C_{14}H_{26}O_2$
226.36
OC(C#CC(O)(CC(C)C)C)(CC(C)C)C
Tetramethyl decynediol
AI3-07159
5-Decyne-4,7-diol, 2,4,7,9-tetramethyl- (9CI)
1,4-Diisobutyl-1,4-dimethylbutynediol
NSC-5630
Surfynol 104
Surfynol 104A
Surfynol 104E
Syrfynol 104
2,4,7,9-Tetramethyl-5-decyne-4,7-diol

126-92-1
$C_8H_{18}O_4S.Na$
233.31
[Na+].CCCCC(CC)COS([O-])(=O)=O
1-Hexanol, 2-ethyl-, hydrogen sulfate, sodium salt
Emersal 6465
2-Ethyl-1-hexanol hydrogen sulfate sodium salt
2-Ethyl-1-hexanol sulfate sodium salt
2-Ethylhexylsiran sodny (Czech)
2-Ethylhexyl sodium sulfate
2-Ethylhexylsulfate sodium
Mono(2-ethylhexyl)sulfate sodium salt
NCI-C50204
Nia Proof 08
Propaste 6708
Sipex Bos
Sodium etasulfate
Sodium ethasulfate
Sodium(2-ethylhexyl)alcohol sulfate
Sodium 2-ethylhexyl sulfate
Sulfuric acid, mono(2-ethylhexyl)ester, sodium salt (8CI)
Tergemist
Tergimist
Tergitol 08
Tergitol Anionic 08
08-Union Carbide

126-96-5
$C_2H_4O_2.1/2Na$
71.55
Sodium diacetate
Acetic acid dimer, sodium salt
Acetic acid, sodium salt (2:1) (9CI)
Acetic acid, sodium salt, Compd with acetic acid (1:1)
Acid acetate
Dykon
Sodium acetate, acid
Sodium acid acetate
Sodium hydrogen diacetate

126-98-7
C_4H_5N
67.10
N#CC(=C)C
2-Propenenitrile, 2-methyl
2-Cyanopropene-1
Isopropene cyanide
Isopropenylnitrile
Methacrylonitrile
Methylacrylonitrile (ACGIH, OSHA)
α-Methylacrylonitrile
2-Methylpropenenitrile
RCRA waste number U152
USAF ST-40

126-99-8
C_4H_5Cl
88.54
C(C=C)(=C)Cl
1,3-Butadiene, 2-chloro
2-Chloor-1,3-butadieen (Dutch)
2-Chlor-1,3-butadien (German)
Chlorobutadiene
2-Chlorobuta-1,3-diene
2-Chloro-1,3-butadiene (OSHA)
Chloropreen (Dutch)
Chloropren (German, Polish)
Chloroprene
β-Chloroprene (ACGIH,OSHA)
Chloroprene, Inhibited [UN 1991]
Chloroprene, Uninhibited (DOT)
2-Cloro-1,3-butadiene (Italian)
Cloroprene (Italian)
Neoprene
UN 1991 [Chloropropene, inhibited]

127-00-4
C_3H_7ClO
94.55
OC(CCl)C
2-Propanol, 1-chloro
1-Chloroisopropyl alcohol
1-Chloro-2-propanol
sec-Propylene chlorohydrin
α-Propylene chlorohydrin

127-06-0
C_3H_7NO
73.11
N(O)=C(C)C
Acetone, oxime
Acetonoxime
Acetoxime
β-Isonitrosopropane
2-Propanone oxime

127-07-1
$CH_4N_2O_2$
76.07
NC(=O)NO
Urea, hydroxy
Biosupressin
Carbamohydroxamic acid
Carbamohydroximic acid
Carbamohydroxyamic acid
Carbamoyl oxime
N-Carbamoylhydroxylamine
Carbamyl hydroxamate
Hidrix
HU
Hydrea
Hydreia
Hydroxycarbamide
Hydroxycarbamine
Hydroxylamine, N-(aminocarbonyl)-
Hydroxylamine, N-carbamoyl-
Hydroxylurea
N-Hydroxymocovina (Czech)
Hydroxyurea
N-Hydroxyurea
Hydroxyurea (D4)
Hydura
Hydurea
Litaler
Litalir
NCI-C04831
NSC-32065
Onco-Carbide
Oxyurea
SK 22591
SQ 1089

127-08-2
$C_2H_3O_2.K$
98.15
[K+].CC(O)=O
Acetic acid, potassium salt
Diuretic salt
Octan draselny (Czech)
Potassium acetate

127-09-3
$C_2H_3O_2.Na$
82.04
[Na+].CC([O-])=O
Acetic acid, sodium salt
Anhydrous sodium acetate
Natriumacetat (German)
Octan sodny (Czech)
Sodium acetate

127-17-3
$C_3H_4O_3$
88.07
O=C(O)C(=O)C
Pyruvic-acid
Acetylformic acid
BTS
α-Ketopropionic acid
2-Oxopropanoic acid
2-Oxopropionic acid
Propanoic acid, 2-oxo- (9CI)
Pyroracemic acid

127-18-4
C_2Cl_4
165.82
C(=C(Cl)Cl)(Cl)Cl
Ethylene, tetrachloro
Ankilostin
Antisol 1
Carbon bichloride
Carbon dichloride

Czterochloroetylen (Polish)
Didakene
Dow-Per
ENT 1,860
Ethene, tetrachloro-
Ethylene tetrachloride
Fedal-UN
NCI-C04580
Nema
PER
Perawin
PERC
Perchloorethyleen, per (Dutch)
Perchlor
Perchloraethylen, per (German)
Perchlorethylene
Perchlorethylene, per (French)
Perchloroethylene (ACGIH, DOT,OSHA)
Perclene
Perclene D
Percloroetilene (Italian)
Percosolve
Perk
Perklone
Persec
RCRA waste number U210
Tetlen
Tetracap
Tetrachlooretheen (Dutch)
Tetrachloraethen (German)
Tetrachlorethylene
Tetrachloroethene
Tetrachloroethylene (OSHA) [UN 1897]
1,1,2,2,-Tetrachloroethylene
Tetracloroetene (Italian)
Tetraleno
Tetralex
Tetravec
Tetroguer
Tetropil
UN 1897 [Tetrachloroethylene]

127-19-5
C_4H_9NO
87.14
O=C(N(C)C)C
Acetamide, N,N-dimethyl
Acetdimethylamide
Acetic acid, dimethylamide
Dimethyl acetamide (ACGIH, OSHA)
Dimethylacetamide
N,N-Dimethylacetamide
Dimethylacetone amide
Dimethylamide acetate
Dimethylamid kyseliny octove (Czech)
DMA
DMAC
NSC-3138
U-5954

127-20-8
$C_3H_3Cl_2O_2.Na$
164.95
[Na+].CC(Cl)(Cl)C([O-])=O
Propionic acid, 2,2-dichloro-, sodium salt
BASFapon B
Dalapon
Dalapon sodium
Dalapon sodium salt
Dichloropropionate
2,2-Dichloropropionic acid, sodium salt
α-α-Dichloropropionic acid sodium salt
2,2-Dichlorpropionsaeure natrium (German)
2,2-DPA
Dowpon
Gramevin
Hico DCPAS
Natriumsalz der 2,2-dichlorpropionsaure
Radapon
Sodium dalapon
Sodium α,α-dichloropropionate
Sodium 2,2-dichloropropionate
Unipon

127-25-3
$C_{21}H_{32}O_2$
316.53
O=C(OC)C(C(C(C(C(=C1)C=C(C2)C(C)C)C2)(CC3)C)C1)(C3)C
Podocarpa-7,13-dien-15-oic acid, 13-isopropyl-, methyl ester
Abietic acid, methyl ester
Methyl abietate
Methyl Ester of Wood Rosin
Wood Rosin, Methyl Ester

127-27-5
$C_{20}H_{30}O_2$
302.46
CC1(CCC2C(=C1)CCC3C2(C)CCCC3(C)C(O)=O)C=C
Pimaric acid
Dextropimaric acid
NSC-2956
1-Phenanthrenecarboxylic acid, 7-ethenyl-1,2,3,4,4a,4b,5,6,7,9,10,10a-dodecahydro-1,4a,7-trimethyl-, (1R-(1α,4aβ,4bα,7β,10aα))- (9CI)
Pimara-8(14),15-dien-19-oic acid
Pimaric acid, d
α-Pimaric acid
d-Pimaric acid
δ8(14)-Pimaric acid
Podocarp-8(14)-en-15-oic acid, 13α-methyl-13-vinyl- (8CI)

127-31-1
$C_{21}H_{29}FO_5$
380.50
CC34CC(O)C1(F)C(CCC2=CC(=O)CCC12C)C3CCC4(O)C(=O)CO
Pregn-4-ene-3,20-dione, 9-fluoro-11-β,17,21-trihydroxy
Alflorone
F-Col
F-Cortef
Florinef
Fludrocortisone
Fludrocortone
Fluodrocortisone
Fluohydrisone
Fluohydrocortisone
9-α-Fluorocortisol
Fluorocortisone
9-α-Fluorohydrocortisone
9-α-Fluoro-17-hydroxycorticosterone
9-Fluoro-11-β,17,21-trihydroxypregn-4-ene-3,20-dione
9-α-Fluoro-11-β,17-α,21-trihydroxy-4-pregnene-3,20-dione
U 5963

127-33-3
$C_{21}H_{21}ClN_2O_8$
464.89
2-Naphthacenecarboxamide, 7-chloro-4-(dimethylamino)-1,4,4a,5,5a,6,11,12a-octahydro- 3,6,10,12,12a-pentahydroxy-1,11-dioxo
7-Chloro-6-demethyltetracycline
Chlortetracycline, 6-demethyl-
Demeclocycline
Declomycin
Demethylchlorotetracycline
6-Demethylchlorotetracycline
6-Demethyl-7-chlorotetracycline
Demethylchlortetracyclin
Demethylchlortetracycline
6-Demethylchlortetracycline
6-Demethyl-7-chlortetracycline
Demethylchlortetracycline Base
DMCT
Ledermycin
Methylchlortetracycline
Mexocine
RP 10192
Tetracycline, 7-chloro-6-demethyl-

127-35-5
$C_{22}H_{27}NO$
321.46
Phenazocine
DEA No. 9715
6,11-Dimethyl-1,2,3,4,5,6-hexahydro-8-hydroxy-3-phenethyl-2,6-methano-3-benzazocine
Fenazocina (Spanish)
1,2,3,4,5,6-Hexahydro-6,11-dimethyl-3-phenethyl-2,6-methano-3-benzazocin-8-ol
1,2,3,4,5,6-Hexahydro-8-hydroxy-6,11-dimethyl-3-phenethyl-2,6-methano-3-benzazocine
2'-Hydroxy-5,9-dimethyl-2-phenethyl-6,7-benzomorphan
(+)2'-Hydroxy-5,9-dimethyl-2-phenethyl-6,7-benzomorphan
2,6-Methano-3-benzazocin-8-ol,

1,2,3,4,5,6-hexahydro-
6,11-dimethyl-3-phenethyl-
NIH 7519
Phenazocinum (Latin)
Prinadol
SKF 6574

127-39-9
$C_{12}H_{22}O_7S.Na$
333.36
Diisobutyl sodium sulfosuccinate
AI3-18859
1,4-Bis(2-methylpropyl)sulfobutanedioate, sodium salt
Butanedioic acid, sulfo-, 1,4-bis(2-methylpropyl) ester, sodium salt

127-41-3
$C_{13}H_{20}O$
192.33
O=C(C=CC(C(=CCC1)C)C1(C)C)C
3-Buten-2-one, 4-(2,6,6-trimethyl-2-cyclohexen-1-yl)
α-Cyclocitrylideneacetone
α-Ionone
4-(2,6,6-Trimethyl-2-cyclohexen-1-yl)-3-buten-2-one

127-42-4
$C_{14}H_{22}O$
206.33
O=C(C=CC(C(=CCC1)C)C1(C)C)CC
1-Penten-3-one, 1-(2,6,6-trimethyl-2-cyclohexen-1-yl)-, (R-(E))- (9CI)
5-(2,6,6-Trimethyl-2-cyclohexenyl)-4-penten-3-one

127-43-5
$C_{14}H_{22}O$
206.33
O=C(C=CC(=C(CCC1)C)C1(C)C)CC
1-Penten-3-one, 1-(2,6,6-trimethyl-1-cyclohexen-1-yl)-(9CI)
AI3-24259
β-Ionone, methyl-
β-Iraldeine
β-Methylionone
Methyl-β-ionone
NSC-163995
5-(2,6,6-Trimethyl-1-cyclohexenyl)-4-penten-3-one

127-47-9
$C_{22}H_{32}O_2$
328.54
O=C(OCC=C(C=CC=C(C=CC(=C(CCC1)C)C1(C)C)C)C)C
Retinol, acetate
Crystalets
Myvak
Myvax
21 CFR 182,5933
Retinyl acetate
all-trans-Retinyl acetate
Vitamin A acetate
trans-Vitamin A acetate
Vitamin A alcohol acetate

127-51-5
$C_{14}H_{22}O$
206.33
O=C(C(=CC(C(=CCC1)C)C1(C)C)C)C
3-Buten-2-one, 3-methyl-4-(2,6,6-trimethyl-2-cyclohexen-1-yl)- (9CI)
AI3-36074
α-Cetone
Cetone α
α-Ionone, isomethyl-
α-Isomethylionone
Isomethyl-α-ionone
Iso-α-methyl ionone
Methyl-α-isoionone
NSC-66432
4-(2,6,6-Trimethyl-2-cyclohexen-1-yl)-3-methyl-3-buten-2-one

127-52-6
$C_6H_6ClNO_2S.Na$
214.62
Chloramine B
AI3-16452
Annogen
Benzene chloramine
Benzenesulfochloramide sodium
Benzenesulfo-sodium chloramide
Benzenesulfonamide, N-chloro-, sodium salt (9CI)
N-Chlorobenzenesulfonamide, sodium salt
(N-Chlorobenzenesulfonamide)-sodium
(N-Chlorobenzenesulfonamido)-sodium
Caswell No. 169
Chloramin B
Chloramine-B
Chlorogen
N-Chloro-N-sodiobenzenesulfonamide
EPA Pesticide Chemical Code 076501
Monochloramine B
Neomagnol
NSC-75446
Sodium benzene sulfochloramide
Sodium benzenesulfochloramine
Sodium benzenesulfonchloramide
Sodium benzenesulfonchloramine
Sodium benzosulfochloramide
Sodium N-chlorobenzenesulfonamide
Sodium, (N-chlorobenzenesulfonamido)-
Sodium, (chloro(phenylsulfonyl)amino)-
Sulfenazone

127-58-2
$C_{11}H_{12}N_4O_2S.Na$
287.32
Sulfanilamide, N[1]-(4-methyl-2-pyrimidinyl)-, monosodium salt
Benzenesulfonamide, 4-amino-N-(4-methyl-2-pyrimidinyl)-, monosodium salt (9CI)
N[1]-(4-Methyl-2-pyrimidinyl)-sulfanilamide sodium salt
Sodium sulfamerazine
Sodium sulphamerazine
Soluble Sulfamerazine
Solumedine
Sulfamerazine sodium

127-63-9
$C_{12}H_{10}O_2S$
218.28
O=S(=O)(c(cccc1)c1)c(cccc2)c2
Phenyl-sulfone
Difenylsulfon (Czech)
Diphenyl sulfone
Diphenyl sulphone
DPS
Phenyl sulphone
Sulfobenzide

127-65-1
$C_7H_8ClNO_2S.Na$
228.66
p-Toluenesulfonamide, N-chloro-, sodium salt
Acti-Chlore
Aktivin
Anexol
Benzenesulfonamide, N-chloro-4-methyl-, sodium salt (9CI)
Berkendyl
Chloralone
Chloramine T
Chlorasan
Chloraseptine
Chlorazan
Chlorazene
Chlorazone
Chlorozone
Chlorseptol
Clorina
Clorosan
Desinfect
Euclorina
Gansil
Gyneclorina
Halamid
Heliogen
Kloramin
Kloramine-T
Multichlor
Sodium chloramine T
Sodium p-toluenesulfonylchloramide
Sodium tosylchloramide
Tampules

Tochlorine
Tolamine
Tosylchloramide sodium

127-68-4
$C_6H_4NO_5S.Na$
225.16
[Na+].[O-]S(=O)(=O)c1cccc(c1)N(=O)=O
Benzenesulfonic acid, m-nitro-, sodium salt
Ludigol F,60
m-Nitrobenzenesulfonic acid sodium salt
Nitrobenzen-m-sulfonan sodny (Czech)
Tiskan (Czech)

127-69-5
$C_{11}H_{13}N_3O_3S$
267.33
O=S(=O)(NC(ON=C1C)=C1C)c(ccc(N)c2)c2
Sulfanilamide, N[1]-(3,4-dimethyl-5-isoxazolyl)
Accuzole
Alphazole
Amidoxal
5-(p-Aminobenzenesulfonamido)-3,4-dimethylisoxazole
5-(p-Aminobenzenesulphonamide)-3,4-dimethylisoxazole
5-(p-Aminobenzenesulphonamido)-3,4-dimethylisoxazole
4-Amino-N-(3,4-dimethyl-5-isoxazolyl)benzenesulphonamide
5-(4-Aminophenylsulfonamido)-3,4-dimethylisoxazole
5-(4-Aminophenylsulphonamido)-3,4-dimethylisoxazole
Astrazolo
Azo Gantrisin
Azosulfizin
Bactesulf
Barazae
Benzenesulfonamide, 4-amino-N-(3,4-dimethyl-5-isoxazolyl)-
Chemouag
3,4-Dimethylisoxale-5-sulfanilamide
3,4-Dimethylisoxazole-5-sulphanilamide
N[1]-(3,4-Dimethyl-5-isoxazolyl)-sulfanilamide
N'-(3,4)Dimethylisoxazol-5-yl-sulphanilamide
N[1]-(3,4-Dimethyl-5-isoxazolyl)-sulphanilamide
3,4-Dimethyl-5-sulfanilamidoisoxazole
3,4-Dimethyl-5-sulphanilamidoisoxazole
3,4-Dimethyl-5-sulphonamidoisoxazole
Dorsulfan
Entusil
Entusul
Ganda
Gantrisin
Gantrisine
Gantrisona
Gantrosan
Isoxamin
J-Sul
Koro-Sulf
NCI-C50022
Neazolin
Neoxazol
Norilgan-S
Novazolo
Novosaxazole
NU 445
Pancid
Renosulfan
Resoxol
Roxosul
Roxosul Tablets
Roxoxol
Saxosozine
SI
SK-Soxazole
Sodizole
Sosol
Soxamide
Soxisol
Soxitabs
Soxo
Soxomide
Stansin
Sulbio
Sulfadimethylisoxazole
Sulfafurazol
Sulfafurazole
Sulfagan
Sulfagen
Sulfaisoxazole
5-Sulfanilamido-3,4-dimethylisoxazole
Sulfapolar
Sulfasoxazole
Sulfalar
Sulfasan
Sulfasol
Sulfazin
Sulfisin
Sulfisonazole
Sulfisoxazol
Sulfisoxazole
Sulfizin
Sulfizol
Sulfizole
Sulfoxol
Suloxsol
Sulphadimethylisoxazole
Sulphafuraz
Sulphafurazol
Sulphafurazole
Sulphafurazolum
Sulphaisoxazole
5-Sulphanilamido-3,4-dimethylisoxazole
Sulphisoxazol
Sulphofurazole
Sulsoxin
Thiasin
TL-Azole
Unisulf
Urisoxin
Uritrisin
Urogan
U.S.-67
Vagilia
V-Sul

127-82-2
$C_{12}H_{12}O_8S_2.Zn$
413.73
Benzenesulfonic acid, p-hydroxy-, zinc salt (2:1)
Benzenesulfonic acid, 4-hydroxy-, zinc salt (2:1)
p-Hydroxybenzenesulfonic acid zinc salt
1-Phenol-4-sulfonic acid zinc salt
Phenozin
Zinc p-hydroxybenzenesulfonate
Zinc phenolsulfonate
Zinc p-phenol sulfonate
Zinc sulfocarbolate
Zinc sulfophenate

127-85-5
$C_6H_7AsNO_3.Na$
239.05
[Na+].Nc1ccc(cc1)[As](O)([O-])=O
Arsanilic acid, monosodium salt
(4-Aminophenyl)arsonic acid sodium salt
Anhydrous sodium arsanilate
Arsamin
Arsanilic acid sodium salt
Arsinosolvin
Arsonic acid, (4-aminophenyl)-, monosodium salt (9CI)
Atoxyl
NCI-C61176
Nuarsol
Piglet Pro-Gen V
Pro-Gen Sodium
Protoxyl
Soamin
Sodium aminarsonate
Sodium p-aminobenzenearsonate
Sodium aminophenol arsonate
Sodium p-aminophenylarsonate
Sodium-analine arsonate
Sodium anilarsonate
Sodium arsanilate [UN 2473]
Sodium p-arsanilate
Sodium arsonilate
Sonate
Trypoxyl
UN 2473 [Sodium arsanilate]

127-90-2
$C_6H_6Cl_8O$
377.72
ClC(COCC(Cl)C(Cl)(Cl)Cl)C(Cl)(Cl)Cl
Ether, bis(2,3,3,3-tetrachloropropyl)
Bis(2,3,3,3-tetrachloropropyl) ether
ENT 25,456
Monsanto CP-16226
Octachlorodipropylether
S 421

127-91-3
$C_{10}H_{16}$
136.26
C(C(CC1C2)C1(C)C)(C2)=C
Bicyclo(3.1.1)heptane, 6,6-dimethyl-2-methylene
Nopinen
Nopinene
β-Pinene
2(10)-Pinene
Pseudopinen
Pseudopinene

127-95-7
$C_2HO_4.K$
128.13
Oxalic acid, monopotassium salt
Kleesalz (German)
Potassium hydrogen oxalate
Potassium salt of sorrel
Sorrel salt

128-03-0
$C_3H_6NS_2.K.H_2O$
177.34
Carbamic acid, dimethyldithio-, potassium salt, hydrate
Potassium dimethyl dithiocarbamate

128-04-1
$C_3H_6NS_2.Na$
143.21
[Na+].CN(C)C(S)=S
Carbamodithioic acid, dimethyl-, sodium salt
Aceto SDD 40
Alcobam NM
Brogdex 555
Carbamic acid, dimethyldithio-, sodium salt
Carbon S
Dibam
Dibam A
Dimethyldithiocarbamic acid, sodium salt
DMDK
Methyl namate
SDDC
Sharstop 204
Sodium N,N-dimethyldithiocarbamate
Sta-Fresh 615
Steriseal Liquid #40
Thiostop N
Vinstop
Vulnopol NM
Wing Stop B

128-09-6
$C_4H_4ClNO_2$
133.54
O=C(N(C(=O)C1)Cl)C1
2,5-Pyrrolidinedione, 1-chloro
1-Chloro-2,5-pyrrolidinedione
N-Chlorosuccinimide
Succinchlorimide
Succinimide, N-chloro-
Succinochlorimide

128-10-9
$C_{12}H_8Cl_6O$
380.91
1,4:5,8-Dimethanonaphthalene, 1,2,3,4,10,10-hexachloro-6,7-epoxy-1,4,4a,5,6,7,8,8a-octahydro- (8CI)
2,7:3,6-Dimethanonaphth-(2,3-b)oxirene, 3,4,5,6,9,9-hexachloro-1a,2,2a,3,6,6a,7,7a-octahydro- (9CI)
NSC-231371

128-37-0
$C_{15}H_{24}O$
220.39
Oc(c(cc(c1)C)C(C)(C)C)c1C(C)(C)C
p-Cresol, 2,6-di-tert-butyl
Advastab 401
Agidol
Agidol 1
Alkofen BP
Antioxidant DBPC
Antioxidant 4
Antioxidant 29
Antioxidant 30
Antioxidant 4K
Antioxidant KB
Antrancine 8
AO 29
AO 4K
BHT
BHT (Food grade)
2,6-Bis(1,1-dimethylethyl)-4-methylphenol
BUKS
Butylated hydroxytoluene
Butylhydroxytoluene
Butylohydroksytoluenu (Polish)
CAO 1
CAO 3
Catalin CAO-3
Chemanox 11
Dalpac
DBMP
DBPC
DBPC (Technical grade)
Deenax
Dibunol
Dibutylated hydroxytoluene
2,6-Di-tert-butyl-p-cresol (ACGIH,OSHA)
2,6-Di-terc.Butyl-p-kresol (Czech)
2,6-Di-tert-butyl-1-hydroxy-4-methylbenzene
3,5-Di-tert-butyl-4-hydroxytoluene
2,6-Di-tert-butyl-p-methylphenol
2,6-Di-tert-butyl-4-methylphenol
4-Hydroxy-3,5-di-tert-butyltoluene
Impruvol
Ionol
Ionol 1
Ionol (Antioxidant)
Ionol CP
Ionole
Kerabit
4-Methyl-2,6-di-terc. butylfenol (Czech)
Methyldi-tert-butylphenol
4-Methyl-2,6-di-tert-butylphenol
NCI-C03598
Nocrac 200
Nonox TBC
P 21
Parabar 441
Paranox 441
Phenol, 2,6-bis(1,1-dimethylethyl)-4-methyl- (9CI)
Stavox
Sumilizer BHT
Sustane
Sustane BHT
Swanox BHT
Tenamene 3
Tenox BHT
Topanol
Topanol O
Topanol OC
Toxolan P
Vanlube PC
Vanlube PCX
Vianol
Vulkanox KB

128-39-2
$C_{14}H_{22}O$
206.36
Oc(c(ccc1)C(C)(C)C)c1C(C)(C)C
Phenol, 2,6-di-tert-butyl
2,6-Bis(tert-butyl)phenol
2,6-Di-tert-butylphenol
Ethanox 701

128-42-7
$C_{14}H_{10}N_2O_{10}S_2$
430.38
O=S(=O)(O)c(c(ccc1N(=O)=O)C=Cc(c(S(=O)(=O)O)cc(N(=O)=O)c2)c2)c1
2,2'-Stilbenedisulfonic acid, 4,4'-dinitro
Benzenesulfonic acid, 2,2'-(1,2-ethenediyl)bis(5-nitro-
Dinitrostilbenedisulfonic acid
4,4'-Dinitro-2,2'-stilbenedisulfonic acid
Kyselina 4,4'-dinitrostilben-2,2'-disulfonova (Czech)

128-44-9
$C_7H_4NO_3S.Na$
205.17
[Na+].O=C1[N-]S(=O)(=O)c2cccc12
1,2-Benzisothiazolin-3-one, 1,1-dioxide, sodium salt
Artificial sweetening substanz gendorf 450
Cristallose

Crystallose
Dagutan
Kristallose
Madhurin
ODA
Saccharine soluble
Saccharinnatrium
Saccharin, sodium
Saccharin, sodium salt
Saccharin soluble
Saccharoidum natricum
Saxin
Sodium 1,2 benzisothiazolin-3-one-1,1-dioxide
Sodium o-benzosulfimide
Sodium benzosulphimide
Sodium o-benzosulphimide
Sodium 2-benzosulphimide
Sodium saccharide
Sodium saccharin
Sodium saccharinate
Sodium saccharine
Soluble Gluside
Soluble Saccharin
Succaril
Sucra
o-Sulfonbenzoic acid imide sodium salt
Sulphobenzoic imide, sodium salt
Sweeta
Sykose
Willosetten

128-46-1
$C_{21}H_{41}N_7O_{12}$
583.69
O(C(C(O)C(O)C1NC)CO)C1OC(C(O)(CO)C(O2)C)C2OC(C(O)C(O)C(NC(=N)N)C3O)C3NC(=N)N
Streptamine, O-β-d-mannopyranosyl-(1-4)-2-deoxy-2-(methylamino)-α-l-glucopyranosyl- (1-2)-5-deoxy-O-3-C-(hydroxymethyl)-α-l-lyxofuranosyl-(1-4)-N,N'-diamidino-, D
Dihydrostreptomycin
DHMS
DST
Streptomycin, dihydro-

128-51-8
$C_{13}H_{20}O_2$
208.33
O=C(OCCC(C(CC1C2)C1(C)C)=C2)C
2-Norpinene-2-ethanol, 6,6-dimethyl-, acetate
Citroviol
6,6-Dimethylbicyclo(3.1.1)-2-heptene-2-ethyl acetate
6,6-Dimethyl-2-norpinene-2-ethanol, acetate
Lignyl acetate
Nopol acetate
Nopyl acetate
2-Pinene-10-methyl acetate

128-56-3
$C_{14}H_7O_5S.Na$
310.26
Sodium anthraquinone-1-sulfonate
1-Anthracenesulfonic acid, 9,10-dihydro-9,10-dioxo-, sodium salt (8CI,9CI)
Anthrachinon-1-sulfonan sodny (Czech)
9,10-Anthraquinone-1-sulfonate sodium salt
1-Anthraquinonesulfonic acid sodium salt
9,10-Dihydro-9,10-dioxo-1-anthracenesulfonic acid sodium salt
Gold Salt
Golden Salt
NSC-7575
Sodium anthraquinone-α-sulfonate

128-58-5
$C_{36}H_{20}O_4$
516.56
O=C(c(c(c(c1c(c(c(c(c(c(c(C2=O)ccc3)c3)cc4OC)c2c5)c46)c5)cc7)c6c8OC)c8)ccc9)c9)c17
Dinaphtho(1,2,3-cd:3',2',1'-lm)perylene-5,10-dione, 16,17-dimethoxy
Ahcovat Jade Green B
Ahcovat Jade Green BDA
Ahcovat Printing Jade Green B
Ahcovat Printing Jade Green BDA
Amanthrene Brilliant Green J
Amanthrene Brilliant Green JP
Amanthrene Green JF
Amanthrene Supra Green JF
Atic Vat Jade Green XBN
Atic Vat Printing Jade Green XBN
Belanthrene Jade Green
Benzadone Jade Green B
Benzadone Jade Green X
Benzadone Jade Green XBN
Benzadone Jade Green XN
Brilliant Green S
Calcoloid Jade Green N
Calcoloid Jade Green NC
Calcoloid Jade Green NP
Caledon Jade Green
Caledon Jade Green XBN
Caledon Jade Green XN
Caledon Printing Jade Green XBN
Caledon Printing Jade Green XN
Carbanthrene Brilliant Green
Carbanthrene Brilliant Green G
C.I. 59825
Cibanone Brilliant Green 2BF
Cibanone Brilliant Green FBF
C.I. Vat Green 1
Dimethoxyviolanthrone
16,17-Dimethoxyviolanthrone
Fenanthren Brilliant Green B
Helanthrene Green B
Indanthren Brilliant Green B
Indanthren Brilliant Green FFB
Indanthren Brilliant Green FFB Extra pure
Indanthrene Brilliant Green B
Indanthrene Brilliant Green BN
Indanthrene Brilliant Green FFB
Jade Green Base
Mayvat Jade Green
Mikethrene Brilliant Green B
Mikethrene Brilliant Green FFB
Navinon Jade Green B
Navinon Jade Green FFB
Nihonthrene Brilliant Green B
Nihonthrene Brilliant Green FFB
Nyanthrene Brilliant Green B
Ostanthren Brilliant Green FFB
Ostanthren Green FFB
Palanthrene Jade Green
Palanthrene Jade Green Supra
Paradone Jade Green B
Paradone Jade Green B New, BX New XS New
Paradone Jade Green BX
Paradone Jade Green XS
Parnithrene Brilliant Green FFB
Pernithrene Brilliant Green GG
Romantrene Brilliant Green FB
Romantrene Brilliant Green FFB
Sandothrene Brilliant Green NBF
Solanthrene Brilliant Green B
Solanthrene Brilliant Green BN
Solanthrene Brilliant Green FF
Tinon Brilliant Green BF
Tinon Brilliant Green 2BF
Tinon Brilliant Green B2F-F
Tinon Brilliant Green BFP
Tyrian Brilliant Green I-B
Tyrian Brilliant hreen I-FFB
Vat Brilliant Green C
Vat Brilliant Green S
Violanthrone, 16,17-dimethoxy-
Zelen Kypova 1 (Czech)
Zelen Ostanthrenova brilantni FFB (Czech)

128-60-9
$C_{34}H_{15}NO_4$
501.49
O=C(c(c(c(c1c(c(c(c(c(c(c(C2=O)ccc3)c3)cc4)c2c5)c46)c5)cc7)c6c8N(=O)=O)c8)cc c9)c9)c17
Anthra(9,1,2-cde)benzo(rst)-pentaphene-5,10-dione, 16-nitro- (9CI)

128-62-1
$C_{22}H_{23}NO_7$
413.46
COc2ccc1C(OC(=O)c1c2OC)C4N(C)CCc5cc3OCOc3c(OC)c45
Noscapine
Capval
Coscopin

Coscotabs
Key-Tusscapine
Longatin
Lyobex
Methoxyhydrastine
Narcompren
Narcosine
Narcotine
1-α-Narcotine
Narcotussin
Nectadon
Nicolane
Nipaxon
Noscapal
Noscapalin
NSC-5366
Opian
Opianine
Terbenol
Tusscapine
Vadebex

128-63-2
$C_{16}H_6Br_4$
517.84
c(c(c(c(cc1Br)Br)cc2)c1cc3)(c2c(cc4Br)Br)c34
Pyrene, 1,3,6,8-tetrabromo- (9CI)
1,3,6,8-Tetrabromopyrene

128-66-5
$C_{24}H_{12}O_2$
332.36
O=C(c(c(c(c1c(c(c(c(C2=O)ccc3)c3)cc4)c2c5)c5)ccc6)c6)c14
Dibenzo(b,def)chrysene-7,14-dione
Ahcovat Printing Golden Yellow
Amanthrene Golden Yellow
Anthravat Golden Yellow
Arlanthrene Golden Yellow
Benzadone Golden Yellow
Calcoloid Golden Yellow
Caledon Golden Yellow
Caledon Printing Yellow
Carbanthrene Golden Yellow
C.I. 59100
Cibanone Golden Yellow
C.I. Vat Yellow
C.I. Vat Yellow 4
Dibenzo(a,b)pyrene-7,14-dione
2,3,7,8-Dibenzopyrene-1,6-quinone
1',2',6',7'-Dibenzpyrene-7,14-quinone
Femanthren Golden Yellow
Golden Yellow
Helanthrene Yellow
Hostavat Golden Yellow
Indanthrene Golden Yellow
Indanthren Golden Yellow
Indanthren Printing Yellow
Leucosol Golden Yellow
Mayvat Golden Yellow
Mikethrene Gold Yellow
NCI-C03565
Nihonthrene Golden Yellow
Nyanthrene Golden Yellow
Palanthrene Golden Yellow
Paradone Golden Yellow
Pharmanthrene Golden Yellow
Romantrene Golden Yellow
Sandothrene Golden Yellow
Sandothrene Printing Yellow
Solanthrene Brilliant Yellow
Tinon Golden Yellow
Tyrion Yellow
Vat Golden Yellow

128-69-8
$C_{24}H_8O_6$
392.32
O=C(OC(=O)c(c1c(c(c(c(c(c2)C(=O)OC3=O)c3cc4)c45)c2)cc6)c5c7)c7)c16
Perylo(3,4-cd:9,10-c'd')di-pyran-1,3,8,10-tetrone (9CI)
NSC-79895
Perylenetetracarboxylic acid dianhydride
Perylenetetracarboxylic anhydride
3,4:9,10-Perylenetetracarboxylic anhydride
3,4,9,10-Perylenetetracarboxylic dianhydride
3,4,9,10-Perylenetetracarboxylic 3,4:9,10-dianhydride (8CI)
Perylene-3,4,9,10-tetracarboxylic dianhydride
Perylene-3,4,9,10-tetracarboxylic 3,4:9,10-dianhydride

128-70-1
$C_{30}H_{14}O_2$
406.44
O=C(c(c(c(c1c(c(c(c(c(C2=O)ccc3)c3)c4)c2cc5)c56)c4cc7)c6)ccc8)c8)c17
8,16-Pyranthrenedione (9CI)
Ahcovat Golden Orange G
Amanthrene Golden Orange G
Benzadone Gold Orange G
C.I. Pigment Orange 40
C.I. Vat Orange 9
C.I. 59700
Calcoloid Golden Orange GD
Calcoloid Golden Orange GFD
Caledon Gold Orange G
Caledon Gold Orange GN
Caledon Paper Gold Orange G
Caledon Printing Orange G
Carbanthrene Golden Orange G
Carbanthrene Golden Orange GD
Carbanthrene Golden Orange GP
Carbanthrene Printing Golden Orange G
Cibanone Golden Orange FG
Cibanone Golden Orange G
Endurol Golden Orange G
Fenanthren Golden Orange G
Indanthren Gold Orange G
Indanthren Golden Orange G
Indanthren Golden Orange GLP
Indanthrene Gold Orange G
Indanthrene Golden Orange G
Indanthrene Golden Orange GA
Mikethren Gold Orange G
Mikethrene Gold Orange G
Monolite Fast Gold Orange GV
Nihonthrene Golden Orange G
NSC-5267
Nyanthrene Golden Orange G
Palanthrene Gold Orange G
Paradone Golden Orange G
Ponsol Golden Orange G
Ponsol Golden Orange GD
Pyranthron
Pyranthrone
Romantrene Golden Orange FG
Sandothrene Golden Orange NG
Solanthrene Orange F-J
Solanthrene Orange J
Tinon Golden Orange G
Tinon Golden Orange GN
Tyrian Golden Orange I-G
Vat Orange 9

128-80-3
$C_{28}H_{22}N_2O_2$
418.52
O=C(c(c(C(=O)c1c(Nc(ccc(c2)C)c2)ccc3Nc(ccc(c4)C)c4)ccc5)c5)c13
Anthraquinone, 1,4-bis(p-tolylamino)
Alizarine Cyanine Green Base
Alizarine Green G Base
Amaplast Green OZ
9,10-Anthracenedione, 1,4-bis-((4-methylphenyl)amino)-(9CI)
Anthcoquinone Cyanine Green Base
Anthraquinone, 1,4-di-p-toluidino- (8CI)
Anthraquinone Green G Base
Arlosol Green B
Arlosol Green BS
Arlosol Green BSS
Bis-1,4-p-tolylaminoanthrchinon (Czech)
C-Green 10
C.I. 61565
C.I. Solvent Green 3
Cyanine Green G Base
D & C Green No. 6
Fat Soluble Anthraquinone Green
Fat Soluble Green Anthraquinone
11091 Green
Green No. 202
Micro-Lex Green 5B
Nitro Fast Green GB
Organol Fast Green J
Organol Green J
Quinazarin Green
Quinizarine Green Base
Quinizarin Green SS
Solvent Green 3
Sudan Green 4B
Toyo Oriental Oil Blue G
Waxoline Green
Waxoline Green G
Zelen Rozpoustedlova 3 (Czech)

Zelen Sudan 4B (Czech)

128-81-4
$C_{16}H_9N_3O_4$
307.25
O=C(NC(=O)c1c(c(c(C(=O)c(c2c cc3)c3)c4N)C2=O)N)c14
1H-Naphth(2,3-f)isoindole-1,3,5,10(2H)-tetrone, 4,11-diamino- (9CI)
2,3-Anthracenedicarboximide, 1,4-diamino-9,10-dihydro-9,10-dioxo- (8CI)
1,4-Diaminoanthraquinone-2,3-dicarboximide
1,4-Diamino-2,3-anthraquinone-dicarboximide
1,4-Diamino-9,10-dihydro-9,10-dioxo-2,3-anthracene-dicarboximide
NSC-115447

128-85-8
$C_{22}H_{18}N_2O_2$
342.39
O=C(c(c(c(C(=O)c1c(NC)ccc2Nc (ccc(c3)C)c3)ccc4)c4)c12
Anthraquinone, 1-(methyl-amino)-4-p-toluidino- (8CI)
Ahcoquinone Blue ASTB Base
9,10-Anthracenedione, 1-(methylamino)-4-((4-methyl-phenyl)amino)- (9CI)
Alizarine Pure Blue B Base
C.I. Solvent Blue 11
C.I. 61525
1-Methylamino-4-(4-methyl-phenylamino)anthraquinone
1-(Methylamino)-4-p-toluidino-anthraquinone
Nitro Fast Blue 3GB
NSC-39906
Organol Blue J
Organol Brilliant Blue J
Solvent Blue 11
Somalia Blue G
Sudan Blue GA
Superlan Astrol B Base
Waxoline Blue GA

128-86-9
$C_{14}H_{10}N_2O_{10}S_2$
430.36
O=C(c(c(c(N)cc1S(=O)(=O)O)C (=O)c2c(O)c(S(=O)(=O)O) cc3N)c1O)c23
Alizarin Brilliant Blue BS
2,6-Anthracenedisulfonic acid, 4,8-diamino-9,10-dihydro-1,5-dihydroxy-9,10-dioxo- (9CI)
4,8-Diamino-1,5-dihydroxy-9,10-dihydro-9,10-dioxo-2,6-anthracenedisulfonic acid, disodium salt

128-89-2
$C_{42}H_{25}N_3O_6$
667.70
O=C(Nc6cccc7C(=O)c5c(Nc3ccc (NC(=O)c1ccccc1)c4C(=O)c 2ccccc2C(=O)c34)cccc5C(=O) c67)c8ccccc8
Anthraquinone, 4,5'-imino-bis(4-benzamido
4,5'-Bis-benzoylamino-1,1'-di-anthrimid (Czech)
4,5'-Iminobis(4-benzamido-anthraquinone)

128-91-6
$C_{14}H_6N_2O_8$
330.20
O=C(c(c(c(N(=O)=O)cc1)C(=O) c2c(O)ccc3N(=O)=O)c1O)c23
Anthraquinone, 1,5-di-hydroxy-4,8-dinitro- (8CI)
9,10-Anthracenedione, 1,5-di-hydroxy-4,8-dinitro- (9CI)
Anthrarufin, 4,8-dinitro-
1,5-Dihydroxy-4,8-dinitro-anthraquinone
4,8-Dihydroxy-1,5-dinitro-anthraquinone
1,5-Dihydroxy-4,8-dinitro-9,10-anthracenedione
1,5-Dinitro-4,8-dihydroxy-anthraquinone
4,8-Dinitro-1,5-dihydroxy-anthraquinone
NSC-37584

128-94-9
$C_{14}H_{10}N_2O_4$
270.26
O=C(c(c(c(N)cc1)C(=O)c2c(N) ccc3O)c1O)c23
Anthraquinone, 1,8-diamino-4,5-dihydroxy
9,10-Anthracenedione, 1,8-di-amino-4,5-dihydroxy-
1,8-Diaminochrysazine
4,5-Diaminochrysazin
1,8-Diamino-4,5-dihydroxy-anthrachinon (Czech)
1,8-Diamino-4,5-dihydroxy-anthraquinone
1,8-Diamino-4,5-dihydroxy-9,10-anthraquinone
4,5-Diamino-1,8-dihydroxy-anthraquinone
1,8-Dihydroxy-4,5-diamino-anthrachinon (Czech)

128-95-0
$C_{14}H_{10}N_2O_2$
238.26
O=C(c(c(C(=O)c1c(N)ccc2N)ccc 3)c3)c12
Anthraquinone, 1,4-diamino
Acetate Red Violet R
Acetoquinone Light Heliotrope NL
Acetylon Fast Red Violet R
Amacel Heliotrope R
Amaplast Red Violet P 2R
9,10-Anthracenedione, 1,4-di-amino- (9CI)
1,4-Anthraquinonyldiamine
Artisil Direct Violet 2RP
Artisil Violet 2RP
Celanthrene Red Violet R
1,4-Diaminoanthrachinon (Czech)
Celliton Fast Red Violet
Celliton Fast Red Violet R
Celliton Fast Red Violet RN
Celliton Fast Red Violet RNA-CF
Celutate Red Violet RH
C.I. 61100
Cibacete Violet 2R
Cibacet Violet E2R
Cibacet Violet 2R
Cilla Fast Red Violet RN
C.I. Disperse Violet 1
C.I. Solvent Violet 11
Diacelliton Fast Violet 5R
1,4-Diaminoanthraquinone
Disperse Violet K
Dispersive Violet K
Duranol Violet 2R
Fenacet Fast Violet 5R
Gracet Violet 2R
Grasol Violet R
Interchem Acetate Red Violet RRLF
Interchem Acetate Violet R
Interchem Hisperse Violet 2RH
Krisolamine
Microsetile Violet 3R
Mideton Fast Red Violet R
Miketon Fast Red Violet R
Nacelan Violet 4R
Nyloquinone Violet R
Oil Violet R
Oracet Violet 2R
Perliton Violet 3R
Resiren Violet TR
Seacyl Violet R
Sectacyl Violet Propyl
Serisol Brilliant Violet 2R
Setacyl Violet P-R
Setacyl Violet R
Setile Violet 3R
Supracet Brilliant Violet 3R
Transetile Violet P 3R

128-99-4
$C_{20}H_{15}N_3O_8S_2$
489.47
O=C(c(c(c(C(=O)c1c(N)c(S(=O) (=O)O)cc2Nc(ccc(S(=O)(=O) O)c3N)c3)ccc4)c4)c12
2-Anthracenesulfonic acid, 1-amino-4-((3-amino-4-sulfo-phenyl)amino)-9,10-dihydro-9,10-dioxo- (9CI)

129-00-0
$C_{16}H_{10}$
202.26
c(c(c(c(cc1)ccc2)c2cc3)(c1ccc4)c34
Pyrene
Benzo(def)phenanthrene
Pyren (German)
β-Pyrene

129-03-3
$C_{21}H_{21}N$
287.43
CN1CCC(CC1)=C3c2ccccc
2C=Cc4ccccc34
Piperidine, 4-(5H-dibenzo-(a,d)cyclohepten-5-ylidene)-1-methyl
Cyproheptadine
4-(5-Dibenzo(a,d)cyclohepten-5-ylidine)-1-methylpiperidine
4-(5H-Dibenzo(a,d)cyclohepten-5-ylidene)-1-methylpiperidine
Dronactin
MK 141
Periactin
Periactine
Periactinol

129-06-6
$C_{19}H_{15}O_4.Na$
330.33
[Na+].CC(=O)CC(c1ccccc1)c3c
([O-])c2ccccc2oc3=O
Coumarin, 3-(α-acetonyl-benzyl)-4-hydroxy-, sodium salt
3-(α-Acetonylbenzyl)-4-hydroxy-coumarin sodium salt
Athrombin
2H-1-Benzopyran-2-one, 4-hydroxy-3-(3-oxo-1-phenyl-butyl)-, sodium salt (9CI)
Coumadin sodium
Coumafene sodium
Cumadin
Marevan
Marevan (sodium salt)
Panivarfin
Panwarfin
Prothromadin
Ratsul Soluble
Sodium, ((3-(α-acetonylbenzyl)-2-oxo-2H-1-benzopyran-4-yl)-oxy)-
Sodium coumadin
Sodium warfarin
Tintorane
Varfine
Waran
Warcoumin
Warfarin sodium
Warfarin, sodium deriv.
Warfarin, sodium salt
Warfilone

129-09-9
$C_{28}H_{14}N_2O_2S_2$
474.56
O=C(c(c(c(SC(=N1)c(cccc2)c2)
c1c3)C(=O)c4ccc(N=C(S5)c
(cccc6)c6)c57)c3)c47
Anthra(2,1-d:6,5-d')bisthi-azole-6,12-dione, 2,8-di-phenyl
C.I. 67300
C.I. Vat Yellow 2
Vat Yellow 2
Zlut Kypova 2 (Czech)
Zlut Ostanthrenova GC (Czech)

129-15-7
$C_{15}H_9NO_4$
267.25
Cc3ccc2C(=O)c1ccccc1C(=O)
c2c3N(=O)=O
Anthraquinone, 2-methyl-1-nitro
9,10-Anthracenedione, 2-methyl-1-nitro- (9CI)
2-Methyl-1-nitro-9,10-anthra-cenedione
2-Methyl-1-nitroanthraquinone
NCI-C01923
1-Nitro-2-methylanthraquinone
1-N-2-MA (Russian)

129-17-9
$C_{27}H_{31}N_2O_6S_2.Na$
566.71
Ammonium, (4-(α-(p-(diethyl-amino)phenyl)-2,4-disulfo-benzylidene)-2,5-cyclo-hexadien-1- ylidene)diethyl-, hydroxide, monosodium salt
Acid Blue 1
Acid Blue V
Acid Bright Azure Z
Acid Brilliant Blue VF
Acid Brilliant Blue Z
Acid Brilliant Sky Blue Z
Acid Brilliant Sky Blue Z
Acid Leather Blue V
Aizen Brilliant Acid Pure Blue VH
Alphazurine 2G
Amacid Blue V
Anhydro-4,4'-bis(diethylamino)-triphenylmethanol-2',4''-di-sulphonic acid, monosodium salt
Bleu Patente V
Blue 1084
1085 Blue
Blue URS
Blue VRS
Brilliant Acid Blue A Export
Brilliant Acid Blue V Extra
Brilliant Acid Blue VS
Brilliant Blue GS
Bucacid Patent Blue VF
Carmin Blue VS
Carmine Blue VF
C.I. 712
C.I. 42045
C.I. Acid Blue 1
C.I. Acid Blue 3
C.I. Acid Blue 1, Sodium salt
C.I. Food Blue 3
Cosmetic Green Blue R25396
4,4'-Di(diethylamino)-4',6'-di-sulphotriphenylmethanol an-hydride, sodium salt
Disulfine Blue VN
Disulphine VN
Disulphine Blue VN 150
E 131
Edicol Supra Blue VR
Erio Brilliant Blue V
Erioglaucine
Erioglaucine Supra
Fenazo Blue XF
Fenazo Blue XV
Food Blue 3
Hexaco Blue VRS
Hexacol Blue VRS
Hidacid Blue V
Intracid Pure Blue V
Kiton Pure Blue V
Kiton Pure Blue V.FQ
L-Blau 3
Leather Blue G
Lissamine Turquoise VN
Merantine Blue VF
Modr Kysela 1 (Czech)
Modr Potravinarska 3 (Czech)
Patentblau V (German)
Patent Blue
Patent Blue V
Patent Blue VF
Patent Blue VF-CF
Patent Blue VF Special
Patent Blue VS
Pontacyl Brilliant Blue
Pontacyl Brilliant Blue V
Schultz Nr. 826 (German)
Sodium Blue VRS
Sodium Patent Blue V
Sulfacid Brilliant Blue 6J
Sulfan Blue
Sulphan Blue
Sumitomo Patent Pure Blue VX
Tetracid Carmine Blue V
Xylene Blue VS

129-20-4
$C_{19}H_{20}N_2O_3$
324.41
CCCCC1C(=O)N(N(C1=O)c2ccc
(O)cc2)c3ccccc3
3,5-Pyrazolidinedione, 4-butyl-1-(p-hydroxyphenyl)-2-phenyl
Artroflog
BM 1
Butaflogin
Butanova
Butapirone
Butilene
4-Butyl-1-(p-hydroxyphenyl)-2-phenyl-3,5-pyrazolidine-dione
4-Butyl-1-(4-hydroxyphenyl)-2-phenyl-3,5-pyrazolidine-dione
4-Butyl-2-(p-hydroxyphenyl)-1-phenyl-3,5-pyrazolidine-dione
4-Butyl-2-(4-hydroxyphenyl)-1-phenyl-3,5-dioxopyrazoli-dine
Crovaril
Deflogin
3,5-Dioxo-1-phenyl-2-(p-hydroxyphenyl)-4-n-butyl-pyrazolidene
Etrozolidina
Flamaril
Flanaril

Flogal
Floghene
Flogistin
Flogitolo
Flogodin
Flogoril
Flogostop
Flopirina
Frabel
G 27202
p-Hydroxyphenylbutazone
1-(p-Hydroxyphenyl)-2-phenyl-4-butyl-3,5-pyrazolidinedione
1-p-Hydroxyphenyl-2-phenyl-3,5-dioxo-4-N-butylpyrazolidine
Idrobutazina
Infamil
Infammil
Ipabutona
Iridil
Isobutazina
Isobutil
Metabolite I
Neo-Farmadol
Neofen
Offitril
Oxalid
Oxazolidin-Geigy
Oxibutol
Oxi-Fenibutol
Oxifenylbutazon
Oxazolidin
Oxiphenbutazone
Oxyphenbutazone
Oxyphenobutazone
Oxyphenylbutazone
1-Phenyl-2-(p-hydroxyphenyl)-3,5-dioxo-4-butylpyrazolidine
Pirabutina
Piraflogin
Poliflogil
Remazin
Reumox
Rumapax
Tandacote
Tandalgesic
Tandearil
Tanderal
Tanderil
Telidal
Tendearil
Valioil
Visubutina
USAF GE-14

129-35-1
$C_{15}H_9ClO_2$
256.69
O=C(c(c(C(=O)c1c(c(cc2)C)Cl)ccc3)c3)c12
1-Chloro-2-methylanthraquinone
9,10-Anthracenedione, 1-chloro-2-methyl- (9CI)
Anthraquinone, 1-chloro-2-methyl- (8CI)
1-Chloro-2-methyl-9,10-anthracenedione
NSC-504

129-40-8
$C_{14}H_6ClNO_4$
287.66
O=C(c(c(c(cc1)Cl)C(=O)c2cccc3N(=O)=O)c1)c23
Anthraquinone, 1-chloro-5-nitro
9,10-Anthracenedione, 1-chloro-5-nitro-
1-Chlor-5-nitroanthrachinon (Czech)
1-Chloro-5-nitroanthraquinone
5-Chloro-1-nitroanthraquinone
1-Nitro-5-chloroanthraquinone

129-42-0
$C_{14}H_{10}N_2O_2$
238.26
O=C(c(c(c(N)cc1)C(=O)c2c(N)ccc3)c1)c23
Anthraquinone, 1,8-diamino
9,10-Anthracenedione, 1,5-diamino- (9CI)
9,10-Anthracenedione, 1,8-diamino-
1,5-Anthraquinonyldiamine
1,8-Anthraquinonyldiamine
1,8-Diaminoanthrachinon (Czech)
1,5-Diaminoanthraquinone
1,8-Diaminoanthraquinone

129-43-1
$C_{14}H_8O_3$
224.22
O=C(c(c(c(O)cc1)C(=O)c2cccc3)c1)c23
Anthraquinone, 1-hydroxy
9,10-Anthracenedione, 1-hydroxy-
1-Hydroxyanthrachinon (Czech)
1-Hydroxyanthraquinone
1-Hydroxy-9,10-anthraquinone

129-44-2
$C_{14}H_{10}N_2O_2$
238.26
O=C(c(c(c(N)cc1)C(=O)c2cccc3N)c1)c23
Anthraquinone, 1,5-diamino
9,10-Anthracenedione, 1,5-diamino-
1,5-Anthraquinonyldiamine
C.I. Disperse Red II
1,5-DAA (Russian)
1,5-Diaminoanthrachinon (Czech)
1,5-Diaminoanthraquinone
1,5-Diamino-9,10-anthraquinone

129-66-8
$C_7H_3N_3O_8$
257.13
OC(=O)c1c(cc(cc1N(=O)=O)N(=O)=O)N(=O)=O
Benzoic acid, trinitro- (10% to 30% water)
Benzoic acid, trinitro- (Dry)
Trinitrobenzoic acid, Dry [UN 0215]
Trinitrobenzoic acid, Wet, At least 10% water, over 25 lbs. in one outside packaging (DOT)
Trinitrobenzoic acid, Wet, Containing less than 30% water [UN 0215]
Trinitrobenzoic acid, Wet, Containing not less than 30% water [UN 1355]
UN 0215 [Trinitrobenzoic acid, Dry or wetted with less than 30 per cent water, by mass]
UN 1355 [Trinitrobenzoic acid, Wetted with not less than 30 per cent water, by mass]

129-67-9
$C_8H_8O_5.2Na$
230.14
7-Oxabicyclo(2.2.1)heptane-2,3-dicarboxylic acid, disodium salt
Accelerate
Aguathol
Des-I-Cate
Dinatrium-(3,6-epoxy-cyclohexaan-1,2-dicarboxylaat) (Dutch)
Dinatrium-(3,6-epoxy-cyclohexan-1,2-dicarboxylat) (German)
Disodium 3,6-endoxohexahydrophthalate
Disodium 3,6-epoxycyclohexane-1,2-dicarboxylate
Disodium 7-oxabicyclo(2.2.1)-heptane-2,3-dicarboxylate
Disodium salt of endothall
Disodium salt of 7-oxabicyclo-(2.2.1)heptane-2,3-dicarboxylic acid
Endotal
Endothal
Endothall
Endothal-natrium (Dutch)
Endothal-sodium
Endothal Weed Killer
3,6-Endoxohexahydrophthalic acid disodium salt
(3,6-Epossi-cicloesan-1,2-di-carbossilato) disodico (Italian)
3,6-Epoxy-cyclohexane 1,2-carboxylate disodique (French)
Herbicide 273
Hydout
Hydrothol
Niagrathal
RCRA waste number P088
Ripenthol
Tri-Endothal

129-79-3
$C_{13}H_5N_3O_7$
315.21
O=C(c(c(c1ccc(N(=O)=O)c2)c(N(=O)=O)cc3N(=O)=O)c3)

c12
Fluoren-9-one, 2,4,7-trinitro
2,4,7-Trinitrofluoren-9-one
2,4,7-Trinitro-9-fluorenone

129-83-9
$C_{17}H_{26}N_2O$
274.39
Phenampromide
DEA No. 9638
Fenampromida (Spanish)
Fenampromide
N-(1-Methyl-2-piperidinoethyl)-propionanilide
Phenampromid
Phenampromidum (Latin)
Propanamide, N-(1-methyl-2-(1-piperidinyl)ethyl)-N-phenyl- (9CI)
Propionanilide, N-(1-methyl-2-piperidinoethyl)-

129-96-4
$C_{10}H_8O_8S_2.2Na$
366.28
2,7-Naphthalenedisulfonic acid, 4,5-dihydroxy-, disodium salt (9CI)
Chromotropic acid disodium salt
Disodium chromotropate
Disodium 1,8-dihydroxylnaphthalene-3,6-disulfonate
Disodium-1,8-dihydroxynaphthalene-3,6-disulfonate
Disodium naphthalene 1,8-dihydroxy-3,6-disulfonate
NSC-4883
Sodium chromotropate
Sodium 1,8-dioxynaphthalene-3,6-disulfonate

130-01-8
$C_{18}H_{25}NO_5$
335.44
Senecionan-11,16-dione, 12-hydroxy
Aureine
12-Hydroxysenecionan-11,16-dione
Senecionin
Senecionine

130-13-2
$C_{10}H_9NO_3S.Na$
246.24
1-Naphthalenesulfonic acid, 4-amino-, monosodium salt (9CI)
AI3-19501
4-Amino-1-naphthalene sulfonic acid, sodium salt
4-Amino-1-napthalene sulfonic acid, sodium salt
Naphthemol
1,4-Naphthionic monosodium salt
Naphthionine
NSC-168
Sodium 1-aminonaphthalene-4-sulfonate
Sodium naphthionate
Sodium α-naphthylamine-4-sulfonate
Sodium 1-naphthylamine-4-sulfonate

130-15-4
$C_{10}H_6O_2$
158.16
O=C(c(c(C(=O)C=1)ccc2)c2)C1
1,4-Naphthoquinone
1,4-Dihydro-1,4-diketo-naphthalene
1,4-Naftochinon (Czech)
1,4-Naphthalenedione
α-Naphthoquinone
RCRA waste number U166
USAF CY-10

130-16-5
C_9H_6ClNO
179.61
n(c(c(c(cc1)Cl)cc2)c1O)c2
8-Quinolinol, 5-chloro
5-Chloro-8-hydroxyquinoline
Chloroxyquinoline
5-Chloro-8-quinolinol

130-17-6
$C_{14}H_{12}N_2O_3S_2$
320.40
O=S(=O)(O)c(c(SC(=N1)c(ccc(N)c2)c2)c1cc3)c3C
7-Benzothiazolesulfonic acid, 2-(p-aminophenyl)-6-methyl
2-(p-Aminophenyl)-6-methyl-benzothiazolyl-7-sulfonic acid

130-20-1
$C_{28}H_{12}Cl_2N_2O_4$
511.31
O=C(c(c(c(Nc(c(N1)c(c(C(=O)c(c2ccc3)c3)c4)C2=O)c4Cl)c1c5Cl)C(=O)c6cccc7)c5)c67
5,9,14,18-Anthrazinetetrone, 7,16-dichloro-6,15-dihydro-(8CI,9CI)
Ahcovat Blue BCF
Alizanthrene Blue RC
Amanthrene Blue BCL
Blue K
C.I. Vat Blue 6
Calcoloid Blue BLC
Calcoloid Blue BLR
Carbanthrene Blue RCS
Cibanone Blue FG
Cibanone Blue GF
D & C Blue No. 9
Dichloroindanthrone
7:16-Dichloro-6:15-indanthrone
Fenan Blue BCS
Fenanthren Blue BC
Harmone B 79
Helanthrene Blue BC
Indanthren Blue BC
Indanthrene Blue BC
Indanthrene Blue BCF
Indo Blue B-I
Intravat Blue GF
Mikethrene Blue BC
Monolite Fast Blue 2RV
Navinon Blue BC
Nihonthrene Brilliant Blue RCL
NSC-74700
Nyanthrene Blue BFP
Ostanthren Blue BCL
Palanthrene Blue BCA
Paradone Blue RC
Pernithrene Blue BC
Ponsol Blue BCS
Romantrene Blue FBC
Sandothrene Blue NGR
Solanthrene Blue B
Solanthrene Blue F-SBA
Tinon Blue GF
Vat Blue 6
Vat Green B
Vat Sky Blue K

130-22-3
$C_{14}H_7O_7S.Na$
342.26
2-Anthracenesulfonic acid, 9,10-dihydro-3,4-dihydroxy-9,10-dioxo-, monosodium salt
Acid Red Alizarine
Ahcoquinone Red S
Alizarin Carmine (Biological stain)
Alizarine Carmine Indicator
Alizarine Red A
Alizarine Red AS
Alizarine Red Indicator
Alizarine Red S (Biological stain)
Alizarine Red S Sodium salt
Alizarine Red SW
Alizarine Red SZ
Alizarine Red W
Alizarine Red WA
Alizarine Red for Wool
Alizarine Red WS
Alizarine S
Alizarine S Extra Conc. A Export
Alizarine S Extra Pure A
Alizarin Red S
Alizarinrot-S (German)
Alizarin S
Alizarinsulfonate
Calcochrome Alizarine Red SC
Carnelio Rubine Lake
Chrome Red Alizarine
C.I. 58005
C.I. Mordant Red 3
Diamond Red W
9,10-Dihydro-3,4-dihydroxy-9,10-dioxo-2-anthracene-sulfonic acid monosodium salt
Ext. D & C Red No. 7
Fenakrom Red W
Mitsui Alizarine Red S
Oxanal Fast Red SW
Sodium alizarinesulfonate
Sodium alizarin-3-sulfonate

130-26-7
C_9H_5ClINO
305.50
n(c(c(c(cc1I)Cl)cc2)c1O)c2
8-Quinolinol, 5-chloro-7-iodo
Alchloquin
Amebil
Amoenol
Bactol
Barquinol
Budoform
Chinoform
5-Chlor-7-jod-8-hydroxy-chinolin (German)
5-Chloro-8-hydroxy-7-iodo-quinoline
5-Chloro-7-iodo-8-hydroxy-quinoline
Chloroiodoquine
5-Chloro-7-iodo-8-quinolinol
Clioquinol
Cliquinol
Eczecidin
Emaform
Entero-Bio Form
Enterum Locorten
Enteroquinol
Enteroseptol
Entero-Vioform
Enterozol
Entrokin
HI-Enterol
Hydriodide-Enterol
Iodochloroxyquinoline
Iodenterol
Iodochlorhydroxyquin
Iodochlorhydroxyquinol
Iodochlorhydroxyquinoline
7-Iodo-5-chloro-8-hydroxy-quinoline
7-Iodo-5-chloroxine
Iodoenterol
Nioform
Quinambicide
Rometin
Vioform
Vioform N.N.R.

130-34-7
$C_{16}H_{10}N_4O_8S_2$
450.39
O=S(=O)(O)c(cc(S(=O)(=O)O)c(c1C(=NN(N=2)c(ccc(N(=O)=O)c3)c3)C2C=4)C4)c1
2H-Naphtho(1,2-d)triazole-6,8-disulfonic acid, 2-(4-nitrophenyl)- (9CI)

130-37-0
$C_{11}H_{10}O_5S.Na$
277.25
[Na+].CC2(CC(=O)c1ccccc1C2=O)S([O-])(=O)=O
2-Naphthalenesulfonic acid, 1,2,3,4-tetrahydro-2-methyl-1,4-dioxo-, sodium salt (9CI)
Bisulfite sodique de menadione (French)
Bisulfito sodico de menadiona (Spanish)
Menachinonum natrium bisulfurosum
Menadione sodio bisolfito
Menadioni natrii bisulfis (Latin)
Menadioni natrii hydrogensulfis
Natrium menadionsulfonicum
Vikasolum

130-59-6
$C_{10}H_5N_3O_6S$
295.21
O=S(=O)(O)c(c(c(c(N=NO1)c12)cc(N(=O)=O)c3)c3)c2
Naphth(1,2-d)(1,2,3)oxadiazole-5-sulfonic acid, 8-nitro-(9CI)

130-60-9
$C_{12}H_{14}N_2O_5.Na$
289.23
Phenol, 2-cyclohexyl-4,6-dinitro-, sodium salt (8CI,9CI)

130-80-3
$C_{24}H_{28}O_4$
380.52
4,4'-Stilbenediol, α,α'-diethyl-, dipropionate, (E)
Clinestrol
Cyren B
DESD
Dibestil
trans-4,4'-(1,2-Diethyl-1,2-ethenediyl)bisphenol dipropionate
α,α'-Diethyl-4,4'-stilbenediol, dipropionate
α,α'-Diethyl-4,4'-stilbenediol trans-dipropionate
trans-α,α'-Diethyl-4,4'-stilbenediol dipropionate
Diethylstilbene dipropionate
α,α'-Diethyl-4,4'-stilbenediol dipropionyl ester
Diethylstilbesterol dipropionate
Diethylstilbestrol dipropionate
Diethylstilbestrol propionate
Dihydroxydiethylstilbene dipropionate
4,4'-Dihydroxy-α,β-diethylstilbene dipropionate
Dipropionato de estilbene (Spanish)
p,p'-Dipropionoxy-trans-α,β-diethylstilbene
Distilbene
Estilben
Estilbin
Estroben
Estroben DF
Estrobene
Estrobene DP
Estrogenin
Estrostilben
Euvestin
Gynolett
Horfemine
Neo-Oestranol II
Neo-Oestronol II
New-Oestranol II
Oestrogynaedron
Orestol
Pabestrol
Pabestrol D
Sinciclan
4,4'-Stilbenediol, α,α'-diethyl-, dipropionate, trans-
Stilbestrol, diethyl dipropionate
Stilbestrol dipropionate
Stilbestrol propionate
Stilbestronate
Stilboestrol
Stilboestrol dipropionate
Stilboestrol DP
Stilbofax
Stilronate
Synestrin
Synoestron
Syntestrin
Syntestrine
Willestrol

130-86-9
$C_{20}H_{19}NO_5$
353.40
CN5CCc2cc1OCOc1cc2C(=O)Cc4ccc3OCOc3c4C5
7,13a-Secoberbin-13a-one, 7-methyl-2,3:9,10-bis-(methylenedioxy)
Biflorine
Bis(1,3)benzodioxolo(4,5-c:5',6'-g)azecin-13(5H)-one, 4,6,7,14-tetrahydro-5-methyl-
Corydinine
Fumarine
Macleyine
Protopine

130-89-2
$C_{20}H_{24}N_2O_2.ClH$
360.92
Quinine, monohydrochloride
Chinimetten
Cinchonan-9-ol, 6'-methoxy-, monohydrochloride, (8-α,9R)- (9CI)
Quinine chloride
Quinine hydrochloride
Quinine muriate

130-95-0
$C_{20}H_{24}N_2O_2$
324.46
O(c(ccc(nccc1C(O)C(N(CCC2C3C=C)C3)C2)c14)c4)C
Quinine
Chinin (German)
Cinchonan-9-ol, 6'-methoxy-, (8-α,9R)- (9CI)
6-Methoxycinchonine
α-(6-Methoxy-4-quinolyl)-5-vinyl-2-quinuclidinemethanol
(-)-Quinine
2-Quinuclidinemethanol, α-(6-methoxy-4-quinolyl)-5-vinyl-

131-08-8
$C_{14}H_7O_5S.Na$
310.26
2-Anthracenesulfonic acid, 9,10-dihydro-9,10-dioxo-, sodium salt
9,10-Anthraquinone-2-sodium sulfonate
2-Anthraquinonesulfonate sodium
Anthraquinone-2-sulfonate sodium salt
2-Anthraquinonesulfonic acid sodium salt
Silver salt
2-Sulfoanthraquinone sodium salt
Sodium-2-anthrachinone-sulphonate
Sodium β-anthraquinone-sulfonate
Sodium 2-anthraquinone-sulfonate
Sodium 9,10-anthraquinone-2-sulfonate

131-09-9
$C_{14}H_7ClO_2$
242.66
O=C(c(c(C(=O)c1cccc2)ccc3Cl)c3)c12
Anthraquinone, 2-chloro
9,10-Anthracenedione, 2-chloro-
2-Chloro-9,10-anthracenedione
2-Chloroanthraquinone

131-11-3
$C_{10}H_{10}O_4$
194.20
O=C(OC)c(c(ccc1)C(=O)OC)c1
Phthalic acid, dimethyl ester
Avolin
1,2-Benzenedicarboxylic acid, dimethyl ester
Dimethyl 1,2-benzenedicarboxylate
Dimethyl benzeneorthodicarboxylate
Dimethylester kyseliny ftalove (Czech)
Dimethyl phthalate (ACGIH, OSHA)
DMP
ENT 262
Fermine
Methyl phthalate
Mipax
NTM
Palatinol M
Phthalic acid methyl ester
Phthalsaeuredimethylester (German)
RCRA waste number U102
Solvanom
Solvarone

131-15-7
$C_{24}H_{38}O_4$
390.62
O=C(OC(CCCCCC)C)c(c(ccc1)C(=O)OC(CCCCCC)C)c1
Phthalic acid, bis(1-methylheptyl) ester
Bis(2-octyl)phthalate
Bis-(2-oktyl)ester kyseliny ftalove (Czech)
Capryl o-phthalate
Dicapryl 1,2-benzenedicarboxylate
Dicapryl phthalate
Dioctanol-2-phthalate
Monoplex DCP
Phthalic acid, bis(2-octyl) ester
Phthalic acid, dicapryl ester
Phthalic acid, di-2-octyl ester

131-16-8
$C_{14}H_{18}O_4$
250.32
O=C(OCCC)c(c(ccc1)C(=O)OCCC)c1
Phthalic acid, dipropyl ester
1,2-Benzenedicarboxylic acid, dipropyl ester
Dipropyl phthalate
Di-n-propyl phthalate

131-17-9
$C_{14}H_{14}O_4$
246.28
O=C(OCC=C)c(c(ccc1)C(=O)OCC=C)c1
1,2-Benzenedicarboxylic acid, di-2-propenyl ester
Dapon 35
Dapon R
Diallylester kyseliny ftalove (Czech)
Diallyl phthalate
NCI-C50657
Phthalic acid, diallyl ester
o-Phthalic acid, diallyl ester

131-18-0
$C_{18}H_{26}O_4$
306.44
O=C(OCCCCC)c(c(ccc1)C(=O)OCCCCC)c1
Phthalic acid, dipentyl ester
Amoil
Amyl phthalate
1,2-Benzenedicarboxylic acid, dipentyl ester
Diamyl phthalate
Dipentyl phthalate
Di-n-pentylphthalate
DPP

131-22-6
$C_{16}H_{13}N_3$
247.32
N(=Nc(cccc1)c1)c(c(c(c(N)c2)ccc3)c3)c2
1-Naphthalenamine, 4-(phenylazo)
C.I. 11350
C.I. Solvent Yellow 4
α-Naphthyl Red
Nubian Yellow TB
Phenylazo α-naphthylamine
4-Phenylazo-1-naphthylamine

131-27-1
$C_{10}H_9NO_6S_2$
303.32
O=S(=O)(O)c(c(c(c(S(=O)(=O)O)cc1)cc2N)c1)c2
1,5-Naphthalenedisulfonic acid, 3-amino
Acid IV
2-Amino-4,8-naphthalenedisulfonic acid
3-Amino-1,5-naphthalenedisulfonic acid
7-Amino-1,5-naphthalenedisulfonic acid
C Acid
4,8-Disulfo-2-naphthalamine
Kyselina C (Czech)
Kyselina 2-naftylamin-4,8-disulfonova (Czech)
β-Naphthylaminedisulfonic acid
β-Naphthylamine-4,8-disulfonic acid
2-Naphthylamine-4,8-disulfonic acid

131-28-2
$C_{23}H_{27}NO_8$
445.51
COc3ccc(C(=O)Cc2c(CCN(C)C)cc1OCOc1c2OC)c(C(O)=O)c3OC
o-Veratric acid, 6-((6-(2-(dimethylamino)ethyl)-2-methoxy-3,4-(methylenedioxy)-phenyl)acetyl)
6-((6-(2-(Dimethylamino)ethyl)-2-methoxy-3,4-(methylenedioxy)phenyl)acetyl)-o-veratric acid
Narcein
Narceine

131-52-2
$C_6Cl_5O.Na$
288.30
[Na+].[O-]c1c(Cl)c(Cl)c(Cl)c(Cl)c1Cl
Phenol, pentachloro-, sodium salt
Dow Dormant Fungicide
Dowicide G-ST
Dowicide G
Napclor-G
Pentachlorophenate sodium
Pentachlorophenol, sodium salt
Pentachlorophenoxy sodium
Pentaphenate
Santobrite
Sodium PCP
Sodium pentachlorophenate [UN 2567]
Sodium pentachlorophenol

Sodium pentachlorophenolate
Sodium pentachlorophenoxide
Sodium, (pentachlorophenoxy)-
Sodium pentachlorphenate
UN 2567 [Sodium pentachlorophenate]
Weedbeads

131-53-3
$C_{14}H_{12}O_4$
244.26
O=C(c(c(O)ccc1)c1)c(c(O)cc(OC)c2)c2
Benzophenone, 2,2'-dihydroxy-4-methoxy
Advastab 47
Benzophenone-8
Cyasorb UV 24
Cyasorb UV 24 Light Absorber
2,2'-Dihydroxy-4-methoxybenzophenone
Dioxybenzon
Dioxybenzone
Methanone, (2-hydroxy-4-methoxyphenyl)(2-hydroxyphenyl)-(9CI)
Spectra-Sorb UV 24
UF 2
UV 24

131-54-4
$C_{15}H_{14}O_5$
274.29
O=C(c(c(O)cc(OC)c1)c1)c(c(O)cc(OC)c2)c2
Benzophenone, 2,2'-dihydroxy-4,4'-dimethoxy
Benzophenone-6
Cyasorb UV 12
Methanone, bis(2-hydroxy-4-methoxyphenyl)- (9CI)
Uvinul D 49

131-55-5
$C_{13}H_{10}O_5$
246.23
O=C(c(c(O)cc(O)c1)c1)c(c(O)cc(O)c2)c2
Benzophenone, 2,2',4,4'-tetrahydroxy
Benzophenone-2
Methanone, bis(2,4-dihydroxyphenyl)- (9CI)
2,2',4,4'-Tetrahydroxy-benzophenone
2,4,2',4'-Tetrahydroxybenzophenone
THBP
Uvinol D-50
Uvinul D-50

131-56-6
$C_{13}H_{10}O_3$
214.23
O=C(c(cccc1)c1)c(c(O)cc(O)c2)c2
Benzophenone, 2,4-dihydroxy
2,4-Dihydroxybenzofenon (Czech)
2,4-Dihydroxybenzophenone
Eastman Inhibitor DHPB
Quinsorb 010
Syntase 100
UF 1
USAF DO-28
USAF ND-54
Uvinul 400

131-57-7
$C_{14}H_{12}O_3$
228.26
O=C(c(cccc1)c1)c(c(O)cc(OC)c2)c2
Benzophenone, 2-hydroxy-4-methoxy
Benzophenone-3
Cyasorb UV 9
2-Hydroxy-4-methoxybenzophenone
(2-Hydroxy-4-methoxyphenyl)-phenylmethanone
Methanone, (2-hydroxy-4-methoxyphenyl)phenyl-
4-Methoxy-2-hydroxybenzophenone
MOB
NCI-C60957
NSC-7778
Oxybenzone
Spectra-Sorb UV 9
Syntase 62
UF 3
USAF CY-9
Uvinul M 40

131-70-4
$C_{12}H_{14}O_4$
222.26
O=C(OCCCC)c(c(ccc1)C(=O)O)c1
Phthalic acid, monobutyl ester
1,2-Benzenedicarboxylic acid, monobutyl ester (9CI)
Butyl hydrogen phthalate
MBP
Monobutyl phthalate
Mono-n-butyl phthalate

131-72-6
$C_{18}H_{24}N_2O_6$
364.40
CCCCCCCCc1cc(cc(N(=O)=O)c1OC(=O)C=CC)N(=O)=O
Dinocap
(E)-2-(1-Methylheptyl)-4,6-dinitrophenyl 2-butenoate (9CI)
2-(1-Methylheptyl)-4,6-dinitrophenyl crotonate

131-73-7
$C_{12}H_5N_7O_{12}$
439.24
O=N(=O)c(cc(N(=O)=O)c(Nc(c(N(=O)=O)cc(N(=O)=O)c1)c1N(=O)=O)c2N(=O)=O)c2
Diphenylamine, 2,2',4,4',6,6'-hexanitro
Aurantia
Bis(2,4,6-trinitro-phenyl)-amin (German)
C.I. 10360
Diphenylamine, hexanitro-
Dipicrylamine
Dipikrylamin (Czech)
DPA
Esanitrodifenilamina (Italian)
2,2',4,4',6,6'-Hexanitrodifenylamin (Czech)
Hexanitrodifenylamine (Dutch)
Hexanitrodiphenylamine (French)
2,2',4,4',6,6'-Hexanitrodiphenylamine
2,4,6,2',4',6'-Hexanitrodiphenylamine
Hexyl (German, Dutch)

131-74-8
$C_6H_6N_4O_7$
246.14
Ammonium-picrate
Ammonium carbazoate
Ammonium picrate, Dry or containing, by weight, less than 10% water [UN 0004]
Ammonium picrate (Wet)
Ammonium picrate, Wet with 10% or more water [UN 1310]
Ammonium picronitrate
Explosive D
Obeline picrate
Phenol, 2,4,6-trinitro-, ammonium salt (9CI)
Picrate of ammonia (DOT)
Picratol
Picric acid, ammonium salt
RCRA waste number P009
2,4,6-Trinitrophenol ammonium salt
UN 0004 [Ammonium picrate, Dry or wetted with less than 10 per cent water, by mass]
UN 1310 [Ammonium picrate, Wetted with not less than 10 per cent water, by mass]

131-79-3
$C_{17}H_{15}N_3$
261.35
2-Naphthylamine, 1-(o-tolylazo)
A.F. Yellow No. 3
C.I. 11390
Cerisol Yellow TB
C.I. Food Yellow 11
C.I. Solvent Yellow 6
Dolkwal Yellow OB
Ext. D & C Yellow No. 10
FD & C Yellow 4
FD & C Yellow No. 4
Jaune OB
1-(2-Methylphenyl)azo-2-naphthalenamine
1-((2-Methylphenyl)azo)-2-naphthalenamine

1-(2-Methylphenyl)azo-2-naphthylamine
Oil Yellow OB
Oil Yellow OB Pure
o-Toluene-1-azo-2-naphthylamine
1-(o-Tolylazo)-2-naphthylamine
Yellow OB
Zlut Maselna OB (Czech)
Zlut Rozpoustedlova 6 (Czech)

131-89-5
$C_{12}H_{14}N_2O_5$
266.28
Phenol, 2-cyclohexyl-4,6-dinitro
6-Cicloesil-2,4-dinitr-fenolo (Italian)
2-Cyclohexyl-4,6-dinitrofenol (Dutch)
2-Cyclohexyl-4,6-dinitrophenol
6-Cyclohexyl-2,4-dinitrophenol
DINEX
Dinitrocyclohexylphenol (DOT)
Dinitro-o-cyclohexylphenol
2,4-Dinitro-6-cyclohexylphenol
4,6-Dinitro-o-cyclohexylphenol
DN
DN 1
DN Dry Mix No. 1
DN Dust No. 12
DNOCHP
Dowspray 17
Dry Mix No. 1
ENT 157
NA 9026 (DOT)
Pedinex (French)
Phenol, 6-cyclohexyl-2,4-dinitro-
RCRA waste number P034
SN 46

131-91-9
$C_{10}H_7NO_2$
173.18
O=Nc(c(c(ccc1)cc2)c1)c2O
2-Naphthol, 1-nitroso
C.I. 10000
α-Nitroso-β-naftol (Czech)
1-Nitroso-2-naftol (Czech)
Nitroso-β-naphthol
α-Nitroso-β-naphthol
1-Nitroso-2-naphthol
Zelen Moridlova 4 (Czech)

131-92-0
$C_{42}H_{23}N_3O_6$
665.64
Benzamide, N,N'-(10,15,16,17-tetrahydro-5,10,15,17-tetraoxo-5H-dinaphtho(2,3-a:2',3'-i)carbazole-4,9-diyl)bis- (9CI)

132-20-7
$C_{16}H_{20}N_2.C_4H_4O_4$
356.46
Pyridine, 2-(α-(2-(dimethylamino)ethyl)benzyl)-, maleate (1:1)
Avil-Retard
Daneral
2-(α-(2-(Dimethylamino)ethyl)-benzyl)pyridine, bimaleate
2-(α-(2-(Dimethylamino)ethyl)-benzyl)pyridine, maleate
1-(N,N-Dimethylamino)-3-(phenyl-3-α-pyridyl)propane maleate
HO 11513
Inhiston
Pheniramine maleate
Phenyl(2-pyridyl)(β-N,N-dimethylaminomethyl) methane maleate
1-Phenyl-1-(2-pyridyl)-3-dimethylaminopropane maleate
Prophenpyridamine maleate
Trimeton maleate
Trimetose

132-22-9
$C_{16}H_{19}ClN_2$
274.82
n(c(ccc1)C(c(ccc(c2)Cl)c2)CCN(C)C)c1
Pyridine, 2-(p-chloro-α-(2-(dimethylamino)ethyl)benzyl)
Allergican
Allergisan
2-(p-Chloro-α-(2-(dimethylamino)ethyl)benzyl)pyridine
4-Chloropheniramine
Chlorophenylpyridamine
1-(p-Chlorophenyl)-1-(2-pyridyl)-3-dimethylaminopropane
1-(p-Chlorophenyl)-1-(2-pyridyl)-3-N,N-dimethylpropylamine
Chloropiril
Chloroprophenpyridamine
Chlorphenamine
Chlorpheniramine
Chlorprophenpyridamine
Chlor-Trimeton
Chlor-Tripolon
Clorfeniramina (Italian)
Cloropiril
Haynon
Histadur
Piriton
Polaronil
2-Pyridinepropanamine, γ-(4-chlorophenyl)-N,N-dimethyl- (9CI)

132-27-4
$C_{12}H_9O.Na$
192.20
[Na+].[O-]c1ccccc1c2ccccc2
2-Biphenylol, sodium salt
Bactrol
(1,1'-Biphenyl)-2-ol, sodium salt
D.C.S.
Dorvicide A
Dowicide
Dowicide A
Dowicide A & A Flakes
Dowizid A
2-Hydroxybiphenyl sodium salt
2-Hydroxydiphenyl sodium
2-Hydroxydiphenyl, sodium salt
Mil-Du-Rid
Mystox WFA
Natriphene
OPP-Na
OPP-Sodium
Orphenol
Phenol, o-phenyl-, sodium deriv.
o-Phenylphenol, sodium salt
2-Phenylphenol sodium salt
Preventol-ON
Preventol ON & ON Extra
Sodium 2-biphenylolate
Sodium (1,1'-biphenyl)-2-olate
Sodium, (2-biphenylyloxy)-
Sodium 2-hydroxydiphenyl
Sodium ortho phenylphenate
Sodium o-phenylphenate
Sodium 2-phenylphenate
Sodium o-phenylphenol
Sodium o-phenylphenolate
Sodium o-phenylphenoxide
SOPP
Stopmold B
Topane

132-29-6
$C_{24}H_{19}O_4P$
402.40
Phosphoric acid, 2-biphenylyl diphenyl ester

132-32-1
$C_{14}H_{14}N_2$
210.30
N(c(c(c1cc(N)cc2)ccc3)c3)(c12)CC
Carbazole, 3-amino-9-ethyl
3-Amino-N-ethylcarbazole
3-Amino-9-ethylcarbazole

132-43-4
$C_{24}H_{47}NO_4S.Na$
468.69
Ethanesulfonic acid, 2-(cyclohexyl(1-oxohexadecyl)amino)-, sodium salt (9CI)

132-53-6
$C_{10}H_7NO_2$
173.18
O=Nc(c(O)c(c(ccc1)c2)c1)c2
1-Naphthol, 2-nitroso
2-Nitroso-1-naphthol

132-60-5
$C_{16}H_{11}NO_2$
249.28
OC(=O)c1cc(nc2ccccc12)c3ccccc3
Cinchoninic acid, 2-phenyl
Aciphenochinoline

Aciphenochinolinium
Agotan
Alutyl
Artam
Artexin
Atigoa
Atocin
Atofan
Atophan
Cinchophen
Cinchophene
Cinchophenic acid
Cinconal
Cincophen
Cincosal
Ikterosan
Mylofanol
Phenophan
Phenoquin
2-Phenylcinchonic acid
2-Phenylcinchoninic acid
2-Phenylquinoline-4-carboxylic acid
2-Phenyl-4-quinolinecarboxylic acid
Polyphlogin
Quinofen
4-Quinolinecarboxylic acid, 2-phenyl- (9CI)
Quinophan
Quinophen
Rhematan
Rheumin
Tervalon
Tophol
Tophosan
Traubofan
Vantyl
Viophan

132-64-9
$C_{12}H_8O$
168.20
O(c(c(c1cccc2)ccc3)c3)c12
Dibenzofuran
2,2'-Biphenylene oxide
2,2'-Biphenylylene oxide
Dibenzo(b,d)furan
Diphenylene oxide

132-65-0
$C_{12}H_8S$
184.26
S(c(c(c1cccc2)ccc3)c3)c12
Dibenzothiophene (9CI)
AI3-00043
(1,1'-Biphenyl)-2,2'-diyl sulfide
2,2'-Biphenylylene sulfide
Dibenzo(b,d)thiophene
Diphenylene sulfide
NSC-2843
α-Thiafluorene
9-Thiafluorene

132-66-1
$C_{18}H_{13}NO_3$
291.32
OC(=O)c1ccccc1C(=O)Nc2cccc3ccccc23
Phthalamic acid, N-1-naphthyl
ACP 322
Alanap
Alanap 1
Alanape
Alanap 10G AT
Ancrack
Dyanap
Kyselina N-1-naftylftalamova (Czech)
Mor-Cran
Naftalam (Czech)
2-((1-Naphthalenylamino)-carbonyl)benzoic acid
N-1-Naphthylphthalamate
α-Naphthylphthalamic acid
N-1-Naphthylphthalamic acid
N-1-Naphthyl-phthalamidsaeure (German)
Naptalam
Naptalame
Naptro
Nip-A-Thin
NPA
NPA-3
PA
Peach-Thin
Premerge Plus
6Q8
Solo

132-67-2
$C_{18}H_{12}NO_3.Na$
313.30
[Na+].OC(=O)c1ccccc1C(=O)Nc2cccc3ccccc23
Phthalamic acid, N-1-naphthyl-, monosodium salt
ACP 322
Alanap-3
α-Naphthylphthalamic acid sodium salt
N-1-Naphthylphthalamic acid sodium salt
Naptalam sodium
NPA
NPA-3
NPA, sodium salt
Phthalamic acid, N-1-naphthyl-, sodium salt
Sodium N-1-naphthylphthalamate
Sodium N-1-naphthylphthalamic acid
Sodium NPA

132-68-3
$C_{21}H_{15}NO_2$
313.35
O=C(Nc(c(c(ccc1)cc2)c1)c2)c(c(O)cc(c3ccc4)c4)c3
2-Naphthalenecarboxamide, 3-hydroxy-N-1-naphthalenyl- (9CI)
Acco Naphthol AS-BO
Acna Naphthol F
Amanil Naphthol AS-BO
Amarthol AS-BO
Anthonaphthol M3B
Azonaphtol AN
Azotol ANF
Brenthol AN
C.I. Azoic Coupling Component 4
C.I. 37560
Celcot RN
Cibanaphthol RN
Dragonthol BO
Hiltonaphthol AS-BO
3-Hydroxy-2-naphthoic-α-naphthalide
1-(2',3'-Hydroxynaphthoyl-amino)naphthalene
Mitsui Naphthozol BO
Naftolo MBO
Naphtanilide BO
Naphtanilide BO Supra
Naphtazol 3B
2-Naphthamide, 3-hydroxy-N-1-naphthyl- (8CI)
Naphthoide BO
Naphthol ACNA F
Naphthol AS-BO
Naphtoelan BO
Naphtol AS-BO
Naphtol AS-BOLL
NSC-37202
Sanatol BO
Solunaptol ANL
Tulathol AS-BO
Ultrazol VII-BO

132-75-2
$C_{12}H_9N$
167.22
N#CCc(c(c(ccc1)cc2)c1)c2
Acetonitrile, (1-naphthyl)
α-Naphthyl acetonitrile
α-(1-Naphthyl)acetonitrile

132-86-5
$C_{10}H_8O_2$
160.18
Oc(c(c(ccc1)cc2O)c1)c2
1,3-Naphthalenediol
Naphthoresorcinol

132-87-6
$C_{17}H_{13}NO_5S$
343.35
O=C(Nc(ccc(c1cc(S(=O)(=O)O)c2)c2O)c1)c(cccc3)c3
2-Naphthalenesulfonic acid, 7-(benzoylamino)-4-hydroxy- (9CI)

132-93-4
$C_{17}H_{19}N_2O_5S.K$
402.54
4-Thia-1-azabicyclo(3.2.0)heptane-2-carboxylic acid, 3,3-dimethyl-7-oxo-6-(2- phenoxypropionamido)-, monopotassium salt
Alfacillin
Alfocillin
Alpen
Astracillin

Bendralan
Bl P 152
BRL 152
Brocsil
Broxil
Chemipen
Chemipen-C
Darcil
Dramcillin-S
K Phenethicillin
Maxipen
Oralopen
PEN 200
Pensig
Phenethicillin K
Phenethicillin K salt
Phenethecillin potassium
Phenethicillin potassium salt
Pheneticillin potassium
Pheno-m-penicillin
Phenoxyaethylpenicillin K-salz (German)
α-Phenoxyethylpenicillin
α-Phenoxyethylpenicillin potassium
α-Phenoxyethylpenicillin potassium salt
Potassium methylphenoxy-methylpenicillin
Potassium phenethicillin
Potassium α-phenoxyethyl penicillin
Potassium (1-phenoxyethyl)-penicillin
Potassium 6-(α-phenoxypro-pionamido)penicillanate
Priospen
Ro-Cillin
Semopen
Synapen
Syncillin
Synerpenin
Synthecillin
Synthecilline

132-98-9
$C_{16}H_{17}N_2O_5S.K$
388.51
[K+].CC3(C)SC2C(N=C(O)CO c1ccccc1)C(=O)N2C3C ([O-])=O
4-Thia-1-azabicyclo(3.2.0)-heptane-2-carboxylic acid, 3,3-dimethyl-7-oxo-6-(2-phenoxy- acetamido)-, monopotassium salt
Antibiocin
Apsin VK
Arcacil
Arcasin
Beromycin
Beromycin 400
Beromycin (Penicillin)
Betapen-Vk
Calciopen K
Cliacil
Compocillin-VK
Distakaps V-K
DISTAQUAINE V-K
Dowpen V-K
DQV-K
Fenoxypen
Isocillin
Icipen
Ispenoral
Ledercillin VK
Megacillin Oral
Oracil-VK
Orapen
Ospeneff
Pedipen
Penagen
Pencompren
Penicillin potassium phenoxy-methyl
Penicillin V potassium
Penicillin V potassium salt
Pen-Vee-K
Pen-Vee-K Powder
Pen-V-K Powder
Penvikal
Pfizerpen VK
D-α-Phenoxymethylpenicil-linate K salt
Phenoxymethylpenicillin potassium
Potassium penicillin V
Potassium penicillin V salt
Potassium phenoxymethyl-penicillin
PVK
Qidpen VK
Robicillin VK
Rocillin-VK
Roscopenin
SK-Penicillin VK
Stabillin VK Syrup 125
Stabillin VK Syrup 62.5
Sumapen VK
Suspen
V-Cil-K
Uticillin VK
V-Cillin K
Veetids
Vepen

133-06-2
$C_9H_8Cl_3NO_2S$
300.59
O=C(N(SC(Cl)(Cl)Cl)C(=O) C1CC=CC2)C12
4-Cyclohexene-1,2-dicar-boximide, N-(trichloro-methyl)thio
Aacaptan
Agrosol S
Agrox 2-Way and 3-Way
Amercide
Bangton
Bean Seed Protectant
Captab
Captaf
Captaf 85W
Captan (OSHA)
Captan-Streptomycin 7.5-0.1 Potato seed piece protectant
Captan 50W
Captancapteneet 26,538
Captane
Captex
ENT 26,538
Essofungicide 406
Flit 406
Fungus Ban Type II
Glyodex 3722
Granox PFM
Gustafson Captan 30-DD
Hexacap
1H-Isoindole-1,3(2H)-dione, 3a,4,7,7a-tetrahydro-2-((trichloromethyl)thio)-
Kaptan
Le captane (French)
Malipur
Merpan
Micro-Check 12
Neracid
NCI-C00077
Orthocide
Orthocide 7.5
Orthocide 50
Orthocide 406
Osocide
SR406
Stauffer Captan
3a,4,7,7a-Tetrahydro-N-(tri-chloromethanesulphenyl)-phthalimide
1,2,3,6-Tetrahydro-N-(trichloro-methylthio)phthalimide
Trichlormethylthioamid kysel-iny 1,2,3,6-tetrahydroftalove (Czech)
N-(Trichlor-methylthio)-phthal-imid (German)
N-Trichloromethylmercapto-4-cyclohexene-1,2-dicarbox-imide
N-(Trichloromethylmercapto)-δ^4-tetrahydrophthalimide
N-Trichloromethylthiocyclohex-4-ene-1,2-dicarboximide
N-(Trichloromethylthio)cyclo-hex-4-ene-1,2-dicarboximide
N-Trichloromethylthio-cis-δ^4-cyclohexene-1,2-dicar-boximide
N-((Trichloromethyl)thio)-4-cyclohexene-1,2-dicarbox-imide
Trichloromethylthio-1,2,5,6-te-trahydrophthalamide
N-((Trichloromethyl)thio)tetra-hydrophthalimide
N-Trichloromethylthio-3a,4,7,7a-tetrahydro-phthalimide
Vancide 89
Vancide 89RE
Vancide P-75
Vangard K
Vanguard K
Vanicide
Vondcaptan

133-07-3
$C_9H_4Cl_3NO_2S$
296.55
O=C(N(SC(Cl)(Cl)Cl)C(=O)c1cc cc2)c12
Phthalimide, N-((trichloro-methyl)thio)
Folpan

Folpel
Folpet
Ftalan
1H-Isoindole-1,3(2H)-dione, 2-((trichloromethyl)thio)-
Orthophaltan
Phaltan
Phthaltan
Thiophal
Trichlormethylthioimid kyseliny ftalove (Czech)
N-(Trichlor-methylthio)-phthalamid (German)
N-(Trichloromethylmercapto)-phthaliamide
N-(Trichlormethylthio)-phthalimide
2-((Trichloromethyl)thio)-1H-isoindole-1,3(2H)-dione
N-(Trichloromethylthio)phthalimide
Troysan Anti-Mildew O

133-13-1
$C_9H_{16}O_4$
188.22
O=C(OCC)C(C(=O)OCC)CC
Malonic acid, ethyl-, diethyl ester (8CI)
AI3-19481
Diethyl ethylmalonate
Diethyl 2-ethylmalonate
Diethyl ethylpropanedioate
NSC-8706
Propanedioic acid, ethyl-, diethyl ester (9CI)

133-14-2
$C_{14}H_6Cl_4O_4$
380.00
O=C(OOC(=O)c(c(cc(c1)Cl)Cl)c1)c(c(cc(c2)Cl)Cl)c2
Peroxide, bis(2,4-dichlorobenzoyl)-, Not more than 52% as a paste or in solution
Bis(2,4-dichlorobenzoyl)peroxide
Cadox TS
Cadox TS 40,50
2,4-Dichlorobenzoylperoxide, More than 75% with water (DOT)
2,4-Dichlorobenzoyl peroxide, Not more than 52% as a paste (DOT)
2,4-Dichlorobenzoyl peroxide, Not more than 52% in solution (DOT)
2,4-Dichlorobenzoyl peroxide, Not more than 75% with water (DOT)
Di-2,4-dichlorobenzoyl peroxide, Maximum concentration 52% as a paste or in solution (DOT)
Di-2,4-dichlorobenzoyl peroxide, Not more than 75% with water (DOT)
Luperco CST
Peroxide, bis(2,4-dichlorobenzoyl)-, More than 75% with water
Peroxide, bis(2,4-dichlorobenzoyl)-, Not more than 75% with water
UN 2137 (DOT)
UN 2138 (DOT)
UN 2139 (DOT)

133-16-4
$C_{13}H_{19}ClN_2O_2$
270.79
CCN(CC)CCOC(=O)c1ccc(N)cc1Cl
Benzoic acid, 4-amino-2-chloro-, 2-(diethylamino)ethyl ester
Chloroprocaine
2-(Diethylamino)ethyl 4-amino-2-chlorobenzoate

133-18-6
$C_{15}H_{15}NO_2$
241.31
O=C(OCCc(cccc1)c1)c(c(N)ccc2)c2
Anthranilic acid, phenethyl ester
Benzoic acid, 2-amino-, 2-phenylethyl ester
Benzylcarbinyl anthranilate
β-Phenethyl-o-aminobenzoate
2-Phenylethyl-o-aminobenzoate
Phenethyl anthranilate
Phenylethyl anthranilate
2-Phenylethyl anthranilate

133-32-4
$C_{12}H_{13}NO_2$
203.26
O=C(O)CCCC(c(c(N1)ccc2)c2)=C1
1H-Indole-3-butanoic acid
Butyric acid, 4-(indolyl)-
Hormex Rooting Powder
Hormodin
IBA
Indole butyric
Indole butyric acid
β-Indolebutyric acid
γ-(Indole-3)-butyric acid
3-Indolebutyric acid
3-Indolyl-γ-butyric acid
γ-(3-Indolyl)butyric acid
γ-(Indol-3-yl)butyric acid
Indolyl-3-butyric acid
4-(Indol-3-yl)butyric acid
4-(3-Indolyl)butyric acid
Jiffy Grow
Kyselina 4-indol-3-ylmaselina (Czech)
Rootone
Rootone F
Seradix

133-35-7
$C_6H_{14}N_2.C_4H_6O_6$
264.27
Piperazine, 1,4-dimethyl-, (R-(R*,R*))-2,3-dihydroxybutanedioate (1:1) (9CI)
Piperazine, 1,4-dimethyl-, tartrate
Piperazine, 1,4-dimethyl-, tartrate (1:1) (8CI)

133-37-9
$C_4H_6O_6$
150.09
O=C(O)C(O)C(O)C(=O)O
Butanedioic acid, 2,3-dihydroxy-, (R*,R*)-(+-)-(9CI)
(R*,R*)-(+-)-2,3-Dihydroxybutanedioic acid
NSC-148314
Paratartaric acid
Racemic acid
Racemic tartaric acid
Resolvable tartaric acid
Tartaric acid D,L
DL-Tartaric acid
Tartaric acid, (+-)- (8CI)
(+-)-Tartaric acid
Traubensaure
Uvic acid

133-49-3
C_6HCl_5S
282.38
Sc(c(c(c(c1Cl)Cl)Cl)Cl)c1Cl
Benzenethiol, pentachloro
PCTP
Pentachloro-benzenethiol
Pentachlorothiophenol
Pentachlorthiofenol (Czech)

USAF B-51

133-53-9
$C_8H_8Cl_2O$
191.06
Oc(c(c(c(c1C)Cl)C)Cl)c1
Dichloro-m-xylenol
AI3-24011
Benzene, 2,4-dichloro-1,3-dimethyl-5-hydroxy-
DCMX
Decasept
2,4-Dichloro-3,5-dimethylphenol
Dichlorometaxylenol
Dichloroxylenol
2,4-Dichloro-m,5-xylenol
2,4-Dichloro-3,5-xylenol
Dichloroxylenolum (Latin)
Dicloroxilenol (Spanish)
Dicloroxilenolo
3,5-Dimethyl-2,4-dichlorophenol
Dixol
Hewsol
NSC-9774
Ottacide
Phenol, 2,4-dichloro-3,5-dimethyl- (9CI)
Prinsyl

3,5-Xylenol, 2,4-dichloro- (8CI)

133-59-5
$C_7H_7ClO_2S$
190.65
O=S(=O)(c(c(ccc1)C)c1)Cl
o-Tolylsulfonyl chloride
Benzenesulfonyl chloride, 2-methyl- (9CI)
2-Methylbenzenesulfonyl chloride
NSC-9354
ortho-Toluenesulfochloride
ortho-Toluenesulfonchloride
o-Toluenesulfonyl chloride (8CI)
ortho-Toluenesulfonyl chloride
o-Tosyl chloride

133-67-5
$C_8H_8Cl_3N_3O_4S_2$
380.66
O=S(=O)(N)c(c(cc(NC(NS1(=O)=O)C(Cl)Cl)c12)Cl)c2
2H-1,2,4-Benzothiadiazine-7-sulfonamide, 6-chloro-3-(dichloromethyl)-3,4-dihydro-, 1,1-dioxide
Achletin
Anatran
Anistadin
Aponorin
Carvacron
6-Chloro-3-(dichloromethyl)-3,4-dihydro-2H-1,2,4-benzothiadiazine-7-sulfonamide-1,1-dioxide
6-Chloro-3-(dichloromethyl)-3,4-dihydro-7-sulfamyl-1,2,4-benzothiadiazine-1,1-dioxide
3-Dichloromethyl-6-chloro-7-sulfamoyl-3,4-dihydro-1,2,4-benzothiadiazine-1,1-dioxide
3-Dichloromethyl-6-chloro-7-sulfamyl-3,4-dihydro-1,2,4-benzothiadiazine 1,1-dioxide
Diurese
Esmarin
Eurinol
Fluitran
Flutra
Gangesol
Hydrotrichlorothiazide
Intromene
Kubacron
Metahydrin
Nakva
Naqua
Salurin
Tachionin
Tolcasone
Trichlormetazid
Trichlormethiazide
Trichloromethiadiazide
Trichloromethiazide
Triclordiuride
Triclormetiazide (Italian)
Triflumen

133-74-4
$C_6H_6ClNO_3S$
207.63
Benzenesulfonic acid, 2-amino-5-chloro- (8CI,9CI)

133-90-4
$C_7H_5Cl_2NO_2$
206.03
O=C(O)c(c(c(N)cc1Cl)Cl)c1
Benzoic acid, 3-amino-2,5-dichloro
ACPM-629
ACP-M-728
Amben
Ambiben
Amiben
Amiben DS
Amibin
3-Amino-2,5-dichlorobenzoic acid
Amoben
Chlorambed
Chloramben
Chlorambene
2,5-Dichloro-3-aminobenzoic acid
Kyselina 3-amino-2,5-dichlorbenzoova (Czech)
NCI-C00055
Ornamental Weeder
Ornamental Weeder 4G
Vegaben
Vegiben

133-91-5
$C_7H_4I_2O_3$
389.91
O=C(O)c(c(O)c(cc1I)I)c1
Salicylic acid, 3,5-diiodo
Benzoic acid, 3,5-diiodo-2-hydroxy-
3,5-Diiodo-2-hydroxybenzoic acid

134-03-2
$C_6H_8O_6.Na$
199.13
L-Ascorbic acid, monosodium salt
Ascorbic acid sodium salt
l-Ascorbic acid sodium salt
Ascorbicin
Ascorbin
Cebitate
Cenolate
Iskia-C
Monosodium ascorbate
Natrascorb
Natri-C
Sodascorbate
Sodium ascorbate
Sodium L-ascorbate
Vitamin C
Vitamin C sodium

134-20-3
$C_8H_9NO_2$
151.18
O=C(OC)c(c(N)ccc1)c1
Anthranilic acid, methyl ester
o-Aminobenzoic acid methyl ester
2-Aminobenzoic acid methyl ester
Benzoic acid, 2-amino-, methyl ester (9CI)
o-Carbomethoxyaniline
2-Carbomethoxyaniline
2-(Methoxycarbonyl)aniline
Methyl o-aminobenzoate
Methyl 2-aminobenzoate
Methyl anthranilate
Methylester kyseliny anthranilove (Czech)
Neroli Oil, Artifical

134-25-8
$C_7H_4Cl_4$
229.92
c(ccc(c1Cl)C(Cl)Cl)(c1)Cl
Benzene, 2,4-dichloro-1-(dichloromethyl)- (9CI)
2,4-Dichloro-1-(dichloromethyl)benzene

134-29-2
$C_7H_9NO.ClH$
159.63
Cl.COc1ccccc1N
o-Anisidine, hydrochloride
o-Aminoanisole hydrochloride
2-Aminoanisole hydrochloride
2-Anisidine hydrochloride
o-Anisylamine hydrochloride
Benzenamine, 2-methoxy-, hydrochloride (9CI)
C.I. 37115
Fast Red BB Base
2-Methoxy-1-aminobenzene hydrochloride
o-Methoxyaniline hydrochloride
2-Methoxyaniline hydrochloride
2-Methoxybenzeneamine hydrochloride
o-Methoxyphenylamine hydrochloride
NCI-C03747

134-30-5
$C_9H_7NO.C_6H_8O_7$
337.28
8-Quinolinol citrate
Caswell No. 719AA
EPA Pesticide Chemical Code 059802
8-Hydroxyquinoline citrate
8-Quinolinol, 2-hydroxy-1,2,3-propanetricarboxylate (1:1) (Salt) (9CI)

134-31-6
$C_{18}H_{14}N_2O_2.H_2O_4S$

388.42
Oc1cccc2cccnc12.OS(O)(=O)=O
8-Quinolinol, sulfate (2:1) (Salt)
Chinosol
Cryptonol
Happy
8-Hydroxy-chinolin-sulfat (German)
8-Hydroxyquinoline sulfate
Octofen
Oxine sulfate
Oxyquinoline sulfate
8-Quinolinol, hydrogen sulfate (2:1)
8-Quinolinol sulfate
Solfato di 8-ossichinolina
Sunoxol
Superol

134-32-7
$C_{10}H_9N$
143.20
c(c(c(N)cc1)ccc2)(c2)c1
1-Naphthylamine
Alfanaftilamina (Italian)
alfa-Naftyloamina (Polish)
1-Aminonaftalen (Czech)
1-Aminonaphthalene
C.I. Azoic Diazo Component 114
Fast Garnet B Base
Fast Garnet Base B
α-Naftalamin (Czech)
1-Naftilamina (Spanish)
α-Naftylamin (Czech)
1-Naftylamin (Czech)
1-Naftylamine (Dutch)
Naphthalidam
Naphthalidine
1-Naphthylamin (German)
α-Naphthylamine (OSHA) [UN 2077]
RCRA waste number U167
UN 2077 [α-Naphthylamine]

134-36-1
$C_{40}H_{71}NO_{14}$
790.00
Erythromycin propionate
Erythromycin 2'-propanoate
Erythromycin 2'-propionate
Propionyl eryhthromycin

134-47-4
$C_{21}H_{16}N_2O_9S_2$
504.49
O=C(Nc(ccc(c1cc(S(=O)(=O)O)c2)c2O)c1)Nc(ccc(c3cc(S(=O)(=O)O)c4)c4O)c3
2-Naphthalenesulfonic acid, 7,7'-(carbonyldiimino)bis-(4-hydroxy- (9CI)
N,N'-Bis(1-hydroxy-3-sulfonaphthyl(6))urea
N,N'-Bis(1-oxy-3-sulfonaphthyl(6))urea
Carbonyl J Acid
I Acid Urea
2-Naphthalenesulfonic acid, 7,7'-ureylenebis(4-hydroxy-(8CI)
NSC-1699
Urea J Acid
6,6'-Ureylenebis(1-naphthol-3-sulfonic acid)

134-49-6
$C_{11}H_{15}NO$
177.27
CC1NCCOC1c2ccccc2
Morpholine, 3-methyl-2-phenyl
A 66
2-Fenyl-3-methylmorfolin (Czech)
3-Methyl-2-phenylmorpholine
Oxazimedrine
Phenmetrazin
Phenmetrazine
2-Phenyl-3-methylmorpholine
dl-2-Phenyl-3-methyltetrahydro-1,4-oxazine
Preludin
Probese-P
Psychamine A 66

134-50-9
$C_{13}H_{10}N_2.ClH$
230.71
[Cl-].Nc2c1ccccc1nc3ccccc23
Acridine, 9-amino-, hydrochloride
Acramine Yellow
9-Acridinamine, monohydrochloride (9CI)
Aminacrine hydrochloride
Aminoacridine hydrochloride
5-Aminoacridine hydrochloride
9-Aminoacridine hydrochloride
9-Aminoacridine monohydrochloride
Monacrin
Monacrin hydrochloride
NSC-7571

134-58-7
$C_4H_4N_6O$
152.14
Nc2nc1nn[nH]c1c(=O)[nH]2
7H-v-Triazolo(4,5-d)pyrimidin-7-one, 5-amino-1,6-dihydro
8 AG
5-Amino-1,6-dihydro-7H-v-triazolo(4,5-d)pyrimidin-7-one
5-Amino-1,4-dihydro-7H-1,2,3-triazolo(4,5-d)pyrimidin-7-one
5-Amino-7-hydroxy-1H-v-triazolo(d)pyrimidine
5-Amino-1H-v-triazolo(d)pyrimidin-7-ol
Azaguanine
Azaguanine-8
8-Azaguanine
Azan
AZG
B-28
Guanazol
Guanazolo
NSC-749
Pathocidin
Pathocidine
SF-337
SK 1150
Triazologuanine
v-Triazolo(4,5-d)pyrimidin-7-ol, 5-amino-
7H-1,2,3-Triazolo(4,5-d)pyrimidin-7-one, 5-amino-1,4-dihydro- (9CI)

134-62-3
$C_{12}H_{17}NO$
191.30
O=C(N(CC)CC)c(cccc1C)c1
m-Toluamide, N,N-diethyl
AI 3-22542
Autan
Baker's Antifol
Benzamide, N,N-diethyl-3-methyl-
Chemform
Deet
Delphene
m-Delphene
DET
DETA
m-DETA
DETA-20
Detamide
Dieltamid
N,N-Diethyl-3-methylbenzamide
Diethyltoluamide
Diethyl-m-toluamide
N,N-Diethyl-m-toluamide
ENT 20,218
ENT 22,542
m-DET
Flypel
Metadelphene
3-Methyl-N,N-diethylbenzamide
MGK Diethyltoluamide
Naugatuck DET
Off
Repel
Repper-Det
Repudin-Special
m-Toluic acid diethylamide

134-72-5
$C_{20}H_{30}N_2O_2.H_2O_4S$
428.60
CNC(C)C(O)c1ccccc1
Ephedrine sulfate (2:1) (Salt), (-)
l-Ephedrine sulfate
Isofedrol
1-α-(1-(Methylamino)ethyl)-benzyl alcohol sulfate
NCI-C55652
1-Phenyl-2-methylamine-propanol-1-sulfate

134-81-6
$C_{14}H_{10}O_2$

210.24
O=C(c(cccc1)c1)C(=O)c(cccc2)c2
Benzil
Dibenzoyl
Diphenyl-α,β-diketone
1,2-Diphenylethanedione
Diphenylglyoxal
Glyoxal, diphenyl-

134-83-8
$C_{13}H_{10}Cl_2$
237.13
c(cccc1)(c1)C(c(ccc(c2)Cl)c2)Cl
4-Chlorobenzhydryl chloride
Benzene, 1-chloro-4-(chlorophenylmethyl)- (9CI)
1-Chloro-4-(chlorophenylmethyl)benzene
Chloro(p-chlorophenyl)phenylmethane
Methane, chloro(p-chlorophenyl)phenyl- (8CI)
NSC-49126

134-84-9
$C_{14}H_{12}O$
196.26
O=C(c(cccc1)c1)c(ccc(c2)C)c2
Benzophenone, 4-methyl
p-Benzophenone, methyl-
4-Methyl benzophenone
Phenyl p-tolyl ketone
USAF DO-54

134-85-0
$C_{13}H_9ClO$
216.67
O=C(c(cccc1)c1)c(ccc(c2)Cl)c2
4-Chlorobenzophenone
AI3-00705
Benzophenone, 4-chloro- (8CI)
p-CBP
p-Chlorobenzophenone
para-Chlorobenzophenone
(4-Chlorophenyl)phenylmethanone
Methanone, (4-chlorophenyl)phenyl- (9CI)
NSC-2872

134-96-3
$C_9H_{10}O_4$
182.19
O=Cc(cc(OC)c(O)c1OC)c1
Benzaldehyde, 3,5-dimethoxy-4-hydroxy
3,5-Dimethoxy-4-hydroxybenzaldehyde
Gallaldehyde 3,5-dimethyl ether
4-Hydroksy-3,5-dwumetoksybenzaldehyd (Polish)
Syringaldehyde
Syringealdehyde
Syringic aldehyde
Syringylaldehyde

135-01-3
$C_{10}H_{14}$
134.24
Benzene, o-diethyl
o-Diethylbenzene

135-02-4
$C_8H_8O_2$
136.16
O=Cc(c(OC)ccc1)c1
o-Anisaldehyde
2-Anisaldehyde
Benzaldehyde, 2-methoxy- (9CI)
o-Methoxybenzaldehyde
2-Methoxybenzaldehyde
6-Methoxybenzaldehyde
2-Methoxybenzenecarboxaldehyde
Salicylaldehyde methyl ether

135-07-9
$C_9H_{11}Cl_2N_3O_4S_2$
360.25
CN2C(CCl)Nc1cc(Cl)c(cc1S2(=O)=O)S(N)(=O)=O
2H-1,2,4-Benzothiadiazine-7-sulfonamide, 6-chloro-3-(chloromethyl)-3,4-dihydro-, 1,1-dioxide
Aquatensen
Enduron
Methychlothiazide
Methyclothiazide
Methycyclothiazide
Methylchlorothiazide
Methylclothiazide
Methylcyclothiazide
NSC-110431

135-09-1
$C_8H_8F_3N_3O_4S_2$
331.31
NS(=O)(=O)c2cc1c(NCNS1(=O)=O)cc2C(F)(F)F
2H-1,2,4-Benzothiadiazine-7-sulfonamide, 3,4-dihydro-6-(trifluoromethyl)-, 1,1-dioxide
Bristab
Bristurin
Di-Ademil
Dihydroflumethazide
Dihydroflumethiazide
3,4-Dihydro-7-sulfamyl-6-trifluoromethyl-2H-1,2,4-benzothiadiazine 1,1-dioxide
3,4-Dihydro-6-trifluoromethyl-2H-1,2,4-benzothiadiazine-7-sulfonamide 1,1-dioxide
3,4-Dihydro-6-trifluoromethyl-7-sulfamoylbenzo-1,2,4-thiadiazine 1,1-dioxide
Diucardin
Elodrin
Finuret
Hydol
Hydrenox
Hydroflumethiazide
Leodrine
Metflorylthiazidine
Methforylthiazidine
Naclex
Olmagran
Rodiuran
Rontyl
Saluron
Sisuril
7-Sulfamyl-6-trifluoromethyl-3,4-dihydro-1,2,4-benzothiadiazine 1,1-dioxide
6-Trifluoromethyl-3,4-dihydro-7-sulfamoyl-2H-1,2,4-benzothiadiazine 1,1-dioxide
Trifluoromethylhydrothiazide
6-Trifluoromethyl-7-sulfamyl-3,4-dihydro-1,2,4-benzothiadiazine-1,1-dioxide
Vergonil

135-12-6
$C_{12}H_7Cl_2NO_3$
284.10
O=N(=O)c(c(Oc(ccc(c1)Cl)c1)ccc2Cl)c2
Ether, 4-chlorophenyl (4'-chloro-2'-nitro)phenyl
Benzene, 4-chloro-1-(4-chlorophenoxy)-2-nitro-
4-Chloro-2-nitrophenyl p-chlorophenyl ether
4,4'-Dichlor-2-nitrodifenylether (Czech)

135-19-3
$C_{10}H_8O$
144.18
Oc(ccc(c1ccc2)c2)c1
2-Naphthol
Azogen Developer A
C.I. 37500
C.I. Azoic Coupling Component 1
C.I. Developer 5
Developer A
Developer AMS
Developer BN
Developer Sodium
β-Hydroxynaphthalene
2-Hydroxynaphthalene
Isonaphthol
β-Monoxynaphthalene
β-Naftol (Dutch)
2-Naftol (Dutch)
β-Naftolo (Italian)
2-Naftolo (Italian)
2-Naphthalenol
Naphthol B
β-Naphthol
2-Naphtol (French)
β-Naphthyl alcohol
β-Naphthyl hydroxide
β-Naphtol (German)

135-20-6
$C_6H_6N_2O_2.H_4N$
156.19
[NH4+].[O-]N(N=O)c1ccccc1
Hydroxylamine, N-nitroso-

N-phenyl-, ammonium salt**
Ammonium N-nitrosophenylhydroxylamine
Cupferon (Czech)
Cupferron
N-Hydroxy-N-nitroso-benzenamine, ammonium salt
Kupferron (Czech)
NCI-C03258
N-Nitrosofenylhydroxylamin amonny (Czech)
N-Nitrosophenylhydroxylamin ammonium salz (German)
N-Nitrosophenylhydroxylamine ammonium salt

135-23-9
$C_{14}H_{19}N_3S.ClH$
297.88
Cl.CN(C)CCN(Cc1cccs1)c2ccccn2
Pyridine, 2-((2-(dimethylamino)ethyl)-2-thenylamino)-, monohydrochloride
Barhist
Capathyn
Compound 01013
Coryzol
2-((2-(Dimethylamino)ethyl)-2-thenyl-amino)pyridine hydrochloride
N,N-Dimethyl-N'-(2-pyridyl)-N'-(2-thenyl)ethylenediamine hydrochloride
N,N-Dimethyl-N'-(2-thenyl)-N'-(2-pyridyl-ethylene-diamine hydrochloride)
Dozar
1,2-Ethanediamine, N,N-dimethyl-N'-2-pyridinyl-N'-(2-thienylmethyl)-, monohydrochloride
Ethylenediamine, N,N-dimethyl-N'-(2-pyridyl)-N'-(2-thenyl)-, hydrochloride
Histafed
Histadyl
Histadyl hydrochloride
Histidyl
Lullamin
Methacon
Methapyrilene hydrochloride
Methoxylene
Pyrathyn
N-(2-Pyridyl)-N-(2-thienyl)-N,N'-dimethyl-ethylenediamine hydrochloride
Semikon
Semikon Hydrochloride
Somnicaps
Tem-Histine
Teralin
Thenyl D.P.E. Hydrochloride
Thenylene
Thenylene hydrochloride
Thenylpyramine hydrochloride
W-53 Hydrochloride
Win 2848 Hydrochloride Salt

135-37-5
$C_6H_{11}NO_5.2Na$
223.13
Glycine, N-(carboxymethyl)-N-(2-hydroxyethyl)-, disodium salt (9CI)
Caswell No. 404
Disodium N-(2-hydroxyethyl)-iminodiacetate
EPA Pesticide Chemical Code 039102

135-48-8
$C_{22}H_{14}$
278.35
c(c(cc(c1ccc2)c2)cc(c3cc(c4ccc5)c5)c4)(c1)c3
Pentacene (9CI)
Benzo(b)naphthacene
lin-Dibenzanthracene
2,3:6,7-Dibenzanthracene
lin-Naphthoanthracene
NSC-90784

135-51-3
$C_{10}H_6O_7S_2.2Na$
348.26
2,7-Naphthalenedisulfonic acid, 3-hydroxy-, disodium salt
Ferricon
2-Naphthol-3,6-disulfonic acid, sodium salt
R Salt

135-57-9
$C_{26}H_{20}N_2O_2S_2$
456.60
O=C(Nc(c(SSc(c(NC(=O)c(cccc1)c1)ccc2)c2)ccc3)c3)c(cccc4)c4
Benzanilide, 2',2'''-dithiobis
Benzamide, N,N'-(dithiodi-2,1-phenylene)bis-
o-(Benzoylamino)phenyl disulfide
Bis(o-benzamidophenyl) disulfide
Bis(2-benzamidophenyl) disulfide
Bis-o-benzoylaminofenyl-disulfid (Czech)
o,o'-Dibenzamidodiphenyl disulfide
Di-o-benzamidophenyl disulphide
2,2'-Dibenzoylaminodiphenyl disulfide
2',2'''-Dithiobisbenzanilide
2',2'''-Dithiodibenzanilide
N,N'-(Dithiodi-2,1-phenylene)-bisbenzamide
Peptazin BAFD
Peptisant 10
Pepton 22

135-61-5
$C_{18}H_{15}NO_2$
277.32
O=C(Nc(c(ccc1)C)c1)c(c(O)cc(c2ccc3)c3)c2
3-Hydroxy-4'-nitro-2-naphthanilide chloroacetate
Acco Naf-Sol AS-D
Acco Naphthol AS-D
Acna Naphthol E
Amanil Naphthol AS-D
Amarthol AS-D
Anthonaphthol AS-D
Azoground D
Azoic Coupling Component 18
Azonaphtol OT
Azotol OT
Brenthol OT
Brentosyn OTN
C.I. Azoic Coupling Component 18
C.I. Azoic Coupling Component 110
C.I. Developer 21
C.I. 37520
Celcot RTO
Cibanaphthol RTO
Daito Grounder D
Dianix Developer ND
Diathol D
Dragonthol D
Hiltonaphthol AS-D
1-(2',3'-Hydroxynaphthoyl-amino)-2-methylbenzene
Miketazol Developer NDF
Mitsui Naphthozol D
Naftolo MD
Naphtanilide D
Naphtanilide D Supra
Naphtazol D
2-Naphthalenecarboxamide, 3-hydroxy-N-(2-methyl-phenyl)- (9CI)
Naphthanil AS-D
Naphthoide AD
Naphthol AS D
Naphthol AS-D Dispersible
Naphthol AS-D Supra
2-Naphtho-o-toluidide, 3-hydroxy- (8CI)
Naphtoelan D
Naphtol AS-D
Naphtol AS-D Supra
NSC-37188
Solunaptol OT
Tulathol AS-D
Ultrazol D

135-62-6
$C_{18}H_{15}NO_3$
293.32
O=C(Nc(c(OC)ccc1)c1)c(c(O)cc(c2ccc3)c3)c2
3-Hydroxy-2-naphthoic acid o-aniside
Acna Naphthol O
Amanil Naphthol AS-OL
Amarthol AS-OL
Anthonaphthol MF
Azoground OL
Azonaphtol OA
Azotol OA
Brentosyn FR
C.I. Azoic Coupling Component 20
C.I. Developer 22

C.I. 37530
Celcot RK
Cibanaphthol RK
Daito Grounder OL
Diathol BO
Diathol OL
Dragonthol OL
Hiltonaphthol AS-OL
3-Hydroxy-2'-methoxy-2-naphthanilide
3-Hydroxy-N-(2-methoxyphenyl)-2-naphthalenecarboxamide
2-(3-Hydroxy-2-naphthamido)-anisole
3-Hydroxy-2-naphthoic o-anisidide
1-(2',3'-Hydroxynaphthoylamino)-2-methoxybenzene
Irganaphthol RK
Kambothol ASOL
2'-Methoxy-2-hydroxy-3-naphthanilide
3-(o-Methoxyphenylaminocarbonyl)-2-naphthol
3-(2-Methoxyphenylcarbamoyl)-2-naphthol
Miketazol Developer NLF
Mitsui Naphthozol OL
Naftolo MOL
Naphtanilide OL
Naphtazol F
2-Naphthalenecarboxamide, 3-hydroxy-N-(2-methoxyphenyl)- (9CI)
Naphthanil OL
2-Naphth-o-anisidide, 3-hydroxy- (8CI)
Naphthoide OL
Naphthol AS-OL
Naphtoelan OL
Naphtol AS-OL
NSC-50680
Solunaptol FRL
Tulathol AS-OL

135-63-7
$C_{18}H_{14}ClNO_2$
311.76
O=C(Nc(c(ccc1Cl)C)c1)c(c(O)cc(c2ccc3)c3)c2
2-Naphthalenecarboxamide, N-(5-chloro-2-methylphenyl)-3-hydroxy- (9CI)
Acco Naf-Sol AS-KB
Acco Naphthol AS-KB
Amanil Naphthol AS-KB
C.I. Azoic Coupling Component 21
C.I. 37526
Hiltonaphthol AS-KB
Naphtanilide KB
Naphtazol C
Naphthanilid KB
Naphthol AS-KB
2-Naphtho-o-toluidide, 5'-chloro-3-hydroxy- (8CI)
Naphtol AS-KB
NSC-37187

135-65-9
$C_{17}H_{12}N_2O_4$
308.28
O=C(Nc(cccc1N(=O)=O)c1)c(c(O)cc(c2ccc3)c3)c2
2-Naphthalenecarboxamide, 3-hydroxy-N-(3-nitrophenyl)- (9CI)
Acco Naphthol AS-BS
Acna Naphthol M
Amanil Naphthol AS-BS
Amarthol AS-BS
Anthonaphthol AS-BS
Azoground BS
Azoic Coupling Component 17
Azonaphtol MNA
Azotol MNA
Azotol NMA
Brenthol MN
C.I. Azoic Coupling Component 17
C.I. 37515
Celcot RM
Cibanaphthol RM
Daito Grounder BS
Diathol BS
Dragonthol BS
Hiltonaphthol AS-BS
Irganaphthol RM
Kambothol ASBS
Mitsui Naphthozol BS
Naftolo MBS
Naphtanilide BS
Naphtazol B
Naphthanil BS
2-Naphthanilide, 3-hydroxy-3'-nitro- (8CI)
Naphthoide BS
Naphthol AS-BS
Naphthol AS-BS Dispersible
Naphthol AS-BS Supra
Naphtoelan BS
Naphtol AS-BS
Naphtol AS-BS Supra
Naptanilide BS Supra
NSC-37168
Solunaptol MNL
Tulathol AS-BS
Ultrazol IV-BS

135-69-3
$C_{14}H_{11}NO_3$
241.24
O=C(c(ccc(c(ccc(N(=O)=O)c1)c1)c2)c2)C
Acetophenone, 4'-(p-nitrophenyl)- (8CI)
4-Acetyl-4'-nitrobiphenyl
Ethanone, 1-(4'-nitro(1,1'-biphenyl)-4-yl)- (9CI)
1-(4'-Nitro(1,1'-biphenyl)-4-yl)-ethanone
4'-(p-Nitrophenyl)acetophenone
NSC-43063

135-70-6
$C_{24}H_{18}$
306.41
c(c(cccc1)c1)(ccc(c(ccc(c(cccc2)c2)c3)c3)c4)c4
p-Quaterphenyl (8CI)
Benzerythrene
1,1'-Biphenyl, 4,4'-diphenyl-
4,4'-Diphenylbiphenyl
NSC-24860
Quadriphenyl
1,1':4',1'':4'',1'''-Quaterphenyl (9CI)
p-Tetraphenyl

135-76-2
$C_{10}H_8O_4S.Na$
247.23
2-Naphthalenesulfonic acid, 6-hydroxy-, monosodium salt (9CI)

135-88-6
$C_{16}H_{13}N$
219.30
N(c(ccc(c1ccc2)c2)c1)c(cccc3)c3
2-Naphthylamine, N-phenyl
Aceto PBN
Agerite
Agerite Powder
Anilinonaphthalene
2-Anilinonaphthalene
Antioxidant 116
Antioxidant PBN
N-Fenyl-2-aminonaftalen (Czech)
Fenyl-β-naftylamin (Czech)
N-(2-Naphthyl)aniline
2-Naphthylphenylamine
β-Naphthylphenylamine
NCI-C02915
Neozon D
Neozone
Neozone D
Nilox PBNA
Nonox D
PBNA
2-Phenylaminonaphthalene
Phenyl-β-naphthylamine
Phenyl-2-naphthylamine
N-Phenyl-β-naphthylamine (ACGIH)
N-Phenyl-2-naphthylamine
Stabilizator AR

135-91-1
$C_{21}H_{30}N_2$
310.47
N(c(ccc(c1)Cc(ccc(N(CC)CC)c2)c2)c1)(CC)CC
Benzenamine, 4,4'-methylenebis(N,N-diethyl- (9CI)
Benzenamine, 4,4'-methylenebis(N,N-diethyl-
4,4'-Methylenebis(N,N-diethylbenzenamine)

135-98-8
$C_{10}H_{14}$
134.24
c(cccc1)(c1)C(CC)C
Benzene, sec-butyl
sec-Butylbenzene [UN 2709]
2-Phenylbutane

UN 2709 [Butyl benzenes]

136-23-2
$C_{18}H_{38}N_2S_4Zn$
476.19
CCCCN(CCCC)C(=S)S[Zn]SC(=S)N(CCCC)CCCC
Zinc, bis(dibutyldithiocarbamato)
Aceto ZDBD
Bis(dibutyldithiocarbamato)zinc
Butazate
Butazate 50-D
Butyl zimate
Butyl ziram
Carbamic acid, dibutyldithio-, zinc complex
Dibutyldithiocarbamic acid zinc salt
USAF GY-5
Vulcacure
Vulkacit LDB/C
Zimate, butyl
Zinc bibutyldithiocarbamate
Zinc dibutyldithiocarbamate
Zinc N,N-dibutyldithiocarbamate

136-24-3
$C_6H_3Cl_3O.1/2Zn$
230.15
Phenol, 2,4,5-trichloro-, zinc salt (9CI)
Zinc, bis(2,4,5-trichlorophenoxy)-

136-25-4
$C_{11}H_9Cl_5O_3$
366.45
CC(Cl)(Cl)C(=O)OCCOc1cc(Cl)c(Cl)cc1Cl
Propionic acid, 2,2-dichloro-, 2-(2,4,5-trichlorophenoxy)-ethyl ester
Baron
2,2-Dichloropropionic acid, 2-(2,4,5-trichlorophenoxy)-ethyl ester
ERBN
Erbon
Ethanol, 2-(2,4,5-trichlorophenoxy)-, 2,2-dichloropropionate
Novege
Novon
Pentanate
2-(2,4,5-Trichlorfenoxy)ethyl-ester kyseliny 2,2-dichlorpropionove (Czech)
2-(2,4,5-Trichlorophenoxy)ethyl 2,2-dichloropropionate
2,4,5-Trichlorophenoxyethyl-α,α-dichloropropionate

136-26-5
$C_{14}H_{29}NO_3$
259.38
O=C(N(CCO)CCO)CCCCCCCCC
Capramide DEA
N,N-Bis(2-hydroxyethyl)decanamide
Capric acid diethanolamide
Decanamide, N,N-bis(2-hydroxyethyl)- (9CI)
Upamide CD

136-30-1
$C_9H_{18}NS_2.Na$
227.39
Carbamic acid, dibutyldithio-, sodium salt
Butyl namate
Dibutyldithiocarbamic acid sodium salt
Dibutyldithiokarbaman sodny (Czech)
Pennac
Sodium DBDT
Sodium dibutyldithiocarbamate
Tepidone
Tepidone Rubber Accelerator
USAF B-35
Vulcacur
Vulcacure

136-32-3
$C_6H_2Cl_3O.Na$
219.42
Phenol, 2,4,5-trichloro-, sodium salt
Dowicide B
Preventol 1
Sodium salt of 2,4,5-trichlorophenol
Sodium 2,4,5-trichlorophenate
Sodium, (2,4,5-trichlorophenoxy)-
2,4,5-Trichlorophenol, sodium salt

136-35-6
$C_{12}H_{11}N_3$
197.26
N(=NNc(cccc1)c1)c(cccc2)c2
Triazene, 1,3-diphenyl
Aniline, N-(phenylazo)-
Cellofor (Czech)
DAAB
Diazoaminobenzen (Czech)
Diazoaminobenzene
p-Diazoaminobenzene
Diazoaminobenzol (German)
1,3-Diphenyltriazene

136-36-7
$C_{13}H_{10}O_3$
214.23
O=C(Oc(cccc1O)c1)c(cccc2)c2
Resorcinol, monobenzoate
Benzoic acid, m-hydroxyphenyl ester
Eastman Inhibitor RMB
3-Hydroxyphenyl benzoate

136-40-3
$C_{11}H_{11}N_5.ClH$
249.73
Cl.Nc2ccc(N=Nc1ccccc1)c(N)n2
Pyridine, 2,6-diamino-3-(phenylazo)-, monohydrochloride
Azodine
Azodium
Azodyne
Azo Gastanol
Azo Gantrisin
Azo-Mandelamine
Azomine
Azo-Standard
Azo-Stat
Azotrex
Baridium
Bisteril
Cystamine (McClung)
Cystopyrin
Cystural
2,6-Diamino-3-phenylazopyridine hydrochloride
2,6-Diamino-3-(phenylazo)-pyridine monohydrochloride
Di-Azo
Diridone
Dolonil
Eucistin
Giracid
Mallofeen
Mallophene
NC 150
NCI-C01672
Nefrecil
Pap
PDP
Phenazo
Phenazodine
Phenazopyridine hydrochloride
Phenazopyridinium chloride
Phenylazo
Phenylazodiaminopyridine hydrochloride
β-Phenylazo-α,α'-diaminopyridine hydrochloride
3-Phenylazo-2,6-diaminopyridine hydrochloride
Phenylazo-α,α'-diaminopyridine monohydrochloride
Phenylazopyridine hydrochloride
3-(Phenylazo)-2,6-pyridinediamine, hydrochloride
Phenylazo Tablets
Phenyl-Idium
Phenyl-Idium 200
Pirid
Piridacil
Pyrazodine
Pyrazofen
Pyredal
Pyridacil
Pyridenal
Pyridene
Pyridiate
2,6-Pyridinediamine, 3-(phenylazo)-, monohydrochloride
Pyridium
Pyridivite
Pyripyridium
Pyrizin
Sedural

Suladyne
Sulodyne
Thiosulfil-A Forte
Urazium
Uridinal
Uriplex
Urobiotic-250
Urodine
Urofeen
Uromide
Urophenyl
Uropyridin
Uropyrine
Utostan
Vestin
W 1655

136-44-7
$C_{10}H_{13}NO_4$
211.24
O=C(OCC(O)CO)c(ccc(N)c1)c1
1,2,3-Propanetriol, p-aminobenzoate
p-Aminobenzoic acid monoglyceryl ester
Escalol 106
Glycerol, 1-p-aminobenzoate
Glyceryl para-aminobenzoate
Monoglycerol p-aminobenzoate
1,2,3-Propanetriol, 1-(4-aminobenzoate)

136-45-8
$C_{13}H_{17}NO_4$
251.28
Di-n-propyl isocinchomeronate
AI3-17591
Caswell No. 400
Dipropylester kyseliny pyridin-2,5-dikarboxylove (Czech)
Dipropyl isocinchomeronate
Di-propylisocinchomeronate
Di-n-propyl-isocinchomeronate (German)
Dipropyl pyridine-2,5-dicarboxylate
Dipropyl 2,5-pyridinedicarboxylate
Di-n-propyl 2,5-pyridinedicarboxylate
ENT 17,591
EPA Pesticide Chemical Code 047201
Isocinchomeronic acid, dipropyl ester
Isocinchomeronyl dipropylester
MGK R-326
MGK Repellent-326
MGK 326
NSC-22364
Pyridin-2,5-dicarbonsaeure-di-n-propylester (German)
2,5-Pyridinedicarboxylic acid, dipropyl ester (8CI,9CI)
R-326
Repper 333

136-47-0
$C_{15}H_{24}N_2O_2.ClH$
300.82
Tetracaine hydrochloride
Amethocaine hydrochloride
Anacel
Anethaine
Benzoic acid, p-(butylamino)-, 2-(dimethylamino)ethyl ester, monohydrochloride
Benzoic acid, 4-(butylamino)-, 2-(dimethylamino)ethyl ester, monohydrochloride
Butethanol
p-(Butylamino)benzoic acid, 2-(dimethylamino)ethyl ester, hydrochloride
p-Butylaminobenzoyl-2-dimethylaminoethanol hydrochloride
Butylocaine
Curtacain
Decicain
Decicaine
Dicaine hydrochloride
Dicainum
Dikain hydrochloride
2-Dimethylaminoethanol 4-N-butylaminobenzoate hydrochloride
Dimethylaminoethyl-p-N-butylaminobenzoate hydrochloride
2-(Dimethylamino)ethyl p-(butylamino)benzoate hydrochloride
2-(Dimethylamino)ethyl p-(butylamino)benzoate monohydrochloride
Gingicain M
Menonasal
Pantocaine hydrochloride
Sterile Tetracaine Hydrochloride
Tetracaine hydrochloride
Tonexol

136-51-6
$C_8H_{16}O_2.1/2Ca$
164.25
Hexanoic acid, 2-ethyl-, calcium salt (9CI)
Calcium 2-ethylhexanoate

136-52-7
$C_8H_{16}O_2.1/2Co$
173.68
Cobalt bis(2-ethylhexanoate)
CO 12
Cobalt(II) 2-ethylhexanoate
Cobalt 2-ethylhexoate
Cobalt(2+) 2-ethylhexanoate
Cobalt octoate
Cobaltous 2-ethylhexanoate
Cobaltous octoate
Hexanoic acid, 2-ethyl-, cobalt(2+) salt (9CI)
NL 49P
NL 51P
NL 51S
Versneller NL 49

136-53-8
$C_8H_{16}O_2.1/2Zn$
176.90
Hexanoic acid, 2-ethyl-, zinc salt (9CI)
Zinc 2-ethylhexanoate

136-60-7
$C_{11}H_{14}O_2$
178.25
O=C(OCCCC)c(cccc1)c1
Benzoic acid, butyl ester
Anthrapole AZ
Benzoic acid n-butyl ester
Butyl benzoate
n-Butyl benzoate
Butylester kyseliny benzoove (Czech)
Dai Cari XBN

136-77-6
$C_{12}H_{18}O_2$
194.30
Oc(c(ccc1O)CCCCCC)c1
Resorcinol, 4-hexyl
Ascaryl
Caprokol
Crystoids
Cystoids anthelmintic
Gelovermin
4-Hexyl-1,3-benzenediol
4-Hexyl-1,3-dihydroxybenzene
Hexylresorcin (German)
4-Hexylresorcine
Hexylresorcinol
p-Hexylresorcinol
4-Hexylresorcinol
4-n-Hexylresorcinol
NCI-C55787
S.T. 37
Sucrets
Worm-Agen

136-78-7
$C_8H_7Cl_2O_5S.Na$
309.10
[Na+].[O-]CCOc1ccc(Cl)cc1Cl.O=S(=O)=O
Ethanol, 2-(2,4-dichlorophenoxy)-, hydrogen sulfate, sodium salt
Crag Herbicide (OSHA)
Crag Herbicide I
Crag Sesone
2,4-DES-NA
2,4-DES-Natrium (German)
2,4-DES Sodium
2-(2,4-Dichlorfenoxy)ethylsiran sodny (Czech)
2-(2,4-Dichlorophenoxy)ethanol hydrogen sulfate sodium salt
2,4-Dichlorophenoxyethyl sulfate, sodium salt
Disul
Disul-Na
Disul-Sodium
Experimental Herbicide I
Herbon 2,4-des-Sodium

Natrium-2,4-dichlorphenoxy-athylsulfat (German)
Ses
Sesone (ACGIH,OSHA)
Sodium 2-(2,4-dichloro-phenoxy)ethyl sulfate
Sodium 2,4-dichlorophenoxy-ethyl sulphate
Sodium 2,4-dichlorophenyl cellosolve sulfate

136-80-1
$C_9H_{13}NO$
151.23
OCCNc(c(ccc1)C)c1
Ethanol, 2-toluidino
Emery 5711
N-β-Hydroxyethyl-o-toluidino-
2-o-Toluidinoethanol
2-o-Tolylaminoethanol
o-Tolyl ethanolamine

136-81-2
$C_{11}H_{16}O$
164.25
Oc(c(ccc1)CCCCC)c1
o-Amyl phenol
AI3-00455
o-Amylphenol
NSC-309965
o-Pentylphenol
2-Pentylphenol
2-Pentyl-phenol
Phenol, o-pentyl- (8CI)
Phenol, 2-pentyl- (9CI)

136-83-4
$C_{15}H_{24}O$
220.35
Oc(c(ccc1)CCCCCCCCC)c1
Phenol, 2-nonyl- (9CI)
Phenol, o-nonyl- (8CI)

136-84-5
$C_5H_{10}N_2O_3$
146.13
O=C(N(CC1)CO)N1CO
2-Imidazolidinone, 1,3-bis-(hydroxymethyl)- (9CI)
Aerotex Reactant No. 100
N,N'-Bis(hydroxymethyl)-ethyleneurea
1,3-Bis(hydroxymethyl)-2-imidazolidinone
BT 324
Calaroc EU
Carbamol TsEM
Cassurit RI
Cyclic dimethylolethyleneurea
1,3-Dihydroxymethyl-2-imida-zolidone
Dimethylol cyclic ethyleneurea
Dimethylolcycloethyleneurea
Dimethylolethyleneurea
N,N'-Dimethylolethyleneurea
1,3-Dimethylolethyleneurea
N,N'-Dimethylol-N,N'-ethyl-eneurea
1,3-Dimethylol-2-imidazolid-inone
Fixapret AH
Karbamol TsEM
Kurbamol TsEM
Neuperm ON
NSC-57546
Prym E
Quecodur AE
Rhonite R 1
Silesian EM
Verapret AN
Zeset S

136-85-6
$C_7H_7N_3$
133.17
N(=NNc1ccc(c2)C)c12
1H-Benzotriazole, 5-methyl
5-Methylbenzotriazole
5-Methyl-1,2,3-benzotriazole

136-92-5
$C_{10}H_{20}N_2S_4.Se$
375.52
Carbamic acid, diethyldithio-, selenium(II) salt
Ethyl selenac
Selenium diethyldithiocarbam-ate

136-95-8
$C_7H_6N_2S$
150.21
N=C(Nc(c1ccc2)c2)S1
Benzothiazole, 2-amino
2-Aminobenzothiazole
2-Aminobenzthiazole
USAF EK-3941
USAF XR-27

136-99-2
$C_{16}H_{32}N_2O$
268.43
N(=C(N(C1)CCO)CCCCCCCCCCC)C1
Lauryl hydroxyethyl imida-zoline
4,5-Dihydro-2-undecyl-1H-imidazole-1-ethanol
1-Hydroxyethyl-2-undecylim-idazoline
1-(2-Hydroxyethyl)-2-undecyl-imidazoline
1H-Imidazole-1-ethanol, 4,5-di-hydro-2-undecyl- (9CI)
2-Imidazoline-1-ethanol, 2-undecyl-
Nalcamine G-11
2-Undecyl-2-imidazoline-1-ethanol

137-00-8
C_6H_9NOS
143.20
N(C(=C(S1)CCO)C)=C1
Thiamine thiazole
AI3-23391
Hemineurine
5-(Hydroxyethyl)-4-methylthia-zole
4-Methyl-5-(β-hydroxyethyl)-thiazole
4-Methyl-5-(2-hydroxyethyl)-thiazole
4-Methyl-5-thiazoleethanol
4-Methyl-5-thiazolethanol
MHT (VAN)
NSC-23262
5-Thiazoleethanol, 4-methyl-(9CI)

137-03-1
$C_{12}H_{22}O$
182.34
O=C(C(CC1)CCCCCCC)C1
Cyclopentanone, 2-n-heptyl
α-Heptyl cyclopentanone
2-n-Heptyl cyclopentanone

137-04-2
$C_6H_6ClN.ClH$
164.03
Benzenamine, 2-chloro-, hydrochloride (9CI)
2-Chlorobenzenamine hydro-chloride

137-05-3
$C_5H_5NO_2$
111.11
O=C(OC)C(C#N)=C
Acrylic acid, 2-cyano-, methyl ester
Adhere
Coapt
α-Cyanoacrylic acid methyl ester
2-Cyanoacrylic acid, methyl ester
Cyanolit
Eastman 910
Eastman 910 Adhesive
Eastman 910 Monomer
Mecrilat
Mecrylate
Methyl cyanoacrylate
Methyl α-cyanoacrylate
Methyl 2-cyanoacrylate (ACGIH,OSHA)

137-06-4
C_7H_8S
124.21
Sc(c(ccc1)C)c1
o-Toluenethiol
o-Mercaptotoluene
o-Methylbenzenethiol
2-Methylbenzenethiol
o-Methylthiophenol
2-Methylthiophenol
o-Thiocresol
2-Toluenethiol
o-Tolyl mercaptan
USAF EK-2676

137-07-5
C_6H_7NS
125.20
Sc(c(N)ccc1)c1
Benzenethiol, o-amino
2-Aminobenzenethiol
o-Aminothiophenol
2-Aminothiophenol
o-Mercaptoaniline
USAF EK-4376

137-08-6
$C_{19}H_{34}N_2O_{10}.Ca$
490.63
[Ca+2].CC(C)(CO)C(O)C(=O)NCCC([O-])=O.CC(C)(CO)C(O)C(=O)NCCC([O-])=O
Pantothenic acid, calcium salt (2:1), (+)
Calcium D(+)-N-(α,γ-di hydroxy-β,β-dimethylbutyryl)-β-alaninate
Calcium panthothenate
Calcium pantothenate
Calcium d-pantothenate
d-Calcium pantothenate
Calpanate
Dextro calcium pantothenate
N-(2,4-Dihydroxy-3,3-dimethylbutyryl)-β-alanine calcium
Pancal
Panthoject
Pantholin
Pantothenate calcium
Pantothenic acid, calcium salt
Pantothenic acid, calcium salt, (+)-
(+)-Pantothenic acid calcium salt
Vitamin B-5

137-09-7
$C_6H_8N_2O.2ClH$
197.08
Phenol, 2,4-diamino-, dihydrochloride
Amidol
2,4-Diaminophenol hydrochloride
NCI-C60026

137-16-6
$C_{15}H_{29}NO_3.Na$
294.38
Sodium lauroyl sarcosinate
Caswell No. 778B
Compound 105
N-Dodecanoyl-N-methylglycine, sodium salt
EPA Pesticide Chemical Code 000174
Gardol
Glycine, N-methyl-N-(1-oxododecyl)-, sodium salt (9CI)
Hamposyl L-30
Lauroylsarcosine sodium salt
N-Lauroylsarcosine, sodium
N-Lauroylsarcosine, sodium salt
Maprosyl 30
Medialan LL-99
N-Methyl-N-(1-oxododecyl)-glycine sodium salt
NSC-117874
Sarcosine, N-lauroyl-, sodium salt (8CI)
Sarcosyl NL
Sarcosyl NL 30
Sarkosyl NL
Sarkosyl NL 30
Sarkosyl NL 35
Sarkosyl NL 97
Sarkosyl NL 100
Sodium lauroylsarcosinate
Sodium N-lauroylsarcosinate
Sodium lauroylsarcosine
Sodium N-lauroylsarcosine

137-17-7
$C_9H_{13}N$
135.23
Cc1cc(C)c(N)cc1C
Aniline, 2,4,5-trimethyl
1-Amino-2,4,5-trimethylbenzene
Benzenamine, 2,4,5-trimethyl-
psi-Cumidine
NCI-C02299
Pseudocumidine
Pseudokumidin (Czech)
1,2,4-Trimethyl-5-aminobenzene
2,4,5-Trimethylanilin (Czech)
2,4,5-Trimethylaniline
2,4,5-Trimethylbenzenamine

137-20-2
$C_{21}H_{40}NO_4S.Na$
425.67
Taurine, N-methyl-N-oleoyl-, sodium salt
Adinol T
Concogel 2 Conc.
Ethanesulfonic acid, 2-(methyl-(1-oxo-91-octadecenyl) amino)-, sodium salt, Z- (9CI)
Hostapon T
Igepon T
Igepon T 33
Igepon T-43
Igepon T 51
Igepon T-71
Igepon T-73
Igepon T 77
Igepon TE
Metaupon Paste
N-Methyl-N-oleoyltaurine sodium salt
Nissan diapion S
Nissan diapon T
Oleoylmethyltaurine sodium salt
OMT
Sodium 2-(N-methyloleamido)-ethane-1-sulfonate
Sodium methyl oleoyl taurate
Sodium N-methyl-N-oleoyl-taurate
Sodium N-oleoyl-N-methyl-taurate
Sodium N-oleoyl-N-methyla-taurine
Sodium oleylmethyltauride

137-26-8
$C_6H_{12}N_2S_4$
240.44
N(C(=S)SSC(N(C)C)=S)(C)C
Disulfide, bis(dimethylthiocarbamoyl)
Aatack
Accelerator thiuram
Aceto TETD
Arasan
Arasan 70
Arasan 75
Arasan-M
Arasan 42-S
Arasan-SF
Arasan-SF-X
Aules
Bis((dimethylamino)carbonothioyl) disulphide
Bis(dimethyl-thiocarbamoyl)-disulfid (German)
Bis(dimethylthiocarbamoyl) disulfide
Bis(dimethylthiocarbamoyl) disulphide
Bis(dimethylthiocarbamyl) disulfide
Chipco thiram 75
Cyuram DS
Disolfuro di tetrametiltiourame (Italian)
Disulfure de tetramethylthiourame (French)
α,α'-Dithiobis(dimethylthio)-formamide
N,N'-(Dithiodicarbonothioyl)-bis(N-methylmethanamine)
Ekagom TB
Falitiram
Fermide
Fernacol
Fernasan
Fernasan A
Fernide
Flo Pro T Seed Protectant
Formamide, 1,1'-dithiobis-(N,N-dimethylthio-
Hermal
Hermat TMT
Heryl
Hexathir
Kregasan
Mercuram
Methyl thiram
Methyl thiuramdisulfide
Methyl Tuads
NA 2771 (DOT)
Nobecutan
Nomersan
Normersan
Panoram 75
Polyram Ultra
Pomarsol
Pomarsol Forte
Pomasol
Puralin
RCRA waste number U244

Rezifilm
Royal TMTD
Sadoplon
Spotrete
Spotrete-F
SQ 1489
Teramethyl thiuram disulfide
Tersan
Tersan 75
Tetramethyldiurane sulphite
Tetramethylenethiuram disulphide
Tetramethylthiocarbamoyldisulphide
Tetramethylthioramdisulfide (Dutch)
Tetramethyl-thiram disulfid (German)
Tetramethylthiuram
Tetramethylthiuram bisulfide
Tetramethylthiuram bisulphide
Tetramethylthiuram disulfide
Tetramethylthiuram disulphide
N,N-Tetramethylthiuram disulphide
N,N,N',N'-Tetramethylthiuram disulfide
Tetramethylthiuran disulphide
Tetramethyl thiurane disulfide
Tetramethyl thiurane disulphide
Tetramethylthiurum disulfide
Tetramethylthiurum disulphide
Tetrapom
Tetrasipton
Tetrathiuram disulfide
Tetrathiuram disulphide
Thillate
Thimer
Thiosan
Thiotex
Thiotox
Thiram (ACGIH,DOT,OSHA)
Thiram 75
Thiramad
Thiram B
Thirame (French)
Thirasan
Thiulix
Thiurad
Thiuram
Thiuram D
Thiuram disulfide, tetramethyl-
Thiuramin
Thiuram M
Thiuram M Rubber Accelerator
Thiuramyl
Thylate
Tirampa
Tiuram (Polish)
Tiuramyl
TMTD
TMTDS
Trametan
Tridipam
Tripomol
TTD
Tuads
Tuex
Tulisan
USAF B-30
USAF EK-2089
USAF P-5
Vancida TM-95
Vancide TM
Vuagt-I-4
Vulcafor TMTD
Vulkacit MTIC
Vulkacit thiuram
Vulkacit thiuram/C

137-29-1
$C_6H_{12}N_2S_4.Cu$
303.98
CN(C)C(=S)S[Cu]SC(=S)N(C)C
Carbamic acid, dimethyldithio-, copper(II) salt
Compound-4018
Copper, bis(dimethyldithiocarbamato)-
Copper dimethyldithiocarbamate
Cumate
Dimethyldithiocarbamic acid copper salt
Wolfen

137-30-4
$C_6H_{12}N_2S_4.Zn$
305.81
CN(C)C(=S)S[Zn]SC(=S)N(C)C
Zinc, bis(dimethyldithiocarbamato)
Aaprotect
Aavolex
Aazira
Accelerator L
Aceto ZDED
Aceto ZDMD
Antene
Amyl Zimate
Alcobam ZM
Bis(dimethylcarbamodithioato-S,S')zinc
Bis-dimethyldithiocarbamate de zinc (French)
Bis(dimethyldithiocarbamato)-zinc
Bis(N,N-dimetil-ditiocarbammato) di zinco (Italian)
Carbamic acid, dimethyldithio-, zinc salt (2:1)
Carbazinc
Ciram
Corona corozate
Corozate
Cuman
Cuman L
Cymate
Dimethylcarbamodithioic acid, zinc complex
Dimethylcarbamodithioic acid, zinc salt
Dimethyldithiocarbamate zinc salt
Dimethyldithiocarbamic acid, zinc salt
Drupina 90
Eptac 1
ENT 988
Fuclasin
Fuclasin Ultra
Fuklasin
Fungostop
Hermat ZDM
Hexazir
Karbam White
Methasan
Methazate
Methyl Zimate
Methyl Zineb
Methyl Ziram
Mexene
Mezene
Milbam
Milban
Molurame
Mycronil
NCI-C50442
Orchard Brand Ziram
Pomarsol Z Forte
Prodaram
Rhodiacid
Soxinal PZ
Soxinol PZ
Tricarbamix Z
Triscabol
Tsimat
Tsiram (Russian)
USAF P-2
Vancide MZ-96
Vulcacure
Vulcacure ZM
Vulkacite L
Vulkacit L
Z 75
Zarlate
ZC
Z-C Spray
Zerlate
Zimate
Zimate, methyl
Zinc bis(diethyldithiocarbamate)
Zinc bis(dimethyldithiocarbamoyl)disulphide
Zinc bis(dimethylthiocarbamoyl)disulfide
Zinc dimethyldithiocarbamate
Zinc N,N-dimethyldithiocarbamate
Zincmate
Zink-bis(N,N-dimethyl-dithiocarbamaat) (Dutch)
Zink-bis(N,N-dimethyl-dithiocarbamat) (German)
Zinkcarbamate
Zink-(N,N-dimethyl-dithiocarbamat) (German)
Ziram
Zirame
Ziram Technical
Ziramvis
Zirasan
Zirasan 90
Zirberk
Zirex 90
Ziride
Zirthane
Zitox

137-32-6
$C_5H_{12}O$
88.17
OCC(CC)C
1-Butanol, 2-methyl

dl-sec-Butyl carbinol
2-Methylbutanol
2-Methyl butanol-1
2-Methyl-1-butanol

137-40-6
$C_3H_5O_2.Na$
96.07
Propionic acid, sodium salt
Impedex
Mycoban
Napropion
Natriumpropionat (German)
Ocuseptine
Propanoic acid, sodium salt
Propionan sodny (Czech)
Sodium propionate

137-41-7
Unknown
Unknown
Potassium N-methyldithio-carbamate
Caswell No. 696
EPA Pesticide Chemical Code 039002

137-42-8
$C_2H_4NS_2.Na$
129.18
[Na+].C=N.S=C=S
Carbamic acid, N-methyldi-thio-, sodium salt
A7 Vapam
Basamid-Fluid
Carbam
Carbathione
Carbation
Karbation
Maposol
Masposol
Metam-Fluid BASF
Metam-sodium (Dutch, French, German, Italian)
Metham (German)
Metham sodium
N-Methyldithiocarbamate de sodium (French)
Methyldithiocarbamic acid, sodium salt
Methyldithiokarbaman sodny (Czech)
N-Metil-ditiocarbammato di sodio (Italian)
N-869
Natrium-N-methyl-dithiocar-bamaat (Dutch)
Natrium-N-methyl-dithiocar-bamat (German)
Sistan
SMDC
Sodium methyldithiocarbamate
Sodium N-methyldithiocar-bamate
Solasan 500
Sometam
Trapex
Trimaton
Vapam
VDM
VPM

137-43-9
C_5H_9Br
149.03
BrC(CCC1)C1
Cyclopentane, bromo- (9CI)
AI3-23448
Bromocyclopentane
Cyclopentyl bromide
NSC-1110

137-47-3
$C_8H_9ClN_2O_4S$
264.68
O=S(=O)(N(C)C)c(ccc(c1N(=O)=O)Cl)c1
Benzenesulfonamide, 4-chloro-N,N-dimethyl-3-nitro- (9CI)
4-Chloro-3-nitro-N,N-dimethyl-benzenesulfonamide
NSC-231629

137-50-8
$C_6H_8N_2O_6S_2$
268.26
O=S(=O)(O)c(c(N)cc(N)c1S(=O)(=O)O)c1
1,3-Benzenedisulfonic acid, 4,6-diamino- (9CI)
4,6-Diamino-1,3-benzenedi-sulfonic acid

137-52-0
$C_{18}H_{14}ClNO_3$
327.76
O=C(Nc(c(OC)ccc1Cl)c1)c(c(O)cc(c2ccc3)c3)c2
2-Naphth-o-anisidide, 5'-chloro-3-hydroxy- (8CI)
Acna Naphthol CA
Amanil Naphthol AS-EL
Azotol KHA
Azotol XA
5'-Chloro-3-hydroxy-2-naphth-o-anisidide
C.I. Azoic Coupling Component 34
C.I. Azoic Coupling Component 41
C.I. 37531
Cibanaphthol RCA
Hiltonaphthol AS-EL
Naphtanilide EL
Naphtazol EL
2-Naphthalenecarboxamide, N-(5-chloro-2-methoxy-phenyl)-3-hydroxy- (9CI)
Naphthanilid EL
Naphthol AS-CA
Naphthol AS-CL
Naphthol AS-EL
Naphthol AS-RC Supra
Naphthol NEL
Naphtol AS-CA
Naphtol AS-CALL
Naphtol AS-RC
NSC-50685

137-58-6
$C_{14}H_{22}N_2O$
234.38
O=C(Nc(c(ccc1)C)c1C)CN(CC)CC
2',6'-Acetoxylidide, 2-(di-ethylamino)
Acetamide, 2-(diethylamino)-N-(2,6-dimethylphenyl)- (9CI)
Anestacon
Diethylaminoaceto-2,6-xylidide
α-Diethylaminoaceto-2,6-xylid-ide
α-Diethylamino-2,6-acetoxyl-idide
2-(Diethylamino)-2',6'-acetoxyl-idide
Diethylaminoacet-2,6-xylidide
α-Diethylamino-2,6-dimethyl-acetanilide
ω-Diethylamino-2,6-dimethyl-acetanilide
alfa-Dietilamino-2,6-dimetil-acetanilide (Italian)
Duncaine
Gravocain
Isicaina
Isicaine
Leostesin
Lida-Mantle
Lidocaine
Lignocaine
Maricaine
Mesocain
Rucaina
Solcain
Xilocaina (Italian)
Xycaine
Xylestesin
Xylocain
Xylocaine
Xylocitin
Xylotox

137-66-6
$C_{22}H_{38}O_7$
414.60
O=C(OCC(O)C(OC(=O)C=1O)C1O)CCCCCCCCCCCCCCC
L-Ascorbic acid, 6-palmitate
Ascorbyl palmitate

137-89-3
$C_{24}H_{38}O_4$
390.62
O=C(OCC(CCCC)CC)c(cccc1C(=O)OCC(CCCC)CC)c1
Isophthalic acid, bis(2-ethyl-hexyl) ester
Bis-(2-ethylhexyl)ester kyseliny isoftalove (Czech)
Bis(2-ethylhexyl) isophthalate
Di-2-ethylhexyl isophthalate
Dioctyl isophthalate
DOIP
Flexol Plasticizer 380
Isophthalic acid, di-(2-ethyl-hexyl)ester

137-99-5
$C_{24}H_{42}O$
346.60
Oc(c(cc(c1)CCCCCCCCC)CCCCCCCCC)c1
2,4-Dinonylphenol
Phenol, 2,4-dinonyl- (9CI)

138-00-1
$C_{16}H_{26}O$
234.38
2,4-Diamylphenol
Phenol, 2,4-dipentyl- (8CI,9CI)

138-15-8
$C_5H_9NO_4.ClH$
183.59
Glutamic acid hydrochloride
Achylin
Acidalin
Acidogen
Acidoride
Acidothyn
Acidulen
Acidulin
Aciglumin
Acigluminum
Aclor
Acridogen
Acridoride
2-Aminopentanedioic acid hydrochloride
Antalka
Flamithin
Flanithin
Gastuloric
Glusatin
Glutamic acid, hydrochloride, L- (8CI)
L-Glutamic acid, hydrochloride (9CI)
Glutamic acid hydrogen chloride
L-Glutamic acid monohydrochloride
Glutamidin
Glutan HCl
Glutan hydrochloric
Glutan hydrochloride
Glutasin
Hydrionic
Hypochylin
Muriamic
NSC-9239
Pepsdol
Pepsidol

138-22-7
$C_7H_{14}O_3$
146.21
O=C(OCCCC)C(O)C
Lactic acid, butyl ester
Butylester kyseliny mlecne (Czech)
Butyl α-hydroxypropionate
Butyl lactate
n-Butyl lactate (ACGIH,OSHA)
Propanoic acid, 2-hydroxy-, butyl ester (9CI)

138-25-0
$C_{10}H_{10}O_7S$
274.25
O=C(OC)c(cc(S(=O)(=O)O)cc1C(=O)OC)c1
1,3-Benzenedicarboxylic acid, 5-sulfo-, 1,3-dimethyl ester (9CI)

138-42-1
$C_6H_5NO_5S$
203.17
O=S(=O)(O)c(ccc(N(=O)=O)c1)c1
Benzenesulfonic acid, p-nitro- (8CI)
Benzenesulfonic acid, 4-nitro- (9CI)
p-Nitrobenzenesulfonic acid
4-Nitrobenzenesulfonic acid
p-Nitrophenylsulfonic acid
NSC-5376

138-52-3
$C_{13}H_{18}O_7$
286.28
O(C(C(C(O)C(O)C1O)CO)C1Oc(c(ccc2)CO)c2
Salicin (8CI)
AI3-19099
Benzyl alcohol, o-hydroxy-, o-glucoside
ß-D-Glucopyranoside, 2-(hydroxymethyl)phenyl
β-D-Glucopyranoside, 2-(hydroxymethyl)phenyl (9CI)
o-(Hydroxymethyl)phenyl β-D-glucopyranoside
NSC-5751
Salicine
Salicoside
Saligenin-β-D-glucopyranoside

138-56-7
$C_{21}H_{28}N_2O_5$
388.45
COc1cc(cc(OC)c1OC)C(=O)NCc2ccc(OCCN(C)C)cc2
Trimethobenzamide
Benzamide, N-(p-(2-(dimethylamino)ethoxy)benzyl)-3,4,5-trimethoxy-
Benzamide, N-((4-(2-(dimethylamino)ethoxy)phenyl)methyl)-3,4,5-trimethoxy-
N-((2-Dimethylaminoethoxy)-benzyl)-3,4,5-trimethoxybenzamide
4-(2-Dimethylaminoethoxy)-N-(3,4,5-trimethoxybenzoyl)-benzylamine
Trimethobenzamidum (Latin)
Trimetobenzamida (Spanish)

138-59-0
$C_7H_{10}O_5$
174.17
OC1CC(=CC(O)C1O)C(O)=O
1-Cyclohexene-1-carboxylic acid, 3,4,5-trihydroxy
Bracken Fern Toxic Component
Shikimate
Shikimic acid
3,4,5-Trihydroxy-1-cyclohexene-1-carboxylic acid

138-60-3
$C_7H_5NO_5$
183.11
Chelidamic acid
NSC-3983
2,6-Pyridinedicarboxylic acid, 1,4-dihydro-4-oxo- (9CI)

138-84-1
$C_7H_7NO_2.K$
176.24
Aminobenzoate potassium
Benzoic acid, 4-amino-, monopotassium salt (9CI)
Monopotassium 4-aminobenzoate
Potaba

138-86-3
$C_{10}H_{16}$
136.26
C(=CCC(C(=C)C)C1)(C1)C
p-Mentha-1,8-diene
Acintene DP
Acintene DP Dipentene
Cajeputene
Cinene
Dipanol
Dipentene [UN 2052]
Inactive Limonene
Kautschin
Limonene
dl-Limonene
p-Mentha-1,8-diene, DL-
1,8(9)-p-Menthadiene
p-Menthane
1-Methyl-4-isopropenyl-1-cyclohexene
Nesol
δ-1,8-Terpodiene
UN 2052 [Dipentene]
Unitene

138-87-4
$C_{10}H_{18}O$
154.28
OC(CCC(C(=C)C)C1)(C1)C
p-Menth-8-en-1-ol
Cyclohexanol, 1-methyl-4-(1-methylethenyl)- (9CI)
t-Menth-1-en-8-ol
β-Terpineol

138-89-6
$C_8H_{10}N_2O$

150.20
O=Nc(ccc(N(C)C)c1)c1
Aniline, N,N-dimethyl-p-nitroso
Accelerine
Benzenamine, N,N-dimethyl-4-nitroso- (9CI)
p-(Dimethylamino)nitroso-benzene
4-(Dimethylamino)nitroso-benzene
Dimethyl-p-nitrosoaniline (DOT)
N,N-Dimethyl-p-nitrosoaniline
Dimethyl(p-nitrosophenyl)amine
NCI-C01821
NDMA
p-Nitrosodimethylaniline [UN 1369]
p-Nitroso-N,N-dimethylaniline
4-Nitrosodimethylaniline
Paranitrosodimethylanilide
Ultra Brilliant Blue P
UN 1369 [p-Nitrosodimethyl-aniline]

138-93-2
Unknown
Unknown
Disodium cyanodithioimido-carbonate

139-02-6
$C_6H_5O.Na$
116.10
Phenol, sodium salt, (Solid)
Phenol sodium
Phenol sodium salt
Sodium carbolate
Sodium phenate
Sodium phenolate, Solid [UN 2497]
Sodium phenoxide
UN 2497 [Sodium phenolate, solid]

139-05-9
$C_6H_{12}NO_3S.Na$
201.24
[Na+].[O-]S(=O)(=O)NC1CCCCC1
Cyclohexanesulfamic acid, monosodium salt
Assugrin
Assurgrin Feinsuss
Assurgrin Vollsuss
Asugryn
Cyclamate
Cyclamate sodium
Cyclamate, sodium salt
Cyclamic acid sodium salt
Cyclohexanesulphamic acid, monosodium salt
Cyclohexylsulphamate sodium
Cyclohexylsulphamic acid, monosodium salt
N-Cyklohexylsulfamat sodny (Czech)
Dulzor-Etas
Hachi-Sugar
Ibiosuc
Natreen
Natriumzyklamate (German)
Sodium cyclamate
Sodium cyclohexanesulfamate
Sodium cyclohexanesulphamate
Sodium cyclohexyl amido-sulphate
Sodium cyclohexyl sulfamate
Sodium cyclohexyl sulphamate
Sodium cyclohexylsulphamidate
Sodium sucaryl
Sucaryl sodium
Succaril
Sucrosa
Sucrun 7
Suessette
Suestamin
Sugarin
Sugaron

139-06-0
$C_{12}H_{24}N_2O_6S_2.Ca$
396.58
Cyclohexanesulfamic acid, calcium salt (2:1)
Calcium cyclamate
Calcium cyclohexanesulfamate
Calcium cyclohexane sulpham-ate
Calcium cyclohexylsulfamate
Calcium cyclohexylsulphamate
Cyclamate calcium
Cyclamate, calcium salt
Cyclan
Cyclohexanesulfamic acid, calcium salt
Cyclohexylsulphamic acid, calcium salt
Cylan
Dietil
Kalziumzyklamate (German)
Sucaryl calcium
Sulfamic acid, cyclohexyl-, calcium salt (2:1) (9CI)

139-07-1
$C_{21}H_{38}N.Cl$
340.05
Ammonium, benzyldimethyl-dodecyl-, chloride
Ammonium, benzyldodecyldi-methyl-, chloride
Benzenemethanaminium, N-do-decyl-N,N-dimethyl-, chloride (9CI)
Benzyl-lauryldimethylammon-ium chloride
Dodecyl-dimethyl-benzyl-ammonium chloride

139-08-2
$C_{23}H_{42}N.Cl$
368.11
[Cl-].CCCCCCCCCCCCCC[N+](C)(C)Cc1ccccc1
Ammonium, benzyldimethyl-tetradecyl-, chloride
Arquad DM14B-90
Benzenemethanaminium, N,N-dimethyl-N-tetradecyl-, chloride (9CI)
Nissan Cation M2-100
Quarton 14 BCL
Tetradecyl-dimethyl-benzyl-ammonium chloride

139-12-8
$C_2H_4O_2.1/3Al$
69.02
Aluminum acetate
Acetic acid, aluminum salt (9CI)
Buro-Sol Concentrate
Domeboro

139-13-9
$C_6H_9NO_6$
191.16
O=C(O)CN(CC(=O)O)CC(=O)O
Acetic acid, nitrilotri
Aminotriacetic acid
N,N-Bis(carboxymethyl)glysine
Chel 300
Complexon I
Glycine, N,N-bis(carboxy-methyl)- (9CI)
Hampshire NTA acid
Komplexon I
Kyselina nitrilotrioctova (Czech)
NCI-C02766
Nitrilotriacetic acid
NTA
Titriplex I
Triglycine
Triglycollamic acid
Trilon A
Versene NTA Acid

139-33-3
$C_{10}H_{14}N_2O_8.2Na$
336.24
[Na+].[Na+].OC(=O)CN(CCN(CC(O)=O)CC([O-])=O)CC([O-])=O
Acetic acid, (ethylenedi-nitrilo)tetra-, disodium salt
Cheladrate
Chelaplex III
Chelaton III
Complexon III
D'e.D.T.A. Disodique (French)
Dinatrium ethylendiamintetra-acetat (Czech)
Disodium diacid ethylenedi-aminetetraacetate
Disodium dihydrogen ethylene-diaminetetraacetate
Disodium dihydrogen(ethylene-dinitrilo)tetraacetate
Disodium edathamil
Disodium edetate
Disodium EDTA
Disodium ethylenediaminetetra-acetate
Disodium ethylenediaminetetra-acetic acid

Disodium (ethylenedinitrilo)-tetraacetate
Disodium (ethylenedinitrilo)-tetraacetic acid
Disodium salt of EDTA
Disodium sequestrene
Disodium tetracemate
Disodium versenate
Disodium versene
Edathamil disodium
Edetate disodium
Edetic acid disodium salt
EDTA disodium
EDTA, disodium salt
Endrate disodium
N,N'-1,2-Ethanediylbis(N-(carboxymethyl)glycine) disodium salt
Ethylenebis(iminodiacetic acid) disodium salt
Ethylenediaminetetraacetate, disodium salt
Ethylenediaminetetraacetic acid, disodium salt
(Ethylenedinitrilo)-tetraacetic acid disodium salt
F 1
F 1 (Complexon)
Glycine, N,N'-1,2-ethanediylbis(N-(carboxymethyl)-, disodium salt (9CI)
Kiresuto B
Komplexon III
Metaquest B
Perma Kleer 50 Crystals Disodium Salt
Perma Kleer Di Crystals
Selekton B 2
Sequestrene sodium 2
Sodium versenate
Tetracemate disodium
Titriplex III
Trilon B
Trilon BD
Triplex III
Veresene disodium salt
Versene Sodium 2

139-40-2
$C_9H_{16}ClN_5$
229.75
CC(C)Nc1nc(Cl)nc(NC(C)C)n1
s-Triazine, 2-chloro-4,6-bis-(isopropylamino)
2,4-Bis(isopropylamino)-6-chloro-s-triazine
2,4-Bis(propylamino)-6-chlor-1,3,5-triazin (German)
2-Chloro-4,6-bis(isopropylamino)-s-triazine
6-Chloro-N,N'-bis(1-methylethyl)-1,3,5-triazine-2,4-diamine
6-Chloro-N,N'-diisopropyl-1,3,5-triazine-2,4-diyldiamine
G-30028
Geigy 30,028
Gesamil
Milogard
Milo-Pro
Plantulin
Primatol P
Propasin
Propazin
Propazine
Propazine (Herbicide)
Prozinex
1,3,5-Triazine-2,4-diamine, 6-chloro-N,N'-bis(1-methylethyl)-

139-41-3
$C_6H_{13}NO_4.Na$
186.16
Sodium dihydroxyethylglycinate
N,N-Bis(2-hydroxyethyl)glycine, monosodium salt
N,N-Bis(2-hydroxyethyl)glycine, sodium salt
Glycine, N,N-bis(2-hydroxyethyl)-, monosodium salt (9CI)
Sodium N,N-bis-2-hydroxyethyl glycinate

139-45-7
$C_{12}H_{20}O_6$
260.32
O=C(OCC(OC(=O)CC)COC(=O)CC)CC
Tripropionin
Glycerine tripropionate
Glycerol tripropionate
Glyceryl tripropionate
Propionin, tri- (8CI)
Tripropionine

139-59-3
$C_{12}H_{11}NO$
185.24
O(c(ccc(N)c1)c1)c(cccc2)c2
Aniline, p-phenoxy
4-Aminobiphenyl ether
4-Aminodifenylether (Czech)
4-Aminodiphenyl ether
p-Aminophenyl phenyl ether
4-Aminophenyl phenyl ether
Benzenamine, 4-phenoxy- (9CI)
p-Phenoxyaniline
4-Phenoxyaniline

139-60-6
$C_{22}H_{40}N_2$
332.64
N(c(ccc(NC(CC)CC(CC)C)c1)c1)C(CC)CC(CC)C
p-Phenylenediamine, N,N'-bis-(1-ethyl-3-methylpentyl)
N,N'-Bis(1-ethyl-3-methylpentyl)-p-phenylenediamine
N,N'-Bis(5-methyl-3-heptyl)-p-phenylenediamine
N,N'-Di(1-ethyl-3-methylpentyl)-p-phenylenediamine
Eastozone 31
Elastozone 31
Santoflex 17
Tenamene 31
UOP 88

139-65-1
$C_{12}H_{12}N_2S$
216.32
S(c(ccc(N)c1)c1)c(ccc(N)c2)c2
Aniline, 4,4'-thiodi
Benzenamine, 4,4'-thiobis-(9CI)
Bis(p-aminophenyl)sulfide
Bis(4-aminophenyl) sulfide
Bis(p-aminophenyl)sulphide
Bis(4-aminophenyl)sulphide
p,p'-Diaminodiphenyl sulfide
4,4'-Diaminodiphenyl sulfide
p,p'-Diaminodiphenyl sulphide
4,4-Diaminodiphenyl sulphide
4,4'-Diaminodiphenylsulphide
Di(p-aminophenyl) sulfide
Di(p-aminophenyl)sulphide
NCI-C01707
Sulfide, bis(p-aminophenyl)
Thioaniline
4,4'-Thioaniline
4,4'-Thiobis(aniline)
4,4'-Thiobisbenzenamine
p,p-Thiodianiline
4,4'-Thiodianiline
Thiodi-p-phenylenediamine

139-66-2
$C_{12}H_{10}S$
186.28
S(c(cccc1)c1)c(cccc2)c2
Phenyl-sulfide
Diphenyl sulfide
Diphenyl sulphide
Diphenyl thioether
Sulfide, diphenyl
1,1'-Thiobis(benzene)

139-85-5
$C_7H_6O_3$
138.13
Protocatechualdehyde
Benzaldehyde, 3,4-dihydroxy-
3,4-Dihydroxybenzaldehyde
3,4-Dihydroxybenzenecarbonal
Protocatechuic aldehyde
Rancinamycin IV

139-87-7
$C_6H_{15}NO_2$
133.22
OCCN(CCO)CC
Ethanol, 2,2'-(ethylimino)di
Diethanolethylamine
Ethanol, 2,2'-(ethylimino)bis-(9CI)
Ethylbis(2-hydroxyethyl)amine
Ethyldiethanolamine
N-Ethyldiethanolamine
2-(N-Ethyl-N-2-hydroxyethylamino)ethanol
N-Ethyl-2,2'-iminodiethanol
2,2'-(Ethylimino)diethanol

139-88-8
$C_{14}H_{29}O_4S.Na$
316.48
4-Undecanol, 7-ethyl-2-methyl-, hydrogen sulfate, sodium salt
4-Ethyl-1-isobutyloktylsiran sodny (Czech)
7-Ethyl-2-methyl-4-hendecanol sulfate sodium salt
7-Ethyl-2-methyl-4-undecanol sulfate sodium salt
Obliterol
Sodium 7-ethyl-2-methyl-4-undecanol sulfate
Sodium 7-ethyl-2-methylundecyl-4-sulfate
Sodium 2-methyl-7-ethylundecanol-4-sulfate
Sodium 2-methyl-7-ethylundecyl sulfate-4
Sodium sotradecol
Sotradecol
STS
Tergitol
Tergitol 4
Tergitol Anionic 4
Tergitol Penetrant 4
Trombovar
Varicol

139-89-9
$C_{10}H_{18}N_2O_7.3Na$
347.22
Trisodium HEDTA
N-(2(Bis(carboxymethyl)-amino)-ethyl)-N-(2-hydroxyethyl)glycine trisodium salt
N-(2-(Bis(carboxymethyl)-amino)ethyl)-N-(2-hydroxyethyl)glycine, trisodium salt
N-(Carboxymethyl)-N'-(2-hydroxyethyl)-N,N'-ethylenediglycine trisodium salt
Caswell No. 487C
CHEL DM 41
Chemcolox 800
Detarol trisodium salt
EPA Pesticide Chemical Code 039109
Glycine, N-(2-(bis(carboxymethyl)amino)ethyl)-N-(2-hydroxyethyl)-, trisodium salt (9CI)
Glycine, N-(carboxymethyl)-N'-(2-hydroxyethyl)-N,N'-ethylenedi-, trisodium salt (8CI)
Hamp-ol Crystals
Hamp-ol 120
(Hydroxyethyl)ethylenediamine-triacetic acid, trisodium salt
N-Hydroxyethylethylenediaminetriacetic acid trisodium salt
N-(Hydroxyethyl)-N,N',N'-ethylenediaminetriacetic acid trisodium
N-(Hydroxyethyl)-N,N',N'-ethylenediaminetriacetic acid trisodium salt
N-(2-Hydroxyethyl)-N,N',N'-ethylenediaminetriacetic acid trisodium salt
(2-Hydroxyethyl)ethylenediaminetriacetic acid, trisodium salt
Monaquest ICA-120
NSC-148338
Perma Kleer 80
Perma Kleer 80 Crystals
Perma Kleer 80, Crystals
Trisodium hydroxyethyl ethylenediaminetriacetate
Trisodium N-hydroxyethylethylenediaminetriacetate
Trisodium N-hydroxyethylethylenediamine-N,N',N'-triacetate
Trisodium N-(hydroxyethyl)-ethylenediaminetriacetate
Trisodium N-(2-hydroxyethyl)-ethylenediaminetriacetate
Trisodium N-(2-hydroxyethyl)-ethylenediamine-N,N',N'-triacetate
Trisodium N-(2-hydroxyethyl)-N,N',N'-ethylenediaminetri acetate
Trisodium (2-hydroxyethyl)-ethylenediaminetriacetate
Trisodium(2-hydroxyethyl)-ethylenediaminetriacetate
Trisodium salt N-hydroxyethyl ethylenediaminetriacetic acid
Trisodium salt N-hydroxyethyl-ethylenediaminetriacetic acid
Trisodium salt of (hydroxyethyl)ethylenediamine
Versen-OL
Versenol
Versenol 120

139-90-2
$C_{13}H_{30}N_2O_4$
278.38
OCCN(CCN(CC(O)C)CC(O)C)CC(O)C
2-Propanol, 1,1'-((2-((2-hydroxyethyl)(2-hydroxypropyl)amino)ethyl)imino)-bis- (9CI)
2-Propanol, 1,1'-((2-((2-hydroxyethyl)(2-hydroxypropyl)amino)ethyl)imino)di- (8CI)
Nalco L-699
NSC-23966

139-91-3
$C_{13}H_{16}N_4O_6$
324.33
2-Oxazolidinone, 5-(morpholinomethyl)-3-((5-nitrofurfurylidene)amino)
Altabactina
Altafur
F-150
Furaltadone
Furazolin
Furazoline
Furmethanol
Furmethonol
Furmetonol
Ibifur
Medifuran
5-Morpholinomethyl-3-(5-nitro-2-furfurylidine-amino)-2-oxazolidinone
NF 260
Nitraldone
Nitrofurmethone
Nitrofurmeton
Otifuril
Sepsinol
Ultrafur
Unifur
Valsyn

139-92-4
$C_{24}H_{48}N_2O_4$
428.64
O=C(NCCN(CC(=O)O)CCO)CCCCCCCCCCCCCCCCC
Glycine, N-(2-hydroxyethyl)-N-(2-((1-oxooctadecyl)-amino)ethyl)- (9CI)

139-94-6
$C_6H_8N_4O_3S$
216.24
CCNC(=O)Nc1ncc(s1)N(=O)=O
Urea, 1-ethyl-3-(5-nitro-2-thiazolyl)
N-Ethyl-N'-(5-nitro-2-thia zolyl)urea
1-Ethyl-3-(5-nitro-2-thiazolyl) urea
Hepzide
Hepzide 30
NCI-C03792
Nithiazid
Nithiazide
Urea, N-ethyl-N'-(5-nitro-2-thiazolyl)-

139-96-8
$C_{12}H_{26}O_4S.C_6H_{15}NO_3$
415.66
CCCCCCCCCCCCOS(O)(=O)=O.OCCN(CCO)CCO
Sulfuric acid, monododecyl ester, Compd. with 2,2',2''-nitrilotriethanol (1:1)
Akyposal TLS
Cycloryl TAWF
Cycloryl WAT
Drene
Elfan 4240 T
Emal T
Emersal 6434
Maprofix TLS
Maprofix TLS 65
Maprofix TLS 500
Melanol LP20T
Propaste T
Rewopol TLS 40
Richonol T
Sipon LT
Sipon LT-6
Sipon LT-40

Standapol TLS 40
Steinapol TLS 40
Stepanol WAT
Sterling WAT
Sulfuric acid, dodecyl ester, triethanolamine salt
Sulfuric acid, monododecyl ester, Compd. with 2,2',2''-nitrilotris(ethanol)
TEA lauryl sulfate
Texapon T-35
Texapon T-42
Texapon TH
Triethanolamine dodecyl sulfate
Triethanolamine lauryl sulfate
Tylorol LT 50

139-99-1
$C_{19}H_{38}O_6S.Na$
417.56
Octadecanoic acid, 9-(sulfooxy)-, 1-methyl ester, sodium salt (9CI)
Methyl oleate, sulfated, sodium salt

140-01-2
$C_{14}H_{23}N_3O_{10}.5Na$
508.28
Pentasodium pentetate
Acetic acid, ((carboxymethylimino)bis(ethylenenitrilo))-tetra-, pentasodium salt
N,N-Bis(2-(bis(carboxymethyl)-amino)ethyl)glycine, pentasodium salt
(((Carboxymethyl)imino)bis-(ethylenenitrilo))tetraacetic acid, pentasodium salt
Caswell No. 642B
CHEL 330
Detapac
Detarex PY
Diethylenetriaminepentaacetate, pentasodium salt
Diethylenetriaminepentaacetic acid pentasodium salt
Diethylenetriaminepentaacetic acid sodium salt
DTPA pentasodium salt
EPA Pesticide Chemical Code 039120
Glycine, N,N-bis(2-(bis(carboxymethyl)amino)ethyl)-, pentasodium salt (9CI)
HAMP-EX 80
Kiresuto P
Pentasodium diethylenetriaminepentaacetate
Pentasodium diethylenetriaminepentaacetic acid
Pentasodium diethylenetriaminepentacetate
Pentasodium DTPA
Perma Kleer 140
Plexene D
Sodium diethylenetriaminepentaacetate
Syntron C
Versenex 80

140-03-4
$C_{21}H_{38}O_4$
354.59
O=C(OC(CCCCCC)CC=CCCCC CCCC(=O)OC)C
Ricinoleic acid, methyl ester, acetate
Flexricin P-4
Methyl 12-acetoxy-9-octadecenoate
Methyl 12-acetoxyoleate
Methyl acetyl ricinoleate
Methylester kyseliny acetylricinolejove (Czech)

140-04-5
$C_{24}H_{44}O_4$
396.61
O=C(OC(CCCCCC)CC=CCCCC CCCC(=O)OCCCC)C
Butyl acetyl ricinoleate
AI3-00400
Bakers P-6
Baryl
Butyl 12-acetoxyoleate
Butyl 12-(acetyloxy)-9-octadecenoate
Flexricin P 6
NSC-2319
Ricinoleic acid, butyl ester, acetate (8CI)
9-Octadecenoic acid, 12-(acetyloxy)-, butyl ester
9-Octadecenoic acid, 12-(acetyloxy)-, butyl ester, (R-(Z))-(9CI)

140-05-6
$C_{23}H_{42}O_5$
398.65
CCCCCCC(CC=CCCCCCCCC (=O)OCCOC)OC(C)=O
Ricinoleic acid, 2-methoxyethyl ester, acetate
Ethylene glycol monomethyl ether acetylricinoleate
Glycol monomethyl ether acetylricinoleate
2-Methoxyethyl 12-acetoxy-9-octadecenoate
2-Methoxyethyl acetyl ricinoleate
2-Methoxyethylester kyseliny acetylricinolove (Czech)
Methyl cellosolve acetylricinoleate

140-07-8
$C_{10}H_{24}N_2O_4$
236.30
OCCN(CCN(CCO)CCO)CCO
Ethanol, 2,2',2'',2'''-(ethylenedinitrilo)tetra- (8CI)
Entol
Ethanol, 2,2',2'',2'''-(1,2-ethanediyldinitrilo)tetrakis- (9CI)
Ethylenediamine tetraethanol
Ethylenediamine, N,N,N',N'-tetrakis(2-hydroxyethyl)-
2,2',2'',2'''-(Ethylenediimino)-tetraethanol
Ethylenedinitrilotetraethanol
NSC-21705
Tetrahydroxyethylethylenediamine
Tetra(hydroxyethyl)ethylenediamine
N,N,N',N'-Tetrakis(2-hydroxyethyl)-1,2-diaminoethane
Tetrakis(hydroxyethyl)ethylenediamine
N,N,N',N'-Tetrakis(hydroxyethyl)ethylenediamine
N,N,N',N'-Tetrakis(2-hydroxyethyl)ethylenediamine
Theed
TKED

140-08-9
$C_6H_{12}Cl_3O_3P$
269.50
O(P(OCCCl)OCCCl)CCCl
Ethanol, 2-chloro-, phosphite (3:1)
2-Chloroethanol phosphite (3:1)
Phosphorous acid, tris(2-chloroethyl) ester
Tris(2-chloroethyl)ester of phosphorus acid
Tris(2-chloroethyl)phosphite

140-10-3
$C_9H_8O_2$
148.17
O=C(O)C=Cc(cccc1)c1
Cinnamic acid, (E)
trans-β-Carboxystyrene
(E)-Cinnamic acid
trans-Cinnamic acid
trans-3-Phenylacrylic acid
2-Propenoic acid, 3-phenyl-, (E)- (9CI)

140-11-4
$C_9H_{10}O_2$
150.19
O=C(OCc(cccc1)c1)C
Acetic acid, benzyl ester
Acetic acid, phenylmethyl ester
α-Acetoxytoluene
Benzyl acetate
Benzylester kyseliny octove (Czech)
Benzyl ethanoate
NCI-C06508
Phenylmethyl acetate

140-18-1
$C_9H_9ClO_2$
184.63
O=C(OCc(cccc1)c1)CCl
Acetic acid, chloro-, benzyl ester
Acetic acid, chloro-, phenylmethyl ester (9CI)

Benzyl chloroacetate
Benzyl-α-chloroacetate
Benzyl monochloracetate
Chloroacetic acid, benzyl ester
Monochloroacetate

140-22-7
$C_{13}H_{14}N_4O$
242.31
O=C(NNc(cccc1)c1)NNc(cccc2)
c2
Carbohydrazide, 1,5-diphenyl
Carbonic dihydrazide, 2,2'-
diphenyl- (9CI)
Diphenylcarbazide
sym-Diphenylcarbazide
1,5-Diphenylcarbazide
2,2'-Diphenylcarbazide
1,5-Diphenylcarbohydrazide
DPC

140-24-9
$C_{24}H_{30}O_4$
382.50
O=C(OCc(cccc1)c1)CCCCCCCC
C(=O)OCc(cccc2)c2
**Decanedioic acid, bis(phenyl-
methyl) ester (9CI)**
AI3-02693
Bis(phenylmethyl) decanedioate
Dibenzyl sebacate
NSC-3896
Sebacic acid, dibenzyl ester
(8CI)

140-29-4
C_8H_7N
117.16
N#CCc(cccc1)c1
Acetonitrile, phenyl
Benzeneacetonitrile (9CI)
Benzyl cyanide
Benzylkyanid (Czech)
Benzyl nitrile
(Cyanomethyl)benzene
α-Cyanotoluene
ω-Cyanotoluene
Phenylacetonitrile
2-Phenylacetonitrile
Phenylacetonitrile, Liquid
[UN 2470]
Phenyl acetyl nitrile
Toluene, α-cyano-
α-Tolunitrile
UN 2470 [Phenylacetonitrile,
Liquid]
USAF KF-21

140-31-8
$C_6H_{15}N_3$
129.24
N(CCNC1)(C1)CCN
Piperazine, 1-(2-aminoethyl)
Aminoethylpiperazine
N-Aminoethylpiperazine
[UN 2815]
1-Aminoethylpiperazine
N-(Aminoethyl)piperazine
N-(β-Aminoethyl)piperazine
N-(2-Aminoethyl)piperazine
1-(2-Aminoethyl)piperazine
AI3-52274
Piperazine, 1-(2-aminoethyl)-
(8CI)
1-Piperazineethanamine (9CI)
UN 2815 [N-Aminoethylpipera-
zine]
USAF DO-46

140-38-5
$C_7H_7ClN_2O$
170.61
NC(=O)Nc1ccc(Cl)cc1
Urea, 1-(p-chlorophenyl)
1-(p-Chlorophenyl)urea
4-Chlorophenylurea

140-39-6
$C_9H_{10}O_2$
150.19
O=C(Oc(ccc(c1)C)c1)C
Acetic acid, p-tolyl ester
Acetic acid, 4-methylphenyl
ester (9CI)
p-Acetoxytoluene
4-Acetoxytoluene
p-Cresol acetate
Cresyl acetate
p-Cresyl acetate
4-Methylbenzoic acid methyl
ester
p-Methylphenyl acetate
4-Methylphenyl acetate
Narceol
Paracresyl acetate
p-Tolyl acetate
p-Tolyl ethanoate

140-41-0
$C_2HCl_3O_2.C_9H_{11}ClN_2O$
362.05
CN(C)C(=O)Nc1ccc(Cl)cc1.
[O-]C(=O)C(Cl)(Cl)Cl
**Acetic acid, trichloro-,
Compd. with 3-(p-chloro-
phenyl)-1,1-dimethylurea
(1:1)**
Acetic acid, trichloro-, Compd.
with N'-(4-chlorophenyl)-
N,N-dimethylurea (1:1) (9CI)
3-(p-Chlorophenyl)-1,1-di-
methylurea, trichloroacetate
GC-2996
Monuron-TCA
3-(Parachlorophenyl)-1,1-di-
methylurea trichloroacetate
Urea, 3-(p-chlorophenyl)-1,1-di-
methyl-, Cmpd. with tri-
chloroacetic acid (1:1)
Urox
Urox 379
Xoru-Ox

140-49-8
$C_{10}H_{10}ClNO_2$
211.66
O=C(Nc(ccc(c1)C(=O)CCl)c1)C
Acetanilide, 4'-(chloroacetyl)
Acetamide, N-(4-(chloro-
acetyl)phenyl)- (9CI)
p-Acetamidophenacyl chloride
p-(Acetylamino)phenacyl
chloride
4'-Chloroacetyl (acetanilide)
NCI-C03770

140-53-4
C_8H_6ClN
151.60
N#CCc(ccc(c1)Cl)c1
Acetonitrile, (p-chlorophenyl)
Benzeneacetonitrile, 4-chloro-
4-Chlor-benzyl-cyanid
(German)
4-Chlorobenzeneacetonitrile
p-Chlorobenzyl cyanide
4-Chlorobenzyl cyanide
p-Chlorophenylacetonitrile
(4-Chlorophenyl)acetonitrile
2-(4-Chlorophenyl)acetonitrile

140-56-7
$C_8H_{10}N_3O_3S.Na$
251.26
[Na+].CN(C)c1ccc(N=NS
([O-])(=O)=O)cc1
**Benzenediazosulfonic acid,
p-(dimethylamino)-, sodium
salt**
Bay 5072
Bay 22555
Bayer 5072
Bayer 22555
DAPA
DAS
Deksonal
Dexon
Dexoxon
Diazenesulfonic acid, (4-(dimethylamino)phenyl)-, sodium
salt
Diazoben
p-Dimethylaminobenzene diazo
sodium sulfonate
p-Dimethylaminobenzenediazosodium sulphonate
p-(Dimethylamino)benzene-
diazosulfonate
p-Dimethylaminobenzene-
diazosulfonic acid, sodium salt
4-Dimethylaminobenzenediazo-
sulfonic acid, sodium salt
p-(Dimethylamino)benzene-
diazosulphonate
p-(Dimethylamino)benzene-
diazosulphonic acid, sodium
salt
4-Dimethylaminobenzene-
diazosulphonic acid, sodium
salt
p-Dimethylaminobenzoldiazosulfonat (natriumsalz)
(German)
(4-(Dimethylamino)phenyl)di-
azenesulfonic acid, sodium
salt

4-((Dimethylamino)phenyl)diazenesulfonic acid, sodium salt
p-(Dimethylamino)-phenyldiazo-natriumsulfonat (German)
N,N-Dimethyl-p-anilinediazosulfonic acid sodium salt
Fenaminosulf
Gold Orange MP
Lesan
NCI-C03010
Pehnaminosulf
Sodium p-(dimethylamino)-benzenediazosulfonate
Sodium 4-(dimethylamino)-benzenediazosulfonate
Sodium p-(dimethylamino)-benzenediazosulphonate
Sodium 4-(dimethylamino)-benzenediazosulphonate
Sodium (4-(dimethylamino)-phenyl)diazenesulfonate
Tropaeolin D

140-57-8
$C_{15}H_{23}ClO_4S$
334.89
CC(COc1ccc(cc1)C(C)(C)C)OS(=O)OCCCl
Sulfurous acid, 2-(p-t-butylphenoxy)-1-methylethyl-2-chloroethyl ester
Acaracide
Aracide
Aramite
Aramite-15W
Aramitearamite-15W
Aratron
2-(p-terc.Butylfenoxy)isopropyl-2'-chlorethylester kyseliny siricite (Czech)
Butylphenoxyisopropyl chloroethyl sulfite
2-(p-Butylphenoxy)isopropyl 2-chloroethyl sulfite
2-(4-t-Butylphenoxy)isopropyl-2-chloroethyl sulfite
2-(p-t-Butylphenoxy)isopropyl 2'-chloroethyl sulphite
2-(p-t-Butylphenoxy)-1-methylethyl 2-chloroethyl ester of sulphurous acid
2-(p-Butylphenoxy)-1-methylethyl 2-chloroethyl sulfite
2-(p-t-Butylphenoxy)-1-methylethyl-2-chloroethyl sulfite
2-(p-t-Butylphenoxy)-1-methylethyl 2'-chloroethyl sulphite
2-(p-t-Butylphenoxy)-1-methylethyl sulphite of 2-chloroethanol
CES
β-Chloroethyl-β'-(p-t-butylphenoxy)-α'-methylethyl sulfite
β-Chloroethyl-β-(p-t-butylphenoxy)-α-methylethyl sulphite
2-Chloroethyl 1-methyl-2-(p-t-butylphenoxy)ethyl sulphate
2-Chloroethyl sulphite of 1-(p-t-butylphenoxy)-2-propanol
Compound 88R
ENT 16,519
Ethanol, 2-chloro-, 2-(p-t-butylphenoxy)-1-methylethyl sulfite
Ethanol, 2-chloro-, ester with 2-(p-tert-butylphenoxy)-1-methylethyl sulfite
Niagaramite
Ortho-Mite
2-Propanol, 1-(p-t-butylphenoxy)-, 2-chloroethyl sulfite
88-R
Sulfurous acid, 2-(p-tert-butylphenoxy)-1-methylethyl-2-chloroethyl ester
Sulfurous acid, 2-chloroethyl-, 2-(4-(1,1-dimethylethyl)-phenoxy)-1-methylethyl ester

140-60-3
$C_{16}H_{26}O_3S$
298.45
O=S(=O)(O)c(ccc(c1)CCCCCCCCCC)c1
4-Decylbenzenesulfonic acid
Benzenesulfonic acid, p-decyl-
Benzenesulfonic acid, 4-decyl-(9CI)
p-n-Decylbenzenesulfonate
p-Decylbenzenesulfonic acid

140-64-7
$C_{19}H_{24}N_4O_2.C_4H_{12}O_8S_2$
592.75
Benzamidine, 4,4'-(pentamethylenedioxy)di-, bis(β-hydroxyethanesulfonate)
Diamidine
4,4'-Diamidinodiphenoxypentane di(β-hydroxyethanesulfonate)
4,4'-Diamidino-α,ω-diphenoxypentane isethionate
Lomidin
Lomidine
Lomidine isoethionate
M&B 800
Pentam 300
p,p'-(Pentamethylenedioxy)dibenzamidine bis(β-hydroxyethanesulfonate)
Pentamidine diisethionate
Pentamidine isethionate
2512 R.P.
R.P. 2512
USAF XR-10

140-66-9
$C_{14}H_{22}O$
206.36
Oc(ccc(c1)C(CC(C)(C)C)(C)C)c1
Phenol, p-(1,1,3,3-tetramethylbutyl)
p-tert-Octylphenol
p-terc.Oktylfenol (Czech)
Phenol, p-(tert-octyl)-
4-(1,1,3,3-Tetramethylbutyl)-phenol

140-67-0
$C_{10}H_{12}O$
148.22
O(c(ccc(c1)CC=C)c1)C
Anisole, p-allyl
p-Allylanisole
p-Allylmethoxybenzene
4-Allyl-1-methoxybenzene
Chavicol methyl ether
Esdragol
Esdragon
Estragole
Isoanethole
p-Methoxyallylbenzene
1-Methoxy-4-(2-propenyl)-benzene
Methyl chavicol
NCI-C60946
Tarragon

140-72-7
$C_{21}H_{38}N.Br$
384.51
Pyridinium, 1-hexadecyl-, bromide
Acetoquat CPB
Bromocet
Cetapharm
Cetasol
Cetazol
Cetylpyridinium bromide
n-Cetylpyridinium bromide
1-Cetylpyridinium bromide
Fixanol C
Hexadecylpyridine bromide
Hexadecylpyridinium bromide
n-Hexadecylpyridinium bromide
1-Hexadecylpyridinium bromide
Morpan CBP
Nitrogenol
Seprisan
Sterogenol
TsPB

140-76-1
C_8H_9N
119.18
n(c(ccc1C=C)C)c1
Pyridine, 5-ethenyl-2-methyl
2-Methyl-5-vinylpyridine
2-Picoline, 5-vinyl-
Pyridine, 2-methyl-5-vinyl-

140-77-2
$C_8H_{14}O_2$
142.20
O=C(O)CCC(CCC1)C1
Cyclopentanepropanoic acid (9CI)
AI3-14247
Cyclopentanepropionic acid (8CI)
Cyclopentylpropionic acid
3-Cyclopentylpropionic acid
NSC-8771

Propionic acid, 3-cyclopentyl-

140-79-4
$C_4H_8N_4O_2$
144.16
O=NN(CCN(N=O)C1)C1
Piperazine, 1,4-dinitroso
Dinitrosopiperazin (German)
Dinitrosopiperazine
N,N'-Dinitrosopiperazine
1,4-Dinitrosopiperazine
DNPZ
NSC-339
USAF DO-36

140-82-9
$C_8H_{19}NO_2$
161.24
O(CCN(CC)CC)CCO
Ethanol, 2-(2-(diethylamino)-ethoxy)- (9CI)
Diethylaminoethoxyethanol
2-β-Diethylaminoethoxyethanol
2-(2-(Diethylamino)ethoxy)-ethanol
NSC-163322

140-87-4
$C_3H_5N_3O$
99.11
NNC(=O)CC#N
Acetic acid, cyano-, hydrazide
AB-42
Armazal
Cianazil
Cyacetacid
Cyacetacide
Cyacetazid
Cyacetazide
Cyanacethydrazide
Cyanacetic acid hydrazide
Cyanacetohydrazide
Cyanacetylhydrazide
Cyanazide
Cyanizide
Cyanoacethydrazide
Cyanoacetic acid hydrazide
Cyanoacetohydrazide
α-Cyanoacetohydrazide
Cyanoacetylhydrazide
Cyanoethydrazide
Cyazid
Cyazide
Dictycide
Dictyzide
Helmox
Hidacian
Hidaciann
Kyanacethydrazid (Czech)
Leandin
Mackreazid
Malonitrile hydrazide
Malononitrile hydrazide
Neohydrazid
Reacid
Reazid
Reazide
Tsiazid
USAF KF-18

140-88-5
$C_5H_8O_2$
100.13
O=C(OCC)C=C
Acrylic acid, ethyl ester
Acrylate d'ethyle (French)
Acrylsaeureaethylester (German)
Aethylacrylat (German)
Akrylanem etylu (Polish)
Carboset 511
Ethoxycarbonylethylene
Ethylacrylaat (Dutch)
Ethyl acrylate (ACGIH,OSHA)
Ethyl acrylate, Inhibited [UN 1917]
Ethylakrylat (Czech)
Ethylester kyseliny akrylove (Czech)
Ethyl propenoate
Ethyl 2-propenoate
Etil acrilato (Italian)
Etilacrilatului (Romanian)
NCI-C50384
2-Propenoic acid, ethyl ester
RCRA waste number U113
UN 1917 [Ethyl acrylate, inhibited]

140-89-6
$C_3H_5OS_2.K$
160.30
CCOC(S)=S.C[N+](C)(C)C
Carbonic acid, dithio-, O-ethyl ester, potassium salt
Carbonodithioic acid, O-ethyl ester, potassium salt (9CI)
(o-Ethyl dithiocarbonato)potassium
o-Ethyl potassium dithiocarbonate
Ethyl potassium xanthate
Ethyl potassium xanthogenate
Ethylxanthic acid potassium salt
Potassium ethyl dithiocarbonate
Potassium O-ethyl dithiocarbonate
Potassium ethylxanthate
Potassium ethyl xanthogenate
Potassium xanthate
Potassium xanthogenate
Xanthic acid, ethyl-, potassium salt
Z3
Z 3 (Pesticide)

140-90-9
$C_3H_6OS_2.Na$
145.20
Carbonodithioic acid, O-ethyl ester, sodium salt (9CI)
O-Ethyl carbonodithioate sodium salt

140-92-1
$C_4H_7OS_2.K$
174.33
Carbonic acid, dithio-, O-isopropyl ester, potassium salt
Carbonodithioic acid, O-(1-methylethyl) ester, potassium salt (9CI)
Dithiocarbonic acid O-isopropyl ester potassium salt
Isopropyl potassium xanthate
Potassium isopropylxanthate
Potassium isopropyl xanthogenate
Xanthic acid, isopropyl-, potassium salt

140-93-2
$C_4H_7OS_2.Na$
158.22
Carbonic acid, dithio-, O-isopropyl ester, sodium salt
Aeroxanthate 343
Carbonodithioic acid, O-(1-methylethyl)ester, sodium salt (9CI)
Good-Rite Nix
Isopropylxanthic acid, sodium salt
Isopropylxanthogenan sodny (Czech)
Natrium-O-isopropyldithiokarbonat (Czech)
Nix
Proxan sodium
Sodium O-isopropyl dithiocarbonate
Sodium isopropylxanthate
Sodium O-isopropyl xanthate
Sodium isopropylxanthogenate
Xanthic acid, isopropyl-, sodium salt
Z 11

140-95-4
$C_3H_8N_2O_3$
120.13
O=C(NCO)NCO
Urea, 1,3-bis(hydroxymethyl)
N,N'-Bis(hydroxymethyl)urea
1,3-Bis(hydroxymethyl)urea
Caurite
Csi Paste
N,N'-Dihydroxymethylurea
Dimethanol urea
Dimethylolurea
N,N'-Dimethylolurea
1,3-Dimethylolurea
DMU
Finish EN
Kaurit S
Knittex ASL
Methural
Methurin (Russian)
Metural
Oxymethurea
Permafresh 477
Protesine DMU
Urea, N,N'-bis(hydroxymethyl)-(9CI)
Ureol P

141-00-4
$C_4H_4O_4.Cd$
228.48
O=C(OC(=O)CC...)
Succinic acid, cadmium salt (1:1)
Cadminate
Cadmium succinate

141-02-6
$C_{20}H_{36}O_4$
340.56
O=C(OCC(CCCC)CC)C=CC(=O)OCC(CCCC)CC
Fumaric acid, bis(2-ethylhexyl) ester
Bis-(2-ethylhexyl)ester kyseliny fumarove (Czech)
Bis(2-ethylhexyl) fumarate
2-Butenedioic acid, bis(2-ethylhexyl) ester
Di(2-ethylhexyl) fumarate
Dioctyl fumarate
DOF
2-Ethylhexyl fumarate
RC Comonomer DOF

141-03-7
$C_{12}H_{22}O_4$
230.34
O=C(OCCCC)CCC(=O)OCCCC
Succinic acid, dibutyl ester
B-9
Butanedioic acid dibutyl ester
Butyl butanedioate
Di-n-butylester kyseliny jantarove (Czech)
Dibutyl succinate
Di-n-butylsuccinate
DNBS
ENT 666
Succinic acid di-n-butyl ester
Tabatrex
Tabutrex

141-04-8
$C_{14}H_{26}O_4$
258.40
O=C(OCC(C)C)CCCCC(=O)OCC(C)C
Adipic acid, diisobutyl ester
DIBA
Diisobutyl adipate
Ftaflex DIBA
Hexanedioic acid, bis(2-methylpropyl) ester (9CI)
Isobutyl adipate

141-05-9
$C_8H_{12}O_4$
172.20
O=C(OCC)C=CC(=O)OCC
Maleic acid, diethyl ester
2-Butenedioic acid (z)-, diethyl ester
Diethylester kyseliny maleinove (Czech)
Diethyl maleate
Ethyl maleate

141-07-1
$C_5H_{12}N_2O_3$
148.15
O=C(NCOC)NCOC
1,3-Bis(methoxymethyl)urea
Bis(methoxymethyl)urea
N,N'-Bis(methoxymethyl)urea
Dimethoxydimethylolurea
N,N'-Dimethoxymethylurea
1,3-Dimethoxymethylurea
Dimethylolurea dimethyl ether
N,N'-Dimethylolurea dimethyl ether
Kaurit W
NSC-75061
Urea, N,N'-bis(methoxymethyl)- (9CI)
Urea, 1,3-bis(methoxymethyl)-(8CI)

141-10-6
$C_{13}H_{20}O$
192.30
O=C(C=CC=C(CCC=C(C)C)C)C
3,5,9-Undecatrien-2-one, 6,10-dimethyl-
AI3-22131
6,10-Dimethyl-3,5,9-undecatrien-2-one

141-12-8
$C_{12}H_{20}O_2$
196.29
O=C(OCC=C(CCC=C(C)C)C)C
2,6-Octadien-1-ol, 3,7-dimethyl-, acetate, (Z)- (9CI)
AI3-35817
Nerol acetate
Neryl acetate
NSC-72031

141-14-0
$C_{13}H_{24}O_2$
212.33
O=C(OCCC(CCC=C(C)C)C)CC
6-Octen-1-ol, 3,7-dimethyl-, propanoate
AI3-24358
3,7-Dimethyl-6-octen-1-ol propanoate

141-17-3
$C_{22}H_{42}O_8$
434.64
O=C(OCCOCCOCCCC)CCCCC(=O)OCCOCCOCCCC
Adipic acid, bis(2-(2-butoxyethoxy)ethyl) ester
Dibutoxyethoxyethyl adipate
Hexanedioic acid, bis(2-(2-butoxyethoxy)ethyl) ester (9CI)
TP-95
Wareflex

141-18-4
$C_{18}H_{34}O_6$
346.52
O=C(OCCOCCCC)CCCCC(=O)OCCOCCCC
Adipic acid, bis(2-butyoxyethyl) ester
Adipic acid, dibutoxyethyl ester
Adipol BCA
Bis(2-butoxyethyl) adipate
Butyl "cellosolve" adipate (BCA)
Dibutoxyethyl adipate
Di(2-butoxyethyl) adipate
Dibutyl cellosolve adipate
Hexanedioic acid, bis(2-butoxyethyl) ester (9CI)
Staflex DBEA

141-19-5
$C_{22}H_{42}O_6$
402.57
O=C(OCCOCCCC)CCCCCCCC(=O)OCCOCCCC
Decanedioic acid, bis(2-butoxyethyl) ester
AI3-03528
Bis(2-butoxyethyl) decanedioate

141-20-8
$C_{16}H_{32}O_4$
288.43
O=C(OCCOCCO)CCCCCCCCCCC
PEG-2 Laurate
AI3-00969
Atlas G-2124
Diethylene glycol laurate
Diethylene glycol lauric acid monoester
Diethylene glycol monolaurate
Diethylene glycol sesquilaurate
Diglycol laurate
Diglycol monolaurate
2,2'-Dihydroxyethyl ether monododecanoate
Dodecanoic acid, 2-(2-hydroxyethoxy)ethyl ester (9CI)
Emcol RDC-D
Ethanol, 2-(2-hydroxyethoxy)-, laurate
G 2124
Glaurin
2-(2-Hydroxyethoxy)ethyl dodecanoate
2-(2-Hydroxyethoxy)ethyl laurate
Lauric acid, 2-(2-hydroxyethoxy)ethyl ester (8CI)
Lauro-Sebum
Nonex 413
NSC-3868
Pegosperse 100L
Pegosperse 100 LN
Polyethylene glycol 100 monolaurate
Polyoxyethylene (2) monolaurate
Unipeg-DGL

141-21-9

$C_{22}H_{46}N_2O_2$
370.61
O=C(NCCNCCO)CCCCCCCCC CCCCCCCC
N-(2-Hydroxyethyl)-N'-stearoylethylenediamine
Ethanolaminoethyl stearamide
Ethanol, 2-(2-stearamidoethyl-amino)-
N-(Ethoxyaminoethyl)stear-amide
N-(2-((2-Hydroxyethyl)amino)-ethyl)octadecanamide
N-(2-((2-Hydroxyethyl)amino)-ethyl)stearamide
N-(2-Hydroxyethyl)-N'-octadec-anoylethylenediamine
Octadecanamide, N-(2-((2-hydroxyethyl)amino)ethyl)- (9CI)
Stearamidoethyl ethanolamine
Stearic N-(aminoethyl)ethan-olamide
N-Stearoyl-N'-(β-hydroxyethyl)-ethylenediamine
N-Stearoyl-N'-(2-hydroxyethyl)-ethylenediamine

141-22-0
$C_{18}H_{34}O_3$
298.52
O=C(O)CCCCCCCC=CCC(O) CCCCCC
Ricinoleic-acid
L'Acide ricinoleique (French)
12-Hydroxy-cis-9-octadecenoic acid
Kyselina 12-hydroxy-9-okta-decenova (Czech)
Kyselina ricinolova (Czech)
9-Octadecenoic acid, 12-hydroxy-, (Z)-
Oleic acid, 12-hydroxy-
Ricinic acid
Ricinolic acid

141-23-1
$C_{19}H_{38}O_3$
314.57
O=C(OC)CCCCCCCCCCC(O) CCCCCC
Stearic acid, 12-hydroxy-, methyl ester
Cenwax ME
12-Hydroxystearic acid, methyl ester
Methyl 12-hydroxystearate

141-24-2
$C_{19}H_{36}O_3$
312.49
O=C(OC)CCCCCCCC=CCC(O) CCCCCC
Methyl ricinoleate
AI3-10523
Castor Oil Acid, Methyl Ester
Flexricin P-1
12-Hydroxy-9-octadecenoic acid, methyl ester
Methyl 12-hydroxy-9-octadec-enoate
NSC-1254
9-Octadecenoic acid, 12-hydroxy-, methyl ester
9-Octadecenoic acid, 12-hydroxy-, methyl ester, (R-(Z))- (9CI)
Ricinoleic acid, methyl ester (8CI)

141-27-5
$C_{10}H_{16}O$
152.26
O=CC=C(CCC=C(C)C)C
2,6-Octadienal, 3,7-dimethyl-, (E)
Citral α
α-Citral
(E)-Citral
trans-Citral
trans-3,7-Dimethyl-2,6-octa-dienal
Geranaldehyde
Geranial

141-28-6
$C_{10}H_{18}O_4$
202.28
O=C(OCC)CCCCC(=O)OCC
Adipic acid, diethyl ester
Diethyl adipate
Diethylester kyseliny adipove (Czech)
Diethyl hexanedioate
Ethyl adipate
Ethyl δ-carboethoxyvalerate
Hexanedioic acid, diethyl ester (9CI)

141-32-2
$C_7H_{12}O_2$
128.19
O=C(OCCCC)C=C
2-Propenoic acid, butyl ester
Acrylic acid, butyl ester
Acrylic acid n-butyl ester
Butylacrylate (OSHA) [UN 2348]
n-Butyl acrylate (ACGIH)
Butylacrylate, Inhibited (DOT)
Butylester kyseliny akrylove (Czech)
Butyl 2-propenoate
UN 2348 [Butylacrylate]

141-37-7
$C_{16}H_{24}O_4$
280.40
7-Oxabicyclo(4.1.0)heptane-3-carboxylic acid, 4-methyl-, (4-methyl-7-oxabicyclo-(4.1.0) Hept-3-yl)methyl ester
Chissonox 201
EP 201
Epoxide-201
3,4-Epoxy-6-methylcyclo-hexenecarboxylic acid (3,4-epoxy-6-methylcyclo-hexylmethyl) ester
3,4-Epoxy-6-methylcyclohexyl-methyl 3,4-epoxy-6-methyl-cyclohexanecarboxylate
3,4-Epoxy-6-methylcyclohexyl-methyl-3',4'-epoxy-6'-methyl-cyclohexane carboxylate
4,5-Epoxy-2-methylcyclohexyl-methyl-4,5-epoxy-2-methyl-cyclohexanecarboxylate
6-Methyl-3,4-epoxycyclohexyl-methyl 6-methyl-3,4-epoxy-cyclohexane carboxylate
Unox 201
Unox Epoxide 201

141-38-8
$C_{26}H_{50}O_3$
410.76
O=C(OCC(CCCC)CC)CCCCCC CC(O1)C1CCCCCCCC
Octadecanoic acid, 9,10-epoxy-, 2-ethylhexyl ester
9,10-Epoxyoctadecanoic acid, 2-ethylhexyl ester
9,10-Epoxystearic acid, 2-ethyl-hexyl ester
2-Ethylhexyl 9,10-epoxyocta-decanoate
2-Ethylhexyl epoxystearate

141-43-5
C_2H_7NO
61.10
OCCN
Ethanol, 2-amino
Aethanolamin (German)
2-Aminoaethanol (German)
2-Aminoetanolo (Italian)
2-Aminoethanol (OSHA)
β-Aminoethyl alcohol
Colamine
Etanolamina (Italian)
Ethanolamine (ACGIH,OSHA) [UN 2491]
Ethanolamine, Solution [UN 2491]
β-Ethanolamine
Ethylolamine
Glycinol
β-Hydroxyethylamine
2-Hydroxyethylamine
Kolamin (Czech)
MEA
Monoaethanolamin (German)
Monoethanolamine (DOT)
Olamine
Thiofaco M-50
UN 2491 [Ethanolamine or Ethanolamine solutions]
USAF EK-1597

141-46-8
$C_2H_4O_2$
60.06
Glycolaldehyde
Acetaldehyde, hydroxy- (9CI)

Diose
Glycolic aldehyde
Hydroxyacetaldehyde
Methylol formaldehyde
Monomethylolformaldehyde

141-52-6
$C_2H_6O.Na$
69.06
[Na+].CC[O-]
Sodium ethanolate
AI3-52660
Caustic Alcohol
Ethanol, sodium salt
Ethoxysodium
Ethyl alcohol, sodium salt
Sodium ethoxide
Sodium ethylate

141-53-7
$CHO_2.Na$
68.01
[Na+].[O-]C=O
Formic acid, sodium salt
Mravencan sodny (Czech)
Salachlor
Sodium formate

141-57-1
$C_3H_7Cl_3Si$
177.54
Silane, propyltrichloro
n-Propyltrichlorosilane
Propyltrichlorosilane [UN 1816]
Silane, trichloropropyl-
UN 1816 [Propyltrichlorosilane]

141-59-3
$C_8H_{18}S$
146.32
SC(CC(C)(C)C)(C)C
2-Pentanethiol, 2,4,4-trimethyl
tert-Octanethiol
t-Octyl mercaptan
terc. Oktanthiol (Czech)
2,4,4-Trimethyl-2-pentanethiol

141-62-8
$C_{10}H_{30}O_3Si_4$
310.69
CSi(C)(C)OSi(C)(C)OSi(C)(C)OSi(C)(C)C
Tetrasiloxane, decamethyl- (9CI)
Decamethyltetrasiloxane

141-63-9
$C_{12}H_{36}O_4Si_5$
384.93
CSi(C)(C)OSi(C)(C)OSi(C)(C)OSi(C)(C)OSi(C)(C)C
Pentasiloxane, dodecamethyl
Dodecamethylpentasiloxane

141-65-1
$C_{16}H_{35}O_4P.Na$
345.47
Tergitol Anionic P-28

141-66-2
$C_8H_{16}NO_5P$
237.22
COP(=O)(OC)OC(C)=CC(=O)N(C)C
Phosphoric acid, dimethyl ester, ester with (E)-3-hydroxy-N,N-dimethylcrotonamide
Bidirl
Bidrin
C 709
Carbicron
Ciba 709
Crotonamide, 3-hydroxy-N,N-dimethyl-, cis-, dimethyl phosphate
Crotonamide, 3-hydroxy-N-N-dimethyl-, dimethyl phosphate, cis-
Crotonamide, 3-hydroxy-N-N-dimethyl-, dimethyl phosphate, (E)-
Diapadrin
Dicrotofos (Dutch)
Dicrotophos (ACGIH,OSHA)
3-(Dimethoxyphosphinyloxy)-N,N-dimethyl-cis-crotonamide
3-(Dimethoxyphosphinyloxy)-N,N dimethylisocrotonamide
3-(Dimethylamino)-1-methyl-3-oxo-1-propenyl dimethyl phosphate
cis-2-Dimethylcarbamoyl-1-methylvinyl dimethylphosphate
O,O-Dimethyl-O-(2-dimethyl-carbamoyl-1-methyl-vinyl)-phosphat (German)
2-Dimethyl cis-2-dimethyl-carbamoyl-1-methylvinyl phosphate
O,O-Dimethyl O-(N,N-dimethylcarbamoyl-1-methylvinyl)phosphate
O,O-Dimethyl-O-(1,4-dimethyl-3-oxo-4-aza-pent-1-enyl)-fosfaat (Dutch)
O,O-Dimethyl-O-(1,4-dimethyl-3-oxo-4-aza-pent-1-enyl)-phosphate
O,O-Dimethyl-O-(1-methyl-2-N,N-dimethyl-carbamoyl)-vinyl-phosphat (German)
Dimethyl phosphate of 3-hydroxy-N,N-dimethyl-cis-crotonamide
Dimethyl phosphate ester with 3-hydroxy-N,N-dimethyl-cis-crotonamide
O,O-Dimetil-O-(1,4-dimetil-3-oxo-4-aza-pent-1-enil)-fosfato (Italian)
Ektafos
ENT 24,482
3-Hydroxydimethyl crotonamide dimethyl phosphate
3-Hydroxy-N,N-dimethyl-cis-crotonamide dimethyl phosphate
OMS 253
Phosphate de dimethyle et de 2-dimethylcarbamoyl 1-methyl vinyle (French)
Phosphoric acid 3-(dimethylamino)-1-methyl-3-oxo-1-propenyl dimethyl ester
Phosphoric acid, dimethyl ester, ester with cis-3-hydroxy-N,N-dimethylcrotonamide
Phosphoric acid, dimethyl 1-methyl-N,N-(dimethylamino)-3-oxo-1-propenyl ester, (E)- (9CI)
SD 3562
Shell SD-3562

141-75-3
C_4H_7ClO
106.56
O=C(CCC)Cl
Butyryl chloride [UN 2353]
n-Butyryl chloride
UN 2353 [Butyryl chloride]

141-76-4
$C_3H_5IO_2$
199.98
O=C(O)CCI
Propionic acid, 3-iodo
3-Iodopropionic acid

141-78-6
$C_4H_8O_2$
88.12
O=C(OCC)C
Acetic acid, ethyl ester
Acetic ether
Acetidin
Acetoxyethane
Aethylacetat (German)
Essigester (German)
Ethylacetaat (Dutch)
Ethyl acetate (ACGIH,OSHA) [UN 1173]
Ethyl acetic ester
Ethyle (acetate d') (French)
Ethylester kyseliny octove (Czech)
Ethyl ethanoate
Etile (acetato di) (Italian)
Octan etylu (Polish)
RCRA waste number U112
UN 1173 [Ethyl acetate]
Vinegar Naphtha

141-79-7
$C_6H_{10}O$
98.16
O=C(C=C(C)C)C
3-Penten-2-one, 4-methyl
Acetone, isopropylidene-
Isobutenyl methyl ketone
Isopropylidene acetone

Mesityloxid (German)
Mesityl oxide (ACGIH,OSHA) [UN 1229]
Mesityloxyde (Dutch)
Methyl isobutenyl ketone
4-Methyl-3-penten-2-on (Dutch, German)
4-Methyl-3-pentene-2-one
2-Methyl-2-pentenone-4
2-Methyl-2-penten-4-one
4-Methyl-3-penten-2-one
4-Metil-3-penten-2-one (Italian)
MIBK
Ossido di mesitile (Italian)
Oxyde de mesityle (French)
UN 1229 [Mesityl oxide]

141-82-2
$C_3H_4O_4$
104.07
O=C(O)CC(=O)O
Malonic-acid
Carboxyacetic acid
Dicarboxymethane
Kyselina malonova (Czech)
Methanedicarboxylic acid
Propanedioic acid
USAF EK-695

141-83-3
$C_2H_6N_4O$
102.07
O=C(NC(=N)N)N
Guanidine carboxamide
(Aminoiminomethyl)urea
Urea, (aminoiminomethyl)- (9CI)

141-84-4
$C_3H_3NOS_2$
133.19
O=C(N=C(S)S1)C1
Rhodanine
4-Oxo-2-thionothiazolidine
4-Oxo-2-thiothiazolidin (Czech)
Rhodanic acid
Rhodanin (Czech)
Rhodaninic acid
Rodanin
2,4-Thiazolidinedione, 2-thio-
4-Thiazolidinone, 2-thioxo-
2-Thio-4-ketothiazolidine
2-Thioxo-4-thiazolidinone
USAF HA-2

141-85-5
$C_6H_6ClN.ClH$
164.03
Aniline, m-chloro-, hydrochloride (8CI)
Amarthol Fast Orange GC Base
Amarthol Fast Orange GC Salt
Ansibases Orange GC
Azofix Orange GC Salt
Azogene Fast Orange GC Base
Azogene Fast Orange GC Salt
Azogene Fast Orange GCN Base
Azogene Fast Orange GEN Salt
Brentamine Fast Orange GC Base
Brentamine Fast Orange GC Salt
Benzenamine, 3-chloro-, hydrochloride (9CI)
m-Chloroaniline hydrochloride
3-Chloroaniline hydrochloride
3-Chlorobenzenamine hydrochloride
C.I. Azoic Diazo Component 2
C.I. 37005
Daito Orange Base GC
Daito Orange Salt GC
Devol Orange C
Devol Orange GC
Devol Orange GC Salt
Diabase Orange GC Base
Diazo Fast Orange GC
Fast Orange Base GC
Fast Orange Base JS
Fast Orange G Base
Fast Orange GC New Salt
Fast Orange JS Salt
Fast Orange MC Base
Fast Orange MC Salt
Fast Orange Salt GC
Fast Orange Salt GCS
Hiltonil Fast Orange GC Base
Hiltosal Fast Orange GC Salt
Hindasol Orange GC Salt
Naphtoelan Fast Orange GC Base
Naphtoelan Fast Orange GC Salt
Natasol Fast Orange GC Salt
NSC-212255
Orange Base Ciba IV
Orange Base Irga IV
Orange Base NGC
Orange GC Salt
Orange GCS Salt
Orange Salt Ciba IV
Orange Salt Irga IV
Orange Salt NGC
Sanyo Fast Orange GC Base
Symulon Orange GC Base

141-86-6
$C_5H_7N_3$
109.15
n(c(N)ccc1)c1N
Pyridine, 2,6-diamino
2,6-Diaminopyridine

141-90-2
$C_4H_4N_2OS$
128.16
n(ccc(n1)O)c1S
Uracil, 2-thio
Antagothyroid
Antagothyroil
Deracil
2,3-Dihydro-2-thioxo-4(1H)-pyrimidinone
6-Hydroxy-2-mercaptopyrimidine
4-Hydroxy-2(1H)-pyrimidinethione
2-Mercapto-4-hydroxypyrimidine
2-Mercapto-4-pyrimidinol
2-Mercapto-4-pyrimidone
2-Mercapto-4(1H)-pyrimidinone
2-Mercaptopyrimid-4-one
Nobilen
4-Pyrimidinol, 2-mercapto-
2-Thio-6-oxypyrimidine
2-Thio-1,3-pyrimidin-4-one
Thiouracil
2-Thiouracil
6-Thiouracil
Tiouracyl (Polish)
TU
2-TU

141-91-3
$C_6H_{13}NO$
115.20
O(C(CNC1)C)C1C
Morpholine, 2,6-dimethyl
2,6-Dimethylmorfolin (Czech)
2,6-Dimethylmorpholine
2,6-Dimethyl-2,3,5,6-tetrahydro-4H-1,4-oxazine

141-93-5
$C_{10}H_{14}$
134.24
c(cccc1CC)(c1)CC
Benzene, m-diethyl
m-Diethylbenzene

141-94-6
$C_{21}H_{45}N_3$
339.59
N(CC(N)(CN1CC(CCCC)CC)C)(C1)CC(CCCC)CC
Hexetidine
AI3-15546
5-Amino-1,3-bis(2-ethylhexyl)-5-methylhexahydropyrimidine
Caswell No. 033BB
Collu Hextril
Duranil Aerosol
Elsix
Glypesin
Hexetidina (Spanish)
Hexetidinum (Latin)
Hexopyrimidine
Hexoral
Hextril
NSC-17764
P 252
5-Pyrimidinamine, 1,3-bis-(2-ethylhexyl)hexahydro-5-methyl- (9CI)
Pyrimidine, 5-amino-1,3-bis-(2-ethylhexyl)hexahydro-5-methyl- (8CI)
Sterilate
Sterisil
Steri/Sol (VAN)
Triocil
Triscol

141-95-7

$C_3H_3O_4.Na$
126.05
Malonic acid, sodium salt
Sodium malonate

141-97-9
$C_6H_{10}O_3$
130.16
O=C(OCC)CC(=O)C
Acetoacetic acid, ethyl ester
Acetoacetic ester
Acetoctan ethylnaty (Czech)
Active acetyl acetate
Butanoic acid, 3-oxo-, ethyl ester
Diacetic ether
EAA
Ethylacetacetat (Czech)
Ethyl acetoacetate
Ethyl acetyl acetate
Ethyl acetylacetonate
Ethylester kyseliny acetoctove (Czech)
Ethyl 3-oxobutanoate
Ethyl 3-oxobutyrate
3-Oxobutanoic acid ethyl ester

141-98-0
$C_6H_{13}NOS$
147.24
O(C(=NCC)S)C(C)C
Z-200
Carbamothioic acid, ethyl-, O-(1-methylethyl) ester (9CI)
Dow Z-200
Ethylcarbamothioic acid, O-(1-methylethyl) ester
N-Ethyl O-isopropyl thiocarbamate
Z 200

142-03-0
$C_4H_7AlO_5$
162.08
Aluminum diacetate
Aluminum, bis(acetato-O)-hydroxy- (9CI)
Aluminum hydroxyacetate
Aluminum subacetate
Basic Aluminum Acetate
Bis(acetato)hydroxyaluminum
Bis(acetato-O)hydroxyaluminum

142-04-1
$C_6H_7N.ClH$
129.60
Cl.Nc1ccccc1
Benzenamine, hydrochloride
Aniline chloride
Aniline hydrochloride [UN 1548]
"Aniline salt"
Anilinium chloride
Chlorhydrate d'aniline (French)
Chlorid anilinu (Czech)
C.I. 76001
Hydrochloride benzenamide
NCI-C03736
Phenylamine hydrochloride
Sul anilinova (Czech)
UN 1548 [Aniline hydrochloride]
USAF EK-442

142-08-5
C_5H_5NO
95.11
n(c(O)ccc1)c1
2(1H)-Pyridone
2-Hydroxypyridine
2-Oxopyridine
2-Pyridinol
2-Pyridinone
2(1H)-Pyridinone (9CI)
α-Pyridone
Pyridone-2 (French)
2-Pyridone

142-09-6
$C_{10}H_{18}O_2$
170.25
O=C(OCCCCCC)C(=C)C
Hexyl methacrylate
AI3-25419
ENT 25,419
n-Hexyl methacrylate
Hexyl 2-methyl-2-propenoate
Methacrylic acid, hexyl ester (8CI)
NSC-24169
2-Propenoic acid, 2-methyl-, hexyl ester (9CI)

142-15-4
$C_{20}H_{38}O_5S.Na$
413.57
Sodium oleoyl isethionate
AI3-00358
9-Octadecenoic acid, 2-sulfoethyl ester, sodium salt
9-Octadecenoic acid (Z)-, 2-sulfoethyl ester, sodium salt (9CI)
2-Sulfoethyl 9-octadecenoate, sodium salt

142-16-5
$C_{20}H_{36}O_4$
340.56
O=C(OCC(CCCC)CC)C=CC(=O)OCC(CCCC)CC
Maleic acid, bis(2-ethylhexyl)ester
Bis-(2-ethylhexyl)ester kyseliny maleinove (Czech)
Bis(2-ethylhexyl)maleate
Di-(2-ethylhexyl)maleate
Dioctyl maleate
DOM
RC Comonomer Dom

142-17-6
$C_{18}H_{34}O_2.1/2Ca$
302.51
9-Octadecenoic acid (Z)-, calcium salt (9CI)
AI3-19804
9-Octadecenoic acid, (Z)-, calcium salt (2:1)

142-22-3
$C_{12}H_{18}O_7$
274.30
O=C(OCC=C)OCCOCCOC(=O)OCC=C
Carbonic acid, oxydiethylenedi-, diallyl ester
Allyl diglycol carbonate
Carbonic acid, allyl ester, diester with diethylene glycol
CR 39
DAGC
Diallyl diglycol carbonate
Diethylene glycol, bis(allyl carbonate)-
01M
Oxydiethylenedicarbonic acid diallyl ester
RAV 7
2,5,8,10-Tetraoxatridec-12-enoic acid, 9-oxo-, 2-propenyl ester (9CI)
Transallyl CR 39

142-26-7
$C_4H_9NO_2$
103.14
O=C(NCCO)C
Acetamide, N-(2-hydroxyethyl)
2-Acetamidoethanol
2-Acetylaminoethanol
Acetylcolamine
N-Acetyl ethanolamine
N-Ethanolacetamide
Hydroxyethyl acetamide
β-Hydroxyethylacetamide
N-β-Hydroxyethylacetamide
N-(2-Hydroxyethyl)acetamide

142-28-9
$C_3H_6Cl_2$
112.99
ClCCCCl
Propane, 1,3-dichloro
1,3-Dichloropropane
Trimethylene dichloride

142-29-0
C_5H_8
68.13
C(=CCC1)C1
UN]
UN 2246 [Cyclopentene]

142-30-3
$C_8H_{14}O_2$
142.20
OC(C#CC(O)(C)C)(C)C
2,5-Dimethyl-3-hexyne-2,5-diol
Acetylenepinacol
AI3-14500
D 43 (VAN)

Dimethylhexynediol
3-Hexyne-2,5-diol, 2,5-dimethyl- (9CI)
Kemitracin-50
NSC-117261
Tetramethylbutynediol

142-31-4
$C_8H_{17}O_4S.Na$
232.30
[Na+].CCCCCCCCOS([O-])(=O)=O
Sulfuric acid, monooctyl ester, sodium salt
Cycloryl OS
Duponol 80
Octyl sodium sulfate
Sipex OLS
Sodium capryl sulfate
Sodium octyl sulfate
Sodium octyl sulphate
SOS

142-45-0
$C_4H_2O_4$
114.06
O=C(O)C#CC(=O)O
2-Butynedioic acid
Acetylenedicarboxylic acid (8CI)
Butynedioic acid
NSC-1903

142-46-1
$C_2H_6N_4S_2$
150.24
N=C(S)NNC(=N)S
Biurea, 2,5-dithio
Bisthiocarbamyl hydrazine
N,N'-Bisthiocarbamyl hydrazine
Bis(thiourea)
2,5-Dithiobiurea
1,2-Hydrazinedicarbothioamide
NCI-C03009
USAF B-55

142-47-2
$C_5H_8NO_4.Na$
169.13
[Na+].NC(CCC([O-])=O)C(O)=O
Glutamic acid, monosodium salt, l-(+)
Accent
Ajinomoto
Chinese seasoning
Glutacyl
Glutamic acid, sodium salt
Glutamat sodny (Czech)
Glutammato monosodico (Italian)
Glutavene
Monosodioglutammato (Italian)
Monosodium glutamate
α-Monosodium glutamate
Monosodium l-glutamate
MSG
Natriumglutaminat (German)
RL-50
Sodium glutamate
Sodium L-glutamate
L(+) Sodium glutamate
Vetsin
Zest

142-50-7
$C_{15}H_{26}O$
222.37
OC(C=C)(CCC=C(CCC=C(C)C)C)C
1,6,10-Dodecatrien-3-ol, 3,7,11-trimethyl-, (S-(Z))- (9CI)
1,6,10-Dodecatrien-3-ol, 3,7,11-trimethyl-, (Z)-(S)-(+)- (8CI)
Nerolidol (VAN)
Nerolidol, cis-(+)-
(+)-Nerolidol
d-Nerolidol
NSC-406963
Peruviol

142-54-1
$C_{15}H_{31}NO_2$
257.41
O=C(NCC(O)C)CCCCCCCCCCC
Lauramide MIPA
Dodecanamide, N-(2-hydroxypropyl)- (9CI)
N-(2-Hydroxypropyl)dodecanamide
Incromide LI
Incromide LMI
Lauroyl isopropanolamide
Monoisopropanolamine lauric acid amide

142-58-5
$C_{16}H_{33}NO_2$
271.44
O=C(NCCO)CCCCCCCCCCCCC
Myristamide MEA
Comperlan MM
N-(2-Hydroxyethyl)tetradecanamide
Loramine MY 228
Monoethanolamine myristic acid condensate
Myristic monoethanolamide
Myristoyl monoethanolamide
Myristyl monoethanolamide
Schercomid MME
Tetradecanamide, N-(2-hydroxyethyl)- (9CI)
Witcamide MM

142-59-6
$C_4H_6N_2S_4.2Na$
256.34
[Na+].[Na+].[S-]C(=S)NCCNC([S-])=S
Carbamic acid, ethylenebis-(dithio-, disodium salt (8CI)
AI3-04473
Carbon D
Caswell No. 585
Chem Bam
Di-natrium-aethylenbisdithiocarbamat (German)
Dinatrium-(N,N'-aethylen-bis-(dithiocarbamat)) (German)
Dinatrium-(N,N'-ethyleen-bis-(dithiocarbamaat)) (Dutch)
Disodium ethylenebis(dithiocarbamate)
Disodium ethylene-1,2-bis-dithiocarbamate
Dithane A-40
Dithane D-14
DSE
1,2-Ethanediylbiscarbamodithioic acid disodium salt
ENT 9106
EPA Pesticide Chemical Code 014503
Ethylen-bis-dithiokarbaman sodny (Czech)
N,N'-Ethylene bis(dithiocarbamate de sodium) (French)
Ethylenebis(dithiocarbamate), disodium salt
Ethylenebis(dithiocarbamic acid) disodium salt
N,N'-Etilen-bis(ditiocarbammato) di sodio (Italian)
Nabam
Nabame (French)
Nabasan
Nafun IPO
Parzate
Parzate Liquid
Spring-Bak

142-62-1
$C_6H_{12}O_2$
116.18
O=C(O)CCCCC
Hexanoic-acid
Butylacetic acid
Caproic acid
n-Caproic acid
Capronic acid
Hexacid 698
Hexanoic acid (DOT)
n-Hexanoic acid
n-Hexoic acid
Kyselina kapronova (Czech)
Pentanecarboxylic acid
Pentiformic acid
Pentylformic acid

142-64-3
$C_4H_{10}N_2.2ClH$
159.08
[Cl-].[Cl-].C1CNCCN1
Piperazine dihydrochloride (ACGIH,OSHA)
Dihydrochloride salt of diethylenediamine
Dowzene DHC
Piperazine hydrochloride

142-68-7

$C_5H_{10}O$
86.13
O(CCCC1)C1
Tetrahydropyran
AI3-16499
NSC-65448
Oxacyclohexane
Oxane (VAN)
Pentamethylene oxide
2H-Pyran, tetrahydro- (9CI)
Tetrahydro-2H-pyran
Tetrahydropyrane
THP

142-71-2
$C_4H_6O_4$.Cu
181.64
Acetic acid, copper(2+) salt
Acetate de cuivre (French)
Acetic acid, cupric salt
Copper(2+) acetate
Copper(II) acetate
Copper acetate (cu(c2H3o2)2)
Copper diacetate
Copper(2+) diacetate
Crystallized Verdigris
Crystals of Venus
Cupric acetate
Cupric diacetate
Neutral Verdigris
Octan mednaty (Czech)

142-72-3
$C_4H_6O_4$.Mg
142.41
Acetic acid, magnesium salt
Cromosan
Magnesium acetate
Magnesium diacetate

142-73-4
$C_4H_7NO_4$
133.12
O=C(O)CNCC(=O)O
Acetic acid, iminodi
Aminodiacetic acid
N-(Carboxymethyl)glycine
Diglycin
Diglycine
Diglykokoll
Glycine, N-(carboxymethyl)-
Hampshire
IDA
Iminobis(acetic acid)
Iminodiacetic acid
2,2'-Iminodiacetic acid
Iminodiethanoic acid
USAF DO-55

142-77-8
$C_{22}H_{42}O_2$
338.64
O=C(OCCCC)CCCCCCCC=C
CCCCCCCC
9-Octadecenoic acid, butyl ester (Z)
Butyl oleate
Oleic acid, butyl ester
Plasthall 503
Uniflex BYO

142-78-9
$C_{14}H_{29}NO_2$
243.38
O=C(NCCO)CCCCCCCCCCC
Lauramide MEA
Amisol LDE
Amisol LME
Comperlan LM
Copramyl
Crillon LME
Cyclomide LM
Dodecanamide, N-(2-hydroxy-ethyl)- (9CI)
2-Dodecanamidoethanol
N-(2-Hydroxyethyl)dodecan-amide
N-(2-Hydroxyethyl)lauramide
Incromide LCL
Incromide LMM
Lauric acid ethanolamide
Lauric acid monoethanolamide
Lauric acid monoethanolamine
Lauric ethylolamide
Lauric N-(2-hydroxyethyl)amide
Lauric monoethanolamide
Lauridit LM
Lauroyl monoethanolamide
Laurylamidoethanol
Laurylethanolamide
Monoethanolamine lauric acid amide
Rewomid L 203
Rolamid CM
Stabilor CMH
Steinamid L 203
Ultrapole H
Vistalan

142-82-5
C_7H_{16}
100.23
C(CCCCC)C
Heptane (ACGIH,OSHA) [UN 1206]
Dipropyl methane
Eptani (Italian)
Gettysolve-C
Heptan (Polish)
n-Heptane (OSHA)
Heptanen (Dutch)
Heptyl hydride
UN 1206 [Heptanes]

142-83-6
C_6H_8O
96.14
O=CC=CC=CC
Sorbaldehyde
Hexa-2,4-dienal
2,4-Hexadienal, (E,E)-
trans,trans-2,4-Hexadienal
1,3-Pentadiene-1-carboxalde-hyde
3-Propyleneacrolein
Sorbic aldehyde

142-84-7
$C_6H_{15}N$
101.22
N(CCC)CCC
Dipropylamine [UN 2383]
Di-n-propylamine
n-Dipropylamine
1-Propanamine, N-propyl-
N-Propyl-1-propanamine
RCRA waste number U110
UN 2383 [Dipropylamine]

142-87-0
$C_{10}H_{21}O_4$S.Na
312.36
Sulfuric acid, decyl ester, sodium salt
Sodium decyl sulfate
Sulfuric acid, monodecyl ester, sodium salt

142-90-5
$C_{16}H_{30}O_2$
254.46
O=C(OCCCCCCCCCCCC)C
(=C)C
Methacrylic acid, dodecyl ester
Acrylic acid, 2-methyl-, dodec-yl ester
Ageflex FM 246
Dodecyl methacrylate
Dodecyl 2-methyl-2-propenoate
Laurylester kyseliny metha-krylove (Czech)
Lauryl methacrylate
Methacrylic acid, lauryl ester

142-91-6
$C_{19}H_{38}O_2$
298.57
O=C(OC(C)C)CCCCCCCCCC
CCCC
Palmitic acid, isopropyl ester
Isopropyl hexadecanoate
Crodamol IPP
Deltyl
Deltyl Prime
Emcol-IP
Emerest 2316
Estol 103
Hexadecanoic acid, isopropyl ester
Isopal
Isopropyl n-hexadecanoate
Isopropyl palmitate
Ja-Fa IPP
Kessco isopropyl palmitate
Plymouth IPP
Propal
Starfol IPP
Stepan d-70
Tegester Isopalm
Unimate IPP
USAF KE-5
Wickenol 111

142-92-7
$C_8H_{16}O_2$
144.24
O=C(OCCCCCC)C
Acetic acid, hexyl ester
Hexyl acetate
n-Hexyl acetate
l-Hexyl acetate
Hexyl alcohol, acetate
Hexylester kyseliny octove (Czech)
Hexyl ethanoate

142-94-9
$C_{18}H_{35}BrO_2$
363.38
O=C(O)C(Br)CCCCCCCCCCCCCCCC
2-Bromostearic acid
2-Bromooctadecanoic acid
α-Bromostearic acid
NSC-58376
Octadecanoic acid, 2-bromo- (9CI)

142-96-1
$C_8H_{18}O$
130.26
O(CCCC)CCCC
Butane, 1,1'-oxybis
1-Butoxybutane
Butyl ether [UN 1149]
n-Butyl ether
Dibutyl ether
Di-n-butyl ether [UN 1149]
n-Dibutyl ether
Dibutyl oxide
Ether butylique (French)
1,1'-Oxybis(butane)
UN 1149 [Dibutyl ethers]

143-00-0
$C_{12}H_{26}O_4S.C_4H_{11}NO_2$
371.53
DEA-lauryl sulfate
Bis(2-hydroxyethyl)ammonium lauryl sulfate
Condanol DLS
Diethanolamine lauryl sulfate
Dodecyl sulfate diethanolamine salt
Lauryl alcohol sulfate, diethanolamine salt
Lauryl sulfate diethanolamine salt
Propaste D
Sipon LD
Stepanol DEA
Sulfuric acid, monododecyl ester, Compd. with 2,2'-iminobis(ethanol) (1:1) (9CI)
Sulfuric acid, monododecyl ester, Compd. with 2,2'-iminodiethanol(1:1)
Texapon DLS
Unipol Conc. 7021
Unipol DEA
Witcolate DLS

143-02-2
$C_{16}H_{34}O_4S$
322.51
O=S(=O)(OCCCCCCCCCCCCCCCC)O
Cetyl sulfate
1-Hexadecanol, hydrogen sulfate (9CI)

143-03-3
$C_{18}H_{38}O_4S$
350.56
O=S(=O)(OCCCCCCCCCCCCCCCCCC)O
Sulfuric acid, monooctadecyl ester (9CI)
Monooctadecyl sulfate

143-06-6
$C_7H_{16}N_2O_2$
160.20
O=C(O)NCCCCCCN
Hexamethylenediamine carbamate
(6-Aminohexyl)carbamic acid
Carbamic acid, (6-aminohexyl)- (9CI)
DIAK 1

143-07-7
$C_{12}H_{24}O_2$
200.36
O=C(O)CCCCCCCCCCC
Lauric-acid
C-1297
Dodecanoic acid
n-Dodecanoic acid
Dodecoic acid
Duodecylic acid
Hydrofol Acid 1255
Hydrofol Acid 1295
Hystrene 9512
Laurostearic acid
Neo-Fat 12
Neo-Fat 12-43
Ninol AA62 Extra
1-Undecanecarboxylic acid
Wecoline 1295

143-08-8
$C_9H_{20}O$
144.29
OCCCCCCCCC
Nonyl-alcohol
Alcohol C-9
Nonalol
1-Nonanol
Nonan-1-ol
n-Nonyl alcohol
Octyl carbinol
Pelargonic alcohol

143-09-9
$C_{10}H_{23}N.ClH$
193.75
1-Decanamine, hydrochloride (9CI)
Decylamine, hydrochloride (8CI)
NSC-5446

143-10-2
$C_{10}H_{22}S$
174.35
SCCCCCCCCCC
1-Decanethiol (9CI)
AI3-06520
Decyl mercaptan
1-Mercaptodecane
NSC-850
Thiols

143-13-5
$C_{11}H_{22}O_2$
186.29
O=C(OCCCCCCCCC)C
Nonyl acetate
Acetic acid, nonyl ester (9CI)
Acetic acid n-nonyl ester
AI3-11583
Nonanol acetate
n-Nonanyl acetate
n-Nonyl acetate
n-Nonyl ethanoate
NSC-82356
Pelargonyl acetate

143-15-7
$C_{12}H_{25}Br$
249.23
BrCCCCCCCCCCCC
Dodecane, 1-bromo- (9CI)
AI3-02166
1-Bromododecane
Dodecyl bromide
n-Dodecyl bromide
Lauryl bromide
NSC-6786

143-16-8
$C_{12}H_{27}N$
185.40
N(CCCCCC)CCCCCC
Dihexylamine
Di-n-hexylamine
1-Hexanamine, n-hexyl-

143-18-0
$C_{18}H_{34}O_2.K$
321.62
Oleic acid, potassium salt
9-Octadecenoic acid (Z)-, potassium salt
Potassium cis-9-octadecenoic acid
Potassium oleate
Trenamine D-200
Trenamine D-201

143-19-1
$C_{18}H_{33}O_2.Na$
304.50
Oleic acid, sodium salt

Eunatrol
Olate Flakes
Sodium oleate

143-22-6
$C_{10}H_{22}O_4$
206.32
O(CCOCCOCCO)CCCC
Ethanol, 2-(2-(2-butoxy-ethoxy)ethoxy)
2-(2-(2-Butoxyethoxy)-ethoxy)ethanol
Butoxytriethylene glycol
Butoxytriglycol
Poly-Solv TB
Triethylene glycol n-butyl ether
Triethylene glycol monobutyl ether
Triglycol monobutyl ether
3,6,9-Trioxa-1-tridecanol

143-23-7
$C_{12}H_{29}N_3$
215.36
N(CCCCCCN)CCCCCCN
N-(6-Aminohexyl)-1,6-hexane-diamine
Bis(6-aminohexyl)amine
Bishexamethylenetriamine
Bis(hexamethylene)triamine
1,13-Diamino-7-azatridecane
6,6'-Diaminodihexylamine
Dihexylamine, 6,6'-diamino-(8CI)
Dihexylenetriamine
1,6-Hexanediamine, N-(6-aminohexyl)- (9CI)
NSC-92231

143-24-8
$C_{10}H_{22}O_5$
222.32
O(CCOCCOCCOC)CCOC
2,5,8,11,14-Pentaoxa-pentadecane
Ansul ether 181AT
Bis(2-methoxyethyl)ether
Bis(2-(2-methoxyethoxy)ethyl) ether
Dimethoxytetraethylene glycol
Dimethoxytetraglycol
Ether, bis(2-(2-methoxyethoxy)-ethyl)
Tetraethylene glycol dimethyl ether
Tetraglyme

143-27-1
$C_{16}H_{35}N$
241.52
NCCCCCCCCCCCCCCCC
1-Hexadecanamine
Alamine 6
Armeen 16D
Cetylamin (German)
Cetylamine
n-Hexadecylamine
Palmitylamine

143-28-2
$C_{18}H_{36}O$
268.54
OCCCCCCCCC=CCCCCCCCC
9-Octadecen-1-ol, (Z)
Adol
Adol 34
Adol 80
Adol 85
Adol 90
Adol 320
Adol 330
Adol 340
Atalco O
Cachalot O-1
Cachalot O-3
Cachalot O-8
Cachalot O-15
Conditioner 1
Crodacol A.10
Crodacol-O
Dermaffine
H.D. Eutanol
HD Oleyl alcohol 70/75
HD Oleyl alcohol 80/85
HD Oleyl alcohol 90/95
HD Oleyl alcohol CG
Lancol
Loxanol 95
Loxanol M
Novol
Ocenol
Oceol
cis-9-Octadecen-1-ol
9-Octadecen-1-ol, cis-
Oleol
Oleyl alcohol
Olive alcohol
Satol
Sipol O
Unjecol 50
Unjecol 70
Unjecol 90
Unjecol 110

143-29-3
$C_{17}H_{36}O_6$
336.53
O(CCOCCOCCCC)COCCOCCO CCCC
Methane, bis(2-(2-butoxy-ethoxy)ethoxy)
Bis(butylcarbitol)formal
Butylcarbitol formal
Cryoflex
Dibutylcarbitolformal
5,8,11,13,16,19-Hexaoxatri-cosane (9CI)
TP 90B

143-33-9
CNNa
49.01
[Na+].[C-]#N
Sodium-cyanide
Cianuro di sodio (Italian)
Cyanide of sodium
Cyanobrik
Cyanogran
Cyanure de sodium (French)
Cymag
Hydrocyanic acid, sodium salt
Kyanid sodny (Czech)
RCRA waste number P106
Sodium cyanide (ACGIH)
Sodium cyanide, Solid [UN 1689]
Sodium cyanide, Solution [UN 1689]
UN 1689 [Sodium cyanide]

143-50-0
$C_{10}Cl_{10}O$
490.60
ClC2(Cl)C4(Cl)C1(Cl)C5(Cl) C(=O)C3(Cl)C1(Cl)C2(Cl) C3(Cl)C45Cl
1,3,4-Metheno-2H-cyclobuta-(cd)pentalen-2-one, 1,1a,3,3a,4,5,5,5a,5b,6-decachloroocta-hydro
Ciba 8514
Chlordecone
Compound 1189
1,2,3,5,6,7,8,9,10,10-Deca-chloro(5.2.1.$0^{2,6}$.$0^{3,9}$.$0^{5,8}$) decano-4-one
Decachloroketone
Decachloro-1,3,4-metheno-2H-cyclobuta(cd)pentalen-2-one
Decachlorooctahydro-1,3,4-metheno-2H-cyclobuta-(cd)pentalen-2-one
1,1a,3,3a,4,5,5,5a,5b,6-Deca-chlorooctahydro-1,3,4-meth-eno-2H-cyclobuta(cd)pentalen-2-one
Decachloropentacyclo-(5.2.1.$0^{2,6}$.$0^{3,9}$.$0^{5,8}$)decan-4-one
Decachloropentacyclo-(5.3.0.$0^{2,6}$.$0^{4,10}$.$0^{5,9}$)decan-3-one
Decachlorotetracyclodecanone
Decachlorotetrahydro-4,7-meth-anoindeneone
ENT 16,391
GC 1189
General Chemicals 1189
Kepone
Kepone-2-one, decachloro-octahydro-
Merex
NCI-C00191
RCRA waste number U142

143-52-2
$C_{18}H_{21}NO_3$
299.36
Metopon
DEA No. 9260
Dihydro-6-methylmorphinone
4,5-α-Epoxy-3-hydroxy-5,17-di-ethylmorphinan-6-one
Methyldihydromorphinone
6-Methyldihydromorphinone
Metopone

Metoponum (Latin)
Morphinan-6-one, 4,5-α-epoxy-3-hydroxy-5-β-17-dimethyl-
Morphinone, dihydro-6-methyl-
Morphinone, methyldihydro-

143-66-8
$C_{24}H_{20}B.Na$
198.09
[Na+].c1ccc(cc1)[B-](c2ccccc2)(c3ccccc3)c4ccccc4
Borate(1-), tetraphenyl-, sodium (9CI)
AI3-60390
Dotite Kalibor
Kalignost
NSC-203323
Sodium tetraphenylborate
Sodium tetraphenylborate(1-)
Sodium tetraphenylboride
Sodium tetraphenylboride(1-)
Sodium tetraphenylboron
Tetraphenylborate(1-) sodium
Tetraphenyl boron sodium salt
Tetraphenyl sodium borate
TPB

143-67-9
$C_{46}H_{58}N_4O_9.H_2O_4S$
909.16
CCC9(O)CC3CN(CCc1c([nH]c2ccccc12)C(C3)(C(=O)OC)c4cc5c(cc4OC)N(C)C6C(O)(C(OC(C)=O)C7(CC)C=CCN8CCC56C78)C(=O)OC)C9
Vincaleukoblastine, sulfate (1:1) (Salt)
Exal
29060 LE
NSC-49842
Velban
Velbe
Vinblastine sulfate
Vincaleukoblastine, sulfate
VLB monosulfate

143-74-8
$C_{19}H_{14}O_5S$
354.39
O=S(=O)(OC(c1cccc2)(c(ccc(O)c3)c3)c(ccc(O)c4)c4)c12
Phenol, 4,4'-(3H-2,1-benzoxathiol-3-ylidene)di-, S,S-dioxide
Fenolipuna
Phenol, 4,4'-(3H-2,1-benzoxathiol-3-ylidene)bis-, S,S-dioxide (9CI)
Phenol Red
Phenolsulfonephthalein
Phenolsulfonphthalein
Phenolsulphonphthalein
PSP
PSP (Indicator)
Sulfonphthal
Sulphental
Sulphonthal

143-81-7
$C_{10}H_{16}N_2O_3.Na$
235.27
CCC(C)C1(CC)C(=O)NC(=O)NC1=O
Barbituric acid, 5-sec-butyl-5-ethyl-, sodium salt
Asturidon
Bubartal
Bubartal TT
Butabarbital
Butabarbital sodium
Butabarbitone sodium
Butabarpal
Butabarpal sodium
Butak
Butasaron
Butased
Butatal sodium
Butazem
Buticaps
Butisol
Butisol sodium
5-sec-Butyl-5-ethylbarbituric acid sodium salt
5-Ethyl-5-(1-methylpropyl)barbituric acid sodium salt
Insolat
Intasedol
Loubarb
Mebutal
Neravan
Noctinal
Paxital
Prelital
2,4,6(1H,3H,5H)-Pyrimidinetrione, 5-ethyl-5-(1-methylpropyl)-, monosodium salt (9CI)
Quiebar
Secbubarbital sodium
Secbutobarbitone sodium
Seda-Bute
Sodium butabarbital
Sodium butobarbitone
Sodium 5-sec-butyl-5-ethylbarbiturate
Sodium 5-ethyl-5-sec-butylbarbiturate
Sodium 5-ethyl-5-(1-methylpropyl)barbiturate

144-11-6
$C_{20}H_{31}NO$
301.52
OC(CCN1CCCCC1)(C2CCCCC2)c3ccccc3
1-Piperidinepropanol, α-cyclohexyl-α-phenyl
Benzhexol
Trihexyphenidyl
Triphenidyl

144-14-9
$C_{22}H_{28}N_2O_2$
352.46
CCOC(=O)C2(CCN(CCc1ccc(N)cc1)CC2)c3ccccc3
Anileridine
Adopol
Alidine
1-(p-Aminophenethyl)-4-phenylisonipecotic acid ethyl ester
1-(p-Aminophenethyl)-4-phenylpiperidine-4-carboxylic acid ethyl ester
N-β-(p-Aminophenyl)ethylnormeperidine
N-(β-(p-Aminophenyl)ethyl)-4-phenyl-4-carbethoxypiperidine
1-(2-(4-Aminophenyl)ethyl)-4-phenyl-4-piperidinecarboxylic acid ethyl ester
Anileridina (Spanish)
Anileridinum (Latin)
DEA No. 9020
Ethyl 1-(p-aminophenethyl)-4-phenylisonipecotate
Ethyl 1-(4-aminophenethyl)-4-phenylisonipecotate
Isonipecotic acid, 1-(p-aminophenethyl)-4-phenyl-, ethyl ester
Leritin
Leritine
Nipecotan
4-Piperidinecarboxylic acid, 1-(2-(4-aminophenyl)ethyl)-4-phenyl-, ethyl ester

144-19-4
$C_8H_{18}O_2$
146.26
OCC(C(O)C(C)C)(C)C
1,3-Pentanediol, 2,2,4-trimethyl
TMPD
2,2,4-Trimethyl-1,3-pentanediol

144-21-8
$CH_3AsO_3.2Na$
183.94
[Na+].[Na+].C[As]([O-])([O-])=O
Methanearsonic acid, disodium salt
Ansar 184
Ansar 8100
Ansar DSMA Liquid
Arrhenal
Arsinyl
Arsonic acid, methyl-, disodium salt
Arsynal
Cacodyl New
Chipco Crab Kleen
Clout
Crab-E-Rad
Crab-3-Rad 100
Cralo-E-Rad
Dal-E-Rad 100
Diarsen
Dimet
Dinate
Disodium methanearsenate
Disodium methanearsonate
Disodium methylarsenate
Disodium methylarsonate
Disodium monomethylarsonate
Disomar

Disomear
Di-Tac
DMA
DMA 100
Drexel DSMA Liquid
DSMA
DSMA Liquid
Jon-Trol
MAA sodium salt
Methar
Methar 30
Metharsan
Metharsinat
Methylarsonat disodny (Czech)
Namate
Neoasycodile
Sodar
Sodium methanearsonate
Sodium metharsonate
Sodium methylarsonate
Somar
Stenosine
Tonarsen
Tonarsin
Versar DSMA LQ
Weed Broom
Weed-E-Rad
Weed-E-Rad 360
Weed-E-Rad DMA Powder
Weed-HOE

144-33-2
$C_6H_6O_7.2Na$
236.10
Citric acid, disodium salt
Alkacitron
Citralka
Disodium citrate
Disodium hydrogen citrate
Disodium monohydrogen citrate
Natrium citricum (German)
1,2,3-Propanetricarboxylic acid, 2-hydroxy-, disodium salt (9CI)
Sodium citrate

144-34-3
$C_{12}H_{24}N_4S_8.Se$
559.84
CN(C)C(=S)S[Se](SC(=S)N(C)C)(SC(=S)N(C)C)SC(=S)N(C)C
Selenium, tetrakis(dimethyldithiocarbamato)
Methyl selenac
Selenium dimethyldithiocarbamate
Tetrakis(dimethylcarbamodithioato-S,S')selenium

144-35-4
$C_{17}H_{18}O_6P_2$
380.27
O(P(OCC1(COP(O2)Oc(cccc3)c3)C2)Oc(cccc4)c4)C1
2,4,8,10-Tetraoxa-3,9-diphosphaspiro(5.5)undecane, 3,9-diphenoxy- (9CI)

144-41-2
$C_8H_{16}NO_4PS_2$
285.34
COP(=S)(OC)SCC(=O)N1CCOCC1
Phosphorodithioic acid, O,O-dimethyl ester, S-ester with 4-(mercaptoacetyl)morpholine
O,O-Dimethyl-S-((morfolino-carbonyl)-methyl)-dithiofosfaat (Dutch)
O,O-Dimethyl S-(morpholinocarbamoylmethyl) dithiophosphate
O,O-Dimethyl morpholinocarbonylmethyl phosphorodithioate
O,O-Dimethyl-S-((morpholino-carbonyl)-methyl)-dithiophosphat (German)
O,O-Dimethyl S-(morpholinocarbonylmethyl) phosphorodithioate
Dimethyl S-(morpholinocarbonylmethyl) phosphorothiolothionate
O,O-Dimetil-S-((morfolino-carbonil)-metil)-ditiofosfato (Italian)
Dithiophosphate de O,O-dimethyle et de S-((morpholino-carbonyle)-methyle) (French)
Ekatin F
Ekatin M
4-(Mercaptoacetyl)morpholine O,O-dimethyl phosphorodithioate
Morfothion (Dutch)
Morpholine, 4-(mercaptoacetyl)-, S-ester with O,O-dimethyl phosphorodithioate
Morphothion
Morphotox
Phosphorodithioic acid, O,O-dimethyl S-(morpholinocarbonylmethyl) ester
Systicide

144-49-0
$C_2H_3FO_2$
78.05
O=C(O)CF
Acetic acid, fluoro
Acide-monofluoracetique (French)
Acido monofluoroacetio (Italian)
Cymonic acid
FAA
Fluoroacetate
Fluoroacetic acid [UN 2642]
2-Fluoroacetic acid
Fluoroethanoic acid
Gifblaar Poison
HFA
MFA
Monofluorazijnzuur (Dutch)
Monofluoressigsaure (German)
Monofluoroacetate
Monofluoroacetic acid
UN 2642 [Fluoroacetic acid]

144-54-7
Unknown
Unknown
Methan
Methylcarbamodithioic acid
Methyldithiocarbamate
Methyldithiocarbanic acid

144-55-8
$CHO_3.Na$
84.01
[Na+].OC([O-])=O
Sodium bicarbonate (1:1)
Baking soda
Bicarbonate of soda
Carbonic acid monosodium salt
Col-Evac
Jusonin
Monosodium carbonate
Neut
Soda Mint
Sodium acid carbonate
Sodium hydrogen carbonate

144-62-7
$C_2H_2O_4$
90.04
O=C(O)C(=O)O
Oxalic-acid
Acide oxalique (French)
Acido ossalico (Italian)
Ethanedioic acid
Ethanedionic acid
Kyselina stavelova (Czech)
NCI-C55209
Oxaalzuur (Dutch)
Oxalic acid (ACGIH,OSHA)
Oxalsaeure (German)

144-74-1
$C_9H_9N_3O_2S_2.Na$
278.32
Sulfanilamide, N^1-2-thiazolyl-, monosodium salt
Monosodium 2-sulfanilamidothiazole
Sodium norsulfazole
Sodium 2-sulfanilamidothiazole
Sodium sulfathiazole
Sodium sulphathiazole
Sodium, (N^1-2-thiazolylsulfanilamido)-
Soluble Sulfathiazole
Sulfanilamide, N^1-2-thiazolyl-, N^1-sodium deriv.
2-Sulfanilamidothiazole sodium salt
Sulfathiazole sodium
N^1-2-Thiazolylsulfanilamide sodium salt

144-79-6
$C_{13}H_{13}ClSi$
232.78
Silane, chloromethyldiphenyl-

(9CI)
Chloromethyldiphenylsilane
Methyldiphenylchlorosilane
Methyldiphenylsilyl chloride
NSC-93961

144-80-9
$C_8H_{10}N_2O_3S$
214.26
O=C(NS(=O)(=O)c(ccc(N)c1)c1)C
Acetamide, N-sulfanilyl
A-500
Acetamide, N-((4-aminophenyl)sulfonyl)- (9CI)
Acetocid
Acetosulfamin
Acetosulfamine
N-Acetyl-4-aminobenzenesulfonamide
N[1]-Acetyl-4-aminophenylsulfonamide
N-Acetylsulfanilamide
N'-Acetylsulfanilamide
N[1]-Acetylsulfanilamide
N-Acetylsulfanilamine
Albamine
Albucid
Alesten
p-Aminobenzenesulfonacetamide
p-Aminobenzenesulfonoacetamide
N-((4-Aminophenyl)sulfonyl)acetamide
Bleph-10
Bleph-10 Liquifilm
Formosulfacetamide
Isopto Cetamide
Ophthel-S
Op-Sulfa 30
Region
Sebizon
Steramide
Sulamyd
Sulf-10
Sulfacet
Sulfacetamide
Sulfacetimide
Sulfacyl
Sulfanilacetamide
Sulfanilamide, N[1]-acetyl-
Sulfanilazetamid (German)
N-Sulfanilylacetamide
Sulphacetamide
Sulphasil
Urosulfon
Urosulfone

144-82-1
$C_9H_{10}N_4O_2S_2$
270.35
O=S(=O)(N=C(SC(=N1)C)N1)c(ccc(N)c2)c2
Sulfanilamide, N[1]-(5-methyl-1,3,4-thiadiazol-2-yl)
2-(p-Aminobenzenesulfonamido)-5-methylthiadiazole
Benzenesulfonamide, 4-amino-N-(5-methyl-1,3,4-thiadiazol-2-yl)-
Lucosil
5-Methyl-2-sulfanilamido-1,3,4-thiadiazole
N[1]-(5-Methyl-1,3,4-thiadiazol-2-yl)-sulfanilamide
Microsul
RP 2145
Rufol
Sulfamethizol
Sulfamethizole
Sulfamethylizole
Sulfamethylthiadiazole
2-Sulfanilamido-5-methyl-1,3,4-thiadiazole
Sulfstat
Sulfurine
Sulphamethizole
Tetracid
Thidicur
Thiosulfil
Ultrasul
Urodiaton
Urolucosil
Utrasul

144-83-2
$C_{11}H_{11}N_3O_2S$
249.31
O=S(=O)(Nc(nccc1)c1)c(ccc(N)c2)c2
Benzenesulfonamide, 4-amino-N-(2-pyridinyl)
Adiplon
2-(p-Aminobenzenesulphonamido)pyridine
Coccoclase
Dagenan
Eubasin
Eubasinum
Haptocil
M+B 693
Piridazol
Plurazol
N-2-Pyridylsulfanilamide
N[1]-2-Pyridylsulfanilamide
Relbapiridina
Ronin
Septipulmon
Streptosilpyridine
Sulfanilamide, N[1]-2-pyridyl-
2-Sulfanilamidopyridin (German)
2-Sulfanilyl aminopyridine
Sulfapyridine
2-Sulfapyridine
Sulfidine
Sulphapyridine
Thioseptal
Trianon

145-39-1
$C_{13}H_{18}N_2O_4$
266.33
O=N(=O)c(c(c(N(=O)=O)c(c1C)C)C(C)(C)C)c1C
Benzene, 1-(1,1-dimethylethyl)-2,6-dinitro-3,4,5-trimethyl
Benzene, 1-tert-butyl-2,6-dinitro-3,4,5-trimethyl-
5-tert-Butyl-1,2,3-trimethyl-4,6-dinitrobenzene
Musk Tibetene

145-42-6
$C_{26}H_{44}NO_7S.Na$
537.76
Taurine, N-choloyl-, sodium salt
Monosodium taurocholic acid
Sodium taurocholate
Taurocholate sodium
Taurocholate sodium salt
Taurocholic acid sodium salt

145-49-3
$C_{14}H_{10}N_2O_4$
270.26
O=C(c(c(c(N)cc1)C(=O)c2c(O)ccc3N)c1O)c23
Anthraquinone, 1,5-diamino-4,8-dihydroxy
9,10-Anthracenedione, 1,5-diamino-4,8-dihydroxy-
Anthrarufin, 4,8-diamino-
Diaminoanthrarufin
1,5-Diaminoanthrarufin
4,8-Diaminoanthrarufin
1,5-Diamino-4,8-dihydroxyanthrachinon (Czech)
4,8-Diamino-1,5-dihydroxyanthraquinone
1,5-Diamino-4,8-dihydroxyanthraquinone
leuco-1,5-Diamino-4,8-dihydroxyanthraquinone
1,5-Dihydroxy-4,8-diaminoanthrachinon (Czech)
1,5-Dihydroxy-4,8-diaminoanthraquinone

145-73-3
$C_8H_{10}O_5$
186.18
O=C(O)C(C(C(=O)O)C(O1)CC2)C12
7-Oxabicyclo(2.2.1)heptane-2,3-dicarboxylic acid
Aquathol
1,2-Cyclohexanedicarboxylic acid, 3,6-endo-epoxy-
Endothal
Endothall
Endothal Technical
Endothal Weed Killer
3,6-Endoxohexahydrophthalic acid
3,6-Endooxohexahydrophthalic acid
Hydout
Hydrothal-47
Hydrothal-191
Hydrothol
Niagrathal
Phthalic acid, hexahydro-3,6-endo-oxy-
RCRA waste number P088
Tri-Endothal

145-94-8
C_9H_9ClO
168.63
4-Indanol, 7-chloro
Chlorindanol
Clorindanol
1H-Inden-4-ol, 2,3-dihydro-7-chloro- (9CI)
Lanesta

146-22-5
$C_{15}H_{11}N_3O_3$
281.29
O=C3CN=C(c1ccccc1)c2cc(ccc2N3)N(=O)=O
2H-1,4-Benzodiazepin-2-one, 1,3-dihydro-7-nitro-5-phenyl
Benzalin
Calsmin
1,3-Dihydro-7-nitro-5-phenyl-2H-1,4-benzodiazepin-2-one
Dormin-5
Eatan
Epibenzalin
Epinelbon
Eunoctin
Eunoktin
Hipnax
Hipsal
LA 1
Mogadan
Mogadon
Mogadone
Nelbon
Neozepam
Neuchlonic
Nitrados
Nitrazepam
Nitrenpax
7-Nitro-5-phenyl-2,3-dihydro-1H-1,4-benzodiazepin-2-one
NSC-58775
Paxisyn
Pelson
Radedorm
Relact
Ro 4-5360
Ro 5-3059
Somnased
Somnibel
Somnite
Sonebon
Sonnolin
Surem
Unisomnia

146-48-5
$C_{21}H_{26}N_2O_3$
354.49
COC(=O)C4C(O)CCC5CN3CCc1c([nH]c2ccccc12)C3CC45
Yohimban-16-α-carboxylic acid, 17-α-hydroxy-, methyl ester
Aphrodine
Aphrosol
Corynine
17-Hydroxyyohimban-16-carboxylic acid methyl ester
Quebrachin
Quebrachine
Yohimbic acid methyl ester
Yohimbin
Yohimbine

146-50-9
$C_{20}H_{30}O_4$
334.46
1,2-Benzenedicarboxylic acid, bis(4-methylpentyl) ester (9CI)
Phthalic acid, diisohexyl ester (8CI)

146-54-3
$C_{18}H_{19}F_3N_2S$
352.45
Phenothiazine, 10-(3-(dimethylamino)propyl)-2-(trifluoromethyl)
10H-Phenothiazine-10-propanamine, N,N-dimethyl-2-(trifluoromethyl)-
Triflupromazine
Vesprin

146-59-8
$C_{21}H_{25}Cl_2N_3O.2ClH$
479.31
CCN(CCCl)CCCNc2c1ccc(Cl)cc1nc3ccc(OC)cc23
1,3-Propanediamine, N-(2-chloroethyl)-N'-(6-chloro-2-methoxy-9-acridinyl)-N-ethyl-, dihydrochloride, hydrate
Acridine mustard
6-Chloro-9-(3-(ethyl-2-chloroethyl)aminopropylamino)-2-methoxyacridine dihydrochloride
9-(3-(Ethyl(2-chloroethyl)-amino)propylamino)-6-chloro-2-methoxyacridine dihydrochloride
ICR 170
2-Methoxy-6-chloro-9-(3-(ethyl-2-chloroethyl)aminopropylamino)acridine dihydrochloride

146-84-9
$C_6H_3N_3O_7.Ag$
336.96
Silver picrate
NSC-168933
Phenol, 2,4,6-trinitro-, silver(1+) salt (9CI)
Picragol
Picrate d'argent (French)
Picrato de plata (Spanish)
Picric acid, silver(1+) salt (8CI)
Picrotol
Silver picrate, Dry
Silver picrate, Wetted with at least 30% water [UN 1347]
Silver, (picryloxy)-
UN 1347 [Silver picrate, Wetted with not less than 30 per cent water, by mass]

147-14-8
$C_{32}H_{16}CuN_8$
576.03
Copper, (29H,31H-phthalocyaninato(2-)-N(29),N(30),N(31),N(32))-, (SP-4-1)- (9CI)
Accosperse Cyan Blue GT
Aqualine Blue
Arlocyanine Blue PS
Bahama Blue BC
Bahama Blue BNC
Bahama Blue Lake NCNF
Bahama Blue WD
Bermuda Blue
Blue GLA
Blue No. 404
Blue Toner GTNF
BT 4651
C.I. Pigment Blue 15
C.I. Pigment Blue 15:1
C.I. Pigment Blue 15:3
C.I. Pigment Blue 15:4
C.I. 74160
C-Ext. Blau 10 (Germany)
Calcotone Blue GP
Ceres Blue BHR
Chromatex Blue BN
Copper phthalocyanin
Copper phthalocyanine
Copper Phthalocyanine Blue
Copper, (phthalocyaninato(2-))-(8CI)
Cromophtal Blue GF
Cromophtal Blue 4G
Cromophtal Blue 4GN
Cyan Blue BNC 55-3745
Cyan Blue BNF 55-3753
Cyan Blue GT 55-3295
Cyan Blue GTNF
Cyan Peacock Blue G
Cyanine Blue BB
Cyanine Blue C
Cyanine Blue LBG
Cyanine Blue LC
Cyanine Blue PRPD
Cyanine Blue S 2100
Cyanine Blue SR 150A
Dainichi Cyanine Blue B
Dainichi Cyanine Blue FPG
Daltolite Fast Blue B
Duratint Blue 1001
Euvinyl Blue 702
Fastogen Blue B
Fastogen Blue FGF
Fastogen Blue FSN
Fastogen Blue TGR
Fastogen Blue 5007
Fastogen Blue 5110
Fastolux Blue
Fenalac Blue B Disp
Germany: C-Ext. Blau 10
Graphtol Blue BL
Graphtol Blue BLF
Graphtol Blue 2GLS
Helio Blue B
Helio Fast Blue BRN

Helio Fast Blue GO
Helio Fast Blue HG
Heliogen Blue A
Heliogen Blue B
Heliogen Blue BG
Heliogen Blue K
Heliogen Blue LBG
Heliogen Blue LBGN
Heliogen Blue WX
Heliogen Blue 6840
Heliogen Blue 6902K
Heliogen Blue 6960
Heliogen Blue 7044T
Heliogen Blue 7080
Heliogen Blue 7100
Hostaperm Blue A 2R
Hostaperm Blue A 3R
Hostaperm Blue AFN
Hostaperm Blue B 2G
Hostaperm Blue B 3G
Hostaperm Blue BG
Irgalite Blue BGL
Irgalite Blue BLP
Irgalite Blue CPV 2
Irgalite Blue CPV 3
Irgalite Blue GLSM
Irgalite Blue LGLD
Irgalite Fast Brilliant Blue BL
Irgaplast Blue RBP
Isol Fast Blue B
Isol Phthalo Blue E 7543
Japan: Blue No. 404
LBX 5
Lionol Blue E
Lionol Blue ER
Lionol Blue ES
Lionol Blue ESP
Lionol Blue GLA
Lionol Blue KL
Lionol Blue NCB Toner
Lionol Blue SM
Lionol Blue SN
Lumatex Blue B
Lutetia Percyanine BRS
Monastral Blue
Monastral Blue B
Monastral Fast Blue
NSC-15976
Nylofil Blue BLL
Peacoline Blue
Phthalocyanine 2ZU
Phthalocyanine Blue
Phthalocyanine VK
Pigment Blue 15
Pigment Fast Blue B
Pigment Sky Blue
Phthalocyanine B 4ZU
Pigment Sky Blue
Polymon Blue LBS
PV Fast Blue A 2R
PV Fast Blue B
PV Fast Blue B 2G
Renol Blue B 2G-H
Sandorin Blue 2GLS
Segnale Light Turquoise NCG
Segnale Light Turquoise NFG
Segnale Light Turquoise NFR
Segnale Light Turquoise PAG
Segnale Light Turquoise SR
Siegle Fast Blue LBGO
Unisperse Blue G-E
Versal Blue B
Versal Blue BG
Vynamon Blue B

147-24-0
$C_{17}H_{21}NO.ClH$
291.85
Cl.CN(C)CCOC(c1ccccc1)c2cccc c2
Ethylamine, 2-(diphenylmethoxy)-N,N-dimethyl-, hydrochloride
Allergan
Allergival
Ambenyl
Bax
Bena
Benadryl
Benadryl hydrochloride
Bendylate
Benocten
Benzehist
Benzhydramine hydrochloride
2-(Benzhydryloxy)-N,N-dimethylethylamine hydrochloride
Dabylen
Denydryl
Difenhydramine hydrochloride
Dimedrol
Dimethylamine benzhydryl ester hydrochloride
β-Dimethylaminoethyl benzhydryl ether hydrochloride
Diphenhydramine hydrochloride
Diphenydramine hydrochloride
Diphenylhydramine hydrochloride
2-Diphenylmethoxy-N,N-dimethylethylamine hydrochloride
Dolestan
Eldadryl
Ethanamine, 2-(diphenylmethoxy)-N,N-dimethyl-, hydrochloride
Ethylamine, N,N-dimethyl-2-(diphenylmethoxy)-, hydrochloride
Felben
Fenylhist
Halbmond
α-Hydroxydiphenylmethane-β-dimethylaminoethyl ether hydrochloride
NCI-C56075
Resmin
Restamin
Rohydra
Sk-Diphenhydramine
Valdrene
Vena
Wehydryl

147-47-7
$C_{12}H_{15}N$
173.28
N(c(c(C(=C1)C)ccc2)c2)C1(C)C
Quinoline, 1,2-dihydro-2,2,4-trimethyl
Acetonanil
Acetonanyl
Acetone anil
Agerite Resin D
1,2-Dihydro-2,2,4-trimethylquinoline
Flectol A
Flectol H
Flectol Pastilles
NCI-C60902
2,2,4-Trimethyl-1,2-dihydrochinolin (Czech)
Trimethyl-1,2-dihydroquinoline
2,2,4-Trimethyl-1,2-dihydroquinoline
Vulkanox HS/LG
Vulkanox HS/Powder

147-52-4
$C_{21}H_{22}N_2O_5S$
414.51
CCOc2ccc1ccccc1c2C(=O)NC4C3SC(C)(C)C(N3C4=O)C(O)=O
4-Thia-1-azabicyclo(3.2.0)-heptane-2-carboxylic acid, 6-(2-ethoxy-1-naphthamido)-3,3- dimethyl-7-oxo
Nafcillin

147-71-7
$C_4H_6O_6$
150.09
O=C(O)C(O)C(O)C(=O)O
Butanedioic acid, 2,3-dihydroxy-, (S-(R*,R*))- (9CI)
(S-(R*,R*))-2,3-Dihydroxybutanedioic acid
D-Tartaric acid

147-81-9
$C_5H_{10}O_5$
150.13
O=CC(O)C(O)C(O)CO
Arabinose (9CI)
L-Arabinose

147-82-0
$C_6H_4Br_3N$
329.84
Nc(c(cc(c1)Br)Br)c1Br
Aniline, 2,4,6-tribromo
Benzenamine, 2,4,6-tribromo- (9CI)
sym-Tribromoaniline
2,4,6-Tribromoaniline
USAF DO-43

147-84-2
$C_5H_{11}NS_2$
149.29
N(C(S)=S)(CC)CC
Carbamic acid, diethyldithio
Dieca
Diethyldithiocarbamic acid
Diethyldithiocarbaminic acid
Diethyldithione

147-85-3
$C_5H_9NO_2$
115.15
O=C(O)C(NCC1)C1
Proline, L
(L)-Proline

147-93-3
$C_7H_6O_2S$
154.19
O=C(O)c(c(S)ccc1)c1
Benzoic acid, o-mercapto
o-Mercaptobenzoesaeure (German)
o-Mercaptobenzoic acid
Thiosalicylic acid
USAF EK-T-2805
USAF KF-2
USAF XR-35

147-94-4
$C_9H_{13}N_3O_5$
243.25
OCC1OC(C(O)C1O)n2ccc(=N)[nH]c2=O
Cytosine, 1-β-d-arabinofuranosyl
AC-1075
Alexan
4-Amino-1-arabinofuranosyl-2-oxo-1,2-dihydropyrimidin (Czech)
4-Amino-1-arabinofuranosyl-2-oxo-1,2-dihydropyrimidine
Arabinocytidine
4-Amino-1-β-D-arabinofuranosyl-2(1H)-pyrimidinon (Czech)
1-β-D-Arabinofuranosyl-4-amino-2(1H)pyrimidinone
1-Arabinofuranosylcytosine
1-β-Arabinofuranosylcytosine
1-(β-D-Arabinofuranosyl)-cytosine
Arabinosylcytosine
β-D-Arabinosylcytosine
Arabitin
Ara-C
Aractidine
Ara-Cytidine
Aracytin
Cylocide
Cytarabin
Cytarabina
Cytarabine
Cytarabinoside
Cytosar
Cytosar-U
Cytosinearabinoside
Cytosine-β-arabinoside
Cytosine β-D-arabinoside
Cytosine, 1-β-D-arabinosyl-
Iretin
NCI-C04728
NSC-63878
2(1H)-Pyrimidinone, 4-amino-1-β-D-arabinofuranosyl- (9CI)
Spongocytidine
U-19,920
U-19920 A
Udicil

148-01-6
$C_8H_7N_3O_5$
225.18
Cc1c(cc(cc1N(=O)=O)N(=O)=O)C(N)=O
o-Toluamide, 3,5-dinitro
Benzamide, 2-methyl-3,5-dinitro-
Coccidine A
Coccidot
Dinitolmid
Dinitolmide (ACGIH,OSHA)
3,5-Dinitro-o-toluamide (OSHA)
D.O.T.
2-Methyl-3,5-dinitrobenzamide
Zoalene
Zoamix

148-18-5
$C_5H_{10}NS_2.Na$
171.27
[Na+].CCN(CC)C([S-])=S
Carbamic acid, diethyldithio-, sodium salt
Carbamodithioic acid, diethyl-, sodium salt (9CI)
Cupral
DDC
DEDC
DEDK
Diethylcarbamodithioic acid, sodium salt
Diethyldithiocarbamate sodium
Diethyldithiocarbamic acid sodium
Diethyldithiocarbamic acid, sodium salt
Diethyl sodium dithiocarbamate
Dithiocarb
Dithiocarbamate
NCI-C02835
Sodium dedt
Sodium diethyldithiocarbamate
Sodium N,N-diethyldithiocarbamate
Sodium salt of N,N-diethyldithiocarbamic acid
Thiocarb
USAF EK-2596

148-24-3
C_9H_7NO
145.17
n(c(c(ccc1)cc2)c1O)c2
8-Quinolinol
Bioquin
8-Chinolinol (Czech)
Fennosan
Hydroxybenzopyridine
8-Hydroxy-chinolin (German)
8-Hydroxyquinoline
NCI-C55298
8-OQ
Oxin
Oxine
Oxybenzopyridine
Oxychinolin
o-Oxychinolin (German)
Oxyquinoline
8-Oxyquinoline
Phenopyridine
8-Quinol
Quinophenol
Tumex
USAF EK-794

148-53-8
$C_8H_8O_3$
152.16
O=Cc(c(O)c(OC)cc1)c1
Benzaldehyde, 2-hydroxy-3-methoxy
m-Anisaldehyde, 2-hydroxy- (8CI)
6-Formylguaiacol
3-Methoxysalicylaldehyde
Orthovanilline
Oxy-2 methoxy-3 benzaldehyde (French)
o-Vanillin

148-56-1
$C_8H_6F_3N_3O_4S_2$
329.29
NS(=O)(=O)c2cc1c(NC=NS1(=O)=O)cc2C(F)(F)F
4H-1,2,4-Benzothiadiazine-7-sulfonamide, 6-(trifluoromethyl)-, 1,1-dioxide
Ademol
Flumethiazide
Rontyl
Routrax
Trifluomethylthiazide
6-(Trifluoromethyl)-1,2,4-benzo-thiadiazine-7-sulfonamide 1,1-dioxide
6-(Trifluoromethyl)-1,4,2-benzothiadiazine-7-sulfonamido 1,1-dioxide
6-Trifluoromethyl-7-sulfamoyl-4H-1,4,2-benzothiadiazine 1,1-dioxide
6-Trifluoromethyl-7-sulfamyl-1,2,4-benzothiadiazine-1,1-dioxide
Trifluoromethylthiazide

148-65-2
$C_{14}H_{18}ClN_3S$
295.86
n(c(N(CC(SC(=C1)Cl)=C1)CCN(C)C)ccc2)c2
Pyridine, 2-((5-chloro-2-thenyl)(2-(dimethylamino)-ethyl)amino)
Chloromethapyrilene
Chloropyrilene
Chlorothen
2-((5-Chloro-2-thenyl)(2-dimethylaminoethyl)amino)-pyridine
Chlorothenylpyramine
N,N-Dimethyl-N'-(2-pyridyl)-

N'-(5-chloro-2-thenyl)ethylenediamine
Ethylenediamine, N-(5-chloro-2-thenyl)-N',N'-dimethyl-N-2-pyridyl-
NCI-C60559
Pyrithen
Tagathen
2-Thenylamine, 5-chloro-N-(2-(dimethylamino)ethyl)-N-2-pyridyl-

148-69-6
$C_{12}H_{16}N_2$
188.26
N#CCCN(c(ccccc1C)c1)CC
Propanenitrile, 3-(ethyl-(3-methylphenyl)amino)-(9CI)
N-(2-Cyanoethyl)-N-ethyl-m-toluidine
3-(Ethyl(3-methylphenyl)-amino)propanenitrile
3-(N-Ethyl-m-toluidino)propionitrite
NSC-93794
Propionitrile, 3-(N-ethyl-m-toluidino)- (8CI)
m-Toluidine, N-cyanoethyl-N-ethyl-

148-71-0
$C_{11}H_{18}N_2$
178.27
N(c(ccc(N)c1C)c1)(CC)CC
CD 2
1,4-Benzenediamine, N4,N4-diethyl-2-methyl- (9CI)
N4,N4-Diethyl-2-methyl-1,4-benzenediamine

148-75-4
$C_{10}H_8O_7S_2$
304.30
O=S(=O)(O)c(ccc(c1cc(S(=O)(=O)O)c2O)c2)c1
2-Naphthol-3,6-disulfonic acid
3-Hydroxy-2,7-naphthalenedisulfonic acid
2,7-Naphthalenedisulfonic acid, 3-hydroxy- (9CI)

148-79-8
$C_{10}H_7N_3S$
201.26
c2ccc1[nH]c(nc1c2)c3cscn3
Benzimidazole, 2-(4-thiazolyl)
Apl-Luster
Arbotect
1H-Benzimidazole, 2-(4-thiazolyl)-
4-(2-Benzimidazolyl)thiazole
Bioguard
Bovizole
Eprofil
Equizole
Lombristop
Mertec
Mertect
Mertect 160
Metasol TK-100
Mintesol
Mintezol
Minzolum
MK 360
Mycozol
Nemapan
Omnizole
Polival
RPH
Storite
TBDZ
TBZ
Tecto
Tecto 60
Tecto RPH
Thiaben
Thiabendazol
Thiabendazole
Thiabenzazole
Thiabenzole
2-(Thiazol-4-yl)benzimidazole
2-(4-Thiazolyl)benzimidazole
2-(4'-Thiazolyl)benzimidazole
2-(4-Thiazolyl)-1H-benzimidazole
Thibenzol
Thibenzole
Thibenzole ATT
Tobaz
Top Form Wormer

148-82-3
$C_{13}H_{18}Cl_2N_2O_2$
305.23
NC(Cc1ccc(cc1)N(CCCl)CCCl)C(O)=O
Alanine, 3-(p-(bis(2-chloroethyl)amino)phenyl)-, L
Alanine nitrogen mustard
Alkeran
AT-290
L-3-(p-(Bis(2-chloroethyl)-amino)phenyl)alanine
p-Bis(β-chloroethyl)aminophenylalanine
p-N-Bis(2-chloroethyl)amino-L-phenylalanine
3-(p-(Bis(2-chloroethyl)amino)-phenyl)-L-alanine
4-(Bis(2-chloroethyl)amino)-L-phenylalanine
CB 3025
3025 C.B.
p-Di-(2-chloroethyl)amino-L-phenylalanine
p-N-Di(chloroethyl)amino-phenylalanine
3-p-(Di(2-chloroethyl)amino)-phenyl-L-alanine
Levofalan
Melfalan
Melphalan
Melphalen
Mephalan
NCI-C04853
NSC-8806
L-PAM
L-Phenylalanine, 4-(bis-(2-chloroethyl)amino)-
Phenylalanine mustard
L-Phenylalanine mustard
Phenylalanine nitrogen mustard
RCRA waste number U150
Sarcoclorin
L-Sarcolysin
p-L-Sarcolysin
Sarcolysine
L-Sarcolysine
Sarkolysin
L-Sarkolysin
SK-15673

148-87-8
$C_{10}H_{14}N_2$
162.26
N#CCCN(c(cccc1)c1)CC
Aniline, N-ethyl-N-(2-cyanoethyl)
Aniline, N-(2-cyanoethyl)-N-ethyl-
N-Ethyl-N-(2-cyanoethyl)aniline
N-Ethyl-N-2-kyanethylanilin (Czech)

149-29-1
$C_7H_6O_4$
154.13
OC1OCC=C2OC(=O)C=C12
4H-Furo(3,2-c)pyran-2(6H)-one, 4-hydroxy
Clairformin
Clavacin
Clavatin
Claviform
Claviformin
2,4-Dihydroxy-2H-pyran-δ-3(6H),α-acetic acid-3,4-lactone
(2,4-Dihydroxy-2H-pyran-3(6H)-ylidene)acetic acid-3,4-lactone
Expansin
Expansine
Gigantin
4-Hydroxy-4H-furo(3,2-c)pyran-2(6H)-one
Leucopin
Mycoin
Mycoin C
Mycoin C3
Mycoine C3
Mycosin
Patulin
Patuline
Penatin
Penicidin
2H-Pyran-$\delta^{3(6H),\alpha}$-acetic acid, 2,4-dihydroxy-, 3,4-lactone
Tercinin
Terinin

149-30-4
$C_7H_5NS_2$
167.25
N(c(c(S1)ccc2)c2)=C1S
2-Benzothiazolethiol
Captax
Kaptax (Czech)

MBT
Mercaptobenzothiazole
2-Mercaptobenzothiazole
2-Merkaptobenzotiazol (Polish)
2-Merkaptobenzthiazol (Czech)
NCI-C56519
Pennac MBT Powder
Rokon
Rotax
Sulfadene
USAF GY-3
USAF XR-29
Vulkacit Mercapto

149-31-5
Unknown
Unknown
2-Methyl-1,3-pentanediol

149-32-6
$C_4H_{10}O_4$
122.14
OCC(O)C(O)CO
Erythritol
Antierythrite
1,2,3,4-Butanetetrol
Erythrite
Erythritol, meso-
meso-Erythritol
L-Erythritol
Erythroglucin
Erythrol
Paycite
Phycitol
Tetrahydroxybutane

149-44-0
$CH_4O_3S.Na$
119.10
Methanesulfinic acid, hydroxy-, monosodium salt
Aldanil
Discolite
Formaldehyde hydrosulfite
Formaldehyde sodium bisulfite adduct
Formaldehyde sodium sulfoxylate
Formaldehydesulfoxylic acid, sodium salt
Formopan
Hydrolit
Hydrosulfite AWC
Hydroxymethanesulfinic acid, sodium salt
Oxymethansulfinsaeuren natrium (German)
Rongalit
Rongalit C
Rongalite
Rongalite C
Sodium formaldehyde sulfoxylate
Sodium hydroxymethanesulfinate
Sodium methanalsulfoxylate
Sodium oxymethanesulfinic acid
Sodium sulfoxylate formaldehyde

149-45-1
$C_6H_4O_8S_2.2Na$
314.20
m-Benzenedisulfonic acid, 4,5-dihydroxy-, disodium salt
1,3-Benzenedisulfonic acid, 4,5-dihydroxy-, disodium salt
3,5-Disulfocatechol disodium salt
SDD
Tiferron
Tiron

149-57-5
$C_8H_{16}O_2$
144.24
O=C(O)C(CCCC)CC
Hexanoic acid, 2-ethyl
Butylethylacetic acid
α-Ethylcaproic acid
2-Ethylhexanoic acid
2-Ethylhexoic acid
Kyselina 2-ethylkapronova (Czech)
Kyselina heptan-3-karboxylova (Czech)

149-73-5
$C_4H_{10}O_3$
106.14
O(C(OC)OC)C
Orthoformic acid, trimethyl ester
Methane, trimethoxy-
Methylester kyseliny orthomravenci (Czech)
Methyl orthoformate
Orthomravencan methylnaty (Czech)
Trimethoxymethane
Trimethylester kyseliny orthomravenci (Czech)
Trimethyl orthoformate

149-74-6
$C_7H_8Cl_2Si$
191.14
Silane, dichloromethylphenyl
Dichlor-fenyl-methylsilane (Czech)
Dichloromethylphenylsilane
Methylphenyldichlorosilane [UN 2437]
Phenylmethyldichlorosilane
UN 2437 [Methylphenyldichlorosilane]

149-91-7
$C_7H_6O_5$
170.13
O=C(O)c(cc(O)c(O)c1O)c1
Gallic-acid
Benzoic acid, 3,4,5-trihydroxy-
Kyselina gallova (Czech)
Kyselina 3,4,5-trihydroxybenzoova (Czech)
3,4,5-Trihydroxybenzoic acid

150-13-0
$C_7H_7NO_2$
137.15
O=C(O)c(ccc(N)c1)c1
Benzoic acid, p-amino
Amben
Aminobenzoic acid
γ-Aminobenzoic acid
p-Aminobenzoic acid
4-Aminobenzoic acid
1-Amino-4-carboxybenzene
Anticanitic Vitamin
Anti-Chromotrichia Factor
Bacterial Vitamin H1
Benzoic acid, 4-amino-
p-Carboxyaniline
4-Carboxyaniline
p-Carboxyphenylamine
Chromotrichia Factor
anti-Chromotrichia Factor
Kyselina p-aminobenzoova (Czech)
PABA
Pabanol
Paraminol
Paranate
Sunbrella
Trichochromogenic Factor
Vitamin BX
Vitamin H

150-19-6
$C_7H_8O_2$
124.15
O(c(cccc1O)c1)C
Phenol, m-methoxy
m-Guaiacol
m-Hydroxyanisole
3-Hydroxyanisole
m-Methoxyphenol
3-Methoxyphenol
Phenol, 3-methoxy- (9CI)
Resorcinol methyl ether
Resorcinol monomethyl ether

150-30-1
$C_9H_{11}NO_2$
165.19
O=C(O)C(N)Cc(cccc1)c1
Alanine, phenyl-, DL- (8CI)
AI3-18436
DL-α-Amino-β-phenylpropionic acid
DL-2-Amino-3-phenylpropanoic acid
NSC-9959
β-Phenylalanine, dl-
DL-Phenylalanine (9CI)
DL-β-Phenylalanine
DL-β-Phenyl-α-alanine
DL-3-Phenylalanine
(+-)-Phenylalanine

150-38-9
$C_{10}H_{13}N_2O_8.3Na$

358.22
[Na+].[Na+].[Na+].OC(=O)CN(CCN(CC([O-])=O)CC([O-])=O)CC([O-])=O
Acetic acid, (ethylenedinitrilo)tetra-, trisodium salt
Edetate trisodium
EDTA trisodium salt
Ethylenediamineacetic acid trisodium salt
Ethylenediaminetetraacetic acid, trisodium salt
Glycine, N,N'-1,2-ethanediylbis(N-(carboxymethyl)-, trisodium salt (9CI)
NCI-C03974
Nevanaid-B Powder
Perma Kleer 50, Trisodium salt
Sequestrene Na3
Sequestrene trisodium
Sequestrene trisodium salt
Trilon AO
Trinatrium ethylendiamintetraacetat (Czech)
Trisodium edetate
Trisodium EDTA
Trisodium ethylenediaminetetraacetate
Trisodium hydrogen ethylenediaminetetraacetate
Trisodium hydrogen (ethylenedinitrilo)tetraacetate
Trisodium versenate
Versene 9

150-39-0
$C_{10}H_{18}N_2O_7$
278.30
O=C(O)CN(CCN(CC(=O)O)CC(=O)O)CCO
Glycine, N-(carboxymethyl)-N'-(2-hydroxyethyl)-N,N'-ethylenedi
N-(Carboxymethyl)-N'-(2-hydroxyethyl)-N,N'-ethylenediglycine
Chel DM Acid
Hamp-ol acid
HEDTA
HEEDTA
N-Hydroxyethylenediaminetriacetic acid
N-(β-Hydroxyethylethylenediamine)-N,N',N'-triacetic acid
N-(2-Hydroxyethyl)ethylenediaminetriacetic acid
(N-Hydroxyethylethylenedinitrilo)triacetic acid
Versenol
Versenol 120

150-46-9
$C_6H_{15}BO_3$
146.02
CCOB(OCC)OCC
Boric acid, triethyl ester
Triethyl borate
Triethylester kyseliny borite (Czech)

150-50-5
$C_{12}H_{27}PS_3$
298.54
CCCCSP(SCCCC)SCCCC
Phosphorotrithious acid, tributyl ester
Chemagro B-1776
Deleaf Defoliant
Easy Off-D
Folex
Merphos
Phosphorotrithious acid, S,S,S-tributyl ester
Tributyl phosphorotrithioite
S,S,S-Tributyl phosphorotrithioite
Tributylthiofosfin (Czech)
S,S,S-Tributyl trithiophosphite

150-59-4
$C_{20}H_{27}N$
281.44
CCN(CCCc1ccccc1)CCCc2ccccc2
Alverine
Alverina (Spanish)
Alverinum (Latin)
Benzenepropanamine, N-ethyl-N-(3-phenylpropyl)-
Bis(γ-phenylpropyl)ethylamine
Di(phenylpropyl)ethylamine
Dipropylamine, N-ethyl-3,3'-diphenyl-
Dipropylin
Dipropyline
N-Ethyl-3,3'-diphenyldipropylamine
N-Ethyl-N-(3-phenylpropyl)benzenepropanamine
Phenopropamine
Phenpropamine
Profenil
Sestron
Sestron Base
Spasmaverine

150-68-5
$C_9H_{11}ClN_2O$
198.67
O=C(N(C)C)Nc(ccc(c1)Cl)c1
Urea, 3-(p-chlorophenyl)-1,1-dimethyl
3-(4-Chloor-fenyl)-1,1-dimethylureum (Dutch)
Chlorfenidim
N-(p-Chlorophenyl)-N',N'-dimethylurea
N'-(4-Chlorophenyl)-N,N-dimethylurea
1-(p-Chlorophenyl)-3,3-dimethylurea
3-(p-Chlorophenyl)-1,1-dimethylurea
3-(4-Chlorophenyl)-1,1-dimethylurea
1-(4-Chloro phenyl)-3,3-dimethyluree (French)
3-(4-Chlor-phenyl)-1,1-dimethyl-harnstoff (German)
3-(4-Cloro-fenil)-1,1-dimetilurea (Italian)
CMU
N,N-Dimethyl-N'-(4-chlorophenyl)urea
1,1-Dimethyl-3-(p-chlorophenyl)urea
Herbicides, Monuron
Karmex Monuron Herbicide
Karmex W. Monuron Herbicide
Lirobetarex
Monurex
Monuron
Monurox
Monuruon
Monuuron
NCI-C02846
Telvar
Telvar Monuron Weedkiller
Telvar W. Monuron Weedkiller
USAF P-8
USAF XR-41

150-69-6
$C_9H_{12}N_2O_2$
180.23
CCOc1ccc(NC(N)=O)cc1
Urea, (p-ethoxyphenyl)
p-Aethoxyphenylharnstoff (German)
Dulcin
Dulcine
p-Ethoxyfenylmocovina (Czech)
p-Ethoxyphenylurea
N-(4-Ethoxyphenyl)urea
4-Ethoxyphenylurea
NCI-C02073
p-Phenetolcarbamid (German)
Phenetolcarbamide
p-Phenetolcarbamide
p-Phenetolecarbamide
Phenethylcarbamid (German)
p-Phenetylurea
Sucrol
Suesstoff
Valzin

150-75-4
C_7H_9NO
123.17
Oc(ccc(NC)c1)c1
Phenol, p-(methylamino)
p-(Methylamino)phenol
4-(Methylamino)phenol

150-76-5
$C_7H_8O_2$
124.15
O(c(ccc(O)c1)c1)C
Phenol, p-methoxy
Hydroquinone monomethyl ether
Mequinol
p-Methoxyphenol
4-Methoxyphenol (ACGIH, OSHA)
MME
Mono methyl ether hydro-

quinone
USAF AN-7

150-78-7
$C_8H_{10}O_2$
138.18
O(c(ccc(OC)c1)c1)C
Benzene, p-dimethoxy
Anisole, p-methoxy-
Benzene, 1,4-dimethoxy-
p-Dimethoxybenzene
1,4-Dimethoxybenzene
Dimethylether hydrochinonu (Czech)
Dimethyl ether hydroquinone
Dimethylhydroquinone
Dimethylhydroquinone ether
DMB
Hydroquinone, dimethyl ether
p-Methoxyanisole
Quinol dimethyl ether
USAF AN-9
USAF UCTL-1791

150-84-5
$C_{12}H_{22}O_2$
198.34
O=C(OCCC(CCC=C(C)C)C)C
6-Octen-1-ol, 3,7-dimethyl-, acetate
Acetic acid, citronellyl ester
Acetic acid, 3,7-dimethyl-6-octen-1-yl ester
Citronellyl acetate
3,7-Dimethyl-6-octen-1-yl acetate
2-Octen-8-ol, 2,6-dimethyl-, acetate

150-86-7
$C_{20}H_{40}O$
296.60
OCC=C(CCCC(CCCC(CCCC(C)C)C)C)C
Phytol
2-Hexadecen-1-ol, 3,7,11,15-tetramethyl-, (R-(R*,R*-(E)))-(9CI)
trans-Phytol
3,7,11,15-Tetramethyl-2-hexadecen-1-ol

151-00-8
C_4H_9NO
87.14
N(O)=CC(C)C
Isobutyraldehyde, oxime
2-Methyl-1-propanal oxime
USAF AM-8

151-01-9
$C_3H_6OS_2$
122.21
O(C(S)=S)CC
Carbonic acid, dithio-, O-ethyl ester
Carbonodithioic acid, O-ethyl ester
o-Ethyl dithiocarbamate
Ethyl xanthate
Ethylxanthic acid
Ethyl xanthogenate
Xanthate

151-05-3
$C_{12}H_{16}O_2$
192.28
O=C(OC(Cc(cccc1)c1)(C)C)C
Phenethyl alcohol, α,α-dimethyl-, acetate
Benzyldimethyl carbinyl acetate
Dimethylbenzyl carbinol acetate
α,α-Dimethylphenethyl acetate
α,α-Dimethylphenethyl alcohol, acetate
DMBCA

151-10-0
$C_8H_{10}O_2$
138.18
O(c(cccc1OC)c1)C
Benzene, m-dimethoxy
Benzene, 1,3-dimethoxy-
m-Dimethoxybenzene
1,3-Dimethoxybenzene
Dimethylether resorcinolu (Czech)
3-Methoxyanisole
Resorcinol dimethyl ether

151-13-3
$C_{22}H_{42}O_3$
354.57
O=C(OCCCC)CCCCCCCC=CCC(O)CCCCCC
9-Octadecenoic acid, 12-hydroxy-, butyl ester, (R-(Z))-
AI3-19737

151-18-8
$C_3H_6N_2$
70.11
N#CCCN
Propionitrile, 3-amino
β-Alaminenitrile
3-Aminopropanenitrile
β-Aminopropionitrile
3-Aminopropionitrile
BAPN
β-Cyanoethylamine

151-19-9
$C_{10}H_{22}O$
158.28
OC(CCC(CC)C)(CC)C
3-Octanol, 3,6-dimethyl- (9CI)
AI3-23412
AR 1
3,6-Dimethyl-3-octanol
NSC-5613

151-21-3
$C_{12}H_{25}O_4S.Na$
288.42
[Na+].CCCCCCCCCCCCOS([O-])(=O)=O
Sulfuric acid, monododecyl ester, sodium salt
AI3-00356
Akyposal SDS
Aquarex ME
Aquarex methyl
Avirol 101
Avirol 118 Conc
Berol 452
Carsonol SLS
Carsonol SLS Paste B
Carsonol SLS Special
Conco Sulfate WA
Conco Sulfate WA-1200
Conco Sulfate WA-1245
Conco Sulfate WAG
Conco Sulfate WAN
Conco Sulfate WAS
Conco Sulfate WN
Cycloryl 21
Cycloryl 31
Cycloryl 580
Cycloryl 585N
Dehydag Sulfate GL Emulsion
Dehydag Sulphate GL Emulsion
Detergent 66
Dodecyl alcohol, hydrogen sulfate, sodium salt
Dodecyl sodium sulfate
Dodecyl sulfate sodium
n-Dodecyl sulfate sodium
Dodecyl sulfate, sodium salt
Dreft
Duponal
Duponal Waqe
Duponol
Duponol C
Duponol ME
Duponol Methyl
Duponol QX
Duponol WA
Duponol WA Dry
Duponol Waq
Duponol Waqa
Duponol Waqe
Duponol Waqm
Emal O
Emal 10
Emersal 6400
Empicol LPZ
Empicol LS 30
Empicol LX 28
Emulsifier No. 104
Finasol OSR_2
Gardinol
Hexamol SLS
Incronol SLS
Irium
Jordanol SL-300
Lanette Wax-S
Laurylsiran sodny (Czech)
Lauryl sodium sulfate
Lauryl sulfate, sodium salt
Maprofix 563
Maprofix LK
Maprofix Neu
Maprofix WAC
Maprofix WAC-LA
Melanol CL

Melanol CL 30
Monododecyl sodium sulfate
Monogen Y 100
Montopol La Paste
NCI-C50191
Neutrazyme
Nikkol SLS
Odoripon Al 95
Orvus Wa Paste
P and G Emulsifier 104
Perlandrol L
Product No. 75
Product No. 161
Quolac Ex-Ub
Rewopol NLS 30
Richonol A
Richonol AF
Richonol C
Sinnopon LS 95
Sinnopon LS 100
Sintapon L
Sipex OP
Sipex SB
Sipex SD
Sipex SP
Sipex UB
Sipon LS
Sipon LS 100
Sipon LSB
Sipon PD
Sipon WD
SLS
Sodium dodecyl sulfate
Sodium n-dodecyl sulfate
Sodium dodecyl sulphate
Sodium lauryl sulfate
Sodium lauryl sulphate
Sodium monododecyl sulfate
Sodium monolauryl sulfate
Solsol Needles
Standapol 112 Conc
Standapol WA-AC
Standapol WAQ
Standapol WAQ Special
Standapol WAS 100
Steinapol NLS 90
Stepanol ME
Stepanol ME Dry
Stepanol ME Dry AW
Stepanol Methyl
Stepanol Methyl Dry AW
Stepanol T 28
Stepanol WA
Stepanol WA-100
Stepanol WAC
Stepanol WA Paste
Stepanol WAQ
Sterling WA Paste
Sterling WAQ-CH
Sterling WAQ-Cosmetic
Sulfetal L 95
Sulfopon WA 1
Sulfopon WA 2
Sulfopon WA 3
Sulfopon WA 1 Special
Sulfotex WA
Sulfotex Wala
Swascol 3L
Swascol 4L
Swascol 1P
Syntapon
Syntapon L
Syntapon L Pasta (Czech)
Tarapon K 12
Texapon DL Conc.
Texapon K12
Texapon K-1296
Texapon L 100
Texapon V HC
Texapon V HC Powder
Texapon ZHC
Texapon Z High Conc. Needles
Trepenol WA
TVM 474
Ultra Sulfate SL-1
Waqe
Witcolate A
Witcolate A Powder
Witcolate C

151-32-6
$C_{24}H_{46}O_4$
398.63
O=C(OCCCCCCCCC)CCCCC
(=O)OCCCCCCCCC
Di-n-nonyl adipate
Adimoll DN
Adipic acid, dinonyl ester (8CI)
Bisoflex DNA
Dinonyl adipate
Dinonyl hexanedioate
Hexanedioic acid, dinonyl ester (9CI)
NSC-11041
Plastomoll Na

151-38-2
$C_5H_{10}HgO_3$
318.74
COCC[Hg]OC(C)=O
Mercury, (acetato)(2-methoxyethyl)
Acetato(2-methoxyethyl)-mercury
Cekusil Universal A
Landisan
Mema
Mercuran
Mercury, acetoxy(2-methoxy-ethyl)-
Methoxyethyl mercuric acetate
Methoxyethylmercury acetate
2-Methoxyethylmerkuriacetat (Czech)
Panogen
Panogen M
Radosan

151-41-7
$C_{12}H_{26}O_4S$
266.44
O=S(=O)(OCCCCCCCCCCCC)O
Sulfuric acid, monododecyl ester
Dodecansulfonic acid, hydroxy-
Dodecyl sulfate
Dodecylsulfuric acid
Lauryl sulfate
Lauryl sulfuric acid
Lauryl sulphate
Monododecyl hydrogen sulfate

151-50-8
CN.K
65.12
[K]C#N
Potassium-cyanide
Cyanide of potassium
Cyanides (OSHA)
Cyanure de potassium (French)
Hydrocyanic acid, potassium salt
Kalium-Cyanid (German)
Potassium cyanide (ACGIH) [UN 1680]
Potassium cyanide, Solid [UN 1680]
Potassium cyanide, Solution [UN 1680]
RCRA waste number P098
UN 1680 [Potassium cyanide]

151-56-4
C_2H_5N
43.08
N(C1)C1
Ethylenimine
Aethylenimin (German)
Aminoethylene
Azacyclopropane
Azirane
Aziridin (German)
Aziridine
1H-Azirine, dihydro-
Dihydroazirene
Dihydro-1H-azirine
Dimethyleneimine
Dimethylenimine
EI
ENT 50,324
Ethyleenimine (Dutch)
Ethyleneimine (ACGIH,OSHA)
Ethylene imine, Inhibited [UN 1185]
Ethylimine
Etilenimina (Italian)
RCRA waste number P054
TL 337
UN 1185 [Ethyleneimine, inhibited]

151-67-7
$C_2HBrClF_3$
197.39
FC(F)(F)C(Cl)Br
Ethane, 2-bromo-2-chloro-1,1,1-trifluoro
Bromochlorotrifluoroethane
1-Bromo-1-chloro-2,2,2-trifluoroethane
2-Bromo-2-chloro-1,1,1-trifluoroethane
Chalothane
Ethane, 1-bromo-1-chloro-2,2,2-trifluoro-
Fluorotane
Fluothane
Ftorotan (Russian)
Halotan
Halothane (ACGIH)

Halsan
Narcotane
Narcotann Ne-Spofa (Russian)
1,1,1-Trifluoro-2-bromo-2-chloroethane
1,1,1-Trifluoro-2-chloro-2-bromoethane
2,2,2-Trifluoro-1-chloro-1-bromoethane

151-83-7
$C_{14}H_{18}N_2O_3$
262.30
Methohexital
(+-)-5-Allyl-1-methyl-5-(1-methyl-2-pentynyl)barbituric acid
DEA No. 2264
Methohexitalum (Latin)
Metoesital
Metohexital (Spanish)
2,4,6(1H,3H,5H)-Pyrimidinetrione, 1-methyl-5-(1-methyl-2-pentynyl)-5-(2-propenyl)-, (+-)-

152-02-3
$C_{19}H_{25}NO$
283.45
Oc4ccc3CC2C1CCCCC1(CCN2CC=C)c3c4
Morphinan-3-ol, 17-allyl-, (-)
N-Allyl-3-hydroxymorphinan
(-)-3-Hydroxy-N-allylmorphinan
levo-3-Hydroxy-N-allyl morphinan
Levallorphan
(-)-Levallorphan
Lorfan
Morphinan, 17-allyl-3-hydroxy-
Naloxiphan
Ro-1-7700

152-16-9
$C_8H_{24}N_4O_3P_2$
286.30
O=P(OP(=O)(N(C)C)N(C)C)(N(C)C)N(C)C
Pyrophosphoramide, octamethyl
Bis(bisdimethylaminophosphonous)anhydride
Bis(dimethylamino)phosphonous anhydride
Bis(dimethylamino)phosphoric anhydride
Bis-N,N,N',N'-tetramethylphosphorodiamidic anhydride
Diphosphoramide, octamethyl-(9CI)
ENT 17,291
Lethalaire G-59
Octamethyl
Octamethyl-difosforzuur-tetramide (Dutch)
Octamethyldiphosphoramide
Octamethyl-diphosphorsaeure-tetramid (German)
Octamethylpyrophosphoramide
Octamethyl pyrophosphortetramide
Octamethyl tetramido pyrophosphate
Octamidophos
Oktamethyl (Czech)
Oktamidofos (Czech)
OMPA
Ompacide
Ompatox
Ompax
Ottometil-pirofosforammide (Italian)
Pestox
Pestox 3
Pestox 66
Pestox III
Pyrophosphoric acid octamethyltetraamide
Pyrophosphoryltetrakisdimethylamide
RCRA waste number P085
Schradan
Schradane (French)
Systam
Systophos
Sytam
Tetrakisdimethylaminophosphonous anhydride

152-18-1
$C_3H_9O_3PS$
156.15
Phosphorothioic acid, O,O,O-trimethyl ester
HC 7900
7900-HC
SD 4741
O,O,O-Trimethylester kyseliny thiofosforecne (Czech)
O,O,O-Trimethyl phosphorothioate
Trimethylthiofosfat (Czech)
O,O,O-Trimethylthiofosfat (Czech)
Trimethylthiophosphate
O,O,O-Trimethyl thiophosphate

152-20-5
$C_3H_9O_3PS$
156.15
COP(=O)(OC)SC
Phosphorothioic acid, O,O,S-trimethyl ester
8000 Bis HC
Dimethylthiomethylphosphate
HC 7901
Methylphosphorothioate-((meo)2(mes)po)
O,O,S-Trimethyl phosphorothioate

152-62-5
$C_{21}H_{28}O_2$
312.49
CC(=O)C3CCC4C2C=CC1=CC(=O)CCC1(C)C2CCC34C
9-β,10-α-Pregna-4,6-diene-3,20-dione
6-Dehydro-retro-progesterone
Diphaston
Dufaston
Duphaston
Duvaron
Dydrogesterone
Gestatron
Gynorest
Hydrogestrone
Isopregnenone
Pregna-4,6-diene-3,20-dione, (9-β,10-α)- (9CI)
Prodel
Retro-6-dehydroprogesterone
Retrone
δ^6-Retroprogesterone
Terolut

152-72-7
$C_{19}H_{15}NO_6$
353.35
CC(=O)CC(c1ccc(cc1)N(=O)=O)c3c(O)c2ccccc2oc3=O
Coumarin, 3-(α-acetonyl-p-nitrobenzyl)-4-hydroxy
Acenocoumarin
Acenocoumarol
Acenocumarol
Acenokumarin (Czech)
3-(α-Acetonyl-p-nitrobenzyl)-4-hydroxy-coumarin
3-(α-Acetonyl-4-nitrobenzyl)-4-hydroxycoumarin
Ascumar
2H-1-Benzopyran-2-one, 4-hydroxy-3-(1-(4-nitrophenyl)-3-oxobutyl)-
G-23350
4-Hydroxy-3-(1-(4-nitrophenyl)-3-oxobutyl)-2H-1-benzopyran-2-one
Nicoumalone
3-(α-(p-Nitrophenol)-β-acetylethyl)-4-hydroxycoumarin
Nitrophenylacetylethyl-4-hydroxycoumarine
3-(α-p-Nitrophenyl-β-acetylethyl)-4-hydroxycoumarin
3-(α-(4'-Nitrophenyl)-β-acetylethyl)-4-hydroxycoumarin
Nitrovarfarian
Nitrowarfarin
Sincoumar
Sinkumar
Sinthrom
Sinthrome
Sintrom
Sintroma
Syncoumar
Syncumar
Syntrom
Zotil

153-18-4
$C_{27}H_{30}O_{16}$
610.57
O(C(C(O)C(O)C1O)C)C1OCC(OC(Oc(c(O)c(c(o2)cc(O)c3)c3O)c2c(ccc(O)c4O)c4)C(O)C5O)C5O
Rutin

Bioflavonoid
Birutan
C.I. 75730
Eldrin
Flavone, 3,3',4',5,7-pentahydroxy-, 3-(O-rhamnosyl-glucoside)
Globulariacitrin
Globularicitrin
Glucopyranoside, quercetin-3 6-O-(6-deoxy-α-l-mannopyranosyl)-, β-D-
Glucopyranoside, quercetin-3 6-O-α-l-rhamnopyranosyl-, β-D
Ilixathin
Melin
Myrticalorin
Myrticolorin
Myticolorin
Osyritin
Osyritrin
Oxyritin
Paliuroside
3,3',4',5,7-Pentahydroxyflavone-3-rutinoside
Phytomelin
Quercetin, 3-(6-O-(6-deoxy-α-l-mannopyranosyl)-β-D-glucopyranoside)
Quercetin 3-rhamnoglucoside
Quercetin rhamnoglucosine
Quercetin, 3-(6-O-α-l-rhamnopyranosyl-β-D-glucopyranoside)
Quercetin 3-rutinoside
Rutabion
Rutinic acid
Rutinoside, 2-(3,4-dihydroxyphenyl)-5,7-dihydroxy-4-oxo-4H-1-benzopyran-3-yl
Rutinoside, quercetin-3, β-
Rutoside
Rutozyd
Sophorin
Tanrutin
USAF CF-5
Violaquercitrin
Vitamin P

153-61-7
$C_{16}H_{16}N_2O_6S_2$
396.46
CC(=O)OCC1=C(N3C(SC1)C(NC(=O)Cc2cccs2)C3=O)C(O)=O
5-Thia-1-azabicyclo(4.2.0)Oct-2-ene-2-carboxylic acid, 3-(hydroxymethyl)-8-oxo-7-(2-(2- thienyl)acetamido)-, acetate (ester)
Cefalotin
Cephalothin
Cephalotin
Cet
CT
7-(2-Thienylacetamido)cephalosporanic acid
7-(Thiophene-2-acetamido)-cephalosporanic acid

153-78-6
$C_{13}H_{11}N$
181.25
Nc3ccc1c(Cc2ccccc12)c3
Fluoren-2-amine
Aminofluoren (German)
2-Aminofluorene
2-Fluorenamine
2-Fluoreneamine
Fluorene, 2-amino-

153-94-6
$C_{11}H_{12}N_2O_2$
204.25
O=C(O)C(N)CC(c(c(N1)ccc2)c2)=C1
Tryptophan, D
D-Tryptophan
D-Trytophane

154-17-6
$C_6H_{12}O_5$
164.18
O=CCC(O)C(O)C(O)CO
D-Arabino-hexose, 2-deoxy
2-Deoxy-D-arabino-hexose
D-2-Deoxyglucose
2-Deoxyglucose
2-Deoxy-D-glucose
2-Desoxy-D-glucose (French)
2 DG
Glucose, 2-deoxy-
D-Glucose, 2-deoxy-
NSC-15193

154-21-2
$C_{18}H_{34}N_2O_6S$
406.60
CCCC1CC(N(C)C1)C(=O)NC(C(C)O)C2OC(SC)C(O)C(O)C2O
D-erythro-D-Galacto-octo-pyranoside, methyl-6,8-dideoxy-6-(1-methyl-4-propyl-l-2- pyrrolidinecarboxamido)-1-thio-, trans-α
Albiotic
Lincocin
Lincolnensin
Lincomycin
Lincomycine (French)
NSC-70731
U-10149

154-23-4
$C_{15}H_{14}O_6$
290.29
OC2Cc1c(O)cc(O)cc1OC2c3ccc(O)c(O)c3
2H-1-Benzopyran-3,5,7-triol, 2-(3,4-dihydroxyphenyl)-3,4-dihydro-, (2R-trans)
Catechin
(+)-Catechin
d-Catechin
D-(+)-Catechin
Catechin (Flavan)
Catechinic acid
Catechol
(+)-Catechol
D-Catechol
Catechol (Flavan)
Catechuic acid
Catergen
Cianidanol
KB-53

154-41-6
$C_9H_{13}NO.ClH$
187.69
Benzyl alcohol, α-(1-aminoethyl)-, hydrochloride, (+-)
α-(1-Aminoethyl)benzenemethanol hydrochloride
α-(1-Aminoethyl)benzyl alcohol hydrochloride
(+-)-2-Amino-1-phenyl-1-propanol hydrochloride
Benzenemethanol, α-(1-aminoethyl)-, hydrochloride, (R*,S*)-, (+-)
α-Hydroxy-β-aminopropylbenzene hydrochloride
Monydrin
Mucorama
Mydriatine
dl-Norephedrine hydrochloride
Obestat
DL-1-Phenyl-2-amino-1-propanol monohydrochloride
Phenylpropanolamine hydrochloride
Propadrine hydrochloride

154-42-7
$C_5H_5N_5S$
167.21
O.Nc2nc1n[nH]cc1c(=S)[nH]2
Purine-6(1H)-thione, 2-amino
2-Amino-1,7-dihydro-6H-purin-6-thion (Czech)
2-Amino-6-mercaptopurine
2-Amino-6-merkaptopurin (Czech)
2-Amino-6-MP
2-Aminopurine-6-thiol
2-Amino-6-purinethiol
2-Aminopurin-6-thiol (Czech)
2-Aminopurine-6(1H)-thione
BW 5071
Guanine, thio-
Lanvis
6-Mercapto-2-aminopurine
6-Mercaptoguanine
NSC-752
Purine-6-thiol, 2-amino-
6H-Purine-6-thione, 2-amino-1,7-dihydro- (9CI)
Tabloid
TG
ThG
Thioguanine
6-Thioguanine
Tioguanin
Tioguanine
Wellcome U3B
X 27

154-69-8
$C_{16}H_{21}N_3.ClH$
291.86
Pyridine, 2-(benzyl(2-(dimethylamino)ethyl)amino)-, monohydrochloride
2-(Benzyl(2-(dimethylamino)-ethyl)amino)pyridine monohydrochloride
N-Benzyl-N-dimethylamino-ethyl α-aminopyridine hydrochloride
2-(Benzyl(2-(dimethylamino)-ethyl)amino)pyridine hydrochloride
N-Benzyl-N',N'-dimethyl-N-2-pyridyl-ethylenediamine hydrochloride
N-Benzyl-N-α-pyridyl-N',N'-dimethyl-aethylendiamin-hydrochlorid (German)
1,2-Ethanediamine, N,N-dimethyl-N'-(phenylmethyl)-N'-2-pyridinyl-, monohydrochloride
Dehistin
Dehistin monohydrochloride
N,N-Dimethyl-N'-(2-pyridyl)-N'-benzylethylenediamine hydrochloride
Ethylenediamine, N-benzyl-N',N'-dimethyl-N-(2-pyridyl)-, hydrochloride
Piristin
Pyrabenzamine
Pyribenzamine hydrochloride
Pyribenzamine monohydrochloride
Pyridine, 2-(benzyl(2-(dimethylamino)ethyl)amino)-, hydrochloride
N^1-α-Pyridyl-N^1-benzyl-N,N-dimethyl ethylenediamine monohydrochloride
Pyrinamine
Resistamine
Stanzamine
Tripelenamine hydrochloride
Tripelennamine hydrochloride
Tripelennamine monohydrochloride
Tri-Tumine

154-93-8
$C_5H_9Cl_2N_3O_2$
214.07
ClCCNC(=O)N(CCCl)N=O
Urea, 1,3-bis(2-chloroethyl)-1-nitroso
BCNU
Bicnu
N,N'-Bis(2-chloroethyl)-N-nitrosourea
Bis(2-chloroethyl)nitrosourea
1,3-Bis(β-chloroethyl)-1-nitrosourea
1,3-Bis-(2-chloroethyl)-1-nitrosourea
Bischloroethylnitrosourea
Carmubris
Carmustin
Carmustine
FDA 0345
NCI-C04773
Nitrumon
NSC-409962
SK 27702
SRI 1720
Urea, N,N'-bis(2-chloroethyl)-N-nitroso- (9CI)

155-04-4
$C_{14}H_8N_2S_4.Zn$
397.85
[Zn].Sc2nc1ccccc1s2
2-Benzothiazolethiol, zinc salt (2:1)
Bis(2-benzothiazolylthio)zinc
Bis(mercaptobenzothiazolato)-zinc
Hermat Zn-MBT
2-Mercaptobenzothiazole zinc salt
Oxaf
Pennac ZT
Tisperse MB-58
USAF GY-7
Vulkacit ZM
Zenite
Zenite Special
Zetax
Zinc 2-benzothiazolethiolate
Zinc benzothiazolyl mercaptide
Zinc benzothiazol-2-ylthiolate
Zinc benzothiazyl-2-mercaptide
Zinc, bis(2-benzothiazolethiolato)-
Zinc mercaptobenzothiazolate
Zinc 2-mercaptobenzothiazole
Zinc mercaptobenzothiazole salt
ZMBT
ZnMB

155-09-9
$C_9H_{11}N$
133.21
NC1CC1c2ccccc2
Cyclopropylamine, 2-phenyl-, trans-(+-)
SKF 385
Transamine
Tranylcypromine

155-41-9
$C_{18}H_{24}NO_4.Br$
398.34
1-α-H,5-α-H-Tropanium, 6-β,7-β-epoxy-3-α-hydroxy-8-methyl-, bromide, (-)-tropate (ester)
Ampyrox
Blocan
Diopal
Epoxymethamine bromide
Epoxytropine tropate methylbromide
Holopan
Hyoscine methyl bromide
Lescopine bromide
Mescopil
Methoscopylamine bromide
Methscopolamine bromide
N-Methylhyoscine bromide
Methylscopolamine bromide
Methylscopolamine hydrobromide
N-Methylscopolammonium bromide
Neo-Avagal
Nutrop
Pamine
Pamine bromide
Paraspan
Proscomide
Restropin
Scopolamine methobromide
Scopolamine methylbromide
(-)-Scopolamine methyl bromide

156-06-9
$C_9H_8O_3$
164.16
O=C(O)C(=O)Cc(cccc1)c1
Phenylpyruvic acid
Benzenepropanoic acid, α-oxo-(9CI)
α-Oxobenzenepropanoic acid

156-08-1
$C_{17}H_{21}N$
239.39
CC(Cc1ccccc1)N(C)Cc2ccccc2
Phenethylamine, N-benzyl-N,α-dimethyl-, (+)
Benzeneethanamine, N,α-dimethyl-N-(phenylmethyl)-, (+)- (9CI)
Benzphetamine

156-10-5
$C_{12}H_{10}N_2O$
198.24
O=Nc(ccc(Nc(cccc1)c1)c2)c2
Diphenylamine, 4-nitroso
Benzenamine, 4-nitroso-N-phenyl- (9CI)
Naugard TKB
NCI-C02244
p-Nitrosodifenylamin (Czech)
p-Nitrosodiphenylamine
4-Nitrosodiphenylamine
p-Nitroso-N-phenylaniline
4-Nitroso-N-phenylaniline
4-Nitroso-N-phenylbenzenamine
N-Phenyl-p-nitrosoaniline
TKB

156-38-7
$C_8H_8O_3$
152.16
O=C(O)Cc(ccc(O)c1)c1
Acetic acid, (p-hydroxyphenyl)
Benzeneacetic acid, 4-hydroxy-(9CI)
4-Hydroxybenzeneacetic acid

(p-Hydroxyphenyl)acetic acid
(4-Hydroxyphenyl)acetic acid

156-39-8
$C_9H_8O_4$
180.17
Pyruvic acid, p-hydroxy-phenyl
Benzenepropanoic acid, 4-hydroxy-α-oxo- (9CI)
p-Hydroxyphenylpyruvic acid
NSC-100738
Pyruvic acid, (p-hydroxy-phenyl)- (8CI)

156-43-4
$C_8H_{11}NO$
137.20
O(c(ccc(N)c1)c1)CC
p-Phenetidine
4-Aminoethoxybenzene
p-Aminofenetol (Czech)
p-Aminophenetole
4-Aminophenetole
Aniline, p-ethoxy-
Benzenamine, 4-ethoxy- (9CI)
p-Ethoxyaniline
4-Ethoxyaniline
p-Fenetidin (Czech)
Phenethidine
p-Phenetidin
p-Phenetidine [UN 2311]
UN 2311 [Phenetidines]

156-51-4
$C_8H_{12}N_2.H_2O_4S$
234.30
NNCCc1ccccc1.OS(O)(=O)=O
Hydrazine, phenethyl-, sulfate (1:1)
Alacine
Alazin
Alazine
EP-411
Estinerval
Felazine
Fenelsin
Fenelzin
Fenelzina
Fenelzine
Fenetsin
Fenizin
Hydrazine, (2-phenylethyl)-, sulfate (1:1)
1-Hydrazino-2-phenylethane hydrogen sulphate
Kalgan
Mao-Rem
Monofen
Monophen
Monoten
N-1544A
Nardelzine
Nardil
P 1531
Phelazin
Phenalzine
Phenalzine dihydrogen sulfate
Phenalzine hydrogen sulphate
Phenelzin
Phenelzine acid sulfate
Phenelzine bisulphate
Phenelzine sulfate
Phenelzine sulphate
Phenethylhydrazine sulfate (1:1)
Phenethylhydrazine sulphate
Phenline
Phenodyn
Phenodyne
Phenylaethyl-hydrazin
Phenylethylhydrazine di-hydrogen sulphate
β-Phenylethylhydrazine di-hydrogen sulfate
2-Phenylethylhydrazine di-hydrogen sulphate
β-Phenylethylhydrazine hydrogen sulphate
2-Phenylethylhydrazine hydrogen sulphate
Phenylethylhydrazine sulphate
β-Phenylethylhydrazine sulfate
2-Phenylethylhydrazine sulphate
S 1544
Stinerval
WL 7

156-54-7
$C_4H_8O_2.Na$
111.11
Butyric acid, sodium salt
Butanoic acid, sodium salt (9CI)
Butyrate sodium
Sodium butanoate
Sodium butyrate
Sodium n-butyrate

156-56-9
$C_7H_{11}NO_2$
141.19
NC(CC1CC1=C)C(O)=O
Cyclopropanepropionic acid, α-amino-2-methylene-, l-(+)
α-Aminomethylenecyclopropane propionic acid
l-α-Amino-β-methylenecyclo-propanepropionic acid
α-Amino-2-methylenecyclo-propanepropionic acid
α-Amino-β-(2-methylenecyclo-propyl)propionic acid
2-Amino-4,5-methylenehex-5-enoic acid
Cyclopropanealanine, 2-methylene, l-
Cyclopropanepropanoic acid, α-amino-2-methylene- (9CI)
Hypoglycin
Hypoglycin A
Hypoglycine
Hypoglycine A
2-Methylenecyclopropaneal-anine
2-Methylenecyclopropanyl-alanine
β-(Methylenecyclopropyl)-alanine

156-59-2
$C_2H_2Cl_2$
96.94
C(=CCl)Cl
Ethylene, 1,2-dichloro-, (Z)
cis-Dichloroethylene
cis-1,2-Dichloroethylene

156-60-5
$C_2H_2Cl_2$
96.94
C(=CCl)Cl
Ethylene, 1,2-dichloro-, (E)
trans-Acetylene dichloride
trans-Dichloroethylene
trans-1,2-Dichloroethylene
RCRA waste number U079

156-62-7
$CN_2.Ca$
80.11
[Ca]=NC#N
Cyanamide, calcium salt (1:1)
Aero-Cyanamid
Aero Cyanamid Granular
Aero Cyanamid Special Grade
Alzodef
Calcium carbimide
Calcium cyanamid
Calcium cyanamide (ACGIH, OSHA)
Calcium cyanamide, Not hydrated (containing more than 0.1% calcium carbide) [UN 1403]
CCC
Cyanamid
Cyanamide
Cyanamide calcique (French)
Cyanamid Granular
Cyanamid Special Grade
CY-L 500
Lime nitrogen (DOT)
NCI-C02937
Nitrogen lime
Nitrolim
Nitrolime
UN 1403 [Calcium cyanamide with more than 0.1 per cent of calcium carbide]
USAF CY-2

156-74-1
$C_{16}H_{38}N_2$
258.48
Decamethonium
Ammonium, decamethylenebis-(trimethyl-
Decamethonum
1,10-Decanediaminium, N,N,N,N',N',N'-hexamethyl-

156-87-6
C_3H_9NO
75.13
OCCCN
1-Propanol, 3-amino

β-Alaninol
γ-Aminopropanol
3-Aminopropanol
3-Amino-1-propanol
3-Aminopropyl alcohol
3-Hydroxypropylamine
Propanolamine
1,3-Propanolamine
3-Propanolamine

157-40-4
C_5H_8
68.12
Spiropentane (6CI,7CI, 8CI,9CI)
Spiro(2.2)pentane

177-10-6
$C_8H_{14}O_2$
142.22
1,4-Dioxaspiro(4.5)decane

177-49-1
$C_{16}H_{32}O_4Si_4$
400.78
6,12,18,24-Tetraoxa-5,7,13,19-tetrasilatetraspiro(4.1.4.1.4.1.4.1)-te tracosane (7CI,8CI, 9CI)
Tetracyclotetramethylene cyclotetrasiloxane
Tetra(silacyclopentyl) cyclotetrasiloxane

181-15-7
$C_{12}H_{22}$
166.31
Spiro(5.6)dodecane (6CI, 7CI,8CI,9CI)

185-94-4
C_5H_8
68.12
Bicyclo(2.1.0)pentane (6CI, 7CI,8CI,9CI)
Housane

187-26-8
C_7H_{10}
94.16
Tricyclo(4.1.0.02,4)heptane (7CI,8CI,9CI)

189-43-5
$C_{32}H_{18}$
402.49
Benzo(xyz)heptaphene (8CI,9CI)
Dinaphtho(2',3',3,4)(2'',3'',9, 10)pyrene

189-45-7
$C_{32}H_{18}$
402.49
Naphthaceno(2,1,12-qra)naphthacene (8CI,9CI)
Dinaphtho(2,3-a:2',3'-h)pyrene

189-55-9
$C_{24}H_{14}$
302.38
c1ccc3c(c1)cc4ccc5cc2ccccc2c6ccc3c4c56
Benzo(rst)pentaphene
DB(a,i)P
Dibenzo(a,i)pyrene
Dibenzo(b,h)pyrene
1,2,7,8-Dibenzopyrene
3,4:9,10-Dibenzopyrene
Dibenz(a,i)pyrene
1,2:7,8-Dibenzpyrene
3,4:9,10-Dibenzpyrene
RCRA waste number U064

189-64-0
$C_{24}H_{14}$
302.38
c1ccc3c(c1)cc4ccc5c2ccccc2cc6ccc3c4c56
Dibenzo(b,def)chrysene
DB(a,h)P
Dibenzo(a,h)pyrene
1,2,6,7-Dibenzopyrene
3,4,8,9-Dibenzopyrene
3,4,8,9-Dibenzpyrene

189-90-2
$C_{19}H_{11}N$
253.31
Anthra(2,1,9-def)isoquinoline (7CI,8CI,9CI)
2-Azabenzo(b)pyrene

189-92-4
$C_{19}H_{11}N$
253.31
Phenaleno(1,9-gh)quinoline
Pyrenoline
Pyrido(2',3':4)pyrene

189-96-8
$C_{24}H_{14}$
302.38
Benzo(pqr)picene (8CI,9CI)
Naphtho(3',4':3,4)pyrene

190-09-0
$C_{42}H_{22}$
526.64
Dinaphtho(2,1,8-c1d1a: 2',1',8'-nop)heptacene-(6CI,8CI,9CI)

190-70-5
$C_{28}H_{14}$
350.42
Benzo(a)coronene (8CI,9CI)

191-07-1
$C_{24}H_{12}$
300.36
Coronene
Hexabenzobenzene

191-20-8
$C_{28}H_{16}$
352.44
Naphtho(1,2,3,4-rst)pentaphene (7CI,8CI,9CI)
Tribenzo(a,i,l)pyrene

191-24-2
$C_{22}H_{12}$
276.34
c1cc2ccc3ccc4ccc5cccc6c(c1)c2c3c4c56
Benzo(ghi)perylene
1,12-Benzoperylene
1,12-Benzperylene

191-26-4
$C_{22}H_{12}$
276.34
c1cc2ccc3cc5cccc6ccc4cc(c1)c2c3c4c56
Dibenzo(def,mno)chrysene
Anthanthren (German)
Anthanthrene
Anthranthrene
Dibenzo(cd,jk)pyrene

191-30-0
$C_{24}H_{14}$
302.38
c1ccc3c(c1)cc4ccc5cccc6c2ccccc2c3c4c56
Dibenzo(def,p)chrysene
BA 51-090462
DB(a,l)P
Dibenzo(a,d)pyrene
Dibenzo(a,l)pyrene
1,2:3,4-Dibenzopyrene
1,2,9,10-Dibenzopyrene
2,3:4,5-Dibenzopyrene
1,2,3,4-Dibenzpyrene
4,5,6,7-Dibenzpyrene

191-79-7
$C_{34}H_{18}$
426.52
Tetrabenzo(de,hi,op,st) pentacene (7CI,8CI,9CI)
1,9:5,10-Di(perinaphthylene) anthracene

192-46-1
$C_{36}H_{20}$
452.56
Dibenzo(j,xyz)heptaphene-(7CI,8CI,9CI)

192-47-2

$C_{28}H_{16}$
352.44
Dibenzo(h,rst)pentaphene
Tribenzo(a,e,i)pyrene
(1,2,4,5,7,8)-Tribenzopyrene
(1,2,4,5,8,9)-Tribenzopyrene
1,2:4,5:8,9-Tribenzopyrene

192-51-8
$C_{24}H_{14}$
302.38
Dibenzo(fg,op)naphthacene (9CI)
Dibenzo(e,l)pyrene
1,2:6,7-Dibenzpyrene (VAN)
NSC-87522

192-54-1
$C_{32}H_{18}$
402.50
Dibenzo(hi,uv)hexacene-(7CI,8CI,9CI)

192-59-6
$C_{28}H_{16}$
352.44
Dibenzo(fg,st)pentacene-(7CI,8CI,9CI)

192-60-9
$C_{32}H_{18}$
402.50
Dibenzo(fg,wx)hexacene-(7CI,8CI,9CI)

192-65-4
$C_{24}H_{14}$
302.38
c1ccc3c(c1)cc4c2ccccc2c5cccc6ccc3c4c56
Naphtho(1,2,3,4-def)chrysene
DB(a,e)P
Dibenzo(a,e)pyrene
1,2,4,5-Dibenzopyrene

192-97-2
$C_{20}H_{12}$
252.32
c1ccc2c(c1)c4cccc5ccc3cccc2c3c45
Benzo(e)pyrene
1,2-Benzopyrene
1,2-Benzpyrene
4,5-Benzopyrene
B(e)P

193-09-9
$C_{24}H_{14}$
302.38
Dibenzo(de,qr)naphthacene (8CI,9CI)
Naphtho(2,3-e)pyrene

193-11-3
$C_{28}H_{16}$
352.44
Dibenzo(de,uv)pentacene-(7CI,8CI,9CI)

193-39-5
$C_{22}H_{12}$
276.34
c(c(c(c(ccc1)c2)c1cc3)c3cc4)(c2c(c5ccc6)c6)c45
Indeno(1,2,3-cd)pyrene
IP
o-Phenylenepyrene
2,3-Phenylenepyrene
2,3-o-Phenylenepyrene
1,10-(ortho-Phenylene)pyrene
1,10-(1,2-Phenylene)pyrene
RCRA waste number U137

194-03-6
$C_{15}H_{9}N$
203.25
Thebenidine
4-Azapyrene
Benzo(lmn)phenanthridine
Dibenzo(c,d,e)quinoline
Pyrenidine

194-59-2
$C_{20}H_{13}N$
267.34
c1ccc4c(c1)ccc5[nH]c3ccc2cccc c2c3c45
7H-Dibenzo(c,g)carbazole
7-Aza-7H-dibenzo(c,g)fluorene
3,4,5,6-Dibenzcarbazol
3,4,5,6-Dibenzcarbazole
3,4,5,6-Dibenzocarbazole
3,4,5,6-Dinaphthacarbazole
7H-DB(c,g)C

194-69-4
$C_{22}H_{14}$
278.36
Benzo(c)chrysene
1,2,5,6-Dibenzphenanthrene

195-19-7
$C_{18}H_{12}$
228.30
Benzo(c)phenanthrene
3,4-Benzophenanthrene
3,4-Benzphenanthrene
Tetrahelicene

196-42-9
$C_{24}H_{14}$
302.38
Naphtho(2,1,8-qra)naphthacene (8CI,9CI)
Naphtho(2,3-a)pyrene
Naphtho(2,3-b)pyrene

196-45-2
$C_{28}H_{16}$
352.44
Naphtho(2,1,8-uva)pentacene (6CI,7CI,8CI,9CI)

196-46-3
$C_{32}H_{18}$
402.50
Naphtho(2,1,8-yza)hexacene (6CI,8CI,9CI)
Naphtho(2,1,8-yxa)hexacene (7CI)

196-78-1
$C_{22}H_{14}$
278.36
Benzo(g)chrysene
1,2,3,4-Dibenzophenanthrene
1,2,3,4-Dibenzphenanthrene

197-70-6
$C_{24}H_{14}$
302.38
Benzo(b)perylene (8CI,9CI)
2,3-Benzoperylene

198-55-0
$C_{20}H_{12}$
252.32
c(c(ccc1)ccc2)(c1c(c(c(cc3)ccc4)c45)c3)c25
Perylene
Dibenz(de,kl)anthracene
Peri-dinaphthalene
Perilene

201-06-9
$C_{16}H_{10}$
202.26
Acephenanthrylene
Benz(e)acenaphthylene
Benzoacenaphthylene
4,5-Benzoacenaphthylene

203-12-3
$C_{18}H_{10}$
226.28
Benzo(ghi)fluoranthene
Benzo(mno)fluoranthene
2,13-Benzofluoranthene
7,10-Benzofluoranthene

203-20-3
$C_{24}H_{14}$
302.38
Dibenz(a,j)aceanthrylene
15,16-Benzdehydrocholanthrene

203-33-8
$C_{20}H_{12}$
252.32
Benz(a)aceanthrylene
1,2-Benzfluoranthene

1,2-Benzfluoranthrene
Benzo(a)fluoranthene
1,2-Benzofluoranthene
Dibenzo(c,lm)fluorene

203-64-5
$C_{15}H_{10}$
190.25
4H-Cyclopenta(def)phenanthrene
Cyclopentaphenanthrene

203-65-6
$C_{14}H_9N$
191.23
4H-Benzo(def)carbazole (8CI, 9CI)
4,5-Iminophenanthrene
Phenanthrene, 4,5-imino-

203-80-5
$C_{13}H_{10}$
166.22
1H-Phenalene (9CI)
1H-Benzonaphthene
peri-Naphthindene
Perinaphthindene
Phenalene (8CI)

205-12-9
$C_{17}H_{12}$
216.29
7H-Benzo(c)fluorene
3,4-Benzofluorene

205-39-0
$C_{16}H_{10}O$
218.25
Benzo(b)naphtho(1,2-d)furan
NSC-109422

205-43-6
$C_{16}H_{10}S$
234.32
c1ccc3c(c1)ccc4sc2ccccc2c34
Benzo(b)naphtho(1,2-d)thiophene
1,2-BNT

205-82-3
$C_{20}H_{12}$
252.32
c1ccc4c(c1)ccc5c2cccc3cccc(c23)c45
Benzo(j)fluoranthene
Benz(j)fluoranthene
10,11-Benzfluoranthene
Benzo(l)fluoranthene
7,8-Benzofluoranthene
10,11-Benzofluoranthene
Benzo-12,13-fluoranthene
B(j)F
Dibenzo(a,jk)fluorene

205-97-0
$C_{24}H_{14}$
302.38
Naphth(2,3-e)acephenanthrylene (9CI)
Dibenzo(b,k)fluoranthene (8CI)

205-99-2
$C_{20}H_{12}$
252.32
Benz(e)acephenanthrylene
3,4-Benz(e)acephenanthrylene
2,3-Benzfluoranthene
3,4-Benzfluoranthene
Benzo(b)fluoranthene
Benzo(e)fluoranthene
2,3-Benzofluoranthene
3,4-Benzofluoranthene
2,3-Benzofluoranthrene
B(b)F

206-44-0
$C_{16}H_{10}$
202.26
c(c(ccc1)ccc2)(c1c(c3ccc4)c4)c23
Fluoranthene
1,2-Benzacenaphthene
Benzene, 1,2-(1,8-naphthylene)-
Benzo(jk)fluorene
Idryl
1,2-(1,8-Naphthalenediyl)-benzene
1,2-(1,8-Naphthylene)benzene
RCRA waste number U120

206-49-5
$C_{15}H_9N$
203.24
Acenaphtho(1,2-b)pyridine (8CI,9CI)
7-Azafluoranthene

206-55-3
$C_{15}H_9N$
203.24
Indeno(1,2,3-de)quinoline (8CI,9CI)
3-Azafluoranthene

206-56-4
$C_{15}H_9N$
203.24
Indeno(1,2,3-ij)isoquinoline (8CI,9CI)
1-Azafluoranthene

207-08-9
$C_{20}H_{12}$
252.32
c2ccc1cc3c(cc1c2)c4cccc5cccc3c45
Benzo(k)fluoranthene
8,9-Benzofluoranthene
11,12-Benzofluoranthene
11,12-Benzo(k)fluoranthene
2,3,1',8'-Binaphthylene
Dibenzo(b,jk)fluorene

207-18-1
$C_{24}H_{14}$
302.38
Naphtho(2,3-k)fluoranthene (8CI,9CI)
Acenaphth(1,2-b)anthracene

207-83-0
$C_{21}H_{14}$
266.35
13H-Dibenzo(a,g)fluorene
1,2,5,6-Dibenzofluorene

208-96-8
$C_{12}H_8$
152.20
c(c(ccc1)ccc2)(c1cc3)c23
Acenaphthylene
Cyclopenta(de)naphthalene

213-46-7
$C_{22}H_{14}$
278.36
Picene
3,4-Benzchrysene
Benzo(a)chrysene
β,β-Binaphthyleneethene
Dibenzo(a,i)phenanthrene
1,2:7,8-Dibenzophenanthrene

214-17-5
$C_{22}H_{14}$
278.36
Benzo(b)chrysene
2,3-Benzochrysene
3,4-Benzotetracene
Benzo(c)tetraphene
3,4-Benzotetraphene
1,2:6,7-Dibenzophenanthrene
2,3:7,8-Dibenzophenanthrene
Dibenzo-2,3,7,8-phenanthrene
Naphth(2,1-a)anthracene

215-58-7
$C_{22}H_{14}$
278.36
c5ccc4cc3c1ccccc1c2ccccc2c3cc4c5
Benzo(b)triphenylene
Db(a,c)A
Dibenz(a,c)anthracene
1,2:3,4-Dibenzanthracene
Dibenzo(a,c)anthracene
1,2:3,4-Dibenzoanthracene

215-62-3
$C_{21}H_{13}N$
279.33
Dibenz(a,c)acridine
NSC-48754

216-00-2
$C_{26}H_{16}$
328.41

Dibenzo(a,c)naphthacene (8CI,9CI)
1,2:3,4-Dibenzotetracene

217-37-8
$C_{26}H_{16}$
328.41
Benzo(c)picene (8CI,9CI)
Fulminene

217-54-9
$C_{26}H_{16}$
328.41
Dibenzo(b,k)chrysene
Anth(2,1-a)anthrene
(Anthra-2',1')-1,2-anthracene
2',1'-Anthra-1,2-anthracene
Anthraceno(2,1-a)anthracene
Anthraceno(2',1':1,2)anthracene

217-59-4
$C_{18}H_{12}$
228.30
Triphenylene
Benzo(l)phenanthrene
9,10-Benzophenanthrene
9,10-Benzphenanthrene
1,2,3,4-Dibenznaphthalene
Isochrysene

218-01-9
$C_{18}H_{12}$
228.30
c(c(c(c(c(c(c(c1)ccc2)c2)c3)c1)ccc4)(c3)c4
Chrysene (OSHA)
1,2-Benzophenanthrene
Benzo(a)phenanthrene
1,2-Benzphenanthrene
Benz(a)phenanthrene
1,2,5,6-Dibenzonaphthalene
RCRA waste number U050

218-02-0
$C_{17}H_{11}N$
229.29
Naphth(2,1-f)isoquinoline
2-Azachrysene

218-08-6
$C_{17}H_{11}N$
229.29
Naphtho(2,1-f)quinoline
1-Azachrysene

222-93-5
$C_{22}H_{14}$
278.35
Pentaphene (8CI,9CI)
Dibenzo(b,h)phenanthrene
2,3:6,7-Dibenzophenanthrene
2,3:6,7-Dibenzphenanthrene

224-41-9
$C_{22}H_{14}$
278.36
c1ccc4c(c1)ccc5cc3ccc2ccccc2c3cc45
Dibenz(a,j)anthracene
1,2:7,8-Dibenzanthracene
3,4,5,6-Dibenzanthracene
Dibenzo(a,j)anthracene
Dibenzo-1,2,7,8-anthracene

224-42-0
$C_{21}H_{13}N$
279.35
c1ccc4c(c1)ccc5nc3ccc2ccccc2c3cc45
Dibenz(a,j)acridine
7-Azadibenz(a,j)anthracene
DB(a,j)AC
Dibenz(a,f)acridine
1,2,7,8-Dibenzacridine
3,4,5,6-Dibenzacridine
Dibenzo(a,j)acridine
3,4,6,7-Dinaphthacridine

224-53-3
$C_{21}H_{13}N$
279.35
Dibenz(c,h)acridine
14-Azadibenz(a,j)anthracene
1,2,7,8-Dibenzacridine (French)
3,4:5,6-Dibenzacridine

225-11-6
$C_{17}H_{11}N$
229.29
c4ccc3nc2ccc1ccccc1c2cc3c4
Benz(a)acridine
7-Azabenz(a)anthracene
1,2-Benzacridine

225-51-4
$C_{17}H_{11}N$
229.29
c4ccc3nc1c(ccc2ccccc12)cc3c4
Benz(c)acridine
12-Azabenz(a)anthracene
B(c)AC
3,4-Benzacridine
7,8-Benzacridine (French)
3,4-Benzoacridine
α-Chrysidine
α-Naphthacridine
RCRA waste number U016

226-36-8
$C_{21}H_{13}N$
279.35
c1ccc4c(c1)ccc5nc2c(ccc3ccccc23)cc45
Dibenz(a,h)acridine
7-Azadibenz(a,h)anthracene
DB(a,h)AC
Dibenz(a,d)acridine
1,2,5,6-Dibenzacridine
1,2,5,6-Dibenzoacridine
1,2,5,6-Dinaphthacridine

226-86-8
$C_{26}H_{16}$
328.41
Dibenzo(a,l)naphthacene (8CI,9CI)

227-04-3
$C_{26}H_{16}$
328.41
Dibenzo(a,j)naphthacene (8CI,9CI)
1,2:7,8-Dibenzotetracene

229-87-8
$C_{13}H_9N$
179.23
c1ccc2c(c1)cnc3ccccc23
Phenanthridine
5-Azaphenanthrene
9-Azaphenanthrene
3,4-Benzoisoquinoline
Benzo(c)quinoline
3,4-Benzoquinoline
6-Phenanthridine

230-17-1
$C_{12}H_8N_2$
180.20
c1ccc2c(c1)nnc3ccccc23
Benzo(c)cinnoline (9CI)
AI3-20950
2,2'-Azobiphenyl
3,4-Benzocinnoline
5,6-Diazaphenanthrene
9,10-Diazaphenanthrene
Diphenylenazone
NSC-86935
5,6-Phenanthroline
Phenazone (VAN)

230-27-3
$C_{13}H_9N$
179.23
c1ccc2c(c1)ccc3cccnc23
Benzo(h)quinoline
4-Azaphenanthrene
α-Benzoquinoline
7,8-Benzoquinoline
α-Naphthoquinoline

238-04-0
$C_{24}H_{14}$
302.38
Naphtho(1,2-k)fluoranthene (8CI,9CI)
Acenaphtho(1,2-b)phenanthrene

238-84-6
$C_{17}H_{12}$
216.29
C3c1ccccc1c4ccc2ccccc2c34
11H-Benzo(a)fluorene
Benzo(a)fluorene
1,2-Benzofluorene

Chrysofluorene
α-Naphthofluorene

239-01-0
$C_{16}H_{11}N$
217.28
11H-Benzo(a)carbazole
1,2-Benzcarbazole

239-35-0
$C_{16}H_{10}S$
234.32
c1ccc3c(c1)ccc4c2ccccc2sc34
Benzo(b)naphtho(2,1-d)thiophene (9CI)
Benzo(a)dibenzothiophene
1,2-Benzo-9-thiafluorene
Naphtho(1,2-b)thianaphthene
NSC-89259

239-64-5
$C_{20}H_{13}N$
267.34
c1ccc4c(c1)ccc5c3ccc2ccccc2c3[nH]c45
7H-Dibenzo(a,i)carbazole
1,2,7,8-Dibenzcarbazole

243-17-4
$C_{17}H_{12}$
216.29
11H-Benzo(b)fluorene
2,3-Benzofluorene

243-28-7
$C_{16}H_{11}N$
217.26
5H-Benzo(b)carbazole (9CI)
2,3-Benzcarbazole
Benzocarbazole
2,3-Benzocarbazole
NSC-59788

243-46-9
$C_{16}H_{10}S$
234.32
Benzo(b)naphtho(2,3-d)thiophene
2,3-BNT

244-36-0
$C_{13}H_{10}$
166.22
1H-Fluorene (8CI,9CI)

244-63-3
$C_{11}H_8N_2$
168.21
c1ccc2c(c1)[nH]c3cnccc23
9H-Pyrido(3,4-b)indole
Norharman

244-99-5
$C_{12}H_9N$
167.20
5H-Indeno(1,2-b)pyridine (8CI,9CI)
4-Azafluorene

253-52-1
$C_8H_6N_2$
130.14
n(ncc(c1ccc2)c2)c1
Phthalazine (9CI)
2,3-Benzodiazine
Benzo(d)pyridazine
2,3-Diazanaphthalene
NSC-62484
β-Phenodiazine

253-66-7
$C_8H_6N_2$
130.14
c2ccc1nnccc1c2
Cinnoline (8CI,9CI)
1,2-Diazanaphthalene
NSC-58374
α-Phenodiazine

253-69-0
$C_8H_6N_2$
130.15
c2cnc1cnccc1c2
1,7-Naphthyridine-(6CI,7CI,8CI,9CI)
1,7-Diazanaphthalene
1,7-Pyridopyridine

253-82-7
$C_8H_6N_2$
130.14
c2ccc1ncncc1c2
Quinazoline (9CI)
1,3-Benzodiazine
Benzo(a)pyrimidine
5,6-Benzopyrimidine
1,3-Diazanaphthalene
NSC-72372
Phenmiazine

256-96-2
$C_{14}H_{11}N$
193.24
Iminostilbene
5-Azadibenzo(a,e)cyclohepta-triene
Dibenz(b,f)azepine
2,3,6,7-Dibenzazepine
5H-Dibenz(b,f)azepine (9CI)
5H-Dibenzo(b,f)azepine
2,2'-Iminostilbene
NSC-123458
Stilbene, 2,2'-imino-

259-79-0
$C_{12}H_8$
152.20
Biphenylene (9CI)
Cyclobutadibenzene
Diphenylene
NSC-101862

260-94-6
$C_{13}H_9N$
179.23
n(c(c(ccc1)cc2cccc3)c1)c23
Acridine [UN 2713]
Akridin (Czech)
9-Azaanthracene
10-Azaanthracene
Benzo(b)quinoline
2,3-Benzoquinoline
Dibenzo(b,e)pyridine
UN 2713 [Acridine]

262-12-4
$C_{12}H_8O_2$
184.20
O2c1ccccc1Oc3ccccc23
Dibenzo-p-dioxin
Dibenzodioxin
Dibenzo(1,4)dioxin
Dibenzo(b,e)(1,4)dioxin
Diphenylene dioxide
NCI-C03656
Oxanthrene
Phenodioxin

262-20-4
$C_{12}H_8OS$
200.26
O(c(c(Sc1cccc2)ccc3)c3)c12
Phenoxathiin
Dibenzothioxin
1,4-Dibenzothioxine
Phenothioxin
Phenoxathine
Phenoxthin
Phenoxathrin
USAF DO-17

271-29-4
$C_7H_6N_2$
118.15
c2cc1cc[nH]c1cn2
1H-Pyrrolo(2,3-c)pyridine
6-Azaindole

271-34-1
$C_7H_6N_2$
118.15
c2cc1[nH]ccc1cn2
1H-Pyrrolo(3,2-c)pyridine
5-Azaindole

271-44-3
$C_7H_6N_2$
118.15
c2ccc1[nH]ncc1c2
1H-Indazole
2-Azaindole
Indazole
Isoindazole

271-89-6
C_8H_6O
118.14
O(c(c(C=1)ccc2)c2)C1
Benzofuran
Benzo(b)furan
2,3-Benzofuran
Benzofurfuran
Coumarone
NCI-C56166
1-Oxindene

272-12-8
C_7H_5NS
135.18
Thieno(2,3-c)pyridine (8CI,9CI)
NSC-152396

272-16-2
C_7H_5NS
135.18
1,2-Benzisothiazole (8CI,9CI)
1-Thia-2-azaindene

272-49-1
$C_7H_6N_2$
118.15
c2cnc1cc[nH]c1c2
1H-Pyrrolo(3,2-b)pyridine
4-Azaindole

272-97-9
$C_6H_5N_3$
119.14
c2cc1[nH]cnc1cn2
1H-Imidazo(4,5-c)pyridine
3,5-Diazaindole

273-53-0
C_7H_5NO
119.13
O(c(c(N=1)ccc2)c2)C1
Benzoxazole
1-Oxa-3-azaindene
USAF EK-5017

274-09-9
$C_7H_6O_2$
122.13
O(c(c(O1)ccc2)c2)C1
Benzene, 1,2-methylenedioxy
Methylenedioxybenzene
1,2-Methylenedioxybenzene

274-40-8
C_8H_7N
117.15
c2cc1cccn1cc2
Indolizine
Pyrrolo(1,2-a)pyridine

275-51-4
$C_{10}H_8$
128.18
c(c(cccc1)cc2)(c1)c2
Azulene
Bicyclo(5.3.0)decapentaene
Bicyclo(0.3.5)deca-1,3,5,7,
9-pentaene
Bicyclo(5.3.0)-deca-2,4,6,8,
10-pentaene
Cyclopentacycloheptene

279-49-2
$C_6H_{10}O$
98.14
7-Oxabicyclo(2.2.1)heptane (8CI,9CI)
Cyclohexane, 3,6-endoxo-
Cyclohexane, 1,4-epoxy-
3,6-Endooxycyclohexane
1,4-Epoxycyclohexane
NSC-54357

280-33-1
C_8H_{14}
110.22
Bicyclo(2.2.2)octane
1,4-Endoethylenecyclohexane

280-57-9
$C_6H_{12}N_2$
112.20
N(CCN(C1)C2)(C1)C2
1,4-Diazabicyclo(2.2.2)octane
Bicyclo(2,2,2)-1,4-diazaoctane
Dabco
Dabco Crystal
Dabco EG
Dabco 33LV
Dabco R-8020
Dabco S-25
D 33LV
1,4-Ethylenepiperazine
Triethylenediamine

281-23-2
$C_{10}H_{16}$
136.24
C(CC(CC1CC23)C3)(C1)C2
Adamantane (8CI)
NSC-527913
Tricyclo(3.3.1.13,7)decane
(9CI)

281-32-3
$C_7H_{10}O_3$
142.16
2,4,10-Trioxatricyclo (3.3.1.13,7)decane (9CI)
1,3,5-Cyclohexanetriol,
cyclic orthoformate
Orthoformic acid, cyclic
1,3,5-cyclohexanetriyl
ester (7CI)
Orthoformic acid, cyclic
ester with 1,3,5-cyclo-
2,4,10-Trioxaadamantane-
(8CI)
hexanetriol
2,8,9-Trioxaadamantane

283-66-9
$C_6H_{12}N_2O_6$
208.16
Hexamethylenetriperoxide-diamine
Hexamethylene triperoxide
diamine, Dry (French)
Hexamethylene triperoxide-
diamine
Hexametilenotriperoxidiamina
(Spanish)
3,4,8,9,12,13-Hexaoxa-1,6-di-
azabicyclo(4.4.4)tetradecane,
Dry

285-67-6
C_5H_8O
84.13
O(C1CCC2)C12
6-Oxabicyclo(3.1.0)hexane
Cyclopentane, 1,2-epoxy-
Cyclopentane oxide
Cyclopentene epoxide
Cyclopenteneoxide
1,2-Epoxycyclopentane

286-08-8
C_7H_{12}
96.17
Bicyclo(4.1.0)heptane (9CI)
cis-Bicyclo(4.1.0)heptane
Norcarane (6CI,8CI)

286-20-4
$C_6H_{10}O$
98.16
O(C1CCCC2)C12
7-Oxabicyclo(4.1.0)heptane
CCHO
Cyclohexene-1-oxide
Cyclohexene epoxide
Cyclohexene oxide
1,2-Cyclohexene oxide
Cyclohexane oxide
Cyclohexane, 1,2-epoxy-
Cyclohexylene oxide
1,2-Epoxycyclohexane
Tetramethyleneoxirane

286-28-2
$C_6H_{10}S$
114.21
C2CCC1SC1C2
7-Thiabicyclo(4.1.0)heptane (9CI)
Cyclohexane, 1,2-epithio-
Cyclohexene episulfide
Cyclohexene sulfide
Cyclohexene, sulfide
1,2-Cyclohexylene sulfide
NSC-59716

286-99-7
$C_{12}H_{22}O$
182.31

13-Oxabicyclo(10.1.0)tridecane (9CI)
AI3-26439
Cyclododecane, 1,2-epoxy-
Cyclododecene epoxide
Epoxycyclododecane
NSC-521077

287-13-8
C_7H_{10}
94.16
Tricyclo(4.1.0.02,7)heptane- (7CI,8CI,9CI)
Tricyclo(3.1.1.06,7)heptane-(6CI)

287-23-0
C_4H_8
56.12
C(CC1)C1
Cyclobutane [UN 2601]
UN 2601 [Cyclobutane]

287-27-4
C_3H_6S
74.15
S(CC1)C1
Thietane (9CI)
NSC-56443
Propane, 1,3-epithio-
Thiacyclobutane
Trimethylene sulfide (8CI)

287-68-3
$H_4Se_2Si_2$
218.12
Cyclodisilaselenane (8CI,9CI)

287-92-3
C_5H_{10}
70.15
C(CCC1)C1
Cyclopentane (ACGIH,OSHA) [UN 1146]
Pentamethylene
UN 1146 [Cyclopentane]

288-13-1
$C_3H_4N_2$
68.09
N(NC=C1)=C1
Pyrazole
1,2-Diazole

288-14-2
C_3H_3NO
69.06
O(N=CC=1)C1
Isoxazole (9CI)
NSC-137774
1-Oxa-2-azacyclopentadiene

288-32-4
$C_3H_4N_2$
68.09
N(C=CN1)=C1
Imidazole
1,3-Diaza-2,4-cyclopentadiene
1,3-Diazole
Formamidine, N,N'-vinylene-
Glyoxalin
Glyoxaline
Imidazol
Iminazole
Imutex
Miazole
Pyrro(b)monazole
USAF EK-4733

288-42-6
C_3H_3NO
69.06
c1cocn1
Oxazole (8CI,9CI)
1,3-Oxazole

288-47-1
C_3H_3NS
85.13
N(C=CS1)=C1
Thiazole

288-88-0
$C_2H_3N_3$
69.08
N(=CNN=1)C1
s-Triazole
TA
1H-1,2,4-Triazole (9CI)

289-16-7
$C_2H_4S_3$
124.25
1,2,4-Trithiolane (8CI,9CI)

289-80-5
$C_4H_4N_2$
80.10
n(nccc1)c1
Pyridazine
1,2-Diazabenzene
1,2-Diazine
Orthodiazine

289-95-2
$C_4H_4N_2$
80.10
n(ccn1)c1
Pyrimidine
1,3-Diazabenzene
m-Diazine
Metadiazine
Miazine

290-37-9
$C_4H_4N_2$
80.10
n(ccnc1)c1
Pyrazine
1,4-Diazabenzene
p-Diazine
1,4-Diazine
Paradiazine
Piazine

290-87-9
$C_3H_3N_3$
81.09
n(cncn1)c1
s-Triazine
Cyanidine
sym-Triazine
1,3,5-Triazine (9CI)
Vedita 250

291-21-4
$C_3H_6S_3$
138.27
S(CSCS1)C1
s-Trithiane
Formaldehyde, thio-, trimer
Thioform (Czech)
Trimethylene trisulfide
Trimethylentrisulfid (Czech)
1,3,5-Trithiacyclohexane
sym-Trithian (Czech)
1,3,5-Trithiane
Trithioformaldehyde

291-64-5
C_7H_{14}
98.21
C1CCCCCC1
Cycloheptane [UN 2241]
UN 2241 [Cycloheptane]

292-64-8
C_8H_{16}
112.22
C1CCCCCCC1
Cyclooctane (9CI)
AI3-26694
NSC-72426

293-96-9
$C_{10}H_{20}$
140.27
C(CCCCCCCC1)C1
Cyclodecane (9CI)

294-62-2
$C_{12}H_{24}$
168.32
C(CCCCCCCCCC1)C1
Cyclododecane (9CI)

294-93-9
$C_8H_{16}O_4$
176.24
C1COCCOCCOCCO1
1,4,7,10-Tetraoxacyclododecane
12-Crown-4
EOCT

Ethylene oxide cyclic tetramer

297-76-7
$C_{24}H_{32}O_4$
384.56
CC(=O)OC4CCC3C2CCC1(C)C(CCC1(OC(C)=O)C#C)C2CCC3=C4
19-Nor-17-α-pregn-4-en-20-yne-3-β,17-diol diacetate
8080 C. B.
Cervicundin
3-β, 17-β-Diacetoxy-17-α-ethynyl-4-oestrene
3-β,17-β-Diacetoxy-19-nor-17-α-pregn-4-en-20-yne
Ethinodiol diacetate
Ethynodiol acetate
Ethynodiol diacetate
β-Ethynodiol diacetate
17-α-Ethynyl-3,17-dihydroxy-4-estrene diacetate
17-α-Ethynylestr-4-ene-3-β,17-β-diol acetate
17-α-Ethynyl-4-estrene-3-β,17-β-diol diacetate
17-α-Ethynyl-4-estrene-3-β,17-diol diacetate
17-α-Ethynyl-19-norandrost-4-ene-3-β,17-β-diol diacetate
Femulen
Luto-Metrodiol
Metrodiol
Metrodiol diacetate
(3-β,17-α)-19-Norpregn-4-en-20-yne-3,17-diol diacetate
Ovulen 50
SC 11800

297-78-9
$C_9H_4Cl_8O$
411.73
ClC1OC(Cl)C2C1C3(Cl)C(=C(Cl)C2(Cl)C3(Cl)Cl)Cl
4,7-Methanoisobenzofuran, 1,3,4,5,6,7,8,8-octachloro-1,3,3a,4,7,7a-hexahydro
CP 14,957
ENT 25,545
ENT 25,545-X
Isobenzan
Octachloro-hexahydro-methanoisobenzofuran
1,3,4,5,6,8,8-Octachloro-1,3,3a,4,7,7a-hexahydro-4,7-methanoisobenzofuran
1,3,4,5,6,7,8,8-Octachloro-2-oxa-3a,4,7,7a-tetrahydro-4,7-methanoindene
1,3,4,5,6,7,10,10-Octachloro-4,7-endo-methylene-4,7,8,9-tetrahydrophthalan
OMS 206
Omtan
R 6700
SD 4402
Shell 4402
Shell WL 1650
Telodrin
WL 1650

297-90-5
$C_{17}H_{23}NO$
257.37
Racemorphan
Cetarin
DEA No. 9733
Dromoran
1,3,4,9,10,10a-Hexahydro-11-methyl-2H-10,4a-iminoethanophenanthren-6-ol, dl-Mixture
DL-3-Hydroxy-N-methylmorphinan
(+-)-3-Hydroxy-N-methylmorphinan
2H-10, 4a-Iminoethanophenanthren-6-ol, 1,3,4,9,10,10a-hexahydro-11-methyl-, dl-
Methorfinan (Czech)
Methorphinan
N-Methyl-3-hydroxymorphinan
Morphinan, 3-hydroxy-N-methyl-, (+-)-
NU 2206
Racemethorphanum
Racemic dromoran
Racemorfano (Spanish)
Racemorphane (French)
Racemorphanum (Latin)
Ro 1-5431

297-97-2
$C_8H_{13}N_2O_3PS$
248.26
O(P(OCC)(Oc(nccn1)c1)=S)CC
Phosphorothioic acid, O,O-diethyl O-pyrazinyl ester
AC 18133
American Cyanamid 18133
CL 18133
Cynem
O,O-Diaethyl-O-(pyrazin-2yl)-monothiophosphat (German)
O,O-Diaethyl-O-(2-pyrazinyl)-thionophosphat (German)
O,O-Diethyl O,2-pyrazinyl phosphorothioate
Diethyl O-2-pyrazinyl phosphorothionate
O,O-Diethyl O-2-pyrazinyl phosphothionate
O,O-Diethyl O-pyrazinyl thiophosphate
EN 18133
ENT 25,580
Ethyl pyrazinyl phosphorothioate
Experimental Nematocide 18,133
Nemafos
Nemaphos
Nematocide
Phosphorothioic acid, O,O-diethyl O-2-pyrazinyl ester
Pyrazinol, O-ester with O,O-diethyl phosphorothioate
RCRA waste number P040
Thionazin
Thionazine
Zinophos

297-99-4
$C_{10}H_{19}ClNO_5P$
299.72
Phosphoric acid, 2-chloro-3-(diethylamino)-1-methyl-3-oxo-1-propenyl dimethyl ester, (E)
Ent 25,515
ML 97
OR 1191
(E)-Phosphamidon
trans-Phosphamidon

298-00-0
$C_8H_{10}NO_5PS$
263.22
COP(=S)(OC)Oc1ccc(cc1)N(=O)=O
Phosphorothioic acid, O,O-dimethyl O-(p-nitrophenyl) ester
A-Gro
Azofos
Azophos
Bay E-601
Bay 11405
Bladan-M
Cekumethion
Dalf
Devithion
O,O-Dimethyl-O-p-nitrofenyl-ester kyseliny thiofosforecne (Czech)
O,O-Dimethyl-O-(4-nitro-fenyl)-monothiofosfaat (Dutch)
O,O-Dimethyl-O-p-nitrofenyl-thiofosfat (Czech)
Dimethyl p-nitrophenyl monothiophosphate
O,O-Dimethyl-O-(4-nitrophenyl)-monothiophosphat (German)
O,O-Dimethyl O-(p-nitrophenyl) phosphorothioate
O,O-Dimethyl-O-(4-nitrophenyl) phosphorothioate
Dimethyl 4-nitrophenyl phosphorothionate
O,O-Dimethyl O-(p-nitrophenyl) thionophosphate
O,O-Dimethyl-O-(p-nitrophenyl)-thionophosphat (German)
O,O-Dimethyl-O-(4-nitrophenyl)-thionophosphat (German)
Dimethyl-p-nitrophenyl thionphosphate
O,O-Dimethyl O-p-nitrophenyl thiophosphate
Dimethyl parathion
Dimethyl p-nitrophenyl thiophosphate
O,O-Dimetil-O-(4-nitro-fenil)-monotiofosfato (Italian)
Drexel Methyl Parathion 4E
E 601
ENT 17,292

Folidol-80
Folidol M
Fosferno M 50
Gearphos
8056HC
M40 & 80
Me-Parathion
Mepaton
Meptox
Metacid
Metacid 50
Metacide
Metafos
Metaphor
Metaphos
Methyl-E 605
Methyl fosferno
Methyl niran
Methyl parathion (ACGIH, OSHA)
Methyl parathion, Liquid [NA 3018]
Methyl parathion mixture, Dry [NA 2783]
Methylthiophos
Metilparation (Hungarian)
Metron
Metyloparation (Polish)
Metylparation (Czech)
NA 2783 [Methyl parathion solid]
NA 3018 [Methyl parathion liquid]
NCI-C02971
p-Nitrophenyldimethylthionophosphate
Nitrox
Nitrox 80
Oleovofotox
Parapest M-50
Parataf
M-Parathion
Parathion methyl
Parathion methyl Homolog
Parathion-metile (Italian)
Paratox
Parton-M
Partron M
Penncap-M
Phenol, p-nitro-, O-ester with O,O-dimethylphosphorothioate
Phosphorothioic acid, O,O-dimethyl O-(4-nitrophenyl) ester
RCRA waste number P071
Sinafid M-48
Sixty-three Special E.C. Insecticide
Tekwaisa
Thiophenit
Thiophosphate de O,O-dimethyle et de O-(4-nitrophenyle) (French)
Thylpar M-50
Toll
Vertac Methyl Parathion Technisch 80%
Vofatox
Wofatos
Wofatox
Wofotox

298-01-1
$C_7H_{13}O_6P$
224.17
Crotonic acid, 3-hydroxy-, methyl ester, dimethyl phosphate, (E)
Apavinphos
2-Butenoic acid, 3-((dimethoxyphosphinyl)oxy)-, methyl ester, (E)- (9CI)
(E)-Mevinphos
trans-Mevinphos
OS 2046
cis-Phosdrin

298-02-2
$C_7H_{17}O_2PS_3$
260.39
CCOP(=S)(OCC)SCSCC
Phosphorodithioic acid, O,O-diethyl S-(ethylthio)-methyl ester
Aastar
AC 3911
American Cyanamid 3,911
O,O-Diaethyl-S-(aethylthio-methyl)-dithiophosphat (German)
O,O-Diethyl S-ethylmercapto-methyl dithiophosphonate
O,O-Diethyl-S-(ethylthio-methyl)-dithiofosfaat (Dutch)
O,O-Diethyl S-ethylthiomethyl dithiophosphonate
O,O-Diethyl ethylthiomethyl phosphorodithioate
O,O-Diethyl S-(ethylthio-methyl phosphorodithioate
O,O-Diethyl S-ethylthiomethyl thiothionophosphate
O,O-Dietil-S-(etiltio-metil)-ditiofosfato (Italian)
Dithiophosphate de O,O-diethyle et d'ethylthiomethyle (French)
EI3911
ENT 24,042
Experimental Insecticide 3911
Foraat (Dutch)
Granutox
L 11/6
Methanethiol, (ethylthio)-, S-ester with O,O-diethyl phosphorodithioate
Phorat (German)
Phorate (ACGIH,OSHA)
Phorate-10G
Rampart
RCRA waste number P094
Thimenox
Thimet
Timet
Vegfru
Vergfru foratox

298-03-3
$C_8H_{19}O_3PS_2$
258.36
CCOP(=S)(OCC)OCCSCC
Phosphorothioic acid, O,O-diethyl O-(2-(ethylthio)ethyl) ester
Bayer 8169
Demeton
Demeton-O
Diaethylthiophosphorsaeureester des aethylthioglykol (German)
O,O-Diethyl O-(2-eththioethyl)-phosphorothioate
Diethyl 2-eththioethyl thionophosphate
O,O-Diethyl O-2-(ethylthio)-ethyl phosphorothioate
O,O-Diethyl-2-ethylthio ethyl phosphorothioate
Diethyl 2-(ethylthio)ethyl phosphorothionate
Di-Septon
E 1059
Ethanethiol, 2-(ethylthio)-, S-ester with O,O-diethyl phosphorodithioate
Ethanol, 2-(ethylthio)-, O-ester with O,O-diethyl phosphorothioate
Ethylthiometon
Mercaptofos (Russian)
Thiodemeton
Thiolmecaptophos

298-04-4
$C_8H_{19}O_2PS_3$
274.42
CCOP(=S)(OCC)SCCSCC
Phosphorodithioic acid, O,O-diethyl S-(2-(ethylthio)-ethyl) ester
Bay 19639
Bayer 19639
Bay S 276
O,O-Diaethyl-S-(3-thia-pentyl)-dithiophosphat (German)
O,O-Diaethyl-S-(2-aethylthio-aethyl)-dithiophosphat (German)
O,O-Diethyl S-(2-eththioethyl) phosphorodithioate
O,O-Diethyl S-(2-eththioethyl) thiothionophosphate
O,O-Diethyl S-(2-ethylmercap-toethyl) dithiophosphate
O,O-Diethyl-S-(2-ethylthio-ethyl)-dithiofosfaat (Dutch)
O,O-Diethyl 2-ethylthioethyl phosphorodithioate
O,O-Diethyl S-2-(ethylthio)-ethyl phosphorodithioate
O,O-Dietil-S-(2-etiltio-etil)-ditiofosfato (Italian)
Dimaz
Disulfaton
Disulfoton (ACGIH,DOT, OSHA)
Disulfoton mixture, Dry (DOT)
Disulfoton mixture, Liquid (DOT)
Di-Syston
Disystox
Dithiodemeton

Dithiophosphate de O,O-diethyle et de S-(2-ethylthioethyle) (French)
Dithiosystox
ENT 23,437
O,O-Ethyl S-2(ethylthio)ethyl phosphorodithioate
Ethylthiodemeton
S-2-(Ethylthio)ethyl O,O-diethyl ester of phosphorodithioic
acid
Frumin AL
Frumin G
Insyst-D
M-74
NA 2783
Phosphorodithionic acid, S-2-(ethylthio)ethyl-O,O-diethyl ester
RCRA waste number P039
S 276
Solvirex
Thiodemeton
Thiodemetron

298-06-6
$C_4H_{11}O_2PS_2$
186.24
O(P(OCC)(S)=S)CC
Phosphorodithioic acid, O,O-diethyl ester
Kyselina O,O-diethyldithiofosforecna (Czech)

298-07-7
$C_{16}H_{35}O_4P$
322.48
O=P(OCC(CCCC)CC)(OCC(CCCC)CC)O
Phosphoric acid, bis(2-ethylhexyl) ester
Bis(2-ethylhexyl)hydrogen phosphate
Bis(2-ethylhexyl)orthophosphoric acid
Bis(2-ethylhexyl)phosphate
Bis(2-ethylhexyl)phosphoric acid
DEHPA Extractant
Di(2-ethylhexyl)orthophosphoric acid
Di(2-ethylhexyl)phosphate
Di-2(ethylhexyl)phosphoric acid
Di(2-ethylhexyl)phosphoric acid (DOT)
2-Ethyl-1-hexanol hydrogen phosphate
HDEHP
Kyselina di-(2-ethylhexyl)fosforecna (Czech)
NA 1902 [Diisooctyl acid phosphate]

298-12-4
$C_2H_2O_3$
74.04
O=CC(=O)O
Glyoxylic-acid
Kyselina glyoxylova (Czech)

298-14-6
$CH_2O_3.K$
101.13
Potassium bicarbonate
Carbonic acid, monopotassium salt (9CI)
K-Lyte
Monopotassium carbonate

298-18-0
$C_4H_6O_2$
86.10
C1OC1C2CO2
Butane, (+-)-1,2:3,4-diepoxy
(R*,R*)-(+-)-2,2'-Bioxirane
dl-Butadiene dioxide
1,2:3,4-Dianhydro-DL-threitol
D,L-Diepoxybutane
(+-)-1,2:3,4-Diepoxybutane
dl-1,2:3,4-Diepoxybutane
dl-1,2:3,4-Diepoxybutane

298-46-4
$C_{15}H_{12}N_2O$
236.29
NC(=O)N2c1ccccc1C=Cc3ccccc23
5H-Dibenz(b,f)azepine-5-carboxamide
Biston
Carbamazepen
Carbamazepine
Carbamezepine
5-Carbamoyl-5H-dibenz(b,f)azepine
5-Carbamoyldibenzo(b,f)azepine
5-Carbamoyl-5H-dibenzo-(b,f)azepine
5-Carbamyldibenzo(b,f)azepine
5-Carbamyl-5H-dibenzo-(b,f)azepine
Carbazepine
Finlepsin
G 32883
Geigy 32883
Stazepin
Tegretal
Tegretol
Telesmin
Timonil

298-50-0
$C_{23}H_{30}NO_3$
368.49
Propantheline
Ammonium, (2-hydroxyethyl)-diisopropylmethyl-, xanthene-9-carboxylate (ester)
2-Propanaminium, N-methyl-N-(1-methylethyl)-N-(2-((9H-xanthen-9-ylcarbonyl)oxy)ethyl)-
Propanthelinium
Propanthelinum

298-59-9
$C_{14}H_{19}NO_2.ClH$
269.80
[Cl-].COC(=O)C(C1CCCCN1)c2ccccc2
2-Piperidineacetic acid, α-phenyl-, methyl ester, hydrochloride
Centedrin
Methylphenidate hydrochloride
Methylphenidylacetate hydrochloride
Methyl α-phenyl-2-piperidine-acetate hydrochloride
Ritalin
Ritalin hydrochloride

298-81-7
$C_{12}H_8O_4$
216.20
O=C(Oc(c(C=1)cc(c2OC=3)C3)c2OC)C1
7H-Furo(3,2-g)(1)benzopyran-7-one, 9-methoxy
Ammoidin
5-Benzofuranacrylic acid, 6-hydroxy-7-methoxy-, δ-lactone
6-Hydroxy-7-methoxy-5-benzofuranacrylic acid δ-lactone
Meladinin
Meladinine
Meloxine
Methoxa-Dome
Methoxalen
Methoxsalen
8-Methoxy-(furano-3'.2': 6.7-coumarin)
9-Methoxy-7H-furo(3,2-g)benzopyran-7-one
8-Methoxy-2',3',6,7-furocoumarin
8-Methoxy-4',5',6,7-furocoumarin
8-Methoxypsoralen
9-Methoxypsoralen
8-Methoxypsoralene
8-MOP
8-MP
NCI-C55903
Oxsoralen
Oxypsoralen
Proralone-MOP
Xanthotoxin
Xanthoxin

299-27-4
$C_6H_{11}O_7.K$
234.27
[K+].OCC(O)C(O)C(O)C(O)C([O-])=O
Gluconic acid, monopotassium salt, D
D-Gluconic acid, monopotassium salt (9CI)
Gluconic acid potassium salt
Gluconsan K
Kalium-β
Kaon Elixir
Katorin
K-IAO

Potalium
Potasoral
Potassium gluconate
Potassium D-gluconate
Potassuril
Sirokal

299-28-5
$C_{12}H_{22}O_{14}.Ca$
430.42
Calcium-gluconate
Gluconate de calcium (French)

299-29-6
$C_{12}H_{22}O_{14}.Fe$
446.19
[Fe].OCC(O)C(O)C(O)C(O)C([O-])=O
Gluconic acid, iron(2+) salt (2:1)
Fergon
Fergon preparations
Ferlucon
Ferronicum
Ferrous gluconate
Gluco-ferrum
Iromin
Iron gluconate
Irox (Gador)
Nionate
Ray-Gluciron

299-39-8
$C_{15}H_{26}N_2.H_2O_4S$
332.51
Sparteine, sulfate
Siarczan sparteiny (Polish)
Spartocin

299-42-3
$C_{10}H_{15}NO$
165.26
OC(c(cccc1)c1)C(NC)C
Ephedrine, L-(-)
Benzenemethanol, α-(1-(methylamino)ethyl)-, (R-(R*,S*))-
Biophedrin
Eciphin
Efedrin
Ephedral
Ephedrate
Ephedremal
Ephedrin
Ephedrine
l-Ephedrine
L(-)-Ephedrine
Ephedrital
Ephedrol
Ephedrosan
Ephedrotal
Ephedsol
Ephendronal
Ephoxamin
Fedrin
α-Hydroxy-β-methyl amine propylbenzene
1-Hydroxy-2-methylamino-1-phenylpropane
I-Sedrin
Isofedrol
Kratedyn
Manadrin
Mandrin
(-)-α-(1-Methylaminoethyl)-benzyl alcohol
l-α-(1-Methylaminoethyl)-benzyl alcohol
l-2-Methylamino-1-phenyl-propanol
Nasol
Norephedrine, N-methyl-
1-Phenyl-2-methylamino-propanol
Sanedrine
Vencipon
Zephrol

299-45-6
$C_{14}H_{17}O_5PS$
328.34
CCOP(=S)(OCC)Oc2ccc1c(C)cc(=O)oc1c2
Coumarin, 7-hydroxy-4-methyl-, O-ester with O,O-diethyphosphorothioate
Bayer E-838
O,O-Diaethyl-O-(4-methyl-coumarin-7-yl)-monothiophosphat (German)
Diethoxy thiophosphoric acid ester of 7-hydroxy-4-methyl coumarin
O,O-Diethyl O-(2-keto-4-methyl-7-α',β'-benzo-α'-pyranyl) thiophosphate
O,O-Diethyl-O-(4-methyl-coumarin-7-yl)-monothio-fosfaat (Dutch)
O,O-Diethyl O-(4-methyl-7-coumarinyl) phosphoro-thioate
O,O-Diethyl O-(4-methyl-7-coumarinyl) thiono-phosphate
Diethyl methylcoumarinyl thio-phosphate
O,O-Diethyl O-(4-methylcoum-arinyl-7) thiophosphate
O,O-Diethyl-O-(4-methyl-7-kumarinyl) ester kyseliny thiofosforescne (Czech)
O,O-Diethyl O-(4-methylumbel-liferone) ester of thio-phosphoric acid
O,O-Diethyl O-(4-methylum-belliferone) phosphorothioate
Diethyl (4-methylumbelliferyl) thionophosphate
O,O-Dietil-O-(4-metilcumarin-7-il)-monotiofosfato (Italian)
O,O-Dietyl-O-4-methylkumar-inyl(7)tiofosfat (Czech)
E 834
E 838
ENT 17,296
Farbenfabriken Bayer
7-Hydroxy-4-methylcoumarin, O,O-diethyl thiophosphoric acid ester
7-Hydroxy-4-methylcoumarin, O-ester with O,O-diethyl phosphorothioate
Hymecromone O,O-diethyl phosphorothioate
4-Methyl-7-hydroxycoumarin diethoxythiophosphate
4-Methylumbelliferone-O,O-diethyl thiophosphate
Phosphorothioic acid, O,O-diethyl ester, O-ester with 7-hydroxy-4-methyl coumarin
Phosphorothioic acid, O,O-di-ethyl O-(4-methyl-2-oxo-2H-1-benzopyran-7-yl) ester (9CI)
Potasan
Potasan-G-Liquid
Thiophosphate de O,O-diethyle et de O-(4-methyl-7-coum-arinyle) (French)

299-75-2
$C_6H_{14}O_6S_2$
246.32
L-Threitol-1,4-bismethane-sulfonate
CB 2562
NSC-39069
Treitol, 1,4-dimethanesulfonate, (2S,3S)- (8CI)
Treosulfan
Tresulfan
Threitol, 1,4-dimethane-sulfonate

299-84-3
$C_8H_8Cl_3O_3PS$
321.54
COP(=S)(OC)Oc1cc(Cl)c(Cl)cc1Cl
Phosphorothioic acid, O,O-dimethyl O-(2,4,5-tri-chlorophenyl) ester
Dermafosu (Polish)
Dermaphos
O,O-Dimethyl O-2,4,5-trichloro-phenyl phosphorothioate
Dimethyl trichlorophenyl thio-phosphate
O,O-Dimethyl O-(2,4,5-tri-chlorophenyl)thiophosphate
O,O-Dimethyl-O-(2,4,5-tri-chlorphenyl)-thionophosphat (German)
Dow ET 14
Dow ET 57
Ectoral
ENT 23,284
ET 14
ET 57
Etrolene
Fenchloorfos (Dutch)
Fenchlorfos
Fenchlorfosu (Polish)
Fenchlorophos
Fenchlorphos
Karlan

Korlan
Korlane
Nanchor
Nanker
Nankor
OMS 123
Phenol, 2,4,5-trichloro-, O-ester with O,O-dimethyl phosphorothioate
Ronnel (ACGIH,OSHA)
Thiophosphate de O,O-dimethyle et de O-(2,4,5-trichlorophenyle) (French)
O-(2,4,5-Trichloor-fenyl)-O,O-dimethyl-monothiofosfaat (Dutch)
Trichlorometafos
O-(2,4,5-Trichlor-phenyl)-O,O-dimethyl-monothiophosphat (German)
O-(2,4,5-Tricloro-fenil)-O,O-dimetil-monotiofosfato (Italian)
Trolen
Trolene
Viozene

299-85-4
$C_{10}H_{14}Cl_2NO_2PS$
314.18
COP(=S)(NC(C)C)Oc1ccc(Cl)cc1Cl
Phosphoramidothioic acid, isopropyl-, O-(2,4-dichlorophenyl) O-methyl ester
O-(2,4-Dichlorophenyl) O-methyl isopropylphosphoramidothioate
O-(2,4-Dichlorophenyl) O-methyl N-isopropylphosphoramidothioate
DMPA
Dow 1329
DOWCO 118
ENT 25,647
(1-Methylethyl)phosphoramidothioic acid O-(2,4-dichlorophenyl) O-methyl ester
Isopropylphosphoramidothioic acid, O-2,4-dichlorophenyl O-methyl ester
K 22023
O-Methyl-O-(2,4-dichlorfenyl)-ester kyseliny isopropylamidothiofosforecne (Czech)
OMS 115
Phenol, 2,4-dichloro-, O-ester with O-methyl isopropylphosphoramidothioate
Zytron

299-86-5
$C_{12}H_{19}ClNO_3P$
291.74
CNP(=O)(OC)Oc1ccc(cc1Cl)C(C)(C)C
Phosphoramidic acid, methyl-, 4-tert-butyl-2-chlorophenyl methyl ester
Amidofos
Amidophos
4-t-Butyl-2-chlorophenyl methyl methylphosphoramidate
O-(4-tert Butyl-2-chloor-fenyl)-O-methyl-fosforzuur-N-methyl-amide (Dutch)
4-tert. Butyl 2-chlorophenyl methylphosphoramidate de methyle (French)
O-(4-tert-Butyl-2-chlorophenyl) O-methyl N-methylamido phosphate
O-(4-tert-Butyl-2-chlor-phenyl)-O-methyl-phosphorsaeure-N-methylamid (German)
O-(4-terz.-Butil-2-cloro-fenil)-O-metil-fosforammide (Italian)
Crufomat
Crufomate (ACGIH,OSHA)
Crufomate A
DOWCO 132
ENT 25,602-X
M-1261
O-Methyl O-2-chloro-4-tert-butylphenyl N-methylamidophosphate
Methylphosphoramidic acid, 4-t-butyl-2-chlorophenyl methyl ester
Montrel
Phenol, 4-t-butyl-2-chloro-, ester with methyl methylphosphoramidate
Phosphoramidic acid, 4-tert-butyl-2-chlorophenylphosphoramidate
Phosphoramidic acid, methyl-, 2-chloro-4-(1,1-dimethylethyl)phenyl methyl ester
Ruelene
Ruelene Drench
Ruelene 25E
Rulene

300-39-0
$C_9H_9I_2NO_3$
432.98
O=C(O)C(N)Cc(cc(c(O)c1I)I)c1
L-Tyrosine, 3,5-diiodo- (9CI)
3,5-Diiodotyrocine
3,5-Diiodo-L-tyrosine
3,5-L-Diiodotyrosine
3,5-Iodo-L-tyrosine
NSC-4143
Tryosine, 3,5-diiodo-
Tyrosine, 3,5-diiodo-, L-(8CI)

300-42-5
$C_{10}H_{15}N.ClH$
185.72
Phenethylamine, N,α-dimethyl-, hydrochloride
A 884
Amdram
Amedrine
Amphedroxy
Amphedroxyn
Apamine
Bombita
C 6379
Corvitin
Daropervamin
Dea Oxo-5
Deofed
Deoxyephedrine
Depoxin
Desamine
Desfedran
Desfedrin
Desossiefedrina
Des-Oxa-D
Desoxedrine
Desoxin
Desoxo-5
Desoxyephedrine hydrochloride
Desoxyfed
Desoxyn
Desoxyphed
Destim
Desyphed
Desyphen
Detrex
Dexophrine
Dexosyn
Dexoval
Dexstim
Dextim
N,α-Dimethylphenethylamine hydrochloride
Doe
Dopidrin
Doxephin
Doxephrin
Doxyfed
Drinalfa
Effroxine
Efroxine
Estimulex
Eufodrin
Eufodrinal
Euphodrin
Euphodrinal
914F
Fenyprin
Gerobit
Gerovit
Heropon
Hiropon
Isophan
Isophen
Kemodrin
Lanazine
Levetamin
Madrine
Mepho-D
Metamfetamina
Metamina
Metamine
Metamphetamin
Metamsustac
Metanfetamina
Methampex
Methamphetamine hydrochloride
Methamphin
Methedrinal
Methedrine
Methedrine hydrochloride
Methoxyn
Methylamphetamine
Methylamphetamine hydro-

chloride
Methylbenzedrin
Methylisamin
Methylisomin
Methylisomyn
N-Methyl-β-phenylisopropyl-aminhydrochlorid (German)
Methylpropamine
Miller Drine
Neodrine
Neopharmedrine
Noradrin
Normadrine
Norodin
Norodrin
Oxydess
Oxydrene
Oxydrin
Oxyfed
Pervitin
Phedoxe
Phedrisox
Philopon
Pisichergina
Premodrin
Psichergina
Psicopan
Psiquergina
Psychergine
Psykoton
Semoxydrine
Soxysympamine
Speed
Stimdex
Stimulex
Syndrox
Tonedrin
Tonedron
Vonedrine

300-57-2
C_9H_{10}
118.19
c(cccc1)(c1)CC=C
Benzene, allyl (8CI)
Allylbenzene
Benzene, 2-propenyl- (9CI)
NSC-18609
1-Propene, 3-phenyl-
2-Propenylbenzene

300-62-9
$C_9H_{13}N$
135.23
NC(C)Cc(cccc1)c1
Phenethylamine, α-methyl, (+-)
Actedron
Adipan
Allodene
dl-Amphetamine
Anorexide
Benzedrine
(+-)-Benzedrine
dl-Benzedrine
(+-)-Desoxynorephedrine
racemic-Desoxynor-ephedrine
Elastonon
Fenylo-izopropylaminyl (Polish)
Isoamycin
Isomyn
Mecodrin
α-Methylbenzeneethaneamine
dl-α-Methylphenethylamine
(+-)-α-Methylphenethylamine
Norephedrane
Norephedrine, deoxy-
Novydrine
Ortedrine
Phenedrine
dl-1-Phenyl-2-aminopropane
Profamina
Propisamine
Psychedrine
Raphetamine
Simpatedrin
Sympamine
Sympatedrine
Weckamine

300-76-5
$C_4H_7Br_2Cl_2O_4P$
380.80
COP(=O)(OC)OC(Br)C(Cl)(Cl)Br
Phosphoric acid, 1,2-dibromo-2,2-dichloroethyl dimethyl ester
Arthodibrom
Bromchlophos
Bromex
Dibrom
O-(1,2-Dibrom-2,2-dichlor-aethyl)-O,O-dimethyl-phosphat (German)
1,2-Dibromo-2,2-dichloroethyl dimethyl phosphate
O-(1,2-Dibromo-2,2-dicloro-etil)-O,O-dimetil-fostato (Italian)
O-(1,2-Dibroom-2,2-dichloor-ethyl)-O,O-dimethyl-fosfaat (Dutch)
O,O-Dimethyl-O-(1,2-dibrom-2,2-dichlor-aethyl)-phosphat (German)
Dimethyl-1,2-dibromo-2,2-dichloroethyl phosphate (OSHA)
O,O-Dimethyl-O-(1,2-dibromo-2,2-dichloroethyl)phosphate
O,O-Dimethyl O-2,2-dichloro-1,2-dibromoethyl phosphate
ENT 24,988
Ethanol, 1,2-dibromo-2,2-dichloro-, dimethyl phosphate
Fosforan O-1,2-dwubromo-2,2-dwuchloroetylo-O,O-dwumetylowy (Polish)
Hibrom
Naled (ACGIH)
Naledu (Polish)
OMS 75
Ortho 4355
Orthodibrom
Orthodibromo
Phosphate de O,O-dimethle et de O-(1,2-dibromo-2,2-dichlorethyle) (French)
RE-4355

300-85-6
$C_4H_8O_3$
104.11
O=C(O)CC(O)C
3-Hydroxybutyric acid
AI3-21675
Butanoic acid, 3-hydroxy- (9CI)
Butyric acid, 3-hydroxy- (8CI)
3 HBA
3-Hydroxybutanoic acid
β-Hydroxy-n-butyric acid
NSC-3806

300-92-5
$C_{36}H_{17}AlO_5$
556.51
Aluminum, hydroxybis-(stearato)
Aluminum distearate (ACGIH)
Aluminum hydroxide distearate
Aluminum, hydroxybis(octadecanoato-O)- (9CI)
Aluminum hydroxydistearate
Special M

301-00-8
$C_{19}H_{32}O_2$
292.46
O=C(OC)CCCCCCCC=CCC=CCC=CCC
9,12,15-Octadecatrienoic acid, methyl ester, (Z,Z,Z)-
AI3-26935

301-02-0
$C_{18}H_{35}NO$
281.48
O=C(N)CCCCCCCC=CCCCCCCCC
(Z)-9-Octadecenamide
Adogen 73
AI3-36742
Armoslip CP
Crodamide O
Crodamide OR
NSC-26987
9-Octadecenamide
9-Octadecenamide, (Z)- (9CI)
9-Octadecenoic acid, amide(cis)
Oleamide
Oleic acid amide
Oleyl amide
Oleylamide
Slip-eze

301-04-2
$C_4H_6O_4.Pb$
325.29
[Pb+2].CC([O-])=O.CC([O-])=O
Acetic acid, lead(2+) salt
Acetate de plomb (French)
Bleiacetat (German)
Dibasic lead acetate
Lead acetate [UN 1616]
Lead(2+) acetate
Lead(II) acetate
Lead diacetate
Lead dibasic acetate

Normal lead acetate
Plumbous acetate
RCRA waste number U144
Salt of Saturn
Sugar of Lead
UN 1616 [Lead acetate]

301-08-6
$C_8H_{16}O_2$.1/2Pb
247.81
Hexanoic acid, 2-ethyl-, lead(2+) salt (9CI)
Lead 2-ethylhexanoate
Lead(2+) 2-ethylhexanoate

301-10-0
$C_8H_{16}O_2$.1/2Sn
203.56
Hexanoic acid, 2-ethyl-, tin-(2+) salt (9CI)
2-Ethylhexanoic acid stannous salt
NSC-75857
Nuocure 28
Stannous 2-ethylhexanoate
Stannous 2-ethylhexoate
Stannous octoate
Tin(II) bis(2-ethylhexanoate)
Tin dioctoate
Tin ethylhexanoate
Tin 2-ethylhexanoate
Tin(II) 2-ethylhexanoate
Tin(2+) 2-ethylhexanoate
Tin(II) 2-ethylhexylate
Tin octoate

301-11-1
$C_{15}H_{27}NO_2S$
285.49
CCCCCCCCCCCC(=O)OCCS C#N
Lauric acid, 2-thiocy-anatoethyl ester
Dodecanoic acid 2-thio-cyanatoethyl ester
ENT 5
Lauric acid ester with 2-hydroxyethyl thiocyanate
Lethane 60
Thiocyanic acid, 2-hydroxy-ethyl ester, laurate
2-Thiocyanoethyl coconate
2-Thiocyanoethyl dodecanoate
β-Thiocyanoethyl laurate
2-Thiocyanoethyl laurate
2-Thiokyanatoethylester kysel-iny laurove (Czech)

301-12-2
$C_6H_{15}O_3PS_2$
230.30
CCS(=O)CCSP(=O)(OC)OC
Phosphorothioic acid, S-(2-(ethylsulfinyl)ethyl) O,O-dimethyl ester
Aimco Systox
Bay 21097
Bayer 21097
Demeton-S-methyl-sulfoxid (German)
Demeton-S-methyl sulfoxide
Demeton-methyl sulphoxide
O,O-Dimethyl-S-(2-aethyl-sulfinyl-aethyl)-thiolphosphat (German)
O,O-Dimethyl S-(2-eththionyl-ethyl) phosphorothioate
Dimethyl S-(2-eththionylethyl) thiophosphate
O,O-Dimethyl S-(2-(ethyl-sulfinyl)ethyl) phosphoro-thioate
O,O-Dimethyl-S-(2-ethylsulfin-yl-ethyl)-monothiofosfaat (Dutch)
O,O-Dimethyl S-(2-ethyl-sulfinyl)ethyl thiophosphate
O,O-Dimethyl S-ethylsulphinyl-ethyl phosphorothiolate
O,O-Dimethyl-S-(3-oxo-3-thia-pentyl)-monothiophosphat (German)
O,O-Dimetil-S-(2-etil-solfinil-etil)-monotiofosfato (Italian)
ENT 24,964
Ethanethiol, 2-(ethylsulfinyl)-, S-ester with O,O-dimethyl phosphorothioate
S-(2-(Ethylsulfinyl)ethyl) O,O-dimethyl phosphoro thioate
Isomethylsystox sulfoxide
Metaisosystoxsulfoxide
Metasystemox
Metasystemox R
Metasystox-R
Methyl demeton-O-sulfoxide
Metilmercaptofosoksid
Oxydemetonmethyl
Oxydemeton-metile (Italian)
Phosphorothioic acid, O,O-di-methyl S-(2-(ethylsulfinyl)-ethyl) ester
R 2170
Thiophosphate de O,O-di-methyle et de S-2-ethylsulf-inylethyle (French)

301-13-3
$C_{24}H_{51}O_3P$
418.72
O(P(OCC(CCCC)CC)OCC(CC CC)CC)CC(CCCC)CC
Phosphorous acid, tris(2-ethyl-hexyl) ester
Tris(2-ethylhexyl)phosphite

301-19-9
$C_{33}H_{40}O_{19}$
740.73
4H-1-Benzopyran-4-one, 3-((6-O-(6-deoxy-α-l-mannopyr-anosyl)-β-d-galactopyran-osyl) oxy)-7-((6-deoxy-α-l-mannopyranosyl)oxy)-5-hydroxy-2-(4-hydroxy-phenyl)
Kaempferol-3-O-gal-rham-7-O-rham
Robinin

302-01-2
H_4N_2
32.06
NN
Hydrazine (ACGIH,OSHA)
Anhydrous hydrazine [UN 2029]
Diamide
Diamine
Hydrazine, Anhydrous [UN 2029]
Hydrazine, Aqueous solution [UN 2030]
Hydrazine, Aqueous solution containing more than 64% hydrazine [UN 2029]
Hydrazine Base
Hydrazyna (Polish)
RCRA waste number U133
UN 2029 [Hydrazine, anhyd-rous or Hydrazine aqueous solutions with more than 64 per cent hydrazine, by mass]
UN 2030 [Hydrazine hydrate or Hydrazine aqueous solutions, with not more than 64 per cent hydrazine, by mass]

302-04-5
CNS
58.08
Thiocyanate
Thiocyanate ion
Thiocyanate ion (1-)

302-17-0
$C_2HCl_3O.H_2O$
165.40
OC(O)C(Cl)(Cl)Cl
Chloral-hydrate
Aquachloral
Bi 3411
Chloraldurat
Dormal
1,1-Ethanediol, 2,2,2-trichloro-(9CI)
Felsules
HS
Hydral
Hydrate de chloral
Kessodrate
Lorinal
Noctec
Nortec
Nycoton
Nycton
Phaldrone
Rectules
Sk-Chloral hydrate
Somni Sed
Somnos
Sontec
Tosyl
Trawotox
Trichloracetaldehyd-hydrat (German)

Trichloroacetaldehyde hydrate
Trichloroacetaldehyde monohydrate
1,1,1-Trichloro-2,2-ethanediol
2,2,2-Trichloro-1,1-ethanediol

302-22-7
$C_{23}H_{29}ClO_4$
404.97
CC(=O)OC3(CCC4C2C=C(Cl)C1=CC(=O)CCC1(C)C2CCC34C)C(C)=O
Pregna-4,6-diene-3,20-dione, 6-chloro-17-hydroxy-, acetate
17-Acetoxy-6-chloro-6-dehydroprogesterone
17-α-Acetoxy-6-chloro-6-dehydroprogesterone
17-α-Acetoxy-6-chloro-6,7-dehydroprogesterone
17-α-Acetoxy-6-chloro-4,6-pregnadiene-3,20-dione
17-α-Acetoxy-6-chloropregna-4,6-diene-3,20-dione
17-(Acetyloxy)-6-chloropregna-4,6-diene-3,20-dione
CAP
Chlormadinon acetate
Chlormadinone acetate
Chlormadinonu (Polish)
6-Chloro-17-α-acetoxy-4,6-pregnadiene-3,20-dione
δ[6]-6-Chloro-17-α-acetoxyprogesterone
6-Chloro-δ[6]-17-acetoxyprogesterone
6-Chloro-δ[6]-(17-α)acetoxyprogesterone
6-Chloro-δ[6]-dehydro-17-acetoxyprogesterone
6-Chloro-6-dehydro-17-α-acetoxyprogesterone
6-Chloro-6-dehydro-17-α-hydroxyprogesterone acetate
6-Chloro-17-α-hydroxypregna-4,6-diene-3,20-dione acetate
6-Chloro-17-α-hydroxy-δ[6]-progesterone acetate
Chloromadinone acetate
6-Chloro-δ[4,6]-pregnadiene-17-α-ol-3,20-dione 17-acetate
6-Chloro-pregna-4,6-dien-17-α-ol-3,20-dione acetate
Clordion
CMA
C-Quens
6-Dehydro-6-chloro-17-α-acetoxyprogesterone
Lormin
Lutinyl
NSC-92338
Pregna-4,6-diene-3,20-dione, 17-(acetoxy)-6-chloro-
RS 1280
Skedule
ST 155

302-27-2
$C_{34}H_{47}NO_{11}$
645.82
CCN3CC1(COC)C(O)CC(OC)C24C5CC6(O)C(OC)C(O)C(OC(C)=O)(C(C(OC)C12)C34)C5C6OC(=O)c
Aconitine (Crystalline)
Aconitane
Aconitin Cristallisat (German)
Aconitine

302-31-8
$C_{17}H_{19}NO_3.1/2C_4H_6O_6$
360.38
Morphinan-3,6-diol, 7,8-didehydro-4,5-epoxy-17-methyl-(5α,6α)-, (R-(R*,R*))-2,3-dihydroxybutanedioate (2:1) (Salt) (9CI)
Morphinan-3,6α-diol, 7,8-didehydro-4,5α-epoxy-17-methyl-, tartrate (2:1) (Salt) (8CI)
Morphine tartrate

302-40-9
$C_{20}H_{25}NO_3$
327.46
CCN(CC)CCOC(=O)C(O)(c1ccccc1)c2ccccc2
Benzilic acid, 2-(diethylamino)ethyl ester
Benactyzin
Benactizina (Italian)
Benactyzine
Benzeneacetic acid, α-hydroxy-α-phenyl-, 2-(diethylamino)ethyl ester (9CI)
Benzilic acid β-diethylaminoethyl ester
Diazil
Diethylaminoethyl benzilate
β-Diethylaminoethyl benzilate
2-(Diethylamino)ethyl benzilate
2-(Diethylamino)ethyl diphenylglycolate
Diphenylglycolic acid 2-(diethylamino)ethyl ester
α-Hydroxy-α-phenylbenzeneacetic acid 2-(diethylamino)ethyl ester

302-41-0
$C_{27}H_{34}N_4O$
430.57
Pirinitramide
A65
(1,4'-Bipiperidine)-4'-carboxamide, 1'-(3-cyano-3,3-diphenylpropyl)-
1'-(3-Cyano-3,3-diphenylpropyl)(1,4'-bipiperidine)-4'-carboxamide
DEA No. 9642
2,2-Diphenyl-4-(4-piperidino-4-carbamoylpiperidino)butyronitrile
Dipidolor
Dipiritramide
Piridolan
Piritramida (Spanish)
Piritramide
Piritramidum (Latin)
R 3365

302-70-5
$C_5H_{11}Cl_2NO.ClH$
208.53
[Cl-].C[N+](O)(CCCl)CCCl
Diethylamine, 2,2'-dichloro-N-methyl-, N-oxide, hydrochloride
Chlormethine-N-oxide hydrochloride
2-Chloro-N-(2-chloroethyl)-N-methylethanamine-N-oxide hydrochloride
2,2'-Dichlorodiethylmethylamine oxide
2,2'-Dichloro-N-methyldiethylamine N-oxide hydrochloride
Ethanamine, 2-chloro-N-(2-chloroethyl)-N-methyl-, N-oxide, hydrochloride (9CI)
HN2 Oxide Hydrochloride
MBAO Hydrochloride
Mechlorethamine oxide hydrochloride
Methyl-bis-(β-chloraethyl)-amin-N-oxyd-hydrochlorid (German)
Methylbis(β-chloroethyl)amine N-oxide hydrochloride
N-Methylbis(2-chloroethyl)-amine N-oxide hydrochloride
N-Methyl-2,2'-dichlorodiethylamine N-oxide hydrochloride
Methyldi(2-chloroethyl)amine N-oxide hydrochloride
Mitomen
Mustron
Nitrogen mustard N-oxide hydrochloride
Nitromin hydrochloride
NSC-10107
Ossiamina
Ossichlorin
Oxyamine
SK-598
XA 2

302-79-4
$C_{20}H_{28}O_2$
300.48
O=C(O)C=C(C=CC=C(C=CC(=C(CCC1)C)C1(C)C)C)C
Retinoic acid, all-trans
Aberel
3,7-Dimethyl-9-(2,6,6-trimethyl-1-cyclohexen-1-yl)-2,4,6,8-nonatetraenoic acid
2,4,6,8-Nonatetranoic acid, 3,7-dimethyl-9-(2,6,6-trimethyl-1-cyclohexen-1-yl)-
NSC-122758
β-RA
Retin-A
Retinoic acid
β-Retinoic acid
trans-Retinoic acid
all-trans-Retinoic acid

Tretinoin
β-all-trans-Retinoic acid
Vitamin A acid

303-07-1
$C_7H_6O_4$
154.12
O=C(O)c(c(O)ccc1)c1O
Benzoic acid, 2,6-dihydroxy- (9CI)
2-Carboxyresorcinol
2,6-Dihydroxybenzoic acid
6-Hydroxysalicylic acid
NSC-49172
γ-Resorcylic acid (8CI)
2,6-Resorcylic acid

303-21-9
$C_{10}H_{12}N_2O_5$
240.24
O=N(=O)c(c(c(N(=O)=O)c(O)c1C(C)C)C)c1
Thymol, 2,6-dinitro
m-Cresol, 2,4-dinitro-6-isobropyl-
2,6-Dinitrothymol
Dinitrothymol 1-2-4 (French)

303-33-3
$C_{16}H_{27}NO_5$
313.44
COC(C)C(O)(C(C)C)C(=O)OC C1=CCN2CCC(O)C12
Heliotrine
Heliotron

303-34-4
$C_{21}H_{33}NO_7$
411.55
COC(C)C(O)(C(=O)OCC1=CCN 2CCC(OC(=O)C(C)=CC)C12) C(C)(C)O
Lasiocarpine
Heliotridine ester with lasiocarpum and angelic acid
NCI-C01478
RCRA waste number U143

303-38-8
$C_7H_6O_4$
154.12
OC(=O)c1cccc(O)c1O
2-Pyrocatechuic acid
Benzoic acid, 2,3-dihydroxy- (9CI)
Catecholcarboxylic acid
2,3 DHB
DOBK
3-Hydroxysalicylic acid
NSC-27435
Pyrocatechuic acid
o-Pyrocatechuic acid (8CI)

303-47-9
$C_{20}H_{18}ClNO_6$
403.84
CC3Cc2c(Cl)cc(C(=O)NC(Cc1cc ccc1)C(O)=O)c(O)c2C(=O)O3
Alanine, N-((5-chloro-8-hydroxy-3-methyl-1-oxo-7-isochromanyl)carbonyl)-3-phenyl-, (-)
N-(((3R)-5-Chloro-8-hydroxy-3-methyl-1-oxo-7-isochromanyl)carbonyl)-3-phenyl-l-alanine
(-)-N-((5-Chloro-8-hydroxy-3-methyl-1-oxo-7-isochromanyl)carbonyl)-3-phenylalanine
NCI-C56586
Ochratoxin A

303-49-1
$C_{19}H_{23}ClN_2$
314.89
CN(C)CCCN2c1ccccc1CCc3ccc (Cl)cc23
5H-Dibenz(b,f)azepine, 10,11-dihydro-3-chloro-5-(3-(dimethylamino)propyl)
Anafranil
Chlorimipramine
3-Chloro-5-(3-(dimethylamino)-propyl)-10,11-dihydro-5H-dibenz(b,f)azepine
3-Chloroimipramine
CIM
Clomipramine
5H-Dibenz(b,f)azepine, 3-chloro-5-(3-(dimethylamino)propyl)-10,11-dihydro-
Monochlorimipramine

304-17-6
$C_{11}H_{11}NO_2$
189.23
O=C(N(C(=O)c1cccc2)C(C)C)c12
Phthalimide, N-isopropyl
N-Isopropilftalimmide (Italian)
N-Isopropylftalimid (Czech)

304-20-1
$C_8H_8N_4.ClH$
196.66
[Cl-].NNc1nncc2ccccc12
Phthalazine, 1-hydrazino-, monohydrochloride
Aiselazine
Appresinum
Aprelazine
Apresazide
Apresine
Apresolin
Apresoline
Apresoline-Esidrix
Apresoline HCl
Apresoline hydrochloride
Apressin
Apressoline
Aprezolin
BA 5968
C-5968
Ciba 5968
Dralzine
Hidralazin
Hipoftalin
Hydralazine chloride
Hydralazine HCl
Hydralazine hydrochloride
Hydralazine monohydrochloride
Hydrallazine hydrochloride
Hydrapress
1-Hydrazinophthalazine hydrochloride
1-Hydrazinophthalazine monohydrochloride
Hyperazin
Hyperazine
Hypophthalin
Hypos
Ipolina
Lopres
Lopress
Nor-Press 25
1(2H)-Phthalazinone, hydrazone hydrochloride
1(2H)-Phthlazinone, hydrazone, monohydrochloride
Praparat 5968
Rolazine
Serpasil Apresoline No. 2

304-59-6
$C_4H_6O_6.K.Na$
212.18
Potassium sodium tartrate
Butanedioic acid, 2,3-dihydroxy-, monopotassium monosodium salt
Butanedioic acid, 2,3-dihydroxy- (R-(R*,R*))-, monopotassium monosodium salt (9CI)
2,3-Dihydroxybutanedioic acid, monopotassium monosodium salt
Monopotassium monosodium tartrate
Rochelle Salt
Seignette Salt
Sodium potassium tartrate
Sodium potassium (dl) tartrate
Tartaric acid, monopotassium monosodium salt

305-01-1
$C_9H_6O_4$
178.15
Coumarin, 6,7-dihydroxy
Aesculetin
Asculetine
2H-1-Benzopyran-2-one, 6,7-dihydroxy- (9CI)
Cichorigenin
Cichoriin aglycon
6,7-Dihydroxycoumarin
Esculetin
Esculetol
Esculin aglycon

305-03-3
$C_{14}H_{19}Cl_2NO_2$
304.24

OC(=O)CCCc1ccc(cc1)N(CCCl)CCCl
Butyric acid, 4-(p-bis-(2-chloroethyl)aminophenyl)
Ambochlorin
Amboclorin
Benzenebutanoic acid, 4-(bis-(2-chloroethyl)amino)-
4-(Bis(2-chloroethyl)amino)-benzenebutanoic acid
γ-(p-Bis(2-chloroethyl)amino-phenyl)butyric acid
4-(p-(Bis(2-chloroethyl)amino)-phenyl)butyric acid
4(p-Bis(β-chloroethyl)amino-phenyl)butyric acid
Butanoic acid, 4-(bis(2-chloro-ethyl)amino)benzene-
CB 1348
Chlorambucil
Chloraminophen
Chloraminophene
Chloroambucil
Chlorobutin
Chlorobutine
N,N-Di-2-chloroethyl-γ-p-aminophenylbutyric acid
p-(N,N-Di-2-chloroethyl)amino-phenyl butyric acid
p-N,N-Di-(β-chloroethyl)amino-phenyl butyric acid
γ-(p-Di(2-chloroethyl)amino-phenyl)butyric acid
Ecloril
Elcoril
Kyselina 4-(N,N-bis-(2-chlor-ethyl)-p-aminofenyl)maselna (Czech)
Leukeran
Leukersan
Leukoran
Linfolizin
Linfolysin
NCI-C03485
NSC-3088
Phenylbuttersaeure-lost (German)
Phenylbutyric acid nitrogen mustard
RCRA waste number U035

305-85-1
$C_6H_3I_2NO_3$
390.90
O=N(=O)c(cc(c(O)c1I)I)c1
Phenol, 2,6-diiodo-4-nitro
Ancylol
2,6-Diiodo-4-nitrophenol
Diisophenol
Disofen
DNP
Disophenol

306-08-1
$C_9H_{10}O_4$
182.18
O=C(O)Cc(ccc(O)c1OC)c1
Homovanillic acid
Acetic acid, (4-hydroxy-3-meth-oxyphenyl)- (8CI)
Benzeneacetic acid, 4-hydroxy-3-methoxy- (9CI)
HMPA
HVA
4-Hydroxy-3-methoxybenzene-acetic acid
(4-Hydroxy-3-methoxyphenyl)-acetic acid
NSC-16682
Vanillacetic acid

306-37-6
$C_2H_8N_2.2ClH$
133.04
Cl.Cl.CNNC
Hydrazine, 1,2-dimethyl-, di-hydrochloride
N,N'-Dimethylhydrazine di-hydrochloride
sym-Dimethylhydrazine di-hydrochloride
1,2-Dimethylhydrazine dihydro-chloride
DMH

306-40-1
$C_{14}H_{30}N_2O_4$
290.46
[OH].[OH].C[N+](C)(C)CCOC(=O)CCC(=O)OCC[N+](C)(C)C
Choline, succinate (2:1) (ester)
Anectine
Choline, succinate (ester)
Choline, succinyl-
Diacetylcholine
Dicholine succinate
Ditilin
Ditiline
Ethanaminium, 2,2'-((1,4-dioxo-1,4-butanediyl)bis(oxy))bis-(N,N,N-trimethyl-
Quelicin
Succinic acid, diester with choline
Succinocholine
Succinoylcholine
Succinylbischoline
Succinyl choline
Succinyldicholine
Suxamethonium
Suxemethonium

306-44-5
$C_3H_5NO_2$
87.09
Pyruvaldehyde, 1-oxime
Acetone, 3-hydroxyimino-
3-Hydroxyiminoacetone
Isonitrosoacetone
1,2-Propanedione, 1-oxime
Pyruvaldehydoxim (Czech)

306-52-5
$C_2H_4Cl_3O_4P$
229.38
O=P(OCC(Cl)(Cl)Cl)(O)O
Ethanol, 2,2,2-trichloro-, di-hydrogen phosphate
Phosphoric acid, 2,2,2-tri-chloroethyl ester
SCH 10159
Trichloroethyl phosphate
2,2,2-Trichloroethyl phosphate
Triclofos
Triclos

306-83-2
$C_2HCl_2F_3$
152.93
FC(F)(F)C(Cl)Cl
Ethane, 2,2-dichloro-1,1,1-tri-fluoro
2,2-Dichloro-1,1,1-trifluoro-ethane
1,1,1-Trifluoro-2,2-dichloro-ethane

307-34-6
C_8F_{18}
438.08
FC(F)(F)C(F)(F)C(F)(F)C(F)(F)C(F)(F)C(F)(F)C(F)(F)C(F)(F)F
Octane, octadecafluoro
Octadecafluorooctane
Perfluorooctane
n-Perfluorooctane

307-35-7
$C_8F_{18}O_2S$
502.12
O=S(=O)(F)C(F)(F)C(F)(F)C(F)(F)C(F)(F)C(F)(F)C(F)(F)C(F)(F)C(F)(F)F
Perfluorooctylsulfonyl fluoride
1,1,2,2,3,3,4,4,5,5,6,6,7,7,8,8,8-Heptadecafluoro-1-octanesulf-onyl fluoride
1-Octanesulfonyl fluoride, 1,1,2,2,3,3,4,4,5,5,6,6,7,7,8,8,8-heptadecafluoro- (9CI)
Perfluorooctanesulfonyl fluoride
N-Perfluorooctanesulfonyl fluoride

309-00-2
$C_{12}H_8Cl_6$
364.90
ClC3=C(Cl)C4(Cl)C2C1CC(C=C1)C2C3(Cl)C4(Cl)Cl
1,4:5,8-Dimethanonaphthal-ene, 1,4,4a,5,8,8a-hexa-hydro-1,2,3,4,10,10-hexa-chloro-, endo, exo mixture (More than 60% aldrin)
Aldrex
Aldrex 30
Aldrex 30 E.C.
Aldrin (ACGIH,DOT,OSHA)
Aldrin, Liquid [NA 2762]
Aldrin, Cast solid (DOT)
Aldrin, Solid [NA 2761]
Aldrine (French)
Aldrite
Aldrosol
Altox

Compound 118
1,4:5,8-Dimethanonaphthalene, 1,2,3,4,10,10-hexachloro-1,4,4a,5,8,8a-hexahydro-, endo,exo
1,4:5,8-Dimethanonaphthalene, 1,4,4a,5,8,8a-hexahydro-1,2,3,4,10,10-hexachloro-, endo, exo mixture (65% or less aldrin)
Drinox
ENT 15,949
Hexachlorohexahydro-endo-exo-dimethanonaphthalene
1,2,3,4,10,10-Hexachloro-1,4,4a,5,8,8a-hexahydro-1,4,5,8-dimethanonaphthalene
1,2,3,4,10,10-Hexachloro-1,4,4a,5,8,8a-hexahydro-exo-1,4-endo-5,8-dimethanonaphthalene
1,2,3,4,10,10-Hexachloro-1,4,4a,5,8,8a-hexahydro-1,4-endo-exo-5,8-dimethanonaphthalene
HHDN
Latka 118 (Czech)
NA 2761 [Aldrin, solid]
NA 2762 [Aldrin, liquid]
NCI-C00044
Octalene
RCRA waste number P004
Seedrin

309-29-5
$C_{24}H_{30}N_2O_2$
378.56
CCN2CC(CCN1CCOCC1)C(C2=O)(c3ccccc3)c4ccccc4
2-Pyrrolidinone, 1-ethyl-4-(2-morpholinoethyl)-3,3-diphenyl
AHR-619
Dopram
Doxapram
1-Ethyl-4-(2-morpholinoethyl)-3,3-diphenyl-2-pyrrolidinone
2-Pyrrolidinone, 1-ethyl-4-(2-(4-morpholinyl)ethyl)-3,3-diphenyl- (9CI)

309-36-4
$C_{14}H_{18}N_2O_3$.Na
285.33
CCC#CC(C)C1(CC=C)C(=O)NC(=O)N(C)C1=O
Barbituric acid, 5-allyl-1-methyl-5-(1-methyl-2-pentynyl)-, sodium salt
5-Allyl-1-methyl-5-(1-methyl-2-pentynyl)barbituric acid sodium salt
Barbituric acid, 5-allyl-1-methyl-5-(1-methyl-2-pentynyl)-sodium deriv.
Brevimytal
Brevital
Brevital sodium
Brietal sodium
Enallynymal sodium
Lilly 22451
Methohexital
Methohexital sodium
Methohexitone sodium
1-Methyl-5-allyl-5-(1-methyl-2-pentynyl)barbituric acid sodium salt
Sodium-dl-5-allyl-1-methyl-5-(1-methyl-2-pentynyl)barbiturate
Sodium methohexital
Sodium methohexitone
Sodium a-dl-1-methyl-5-allyl-5-(1-methyl-2-pentynyl)-barbiturate

311-45-5
$C_{10}H_{14}NO_6P$
275.22
CCOP(=O)(OCC)Oc1ccc(cc1)N(=O)=O
Phosphoric acid, diethyl p-nitrophenyl ester
Chinorta
O,O'Diaethyl-p-nitrophenylphosphat (German)
Diaethyl-p-nitrophenylphosphorsaeureester (German)
Diethyl-p-nitrofenyl ester kyseliny fosforecne (Czech)
Diethyl p-nitrophenyl phosphate
O,O-Diethyl O-p-nitrophenyl phosphate
Diethyl paraoxon
O,O-Diethyl phosphoric acid O-p-nitrophenyl ester
O,O-Dietyl-O-p-nitrofenylfosfat (Czech)
E 600
ENT 16,087
Ester 25
Ethyl p-nitrophenyl ethylphosphate
Ethyl paraoxon
Eticol
Fosfakol
HC 2072
Mintaco
Mintacol
Miotisal
Miotisal A
p-Nitrophenyl diethylphosphate
Oxyparathion
Paraoxon
Paraoxone
Paroxan
Pestox 101
Phenol, p-nitro-, ester with diethyl phosphate
Phosphacol
Phosphonothioic acid, diethyl-paranitrophenyl ester
Phosphoric acid diethyl 4-nitrophenyl ester
RCRA waste number P041
Soluglacit
TS 219

311-47-7
$C_6H_{12}ClO_4P$
214.60
CCOP(=O)(OCC)OC=CCl
Phosphoric acid, 2-chlorovinyl diethyl ester
2-Chlorovinyl diethyl phosphate
2-Chlorvinyl-diethylfosfat (Czech)
Compound 1836
Diethyl 2-chlorovinyl phosphate
O,O-Diethyl O-(2-chlorovinyl) phosphate
Ethenol, 2-chloro-, diethyl phosphate
Latka 1836 (Czech)
OS 1836
SD 1836
Shell OS 1836

311-89-7
$C_{12}F_{27}N$
671.13
FC(F)(N(C(F)(F)C(F)(F)C(F)(F)C(F)(F)F)C(F)(F)C(F)(F)C(F)(F)C(F)(F)F)C(F)(F)C(F)(F)C(F)(F)F
Tributylamine, heptacosafluoro
1-Butanamine, 1,1,2,2,3,3,4,4,4-nonafluoro-N,N-bis-(nonafluorobutyl)- (9CI)
FC 43
FC 47
Fluorinert FC 43
Fluorocarbon FC 43
Fluosol 43
Heptacosafluorotributylamine
Mediflor FC 43
Perfluorotributylamine
Tri(perfluorobutyl)amine
Tris(nonafluorobutyl)amine

313-71-3
$C_{32}H_{18}$
402.50
Dibenzo(de,yz)hexacene-(7CI,8CI,9CI)

313-80-4
$C_{15}H_9N$
203.25
Naphtho(2,1,8-def)quinoline
1-Azapyrene

314-13-6
$C_{34}H_{24}N_6O_{14}S_4$.4Na
960.84
[Na+].[Na+].[Na+].[Na+].Cc1cc(ccc1N=Nc3ccc2c(cc(c(N)c2c3O)S([O-])(=O)=O)S([O-])(=O)=O)c6ccc(N=Nc5ccc4c(cc(c(N)c4c5O)S([O
1,3-Naphthalenedisulfonic acid, 6,6'-((3,3'-dimethyl-4,4'-biphenylylene)bis(azo))-bis (4-amino-5-hydroxy-, tetrasodium salt
Azovan Blue
4,4'-Bis(1-amino-8-hydroxy-2,4-disulfo-7-naphthylazo)-

3,3'-bitolyl, tetrasodium salt
4,4'-Bis(7-(1-amino-8-hydroxy-2,4-disulfo)naphthylazo)-3,3'-bitolyl, tetrasodium salt
4,4'-Bis(1-amino-8-hydroxy-2,4-disulpho-7-naphthylazo)-3,3'-bitolyl, tetrasodium salt
4,4'-Bis(7-(1-amino-8-hydroxy-2,4-disulpho)naphthylazo)-3,3'-bitolyl, tetrasodium salt
Blekit evansa (Polish)
Chlorazol Sky Blue ff
C.I. 23860
C.I. Direct Blue 53
C.I. Direct Blue 53, Tetrasodium salt
Diamine Sky Blue FF
Diazobleu
Diazol Pure Blue BF
Direct Blue 53
Dye Evans Blue
EB
Evablin
Evans Blue
Evans Blue Dye
Evans Blue, Sodium salt
Geigy-Blau 536
Geigy Blue 536, Med
Modr Evansova (Czech)
Modr Prima 53 (Czech)
T 1824

314-19-2
$C_{17}H_{17}NO_2.ClH$
303.81
4H-Dibenzo(de,g)quinoline-10,11-diol, 5,6,6a,7-tetrahydro-6-methyl-, hydrochloride, (R)
Apomorphine chloride
Apomorphine hydrochloride
(-)-Apomorphine hydrochloride
Apomorphinium chloride
(-)-Apomorphinium hydrochloride
6a-β-Aporphine-10,11-diol, hydrochloride
N-Methylnorapomorphine hydrochloride
6a-β-Noraporphine-10,11-diol, 6-methyl-, hydrochloride

314-40-9
$C_9H_{13}BrN_2O_2$
261.15
CCC(C)n1c(=O)[nH]c(C)c(Br)c1=O
Uracil, 5-bromo-3-sec-butyl-6-methyl
Borea
Bromacil (ACGIH,OSHA)
Bromax
Bromazil
5-Bromo-3-sec-butyl-6-methyluracil
5-Bromo-6-methyl-3-(1-methylpropyl)-2,4(1H,3H)-pyrimidinedione
5-Bromo-6-methyl-3-(1-methylpropyl)uracil
3-sek.Butyl-5-brom-6-methyluracil (German)
Cynogan
Du Pont Herbicide 976
Eerex Granular Weed Killer
Eerex Water soluble concentrate weed killer
Herbicide 976
Hyvarex
Hyvar X
Hyvar X-L
Hyvar X Bromacil
Hyvar X Weed Killer
Hyvar X-WS
Krovar II
Nalkil
Uragan
Uragon
Urox B
Urox B Water soluble concentrate weed killer
Urox-HX
Urox HX Granular Weed Killer

314-42-1
$C_8H_{11}BrN_2O_2$
247.12
CC(C)n1c(=O)[nH]c(C)c(Br)c1=O
Uracil, 5-bromo-3-isopropyl-6-methyl
5-Brom-3-isopropyl-6-methyluracil (German)
5-Bromo-3-isopropyl-6-methyl, 2,4-pyrimidinedione (French)
5-Bromo-3-isopropyl-6-methyluracil
5-Bromo-3-isopropyl-6-metiluracil (Italian)
5-Broom-3-isopropyl-6-methyluracil-(Dutch)
5-Bromo-6-methyl-3-(1-methylethyl)-2,4-(1H,3H)-pyrimidinedione
Du Pont Herbicide 82
H-82
Herbicide 82
Hyvar
Isocil
Isoprocil (French)
3-Isopropyl-5-bromo-6-methyluracil
Lorox
2,4(1H,3H)-Pyrimidinedione, 5-bromo-6-methyl-3-(1-methylethyl)- (9CI)

315-18-4
$C_{12}H_{18}N_2O_2$
222.32
CNC(=O)Oc1cc(C)c(N(C)C)c(C)c1
Carbamic acid, methyl-, 4-dimethylamino-3,5-xylyl ester
Carbamate, 4-dimethylamino-3,5-xylyl N-methyl-
Carbamic acid, methyl-, 4-(dimethylamino)-3,5-dimethylphenyl ester
4-(Dimethylamine)-3,5-xylyl N-methylcarbamate
4-(Dimethylamino)-3,5-dimethylphenol methylcarbamat (ester)
4-(Dimethylamino)-3,5-dimethylphenyl N-methylcarbamate
4-(Dimethylamino)-3,5-xylenol, methylcarbamate (ester)
4-Dimethylamino-3,5-xylyl methylcarbamate
4-Dimethylamino-3,5-xylyl N-methylcarbamate
4-(N,N-dimethylamino)-3,5-xylyl N-methylcarbamate
DOWCO 139
ENT 25,766
Methylcarbamic acid, 4-(dimethylamino)-3,5-xylyl ester
Methyl-4-dimethylamino-3,5-xylyl carbamate
Methyl-4-dimethylamino-3,5-xylyl ester of carbamic acid
Mexacarbate (DOT)
NA 2757 (DOT)
NCI-C00544
OMS-47
OMS 639
3,5-Xylenol, 4-(dimethylamino)-, methylcarbamate
Zactran
Zectane
Zectran
Zextran

315-22-0
$C_{16}H_{23}NO_6$
325.40
CC1C(=O)OC2CCN3CC=C(COC(=O)C(C)(O)C1(C)O)C23
Monocrotaline
Crotaline
(13-α,14-α)-14,19-Dihydro-12,13-dihydroxy-20-norcrotalanan-11,15-dione
Monocrotalin
NCI-C56462
20-Norcrotalanan-11,15-dione, 14,19-dihydro-12,13-dihydroxy-, (13-α,14-α)- (9CI)
NSC-28693

315-30-0
$C_5H_4N_4O$
136.13
n(c(O)c(c(n1)NN=2)C2)c1
H-Pyrazolo(3,4-d)pyrimidin-4-ol
Adenock
AL-100
Allopurinol
Allozym
Allural
Alositol
Aluline
Anoprolin
Anzief
Apurin

Apurol
Bleminol
Bloxanth
BW 56-158
Caplenal
Cellidrin
Dabrosin
Embarin
Epidropal
1,5-Dihydro-4H-pyrazolo-(3,4-d)pyrimidin-4-one
Foligan
Gichtex
HPP
4'-Hydroxypyrazolol(3,4-d)-pyrimidine
4-Hydroxypyrazolopyrimidine
4-Hydroxy-1H-pyrazolo(3,4-d)-pyrimidine
4-Hydroxy-3,4-pyrazolo-pyrimidine
4-Hydroxypyrazolo(3,4-d)pyr-imidine
4-Hydroxypyrazolyl(3,4-d)pyr-imidine
Ketanrift
Ketobun-A
Lopurin
Lysuron
Miniplanor
Monarch
Nektrohan
NSC-1390
4H-Pyrazolo(3,4-d)pyrimidin-4-one
Remid
Riball
Suspendol
Takanarumin
Urbol
Uricemil
Uritas
Urobenyl
Urosin
Xanturat
Zyloprim
Zyloric

315-37-7
$C_{26}H_{40}O_3$
400.66
CCCCCCC(=O)OC3CCC4C2CC C1=CC(=O)CCC1(C)C2CC C34C
Testosterone, heptanoate
Atlatest
Androst-4-en-3-one, 17-((1-oxo-heptyl)oxy)-, (17-β)-
Androtardyl
Delatestryl
Heptanoic acid, ester with tes-tosterone
Reposo-TMD
Malogen L.A.200
NSC-17591
TE
Testosterone enantate
Testosterone enanthate
Testosterone ethanate
Testosterone heptoate
Testosterone heptylate
Testosterone oenanthate
Testate
Testostroval

316-41-6
$C_{40}H_{36}N_2O_8.O_4S$
768.84
Berbinium, 7,8,13,13a-tetra-dehydro-9,10-dimethoxy-2,3-(methylenedioxy)-, sulfate (2:1)
Benzo(g)-1,3-benzodioxolo-(5,6-a)quinolizinium, 5,6-di-hydro-9,10-dimethoxy-, sulfate (2:1)
Berberine sulfate
Berberine sulfate (2:1)
Berberin sulfate
Neutral Berberine Sulfate

316-42-7
$C_{29}H_{40}N_2O_4.2ClH$
553.63
CCC3CN2CCc1cc(OC)c(OC)cc1 C2CC3CC4NCCc5cc(OC)c (OC)cc45
Emetine, dihydrochloride
Amebicide
(-)-Emetine dihydrochloride
l-Emetine dihydrochloride
Emetine, hydrochloride
NSC-33669

316-46-1
$C_9H_{11}FN_2O_6$
262.22
Uridine, 5-fluoro
5-Fluorouridine
FUR
5-FUR

316-49-4
$C_{19}H_{14}$
242.33
Cc1cccc3c1ccc4cc2ccccc2cc34
Benz(a)anthracene, 4-methyl
4-Methylbenz(a)anthracene
4'-Methyl-1:2-benzanthracene

317-34-0
$C_7H_8N_4O_2.1/2C_2H_8N_2$
210.25
Theophylline, Compd. with ethylenediamine (2:1)
Aminocardol
Aminodur
Aminofilina (Spanish)
Aminophylline
Ammophyllin
Cardophylin
Cardophyllin
Cardiofilina
Cardiomin
Carena
Cariomin
Diaphylline
3,7-Dihydro-1,3-dimethyl-1H-purine-2,6-dione Compd. with 1,2-ethanediamine (2:1)
Diophyllin
Diuxanthine
DOBO
Dura-Tab S.M. Aminophylline
Ethophylline
Ethylenediamine, Compd. with theophylline (1:2)
Etilen-Xantisan Tabl.
Eufilina (Polish)
Euphyllin
Euphylline
Euufilin
Eurphyllin
Genophyllin
Grifomin
Inophylline
Lasodex
Linampheta
Metaphyllin
Metaphylline
Methophylline
Minaphil
Miofilin
Neophyiline
Peterphyllin
Phylcardin
Phyllindon
Phyllocontin
1H-Purine-2,6-dione, 3,7-di-hydro-1,3-dimethyl-, Compd. with 1,2-ethanediamine (2:1) (9CI)
Rectalad-Aminophylline
Somophyllin
Somophyllin O
Stenovasan
Tad
Tefamin
Theodrox
Theolamine
Theolone
Theomin
Theophyldine
Theophyllamine
Theophyllaminium
Theophyllin aethylendiamin (German)
Theophylline ethylenediamine
Theophyllin ethylenediamine
Thephyldine
TH/100
Vasofilina

319-78-8
$C_6H_{13}NO_2$
131.20
Isoleucine, D
D-Isoleucine

319-84-6
$C_6H_6Cl_6$
290.82
C(C(C(C(C1Cl)Cl)Cl)Cl)(C1Cl) Cl
Cyclohexane, 1,2,3,4,5,6-hexa-chloro-, α-isomer
Benzene hexachloride-α-isomer
α-Benzenehexachloride

α-BHC
Cyclohexane, 1,2,3,4,5,6-hexachloro-, α-
Cyclohexane, α-1,2,3,4,5,6-hexachloro-
ENT 9,232
α-HCH
α-Hexachloran
α-Hexachlorane
Hexachlorcyclohexan (German)
α-Hexachlorcyclohexane
α-Hexachlorocyclohexane
α-1,2,3,4,5,6-Hexachlorocyclohexane
1-α,2-α,3-β,4-α,5-β,6-β-Hexachlorocyclohexane
α-Lindane

319-85-7
$C_6H_6Cl_6$
290.82
C(C(C(C(C1Cl)Cl)Cl)Cl)(C1Cl)Cl
Cyclohexane, 1,2,3,4,5,6-hexachloro-, β-isomer
trans-α-Benzenehexachloride
Beta-Isomer
β-BHC
Cyclohexane, β-1,2,3,4,5,6-hexachloro-
Cyclohexane, 1,2,3,4,5,6-hexachloro-, β-
Cyclohexane, 1,2,3,4,5,6-hexachloro-, trans-
ENT 9,233
β-HCH
β-Hexachlorobenzene
1-α,2-β,3-α,4-β,5-α,6-β-Hexachlorocyclohexane
β-Hexachlorocyclohexane
β-1,2,3,4,5,6-Hexachlorocyclohexane
β-Lindane

319-86-8
$C_6H_6Cl_6$
290.82
C(C(C(C(C1Cl)Cl)Cl)Cl)(C1Cl)Cl
Cyclohexane, 1,2,3,4,5,6-hexachloro-, δ-isomer
δ-BHC
Cyclohexane, δ-1,2,3,4,5,6-hexachloro-
ENT 9,234
δ-HCH
1-α,2-α,3-α,4-β,5-α,6-β-Hexachlorocyclohexane
δ-Hexachlorocyclohexane
δ-1,2,3,4,5,6-Hexachlorocyclohexane
δ-Lindane

319-87-9
C_6Cl_5F
268.33
Benzene, pentachlorofluoro- (8CI,9CI)
NSC-146405

319-94-8
$C_6H_5Cl_5$
254.36
ClC1C=C(Cl)C(Cl)C(Cl)C1Cl
Cyclohexene, 1,3,4,5,6-pentachloro-, γ
Cyclohexene, 1,3,4,5,6-pentachloro-, (3-α,4-β,5-β,6-α)-(9CI)
γ-PCCH
γ-Pentachlorocyclohexene

320-60-5
$C_7H_3Cl_2F_3$
215.00
FC(F)(F)c(c(cc(c1)Cl)Cl)c1
Benzene, 2,4-dichloro-1-(trifluoromethyl)- (9CI)
2,4-Dichlorobenzotrifluoride
2,4-Dichloro-1-(trifluoromethyl)benzene

320-67-2
$C_8H_{12}N_4O_5$
244.24
Nc2ncn(C1OC(CO)C(O)C1O)c(=O)n2
s-Triazin-2(1H)-one, 4-amino-1-β-D-ribofuranosyl
5-AC
4-Amino-1-β-D-ribofuranosyl-D-triazin-2(1H)-one
Antibiotic U 18496
Azacitidine
Azacytidine
5-Azacytidine
5'-Azacytidine
5 AZC
5-AZCR
Ladakamycin
Mylosar
NCI-C01569
NSC-102816
1,3,5-Triazin-2(1H)-one, 4-amino-1-β-D-ribofuranosyl-
U 18496

320-72-9
$C_7H_4Cl_2O_3$
207.01
O=C(O)c(c(O)c(cc1Cl)Cl)c1
Salicylic acid, 3,5-dichloro
3,5-Dichlorosalicylic acid
USAF DO-68

321-14-2
$C_7H_5ClO_3$
172.57
O=C(O)c(c(O)ccc1Cl)c1
Salicylic acid, 5-chloro
Benzoic acid, 5-chloro-2-hydroxy-
5-Chlorosalicylic acid

321-38-0
$C_{10}H_7F$
146.17
Fc(c(c(ccc1)cc2)c1)c2
Naphthalene, 1-fluoro
1-Fluornaftalen (Czech)
α-Fluoronaphthalene

321-60-8
$C_{12}H_9F$
172.20
Fc(c(c(cccc1)c1)ccc2)c2
2-Fluorobiphenyl
Biphenyl, 2-fluoro-
1,1'-Biphenyl, 2-fluoro- (9CI)
2-Fluoro-1,1'-biphenyl
o-Fluorodiphenyl
NSC-10366

321-64-2
$C_{13}H_{14}N_2$
198.29
Acridine, 1,2,3,4-tetrahydro-9-amino
9-Acridinamine, 1,2,3,4-tetrahydro- (9CI)
5-Amino-6,7,8,9-tetrahydroacridine (European)
9-Amino-1,2,3,4-tetrahydroacridine
CS 12602
Romotal
Tacrine
1,2,3,4-Tetrahydro-9-acridinamine
Tetrahydroaminacrine
Tetrahydroaminoacridine
Tetrahydroaminocrin
1,2,3,4-Tetrahydro-5-aminoacridine
Tetrahydroaminocrine
THA

324-74-3
$C_{12}H_9F$
172.20
4-Fluorobiphenyl
Biphenyl, 4-fluoro- (8CI)
1,1'-Biphenyl, 4-fluoro- (9CI)
p-Fluorodiphenyl
NSC-56686

326-61-4
$C_{10}H_{10}O_4$
194.20
O=C(OCc(ccc(OCO1)c12)c2)C
Acetic acid, (3,4-methylenedioxy)benzyl ester
Heliotropyl acetate
3,4-Methylenedioxybenzyl acetate
Piperonyl acetate

326-91-0
$C_8H_5F_3O_2S$
222.19
O=C(C(SC=C1)=C1)CC(=O)C(F)(F)F
Thenoyltrifluoroacetone

AI3-31295
1,3-Butanedione, 4,4,4-trifluoro-1-(2-thienyl)- (9CI)
NSC-66544
Perfluoroacetyl(2-thenoyl)-methane
α-Thenoyltrifluoroacetone
2-Thenoyltrifluoroacetone
1-Thenoyl-3,3,3-trifluoroacetone
1,1,1-Trifluoro-3-(2-thenoyl)-acetone
TTA
TTB

327-54-8
$C_6H_2F_4$
150.08
Fc1cc(F)c(F)cc1F
Benzene, 1,2,4,5-tetrafluoro- (8CI,9CI)
NSC-10249

327-57-1
$C_6H_{13}NO_2$
131.20
O=C(O)C(N)CCCC
Norleucine, l
α-Aminocaproic acid
2-Aminocaproic acid
2-Aminohexanoic acid
(S)-2-Aminohexanoic acid
Caprine
Glycoleucine
Norleucine
L-Norleucine (9CI)
L-(+)-Norleucine

327-97-9
$C_{16}H_{18}O_9$
354.34
Cyclohexanecarboxylic acid, 3-((3-(3,4-dihydroxyphenyl)-1-oxo-2-propenyl)oxy)-1,4,5-trihydroxy-, (1s-(1-α,3-β,4-α,5-α))
3-Caffeoylquinic acid
3-o-Caffeoylquinic acid
Chlorogenic acid

327-98-0
$C_{10}H_{12}Cl_3O_2PS$
333.60
CCOP(=S)(CC)Oc1cc(Cl)c(Cl)cc1Cl
Phosphonothioic acid, ethyl-, O-ethyl O-(2,4,5-trichlorophenyl) ester
5082A
O-Aethyl-O-(2,4,5-trichlor-phenyl)-aethylthionophosphonat (German)
Agrisil
Agritox
Bay 37289
Bayer 37289
Bayer 5081
Bayer S 4400
Chemagro 37289
ENT 25,712
Ethyl trichlorophenylethylphosphonothioate
O-Ethyl O-2,4,5-trichlorophenyl ethylphosphonothioate
Fenophosphon
OMS 412
OMS 578
Phenol, 2,4,5-trichloro-, O-ester with O-ethyl ethylphosphonothioate
Phytosol
S 4400
Stauffer N-3049
Trichloronat
Trichloronate
Wirkstoff 37289

328-38-1
$C_6H_{13}NO_2$
131.20
O=C(O)C(N)CC(C)C
Leucine, D
D-Leucine

328-39-2
$C_6H_{13}NO_2$
131.17
O=C(O)C(N)CC(C)C
DL-Leucine (9CI)
AI3-26709
Leucine, DL- (8CI)
(+-)-Leucine
NSC-9252

328-42-7
$C_4H_4O_5$
132.07
OC(=O)CC(=O)C(O)=O
Oxalacetic acid
Butanedioic acid, oxo- (9CI)
Ketosuccinic acid
2-Ketosuccinic acid
NSC-77688
Oxobutanedioic acid
Oxosuccinic acid

328-50-7
$C_5H_6O_5$
146.10
O=C(O)C(=O)CCC(=O)O
α-Ketoglutaric acid
AI3-26938
Glutaric acid, α keto
Glutaric acid, 2-oxo- (8CI)
NSC-17391
α-Oxoglutaric acid
2-Oxopentanedioic acid
2-Oxo-1,5-pentanedioic acid
Pentanedioic acid, 2-oxo- (9CI)

328-57-4
$C_7H_8F_2Si$
158.22
Silane, difluoromethyl phenyl- (6CI,7CI,8CI, 9CI)
Difluoromethylphenylsilane
Methylphenylsilane difluoride

328-74-5
$C_8H_5F_6N$
229.14
FC(F)(F)c(cc(N)cc1C(F)(F)F)c1
3,5-Xylidine, α,α,α,α',α',α'-hexafluoro
Aniline, 3,5-bis(trifluoro-methyl)-
Benzenamine, 3,5-bis(trifluoro-methyl)- (9CI)
3,5-Bis(trifluoromethyl)aniline

328-84-7
$C_7H_3Cl_2F_3$
215.00
FC(F)(F)c(ccc(c1Cl)Cl)c1
3,4-Dichlorobenzotrifluoride
Benzene, 1,2-dichloro-4-(tri-fluoromethyl)- (9CI)
3,4-Dichlorophenyltrifluoro-methane
1,2-Dichloro-4-(trifluoro-methyl)benzene
3,4-Dichloro-α,α,α-trifluoro-toluene
Toluene, 3,4-dichloro-α,α,α-trifluoro-

329-01-1
$C_8H_4F_3NO$
187.13
O=C=Nc(cccc1C(F)(F)F)c1
Isocyanic acid, (m-trifluoro-methylphenyl) ester
(α,α,α-Trifluoro-m-tolyl) isocyanate
TIC

329-63-5
$C_9H_{13}NO_3.ClH$
219.69
Benzyl alcohol, 3,4-dihydroxy-α-((methylamino)methyl)-, hydrochloride, (+-)
(+-)-Adrenaline hydrochloride
dl-Adrenaline hydrochloride
1,2-Benzenediol, 4-(1-hydroxy-2-(methylamino)ethyl)-, hydrochloride, (+-)- (9CI)
(+-)-3,4-Dihydroxy-α-((methyl-amino)methyl)benzyl alcohol hydrochloride
(+-)-Epinephrine hydrochloride
dl-Epinephrine hydrochloride

329-65-7
$C_9H_{13}NO_3$
183.23
Benzyl alcohol, 3,4-dihydroxy-α-((methylamino)methyl)-, (+-)
DL-Adrenaline
dl-Epinephrine
Epinephrine Racemic

329-71-5
$C_6H_4N_2O_5$
184.12
Oc1cc(ccc1N(=O)=O)N(=O)=O
Phenol, 2,5-dinitro
2,5-Dinitrofenol (Czech)
γ-Dinitrophenol
2,5-Dinitrophenol
2,5-DNP
Phenol, γ-dinitro-

330-13-2
$C_6H_6NO_6P$
219.08
O=P(Oc(ccc(N(=O)=O)c1)c1)(O)O
Nitrophenylphosphate
Mono(4-nitrophenyl) phosphate
p-Nitrophenyl dihydrogen phosphate
4-Nitrophenyl dihydrogen phosphate
NSC-404086
Phenol, p-nitro-, dihydrogen phosphate
Phosphoric acid, mono(p-nitrophenyl) ester (8CI)
Phosphoric acid, mono(4-nitrophenyl) ester (9CI)

330-39-2
$C_9H_{11}FN_2O$
182.19
CN(C)C(=O)Nc1cccc(F)c1
Urea, N'-(3-fluorophenyl)-N,N-dimethyl- (9CI)
Urea, 3-(m-fluorophenyl)-1,1-dimethyl- (8CI)

330-54-1
$C_9H_{10}Cl_2N_2O$
233.11
O=C(N(C)C)Nc(ccc(c1Cl)Cl)c1
Urea, 3-(3,4-dichlorophenyl)-1,1-dimethyl
AF 101
Cekiuron
Crisuron
Dailon
DCMU
Diater
Dichlorfenidim
3-(3,4-Dichlorophenol)-1,1-dimethylurea
N'-(3,4-Dichlorophenyl)-N,N-dimethylurea
3-(3,4-Dichlorophenyl)-1,1-dimethylurea
1-(3,4-Dichlorophenyl)-3,3-dimethyluree (French)
3-(3,4-Dichlor-phenyl)-1,1-dimethyl-harnstoff (German)
3-(3,4-Dichloor-fenyl)-1,1-dimethylureum (Dutch)
3-(3,4-Dicloro-fenyl)-1,1-dimetil-urea (Italian)
1,1-Dimethyl-3-(3,4-dichlorophenyl)urea
Di-On
Direx 4L
Diumate
Diurex
Diurol
Diuron (OSHA)
Diuron 4L
DMU
Drexel
Drexel Diuron 4L
Duran
Dynex
Farmco Diuron
Herbatox
Herbixol
HW 920
Karmex
Karmex Diuron Herbicide
Karmex DW
Marmer
Sup'r Flo
Telvar
Telvar Diuron Weed Killer
Tigrex
Unidron
Urox D
USAF P-7
USAF XR-42
Vonduron

330-55-2
$C_9H_{10}Cl_2N_2O_2$
249.11
O=C(N(OC)C)Nc(ccc(c1Cl)Cl)c1
Urea, 3-(3,4-dichlorophenyl)-1-methoxy-1-methyl
Afalon
Afalon Inuron
Afalonu (Polish)
Aphalon
Cephalon
3-(3,4-Dichloor-fenyl)-1-methoxy-1-methylureum (Dutch)
3-(3,4-Dichloro-fenil)-1-metossi-1-metil-urea (Italian)
3-(3,4-Dichlorophenyl)-1-methoxymethylurea
3-(3,4-Dichlorophenyl)-1-methoxy-1-methylurea
N'-(3,4-Dichlorophenyl)-N-methoxy-N-methylurea
1-(3,4-Dichlorophenyl)-methoxy-3-methyluree (French)
N-(3,4-Dichlorophenyl)-N'-methyl-N'-methoxyurea
3-(3,4-Dichlor-phenyl)-1-methoxy-1-methyl-harnstoff (German)
3-(4,5-Dichlorphenyl)-1-methoxy-1-methylharnstoff (German)
Du Pont 326
Dupont Herbicide 326
N-(3,4-Dwuchlorofenylo)-N'-metoksy-N'-metylomocznik (Polish)
Garnitan
Herbicide 326
HOE 2810
Laroks (Polish)
Linex 4L
Linorox
Linurex
Linuron
Linuron (Herbicide)
Lorex
Lorox
Lorox Linuron Weed Killer
Methoxydiuron
1-Methoxy-1-methyl-3-(3,4-dichlorophenyl)urea
Premalin
Sarclex
Scarclex
Sinuron

331-39-5
$C_9H_8O_4$
180.17
Cinnamic acid, 3,4-dihydroxy
Caffeic acid
3,4-Dihydroxybenzeneacrylic acid
3,4-Dihydroxycinnamic acid
2-Propenoic acid, 3-(3,4-dihydroxyphenyl)- (9CI)

332-33-2
$C_9H_{11}FN_2O$
182.22
CN(C)C(=O)Nc1ccc(F)cc1
Urea, 1,1-dimethyl-3-(p-fluorophenyl)
1,1-Dimethyl-3-(p-fluorophenyl)urea
N'-(4-Fluorophenyl)-N,N-dimethylurea

332-77-4
$C_6H_{10}O_3$
130.14
O(C(OC)C=C1)C1OC
Furan, 2,5-dihydro-2,5-dimethoxy-
2,5-Dihydro-2,5-dimethoxyfuran
2,5-Dimethoxy-2,5-dihydrofuran
2,5-Dimethoxyfuran
Furan, 2,5-dimethoxy-
NSC-43243

333-18-6
$C_2H_8N_2.2ClH$
133.04
[Cl-].[Cl-].NCCN
Ethylenediamine, dihydrochloride
Chlor-Ethamine
Ethylenediamine hydrochloride

333-20-0
CNS.K
97.18
[K+].[S-]C#N
Thiocyanic acid, potassium salt
Arterocyn
Aterocyn

Kyonate
Potassium isothiocyanate
Potassium rhodanate
Potassium rhodanide
Potassium sulfocyanate
Potassium thiocyanate
Potassium thiocyanide
Rhocya
Rodanca
Rhodanide
Thio-Cara

333-27-7
$C_2H_3F_3O_3S$
164.11
O=S(=O)(OC)C(F)(F)F
Methyl trifluoromethane sulfonate
AI3-62911
Methanesulfonic acid, trifluoro-, methyl ester (9CI)
Methyl triflate
Methyl trifluoromethanesulfonate
NSC-270679

333-41-5
$C_{12}H_{21}N_2O_3PS$
304.38
O(P(OCC)(Oc(nc(nc1C)C(C)C)c1)=S)CC
Phosphorothioic acid, O,O-diethyl O-(2-isopropyl-6-methyl-4-pyrimidinyl) ester
Alfa-Tox
Basudin
Basudin 10 G
Bazinon
Bazuden
Dazzel
Desapon
O,O-Diaethyl-O-(2-isopropyl-4-methyl-pyrimidin-6-yl)-monothiophosphat (German)
O,O-Diaethyl-O-(2-isopropyl-4-methyl)-6-pyrimidyl-thionophosphat (German)
Diagran
Dianon
Diaterr-Fos
Diazajet
Diazatol
Diazide
Diazinon (ACGIH,OSHA) [UN 2783]
Diazinone
Diazitol
Diazol
O,O-Diethyl-o-(2-isopropyl-4-methyl-pyrimidin-6-yl)-monothiofosfaat (Dutch)
O,O-Diethyl-O-(2-isopropyl-4-methyl-6-pyrimidinyl)-phosphorothioate
O,O-Diethyl O-(2-isopropyl-6-methyl-4-pyrimidinyl) phosphorothioate
O,O-Diethyl-O-(2-isopropyl-4-methyl-6-pyrimidyl)phosphorothioate
O,O-Diethyl O-(2-isopropyl-4-methyl-6-pyrimidyl) thionophosphate
O,O-Diethyl 2-isopropyl-4-methylpyrimidyl-6-thiophosphate
Diethyl 4-(2-isopropyl-6-methylpyrimidinyl)phosphorothionate
O,O-Dietil-O-(2-isopropil-4-metil-pirimidin-6-il)-monotiofosfato (Italian)
O,O-Diethyl O-6-methyl-2-isopropyl-4-pyrimidinyl phosphorothioate
Dimpylate
Dizinon
Dizzitol
Dyzol
ENT 19,507
Exodin
Fezudin
G 301
G-24480
Gardentox
Geigy 24480
O-2-Isopropyl-4-methyl-pyrimidyl-O,O-diethyl phosphorothioate
Isopropylmethylpyrimidyl diethyl thiophosphate
Kayazinon
Kayazol
NA 2783 [Organophosphorus pesticides, solid, toxic, N.O.S.]
NCI-C08673
Nedcidol
Neocidol
Nipsan
Nucidol
OMS 469
Phosphorothioate, O,O-diethyl O-6-(2-isopropyl-4-methyl-pyrimidyl)
4-Pyrimidinol, 2-isopropyl-6-methyl-, O-ester with O,O-diethyl phosphorothioate
Sarolex
Spectracide
Srolex
Thiophosphate de O,O-diethyle et de O-2-isopropyl-4-methyl-6-pyrimidyle (French)

333-43-7
$C_{11}H_{17}OPS_2$
260.37
CCOP(=S)(CC)Sc1ccc(C)cc1
Phosphonodithioic acid, ethyl-, O-ethyl S-(p-tolyl) ester
Bay 38156
Bayer 38156
ENT 25,713
O-Ethyl S-(4-methylphenyl) ethylphosphonodithioate
O-Ethyl-S-p-tolylester kyseliny ethyldithiofosfonove (Czech)
O-Ethyl S-p-tolyl ethylphosphonodithioate
N 2788
S 4706
Stauffer N-2788
p-Toluenethiol, S-ester with O-ethyl ethylphosphonodi thioate

334-48-5
$C_{10}H_{20}O_2$
172.30
O=C(O)CCCCCCCCC
Decanoic-acid
Capric acid
n-Capric acid
Caprinic acid
Caprynic acid
n-Decanoic acid
n-Decoic acid
Decylic acid
n-Decylic acid
Hexacid 1095
Neo-Fat 10
1-Nonanecarboxylic acid

334-56-5
$C_{10}H_{21}F$
160.31
FCCCCCCCCCC
Decane, 1-fluoro
1-Fluorodecane

334-68-9
$C_{12}H_{25}F$
188.37
Dodecane, 1-fluoro
1-Fluorododecane

334-88-3
CH_2N_2
42.05
Methane, diazo
Azimethylene
Diazirine
Diazomethane (ACGIH,OSHA)

335-02-4
CF_3NO_2
115.01
Methane, trifluoronitro- (8CI, 9CI)
Fluoropicrin

335-05-7
CF_4O_2S
152.07
O=S(=O)(F)C(F)(F)F
Methanesulfonyl fluoride, trifluoro- (9CI)
Trifluoromethanesulfonyl fluoride

335-42-2
C_4F_8O
216.03
O=C(F)C(F)(F)C(F)(F)C(F)(F)F

Butanoyl fluoride, heptafluoro- (9CI)
Heptafluorobutanoyl fluoride

335-57-9
C_7F_{16}
388.07
FC(F)(F)C(F)(F)C(F)(F)C(F)(F)C(F)(F)C(F)(F)C(F)(F)F
Heptane, hexadecafluoro
Hexadecafluoroheptane
Perfluoroheptane
Perfluoro-n-heptane

335-66-0
$C_8F_{16}O$
416.06
O=C(F)C(F)(F)C(F)(F)C(F)(F)C(F)(F)C(F)(F)C(F)(F)C(F)(F)F
Octanoyl fluoride, pentadecafluoro- (9CI)
Pentadecafluorooctanoyl fluoride

335-67-1
$C_8HF_{15}O_2$
414.09
O=C(O)C(F)(F)C(F)(F)C(F)(F)C(F)(F)C(F)(F)C(F)(F)C(F)(F)F
Octanoic acid, pentadecafluoro
Pentadecafluorooctanoic acid
Pentadecafluoro-n-octanoic acid
Perfluorocaprylic acid
Perfluoroheptanecarboxylic acid
Perfluoroctanoic acid
Perfluorooctanoic acid
PFOA

335-71-7
$C_7F_{16}O_2S$
452.12
O=S(=O)(F)C(F)(F)C(F)(F)C(F)(F)C(F)(F)C(F)(F)C(F)(F)C(F)(F)F
1-Heptanesulfonyl fluoride, 1,1,2,2,3,3,4,4,5,5,6,6,7,7,7-pentadecafluoro- (9CI)
1,1,2,2,3,3,4,4,5,5,6,6,7,7,7-Pentadecafluoro-1-heptanesulfonyl fluoride

338-45-4
$C_7H_{13}O_6P$
224.17
Crotonic acid, 3-hydroxy-, methyl ester, dimethyl phosphate, (Z)
2-Butenoic acid, 3-((dimethoxyphosphinyl)oxy)-, methyl ester, (Z)- (9CI)
cis-Mevinphos
trans-Phosdrin

338-84-1
$C_{15}F_{33}N$
821.11
FC(F)(N(C(F)(F)C(F)(F)C(F)(F)C(F)(F)C(F)(F)F)C(F)(F)C(F)(F)C(F)(F)C(F)(F)C(F)(F)F)C(F)(F)C(F)(F)C(F)(F)C(F)(F)F
Fluorocarbon FC 70
Fluorinert FC 70
1-Pentanamine, 1,1,2,2,3,3,4,4,5,5,5-undecafluoro-N,N-bis(undecafluoropentyl)-(9CI)
Tris(undecafluoropentyl)amine

339-43-5
$C_{11}H_{17}N_3O_3S$
271.37
O=C(NCCCC)NS(=O)(=O)c(ccc(N)c1)c1
Urea, 1-butyl-3-sulfanilyl
Alentin
N-(4-Aminobenzenesulfonyl)-N'-butylurea
4-Amino-N-((butylamino)carbonyl)benzenesulfonamide
Aminophenurobutane
Bucarban
Bucrol
Bukarban
Burcol
Butisulfina
N'-(Butylcarbamoyl)sulfanilamide
N[1]-(Butylcarbamoyl)sulfanilamide
N-Butylsulfanilylurea
1-Butyl-3-sulfanilylurea
BZ 55
Carbutamide
Carbutamid
Cicloral
Diaboral
Emedan
Glucidoral
Glucofren
Glybutamide
Inbuton
Invenol
Nadisan
Nadizan
Norboral
Oranil
Oranyl
Orasulin
N[1]-Sulfanilyl-N[2]-butylcarbamide
N[1]-Sulfanilyl-N[2]-butylurea
N-Sulfanilyl N'butyluree (French)
U 6987

340-57-8
$C_{15}H_{11}ClN_2O$
270.71
Mecloqualone
B 208
B 208-Tropon
Casfen
CHI 8
3-(o-Chlorophenyl)-2-methyl-4(3H)-quinazolinone
3-(o-Chlorophenyl)-2-methyl-4-quinazolone
DEA No. 2572
Meclocualona (Spanish)
Mecloqualon
Mecloqualonum (Latin)
2-Methyl-3-(2-chlorophenyl)-chinazolon-4
2-Methyl-3-(o-chlorophenyl)-4-quinazolinone
NSC-142005
Nubarene
4(3H)-Quinazolinone, 3-(o-chlorophenyl)-2-methyl-(8CI)
4(3H)-Quinazolinone, 3-(2-chlorophenyl)-2-methyl-(9CI)
W 4744

342-69-8
$C_{11}H_{14}N_4O_4S$
298.35
9H-Purine, 6-(methylthio)-9-β-d-ribofuranosyl
Inosine, 6-(methylthio)-
6-Methylmercaptopurine ribonucleoside
6-Methylmercaptopurine riboside
6-Methyl MP-Riboside
6-Methyl-9-ribofuranosylpurine-6-thiol
Methylthioinosine
6-Methylthioinosine
6-(Methylthio)purine ribonucleoside
6-Methylthiopurine riboside
NCI-C04784
NSC-40774
Purine-6-thiol, 6-methyl-9-ribofuranosyl-
β-D-Ribosyl-6-methylthiopurine
SQ 21977

343-65-7
$C_{10}H_{12}N_2O_3$
208.22
NC(CC(=O)c1ccccc1N)C(O)=O
Kynurenine

344-04-7
C_6BrF_5
246.96
Fc(c(c(F)c(F)c1F)Br)c1F
Benzene, bromopentafluoro-(9CI)
Bromopentafluorobenzene
Bromoperfluorobenzene
NSC-21630
Pentafluorobromobenzene
Pentafluorophenyl bromide

344-07-0
C_6ClF_5
202.51
Fc(c(c(F)c(F)c1F)Cl)c1F

Benzene, chloropentafluoro
Chloropentafluorobenzene
Chloroperfluorobenzene
Pentafluorochlorobenzene
Pentafluorophenyl chloride

344-72-9
$C_7H_7F_3N_2O_2S$
240.22
5-Thiazolecarboxylic acid, 2-amino-4-(trifluoromethyl)-, ethyl ester
2-Amino-4-(trifluoromethyl)-5-thiazolecarboxylic acid ethyl ester

345-35-7
C_7H_6ClF
144.58
Fc(c(ccc1)CCl)c1
Benzene, 1-(chloromethyl)-2-fluoro- (9CI)
α-Chloro-o-fluorotoluene
α-Chloro-2-fluorotoluene
1-(Chloromethyl)-2-fluorobenzene
o-Fluorobenzyl chloride
2-Fluorobenzyl chloride
NSC-88295
Toluene, α-chloro-o-fluoro-

346-18-9
$C_{11}H_{13}ClF_3N_3O_4S_3$
439.90
CN2C(CSCC(F)(F)F)Nc1cc(Cl)c(cc1S2(=O)=O)S(N)(=O)=O
2H-1,2,4-Benzothiadiazine-7-sulfonamide, 6-chloro-3,4-dihydro-2-methyl-3-(((2,2,2-trifluoroethyl)-thio)methyl)-, 1,1-dioxide
Drenusil
P 2525
Polythiazide
Renese

348-51-6
C_6H_4ClF
130.55
Fc(c(ccc1)Cl)c1
Benzene, 1-chloro-2-fluoro- (9CI)
o-Chlorofluorobenzene
1-Chloro-2-fluorobenzene
o-Fluorochlorobenzene
1-Fluoro-2-chlorobenzene
NSC-10270

348-54-9
C_6H_6FN
111.13
Fc(c(N)ccc1)c1
Aniline, 2-fluoro
o-Fluoroaniline
2-Fluoroaniline [UN 2941]
2-Fluorobenzenamine
UN 2941 [Fluoroanilines]

348-67-4
$C_5H_{11}NO_2S$
149.23
O=C(O)C(N)CCSC
Methionine, D
D-Methionine
D-Metionien (Australian)

349-88-2
$C_6H_4ClFO_2S$
194.61
O=S(=O)(c(ccc(F)c1)c1)Cl
4-Fluorobenzenesulfonyl chloride
Benzenesulfonyl chloride, p-fluoro-
Benzenesulfonyl chloride, 4-fluoro- (9CI)
NSC-140128
p-Fluorobenzenesulfonyl chloride
4-Fluorophenylsulfonyl chloride

350-03-8
C_7H_7NO
121.15
O=C(c(cccn1)c1)C
Ketone, methyl 3-pyridyl
3-Acetopyridine
β-Acetylpyridine
3-Acetylpyridine
Methyl pyridyl ketone
Methyl β-pyridyl ketone
Methyl 3-pyridyl ketone
Pyridine, 3-acetyl-

350-30-1
$C_6H_3ClFNO_2$
175.55
Benzene, 2-chloro-1-fluoro-4-nitro
2-Chloro-1-fluoro-4-nitrobenzene
3-Chloro-4-fluoronitrobenzene

350-46-9
$C_6H_4FNO_2$
141.11
O=N(=O)c(ccc(F)c1)c1
Benzene, 1-fluoro-4-nitro
p-Fluoronitrobenzene
4-Fluoronitrobenzene
p-Nitrofluorobenzene
4-Nitrofluorobenzene

350-50-5
C_7H_7F
110.13
FCc(cccc1)c1
Benzene, (fluoromethyl)- (9CI)
Benzyl fluoride
(Fluoromethyl)benzene
α-Fluorotoluene

351-05-3
C_8H_7BrFNO
232.07
Acetanilide, 4'-bromo-2-fluoro
Acetamide, N-(4-bromophenyl)-2-fluoro- (9CI)
4'-Bromo-2-fluoroacetanilide
FABA
Fluoroaceto-p-bromoanilide
Yanomite

351-28-0
C_8H_8FNO
153.15
CC(=O)Nc1cccc(F)c1
Acetamide, N-(3-fluorophenyl)- (9CI)
Acetanilide, 3'-fluoro- (8CI)
NSC-10348

351-36-0
$C_9H_8F_3NO$
203.18
CC(=O)Nc1cccc(c1)C(F)(F)F
m-Acetotoluidide, α,α,α-trifluoro
Acetamide, N-(3-(trifluoromethyl)phenyl)- (9CI)
Acetanilide, 3-(trifluoromethyl)-
m-Trifluoromethyl acetanilide
3-Trifluoromethylacetanilide
USAF MA-14

351-83-7
C_8H_8FNO
153.17
CC(=O)Nc1ccc(F)cc1
Acetanilide, 4'-fluoro
Acetamide, N-(4-fluorophenyl)- (9CI)
p-Fluoroacetanilide
4-Fluoroacetanilide
4'-Fluoroacetanilide

352-11-4
C_7H_6ClF
144.58
Fc(ccc(c1)CCl)c1
Benzene, 1-(chloromethyl)-4-fluoro- (9CI)
α-Chloro-p-fluorotoluene
α-Chloro-4-fluorotoluene
1-(Chloromethyl)-4-fluorobenzene
p-Fluorobenzyl chloride
4-Fluorobenzyl chloride
NSC-25084
Toluene, α-chloro-p-fluoro- (8CI)

352-32-9
C_7H_7F
110.14
Fc(ccc(c1)C)c1
Toluene, p-fluoro

Benzene, 1-fluoro-4-methyl- (9CI)
1-Fluoro-4-methylbenzene
p-Fluorotoluene [UN 2388]
4-Fluorotoluene
NSC-8861
Toluene, p-fluoro- (8CI)
UN 2388 [Fluorotoluenes]

352-33-0
C_6H_4ClF
130.55
Fc(ccc(c1)Cl)c1
Benzene, 1-chloro-4-fluoro- (9CI)
p-Chlorofluorobenzene
1-Chloro-4-fluorobenzene
p-Fluorochlorobenzene
1,4-Fluorochlorobenzene
1-Fluoro-4-chlorobenzene
NSC-10272

352-34-1
C_6H_4FI
222.00
Fc(ccc(c1)I)c1
Benzene, 1-fluoro-4-iodo- (9CI)
p-Fluoroiodobenzene
1-Fluoro-4-iodobenzene
p-Iodofluorobenzene
4-Iodofluorobenzene
NSC-10280

352-70-5
C_7H_7F
110.14
Fc(cccc1C)c1
Toluene, m-fluoro
Benzene, 1-fluoro-3-methyl- (9CI)
1-Fluoro-3-methylbenzene
m-Fluorotoluene [UN 2388]
3-Fluorotoluene
NSC-8860
Toluene, m-fluoro- (8CI)
UN 2388 [Fluorotoluenes]

352-87-4
$C_6H_7F_3O_2$
168.12
O=C(OCC(F)(F)F)C(=C)C
2,2,2-Trifluoroethyl methacrylate
Methacrylic acid, 2,2,2-trifluoroethyl ester
NSC-32617
2-Propenoic acid, 2-methyl-, 2,2,2-trifluoroethyl ester (9CI)
Trifluoroethyl methacrylate

352-93-2
$C_4H_{10}S$
90.20
S(CC)CC
Ethyl-sulfide
Diethylsulfid (Czech)
Diethyl sulfide [UN 2375]
Diethylthioether
Ethane, 1,1'-thiobis-
Ethyl monosulfide
Ethylthioethane
Ethyl thioether
Sulfodor (Czech)
3-Thiapentane
Thioethyl ether
UN 2375 [Diethyl sulfide]

353-36-6
C_2H_5F
48.07
FCC
Ethane, fluoro
Ethyl fluoride [UN 2453]
Fluoroethane
Monofluoroethane
R 161
UN 2453 [Ethyl fluoride]

353-42-4
$C_2H_6O.BF_3$
113.89
Boron trifluoride-dimethyl ether
Boron trifluoride dimethyl etherate [UN 2965]
Bortrifluorid-dimethylether (Czech)
Fluorid bority-dimethylether (1:1) (Czech)
UN 2965 [Boron trifluoride dimethyl etherate]

353-50-4
CF_2O
66.01
O=C(F)F
Carbonic difluoride (9CI)
Carbon difluoride oxide
Carbon fluoride oxide
Carbon oxyfluoride
Carbonyl difluoride
Carbonyl fluoride (ACGIH, OSHA) [UN 2417]
Difluoroformaldehyde
Fluophosgene
Fluoroformyl fluoride
Fluorophosgene
RCRA waste number U033
UN 2417 [Carbonyl fluoride]

353-58-2
$CBrCl_2F$
181.82
Methane, bromodichlorofluoro- (8CI,9CI)

353-59-3
$CBrClF_2$
165.37
FC(F)(Br)Cl
Methane, bromochlorodifluoro
BCF
Bromochlorodifluoromethane
Chlorodifluorobromomethane [UN 1974]
Chlorodifluoromonobromomethane
Flugex 12B1
Fluorocarbon 1211
Freon 12B1
Halon 1211
UN 1974 [Chlorodifluorobromomethane]

353-61-7
C_4H_9F
76.11
FC(C)(C)C
Propane, 2-fluoro-2-methyl- (9CI)
tert-Butyl fluoride
2-Fluoro-2-methylpropane

353-66-2
$C_2H_6F_2Si$
96.15
Silane, difluorodimethyl- (9CI)
Difluorodimethylsilane
Dimethyldifluorosilane

354-06-3
$C_2HBrClF_3$
197.39
FC(F)(Br)C(F)Cl
Ethane, 1-bromo-2-chloro-1,1,2-trifluoro
1-Bromo-2-chloro-1,1,2-trifluoroethane
1,1,2-Trifluoro-1-bromo-2-chloroethane

354-23-4
$C_2HCl_2F_3$
152.93
FC(Cl)C(F)(F)Cl
Ethane, 1,2-dichloro-1,1,2-trifluoro
1,2-Dichloro-1,1,2-trifluoroethane
1,1,2-Trifluoro-1,2-dichloroethane

354-32-5
C_2ClF_3O
132.47
O=C(C(F)(F)F)Cl
Acetyl chloride, trifluoro- (9CI)
Trifluoroacetyl chloride

354-33-6
C_2HF_5
120.02
Ethane, pentafluoro- (8CI,9CI)
R 125

354-58-5
$C_2Cl_3F_3$
187.37
FC(F)(F)C(Cl)(Cl)Cl
Ethane, 1,1,1-trichloro-2,2,2-trifluoro
FC 113
FC133a
Freon FT
Precision Cleaning Agent
TF
T-WD602
Trichlorotrifluoroethane

354-64-3
C_2F_5I
245.92
FC(F)(F)C(F)(F)I
Ethane, pentafluoroiodo- (9CI)
Pentafluoroethyl iodide
Pentafluoroiodoethane

355-02-2
C_7F_{14}
350.07
FC(F)(F)C(F)(C(F)(F)C(F)(F)C(F)(F)C1(F)F)C1(F)F
Cyclohexane, 1-trifluoromethyl-1,2,2,3,3,4,4,5,5,6,6-undecafluoro
Perfluoromethylcyclohexane

355-25-9
C_4F_{10}
238.03
FC(F)(F)C(F)(F)C(F)(F)C(F)(F)F
Butane, decafluoro- (9CI)
Decafluorobutane
Perfluorobutane

355-42-0
C_6F_{14}
338.04
FC(F)(F)C(F)(F)C(F)(F)C(F)(F)C(F)(F)C(F)(F)F
Hexane, tetradecafluoro- (9CI)
Perfluorohexane
Perfluoro-n-hexane
Tetradecafluorohexane

355-68-0
C_6F_{12}
300.05
Cyclohexane, dodecafluoro- (9CI)
Dodecafluorocyclohexane
NSC-68382
Perfluorocyclohexane

357-56-2
$C_{25}H_{32}N_2O_2$
392.59
CC(CN1CCOCC1)C(C(=O)N2CCCC2)(c3ccccc3)c4ccccc4
Pyrrolidine, 1-(2,2-diphenyl-3-methyl-4-morpholino-butyryl)-, (+)
Alcoid
Dauran
Dextromoramide
Dimorlin
(+)-2,2-Diphenyl-3-methyl-4-morpholinobutyrylpyrrolidine
Jetrium
Jetrium R
Linfadol
MCP 875
Moramide
Narcolo
Palfadonna
Pyrrolamidol
Pyrrolamidolum
R 875
SKF 5137
Troxilan
Yetrium

357-57-3
$C_{23}H_{26}N_2O_4$
394.51
O=C(N(c(c(C1(C(N(C2)CC(C3C4C5OC6)=C6)C3)C2)cc(OC)c7OC)c7)C14)C5
Brucine
Brucin (German)
Brucina (Italian)
Brucine [UN 1570]
Brucine, Solid (DOT)
(-)-Brucine
Brucine alkaloid
Dimethoxy strychnine
2,3-Dimethoxy-strychnine
Dimethoxy strychnine (DOT)
10,11-Dimethystrychnine
RCRA waste number P018
Strychnidin-10-one, 2,3-dimethoxy- (9CI)
Strychnine, 2,3-dimethoxy-
UN 1570 [Brucine]

358-43-0
$C_6H_{15}FSi$
134.27
Silane, triethylfluoro- (8CI,9CI)

359-06-8
C_2H_2ClFO
96.49
O=C(Cl)CF
Acetyl chloride, fluoro
Chlorid kyseliny fluoroctove (Czech)
Fluoroacetyl chloride
TL 670

359-10-4
C_2HClF_2
98.48
Ethylene, 2-chloro-1,1-difluoro
Chloro-1,1-difluoroethylene
2-Chloro-1,1-difluoroethylene
1,1-Difluorochloroethylene
Ethene, 2-chloro-1,1-difluoro-

359-11-5
C_2HF_3
82.03
FC=C(F)F
Trifluoroethene
Ethene, trifluoro- (9CI)
Ethylene, trifluoro-
Trifluoroethylene

359-28-4
$C_2H_2Cl_3F$
151.40
Ethane, 1,1,2-trichloro-2-fluoro- (6CI,7CI,8CI,9CI)
1-Fluoro-1,2,2-trichloroethane
2-Fluoro-1,1,2-trichloroethane
1,1,2-Trichloro-2-fluoroethane

359-35-3
$C_2H_2F_4$
102.03
Ethane, 1,1,2,2-tetrafluoro- (8CI,9CI)
Freon 134
R 134

359-83-1
$C_{19}H_{27}NO$
285.47
CC2C3Cc1ccc(O)cc1C2(C)CCN3CC=C(C)C
2,6-Methano-3-benzazocin-8-ol, 1,2,3,4,5,6-hexahydro-6,11-dimethyl-3-(3-methyl-2-butenyl)
II-C-2
2-(3,3-Dimethylallyl)cyclazocine
2-Dimethylallyl-5,9-dimethyl-2'-hydroxybenzomorphan
2-(3,3-Dimethylallyl)-2',2'-hydroxy-5,9-dimethyl-6,7-benzomorphan
Fortalgesic
Fortalin
Fortral
2'-Hydroxy-5,9-dimethyl-2-(3,3-dimethylallyl)-6,7-benzomorphan
dl-2'-Hydroxy-5,9-dimethyl-2-(3,3-dimethylallyl)-6,7-benzomorphan
KF-1820
Liticon
3-(3-Methyl-2-butenyl)-1,2,3,4,5,6-hexahydro-6,11-dimethyl-2,6-methano-3-benzazocin-8-ol
NIH 7958
NSC-107430
Pentagin
Pentazocine

Sosigon
Talwan
Talwin
WIN 20228

360-46-3
C_4F_9NO
249.03
1,2-Oxazetidine, 3,3,4,4-tetrafluoro-2-(pentafluoroethyl)- (8CI)

360-68-9
$C_{27}H_{48}O$
388.75
5-β-Cholestan-3-β-ol
Cholestan-3-ol, (3-β,5-β)- (9CI)
3-β-Cholestanol
Coprostanol
Coprostan-3-β-ol
Coprosterol
Dihydrocholesterol
3-β-Hydroxycholestane
Koprosterin (German)
Stercorin
Zymostanol

360-89-4
C_4F_8
200.04
FC(F)(F)C(F)=C(F)C(F)(F)F
2-Butene, 1,1,1,2,3,4,4,4-octafluoro
FC-1318
Octafluorobutene-2
Octafluorobut-2-ene [UN 2422]
Perfluorobut-2-ene
Perfluoro-2-butene (DOT)
UN 2422 [Octafluorobut-2-ene]

361-37-5
$C_{21}H_{27}N_3O_2$
353.51
CCC(CO)NC(=O)C2CN(C)C1Cc3cn(C)c4cccc(C1=C2)c34
Ergoline-8-β-carboxamide, 9,10-didehydro-N-(1-(hydroxymethyl)propyl)-1,6-dimethyl
Deseril
Desernyl
Deseryl
Methyllysergic acid butanolamide
1-Methyllysergic acid butanolamide
Methysergid
Methysergide
UML 491

362-29-8
$C_{20}H_{24}N_2OS$
340.48
CCC(=O)c3ccc2Sc1ccccc1N(CC(C)N(C)C)c2c3
Propiomazine
CB 1678
10-Dimethylaminoisopropyl-2-propionylphenothiazine
1-(10-(2-(Dimethylamino)-propyl)phenothiazin-2-yl)-1-propanone
10-(2-Dimethylaminopropyl)-2-propionylphenothiazine
Phenoctyl
Propiomazina (Spanish)
Propiomazinum (Latin)
1-Propanone, 1-(10-(2-(dimethylamino)propyl)phenothiazin-2-yl)-
1-Propanone, 1-(10-(2-(dimethylamino)propyl)-10H-phenothiazin-2-yl)-
3-Propionyl-10-dimethylaminoisopropylphenothiazine
2-Propionyl-10-(2-(dimethylamino)propyl)phenothiazine
Propionylpromethazine
Wy-1359

363-72-4
C_6HF_5
168.07
Fc(c(F)cc(F)c1F)c1F
Benzene, pentafluoro- (9CI)
NSC-88293
Pentafluorobenzene
1,2,3,4,5-Pentafluorobenzene

364-62-5
$C_{14}H_{22}ClN_3O_2$
299.84
CCN(CC)CCNC(=O)c1cc(Cl)c(N)cc1OC
n-Anisamide, 4-amino-5-chloro-N-(2-(diethylamino)ethyl)
4-Amino-5-chloro-N-(2-(diethylamino)ethyl)-2-methoxybenzamide
o-Anisamide, 4-amino-5-chloro-N-(2-(diethylamino)ethyl)- (8CI)
Benzamide, 4-amino-5-chloro-N-(2-(diethylamino)ethyl)-2-methoxy- (9CI)
5-Chloro-2-methoxyprocainamide
DEL
DEL 1267
N-(Diethylaminoethyl)-2-methoxy-4-amino-5-chlorobenzamide
Maxolon
Metaclopramide
Metaclopromide
Methochlopramide
Methoclopramide
2-Methoxy-5-chloroprocainamide
Metochlopramide
Metoclol
Metoclopramide
Moriperan
Plasil
Primperan
Reliveran

364-76-1
$C_6H_5FN_2O_2$
156.13
O=N(=O)c(c(F)ccc1N)c1
Aniline, 4-fluoro-3-nitro
Benzenamine, 4-fluoro-3-nitro-
4-Fluoro-3-nitroaniline
4-F-3NA

365-07-1
$C_{10}H_{17}N_2O_8P$
324.26
5'-Thymidylic acid
Deoxyribosylthymine monophosphate
Deoxythymidine monophosphate
Deoxythymidine 5'-monophosphate
Deoxythymidine 5'-phosphate
Deoxythymydilic acid
Deoxy TMP
DPT
DTMP
5'-DTMP
Thymidine mononucleotide
Thymidine monophosphate
Thymidine 5'-monophosphate
Thymidine-5'-monophosphoric acid
Thymidine phosphate
Thymidine 5'-phosphate
Thymidine 5'-phosphoric acid
Thymidylic acid
TMP
TMP (Nucleotide)
5'-TMP

366-18-7
$C_{10}H_8N_2$
156.20
n(c(ccc1)c(nccc2)c2)c1
2,2'-Bipyridine
2,2'-Bipyridin
Bipyridine
α,α'-Bipyridine
α,α'-Bipyridyl
2,2'-Bipyridyl
C.I. 588
α,α'-Dipyridyl
2,2'-Dipyridyl

366-29-0
$C_{16}H_{20}N_2$
240.38
N(c(ccc(c(ccc(N(C)C)c1)c1)c2)c2)(C)C
Benzidine, N,N,N',N'-tetramethyl
(1,1'-Biphenyl)-4,4'-diamine, N,N,N',N'-tetramethyl-
4,4'-Bis(N,N-dimethylamino)-biphenyl
N,N,N',N'-Tetramethylbenzidine
N,N,N',N'-Tetramethyl-p,p'-benzidine

366-70-1
$C_{12}H_{19}N_3O.ClH$
257.80
Cl.CNNCc1ccc(cc1)C(=O)NC(C)C
p-Toluamide, N-isopropyl-α-(2-methylhydrazino)-, monohydrochloride
Benzamide, N-(1-methylethyl)-4-((2-methylhydrazino)-methyl)-, monohydrochloride
Benzamide, (p-(N'-methyl-hydrazinomethyl)-N-iso-propyl)-
Chlorhydrate de 1-methyl 2 p-(isopropylcarbamoyl)-benzyl-hydrazine (French)
Ibenzmethyzin hydrochloride
Ibenzmethyzine hydrochloride
IBZ
1-(p-Isopropylcarbamoyl-benzyl)-2-methylhydrazine hydrochloride
2-(p-(Isopropylcarbamoyl)-benzyl)-1-methylhydrazine hydrochloride
N-Isopropyl-α-(2-methylhydra-zino)-p-toluamide hydro-chloride
N-Isopropyl-p-(2-methylhydra-zinomethyl)benzamide hydro-chloride
Matulane
MBH
p-(N'-Methylhydrazinomethyl)-N-isopropylbenzamide hydro-chloride
1-Methyl-2-p-(isopropylcar-bamyol)benzohydrazine hydrochloride
1-Methyl-2-(p-isopropylcar-bamoylbenzyl)hydrazine hydrochloride
MIH
MIH Hydrochloride
Nathulane
Natulan
Natulanar
Natulan Hydrochloride
NCI-C01810
NSC-77213
PCB Hydrochloride
Procarbazin (German)
Procarbazine hydrochloride
Ro 4-6467
p-Toluamide, N-isopropyl-α-(2-methylhydrazino)-, hydrochloride

367-11-3
$C_6H_4F_2$
114.10
Fc1ccccc1F
Benzene, o-difluoro
o-Difluorobenzene
Orthodifluorobenzene

367-12-4
C_6H_5FO
112.11
Fc(c(O)ccc1)c1
Phenol, o-fluoro
o-Fluorophenol
2-Fluorophenol

367-21-5
C_6H_5ClFN
145.56
Fc(c(cc(N)c1)Cl)c1
3-Chloro-4-fluoroaniline
Aniline, 3-chloro-4-fluoro-
Benzenamine, 3-chloro-4-fluoro- (9CI)
3-Chloro-4-fluorobenzenamine
NSC-10290

367-25-9
$C_6H_5F_2N$
129.12
Fc(c(N)ccc1F)c1
Aniline, 2,4-difluoro
Benzenamine, 2,4-difluoro- (9CI)
2,4-Difluoroaniline

367-51-1
$C_2H_3O_2S.Na$
114.10
[Na+].OC(=O)CS
Acetic acid, mercapto-, mono-sodium salt
Mercaptoacetic acid sodium salt
Sodium mercaptoacetate
Sodium thioglycolate
Sodium thioglycollate
Thioglycolatesodium
Thioglycollic acid, sodium salt
USAF EK-5199

368-55-8
$C_4F_5N_3$
185.06
1,3,5-Triazine, 2,4-di fluoro-6-(trifluoro-methyl)- (9CI)
s-Triazine, 2,4-difluoro-6-(trifluoromethyl)- (6CI,8CI)

368-61-6
$C_{10}H_8F_3N_5$
255.20
1,3,5-Triazine-2,4-diamine, N-phenyl-6-(trifluoro-methyl)- (9CI)
s-Triazine, 2-amino-4-ani lino-6-(trifluoromethyl)- (6CI,7CI,8CI)

368-66-1
$C_6F_9N_3$
285.09
FC(F)(F)c1nc(nc(n1)C(F)(F)F)C(F)(F)F
s-Triazine, 2,4,6-tris(trifluoro-methyl)
2,4,6-Tris(trifluoromethyl)-s-triazine
TTT

369-58-4
$C_6H_5N_2.F_6P$
250.07
Benzenediazonium, hexa-fluorophosphate(1-) (9CI)
Phenyldiazonium hexafluoro-phosphate

369-77-7
$C_{14}H_9Cl_2F_3N_2O$
349.13
O=C(Nc(ccc(c1)Cl)c1)Nc(ccc(c2C(F)(F)F)Cl)c2
Cloflucarban
Carbanilide, 4,4'-dichloro-3-(trifluoromethyl)- (8CI)
N-(4-Chlorophenyl)-N'-(4-chloro-3-(trifluoro-methyl)phenyl)urea
Cloflucarbon
4,4'-Dichloro-3-(trifluoro-methyl)carbanilide
4,4'-Dichloro-3-(trifluoro-methyl)-carbanilide
Halocarban
Halocarbano (Spanish)
Halocarbanum (Latin)
NSC-114133
TFC
Trifluoromethyldichlorocar banilide
3-(Trifluoromethyl)-4,4'-di-chlorocarbanilide
3-Trifluoromethyl-4,4'-dichloro-N,N'-diphenylurea
Urea, N-(4-chlorophenyl)-N'-(4-chloro-3-(trifluoromethyl)-phenyl)- (9CI)

369-90-4
$C_{13}H_9F_3N_2O_2$
282.22
Benzenamine, N-(4-nitro-phenyl)-3-(trifluoro-methyl)- (9CI)
m-Toluidine, α,α,α-tri fluoro-N-(p-nitrophenyl)- (8CI)

370-32-1
$C_9H_8F_3NO_2$
219.16
Ethanol, 2,2,2-trifluoro-, phenylcarbamate (9CI)
Ethanol, 2,2,2-trifluoro-, carbanilate (7CI,8CI)
2,2,2-Trifluoroethyl carbanilate
2,2,2-Trifluoroethyl N-phenylcarbamate

371-40-4
C_6H_6FN
111.13
Fc(ccc(N)c1)c1
Aniline, 4-fluoro
Benzenamine, 4-fluoro- (9CI)
4-Fluoranilin (Czech)
p-Fluoroaniline
4-Fluoroaniline (DOT)
4-Fluorobenzenamine
p-Fluorophenylamine
UN 2944 (DOT)

371-41-5
C_6H_5FO
112.11
Fc(ccc(O)c1)c1
Phenol, p-fluoro
4-Fluorophenol

371-62-0
C_2H_5FO
64.07
FCCO
Ethanol, 2-fluoro
β-Fluoroethanol
2-Fluoroethanol
TL 741

371-86-8
$C_6H_{16}FN_2OP$
182.21
Phosphorodiamidic fluoride, N,N'-diisopropyl
Bis(isopropylamido) fluorophosphate
Bisisopropylaminofluorophosphine oxide
Bis(monoisopropylamino)-fluorophosphate
Bis(monoisopropylamino)-fluorophosphine oxide
N,N'-Diisopropil-fosforodi-ammido-fluoruro (Italian)
Di(isopropylamido)phosphoryl fluoride
N,N'-Diisopropyl-diamido-fosforzuur-fluoride (Dutch)
N,N'-Diisopropyl-diamido-phosphorsaeure-fluorid (German)
N,N'-Diisopropyldiamidophosphoryl fluoride
N,N'-Diisopropylphosphorodiamidic fluoride
Fluorobisisopropylamino-phosphine oxide
Fluorure de N,N'-diisopropyle phosphorodiamide (French)
Isopestox
Mipafox [UN 2783]
Pestox 15
Peston XV
Pestox XV
Phosphine oxide, fluorobis-(isopropylamino)-
Phosphorodi(isopropylamidic) fluoride
UN 2783 [Organophosphorus pesticides, solid, toxic, N.O.S.]

372-09-8
$C_3H_3NO_2$
85.07
O=C(O)CC#N
Acetic acid, cyano
Acide cyanacetique (French)
CAA
Cyanessigsaeure (German)
Cyanoacetic acid
Kyselina kyanoctova (Czech)
Malonic mononitrile
Monocyanoacetic acid
USAF KF-17

372-18-9
$C_6H_4F_2$
114.10
Fc1cccc(F)c1
Benzene, m-difluoro
m-Difluorobenzene
Metadifluorobenzene

372-19-0
C_6H_6FN
111.13
Nc1cccc(F)c1
Aniline, 3-fluoro
3-Fluoranilin (Czech)
3-Fluoroaniline

372-20-3
C_6H_5FO
112.10
Oc1cccc(F)c1
Phenol, m-fluoro- (8CI)
m-Fluorophenol
3-Fluorophenol
NSC-87078
Phenol, 3-fluoro- (9CI)

372-31-6
$C_6H_7F_3O_3$
184.11
O=C(OCC)CC(=O)C(F)(F)F
Butanoic acid, 4,4,4-trifluoro-3-oxo-, ethyl ester (9CI)
Acetoacetic acid, 4,4,4-trifluoro-, ethyl ester
AI3-52657
Ethyl trifluoroacetoacetate
Ethyl 4,4,4-trifluoroacetoacetate
Ethyl (trifluoroacetyl)acetate
NSC-42739

372-38-3
$C_6H_3F_3$
132.09
Fc1cc(F)cc(F)c1
Benzene, 1,3,5-trifluoro- (8CI, 9CI)
NSC-10264

373-02-4
$C_4H_6O_4.Ni$
176.81
[Ni+2].CC([O-])=O.CC([O-])=O
Nickel(II) acetate (1:2)
Acetic acid, nickel(2+) salt
Nickelous acetate

373-14-8
$C_6H_{13}F$
104.19
FCCCCCC
Hexane, 1-fluoro
Fluorohexane
1-Fluorohexane

373-49-9
$C_{16}H_{30}O_2$
254.41
O=C(O)CCCCCCCC=CCCCCCC
9-Hexadecenoic acid, (Z)-(9CI)
AI3-36443
cis-9-Hexadecenoic acid
9-cis-Hexadecenoic acid
NSC-277452
Palmitoleic acid
Palmitolinoleic acid

374-01-6
$C_3H_5F_3O$
114.07
FC(F)(F)C(O)C
2-Propanol, 1,1,1-trifluoro-(9CI)
1-Methyl-2,2,2-trifluoroethanol
NSC-3637
1,1,1-Trifluoro-2-propanol

374-07-2
$C_2Cl_2F_4$
170.92
FC(F)(F)C(F)(Cl)Cl
1,1-Dichloro-1,2,2,2-tetrafluoroethane
Dichlorotetrafluoroethane
1,1-Dichlorotetrafluoroethane
Ethane, 1,1-dichlorotetrafluoro-
Ethane, 1,1-dichloro-1,2,2,2-tetrafluoro- (9CI)
Frigen 114a
1,1,1,2-Tetrafluoro-2,2-dichloroethane

375-01-9
$C_4H_3F_7O$
200.07
FC(F)(F)C(F)(F)C(F)(F)CO
1-Butanol, 2,2,3,3,4,4,4-heptafluoro
α,α-Dihydroperfluorobutanol
1,1-Dihydroperfluorobutanol
1,1-H,H-Heptafluorobutanol
2,2,3,3,4,4,4-Heptafluorobutanol

375-22-4

$C_4HF_7O_2$
214.05
O=C(O)C(F)(F)C(F)(F)C(F)(F)F
Butyric acid, heptafluoro
Heptafluorobutyric acid
Kyselina heptafluormaselna (Czech)

375-72-4
$C_4F_{10}O_2S$
302.09
O=S(=O)(F)C(F)(F)C(F)(F)C(F)(F)C(F)(F)F
1-Butanesulfonyl fluoride, 1,1,2,2,3,3,4,4,4-nonafluoro- (9CI)
1,1,2,2,3,3,4,4,4-Nonafluoro-1-butanesulfonyl fluoride

375-81-5
$C_5F_{12}O_2S$
352.10
O=S(=O)(F)C(F)(F)C(F)(F)C(F)(F)C(F)(F)C(F)(F)F
1-Pentanesulfonyl fluoride, 1,1,2,2,3,3,4,4,5,5,5-undecafluoro- (8CI)
1,1,2,2,3,3,4,4,5,5,5-Undecafluoro-1-pentanesulfonyl fluoride

375-83-7
C_7HF_{15}
370.06
Heptane, 1,1,1,2,2,3,3,4,4,5,5,6,6,7,7-pentadecafluoro- (6CI,7CI,8CI,9CI)
1-Hydroperfluoroheptane
1H-Pentadecafluoroheptane
1H-Perfluoroheptane

376-14-7
$C_{16}H_{14}F_{17}NO_4S$
639.32
O=C(OCCN(S(=O)(=O)C(F)(F)C(F)(F)C(F)(F)C(F)(F)C(F)(F)C(F)(F)C(F)(F)C(F)(F)F)CC)C(=C)C
2-Propenoic acid, 2-methyl-, 2-(Ethyl((heptadecafluorooctyl)sulfonyl)amino)ethyl ester (9CI)

378-44-9
$C_{22}H_{29}FO_5$
392.51
OCC(=O)C3(O)CCC4C2CCC1=CC(=O)C=CC1C2(F)C(O)CC34
Pregna-1,4-diene-3,20-dione, 9-fluoro-11-β,17,21-trihydroxy-16-β-methyl
Betamethasone
Betamethazone
Betnelan
Betsolan
Celestone
9-α-Fluoro-16-β-methylprednisolone
9-α-Fluoro-16-β-methyl-1,4-pregnadiene-11-β,17-α,21-triol-3,20-dione
9-Fluoro-11-β,17,21-trihydroxy-16-β-methylpregna-1,4-diene-3,20-dione
9-α-Fluoro-11-β,17,21-trihydroxy-16-β-methylpregna-1,4-diene-3,20-dione
9-α-Fluoro-11-β,17-α,21-trihydroxy-16-β-methylpregna-1,4-diene-3,20-dione
16-β-Methyl-1,4-pregnadiene-9-α-fluoro-11-β,17-α,21-triol-3,20-dione
NSC-39470
Sch 4831

379-34-0
$C_{12}H_{12}N_2O_4$
248.23
4-Hydroxyphenobarbital
5-Ethyl-5-(4-hydroxyphenyl)-2,4,6(1H,3H,5H)-pyrimidinetrione
NSC-159266

379-52-2
$C_{18}H_{15}FSn$
369.02
Stannane, fluorotriphenyl
Biomet 204
Fluorotriphenylstannane
Tin, fluorotriphenyl-
Triphenyltin fluoride

379-79-3
$C_{66}H_{70}N_{10}O_{10}.C_4H_6O_6$
1313.56
Ergotamine-tartrate
Ergam
Ergate
Ergomar
Ergostat
Ergotamine bitartrate
Ergotartrate
Etin
Exmigra
Femergin
Gotamine tartrate
Gynergen
Lingraine
Lingran
Neo-Ergotin
Rigetamin
Secagyn
Secupan

381-73-7
$C_2H_2F_2O_2$
96.04
O=C(O)C(F)F
Acetic acid, difluoro
Difluoroacetic acid

382-10-5
$C_4H_2F_6$
164.06
FC(F)(F)C(C(F)(F)F)=C
Propene, 3,3,3-trifluoro-2-(trifluoromethyl)
1,1-Bis(trifluoromethyl)ethene
Hexafluoroisobutylene
3,3,3,4,4,4-Hexafluoroisobutylene
3,3,3-Trifluoro-2-(trifluoromethyl)propene

382-21-8
C_4F_8
200.04
FC(F)=C(C(F)(F)F)C(F)(F)F
1-Propene, 1,1,3,3,3-pentafluoro-2-trifluoromethyl
Isobutene, octafluoro-
Octafluoroisobutylene
Octafluoro-sec-butene
Perfluoroisobutylene
PFIB

383-07-3
$C_{17}H_{16}F_{17}NO_4S$
653.35
O=C(OCCN(S(=O)(=O)C(F)(F)C(F)(F)C(F)(F)C(F)(F)C(F)(F)C(F)(F)C(F)(F)C(F)(F)F)CCCC)C=C
2-Propenoic acid, 2-(butyl((heptadecafluorooctyl)sulfonyl)amino)ethyl ester (9CI)
2-Propenoic acid, 2(((heptadecafluorooctyl)sulfonyl)butylamino)ethyl ester

383-63-1
$C_4H_5F_3O_2$
142.08
O=C(OCC)C(F)(F)F
Acetic acid, trifluoro-, ethyl ester (9CI)
AI3-52221
Ethyl trifluoroacetate
Ethyl trifluoroethanoate
NSC-220215
Trifluoroacetic acid, ethyl ester

384-22-5
$C_7H_4F_3NO_2$
191.12
FC(F)(F)c1ccccc1N(=O)=O
Toluene, 2-nitro-α,α,α-trifluoro
Benzene, 1-nitro-2-(trifluoromethyl)- (9CI)
o-Nitrobenzotrifluoride [UN 2306]
o-(Trifluoromethyl)nitrobenzene
UN 2306 [Nitrobenzotrifluorides]

385-00-2

$C_7H_4F_2O_2$
158.11
Benzoic acid, 2,6-difluoro
2,6-Difluorobenzoic acid

385-13-7
$C_{30}H_{16}$
376.46
Tetrabenzo(de,hi,mn,qr) naphthacene (7CI,8CI, 9CI)
Pyreno(1',2':1,2)pyrene
Pyreno(4,5-e)pyrene

385-14-8
$C_{28}H_{16}$
352.44
Benzo(p)naphtho-(1,8,7-ghi)chrysene-(7CI,8CI,9CI)

389-08-2
$C_{12}H_{12}N_2O_3$
232.26
CCn1cc(C(O)=O)c(=O)c2ccc(C)nc12
1,8-Naphthyridine-3-carboxylic acid, 1-ethyl-1,4-dihydro-7-methyl-4-oxo
Acide 1-etil-7-metil-1,8-naftiridin-4-one-3-carbossilico (Italian)
Acide nalidixico (Italian)
Acide nalidixique (French)
1-Aethyl-7-methyl-1,8-naphthyridin-4-on-3-karbonsaeure (German)
Betaxina
3-Carboxy-1-ethyl-7-methyl-1,8-naphthidin-4-one
Chinoin
CYBIS
1,4-Dihydro-1-ethyl-7-methyl-4-oxo-1,8-naphthyridine-3-carboxylic acid
Dixiben
1-Ethyl-1,4-dihydro-7-methyl-4-oxo-1,8-naphthyridine-3-carboxilic acid
1-Ethyl-7-methyl-1,4-dihydro-1,8-naphthyridin-4-one-3-carboxylic acid
1-Ethyl-7-methyl-4-oxo-1,4-dihydro-1,8-naphthyridine-3-carboxylic acid
Eucisten
Innoxalon
Kusnarin
NA
Nalidic acid
Nalidicron
Nalidixic acid
Nalidixin
Nalidixinic acid
Nalitucsan
Nalix
Nalurin
Narigix
Naxuril
NCI-C56199
Neggram
Negram
Nevigramon
Nicelate
Nogram
NSC-82174
Poleon
Specifen
Uralgin
Uriben
Uriclar
Urisal
Urodixin
Uroman
Uroneg
Uropan
WIN 18,320
Wintomylon

390-64-7
$C_{24}H_{27}N$
329.52
Phenethylamine, N-(3,3-diphenylpropyl)-α-methyl
B-436
Benzenepropanamine, N-(1-methyl-2-phenylethyl)-γ-phenyl-
Bismethin
Carditin
Corontin
Corpax
N-(3,3-Diphenylpropyl)-α-methylphenaethylamin (German)
N-(3,3-Diphenylpropyl)-α-methylphenethylamine
Elecor
Falliocor
Hostaginan
N-(1-Methyl-2-phenylethyl)-γ-phenylbenzenepropanamine
1-Phenyl-2-(1',1'-diphenylpropyl-3'-amino)propane
N-(3'-Phenylo-2-propylo)-1,1-diphenylo-3-propyloamine (Polish)
Prenylamine
Segontin
Synadrin
Valecor

392-56-3
C_6F_6
186.06
Fc(c(F)c(F)c(F)c1F)c1F
Benzene, hexafluoro
Hexafluorobenzene
Perfluorobenzene

392-85-8
$C_7H_4F_4$
164.10
FC(F)(F)c(c(F)ccc1)c1
Benzene, 1-fluoro-2-(trifluoromethyl)- (9CI)
o-Fluorobenzotrifluoride
2-Fluorobenzotrifluoride
1-Fluoro-2-(trifluoromethyl)-benzene
NSC-10314
o,α,α,α-Tetrafluorotoluene

393-52-2
C_7H_4ClFO
158.56
O=C(c(c(F)ccc1)c1)Cl
Benzoyl chloride, o-fluoro
o-Fluorobenzoyl chloride
2-Fluorobenzoyl chloride

393-75-9
$C_7H_2ClF_3N_2O_4$
270.56
O=N(=O)c(c(c(N(=O)=O)cc1C(F)(F)F)Cl)c1
Toluene, 4-chloro-3,5-dinitro-α,α,α-trifluoro
Benzene, 2-chloro-1,3-dinitro-5-(trifluoromethyl)-
Benzotrifluoride, 4-chloro-3,5-dinitro-
4-Chloro-3,5-dinitrobenzotrifluoride
4-Chloro-3,5-dinitro-α,α,α-trifluorotoluene
3,5-Dinitro-4-chloro-α,α,α-trifluorotoluene

395-28-8
$C_{18}H_{23}NO_3$
301.42
CC(COc1ccccc1)NC(C)C(O)c2ccc(O)cc2
Benzyl alcohol, p-hydroxy-α-(1-((1-methyl-2-phenoxyethyl)amino)ethyl)
Dilavase
Duvadilan
p-Hydroxy-N-(1-methyl-2-phenoxyethyl)norephedrine
1-(4-Hydroxyphenyl)-2-(1-methyl-2-phenoxyethylamino)-propanol
1-(p-Hydroxyphenyl)-2-(1'-methyl-2'-phenoxyethylamino)propanol-2-hydrochloride
Isoxsuprine
2-(Phenoxy-2-propylamino)-1-(p-hydroxyphenyl)-1-propanol hydrochloride
Vasodilan
Vasodilian

396-01-0
$C_{12}H_{11}N_7$
253.30
Nc3nc(N)c2nc(c1ccccc1)c(N)nc2n3
Pteridine, 2,4,7-triamino-6-phenyl
Ademine
Diren
Ditak
Dyrenium

Dytac
Jatropur
NCI-C56042
Noridil
6-Phenyl-2,4,7-pteridinetriamine
6-Phenyl-2,4,7-triaminopteridine
2,4,7-Pteridinetriamine, 6-phenyl- (9CI)
Pterofen
Pterophene
SKF 8542
Taturil
Teriam
Teridin
2,4,7-Triamino-6-fenilpteridina (Italian)
2,4,7-Triamino-6-phenylpteridine
Triampur
Triamteren
Triamterene
Triamteril
Triamteril Complex
Tri-Span
Triteren

398-23-2
$C_{12}H_8F_2$
190.19
Biphenyl, 4,4'-difluoro- (8CI)
AI3-19148
1,1'-Biphenyl, 4,4'-difluoro- (9CI)
4,4'-Difluorodiphenyl
NSC-4173

400-99-7
$C_7H_4F_3NO_3$
207.12
O=N(=O)c(c(O)ccc1C(F)(F)F)c1
p-Cresol, 2-nitro-α,α,α-trifluoro
2-Nitro-α,α,α-trifluoro-p-cresol

401-78-5
$C_7H_4BrF_3$
225.02
FC(F)(F)c(cccc1Br)c1
Toluene, m-bromo-α,α,α-trifluoro
Benzene, 1-bromo-3-(trifluoromethyl)-
3-Brombenzotrifluorid (Czech)
m-Bromobenzotrifluoride
3-Bromobenzotrifluoride
3-Bromobenzyltrifluoride
m-Bromo(trifluoromethyl)benzene
3-Bromotrifluoromethylbenzene
Toluene, α,α,α-trifluoro-3-bromo-
m-(Trifluoromethyl)bromobenzene
3-(Trifluoromethyl)bromobenzene
m-(Trifluoromethyl)phenyl bromide
3-(Trifluoromethyl)phenyl bromide

402-31-3
$C_8H_4F_6$
214.11
FC(F)(F)c(cccc1C(F)(F)F)c1
Benzene, 1,3-bis(trifluoromethyl)- (9CI)
AI3-52239
m-Bis(trifluoromethyl)benzene
1,3-Bis(trifluoromethyl)benzene
Hexafluoro-m-xylene
α,α,α,α',α',α'-Hexafluoro-m-xylene
NSC-10342
m-Xylene, α,α,α,α',α',α'-hexafluoro- (8CI)

402-54-0
$C_7H_4F_3NO_2$
191.12
Toluene, 4-nitro-α,α,α-trifluoro
Benzene, 1-nitro-4-(trifluoromethyl)- (9CI)
p-Nitrobenzotrifluoride [UN 2306]
p-Nitro(trifluoromethyl)benzene
4-(Trifluoromethyl)nitrobenzene
UN 2306 [Nitrobenzotrifluorides]

402-67-5
$C_6H_4FNO_2$
141.10
Fc1cccc(c1)N(=O)=O
Benzene, 1-fluoro-3-nitro- (9CI)
m-Fluoronitrobenzene
3-Fluoronitrobenzene
1-Fluoro-3-nitrobenzene
m-Nitrofluorobenzene
NSC-60651

403-19-0
$C_6H_4FNO_3$
157.10
Phenol, 2-fluoro-4-nitro- (8CI, 9CI)

403-24-7
$C_7H_4FNO_4$
185.11
Benzoic acid, 2-fluoro-4-nitro- (8CI,9CI)
NSC-190361

403-42-9
C_8H_7FO
138.14
O=C(c(ccc(F)c1)c1)C
Acetophenone, 4'-fluoro- (8CI)
Ethanone, 1-(4-fluorophenyl)- (9CI)
p-Fluoroacetophenone
4-Fluoroacetophenone
4'-Fluoroacetophenone
1-(4-Fluorophenyl)ethanone
NSC-30635

404-86-4
$C_{18}H_{27}NO_3$
305.46
O=C(NCc(ccc(O)c1OC)c1)CCCCC=CC(C)C
6-Nonenamide, N-((4-hydroxy-3-methoxyphenyl)methyl)-8-methyl-, (E)
Capsaicin
Capsaicine
N-((4-Hydroxy-3-methoxyphenyl)methyl)-8-methyl-6-nonenamide
trans-N-((4-Hydroxy-3-methoxyphenyl)methyl)-8-methyl-6-nonenamide
trans-8-Methyl-N-vanillyl-6-nonenamide
NCI-C56564
6-Nonenamide, 8-methyl-N-vanillyl-, (E)- (8CI)

405-30-1
$C_{13}H_{10}ClFS$
252.74
Fc2ccc(SCc1ccc(Cl)cc1)cc2
Sulfide, p-chlorobenzyl p-fluorophenyl
p-Chlorobenzyl p-fluorophenyl sulfide
4-Chlorobenzyl 4'-fluorophenyl sulfide
p-Chlorobenzyl p-fluorophenyl sulphide
1-Chloro-4-(((4-fluorophenyl)-thio)methyl)benzene
Fluorbenside
Fluoroparacide
Fluorparacide
Fluorosulfacide
Fluorosulphacide
HRS 942
RD 2454

405-50-5
$C_8H_7FO_2$
154.14
O=C(O)Cc(ccc(F)c1)c1
Acetic acid, (p-fluorophenyl)- (8CI)
AI3-52627
Ba 2821
Benzeneacetic acid, 4-fluoro- (9CI)
4-Fluorobenzeneacetic acid
(p-Fluorophenyl)acetic acid
(4-Fluorophenyl)acetic acid
NSC-402

406-90-6
$C_4H_5F_3O$
126.09
FC(F)(F)COC=C
Ether, 2,2,2-trifluoroethyl

vinyl
Ethene, (2,2,2-trifluoroethoxy)-
Floroxene
Fluooxene
Fluoromar
Fluoroxene
Fluorxene
Fluroxene
2,2,2-Trifluoroethyl vinyl ether

407-25-0
$C_4F_6O_3$
210.04
O=C(OC(=O)C(F)(F)F)C(F)(F)F
Acetic acid, trifluoro-, anhydride
Anhydrid kyseliny trifluoroctove (Czech)
Bis(trifluoroacetic) anhydride
Hexafluoroacetic anhydride
Perfluoroacetic anhydride
Trifluoroacetic acid anhydride
Trifluoroacetic anhydride
Trifluoroacetyl anhydride

407-96-5
$C_7H_{14}F_2$
136.19
Heptane, 1,1-difluoro- (6CI,7CI,8CI,9CI)
1,1-Difluoroheptane

408-35-5
$C_{16}H_{32}O_2.Na$
279.47
Palmitic acid, sodium salt
Hexadecanoic acid, sodium salt (9CI)
Sodium hexadecanoate
Sodium palmitate
Sodium pentadecanecarboxylate

409-02-9
$C_8H_{14}O$
126.22
CC(C)C=CCCC(C)=O
5-Hepten-2-one, 6-methyl
Methyl heptenone
6-Methyl-5-hepten-2-one

409-21-2
CSi
40.10
Silicon-carbide
Carbolon
Carbon silicide
Carborundeum
Carborundum
KZ 3M
KZ 5M
KZ 7M
Silicon carbide (ACGIH,OSHA)
Silicon monocarbide
Silundum

414-29-9
$C_{22}H_{14}$
278.36
Dibenzanthracene

420-04-2
CH_2N_2
42.05
N#CN
Cyanamide (ACGIH,OSHA)
Amidocyanogen
Carbamonitrile
Carbimide
Carbodiimide
Cyanoamine
N-Cyanoamine
Cyanogenamide
Cyanogen nitride
Hydrogen Cyanamide
USAF EK-1995

420-05-3
CHNO
43.02
Cyanic acid (8CI,9CI)

420-12-2
C_2H_4S
60.12
S(C1)C1
Ethylene-sulfide
Aethylensulfid (German)
2,3-Dihydrothiirene
Ethylene episulfide
Ethylene episulphide
Ethylene sulphide
Thiacyclopropane
Thiirane
Thiirene, 2,3-dihydro-

420-26-8
C_3H_7F
62.09
Propane, 2-fluoro- (8CI,9CI)

420-46-2
$C_2H_3F_3$
84.05
Ethane, 1,1,1-trifluoro
FC143a
Fluorocarbon FC143a
Methylfluoroform
R 143a
1,1,1-Trifluoroethane
1,1,1-Trifluoroform

420-56-4
C_3H_9FSi
92.19
Trimethylfluorosilane
Silane, fluorotrimethyl- (9CI)

421-20-5
CH_3FO_3S
114.10
COS(F)(=O)=O
Fluorosulfuric acid, methyl ester
Magic Methyl
Methyl fluorosulfate
Methyl fluorosulfonate
Methyl fluosulfonate

421-50-1
$C_3H_3F_3O$
112.05
O=C(C(F)(F)F)C
2-Propanone, 1,1,1-trifluoro- (9CI)
Methyl trifluoromethyl ketone
NSC-66412
1,1,1-Trifluoroacetone
3,3,3-Trifluoroacetone
Trifluoromethyl methyl ketone
1,1,1-Trifluoro-2-propanone

421-75-0
$C_3H_3ClF_4$
150.50
Propane, 1-chloro-1,1,2,2-tetrafluoro- (8CI)

422-02-6
$C_3H_2ClF_5$
168.49
Propane, 3-chloro-1,1,1,2,2-pentafluoro- (8CI)

422-05-9
$C_3H_3F_5O$
150.06
FC(F)(F)C(F)(F)CO
1-Propanol, 2,2,3,3,3-pentafluoro
2,2,3,3,3-Pentafluoropropanol
2,2,3,3,3-Pentafluoro-1-propanol

422-55-9
C_3HClF_6
186.49
Propane, 1-chloro-1,1,2,2,3,3-hexafluoro
1-Chloro-1,1,2,2,3,3-hexafluoropropane

423-50-7
$C_6F_{14}O_2S$
402.11
O=S(=O)(F)C(F)(F)C(F)(F)C(F)(F)C(F)(F)C(F)(F)C(F)(F)F
1-Hexanesulfonyl fluoride, 1,1,2,2,3,3,4,4,5,5,6,6,6-tridecafluoro- (9CI)
1,1,2,2,3,3,4,4,5,5,6,6,6-Tridecafluoro-1-hexanesulfonyl fluoride

427-00-9
$C_{17}H_{21}NO_2$
271.35
Desomorphine

DEA No. 9055
Desomorfin (Czech)
Desomorfina (Spanish)
Desomorphinum (Latin)
Dihydrodeoxymorphine
Dihydrodesoxymorphine-d
4,5-Epoxy-3-hydroxy-N-methyl-morphinan
Morphinan-3-ol, 4,5-α-epoxy-17-methyl-
Morphine, 6-deoxy-7,8-dihydro-
Permonid

428-59-1
C_3F_6O
166.03
FC(F)(O1)C1(F)C(F)(F)F
Propane, 1,2-epoxy-1,1,2,3,3,3-hexafluoro
Hexafluoroepoxypropane
Hexafluoro-1,2-epoxypropane
Hexafluoropropene epoxide
Hexafluoropropene oxide
Hexafluoropropylene oxide [NA 1956]
NA 1956 [Hexafluoropropylene oxide]
Oxirane, trifluoro(trifluoromethyl)-
Perfluoro(methyloxirane)
Perfluoropropylene oxide
Propylene oxide hexafluoride
(Trifluoromethyl)trifluorooxirane
Trifluoro(trifluoromethyl)-oxirane

430-51-3
C_3H_5FO
76.07
CC(=O)CF
2-Propanone, 1-fluoro- (8CI, 9CI)
NSC-21302

430-66-0
$C_2H_3F_3$
84.04
Ethane, 1,1,2-trifluoro- (8CI, 9CI)
Freon 143

431-03-8
$C_4H_6O_2$
86.10
O=C(C(=O)C)C
2,3-Butanedione [UN 2346]
Biacetyl
Butadione
2,3-Butadione
Diacetyl (DOT)
2,3-Diketobutane
Dimethyl diketone
Dimethylglyoxal
Glyoxal, dimethyl-
UN 2346 [Butanedione]

431-47-0
$C_3H_3F_3O_2$
128.05
O=C(OC)C(F)(F)F
Acetic acid, trifluoro-, methyl ester (9CI)
Methyl trifluoroacetate

432-08-6
$C_{18}F_{39}N$
971.14
1-Hexanamine, 1,1,2,2,3,3,4,4,5,5,6,6,6-tridecafluoro-N,N-bis(tridecafluorohexyl)-(9CI)

432-25-7
$C_{10}H_{16}O$
152.24
O=CC(=C(CCC1)C)C1(C)C
1-Cyclohexene-1-carboxaldehyde, 2,6,6-trimethyl-
AI3-37227

434-07-1
$C_{21}H_{32}O_3$
332.53
5-α,17-β-Androstan-3-one, 17-hydroxy-2-(hydroxymethylene)-17-methyl
Adroidin
Adroyd
Anadrol
Anadroyd
Anapolon
Anasteron
Anasteronal
Anasterone
Androstan-3-one, 17-hydroxy-2-(hydroxymethylene)-17-methyl-, (5-α,17-β)-
5-α-Androstan-3-one, 17-β-hydroxy-2-(hydroxymethylene)-
17-methyl- (8CI)
Androstano(2,3-c)(1,2,5)oxadiazol-17-ol, 17-methyl-, (5-α,17-β)-
Becorel
C.I. 406
4,5-Dihydro-2-hydroxymethylene-17-α-methyltestosterone
Dynasten
HMD
17-Hydroxy-2-(hydroxymethylene)-17-methyl-5-α-17-β-androst-3-one
17-β-Hydroxy-2-hydroxymethylene-17-α-methyl-3-androstanone
17-β-Hydroxy-2-(hydroxymethylene)-17-α-methyl-5-α-androstan-3-one
17-β-Hydroxy-2-(hydroxymethylene)-17-methyl-5-α-androstan-3-one
2-Hydroxymethylene-17-α-methyl-5-α-androstan-17-β-ol-3-one
2-Hydroxymethylene-17-α-methyl-dihydrotestosterone
2-(Hydroxymethylene)-17-α-methyldihydrotestosterone
2-Hydroxymethylene-17-α-methyl-17-β-hydroxy-3-androstanone
Methabol
17-α-Methyl-2-hydroxymethylene-17-hydroxy-5-α-androstan-3-one
Nastenon
NSC-26198
Oximetholonum
Oximetolona
Oxitosona-50
Oxymethalone
Oxymethenolone
Oxymetholone
Pavisoid
Plenastril
Protanabol
Roboral
Synasteron
Zenalosyn

434-13-9
$C_{24}H_{40}O_3$
376.64
CC(CCC(O)=O)C3CCC4C2CCC1CC(O)CCC1(C)C2CCC34C
5-β-Cholan-24-oic acid, 3-α-hydroxy
5-β-Cholanic acid, 3-α-hydroxy-
Cholan-24-oic acid, 3-hydroxy-, (3-α,5-β)- (9CI)
3-α-Hydroxycholanic acid
3-α-Hydroxy-5-β-cholanic acid
Lithocholic acid
17-β-(1-Methyl-3-carboxypropyl)ethiocholan-3-α-ol
NCI-C03861

434-16-2
$C_{27}H_{44}O$
384.71
OC(CCC(C1=CC=C2C(C(C(C3)C(CCCC(C)C)C)(CC4)C)C3)(C24)C)C1
Cholesta-5,7-dien-3-β-ol
Cholesta-5,7-dien-3-ol, (3-β)-(9CI)
5,7-Cholestadien-3-β-ol
δ^7-Cholesterol
$\delta^{5,7}$-Cholesterol
Cholesterol, 7-dehydro-
Dehydrocholesterin (German)
7-Dehydrocholesterin
Dehydrocholesterol
7-Dehydrocholesterol
7,8-Didehydrocholesterol
Provitamin D_3

434-22-0
$C_{18}H_{26}O_2$
274.44
CC34CCC1C(CCC2=CC(=O)CCC12)C3CCC4O

Estr-4-en-3-one, 17-β-hydroxy
Estr-4-en-3-one, 17-hydroxy-, (17-β)- (9CI)
17-β-Hydroestr-4-en-3-one
Menidrabol
Nandrolon
Nandrolone
Norandrostenolon
Norandrostenolone
19-Norandrostenolone
Nortestonate
Nortestosterone
19-Nortestosterone
(+)-19-Nortestosterone
Oestrenolon

434-45-7
$C_8H_5F_3O$
174.12
O=C(c(cccc1)c1)C(F)(F)F
Acetophenone, 2,2,2-trifluoro- (8CI)
Ethanone, 2,2,2-trifluoro-1-phenyl- (9CI)
NSC-42752
Phenyl trifluoromethyl ketone
Trifluoroacetophenone
α,α,α-Trifluoroacetophenone
1,1,1-Trifluoroacetophenone
2,2,2-Trifluoroacetophenone
Trifluoromethyl phenyl ketone
2,2,2-Trifluoro-1-phenylethanone

434-64-0
C_7F_8
236.07
FC(F)(F)c(c(F)c(F)c(F)c1F)c1F
Toluene, octafluoro
Perfluorotoluene

435-02-9
$C_{38}H_{20}$
476.58
Dibenzo(de,yz)naphtho-(8,1,2-hij)hexaphene-(7CI,8CI,9CI)

435-97-2
$C_{18}H_{16}O_3$
280.34
CCC(c1ccccc1)c3c(O)c2ccccc2oc3=O
Coumarin, 3-(α-ethylbenzyl)-4-hydroxy
2H-1-Benzopyran-2-one, 4-hydroxy-3-(1-phenylpropyl)-(9CI)
Falithrom
Fencumar
Liquamar
Marcoumar
Marcumar
Phenprocoumarol
Phenprocoumarole
Phenprocoumon
3-(1'-Phenyl-propyl)-4-oxy-coumarin (German)
Ro 1-4849

437-38-7
$C_{22}H_{28}N_2O$
336.52
CCC(=O)N(C1CCN(CC1)CCc2ccccc2)c3ccccc3
Propionanilide, N-(1-phenethyl-4-piperidyl)
Fentanest
Fentanil
Fentanyl
Pentanyl
Phentanyl
N-Phenethyl-4-(n-propionyl-anilino)piperidine
1-Phenethyl-4-n-propionyl-anilinopiperidine
Propanamide, N-phenyl-N-(1-(2-phenylethyl)-4-piperidinyl)- (9CI)
R 4263
Sentonil

438-60-8
$C_{19}H_{21}N$
263.41
CNCCCC2c1ccccc1C=Cc3ccccc23
5H-Dibenzo(a,d)cycloheptene-5-propylamine, N-methyl
N-3-(5H-Dibenzo(a,d)cyclohepten-5-yl)propyl-N-methylamine
5-(3-Methylaminopropyl)-5H-dibenzo(a,d)cycloheptene
MK 240
Protriptyline
Protryptyline
Triptil
Vivactil

438-67-5
$C_{18}H_{22}O_5S.Na$
373.45
Estrone, hydrogen sulfate, sodium salt
Conestoral
Estra-1,3,5(10)-trien-17-one, 3-(sulfooxy)-, sodium salt (9CI)
Estrone sodium sulfate
Estrone sulfate sodium
Estrone sulfate sodium salt
Estrone-3-sulfate sodium salt
Evex
Morestin
Oestrone-3-sulphate sodium salt
Sodium estrone sulfate
Sodium estrone-3-sulfate

439-14-5
$C_{16}H_{13}ClN_2O$
284.76
O=C(N(c(c(cc(c1)Cl)C(=N2)c(cccc3)c3)c1)C)C2
2H-1,4-Benzodiazepin-2-one, 7-chloro-1,3-dihydro-1-methyl-5-phenyl
Alboral
Aliseum
Amiprol
Ansiolin
Ansiolisina
Apaurin
Apozepam
Assival
Atensine
Atilen
Bialzepam
Calmocitene
Calmpose
Cercine
Ceregulart
7-Chloro-1,3-dihydro-1-methyl-5-phenyl-2H-1,4-benzodiazepin-2-one
7-Chloro-1-methyl-5-3H-1,4-benzodiazepin-2(1H)-one
7-Chloro-1-methyl-2-oxo-5-phenyl-3H-1,4-benzodiazepine
7-Chloro-1-methyl-5-phenyl-2H-1,4-benzodiazepin-2-one
7-Chloro-1-methyl-5-phenyl-3H-1,4-benzodiazepin-2(1H)-one
7-Chloro-1-methyl-5-phenyl-1,3-dihydro-2H-1,4-benzodiazepin-2-one
Condition
DAP
Diacepan
Diapam
Diazemuls
Diazepam
Diazepamu (Polish)
Diazepan
Diazetard
Dienpax
Dipam
Dipezona
Domalium
Duksen
Duxen
E-Pam
Eridan
Faustan,
Freudal
Frustan
Gihitan
Horizon
Kabivitrum
Kiatrium
LA-III
Lembrol
Levium
Liberetas
Methyl diazepinone
1-Methyl-5-phenyl-7-chloro-1,3-dihydro-2H-1,4-benzodiazepin-2-one
Morosan
Noan
NSC-77518
Pacitran
Paranten
Paxate
Paxel
Plidan

Quetinil
Quiatril
Quievita
Relaminal
Relanium
Relax
Renborin
Ro 5-2807
S.A. R.L.
Saromet
Sedipam
Seduksen
Seduxen
Serenack
Serenamin
Serenzin
Setonil
Sibazon
Sonacon
Stesolid
Stesolin
Tensopam
Tranimul
Tranqdyn
Tranquirit
Umbrium
Unisedil
Usempax AP
Valeo
Valitran
Valium
Vatran
Velium
Vival
Vivol
WY-3467
Zipan

440-58-4
$C_{12}H_{11}I_3N_2O_4$
627.95
Benzoic acid, 3-(acetylamino)-5-((acetylamino)methyl)-2,4,6-triiodo
3-Acetamido-5-(acetamidomethyl)-2,4,6-triiodobenzoic acid
3-(Acetylamino)-5-((acetylamino)methyl)-2,4,6-triiodobenzoic acid
Amet (German)
Ametriodinic acid
B-4130
α,5-Diacetamido-2,4,6-triiodo-m-toluic acid
Iodamide
Jodamid (German)
Jodomiron
SH 926
Uromiro
Uromiron

441-38-3
$C_{14}H_{13}NO_2$
227.28
OC(c(cccc1)c1)C(=NO)c(cccc2)c2
Benzoin, oxime
Benzoinoxim (Czech)
Benzoin, α-oxime
α-Benzoin oxime
Cupron (Czech)
Cuprone
USAF FA-5

441-61-2
Unknown
Unknown
DEA No. 9623
Ethylmethylthiambutene

441-91-8
$C_{19}H_{22}N_2O$
294.40
Benanserin

443-48-1
$C_6H_9N_3O_3$
171.18
Cc1ncc(N(=O)=O)n1CCO
Imidazole-1-ethanol, 2-methyl-5-nitro
Acromona
Anagiardil
Atrivyl
Bayer 5360
Bexon
Clont
Cont
Danizol
Deflamon-wirkstoff
Efloran
Elyzol
Entizol
1-(β-Ethylol)-2-methyl-5-nitro-3-azapyrrole
Eumin
Flagemona
Flagesol
Flagil
Flagyl
Flegyl
Giatricol
Gineflavir
1-(β-Hydroxyethyl)-2-methyl-5-nitroimidazole
1-(2-Hydroxyethyl)-2-methyl-5-nitroimidazole
1-Hydroxyethyl-2-methyl-5-nitroimidazole
1-(2-Hydroxy-1-ethyl)-2-methyl-5-nitroimidazole
Klion
Meronidal
2-Methyl-1-(2-hydroxyethyl)-5-nitroimidazole
2-Methyl-3-(2-hydroxyethyl)-4-nitroimidazole
2-Methyl-5-nitroimidazole-1-ethanol
Metronidaz
Metronidazol
Metronidazole
Metronidazolo
Monagyl
Nalox
Neo-Tric
Nida
Novonidazol
NSC-50364
Orvagil
1-(β-Oxyethyl)-2-methyl-5-nitroimidazole
RP 8823
Sanatrichom
SC 10295
Takimetol
Trichazol
Trichex
Trichocide
Trichomol
Trichomonacid (pharmaceutical)
Trichopal
Trichopol
Tricocet
Tricom
Tricowas B
Trikacide
Trikamon
Trikojol
Trikozol
Trimeks
Trivazol
Vagilen
Vagimid
Vertisal

443-83-4
C_7H_6ClF
144.58
Fc(c(c(cc1)Cl)C)c1
Benzene, 1-chloro-3-fluoro-2-methyl- (9CI)
1-Chloro-3-fluoro-2-methylbenzene
2-Chloro-6-fluorotoluene

445-03-4
$C_7H_5ClF_3N$
195.57
FC(F)(F)c(c(N)ccc1Cl)c1
Benzenamine, 4-chloro-2-(trifluoromethyl)- (9CI)
4-Chloro-2-(trifluoromethyl)benzenamine

445-29-4
$C_7H_5FO_2$
140.12
O=C(O)c(c(F)ccc1)c1
Benzoic acid, o-fluoro
o-Fluorbenzoesaeure (German)
o-Fluorobenzoic acid

445-66-9
$C_7H_4F_3N_3O_4$
251.10
Nc1c(cc(cc1N(=O)=O)C(F)(F)F)N(=O)=O
Benzenamine, 2,6-dinitro-4-(trifluoromethyl)- (9CI)
p-Toluidine, α,α,α-trifluoro-2,6-dinitro- (8CI)

446-36-6
$C_6H_4FNO_3$

157.10
NSC-10284

446-86-6
$C_9H_7N_7O_2S$
277.29
CN1C=NC(C1Sc2ncnc3nc[nH]c23)N(=O)=O
Purine, 6-((1-methyl-4-nitro-imidazol-5-yl)thio)
Azamun (Czech)
Azanin
Azatioprin
Azothioprin (Czech)
Azathioprine
Azothioprine
BW 57-322
Ccucol
Imuran
Imurek
Imurel
Methylnitroimidazolylmercaptopurine
6-(1'-Methyl-4'-nitro-5'-imidazolyl)-mercaptopurine
6-(Methyl-p-nitro-5-imidazolyl)-thiopurine
6-((1-Methyl-4-nitroimidazol-5-yl)thio)purine
6-(1-Methyl-p-nitro-5-imidazolyl)-thiopurine
6-(1-Methyl-4-nitroimidazol-5-ylthio)purin (Czech)
6-(1-Methyl-4-nitroimidazol-5-ylthio)purine
6-((1-Methyl-4-nitro-1H-imidazol-5-yl)thio)-1H-purine
NCI-C03474
NSC-39084
1H-Purine, 6-((1-methyl-4-nitro-1H-imidazol-5-yl)thio)-
Rorasul

447-05-2
$C_8H_{12}NO_6P$
249.18
Pyridoxol, 5-(dihydrogen phosphate)
3,4-Pyridinedimethanol, 5-hydroxy-6-methyl-, α^3-(dihydrogen phosphate) (9CI)
Pyridoxine phosphate
Pyridoxine 5-phosphate
Pyridoxine 5'-phosphate
Pyridoxol 5-phosphate
Pyridoxol 5'-phosphate

447-41-6
$C_{19}H_{25}NO_2$
299.41
Nylidrin
Benzenemethanol, 4-hydroxy-α-(1-((1-methyl-3-phenylpropyl)amino)ethyl)-
Benzyl alcohol, p-hydroxy-α-(1-((1-methyl-3-phenylpropyl)amino)ethyl)-
Bufenina (Spanish)
Buphenin
Buphenine
Bupheninum (Latin)
4-Hydroxy-α-(1-((1-methyl-3-phenylpropyl)amino)ethyl)-benzenemethanol
p-Hydroxy-α-(1-((1-methyl-3-phenylpropyl)amino)ethyl)-benzyl alcohol
p-Hydroxy-n-(1-methyl-3-phenylpropyl)norephedrine
1-(p-Hydroxyphenyl)-2-(1'-methyl-3'-phenylpropyl-amino)-1-propanol
Nilidrine
Nylidrinum
Phenyl-sec-butyl norsuprifen
Suprifen-PSB

447-53-0
$C_{10}H_{10}$
130.20
C2Cc1ccccc1C=C2
Naphthalene, 1,2-dihydro
Dialin
1,2-Dihydronaphthalene

451-13-8
$C_8H_8O_4$
168.15
O=C(O)Cc(c(O)ccc1O)c1
Homogentisic acid
Alcapton
Acetic acid, (2,5-dihydroxy-phenyl)- (8CI)
Benzeneacetic acid, 2,5-di-hydroxy- (9CI)
2,5-Dihydroxybenzeneacetic acid
2,5-Dihydroxyphenylacetic acid
2,5-Dihydroxy-α-toluic acid
Homogentisate acid
Homogentisinic acid
NSC-88940

451-40-1
$C_{14}H_{12}O$
196.26
O=C(c(cccc1)c1)Cc(cccc2)c2
Acetophenone, 2-phenyl
Benzyl phenyl ketone
Deoxybenzoin
Desoxybenzoin
1,2-Diphenylethanone
Ethanone, 1,2-diphenyl- (9CI)
2-Phenylacetophenone
Phenyl benzyl ketone

452-06-2
$C_5H_5N_5$
135.15
Nc2ncc1[nH]cnc1n2
Purine, 2-amino
2-Aminopurine
1H-Purin-2-amine
SQ 22451

452-35-7
$C_9H_{10}N_2O_3S_2$
258.33
CCOc2ccc1nc(sc1c2)S(N)(=O)=O
2-Benzothiazolesulfonamide, 6-ethoxy
Cardrase
Diuretic C
Ethoxazolamide
6-Ethoxy-2-benzothiazole-sulfonamide
Ethoxyzolamide
Ethoxzolamide
Etoxzolamide

452-58-4
$C_5H_7N_3$
109.15
Pyridine, 2,3-diamino
2,3-Diaminopyridine

452-86-8
$C_7H_8O_2$
124.15
Oc(c(O)cc(c1)C)c1
Pyrocatechol, 4-methyl
1,2-Benzenediol, 4-methyl- (9CI)
3,4-Dihydroxytoluene
Homocatechol
Homopyrocatechol
4-Methylcatechol
p-Methylpyrocatechol
4-Methylpyrocatechol
Toluene-3,4-diol

454-29-5
$C_4H_9NO_2S$
135.20
DL-Homocysteine (Free base)
Butyric acid, 2-amino-4-mercapto-, DL-
Homocysteine, DL-
USAF B-12

454-31-9
$C_4H_6F_2O_2$
124.09
O=C(OCC)C(F)F
Acetic acid, difluoro-, ethyl ester (9CI)
Ethyl difluoroacetate

454-57-9
$C_8H_{11}FSi$
154.26
Silane, fluorodimethyl-phenyl- (7CI,8CI,9CI)
Fluorodimethylphenylsilane
Phenyldimethylfluorosilane

454-92-2
$C_8H_5F_3O_2$
190.12
OC(=O)c1cccc(c1)C(F)(F)F
3-Trifluoromethylbenzoate
Benzoic acid, 3-(trifluoro-

methyl)- (9CI)
NSC-43025
m-Toluic acid, α,α,α-trifluoro- (8CI)
m-(Trifluoromethyl)benzoic acid
3-(Trifluoromethyl)benzoic acid

455-19-6
$C_8H_5F_3O$
174.12
O=Cc(ccc(c1)C(F)(F)F)c1
4-(Trifluoromethyl)benzaldehyde
Benzaldehyde, 4-(trifluoro-methyl)- (9CI)
p-(Trifluoromethyl)benzaldehyde

455-20-9
$C_8H_9FO_2S$
188.22
O=S(=O)(F)c(ccc(c1)CC)c1
Benzenesulfonyl fluoride, 4-ethyl- (9CI)
4-Ethylbenzenesulfonyl fluoride

455-24-3
$C_8H_5F_3O_2$
190.12
O=C(O)c(ccc(c1)C(F)(F)F)c1
4-(Trifluoromethyl)benzoic acid
Benzoic acid, 4-(trifluoro-methyl)- (9CI)
NSC-88327

455-36-7
C_8H_7FO
138.14
CC(=O)c1cccc(F)c1
Ethanone, 1-(3-fluorophenyl)- (9CI)
Acetophenone, 3'-fluoro- (8CI)
NSC-88301

455-38-9
$C_7H_5FO_2$
140.11
O=C(O)c(cccc1F)c1
3-Fluorobenzoic acid
Benzoic acid, m-fluoro- (8CI)
Benzoic acid, 3-fluoro- (9CI)
m-Fluorobenzoic acid
NSC-10320

456-22-4
$C_7H_5FO_2$
140.11
O=C(O)c(ccc(F)c1)c1
4-Fluorobenzoic acid
Benzoic acid, p-fluoro- (8CI)
Benzoic acid, 4-fluoro- (9CI)
p-Fluorobenzoic acid
NSC-10321

456-23-5
$C_7H_5O_3$
137.11
Benzoic acid, 4-hydroxy-, ion-(1-) (9CI)
Benzoic acid, p-hydroxy-, ion-(1-) (8CI)

456-42-8
C_7H_6ClF
144.58
Fc(cccc1CCl)c1
Benzene, 1-(chloromethyl)-3-fluoro- (9CI)
α-Chloro-m-fluorotoluene
α-Chloro-3-fluorotoluene
1-(Chloromethyl)-3-fluoro-benzene
m-Fluorobenzyl chloride
3-Fluorobenzyl chloride
NSC-60720
Toluene, α-chloro-m-fluoro-(8CI)

456-49-5
C_7H_7FO
126.13
Benzene, 1-fluoro-3-methoxy-(9CI)
Anisole, m-fluoro- (8CI)
NSC-88277

456-59-7
$C_{17}H_{24}O_3$
276.41
O=C(OC(CC(CC1(C)C)C)C1)C(O)c(cccc2)c2
Mandelic acid, 3,3,5-trimethylcyclohexyl ester
Arto-Espasmol
Benzeneacetic acid, α-hydroxy-, 3,3,5-trimethylcyclohexyl ester (9CI)
BS 572
Capilan
Ciclospasmol
Clandilon
Cyclandelate
Cyclolyt
Cyclomandol
Cyclospasmol
Dilatan
α-Hydroxybenzeneacetic acid 3,3,5-trimethylcyclohexyl ester
Perebral
Saiclate
Sancyclan
Sepyron
Spasmione
Spasmocyclon
Spasmocyclone
3,3,5-Trimethylcyclohexanol α-phenyl-α-hydroxyacetate
3,5,5-Trimethylcyclohexyl amygdalate
3,3,5-Trimethylcyclohexyl mandelate

457-87-4
$C_{11}H_{17}N$
163.26
Ethylamphetamine
DEA No. 1475
N-Ethylamphetamine
Etilamfetamina (Spanish)
Etilamfetamine
Etilamfetaminum (Latin)
Phenethylamine, N-ethyl-α-methyl-
1-Phenyl-2-aethylamino-propan (German)
α-Phenyl-β-ethylaminopropane
1-Phenyl-2-ethylaminopropane

458-24-2
$C_{12}H_{16}F_3N$
231.29
CCNC(C)Cc1cccc(c1)C(F)(F)F
Phenethylamine, N-ethyl-α-methyl-m-(trifluoromethyl)
N-Ethyl-α-methyl-3-trifluoro-methylphenethylamine
Fenfluramine
3-(Trifluoromethyl)-N-ethyl-α-methylphenethylamine
1-(meta-Trifluoromethyl-phenyl)-2 ethylamino-propane

458-35-5
$C_{10}H_{12}O_3$
180.20
O(c(c(O)ccc1C=CCO)c1)C
Coniferyl alcohol
AI3-36149
3-(4-Hydroxy-3-methoxy-phenyl)-2-propen-1-ol
4-(3-Hydroxy-1-propenyl)-2-methoxyphenol
Phenol, 4-(3-hydroxy-1-propen-yl)-2-methoxy-

458-36-6
$C_{10}H_{10}O_3$
178.19
2-Propenal, 3-(4-hydroxy-3-methoxyphenyl)- (9CI)
Cinnamaldehyde, 4-hydroxy-3-methoxy- (8CI)
Coniferaldehyde
p-Coniferaldehyde
Coniferyl aldehyde
Ferulaldehyde

458-37-7
$C_{21}H_{20}O_6$
368.41
O=C(C=Cc(ccc(O)c1OC)c1)CC(=O)C=Cc(ccc(O)c2OC)c2
1,6-Heptadiene-3,5-dione, 1,7-bis(4-hydroxy-3-methoxyphenyl)
C.I. 75300
C.I. Natural Yellow 3

Curcuma
Curcumin
Diferuloylmethane
Golden Seal
Haidr
Halad
Haldar
Halud
Hydrastis
Indian Saffron
Indian Turmeric
Kacha Haldi
Kurkumin (Czech)
Merita Earth
NCI-C61325
Orange Root
Souchet
Yellow Ginger
Yellow Puccoon
Yellow Root
Yo-Kin
Zlut Prirodni 3 (Czech)

458-88-8
$C_8H_{17}N$
127.26
CCCC1CCCCN1
Piperidine, 2-propyl-, (S)
Cicutin
Cicutine
d-Conicine
Coniin
Coniine
(+)-Coniine
α-Conine
Conine
β-Propylpiperidine
2-Propylpiperidine

459-60-9
C_7H_7FO
126.13
O(c(ccc(F)c1)c1)C
Anisole, p-fluoro- (8CI)
AI3-10595
Benzene, 1-fluoro-4-methoxy- (9CI)
p-Fluoroanisole
4-Fluoroanisole
p-Fluoromethoxybenzene
1-Fluoro-4-methoxybenzene
p-Fluorophenyl methyl ether
p-Methoxyfluorobenzene
NSC-4672

459-72-3
$C_4H_7FO_2$
106.11
O=C(OCC)CF
Acetic acid, fluoro-, ethyl ester
Ethylester kyseliny fluoroctove (Czech)
Ethyl fluoroacetate

459-80-3
$C_{10}H_{16}O_2$
168.26
O=C(O)C=C(CCC=C(C)C)C
2,6-Octadienoic acid, 3,7-dimethyl
3,7-Dimethyl-2,6(and 2,7)-octadienoic acid
Geranic acid

459-99-4
$C_4H_6F_2O_2$
124.10
FCCOC(=O)CF
Acetic acid, fluoro-, (2-fluoroethyl) ester
2-Fluorethylester kyseliny fluoroctove (Czech)
β-Fluoroethyl fluoroacetate
2-Fluoroethyl fluoroacetate
TL 855

460-00-4
C_6H_4BrF
175.00
Fc(ccc(c1)Br)c1
Benzene, 1-bromo-4-fluoro- (9CI)
p-Bromofluorobenzene
4-Bromofluorobenzene
1-Bromo-4-fluorobenzene
p-Fluorobromobenzene
4-Fluorobromobenzene
1-Fluoro-4-bromobenzene
4-Fluoro-1-bromobenzene
p-Fluorophenyl bromide
4-Fluorophenyl bromide
NSC-10268

460-12-8
C_4H_2
50.06
C(C#C)#C
1,3-Butadiyne (9CI)

460-13-9
C_3H_7F
62.09
1-Fluoropropane
n-Fluoropropane

460-19-5
C_2N_2
52.04
N#CC#N
Cyanogen (ACGIH,DOT, OSHA)
Carbon nitride
Cyanogen gas (DOT)
Cyanogen, Liquefied [UN 1026]
Cyanogene (French)
Dicyan
Dicyanogen
Ethanedinitrile
Nitriloacetonitrile
Oxalic acid dinitrile
Oxalic nitrile
Oxalonitrile
Oxalyl cyanide
Prussite
RCRA waste number P031
UN 1026 [Cyanogen, liquefied]

460-35-5
$C_3H_4ClF_3$
132.52
FC(F)(F)CCCl
Propane, 3-chloro-1,1,1-trifluoro
1-Chloro-3,3,3-trifluoropropane
3-Chloro-1,1,1-trifluoropropane
Freon 253
1,1,1-Trifluoro-3-chloropropane

460-40-2
$C_3H_3F_3O$
112.06
Propionaldehyde, 3,3,3-trifluoro
Trifluoropropionaldehyde
3,3,3-Trifluoropropionaldehyde

461-56-3
$C_3H_5FO_2$
92.08
O=C(O)CCF
Propionic acid, 3-fluoro
ω-Fluoropropionic acid
3-Fluoropropionic acid

461-58-5
$C_2H_4N_4$
84.10
N#CNC(=N)N
Guanidine, cyano
Cyanoguanidine
Dicyandiamide
Dicyandiamin (German)

461-72-3
$C_3H_4N_2O_2$
100.09
O=C(NCC1=O)N1
Hydantoin
Glycolylurea
2,4-Imidazolidinedione (9CI)

461-78-9
$C_{10}H_{14}ClN$
183.70
CC(C)(N)Cc1ccc(Cl)cc1
Phenethylamine, p-chloro-α,α-dimethyl
p-Chloro-α,α-dimethylphenethylamine
Chlorophentermine
β-(p-Chlorophenyl)-α,α-dimethylethylamine
1-(p-Chlorophenyl)-2-methyl-2-aminopropane
Chlorphenteramine
Chlorphentermine

462-06-6

C_6H_5F
96.11
Fc(cccc1)c1
Benzene, fluoro
Fluorobenzene [UN 2387]
Phenyl fluoride
UN 2387 [Fluorobenzene]

462-08-8
$C_5H_6N_2$
94.13
n(cccc1N)c1
Pyridine, 3-amino
β-Aminopyridine
m-Aminopyridine [UN 2671]
3-Aminopyridine
Amino-3 pyridine
3-Pyridinamine
3-Pyridylamine
UN 2671 [Aminopyridines (o-; m-; p-)]

462-18-0
$C_{13}H_{26}O$
198.39
7-Tridecanone
Dihexyl ketone
Di-n-hexyl ketone
Enanthone
Hexyl ketone

462-27-1
$C_3H_4ClFO_2$
126.52
Formic acid, chloro-, 2-fluoroethyl ester
Chloroformic acid 2-fluoroethyl ester
2-Fluoroethyl chloroformate
2-Fluorethylester kyseliny chlormravenci (Czech)
TL 751

462-60-2
$C_3H_6N_2O_3$
118.08
O=C(NCC(=O)O)N
Hydantoic acid
Acetic acid, ((aminocarbonyl)-amino)-
Acetic acid, ureido-
N-(Aminocarbonyl)glycine
Carbamoylglycine
N-Carbamoylglycine
Glycine, N-(aminocarbonyl)- (9CI)
Glycine, N-carbamoyl- (8CI)
Glycoluric acid
NSC-49417
Urea, (carboxymethyl)-

462-73-7
C_4H_8ClF
110.57
Butane, 1-chloro-4-fluoro
4-Fluorobutyl chloride

462-94-2
$C_5H_{14}N_2$
102.21
NCCCCCN
1,5-Pentanediamine
Animal Coniine
Cadaverin
Cadaverine
1,5-Diaminopentane
Pentamethylenediamine
1,5-Pentamethylenediamine

462-95-3
$C_5H_{12}O_2$
104.17
O(CC)COCC
Methane, diethoxy
Diethoxymethane [UN 2373]
Diethylformal
Ethylal
UN 2373 [Diethoxymethane]

463-04-7
$C_5H_{11}NO_2$
117.17
O=NOCCCCC
Nitrous acid, pentyl ester
Amyl nitrite [UN 1113]
n-Amyl nitrite
Nitramyl
1-Nitropentane
Pentyl alcohol, nitrite
Pentyl nitrite
UN 1113 [Amyl nitrites]

463-11-6
$C_8H_{17}F$
132.25
Octane, 1-fluoro
1-Fluorooctane

463-18-3
$C_9H_{19}F$
146.28
Nonane, 1-fluoro
1-Fluorononane

463-40-1
$C_{18}H_{30}O_2$
278.44
O=C(O)CCCCCCCC=CCC=CC C=CCC
Linolenic acid (8CI)
AI3-23986
α-Linolenic acid
NSC-2042
all-cis-9,12,15-Octadecatrienoic acid
cis,cis,cis-9,12,15-Octadecatri-enoic acid
(Z,Z,Z)-9,12,15-Octadecatri-enoic acid
9,12,15-Octadecatrienoic acid, (Z,Z,Z)- (9CI)

463-49-0
C_3H_4
40.07
C(=C)=C
Allene
sym-Allylene
Dimethylenemethane
Propadiene
1,2-Propadiene (9CI)
Propadiene, Inhibited [UN 2200]
UN 2200 [Propadiene, Inhibited.]

463-51-4
C_2H_2O
42.04
O=C=C
Ketene (ACGIH,OSHA)
Carbomethene
Ethenone
Keto-Ethylene

463-56-9
CHNS
59.09
N#CS
Ammonium thiocyanate
Thiocyanic acid (9CI)

463-58-1
COS
60.07
O=C=S
Carbonyl sulfide [UN 2204]
Carbon oxide sulfide (9CI)
Carbon oxysulfide
Carbonyl sulfide-^{32}S
Oxycarbon sulfide
UN 2204 [Carbonyl sulfide]

463-71-8
CCl_2S
114.97
S=C(Cl)Cl
Thiophosgene
Carbon chlorosulfide
Carbonothioic dichloride (9CI)
Carbonyl chloride, thio-
Dichlorothiocarbonyl
Phosgene, thio-
Thiocarbonic dichloride
Thiocarbonyl chloride
Thiocarbonylchloride (DOT)
Thiocarbonyl dichloride
Thiofosgen (Czech)
Thiokarbonylchlorid (Czech)
Thiophosgene [UN 2474]
UN 2474 [Thiophosgene]

463-77-4
CH_3NO_2
61.03
O=C(O)N
Carbamic acid (8CI,9CI)
Aminoformic acid
Formic acid, amino-

463-82-1
C_5H_{12}
72.17
C(C)(C)(C)C
Propane, 2,2-dimethyl
2,2-Dimethylpropane
[UN 2044]
Neopentane
tert-Pentane (DOT)
UN 2044 [2,2-Dimethylpropane other than pentane and isopentane]

463-88-7
$C_5H_{12}N.HO$
103.19
Ammonium, trimethylvinyl-, hydroxide
Ethenaminium, N,N,N-trimethyl-, hydroxide (9CI)
Neirine
Neurin
Neurine
N,N,N-Trimethylethenaminium hydroxide
Trimethyl vinyl ammonium hydroxide
Vinyltrimethylammonium hydroxide
Vitaloid

464-06-2
C_7H_{16}
100.20
C(C(C)C)(C)(C)C
2,2,3-Trimethylbutane
Butane, 2,2,3-trimethyl- (9CI)
NSC-73938
Triptan
Triptane

464-07-3
$C_6H_{14}O$
102.20
OC(C(C)(C)C)C
2-Butanol, 3,3-dimethyl
tert-Butyl methyl carbinol
3,3-Dimethyl-2-butanol
Pinacolyl alcohol (6CI)

464-10-8
CBr_3NO_2
297.75
O=N(=O)C(Br)(Br)Br
Methane, nitrotribromo
Nitrotribromomethane

464-17-5
$C_{10}H_{16}$
136.24
Bicyclo(2.2.1)hept-2-ene, 1,7,7-trimethyl- (9CI)
2-Bornene (8CI)
Bornylene
NSC-193373
1,7,7-Trimethylbicyclo(2.2.1)-hept-2-ene
1,7,7-Trimethylnorbornene
1,7,7-Trimethyl-2-norbornene

464-48-2
$C_{10}H_{16}O$
152.26
O=C(C(C(C1C2)(C)C)(C2)C)C1
Camphor, l-, (-)
l-Camphor
l-(-)-Camphor

464-49-3
$C_{10}H_{16}O$
152.26
O=C(C(C(C1C2)(C)C)(C2)C)C1
Camphor, (1R,4R)-(+)
Alcanfor
Bicyclo(2.2.1)heptan-2-one, 1,7,7-trimethyl-, (1R)-
(+)-2-Bornanone
d-2-Bornanone
d-2-Camphanone
Camphor, (+)-
(+)-Camphor
d-Camphor
D-(+)-Camphor
Camphor USP
Japanese Camphor

465-16-7
$C_{32}H_{48}O_9$
576.80
Card-20(22)-enolide, 16-(acetyloxy)-3-((2,6-dideoxy-3-O-methyl-l-arabino-hexopyranosyl)oxy)- 14-hydroxy-, (3-β,16-β)
Corrigen
Foliandrin
Folinerin
Folinevin
Neriol
Neriolin
Neriostene
Oleandrin
Oleandrina (Spanish)
Oleandrine

465-65-6
$C_{19}H_{21}NO_4$
327.41
Oc1ccc2CC4N(CCC35C(Oc1c23)C(=O)CCC45O)CC=C
Morphinan-6-one, 4,5-α-epoxy-3,14-dihydroxy-17-(2-propenyl)
l-N-Allyl-7,8-dihydro-14-hydroxynormorphinone
17-Allyl-4,5-α-epoxy-3,14-dihydroxymorphinan-6-one
l-N-Allyl-14-hydroxynordihydromorphinone
12-Allyl-7,7a,8,9-tetrahydro-3,7a-dihydroxy-4ah-8,9c-iminoethanophenanthro(4,5-bcd)-furanone
Naloxone
l-Naloxone
Normorphinone, N-allyl-7,8-dihydro-14-hydroxy-, (-)-

465-73-6
$C_{12}H_8Cl_6$
364.90
C(=C(C(C1(Cl)Cl)(C(C(C=CC23)C2)C34)Cl)Cl)(C14Cl)Cl
1,4:5,8-Dimethanonaphthalene, 1,2,3,4,10,10-hexachloro-1,4,4a,5,8,8a-hexahydro-, endo, endo
Compound 711
ENT 19,244
Experimental Insecticide 711
1,2,3,4,10,10-Hexachloro-1,4,4a,5,8,8a-hexahydro-1,4:5,8-endo-endo-dimethanonaphthalene
1,2,3,4,10,10-Hexachloro-1,4,4a,5,8,8a-hexahydro-1,4-endo,endo-5,8-dimethanonaphthalene
Isodrin
Latka 711 (Czech)
RCRA waste number P060

466-40-0
$C_{21}H_{27}NO$
309.45
Isomethadone
DEA No. 9226
6-Dimethylamino-5-methyl-4,4-diphenyl-3-hexanone
6-(Dimethylamino)-5-methyl-4,4-diphenyl-3-hexanone
1,1-Diphenyl-1-(dimethylamino-isopropyl)butanone-2
3-Hexanone, 6-(dimethyl-amino)-4,4-diphenyl-5-methyl-
Isoamidone II
Isometadona (Spanish)
Isometadone
Isomethadone II
Isomethadonum (Latin)
Win 1783

466-90-0
$C_{20}H_{23}NO_4$
341.40
Thebacon
Acedicon
Acedikon (Czech)
Acetyldihydrocodeinone
Acetyldihydrokodeinon (Czech)
DEA No. 9315
Demethyldihydrothebaine, acetate
Demethyldihydrothebanine acetate
Dihydrocodeinone enol acetate
Morphinan-6-ol, 6,7-didehydro-4,5-epoxy-3-methoxy-17-methyl-, acetate (ester), (5α)- (9CI)
Morphinan-6-ol, 6,7-didehydro-4,5-α-epoxy-3-methoxy-17-methyl-, acetate

Morphinan-6-ol, 6,7-didehydro-4,5α-epoxy-3-methoxy-17-methyl-, acetate (ester) (8CI)
NSC-117860
Tebacon
Tebacona (Spanish)
Tebacone
Thebacone (French)
Thebaconum (Latin)
Thebaine, demethyldihydro-, acetate

466-97-7
$C_{16}H_{17}NO_3$
271.31
Normorphine
DEA No. 9313
Demethylmorphine
Desmethylmorphine
4,5-Epoxy-3,6-dihydroxy-morphin-7-ene
Morphinan-3,6-diol, 7,8-didehydro-4,5-epoxy-, (5α,6α)- (9CI)
Morphinan-3,6α-diol, 7,8-didehydro-4,5α-epoxy- (8CI)
Normorfina (Spanish)
N-Normorphine
(-)-Normorphine
Normorphinum (Latin)
NSC-270042

466-99-9
$C_{17}H_{19}NO_3$
285.37
CN2CCC14C3CCC(=O)C1Oc5c(O)ccc(CC23)c45
Morphinan-6-one, 4,5-α-epoxy-3-hydroxy-17-methyl
Dihydromorfinon (Czech)
Dihydromorphinone
Dilaudid
DIMO
Hydromorphone
Hymorphan
Laudicon
Morphinone, dihydro-
Paramorphan

467-15-2
$C_{17}H_{19}NO_3$
285.37
Morphinan-6-α-ol, 7,8-didehydro-4,5-α-epoxy-3-methoxy
N-Desmethylcodeine
7,8-Didehydro-4,5-α-epoxy-3-methoxymorphinan-6-α-ol
Morphinan-6-ol, 7,8-didehydro-4,5-epoxy-3-methoxy-, (5-α,6-α)- (9CI)
Norcodeine
N-Norcodeine
Normorphine 3-methyl ether

467-18-5
Unknown
Unknown
DEA No. 9308
Myrophine

467-60-7
$C_{18}H_{21}NO$
267.36
Pipradrol
DEA No. 1750
Detaril
α,α-Diphenyl-2-piperidine-methanol
Gerodyl
Meratran
MRD 108
2-Piperidinemethanol, α,α-diphenyl-
α-(2-Piperidyl)benzhydrol
Pipradol (Spanish)
α-Pipradol
Pipradrolo
Pipradrolum (Latin)
Piridrol
Pyridrol
Pyridrole

467-62-9
$C_{19}H_{19}N_3O$
305.36
OC(c(ccc(N)c1)c1)(c(ccc(N)c2)c2)c(ccc(N)c3)c3
Tris(p-aminophenyl)methanol
Benzenemethanol, 4-amino-α,α-bis(4-aminophenyl)- (9CI)
Fuchsin dye base
Methanol, tris(4-aminophenyl)-
Pararosaniline base
4,4',4''-Triaminotriphenyl-carbinol
Tris (4-aminophenyl) carbinol

467-81-2
$C_{35}H_{52}O_5$
552.87
Olean-12-en-28-oic acid, 22-β-hydroxy-3-oxo-, 2-methylcrotonate, (Z)
22-β-Angeloyloxyoleanolic acid
Lantadene A
22-((2-Methyl-1-oxo-2-butenyl)-oxy)-3-oxo-olean-12-en-28-oic acid, 22-β(Z)-
Rehmannic acid

467-82-3
$C_{35}H_{52}O_5$
552.87
Olean-12-en-28-oic acid, 22-β-hydroxy-3-oxo-, 3-methylcrotonate
Lantadene B
22-((3-Methyl-1-oxo-2-butenyl)-oxy)-3-oxo-olean-12-en-28-oic acid, (22,β)-

467-83-4
Unknown
Unknown
Dipipanone
DEA No. 9622
Dipipanona (Spanish)
Dipipanonum (Latin)

467-84-5
$C_{23}H_{29}NO_2$
351.48
Phenadoxone
CB 11
DEA No. 9637
4,4-Diphenyl-6-morpholino-3-heptanone
Fenadossone
Fenadoxona (Spanish)
3-Heptanone, 4,4-diphenyl-6-morpholino-
3-Heptanone, 6-(4-morpholinyl)-4,4-diphenyl- (9CI)
6-(4-Morpholinyl)-4,4-diphenyl-3-heptanone
Phenodoxone
Phenadoxonum (Latin)

467-85-6
$C_{20}H_{25}NO$
295.42
Normethadone
AI3-23988
Dacartil
DEA No. 9635
Desmethylmethadone
Eucopon
3-Hexanone, 6-(dimethyl-amino)-4,4-diphenyl- (8CI, 9CI)
Hoechst 10582
Isoamidone I
Isomethadone I
Mepidon
Noramidone
Normedon
Normetadona (Spanish)
Normetadone
Normethadonum (Latin)
NSC-19598
Phenyldimazone
Thiofentanyl

467-86-7
Unknown
Unknown
DEA No. 9621
Dioxaphetyl butyrate

468-07-5
$C_{24}H_{29}NO$
347.49
Phenomorphan
DEA No. 9647
Fenomorfano (Spanish)
Morphinan-3-ol, 17-phenethyl-, (-)-
Phenomorphane (French)
Phenomorphanum (Latin)

468-50-8
Unknown
Unknown
DEA No. 9608
Betameprodine

468-51-9
Unknown
Unknown
DEA No. 9604
Alphameprodine

468-56-4
$C_{15}H_{21}NO_3$
263.33
Hydroxypethidine
DEA No. 9627
Hidroxipetidina (Spanish)
Hydropetidine
Hydroxypethidinum (Latin)
Idrossipetidina
Isonipecotic acid, 4-(m-hydroxyphenyl)-1-methyl-, ethyl ester
Oxipethidine
Oxipethidinum
4-Piperidinecarboxylic acid, 4-(m-hydroxyphenyl)-1-methyl-, ethyl ester
Win 771

468-59-7
$C_{16}H_{23}NO_2$
261.36
Betaprodine
Betaprodina (Spanish)
Betaprodinum (Latin)
DEA No. 9611
Nu-1779
4-Piperidinol, 1,3-dimethyl-4-phenyl-, propanoate (ester), trans- (9CI)
4-Piperidinol, 1,3-dimethyl-4-phenyl-, propionate (ester), stereoisomer
β-Prodine
β-Prodinol

468-76-8
$C_{24}H_{39}NO_4$
405.57
Cassaine
Acetic acid, (dodecahydro-7-hydroxy-1,4b,8,8-tetramethyl-10-oxo-2(1H)-phenanthrenylidene)-, 2-(dimethylamino)ethyl ester, (1R-(1α, 2e,4aα,4bβ,7β,8aα,10aβ))-
(Dodecahydro-7β-hydroxy-1α,4bβ,8,8-tetramethyl-10-oxo-2- (1H) -phenanthrenylidene)acetic acid 2-(dimethylamino)ethyl ester
(E)-3β-Hydroxy-14α-methyl-7-oxopodocarpane-δ13α-acetic acid 2-(dimethylamino)-ethyl ester
Nervocidine
Podocarpane-δ(13,α)-acetic acid, 3β-hydroxy-14α-methyl-7-oxo-, 2-(dimethylamino)ethyl ester, (E)-

469-21-6
$C_{17}H_{22}N_2O$
270.41
CN(C)CCOC(C)(c1ccccc1)c2ccccn2
Pyridine, 2-(α-(2-(dimethylamino)ethoxy)-α-methylbenzyl)
2-(α-(2-(Dimethylamino)ethoxy)-α-methylbenzyl)pyridine
2-Dimethylainoethoxyphenylmethyl-2-picoline
Doxylamine
Ethanamine, N,N-dimethyl-2-(1-phenyl-1-(2-pyridinyl)ethoxy)- (9CI)
NCI-C60684
Phenyl-2-pyridylmethyl-β-N,N-dimethylaminoethyl ether

469-59-0
$C_{27}H_{39}NO_3$
425.67
CC1CNC6C(C1)OC5(CCC4C3CC=C2CC(O)CCC2(C)C3C(=O)C4=C5C)C6C
Spiro(9H-benzo(a)fluorene-9,2'(3'H)-furo(3,2-b)-pyridin)-11(1H)-one, 2,3,3'a, 4,4',5',6,6',6a,6b,7,7',7'a, 8,11a,11b-hexadecahydro-3-hydroxy- 3',6',10,11b-tetramethyl
Jervine

469-61-4
$C_{15}H_{24}$
204.39
C(C(C(C1C(=C2)C)(C)C)CC3)(C1)(C3C)C2
1H-3a,7-Methanoazulene, 2,3,4,7,8,8a-hexahydro-3,6,8,8-tetramethyl-, (3R-(3-α,3a-β, 7-β,8a-α))
Cedr-8-ene
α-Cedrene

469-62-5
$C_{22}H_{29}NO_2$
339.52
CCC(=O)OC(Cc1ccccc1)(C(C)CN(C)C)c2ccccc2
2-Butanol, 4-(dimethylamino)-3-methyl-1,2-diphenyl-, propionate, (+)
Darvon
Dextropropoxyphene
Dextroproxifeno (Spanish)
α-(+)-4-Dimethylamino-1,2-diphenyl-3-methyl-2-butanol propionate ester
Dolene
Doloxene
Propoxyphene, (+)-
d-Propoxyphene
Proxagesic
SK 65

469-79-4
$C_{15}H_{21}NO_2$
247.33
Ketobemidone
A 21 Lundbeck
Cetobemidon
Cetobemidona (Spanish)
Cetobemidone (French)
Cetobemidonum (Latin)
Ciba 7115
Cliradon
Cliradone
Cymidon (VAN)
DEA No. 9628
Ethyl (4-(m-hydroxyphenyl)-1-methyl)-4-piperidyl ketone
Hoechst 10720
1-(4-(3-Hydroxyphenyl)-1-methyl-4-piperidinyl)-1-propanone
K 4710
Ketobemidonum
Ketone, ethyl 4-(m-hydroxyphenyl)-1-methylpiperidyl
NSC-117863
1-Propanone, 1-(4-(3-hydroxyphenyl)-1-methyl-4-piperidinyl)- (9CI)
1-Propanone, 1-(4-(m-hydroxyphenyl)-1-methyl-4-piperidyl)- (8CI)
Win 1539

469-81-8
$C_{20}H_{30}N_2O_3$
346.47
4-Piperidinecarboxylic acid, 1-[2-(4-morpholinyl)ethyl]-4-phenyl-, ethyl ester (9CI)
Isonipecotic acid, 1-(2-morpholinoethyl)-4-phenyl-, ethyl ester (6CI, 8CI)
Morpheridine
Morpheridin
TA 1

469-82-9
$C_{18}H_{27}NO_4$
321.42
Etoxeridine
Carbetidine
DEA No. 9625
Etosseridina
Etoxeridina (Spanish)
Etoxeridinum (Latin)
Isonipecotic acid, 1-(2-(2-hydroxyethoxy)ethyl)-4-phenyl-, ethyl ester (8CI)
UcB 2073
Wy-2039

469-92-1
$C_{15}H_{24}$

204.36
3a,7-Methano-3aH-cyclopenta-cyclooctene, 1,4,5,6,7,8, 9,9a-octahydro-1,1,7-tri-methyl-, (3aR-(3aα,7α,9aβ))-(9CI)
Clovene
3a,7-Methano-3aH-cyclopenta-cyclooctene, 1,4,5,6,7,8, 9,9a-octahydro-1,1,7-trimethyl-
3a,7-Methano-3aH-cyclopenta-cyclooctene, 1,4,5,6,7,8, 9,9a-octahydro-1,1,7β-tri-methyl- (8CI)

470-67-7
$C_{10}H_{18}O$
154.28
O(C(CC1)(CC2)C(C)C)C12C
p-Menthane, 1,4-epoxy
1,4-Cineol
1,4-Cineole
1,4-Epoxy-p-menthane
Isocineole
7-Oxabicyclo(2.2.1)heptane, 1-isopropyl-4-methyl- (6CI)
7-Oxabicyclo(2.2.1)heptane, 1-methyl-4-(1-methylethyl)-(9CI)

470-82-6
$C_{10}H_{18}O$
154.28
O(C(CCC1C2)(C2)C)C1(C)C
p-Menthane, 1,8-epoxy
Cajeputol
1,8-Cineol
Cineole
1,8-Cineole
1,8-Epoxy-p-menthane
Eucalyptol
Eucalyptole
Eukalyptol (Czech)
Limonene oxide
NCI-C56575
1,8-Oxido-p-menthane
2-Oxabicyclo(2.2.2)Octane, 1,3,3-trimethyl-

470-90-6
$C_{12}H_{14}Cl_3O_4P$
359.58
CCOP(=O)(OCC)OC(=CCl)c1ccc(Cl)cc1Cl
Phosphoric acid, 2-chloro-1-(2,4-dichlorophenyl)vinyl diethyl ester
Apachlor
Benzyl alcohol, 2,4-dichloro-α-(chloromethylene)-, diethyl phosphate
Birlane
Birlane 24
C8949
C-10015
CFV
CGA 26351
Chlofenvinphos
O-2-Chloor-1-(2,4-dichloor-fenyl)-vinyl-O,O-diethylfosfaat (Dutch)
O-2-Chlor-1-(2,4-dichlor-phenyl)-vinyl-O,O-diaethyl-phosphat (German)
Chlorfenvinfos
Chlorfenvinphos
Chlorfenwinfosem (Polish)
2-Chloro-1-(2,4-dichloro-phenyl)vinyl diethyl phosphate
β-2-Chloro-1-(2',4'-dichloro-phenyl) vinyl diethylphosphate
Chlorofenvinphos
Chlorphenvinfos
Chlorphenvinphos
Clofenvinfos
O-2-Cloro-1-(2,4-dicloro-fenil)-vinil-O,O-dietilfosfato (Italian)
Compound 4072
CVP
Dermaton
O,O-Diaethyl-O-1-(4,5-dichlor-phenyl)-2-chlor-vinyl-phosphat (German)
O,O-Diethyl O-(2-chloro-1-(2',4'-dichlorophenyl)vinyl) phosphate
Diethyl 1-(2,4-dichlorophenyl)-2-chlorovinyl phosphate
O,O-Dwuetylo-O-1-(2,4-dwu-chlorofenylo)-2-chlorowinyl-ofosforan (Polish)
ENT 24,969
GC 4072
OMS 1328
Phosphate de O,O-diethyle et de O-2-chloro-1-(2,4-dichloro-phenyl) vinyle (French)
Sapecron
Saprecon C
SD 4072
SD 7859
Shell 4072
Steladone
Supona
Supone
Unitox
Vinylphate

471-01-2
$C_9H_{14}O$
138.21
3-Cyclohexen-1-one, 3,5,5-tri-methyl- (8CI,9CI)
β-Isophorone
β-Phorone

471-03-4
$C_4H_8Cl_2O_2S$
191.08
Sulfone, bis(2-chloroethyl)
Bis(2-chloroethyl)sulfone
Bis(β-chloroethyl)sulfone
H Sulfone
Mustard Gas Sulfone
Mustard Sulfone
Yperite Sulfone

471-25-0
$C_3H_2O_2$
70.05
O=C(O)C#C
Propiolic-acid
Acetylenecarboxylic acid
Carboxyacetylene
Propargylic acid
Propynoic acid
2-Propynoic acid

471-29-4
$C_2H_7N_3$
73.12
N=C(NC)N
Guanidine, methyl
Methylguanidin (German)
Methylguanidine
Monomethyl guanidin (German)
Monomethylguanidine

471-34-1
$CO_3.Ca$
100.09
Carbonic acid, calcium salt (1:1)
Atomit
Calcium carbonate

471-46-5
$C_2H_4N_2O_2$
88.08
O=C(N)C(=O)N
Oxamide
Amid kyseliny stavelove (Czech)
Ethanediamide
Formimidic acid, 1-carbamoyl-
Oxalamide
Oxalic acid diamide
Oxamid (Czech)
Oxamimidic acid

471-47-6
$C_2H_3NO_3$
89.04
O=C(O)C(=O)N
Oxamic acid
Acetic acid, aminooxo- (9CI)
AI3-52243
Formic acid, (aminocarbonyl)-
Formic acid, carbamoyl-
Glycine, 2-oxo-
Glyoxylic acid, amino-
NSC-47001
Oxalic acid monoamide
Oxamate, (aminocarbonyl)-

471-62-5
$C_{30}H_{52}$
412.75
Hopane
A'-Neogammacerane

471-74-9
$C_{20}H_{30}O_2$
302.46

Podocarp-8(14)-en-15-oic acid, 13β-methyl-13-vinyl- (8CI)
AI3-24620
Cryptopimaric acid
Isodextropimaric acid
Isopimaric acid (D)
δ8(14)-Isopimaric acid
NSC-6435
1-Phenanthrenecarboxylic acid, 7-ethenyl-1,2,3,4,4a,4b,5,6,7, 9,10,10a-dodecahydro-1,4a, 7-trimethyl-, (1R-(1α,4a β,4b α,7α,10a α))-
Sandaracopimaric acid
(-)-Sandaracopimaric acid

471-77-2
$C_{20}H_{30}O_2$
302.50
Podocarp-8(14)-en-15-oic acid, 13-isopropylidene
Neoabietic acid

473-03-0
$C_{30}H_{52}O$
428.74
OC(C(C(C(C(CC1)(C)C)C2)(C1)C)CCC=C(CCC(C(CCC3)=C)C3(C)C)C)(C2)C
2-Naphthalenol, 1-(6-(2,2-dimethyl-6-methylenecyclohexyl)-4-methyl-3-hexenyl)-decahydro-2,5,5,8a-tetramethyl-, (1R-(1α(3E,6(S*)), 2β,4aβ,8aα))- (9CI)

473-13-2
$C_{15}H_{24}$
204.36
Naphthalene, 1,2,3,4,4a,5,6, 8a-octahydro-4a,8-dimethyl-2-(1-methylethenyl)-, (2R-(2α,4aα,8aβ))- (9CI)
Eudesma-3,11-diene (8CI)
α-Selinene
α-Selinene, (-)-
(-)-α-Selinene

473-54-1
$C_{10}H_{18}O$
154.25
OC(C(CC1C2)C1(C)C)(C2)C
2-Pinanol
Bicyclo(3.1.1)heptan-2-ol, 2,6,6-trimethyl- (9CI)
2,6,6-Trimethylbicyclo(3.1.1)-heptan-2-ol

473-55-2
$C_{10}H_{18}$
138.25
C(CC1C(C2)C)(C1(C)C)C2
Dihydropinene
Bicyclo(3.1.1)heptane, 2,6,6-trimethyl- (9CI)
NSC-76674
Pinane (8CI)
2,6,6-Trimethylbicyclo(3.1.1)-heptane

473-72-3
$C_{10}H_{16}O_3$
184.24
Cyclobutaneacetic acid, 3-acetyl-2,2-dimethyl- (8CI, 9CI)
NSC-29469
NSC-46248
NSC-96748
Pinonic acid

473-81-4
$C_3H_6O_4$
106.08
OCC(O)C(O)=O
Glyceric acid

474-25-9
$C_{24}H_{40}O_4$
392.64
CC(CCC(O)=O)C3CCC4C2C(O)CC1CC(O)CCC1(C)C2CCC34C
5-β-Cholan-24-oic acid, 3-α, 7-α-dihydroxy
Anthropodeoxycholic acid
Anthropodesoxycholic acid
Anthropododesoxycholic acid
CDC
CDCA
Chendal
Chendol
Chenic acid
Chenodeoxycholic acid
Chenodesoxycholic acid
Chenodesoxycholsaeure (German)
Chenodiol
Cholan-24-oic acid, 3,7-dihydroxy-, (3-α,5-β,7-α)- (9CI)
3-α,7-α-Dihydroxycholanic acid
3-α,7-α-Dihydroxy-5-β-cholan-24-oic acid
3-α,7-α-Dihydroxycholansaeure (German)
Gallodesoxycholic acid

474-62-4
$C_{28}H_{48}O$
400.69
Campesterol
Campesterin
Campestrol
Ergost-5-en-3-ol, (3β,24R)- (9CI)
Ergost-5-en-3β-ol, (24R)- (8CI)
(24R)-5-Ergosten-3β-ol
24α-Methylcholesterol
NSC-224330

474-67-9
$C_{28}H_{46}O$
398.68
Brassicasterol
Ergosta-5,22-dien-3-ol, (3β,22E)-
24 β-Methylcholesta-5,22-dien-3 β-ol

474-86-2
$C_{18}H_{20}O_2$
268.38
Estra-1,3,5(10),7-tetraen-17-one, 3-hydroxy
Equilin
1,3,5,7-Estratetraen-3-ol-17-one
3-Hydroxyestra-1,3,5(10),7-tetraen-17-one

475-00-3
$C_{33}H_{49}NO_7$
571.83
CC1CNC(C(O)C1)C(C)c5ccc4C3CC=C2CC(CCC2(C)C3Cc4c5C)OC6OC(CO)C(O)C(O)C6O
Veratrosine
Veratramine 3-glycoside

475-20-7
$C_{15}H_{24}$
204.36
C(C(C(C1=C)C2)C(CCC3)(C)C)(C13C)C2
Longifolene
1,4-Methanoazulene, decahydro-4,8,8-trimethyl-9-methylene-
1,4-Methanoazulene, decahydro-4,8,8-trimethyl-9-methylene-, (1S-(1α,3aβ,4α, 8aβ))- (9CI)
1,4-Methanoazulene, decahydro-4,8,8-trimethyl-9-methylene-, (1S,3aR,4S,8aS)-(+)- (8CI)
Junipen
Junipene
Kuromatsuen
Kuromatsuene
Longifolen
d-Longifolene
(+)-Longifolene
NSC-150808

475-26-3
$C_{14}H_9Cl_3F_2$
321.58
Fc1ccc(cc1)C(c2ccc(F)cc2)C(Cl)(Cl)Cl
Ethane, 1,1,1-trichloro-2,2-bis-(p-fluorophenyl)
Benzene, 1,1'-(2,2,2-trichloroethylidene)bis(4-fluoro-
1,1-Bis(p-fluorophenyl)-2,2,2-trichloroethane
2,2-Bis(p-fluorophenyl)-1,1,1-trichloroethane
2,2-Bis(4-fluorophenyl)-1,1,1-trichloroethane
DFDT
Difluorodiphenyltrichloroethane
p,p'-Difluorodiphenyltrichloro-

ethane
Ethane, 2,2-bis(p-fluorophenyl)-1,1,1-trichloro-
Fluoro-DDT
Fluorogesarol
GIX
HO-2,474
1,1,1-Trichloro-2,2-bis(p-fluorophenyl) ethane
1,1,1-Trichloroethane, 2,2-bis-(p-fluorophenyl-

476-45-9
$C_{15}H_{14}O_6$
290.29
1,4-Naphthoquinone, 3-acetonyl-5,8-dihydroxy-6-methoxy-2-methyl
Javanicin
1,4-Naphthalenedione, 5,8-dihydroxy-6-methoxy-2-methyl-3-(2-oxopropyl)-

476-66-4
$C_{14}H_6O_8$
302.20
(1)Benzopyrano(5,4,3-cde)-(1)benzopyran-5,10-dione, 2,3,7,8-tetrahydroxy
Alizarine Yellow
Benzoaric acid
C.I. 55005
C.I. 75270
Elagostasine
Eleagic acid
Ellagic acid
Gallogen
Gallogen (Astringent)
Lagistase
2,3,7,8-Tetrahydroxy(1)benzopyrano(5,4,3-cde)(1)benzopyran-5,10-dione

476-73-3
$C_{10}H_6O_8$
254.15
OC(=O)c1ccc(C(O)=O)c(C(O)=O)c1C(O)=O
1,2,3,4-Benzenetetracarboxylic acid (8CI,9CI)
Mellophanic acid

477-30-5
$C_{21}H_{25}NO_5$
371.47
CNC2CCc1cc(OC)c(OC)c(OC)c1c3ccc(OC)c(=O)cc23
Colchicine, N-deacetyl-N-methyl
Alkaloid H 3, From colchicum antumnale
Benzo(a)heptalen-9(5H)-one, 6,7-dihydro-1,2,3,10-tetramethoxy-7-(methylamino)-, (S)-
C-12669
Ciba 12669A
Colcemid
Colcemide
Colchamin
Colchamine
Colchicine, 7-deacetamido-7-(methylamino)-
Colchicine, deacetyl-N-methyl-
Colchine, N-deacetyl-N-methyl
Colemid
Deacetylmethylcolchicine
Deacetyl-N-methylcolchicine
N-Deacetyl-N-methylcolchicine
Demecolcin
Demecolcine
Desacetylmethylcolchicine
N-Desacetylmethylcolchicine
N-Desacetyl-N-methylcolchicine
Desmecolchine
Desmecolcine
6,7-Dihydro-1,2,3,10-tetramethoxy-7-(methylamino)-benzo(α)heptalen-9(5H)-one
Kolchamin
Kolchicin (Czech)
Kolkamin
Methylcolchicine
N-Methyl-N-deacetylcolchicine
N-Methyldemecolcine
N-Methyl-N-desacetylcolchicine
NSC-3096
Omain
Omaine
Reichstein's F
Santavy's Substance F
Substance F

477-73-6
$C_{20}H_{19}N_4.Cl$
350.88
Phenazinium, 3,7-diamino-2,8-dimethyl-5-phenyl-, chloride
Basic Red 2
Brilliant Safranine BR
Brilliant Safranine G
Brilliant Safranine GR
Calcozine Red Y
Cerven zasadita 2 (Czech)
C.I. 50240
C.I. Basic Red 2
2,8-Dimethylphenosafranine
Gossypimine
Hidaco Safranine
Leather Red HT
Mitsui Safranine
Nippon Kagaku Safranine GK
Nippon Kagaku Safranine T
Safranin
Safranine
Safranine A
Safranine B
Safranine G
Safranine GF
Safranine J
Safranine O
Safranine OK
Safranine Superfine G
Safranine T
Safranine TH
Safranine TN
Safranine Y
Safranine YN
Safranine ZH
Safranin T
Tolusafranine

478-42-2
$C_{13}H_{10}O_5$
246.22
5H-Furo(3,2-g)(1)benzopyran-5-one, 4-hydroxy-9-methoxy-7-methyl-(6CI,7CI,8CI,9CI)
Khellinol
5-Norkhellin

478-43-3
$C_{15}H_8O_6$
284.22
Rhein
2-Anthracenecarboxylic acid, 9,10-dihydro-4,5-dihydroxy-9,10-dioxo- (9CI)
2-Anthroic acid, 9,10-dihydro-4,5-dihydroxy-9,10-dioxo-(8CI)
Cassic acid
Chrysazin-3-carboxylic acid
9,10-Dihydro-4,5-dihydroxy-9,10-dioxo-2-anthracene-carboxylic acid
1,8-Dihydroxyanthraquinone-3-carboxylic acid
4,5-Dihydroxy-2-anthraquinone-carboxylic acid
Monorhein
NSC-38629
Rheic acid
Rhubarb Yellow

478-94-4
$C_{16}H_{17}N_3O$
267.31
Lysergamide
DEA No. 7310
9,10-Didehydro-6-methyl-ergoline-8-β-carboxamide
Ergine
LA
LA-III
Lysergic acid amide
Ololiuqui

479-18-5
$C_{10}H_{14}N_4O_4$
254.28
O=C(N(C(=O)C(N(C=N1)CC(O)CO)=C12)C)N2C
Theophylline, 7-(2,3-dihydroxypropyl)
Afi-Phyllin
Aristophyllin
Astmamasit
Astrophyllin
Circain
Circair
Coronal
Coronarin
Corphyllin
Cor-theophylline
7-(2,3-Dihydroxypropyl)-

3,7-dihydro-1,3-dimethyl-1H-purine-2,6-dione
Dihydroxypropyl theopylin (German)
Dihydroxypropyl theophylline
(1,2-Dihydroxy-3-propyl)thiophyllin
7-(2,3-Dihydroxypropyl)theophylline
Dilor
1,3-Dimethyl-7-(2,3-dihydroxypropyl)xanthine
7-(2,3-Dioxypropyl)theophylline
Diphyllin
Diprofillin
Diprofilline
Diprophyllin
Diprophylline
DT
Dyphylline
Glyfyllin
Glyphyllin
Glyphylline
Hidroxiteofillina
Hiphyllin
Hyphylline
Lufyllin
Neophyllin
Neophylline
Neophyllin M
Neostenovasan
Neothylline
Neotilina
Neo-Vasophylline
Neufil
Neutrafil
Neutrafillina
Neutraphyllin
Neutraphylline
Neutroxantina
Propyphyllin
Protheophylline
Purifilin
1H-Purine-2,6-dione, 7-(2,3-dihydroxypropyl)-3,7-dihydro-1,3-dimethyl
Sibephyllin
Sibephylline
Silbephylline
Solufilin
Solufyllin
Soluphyllin
Synthophylline
Tefilan
Theal
Theal Ampules
Thefylan

479-23-2
$C_{20}H_{14}$
254.34
Cholanthrene
Benz(j)aceanthrylene, 1,2-dihydro-
7,8-Dimethylenebenz(a)anthracene

479-27-6
$C_{10}H_{10}N_2$
158.19
c(c(c(N)cc1)c(N)cc2)(c1)c2
1,8-Naphthalenediamine (9CI)
AI3-03804
1,8-Diaminonaphthalene
Naphthalene-1,8-diamine
1,8-Naphthylenediamine
NSC-6081

479-45-8
$C_7H_5N_5O_8$
287.17
CN(N(=O)=O)c1c(cc(cc1N(=O)=O)N(=O)=O)N(=O)=O
Aniline, N-methyl-N,2,4,6-tetranitro
Benzenamine, N-methyl-N,2,4,6-tetranitro- (9CI)
CE
N-Methyl-n,2,4,6-tetranitroaniline
Nitramine
Picrylmethylnitramine
Picrylnitromethylamine
Tetralit
Tetralite
Tetril
Tetryl (ACGIH,OSHA) [UN 0208]
2,4,6-Tetryl
Trinitrophenylmethylnitramine [UN 0208]
2,4,6-Trinitrophenylmethylnitramine (OSHA)
2,4,6-Trinitrophenyl-N-methylnitramine
UN 0208 [Trinitrophenylmethylnitramine (tetryl)]

479-47-0
$C_{10}H_6O_8$
254.15
OC(=O)c1cc(C(O)=O)c(C(O)=O)c(c1)C(O)=O
1,2,3,5-Benzenetetracarboxylic acid (8CI,9CI)
Prehnitic acid

479-61-8
$C_{55}H_{72}MgN_4O_5$
893.48
Chlorophyll A
Chlorophyll A2
Magnesium, (3,7,11,15-tetramethyl-2-hexadecenyl 9-ethenyl-14-ethyl-21-(methoxycarbonyl)-4,8,13,18-tetramethyl-20-oxo-3-phorbinepropanato(2-)-N23,N24,N25,N26)-, (SP-4-2-(3S-(3α(2E,7S*, 11S*),4β,21β)))- (9CI)

479-66-3
$C_{14}H_{12}O_8$
308.25
Fulvic acid
3,4-Dihydro-3,7,8-trihydroxy-3-methyl-10-oxo-1H, 10H-pyrano(4,3-b)(1)benzopyran-9-carboxylic acid

479-73-2
$C_{19}H_{17}N_3$
287.37
N=C(C=CC(C=1)=C(c(ccc(N)c2)c2)c(ccc(N)c3)c3)C1
Pararosaniline
α-(4-Aminophenyl)-α-(4-imino-2,5-cyclohexadien-1-ylidene)-4-toluidine monohydrochloride

479-79-8
$C_{17}H_{10}O$
230.27
11H-Benzo(a)fluoren-11-one (8CI,9CI)

480-18-2
$C_{15}H_{12}O_7$
304.27
O=C(c(c(OC1c(ccc(O)c2O)c2)cc(O)c3)c3O)C1O
Flavanone, 3,3',4',5,7-pentahydroxy
4H-1-Benzopyran-4-one, 2-(3,4-dihydroxyphenyl)-2,3-dihydro-3,5,7-trihydroxy-
4H-1-Benzopyran-4-one, 2-(3,4-dihydroxyphenyl)-2,3-dihydro-3,5,7-trihydroxy-, (2R-trans)
Catechin hydrate
Dihydroquercetin
(+)-Dihydroquercetin
2,3-Dihydroquercetin
(2R,3R)-Dihydroquercetin
Distylin
Flavone, 2,3-dihydro-3,3',4',5,7-pentahydroxy-
3,3',4',5,7-Pentahydroxyflavanone
Quercetin, dihydro-
Taxifolin
Taxifoliol

480-22-8
$C_{14}H_{10}O_3$
226.24
Oc2cccc3cc1cccc(O)c1c(O)c23
1,8,9-Anthracenetriol
1,8,9-Anthratriol
Antraderm
Batidrol
Cignolin
Cigthranol
Cygnolin
Dermaline
Derobin
Dihydroxy-anthranol
1,8-Dihydroxyanthranol
1,8-Dihydroxy-9-anthranol
1,8-Dihydroxyanthrone
1,8-Dihydroxy-9-anthrone
Dioxyanthranol
Dithranol
Dithrocream
Lasan

Psoradrate
Psoriacide
Psoriacid-Stift
1,8,9-Trihydroxyanthracene

480-34-2
$C_{11}H_{10}O_4$
206.20
4H-1-Benzopyran-4-one, 5-hydroxy-7-methoxy-2-methyl- (9CI)
Chromone, 5-hydroxy-7-methoxy-2-methyl- (8CI)
Eugenin

480-54-6
$C_{18}H_{25}NO_6$
351.44
CC=C1CC(C)C(O)(CO)C(=O)OCC2=CCN3CCC(OC1=O)C23
Retrorsine
12,18-Dihydroxy-senecionan-11,16-dione
β-Longilobine
cis-Retronecic acid ester of retronecine
Senecionan-11,16-dione, 12,18-dihydroxy-

480-63-7
$C_{10}H_{12}O_2$
164.22
O=C(O)c(c(cc(c1)C)C)c1C
Benzoic acid, 2,4,6-trimethyl
β-Isodurylic acid
Mesitoic acid
2,4,6-Trimethylbenzoic acid

480-81-9
$C_{18}H_{23}NO_5$
333.42
CC=C1CC(=C)C(C)(O)C(=O)OCC2=CCN3CCC(OC1=O)C23
Seneciphylline
Jacodine
Senecionan-11,16-dione, 13,19-didehydro-12-hydroxy- (9CI)
Seneciphyllin

480-82-0
$C_{15}H_{25}NO_5$
299.36
2,3-Dihydroxy-2-(1-methylethyl)-butanoic acid) 2,3,5,7a-tetrahydro-1-hydroxy-1H-pyrrolizin-7-yl) methyl ester
Indicine
NSC-136052

480-85-3
$C_8H_{13}NO_2$
155.22
OCC1=CCN2CCC(O)C12
Retronecine
1H-Pyrrolizine-7-methanol, 2,3,5,7a-tetrahydro-1-hydroxy-, (1R-trans)- (9CI)
Retronecin
(+)-Retronecine

480-91-1
C_8H_7NO
133.14
1H-Isoindol-1-one, 2,3-dihydro- (9CI)
NSC-3689
Phthalimidine (8CI)

480-96-6
$C_6H_4N_2O_2$
136.10
O=N(ON=C1C=CC=C2)=C12
Benzofuroxan
AI3-62099
Benzofurazan oxide
Benzofurazan N-oxide
Benzofurazan, 1-oxide (9CI)
Benzofuroxane
NSC-19930

481-06-1
$C_{15}H_{18}O_3$
246.33
O=C(OC(C(C(C=CC1=O)(CC2)C)=C1C)C23)C3C
Eudesma-1,4-dien-12-oic acid, 6-α-hydroxy-3-oxo-, γ-lactone, (11S)
11-Epiisoeusantona-1,4-dienic acid, 6α-hydroxy-3-oxo-, γ-lactone
Eudesma-1,4-dien-12-oic acid, 6-α-hydroxy-3-oxo-, γ-lactone, (11S)-(-)-
Naphtho(1,2-b)puran-2,8(3H,-4H)-dione, 3a,5,5a,9b-tetrahydro-3,5a,9-trimethyl-
Santonin
α-Santonin
l-α-Santonin
Santoninic anhydride

481-21-0
$C_{27}H_{48}$
372.68
Cholestane, (5α)- (9CI)
Cholestane (VAN)
α-Cholestane
5α-Cholestane (8CI)
NSC-224419
28,29,30-Trinorlanostane

481-37-8
Unknown
Unknown
Ecgonine
DEA No. 9180
3 β-Hydroxy-2 β-tropanecarboxylic acid

481-39-0
$C_{10}H_6O_3$
174.16
Oc1cccc2C(=O)C=CC(=O)c12
1,4-Naphthalenedione, 5-hydroxy
Akhnot
C.I. 75500
C.I. Natural Brown 7
5-Hydroxy-1,4-naftochinon (Czech)
5-Hydroxy-1,4-naphthoquinone
Iuglon
Juglane
Juglon
Juglone
Lawsone
1,4-Naphthoquinone, 5-hydroxy-
1,4-Naphthoquinone, 8-hydroxy-
Nucin
Oil Red BS
Regianin
Walnut Extract

481-74-3
$C_{15}H_{10}O_4$
254.25
Anthraquinone, 1,8-dihydroxy-3-methyl
9,10-Anthracenedione, 1,8-dihydroxy-3-methyl- (9CI)
Chrysophanic acid
Chrysophanol
C.I. 75400
C.I. Natural Yellow 23
1,8-Dihydroxy-3-methylanthraquinone
3-Methylchrysazin
Turkey Rhubarb

481-97-0
$C_{18}H_{22}O_5S$
350.44
Estrone sulfate
Estrone hydrogen sulfate

482-05-3
$C_{14}H_{10}O_4$
242.23
O=C(O)c(c(c(c(ccc1)C(=O)O)c1)ccc2)c2
Diphenic acid
AI3-23779
O,O'-Bibenzoic acid
2,2'-Bibenzoic acid
2,2'-Biphenyldicarboxylic acid
(1,1'-Biphenyl)-2,2'-dicarboxylic acid (9CI)
2,2'-Dicarboxybiphenyl
O,O'-Diphenic acid
2,2'-Diphenic acid
NSC-1966

482-44-0

$C_{16}H_{14}O_4$
270.30
7H-Furo(3,2-g)(1)benzopyran-7-one, 9-((3-methyl-2-butenyl)oxy)
Ammidin
5-Benzofuranacrylic acid, 6-hydroxy-7-((3-methyl-2-butenyl)oxy)-, δ-lactone
Imperatorin
8-Isoamylenoxypsoralen
8-Isopentenyloxypsoralene
Marmelosin
9-((3-Methyl-2-butenyl)oxy)-7H-furo(3,2-g)(1)benzopyran-7-one

482-54-2
$C_{14}H_{22}N_2O_8$
346.38
O=C(O)CN(C(C(N(CC(=O)O)CC(=O)O)CCC1)C1)CC(=O)O
Acetic acid, (1,2-cyclohexylenedinitrilo)tetra
CDTA
CGTA
CHEL 600
Complexon IV
1,2-Cyclohexanediaminetetraacetic acid
1,2-Cyclohexanediamine-N,N,N',N'-tetraacetic acid
1,2-Cyclohexylenediaminetetraacetic acid
(1,2-Cyclohexylenedinitrilo)tetraacetic acid
CYDTA
DCTA
1,2-Diaminocyclohexanetetraacetic acid
1,2-Diaminocyclohexane-N,N'-tetraacetic acid
Glycine, N,N[1]-1,2-cyclohexanediylbis(N-(carboxymethyl)-(9CI)
Komplexon IV
Kyselina 1,2-cyklohexylendiamintetraoctova (Czech)
OCTA

482-89-3
$C_{16}H_{10}N_2O$
246.28
O=C(c(c(N1)ccc2)c2)C1=C(Nc(c3ccc4)c4)C3=O
($\delta^{2,2'}$-Biindoline)-3,3'-dione
$\delta^{2,2'}$-Bipseudoindoxyl
11669 Blue
Blue No. 201
C.I. 73000
C.I. Vat Blue 1
Cystoceva
D & C Blue No. 6
Diindogen
(2,2'-Biindoline)-3,3'-dione
Indigo
Indigo Blue
Indigo Ciba
Indigo Ciba SL
Indigo J
Indigo N
Indigo Nac
Indigo Nacco
Indigo P
Indigo PLN
Indigo Powder W
Indigo Pure BASF
Indigo Pure BASF Powder K
Indigo Synthetic
Indigotin
Indigo VS
3H-Indol-3-one, 2(1,3-dihydro-3-oxo-2H-indol-2-ylidene)-1,2-dihydro- (9CI)
Lithosol Deep Blue V
Mitsui Indigo Paste
Mitsui Indigo Pure
Modr Kypova 1 (Czech)
Monolite Fast Navy Blue BV
NCI-C61392
Synthetic Indigo
Synthetic Indigo TS
Vat Blue 1
Vulcafix Blue R
Vulcafor Blue A
Vulcanosine Dark Blue L
Vynamon Blue A

483-18-1
$C_{29}H_{40}N_2O_4$
480.71
Cl.Cl.CCC3CN2CCc1cc(OC)c(OC)cc1C2CC3CC4NCCc5cc(OC)c(OC)cc45
2H-Benzo(a)quinolizine, 3-ethyl-1,3,4,6,7,11b-hexahydro-9,10-dimethoxy-2-((1,2,3,4- tetrahydro-6,7-dimethoxy-1-isoquinolyl)methyl)
Cephaeline methyl ether
Emetine
(-)-Emetine
NSC-33669
6',7',10,11-Tetramethoxyemetan

483-20-5
$C_{16}H_{10}N_2O_8S_2$
422.38
1H-Indole-5-sulfonic acid, 2-(1,3-dihydro-3-oxo-5-sulfo-2H-indol-2-ylidene)-2,3-dihydro-3-oxo- (9CI)
(δ(2,2')-Biindoline)-5,5'-disulfonic acid, 3,3'-dioxo- (8CI)
Blue X
5,5'-Indigotindisulfonic acid

483-55-6
$C_{11}H_8O_3$
188.19
CC2=C(O)C(=O)c1ccccc1C2=O
1,4-Naphthoquinone, 2-hydroxy-3-methyl
2-Hydroxy-3-methyl-1,4-naphthoquinone
3-Hydroxy-2-methyl-1,4-naphthoquinone
2-Methyl-3-hydroxy-1,4-naphthoquinone
Phthiocol

483-65-8
$C_{18}H_{18}$
234.34
Phenanthrene, 1-methyl-7-(1-methylethyl)- (9CI)
AI3-00840
7-Isopropyl-1-methylphenanthrene
1-Methyl-7-isopropylphenanthrene
1-Methyl-7-(1-methylethyl)-phenanthrene
NSC-26317
Phenanthrene, 7-isopropyl-1-methyl- (8CI)
Reten
Retene

483-76-1
$C_{15}H_{24}$
204.36
Naphthalene, 1,2,3,5,6,8a-hexahydro-4,7-dimethyl-1-(1-methylethyl)-, (1S-cis)- (9CI)
Cadina-1(10),4-diene (8CI)
δ-Cadinene
δ-Cadinene, (+)-
(+)-δ-Cadinene

483-77-2
$C_{15}H_{22}$
202.34
Naphthalene, 1,2,3,4-tetrahydro-1,6-dimethyl-4-(1-methylethyl)-, (1S-cis)- (9CI)
Cadina-1,3,5-triene (8CI)
Calamenene
Calamenene, (-)-
(-)-Calamenene

483-78-3
$C_{15}H_{18}$
198.31
Naphthalene, 1,6-dimethyl-4-(1-methylethyl)- (9CI)
Cadalene
Naphthalene, 4-isopropyl-1,6-dimethyl- (8CI)

483-87-4
$C_{16}H_{14}$
206.29
Phenanthrene, 1,7-dimethyl- (8CI,9CI)
Pimanthrene

484-17-3
$C_{14}H_{10}O$
194.24
Oc2cc1ccccc1c3ccccc23
9-Phenanthrol
9-Hydroxyphenanthrene

9-Phenanthrenol

484-20-8
$C_{12}H_8O_4$
216.20
7H-Furo(3,2-g)(1)benzopyran-7-one, 4-methoxy
Bergaptan
Bergapten
Bergaptene
Heraclin
6-Hydroxy-4-methoxy-5-benzofuranacrylic acid, γ-lactone
Majudin
5-Methoxy-6,7-furanocoumarin
4-Methoxy-7H-furo(3,2-g)(1)-benzopyran-7-one
5-Methoxypsoralen
5-MOP

485-31-4
$C_{15}H_{18}N_2O_6$
322.35
CCC(C)c1cc(cc(N(=O)=O)c1OC(=O)C=C(C)C)N(=O)=O
Crotonic acid, 3-methyl-, 2-sec-butyl-4,6-dinitrophenyl ester
Acricid
Ambox
Binapacryl
2-Butenoic acid, 3-methyl-, 2-(1-methylpropyl)-4,6-dinitrophenyl ester (9CI)
2-sek.Butyl-4,6-dinitrofenyl-ester kyseliny 3-methylkrotonove (Czech)
2-sec-Butyl-4,6-dinitrophenyl-3,3-dimethylacrylate
2-sec-Butyl-4,6-dinitrophenyl 3-methyl-2-butenoate
2-sec Butyl-4,6-dinitrophenyl 3-methylcrotonate
2-sec-Butyl-4,6-dinitrophenyl senecioate
Dapacryl
3,3 Dimethyl-acrylate de 2,4-dinitro-6-(1-methylpropyle) phenyle (French)
3,3-Dimethylacrylic acid 2-sec-butyl-4,6-dinitrophenyl ester
Dinapacryl
4,6-Dinitro-2-sec-butylphenyl β,β-dimethylacrylate
2,4-Dinitro-6-sec-butylphenyl 2-methylcrotonate
4,6-Dinitrophenyl-2-sec-butyl-3-methyl-2-butenonate
Dinoseb, 3,3-dimethylacryl ester
Dinoseb methacrylate
Endosan
ENT 25,793
FMC 9044
HOE 2784
HOE 2784 OA
3-Methyl-2-butenoic acid 2-sec-butyl-4,6-dinitrophenyl ester
3-Methyl-2-butenoic acid 2-(1-methylpropyl)-4,6-dinitrophenyl ester
3-Methylcrotonic acid 2-sec-butyl-4,6-dinitrophenyl ester
(6-(1-Methyl-propyl)-2,4-dinitro-fenyl)-3,3-dimethyl-acrylaat (Dutch)
2-(1-Methylpropyl)-4,6-dinitrophenyl β,β-dimethacrylate
(6-(1-Methyl-propyl)-2,4-dinitro-phenyl)-3,3-dimethyl-acrylat (German)
(6-(1-Metil-propil)-2,4-dinitro-fenil)-3,3-dimetil-acrilato (Italian)
Morocide
Morrocid
NIA 9044
Niagara 9044
Phenol, 2-sec-butyl-4,6-dinitro-, 3-methylcrotonate
Senecioic acid 2-sec butyl-4,6-dinitrophenyl ester

485-35-8
$C_{11}H_{14}N_2O$
190.27
O=C(N(C(C(C(CC1CN2)C2)=CC=3)C1)C3
Cytisine
Baptitoxin
Baptitoxine
Cystisine
Cytiton
Cytitone
1,5-Methano-8H-pyrido-(1,2-a)(1,5)diazocin-8-one, 1,2,3,4,5,6-hexahydro-
Sophorine
Ulexine

485-47-2
$C_9H_6O_4$
178.15
O=C(c(c(c(C1=O)ccc2)c2)C1(O)O
1,3-Indandione, 2,2-dihydroxy
2,2-Dihydroxy-1,3-indandione
2,2-Dihydroxy-1H-indene-1,3(2H)-dione
1,2,3-Indantrione, 2-hydrate
1,2,3-Indantrione monohydrate
Ninhydrin
Ninhydrin hydrate
Triketohydrindene hydrate

485-71-2
$C_{19}H_{22}N_2O$
294.43
OC(c(c(c(nc1)ccc2)c2)c1)C(N(CCC3C4C=C)C4)C3
Cinchonidine
Cinchonan-9-ol, (8-α,9R)- (9CI)
(-)-Cinchonidine
Cinchovatine
α-Quinidine
2-Quinuclidinemethanol, α-4-quinolyl-5-vinyl-

486-12-4
$C_{19}H_{22}N_2$
278.43
Cc1ccc(cc1)C(=CCN2CCCC2)c3ccccn3
Pyridine, 2-(3-(1-pyrrolidinyl)-1-p-tolylpropenyl)-, (E)
NCI-C61450
Pyridine, 2-(1-(4-methylphenyl)-3-(1-pyrrolidinyl)-1-propenyl)-, (E)- (9CI)
Triprolidin
Triprolidine
Tripyrolidine

486-25-9
$C_{13}H_8O$
180.21
O=C(c(c(c(c1cccc2)ccc3)c3)c12
9H-Fluoren-9-one
Fluoren-9-one
9-Fluorenone

486-56-6
$C_{10}H_{12}N_2O$
176.24
Cotinine
(-)-Cotinine
(S)-Cotinine
2-Pyrrolidinone, 1-methyl-5-(3-pyridinyl)-, (S)- (9CI)

486-84-0
$C_{12}H_{10}N_2$
182.24
9H-Pyrido(3,4-b)indole, 1-methyl
Aribine
Harman
Harmane
Locuturine
Loturine
2-Methyl-β-carboline
3-Methyl-4-carboline
1-Methylnorharman
1-Methyl-9H-pyrido(3,4-b)-indole
Passiflorin

487-19-4
$C_{10}H_{10}N_2$
158.22
Cn1cccc1c2cccnc2
Pyridine, 3-(1-methyl-2-pyrrolyl)
1-Methyl-2-(3-pyridyl)pyrrole
3-(1-Methyl-2-pyrrolyl)pyridine
Nicotyrine
β-Nicotyrine
3,2'-Nicotyrine
Pyridine, 3-(1-methyl-1H-pyrrol-2-yl)- (9CI)

487-52-5
$C_{15}H_{12}O_5$
272.26
Butein
2',3,4,4'-Tetrahydroxychalcone

487-68-3
$C_{10}H_{12}O$
148.22
O=Cc(c(cc(c1)C)C)c1C
Benzaldehyde, 2,4,6-trimethyl
Mesitaldehyde
Mesitylaldehyde
Mesitylenecarboxaldehyde
2-Mesitylenecarboxaldehyde
2,4,6-Trimethylbenzaldehyde

487-93-4
$C_{12}H_{16}N_2O$
204.26
Bufotenin
Bufotenine
Cinobufotenine
Cohoba
DEA No. 7433
3-(β-Dimethylaminoethyl)-5-hydroxyindole
3-((β-Dimethylamino)ethyl)-5-hydroxyindole
3-(2-Dimethylaminoethyl)indol-5-ol
3-(2-Dimethylaminoethyl)-5-indolol
3-((2-Dimethylamino)ethyl)-5-indolol
N,N-Dimethyl-5-hydroxytryptamine
N,N-Dimethylserotonin
5-Hydroxy-N,N-dimethyltryptamine
Indol-5-ol, 3-(2-(dimethylamino)ethyl)- (8CI)
1H-Indol-5-ol, 3-(2-(dimethylamino)ethyl)- (9CI)
Mapine
Mappine
NSC-89593

488-10-8
$C_{11}H_{16}O$
164.27
O=C(C(=C(C1)C)CC=CCC)C1
2-Cyclopenten-1-one, 3-methyl-2-(2-pentenyl)-, (Z)
Jasmone
cis-Jasmone
(Z)-Jasmone
3-Methyl-2-(cis-2-penten-1-yl)-2-cyclopenten-1-one

488-17-5
$C_7H_8O_2$
124.15
Oc(c(ccc1)C)c1O
Pyrocatechol, 3-methyl
1,2-Benzenediol, 3-methyl-
2,3-Dihydroxytoluene
3-Methylcatechol
3-Methylpyrocatechol
2,3-Toluenediol

488-21-1
$C_6H_8O_4$
144.13
2-Butenedioic acid, 2,3-dimethyl-, (Z)- (9CI)
Maleic acid, dimethyl- (8CI)
Pyrocinchonic acid

488-23-3
$C_{10}H_{14}$
134.24
c(c(c(c(c1)C)C)C)(c1)C
Benzene, 1,2,3,4-tetramethyl
Prehnitene
Prehnitol
1,2,3,4-Tetramethylbenzene

488-41-5
$C_6H_{12}Br_2O_4$
308.00
OC(CBr)C(O)C(O)C(O)CBr
Mannitol, 1,6-dibromo-1,6-dideoxy-, D
DBM
1,6-Dibrom-1,6-didesoxy-d-mannit (German)
Dibromannit
1,6-Dibromo-1,6-dideoxy-d-mannitol
1,6-Dibromo-1,6-d-didesoxy-mannitol
Dibromomannitol
D-Dibromomannitol
1,6-Dibromomannitol
Mieobromol
Mitobronitol
Myebrol
Myelobromol
NCI-C04762
NSC-94100
R 54

488-47-1
$C_6H_2Br_4O_2$
425.72
Pyrocatechol, tetrabromo
1,2-Benzenediol, 3,4,5,6-tetrabromo-
Tetrabromocatechol
3,4,5,6-Tetrabromocatechol
Tetrabromopyrocatechol

488-81-3
$C_5H_{12}O_5$
152.17
OCC(O)C(O)C(O)CO
Ribitol
Adonite
Adonitol
1,2,3,4,5-Pentanepentol
Pentitol

488-93-7
$C_5H_4O_3$
112.08
OC(=O)c1ccoc1
3-Furancarboxylic acid (9CI)
3-Furoic acid (8CI)
NSC-349941

489-01-0
$C_{15}H_{24}O_2$
236.35
Topanol 354
2,6-Bis(1,1-dimethylethyl)-4-methoxyphenol
2,6-Di-tert-butyl-4-methoxyphenol
NSC-14451
Phenol, 2,6-bis(1,1-dimethylethyl)-4-methoxy- (9CI)
Phenol, 2,6-di-tert-butyl-4-methoxy- (8CI)

489-39-4
$C_{15}H_{24}$
204.36
1H-Cycloprop(e)azulene, decahydro-1,1,7-trimethyl-4-methylene-, (1aR-(1aα,4aα,7α,7aβ,7bα))- (9CI)
Aromadendrene (VAN)
Aromadendrene, (+)-
(+)-Aromadendrene
1H-Cycloprop(e)azulene, decahydro-1,1,7-trimethyl-4-methylene-, (1aR,4aR,7R,7aR,7bS)-(+)- (8CI)

489-98-5
$C_6H_4N_4O_6$
228.10
Nc1c(cc(cc1N(=O)=O)N(=O)=O)N(=O)=O
Aniline, 2,4,6-trinitro- (8CI)
AI3-28913
Benzenamine, 2,4,6-trinitro-(9CI)
NSC-4860
Picramide
2,4,6-Trinitroaniline

490-11-9
$C_7H_5NO_4$
167.11
OC(=O)c1ccncc1C(O)=O
3,4-Pyridinedicarboxylic acid (9CI)
Cinchomeronic acid
NSC-178

490-55-1
$C_9H_9N_3S$
191.27
Thiazole, 2,4-diamino-5-phenyl
Amifenazol
Amiphenazol
Amiphenazole
Daftazol
DAPT
Daptazile
Daptazole
DHA-245
2,4-Diamino-5-phenylthiazole
Dizol

Fenamizol
Phenamizol
Phenamizole
5-Phenyl-2,4-thiazolediamine

490-64-2
$C_{10}H_{12}O_5$
212.20
O=C(O)c(c(OC)cc(OC)c1OC)c1
Benzoic acid, 2,4,5-tri-methoxy-
AI3-38428
2,4,5-Trimethoxybenzoic acid

490-65-3
$C_{14}H_{16}$
184.28
Naphthalene, 1-methyl-7-(1-methylethyl)- (9CI)
Eudalene (6CI)
Eudalin
Naphthalene, 7-isopropyl-1-methyl- (7CI,8CI)

490-67-5
$C_{19}H_{26}O_{12}$
446.41
Benzoic acid, 2-((6-O-β-D-xylo-pyranosyl-β-D-glucopyrano-syl)oxy)-, methyl4 ester (9CI)
Gaultherin (8CI)
Monotropitin
Monotropitoside

490-78-8
$C_8H_8O_3$
152.16
O=C(c(c(O)ccc1O)c1)C
Acetophenone, 2',5'-dihydroxy
Acetylhydroquinone
2-Acetylhydroquinone
2,5-Dihydroxyacetophenone
2',5'-Dihydroxyacetophenone
Ethanone, 1-(2,5-dihydroxy-phenyl)- (9CI)
Quinacetophenone

490-79-9
$C_7H_6O_4$
154.13
O=C(O)c(c(O)ccc1O)c1
Gentisic-acid
Benzoic acid, 2,5-dihydroxy-
2,5-DHBA
2,5-Dihydroxybenzoic acid
Gentisate
Hydroquinonecarboxylic acid
5-Hydroxysalicylic acid
Kyselina 2,5-dihydroxy-benzoova (Czech)
Kyselina gentisinova (Czech)
Salicylic acid, 5-hydroxy-

491-04-3
$C_{10}H_{18}O$
154.25
OC(C=C(CC1)C)C1C(C)C
3-Hydroxy-4-isopropyl-1-methylcyclohexene
3-Carvomenthenol
2-Cyclohexen-1-ol, 3-methyl-6-(1-methylethyl)- (9CI)
p-Menth-1-en-3-ol
1-Methyl-4-isopropyl-1-cyclo-hexen-3-ol
3-Methyl-6-(1-methylethyl)-2-cyclohexen-1-ol
Piperitol
Piperitol (monoterpene)

491-07-6
$C_{10}H_{18}O$
154.28
O=C(C(CCC1C)C(C)C)C1
p-Menthan-3-one, (Z)
Cyclohexanone, 5-methyl-2-(1-methylethyl)-, (Z)-
Isomenthone
2-Isopropyl-5-methyl-cyclo-hexanone
5-Methyl-2-(1-methylethyl)-cyclohexanone

491-11-2
$C_6H_4ClNO_3$
173.56
Phenol, 3-chloro-4-nitro
3-Chloro-4-nitrophenol

491-30-5
C_9H_7NO
145.15
O=c1[nH]ccc2ccccc12
Isocarbostyril (VAN) (8CI)
AI3-62131
1-Hydroxyisoquinoline
1-Isoquinolinol
1(2H)-Isoquinolinone (9CI)
Isoquinolin-1-one
1(2H)-Isoquinolone
NSC-27273

491-35-0
$C_{10}H_9N$
143.20
n(c(c(c(c1)C)ccc2)c2)c1
Lepidine
Cincholepidine
Lepidin
4-Lepidine
γ-Methylquinoline
4-Methylquinoline
Quinoline, 4-methyl-

491-59-8
$C_{15}H_{12}O_3$
240.27
1,8,9-Anthracenetriol, 3-methyl
Chrysarobin
Chrysophanic acid anthranol
3-Methyl-1,8,9-anthracenetriol
3-Methylanthralin
1,8,9-Trihydroxy-3-methyl-anthracene

492-17-1
$C_{12}H_{12}N_2$
184.26
Nc1ccc(cc1)c2ccccc2N
2,4'-Biphenyldiamine
o,p'-Bianiline
(1,1'-Biphenyl)-2,4'-diamine
2,4'-Diaminobifenyl (Czech)
o,p'-Diaminobiphenyl
2,4'-Diaminobiphenyl
2,4'-Diaminodiphenyl
o,p'-Dianiline
Difenylin
2,4'-Diphenyldiamine
Diphenyline

492-22-8
$C_{13}H_8OS$
212.27
Oc(c(c(sc1cccc2)ccc3)c3)c12
Thioxanthen-9-one (8CI)
NSC-15912
Thiaxanthenone
Thiaxanthon
Thiaxanthone
Thioxanthene-9-one
9H-Thioxanthen-9-one (9CI)
Thioxanthene, 9-oxo-
9H-Thioxanthene, 9-oxo-
Thioxanthenone
Thioxanthone
9-Thioxanthone

492-27-3
$C_{10}H_7NO_3$
189.16
OC(=O)c2cc(O)c1ccccc1n2
Kynurenic acid
4-Hydroxyquinaldic acid
4-Hydroxyquinaldinic acid
4-Hydroxyquinoline-2-carboxyl-ic acid
Kinurenic acid
Kynuronic acid
NSC-58973
Quinaldic acid, 4-hydroxy- (8CI)
2-Quinolinecarboxylic acid, 4-hydroxy- (9CI)
Quinurenic acid

492-37-5
$C_9H_{10}O_2$
150.18
O=C(O)C(c(cccc1)c1)C
Hydratropic acid
Benzeneacetic acid, α-methyl- (9CI)
α-Methylbenzeneacetic acid
α-Methylphenylacetic acid
NSC-42872
2-Phenylpropanoic acid
α-Phenylpropionic acid
2-Phenylpropionic acid

492-39-7
$C_9H_{13}NO$
151.20
OC(c(cccc1)c1)C(N)C
Cathine
threo-2-Amino-1-hydroxy-1-phenylpropane
Benzenemethanol, α-(1-aminoethyl)-, (S-(R*,R*))- (9CI)
Cathina (Spanish)
Cathinum (Latin)
DEA No. 1230
Katine
psi-Norephedrine
Norpseudoephedrine, (+)-
d-Norpseudoephedrine
Nor-psi-ephedrine
d-Nor-psi-ephedrine
threo-1-Phenyl-1-hydroxy-2-aminopropane
Pseudonorephedrine

492-80-8
$C_{17}H_{21}N_3$
267.41
N=C(c(ccc(N(C)C)c1)c1)c(ccc(N(C)C)c2)c2
Aniline, 4,4'-(imidocarbonyl)bis(N,N-dimethyl
Apyonine Auramine Base
Auramine
Auramine Base
Auramine (Free base)
Auramine N Base
Auramine OAF
Auramine O Base
Auramine OO
Auramine SS
Auremine
Benzenamine, 4,4'-carbonimidoylbis(N,N-dimethyl-(9CI)
Bis(p-dimethylaminophenyl)-methyleneimine
Brilliant Oil Yellow
C.I. 41000B
C.I. Basic Yellow 2, Free base
C.I. Solvent Yellow 34
4,4'-Dimethylaminobenzophenonimide
Glauramine
4,4'-(Imidocarbonyl)bis(N,N-dimethylaniline)
RCRA waste number U014
Tetramethyldiaminodiphenylacetimine
Waxoline Yellow O
Yellow Pyoctanine

492-86-4
$C_8H_7ClO_3$
186.59
O=C(O)C(O)c(ccc(c1)Cl)c1
Benzeneacetic acid, 4-chloro-α-hydroxy- (9CI)
AI3-16648
4-Chloro-α-hydroxybenzeneacetic acid
p-Chloromandelic acid
4-Chloromandelic acid
Mandelic acid, p-chloro- (8CI)
NSC-31400

493-01-6
$C_{10}H_{18}$
138.25
C(C(CCC1)CCC2)(C1)C2
Naphthalene, decahydro-, cis-(9CI)
cis-Bicyclo(4.4.0)decane
cis-Decahydronaphthalene
cis-Decalin
cis-Perhydronaphthalene
NSC-77452

493-02-7
$C_{10}H_{18}$
138.25
C(C(CCC1)CCC2)(C1)C2
Naphthalene, decahydro-, trans- (9CI)
trans-Bicyclo(4.4.0)decane
Decahydronaphthalene, trans-
trans-Decahydronaphthalene
trans-Decalin
NSC-77453
trans-Perhydronaphthalene

493-09-4
$C_8H_8O_2$
136.15
C2COc1ccccc1O2
1,4-Benzodioxan (8CI)
AI3-05084
Benzene, 1,2-(1,2-ethanediylbis(oxy))-
1,4-Benzodioxane
1,4-Benzodioxin, 2,3-dihydro-(9CI)
1,2-(Ethylenedioxy)benzene
Ethylene o-phenylene dioxide
NSC-406705
Pyrocatechol ethylene ether

493-52-7
$C_{15}H_{15}N_3O_2$
269.33
O=C(O)c(c(N=Nc(ccc(N(C)C)c1)c1)ccc2)c2
Benzoic acid, 2-((4-dimethylamino)phenylazo)
2-Carboxy-4'-(dimethylamino)-azobenzene
Cerven kysela 2 (Czech)
Cerven methylova (Czech)
C.I. 13020
C.I. Acid Red 2
p-(Dimethylamino)azobenzene-o-carboxylic acid
4'-Dimethylaminoazobenzene-2-carboxylic acid
o-((p-(Dimethylamino)phenyl)-azo)benzoic acid
2-((4-Dimethylamino)phenyl-azo)benzoic acid
Methyl Red

493-77-6
$C_{21}H_{15}N_3$
309.35
n(c(nc(n1)c(cccc2)c2)c(cccc3)c3)c1c(cccc4)c4
s-Triazine, 2,4,6-triphenyl-(8CI)
AI3-61032
Cyaphenine
Kyaphenine
NSC-46521
1,3,5-Triazine, 2,4,6-triphenyl-(9CI)
Triphenyl-s-triazine
s-Triphenyltriazine
2,4,6-Triphenyltriazine
2,4,6-Triphenyl-s-triazine
2,4,6-Triphenyl-1,3,5-triazine

494-03-1
$C_{14}H_{15}Cl_2N$
268.20
ClCCN(CCCl)c2ccc1ccccc1c2
2-Naphthylamine, N,N-bis-(2-chloroethyl)
N,N-Bis-(2-chlorethyl)-2-naftylamin (Czech)
2-Bis(2-chloroethyl)aminonaphthalene
N,N,-Bis(2-chloroethyl)-2-naphthylamine
Bis(2-chloroethyl)-β-naphthylamine
Chlornaftina
Chlornaphazin
Chlornaphazine
Chlornaphthin
Chloronaftina
Chloronaphthine
Dichloroethyl-β-naphthylamine
Di(2-chloroethyl)-β-naphthylamine
NN-Di(2-chloroethyl)-β-naphthylamine
2-N,N-Di(2-chloroethyl)naphthylamine
Erysan
Naphthylamine mustard
β-Naphthyl-bis-(β-chloroethyl)-amine
2-Naphthylbis(2-chloroethyl)-amine
β-Naphthyl-di-(2-chloroethyl)-amine
NSC-62209
R48
RCRA waste number U026

494-04-2
$C_{15}H_{11}N_3$
233.25
3,2':4',3''-Terpyridine (8CI,9CI)
Nicotelline

494-19-9
$C_{14}H_{13}N$
195.28
N(c(c(ccc1)CCc2cccc3)c1)c23

5H-Dibenz(b,f)azepine, 10,11-dihydro
10,11-Dihydro-5-dibenz(b,f)-azepine
Iminodibenzyl

494-38-2
$C_{17}H_{19}N_3$
265.39
Acridine, 3,6-bis(dimethyl-amino)
Acridinediamine, N,N,N[1],N[1]-tetramethyl- (9CI)
Acridine Orange
Acridine Orange Base
Acridine Orange Free Base
Acridine Orange NO
Acridinorange
Basic Orange 3rn
2,8-Bisdimethylaminoacridine
3,6-Bis(dimethylamino)acridine
3,6-Bis-(dimethylamino)akridin (Czech)
Brilliant Acridine Orange E
C.I. Basic Orange 14
C.I. 46005
C.I. 46005B
C.I. No. 46005:1
C.I. Solvent Orange 15
3,6-Di(dimethylamino)acridine
Euchrysine
Oranz Akridinova (Czech)
Oranz Rozpoustedlova 15 (Czech)
Rhoduline Orange
Rhoduline Orange N
Rhoduline Orange NO
Solvent Orange 15
N,N,N',N'-Tetramethyl-3,6-acridinediamine
Waxoline Orange A

494-52-0
$C_{10}H_{14}N_2$
162.26
C1CCC(NC1)c2cccnc2
Anabasine
Anabasin
(-)-Anabasin
Anabazin
Neonicotine
Neonikotin
Piperidine, 2-(3-pyridyl)-
1-3-(2'-Piperidyl)pyridine
Pyridine, 3-(2-piperidyl)-
3-(2-Piperidinyl)pyridine
2-(3'-Pyridyl) piperidine
(-)-2-(3'-Pyridyl)piperidine

494-99-5
$C_9H_{12}O_2$
152.19
O(c(c(OC)cc(c1)C)c1)C
Benzene, 1,2-dimethoxy-4-methyl- (9CI)
1,2-Dimethoxy-4-methylbenzene
3,4-Dimethoxytoluene
Homoveratrole
4-Methyl-1,2-dimethoxybenzene
4-Methylveratrol
4-Methylveratrole
NSC-7378
Toluene, 3,4-dimethoxy- (8CI)

495-10-3
C_9H_7N
132.16
Benzeneacetonitrile, α-methylene- (9CI)
Atroponitrile (6CI,7CI,8CI)
α-Cyanostyrene
α-Phenylacrylonitrile
2-Phenylacrylonitrile

495-18-1
$C_7H_7NO_2$
137.15
O=C(NO)c(cccc1)c1
Benzohydroxamic-acid
Benzamide, N-hydroxy-
Benzohydroxamate
Benzoylhydroxamic acid
Phenylhydroxamic acid

495-40-9
$C_{10}H_{12}O$
148.20
O=C(c(cccc1)c1)CCC
Butyrophenone (8CI)
AI3-02062
1-Butanone, 1-phenyl- (9CI)
n-Butyrophenone
NSC-8463
1-Phenyl-1-butanone
Phenyl propyl ketone
Propyl phenyl ketone

495-48-7
$C_{12}H_{10}N_2O$
198.24
O=N(=Nc(cccc1)c1)c(cccc2)c2
Azoxybenzene
Azobenzene, oxide
Azossibenzene (Italian)
Azoxybenzeen (Dutch)
Azoxybenzide
Azoxybenzol (German)
Azoxydibenzene
Benzene, azoxydi-
Diazene, diphenyl-, 1-oxide (9CI)
Fenazox (German)
Ordinary azoxybenzene

495-54-5
$C_{12}H_{12}N_4$
212.28
N(=Nc(cccc1)c1)c(c(N)cc(N)c2)c2
m-Phenylenediamine, 4-(phenylazo)
Azobenzene-2,4-diamine
Chrysoidin A
Chrysoidine Base
Chrysoidine Base A
Chrysoidine Base B
Chrysoidine G Base
Chrysoidine J Base
Chrysoidine Y Base
Chrysoidine Y Base New
Chrysoidine YD Base
C.I. 11270
C.I. Solvent Orange 3
2,4-Diaminoazobenzen (Czech)
Diaminoazobenzene
2,4-Diaminoazobenzene
Fat Brown GG
Grasan Chrysoidine
Oranz Rozpoustedlova 3 (Czech)
Oranz Zasadita 2 (Czech)
Waxoline Orange Y

495-69-2
$C_9H_9NO_3$
179.19
O=C(NCC(=O)O)c(cccc1)c1
Hippuric-acid
Acido ippurico (Italian)
Benzamidoacetic acid
Benzoylglycine
Glycine, N-benzoyl-
Phenylcarbonylaminoacetic acid

495-73-8
$C_{13}H_{11}N_3O_2$
241.27
ON=C1C=CC(C=C1)=NNC(=O)c2ccccc2
Benzoic acid, (4-oxo-2,5-cyclohexadien-1-ylidene)hydrazide, oxime
Bay 15080
Bayer 15080
Benchinox
Benguinox
Benquinox
Benzoic acid, (4-(hydroxyimino)-2,5-cyclohexadien-1-ylidene) hydrazide
1,4-Benzoquinone N'-benzoylhydrazone oxime
p-Benzoquinone oxime benzoylhydrazone
Cereden
Ceredon
Cereline
Cerenox
Chinonoxim-benzoylhydrazon (German)
Chinonoxime-benzoylhydrazone
COBH
GBH
Lerenox
QBH
Quinone oxime benzoylhydrazone
Tillantox
Tserenox

495-76-1
$C_8H_8O_3$
152.15
O(c(c(O1)cc(c2)CO)c2)C1

Piperonyl alcohol (8CI)
AI3-05702
1,3-Benzodioxole-5-methanol (9CI)
Benzyl alcohol, 3,4-(methylenedioxy)-
5-Hydroxymethyl-1,3-benzodioxole
1-Hydroxymethyl-3,4-methylenedioxybenzene
NSC-26265
Piperonol

496-03-7
$C_8H_{16}O_2$
144.21
O=CC(C(O)CCC)CC
2-Ethyl-3-hydroxyhexanal
AI3-01508
Butyraldol
Hexanal, 2-ethyl-3-hydroxy-
3-Hydroxy-2-ethylhexanal

496-10-6
C_9H_{16}
124.23
1H-Indene, octahydro- (9CI)
Bicyclo(4.3.0)nonane
Hydrindan
Indan, hexahydro- (8CI)

496-11-7
C_9H_{10}
118.19
c(c(ccc1)CC2)(c1)C2
Indan
2,3-Dihydroindene
Hydrindene
1,2-Hydrindene
Hydrindonaphthene
Indane
Indene, 2,3-dihydro-

496-14-0
C_8H_8O
120.15
C1OCc2ccccc12
Isocoumarin

496-15-1
C_8H_9N
119.16
N(c(c(ccc1)C2)c1)C2
Indoline
AI3-39164
2,3-Dihydroindole
1H-Indole, 2,3-dihydro-

496-16-2
C_8H_8O
120.15
O(c(c(ccc1)C2)c1)C2
Coumaran
Benzofuran, 2,3-dihydro- (9CI)
2,3-Dihydrobenzofuran

496-46-8
$C_4H_6N_4O_2$
142.09
O=C(NC(NC(=O)N1)C12)N2
Glycoluril (8CI)
Acetylene carbamide
Acetylenediurea
Acetylenediureine
Acetyleneurea
AI3-08271
Diurea glyoxalate
Glyoxalbiuret
Glyoxaldiureine
Glyoxaldiurene
Imidazo(4,5-d)imidazole-2,5(1H,3H)-dione, tetrahydro- (9CI)
NSC-2765
Tetrahydroimidaz(d)imidazole-2,5(1H,3H)-dione
Tetrahydroimidazo(4,5-d)imidazole-2,5(1H,3H)-dione

496-72-0
$C_7H_{10}N_2$
122.19
Nc(c(N)cc(c1)C)c1
Toluene-3,4-diamine
1,2-Benzenediamine, 4-methyl-
3,4-Diaminotoluene
3,4-Toluylenediamine
3,4-Tolylenediamine

496-73-1
$C_7H_8O_2$
124.14
Oc(c(ccc1O)C)c1
1,3-Benzenediol, 4-methyl- (9CI)
4-Methyl-1,3-benzenediol
4-Methylresorcinol

496-78-6
$C_9H_{12}O$
136.19
Oc(c(cc(c1C)C)C)c1
Phenol, 2,4,5-trimethyl- (9CI)
1-Hydroxy-2,4,5-trimethylbenzene
NSC-38776
Pseudocumenol
2,4,5-Trimethylphenol

497-03-0
C_5H_8O
84.12
O=CC(=CC)C
Crotonaldehyde, 2-methyl-, (E)- (8CI)
AI3-24379
2-Butenal, 2-methyl-, (E)- (9CI)
E-2-Methyl-2-butenal
(E)-2-Methyl-2-butenal
trans-2-Methyl-2-butenal
NSC-2179
Tiglaldehyde
trans-Tiglaldehyde
Tiglic acid aldehyde
Tiglic aldehyde

497-04-1
$C_3H_7ClO_2$
110.54
1,3-Propanediol, 2-chloro- (8CI,9CI)
Glycerol β-chlorohydrin

497-06-3
$C_4H_8O_2$
88.11
3-Butene-1,2-diol
AI3-07552

497-18-7
CH_6N_4O
90.11
O=C(NN)NN
Carbohydrazide
Carbazic acid, hydrazide
Carbazide (DOT)
Carbodihydrazide
Carbonic acid, dihydrazide
Carbonic dihydrazide (9CI)
Carbonohydrazide
1,3-Diaminomocovina (Czech)
1,3-Diaminourea
Hydrazine, carbonyldi-
Karbazid (Czech)
Semicarbazide, 4-amino-
Urea, 1,3-diamino-

497-19-8
$CO_3.2Na$
105.99
[Na+].[Na+].[O-]C([O-])=O
Sodium carbonate (2:1)
Carbonic acid, disodium salt
Crystol carbonate
Soda ash
Disodium carbonate
Trona

497-23-4
$C_4H_4O_2$
84.08
2(5H)-Furanone
2-Butenoic acid, 4-hydroxy-, γ-lactone
2-Butenoic acid γ-lactone
δ,α,β-Butenolide
2-Buten-4-olide
γ-Crotolactone
α,β-Crotonolactone
Crotonic acid, 4-hydroxy-, γ-lactone (6CI,7CI)
γ-Crotonolactone
4-Hydroxy-2-butenoic acid lactone
4-Hydroxy-2-butenoic acid γ-lactone
γ-Hydroxycrotonic acid lactone
Isocrotonolactone
2-Oxo-2,5-dihydrofuran

497-26-7
$C_4H_8O_2$
88.12
O(CCO1)C1C
1,3-Dioxolane, 2-methyl
Methyldioxolane
2-Methyl-1,3-dioxolane

497-38-1
$C_7H_{10}O$
110.17
O=C(C(CC1C2)C2)C1
2-Norbornanone
Norcamphor

497-39-2
$C_{15}H_{24}O$
220.39
Oc(c(cc(c1C)C(C)(C)C)C(C)(C)C)c1
m-Cresol, 4,6-di-tert-butyl
DBMC
4,6-Di-t-butyl-m-cresol
2,4-Di-t-butyl-5-methylphenol
Phenol, 2,4-bis(1,1-dimethylethyl)-5-methyl- (9CI)

497-56-3
$C_7H_6N_2O_5$
198.15
o-Cresol, 3,5-dinitro
3,5-Dinitro-o-cresol

497-59-6
$C_7H_4O_7$
200.10
O=C(O)c(oc(c(O)c1O)C(=O)O)c1
Meconic acid
3-Hydroxy-4-oxo-4H-pyran-2,6-dicarboxylic acid
NSC-805
4H-Pyran-2,6-dicarboxylic acid, 3-hydroxy-4-oxo- (9CI)

498-02-2
$C_9H_{10}O_3$
166.19
O=C(c(ccc(O)c1OC)c1)C
Acetophenone, 4'-hydroxy-3'-methoxy
Acetoguaiacon
Acetoguaiacone
Acetovanillone
Acetovanilone
Acetovanyllon
Apocynin
Apocynine
Ethanone, 1-(4-hydroxy-3-methoxyphenyl)- (9CI)
4'-Hydroxy-3'-methoxyacetophenone
3-Metoksy-4-hydroksyacetofenon (Polish)

498-21-5
$C_5H_8O_4$
132.12
O=C(O)C(CC(=O)O)C
Methylsuccinic acid
Butanedioic acid, methyl- (9CI)
Methylbutanedioic acid
2-Methylsuccinic acid
NSC-5276
1,2-Propanedicarboxylic acid
Pyrotartaric acid
Succinic acid, methyl- (8CI)

498-23-7
$C_5H_6O_4$
130.11
CC(=CC(O)=O)C(O)=O
Citraconic-acid
2-Butenedioic acid, 2-methyl-, (Z)- (9CI)
Kyselina citrakonova (Czech)
Maleic acid, methyl-
cis-Methylbutenedioic acid
2-Methyl-2-butenedioic acid
Methylmaleic acid

498-43-1
$C_6H_{12}O_6$
180.16
D-Ribo-hexonic acid, 3-deoxy- (9CI)
Glucometasaccharinic acid, α-D- (8CI)
α-D-Glucometasaccharinic acid

498-60-2
$C_5H_4O_2$
96.09
O=Cc1ccoc1
3-Furancarboxaldehyde (9CI)
3-Furaldehyde (8CI)

498-66-8
C_7H_{10}
94.17
C(C=CC1C2)(C1)C2
2-Norbornene
Norbornylene
Norcamphene

498-81-7
$C_{10}H_{20}O$
156.30
OC(C(CCC(C1)C)C1)(C)C
p-Menthan-8-ol
Cyclohexanemethanol, α,α,4-trimethyl- (9CI)
Dihydro-α-terpineol
1-Methyl-4-isopropylcyclohexane-8-ol

499-03-6
$C_{10}H_{16}$
136.24
Cyclohexene, 1-methyl-3-(1-methylethenyl)-, (+-)- (9CI)
m-Mentha-1,8-diene, (+-)- (8CI)

499-06-9
$C_9H_{10}O_2$
150.19
O=C(O)c(cc(cc1C)C)c1
Benzoic acid, 3,5-dimethyl
3,5-Dimethylbenzoic acid
Mesitylenic acid

499-12-7
$C_6H_6O_6$
174.12
O=C(O)C(=CC(=O)O)CC(=O)O
1-Propene-1,2,3-tricarboxylic acid
Achilleic acid
Aconitic acid
Citridic acid
Equisetic acid

499-48-9
$C_{13}H_{26}N_2O_3$
258.36
Elaiomycin
4-Methoxy-3-(1-octenylazoxy)-2-butanol

499-70-7
$C_{10}H_{18}O$
154.25
Cyclohexanone, 2-methyl-5-(1-methylethyl)-, trans- (9CI)
Carvomenthone

499-74-1
$C_{10}H_{16}O$
152.24
2-Cyclohexen-1-one, 6-methyl-3-(1-methylethyl)-
AI3-27537

499-75-2
$C_{10}H_{14}O$
150.24
Oc(c(ccc1C(C)C)C)c1
Carvacrol
Antioxine
o-Cresol, 5-isopropyl-
p-Cymene, 2-hydroxy-
2-p-Cymenol
2-Hydroxy-p-cymene
Isopropyl-o-cresol
3-Isopropyl-6-methylphenol
Isothymol
2-Methyl-5-isopropylphenol
2-Methyl-5-(1-methylethyl)-phenol
Phenol, 3-isopropyl-6-methyl-
Phenol, 5-isopropyl-2-methyl-
Phenol, 2-methyl-5-(1-methylethyl)- (9CI)
o-Thymol

499-80-9
$C_7H_5NO_4$
167.11

O=C(O)c(ccnc1C(=O)O)c1
2,4-Pyridinedicarboxylic acid (9CI)
NSC-403248

499-81-0
$C_7H_5NO_4$
167.11
3,5-Pyridinedicarboxylic acid (9CI)
AI3-52669
5-Carboxynicotinic acid
Dinicotinic acid
NSC-6497
Pyridine-3,5-dicarboxylic acid

499-83-2
$C_7H_5NO_4$
167.11
O=C(O)c(nc(cc1)C(=O)O)c1
Dipicolinic acid
Dipicolinate
2,6-Dipicolinic acid
NSC-176
2,6-Pyridinedicarboxylic acid (9CI)

500-22-1
C_6H_5NO
107.12
O=Cc(cccn1)c1
Nicotinaldehyde
3-Formylpyridine
Nicotinealdehyde
Nicotinic aldehyde
3-Pyridinaldehyde
3-Pyridinealdehyde
Pyridine-3-carbaldehyde
β-Pyridinecarbonaldehyde
3-Pyridinecarboxaldehyde (9CI)
3-Pyridylaldehyde
3-Pyridylcarboxaldehyde
Rowalind

500-28-7
$C_8H_9ClNO_5PS$
297.66
COP(=S)(OC)Oc1ccc(N(=O)=O)c(Cl)c1
Phosphorothioic acid, O-(3-chloro-4-nitrophenyl) O,O-dimethyl ester
Bayer 22/190
O-(3-Chloor-4-nitro-fenyl)-O,O-dimethyl-monothiofosfaat (Dutch)
Chloorthion (Dutch)
O-(3-Chlor-4-nitro-phenyl)-O,O-dimethyl-monothiophosphat (German)
O-(3-Chloro-4-nitrophenyl) O,O-dimethyl phosphorothioate
Chlorothion
Chlorthion
Chlorthion methyl
Chlortion (Czech)
O-(3-Cloro-4-nitro-fenil)-O,O-dimetil-monotiofosfato (Italian)
Compound 22/190
O,O-Dimethyl-O-3-chlor-4-nitrofenylester kyseliny thiofosforecne (Czech)
O,O-Dimethyl-O-3-chlor-4-nitrofenyltiofosfat (Czech)
O,O-Dimethyl-O-(3-chlor-4-nitrophenyl)-monothiophosphat (German)
O,O-Dimethyl O-(3-chloro-4-nitrophenyl) phosphoro thioate
Dimethyl 3-chloro-4-nitrophenyl thionophosphate
O,O-Dimethyl O-(3-chloro-4-nitrophenyl) thiophosphate
O,O-Dimethyl-O-(4-nitro-5-chlorphenyl)-thionophosphat (German)
O,O-Dimethyl p-nitro-m-chlorophenyl thiophosphate
O,O-Dimethyl O-4-nitro-3-chlorophenyl thiophosphate
ENT 18,861
Methylchlorothion
p-Nitro-m-chlorophenyl dimethyl thionophosphate
OMS 217
Phenol, 3-chloro-4-nitro-, O-ester with O,O-dimethyl phosphorothioate
Thiophosphate de O,O-dimethyle et de O-3-chloro-4-nitrophenyle (French)

500-34-5
$C_{15}H_{21}NO_2$
247.37
4-Piperidinol, 2,2,6-trimethyl-, benzoate (ester)
β-Eucaine
Eucaine B
Eukain B
Trimethylbenzoyloxypiperidine

500-38-9
$C_{18}H_{22}O_4$
302.40
Pyrocatechol, 4,4'-(2,3-dimethyltetramethylene)di
2,3-Bis(3,4-dihydroxyphenylmethyl)butane
Butane, 1,4-bis(3,4-dihydroxyphenyl)-2,3-dimethyl-
Dihydronorguaiaretic acid
β,γ-Dimethyl-α,δ-bis(3,4-dihydroxyphenyl)butane
4,4'-(2,3-Dimethyltetramethylene)dipyrocatechol
Dinorguaiaretic acid, dihydro-
NDGA
Nordihydroguaiaretic acid
Nordihydroguairaretic acid
Norguaiaretic acid, dihydro-

500-42-5
$C_9H_8ClN_5$
221.67
s-Triazine, 2-amino-4-(p-chloroanilino)
2-Amino-4-(p-chloroanilino)-s-triazine
ASA 226
Chlorazanil
Chlorazinil
Chloroazinal
2-(4-Chlorophenylamino)-4-amino-1,3,5-triazine
N-(4-Chlorophenyl)-1,3,5-triazine-2,4-diamine
Daquin
Diurazine
neo-Urofort
Neurofort
Orpizin
Riker 545
SKF 3195
Triazurol
Neo-Urofort

500-44-7
$C_8H_{10}N_2O_4$
198.17
(L)-Mimosine
AI3-51821
Leucaenine
Leucaenol
Leucena glauca α-amino acid
Leucenine
Leucenol
Mimosin
Mimosine
L-Mimosine
NSC-69188
1(4H)-Pyridinepropanoic acid, α-amino-3-hydroxy-4-oxo-, (S)- (9CI)

500-66-3
$C_{11}H_{16}O_2$
180.27
Resorcinol, 5-pentyl
5-n-Amylresorcinol
1,3-Benzenediol, 5-pentyl-(9CI)
3,5-Dihydroxyamylbenzene
Olivetol
5-Pentylresorcinol
5-n-Pentylresorcinol

500-99-2
$C_8H_{10}O_3$
154.17
COc1cc(O)cc(OC)c1
Phloroglucinol dimethyl ether
3,5-Dimethoxyphenol
1-Hydroxy-3,5-dimethoxybenzene
NSC-70955
Phenol, 3,5-dimethoxy- (9CI)
Taxicatigenin

501-24-6
$C_{21}H_{36}O$
304.52
Oc(cccc1CCCCCCCCCCCCCCC

C)c1
Phenol, m-pentadecyl- (8CI)
Anacardol, tetrahydro-
Cyclogallipharaol
Hydrocardanol
Hydroginkgol
NSC-9781
m-Pentadecylphenol
3-Pentadecylphenol
3-n-Pentadecylphenol
Phenol, 3-pentadecyl- (9CI)
Tetrahydroanacardol

501-52-0
$C_9H_{10}O_2$
150.18
O=C(O)CCc(cccc1)c1
3-Phenylpropionic acid
AI3-00892
Benzenepropanoic acid (9CI)
Benzenepropionic acid
Benzylacetic acid
Dihydrocinnamic acid
Hydrocinnamic acid (8CI)
NSC-9272
Phenylpropanoic acid
3-Phenylpropanoic acid
Phenylpropionic acid
β-Phenylpropionic acid

501-53-1
$C_8H_7ClO_2$
170.60
O=C(OCc(cccc1)c1)Cl
Formic acid, chloro-, benzyl ester
Benzylcarbonyl chloride
Benzyl chlorocarbonate (DOT)
Benzyl chloroformate [UN 1739]
Benzyloxycarbonyl chloride
BZCF
Carbobenzoxy chloride
Carbobenzyloxy chloride
Chloroformic acid benzyl ester
UN 1739 [Benzyl chloroformate]

501-58-6
$C_{14}H_{14}N_2O_2$
242.27
Azobenzene, 4,4'-dimethoxy- (8CI)
4,4'-Azodianisole
Diazene, bis(4-methoxyphenyl)- (9CI)
4,4'-Dimethoxyazobenzene
NSC-31011

501-60-0
$C_{14}H_{14}N_2$
210.27
Diazene, bis(4-methylphenyl)- (9CI)
AI3-08890
p,p'-Azotoluene (8CI)
4,4'-Azotoluene
4,4'-Dimethylazobenzene
NSC-31008

501-65-5
$C_{14}H_{10}$
178.23
C(#Cc(cccc1)c1)c(cccc2)c2
Biphenylacetylene
Acetylene, diphenyl- (8CI)
AI3-04360
Benzene, 1,1'-(1,2-ethynediyl)-bis- (9CI)
1,2-Diphenylacetylene
Diphenylethyne
Ethyne, diphenyl-
1,1'-(1,2-Ethynediyl)bisbenzene
NSC-5185
Tolan
Tolane

501-82-6
$C_7H_7NO_2$
137.14
Phenylcarbamic acid
Carbanilic acid

501-94-0
$C_8H_{10}O_2$
138.17
OCCc(ccc(O)c1)c1
4-Hydroxyphenylethanol
Benzeneethanol, 4-hydroxy- (9CI)
4-Hydroxybenzeneethanol
p-Hydroxyphenethyl alcohol
NSC-59876

501-97-3
$C_9H_{10}O_3$
166.19
OC(=O)CCc1ccc(O)cc1
Hydrocinnamic acid, p-hydroxy
Benzenepropanoic acid, 4-hydroxy- (9CI)
Dihydro-p-coumaric acid
Hydro-p-coumaric acid
4-Hydroxybenzenepropanoic acid
p-Hydroxyhydrocinnamic acid
3-(4-Hydroxyphenyl)propanoic acid
β-(p-Hydroxyphenyl)propionic acid
p-Hydroxyphenylpropionic acid
3-(p-Hydroxyphenyl)propionic acid
3-(4-Hydroxyphenyl)propionic acid
3-(4'-Hydroxyphenyl)propionic acid
4-Hydroxyphenylpropionic acid
Phloretic acid

501-98-4
$C_9H_8O_3$
164.17
Cinnamic acid, p-hydroxy-, (E)
(E)-p-Coumaric acid
trans-p-Coumaric acid
trans-p-Coumarinic acid
(E)-p-Hydroxycinnamic acid
trans-p-Hydroxycinnamic acid
trans-4-Hydroxycinnamic acid
Naringeninic acid
2-Propenoic acid, 3-(4-hydroxyphenyl)-, (E)- (9CI)

502-39-6
$C_3H_6HgN_4$
298.72
C[Hg]NC(=N)NC#N
Mercury, (3-cyanoguanidino)methyl
Agrosol
Cyano(methylmercuri)guanidine
Guanidine, cyano-, methylmercury deriv.
MEMA
Methylmercuric cyancguanidine
Methylmercuric dicyandiamide
Methylmercury dicyandiamide
Methylmerkuridikyandiamid (Czech)
MMD
Morsodren
Morton EP-227
Morton Soil Drench
Morton Soil-Drench-C
Pandrinox
Pano-Drench 4
Panodrin A-13
Panogen
Panogen 15
Panogen 43
Panogen PX
Panogen Turf Fungicide
Panogen Turf Spray
Panospray 30
R 8
R 8 (Fungicide)
Zaprawa Nasienna Plynna

502-41-0
$C_7H_{14}O$
114.19
OC1CCCCCC1
Cycloheptanol (8CI,9CI)
NSC-52221

502-42-1
$C_7H_{12}O$
112.19
O=C(CCCCC1)C1
Cycloheptanone
Ketocycloheptane
Ketoheptamethylene
Suberon
Suberone

502-44-3
$C_6H_{10}O_2$
114.16
O=C(OCCCC1)C1
Hexanoic acid, ε-lactone

Caprolactone
ε-Caprolactone
6-Hexanolactone
1,6-Hexanolide
6-Hydroxyhexanoic acid lactone
ε-Kaprolakton (Czech)
2-Oxepanone (8CI,9CI)

502-49-8
$C_8H_{14}O$
126.22
O=C(CCCCCC1)C1
Cyclooctanone

502-55-6
$C_6H_{10}O_2S_4$
242.40
O(C(=S)SSC(OCC)=S)CC
Formic acid, dithiobis(thio-, O,O-diethyl ester
Auligen
Aulinogen
BEK
BEXIDE
BEXT
Biethylxanthogentrisulfide
Bis(ethylxanthic)disulfide
Bis(ethylxanthogen)
Bisethylxanthogen disulfide
DEX
Di-ethoxythiokarbonyl-disulfid (Czech)
Diethyldithio bis(thionoformate)
Diethyl dixanthogen
Diethyl xanthogenate
Diethylxanthogen disulfide
Dithiobis(thioformic acid) O,O-diethyl ester
Dixanthogen
Ethylxanthic disulfide
Ethyl xanthogen disulfide
EXD
Herbisan
K Preparation
Lenisarin
Preparation K
Skabilan
Sulfasan
Thioperoxydicarbonic acid diethyl ester
Xanthogen, bis(ethyl-

502-56-7
$C_9H_{18}O$
142.27
O=C(CCCC)CCCC
5-Nonanone
Butyl ketone
Dibutyl ketone
Nonan-5-one
5-Oxononane

502-61-4
$C_{15}H_{24}$
204.36
1,3,6,10-Dodecatetraene, 3,7,11-trimethyl-, (E,E)- (9CI)
3,7,11-Trimethyl-1,3,6,10-dodecatetraene

502-69-2
$C_{18}H_{36}O$
268.48
O=C(CCCC(CCCC(CCCC(C)C)C)C)C
Hexahydrofarnesylacetone
2-Pentadecanone, 6,10,14-trimethyl- (VAN)(9CI)
6,10,14-Trimethylpentadecan-2-one
6,10,14-Trimethyl-2-pentadecanone

502-72-7
$C_{15}H_{28}O$
224.43
O=C(CCCCCCCCCCCCC1)C1
Cyclopentadecanone
Normuscone

502-98-7
$C_2H_4Cl_2N_6$
182.96
Chlorazodin
1,1'-Azobis(N-chloroformamidine)
Chlorazodine (French)
Chlorazodinum (Latin)
Chloroazodin
Clorazodina (Spanish)
Diazenedicarboximidamide, N,N''-dichloro
N,N'-Dichloroazodicarbonamidine (Salts of) (Dry)
N,N'-Dicloroazodicarbonamidina (Spanish)

503-05-9
$C_{18}H_{32}O_2$
280.45
Malvalic acid
1-Cyclopropene-1-heptanoic acid, 2-octyl-
Malvic acid
2-Octyl-1-cyclopropene-1-heptanoic acid

503-09-3
C_3H_5FO
76.08
Propane, 1,2-epoxy-3-fluoro
1,2-Epoxy-3-fluoropropane
Epifluorohydrin

503-17-3
C_4H_6
54.10
C(#CC)C
Crotonylene [UN 1144]
2-Butyne (8CI,9CI)
Dimethylacetylene
UN 1144 [Crotonylene]

503-28-6
$C_2H_6N_2$
58.08
Azomethane

503-29-7
C_3H_7N
57.09
N(CC1)C1
Azetidine
AI3-61395

503-30-0
C_3H_6O
58.09
O(CC1)C1
Oxetane
Cyclooxabutane
1,3-Epoxypropane
Oxacyclobutane
Oxetan
α,γ-Propane oxide
1,3-Propylene oxide
Trimethylenoxid (German)
Trimethylene oxide

503-38-8
$C_2Cl_4O_2$
197.82
ClC(=O)OC(Cl)(Cl)Cl
Formic acid, chloro-, trichloromethyl ester
Carbonochloridic acid trichloromethyl ester
Difosgen (Czech)
Diphosgen
Diphosgene [UN 1076]
Methanol, trichloro-, chloroformate
Trichloromethyl chloroformate
Trichlormethylester kyseliny chlormravenci (Czech)
UN 1076 [Phosgene]

503-40-2
$CH_4O_6S_2$
176.17
O=S(=O)(O)CS(=O)(=O)O
Methanedisulfonic acid (9CI)

503-60-6
C_5H_9Cl
104.58
C(=CCCl)(C)C
1-Chloro-3-methyl-2-butene
2-Butene, 1-chloro-3-methyl- (9CI)
γ,γ-Dimethylallyl chloride
3,3-Dimethylallyl chloride
3-Methyl-2-butenyl chloride
3-Methylcrotyl chloride
Prenyl chloride

503-64-0
$C_4H_6O_2$
86.09

2-Butenoic acid, (Z)- (9CI)
Crotonic acid, (Z)- (8CI)
Isocrotonic acid

503-74-2
$C_5H_{10}O_2$
102.15
O=C(O)CC(C)C
Isovaleric-acid
Acetic acid, isopropyl-
Butyric acid, 3-methyl-
Delphinic acid
Isopentanoic acid (DOT)
Isopropylacetic acid
Isovalerianic
Isovalerianic acid
Kyselina isovalerova (Czech)
3-Methylbutanoic acid
β-Methylbutyric acid
3-Methylbutyrate
3-Methylbutyric acid
NA 1760

504-01-8
$C_6H_{12}O_2$
116.16
Cyclohexane-1,3-diol
AI3-06234
1,3-Benzenediol, hexahydro-
1,3-Cyclohexanediol (9CI)
1,3-Dihydroxycyclohexane
NSC-30235
Resorcitol

504-02-9
$C_6H_8O_2$
112.13
O=C(CCCC1=O)C1
1,3-Cyclohexanedione (9CI)
AI3-11062
1,3-Benzenediol, dihydro-
1,3-Cyclohexandione
Cyclohexane-1,3-dione
1,3 Cyclohexanedione
1,3-Cyclohexanone
Dihydroresorcinol
Hydroresorcinol
NSC-57477
Resorcinol, dihydro-

504-08-5
$C_3H_5N_5$
111.08
Nc1ncnc(N)n1
Formoguanamine
AI3-51263
Diamino-s-triazine
2,6-Diamino-s-triazine
4,6-Diamino-s-triazine
Guanamine
NSC-251
1,3,5-Triazine-2,4-diamine (9CI)
s-Triazine, 2,4-diamino- (8CI)

504-15-4
$C_7H_8O_2$
124.15
Oc(cc(O)cc1C)c1
Resorcinol, 5-methyl
1,3-Dihydroxy-5-methylbenzene
3,5-Dihydroxytoluene
5-Methyl-1,3-benzenediol
5-Methylresorcinol
Orcin
Orcinol
Orcinol, 5-methylresorcinol

504-17-6
$C_4H_4N_2O_2S$
144.16
O=C(N=C(S)NC1=O)C1
Barbituric acid, 2-thio
Austranal
Bathyran
2-Mercaptobarbituric acid
4,6(1H,5H)-Pyrimidinedione, dihydro-2-thioxo- (9CI)
Thiobarbituric acid
2-Thiobarbituric acid
USAF EK-660

504-20-1
$C_9H_{14}O$
138.23
O=C(C=C(C)C)C=C(C)C
2,5-Heptadien-4-one, 2,6-dimethyl
Diisopropylidene acetone
sym-Diisopropylidene acetone
2,6-Dimethyl-2,5-heptadien-4-one
Phoron (German)
Phorone

504-24-5
$C_5H_6N_2$
94.13
n(ccc(N)c1)c1
Pyridine, 4-amino
4-Aminopyridine
Amino-4 pyridine
γ-Aminopyridine
p-Aminopyridine [UN 2671]
4-AP
Avitrol
Avitrol 200
RCRA waste number P008
4-Pyridylamine
4-Pyridinamine
UN 2671 [Aminopyridines (o-; m-; p-)]
VMI 10-3

504-29-0
$C_5H_6N_2$
94.13
n(c(N)ccc1)c1
Pyridine, 2-amino
o-Aminopyridine [UN 2671]
α-Aminopyridine
Amino-2 pyridine
2-Aminopyridine (ACGIH, OSHA)
α-Pyridinamine
α-Pyridylamine
2-Pyridylamine
UN 2671 [Aminopyridines (o-; m-; p-)]

504-31-4
$C_5H_4O_2$
96.09
2H-Pyran-2-one (8CI,9CI)
Coumalin
2,4-Pentadienoic acid, 5-hydroxy-, δ-lactone
2H-Pyran, 2-oxo-
α-Pyrone
2-Pyrone

504-53-0
$C_{35}H_{70}O$
506.94
O=C(CCCCCCCCCCCCCCCC C)CCCCCCCCCCCCCCCCC
Stearone
AI3-15184
Diheptadecyl ketone
Di-n-heptadecyl ketone
Heptadecyl ketone
NSC-83612
18-Pentatriacontanone (9CI)

504-57-4
$C_{19}H_{38}O$
282.51
O=C(CCCCCCCCC)CCCCC CCCC
10-Nonadecanone (9CI)
AI3-05050
Dinonyl ketone
Di-n-nonyl ketone
NSC-1773

504-60-9
C_5H_8
68.13
C(=CC=C)C
1,3-Pentadiene
1-Methylbutadiene
Piperylene
RCRA waste number U186

504-63-2
$C_3H_8O_2$
76.11
OCCCO
1,3-Propanediol
2-Deoxyglycerol
1,3-Dihydroxypropane
2-(Hydroxymethyl)ethanol
NSC-65426
PG
Propane-1,3-diol
β-Propylene glycol
1,3-Propylene glycol
Trimethylene glycol

504-66-5
C_2HN_3

67.03
N#CNC#N
Dicyanamide
Cyanamide, cyano- (9CI)
Cyanocyanamide
Dicyanimide
Ditsianamid
Imidodicarbonitrile

504-88-1
$C_3H_5NO_4$
119.09
OC(=O)CCN(=O)=O
Propionic acid, 3-nitro
BNP
Bovinocidin
Hiptagenic acid
NCI-C03076
β-Nitropropionic acid
3-Nitropropionic acid
Propanoic acid, 3-nitro- (9CI)

504-90-5
$C_2H_4N_2S_4$
184.32
Disulfide, bis(thiocarbamoyl)
Dithiocarbamoyl disulfide
Thiuram disulfide

504-96-1
$C_{20}H_{38}$
278.52
Neophytadiene
1,3-Butadiene, 2-(4,8,12-trimethyltridecyl)-
1-Hexadecene, 7,11,15-trimethyl-3-methylene-
2-(4,8,12-Trimethyltridecyl)-buta-1,3-diene

504-99-4
$C_9H_{18}O_2$
158.24
Octanoic acid, 6-methyl- (8CI, 9CI)

505-10-2
$C_4H_{10}OS$
106.19
OCCCSC
1-Propanol, 3-(methylthio)- (9CI)
AI3-17420
3-Hydroxypropyl methyl sulfide
Methionol
3-Methylmercapto-1-propanol
γ-Methylmercaptopropyl alcohol
3-(Methylthio)-1-propanol
NSC-2859

505-22-6
$C_4H_8O_2$
88.12
m-Dioxane
1,3-Dioxacyclohexane
1,3-Dioxane
1,3-Propanediol formal

505-29-3
$C_4H_8S_2$
120.24
S(CCSC1)C1
p-Dithiane
1,4-Dithiacyclohexane
1,4-Dithiane

505-32-8
$C_{20}H_{40}O$
296.54
OC(C=C)(CCCC(CCCC(CCCC(C)C)C)C)C
Isophytol
AI3-25090
1-Hexadecen-3-ol, 3,7,11,15-tetramethyl- (VAN) (9CI)
NSC-93744
2,6,10,14-Tetramethylhexadec-15-en-14-ol
3,7,11,15-Tetramethyl-1-hexadecen-3-ol
2,6,10-Trimethyl-14-vinylpentadecan-14-ol

505-47-5
$C_6H_{11}NO_4$
161.15
OC(=O)CCNCCC(O)=O
β-Alanine, N-(2-carboxyethyl)- (9CI)
NSC-41820
Propanoic acid, 3,3'-iminobis-
Propionic acid, 3,3'-iminodi- (8CI)

505-48-6
$C_8H_{14}O_4$
174.20
O=C(O)CCCCCCC(=O)O
Suberic acid
AI3-52672
Hexamethylenedicarboxylic acid
1,6-Hexanedicarboxylic acid
NSC-25952
Octanedioic acid (9CI)
1,8-Octanedioic acid

505-52-2
$C_{13}H_{24}O_4$
244.33
O=C(O)CCCCCCCCCCCC(=O)O
Tridecanedioic acid (9CI)
AI3-18168
Brassylic acid
NSC-9498
1,13-Tridecanedioic acid
1,11-Undecanedicarboxylic acid

505-54-4
$C_{16}H_{30}O_4$
286.41
O=C(O)CCCCCCCCCCCCCCC(=O)O
Hexadecanedioic acid (9CI)
1,16-Hexadecanedioic acid
Hexadecane-1,16-dioic acid
NSC-15164
Thapsic acid
1,14-Tetradecanedicarboxylic acid
n-Tetradecane-ω,ω'-dicarboxylic acid

505-57-7
$C_6H_{10}O$
98.16
O=CC=CCCC
2-Hexenal
Hex-2-enal
Hex-2-en-1-al
Hexylenic aldehyde
Leaf Aldehyde

505-60-2
$C_4H_8Cl_2S$
159.08
ClCCSCCCl
Sulfide, bis(2-chloroethyl)
Bis(β-chloroethyl)sulfide
Bis(2-chloroethyl)sulfide
Bis(2-chloroethyl)sulphide
1-Chloro-2-(β-chloroethylthio)-ethane
2,2'-Dichlorodiethyl sulfide
Di-2-chloroethyl sulfide
β,β'-Dichloroethyl sulfide
β,β-Dichlor-ethyl-sulphide
2,2'-Dichloroethyl sulphide
Distilled Mustard
Gelbkreuz (Czech)
HD
Kampstoff "Lost"
Mustard HD
Mustard Gas
Mustard, Sulfur
Mustard Vapor
Schwefel-Lost
S-Lost
S Mustard
Sulfur Mustard Gas
Sulfur Mustard
Sulphur Mustard
Sulphur Mustard Gas
1,1'-Thiobis(2-chloroethane)
Yellow Cross Liquid
Yperite

505-70-4
$C_6H_6O_4$
142.11
Muconic acid
1,3-Butadiene-1,4-dicarboxylic acid
2,4-Hexadienedioic acid (9CI)
NSC-16627

505-75-9
$C_{17}H_{22}O_2$
258.39

8,10,12-Heptadecatriene-4,6-diyne-1,14-diol, (E,E,E)-(-)
Cicutoxin

506-05-8
$C_{11}H_{23}F$
174.34
Undecane, 1-fluoro
1-Fluoroundecane

506-12-7
$C_{17}H_{34}O_2$
270.51
O=C(O)CCCCCCCCCCCCCCC
Heptadecanoic-acid
n-Heptadecoic acid
n-Heptadecylic acid
Margaric acid

506-13-8
$C_{16}H_{32}O_3$
272.43
O=C(O)CCCCCCCCCCCCCCO
Hexadecanoic acid, 16-hydroxy- (9CI)
16-Hydroxyhexadecanoic acid
ω-Hydroxypalmitic acid
16-Hydroxypalmitic acid
Juniperic acid
NSC-159292
Palmitic acid, 16-hydroxy-

506-21-8
$C_{18}H_{32}O_2$
280.45
Linolelaidic acid
AI3-36448
9,12-Octadecadienoic acid, (E,E)-

506-30-9
$C_{20}H_{40}O_2$
312.60
O=C(O)CCCCCCCCCCCCCCCCCCC
Eicosanoic-acid
Arachic acid
Arachidic acid

506-32-1
$C_{20}H_{32}O_2$
304.52
Arachidonic-acid
Arachidonate
(all-Z)-5,8,11,14-Eicosatetraenoic acid

506-38-7
$C_{25}H_{50}O_2$
382.67
NSC-89289
Pentacosanoic acid

506-46-7
$C_{26}H_{52}O_2$
396.70
Hexacosanoic acid (9CI)
Ceratinic acid
Ceric acid
Cerinic acid
Cerotic acid
Hexacosanic acid
NSC-4205

506-48-9
$C_{28}H_{56}O_2$
424.75
O=C(O)CCCCCCCCCCCCCCCCCCCCCCCCCC
NSC-407311

506-50-3
$C_{30}H_{60}O_2$
452.81
Triacontanoic acid (9CI)
Melissic acid
NSC-53832
n-Triacontanoic acid
1-Triacontanoic acid

506-51-4
$C_{24}H_{50}O$
354.66
OCCCCCCCCCCCCCCCCCCCCCCCC
Lignoceryl alcohol
Lignoceric alcohol
Lignocerol
NSC-93768
n-Tetracosanol
1-Tetracosanol (9CI)
Tetracosyl alcohol

506-52-5
$C_{26}H_{54}O$
382.71
OCCCCCCCCCCCCCCCCCCCCCCCCCC
1-Hexacosanol (9CI)
n-Hexacosanol
Hexacosyl alcohol
NSC-4058

506-59-2
$C_2H_7N.ClH$
81.56
[Cl-].CNC
Dimethylamine, hydrochloride
Dimethylammonium chloride
Hydrochloric acid dimethylamine
Methanamine, N-methyl-, hydrochloride (9CI)

506-61-6
$C_2AgN_2.K$
199.01
Potassium-silver-cyanide
Kyanostribrnan draselny (Czech)
RCRA waste number P099
Silver potassium cyanide

506-64-9
CAgN
133.89
Silver-cyanide
Cyanure d'argent (French)
Kyanid stribrny (Czech)
RCRA waste number P104
Silver cyanide [UN 1684]
UN 1684 [Silver cyanide]

506-68-3
CBrN
105.93
N#CBr
Cyanogen-bromide
Bromine cyanide
Bromocyan
Bromocyanide
Bromocyanogen
Bromure de cyanogen (French)
Campilit
Cyanobromide
Cyanogen bromide [UN 1889]
Cyanogen monobromide
RCRA waste number U246
TL 822
UN 1889 [Cyanogen bromide]

506-77-4
CClN
61.47
N#CCl
Cyanogen-chloride
Chlorcyan
Chlorine cyanide
Chlorocyan
Chlorocyanide
Chlorocyanogen
Chlorure de cyanogene (French)
Cyanogen chloride (ACGIH, OSHA)
Cyanogen chloride, Containing less than 0.9% water (DOT)
Cyanogen chloride, Inhibited [UN 1589]
RCRA waste number P033
UN 1589 [Cyanogen chloride, inhibited]

506-78-5
CIN
152.92
N#CI
Iodine-cyanide
Cyanogen iodide
Jodcyan

506-82-1
C_2H_6Cd
142.48
Cadmium, dimethyl- (8CI,9CI)

Dimethylcadmium

506-85-4
CNO
42.01
Fulminic acid
Acide fulminique (French)
Acido fulminico (Spanish)

506-87-6
$CH_2O_3.2H_3N$
96.11
Ammonium-carbonate
Ammoniumcarbonat (German)
Ammonium carbonate (DOT)
Carbonate d'ammoniaque (French)
Carbonic acid, ammonium salt
Carbonic acid, diammonium salt (8CI,9CI)
Diammonium carbonate
NA 9084 (DOT)

506-93-4
$CH_5N_3.HNO_3$
122.06
Guanidine, mononitrate (8CI, 9CI)
Guanidine nitrate
Guanidine nitrate (1:1)
Guanidinium nitrate [UN 1467]
Nitrate de guanidine (French)
Nitrato de guanidina (Spanish)
NSC-7295
UN 1467 [Guanidine nitrate]

506-96-7
C_2H_3BrO
122.96
O=C(Br)C
Acetyl bromide [UN 1716]
UN 1716 [Acetyl bromide]

507-02-8
C_2H_3IO
169.95
O=C(C)I
Acetyl iodide [UN 1898]
UN 1898 [Acetyl iodide]

507-09-5
C_2H_4OS
76.12
O=C(S)C
Acetic acid, thio
Acetyl mercaptan
Ethanethioic acid
Ethanethiolic acid
Kyselina thiooctova (Czech)
Methanecarbothiolic acid
Thiacetic acid
Thioacetic acid [UN 2436]
Thiolacetic acid
Thionoacetic acid
UN 2436 [Thioacetic acid]
USAF EK-P-737

507-19-7
C_4H_9Br
137.04
BrC(C)(C)C
Propane, 2-bromo-2-methyl
2-Bromoisobutane
2-Bromo-2-methylpropane [UN 2342]
tert-Butyl bromide
Trimethylbromomethane
UN 2342 [Bromomethylpropanes]

507-20-0
C_4H_9Cl
92.58
C(Cl)(C)(C)C
Propane, 2-chloro-2-methyl
tert-Butyl chloride
2-Chloroisobutane
2-Chloro-2-methylpropane
Trimethylchloromethane

507-25-5
CI_4
519.61
C(I)(I)(I)I
Carbon-tetraiodide
Carbon iodide
Methane, tetraiodo-
Tetraiodomethane

507-45-9
$C_5H_{10}Cl_2$
141.04
Butane, 2,3-dichloro-2-methyl- (8CI,9CI)
Amylene dichloride

507-60-8
$C_{32}H_{44}O_{12}$
620.69
Scilliroside
6 β-(Acetyloxy)-3-β-(β-D-glucopyranosyloxy)-8,14-dihydroxybufa-4,20,22-trienolide
3-β,6-β-6-Acetyloxy-3-(β-D-glucopyranosyloxy)-8,14-dihydroxybufa-4,20,22-trienolide
(3β,6β)-6-(Acetyloxy)-3-(β-D-glucopyranosyloxy)-8,14-dihydroxybufa-4,20,22-trienolide
Bufa-4,20,22-trienolide, 6-(acetyloxy)-3-(β-D-glucopyranosyloxy)-8,14-dihydroxy-, (3β,6β)-
Caswell No. 722
EPA Pesticide Chemical Code 070801
3β-(β-D-Glucopyropyanosyloxy)-6β,8,14-trihydroxybufa-4,20,22-trienolide 6-accetate
NSC-7523
Red Squill
Scillirosid
Scilliroside, (3-β,6-β)-
Scilliroside, (3β,6β)-
Scillirosidin + glucose (German)
Silmurin

507-70-0
$C_{10}H_{18}O$
154.28
OC(C(C(C1C2)(C)C)(C2)C)C1
Borneol [UN 1312]
2-Bornanol, endo-
Baros Camphor
Bhimsaim Camphor
Bicyclo(2.2.1)heptan-2-ol, 1,7,7-trimethyl-, endo- (9CI)
Borneo Camphor
trans-Borneol
Bornyl alcohol
Camphane, 2-hydroxy-
2-Camphanol
Camphol
Dryobalanops camphor
2-Hydroxycamphane
Malayan Camphor
Sumatra Camphor
1,7,7-Trimethyl-bicyclo(2.2.1)-heptan-2-ol, endo-
UN 1312 [Borneol]

508-32-7
$C_{10}H_{16}$
136.24
C(C1(C(C2C3)(C)C)C)(C13)C2
Tricyclo(2.2.1.0(2.6))heptane, 1,7,7-trimethyl-
AI3-26465
NSC-86978
Tricyclene
Tricyclo(2.2.1.02,6)heptane, 1,7,7-trimethyl- (9CI)
1,7,7-Trimethyltricyclo-(2.2.1.$0^{2,6}$)heptane

508-75-8
$C_{29}H_{42}O_{10}$
550.71
CC6OC(OC4CCC3(C=O)C2CCC1(C)C(CCC1(O)C2CCC3(O)C4)C5=CC(=O)OC5)C(O)C(O)C6O
Convallatoxin
Convallaotoxin
Convallaton
Convallatoxoside
Convallotoxin
Corglycon
Corglycone
Corglykon
Korglykon
Mannopyranoside, strophanthidin-3 6-deoxy-, α-l-
Rhamnoside, strophanthidin-3, α-l-
Strophanthidin, 3-(6-deoxy-α-l-mannopyranoside)
Strophanthidin α-l-rhamnoside

509-14-8
CN_4O_8
196.05
O=N(=O)C(N(=O)=O)(N(=O)=O)N(=O)=O
Methane, tetranitro
NCI-C55947
RCRA waste number P112
Tetranitromethane (ACGIH, OSHA) [UN 1510]
TNM
UN 1510 [Tetranitromethane]

509-15-9
$C_{20}H_{22}N_2O_2$
322.44
CN2CC3(C=C)C1CC4OCC1C2C3C45C(=O)Nc6ccccc56
Gelsemine
Gelsemin

509-34-2
$C_{28}H_{30}N_2O_3$
442.55
O=C(OC(c(c(Oc1cc(N(CC)CC)cc2)cc(N(CC)CC)c3)c3)(c12)c4cccc5)c45
Fluoran, 3',6'-bis(diethylamino)- (8CI)
Aizen Rhodamine B Base
Certiqual Rhodamine
C.I. Solvent Red 49
Eljon Magenta Toner
Fast Oil Pink B
Lacquer Pink S
NSC-43944
Rhodamine B Base
Rhodamine B Base Extra
Rhodamine B Extra Base
Rhodamine B Lactone
Rhodamine Base B Extra
Rhodamine S Lactone
Solvent Red 49
Spiro(isobenzofuran-1(3H), 9'-(9H)xanthen)-3-one, 3',6'-bis(diethylamino)- (9CI)
Waxoline Rhodamine B
Waxoline Rhodamine BS

509-60-4
$C_{17}H_{21}NO_3$
287.35
Dihydromorphine
DEA No. 9145
Dihydromorfin (Czech)
Hydromorphine
Morphinan-3,6-diol, 4,5-epoxy-17-methyl-, (5α,6α)- (9CI)
Morphinan-3,6-α-diol, 4,5-α-epoxy-17-methyl-
Morphine, dihydro-
NSC-117865
Paramorfan

509-67-1
$C_{23}H_{30}N_2O_4$
398.49
Pholcodine
Codylin
DEA No. 9314
Dia-Tuss
7,8-Didehydro-4,5-α-epoxy-17-methyl-3-(2-morpholinoethoxy)morphinan-6-α-ol
Ethnine
Ethnine Simplex
Folcodina (Spanish)
Folcodine
Folkodin (Czech)
Glycodine
Hibernyl
Homocodeine
Memine
Morphinan-6-α-ol, 7,8-didehydro-4,5-α-epoxy-17-methyl-3-(2-morpholinoethoxy)-
Morphine, O^3-(2-morpholinoethyl)-
Morphine, 3-O-(2-morpholinoethyl)-
β-Morpholinoethylmorphine
O^3-(2-Morpholinoethyl)-morphine
3-(2-Morpholinoethyl)morphine
Morpholinylethylmorphine
3-(2-(4-Morpholinyl)ethyl)-morphine
3-Morpholylaethylmorphin (German)
Neocodin
Pectolin
Pholcodin
Pholcodinum (Latin)
Prodromine
Tetrahydro-1,4-oxazinylmethyl-codeine
Weifacodine

509-74-0
$C_{23}H_{31}NO_2$
353.50
Acetylmethadol
Acemethadone
Acetilmetadol (Spanish)
3-Acetoxy-6-dimethylamino-4,4-diphenylheptane
5-Acetoxy-2-dimethylamino-4,4-diphenylheptane
O-Acetyl-6-dimethylamino-4,4-diphenyl-3-heptanol
Acetylmethadol
Acetylmethadolum (Latin)
Amidolacetate
Benzeneethanol, β-(2-(dimethylamino)propyl)-α-ethyl-β-phenyl-, acetate (ester)
DEA No. 9601
6-(Dimethylamino)-4,4-diphenyl-3-heptanol acetate (ester)
β-(2-(Dimethylamino)propyl)-α-ethyl-β-phenylbenzene-ethanol acetate hydrochloride
3-Heptanol, 6-(dimethylamino)-4,4-diphenyl-, acetate
3-Heptanol, 6-(dimethylamino)-4,4-diphenyl-, acetate (ester)
Methadyl acetate
Race-acetylmethadol

509-78-4
$C_{20}H_{25}NO_3$
327.42
Dimenoxadol
Dimenossadolo
Dimenoxadolum (Latin)
2-(Dimethylamino)ethyl ethoxy-diphenylacetate
DEA No. 9617

509-86-4
$C_{13}H_{18}N_2O_3$
250.33
CCC1(C(=O)NC(=O)NC1=O)C2=CCCCCC2
Barbituric acid, 5-(1-cyclohepten-1-yl)-5-ethyl
Cycloheptenylethylbarbituric acid
5-(1-Cyclohepten-1-yl)-5-ethy barbituric acid
Cycloheptenylethylmalonylur
5-(1-Cyclohepten-1-yl)-5-ethy 2,4,6(1H,3H,5H)-pyrimidine trione
5-Ethyl-5-cycloheptenylbarbituric acid
5-Ethyl-5-(1'-cycloheptenyl)-barbituric acid
G 475
Heptabarb
Heptabarbital
Heptabarbitone
Heptabarbum
Heptadorm
Heptamal
Heptbarbital
Medapan
Medomin
Medomine
Noctyn
2,4,6(1H,3H,5H)-Pyrimidinetrione, 5-(1-cyclohepten-1-yl)-5-ethyl- (9CI)

510-15-6
$C_{16}H_{14}Cl_2O_3$
325.20
O=C(OCC)C(O)(c(ccc(c1)Cl)c1)c(ccc(c2)Cl)c2
Benzilic acid, 4,4'-dichloro-, ethyl ester
ACAR
Acaraben
Acaraben 4E
Akar
Akar 50
Akar 338
Benzeneacetic acid, 4-chloro-α-(4-chlorophenyl)-α-hydroxy-, ethyl ester
Benzilan
Benz-O-Chlor
Chlorbenzilat
Chlorbenzilate
Chlorbenzylate
Chlorobenzilate

Chlorobenzylate
Compound 338
4,4'-Dichlorobenzilate
4,4'-Dichlorobenzilic acid ethyl ester
4,4'-Dichlorbenzilsaeureaethylester (German)
ENT 18,596
Ethyl 4-chloro-α-(4-chlorophenyl)-α-hydroxybenzeneacetate
Ethyl p,p'-dichlorobenzilate
Ethyl 4,4'-dichlorobenzilate
Ethyl 4,4'-dichlorodiphenyl glycollate
Ethyl 4,4'-dichlorophenyl glycollate
Ethyl ester of 4,4'-dichlorobenzilic acid
Ethylester kyseliny 4,4-dichlorbenzilove (Czech)
Ethyl-2-hydroxy-2,2-bis-(4-chlorophenyl)acetate
Folbex
Folbex Smoke-Strips
G 338
G 23992
Geigy 338
Kop-Mite
NCI-C00408
NCI-C60413
RCRA waste number U038

510-53-2
$C_{18}H_{25}NO$
271.40
Racemethorphan
DEA No. 9732
Morphinan, 3-methoxy-17-methyl-, (+-)- (8CI,9CI)
Racemethorphane (French)
Racemethorphanum (Latin)
Racemetorfano (Spanish)

511-45-5
$C_{20}H_{25}NO$
295.46
1-Piperidinepropanol, α,α-diphenyl
238 C
α,α-Diphenyl-1-piperidinepropanol
HH 212
Nonplesin
Parks
Parks 12
Parks 12 Hommel
PDP
Pridinol

511-59-1
$C_{15}H_{24}$
204.36
C(C(CC1C2)C2)(C1(CCC=C(C)C)C)=C
Bicyclo(2.2.1)heptane, 2-methyl-3-methylene-2-(4-methyl-3-pentenyl)-, (1S-exo)- (9CI)
β-Santalene

512-13-0
$C_{10}H_{18}O$
154.28
OC(C(CC1C2)(C2)C)C1(C)C
2-Norbornanol, 1,3,3-trimethyl-, (-)-endo
Bicyclo(2.2.1)heptan-2-ol, 1,3,3-trimethyl-, (1R-endo)-
α-Fenchol
endo-Fenchol
α-Fenchyl alcohol

512-26-5
$C_6H_8O_7.3/2Pb$
502.93
Lead citrate
Citric acid, lead(2+) salt (2:3) (8CI)
Lead citrate (VAN)
NSC-112228
1,2,3-Propanetricarboxylic acid, 2-hydroxy-, lead(2+) salt (2:3) (9CI)
Trilead dicitrate

512-42-5
$CH_4O_4S.Na$
135.10
Sulfuric acid, monomethyl ester, sodium salt (9CI)
Monomethyl sulfate sodium salt

512-56-1
$C_3H_9O_4P$
140.09
O=P(OC)(OC)OC
Phosphoric acid, trimethyl ester
Methyl phosphate
NCI-C03781
TMP
Trimethylfosfat (Czech)
Trimethyl phosphate
O,O,O-Trimethyl phosphate

512-69-6
$C_{18}H_{32}O_{16}$
504.44
O(C(C(O)C(O)C1O)CO)C1OCC(OC(OC(OC(C2O)CO)(C2O)CO)C(O)C3O)C3O
Raffinose (8CI)
AI3-19427
α-D-Glucopyranoside, β-D-fructofuranosyl O-α-D-galactopyranosyl-(1-6)- (9CI)
α-D-Glucopyranoside, β-D-fructofuranosyl O-α-D-galactopyranosyl (1 to 6)-, hydrate
NSC-170228
D-Raffinose
d-(+)-Raffinose

512-82-3
$C_4H_{12}O_4P.HO$
172.14
Phosphonium, tetrakis-(hydroxymethyl)-, hydroxide
Tetrakis(hydroxymethyl) phosphonium hydroxide
THPOH

512-85-6
$C_{10}H_{16}O_2$
168.26
CC(C)C12CCC(C)(OO1)C=C2
p-Menth-2-ene, 1,4-epidioxy
Ascaridol
Ascaridole (DOT)
Ascarisin
2,3-Dioxabicyclo(2.2.2)oct-5-ene, 1-isopropyl-4-methyl-
1-Methyl-4-(1-methylethyl)-2,3-dioxabicyclo(2.2.2)Oct-5-ene
1,4-Peroxido-p-menthene-2

513-08-6
$C_9H_{21}O_4P$
224.24
CCCOP(=O)(OCCC)OCCC
Phosphoric acid, tripropyl ester
AI3-07848

513-10-0
$C_9H_{23}NO_3PS.I$
383.26
Ammonium, (2-mercaptoethyl)trimethyl-, iodide, S-este with O,O-diethylphosphorothioate
Ammonium, (2-(O,O-diethylphosphorothio)ethyl)trimethyl-, iodide
2-Diaethoxyphosphinyl-thioaethyl-trimethyl-ammonium-jodid (German)
N-(2-(Diethoxyphosphinylthio)-ethyl)trimethylammonium iodide
2-Diethoxy-phosphinylthioethyltrimethylammonium iodide
Diethoxyphosphoryl-thiocholine iodide
O,O-Diethyl S-2-trimethylammonium ethylphosphonothiolate iodide
S-(2-Dimethylaminoethyl)-O.O-diethylphosphorothioate methiodide
Echodide
Echothiophate
Echothiophate iodide
Ecothiopate
Ecothiopate iodide
Ecothiophate iodide
S-Ester of (2-mercaptoethyl)trimethylammonium iodide with O,O-diethyl phosphorothioate
Ethanaminium, 2-((diethoxyphosphinyl)thio)-N,N,N-trimethyl-, iodide
(2-Mercaptoethyl)trimethyl-

ammonium iodide S-ester with O,O-diethyl phosphorothioate
217 MI
Phospholine iodide
Phospholine (The pharmaceutical)
S-(2-(N,N,N-Trimethylammonio)ethyl) O,O-diethylphosphorothiolate iodide

513-23-5
$C_{10}H_{18}O$
154.25
Bicyclo(3.1.0)hexan-3-ol, 4-methyl-1-(1-methylethyl)-(9CI)
3-Thujanol

513-31-5
$C_3H_4Br_2$
199.89
BrCC(Br)=C
1-Propene, 2,3-dibromo
2,3-Dibromopropene

513-35-9
C_5H_{10}
70.15
C(=CC)(C)C
2-Butene, 2-methyl
Amylene
Ethylene, trimethyl-
β-Iso-amylene
Isopentene [UN 2371]
2-Methyl-2-butene
Methyl butene (DOT)
2-Methyl-2-butene [UN 2460]
Trimethylethylene
UN 2371 [Isopentenes]
UN 2460 [2-Methyl-2-butene]

513-36-0
C_4H_9Cl
92.57
ClCC(C)C
Isobutyl chloride
1-Chloro-2-methylpropane
Propane, 1-chloro-2-methyl-(9CI)

513-37-1
C_4H_7Cl
90.56
CC(C)=CCl
Propene, 1-chloro-2-methyl
α-Chloroisobutylene
1-Chloroisobutylene
1-Chloro-2-methylpropene
1-Chloro-2-methyl-1-propene
Dimethylvinylchloride
β, β-Dimethylvinyl chloride
2,2-Dimethylvinyl chloride
Isocrotyl chloride
NCI-C54819
1-Propene, 1-chloro-2-methyl-

513-38-2
C_4H_9I
184.03
C(CI)(C)C
Propane, 1-iodo-2-methyl
1-Iodo-2-methylpropane
Isobutyl iodide
Isobutyljodid (Czech)
1-Jod-2-methylpropan (Czech)
Primary isobutyl iodide

513-42-8
C_4H_8O
72.12
OCC(=C)C
2-Propen-1-ol, 2-methyl
Isopropenyl carbinol
Methallyl alcohol [UN 2614]
UN 2614 [Methallyl alcohol]

513-44-0
$C_4H_{10}S$
90.20
SCC(C)C
1-Propanethiol, 2-methyl
Isobutanethiol
Isobutyl mercaptan

513-48-4
C_4H_9I
184.03
C(CC)(C)I
Butane, 2-iodo
sec-Butyl iodide
2-Iodobutane [UN 2390]
2-Jodbutan (Czech)
UN 2390 [2-Iodobutane]

513-49-5
$C_4H_{11}N$
73.16
NC(CC)C
sec-Butylamine, (S)
(+)-2-Butylamine
S-2-Butylamine

513-53-1
$C_4H_{10}S$
90.19
SC(CC)C
sec-Butyl mercaptan
sec-Butanethiol
2-Butanethiol (9CI)
secondary Butylmercaptan
2-Butyl mercaptan
sec-Butyl thioalcohol
sec-Butyl thiol
2-Mercaptobutane
1-Methyl-1-propanethiol
NSC-78417

513-74-6
$CH_3NS_2.H_3N$
110.19
Ammonium dithiocarbamate
Ammonium sulfocarbamate
Carbamic acid, dithio-, monoammonium salt (8CI)
Carbamodithioic acid, monoammonium salt (9CI)
Dithiocarbamic acid monoammonium salt
Monoammonium carbamodithioate
NSC-202959

513-77-9
$CO_3.Ba$
197.35
[Ba+2].OC(O)=O
Barium carbonate (1:1)
Barium carbonate
Carbonic acid, barium salt (1:1)
C.I. 77099
C.I. Pigment White 10

513-78-0
$CO_3.Cd$
172.41
[Cd].[O-]C([O-])=O
Carbonic acid, cadmium salt
Cadmium carbonate
Cadmium monocarbonate
Chemcarb

513-79-1
$CO_3.Co$
118.94
Carbonic acid, cobalt(2+) salt (1:1)
Cobaltous carbonate

513-81-5
C_6H_{10}
82.15
C(C(=C)C)(=C)C
1,3-Butadiene, 2,3-dimethyl-(9CI)
Biisopropenyl
Diisopropenyl
2,3-Dimethylbutadiene
2,3-Dimethylbuta-1,3-diene
2,3-Dimethyl-1,3-butadiene
2,3-Dimethylenebutane
NSC-8656

513-85-9
$C_4H_{10}O_2$
90.14
OC(C(O)C)C
2,3-Butanediol
2,3-Butylene glycol
2,3-Dihydroxybutane
Dimethylene glycol

513-86-0
$C_4H_8O_2$
88.12
O=C(C(O)C)C
2-Butanone, 3-hydroxy
Acetoin
Acetyl methyl carbinol [UN 2621]

2,3-Butanolone
2-Butanol-3-one
Dimethylketol
3-Hydroxy-2-butanone
1-Hydroxyethyl methyl ketone
γ-Hydroxy-β-oxobutane
UN 2621 [Acetyl methyl carbinol]

513-88-2
$C_3H_4Cl_2O$
126.97
2-Propanone, 1,1-dichloro
α,α-Dichloroacetone
DCP
1,1-Dichloroacetone
Dichloromethyl methyl ketone
1,1-Dichloropropanone

514-10-3
$C_{20}H_{30}O_2$
302.50
O=C(O)C(C(C(C(C(=C1)C=C(C2)C(C)C)C2)(CC3)C)C1)(C3)C
Podocarpa-7,13-dien-15-oic acid, 13-isopropyl
Abietic acid
13-Isopropylpodocarpa-7,13-dien-15-oic acid
Kyselina abietova (Czech)
Sylvic acid

514-73-8
$C_{23}H_{23}N_2S_2.I$
518.47
Dithiazanine iodide
Abminthic
AI3-50132
Anelmid
Anguifugan
Benzothiazolium, 3-ethyl-2-(5-(3-ethyl-2-benzothiazolinylidene)-1,3-pentadienyl)-, iodide (8CI)
Benzothiazolium, 3-ethyl-2-(5-(3-ethyl-2(3H)-benzothiazolylidene)-1,3-pentadienyl)-, iodide (9CI)
(2-Bis(3-ethylbenzothiazolyl))-pentamethine cyanine iodide
Compound 01748
Dejo
Delvex
3,3'-Diethyldithiacarbodicyanine iodide
3,3'-Diethylpentamethinethiacyanine iodide
Diethylthiadicarbocyanine iodide
3,3-Diethylthiadicarbocyanine iodide
3,3'-Diethylthiadicarbocyanine iodide
3,3'-Diethyl-2,2'-thiadicarbocyanine iodide
Dilombrin
Dilombrine
Dithiazanin iodide
Dithiazanine iodide
Dithiazanini iodidum (Latin)
Dithiazine
Dithiazinine
Dizan
Dqoci
Eastman 7663
3-Ethyl-2-(5-(3-ethyl-2-benzothiazolinylidene)-1,3-pentadienyl)benzothiazolium iodide
Iodure de dithiazanine (French)
Ioduro de ditiazanina (Spanish)
L-01748
Netocyd
NK 136
NSC-221154
Omni-Passin
Partel
Telmicid
Telmid
Telmide
Vercidon

514-94-3
$C_{10}H_{16}$
136.24
1,3-Cyclohexadiene, 1,5,5,6-tetramethyl- (7CI,8CI,9CI)
α-Pyronene (6CI)
1,5,5,6-Tetramethyl-1,3-cyclohexadiene

515-00-4
$C_{10}H_{16}O$
152.24
OCC(C(CC1C2)C1(C)C)=C2
Bicyclo(3.1.1)hept-2-ene-2-methanol, 6,6-dimethyl- (9CI)

515-30-0
$C_9H_{10}O_3$
166.18
O=C(O)C(O)(c(cccc1)c1)C
Atrolactic acid
Benzeneacetic acid, α-hydroxy-α-methyl- (9CI)
α-Hydroxy-α-methylbenzeneacetic acid
α-Hydroxy-α-phenylpropionic acid
2-Hydroxy-2-phenylpropionic acid
Mandelic acid, α-methyl- (8CI)
α-Methylmandelic acid
NSC-401846
2-Phenyl-2-hydroxypropionic acid
α-Phenyllactic acid

515-40-2
$C_{10}H_{13}Cl$
168.67
c(cccc1)(c1)C(CCl)(C)C
Benzene, (2-chloro-1,1-dimethylethyl)- (9CI)
β-Chloro-tert-butylbenzene
(β-Chloro-tert-butyl)benzene
(β-Chloro-α,α-dimethyl)ethylbenzene
(2-Chloro-1,1-dimethylethyl)-benzene
β,β-Dimethylphenethyl chloride
2-Methyl-2-phenylpropyl chloride
Neophyl chloride
NSC-54159

515-42-4
$C_6H_5O_3S.Na$
180.16
[Na+].OS(=O)(=O)c1ccccc1
Benzenesulfonic acid, sodium salt
Sodium benzenesulfonate
Sodium benzosulfonate
Sodium phenylsulfonate

515-46-8
$C_8H_{10}O_3S$
186.23
Benzenesulfonic acid, ethyl ester (8CI,9CI)
NSC-3217

515-84-4
$C_4H_5Cl_3O_2$
191.44
O=C(OCC)C(Cl)(Cl)Cl
Ethyl trichloroacetate
Acetic acid, ester with trichloroethanol
Acetic acid, trichloro-, ethyl ester (9CI)
Acetic acid, 2,2,2-trichloroethyl ester
AI3-28577
NSC-8829
Trichloroethyl acetate

516-03-0
$C_2H_2O_4.Fe$
145.88
Ferrous oxalate
Ethanedioic acid, iron(2+) salt (1:1) (9CI)
Ferrox
Iron oxalate
Iron(II) oxalate
Iron(2+) oxalate
Iron protoxalate
Oxalic acid, iron(2+) salt (1:1)

516-05-2
$C_4H_6O_4$
118.10
CC(C(O)=O)C(O)=O
Malonic acid, methyl
Methylmalonic acid

516-06-3
$C_5H_{11}NO_2$

117.14
O=C(O)C(N)C(C)C
DL-Valine (9CI)
AI3-18308
DL-α-Aminoisovaleric acid
NSC-9755
Valine, DL- (8CI)

516-21-2
$C_{11}H_{14}ClN_5$
251.75
s-Triazine, 1,2-dihydro-1-(p-chlorophenyl)-4,6-diamino-2,2-dimethyl
4-Amino-6-p-chloroanilino-1,2-dwuhydro-2,2-dwumethylo-1,3,5-trojazyna (Polish)
Chlorguanide triazine
CGT
1-(p-Chlorophenyl)-4,6-diamino-2,2-dimethyl-1,2-dihydro-s-triazine
1-p-Chlorophenyl-1,2-dihydro-2,2-dimethyl-4,6-diamino-s-triazine
Cycloguanil
Cycloguanyl
WR 5473

517-16-8
$C_{15}H_{17}HgNO_2S$
475.98
CC[Hg]N(c1ccccc1)S(=O)(=O)c2ccc(C)cc2
Mercury, ethyl(p-toluenesulfonanilidato)
Ceresan M
Ceresan M-2X
Ceresan M-DB
Compound-1452-F
EMTS
N-Ethylmercuri-N-phenyl-p-toluenesulfonamide
N-(Ethylmercuri)-p-toluenesulfonanilide
N-(Ethylmercuri)-p-toluenesulphonanilide
Ethylmercury p-toluenesulfanilide
Ethylmercury p-toluene sulfonamide
Ethylmercury p-toluenesulfonanilide
N-Ethylmerkuri-p-toluensulfoanilid (Czech)
Granosan
Granosan M
Granosan MDB
Mercury, ethyl(4-methyl-N-phenylbenzenesulfonamidato-n)- (9CI)
Mercury, ethyl(N-phenyl-p-toluenesulfonamidato)-
Mercury, ethyl(N-phenyl-p-toluenesulfonamido)-
Mergon
Mergon D
(N-Phenyl-p-toluenesulfonamido)ethylmercury
Seed Dressing Universal
p-Toluenesulfonamide, N-(ethylmercuri)-N-phenyl-+
p-Toluenesulfonanilide, N-(ethylmercuri)-
Zaprawa Nasienna Universal
Zaprawa Nasienna Universal R
Zaprawa Nasienna Uniwersalna

517-18-0
$C_{18}H_{22}O_3$
286.40
2-Naphthalenepropionic acid, β-ethyl-6-methoxy-α,α-dimethyl
Methallenestril
Methallenestrol
3-(6-Methoxy-2-naphthyl)-2,2-dimethylpentanoic acid
2-Naphthalenepropanoic acid, β-ethyl-6-methoxy-α,α-dimethyl- (9CI)
Novestrine
Vallestril

517-23-7
$C_6H_8O_3$
128.13
O=C(OCC1)C1C(=O)C
2(3H)-Furanone, 3-acetyldihydro- (9CI)
AI3-05827
α-Acetobutyrolactone
α-Acetylbutyrolactone
α-Acetyl-γ-butyrolactone
2-Acetylbutyrolactone
2-Acetyl-γ-butyrolactone
3-Acetyldihydro-2(3H)-furanone
3-Acetyl-2(3H)-4,5-dihydrofuranone
α-Acetyl-γ-hydroxybutyric acid γ-lactone
2-Acetyl-4-hydroxybutyric acid γ-lactone
3-Acetyltetrahydro-2-furanone
α-(2-Hydroxyethyl)acetoacetic acid γ-lactone
NSC-2019
2-Oxo-3-acetyltetrahydrofuran

517-25-9
CHN_3O_6
151.05
O=N(=O)C(N(=O)=O)N(=O)=O
Methane, trinitro
Nitroform
Trinitromethane (DOT)

517-28-2
$C_{16}H_{14}O_6$
302.30
O(c(c(C(c(c(cc(O)c1O)C2)c1)C23O)ccc4O)c4O)C3
Hematoxylin
NCI-C55889

517-60-2
$C_{12}H_6O_{12}$
342.17
OC(=O)c1c(C(O)=O)c(C(O)=O)c(C(O)=O)c(C(O)=O)c1C(O)=O
Mellitic acid
Benzenehexacarboxylic acid (9CI)
1,2,3,4,5,6-Benzenehexacarboxylic acid
Hexacarboxybenzene
Mellic acid
NSC-229358

517-92-0
$C_{14}H_4N_4O_{12}$
420.18
1,8-Dihydroxy-2,4,5,7-tetranitroanthraquinone
Acide chrysaminique (French)
Acido crisaminico (Spanish)
9,10-Anthracenedione, 1,8-dihydroxy-2,4,5,7-tetranitro- (9CI)
Anthraquinone, 1,8-dihydroxy-2,4,5,7-tetranitro- (8CI)
Chrysammic acid
Chrysamminic acid
1,8-Dihidroxi-2,4,5,7-tetranitroantraquinona (Spanish)
Dihydroxy-1,8 tetranitro-2,4,5,7 anthraquinone (French)
NSC-9046
2,4,5,7-Tetranitrochrysazin

518-28-5
$C_{22}H_{22}O_8$
414.44
COc1cc(cc(OC)c1OC)C4C2C(COC2=O)C(O)c5cc3OCOc3cc45
Furo(3',4':6,7)naphtho(2,3-d)-1,3-dioxol-6(5ah)-one, 5,8,8a,9-tetrahydro-9-hydroxy-5- (3,4,5-trimethoxyphenyl)
NSC-24818
Podophyllinic acid lactone
Podophyllotoxin

518-45-6
$C_{20}H_{12}O_5$
332.31
Benzoic acid, 2-(6-hydroxy-3-oxo-3H-xanthen-9-yl)- (9CI) (VAN)
Benzoic acid, o-(6-hydroxy-3-oxo-3H-xanthen-9-yl)- (8CI) (VAN)

518-47-8
$C_{20}H_{10}O_5.2Na$
376.28
[Na+].[Na+].[O-]C(=O)c1ccccc1c3c2ccc([O-])cc2oc4cc(=O)ccc34
Fluorescein, disodium salt
Aizen Uranine

Calcocid Uranine B4315
9-o-Carboxyphenyl-6-hydroxy-3-isoxanthone, disodium salt
Certiqual Fluoresceine
C.I. 766
C.I. 45350 Disodium salt
C.I. 45350 Sodium salt
C.I. Acid Yellow 73
D & C Yellow No. 8
Disodium 6-hydroxy-3-oxo-9-xanthene-o-benzoate
Fluorescein sodium
Fluorescein sodium B.P
Fluorescein, Soluble
Fluor-I-Strip A.T.
Ful-Glo
Funduscein
Furanium
Hidacid Uranine
NCI-C54706
Resorcinol phthalein sodium
Sodium fluorescein
Sodium fluoresceinate
Sodium salt of hydroxy-o-carboxy-phenyl-fluorone
Soluble Fluorescein
Soluble Fluoresceine
Spiro(isobenzofuran-1(3H), 9'-(9H)xanthen)-3-one, 3',6'-dihydroxy-, disodium salt
Uranin
Uranine
Uranine A Extra
Uranine O
Uranine SS
Uranine USP XII
Uranine WSS
Uranine Yellow
11824 Yellow
12417 Yellow

518-75-2
$C_{13}H_{14}O_5$
250.27
CC2OC=C1C(=C(C(O)=O)C(=O)C(=C1C2C)C)O
3H-2-Benzopyran-7-carboxylic acid, 4,6-dihydro-8-hydroxy-3,4,5-trimethyl-6-oxo-, (3R-trans)
Antimycin
Citrinin
(3R,4S)-4,6-Dihydro-8-hydroxy-3,4,5-trimethyl-6-oxo-3H-2-benzopyran-7-carboxylic acid

518-82-1
$C_{15}H_{10}O_5$
270.25
Anthraquinone, 6-methyl-1,3,8-trihydroxy
9,10-Anthracenedione, 1,3,8-trihydroxy-6-methyl- (9CI)
Anthraquinone, 1,3,8-tri-hydroxy-6-methyl-
C.I. 75440
C.I. Natural Yellow 14
Emodin
Emodol
Frangula emodin
6-Methyl-1,3,8-trihydroxy-anthraquinone
Persian Berry Lake
Rheum Emodin
Schuttgelb

519-09-5
$C_{16}H_{19}NO_4$
289.33
Benzoyl ecgonine
Benzoylecgonine
DEA No. 9180

519-44-8
$C_6H_4N_2O_6$
200.10
Oc1ccc(N(=O)=O)c(O)c1N(=O)=O
2,4-Dinitroresorcinol (Heavy metal salts of) (Dry)
1,3-Benzenediol, 2,4-dinitro- (9CI)
2,4-Dinitroresorcinol (Spanish)
Dinitro-2,4 resorcinol (French)
NSC-243680
Resorcinol, 2,4-dinitro- (8CI)

519-73-3
$C_{19}H_{16}$
244.34
c(cccc1)(c1)C(c(cccc2)c2)c(cccc3)c3
Triphenylmethane
AI3-02337
Benzene, 1,1',1''-methylidyne-tris- (9CI)
Methane, triphenyl- (8CI)
1,1',1''-Methylidynetrisbenzene
NSC-4049
Tritane

519-95-9
$C_{20}H_{14}O_3$
302.33
Florantyrone
Anchol
Bilyn
Cistoplex
Florantirona (Spanish)
Florantyron
Florantyronum (Latin)
Fluochol
8-Fluoranthenebutyric acid, γ-oxo-
4-(8-Fluoranthenyl)-4-oxo-butyric acid
β-(8-Fluoranthoyl)propionic acid
β-(8-Fluoranthyloyl)propionic acid
Fluorantyrone
Idrobil
Idroepar
γ-Oxo-8-fluoranthenebutanoic acid
γ-Oxo-8-fluoranthenebutyric acid
SC 1674
Zanchol

520-18-3
$C_{15}H_{10}O_6$
286.25
Oc1ccc(cc1)c3oc2cc(O)cc(O)c2c(=O)c3O
Flavone, 3,4',5,7-tetrahydroxy
4H-1-Benzopyran-4-one, 3,5,7-trihydroxy-2-(4-hydroxy-phenyl)-
Campherol
C.I. 75640
Indigo Yellow
Kaempferol
Kaempherol
Kampherol
Kempferol
Nimbecetin
Pelargidenolon
Pelargidenolon 1497
Populnetin
Rhamnolutein
Rhamnolutin
Robigenin
Swartziol
3,4',5,7-Tetrahydroxyflavone
Trifolitin
5,7,4'-Trihydroxyflavonol

520-45-6
$C_8H_8O_4$
168.16
O=C(OC(=CC1=O)C)C1C(=O)C
2H-Pyran-2,4(3H)-dione, 3-acetyl-6-methyl
Acetic acid, dehydro-
3-Acetyl-4-hydroxy-6-methyl-2H-pyran-2-one
3-Acetyl-6-methyl-2,4-pyran-dione
3-Acetyl-6-methylpyrandione-2,4
3-Acetyl-6-methyl-2H-pyran-2,4(3H)-dione
3-Acetyl-6-methyl-2H-pyran-2,4(3H)-dione. enol form
Dehydracetic acid
Dehydroacetic acid
DHA
DHS
4-Hexenoic acid, 2-acetyl-5-hydroxy-3-oxo-, δ-lactone
Kyselina dehydroacetova (Czech)
Methylacetopyronone
2H-Pyran-2-one, 3-acetyl-4-hydroxy-6-methyl-

520-52-5
$C_{12}H_{17}N_2O_4P$
284.24
Psilocybine
CY 39
DEA No. 7437
DEA No. 7438

3-(2-Dimethylaminoethyl)indol-4-yl dihydrogen phosphate
3-(2-(Dimethylamino)ethyl)-1H-indol-4-ol dihydrogen phosphate ester
3-2'-Dimethylaminoethylindol-4-phosphate
Indocybin
Indol-4-ol, 3-(2-(dimethylamino)ethyl)-, dihydrogen phosphate
O-Phosphoryl-4-hydroxy-N,N-dimethyltryptamine
4-Phosphoryloxy-ω-N,N-dimethyltryptamine
Psilocibin
Psilocibina (Spanish)
Psilocin phosphate ester
Psilocybin
Psilocybinum (Latin)
Psilocyn
Psilotsibin
Teonanacatl

520-63-8
$C_8H_{13}NO_2$
155.22
1H-Pyrrolizine-7-methanol, 2,3,5,7a-tetrahydro-1-hydroxy-, (1S-cis)
Heliotridine

520-68-3
$C_{20}H_{31}NO_7.ClH$
433.98
2-Butenoic acid, 2-methyl-, 7-((2,3-dihydroxy-2-(1-hydroxyethyl)-3-methyl-1-oxobutoxy) methyl)-2,3,5,7a-tetrahydro-1H-pyrrolizin-1-ylester, hydrochloride
Echimidine hydrochloride

520-69-4
$C_9H_{15}N$
137.22
1H-Pyrrole, 3-ethyl-2,4,5-trimethyl- (9CI)
Phyllopyrrole
Pyrrole, 3-ethyl-2,4,5-trimethyl-(8CI)

520-85-4
$C_{22}H_{32}O_3$
344.54
CC3CC1C(CCC2(C)C1CCC2(O)C(C)=O)C4(C)CCC(=O)C=C34
Pregn-4-ene-3,20-dione, 17-hydroxy-6-α-methyl
Farlutal
Medroxyprogesteron
Medroxyprogesterone
6-α-Methyl-17-α-hydroxyprogesterone
Pregn-4-ene-3,20-dione, 17-hydroxy-6-methyl-, (6-α)-(9CI)
Provera
U 8840

521-31-3
$C_8H_7N_3O_2$
177.14
Oc(nnc(O)c1c(N)ccc2)c12
Luminol
AI3-52555
5-Amino-2,3-dihydro-1,4-phthalazinedione
3-Aminophthalhydrazide
3-Aminophthalic acid hydrazide
3-Aminophthalic hydrazide
1,2-Benzenedicarboxylic acid, 3-amino-, cyclic hydrazide
NSC-5064
1,4-Phthalazinedione, 5-amino-2,3-dihydro- (9CI)

521-35-7
$C_{21}H_{26}O_2$
310.47
6H-Dibenzo(b,d)pyran-1-ol, 6,6,9-trimethyl-3-pentyl
3-Amyl-1-hydroxy-6,6,9-trimethyl-6H-dibenzo(b,d)pyran
Cannabinol
CBN

522-12-3
$C_{21}H_{20}O_{11}$
448.41
Quercitrin
C.I. 75720
Flavone, 3,3',4',5,7-pentahydroxy-, 3-(6-deoxy-α-l-mannopyranoside)
Flavone, 3,3',4',5,7-pentahydroxy-, 3-rhamnoside
Mannopyranoside, quercetin-3 6-deoxy-, α-l-
NCI-C60102
3,3',4',5,7-Pentahydroxyflavone-3-l-rhamnoside
Quercetin, 3-(6-deoxy-α-l-mannopyranoside)
Quercetrin-3-O-rham
Quercetin-3-l-rhamnoside
Quercetrin
Quercimelin
Quercitroside
Rhamnoside, quercetin-3
USAF CF-2

522-40-7
$C_{18}H_{22}O_8P_2$
428.34
CCC(=C(CC)c1ccc(OP(O)(O)=O)cc1)c2ccc(OP(O)(O)=O)cc2
4,4'-Stilbenediol, α,α'-diethyl-, bis(dihydrogen phosphate), (E)
DESdp
α,α'-Diethyl-(E)-4,4'-stilbenediol bis(dihydrogen phosphate)
Diethylstilbesterol diphosphate
Diethylstilbestrol diphosphate
Diethylstilbestrol phosphate
Diethylstilbestryl diphosphate
Fosfestrol
Honvan
Phosphestrol
Phenol, 4,4'-(1,2-diethyl-1,2-ethenediyl)bis-, bis(dihydrogen phosphate), (E)-
ST52-ASTA
Stilbestrol diphosphate
Stilphostrol

522-66-7
$C_{20}H_{26}N_2O_2$
326.43
CCC1CN2CCC1CC2C(O)c3ccnc4ccc(OC)cc34
Cinchonan-9-ol, 10,11-dihydro-6'-methoxy-, (8α,9R)-(9CI)
Dihydroquinine
Hydroquinine (8CI)
NSC-41799
Quinine, 10,11-dihydro-

523-31-9
$C_{22}H_{18}O_4$
346.38
1,2-Benzenedicarboxylic acid, bis(phenylmethyl) ester (9CI)
NSC-4057
Phthalic acid, dibenzyl ester (8CI)

523-44-4
$C_{16}H_{11}N_2O_4S.Na$
350.34
[Na+].Oc2ccc(N=Nc1ccc(cc1)S([O-])(=O)=O)c3ccccc23
Benzenesulfonic acid, p-((4-hydroxy-1-naphthyl)-azo)-, sodium salt
Acid Leather Orange I
Acid Orange I
Acid Phosphine CL
A.F. Orange No. 1
Aizen Food Orange No. 1
Aizen Naphthol Orange I
Aizen Orange I
Certiqual Orange I
C.I. 150
C.I. 14600
C.I. Acid Orange 20
C.I. Acid Orange 20, Monosodium salt
D & C Orange No. 3
Dye Orange No. 1
Egacid Orange GG
Elgacid Orange 2G
Eniacid Orange I
Ext. D & C Orange No. 3
External D and C Orange No. 3
FD & C Orange No. 1
Hispacid Orange 1
4-((4-Hydroxy-1-naphthalenyl)-azo)benzenesulfonic acid, monosodium salt
4-((4-Hydroxy-1-naphthalenyl)-azo)benzenesulphonic acid,

monosodium salt
p-((4-Hydroxy-1-naphthyl)azo)-benzenesulfonic acid, monosodium salt
p-((4-Hydroxy-1-naphthyl)azo)-benzenesulfonic acid, sodium salt
p-((4-Hydroxy-1-naphthyl)azo)-benzenesulphonic acid, monosodium salt
p-((4-Hydroxy-1-naphthyl)azo)-benzenesulphonic acid, sodium salt
Java Orange I
Nankai Acid Orange I
Naphthalene Orange I
Naphthol Orange
α-Naphthol Orange
Neklacid Orange 1
1333 Orange
Orange I
Orange I Extra conc. A export
Orange IM
Orange I, Sodium salt
Orange S
Oranz I (Czech)
Oranz Kysela 20 (Czech)
Schultz No. 185
Sodium azo-α-naphtholsulfanilate
Sodium azo-α-naphtholsulphanilate
4-p-Sulfophenylazo-1-naphthol monosodium salt
4-p-Sulphophenylazo-1-naphthol, monosodium salt
Tertracid Orange I
Tropaeolin 1
Tropaeolin G
Tropaeolin OOO No. 1
Tropeolin (Czech)

523-47-7
$C_{15}H_{24}$
204.36
CC1=CCC2C(C1)C(CC=C2C)C(C)(C)C
β-Cadinene
3,9-Cadinadiene
Cadina-3,9-diene (8CI)
Cadinene
β-Cadinene, (-)-
(-)-β-Cadinene
Naphthalene, 1,2,4a,5,8,8a-hexahydro-4,7-dimethyl-1-(1-methylethyl)-, (1S-(1α,-4aβ,8aα))- (9CI)
NCI-C56008
NSC-46152

523-50-2
$C_{11}H_6O_3$
186.17
O=c3ccc2ccc1occc1c2o3
2H-Furo(2,3-h)(1)benzopyran-2-one
Angecin
Angelecin
Angelicin
Angelicin (coumarin deriv)
Furo(2,3-h)coumarin
Furo(5',4',7,8)coumarin
4-Hydroxy-5-benzofuranacrylic acid γ-lactone
3-(4-Hydroxy-5-benzofuranyl)-2-propenoic acid γ-lactone
Isopsoralin

523-87-5
$C_{17}H_{21}NO.C_7H_7ClN_4O_2$
470.02
Cn2c1N=C(Cl)[N-]c1c(=O)n(C)c2=O
Theophylline, 8-chloro-, Compd. with 2-(diphenylmethoxy)-N,N-dimethylethylamine (1:1)
Amosyt
Anautine
Andramine
Aviomarin
o-Benzhydryldimethylaminoethanol 8-chlorotheophyllinate
Benzhydryl-β-dimethylaminoethylether 8-chlorotheophylline
2-(Benzhydryloxy)-N,N-dimethylethylamine Compd. with 8-chlorotheophylline
Chloranautine
8-Chlorotheophylline, Compd. with 2-(diphenylmethoxy)-N,N-dimethylethylamine (1:1)
Diamarin
Dimenhydrinate
Diphenhydramine 8-chlorotheophylline
Diphenhydrinate
Dramamin
Dramamine
Dramarin
Dramyl
Dromyl
Eldodram
Ethylamine, N,N-dimethyl-2-(diphenylmethoxy)-, Compd. with 8-chlorotheophylline
Gravinol
Gravol
Menhydrinate
NCI-C60639
Neo-Navigan
Novamin
Novamine
Permital
Reise-Engletten
Supremal
Teodramin
Travelin
Travelmin
Vomex A
Xamamina

523-88-6
$C_{14}H_8Br_2O_4$
400.04
Salicil, 5,5'-dibromo
DBS
5,5'-Dibrom-2,2'-dioxybenzil (German)
5,5'-Dibromo-2,2'-dihydroxybenzil
5,5'-Dibromo-2,2'-dihydroxybibenzoyl
Dibromosalicil
5,5'-Dibromosalicil
Dibromosalicyl
Dibrosal

524-36-7
$C_8H_{12}N_2O_2.2ClH$
241.14
Cc1ncc(CO)c(CN)c1O
Pyridoxamine-dihydrochloride
4-(Aminomethyl)-5-hydroxy-6-methyl-3-pyridinemethanol dihydrochloride
2-Methyl-3-hydroxy-4-aminomethyl-5-hydroxymethylpyridene dihydrochloride
3-Pyridinemethanol, 4-(aminomethyl)-5-hydroxy-6-methyl-, dihydrochloride
Pyridoxylamine dihydrochloride

524-38-9
$C_8H_5NO_3$
163.14
O=C(N(O)C(=O)c1cccc2)c12
Phthalimide, N-hydroxy
N-Hydroxyphthalimide

524-40-3
$C_7H_6N_2O$
134.15
Nicotinonitrile, 1,2-dihydro-4-methyl-2-oxo
1,2-Dihydro-4-methoxy-1-methyl-2-oxonicotinonitrile
Ricidine
Ricinine

524-42-5
$C_{10}H_6O_2$
158.16
O=C(c(c(C=C1)ccc2)c2)C1=O
1,2-Naphthoquinone
1,2-Naftochinon (Czech)
1,2-Naphthalenedione
1,2-Naphthaquinone
β-Naphthoquinone

524-84-5
$C_{14}H_{17}NS_2$
263.42
Dimethylthiambutene
Allylamine, 3,3-di-2-thienyl-N,N,1-trimethyl-
Allylamine, N,N,1-trimethyl-3,3-di-2-thienyl-
3-Buten-2-amine, N,N-dimethyl-4,4-di-2-thienyl- (9CI)
DEA No. 9619
Dimethibutin
3-Dimethylamino-1,1-bis-(2-thienyl)-1-butene
3-Dimethylamino-1,1-di-

(2'-thienyl)but-1-ene
3-Dimethylamino-1,1-di-(2'-thienyl)-1-butene
N,N-Dimethyl-4,4-di-2-thienyl-3-buten-2-amine
Dimethylthiambutenum (Latin)
Dimetiltiambutene
Dimetiltiambuteno (Spanish)
Ohton
N,N,1-Trimethyl-3,3-di-2-thienylallylamine
N,N,1-Trimethyl-3,3-di(2-thienyl)-2-propenylamine
338C48

525-37-1
$C_{10}H_8O_6S_2$
288.30
Naphthalene-1,6-disulfonic acid
N-1,6-DSA

525-47-3
$C_{10}H_6N_2O_4$
218.18
OC(=O)c2ccc1c(cccc1n2)N(=O)=O
Quinaldic acid, 5-nitro
5-Nitroquinaldic acid
5-Nitroquinaldinic acid
5-Nitro-2-quinolinecarboxylic acid

525-52-0
$C_{12}H_{12}O_6$
252.22
O=C(Oc(c(OC(=O)C)c(OC(=O)C)cc1)c1)C
Pyrogallol, triacetate (8CI)
Acetpyrogall
AI3-03113
1,2,3-Benzenetriol, triacetate (9CI)
Lenigallol
NSC-24068
Pyracetol
Pyrogallol triacetate
Triacetylpyrogallol

525-66-6
$C_{16}H_{21}NO_2$
259.38
CC(C)NCC(O)COc1cccc2ccccc12
2-Propanol, 1-(isopropylamino)-3-(1-naphthyloxy)
AY 64043
Dociton
ICI 45520
Inderal
1-Isopropylamine-3-(1-naphthyloxy)-2-propanol
1-Isopropylamino-3-(1-naphthyloxy)-2-propanol
NSC-91523
Propanalol
Propanolol
Propranalol
Propranolol

525-79-1
$C_{10}H_9N_5O$
215.24
O(C(=CC=1)CNc(ncnc2NC=N3)c23)C1
Adenine, N-furfuryl
FAP
N-Furfuryladenine
N^6-Furfuryladenine
6-Furfuryladenine
N^6-(Furfurylamino)purine
6-(Furfurylamino)purine
Kinetin
Kinetin (Plant hormone)

525-82-6
$C_{15}H_{10}O_2$
222.25
o(c(c(c(O)c1)ccc2)c2)c1c(cccc3)c3
Asmacoril
Chromocor
Cromaril
Flavone
2-Phenyl-4H-1-benzopyran-4-one
Phenylchromone
2-Phenylchromone

526-08-9
$C_{15}H_{14}N_4O_2S$
314.39
Nc1ccc(cc1)S(=O)(=O)Nc2ccnn2c3ccccc3
Benzenesulfonamide, 4-amino-N-(1-phenyl-1H-pyrazol-5-yl)
3-(p-Aminobenzenesulfonamido)-2-phenylpyrazole
4-Amino-N-(1-phenyl-1H-pyrazol-5-yl)benzenesulfonamide
Depocid
Depotsulfonamide
Eftolon
Firmazolo
Inamil
Isarol
Merian
Microtan Pirazolo
Orisul
Orisulf
Paidazolo
N'-(1-Phenylpyrazol-5-yl)sulfanilamide
1-Phenyl-5-sulfanilamidopyrazole
Plisulfan
Raziosulfa
SP
SPP
Sulfabid
Sulfafenazolo (Italian)
Sulfanilamide, N^1-(1-phenylpyrazol-5-yl)- (8CI)
5-Sulfanilamido-1-phenylpyrazole
Sulfaphenazole
Sulfaphenazon
Sulfaphenylpipazol
Sulfaphenylpyrazole
Sulphenazole

526-26-1
$C_7H_5O_3.1/2Sr$
180.93
Benzoic acid, 2-hydroxy-, strontium salt (2:1)
2-Hydroxybenzoic acid strontium salt (2:1)
Stroncylate
Strontium salicylate

526-73-8
C_9H_{12}
120.21
c(c(c(cc1)C)C)(c1)C
Benzene, 1,2,3-trimethyl
Hemimellitene
1,2,3-Trimethylbenzene

526-75-0
$C_8H_{10}O$
122.18
Oc(c(c(cc1)C)C)c1
2,3-Xylenol
2,3-Dimethylphenol
Phenol, 2,3-dimethyl-
UN 2261 [Xylenols]
o-Xylenol [UN 2261]

526-94-3
$C_4H_4O_6.Na$
171.07
Tartaric acid, monosodium salt
Butanedioic acid, 2,3-dihydroxy-, (R-(R*,R*))-, monosodium salt
Natriumtartrat (German)
Sodium tartrate

526-95-4
$C_6H_{12}O_7$
196.16
O=C(O)C(O)C(O)C(O)C(O)CO
Gluconic acid
Dextronic acid
Glosanto
Gluconic acid, D- (8CI)
D-Gluconic acid (9CI)
Glycogenic acid
Glyconic acid
Maltonic acid
NSC-77381
Pentahydroxycaproic acid
2,3,4,5,6-Pentahydroxyhexanoic acid

526-99-8
$C_6H_{10}O_8$
210.16
O=C(O)C(O)C(O)C(O)C(O)C(=O)O
Galactaric-acid
Galactosaccharic acid

Mucic acid
Saccharolactic acid
Schleimsaure
Tetrahydroxyadipic acid

527-07-1
$C_6H_{11}O_7.Na$
218.16
Gluconic acid, monosodium salt, D
Glonsen
Gluconato di sodio (Italian)
Gluconic acid sodium salt
Monosodium gluconate
Pasexon 100T
PMP Sodium gluconate
Sodium gluconate
Sodium D-gluconate

527-09-3
$C_{12}H_{22}CuO_{14}$
453.84
Copper gluconate
Bis(D-gluconato-O1,O2)copper
Chelates of copper gluconate
Copper, bis(D-gluconato)-
Copper, bis(D-gluconato-O1,O2)- (9CI)
Copper D-gluconate (1:2)
Cupric gluconate
Gluconic acid, copper(2+) salt
Gluconic acid, copper(2+) salt (2:1), D-

527-20-8
$C_6H_2Cl_5N$
265.34
Nc1c(Cl)c(Cl)c(Cl)c(Cl)c1Cl
Aniline, 2,3,4,5,6-pentachloro
Benzenamine, 2,3,4,5,6-pentachloro- (9CI)
PCA
Pentachloroaminobenzene
Pentachloroaniline
2,3,4,5,6-Pentachloroaniline
2,3,4,5,6-Pentachlorobenzenamine

527-35-5
$C_{10}H_{14}O$
150.22
Oc(c(c(cc1C)C)C)c1C
Phenol, 2,3,5,6-tetramethyl- (9CI)
Durenol
NSC-65612
Phenol, tetramethyl-
2,3,5,6-Tetramethylphenol

527-53-7
$C_{10}H_{14}$
134.24
c(cc(c(c1C)C)C)(c1)C
Benzene, 1,2,3,5-tetramethyl
Isodurene
1,2,3,5-Tetramethylbenzene
1,2,4,5-Tetramethylbenzene

527-54-8
$C_9H_{12}O$
136.19
Oc(cc(c(c1C)C)C)c1
Phenol, 3,4,5-trimethyl- (9CI)
3,4,5-Hemimellitenol
1-Hydroxy-3,4,5-trimethylbenzene
NSC-65648
3,4,5-Trimethylphenol

527-55-9
$C_8H_{10}O_2$
138.17
Cc1cc(O)cc(O)c1C
4-Methylorcinol
1,3-Benzenediol, 4,5-dimethyl- (9CI)
1,3-Dihydroxy-4,5-dimethylbenzene
3,5-Dihydroxy-1,2-dimethylbenzene
1,2-Dimethyl-3,5-dihydroxybenzene
NSC-243682
Resorcinol, 4,5-dimethyl- (8CI)
o-Xylorcinol

527-60-6
$C_9H_{12}O$
136.21
Oc(c(cc(c1)C)C)c1C
Mesitol
2-Hydroxymesitylene
Mesityl alcohol
Phenol, 2,4,6-trimethyl- (9CI)
2,4,6-Trimethylphenol
2,4,6-Trimetylofenol (polish)

527-62-8
$C_6H_5Cl_2NO$
178.02
2-Amino-4,6-dichlorophenol

527-72-0
$C_5H_4O_2S$
128.15
O=C(O)C(SC=C1)=C1
2-Carboxythiophene
2-Thenoic acid
α-Thiophenecarboxylic acid
2-Thiophenic acid

527-73-1
$C_3H_3N_3O_2$
113.09
O=N(=O)C(=NC=C1)N1
Imidazole, 2-nitro
11A
Amicin
Azomycin
2-Nitro-1H-imidazole
2-Nitroimidazole
Ro 05-9129

527-84-4
$C_{10}H_{14}$
134.24
c(c(ccc1)C)(c1)C(C)C
o-Cymene
UN 2046 [Cymenes]

527-85-5
C_8H_9NO
135.16
O=C(N)c(c(ccc1)C)c1
Benzamide, 2-methyl- (9CI)
o-Methylbenzamide
2-Methylbenzamide
NSC-2169
o-Toluamide (8CI)
o-Toluic amide
o-Tolylamide

528-21-2
$C_8H_8O_4$
168.16
O=C(c(c(O)c(O)c(O)c1)c1)C
Acetophenone, 2',3',4'-trihydroxy
Alizarine Yellow C
Alizarin Yellow C
C.I. 57000
Ethanone, 1-(2,3,4-trihydroxyphenyl)- (9CI)
Gallacetophenone
Galloacetophenone
2,3,4-Trihydroxyacetophenone

528-29-0
$C_6H_4N_2O_4$
168.12
O=N(=O)c1ccccc1N(=O)=O
Benzene, o-dinitro
Benzene, 1,2-dinitro- (9CI)
o-Dinitrobenzene (ACGIH, OSHA) [UN 1597]
UN 1597 [Dinitrobenzenes, solid]

528-44-9
$C_9H_6O_6$
210.15
O=C(O)c(ccc(c1C(=O)O)C(=O)O)c1
1,2,4-Benzenetricarboxylic acid
TMA
1,2,4-Tricarboxybenzene
Trimellitic acid

528-45-0
$C_7H_4N_2O_6$
212.13
O=C(O)c(ccc(N(=O)=O)c1N(=O)=O)c1
Benzoic acid, 3,4-dinitro
3,4-Dinitrobenzoic acid

528-46-1

$C_9H_6O_5$
194.15
Phthalonic acid
2-Carboxy-α-oxobenzeneacetic acid

528-50-7
$C_{12}H_{22}O_{11}$
342.30
O=CC(O)C(O)C(OC(OC(C(O)C1O)CO)C1O)C(O)CO
D-Glucose, 4-O-β-D-glucopyranosyl-
AI3-18877
4-O-β-D-Glucopyranosyl-D-glucose

528-74-5
$C_{20}H_{20}Cl_2N_8O_5$
523.38
CN(Cc2cnc1nc(N)nc(N)c1n2)c3c(Cl)cc(cc3Cl)C(=O)NC(CCC(O)=O)C(O)=O
Glutamic acid, N-(3,5-dichloro-4-(((2,4-diamino-6-pteridinyl)methyl)methylamino)benzoyl)
Amethopterin, 3',5'-dichloro-
DCM
Dichloroamethopterin
3',5'-Dichloroamethopterin
3',5'-Dichloro-4-amino-4-deoxy-N^{10}-methylpteroglutamic acid
N-(3,5-Dichloro-4-((2,4-diamino-6-pteridinylmethyl)methylamino)benzoyl)glutamic acid
Dichloromethotrexate
3',5'-Dichloromethotrexate
Methotrexate, dichloro-
NCI-C04875
NSC-29630

528-90-5
$C_{10}H_{12}O_2$
164.22
Benzoic acid, 2,4,5-trimethyl
Durylic acid
2,4,5-Trimethylbenzoic acid

529-16-8
C_9H_{14}
122.21
Bicyclo(2.2.1)hept-2-ene, 2,3-dimethyl- (9CI)
2-Norbornene, 2,3-dimethyl- (8CI)
Santene

529-19-1
C_8H_7N
117.16
N#Cc(c(ccc1)C)c1
2-Cyanotoluene
o-Cyanotoluene
2-Methylbenzenecarbonitrile
o-Methylbenzonitrile
2-Methylbenzonitrile
2-Toluenkarbonitril (Czech)
o-Toluic nitrile
o-Toluonitrile
o-Tolylnitrile

529-20-4
C_8H_8O
120.15
Cc1ccccc1C=O
Benzaldehyde, 2-methyl- (9CI)
AI3-21918
2-Formyltoluene
o-Methylbenzaldehyde
2-Methylbenzaldehyde
NSC-103152
o-Tolualdehyde (8CI)
2-Tolualdehyde
o-Toluic aldehyde
o-Toluylaldehyde
o-Tolylaldehyde

529-34-0
$C_{10}H_{10}O$
146.20
O=C(c(c(ccc1)CC2)c1)C2
1(2H)-Naphthalenone, 3,4-dihydro
α-Tetralone
1-Tetralone

529-64-6
$C_9H_{10}O_3$
166.18
OCC(C(O)=O)c1ccccc1
Tropic acid
Benzeneacetic acid, α-(hydroxymethyl)- (9CI)
Hydracrylic acid, 2-phenyl-
Hydratropic acid, β-hydroxy-
α-(Hydroxymethyl)benzeneacetic acid
NSC-20990
2-Phenylhydracrylic acid
α-Toluic acid, α-(hydroxymethyl)-

529-69-1
$C_6H_5N_5O_2$
179.16
4,7(3H,8H)-Pteridinedione, 2-amino
2-Amino-4,7(3H,8H)-pteridinedione
Isoxanthopterin
4,7(1H,8H)-Pteridinedione, 2-amino- (9CI)
Ranachrome 4

529-84-0
$C_{10}H_8O_4$
192.18
O=C(Oc(c(C=1C)cc(O)c2O)c2)C1
Coumarin, 6,7-dihydroxy-4-methyl
2H-1-Benzopyran-2-one, 6,7-dihydroxy-4-methyl- (9CI)
6,7-Dihydroxy-4-methylcoumarin
4-Methylaesculetin
Methylesculetin
4-Methylesculetin
4-Methylesculetol

529-96-4
$C_8H_{13}N_2O_5P$
248.18
Cc1ncc(COP(O)(O)=O)c(CN)c1O
Pyridoxamine phosphate
Pyridoxamine-P

530-44-9
$C_{15}H_{15}NO$
225.28
O=C(c(ccc(N(C)C)c1)c1)c(cccc2)c2
Benzophenone, 4-(dimethylamino)- (8CI)
p-Benzoyl-N,N-dimethylaniline
p-Dimethylaminobenzophenone
4-Dimethylaminobenzophenone
4-(Dimethylamino)benzophenone
4-N,N-Dimethylaminobenzophenone
(4-(Dimethylamino)phenyl)phenylmethanone
Methanone, (4-(dimethylamino)phenyl)phenyl- (9CI)
NSC-15962

530-47-2
$C_{12}H_{12}N_2.ClH$
220.69
NN(c1ccccc1)c2ccccc2
Hydrazine, 1,1-diphenyl-, monohydrochloride (9CI)
AI3-52355
N,N-Diphenylhydrazine hydrochloride
1,1-Diphenylhydrazine hydrochloride (VAN)
1,1-Diphenylhydrazine monohydrochloride
NSC-5937

530-48-3
$C_{14}H_{12}$
180.25
c(C(c(cccc1)c1)=C)(cccc2)c2
1,1-Diphenylethylene
AI3-06164
Benzene, 1,1'-ethenylidenebis- (9CI)
DDNU (VAN)
1,1-Diphenylethene
Diphenylethylene
α,α-Diphenylethylene
as-Diphenylethylene
1,1'-Ethenylidenebisbenzene
Ethylene, 1,1-diphenyl- (8CI)
α-Methylene-diphenylmethane
NSC-57645
α-Phenylstyrene

530-50-7
$C_{12}H_{12}N_2$
184.23
1,1-Diphenylhydrazine
AI3-23023
α,α-Diphenylhydrazine
N,N-Diphenylhydrazine
Hydrazine, 1,1-diphenyl-
Hydrozobenzene
NCI-C01854

530-56-3
$C_9H_{12}O_4$
184.19
COc1cc(CO)cc(OC)c1O
Benzenemethanol, 4-hydroxy-3,5-dimethoxy- (9CI)
Benzyl alcohol, 4-hydroxy-3,5-dimethoxy- (8CI)
Syringyl alcohol

530-57-4
$C_9H_{10}O_5$
198.19
O=C(O)c(cc(OC)c(O)c1OC)c1
Benzoic acid, 4-hydroxy-3,5-dimethoxy
4-Hydroxy-3,5-dimethoxybenzoic acid
Syringic acid

530-59-6
$C_{11}H_{12}O_5$
224.21
Sinapic acid
NSC-59261

530-66-5
$C_9H_7N.H_2O_4S$
227.23
Quinoline, sulfate (1:1) (8CI,9CI)

530-75-6
$C_{16}H_{12}O_6$
300.27
Acetylsalicylic acid
Acetylsalicylsalicylic acid

530-91-6
$C_{10}H_{12}O$
148.22
OC(CCc(c1ccc2)c2)C1
2-Naphthol, 1,2,3,4-tetrahydro
1,2,3,4-Tetrahydro-2-naphthol
β-Tetralol

531-73-7
$C_{13}H_{11}N_3.2ClH$
282.19
Acridine, 3,6-diamino-, dihydrochloride
3,6-Acridinediamine, dihydrochloride (9CI)
3,6-Diaminoacridine dihydrochloride
2,8-Diaminoacridinium chloride hydrochloride
3,6-Diaminoacridinium chloride hydrochloride
Proflavine dihydrochloride

531-76-0
$C_{13}H_{18}Cl_2N_2O_2$
305.23
NC(Cc1ccc(cc1)N(CCCl)CCCl)C(O)=O
Alanine, 3-(p-(bis(2-chloroethyl)amino)phenyl)-, DL
3-(p-(Bis(2-chloroethyl)amino)phenyl)alanine
DL-3-(p-(Bis(2-chloroethyl)amino)phenyl)alanine
4-(Bis(2-chloroethyl)amino)-DL-phenylalanine
CB-3307
p-Di-(2-chloraethyl)-amino-DL-phenyl-alanin (German)
p-Di(2-chloroethyl)amino-DL-phenylalanine
Merfalan
Merphalan
o-Merphalan
NCI-C04944
NSC-14210
DL-Phenylalanine, 4-(bis(2-chloroethyl)amino)-
DL-Phenylalanine mustard
Phenylalanin-lost (German)
DL-Phenylalanin-lost (German)
Sarcoclorin
DL-Sarcolysin
DL-Sarcolysine

531-82-8
$C_9H_7N_3O_4S$
253.25
CC(=O)Nc1nc(cs1)c2ccc(o2)N(=O)=O
Acetamide, N-(4-(5-nitro-2-furyl)-2-thiazolyl)
2-Acetamido-4-(5-nitro-2-furyl)-thiazole
2-Acetamino-4-(5-nitro-2-furyl)-thiazole
2-Acetylamino-4-(5-nitro-2-furyl)thiazole
AS17665
Furathiazole
Furium
Furothiazole
NFTA
N-(4-(5-Nitro-2-furanyl)-2-thiazolyl)acetamide
N-(4-(5-Nitro-2-furyl)-2-thiazolyl)acetamide
N-(4-(5-Nitro-2-furyl)thiazol-2-yl)acetamide
Thiazole, 2-acetamido-4-(5-nitro-2-furyl)-
Thiazole, 2-acetamino-4-(5-nitro-2-furyl)-
Thiazole, 2-acetylamino-4-(5-nitro-2-furyl)-

531-85-1
$C_{12}H_{12}N_2.2ClH$
257.18
Cl.Cl.Nc1ccc(cc1)c2ccc(N)cc2
Benzidine, dihydrochloride
Benzidine hydrochloride
(1,1'-Biphenyl)-4,4'-diamine, dihydrochloride
Dihidrocloruro de benzidina (Spanish)

531-86-2
$C_{12}H_{12}N_2.H_2O_4S$
282.34
Benzidine-sulfate
(1,1'-Biphenyl)-4,4'-diamine, sulfate (1:1)

532-02-5
$C_{10}H_8O_3S.Na$
231.23
[Na+].[O-]S(=O)(=O)c2ccc1ccccc1c2
2-Naphthalenesulfonic acid, sodium salt (9CI)
β-Naphthalenesulfonic sodium salt
NSC-7415
Sodium naphthalene-2-sulfonate
Sodium naphthalene-6-sulfonate
Sodium β-naphthalenesulfonate
Sodium 2-naphthalenesulfonate
Sodium naphthalene-2-sulphonate
Sodium salt of β-naphthalenesulfonic acid

532-03-6
$C_{11}H_{15}NO_5$
241.27
O=C(OCC(O)COc(c(OC)ccc1)c1)N
1,2-Propanediol, 3-(o-methoxyphenoxy)-, 1-carbamate
AHR 85
Carbamic acid, 2-hydroxy-3-(o-methoxyphenoxy)propyl ester
Etroflex
Glycerylguaiacolate carbamate
Glycerylguajacol-carbamat
Guaiacol glyceryl ether carbamate
Guiacol-Gliceriletere Monocarbammato
2-Hydroxy-3-(o-methoxyphenoxy)propyl 1-carbamate
Lumirelax
Methocarbamol
3-(2-Methoxyphenoxy)-1-glyceryl carbamate
3-(o-Methoxyphenoxy)-2-hydroxypropyl carbamate
3-(o-Methoxyphenoxy)-1,2-propanediol 1-carbamate
Metocarbamol
Metocarbamolo
Metofenina
Miolaxene

Miorilas
Miowas
Myolaxene
Neuraxin
Perilax
Reflexyn
Relax
Relestrid
Robaxan
Robaxin
Robaxine
Robaxon
Robinax
Tresortil

532-12-7
$C_9H_{10}N_2$
146.21
C1CC=C(N1)c2cccnc2
Pyridine, 3-(1-pyrrolin-2-yl)
Myosmine

532-27-4
C_8H_7ClO
154.60
O=C(c(cccc1)c1)CCl
Acetophenone, 2-chloro
CAF
CAP
Chemical Mace
Chloracetophenone
Chloroacetophenone (DOT)
Chloroacetophenone, Gas, liquid, or solid [UN 1697]
α-Chloroacetophenone (ACGIH,OSHA)
ω-Chloroacetophenone
1-Chloroacetophenone
2-Chloroacetophenone
Chloromethyl phenyl ketone
2-Chloro-1-phenylethanone
CN
Ethanone, 2-chloro-1-phenyl-
Mace (Lacrimator)
NCI-C55107
Phenacyl chloride (OSHA)
Phenylchloromethylketone
UN 1697 [Chloroacetophenone (CN) liquid or solid]

532-28-5
C_8H_7NO
133.16
N#CC(O)c(cccc1)c1
Mandelonitrile
Acetonitrile, hydroxyphenyl-
Amygdalonitrile
Benzaldehyde cyanohydrin
Benzaldehydkyanhydrin (Czech)
Glycolonitrile, phenyl-
Mandelic acid nitrile
Nitril kyseliny mandlove (Czech)

532-32-1
$C_7H_5O_2.Na$
144.11
[Na+].[O-]C(=O)c1ccccc1
Benzoic acid, sodium salt
Antimol
Benzoan sodny (Czech)
Benzoate of soda
Benzoate sodium
Benzoesaeure (na-salz) (German)
Sobenate
Sodium benzoate
Sodium benzoic acid

532-34-3
$C_{12}H_{18}O_4$
226.30
O=C(OCCCC)C(OC(CC1=O)(C)C)=C1
2H-Pyran-6-carboxylic acid, 3,4-dihydro-2,2-dimethyl-4-oxo-, butyl ester
BMOO
Butopyronoxyl
Butyl 3,4-dihydro-2,2-dimethyl-4-oxo-2h-pyran-6-carboxylate
n-Butyl ester of 3,4-dihydro-2,2-dimethyl-4-oxo-2H-pyran-6-carboxylic acid
Butyl mesityl oxide
n-Butyl mesityl oxide oxalate
n-Butylmesityloxid oxalate
2-Carbo-n-butoxy-6,6-dimethyl-5,6-dihydro-1,4-pyrone
3,4-Dihydro-2,2-dimethyl-4-oxo-2h-pyran-6-carboxylic acid, n-butyl ester
Dihydropyrone
α,α-Dimethyl-α'-carbobutoxy-dihydro-γ-pyrone
2,2-Dimethyl-6-carbobutoxy-2,3-dihydro-4-pyrone
ENT 9
Indalone

532-82-1
$C_{12}H_{12}N_4.ClH$
248.74
Cl.Nc2ccc(N=Nc1ccccc1)c(N)c2
m-Phenylenediamine, 4-(phenylazo)-, hydrochloride
Astra Chrysoidine R
Brasilazina Orange Y
Brilliant Oil Orange Y Base
Calcozine Chrysoidine Y
Calcozine Orange YS
Chrysoidin
Chrysoidine
Chrysoidine A
Chrysoidine B
Chrysoidine C Crystals
Chrysoidine Crystals
Chrysoidine G
Chrysoidine GN
Chrysoidine GS
Chrysoidine HR
Chrysoidine (II)
Chrysoidine J
Chrysoidine M
Chrysoidine Orange
Chrysoidine PRL
Chrysoidine PRR
Chrysoidine SL
Chrysoidine Special (Biological stain and indicator)
Chrysoidine SS
Chrysoidine Y
Chrysoidine Y Base New
Chrysoidine Y Crystals
Chrysoidine Y EX
Chrysoidine YGH
Chrysoidine YL
Chrysoidine YN
Chrysoidine Y Special
Chrysoidin FB
Chrysoidin Y
Chrysoidin YN
Chryzoidyna F.B. (Polish)
C.I. 11270
C.I. Basic Orange 2
C.I. Basic Orange 3
C.I. Basic Orange 2, Monohydrochloride
C.I. Solvent Orange 3
2,4-Diaminoazobenzene hydrochloride
Diazocard Chrysoidine G
Elcozine Chrysoidine Y
Leather Orange HR
Nippon kagaku chrysoidine
4-(Phenylazo)-1,3-benzenediamine, monohydrochloride
4-Phenylazo-m-phenylenediamine hydrochloride
4-(Phenylazo)-m-phenylenediamine, monohydrochloride
Pure Chrysoidine YBH
Pure Chrysoidine YD
Pyracryl Orange Y
Sugai Chrysoidine
Tertrophene Brown CG

533-00-6
$C_7H_6O_2.1/2Ba$
190.79
Benzoic acid, barium salt (9CI)
Barium benzoate

533-15-3
$C_8H_{17}NO$
143.22
OCCC(N(CCC1)C)C1
2-Piperidineethanol, 1-methyl-(9CI)
1-Methyl-2-piperidineethanol
NSC-75616

533-17-5
C_8H_8ClNO
169.60
O=C(Nc(c(ccc1)Cl)c1)C
o-Chloroacetanilide
Acetamide, N-(2-chlorophenyl)-(9CI)
Acetanilide, 2'-chloro- (8CI)
Acetic acid, amide, N(2-chlorophenyl)-
2'-Chloroacetanilide
N-(2-Chlorophenyl)acetamide

NSC-40562

533-18-6
$C_9H_{10}O_2$
150.18
O=C(Oc(c(ccc1)C)c1)C
Acetic acid, o-tolyl ester (8CI)
Acetic acid, 2-methylphenyl ester (9CI)
o-Acetoxytoluene
Acetyl-o-cresol
AI3-04168
o-Cresyl acetate
o-Cresylic acetate
o-Methylphenyl acetate
2-Methylphenyl acetate
NSC-58961
o-Tolyl acetate

533-23-3
$C_{10}H_{10}Cl_2O_3$
249.10
Acetic acid, (2,4-dichlorophenoxy)-, ethyl ester
Benzeneacetic acid, 2,4-dichloro-, ethyl ester (9CI)
Dicotox
2,4-D Ethyl ester
(2,4-Dichlorophenoxy)acetic acid ethyl ester
Ethyl (2,4-dichlorophenoxy)-acetate
Weedone 40
Weedone Concentrate 48

533-45-9
C_6H_8ClNS
161.66
Cc1ncsc1CCCl
Thiazole, 5-(2-chloroethyl)-4-methyl
Chlorethiazol
Chlorethiazole
Chlormethiazol
Chlormethiazole
Chlormithiazole
5-(2-Chloroethyl)-4-methyl-thiazole
Chloro-S.C.T.Z.
Clomethiazole
Clomethiazolum
Distraneurin
Emineurina
Heminevrin
4-Methyl-5-(β-chloroethyl)-thiazole
SCTZ
Somnevrin
WY 1485

533-51-7
$C_2H_2O_4.2Ag$
305.77
Silver oxalate (Dry)
Disilver oxalate
Ethanedioic acid, disilver(1+) salt
Oxalate d'argent (French)
Oxalato de plata (Spanish)
Oxalic acid disilver salt
Oxalic acid, disilver(1+) salt
Oxalic acid silver salt (1:2)
Silver oxalate
Silver oxalate, Dry

533-58-4
C_6H_5IO
220.01
Oc(c(ccc1)I)c1
Phenol, o-iodo
o-Iodophenol
2-Iodophenol
o-Jodfenol (Czech)
2-Jodfenol (Czech)

533-60-8
$C_6H_{10}O_2$
114.14
O=C(C(O)CCC1)C1
Cyclohexanone, 2-hydroxy-(9CI)
Adipoin
AI3-06549
2-Hydroxycyclohexanone
2-Hydroxy-1-cyclohexanone
NSC-298536

533-67-5
$C_5H_{10}O_4$
134.13
O=CCC(O)C(O)CO
Deoxyribose
AI3-52228
2-Deoxy-D-erythro-pentose
D-Erythro-pentose, 2-deoxy-

533-73-3
$C_6H_6O_3$
126.12
Oc(c(O)cc(O)c1)c1
1,2,4-Benzenetriol
Hydroquinone, hydroxy-
Hydroxyhydroquinone
Hydroxyquinol
Oxyhydrochinon (German)
Oxyhydroquinone
1,2,4-Trihydroxybenzene

533-74-4
$C_5H_{10}N_2S_2$
162.29
N(CSC(N1C)=S)(C1)C
2H-1,3,5-Thiadiazine-2-thione, tetrahydro-3,5-dimethyl
Basamid
Basamid G
Basamid-Granular
Basamid P
Basamid-Puder
Carbothialdin
Carbothialdine
CRAG
CRAG 974
CRAG Fungicide 974
CRAG Nemacide
CRAG 85W
Dazomet
Dazomet-Powder BASF
Dimethylformocarbothialdine
3,5-Dimethylperhydro-1,3,5-thiadiazin-2-thion (Czech, German)
3,5-Dimethyltetrahydro-1,3,5-thiadiazine-2-thione
3,5-Dimethyl-1,2,3,5-tetrahydro-1,3,5-thiadiazinethione-2
3,5-Dimethyltetrahydro-1,3,5-2h-thiadiazine-2-thione
3,5-Dimethyl-1,3,5-2h-tetrahydrothiadiazine-2-thione
3,5-Dimethyltetrahydro-2h-1,3,5-thiadiazine-2-thione
3,5-Dimethyl-2-thionotetrahydro-1,3,5-thiadiazine
3,5-Dimetil-peridro-1,3,5-tiadiazin-2-tione (Italian)
DMTT
Fennosan B 100
Micofume
Mylon (Czech)
Mylone
Mylone 85
N 521
Nalcon 243
Nefusan
Prezervit
Salvo (Czech)
Stauffer N 521
Tetrahydro-2H-3,5-dimethyl-1,3,5-thiadiazine-2-thione
Tetrahydro-3,5-dimethyl-2H-1,3,5-thiadiazine-2-thione
Thiazon
Thiazone
2-Thio-3,5-dimethyltetrahydro-1,3,5-thiadiazine
Tiazon
Troysan 142
UCC 974

533-75-5
$C_7H_6O_2$
122.13
Oc1cccccc1=O
2,4,6-Cycloheptatrien-1-one, 2-hydroxy
Purpurocatechol
Tropolone

533-96-0
$CH_2O_3.3/2Na$
96.51
Magadi soda
Carbonic acid, sodium salt (2:3) (9CI)
Snowflake crystals
Sodium Sesquicarbonate
Trona
Urao

533-98-2
$C_4H_8Br_2$
215.92
BrCC(Br)CC

Butane, 1,2-dibromo- (9CI)
AI3-14677
α-Butylene dibromide
1,2-Dibromobutane
NSC-6181

534-07-6
$C_3H_4Cl_2O$
126.97
O=C(CCl)CCl
2-Propanone, 1,3-dichloro
Bis(chloromethyl)ketone
sym-Dichloroacetone
α,α'-Dichloroacetone
α,γ-Dichloroacetone
1,3-Dichloroacetone [UN 2649]
1,3-Dichloro-2-propanone
UN 2649 [1,3-Dichloroacetone]

534-13-4
$C_3H_8N_2S$
104.19
N(=C(S)NC)C
Urea, 1,3-dimethyl-2-thio
1,3-Dimethylisothiourea
Dimethylthiocarbamide
N,N'-Dimethylthiourea
sym-Dimethylthiourea
1,3-Dimethylthiourea
Thiourea, N,N'-dimethyl- (9CI)

534-15-6
$C_4H_{10}O_2$
90.14
O(C(OC)C)C
Acetaldehyde, dimethyl acetal
1,1-Dimethoxyethane [UN 2377]
Dimethylacetal
Dimethyl aldehyde
Ethane, 1,1-dimethoxy-
Ethylidene dimethyl ether
Methyl formyl
UN 2377 [1,1-Dimethoxy-ethane]

534-22-5
C_5H_6O
82.11
O(C(=CC=1)C)C1
Furan, 2-methyl
Methylfuran (DOT)
2-Methylfuran [UN 2301]
Silvan (Czech)
Sylvan
UN 2301 [2-Methylfuran]

534-26-9
$C_4H_8N_2$
84.11
N(=C(NC1)C)C1
1H-Imidazole, 4,5-dihydro-2-methyl-
AI3-16866
4,5-Dihydro-2-methyl-1H-imidazole

534-52-1
$C_7H_6N_2O_5$
198.15
O=N(=O)c(cc(N(=O)=O)c(O)c1C)c1
o-Cresol, 4,6-dinitro
Antinonin
Antinonnin
Arborol
Capsine
Chemsect DNOC
C.I. 10310
Degrassan
Dekrysil
Detal
Detol
Dinitrocresol
Dinitro-o-cresol (ACGIH, OSHA)
2,4-Dinitro-o-cresol
4,6-Dinitro-o-cresol
4,6-Dinitro-o-cresolo (Italian)
Dinitrodendtroxal
3,5-Dinitro-2-hydroxytoluene
4,6-Dinitrokresol (Dutch)
4,6-Dinitro-o-kresol (Czech)
Dinitrol
Dinitromethyl cyclohexyltrienol
2,4-Dinitro-6-methylphenol
DINOC
DINOK
Dinurania
Ditrosol
DN
DNC
DN-Dry Mix No. 2
DNOC
DNOK (Czech)
Dwunitro-o-krezol (Polish)
Effusan
Effusan 3436
Elgetol
Elgetol 30
Elipol
ENT 154
Extrar
Hedolit
Hedolite
K III
K IV
Krenite (Obs.)
Kresamone
Kresonite-E
Krezotol 50
Le dinitrocresol-4,6 (French)
Lipan
2-Methyl-4,6-dinitrophenol
Nitrador
Nitrofan
Oranz Viktoria (Czech)
Phenol, 2-methyl-4,6-dinitro- (9CI)
Prokarbol
Rafex
Rafex 35
Raphatox
RCRA waste number P047
Sandolin
Sandolin A
Selinon
Sinox
Trifocide
Trifrina
Winterwash
Zahlreiche bezeichnungen (German)

534-59-8
$C_7H_{12}O_4$
160.17
2-n-Butylmalonate
Butylmalonate
2-Butylmalonate
n-Butylmalonic acid
2-Butylmalonic acid
2-n-Butylmalonic acid
α-Carboxycaproic acid
Malonic acid, butyl- (8CI)
NSC-791
1,1-Pentanedicarboxylic acid
Propanedioic acid, butyl- (9CI)

535-13-7
$C_5H_9ClO_2$
136.59
O=C(OCC)C(Cl)C
Propionic acid, 2-chloro-, ethyl ester
Ethyl-2-chloropropionate [UN 2935]
Propanoic acid, 2-chloro-, ethyl ester (9CI)
UN 2935 [Ethyl-2-chloropropionate]

535-15-9
$C_4H_6Cl_2O_2$
157.00
O=C(OCC)C(Cl)Cl
Acetic acid, dichloro-, ethyl ester (9CI)
AI3-28936
Dichloroacetic acid, ethyl ester
Ethyl dichloroacetate
Ethyl 2,2-dichloroacetate
Ethyl dichloroethanoate
NSC-27788

535-32-0
$C_6H_{13}NO_2S$
163.26
Butyric acid, 2-amino-4-(ethylthio)-, d
D-2-Amino-4-(ethylthio)butyric acid
D-Ethionine

535-75-1
$C_6H_{11}NO_2$
129.18
O=C(O)C(NCCC1)C1
Pipecolic-acid
Acide pipecolique (French)
Acide piperidine-carboxylique-2 (French)
Dihydrobaikiane
Hexahydropicolinic acid
Homoproline

Pipecolate
Pipecolinic acid
α-Pipecolinic acid
2-Piperidinecarboxylic acid (9CI)
Piperolinic acid

535-77-3
$C_{10}H_{14}$
134.24
c(cccc1C(C)C)(c1)C
m-Cymene [UN 2046]
UN 2046 [Cymenes]

535-80-8
$C_7H_5ClO_2$
156.57
O=C(O)c(cccc1Cl)c1
Benzoic acid, m-chloro
Acido m-clorobenzoico (Italian)
Benzoic acid, 3-chloro-
m-Chlorobenzoic acid
3-Chlorobenzoic acid

535-83-1
$C_7H_7NO_2$
137.15
Trigonelline
Betain nicotinate
Betaine nicotinate
Caffearine
3-Carboxy-1-methylpyridinium hydroxide inner salt
Coffearin
Coffearine
Gynesine
N-Methylnicotinate
N-Methylnicotinic acid
N'-Methylnicotinic acid
Nicotinic acid N-methylbetaine
Pyridinium, 3-carboxy-1-methyl-, hydroxide, inner salt (8CI)
Trigenolline
Trigonellin

535-89-7
$C_7H_{10}ClN_3$
171.65
Pyrimidine, 2-chloro-4-(dimethylamino)-6-methyl
Castrix
2-Chloor-4-dimethylamino-6-methyl-pyrimidine (Dutch)
2-Chlor-4-dimethylamino-6-methylpyrimidin (German)
2-Chloro-4-dimethylamino-6-methyl-pyrimidine
2-Chloro-4-methyl-6-dimethyl-aminopyrimidine
2-Chloro-n,n-6-trimethyl-4-pyrimidinamine
2-Cloro-4-dimetilamino-6-metil-pirimidina (Italian)
Crimidin (German)
Crimidina (Italian)
Crimidine
W 491

536-33-4
$C_8H_{10}N_2S$
166.26
CCc1cc(ccn1)C(N)=S
Isonicotinamide, 2-ethylthio
Aetina
Aetiva
Amidazin
Amidazine
Bayer 5312
ETH
Ethina
Ethinamide
Ethimide
Ethionamide
Ethioniamide
α-Ethylisonicotinic acid thioamide
2-Ethylisonicotinic acid thioamide
2-Ethylisonicotinic thioamide
α-Ethylisonicotinoylthioamide
Ethylisothiamide
α-Ethylisothionicotinamide
2-Ethylisothionicotinamide
2-Ethyl-4-pyridinecarbothioamide
2-Ethyl-4-thioamidylpyridine
2-Ethyl-4-thiocarbamoylpyridine
α-Ethylthioisonicotinamide
2-Ethylthioisonicotinamide
Ethyonomide
Etimid
EtioCIdan
Etionamid
Etionid
Etionizina
Etionizine
ETP
Fatoliamid
F.I. 58-30
Iridocin
Iridozin
Isothin
Isotiamida
Itiocide
Nicotion
Nisotin
Nizotin
NCI-C01694
Rigenicid
Sertinon
Teberus
1314 TH
TH 1314
Thianid
Thianide
Thioamide
Thiomid
Thioniden
Tio-Mid
Trecator
Trescatyl
Trescazide
Tubermin
Tuberoid
Tuberoson

536-40-3
$C_7H_7ClN_2O$
170.59
Benzoic acid, p-chloro-, hydrazide (8CI)
Benzoic acid, 4-chloro-, hydrazide (9CI)
p-Chlorobenzhydrazide
4-Chlorobenzhydrazide
p-Chlorobenzohydrazide
4-Chlorobenzohydrazide
p-Chlorobenzoic acid, hydrazide
4-Chlorobenzoic acid, hydrazide
p-Chlorobenzoic hydrazide
p-Chlorobenzoyl hydrazide
4-Chlorobenzoyl hydrazide
p-Chlorobenzoylhydrazine
4-Chlorobenzoylhydrazine
NSC-54990

536-50-5
$C_9H_{12}O$
136.21
OC(c(ccc(c1)C)c1)C
Benzyl alcohol, p,α-dimethyl
Bilagen
α,4-Dimethylbenzenemethanol (9CI)
p,α-Dimethylbenzyl alcohol
Ethanol, 1-(p-tolyl)-
4-(α-Hydroxyethyl)toluene
1-(p-Methylphenyl)ethanol
1-(4-Methylphenyl)ethanol
Methyl-p-tolylcarbinol
Norbilan
1-(p-Tolyl)ethanol
p-Tolylmethylcarbinol (German)
Tomobil

536-57-2
$C_7H_8O_2S$
156.21
O=S(O)c(ccc(c1)C)c1
Benzenesulfinic acid, 4-methyl- (9CI)
4-Methylbenzenesulfinic acid

536-59-4
$C_{10}H_{16}O$
152.26
OCC(=CCC(C(=C)C)C1)C1
p-Mentha-1,8-dien-7-ol
Cyclohex-1-ene-1-methanol, 4-(1-methylethenyl)-
Dihydrocuminyl alcohol
4-Isopropenyl-cyclohex-1-ene-1-methanol
Perilla alcohol
Perillol
Perillyl alcohol

536-60-7
$C_{10}H_{14}O$
150.24
OCc(ccc(c1)C(C)C)c1
p-Cymen-7-ol
Benzyl alcohol, p-isopropyl-
Cumic alcohol
Cuminic alcohol
Cuminol

Cuminyl alcohol
Cumyl alcohol
p-Isopropylbenzyl alcohol

536-66-3
$C_{10}H_{12}O_2$
164.20
O=C(O)c(ccc(c1)C(C)C)c1
4-Isopropylbenzoate
AI3-17970
Benzoic acid, p-isopropyl- (8CI)
Benzoic acid, 4-(1-methylethyl)- (9CI)
Cuminic acid
p-Isopropylbenzoic acid
4-Isopropylbenzoic acid
4-(1-Methylethyl)benzoic acid
NSC-1907

536-69-6
$C_{10}H_{13}NO_2$
179.24
O=C(O)c(ncc(c1)CCCC)c1
2-Pyridinecarboxylic acid, 5-butyl
5-Butylpicolinic acid
5-Butyl-2-pyridinecarboxylic acid
Fusaric acid
Fusarinic acid
Picolinic acid, 5-butyl-

536-74-3
C_8H_6
102.14
C(c(cccc1)c1)#C
Benzene, ethynyl
Acetylene, phenyl-
Ethinylbenzene
Ethynylbenzene
Phenylacetylene

536-75-4
C_7H_9N
107.15
n(ccc(c1)CC)c1
Pyridine, 4-ethyl- (9CI)
γ-Ethylpyridine
4-Ethylpyridine
NSC-822

536-78-7
C_7H_9N
107.15
n(cccc1CC)c1
3-Ethylpyridine
Pyridine, 3-ethyl- (9CI)

536-90-3
C_7H_9NO
123.17
O(c(cccc1N)c1)C
m-Anisidine
m-Aminoanisole
3-Aminoanisole
m-Anisylamine
Benzenamine, 3-methoxy- (9CI)
3-Methoxyaniline
3-Methoxybenzenamine

537-00-8
$C_6H_9O_6.Ce$
317.27
Cerium-acetate
Cerium(III) acetate
Cerium triacetate
Cerous acetate
Cetic acid, cerium(3+) salt (8CI,9CI)

537-01-9
$CH_2O_3.2/3Ce$
155.44
Carbonic acid, cerium(3+) salt (3:2) (9CI)
Cerium carbonate (VAN)
Cerium(III) carbonate
Cerium tricarbonate
Cerous carbonate
NSC-253011

537-03-1
$C_6La_2O_{12}$
541.87
Lanthanum, tris(ethanedioato-(2-))di- (9CI)
Tris(ethanedioato(2-))dilanthanum

537-17-7
$C_9H_9N_5$
187.17
1,3,5-Triazine-2,4-diamine, N-phenyl- (9CI)
Amanozina (Spanish)
Amanozine
Amanozinum (Latin)
s-Triazine, 2-amino-4-anilino- (8CI)

537-24-6
$C_9H_9Br_2NO_3$
338.98
NC(Cc1cc(Br)c(O)c(Br)c1)C(O)=O
Tyrosine, 3,5-dibromo- (8CI, 9CI)
NSC-39452

537-26-8
$C_{15}H_{19}NO_2$
245.35
CN1C2CCC1CC(C2)OC(=O)c3ccccc3
1-α-H,5-α-H-Tropan-3-β-ol, benzoate (ester)
Benzilate of pseudotropanol
Benzoylpseudotropeine
Benzoylpseudotropine
Benzoyl-psi-tropeine
o-Benzoyltropine
exo-8-Methyl-8-azabicyclo-(3.2.1)-octan-3-ol benzoate
Pseudotropanol benzilate
Pseudotropine benzoate
Pseudotropine, benzoate (ester)
Tropacaine
Tropacocain
Tropacocaine
psi-Tropine benzoate

537-45-1
$C_6H_2Br_2ClNO$
299.36
O=C(C(=CC(=NCl)C=1)Br)C1Br
2,5-Cyclohexadien-1-one, 4-chloroimino-2,6-dibromo
BQC Reagent
4-Chloroimino-2,6-dibromo-2,5-cyclohexadiene-1-one
2,5-Cyclohexadien-1-one, 2,6-dibromo-4-(chloroimino)- (8CI,9CI)
2,6-Dibromoquinone chlorimide
2,6-Dibromoquinone chloroimide
2,6-Dibromoquinone chloroimine

537-46-2
$C_{10}H_{15}N$
149.26
CNC(C)Cc1ccccc1
d-1-Phenyl-2-methylaminopropane
d-1-Phenyl-2-methylaminopropan (German)
Phenethylamine, N,α-dimethyl-, (+)

537-47-3
$C_7H_9N_3O$
151.15
Hydrazinecarboxamide, N-phenyl- (9CI)
NSC-231527
4-Phenylsemicarbazide
Semicarbazide, 4-phenyl- (8CI)

537-64-4
$C_{14}H_{14}Hg$
382.86
Mercury, bis(4-methylphenyl)- (9CI)
Bis(4-methylphenyl)mercury
Di-p-tolyl mercury
Mercury, di-p-tolyl- (8CI)
NSC-33535

537-65-5
$C_{12}H_{13}N_3$
199.24
N(c(ccc(N)c1)c1)c(ccc(N)c2)c2
4,4'-Diaminodiphenylamine
AI3-12116
N-(4-Aminophenyl)-1,4-benzenediamine
Aniline, 4,4'-iminodi-
Benzenamine, 4,4'-iminobis-
1,4-Benzenediamine, N-

(4-aminophenyl)- (9CI)
Bis(p-aminophenyl)amine
C.I. 76120
p,p'-Diaminodiphenylamine
Di(p-aminophenyl)amine
Diazol Black C
Diphenylamine, 4,4'-diamino- (8CI)
4,4'-Iminodianiline
Indamine
NSC-33417
p-Phenylenediamine, N-(p-aminophenyl)-

537-92-8
$C_9H_{11}NO$
149.21
O=C(Nc(cccc1C)c1)C
m-Acetotoluidide
Acetamide, N-(3-methylphenyl)- (9CI)
3-Acetamidotoluene
Aceto-m-aminotoluene
Acetotoluide
N-Acetyl-m-toluidine
m-Methylacetanilide
3-Methylacetanilide
3'-Methylacetanilide
m-Tolylacetamide
N-m-Tolylacetamide

537-98-4
$C_{10}H_{10}O_4$
194.20
Cinnamic acid, 4-hydroxy-3-methoxy-, (E)
Cinnamic acid, 4-hydroxy-3-methoxy-, trans-
Ferulic acid
Ferulic acid, trans-
4-Hydroxy-3-methoxycinnamic acid
3-(4-Hydroxy-3-methoxyphenyl)propenoic acid
2-Propenoic acid, 3-(4-hydroxy-3-methoxyphenyl)-, (E)- (9CI)

538-07-8
$C_6H_{13}Cl_2N$
170.10
CCN(CCCl)CCCl
Triethylamine, 2,2'-dichloro
Bis(2-chloroethyl)ethylamine
2,2'-Dichlorotriethylamine
Ethylbis(β-chloroethyl)amine
Ethylbis(2-chloroethyl)amine
Ethyl-S
HN1
TL 329
TL 1149

538-08-9
$C_7H_{10}N_2$
122.19
N#CN(CC=C)CC=C
Cyanamide, diallyl
Cyanamide, di-2-propenyl- (9CI)
Diallylcyanamide
Diallylkyanamid (Czech)

538-23-8
$C_{27}H_{50}O_6$
470.77
O=C(OCC(OC(=O)CCCCCCC)COC(=O)CCCCCCC)CCCCCCC
Trioctanoin
Caprylic acid triglyceride
Glycerol tricaprylate
Glycerol trioctanoate
Glyceryl trioctanoate
MCT
Octanoic acid, 1,2,3-propanetriyl ester
Octanoic acid triglyceride
Octanoin, tri- (8CI)
Rato
Tricaprylic glyceride
Tricaprylin
Trioctanoylglycerol

538-24-9
$C_{39}H_{74}O_6$
639.01
O=C(OCC(OC(=O)CCCCCCCCCCC)COC(=O)CCCCCCCCCCC)CCCCCCCCCCC
Trilaurin
AI3-11124
Dodecanoic acid, 1,2,3-propanetriyl ester (9CI)
Dodecanoic acid, 1,2,3-propantriyl ester
Glycerin trilaurate
Glycerol trilaurate
Glyceryl tridodecanoate
Glyceryl trilaurate
Lauric acid triglyceride
Lauric acid triglycerin ester
Laurin, tri- (8CI)
NSC-4061
1,2,3-Propanetriol tridodecanoate
1,2,3-Propanetriyl dodecanoate

538-28-3
$C_8H_{10}N_2S.ClH$
202.72
[Cl-].NC(=N)SCc1ccccc1
Pseudourea, 2-benzyl-2-thio-, monohydrochloride
Benzylisothiourea hydrochloride
Benzylisothiouronium chloride
2-Benzylisothiouronium chloride
Benzyl thiopseudourea hydrochloride
Benzylthiuronium chloride
S-Benzylthiuronium chloride
BTKH
Isothiouronium chloride, benzyl-
Pseudourea, 2-benzyl-2-thio-, hydrochloride
Pseudourea, 2-thio-2-benzyl-, hydrochloride
TL 944
USAF EK-2124

538-41-0
$C_{12}H_{12}N_4$
212.28
N(=Nc(ccc(N)c1)c1)c(ccc(N)c2)c2
Aniline, 4,4'-azodi
4,4'-Azodianiline

538-43-2
$C_9H_{12}O_3$
168.21
OCC(O)COc1ccccc1
1,2-Propanediol, 3-phenoxy
Antodyn
Antodyne
1-Fenoxy-2,3-propandiol (Czech)
Glycerol α-monophenyl ether
Phenol-glycerinaether (German)
Phenol glycerol ether
Phenol glyceryl ether
1-Phenoxy-2,3-propanediol
3-Phenoxy-1,2-propanediol
Phenyl-α-glycerol ether
Phenylglyceryl ether
U 27,462

538-51-2
$C_{13}H_{11}N$
181.23
C(=Nc1ccccc1)c2ccccc2
Aniline, N-benzylidene- (8CI)
AI3-01538
Benzalaniline
N-Benzalaniline
Benzaldehyde anil
Benzenamine, N-(phenylmethylene)- (9CI)
N-Benzylidenaniline
Benzylideneaniline
N-Benzylideneaniline
NSC-736

538-65-8
$C_{13}H_{16}O_2$
204.29
CCCCOC(=O)C=Cc1ccccc1
Cinnamic acid, butyl ester
Butyl cinnamate
n-Butyl cinnamate
n-Butyl phenylacrylate
Cinnamate de n-butyle (French)
Cinnamic acid n-butyl ester
Eliminoxy
2-Propenoic acid, 3-phenyl-, butyl ester

538-68-1
$C_{11}H_{16}$
148.25
c(cccc1)(c1)CCCCC
Benzene, pentyl- (9CI)
AI3-00452
Amylbenzene

n-Amylbenzene
NSC-73982
Pentane, 1-phenyl-
Pentylbenzene
n-Pentylbenzene
Phenylpentane
1-Phenylpentane
1-Phenyl-n-pentane

538-74-9
$C_{14}H_{14}S$
214.33
S(Cc(cccc1)c1)Cc(cccc2)c2
Benzyl sulfide
AI3-00842
Benzene, 1,1'-(thiobis(methylene))bis- (9CI)
Benzyl monosulfide
Benzyl thioether
Dibenzyl monosulfide
Dibenzyl sulfide
1,3-Diphenyl-2-thiapropane
NSC-212544
Sulfide, dibenzyl
1,1'-(Thiobis(methylene))bisbenzene

538-75-0
$C_{13}H_{22}N_2$
206.32
N(=C=NC(CCCC1)C1)C(CCCC2)C2
Dicyclohexylcarbodiimide
AI3-08191
Carbodicyclohexylimide
Carbodiimide, dicyclohexyl- (8CI)
Cyclohexanamine, N,N'-methanetetraylbis- (9CI)
DCC
DCCD
DCCI
Dicylcohexylcarbodiimide
N,N'-Dicyclohexylcarbodiimide
1,3-Dicyclohexylcarbodiimide
N,N'-Methanetetraylbiscyclohexanamine
NSC-30022

538-86-3
$C_8H_{10}O$
122.18
O(Cc(cccc1)c1)C
Ether, benzyl methyl
Benzene, (methoxymethyl)- (9CI)
Benzyl methyl ether
α-Methoxytoluene
Methyl benzyl ether
α-Methylbenzyl ether

538-93-2
$C_{10}H_{14}$
134.24
c(cccc1)(c1)CC(C)C
Benzene, isobutyl
Isobutylbenzene
2-Methyl-1-phenylpropane

539-03-7
C_8H_8ClNO
169.62
O=C(Nc(ccc(c1)Cl)c1)C
Acetanilide, 4'-chloro
Acetamide, N-(4-chlorophenyl)-
Acetic-4-chloroanilide
4-Chloroacetanilide
4'-Chloroacetanilide
N-(p-Chlorophenyl)acetamide
N-(4-Chlorophenyl)acetamide

539-17-3
$C_{14}H_{16}N_4$
240.34
N(=Nc(ccc(N)c1)c1)c(ccc(N(C)C)c2)c2
Aniline, 4-(p-dimethylaminophenylazo)
Acetile Diazo Black N
Acetile Diazo Black R
ADAB
p-Aminobenzeneazodimethylaniline
4'-Amino-Dab
4'-Amino-N,N-dimethyl-4-aminoazobenzene
Aniline, N,N-dimethyl-4,4'-azodi- (7CI)
Azobenzene, 4-amino-4'-dimethylamino-
Benzenamine, 4-((4-aminophenyl)azo)-N,N-dimethyl- (9CI)
C.I. 11025
C.I. Disperse Black 3
Diazo Nero Microsetile G
4-Dimethylaminoazobenzene
Interchem Acetate Developed Black
Meisei Acemyl Diazo Black B
Meisei Teryl Diazo Black CR
Microsetile Diazo Black G
Supracet Diazo Black A

539-21-9
$C_8H_{11}N_7S$
237.32
NC(=N)NN=C1C=CC(C=C1)=NNC(N)=S
Guanidine, ((4-oxo-2,5-cyclohexadien-1-ylidene)amino)-, thiosemicarbazone
Ambazon
Ambazone
Anginon
1-Amidinohydrazono-4-thiosemicarbazono-2,5-cyclohexadiene
p-Benzoquinone amidinohydrazone thiosemicarbazone
Benzoquinone guanylhydrazone thiosemicarbazone
DC 0572
Faringosept
Guanothiazon
Hydrazinecarbothioamide, 2-(4-((aminoiminomethyl)-hydrazono)-2,5-cyclohexadien-1-ylidene)-
Inversal
Iversal
Ivertol
((4-Oxo-2,5-cyclohexadien-1-ylidene)amino)guanidine thiosemicarbazone
Primal
Promassol

539-30-0
$C_9H_{12}O$
136.19
O(Cc(cccc1)c1)CC
Benzene, (ethoxymethyl)- (9CI)
AI3-02270
Benzyl ethyl ether
Benzyl ethyl oxide
Ether, benzyl ethyl (8CI)
(Ethoxymethyl)benzene
NSC-8066

539-32-2
$C_9H_{13}N$
135.23
n(cccc1CCCC)c1
Pyridine, 3-butyl
3-Butylpyridine
3-n-Butylpyridine
1-(3-Pyridyl)butane

539-66-2
$C_5H_{10}O_2.Na$
125.12
Butanoic acid, 3-methyl-, sodium salt (9CI)
Isovaleric acid, sodium salt (8CI)

539-80-0
C_7H_6O
106.13
Tropone
2,4,6-Cycloheptatrien-1-one

539-82-2
$C_7H_{14}O_2$
130.19
O=C(OCC)CCCC
Pentanoic acid, ethyl ester (9CI)
AI3-01270
Ethyl pentanoate
Ethyl valerate
Ethyl n-valerate
NSC-8868
Valeric acid, ethyl ester (8CI)

539-88-8
$C_7H_{12}O_3$
144.19
O=C(OCC)CCC(=O)C
Levulinic acid, ethyl ester
Ethyl 3-acetylpropionate

Ethyl ketovalerate
Ethyl 4-ketovalerate
Ethyl laevulinate
Ethyl levulate
Ethyl 4-oxopentanoate
Ethyl 4-oxovalerate
Pentanoic acid, 4-oxo-, ethyl ester (9CI)

539-90-2
$C_8H_{16}O_2$
144.24
O=C(OCC(C)C)CCC
Butyric acid, isobutyl ester
Isobutyl butanoate
Isobutyl butyrate
Isobutyl-n-butyrate
2-Methylpropyl butyrate

540-07-8
$C_{11}H_{22}O_2$
186.29
O=C(OCCCCC)CCCCC
Hexanoic acid, pentyl ester (9CI)
AI3-06030
Amyl caproate
n-Amyl caproate
Amyl capronate
Amyl hexanoate
Amyl hexoate
NSC-46119
Pentyl caproate
Pentyl hexanoate

540-09-0
$C_{23}H_{46}O$
338.69
O=C(CCCCCCCCCCC)CCCCCCCCCCC
12-Tricosanone
Di-n-undecyl ketone

540-10-3
$C_{32}H_{64}O_2$
480.96
O=C(OCCCCCCCCCCCCCCCC)CCCCCCCCCCCCCCC
Palmitic acid, hexadecyl ester
Cetin
Cetyl palmitate
Hexadecanoic acid, hexadecyl ester
Hexadecyl palmitate
Palmityl palmitate
Standamul 1616

540-18-1
$C_9H_{18}O_2$
158.27
O=C(OCCCCC)CCC
Butyric acid, pentyl ester
Amyl butyrate
n-Amyl butyrate [UN 2620]
Butanoic acid pentyl ester
Pentyl butyrate
UN 2620 [Amyl butyrates]

540-23-8
$C_7H_9N.ClH$
143.63
Cl.Cc1ccc(N)cc1
p-Toluidine, hydrochloride
4-Aminotoluene hydrochloride
Benzenamine, 4-methyl-, hydrochloride (9CI)
4-Methylaniline hydrochloride
p-Toluidinium chloride

540-24-9
$C_6H_8N_2.ClH$
144.62
p-Phenylenediamine, monohydrochloride
4-Dimethylaminoaniline hydrochloride

540-36-3
$C_6H_4F_2$
114.10
Fc1ccc(F)cc1
Benzene, p-difluoro
Benzene, 1,4-difluoro-
para-Difluorobenzene

540-37-4
C_6H_6IN
219.03
Nc(ccc(c1)I)c1
Aniline, p-iodo
p-Aminophenyl iodide
Aniline, 4-iodo-
Benzenamine, 4-iodo- (9CI)
p-Iodoaniline
4-Iodoaniline

540-38-5
C_6H_5IO
220.01
Oc(ccc(c1)I)c1
Phenol, p-iodo
p-Iodophenol
4-Iodophenol

540-42-1
$C_7H_{14}O_2$
130.21
O=C(OCC(C)C)CC
Propionic acid, isobutyl ester
Isobutyl propionate [UN 2394]
2-Methylpropyl propionate
Propanoic acid, 2-methylpropyl ester (9CI)
UN 2394 [Isobutyl propionate]

540-49-8
$C_2H_2Br_2$
185.86
BrC=CBr
Ethylene, 1,2-dibromo
1,2-Dibromoethylene

540-51-2
C_2H_5BrO
124.98
OCCBr
Ethanol, 2-bromo
BE
Bromoethanol
2-Bromoethanol
Ethylenebromohydrin
Glycol bromohydrin

540-54-5
C_3H_7Cl
78.55
ClCCC
Propane, 1-chloro
1-Chloropropane
Propyl chloride [UN 1278]
UN 1278 [Propyl chloride]

540-59-0
$C_2H_2Cl_2$
96.94
C(=CCl)Cl
Ethylene, 1,2-dichloro
Acetylene dichloride (OSHA)
1,2-Dichlor-aethen (German)
1,2-Dichloroethene
Dichloro-1,2-ethylene (French)
sym-Dichloroethylene
1,2-Dichloroethylene (ACGIH, OSHA)
Dioform
NCI-C56031

540-63-6
$C_2H_6S_2$
94.20
SCCS
1,2-Ethanedithiol
1,2-Dimercaptoethane
Dithioethyleneglycol
Dithioglycol
Ethylene dimercaptan
α-Ethylene dimercaptan
Ethylene dithioglycol
Ethylenedithiol
Ethyl hydropersulfide

540-67-0
C_3H_8O
60.11
Ethane, methoxy-
Ether, ethyl methyl
Ethyl methyl ether [UN 1039]
Methane, ethoxy
Methyl ethyl ether (DOT)
UN 1039 [Ethyl methyl ether]

540-69-2
$CH_2O_2.H_3N$
63.07
N.[O-]C=O
Ammonium-formate
Formic acid ammonium salt
Mravencan amonny (Czech)

540-72-7
CNS.Na
81.07
[Na+].[S-]C#N
Thiocyanic acid, sodium salt
Haimased
Natriumrhodanid (German)
Scyan
Sodium isothiocyanate
Sodium rhodanate
Sodium rhodanide
Sodium sulfocyanate
Sodium sulfocyanide
Sodium thiocyanate
Sodium thiocyanide
Thiocyanate sodium
USAF EK-T-434

540-73-8
$C_2H_8N_2$
60.12
CNNC
Hydrazine, 1,2-dimethyl
1,2-Dimethylhydrazin (German)
Dimethylhydrazine, symmetrical [UN 2382]
N,N'-Dimethylhydrazine
sym-Dimethylhydrazine
1,2-Dimethyl-hydrazine
DMH
Hydrazomethane
RCRA waste number U099
SDMH
Symetryczna dwumetylohydrazyna (Polish)
UN 2382 [Dimethylhydrazine, symmetrical]

540-84-1
C_8H_{18}
114.26
C(CC(C)C)(C)(C)C
Pentane, 2,2,4-trimethyl
Isobutyltrimethylmethane
Isooctane [UN 1262]
2,2,4-Trimethylpentane
UN 1262 [Octanes]

540-88-5
$C_6H_{12}O_2$
116.18
O=C(OC(C)(C)C)C
Acetic acid, tert-butyl ester
Acetic acid, 1,1-dimethylethyl ester (9CI)
t-Butyl acetate
tert-Butyl acetate (ACGIH,-OSHA) [UN 1123]
Texaco Lead Appreciator
TLA
UN 1123 [Butyl acetates]

540-97-6
$C_{12}H_{36}O_6Si_6$
444.93
Cyclohexasiloxane, dodecamethyl- (9CI)
Dodecamethylcyclohexasiloxane

541-01-5
$C_{16}H_{48}O_6Si_7$
533.15
CSi(C)(C)OSi(C)(C)OSi(C)(C)OSi(C)(C)OSi(C)(C)OSi(C)(C)OSi(C)(C)C
Heptasiloxane, hexadecamethyl- (9CI)
Hexadecamethylheptasiloxane

541-02-6
$C_{10}H_{30}O_5Si_5$
370.85
Cyclopentasiloxane, decamethyl
Decamethylcyclopentasiloxane
Dekamethylcyklopentasiloxan (Czech)

541-05-9
$C_6H_{18}O_3Si_3$
222.46
CSi1(C)OSi(C)(C)OSi(C)(C)O1
Hexamethylcyclotrisiloxane
AI3-62005
Cyclotrisiloxane, hexamethyl-

541-09-3
$C_4H_6O_6U$
388.10
Uranium, bis(aceto)dioxo
NA 9180 (DOT)
Uranium acetate
Uranium oxyacetate
Uranyl acetate (DOT)

541-15-1
$C_7H_{15}NO_3$
161.23
Ammonium, (3-carboxy-2-hydroxypropyl)trimethyl-, hydroxide, inner salt, l
(3-Carboxy-2-hydroxypropyl)trimethyl-ammonium hydroxide, inner salt
3-Carboxy-2-hydroxy-N,N,N-trimethyl-1-propanaminium hydroxide, inner salt
Carnitine
(-)-Carnitine
l-Carnitine
(R)-Carnitine
Karnitin
1-Propanaminium, 3-carboxy-2-hydroxy-N,N,N-trimethyl-, hydroxide, inner salt, (R)-(9CI)
γ-Trimethyl-ammonium-β-hydroxybutirate
γ-Trimethyl-β-hydroxybutyrobetaine
Vitamin BT

541-23-1
$C_5H_{13}N.ClH$
123.65
Isopentylamine, hydrochloride
Isoamylamine hydrochloride

541-25-3
$C_2H_2AsCl_3$
207.31
ClC=C[As](Cl)Cl
Arsine, dichloro(2-chlorovinyl)
Arsine, (2-chlorovinyl)dichloro-
Arsonous dichloride, (2-chloroethenyl)- (9CI)
Chlorovinylarsine dichloride
β-Chlorovinylbichloroarsine
2-Chlorovinyldichloroarsine
Dichloro(2-chlorovinyl)arsine
Lewisite
Lewisite (Arsenic compound)
M-1

541-31-1
$C_5H_{12}S$
104.22
SCCC(C)C
1-Butanethiol, 3-methyl- (9CI)
3-Methyl-1-butanethiol

541-33-3
$C_4H_8Cl_2$
127.02
C(Cl)(Cl)CCC
Butane, 1,1-dichloro
Butylidene chloride
1,1-Dichlorobutane

541-35-5
C_4H_9NO
87.12
O=C(N)CCC
Butanamide (9CI)
AI3-24199
Butanimidic acid
Butanoic acid, amide
n-Butylamide
Butyramide (8CI)
n-Butyramide
NSC-8424

541-41-3
$C_3H_5ClO_2$
108.53
O=C(OCC)Cl
Formic acid, chloro-, ethyl ester
Chlorocarbonate D'ethyle (French)
Chloroformic acid ethyl ester
Clhorameisensaeureaethylester (German)
ECF
Ethylchloorformiaat (Dutch)
Ethyl chlorocarbonate (DOT)
Ethyl chloroformate [UN 1182]
Ethyle, chloroformiat d'

(French)
Ethylester kyseliny chlormravenci (Czech)
Etil clorocarbonato (Italian)
Etil cloroformiato (Italian)
TL 423
UN 1182 [Ethyl chloroformate]

541-42-4
$C_3H_7NO_2$
89.11
O=NOC(C)C
Nitrous acid, isopropyl ester
Isopropylester kyseliny dusite (Czech)
Isopropyl nitrite
Nitrous acid, 1-methylethyl ester (9CI)
2-Propanol nitrite

541-46-8
$C_5H_{11}NO$
101.14
O=C(N)CC(C)C
Isovaleramide
Butanamide, 3-methyl- (9CI)
Isovaleric amide
3-Methylbutanamide
β-Methylbutyramide
NSC-402555

541-47-9
$C_5H_8O_2$
100.13
O=C(O)C=C(C)C
Crotonic acid, 3-methyl
2-Butenoic acid, 3-methyl- (9CI)
β,β-Dimethacrylic acid
β,β-Dimethylacrylic acid
3,3-Dimethylacrylic acid
Kyselina 3-methyl-2-butenova (Czech)
3-Methyl-2-butenoic acid
β-Methylcrotonic acid
3-Methylcrotonic acid
Senecioic acid

541-48-0
$C_4H_9NO_2$
103.11
O=C(O)CC(N)C
3-Aminobutyric acid
AI3-26821
3-Aminobutanoic acid
Butanoic acid, 3-amino-

541-50-4
$C_4H_6O_3$
102.09
CC(=O)CC(O)=O
Acetoacetic acid
Acetoacetate
3-Ketobutyrate
Oxobutyrate
3-Oxobutyric acid

541-53-7
$C_2H_5N_3S_2$
135.22
N=C(S)NC(=N)S
Biuret, 2,4-dithio
Dithiobiuret
DTB
RCRA waste number P049
Urea, 2-thio-1-(thiocarbamoyl)-
USAF B-44
USAF EK-P-6281

541-58-2
C_5H_7NS
113.19
N(C(=CS1)C)=C1C
Thiazole, 2,4-dimethyl
2,4-Dimethylthiazole

541-69-5
$C_6H_8N_2.2ClH$
181.08
Cl.Cl.Nc1cccc(N)c1
m-Phenylenediamine, dihydrochloride
m-Aminoaniline dihydrochloride
3-Aminoaniline dihydrochloride
1,3-Benzenediamine hydrochloride
m-Benzenediamine dihydrochloride
m-Diaminobenzene dihydrochloride
1,3-Diaminobenzene dihydrochloride
m-Phenylenediamine hydrochloride
1,3-Phenylenediamine dihydrochloride
USAF EK-206

541-70-8
$C_6H_8N_2.H_2O_4S$
206.21
m-Phenylenediamine sulfate
1,3-Benzenediamine, sulfate (1:1) (9CI)
m-Phenylenediamine, sulfate (1:1)
1,3-Phenylenediamine sulfate

541-73-1
$C_6H_4Cl_2$
147.00
c(cccc1Cl)(c1)Cl
Benzene, m-dichloro
Benzene, 1,3-dichloro- (9CI)
m-Dichlorobenzene
1,3-Dichlorobenzene
m-Dichlorobenzol
m-Phenylenedichloride
RCRA waste number U071

541-85-5
$C_8H_{16}O$
128.24
CCC(C)CC(=O)CC
3-Heptanone, 5-methyl
Ethyl sec-amyl ketone
Ethyl amyl ketone (ACGIH)
3-Methyl-5-heptanone
5-Methyl-3-heptanone

541-88-8
$C_4H_4Cl_2O_3$
170.98
O=C(OC(=O)CCl)CCl
Chloracetic anhydride
Acetic acid, chloro-, anhydride (9CI)
AI3-52321
Chloroacetic acid anhydride
Chloroacetic anhydride
2-Chloroacetic anhydride
Chloroacetyl anhydride
Chloroethanoic anhydride
α,α'-Dichloroacetic anhydride
sym-Dichloroacetic anhydride
Monochloroacetic acid anhydride
NSC-71207

541-91-3
$C_{16}H_{30}O$
238.46
O=C(CCCCCCCCCCCCC1C)C1
Cyclopentadecanone, 3-methyl
3-Methylcyclopentadecanone
3-Methyl-1-cyclopentadecanone
Moschus ketone
Muscone
Muskone

542-02-9
$C_4H_7N_5$
125.10
n(c(nc(n1)C)N)c1N
s-Triazine, 2,4-diamino-6-methyl- (8CI)
Acetoguanamine
AI3-50715
2,4-Diamino-6-methyl-1,3,5-triazine
ENT 50,715
2-Methyl-4,6-diamino-1,3,5-triazine
6-Methyl-1,3,5-triazine-2,4-diamine
NSC-257
1,3,5-Triazine-2,4-diamine, 6-methyl- (9CI)

542-10-9
$C_6H_{10}O_4$
146.14
O=C(OC(OC(=O)C)C)C
1,1-Ethanediol, diacetate (9CI)
AI3-24218
1,1-Diacetoxyethane
Ethylidene acetate
Ethylidene diacetate
NSC-8852

542-11-0
$C_6H_7N.BrH$
174.03
Benzenamine, hydrobromide (9CI)

542-16-5
$C_6H_7N.1/2H_2O_4S$
142.16
Benzenamine, sulfate (2:1) (9CI)
Aniline sulfate

542-18-7
$C_6H_{11}Cl$
118.62
C(CCCC1)(C1)Cl
Cyclohexane, chloro
Chlorocyclohexane
Cyclohexyl chloride
Monochlorocyclohexane

542-28-9
$C_5H_8O_2$
100.12
O=C1CCCCO1
δ-Valerolactone
AI3-25024
NSC-6247
Pentanoic acid, 5-hydroxy-, δ-lactone
2H-Pyran-2-one, tetrahydro- (9CI)
Tetrahydro-2-pyranone
Valeric acid, δ-hydroxy-, δ-lactone
5-Valerolactone
γ-Valerolactone
δ-Valeryllactone

542-42-7
$C_{16}H_{32}O_2.1/2Ca$
276.74
Hexadecanoic acid, calcium salt (9CI)
Calcium hexadecanoate

542-50-7
$C_{27}H_{54}O$
394.73
14-Heptacosanone (8CI,9CI)
Myristone
NSC-15184

542-54-1
$C_6H_{11}N$
97.18
N#CCCC(C)C
Valeronitrile, 4-methyl
Isoamyl cyanide
Isoamylkyanid (Czech)
Isocapronitrile
Isopentyl cyanide
4-Methylpentanenitrile
4-Methylvaleronitrile

542-55-2
$C_5H_{10}O_2$
102.15
O=COCC(C)C
Formic acid, isobutyl ester
Isobutylester kyseliny mravenci (Czech)
Isobutyl formate [UN 2393]
Iso-butyl formate
Tetryl formate
UN 2393 [Isobutyl formate]

542-56-3
$C_4H_9NO_2$
103.14
O=NOCC(C)C
Nitrous acid, isobutyl ester
IBN
Isobutyl nitrite
NCI-C61052
Nitrous acid, 2-methylpropyl ester (9CI)

542-58-5
$C_4H_7ClO_2$
122.56
O=C(OCCCl)C
Ethanol, 2-chloro-, acetate
Acetoxyethyl chloride
2-Chlorethylacetat (German)
2-Chloroethanol acetate
β-Chloroethyl acetate
2-Chloroethyl acetate

542-59-6
$C_4H_8O_3$
104.12
O=C(OCCO)C
Ethylene glycol, monoacetate
Acetic acid 2-hydroxyethyl ester
1,2-Ethanediol, monoacetate
Ethylene glycol acetate
Glycol monoacetate
Glycol-monoacetin
2-Hydroxyethyl acetate
2-Hydroxyethylester kyseliny octove (Czech)

542-62-1
C_2BaN_2
189.38
Barium-cyanide
Barium cyanide [UN 1565]
Barium cyanide, Solid (DOT)
Barium dicyanide
RCRA waste number P013
UN 1565 [Barium cyanide]

542-69-8
C_4H_9I
184.03
C(CCC)I
Butane, 1-iodo
Butyl iodide
n-Butyl iodide
1-Iodobutane
1-Jodbutan (Czech)

542-75-6
$C_3H_4Cl_2$
110.97
C(=CCl)CCl
Propene, 1,3-dichloro
α-Chloroallyl chloride
γ-Chloroallyl chloride
3-Chloroallyl chloride
3-Chloropropenyl chloride
DCP
D-D92
Dichloropropene
1,3-Dichloropropene (ACGIH, OSHA)
1,3-Dichloropropene-1
1,3-Dichloro-2-propene
α,γ-Dichloropropylene
1,3-Dichloropropylene
Dorlone II
NCI-C03985
RCRA waste number U084
Telone
Telone II
Telone II Soil Fumigant
Vidden D

542-76-7
C_3H_4ClN
89.53
N#CCCCl
Propionitrile, 3-chloro
3-Chloropropanenitrile
3-Chloropropanonitrile
β-Chloropropionitrile
3-Chloropropionitrile
3-Chlorpropannitril (Czech)
RCRA waste number P027
USAF A-8798

542-78-9
$C_3H_4O_2$
72.07
O=CCC=O
Propanedial
Malonaldehyde
Malondialdehyde
Malonic aldehyde
Malonic dialdehyde
Malonodialdehyde
Malonyldialdehyde
NCI-C54842
1,3-Propanedial
1,3-Propanedialdehyde
1,3-Propanedione

542-81-4
C_3H_7ClS
110.61
Ethane, 1-chloro-2-(methylthio)- (9CI)
2-Chloroethyl methyl sulfide
1-Chloro-2-(methylthio)ethane
Hemisulfur mustard
2-Methylthioethyl chloride
NSC-91724

Sulfide, 2-chloroethyl methyl (8CI)

542-85-8
C_3H_5NS
87.15
N(=C=S)CC
Isothiocyanic acid, ethyl ester
Ethane, isothiocyanato- (9CI)
Ethyl isothiocyanate
Ethyl Mustard Oil
Isothiocyanatoethane

542-88-1
$C_2H_4Cl_2O$
114.96
O(CCl)CCl
Ether, bis(chloromethyl)
BCME
Bis(chloromethyl) ether (ACGIH,OSHA)
Bis-CME
Chloro(chloromethoxy)methane
Chloromethyl ether
Dichlordimethylaether (German)
sym-Dichloro-dimethyl ether
1,1'-Dichlorodimethyl ether
Dichlorodimethyl ether, symmetrical [UN 2249]
sym-Dichloromethyl ether
Dimethyl-1,1'-dichloroether
Methane, oxybis(chloro-
Oxybis(chloromethane)
RCRA waste number P016
UN 2249 [Dichlorodimethyl ether, symmetrical]

542-90-5
C_3H_5NS
87.14
N#CSCC
Ethylthiocyanate
Aethylrhodanid (German)
AI3-18429
Ethane, thiocyanato-
Ethyl rhodanate
Ethyl sulfocyanate
Ethyl thiocyanate
Ethylthiokyanat (Czech)
NSC-8827
Thiocyanic acid, ethyl ester (8CI,9CI)

542-92-7
C_5H_6
66.11
C(=CC=C1)C1
1,3-Cyclopentadiene
Cyclopentadiene (ACGIH, OSHA)
Pentole
Pyropentylene
R-Pentine

543-20-4
$C_4H_4Cl_2O_2$
154.98
O=C(CCC(=O)Cl)Cl
Succinyl-chloride
Succinic acid dichloride
Succinic chloride
Succinoyl chloride
Succinyl dichloride

543-24-8
$C_4H_7NO_3$
117.10
O=C(NCC(=O)O)C
Aceturic acid
Acetamidoacetic acid
Acetylaminoacetic acid
Acetylglycine
N-Acetylglycine
Acetylglycocoll
AI3-17738
Ethanoylaminoethanoic acid
Glycine, N-acetyl- (9CI)
NSC-7605

543-27-1
$C_5H_9ClO_2$
136.58
O=C(OCC(C)C)Cl
Isobutyl chloroformate
Carbonochloridic acid, 2-methylpropyl ester (9CI)
Chlorocarbonic acid isobutyl ester
Formic acid, chloro-, isobutyl ester (8CI)
Isobutyl chlorocarbonate
2-Methylpropyl carbonochloridate
2-Methylpropyl chloroformate
NSC-8429

543-28-2
$C_5H_{11}NO_2$
117.14
O=C(OCC(C)C)N
Carbamic acid, isobutyl ester (8CI)
Carbamic acid, 2-methylpropyl ester (9CI)
Isobutyl carbamate
2-Methylpropyl carbamate
NSC-31187

543-29-3
$C_4H_9NO_3$
119.11
Nitric acid, 2-methylpropyl ester (9CI)
Isobutyl nitrate
Nitric acid, isobutyl ester (8CI)

543-38-4
$C_5H_{12}N_4O_3$
176.21
O=C(O)C(N)CCONC(=N)N
Butyric acid, 2-amino-4-(guanidinooxy)-, L
Canavanin
Canavanine
L-Canavanine
L-Homoserine, O-((aminoiminomethyl)amino)- (9CI)

543-39-5
$C_{10}H_{18}O$
154.28
OC(CCCC(C=C)=C)(C)C
7-Octen-2-ol, 2-methyl-6-methylene
3-Methylene-7-methyl-1-octen-7-ol
Myrcenol

543-49-7
$C_7H_{16}O$
116.23
OC(CCCCC)C
2-Heptanol
Amyl methyl carbinol
Heptanol-2
2-Hydroxyheptane
Methyl amyl carbinol

543-59-9
$C_5H_{11}Cl$
106.61
ClCCCCC
Pentane, 1-chloro
Amyl chloride [UN 1107]
n-Amyl chloride
1-Chloropentane
Pentyl chloride
UN 1107 [Amyl chlorides]

543-67-9
$C_3H_7NO_2$
89.11
Nitrous acid, propyl ester
Nitrous acid, n-propyl ester
Propanol nitrite
Propyl nitrite
n-Propyl-nitrite

543-80-6
$C_4H_6O_4.Ba$
255.44
[Ba+2].CC([O-])=O.CC([O-])=O
Acetic acid, barium salt
Barium acetate
Barium diacetate
Octan barnaty (Czech)

543-81-7
$C_4H_6O_4.Be$
127.11
Acetic acid, beryllium salt
Beryllium acetate
Beryllium acetate normal

543-86-2
$C_6H_{13}NO_2$
131.20

Carbamic acid, isopentyl ester
1-Butanol, 3-methyl-, carbamate
Carbamate (isoamyl)
Isoamyl aminoformate
Isoamyl carbamate

543-90-8
$C_2H_4O_2$.1/2Cd
116.25
[Cd+2].CC([O-])=O.CC([O-])=O
Cadmium(II) acetate
Acetic acid, cadmium salt
Bis(acetoxy)cadmium
Cadmium acetate
Cadmium diacetate
C.I. 77185

543-94-2
$C_4H_6O_4$.Sr
205.72
Acetic acid, strontium salt
Strontium acetate

544-00-3
$C_{10}H_{23}N$
157.29
N(CCC(C)C)CCC(C)C
Diisopentylamine (8CI)
AI3-35092
1-Butanamine, 3-methyl-N-(3-methylbutyl)- (9CI)
Diisoamylamine
3-Methyl-N-(3-methylbutyl)-1-butanamine
NSC-6261

544-01-4
$C_{10}H_{22}O$
158.28
O(CCC(C)C)CCC(C)C
Di-iso-amyl ether
AI3-02268
Butane, 1,1'-oxybis(3-methyl- (9CI)
Diisoamyl ether
Diisopentyl ether
Isoamyl ether
Isoamyl oxide
Isopentyl ether (8CI)
NSC-9281
1,1'-Oxybis(3-methylbutane)

544-02-5
$C_{10}H_{22}S$
174.35
S(CCC(C)C)CCC(C)C
Butane, 1,1'-thiobis(3-methyl-
AI3-18864
1,1'-Thiobis(3-methylbutane)

544-10-5
$C_6H_{13}Cl$
120.62
ClCCCCCC
1-Chlorohexane
AI3-28589
Chlorhexane
Hexane, 1-chloro-

544-12-7
$C_6H_{12}O$
100.16
OCCC=CCC
3-Hexen-1-ol
AI3-25091

544-13-8
$C_5H_6N_2$
94.13
N#CCCCC#N
1,3-Trimethylenedinitrile
1,3-Dicyanopropane
Glutaric acid dinitrile
Glutarodinitrile
Glutaronitrile (8CI)
Pentanedinitrile (9CI)
Pyrotartaric acid nitrile

544-16-1
$C_4H_9NO_2$
103.14
O=NOCCCC
Nitrous acid, butyl ester
Butyl nitrite [UN 2351]
n-Butyl nitrite
NBN
NCI-C56553
Nitrous acid, n-butyl ester
UN 2351 [Butyl nitrites]

544-17-2
$C_2H_2O_4$.Ca
130.12
[Ca+2].[O-]C=O.[O-]C=O
Formic acid, calcium salt
Calcium formate
Mravencan vapenaty (Czech)

544-18-3
CH_2O_2.1/2Co
75.49
Cobaltous formate
Cobalt diformate
Cobalt formate (VAN)
Cobalt(2+) formate
Formic acid, cobalt(2+) salt (9CI)
NSC-112231

544-19-4
CH_2O_2.1/2Cu
77.80
Cupric formate
Copper formate (VAN)
Copper(2+) formate
Cupric diformate
Formic acid, copper(2+) salt (9CI)
Formic acid, copper(2+) salt (1:1)
NSC-112232
Tubercuprose

544-25-2
C_7H_8
92.15
C1C=CC=CC=C1
1,3,5-Cycloheptatriene
Cycloheptatriene [UN 2603]
Tropilidene
Tropilidin
UN 2603 [Cycloheptatriene]

544-40-1
$C_8H_{18}S$
146.32
S(CCCC)CCCC
Butyl-sulfide
Butyl monosulfide
n-Butyl-sulfide
Butylthiobutane
n-Dibutyl sulfide
Di-n-butylsulfide
Dibutyl sulphide
Dibutyl thioether
5-Thianonane
Thianonane-5

544-60-5
$C_{18}H_{34}O_2.H_3N$
299.49
Ammonium oleate
AI3-36578
Ammonium 9-octadecenoate
Ammonium oleate
Caswell No. 046C
9-Octadecenoic acid, ammonium salt
9-Octadecenoic acid (Z)-, ammonium salt (9CI)

544-63-8
$C_{14}H_{28}O_2$
228.42
O=C(O)CCCCCCCCCCCCC
Myristic-acid
Crodacid
Emery 655
Hydrofol Acid 1495
Hystrene 9014
n-Tetradecoic acid
Neo-Fat 14
Tetradecanoic acid
n-Tetradecanoic acid
1-Tridecanecarboxylic acid
Univol U 316S

544-76-3
$C_{16}H_{34}$
226.50
C(CCCCCCCCCCCCCC)C
Hexadecane
Cetane
n-Cetane
n-Hexadecane

544-77-4
$C_{16}H_{33}I$
352.34

C(CCCCCCCCCCCCCCCC)I
Hexadecane, 1-iodo- (9CI)
1-Iodohexadecane

544-85-4
$C_{32}H_{66}$
450.98
C(CCCCCCCCCCCCCCCCCCCC CCCCCCCCCCC)C
Dotriacontane

544-92-3
CCuN
89.56
Copper cyanide [UN 1587]
Copper(I) cyanide
Cupricin
Cuprous cyanide
RCRA waste number P029
UN 1587 [Copper cyanide]

544-97-8
C_2H_6Zn
95.45
Zinc, dimethyl
Dimethyl zinc
Dimethylzinc [UN 1370]
UN 1370 [Dimethylzinc]

545-06-2
C_2Cl_3N
144.38
N#CC(Cl)(Cl)Cl
Acetonitrile, trichloro
Cyanotrichloromethane
Nitrile trichloracetique (French)
Trichlor-acetonitril (German)
Trichlormethylkyanid (Czech)
Trichloroacetonitrile
Trichloromethyl cyanide
Trichloromethylnitrile
Trichlouracetonitril (Dutch)
Tritox

545-55-1
$C_6H_{12}N_3OP$
173.18
O=P(N1CC1)(N2CC2)N3CC3
Phosphine oxide, tris-(1-aziridinyl)
A 6366
Aphoxide
APO
Aziridine, 1,1',1''-phosphinylidynetris-
1-Aziridinyl phosphine oxide (tris) (DOT)
Cbc 906288
ENT 24,915
Imperon Fixer T
NSC-9717
1,1',1''-Phosphinylidynetris-aziridine
Phosphoramide, N,N',N''-tri-ethylene-
Phosphoric acid triethylene imide
Phosphoric acid triethylene-imine (DOT)
Phosphoric triamide, N,N',N''-tri-1,2-ethanediyl-
Phosphoric triamide, N,N',N''-triethylene-
SK-3818
TAPO
TEF
TEPA
Triaethylenphosphorsaeureamid (German)
Triaziridinophosphine oxide
Triaziridinylphosphine oxide
Tri(aziridinyl)phosphine oxide
Tri-1-aziridinylphosphine oxide
Tri(1-aziridinyl)phosphine oxide
N,N',N''-Tri-1,2-ethanediyl-phosphoric triamide
Triethylenephosphoramide
N,N',N''-Triethylenephos-phoramide
Triethylenephosphoric triamide
N,N',N''-Triethylenephosphoric triamide
Triethylenephosphorotriamide
Triethylenfosforamid (Czech)
Tris(1-aziridine)phosphine oxide
Tris-(1-aziridinyl)fosfinoxid (Czech)
Tris(aziridinyl)phosphine oxide
Tris(1-aziridinyl)phosphine oxide
Tris-(1-aziridinyl)phosphine oxide (DOT)
Tris-(1-aziridinyl)phosphine oxide, Solution [UN 2501]
Tris(N-ethylene)phosphorotri-amidate
UN 2501 [Tris-(1-aziridinyl)-phosphine oxide, solution]

545-59-5
C25H32N2O2
392.54
Racemoramide
DEA No. 9645
Pyrrolidine, 1-(3-methyl-4-(4-morpholinyl)-1-oxo-2,2-di-phenylbutyl)-, (+-)- (9CI)
R 610
Racemoramida (Spanish)
Racemoramidum (Latin)

545-90-4
$C_{21}H_{29}NO$
311.47
Dimepheptanol
Amidol
Benzeneethanol, β-(2-(dimethyl-amino)propyl)-α-ethyl-β-phenyl- (9CI)
Bimethadol
Bimethadolum
DEA No. 9618
Dimefeptanol (Spanish)
Dimefeptanolo
Dimepheptanolum (Latin)
3-Heptanol, 6-(dimethylamino)-4,4-diphenyl- (8CI)
Methadol
NIH 2933
Racemethadol

546-46-3
$C_6H_5O_7.3/2Zn$
290.18
Citric acid, zinc salt (2:3)
1,2,3-Propanetricarboxylic acid, 2-hydroxy-, zinc salt (2:3) (9CI)
Zinc citrate

546-56-5
$C_{48}H_{40}O_4Si_4$
793.19
O1Si(OSi(OSi(OSi1(c2ccccc2) c3ccccc3)(c4ccccc4)c5ccccc5) (c6ccccc6)c7ccccc7)(c8ccc cc8)c9ccccc9
Cyclotetrasiloxane, octa-phenyl- (9CI)
NSC-293057
Octaphenylcyclotetrasiloxane
1,1,3,3,5,5,7,7-Octaphenylcyclo-tetrasiloxane
Octaphenyltetracyclosiloxane

546-67-8
$C_8H_{16}O_8.Pb$
447.43
Acetic acid, lead(4+) salt
Lead tetraacetate

546-68-9
$C_3H_8O.1/4Ti$
72.07
CCC[O-].CCC[O-].CCCO[Ti] OCCC
Isopropyl alcohol, titan-ium(4+) salt
Isopropyl orthotitanate
Isopropyl titanate(IV)
Tetraisopropoxide titanium
Tetraisopropoxytitanium
Tetraisopropyl orthotitanate
Tetraisopropyl titanate
Tetrakis(isopropoxy)titanium
Titanium(4+) isopropoxide
Titanium isopropylate
Titanium tetraisopropoxide
Titanium tetraisopropylate
Titanium tetra-n-propoxide
Tyzor TPt

546-80-5
$C_{10}H_{16}O$
152.26
O=C(C(C(C12C(C)C)C1)C)C2
3-Thujanone, (1S,4R,5R)-(-)
Bicyclo(3.1.0)hexan-3-one, 4-methyl-1-(1-methylethyl)-, (1S-1-α,4-α,5-α)- (9CI)
(-)-3-Isothujone
3-Thujanone, (-)-
Thujon
Thujone

(-)-Thujone
α-Thujone
l-Thujone

546-88-3
$C_2H_5NO_2$
75.06
Acetohydroxamic acid
Acetamide, N-hydroxy- (9CI)
Acethydroxamsaure (German)
Acetic acid, oxime
Acetohydroximic acid
Acetylhydroxamic acid
AHA
AI3-62232
Hydroxylamine, N-acetyl-
Lithostat
Methylhydroxamic acid
NSC-176136

546-89-4
$C_2H_4O_2.Li$
67.00
Acetic acid, lithium salt
Lithium acetate
Quilone

546-93-0
$CO_3.Mg$
84.32
Magnesium(II) carbonate (1:1)
Carbonate magnesium
Carbonic acid, magnesium salt
C.I. 77713
DCI Light Magnesium carbonate
Hydromagnesite
Magmaster
Magnesite (OSHA)
Stan-Mag Magnesium carbonate

546-99-6
$C_{30}H_{50}$
410.73
A'-Neogammacer-17(21)-ene (8CI,9CI)

547-57-9
$C_{12}H_{10}N_2O_5S.Na$
317.27
Benzenesulfonic acid, 4-((2,4-dihydroxyphenyl)azo)-, monosodium salt (9CI)
Acid Leather Yellow PGW
Acid Phosphine G New
Acid Yellow
Acme Yellow Acid Yellow RS
Benzenesulfonic acid, p-((2,4-dihydroxyphenyl)azo)-, sodium salt
C.I. Acid Orange 6
C.I. Acid Orange 6, Monosodium salt (8CI)
C.I. Food Yellow 8
C.I. 14270
Cetil Chromine Yellow GR
Chrysoin G
Chrysoin S
Chrysoin S Specially Pure
Chrysoine
Chrysoine Extra
Chrysoine Extra Pure A
Chrysoine N
Chrysoine S
Chrysoine S Extra Pure
Chrysonine S
Chrysonis S
Curol Orange G
Dermina Yellow G
2,4-Dihydroxyazobenzene-4'-sulfonate sodium salt
p-(2,4-Dihydroxyphenylazo)-benzenesulfonic acid, sodium salt
p-((2,4-Dihydroxyphenyl)azo)-benzenesulfonic acid, sodium salt
Eniacid Yellow RS
Eurocert Chrysoine S
Gold Yellow
Hispacid Yellow CG
Naphthazine Yellow RP
Neklacid Yellow G
NSC-10441
Orange Acid G
Resorcin Yellow
Resorcine Yellow
Resorcine Yellow O Extra
Resorcinol yellow
Resorcinol Yellow A
Sodium azoresorcinolsulfanilate
Tertracid Yellow TRO
Tropaelin-O
Tropaeolene (Biological stain)
Tropaeolin O
Tropaeolin R
Tropaeoline
Tropaeoline O
Tropeolin O
Yellow T

547-58-0
$C_{14}H_{14}N_3O_3S.Na$
327.36
[Na+].CN(C)c2ccc(N=Nc1ccc(cc1)S([O-])(=O)=O)cc2
Benzenesulfonic acid, p-((p-(dimethylamino)phenyl)-azo)-, sodium salt
C.I. 13025
C.I. Acid Orange 52
Dexon
Diazoben
4-Dimethylaminoazobenzene-4'-sulphonic acid sodium salt
p-((p-(Dimethylamino)phenyl)-azo)benzenesulfonic acid sodium salt
Eniamethyl Orange
Gold Orange
Helianthine
Helianthine B
KCA Methyl Orange
Methyl Orange
Methyl Orange B
Methyloranz (Czech)
Orange 3
Orange III
Oranz III (Czech)
Oranz Kysela 52 (Czech)
Oranz Methylova (Czech)
Tropaeolin

547-60-4
$C_{10}H_{16}O$
152.24
O=C(C(C(CC12)C2(C)C)C)C1
Pinocamphone
Bicyclo(3.1.1)heptan-3-one, 2,6,6-trimethyl-, (1α,2α,5α)-(9CI)
(1 α,2 α,5 α)-2,6,6-Trimethyl-bicyclo(3.1.1)heptan-3-one

547-63-7
$C_5H_{10}O_2$
102.15
O=C(OC)C(C)C
Isobutyric acid, methyl ester
Methylester kyseliny isomaselne (Czech)
Methyl isobutyrate

547-64-8
$C_4H_8O_3$
104.11
O=C(OC)C(O)C
Methyl lactate
AI3-00584
2-Hydroxypropanoic acid methyl ester
Lactic acid, methyl ester (8CI)
Methyl 2-hydroxypropanoate
Methyl 2-hydroxypropionate
NSC-406248
Propanoic acid, 2-hydroxy-, methyl ester (9CI)

548-26-5
$C_{20}H_8Br_4O_5.2Na$
693.90
OC(=O)c1ccccc1c3c2cc(Br)c(O)c(Br)c2oc4c(Br)c(=O)c(Br)cc34
Benzoic acid, 2-(2,4,5,7-tetra-bromo-6-hydroxy-3-oxo-3H-xanthen-9-yl)-, disodium salt
Aizen Eosine GH
Bromo acid
Bromo B
Bromo 4D
Bromo 4DC
Bromo 4DL
Bromoeosine
Bromo FL
Bromofluoresceic acid
Bromo fluorescein
Bromo JPS
Bromo TS
Bromo X-100
Bromo XX
Bronze Bromo
Certiqual Eosine
C.I. 45380
C.I. Acid Red 87

Eosin
Eosine B
Eosine BPC
Eosine BS
Eosine BS-SF
Eosine DA
Eosine DWC 73
Eosine Extra conc. A. export
Eosine FA
Eosine 3G
Eosine GF
Eosine J
Eosine K Salt Free
Eosine Lake Red Y
Eosine Salt Free
Eosine W/S
Eosine YB
Eosine YS
Eosin YS
Fenazo Eosine XG
Fluorescein, 2',4',5',7'-tetrabromo-, disodium salt
Hidacid Boiling Bromo
Hidacid Bromo Acid Regular
Hidacid Dibromo Fluorescein
Hidacid Eosine Soda Salt
Hidacid White Bromo
Irgalite Bronze Red CL
Phloxine Red 20-7600

548-39-0
$C_{13}H_8O$
180.21
Phenalen-1-one (8CI)
NSC-150161
Perinaphthenone
7-Perinaphthenone
Phenalenone
1H-Phenalen-1-one (9CI)

548-62-9
$C_{25}H_{30}N_3.Cl$
408.03
[Cl-].CN(C)c1ccc(cc1)C(=C2C=C C(C=C2)=[N+](C)C)c3ccc (cc3)N(C)C
Ammonium, (4-(bis(p-(dimethylamino)phenyl)methylene)-2,5-cyclohexadien-1-ylidene)dimethyl-, chloride
Adergon
Aizen Crystal Violet
Aizen Crystal Violet Extra Pure
Aniline Violet
Aniline Violet Pyoktanine
Atmonil
Avermin
Axuris
Badil
Basic Violet 3
Basic Violet BN
Bismuth Violet
Blaues Pyoktanin (German)
Brilliant Violet 5B
Calcozine Violet C
Calcozine Violet 6BN
C.I. 42555
C.I. Basic Violet 3
Crystal Violet
Crystal Violet O
Crystal Violet 5BO
Crystal Violet 6B
Crystal Violet 6BO
Crystal Violet 10B
Crystal Violet AO
Crystal Violet AON
Crystal Violet Base
Crystal Violet BP
Crystal Violet BPC
Crystal Violet Chloride
Crystal Violet Extra Pure
Crystal Violet Extra Pure APN
Crystal Violet Extra Pure APNX
Crystal Violet FN
Crystal Violet HL2
Crystal Violet Pure DSC
Crystal Violet Pure DSC Brilliant
Crystal Violet SS
Crystal Violet Technical
Crystal Violet USP
Gentersal
Gentian Violet
Gentianaviolett (German)
Gentiaverm
Genticid
Gentioletten
Hectograph Violet SR
Hecto Violet R
Hexamethylparaosaniline chloride
Hexamethyl-p-rosaniline chloride
Hexamethyl p-rosaniline hydrochloride
Hexamethyl Violet
Hidaco Brilliant Crystal Violet
Hidaco Crystal Violet
Kristall-Violett (German)
Meroxyl
Meroxylan
Meroxylan-Wander
Meroxyl-Wander
Methylrosanilinchlorid (German)
Methylrosaniline chloride
Methylrosanilinum chloratum
Methyl Violet 5BNO
Methyl Violet 5BO
Methyl Violet 10B
Methyl Violet 10BD
Methyl Violet 10BK
Methyl Violet 10BN
Methyl Violet 10BNS
Methyl Violet 10BO
Methylviolett (German)
Mitsui Crystal Violet
NCI-C55969
Oxiuran
Oxycolor
Oxyozyl
Paper Blue R
Pararosaniline, N,N,N',N',N'',N''-hexamethyl-, chloride
Plastoresin Violet 5BO
Pyoktanin
Pyoverm
Vermicid
Vianin
Viocid
12416 Violet
Violet 6BN
Violet 5BO
Violet CP
Violet Gencianova (Czech)
Violet Krystalova (Czech)
Violet XXIII
Violet Zasadita 3 (Czech)

548-73-2
$C_{22}H_{22}FN_3O_2$
379.47
Fc1ccc(cc1)C(=O)CCCN2CCC (=CC2)n3c(=O)[nH]c4ccccc34
2-Benzimidazolinone, 1-(1-(3-(p-fluorobenzoyl)propyl)-1,2,3,6-tetrahydro-4-pyridyl)
Dehidrobenzperidol
Dehydrobenzperidol
Deidrobenzperidolo
DHBP
Dihidrobenzperidol
Dridol
Droleptan
Droperidol
1-(1-(3-(p-Fluorobenzoyl)-propyl)-1,2,3,6-tetrahydro-4-pyridyl)-2-benzimidazolinone
1-(1-(4-(p-Fluorophenyl)-4-oxo-butyl)-1,2,3,6-tetrahydro-4-pyridyl)-2-benzimidazolinone
Halkan
Inappin
Inapsin
Inapsine
Innovan
Innovar
Innovar-Vet
Inopsin
Inoval
Leptanal
Leptofen
MCN-JR-4749
Properidol
R 4749
Sintosian
Thalamonal
Vetkalm

548-93-6
$C_7H_7NO_3$
153.15
Nc1c(O)cccc1C(O)=O
Benzoic acid, 2-amino-3-hydroxy
2-Amino-3-hydroxybenzoic acid
Anthranilic acid, 3-hydroxy-
3-Hydroxyanthranilic acid
3-Hydroxy-anthranilsaeure (German)
3-OHAA
3-Oxyanthranilic acid

549-49-5
$C_{20}H_{24}N_2O_2.BrH$
405.38
Quinine, monohydrobromide

Bromoquinine
Chinin hydrobromid (German)
Quinine hydrobromide

549-56-4
$C_{20}H_{24}N_2O_2.H_2O_4S$
422.49
Cinchonan-9-ol, 6'-methoxy-, (8α,9R)-, sulfate (1:1) (Salt) (9CI)
8α,9R-6'-Methoxycinchonan-9-ol, sulfate (1:1) salt
Quinine bisulfate

550-44-7
$C_9H_7NO_2$
161.15
O=C(N(C(=O)c1cccc2)C)c12
Phthalimide, N-methyl- (8CI)
AI3-01393
1H-Isoindole-1,3(2H)-dione, 2-methyl- (9CI)
2-Methyl-1H-isoindole-1,3(2H)-dione
N-Methylphthalimide
NSC-44059

550-82-3
$C_{12}H_7NO_4$
229.20
O=C(C=CC(=N(=O)c(c(O1)cc(O)c2)c2)C1=3)C3
3H-Phenoxazin-3-one, 7-hydroxy-, 10-oxide
Azoresorcin
Diazoresorcinol
7-Hydroxy-3H-fenoxazin-3-on-10-oxid (Czech)
7-Hydroxy-3H-phenoxazin-3-one 10-oxide
Resazoin
Resazurin
Resazurine

550-99-2
$C_{14}H_{14}N_2.ClH$
246.76
2-Imidazoline, 2-(1-naphthylmethyl)-, monohydrochloride
Albalon Liquifilm
Clera
Coldan
1H-Imidazole, 4,5-dihydro-2-(1-naphthalenylmethyl)-, monohydrochloride
Naphazoline hydrochloride
Naphcon
Naphcon Forte
2-(1-Naphthylmethyl)imidazoline hydrochloride
2-(1-Naphthylmethyl)-2-imidazoline hydrochloride
Niazol
Privine hydrochloride
Prizole hydrochloride
Rhinantin
Rhinoperd
Sanorin
Sanorin-Spofa
Stricylon
Vasocon

551-06-4
$C_{11}H_7NS$
185.25
N(c(c(c(ccc1)cc2)c1)c2)=C=S
Isothiocyanic acid, 1-naphthyl ester
ANI
ANIT
1-Isothiocyanate-naphthalene
1-Isothiocyanatonaphthalene
Kesscocide
1-Naftylisothiokyanat (Czech)
Naphthalene, isothiocyanato-
α-Naphthyl isothiocyanate
1-Naphthyl isothiocyanate

551-11-1
$C_{20}H_{34}O_5$
354.54
CCCCCC(O)C=CC1C(O)CC(O)C1CC=CCCCC(O)=O
Prosta-5,13-dien-1-oic acid, (5Z,9-α,11-α,13E,15S)-9,11,15-trihydroxy
Amoglandin
7-(3,5-Dihydroxy-2-(3-hydroxy-1-octenyl)cyclopentyl)-5-heptenoic acid
Dinoprost
Enzaprost
Enzaprost F
5-Heptenoic acid, 7-(3,5-dihydroxy-2-(3-hydroxy-1-octenyl)cyclopentyl)-
5-Heptenoic acid, 7-(3,5-dihydroxy-2-(3-hydroxy-1-octenyl)cyclopentyl)-, l-
Panacelan
PGF2-α
l-PGF2-α
Prosta-5,13-dien-1-oic acid, 9,11,15-trihydroxy-, (5Z,9-α,11-α,13E,15S)- (9CI)
Prostaglandin F2-α
l-Prostaglandin F2-α
Prostaglandin F2a
Prostalmon F
Prostarmon F
Prostin F2-α
(5Z,9-α,11-α,13E,15S)-9,11,15-Trihydroxyprosta-5,13-dien-1-oic acid
9,11,15-Trihydroxyprosta-5,13-dien-1-oic acid
U-14583

551-45-1
$C_9H_{12}O_2$
152.19
2-Cyclopenten-1-one, 2-allyl-4-hydroxy-3-methyl- (VAN) (8CI)
AI3-20047
Allethrolon
Allethrolone (VAN)
2-Cyclopenten-1-one, 4-hydroxy-3-methyl-2-(2-propenyl)- (9CI)
NSC-42192

551-62-2
$C_6H_2F_4$
150.08
Benzene, 1,2,3,4-tetrafluoro- (8CI,9CI)
NSC-21635

551-74-6
$C_{10}H_{22}Cl_2N_2O_4.2ClH$
378.16
Mannitol, 1,6-bis((2-chloroethyl)amino)-1,6-dideoxy-, dihydrochloride, D
BCM
1,6-Bis-(chloroethylamino)-1,6-desoxy-D-mannitol dihydrochloride
1,6-Bis-(chloroethylamino)-1,6-dideoxy-D-mannite dihydrochloride
1,6-Bis(β-chloroethylamino)-1,6-dideoxy-D-mannitol dihydrochloride
Degranol
Degranol chinoin
1,6-Dideoxy-1,6-di(2-chloroethylamino)-D-mannitol dihydrochloride
Dimesylmannitol
Mannitol Mustard
Mannitol Mustard Dihydrochloride
Mannitol Nitrogen Mustard
Mannogranol
Mannomustine
Mannomustine dihydrochloride
NSC-9698

551-76-8
$C_7H_5Cl_3O$
211.47
Phenol, 2,4,6-trichloro-3-methyl- (9CI)
m-Cresol, 2,4,6-trichloro- (8CI)
2,4,6-Trichloro-m-cresol

551-92-8
$C_5H_7N_3O_2$
141.15
Cc1ncc(N(=O)=O)n1C
Imidazole, 1,2-dimethyl-5-nitro
1,2-Dimethyl-5-nitroimidazole
Dimetridazole
Emtryl
Emtrylvet
Emtrymix
1H-Imidazole, 1,2-dimethyl-5-nitro- (9CI)
8595 R.P.

551-93-9

C_8H_9NO
135.16
O=C(c(c(N)ccc1)c1)C
o-Aminoacetophenone
Acetophenone, 2'-amino- (8CI)
1-Acetyl-2-aminobenzene
o-Acetylaniline
2-Acetylaniline
AI3-04095
2-Aminoacetophenone
2'-Aminoacetophenone
o-Aminoacetylbenzene
1-(2-Aminophenyl)ethanone
Ethanone, 1-(2-aminophenyl)-
(9CI)
NSC-8820

552-16-9
$C_7H_5NO_4$
167.13
O=C(O)c(c(N(=O)=O)ccc1)c1
Benzoic acid, o-nitro
o-Nitrobenzoic acid
2-Nitrobenzoic acid

552-22-7
$C_{20}H_{24}I_2O_2$
550.22
O(c(c(cc(c1C)c(c(cc(OI)c2C(C)
C)C)c2)C(C)C)c1)I
Thymol iodide
Annidalin
Aristol
4,4'-Bis(iodooxy)-2,2'-di-
methyl-5,5'-bis(1-methylethyl-
1,1'-biphenyl
Bithymol diiodide
Diiododithymol
Dithymol diiodide
Hypoiodous acid, 2,2'-dimethyl-
5,5'-bis(1-methylethyl)-
(1,1'-biphenyl)-4,4'-diyl ester
(9CI)
Iodistol
Iodohydromol
Iodosol
Iodothymol (VAN)
Iosol
Iothymol
Lothymol
NSC-2222
Thymiode
Thymiodol
Thymodin
Thymotol

552-25-0
$C_{21}H_{28}N_2O$
324.45
Diampromide
DEA No. 9615
Diampromid
Diampromida (Spanish)
Diampromidum (Latin)
N-(2-(Methylphenethylamino)-
propyl)propionanilide
Propanamide, N-(2-(methyl-
(2-phenylethyl)amino)propyl)-
N-phenyl- (9CI)
Propionanilide, N-(2-(methyl-
phenethylamino)propyl)-

552-30-7
$C_9H_4O_5$
192.13
O=C(OC(=O)c1ccc(C(=O)O)c2)
c12
1,2,4-Benzenetricarboxylic acid 1,2-anhydride
Anhydro trimellic acid
1,2,4-Benzenetricarboxylic acid
anhydride
1,2,4-Benzenetricarboxylic acid,
cyclic 1,2-anhydride
1,2,4-Benzenetricarboxylic anhydride
4-Carboxyphthalic anhydride
1,3-Dihydro-1,3-dioxo-5-isoben-
zofurancarboxylic acid
1,3-Dioxo-5-phthalancarboxylic
acid
Diphenylmethane-4,4'-diisocy-
anate-trimellic anhydride-
ethomid HT polymer
NCI-C56633
5-Phthalancarboxylic acid,
1,3-dioxo-
TMA
TMAN
Trimellic acid anhydride
Trimellic acid 1,2-anhydride
Trimellitic acid cyclic 1,2-
anhydride
Trimellitic anhydride (ACGIH,
OSHA)

552-32-9
$C_8H_8N_2O_3$
180.15
O=C(Nc(c(N(=O)=O)ccc1)c1)C
**Acetamide, N-(2-nitrophenyl)-
(9CI)**
Acetanilide, 2'-nitro- (8CI)
AI3-08843
o-Nitroacetanilide
2'-Nitroacetanilide
N-(2-Nitrophenyl)acetamide
NSC-1313

552-45-4
C_8H_9Cl
140.61
c(c(ccc1)C)(c1)CCl
α-Chloro-o-xylene
Benzene, 1-(chloromethyl)-
2-methyl- (9CI)
1-(Chloromethyl)-2-methyl-
benzene
2-(Chloromethyl)toluene
ω-Chloro-o-xylene
o-Methylbenzyl chloride
2-Methylbenzyl chloride
o-Xylene, α-chloro-
o-Xylyl chloride
o-Xylyl-α-chloride

552-46-5
$C_{10}H_9N.ClH$
179.66
**1-Naphthylamine hydro-
chloride**
1-Amino-naphthalene hydrochloride
1-Naphthalenamine hydro-
chloride
α-Naphthylamine hydrochloride

552-62-5
$C_6H_6N_4O_2$
166.16
Cn1cnc2[nH]c(=O)[nH]c(=O)c12
Xanthine, 7-methyl
Heteroxanthin
7-Methylxanthin

552-82-9
$C_{13}H_{13}N$
183.25
N(c(cccc1)c1)(c(cccc2)c2)C
**Benzenamine, N-methyl-
N-phenyl- (9CI)**
AI3-02479
Demelverine
Diphenylamine, N-methyl-
(8CI)
Diphenylmethylamine
N,N-Diphenylmethylamine
N-Methyldiphenethylamine
Methyldiphenylamine
N-Methyldiphenylamine
N-Methyl-N-phenylaniline
N-Methyl-N-phenylbenzen-
amine
NSC-3790

552-86-3
$C_{10}H_8O_4$
192.17
O=C(C(OC=C1)=C1)C(O)C
(OC=C2)=C2
Furoin
AI3-02545
1,2-Di-2-furanyl-2-hydroxy-
ethanone
Ethanone, 1,2-di-2-furanyl-
2-hydroxy- (9CI)
Ethanone, 1,2-di-2-furyl-
2-hydroxy-
α-Furoin
Furoylfurylcarbinol
Ketone, 2-furyl α-hydroxyfur-
furyl
NSC-18522

552-89-6
$C_7H_5NO_3$
151.13
O=Cc(c(N(=O)=O)ccc1)c1
Benzaldehyde, o-nitro
o-Nitrobenzaldehyde
2-Nitrobenzaldehyde

552-94-3
$C_{14}H_{10}O_5$
258.24

O=C(Oc(c(ccc1)C(=O)O)c1)
c(c(O)ccc2)c2
Benzoic acid, 2-hydroxy-, 2-carboxyphenyl ester
Diacesal
Diplosal
Disalcid
Disalicylic acid
Disalyl
NSC-49171
Nobacid
SAA
Salical
Salicylic acid, bimolecular ester
Salicyloylsalicylic acid
Salicylsalicylic acid
o-Salicylsalicylic acid
Salina
Saloxium
Salsalate
Salysal
Sasapirin
Sasapyrin
Sasapyrine
Sasapyrinum

553-12-8
$C_{34}H_{34}N_4O_4$
562.64
O=C(O)CCc(c(nc1cc(Nc(c2C=C)
cc(nc(c3C=C)cc(Nc4c5CC-
C(=O)O)c5C)c3C)c2C)c4)c1C
Protoporphyrin IX
NSC-2632
Ooporphyrin
21H,23H-Porphine-2,18-dipropanoic acid, 7,12-diethenyl-3,8,13,17-tetramethyl- (9CI)
2,18-Porphinedipropionic acid, 3,8,13,17-tetramethyl-7,12-divinyl- (8CI)
Protoporphyrin
Protoporphyrin IX (VAN)

553-26-4
$C_{10}H_8N_2$
156.20
n(ccc(c1)c(ccnc2)c2)c1
4,4'-Bipyridine
γ,γ'-Bipyridyl
4,4-Bipyridyl
4,4'-Bipyridyl
4,4'-Dipyridine
γ,γ'-Dipyridyl
4,4-Dipyridyl
4,4'-Dipyridyl
4-(4-Pyridyl)pyridine

553-27-5
$C_{10}H_{13}Cl_2N$
218.14
ClCCN(CCCl)c1ccccc1
Aniline, N,N-bis(2-chloroethyl)
Aniline mustard
Anilinlost (German)
N,N-Bis(2-chloroethyl)aniline
N,N-Bis(2-chloroethyl)benzenamine
β,β'-Dichlorodiethylaniline
N,N,-Di(2-chloroethyl)aniline
Lymphchin
Lymphocin
Lymphoquin
NSC-18429
Phenylbis(2-chloroethylamine)
TL 476

553-30-0
$C_{13}H_{11}N_3.H_2O_4S$
307.35
Nc3ccc2C=c1ccc(N)cc1=[N+]
c2c3.OS([O-])(=O)=O
Acridine, 3,6-diamino-, sulfate (1:1)
3,6-Acridinediamine sulfate
3,6-Acridinediamine, sulfate (1:1) (9CI)
3,6-Acridinediamine sulphate
3,6-Diaminoacridine bisulphate
3,6-Diaminoacridine sulphate (1:1)
3,6-Diaminoacridinium monohydrogen sulphate
2,8-Diaminoacridinium sulphate
Flavine
Flavin sulphate
Isoflav
Neutral Proflavine Sulphate
Pancridine
Proflavine (sulfate)
Proflavine sulphate
Proflavin sulfate
Sanoflavin

553-70-8
$C_7H_6O_2.1/2Mg$
134.28
Benzoic acid, magnesium salt (9CI)
Magnesium benzoate

553-72-0
$C_7H_6O_2.1/2Zn$
154.81
Benzoic acid, zinc salt (9CI)
NSC-176118
Zinc benzoate
Zinc dibenzoate

553-82-2
$C_7H_6Cl_2O$
177.03
O(c(c(cc(c1)Cl)Cl)c1)C
2,4-Dichloroanisole
AI3-16644
Anisole, 2,4-dichloro- (8CI)
Benzene, 2,4-dichloro-1-methoxy- (9CI)
1,5-Dichloro-2-methoxybenzene
2,4-Dichloro-1-methoxybenzene
NSC-6077

553-90-2
$C_4H_6O_4$
118.09
O=C(OC)C(=O)OC
Ethanedioic acid, dimethyl ester (9CI)
AI3-21214
Dimethyl ethanedioate
Dimethyl oxalate
Methyl oxalate
NSC-9374
Oxalic acid, dimethyl ester (8CI)

553-94-6
C_8H_9Br
185.06
c(ccc(c1Br)C)(c1)C
Benzene, 2-bromo-1,4-dimethyl- (9CI)
1-Bromo-2,5-dimethylbenzene
2-Bromo-1,4-dimethylbenzene
Bromo-p-xylene
2-Bromo-p-xylene
2-Bromo-1,4-xylene
2,5-Dimethylbromobenzene
2,5-Dimethylphenyl bromide
NSC-8051
p-Xylene, 2-bromo- (8CI)
2,5-Xylyl bromide

553-97-9
$C_7H_6O_2$
122.13
O=C(C(=CC(=O)C=1)C)C1
p-Benzoquinone, 2-methyl
2-Methyl-1,4-benzochinon (Czech)
Methyl-p-benzoquinone
Methyl-1,4-benzoquinone
2-Methyl-p-benzoquinone
2-Methylbenzoquinone-1,4
2-Methyl-1,4-benzoquinone
2-Methyl-1,4-quinone
1,4-Toluchinon (Czech)
Toluquinone
p-Toluquinone
1,4-Toluquinone

554-00-7
$C_6H_5Cl_2N$
162.02
Nc(c(cc(c1)Cl)Cl)c1
Aniline, 2,4-dichloro
Benzenamine, 2,4-dichloro- (9CI)
2,4-Dichloranilin (German)
2,4-Dichloroaniline

554-12-1
$C_4H_8O_2$
88.12
O=C(OC)CC
Propionic acid, methyl ester
Methylester kyseliny propionove (Czech)
Methyl propanoate
Methyl propionate [UN 1248]
Methyl propylate
Propanoic acid, methyl ester
Propionate de methyle (French)
UN 1248 [Methyl propionate]

554-13-2
$CO_3.2Li$
73.89
Lithium carbonate (2:1)
Camcolit
Candamide
Carbolith
Carbonic acid, dilithium salt
Carbonic acid lithium salt
Ceglution
CP-15467-61
Dilithium carbonate
Eskalith
Hypnorex
Limas
Liskonum
Lithane
Lithicarb
Lithinate
Lithium carbonate
Lithobid
Lithonate
Lithotabs
NSC-16895
Plenur
Priadel
Quilonum retard

554-14-3
C_5H_6S
98.17
S(C(=CC=1)C)C1
Thiophene, 2-methyl
2-Methylthiophene

554-35-8
$C_{10}H_{17}NO_6$
247.28
CC(C)(OC1OC(CO)C(O)C(O)C1O)C#N
Propanenitrile, 2-(β-D-glucopyranosyloxy)-2-methyl
2-(β-D-Glucopyranosyloxy)isobutyronitrile
2-(β-D-Glucopyranosyloxy)-2-methylpropanenitrile
Linamarin
Phaseolunatin

554-57-4
$C_5H_8N_4O_3S_2$
236.29
CC(=O)N=c1sc(nn1C)S(N)(=O)=O
Acetamide, N-(4-methyl-2-sulfamoyl-δ²-1,3,4-thiadiazolin-5-ylidene)
Acetamide, N-(5-(aminosulfonyl)-3-methyl-1,3,4-thiadiazol-2(3H)-ylidene)- (9CI)
2-Acetylimino-3-methyl-δ⁴-1,3,4-thiadiazoline-5-sulfonamide
5-Acetylimino-4-methyl-δ²-1,3,4-thiadiazoline-2-sulfonamide
Methazolamide
Methenamide
Naptazane
Neptazane
Neptazaneat

554-60-9
$C_{10}H_{16}$
136.24
Pseudocarene
Bicyclo(4.1.0)heptane, 7,7-dimethyl-3-methylene-
β-Carene
psi-Carene
3(10)-Carene
Norcarane, 7,7-dimethyl-3-methylene-

554-68-7
$C_6H_{15}N.ClH$
137.68
Triethylamine, hydrochloride

554-70-1
$C_6H_{15}P$
118.16
P(CC)(CC)CC
Phosphine, triethyl- (9CI)
Triethylphosphine

554-73-4
$C_{18}H_{15}N_3O_3S.Na$
376.37
Benzenesulfonic acid, 4-((4-(phenylamino)phenyl)azo)-, monosodium salt (9CI)
Acid Yellow D
Benzenesulfonic acid, p-((p-anilinophenyl)azo)-, monosodium salt
Benzenesulfonic acid, p-((p-anilinophenyl)azo)-, sodium salt
C.I. Acid Orange 5
C.I. Acid Orange 5, Monosodium salt (8CI)
C.I. 13080
Diphenylamine Orange
Hispacid Orange IV
NSC-10456
Orange GS
Orange IV
Orange N
Orange 4 Lake
Sodium p-diphenylamino-azobenzenesulfonate
Solar Orange IV
Tertracid Orange IV
Tropaeolin OO
Tropeolin OO

554-77-8
$C_6H_5ClHgO_3S$
393.21
Chloromercuriphenylsulfonate
AI3-51823
Mercury, chloro(4-sulfophenyl)-

554-84-7
$C_6H_5NO_3$
139.12
O=N(=O)c(cccc1O)c1
Phenol, m-nitro
m-Hydroxynitrobenzene
3-Hydroxynitrobenzene
m-Nitrofenol (Czech)
m-Nitrophenol [UN 1663]
3-Nitrophenol
UN 1663 [Nitrophenols (o-; m-; p-;)]

554-95-0
$C_9H_6O_6$
210.14
O=C(O)c(cc(cc1C(=O)O)C(=O)O)c1
1,3,5-Benzenetricarboxylic acid (9CI)
AI3-06468
5-Carboxyisophthalic acid
NSC-3998
1,3,5-Tricarboxybenzene
Trimesic acid
Trimesinic acid
Trimesitinic acid

555-03-3
$C_7H_7NO_3$
153.15
O=N(=O)c(cccc1OC)c1
Anisole, m-nitro
m-Methoxynitrobenzene
3-Methoxynitrobenzene
Methyl m-nitrophenyl ether
m-Nitroanisole [UN 2730]
3-Nitroanisole
UN 2730 [Nitroanisole]

555-06-6
$C_7H_6NO_2.Na$
159.13
Benzoic acid, p-amino-, monosodium salt
p-Aminobenzoic acid sodium salt
Antergyl
Benzoic acid, 4-amino-, monosodium salt (9CI)
PAB
PABAVJT
Sodium p-aminobenzoate
Sodium 4-aminobenzoate

555-10-2
$C_{10}H_{16}$
136.24
C(C=CC(C1)C(C)C)(C1)=C
β-Phellandrene
Cyclohexene, 3-methylene-6-(1-methylethyl)- (9CI)
3-Isopropyl-6-methylene-1-cyclohexene
4-Isopropyl-1-methylene-2-cyclohexene
2-p-Menthadiene

p-Mentha-1(7),2-diene (8CI)
3-Methylene-6-(1-methylethyl)-cyclohexene
NSC-53044
Phellandrene, β

555-16-8
$C_7H_5NO_3$
151.13
O=Cc(ccc(N(=O)=O)c1)c1
Benzaldehyde, p-nitro
Benzaldehyde, 4-nitro- (9CI)
p-Formylnitrobenzene
p-Nitrobenzaldehyde
4-Nitrobenzaldehyde

555-30-6
$C_{10}H_{13}NO_4$
211.24
O=C(O)C(N)(Cc(ccc(O)c1O)c1)C
Alanine, 3-(3,4-dihydroxyphenyl)-2-methyl-, l-(-)
Aldomet
Aldometil
Aldomin
Aelpha Medopa
AMD
Bayer 1440 L
Baypresol
L(-)-β-(3,4-Dihydroxyphenyl)-α-methylalanine
L-(-)-3-(3,4-Dihydroxyphenyl)-2-methylalanine
L-3-(3,4-Dihydroxyphenyl)-2-methylalanine
Dopamet
Dopatec
Dopegyt
Hyperpax
L-(α-MD)
Medomet
Medopren
Methoplain
α-Methyl-l-3,4-dihydroxyphenylalanine
L-α-Methyl-3,4-dihydroxyphenylalanine
α-Methyl-β-(3,4-dihydroxyphenyl)-l-alanine
L-(-)-α-Methyl-β-(3,4-dihydroxyphenyl)alanine
Methyldopa
(-)-Methyldopa
L-Methyldopa
L-α-Methyldopa
α-Methyl dopa
α-Methyldopa, L-
α-Methyl-L-DOPA
MK. B51
MK 351
NCI-C55721
NR.C 2294
Presinol
Presolisin
Sedometil
Sembrina
L-Tyrosine, 3-hydroxy-α-methyl- (9CI)

555-31-7
$C_9H_{21}AlO_3$
204.28
C.[Al].CC.CC=[O-].CC=[O-].CC=[O-]
Aluminum(II) isopropylate
Aluminum isopropoxide
Triisopropoxyaluminum

555-32-8
$C_7H_6O_2.1/3Al$
131.12
Benzoic acid, aluminum salt (9CI)
Aluminum benzoate

555-34-0
$C_2H_2O_4.1/3Fe.Na$
116.31
Ethanedioic acid, iron(3+) sodium salt (3:1:3) (9CI)

555-35-1
$C_{16}H_{32}O_2.1/3Al$
265.42
Aluminum palmitate
AI3-19807
Aluminum hexadecanoate
Aluminum stearate palmitate
Hexadecanoic acid, aluminum salt (9CI)
Hexadecanoic acid, aluminum salt (3:1)
NSC-115894
Palmitic acid, aluminum salt (8CI)

555-36-2
$C_{18}H_{36}O_2.1/3Fe$
303.10
Octadecanoic acid, iron(3+) salt (9CI)
Iron(3+) octadecanoate

555-37-3
$C_{12}H_{16}Cl_2N_2O$
275.20
CCCCN(C)C(=O)Nc1ccc(Cl)c(Cl)c1
Urea, 1-Butyl-3-(3,4-dichlorophenyl)-1-methyl
N-Butyl-N'-(3,4-dichlorophenyl)-N-methylurea
1-Butyl-3-(3,4-dichlorophenyl)-1-methylurea
3-(3,4-Dichlorphenyl)-1-n-butylharnstoff (German)
3-(3,4-Dichlorophenyl)-1-methyl-1-butylurea
Granurex
Kloben
Kloben Neburon
Neburea
Neburex
Neburon

555-43-1
$C_{57}H_{110}O_6$
891.50
Tristearin
AI3-01633
Dynasan 118
Glycerol trioctadecanoate
Glycerol, trioctadecanoate
Glycerol tristearate
Glyceryl tristearate
Glycowax S 932
Hardened Oil
Octadecanoic acid, 1,2,3-propanetriyl ester
1,2,3-Propanetriol trioctadecanoate
1,2,3-Propanetriyl octadecanoate
Spezialfett 118
Stearic acid triglyceride
Stearic acid triglycerin ester
Stearic triglyceride
Stearin
Stearin, tri-
Stearoyl triglyceride
Trioctadecanoin

555-45-3
$C_{45}H_{86}O_6$
723.17
Trimyristin
AI3-18153
Dynasan 114
Glycerol trimyristate
Glyceryl trimyristate
Myristic acid triglyceride
Myristin
Myristin, tri- (8CI)
NSC-4062
1,2,3-Propanetriyl tetradecanoate
Tetradecanoic acid, 1,2,3-propanetriyl ester (9CI)
VP 114

555-48-6
$C_8H_{10}N_2O$
150.17
O=C(Nc(cccc1)c1)CN
Acetamide, 2-amino-N-phenyl- (9CI)
Acetanilide, 2-amino- (8CI)
2-Aminoacetanilide
Aminoacetic anilide
2-Amino-N-phenylacetamide
Benzokoll
Glycine anilide
Glycocollanilide
N-Glycylaniline
NSC-226561

555-75-9
$C_2H_6O.1/3Al$
55.06
CCO[Al](OCC)OCC
Aluminum ethylate
Aluminum ethoxide
Aluminum triethoxide
Ethanol, aluminum salt (9CI)
Ethyl alcohol, aluminum salt

Triethoxyaluminum

555-77-1
$C_6H_{12}Cl_3N$
204.54
ClCCN(CCCl)CCCl
Triethylamine, 2,2',2''-trichloro
HN3
TL 145
Trichlormethine
Tri-(2-chloroethyl)amine
2,2',2''-Trichlorotriethylamine
Tris(β-chloroethyl)amine
Tris(2-chloroethyl)amine
TS 160

555-84-0
$C_8H_8N_4O_4$
224.20
O=C1NCCN1N=Cc2ccc(o2)N(=O)=O
2-Imidazolidinone, 1-((5-nitrofurfurylidene)amino)
2-Imidazolidinone, 1-(((5-nitro-2-furanyl)methylene)amino)-
NF 246
Nifuradene
N-(5-Nitro-2-furfurylidene)-1-amino-2-imidazolidone
N-(5-Nitro-2-furfurylidene-amino)-2-imidazoaidinone
1-((5-Nitrofurfurylidene)-amino)-2-imidazolidinone
Nifuradine
NSC-6470
Oxafuradene
Oxifuradene
Oxyfuradene
Renafur

555-89-5
$C_{13}H_{10}Cl_2O_2$
269.13
Clc2ccc(OCOc1ccc(Cl)cc1)cc2
Methane, bis(p-chlorophenoxy)
Bis(p-chlorophenoxy)methane
DCPM
Di-p-chlorodiphenoxymethane
Di-(p-chlorophenoxy)methane
Di-(4-chlorophenoxy)methane
ENT 15,208
K 1875
1,1'-(Methylenebis(oxy))bis-(4-chloro)benzene
Neotran
Oxythane

556-08-1
$C_9H_9NO_3$
179.17
O=C(Nc(ccc(c1)C(=O)O)c1)C
Acedoben
Acedoben (Spanish)
Acedobene (French)
Acedobenum (Latin)
4-Acetamidobenzoic acid
p-Acetaminobenzoic acid
p-Acetoaminobenzoic acid
p-Acetylaminobenzoic acid
4-Acetylaminobenzoic acid
4-(Acetylamino)benzoic acid
N-Acetyl-p-aminobenzoic acid
AI3-16506
Benzoic acid, p-acetamido-(8CI)
Benzoic acid, 4-(acetylamino)-(9CI)
4-Carboxyacetanilide
4'-Carboxyacetanilide
NSC-4002
PAAB

556-18-3
C_7H_7NO
121.15
O=Cc(ccc(N)c1)c1
Benzaldehyde, 4-amino
p-Aminobenzaldehyde
4-Aminobenzaldehyde

556-22-9
$C_{20}H_{40}N_2.C_2H_4O_2$
368.68
CCCCCCCCCCCCCCCCCC1=NCCN1.CC([O-])=O
2-Imidazoline, 2-heptadecyl-, monoacetate
Crag 341
Crag Fruit Fungicide 341
Crag Fungicide 341
Experimental Fungicide 341
Glyodin acetate
Glyoxide Dry
2-Heptadecyl-4,5-dihydro-1H-imidazolyl monoacetate
2-Heptadecyl glyoxalidine acetate
2-Heptadecyl-2-imidazoline acetate
2-Imidazoline, 2-heptadecyl-, acetate

556-24-1
$C_6H_{12}O_2$
116.18
O=C(OC)CC(C)C
Isovaleric acid, methyl ester
Butanoic acid, 3-methyl-, methyl ester (9CI)
Methyl isopentanoate
Methyl isovalerate [UN 2400]
Methylisovalerate
Methyl 3-methylbutanoate
Methyl 3-methylbutyrate
UN 2400 [Methyl isovalerate]

556-33-2
$C_6H_{11}N_3O_4$
189.15
NCC(=O)NCC(=O)NCC(O)=O
Glycyl-glycyl-glycine
Diglycylglycine
Glycine, N-(N-glycylglycyl)-(9CI)
Glycylglycylglycine
N-(N-Glycylglycyl)glycine
NSC-46707

556-48-9
$C_6H_{12}O_2$
116.16
OC1CCC(O)CC1
1,4-Cyclohexanediol (8CI,9CI)
NSC-5651
Quinitol

556-50-3
$C_4H_8N_2O_3$
132.11
O=C(NCC(=O)O)CN
Glycylglycine
AI3-62521
Glycine dipeptide
Glycine, N-glycyl- (9CI)
α-Glycylglycine
N-Glycylglycine
NSC-49346

556-52-5
$C_3H_6O_2$
74.09
O(C1CO)C1
1-Propanol, 2,3-epoxy
Epihydrin alcohol
2,3-Epoxypropanol
2,3-Epoxy-1-propanol (OSHA)
Glycide
Glycidol (ACGIH,OSHA)
Glycidyl alcohol
3-Hydroxy-1,2-epoxypropane
Methanol, oxiranyl-
NCI-C55549

556-53-6
$C_3H_9N.ClH$
95.57
Propylamine, hydrochloride (8CI)
NSC-210913
1-Propanamine, hydrochloride (9CI)
Propylamine hydrochloride
n-Propylamine hydrochloride
Propylammonium chloride
1-Propylammonium chloride

556-56-9
C_3H_5I
167.98
C(=C)CI
Propene, 3-iodo
Allyl iodide [UN 1723]
3-Iodopropene
3-Iodopropylene
1-Propene, 3-iodo- (9CI)
UN 1723 [Allyl iodide]

556-61-6
C_2H_3NS
73.12

556-63-8

N(=C=S)C
Methane, isothiocyanato
EP-161E
Isothiocyanate de methyle (French)
Isothiocyanatomethane
Isothiocyanic acid, methyl ester
Isotiocianato di metile (Italian)
Methylisothiocyanaat (Dutch)
Methyl-isothiocyanat (German)
Methyl isothiocyanate [UN 2477]
Methylisothiokyanat (Czech)
Methyl Mustard Oil
Methylsenfoel (German)
MIC
MIT
MITC
Morton EP-161E
Trapex
Trapexide
UN 2477 [Methyl isothiocyanate]
Vorlex
Vortex
WN 12

556-63-8
$CH_2O_2.Li$
52.97
Formic acid, lithium salt (9CI)
Lithium formate

556-64-9
C_2H_3NS
73.12
N#CSC
Thiocyanic acid, methyl ester
Methane, thiocyanato-
Methylrhodanid (German)
Methyl sulfocyanate
Methyl thiocyanate
Methylthiokyanat (Czech)

556-67-2
$C_8H_{24}O_4Si_4$
296.68
CSi1(C)OSi(C)(C)OSi(C)(C)OSi(C)(C)O1
Cyclotetrasiloxane, octamethyl
Octamethylcyclotetrasiloxane
Oktamethylcyklotetrasiloxan (Czech)

556-68-3
$C_{16}H_{48}O_8Si_8$
593.24
Cyclooctasiloxane, hexadecamethyl- (8CI,9CI)

556-69-4
$C_{18}H_{54}O_7Si_8$
607.31
Octasiloxane, octadecamethyl- (9CI)
Octadecamethyloctasiloxane

556-71-8
$C_{18}H_{54}O_9Si_9$
667.39
Cyclononasiloxane, octadecamethyl- (8CI,9CI)

556-82-1
$C_5H_{10}O$
86.15
OCC=C(C)C
2-Buten-1-ol, 3-methyl
Dimethylallyl alcohol
γ,γ-Dimethylallyl alcohol
3,3-Dimethylallyl alcohol
3-Methyl-2-buten-1-ol
Prenol
Prenyl alcohol

556-88-7
$CH_4N_4O_2$
104.09
O=N(=O)NC(=N)N
Guanidine, 1-nitro
Guanidine, nitro-
Nitroguanidine
α-Nitroguanidine
2-Nitroguanidine
Nitroguanidine, Containing less than 20% water [UN 0282]
Nitroguanidine, Dry [UN 0282]
Nitroguanidine, Wetted with not less than 20 per cent water, by mass [UN 1336]
Picrite (The explosive)
UN 0282 [Nitroguanidine; (Picrite), Dry or wetted with less than 20 per cent water, by mass]
UN 1336 [Nitroguanidine; (Picrite) wetted with not less than 20 per cent water, by mass]

557-00-6
$C_8H_{16}O_2$
144.24
Isovaleric acid, propyl ester
Butanoic acid, 3-methyl-, propyl ester (9CI)
Propyl isovalerate
Propyl 3-methylbutyrate

557-04-0
$C_{36}H_{70}O_4.Mg$
591.37
Stearic acid, magnesium salt
Magnesium stearate (ACGIH)
Octadecanoic acid, magnesium salt

557-05-1
$C_{36}H_{70}O_4.Zn$
632.43
Zinc-stearate
Coad
Dermarone
Dibasic zinc stearate
Hydense
Hytech
Metallac
Metasap 576
Mathe
Octadecanoic acid, zinc salt
Petrac ZN-41
Stavinor ZN-E
Stearic acid, zinc salt
Synpro stearate
Talculin Z
Zinc distearate
Zinc octadecanoate
Zinc stearate (ACGIH,OSHA)

557-07-3
$C_{36}H_{68}O_4.Zn$
630.41
Zinc oleate (1:2)
Oleic acid, zinc salt
Zinc oleate

557-08-4
$C_{11}H_{20}O_2.1/2Zn$
216.97
Zinc undecylenate
NSC-402438
10-Undecenoic acid, zinc salt (9CI)
10-Undecenoic acid, zinc (2+) salt
Undecylenic acid, zinc salt
Zinc undecenoate
Zinc 10-undecenoate

557-11-9
$C_4H_8N_2O$
100.14
O=C(NCC=C)N
Urea, allyl
Allylcarbamide
Allylurea
N-Allylurea
1-Allylurea
Monoallylurea
N-2-Propenylurea
Urea, 2-propenyl- (9CI)

557-17-5
$C_4H_{10}O$
74.14
CCCOC
Ether, methyl propyl
α-Methoxy propane
1-Methoxypropane
Methyl propyl ether [UN 2612]
Methyl n-propyl ether
Metopryl
Neothyl
Propane, 1-methoxy- (9CI)
UN 2612 [Methyl propyl ether]

557-19-7
C_2N_2Ni
110.75

Nickel cyanide, [UN 1653]
Nickel cyanide, Solid (DOT)
RCRA waste number P074
UN 1653 [Nickel cyanide]

557-20-0
$C_4H_{10}Zn$
123.51
Zinc, diethyl
Diethylzinc [UN 1366]
Diethyl zinc
UN 1366 [Diethylzinc]
Zinc ethide
Zinc ethyl (DOT)

557-21-1
C_2N_2Zn
117.41
Zinc cyanide [UN 1713]
Cyanure de zinc (French)
RCRA waste number P121
UN 1713 [Zinc cyanide]
Zinc dicyanide

557-24-4
$C_4H_5NO_3$
115.08
O=C(O)C=CC(=O)N
Maleamic acid (8CI)
Acrylic acid, 3-carbamoyl-, (Z)-
AI3-16135
2-Butenoic acid, 4-amino-4-oxo-, (Z)- (9CI)
Maleamate
Maleic monoamide
NSC-8155

557-25-5
$C_7H_{14}O_4$
162.19
O=C(OCC(O)CO)CCC
Butanoic acid, 2,3-dihydroxypropyl ester (9CI)
Butyrin, 1-mono- (8CI)
2,3-Dihydroxypropyl butanoate
Glycerol-α-mono-n-butyrate
α-Monobutyrin
NSC-8451

557-28-8
$C_3H_6O_2.1/2Zn$
106.77
Zinc propionate
Propanoic acid, zinc salt (9CI)
Propionic acid, zinc salt
Zinc dipropionate
Zinc propanoate

557-31-3
$C_5H_{10}O$
86.15
O(CC=C)CC
Ether, allyl ethyl
Allyl ethyl ether [UN 2335]
3-Ethoxy-1-propene
Ethyl allyl ether
1-Propene, 3-ethoxy- (9CI)
UN 2335 [Allyl ethyl ether]

557-34-6
$C_4H_6O_4.Zn$
183.47
[Zn+2].CC([O-])=O.CC([O-])=O
Acetic acid, zinc(II) salt
Acetic acid, zinc salt (8CI,9CI)
Dicarbomethoxyzinc
Zinc acetate
Zinc diacetate

557-36-8
$C_8H_{17}I$
240.13
C(CCCCCC)(C)I
Octane, 2-iodo- (9CI)
2-Iodooctane

557-40-4
$C_6H_{10}O$
98.16
O(CC=C)CC=C
Ether, diallyl
Allylether
Diallylether [UN 2360]
Ether, propenyl
Propenyl ether
3,3'-Oxybis(1-propene)
UN 2360 [Diallylether]

557-41-5
$CH_2O_2.1/2Zn$
78.72
Zinc formate
Caswell No. 915
EPA Pesticide Chemical Code 087802
Formic acid, zinc salt (9CI)
Zinc diformate

557-59-5
$C_{24}H_{48}O_2$
368.64
O=C(O)CCCCCCCCCCCCCCCCCCCCCCC
Lignoceric acid
Tetracosanoic acid (9CI)

557-61-9
$C_{28}H_{58}O$
410.77
OCCCCCCCCCCCCCCCCCCCCCCCCCCCC
1-Octacosanol (9CI)2
Cluytyl alcohol
Montanyl alcohol
NSC-10770
Octacosanol
Octacosanol-1
n-Octacosanol
Octacosyl alcohol

557-66-4
$C_2H_7N.ClH$
81.56
Ethylamine, hydrochloride
Ethyl ammonium chloride

557-91-5
$C_2H_4Br_2$
187.88
BrC(Br)C
Ethane, 1,1-dibromo
1,1-Dibromoethane
Ethylidene bromide
Ethylidene dibromide

557-93-7
C_3H_5Br
120.99
C(Br)(=C)C
Propene, 2-bromo
2-Bromopropene
2-Bromopropylene
Isopropylene bromide
α-Methylvinyl bromide
1-Propene, 2-bromo- (9CI)

557-98-2
C_3H_5Cl
76.53
C(=C)(C)Cl
Propene, 2-chloro
2-Chloropropene [UN 2456]
2-Chloro-1-propene
UN 2456 [2-Chloropropene]

558-13-4
CBr_4
331.65
BrC(Br)(Br)Br
Carbon-tetrabromide
Bromid uhlicity (Czech)
Carbon bromide
Carbon tetrabromide (ACGIH, OSHA) [UN 2516]
Methane, tetrabromide
Methane, tetrabromo-
Tetrabromomethane
UN 2516 [Carbon tetrabromide]

558-25-8
CH_3FO_2S
98.10
O=S(=O)(F)C
Methanesulfonyl-fluoride
Fumette
Methanesulphonyl fluoride
MSF

558-30-5
C_4H_8O
72.12
O(C1(C)C)C1
Propane, 1,2-epoxy-2-methyl
Isobutyleneoxide

558-37-2

C_6H_{12}
84.16
C(C(C)(C)C)=C
1-Butene, 3,3-dimethyl- (9CI)
tert-Butylethene
tert-Butylethylene
3,3-Dimethylbutene
2,2-Dimethyl-3-butene
3,3-Dimethyl-1-butene
tert-Hexene
Neohexene
NSC-74119
Trimethylvinylmethane

560-21-4
C_8H_{18}
114.23
C(C(C)C)(CC)(C)C
2,3,3-Trimethylpentane
NSC-77447
Pentane, 2,3,3-trimethyl- (9CI)

560-23-6
C_8H_{16}
112.22
1-Pentene, 2,3,3-trimethyl- (8CI,9CI)
2,3,3-Trimethyl-1-pentene

561-27-3
$C_{21}H_{23}NO_5$
369.45
CN1CCC24C3Oc5c(OC(C)=O)ccc(CC1C2C=CC3OC(C)=O)c45
Morphinan-3,6-α-diol, 7,8-didehydro-4,5-α-epoxy-17-methyl-, diacetate (ester)
Acetomorfine
Acetomorphine
Aspron
Boy
Diacephin
Diacetylmorfin
Diacetylmorphine
Diamorfina
Diamorphine
Diaphorm
Diasetielmorfien
Diasetilmorfin
Diasetylmorfiimi
Diazetylmorphine
7,8-Dihydro-4,5-α-epoxy-17-methylmorphinan-3,6-α-diol diacetate
Dooje
Eclorion
Eroina
"H"
Hairy
Harry
Heroien
Heroiin
Heroin
Herolan
Horse
Ieroin
Iroini
Joy Powder
Morphacetin
Morphine diacetate
Preza
Scot
White Stuff

561-48-8
$C_{23}H_{29}NO$
335.49
Norpipanone
DEA No. 9636
3-Hexanone, 4,4-diphenyl-6-piperidino- (8CI)
3-Hexanone, 4,4-diphenyl-6-(1-piperidinyl)- (9CI)
Hoechst 10495
Norpipanona (Spanish)
Norpipanonum (Latin)

561-76-2
$C_{16}H_{23}NO_2$
261.37
Properidine
DEA No. 9644
Ipropethidine
Isonipecotic acid, 1-methyl-4-phenyl-, isopropyl ester (8CI)
4-Piperidinecarboxylic acid, 1-methyl-4-phenyl-, 1-methylethyl ester (9CI)
Properidina (Spanish)
Properidinum (Latin)

562-10-7
$C_{17}H_{22}N_2O.C_4H_6O_4$
388.51
Pyridine, 2-(α-(2-(dimethylamino)ethoxy)-α-methylbenzyl)-, succinate (1:1)
Decapryn
Decapryn succinate
Decarpyn succinate (1:1)
Dimethylaminoethoxy-methylbenzyl-pyridine succinate
2-(α-(2-Dimethylaminoethoxy)-α-methylbenzyl)pyridine succinate
2-Dimethylaminoethoxyphenylmethyl-2-picoline succinate
Doxylamine succinate
Doxylamine succinate (1:1)
Hoggar N
Mereprine
Phenyl2-pyridylmethyl-β-N,N-dimethylaminoethyl ether succinate
Unisom

562-26-5
$C_{23}H_{29}NO_3$
367.48
Phenoperidine
DEA No. 9641
Fenoperidina (Spanish)
Phenoperidinum (Latin)
R 1406

562-49-2
C_7H_{16}
100.20
C(CC)(CC)(C)C
Pentane, 3,3-dimethyl- (9CI)
3,3-Dimethylpentane
NSC-74150

562-73-2
$C_7H_{12}O_6$
192.17
Cyclohexanecarboxylic acid, 1,3,4,5-tetrahydroxy- (8CI, 9CI)
NSC-243743

562-74-3
$C_{10}H_{18}O$
154.28
OC(CCC(=C1)C)(C1)C(C)C
para-Menth-1-en-4-ol
4-Carvomenthenol
3-Cyclohexen-1-ol, 4-methyl-1-(1-methylethyl)- (9CI)
1-para-Menthen-4-ol
Terpinenol-4
Terpinen-4-ol
4-Terpinenol
Terpinenolu-4 (Czech)
4-Terpineol

563-04-2
$C_{21}H_{21}O_4P$
368.37
O=P(Oc(cccc1C)c1)(Oc(cccc2C)c2)Oc(cccc3C)c3
Tri-m-cresyl phosphate
NSC-4055
Phosphoric acid, tris(3-methylphenyl) ester (9CI)
Phosphoric acid tri-m-tolyl ester
Phosphoric acid, tri(3-tolyl)ester
Tri-m-cresyl phosphite
Tri-m-tolyl phosphate
Tris-m-cresyl phosphate
Tris(3-methylphenyl) phosphate
Tris(m-tolyl) phosphate

563-12-2
$C_9H_{22}O_4P_2S_4$
384.49
CCOP(=S)(OCC)SCSP(=S)(OCC)OCC
Phosphorodithioic acid, S,S'-methylene O,O,O',O'-tetraethyl ester
AC 3422
Bis(S-(diethoxyphosphinothioyl)mercapto)methane
Bladan
Diethion
Embathion
ENT 24,105
Ethanox
Ethiol
Ethion (ACGIH,DOT,OSHA) [NA 2783]
Ethion mixture, Dry (DOT)

Ethodan
Ethyl methylene phosphorodithioate
FMC-1240
Fosfono 50
Hylemox
Itopaz
Kwit
Methyleen-S,S'-bis(O,O-diethyl-dithiofosfaat) (Dutch)
S,S'-Methylen-bis(O,O-diaethyl-dithiophosphat) (German)
Methylene-S,S'-bis(O,O-diaethyl-dithiophosphat) (German)
S,S'-Methylene O,O,O',O'-tetraethyl phosphorodithioate
NA 2783 [Organophosphorus pesticides, solid, toxic, N.O.S.]
Nia 1240
Niagara 1240
Nialate
Phosphorodithioic acid, O,O-diethyl ester, S,S-diester with methanedithiol
Phosphotox E
Rhodiacide
Rhodocide
Rodocid
RP 8167
Soprathion
O,O,O',O'-Tetraaethyl-bis(dithiophosphat) (German)
O,O,O',O'-Tetraethyl S,S'-methylenebisphosphordithioate
O,O,O',O'-Tetraethyl-S,S'-methylenebisphosphorodithioate
Tetraethyl S,S'-methylene bis-(phosphorothiolothionate)
O,O,O',O'-Tetraethyl S,S'-methylene di(phosphorodithioate)
Vegfru fosmite

563-16-6
C_8H_{18}
114.23
Hexane, 3,3-dimethyl- (9CI)
3,3-Dimethylhexane
NSC-74174

563-25-7
$C_8H_{18}F_2Sn$
270.95
Stannane, dibutyldifluoro
Dibutyldifluorostannane
Dibutyltin difluoride
Tin, dibutyl-, difluoride

563-41-7
$CH_5N_3O.ClH$
111.51
Cl.NNC(N)=O
Hydrazinecarboxamide, monohydrochloride (9CI)
Amidourea hydrochloride
Aminourea hydrochloride
Carbamylhydrazine hydrochloride
CH
Hydrazinecarboxamide, hydrochloride
NSC-4732
Semicarbazide chloride
Semicarbazide hydrochloride
Semicarbazide, monohydrochloride (8CI)

563-43-9
$C_2H_5AlCl_2$
126.95
Aluminum, dichloroethyl
Dichloroethylaluminum
Dichloromonoethylaluminum
Ethyl aluminum dichloride (DOT)
Ethyldichloroaluminum
UN 1924 (DOT)

563-45-1
C_5H_{10}
70.15
C(C(C)C)=C
1-Butene, 3-methyl
Isopentene [UN 2371]
3-Methyl-1-butene [UN 2561]
UN 2371 [Isopentenes]
UN 2561 [3-Methyl-1-butene]

563-46-2
C_5H_{10}
70.15
C(=C)(CC)C
1-Butene, 2-methyl
Isopentene [UN 2371]
2-Methyl-1-butene [UN 2459]
UN 2371 [Isopentenes]
UN 2459 [2-Methyl-1-butene]

563-47-3
C_4H_7Cl
90.56
C(=C)(CCl)C
Propene, 3-chloro-2-methyl
γ-Chloroisobutylene
3-Chlor-2-methyl-prop-1-en (German)
1-Chloro-2-methyl-2-propene
3-Chloro-2-methylpropene
3-Chloro-2-methyl-1-propene
3-Cloro-2-metil-prop-1-ene (Italian)
Chlorure de methallyle (French)
Cloruro di metallile (Italian)
Isobutenyl chloride
Methallyl chloride
β-Methallyl chloride
2-Methyl-allylchlorid (German)
β-Methylallyl chloride
2-Methylallyl chloride
Methyl allyl chloride [UN 2554]
NCI-C54820
UN 2554 [Methyl allyl chloride]

563-52-0
C_4H_7Cl
90.56
C(=C)C(Cl)C
1-Butene, 3-chloro
3-Chloro-1-butene
α-Methallyl chloride
α-Methylallyl chloride
1-Methylallyl chloride

563-54-2
$C_3H_4Cl_2$
110.97
CC(Cl)=CCl
Propene, 1,2-dichloro
Dichlor
1,2-Dichloropropene
1,2-Dichloropropylene
Dichlorpropen-gemisch (German)
PDC
Propylene dichloride
RCRA waste number U083

563-57-5
$C_3H_4Cl_2$
110.97
1-Propene, 3,3-dichloro- (9CI)
Propene, 3,3-dichloro- (8CI)

563-58-6
$C_3H_4Cl_2$
110.97
Propene, 1,1-dichloro
1,1-Dichloropropene
1,1-Dichloropropylene
1-Propene, 1,1-dichloro- (9CI)

563-68-8
$C_2H_3O_2.Tl$
263.42
[Tl].CC(O)=O
Acetic acid, thallium(i) salt
RCRA waste number U214
Thallium acetate
Thallium(1+) acetate
Thallium(I) acetate
Thallium monoacetate
Thallous acetate

563-71-3
$CH_2O_3.Fe$
117.87
Ferrous carbonate
Blaud's Mass
Carbonic acid, iron(2+) salt (1:1) (9CI)
Iron carbonate
Iron II carbonate
Iron (2+) carbonate

563-72-4
$C_2H_2O_4.Ca$
130.11
Ethanedioic acid, calcium salt

(1:1) (9CI)

563-76-8
$C_3H_4Br_2O$
215.87
O=C(Br)C(Br)C
Propanoyl bromide, 2-bromo- (9CI)
2-Bromopropanoyl bromide

563-78-0
C_6H_{12}
84.16
2,3-Dimethyl-1-butene
1-Butene, 2,3-dimethyl- (9CI)
NSC-73906

563-79-1
C_6H_{12}
84.16
C(=C(C)C)(C)C
2,3-Dimethyl-2-butene
AI3-37707
2-Butene, 2,3-dimethyl- (9CI)
2,3-Dimethylbutene-2
2,3-Dimethylbut-2-ene
NSC-73907
1,1,2,2-Tetramethylethylene

563-80-4
$C_5H_{10}O$
86.15
O=C(C(C)C)C
2-Butanone, 3-methyl
2-Acetyl propane
Isopropyl methyl ketone
3-Methyl-2-butanone
3-Methylbutan-2-one [UN 2397]
Methyl isopropyl ketone (ACGIH,OSHA)
MIPK
UN 2397 [3-Methylbutan-2-one]

563-83-7
C_4H_9NO
87.12
O=C(N)C(C)C
Isobutyramide (8CI)
Isobutylamide
Isobutyrimidic acid
Isopropylformamide
2-Methylpropanamide
2-Methylpropionamide
NSC-8423
Propanamide, 2-methyl- (9CI)

563-84-8
C_4H_9ClO
108.57
2-Butanol, 3-chloro- (8CI,9CI)
NSC-239709

564-00-1
$C_4H_6O_2$
86.10
C1OC1C2CO2
Butane, 1,2:3,4-diepoxy-, meso
(R*,S*)-2,2'-Bioxirane
1,2:3,4-Dianhydroerythritol
meso-Diepoxybutane
meso-1,2,3,4-Diepoxybutane
(R*,S*)-Diepoxybutane
Erythritol anhydride
Erythritol, 1,2:3,4-dianhydro-

564-02-3
C_8H_{18}
114.23
2,2,3-Trimethylpentane
NSC-73954
Pentane, 2,2,3-trimethyl- (9CI)

564-25-0
$C_{22}H_{24}N_2O_8$
444.48
CC3C2C(O)C1C(N(C)C)C(=C(C(N)=O)C(=O)C1(O)C(=C2C(=O)c4c(O)cccc34)O)O
2-Naphthacenecarboxamide, 4-α-S-(dimethylamino)-1,4,4a-α-5,5a-α,6,11,12a-octahydro-3,5-α,10,12,12a-α-pentahydroxy-6-α-methyl-1,11-dioxo
Doxiciclina (Italian)
α-6-Deoxy-5-hydroxytetracycline
α-6-Deoxyoxytetracycline
6-α-Deoxy-5-oxytetracycline
Doxycycline
GS-3065
5-Hydroxy-α-6-deoxytetracycline
Liviatin
Vibramycin

565-33-3
$C_{14}H_{21}N_3O_3S$
311.44
Urea, 1-cyclohexyl-3-(4-methylmetanilyl)
3-Amino-4-methylbenzenesulfonylcyclohexylurea
1-(3-Amino-p-tolylsulfonyl)-3-cyclohexylurea
Benzenesulfonamide, 3-amino-N-((cyclohexylamino)carbonyl)-4-methyl-
G 25,804
G 444E
N-Cyclohexyl-N'-(3-amino-4-methylbenzenesulfonyl)urea
Geigy 444E
Géigy Herbicide 444E
Euglycin
Glyhexylamide
Glyhexylamine isodiane
Melanex
Metahexamide
Methahexamide
Methexamide
Melonex
1,3,5-Triazine-2,4-diamine, 6-chloro-N,N,N',N'-tetraethyl-
S 1600
WP 40

565-59-3
C_7H_{16}
100.20
C(C(CC)C)(C)C
2,3-Dimethylpentane
3,4-Dimethylpentane
NSC-23696
Pentane, 2,3-dimethyl- (9CI)

565-60-6
$C_6H_{14}O$
102.18
2-Pentanol, 3-methyl- (9CI)
3-Methyl-2-pentanol
NSC-92741

565-61-7
$C_6H_{12}O$
100.16
2-Pentanone, 3-methyl- (9CI)
sec-Butyl methyl ketone
Methyl sec-butyl ketone
Methyl 1-methylpropyl ketone
3-Methyl-2-pentanone
NSC-66492

565-63-9
$C_5H_8O_2$
100.12
Crotonic acid, 2-methyl-, (Z)- (8CI)
Angelic acid
2-Butenoic acid, 2-methyl-, (Z)- (9CI)
cis-2-Methyl-2-butenoic acid
NSC-96885

565-64-0
$C_3H_4Cl_2O_2$
142.97
OC(=O)C(Cl)CCl
Propionic acid, 2,3-dichloro
2,3-Dichloropropionic acid

565-67-3
$C_6H_{14}O$
102.20
CCC(O)C(C)C
3-Pentanol, 2-methyl
2-Methyl-3-pentanol
Propanol, 1-isopropyl-

565-69-5
$C_6H_{12}O$
100.16
CCC(=O)C(C)C
3-Pentanone, 2-methyl- (8CI, 9CI)

565-70-8
$C_4H_8O_3$
104.11
CCC(O)C(O)=O
2-Hydroxybutyric acid
Butanoic acid, 2-hydroxy- (9CI)
Butyric acid, 2-hydroxy- (8CI)
2-Hydroxybutanoic acid
α-Hydroxybutyric acid
α-Hydroxy-n-butyric acid
NSC-6495

565-71-9
$C_3H_7NO_3$
105.09
NCC(O)C(O)=O
Propanoic acid, 3-amino-2-hydroxy- (9CI)
Isoserine (8CI)

565-74-2
$C_5H_9BrO_2$
181.05
O=C(O)C(Br)C(C)C
Butyric acid, 2-bromo-3-methyl
α-Bromoisovaleric acid
2-Bromoisovaleric acid
2-Bromo-3-methylbutanoic acid
2-Bromo-3-methylbutyric acid
Butanoic acid, 2-bromo-3-methyl- (9CI)

565-75-3
C_8H_{18}
114.23
C(C(C(C)C)C)(C)C
2,3,4-Trimethylpentane
NSC-24846
Pentane, 2,3,4-trimethyl- (9CI)

565-76-4
Unknown
Unknown
2,3,4-Trimethyl-1-pentene

565-77-5
C_8H_{16}
112.22
2-Pentene, 2,3,4-trimethyl- (8CI,9CI)
NSC-73943

565-80-0
$C_7H_{14}O$
114.19
O=C(C(C)C)C(C)C
3-Pentanone, 2,4-dimethyl- (9CI)
Diisopropyl ketone
2,4-Dimethylpentan-3-one
2,4-Dimethyl-3-pentanone
Isobutyrone
Isopropyl ketone
NSC-14662

567-18-0
$C_{10}H_8O_4S$
224.24
O=S(=O)(O)c(ccc(c1ccc2)c2)c1O
2-Naphthalenesulfonic acid, 1-hydroxy- (9CI)
1-Hydroxy-2-naphthalene-sulfonic acid

567-61-3
$C_8H_8O_3$
152.16
Cc1cccc(O)c1C(O)=O
2,6-Cresotic acid
Benzoic acid, 2-hydroxy-6-methyl- (9CI)
6-Methylsalicylic acid
6-MS
6-MSA

567-72-6
$C_{27}H_{42}O$
382.63
Cholesta-3,5-dien-7-one (9CI)
AI3-52863
3,5-Cholestadien-7-one
δ3,5-Cholestadien-7-one
NSC-134914

568-81-0
$C_{20}H_{16}$
256.36
Benz(a)anthracene, 6,12-dimethyl
4,9-Dimethyl-1,2-benzanthracene
6,12-Dimethylbenz(a)anthracene

569-41-5
$C_{12}H_{12}$
156.23
Cc1cccc2cccc(C)c12
Naphthalene, 1,8-dimethyl- (8CI,9CI)

569-51-7
$C_9H_6O_6$
210.14
OC(=O)c1cccc(C(O)=O)c1C(O)=O
Benzene 1,2,3-tricarboxylic acid
1,2,3-Benzenetricarboxylic acid (9CI)
Hemimellitic acid
NSC-401092
1,2,3-Tricarboxybenzene

569-57-3
$C_{23}H_{21}ClO_3$
380.89
COc1ccc(cc1)C(Cl)=C(c2ccc(OC)cc2)c3ccc(OC)cc3
Ethylene, chlorotris(p-methoxyphenyl)
Anisene
Benzene, 1,1',1''-(1-chloro-1-ethenyl-2-ylidene)tris-(4-methoxy)-
Chlorestrolo
1,1',1''-(1-Chloro-1-ethenyl-2-ylidene)-tris(4-methoxybenzene)
Chlorotrianisene
Chlorotrianizen
Chlorotrisin
Chlorotris(p-methoxyphenyl)-ethylene
Chlortrianisen
Clorestrolo
Clorotrisin
CTA
Hormonisene
Khlortrianizen
Merbentul
Metace
NSC-10108
Rianil
TACE
TACE-FN
Tri-p-anisylchloroethylene
Tris(p-methoxyphenyl)chloroethylene

569-61-9
$C_{19}H_{17}N_3.ClH$
323.85
[Cl-].Nc1ccc(cc1)C(=C2C=CC(=[NH2+])C=C2)c3ccc(N)cc3
Benzenamine, 4-((4-aminophenyl)(4-imino-2,5-cyclohexadien-1-ylidene)methyl), monohydrochloride
Basic Parafuchsine
Calcozine Magenta N
Cerven Zasadita 9 (Czech)
C.I. 42500
C.I. Basic Red 9, Monohydrochloride
p-Fuchsin
Fuchsine DR-001
Fuchsine SPC
Fuchsin SP
4,4'-((4-Imino-2,5-cyclohexadien-1-ylidene)methylene)dianiline monohydrochloride
NCI-C54739
Parafuchsin (German)
Parafuchsine
Para-Magenta
Pararosaniline
Pararosaniline chloride
Pararosaniline hydrochloride
p-Rosaniline HCL
Schultz-Tab. No. 779 (German)
4,4'4''-Triaminotriphenylmethan-hydrochlorid (German)

569-64-2
$C_{23}H_{25}N_2.Cl$
364.95
[Cl-].CN(C)c1ccc(cc1)C(=C2C=CC(C=C2)=[N+](C)C)c3c

cccc3
Ammonium, (4-(p-(dimethyl-amino)-α-phenylbenzylidene)-2,5-cyclohexadien-1-ylidene)- dimethyl-, chloride
Acryl Brilliant Green B
ADC Malachite Green Crystals
Aizen Malachite Green
Aizen Malachite Green Crystals
Aniline Green
Astra Malachite Green
Astra Malachite Green B
Astra Malachite Green BXX
Basic Green 4
Benzaldehyde Green
Benzal Green
Bronze Green Toner A-8002
Burma Green B
Calcozine Green V
China Green
China Green (Biological stain)
C.I. 42000
C.I. Basic Green 4
Diabasic Malachite Green
Diamond Green B
Diamond Green B Extra
Diamond Green BX
Diamond Green P Extra
Fast Green
Fast Green O
Green MX
Grenoble Green
Hidaco Malachite Green Base
Hidaco Malachite Green LC
Hidaco Malachite Green SC
Light Green N
Light Green N (Biological stain)
Lincoln Green Toner B 15-2900
Malachite Green
Malachite Green A
Malachite Green AN
Malachite Green B
Malachite Green Chloride
Malachite Green CP
Malachite Green Crystals
Malachite Green Crystals BPC
Malachite Green Hydrochloride
Malachite Green (Indicator)
Malachite Green J3E
Malachite Green Powder
Malachite Green WS
Malachite Lake Green A
Malachit-Grun (German)
Mitsui Malachite Green
New Victoria Green Extra I
New Victoria Green Extra II
New Victoria Green Extra O
OJI Malachite Green
Solid Green Crystals O
Solid Green O
Tertrophene Green M
Tetramethyl diapara-amido-triphenyl carbinol
Tokyo Aniline Malachite Green
Victoria Green
Victoria Green B
Victoria Green S
Victoria Green WB
Victoria Green WPB
Zelen Malachitova (Czech)
Zelen Zasadita 4 (Czech)

569-65-3
$C_{25}H_{27}ClN_2$
390.99
N(CCN(C1)Cc(cccc2C)c2)(C1)C(c(cccc3)c3)c(ccc(c4)Cl)c4
Piperazine, 1-(p-chloro-α-phenylbenzyl)-4-(m-methylbenzyl)
Ancolan
Ancolon
Bonadettes
Bonadoxin
Bonamine
Calmonal
Bonine
Chiclida
1-(p-Chlorobenzhydryl)-4-(m-methylbenzyl)diethylenediamine
1-p-Chlorobenzhydryl-4-m-methylbenzylpiperazine
1-(p-Chloro-α-phenylbenzyl)-4-(m-methylbenzyl)piperazine
Histamethine
Histamethizine
Histametizine
Histametizyne
Itinerol
Longifene
Marex
Meclizine
Meclozine
Monamine
Navicalm
Neo-Istafene
Neo-Suprimal
Neo-Suprimel
Parachloramine
Peremesin
Postafen
Postafene
Sabari
Sea-Legs
Siguran
Subari
Suprimal
Travelon
UCB 170
UCB 5052
UCB 5062
Vibazine
Vomissels

570-24-1
$C_7H_8N_2O_2$
152.14
Cc1cccc(N(=O)=O)c1N
Benzenamine, 2-methyl-6-nitro- (9CI)
NSC-286

570-74-1
$C_{27}H_{46}$
370.66
Cholest-5-ene (9CI)
NSC-118131

571-58-4
$C_{12}H_{12}$
156.23
Cc1ccc(C)c2ccccc12
1,4-Dimethylnaphthalene
1,4-Dimethylnapthalene
Naphthalene, 1,4-dimethyl- (9CI)
NSC-61779

571-60-8
$C_{10}H_8O_2$
160.18
Oc(c(c(c(O)c1)ccc2)c2)c1
1,4-Naphthalenediol

571-61-9
$C_{12}H_{12}$
156.23
Cc1cccc2c(C)cccc12
Naphthalene, 1,5-dimethyl- (9CI)
1,5-Dimethylnaphthalene
NSC-59388

572-89-4
$C_{20}H_{16}$
256.35
Benz(a)anthracene, 2,9-dimethyl- (9CI)

573-11-5
$C_{10}H_{12}O_5$
212.20
Benzoic acid, 2,3,4-trimethoxy- (8CI,9CI)

573-35-3
$C_6H_{13}O_9P$
260.14
myo-Inositol, 1-(dihydrogen phosphate) (9CI)
Inositol, 1-(dihydrogen phosphate), myo- (8CI)
Inositol monophosphate

573-56-8
$C_6H_4N_2O_5$
184.12
Oc1c(cccc1N(=O)=O)N(=O)=O
Phenol, 2,6-dinitro
β-Dinitrophenol
2,6-Dinitrofenol (Czech)
2,6-Dinitrophenol

573-58-0
$C_{32}H_{24}N_6O_6S_2.2Na$
698.72
[Na+].[Na+].Nc5c(N=Nc1ccc(cc1)c4ccc(N=Nc3cc(c2ccccc2c3N)S([O-])(=O)=O)cc4)cc(c6ccccc56)S([O-])(=O)=O
1-Naphthalenesulfonic acid, 3,3'-(4,4'-biphenylenebis-(azo))bis(4-amino-, disodium

salt
Atlantic Congo Red
Atul Congo Red
Azocard Red Congo
Benzo Congo Red
Brasilamina Congo 4B
Cerven Kongo (Czech)
Cerven Prima 28 (Czech)
C.I. 22120
C.I. Direct Red 28
C.I. Direct Red 28, Disodium salt
Congo Red
Congo Red 4B
Congo Red 4BX
Congo Red CR
Congo Red H
Congo Red ICI
Congo Red L
Congo Red M
Congo Red N
Congo Red R
Congo Red RS
Congo Red W
Cotton Red 5B
Cotton Red 4BC
Cotton Red L
CR
Diacotton Congo Red
Direct Red 28
Direct Red C
Direct Red DC-CF
Direct Red K
Erie Congo 4B
Hispamin Congo 4B
Kayaku Congo Red
Mitsui Congo Red
Peeramine Congo Red
Sodium diphenyldiazo-bis-(α-naphthylaminesulfonate)
Sugai Congo Red
Tertrodirect Red C
Trisulfon Congo Red
Vondacel Red CL

573-89-7
$C_{16}H_{12}N_2O_4S$
328.34
O=S(=O)(O)c(ccc(N=Nc(c(c(ccc1)cc2)c1)c2O)c3)c3
Benzenesulfonic acid, 4-((2-hydroxy-1-naphthalenyl)azo)-(9CI)
C.I. Acid Orange 7, Free acid

573-98-8
$C_{12}H_{12}$
156.23
Cc2ccc1ccccc1c2C
Naphthalene, 1,2-dimethyl-(9CI)
1,2-Dimethylnaphthalene
NSC-59832

574-00-5
$C_{10}H_8O_2$
160.17
1,2-Dihydroxynaphthalene
1,2-Naphthalenediol
NSC-401609

574-42-5
$C_{26}H_{22}O$
350.46
Benzene, 1,1',1'',1'''-(oxydimethylidyne)tetrakis- (9CI)
AI3-15103
Benzhydrol ether
Benzhydryl ether
Benzohydrol ether
Bis(benzhydryl) ether
Bis(diphenylmethyl) ether
Dibenzhydryl ether
Dibenzohydryl ether
Diphenylmethyl ether
Ether, bis(diphenylmethyl) (8CI)
NSC-2438
1,1,1',1'-Tetraphenyldimethyl ether

574-69-6
$C_{16}H_{12}N_2O_4S$
328.34
O=S(=O)(O)c(ccc(N=Nc(c(c(c(O)c1)ccc2)c2)c1)c3)c3
Benzenesulfonic acid, 4-((4-hydroxy-1-naphthalenyl)azo)- (9CI)

574-93-6
$C_{32}H_{18}N_8$
514.50
n(c(nc(nc(nc(nc(nc(nc(n1)c2ccc3)c23)c4cccc5)c45)c6cccc7)c67)c(c8ccc9)c9)c18
Phthalocyanine
CI Pigment Blue 16
5,28:14,19-Diimino-7,12:26,21-dinitrilotetrabenzo(c,h,m,r)-(1,6,11,16) tetraazacycloeicosine
Heliogen Blue g
Heliogen Blue 7560
Heliogen Blue 7800
Irgazin Blue 3GT
Lionol Blue KW
Monolite Fast Blue GS
29H,31H-Phthalocyanine (9CI)
Pigment Blue Green Phthalocyanine U
Polymon Blue G
Tetrabenzo(b,g,l,q)porphyrazine

574-98-1
$C_{10}H_8BrNO_2$
254.08
O=C(N(C(=O)c1cccc2)CCBr)c12
Phthalimide, N-(2-bromoethyl)- (8CI)
2-(2-Bromoethyl)-1H-isoindole-1,3(2H)-dione
β-Bromoethylphthalimide
2-(Bromoethyl)phthalimide
N-(2-Bromoethyl)phthalimide
1-Bromo-2-phthalimidoethane
1H-Isoindole-1,3(2H)-dione, 2-(2-bromoethyl)- (9CI)
NSC-2688
β-Phthalimidoethyl bromide

575-37-1
$C_{12}H_{12}$
156.23
Cc2ccc1cccc(C)c1c2
Naphthalene, 1,7-dimethyl-(9CI)
1,7-Dimethylnaphthalene
NSC-60773

575-41-7
$C_{12}H_{12}$
156.24
Cc2cc(C)c1ccccc1c2
Naphthalene, 1,3-dimethyl
1,3-Dimethylnaphthalene

575-43-9
$C_{12}H_{12}$
156.23
Cc2ccc1c(C)cccc1c2
1,6-Dimethylnaphthalene
AI3-17608
Dimethylnaphthalene
Naphthalene, 1,6-dimethyl-(9CI)
NSC-52966

575-44-0
$C_{10}H_8O_2$
160.17
Oc(c(c(cc1)cc(O)c2)c2)c1
1,6-Naphthalenediol (9CI)
C.I. 76630
1,6-Dihydroxynaphthalene
6-Hydroxy-1-naphthol
Naphthalene, 1,6-dihydroxy-
2,5-Naphthalenediol
NSC-7201

575-89-3
$C_8H_5Cl_3O_3$
255.48
Acetic acid, (2,4,6-trichlorophenoxy)- (9CI)
AI3-09412
NSC-61993
2,4,6-T
(2,4,6-Trichlorophenoxy)acetic acid

575-90-6
$C_8H_6Cl_2O_3$
221.04
Acetic acid, (2,6-dichlorophenoxy)- (9CI)
2,6-D
2,6-Dichlorophenoxyacetate
(2,6-Dichlorophenoxy)acetic acid
NSC-409411

576-24-9
$C_6H_4Cl_2O$
163.00
Oc(c(c(cc1)Cl)Cl)c1
Phenol, 2,3-dichloro

576-26-1
$C_8H_{10}O$
122.18
Oc(c(ccc1)C)c1C
2,6-Xylenol
2,6-Dimethylphenol
2,6-DMP

576-55-6
$C_7H_4Br_4O$
423.72
Oc(c(c(c(c1Br)Br)Br)C)c1Br
Tetrabromo-o-cresol
AI3-01565
o-Cresol, tetrabromo- (8CI)
o-Cresol, 3,4,5,6-tetrabromo-
2-Methyl-3,4,5,6-tetrabromophenol
NSC-4866
Phenol, 2,3,4,5-tetrabromo-6-methyl- (9CI)
3,4,5,6-Tetrabromo-o-cresol
2,3,4,5-Tetrabromo-6-methylphenol
3,4,5,6-Tetrabromo-2-methylphenol

576-68-1
$C_{10}H_{22}Cl_2N_2O_4$
305.24
OC(CNCCCl)C(O)C(O)C(O)CNCCCl
Mannitol, 1,6-bis((2-chloroethyl)amino)-1,6-dideoxy-, D-,
BCM
1,6-Bis(chloroethylamino)-1,6-bis-deoxy-D-mannitol
1,6-Bis(chloroethylamino)-1,6-dideoxy-D-mannite
1,6-Bis((β-chloroethyl)amino)-1,6-dideoxy-D-mannitol
1,6-Bis((2-chloroethyl)amino)-1,6-dideoxy-D-mannitol
Degranol
Mannit-Lost (German)
Mannit-Mustard (German)
Mannitol Mustard
Mannitol Nitrogen Mustard
Mannomustine

577-11-7
$C_{20}H_{38}O_7S.Na$
445.63
[Na+].CCCCC(CC)COC(=O)CC(C(=O)OCC(CC)CCCC)S([O-])(=O)=O
Succinic acid, sulfo-, 1,4-bis-(2-ethylhexyl) ester, sodium salt
Aerosol GPG
Aerosol OT
Aerosol OT 75
Aerosol OT-B
Alcopol O
Alphasol OT
Berol 478
Bis(ethylhexyl) ester of sodium sulfosuccinic acid
Bis-2-ethylhexylester sulfo-jantaranu sodneho (Czech)
Bis(2-ethylhexyl)sodium sulfosuccinate
Bis(2-ethylhexyl) S-sodium sulfosuccinate
1,4-Bis(2-ethylhexyl) sodium sulfosuccinate
Butanedioic acid, sulfo-, 1,4-bis(2-ethylhexyl) ester, sodium salt (9CI)
Celanol DOS 75
Clestol
Colace
Complemix
Constonate
Coprol
Defilin
Di-(2-ethylhexyl) sodium sulfosuccinate
Dioctlyn
Dioctylal
Dioctyl ester of sodium sulfosuccinate
Dioctyl ester of sodium sulfosuccinic acid
Dioctyl-Medo Forte
Dioctyl sodium sulfosuccinate
Dioctyl sulfosuccinate sodium
Dioctyl sulfosuccinate sodium salt
Diomedicone
Diosuccin
Diotilan
Diovac
Docusate sodium
Doxinate
Doxol
D-S-S
Dulsivac
Duosol
2-Ethylhexyl sulfosuccinate sodium
Humifen WT 27G
Konlax
Kosate
Laxinate
Laxinate 100
Manoxal OT
Manoxol OT
Mervamine
Modane Soft
Molatoc
Molcer
Molofac
Monawet MD 70E
Monawet MO-70
Monawet MO-70 RP
Monawet MO-84 R2W
Monoxol OT
Nekal WT-27
Nevax
Nikkol OTP 70
Norval
Obston
Rapisol
Regutol
Requtol
Revac
Sanmorin OT 70
SBO
Sobital
Sodium bis(2-ethylhexyl) sulfosuccinate
Sodium di-(2-ethylhexyl) sulfosuccinate
Sodium dioctyl sulfosuccinate
Sodium dioctyl sulphosuccinate
Sodium 2-ethylhexylsulfosuccinate
Sodium sulfodi-(2-ethylhexyl)-sulfosuccinate
Softil
Soliwax
Solusol-75%
Solusol-100%
Sulfimel DOS
Sulfosuccinic acid, bis(2-ethylhexyl)ester sodium salt
SV 102
Tex-Wet 1001
Triton GR-5
Triton GR 7
Vatsol OT
Velmol
Waxsol
Wetaid SR

577-16-2
$C_9H_{10}O$
134.18
Acetophenone, 2'-methyl- (8CI)
o-Acetyltoluene
2-Acetyltoluene
Ethanone, 1-(2-methylphenyl)- (9CI)
o-Methylacetophenone
2-Methylacetophenone
2'-Methylacetophenone
2'-Methylacetylphenone
NSC-84233

577-19-5
$C_6H_4BrNO_2$
202.02
O=N(=O)c(c(ccc1)Br)c1
Benzene, 1-bromo-2-nitro
o-Bromonitrobenzene
2-Bromonitrobenzene
1-Bromo-2-nitrobenzene (DOT)
o-Nitrobromobenzene
o-Nitrobromobenzene, Liquid [UN 2732]
2-Nitrobromobenzene
UN 2732 [Nitrobromobenzenes liquid]

577-55-9
$C_{12}H_{18}$
162.30
c(c(ccc1)C(C)C)(c1)C(C)C
Benzene, o-diisopropyl

o-Diisopropylbenzene

577-59-3
$C_8H_7NO_3$
165.16
CC(=O)c1ccccc1N(=O)=O
Acetophenone, 2'-nitro
Ethanone, 1-(2-nitrophenyl)- (9CI)
o-Nitroacetophenone
2'-Nitroacetophenone
1-(2-Nitrophenyl)ethanone

577-71-9
$C_6H_4N_2O_5$
184.12
Oc1ccc(N(=O)=O)c(c1)N(=O)=O
Phenol, 3,4-dinitro
3,4-Dinitrofenol (Czech)
3,4-Dinitrophenol

578-46-1
$C_7H_8N_2O_2$
152.17
m-Toluidine, 6-nitro
3-Amino-4-nitrotoluene
6-Nitro-m-toluidine

578-54-1
$C_8H_{11}N$
121.20
Nc(c(ccc1)CC)c1
Aniline, 2-ethyl
o-Aminoethylbenzene
Aniline, o-ethyl- (8CI)
Benzenamine, 2-ethyl- (9CI)
2-Ethyl aniline
o-Ethylaniline
2-Ethylaniline [UN 2273]
2-Ethylbenzenamine
UN 2273 [2-Ethylaniline]

578-57-4
C_7H_7BrO
187.05
O(c(c(ccc1)Br)c1)C
Anisole, o-bromo
Anisyl bromide
Benzene, 1-bromo-2-methoxy- (9CI)
o-Bromoanisole
2-Bromoanisole
o-Bromophenyl methyl ether
o-Methoxybromobenzene
2-Methoxybromobenzene
o-Methoxyphenyl bromide
2-Methoxyphenyl bromide

578-66-5
$C_9H_8N_2$
144.19
n(c(c(ccc1)cc2)c1N)c2
Quinoline, 8-amino
8-Aminoquinoline

578-94-9
$C_{12}H_9AsClN$
277.59
Phenarsazine, 10-chloro-5,10-dihydro
Adamsite
5-Aza-10-arsenaanthracene chloride
10-Chloro-5,10-dihydro-arsacridine
10-Chloro-5,10-dihydrophen-arsazine
Diphenylaminechlorarsine
Diphenylamine chloroarsine [UN 1698]
DM
Fenarsazinchlorid (Czech)
Phenarsazine chloride
Phenazarsine chloride
UN 1698 [Diphenylamine chloroarsine]

578-95-0
$C_{13}H_9NO$
195.23
Oc(c(c(nc1cccc2)ccc3)c3)c12
9-Acridanone
Acridanone
9(10H)-Acridinone (9CI)
Acridone
9-Acridone
9(10H)-Acridone

579-07-7
$C_9H_8O_2$
148.16
O=C(c(cccc1)c1)C(=O)C
1,2-Propanedione, 1-phenyl- (9CI)
Acetylbenzoyl
AI3-23868
Benzoylacetyl
Benzoyl methyl ketone
Methylphenylglyoxal
NSC-7643
Phenylmethyldiketone
1-Phenyl-1,2-propanedione
3-Phenyl-2,3-propanedione
Pyruvophenone

579-23-7
$C_{22}H_{22}O_5$
366.44
Cyclohexanone, 2,6-vanillylidene
2,6-Divanillylidenecyclohexan-one

579-66-8
$C_{10}H_{15}N$
149.26
Nc(c(ccc1)CC)c1CC
Aniline, 2,6-diethyl
Benzenamine, 2,6-diethyl- (9CI)
2,6-Diethylaniline

579-75-9
$C_8H_8O_3$
152.16
O=C(O)c(c(OC)ccc1)c1
o-Anisic acid
2-Anisic acid
Benzoic acid, 2-methoxy- (9CI)
Kyselina 2-methoxybenzoova (Czech)
o-Methoxybenzoic acid
2-Methoxybenzoic acid
o-Methylsalicylic acid
Salicylic acid methyl ether

580-13-2
$C_{10}H_7Br$
207.07
c(c(ccc1Br)ccc2)(c2)c1
Naphthalene, 2-bromo- (9CI)
AI3-19928
2-Bromonaphthalene

580-15-4
$C_9H_8N_2$
144.16
Nc2ccc1ncccc1c2
6-Aminoquinoline
6-Quinolinamine (9CI)
Quinoline, 6-amino- (8CI)

580-17-6
$C_9H_8N_2$
144.19
n(c(c(ccc1)cc2N)c1)c2
Quinoline, 3-amino
3-Aminoquinoline
3-Quinolineamine

580-22-3
$C_9H_8N_2$
144.19
n(c(c(ccc1)cc2)c1)c2N
Quinoline, 2-amino
2-Aminoquinoline
2-Quinolinamine (9CI)

580-48-3
$C_{11}H_{20}ClN_5$
257.81
CCN(CC)c1nc(Cl)nc(n1)N(CC)CC
s-Triazine, 2-chloro-4,6-bis-(diethylamino)
Chlorazine
2-Chloro-4,6-bis(diethylamino)-s-triazine
6-Chloro-N,N,N',N'-tetraethyl-1,3,5-triazine-2,4-diamine

580-51-8
$C_{12}H_{10}O$
170.21
Oc(cccc1c(cccc2)c2)c1
3-Hydroxybiphenyl
(1,1'-Biphenyl)-3-ol (9CI)
3-Biphenylol (8CI)

581-28-2
$C_{13}H_{10}N_2$
194.25
Nc3ccc2nc1ccccc1cc2c3
Acridine, 2-amino
2-Acridinamine (9CI)
2-Aminoacridine
3-Aminoacridine (European)
2-Aminoakridin (Czech)

581-29-3
$C_{13}H_{10}N_2$
194.25
Nc3ccc2cc1ccccc1nc2c3
Acridine, 3-amino
3-Acridinamine (9CI)
2-Aminoacridine (European)
3-Aminoacridine
3-Aminoakridin (Czech)

581-30-6
$C_{12}H_7NOS$
213.25
3H-Phenothiazin-3-one (9CI)
AI3-03564

581-40-8
$C_{12}H_{12}$
156.23
Cc2cc1ccccc1cc2C
Naphthalene, 2,3-dimethyl- (9CI)
AI3-17609

581-42-0
$C_{12}H_{12}$
156.23
Cc2ccc1cc(C)ccc1c2
2,6-Dimethylnaphthalene
AI3-01876
Naphthalene, 2,6-dimethyl- (9CI)

581-49-7
$C_{10}H_{12}N_2$
160.24
Anatabine
(-)-Anatabine
2,3'-Bipyridine, 1,2,3,6-tetrahydro-, (S)- (9CI)

581-50-0
$C_{10}H_8N_2$
156.20
c1ccc(nc1)c2cccnc2
2,3'-Bipyridine
α,β'-Bipyridine
2,3'-Bipyridyl
2,3'-Dipyridine
α,β-Dipyridyl
2,3'-Dipyridyl
Isonicoteine

581-75-9
$C_{10}H_8O_6S_2$
288.30
O=S(=O)(O)c(ccc(c1ccc2S(=O)(=O)O)c2)c1
Naphthalene-2,6-disulfonic acid
N-2,6-DSA
2,6-Naphthalenedisulfonic acid (9CI)

581-89-5
$C_{10}H_7NO_2$
173.18
O=N(=O)c2ccc1ccccc1c2
Naphthalene, 2-nitro
β-Nitronaphthalene
2-Nitronaphthalene

581-96-4
$C_{12}H_{10}O_2$
186.22
OC(=O)Cc2ccc1ccccc1c2
2-Naphthaleneacetic acid
Acide β-naphthylacetique (French)
Betoxan
β-Naphthaleneacetate
β-Naphthaleneacetic acid
β-Naphthylacetic acid
2-(2-Naphthyl)acetic acid

582-16-1
$C_{12}H_{12}$
156.23
Naphthalene, 2,7-dimethyl- (9CI)
AI3-17610

582-17-2
$C_{10}H_8O_2$
160.18
Oc(ccc(c1cc(O)c2)c2)c1
2,7-Naphthalenediol
C.I. 76645
2,7-Dihydroxynaphthalene
Naphthalenediol-2,7 (French)
Naphthalene-2,7-diol

582-25-2
$C_7H_6O_2.K$
161.23
Potassium benzoate
Benzoic acid, potassium salt (9CI)

582-33-2
$C_9H_{11}NO_2$
165.19
O=C(OCC)c(cccc1N)c1
Tricaine
AI3-02743
m-Aminobenzoic acid, ethyl ester
3-Aminobenzoic acid, ethyl ester, methanesulfonate
Benzoic acid, m-amino-, ethyl ester (8CI)
Benzoic acid, 2-amino-, ethyl ester
Benzoic acid, 3-amino-, ethyl ester (9CI)
Ethyl 3-aminobenzoate

582-60-5
$C_9H_{10}N_2$
146.21
N(c(c(N1)cc(c2C)C)c2)=C1
Benzimidazole, 5,6-dimethyl
5,6-Dimethylbenzimidazole

582-61-6
$C_7H_5N_3O$
147.12
Benzoyl azide
Azida de benzoilo (Spanish)
Azoture de benzoyle (French)
Benzazide
Benzoic acid azide

583-04-0
$C_{10}H_{10}O_2$
162.19
O=C(OCC=C)c(cccc1)c1
Benzoic acid, allyl ester (8CI)
AI3-07823
Benzoic acid, 2-propenyl ester (9CI)

583-08-4
$C_8H_8N_2O_3$
180.15
Nicotinuric acid
Glycine, N-nicotinoyl- (8CI)
Glycine, N-(3-pyridinylcarbonyl)- (9CI)

583-15-3
$C_{14}H_{10}O_4.Hg$
442.83
Mercury(II) benzoate
Mercuric benzoate
Mercuric benzoate, Solid (DOT)
Mercury benzoate [UN 1631]
UN 1631 [Mercury benzoate]

583-39-1
$C_7H_6N_2S$
150.21
N(c(c(N1)ccc2)c2)=C1S
2-Benzimidazolethiol
Antiegene MB
Antioxidant MB (Czech)
AOMB
ASM MB
2-Benzimidazolinthion (Czech)
2-Mercaptobenzimidazole
Mercaptobenzoimidazole
2-Mercaptobenzoimidazole
Merkaptobenzimidazol (Czech)
2-Merkaptobenzimidazol (Czech)
NCI-C60980

o-Phenylenethiourea
USAF EK-6540
USAF XF-21

583-48-2
C_8H_{18}
114.23
C(C(CC)C)(CC)C
Hexane, 3,4-dimethyl- (9CI)
3,4-Dimethylhexane

583-53-9
$C_6H_4Br_2$
235.91
c(c(ccc1)Br)(c1)Br
Benzene, o-dibromo- (8CI)
AI3-10009
Benzene, 1,2-dibromo- (9CI)
1,2-Dibromobenzene

583-55-1
C_6H_4BrI
282.91
c(c(ccc1)Br)(c1)I
Benzene, 1-bromo-2-iodo- (9CI)
Benzene, 1-bromo-2-iodo-
1-Bromo-2-iodobenzene

583-57-3
C_8H_{16}
112.24
Cyclohexane, 1,2-dimethyl
o-Dimethylcyclohexane
1,2-Dimethylcyclohexane [UN 2263]
UN 2263 [Dimethylcyclohexanes]

583-58-4
C_7H_9N
107.17
n(ccc(c1C)C)c1
3,4-Lutidine
3,4-Dimethylpyridine
Pyridine, 3,4-dimethyl- (9CI)

583-59-5
$C_7H_{14}O$
114.21
OC(C(CCC1)C)C1
Cyclohexanol, o-methyl
o-Methylcyclohexanol

583-60-8
$C_7H_{12}O$
112.19
O=C(C(CCC1)C)C1
Cyclohexanone, 2-methyl
2-Methyl-cyclohexanon (German, Dutch)
o-Methylcyclohexanone (ACGIH,OSHA)
2-Methylcyclohexanone
2-Metilcicloesanone (Italian)

583-61-9
C_7H_9N
107.15
n(c(c(cc1)C)C)c1
Pyridine, 2,3-dimethyl- (9CI)
AI3-24280
2,3-Dimethylpyridine
2,3-Lutidine (8CI)

583-63-1
$C_6H_4O_2$
108.10
o-Benzoquinone
Benzoquinone [UN 2587]
1,2-Benzoquinone
3,5-Cyclohexadiene-1,2-dione (9CI)
o-Quinone
UN 2587 [Benzoquinone]

583-78-8
$C_6H_4Cl_2O$
163.00
Oc(c(ccc1Cl)Cl)c1
Phenol, 2,5-dichloro
2,5-Dichlorophenol

583-91-5
$C_5H_{10}O_3S$
150.21
O=C(O)C(O)CCSC
Butyric acid, 2-hydroxy-4-(methylthio)
Alimet
Butanoic acid, 2-hydroxy-4-(methylthio)- (9CI)
2-Hydroxy-4-(methylthio)-butanoic acid
Methionine hydroxy analog
MHA acid
MHA-FA

583-92-6
$C_5H_8O_3S$
148.18
2-Keto-4-methylthiobutyric acid
α-Keto-γ-methiolbutyrate
α-Ketomethionine
2-Keto-4-thiomethylbutyrate
2-Ketothiomethylbutyric acid
S-Methyl-α-ketobutyric acid
4-Methylmercapto-2-oxobutyrate
4-Methylthio-2-ketobutyric acid
4-Methylthio-2-oxobutanoate
4-(Methylthio)-2-oxobutanoic acid
α-Oxomethionine
α-Oxo-γ-methylthiobutyric acid
2-Oxo-4-thiomethylbutyric acid

584-02-1
$C_5H_{12}O$
88.17
OC(CC)CC
3-Pentanol
Diethyl carbinol
Diethylcarbinol (DOT)
Isoamyl alcohol (ACGIH)
Pentanol-3
Pentan-3-ol
UN 2706 (DOT)

584-03-2
$C_4H_{10}O_2$
90.14
OCC(O)CC
1,2-Butanediol
1,2-Butylene glycol

584-08-7
$CO_3.2K$
138.21
Potassium carbonate (2:1)
Carbonic acid, dipotassium salt
Kaliumcarbonat (German)
K-Gran
Pearl Ash
Potash

584-79-2
$C_{19}H_{26}O_3$
302.45
O=C(OC(C(=C(C1=O)CC=C)C)C1)C(C2(C)C)C2C=C(C)C
Cyclopropanecarboxylic acid, 2,2-dimethyl-3-(2-methyl-1-propenyl)-, 2-methyl-4-oxo- 3-(2-propenyl)-2-cyclopenten-1-yl ester
(+)-Allelrethonyl (+)-cis,trans-chrysanthemate
Allethrin
d-Allethrin
d-trans Allethrin
Allethrin (DOT)
Allethrin I
Allyl cinerin
Allyl homolog of cinerin I
d,l-2-Allyl-4-hydroxy-3-methyl-2-cyclopenten-1-one-d,l-chrysanthemum monocarboxylate
3-Allyl-4-keto-2-methylcyclopentenyl chrysanthemum-monocarboxylate
3-Allyl-2-methyl-4-oxo-2-cyclopenten-1-yl chrysanthemate
dl-3-Allyl-2-methyl-4-oxocyclopent-2-enyl DL-cis trans chrysanthemate
Allylrethronyl dl-cis-trans-chrysanthemate
Bioaltrina (Portuguese)
Bioallethrin
Cinerin I Allyl Homolog
Depallethrin
ENT 17,510
Exthrin
FDA 1446
FMC 249
NA 2902 (DOT)
Necarboxylic acid
NIA 249

OMS 468
Pallethrine
Pynamin
Pynamin-Forte
Pyresin
Pyresyn
Synthetic Pyrethrins

584-84-9
$C_9H_6N_2O_2$
174.17
O=C=Nc(c(ccc1N=C=O)C)c1
Benzene, 2,4-diisocyanato-1-methyl
Cresorcinol diisocyanate
Desmodur T80
Di-isocyanate de toluylene (French)
Di-iso-cyanatoluene
2,4-Diisocyanato-1-methylbenzene (9CI)
2,4-Diisocyanatotoluene
Diisocyanat-toluol (German)
Isocyanic acid, methylphenylene ester
Isocyanic acid, 4-methyl-m-phenylene ester
Hylene T
Hylene TCPA
Hylene TLC
Hylene TM
Hylene TM-65
Hylene TRF
4-Methyl-phenylene diisocyanate
4-Methyl-phenylene isocyanate
Mondur TD
Mondur TD-80
Mondur TDS
Nacconate IOO
NCI-C50533
Niax TDI
Niax TDI-P
RCRA waste number U223
Rubinate TDI 80/20
TDI (OSHA)
2,4-TDI
TDI-80
Tolueen-diisocyanaat (Dutch)
Toluen-disocianato (Italian)
Toluene diisocyanate
Toluene-2,4-diisocyanate (ACGIH,OSHA)
2,4-Toluenediisocyanate
Toluilenodwuizocyjanian (Polish)
Tuluylendiisocyanat (German)
Toluylene-2,4-diisocyanate
Tolyene 2,4-diisocyanate
meta-Tolylene diisocyanate
Tolylene-2,4-diisocyanate
2,4-Tolylenediisocyanate

584-90-7
$C_{14}H_{14}N_2$
210.27
Diazene, bis(2-methylphenyl)-(9CI)
o,o'-Azotoluene (8CI)
NSC-31007

584-93-0
$C_5H_9BrO_2$
181.05
O=C(O)C(Br)CCC
Valeric acid, 2-bromo
α-Bromovaleric acid
Valeric acid, α-bromo-

584-94-1
C_8H_{18}
114.23
2,3-Dimethylhexane
Hexane, 2,3-dimethyl- (9CI)

585-07-9
$C_8H_{14}O_2$
142.22
CC(=C)C(=O)OC(C)(C)C
Methacrylic acid, tert-butyl ester
tert-Butyl methacrylate
2-Propenoic acid, 2-methyl-, 1,1-dimethylethyl ester (9CI)

585-25-1
$C_8H_{14}O_2$
142.20
2,3-Octanedione (8CI,9CI)
NSC-7642

585-34-2
$C_{10}H_{14}O$
150.22
Oc(cccc1C(C)(C)C)c1
Phenol, 3-(1,1-dimethylethyl)-(9CI)
3-(1,1-Dimethylethyl)phenol

585-71-7
C_8H_9Br
185.06
BrC(c(cccc1)c1)C
(1-Bromoethyl)benzene
Benzene, (1-bromoethyl)- (9CI)
(α-Bromoethyl)benzene
1-Bromo-1-phenylethane
α-Methylbenzyl bromide
α-Phenethyl bromide
1-Phenethyl bromide
1-Phenyl-1-bromoethane
α-Phenylethyl bromide
1-Phenylethyl bromide

585-74-0
$C_9H_{10}O$
134.18
CC(=O)c1cccc(C)c1
Acetophenone, 3'-methyl-(8CI)
Ethanone, 1-(3-methylphenyl)-(9CI)

585-76-2
$C_7H_5BrO_2$
201.02
O=C(O)c(cccc1Br)c1
Benzoic acid, m-bromo- (8CI)
AI3-08854
Benzoic acid, 3-bromo- (9CI)
3-Bromobenzoic acid

585-79-5
$C_6H_4BrNO_2$
202.02
O=N(=O)c(cccc1Br)c1
Benzene, 1-bromo-3-nitro
m-Bromonitrobenzene
1-Bromo-3-nitrobenzene (DOT)
3-Bromonitrobenzene
m-Nitrobromobenzene, Solid [UN 2732]
3-Nitrobromobenzene
UN 2732 [Nitrobromobenzenes solid]

585-84-2
$C_6H_6O_6$
174.11
O=C(O)C(=CC(=O)O)CC(=O)O
1-Propene-1,2,3-tricarboxylic acid, (Z)- (9CI)
(Z)-1-Propene-1,2,3-tricarboxylic acid

586-11-8
$C_6H_4N_2O_5$
184.12
Oc1cc(cc(c1)N(=O)=O)N(=O)=O
Phenol, 3,5-dinitro
3,5-Dinitrofenol (Czech)
3,5-Dinitrophenol

586-37-8
$C_9H_{10}O_2$
150.18
COc1cccc(c1)C(C)=O
Acetophenone, 3'-methoxy-(8CI)
AI3-26011
Ethanone, 1-(3-methoxyphenyl)-(9CI)

586-38-9
$C_8H_8O_3$
152.16
O=C(O)c(cccc1OC)c1
m-Anisic acid
Benzoic acid, 3-methoxy- (9CI)
m-Methoxybenzoic acid
3-Methoxybenzoic acid

586-61-8
$C_9H_{11}Br$
199.09
c(ccc(c1)Br)(c1)C(C)C
Benzene, 1-bromo-4-(1-methylethyl)- (9CI)
p-Bromoisopropylbenzene
1-Bromo-4-(1-methylethyl)-

benzene
Cumene, p-bromo- (8CI)

586-62-9
$C_{10}H_{16}$
136.26
C(=C(C)C)(CCC(=C1)C)C1
Terpinolene [UN 2541]
Cyclohexene, 1-methyl-4-(1-methylethylidene)-
UN 2541 [Terpinolene]

586-76-5
$C_7H_5BrO_2$
201.03
O=C(O)c(ccc(c1)Br)c1
Benzoic acid, p-bromo
Benzoic acid, 4-bromo- (9CI)
p-Bromobenzoic acid
4-Bromobenzoic acid
p-Carboxybromobenzene

586-78-7
$C_6H_4BrNO_2$
202.01
O=N(=O)c(ccc(c1)Br)c1
Benzene, 1-bromo-4-nitro
p-Bromonitrobenzene
4-Bromonitrobenzene
p-Nitrobromobenzene, Liquid [UN 2732]
4-Nitrobromobenzene
UN 2732 [Nitrobromobenzenes liquid]

586-82-3
$C_{10}H_{18}O$
154.25
3-Cyclohexen-1-ol, 1-methyl-4-(1-methylethyl)- (9CI)
p-Menth-3-en-1-ol (8CI)
1-Terpineol

586-89-0
$C_9H_8O_3$
164.16
Benzoic acid, 4-acetyl- (9CI)
Benzoic acid, p-acetyl- (8CI)
NSC-16644

586-91-4
$C_{14}H_{10}N_2O_4$
270.25
4-Azobenzoate
p-Azobenzoate
para-Azobenzoate

586-95-8
C_6H_7NO
109.14
n(ccc(c1)CO)c1
4-Pyridinemethanol
4-(Hydroxymethyl)pyridine
γ-Picolyl alcohol
4-Pyridylcarbinol
4-Pyridylmethanol

586-96-9
C_6H_5NO
107.11
O=Nc(cccc1)c1
Nitrosobenzene
Benzene, nitroso- (9CI)

586-98-1
C_6H_7NO
109.14
OCc1ccccn1
2-Pyridinemethanol
2-(Hydroxymethyl)pyridine
α-Picolyl alcohol
Pyridine-2-carbinol
2-Pyridinylmethanol
2-Pyridylcarbinol
2-Pyridylmethanol

587-02-0
$C_8H_{11}N$
121.20
CCc1cccc(N)c1
Aniline, m-ethyl
Benzenamine, 3-ethyl- (9CI)
m-Ethylaniline
3-Ethylaniline

587-03-1
$C_8H_{10}O$
122.17
Cc1cccc(CO)c1
3-Methylbenzyl alcohol
AI3-21575
Benzenemethanol, 3-methyl-

587-04-2
C_7H_5ClO
140.57
Clc1cccc(C=O)c1
Benzaldehyde, m-chloro- (8CI)
AI3-10532
Benzaldehyde, 3-chloro- (9CI)

587-26-8
$CH_2O_3.2/3La$
154.63
Carbonic acid, lanthanum(3+) salt (3:2) (9CI)

587-33-7
$C_9H_{11}NO_3$
181.19
L-Phenylalanine, 3-hydroxy- (9CI)
m-Tyrosine
m-Tyrosine, L- (8CI)

587-34-8
$C_9H_{11}ClN_2O$
198.67
CN(C)C(=O)Nc1cccc(Cl)c1
Urea, 1-(m-chlorophenyl)-3,3-dimethyl
1-(m-Chlorophenyl)-3,3-dimethylurea
N,N-Dimethyl-N'-(3-chlorophenyl)-urea

587-56-4
$C_9H_9Cl_2NO_2$
234.08
Carbamic acid, (3-chlorophenyl)-, 2-chloroethyl ester (9CI)
AI3-22781
Carbanilic acid, m-chloro-, 2-chloroethyl ester (8CI)

587-64-4
$C_8H_6Cl_2O_3$
221.04
Acetic acid, (3,5-dichlorophenoxy)- (8CI,9CI)
NSC-190561

587-65-5
C_8H_8ClNO
169.62
CC(=O)Nc1ccccc1Cl
Acetanilide, 2-chloro
Acetamide, 2-chloro-N-phenyl- (9CI)
Chloroacetanilide
α-Chloroacetanilide
2-Chloroacetanilide
N-Phenylchloroacetamide

587-84-8
$C_{12}H_{11}N.H_2O_4S$
267.32
Diphenylamine, hydrogen sulfate
Diphenylamine sulfate
USAF EK-743

587-85-9
$C_{12}H_{10}Hg$
354.81
Mercury, diphenyl
Difenylrtut (Czech)
Diphenylmercury

587-98-4
$C_{18}H_{14}N_3O_3S.Na$
375.40
Benzenesulfonic acid, 3-((4-(phenylamino)-phenyl)azo)-, monosodium salt
Acidic Metanil Yellow
Acid Leather Yellow PRW
Acid Leather Yellow R
Acid metanil Yellow
Acid Yellow 36
Aizen Metanil Yellow
Amacid Yellow M
Brasilan Metanil Yellow
Bucacid Metanil Yellow

Calcocid Yellow MXXX
C.I. 13065
C.I. Acid Yellow 36
C.I. Acid Yellow 36 monosodium salt
Diacid Metanil Yellow
Eniacid Metanil Yellow GN
Ext. D & C Yellow No. 1
Fenazo Yellow M
Hidacid Metanil Yellow
Hispacid Yellow MG
Java Metanil Yellow G
Kiton Orange MNO
Kiton Yellow MS
Metanile Yellow O
Metanil Yellow
Metanil Yellow 1955
Metanil Yellow C
Metanil Yellow E
Metanil Yellow Extra
Metanil Yellow F
Metanil Yellow G
Metanil Yellow Griesbach
Metanil Yellow K
Metanil Yellow KRSU
Metanil Yellow M3X
Metanil Yellow O
Metanil Yellow PL
Metanil Yellow S
Metanil Yellow Supra P
Metanil Yellow VS
Metanil Yellow WS
Metanil Yellow Y
Metanil Yellow YK
Mitsui Metanil Yellow
Monoazo
Remaderm Yellow HPR
Shikiso Metanil Yellow
Symulon Metanil Yellow
Takaoka Metanil Yellow
Tertracid Yellow M
Tropaeolin G
Vondacid Metanil Yellow G
11363 Yellow
Yodochrome Metanil Yellow
Zlut Kysela 36 (Czech)
Zlut Metanilova (Czech)

588-04-5
$C_{14}H_{14}N_2$
210.27
Diazene, bis(3-methylphenyl)- (9CI)

588-05-6
$C_8H_{11}NO$
137.18
3-Tyramine
3-Hydroxyphenylethylamine
m-Tyramine
meta-Tyramine

588-07-8
C_8H_8ClNO
169.62
O=C(Nc(cccc1Cl)c1)C
Acetanilide, 3'-chloro
m-Chloroacetanilide

588-22-7
$C_8H_6Cl_2O_3$
221.04
OC(=O)COc1ccc(Cl)c(Cl)c1
Acetic acid, (3,4-dichlorophenoxy)
3,4-D
3,4-DA
3,4-Dichlorophenoxyacetic acid

588-30-7
$C_9H_8O_3$
164.16
OC(=O)C=Cc1cccc(O)c1
3-Coumaric acid
AI3-32389
2-Propenoic acid, 3-(3-hydroxyphenyl)- (9CI)

588-32-9
$C_8H_7ClO_3$
186.59
OC(=O)COc1cccc(Cl)c1
3-Chlorophenoxyacetic acid
Acetic acid, (m-chlorophenoxy)- (8CI)
Acetic acid, (3-chlorophenoxy)- (9CI)

588-46-5
$C_9H_{11}NO$
149.21
CC(=O)NCc1ccccc1
Acetamide, N-benzyl
Acetamide, N-(phenylmethyl)- (9CI)
N-Acetylbenzylamine
Benzylacetamide
N-Benzylacetamide

588-59-0
$C_{14}H_{12}$
180.26
c(cccc1)(c1)C=Cc(cccc2)c2
Stilbene
Benzene, 1,1'-(1,2-ethenediyl)bis- (9CI)
Bibenzal
Bibenzylidene
Bibenzylidine
α,β-Diphenylethylene
1,2-Diphenylethylene
Stilben (German)

589-16-2
$C_8H_{11}N$
121.20
Nc(ccc(c1)CC)c1
Aniline, 4-ethyl
1-Amino-4-ethylbenzene
p-Ethylaniline

589-18-4
$C_8H_{10}O$
122.18
OCc(ccc(c1)C)c1
Benzyl alcohol, p-methyl
4-Methyl-benzenemethanol (9CI)
p-Methylbenzylalcohol
4-Methylbenzyl alcohol
p-Methylbenzylalkohol (German)
p-Tolylcarbinol
4-Tolylcarbinol

589-29-7
$C_8H_{10}O_2$
138.17
OCc(ccc(c1)CO)c1
p-Xylene-α,α'-diol (8CI)
AI3-25222
1,4-Benzenedimethanol (9CI)

589-34-4
C_7H_{16}
100.20
C(CCC)(CC)C
3-Methylhexane
Hexane, 3-methyl- (9CI)

589-35-5
$C_6H_{14}O$
102.18
OCCC(CC)C
1-Pentanol, 3-methyl- (9CI)
AI3-38563
3-Methyl-1-pentanol

589-38-8
$C_6H_{12}O$
100.18
O=C(CCC)CC
3-Hexanone
Aethylpropylketon (German)
Ethyl propyl ketone

589-43-5
C_8H_{18}
114.23
2,4-Dimethylhexane
Hexane, 2,4-dimethyl- (9CI)

589-53-7
C_8H_{18}
114.23
Heptane, 4-methyl- (9CI)

589-55-9
$C_7H_{16}O$
116.20
OC(CCC)CCC
4-Heptanol (9CI)

589-59-3
$C_9H_{18}O_2$
158.27
O=C(OCC(C)C)CC(C)C
Isovaleric acid, isobutyl ester
Butanoic acid, 3-methyl-,

2-methylpropyl ester (9CI)
Isobutyl isopentanoate
Isobutyl isovalerate
2-Methylpropyl isovalerate
2-Methylpropyl 3-methylbutyrate

589-62-8
$C_8H_{18}O$
130.23
CCCCC(O)CCC
4-Octanol (9CI)
AI3-37214

589-63-9
$C_8H_{16}O$
128.21
4-Octanone (8CI,9CI)
NSC-43245

589-75-3
$C_{12}H_{24}O_2$
200.32
O=C(OCCCC)CCCCCCC
Octanoic acid, butyl ester (9CI)
AI3-30983
Butyl caprylate
n-Butylcaprylate
Butyl octanoate

589-81-1
C_8H_{18}
114.23
C(CCCC)(CC)C
Heptane, 3-methyl- (9CI)
3-Methylheptane

589-82-2
$C_7H_{16}O$
116.23
OC(CCCC)CC
3-Heptanol
3-Hydroxyheptane

589-87-7
C_6H_4BrI
282.91
c(ccc(c1)Br)(c1)I
Benzene, 1-bromo-4-iodo- (9CI)
AI3-09032
1-Bromo-4-iodobenzene

589-90-2
C_8H_{16}
112.24
Cyclohexane, 1,4-dimethyl
1,4-Dimethylcyclohexane [UN 2263]
UN 2263 [Dimethylcyclohexanes]

589-91-3
$C_7H_{14}O$
114.19
OC(CCC(C1)C)C1
4-Methylcyclohexanol
AI3-01169
Cyclohexanol, 4-methyl- (9CI)

589-92-4
$C_7H_{12}O$
112.19
O=C(CCC(C1)C)C1
Cyclohexanone, 4-methyl
4-Methylcyclohexanone
Methyl-4 cyclohexanone-1 (French)

589-93-5
C_7H_9N
107.17
n(c(ccc1C)C)c1
2,5-Lutidine
2,5-Dimethylpyridine
Pyridine, 2,5-dimethyl- (9CI)

589-98-0
$C_8H_{18}O$
130.26
OC(CCCCC)CC
3-Octanol
Amylethylcarbinol
Ethylamylcarbinol
Ethyl-n-amylcarbinol
Octanol-3

590-00-1
$C_6H_7O_2.K$
150.23
Sorbic acid, potassium salt
2,4-Hexadienoic acid potassium salt
Potassium sorbate
Sorbistat-K
Sorbistat-potassium

590-01-2
$C_7H_{14}O_2$
130.21
O=C(OCCCC)CC
Propionic acid, butyl ester
Butyl propanoate
Butylpropionate [UN 1914]
n-Butyl propionate
Propanoic acid, butyl ester (9CI)
UN 1914 [Butylpropionate]

590-02-3
$C_6H_{11}ClO_2$
150.60
O=C(OCCCC)CCl
n-Butyl chloroacetate
Acetic acid, chloro-, butyl ester (9CI)
AI3-18528
Butyl chloroacetate
n-Butyl-chloroacetate
Caswell No. 125J

590-17-0
C_2H_2BrN
119.94
Acetonitrile, bromo- (8CI,9CI)

590-18-1
C_4H_8
56.11
C(=CC)C
cis-2-Butene
2-Butene, (Z)- (9CI)
cis-Butene
(Z)-2-Butene
2-Butene-cis
cis-Butylene
β-cis-Butylene
cis-1,2-Dimethylethylene
High-boiling Butene-2

590-19-2
C_4H_6
54.09
C(=CC)=C
1,2-Butadiene (9CI)
Allene, methyl-
Methylallene
1-Methylallene

590-21-6
C_3H_5Cl
76.53
C(=CCl)C
Propene, 1-chloro
1-Chloropropene
1-Chloro-1-propene
Propenyl chloride

590-26-1
$C_2H_2I_2$
279.85
Ethene, 1,2-diiodo-, (Z)- (9CI)
cis-1,2-Diiodoethene
cis-Diiodoethylene
cis-1,2-Diiodoethylene
Ethylene, 1,2-diiodo-, (Z)- (8CI)

590-27-2
$C_2H_2I_2$
279.85
Ethene, 1,2-diiodo-, (E)- (9CI)
trans-Diiodoethene
trans-1,2-Diiodoethene
trans-Diiodoethylene
(E)-1,2-Diiodoethylene
trans-1,2-Diiodoethylene
Ethylene, 1,2-diiodo-, (E)- (8CI)

590-28-3
CNO.K
81.12

Cyanic acid, potassium salt
Aero Cyanate
Aero Cyanate Weedkiller
Alicyanate
Bonide Krab Crabgrass Killer
Bulpur
Ded-Weed Crabgrass Killer
D & P Double O Crabgrass Killer
Dupont PC Crabgrass Killer
Green Cross Crabgrass Killer
Kaliumcyanat (German)
Miller P.C. Weedkiller
P.C. 80 Crabgrass Killer
Potassium cyanate
Potassium isocyanate
Weedanol Cyanol
Weedone Crab Grass Killer

590-29-4
$CHO_2.K$
84.12
Formic acid, potassium salt
Mravencan draselny (Czech)
Potassium formate

590-35-2
C_7H_{16}
100.20
C(CCC)(C)(C)C
Pentane, 2,2-dimethyl- (9CI)
2,2-Dimethylpentane

590-36-3
$C_6H_{14}O$
102.20
OC(CCC)(C)C
2-Pentanol, 2-methyl
2-Methyl-2-pentanol
2-Methylpentan-2-ol [UN 2560]
UN 2560 [2-Methylpentan-2-ol]

590-42-1
C_5H_9NS
115.19
N(=C=S)C(C)(C)C
Propane, 2-isothiocyanato-2-methyl- (9CI)
2-Isothiocyanato-2-methylpropane

590-46-5
$C_5H_{12}NO_2.Cl$
153.63
[Cl-].C[N+](C)(C)CC(O)=O
Ammonium, (carboxymethyl)-trimethyl-, chloride
(Carboxymethyl)trimethylammonium chloride
Glykokollbetain-chlorid (German)
Methanaminium, 1-carboxy-N,N,N-trimethyl-, chloride

590-50-1
$C_7H_{14}O$
114.19
2-Pentanone, 4,4-dimethyl- (8CI,9CI)
NSC-944

590-66-9
C_8H_{16}
112.22
Cyclohexane, 1,1-dimethyl- (9CI)
AI3-28793

590-67-0
$C_7H_{14}O$
114.19
Cyclohexanol, 1-methyl- (9CI)
AI3-15917

590-73-8
C_8H_{18}
114.23
Hexane, 2,2-dimethyl- (9CI)

590-86-3
$C_5H_{10}O$
86.15
O=CCC(C)C
Butyraldehyde, 3-methyl
1-Butanal, 3-methyl-
Isoamyl aldehyde
Isopentaldehyde
Isovaleral
Isovaleraldehyde
Isovaleric aldehyde
2-Methylbutanal-4
3-Methylbutanal
3-Methylbutyraldehyde

590-88-5
$C_4H_{12}N_2$
88.14
1,3-Butanediamine (8CI,9CI)
AI3-52308
1,3-Diaminobutane

590-90-9
$C_4H_8O_2$
88.11
O=C(CCO)C
2-Butanone, 4-hydroxy- (9CI)
AI3-11747
4-Hydroxy-2-butanone

590-92-1
$C_3H_5BrO_2$
152.99
O=C(O)CCBr
Propionic acid, 3-bromo
β-Bromopropionic acid
3-Bromopropionic acid

590-96-5
$C_2H_6N_2O_2$
90.10
CN(=O)=NCO
Methanol, (methyl-ONN-azoxy)
1-Hydroxymethyl-2-methyl-ditmide-2-oxide
MAM
Methylazoxymethanol

591-01-5
$C_2H_6N_4O.1/2H_2O_4S$
151.12
Urea, (aminoiminomethyl)-, sulfate (2:1) (9CI)
(Aminoiminomethyl)urea sulfate (2:1)
Guanylurea sulfate

591-08-2
$C_3H_6N_2OS$
118.17
O=C(NC(=N)S)C
Urea, 1-acetyl-2-thio
Acetyl thiourea
1-Acetyl-2-thiourea
RCRA waste number P002
USAF EK-4890

591-17-3
C_7H_7Br
171.05
c(cccc1Br)(c1)C
Toluene, m-bromo
Benzene, 1-bromo-3-methyl-
m-Bromotoluene
3-Bromotoluene
5-Bromotoluene
m-Methylbromobenzene
3-Methylbromobenzene
m-Tolyl bromide

591-18-4
C_6H_4BrI
282.91
c(cccc1Br)(c1)I
Benzene, 1-bromo-3-iodo- (9CI)
1-Bromo-3-iodobenzene

591-19-5
C_6H_6BrN
172.02
Nc(cccc1Br)c1
Aniline, m-bromo- (8CI)
Benzenamine, 3-bromo- (9CI)
m-Bromoaniline
3-Bromobenzenamine

591-20-8
C_6H_5BrO
173.01
Oc(cccc1Br)c1
3-Bromophenol
Phenol, 3-bromo- (9CI)

591-21-9
C_8H_{16}

112.24
Cyclohexane, 1,3-dimethyl
m-Dimethylcyclohexane
1,3-Dimethylcyclohexane [UN 2263]
UN 2263 [Dimethylcyclohexanes]

591-22-0
C_7H_9N
107.15
n(cc(cc1C)C)c1
Pyridine, 3,5-dimethyl- (9CI)
3,5-Dimethylpyridine

591-23-1
$C_7H_{14}O$
114.21
OC(CCCC1C)C1
Cyclohexanol, m-methyl
m-Methylcyclohexanol

591-24-2
$C_7H_{12}O$
112.19
O=C(CCCC1C)C1
Cyclohexanone, 3-methyl
3-Methylcyclohexanone
Methyl-3 cyclohexanone-1 (French)

591-27-5
C_6H_7NO
109.14
Oc(cccc1N)c1
Phenol, m-amino
m-Aminofenol (Czech)
3-Amino-1-hydroxybenzene
m-Aminophenol [UN 2512]
3-Aminophenol
BASF Ursol EG
C.I. 76545
C.I. Oxidation Base 7
Fouramine EG
Fourrine 65
Fourrine EG
Furro EG
Futramine EG
3-Hydroxyaniline
Nako TEG
Pelagol EG
Renal EG
Tertral eg
UN 2512 [Aminophenols (o-; m-; p-)]
Ursol EG
Zoba EG

591-31-1
$C_8H_8O_2$
136.16
O=Cc(cccc1OC)c1
m-Anisaldehyde
Benzaldehyde, 3-methoxy- (9CI)
m-Methoxybenzaldehyde
3-Methoxybenzaldehyde

591-33-3
$C_{10}H_{13}NO_2$
179.24
O=C(Nc(cccc1OCC)c1)C
m-Acetophenetidide
Acetamide, N-(3-ethoxyphenyl)- (9CI)
Acetanilide, 3'-ethoxy-
m-Ethoxyacetanilide
3-Ethoxyacetanilide
3'-Ethoxyacetanilide

591-35-5
$C_6H_4Cl_2O$
163.00
Oc1cc(Cl)cc(Cl)c1
Phenol, 3,5-dichloro
3,5-Dichlorophenol

591-47-9
C_7H_{12}
96.17
C(=CCCC1C)C1
4-Methylcyclohexene
Cyclohexene, 4-methyl- (9CI)

591-48-0
C_7H_{12}
96.17
C(=CC(CC1)C)C1
Cyclohexene, 3-methyl- (9CI)
3-Methylcyclohexene
3-Methyl-1-cyclohexene

591-49-1
C_7H_{12}
96.17
C(=CCCC1)(C1)C
Cyclohexene, 1-methyl- (9CI)
AI3-52478
1-Methylcyclohexene

591-50-4
C_6H_5I
204.01
c(cccc1)(c1)I
Benzene, iodo
Benzeneiodide
Iodobenzene
Phenyl iodide

591-60-6
$C_8H_{14}O_3$
158.22
O=C(OCCCC)CC(=O)C
Acetoacetic acid, butyl ester
Butanoic acid, 3-oxo-, butyl ester
Butyl acetoacetate
Butylester kyseliny acetoctove (Czech)

591-76-4
C_7H_{16}
100.20
C(CCCC)(C)C
Isoheptane
Hexane, 2-methyl- (9CI)
2-Methylhexane

591-78-6
$C_6H_{12}O$
100.18
O=C(CCCC)C
2-Hexanone (OSHA)
Butyl methyl ketone
n-Butyl methyl ketone
Hexanone-2
Ketone, butyl methyl
MBK
Methyl n-butyl ketone (ACGIH, OSHA)
MNBK

591-80-0
$C_5H_8O_2$
100.13
O=C(O)CCC=C
4-Pentenoic acid
Allylacetic acid

591-87-7
$C_5H_8O_2$
100.13
O=C(OCC=C)C
Acetic acid, allyl ester
Acetic acid, 2-propenyl ester (9CI)
3-Acetoxypropene
Allyl acetate [UN 2333]
UN 2333 [Allyl acetate]

591-89-9
$C_2HgN_2.2CKN$
382.87
Potassium tetracyanomercurate (II)
Mercuric potassium cyanide
Mercuric potassium cyanide [UN 1626]
Mercuric potassium cyanide, Solid (DOT)
UN 1626 [Mercuric potassium cyanide]

591-93-5
C_5H_8
68.12
C(=C)CC=C
1,4-Pentadiene (9CI)

591-95-7
C_5H_8
68.12
1,2-Pentadiene (8CI,9CI)
Ethylallene

1-Methyl-2,3-butadiene

591-97-9
C_4H_8Cl
91.57
C(=CC)CCl
2-Butene, 1-chloro
2-Butenyl chloride
1-Chloro-2-butene
Crotyl chloride
Krotylchlorid (Czech)
γ-Methallyl chloride
γ-Methylallyl chloride

592-01-8
C_2CaN_2
92.12
Calcium cyanide [UN 1575]
Calcium cyanide, Solid (DOT)
Calcyanide
Cyanogas
Cyanure de calcium (French)
RCRA waste number P021
UN 1575 [Calcium cyanide]

592-04-1
C_2HgN_2
252.63
Mercury(II) cyanide
Cyanure de mercure (French)
Mercuric cyanide (DOT)
Mercuric cyanide, Solid (DOT)
Mercury cyanide [UN 1636]
UN 1636 [Mercury cyanide]

592-05-2
C_2N_2Pb
259.22
Lead cyanide
C.I. Pigment Yellow 48
C.I. 77610
Cianuro de plomo (Spanish)
Cyanure de plomb (French)
Lead(II)cyanide
UN 1620 [Lead cyanide]

592-13-2
C_8H_{18}
114.23
CC(C)CCC(C)C
Hexane, 2,5-dimethyl- (9CI)

592-20-1
$C_5H_8O_3$
116.12
O=C(OCC(=O)C)C
2-Propanone, 1-(acetyloxy)- (9CI)
1-Acetoxy-2-propanone
1-(Acetyloxy)-2-propanone
AI3-08537
2-Propanone, 1-hydroxy-, acetate (8CI)

592-27-8
C_8H_{18}
114.23
Heptane, 2-methyl- (9CI)

592-31-4
$C_5H_{12}N_2O$
116.19
O=C(NCCCC)N
Urea, butyl
n-Butylurea
NCI-C02131

592-35-8
$C_5H_{11}NO_2$
117.17
O=C(OCCCC)N
Carbamic acid, butyl ester
Butyl carbamate
USAF FO-1
USAF EL-101

592-41-6
C_6H_{12}
84.18
C(=C)CCCC
1-Hexene [UN 2370]
Hex-1-ene (DOT)
UN 2370 [1-Hexene]

592-42-7
C_6H_{10}
82.15
C(=C)CCC=C
1,5-Hexadiene (9CI)

592-43-8
C_6H_{12}
84.16
C(=CC)CCC
2-Hexene (Mixed cis & trans)
AI3-28402
2-Hexene

592-45-0
C_6H_{10}
82.15
C(=C)CC=CC
1,4-Hexadiene (9CI)
1-Allylpropene
Allylpropenyl

592-46-1
C_6H_{10}
82.15
CC=CC=CC
2,4-Hexadiene (9CI)

592-47-2
C_6H_{12}
84.16
3-Hexene (8CI,9CI)

592-48-3
C_6H_{10}
82.15
CCC=CC=C
1,3-Hexadiene (8CI,9CI)

592-50-7
$C_5H_{11}F$
90.14
FCCCCC
1-Fluoropentane
n-Amyl fluoride
Pentane, 1-fluoro- (9CI)

592-51-8
C_5H_7N
81.11
N#CCCC=C
4-Pentenenitrile (9CI)
Allylacetonitrile
Allylmethyl cyanide
3-Butenyl cyanide
4-Cyano-1-butene
4-Pentenoic acid, nitrile
4-Pentenonitrile

592-57-4
C_6H_8
80.13
C(=CC=CC1)C1
1,3-Cyclohexadiene (9CI)

592-62-1
$C_4H_8N_2O_3$
132.14
CC(=O)OCN=N(C)=O
Methanol, (methyl-ONN-azoxy)-, acetate (ester)
Acetic acid, (methyl-ONN-azoxy)methyl ester
Cycasin acetate
MAM AC
MAM acetate
Methylazoxymethanol acetate
Methylazoxymethyl acetate
Methylazoxymethylester kyseliny octove (Czech)
(Methyl-ONN-azoxy)methanol, acetate (ester)

592-76-7
C_7H_{14}
98.19
C(=C)CCCCC
1-Heptene (9CI)
Heptene
n-Heptene
n-Hept-1-ene
1-n-Heptene
n-Hepteno (Spanish)
α-Heptylene

592-77-8
C_7H_{14}
98.19
2-Heptene (8CI,9CI)

592-78-9
C_7H_{14}
98.19
3-Heptene (8CI,9CI)

592-84-7
$C_5H_{10}O_2$
102.15
O=COCCCC
Formic acid, butyl ester
Butylester kyseliny mravenci (Czech)
Butyl formate (DOT)
n-Butyl formate [UN 1128]
UN 1128 [n-Butyl formate]

592-85-8
$C_2N_2S_2.Hg$
316.75
Thiocyanic acid, mercury(II) salt
Mercuric sulfocyanate
Mercuric sulfo cyanate, Solid (DOT)
Mercuric sulfocyanide
Mercuric thiocyanate
Mercuric thiocyanate, Solid (DOT)
Mercury, bis(thiocyanato)-
Mercury dithiocyanate
Mercury thiocyanate [UN 1646]
Mercury(II)thiocyanate
Thiocyanic acid, mercury(2+) salt
UN 1646 [Mercury thiocyanate]

592-87-0
CHNS.1/2Pb
162.69
Lead thiocyanate
Isothiocyanic acid, lead(2+) salt
Lead dithiocyanate
Lead isothiocyanate
Lead sulfocyanate
Lead(II) thiocyanate
Lead(2+) thiocyanate
Thiocyanic acid, lead(2+) salt (9CI)

592-88-1
$C_6H_{10}S$
114.22
S(CC=C)CC=C
Allyl-sulfide
Allyl monosulfide
Diallyl monosulfide
Diallyl sulfide
Diallyl thioether
Oil Garlic
1-Propene, 3,3'-thiobis- (9CI)
2-Propenyl sulphide
Thioallyl ether
3,3-Thiobis(1-propene)

592-90-5
$C_6H_{12}O$
100.16
O(CCCCC1)C1
Oxepane (9CI)
Oxacycloheptane

592-99-4
C_8H_{16}
112.22
4-Octene (8CI,9CI)

593-08-8
$C_{13}H_{26}O$
198.35
O=C(CCCCCCCCCCC)C
2-Tridecanone (9CI)
AI3-04238
Methyl undecyl ketone

593-26-0
$C_{16}H_{32}O_2.H_3N$
273.45
Hexadecanoic acid, ammonium salt (9CI)
Ammonium hexadecanoate

593-29-3
$C_{18}H_{35}O_2.K$
322.63
Stearic acid, potassium salt
Octadecanoic acid, potassium salt
Potassium stearate (ACGIH)

593-39-5
$C_{18}H_{34}O_2$
282.47
6-Octadecenoic acid, (Z)- (8CI,9CI)
Petroselinic acid

593-45-3
$C_{18}H_{38}$
254.50
C(CCCCCCCCCCCCCCCC)C
n-Octadecane
AI3-06523
Octadecane (9CI)

593-49-7
$C_{27}H_{56}$
380.74
Heptacosane
AI3-36283

593-50-0
$C_{30}H_{62}O$
438.82
OCCCCCCCCCCCCCCCCCCCCCCCCCCCCCC
1-Triacontanol (9CI)
AI3-20480

593-51-1
$CH_5N.ClH$
67.51
Mathylamine, hydrochloride (9CI)
AI3-52469

593-53-3
CH_3F
34.04
FC
Methane, fluoro
Fluoromethane
Freon 41
Methyl fluoride [UN 2454]
UN 2454 [Methyl fluoride]

593-56-6
$CH_5NO.ClH$
83.53
Hydroxylamine, o-methyl-, hydrochloride
Methoxyamine, hydrochloride
Methoxylamine hydrochloride
o-Methylhydroxylamine
Methyloxyammonium chloride
o-Methylhydroxylamine hydrochloride

593-60-2
C_2H_3Br
106.96
BrC=C
Ethylene, bromo
Bromoethene (9CI)
Bromoethylene
Bromure de vinyle (French)
Ethene, bromo-
NCI-C50373
UN 1085 [Vinyl bromide, inhibited]
Vinile (bromuro di) (Italian)
Vinylbromid (German)
Vinyl bromide (ACGIH,OSHA)
Vinyl bromide, Inhibited [UN 1085]
Vinyle (bromure de) (French)

593-70-4
CH_2ClF
68.48
FCCl
Methane, chlorofluoro
CFC 31
Chlorofluoromethane
FC 31
Freon 31
Monochloromonofluoromethane
R 31
R 31 (Refrigerant)

593-71-5
CH_2ClI
176.38
Methane, chloroiodo- (8CI,9CI)

593-74-8
C_2H_6Hg

230.67
C[Hg]C
Mercury, dimethyl
Dimethyl mercury

593-75-9
C_2H_3N
41.05
Methylisonitrile
Methyl isocyanide

593-77-1
CH_5NO
47.07
CNO
Hydroxylamine, N-methyl
Methanamine, N-hydroxy-
Methylhydroxylamine
NCI-C60066

593-79-3
C_2H_6Se
109.04
Selenide, dimethyl
Dimethyl selenide
Dimethylselenium
Methyl selenide (8CI)
Methyl selenium

593-81-7
$C_3H_9N.ClH$
95.59
Trimethylamine, hydrochloride
Trimethylammonium chloride

593-82-8
$C_2H_8N_2.ClH$
96.58
[Cl-].CN(C)N
Hydrazine, 1,1-dimethyl-, hydrochloride
1,1-Dimethylhydrazine hydrochloride

593-85-1
$CH_5N_3.I/2CH_2O_3$
493.30
Carbonic acid, Compd. with guanidine (1:2)
AI3-14631
Guanidine carbonate

593-89-5
CH_3AsCl_2
160.86
Methyldichloroarsine
Arsine, dichloromethyl-
Arsonous dichloride, methyl- (9CI)
Dichloromethylarsine
Methylarsine dichloride
Methylarsonous dichloride
Methyl-arsonous dichloride
Methyldichlorarsine
NA 1556 [Methyldichloroarsine]
TL 294

593-92-0
$C_2H_2Br_2$
185.86
1,1-Dibromoethylene

593-94-2
$CHBr_2I$
299.73
Methane, dibromoiodo- (9CI)

593-96-4
C_2H_4BrCl
143.41
Ethane, 1-bromo-1-chloro- (8CI,9CI)

594-04-7
$CHCl_2I$
210.83
Methane, dichloroiodo- (8CI, 9CI)

594-11-6
C_4H_8
56.11
C(C1)(C1)C
Cyclopropane, methyl- (9CI)
Methylcyclopropane

594-15-0
CBr_3Cl
287.18
Methane, tribromochloro- (8CI,9CI)

594-16-1
$C_3H_6Br_2$
201.89
BrC(Br)(C)C
Propane, 2,2-dibromo- (9CI)
2,2-Dibromopropane

594-18-3
CBr_2Cl_2
242.72
Methane, dibromodichloro- (9CI)

594-19-4
C_4H_9Li
64.06
[Li]C(C)(C)C
Lithium, (1,1-dimethylethyl)- (9CI)

594-20-7
$C_3H_6Cl_2$
112.99
C(Cl)(Cl)(C)C
Propane, 2,2-dichloro- (9CI)
2,2-Dichloropropane

594-27-4
$C_4H_{12}Sn$
178.85
C[Sn](C)(C)C
Stannane, tetramethyl
Tetramethylcin (Czech)
Tetramethylstannane
Tetramethyl tin
Tin, tetramethyl-

594-34-3
$C_4H_8Br_2$
215.94
BrCC(Br)(C)C
Propane, 1,2-dibromo-2-methyl
1,2-Dibromo-2-methylpropane

594-36-5
$C_5H_{11}Cl$
106.60
C(CC)(Cl)(C)C
tert-Amyl chloride
Amyl chloride
Butane, 2-chloro-2-methyl- (8CI,9CI)
2-Chloro-2-methylbutane
Chlorure d'amyle (French)
Cloruro de amilo (Spanish)

594-37-6
$C_4H_8Cl_2$
127.02
Propane, 1,2-dichloro-2-methyl
1,2-Dichloro-2-methylpropane
Isobutylene dichloride

594-38-7
$C_5H_{11}I$
198.05
Butane, 2-iodo-2-methyl- (8CI, 9CI)

594-42-3
CCl_4S
185.87
S(C(Cl)(Cl)Cl)Cl
Methanesulfenyl chloride, trichloro
Clairsit
Mercaptan methylique perchlore (French)
PCM
Perchlormethylmerkaptan (Czech)
Perchloromethyl mercaptan (ACGIH,OSHA)
Perchloromethylmercaptan [UN 1670]
RCRA waste number P118

Trichloromethane sulfenyl chloride
Trichloromethanesulphenyl chloride
Trichloromethylsulfenyl chloride
Trichloromethylsulphenyl chloride
UN 1670 [Perchloromethylmercaptan]

594-45-6
$C_2H_6O_3S$
110.13
O=S(=O)(O)CC
Ethane sulfonate
Ethanesulfonic acid (9CI)

594-56-9
C_7H_{14}
98.19
2,3,3-Trimethyl-1-butene

594-60-5
$C_6H_{14}O$
102.18
2-Butanol, 2,3-dimethyl- (9CI)
AI3-17876

594-61-6
$C_4H_8O_3$
104.11
O=C(O)C(O)(C)C
2-Hydroxyisobutyric acid
AI3-31313
2-Hydroxy-2-methylpropanoic acid
Lactic acid, 2-methyl- (8CI)
Propanoic acid, 2-hydroxy-2-methyl- (9CI)

594-65-0
$C_2H_2Cl_3NO$
162.40
O=C(N)C(Cl)(Cl)Cl
Acetamide, 2,2,2-trichloro
Acetamide, α-trichloro-
Amid kyseliny trichloroctove (Czech)
Trichloroacetamide
α,α,α-Trichloroacetamide
2,2,2-Trichloroacetamide

594-70-7
$C_4H_9NO_2$
103.11
O=N(=O)C(C)(C)C
Propane, 2-methyl-2-nitro- (9CI)
2-Methyl-2-nitropropane

594-71-8
$C_3H_6ClNO_2$
123.55
CC(C)(Cl)N(=O)=O
Propane, 2-chloro-2-nitro
2-Chloro-2-nitropropane

594-72-9
$C_2H_3Cl_2NO_2$
143.96
CC(Cl)(Cl)N(=O)=O
Ethane, 1,1-dichloro-1-nitro
1,1-Dichloor-1-nitroethaan (Dutch)
1,1-Dichlor-1-nitroaethan (German)
Dichloronitroethane
1,1-Dichloro-1-nitroethane (ACGIH,OSHA) [UN 2650]
1,1-Dicloro-1-nitroetano (Italian)
Ethide
UN 2650 [1,1-Dichloro-1-nitroethane]

594-82-1
C_8H_{18}
114.23
CC(C)(C)C(C)(C)C
Butane, 2,2,3,3-tetramethyl- (9CI)

594-83-2
$C_7H_{16}O$
116.20
2-Butanol, 2,3,3-trimethyl- (9CI)

594-89-8
C_3HCl_7
285.21
Propane, 1,1,1,2,2,3,3-heptachloro- (8CI,9CI)
NSC-7298

594-90-1
C_3Cl_8
319.66
Propane, octachloro- (6CI,7CI,8CI,9CI)
Octachloropropane
Perchloropropane

595-33-5
$C_{24}H_{32}O_4$
384.56
Pregna-4,6-diene-3,20-dione, 17-hydroxy-6-methyl-, acetate
17-α-Acetoxy-6-dehydro-6-methylprogesterone
17-Acetoxy-6-methylpregna-4,6-diene-3,20-dione
17-α-Acetoxy-6-methylpregna-4,6-diene-3,20-dione
17-α-Acetoxy-6-methyl-4,6-pregnadiene-3,20-dione
BDH 1298
6-Dehydro-6-methyl-17-α-acetoxyprogesterone
DMAP
17-Hydroxy-6-methylpregna-4,6-diene-3,20-dione acetate
Megace
Megestrol acetate
Megestryl acetate
6-Methyl-17-α-acetoxypregna-4,6-diene-3,20-dione
6-Methyl-6-dehydro-17-α-acetoxyprogesterone
6-Methyl-δ^6-dehydro-17-α-acetoxyprogesterone
6-Methyl-6-dehydro-17-α-acetylprogesterone
6-Methyl-17-α-hydroxy-δ^6-progesterone acetate
6-Methyl-$\delta^{4,6}$-pregnadien-17-α-ol-3,20-dione acetate
NSC-71423
Ovaban
SC10363
Volidan

595-37-9
$C_6H_{12}O_2$
116.16
CCC(C)(C)C(O)=O
Butanoic acid, 2,2-dimethyl- (9CI)
Butyric acid, 2,2-dimethyl- (8CI)
NSC-16045

595-41-5
$C_7H_{16}O$
116.20
3-Pentanol, 2,3-dimethyl- (8CI,9CI)

595-44-8
Unknown
Unknown
1,1-Dichloro-1-nitropropane

595-46-0
$C_5H_8O_4$
132.12
O=C(O)C(C(=O)O)(C)C
Malonic acid, dimethyl- (8CI)
Dimethylmalonic acid
Dimethylpropanedioic acid
Propanedioic acid, dimethyl- (9CI)

595-90-4
$C_{24}H_{20}Sn$
427.13
c1ccc(cc1)[Sn](c2ccccc2)(c3ccccc3)c4ccccc4
Tetraphenyl tin
AI3-14319
Stannane, tetraphenyl- (9CI)
Tetraphenylstannane

596-15-6
$C_{17}H_{19}NO_3.C_2H_4O_2$
345.39

Morphinan-3,6-diol, 7,8-didehydro-4,5-epoxy-17-methyl-(5α,6α)-, acetate (Salt) (9CI)
Morphine acetate

597-05-7
$C_8H_{18}O$
130.23
3-Pentanol, 3-ethyl-2-methyl- (8CI,9CI)
NSC-900

597-09-1
$C_5H_{11}NO_4$
149.14
O=N(=O)C(CC)(CO)CO
2-Nitro-2-ethyl-1,3-propanediol
AI3-02257
2-Ethyl-2-nitro-1,3-propanediol
1,3-Propanediol, 2-ethyl-2-nitro- (9CI)

597-25-1
$C_6H_{14}NO_4P$
195.18
COP(=O)(OC)N1CCOCC1
Phosphonic acid, morpholino-, dimethyl ester
Dimethylmorpholinophosphonate
Dimethyl morpholinophosphoramidate
DMMPA
HC 1717
Morpholinophosphonic acid dimethyl ester
NCI-C54740
Phosphonic acid, 4-morpholinyl-, dimethyl ester (9CI)

597-31-9
$C_5H_{10}O_2$
102.13
O=CC(CO)(C)C
3-Hydroxy-2,2-dimethylpropionaldehyde
2,2-Dimethyl-3-hydroxypropanal
α,α-Dimethyl-β-hydroxypropionaldehyde
2,2-Dimethyl-β-hydroxypropionaldehyde
Hydracrylaldehyde, 2,2-dimethyl-
3-Hydroxy-2,2-dimethylpropanal
2-(Hyrdoxymethyl)-2-methylpropanal
Hydroxypivalaldehyde
3-Hydroxypivalaldehyde
Hydroxypivaldehyde
Pentaaldol
Pentaldol
Propanal, 3-hydroxy-2,2-dimethyl- (9CI)
Propionaldehyde, 3-hydroxy-2,2-dimethyl-

597-43-3
$C_6H_{10}O_4$
146.14
Butanedioic acid, 2,2-dimethyl- (9CI)
NSC-408419
Succinic acid, 2,2-dimethyl- (8CI)

597-49-9
$C_7H_{16}O$
116.23
OC(CC)(CC)CC
3-Pentanol, 3-ethyl
3-Aethyl-pentanol-(3) (German)
3-Ethyl-3-pentanol
Triethylcarbinol
Triethylmethanol

597-64-8
$C_8H_{20}Sn$
234.97
CC[Sn](CC)(CC)CC
Stannane, tetraethyl
TET
Tetraethylstannane
Tetraethyl tin
Tin, tetraethyl-

597-76-2
$C_8H_{18}O$
130.23
CCCC(O)(CC)CC
3-Hexanol, 3-ethyl- (9CI)
NSC-28044

597-88-6
$C_{10}H_{14}NO_5PS$
291.28
CCOP(=O)(Oc1ccc(cc1)N(=O)=O)SCC
Phosphorothioic acid, O,S-diethyl-O-(p-nitrophenyl) ester
O,S-Diaethyl-O-(p-nitrophenyl)-phosphat (German)
O,S-Diethyl-O-p-nitrofenyl ester kyseliny thiofosforecne (Czech)
O,S-Diethyl-O-p-nitrofenyl-thiofosfat (Czech)
O,S-Diethyl O-(p-nitrophenyl) phosphorothioate
O,S-Diethyl O-(4-nitrophenyl)-phosphorothioate
O,S-Diethyl O-(4-nitrophenyl)-thiophosphate
S-Ethyl parathion
Isoparathion
Phosphorothioic acid, O,S-diethyl O-(4-nitrophenyl) ester (9CI)

597-89-7
$C_8H_{10}NO_5PS$
263.22
Phosphorothioic acid, O,S-dimethyl O-(p-nitrophenyl) ester
O,S-Dimethyl O-(p-nitrophenyl) phosphorothioate
O,S-Dimethyl O-(4-nitrophenyl)thiophosphate

597-96-6
$C_7H_{16}O$
116.23
CCCC(C)(O)CC
3-Hexanol, 3-methyl
3-Methyl-hexanol-(3) (German)
3-Methyl-3-hexanol

598-01-6
$C_8H_{18}O$
130.23
4-Heptanol, 4-methyl- (6CI,8CI,9CI)
4-Hydroxy-4-methylheptane
4-Methyl-4-heptanol

598-02-7
$C_4H_{11}O_4P$
154.10
O=P(OCC)(OCC)O
Diethyl phosphate
AI3-17002
Phosphoric acid, diethyl ester (9CI)

598-06-1
$C_8H_{18}O$
130.23
3-Heptanol, 3-methyl-, (.+-.)- (8CI,9CI)
(.+-.)-3-Methyl-3-heptanol

598-14-1
$C_2H_5AsCl_2$
174.89
CC[As](Cl)Cl
Arsine, dichloroethyl
Arsenic dichloroethane
Arsonous dichloride, ethyl- (9CI)
Dichloroethylarsine
Dick (German)
ED
Ethyldichlorarsine
Ethyldichloroarsine [UN 1892]
TL 214
UN 1892 [Ethyldichloroarsine]

598-16-3
C_2HBr_3
264.76
BrC=C(Br)Br
Ethylene, tribromo
Ethene, tribromo-
Tribromoethene
Tribromoethylene

1,1,2-Tribromoethylene

598-18-5
C_3H_7BrO
138.99
1-Propanol, 2-bromo- (8CI,9CI)

598-20-9
$C_2H_3Br_2Cl$
222.32
1-Chloro-1,2-dibromoethane
Ethane, 1,2-dibromo-1-chloro-

598-25-4
C_5H_8
68.12
1,2-Butadiene, 3-methyl- (9CI)

598-30-1
C_4H_9Li
64.06
Lithium, (1-methylpropyl)- (9CI)
(1-Methylpropyl)lithium

598-31-2
C_3H_5BrO
136.99
2-Propanone, bromo
Acetonyl bromide
Acetyl methyl bromide
Bromoacetone [UN 1569]
Bromoacetone, Liquid (DOT)
Bromomethyl methyl ketone
Bromo-2-propanone
1-Bromo-2-propanone
B-Stoff
Martonite
Monobromoacetone
2-Propanone, 1-bromo- (9CI)
RCRA waste number P017
UN 1569 [Bromoacetone]

598-32-3
C_4H_8O
72.12
OC(C=C)C
3-Buten-2-ol
Methyl vinyl carbinol
Propenol, 1-methyl

598-36-7
C_3H_9NS
91.17
β-Mercaptopropylamine
Methyl-2-cysteamine
2-Propanethiol, 1-amino- (9CI)

598-38-9
$C_2H_4Cl_2O$
114.96
OCC(Cl)Cl
Ethanol, 2,2-dichloro
2,2-Dichloroethanol

598-50-5
$C_2H_6N_2O$
74.10
O=C(NC)N
Urea, methyl
Methylmocovina (Czech)
Methylurea
n-Methylurea
1-Methylurea

598-52-7
$C_2H_6N_2S$
90.16
CNC(N)=S
Urea, 1-methyl-2-thio
Methyl thiourea
1-Methylthiourea

598-53-8
$C_4H_{10}O$
74.12
COC(C)C
Propane, 2-methoxy- (9CI)
Ether, isopropyl methyl (8CI)

598-55-0
$C_2H_5NO_2$
75.08
O=C(OC)N
Carbamic acid, methyl ester
Methyl carbamate
Methylester kyseliny karbaminove (Czech)
Methylkarbamat (Czech)
Methylurethan
Methylurethane
NCI-C55594
Urethylane

598-56-1
$C_4H_{11}N$
73.13
N(CC)(C)C
N,N-Dimethylethylamine
AI3-52225
N,N-Dimethylethanamine
Dimethylethylamine
Ethanamine, N,N-dimethyl-
Ethylamine, N,N-dimethyl-
N-Ethyldimethylamine
Methanamine, N-ethyl-N-methyl-

598-58-3
CH_3NO_3
77.05
O=N(=O)OC
Nitric acid, methyl ester
Methylester kyseliny dusicne (Czech)
Methyl nitrate (DOT)

598-61-8
C_5H_{10}
70.13
Cyclobutane, methyl- (8CI,9CI)

598-62-9
$CH_2O_3.Mn$
116.96
Manganese carbonate
Carbonic acid, manganese(2+) salt (1:1) (9CI)
Manganese carbonate (1:1)
Manganese(II) carbonate
Manganese(2+) carbonate
Manganese(2+) carbonate (1:1)
Manganous carbonate
Natural Rhodochrosite

598-63-0
$CO_3.Pb$
267.20
Lead-carbonate
Carbonic acid, lead(2+) salt (1:1)
Cerussete
Dibasic lead carbonate
Lead(2+) carbonate
White Lead

598-72-1
$C_3H_5BrO_2$
152.99
O=C(O)C(Br)C
Propionic acid, 2-bromo
α-Bromopropionic acid

598-73-2
C_2BrF_3
160.93
FC(F)=C(F)Br
Ethylene, bromotrifluoro
Bromotrifluoroethene
Bromotrifluoroethylene [UN 2419]
Ethene, bromotrifluoro- (9CI)
Trifluorobromoethylene
Trifluorovinyl bromide
UN 2419 [Bromotrifluoroethylene]

598-75-4
$C_5H_{12}O$
88.15
OC(C(C)C)C
3-Methyl-2-butanol
2-Butanol, 3-methyl- (9CI)

598-77-6
$C_3H_5Cl_3$
147.43
C(C(Cl)Cl)(Cl)C
Propane, 1,1,2-trichloro
1,1,2-Trichloropropane

598-78-7
$C_3H_5ClO_2$
108.53

O=C(O)C(Cl)C
Propionic acid, 2-chloro
α-Chloropropionic acid
Propionic acid, α-chloro-

598-79-8
$C_3H_3ClO_2$
106.51
O=C(O)C(=C)Cl
Acrylic acid, 2-chloro
Chloroacrylic acid
α-Chloroacrylic acid
2-Chloroacrylic acid
2-Propenoic acid, 2-chloro-

598-82-3
$C_3H_6O_3$
90.08
O=C(O)C(O)C
Lactic acid
Ethylidenelactic acid
2-Hydroxypropanoic acid
(+-)-2-Hydroxypropanoic acid
α-Hydroxypropionic acid
DL-Lactate
DL-Lactic acid
Milchsaure (German)
Ordinary Lactic Acid
Propanoic acid, 2-hydroxy-, (+-)- (9CI)
Racemic lactic acid
Tonsillosan

598-88-9
$C_2Cl_2F_2$
132.92
Ethene, 1,2-dichloro-1,2-difluoro- (9CI)
Ethylene, 1,2-dichloro-1,2-difluoro- (8CI)

598-92-5
$C_2H_4ClNO_2$
109.51
O=N(=O)C(Cl)C
1-Chloro-1-nitroethane
AI3-15633
Ethane, 1-chloro-1-nitro- (8CI, 9CI)

598-94-7
$C_3H_8N_2O$
88.13
O=C(N(C)C)N
Urea, 1,1-dimethyl
1,1-Dimethylurea

598-96-9
C_8H_{16}
112.22
C(=CC)(C(C)(C)C)C
3,4,4-Trimethyl-2-pentene
2-Pentene, 3,4,4-trimethyl- (9CI)
3,4,4-Trimethylpentene-2

598-98-1
$C_6H_{12}O_2$
116.16
Propanoic acid, 2,2-dimethyl-, methyl ester (9CI)
Methyl 2,2-dimethylpropanoate
Methyl pivalate

598-99-2
$C_3H_3Cl_3O_2$
177.41
O=C(OC)C(Cl)(Cl)Cl
Acetic acid, trichloro-, methyl ester
Methyl trichloroacetate
Methyl trichloroacetate [UN 2533]
UN 2533 [Methyl trichloroacetate]

599-64-4
$C_{15}H_{16}O$
212.31
Oc(ccc(c1)C(c(cccc2)c2)(C)C)c1
Phenol, p-(α,α-dimethylbenzyl)
p-Cumylphenol
p-(α-Cumyl)phenol
p-(α,α-Dimethylbenzyl)phenol
4-(Dimethylphenylmethyl)-phenol
4-Hydroxydiphenyldimethylmethane
Phenol, 4-(1-methyl-1-phenethyl)- (9CI)

599-69-9
$C_9H_{13}NO_2S$
199.29
O=S(=O)(N(C)C)c(ccc(c1)C)c1
p-Toluenesulfonamide, N,N-dimethyl
N,N-Dimethyl p-toluene sulfonamide

599-79-1
$C_{18}H_{14}N_4O_5S$
398.42
OC(=O)c3cc(N=Nc1ccc(cc1)S(=O)(=O)Nc2ccccn2)ccc3O
Salicylic acid, 5-((p-(2-pyridylsulfamoyl)phenyl)azo)
Accucol
Asulfidine
Azopyrin
Azulfidine
4-(Pyridyl-2-amidosulfonyl)-3'-carboxy-4'-hydroxyazobenzene
5-(4-(2-Pyridylsulfamoyl)-phenylazo)-2-hydroxybenzoic acid
5-(p-(2-Pyridylsulfamoyl)-phenylazo)salicylic acid
Rorasul
Salazopyrin
Salazosulfapyridine
Salicylazosulfapyridine
S.A.S.-500
SASP
Sulcolon
Sulfasalazine
Sulphasalazine
W-T SASP Oral

600-00-0
$C_6H_{11}BrO_2$
195.06
O=C(OCC)C(Br)(C)C
Propanoic acid, 2-bromo-2-methyl-, ethyl ester (9CI)
Ethyl α-bromoisobutyrate
Ethyl 2-bromo-2-methylpropanoate

600-05-5
$C_3H_4Br_2O_2$
231.89
O=C(O)C(Br)CBr
Propionic acid, 2,3-dibromo
2,3-Dibromopropanoic acid
α,β-Dibromopropionic acid
2,3-Dibromopropionic acid
Propanoic acid, 2,3-dibromo- (9CI)

600-07-7
$C_5H_{10}O_2$
102.13
O=C(O)C(CC)C
Butanoic acid, 2-methyl-, (+-)- (9CI)
(+-)-2-Methylbutanoic acid
DL-2-Methylbutyric acid

600-14-6
$C_5H_8O_2$
100.13
O=C(C(=O)CC)C
2,3-Pentanedione
Acetylpropionyl

600-22-6
$C_4H_6O_3$
102.09
Propanoic acid, 2-oxo-, methyl ester (9CI)
NSC-65430
Pyruvic acid, methyl ester (8CI)

600-24-8
$C_4H_9NO_2$
103.14
Butane, 2-nitro
2-Nitrobutane

600-25-9
$C_3H_6ClNO_2$
123.55
CCC(Cl)N(=O)=O
Propane, 1-chloro-1-nitro
Chloronitropropane
1-Chloro-1-nitropropane

(ACGIH,OSHA)
Korax
Lanstan

600-33-9
$C_3H_3ClO_4$
138.51
Propanedioic acid, chloro- (9CI)
Malonic acid, chloro- (8CI)

600-36-2
$C_7H_{16}O$
116.20
OC(C(C)C)C(C)C
3-Pentanol, 2,4-dimethyl- (9CI)
2,4-Dimethyl-3-pentanol

600-40-8
$C_2H_4N_2O_4$
120.05
1,1-Dinitroethane (Dry)
1,1-Dinitroetano (Spanish)
1,1-Dinitroethane
Dinitro-1,1 ethane (French)
Ethane, 1,1-dinitro- (8CI,9CI)

601-54-7
$C_{27}H_{44}O$
384.71
Cholest-5-en-3-one
Cholestenone
δ^5-Cholestenone
5-Cholesten-3-one
Cholesterone

601-75-2
$C_5H_8O_4$
132.12
CCC(C(O)=O)C(O)=O
Ethylmalonic acid
Ethylmalonate
Malonic acid, ethyl- (8CI)
Propanedioic acid, ethyl- (9CI)

601-77-4
$C_6H_{14}N_2O$
130.22
CC(C)N(N=O)C(C)C
Diisopropylamine, N-nitroso
Diisopropylnitrosamin (German)
N-Nitrosodiisopropylamine

601-88-7
$C_6H_3Cl_2NO_2$
191.99
Benzene, 1,3-dichloro-2-nitro- (8CI,9CI)

601-89-8
$C_6H_5NO_4$
155.10
O=N(=O)c(c(O)ccc1)c1O
1,3-Benzenediol, 2-nitro- (9CI)
AI3-52603
2-Nitro-1,3-benzenediol

602-01-7
$C_7H_6N_2O_4$
182.15
O=N(=O)c(c(N(=O)=O)c(cc1)C)c1
Toluene, 2,3-dinitro
Benzene, 1-methyl-2,3-dinitro- (9CI)
2,3-Dinitrotoluene
2,3-DNT

602-29-9
$C_7H_5N_3O_6$
227.15
Toluene, 2,3,4-trinitro
Benzene, 1-methyl-2,3,4-tri-nitro-
2,3,4-Trinitrotoluene

602-38-0
$C_{10}H_6N_2O_4$
218.18
O=N(=O)c(c(c(cc1)ccc2)c2N(=O)=O)c1
Naphthalene, 1,8-dinitro
1,8-Dinitronaphthalene

602-55-1
$C_{20}H_{14}$
254.33
Anthracene, 9-phenyl- (9CI)

602-60-8
$C_{14}H_9NO_2$
223.24
O=N(=O)c2c1ccccc1cc3ccccc23
Anthracene, 9-nitro
5-Nitroanthracene
9-Nitroanthracene

602-87-9
$C_{12}H_9NO_2$
199.22
O=N(=O)c(c(c(c(cc1)CC2)c2c3)c1)c3
Acenaphthene, 5-nitro
Acenaphthylene, 1,2-dihydro-5-nitro-
1,2-Dihydro-5-nitro-acenaphthylene
5-NAN
NCI-C01967
5-Nitroacenaphthene
5-Nitroacenapthene
5-Nitronaphthalene ethylene

602-99-3
$C_7H_5N_3O_7$
243.15
m-Cresol, 2,4,6-trinitro
Cresylite
Phenol, 3-methyl-2,4,6-trinitro- (9CI)
Trinitro-meta-cresol [UN 0216]
Trinitro-m-cresol
Trinitro-m-cresolic acid
UN 0216 [Trinitro-meta-cresol]

603-32-7
$C_{18}H_{15}As$
306.24
c1ccc(cc1)[As](c2ccccc2)c3ccccc3
Arsine, triphenyl- (9CI)
AI3-28453
Triphenylarsene
Triphenylarsine

603-34-9
$C_{18}H_{15}N$
245.34
N(c(cccc1)c1)(c(cccc2)c2)c(cccc3)c3
Triphenylamine
Benzenamine, N,N-diphenyl- (9CI)
N,N-Diphenylaniline
Triphenyl amine (ACGIH, OSHA)

603-35-0
$C_{18}H_{15}P$
262.30
c(P(c(cccc1)c1)c(cccc2)c2)(cccc3)c3
Phosphine, triphenyl
Trifenylfosfin (Czech)
Triphenylphosphine

603-36-1
$C_{18}H_{15}Sb$
353.08
c1ccc(cc1)[Sb](c2ccccc2)c3ccccc3
Stibine, triphenyl
Trifenylstibin (Czech)
Triphenylantimony
Triphenyl stibine

603-45-2
$C_{19}H_{14}O_3$
290.32
Oc1ccc(cc1)C(=C2C=CC(=O)C=C2)c3ccc(O)cc3
Aurin
AI3-18245
4(Bis(p-hydroxyphenyl)methylene)-2,5-cyclohexadien-1-one
2,5-Cyclohexadien-1-one, 4-(bis(p-hydroxyphenyl)-methylene)- (8CI)
2,5-Cyclohexadien-1-one, 4-(bis(4-hydroxyphenyl)-methylene)- (9CI)

603-48-5
$C_{25}H_{31}N_3$
373.52

N(c(ccc(c1)C(c(ccc(N(C)C)c2)
c2)c(ccc(N(C)C)c3)c3)c1)(C)C
Leucogentian violet
AI3-19978
Aniline, 4,4',4''-methylidyne-tris(N,N-dimethyl- (8CI)
Benzenamine, 4,4',4''-methyl-idynetris(N,N-dimethyl- (9CI)

603-50-9
$C_{22}H_{19}NO_4$
361.42
O=C(Oc(ccc(c1)C(c(ccc(OC(=O)
C)c2)c2)c(nccc3)c3)c1)C
Phenol, 4,4'-(2-pyridylmethyl-ene)di-, diacetate (ester)
Bicol
Bis(p-acetoxyphenyl)-2-pyridyl-methane
Bisacodyl
2-(4,4'-Diacetoxydiphenyl-methyl)pyridine
(4,4'-Diacetoxydiphenyl)(2-pyridyl)methane
4,4'-Diacetoxydiphenylpyrid-2-ylmethane
Di-(p-acetoxyphenyl)-2-pyridylmethane
Di-(4-acetoxyphenyl)-2-pyridyl-methane
Dulcolan
Dulcolax
Laxans
LA 96A
Phenol, 4,4'-(2-pyridinyl-methylene)bis-, diacetate (ester)
Pyrilax
4,4'-(2-Pyridylmethylene)-diphenol diacetate
SK-Bisacodyl
Theralax

603-54-3
$C_{13}H_{12}N_2O$
212.24
O=C(N(c(cccc1)c1)c(cccc2)c2)N
Urea, N,N-diphenyl- (9CI)
AI3-61314
N,N-Diphenylurea

603-62-3
$C_8H_4N_2O_4$
192.12
O=C(NC(=O)c1c(N(=O)=O)ccc2)
c12
3-Nitrophthalimide
AI3-08841
1H-Isoindole-1,3(2H)-dione, 4-nitro- (9CI)
4-Nitro-1H-isoindole-1,3(2H)-dione

603-65-6
$C_{11}H_{11}BrN_2O$
267.13
3H-Pyrazol-3-one, 2-(4-bromophenyl)-1,2-di-hydro-1,5-dimethyl- (9CI)
p-Bromoantipyrine
Bromopyrine
3-Pyrazolin-5-one, 1-(p-bro-mophenyl)-2,3-dimethyl- (6CI,8CI)

603-76-9
C_9H_9N
131.17
N(c(c(C=1)ccc2)c2)(C1)C
1-Methylindole
Indole, 1-methyl- (8CI)
1H-Indole, 1-methyl- (9CI)
1-Methyl-1H-indole

603-79-2
$C_9H_{10}O_2$
150.19
Cc1cccc(C(O)=O)c1C
Benzoic acid, 2,3-dimethyl
2,3-Dimethylbenzoic acid
Hemellitic acid

603-83-8
$C_7H_8N_2O_2$
152.17
o-Toluidine, 3-nitro
2-Amino-6-nitrotoluene
Benzenamine, 2-methyl-3-nitro-
3-Nitro-o-toluidine

603-86-1
$C_6H_4ClNO_3$
173.55
O=N(=O)c(c(O)c(cc1)Cl)c1
Phenol, 2-chloro-6-nitro- (9CI)
2-Chloro-6-nitrophenol

604-35-3
$C_{29}H_{48}O_2$
428.70
O=C(OC(CCC(C1=CCC2C(C(C
(C3)C(CCCC(C)C)C)(CC4)C)
C3)(C24)C)C1)C
Cholesterol acetate
AI3-24120
Cholest-5-en-3-ol (3β)-, acetate (9CI)
Cholesterol, acetate (8CI)

604-44-4
$C_{10}H_7ClO$
178.62
Oc(c(c(c(c1)Cl)ccc2)c2)c1
4-Chloro-1-naphthol
4-Chloro-1-naphthalenol
1-Naphthalenol, 4-chloro- (9CI)
1-Naphthol, 4-chloro- (8CI)

604-53-5
$C_{20}H_{14}$
254.33
c(c(c(c(c(c(c(ccc1)cc2)c1)c2)cc3)
ccc4)(c4)c3
1,1'-Binaphthalene (9CI)
AI3-18146
1,1'-Binaphthyl (8CI)

604-75-1
$C_{15}H_{11}ClN_2O_2$
286.73
OC3N=C(c1ccccc1)c2cc(Cl)ccc2
NC3=O
2H-1,4-Benzodiazepin-2-one, 7-chloro-1,3-dihydro-3-hydroxy-5-phenyl
Adumbran
Ansiolisina
Ansioxacepam
Anxiolit
Aplakil
Astress
Bonare
7-Chloro-1,3-dihydro-3-hydroxy-5-phenyl-2H-1,4-benzodiazepine-2-one
7-Chloro-3-hydroxy-5-phenyl-1,3-dihydro-2H-1,4-benzo-diazepin-2-one
Enidrel
Noctazepam
Hilong
Isodin
Limbial
Nesontil
Notaral
OX
Oxazepam
Pacienx
Praxiten
Propax
Psicopax
Quen
Quilibrex
Ro 5-6789
Rondar
Serax
Serenal
Serenid
Serenid-D
Serepax
Seresta
Serpax
Sigacalm
Sobril
Tazepam
Tranquo-buscopan-wirkstoff
Vaben
WY-3498
Z10-TR

604-88-6
$C_{18}H_{30}$
246.48
c(c(c(c(c1CC)CC)CC)CC)(c1CC)
CC
Benzene, hexaethyl
Hexaethylbenzene

605-01-6
$C_{16}H_{26}$
218.38
c(c(c(c(c1CC)CC)CC)CC)(c1)CC

Benzene, pentaethyl- (9CI)
AI3-04220
Pentaethylbenzene

605-02-7
$C_{16}H_{12}$
204.27
Naphthalene, 1-phenyl- (9CI)

605-32-3
$C_{14}H_8O_3$
224.22
O=C(c(c(C(=O)c1cccc2)ccc3O)c3)c12
Anthraquinone, 2-hydroxy
9,10-Anthracenedione, 2-hydroxy- (9CI)
β-Hydroxyanthraquinone
2-Hydroxyanthraquinone

605-39-0
$C_{14}H_{14}$
182.27
o,o'-Bitolyl (8CI)
1,1'-Biphenyl, 2,2'-dimethyl- (9CI)

605-45-8
$C_{14}H_{18}O_4$
250.32
O=C(OC(C)C)c(c(ccc1)C(=O)OC(C)C)c1
Phthalic acid, diisopropyl ester
1,2-Benzenedicarboxylic acid, bis(1-methylethyl) ester
Diisopropyl phthalate

605-48-1
$C_{14}H_8Cl_2$
247.12
c(c(c(c(c1ccc2)c2)Cl)ccc3)(c1Cl)c3
Anthracene, 9,10-dichloro- (9CI)
9,10-Dichloroanthracene

605-69-6
$C_{10}H_6N_2O_5$
234.18
O=N(=O)c(c(O)c(c(c1N(=O)=O)ccc2)c2)c1
1-Naphthol, 2,4-dinitro
C.I. 10315
2,4-Dinitro-1-naftol (Czech)
2-4 Dinitro-α-naphtol (French)
2,4-Dinitro-1-naphthol
Golden Yellow
Manchester Yellow
Maritus Yellow
Naphthol Yellow
Naphthylene Yellow
Saffron Yellow
Zlut Marciova (Czech)
Zlut Naftolova (Czech)

605-71-0
$C_{10}H_6N_2O_4$
218.18
Naphthalene, 1,5-dinitro
1,5-Dinitronaphthalene

605-99-2
$C_6H_6N_4O_3$
182.11
1H-Purine-2,6,8(3H)-trione, 4,9-dihydro-3-methyl- (9CI)
Uric acid, 3-methyl- (8CI)

606-03-1
$C_{24}H_{21}N_3O_3$
399.43
1,3,5-Triazine-2,4,6(1H,3H,5H)-trione, 1,3,5-tris(phenylmethyl)- (9CI)
s-Triazine-2,4,6(1H,3H,5H)-trione, 1,3,5-tribenzyl- (8CI)
s-'Triazine-2,4,6(1H,3H,5H)-trione, tribenzyl-

606-07-5
$C_8H_5Cl_5$
278.39
Benzene, pentachloroethyl- (6CI,7CI,9CI)
Alkazene 32

606-20-2
$C_7H_6N_2O_4$
182.15
O=N(=O)c(c(c(N(=O)=O)cc1)C)c1
Toluene, 2,6-dinitro
Benzene, 2-methyl-1,3-dinitro- (9CI)
2,6-Dinitrotoluene
2,6-DNT
RCRA waste number U106

606-21-3
$C_6H_3ClN_2O_4$
202.56
O=N(=O)c(c(c(N(=O)=O)cc1)Cl)c1
Benzene, 2-chloro-1,3-dinitro
2,6-Dinitrochlorobenzene

606-22-4
$C_6H_5N_3O_4$
183.14
Aniline, 2,6-dinitro
Benzenamine, 2,6-dinitro- (9CI)
2,6-Dinitroaniline
NCI-C60753

606-23-5
$C_9H_6O_2$
146.15
O=C(c(c(C1=O)ccc2)c2)C1
1,3-Indandione
1,3-Diketohydrindene
1H-Indene-1,3(2H)-dione

606-28-0
$C_{15}H_{12}O_3$
240.26
O=C(OC)c(c(ccc1)C(=O)c(cccc2)c2)c1
Benzoic acid, o-benzoyl-, methyl ester (8CI)
AI3-00516
Benzoic acid, 2-benzoyl-, methyl ester (9CI)
2-Benzoylbenzoic acid, methyl ester
Methyl 2-benzoylbenzoate

606-35-9
$C_7H_5N_3O_7$
243.15
O=N(=O)c(cc(N(=O)=O)c(OC)c1N(=O)=O)c1
Anisole, 2,4,6-trinitro
Benzene, 2-methoxy-1,3,5-trinitro- (9CI)
Methyl picrate
2,4,6-Trinitroanisole

606-37-1
$C_{10}H_6N_2O_4$
218.18
Naphthalene, 1,3-dinitro
1,3-Dinitronaphthalene

606-40-6
$C_{10}H_7ClO$
178.62
1-Naphthalenol, 2-chloro- (9CI)
1-Naphthol, 2-chloro- (8CI)

606-43-9
$C_{10}H_9NO$
159.18
O=C(N(c(c(C=1)ccc2)c2)C)C1
Carbostyril, 1-methyl- (8CI)
AI3-24119
1-Methyl-2(1H)-quinolinone
1-Methyl-2-quinolone
2(1H)-Quinolinone, 1-methyl- (9CI)

606-58-6
$C_{12}H_{13}N_5O_4$
291.30
7H-Pyrrolo(2,3-d)pyrimidine-5-carbonitrile, 4-amino-7-β-D-ribofuranosyl
Ahygroscopin-B
4-Amino-5-cyano-7-(D-ribofuranosyl)-7H-pyrrolo(2,3-d)-pyrimidine
4-Amino-7-β-D-ribofuranosyl-7H-pyrrolo(2,3-d)pyrimidine-5-carbonitrile
Antibiotic 1037
Antibiotic A-399-Y4
Antibiotic E212

A-399-Y4
Cyanotubericidin
E-212
E-212-1
Naritheracin
NSC-63701
Siromycin
Toyocamycin
Toyocamycin Nucleoside
Unamycin-B
Uramycin B
Vengicide

607-12-5
$C_{12}H_9ClO$
204.66
4-Chloro-2-phenylphenol
AI3-09046
2-Biphenylol, 5-chloro- (8CI)
(1,1'-Biphenyl)-2-ol, 5-chloro- (9CI)
Caswell No. 210
5-Chloro-2-biphenylol
5-Chloro-(1,1'-biphenyl)-2-ol
EPA Pesticide Chemical Code 062208

607-30-7
$C_{10}H_9NO$
159.20
ONc1cccc2ccccc12
Hydroxylamine, N-1-naphthyl
N-Hydroxy-1-aminonaphthalene
N-Hydroxy-1-naphthylamine
1-Naphthylhydroxylamine
N-1-Naphthylhydroxylamine

607-34-1
$C_9H_6N_2O_2$
174.17
O=N(=O)c1cccc2ncccc12
Quinoline, 5-nitro
5-Nitroquinoline

607-35-2
$C_9H_6N_2O_2$
174.17
O=N(=O)c(c(nccc1)c1cc2)c2
Quinoline, 8-nitro
8-Nitroquinoline

607-57-8
$C_{13}H_9NO_2$
211.23
O=N(=O)c(ccc(c1Cc2cccc3)c23)c1
9H-Fluorene, 2-nitro
Fluorene, 2-nitro-
2-Nitrofluorene

607-85-2
$C_{10}H_{12}O_3$
180.20
O=C(OC(C)C)c(c(O)ccc1)c1
Benzoic acid, 2-hydroxy-, 1-methylethyl ester
AI3-00511
1-Methylethyl 2-hydroxybenzoate

607-91-0
$C_{11}H_{12}O_3$
192.23
COc1cc(CC=C)cc2OCOc12
Benzene, 5-allyl-1-methoxy-2,3-(methylenedioxy)
5-Allyl-1-methoxy-2,3-(methylenedioxy)benzene
Myristicin

607-99-8
$C_7H_5Br_3O$
344.83
O(c(c(cc(c1)Br)Br)c1Br)C
Benzene, 1,3,5-tribromo-2-methoxy- (9CI)
AI3-00641
1,3,5-Tribromo-2-methoxybenzene

608-21-9
$C_6H_3Br_3$
314.80
Benzene, 1,2,3-tribromo- (8CI, 9CI)

608-25-3
$C_7H_8O_2$
124.14
Oc(c(c(O)cc1)C)c1
2-Methylresorcinol
AI3-61050
1,3-Benzenediol, 2-methyl- (9CI)
2,6-Dihydroxytoluene
2-Methyl-1,3-benzenediol
Resorcinol, 2-methyl- (8CI)

608-27-5
$C_6H_5Cl_2N$
162.01
Nc(c(c(cc1)Cl)Cl)c1
2,3-Dichloroaniline
Aniline, 2,3-dichloro- (8CI)
Benzenamine, 2,3-dichloro- (9CI)
2,3-Dichlorobenzenamine

608-29-7
$C_6H_3I_3$
455.80
Benzene, 1,2,3-triiodo- (7CI,8CI,9CI)
1,2,3-Triiodobenzene

608-31-1
$C_6H_5Cl_2N$
162.01
Nc(c(ccc1)Cl)c1Cl
Benzenamine, 2,6-dichloro- (9CI)
2,6-Dichloroaniline
2,6-Dichlorobenzenamine

608-33-3
$C_6H_4Br_2O$
251.91
Oc(c(ccc1)Br)c1Br
2,6-Dibromophenol
Phenol, 2,6-dibromo- (9CI)

608-44-6
$C_6H_4Cl_2O_2$
179.00
Hydroquinone, 2,3-dichloro- (8CI)
1,4-Benzenediol, 2,3-dichloro- (9CI)

608-50-4
$C_9H_{10}N_2O_4$
210.18
2,4-Dinitro-1,3,5-trimethylbenzene
Benzene, 1,3,5-trimethyl-2,4-dinitro- (9CI)
2,4-Dinitromesitylene
Mesitylene, 2,4-dinitro- (8CI)

608-66-2
$C_6H_{14}O_6$
182.17
OCC(O)C(O)C(O)C(O)CO
Dulcitol
AI3-19423
Galactitol (9CI)

608-68-4
$C_6H_{10}O_6$
178.14
O=C(OC)C(O)C(O)C(=O)OC
Butanedioic acid, 2,3-dihydroxy- (R-(R*,R*))-, dimethyl ester (9CI)

608-71-9
C_6HBr_5O
488.62
Oc(c(c(c(c1Br)Br)Br)Br)c1Br
Phenol, pentabromo
Pentabromfenol (Czech)
Pentabromophenol

608-73-1
$C_6H_6Cl_6$
290.82
ClC1C(Cl)C(Cl)C(Cl)C(Cl)C1Cl
Cyclohexane, 1,2,3,4,5,6-hexachloro
Benzene hexachloride
BHC
Compound-666
DBH
ENT 8,601
Gammexane
HCCH
HCH
Hexa

Hexachlor
Hexachloran
Hexachlorocyclohexane
1,2,3,4,5,6-Hexachloro-cyclohexane
Hexaklor
Hexylan
Jacutin
Latka 666 (Czech)

608-74-2
C_6I_6
833.49
c(c(c(c(c1I)I)I)I)(c1I)I
Benzene, hexaiodo- (9CI)
Hexaiodobenzene

608-90-2
C_6HBr_5
472.59
Benzene, pentabromo- (8CI, 9CI)

608-93-5
C_6HCl_5
250.32
c(c(c(c(c1Cl)Cl)Cl)Cl)(c1)Cl
Benzene, pentachloro
PCP
Pentachlorobenzene
QCB
RCRA waste number U183

608-94-6
$C_6H_3Cl_3O_2$
213.45
1,4-Benzenediol, 2,3,5-trichloro- (9CI)
Hydroquinone, trichloro- (8CI)

608-96-8
C_6HI_5
707.60
Benzene, pentaiodo- (8CI, 9CI)
Pentaiodobenzene

609-08-5
$C_8H_{14}O_4$
174.20
O=C(OCC)C(C(=O)OCC)C
Malonic acid, methyl-, diethyl ester (8CI)
Diethyl methylpropanedioate
Propanedioic acid, methyl-, diethyl ester (9CI)

609-15-4
$C_6H_9ClO_3$
164.59
O=C(OCC)C(C(=O)C)Cl
A 21960
Acetoacetic acid, 2-chloro-, ethyl ester (8CI)
Butanoic acid, 2-chloro-3-oxo-, ethyl ester (9CI)
2-Chloro-3-oxobutanoic acid, ethyl ester
Ethyl 2-chloroacetoacetate
Ethyl 2-chloro-3-oxobutanoate

609-19-8
$C_6H_3Cl_3O$
197.44
Oc1cc(Cl)c(Cl)c(Cl)c1
Phenol, 3,4,5-trichloro
3,4,5-Trichlorophenol

609-20-1
$C_6H_6Cl_2N_2$
177.04
Nc(c(cc(N)c1)Cl)c1Cl
p-Phenylenediamine, 2,6-dichloro
1,4-Benzenediamine, 2,6-dichloro-
C.I. 37020
Daito Brown Salt RR
1,4-Diamino-2,6-dichlorobenzene
2,5-Diamino-1,3-dichlorobenzene
2,6-Dichloro-1,4-benzenediamine
2,6-Dichloro-p-phenylenediamine
2,6-Dichloro-1,4-phenylenediamine
Fast Brown RR Salt
NCI-C50260

609-22-3
$C_7H_6Br_2O$
265.93
Oc(c(cc(c1)Br)C)c1Br
Phenol, 2,4-dibromo-6-methyl- (9CI)
AI3-00101
2,4-Dibromo-6-methylphenol

609-23-4
$C_6H_3I_3O$
471.79
Oc(c(cc(c1)I)I)c1I
Phenol, 2,4,6-triiodo
2,4,6-Triiodophenol
2,4,6-Trijodfenol (Czech)

609-26-7
C_8H_{18}
114.23
2-Methyl-3-ethylpentane

609-31-4
$C_4H_9NO_3$
119.11
O=N(=O)C(CC)CO
1-Butanol, 2-nitro- (9CI)
AI3-04493
Caswell No. 601
EPA Pesticide Chemical Code 056901
2-Nitro-1-butanol

609-36-9
$C_5H_9NO_2$
115.13
O=C(O)C(NCC1)C1
DL-Proline (9CI)
Proline, DL- (8CI)

609-39-2
$C_4H_3NO_3$
113.08
Furan, 2-nitro
Nitrofuran
2-Nitrofuran
5-Nitrofuran

609-40-5
$C_4H_3NO_2S$
129.14
O=N(=O)C(SC=C1)=C1
Thiophene, 2-nitro
2-Nitrothiophene

609-46-1
$C_6H_6O_4S$
174.18
O=S(=O)(O)c(c(O)ccc1)c1
Benzenesulfonic acid, 2-hydroxy- (9CI)
Benzenesulfonic acid, o-hydroxy- (8CI)
NSC-243747

609-54-1
$C_8H_{10}O_3S$
186.23
O=S(=O)(O)c(c(ccc1C)C)c1
Benzenesulfonic acid, 2,5-dimethyl- (9CI)
2,5-Dimethylbenzenesulfonic acid

609-60-9
$C_8H_9ClO_2S$
204.68
O=S(=O)(c(c(cc(c1)C)C)c1)Cl
Benzenesulfonyl chloride, 2,4-dimethyl- (9CI)
2,4-Dimethylbenzenesulfonyl chloride

609-66-5
C_7H_6ClNO
155.58
Benzamide, o-chloro- (8CI)
AI3-09664
Benzamide, 2-chloro- (9CI)
2-Chlorobenzamide

609-72-3
$C_9H_{13}N$
135.23

N(c(c(ccc1)C)c1)(C)C
o-Toluidine, N,N-dimethyl
Benzamine, N,N,2-trimethyl- (9CI)
Dimethyl-o-toluidine
N,N-Dimethyl-o-toluidine
o-Methyldimethylaniline
N,N,2-Trimethylaniline

609-89-2
$C_6H_3Cl_2NO_3$
208.00
Oc1c(Cl)cc(Cl)cc1N(=O)=O
Phenol, 2,4-dichloro-6-nitro
2,4-Dichlor-6-nitrofenol (Czech)
2,4-Dichloro-6-nitrophenol

609-93-8
$C_7H_6N_2O_5$
198.15
O=N(=O)c(c(O)c(N(=O)=O)cc1C)c1
p-Cresol, 2,6-dinitro
Dinitro-p-cresol
DNPC
Victoria Orange
Victoria Yellow

609-99-4
$C_7H_4N_2O_7$
228.13
O=C(O)c(c(O)c(N(=O)=O)cc1N(=O)=O)c1
Salicylic acid, 3,5-dinitro
Benzoic acid, 2-hydroxy-3,5-dinitro- (9CI)
3,5-Dinitrosalicylic acid

610-02-6
$C_7H_6O_5$
170.12
OC(=O)c1ccc(O)c(O)c1O
Benzoic acid, 2,3,4-trihydroxy- (9CI)

610-14-0
$C_7H_4ClNO_3$
185.57
O=C(c(c(N(=O)=O)ccc1)c1)Cl
Benzoyl chloride, o-nitro
Benzoyl chloride, 2-nitro- (9CI)
o-Nitrobenzoyl chloride
2-Nitrobenzoyl chloride

610-15-1
$C_7H_6N_2O_3$
166.15
Benzamide, o-nitro
Benzmide, 2-nitro- (9CI)
o-Nitrobenzamide
2-Nitrobenzamide

610-25-3
$C_7H_5N_3O_6$
227.15
Toluene, 2,4,5-trinitro
Benzene, 1-methyl-2,4,5-tri-nitro-
2,4,5-Trinitrotoluene

610-28-6
$C_7H_4N_2O_6$
212.13
O=C(O)c(c(N(=O)=O)ccc1N(=O)=O)c1
Benzoic acid, 2,5-dinitro
2,5-Dinitrobenzoic acid

610-30-0
$C_7H_4N_2O_6$
212.13
O=C(O)c(c(N(=O)=O)cc(N(=O)=O)c1)c1
Benzoic acid, 2,4-dinitro
2,4-Dinitrobenzoic acid

610-35-5
$C_8H_6O_5$
182.13
O=C(O)c(c(ccc1O)C(=O)O)c1
1,2-Benzenedicarboxylic acid, 4-hydroxy- (9CI)
4-Hydroxy-1,2-benzenedi-carboxylic acid

610-38-8
$C_6H_3BrN_2O_4$
246.99
4-Bromo-1,2-dinitrobenzene
Benzene, 4-bromo-1,2-dinitro-

610-39-9
$C_7H_6N_2O_4$
182.15
O=N(=O)c(c(N(=O)=O)ccc1C)c1
Toluene, 3,4-dinitro
Benzene, 4-methyl-1,2-dinitro- (9CI)
3,4-Dinitrotoluene
3,4-DNT

610-48-0
$C_{15}H_{12}$
192.26
1-Methylanthracene
Anthracene, 1-methyl- (9CI)

610-54-8
$C_8H_8N_2O_5$
212.18
O=N(=O)c(ccc(OCC)c1N(=O)=O)c1
Phenetole, 2,4-dinitro
2,4-Dinitrofenetol (Czech)
2,4-Dinitrophenetole

610-66-2
$C_8H_6N_2O_2$
162.16
O=N(=O)c(c(ccc1)CC#N)c1
Acetonitrile, (o-nitrophenyl)
Benzeneacetonitrile, 2-nitro- (9CI)
o-Nitrobenzacetonitrile
2-Nitrobenzeneacetonitrile
2-Nitrobenzyl nitrile
(o-Nitrophenyl)acetonitrile
(2-Nitrophenyl)acetonitrile

610-69-5
$C_8H_7NO_4$
181.14
O=C(Oc(c(N(=O)=O)ccc1)c1)C
Acetic acid, 2-nitrophenyl ester (9CI)
AI3-21066
2-Nitrophenyl acetate

610-72-0
$C_9H_{10}O_2$
150.18
O=C(O)c(c(ccc1C)C)c1
Benzoic acid, 2,5-dimethyl- (9CI)
2,5-Dimethylbenzoic acid

610-94-6
$C_8H_7BrO_2$
215.05
O=C(OC)c(c(ccc1)Br)c1
Benzoic acid, o-bromo-, methyl ester (8CI)
Benzoic acid, 2-bromo-, methyl ester (9CI)
Methyl 2-bromobenzoate

610-96-8
$C_8H_7ClO_2$
170.60
O=C(OC)c(c(ccc1)Cl)c1
Benzoic acid, o-chloro-, methyl ester (8CI)
AI3-23031
Benzoic acid, 2-chloro-, methyl ester (9CI)
Methyl 2-chlorobenzoate

611-00-7
$C_7H_4Br_2O_2$
279.92
Benzoic acid, 2,4-dibromo- (8CI,9CI)

611-01-8
$C_9H_{10}O_2$
150.18
O=C(O)c(c(cc(c1)C)C)c1
2,4-Dimethylbenzoic acid
Benzoic acid, 2,4-dimethyl- (9CI)

611-05-2
$C_7H_8N_2O_2$
152.14

Cc1cc(N)ccc1N(=O)=O
Benzenamine, 3-methyl-4-nitro- (9CI)
NSC-17041
m-Toluidine, 4-nitro- (8CI)

611-06-3
$C_6H_3Cl_2NO_2$
192.00
O=N(=O)c(c(cc(c1)Cl)Cl)c1
Benzene, 2,4-dichloro-1-nitro
2,4-Dichloronitrobenzene

611-07-4
$C_6H_4ClNO_3$
173.55
O=N(=O)c(c(O)cc(c1)Cl)c1
Phenol, 5-chloro-2-nitro- (9CI)
5-Chloro-2-nitrophenol

611-14-3
C_9H_{12}
120.21
c(c(ccc1)C)(c1)CC
Toluene, o-ethyl
Benzene, 1-ethyl-2-methyl- (9CI)
1-Ethyl-2-methylbenzene
o-Ethyltoluene
2-Ethyltoluene
o-Ethyl methylbenzene
o-Methylethylbenzene

611-15-4
C_9H_{10}
118.19
c(c(ccc1)C=C)(c1)C
Styrene, o-methyl
Benzene, 1-ethenyl-2-methyl- (9CI)
o-Methylstyrene
2-Methylstyrene
o-Vinyltoluene
2-Vinyltoluene

611-17-6
C_7H_6BrCl
205.48
α-Bromo-o-chlorotoluene
Benzene, 1-(bromomethyl)-2-chloro-
Benzyl bromide, 2-chloro
o-Chlorobenzyl bromide
2-Chlorobenzyl bromide
Toluene, α-bromo-o-chloro-

611-19-8
$C_7H_6Cl_2$
161.03
c(c(ccc1)Cl)(c1)CCl
2-Chlorobenzyl chloride
AI3-14885
Benzene, 1-chloro-2-(chloromethyl)- (9CI)
1-Chloro-2-(chloromethyl)-benzene
Toluene, o,α-dichloro- (8CI)

611-20-1
C_7H_5NO
119.12
N#Cc(c(O)ccc1)c1
Benzonitrile, 2-hydroxy- (9CI)
o-Cyanophenol
o-Hydroxybenzonitrile
2-Hydroxybenzonitrile
Salicylonitrile (8CI)

611-32-5
$C_{10}H_9N$
143.20
n(c(c(ccc1)cc2)c1C)c2
Quinoline, 8-methyl
8-Methylquinoline

611-34-7
$C_9H_8N_2$
144.19
Nc1cccc2ncccc12
Quinoline, 5-amino
5-Aminoquinoline
5-Quinolinamine

611-59-6
$C_7H_8N_4O_2$
180.19
Cn1cnc2[nH]c(=O)n(C)c(=O)c12
Paraxanthine
1,7-Dimethylxanthine
1H-Purine-2,6-dione, 3,7-dihydro-1,7-dimethyl- (9CI)

611-70-1
$C_{10}H_{12}O$
148.20
O=C(c(cccc1)c1)C(C)C
Isobutyrophenone (8CI)
AI3-11204
2-Methyl-1-phenyl-1-propanone
1-Propanone, 2-methyl-1-phenyl- (9CI)

611-72-3
$C_8H_8O_3$
152.15
O=C(O)C(O)c(cccc1)c1
Benzeneacetic acid, α-hydroxy-, (+-)- (9CI)
α-Hydroxybenzeneacetic acid, (+-)-
(+-)-α-Hydroxybenzeneacetic acid

611-73-4
$C_8H_6O_3$
150.14
OC(=O)C(=O)c1ccccc1
Benzeneglyoxylic-acid
Glyoxylic acid, phenyl-

611-74-5
$C_9H_{11}NO$
149.21
O=C(N(C)C)c(cccc1)c1
Benzamide, N,N-dimethyl
Dimethylbenzamide
N,N-Dimethylbenzamide

611-92-7
$C_{15}H_{16}N_2O$
240.33
O=C(N(c(cccc1)c1)C)N(c(cccc2)c2)C
Carbanilide, N,N'-dimethyl
Centralite II
N,N'-Dimethyl carbanilide
Methyl centralite
Urea, N,N'-dimethyl-N,N'-diphenyl- (9CI)

612-00-0
$C_{14}H_{14}$
182.27
1,1-Diphenylethane
AI3-04341
Benzene, 1,1'-ethylidenebis- (9CI)
Ethane, 1,1-diphenyl- (8CI)

612-12-4
$C_8H_8Cl_2$
175.06
c(c(ccc1)CCl)(c1)CCl
o-Xylene, α,α'-dichloro
Benzene, 1,2-bis(chloromethyl)-
1,2-Bis(chloromethyl)benzene
α,α'-Dichloro-o-xylene
o-Xylylene dichloride

612-20-4
$C_8H_8O_3$
152.15
Benzoic acid, 2-(hydroxymethyl)- (9CI)
o-Toluic acid, α-hydroxy- (8CI)

612-23-7
$C_7H_6ClNO_2$
171.59
ClCc1ccccc1N(=O)=O
Toluene, α-chloro-o-nitro
Benzene, 1-(chloromethyl)-2-nitro- (9CI)
1-(Chloromethyl)-2-nitrobenzene
α-Chloro-o-nitrotoluene
o-Nitrobenzyl chloride
2-Nitrobenzyl chloride

612-25-9
$C_7H_7NO_3$
153.13
2-Nitrobenzyl alcohol
AI3-16131
Benzenemethanol, 2-nitro- (9CI)

Benzyl alcohol, o-nitro- (8CI)

612-35-1
$C_{14}H_{12}O_2$
212.25
Benzoic acid, 2-(phenylmethyl)- (9CI)
NSC-74872
o-Toluic acid, α-phenyl- (8CI)

612-52-2
$C_{10}H_9N.ClH$
179.66
2-Naphthylamine, hydrochloride
2-Aminonaphthalene hydrochloride
β-Naphthylamine hydrochloride
2-Naphthalenamine, hydrochloride

612-58-8
$C_{10}H_9N$
143.20
Cc2cnc1ccccc1c2
Quinoline, 3-methyl
3-Methylquinoline

612-60-2
$C_{10}H_9N$
143.20
n(c(c(ccc1C)cc2)c1)c2
Quinoline, 7-methyl
7-Methylquinoline
m-Toluquinoline

612-64-6
$C_8H_{10}N_2O$
150.20
CCN(N=O)c1ccccc1
Aniline, N-ethyl-N-nitroso
Benzenamine, N-ethyl-N-nitroso- (9CI)
Ethylnitrosoaniline
NEA
Nitrosoethylaniline
N-Nitroso-N-ethyl aniline

612-71-5
$C_{24}H_{18}$
306.42
1,3,5-Triphenylbenzene
5'-Phenyl-m-terphenyl
1,1':3',1''-Terphenyl- (9CI)
m-Terphenyl, 5'-phenyl
Triphenylbenzene
symmetrical Triphenylbenzene

612-75-9
$C_{14}H_{14}$
182.27
m,m'-Bitolyl (8CI)
1,1'-Biphenyl, 3,3'-dimethyl- (9CI)

612-82-8
$C_{14}H_{16}N_2.2ClH$
285.24
Cl.Cl.Cc1cc(ccc1N)c2ccc(N)c(C)c2
Benzidine, 3,3'-dimethyl-, dihydrochloride
3,3'-Dimethylbenzidine dihydrochloride

612-83-9
$C_{12}H_{10}Cl_2N_2.2ClH$
326.06
[Cl-].[Cl-].Nc1ccc(cc1Cl)c2ccc(N)c(Cl)c2
Benzidine, 3,3'-dichloro-, dihydrochloride
(1,1'-Biphenyl)-4,4'-diamine, 3,3'-dichloro-, dihydrochloride
3,3'-Dichlorobenzidine dihydrochloride

612-94-2
$C_{16}H_{12}$
204.27
Naphthalene, 2-phenyl- (9CI)

613-12-7
$C_{15}H_{12}$
192.27
Anthracene, 2-methyl
2-Methylanthracene

613-13-8
$C_{14}H_{11}N$
193.26
Nc3ccc2cc1ccccc1cc2c3
2-Anthracenamine
β-Aminoanthracene
2-Aminoanthracene
2-Anthracylamine
2-Anthrylamine
2-Anthramine

613-26-3
$C_{16}H_{14}$
206.29
Anthracene, 2,6-dimethyl- (8CI,9CI)

613-29-6
$C_{14}H_{23}N$
205.34
N(c(cccc1)c1)(CCCC)CCCC
N,N-Dibutylaniline
Benzenamine, N,N-dibutyl- (9CI)
N,N-Dibutylbenzenamine

613-31-0
$C_{14}H_{12}$
180.25
C2c1ccccc1Cc3ccccc23
Anthracene, 9,10-dihydro- (9CI)
AI3-09026

613-33-2
$C_{14}H_{14}$
182.27
p,p'-Bitolyl (8CI)
AI3-21616
1,1'-Biphenyl, 4,4'-dimethyl- (9CI)

613-35-4
$C_{16}H_{16}N_2O_2$
268.34
O=C(Nc(ccc(c(ccc(NC(=O)C)c1)c1)c2)c2)C
4',4'''-Biacetanilide
N,N'-(1,1'-Biphenyl)-4,4'-diyl-bis-acetamide 4',4'''-biacetanilide
N,N'-4,4'-Biphenylylenebis-acetamide
4,4'-Diacetamidobiphenyl
4,4'-Diacetylaminobiphenyl
Diacetylbenzidine
N,N'-Diacetyl benzidine
4,4'-Diacetylbenzidine

613-37-6
$C_{13}H_{12}O$
184.25
O(c(ccc(c(cccc1)c1)c2)c2)C
Anisole, p-phenyl
1,1'-Biphenyl, 4-methoxy- (9CI)
p-Methoxybiphenyl
4-Methoxybiphenyl

613-46-7
$C_{11}H_7N$
153.19
N#Cc(ccc(c1ccc2)c2)c1
2-Naphthonitrile
β-Cyanonaphthalene
2-Cyanonaphthalene
2-Naphthalenecarbonitrile (9CI)
β-Naphthonitrile
2-Naphthalenenitrile

613-50-3
$C_9H_6N_2O_2$
174.17
O=N(=O)c2ccc1ncccc1c2
Quinoline, 6-nitro
6-Nitroquinoline

613-55-8
$C_{14}H_{14}N_2O_2$
242.28
Diazene, bis(2-methoxyphenyl)- (9CI)
Anisole, 2,2'-azodi- (6CI, 7CI)
Azobenzene, 2,2'-dimethoxy- (8CI)

2,2'-Dimethoxyazobenzene

613-59-2
$C_{17}H_{14}$
218.30
Naphthalene, 2-(phenyl-methyl)- (9CI)
Naphthalene, 2-benzyl- (8CI)

613-69-4
$C_9H_{10}O_2$
150.18
O=Cc(c(OCC)ccc1)c1
Benzaldehyde, o-ethoxy- (8CI)
AI3-01358
Benzaldehyde, 2-ethoxy- (9CI)
2-Ethoxybenzaldehyde

613-90-1
C_8H_5NO
131.13
O=C(C#N)c(cccc1)c1
Benzeneacetonitrile, α-oxo- (9CI)
AI3-13150
Glyoxylonitrile, phenyl- (8CI)
α-Oxobenzeneacetonitrile

613-93-4
C_8H_9NO
135.18
O=C(NC)c(cccc1)c1
Benzamide, N-methyl
N-Methylbenzamide
N-Methylbenzenamide

613-94-5
$C_7H_8N_2O$
136.17
O=C(NN)c(cccc1)c1
Benzoic acid, hydrazide
Hydrazine, benzoyl-
Benzhydrazide
Benzohydrazide
Benzohydrazine
Benzoic hydrazide
Benzoyl hydrazide
Benzoyl hydrazine
Hydrazid kyseliny benzoove
(Czech)

613-97-8
$C_9H_{13}N$
135.20
N(c(cccc1)c1)(CC)C
N-Ethyl-N-methylaniline
Aniline, N-ethyl-N-methyl- (8CI)
Benzenamine, N-ethyl-N-methyl- (9CI)
N-Ethyl-N-methylbenzenamine

614-00-6
$C_7H_8N_2O$
136.17
O=NN(c(cccc1)c1)C
Aniline, N-methyl-N-nitroso
Benzenamine, N-methyl-N-nitroso-
Methylnitrosoaniline
N-Methyl-N-nitrosoaniline
N-Methyl-N-nitrosobenzenamine
Methylphenylnitrosamine
MNA
Nitrosomethylaniline
N-Nitroso-N-methylaniline
N-Nitrosomethylphenylamine
NMA
Phenylmethylnitrosamine

614-26-6
$C_{12}H_8Cl_2N_2O$
267.12
Azoxybenzene, 4,4'-dichloro
DCAOB
Diazene, bis(4-chlorphenyl)-, 1-oxide (9CI)
p,p'-Dichloroazoxybenzene
4,4'-Dichloroazoxybenzene

614-27-7
$C_{10}H_{10}O_3$
178.19
Benzenepropanoic acid, β-oxo-, methyl ester (9CI)
Acetic acid, benzoyl-, methyl ester (8CI)
NSC-407764

614-33-5
$C_{24}H_{20}O_6$
404.42
O=C(OCC(OC(=O)c(cccc1)c1)COC(=O)c(cccc2)c2)c(cccc3)c3
Glycerol, tribenzoate (8CI)
AI3-08196
1,2,3-Propanetriol, tribenzoate (9CI)

614-45-9
$C_{11}H_{14}O_3$
194.25
O=C(OOC(C)(C)C)c(cccc1)c1
Peroxybenzoic acid, tert-butyl ester
terc.Butylester kyseliny peroxybenzoove (Czech)
terc.Butylperbenzoan (Czech)
t-Butyl perbenzoate
t-Butyl peroxy benzoate
tert-Butyl peroxybenzoate, Not more than 50% with inert inorganic solid (DOT)
tert-Butyl peroxybenzoate, Not more than 75% in solution (DOT)
tert-Butyl peroxybenzoate, Technical pure or in concentration of more than 75% (DOT)
Esperox 10
Novox
Perbenzoate de butyle tertiaire (French)
Peroxybenzoic acid, tert-butyl ester, Not more than 50% with inert inorganic solid
Peroxybenzoic acid, tert-butyl ester, Not more than 75% in solution
Trigonox C
UN 2097 (DOT)
UN 2098 (DOT)
UN 2890 (DOT)

614-60-8
$C_9H_8O_3$
164.17
O=C(O)C=Cc(c(O)ccc1)c1
Cinnamic acid, o-hydroxy-, (E)
trans-o-Coumaric acid
trans-o-Hydroxycinnamic acid
trans-2-Hydroxycinnamic acid
2-Propenoic acid, 3-(2-hydroxyphenyl)-, (E)- (9CI)

614-61-9
$C_8H_7ClO_3$
186.60
OC(=O)COc1ccccc1Cl
Acetic acid, o-chlorophenoxy
Acetic acid, (2-chlorophenoxy)- (9CI)
Acide o-chlorophenoxyacetique (French)
o-Chlorophenoxyacetic acid
(2-Chlorophenoxy)acetic acid

614-68-6
C_8H_7NO
133.14
O=C=Nc(c(ccc1)C)c1
Benzene, 1-isocyanato-2-methyl-
AI3-28255
1-Isocyanato-2-methylbenzene

614-75-5
$C_8H_8O_3$
152.15
OC(=O)Cc1ccccc1O
2-Hydroxyphenylacetic acid
Acetic acid, (o-hydroxyphenyl)- (8CI)
Benzeneacetic acid, 2-hydroxy- (9CI)
2-Hydroxybenzeneacetic acid

614-78-8
$C_8H_{10}N_2S$
166.23
N=C(S)Nc(c(ccc1)C)c1
Thiourea, (2-methylphenyl)- (9CI)
AI3-24571
(2-Methylphenyl)thiourea

o-Tolyl thiourea
1-o-Tolyl-2-thiourea
Urea, 2-thio-1-o-tolyl- (8CI)

614-80-2
$C_8H_9NO_2$
151.18
O=C(Nc(c(O)ccc1)c1)C
Acetanilide, 2'-hydroxy
Acetamide, N-(2-hydroxy-phenyl)- (9CI)
o-Acetamidophenol
2-Acetamidophenol
Acet-o-aminofenol (Czech)
2-Acetaminophenol
o-Acetylaminofenol (Czech)
N-Acetyl-2-aminophenol
o-(Acetylamino)phenol
2-(Acetylamino)phenol
o-Hydroxyacetanilide
2'-Hydroxyacetanilide
2-Hydroxyanilid kyseliny octove (Czech)
Phenol, 2-acetamido-

614-94-8
$C_7H_{10}N_2O.2ClH$
211.11
m-Phenylenediamine, 4-methoxy-, dihydrochloride
1,3-Benzenediamine, 4-methoxy-, dihydrochloride
2,4-Diaminoanisole dihydrochloride

614-95-9
$C_5H_{10}N_2O_3$
146.17
CCOC(=O)N(CC)N=O
Carbamic acid, ethylnitroso-, ethyl ester
Aethylnitrosourethan (German)
ENU
Ethylester kyseliny N-ethyl-N-nitrosokarbaminove (Czech)
Ethylnitrosocarbamic acid, ethyl ester
N-Ethyl-N-nitrosocarbamic acid ethyl ester
N-Ethyl-N-nitrosourethan
N-Ethyl-N-nitrosourethane
NEU
Nitrosoethylurethan
N-Nitroso-N-ethylurethan
NSC-24890

614-96-0
C_9H_9N
131.17
Cc2ccc1[nH]ccc1c2
Indole, 5-methyl- (8CI)
1H-Indole, 5-methyl- (9CI)

614-99-3
$C_7H_8O_3$
140.15
O=C(OCC)C(OC=C1)=C1
2-Furoic acid, ethyl ester
Ethyl furoate

615-05-4
$C_7H_{10}N_2O$
138.19
O(c(c(N)cc(N)c1)c1)C
Anisole, 2,4-diamino
1,3-Benzenediamine, 4-methoxy- (9CI)
C.I. 76050
C.I. Oxidation Base 12
2,4 DAA
2,4-Diamineanisole
2,4-Diaminoanisol
2,4-Diaminoanisole
2,4-Diaminoanisole Base
m-Diaminoanisole 1,3-diamino-4-methoxybenzene
2,4-Diamino-1-methoxybenzene
Furro L
4-Methoxy-1,3-benzenediamine
p-Methoxy-m-phenylenediamine
4-Methoxy-m-phenylenediamine
4-MMPD
Pelagol DA
Pelagol Grey L
Pelagol L
m-Phenylenediamine, 4-methoxy- (8CI)

615-13-4
C_9H_8O
132.17
O=C(Cc(c1ccc2)c2)C1
2-Indanone
1,3-Dihydro-2H-inden-2-one
2H-Inden-2-one, 1,3-dihydro-

615-15-6
$C_8H_8N_2$
132.18
N(c(c(N1)ccc2)c2)=C1C
Benzimidazole, 2-methyl
Acetamidine, N-N'-o-phenylene-
Methyl-2-benzimidazole
2-Methylbenzimidazole

615-21-4
$C_7H_7N_3S$
165.23
N(N)=C(Nc(c1ccc2)c2)S1
Benzothiazole, 2-hydrazino
2-Hydrazinobenzothiazole
USAF EK-3967

615-22-5
$C_8H_7NS_2$
181.28
N(c(c(S1)ccc2)c2)=C1SC
2-(Methylthio)benzothiazole
AI3-51102
Benzothiazole, 2-(methylthio)- (9CI)

615-28-1
$C_6H_8N_2.2ClH$
181.08
Cl.Cl.Nc1ccccc1N
o-Phenylenediamine, dihydrochloride
USAF EK-678

615-29-2
$C_7H_{16}O$
116.20
3-Hexanol, 4-methyl- (8CI,9CI)
NSC-91500

615-36-1
C_6H_6BrN
172.02
Nc(c(ccc1)Br)c1
Aniline, o-bromo- (8CI)
Benzenamine, 2-bromo- (9CI)
2-Bromobenzenamine

615-37-2
C_7H_7I
218.04
c(c(ccc1)C)(c1)I
Benzene, 1-iodo-2-methyl- (9CI)
1-Iodo-2-methylbenzene
Toluene, o-iodo- (8CI)

615-41-8
C_6H_4ClI
238.45
c(c(ccc1)Cl)(c1)I
Benzene, 1-chloro-2-iodo- (9CI)
AI3-16897
1-Chloro-2-iodobenzene

615-42-9
$C_6H_4I_2$
329.91
c(c(ccc1)I)(c1)I
Benzene, o-diiodo- (8CI)
Benzene, 1,2-diiodo- (9CI)
1,2-Diiodobenzene

615-43-0
C_6H_6IN
219.02
Nc(c(ccc1)I)c1
Benzenamine, 2-iodo- (9CI)
AI3-00493
o-Iodoaniline
2-Iodobenzenamine

615-45-2
$C_7H_{10}N_2.2ClH$
195.11
Toluene-2,5-diamine, dihydrochloride
1,4-Benzenediamine, 2-methyl-, dihydrochloride (9CI)
2,5-Diaminotoluene dihydro-

chloride
p-Toluenediamine dihydro-
chloride

615-46-3
$C_6H_7ClN_2.2ClH$
214.50
2-Chloro-1,4-benzenediamine dihydrochloride
1,4-Benzenediamine, 2-chloro-, dihydrochloride (9CI)
p-Phenylenediamine, 2-chloro-, dihydrochloride

615-50-9
$C_7H_{10}N_2.H_2O_4S$
220.27
Toluene-2,5-diamine, sulfate (1:1)
1,4-Benzenediamine, 2-methyl-, sulfate (1:1) (9CI)
C.I. 76043
p-Diaminotoluene sulfate
2,5-Diaminotoluene sulfate
2,5-Diaminotoluene sulphate
Fouramine Standard
2-Methyl-1,4-benzenediamine sulfate
2-Methyl-p-phenylenediamine sulphate
Toluene-2,5-diamine sulphate
p-Toluenediamine sulphate
Toluylene-2,5-diamine sulphate
p-Toluylenediamine sulphate
p-Tolylenediamine sulphate

615-53-2
$C_4H_8N_2O_3$
132.14
O=C(OCC)N(N=O)C
Carbamic acid, N-methyl-N-nitroso-, ethyl ester
Ethylester kyseliny N-methyl-N-nitrosokarbaminove (Czech)
Ethyl ester of methylnitroso-carbamic acid
N-Methyl-N-nitrosocarbamic acid, ethyl ester
N-Methyl-N-nitrosoethylcar-bamate
Methylnitrosourethan (German)
Methylnitrosourethane
N-Methyl-N-nitroso-urethane
MNU
MNUN
Nitrosomethylurethan (German)
Nitrosomethylurethane
N-Nitroso-N-methylurethane
NMUT
NSC-2860
RCRA waste number U178

615-54-3
$C_6H_3Br_3$
314.80
c(ccc(c1Br)Br)(c1)Br
1,2,4-Tribromobenzene
AI3-18131
Benzene, 1,2,4-tribromo- (9CI)

615-56-5
$C_6H_4Br_2O$
251.92
Phenol, 3,4-dibromo- (9CI)
3,4-Dibromophenol

615-57-6
$C_6H_5Br_2N$
250.91
Nc(c(cc(c1)Br)Br)c1
Benzenamine, 2,4-dibromo- (9CI)
2,4-Dibromoaniline
2,4-Dibromobenzenamine

615-58-7
$C_6H_4Br_2O$
251.92
Oc(c(cc(c1)Br)Br)c1
Phenol, 2,4-dibromo
2,4-Dibromophenol

615-59-8
$C_7H_6Br_2$
249.93
c(ccc(c1C)Br)(c1)Br
Benzene, 1,4-dibromo-2-methyl- (9CI)
1,4-Dibromo-2-methyl-benzene

615-65-6
C_7H_8ClN
141.61
Nc(c(cc(c1)C)Cl)c1
p-Toluidine, 2-chloro
Benzenamine, 2-chloro-4-methyl-
2-Chloro-4-methylaniline
2-Chloro-p-toluidine
2-Chlor-4-toluidin (Czech)
4-Methyl-2-chloroaniline

615-66-7
$C_6H_7ClN_2$
142.60
Nc(c(cc(N)c1)Cl)c1
p-Phenylenediamine, 2-chloro
3-Chloro-4-aminoaniline
2-Chlor-p-fenylendiamin (Czech)
3-Chlor-p-fenylendiamin (Czech)
o-Chloro-p-phenylenediamine
2-Chloro-p-phenylenediamine
C.I. 76065
Ursol Brown O

615-67-8
$C_6H_5ClO_2$
144.56
Oc(c(cc(O)c1)Cl)c1
Hydroquinone, chloro
Chlorohydroquinone
Monochlorohydroquinone

615-74-7
C_7H_7ClO
142.59
Oc(c(ccc1C)Cl)c1
Phenol, 2-chloro-5-methyl
6-Chloro-m-cresol
2-Chloro-5-methylphenol

616-02-4
$C_5H_4O_3$
112.09
O=C(OC(=O)C=1)C1C
Citraconic-anhydride
Anhydrid kyseliny citrakonove (Czech)
Citraconic acid anhydride
2,5-Furandione, 3-methyl-
Maleic anhydride, methyl-
Methylmaleic anhydride
α-Methylmaleic anhydride
2-Methylmaleic anhydride
3-Methylmaleic anhydride
Monomethylmaleic anhydride

616-05-7
$C_6H_{11}BrO_2$
195.08
O=C(O)C(Br)CCCC
Hexanoic acid, 2-bromo
α-Bromocaproic acid
α-Bromo-n-caproic acid
2-Bromohexanoic acid

616-06-8
$C_6H_{13}NO_2$
131.17
O=C(O)C(N)CCCC
DL-Norleucine (9CI)
Norleucine, DL- (8CI)

616-12-6
C_6H_{12}
84.16
2-Pentene, 3-methyl-, (E)- (9CI)

616-13-7
$C_5H_{11}Cl$
106.60
ClCC(CC)C
Butane, 1-chloro-2-methyl- (9CI)
1-Chloro-2-methylbutane

616-20-6
$C_5H_{11}Cl$
106.60
C(Cl)(CC)CC
Pentane, 3-chloro- (9CI)
3-Chloropentane

616-21-7

$C_4H_8Cl_2$
127.01
ClCC(Cl)CC
1,2-Dichlorobutane
Butane, 1,2-dichloro- (9CI)

616-23-9
$C_3H_6Cl_2O$
128.99
OCC(CCl)Cl
1-Propanol, 2,3-dichloro
2,3-Dichloropropanol
1,2-Dichloro-3-propanol
1,2-Dichloropropanol-3
2,3-Dichloro-1-propanol
Glycerol α,β-dichlorohydrin

616-24-0
$C_5H_{13}N$
87.16
NC(CC)CC
Propylamine, 1-ethyl- (8CI)
3-Aminopentane
3-Pentanamine (9CI)

616-25-1
$C_5H_{10}O$
86.13
OC(C=C)CC
1-Penten-3-ol (9CI)
AI3-28606

616-29-5
$C_3H_{10}N_2O$
90.11
OC(CN)CN
1,3-Diamino-2-propanol
AI3-15325
1,3-Diamino-2-hydroxypropane
2-Propanol, 1,3-diamino- (8CI, 9CI)

616-30-8
$C_3H_9NO_2$
91.13
OCC(O)CN
1,2-Propanediol, 3-amino
1-Aminoglycerol
3-Amino-1,2-propanediol
2,3-Dihydroxypropylamine

616-38-6
$C_3H_6O_3$
90.09
O=C(OC)OC
Carbonic acid, dimethyl ester
Dimethyl carbonate [UN 1161]
Methyl carbonate
UN 1161 [Dimethyl carbonate]

616-39-7
$C_5H_{13}N$
87.16
N(CC)(CC)C
Ethanamine, N-ethyl-N-methyl- (9CI)
N-Ethyl-N-methylethanamine

616-43-3
C_5H_7N
81.11
Cc1cc[nH]c1
1H-Pyrrole, 3-methyl- (9CI)
Pyrrole, 3-methyl- (8CI)

616-44-4
C_5H_6S
98.17
S(C=CC=1C)C1
Thiophene, 3-methyl
3-Methylthiophene
3-Thiotolene

616-45-5
C_4H_7NO
85.12
O=C(NCC1)C1
2-Pyrrolidinone
γ-Aminobutyric acid lactam
4-Aminobutyric acid lactam
γ-Aminobutyric lactam
γ-Aminobutyrolactam
Butyrolactam
γ-Butyrolactam
Lam
2-Oxopyrrolidine
2-Pyrol
α-Pyrrolidinone
Pyrrolidon (German)
Pyrrolidone
α-Pyrrolidone
2-Pyrrolidone

616-47-7
$C_4H_6N_2$
82.12
N(C=CN1C)=C1
Imidazole, 1-methyl
1-Methylimidazole

616-52-4
$C_7H_{17}O_3P$
180.19
Pinacolyl methylphosphonic acid
PMPA

616-62-6
$C_6H_{10}O_4$
146.14
CCCC(C(O)=O)C(O)=O
Propanedioic acid, propyl- (9CI)
1,1-Butanedicarboxylic acid
Malonic acid, propyl- (8CI)

616-72-8
$C_8H_8N_2O_4$
196.18
Cc1cc(C)c(cc1N(=O)=O)N(=O)=O
m-Xylene, 4,6-dinitro
1,3-Dimethyl-4,6-dinitrobenzene
1,5-Dimethyl-2,4-dinitrobenzene
4,6-Dinitro-m-xylene
4,6-Dinitro-1,3-xylene

616-73-9
$C_7H_6N_2O_5$
198.15
Cc1cc(O)c(cc1N(=O)=O)N(=O)=O
m-Cresol, 4,6-dinitro
4,6-Dinitro-m-cresol
3-Methyl-4,6-dinitrophenol

616-74-0
Unknown
Unknown
4,6-Dinitroresorcinol (Heavy metal salts of) (Dry)

616-79-5
$C_7H_6N_2O_4$
182.12
O=C(O)c(c(N)ccc1N(=O)=O)c1
Anthranilic acid, 5-nitro- (8CI)
2-Amino-5-nitrobenzoic acid
Benzoic acid, 2-amino-5-nitro- (9CI)

616-82-0
$C_7H_5NO_5$
183.11
OC(=O)c1ccc(O)c(c1)N(=O)=O
Benzoic acid, 4-hydroxy-3-nitro- (9CI)
4-Hydroxy-3-nitrobenzoic acid

616-88-6
$C_8H_{14}O_4$
174.20
Propanedioic acid, pentyl- (9CI)
1,1-Hexanedicarboxylic acid
Malonic acid, pentyl- (8CI)
NSC-521917

616-91-1
$C_5H_9NO_3S$
163.21
O=C(NC(C(=O)O)CS)C
Cysteine, N-acetyl-, L
L-α-Acetamido-β-mercaptopropionic acid
Acetein
Acetylcysteine
N-Acetylcysteine
N-Acetyl-L-cysteine
N-Acetyl-3-mercaptoalanine
Airbron
Broncholysin
l-Cysteine, N-acetyl- (9CI)
Fluimucetin
Fluimucil

Flumicil
Inspir
Mercapturic acid
Mercapturic acid, (R)-
Mucolyticum
Mucolyticum lappe
Mucolytikum lappe
Mucomyst
Mucosolvin
NAC
NAC-TB
NSC-111180
Parvolex
Respaire

617-29-8
$C_7H_{16}O$
116.20
3-Hexanol, 2-methyl- (8CI,9CI)
NSC-91501

617-35-6
$C_5H_8O_3$
116.12
O=C(OCC)C(=O)C
Ethyl pyruvate
AI3-05636
Ethyl 2-oxopropanoate
Propanoic acid, 2-oxo-, ethyl ester (9CI)
Pyruvic acid, ethyl ester (8CI)

617-42-5
$C_4H_3ClO_4$
150.52
2-Chloromaleic acid

617-43-6
$C_4H_3ClO_4$
150.52
2-Chlorofumaric acid

617-48-1
$C_4H_6O_5$
134.09
O=C(O)CC(O)C(=O)O
Butanedioic acid, hydroxy-, (+-)- (9CI)
AI3-06292
(+-)-Hydroxybutanedioic acid
Malic acid, DL- (8CI)

617-50-5
$C_7H_{14}O_2$
130.21
Isobutyric acid, isopropyl ester
Isopropyl isobutyrate [UN 2406]
Isopropyl 2-methylpropanoate
Propanoic acid, 2-methyl-, 1-methylethyl ester (9CI)
UN 2406 [Isopropyl isobutyrate]

617-51-6
$C_6H_{12}O_3$
132.16
Isopropyl lactate
AI3-00586
2-Hydroxy-propanoic acid 1-methylethyl ester
Lactic acid, isopropyl ester
Propanoic acid, 2-hydroxy-, 1-methylethyl ester

617-62-9
$C_6H_{10}O_4$
146.14
Glutaric acid, 2-methyl- (8CI)
AI3-23452
Pentanedioic acid, 2-methyl- (9CI)

617-65-2
$C_5H_9NO_4$
147.15
O=C(O)C(N)CCC(=O)O
Glutamic acid, DL
(+-)-Glutamic acid
DL-Glutamic acid (9CI)

617-68-5
$C_6H_{14}N_2O_2.2ClH$
219.10
DL-Lysine, dihydrochloride (9CI)
Lysine, dihydrochloride, DL- (8CI)

617-78-7
C_7H_{16}
100.20
Pentane, 3-ethyl- (9CI)

617-79-8
$C_6H_{15}N$
101.22
CCC(CC)CN
Butylamine, 2-ethyl
1-Amino-2-ethylbutane
1-Butanamine, 2-ethyl-
2-Ethylbutylamine

617-84-5
$C_5H_{11}NO$
101.17
O=CN(CC)CC
Formamide, N,N-diethyl
Diethylamid kyseliny mravenci (Czech)
Diethyl formamide

617-86-7
$C_6H_{16}Si$
116.28
CC[SiH](CC)CC
Silane, triethyl- (9CI)
Triethylsilane

617-89-0
C_5H_7NO
97.13
O(C(=CC=1)CN)C1
Furfurylamine [UN 2526]
2-Furanmethylamine
(2-Furylmethyl)amine
Methylamine, 1-(2-furyl)-
UN 2526 [Furfurylamine]
USAF Q-1

617-94-7
$C_9H_{12}O$
136.21
OC(c(cccc1)c1)(C)C
Benzyl alcohol, α,α-dimethyl
Benzenemethanol, α,α-dimethyl- (9CI)
α-Cumyl alcohol
α,α-Dimethylbenzenemethanol
α,α-Dimethylbenzyl alcohol
Dimethylphenylcarbinol
Dimethylphenylmethanol
1-Hydroxycumene
Phenyldimethylcarbinol
2-Phenylisopropanol

618-32-6
C_7H_5BrO
185.02
O=C(c(cccc1)c1)Br
Benzoyl bromide (9CI)

618-36-0
$C_8H_{11}N$
121.18
NC(c(cccc1)c1)C
Benzenemethanamine, α-methyl-, (+-)- (9CI)
(+-)-α-Methylbenzenemethanamine

618-41-7
$C_6H_6O_2S$
142.18
Benzenesulfinic acid (8CI,9CI)
NSC-227915
Phenylsulfinic acid

618-42-8
$C_7H_{13}NO$
127.21
O=C(N(CCCC1)C1)C
Piperidine, 1-acetyl
n-Acetylpiperidin (German)
1-Acetylpiperidine

618-45-1
$C_9H_{12}O$
136.21
Oc(cccc1C(C)C)c1
Phenol, m-isopropyl
m-Cumenol
m-Isopropylphenol
3-Isopropylphenol

Phenol, 3-(1-methylethyl)-

618-46-2
$C_7H_4Cl_2O$
175.01
O=C(c(cccc1Cl)c1)Cl
Benzoyl chloride, m-chloro- (8CI)
Benzoyl chloride, 3-chloro- (9CI)
3-Chlorobenzoyl chloride

618-51-9
$C_7H_5IO_2$
248.02
O=C(O)c(cccc1I)c1
3-Iodobenzoic acid
Benzoic acid, m-iodo- (8CI)
Benzoic acid, 3-iodo- (9CI)

618-62-2
$C_6H_3Cl_2NO_2$
191.99
O=N(=O)c(cc(cc1Cl)Cl)c1
Benzene, 1,3-dichloro-5-nitro- (9CI)
1,3-Dichloro-5-nitrobenzene

618-80-4
$C_6H_3Cl_2NO_3$
207.99
O=N(=O)c(cc(c(O)c1Cl)Cl)c1
2,6-Dichloro-4-nitrophenol
AI3-03649
Phenol, 2,6-dichloro-4-nitro- (9CI)

618-85-9
$C_7H_6N_2O_4$
182.15
Cc1cc(cc(c1)N(=O)=O)N(=O)=O
Toluene, 3,5-dinitro
Benzene, 1-methyl-3,5-dinitro- (9CI)
3,5-Dinitrotoluene
3,5-DNT
1-Methyl-3,5-dinitro-benzene

618-87-1
$C_6H_5N_3O_4$
183.14
Nc1cc(cc(c1)N(=O)=O)N(=O)=O
Aniline, 3,5-dinitro
Benzeneamine, 3,5-dinitro-
3,5-Dinitroaniline

618-88-2
$C_8H_5NO_6$
211.12
O=C(O)c(cc(N(=O)=O)cc1C(=O)O)c1
Isophthalic acid, 5-nitro- (8CI)
1,3-Benzenedicarboxylic acid, 5-nitro- (9CI)
5-Nitro-1,3-benzenedicarboxylic acid
5-Nitroisophthalic acid

618-89-3
$C_8H_7BrO_2$
215.05
O=C(OC)c(cccc1Br)c1
Benzoic acid, 3-bromo-, methyl ester (9CI)
Methyl 3-bromobenzoate

619-04-5
$C_9H_{10}O_2$
150.19
O=C(O)c(ccc(c1C)C)c1
Benzoic acid, 3,4-dimethyl
3,4-Dimethylbenzoic acid

619-08-9
$C_6H_4ClNO_3$
173.56
O=N(=O)c(ccc(O)c1Cl)c1
Phenol, 2-chloro-4-nitro
2-Chloro-4-nitrophenol

619-14-7
$C_7H_5NO_5$
183.11
OC(=O)c1ccc(cc1O)N(=O)=O
Benzoic acid, 3-hydroxy-4-nitro- (8CI,9CI)
NSC-46823

619-15-8
$C_7H_6N_2O_4$
182.15
O=N(=O)c(ccc(N(=O)=O)c1C)c1
Toluene, 2,5-dinitro
Benzene, 2-methyl-1,4-dinitro- (9CI)
2,5-Dinitrotoluene
2,5-DNT

619-17-0
$C_7H_6N_2O_4$
182.15
O=C(O)c(c(N)cc(N(=O)=O)c1)c1
Anthranilic acid, 4-nitro
2-Amino-4-nitro-benzoic acid
Benzoic acid, 2-amino-4-nitro- (9CI)
NCI-C01945
4-Nitroanthranilic acid

619-21-6
$C_8H_6O_3$
150.13
O=Cc(cccc1C(=O)O)c1
Benzoic acid, 3-formyl- (9CI)
3-Carboxybenzaldehyde
3-Formylbenzoic acid

619-23-8
$C_7H_6ClNO_2$
171.59
O=N(=O)c(cccc1CCl)c1
Toluene, α-chloro-m-nitro
Benzene, 1-(chloromethyl)-3-nitro- (9CI)
α-Chloro-m-nitrotoluene
m-Nitrobenzyl chloride
3-Nitrobenzyl chloride

619-24-9
$C_7H_4N_2O_2$
148.13
N#Cc(cccc1N(=O)=O)c1
Benzonitrile, m-nitro
Benzonitrile, 3-nitro-
m-Cyanonitrobenzene
3-Cyanonitrobenzene
m-Nitrobenzonitrile
3-Nitrobenzonitrile

619-33-0
$C_6H_{12}Cl_2O_2$
187.07
Ethane, 1,1-dichloro-2,2-diethoxy- (9CI)
Acetaldehyde, dichloro-, diethyl acetal (8CI)

619-42-1
$C_8H_7BrO_2$
215.05
O=C(OC)c(ccc(c1)Br)c1
Benzoic acid, p-bromo-, methyl ester (8CI)
Benzoic acid, 4-bromo-, methyl ester (9CI)
Methyl 4-bromobenzoate

619-45-4
$C_8H_9NO_2$
151.18
O=C(OC)c(ccc(N)c1)c1
Benzoic acid, p-amino-, methyl ester
p-Aminobenzoic acid methyl ester
Methyl p-aminobenzoate

619-50-1
$C_8H_7NO_4$
181.16
O=C(OC)c(ccc(N(=O)=O)c1)c1
Benzoic acid, p-nitro-, methyl ester
Methyl-p-nitrobenzoate

619-55-6
C_8H_9NO
135.16
O=C(N)c(ccc(c1)C)c1
Benzamide, 4-methyl- (9CI)
4-Methylbenzamide

619-56-7

C_7H_6ClNO
155.58
NC(=O)c1ccc(Cl)cc1
Benzamide, 4-chloro- (9CI)
Benzamide, p-chloro- (8CI)
NSC-406894
NSC-74687

619-58-9
$C_7H_5IO_2$
248.02
O=C(O)c(ccc(c1)I)c1
Benzoic acid, p-iodo
p-Iodobenzoic acid
4-Jodbenzoesaeure (German)

619-64-7
$C_9H_{10}O_2$
150.18
CCc1ccc(cc1)C(O)=O
4-Ethylbenzoic acid
4-Ethylbenzoate

619-65-8
$C_8H_5NO_2$
147.13
O=C(O)c(ccc(C#N)c1)c1
Benzoic acid, p-cyano- (8CI)
Benzoic acid, 4-cyano- (9CI)
4-Cyanobenzoic acid

619-66-9
$C_8H_6O_3$
150.14
O=Cc(ccc(c1)C(=O)O)c1
Terephthalaldehydic-acid
4-Formylbenzoic acid

619-72-7
$C_7H_4N_2O_2$
148.13
N#Cc(ccc(N(=O)=O)c1)c1
Benzonitrile, p-nitro
Benzonitrile, 4-nitro- (9CI)
p-Cyanonitrobenzene
4-Cyanonitrobenzene
p-Nitrobenzonitrile
4-Nitrobenzonitrile

619-80-7
$C_7H_6N_2O_3$
166.15
O=C(N)c(ccc(N(=O)=O)c1)c1
Benzamide, p-nitro

619-84-1
$C_9H_{11}NO_2$
165.21
O=C(O)c(ccc(N(C)C)c1)c1
Benzoic acid, p-(dimethyl-amino) (8CI)
Benzoic acid, 4-(dimethyl-amino)- (9CI)
N,N-Dimethyl-4-aminobenzoic acid
p-Dimethylaminobenzoic acid
4-(Dimethylamino)benzoic acid

619-86-3
$C_9H_{10}O_3$
166.18
O=C(O)c(ccc(OCC)c1)c1
Benzoic acid, p-ethoxy- (8CI)
AI3-20152
Benzoic acid, 4-ethoxy- (9CI)
4-Ethoxybenzoic acid

619-99-8
C_8H_{18}
114.23
C(CCC)(CC)CC
Hexane, 3-ethyl- (9CI)
3-Ethylhexane

620-02-0
$C_6H_6O_2$
110.12
O=CC(OC(=C1)C)=C1
2-Furaldehyde, 5-methyl
2-Formyl-5-methylfuran
5-Methyl-2-furaldehyde
5-Methyl-2-furancarbox-aldehyde
5-Methyl furfural
5-Methyl-2-furfural
5-Methylfurfuraldehyde

620-05-3
C_7H_7I
218.04
c(cccc1)(c1)CI
Toluene, α-iodo
Benzene, (iodomethyl)- (9CI)
Benzyl iodide [UN 2653]
(Iodomethyl)benzene
Iodophenylmethane
α-Iodotoluene
UN 2653 [Benzyl iodide]

620-13-3
C_8H_9Br
185.06
BrCc(cccc1C)c1
m-Xylyl bromide
AI3-20884
Benzene, 1-(bromomethyl)-3-methyl- (9CI)
1-(Bromomethyl)-3-methyl-benzene
α-Bromo-m-xylene
ω-Bromo-m-xylene
m-Methylbenzyl bromide
3-Methylbenzyl bromide
m-Xylene, α-bromo- (8CI)

620-14-4
C_9H_{12}
120.19
CCc1cccc(C)c1
3-Ethyltoluene
Benzene, 1-ethyl-3-methyl- (9CI)
Toluene, m-ethyl- (8CI)

620-17-7
$C_8H_{10}O$
122.17
Oc(cccc1CC)c1
3-Ethylphenol
AI3-19938
Benzene, 1-ethyl-3-hydroxy-
1-Ethyl-3-hydroxybenzene
m-Ethylphenol
meta-Ethylphenol
Phenol, m-ethyl- (8CI)
Phenol, 3-ethyl- (9CI)

620-19-9
C_8H_9Cl
140.61
Benzene, 1-(chloromethyl)-3-methyl- (9CI)
m-Xylene, α-chloro- (8CI)

620-20-2
$C_7H_6Cl_2$
161.03
c(cccc1Cl)(c1)CCl
Benzene, 1-chloro-3-(chloro-methyl)- (9CI)
1-Chloro-3-(chloromethyl)-benzene
Toluene, m,α-dichloro- (8CI)

620-22-4
C_8H_7N
117.16
N#Cc(cccc1C)c1
m-Tolunitrile
MTN
Nitril kyseliny m-toluylove (Czech)
3-Toluenkarbonitril (Czech)
m-Tolynitrile

620-23-5
C_8H_8O
120.15
O=Cc(cccc1C)c1
Benzaldehyde, 3-methyl- (9CI)
AI3-02278
3-Methylbenzaldehyde
m-Tolualdehyde (8CI)

620-24-6
$C_7H_8O_2$
124.14
OCc1cccc(O)c1
3-Hydroxybenzyl alcohol
AI3-31880
Benzenemethanol, 3-hydroxy- (9CI)
Benzyl alcohol, m-hydroxy- (8CI)

620-42-8

$C_{21}H_{21}O_6P$
400.39
O(c(ccc(c1)C)c1)P(Oc(ccc(c2)C)c2)Oc(ccc(c3)C)c3
Phosphorous acid, tri-p-cresyl ester
Tri-p-cresyl phosphite

620-71-3
$C_9H_{11}NO$
149.21
O=C(Nc(cccc1)c1)CC
Propionanilide

620-73-5
$C_8H_7ClO_2$
170.60
O=C(Oc(cccc1)c1)CCl
Acetic acid, chloro-, phenyl ester (9CI)
AI3-23257
Phenyl chloroacetate

620-84-8
$C_{13}H_{13}N$
183.25
N(c(ccc(c1)C)c1)c(cccc2)c2
Benzenamine, 4-methyl-N-phenyl- (9CI)
4-Methyl-N-phenylbenzenamine

620-88-2
$C_{12}H_9NO_3$
215.22
O=N(=O)c(ccc(Oc(cccc1)c1)c2)c2
Ether, 4-nitrophenyl phenyl
Benzene, 1-nitro-4-phenoxy- (9CI)
4-Nitrobiphenyl ether
4-Nitrodifenylether (Czech)
p-Nitrodiphenyl ether
4-Nitrodiphenyl ether
p-Nitrophenylphenyl ether
4-Nitrophenylphenyl ether

620-92-8
$C_{13}H_{12}O_2$
200.25
Oc(ccc(c1)Cc(ccc(O)c2)c2)c1
Phenol, 4,4'-methylenebis
Bis(p-hydroxyphenyl)methane
Bis(4-hydroxyphenyl)methane
p,p'-Bis(hydroxyphenyl)-methane
4,4'-Methylenebisphenol
4,4'-Methylene diphenol
Phenol, 4,4'-methylenedi-

620-95-1
$C_{12}H_{11}N$
169.22
Pyridine, 3-(phenylmethyl)- (9CI)
Pyridine, 3-benzyl- (8CI)

621-09-0
$C_{14}H_{14}N_2$
210.30
N(c(cccc1)c1)=C(Nc(cccc2)c2)C
Acetamidine, N,N'-diphenyl
N,N'-Diphenylacetamidine

621-23-8
$C_9H_{12}O_3$
168.21
O(c(cc(OC)cc1OC)c1)C
Benzene, 1,3,5-trimethoxy
Phloroglucinol trimethyl ether
1,3,5-Trimethoxybenzene

621-31-8
$C_8H_{11}NO$
137.20
CCNc1cccc(O)c1
Phenol, m-(ethylamino)
3-(Ethylamino)phenol

621-32-9
$C_9H_{12}O$
136.19
O(c(cccc1C)c1)CC
Benzene, 1-ethoxy-3-methyl- (9CI)
AI3-24476
1-Ethoxy-3-methylbenzene
Phenetole, m-methyl- (8CI)

621-33-0
$C_8H_{11}NO$
137.20
O(c(cccc1N)c1)CC
m-Phenetidine
Benzenamine, 3-ethoxy- (9CI)
m-Ethoxyaniline
3-Ethoxyaniline
3-Ethoxybenzenamine

621-35-2
$C_8H_{10}N_2O.ClH$
186.63
Acetamide, N-(3-amino-phenyl)-, monohydrochloride (9CI)
3-Aminoacetanilide, hydro-chloride
N-(3-Aminophenyl)acetamide monohydrochloride

621-36-3
$C_9H_{10}O_2$
150.18
Cc1cccc(CC(O)=O)c1
Benzeneacetic acid, 3-methyl- (9CI)
Acetic acid, m-tolyl- (8CI)
NSC-76090

621-37-4
$C_8H_8O_3$
152.15
3-Hydroxyphenylacetic acid
Acetic acid, (m-hydroxy-phenyl)- (8CI)
Benzeneacetic acid, 3-hydroxy- (9CI)
3-Hydroxybenzeneacetic acid
meta-Hydroxyphenylacetic acid

621-38-5
C_8H_8BrNO
214.07
CC(=O)Nc1cccc(Br)c1
Acetamide, N-(3-bromo-phenyl)- (9CI)
Acetanilide, 3'-bromo- (6CI,7CI,8CI)
3-Bromoacetanilide
3'-Bromoacetanilide
m-Bromoacetanilide

621-42-1
$C_8H_9NO_2$
151.18
O=C(Nc(cccc1O)c1)C
Acetanilide, 3'-hydroxy
Acetamide, N-(3-hydroxy-phenyl)- (9CI)
m-Acetamidophenol
3-Acetamidophenol
m-(Acetylamino)phenol
3-(Acetylamino)phenol
BS 479
m-Hydroxyacetanilide
3-Hydroxyacetanilide
3'-Hydroxyacetanilide
Metacetamol
Metalid
Nebs
Pedituss
Pyrapap
Rystal

621-51-2
$C_9H_{10}O_3$
166.18
CCOc1cccc(c1)C(O)=O
Benzoic acid, 3-ethoxy- (9CI)
Benzoic acid, m-ethoxy- (8CI)

621-59-0
$C_8H_8O_3$
152.16
O=Cc(ccc(OC)c1O)c1
Benzaldehyde, 3-hydroxy-4-methoxy
p-Anisaldehyde, 3-hydroxy- (8CI)
5-Formylguaiacol
Isovanillin
Isovanilline
Oxy-3 methoxy-4 benzaldehyde (French)

621-62-5
$C_6H_{13}ClO_2$
152.62
O(C(OCC)CCl)CC

Acetaldehyde, chloro-, diethyl acetal (8CI)
AI3-08039
2-Chloro-1,1-diethoxyethane
Ethane, 2-chloro-1,1-diethoxy- (9CI)

621-63-6
$C_6H_{14}O_3$
134.18
Ethanol, 2,2-diethoxy- (9CI)
Glycolaldehyde, diethyl acetal (8CI)
NSC-9255

621-64-7
$C_6H_{14}N_2O$
130.22
O=NN(CCC)CCC
Dipropylamine, N-nitroso
Dipropylnitrosamine
Di-n-propylnitrosamine
DPN
DPNA
NDPA
N-Nitrosodipropylamine
N-Nitrosodi-n-propylamine
N-Nitroso-N-propyl-1-propanamine
Propanamine, N-nitroso-N-propyl-
Propylamine, N-nitroso-N-di-
RCRA waste number U111

621-68-1
$C_7H_{15}Cl_2N$
184.13
CCCN(CCCl)CCCl
Propylamine, N,N-bis-(2-chloroethyl)
N,N-Bis(2-chloroethyl)-propylamine
Propylbis(β-chloroethyl)-amine

621-70-5
$C_{21}H_{38}O_6$
386.59
O=C(OCC(OC(=O)CCCCC)COC(=O)CCCCC)CCCCC
Glycerol-trihexanoate
Tricapronin

621-77-2
$C_{15}H_{33}N$
227.43
N(CCCCC)(CCCCC)CCCCC
Triamylamine
AI3-11530
N,N-Dipentyl-1-pentanamine
1-Pentanamine, N,N-dipentyl- (9CI)
Tripentylamine (8CI)

621-82-9
$C_9H_8O_2$
148.17
O=C(O)C=Cc(cccc1)c1
Cinnamic-acid
Kyselina skoricove (Czech)
Phenylacrylic acid
tert-β-Phenylacrylic acid
3-Phenylacrylic acid
3-Phenylpropenoic acid
3-Phenyl-2-propenoic acid
2-Propenoic acid, 3-phenyl-
Zimtsaeure (German)

622-08-2
$C_9H_{12}O_2$
152.21
O(Cc(cccc1)c1)CCO
Ethanol, 2-(benzyloxy)
Benzyl cellosolve
Benzylcelosolv (Czech)
2-Benzyloxyethanol
Ethylene glycol monobenzyl ether

622-24-2
C_8H_9Cl
140.61
c(cccc1)(c1)CCCl
Benzene, (2-chloroethyl)- (9CI)
(2-Chloroethyl)benzene

622-25-3
C_8H_7Cl
138.60
c(cccc1)(c1)C=CCl
Benzene, (2-chloroethenyl)- (9CI)
(2-Chloroethenyl)benzene

622-32-2
C_7H_7NO
121.13
Benzaldehyde, oxime, (Z)- (8CI,9CI)
NSC-154850

622-38-8
$C_8H_{10}S$
138.23
CCSc1ccccc1
Benzene, (ethylthio)- (9CI)
AI3-17095
Sulfide, ethyl phenyl (8CI)

622-40-2
$C_6H_{13}NO_2$
131.20
O(CCN(C1)CCO)C1
4-Morpholineethanol
Ethanol, 2-morpholino-
Ethanol, 2-(morpholinyl)-
N-β-Hydroxyethylmorpholine
N-(2-Hydroxyethyl)morfolin (Czech)
N-(2-Hydroxyethyl)morpholine
Morpholine ethanol
β-Oxyaethyl-morpholin (German)

622-44-6
$C_7H_5Cl_2N$
174.03
ClC(Cl)=Nc1ccccc1
Imidocarbonyl chloride, phenyl
Carbonimidic dichloride, phenyl- (9CI)
N-(Dichloromethylene)aniline
N-Phenylcarbonimidic dichloride
Phenyl carbylamine chloride
Phenylcarbylamine chloride [UN 1672]
Phenylimidocarbonyl chloride
N-Phenylimidophosgene
Phenyliminocarbonyl dichloride
N-Phenyliminocarbonyl dichloride
Phenylisonitrile dichloride
UN 1672 [Phenylcarbylamine chloride]

622-45-7
$C_8H_{14}O_2$
142.22
O=C(OC(CCCC1)C1)C
Acetic acid, cyclohexyl ester
Cyclohexanol, acetate
Cyclohexanolazetat (German)
Cyclohexanyl acetate
Cyclohexyl acetate [UN 2243]
Cyclohexylester kyseliny octove (Czech)
UN 2243 [Cyclohexyl acetate]

622-46-8
$C_7H_7NO_2$
137.13
NC(=O)Oc1ccccc1
Carbamic acid, phenyl ester (9CI)
AI3-50866

622-47-9
$C_9H_{10}O_2$
150.19
Cc1ccc(CC(O)=O)cc1
Acetic acid, (p-tolyl)
Benzeneacetic acid, 4-methyl- (9CI)
4-Methylbenzeneacetic acid
p-Methylphenylacetic acid
(4-Methylphenyl)acetic acid
(p-Tolyl)acetic acid

622-51-5
$C_8H_{10}N_2O$
150.20
Cc1ccc(NC(N)=O)cc1
Urea, p-tolyl
NCI-C02153
4-Methylphenylurea
p-Tolycarbamide

p-Tolylurea
p-Tolyurea

622-62-8
$C_8H_{10}O_2$
138.18
O(c(ccc(O)c1)c1)CC
Phenol, p-ethoxy
Ether monoethylique de l' hydroquinone (French)
p-Ethoxyphenol
4-Ethoxyphenol
4-Ethyloxyphenol
Hydroquinone monoethyl ether
p-Hydroxyphenetole
Phenol, 4-ethoxy- (9CI)

622-78-6
C_8H_7NS
149.22
N(=C=S)Cc(cccc1)c1
Isothiocyanic acid, benzyl ester
Benzyl-isothiocyanate
Benzyl Mustard Oil
Benzylsenfoel (German)

622-80-0
$C_9H_{13}N$
135.20
N(c(cccc1)c1)CCC
Benzenamine, N-propyl-
AI3-08884
N-Propylbenzenamine

622-85-5
$C_9H_{12}O$
136.21
O(c(cccc1)c1)CCC
Ether, propyl phenyl
Propoxyphenyl
Propyl phenyl ether

622-86-6
C_8H_9ClO
156.61
Benzene, (2-chloroethoxy)- (9CI)
NSC-1073
Phenetole, β-chloro- (8CI)

622-96-8
C_9H_{12}
120.21
c(ccc(c1)C)(c1)CC
Toluene, p-ethyl
Benzene, 1-ethyl-4-methyl- (9CI)
1-Ethyl-4-methylbenzene
p-Ethylmethylbenzene
p-Ethyltoluene
4-Ethyltoluene
p-Methylethylbenzene
4-Methylethylbenzene

622-97-9
C_9H_{10}
118.19
c(ccc(c1)C=C)(c1)C
Styrene, p-methyl
Benzene, 1-ethenyl-4-methyl- (9CI)
p-Methylstyrene
1-p-Tolylethene
p-Vinyltoluene
4-Vinyltoluene

623-00-7
C_7H_4BrN
182.03
Brc1ccc(C#N)cc1
Benzonitrile, 4-bromo
4-Bromobenzonitrile

623-03-0
C_7H_4ClN
137.57
N#Cc(ccc(c1)Cl)c1
Benzonitrile, p-chloro
p-Chlorobenzonitrile
Nitril kyseliny p-chlorbenzoove (Czech)

623-05-2
$C_7H_8O_2$
124.14
OCc1ccc(O)cc1
4-Hydroxybenzyl alcohol
Benzenemethanol, 4-hydroxy- (9CI)
Benzyl alcohol, p-hydroxy- (8CI)
4-Hydroxybenzenemethanol
p-Hydroxybenzyl alcohol
4-Methylol phenol

623-08-5
$C_8H_{11}N$
121.18
N(c(ccc(c1)C)c1)C
Benzenamine, N,4-dimethyl- (9CI)
N,4-Dimethylbenzenamine

623-12-1
C_7H_7ClO
142.58
O(c(ccc(c1)Cl)c1)C
Anisole, p-chloro- (8CI)
AI3-07211
Benzene, 1-chloro-4-methoxy- (9CI)
1-Chloro-4-methoxybenzene

623-15-4
$C_8H_8O_2$
136.15
O=C(C=CC(OC=C1)=C1)C
Monofurfurylideneacetone
AI3-05777
3-Butene-2-one, 4-(2-furanyl)-
3-Buten-2-one, 4-(2-furanyl)- (9CI)
3-Buten-2-one, 4-(2-furyl)- (8CI)
4-(2-Furanyl)-3-buten-2-one

623-17-6
$C_7H_8O_3$
140.15
O=C(OCC(OC=C1)=C1)C
Furfuryl alcohol, acetate
Acetic acid furfuryl ester
2-Acetoxymethylfuran
2-Furanmethanol, acetate (9CI)
2-Furanmethyl acetate
Furfuryl acetate

623-27-8
$C_8H_6O_2$
134.14
O=Cc(ccc(c1)C=O)c1
Terephthalaldehyde
p-Benzenedicarboxaldehyde
1,4-Benzenedicarboxaldehyde (9CI)
1,4-Diformylbenzene
p-Formylbenzaldehyde
4-Formylbenzaldehyde
p-Phthalaldehyde
Terephtaldehyde
Terephtaldehydes (French)
Terephthalic aldehyde

623-30-3
$C_7H_6O_2$
122.12
O=CC=CC(OC=C1)=C1
2-Furanacrolein (8CI)
AI3-00703
3-(2-Furanyl)-2-propenal
2-Propenal, 3-(2-furanyl)- (9CI)

623-33-6
$C_4H_9NO_2.ClH$
139.60
Glycine, ethyl ester, hydrochloride
USAF DO-10

623-36-9
$C_6H_{10}O$
98.16
O=CC(=CCC)C
2-Pentenal, 2-methyl
2-Methyl-2-pentenal
2-Methyl-2-penten-1-al
2-Methyl-2-pentene-1-al

623-37-0
$C_6H_{14}O$
102.18
OC(CCC)CC
3-Hexanol (9CI)

623-38-1
$C_5H_8O_2$

100.12
Pentanal, 3-oxo- (9CI)
Valeraldehyde, 3-oxo- (8CI)

623-42-7
$C_5H_{10}O_2$
102.15
O=C(OC)CCC
Butyric acid, methyl ester
Methyl n-butanoate
Methyl butyrate [UN 1237]
Methyl-n-butyrate
UN 1237 [Methyl butyrate]

623-47-2
$C_5H_6O_2$
98.10
2-Propynoic acid, ethyl ester (9CI)
NSC-60551
Propiolic acid, ethyl ester (8CI)

623-48-3
$C_4H_7IO_2$
214.01
O=C(OCC)CI
Acetic acid, iodo-, ethyl ester
Ethylester kyseliny jodoctove (Czech)
Ethyl iodoacetate
Ethyl monoiodoacetate
S 9

623-51-8
$C_4H_8O_2S$
120.18
Acetic acid, mercapto-, ethyl ester
Ethyl mercaptoacetate
Ethyl α-mercaptoacetate
Ethyl 2-mercaptoacetate
Ethyl mercaptoacetic acid
Ethyl thioglycolate
Mercaptoacetic acid ethyl ester
Thioglycolic acid ethyl ester
Thioglykolsaeure-aethylester (German)
USAF EK-2070

623-55-2
$C_7H_{16}O$
116.20
3-Hexanol, 5-methyl- (8CI,9CI)
NSC-25530

623-56-3
$C_7H_{14}O$
114.19
3-Hexanone, 5-methyl- (8CI, 9CI)

623-68-7
$C_8H_{10}O_3$
154.18
O=C(OC(=O)C=CC)C=CC
Crotonic-anhydride
Anhydrid kyseliny krotonove (Czech)
2-Butenoic acid, anhydride (9CI)
Crotonic acid anhydride

623-69-8
$C_5H_{12}O_3$
120.15
2-Propanol, 1,3-dimethoxy- (8CI,9CI)
NSC-263483

623-70-1
$C_6H_{10}O_2$
114.16
O=C(OCC)C=CC
Crotonic acid, ethyl ester, (E)
2-Butenoic acid, ethyl ester, (E)- (9CI)
trans-2-Butenoic acid ethyl ester
Crotonate d'ethyle (French)
α-Crotonic acid ethyl ester
Ethyl crotonate [UN 1862]
Ethyl (E)-crotonate
Ethyl trans-crotonate
Ethylcrotonate
UN 1862 [Ethyl crotonate]

623-78-9
$C_5H_{11}NO_2$
117.17
O=C(OCC)NCC
Carbamic acid, ethyl-, ethyl ester
Ethylcarbamic acid, ethyl ester
Ethyl-N-ethyl carbamate

623-80-3
$C_5H_{10}OS_2$
150.27
Carbonodithioic acid, S,S-diethyl ester (9CI)
Carbonic acid, dithio-, S,S-diethyl ester (8CI)

623-81-4
$C_4H_{10}O_3S$
138.19
O=S(OCC)OCC
Sulfurous acid, diethyl ester (8CI,9CI)
NSC-8838

623-84-7
$C_7H_{12}O_4$
160.19
O=C(OCC(OC(=O)C)C)C
1,2-Propanediol, diacetate
Propylene glycol diacetate
α-Propylene glycol diacetate

623-87-0
$C_3H_6N_2O_7$
182.08
Glycerol-1,3-dinitrate
1,3-Dinitroglycerin
1,3-DNG
Glycerol, 1,3-dinitrate
1,3-Glyceryl dinitrate
1,2,3-Propanetriol, 1,3-dinitrate (9CI)

623-91-6
$C_8H_{12}O_4$
172.20
O=C(OCC)C=CC(=O)OCC
2-Butenedioic acid, diethyl ester, (E)
Diethylester kyseliny fumarove (Czech)
Diethyl fumarate
Ethyl fumarate
Fumaric acid, diethyl ester

623-93-8
$C_9H_{20}O$
144.26
5-Nonanol (8CI,9CI)
NSC-4552

624-00-0
$C_{10}H_{20}O_3$
188.27
O=C(O)CCCC(O)CCCCC
Decanoic acid, 5-hydroxy- (8CI,9CI)
5-Hydroxydecanoic acid

624-15-7
$C_{10}H_{18}O$
154.25
OCC=C(CCC=C(C)C)C
Geraniol
3,7-Dimethyl-2,6-octadien-1-ol
trans-3,7-Dimethyl-2,6-octadien-1-ol
2,6-Octadien-1-ol, 3,7-dimethyl- (9CI)

624-16-8
$C_{10}H_{20}O$
156.27
4-Decanone (6CI,7CI, 8CI,9CI)
Hexyl propyl ketone
Propyl hexyl ketone

624-17-9
$C_{13}H_{24}O_4$
244.33
O=C(OCC)CCCCCCCC(=O)OCC
Nonanedioic acid, diethyl ester (9CI)
AI3-06279
Azelaic acid, diethyl ester
Diethyl nonanedioate

624-18-0
$C_6H_8N_2.2ClH$
181.08
Cl.Cl.Nc1ccc(N)cc1
p-Phenylenediamine, dihydrochloride
p-Aminoaniline dihydrochloride
4-Aminoaniline dihydrochloride
p-Benzenediamine dihydrochloride
1,4-Benzenediamine dihydrochloride
C.I. 76061
C.I. Oxidation Base 10A
p-Diaminobenzene dihydrochloride
1,4-Diaminobenzene dihydrochloride
Durafur Black RC
Fourrine 64
Fourrine DS
NCI-C03930
Oxidation Base 10A
p-PD HCl
p-PDA HCl
Pelagol CD
Pelagol Grey CD
p-Phenylenediamine hydrochloride
1,4-Phenylenediamine dihydrochloride

624-19-1
$C_6H_6BrN.ClH$
208.48
Benzenamine, 4-bromo-, hydrochloride (9CI)
4-Bromobenzenamine hydrochloride

624-24-8
$C_6H_{12}O_2$
116.18
O=C(OC)CCCC
Valeric acid, methyl ester
Methyl pentanoate
Methyl valerate
Methyl n-valerate
Methyl valerianate
Pentanoic acid, methyl ester (9CI)

624-29-3
C_8H_{16}
112.22
1,4-Dimethylcyclohexane-cis

624-31-7
C_7H_7I
218.04
c(ccc(c1)C)(c1)I
Benzene, 1-iodo-4-methyl- (9CI)
AI3-16899
1-Iodo-4-methylbenzene
Toluene, p-iodo- (8CI)

624-38-4
$C_6H_4I_2$
329.91
c(ccc(c1)I)(c1)I
Benzene, p-diiodo- (8CI)
AI3-08885
Benzene, 1,4-diiodo- (9CI)
1,4-Diiodobenzene

624-41-9
$C_7H_{14}O_2$
130.19
O=C(OCC(CC)C)C
1-Butanol, 2-methyl-, acetate (9CI)
2-Methyl-1-butanol acetate

624-42-0
$C_8H_{16}O$
128.21
Ethyl iso-amyl ketone
Ethyl isoamyl ketone
3-Heptanone, 6-methyl-
6-Methyl-3-heptanone

624-43-1
$C_3H_7NO_5$
137.11
OCC(O)CON(=O)=O
Glycerol, 1-nitrate
Glycerol 1-mononitrate
Glycerol-1-nitrat (German)
Glyceryl 1-mononitrate
1-MNG
1-Mononitroglycerin
1,2,3-Propanetriol, 1-nitrate (9CI)

624-47-5
$C_6H_{12}O_4$
148.16
1,2,3-Propanetriol, 1-propanoate (9CI)
NSC-35011
Propionin, 1-mono- (8CI)

624-48-6
$C_6H_8O_4$
144.14
O=C(OC)C=CC(=O)OC
2-Butenedioic acid, dimethyl ester, (Z)
Dimethylester kyseliny maleinove (Czech)
Dimethyl maleate
Maleic acid, dimethyl ester
Methyl maleate
Sipomer DMM

624-51-1
$C_9H_{20}O$
144.26
OC(CCCCCC)CC
3-Nonanol
AI3-37211
Hexyl ethyl carbinol

624-54-4
$C_8H_{16}O_2$
144.21
O=C(OCCCCC)CC
Propanoic acid, pentyl ester (9CI)
AI3-24356
Pentyl propanoate
n-Pentyl propionate
Propionic acid, pentyl ester (8CI)

624-60-2
$C_3H_9N.ClH$
95.57
Ethanamine, N-methyl-, hydrochloride (9CI)
N-Methylethanamine hydrochloride

624-61-3
C_2Br_2
183.8
Dibromoacetylene
Acetylene, dibromo-
Dirbomoacetileno (Spanish)
Dibromoethylene

624-64-6
C_4H_8
56.11
C(=CC)C
trans-2-Butene
trans-Butene
(E)-2-Butene
2-Butene, (E)- (9CI)
2-Butene-trans
2-trans-Butene
β-trans-Butylene
trans-1,2-Dimethylethylene
Low-boiling Butene-2

624-65-7
C_3H_3Cl
74.51
C(#C)CCl
Propyne, 3-chloro- (8CI)
3-Chloro-1-propyne
1-Propyne, 3-chloro- (9CI)

624-67-9
C_3H_2O
54.05
O=CC#C
Propiolaldehye
Propargylaldehyde
Propioaldehyde
2-Propynal (9CI)
Propynal

624-72-6
$C_2H_4F_2$
66.06
Ethane, 1,2-difluoro
1,2-Difluoroethane

FC143
Fluorocarbon FC143
Freon 152

624-73-7
$C_2H_4I_2$
281.86
C(CI)I
Ethane, 1,2-diiodo- (9CI)
1,2-Diiodoethane

624-74-8
C_2I_2
277.83
Diiodoacetylene
Acetylene, diiodo-
Diioacetylene
Diyodoacetileno (Spanish)
Ethyne, diiodo-

624-76-0
C_2H_5IO
171.97
OCCI
Ethanol, 2-iodo
Ethylene iodohydrin
Iodoethanol
2-Iodoethanol

624-78-2
C_3H_9N
59.11
N(CC)C
Ethanamine, N-methyl- (9CI)
N-Methylethanamine

624-79-3
C_3H_5N
55.09
Carbylamine, ethyl
Ethane, isocyano- (9CI)
Ethylcarbylamine
Ethyl isocyanide (8CI)
Ethylisokyanid (Czech)
Ethyl isonitrile
Ethylkarbylamin (Czech)
Isocyanoethane

624-83-9
C_2H_3NO
57.06
O=C=NC
Isocyanic acid, methyl ester
Isocyanate de methyle (French)
Iso-cyanatomethane
MIC
Methylisocyanaat (Dutch)
Methyl isocyanat (German)
Methyl isocyanate (ACGIH, OSHA) [UN 2480]
Methyl isocyanate, Solutions (DOT)
Methylisokyanat (Czech)
Metil isocianato (Italian)
RCRA waste number P064
TL 1450
UN 2480 [Methyl isocyanate]

624-84-0
CH_4N_2O
60.07
O=CNN
Formic acid, hydrazide
Carbazaldehyde
Formal hydrazine
Formhydrazid (German)
Formhydrazide
Formic hydrazide
Formohydrazide
Formylhydrazide
Formylhydrazine
N-Formylhydrazine
Hydrazinecarboxaldehyde (9CI)
Hydrazine, formyl-

624-89-5
C_3H_8S
76.16
S(CC)C
Ethane, (methylthio)-
AI3-18786
(Methylthio)ethane

624-91-9
CH_3NO_2
61.05
O=NOC
Nitrous acid, methyl ester
Methylester kyseliny dusite (Czech)
Methyl nitrite (DOT)

624-92-0
$C_2H_6S_2$
94.20
S(SC)C
Disulfide, dimethyl
Dimethyldisulfide
Dimethyl disulfide [UN 2381]
UN 2381 [Dimethyl disulfide]

624-95-3
$C_6H_{14}O$
102.18
OCCC(C)(C)C
3,3-Dimethylbutan-1-ol
1-Butanol, 3,3-dimethyl- (9CI)
3,3-Dimethyl-1-butanol

625-06-9
$C_7H_{16}O$
116.20
2-Pentanol, 2,4-dimethyl- (9CI)

625-16-1
$C_7H_{14}O_2$
130.19
tert-Amyl acetate

625-23-0
$C_7H_{16}O$
116.20
CCCCC(C)(C)O
2-Hexanol, 2-methyl- (9CI)

625-25-2
$C_8H_{18}O$
130.23
2-Heptanol, 2-methyl- (9CI)
AI3-37268

625-27-4
C_6H_{12}
84.18
CCC=C(C)C
2-Pentene, 2-methyl
2-Methyl-2-pentene
2-Methyl-pentene-2

625-29-6
$C_5H_{11}Cl$
106.60
C(Cl)(CCC)C
Pentane, 2-chloro- (9CI)
2-Chloropentane

625-30-9
$C_5H_{13}N$
87.16
sec-Amylamine
Butylamine, 1-methyl- (8CI)
2-Pentanamine (9CI)

625-31-0
$C_5H_{10}O$
86.13
OC(CC=C)C
4-Penten-2-ol (9CI)
AI3-28609

625-33-2
C_5H_8O
84.13
O=C(C=CC)C
3-Penten-2-one
Ethylidene acetone
Methyl propenyl ketone

625-38-7
$C_4H_6O_2$
86.09
OC(=O)CC=C
3-Butenoic acid (9CI)
AI3-32117

625-43-4
$C_5H_{13}N$
87.16
N(CC(C)C)C
1-Propanamine, N,2-dimethyl- (9CI)
N,2-Dimethyl-1-propanamine
N,2-Dimethylpropylamine

N-Methylisobutylamine

625-44-5
$C_5H_{12}O$
88.15
Propane, 1-methoxy-2-methyl- (9CI)
Ether, isobutyl methyl (8CI)

625-45-6
$C_3H_6O_3$
90.09
O=C(O)COC
Acetic acid, methoxy
Methoxyacetic acid
2-Methoxyacetic acid

625-48-9
$C_2H_5NO_3$
91.08
OCCN(=O)=O
Ethanol, 2-nitro
β-Nitroalcohol
2-Nitroethanol

625-50-3
C_4H_9NO
87.14
O=C(NCC)C
Acetamide, N-ethyl
Acetamidoethane
N-Aethylacetamid (German)
Ethylacetamide
N-Ethylacetamide

625-52-5
$C_3H_8N_2O$
88.13
O=C(NCC)N
Urea, ethyl
Ethylurea
n-Ethylurea
1-Ethylurea
Urea, 1-ethyl-

625-53-6
$C_3H_8N_2S$
104.19
N=C(S)NCC
Urea, 1-ethyl-2-thio
Ethyl thiourea
1-Ethylthiourea

625-54-7
$C_5H_{12}O$
88.15
O(C(C)C)CC
Propane, 2-ethoxy- (9CI)
2-Ethoxypropane
Ethyl isopropyl ether

625-55-8
$C_4H_8O_2$
88.12
O=COC(C)C
Formic acid, isopropyl ester
Isopropyl formate [UN 1281]
UN 1281 [Propyl formates]

625-58-1
$C_2H_5NO_3$
91.08
Nitric acid, ethyl ester
Ethylester kyseliny dusicne (Czech)
Ethyl nitrate (DOT)
Nitric ether (DOT)

625-60-5
C_4H_8OS
104.17
O=C(SCC)C
Ethanethioic acid, S-ethyl ester
AI3-14852
S-Ethyl ethanethioate

625-65-0
C_7H_{14}
98.19
2-Pentene, 2,4-dimethyl- (8CI, 9CI)
NSC-74139

625-67-2
$C_5H_{10}Cl_2$
141.04
Pentane, 2,4-dichloro- (6CI,7CI,8CI,9CI)
2,4-Dichloropentane

625-69-4
$C_5H_{12}O_2$
104.17
CC(O)CC(C)O
2,4-Pentanediol
2,4-Amyleneglycol
Isoamylene alcohol
Pentanediol-2,4

625-76-3
$CH_2N_2O_4$
106.03
Dinitromethane
Dinitrometano (Spanish)
Methane, dinitro-

625-77-4
$C_4H_7NO_2$
101.10
Acetamide, N-acetyl- (9CI)
Diacetamide (8CI)
NSC-405639

625-80-9
$C_6H_{14}S$
118.24
S(C(C)C)C(C)C
Isopropyl sulfide (8CI)
Diisopropyl sulfide
Propane, 2,2'-thiobis- (9CI)
2,2'-Thiobispropane

625-84-3
C_6H_9N
95.14
Cc1ccc(C)[nH]1
Pyrrole, 2,5-dimethyl- (8CI)
1H-Pyrrole, 2,5-dimethyl- (9CI)

625-86-5
C_6H_8O
96.14
O(C(=CC=1)C)C1C
Furan, 2,5-dimethyl
2,5-Dimethylfuran

625-92-3
$C_5H_3Br_2N$
236.89
Pyridine, 3,5-dibromo- (8CI, 9CI)
NSC-6209

625-95-6
C_7H_7I
218.04
c(cccc1C)(c1)I
Benzene, 1-iodo-3-methyl- (9CI)
1-Iodo-3-methylbenzene
m-Iodotoluene
Toluene, m-iodo- (8CI)

625-98-9
C_6H_4ClF
130.55
Fc(cccc1Cl)c1
Benzene, 1-chloro-3-fluoro- (9CI)
1-Chloro-3-fluorobenzene

625-99-0
C_6H_4ClI
238.45
c(cccc1Cl)(c1)I
Benzene, 1-chloro-3-iodo- (9CI)
AI3-22030
1-Chloro-3-iodobenzene

626-00-6
$C_6H_4I_2$
329.91
c(cccc1I)(c1)I
Benzene, 1,3-diiodo- (9CI)
1,3-Diiodobenzene

626-01-7
C_6H_6IN
219.03
Nc(cccc1I)c1

Aniline, m-iodo
m-Aminoiodobenzene
3-Aminonitrobenzene
Benzenamine, 3-iodo- (9CI)
m-Iodoaniline
3-Iodoaniline

626-02-8
C_6H_5IO
220.01
Oc(cccc1I)c1
Phenol, m-iodo
m-Hydroxyiodobenzene
m-Iodophenol
3-Iodophenol
3-Jodphenol (German)
Phenol, 3-iodo-

626-16-4
$C_8H_8Cl_2$
175.06
c(cccc1CCl)(c1)CCl
m-Xylene, α,α'-dichloro
Benzene, 1,3-bis(chloromethyl)-
1,3-Bis(chloromethyl)benzene
α,α'-Dichloro-m-xylene
m-Xylylene dichloride

626-17-5
$C_8H_4N_2$
128.14
N#Cc(cccc1C#N)c1
1,3-Benzenedicarbonitrile
1,3-Benzendikarbonitril (Czech)
m-Dicyanobenzene
1,3-Dicyanobenzene
Dinitrile of isophthalic acid
IPN
Isoftalodinitril (Czech)
Isoftalonitril (Czech)
Isophthalodinitrile
Isophthalonitrile
Nitril kyseliny isoftalove (Czech)
m-PDN
m-Phthalodinitrile (ACGIH, OSHA)

626-23-3
$C_8H_{19}N$
129.24
N(C(CC)C)C(CC)C
Di-sec-butylamine
AI3-28514
2-Butanamine, N-(1-methylpropyl)- (9CI)
Di-sec-butylamine (8CI)
N-(1-Methylpropyl)-2-butanamine

626-26-6
$C_8H_{18}S$
146.30
Butane, 2,2'-thiobis- (9CI)
sec-Butyl sulfide (8CI)

626-36-8
$C_4H_8N_2O_3$
132.11
O=C(OCC)NC(=O)N
Carbamic acid, (aminocarbonyl)-, ethyl ester (9CI)
Ethyl (aminocarbonyl)carbamate

626-38-0
$C_7H_{14}O_2$
130.21
O=C(OC(CCC)C)C
Acetic acid, 2-pentyl ester
2-Acetoxypentane
sec-Amyl acetate (ACGIH, OSHA) [UN 1104]
sek.Amylester kyseliny octove (Czech)
2-Amylester kyseliny octove (Czech)
1-Methylbutyl acetate
2-Pentanol, acetate (8CI, 9CI)
2-Pentyl acetate
UN 1104 [Amyl acetates]

626-39-1
$C_6H_3Br_3$
314.80
c(cc(cc1Br)Br)(c1)Br
1,3,5-Tribromobenzene
AI3-15483
Benzene, 1,3,5-tribromo- (9CI)

626-41-5
$C_6H_4Br_2O$
251.91
Phenol, 3,5-dibromo- (8CI,9CI)

626-43-7
$C_6H_5Cl_2N$
162.01
Nc(cc(cc1Cl)Cl)c1
3,5-Dichloroaniline
Aniline, 3,5-dichloro-
Benzenamine, 3,5-dichloro- (9CI)
3,5-Dichlorobenzenamine

626-44-8
$C_6H_3I_3$
455.80
Benzene, 1,3,5-triiodo- (6CI,7CI,8CI,9CI)
1,3,5-Triiodobenzene

626-48-2
$C_5H_6N_2O_2$
126.13
n(c(O)cc(n1)C)c1O
Uracil, 6-methyl
4-Methyluracil
6-Methyluracil
Pseudothymine
2,4(1H,3H)-Pyrimidinedione, 6-methyl- (9CI)

626-51-7
$C_6H_{10}O_4$
146.14
CC(CC(O)=O)CC(O)=O
3-Methylglutaric acid
Pentanedioic acid, 3-methyl- (9CI)

626-55-1
C_5H_4BrN
157.99
n(cccc1Br)c1
Pyridine, 3-bromo- (9CI)
AI3-17853
3-Bromopyridine

626-60-8
C_5H_4ClN
113.55
n(cccc1Cl)c1
Pyridine, 3-chloro
m-Chloropyridine
3-Chloropyridine

626-61-9
C_5H_4ClN
113.55
Clc1ccncc1
Pyridine, 4-chloro
4-Chloropyridine

626-62-0
$C_6H_{11}I$
210.06
C(CCCC1)(C1)I
Cyclohexane, iodo- (9CI)
Cyclohexyliodide
Iodocyclohexane

626-64-2
C_5H_5NO
95.11
O=c1cc[nH]cc1
4-Pyridinol
γ-Hydroxypyridine
4-Hydroxypyridine

626-67-5
$C_6H_{13}N$
99.20
N(CCCC1)(C1)C
Piperidine, 1-methyl
n-Methylpiperidine
1-Methylpiperidine [UN 2399]
UN 2399 [1-Methylpiperidine]

626-70-0
$C_7H_{12}O_4$
160.17
Hexanedioic acid, 2-methyl- (8CI,9CI)

626-77-7
$C_9H_{18}O_2$

158.24
Hexanoic acid, propyl ester (9CI)
AI3-06014
Propyl caproate
Propyl hexanoate

626-82-4
$C_{10}H_{20}O_2$
172.30
O=C(OCCCC)CCCCC
Hexanoic acid, butyl ester
Butyl caproate
Butyl hexanoate
n-Butyl hexanoate

626-89-1
$C_6H_{14}O$
102.20
Isohexyl-alcohol
Iso-hexanol
Isohexanol
1-Pentanol, 4-methyl- (9CI)

626-93-7
$C_6H_{14}O$
102.18
OC(CCCC)C
2-Hexanol (9CI)

626-97-1
$C_5H_{11}NO$
101.14
O=C(N)CCCC
Valeramide (8CI)
AI3-24387
Pentanamide (9CI)

626-98-2
$C_5H_8O_2$
100.13
CCC=CC(O)=O
2-Pentenoic acid

627-00-9
$C_4H_7ClO_2$
122.55
O=C(O)CCCCl
Butanoic acid, 4-chloro- (9CI)
Butyric acid, 4-chloro- (8CI)
4-Chlorobutanoic acid

627-03-2
$C_4H_8O_3$
104.12
O=C(O)COCC
Acetic acid, ethoxy
Ethoxyacetic acid
2-Ethoxyacetic acid

627-05-4
$C_4H_9NO_2$
103.14
CCCCN(=O)=O
Butane, 1-nitro
1-Nitrobutane

627-08-7
$C_6H_{14}O$
102.18
O(C(C)C)CCC
Propane, 1-(1-methylethoxy)- (9CI)
1-(1-Methylethoxy)propane

627-11-2
$C_3H_4Cl_2O_2$
142.97
O=C(OCCCl)Cl
Formic acid, chloro-, 2-chloroethyl ester
2-Chloroethyl chloroformate
2-Chlorethylester kyseliny chlormravenci (Czech)
Chloroformic acid 2-chloroethyl ester
TL 207

627-12-3
$C_4H_9NO_2$
103.14
O=C(OCCC)N
Carbamic acid, propyl ester
Propyl carbamate
n-Propyl carbamate
Propylester kyseliny karbaminove (Czech)
Propyl urethane

627-13-4
$C_3H_7NO_3$
105.11
n-Propyl nitrate (ACGIH, OSHA) [UN 1865]
Nitrate de propyle normal (French)
Nitric acid, propyl ester
Propylester kyseliny dusicne (Czech)
Propyl nitrate
UN 1865 [n-Propyl nitrate]

627-18-9
C_3H_7BrO
139.01
OCCCBr
1-Propanol, 3-bromo
3-Bromopropanol
3-Bromo-1-propanol
3-Hydroxypropyl bromide

627-19-0
C_5H_8
68.12
C(#C)CCC
1-Pentyne
AI3-37709

627-20-3
C_5H_{10}
70.13
C(=CC)CC
cis-2-Pentene
β-Amylene-cis
cis-β-Amylene
cis-β-n-Amylene
cis-Pentene
2-Pentene, (Z)- (9CI)
(Z)-2-Pentene

627-21-4
C_5H_8
68.12
2-Pentyne (8CI,9CI)

627-22-5
C_4H_5Cl
88.54
1,3-Butadiene, 1-chloro
1-Chlorobutadiene
1-Chloro-1,3-butadiene

627-23-6
C_4H_5Cl
88.54
1,2-Butadiene, 1-chloro- (6CI,8CI,9CI)

627-30-5
C_3H_7ClO
94.55
OCCCCl
1-Propanol, 3-chloro
3-Chlorpropan-1-ol (German)
3-Chloropropanol
3-Chloropropanol-1 [UN 2849]
Trimethylene chlorohydrin
UN 2849 [3-Chloropropanol-1]

627-35-0
$C_4H_{11}N$
73.13
CCCNC
1-Propanamine, N-methyl- (9CI)
NSC-165652
Propylamine, N-methyl- (8CI)

627-42-9
C_3H_7ClO
94.54
O(CCCl)C
β-Chloroethyl methyl ether
AI3-61817
2-Chloroethyl methyl ether
1-Chloro-2-methoxyethane
Ethane, 1-chloro-2-methoxy
Ether, 2-chloroethyl methyl
2-Methoxyethyl chloride

627-44-1
$C_4H_{10}Hg$
258.73

CC[Hg]CC
Mercury, diethyl
Diethyl mercury

627-45-2
C_3H_7NO
73.11
O=CNCC
Formamide, N-ethyl
N-Aethylformamid (German)
Ethylformamide
N-Ethylformamide
N-Formylethylamine

627-51-0
C_4H_6S
86.16
Ethene, 1,1'-thiobis- (9CI)
Vinyl sulfide (8CI)

627-53-2
$C_4H_{10}Se$
137.08
Diethyl selenide
Diethylselenide
Diethylselenium
Selenide, diethyl-
TL 264

627-58-7
C_8H_{14}
110.20
C(CCC(=C)C)(=C)C
1,5-Hexadiene, 2,5-dimethyl- (9CI)
AI3-30528
2,5-Dimethylhexa-1,5-diene
2,5-Dimethyl-1,5-hexadiene

627-59-8
$C_7H_{16}O$
116.20
2-Hexanol, 5-methyl- (8CI,9CI)

627-63-4
$C_4H_2Cl_2O_2$
152.96
O=C(C=CC(=O)Cl)Cl
Fumaryl chloride [UN 1780]
Chlorure de fumaryle (French)
Dichlorid kyseliny fumarove (Czech)
Fumaroyl chloride
Fumaroyl dichloride
Fumarylchlorid (Czech)
TL 189
UN 1780 [Fumaryl chloride]

627-83-8
$C_{38}H_{74}O_4$
595.00
O=C(OCCOC(=O)CCCCCCCCCCCCCCCCC)CCCCCCCCCCCCCCCCC
Glycol distearate
1,2-Ethanediyl octadecanoate
Ethylene glycol distearate
Felsapon-GDS
Octadecanoic acid, 1,2-ethanediyl ester (9CI)
Radia 7267
Stearic acid, ethylene ester (8CI)
Unipeg-EGDS
Unitina AGS

627-90-7
$C_{13}H_{26}O_2$
214.35
O=C(OCC)CCCCCCCCCC
Undecanoic acid, ethyl ester
AI3-04250
Ethyl undecanoate

627-91-8
$C_7H_{12}O_4$
160.17
O=C(OC)CCCCC(=O)O
Adipic acid, monomethyl ester (8CI)
AI3-01772
Hexanedioic acid, monomethyl ester (9CI)
Methyl hydrogen adipate
Monomethyl hexanedioate

627-93-0
$C_8H_{14}O_4$
174.22
O=C(OC)CCCCC(=O)OC
Adipic acid, dimethyl ester
Dimethyl adipate
Dimethyl hexanedioate
Hexanedioic acid, dimethyl ester (9CI)
Methyl adipate

627-97-4
C_8H_{16}
112.22
2-Heptene, 2-methyl- (8CI,9CI)
NSC-102776

627-98-5
$C_7H_{16}O$
116.20
1-Hexanol, 5-methyl- (8CI,9CI)

628-02-4
$C_6H_{13}NO$
115.20
O=C(N)CCCCC
Hexanamide
Caproamide
Capronamide
NCI-C02142

628-13-7
$C_5H_5N.ClH$
115.57
Pyridine, hydrochloride
Pyridinium chloride
Pyridinium monochloride

628-17-1
$C_5H_{11}I$
198.06
C(CCCC)I
Pentane, 1-iodo
Amyl iodide
n-Amyl iodide
1-Iodopentane
1-Jodpentan (Czech)
Pentyl iodide
n-Pentyl iodide
1-Pentyl iodide

628-28-4
$C_5H_{12}O$
88.17
Ether, butyl methyl
Butane, 1-methoxy- (9CI)
Butyl methyl ether [UN 2350]
α-Methoxybutane
1-Methoxybutane
Methyl butyl ether
Methyl n-butyl ether
UN 2350 [Butyl methyl ether]

628-29-5
$C_5H_{12}S$
104.22
S(CCCC)C
Butane, 1-(methylthio)- (9CI)
1-(Methylthio)butane

628-32-0
$C_5H_{12}O$
88.17
CCCOCC
Ether, ethyl propyl
1-Ethoxypropane
Ethyl propyl ether [UN 2615]
Ethyl n-propyl ether
Propane, 1-ethoxy- (9CI)
Propyl ethyl ether
UN 2615 [Ethyl propyl ether]

628-34-2
C_4H_9ClO
108.58
CCOCCCl
Ether, 2-chloroethyl ethyl
2-Chloroethoxyethane
β-Chloroethyl ethyl ether
2-Chloroethyl ethyl ether
Ethane, 1-chloro-2-ethoxy- (9CI)
2-Ethoxyethyl chloride
Ethyl β-chloroethyl ether

628-36-4
$C_2H_4N_2O_2$
88.08
O=CNNC=O
Formic acid, hydrazodi
1,2-Diformylhydrazin (German)

1,2-Diformylhydrazine
Hydrazine, 1,2-diformyl-

628-37-5
Unknown
Unknown
Diethyl peroxide

628-39-7
$C_4H_{10}Se_2$
216.04
Diselenide, diethyl
AI3-10600

628-41-1
C_6H_8
80.13
C(=CCC=C1)C1
1,4-Cyclohexadiene (9CI)

628-46-6
$C_7H_{14}O_2$
130.19
CC(C)CCCC(O)=O
Hexanoic acid, 5-methyl- (8CI,9CI)
Isoenanthic acid
NSC-1075

628-55-7
$C_8H_{18}O$
130.23
O(CC(C)C)CC(C)C
Propane, 1,1'-oxybis(2-methyl- (9CI)
1,1'-Oxybis(2-methylpropane)

628-61-5
$C_8H_{17}Cl$
148.68
C(CCCCCC)(Cl)C
Octane, 2-chloro- (9CI)
2-Chlorooctane

628-63-7
$C_7H_{14}O_2$
130.21
O=C(OCCCCC)C
Acetic acid, pentyl ester
Acetate d'amyle (French)
Acetic acid, amyl ester
Amyl acetate [UN 1104]
n-Amyl acetate (ACGIH, OSHA)
Amyl acetic ester
Amyl acetic ether
Amylazetat (German)
Amylester kyseliny octove (Czech)
Birnenoel
Octan amylu (Polish)
Pear Oil
Pent-Acetate
Pent-Acetate 28
1-Pentanol acetate
Pentyl acetate
n-Pentyl acetate
1-Pentyl acetate
Primary amyl acetate
UN 1104 [Amyl acetates]

628-71-7
C_7H_{12}
96.17
C(#C)CCCCC
1-Heptyne (9CI)

628-73-9
$C_6H_{11}N$
97.18
N#CCCCCC
Hexanenitrile
Capronitrile
n-Capronitrile

628-76-2
$C_5H_{10}Cl_2$
141.05
ClCCCCCCl
Pentane, 1,5-dichloro
1,5-Dichloropentane

628-81-9
$C_6H_{14}O$
102.20
O(CCCC)CC
Ether, butyl ethyl
Butyl ethyl ether
Ether ethylbutylique (French)
Ethyl butyl ether [UN 1179]
UN 1179 [Ethyl butyl ether]

628-82-0
$C_4H_8O_3$
104.11
Ethanol, 2-methoxy-, formate (7CI,8CI,9CI)
2-Methoxyethyl formate

628-86-4
$C_2HgN_2O_2$
284.63
Mercury fulminate (DOT)
Fulminate of mercury, Dry (DOT)
Fulminate of mercury, Wet
Fulminate of mercury, Wet (DOT)
Initiating explosive fulminate of mercury (DOT)
Mercury fulminate, Containing, by weight, at least 20% water [UN 0135]
Mercury fulminate (Dry)
Mercury fulminate (Wet)
RCRA waste number P065
UN 0135 [Mercury fulminate, Wetted with not less than 20 per cent water, or mixture of alcohol and water, by mass]

628-89-7
$C_4H_9ClO_2$
124.58
O(CCO)CCCl
Ethanol, 2-(2-chloroethoxy)
2-(2-Chloroethoxy)ethanol
2-(2'-Chloroethoxy)ethanol
Diglycol chlorhydrin

628-90-0
$C_4H_{10}O_3$
106.12
Methane, oxybis(methoxy- (9CI)
Bis(methoxymethyl) ether
α,α'-Dimethoxydimethyl ether
Ether, bis(methoxymethyl)- (6CI,7CI,8CI)
2,4,6-Trioxaheptane

628-92-2
C_7H_{12}
96.19
C1CCC=CCC1
Cycloheptene [UN 2242]
UN 2242 [Cycloheptene]

628-94-4
$C_6H_{12}N_2O_2$
144.20
O=C(N)CCCCC(=O)N
Adipamide
Adipic acid amide
Adipic acid diamide
Adipic diamide
1,4-Butanedicarboxamide
Hexanediamide (9CI)
NCI-C02095

628-96-6
$C_2H_4N_2O_6$
152.08
O=N(=O)OCCON(=O)=O
Ethylene glycol dinitrate (ACGIH,DOT,OSHA)
Dinitroglicol (Italian)
Dinitroglycol
EGDN
Ethanediol dinitrate
Ethylene dinitrate
Ethylene nitrate
Ethylenglykoldinitrat (Czech)
Glycoldinitraat (Dutch)
Glycol dinitrate
Glycol (dinitrate de) (French)
Glykoldinitrat (German)
Nitroglycol
Nitroglykol (Czech)

628-97-7
$C_{18}H_{36}O_2$
284.48
O=C(OCC)CCCCCCCCCCCCCC
Ethyl palmitate

AI3-06331
Ethyl hexadecanoate
Hexadecanoic acid, ethyl ester (9CI)
Palmitic acid, ethyl ester (8CI)

628-99-9
$C_9H_{20}O$
144.26
OC(CCCCCCC)C
2-Nonanol (9CI)
AI3-37210

629-01-6
$C_8H_{17}NO$
143.22
O=C(N)CCCCCCC
Octanamide (9CI)

629-04-9
$C_7H_{15}Br$
179.13
BrCCCCCCC
Heptane, 1-bromo
1-Bromoheptane
Heptyl bromide
n-Heptyl bromide

629-05-0
C_8H_{14}
110.22
C(#C)CCCCCC
1-Octyne
Hexylacetylene

629-06-1
$C_7H_{15}Cl$
134.65
ClCCCCCCC
Heptane, 1-chloro- (9CI)
1-Chloroheptane

629-08-3
$C_7H_{13}N$
111.18
N#CCCCCCC
Heptanenitrile (9CI)
AI3-28301

629-11-8
$C_6H_{14}O_2$
118.20
OCCCCCCO
1,6-Hexanediol
HDO
Hexamethylene glycol

629-13-0
$C_2H_4N_6$
112.06
1,2-Diazidoethane
1,2-Diazidoetano (Spanish)
Diazido-1,2 ethane (French)
Ethane, 1,2-diazido-

629-14-1
$C_6H_{14}O_2$
118.20
O(CCOCC)CC
Ethane, 1,2-diethoxy
1,2-Diethoxyethane
Diethyl cellosolve (DOT)
Diethylether ethylenglykolu (Czech)
Ethylene glycol diethyl ether [UN 1153]
Ethyl glyme
UN 1153 [Ethylene glycol diethyl ether]

629-19-6
$C_6H_{14}S_2$
150.31
S(SCCC)CCC
n-Propyl disulfide
AI3-37201
Dipropyl disulfide
Disulfide, dipropyl (9CI)
Propyl disulfide (8CI)

629-20-9
C_8H_8
104.16
c(cccccc1)c1
1,3,5,7-Cyclooctatetraene
(8)Annulene
Cyclooctatetraene [UN 2358]
UN 2358 [Cyclooctatetraene]

629-25-4
$C_{12}H_{23}O_2.Na$
222.34
Lauric acid, sodium salt
Lauran sodny (Czech)
Sodium dodecanoate
Sodium laurate

629-27-6
$C_8H_{17}I$
240.15
C(CCCCCCC)I
Octane, 1-iodo
1-Iodooctane
1-Jodoktan (Czech)
Octyl iodide
n-Octyl iodide
1-Octyl iodide
1-n-Octyl iodide

629-30-1
$C_7H_{16}O_2$
132.20
OCCCCCCCO
1,7-Heptanediol (9CI)
AI3-11199

629-36-7
$C_6H_{12}Cl_2O$
171.07
O(CCCCl)CCCCl
Propane, 1,1'-oxybis(3-chloro- (9CI)
1,1'-Oxybis(3-chloropropane)

629-38-9
$C_7H_{14}O_4$
162.21
CCOCCOCOC(C)=O
Ethanol, 2-(2-methoxyethoxy)-, acetate
Acetic acid, 2-(2-methoxyethoxy)ethyl ester
Diethylene glycol monomethyl ether acetate
2-(2-Methoxyethoxy)ethanol acetate
2-(2-Methoxyethoxy)ethyl acetate
2-(2-Methoxyethoxy)ethylester kyseliny octove (Czech)
Methyl carbitol acetate
Methylkarbitolacetat (Czech)

629-39-0
$C_8H_{17}NO_3$
175.22
O=N(=O)OCCCCCCCC
Octyl nitrate
Nitric acid, octyl ester (8CI, 9CI)

629-40-3
$C_8H_{12}N_2$
136.22
N#CCCCCCCC#N
Suberonitrile
1,6-Dicyanohexane
Octanedinitrile (9CI)

629-45-8
$C_8H_{18}S_2$
178.36
S(SCCCC)CCCC
n-Butyl disulfide
AI3-32578
Butyl disulfide (8CI)
Disulfide, dibutyl (9CI)

629-50-5
$C_{13}H_{28}$
184.41
C(CCCCCCCCCCC)C
Tridecane
n-Tridecane

629-59-4
$C_{14}H_{30}$
198.44
C(CCCCCCCCCCCC)C
Tetradecane

629-60-7
$C_{13}H_{25}N$
195.34
N#CCCCCCCCCCCCC
Tridecanenitrile (9CI)

AI3-16570

629-62-9
$C_{15}H_{32}$
212.47
C(CCCCCCCCCCCCC)C
Pentadecane
n-Pentadecane

629-63-0
$C_{14}H_{27}N$
209.37
N#CCCCCCCCCCCCCC
Myristonitrile (8CI)
AI3-00310
Tetradecanenitrile (9CI)

629-70-9
$C_{18}H_{36}O_2$
284.54
O=C(OCCCCCCCCCCCCC
CCC)C
1-Hexadecanol, acetate
1-Acetoxyhexadecane
Cetyl acetate
ENT 1,025
Hexadecyl acetate
Palmityl acetate

629-72-1
$C_{15}H_{31}Br$
291.32
BrCCCCCCCCCCCCCCC
Pentadecane, 1-bromo- (9CI)
1-Bromopentadecane

629-73-2
$C_{16}H_{32}$
224.43
C(=C)CCCCCCCCCCCCCC
Cetene
AI3-06556
α-Hexadecene
n-Hexadec-1-ene
1-Hexadecene (9CI)
α-Hexadecylene
1-Cetene
Cetylene

629-74-3
$C_{16}H_{30}$
222.41
C(#C)CCCCCCCCCCCCCC
1-Hexadecyne (9CI)

629-76-5
$C_{15}H_{32}O$
228.42
OCCCCCCCCCCCCCCC
1-Pentadecanol (9CI)
AI3-33881
Pentadecanol

629-78-7
$C_{17}H_{36}$
240.53
C(CCCCCCCCCCCCCCC)C
Heptadecane
n-Heptadecane

629-79-8
$C_{16}H_{31}N$
237.42
N#CCCCCCCCCCCCCCCC
Hexadecanenitrile (9CI)
AI3-11234
Palmitonitrile (8CI)

629-80-1
$C_{16}H_{32}O$
240.43
O=CCCCCCCCCCCCCCCC
Hexadecanal
AI3-24252

629-82-3
$C_{16}H_{34}O$
242.50
O(CCCCCCCC)CCCCCCCC
Octyl-ether
Antar
Caprylic ether
Dioctyl ether
Ether, di-n-octyl-
Octane, 1,1'-oxybis- (9CI)

629-83-4
$C_{31}H_{62}O_2$
466.83
Triacontanoic acid, methyl ester (8CI,9CI)
NSC-20743

629-92-5
$C_{19}H_{40}$
268.53
C(CCCCCCCCCCCCCCCCC)C
Nonadecane (9CI)
AI3-36122

629-93-6
$C_{18}H_{37}I$
380.40
C(CCCCCCCCCCCCCCCCC)I
Octadecane, 1-iodo- (9CI)
1-Iodooctadecane
Octadecyl iodide

629-94-7
$C_{21}H_{44}$
296.58
C(CCCCCCCCCCCCCCC
CCC)C
Heneicosane
AI3-36479

629-96-9
$C_{20}H_{42}O$
298.55
OCCCCCCCCCCCCCCC
CCCCC
1-Eicosanol (9CI)
AI3-36485
Arachic alcohol
Arachidic alcohol
Arachidyl alcohol
n-Eicosanol
n-1-Eicosanol
Eicosyl alcohol
Pri-n-eicosyl alcohol

629-97-0
$C_{22}H_{46}$
310.61
C(CCCCCCCCCCCCCCC
CCCCC)C
Docosane (9CI)

629-99-2
$C_{25}H_{52}$
352.69
C(CCCCCCCCCCCCCCC
CCCCCCC)C
Pentacosane (9CI)
AI3-36478

630-01-3
$C_{26}H_{54}$
366.71
C(CCCCCCCCCCCCCCCCC
CCCCCC)C
Hexacosane (9CI)

630-02-4
$C_{28}H_{58}$
394.77
C(CCCCCCCCCCCCCCCCC
CCCCCCCC)C
n-Octacosane
AI3-52615
Octacosane (9CI)

630-03-5
$C_{29}H_{60}$
408.80
C(CCCCCCCCCCCCCCCCC
CCCCCCCCC)C
Nonacosane
AI3-36284

630-04-6
$C_{31}H_{64}$
436.86
Hentriacontane

630-05-7
$C_{33}H_{68}$
464.90
Tritriacontane (8CI,9CI)

630-06-8
$C_{36}H_{74}$
506.98

C(CCCCCCCCCCCCCCCCCC CCCCCCCCCCCCCCCC)C
Hexatriacontane (9CI)
AI3-52389

630-07-9
$C_{35}H_{72}$
492.96
Pentatriacontane (9CI)

630-08-0
CO
28.01
O#C
Carbon-monoxide
Carbone (oxyde de) (French)
Carbonic oxide
Carbonio (ossido di) (Italian)
Carbon monoxide (ACGIH, OSHA) [UN 1016]
Carbon monoxide, Cryogenic liquid [NA 9202]
Carbon oxide (co)
Exhaust Gas
Flue Gas
Kohlenmonoxid (German)
Kohlenoxyd (German)
Koolmonoxyde (Dutch)
NA 9202 [Carbon monoxide, refrigerated liquid (cryogenic liquid)]
Oxyde de carbone (French)
UN 1016 [Carbon monoxide]
Wegla tlenek (Polish)

630-10-4
CH_4N_2Se
123.03
Urea, seleno
RCRA waste number P103
Selenourea

630-17-1
$C_5H_{11}Br$
151.05
BrCC(C)(C)C
Neopentyl bromide
1-Bromo-2,2-dimethylpropane
Propane, 1-bromo-2,2-dimethyl- (9CI)

630-18-2
C_5H_9N
83.13
N#CC(C)(C)C
Pivalonitrile (8CI)
AI3-33242
2,2-Dimethylpropanenitrile
Propanenitrile, 2,2-dimethyl- (9CI)

630-19-3
$C_5H_{10}O$
86.13
O=CC(C)(C)C
Pivaldehyde
AI3-33229
2,2-Dimethylpropanal
Pivalaldehyde (8CI)
Propanal, 2,2-dimethyl- (9CI)

630-20-6
$C_2H_2Cl_4$
167.84
C(CCl)(Cl)(Cl)Cl
Ethane, 1,1,1,2-tetrachloro
NCI-C52459
RCRA waste number U208
1,1,1,2-Tetrachloroethane

630-24-0
$C_2Br_2Cl_4$
325.64
Ethane, 1,1-dibromo-1,2,2,2-tetrachloro- (9CI)

630-56-8
$C_{27}H_{40}O_4$
428.67
CCCCCC(=O)OC3(CCC4C2CC C1=CC(=O)CCC1(C)C2CC C34C)C(C)=O
Pregn-4-ene-3,20-dione, 17-hydroxy-, hexanoate
Capron
Corlutin L.A.
Delalutin
Depo-Proluton
Duraluton
Estralutin
Gesterol L.A.
17-α-Hexanoyloxypregn-4-ene-3,20-dione
Hormofort
HPC
17-Hydroxypregn-4-ene-3,20-dione hexanoate
Hydroxyprogesterone caproate
17-α-Hydroxyprogesterone caproate
17-α-Hydroxy progesterone n-caproate
17-α-Hydroxyprogesterone hexanoate
Hylutin
Hyproval-PA
Hyroxon
Idrogestene
Lutate
Luteocrin
Luteocrin Depot
Lutopron
Neolutin
NSC-17592
17-((1-Oxohexyl)oxy)pregn-4-ene-3,20-dione
Pregn-4-ene-3,20-dione, 17-((1-oxohexyl)oxy)-
Primolut Depot
Progesterone caproate
Progesterone Retard Pharlon
Proluton Depot
Relutin
Squibb
Syngynon
Teralutil

630-60-4
$C_{29}H_{44}O_{12}$
584.73
O=C(OCC=1C(C(C(O)(C(C(C(C(C (O)(C2)CC(OC(OC(C(O) C3O)C)C3O)C4)(C4O)CO) C5O)C2)C6)(C5)C)C6)C1
Ouabain
Acocantherin
Astrobain
Gratibain
Gratus strophanthin
g-Strophanthin
Ouabagenin-l-rhamnosid (German)
Ouabagenin l-rhamnoside
Oubain
Ouabaine
Purostrophan
Strophanthin G
G-Strophanthin
Strophoperm

630-72-8
$C_2N_4O_6$
176.02
Trinitroacetonitrile
Acetonitrile, trinitro-
Trinitroacetonitrile (French)
Trinitroacetonitrilo (Spanish)

630-93-3
$C_{15}H_{11}N_2O_2.Na$
274.27
[Na+].O=C1NC(C(=O)[N-]1)(c2ccccc2)c3ccccc3
Hydantoin, 5,5-diphenyl-, monosodium salt
Alepsin
Antilepsin
Antisacer
Auranile
Citrullamon
Danten
Dantoin
Denyl
Denyl sodium
Derizene
Difenin
Difetoin
Difhydan
Di-Hydan
Dihydantoin
Dilantin
Dilantin sodium
Di-Len
Dintoina
Diphantoine sodium
Diphedan
Diphenate
Diphenin
Diphenine sodium
Diphentoin
Diphenylhydantoin sodium
5,5-Diphenylhydantoin sodium
5,5-Diphenyl-2,4-imidazolidinedione, monosodium salt

Diphenylan sodium
Di-Phetine
Ditoin
Divulsan
DPH
Enkefal
Epamin
Epanutin
Epelin
Epifenyl
Epihydan
Epilan-D
Epilantin
Epinat
Eptoin
Fenantoin
Fenitoin
Fenytoine
Hidantal sodium
Hydantin sodium
Hydantoinal sodium
Hydantoin, 5,5-diphenyl-, sodium salt
Hydantoin sodium
Idantoil
Idantoinal
2,4-Imidazolidinedione, 5,5-diphenyl-, monosodium salt
Lepitoin
Lepitoin sodium
Minetoin
Novantoina
Novodiphenyl
OM-Hydantoine sodium
Phenhydan
Phenytoin sodium
Phenytoinum sodium
Saceril
SDPH
Sodanton
Sodium diphenylhydantoin
Sodium diphenyl hydantoinate
Sodium 5,5-diphenylhydantoinate
Sodium 5,5-diphenyl-2,4-imidazolidinedione
Sodium phenytoin
Solantoin
Solantyl
Soluble Phenytoin
Sylantoic
Tacosal
Thilophenyt
Zentropil

631-36-7
$C_8H_{20}Si$
144.33
Silane, tetraethyl- (9CI)
Tetraethylsilane

631-40-3
$C_{12}H_{28}N.I$
313.31
[I-].CCC[N+](CCC)(CCC)CCC
Ammonium, tetrapropyl-, iodide
1-Propanaminium, N,N,N-tripropyl-, iodide (9CI)
Tetrapropylammonium iodide
Tetra-n-propylammonium iodide

631-60-7
$C_2H_3O_2.Hg$
259.64
[Hg].CC([O-])=O
Acetic acid, mercury (1+) salt (8CI,9CI)
Mercurous acetate (DOT)
Mercurous acetate, Solid (DOT)
Mercury acetate [UN 1629]
Mercury monoacetate
UN 1629 [Mercury acetate]

631-61-8
$C_2H_3O_2.H_4N$
77.10
Acetic acid, ammonium salt
Ammonium acetate

631-64-1
$C_2H_2Br_2O_2$
217.84
O=C(O)C(Br)Br
Acetic acid, dibromo- (9CI)
Dibromoacetic acid

632-21-3
$C_3H_2Cl_4O$
195.85
2-Propanone, 1,1,3,3-tetrachloro
TCA
1,1,3,3-Tetrachloroacetone
1,1,3,3-Tetrachloropropanone
1,1,3,3-Tetrachloro-2-propanone

632-22-4
$C_5H_{12}N_2O$
116.19
O=C(N(C)C)N(C)C
Urea, 1,1,3,3-tetramethyl
TEMUR
Tetramethylurea
1,1,3,3-Tetramethylurea
Tetramethyluree (French)
TMU

632-25-7
$C_7H_6O_5S$
202.19
O=C(O)c(c(S(=O)(=O)O)ccc1)c1
2-Sulfobenzoic acid
Benzoic acid, 2-sulfo- (9CI)
o-Sulfobenzoic acid

632-46-2
$C_9H_{10}O_2$
150.19
Cc1cccc(C)c1C(O)=O
Benzoic acid, 2,6-dimethyl
2,6-Dimethylbenzoic acid

632-51-9
$C_{26}H_{20}$
332.44
c(C(c(cccc1)c1)=C(c(cccc2)c2)c(cccc3)c3)(cccc4)c4
Benzene, 1,1',1'',1'''-(1,2-ethendiylidene)tetrakis-
AI3-19250
Benzene, 1,1',1'',1'''-(1,2-ethenediylidene)tetrakis- (9CI)
Ethylene, tetraphenyl- (8CI)

632-58-6
$C_8H_2Cl_4O_4$
303.90
O=C(O)c(c(c(c(c(c1Cl)Cl)Cl)C(=O)O)c1Cl
Phthalic acid, tetrachloro
1,2-Benzenedicarboxylic acid, 3,4,5,6-tetrachloro- (9CI)
Tetrachlorophthalic acid

632-79-1
$C_8Br_4O_3$
463.70
O=C(OC(=O)c1c(c(c(c2Br)Br)Br)Br)c12
1,3-Isobenzofurandione, 4,5,6,7-tetrabromo- (9CI)
Bromphthal
FG 4000
Fire Master PHT 4
NSC-4874
Phthalic anhydride, tetrabromo- (8CI)
4,5,6,7-Tetrabromo-1,3-isobenzofurandione
Tetrabromophthalic anhydride
3,4,5,6-Tetrabromophthalic anhydride

632-83-7
$C_{14}H_7BrO_2$
287.11
9,10-Anthracenedione, 1-bromo- (9CI)
Anthraquinone, 1-bromo- (8CI)

632-99-5
$C_{20}H_{19}N_3.ClH$
337.88
Cl.Cc1cc(ccc1N)C(=C2C=CC(N)C=C2)c3ccc(N)cc3
Benzenamine, 4-((4-aminophenyl)(4-imino-2,5-cyclohexadien-1-ylidene)methyl)-2-methyl-, monohydrochloride
Aizen Magenta
Aniline Red
Astra Fuchsine B
Basic Fuchsine
Basic Magenta
Basic Magenta E-200
Calcozine Fuchsine HO
Calcozine Magenta RIN
Calcozine Magenta XX
Cerise B

C.I. 42510
C.I. Basic Violet 14
C.I. Basic Violet 14, monohydrochloride (8CI)
Diabasic Magenta
Diamond Fuchsine
Fuchsine
Fuchsine A
Fuchsine CS
Fuchsine G
Fuchsine HO
Fuchsine N
Fuchsine RTN
Fuchsine Y
Magenta
Magenta DP
Magenta E
Magenta G
Magenta I
Magenta PN
Magenta Powder N
Magenta S
Magenta Supertine
Orient Basic Magenta
12418 Red
Rosaniline
Rosaniline chloride
Rosaniline hydrochloride
Rosanilinium chloride
Violet Zasadita 14 (Czech)

633-03-4
$C_{27}H_{33}N_2.HO_4S$
482.69
Ammonium, (4-(p-(diethylamino)-α-phenylbenzylidene)-2,5-cyclohexadien-1-yl idene) diethyl-, sulfate (1:1)
ADC Brilliant Green Crystals
Aizen Diamond Green GH
Azien Malachite Green GH
Aniline Green
Astra Diamond Green GX
Avon Green A-4379
Basic Bright Green
Basic Bright-Green Sulfate
Basic Brilliant Green
Basic Green 1
Basic Green V
Benzaldehyde Green
Brilliant Green
Brilliant Green Aseptic
Brilliant Green B
Brilliant Green B.P.
Brilliant Green BP Crystals
Brilliant Green BPc
Brilliant Green Crystals
Brilliant Green Crystals H
Brilliant Green DSC
Brilliant Green G
Brilliant Green GX
Brilliant Green Lake
Brilliant Green P
Brilliant Green Special
Brilliant Green Sulfate
Brilliant Green WP Crystals
Brilliant Green Y
Brilliant Green YN
Brilliant Green YNS
Brillant-Grun (German)
Brilliant Lake Green Y
Brilliant Tungstate Green Toner GT-288
Calcozine Brilliant Green G
C.I. 42040
C.I. Basic Green 1
C.I. Basic Green 1, Sulfate (1:1)
Deorlene Green JJO
Diamond Green G
Emerald Green
Ethyl Green
Fast Green J
Fast Green JJO
12415 Green
Green EN
Hidaco Brilliant Green
Malachite Green G
Mitsui Brilliant Green GX
Solid Green
Tertrophene Brilliant Green G
Tokyo Aniline Brilliant Green
Zelen Brilantni (Czech)
Zelen Malachitova G (Czech)
Zelen Smaragdova (Czech)
Zelen Zasadita 1 (Czech)

633-31-8
$C_{30}H_{50}O_2$
442.73
O=C(OC(CCC(C1=CCC2C(C(C(C3)C(CCCC(C)C)C)(CC4)C)C3)(C24)C)C1)CC
Cholesteryl n-propionate
Cholest-5-en-3-ol (3β)-, propanoate (9CI)

633-96-5
$C_{16}H_{12}N_2O_4S.Na$
351.35
[Na+].Oc2ccc1ccccc1c2N=Nc3ccc(cc3)S([O-])(=O)=O
Benzenesulfonic acid, 4-((2-hydroxy-1-naphthalenyl)-azo)-, monosodium salt
Acid Leather Orange Extra
Acid Leather Orange Extra G
Acid Leather Orange Extra PRW
Acid Orange
Acid Orange 7
Acid Orange A
Acid Orange II
Acilan Orange II
Airedale Orange II
Amacid Orange Y
Atul Acid Orange II
Betanaphthol Orange
Brasilan Orange A
Bucacid Orange A
Calcocid Orange Y
Certiqual Orange II
C.I. 15510
C.I. Acid Orange 7
C.I. Acid Orange 7, Monosodium salt
Colacid Orange
Curol Orange
D & C Orange No. 4
Diacid Orange II
Erio Orange II
Fenazo Orange
Hidacid Orange II
Hispacid Orange AF
p-((2-Hydroxy-1-naphthyl)azo)-benzenesulfonic acid sodium salt
Java Orange II
Kiton Orange II
Kromon Lake Orange Toner
Lake Orange A
Lake Orange II YS
Leather Orange Extra
Lurazol Orange E
Lutetia Orange 3JR
Mandarin G
Naphthalene Lake Orange G
Naphthalene Orange G
Naphthol Orange
β-Naphthol Orange
2-Naphthol Orange II
β-Naphthyl Orange
Naphtocard Orange II
Neklacid Orange II
No. 177 Orange Lake
Nubilon Orange R
11550 Orange
Orange Extra N
Orange Extra P
Orange II
Orange IIC
Orange II for Lakes
Orange IIP
Orange IIS
Orange IISM
Orange II Special for Lacquer
Orange No. 205
Orange 2 Sodium salt
Orange Toner GRT
Orange Y
Orange YA
Orange YZ
Oranz Kysela 7 (Czech)
Peeracid Orange II
Perca Orange gr
Persian Orange
Persian Orange Lake
Persian Orange X
Pure Orange II S
Rifa acid Orange II
Sanyo Gum Orange A
Solar Orange
Special Orange GR
Special Orange H
Symuler Orange Lake 43
Symulon Acid Orange II
Tangarine Lake X-917
Tertracid Orange II
Tropaeolin OOO
Tropaeolin OOO 2
Vondacid Orange II
Wool Orange A

634-03-7
$C_{12}H_{17}NO$
191.30
CC1C(OCCN1C)c2ccccc2
Morpholine, 3,4-dimethyl-2-phenyl-, (+)
Antapentan
Mephenmetrazine
Morpholine, 3,4-dimethyl-2-phenyl-, (2-R-trans)- (9CI)

Phendimetrazine
(+)-Phendimetrazine
Sedafamen

634-36-6
$C_9H_{12}O_3$
168.19
O(c(c(OC)ccc1)c1OC)C
Benzene, 1,2,3-trimethoxy- (9CI)
AI3-02077
1,2,3-Trimethoxybenzene

634-66-2
$C_6H_2Cl_4$
215.88
c(c(c(c(c1)Cl)Cl)Cl)(c1)Cl
Benzene, 1,2,3,4-tetrachloro
1,2,3,4-Tetrachlorobenzene

634-67-3
$C_6H_4Cl_3N$
196.46
Nc1ccc(Cl)c(Cl)c1Cl
Aniline, 2,3,4-trichloro- (8CI)
Benzenamine, 2,3,4-trichloro- (9CI)

634-68-4
$C_6H_2I_4$
581.70
Benzene, 1,2,3,4-tetraiodo- (8CI,9CI)

634-83-3
$C_6H_3Cl_4N$
230.90
Nc1cc(Cl)c(Cl)c(Cl)c1Cl
Benzenamine, 2,3,4,5-tetra-chloro- (9CI)
Aniline, 2,3,4,5-tetrachloro- (8CI)

634-89-9
$C_6H_2Br_4$
393.70
Benzene, 1,2,3,5-tetrabromo- (8CI,9CI)

634-90-2
$C_6H_2Cl_4$
215.88
c(cc(c(c1Cl)Cl)Cl)(c1)Cl
Benzene, 1,2,3,5-tetrachloro
1,2,3,5-Tetrachlorobenzene

634-91-3
$C_6H_4Cl_3N$
196.46
Nc1cc(Cl)c(Cl)c(Cl)c1
Benzenamine, 3,4,5-trichloro- (9CI)
Aniline, 3,4,5-trichloro- (8CI)
NSC-115260

634-92-4
$C_6H_2I_4$
581.70
Benzene, 1,2,3,5-tetraiodo- (7CI,8CI,9CI)

634-93-5
$C_6H_4Cl_3N$
196.46
Nc(c(cc(c1)Cl)Cl)c1Cl
Aniline, 2,4,6-trichloro
Benzenamine, 2,4,6-trichloro- (9CI)
sym-Trichloroaniline
2,4,6-Trichloroaniline
2,4,6-Trichlorobenzenamine

634-95-7
$C_5H_{12}N_2O$
116.19
O=C(N(CC)CC)N
Urea, 1,1-diethyl
asym-Diethylurea
N,N-Diethylurea
1,1-Diethylurea
Urea, N,N-diethyl- (9CI)

634-97-9
$C_5H_5NO_2$
111.09
OC(=O)c1ccc[nH]1
2-Pyrrolecarboxylic acid
Pyrrole-2-carboxylic acid (8CI)
1H-Pyrrole-2-carboxylic acid (9CI)

635-08-5
$C_8H_{10}O_4$
170.17
1-Cyclohexene-1,2-dicarboxylic acid (8CI,9CI)

635-21-2
$C_7H_6ClNO_2$
171.58
O=C(O)c(c(N)ccc1Cl)c1
Anthranilic acid, 5-chloro- (8CI)
AI3-15229
2-Amino-5-chlorobenzoic acid
Benzoic acid, 2-amino-5-chloro- (9CI)

635-22-3
$C_6H_5ClN_2O_2$
172.58
O=N(=O)c(c(ccc1N)Cl)c1
Aniline, 4-chloro-3-nitro
4-Chloro-3-nitroaniline

635-39-2
$C_5H_5N_5O.ClH$
187.61
Guanine, hydrochloride
USAF S-1

635-46-1
$C_9H_{11}N$
133.19
N(c(c(ccc1)CC2)c1)C2
Quinoline, 1,2,3,4-tetrahydro- (9CI)
AI3-10034
1,2,3,4-Tetrahydroquinoline

635-65-4
$C_{33}H_{36}N_4O_6$
584.73
O=C(NC(C=1C=C)=CC(NC(=C2CCC(=O)O)CC(NC(=C3C)C=C(NC(=O)C=4C=C)C4C)=C3CCC(=O)O)=C2C)C1C
Biline-8,12-dipropionic acid, 1,10,19,22,23,24-hexahydro-2,7,13,17-tetramethyl-1,19-dioxo- 3,18-divinyl
Bilirubin
Bilirubin IX-α
Hematoidin
Hemetoidin
Principal Bile Pigment

635-90-5
$C_{10}H_9N$
143.18
c1ccc(cc1)n2cccc2
Pyrrole, 1-phenyl- (8CI)
AI3-11735
1H-Pyrrole, 1-phenyl- (9CI)

636-09-9
$C_{12}H_{14}O_4$
222.26
O=C(OCC)c(ccc(c1)C(=O)OCC)c1
Terephthalic acid, diethyl ester
1,4-Benzenedicarboxylic acid, diethyl ester
Diethyl terephthalate

636-21-5
$C_7H_9N.ClH$
143.63
Cl.Cc1ccccc1N
o-Toluidine, hydrochloride
1-Amino-2-methylbenzene hydrochloride
2-Amino-1-methylbenzene hydrochloride
2-Aminotoluene hydrochloride
o-Aminotoluene hydrochloride
Benzenamine, 2-methyl-, hydro-chloride
1-Methyl-2-aminobenzene hydrochloride
2-Methyl-1-aminobenzene hydrochloride
o-Methylaniline hydrochloride
2-Methylaniline hydrochloride
o-Methylbenzenamine hydro-

chloride
2-Methylbenzenamine hydrochloride
NCI-C02335
RCRA waste number U222
o-Toluidin hydrochloride
2-Toluidine hydrochloride
o-Tolylamine hydrochloride

636-23-7
$C_7H_{10}N_2.2ClH$
195.11
Cl.Cl.Cc1ccc(N)cc1N
Toluene-2,4-diamine, dihydrochloride
2,4-Diaminotoluene dihydrochloride
Metatolylenediamine dihydrochloride
2,4-Toluenediamine dihydrochloride

636-26-0
$C_5H_6N_2OS$
142.19
n(cc(c(n1)O)C)c1S
Thymine, 2-thio
4(1H)-Pyrimidinone, 2,3-dihydro-5-methyl-2-thioxo- (9CI)
Thiothymine
2-Thiothymine

636-28-2
$C_6H_2Br_4$
393.70
c(c(cc(c1Br)Br)Br)(c1)Br
Benzene, 1,2,4,5-tetrabromo- (9CI)
1,2,4,5-Tetrabromobenzene

636-30-6
$C_6H_4Cl_3N$
196.46
Nc(c(cc(c1Cl)Cl)Cl)c1
Aniline, 2,4,5-trichloro- (8CI)
Benzenamine, 2,4,5-trichloro- (9CI)
2,4,5-Trichlorobenzenamine

636-31-7
$C_6H_2I_4$
581.70
Benzene, 1,2,4,5-tetraiodo- (7CI,8CI,9CI)
1,2,4,5-Tetraiodobenzene

636-41-9
C_5H_7N
81.11
Cc1ccc[nH]1
Pyrrole, 2-methyl- (8CI)
1H-Pyrrole, 2-methyl- (9CI)

636-46-4
$C_8H_6O_5$
182.14
Isophthalic acid, 4-hydroxy
1,3-Benzenedicarboxylic acid, 4-hydroxy- (9CI)
Eupirina
4-HIPA
4-Hydroxyisophthalic acid

636-53-3
$C_{12}H_{14}O_4$
222.26
O=C(OCC)c(cccc1C(=O)OCC)c1
Isophthalic acid, diethyl ester
Diethyl isophthalate

636-70-4
$C_6H_{15}N.BrH$
182.10
Ethanamine, N,N-diethyl-, hydrobromide (9CI)
N,N-Diethylethanamine hydrobromide

636-78-2
$C_7H_6O_5S$
202.19
4-Sulfobenzoic acid
para-Sulfobenzoic acid
para-Sulphobenzoic acid

636-98-6
$C_6H_4INO_2$
249.00
O=N(=O)c(ccc(c1)I)c1
Benzene, 1-iodo-4-nitro- (9CI)
AI3-08878
1-Iodo-4-nitrobenzene

637-07-0
$C_{12}H_{15}ClO_3$
242.72
CCOC(=O)C(C)(C)Oc1ccc(Cl)cc1
Propionic acid, 2-(p-chlorophenoxy)-2-methyl-, ethyl ester
Acetic acid, (p-chlorophenoxy)-dimethyl-, ethyl ester
Amotril
Amotril S
Angiokapsul
Anparton
Antilipid
Antilipide
Apolan
Arterioflexin
Artes
Arterosol
Artevil
Ateculon
Ateriosan
Athebrate
Atheromide
Atheropront
Athranid-wirkstoff
Atrolen
Atromid
Atromida
Atromidin
Atromid S
Atrovis
AY 61123
Azionyl
Bioscleran
Bresit
Cartagyl
α-(p-Chlorophenoxy)isobutyric acid, ethyl ester
α-p-Chlorophenoxyisobutyryl ethyl ester
2-(4-Chlorophenoxy)-2-methylpropanoic acid ethyl ester
2-(p-Chlorophenoxy)-2-methylpropionic acid ethyl ester
Cinnarizin
Cinnarizine
Citiflus
Claripex
Claripex CPIB
Cloberat
Clobrat
Clobren-SF
Clofar
Clofibram
Clofibrat
Clofibrate
Clofibrato (Spanish)
Clofinit
Clofipront
CPIB
Deliva
Dura Clofibrat
ELPI
EPIB
Ethyl-α-p-chlorophenoxy-isobutyrate
Ethyl chlorophenoxyisobutyrate
Ethyl α-(4-chlorophenoxy)isobutyrate
Ethyl para-chlorophenoxyisobutyrate
Ethyl 2-(p-chlorophenoxy)isobutyrate
Ethyl α-(p-chlorophenoxy)-α-methylpropionate
Ethyl α-(4-chlorophenoxy)-α-methylpropionate
Ethyl 2-(p-chlorophenoxy)-2-methylpropionate
Ethyl 2-(4-chlorophenoxy)-2-methylpropionate
Ethyl clofibrate
Fibralem
Gerastop
Hyclorate
ICI 28257
Isobutyric acid, α-(p-chlorophenoxy)-, ethyl ester
Klofiran
Levatrom
Lipamid
Lipavil
Lipavlon
Lipide 500
Lipidsenker
Lipofacton
Lipomid
Liponorm
Lipore duct

Liporil
Liposid
Liprin
Liprinal
Lobetrin
Miscleron
Negalip
Neo-Atomid
Neo-Atromid
Normalip
Normat
Normet
Normolipol
NSC-79389
Oxan 600
Persantinat
Propanoic acid, 2-(4-chloro-phenoxy)-2-methyl-, ethyl ester
Propionic acid, 2-(4-chlorophen-oxy)-2-methyl-, ethyl ester
Recolip
Regardin
Regelan
Regelan N
Robigram
Scrobin
Serofinex
Serotinex
Skerolip
Skleromex
Skleromexe
Sklero-Tablinen
Sklero-Tabuls
Ticlobran
Vincamin compositum
Xyduril
Yoclo

637-12-7
$C_{54}H_{105}O_6$.Al
877.57
Stearic acid, aluminum salt
Alugel 34TN
Aluminium stearate
Aluminum stearate
Aluminum tristearate (ACGIH)
Metasap XX
Octadecanoic acid, aluminum salt
Rofob 3
SA 1500
Tribasic aluminum stearate

637-27-4
$C_9H_{10}O_2$
150.18
O=C(Oc(cccc1)c1)CC
Propanoic acid, phenyl ester (9CI)
AI3-04253
Phenyl propanoate

637-31-0
$C_{16}H_{21}N_3$
255.34
Diphenylamine, 4,4'-bis(di-methylamino)- (8CI)
1,4-Benzenediamine, N'-(4-(di-methylamino)phenyl)-N,N-di-methyl- (9CI)

637-50-3
C_9H_{10}
118.19
c(cccc1)(c1)C=CC
Benzene, propenyl
β-Methylstyrene
ω-Methylstyrene
3-Phenyl-1-propene
Propenyl benzene

637-59-2
$C_9H_{11}Br$
199.09
BrCCCc(cccc1)c1
Benzene, (3-bromopropyl)- (9CI)
1-Bromo-3-phenylpropane
(3-Bromopropyl)benzene

637-61-6
C_6H_4ClNO
141.56
O=C(C=CC(=NCl)C=1)C1
2,5-Cyclohexadien-1-one, 4-chloroimino
4-Chloroimino-2,5-cyclohexa-diene-1-one
Quinone chlorimide

637-62-7
$C_6H_5NO_2$
123.12
ON=C1C=CC(=O)C=C1
2,5-Cyclohexadien-1-one, 4-hydroxyimino
p-Benzoquinone, monooxime (8CI)
2,5-Cyclohexadiene-1,4-dione, monooxime (9CI)
4-Hydroxyimino-2,5-cyclohexa-diene-1-one
Quinone monooxime
p-Quinone monooxime
p-Quinone monoxime
Quinone 4-oxime

637-64-9
$C_7H_{12}O_3$
144.17
O=C(OCC(OCC1)C1)C
Tetrahydrofurfuryl acetate
AI3-21209
2-Furanmethanol, tetrahydro-, acetate (9CI)
Furfuryl alcohol, tetrahydro-, acetate (8CI)
Tetrahydro-2-furanmethanol acetate

637-69-4
$C_9H_{10}O$
134.18
Anisole, p-vinyl- (8CI)
AI3-05526
Benzene, 1-ethenyl-4-methoxy- (9CI)

637-78-5
$C_6H_{12}O_2$
116.18
O=C(OC(C)C)CC
Propionic acid, isopropyl ester
Isopropyl propionate [UN 2409]
Propanoic acid, 1-methylethyl ester (9CI)
Iso-propyl propanoate
Iso-propyl propionate
UN 2409 [Isopropyl propionate]

637-87-6
C_6H_4ClI
238.45
c(ccc(c1)Cl)(c1)I
Benzene, 1-chloro-4-iodo- (9CI)
1-Chloro-4-iodobenzene

637-88-7
$C_6H_8O_2$
112.13
O=C(CCC(=O)C1)C1
1,4-Cyclohexanedione (9CI)
AI3-22410

637-92-3
$C_6H_{14}O$
102.18
O(C(C)(C)C)CC
Ether, tert-butyl ethyl (8CI)
2-Ethoxy-2-methylpropane
Propane, 2-ethoxy-2-methyl- (9CI)

637-97-8
$C_5H_{11}I$
198.05
C(CCC)(C)I
Pentane, 2-iodo- (9CI)
sec-Amyl iodide
2-Iodopentane

638-02-8
C_6H_8S
112.20
Thiophene, 2,5-dimethyl- (8CI,9CI)
NSC-60689

638-03-9
C_7H_9N.ClH
143.63
Cl.Cc1cccc(N)c1
m-Toluidine, hydrochloride

638-04-0
C_8H_{16}
112.22
CC1CCCC(C)C1

Cyclohexane, 1,3-dimethyl-, cis- (9CI)

638-16-4
$C_3H_3N_3S_3$
177.27
n(c(nc(n1)S)S)c1S
s-Triazine-2,4,6-trithiol
Cyanuric acid, trithio-
1,3,5-Triazine-2,4,6-trimercaptan
2,4,6-Triazinetrithiol
1,3,5-Triazine-2,4,6(1H,3H,5H)-trithione
1,3,5-Trimercaptotriazine
2,4,6-Trimercapto-s-triazine
Trithiocyanuric acid
USAF TH-3

638-17-5
$C_6H_{13}NS_2$
163.32
4H-1,3,5-Dithiazine, dihydro-2,4,6-trimethyl-, (2-α,4-α, 6-α)
Thialdine

638-21-1
C_6H_7P
110.10
Pc1ccccc1
Phosphine, phenyl
Fenylfosfin (Czech)
Phenylphosphine (ACGIH, OSHA)

638-23-3
$C_5H_9NO_4S$
179.21
O=C(O)C(N)CSCC(=O)O
Alanine, 3-((carboxymethyl)-thio)-, L
Carbocisteine
Carbocysteine
l-Carboxymethylcysteine
s-(Carboxymethyl)cysteine
3-(Carboxymethylthio)alanine
l-3-((Carboxymethyl)thio)-alanine
l-Cysteine, S-(carboxymethyl)-(9CI)
L.J. 206
Mucodyne
Rhinathiol
Rinatiol
Thiodril

638-26-6
$C_{18}H_{36}O_3$
300.48
Octadecanoic acid, 10-hydroxy-(8CI,9CI)
NSC-79060

638-28-8
$C_6H_{13}Cl$
120.62
C(CCCC)(Cl)C
Hexane, 2-chloro- (9CI)
2-Chlorohexane

638-29-9
C_5H_9ClO
120.58
O=C(CCCC)Cl
Chlorure de valeryle (French)
Cloruro de valerilo (Spanish)
Pentanoyl chloride (9CI)
Valeryl chloride
UN 2502 [Valeryl chloride]

638-32-4
$C_4H_7NO_3$
117.10
NC(=O)CCC(O)=O
Butanoic acid, 4-amino-4-oxo-(9CI)
NSC-78416
Succinamic acid (8CI)

638-36-8
$C_{20}H_{42}$
282.56
Phytane
2,6,10,14-Tetramethylhexadecane

638-38-0
$C_4H_6O_4.Mn$
173.04
[Mn].[O-]C(=O)CCC([O-])=O
Acetic acid, manganese(II) salt (2:1)
Diacetylmanganese
Manganese acetate
Manganese(2+) acetate
Manganese(II) acetate
Manganese acetate (Mn(OAc)$_2$)
Manganese diacetate
Manganous acetate
Octan manganaty (Czech)

638-45-9
$C_6H_{13}I$
212.07
C(CCCCC)I
Hexane, 1-iodo- (9CI)
1-Iodohexane

638-46-0
$C_6H_{14}S$
118.24
Butane, 1-(ethylthio)- (9CI)
NSC-19
Sulfide, butyl ethyl (8CI)

638-49-3
$C_6H_{12}O_2$
116.18
O=COCCCCC
Formic acid, pentyl ester
n-Amyl formate
Amyl formate [UN 1109]
Pentyl formate
m-Pentyl formate
UN 1109 [Amyl formates]

638-53-9
$C_{13}H_{26}O_2$
214.39
O=C(O)CCCCCCCCCCCC
Tridecanoic-acid
n-Tridecoic acid
Tridecylic acid

638-56-2
$C_8H_{16}Cl_2O_3$
231.12
Bis(2-(2-chloroethoxy)ethyl) ether
Ethane, 1,1'-oxybis(2-(2-chloroethoxy)- (9CI)
Ether, bis(2-(2-chloroethoxy)-ethyl) (8CI)
1,11-Dichloro-3,6,9-trioxaundecane

638-59-5
$C_{16}H_{32}O_2$
256.43
O=C(OCCCCCCCCCCCCCC)C
1-Tetradecanol, acetate
AI3-11586
Myristyl acetate
1-Tetradecanol acetate
Tetradecyl acetate

638-65-3
$C_{18}H_{35}N$
265.54
N#CCCCCCCCCCCCCCCCCC
Stearonitrile
Heptadecyl cyanide
Nitril kyseliny stearove (Czech)
Octadecanenitrile
Octadecanonitrile
Oktadekannitril (Czech)

638-66-4
$C_{18}H_{36}O$
268.48
Octadecanal (9CI)
Stearaldehyde (8CI)

638-67-5
$C_{23}H_{48}$
324.63
C(CCCCCCCCCCCCCCCCCC
CCCC)C
Tricosane (9CI)
AI3-35917

638-68-6
$C_{30}H_{62}$
422.82
C(CCCCCCCCCCCCCCCCCCC

CCCCCCCCCC)C
Triacontane (9CI)

639-13-4
$C_{29}H_{42}O_{11}$
566.71
5-β-Card-20(22)-enolide, 19-oxo-3-β-((l-rhamnopyranosyl)oxy)-5,12-β,14-trihydroxy
Antiarigenin + L-rhamnose (German)
β-Antiarin

639-46-3
$C_{17}H_{19}NO_4$
301.34
DEA No. 9307
Genomorphin
Genomorphine
Morphinan-3,6-diol, 7,8-didehydro-4,5-epoxy-17-methyl-, (5-α,6-α)-, 17-oxide (9CI)
Morphinan-3,6-α-diol, 7,8-didehydro-4,5-α-epoxy-17-methyl-, 17-oxide
Morphine oxide
Morphine-N-oxide

639-48-5
$C_{29}H_{25}N_3O_5$
495.52
Nicomorphine
DEA No. 9312
Nicomorfina
Nicomorphinum (Latin)

639-58-7
$C_{18}H_{15}ClSn$
385.47
Cl[Sn](c1ccccc1)(c2ccccc2)c3ccccc3
Stannane, chlorotriphenyl
Aquatin
Aquatin 20 EC
Brestanol
Chlorotriphenylstannane
Chlorotriphenyltin
Fentin chloride
GC 8993
General Chemicals 8993
HOE 2872
LS 4442
Phenostat-C
TPTC
Tinmate
Triphenylchlorostannane
Triphenylchlorotin
Triphenyltin chloride

640-15-3
$C_6H_{15}O_2PS_3$
246.36
CCSCCSP(=S)(OC)OC
Phosphorodithioic acid, S-(2-(ethylthio)ethyl) O,O-dimethyl ester
Bay 23129
Bayer 23129
Compound M-81
O,O-Dimethyl-S-(2-aethylthio-aethyl)-dithio phosphat (German)
O,O-Dimethyl-S-(2-ethylmercaptoethyl) dithiophosphate
O,O-Dimethyl-S-2-ethylmerkaptoethylester kyseliny dithiofosforecne (Czech)
O,O-Dimethyl-S-(2-ethylthio-ethyl)-dithiofosfaat (Dutch)
O,O-Dimethyl S-(2-(ethylthio)-ethyl) phosphorodithioate
O,O-Dimetil-S-(etiltio-etil)-ditiofosfato (Italian)
Dithiometasystox
Dithiomethon
Dithiometon (French)
Dithiophosphate de O,O-dimethyle et de S-(2-ethylthio-ethyle) (French)
Ekatin
Ekatin aerosol
Ekatine-25
Ekatin ULV
Ethanethiol, 2-(ethylthio)-, S-ester with O,O-dimethyl phosphorodithioate
2-Ethylthioethyl O,O-dimethyl phosphorodithioate
S-(2-(Ethylthio)ethyl) O,O-dimethylphosphorodithionate
S-(2-(Ethylthio)ethyl)dimethyl phosphorothiolothionate
Intrathion
Intration
Luxistelm
M 81
Phosphorodithioic acid, O,O-dimethyl S-(2-ethylthio)ethyl ester
San 230
Thiameton
Thiometon
Thiotox

640-19-7
C_2H_4FNO
77.07
O=C(N)CF
Acetamide, 2-fluoro
AFL 1081
Amid kyseliny fluoroctove (Czech)
Compound 1081
FAA
Fluorakil 100
Fluoroacetamide
2-Fluoroacetamide
Fluoroacetic acid amide
Fussol
Megatox
Monofluoroacetamide
Navron
RCRA waste number P057
Rodex
Yanock

640-61-9
$C_8H_{11}NO_2S$
185.24
O=S(=O)(NC)c(ccc(c1)C)c1
Benzenesulfonamide, N,4-dimethyl- (9CI)
AI3-00115
N,4-Dimethylbenzenesulfonamide
N-Methyl-p-toluenesulfonamide
p-Toluenesulfonamide, N-methyl- (8CI)

640-68-6
$C_5H_{11}NO_2$
117.17
O=C(O)C(N)C(C)C
Valine, D
D-Valine

641-16-7
$C_6H_2N_4O_9$
274.08
2,3,4,6-Tetranitrophenol
Phenol, 2,3,4,6-tetranitro-
2,3,4,6-Tetranitrofenol (Spanish)
Tetranitro-2,3,4,6 phenol (French)

641-85-0
$C_{21}H_{36}$
288.52
Pregnane, (5α)- (9CI)
Allopregnane
5α-Pregnane (8CI)

641-96-3
$C_{24}H_{18}$
306.41
1,1':2',1'':2'',1'''-Quaterphenyl (9CI)
1,1'-Biphenyl, 2,2'-diphenyl-
2,2'-Diphenylbiphenyl
NSC-90717
o-Quaterphenyl (8CI)

643-20-9
$C_7H_{13}N$
111.18
Pyrrolizidine
1-Azabicyclo(3.3.0)octane
Hexahydropyrrolizine
1H-Pyrrolizine, hexahydro- (9CI)

643-22-1
$C_{37}H_{67}NO_{13}.C_{18}H_{36}O_2$
1018.59
CN(C)C2C1CC4C(=C(O)C1(O)C(=O)C(=C2O)C(N)=O)C(=O)c3c(O)cccc3C4(C)O
Erythromycin, stearate (Salt)
Bristamycin
Dowmycin E
Erypar
Ethril

Erythrocin stearate
Erythromycin, octadecanoate (Salt)
Erythromycin stearic acid salt
Gallimycin
NCI-C55674
Pfizer-E
Qidmycin
SK-Erythromycin

643-28-7
$C_9H_{13}N$
135.23
Nc(c(ccc1)C(C)C)c1
Aniline, o-isopropyl
o-Aminoisopropylbenzene
2-Aminoisopropylbenzene
Benzenamine, 2-(1-methylethyl)- (9CI)
o-Cumidine
2-Isopropyl aniline
N-Isopropylaniline (ACGIH)
o-Isopropylaniline

643-43-6
$C_8H_6N_2O_6$
226.13
O=C(O)Cc(c(N(=O)=O)cc(N(=O)=O)c1)c1
2,4-Dinitrophenylacetic acid
AI3-01812
Benzeneacetic acid, 2,4-dinitro- (9CI)
2,4-Dinitrobenzeneacetic acid

643-58-3
$C_{13}H_{12}$
168.24
Cc1ccccc1c2ccccc2
2-Methylbiphenyl
Biphenyl, 2-methyl- (8CI)
1,1'-Biphenyl, 2-methyl- (9CI)

643-79-8
$C_8H_6O_2$
134.14
O=Cc(c(ccc1)C=O)c1
Phthalaldehyde
Phtalaldehydes (French)

643-93-6
$C_{13}H_{12}$
168.24
c(cccc1c(cccc2)c2)(c1)C
1,1'-Biphenyl, 3-methyl- (9CI)
3-Methyl-1,1'-biphenyl

644-08-6
$C_{13}H_{12}$
168.25
Cc1ccc(cc1)c2ccccc2
Biphenyl, 4-methyl
FEMA 3186
p-Methylbiphenyl
4-Methyl-1,1'-biphenyl
p-Methyldiphenyl
4-Methyldiphenyl

644-26-8
$C_{14}H_{21}NO_2$
235.36
2-Butanol, 1-(dimethylamino)-2-methyl-, benzoate (ester)
Amyleine
Amylocaine
Stovaine
Stovine

644-31-5
$C_9H_8O_4$
180.17
Peroxide, acetyl benzoyl
Acetyl benzoyl peroxide, More than 40% in solution (DOT)
Acetyl benzoyl peroxide, Maximum concentration 45% in solution (DOT)
Acetyl benzoyl peroxide, Solid (DOT)
Acetyl benzoyl peroxide, Solution, Not over 40% peroxide (DOT)
Peroxide, acetyl benzoyl, Maximum concentration 45% in solution
UN 2081 (DOT)

644-35-9
$C_9H_{12}O$
136.21
Oc(c(ccc1)CCC)c1
Phenol, o-propyl
o-Propylphenol

644-49-5
$C_7H_{14}O_2$
130.19
Propanoic acid, 2-methyl-, propyl ester (9CI)
Isobutyric acid, propyl ester (8CI)
NSC-406702

644-64-4
$C_{10}H_{16}N_4O_3$
240.30
CN(C)C(=O)Oc1cc(C)n(n1)C(=O)N(C)C
Carbamic acid, dimethyl-, 1-((dimethylamino)carbonyl)-5-methyl-1H-pyrazol-3-yl ester
Carbamic acid, dimethyl-, 5-methyl-1H-pyrazol-3-yl ester
Dimethylcarbamic acid 1-((dimethylamino)carbonyl)-5-methyl-1H-pyrazol-3-yl ester
Dimethylcarbamic acid ester with 3-hydroxy-N,N,5-trimethylpyrazole-1-carboxamide
2-Dimethylcarbamoyl-3-methylpyrazolyl-(5)-N,N-dimethylcarbamat (German)
1-Dimethylcarbamoyl-5-methyl-3-pyrazolyl dimethylcarbamate
2-Dimethylcarbamoyl-3-methyl-5-pyrazolyl dimethylcarbamate
Dimethyl 2-carbamyl-3-methylpyrazolyldimethylcarbamate
2-(N,N-Dimethylcarbamyl)-3-methylpyrazolyl-5 N,N-dimethylcarbamate
Dimetilan
Dimetilane
ENT 25,595-X
ENT 25,922
G-22870
Geigy 22870
Geigy GS-13332
GS-13332
5-Methyl-1H-pyrazol-3-yl dimethylcarbamate
OMS 479
Pyrazole-1-carboxamide, 3-hydroxy-N,N,5-trimethyl-, dimethylcarbamate (ester)
Snip
Snip Fly
Snip Fly Bands

644-97-3
$C_6H_5Cl_2P$
178.98
c(P(Cl)Cl)(cccc1)c1
Phosphonous dichloride, phenyl
Benzene phosphorus dichloride (DOT)
Dichlorophenylphosphine
Phenyldichlorophosphine
Phenylphosphine dichloride
Phenylphosphonous acid dichloride
Phenylphosphonous dichloride
Phenyl phosphorus dichloride [UN 2798]
UN 2798 [Phenyl phosphorus dichloride]

645-00-1
$C_6H_4INO_2$
249.00
O=N(=O)c(cccc1I)c1
Benzene, 1-iodo-3-nitro- (9CI)
AI3-15390
1-Iodo-3-nitrobenzene

645-05-6
$C_9H_{18}N_6$
210.33
CN(C)c1nc(nc(n1)N(C)C)N(C)C
Melamine, hexamethyl
Altretamine
ENT 50,852
Hemel
Hexamethylmelamine
Hexastat
HMM
NCI-C50259

NSC-13875
1,3,5-Triazine-2,4,6-triamine, N,N,N',N',N'',N''-hexamethyl- (9CI)
s-Triazine, 2,4,6-tris(dimethylamino)-
2,4,6-Tris(dimethylamino)-s-triazine
2,4,6-Tris(dimethylamino)-1,3,5-triazine

645-08-9
$C_8H_8O_4$
168.16
p-Anisic acid, 3-hydroxy
Acide isovanillique (French)
Benzoic acid, 3-hydroxy-4-methoxy-
3-Hydroxyanisic acid
3-Hydroxy-4-methoxybenzoic acid
Isovanillic acid

645-09-0
$C_7H_6N_2O_3$
166.12
O=C(N)c(cccc1N(=O)=O)c1
Benzamide, m-nitro- (8CI)
Benzamide, 3-nitro- (9CI)
3-Nitrobenzamide

645-12-5
$C_5H_3NO_5$
157.09
OC(=O)c1ccc(o1)N(=O)=O
2-Furoic acid, 5-nitro
2-Furancarboxylic acid, 5-nitro (9CI)
5-Nitrofurancarboxylic acid
Nitrofurate
5-Nitrofuroic acid
5-Nitropyromucate

645-13-6
$C_{11}H_{14}O$
162.23
O=C(c(ccc(c1)C(C)C)c1)C
Acetophenone, 4'-isopropyl- (8CI)
AI3-15527
Ethanone, 1-(4-(1-methylethyl)-phenyl)- (9CI)
1-(4-(1-Methylethyl)phenyl)-ethanone

645-15-8
$C_{12}H_9N_2O_8P$
340.17
O=P(Oc(ccc(N(=O)=O)c1)c1)(Oc(ccc(N(=O)=O)c2)c2)O
Bis(4-nitrophenyl)phosphate
Bis(4-nitrophenyl) phosphate
Phosphoric acid, bis(p-nitrophenyl) ester (8CI)
Phosphoric acid, bis(4-nitrophenyl) ester (9CI)

645-18-1
$C_{10}H_{10}ClN_5$
235.65
1,3,5-Triazine-2,4-diamine, N-(4-chlorophenyl)-6-methyl- (9CI)
s-Triazine, 2-amino-4-(p-chloroanilino)-6-methyl- (8CI)

645-49-8
$C_{14}H_{12}$
180.25
C(=Cc1ccccc1)c2ccccc2
cis-1,2-Diphenylethylene
Benzene, 1,1'-(1,2-ethenediyl)bis-, (Z)- (9CI)
cis-Diphenylethene
Isostilbene
Stilbene, (Z)- (8CI)
cis-Stilbene
(Z)-Stilbene

645-55-6
$C_6H_6N_2O_2$
138.11
N-Nitroaniline
Aniline, N-nitro-
Benzenamine, N-nitro-
N-Nitroanilina (Spanish)
N-Nitroaniline (French)
Phenylnitramine
N-Phenylnitramine

645-56-7
$C_9H_{12}O$
136.21
Oc(ccc(c1)CCC)c1
Phenol, p-propyl
p-Propylphenol

645-62-5
$C_8H_{14}O$
126.22
O=CC(=CCCC)CC
2-Hexenal, 2-ethyl
Acrolein, 2-ethyl-3-propyl-
2-Ethylhexenal
2-Ethyl-2-hexenal
α-Ethyl-β-n-propylacrolein
2-Ethyl-3-propyl acrolein

645-92-1
$C_3H_5N_5O$
127.08
Ammeline
AI3-50983
4,6-Diamino-1,3,5-triazin-2(1H)-one
s-Triazin-2-ol, 4,6-diamino- (8CI)
1,3,5-Triazin-2(1H)-one, 4,6-diamino- (9CI)

645-93-2
$C_3H_4N_4O_2$
128.07
1,3,5-Triazine-2,4(1H,3H)-dione, 6-amino-
AI3-61038

646-04-8
C_5H_{10}
70.13
C(=CC)CC
trans-2-Pentene
β-Amylene-trans
trans-β-Amylene
trans-β-n-Amylene
2-Pentene, (E)- (9CI)
(E)-2-Pentene
2-trans-Pentene

646-06-0
$C_3H_6O_2$
74.09
O(CCO1)C1
1,3-Dioxolane
Dioxolan (Czech)
1,3-Dioxolan
1,3-Dioxacyclopentane
Ethylene glycol formal
Formal glycol
Glycol formal

646-07-1
$C_6H_{12}O_2$
116.18
O=C(O)CCC(C)C
Isohexanoic acid (Mixed isomers)

646-13-9
$C_{22}H_{44}O_2$
340.66
O=C(OCC(C)C)CCCCCCCCCCCCCCCCC
Stearic acid, isobutyl ester
Isobutyl stearate

646-14-0
$C_6H_{13}NO_2$
131.17
Hexane, 1-nitro- (8CI,9CI)

646-20-8
$C_7H_{10}N_2$
122.19
N#CCCCCCC#N
Heptanedinitrile
1,5-Dicyanopentane
Pimelic acid dinitrile
Pimelonitrile

646-30-0
$C_{19}H_{38}O_2$
298.51
O=C(O)CCCCCCCCCCCCCCCCC
Nonadecanoic acid (9CI)
AI3-36442

646-31-1
$C_{24}H_{50}$
338.66
C(CCCCCCCCCCCCCCCCCCC CCCC)C
Tetracosane (9CI)
AI3-52698

647-42-7
$C_8H_5F_{13}O$
364.11
FC(F)(F)C(F)(F)C(F)(F)C(F)(F) C(F)(F)C(F)(F)CCO
1-Octanol, 3,3,4,4,5,5,6,6, 7,7,8,8,8-tridecafluoro- (9CI)
1H,1H,2H,2H-Perfluorooctanol
1,1,2,2-Tetrahydroperfluoro-1-octanol

650-20-4
$C_4H_{10}FO_2P$
140.09
Phosphonofluoridic acid, ethyl-, ethyl ester (6CI, 7CI,8CI,9CI)

650-51-1
$C_2Cl_3O_2.Na$
185.36
[Na+].OC(=O)C(Cl)(Cl)Cl
Acetic acid, trichloro-, sodium salt
ACP Grass Killer
Antiperz
Antyperz
Dow Sodium TCA Inhibited
Green Cross Couch Grass Killer
NaTA
NaTCA
Natriumtrichlooracetaat (Dutch)
Natriumtrichloracetat (German)
Sodio(tricloroacetato di) (Italian)
Sodium TCA
Sodium TCA Inhibited
Sodium (trichloracetate de) (French)
Sodium trichloroacetate
STCA
TCA
TCA Sodium
Trichloressigsaures natrium (German)
Trichloroacetic acid sodium salt
Trichloroctan sodny (Czech)
Varitox
Weedmaster Grass Killer

651-55-8
$C_{18}H_{22}O_2$
270.37
Dihydroequilin
α-Dihydroequilin
β-Dihydroequilin
17 α-Dihydroequilin

652-67-5
$C_6H_{10}O_4$
146.16
OC1CC2C(O)CC12
Glucitol, 1,4:3,6-dianhydro-, D
AT 101
Devicoran
D-1,4:3,6-Dianhydroglucitol
1,4:3,6-Dianhydrosorbitol
Hydronol
Isobide
Isosorbide
(+)-D-Isosorbide

653-03-2
$C_{24}H_{31}N_3OS$
409.64
CCCC(=O)c4ccc3Sc1ccccc1N (CCCN2CCN(C)CC2)c3c4
1-Butanone, 1-(10-(3-(4-methyl-1-piperazinyl)-propyl)phenothiazin-2-yl)
AHR-712
Bayer 1362
1-Butanone, 1-(10-(3-(4-methyl-1-piperazinyl)propyl)-10H-phenothiazin-2-yl)- (9CI)
Butaperazine
3-n-Butyryl-10-(3'-N-methyl-piperazino-N'-propyl)phenothiazin
Butyrylperazine
Emerex
Megalectil
1-(10-(3-(4-Methyl-1-piperazin-yl)propyl)phenothiazin-2-yl)-1-butanone
Randolectil
Repoise
Riker 595
Tyrylen

653-63-4
$C_{10}H_{14}N_5O_6P$
331.19
Nc1ncnc2n(cnc12)C3CC(O)C (COP(O)(O)=O)O3
2'-Deoxy-5'-adenosine monophosphate
Adenosine, 2'-deoxy-, 5'-(dihydrogen phosphate) (8CI)
5'-Adenylic acid, 2'-deoxy- (9CI)
Deoxyadenosine monophosphate
Deoxyadenylate
2'-Deoxy-5'-adenylic acid
2'Deoxy-5'-AMP

655-86-7
C12H10N4
MW:
2,3-Phenazinediamine (9CI)
2,3-Diaminophenazine
Phenazine, 2,3-diamino- (8CI)

656-31-5
$C_7H_7FN_2O$
154.13
NC(=O)Nc1ccccc1F
Urea, (2-fluorophenyl)- (9CI)
NSC-204323
Urea, (o-fluorophenyl)- (8CI)

657-27-2
$C_6H_{14}N_2O_2.ClH$
182.68
[Cl-].NCCCCC(N)C(O)=O
Lysine, monohydrochloride, L
Darvyl
Lyamine
Lysine hydrochloride
L-Lysine hydrochloride
Lysine monohydrochloride
L-Lysine, monohydrochloride

657-37-4
C_7H_8FOP
158.11
Phosphinic fluoride, methyl phenyl- (7CI,8CI, 9CI)

657-84-1
$C_7H_7O_3S.Na$
194.19
p-Toluenesulfonic acid, sodium salt
Benzenesulfonic acid, 4-methyl-, sodium salt (9CI)
Naxonate hydrotrope
Sodium p-methylbenzene-sulfonate
Sodium paratoluene sulphonate
Sodium p-toluenesulfonate
Sodium p-tolyl sulfonate
Sodium tosylate
4-Toluenesulfonic acid sodium salt

658-65-1
$C_7H_{12}FN_5$
185.17
1,3,5-Triazine-2,4-diamine, N,N'-diethyl-6-fluoro- (9CI)
s-Triazine, 2,4-bis(ethylamino)-6-fluoro- (8CI)

659-30-3
$C_7H_7FN_2O$
154.16
NC(=O)Nc1ccc(F)cc1
Urea, 1-(p-fluorophenyl)
1-(p-Fluorophenyl)urea
(4-Fluorophenyl)urea

659-70-1
$C_{10}H_{20}O_2$
172.30
O=C(OCCC(C)C)CC(C)C
Isovaleric acid, isopentyl ester
Isoamyl isovalerate
Isopentyl isovalerate

659-84-7
C_8H_{12}
108.18
Bicyclo(5.1.0)oct-3-ene (7CI,9CI)

660-60-6
$C_{18}H_{36}O_2.1/2Cu$
316.25
Octadecanoic acid, copper(2+) salt
AI3-00903
Copper(2+) octadecanoate

660-68-4
$C_4H_{11}N.ClH$
109.62
Diethylamine, hydrochloride
Diethylammonium chloride
Ethanamine, N-ethyl-, hydrochloride (9CI)

661-11-0
$C_7H_{15}F$
118.22
FCCCCCCC
Heptane, 1-fluoro
1-Fluoroheptane

661-19-8
$C_{22}H_{46}O$
326.61
OCCCCCCCCCCCCCCCCCCCCCC
Behenyl alcohol
AI3-36489
Behenic alcohol
1-Docosanol (9CI)
Docosyl alcohol
IK 2
Tadenan

666-52-4
C_3D_6O
52.03
2-Propanone-1,1,1,3,3,3-d6 (9CI)
Acetone-d6 (8CI)

669-04-5
$C_9H_8O_2$
148.16
3(2H)-Benzofuranone, 7-methyl- (7CI,8CI,9CI)
7-Methylbenzo(b)furan-3(2H)-one

669-90-9
$C_6H_{10}O_7$
194.14
O=C(O)C(=O)C(O)C(O)C(O)CO
D-Arabino-2-hexulosonic acid (9CI)

670-54-2
C_6N_4
128.10
N#CC(C#N)=C(C#N)C#N
Ethenetetracarbonitrile
Tetracyanoethylene
Tetrakyanethylen (Czech)

671-04-5
$C_{10}H_{12}ClNO_2$
213.68
CNC(=O)Oc1cc(C)c(C)cc1Cl
Carbamic acid, methyl-, 6-chloro-3,4-xylyl ester
Banol
Banol Tuco Sok
Carbamic acid, methyl-, (2-chloro-4,5-dimethyl)phenyl ester
Carbamic acid, methyl-, 2-chloro-4,5-xylyl ester
Carbanolate
2-Chloro-4,5-dimethylphenyl methylcarbamate
Chloroxylam
6-Chloro-3,4-xylenyl N-methylcarbamate
2-Chloro-4,5-xylyl N-methylcarbamate
6-Chloro-3,4-xylyl methylcarbamate
6-Chloro-3,4-xylyl N-methylcarbamate
OMS-174
Phenol, 2-chloro-4,5-dimethyl-, methyl carbamate (9CI)
SOK
U 12927
U-17004
Upjohn U-12,927

671-16-9
$C_{12}H_{19}N_3O$
221.34
CNNCc1ccc(cc1)C(=O)NC(C)C
p-Toluamide, N-isopropyl-α-(2-methylhydrazino)
Benzamide, N-(1-methylethyl)-4-((2-methylhydrazino)-methyl)- (9CI)
Ibenzmethyzine
2-(p-Isopropylcarbamoylbenzyl)-1-methylhydrazine
n-Isopropyl-α-(2-methylhydrazino)-p-toluamide
Matulane
4-((2-Methylhydrazino)methyl)-N-isopropylbenzamide
1-Methyl-2-(p-(isopropylcarbamoyl)benzyl)hydrazine
MIH
Natulan
NSC-77213
PCB
Procarbazin (German)
Procarbazine
Ro 4-6467

671-36-3
$C_7H_{15}N$
113.20
CC1CCCCN1C
Piperidine, 1,2-dimethyl- (8CI, 9CI)
NSC-363752

671-56-7
C_4H_7ClO
106.55
3-Buten-2-ol, 1-chloro- (6CI,7CI,8CI,9CI)
1-Chloro-3-buten-2-ol
4-Chloro-3-hydroxy-1-butene

672-04-8
$C_{13}H_{19}NO_2$
221.29
m-(1-Ethylpropyl)phenyl methylcarbamate
BUX
Carbamic acid, methyl-, m-(1-ethylpropyl)phenyl ester
Phenol, m-(1-ethylpropyl)-, methylcarbamate (8CI)
Phenol, 3-(1-ethylpropyl)-, methylcarbamate

672-15-1
$C_4H_9NO_3$
119.11
Butyric acid, 2-amino-4-hydroxy-, L- (8CI)
L-Homoserine (9CI)

672-65-1
C_8H_9Cl
140.61
c(cccc1)(c1)C(Cl)C
Benzene, (1-chloroethyl)- (9CI)
(1-Chloroethyl)benzene
α-Phenylethyl chloride

672-76-4
$C_{10}H_{12}O_2$
164.22
O=C(C(O)=CC=C(C=1)C(C)C)C1
2,4,6-Cycloheptatrien-1-one, 2-hydroxy-5-(1-methylethyl)
2,4,6-Cycloheptatrien-1-one, 2-hydroxy-5-isopropyl- (8CI)
γ-Thujaplicin
γ-Thujaplicine

673-04-1
$C_8H_{15}N_5O$
197.28
CCNc1nc(NCC)nc(OC)n1
s-Triazine, 2,4-bis(ethylamino)-6-methoxy
2,4-Bis(ethylamino)-6-methoxy-s-triazine
4,6-Bis(ethylamino)-2-methoxy-s-triazine

G-30044
Geigy 30,044
Gesadural
2-Methoxy-4,6-bis(ethylamino)-s-triazine
Methoxy Simazine
Pimeton
Simeton
Simetone

673-06-3
$C_9H_{11}NO_2$
165.21
O=C(O)C(N)Cc(cccc1)c1
Alanine, phenyl-, D
NCI-C60195
D-Phenylalanine
D-β-Phenylalanine

673-84-7
$C_{10}H_{16}$
136.24
C(=CC=CC(=CC)C)(C)C
Alloocimene
AI3-00737
2,6-Dimethyl-2,4,6-octatriene
allo-Ocimene
2,4,6-Octatriene, 2,6-dimethyl-(VAN) (9CI)

674-76-0
C_6H_{12}
84.16
2-Pentene, 4-methyl-, (E)-(8CI,9CI)
NSC-73914

674-81-7
CH_4N_4O
88.09
NC(=N)NN=O
Guanidine, nitroso
NA 0473 [Nitrosoguanidine]
Nitrosoguanidin (German)
Nitrosoguanidine [NA 0473]
Nitrosoguanidine, Initiating explosive (DOT)
N-Nitrosoguanidine

674-82-8
$C_4H_4O_2$
84.08
O=C(OC1=C)C1
2-Oxetanone, 4-methylene
3-Buteno-β-lactone
Diketene
Diketene, Inhibited [UN 2521]
Ketene, dimer
4-Methylene-2-oxetanone
UN 2521 [Diketene, inhibited]

675-14-9
$C_3F_3N_3$
135.06
Fc(nc(F)nc1F)n1
s-Triazine, 2,4,6-trifluoro
Cyanuric fluoride
2,4,6-Trifluoro-s-triazine

675-20-7
C_5H_9NO
99.15
O=C(NCCC1)C1
2-Piperidone
Piperidon (German)
Piperidone-2 (French)

675-62-7
$C_4H_7Cl_2F_3Si$
211.10
Silane, dichloromethyl-(3,3,3-trifluoropropyl)
Dichloromethyl-3,3,3-trifluoropropylsilane

676-22-2
$C_{12}H_{18}$
162.27
1,5,9-Cyclododecatriene, (E,E,E)- (8CI,9CI)

676-49-3
CH_3D
15.03
Methane-d (8CI,9CI)

676-58-4
CH_3ClMg
74.79
C[Mg]Cl
Methylmagnesium chloride
Chloromethylmagnesium
Magnesium, chloromethyl-(9CI)

676-97-1
CH_3Cl_2OP
132.91
CP(Cl)(Cl)=O
Phosphonic dichloride, methyl
Methyl phosphonic dichloride [NA 9206]
NA 9206 [Methyl phosphonic dichloride]

676-98-2
CH_3Cl_2PS
148.98
Methyl phosphonothioic dichloride, Anhydrous
NA 1760
Phosphonothioic dichloride, methyl-, Anhydrous

677-21-4
$C_3H_3F_3$
96.06
FC(F)(F)C=C
Propene, 3,3,3-trifluoro
1-Propene, 3,3,3-trifluoro- (9CI)
Trifluoromethylethylene
1,1,1-Trifluoropropene
3,3,3-Trifluoropropene
3,3,3-Trifluoropropylene

677-67-8
$C_2F_4O_3S$
180.08
O=C(F)C(F)(F)S(=O)(=O)F
Acetyl fluoride, difluoro-(fluorosulfonyl)- (9CI)
Difluoro(fluorosulfonyl)acetyl fluoride

678-26-2
C_5F_{12}
288.04
FC(F)(F)C(F)(F)C(F)(F)C(F)(F)C(F)(F)F
Perfluoropentane
Dodecafluoropentane
Pentane, dodecafluoro- (9CI)

678-39-7
$C_{10}H_5F_{17}O$
464.12
FC(F)(F)C(F)(F)C(F)(F)C(F)(F)C(F)(F)C(F)(F)C(F)(F)C(F)(F)CCO
1H,1H,2H,2H-Perfluorodecanol
1-Decanol, 3,3,4,4,5,5,6,6,7,7,8,8,9,9,10,10,10-heptadecafluoro- (9CI)
1,1,2,2-Tetrahydroperfluoro-1-decanol

680-31-9
$C_6H_{18}N_3OP$
179.24
O=P(N(C)C)(N(C)C)N(C)C
Phosphoric triamide, hexamethyl
Eastman Inhibitor HPT
ENT 50,882
Hempa
Hexametapol
Hexamethylfosforamid (Czech)
Hexamethyl phosphoramide (ACGIH)
Hexamethylphosphoric acid triamide
Hexamethylphosphoric triamide
N,N,N,N,N,N-Hexamethylphosphoric triamide
Hexamethylphosphorotriamide
Hexamethylphosphotriamide
Hexmethylphosphoramide
HMPA
HMPT
HPT
Phosphoric acid hexamethyltriamide
Phosphoric tris(dimethylamide)
Phosphoryl hexamethyltriamide
Tri(dimethylamino)phosphineoxide
Tris-(dimethylamid) kyseliny

fosforecne (Czech)
Tris(dimethylamino)phosphine oxide
Tris(dimethylamino)phosphorus oxide

681-57-2
$C_7H_{12}O_4$
160.17
Pentanedioic acid, 2,2-dimethyl- (9CI)
Glutaric acid, 2,2-dimethyl- (8CI)
NSC-61979

681-84-5
$C_4H_{12}O_4Si$
152.25
COSi(OC)(OC)OC
Silicic acid, tetramethyl ester
Methyl orthosilicate [UN 2606]
Methyl silicate (ACGIH,OSHA)
Silicic acid, methyl ester of ortho-
Tetramethoxysilane
Tetramethyl silicate
Tetramethylsilikat (Czech)
TL 190
UN 2606 [Methyl orthosilicate]

682-01-9
$C_{12}H_{28}O_4Si$
264.44
CCCOSi(OCCC)(OCCC)OCCC
Silicic acid (H4SiO4), tetrapropyl ester (9CI)
Tetrapropoxysilane
Tetrapropyl silicate

682-09-7
$C_{12}H_{22}O_3$
214.34
O(CC(CC)(CO)COCC=C)CC=C
1,3-Propanediol, 2-ethyl-2-(hydroxymethyl)-, diallyl ether
Trimethylolpropane diallyl ether

683-08-9
$C_5H_{13}O_3P$
152.15
Phosphonic acid, methyl-, diethyl ester
Diethyl methylphosphonate

683-10-3
$C_{16}H_{33}NO_2$
271.44
O=C(O)CN(CCCCCCCCCCCC)(C)C
Lauryl betaine
1-Dodecanaminium, N-(carboxymethyl)-N,N-dimethyl-, hydroxide, inner salt (9CI)
Laurylbetain
Lauryldimethylbetaine
Lauryl dimethyl glycine
Unibetaine LB

683-18-1
$C_8H_{18}Cl_2Sn$
303.85
CCCC[Sn](Cl)(Cl)CCCC
Stannane, dibutyldichloro
Chlorid di-n-butylcinicity (Czech)
D.B.T.C.
Dibutyldichlorostannane
Dibutyldichlorotin
Dibutylstannium dichloride
Dibutyltin chloride
Dibutyltin dichloride
Di-n-butyltin dichloride
Di-n-butyl-zinn-dichlorid (German)
Dichlorodibutylstannane
Dichlorodibutyltin
Tin, dibutyl-, dichloride

683-51-2
C_3H_3ClO
90.51
Acrolein, 2-chloro
α-Chloroacrolein
2-Chloroacrolein
2-Chloropropenaldehyde
2-Propenal, 2-chloro- (9CI)

683-53-4
$C_2H_3BrCl_2$
177.86
Ethane, 2-bromo-1,1-dichloro- (8CI,9CI)

683-68-1
$C_2H_2Br_2Cl_2$
256.75
1,2-Dibromo-1,2-dichloroethane
1,2-DBDCE
Ethane, 1,2-dibromo-1,2-dichloro- (9CI)

683-70-5
C_4H_5ClO
104.54
3-Buten-2-one, 3-chloro- (8CI, 9CI)

684-16-2
C_3F_6O
166.03
O=C(C(F)(F)F)C(F)(F)F
2-Propanone, 1,1,1,3,3,3-hexafluoro
Acetone, hexafluoro-
6FK
Hexafluoroacetone (ACGIH, OSHA) [UN 2420]
NCI-C56440
UN 2420 [Hexafluoroacetone]

684-19-5
$C_8H_{19}OP$
162.21
Phosphine oxide, bis-(1,1-dimethylethyl)- (9CI)
Bis(tert-butyl)phosphine oxide
Di-tert-butylphosphine oxide
Phosphine oxide, di-tert-butyl- (7CI,8CI)

684-93-5
$C_2H_5N_3O_2$
103.10
O=C(N(N=O)C)N
Urea, N-methyl-N-nitroso
Carbamide, N-methyl-N-nitroso-
Methylnitroso-harnstoff (German)
N-Methyl-N-nitroso-harnstoff (German)
1-Methyl-1-nitrosomocovina (Czech)
Methylnitrosourea
N-Methyl-N-nitrosourea
1-Methyl-1-nitrosourea
Methylnitrosouree (French)
MNU
N-Nitroso-N-methylcarbamide
N-Nitroso-N-methyl-harnstoff (German)
Nitrosomethylurea
N-Nitroso-N-methylurea
1-Nitroso-1-methylurea
NMH
NMU
NSC-23909
RCRA waste number U177
SKI 24464
SRI 859
Urea, 1-methyl-1-nitroso-

684-94-6
$C_7H_{12}O$
112.17
3-Penten-2-one, 3,4-dimethyl- (8CI,9CI)

685-91-6
$C_6H_{13}NO$
115.20
O=C(N(CC)CC)C
Acetamide, N,N-diethyl
N,N-Diethylacetamide

686-31-7
$C_{13}H_{26}O_3$
230.39
O=C(OOC(CC)(C)C)C(CCCC)CC
Peroxyhexanoic acid, 2-ethyl-tert-pentyl ester
tert-Amyl peroxy-2-ethylhexanoate, Technically pure (DO
2-Ethylperoxyhexanoic acid tert-pentyl ester
UN 2898 (DOT)

686-68-0
$CH_4N_2O_3$
92.07
Urea, 1,3-dihydroxy

687-48-9
$C_5H_{11}NO_2$
117.17
Carbamic acid, dimethyl-, ethyl ester
Ethyl N,N-dimethyl carbamate
Di-N-methyl ethyl carbamate

688-37-9
$C_{18}H_{34}O_2.1/3Al$
291.46
Aluminum oleate
AI3-19803
Aluminum trioleate
9-Octadecenoic acid (Z)-, aluminum salt
9-Octadecenoic acid, (Z)-, aluminum salt (3:1)
Oleic acid aluminum salt

688-73-3
$C_{12}H_{28}Sn$
291.09
Stannane, tri-n-butyl-, hydride
Stannane, tributyl- (8CI,9CI)
Tin, tri-n-butyl-, hydride
Tributylstannane
Tributylstannic hydride
Tributyltin
Tributyltin hydride
Tri-n-butyltin hydride

688-74-4
$C_{12}H_{27}BO_3$
230.20
CCCCOB(OCCCC)OCCCC
Boric acid, tributyl ester
Borane, tributoxy-
Borester 2
Butyl borate
n-Butyl borate
Butyl borate, tri-
Tributoxyborane
Tri-n-butoxyborane
Tributyl borate
Tri-n-butyl borate
Tributylester kyseliny borite (Czech)

688-84-6
$C_{12}H_{22}O_2$
198.34
O=C(OCC(CCCC)CC)C(=C)C
Methacrylic acid, 2-ethylhexyl ester
2-Ethylhexyl methacryate
2-Ethyl-1-hexyl methacrylate

689-11-2
$C_5H_{12}N_2O$
116.19
O=C(NC(CC)C)N
Urea, sec-butyl
sec-Butylurea

689-12-3
$C_6H_{10}O_2$
114.14
O=C(OC(C)C)C=C
2-Propenoic acid, 1-methylethyl ester (9CI)
Acrylic acid, isopropyl ester (8CI)
AI3-15706
Isopropyl acrylate
Isopropyl 2-propenoate
1-Methylethyl 2-propenoate

689-31-6
$C_6H_8O_5$
160.13
3-Oxoadipic acid
β-Ketoadipate
3-Ketoadipic
β-Ketoadipic acid
β-Oxoadipic acid

689-89-4
$C_7H_{10}O_2$
126.16
O=C(OC)C=CC=CC
2,4-Hexadienoic acid, methyl ester, (E,E)-
AI3-30202
(E,E)-Methyl 2,4-hexadienoate
Methyl sorbate

689-97-4
C_4H_4
52.08
C(#C)C=C
1-Buten-3-yne
Vinylacetylene

690-08-4
C_7H_{14}
98.19
2-Pentene, 4,4-dimethyl-, (E)- (9CI)

691-37-2
C_6H_{12}
84.16
C(=C)CC(C)C
4-Methyl-1-pentene
1-Pentene, 4-methyl- (9CI)

691-38-3
C_6H_{12}
84.16
CC=CC(C)C
2-Pentene, 4-methyl-, (Z)- (9CI)

691-60-1
$C_4H_{10}N_2O$
102.16
O=C(NC(C)C)N
Urea, isopropyl
Isopropylurea
n-Isopropylurea
1-Isopropylurea
Urea, (1-methylethyl)- (9CI)

692-29-5
$C_4H_6O_3$
102.10
Succinaldehydic-acid
Butanoic acid, 4-oxo- (9CI)
Butryaldehydic acid
3-Formylpropanoic acid
β-Formylpropionic acid
3-Formylpropionic acid
γ-Oxybutyric acid
Succinic acid semialdehyde
Succinic semialdehyde

692-42-2
$C_4H_{11}As$
134.05
Diethylarsine
Arsine, diethyl-
RCRA waste No. P038

692-45-5
$C_3H_4O_2$
72.07
C=COC=O
Formic acid, vinyl ester
Vinylester kyseliny mravenci (Czech)
Vinyl formate

692-86-4
$C_{13}H_{24}O_2$
212.37
O=C(OCC)CCCCCCCCC=C
10-Undecenoic acid, ethyl ester
Ethyl 10-hendecenoate
Ethyl undecenoate
Ethyl 10-undecenoate
Ethyl undecylenate

693-02-7
C_6H_{10}
82.15
C(#C)CCCC
1-Hexyne (9CI)

693-07-2
C_4H_9ClS
124.64
S(CCCl)CC
Sulfide, chloroethyl ethyl
Chlordiethylsulfid (Czech)
2-Chloroethyl ethyl sulfide
1-Chloro-2-(ethylthio)ethane
2-Chloroethyl ethyl thioether
Ethane, 1-chloro-2-(ethylthio)-

Ethyl β-chloroethyl sulfide
Ethyl 2-chloroethyl sulfide
β-Ethylmerkaptoethylchlorid (Czech)
2-(Ethylthio)chloroethane
2-Ethylthioethyl chloride
Half-Mustard Gas
H-MG

693-13-0
$C_7H_{14}N_2$
126.23
N(=C=NC(C)C)C(C)C
Carbodiimide, diisopropyl
Diisopropylcarbodiimide
2-Propanamine, N,N'-methanetetraylbis- (9CI)

693-16-3
$C_8H_{19}N$
129.28
NC(CCCCCC)C
Heptylamine, 1-methyl
2-Aminooctane
Caprylamine
1-Methylheptylamine
2-Octanamine
2-Octylamine

693-21-0
$C_4H_8N_2O_7$
196.14
O=N(=O)OCCOCCON(=O)=O
Diethylene glycol dinitrate (DOT)
Bis(hydroxyaethyl)-aether-dinitrat (German)
Diethyleneglycol dinitrate, desensitized [UN 0075]
Diethylenglykoldinitrate (Czech)
Diglycoldinitraat (Dutch)
Diglycol (dinitrate de) (French)
Diglykoldinitrat (German)
Di(hydroxyethyl) ether dinitrate
Dinitrate de diethylene-glycol (French)
Dinitrodiglicol (Italian)
Dinitrodiglykol (Czech)
UN 0075 [Diethyleneglycol dinitrate, desensitized with not less than 25 percent nonvolatile water- insoluble phlegmatizer, by mass]

693-23-2
$C_{12}H_{22}O_4$
230.30
O=C(O)CCCCCCCCCCC(=O)O
Dodecanedioic acid (9CI)
Decamethylenedicarboxylic acid
1,10-Decanedicarboxylic acid
1,10-Dicarboxydecane
1,12-Dodecanedioic acid

693-30-1
C_4H_9ClOS
140.64
Ethanol, 2-((2-chloroethyl)-thio)
CH
β-Chloroethyl β-hydroxyethyl sulfide
2-Chloroethyl 2-hydroxyethyl sulfide
2-((2-Chloroethyl)thio)ethanol
Half Mustard Gas
Half Sulfur Mustard
2-Hydroxyethyl 2-chloroethyl sulfide
Mustard chlorohydrin
Sulfide, 2-chloroethyl 2-hydroxyethyl
Sulfur Half-Mustard

693-33-4
$C_{20}H_{41}NO_2$
327.55
O=C(O)CN(CCCCCCCCCCCCCCCC)(C)C
Cetyl Betaine
Ammonium, (carboxymethyl)-hexadecyldimethyl-, hydroxide, inner salt
N-(Carboxymethyl)-N,N-dimethyl-1-hexadecanaminium hydroxide, inner salt
1-Hexadecanaminium, N-(carboxymethyl)-N,N-dimethyl-, hydroxide, inner salt (9CI)
Palmityldimethylbetaine

693-36-7
$C_{42}H_{82}O_4S$
683.18
O=C(OCCCCCCCCCCCCCCCCCC)CCSCCC(=O)OCCCCCCCCCCCCCCCCCC
Distearyl thiodipropionate
Advastab PS 802
Advastab 802
Antiok S
Cyanox STDP
Dioctadecyl 3,3'-thiobis-propanoate
Dioctadecyl thiodipropionate
Dioctadecyl 3,3'-thiodipropionate
3,3'-Dioctadecyl thiodipropionate
Distearyl β-thiodipropionate
Distearyl β,β'-thiodipropionate
Distearyl 3,3'-thiodipropionate
DSTDP
Plastanox STDP
Plastanox STDP Antioxidant
Propanoic acid, 3,3'-thiobis-, dioctadecyl ester (9CI)
Propionic acid, 3,3'-thiodi-, dioctadecyl ester (8CI)
PS 802
3,3'-Thiobispropanoic acid, dioctadecyl ester
Thiodipropionic acid, distearyl ester3

693-54-9
$C_{10}H_{20}O$
156.30
O=C(CCCCCCCC)C
2-Decanone
Methyl octyl ketone
Methyl n-octyl ketone
Octyl methyl ketone

693-58-3
$C_9H_{19}Br$
207.15
BrCCCCCCCCC
n-Nonyl bromide
1-Bromononane
Nonane, 1-bromo- (9CI)

693-65-2
$C_{10}H_{22}O$
158.32
O(CCCCC)CCCCC
Pentyl-ether
Amyl ether
n-Amyl ether
Diamyl ether
Di-n-amyl ether
Dipentyl ether
Ether, di-n-pentyl-
Pentane, 1,1'-oxybis- (9CI)

693-67-4
$C_{11}H_{23}Br$
235.21
BrCCCCCCCCCCC
Undecane, 1-bromo- (9CI)
1-Bromoundecane

693-72-1
$C_{18}H_{34}O_2$
282.47
11-Octadecenoic acid, (E)-
AI3-36458

693-89-0
C_6H_{10}
82.15
Cyclopentene, 1-methyl- (9CI)

693-92-5
C_4H_5NS
99.15
Isothiazole, 3-methyl- (8CI,9CI)

693-95-8
C_4H_5NS
99.15
N(C(=CS1)C)=C1
4-Methylthiazole
Thiazole, 4-methyl- (9CI)

693-98-1
$C_4H_6N_2$
82.12
N(C=CN1)=C1C
Imidazole, 2-methyl

2-Methylimidazole

694-05-3
C_5H_9N
83.15
Pyridine, 1,2,3,6-tetrahydro
δ^3-Piperidine
1,2,3,6-Tetrahydropyridine [UN 2410]
1,2,5,6-Tetrahydropyridine
UN 2410 [1,2,3,6-Tetrahydropyridine]

694-59-7
C_5H_5NO
95.11
O=n(cccc1)c1
Pyridine-1-oxide
Pyridine N-oxide

694-80-4
C_6H_4BrCl
191.45
c(c(ccc1)Br)(c1)Cl
Benzene, 1-bromo-2-chloro- (9CI)
AI3-31290
1-Bromo-2-chlorobenzene

694-83-7
$C_6H_{14}N_2$
114.18
NC(C(N)CCC1)C1
1,2-Cyclohexanediamine (9CI)
1,2-Cylohexanediamine
1,2-Diaminocyclohexane

694-87-1
C_8H_8
104.15
Bicyclo(4.2.0)octa-1,3,5-triene (8CI,9CI)
Benzocyclobutene, 1,2-dihydro-

695-06-7
$C_6H_{10}O_2$
114.16
O=C(OC(C1)CC)C1
2(3H)-Furanone, 5-ethyldihydro
γ-Caprolactone
6-Caprolactone
γ-Ethylbutyrolactone
γ-Ethyl-n-butyrolactone
γ-Hexalactone
γ-Hexanolactone
Hexanolide-1,4
4-Hydroxyhexanoic acid lactone
Tonkalide
Toukalide

695-77-2
$C_5H_2Cl_4$
203.87
1,3-Cyclopentadiene, 1,2,3,4-tetrachloro
Tetrachlorocyclopentadiene
1,2,3,4-Tetrachlorocyclopentadiene

695-98-7
$C_8H_{11}N$
121.18
Pyridine, 2,3,5-trimethyl- (8CI,9CI)

696-07-1
$C_4H_3IN_2O_2$
237.99
n(cc(c(n1)O)I)c1O
Uracil, 5-iodo
5-Iodouracil

696-23-1
$C_4H_5N_3O_2$
127.12
O=N(=O)C(N=C(N1)C)=C1
Imidazole, 2-methyl-5-nitro
Imidazole, 2-methyl-4-nitro-
Menidazole
2-Methyl-4-nitroimidazole
2-Methyl-5-nitro-1H-imidazole (9CI)
RP 8532

696-28-6
$C_6H_5AsCl_2$
222.93
Cl[As](Cl)c1ccccc1
Arsine, dichlorophenyl
Arsonous dichloride, phenyl- (9CI)
Dichlor-fenylarsin (Czech)
Dichlorophenylarsine
FDA
Fenildicloroarsina (Italian)
Fenyldichlorarsin (Czech)
NA 1556 (DOT)
Phenylarsinedichloride
Phenyl dichlorarsine
Phenyl dichloroarsine (DOT)
Phenyldichloroarsine
RCRA waste number P036
TL 69

696-29-7
C_9H_{18}
126.24
C(CCCC1)(C1)C(C)C
Isopropylcyclohexane
Cyclohexane, isopropyl- (8CI)
Cyclohexane, (1-methylethyl)- (9CI)
Iso-propylcyclohexane
(1-Methylethyl)cyclohexane

696-54-8
$C_6H_6N_2O$
122.14
ON=Cc1ccncc1
Isonicotinaldehyde, oxime
4-(Hydroxyiminomethyl)-pyridine
Pyridine-4-aldoxime
4-Pyridinealdoxime
Pyridine, 4-formyl-, oxime

696-71-9
$C_8H_{16}O$
128.21
OC1CCCCCCC1
Cyclooctanol (9CI)

697-18-7
$C_2F_4O_3S$
180.08
O=S(=O)(OC1(F)F)C1(F)F
1,2-Oxathietane, 3,3,4,4-tetrafluoro-, 2,2-dioxide (9CI)

697-40-5
$C_9H_{15}Br$
203.13
Bicyclo(2.2.2)octane, 1-bromo-4-methyl- (7CI, 8CI,9CI)

697-82-5
$C_9H_{12}O$
136.19
Oc(c(c(cc1C)C)C)c1
Phenol, 2,3,5-trimethyl- (9CI)
2,3,5-Trimethylphenol

697-86-9
$C_6H_4BrCl_2N$
240.91
Benzenamine, 2-bromo-4,6-dichloro- (9CI)
Aniline, 2-bromo-4,6-dichloro- (8CI)

698-01-1
$C_8H_{10}ClN$
155.63
Aniline, o-chloro-N,N-dimethyl-
Benzenamine, 2-chloro-N,N-dimethyl-
o-Chloro-N,N-dimethylaniline
2-Chloro-N,N-dimethylaniline

698-63-5
$C_5H_3NO_4$
141.09
2-Furaldehyde, 5-nitro
5-Furancarboxaldehyde, 5-nitro- (9CI)
5-Nitro-2-furaldehyde
Nitrofurfural
5-Nitrofurfural
5-Nitrofurfuraldehyde

698-71-5
$C_9H_{12}O$

136.19
CCc1cc(C)cc(O)c1
m-Cresol, 5-ethyl- (8CI)
AI3-24210
Phenol, 3-ethyl-5-methyl- (9CI)

698-72-6
$C_5H_5Cl_2N_3$
178.02
1,3,5-Triazine, 2,4-dichloro-6-ethyl- (9CI)
s-Triazine, 2,4-dichloro-6-ethyl- (7CI,8CI)

698-76-0
$C_8H_{14}O_2$
142.20
O=C(OC(CC1)CCC)C1
2H-Pyran-2-one, tetrahydro-6-propyl- (9CI)
δ-Octalactone
Tetrahydro-6-propyl-2H-pyran-2-one

700-12-9
$C_{11}H_{16}$
148.25
c(c(c(c(c1C)C)C)C)(c1)C
Pentamethylbenzene
Benzene, pentamethyl- (9CI)

700-13-0
$C_9H_{12}O_2$
152.21
Oc(c(cc(O)c1C)C)c1C
Hydroquinone, trimethyl
1,4-Benzenediol, 2,3,5-trimethyl- (9CI)
psi-Cumohydroquinone
Pseudocumohydroquinone
Trimethylhydroquinone
2,3,5-Trimethylhydroquinone

700-17-4
$C_6H_3F_4N$
165.08
Benzenamine, 2,3,5,6-tetrafluoro- (9CI)
Aniline, 2,3,5,6-tetrafluoro- (8CI)
NSC-88347

700-19-6
$C_7H_8O_4$
156.14
1,2,4,5-Benzenetetrol, 3-methyl- (9CI)
Toluene-2,3,5,6-tetrol (8CI)

700-38-9
$C_7H_7NO_3$
153.13
Cc1ccc(N(=O)=O)c(O)c1
Phenol, 5-methyl-2-nitro- (9CI)
AI3-19031

701-64-4
$C_6H_7O_4P$
174.09
O=P(Oc(cccc1)c1)(O)O
Phosphoric acid, monophenyl ester (9CI)
Monophenyl phosphate

701-77-9
$C_7H_{10}ClN_3$
171.63
1,3,5-Triazine, 2-chloro-4,6-diethyl- (9CI)
s-Triazine, 2-chloro-4,6-diethyl- (7CI,8CI)

701-82-6
$C_7H_8N_2O_2$
152.14
O=C(Nc(cccc1O)c1)N
Urea, (m-hydroxyphenyl)- (8CI)
AI3-61347
(3-Hydroxyphenyl)urea
Urea, (3-hydroxyphenyl)- (9CI)

702-03-4
Unknown
Unknown
N-(2-Cyanoethyl)cyclohexylamine

702-54-5
$C_8H_{13}NO_3$
171.22
O=C(OCC(C1=O)(CC)CC)N1
2H-1,3-Oxazine-2,4(3H)-dione, 5,5-diethyldihydro
Dietadione (Italian)
Diethadion
Diethadione
5,5-Diethyldihydro-2H-1,3-oxazine-2,4(3H)-dione
5,5-Diethyl-1,3-oxazin-2,4-dione
5,5-Diethyl-1,3-oxazine-2,4-dione
5,5-Dietildiidro-1,3-ossazin-2,4-dione (Italian)
5,5-Diethyltetrahydro-2H-1,3-oxazine-2,4(3H)-dione
Dietroxine
Dihydro-5,5-diethyl-2H-1,3-oxazine-2,4(3H)-dione
Diidro-5,5-dietil-2H-1,3-ossazin-2,4(3H)-dione (Italian)
Dioxone
L 1811
Ledosten
Lepton
Persisten
TOCE
TOCEN

702-79-4
$C_{12}H_{20}$
164.29
Tricyclo(3.3.1.1(3,7))decane, 1,3-dimethyl- (9CI)
Adamantane, 1,3-dimethyl- (8CI)

703-59-3
$C_9H_6O_3$
162.14
Homophthalic acid anhydride
1H-2-Benzopyran-1,3(4H)-dione (9CI)

704-76-7
$C_{10}H_{16}O_2$
168.24
Cyclohexanone, 2-(ethoxymethylene)-4-methyl- (7CI,8CI,9CI)

705-25-9
$C_4H_4F_3N_5$
179.11
1,3,5-Triazine-2,4-diamine, 6-(trifluoromethyl)- (9CI)
s-Triazine, 2,4-diamino-6-(trifluoromethyl)- (6CI,7CI,8CI)

705-78-2
$C_7H_{11}N_3O_2$
169.18
1,3,5-Triazine, 2-ethyl-4,6-dimethoxy- (9CI)
s-Triazine, 2-ethyl-4,6-dimethoxy- (7CI,8CI)

705-86-2
$C_{10}H_{18}O_2$
170.28
O=C(OC(CC1)CCCCC)C1
2H-Pyran-2-one, tetrahydro-6-pentyl
AI3-36028
Amyl-δ-valerolactone
Decanolide-1,5
δ-Decalactone

706-14-9
$C_{10}H_{18}O_2$
170.28
O=C(OC(C1)CCCCCC)C1
2(3H)-Furanone, 5-hexyldihydro
γ-n-Decalactone
Decanolide-1,4
γ-n-Hexyl-γ-butyrolactone
Hydroxydecanoic acid γ-lactone

706-78-5
C_5Cl_8
343.68
Octachlorocyclopentene

AI3-27067
Cyclopentene, octachloro- (9CI)

708-53-2
$C_9H_{10}O_4$
182.18
Ethanone, 1-(2,3-dihydroxy-4-methoxyphenyl)- (9CI)
Acetophenone, 2',3'-dihydroxy-4'-methoxy- (6CI, 7CI,8CI)
2,3-Dihydroxy-4-methoxyacetophenone
Gallacetophenone 4-O-methyl ether

709-93-3
$C_9H_{11}NO_3$
181.19
1,2-Ethanediol, mono (phenylcarbamate) (9CI)
Ethylene glycol, carbanilate (7CI)
Ethylene glycol, monocarbanilate (8CI)
2-Hydroxyethyl N-phenyl carbamate

709-98-8
$C_9H_9Cl_2NO$
218.09
CCC(=O)Nc1ccc(Cl)c(Cl)c1
Propionanilide, 3',4'-dichloro
Bay 30130
Chem Rice
Crystal Propanil-4
DCPA
3,4-Dichloranilid kyseliny propionove (Czech)
N-(3,4-Dichlorophenyl)propanamide
N-(3,4-Dichlorophenyl)propionamide
3',4'-Dichlorophenylpropionanilide
Dichloropropionanilide
3,4-Dichloropropionanilide
3',4'-Dichloropropionanilide
DIPRAM
DPA
Farmco propanil
FW 734
Grascide
Herbax Technical
Montrose Propanil
Propanamide, N-(3,4-dichlorophenyl)-
Propanex
Propanid
Propanide
Propanil
Propionic acid 3,4-dichloroanilide
Prop-Job
Riselect
Rogue
Rosanil
S 10165
Stam
Stam F 34
Stam LV 10
Stam M-4
Stampede
Stampede 3E
Stam Supernox
Strel
Supernox
Surcopur
Surpur
Vertac

712-48-1
$C_{12}H_{10}AsCl$
264.59
Arsine, chlorodiphenyl
Arsinous chloride, diphenyl- (9CI)
Blue Cross
Chlor-difenylarsin (Czech)
Chlorodiphenylarsine
Clark I
DA
Difenylchlorarsin (Czech)
Diphenylchloorarsine (Dutch)
Diphenylchloroarsine, Solid or liquid [UN 1699]
Sneezing Gas
UN 1699 [Diphenylchloroarsine, solid or liquid]

712-68-5
$C_6H_4N_4O_3S$
212.20
Nc1nnc(s1)c2ccc(o2)N(=O)=O
1,3,4-Thiadiazole, 2-amino-5-(5-nitro-2-furyl)
2-Amino-5-(5-nitro-2-furyl)-1,3,4-thiadiazole
5-Amino-2-(5-nitro-2-furyl)-1,3,4-thiadiazole
ASA-140
Furidazina
Furidiazina
Furidiazine
NF-475
2-(5-Nitro-2-furyl)-5-amino-1,3,4-thiadiazole
5-(5-Nitro-2-furyl)-2-amino-1,3,4-thiadiazole
Ph/778
1,3,4-Thiadiazol-2-amine, 5-(5-nitro-2-furanyl)- (9CI)
Thiafur
Tiafur
Triafur

713-46-2
$C_{12}H_{18}O_2$
194.27
Ethanol, 2-(p-tert-butylphenoxy)- (8CI)
AI3-00033
Ethanol, 2-(4-(1,1-dimethylethyl)phenoxy)- (9CI)

713-68-8
$C_{12}H_{10}O_2$
186.22
O(c(cccc1O)c1)c(cccc2)c2
Phenol, m-phenoxy
m-Phenoxyphenol

713-95-1
$C_{12}H_{22}O_2$
198.34
O=C(OC(CC1)CCCCCCC)C1
2H-Pyran-2-one, 6-heptyltetrahydro
δ-Dodecalactone
n-Heptyl-δ-valerolactone
5-Hydroxydodecanoic acid lactone
5-Hydroxydodecanoic acid δ-lactone

715-99-1
$C_{11}H_{11}N_3O$
201.25
Cc1nn(c(C)c1N=O)c2ccccc2
Pyrazole, 3,5-dimethyl-4-nitroso-1-phenyl
3,5-Dimethyl-4-nitroso-1-phenylpyrazole
1-Phenyl-3,5-dimethyl-4-nitroso-pyrazol (German)

717-62-4
$C_5F_7N_3$
235.06
1,3,5-Triazine, 2-fluoro-4,6-bis(trifluoromethyl)- (9CI)
s-Triazine, 2-fluoro-4,6-bis (trifluoromethyl)- (6CI, 8CI)

717-74-8
$C_{15}H_{24}$
204.36
c(cc(cc1C(C)C)C(C)C)(c1)C(C)C
Benzene, 1,3,5-triisopropyl- (8CI)
AI3-51088
Benzene, 1,3,5-tris(1-methylethyl)- (9CI)
1,3,5-Triisopropylbenzene
1,3,5-Tris(1-methylethyl)benzene

717-90-8
$C_9H_6F_2N_4$
208.17
1,3,5-Triazin-2-amine, 4,6-difluoro-N-phenyl- (9CI)
2,4-Difluoro-6-(phenylamino)-s-triazine
s-Triazine, 2-anilino-4,6-difluoro- (6CI,8CI)

717-91-9
$C_9H_5F_2N_3S$

225.22
1,3,5-Triazine, 2,4-difluoro-6-(phenylthio)-(9CI)
s-Triazine, 2,4-difluoro-6-(phenylthio)- (6CI,8CI)

719-22-2
$C_{14}H_{20}O_2$
220.34
O=C(C(=CC(=O)C=1)C(C)(C)C)C1C(C)(C)C
p-Benzoquinone, 2,6-di-tert-butyl
DBQ
2,6-Di-tert-butyl-p-benzoquinone

719-32-4
$C_8Cl_6O_2$
340.80
O=C(c(c(c(c(c1Cl)C(=O)Cl)Cl)Cl)c1Cl)Cl
2,3,5,6-Tetrachloroterephthaloyl chloride
1,4-Benzenedicarbonyl dichloride, 2,3,5,6-tetrachloro-(9CI)
Perchloroterephthaloyl chloride
Terephthaloyl chloride, tetrachloro-
Tetrachloroterephthalic dichloride
Tetrachloroterephthaloyl chloride
Tetrachloroterephthaloyl dichloride
2,3,5,6-Tetrachloroterephthaloyl dichloride

719-59-5
$C_{13}H_{10}ClNO$
231.69
O=C(c(cccc1)c1)c(c(N)ccc2Cl)c2
Methanone, (2-amino-5-chlorophenyl)phenyl
2-Amino-5-chlorobenzophenone
(2-Amino-5-chlorophenyl)-phenylmethanone

721-50-6
$C_{13}H_{20}N_2O$
220.35
CCCNC(C)C(=O)Nc1ccccc1C
o-Propionotoluidide, 2-(propylamino)
Astra 1512
Astra 1515
Citanest
L 67
o-Methyl-α-propylamino-propionanilide
Prilocaine
Propanamide, N-(2-methyl-phenyl)-2-(propylamino)-(9CI)
Propitocaine
α-n-Propyl-amino-2-methylpropionanilide

721-61-9
$C_8H_{12}F_3N_5$
235.18
1,3,5-Triazine-2,4-diamine, N,N'-diethyl-6-(trifluoro-methyl)- (9CI)
s-Triazine, 2,4-bis(ethylamino)-6-(trifluoromethyl)- (8CI)

723-46-6
$C_{10}H_{11}N_3O_3S$
253.30
O=S(=O)(N=C(NOC=1C)C1)c(ccc(N)c2)c2
Sulfanilamide, N'-(5-methyl-3-isoxazolyl)
A047
4-Amino-N-(5-methyl-3-isoxazolyl)benzenesulfonamide
3-(para-Aminophenyl-sulphonamido)-5-methylisoxazole
Azo-Gantanol
Bactrim
Benzenesulfonamide, 4-amino-N-(5-methyl-3-isoxazolyl)-
Co-Trimoxazole
Eusaprim
Fectrim
Gantanol
N'-(5-Methyl-3-isoxazole)sulfanilamide
N'-(5-Methyl-3-isoxazolyl)sulfanilamide
5-Methyl-3-sulfanilamidoisoxazole
5-Methyl-3-sulfanylamidoisoxazole
N'-(5-Methylisoxazol-3-yl)-sulphanilamide
N[1]-(5-Methyl-3-isoxazolyl)-sulphanilamide
MS 53
5-Methyl-3-sulphanil-amidoisoxazole
Metoxal
Radonil
Ro 4-2130
SIM
Sinomin
Septra
Septran
Septrin
Sulfamethalazole
Sulfamethoxazol
Sulfamethoxazole
Sulfamethylisoxazole
3-Sulfanilamido-5-methylisoxazole
Sulfisomezole
Sulphamethalazole
Sulphamethoxazol
Sulphamethoxazole
Sulpha-methoxizole
Sulphamethylisoxazole
3-Sulphanilamido-5-methylisoxazole
Sulphisomezole
TRIB
Trimetoprim-Sulfa

723-62-6
$C_{15}H_{10}O_2$
222.25
O=C(O)c(c(c(ccc1)cc2cccc3)c1)c23
9-Anthroic acid
Anthracene-9-carboxylic acid
9-Carboxyanthracene

730-40-5
$C_{12}H_{10}N_4O_2$
242.21
O=N(=O)c(ccc(N=Nc(ccc(N)c1)c1)c2)c2
Disperse Orange 3
Benzenamine, 4-((4-nitro phenyl)azo)- (9CI)
4-((4-Nitrophenyl)azo)benzenamine

731-27-1
$C_{10}H_{13}Cl_2FN_2O_2S_2$
347.27
CN(C)S(=O)(=O)N(SC(F)(Cl)Cl)c1ccc(C)cc1
Sulfamide, N-((dichlorofluoro-methyl)thio)-N',N'-dimethyl-N-(p-tolyl)
Bay 49854
Bay 5212
Bay 5712a
Bayer 49854
Bayer 5712a
Dichlofluanid-methyl
N'-Dichlorofluoromethylthio-N,N-dimethyl-N'-(4-tolyl)sulfamide
N,N-Dimethyl-N-(4-tolyl)-N-(dichlorofluor-methylthio)-sulfamide
N,N-Dimethyl-N'-(4-tolyl)-N'-(dichlorfluormethylthio)-sulfamid (German)
Euparen M
KUE 13183b
Methanesulfanenamide, 1,1-dichloro-N-((dimethylamino)-sulfonyl)-1-fluoro-N-(4-methylphenyl)-
Tolylfluanid
Tolyfluanide

731-92-0
$C_{12}H_8N_2O_5$
260.22
2-Biphenylol, 3,5-dinitro
2,4-Dinitro-6-phenylphenol
Phenol, 2,4-dinitro-6-phenyl-

732-11-6
$C_{11}H_{12}NO_4PS_2$
317.33
COP(=S)(OC)SCN2C(=O)c1ccccc1C2=O

Phosphorodithioic acid, O,O-dimethyl ester, S-ester with N-(mercaptomethyl)-phthalimide
APPA
Decemthion
Decemthion p-6
(O,O-Dimethyl-phthalimidio-methyl-dithiophosphate)
O,O-Dimethyl 5-(phthalimido-methyl)dithiophosphate
O,O-Dimethyl S-(N-phthalimi-domethyl) dithiophosphate
O,O-Dimethyl S-phthalimido-methyl phosphorodithioate
ENT 25,705
Fosmet
Ftalophos
Imidan
Kemolate
N-(Mercaptomethyl)phthalimide S-(O,O-dimethyl phosphorodi-thioate)
Percolate
Phosphorodithioic acid, S-((1,3-dihydro-1,3-dioxo-iso-indol-2-yl)methyl) O,O-di-methyl ester
Phosmet
Phthalimide, N-(mercapto-methyl)-, S-ester with O,O-di-methyl phosphorodithioate
Phthalimido O,O-dimethyl phosphorodithioate
Phthalimidomethyl O,O-di-methyl phosphorodithioate
Phthalophos
PMP
Prolate
R 1504
Smidan
Stauffer R 1504

732-26-3
$C_{18}H_{30}O$
262.44
Oc(c(cc(c1)C(C)(C)C)C(C)(C)C)c1C(C)(C)C
2,4,6-Tri-tert-butylphenol
Phenol, 2,4,6-tri-tert-butyl-(8CI)
Phenol, 2,4,6-tris(1,1-dimethyl-ethyl)- (9CI)
2,4,6-Tris(1,1-dimethylethyl)-phenol

735-69-3
$C_{16}H_{13}N_3S$
279.37
1,3,5-Triazine, 2-(methyl-thio)-4,6-diphenyl- (9CI)
s-Triazine, 2-(methylthio)-4,6-diphenyl- (7CI,8CI)

738-70-5
$C_{14}H_{18}N_4O_3$
290.36
COc2cc(Cc1cnc(N)nc1N)cc(OC)c2OC
Pyrimidine, 2,4-diamino-5-(3,4,5-trimethoxybenzyl)
BW 56-72
2,4-Diamino-5-(3,4,5-tri-methoxybenzyl)pyrimidine
Monoprim
NIH 204
NSC-106568
Proloprim
2,4-Pyrimidinediamine, 5-((3,4,5-trimethoxyphenyl)-methyl)-
Syraprim
Tiempe
Trimanyl
Trimethoprim
Trimethopriom
5-(3,4,5-Trimethoxybenzyl)-2,4-diaminopyrimidine
Trimopan
Trimpex
Wellcoprim

738-71-6
$C_{16}H_{13}N_3O_2$
279.30
1,3,5-Triazine, 2-methyl-4,6-diphenoxy- (9CI)
s-Triazine, 2-methyl-4,6-di-phenoxy- (7CI,8CI)

738-87-4
$C_{19}H_{34}O_2$
294.48
Sterculic acid
1-Cyclopropene-1-octanoic acid, 2-octyl-
Sterculinic acid

739-71-9
$C_{20}H_{26}N_2$
294.48
CC(CN(C)C)CN2c1ccccc1CCc3ccccc23
5H-Dibenz(b,f)azepine, 5-(3-(dimethylamino)-2-methylpropyl)-10,11-di-hydro
5H-Dibenz(b,f)azepine, 10,11-dihydro-5-(3-(di-methylamino)-2-methyl-propyl)-
10,11-Dihydro-5-(3-dimethyl-amino-2-methylpropyl)-5H-di-benz(b,f)azepine
5H-Dibenz(b,f)azepine-5-pro-panamine, 10,11-dihydro-N,N,β-trimethyl-
1-(3-Dimethylamino-2-methyl-propyl)-4,5-dihydro-2,3:6,7-di-benzazepine
5-(3-(Dimethylamino)-2-methyl-propyl)-10,11-dihydro-5H-di-benz(b,f)azepine
5-(γ-Dimethylamino-β-methyl-propyl)-10,11-dihydro-5H-di-benzo(b,f)azepine
FI 6120
IL 6001
β-Methylimipramine
2'-Metil-3'-dimetilamino-propil-5-iminodibenzile (Italian)
7162 RP
Sapilent
Surmontil
Surmontil 315695
Trimeprimina (Italian)
Trimeprimine
Trimeproprimine
Trimipramine

740-77-2
$C_{17}H_{15}N_3O_2$
293.33
s-Triazine, 2-ethyl-4,6-di-phenoxy- (7CI,8CI)

741-58-2
$C_{14}H_{24}NO_4PS_3$
397.54
CC(C)OP(=S)(OC(C)C)SCCNS(=O)(=O)c1ccccc1
Phosphorodithioic acid, O,O-diisopropyl ester, S-ester with N-(2-mercapto-ethyl) benzenesulfonamide
Bensulide
Benzenesulfonamide, N-(2-mer-captoethyl)-, S-ester with O,O-diisopropylphosphorodi-thioate
Benzulfide
Betamec
Betasan
O,O-Bis(1-methylethyl) S-(2-((phenylsulfonyl)amino)-ethyl)pheosphorodithioate
N-(β-O,O-Diisopropyldithio-phosphorylethyl)benzene sulfonamide
N-(2-(O,O-Diisopropyldithio-phosphoryl)ethyl)benzene-sulfonamide
S-(O,O-Diisopropyl phosphoro-dithioate) ester of N-(2-mer-captoethyl)benzenesulfon-amide
Disan
Exporsan
N-(2-Mercaptoethyl)benzene-sulfonamide S-(O,O-diiso-propylphosphorodithioate)
Phosphorodithioic acid, O,O-bis(1-methylethyl) S-(2-((phenylsulfonyl)amino)-ethyl) ester
Prefar
Pre-San
R-4461

744-45-6
$C_{20}H_{14}O_4$
318.33
O=C(Oc(cccc1)c1)c(cccc2C(=O)Oc(cccc3)c3)c2
Isophthalic acid, diphenyl ester (8CI)
1,3-Benzenedicarboxylic acid, diphenyl ester (9CI)

Diphenyl 1,3-benzenedicarboxylate

746-53-2
$C_{13}H_4N_4O_9$
360.17
O=C(c(c(c1c(N(=O)=O)cc(N(=O)=O)c2)c(N(=O)=O)cc3N(=O)=O)c3)c12
Fluoren-9-one, 2,4,5,7-tetranitro- (8CI)
9H-Fluoren-9-one, 2,4,5,7-tetranitro- (9CI)
2,4,5,7-Tetranitrofluoren-9-one
2,4,5,7-Tetranitro-9H-fluoren-9-one

747-45-5
$C_{20}H_{24}N_2O_2.H_2O_4S$
422.54
Quinidine, sulfate (1:1) (Salt)
Biquin durules
Chinidin duriles
Chinidin vufb
Cinchonan-9-ol, 6'-methoxy-, (9S)-, sulfate (1:1) (Salt) (9CI)
(9S)-6'-Methoxycinchonan-9-ol sulfate (1:1) (Salt)
Kinidin duretter
Kinidin durules
Kinilentin
Optochinidin
Quinidine bisulfate

747-90-0
$C_{27}H_{44}$
368.71
CC(C)CCCC(C)C3CCC4C2CC=C1C=CCCC1(C)C2CCC34C
Cholesta-3,5-diene
δ-[3,5]-Cholestadien (German)
$\delta^{3,5}$-Cholestadiene
δ-3,5-Cholestadiene
Cholesterilene

749-49-5
$C_{20}H_{24}N_2O_2.C_3H_6O_3$
414.49
Cinchonan-9-ol, 6'-methoxy-, (8α,9R)-, mono(2-hydroxypropanoate) (Salt) (9CI)
Quinine lactate
Quinine, monolactate (Salt) (8CI)

751-38-2
$C_{34}H_{24}$
432.56
Naphthalene, 1,2,3,4-tetraphenyl- (8CI,9CI)
NSC-63121

752-74-9
$C_{57}H_{45}N_3O_3Sn_3$
1176.08
Tin, (2,4,6-trioxo-s-triazine-1,3,5(2H,4H,6H)-triyl)tris(triphenyl- (7CI,8CI)
s-Triazine-2,4,6(1H,3H,5H)-trione, tris(triphenylstannyl)-

753-73-1
$C_2H_6Cl_2Sn$
219.67
C[Sn](C)(Cl)Cl
Stannane, dichlorodimethyl
Dichlorid dimethylcinicity (Czech)
Dichlorodimethylstannane
Dichlorodimethyltin
Dimethyldichlorostannane
Dimethyldichlorotin
Dimethyltin dichloride
Tin, dimethyl-, dichloride

753-89-9
$C_5H_{11}Cl$
106.60
C(CCl)(C)(C)C
Propane, 1-chloro-2,2-dimethyl- (9CI)
1-Chloro-2,2-dimethylpropane
Neopentyl chloride

754-10-9
$C_5H_{11}NO$
101.14
O=C(N)C(C)(C)C
Pivalamide (8CI)
2,2-Dimethylpropanamide
Propanamide, 2,2-dimethyl- (9CI)

755-73-7
$C_5H_6F_4O_3$
190.09
O=C(OC)C(F)(F)C(F)(F)OC
Propanoic acid, 2,2,3,3-tetrafluoro-3-methoxy-, methyl ester (9CI)
2,2,3,3-Tetrafluoro-3-methoxypropanoic acid, methyl ester

756-79-6
$C_3H_9O_3P$
124.09
O=P(OC)(OC)C
Phosphonic acid, methyl-, dimethyl ester
Dimethyl methylphosphonate
DMMP
Methanephosphonic acid dimethyl ester
Methyl phosphonic acid, dimethyl ester
NCI-C54762

756-80-9
$C_2H_7O_2PS_2$
158.18
O(P(OC)(S)=S)C
Phosphorodithioic acid, O,O-dimethyl ester
O,O-Dimethylphosphorodithioate
Kwas dwumetylo-dwutiofosforowy (Polish)
Kyselina O,O-dimethyldithiofosforcna (Czech)

757-58-4
$C_{12}H_{30}O_{13}P_4$
506.30
Tetraphosphoric acid, hexaethyl ester, 0% to 25%
Bladan
Bladan Base
Ethyl tetraphosphate
Ethyl tetraphosphate, hexa-
HET
HETP
Hexaethyltetrafosfat (Czech)
Hexaethyl tetraphosphate (DOT)
Hexaethyl tetraphosphate and compressed gas mixture [UN 1612]
Hexaethyl tetraphosphate, Liquid [UN 1611]
Hexaethyl tetraphosphate, Liquid, Containing not more than 25% hexaethyl tetraphosphate (DOT)
Hexaethyl tetraphosphate Mixture, Dry containing more than 2% hexaethyl tetraphosphate (DOT)
Hexaethyl tetraphosphate Mixture, Dry containing not more than 2% hexaethyl tetraphosphate (DOT)
Hexaethyl tetraphosphate Mixture, Liquid containing > 25% hexaethyl tetraphosphate (DOT)
HTP
NA 2783 (DOT)
RCRA waste number P062
Tetraphosphate hexaethylique (French)
Tetraphosphoric acid, hexaethyl ester
Tetraphosphoric acid, hexaethyl ester, Mixed with compressed gas
UN 1611 [Hexaethyl tetraphosphate liquid]
UN 1612 [Hexaethyl tetraphosphate and compressed gas mixtures]
UN 2783 (DOT)

758-17-8
$C_2H_6N_2O$
74.10
CN(N)C=O
Formic acid, methylhydrazide
N-Formyl-N-methylhydrazine

Hydrazine, 1-formyl-1-methyl-
1-Methyl-1-formylhydrazide
N-Methyl-N-formylhydrazine
MFH

758-87-2
$C_6H_{10}O$
98.15
4-Penten-2-one, 3-methyl- (8CI,9CI)
3-Methyl-4-penten-2-one

759-36-4
$C_{10}H_{18}O_4$
202.25
Propanedioic acid, (1-methylethyl)-, diethyl ester (9CI)
Malonic acid, isopropyl-, diethyl ester (8CI)
NSC-1007

759-73-9
$C_3H_7N_3O_2$
117.13
O=C(N(N=O)CC)N
Urea, 1-ethyl-1-nitroso
Aenh (German)
Aethylnitroso-harnstoff (German)
Carbamide, N-ethyl-N-nitroso-
ENU
1-Ethyl-1-nitrosomocovina (Czech)
Ethylnitrosourea
n-Ethylnitrosourea
n-Ethyl-N-nitroso-urea
1-Ethyl-1-nitrosourea
NEU
Nitrosoethylurea
NSC-45403
RCRA waste number U176

759-94-4
$C_9H_{19}NOS$
189.35
O=C(N(CCC)CCC)SCC
Carbamic acid, dipropylthio-, S-ethyl ester
S-Aethyl-N,N-dipropylthiol-carbamat (German)
Dipropylcarbamothioic acid S-ethyl ester
N,N-Dipropylthiocarbamic acid S-ethyl ester
EPTAM
EPTC
Eradicane
S-Ethyl-N,N-dipropylthio-carbamate
S-Ethyl N,N-di-n-propylthiocar-bamate
Ethyl di-n-propylthiolcarbamate
Ethyl-N,N-dipropylthiolcar-bamate
Ethyl N,N-di-n-propylthiolcar-bamate
FDA 1541
Genep EPTC
Knoxweed
R-1608
Stauffer R 1608
Torbin

760-20-3
C_6H_{12}
84.16
1-Pentene, 3-methyl- (9CI)

760-21-4
C_6H_{12}
84.16
C(=C)(CC)CC
2-Ethyl-1-butene
1-Butene, 2-ethyl- (8CI)
3-Methylenepentane
Pentane, 3-methylene- (9CI)

760-23-6
$C_4H_6Cl_2$
125.00
C(C(Cl)CCl)=C
1-Butene, 3,4-dichloro
3,4-Dichloro-1-butene

760-55-4
$C_{11}H_{20}O_4$
216.28
Propanedioic acid, octyl- (9CI)
Malonic acid, octyl- (8CI)

760-67-8
$C_8H_{15}ClO$
162.68
O=C(C(CCCC)CC)Cl
Hexanoyl chloride, 2-ethyl
2-Ethylcaproyl chloride
2-Ethylhexanoic acid chloride
2-Ethylhexanoyl chloride

760-78-1
$C_5H_{11}NO_2$
117.14
O=C(O)C(N)CCC
Norvaline
AI3-18299
2-Aminovaleric acid
Norvaline, DL- (8CI)
DL-Norvaline (9CI)

760-93-0
$C_8H_{10}O_3$
154.17
O=C(OC(=O)C(=C)C)C(=C)C
Methacrylic anhydride
Methacrylic acid anhydride
Methacryloyl anhydride
2-Methyl-2-propenoic acid anhydride
2-Propenoic acid, 2-methyl-, anhydride (9CI)

761-65-9
$C_9H_{19}NO$
157.29
O=CN(CCCC)CCCC
Formamide, N,N-di-n-butyl
DBF
Dibutylamid kyseliny mravenci (Czech)
N,N-Di-n-butylformamide

762-04-9
$C_4H_{11}O_3P$
138.12
O=P(OCC)OCC
Phosphorous acid, diethyl ester
Diethylfosfit (Czech)
Diethyl hydrogen phosphite
Diethyl phosphite

762-12-9
$C_{20}H_{38}O_4$
342.58
O=C(OOC(=O)CCCCCCCCC)CCCCCCCCC
Decanoyl-peroxide
Decanox
Decanoyl peroxide, Technically pure (DOT)
Didecanoyl peroxide
Didecanoyl peroxide, Technically pure (DOT)
Perkadox SE 10
Peroxide, bis(1-oxodecyl)
UN 2120 (DOT)

762-13-0
$C_{18}H_{34}O_4$
314.47
O=C(OOC(=O)CCCCCCCC)CCCCCCCC
Pelargonyl peroxide (Technically pure)
Di-n-nonanoyl peroxide, Technically pure
Dipelargonyl peroxide
Nonanoyl peroxide
Pelargonoyl peroxide
Pelargonyl peroxide
Pelargonyl peroxide, Technically pure
Peroxide, bis(1-oxononyl) (9CI)
UN 2130

762-16-3
$C_{16}H_{30}O_4$
286.46
O=C(OOC(=O)CCCCCCC)CCCCCCC
Peroxide, octanoyl
Caprolyl peroxide
Capryloyl peroxide (DOT)
Capryl peroxide
Caprylyl peroxide
Caprylyl peroxide, Solution (DOT)
Dicaprylyl peroxide
Dioctanoyl peroxide
Di-n-octanoyl peroxide, Technically pure (DOT)
NA 2129 (DOT)

n-Octanoyl peroxide (DOT)
n-Octanoyl peroxide, Technically pure (DOT)
Perkadox SE 8
Peroxide, bis(1-oxooctyl) (9CI)
UN 2129 (DOT)

762-42-5
$C_6H_6O_4$
142.12
O=C(OC)C#CC(=O)OC
2-Butynedioic acid, dimethyl ester
Acetylenedicarboxylic acid, dimethyl ester
Dimethyl acetylenedicarboxylic acid

762-46-9
$C_7H_{16}O_2$
132.20
Hydroperoxide, 1-methylhexyl (8CI,9CI)

762-62-9
C_7H_{14}
98.19
1-Pentene, 4,4-dimethyl- (9CI)

762-63-0
C_7H_{14}
98.19
2-Pentene, 4,4-dimethyl-, (Z)- (8CI,9CI)
NSC-74142

762-84-5
$C_6H_{13}NO$
115.18
Acetamide, N-(1,1-dimethylethyl)- (9CI)
Acetamide, N-tert-butyl- (6CI,7CI,8CI)
2-Acetamido-2-methylpropane
N-tert-Butylacetamide
N-(1,1-Dimethylethyl)-acetamide

763-29-1
C_6H_{12}
84.18
C(=C)(CCC)C
1-Pentene, 2-methyl
2-Methylpentene
2-Methyl-pentene-1

763-30-4
C_6H_{10}
82.16
1,4-Pentadiene, 2-methyl
2-Methyl-1,4-pentadiene

763-32-6
$C_5H_{10}O$
86.13
OCCC(=C)C
Isobutenylcarbinol
3-Buten-1-ol, 3-methyl- (9CI)
3-Methyl-3-buten-1-ol

763-69-9
$C_7H_{14}O_3$
146.21
O=C(OCC)CCOCC
Propionic acid, 3-ethoxy-, ethyl ester
Ethoxypropionic acid, ethyl ester
Ethylester kyseliny 3-ethoxypropionove (Czech)
Ethyl β-ethoxypropionate

763-93-9
$C_6H_{10}O$
98.14
3-Hexen-2-one (8CI,9CI)

764-01-2
C_4H_6O
70.09
CC#CCO
2-Butyn-1-ol
AI3-37252

764-13-6
C_8H_{14}
110.20
C(=CC=C(C)C)(C)C
2,4-Hexadiene, 2,5-dimethyl- (9CI)
AI3-26989
2,5-Dimethyl-2,4-hexadiene

764-35-2
Unknown
Unknown
Methyl propyl acetylene

764-39-6
C_5H_8O
84.13
O=CC=CCC
2-Pentenal

764-41-0
$C_4H_6Cl_2$
125.00
C(=CCCl)CCl
2-Butene, 1,4-dichloro
DCB
1,4-DCB
1,4-Dichlorobutene-2
1,4-Dichloro-2-butene
RCRA waste number U074

764-42-1
$C_4H_2N_2$
78.08
N#CC=CC#N
Fumaronitrile

764-71-6
$C_8H_{16}O_2.K$
183.34
Octanoic acid, potassium salt
Potassium caprylate
Potassium octanoate

764-85-2
$C_9H_{17}ClO$
176.69
O=C(CCCCCCCC)Cl
Nonanoyl chloride (9CI)

764-93-2
$C_{10}H_{18}$
138.25
C(#C)CCCCCCCC
1-Decyne (9CI)

765-05-9
$C_{12}H_{24}O$
184.32
O(CCCCCCCCCC)C=C
Decane, 1-(ethenyloxy)- (9CI)
Decyl vinyl ether
1-(Ethenyloxy)decane

765-09-3
$C_{13}H_{27}Br$
263.31
BrCCCCCCCCCCCCC
Tridecane, 1-bromo
1-Bromotridecane

765-27-5
$C_{20}H_{38}$
278.52
1-Eicosyne (8CI,9CI)

765-30-0
C_3H_7N
57.09
NC(C1)C1
Cyclopropanamine (9CI)
Cyclopropylamine (8CI)

765-34-4
$C_3H_4O_2$
72.07
O=CC1CO1
Glycidaldehyde [UN 2622]
Epihydrinaldehyde
Epihydrine aldehyde
2,3-Epoxypropanal
2,3-Epoxy-1-propanal
2,3-Epoxypropionaldehyde
Formyloxiran
Glycidal
Glycidylaldehyde
Oxirane-carboxaldehyde
Propionaldehyde, 2,3-epoxy-
RCRA waste number U126

UN 2622 [Glycidaldehyde]

765-38-8
$C_5H_{11}N$
85.14
N(C(CC1)C)C1
Pyrrolidine, 2-methyl- (9CI)
2-Methylpyrrolidine

765-43-5
C_5H_8O
84.12
Ethanone, 1-cyclopropyl- (9CI)
AI3-08707
Ketone, cyclopropyl methyl (8CI)

765-69-5
$C_6H_8O_2$
112.13
1,3-Cyclopentanedione, 2-methyl- (8CI,9CI)
NSC-54458

765-87-7
$C_6H_8O_2$
112.14
O=C(C(=O)CCC1)C1
1,2-Cyclohexanedione
1,2-Dioxocyclohexane

766-09-6
$C_7H_{15}N$
113.23
N(CCCC1)(C1)CC
Piperidine, 1-ethyl
N-Aethylpiperidin (German)
1-Ethyl piperidine [UN 2386]
1-Ethylpiperidine
UN 2386 [1-Ethylpiperidine]

766-15-4
$C_6H_{12}O_2$
116.18
CC1(C)CCOCO1
m-Dioxane, 4,4-dimethyl
4,4-Dimethyl-m-dioxane
4,4-Dimethyldioxane-1,3
4,4-Dimethyl-1,3-dioxane

766-39-2
$C_6H_6O_3$
126.12
O=C(OC(=O)C=1C)C1C
Maleic anhydride, dimethyl
α,β-Dimethylmaleic anhydride

766-51-8
C_7H_7ClO
142.58
O(c(c(ccc1)Cl)c1)C
Anisole, o-chloro- (8CI)
Benzene, 1-chloro-2-methoxy- (9CI)
1-Chloro-2-methoxybenzene

766-76-7
$C_7H_5O_2$
121.12
O=C(O)c(cccc1)c1
Benzoic acid, ion(1-) (8CI,9CI)

766-77-8
$C_8H_{12}Si$
136.27
CSi(C)c1ccccc1
Silane, dimethylphenyl- (9CI)
Dimethylphenylsilane
Phenyldimethylsilane

766-92-7
$C_8H_{10}S$
138.23
S(Cc(cccc1)c1)C
Benzene, ((methylthio)-methyl)- (9CI)
((Methylthio)methyl)benzene
Sulfide, benzyl methyl (8CI)

767-00-0
C_7H_5NO
119.13
N#Cc(ccc(O)c1)c1
Benzonitrile, p-hydroxy
4-Hydroxybenzonitrile

767-15-7
$C_6H_9N_3$
123.14
Cc1cc(C)nc(N)n1
Pyrimidine, 2-amino-4,6-dimethyl- (8CI)
AI3-08092
2-Amino-4,6-dimethylpyrimidine
4,6-Dimethyl-2-pyrimidinamine
2-Pyrimidinamine, 4,6-dimethyl- (9CI)

767-17-9
$C_3H_5N_5S$
143.19
n(c(nc(n1)N)N)c1S
s-Triazine-2-thione, 4,6-diamino
4,6-Diamino-1,3,5-triazine 2-thione
s-Triazine-2-thiol, 4,6-diamino-
1,3,5-Triazine-2(1H)-thione, 4,6-diamino- (9CI)
s-Triazine-2(1H)-thione, tetrahydro-4,6-diimino-
USAF B-45
USAF CY-14

767-58-8
$C_{10}H_{12}$
132.21
Indan, 1-methyl- (8CI)
1H-Indene, 2,3-dihydro-1-methyl- (9CI)

767-59-9
$C_{10}H_{10}$
130.19
CC1C=Cc2ccccc12
1H-Indene, 1-methyl- (9CI)
Indene, 1-methyl- (8CI)
NSC-38857

767-60-2
$C_{10}H_{10}$
130.19
1H-Indene, 3-methyl- (9CI)
Indene, 3-methyl- (8CI)

767-81-7
$C_7H_8O_3$
140.14
6-Methyl-1,2,4-benzenetriol

767-98-6
$C_7H_6N_2O$
134.14
Mallorepine
3-Cyano-1-methyl-4-pyridone
3-Pyridinecarbonitrile, 1,4-dihydro-1-methyl-4-oxo-

768-33-2
$C_8H_{11}ClSi$
170.71
Silane, chlorodimethylphenyl- (9CI)
Chlorodimethylphenylsilane

768-49-0
$C_{10}H_{12}$
132.21
c(cccc1)(c1)C=C(C)C
Benzene, (2-methyl-1-propenyl)- (9CI)
Benzene, (2-methylpropenyl)- (8CI)
(2-Methylpropenyl)benzene
(2-Methyl-1-propenyl)benzene

768-50-3
$C_9H_{14}O$
138.21
O=C(CC(=CCCC1)C1)C
2-Propanone, 1-(1-cyclohexen-1-yl)- (9CI)
AI3-21374
1-(1-Cyclohexen-1-yl)-2-propanone

768-52-5
$C_9H_{13}N$
135.23
N(c(cccc1)c1)C(C)C
Aniline, n-isopropyl
n-Isopropylaniline (OSHA)

768-56-9
$C_{10}H_{12}$
132.22
1-Butene, 4-phenyl
Benzene, 3-butenyl-
4-Phenylbutene-1
4-Phenyl-1-butene

768-59-2
$C_9H_{12}O$
136.19
Benzenemethanol, 4-ethyl-
AI3-21535

768-94-5
$C_{10}H_{17}N$
151.28
NC(CC(CC1CC23)C2)(C1)C3
Tricyclo(3.3.1.1^{3,7})decan-1-amine
1-Adamantamine
1-Adamantanamine
Amantadine
1-Aminoadamantane
1-Aminoadamatane
1-Aminodiamantane
1-Aminotricyclo(3.3.1.1^{3,7})decane
EXP-105-1
PK-Merz
Symmetrel

769-11-9
$C_6H_5ClN_2O_2$
172.56
Benzenamine, 2-chloro-6-nitro- (9CI)
Aniline, 2-chloro-6-nitro- (8CI)
NSC-37398

769-25-5
$C_{11}H_{14}$
146.23
Benzene, 2-ethenyl-1,3,5-trimethyl- (9CI)
NSC-51012
Styrene, 2,4,6-trimethyl- (8CI)

770-12-7
$C_6H_5Cl_2O_2P$
210.98
O=P(Oc(cccc1)c1)(Cl)Cl
Phenyl dichlorophosphate
Phenyl phosphorodichloridate
Phosphorodichloridic acid, phenyl ester (9CI)

770-19-4
$C_7H_7FN_2O$
154.13
NC(=O)Nc1cccc(F)c1
Urea, (3-fluorophenyl)- (9CI)
Urea, (m-fluorophenyl)- (8CI)

770-35-4
$C_9H_{12}O_2$
152.19
O(c(cccc1)c1)CC(O)C
Propylene phenoxetol
AI3-14682
1-Phenoxy-2-propanol
2-Propanol, 1-phenoxy- (9CI)

771-29-9
$C_{10}H_{12}O_2$
164.20
Tetralin hydroperoxide, Technically pure
AI3-12032
Hidroperoxido de tetrahidronaftilo (Spanish)
Hydroperoxide, 1,2,3,4-tetrahydro-1-naphthalenyl
Hydroperoxide, 1,2,3,4-tetrahydro-1-naphthyl-
Hydroperoxyde de tetrahydronaphtyle (French)
Hydroperoxyde de tetraline (French)
Idroperossido di tetralina (Italian)
Tetrahydronaphthyl hydroperoxide
Tetralin hydroperoxide
Tetralin hydroperoxide, Technically pure
Tetralinehydroperoxyde (Dutch)
Tetralinhydroperoxid (German)
UN 2136

771-60-8
$C_6H_2F_5N$
183.09
Fc(c(N)c(F)c(F)c1F)c1F
Aniline, 2,3,4,5,6-pentafluoro
Aminopentafluorobenzene
Benzenamine, 2,3,4,5,6-pentafluoro- (9CI)
Pentafluoroaniline
2,3,4,5,6-Pentafluoroaniline
Pentafluorophenylamine

771-61-9
C_6HF_5O
184.07
Fc(c(F)c(O)c(F)c1F)c1F
Phenol, pentafluoro
Pentafluorophenol

771-62-0
C_6HF_5S
200.13
Benzenethiol, pentafluoro
Pentafluorobenzenethiol-
Pentafluorothiophenol

772-54-3
Unknown
Unknown
N-Benzyldiethylamine

772-65-6
$C_9H_{12}Cl_2Si$
219.19
Silane, dichloromethyl(2-phenylethyl)- (9CI)
Dichloromethyl(2-phenylethyl)-silane
Methyl(β-phenethyl)dichlorosilane

773-64-8
$C_9H_{11}ClO_2S$
218.71
Benzenesulfonyl chloride, 2,4,6-trimethyl
2-Mesitylenesulfonyl chloride
2,4,6-Trimethylbenzenesulfonyl chloride

773-99-9
$C_{12}H_{12}O$
172.23
1-Naphthaleneethanol (8CI, 9CI)
NSC-28342

774-40-3
$C_{10}H_{11}O_2$
163.21
CCOC(=O)C(O)c1ccccc1
Mandelic acid, ethyl ester
Ethyl mandelate
Mandelsaeureaethylester (German)

776-19-2
$C_4H_2Cl_4N_4O_2$
279.87
O=C(N(C(N(C(=O)N1Cl)Cl)Cl)C12)Cl)N2Cl
1,3,4,6-Tetrachloroglycoluril
AI3-60335
Caswell No. 829
EPA Pesticide Chemical Code 078401
Imidazo(4,5-d)imidazole-2,5(1H,3H)-dione, 1,3,4,6-tetrachlorotetrahydro- (9CI)

776-35-2
$C_{14}H_{12}$
180.25
Phenanthrene, 9,10-dihydro- (9CI)

776-74-9
$C_{13}H_{11}Br$
247.15
BrC(c(cccc1)c1)c(cccc2)c2
Methane, bromodiphenyl
Diphenylmethyl bromide [UN 1770]
Diphenyl methyl bromide, Solid (DOT)
Diphenyl methyl bromide, Solution (DOT)

UN 1770 [Diphenylmethyl bromide]

777-11-7
$C_9H_4Cl_3IO$
361.38
Clc1cc(Cl)c(OCC#CI)cc1Cl
Ether, 3-iodo-2-propynyl 2,4,5-trichlorophenyl
Haloprogin
Halotex
3-Iodo-2-propynyl 2,4,5-trichlorophenyl ether
M 1028
Mycanden
Mycilan
Polik
2,4,5-Trichlorophenyl iodopropargyl ether
2,4,5-Trichlorophenyl-γ-iodopropargyl ether

777-37-7
$C_7H_3ClF_3NO_2$
225.55
O=N(=O)c(ccc(c1C(F)(F)F)Cl)c1
2-Chloro-5-nitrobenzotrifluoride
AI3-28914
Benzene, 1-chloro-4-nitro-2-(trifluoromethyl)- (9CI)
1-Chloro-4-nitro-2-(trifluoromethyl)benzene
4-Nitro-2-(trifluoromethyl)-chlorobenzene
Toluene, 2-chloro-α,α,α-trifluoro-5-nitro- (8CI)

777-56-0
$C_8H_{14}FN_5$
199.23
s-Triazine, 2-(ethylamino)-4-fluoro-6-(isopropyl amino)- (7CI,8CI)

777-95-7
$C_{10}H_{16}O_4$
200.23
1,6-Dioxacyclododecane-7,12-dione (9CI)
Adipic acid, cyclic tetramethylene ester

778-22-3
$C_{15}H_{16}$
196.29
Benzene, 1,1'-(1-methylethylidene)bis- (9CI)
Propane, 2,2-diphenyl- (8CI)

778-25-6
$C_{13}H_{14}OSi$
214.36
Silanol, diphenylmethyl
Diphenylmethylsilanol

779-02-2
$C_{15}H_{12}$
192.27
c(c(c(c(c1ccc2)c2)C)ccc3)(c3)c1
Anthracene, 9-methyl
9-Methylanthracene

779-51-1
$C_{15}H_{14}$
194.28
Benzene, 1,1'-(1-methyl-1,2-ethenediyl)bis- (9CI)
Stilbene, α-methyl- (8CI)

780-69-8
$C_{12}H_{20}O_3Si$
240.41
Silane, phenyltriethoxy
Phenyltriethoxysilane
Silane, triethoxyphenyl-
Triethoxyfenylsilan (Czech)
Triethoxyphenylsilane

781-43-1
$C_{16}H_{14}$
206.30
Cc2c1ccccc1c(C)c3ccccc23
Anthracene, 9,10-dimethyl
9,10-Dimethylanthracene

781-74-8
$C_{14}H_{14}O_2$
214.27
1-Naphthalenebutanoic acid (9CI)
1-Naphthalenebutyric acid (6CI,7CI,8CI)
γ-(1-Naphthyl)butyric acid
4-(1-Naphthyl)butyric acid

782-74-1
$C_{12}H_{10}Cl_2N_2$
253.12
N(Nc(c(ccc1)Cl)c1)c(c(ccc2)Cl)c2
1,2-Bis(2-chlorophenyl)-hydrazine
1,2-Bis(2-chlorophenyl)hydrazine
2,2'-Dichlorohydrazobenzene
Hydrazine, 1,2-bis(2-chlorophenyl)- (9CI)
Hydrazobenzene, 2,2'-dichloro-

785-30-8
$C_{13}H_{13}N_3O$
227.25
O=C(Nc(ccc(N)c1)c1)c(ccc(N)c2)c2
4,4'-Diaminobenzanilide
4-Amino-N-(4-aminophenyl)-benzamide
Benzamide, 4-amino-N-(4-aminophenyl)- (9CI)
Benzanilide, 4,4'-diamino- (8CI)

786-19-6
$C_{11}H_{16}ClO_2PS_3$
342.87
CCOP(=S)(OCC)SCSc1ccc(Cl)cc1
Phosphorodithioic acid, S-(((p-chlorophenyl)thio)-methyl) O,O-diethyl ester
Acarithion
Akarithion
Carbofenothion (Dutch)
Carbophenothion
S-((p-Chlorophenylthio)methyl) O,O-diethyl phosphorodithioate
S-(4-Chlorophenylthiomethyl)-diethyl phosphorothiolothionate
Dagadip
O,O-Diaethyl-S-((4-chlor-phenyl-thio)-methyl)dithiophosphat (German)
O,O-Diethyl-S-((4-chloor-fenyl-thio)-methyl)-dithiofosfaat (Dutch)
O,O-Diethyl-S-p-chlorfenylthiomethylester kyseliny dithiofosforecne (Czech)
O,O-Diethyl p-chlorophenyl-mercaptomethyl dithiophosphate
O,O-Diethyl S-p-chlorophenyl-thiomethyl dithiophosphate
O,O-Diethyl S-(4-chlorophenyl-thiomethyl) dithiophosphate
O,O-Diethyl S-(p-chlorophenyl-thiomethyl) phosphorodithioate
O,O-Diethyl dithiophosphoric acid p-chlorophenylthiomethyl ester
O,O-Dietil-S-((4-cloro-fenil-tio)-metile)-ditiofosfato (Italian)
Dithiophosphate de O,O-diethyle et de (4-chloro-phenyl) thiomethyle (French)
Endyl
ENT 23,708
Garrathion
Karbofenothion (Czech)
Lethox
Nephocarp
Oleoakarithion
OMS 244
R-1303
Stauffer R-1,303
Trithion
Trithion Miticide

788-17-0
$C_{16}H_{20}N_2$
240.34
N(c(ccc(Nc(cccc1)c1)c2)c2)C(CC)C
1,4-Benzenediamine, N-(1-methylpropyl)-N'-phenyl-

(9CI)
N-(1-Methylpropyl)-N'-phenyl-1,4-benzenediamine

789-02-6
$C_{14}H_9Cl_5$
354.48
Clc1ccc(cc1)C(c2ccccc2Cl)C(Cl)(Cl)Cl
Ethane, 2-(o-chlorophenyl)-2-(p-chlorophenyl)-1,1,1-trichloro
2,2-Bis(o,p-chlorophenyl)-1,1,1-trichloroethane
o,p'-DDT
1,1,1-Trichloro-2-(o-chlorophenyl)-2-(p-chlorophenyl)ethane

789-07-1
$C_{16}H_9NO_2$
247.26
Pyrene, 2-nitro
2-Nitropyrene

789-24-2
$C_{19}H_{14}$
242.32
9H-Fluorene, 9-phenyl- (9CI)
Fluorene, 9-phenyl- (8CI)
NSC-27929

789-61-7
$C_{10}H_{13}N_5O_3S$
283.34
Nc3nc(S)c2NCN(C1CC(O)C(CO)O1)c2n3
9H-Purine-6(1H)-thione, 2-amino-9-(2-deoxy-β-D-erythro-pentofuranosyl)
B-TGDR
β-2'-Deoxythioguanosine
Guanosine, 2'-deoxy-6-thio-(9CI)
NCI-C01581
TGDR
β-TGDR
Thioguanine deoxyriboside
β-Thioguanine deoxyriboside

790-60-3
$C_{18}H_{10}O$
242.28
Benz(a)oxireno(c)anthracene
Benz(a)anthracene 5,6-oxide
Benz(3,4)anthra(1,2-b)oxirene (9CI)
Benz(a)anthracene-5,6-epoxide

791-28-6
$C_{18}H_{15}OP$
278.29
O=P(c(cccc1)c1)(c(cccc2)c2)c(cccc3)c3
Phosphine oxide, triphenyl- (9CI)
Triphenylphosphine oxide

791-31-1
$C_{18}H_{16}OSi$
276.43
Silane, hydroxytriphenyl
Hydroxytriphenylsilane

792-74-5
$C_{16}H_{14}O_4$
270.28
O=C(OC)c(ccc(c(ccc(c1)C(=O)OC)c1)c2)c2
Dimethyl 4,4'-biphenyldicarboxylate
4,4'-Biphenyldicarboxylic acid, dimethyl ester
(1,1'-Biphenyl)-4,4'-dicarboxylic acid, dimethyl ester (9CI)

793-24-8
$C_{18}H_{24}N_2$
268.39
N(c(ccc(Nc(cccc1)c1)c2)c2)C(CC(C)C)C
Santoflex 13
Antioxidant 4020
1,4-Benzenediamine, N-(1,3-dimethylbutyl)-N'-phenyl- (9CI)
Diafen 13
N-(1,3-Dimethylbutyl)-N'-phenyl-p-phenylenediamine
NCI-C56315
Permanax 120
N-Phenyl-N'-(1,3-dimethyl butyl)-para-phenylenediamine
p-Phenylenediamine, N-(1,3-dimethylbutyl)-N'-phenyl-
Vulkanox 4020
Wingstay 300

794-93-4
$C_{11}H_{11}N_5O_5$
293.27
OCN(CO)c2ncc(C=Cc1ccc(o1)N(=O)=O)nn2
Methanol, ((6-(2-(5-nitro-2-furyl)vinyl)-as-triazin-3-yl)imino)di
3-Bis(hydroxymethyl)amino-6-(5-nitro-2-furylethenyl)-1,2,4-triazine
Bis(hydroxymethyl)furatrizine
DHNT
3-Di(hydroxymethyl)amino-6-(2-(5-nitro-2-furyl)vinyl)-1,2,4-triazine
Dihydroxymethyl furatrizine
FT
Furatone
Furatone-S
Methanol, ((6-(2-(5-nitro-2-furanyl)ethenyl)-1,2,4-triazin-3-yl)imino)bis-
N-(6-(5-Nitrofurfurylidene-methyl)-1,2,4-triazin-3-yl)-iminodimethanol
6-(5-Nitro-2-furylvinyl)-3-(dihydroxydimethylamino)-1,2,4-triazene
((6-(2-(5-Nitro-2-furyl)vinyl)-as-triazin-3-yl)imino)dimethanol
N-(6-(2-(5-Nitro-2-furyl)vinyl)-1,2,4-triazin-3-yl)iminodimethanol
Panfuran-S
1,2,4-Triazine, 3-di(hydroxymethyl)amino-6-(5-nitro-2-furylethenyl)-

795-38-0
$C_{18}H_{23}NO_3$
301.42
Morphinan-6-β-ol, 4,5-α-epoxy-3-methoxy-17-methyl
DHIC
Dihydroisocodeine
Isocodeine, dihydro-

797-63-7
$C_{21}H_{28}O_2$
312.49
OC3(CCC4C2CCC1=CC(=O)CCC1C2CCC34)C#C
18,19-Dinor-17-α-pregn-4-en-20-yn-3-one, 13-ethyl-17-hydroxy-, (+)
17-α-Ethinyl-13-β-ethyl-17-β-hydroxy-4-estren-3-one
13-Ethyl-17-α-ethynylgon-4-en-17-β-ol-3-one
13-Ethyl-17-α-ethynyl-17-β-hydroxy-4-gonen-3-one
17-Ethynyl-18-methyl-19-nortestosterone
17-β-Hydroxy-18-methyl-19-nor-17-α-pregn-4-en-20-yn-3-one
18-Methyl-17-α-ethynyl-19-nortestosterone
d-Norgestrel
d(-)-Norgestrel
D-Norgestrel
Postinor

799-87-1
$C_{13}H_{11}N_2O_8P$
354.22
Bis(4-nitrophenyl)methyl phosphate
Bis(p-nitrophenyl)methyl phosphate
Phosphoric acid, methyl bis-(4-nitrophenyl) ester

800-24-8
$C_{16}H_{22}N_2O_6$
338.40
p-Benzoquinone, 2,5-bis-(1-aziridinyl)-3,6-bis-(2-methoxyethoxy)
A-139
Aziridyl benzoquinone
Bay A 139
Bayer A 139
Bayer E39 Soluble

Benzoquinone aziridine
2,5-Bis(1-aziridinyl)-3,6-bis-(2-methoxyethoxy)-2,5-cyclohexadiene-1,4-dione
2,5-Bis(1-aziridinyl)-3,6-bis-(2-methoxyethoxy)-p-benzoquinone
3,6-Bis(β-methoxyethoxy)-2,5-bis(ethylenimino)-p-benzoquinone
2,5-Bismethoxyethoxy-3,6-bis-ethyleneimino-1,4-benzoquinone
3,6-Bis(β-methoxyethoxy)-2,5-bis(ethyleneimino)-p-benzoquinone
E 39 Soluble
NSC-17262

804-36-4
$C_{14}H_{12}N_6O_6$
360.32
Guanidine, ((3-(5-nitro-2-furyl)-1-(2-(5-nitro-2-furyl)-vinyl)allylidene)amino)
1,5-Bis(5-nitro-2-furanyl)-1,4-pentadien-3-one, (amino-iminomethyl)hydrazone
Bis(5-nitrofurfurylidene)acetone guanylhydrazone
sym-Bis(5-nitro-2-furfurylidene) acetone guanylhydrazone
1,5-Bis(5-nitro-2-furyl)-3-pentadienone amidinonhydrazone
1,5-Bis(5-nitro-2-furyl)-3-pentadienone guanylhydrazone
Difuran
Difurazone
((3-(5-Nitro-2-furyl)-1-(2-(5-nitro-2-furyl)vinyl)allylidene)-amino)guanidine
Nitrovin
Panazon
Panazone
Payzone

804-63-7
$C_{20}H_{24}N_2O_2.O_4S$
420.52
Quinine, sulfate
Quinine bisulfate
Quinine hydrogen sulfate

807-28-3
$C_{26}H_{26}OSi_2$
410.66
Disiloxane, 1,3-dimethyl-1,1,3,3-tetraphenyl- (9CI)
1,1,3,3-Tetraphenyl-1,3-dimethyldisiloxane

811-54-1
$CH_2O_2.1/2Pb$
149.62
Lead formate
Formic acid, lead(2+) salt (8CI,9CI)
Lead(2+) formate

811-95-0
$C_2H_2Cl_3F$
151.39
Ethane, 1,1,2-trichloro-1-fluoro- (8CI,9CI)

811-97-2
$C_2H_2F_4$
102.03
Norflurane
Ethane, 1,1,1,2-tetrafluoro- (9CI)
Norflurano (Spanish)
Norfluranum (Latin)
1,1,1,2-Tetrafluoroethane

812-00-0
CH_5O_4P
112.03
O=P(OC)(O)O
Phosphoric acid, monomethyl ester
Dihydrogen methyl phosphate
Methyl phosphate
O-Methylphosphate
Monomethyl dihydrogen phosphate
Monomethyl phosphate

812-03-3
$C_3H_4Cl_4$
181.88
C(C(Cl)C)(Cl)(Cl)Cl
Propane, 1,1,1,2-tetrachloro- (9CI)
1,1,1,2-Tetrachloropropane

813-77-4
$C_2H_6ClO_3P$
144.49
Phosphorochloridic acid, dimethyl ester (8CI,9CI)

813-78-5
$C_2H_7O_4P$
126.06
O=P(OC)(OC)O
Phosphoric acid, dimethyl ester
Dimethyl hydrogen phosphate
O,O-Dimethyl hydrogen phosphate
Dimethyl phosphate

813-94-5
$C_6H_8O_7.3/2Ca$
252.25
Tricalcium citrate
Acicontral
Calcium citrate
Citric acid, calcium salt (2:3)
Citrical
2-Hydroxy-1,2,3-propanetricarboxylic acid calcium salt (2:3)
1,2,3-Propanetricarboxylic acid, 2-hydroxy-, calcium salt (2:3) (9CI)
Tribasic calcium citrate

814-29-9
$C_{12}H_{27}OP$
218.36
O=P(CCCC)(CCCC)CCCC
Phosphine oxide, tributyl
Tributylfosfinoxid (Czech)
Tributylphosphine oxide
Trisbutylphosphine oxide

814-49-3
$C_4H_{10}ClO_3P$
172.56
O=P(OCC)(OCC)Cl
Phosphorochloridic acid, diethyl ester
Chlorophosphoric acid, diethyl ester
Diethoxyphosphorus oxychloride
Diethylchlorfosfat (Czech)
Diethyl chlorophosphate

814-68-6
C_3H_3ClO
90.51
O=C(C=C)Cl
Acryloyl-chloride
Acrylic acid chloride
Acrylyl chloride
Chlorid kyseliny akrylove (Czech)
Propenoyl chloride
2-Propenoyl chloride (9CI)

814-71-1
$C_2H_4O_2S.1/2Ca$
112.16
Calcium thioglycolate
Acetic acid, mercapto-, calcium salt (2:1) (9CI)
Calcium (2:1)
Calcium mercaptoacetate
Calcium thioglycollate
DEPIL
Ebacream
Jully
Mercaptoacetic acid calcium derivative
Surgex
Vikor

814-75-5
C_4H_7BrO
151.00
O=C(C(Br)C)C
2-Butanone, 3-bromo- (9CI)
3-Bromo-2-butanone

814-78-8
C_5H_8O
84.13

O=C(C(=C)C)C
3-Buten-2-one, 3-methyl
Isopropenyl methyl ketone
Ketone, methyl isopropenyl
3-Methyl-3-buten-2-on (German)
2-Methyl-1-buten-3-one
3-Methyl-3-buten-2-one
Methyl isopropenyl ketone
Methyl isopropenyl ketone, Inhibited [UN 1246]
UN 1246 [Methyl isopropenyl ketone, inhibited]

814-80-2
$C_6H_{10}O_6.Ca$
218.24
Lactic acid, calcium salt (2:1)
Calcium lactate
Calphosan
Conclyte calcium
2-Hydroxypropanoic acid calcium salt
Propanoic acid, 2-hydroxy-, calcium salt

814-81-3
$C_3H_6O_3.1/2Cu$
121.85
Cupric lactate
Lactic acid, copper(2+) salt (2:1) (8CI)

814-90-4
$C_2H_2O_4.Cr$
142.03
Chromous oxalate
Ethanedioic acid, chromium(2+) salt (1:1)
Oxalic acid, chromium(2+) salt (1:1)

814-91-5
$C_2H_2O_4.Cu$
153.58
Copper(II) oxalate
Caswell No. 248A
Copper oxalate
Cupric oxalate
EPA Pesticide Chemical Code 023305
Ethanedioic acid copper salt
Ethanedioic acid, copper(2+) salt (1:1) (9CI)
Oxalic acid, copper(2+) salt (1:1) (8CI)

814-93-7
$C_2H_2O_4.Pb$
297.24
Lead oxalate
Ethanedioic acid, lead(2+) salt (1:1) (9CI)

814-94-8
$C_2H_2O_4.Sn$
208.73
Tin(II) oxalate
Ethanedioic acid, tin(2+) salt (1:1) (9CI)
Oxalic acid, tin(2+) salt (1:1) (8CI)
Stannous oxalate
Stavelan cinaty (Czech)
Tin(2+) oxalate
Tin oxalate

814-95-9
$C_2H_2O_4.Sr$
177.66
Ethanedioic acid, strontium salt (1:1) (9CI)
Oxalic acid, strontium salt (1:1) (8CI)

815-17-8
$C_6H_{10}O_3$
130.14
O=C(O)C(=O)C(C)(C)C
Terimethylpyruvic acid
AI3-11509
Butanoic acid, 3,3-dimethyl-2-oxo- (9CI)
tert-Butylglyoxylic acid
Butyric acid, 3,3-dimethyl-2-oxo-
3,3-Dimethyl-2-oxobutanoic acid
3,3-Dimethyl-2-oxobutyric acid

815-24-7
$C_9H_{18}O$
142.24
O=C(C(C)(C)C)C(C)(C)C
3-Pentanone, 2,2,4,4-tetramethyl- (9CI)
AI3-11096
2,2,4,4-Tetramethyl-3-pentanone

815-82-7
$C_4H_6O_6.Cu$
213.63
Cupric tartrate
Butanedioic acid, 2,3-dihydroxy- (R-(R*,R*))-, copper(2+) salt (1:1) (9CI)
2,3-Dihydroxybutanedioic acid copper salt
Tartaric acid, copper(2+) salt (1:1), (+)- (8CI)

815-84-9
$C_4H_4O_6.Pb$
355.27
Lead tartrate
Butanedioic acid, 2,3-dihydroxy- (R-(R*,R*))-, lead-(2+)
salt (1:1) (9CI)
Lead(II) tartrate (1:1)
Tartaric acid, lead(2+) salt (1:1) (8CI)

816-40-0
C_4H_7BrO
151.02
2-Butanone, 1-bromo
1-Bromo-2-butanone
TL 819

816-66-0
$C_6H_{10}O_3$
130.14
O=C(O)C(=O)CC(C)C
α-Ketoisocaproic acid
4-Methyl-2-oxopentanoic acid
Pentanoic acid, 4-methyl-2-oxo- (9CI)

816-79-5
C_7H_{14}
98.19
C(=CC)(CC)CC
2-Pentene, 3-ethyl- (9CI)
3-Ethyl-2-pentene

817-09-4
$C_6H_{12}Cl_3N.ClH$
241.00
Triethylamine, 2,2',2''-trichloro-, hydrochloride
HN3 HCl
HN3 Hydrochloride
Lekamin
NSC-30211
R-47
Sinalost
SK-100
Tri(β-chloroethyl)amine hydrochloride
Trichlormethine
Tri-(2-chloroethyl)amine hydrochloride
Trillekamin
2,2',2''-Trichlorotriethylamine hydrochloride
Trichlormethinium chloride
Trichlor-triaethylamin-hydrochlorid (German)
Trimitan
Trimustine
Trimustine hydrochloride
Tris(β-chloroethyl)amine hydrochloride
Tris(2-chloroethyl)amine hydrochloride
Tris(2-chloroethyl)amine monohydrochloride
Tris(2-chloroethyl)ammonium chloride
Tris-N-Lost (German)
TS-160

818-08-6
$C_8H_{18}OSn$
248.95
CCCC[Sn](=O)CCCC
Stannane, dibutyloxo
DBOT
Dibutyloxide of tin
Dibutyloxostannane

Dibutyloxotin
Dibutylstannane oxide
Dibutylstannium oxide
Dibutyltin oxide
Di-n-butyltin oxide
Di-n-butyl-zinn-oxyd (German)
Kyslicnik di-n-butylcinicity (Czech)
Tin, dibutyloxo-
Tin, dibutyl-, oxide

818-23-5
$C_{15}H_{30}O$
226.45
8-Pentadecanone
Caprylone
Diheptyl ketone
Heptyl ketone
8-Oxopentadecane
Pentadecan-8-one

818-38-2
$C_9H_{16}O_4$
188.22
O=C(OCC)CCCC(=O)OCC
Glutaric acid, diethyl ester (8CI)
AI3-06007
Diethyl glutarate
Diethyl pentanedioate
Pentanedioic acid, diethyl ester (9CI)

818-61-1
$C_5H_8O_3$
116.13
O=C(OCCO)C=C
Acrylic acid, 2-hydroxyethyl ester
2-(Acryloyloxy)ethanol
Bisomer 2HEA
Ethylene glycol, acrylate
Ethylene glycol, monoacrylate
Hydroxyethyl acrylate
β-Hydroxyethyl acrylate
2-Hydroxyethyl acrylate
2-Hydroxyethylester kyseliny akrylove (Czech)
2-Propenoic acid, 2-hydroxyethyl ester (9CI)

818-92-8
C_3H_5F
60.07
1-Propene, 3-fluoro- (9CI)
Propene, 3-fluoro- (8CI)

819-97-6
$C_8H_{16}O_2$
144.21
Butanoic acid, 1-methylpropyl ester (9CI)
Butyric acid, sec-butyl ester (8CI)

820-29-1
$C_{10}H_{20}O$
156.27
5-Decanone (8CI,9CI)
NSC-244937

820-69-9
$C_6H_8O_2$
112.13
3-Hexene-2,5-dione, (E)- (8CI, 9CI)

821-08-9
Unknown
Unknown
Divinyl acetylene

821-09-0
$C_5H_{10}O$
86.13
4-Penten-1-ol (8CI,9CI)
NSC-97503

821-10-3
$C_4H_4Cl_2$
122.98
C(#CCCl)CCl
2-Butyne, 1,4-dichloro
1,4-Dichlorobutyne
1,4-Dichloro-2-butyne

821-11-4
$C_4H_8O_2$
88.11
2-Butene-1,4-diol, (E)- (8CI, 9CI)

821-25-0
$C_7H_{14}Cl_2$
169.10
Heptane, 1,1-dichloro- (6CI,7CI,8CI,9CI)
1,1-Dichloroheptane

821-38-5
$C_{14}H_{26}O_4$
258.36
O=C(O)CCCCCCCCCCCCC(=O)O
Tetradecanedioic acid (9CI)

821-55-6
$C_9H_{18}O$
142.27
O=C(CCCCCCC)C
2-Nonanone
Ketone, heptyl methyl
Methyl heptyl ketone
Nonan-2-one

821-88-5
$C_9H_{18}Cl_2$
197.15
Nonane, 1,1-dichloro- (7CI,8CI,9CI)
1,1-Dichlorononane

821-91-0
$C_{15}H_{33}NO_2$
259.43
OCC(O)CNCCCCCCCCCCCC
1,2-Propanediol, 3-(dodecylamino)- (9CI)
N-(2,3-Dihydroxypropyl)dodecylamine
3-(Dodecylamino)-1,2-propanediol

821-95-4
$C_{11}H_{22}$
154.30
C(=C)CCCCCCCCC
1-Undecene (9CI)
1-Hendecene
α-Nonylethylene
α-Undecene
n-1-Undecene
α-Undecylene

821-99-8
$C_9H_{18}Cl_2$
197.15
ClCCCCCCCCCCl
Nonane, 1,9-dichloro- (9CI)
1,9-Dichlorononane

822-06-0
$C_8H_{12}N_2O_2$
168.22
O=C=NCCCCCCN=C=O
Hexane, 1,6-diisocyanato
Desmodur H
Desmodur N
1,6-Diisocyanatohexane
Hexamethylendiisokyanat (Czech)
Hexamethylene diisocyanate
Hexamethylenediisocyanate [UN 2281]
Hexamethylene-1,6-diisocyanate
1,6-Hexamethylene diisocyanate
1,6-Hexanediol diisocyanate
HMDI
Isocyanic acid, diester with 1,6-hexanediol
Isocyanic acid, hexamethylene ester
Metyleno-bis-fenyloizocyjanian (Polish)
Szesciometylenodwuizocyjanian (Polish)
TL 78
UN 2281 [Hexamethylenediisocyanate]

822-13-9
$C_{13}H_{27}Cl$
218.81
ClCCCCCCCCCCCCC
Tridecane, 1-chloro- (9CI)
1-Chlorotridecane
Tridecylchloride

822-16-2
$C_{18}H_{35}O_2$.Na
306.52
Stearic acid, sodium salt
Bonderlube 235
Flexichem B
Octadecanoic acid, sodium salt (9CI)
Prodhygine
Sodium octadecanoate
Sodium stearate (ACGIH)

822-50-4
C_7H_{14}
98.19
Cyclopentane, 1,2-dimethyl-, trans- (8CI,9CI)
NSC-74147

822-67-3
$C_6H_{10}O$
98.14
2-Cyclohexen-1-ol (9CI)

822-68-4
$C_6H_{13}P$
116.14
Phosphine, cyclohexyl- (8CI, 9CI)

822-80-0
C_7H_8O
108.14
Bicyclo(2.2.1)hepta-2,5-dien-7-ol (9CI)
7-Hydroxynorbornadiene

822-83-3
$C_6H_{12}O_2$
116.16
1,3-Dioxolane, 2-(1-methylethyl)- (9CI)
1,3-Dioxolane, 2-isopropyl- (8CI)
NSC-139442

822-86-6
$C_6H_{10}Cl_2$
153.05
Cyclohexane, 1,2-dichloro-, trans- (8CI,9CI)

823-22-3
$C_6H_{10}O_2$
114.14
2H-Pyran-2-one, tetrahydro-6-methyl- (8CI,9CI)
Hexanoic acid, 5-hydroxy-, lactone
Hexanoic acid, 5-hydroxy-, δ-lactone
NSC-32863
NSC-134774

823-40-5
$C_7H_{10}N_2$
122.19
Nc(c(c(N)cc1)C)c1
Toluene-2,6-diamine
1,3-Benzenediamine, 2-methyl-
2,6-Diaminotoluene
2,6-Toluylenediamine
2,6-Tolylenediamine

823-76-7
$C_8H_{14}O$
126.20
Ethanone, 1-cyclohexyl- (9CI)
Ketone, cyclohexyl methyl (8CI)

823-87-0
$C_6H_5NO_2$.Na
146.09
Phenol, 4-nitroso-, sodium salt
AI3-09061
4-Nitrosophenol sodium salt

823-94-9
$C_6H_9N_3$
123.14
1,3,5-Triazine, 2,4,6-trimethyl- (9CI)
s-Triazine, 2,4,6-trimethyl- (8CI)

823-95-0
$C_3H_4FN_5$
129.12
s-Triazine, 2,4-diamino-6-fluoro
2,4-Diamino-6-fluoro-s-triazine

824-11-3
$C_6H_{11}O_3P$
162.13
O(P(OCC1(CC)C2)O2)C1
Trimethylolpropane phosphite
4-Aethyl-1-phospha-2,6,7-trioxabicyclo(2.2.2)octan (German)
4-Ethyl-1-phospha-2,6,7-trioxabicyclo(2.2.2)octane
4-Ethyl-2,6,7-trioxa-1-phosphabicyclo(2.2.2)octane
2-(Hydroxymethyl)-2-ethyl-1,3-propanediol, cyclic phosphite (1:1)
1,3-Propanediol, 2-ethyl-2-(hydroxymethyl)-, cyclic phosphite (1:1) (8CI)
2,6,7-Trioxa-1-phosphabicyclo-(2.2.2)octane, 4-ethyl- (9CI)
1,1,1-Trishydroxymethylpropane bicyclic phosphite

824-22-6
$C_{10}H_{12}$
132.21
1H-Indene, 2,3-dihydro-4-methyl- (9CI)
Indan, 4-methyl- (8CI)

824-63-5
$C_{10}H_{12}$
132.21
1H-Indene, 2,3-dihydro-2-methyl- (9CI)
Indan, 2-methyl- (8CI)

824-69-1
$C_6H_4Cl_2O_2$
179.00
Oc(c(cc(O)c1Cl)Cl)c1
Hydroquinone, 2,5-dichloro- (8CI)
1,4-Benzenediol, 2,5-dichloro- (9CI)
2,5-Dichloro-1,4-benzenediol

824-72-6
$C_6H_5Cl_2OP$
194.98
O=P(c(cccc1)c1)(Cl)Cl
Benzene phosphorus dichloride
AI3-15064
Phenyl phosphorus dichloride [UN 2798]
Phenylphosphonic dichloride
Phosphonic dichloride, phenyl- (9CI)
UN 2798 [Phenyl phosphorus dichloride]

824-78-2
$C_6H_5NO_3$.Na
162.10
[Na+].[O-]c1ccc(cc1)N(=O)=O
p-Nitrophenol sodium salt
AI3-09021
para-Nitro sodium phenolate
4-Nitrophenol sodium salt
Phenol, p-nitro-, sodium salt
Phenol, 4-nitro-, sodium salt
Sodium nitrophenate
Sodium p-nitrophenol
Sodium p-nitrophenolate
Sodium 4-nitrophenolate
Sodium p-nitrophenoxide
Sodium 4-nitrophenoxide

824-79-3
$C_7H_7O_2S$.Na
178.19
p-Toluenesulfinic acid, sodium salt
Benzenesulfinic acid, 4-methyl-, sodium salt (9CI)
Sodium 4-methylbenzenesulfinate
Sodium p-toluenesulfinate
Sodium 4-toluenesulfinate
Sodium p-tolylsulfinate
p-Toluensulfinan sodny (Czech)

825-41-2
$C_6H_5ClN_2O_2$
172.56
Benzenamine, 3-chloro-4-nitro- (9CI)
Aniline, 3-chloro-4-nitro- (8CI)
NSC-39966

825-51-4
$C_{10}H_{18}O$
154.28
OC(CCC(C1CCC2)C2)C1
2-Naphthalenol, decahydro
Decahydronaphthol-2
Decahydro-β-naphthol
trans-Decahydro-β-naphthol
2-Decalinol
2-Decalol
2-Hydroxydecalin
Naphthalen-2-ol, decahydro-

825-52-5
$C_8H_6N_2O$
146.14
N#CC(=NO)c(cccc1)c1
Benzeneacetonitrile, α-(hydroxyimino)- (9CI)
α-(Hydroxyimino)benzeneacetonitrile
Oximinophenylacetonitrile

825-55-8
$C_{10}H_8S$
160.24
c1ccc(cc1)c2cccs2
Thiophene, 2-phenyl- (8CI,9CI)

825-86-5
$C_6H_6INO_2S$
283.09
Benzenesulfonamide, p-iodo
p-Iodobenzenesulfonamide

826-36-8
$C_9H_{17}NO$
155.27
O=C(CC(NC1(C)C)(C)C)C1
4-Piperidone, 2,2,6,6-tetramethyl
4-Oxo-2,2,6,6-tetramethylpiperidine
4-Piperidinone, 2,2,6,6-tetramethyl- (9CI)
2,2,6,6-Tetramethylpiperidinone
2,2,6,6-Tetramethylpiperidone
2,2,6,6-Tetramethyl-4-piperidone
Triacetonamin
Triacetonamine
Triacetone amine
Trojacetonoaminy (Polish)
Vincubina
Vincubine

826-55-1
$C_{10}H_{12}O_2$
164.20
Benzeneacetic acid, α,α-dimethyl- (9CI)
Hydratropic acid, α-methyl- (8CI)
NSC-28952
NSC-29095

826-81-3
$C_{10}H_9NO$
159.20
n(c(c(ccc1)cc2)c1O)c2C
8-Quinolinol, 2-methyl
8-Hydroxy-2-methylquinoline
Hydroxyquinaldine
8-Hydroxyqinaldine
2-Methyl-8-hydroxyquinoline
2-Methyloxine
2-Methyl-8-quinolinol

827-16-7
$C_6H_9N_3O_3$
171.14
s-Triazine-2,4,6(1H,3H,5H)-trione, 1,3,5-trimethyl- (8CI)
1,3,5-Triazine-2,4,6(1H,3H,5H)-trione, 1,3,5-trimethyl- (9CI)

827-19-0
$C_8H_{10}O_3S.Na$
209.22
Benzenesulfonic acid, 2,5-dimethyl-, sodium salt (9CI)

827-52-1
$C_{12}H_{16}$
160.28
c(cccc1)(c1)C(CCCC2)C2
Benzene, cyclohexyl
Phenylcyclohexane

827-54-3
$C_{12}H_{10}$
154.21
Naphthalene, 2-ethenyl- (9CI)
Naphthalene, 2-vinyl- (8CI)

827-94-1
$C_6H_4Br_2N_2O_2$
295.91
O=N(=O)c(cc(c(N)c1Br)Br)c1
2,6-Dibromo-4-nitroaniline
Aniline, 2,6-dibromo-4-nitro- (8CI)
Benzenamine, 2,6-dibromo-4-nitro- (9CI)
2,6-Dibromo-4-nitrobenzenamine

827-95-2
$C_7H_5NO_4.Na$
190.10
Benzoic acid, 3-nitro-, sodium salt (9CI)
Sodium 3-nitrobenzoate

828-00-2
$C_8H_{14}O_4$
174.22
O=C(OC(OC(OC1C)C)C1)C
Acetic acid, ester with 2,6-dimethyl-m-dioxan-4-ol
Acetic acid, 2,6-dimethyl-m-dioxan-4-yl ester
Acetomethoxan
Acetomethoxane
6-Acetoxy-2,4-dimethyl-m-dioxane
DDOA
Dimethoxane
2,6-Dimethyl-m-dioxan-4-ol acetate
2,6-Dimethyl-m-dioxan-4-yl acetate
m-Dioxan-4-ol, 2,6-dimethyl-, acetate
1,3-Dioxan-4-ol, 2,6-dimethyl-, acetate
Dioxin (Bactericide) (Obs.)
G1V Gard DXN
NCI-C56213

829-26-5
$C_{13}H_{14}$
170.25
Cc2ccc1cc(C)c(C)cc1c2
Naphthalene, 2,3,6-trimethyl- (9CI)
AI3-17611

829-84-5
$C_{12}H_{23}P$
198.29
P(C(CCCC1)C1)C(CCCC2)C2
Phosphine, dicyclohexyl- (9CI)
Dicyclohexylphosphine

829-99-2
$C_{13}H_{18}$
174.29
Benzene, 1-heptenyl- (9CI)
1-Heptene, 1-phenyl- (7CI, 8CI)

830-03-5
$C_8H_7NO_4$
181.16
O=C(Oc(ccc(N(=O)=O)c1)c1)C
Acetic acid, p-nitrophenyl ester
Acetic acid, 4-nitrophenyl ester (9CI)
p-Acetoxynitrobenzene
p-Nitrophenol acetate
p-Nitrophenyl acetate
4-Nitrophenyl acetate

830-09-1

$C_{10}H_{10}O_3$
178.19
O=C(O)C=Cc(ccc(OC)c1)c1
2-Propenoic acid, 3-(4-methoxyphenyl)- (9CI)
AI3-23399
3-(4-Methoxyphenyl)-2-propenoic acid

830-13-7
$C_{12}H_{22}O$
182.31
O=C(CCCCCCCCCC1)C1
Cyclododecanone (9CI)

830-81-9
$C_{12}H_{10}O_2$
186.21
O=C(Oc(c(c(ccc1)cc2)c1)c2)C
α-Naphthyl acetate
AI3-17246
1-Naphthalenol, acetate (9CI)
1-Naphthol, acetate (8CI)
a-Naphthyl acetate

830-96-6
$C_{11}H_{11}NO_2$
189.23
O=C(O)CCC(c(c(N1)ccc2)c2)=C1
1H-Indole-3-propionic acid
Indolepropionic acid
β-Indolepropionic acid
3-(3-Indolyl)propanoic acid

831-52-7
$C_6H_4N_3O_5$.Na
221.12
Picramic acid, sodium salt
Picramic acid, sodium salt, dry
Sodium picramate, Dry or containing less than 20% water [UN 0235]
Sodium picramate, Wet (with at least 20% water) [UN 1349]
UN 0235 [Sodium picramate, Dry or wetted with less than 20 per cent water, by mass]
UN 1349 [Sodium picramate, Wetted with not less than 20 per cent water, by mass]

831-61-8
$C_9H_{10}O_5$
198.19
O=C(OCC)c(cc(O)c(O)c1O)c1
Gallic acid, ethyl ester
Benzoic acid, 3,4,5-trihydroxy-, ethyl ester (9CI)
Ethylester kyseliny gallove (Czech)
Ethyl gallate
Ethyl 3,4,5-trihydroxybenzoate
Nipagallin A
Nipa No. 48
Phyllemblin
Progallin A

831-81-2
$C_{13}H_{11}Cl$
202.68
(4-Chlorophenyl)phenylmethane
AI3-22090
Benzene, 1-chloro-4-(phenylmethyl)- (9CI)
p-Chlorobenzylbenzene
p-Chlorodiphenylmethane
4-Chloroditan
(p-Chlorophenyl)phenylmethane
Methane, (p-chlorophenyl)-phenyl- (8CI)
Methane, (4-chlorophenyl)-phenyl-

831-82-3
$C_{12}H_{10}O_2$
186.21
O(c(ccc(O)c1)c1)c(cccc2)c2
Phenol, p-phenoxy- (8CI)
Phenol, 4-phenoxy- (9CI)
4-Phenoxyphenol

831-91-4
$C_{13}H_{12}S$
200.30
Benzene, ((phenylmethyl)thio)- (9CI)
Sulfide, benzyl phenyl (8CI)

832-64-4
$C_{15}H_{12}$
192.26
Phenanthrene, 4-methyl- (9CI)

832-69-9
$C_{15}H_{12}$
192.27
Phenanthrene, 1-methyl
1-Methylphenanthrene

832-71-3
$C_{15}H_{12}$
192.26
Phenanthrene, 3-methyl- (9CI)

833-43-2
$C_{10}H_{11}NO_4$
209.20
O=C(OCCc(c(N(=O)=O)ccc1)c1)C
Benzeneethanol, 2-nitro-, acetate (ester)
AI3-35599
o-Nitrophenethyl alcohol, acetate
2-Nitrobenzeneethanol acetate (ester)

833-66-9
$C_{10}H_8O_4$S.K
263.34
2-Naphthalenesulfonic acid, 6-hydroxy-, monopotassium salt (9CI)
Schaeffer's acid, monopotassium salt

833-81-8
$C_{15}H_{14}$
194.28
Benzene, 1,1'-(1-methyl-1,2-ethenediyl)bis-, (E)- (9CI)
trans-1,2-Diphenylpropene
trans-1,2-Diphenyl-1-propene
(E)-1,2-Diphenyl-1-propene
(E)-α-Methylstilbene
trans-α-Methylstilbene
Stilbene, α-methyl-, (E)- (8CI)

834-12-8
$C_9H_{17}N_5S$
227.37
CCNc1nc(NC(C)C)nc(SC)n1
s-Triazine, 2-ethylamino-4-isopropylamino-6-methylthio
A 1093
Ametrex
Ametryn
Ametryne
Crisatrine
2-Ethylamino-4-isopropylamino-6-methylmercapto-s-triazine
2-Ethylamino-4-isopropylamino-6-methylthio-s-triazine
2-Ethylamino-4-isopropylamino-6-methylthio-1,3,5-triazine
N-Ethyl-N'-isopropyl-6-methylthio-1,3,5-triazine-2,4-diyldiamine
Evik
G-34162
Gesapax
2-Methylmercapto-4-ethylamino-6-isopropylamino-s-triazine
2-Methylmercapto-4-isopropylamino-6-ethylamino-s-triazine
2-Methylthio-4-ethylamino-6-isopropylamino-s-triazine
Trinatox D

834-14-0
$C_{14}H_{14}O$
198.26
Benzene, 1-methoxy-4-(phenylmethyl)- (9CI)
Anisole, p-benzyl- (8CI)
NSC-2413

834-28-6
$C_{10}H_{15}N_5$.ClH
241.76
Cl.NC(=N)NC(=N)NCCc1ccccc1
Biguanide, 1-phenethyl-, monohydrochloride
Biguanide, 1-phenethyl-, hydro-

chloride
DBI-TD
DIPAR
1-Fenetilbiguanide cloridrato (Italian)
Meltrol
Phenethylbiguanide hydrochloride
N'-β-Phenethylbiguanide hydrochloride
1-Phenethylbiguanide hydrochloride
N[1]-β-Phenethylbiguanide hydrochloride
1-Phenylaethylbiguanid hydrochlorid (German)
Phenformin, hydrochloride
Phenformin HCl No. 9113
Phenoformine hydrochloride
N-β-Phenylethyl biguanide hydrochloride
USAF VI-6

835-31-4
$C_{14}H_{14}N_2$
210.30
C3CN=C(Cc1cccc2ccccc12)N3
2-Imidazoline, 2-(1-naphthylmethyl)
Antan
Imidin
Ciba 2020/r
Naphazoline
Naphthizine
2-(Naphthyl-(1')-methyl)imidazolin (German)
α-Naphthylmethyl imidazoline
2-(α-Naphthylmethyl)-imidazoline
2-(1-Naphthylmethyl)-2-imidazoline
Privine
Rhinazine
Sanorin

835-64-3
$C_{13}H_9NO_2$
211.23
O(c(c(N=1)ccc2)c2)C1c(c(O)ccc3)c3
Phenol, 2-(2-benzoxazolyl)
2-(o-Hydroxyphenyl)-benzoxazole
USAF EK-6754

836-24-8
$C_{10}H_{19}N_5S$
241.33
1,3,5-Triazine-2,4-diamine, N-ethyl-N'-(1-methylpropyl)-6-(methylthio)- (9CI)
GS 14253
s-Triazine, 2-(sec-butylamino)-4-(ethylamino)-6-(methylthio)- (8CI)

836-30-6
$C_{12}H_{10}N_2O_2$
214.24
O=N(=O)c(ccc(Nc(cccc1)c1)c2)c2
Diphenylamine, 4-nitro
Benzenamine, 4-nitro-N-phenyl-
4-Nitrodifenylamin (Czech)
p-Nitrodiphenylamine
4-Nitrodiphenylamine

838-85-7
$C_{12}H_{11}O_4P$
250.19
O=P(Oc(cccc1)c1)(Oc(cccc2)c2)O
Phosphoric acid, diphenyl ester (9CI)
Diphenyl phosphate

838-88-0
$C_{15}H_{18}N_2$
226.35
Cc2cc(Cc1ccc(N)c(C)c1)ccc2N
Aniline, 4,4'-methylenebis-(2-methyl
Benzenamine, 4,4'-methylenebis(2-methyl-
Bis-4-amino-3-methylfenylmethan (Czech)
Bis(4-amino-3-methylphenyl)-methane
3,3'-Dimethyl-4,4'-diaminodiphenylmethane
MBOT
ME-MDA
Methane, bis(4-amino-3-methylphenyl)-
4,4'-Methylenebis(2-methylaniline)
4,4'-Methylenebis(2-methylbenzenamine)
4,4'-Methylenebis(o-toluidine)
4,4'-Methylene di-o-toluidine
o-Toluidine, 4,4'-methylenedi-

839-90-7
$C_9H_{15}N_3O_6$
261.22
O=C(N(C(=O)N(C1=O)CCO)CCO)N1CCO
1,3,5-Tris(2-hydroxyethyl) isocyanurate
AI3-60291
Isocyanuric acid tris(2-hydroxyethyl) ester
THEIC
s-Triazine-2,4,6(1H,3H,5H)-trione, 1,3,5-tris(2-hydroxyethyl)-
1,3,5-Triazine-2,4,6(1H,3H,5H)-trione, 1,3,5-tris(2-hydroxyethyl)- (9CI)
Tris(hydroxyethyl) isocyanurate
Tris(β-hydroxyethyl) isocyanurate
Tris(2-hydroxyethyl) isocyanurate
N,N',N''-Tris(2-hydroxyethyl) isocyanurate
1,3,5-Tris(2-hydroxyethyl) isocyanuric acid
Tris(2-hydroxyethyl)-s-triazine-2,4,6-trione

840-65-3
$C_{14}H_{12}O_4$
244.25
O=C(OC)c(ccc(c1ccc2C(=O)OC)c2)c1
2,6-Naphthalenedicarboxylic acid, dimethyl ester (9CI)
Dimethyl 2,6-naphthalenedicarboxylate
2,6-Naphthalene dicarboxylic acid, dimethyl ester

841-06-5
$C_{11}H_{21}N_5OS$
271.43
COCCCNc1nc(NC(C)C)nc(SC)n1
s-Triazine, 2-(isopropylamino)-4-((3-methoxypropyl)amino)-6-(methylthio)
G 36393
Gesaran
2-Isopropylamino-4-(3-methoxypropylamino)-6-methylthio-1,3,5-triazin (German)
4-Isopropylamino-6-(3'-methoxypropylamino)-2-methythio 1,3,5-triazin (German)
2-Isopropylamino-4-(3-methoxypropylamino)-6-methylthio-s-triazine
Methoproptryne
Methoprotryn
Metoprotryn
Metoprotryne

842-07-9
$C_{16}H_{12}N_2O$
248.30
Oc(ccc(c1ccc2)c2)c1N=Nc(ccc3)c3
2-Naphthol, 1-(phenylazo)
Atul Orange R
Benzeneazo-β-naphthol
Benzene-1-azo-2-naphthol
1-Benzeneazo-2-naphthol
Benzene-1-azo-2-naphthol
1-Benzoazo-2-naphthol
Brasilazina Oil Orange
Brilliant Oil Orange R
Calcogas M
Calcogas Orange NC
Calco Oil Orange 7078
Calco Oil Orange 7078-Y
Calco Oil Orange Z-7078
Campbelline Oil Orange
Carminaph
Ceres Orange R
Cerotinorange G
C.I. 12055
C.I. Solvent Yellow 14
Dispersol Yellow PP
Dunkelgelb
Enial Orange I
Fast Oil Orange
Fast Oil Orange I

Fast Orange
Fat Orange 4A
Fat Orange G
Fat Orange I
Fat Orange R
Fat Orange RS
Fettorange 4A
Fettorange LG
Fettorange R
Grasal Orange
Grasan Orange R
Hidaco Oil Orange
Lacquer Orange VG
Motiorange R
NCI-C53929
Oil Orange
Oil Orange 31
Oil Orange 2311
Oil Orange 2B
Oil Orange E
Oil Orange EP
Oil Orange G
Oil Orange PEL
Oil Orange R
Oil Orange R-14
Oil Orange 7078-V
Oil Orange Z-7078
Oleal Orange R
Orange 2 Insoluble
Orange Insoluble OLG
Orange PEL
Orange A L'huile
Orange 3RA Soluble in Grease
Orange Resenole No. 3
Orange R Fat Soluble
Organol Orange
Orange Soluble A L'huile
Orient Oil Orange PS
Petrol Orange Y
1-(Phenylazo)-2-naphthalenol
1-Phenylazo-β-naphthol
1-Phenylazo-2-naphthol
Plastoresin Orange F4A
Pyronalorange
Resoform Orange G
Resinol Orange R
Sansel Orange G
Scharlach B
Silotras Orange TR
Solvent Yellow 14
Somalia Orange I
Soudan I
Spirit Orange
Spirit Yellow I
Stearix Orange
Sudan I
Sudan J
Sudan 1
Sudan Orange R
Sudan Orange RA
Sudan Orange RA New
Tertrogras Orange SV
Toyo Oil Orange
Waxakol Orange GL
Waxoline Yellow I
Waxoline Yellow IM
Waxoline Yellow IP
Waxoline Yellow IS
Zlut Rozpoustedlova 14 (Czech)

842-15-9
$C_{10}H_9NO_6S_2.K$
342.41
1,3-Naphthalenedisulfonic acid, 7-amino-, monopotassium salt (9CI)

842-18-2
$C_{10}H_8O_7S_2.2K$
382.50
1,3-Naphthalenedisulfonic acid, 7-hydroxy-, dipotassium salt (9CI)

845-52-3
$C_{12}H_{23}N_5O_2S$
301.46
COCCCNc1nc(NCCCOC)nc(SC)n1
s-Triazine, 2,4-bis((3-methoxypropyl)amino)-6-(methylthio)
2,4-Bis(3-methoxypropylamino)-6-methylthio-s-triazine
CP 17029
Lambast
2-Methylmercapto-4,6-bis-(3-methoxypropylamino)-s-triazine
MPMT

846-49-1
$C_{15}H_{10}Cl_2N_2O_2$
321.17
OC3N=C(c1ccccc1Cl)c2cc(Cl)ccc2NC3=O
2H,1,4-Benzodiazepin-2-one, 7-chloro-5-(o-chlorophenyl)-1,3-dihydro-3-hydroxy
Almazine
Ativan
7-Chloro-5-(o-chlorophenyl)-1,3-dihydro-3-hydroxy-2H-1,4-benzodiazepin-2-one
7-Chloro-5-(2-chlorophenyl)-1,3-dihydro-3-hydroxy-2H-1,4-benzodiazepin-2-one
7-Chloro-5-(2-chlorophenyl)-3-hydroxy-1H-1,4-benzodiazepin-2(3H)-one
Emotival
Lorax
Lorazepam
Lorsilan
Psicopax
Tavor
Temesta
WY 4036
WYPAX

846-50-4
$C_{16}H_{13}ClN_2O_2$
300.76
CN2C(=O)C(O)N=C(c1ccccc1)c3cc(Cl)ccc23
2H-1,4-Benzodiazepin-2-one, 7-chloro-1,3-dihydro-3-hydroxy-1-methyl-5-phenyl
2H-1,4-Benzodiazepin-2-one, 1,3-dihydro-7-chloro-3-hydroxy-1-methyl-5-phenyl-
Cerepax
7-Chloro-1,3-dihydro-3-hydroxy-1-methyl-5-phenyl-2H-1,4-benzodiazepin-2-one
Crisonar
1,3-Dihydro-7-chloro-3-hydroxy-1-methyl-5-phenyl-2H-1,4-benzodiazepin-2-one
ER 115
Euhypnos
Hydroxydiazepam
3-Hydroxydiazepam
K3917
Levanxene
Levanxol
Mabertin
Methyloxazepam
N-Methyloxazepam
Normison
Oxydiazepam
Planum
Remestan
Restoril
Ro 5-5345
Signopam
Temazepam
WY 2917
WY 3917

847-51-8
$C_{10}H_7F_7N_2O_2$
320.15
O=C(Nc(c(O)cc(N)c1)c1)C(F)(F)C(F)(F)C(F)(F)F
Butanamide, N-(4-amino-2-hydroxyphenyl)-2,2,3,3,4,4,4-heptafluoro- (9CI)
Butyranilide, 4'-amino-2,2,3,3,4,4,4-heptafluoro-2'-hydroxy-

848-75-9
$C_{16}H_{12}Cl_2N_2O_2$
335.18
Lormetazepam
2H-1,4-Benzodiazepin-2-one, 7-chloro-5-(2-chlorophenyl)-1,3-dihydro-3-hydroxy-1-methyl-
2H-1,4-Benzodiazepin-2-one, 1,3-dihydro-7-chloro-5-(o-chlorophenyl)-3-hydroxy-1-methyl-
7-Chloro-5-(o-chlorophenyl)-1,3-dihydro-3-hydroxy-1-methyl-2H-1,4-benzodiazepin-2-one
7-Chloro-5-(2-chlorophenyl)-3-hydroxy-1-methyl-2,3-dihydro-1H-1,4-benzodiazepin-2-one
DEA No. 2774
Lormetazepamum (Latin)
Methyllorazepam
N-Methyllorazepam
Ro 5-5516
WY-4082

849-99-0
$C_{18}H_{30}O_4$
310.48
O=C(OC(CCCC1)C1)CCCCC
(=O)OC(CCCC2)C2
Adipic acid, dicyclohexyl ester
Dicyclohexyl adipate
Ergoplast ADC
Hexanedioic acid, dicyclohexyl ester (9CI)

853-35-0
$C_{14}H_6O_8S_2.2Na$
412.30
1,5-Anthracenedisulfonic acid, 9,10-dihydro-9,10-dioxo-, disodium salt
Anthrachinon-1,5-disulfonan disodny (Czech)
Anthrachinon-1,5-disulfonan sodny (Czech)
Disodium anthraquinone-1,5-disulfonate
Sodium anthraquinone-1,5-disulfonate

853-67-8
$C_{14}H_8O_8S_2.2Na$
414.33
2,7-Anthracenedisulfonic acid, 9,10-dihydro-9,10-dioxo-, disodium salt (9CI)
2,7-Anthraquinonedisulfonic acid, disodium salt
Anthraquinone-2,7-disulfonic acid, disodium salt

860-22-0
$C_{16}H_8N_2O_8S_2.2Na$
466.36
[Na+].[Na+].[O-]S(=O)(=O)c4cc
c3NC(=C2Nc1ccc(cc1C2=O)
S([O-])(=O)=O)C(=O)c3c4
($\delta^{2,2'}$-Biindoline)-5,5'-disulfonic acid, 3,3'-dioxo-, disodium salt
Acid Blue W
Acid Leather Blue IC
A.F. Blue No. 2
Airedale Blue IN
Amacid Brilliant Blue
Aniline Carmine Powder
Atul Indigo Carmine
1311 Blue
12070 Blue
Bucacid Indigotine B
Canacert Indigo Carmine
Carmine Blue (Biological stain)
C.I. 73015
C.I. 75781
C.I. Acid Blue 74
C.I. Food Blue 1
C.I. Food Blue 1, Disodium salt
C.I. Natural Blue 2
Cilefa Blue R
Disodium indigo-5,5-disulfonate
Disodium salt of 1-indigotin-S,S'-disulphonic acid
Dolkwal Indigo Carmine
E 132
Edicol Supra Blue X
FD & C Blue No. 2
Food Blue 2
Grape Blue A Geigy
HD Indigo Carmine
HD Indigo Carmine Supra
Hexacert Blue No. 2
Hexacol Indigo Carmine Supra
Indigo Carmine
Indigo Carmine A
Indigo Carmine AC
Indigo Carmine (Biological stain)
Indigo Carmine BP
Indigo Carmine Conc. FQ
Indigo Carmine Disodium salt
Indigo Carmine Powder
Indigo Carmine X
Indigo Disulfonate (Biological stain)
Indigo Extract
Indigo-Karmin (German)
Indigotin I
5,5'-Indigotindisulfonic acid
Indigotin-5,5'-disulfonic acid disodium salt
Indigotine
Indigotine B
Indigotine Blue LZ
Indigotine Conc. Powder
Indigotine Disodium salt
Indigotine Extra Pure A
Indigotine I
Indigotine Lake
Indigotine N
Indocarmine F
Intense Blue
L-Blau 2 (German)
Maple Indigo Carmine
Mitsui Indigo Carmine
Modr Kysela 74 (Czech)
Modr Pigment 63 (Czech)
Modr Potravinarska 1 (Czech)
Murabba
Sachsischblau
San-EI Indigo Carmine
Schultz Nr. 1309 (German)
Sodium indigotindisulfonate
Sodium 5,5'-indigotidisulfonate
Soluble Indigo
Sumitomo Wool Blue SBC
USACERT Blue No. 2

865-21-4
$C_{46}H_{58}N_4O_9$
811.08
CCC9(O)CC8CC(c1cc2c(cc1OC)
N(C)C3C(O)(C(O)OC)C(OC
(C)=O)C4(CC)C=CCN5CC
C23C45)c7[nH]c6ccccc6c7C
CN(C8)C9
Vincaleukoblastine
29060-LE
NCI-C04842
NDC 0002-1452-01
Nincaluicolflastine
NSC-47842
Rozevin
Vinblastin
Vinblastine
Vincaleucoblastin
Vincaleucoblastine
Vincoblastine
VLB
VR-8

865-47-4
$C_4H_{10}O.K$
113.22
[K+].CC(C)(C)[O-]
2-Propanol, 2-methyl-, potassium salt (9CI)
2-Methyl-2-propanol potassium salt

865-86-1
$C_{12}H_5F_{21}O$
564.14
FC(F)(F)C(F)(F)C(F)(F)C(F)(F)
C(F)(F)C(F)(F)C(F)(F)C(F)(F)
C(F)(F)C(F)(F)CCO
1-Dodecanol, 3,3,4,4,5,5,6,6,7,7,8,8,9,9,10,10,11,11,12,12,12-heneicosafluoro-(9CI)
1,1,2,2-Tetrahydroperfluoro dodecanol

866-81-9
$C_6H_8O_7.3/2Co$
280.52
Cobaltous citrate
Citric acid, cobalt(2+) salt (2:3)
Cobalt citrate
1,2,3-Propanetricarboxylic acid, 2-hydroxy-, cobalt(2+) salt (2:3) (9CI)

866-82-0
$C_6H_4O_7.2Cu$
315.18
CC([O-])=O.[O-]C(=O)CC
(=[O-])C([O-])=O.[Cu][Cu]
Copper(I) citrate
Citric acid, copper(2+) salt (8CI)
Copper citrate
Cupric citrate
Cuprocitrol
2-Hydroxy-1,2,3-propanetricarboxylic acid, copper salt (1:2)
1,2,3-Propanetricarboxylic acid, 2-hydroxy-, copper(2+) salt (1:2) (9CI)

866-84-2
$C_6H_5O_7.3K$
306.41
Citric acid, tripotassium salt
Kajos
Kaliksir
Porekal
Potassium citrate
1,2,3-Propanetricarboxylic acid, 2-hydroxy-, tripotassium salt (9CI)
Seltz-K

Tripotassium citrate
Tripotassium citrate monohydrate

867-13-0
$C_8H_{17}O_5P$
224.22
O=P(OCC)(OCC)CC(=O)OCC
Acetic acid, diethylphos phono-, ethyl ester
Acetic acid, (diethoxyphosphinyl)-, ethyl ester
Acetic acid, phosphono-, triethyl ester
Ethyl (diethoxyphosphinyl)-acetate
Ethyl diethoxyphosphoryl acetate
Ethyl (diethylphosphono)acetate
TL 465
Triethyl phosphonoacetate

867-27-6
$C_6H_{15}O_3PS_2$
230.30
CCSCCOP(=S)(OC)OC
Phosphorothioic acid, O-(2-(ethylthio)ethyl) O,O-dimethyl ester
Bay 15203
Demeton-O-Methyl
Demeton-O-Metile (Italian)
O,O-Dimethyl-O-(2-aethylthio-aethyl)monothiophosphat (German)
O,O-Dimethyl 2-ethylmercapto-ethyl thiophosphate, thiono isomer
O,O-Dimethyl-O-(2-ethyl-thio-ethyl)-monothiofosfaat (Dutch)
O,O-Dimethyl O-ethylmercapto-ethyl thiophosphate
O,O-Dimethyl O-2-(ethylthio)-ethyl phosphorothioate
O,O-Dimetil-O-(2-etiltio-etil)-monotiofosfato (Italian)
ENT 18,862
Ethanol, 2-(ethylthio)-, O-ester with O,O-dimethyl phosphorothioate
β-Ethylmercaptoethyl dimethyl thionophosphate
O-(2-(Ethylthio)ethyl) O,O-dimethyl phosphorothioate
2-(Ethylthio)ethyl dimethyl phosphorothionate
Methylcistox
Methyl-Demeton-O
O-Methyldemeton
Methylmercaptophos
Methylsystox
Thiophosphate de O,O-dimethyle et de O-2-ethylthio-ethyle (French)

867-44-7
$C_4H_{12}N_4S_2.H_2O_4S$
278.40
Pseudourea, 2-methyl-2-thio-, sulfate (2:1)
Carbamimidothioic acid, methyl ester, sulfate (2:1) (9CI)
S-Methylisothiourea hemisulfate
S-Methylisothiourea sulfate (2:1)
S-Methylthiouronium sulfate (2:1)

867-81-2
$C_9H_{17}NO_5.Na$
242.22
Sodium pantothenate
β-Alanine, N-(2,4-dihydroxy-3,3-dimethyl-1-oxobutyl)-, monosodium salt, (R)- (9CI)
Panthoject
Pantothenic acid, monosodium salt, D-
Pantothenic acid, sodium salt
Sodium D-pantothenate
Sodium pantothenic acid

868-14-4
$C_4H_5O_6.K$
188.19
Tartaric acid, monopotassium salt
Acid potassium tartrate
Butanedioic acid, 2,3-dihydroxy-, (R-(R*,R*))-, monopotassium salt (9CI)
Cream of Tartar
Cremor Tartari
Faccla
Faccula
Faecla
Faecula
Monopotassium tartrate
Potassium acid tartrate
Potassium bitartrate
Potassium hydrogen tartrate
Potassium tartrate
Tartar
Tartar Cream

868-18-8
$C_4H_4O_6.2Na$
194.06
Tartaric acid, disodium salt
Bisodium tartrate
Butanedioic acid, 2,3-dihydroxy-(R-(R*,R*))-, disodiu salt (9CI)
Disodium tartrate
Disodium L-(+)-tartrate
Sal Tartar
Sodium tartrate
Sodium L-(+)-tartrate

868-19-9
$C_4H_6O_6.Sr$
237.71
Butanedioic acid, 2,3-dihydroxy- (R-(R*,R*))-, strontium salt (1:1) (9CI)
Tartaric acid, strontium salt (1:1) (8CI)

868-57-5
$C_6H_{12}O_2$
116.16
O=C(OC)C(CC)C
Butanoic acid, 2-methyl-, methyl ester
AI3-34461
Methyl 2-methylbutanoate

868-77-9
$C_6H_{10}O_3$
130.16
O=C(OCCO)C(=C)C
Methacrylic acid, 2-hydroxyethyl ester
Ethylene glycol methacrylate
Ethylene glycol, monomethacrylate
Glycol methacrylate
Glycol monomethacrylate
Hydroxyethyl methacrylate
β-Hydroxyethyl methacrylate
2-Hydroxyethyl methacrylate
Mhoromer
Monomer MG-1

868-85-9
$C_2H_7O_3P$
110.06
O=P(OC)OC
Phosphonic acid, dimethyl ester
Dimethylester kyseliny fosforite (Czech)
Dimethylfosfit (Czech)
Dimethylfosfonat (Czech)
Dimethylhydrogenphosphite
Dimethyl phosphite
Dimethyl phosphonate
NCI-C54773

869-01-2
$C_5H_{11}N_3O_2$
145.19
CCCCN(N=O)C(N)=O
Urea, 1-butyl-1-nitroso
BNU
Butylnitrosoharnstoff (German)
n-Butylnitrosourea
n-Butyl-N-nitrosourea
1-Butyl-1-nitrosourea
n-Nitrosobutylurea
1-Nitroso-1-butylurea
Urea, n-butyl-N-nitroso-

869-24-9
$C_6H_{14}ClN.ClH$
172.12
Triethylamine, 2-chloro-, hydrochloride
Bis(ethyl)-2-chloroethylamine hydrochloride
β-Chloroethyldiethylamine hydrochloride
(2-Chloroethyl)diethylamine

hydrochloride
2-Chlorotriethylamine hydrochloride
Diethylaminoethyl chloride hydrochloride
β-Diethylamino-ethyl chloride hydrochloride
Diethyl-β-chloroethylamine hydrochloride

869-29-4
$C_7H_{10}O_4$
158.15
Allylidene diacetate
Acrolein diacetate
AI3-24349
Allylidene acetate
Caswell No. 706
Diacetoxypropene
1,1-Diacetoxypropene-2
3,3-Diacetoxypropene
EPA Pesticide Chemical Code 068402
2-Propene-1,1-diol, diacetate (8CI,9CI)
SD-345
Shell SD 345
Shell 345

869-79-4
$C_9H_{20}N_2O$
172.27
Urea, N,N'-bis(1-methylpropyl)- (9CI)
1,3-Di-sec-butylurea
Urea, 1,3-di-sec-butyl- (6CI,7CI,8CI)

870-08-6
$C_{16}H_{34}OSn$
361.19
Stannane, dioctyloxo
Dioctyloxostannane
Dioctyltin oxide
Di-n-octyltin oxide
Di-n-octyl-zinn oxyd (German)
Stannane, oxodioctyl-
Tin, dioctyl-, oxide

870-23-5
C_3H_6S
74.15
SCC=C
2-Propene-1-thiol (9CI)
AI3-23286

870-72-4
$CH_3O_4S.Na$
134.09
Methanesulfonic acid, hydroxy-, sodium salt
Sodium formaldehyde bisulfite

871-27-2
$C_4H_{11}Al$
86.11
Diethylaluminum hydride
Aluminum, diethylhydro- (8CI, 9CI)
Diethylhydroaluminum

871-37-4
$C_{22}H_{43}NO_2$
353.58
O=C(O)CN(CCCCCCCCC=CCCCCCCCC)(C)C
Oleyl betaine
Ammonium, (carboxymethyl)dimethyl-cis-9-octadecenyl-, hydroxide, inner salt
N-(Carboxymethyl)-N,N-dimethyl-9-octadecen-1-aminium hydroxide, inner salt
Jortaine OB
9-Octadecen-1-aminium, N-(carboxymethyl)-N,N-dimethyl-, hydroxide, inner salt, (Z)- (9CI)
Oleyl dimethyl glycine
Oleyldimethylbetaine
Unibetaine OLB-30
Unibetaine OLB-50

871-83-0
$C_{10}H_{22}$
142.28
Nonane, 2-methyl- (9CI)

872-05-9
$C_{10}H_{20}$
140.27
C(=C)CCCCCCCC
1-Decene (9CI)
α-Decene
n-1-Decene
1-n-Decene
Decylene
n-Decylene

872-10-6
$C_{10}H_{22}S$
174.35
Diamyl sulfide

872-31-1
C_4H_3BrS
163.04
S(C=CC=1Br)C1
Thiophene, 3-bromo- (9CI)
3-Bromothiophene

872-36-6
$C_3H_2O_3$
86.05
Carbonic acid, cyclic vinylene ester
1,3-Dioxol-2-one (9CI)
Vinylene carbonate

872-50-4
C_5H_9NO
99.15
O=C(N(CC1)C)C1
2-Pyrrolidinone, 1-methyl
N-Methylpyrrolidinone
N-Methyl-2-pyrrolidinone
1-Methyl-2-pyrrolidinone
1-Methyl-5-pyrrolidinone
Methylpyrrolidone
N-Methylpyrrolidone
N-Methyl-α-pyrrolidone
N-Methyl-2-pyrrolidone
1-Methyl-2-pyrrolidone
M-Pyrol
NMP

872-55-9
C_6H_8S
112.20
CCc1cccs1
Thiophene, 2-ethyl- (8CI,9CI)

872-85-5
C_6H_5NO
107.12
O=Cc(ccnc1)c1
Isonicotinaldehyde
4-Formylpyridine
Isonicotinic aldehyde
p-Pyridinealdehyde
4-Pyridinealdehyde
Pyridine-4-carbaldehyde
4-Pyridinecarboxaldehyde (9CI)

872-93-5
$C_5H_{10}O_2S$
134.20
O=S(=O)(CCC1C)C1
Thiophene, tetrahydro-3-methyl-, 1,1-dioxide (9CI)
Tetrahydro-3-methylthiophene 1,1-dioxide

873-25-6
$C_4H_3Cl_3N_2$
185.43
1H-Imidazole, 2,4,5-trichloro-1-methyl- (9CI)
Imidazole, 2,4,5-trichloro-1-methyl- (8CI)

873-32-5
C_7H_4ClN
137.57
N#Cc(c(ccc1)Cl)c1
Benzonitrile, o-chloro
o-Chlorbenzonitril (Czech)
o-Chlorobenzonitrile
Nitril kyseliny o-chlorbenzoove (Czech)

873-55-2
$C_6H_6O_2S.Na$
165.17
OS(=O)c1ccccc1
Benzenesulfinic acid, sodium

salt (9CI)
AI3-52287
Sodium benzenesulfinate

873-62-1
C_7H_5NO
119.12
Oc1cccc(C#N)c1
Benzonitrile, 3-hydroxy- (9CI)
Benzonitrile, m-hydroxy- (8CI)
NSC-60108

873-69-8
$C_6H_6N_2O$
122.14
ON=Cc1ccccn1
Picolinaldehyde, oxime
2-Formylpyridine ketoxime
2-Formylpyridine oxime
2-Hydroxyiminomethyl pyridine
Picolinaldoxime
α-Picolinealdoxime
2-Pyridinaldoxime
Pyridine-2-aldoximate
Pyridine-2-aldoxime
2-Pyridinealdoxime
2-Pyridinecarboxaldehyde, oxime
2-Pyridylaldoxime
Pyrine-2-aldoximate

873-74-5
$C_7H_6N_2$
118.15
N#Cc(ccc(N)c1)c1
Aniline, p-cyano
p-Aminobenzonitrile
4-Aminobenzonitrile
Benzonitrile, p-amino- (8CI)
Benzonitrile, 4-amino- (9CI)
p-Cyanoaniline
4-Cyanoaniline

873-76-7
C_7H_7ClO
142.58
OCc1ccc(Cl)cc1
Benzenemethanol, 4-chloro- (9CI)
AI3-20628
Benzyl alcohol, p-chloro- (8CI)

873-83-6
$C_4H_5N_3O_2$
127.12
n(c(O)cc(n1)N)c1O
Uracil, 6-amino
6-Aminouracil

873-94-9
$C_9H_{16}O$
140.23
O=C(CC(CC1(C)C)C)C1
Dihydro-isophorone
AI3-33978
Cyclohexanone, 3,3,5-trimethyl-
3,3,5-Trimethylcyclohexanone

874-05-5
$C_7H_7N_3$
133.13
N=C(NNc1cccc2)c12
1H-Indazol-3-amine (9CI)

874-23-7
$C_8H_{12}O_2$
140.18
O=C(C(C(=O)CCC1)C1)C
Cyclohexanone, 2-acetyl- (9CI)
2-Acetylcyclohexanone
AI3-19261

874-24-8
$C_6H_5NO_3$
139.11
OC(=O)c1ncccc1O
3-Hydroxypicolinic acid
3-Hydroxy-2-pyridinecarboxylic acid

874-35-1
$C_{10}H_{12}$
132.21
1H-Indene, 2,3-dihydro-5-methyl- (9CI)
Indan, 5-methyl- (8CI)

874-41-9
$C_{10}H_{14}$
134.22
c(ccc(c1C)CC)(c1)C
Benzene, 1-ethyl-2,4-dimethyl- (9CI)
1-Ethyl-2,4-dimethylbenzene
m-Xylene, 4-ethyl- (8CI)

874-42-0
$C_7H_4Cl_2O$
175.01
O=Cc(c(cc(c1)Cl)Cl)c1
Benzaldehyde, 2,4-dichloro- (9CI)
AI3-16063
2,4-Dichlorobenzaldehyde

874-60-2
C_8H_7ClO
154.60
O=C(c(ccc(c1)C)c1)Cl
p-Toluoyl chloride
Benzoyl chloride, 4-methyl- (9CI)
4-Methylbenzoic acid chloride
p-Methylbenzoyl chloride
4-Methylbenzoyl chloride
p-Toluic acid chloride
4-Toluoyl chloride
p-Toluyl chloride

874-68-0
$C_9H_{14}O$
138.21
2-Propanone, 1-cyclohexylidene-
AI3-26201

875-51-4
$C_6H_5BrN_2O_2$
217.01
O=N(=O)c(c(N)ccc1Br)c1
Benzenamine, 4-bromo-2-nitro- (9CI)
AI3-15013
4-Bromo-2-nitroaniline
4-Bromo-2-nitrobenzenamine
4-Bromo-2-nitrobenzeneamine
2-Nitro-4-bromoaniline

875-74-1
$C_8H_9NO_2$
151.16
O=C(O)C(N)c(cccc1)c1
Benzeneacetic acid, α-amino-, (R)- (9CI)
(R)-α-Aminobenzeneacetic acid

875-79-6
$C_{10}H_{11}N$
145.20
Cc2cc1ccccc1n2C
Indole, 1,2-dimethyl- (8CI)
1H-Indole, 1,2-dimethyl- (9CI)

875-99-0
$C_{10}H_{14}O$
150.22
β-Pericyclocamphanone
Pericyclocamphanone
4,7,7-Trimethyltricyclo(2.2.1.0(2,6))heptan-3-one

876-27-7
$C_8H_7ClO_2$
170.60
Acetic acid, p-chlorophenyl ester
Acetic acid, 4-chlorophenyl ester (9CI)
p-Acetoxychlorobenzene
p-Chlorfenylester kyseliny octove (Czech)
p-Chlorophenyl acetate
4-Chlorophenyl acetate
Phenol, p-chloro-, acetate

877-10-1
$C_8H_6Cl_4$
243.95
Benzene, 1,2,4,5-tetrachloro-3,6-dimethyl- (9CI)
AI3-52685
p-Xylene, 2,3,5,6-tetrachloro- (8CI)

877-11-2
$C_7H_3Cl_5$

264.36
2,3,4,5,6-Pentachlorotoluene
Benzene, pentachloromethyl- (9CI)
Pentachlorotoluene
Toluene, 2,3,4,5,6-pentachloro- (8CI)

877-22-5
$C_8H_8O_4$
168.16
COc1cccc(C(O)=O)c1O
m-Anisic acid, 2-hydroxy
Acide orthovanillique
Benzoic acid, 2-hydroxy-3-methoxy-
3-Hydroxy-m-anisic acid
2-Hydroxy-3-methoxybenzoic acid
3-Methoxysalicylic acid
o-Vanillic acid

877-24-7
$C_8H_6O_4$.K
205.23
Potassium biphthalate
AI3-51765
1,2-Benzenedicarboxylic acid, monopotassium salt
Monopotassium 1,2-benzenedicarboxylate
Potassium acid phthalate

877-43-0
$C_{11}H_{11}N$
157.21
n(c(c(cc(c1)C)cc2)c1)c2C
Quinoline, 2,6-dimethyl- (9CI)
AI3-03277
2,6-Dimethylquinoline

877-44-1
$C_{12}H_{18}$
162.27
Benzene, 1,2,4-triethyl- (8CI, 9CI)

877-65-6
$C_{11}H_{16}O$
164.25
OCc(ccc(c1)C(C)(C)C)c1
Benzenemethanol, 4-(1,1-dimethylethyl)-
AI3-21419
p-tert-Butylbenzyl alcohol
4-(1,1-Dimethylethyl)benzenemethanol

877-83-8
$C_4Cl_4N_4$
245.88
Carbonimidic dichloride, (4,6-dichloro-1,3,5-triazin-2-yl)- (9CI)
Cyanogen chloride, tetramer (6CI)
Imidocarbonyl chloride, (4,6-dichloro-s-triazin-2-yl)- (7CI,8CI)

877-89-4
$C_6H_9N_3O_3$
171.14
1,3,5-Triazine, 2,4,6-trimethoxy- (9CI)
s-Triazine, 2,4,6-trimethoxy- (8CI)

878-13-7
$C_{11}H_{20}O$
168.28
Cycloundecanone (8CI,9CI)
NSC-96907

879-39-0
$C_6HCl_4NO_2$
260.88
Clc1cc(N(=O)=O)c(Cl)c(Cl)c1Cl
Benzene, 1,2,3,4-tetrachloro-5-nitro
Benzene, 5-nitro-1,2,3,4-tetrachloro-
DB-905
Folosan DB-905 Fumite
Folsan
Fusarex
TCBN
1,2,3,4-Tetrachloro-5-nitrobenzene
2,3,4,5-Tetrachloronitrobenzene

879-97-0
$C_{12}H_{18}O$
178.27
Oc(c(cc(c1)C(C)(C)C)c1C
Phenol, 4-(1,1-dimethylethyl)-2,6-dimethyl- (9CI)
4-(1,1-Dimethylethyl)-2,6-dimethylphenol

880-93-3
$C_{11}H_9NO_2$
187.19
Naphthalene, 1-methyl-4-nitro- (8CI,9CI)
NSC-144478

881-03-8
$C_{11}H_9NO_2$
187.21
O=N(=O)c(c(c(ccc1)cc2)c1)c2C
Naphthalene, 2-methyl-1-nitro (8CI,9CI)
2-Methyl-1-nitronaphthalene
NSC-7516

882-09-7
$C_{10}H_{11}ClO_3$
214.66
CC(C)(Oc1ccc(Cl)cc1)C(O)=O
Propionic acid, 2-(p-chlorophenoxy)-2-methyl
Acetic acid, (p-chlorophenoxy)dimethyl-
Acide (p-chlorophenoxy)-2 methyl-2 propionique (French)
α-(p-Chlorophenoxy)isobutyric acid
2-(4-Chlorophenoxy)-2-methylpropanoic acid
2-(p-Chlorophenoxy)-2-methylpropionic acid
Chlorophibrinic acid
Clofibric acid
Clofibrinic acid
Clofibrinsaeure (German)

882-33-7
$C_{12}H_{10}S_2$
218.34
S(Sc(cccc1)c1)c(cccc2)c2
Phenyl-disulfide
Diphenyl disulfide
Disulfide diphenyl
USAF E-1

882-71-3
$C_9H_{10}F_2O_3S$
236.24
2-Propanol, 1,3-difluoro-, benzenesulfonate (7CI, 8CI)

883-20-5
$C_{15}H_{12}$
192.27
Phenanthrene, 9-methyl
9-Methylphenanthrene

883-40-9
$C_{13}H_{10}N_2$
194.22
Diazodiphenylmethane
Benzene, 1,1'-(diazomethylene)-bis-
Diazodifenilmetano (Spanish)
Diazodiphenylmethane (French)
1,1'-Diphenyldiazomethane
Methane, diazodiphenyl-

883-99-8
$C_{12}H_{10}O_3$
202.21
O=C(OC)c(c(O)cc(c1ccc2)c2)c1
2-Naphthalenecarboxylic acid, 3-hydroxy-, methyl ester (9CI)
AI3-30205
2-Naphthoic acid, 3-hydroxy-, methyl ester (8CI)

884-43-5
$C_9H_{15}N_3O_3$
213.22
1,3,5-Triazine, 2,4,6-triethoxy- (9CI)

s-Triazine, 2,4,6-triethoxy- (8CI)

885-82-5
$C_{12}H_9NO_3$
215.20
(1,1'-Biphenyl)-4-ol, 3-nitro- (9CI)
AI3-24008
4-Biphenylol, 3-nitro- (8CI)
NSC-95810
Phenol, 2-nitro-4-phenyl-

886-45-3
$C_9H_4ClF_2N_3O$
243.60
s-Triazine, 2-(o-chlorophenoxy)-4,6-difluoro- (6CI,8CI)

886-50-0
$C_{10}H_{19}N_5S$
241.40
CCNc1nc(NC(C)(C)C)nc(SC)n1
s-Triazine, 2-(tertbutylamino)-4-(ethylamino)-6-(methylthio)
4-Aethylamino-2-tert-butylamino-6-methylthio-s-triazin (German)
2-tert.Butylamino-4-aethylamino-6-methylthio-1,3,5-triazin (German)
2-tert-Butylamino-4-ethylamino-6-methylmercapto-s-triazine
2-tert-Butylamino-4-ethylamino-6-methylthio-s-triazine
Clarosan
GS 14260
HS-14260
Igran
Igran 50
2-Methylthio-4-ethylamino-6-tert-butylamino-s-triazine
Prebane
Shortstop
Short-Stop E
Terbutrex
Terbutryn
Terbutryne

886-59-9
$C_{13}H_{18}N_2O$
218.29
O=C(NC1CCCCC1)Nc2ccccc2
Urea, N-cyclohexyl-N'-phenyl- (9CI)
NSC-80588
Urea, 1-cyclohexyl-3-phenyl- (8CI)

886-65-7
$C_{16}H_{14}$
206.29
c(cccc1)(c1)C=CC=Cc(cccc2)c2
1,4-Diphenylbutadiene
Benzene, 1,1'-(1,3-butadiene-1,4-diyl)bis- (9CI)
1,3-Butadiene, 1,4-diphenyl- (8CI)
1,1'-(1,3-Butadiene-1,4-diyl)bisbenzene

886-74-8
$C_{10}H_{12}ClNO_4$
245.68
NC(=O)OCC(O)COc1ccc(Cl)cc1
Carbamic acid, 3-(p-chlorophenoxy)-2-hydroxypropyl ester
3-(p-Chlorophenoxy)-2-hydroxypropyl carbamate
3-(p-Chlorophenoxy)-1,2-propanediol-1-carbamate
3-(4-Chlorophenoxy)-1,2-propanediol-1-carbamate
Chlorphenesin carbamate
Maolate
1,2-Propanediol, 3-(p-chlorophenoxy)-, 1-carbamate
U-19,646

886-77-1
$C_{13}H_{10}O_3$
214.22
O=C(C=CC(OC=C1)=C1)C=CC(OC=C2)=C2
Difurfurylideneacetone
AI3-09311
1,5-Di-2-furanyl-1,4-pentadien-3-one
1,5-Difuryl-1,4-pentadien-3-one
1,4-Pentadien-3-one, 1,5-di-2-furanyl- (9CI)

887-54-7
$C_9H_4Cl_4O_4$
317.94
1,4-Benzenedicarboxylic acid, 2,3,5,6-tetrachloro-, monomethyl ester (9CI)
Terephthalic acid, tetrachloro-, methyl ester
Terephthalic acid, tetrachloro-, monomethyl ester (8CI)

892-20-6
$C_{18}H_{16}Sn$
351.03
Stannane, triphenyl
Triphenylstannane
Triphenylstannyl hydride
Triphenyltin
Triphenyltin hydride

892-21-7
$C_{16}H_9NO_2$
247.26
Fluoranthene, 3-nitro
3-Nitrofluoranthene
4-Nitrofluoranthene

895-85-2
Unknown
Unknown
Di-(2-methylbenzoyl) peroxide (Not more than 85% with water)
UN 2593

897-55-2
$C_{19}H_{18}N_2$
274.39
n(c(c(c(c1)C=Cc(ccc(N(C)C)c2)c2)ccc3)c3)c1
Quinoline, 4-(p-(dimethylamino)styryl)
2-(4-N,N-Dimethylaminostyryl)-quinoline
4-(p-(Dimethylamino)styryl)-quinoline
4-(4-Dimethylaminostyryl)-quinoline
4M20

898-22-6
$C_{12}H_{15}N_3O_6$
297.27
1,3,5-Triazine-2,4,6-tricarboxylic acid, triethyl ester (9CI)
s-Triazine-2,4,6-tricarboxylic acid, triethyl ester (6CI,7CI,8CI)
Triethyl 1,3,5-triazine-2,4,6-tricarboxylate
2,4,6-Tris(ethoxycarbonyl)-1,3,5-triazine

900-38-9
$C_{16}H_{19}N_3O_2S$
317.41
Urea, 1-(p-ethoxyphenyl)-3-(6-ethoxy-2-pyridyl)-2-thio- (7CI,8CI)

900-95-8
$C_{20}H_{18}O_2Sn$
409.07
CC(=O)O[Sn](c1ccccc1)(c2ccccc2)c3ccccc3
Stannane, acetoxytriphenyl
Acetatotriphenylstannane
Acetate de triphenyl-etain (French)
Acetato di stagno trifenile (Italian)
Acetoxy-triphenyl-stannan (German)
Acetoxy-triphenylstannane
Batasan
Brestan
Brestan 60
ENT 25,208
Fenolovo acetate
Fentin acetaat (Dutch)
Fentin acetat (German)
Fentin acetate
Fentin azetat (German)
Fentine acetate (French)
Fintin acetato (Italian)
GC 6936

HOE-2824
Liromatin
Lirostanol
OMS 1020
Phenostat A
Phentinoacetate
Phentin acetate
Stannane, (acetyloxy)triphenyl- (9CI)
Suzu
Tin, acetoxytriphenyl-
Tinestan
Tinestan 60 WP
Tin triphenyl acetate
TPTA
TPZA
Trifenil stagno acetato (Italian)
Trifenyl-tinacetaat (Dutch)
Triphenylaceto stannane
Triphenylstannium acetate
Triphenyltin acetate
Triphenyl-zinnacetat (German)
Tubotin
VP 1940

901-44-0
$C_{19}H_{24}O_4$
316.40
O(c(ccc(c1)C(c(ccc(OCCO)c2)c2)(C)C)c1)CCO
Ethanol, 2,2'-((1-methylethylidene)bis(4,1-phenyleneoxy))-bis- (9CI)
AI3-15587
2,2'-(Isopropylidenebis(p-phenyleneoxy))diethanol

905-14-6
$C_{14}H_{13}N_2O_8P$
368.23
Phosphoric acid, ethyl bis-(4-nitrophenyl) ester (9CI)
Phosphoric acid, ethyl bis-(p-nitrophenyl) ester (8CI)

911-45-5
$C_{26}H_{28}ClNO$
406.00
CCN(CC)CCOc1ccc(cc1)C(=C(Cl)c2ccccc2)c3ccccc3
Triethylamine, 2-(p-(2-chloro-1,2-diphenylvinyl)phenoxy)
Chlomaphene
Chloramifene
Chloramiphene
2-(4-(2-Chloro-1,2-diphenylethenyl)phenoxy)-N,N-diethylethanamine
2-(p-(β-Chloro-α-phenylstyryl)phenoxy)-triethylamine
Cisclomiphene
Clomifene
Clomiphene
Clomiphene B
1-(p-(β-Diethylaminoethoxy)-phenyl)-1,2-diphenylchloroethylene

911-65-9
$C_{22}H_{28}N_4O_3$
396.47
Etonitazene
Arc 1G2
Ba-20684
Benzimidazole, 1-(2-(diethylamino)ethyl)-2-(p-ethoxybenzyl)-5-nitro-
1H-Benzimidazole-1-ethanamine, 2-((4-ethoxyphenyl)-methyl)-N,N-diethyl-5-nitro-(9CI)
Ciba 20-684BA
DEA No. 9624
1-(2-(Diethylamino)ethyl)-2-(p-ethoxybenzyl)-5-nitrobenzimidazole
Etobedolum
Etonitazeno (Spanish)
Etonitazenum (Latin)
Etonitazine
Etonitazinum
NIH 7607

914-00-1
$C_{22}H_{22}N_2O_8$
442.46
CN(C)C4C3C(O)C2C(=C)c1cccc(O)c1C(=O)C2=C(O)C3(O)C(=O)C(=C4O)C(N)=O
2-Naphthacenecarboxamide, 4-(dimethylamino)-1,4,4a,5,5a,6,11,12a-octahydro- 3,5,10,12,12a-pentahydroxy-6-methylene-1,11-dioxo
6-Deoxy-6-demethyl-6-methylene-5-oxytetracycline
GS-2876
Methacycline
Methacycline amphoteric
Methacycline base
6-Methylene-5-hydroxytetracycline
6-Methylene-5-oxytetracycline
Oxytetracycline, 6-methylene-
Rondomycin

915-30-0
$C_{30}H_{32}N_2O_2$
452.64
CCOC(=O)C1(CCN(CC1)CCC(C#N)(c2ccccc2)c3ccccc3)c4ccccc4
Isonipecotic acid, 1-(3-cyano-3,3-diphenylpropyl)-4-phenyl-, ethyl ester
1-(3-Cyano-3,3-diphenylpropyl)-4-phenyl-isonipecotic acid ethyl ester
Diphenoxylate
4-Piperidinecarboxylic acid, 1-(3-cyano-3,3-diphenyl-propyl)-4-phenyl-, ethyl ester (9CI)

915-67-3
$C_{20}H_{11}N_2O_{10}S_3.3Na$
604.48
[Na+].[Na+].[Na+].Oc4c(N=Nc1ccc(c2ccccc12)S([O-])(=O)=O)c3ccc(cc3cc4S([O-])(=O)=O)S([O-])(=O)=O
2,7-Naphthalenedisulfonic acid, 3-hydroxy-4-((4-sulfo-1-naphthyl)azo)-, trisodium salt
Acetacid Red 2BR
Acid Amaranth
Acid Amaranth I
Acid Amaranth N
Acid Leather Red 12BW
Acid Leather Rubine S
Acid Red 37
Acilan Red SE
Aizen Amaranth
Amacid Amaranth
Amarant (Czech)
Amaranth
Amaranth A
Amaranth B
Amaranth BPC
Amaranthe
Amaranthe USP (Biological stain)
Amaranth Extra
Amaranth Lake
Amaranth S
Amaranth S Specially Pure
Amaranth USP
Amaranth WD
Azo Red R
Azo Rubine S
Azo Rubine SF
Azo Rubine S.Fq
Azorubin S
Azo Ruby S
Bordeaux
Bordeaux S
Bordeaux S Extra Conc. A. Export
Bordeaux S Extra Pure A
Calcocid Amaranth
Canacert Amaranth
Certicol Amaranth S
Cerven Kysela 27 (Czech)
Cerven Potravinarska 9 (Czech)
C.I. 184
C.I. 16185
C.I. Acid Red 27
C.I. Acid Red 27, Trisodium salt
C.I. Food Red 9
Cilefa Rubine 2B
Daishiki Amaranth
D & C Red 2
Dolkwal Amaranth
Dye FD & C Red 2
Dye Red Raspberry
E 123
Edicol Amaranth
Edicol Supra Amaranth A
EEC No. 123
Eurocert Amaranth
Fast Red
FD & C Red No. 2
FD & C Red No. 2-Aluminium Lake
Food Red 2
Food Red 9

Fruit Red A Geigy
HD Amaranth B
HD Amaranth Supra
Hexacert Red No. 2
Hexacol Amaranth B Extra
Hidacid Amaranth
Hispacid Red AM
2-Hydroxy-1,1'-azonaphthalene-3,6,4'-trisulfonic acid trisodium salt
3-Hydroxy-4-((4-sulfo-1-naphthalenyl)azo)-2,7-naphthlene-disulfonic acid, trisodium salt
3-Hydroxy-4-((4-sulfo-1-naphthyl)azo)-2,7-naphthalenedisulfonic acid, trisodium salt
3-Hydroxy-4-((4-sulpho-1-naphthalenyl)azo)-2,7-naphthalenedisulphonic acid, trisodium salt
3-Hydroxy-4-((4-sulpho-1-naphthyl)azo)-2,7-naphthalenedisulphonic acid, trisodium salt
Java Amaranth
Kayaku Amaranth
Kayaku Food Colour Red No. 2
KCA Foodcol Amaranth A
Kiton Rubine S
Lissamine Amaranth AC
L-Red 3
L-Rot 3 (German)
Maple Amaranth
Naphthol Red B
Naphthol Red C
Naphthol Red LZS
Naphthol Red O
Naphthol Red S
Naphthol Red S Conc. Specially Pure
Naphthol Red SI
Naphthol Red S Specially Pure
Naptholrot S (German)
Neklacid Red A
Rakuto Amaranth
Raspberry Red for Jellies
1302 Red
1508 Red
Red Dye No. 2
Red No. 2
San-EI Amaranth
S-Azo Rubine
Schultz Nr. 212 (German)
Shikiso Amaranth
Solar Red O
1-(4-Sulfo-1-naphthylazo)-2-naphthol-3,6-disulfonic acid trisodium salt
1-(4-Sulpho-1-naphthylazo)-2-naphthol-3,6-disulphonic acid, trisodium salt
Takaoka Amaranth
Tertracid Red A
Toyo amaranth
Trisodium salt of 1-(4-sulfo-1-naphthylazo)-2-naphthol-3,6-disulfonic acid
Trisodium salt of 1-(4-sulpho-1-naphthylazo)-2-naphthol-3,6-disulphonic acid
USACERT Red No. 2
Victoria Rubine O
Victoria Rubine O for Food
Whortleberry Red
Wool Bordeaux 6RK
Wool Red
Wool Red 40F

917-61-3
CNO.Na
65.01
Cyanic acid, sodium salt
Cyansan
San-Cyan
Sodium cyanate
Sodium isocyanate
Weecon
Zassol

917-92-0
C_6H_{10}
82.15
1-Butyne, 3,3-dimethyl- (8CI, 9CI)

917-93-1
C_4H_7ClO
106.55
O=CC(C)(C)Cl
Propanal, 2-chloro-2-methyl- (9CI)
2-Chloroisobutyraldehyde
2-Chloro-2-methylpropanal

918-00-3
$C_3H_3Cl_3O$
161.41
2-Propanone, 1,1,1-trichloro
α,α,α-Trichloroacetone
1,1,1-Trichloroacetone
1,1,1-Trichloropropanone

918-20-7
C_4ClD_9
83.50
Propane-1,1,1,3,3,3-d6, 2-chloro-2-(methyl-d3)- (8CI,9CI)

918-37-6
$C_2N_6O_{12}$
300.02
O=N(=O)C(N(=O)=O)(N(=O)=O)C(N(=O)=O)(N(=O)=O)N(=O)=O
Hexanitroethane
Ethane, hexanitro- (8CI,9CI)
Hexanitroetano (Spanish)
Hexanitroethane (French)

918-54-7
$C_2H_3N_3O_7$
181.04
Trinitroethanol
Ethanol, 2,2,2-trinitro-
Trinitroetanol (Spanish)
Trinitroethanol (French)
2,2,2-Trinitroethanol

919-30-2
$C_9H_{23}NO_3Si$
221.42
CCOSi(CCCN)(OCC)OCC
1-Propanamine, 3-(triethoxysilyl)
A 1100
AGM-9
(γ-Aminopropyl)triethoxysilane
(3-Aminopropyl)triethoxysilane
Propylamine, 3-(triethoxysilyl)-
Silane, γ-aminopropyltriethoxy-
Silane, (3-aminopropyl)triethoxy-
Silicone A-1100
Triethoxy(3-aminopropyl)silane
3-(Triethoxysilyl)propylamine

919-31-3
$C_9H_{19}NO_3Si$
217.38
CCOSi(CCC#N)(OCC)OCC
Silane, (2-cyanoethyl)triethoxy
β-Cyanoethyltriethoxysilane
(2-Cyanoethyl)triethoxysilane
Propionitrile, 3-(triethoxysilyl)-
Triethoxy-2-kyanethylsilan (Czech)

919-44-8
$C_7H_{14}NO_5P$
223.19
Phosphoric acid, dimethyl 1-methyl-3-(methylamino)-2-oxo-1-propenyl ester, (Z)
Monocrotophos

919-54-0
$C_8H_{17}O_4PS_2$
272.34
CCOC(=O)CSP(=S)(OCC)OCC
Acetic acid, mercapto-, ethyl ester, S-ester with O,O-diethyl phosphorodithioate
Acethion
Acethione
Acetic acid, ((diethoxyphosphinothioyl)thio)-, ethyl ester (9CI)
O,O-Diethyl S-carboethoxymethyl dithiophosphate
O,O-Diethyl S-carboethoxymethyl phosphorodithioate
ENT 25,650
Ethoxyphas
Ethoxyphos
Ethyl ((diethoxyphosphinothioyl)thio)acetate
Hercules 4580
Phosphorodithioic acid, O,O-diethyl ester, S-ester with ethyl mercaptoacetate

919-76-6
$C_7H_{16}NO_4PS_2$

273.33
COCCNC(=O)CSP(=S)(OC)OC
Phosphorodithioic acid, O,O-dimethyl ester, S-ester with 2-mercapto-N-(2-methoxyethyl) acetamide
Acetamide, 2-mercapto-N-(2-methoxyethyl)-, S-ester with O,O-dimethyl phosphorodithioate
Amidiphos
Amidithion
C 2446
Ciba 2446
Ciba Thiocron
O,O-Dimethyl-S-(2-methoxy-ethylcarbamoylmethyl)dithiophosphate
O,O-Dimethyl S-(2-methoxy-ethylcarbamoyl methyl) phosphorodithioate
ENT 27,160
Medithionat
S-(2-((2-Methoxyethyl)amino-2-oxoethyl) O,O-dimethyl) phosphorodithioate
S-(N-2-Methoxyethylcarbamoyl-methyl)dimethyl phophorothiolothionate
Thiocron
Thiocron 30

919-86-8
$C_6H_{15}O_3PS_2$
230.30
O=P(OC)(OC)SCCSCC
Phosphorothioic acid, S-(2-(ethylthio)ethyl) O,O-dimethyl ester
Bay 18436
Bayer 25/154
Demeton-S-methyl
Demeton-S-metile (Italian)
O,O-Dimethyl-S-(2-aethylthio-aethyl)-monothiophosphat (German)
O,O-Dimethyl S-(2-eththio-ethyl)phosphorothioate
Dimethyl S-(2-eththioethyl)-thiophosphate
O,O-Dimethyl S-ethylmercapto-ethyl thiophosphate
O,O-Dimethyl 2-ethylmercapto-ethyl thiophosphate, thiolo isomer
O,O-Dimethyl-S-(2-ethylthio-ethyl)-monothiofosfaat (Dutch)
O,O-Dimethyl S-(2-(ethylthio)-ethyl)phosphorothioate
O,O-Dimethyl-S-(3-thia-pentyl)-monothiophosphat (German)
O,O-Dimetil-S-(2-etiltio-etil)-monotiofosfato (Italian)
Duratox
Ethanethiol, 2-(ethylthio)-, S-ester with O,O-dimethyl phosphorothioate
S-(2-(Ethylthio)ethyl) O,O-dimethyl phosphorothioate
S-(2-(Ethylthio)ethyl) dimethyl phosphorothiolate
S-(2-(Ethylthio)ethyl) O,O-dimethyl thiophosphate
Isometasystox
Isomethylsystox
Metaisoseptox
Metaisosystox
Metasystox Forte
Metasystox (I)
Metasystox J
Methyl demeton thioester
Methyl isosystox
Methyl-mercaptofos teolery
Phosphorothioic acid, O,O-dimethyl S-(2-(ethylthio)ethyl) ester
Thiophosphate de O,O-dimethyle et de S-2-ethylthio-ethyle (French)

919-94-8
$C_7H_{16}O$
116.20
O(C(CC)(C)C)CC
Butane, 2-ethoxy-2-methyl- (9CI)
2-Ethoxy-2-methylbutane
Ethyl tert-amyl ether

920-46-7
C_4H_5ClO
104.54
O=C(C(=C)C)Cl
Methacryloyl-chloride
Methacryl chloride
Methacrylic acid chloride
Methacrylic chloride
Methacrylyl chloride
Methylacryloyl chloride
α-Methylacryloyl chloride
2-Methylpropenoic acid chloride
2-Methylpropenoyl chloride
2-Propenoyl chloride, 2-methyl- (9CI)

920-66-1
$C_3H_2F_6O$
168.05
FC(F)(F)C(O)C(F)(F)F
2-Propanol, 1,1,1,3,3,3-hexafluoro
Hexafluoroisopropanol
HFIP
1,1,1,3,3,3-Hexafluoro-2-propanol

921-03-9
$C_3H_3Cl_3O$
161.41
ClCC(=O)C(Cl)Cl
2-Propanone, 1,1,3-trichloro
α,α',α'-Trichloroacetone
1,1,3-Trichloroacetone
1,1,3-Trichloro-2-propanone

921-09-5
$C_4Cl_4H_2$
191.86
ClC(Cl)=C(Cl)C(Cl)=C
1,3-Butadiene, 1,1,2,3-tetrachloro
1,1,2,3-Tetrachloro-1,3-butadiene

921-47-1
C_9H_{20}
128.26
Hexane, 2,3,4-trimethyl- (8CI,9CI)

922-28-1
C_9H_{20}
128.26
Heptane, 3,4-dimethyl- (9CI)

922-61-2
C_6H_{12}
84.16
2-Pentene, 3-methyl- (8CI,9CI)

922-62-3
C_6H_{12}
84.16
C(=CC)(CC)C
2-Pentene, 3-methyl-, (Z)- (9CI)
cis-3-Methyl-2-pentene
(Z)-3-Methyl-2-pentene

922-67-8
$C_4H_4O_2$
84.08
Propiolic acid, methyl ester
Acetylenecarboxylic acid methyl ester
Methyl propiolate
Propynoic acid, methyl ester

922-68-9
$C_3H_4O_3$
88.06
O=C(OC)C=O
Acetic acid, oxo-, methyl ester (9CI)
Methyl glyoxylate
Methyl oxoacetate

922-80-5
$C_{14}H_{26}O_7S.Na$
361.41
Diamyl sodium sulfosuccinate
AI3-18857
Butanedioic acid, sulfo-, 1,4-dipentyl ester, sodium salt
1,4-Dipentylsulfobutanedioic acid, sodium salt

923-02-4
$C_5H_9NO_2$

115.13
O=C(NCO)C(=C)C
2-Propenamide, N-(hydroxymethyl)-2-methyl- (9CI)
N-(Hydroxymethyl)-2-methyl-2-propenamide
Methylolmethacrylamide

923-26-2
$C_7H_{12}O_3$
144.19
O=C(OCC(O)C)C(=C)C
Methacrylic acid, 2-hydroxypropyl ester
β-Hydroxypropyl methacrylate
2-Hydroxypropyl methacrylate
2-Propenoic acid, 2-methyl-, 2-hydroxypropyl ester (9CI)

924-06-1
$C_{15}H_{32}O_2$
244.42
Ethanol, 2-((1-hexylheptyl)oxy)- (7CI, 8CI,9CI)
Ethylene glycol mono(di-n-hexylcarbonyl) ether
Ethylene glycol mono-7-tridecyl ether

924-16-3
$C_8H_{18}N_2O$
158.28
O=NN(CCCC)CCCC
1-Butanamine, n-butyl-N-nitroso
Butylamine, N-nitrosodi-
n-Butyl-N-nitroso-1-butamine
DBN
DBNA
Dibutylamine, N-nitroso- (6CI)
Di-n-butylnitrosamin (German)
Dibutylnitrosamine
Di-n-butylnitrosamine
N,N-Di-n-butylnitrosamine
N,N-Dibutylnitrosoamine
NDBA
Nitrosodibutylamine
N-Nitrosodibutylamine
N-Nitroso-di-n-butylamine
RCRA waste number U172

924-42-5
$C_4H_7NO_2$
101.12
O=C(NCO)C=C
Acrylamide, N-(hydroxymethyl)
N-(Hydroxymethyl)acrylamide
N-(Hydroxymethyl)-2-propenamide
N-Methanolacrylamide
N-Methylolacrylamide
Monomethylolacrylamide
NCI-C60333
2-Propenamide, N-(hydroxymethyl)-
Uramine T 80

924-43-6
$C_4H_9NO_2$
103.14
O=NOC(CC)C
Nitrous acid, sec-butyl ester
sec-Butyl nitrite
Nitrous acid, 1-methyl propyl ester

924-50-5
$C_6H_{10}O_2$
114.14
O=C(OC)C=C(C)C
2-Butenoic acid, 3-methyl-, methyl ester (9CI)
Methyl 3-methyl-2-butenoate

924-83-4
$C_7H_{10}O_4$
158.15
O=C(OC(C)C)C=CC(=O)O
2-Butenedioic acid (Z)-, mono (1-methylethyl) ester (9CI)
Isopropyl maleate half ester
Monoisopropyl maleate

925-21-3
$C_8H_{12}O_4$
172.18
O=C(OCCCC)C=CC(=O)O
2-Butenedioic acid (Z)-, monobutyl ester (9CI)

925-60-0
$C_6H_{10}O_2$
114.14
O=C(OCCC)C=C
2-Propenoic acid, propyl ester (9CI)
Acrylic acid, propyl ester (8CI)
Propyl acrylate
n-Propyl acrylate
Propyl 2-propenoate

925-78-0
$C_9H_{18}O$
142.24
O=C(CCCCCC)CC
3-Nonanone
AI3-36117

926-06-7
$C_4H_{10}O_3S$
138.20
O=S(=O)(OC(C)C)C
Methanesulfonic acid, isopropyl ester
IMS
Isopropyl mesylate
Isopropylmethanesulfonate
Isopropyl methane sulphonate
Methanesulfonic acid, 1-methylethyl ester
2-Propyl methanesulphonate

926-39-6
$C_2H_7NO_4S$
141.14
O=S(=O)(OCCN)O
Ethanolamine O-sulfate
AI3-16953
2-Aminoethanol, hydrogen sulfate (ester)
Ethanol, 2-amino-, hydrogen sulfate (ester) (9CI)

926-41-0
$C_7H_{14}O_2$
130.19
1-Propanol, 2,2-dimethyl-, acetate (8CI,9CI)
NSC-945

926-42-1
$C_5H_{11}NO_3$
133.14
1-Propanol, 2,2-dimethyl-, nitrate (8CI,9CI)

926-56-7
C_6H_{10}
82.15
C(=CC=C)(C)C
4-Methyl-1,3-pentadiene
Methylpentadiene
Metilpentadieno (Spanish)
1,3-Pentadiene, 4-methyl- (9CI)

926-57-8
$C_4H_5Cl_2$
123.99
C(=CCCl)(C)Cl
2-Butene, 1,3-dichloro
1,3-Dichloro-2-butene

926-64-7
$C_4H_8N_2$
84.14
CN(C)CC#N
Acetonitrile, (dimethylamino)
N-(Cyanomethyl)dimethylamine
Dimethylaminoacetonitrile
2-Dimethylaminoacetonitrile [UN 2378]
Glycinonitrile, N,N-dimethyl-
UN 2378 [2-Dimethylaminoacetonitrile]

926-65-8
$C_5H_{10}O$
86.13
O(C=C)C(C)C
Vinyl isopropyl ether
2-(Ethenyloxy)propane
Propane, 2-(ethenyloxy)- (9CI)

926-82-9
C_9H_{20}
128.26
C(CC)(CC(CC)C)C

Heptane, 3,5-dimethyl- (9CI)
3,5-Dimethylheptane

927-07-1
$C_9H_{18}O_3$
174.27
O=C(OOC(C)(C)C)C(C)(C)C
Peroxypivalic acid, tert-butyl ester
t-Butyl peroxypivalate
tert-Butyl peroxypivalate, Not more than 77% in solution (DOT)
tert-Butyl perpivalate
tert-Butyl trimethylperoxy-acetate
Esperox 31M
Lupersol 11
Peroxypivalic acid, tert-butyl ester, Not more than 77% in solution
Trigonox 25/75
UN 2110 (DOT)

927-49-1
$C_{11}H_{22}O$
170.33
O=C(CCCCC)CCCCC
6-Undecanone
Amyl ketone
Diamyl ketone
Dipentyl ketone
6-Oxoundecane
Pentyl ketone
Undecan-6-one

927-60-6
$C_4H_9NO_2$
103.11
O=C(N)CCCO
4-Hydroxybutyramide
AI3-26368
Butanamide, 4-hydroxy- (9CI)
Butyramide, 4-hydroxy-
4-Hydroxybutanamide
Hydroxybutyramide
γ-Hydroxybutyramide
4-Hydroxybutyric acid amide
PC 621

927-62-8
$C_6H_{15}N$
101.19
N(CCCC)(C)C
1-Butanamine, N,N-dimethyl- (9CI)
N,N-Dimethyl-1-butanamine
Dimethylbutylamine

927-68-4
$C_4H_7BrO_2$
167.02
O=C(OCCBr)C
Ethanol, 2-bromo-, acetate
Bromoethyl acetate
2-Bromoethylacetate

927-73-1
C_4H_7Cl
90.55
C(=C)CCCl
1-Butene, 4-chloro- (9CI)
4-Chloro-1-butene

927-80-0
C_4H_6O
70.09
Ethoxyacetylene
Ether, ethyl ethynyl
Ethoxyethyne
Ethoxy-ethyne
Ethyl ethynyl ether
Ethyne, ethoxy- (9CI)

928-45-0
$C_4H_9NO_3$
119.12
CCCCON(=O)=O
Butyl nitrate

928-51-8
C_4H_9ClO
108.58
OCCCCCl
1-Butanol, 4-chloro
4-Chlorbutan-1-ol (German)
4-Chloro-1-butane-ol
4-Chlorobutanol
4-Chloro-1-butanol
Tetramethylene chlorohydrin

928-55-2
$C_5H_{10}O$
86.15
CCOC=CC
Ether, ethyl propenyl
Ethyl propenyl ether
Ethyl 1-propenyl ether

928-65-4
$C_6H_{13}Cl_3Si$
219.63
Silane, hexyltrichloro
Hexyltrichlorosilane [UN 1784]
Silane, trichlorohexyl-
UN 1784 [Hexyltrichloro-silane]

928-68-7
$C_8H_{16}O$
128.21
O=C(CCCC(C)C)C
2-Heptanone, 6-methyl- (9CI)
6-Methyl-2-heptanone

928-72-3
$C_4H_5NO_4.2Na$
177.08
[Na+].[Na+].[O-]C(=O)CNCC([O-])=O
Acetic acid, iminodi-, disodium salt
Disodium iminodiacetate
Iminodioctan disodny (Czech)
Iminodioctan sodny (Czech)

928-80-3
$C_{10}H_{20}O$
156.27
3-Decanone (9CI)
Ethyl heptyl ketone

928-95-0
$C_6H_{12}O$
100.18
OCC=CCCC
2-Hexen-1-ol, (E)
trans-2-Hexenol
2-Hexenol
2-Hexen-1-ol, trans-

928-96-1
$C_6H_{12}O$
100.18
OCCC=CCC
3-Hexen-1-ol, (Z)
Blatteralkohol
β-γ-Hexenol
3-Hexen-1-ol, cis-
cis-3-Hexenol
cis-3-Hexen-1-ol
Leaf Alcohol

928-97-2
$C_6H_{12}O$
100.16
OCCC=CCC
3-Hexen-1-ol, (E)-
AI3-34794
trans-3-Hexenol
trans-3-Hexen-1-ol
(E)-3-Hexen-1-ol

929-06-6
$C_4H_{11}NO_2$
105.16
O(CCO)CCN
Ethanol, 2-(2-aminoethoxy)
2-Aminoethoxyethanol
2-(2-Aminoethoxy)ethanol (DOT)
Diglycolamine
NA 1760 (DOT)

929-16-8
$C_{18}H_{35}ClO_2$
318.93
2-Chloroethyl palmitate
AI3-03502
Hexadecanoic acid, 2-chloro-ethyl ester

929-73-7
$C_{12}H_{27}N.ClH$
221.86
Dodecylamine, hydrochloride

Dodecanamine hydrochloride
1-Dodecanamine, hydrochloride (9CI)
n-Dodecylamine hydrochloride
Dodecylammonium chloride
n-Dodecylammonium chloride
Laurylamine hydrochloride
Laurylammonium hydrochloride

929-77-1
$C_{23}H_{46}O_2$
354.62
O=C(OC)CCCCCCCCCCCCCCCCCCCC
Behenic acid, methyl ester
AI3-36456
Docosanoic acid, methyl ester (9CI)
Methyl behenate
Methyl docosanoate

930-02-9
$C_{20}H_{40}O$
296.54
O(CCCCCCCCCCCCCCCCCC)C=C
Ether, octadecyl vinyl (8CI)
1-(Ethenyloxy)octadecane
Octadecane, 1-(ethenyloxy)- (9CI)

930-22-3
C_4H_6O
70.10
O(C1C=C)C1
1-Butene, 3,4-epoxy
Butadiene monoxide
1,2-Epoxybutene-3
3,4-Epoxy-1-butene
Vinyloxirane

930-27-8
C_5H_6O
82.11
Cc1ccoc1
Furan, 3-methyl
3-Methylfuran

930-30-3
C_5H_6O
82.10
Cyclopentenone
2-Cyclopenten-1-one (9CI)

930-37-0
$C_4H_8O_2$
88.12
O(C1COC)C1
Propane, 1,2-epoxy-3-methoxy
1,2-Epoxy-3-methoxypropane
Glycidol methyl ether
Glycidyl methyl ether
(Methoxymethyl)oxirane
3-Methoxypropylene oxide
Methyl glycidyl ether
Oxirane, (methoxymethyl)- (9CI)

930-55-2
$C_4H_8N_2O$
100.14
O=NN(CCC1)C1
Pyrrolidine, 1-nitroso
N-Nitrosopyrrolidin (German)
N-Nitrosopyrrolidine
1-Nitrosopyrrolidine
No-Pyr
N-N-PYR
NPYR
Pyrrole, tetrahydro-N-nitroso-
RCRA waste number U180

930-65-4
C_6H_9Cl
116.59
Cyclohexene, 4-chloro- (6CI,7CI,8CI,9CI)
4-Chlorocyclohexene

930-66-5
C_6H_9Cl
116.59
Cyclohexene, 1-chloro- (9CI)

930-68-7
C_6H_8O
96.14
O=C(C=CCC1)C1
2-Cyclohexen-1-one
Cyclohexenone
2-Cyclohexenone

930-69-8
$C_6H_6S.Na$
133.17
Sodium benzenethiolate
Benzenethiol, sodium salt (9CI)
Sodium phenylmercaptide
Sodium phenylsulfide
Sodium phenylthiolate
Sodium thiophenate
Sodium thiophenolate
Sodium thiophenoxide
Sodium thiophenylate
Thiophenyl sodium salt

930-90-5
C_8H_{16}
112.22
Cyclopentane, 1-ethyl-2-methyl-, trans- (8CI,9CI)

931-17-9
$C_6H_{12}O_2$
116.18
1,2-Cyclohexanediol
Brenzcatechin
Brenzkatechin (German)
Pyrocatechitol

931-19-1
C_6H_7NO
109.12
O=n(c(ccc1)C)c1
Pyridine, 2-methyl-, 1-oxide (9CI)
AI3-24502
2-Methylpyridine 1-oxide
2-Picoline, 1-oxide (8CI)

931-20-4
$C_6H_{11}NO$
113.15
O=C(N(CCC1)C)C1
2-Piperidinone, 1-methyl- (9CI)
AI3-11535
1-Methyl-2-piperidinone
1-Methyl-2-piperidone
2-Piperidone, 1-methyl- (8CI)

931-35-1
$C_6H_{12}N_2$
112.18
1H-Imidazole, 2-ethyl-4,5-dihydro-4-methyl- (9CI)
2-Ethyl-5-methyl-2-imidazoline
2-Imidazoline, 2-ethyl-4-methyl- (7CI,8CI)

931-36-2
$C_6H_{10}N_2$
110.15
N(C(=CN1)C)=C1CC
Imidazole, 2-ethyl-4-methyl- (8CI)
2-Ethyl-4-methylimidazole
2-Ethyl-4-methyl-1H-imidazole
1H-Imidazole, 2-ethyl-4-methyl- (9CI)

931-51-1
$C_6H_{11}ClMg$
142.91
Cl[Mg]C1CCCCC1
Cyclohexylmagnesium chloride
Chlorocyclohexylmagnesium
Magnesium, chlorocyclohexyl- (9CI)

931-54-4
C_7H_5N
103.12
Phenylisocyanide

931-56-6
$C_7H_{14}O$
114.19
Cyclohexane, methoxy- (9CI)
Cyclohexyl methyl ether
Ether, cyclohexyl methyl (6CI,7CI,8CI)

Methoxycyclohexane
1-Methoxycyclohexane
Methyl cyclohexyl ether

931-64-6
C_8H_{12}
108.18
Bicyclo(2.2.2)oct-2-ene (8CI, 9CI)
Cyclohexene, 3,6-endo-(1,2-ethanediyl)-
3,6-Endoethylenecyclohexene

931-86-2
$C_3H_4N_4O$
112.07
1,3,5-Triazin-2(1H)-one, 4-amino- (9CI)
5-Azacytosine
NSC-51100
NSC-54006
s-Triazin-2-ol, 4-amino-
s-Triazin-2(1H)-one, 4-amino- (8CI)

931-88-4
C_8H_{14}
110.20
Cyclooctene (9CI)
AI3-26693

931-97-5
$C_7H_{11}NO$
125.19
N#CC(O)(CCCC1)C1
Cyclohexanecarbonitrile, 1-hydroxy
1-Hydroxycycloheptanecarbonitrile
1-Hydroxycyclohexanecarbonitrile

932-66-1
$C_8H_{12}O$
124.18
Ethanone, 1-(1-cyclohexen-1-yl)- (9CI)
Ketone, 1-cyclohexen-1-yl methyl (8CI)

NSC-12216

932-83-2
$C_6H_{12}N_2O$
128.20
O=NN1CCCCCC1
1H-Azepine, hexahydro-1-nitroso
Hexahydro-1-nitroso-1H-azepine
N-6-MI
N-Nitrosoazacycloheptane
N-Nitrosohexahydroazepine
N-Nitrosohexamethyleneimine
Nitrosohexamethylenimine
N-Nitrosoperhydroazepine

932-98-9
C_6H_4ClNO
141.55
4-Chloronitrosobenzene
Benzene, 1-chloro-4-nitroso- (9CI)
1-Chloro-4-nitrosobenzene

933-19-7
$C_4H_5N_3O_2$
127.08
1,3,5-Triazine-2,4(1H,3H)-dione, 6-methyl- (9CI)
s-Triazine-2,4-diol, 6-methyl- (6CI,7CI)
s-Triazine-2,4(1H,3H)-dione, 6-methyl- (8CI)

933-20-0
$C_3H_2Cl_2N_4$
164.99
s-Triazine, 2-amino-4,6-dichloro
2-Amino-4,6-dichloro-s-triazine
1,3,5-Triazin-2-amine, 4,6-dichloro-

933-40-4
$C_8H_{16}O_2$
144.21
Cyclohexane, 1,1-dimethoxy- (9CI)
Cyclohexanone, dimethyl acetal (8CI)

933-52-8
$C_8H_{12}O_2$
140.20
O=C(C(C1=O)(C)C)C1(C)C
1,3-Cyclobutanedione, 2,2,4,4-tetramethyl
Tetramethylcyclobuta-1,3-dione
Tetramethyl-1,3-cyclobutanedione

933-75-5
$C_6H_3Cl_3O$
197.44
Oc(c(ccc1Cl)Cl)c1Cl
Phenol, 2,3,6-trichloro
2,3,6-Trichlorophenol

933-78-8
$C_6H_3Cl_3O$
197.45
Oc1cc(Cl)cc(Cl)c1Cl
2,3,5-Trichlorophenol

933-87-9
$C_4H_4BrN_3O_2$
205.98
1-Methyl-5-bromo-4-nitroimidazole
MTR 1-80

933-98-2
$C_{10}H_{14}$
134.22
c(c(c(cc1)C)C)(c1)CC
Benzene, 1-ethyl-2,3-dimethyl- (9CI)
1-Ethyl-2,3-dimethylbenzene

933-99-3
$C_8H_{10}O_2$
138.17
3-Ethylcatechol
1,2-Benzenediol, 3-ethyl-

934-00-9
$C_7H_8O_3$
140.14
Pyrocatechol, 3-methoxy- (8CI)
AI3-21349
1,2-Benzenediol, 3-methoxy- (9CI)

934-32-7
$C_7H_7N_3$
133.17
N=C(Nc(c1ccc2)c2)N1
Benzimidazole, 2-amino
2-Aminobenzimidazole
USAF EK-4037

934-34-9
C_7H_5NOS
151.18
O=C(Nc(c1ccc2)c2)S1
2(3H)-Benzothiazolone (9CI)
AI3-24484
2-Benzothiazolinone (8CI)

934-73-6
C_7H_7ClOS
174.65
CS(=O)c1ccc(Cl)cc1
Sulfoxide, p-chlorophenyl methyl
Benzene, 1-chloro-4-(methylsulfinyl)- (9CI)
p-Chlorophenyl methyl sulfoxide
4-Chlorophenyl methyl sulfoxide
Methyl 4-chlorophenyl sulfoxide

934-74-7
$C_{10}H_{14}$
134.22
c(cc(cc1C)CC)(c1)C
Benzene, 1-ethyl-3,5-dimethyl- (9CI)
1-Ethyl-3,5-dimethylbenzene
m-Xylene, 5-ethyl- (8CI)

934-75-8
$C_5H_9N_5$
139.13
1,3,5-Triazine-2,4-diamine, 6-ethyl-
AI3-51294

934-80-5
$C_{10}H_{14}$
134.22
c(ccc(c1C)C)(c1)CC
Benzene, 4-ethyl-1,2-dimethyl- (9CI)
4-Ethyl-1,2-dimethylbenzene
o-Xylene, 4-ethyl- (8CI)

935-31-9
$C_{10}H_{18}$
138.25
Cyclodecene, (Z)- (8CI,9CI)
NSC-105776

935-45-5
$C_9H_{18}O_2$
158.24
1,3-Dioxolane, 2-ethyl-2-isobutyl- (7CI,8CI)

935-92-2
$C_9H_{10}O_2$
150.18
O=C(C(=CC(=O)C=1C)C)C1C
p-Benzoquinone, 2,3,5-trimethyl- (8CI)
2,5-Cyclohexadiene-1,4-dione, 2,3,5-trimethyl- (9CI)
2,3,5-Trimethyl-2,5-cyclohexadiene-1,4-dione

935-95-5
$C_6H_2Cl_4O$
231.88
Oc1c(Cl)c(Cl)cc(Cl)c1Cl
Phenol, 2,3,5,6-tetrachloro
2,3,5,6-Tetrachlorophenol

936-16-3
$C_7H_7NO_2S$
169.20
1,2-Benzoisothiazoline 1,1-dioxide
1,2-BITDO

936-49-2
$C_9H_{10}N_2$
146.21
N(=C(NC1)c(cccc2)c2)C1
2-Imidazoline, 2-phenyl
1H-Imidazole, 4,5-dihydro-2-phenyl-
2-Phenyl-2-imidazoline

937-14-4
$C_7H_5ClO_3$
172.57
O=C(OO)c(cccc1Cl)c1
m-Chloroperbenzoic acid
Acide chloro-3 peroxybenzoique (French)
Acido 3-cloroperoxibenzoico (Spanish)
Benzenecarboperoxoic acid, 3-chloro- (9CI)
3-Chlorobenzenecarboperoxoic acid
3-Chloro-benzenecarboperoxoic acid
m-Chlorobenzoyl hydroperoxide
3-Chloroperbenzoic acid
m-Chloroperoxybenzoic acid
3-Chloroperoxybenzoic acid
3-Chloroperoxybenzoic acid, Maximum concentration 86%
3-Chloroperoxybenzoic acid (Not more than 86% with 3-chlorobenzoic acid)
Peroxybenzoic acid, m-chloro- (8CI)
Peroxybenzoic acid, m-chloro-, Maximum concentration 86%
UN 2755

937-30-4
$C_{10}H_{12}O$
148.20
O=C(c(ccc(c1)CC)c1)C
Acetophenone, 4'-ethyl- (8CI)
AI3-08507
Ethanone, 1-(4-ethylphenyl)- (9CI)
1-(4-Ethylphenyl)ethanone

937-40-6
$C_8H_{10}N_2O$
150.20
CN(Cc1ccccc1)N=O
Benzylamine, N-methyl-N-nitroso
Methyl-benzyl-nitrosoamin (German)
Methylbenzylnitrosamine
N-Methyl-N-benzylnitrosamine
N-Methyl-N-nitrosobenzylamine
N-Nitrosobenzylmethylamine
N-Nitrosomethylbenzylamine

938-09-0
$C_8H_9ClO_2S$
204.68
Benzene, ((2-chloroethyl)sulfonyl)- (9CI)
NSC-145234
NSC-207409
NSC-236821
Sulfone, 2-chloroethyl phenyl (8CI)

938-22-7
$C_7H_4Cl_4O$
245.92
Benzene, 1,2,3,5-tetrachloro-4-methoxy- (9CI)
Anisole, 2,3,4,6-tetrachloro- (8CI)

938-23-8
$C_7H_5Cl_3O_2$
227.47
Phenol, 2,3,5-trichloro-6-methoxy- (8CI,9CI)

938-73-8
$C_9H_{11}NO_2$
165.21
O=C(N)c(c(OCC)ccc1)c1
Benzamide, o-ethoxy
Anovigam
Benzamide, 2-ethoxy- (9CI)
Etamide
Ethbenzamide
Ethenzamid
Ethenzamide
Ethosalicyl
o-Ethoxybenzamide
2-Ethoxybenzamide
Etocil
Etosalicil
Etosalicyl
Eusal
H.P. 209
Katagrippe
Lindatox
Lucamide
Pirosolvina
Protopyrin
Trancalgyl

938-86-3
$C_7H_4Cl_4O$
245.92
Benzene, 1,2,3,4-tetrachloro-5-methoxy- (9CI)
Anisole, 2,3,4,5-tetrachloro- (8CI)

939-23-1
$C_{11}H_9N$
155.21
n(ccc(c1)c(cccc2)c2)c1
Pyridine, 4-phenyl
p-Phenylpyridine
4-Phenylpyridine

939-27-5
$C_{12}H_{12}$
156.24
Naphthalene, 2-ethyl
2-Ethylnaphthalene

939-48-0
$C_{10}H_{12}O_2$
164.22
O=C(OC(C)C)c(cccc1)c1
Benzoic acid, isopropyl ester
Isopropyl benzoate
Isopropylester kyseliny benzoove (Czech)

939-97-9
$C_{11}H_{14}O$
162.23
O=Cc(ccc(c1)C(C)(C)C)c1
Benzaldehyde, 4-(1,1-dimethylethyl)-
AI3-37199
4-(1,1-Dimethylethyl)benzaldehyde

940-41-0
$C_8H_9Cl_3Si$
239.61
ClSi(Cl)(Cl)CCc1ccccc1
Silane, trichlorophenethyl
(2-Phenylethyl)trichlorosilane
Trichlor-2-fenylethylsilan (Czech)
Trichlorophenethylsilane
Trichloro-2-phenylethylsilane

940-71-6
$Cl_6N_3P_3$
347.64
ClP1(=NP(=NP(=N1)(Cl)Cl)(Cl)Cl)Cl
Hexachlorocyclotriphosphazene
AI3-24917
Hexachlorocyclotriphosphazine
NSC-209799
NSC-2667
Phosnic 390
Phosphonitrile chloride, cyclic trimer
Triphosphonitrilic chloride
1,3,5,2,4,6-Triazatriphosphorine, 2,2,4,4,6,6-hexachloro-2,2,4,4,6,6-hexahydro- (8CI,9CI)

941-69-5
$C_{10}H_7NO_2$
173.18
O=C(N(c(cccc1)c1)C(=O)C=2)C2
Maleimide, N-phenyl
N-Fenylimid kyseliny maleinove (Czech)
N-Phenylmaleimide

941-98-0
$C_{12}H_{10}O$
170.22
O=C(c(c(c(ccc1)cc2)c1)c2)C
1'-Acetonaphthone
1-Acetonaphthalene
α-Acetonaphthone
1-Acetonaphthone
1-Acetylnaphthalene
Ethanone, 1-(1-naphthalenyl)- (9CI)
α-Methyl naphthyl ketone
Methyl α-naphthyl ketone
Methyl 1-naphthyl ketone
1-(1-Naphthalenyl)ethanone
α-Naphthyl methyl ketone
1-Naphthyl methyl ketone

943-27-1
$C_{12}H_{16}O$
176.26
O=C(c(ccc(c1)C(C)(C)C)c1)C
Acetophenone, 4'-tert-butyl- (8CI)
4-tert-Butylacetophenone
1-(4-(1,1-Dimethylethyl)-phenyl)ethanone
Ethanone, 1-(4-(1,1-dimethylethyl)phenyl)- (9CI)

943-89-5
$C_{10}H_{10}O_3$
178.19
O=C(O)C=Cc(ccc(OC)c1)c1
2-Propenoic acid, 3-(4-methoxyphenyl)-, (E)- (9CI)
(E)-3-(4-Methoxyphenyl)-2-propenoic acid
trans-2-Propenoic acid, 3-(4-methoxyphenyl)-

944-22-9
$C_{10}H_{15}OPS_2$
246.34
CCOP(=S)(CC)Sc1ccccc1
Phosphonodithioic acid, ethyl-, O-ethyl S-phenyl ester
O-Aethyl-S-phenyl-aethyl-dithiophosphonat (German)
Difonate
Dyfonate
Dyphonate
ENT 25,796
O-Ethyl-S-phenyl ethylphosphonodithioate
O-Ethyl S-phenyl ethyldithiophosphonate
Fonofos (ACGIH,OSHA)
Fonophos
N 2790
OMS 410
Stauffer N 2790

944-61-6
$C_8H_6Cl_4O_2$
275.95
Benzene, 1,2,3,4-tetrachloro -5,6-dimethoxy- (8CI,9CI)

944-77-4
$C_8H_6Cl_4O_2$
275.95
Benzene, 1,2,3,5-tetrachloro-4,6-dimethoxy- (8CI,9CI)

944-78-5
$C_8H_6Cl_4O_2$
275.95
Benzene, 1,2,4,5-tetrachloro-3,6-dimethoxy- (8CI,9CI)

946-31-6
$C_6H_3ClN_2O_5$
218.56
Phenol, 2-chloro-4,6-dinitro
2-Chloro-4,6-dinitrophenol
USAF DO-60

947-02-4
$C_7H_{14}NO_3PS_2$
255.31
O=P(OCC)(OCC)N=C(SCC1)S1
Imidocarbonic acid, phosphonodithio-, cyclic ethylene p,p-diethyl ester
AC 47031
American Cyanamid 47031
American Cyanamid AC 47,031
American Cyanamid CL-47031
C.I. 47031
Cyalane
Cyclic ethylene(diethoxyphosphinothioyl)dithioimidocarbonate
Cyclic ethylene p,p-diethyl phosphonodithioimidocarbonate
Cylan
Cyolan
Cyolane
Cyolane Cylan
Cyolane Insecticide
(Diethoxyphosphinyl)dithioimidocarbonic acid cyclic ethylene ester
2-(Diethoxyphosphinylimino)-1,3-dithiolane
p,p-Diethyl cyclic ethylene ester of phosphonodithioimidocarbonic acid
EI 47031
ENT 25,830
1,2-Ethanedithiol, cyclic ester with p,p-diethyl phosphonodithioimidocarbonate
1,2-Ethanedithiol, cyclic S,S-ester with phosphonodithioimidocarbonic acid p,p-diethyl ester
Imidocarbonic acid, (diethoxyphosphinyl)dithio-, cyclic ethylene ester
Imidocarbonic acid, phosphonodithio-, p,p-diethyl cyclic ethylene ester
OMS 646
Phosfolan
Phospholan
Phosphoramidic acid, 1,3-dithiolan-2-ylidene-, diethyl ester

947-04-6
$C_{12}H_{23}NO$
197.36
O=C(NCCCCCCCCCC1)C1
Azacyclotridecan-2-one
2-Azacyclotridecanone
Cyclododecalactam

947-19-3

$C_{13}H_{16}O_2$
204.27
O=C(c(cccc1)c1)C(O)(CCC
C2)C2
Methanone, (1-hydroxycyclo-hexyl)phenyl- (9CI)
1-Hydroxycyclohexyl phenyl ketone
(1-Hydroxycyclohexyl)phenyl-methanone

947-72-8
$C_{14}H_9Cl$
212.68
Phenanthrene, 9-chloro
9-Chlorophenanthrene
10-Chlorophenanthrene

947-91-1
$C_{14}H_{12}O$
196.26
Acetaldehyde, diphenyl
Benzeneacetaldehyde, α-phenyl- (9CI)
Diphenylacetaldehyde
2,2-Diphenylacetaldehyde
Diphenylketen

947-92-2
$C_{12}H_{22}N_2O$
210.36
O=NN(C1CCCCC1)C2CCCCC2
Dicyclohexylamine, N-nitroso
Dicyclohexylnitrosamin (German)
Dicyclohexylnitrosamine
N-Nitrosodicyclohexylamine

948-03-8
$C_{12}H_{10}O_3$
202.21
O=C(OC)c(ccc(c1ccc2)c2)c1O
2-Naphthalenecarboxylic acid, 1-hydroxy-, methyl ester (9CI)

948-65-2
$C_{14}H_{11}N$
193.24
N(c(c(C=1)ccc2)c2)C1c(cccc3)c3
α-Phenylindole
Indole, 2-phenyl- (8CI)
1H-Indole, 2-phenyl- (9CI)
2-Phenyl-1H-indole

949-13-3
$C_{14}H_{22}O$
206.36
Phenol, o-octyl
o-Octylphenol
2-Octylphenol
Phenol, 2-octyl- (9CI)

949-38-2
$C_{13}H_{11}ClO$
218.68
Benzyl o-chlorophenyl ether
Benzene, 1-chloro-2-(phenyl-methoxy)-
Ether, benzyl o-chlorophenyl-

949-42-8
$C_6H_3Cl_6N_3$
329.81
s-Triazine, 2-methyl-4,6-bis-(trichloromethyl)- (8CI)
2-Methyl-4,6-bis(trichloro-methyl)-1,3,5-triazine
1,3,5-Triazine, 2-methyl-4,6-bis-(trichloromethyl)- (9CI)

949-43-9
$C_8H_8Cl_3N_3O$
268.53
s-Triazine, 2-(allyloxy)-4-methyl-6-(trichloro-methyl)- (7CI,8CI)

949-87-1
$C_{13}H_{12}N_2$
196.24
Diazene, (4-methylphenyl)-phenyl- (9CI)
Azobenzene, 4-methyl- (8CI)
NSC-31005

950-10-7
$C_8H_{16}NO_3PS_2$
269.34
O=P(OCC)(OCC)N=C(SCC1C)
S1
1,3-Dithiolane, 2-(diethoxy-phosphinylimino)-4-methyl
AC 47470
American Cyanamid CL-47470
CL-47,470
Cyclic propylene (diethoxy-phosphinyl)dithioimido-carbonate
Cytrolane
p,p-Diethyl cyclic propylene ester of phosphonodithioimi-docarbonic acid
2-(Diethoxyphosphinylimino)-4-methyl-1,3-dithiolane
Diethyl (4-methyl-1,3-dithiolan-2-ylidene)phosphoroamidate
EI-47470
ENT 25,991
Mephosfolan
Mephospholan
Phosphoramidic acid, (4-methyl-1,3-dithiolan-2-ylidene)-, diethyl ester

950-35-6
$C_8H_{10}NO_6P$
247.16
COP(=O)(OC)Oc1ccc(cc1)
N(=O)=O
Phosphoric acid, dimethyl p-nitrophenyl ester
Bay 11678
Desmethylnitrophos
O,O-Dimetyl-O-p-nitrofenyl-fosfat (Czech)
Dimethyl-p-nitrofenylester ky-seliny fosforecne (Czech)
Dimethyl-p-nitrophenyl phosph-ate
Dimethyl-4-nitrophenyl phosph-ate
Dimethyl paraoxon
Methyl-E-600
Methyl paraoxon
Paraoxon-methyl
Phosphoric acid, dimethyl-4-nitrophenyl ester (9CI)

950-37-8
$C_6H_{11}N_2O_4PS_3$
302.34
COc1nn(CSP(=S)(OC)OC)c
(=O)s1
Phosphorodithioic acid, O,O-dimethyl ester, S-ester with 4-(mercaptomethyl)-2-methoxy- δ^2-1,3,4-thiadia-zolin-5-one
Ciba-Geigy GS 13005
(O,O-Dimethyl)-S-(-2-methoxy-δ^2-1,3,4-thiadiazolin-5-on-4-ylmethyl)dithiophosphate
S-(2,3-Dihydro-5-methoxy-2-oxo-1,3,4-thiadiazol-3-methyl) dimethyl phosphoro-thiolothionate
S-2,3-Dihydro-5-methoxy-2-oxo-1,3,4-thiadiazol-3-yl-methyl O,O-dimethylphos-phorodithioate
O,O-Dimethyl S-(5-methoxy-1,3,4-thiadiazolinyl-3-methyl) dithiophosphate
O,O-Dimethyl-S-(2-methoxy-1,3,4-thiadiazol-5(4H)-onyl-(4)-methyl) phosphorodithio-ate
O,O-Dimethyl-S-(2-methoxy-1,3,4-thiadiazol-5-on-4-ly)-methyl-dithiophosphat (German)
O,O-Dimethyl-S-(2-methoxy-1,3,4-thiadiazol-5-(4H)-onyl-(4)-methyl)-dithiophosphat (German)
O,O-Dimethyl-S-((2-methoxy-1,3,4 (4H)-thiodiazol-5-on-4-yl)-methyl)-dithiofosfaat (Dutch)
O,O-Dimetil-S-((2-metossi-1,3,4-(4H)-tiadiazol-5-on-4-il)-metil)-ditifosfato (Italian)
DMTP(Japan)
ENT 27,193
Fisons NC 2964
Geigy 13005
Geigy GS-13005
GS 13005
Methidathion
Methidathion 50S
S-2-Metoksy-1,3,4-tiadiazolo-5-on-N-metylo-O,O-dwume-

tylowy (Polish)
S-((5-Methoxy-2-oxo-1,3,4-thiadiazol-3(2H)-yl)methyl) O,O-dimethyl phosphorodithioate
OMS 844
Somonil
Supracid
Supracide
Ultracid 40
Ultracide

951-86-0
$C_{14}H_{10}Cl_2$
249.14
Benzene, 1,1'-(1,2-dichloro-1,2-ethenediyl)-bis-, (E)- (9CI)
(E)-1,2-Dichloro-1,2-diphenylethene
trans-1,2-Dichlorodiphenylethylene
trans-Dichlorostilbene
trans-α,α'-Dichlorostilbene
Stilbene, α,α'-dichloro-, (E)- (8CI)

952-23-8
$C_{13}H_{11}N_3.ClH$
245.73
Cl.Nc3ccc2cc1ccc(N)cc1nc2c3
Acridine, 3,6-diamino-, monohydrochloride
3,6-Acridinediamine, monohydrochloride (9CI)
3,6-Diaminoacridine monohydrochloride
3,6-Diaminoacridinium chloride
3,6-Diaminoacridinium chloride hydrochloride
2,8-Diaminoacridinium chloride monohydrochloride
Proflavine hydrochloride
Proflavine monohydrochloride

952-47-6
$C_{13}H_{12}N_2O$
212.25
Phenol, 4-methyl-2-(phenylazo)- (9CI)
C.I. 11850
p-Cresol, 2-(phenylazo)- (6CI,7CI,8CI)
2-Hydroxy-5-methylazobenzene
4-Methyl-2-(phenylazo)phenol
Oil Yellow APC
2-Phenylazo-4-methylphenol
2-Phenylazo-4-methyl-1-phenol

953-17-3
$C_9H_{12}ClO_2PS_3$
314.81
COP(=S)(OC)SCSc1ccc(Cl)cc1
Phosphorodithioic acid, S-(((p-chlorophenyl)thio)-methyl) O,O-dimethyl ester
S-(((p-Chlorophenyl)thio)-methyl) O,O-dimethyl phosphorodithioate
S-(((4-Chlorophenyl)thio)-methyl) O,O-dimethylphosphorodithioate
Dimethyl p-chlorophenylthiomethyl dithiophosphate
O,O-Dimethyl-S-(p-chlorophenylthiomethyl)phosphorodithioate
O,O-Dimethylthiophosphoric acid p-chlorophenyl ester
ENT 25,586
ENT 25,599
G-29288
Geigy G-29288
Methanethiol, ((p-chlorophenyl)thio)-, s-ester with O,O-dimethyl phosphorodithioate
Methylcarbophenothion
Methyl trithion
R-1492
Stauffer R-1492
Tri-Me

953-91-3
$C_{13}H_{18}O_3S$
254.35
O=S(=O)(OC(CCCC1)C1)c(ccc(c2)C)c2
Benzenesulfonic acid, 4-methyl-, cyclohexyl ester (9CI)
Cyclohexyl 4-methylbenzenesulfonate
Cyclohexyl tosylate
p-Toluenesulfonic acid, cyclohexyl ester (8CI)

956-38-7
$C_7H_4BrCl_6N_3$
422.76
s-Triazine, 2-(2-bromo ethyl)-4,6-bis(tri chloromethyl)- (7CI, 8CI)

956-90-1
$C_{17}H_{25}N.ClH$
279.89
[Cl-].C1CCN(CC1)C2(CCCCC2)c3ccccc3
Piperidine, 1-(1-phenylcyclohexyl)-, hydrochloride
Angel Dust
C.I. 395
CN-25,253-2
DOA
Elephant tranquilizer
Elysion
GP-121
HOG
NSC-40902
PCP hydrochloride
Peace Pill
Phencyclidine hydrochloride
1-(1-Phenylcyclohexyl)piperidine hydrochloride
Sernyl
Sernylan
Sernyl hydrochloride
Trank

957-51-7
$C_{16}H_{17}NO$
239.34
CN(C)C(=O)C(c1ccccc1)c2ccccc2
Acetamide, N,N-dimethyl-2,2-diphenyl
Benzeneacetamide, N,N-dimethyl-α-phenyl- (9CI)
DIF 4
Diamide
Difenamid (Czech)
Dimethylamid kyseliny difenyloctove (Czech)
N,N-Dimethyldiphenylacetamide
N,N-Dimethyl-α,α-diphenylacetamide
N,N,-Dimethyl-2,2-diphenylacetamide
N,N-Dimethyl-α-phenylbenzeneacetamide
DIMID
Diphenamid
Diphenamide
2,2-Diphenyl-N,N-dimethylacetamide
Diphenylamide
DYMID
ENIDE
ENIDE 50
FDN
Fenam
L-34314
Lilly 34,314
U 4513
80W

958-93-0
$C_{14}H_{19}N_3S.ClH$
297.88
Pyridine, 2-((2-(dimethylamino)ethyl)-3-thenyl amino)-, monohydrochloride
Dethylandiamine
2-((2-(Dimethylamino)ethyl)-3-thenyl-amino)-pyridine hydrochloride
N,N-Dimethyl-N'-(3-thenyl)-N'-(2-pyridyl) ethylenediamine hydrochloride
Thenfadil hydrochloride
Thenyldiamine chloride
Thenyldiamine hydrochloride
WIN-2848

959-26-2
$C_{12}H_{14}O_6$
254.24
O=C(OCCO)c(ccc(c1)C(=O)OCCO)c1
Bis(2-hydroxyethyl) tere-

phthalate
1,4-Benzenedicarboxylic acid, bis(2-hydroxyethyl) ester (9CI)
Bis(ethylene glycol) terephthalate
Bis(hydroxyethyl) terephthalate
Bis(β-hydroxyethyl) terephthalate
Terephthalic acid, bis(2-hydroxyethyl) ester

959-52-4
$C_{12}H_{15}N_3O_3$
249.30
O=C(N(CN(C(=O)C=C)CN1C(=O)C=C)C1)C=C
s-Triazine, hexahydro-1,3,5-triacryloyl
Fixierer P
Triacrylformal
Triacryloylhexahydrotriazine
Tri(N-acryloyl)hexahydrotriazine
Triacryloylhexahydro-s-triazine
1,3,5-Triacryloylhexahydrotriazine
Triacryloylperhydrotriazine
1,3,5-Triazine, hexahydro-1,3,5-tris(1-oxo-2-propenyl)-(9CI)
Tris(N-acryloyl)hexahydrotriazine
Tris(acryloyl)hexahydro-s-triazine

959-55-7
$C_{17}H_{30}N.Cl$
283.93
Ammonium, benzyldimethyloctyl-, chloride
Benzenemethanaminium, N,N-dimethyl-N-octyl-, chloride (9CI)
Octyl-dimethyl-benzylammonium chloride

959-60-4
$C_9H_{12}Cl_3N_3O_3$
316.57
s-Triazine, 2,4,6-tris(2-chloroethoxy)- (7CI, 8CI)

959-98-8
$C_9H_6Cl_6O_3S$
406.91
5-Norbornene-2,3-dimethanol, 1,4,5,6,7,7-hexachloro-, cyclic sulfite, endo
α-Benzoepin
Endosulfan A
a-Endosulfan-α
α-Endosulfan
Endosulfan 1
α-Thiodan
β-Thionex

960-71-4
$C_{18}H_{15}B$
242.13
c1ccc(cc1)B(c2ccccc2)c3ccccc3
Borane, triphenyl-
AI3-60391
Triphenylborane

961-07-9
$C_{10}H_{13}N_5O_4$
267.28
O(C(C(O)C1)CO)C1N(c(nc(nc2O)N)c2N=3)C3
Guanosine, 2'-deoxy
Deoxyguanosine
2'-Deoxyguanosine
Guanine deoxyriboside

961-11-5
$C_{10}H_9Cl_4O_4P$
365.96
COP(=O)(OC)OC(=CCl)c1cc(Cl)c(Cl)cc1Cl
Phosphoric acid, 2-chloro-1-(2,4,5-trichlorophenyl)-vinyl dimethyl ester
Benzyl alcohol, 2,4,5-trichloro-α-(chloromethylene)-, dimethyl phosphate
2-Chloro-1-(2,4,5-trichlorophenyl)vinyl dimethyl phosphate
2-Chloro-1-(2,4,5-trichlorophenyl)vinyl phosphoric acid dimethyl ester
O,O-Dimethyl-O-2-chlor-1-(2,4,5-trichlorphenyl)-vinylphosphat (German)
IPO 8
NCI-C00168
Phosphoric acid, 2-chloro-1-(2,4,5-trichlorophenyl)ethenyl dimethyl ester
2,4,5-Trichloro-α-(chloromethylene)benzyl phosphate

961-22-8
$C_{10}H_{12}N_3O_4PS$
301.26
COP(=O)(OC)SCN2N=Nc1ccccc1C2-O
Azinphosmethyl oxon

961-68-2
$C_{12}H_9N_3O_4$
259.24
O=N(=O)c(ccc(Nc(cccc1)c1)c2N(=O)=O)c2
Diphenylamine, 2,4-dinitro
2,4-Dinitrodiphenylamine

961-69-3
$C_{14}H_{17}NO_4.K$
302.39
Benzeneacetic acid, α-((3-ethoxy-1-methyl-3-oxo-1-propenyl)amino)-, monopotassium salt, (R)- (9CI)
Dane Salt

961-71-7
$C_{17}H_{22}N_2$
254.41
Ethylenediamine, N-benzyl-N',N'-dimethyl-N-phenyl
Antergan
N-Benzyl-N',N'-dimethyl-N-phenylethylenediamine
Bridal
Dimetina
1,2-Ethanediamine, N,N-dimethyl-N'-phenyl-N'-(phenylmethyl)- (9CI)
Lergitin
NCI-C60719
Phenbenzamine
PM245
2339 RP

962-32-3
$C_{18}H_{12}O$
244.30
O1c4c1c3cc2ccccc2cc3c5ccccc45
Benz(a)anthracene, 5,6-epoxy-5,6-dihydro
Benz(a)anthracene, 5,6-oxide
Benz(a)anthra-5,6-oxide
Benz(3,4)anthra(1,2-b)oxirene, 1a,11b-dihydro- (9CI)
Benzo(a)anthracene 5,6-oxide
3,4-Dihydro-3,4-epoxy-1,2-benzanthracene
5,6-Epoxy-5,6-dihydrobenz(a)-anthracene

962-58-3
$C_{12}H_{21}N_2O_4P$
288.29
CCOP(=O)(OCC)Oc1cc(C)nc(n1)C(C)C
Diazoxon
O,O-Diethyl O-(2-isopropyl-4-methylpyrimid-6-yl)phosphate

963-89-3
$C_{19}H_{15}N$
257.35
Cc4ccc3nc1c(ccc2ccccc12)c(C)c3c4
Benz(c)acridine, 7,9-dimethyl
3,10-Dimethyl-7,8-benzacridine (French)
7,9-Dimethylbenz(c)acridine

964-79-4
$C_{13}H_{11}N_3O_4$
273.23
Benzenamine, N-(3-methylphenyl)-2,4-dinitro- (9CI)
m-Toluidine, N-(2,4-dinitrophenyl)- (8CI)

965-32-2
$C_{19}H_{34}N.Cl$
311.93
Benzenemethanaminium, N-decyl-N,N-dimethyl-, chloride (9CI)
N-Capryl-N,N-dimethylbenzyl-ammonium chloride

965-90-2
$C_{20}H_{32}O$
288.52
19-Nor-17-α-pregn-4-en-17-ol
Durabolin-O
Duraboral
Ethylestrenol
Ethylnandrol
Maxibalin
Maxibolin
Neodurabolin
19-Norpregn-4-en-17-ol, (17-α)-(9CI)
Orabolin
ORG-483
Orgabolin
Orgaboral

968-81-0
$C_{15}H_{20}N_2O_4S$
324.43
O=C(NC(CCCC1)C1)NS(=O)(=O)c(ccc(c2)C(=O)C)c2
Urea, 1-((p-acetylphenyl)-sulfonyl)-3-cyclohexyl
Acetohexamide
1-(p-Acetylbenzenesulfonyl)-3-cyclohexylurea
1-((p-Acetylphenyl)sulfonyl)-3-cyclohexylurea
Benzenesulfonamide, 4-acetyl-N-((cyclohexylamino)carbonyl)-
Cyclamide
Dimelin
Dimelor
Dymelor
NCI-C03247
Ordimel
Tsiklamid

968-93-4
$C_{19}H_{24}O_3$
300.40
Testolactone
1,2,3,4,4a,4b,7,9,10,10a-Decahydro-2-hydroxy-2,4b-dimethyl-7-oxo-1-phenanthrene-propionic acid δ-lactone
δ(1)-Dehydrotestolactone
1-Dehydrotestololactone
δ(1)-Dehydrotestololactone
1,2-Dehydrotestololactone
1,2-Didehydrotestololactone
Fludestrin
13-Hydroxy-3-oxo-13,17-seco-androsta-1,4-dien-17-oic acid δ-lactone
NSC-12173
NSC-23759
D-homo-17a-Oxaandrosta-1,4-diene-3,17-dione (9CI)
17α-Oxo-d-homo-1,4-andro-stadiene-3,17-dione
13,17-Secoandrosta-1,4-dien-17-oic acid, 13-hydroxy-3-oxo-, δ-lactone (8CI)
SQ 9538
Teolit
Teslac
Teslak
Testolacton
Testolactona (Spanish)
δ(1)-Testolactone
Testolactonum (Latin)
Testolattone
δ-1-Testololactone
δ(1)-Testololactone
Testololactone, 1-dehydro-
Testololactone, 1,2-didehydro-

970-06-9
$C_{22}H_{16}$
280.37
Naphthalene, 1,7-diphenyl-(8CI,9CI)

970-76-3
$C_{12}H_8N_4O_6$
304.19
Benzenamine, 2,4-dinitro-N-(4-nitrophenyl)- (9CI)
Diphenylamine, 2,4,4'-trinitro-(8CI)

970-91-2
$C_{12}H_8N_4O_6$
304.22
Benzenamine, 2,4-dinitro-N-(3-nitrophenyl)- (9CI)
Diphenylamine, 2,3',4-tri-nitro- (7CI,8CI)
2,4,3'-Trinitrodiphenylamine
2,3',4-Trinitrodiphenyl-amine

971-15-3
$C_{12}H_{20}N_2S_8$
448.82
N(C(=S)SSSSSSC(N(CCCC1)C1)=S)(CCCC2)C2
Piperidine, 1,1'-(hexathiodi-carbonothioyl)bis- (9CI)
1,1'-(Hexathiodicarbonothioyl)-bispiperidine

972-02-1
$C_{21}H_{27}NO$
309.49
OC(CCCN1CCCCC1)(c2ccccc2)c3ccccc3
1-Piperidinebutanol, α,α-di-phenyl
Benzhydrol, α-(3-piperidino-propyl)-
Difenidol
Diphenidol
α,α-Diphenyl-1-piperidine-butanol
Nometic
SKF 478
SK&F No. 478-A
Vontrol

973-21-7
$C_{14}H_{18}N_2O_7$
326.34
CCC(C)c1cc(cc(N(=O)=O)c1OC(=O)OC(C)C)N(=O)=O
Carbonic acid, 2-sec-butyl-4,6-dinitrophenyl isopropyl ester
Acrec
Acrex
(2-sek.Butyl-4,6-dinitrofenyl)-isopropylkarbonat (Czech)
2-sec-Butyl-4,6-dinitrophenyl isopropyl carbonate
Carbonic acid, 1-methylethyl 2-(1-methylpropyl)-4,6-di-nitrophenylester (9CI)
Dessin
2,4-Dinitro-6-sek.butyl-iso-propylphenylcarbonat (German)
2,4-Dinitro-6-sec-butylphenyl isopropyl carbonate
Dinobuton
Dinofen
Drawinol
DS 18302
ENT 27,244
Isophen
Isophen (Pesticide)
Isopropyl 2,4-dinitro-6-sec-butyl phenyl carbonate
Isopropyl-2-(1-methyl-n-propyl)-4,6-dinitrophenyl carbonate
Kasebon
MC 1053
1-Methylethyl 2-(1-ethyl-propyl)-4,6-dinitrophenyl carbonate
2-(1-Methyl-2-propyl)-4,6-di-nitrophenyl isopropylcarbonate
OMS 1056
Phenol, 2-sec-butyl-4,6-dinitro-, isopropylcarbonate
Sytasol
Talan
UC 19786
Union Carbide 19786

979-32-8
$C_{24}H_{32}O_3$
368.56
CCCCC(=O)OC3CCC4C2CCc1cc(O)ccc1C2CCC34C
Estradiol, 17-valerate
Atladiol
Deladiol
Delahormone Unimatic
Delestrogen
Delestrogen 4X
Dura-estradiol

Estradiol valerate
Estradiol 17-β-valerate
Estradiol valerianate
Estra-1,3,5(10)-triene-3,17-diol (17-β)-, 17-pentanoate (9CI)
Estraval
Femogex
Neofollin
Oestradiol valerate
Pharlon
Progynon
Progynon-Depot
Progynova

980-26-7
$C_{22}H_{16}N_2O_2$
340.37
Oc(c(c(nc1cc(c(nc(c2cc(c3)C)c3)c4)c2O)ccc5C)c5)c14
Quino(2,3-b)acridine-7,14-dione, 5,12-dihydro-2,9-dimethyl- (9CI)
C.I. Pigment Red 122

981-40-8
$C_{22}H_{23}O_4P$
382.42
Phosphoric acid, (p-tert-butylphenyl) diphenyl ester

982-57-0
$C_{15}H_{15}Cl_2N_2O_8.Na$
445.21
Acetamide, 2,2-dichloro-N-(β-hydroxy-α-(hydroxymethyl)-p-nitrophenethyl)-, α-ester with sodium succinate
Chloramphenicol monosuccinate sodium salt
Chloramphenicol sodium monosuccinate
Chloramphenicol sodium succinate
Chloramphenicol succinate sodium
Chloramphenicol-sukzinat-natrium (German)
Protophenicol
Sodium chloramphenicol succinate

989-38-8
$C_{28}H_{30}N_2O_3.ClH$
479.06
Benzoic acid, o-(6-(ethylamino)-3-(ethylimino)-2,7-dimethyl-3H-xanthen-9-yl)-, ethyl ester, monohydrochloride
Basic Red 1
Basic Rhodamine Yellow
Basic Rhodaminic Yellow
Calcozine Red 6G
Calcozine Rhodamine 6GX
Cerven Zasadita 1 (Czech)
C.I. 45160
C.I. Basic Red 1
C.I. Basic Red 1, monohydrochloride
Elcozine Rhodamine 6GDN
Eljon Pink Toner
Fanal Pink GFK
Fanal Red 25532
Heliostable Brilliant Pink B Extra
Mitsui Rhodamine 6GCP
NCI-C56122
Nyco Liquid Red GF
Rhodamine 69DN Extra
Rhodamine F4G
Rhodamine F5G
Rhodamine F5G Chloride
Rhodamine 6G
Rhodamine 6GB
Rhodamine 6G (Biological stain)
Rhodamine 6GBN
Rhodamine 6GCP
Rhodamine 4GD
Rhodamine 6GD
Rhodamine GDN
Rhodamine 5GDN
Rhodamine 6 GDN
Rhodamine 6 GDN Extra
Rhodamine 6GEX Ethyl ester
Rhodamine 6G Extra
Rhodamine 6G Extra base
Rhodamine 4GH
Rhodamine 6GH
Rhodamine 5GL
Rhodamine 6G Lake
Rhodamine 6GX
Rhodamine J
Rhodamine 6JH
Rhodamine 7JH
Rhodamine Lake Red 6G
Rhodamine Y 20-7425
Rhodamine ZH
Rhodamine 6ZH
Silosuper Pink B

991-42-4
$C_{33}H_{25}N_3O_3$
511.61
5-Norbornene-2,3-dicarboximide, 5-(α-hydroxy-α-2-pyridylbenzyl)-7-(α-2- pyridylbenzylidene)
Compound S-6,999
ENT 51,762
MCN 1025
Nobormide
Norbormide
Raticate
S-6,999
Shoxin

992-59-6
$C_{34}H_{26}N_6O_6S_2.2Na$
724.76
1-Naphthalenesulfonic acid, 3,3'-((3,3'-dimethyl(1,1'-biphenyl)-4,4'-diyl)bis(azo))bis (4-amino-, disodium salt
Amanil Purpurine 4B
Atul Direct Red 4B
Azamin 4B
Azocard Red 4B
Bencidal Purple 4B
Benzanil Purpurine 4B
Benzopurpurin 4B
Benzopurpurine 4B
Benzopurpurine 4BKX
Benzopurpurine 4BX
Brasilamina Red 4B
Calcomine Red 4BX
Cerven prima 2 (Czech)
Chrome Leather Red 4B
C.I. 23500
C.I. Direct Red 2
Cotton Red 4B
Diacotton Benzopurpurine 4B
Diamine Purpurine 4B
Diaphtamine Purpurine
Diazamine Purpurine 4B
Diazine Red 4B
Diazol Purpurine 4B
Diphenyl Red 4B
Diphenyl Red 4BS
Direct Purpurine 4B
Direct Purpurine M4B
Direct Red 2
Direct Red 4A
Direct Red 4B
Direct Red DCB
Ditolylbis(azonaphthionic acid)
Eclipse Red
Erie Benzo 4BP
Erie Red 4B
Fast Scarlet
Hispamin Red 4B
Kayaku Benzopurpurine 4B
Mitsui Benzopurpurine 4BX
Paper Red 4B
Phenamine Purpurine 4B
Purpurin 4B
Purpurine 4B
Tertrodirect Red 4B

993-12-4
C_2H_6ClPS
128.56
Dimethyl phosphorochloridothioate
Chlorodimethylphosphine sulfide
Dimethylphosphinothioic chloride
Dimethylphosphinothioyl chloride
Dimethyl thiophosphoryl chloride
Dimethylthiophosphoryl chloride
NA 2267
Phosphinothioic chloride, dimethyl-
UN 2267 [Dimethyl thiophosphoryl chloride]

993-13-5
CH_5O_3P
96.02
CP(O)(O)=O
Methylphosphonate
AI3-51156
Phosphonic acid, methyl- (9CI)

993-16-8
CH_3Cl_3Sn
240.08
Stannane, methyltrichloro
Methyltin trichloride
Methyltrichlorostannane
Methyltrichlorotin
Monomethyltin trichloride
Stananne, trichloromethyl- (8CI,9CI)
Tin, methyl-, trichloride
Trichloromethylstannane
Trichloromethyltin

993-43-1
$C_2H_5Cl_2PS$
163.00
S=P(CC)(Cl)Cl
Phosphonothioic dichloride, ethyl
Dichloroethylphosphine sulfide
Ethyl phosphonothioic dichloride, Anhydrous (DOT)
Ethylphosphonothioic dichloride
Ethylphosphonothionic dichloride
Ethyl phosphonothioyl dichloride
Ethylthionophosphonyl dichloride
Ethylthiophosphonic dichloride
NA 1760 (DOT)

994-05-8
$C_6H_{14}O$
102.18
Butane, 2-methoxy-2-methyl- (9CI)
Ether, methyl tert-pentyl (8CI)
TAME (Ether)

994-30-9
$C_6H_{15}ClSi$
150.72
CCSi(Cl)(CC)CC
Silane, chlorotriethyl- (9CI)
Chlorotriethylsilane
Triethylchlorosilane

994-31-0
$C_6H_{15}ClSn$
241.35
CC[Sn](Cl)(CC)CC
Stannane, chlorotriethyl
Chlorotriethylstannane
Chlorotriethyltin
Tin, triethyl-, chloride
Triaethylzinnchlorid (German)
Triethylchlorostannane
Triethylchlorotin
Triethylstannium chloride
Triethylstannyl chloride
Triethyltin chloride

994-32-1
$C_6H_{16}OSn$
222.90
Stannane, triethylhydroxy- (8CI,9CI)

994-36-5
$C_6H_8O_7.xNa$
Unknown
1,2,3-Propanetricarboxylic acid, 2-hydroxy-, sodium salt (9CI)

995-33-5
$C_{17}H_{34}O_6$
334.51
O=C(OCCCC)CCC(OOC(C)(C)C)(OOC(C)(C)C)C
Valeric acid, 4,4-bis(tert-butylperoxy)-, butyl ester
4,4-Bis(tert-butylperoxy)valeric acid butyl ester
n-Butyl-4,4-di(tert-butylperoxy)-valerate, Not more than 52% with inert solid (DOT)
n-Butyl-4,4-di(tert-butylperoxy)-valerate, Technically pure (DOT)
Pentanoic acid, 4,4-bis((1,1-dimethylethyl)dioxy)-, butyl ester
Trigonox 17/40
UN 2140 (DOT)
UN 2141 (DOT)

996-98-5
$C_2H_6N_4O_2$
118.12
O=C(NN)C(=O)NN
Oxalic acid, dihydrazide
Ethanedioic acid, dihydrazide (9CI)
Oxaldihydrazide
Oxalhydrazide
Oxalic acid bishydrazide
Oxalic acid hydrazide
Oxalic dihydrazide
Oxalic hydrazide
Oxaloyl dihydrazide
Oxaloylhydrazide
Oxalyl dihydrazide
Oxalyl hydrazide
Oxalylhydrazine

997-50-2
$C_6H_{16}Sn$
206.91
Stannane, triethyl
Monohydrotriethyletain (French)
TET
Triethylstannane
Triethyltin
Triethyltin hydride

998-30-1
$C_6H_{16}O_3Si$
164.31
CCOSi(OCC)OCC
Silane, triethoxy
Triethoxysilane

998-40-3
$C_{12}H_{27}P$
202.36
P(CCCC)(CCCC)CCCC
Phosphine, tributyl
Tributylfosfin (Czech)
Tributylphosphine
Tri-n-butylphosphine

999-21-3
$C_{10}H_{12}O_4$
196.22
O=C(OCC=C)C=CC(=O)OCC=C
Maleic acid, diallyl ester
Diallylester kyseliny maleinove (Czech)
Diallyl maleate
Sipomer DAM

999-33-7
$C_8H_{18}ClN$
163.69
1-Butanamine, N-butyl-N-chloro- (9CI)
Chlorodibutylamine
N-Chlorodibutylamine
Dibutylamine, N-chloro- (6CI,7CI,8CI)
Dibutylchloramine

999-55-3
$C_6H_8O_2$
112.13
O=C(OCC=C)C=C
Acrylic acid, allyl ester (8CI)
AI3-15698
Allyl acrylate
2-Propenoic acid, 2-propenyl ester (9CI)
2-Propenyl 2-propenoate

999-61-1
$C_6H_{10}O_3$
130.16
O=C(OCC(O)C)C=C
Acrylic acid, 2-hydroxypropyl ester
HPA
β-Hydroxypropyl acrylate
2-Hydroxypropyl acrylate (ACGIH,OSHA)
1,2-Propanediol, 1-acrylate
2-Propenoic acid, 2-hydroxypropyl ester (9CI)
Propylene glycol monoacrylate

999-81-5
$C_5H_{13}ClN.Cl$
158.09
[Cl-].C[N+](C)(C)CCCl
Ammonium, (2-chloroethyl)-trimethyl-, chloride

AC 38555
Antywylegacz
CCC
CCC Plant Growth Regulant
CE CE CE
2-Chloraethyl-trimethyl-ammoniumchlorid (German)
Chlorcholinchlorid (Czech, German)
Chlorcholine chloride
Chlormequat
Chlormequat chloride
Chlorocholine chloride
(β-Chloroethyl)trimethyl-ammonium chloride
(2-Chloroethyl)trimethyl-ammonium chloride
2-Chloro-N,N,N-trimethylethan-aminium chloride
Choline dichloride
60-CS-16
Cyclocel
Cycocel
Cycocel-Extra
Cycogan
Cycogan Extra
Cyocel
EI 38,555
Ethanaminium, 2-chloro-N,N,N-trimethyl-, chloride (9CI)
HICO CCC
Hormocel-2CCC
Increcel
Lihocin
NCI-C02960
Retacel
Stabilan
Trimethyl-β-chlorethylammon-iumchlorid (Czech)
TUR

999-97-3
$C_6H_{19}NSi_2$
161.44
Disilazane, 1,1,1,3,3,3-hexa-methyl
Bis(trimethylsilyl)amine
Hexamethyldisilazane
Hexamethylsilazane
HMDS
OAP
Silanamine, 1,1,1-trimethyl-N-(trimethylsilyl)- (9CI)

1000-36-8
$C_{11}H_{25}O_3P$
236.29
Phosphonic acid, methyl-, di-pentyl ester (8CI,9CI)

1000-82-4
$C_2H_6N_2O_2$
90.07
O=C(NCO)N
(Hydroxymethyl)urea
N-(Hydroxymethyl)urea
1-(Hydroxymethyl)urea
Methylolurea
N-Methylolurea
Methylolureas
Mono(hydroxymethyl)urea
Monomethylolurea
Urea, (hydroxymethyl)- (9CI)

1000-90-4
$C_8H_{14}O_2S_4Zn$
270.46
Zinc, bis(O-(1-methylethyl) carbonodithioato-S,S')-, (T-4)- (9CI)
Carbonodithioic acid, O-(1-methylethyl) ester, zinc salt

1001-55-4
$C_4H_5NO_2$
99.10
CC(=O)OCC#N
Glycolonitrile, acetate
Acetic acid cyanomethyl ester
Cyanomethyl acetate
Kyanmethylester kyseliny octove (Czech)

1002-11-5
$C_{10}H_{21}Cl$
176.73
Decane, 3-chloro- (6CI, 7CI,8CI,9CI)
3-Chlorodecane

1002-16-0
$C_5H_{11}NO_3$
133.14
Amyl nitrate
AI3-15292
Amylester kyseliny dusicne (Czech)
Nitrate d'amyle (French)
Nitrato de amilo (Spanish)
Nitric acid, pentyl ester
UN 1112 [Amyl nitrate]

1002-43-3
$C_{12}H_{26}$
170.34
Undecane, 3-methyl- (8CI,9CI)

1002-53-5
$C_8H_{20}Sn$
234.97
Stannane, dibutyl
Dibutylstannane
Dibutyltin
Dibutyltin dihydride (6CI)
Dibutyltin hydride
Dibutyl-zinn (German)

1002-62-6
$C_{10}H_{19}O_2.Na$
194.28
Decanoic acid, sodium salt
Caprinic acid sodium salt
Sodium caprinate
Sodium caprate
Sodium decanoate
Sodium-n-decanoate
Sodium decanoic acid

1002-67-1
$C_7H_{16}O_3$
148.23
CCOCCOCCOC
Ether, 2-ethoxyethyl 2-meth-oxyethyl
Diethylene glycol ethyl methyl ether
2-Ethoxyethyl 2-methoxyethyl ether
Ethylmethylether diethylen-glykolu (Czech)

1002-69-3
$C_{10}H_{21}Cl$
176.73
ClCCCCCCCCCC
1-Chlorodecane
Decane, 1-chloro- (9CI)

1002-84-2
$C_{15}H_{30}O_2$
242.45
O=C(O)CCCCCCCCCCCCCC
Pentadecanoic-acid
Pentadecyclic acid

1002-89-7
$C_{18}H_{35}O_2.H_4N$
301.58
Stearic acid, ammonium salt
Ammonium stearate (ACGIH)
Octadecanoic acid, ammonium salt

1003-03-8
$C_5H_{11}N$
85.14
NC(CCC1)C1
Cyclopentanamine (9CI)
Cyclopentylamine (8CI)

1003-09-4
C_4H_3BrS
163.04
S(C(=CC=1)Br)C1
Thiophene, 2-bromo- (9CI)
2-Bromothiophene

1003-14-1
$C_5H_{10}O$
86.15
Pentane, 1,2-epoxy
1,2-Epoxypentane
Propyloxirane

1003-29-8
C_5H_5NO
95.09
O=CC(NC=C1)=C1

Pyrrole-2-carboxaldehyde (8CI)
AI3-35104
1H-Pyrrole-2-carboxaldehyde (9CI)

1003-38-9
$C_6H_{12}O$
100.16
2,5-Dimethyltetrahydrofuran
Furan, tetrahydro-2,5-dimethyl- (9CI)
2,5-Me2-THF

1003-67-4
C_6H_7NO
109.14
O=n(ccc(c1)C)c1
Pyridine, 4-methyl-, 1-oxide
4-Methylpyridine 1-oxide

1003-73-2
C_6H_7NO
109.14
O=n(cccc1C)c1
Pyridine, 3-methyl-, 1-oxide
3-Methylpyridine-1-oxide

1004-24-6
$C_8H_{14}O$
126.20
Cyclohexanemethanol, 4-methylene- (7CI,8CI, 9CI)
4-Methylenecyclohexane methanol

1004-29-1
$C_8H_{16}O$
128.21
Furan, 2-butyltetrahydro- (8CI,9CI)
Octane, 1,4-epoxy-

1004-38-2
$C_4H_7N_5$
125.10
n(c(N)cc(n1)N)c1N
2,4,6-Triaminopyrimidine
AI3-60016
2,4,6-Pyrimidinetriamine (9CI)
Pyrimidine, 2,4,6-triamino- (8CI)

1005-35-2
$C_8H_{11}OS.Cl$
190.69
(4-Hydroxyphenyl)dimethylsulfonium chloride
(p-Hydroxyphenyl)dimethylsulfonium chloride
Sulfonium, (p-hydroxyphenyl)-dimethyl-, chloride
Sulfonium, (4-hydroxyphenyl)-dimethyl-, chloride (9CI)

1005-64-7
$C_{10}H_{12}$
132.21
Benzene, 1-butenyl-, (E)- (9CI)
1-Butene, 1-phenyl-, (E)- (8CI)

1005-67-0
$C_8H_{17}NO$
143.22
Morpholine, 4-butyl- (9CI)
4-Butylmorpholine

1005-93-2
$C_6H_{11}O_4P$
178.14
CC1(CCO)C(O)[P](=O)C1O
1,3-Propanediol, 2-ethyl-2-(hydroxymethyl)-, cyclic phosphate
4-Aethyl-1-phospha-2,6,7-trioxabicyclo(2.2.2)octan-1-oxid (German)
2-Ethyl-2-(hydroxymethyl)-1,3-propanediol, cyclic phosphate (1:1)
4-Ethyl-1-phospha-2,6,7-trioxabicyclo(2.2.2)octane-1-oxide
4-Ethyl-2,6,7-trioxa-1-phosphabicyclo(2.2.2)octane-1-oxide
2-(Hydroxymethyl)-2-ethyl-1,3-propanediol, cyclic phosphate (1:1)
TMP-P
Trimethyolpropane phosphate
Trimethylopropane phosphate
2,6,7-Trioxa-1-phosphabicyclo-(2.2.2)octane, 4-ethyl-, 1-oxide (9CI)

1006-31-1
$C_7H_4Cl_4$
229.92
Benzene, 1,2,4,5-tetrachloro-3-methyl- (9CI)
2,3,5,6-Tetrachlorotoluene
Toluene, 2,3,5,6-tetrachloro- (6CI,7CI,8CI)

1006-32-2
$C_7H_4Cl_4$
229.92
Benzene, 1,2,3,4-tetrachloro-5-methyl- (9CI)
2,3,4,5-Tetrachlorotoluene
Toluene, 2,3,4,5-tetrachloro (6CI,7CI,8CI)

1006-94-6
C_9H_9NO
147.19
COc2ccc1[nH]ccc1c2
Indole, 5-methoxy
Methoxy-5 indole (French)
5-Methoxyindole

1007-26-7
$C_{11}H_{16}$
148.25
CC(C)(C)Cc1ccccc1
Benzene, (2,2-dimethylpropyl)- (9CI)
Benzene, neopentyl- (8CI)

1007-28-9
$C_5H_8ClN_5$
173.57
6-Deisopropylatrazine
2-Amino-4-chloro-6-ethylamino-s-triazine
Caswell No. 033F
G-28279
s-Triazine, 2-amino-4-chloro-6-(ethylamino)- (8CI)
1,3,5-Triazine-2,4-diamine, 6-chloro-N-ethyl- (9CI)

1007-32-5
$C_{10}H_{12}O$
148.20
2-Butanone, 1-phenyl- (8CI, 9CI)
NSC-133447

1007-36-9
$C_8H_{10}N_2O$
150.20
CNC(=O)Nc1ccccc1
Urea, 1-methyl-3-phenyl
N-Methyl-N'-phenylurea
1-Methyl-3-phenylurea

1008-72-6
$C_7H_6O_4S.Na$
209.18
Benzenesulfonic acid, o-formyl-, sodium salt (8CI)
Benzenesulfonic acid, 2-formyl-, sodium salt (9CI)
Sodium 2-formylbenzenesulfonate

1008-80-6
$C_{12}H_{22}$
166.31
Naphthalene, decahydro-2,3-dimethyl- (8CI,9CI)

1008-88-4
$C_{11}H_9N$
155.19
n(cccc1c(cccc2)c2)c1
3-Phenylpyridine
Pyridine, 3-phenyl- (9CI)

1008-89-5
$C_{11}H_9N$
155.21

c1ccc(cc1)c2ccccn2
Pyridine, 2-phenyl
o-Phenylpyridine
2-Phenylpyridine

1009-61-6
$C_{10}H_{10}O_2$
162.19
O=C(c(ccc(c1)C(=O)C)c1)C
Benzene, p-diacetyl- (8CI)
Ethanone, 1,1'-(1,4-phenylene)-bis- (9CI)
1,1'-(1,4-Phenylene)bisethanone

1009-74-1
$C_9H_{15}N_3$
165.24
1,3,5-Triazine, 2,4,6-triethyl- (9CI)
s-Triazine, 2,4,6-triethyl- (6CI,7CI,8CI)
Triethyl-s-triazine
2,4,6-Triethyl-1,3,5-triazine

1010-48-6
$C_{11}H_{14}O_2$
178.25
Hydrocinnamic acid, β,β-di methyl
Benzenepropanoic acid, β,β-dimethyl- (9CI)
β,β-Dimethylhydrocinnamic acid
3-Phenylisopentanoic acid
β-Phenylisovaleric acid

1011-12-7
$C_{12}H_{18}O$
178.27
O=C(C(=C(CCCC1)C1)CCC2)C2
Cyclohexanone, 2-cyclohexylidene- (9CI)
AI3-04026
2-Cyclohexylidenecyclohexanone

1011-91-2
$C_7H_{13}N_5S$
199.25
1,3,5-Triazine-2(1H)-thione, 4,6-bis(ethylamino)- (9CI)
s-Triazine-2-thiol, 4,6-bis-(ethylamino)- (8CI)

1012-72-2
$C_{14}H_{22}$
190.33
c(ccc(c1)C(C)(C)C)(c1)C(C)(C)C
1,4-Di-tert-butylbenzene
AI3-11248
Benzene, 1,4-bis(1,1-dimethylethyl)- (9CI)
Benzene, p-di-tert-butyl- (8CI)
1,4-Bis(1,1-dimethylethyl)-benzene
p-Di-tert-butylbenzene

1013-23-6
$C_{12}H_8OS$
200.26
Dibenzothiophene, 5-oxide (9CI)
Dibenzothiophene sulfone

1013-75-8
$C_{11}H_{15}NO_2$
193.24
Carbanilic acid, N-ethyl-, ethyl ester (8CI)
AI3-06187
Carbamic acid, ethylphenyl-, ethyl ester (9CI)
Ethyl N-ethyl-N-phenylcarbamate
Ethyl phenylurethane
N-Ethyl-N-phenylurethane

1014-41-1
$C_{14}H_{22}$
190.33
Benzene, 1,4-bis(1-methylpropyl)- (9CI)
Benzene, p-di-sec-butyl- (8CI)

1014-69-3
$C_8H_{15}N_5S$
213.34
CNc1nc(NC(C)C)nc(SC)n1
s-Triazine, 2-(isopropylamino)-4-(methylamino)-6-(methylthio)
Desmethryn
Desmetryn (German, Dutch)
Desmetryne
G 34360
GS 34360
2-Isopropilamino-4-metilamino-6-metiltio-1,3,5-triazina (Italian)
2-Isopropylamino-4-methylamino-6-methylmercapto-s-triazine
2-Isopropylamino-4-methylamino-6-methylthio-1,3,5-triazine
2-Methylamino-4-methylthio-6-isopropylamino-1,3,5-triazine
Methylmercapto-4-isopropylamino-6-methylamino-s-triazine
2-Methylthio-4-isopropylamino-6-methylamino-s-triazine
2-(Methylthio)-4-(methylamino)-6-(isopropylamino)-s-triazine
Samuron
Semeron
Topusyn

1014-70-6
$C_8H_{15}N_5S$
213.34
CCNc1nc(NCC)nc(SC)n1
s-Triazine, 2,4-bis(ethylamino)-6-(methylthio)
2,4-Bis(ethylamino)-6-methylmercapto-s-triazine
4,6-Bis(ethylamino)-2-methylthio-1,3,5-triazine
2,4-Bis(ethylamino)-6-(methylthio)-s-triazine
Cymetrin
N,N'-Diethyl-6-(methylthio)-1,3,5-triazine-2,4-diamine
G 32911
Gy-Bon
2-Methylmercapto-4,6-bis(ethylamino)-s-triazine
2-Methylthio-4,6-bis(ethylamino)-s-triazine
2-Methylthio-4,6-bis(monoethylamino)-2-triazine
Simetryn
Simetryne
Symetryne

1014-83-1
$C_{10}H_{10}O_4$
194.19
2-Propenoic acid, 3-(4-hydroxy-3-methoxyphenyl)-, (Z)- (9CI)
Cinnamic acid, 4-hydroxy-3-methoxy-, (Z)- (8CI)

1016-05-3
$C_{12}H_8O_2S$
216.26
Dibenzothiophene, 5,5-dioxide
AI3-03709

1017-55-6
$C_{12}H_{21}N_3$
207.32
1,3,5-Triazine, 2,4,6-tripropyl- (9CI)
s-Triazine, 2,4,6-tripropyl- (7CI,8CI)
Tripropyl-s-triazine
2,4,6-Tripropyl-1,3,5-triazine

1017-56-7
$C_6H_{12}N_6O_3$
216.24
n(c(nc(n1)NCO)NCO)c1NCO
Methanol, (s-triazine-2,4,6-triyltriimino)tri
Melamine, N^2,N^4,N^6-tris-(hydroxymethyl)-
N,N',N''-Trihydroxymethylmelamine
N,N',N''-Tris(hydroxymethyl)-1,3,5-triazine-2,4,6-triamine
Trimethylolmelamine

1018-71-9
$C_{10}H_6Cl_2N_2O_2$

257.08
Pyrrole, 3-chloro-4-(3-chloro-2-nitrophenyl)
3-Chloro-4-(3-chloro-2-nitrophenyl)pyrrole
3-Chloro-4-(2'-nitro-3'-chlorophenyl)pyrrole
NSC-107654
PN
Pyroace
Pyrollnitrin
Pyrrolnitrin

1020-39-9
$C_{13}H_{10}ClNO$
231.69
Clc1ccccc1NC(=O)c2ccccc2
Benzamide, N-(2-chlorophenyl)
Benzanilide, 2'-chloro- (8CI)
2'-Chlorobenzanilide
N-(2-Chlorophenyl)benzamide

1020-53-7
$C_{10}H_{15}N_5S$
237.33
s-Triazine, 2,4-bis(allyl amino)-6-(methylthio)- (7CI,8CI)

1022-22-6
$C_{14}H_9Cl_3$
283.58
ClC=C(c1ccc(Cl)cc1)c2ccc(Cl)cc2
Ethylene, 1,1-bis(p-chlorophenyl)-2-chloro
1,1-Bis(p-chlorophenyl)-2-chloroethylene

1024-57-3
$C_{10}H_5Cl_7O$
389.30
ClC2C1OC1C3C2C4(Cl)C(=C(Cl)C3(Cl)C4(Cl)Cl)Cl
4,7-Methanoindan, 1,4,5,6,7,8,8-heptachloro-2,3-epoxy-3a,4,7,7a-tetrahydro
ENT 25,584
Epoxyheptachlor
HCE
Heptachlor epoxide
1,4,5,6,7,8,8-Heptachloro-2,3-epoxy-2,3,3a,4,7,7a-hexahydro-4,7-methanoindene
1,4,5,6,7,8,8-Heptachloro-2,3-epoxy-3a,4,7,7a-tetrahydro-4,7-methanoindan
2,3,4,5,6,7,7-Heptachloro-1a,1b,5,5a,6,6a-hexahydro-2,5-methano-2H-indeno-(1,2-b)oxirene
Hiptachlor epoxide
2,5-Methano-2H-oxireno-(a)indene, 2,3,4,5,6,7,7-heptachloro-1a,1b,5,5a,6,6a-hexahydro-
Velsicol 53-CS-17

1025-15-6
$C_{12}H_{15}N_3O_3$
249.25
O=C(N(C(=O)N(C1=O)CC=C)CC=C)N1CC=C
1,3,5-Triazine-2,4,6(1H,3H,5H)-trione, 1,3,5-tri-2-propenyl- (9CI)
AI3-60290
Triallyl isocyanurate

1031-07-8
Unknown
Unknown
ClC2=C(Cl)C3(Cl)C1COS(=O)(=O)OCC1C2(Cl)C3(Cl)Cl
5-Norbornene-2,3-dimethanol, 1,4,5,6,7,7-hexachloro-, cyclic sulfate
Endosulfan sulfate

1031-47-6
$C_{12}H_{19}N_6OP$
294.26
Triamiphos
AI3-27223
5-Amino-1-bis(dimethylamide)-phosphoryl-3-phenyl-1,2,4-triazole
5-Amino-1-bis(dimethylamido)-phosphoryl-3-phenyl-1,2,4-triazole
5-Amino-1-(bis(dimethyl-amino)phosphinyl)-3-phenyl-1,2,4-triazole
5-Amino-3-fenyl-1-bis(di-methyl-amino)-fosforyl-1,2,4-triazool (Dutch)
5-Amino-3-fenil-1-bis(-dimetil-amino)-fosforil-1,2,4-triazolo (Italian)
5-Amino-3-phenyl-1-bis (di-methyl-amino)-phosphoryle-1,2,4-triazole (French)
5-Amino-3-phenyl-1-bis(di-methylamino)-phosphoryl-1H-1,2,4-triazol (German)
5-Amino-3-phenyl-1,2,4-tri-azole-1-yl-N,N,N',N'-tetra-methylphosphodiamide
5-Amino-3-phenyl-1,2,4-tri-azolyl-1-bis(dimethylamido)-pheosphate
5-Amino-3-phenyl-1,2,4-tri-azolylbis(dimethylamino)-phosphinoxid (German)
5-Amino-3-phenyl-1,2,4-tria-zolyl-N,N,N'N'-tetramethyl-phosphonamide
p-(5-Amino-3-phenyl-1H-1,2,4-triazol-1-yl)-N,N,N',N'-tetramethylphosphonic diamide
Bis(dimethylamino)-3-amino-5-phenyltriazolyl phosphine oxide
Caswell No. 037A
EPA Pesticide Chemical Code 237200
Niagara 5943
3-Phenyl-5-amino-1,2,4-tria-zolyl-(1)-(N,N'-tetramethyl) diamidophosphonate
Phosphonic diamide, P-(5-am-ino-3-phenyl-1H-1,2,4-triazol-1-yl)-N,N,N',N'-tetramethyl-(8CI,9CI)
Triamifos (German, Dutch, Italian)
Triamphos
Wepsin
Wepsyn
Wepsyn 155
WP 155

1034-41-9
$C_{10}H_2Cl_{10}O$
492.65
Chlordecone alcohol
AI3-24875
1,3,4-Metheno-1H-cyclobuta-(cd)pentalen-2-ol, 1,1a,3,3a,4,5,5,5a,5b,6-decachloroocta-hydro-

1037-57-6
$C_{10}H_{10}N_6O_5$
294.19
2-Furaldehyde, 5-nitro-, (4,6-dimethoxy-s-triazin-2-yl)-hydrazone (8CI)

1041-00-5
$C_{18}H_{14}N_2O_2$
290.31
O(c(c(N=1)cc(c2)C)c2)C1C=CC(Oc(c3cc(c4)C)c4)=N3
Benzoxazole, 2,2'-(1,2-ethenediyl)bis(5-methyl- (9CI)
α,β-Di(5-methylbenzoxazol-2-yl)ethene
2,2'-(1,2-Ethenediyl)bis-(5-methylbenzoxazole)

1047-16-1
$C_{20}H_{12}N_2O_2$
312.32
Oc(c(c(nc1cc(c(nc(c2ccc3)c3)c4)c2O)ccc5)c5)c14
Cinquasia Red
C.I. Pigment Red 122
C.I. Pigment Violet 19
C.I. 46500
Cinquasia Red B
Cinquasia Red Y
Cinquasia Violet
Cinquasia Violet R
5,12-Dihydroquino(2,3-b)acridine-7,14-dione
Hostaperm Red E 3B
Hostaperm Red E 5B
Hostaperm Red Violet ER
Linear trans quinacridone
Monastral Red
Monastral Red B
Monastral Red Y

Monastral Violet R
Paliogen Red BG
Permanent Red E3B
Permanent Red E5B
Pigment Quinacridone Red
Pigment Violet 19
Pigment Violet #19
PV Fast Red E 3B
PV Fast Red E 5B
Quinacridone
Quinacridone Red
Quinacridone Red MC
Quinacridone Violet
Quinacridone Violet MC
Quino(2,3-b)acridine-7,14-dione, 5,12-dihydro- (9CI)

1052-38-6
$C_{18}H_{18}N_8$
346.44
N(=Nc(c(N)cc(N)c1)c1)c(cccc2N=Nc(c(N)cc(N)c3)c3)c2
1,3-Benzenediamine, 4,4'-(1,3-phenylenebis(azo))bis

1058-92-0
$C_{16}H_{11}ClN_2O_9S_2.2Na$
520.82
Eriochrome Blue SE
3-((5-Chloro-2-hydroxyphenyl)-azo)-4,5-dihydroxy-2,7-naphthalenedisulfonic acid disodium salt
2,7-Naphthalenedisulfonic acid, 3-((5-chloro-2-hydroxy phenyl)azo)-4,5-dihydroxy-, disodium salt (9CI)

1064-48-8
$C_{22}H_{16}N_6O_9S_2.2Na$
618.54
2,7-Naphthalenedisulfonic acid, 4-amino-5-hydroxy-3-((4-nitrophenyl)azo)-6-(phenylazo)-, disodium salt
Acidal Black 10B
Acidal Navy Blue 3BR
Acid Black 10A
Acid Black 10B
Acid Black 12B
Acid Black 10BA
Acid Black Base M
Acid Black 4BN
Acid Black 4BNU
Acid Black 10BN
Acid Black BRX
Acid Black BX
Acid Black H
Acid Black 1
Acid Black JVS
Acid Blue Black B
Acid Blue Black 10B
Acid Blue Black BG
Acid Blue Black Double 600
Acid Leather Blue IGW
Acid Leather Dark Blue G
Acid Leather Fast Blue Black G
Airedale Black 2BG
Amacid Black 10BR
Atul Acid Black 10BX
Atul Acid Black BX
Azanol Fast Acid Black 10B
Azo Dark Blue C 2B
Azo Dark Blue HR
Azo Dark Blue S
Azo Dark Blue SH
Blue Black 12B
Blue Black SX
Boruta Black A
Brasilan Black BS
Bucacid Blue Black
Calcocid Blue Black
Calcocid Blue Black 2R
Cern Kysela 1 (Czech)
C.I. 20470
C.I. Acid Black 1 (7CI)
C.I. Acid Black 1, Disodium salt (8CI)
Colacid Black 10A
Comacid Blue Black B
Diacid Blue Black 10B
Eniacid Black IVS
Eniacid Black SH
Eriosin Blue Black B
Fast Sulon Black BN
Fenazo Blue Black

1066-17-7
$C_{45}H_{85}N_{13}O_{10}$
968.43
Colistin
Colimycin
Colisticina
Coly-Mycin
Colymysin S
Polymyxin E
Totazina

1066-30-4
$C_6H_9O_6.Cr$
229.15
[Cr+3].CC([O-])=O.CC([O-])=O.CC([O-])=O
Acetic acid, chromium(3+) salt
Chromic acetate
Chromic acetate (III)
Chromium acetate
Chromium(III) acetate
Chromium triacetate

1066-33-7
$CO_3.2H_4N$
96.11
Ammonium bicarbonate (1:1)
Acid ammonium carbonate
Ammonium carbonate
Ammonium hydrogen carbonate
Carbonic acid, monoammonium salt
Monoammonium carbonate

1066-35-9
C_2H_7ClSi
94.62
Silane, chlorodimethyl- (9CI)
Chlorodimethylsilane
Dimethylchlorosilane

1066-42-8
$C_2H_8O_2Si$
92.17
Silanediol, dimethyl- (9CI)
Dimethylsilanediol

1066-45-1
C_3H_9ClSn
199.26
C[Sn](C)(C)Cl
Stannane, chlorotrimethyl
Chlorotrimethylstannane
Chlorotrimethyltin
M&T Chemicals 1222-45
Trimethylchlorostannane
Trimethylchlorotin
Trimethylstannyl chloride
Trimethyltin chloride

1066-51-9
CH_6NO_3P
111.03
1-Aminomethylphosphonic acid
Caswell No. 037C
EPA Pesticide Chemical Code 207800
Phosphonic acid, (aminomethyl)- (9CI)

1067-08-9
C_8H_{18}
114.23
Pentane, 3-ethyl-3-methyl- (8CI,9CI)
NSC-73956

1067-14-7
$C_6H_{15}ClPb$
329.85
Plumbane, chlorotriethyl
Lead, triethyl-, chloride
Plumbane, triethylchloro-
Triethyl lead chloride
Triethylplumbium chloride

1067-20-5
C_9H_{20}
128.26
3,3-Diethylpentane

1067-33-0
$C_{12}H_{24}O_4Sn$
351.05
CCCC[Sn](CCCC)(OC(C)=O)OC(C)=O
Stannane, diacetoxydibutyl
BA 2726
Bis(acetyloxy)dibutylstannane
Diacetoxybutyltin
Diacetoxydibutylstannane
Diacetoxydibutyltin
Dibutylstannium diacetate

Dibutyl tin diacetate
Fomrez Sul-3
NCI-C02028
T 1 (Catalyst)
Tin, dibutyl-, diacetate

1067-53-4
$C_{11}H_{24}O_6Si$
280.44
Silane, tris(2-methoxyethoxy)-vinyl
A 172
6-Ethenyl-6-(methoxyethoxy)-2,5,7,10-tetraoxa-6-silaundecane (9CI)
GF 58
Silicone A-172
Tris(methoxyethoxy)vinylsilane
(Tris(β-methoxyethoxy))vinylsilane
Tris(2-methoxyethoxy)vinylsilane
Vinyltris(methoxyethoxy)silane
Vinyltris(β-methoxyethoxy)silane
Vinyltris(2-methoxyethoxy)silane

1067-66-9
$C_{12}H_{28}N_2O_5Si$
308.44
2-Oxa-7,10-diaza-3-silatridecan-13-oic acid, 3,3-dimethoxy-, methyl ester (9CI)

1067-71-6
$C_7H_{15}O_4P$
194.17
Phosphonic acid, (2-oxopropyl)-, diethyl ester (9CI)
NSC-408852
Phosphonic acid, acetonyl-, diethyl ester (8CI)

1067-90-9
$C_8H_{17}O_4P$
208.20
Phosphonic acid, (3-oxobutyl)-, diethyl ester (7CI,8CI,9CI)
Diethyl (3-oxobutyl)-phosphonate

1067-98-7
$C_9H_{18}Cl_3O_4P$
327.57
1-Propanol, 3-chloro-, phosphate (3:1) (8CI,9CI)

1068-27-5
$C_{16}H_{30}O_4$
286.46
O(OC(C)(C)C)C(C#CC(OOC(C)(C)C)(C)C)(C)C
3-Hexyne, 2,5-dimethyl-2,5-di(t-butylperoxy)
2,5-Dimethyl-2,5-bis-(tert-butylperoxy)hexyne-3, Max. concent. 52% with inert solid (DOT)
2,5-Dimethyl-2,5-di(t-butylperoxy)hexyne-3
2,5-Dimethyl-2,5-di-(tert-butylperoxy)hexyne-3, Technically pure (DOT)
3-Hexyne, 2,5-dimethyl-2,5-di-(t-butylperoxy)-, Maximum concentration 52% with inert solid
UN 2158 (DOT)
UN 2159 (DOT)

1068-61-7
$C_4H_6O_2.1/2Pb$
189.69
Lead methacrylate
Lead(2+) 2-methyl-2-propenoate
2-Propenoic acid, 2-methyl-, lead(2+) salt (9CI)

1068-87-7
C_9H_{20}
128.26
2,4-Dimethyl-3-ethyl pentane
Pentane, 3-ethyl-2,4-dimethyl-(9CI)

1068-90-2
$C_9H_{15}NO_5$
217.25
O=C(NC(C(=O)OCC)C(=O)OCC)C
Malonic acid, acetamido-, diethyl ester
Acetamidomalonic acid diethyl ester
Diethylester kyseliny acetylaminomalonove (Czech)

1069-53-0
C_9H_{20}
128.26
Hexane, 2,3,5-trimethyl-(8CI,9CI)

1069-66-5
$C_8H_{15}O_2.Na$
166.22
Valeric acid, 2-propyl-, sodium salt
Convulex
Depakene
Depakine
Depekane
Dipropylacetate sodium
DPA sodium
Epilim
Ergenyl
Eurekene
KW-066
Labazene
Pentanoic acid, 2-propyl-, sodium salt
2-Propylpentanoic acid sodium salt
2-Propylvaleric acid sodium salt
Sodium bispropylacetate
Sodium dipropylacetate
Sodium α,α-dipropylacetate
Sodium n-dipropylacetate
Sodium 2-propylpentanoate
Sodium 2-propylvalerate
Sodium valproate
Valproate sodium
Valproic acid sodium salt

1070-00-4
$C_{24}H_{51}Al$
366.65
Tri-n-octylaluminum
Aluminum, trioctyl- (9CI)
Trioctylaluminum

1070-03-7
$C_8H_{19}O_4P$
210.21
O=P(OCC(CCCC)CC)(O)O
2-Ethylhexyl dihydrogen phosphate
Mono(2-ethylhexyl) phosphate
Mono(2-ethylhexyl)phosphate
Phosphoric acid, mono(2-ethylhexyl) ester (9CI)

1070-10-6
$C_8H_{18}O.1/4Ti$
142.21
1-Hexanol, 2-ethyl-, titanium(4+) salt (9CI)
2-Ethyl-1-hexanol titanium(4+) salt

1070-19-5
$C_5H_9N_3O_2$
143.13
O=C(OC(C)(C)C)N=N=N
tert-Butoxycarbonyl azide
Azida de terc-butoxicarbonilo (Spanish)
Azoture de tert-butoxycarbonyle (French)
t-Butoxycarbonyl azide
tert-Butyl azidoformate
tert-Butyloxycarbonyl azide
Carbonazidic acid, 1,1-dimethylethyl ester (9CI)
1,1-Dimethylethyl carbonazidate
Formic acid, azido-, tert-butyl ester (8CI)

1070-78-6
$C_3H_4Cl_4$
181.87
C(CCCl)(Cl)(Cl)Cl
Propane, 1,1,1,3-tetrachloro
1,1,1,3-Tetrachloropropane

1070-83-3
$C_6H_{12}O_2$

116.16
O=C(O)CC(C)(C)C
Butanoic acid, 3,3-dimethyl- (9CI)
AI3-19252
Butyric acid, 3,3-dimethyl- (8CI)
3,3-Dimethylbutanoic acid

1070-87-7
C_9H_{20}
128.26
Pentane, 2,2,4,4-tetramethyl- (9CI)

1071-22-3
$C_3H_4Cl_3NSi$
188.52
ClSi(Cl)(Cl)CCC#N
Propionitrile, 3-(trichlorosilyl)
β-Cyanoethyltrichlorosilane
2-Cyanoethyltrichlorosilane
Silane, (2-cyanoethyl)trichloro-
Trichlor-2-kyanethylsilan (Czech)

1071-23-4
$C_2H_8NO_4P$
141.08
O=P(OCCN)(O)O
Ethanol, 2-amino-, dihydrogen phosphate (ester)
Ethanolamine phosphate
Ethanol, 2-amino-, phosphate
Phosphoethanolamine
O-Phosphoethanolamine
Phosphoric acid, 2-aminoethyl phenyl ester
Phosphoryethanolamine
Phosphorylethanolamine

1071-26-7
C_9H_{20}
128.26
Heptane, 2,2-dimethyl- (9CI)

1071-81-4
$C_{10}H_{22}$
142.28
Hexane, 2,2,5,5-tetramethyl- (8CI,9CI)
Bineopentyl

1071-83-6
$C_3H_8NO_5P$
169.09
OC(=O)CNCP(O)(O)=O
Glycine, N-(phosphonomethyl)
Glyphosate
MON 0573
N-(Phosphonomethyl)glycine

1072-05-5
C_9H_{20}
128.26
Heptane, 2,6-dimethyl- (8CI, 9CI)

1072-16-8
$C_{10}H_{22}$
142.28
Octane, 2,7-dimethyl- (8CI,9CI)
Diisoamyl
NSC-6232

1072-33-9
$C_{15}H_{30}O_2$
242.40
1-Tridecanol, acetate
AI3-35251

1072-35-1
$C_{18}H_{36}O_2.1/2Pb$
388.08
Lead stearate
Lead(2+) octadecanoate
Octadecanoic acid, lead(2+) salt (9CI)

1072-44-2
C_3H_7N
57.09
CN1CC1
Aziridine, 1-methyl- (8CI,9CI)

1072-52-2
C_4H_9NO
87.14
OCCN(C1)C1
1-Aziridineethanol
2-(1-Aziridinyl)ethanol
β-Hydroxy-1-ethylaziridine
2-Hydroxy-1-ethylaziridine
N-(β-Hydroxyethyl)aziridine
N-(2-Hydroxyethyl)aziridine
N-Hydroxyethyl ethylene imine
N-(2-Hydroxyethyl)ethylenimine
1-(2-Hydroxyethyl)ethylenimine

1072-63-5
$C_5H_6N_2$
94.10
N(C=CN1C=C)=C1
1H-Imidazole, 1-ethenyl- (9CI)
1-Ethenyl-1H-imidazole

1072-71-5
$C_2H_2N_2S_3$
150.24
N(N=C(S)S1)=C1S
1,3,4-Thiadiazole-2,5-dithiol
Bismuthiol I
2,5-Dimercaptothiadiazole
2,5-Dimercapto-1,3,4-thiadiazole
PY 61H
1,3,4-Thiadiazol-dithiol-(2,5) (German)
USAF A-8354

1072-83-9
C_6H_7NO
109.12
O=C(C(NC=C1)=C1)C
2-Acetylpyrrole
Ethanone, 1-(1H-pyrrol-2-yl)- (9CI)
Ketone, methyl pyrrol-2-yl (8CI)
1-(1H-Pyrrol-2-yl)ethanone

1072-91-9
$C_6H_{10}N_2$
110.15
1H-Pyrazole, 1,3,5-trimethyl- (9CI)
Pyrazole, 1,3,5-trimethyl- (8CI)

1072-98-6
$C_5H_5ClN_2$
128.55
Nc1ccc(Cl)cn1
2-Pyridinamine, 5-chloro-
AI3-52448

1073-06-9
C_6H_4BrF
175.01
Fc(cccc1Br)c1
Benzene, 1-fluoro-3-bromo
3-Bromfluorbenzen (Czech)
3-Bromofluorobenzene

1073-23-0
C_7H_9NO
123.15
O=n(c(ccc1)C)c1C
2,6-Lutidine N-oxide
AI3-60115
2,6-Dimethylpyridine 1-oxide
2,6-Lutidine, 1-oxide (8CI)
Pyridine, 2,6-dimethyl-, 1-oxide (9CI)

1073-67-2
C_8H_7Cl
138.60
c(ccc(c1)C=C)(c1)Cl
p-Chlorostyrene
Benzene, 1-chloro-4-ethenyl- (9CI)
1-Chloro-4-ethenylbenzene
4-Chlorostyrene
Parachlorostyrene
Styrene, p-chloro- (8CI)
Styrene, 4-chloro-

1073-72-9
C_7H_8OS
140.21
Oc(ccc(SC)c1)c1
Phenol, p-(methylthio)
4-Methylmercaptophenol

p-(Methylthio)phenol
4-(Methylthio)phenol

1073-75-2
$C_4H_9O_4P$
152.09
O(P(OC1)OCCO)C1
Ethanol, 2-(1,3,2-dioxaphospholan-2-yloxy)- (9CI)
2-(1,3,2-Dioxaphospholan-2-yloxy)ethanol
Phosphorus acid, cyclic ethylene 2-hydroxyethyl ester

1074-11-9
$C_8H_8Cl_2$
175.06
Benzene, (1,2-dichloroethyl)- (8CI,9CI)

1074-17-5
$C_{10}H_{14}$
134.22
c(c(ccc1)C)(c1)CCC
Benzene, 1-methyl-2-propyl- (9CI)
1-Methyl-2-propylbenzene
Toluene, o-propyl- (8CI)

1074-43-7
$C_{10}H_{14}$
134.22
c(cccc1C)(c1)CCC
Benzene, 1-methyl-3-propyl- (9CI)
1-Methyl-3-propylbenzene
Toluene, m-propyl- (8CI)

1074-55-1
$C_{10}H_{14}$
134.22
c(ccc(c1)C)(c1)CCC
Benzene, 1-methyl-4-propyl- (9CI)
1-Methyl-4-propylbenzene
Toluene, p-propyl- (8CI)

1074-82-4
$C_8H_5NO_2.K$
186.23
[K+].[O-]C1=NC(=O)c2ccccc12
Potassium phthalimide
1H-Isoindole-1,3(2H)-dione, potassium salt (9CI)
1H-Isoindole-1,3(2h)-dione, potassium salt
Phthalimide, potassium salt (8CI)
n-Potassiophthalimide
Potassium phthalimidate

1075-49-6
$C_9H_8O_2$
148.16
Benzoic acid, 4-ethenyl- (9CI)
Benzoic acid, p-vinyl- (8CI)
NSC-176003

1075-59-8
$C_5H_7N_3O_3$
157.11
1,3,5-Triazin-2(1H)-one, 4,6-dimethoxy- (9CI)
s-Triazin-2-ol, 4,6-dimethoxy- (8CI)

1076-22-8
$C_6H_6N_4O_2$
166.16
Cn1c(=O)[nH]c(=O)c2[nH]cnc12
Xanthine, 3-methyl
3-Methylxanthine

1076-46-6
$C_7H_8Cl_2N_2O_2$
223.06
Ammonium chloramben
3-Amino-2,5-dichlorobenzoic acid, ammonium salt
Caswell No. 168B
Chloramben ammonium salt
EPA Pesticide Chemical Code 029902

1076-47-7
$C_{10}H_{12}O_2$
164.20
Benzoic acid, 2,3,4-trimethyl- (8CI,9CI)
Prehnitylic acid

1076-61-5
$C_{12}H_{16}$
160.26
Naphthalene, 1,2,3,4-tetrahydro-6,7-dimethyl- (8CI,9CI)

1077-16-3
$C_{12}H_{18}$
162.27
c(cccc1)(c1)CCCCCC
n-Hexylbenzene
Benzene, hexyl- (9CI)
Hexane, 1-phenyl- (8CI)
Hexylbenzene

1077-56-1
$C_9H_{13}NO_2S$
199.27
O=S(=O)(NCC)c(c(ccc1)C)c1
N-Ethyl-o-toluenesulfonamide
Benzenesulfonamide, N-ethyl-2-methyl- (9CI)
N-Ethyl-2-methylbenzenesulfonamide
o-Toluenesulfonamide, N-ethyl-

1078-19-9
$C_{11}H_{12}O_2$
176.22
O=C(c(cc(cc(OC)c1)CC2)c1)C2
1(2H)-Naphthalenone, 3,4-dihydro-6-methoxy- (9CI)
3,4-Dihydro-6-methoxy-1(2H)-naphthalenone

1078-61-1
$C_9H_{10}O_4$
182.19
Hydrocinnamic acid, 3,4-dihydroxy
Benzenepropanoic acid, 3,4-dihydroxy- (9CI)
Dihydrocaffeic acid
3,4-Dihydroxyhydrocinnamic acid
Hydrocaffeic acid

1078-71-3
$C_{13}H_{20}$
176.30
c(cccc1)(c1)CCCCCCC
Benzene, heptyl- (9CI)
Heptane, 1-phenyl- (8CI)
Heptylbenzene

1079-21-6
$C_{12}H_{10}O_2$
186.22
Oc(c(c(cccc1)c1)cc(O)c2)c2
2,5-Biphenyldiol
2,5-Dihydroxybiphenyl
Hydroquinone, phenyl-
Phenylhydroquinone

1079-33-0
$C_{10}H_9NO_2S$
207.26
CNC(=O)Oc1cccc2sccc12
Carbamic acid, methyl-, benzo(b)thien-4-yl ester
4-Benzothienylester kyseliny methylkarbaminove (Czech)
Benzo(b)thien-4-yl methylcarbamate
4-Benzothienyl methylcarbamate
Benzo(b)thiophene-4-ol, methylcarbamate
ENT 27,041
MCA-600
Mobam
Mobam phenol
Mobil MC-A-600
MOS-708
OMS-708

1079-66-9
$C_{12}H_{10}ClP$
220.64
c(P(c(cccc1)c1)Cl)(cccc2)c2
Phosphinous chloride, diphenyl- (9CI)
Diphenylphosphinous chloride

1079-96-5
$C_{14}H_{22}$
190.33
Benzene, 1,3-bis(1-methyl propyl)- (9CI)
Benzene, m-di-sec-butyl- (6CI,7CI,8CI)

1081-71-6
$C_{10}H_{10}O_5$
210.19
Vanilpyruvic acid
4-Hydroxy-3-methoxy-α-oxo-benzenepropanoic acid
3-Methoxy-4-hydroxyphenyl-pyruvic acid

1081-75-0
$C_{15}H_{16}$
196.29
Benzene, 1,1'-(1,3-propane-diyl)bis- (9CI)
Propane, 1,3-diphenyl- (8CI)

1081-77-2
$C_{15}H_{24}$
204.36
c(cccc1)(c1)CCCCCCCCC
Benzene, nonyl- (9CI)
Nonylbenzene

1082-88-8
$C_{12}H_{19}NO_3$
225.28
DEA No. 7390
α-Methylmescaline
Phenethylamine, α-methyl-3,4,5-trimethoxy-
Phenethylamine, 3,4,5-tri-methoxy-α-methyl-
3,4,5-Trimethoxyamphetamine
Trimethoxyphenyl-β-aminopropane
3,4,5-Trimethoxyphenyl-β-aminopropane

1085-98-9
$C_9H_{11}Cl_2FN_2O_2S_2$
333.24
CN(C)S(=O)(=O)N(SC(F)(Cl)Cl)c1ccccc1
Sulfamide, N-((dichlorofluoro-methyl)thio)-N',N'-dimethyl-N-phenyl
Aniline, N-((dichlorofluoro-methyl)thio)-N-((dimethyl-amino)sulfonyl)-
Bay 47531
Bayer 47531
Dichlofluanid
Dichlofluanide
Dichlorfluanid (Czech)
N-Dichlorfluormethylthio-N',N'-dimethylaminosulfon-saeureanilid (German)
N-(Dichlor-fluor-methyl-thio)-N',N'-dimethyl-N-phenyl-schwefel-saeurediamid (German)
1,1-Dichloro-N-((dimethyl-amino)sulfonyl)-1-fluoro-N-phenylmethane sulfenamide
N-(Dichlorofluoromethylthio)-N',N'-dimethyl-N-phenylsulf-amide
N-(Dichlorofluoromethylthio)-N-(dimethylsulfamoyl)-aniline
N,N-Dimethyl-N'-phenyl-N'-fluorodichloromethylthio-sulfamide
Elvaron
Eparen
Euparen
Euparene
KU 13-032-C
KUE 13032c

1087-21-4
$C_{14}H_{14}O_4$
246.26
O=C(OCC=C)c(cccc1C(=O)OCC=C)c1
Isophthalic acid, diallyl ester (8CI)
AI3-16904
1,3-Benzenedicarboxylic acid, di-2-propenyl ester (9CI)
Di-2-propenyl 1,3-benzenedi-carboxylate

1088-11-5
$C_{15}H_{11}ClN_2O$
270.73
Clc3ccc2NC(=O)CN=C(c1ccccc1)c2c3
2H-1,4-Benzodiazepin-2-one, 7-chloro-1,3-dihydro-5-phenyl
A-101
2H-1,4-Benzodiazepin-2-one, 1,3-dihydro-7-chloro-5-phenyl-
7-Chloro-1,3-dihydro-5-phenyl-2H-1,4-benzodiazepin-2-one
Calmday
Dealkylprazepam
Demethyldiazepam
N-Demethyldiazepam
1-Demethyldiazepam
N-Deoxydemoxapam
Desalkylprazepam
N-Descyclopropylmethyl prazepam
Desmethyldiazepam
N-Desmethyldiazepam
DMDZ
Madar
NDD
Nordiazepam
Norprazepam
Ro 5-2180
Stilny

1090-13-7
$C_{18}H_{10}O_2$
258.28
O=C(c(c(c(C(=O)c1cccc2)cc(c3ccc4)c4)c3)c12
5,12-Naphthacenedione (9CI)

1095-66-5
$C_{18}H_{34}O_2.C_4H_9NO$
369.58
9-Octadecenoic acid (Z)-, Compd. with morpholine (1:1) (9CI)
Morpholine 9-octadecenoate
Morpholine oleate
Oleic acid, morpholine salt
Oleic acid, morpholine soap

1096-48-6
$C_{20}H_{19}NO_2$
305.37
O=C(c(c(c(NC(CCCC1)C1)cc2)C(=O)c3cccc4)c2)c34
9,10-Anthracenedione, 1-(cyclohexylamino)- (9CI)
1-Cyclohexanaminoanthracene-9,10-dione
1-(Cyclohexylamino)-9,10-anthracenedione

1096-84-0
$C_{21}H_{16}O_2$
300.37
2-Naphthol, 1,1'-methylenedi
1,1'-Methylenebis(2-naphthol)

1100-88-5
$C_{25}H_{22}P.Cl$
388.88
[Cl-].C(c1ccccc1)[P+](c2ccccc2)(c3ccccc3)c4ccccc4
Phosphonium, benzyltri-phenyl-, chloride (8CI)
Phosphonium, triphenyl(phenyl-methyl)-, chloride (9CI)
Triphenyl(phenylmethyl)phos-phonium chloride

1103-38-4
$C_{20}H_{14}N_2O_4S.1/2Ba$
447.07
Barium Lithol Red
Barium Lithol
Calcotone Red B
C.I. Pigment Red 49, Barium salt (2:1)
C.I. Pigment Red 49:1
D & C Red No. 12
Dainichi Lithol Red R
Eljon Lithol Red MS
Irgalite Red BRL
Isol Red Toner GB
Isol Red Toner RB
Isol Red 3BK
Isol Tobias Red GB
Isol Tobias Red RB
Isol Tobias Red 3BK
Light Red RB
Light Red RCN
Lithol Red Barium Toner

Lithol Red 18959
Lithol Red 22060
Lithol Red 27965
1-Naphthalenesulfonic acid, 2-((2-hydroxy-1-naphthalenyl)azo)-, barium salt (2:1) (9CI)
Pigment Red 49:1
Poster Red
Red No. 207
Red Toner YTA
1883 Red
Sanyo Fast Red NN
Sanyo Lacquer Red RN
Sanyo Lithol Red R
Symuler Red 2R Ba Salt
Vulcanosine Red RBKX

1103-39-5
$C_{20}H_{14}N_2O_4S.1/2Ca$
398.45
Calcium Lithol Red
Brilliant Toner RB
Calcium lithol
Calcotone Red 2B
C.I. Pigment Red 49 Ca Salt
C.I. Pigment Red 49, calcium salt (2:1) (8CI)
C.I. Pigment Red 49:2
D & C Red No. 11
Eljon Bordeaux
Eljon Lithol Red BS
Isol Red Toner RC
Isol Tobias Red RC
Light Red RCA
Lithol Red CA
Lithol Red Calcium Toner
Lithol Red RC Extra
Lithol Red Toner 3BX
Lithol Red 19592
1-Naphthalenesulfonic acid, 2-((2-hydroxy-1-naphthalenyl)azo)-, calcium salt (2:1) (9CI)
No 66 Conc Lithol Toner
Pigment Red 49:2
Red No. 206
Red Toner EBA
1793 Red
Vulcanosine Red RCKX
Winthrop Red X 1666

1111-67-7
CHNS.Cu
122.63
Cuprous thiocyanate
Caswell No. 266A
Copper(1+) thiocyanate
EPA Pesticide Chemical Code 025602
Thiocyanic acid, copper(1+) salt (8CI,9CI)

1111-78-0
$CH_2NO_2.H_4N$
78.09
Carbamic acid, ammonium salt
Ammonium aminoformate
Ammonium carbamate (DOT)
NA 9083 (DOT)

1112-38-5
$C_2H_7O_3PS$
142.12
O,O-Dimethyl phosphorothionate
AI3-21247
Phosphorothioic acid, O,O-dimethyl ester

1112-63-6
$C_{12}H_{30}OSn_2$
427.80
Distannoxane, hexaethyl
Hexaethyldistannoxane
1,1,1,3,3,3-Hexaethyldistannoxane

1113-01-5
$C_4H_9O_4PS_2$
216.22
COP(=S)(OC)SCC(O)=O
Phosphorodithioic acid, O,O-dimethyl S-carboxymethyl ester
Acetic acid, ((dimethoxyphosphinothioyl)thio)-
Carboxy derivative of dimethoate
O,O-Dimethyl S-carboxymethyl phosphorodithioate
O,O,-Dimethyldithiophosphorylacetic acid

1113-02-6
$C_5H_{12}NO_4PS$
213.21
Phosphorothioic acid, O,O-dimethyl ester, S-ester with 2-mercapto-N-methylacetamide
Acetamide, 2-mercapto-N-methyl-, S-ester with O,O-dimethylphosphorothioate
Bay 45432
Bayer 45,432
Dimethoate O-Analog
Dimethoate Oxygen Analog
Dimethoate Po Isologue
Dimethoxon
O,O-Dimethyl S-(2-(methylamino)-2-oxoethyl)phosphorothioate
O,O-Dimethyl-S-((N-methylcarbamoyl)-methyl)-monothiofosfaat (Dutch)
O,O-Dimethyl-S-(N-methyl-carbamoyl)-methyl-monothiophosphat (German)
O,O-Dimethyl S-((methylcarbamoyl)methyl)phosphorothioate
O,O-Dimethyl-S-(N-methylcarbamoylmethyl)phosphorothioate
O,O-Dimethyl S-(N-methylcarbamoylmethyl) phosphorothiolate
Dimethyl-S-(N-methyl-carbamoyl-methyl)phosphorothiolate
O,O-Dimethyl-S-(N-methyl-carbamoyl-methyl)-thiolphosphat (German)
O,O-Dimethyl S-(N-methylcarbamoylmethyl) thiophosphate
O,O-Dimethyl-S-(2-oxo-3-azabutyl)-monothiophosphate
O,O-Dimetil-S-(N-metil-carbamoil)-metil-monotiofosfato (Italian)
ENT 25,776
Folimat
O-Analog of Dimethoate
Omethoat
Omethoate
Phosphorothioic acid, O,O-dimethyl S-(2-(methylamino)-2-oxoethyl) ester
PO-Dimethoate
P=O-Rogor
S6876
Thiophosphate de O,O-dimethyle et de S-(N-methylcarbamoyl) methyle (French)

1113-14-0
$C_8H_{16}O_4S_2$
240.34
trans-1,2-Bis(propylsulfonyl)-ethene
B-1843
(E)-1,2-Bis(propylsulfonyl)-ethylene
trans-1,2-Bis(propylsulfonyl)-ethylene
trans-1,2-Bis(n-propylsulfonyl)-ethylene
C-272
Caswell No. 099
CHE 1843
Chemagro B-1843
Ethylene, 1,2-bis(propylsulfonyl)-, (E)- (8CI)
(E)-1,1'-(1,2-Ethenediylbis(sulfonyl)bis(propane))
EPA Pesticide Chemical Code 011701
Preseed
Propane, 1,1'-(1,2-ethenediylbis(sulfonyl))bis-, (E)- (9CI)
Vancide PA
Vancide PA Dispersion

1113-38-8
$C_2H_2O_4.2H_3N$
124.12
Oxalic acid, diammonium salt
Ammonium oxalate [NA 2449]
Ethanedioic acid diammonium salt
NA 2449 [Oxalates, water soluble]

1113-68-4
$C_5H_9NO_2$

115.13
O=C(N(C(=O)C)C)C
Acetamide, N-acetyl-N-methyl-
N-Acetyl-N-methylacetamide
AI3-28901

1114-71-2
$C_{10}H_{21}NOS$
203.38
CCCCN(CC)C(=O)SCCC
Carbamic acid, butylethylthio-, S-propyl ester
Butylethylthiocarbamic acid S-propyl ester
Carbamothioic acid, butyl-ethyl-, S-propyl ester (9CI)
PEBC
Pebulate
S-Propyl-N-aethyl-N-butyl-thiocarbamat (German)
S-Propyl butylethylthiocarbamate
Propyl-ethylbutylthiocarbamate
Propyl ethylbutylthiolcarbamate
n-Propyl-N-ethyl-N-(n-butyl)-thiocarbamate
Propylethyl-n-butylthiocarbamate
Propyl N-ethyl-n-butylthiocarbamate
S-(n-Propyl)-N-ethyl-N-n-butylthiocarbamate
N-Propyl-N-ethyl-N-(n-butyl)-thiolcarbamate
R-2061
Stauffer R-2061
Tillam (Russian)
Tillam-6-E

1115-08-8
C_6H_{10}
82.15
C(=C)C(C=C)C
1,4-Pentadiene, 3-methyl- (9CI)
Methylpentadiene
3-Methyl-1,4-pentadiene

1115-11-3
C_5H_8O
84.12
2-Butenal, 2-methyl- (9CI)
Crotonaldehyde, 2-methyl- (8CI)

1115-15-7
C_4H_6OS
102.16
Vinyl-sulfoxide
Divinyl sulfoxide
Ethene, 1,1-sulfinylbis-
TL 907

1115-20-4
$C_{10}H_{20}O_4$
204.27
O=C(OCC(CO)(C)C)C(CO)(C)C
3-Hydroxy-2,2-dimethylpropyl hydroxypivalate
Esterdiol 204
Hydracrylic acid, 2,2-dimethyl-, 3-hydroxy-2,2-dimethylpropyl ester
3-Hydroxy-2,2-dimethylpropyl 2,2-dimethylhydracrylate
Hydroxyneopentyl hydroxypivalate
Hydroxypivalic acid neopentyl glycol ester
3-(Hydroxypivaloyloxy)-2,2-dimethylpropanol
Hydroxypivalyl hydroxypivalate
Neopentyl glycol monohydroxypivalate
Propanoic acid, 3-hydroxy-2,2-dimethyl-, 3-hydroxy-2,2-dimethylpropyl ester (9CI)

1115-30-6
$C_{10}H_{16}O_5$
216.23
O=C(OCC)C(C(=O)C)CC(=O)OCC
Butanedioic acid, acetyl-, diethyl ester (9CI)
AI3-05620
Diethyl acetylbutanedioate

1115-96-4
$C_9H_{19}NO$
157.25
Heptanamide, N,N-dimethyl- (8CI,9CI)

1116-40-1
$C_{12}H_{27}N$
185.35
N(CC(C)C)(CC(C)C)CC(C)C
1-Propanamine, 2-methyl-N,N-bis(2-methylpropyl)- (9CI)
2-Methyl-N,N-bis(2-methylpropyl)-1-propanamine

1116-54-7
$C_4H_{10}N_2O_3$
134.16
O=NN(CCO)CCO
Ethanol, N-nitrosoiminodi
Bis(β-hydroxyaethyl)nitrosamin (German)
Bis(β-hydroxyethyl)nitrosamine
Diaethanolnitrosamin (German)
Diethanolnitrosoamine
Diethylamine, 2,2'-dihydroxy-N-nitroso-
2,2'-Iminodi-N-nitrosoethanol
NCI-C55583
NDELA
N-Nitrosoaminodiethanol
N-Nitrosobis(2-hydroxyethyl)-amine
N-Nitrosodiaethanolamin (German)
N-Nitrosodiethanolamine
2,2'-(Nitrosoimino)bisethanol
Nitrosoimino diethanol
RCRA waste number U173

1116-70-7
$C_{12}H_{27}Al$
198.33
Tri-n-butylaluminum
Aluminum, tributyl- (9CI)
Tributylaluminum

1116-73-0
$C_{18}H_{39}Al$
282.49
Tri-n-hexylaluminum
Aluminum, trihexyl- (9CI)
Trihexylalane
Trihexylaluminum

1116-76-3
$C_{24}H_{51}N$
353.76
N(CCCCCCCC)(CCCCCCCC)CCCCCCCC
1-Octanamine, N,N-dioctyl
Alamine 308
Alamine 336
Tricaprylylamine
Trioctylamine
Tri-n-octylamine

1117-86-8
$C_8H_{18}O_2$
146.23
1,2-Octanediol (9CI)
AI3-13058

1117-97-1
C_2H_7NO
61.08
O(NC)C
N-Methoxymethylamine
N,O-Dimethylhydroxylamine
O,N-Dimethylhydroxylamine
Hydroxylamine, N,O-dimethyl-
Methanamine, N-methoxy- (9CI)
Methoxyamine, N-methyl-
N-Methoxymethanamine
Methoxymethylamine
N-Methoxy-N-methylamine
Methylamine, N-methoxy-
Methylmethoxyamine
N-Methyl-O-methylhydroxylamine
O-Methyl-N-methylhydroxylamine

1118-39-4
$C_{12}H_{20}O_2$
196.32
O=C(OC(CCCC(C=C)=C)(C)C)C
7-Octen-2-ol, 2-methyl-6-methylene-, acetate
Acetic acid, 2-methyl-6-methylene-7-octen-2-yl ester

Acetic acid, myrcenyl ester
3-Methylene-7-methyl-1-octen-7-yl acetate
Myrcenyl acetate
2-Methyl-6-methylene-7-octen-2-yl acetate

1118-41-8
$C_{36}H_{76}N.Cl$
558.45
1-Heptadecanaminium, N-heptadecyl-N,N-di methyl-, chloride (9CI)
Ammonium, diheptadecyldimethyl-, chloride (8CI)
Diheptadecyldimethylammonium chloride (6CI,7CI)

1118-46-3
$C_4H_9Cl_3Sn$
282.17
CCCC[Sn](Cl)(Cl)Cl
Stannane, butyltrichloro
Butylstannium trichloride
Chlorid n-butylcinicity (Czech)
Monobutyltin trichloride
Tin, n-butyl-, trichloride

1118-47-4
$C_7H_{14}O_2$
130.19
CC(C)(C)CCC(O)=O
Pentanoic acid, 4,4-dimethyl- (9CI)
4,4-Dimethylpentanoic acid

1118-58-7
C_6H_{10}
82.15
C(C=CC)(=C)C
2-Methyl-1,3-pentadiene
AI3-14672-X
2,4-Dimethyl-1,3-butadiene
Methylpentadiene
1,3-Pentadiene, 2-methyl- (9CI)
1,3-Pentadiene, 2-methyl- (85%), and 4-methyl-1,3-pentadiene (15%)

1118-92-9
$C_{10}H_{21}NO$
171.32
O=C(N(C)C)CCCCCCC
Octanamide, N,N-dimethyl
N,N-dimethylcaprylamide
N,N-Dimethyloctanamide

1119-16-0
$C_6H_{12}O$
100.18
O=CCCC(C)C
Valeraldehyde, 4-methyl
Isocaproaldehyde
Isohexanal
4-Methylpentanal
4-Methyl valeraldehyde
Pentanal, 4-methyl- (9CI)

1119-33-1
$C_7H_{13}NO_4$
175.18
CCOC(=O)CCC(N)C(O)=O
Ethyl Glutamate
γ-Ethylglutamate
5-Ethyl-L-glutamate
Glutamic acid, 5-ethyl ester, L- (8CI)
L-Glutamic acid, 5-ethyl ester (9CI)

1119-34-2
$C_6H_{14}N_4O_2.ClH$
210.70
[Cl-].NC(CCCNC(N)=N)C(O)=O
Arginine, monohydrochloride, L
Argamine
Arginine hydrochloride
Arginine, hydrochloride, L-
L-Arginine hydrochloride
Arginine monohydrochloride
L-Arginine, monohydrochloride
Argivene
Detoxargin
Levargin
Minophagen A
R-Gene

1119-40-0
$C_7H_{12}O_4$
160.17
O=C(OC)CCCC(=O)OC
Dimethyl glutarate
AI3-06026
Dimethyl pentanedioate
Glutaric acid, dimethyl ester (8CI)
Methyl glutarate
Pentanedioic acid, dimethyl ester (9CI)

1119-44-4
$C_7H_{12}O$
112.17
O=C(C=CCCC)C
3-Hepten-2-one
AI3-22032

1119-46-6
$C_5H_9ClO_2$
136.58
Pentanoic acid, 5-chloro- (9CI)
Valeric acid, 5-chloro- (8CI)

1119-49-9
$C_6H_{13}NO$
115.17
O=C(NCCCC)C
N-Butyl acetamide
Acetamide, N-butyl- (8CI,9CI)
AI3-02183
N-Butylacetamide

1119-72-8
$C_6H_6O_4$
142.11
2,4-Hexadienedioic acid, (Z,Z)- (9CI)

1119-85-3
$C_6H_6N_2$
106.14
N#CCC=CCC#N
3-Hexenedinitrile
1,4-Dicyano-2-butene
3-Hexenedinitrile (8CI,9CI)
NSC-11685

1119-97-7
$C_{17}H_{38}N.Br$
336.47
[Br-].CCCCCCCCCCCCCC[N+](C)(C)C
Ammonium, tetradecyltrimethyl-, bromide
Ammonium, trimethyltetradecyl-, bromide
Morpan T
Myristyltrimethylammonium bromide
MYTAB
Quaternium-13
1-Tetradecanaminium, N,N,N-trimethyl-, bromide (9CI)
Tetradecyltrimethylammonium bromide
Tetradonium bromide
Trimethylmyristylammonium bromide
N,N,N-Trimethyl-1-tetradecanaminium bromide
Trimethyltetradecylammonium bromide

1120-01-0
$C_{16}H_{33}O_4S.Na$
344.54
1-Hexadecanol, hydrogen sulfate, sodium salt
Avitex C
Avitex SF
Cetyl sodium sulfate
Cetyl sulfate sodium salt
Conco Sulfate C
Hexadecyl sodium sulfate
Nikkol S.C.S
SHS
Sodium cetyl sulfate
Sodium hexadecyl sulfate
Sodium monohexadecyl sulfate
Sodium palmityl sulfate
Tergitol anionic 7

1120-04-3
$C_{18}H_{37}O_4S.Na$
372.60
Sulfuric acid, monooctadecyl ester, sodium salt
Octadecyl sodium sulfate

Sodium monooctadecyl sulfate
Sodium monostearyl sulfate
Sodium octadecyl sulfate
Sodium stearyl sulfate

1120-06-5
$C_{10}H_{22}O$
158.28
2-Decanol (9CI)
AI3-11545

1120-16-7
$C_{12}H_{25}NO$
199.33
O=C(N)CCCCCCCCCCC
Lauramide
Dodecanamide (9CI)
Lauric acid amide

1120-21-4
$C_{11}H_{24}$
156.35
C(CCCCCCCCC)C
Undecane [UN 2330]
Hendecane
n-Undecane
UN 2330 [Undecane]

1120-23-6
$C_8H_{17}ClO_2$
180.67
O(CCOCCCl)CCCC
2-β-Butoxyethoxyethyl chloride
AI3-09117
Butane, 1-(2-(2-chloroethoxy)-ethoxy)-
1-(2-(2-Chloroethoxy)ethoxy)-butane
Ethane, 1-butoxy-2-(2-chloro-ethoxy)-

1120-24-7
$C_{12}H_{27}N$
185.35
N(CCCCCCCCCC)(C)C
Decylamine, N,N-dimethyl- (8CI)
AI3-27531
1-Decanamine, N,N-dimethyl- (9CI)
N,N-Dimethyl-1-decanamine
N,N-Dimethyldecylamine

1120-25-8
$C_{17}H_{32}O_2$
268.44
9-Hexadecenoic acid, methyl ester, (Z)-
AI3-36450

1120-28-1
$C_{21}H_{42}O_2$
326.56
O=C(OC)CCCCCCCCCCCCCCCCCCC
Eicosanoic acid, methyl ester
AI3-36455
Methyl eicosanoate

1120-36-1
$C_{14}H_{28}$
196.38
C(=C)CCCCCCCCCCCC
1-Tetradecene (9CI)
AI3-10509
α-Tetradecene
n-Tetradec-1-ene
1-Tetradecylene

1120-46-3
$C_{18}H_{34}O_2.1/2Pb$
386.06
Lead oleate
Lead(II) oleate (1:2)
9-Octadecenoic acid (Z)-, lead-(2+) salt (9CI)
Oleic acid lead salt
Oleic acid, lead(2+) salt (2:1)

1120-48-5
$C_{16}H_{35}N$
241.45
N(CCCCCCCC)CCCCCCCC
Dioctylamine (8CI)
AI3-15029
1-Octanamine, N-octyl- (9CI)
N-Octyl-1-octanamine

1120-49-6
$C_{20}H_{43}N$
297.56
N(CCCCCCCCCC)CCCCCCCCCC
1-Decanamine, N-decyl- (9CI)
N-Decyl-1-decanamine

1120-56-5
C_5H_8
68.12
C(CC1)(C1)=C
Cyclobutane, methylene- (9CI)
Methylenecyclobutane

1120-62-3
C_6H_{10}
82.15
Cyclopentene, 3-methyl- (8CI, 9CI)

1120-71-4
$C_3H_6O_3S$
122.15
O=S(=O)(OCC1)C1
1,2-Oxathiolane 2,2-dioxide
3-Hydroxy-1-propanesulfonic acid γ-sultone
3-Hydroxy-1-propanesulphonic acid sulfone
3-Hydroxy-1-propanesulphonic acid sultone
1,2-Oxathrolane 2,2-dioxide
1-Propanesulfonic acid-3-hydroxy-γ-sultone
Propane sultone (ACGIH)
1,3-Propane sultone
Propanesultone
RCRA waste number U193

1120-72-5
$C_6H_{10}O$
98.14
O=C(C(CC1)C)C1
Cyclopentanone, 2-methyl- (9CI)
AI3-39196
2-Methylcyclopentanone

1120-73-6
C_6H_8O
96.13
2-Cyclopenten-1-one, 2-methyl- (8CI,9CI)

1120-85-0
$C_6H_{13}N$
99.20
CCCCN1CC1
Aziridine, 1-butyl
n-Butylaziridine
1-Butylaziridine
n-Butylethylenimine

1120-87-2
C_5H_4BrN
157.99
n(ccc(c1)Br)c1
Pyridine, 4-bromo- (9CI)
4-Bromopyridine

1120-97-4
$C_5H_{10}O_2$
102.15
1,3-Dioxane, 4-methyl
4-Methyl-m-dioxane
4-Methyl-1,3-dioxane

1121-05-7
$C_7H_{10}O$
110.16
2-Cyclopenten-1-one, 2,3-dimethyl- (8CI,9CI)

1121-21-7
$C_6H_{10}Cl_2$
153.05
1,2-Dichlorocyclohexane
Cyclohexane, 1,2-dichloro- (9CI)

1121-31-9
C_5H_5NOS
127.17
O=n(c(S)ccc1)c1
2-Pyridinethiol, 1-oxide
2-Mercaptopyridine monoxide

Omadine

1121-55-7
C_7H_7N
105.13
Pyridine, 3-ethenyl-
AI3-18209

1121-60-4
C_6H_5NO
107.11
O=Cc(nccc1)c1
Picolinaldehyde (8CI)
AI3-33230
2-Pyridinecarboxaldehyde (9CI)
2-Pyridylaldehyde

1121-70-6
$C_7H_7O.Na$
130.13
p-Cresol, sodium salt
4-Methylphenol sodium salt
Phenol, 4-methyl-, sodium salt (9CI)
Sodium p-cresolate
Sodium p-cresoxide
Sodium p-methylphenolate
Sodium 4-methylphenolate
Sodium p-methylphenoxide
Sodium 4-methylphenoxide

1121-78-4
C_6H_7NO
109.14
3-Pyridinol, 6-methyl
3-Hydroxy-6-methylpyridine
6-Methyl-3-pyridinol

1121-92-2
$C_7H_{13}NO$
127.21
N(CCCCCC1)C1
Azocine, octahydro
Suberone oxime
Suberonisoxim (German)

1122-39-0
$C_8H_{11}N$
121.18
Pyridine, 2,4,5-trimethyl- (8CI,-9CI)

1122-54-9
C_7H_7NO
121.15
O=C(c(ccnc1)c1)C
Ketone, methyl 4-pyridyl
4-Acetylpyridine
Methyl 4-pyridyl ketone
Pyridine, 4-acetyl-

1122-58-3
$C_7H_{10}N_2$
122.19
n(ccc(N(C)C)c1)c1
Pyridine, 4-(dimethylamino)
4-Dimethylaminepyridine
4-Dimethylaminopyridine
γ-(Dimethylamino)pyridine
p-Dimethylaminopyridine

1122-60-7
$C_6H_{11}NO_2$
129.18
O=N(=O)C(CCCC1)C1
Cyclohexane, nitro
Nitrocyclohexane

1122-61-8
$C_5H_4N_2O_2$
124.09
O=N(=O)c1ccncc1
4-Nitropyridine
AI3-25255
Pyridine, 4-nitro- (9CI)

1122-62-9
C_7H_7NO
121.13
O=C(c(nccc1)c1)C
Ethanone, 1-(2-pyridinyl)- (9CI)
AI3-52210
Ketone, methyl 2-pyridyl (8CI)
1-(2-Pyridinyl)ethanone

1122-69-6
$C_8H_{11}N$
121.18
Pyridine, 2-ethyl-6-methyl- (9CI)
2-Picoline, 6-ethyl- (8CI)

1122-73-2
$C_4H_6N_4O$
126.09
1,3,5-Triazin-2-amine, 4-methoxy- (9CI)
s-Triazine, 2-amino-4-methoxy- (8CI)

1122-81-2
$C_8H_{11}N$
121.18
n(ccc(c1)CCC)c1
Pyridine, 4-propyl- (9CI)
4-Propylpyridine

1122-82-3
$C_7H_{11}NS$
141.25
N(=C=S)C(CCCC1)C1
Isothiocyanic acid, cyclohexyl ester
Cyclohexyl-isothiocyanat (German)

1123-09-7
$C_8H_{12}O$
124.18
2-Cyclohexen-1-one, 3,5-dimethyl- (8CI,9CI)
NSC-845
NSC-10113

1123-84-8
$C_8H_6Cl_2$
173.04
Benzene, 1,4-dichloro-2-ethenyl- (9CI)
NSC-20953
Styrene, 2,5-dichloro- (8CI)

1123-85-9
$C_9H_{12}O$
136.21
OCC(c(cccc1)c1)C
Phenethyl alcohol, β-methyl
Benzeneethanol, β-methyl- (9CI)
Hydratropic alcohol
Hydratropyl alcohol
β-Methylphenethyl alcohol
α-Methyl phenylethyl alcohol
2-Phenylpropan-1-ol
β-Phenylpropyl alcohol
2-Phenylpropyl alcohol

1124-05-6
$C_8H_8Cl_2$
175.06
c(c(cc(c1Cl)C)Cl)(c1)C
Benzene, 1,4-dichloro-2,5-dimethyl- (9CI)
1,4-Dichloro-2,5-dimethylbenzene
2,5-Dichloro-p-xylene
p-Xylene, 2,5-dichloro- (8CI)

1124-11-4
$C_8H_{12}N_2$
136.22
n(c(c(nc1C)C)C)c1C
Pyrazine, tetramethyl
Tetramethylpyrazine
2,3,5,6-Tetramethyl pyrazine

1124-14-7
C_9H_9Br
168.33
Benzene, 1-bromo-4-cyclo propyl- (6CI,7CI,8CI, 9CI)
(p-Bromophenyl)cyclopropane
1-Bromo-4-cyclopropylbenzene
(4-Bromophenyl)cyclopropane
4-Cyclopropylbromobenzene
4-Cyclopropylphenyl bromide

1124-18-1

$C_7H_5D_3$
89.12
Benzene, methyl-d3- (9CI)
Toluene-α,α,α-d3 (8CI)

1124-19-2
$C_6H_5Cl_3Sn$
302.17
Phenyltin trichloride
NSC-92617
Stannane, trichlorophenyl- (8CI, 9CI)

1124-27-2
$C_{10}H_{18}$
138.25
C(=C(C)C)(CCC(C1)C)C1
Cyclohexane, 1-methyl-4-(1-methylethylidene)-(9CI)
1-Isopropylidene-4-methyl-cyclohexane
4(8)-p-Menthene
p-Menth-4(8)-ene (7CI,8CI)
4-Methyl-1-isopropylidene-cyclohexane

1124-33-0
$C_5H_4N_2O_3$
140.09
O=N(=O)c(ccn(=O)c1)c1
4-Nitropyridine 1-oxide
AI3-25256
Avitrol 100
Caswell No. 604AA
EPA Pesticide Chemical Code 597300
4-Nitropyridine-N-oxide
Pyridine, 4-nitro-, 1-oxide (8CI,9CI)

1124-48-7
$C_6H_6O_4$
142.11
2-Furanacetic acid, 2,5-dihydro-5-oxo-, (+)- (8CI,9CI)

1124-53-4
$C_8H_{15}NO$
141.24
Acetamide, N-cyclohexyl
Acetamidocyclohexane
N-Acetylcyclohexylamine
N-Cyclohexylacetamide

1125-27-5
$C_8H_{10}Cl_2Si$
205.17
Silane, dichloroethylphenyl
Dichloroethylphenylsilane
Ethyl phenyl dichlorosilane
Ethylphenyldichlorosilane [UN 2435]
UN 2435 [Ethylphenyldichlorosilane]

1125-80-0
$C_{10}H_9N$
143.20
Cc2cc1ccccc1cn2
Isoquinoline, 3-methyl
3-Methylisoquinoline

1126-46-1
$C_8H_7ClO_2$
170.60
O=C(OC)c(ccc(c1)Cl)c1
Benzoic acid, p-chloro-, methyl ester (8CI)
AI3-23578
Benzoic acid, 4-chloro-, methyl ester (9CI)
Methyl 4-chlorobenzoate

1126-78-9
$C_{10}H_{15}N$
149.26
N(c(cccc1)c1)CCCC
Aniline, n-butyl
Benzenamine, n-butyl- (9CI)
N-Butylaniline [UN 2738]
N-(n-Butyl)aniline
N-n-Butylaniline
4-(Phenylamino)butane
UN 2738 [N-Butylaniline]

1126-79-0
$C_{10}H_{14}O$
150.24
O(c(cccc1)c1)CCCC
Ether, butyl phenyl
Butoxyphenyl
Butyl phenyl ether

1127-75-9
$C_7H_8N_4S$
180.21
9H-Purine, 9-methyl-6-(methylthio)- (8CI,9CI)
NSC-38302

1127-76-0
$C_{12}H_{12}$
156.24
CCc1cccc2ccccc12
Naphthalene, 1-ethyl
1-Ethylnaphthalene

1128-16-1
$C_6H_2Cl_5N$
265.35
n(c(c(cc1Cl)Cl)C(Cl)(Cl)Cl)c1
Pyridine, 3,5-dichloro-2-(trichloromethyl)- (9CI)
3,5-Dichloro-2-(trichloromethyl)pyridine
3,5-Trichloro-2-(trichloromethyl)pyridine

1129-41-5
$C_9H_{11}NO_2$
165.21
O=C(Oc(cccc1C)c1)NC
Carbamic acid, methyl-, m-tolyl ester
Carbamic acid, methyl-, 3-tolyl ester
m-Cresyl methylcarbamate
m-Cresyl ester of N-methylcarbamic acid
Dicresyl
Dicresyl N-methylcarbamate
DRC 3341
Kumiai
Metacrate
Metholcarb
Methylcarbamic acid m-tolyl ester
3-Methylphenyl N-methylcarbamate
Metolcarb
MTMC
m-Tolylester kyseliny methylkarbaminove (Czech)
m-Tolyl N-methylcarbamate
3-Tolyl-N-methylcarbamate
Tsumacide

1129-50-6
$C_{10}H_{13}NO$
163.24
O=C(Nc(cccc1)c1)CCC
n-Butyranilide
AI3-01391
Butanamide, N-phenyl- (9CI)
Butyranilide (8CI)
N-Phenylbutanamide
NSC-6123

1131-18-6
$C_{10}H_{11}N_3$
173.24
N(N(c(cccc1)c1)C(N)=C2)=C2C
1H-Pyrazole-5-amine, 3-methyl-1-phenyl
5-Amino-3-methyl-1-phenylpyrazole
1-Phenyl-3-methyl-5-aminopyrazole
Pyrazole, 5-amino-3-methyl-1-phenyl-

1131-60-8
$C_{12}H_{16}O$
176.26
Oc(ccc(c1)C(CCCC2)C2)c1
Phenol, p-cyclohexyl- (8CI)
AI3-09330
4-Cyclohexylphenol
Phenol, 4-cyclohexyl- (9CI)

1131-62-0
$C_{10}H_{12}O_3$
180.20
O=C(c(ccc(OC)c1OC)c1)C
Acetophenone, 3',4'-dimethoxy- (8CI)
AI3-11163

1-(3,4-Dimethoxyphenyl)-ethanone
Ethanone, 1-(3,4-dimethoxy-phenyl)- (9CI)

1132-21-4
$C_9H_{10}O_4$
182.18
O=C(O)c(cc(OC)cc1OC)c1
Benzoic acid, 3,5-dimethoxy- (9CI)
AI3-52341
3,5-Dimethoxybenzoic acid

1133-03-5
$C_{11}H_{16}O_3$
196.25
2(4H)-Benzofuranone, 5,6,7,7a-tetrahydro-6-hydroxy-4,4,7a-tri methyl-, cis- (8CI,9CI)

1133-63-7
$C_{12}H_{10}O_2$
186.21
(1,1'-Biphenyl)-2,3-diol (9CI)
1,2-Benzenediol, 3-phenyl-
2,3-Biphenyldiol (8CI)
Pyrocatechol, 3-phenyl-

1133-64-8
$C_{10}H_{13}N_3O$
191.26
O=NN1CCCCC1c2cccnc2
Anabasine, 1-nitroso
NAB
N-Nitrosoanabasine
1-Nitrosoanabasine
N'-Nitrosoanabasine
N-Nitroso-2-(3'-pyridyl)piperidine
Pyridine, 3-(1-nitroso-2-piperidinyl)-, (S)- (9CI)

1133-80-8
$C_{13}H_9Br$
245.12
9H-Fluorene, 2-bromo- (9CI)
Fluorene, 2-bromo- (8CI)
NSC-1463

1134-04-9
C_6Cl_7N
334.24
n(c(c(c(c1Cl)Cl)Cl)C(Cl)(Cl)Cl)c1Cl
Pyridine, 2,3,4,5-tetrachloro-6-(trichloromethyl)- (9CI)
Heptachloro-2-picoline

1134-23-2
$C_{11}H_{21}NOS$
215.39
CCSC(=O)N(CC)C1CCCCC1
Cyclohexanecarbamic acid, N-ethylthio-, S-ethyl ester
Carbamic acid, cyclohexylethylthio-, S-ethyl ester
Carbamothioic acid, cyclohexylethyl-, S-ethyl ester
Cycloate
S-Ethyl cyclohexylethylthiocarbamate
S-Ethyl N-ethyl N-cyclohexylthiolcarbamate
Etsan
Eurex
Hexylthiocarbam
R 2063
Ro-Neet
Ronit

1134-36-7
$C_{12}H_{11}NO$
185.22
Oc(c(N)cc(c(cccc1)c1)c2)c2
(1,1'-Biphenyl)-4-ol, 3-amino- (9CI)
3-Amino-(1,1'-biphenyl)-4-ol

1134-62-9
$C_{14}H_{16}$
184.28
Naphthalene, 2-butyl- (8CI,9CI)

1135-23-5
$C_{10}H_{12}O_4$
196.20
Benzenepropanoic acid, 4-hydroxy-3-methoxy- (9CI)
Ferulic acid, α,β-dihydro-
Hydrocinnamic acid, 4-hydroxy-3-methoxy- (8CI)
Hydroferulic acid
Shorbic acid

1135-24-6
$C_{10}H_{10}O_4$
194.19
O=C(O)C=Cc(ccc(O)c1OC)c1
Ferulic acid
Cinnamic acid, 4-hydroxy-3-methoxy- (8CI)
3-(4-Hydroxy-3-methoxyphenyl)-2-propenoic acid
2-Propenoic acid, 3-(4-hydroxy-3-methoxyphenyl)- (9CI)

1135-99-5
$C_{12}H_{10}Cl_2Sn$
343.81
Stannane, dichlorodiphenyl
Difenylstanniumdichlorid (Czech)
Diphenylstannium dichloride
Diphenyltin dichloride
Tin, diphenyl-, dichloride

1136-86-3
$C_{11}H_{14}O_4$
210.23
Ethanone, 1-(3,4,5-trimethoxyphenyl)- (9CI)

1137-41-3
$C_{13}H_{11}NO$
197.25
O=C(c(cccc1)c1)c(ccc(N)c2)c2
Benzophenone, 4-amino
p-Aminobenzophenone
USAF A-233

1137-42-4
$C_{13}H_{10}O_2$
198.22
O=C(c(cccc1)c1)c(ccc(O)c2)c2
4-Hydroxybenzophenone
AI3-00862
Benzophenone, 4-hydroxy- (8CI)
(4-Hydroxyphenyl)phenylmethanone
Methanone, (4-hydroxyphenyl)phenyl- (9CI)

1138-52-9
$C_{14}H_{22}O$
206.33
Oc(cc(cc1C(C)(C)C)C(C)(C)C)c1
Phenol, 3,5-bis(1,1-dimethylethyl)- (9CI)
3,5-Bis(1,1-dimethylethyl)phenol
Phenol, 3,5-di-tert-butyl- (8CI)

1139-30-6
$C_{15}H_{24}O$
220.39
O(C1(CCC(C(C(CC2)=C)C3)C3(C)C)C)C12
5-Oxatricyclo(8.2.0.0^{4,6})dodecane, 4,12,12-trimethyl-9-methylene-, (1R,4R,6R,10S)
Caryophylene oxide
Caryophyllene epoxide
(-)-β-Caryophyllene epoxide
β-Caryophyllene epoxide
Caryophyllene oxide
(-)-Caryophyllene oxide
β-Caryophyllene oxide
Epoxycaryophyllene
(-)-Epoxydihydrocaryophyllene
4,11,11-Trimethyl-8-methylene-5-oxatricyclo(8.2.0.0(4,6))dodecane

1139-82-8
$C_{15}H_{12}O$
208.26
6H-Dibenzo(a,c)cyclohepten-6-one, 5,7-di hydro- (7CI,8CI,9CI)

1141-38-4
$C_{12}H_8O_4$

216.19
2,6-Naphthalenedicarboxylic acid (9CI)

1142-15-0
$C_{15}H_{14}O$
210.28
Benzene, 1-methoxy-4-(2-phenylethenyl)- (9CI)
Anisole, p-styryl- (8CI)
NSC-2139

1142-19-4
$C_{12}H_{8}Cl_{2}S_{2}$
287.23
S(Sc(ccc(c1)Cl)c1)c(ccc(c2)Cl)c2
Bis(p-chlorophenyl)disulfide
AI3-16362
Bis(4-chlorophenyl) disulfide
p-Chlorophenyl disulfide
DDDS (Pesticide)
p,p'-Dichlorodiphenyl disulfide
4,4'-Dichlorodiphenyl disulfide
Di(p-chlorophenyl) disulfide
Disulfide, bis(p-chlorophenyl) (8CI)
Disulfide, bis(4-chlorophenyl) (9CI)

1143-38-0
$C_{14}H_{10}O_{3}$
226.24
Oc2cccc3Cc1cccc(O)c1C(=O)c23
Anthrone, 1,8-dihydroxy
9(10H)-Anthracenone, 1,8-dihydroxy-
Anthra-Derm
Anthralin
Batridol
Chrysodermol
Cignolin
Cigthranol
1,8-Dihydroxyanthrone
1,8-Dihydroxy-9-anthrone
Dithranol
Psoriacid-Stift

1143-72-2
$C_{13}H_{10}O_{4}$
230.22
O=C(c(cccc1)c1)c(c(O)c(O)c(O)c2)c2
Benzophenone, 2,3,4-trihydroxy- (8CI)
Methanone, phenyl(2,3,4-trihydroxyphenyl)- (9CI)
Phenyl(2,3,4-trihydroxyphenyl)-methanone

1144-74-7
$C_{13}H_{9}NO_{3}$
227.21
O=C(c(cccc1)c1)c(ccc(N(=O)=O)c2)c2
Methanone, (4-nitrophenyl)-phenyl- (9CI)
Benzophenone, 4-nitro-
(4-Nitrophenyl)phenylmethanone

1145-44-4
$C_{7}H_{4}Cl_{7}N_{3}$
378.28
1,3,5-Triazine, 2-(2-chloroethyl)-4,6-bis(trichloromethyl)- (9CI)
NSC-372149
s-Triazine, 2-(2-chloroethyl)-4,6-bis(trichloromethyl)- (8CI)

1152-61-0
$C_{12}H_{13}NO_{6}$
267.23
O=C(OCc(cccc1)c1)NC(C(=O)O)CC(=O)O
Aspartic acid, N-carboxy-, N-benzyl ester, L- (8CI)
L-Aspartic acid, N-((phenylmethoxy)carbonyl)- (9CI)
N-((Phenylmethoxy)carbonyl)-L-aspartic acid

1153-05-5
$C_{18}H_{15}AsO$
322.24
Arsine oxide, triphenyl- (9CI)
Triphenylarsine oxide

1155-00-6
$C_{12}H_{8}N_{2}O_{4}S_{2}$
308.33
O=N(=O)c(c(SSc(c(N(=O)=O)ccc1)c1)ccc2)c2
Disulfide, bis(o-nitrophenyl) (8CI)
AI3-08905
Disulfide, bis(2-nitrophenyl) (9CI)

1156-19-0
$C_{14}H_{21}N_{3}O_{3}S$
311.44
Cc1ccc(cc1)S(=O)(=O)NC(=O)NN2CCCCCC2
Urea, 1-(hexahydro-1H-azepin-1-yl)-3-(p-tolylsulfonyl)
Benzenesulfonamide, N-(((hexahydro-1H-azepin-1-yl)-amino)-carbonyl)-4-methyl-
1-(Hexahydro-1-azepinyl)-3-p-tolylsulfonylurea
1-(Hexahydro-1H-azepin-1-yl)-3-(p-tolylsulfonyl)urea
NSC-70762
NCI-C03327
Tolazamide
Tolinase
N-(p-Toluenesulfonyl)-N'-hexamethyleniminourea
4-(p-Tolylsulfonyl)-1,1-hexamethylenesemicarbazide
U 17835

1162-06-7
$C_{20}H_{18}O_{2}Pb$
497.57
CC(=O)O[Pb](c1ccccc1)(c2cccc c2)c3ccccc3
Plumbane, acetoxytriphenyl
Acetoxytriphenyllead
Plumbane, (acetyloxy)triphenyl- (9CI)
Triphenyllead acetate

1162-65-8
$C_{17}H_{12}O_{6}$
312.29
COc4cc2OC1OC=CC1c2c5oc(=O)c3C(=O)CCc3c45
Cyclopenta(c)furo(3',2':4,5)-furo(2,3-h)(1)benzopyran-1,11-dione, 2,3,6a,9a-tetrahydro- 4-methoxy
AFB1
AFBI
Aflatoxin B
Aflatoxin B1

1163-19-5
$C_{12}Br_{10}O$
959.22
O(c(c(c(c(c1Br)Br)Br)Br)c1Br)c(c(c(c(c2Br)Br)Br)Br)c2Br
Ether, bis(pentabromophenyl)
Benzene, 1,1'-oxybis(2,3,4,5,6-pentabromo- (9CI)
Berkflam B 10E
BR 55N
Bromkal 83-10DE
Bromkal 82-ODE
DBDPO
Decabromobiphenyl ether
Decabromobiphenyl oxide
Decabromodiphenyl oxide
Decabromophenyl ether
DE 83R
Ether, decabromodiphenyl
FR 300
FR 300BA
FRP 53
NCI-C55287
Pentabromophenyl ether
Saytex 102
Saytex 102E
Tardex 100

1165-39-5
$C_{17}H_{12}O_{7}$
328.29
COc4cc2OC1OC=CC1c2c5oc(=O)c3C(=O)OCCc3c45
1H,12H-Furo(3',2':4,5)furo-(2,3-h)pyrano(3,4-c)(1)benzopyran-1,12-dione, 3,4,7a,10a-tetrahydro-5-methoxy
Aflatoxin G1

1165-48-6
$C_{20}H_{21}NO_{3}$
323.42

COc2ccc1c(=O)c(C)c(oc1c2CN(C)C)c3ccccc3
Flavone, 8-(dimethylaminomethyl)-7-methoxy-3-methyl
Dimefline
8-(Dimethylaminomethyl)-7-methoxy-3-methylflavone
8-((Dimethylamino)methyl)-7-methoxy-3-methyl-2-phenylflavone
DW 62
Malivan
N-(7-Methoxy-3-methyl-4-oxo-2-phenyl-4H-chromen-8-yl)-methyl-N,N-dimethylamine
Reanimil
REC 7/0267
Remeflin

1166-18-3
$C_{24}H_{18}$
306.41
m-Quaterphenyl (8CI)
1,1':3',1'':3'',1'''-Quaterphenyl (9CI)

1172-18-5
$C_{21}H_{23}ClFN_3O.2ClH$
460.84
[Cl-].[Cl-].CCN(CC)CCN2C(=O)CN=C(c1ccccc1F)c3cc(Cl)ccc23
2H-1,4-Benzodiazepin-2-one, 7-chloro-1-(2-(diethylamino)-ethyl)-5-(o-fluorophenyl)-, 1,3-dihydro-, dihydrochloride
Benozil
Dalmadorm
Dalmadorm hydrochloride
Dalmane
Dalmate
Dormodor
Felison
Flurazepam dihydrochloride
Flurazepam hydrochloride
ID 480
ID 480 Dihydrochloride
Insumin
Lunipax
NSC-78559
Ro 5-6901
Somlan

1172-82-3
$C_{24}H_{33}O_4$
385.52
Medroxyprogesterone 17-acetate
17-Hydroxy-6-methylpregn-4-ene-3,20-dione acetate
Medroxyprogesterone acetate
Pregn-4-ene-3,20-dione, 17-hydroxy-6-methyl-, acetate

1176-44-9
$C_{30}H_{52}$
412.74
A'-Neogammacerane, (21β)-(9CI)
Moretane
A'-Neo-21αH-gammacerane (8CI)
Zeorinane

1182-65-6
$C_{34}H_{52}O_3S$
540.85
O=S(=O)(OC(CCC(C1=CCC2C(C(C(C3)C(CCCC(C)C)C)(CC4)C)C3)(C24)C)C1)c(ccc(c5)C)c5
Cholest-5-en-3-ol (3β)-, 4-methylbenzenesulfonate (9CI)
Cholesterol, p-toluenesulfonate (8CI)

1184-57-2
CH_4HgO
232.64
C[Hg]O
Mercury, hydroxymethyl
Methylhydrargyriumhydroxid (Czech)
Methylmercuric hydroxide
Methylmercury hydroxide
Methylmerkurihydroxid (Czech)

1184-60-7
C_3H_5F
60.07
1-Propene, 2-fluoro- (9CI)
Propene, 2-fluoro- (8CI)

1184-64-1
$CO_3.Cu$
123.55
Carbonic acid, copper(2+) salt (1:1)
Copper carbonate (1:1)
Copper (II) carbonate
Cupric carbonate (1:1)

1184-78-7
C_3H_9NO
75.10
O=N(C)(C)C
Trimethyloxamine
AI3-60110
N,N-Dimethylmethanamine N-oxide
Methanamine, N,N-dimethyl-, N-oxide (9CI)
Trimethylamine, N-oxide (8CI)

1185-09-7
C_2HCl_5S
234.36
S(C(C(Cl)Cl)(Cl)Cl)Cl
1,1,2,2-Tetrachloroethylsulfenyl chloride
Ethanesulfenyl chloride, 1,1,2,2-tetrachloro- (9CI)
1,1,2,2-Tetrachloroethanesulfenyl chloride

1185-33-7
$C_6H_{14}O$
102.20
CCC(C)(C)CO
1-Butanol, 2,2-dimethyl
2,2-Dimethylbutanol

1185-39-3
$C_7H_{14}O_2$
130.21
Valeric acid, 2,2-dimethyl
2,2-Dimethylpentanoic acid
2,2-Dimethylvaleric acid

1185-55-3
$C_4H_{12}O_3Si$
136.25
COSi(C)(OC)OC
Silane, methyltrimethoxy
Methyltrimethoxysilane
Silane, trimethoxymethyl-
Trimethoxymethylsilane
Union Carbide A-163

1185-57-5
$C_6H_8O_7.xFe.xH_4N$
709.44
Citric acid, ammonium iron-(3+) salt
Ammonium iron(III) citrate
FAC
Ferric ammonium citrate

1185-81-5
$C_{32}H_{68}S_2Sn$
635.73
Stannane, dibutylbis(dodecylthio)- (9CI)
Dibutylbis(dodecylthio)stannane

1186-09-0
$C_6H_{15}O_3PS$
198.24
CCOP(=O)(OCC)SCC
Phosphorothioic acid, O,O,S-triethyl ester
O,O-Diethyl S-ethyl phosphorothioate
O,O,S-Triethyl phosphorothioate
O,O,S-Triethyl thiophosphate

1186-53-4
C_9H_{20}
128.26
2,2,3,4-Tetramethylpentane

1187-42-4
$C_4H_4N_4$
108.08
N#CC(N)=C(C#N)N

Diaminomaleonitrile
2-Butenedinitrile, 2,3-diamino-, (Z)- (9CI)
(Z)-2,3-Diamino-2-butenedinitrile
Maleonitrile, diamino- (8CI)

1187-58-2
C_4H_9NO
87.14
O=C(NC)CC
Propionamide, N-methyl
N-Methylpropanamide
N-Methylpropionamide
N-Methylpropionic acid amide
N-Methylpropionsaureamid (German)
Propanamide, N-methyl- (9CI)

1187-59-3
C_4H_7NO
85.12
O=C(NC)C=C
Acrylamide, N-methyl
N-Methylacrylamide
2-Propenamide, N-methyl- (9CI)

1187-87-7
$C_8H_{14}O$
126.20
2-Hexanone, 5-methyl-3-methylene- (6CI,7CI, 8CI,9CI)

1187-93-5
C_3F_6O
166.03
FC(F)(F)OC(F)=C(F)F
Ether, trifluoromethyl trifluorovinyl
Ethene, trifluoro(trifluoromethoxy)-
Trifluoromethyl trifluorovinyl ether

1188-02-9
$C_8H_{16}O_2$
144.21
O=C(O)C(CCCCC)C
Heptanoic acid, 2-methyl- (9CI)
2-Methylheptanoic acid

1188-47-2
Unknown
Unknown
Nitrilotriacetic acid and its salts
NTA, copper(2+) salt (1:1)

1188-48-3
Unknown
Unknown
Nitrilotriacetic acid and its salts
NTA, magnesium salt (1:1)

1189-08-8
$C_{12}H_{18}O_4$
226.27
O=C(OCCC(OC(=O)C(=C)C)C)C(=C)C
2-Propenoic acid, 2-methyl-, 1-methyl-1,3-propanediyl ester (9CI)

1189-85-1
$C_8H_{18}CrO_4$
230.26
Chromic acid, di-t-butyl ester
t-Butyl chromate
tert-Butyl chromate (ACGIH, OSHA)

1189-99-7
$C_{10}H_{22}$
142.28
Heptane, 2,5,5-trimethyl- (8CI,9CI)

1190-16-5
$C_5H_9Cl_2NSi$
182.14
CSi(Cl)(Cl)CCCC#N
Butyronitrile, 4-(dichloromethylsilyl)
3-Cyanopropyldichloromethylsilane
Dichlor-3-kyanpropyl-methylsilan (Czech)

1190-22-3
$C_4H_8Cl_2$
127.01
C(Cl)(CCCl)C
Butane, 1,3-dichloro- (9CI)
1,3-Dichlorobutane

1190-28-9
$C_6H_{11}O_6PS_2$
274.26
Butanedioic acid, ((dimethoxyphosphinothioyl)thio)-
AI3-10627

1191-04-4
$C_6H_{10}O_2$
114.14
O=C(O)C=CCCC
2-Hexenoic acid (9CI)

1191-15-7
$C_8H_{19}Al$
142.25
Aluminum, diisobutylhydro
Al-Alchili (Italian)
Al-Diisobutyl
Aluminum, hydrobis(2-methylpropyl)- (9CI)
Aluminum, hydrodiisobutyl- (8CI)
Bis(isobutyl)hydroaluminum
Diisobutylaluminium hydride
Diisobutylaluminum
Diisobutylaluminum hydride
Diisobutylhydroaluminum
Hydrodiisobutylaluminum

1191-17-9
Unknown
Unknown
2,2'-Dichloroethyl ether

1191-25-9
$C_6H_{12}O_3$
132.16
6-Hydroxyhexanoic acid
ε-Hydroxycaproic acid
6-Hydroxyhexanoate

1191-50-0
$C_{14}H_{29}O_4S.Na$
316.48
1-Tetradecanol, hydrogen sulfate, sodium salt
7-Ethyl-2-methyl-4-hendecanol sulfate sodium salt
Myristyl sulfate, sodium salt
Niaproof 4
Sodium myristyl sulfate
Sodium sotradecol
Sodium tetradecyl sulfate
STS
Sulfuric acid, monotetradecyl ester, sodium salt
Sulfuric acid, myristyl ester, sodium salt
Tetradecyl sodium sulfate
Tetradecyl sulfate, sodium salt
Trombavar
Trombovar

1191-79-3
$C_{72}H_{140}O_8.Ba.Cd$
1383.86
Stearic acid, barium cadmium salt (4:1:1)
Barium cadmium stearate
Cadmium barium stearate
Octadecanoic acid, barium cadmium salt (4:1:1) (9CI)

1191-80-6
$C_{36}H_{66}O_4.Hg$
763.61
Mercury-oleate
Mercuric oleate
Mercuric oleate, Solid (DOT)
Mercury oleate [UN 1640]
Oleate of mercury
UN 1640 [Mercury oleate]

1191-95-3
C_4H_6O

70.09
O=C(CC1)C1
Cyclobutanone (9CI)
AI3-37787

1192-18-3
C_7H_{14}
98.19
Cyclopentane, 1,2-dimethyl-, cis- (9CI)

1192-22-9
$C_6H_{12}O$
100.18
CCC1OC1(C)C
Pentane, 2,3-epoxy-2-methyl
2,3-Epoxy-2-methylpentane
2-Methyl-2,3-pentylene oxide

1192-28-5
C_5H_9NO
99.15
N(O)=C(CCC1)C1
Cyclopentanone, oxime (8CI, 9CI)

1192-37-6
C_7H_{12}
96.17
C(CCCC1)(C1)=C
Cyclohexane, methylene- (9CI)
Methylenecyclohexane

1192-58-1
C_6H_7NO
109.14
1H-Pyrrole-2-carboxaldehyde, 1-methyl
1-Methyl-1H-pyrrole-2-carboxaldehyde

1192-62-7
$C_6H_6O_2$
110.12
O=C(C(OC=C1)=C1)C
Ketone, 2-furyl methyl
Acetylfuran
2-Acetylfuran
Ethanone, 1-(2-furanyl)- (9CI)
Furan, 2-acetyl-
1-(2-Furanyl)ethanone
2-Furyl methyl ketone
Methyl 2-furyl ketone

1192-79-6
C_6H_7NO
109.12
1H-Pyrrole-2-carboxaldehyde, 5-methyl- (9CI)
NSC-81349
Pyrrole-2-carboxaldehyde, 5-methyl- (8CI)

1193-02-8
C_6H_7NS
125.20
Benzenethiol, p-amino
p-Aminobenzenethiol
4-Aminobenzenethiol
p-Aminophenylmercaptan
p-Aminothiophenol
4-Aminothiophenol
Benzenethiol, 4-amino- (9CI)
p-Mercaptoaniline
4-Mercaptoaniline

1193-18-6
$C_7H_{10}O$
110.17
O=C(C=C(CC1)C)C1
2-Cyclohexen-1-one, 3-methyl
3-Methyl-2-cyclohexen-1-one
Seudenone

1193-79-9
$C_7H_8O_2$
124.15
CC(=O)c1ccc(C)o1
Furan, 2-acetyl-5-methyl
2-Acetyl-5-methyl-furan

1193-81-3
$C_8H_{16}O$
128.21
OC(C(CCCC1)C1)C
Cyclohexanemethanol, α-methyl- (9CI)
AI3-36516
Cyclohexylmethylcarbinol
α-Methylcyclohexanemethanol

1193-82-4
C_7H_8OS
140.21
CS(=O)c1ccccc1
Benzene, (methylsulfinyl)- (9CI)
Sulfoxide, methyl phenyl (8CI)

1194-65-6
$C_7H_3Cl_2N$
172.01
Clc1cccc(Cl)c1C#N
Benzonitrile, 2,6-dichloro
Carsoron
Casaron
Casoron
Casoron 133
Code H 133
2,6-DBN
DBN (The herbicide)
DCB
Decabane
Dichlobenil
2,6-Dichlorbenzonitril (German)
2,6-Dichlorobenzonitrile
Du-Sprex
H 133
H 1313
NIA 5996
Niagara 5006
Niagara 5996

1195-16-0
$C_6H_9NO_2S$
159.22
O=C(NC(C(=O)SC1)C1)C
Acetamide, N-(tetrahydro-2-oxo-3-thienyl)
2-Acetamido-4-mercaptobutyric acid thiolactone
2-Acetamide-4-mercaptobutyric acid γ-thiolactone
α-Acetamido-γ-thiobutyrolactone
N-Acetylhomocysteine thiolactone
N-Acetylhomocysteinthiolakton (German)
ACHTL
AHCTL
BO 714
Citiolase
Citiolone
N-(Tetrahydro-2-oxo-3-thienyl)-acetamide
Thioxidrene

1195-42-2
$C_9H_{19}N$
141.25
Cyclohexanamine, N-(1-methylethyl)- (9CI)
Isopropyl cyclohexylamine
N-Isopropylcyclohexylamine

1195-79-5
$C_{10}H_{16}O$
152.24
O=C(C(CC1C2)(C2)C)C1(C)C
Fenchone
AI3-00736
Bicyclo(2.2.1)heptan-2-one, 1,3,3-trimethyl- (9CI)
Fenchon (German)
2-Norbornanone, 1,3,3-tri-methyl- (8CI)
1,3,3-Trimethylbicyclo(2.2.1)-heptan-2-one
1,3,3-Trimethyl-2-norbornanone
1,3,3-Trimethyl-2-norcamphan-one

1196-58-3
$C_{11}H_{16}$
148.25
Benzene, (1-ethylpropyl)- (9CI)

1197-01-9
$C_{10}H_{14}O$
150.22
OC(c(ccc(c1)C)c1)(C)C
Benzenemethanol, α,α,4-tri-methyl-
AI3-00732
p-Cymen-8-ol

α,α,4-Trimethylbenzenemethan-ol

1197-09-7
$C_8H_8O_3$
152.15
3,4-Dihydroxyacetophenone
Qingxintong

1197-37-1
$C_8H_{12}N_2O$
152.18
O(c(ccc(N)c1N)c1)CC
4-Ethoxy-1,2-benzenediamine
1,2-Benzenediamine, 4-ethoxy- (9CI)
o-Phenylenediamine, 4-ethoxy-

1197-55-3
$C_8H_9NO_2$
151.18
O=C(O)Cc(ccc(N)c1)c1
Acetic acid, (p-aminophenyl)
4-Aminobenzeneacetic acid
(p-Aminophenyl)acetic acid
4-Aminophenylacetic acid
p-Amino-α-toluic acid
Benzeneacetic acid, 4-amino- (9CI)
4-Carboxymethylaniline

1198-27-2
$C_{10}H_9NO.ClH$
195.66
Cl.Nc1c(O)ccc2ccccc12
2-Naphthol, 1-amino-, hydrochloride
1-Amino-2-naphthol hydrochloride
2-Hydroxy-1-naphthylamine hydrochloride

1198-37-4
$C_{11}H_{11}N$
157.21
n(c(c(c(c1)C)ccc2)c2)c1C
Quinoline, 2,4-dimethyl- (9CI)
AI3-08881
2,4-Dimethylquinoline

1198-55-6
$C_6H_2Cl_4O_2$
247.88
Oc(c(c(c(c1Cl)Cl)Cl)Cl)c1O
Pyrocatechol, tetrachloro
Tetrachloro-1,2-benzenediol
3,4,5,6-Tetrachloro-1,2-benzenediol
Tetrachlorocatechol
Tetrachloropyrocatechol
Tetrachlorpyrokatechin (Czech)
Tetrachlorpyrokatechol (Czech)

1199-77-5
$C_{10}H_{10}O_2$
162.19
CC(=Cc1ccccc1)C(O)=O
2-Propenoic acid, 2-methyl-3-phenyl- (9CI)
Cinnamic acid, α-methyl- (8CI)
NSC-401113

1200-14-2
$C_{11}H_{14}O$
162.23
O=Cc(ccc(c1)CCCC)c1
Benzaldehyde, 4-butyl- (9CI)
4-Butylbenzaldehyde

1201-30-5
C_6HCl_6N
299.79
n(c(c(c(c1Cl)Cl)Cl)C(Cl)(Cl)Cl)c1
Pyridine, 3,4,5-trichloro-2-(trichloromethyl)- (9CI)

1201-99-6
$C_9H_6Cl_2O_2$
217.05
Cinnamic acid, 2,4-dichloro
2,4-Dichlorocinnamic acid
2-Propenoic acid, 3-(2,4-dichlorophenyl)- (9CI)

1202-34-2
$C_{10}H_9N_3$
171.18
n(c(Nc(nccc1)c1)ccc2)c2
2,2'-Dipyridylamine
AI3-52643
2-Pyridinamine, N-2-pyridinyl- (9CI)
Pyridine, 2,2'-iminodi- (8CI)
N-2-Pyridinyl-2-pyridinamine

1203-17-4
$C_{14}H_{20}$
188.31
c(c(ccc1)C(C2C)(C)C)(c1)C2(C)C
1,1,2,3,3-Pentamethylindan
2,3-Dihydro-1,1,2,3,3-pentamethyl-1H-indene
Indan, 1,1,2,3,3-pentamethyl-
1H-Indene, 2,3-dihydro-1,1,2,3,3-pentamethyl- (9CI)
1,1,2,3,3-Pentamethylindane

1203-86-7
$C_8H_3Cl_5O$
292.37
O=C(c(c(cc(c1Cl)Cl)Cl)c1)C(Cl)Cl
2,2,2',4',5'-Pentachloroacetophenone
Acetophenone, 2,2,2',4',5'-pentachloro-
2,2-Dichloro-1-(2,4,5-trichlorophenyl)ethanone
Ethanone, 2,2-dichloro-1-(2,4,5-trichlorophenyl)- (9CI)
Pentachloroacetophenone
α,α,2,4,5-Pentachloroacetophenone
2,4,5-Trichlorophenacylidene chloride
2,4,5-Trichlorophenacylidene dichloride

1204-72-4
$C_{11}H_9NO_2$
187.21
Naphthalene, 2-methyl-3-nitro
3-Methyl-2-nitronaphthalene
2-Methyl-3-nitronaphthalene (1204-72-4)

1205-08-9
$C_{10}H_{11}NO_3$
193.20
COC(=O)CNC(=O)c1ccccc1
Glycine, N-benzoyl-, methyl ester (9CI)
Hippuric acid, methyl ester (8CI)

1205-64-7
$C_{13}H_{13}N$
183.25
N(c(cccc1C)c1)c(cccc2)c2
Benzenamine, 3-methyl-N-phenyl- (9CI)
3-Methyldiphenylamine
3-Methyl-N-phenylbenzenamine

1205-71-6
$C_{12}H_{10}ClN$
203.66
Benzenamine, 4-chloro-N-phenyl- (9CI)
Diphenylamine, 4-chloro- (8CI)
NSC-231508

1208-42-0
$C_{10}H_{14}O_5$
214.22
1,2,3-Propanetriol, 1-(4-hydroxy-3-methoxyphenyl)- (8CI,9CI)
Guaiacylglycerol

1208-52-2
$C_{13}H_{14}N_2$
198.29
Nc(c(ccc1)Cc(ccc(N)c2)c2)c1
Aniline, 2',4-methylenedi
Benzenamine, 2-((4-aminophenyl)methyl)- (9CI)
2',4-Bis(aminophenyl)methane
o,p'-Diaminodiphenylmethane
2,4'-Diaminodiphenylmethan (German)
2,4'-Diaminodiphenylmethane
2,4'-Diphenylmethanediamine
2,4'-Methylenebis(aniline)
2,4'-Methylenedianiline

1208-67-9
$C_{10}H_{12}O_4S.Na$
251.26
4-Sulfophenylmethallyl ether
Benzenesulfonic acid, 4-((2-methyl-2-propenyl)oxy)-, sodium salt (9CI)
p-Sulfophenyl methallyl ether, sodium salt

1208-86-2
$C_{13}H_{13}NO$
199.25
Benzenamine, 4-methoxy-N-phenyl- (9CI)
p-Anisidine, N-phenyl- (8CI)
NSC-31630

1209-98-9
$C_{15}H_{21}N$
215.33
Fencamfamine
Bicyclo(2.2.1)heptan-2-amine, N-ethyl-3-phenyl-
DEA No. 1760
2-Ethylamino-3-phenylnorbornane
Euvitol
Fenacamfamin
Fencamfamin
Fencamfamine (French)
Fencamfaminum (Latin)
Fencanfamina (Spanish)
2-Norbornanamine, N-ethyl-3-phenyl-
2-Phenyl-3-ethylaminobicyclo-(2.2.1)heptane
3-Phenyl-N-ethyl-2-norbornanamine
Reactivan

1210-05-5
$C_{14}H_{10}O_2$
210.24
O=Cc1ccccc1c2ccccc2C=O
Diphenaldehyde
2,2'-Biphenyldicarboxaldehyde
2,2'-Diformylbiphenyl

1210-12-4
$C_{15}H_9N$
203.24
N#Cc(c(c(ccc1)cc2cccc3)c1)c23
9-Anthracenecarbonitrile (9CI)
9-Anthronitrile (8CI)

1210-35-1
$C_{15}H_{12}O$
208.26
O=C(c(c(c(ccc1)CCc2cccc3)c1)c23
Dibenzsuberone
Dibenzocycloheptanone
5H-Dibenzo(a,d)cyclohepten-5-one, 10,11-dihydro- (9CI)
Dibenzosuberone
10,11-Dihydro-5H-dibenzo(a,d)-cyclohepten-5-one

1212-29-9
$C_{13}H_{24}N_2S$
240.45
N(=C(S)NC(CCCC1)C1)C(CCCC2)C2
Urea, 1,3-dicyclohexyl-2-thio
Dicyclohexyl thiourea
N,N'-Dicyclohexylthiourea
NCI-C04524

1214-39-7
$C_{12}H_{11}N_5$
225.22
n(c(NCc(cccc1)c1)c(N=CN2)c2n3)c3
N-benzyladine
ABG 3034
Adenine, N-benzyl- (8CI)
BA (Growth stimulant)
6-BA
BAP (Growth stimulant)
6-BAP
Benzyladenine
N-Benzyladenine
N6-Benzyladenine
N^6-Benzyladenine
6-Benzyladenine
Benzylaminopurine
N^6-(Benzylamino)purine
6-(Benzylamino)purine
6-(N-Benzylamino)purine
Caswell No. 081EE
EPA Pesticide Chemical Code 116901
N-(Phenylmethyl)-1H-purin-6-amine
1H-Purin-6-amine, N-(phenylmethyl)- (9CI)
SD 4901
SQ 4609

1215-57-2
$C_{15}H_{14}N_2$
222.28
N(=C=Nc(c(ccc1)C)c1)c(c(ccc2)C)c2
Benzenamine, N,N'-methanetetraylbis(2-methyl- (9CI)
Bis(o-tolylcarbodiimide)
N,N'-Methanetetraylbis-(2-methylbenzenamine)

1217-08-9
$C_{17}H_{26}O$
246.39
OCC(c(ccc(c1C(C2C)(C)C)C2(C)C)c1)C
1H-Indene-5-ethanol, 2,3-dihydro-β,1,1,2,3,3-hexamethyl- (9CI)

1220-00-4
$C_{13}H_{10}Cl_2N_2S$
297.21
Urea, 1,3-bis(p-chlorophenyl)-2-thio
Di-p-chlorophenylthiourea
Di-4-chlorophenyl thiourea

1220-94-6
$C_{15}H_{12}N_2O_2$
252.29
O=C(c(c(C(=O)c1c(N)ccc2NC)ccc3)c3)c12
Anthraquinone, 1-amino-4-methylamino
Acetoquinone Light Violet N
Amacel Violet 6B
4-Amino-1-methylaminoanthraquinone
9,10-Anthracenedione, 1-amino-4-(methylamino)- (9CI)
Celliton Fast Violet 6B
Celliton Fast Violet 6BA-CF
Cilla Fast Violet 6B
C.I. 61105
C.I. Disperse Violet 4
C.I. Solvent Violet 12
Diacelliton Fast Violet BF
Disperse Fast Violet B
Disperse Violet 4S
Dispersol Violet B
Duranol Brilliant Violet B
Fenacet Fast Violet 6B
Interchem Acetate Violet 6B
Kayalon Fast Violet BB
1-MA-4-AA (Russian)
Microsetile Violet B
Nacelan Violet 4B
Oracet Violet B
Oracet Violet BN
Serisol Fast Violet 6B
Supracet Violet 2B
Violet Disperzni 4 (Czech)
Violet Rozpoustedlova 12 (Czech)

1222-05-5
$C_{18}H_{26}O$
258.44
O(CC(c(c1cc(c2C(C3C)(C)C)C3(C)C)c2)C)C1
Cyclopenta(g)-2-benzopyran, 1,3,4,6,7,8-hexahydro-4,6,6,7,8,8-hexamethyl
Galoxolide
1,3,4,6,7,8-Hexahydro-4,6,6,7,8,8-hexamethyl-cyclopenta-γ-2-benzopyran

1224-63-1
$C_{12}H_{14}Cl_3O_3PS$
375.64
CCOP(=S)(OCC)OC(=CCl)c1ccc(Cl)cc1Cl
Phosphorothioic acid, O-(2-chloro-1-(2,4-dichlorophenyl)vinyl) O,O-diethyl ester
O'-(2-Chloro-1-(2,4-dichlorophenyl)ethenyl) O,O-diethyl phosphorothioate
ENT 27,097

OMS-774
Phosphorothioic acid, O-(2-chloro-1-(2,4-dichlorophenyl)ethenyl) O,O-diethyl ester
SD-8803
Shell SD-8803

1229-35-2
$C_{18}H_{20}N_2S.ClH$
332.92
Phenothiazine, 10-((1-methyl-3-pyrrolidinyl)methyl)-, hydrochloride
Dilosyn
Disyncran
Methdilazine hydrochloride
10-((1-Methyl-3-pyrrolidinyl)-methyl)phenothiazine, hydrochloride
MJ 5022
Tacaryl

1229-55-6
$C_{17}H_{14}N_2O_2$
278.33
O(c(c(N=Nc(c(c(ccc1)cc2)c1)c2O)ccc3)c3)C
C.I. Solvent Red
C.I. 12150
Oil Pink
Solvent Red 1

1230-80-4
$C_{15}H_{12}N_4O_2$
280.29
1,3,5-Triazin-2-amine, 4,6-diphenoxy- (9CI)
s-Triazine, 2-amino-4,6-diphenoxy- (6CI,7CI,8CI)

1235-74-1
$C_{21}H_{30}O_2$
314.47
Podocarpa-8,11,13-trien-15-oic acid, 13-isopropyl-, methyl ester (8CI)
1-Phenanthrenecarboxylic acid, 1,2,3,4,4a,9,10,10a-octahydro-1,4a-dimethyl-7-(1-methylethyl)-,methyl ester, (1R-(1α,4aβ,10aα))- (9CI)

1239-45-8
$C_{21}H_{20}N_3.Br$
394.35
[Br-].CC[n+]3c(c1ccccc1)c2cc(N)ccc2c4ccc(N)cc34
Phenanthridinium, 3,8-diamino-5-ethyl-6-phenyl-, bromide
2,7-Diamino-10-ethyl-9-phenylphenanthridinium bromide
3,8-Diamino-5-ethyl-6-phenylphenanthridinium bromide
2,7-Diamino-9-phenyl-10-ethylphenanthridinium bromide
2,7-Diamino-9-phenylphenanthridine ethobromide
Dromilac
Ethidium bromide
Homidium bromide
RD 1572

1241-94-7
$C_{20}H_{27}O_4P$
362.44
O=P(OCC(CCCC)CC)(Oc(cccc1)c1)Oc(cccc2)c2
Phosphoric acid, 2-ethylhexyl diphenyl ester
Diphenyl 2-ethylhexyl phosphate
(2-Ethylhexyl)-difenylfosfat (Czech)
2-Ethylhexyl diphenylphosphate
1-Hexanol, 2-ethyl-, ester with diphenyl phosphate
Santicizer 141
Santicizer 141 (Monsanto)

1242-76-8
$C_{25}H_{24}$
324.47
Picene, 1,2,3,4-tetrahydro-2,2,9-trimethyl- (8CI,9CI)

1248-18-6
$C_{20}H_{13}N_2O_4S.Na$
400.40
[Na+].Oc2ccc1ccccc1c2N=Nc4ccc3ccccc3c4S([O-])(=O)=O
1-Naphthalenesulfonic acid, 2-((2-hydroxy-1-naphthalenyl)azo)-, monosodium salt
Atul Pigment Red RS Sodium salt
Brasilaca Red R
Britone Red Y
Carnelio Pale Lithol Red
Certiqual Lithol Red
C.I. 15630
C.I. Pigment Red 49
C.I. Pigment Red 49, Monosodium salt
D & C Red No. 10
Eljon Lithol Red No. 10
Graphic Red Y
Irgalite Red RL
Kromon Sodium Lithol
Lake Red R
Lake Red RL
Light Red RS
Lithol Red
Lithol Red 3580
Lithol Red 17676
Lithol Red B
Lithol Red 3GS
Lithol Red Lake
Lithol Red R
Lithol Red RB Extra
Lithol Red RL 151
Lithol Red RS
Lithol Red Sodium salt
Lithol Red Toner
Lithol Toner
Lithol Toner Extra Light 5000
Lithol Toner Sodium salt RT 314
Lithol Toner YA 8003
Lutetia Red R
Monolite Red R
New York Red
Ohio Red
Oralith Red SR Water Soluble
Plastoresin Red SR
Recolite Red LYS
11935 Red
Red No. 205
Resamine Red RB
Resamine Red RC
Segnale Red R
Signal Red
Siloton Red R
Sodium Lithol
Sodium Lithol Red
Sodium Lithol Red 20-4018
Sunburst Red
Tertropigment Red LR
Undeveldoped Lithol Toner
Vulcafix Red R
Uulcol Scarlet 2G

1249-84-9
$C_{25}H_{44}N_2O.2ClH$
461.55
Azacosterol hydrochloride
AI3-52592
Androst-5-en-3-ol, 17-((3-(dimethylamino)propyl)methylamino)-, dihydrochloride, (3β,17β)-
Androst-5-en-3-β-ol, 17-β-((3-(dimethylamino)propyl)-methylamino)-, dihydrochloride
Azacosterol dihydrochloride
Azacosterol hydrochloride
Azasterol
Caswell No. 286A
20,25-Diazachlolestenol dihydrochloride
Diazacholesterol dihydrochloride
20,25-Diazacholesterol dihydrochloride
Diazacosterol hydrochloride
17-β-((3-(Dimethylamino)-propyl)methylamino)androst-5-en-3-β-ol dihydrochloride
EPA Pesticide Chemical Code 098101
IMD 760
Ornitrol
SC 12937

1250-95-9
$C_{27}H_{46}O_2$
402.73
5-α-Cholestan-3-β-ol, 5,6-α-epoxy
Cholestan-3-ol, 5,6-epoxy-, (3-β,5-α,6-α)-
Cholesterol-α-epoxide
Cholesterol 5-α,6-α-epoxide
Cholesterol oxide

Cholesterol α-oxide
5-α,6-α-Epoxycholestanol
Epoxycholesterol

1254-78-0
$C_{26}H_{47}O_3P$
438.63
O(P(OCCCCCCCCCC)Oc(cccc1)c1)CCCCCCCCCC
Didecyl phenyl phosphite
Irgaplast CH 300
Phenyl didecyl phosphite
Phosphorous acid, didecyl phenyl ester (9CI)

1256-83-3
$C_{29}H_{48}O_3$
444.70
Cholestan-6-one, 3-(acetyloxy)-, (3β,5α)- (9CI)
5α-Cholestan-6-one, 3β-hydroxy-, acetate (8CI)

1260-17-9
$C_{22}H_{20}O_{13}$
492.39
O=C(O)c(c(O)cc(c1C(=O)c(c2c(O)c(O)c3C(OC(C(O)C4O)CO)C4O)c3O)C2=O)c1C
Carmine
AI3-18242
2-Anthracenecarboxylic acid, 7-α-D-glucopyranosyl-9,10-dihydro-3,5,6,8-tetrahydroxy-1-methyl-9,10-dioxo-
2-Anthracenecarboxylic acid, 7-β-D-glucopyranosyl-9,10-dihydro-3,5,6,8-tetrahydroxy-1-methyl-9,10-dioxo- (9CI)
1-Anthroic acid, 9,10-dihydro-2,5,7,8-tetrahydroxy-4-methyl-9,10-dioxo-6-(2,3,4,5-tetrahydroxyhexanoyl)- (8CI)
Carminic acid
C.I. Natural Red 4
C.I. 75470
Cochineal Tincture
E 120
7-α-D-Glucopyranosyl-9,10-dihydro-3,5,6,8-tetrahydr oxy-1-methyl-9,10-dioxo-2-anthracene-carboxylic acid

1264-62-6
$C_{43}H_{75}NO_{16}$
862.19
Succinic acid, monoethyl ester, monoester with erythromycin
Erythrocin ethyl succinate
Erythromycin, ethyl succinate
Erythromycin, mono(ethyl succinate)
Pediamycin

1271-19-8
$C_{10}H_{10}Cl_2Ti$
249.00
Titanium, dichloro-di-pi-cyclopentadienyl
Dicyclopentadienyltitanium dichloride
Dichlorobis(eta^5-2,4-cyclopentadien-1-yltitanium)
Dichlorodicyclopentadienyltitanium
Dichlorodi-pi-cyclopentadienyltitanium
Dichlorotitanocene
Dicyclopentadienyldichlorotitanium
NCI-C04502
Titanium, dichlorobis(eta^5-2,4-cyclopentadien-1-yl)- (9CI)
Titanium ferrocene
Titanocene
Titanocene, dichloride

1271-24-5
$C_{10}H_{10}Cr$
182.19
Chromocene (9CI)
Chromium, bis(eta(5)-2,4-cyclopentadien-1-yl)-
Chromium, di-pi-cyclopentadienyl- (8CI)

1271-28-9
$C_{10}H_{10}.Ni$
188.91
Nickel, Compd with pi-cyclopentadienyl (1:2)
pi-Cyclopentadienyl Compd. with nickel
Di-pi-cyclopentadienylnickel
Nickel biscyclopentadiene
Nickelocene
Nikelocen (Czech)

1271-55-2
$C_{12}H_{12}FeO$
228.09
Ketone, ferrocenyl methyl
Acetoferrocene
Acetylferrocene
1-Acetylferrocene
Ferrocene, acetyl-
Monacetylferrocene

1277-43-6
$C_{10}H_{10}Co$
189.13
Cobalt, di-pi-cyclopentadienyl
Bis(cyclopentadienyl)cobalt
Cobaltocene
Dicyclopentadienylcobalt
Kobaltocen (Czech)

1291-32-3
$C_{10}H_{10}Cl_2Zr$
292.32
Zirconium, dichloro-di-pi-cyclopentadienyl
Zirconocene, dichloride

1295-35-8
$C_{16}H_{24}Ni$
275.06
Nickel, bis((1,2,5,6-eta)-1,5-cyclooctadiene)- (9CI)
Nickel, bis(1,5-cyclooctadiene)- (8CI)

1300-21-6
$C_2H_4Cl_2$
98.96
Ethane, dichloro
Dichloroethane

1300-51-2
$C_6H_6O_4S.Na$
197.17
Sodium phenolsulfonate
Benzenesulfonic acid, hydroxy-, monosodium salt (9CI)
Hydroxybenzenesulfonic acid, monosodium salt

1300-71-6
$C_8H_{10}O$
122.18
Xylenol [UN 2261]
Dimethylphenol
Phenol, dimethyl-
Stericol
UN 2261 [Xylenols]
Xilenoli (Italian)
Xylenolen (Dutch)

1300-72-7
$C_8H_{10}O_3S.Na$
209.23
Xylenesulfonic acid, sodium salt
Benzenesulfonic acid, dimethyl-, sodium salt
Conco SXS
Cyclophil SXS30
Eltesol SX 30
Hydrotrope
Naxonate
Naxonate G
NCI-C55403
Sodium dimethylbenzenesulfonate
Sodium xylenesulfonate
Stepanate X
Surco SXS
Ultrawet 40SX

1300-73-8
$C_8H_{11}N$
121.20
Xylidine
Acid Leather Brown 2G
Acid Orange 24
Aminodimethylbenzene
11460 Brown
Dimethylaminobenzene (OSHA)

Dimethylaniline
Dimethylphenylamine
Resorcine Brown J
Resorcine Brown R
UN 1711 [Xylidines, solid or solution]
Xilidine (Italian)
Xylidine (ACGIH,OSHA) [UN 1711]
Xylidinen (Dutch)

1300-78-3
$C_{27}H_{30}Hg_3O_6$
1052.31
Caswell No. 859B
EPA Pesticide Chemical Code 459300
Tolylmercuric acetate

1302-42-7
$AlO_2.Na$
81.97
Sodium aluminate
Aluminate, sodium (9CI)
Aluminum sodium oxide
Sodium aluminate solution
Sodium aluminum dioxide
Sodium aluminum oxide
Sodium metaaluminate

1302-52-9
$Al_2O_{18}Si_6.3Be$
537.53
Beryl
Beryl Ore
Beryllium aluminium silicate
Beryllium aluminosilicate
Beryllium aluminum silicate

1302-74-5
Al_2O_3
101.96
Corundum
Aluminum oxide
α-Corundum
Electrocorundum
EN 237
KER 710
KO 7
Korund
KU 10
KU 5-3
MP 1
MP 1 (Refractory)

1302-76-7
$O_5Si.2Al$
162.05
Aluminum(III) silicate (2:1)
Aluminosilcate
Aluminum oxide silicate
Ceramic Fibre
Cyanite
Disthene
Kaopolite
Kyanite
Mullite
Oil-Dri
Safe-N-Dri
Silicic acid aluminum salt
Snow Tex
Valfor

1302-78-9
Unknown
Unknown
Bentonite
Albagel Premium USP 4444
Bentonite 2073
Bentonite Magma
Hi-Jel
Imvite I.G.B.A
Magbond
Montmorillonite
Panther Creek Bentonite
Southern Bentonite
Tixoton
Volcaly Bentonite BC
Volclay
Wilkinite

1303-00-0
AsGa
144.64
Gallium-arsenide
Gallium monoarsenide

1303-28-2
As_2O_5
229.84
Arsenic-pentoxide
Anhydride arsenique (French)
Arsenic acid
Arsenic acid anhydride
Arsenic anhydride
Arsenic oxide
Arsenic pentaoxide
Arsenic pentoxide [UN 1559]
Arsenic pentoxide, Solid (DOT)
Arsenic (V) oxide
Diarsenic pentoxide
RCRA waste number P011
UN 1559 [Arsenic pentoxide]
Zotox

1303-32-8
Unknown
Unknown
Arsenic disulfide

1303-33-9
As_2S_3
246.02
Arsenic sulfide, Solid [UN 1557]
Arsenic sesquisulfide
Arsenic sesquisulphide
Arsenic sulfide Yellow
Arsenic sulphide [NA 1557]
Arsenic tersulphide
Arsenic trisulfide (DOT)
Arsenic Yellow
Arsenious sulphide
Arsenous sulfide
Auripigment
C.I. 77086
C.I. Pigment Yellow
Diarsenic trisulfide
Diarsenic trisulphide
King's Yellow
King's Gold
NA 1557 [Arsenic sulfide]
Orpiment
UN 1557 [Arsenic compounds, solid, N.O.S. including arsenates, N.O.S.; arsenites, N.O.S.; arsenic sulfides, N.O.S.; and organic compounds of arsenic, N.O.S.]

1303-36-2
As_2Se_3
386.72
Arsenic selenide (9CI)

1303-39-5
$As_4O_{15}.5Zn$
866.63
Arsenic acid, zinc salt
UN 1712 [Zinc arsenate or Zinc arsenite or Zinc arsenate and Zinc arsenite mixtures.]
Zinc arsenate
Zinc arsenate, Basic
Zinc arsenate, Solid

1303-86-2
B_2O_3
69.62
Boron-oxide
Boric anhydride
Boron oxide (ACGIH,OSHA)
Boron sesquioxide
Boron trioxide
Fused Boric Acid

1303-96-4
$B_4O_7.2Na.10H_2O$
381.42
Sodium borate, decahydrate
Antipyonin
Borascu
Borates, tetra, sodium salt, decahydrate (ACGIH,OSHA)
Borax (8CI)
Borax decahydrate
Boricin
Gertley borate
Jaikin
Neobor
Polybor
Sodium biborate
Sodium biborate decahydrate
Sodium pyroborate
Sodium pyroborate decahydrate
Sodium tetraborate
Sodium tetraborate decahydrate
Tronabor

1304-28-5
BaO

153.34
Barium oxide [UN 1884]
Barium monoxide
Barium protoxide
Baryta
Calcined baryta
Oxyde de baryum (French)
UN 1884 [Barium oxide]

1304-29-6
BaO_2
169.34
Barium peroxide [UN 1449]
Bario (perossido di) (Italian)
Barium binoxide
Barium dioxide
Bariumperoxid (German)
Bariumperoxyde (Dutch)
Barium superoxide
Dioxyde de baryum (French)
Peroxyde de baryum (French)
UN 1449 [Barium peroxide]

1304-56-9
BeO
25.01
Beryllium-oxide
Beryllia
Beryllium monoxide
Thermalox

1304-82-1
Bi_2Te_3
800.76
Bismuth-telluride
Bismuth sesquitelluride
Bismuth telluride (ACGIH, OSHA)
Bismuth telluride, Undoped (OSHA)

1304-85-4
$Bi_5H_9N_4O_{22}$
1462.03
Bismuth-hydroxide-nitrate-oxide
Basic Bismuth Nitrate
Bismuth Magistery
Bismuth subnitrate
Bismuth subnitricum
Bismuth White
Bismuthyl nitrate
Blanc de fard
C.I. 77169
C.I. Pigment White 17
Cosmetic White
Flake White
Magistery of Bismuth
Novismuth
Paint White
Snowcal 5SW
Spanish White
Vicalin

1305-62-0
CaH_2O_2
74.10
Calcium-hydroxide
Bell Mine
Calcium hydrate
Calcium hydroxide (ACGIH, OSHA)
Hydrated lime
Kemikal
Lime Water
Slaked Lime

1305-78-8
CaO
56.08
Calcium oxide (ACGIH, OSHA) [UN 1910]
Burnt Lime
Calcia
CALX
Lime
Lime, Burned
Lime, Unslaked (DOT)
Oxyde de calcium (French)
Quicklime (DOT)
UN 1910 [Calcium oxide]
Wapniowy tlenek (Polish)

1305-79-9
CaO_2
72.08
Calcium peroxide [UN 1457]
UN 1457 [Calcium peroxide]

1305-99-3
Ca_3P_2
182.18
Calcium phosphide [UN 1360]
UN 1360 [Calcium phosphide]

1306-06-5
$Ca_5HO_{13}P_3$
502.31
Durapatite
Alveograf
Apatite, hydroxy
Calcium orthophosphate, basic
Calcium phosphate hydroxide
Hydroxyapatite
Monite
Pentacalcium monohydroxyorthophosphate
Periograf
Supertite 10
Win 40350

1306-19-0
CdO
128.40
O=[Cd]
Cadmium-oxide
Kadmu tlenek (Polish)
NCI-C02551

1306-23-6
CdS
144.46
Cadmium-sulfide
Aurora Yellow
Cadmium Golden 366
Cadmium Lemon Yellow 527
Cadmium Orange
Cadmium Primrose 819
Cadmium sulphide
Cadmium Yellow
Cadmium Yellow 000
Cadmium Yellow Conc. Deep
Cadmium Yellow Conc. Golden
Cadmium Yellow Conc. Lemon
Cadmium Yellow Conc. Primrose
Cadmium Yellow OZ Dark
Cadmium Yellow Primrose 47-4100
Cadmium Yellow 10G Conc
Cadmium Yellow 892
Cadmopur Golden Yellow N
Cadmopur Yellow
Capsebon
C.I. 77199
C.I. Pigment Orange 20
C.I. Pigment Yellow 37
Ferro Lemon Yellow
Ferro Orange Yellow
Ferro Yellow
Greenockite
NCI-C02711

1306-24-7
CdSe
191.37
Cadmium selenide (9CI)

1306-38-3
CeO_2
172.11
Ceric oxide
Cerium dioxide
Cerium oxide (9CI)

1307-96-6
CoO
74.93
Cobalt(2+) oxide
C.I. 77322
C.I. Pigment Black 13
Cobalt Black
Cobalt monooxide
Cobalt monoxide
Cobaltous oxide
Cobalt oxide
Cobalt(II) oxide
Monocobalt oxide
Zaffre

1308-06-1
Co_3O_4
240.80
Cobalt oxide (9CI)

1308-14-1
CrH_3O_3
103.03
Chromic (III) hydroxide
Chromic acid Solution

[UN 1755]
UN 1755 [Chromic acid solution]

1308-31-2
Cr_2FeO_4
223.85
Chromite (Mineral)
Chrome ore
Chromite
Chromite ore
Iron chromite

1308-38-9
Cr_2O_3
152.00
Chromium(III) oxide (2:3)
Anadomis Green
Anadonis Green
Casalis Green
Chrome Green
Chrome ochre
Chrome oxide
Chrome Oxide Green
Chrome Oxide Green GN-M
Chrome Oxide Pigment
Chrome Oxide X1134
Chromia
Chromic Acid Green
Chromic oxide
Chromium III oxide
Chromium oxide
Chromium(3+) oxide
Chromium Oxide Green
Chromium Oxide Pigment
Chromium Oxide X1134
Chromium sesquioxide
Chromium(3+) trioxide
C.I. 77288
C.I. No.77278
C.I. Pigment Green 17
Dichromium trioxide
11661 Green
Green Chrome Oxide
Green Chromic Oxide
Green Chromium Oxide
Green Cinnabar
Green Oxide of Chromium
Green Oxide of Chromium OC-31
Green Rouge
Guignet's Green
Leaf Green
Levanox Green GA
Levanox Green GA (Hydrated chromic oxide)
Oil Green
Oxide of chromium
Pure Chromium Oxide Green 59
Ultramarine Green

1309-36-0
FeS_2
119.98
Pyrite (9CI)

1309-37-1
Fe_2O_3
159.70
Iron(III) oxide
Anchred Standard
Anhydrous Iron oxide
Anhydrous Oxide of iron
Armenian Bole
Bauxite Residue
Black Oxide of Iron
Blended Red Oxides of Iron
Burnt Island Red
Burnt Sienna
Burnt Umber
Calcotone Red
Caput Mortuum
Colcothar
Colloidal Ferric Oxide
C.I. 77491
C.I. Pigment Red 101
C.I. Pigment Red 102
C.I. Pigment Red 101 and 102
Crocus martis adstringens
Deanox
Deanox DNX Pigments
Eisenoxyd
English Red
Ferric oxide
Ferric oxide (Colloidal)
Ferrugo
Indian Red
Iron oxide (ACGIH,OSHA)
Iron Oxide Red
Iron Oxide Pigments
Iron Red
Iron Sesquioxide
Jeweler's Rouge
Levanox Red 130A
Light Red
Manufactured Iron Oxides
Mars Brown
Mars Red
Natural Iron Oxides
Natural Red Oxide
Ochre
Prussian Brown
Raddle
11554 Red
Red Iron Oxide
Red Ochre
Red Oxide
Red Oxide D3452
Red Oxide D6984
Red Oxide of Iron
Rouge (OSHA)
Rubigo
Sienna
Specular Iron
Stone Red
Supra
Synthetic Iron oxide
Venetian Red
Vitriol Red
Vogel's Iron Red
Yellow Ferric oxide
Yellow Oxide of Iron

1309-42-8
H_2MgO_2
58.31
O[Mg]O
Magnesia
Caustic Magnesite
Magnesia Magma
Magnesium dihydroxide
Magnesium hydrate
Magnesium hydroxide (9CI)
Magnesium Hydroxide Gel
Magnesium oxide
Marinco H
Marinco H 1241
Milk Of Magnesia
Mint-O-Mag
Nemalite
S/G 84

1309-48-4
MgO
40.31
Magnesium-oxide
Akro-Mag
Animag
Calcined brucite
Calcined magnesia
Calcined magnesite
Granmag
Magcal
Magchem 100
Maglite
Magnesia
Magnesia USTA
Magnesium oxide (ACGIH, OSHA)
Magnezu Tlenek (Polish)
Magox
Magox 85
Magox 90
Magox 95
Magox 98
Magox op
Marmag
Oxymag
Periclase
Seawater Magnesia

1309-60-0
O_2Pb
239.19
Lead dioxide [UN 1872]
Bioxyde de plomb (French)
C.I. 77580
Lead Brown
Lead oxide Brown
Lead peroxide (DOT)
Lead superoxide
Peroxyde de plomb (French)
UN 1872 [Lead dioxide]

1309-64-4
O_3Sb_2
291.50
Antimony-oxide
A 1530
A 1582
A 1588LP
Antimonious oxide
Antimony(3+) oxide
Antimony peroxide
Antimony sesquioxide
Antimony trioxide
Antimony white

Antox
AP 50
Chemetron Fire Shield
C.I. 77052
C.I. Pigment White 11
Dechlorane A-O
Diantimony trioxide
Exitelite
Extrema
Flowers of antimony
NCI-C55152
Nyacol A 1530
Senarmontite
Thermoguard B
Thermoguard S
Timonox
Valentinite
Weisspiessglanz (German)

1310-32-3
FeSe
134.81
Iron selenide (9CI)

1310-43-6
Fe_2P
142.67
Diiron monophosphide
Di-iron phosphide
Ferrous phosphide
Iron phosphide (9CI)

1310-53-8
GeO_2
104.59
Germanium-dioxide
Germania
Germanic acid
Germanium oxide
Germanium oxide (geo2)

1310-58-3
HKO
56.11
Potassium-hydroxide
Caustic potash
Caustic potash, Dry, Solid, Flake, Bead, or Granular (DOT)
Caustic potash, Liquid or solution (DOT)
Hydroxyde de potassium (French)
Kaliumhydroxid (German)
Kaliumhydroxyde (Dutch)
Lye
Potassa
Potasse Caustique (French)
Potassio (idrossido di) (Italian)
Potassium hydrate (DOT)
Potassium hydroxide (ACGIH, OSHA)
Potassium hydroxide, Dry, Solid, Flake, Bead, or Granular [UN 1813]
Potassium hydroxide, Liquid or solution [UN 1814]
Potassium (hydroxyde de) (French)
UN 1813 [Potassium hydroxide, solid]
UN 1814 [Potassium hydroxide, solution]

1310-65-2
HLiO
23.95
Lithium hydroxide (Li(OH)) (9CI)
Lithium hydroxide, Solution [UN 2679]
UN 2679 [Lithium hydroxide, solution]

1310-73-2
HNaO
40.00
Sodium-hydroxide
Caustic soda
Caustic soda (DOT)
Caustic soda, Liquid (DOT)
Caustic soda, Solid (DOT)
Caustic soda, Solution (DOT)
Hydroxyde de sodium (French)
Lewis-Red Devil Lye
Lye
Lye (DOT)
Natriumhydroxid (German)
Natriumhydroxyde (Dutch)
Soda lye
Sodio(idrossido di) (Italian)
Sodium hydrate (DOT)
Sodium hydroxide (ACGIH, DOT,OSHA)
Sodium hydroxide, Liquid [UN 1824]
Sodium hydroxide, Solid [UN 1823]
Sodium hydroxide, Solution [UN 1824]
Sodium(hydroxyde de) (French)
UN 1823 [Sodium hydroxide, solid]
UN 1824 [Sodium hydroxide solution]
White Caustic

1312-03-4
Hg_3O_6S
729.83
Mercury-oxide-sulfate
Basic Mercuric Sulfate
Mercuric basic sulfate
Mercuric subsulfate, Solid [NA 2025]
NA 2025 [Mercury compounds, solid, N.O.S.]
Turpeth mineral

1312-73-8
K_2S
110.26
Potassium sulfide (DOT)
Potassium monosulfide
Potassium sulfide (2:1)
Potassium sulfide (2:1), Hydrated, Containing at least 30% water
Potassium sulphide, Anhydrous or containing less than 30% water [UN 1382]
Potassium sulphide, Hydrated, Containing not less than 30% water [UN 1847]
UN 1382 [Potassium sulfide, Anhydrous or potassium sulfide with less than 30 per 1 cent water of crystallization]
UN 1847 [Potassium sulfide, Hydrated with not less than 30 per cent water of crystallization]

1312-76-1
Unknown
Unknown
Potassium polysilicate
Caswell No. 701B
EPA Pesticide Chemical Code 072606
Kasil
Kasil 6
Potassium metasilicate
Potassium silicate
Potassium silicate solution
Potassium Water Glass
PS 7
Pyramid 120
Silicic acid, potassium salt (9CI)
Soluble Potash Glass
Soluble Potash Water Glass

1312-81-8
La_2O_3
325.81
Lanthanum oxide (9CI)

1313-13-9
MnO_2
86.94
Manganese-dioxide
Black manganese oxide
Bog manganese
Braunstein (German)
Bruinsteen (Dutch)
Cement Black
C.I. 77728
C.I. Pigment Black 14
C.I. Pigment Brown 8
Mangaanbioxyde (Dutch)
Mangaandioxyde (Dutch)
Mangandioxid (German)
Manganese binoxide
Manganese (biossido di) (Italian)
Manganese (bioxyd de) (French)
Manganese Black
Manganese (diossido di) (Italian)
Manganese (dioxyde de) (French)
Manganese oxide
Manganese (IV) oxide

Manganese peroxide
Manganese superoxide
Pyrolusite Brown

1313-27-5
MoO_3
143.94
Molybdenum-trioxide
Molybdic anhydride
Molybdic trioxide

1313-60-6
Na_2O_2
77.98
Sodium peroxide [UN 1504]
Disodium dioxide
Disodium peroxide
Flocool 180
Sodium dioxide
Sodium oxide (Na2-O2)
Solozone
UN 1504 [Sodium peroxide]

1313-82-2
Na_2S
78.04
Sodium sulfide (Anhydrous) [UN 1385]
Sodium monosulfide
Sodium sulphide
Sodium sulphide, Anhydrous or containing less than 30% water [UN 1385]
UN 1385 [Sodium sulfide, anhydrous or sodium sulfide with less than 30 per cent water of crystallization]

1313-97-9
Nd_2O_3
336.48
Neodymium oxide (9CI)

1313-99-1
NiO
74.71
Nickel(II) oxide (1:1)
Black Nickel Oxide
Bunsenite
C.I. 77777
Green Nickel Oxide
Mononickel oxide
Nickel monoxide
Nickelous oxide
Nickel protoxide
Nickel oxide
Nickel Oxide Sinter 75

1314-06-3
Ni_2O_3
165.42
Nickel(III) oxide
Dinickel trioxide
Nickelic oxide
Nickel oxide
Nickel oxide peroxide
Nickel peroxide
Nickel sisquioxide
Nickel trioxide

1314-13-2
OZn
81.37
Zinc-oxide
Actox 14
Actox 16
Actox 216
AI3-00277
Akro-Zinc Bar 85
Akro-Zinc Bar 90
Amalox
Azo-33
Azo-55
Azo-66
Azo-77
Azodox-55
Azodox-55TT
Azo-55TT
Azo-66TT
Azo-77TT
Cadox XX 78
Calamine
Calamine (Spray)
Chinese White
C.I. 77947
C.I. Pigment White 4
Cynku tlenek (Polish)
Eledtrox 2500
Emanay Zinc Oxide
Emar
Felling Zinc Oxide
Flowers of Zinc
Giap 10
Green Seal-8
Hubbuck's White
Kadox 15
Kadox-25
Kadox 72
K-Zinc
Outmine
Ozide
Ozlo
Pasco
Permanent White
Philosopher's Wool
Powder Base 900
Protox Type 166
Protox Type 167
Protox Type 168
Protox Type 169
Protox Type 267
Protox Type 268
Red-Seal-9
Snow White
Unichem ZO
Vandem VAC
Vandem VOC
White Seal-7
XX 78
XX 203
XX 601
Zinca 20
Zincite
Zincoid
Zinc oxide (ACGIH,OSHA)
Zinc oxide Fume
Zinc white
Zn-0401 E 3/16''
Zn 0701T

1314-18-7
O_2Sr
119.62
Strontium dioxide
Peroxido de estroncio (Spanish)
Peroxyde de strontium (French)
Strontium peroxide [UN 1509] (9CI)
UN 1509 [Strontium peroxide]

1314-20-1
O_2Th
264.00
Thorium-oxide
Thoria
Thorium dioxide
Thorotrast
Thortrast
Umbrathor

1314-22-3
O_2Zn
97.37
Zinc peroxide [UN 1516]
UN 1516 [Zinc peroxide]
Zinc superoxide

1314-23-4
O_2Zr
123.22
O=[Zr]=O
Zirconium oxide (9CI)
AI3-29087

1314-24-5
O_3P_2
109.94
Phosphorus trioxide [UN 2578]
Diphosphorus trioxide
UN 2578 [Phosphorus trioxide]

1314-27-8
O_3Pb_2
462.40
Lead trioxide
Lead oxide (8CI,9CI)
Lead sesquioxide

1314-32-5
O_3Tl_2
456.74
Thallic-oxide
Dithallium trioxide
RCRA waste number P113
Thallium oxide (8CI,9CI)
Thallium(111) oxide
Thallium(3+) oxide
Thallium peroxide
Thallium sesquioxide

1314-34-7
O_3V_2
149.88
Vanadium-trioxide
UN 2860 [Vanadium trioxide, nonfused form]
Vanadic oxide
Vanadium oxide
Vanadium sesquioxide
Vanadium trioxide, Nonfused form [UN 2860]

1314-35-8
O_3W
231.85
Tungsten-oxide
C.I. 77901
Tungsten Blue
Tungsten trioxide
Tungstic anhydride
Tungstic oxide
Wolframite

1314-36-9
O_3Y_2
225.82
Yttrium-oxide
Yttria

1314-41-6
O_4Pb_3
685.57
Lead-tetroxide
C.I. 77578
C.I. Pigment Red 105
Gold Satinobre
Lead orthoplumbate
Lead oxide
Lead Oxide Red
Lead tetraoxide
Mineral Orange
Mineral Red
Minium
Minium Non Setting RL-95
Orange lead
Paris Red
Plumboplumbic oxide
Red Lead
Red Lead Oxide
Sandix
Saturn Red
Trilead tetroxide

1314-56-3
O_5P_2
141.94
Phosphorus-oxide
Diphosphorus pentoxide
NA 1807 (DOT)
Phosphoric anhydride (DOT)
Phosphorus(V) oxide
Phosphorus pentaoxide
Phosphorus pentoxide [UN 1807]
UN 1807 [Phosphorus pentoxide]

1314-60-9
O_5Sb_2
323.50
Antimony-pentoxide
A 1530
Antimonic "acid"
Antimonic oxide
Antimony pentaoxide
Diantimony pentaoxide
Diantimony pentoxide
Stibic anhydride

1314-61-0
O_5Ta_2
441.89
Tantalum oxide dust
Tantalic acid anhydride
Tantalum oxide (ACGIH) (8CI, 9CI)
Tantalum penta oxide
Tantalum pentoxide

1314-62-1
O_5V_2
181.88
Vanadium pentoxide (Dust)
Anhydride vanadique (French)
C.I. 77938
RCRA waste number P120
UN 2862 [Vanadium pentoxide, nonfused form]
Vanadic anhydride
Vanadio, pentossido di (Italian)
Vanadium Dust and Fume (ACGIH)
Vanadium oxide (V2O5)
Vanadium pentaoxide
Vanadium pentoxide (Fume)
Vanadiumpentoxid (German)
Vanadium pentoxide, Nonfused form [UN 2862]
Vanadiumpentoxyde (Dutch)
Vanadium, pentoxyde de (French)
Wanadu pieciotlenek (Polish)

1314-64-3
O_6SU
366.09
Uranyl sulfate
Dioxo(sulfato(2-)-O)uranium
Uranium, dioxosulfato-
Uranium, dioxo(sulfato(2-)-O)- (9CI)
Uranium oxide sulfate
Uranium oxysulfate

1314-80-3
P_2S_5
222.24
S=P(=S)SP(=S)=S
Phosphorus-sulfide
Pentasulfure de phosphore (French)
Phosphoric sulfide
Phosphorus pentasulfide (ACGIH,DOT,OSHA)
Phosphorus pentasulphide, Free from yellow or white phosphorus [UN 1340]
Phosphorus persulfide
RCRA waste number U189
Sirnik fosforecny (Czech)
Sulfur phosphide
Thiophosphoric anhydride
UN 1340 [Phosphorus pentasulfide, Free from yellow or white phosphorus]

1314-84-7
P_2Zn_3
258.05
Zinc phosphide [UN 1714]
Blue-Ox
Kilrat
Mous-Con
Phosphure de zinc (French)
Phosvin
RCRA waste number P122
Rumetan
UN 1714 [Zinc phosphide]
Zinco(fosfuro di) (Italian)
Zinc(phosphure de) (French)
Zinc-Tox
Zinkfosfide (Dutch)
Zinkphosphid (German)
ZP

1314-85-8
P_4S_3
220.06
Phosphorus-sesquisulfide
Phosphorous sesquisulfide
Phosphorus sesquisulfide (DOT)
Phosphorus sesquisulphide, Free from yellow or white phosphorus [UN 1341]
Phosphorus (III) sulfide (IV)
Sesquisulfure de phosphore (French)
Tetraphosphorus trisulfide
Trisulfurated phosphorus
UN 1341 [Phosphorus sesquisulfide, Free from yellow or white phosphorous]

1314-87-0
PbS
239.25
Lead-sulfide
C.I. 77640
Galena
Natural lead sulfide
Plumbous sulfide

1314-91-6
PbTe
334.79
Lead telluride (8CI,9CI)

1314-96-1
SSr
119.69
Strontium sulfide (SrS) (9CI)
CI Pigment White 8

C.I. 77847
RCRA waste No. P107
Strontium monosulfide
Strontium sulphide

1314-98-3
SZn
97.46
Zinc sulfide (9CI)
Albalith
C.I. Pigment White 7
Irtran 2
Sachtolith
Zinc blende
Zinc monosulfide

1315-04-4
S_5Sb_2
MW:
Antimony pentasulfide
Antimonial saffron
Antimonic sulfide
Antimony Red
Antimony sulfide (Sb_2S_5) (8CI, 9CI)
Antimony Sulfide Golden
C.I. 77061
Golden Antimony Sulfide

1317-25-5
$C_4H_9Al_2ClN_4O_7$
314.53
Alcloxa
ALCA
Alcloxum (Latin)
Aluminium chlorohydroxy allantoinate
Aluminum chlorhydroxy allantoinate
Aluminum, chloro((2,5-dioxo-4-imidazolidinyl)ureato)tetrahydroxydi- (9CI)
Aluminum chlorohydroxy allantoinate
Aluminum, chlorotetrahydroxy((4,5-dihydro-2-hydroxy-5-oxo-1H-imidazol-4-yl)ureato)di-
Aluminum, chlorotetrahydroxy((2-hydroxy-5-oxo-2-imidazolin-4-yl)ureato)di-
Caswell No. 029A
Chlorhydroxyaluminum allantoinate
Chloro((2,5-dioxo-4-imidazolidinyl)ureato)tetrahydroxydialuminum
Chlorotetrahydroxy((2-hydroxy-5-oxo-2-imidazolin-4-yl)ureato)dialuminum
Dialuminum tetrahydroxychloro allantoinate
RC-173

1317-33-5
MoS_2
160.06
Molybdenum-sulfide
DAG 325
Molybdenum disulfide
Molybdenum (IV) sulfide
Molykote
Mopol M
Mopol S

1317-34-6
Mn_2O_3
157.88
Manganese(III) oxide
Cassel Brown
C.I. Natural Brown 8
C.I. 77727
Cologne Earth
Cologne Umber
Cullen Earth
Dimanganese trioxide
Manganese manganate
Manganese(3+) oxide
Manganese sisquioxide
Manganese trioxide
Manganic oxide
Rubens Brown
Soluble Vandyke Brown
Vandyke Brown
Walnut Stain

1317-35-7
Mn_3O_4
228.82
Manganese-oxide
Manganese tetroxide (ACGIH,-OSHA)
Manganomanganic oxide
Trimanganese tetraoxide
Trimanganese tetroxide

1317-36-8
OPb
223.19
Lead-monoxide
C.I. 77577
C.I. Pigment Yellow 46
Lead monooxide
Lead oxide
Lead Oxide Yellow
Lead protoxide
Lead(II) oxide
Litharge
Litharge Pure
Litharge Yellow L-28
Massicot
Massicotite
Plumbous oxide
Yellow Lead Ocher

1317-37-9
FeS
87.91
S=[Fe]
Ferrous sulfide
Black Iron Sulfide
C.I. 77540
Ferrous monosulfide
Iron monosulfide
Iron protosulfide
Iron sulfide (9CI)
Iron sulfuret

1317-38-0
CuO
79.54
O=[Cu]
Copper-oxide
Black Copper Oxide
C.I. 77403
C.I. Pigment Black 15
Copper Brown
Copper(II) oxide
Copper monooxide
Copper monoxide
Copper(2+) oxide
Cupric oxide

1317-39-1
Cu_2O
143.08
Copper(I) oxide
Brown Copper Oxide
Caocobre
C.I. 77402
Copox
Copper nordox
Copper oxide (8CI,9CI)
Copper (1+) oxide
Copper-Sandoz
Copper Sardez
Copper suboxide
Cupper oxide (Russian)
Cupramar
Cuprocide
Cuprous oxide
Dicopper monoxide
Fungimar
Kupferoxydul (German)
Kuprite
Nordox
Oleocuivre
Oleo nordox
Oxyde cuivreux (French)
Perecot
Perenox
Red Copper Oxide
Yellow Cuprocide

1317-40-4
CuS
95.60
Copper (II) sulfide
C.I. 77450
C.I. Pigment Blue 34
Copper Blue
Copper monosulfide
Copper sulfide
Copper(2+) sulfide
Cupric sulfide
Horace Vernet's Blue
Monocopper monosulfide
Oil Blue

1317-42-6
CoS
90.99
Cobalt(2+) sulfide
Cobalt monosulfide
Cobaltous sulfide

Cobalt sulfide
Cobalt sulfide (Amorphous)

1317-60-8
Fe_2O_3
159.70
Hematite
Blood Stone
Haematite
Iron Ore
Red Iron Ore

1317-61-9
Fe_3O_4
231.54
Ferrosoferric oxide
Iron oxide (9CI)
Iron(III) oxide

1317-65-3
$CO_3.Ca$
100.09
Calcium-carbonate
Agricultural limestone
Agstone
Bell Mine Pulverized Limestone
Calcium carbonate (ACGIH, OSHA)
Chalk
Domolite
Franklin
Limestone (OSHA)
Lithograpic Stone
Marble (OSHA)
Natural calcium carbonate
Portland Stone
Sohnhofen Stone

1317-70-0
O_2Ti
79.88
Anatase titanium dioxide
Anatase (TiO_2) (9CI)

1317-80-2
O_2Ti
79.88
Rutile titanium dioxide
Rutile (TiO_2) (9CI)
Titanium oxide

1317-86-8
Unknown
Unknown
Stibnite

1317-95-9
Unknown
Unknown
Silica, Crystalline-tripoli (ACGIH,OSHA)
Tripoli

1317-98-2
Unknown
Unknown
Valentinite

1318-00-9
Unknown
Unknown
Vermiculite

1318-02-1
Unknown
Unknown
Zeolite
Calcium aluminum silicate
Clinoptilolite
Erionite
Hydrated alkali-aluminum silicate
MS 4A
MS 5A
Silver sodium zeolite
Sodium aluminosilicate
Zeolite A
Zeolite NaA
Zeolites
Zeolithe A

1318-09-8
Unknown
Unknown
Amphiboles

1318-16-7
$Al_2O_3.xH_2O$
228.10
Bauxite
Florite
Porocel
Porocel O

1319-41-1
Unknown
Unknown
Saponite
Saponite ((Mg0.5-1Fe0-0.5)3(Si3.67Al0.33)(Na0-0.33Ca0-0.17)(OH)2O10.4H2O) (9CI)
Zebedassite

1319-46-6
$C_2H_2O_8Pb_3$
775.63
Lead hydroxide carbonate
Basic lead carbonate
Berlin White
Bis(carbonato(2-))dihydroxytrilead
Carbonic acid, lead salt, basic
Ceruse
Cerussa
C.I. Pigment White 1
Flake Lead
Lead, bis(carbonato)dihydroxytri-
Lead, bis(carbonato(2-))dihydroxytri- (9CI)
Lead carbonate hydroxide
Lead subcarbonate
Silver White
White Lead
White Lead, Hydrocerussite

1319-72-8
$C_8H_5Cl_3O_3.C_3H_9NO$
330.60
2-Propanol, 1-amino-, (2,4,5-trichlorophenoxy)acetate (Salt) (9CI)
Acetic acid, (2,4,5-trichlorophenoxy)-, Compd. with 1-amino-2-propanol (1:1) (8CI)
2-Propanol, 1-amino-, Compd. with (2,4,5-trichlorophenoxy)acetic acid (1:1)

1319-73-9
C_9H_{10}
118.18
Benzene, ethenyl-, monomethyl deriv. (9CI)
Styrene, methyl- (8CI)

1319-77-3
C_7H_8O
108.15
Cresol (ACGIH,OSHA) [UN 2076]
Acede cresylique (French)
Bacillol
Cresoli (Italian)
Cresylic acid [UN 2022]
Hydroxytoluole (German)
Kresole (German)
Kresolen (Dutch)
Krezol (Polish)
Phenol, methyl- (9CI)
RCRA waste number U052
Tekresol
ar-Toluenol
Tricresol
UN 2022 [Cresylic acid]
UN 2076 [Cresols (o-; m-; p-)]

1319-80-8
$C_{10}H_8Cl_8$
411.78
Norbornane, 2,2-dimethyl-3-methylene-, octachloro deriv.
Octachlorocamphene

1319-85-3
$C_7H_3Cl_3O_2$
225.46
Benzoic acid, trichloro- (8CI, 9CI)

1319-86-4
Unknown
Unknown

Acetato(chloromethoxy-propyl)mercury

1319-91-1
$C_{22}H_{43}IO_2.1/2Ca$
486.53
Calcium iodobehenate
Calioben
Docosanoic acid, iodo-, calcium salt
Iododocosanoic acid calcium salt
Saiodin
Sajodin

1319-96-6
$C_{10}H_{17}ClN_2O$
216.74
Urea, 3-(chloro-2-norbornyl)-1,1-dimethyl
1-(Chloro-2-norbornyl)-3,3-dimethylurea
Hercules 7175

1320-01-0
$C_{12}H_{18}$
162.28
Amyl toluene
Amyltoluene
Benzene, methylpentyl- (9CI)
Toluene, pentyl- (6CI, 8CI)

1320-07-6
$C_{20}H_{18}N_4O_5S.Na$
449.42
Acid Orange 24
Benzenesulfonic acid, 4-((3-((dimethylphenyl)azo)-2,4-dihydroxyphenyl)azo)-, monosodium salt (9CI)
Brown No. 201 (Japan)
C-Ext. Braun 4 (Germany)
C.I. 20170
D & C Brown No. 1
4-((3-((2,4-Dimethylphenyl)-azo)-2,4-dihydroxyphenyl)-azo)benzenesulfonic acid, monosodium salt
Germany: C-Ext. Braun 4
Japan: Brown No. 201
Resorcin Brown

1320-15-6
$C_{18}H_{26}Cl_2O_3$
361.31
Isooctyl 4-(2,4-dichlorophenoxy)butyrate
Butanoic acid, 4-(2,4-dichlorophenoxy)-, isooctyl ester
Butyric acid, 4-(2,4-dichlorophenoxy)-, isooctyl ester
Caswell No. 316D
2,4-DB isoocytl ester
4-(2,4-Dichlorophenoxy)butyric acid isooctyl ester
EPA Pesticide Chemical Code 030863

1320-16-7
$C_{11}H_{14}O_2$
178.23
Benzoic acid, (1,1-dimethylethyl)- (9CI)
(1,1-Dimethylethyl)benzoic acid

1320-18-9
$C_{15}H_{20}Cl_2O_4$
351.23
2,4-D, Propylene glycol butyl ether ester
Acetic acid, (2,4-dichlorophenoxy)-, 2-butoxymethylethyl ester (9CI)
Acetic acid, (2,4-dichlorophenoxy)-, butoxy propylene deriv (8CI)
Acetic acid, 2,4-dichlorophenoxy-, butoxypropyl ester
2,4-Dichlorophenoxyacetic acid propylene glycol butyl ether ester
2,4-Dichlorophenoxy), 2-butoxymethylethyl ester
2,4-D, propylene glycol butyl ether ester
2,4-D PGBE
Propylene glycol butyl ether 2,4-dichlorophenoxyacetate

1320-21-4
$C_{13}H_{20}O$
192.30
Amyl xylyl ether
Benzene, dimethyl(pentyloxy)-(9CI)
Ether, pentyl xylyl (8CI)

1320-37-2
$C_2Cl_2F_4$
170.92
Ethane, dichlorotetrafluoro
Dichlorotetrafluoroethane [UN 1958]
Dwuchloroczterofluoroetan (Polish)
Tetrafluorodichloroethane
UN 1958 [Dichlorotetrafluoroethane]

1320-41-8
$C_2H_2F_2$
64.03
Ethene, difluoro- (9CI)
Ethylene, difluoro- (8CI)

1320-66-7
C_4H_9ClO
108.57
Butylene chlorohydrin
Butanol, chloro- (VAN) (9CI)
Butene chlorohydrin
Chlorobutanol

1320-67-8
$C_4H_{10}O_2$
90.14
1,2-Propanediol, monomethyl ether
Methyl ether of propylene glycol

1320-98-5
$C_6H_{14}O$
102.20
CC(C)CCCO
Pentanol, 4-methyl
4-Methylpentanol

1321-04-6
$C_7H_4I_2O_3$
389.92
Benzoic acid, 2-hydroxydiiodo-(9CI)
Salicylic acid, diiodo- (8CI)

1321-10-4
C_7H_7ClO
142.58
Phenol, chloromethyl- (9CI)
Cresol, chloro- (8CI)

1321-11-5
$C_7H_7NO_2$
137.13
Benzoic acid, amino- (9CI)

1321-12-6
$C_7H_7NO_2$
137.13
Nitrotoluenes
Benzene, methylnitro-
Methylnitrobenzene
Mononitrotoluene
Nitrophenylmethane
Nitrotoluene
Nitrotoluol

1321-16-0
$C_7H_{10}O$
110.17
O=CC1=CCCCC1
Cyclohexenecarboxaldehyde
Benzaldehyde, tetrahydro-
Tetrahydrobenzaldehyde

1321-38-6
$C_9H_6N_2O_2$
174.15
[O-][C][N]C([N][C][O-])c1ccccc1
Benzene, diisocyanatomethyl-(9CI)
Isocyanic acid, methylphenylene ester (8CI)

1321-40-0
C_9H_7NO

145.17
Quinolinol
Hydroxyquinoleine (French)

1321-60-4
$C_9H_{18}O$
142.24
Trimethylcyclohexanol
Cyclohexanol, trimethyl- (8CI, 9CI)

1321-64-8
$C_{10}H_3Cl_5$
300.38
Naphthalene, pentachloro
Pentachloronaphthalene (ACGIH,OSHA)

1321-65-9
$C_{10}H_5Cl_3$
231.50
Naphthalene, trichloro
Halowax
Nibren Wax
Seekay Wax
Trichloronaphthalene (ACGIH, OSHA)

1321-67-1
$C_{10}H_8O$
144.18
Naphthol
Naphthalenol (9CI)

1321-69-3
$C_{10}H_8O_3S.Na$
231.23
Naphthalene sulfonic acid, sodium salt solution
Naphthalenesulfonic acid, sodium salt (8CI,9CI)
Sodium naphthalene sulfonate solution
Sodium naphthalenesulfonate

1321-74-0
$C_{10}H_{10}$
130.20
Benzene, divinyl
Divinylbenzene
Vinylstyrene

1321-87-5
$C_{10}H_{18}O_2$
170.25
Octynediol, dimethyl- (6CI,8CI,9CI)
Dimethyloctynediol
Syrfynol 82

1321-89-7
$C_{10}H_{20}O$
156.27
Isodecylaldehyde
2,6-Dimethyl octanal
2,6-Dimethyl octanoic aldehyde
Isoaldehyde C-10
Isodecanal (9CI)

1321-94-4
$C_{11}H_{10}$
142.21
Naphthalene, methyl
Methylnaftalen (Czech)
Methylnaphthalene

1322-14-1
$C_{11}H_{20}O_2.1/2Ca$
204.32
10-Undecenoic acid, calcium salt (9CI)
Calcium 10-undecenoate

1322-17-4
$C_{11}H_{22}O_3$
202.29
1,3-Nonanediol, monoacetate (9CI)
AI3-36032-X
1,3-Nonanediol acetate
Nonane-1,3-diol monoacetate

1322-20-9
$C_{12}H_{10}O$
170.21
(1,1'-Biphenyl)ol (9CI)
Biphenylol (8CI)
Phenol, phenyl-

1322-38-9
$C_{13}H_8Br_3NO_2$
449.95
Salicylanilide, tribromo
Tribromosalicylanilide

1322-60-7
$C_{14}H_8O_4$
240.22
9,10-Anthracenedione, dihydroxy- (9CI)
Anthraquinone, dihydroxy- (8CI)

1322-93-6
$C_{16}H_{20}O_3S.Na$
315.39
Naphthalenesulfonic acid, bis-(1-methylethyl)-, sodium salt (9CI)
Naphthalenesulfonic acid, diisopropyl-, sodium salt (8CI)

1322-97-0
$C_{16}H_{26}O_2$
250.42
Ethanol, 2-(octylphenoxy)- (8CI,9CI)
Octyl phenol condensed with 1 mole ethylene oxide
Octyl phenoxy ethanol
Octylphenoxyethanol
PE-1

1322-98-1
$C_{16}H_{25}O_3S.Na$
320.46
Benzenesulfonic acid, decyl-, sodium salt
Decyl benzene sodium sulfonate
Santomerse D
Sodium decylbenzenesulfonamide
Sodium decylbenzenesulfonate
Ultrawet DS

1323-38-2
$C_{21}H_{40}O_5$
372.55
9-Octadecenoic acid, 12-hydroxy-, (R-(Z))-, monoester with 1,2,3-propanetriol
AI3-00967
9-Octadecenoic acid, 12-hydroxy-, monoester with 1,2,3-propanetriol, (R-(Z))-

1323-39-3
$C_{21}H_{42}O_3$
342.63
Stearic acid, monoester with 1,2-propanediol
Atlas G 924
Cerasynt PA
Cerasynt PN
Crill 26
Dragil-P
Emcol PS-50 RHP
Emerest 2381
Monosteol
Monosteol TG
Noca
Nonex 32
1,2-Propanediol, monostearate
Propylene glycol monostearate
Prostearin
Tegin P
Octadecanoic acid, monoester with 1,2-propanediol
USAF KE-13

1323-65-5
$C_{24}H_{42}O$
346.60
Dinonyl phenol
AI3-22033-X
Dinonylphenol
Phenol, dinonyl- (9CI)
Phenol, dinonyl- (Mixture of isomers)

1324-04-5
$C_{22}H_{13}NO_8S_2$
483.47
1H-Indene-1,3(2H)-dione, 2-benzo(f)quinolin-3-yl-, disulfo deriv. (9CI)

1324-17-0
$C_{34}H_{15}BrO_2$
535.40
Benzo(rst)phenanthro(10,1,2-cde)pentaphene-9,18-dione, bromo- (9CI)

1324-27-2
$C_{28}H_{13}ClN_2O_4$
476.86
5,9,14,18-Anthrazinetetrone, chloro-6,15-dihydro- (9CI)

1324-35-2
$C_{30}H_{12}Br_2O_2$
564.23
8,16-Pyranthrenedione, dibromo- (9CI)
Dibromo-8,16-pyranthrenedione

1324-54-5
Unknown
Unknown
C.I. Vat Blue 18
Ahcovat Navy Blue BR
Ahcovat Printing Navy Blue XSA
Alizanthrene Navy Blue R
Alizanthrene Navy Blue RT
Amanthrene Navy Blue BN
Amanthrene Supra Navy Blue BN
Amanthrene Supra Navy Blue BNR
Anthravat Navy Blue BR
Calcoloid Navy Blue
Calcoloid Navy Blue NTC
Caldedon Navy Blue AR
Caledon Navy Blue ART
Caledon Navy Blue 2R
Carbanthrene Navy Blue RA
C.I. 59815
Cibanone Navy Blue FRA
Cibanone Navy Blue RA
Mikethren Navy Blue FRA
Modr Kypova 18 (Czech)
Paradone Dark Blue RFW
Ponsol Jade Green Supra D
Ponsol Navy Blue RA
Ponsol Navy Blue RAD
Romantrene Navy Blue FRA
Sandothrene Dark Blue NR
Tinon Navy Blue RA
Vat Blue 18

1324-55-6
$C_{34}H_{14}Cl_2O_2$
525.38
Benzo(rst)phenanthro-(10,1,2-cde)pentaphene-9,18-dione, dichloro
Ahcovat Brilliant Violet 2R
Ahcovat Brilliant Violet 4R
Amanthrene Brilliant Violet RR
Arlanthrene Violet 4R
Atic Vat Brilliant Purple 4R
Benzadone Brilliant Purple 2R
Benzadone Brilliant Purple 4R
Brilliant Violet K
Calcoloid Violet 4RD
Calcoloid Violet 4RP
Caledon Brilliant Purple 4R
Caledon Brilliant Purple 4RP
Caledon Printing Purple 4R
Carbanthrene Brilliant Violet 4R
Carbanthrene Violet 2R
Carbanthrene Violet 2RP
C.I. 60010
Cibanone Violet F 4R
Cibanone Violet F 2RB
Cibanone Violet 2R
Cibanone Violet 4R
C.I. Pigment Violet 31
C.I. Vat Violet 1 (8CI)
Dichloroisoviolanthrone
Fenanthren Brilliant Violet 2R
Fenanthren Brilliant Violet 4R
Indanthren Brilliant Violet 4R
Indanthren Brilliant Violet RR
Indanthrene Brilliant Violet 4R
Indanthrene Brilliant Violet RR
Indanthren Printing Violet F 4R
Indofast Violet Lake
Nihonthrene Brilliant Violet 4R
Nihonthrene Brilliant Violet RR
Nyanthrene Brilliant Violet 4R
Ponolith Fast Violet 4RN
Sandothrene Violet N 4R
Sandothrene Violet N 2RB
Sandothrene Violet 4R
Solanthrene Brilliant Violet F 2R
Symuler Fast Violet R
Tinon Violet B 4RP
Tinon Violet 4R
Tinon Violet 2RB
Vat Bright Violet K
Vat Brilliant Violet K
Vat Brilliant Violet KD
Vat Brilliant Violet KP
Violet Kypova 1 (Czech)
Violet Pigment 31

1324-58-9
$C_{34}H_{22}Cl_2N_4O_{11}S_3.3Na$
898.62
Diindolo(3,2-b:3',2'-m)triphenodioxazinetrisulfonic acid, 8,18-dichloro-5,15-diethyl-5,15-dihydro-, trisodium salt (9CI)

1324-76-1
$C_{37}H_{29}N_3O_3S$
595.70
CI Pigment Blue 61
Benzenesulfonic acid, ((4-((4-(phenylamino)phenyl)(4-(phenylimino)-2,5-cyclohexadien-1-ylidene)methyl)-phenyl)amino)- (9CI)
C.I. 42765
Pigment Blue 61
Reflex Blue AGL
Reflex Blue AGM
Reflex Blue R

1325-16-2
$C_{17}H_{12}N_2O_9S_2.2Ba$
727.09
Benzoic acid, 2-((2-hydroxy-3,6-disulfo-1-naphthalenyl)-azo)-, barium salt (1:2) (9CI)

1325-35-5
Unknown
Unknown
C.I. Direct Orange 15 (9CI)

1325-37-7
Unknown
Unknown
C.I. Direct Yellow 11 (9CI)
Airedale Yellow RD
Amanil Fast Yellow AN
Atlantic Stilbene Yellow GA
Azine Fast Yellow A
Azomine Yellow R
Bencidal Fast Yellow X
Benzanil Yellow R
Benzo Fast Yellow A
Calcomine Yellow 2G
Chlorazol Paper Yellow R
Chrome Leather Yellow A
C.I. Yellow 11
C.I. 40000
Cresotine Yellow A
Diamine Fast Yellow A
Diazine Yellow R
Diazol Fast Yellow A
Diphenyl Fast Yellow FA
Diphenyl Yellow G
Direct Orange G
Direct Yellow F
Erie Yellow FP
Erie Yellow SR
Fast Yellow G
Fenamin Yellow TP
Fixanol Yellow GS
Hispamin Yellow F
Iredale Yellow RD
Kansai Direct Fast Yellow A
Manil Fast Yellow AN
Nankai Direct Fast Yellow A
Naphthalene Leather Yellow GL
NCI-C60888
Nippon Fast Yellow A
Paper Yellow RF
Paramine Fast Yellow GR
Paramine Yellow 2R
Peeramine Stilbene Yellow GA
Pheno Fast Yellow 95
Polycor Yellow R
Stilbene Yellow TK
Stilbene Yellow TR
Sun Yellow G
Tetrodirect Yellow A
Trisulfon Yellow G
Trisulfon Yellow GF
Vondacel Yellow RN
Wogenal Yellow CG
Yellow EMBL

Zomine Yellow R

1325-38-8
Unknown
Unknown
C.I. Direct Yellow 6 (9CI)

1325-82-2
Unknown
Unknown
C.I. Pigment Violet 3
Benzoic Methyl Violet Lake
Brillfast Violet
Calcotone Violet RP
C.I. 42535:2
C.I. Basic Violet 1, molybdate-tungstatephosphate (9CI)
C.I. 42535 Lake
Conc. Violet R
Consol Violet
Dainichi Fast Violet M toner
Dainichi Fast Violet MX
Eljon Violet Toner
Fanal Violet RA Supra
Fanal Violet R Supra
Fanatone Violet
Fast Bronze Violet
Fastel Violet R
Fastel Violet R Supra
Fast Violet Toner R
Federal Fast Violet 7001
Halopont Violet NM
Helmerco Violet MR
Irgalite Violet TCR
Kromal Violet R
Lake Basic Violet
Methyl Violet Lake
Methyl Violet PMA Lake VA-4150
NCI-C54659
No. 34 Forthbrite Fast Violet
No. 48 Forthbrite Fast Violet
Nyco Liquid Violet RF
Nyco Super Violet 4R
Permanent Purple
Permanent Purple Toner
PMA Violet 3
Purple Lake
Pyramid Violet Toner NX
Recolite Violet RDS
Recolite Violet RTS
Siegle Bluish Violet Extract DH47
Silosuper Violet R
Solar Coating Violet RMN47-3012
Solar Violet RCL
Solar Violet RMN47-3612
Syton Violet R
Tintofen Violet R
Tintofen Violet R Supra
Toning Blue MV
Tropical Violet Toner
Violet Lake
Violet Toner PTMA 55-2925

1325-87-7
$C_{33}H_{40}N_3.x$
Unknown
CI Pigment Blue 1
Cascade Blue
C.I. Pigment Blue 1
C.I. 42595 Phosphotungsto-molybdic acid salt
C.I. 42595:2
Conc Blue B
Dainichi Fast Blue EX
Dainichi Fast Blue Toner
Ethanaminium, N-(4-((4-(di-ethylamino)phenyl)(4-(ethyl-amino)-1-naphthalenyl)methyl-ene)-2,5-cyclohexadien-1-yl-idene)-N-ethyl-, molybdate-tungstatephosphate (9CI)
Fanal Blue B Supra
Fanal Blue BG Supra Powder
Fanatone Blue B
Fast Blue B Supra
Fast Blue Lake
Fast Blue Toner B
Halopont Blue BGM
Heliostable Blue B
Heliostable Brilliant Blue B Extra
Helmerco Blue M 4G
Irgalite Brilliant Blue MRS
Irgalite Victoria Blue TRCN
Kromal Blue OB
Kromal Blue RBS
Marine Blue A 8021
Nyco Liquid Blue BF
Nyco Super Blue B
Permanent Victoria Blue Toner
Pigment Blue 1
Pyramid Royal Blue Toner
Recolite Royal Blue BDS
Recolite Royal Blue BTS
Royal Victoria Blue CP 637
Sicilian Blue A 7021
Siegle Blue Extract D 448
Solar Blue UMN 57-6692
Solfast Victoria Blue CP 476
Symulex Blue BF
Syton Blue B
Tropical Royal Blue Toner
Ultra Blue B
Victoria Pure Blue B

1326-03-0
$C_{28}H_{31}N_2O_3.x$
Unknown
Xanthylium, 9-(2-carboxy-phenyl)-3,6-bis(diethyl-amino)-, molybdatetungs-tatephosphate (9CI)

1326-82-5
Unknown
Unknown
C.I. Sulphur Black 1 (9CI)
Asphalen Sulphur Black C
Asphalen Sulphur Black S
Atul Sulphur Black GXE
Atul Sulphur Black GXR
Atul Sulphur Black RP
Calcogene Black GX-CF
C.I. 53185
Diresul Black P
Eclipse Black BG
Eclipse Deep Black S
Fenoxyl Black RD
Immedial Black GN
Immedial Black AT
Immedial Black MF
Immedial Carbon B
Immedial Carbon BO
Immedial Carbon BR
Immedial Carbon CBO
Immedial Carbon CM
Immedial Carbon CMR
Immedial Carbon LP
Immedial Carbon MLB
Immedial Carbon NGD
Immedial Carbon RRC
Immedial MFS Grains
Immedial Pinting Black B Paste
Katigen Deep Black NND-CF
Katigen Deep Black RND-CF
Kayaku Sulphur Black BBR 200
Kayaku Sulphur Black BBX
Kayaku Sulphur Black BDX
Kayaku Sulphur Black BNX
Kayaku Sulphur Black BX
Kayaku Sulphur Black BX 200
Kayaku Sulphur Black BX 200 Flakes
Kayaku Sulphur Black G
Kayaku Sulphur Black TB
Kayaku Sulphur Black 3BX
Kayasol Black B
Mitsui Sulphur Black B
Mitsui Sulphur Black BC
Mitsui Sulphur Black BF
Mitsui Sulphur Black BO
Mitsui Sulphur Black BS
Mitsui Sulphur Black G
Mitsui Sulphur Black GF
Mundial Black MO
Nissen Black BGL
Nissen Black BK
Nissen Black BX

1326-83-6
Unknown
Unknown
C.I. Solubilised Sulphur Black 1 (9CI)
C.I. Solubilized Sulfur Black 1
C.I. 53186

1326-85-8
Unknown
Unknown
C.I. Sulphur Black 2 (9CI)

1326-86-9
Unknown
Unknown
C.I. Solubilised Sulphur Black 2 (9CI)

1326-96-1
Unknown
Unknown
C.I. Sulphur Red 10 (9CI)
C.I. Leuco Sulfur Red 10

C.I. Sulfur Red 10
C.I. 53228

1327-14-6
Unknown
Unknown
C.I. Sulphur Black 11 (9CI)
C.I. Sulfur Black 11, C.I. 53290

1327-33-9
Unknown
Unknown
Antimony oxide (VAN)(9CI)

1327-36-2
Unknown
Unknown
Aluminosilicate
Aluminatesilicate (9CI)
Aluminum silicate
C.I. 77004
Pyrophyllite

1327-41-9
Unknown
Unknown
Aluminum-chloride-hydroxide
Aluminum chloride, Basic (9CI)
Polyaluminium chloride

1327-53-3
As_2O_3
197.84
O=[As]O[As]=O
Arsenic trioxide (ACGIH) [UN 1561]
Acide arsenieux (French)
Anhydride arsenieux (French)
Arsenic blanc (French)
Arsenic oxide
Arsenic (III) oxide
Arsenic sesquioxide
Arsenic trioxide, Solid (DOT)
Arsenicum album
Arsenic, White, Solid (DOT)
Arsenigen saure (German)
Arsenious acid
Arsenious acid, Solid (DOT)
Arsenious oxide
Arsenious trioxide
Arsenite
Arsenolite
Arsenous acid
Arsenous acid anhydride
Arsenous anhydride
Arsenous oxide
Arsenous oxide anhydride
Arsentrioxide
Arsodent
Claudelite
Claudetite
Crude Arsenic
Diarsenic trioxide
RCRA waste number P012
UN 1561 [Arsenic trioxide]
White Arsenic

1327-57-7
Unknown
Unknown
C.I. Sulphur Blue 7 (9CI)
Acco Sulfur Blue B-CF
Acco Sulfur Blue GLP-CF
Acco Sulfur Blue GLR-CF
Acco Sulfur Blue R-CF
Acco Sulfur Blue 2R-CF
Acco Sulfur Blue 4R-CF
Acco Sulfur Blue 6R-CF
Atul Sulfur Navy Blue
Calcogene Blue 2B-CF
Calcogene Blue 2R-CF
Calcogene Blue 6R-CF
Calcogene Navy Blue 2GS-CF
C.I. Sulfur Blue 7
C.I. 53440
Dark Blue Z
Diresul Blue 8RS
Diresul Blue 9RS
Diresul Navy Blue GIS
Eclipse Dark Blue B
Fenoxyl Blue L
Fenoxyl Blue LC
Fenoxyl Blue LCR
Fenoxyl Blue 9R
Immedial Indone RR
Immedial Indone Violet B
Katigen Blue BCR Conc CF
Katigen Blue BC3R Conc CF
Katigen Blue BC8R Conc CF
Kayaku Indone Violet BC
Kayaku Sulphur Blue BK
Kayaku Sulphur Blue BN
Kayaku Sulphur Blue BP
Kayaku Sulphur Blue FBB
Kayaku Sulphur Blue RC
Kayaku Sulphur Blue RN
Kayaku Sulphur Blue RS
Kayaku Sulphur Blue TFB
Kayaku Sulphur Blue TFR
Kayaku Sulphur Blue 4R
Kayaku Sulphur Indone Violet
Mitsui Sulphur Blue BC
Mitsui Sulphur Blue FB
Mitsui Sulphur Blue LC
Mitsui Sulphur Blue LCR
Mitsui Sulphur Blue R
Mitsui Sulphur Blue RC
Mitsui Sulphur Blue RCP
Mitsui Sulphur Blue TFA
Mitsui Sulphur Blue TFB
Mitsui Sulphur Blue 3BN
Mitsui Sulphur Blue 4R

1327-59-9
Unknown
Unknown
C.I. Sulphur Blue 13 (9CI)

1327-79-3
Unknown
Unknown
C.I. Vat Blue 43 (9CI)

1328-04-7
Unknown
Unknown
C.I. Pigment Violet 5:1 (9CI)
C.I. 58055:1

1328-24-1
$C_{28}H_{18}N_3NaO_7S$
563.53
C.I. Acid Black 48 (9CI)

1328-51-4
Unknown
Unknown
Luxol Fast Blue MBS
C.I. Solvent Blue 38 (9CI)
Luxol-Fast-Blue
Methasol Fast Blue

1328-53-6
Unknown
Unknown
C.I. Pigment Green 7
Accosperse Cyan Green G
Brilliant Green Phthalocyanine
Calcotone Green G
Ceres Green 3B
Chromatex Green G
Colanyl Green GG
C.I. 74260
C.I. Pigment Green 42
Copper Phthalocyanine Green
Cromophtal Green GF
Cyan Green 15-3100
Cyanine Green GP
Cyanine Green NB
Cyanine Green T
Cyanine Green Toner
Dainichi Cyanine Green FG
Dainichi Cyanine Green FGH
Daltolite Fast Green GN
Duratint Green 1001
Fastogen Green 5005
Fastogen Green B
Fastolux Green
Fenalac Green G
Fenalac Green G Disp
Granada Green Lake GL
Graphtol Green 2GLS
Heliogen Green 8680
Heliogen Green 8730
Heliogen Green A
Heliogen Green G
Heliogen Green GA
Heliogen Green GN
Heliogen Green GNA
Heliogen Green GTA
Heliogen Green GV
Heliogen Green GWS
Heliogen Green 8681K
Heliogen Green 8682T
Hostaperm Green GG
Irgalite Fast Brilliant Green GL
Irgalite Fast Brilliant Green 3GL
Irgalite Green GLN
Klondike Yellow X-2261
Lutetia Fast Emerald J
Microlith Green G-FP
Monarch Green WD
Monastral Fast Green BGNA
Monastral Fast Green G

Monastral Fast Green GD
Monastral Fast Green GF
Monastral Fast Green GFNP
Monastral Fast Green GN
Monastral Fast Green GNA
Monastral Fast Green GTP
Monastral Fast Green GV
Monastral Fast Green GWD
Monastral Fast Green 2GWD
Monastral Fast Green GX
Monastral Fast Green GXB
Monastral Fast Green GYH
Monastral Fast Green LGNA
Monastral Green B
Monastral Green B Pigment
Monastral Green G
Monastral Green GFN
Monastral Green GH
Monastral Green GN
Monolite Fast Green GVSA
NCI-C54637
Non-Flocculating Green G 25
Opaline Green G 1
Permanent Green Toner GT-376
Phthalocyanine Brilliant Green
Phthalocyanine Green
Phthalocyanine Green LX
Phthalocyanine Green V
Phthalocyanine Green VFT 1080
Phthalocyanine Green WDG 47
Pigment Fast Green G
Pigment Fast Green GN
Pigment Green 7
Pigment Green Phthalocyanine
Pigment Green Phthalocyanine V
Polychloro copper phthalocyanine
Polymo Green FBH
Polymo Green FGH
Polymon Green G
Polymon Green 6G
Polymon Green GN
PV-Fast Green G
Ramapo
Sanyo Cyanine Green
Sanyo Phthalocyanine Green FB Pure
Sanyo Phthalocyanine Green F6G
Segnale Light Green G
Sherwood Green A 4436
Siegle Fast Green G
Solfast Green
Solfast Green 63102
Synthaline Green
Termosolido Green FG Supra
Thalo Green No.1
Versal Green G
Vulcal Fast Green F2G
Vulcanosine Fast Green G
Vulcol Fast Green F2G
Vynamon Green BE
Vynamon Green BES
Vynamon Green GNA

1330-16-1
$C_{10}H_{16}$
136.26
Bicyclo(3.1.1)heptane, 2,6,6-trimethyl-, didehydro deriv.
Pinene (DOT)
α-Pinene [UN 2368]
UN 2368 [α-Pinene]

1330-19-4
$C_7H_{14}O_2$
130.19
Isoheptanoic acid (8CI,9CI)

1330-20-7
C_8H_{10}
106.18
Xylene (ACGIH,OSHA) [UN 1307]
Benzene, dimethyl-
Dimethylbenzene (OSHA)
Ksylen (Polish)
Methyl toluene
NCI-C55232
RCRA waste number U239
UN 1307 [Xylenes]
Violet 3
Xiloli (Italian)
Xylenen (Dutch)
Xylol (DOT)
Xylole (German)

1330-37-6
$C_{32}H_8Cl_8CuN_8$
851.59
Copper, (octachloro-29H,31H-phthalocyaninato(2-)-N29, N30,N31,N32)- (9CI)

1330-38-7
$C_{32}H_{14}CuN_8O_6S_2.2Na$
780.12
Heliogen Blue SBL
Copper(dihydrogen phthalocyaninedisulfonato(2)) disodium salt
Cuprate(2-), (29H,31H-phthalocyaninedisulfonato(4-)-N29, N30,N31,N32)-, disodium (9CI)

1330-39-8
$C_{32}H_{13}CuN_8O_9S_3.3Na$
882.22
Copper phthalocyanine trisulfonic acid
Copper, (29H,31H-phthalocyanine-ar,ar',ar''-trisulfonato(2-)-N29,N30,N31,N32)-, trisodium salt
Copper phthalocyaninetrisulfonic acid, sodium salt
Cuprate(3-), (29H,31H-phthalocyaninetrisulfonato(5-)-N29,N30,N31,N32)-, trisodium (9CI)

1330-40-1
$C_{56}H_{32}CuN_8$
880.42
Copper, (tetraphenyl-29H, 31H-phthalocyaninato(2-)-N29,N30,N31,N32)- (9CI)
(Tetraphenyl-29H,31H-phthalocyaninato(2-)-N29,N30,N31, N32)copper

1330-43-4
$B_4Na_2O_7$
201.22
Boric acid, disodium salt
Anhydrous borax
Borates, tetra, sodium salt, Anhydrous (OSHA)
Borax Glass
Disodium tetraborate
FR 28
Fused Borax
Rasorite 65
Sodium biborate
Sodium tetraborate
Sodium tetraborate ($Na_2B_4O_7$)

1330-61-6
$C_{13}H_{24}O_2$
212.37
CCCCCCCCC(C)OC(=O)C=C
Acrylic acid, isodecyl ester
Ageflex FA-10
Isodecyl acrylate
Isodecyl alcohol, acrylate
Isodecyl propenoate
2-Propenoic acid, isodecyl ester (9CI)

1330-70-7
$C_{18}H_{36}O_3$
300.48
Octadecanoic acid, hydroxy- (9CI)
Hydroxyoctadecanoic acid

1330-76-3
$C_{20}H_{36}O_4$
340.50
Diisooctyl maleate
2-Butenedioic acid (Z)-, diisooctyl ester (9CI)
DIOM
Isooctyl alcohol, maleate (2:1)
Maleic acid, diisodecyl ester
Maleic acid, diisooctyl ester
RC Comonomer DIOM

1330-78-5
$C_{21}H_{21}O_4P$
368.37
Cc1ccccc1OP(=O)(Oc2ccccc2C)Oc3ccccc3C
Tricresyl phosphate
AI3-16771
Caswell No. 884
Celluflex 179C
Cresyl phosphate
Disflamoll TKP
Durad

EPA Pesticide Chemical Code 083401
Flexol Plasticizer TCP
Fosfato de tricresilo (Spanish)
Fyrquel 150
IMOL S 140
Kronitex
Lindol
NCI-C61041
Phosphate de tricresyle (French)
Phosphoric acid, tris(methylphenyl) ester
Phosphoric acid, tritolyl ester
TCP
Tricresilfosfati (Italian)
Tricresylfosfaten (Dutch)
Tricresylphosphate, With more than 3% ortho isomer [UN 2574]
Trikresylfosfat (Czech)
Trikresylphosphate (German)
Tris(methylphenyl) phosphate
Tris(tolyloxy)phosphine oxide
Tritolylfosfat (Czech)
Tritolyl phosphate
UN 2574 [Tricresylphosphate, With more than 3% ortho isomer]

1330-80-9
$C_{21}H_{40}O_3$
340.55
Propylene glycol oleate
AI3-00972
9-Octadecenoic acid, monoester with 1,2-propanediol
9-Octadecenoic acid (Z)-, monoester with 1,2-propanediol (9CI)

1330-85-4
$C_{22}H_{40}N.Cl$
354.01
Dodecylbenzyltrimonium chloride
Benzenemethanaminium, ar-dodecyl-N,N,N-trimethyl-, chloride (9CI)
Benzenemethanaminium, 4-dodecyl-N,N,N-trimethyl-, chloride
Caswell No. 414
Dodecylbenzyl trimethyl ammonium chloride
(Dodecylbenzyl)trimethylammonium chloride
4-Dodecyl-N,N,N-trimethylbenzenemethanaminium chloride
EPA Pesticide Chemical Code 069125
Quaternium-28
Trimethyl dodecylbenzyl ammonium chloride

1330-86-5
$C_{22}H_{42}O_4$
370.57
Diisooctyl adipate
Adipic acid, diisooctyl ester
Adipol 10A
Di-iso-octyl adipate
Diisooctyl hexanedioate
Diisooctyl phthalate
Dimethyl heptyladipate
DIOA
Hexanedioic acid, diisooctyl ester (9CI)
Isooctyl adipate
PX 208

1330-96-7
$C_{26}H_{42}O_4$
418.62
Isodecyl octyl phthalate
1,2-Benzenedicarboxylic acid, isodecyl octyl ester (9CI)
Dinopol IDO
Octyl isodecyl phthalate
ODP
Phthalic acid, isodecyl octyl ester
PX 118

1331-09-5
$C_5H_{10}O_2$
102.15
CC1COCCO1
Dioxolane, methyl
Methyl dioxolane

1331-11-9
$C_5H_{10}O_3$
118.13
3-Ethoxypropionic acid
Ethoxypropionic acid
Propionic acid, ethoxy-

1331-17-5
$C_6H_{12}O_2$
116.18
OCCCOCC=C
Propanol, allyloxy
Allyl ether of propylene glycol
Allylether propylenglykolu (Czech)
Propylene glycol, allyl ether

1331-22-2
$C_7H_{12}O$
112.19
Cyclohexanone, methyl
Methylcyclohexanone
Methyl cyclohexanone [UN 2297]
Metylocykloheksanon (Polish)
UN 2297 [Methyl cyclohexanone]

1331-28-8
C_8H_7Cl
138.60
Styrene, chloro
Chlorostyrene

1331-43-7
$C_{10}H_{20}$
140.30
CCC1(CC)CCCCC1
Cyclohexane, diethyl- (Mixed isomers)
Cyclohexane, diethyl-
Diethylcyclohexane
Diethylcyclohexane (Mixed isomers)

1331-50-6
$C_{12}H_{24}O$
184.36
Nonanone, trimethyl
Trimethyl nonanone

1331-61-9
$C_{18}H_{30}O_3S.H_3N$
343.52
Ammonium dodecylbenzenesulfonate
Ammonium lauryl benzene sulfonate
Benzenesulfonic acid, dodecyl-, ammonium salt (9CI)
Dodecylbenzenesulfonic acid, ammonium salt

1331-81-3
$C_8H_{10}O_2$
138.17
Benzenemethanol, ar-methoxy- (9CI)
ar-Methoxybenzenemethanol

1332-03-2
$Cu_4H_6O_{10}S.H_2O$
470.31
Copper (II) hydroxide sulfate
Basic Copper Sulfate
Basic Copper TS-53
Caswell No. 073
Chemform Brand Fixed Copper Fungicide
Citco Tri-basic Copper Sulfate-50XF
Copper hydroxide sulfate, monohydrate (8CI,9CI)
Copper(2+) hydroxide sulfate, monohydrate
Copper Pride
Copper sulfate, tribasic
CP Basic Copper TS-53 WP
CP Basic Sulfate
EPA Pesticide Chemical Code 008101
Griffin Super Cu
Kilcop 53
Kobasic
Lilly/Miller Microcop Fungicide
Nutri-Sperse Copper 50
Phyto-Bordeaux
SA-50 Brand Neutral Copper Fungicide
TBC5 53
Tennessee Brand Tri-Basic Copper Sulfate

TNCS 53
Top Cop Tri Basic
Triangle
Tri-basic Copper Fungicide
Tribasic copper sulfate
Tribasic copper sulfate monohydrate
53WP

1332-07-6
$3ZnO.2B_2O_3$
383.35
Zinc borate
Borax 2335
Boric acid, zinc salt (9CI)
ZB 112
ZB 237
ZN 100

1332-21-4
Unknown
Unknown
Asbestos (ACGIH,DOT, OSHA)
Amianthus
Amosite (Obs.)
Amphibole
Asbest (German)
Asbestos Fiber
Asbestos Fibre
Fibrous Grunerite
NCI-C08991
Serpentine

1332-37-2
Unknown
Unknown
Iron-oxide
Ferrous ferrite
Iron Mass, Spent (DOT)
Iron oxide, Spent [UN 1376]
Iron Oxide Red 130B
Iron Sponge, Spent [UN 1376]
MIO 40GN
Siferrit
UN 1376 [Iron oxide, spent, or Iron sponge, spent obtained from coal gas purification]

1332-40-7
$Cl_2Cu_4H_6O_6$
427.12
Copper chloride, Mixed with copper oxide, hydrate
Agrizan
BASF-Grunkupfer
Basic Copper Chloride
Blitox
Blitox 50
Blue Copper
Blue Copper-50
Chemocin
Chempar
Cobox
Cobox Blue
Colloidox
Copper chloride, Basic
Copper chloride oxide
Copper chloride oxide, hydrate (9CI)
Copper OC Fungicide
Copper oxychloride
Coppersan
Coppesan
Coppesan Blue
Coprantol
Coprex
Coprosan Blue
Cop-Tox
Coxysan
Cu-56
Cupral 45
Cupramar
Cupramer
Cuprantol
Cupravet
Cupravit
Cupravit Forte
Cupravit Green
Cupricol
Cupric oxide chloride
Cupritox
Cuprokylt
Cuprol
Cuprosan
Cuprosana
Cuprosan Blue
Cuprovinol
Cuprox
Cuproxol
Devicopper
Faligruen
Fycol 8
Fytolan
Kauritil
Kilex
KT 35
Kupferoxychlorid (German)
Kupricol
Kuprikol
Microcop
Miedzian
Miedzian 50
Neoram Blu
Oxicob
Oxivor
Oxychlorue de cuivre (French)
Oxyclor
Oxycur
Parrycop
Peprosan
Recop
Rhodiacuivre
Tamraghol
Tricop 50
Viricuivre
Vitigran
Vitigran Blue

1332-58-7
Unknown
Unknown
Clay (Kaolin)
Altowhites
Bentone
Continental
Dixie
Emathlite
Fitrol
Fitrol desiccite 25
Glomax
Hydrite
Kaolin (OSHA)
Kaopaous
Kaophills-2
Langford
Mcnamee
Parclay
Peerless
Snow Tex

1332-65-6
$ClCu_2H_3O_3$
213.57
Copper chloride hydroxide (9CI)
Aviocaffaro
Aviocaffaro PF
Basic Copper Chloride
Caswell No. 072
Chempar
Cop Tox
Copper hydroxide chloride
Copper oxychloride
Copper(II) oxychloride
Criscobre
Cudrox
Cuidrox
Cupravit
Cuprocaffaro
Cuprokylt
Cuprosana
Cuprovinol
Devicopper
Dicopper chloride trihydroxide
EPA Pesticide Chemical Code 008001
Fernacot
Kilex
Natural atacamite, Natural paratacamite (γ) and (β)
Neoram
Neoram Blu
Oxy COC
Pasta Caffaro
Pere-Col
Recop
Rhodiacuivre

1332-81-6
O_4Sb_2
307.50
Antimony oxide (9CI)

1333-07-9
$C_7H_9NO_2S$
171.21
Toluenesulfonamide
AI3-04488-X
Benzenesulfonamide, ar-methyl- (9CI)
Benzenesulfonamide, 2(and 4)-methyl-
ar-Methylbenzenesulfonamide
ar-Toluenesulfonamide

1333-13-7

$C_{11}H_{16}O$
164.25
tert-Butyl-m-cresol
AI3-17283
m-Cresol, tert-butyl- (8CI)
(1,1-Dimethylethyl)-3-methyl-phenol
Phenol, (1,1-dimethylethyl)-3-methyl- (9CI)

1333-17-1
$C_{14}H_{26}O_2$
226.36
Decynediol, tetramethyl- (8CI, 9CI)

1333-21-7
$C_{72}H_{123}O_3P$
1067.74
Phenol, dinonyl-, phosphite (3:1) (9CI)
Dinonylphenol phosphite (3:1)

1333-38-6
$C_5H_6O_2$
98.10
Angelica lactone
2-Furanone, 5-methyl- (8CI)
2(?H)-Furanone, 5-methyl- (9CI)
5-Methyl-2(?H)-furanone

1333-39-7
$C_6H_6O_4S$
174.18
Benzenesulfonic acid, hydroxy
Hydroxybenzenesulfonic acid
Phenolsulfonic acid
Phenolsulphonic acid, Liquid [UN 1803]
Sulfocarbolic acid
UN 1803 [Phenolsulfonic acid, Liquid]

1333-41-1
C_6H_7N
93.14
Picoline [UN 2313]
Methylpyridine
UN 2313 [Picolines]

1333-52-4
$C_{12}H_{10}O$
170.21
Ethanone, 1-(naphthalenyl)- (9CI)
1-(Naphthalenyl)ethanone

1333-73-9
Unknown
Unknown
Boric acid, sodium salt (9CI)
Polybor 3
Sodium borate

1333-74-0
H_2
2.02
[H][H]
Hydrogen (DOT)
Hydrogen, Compressed [UN 1049]
Hydrogen, Refrigerated liquid [UN 1966]
UN 1049 [Hydrogen, compressed]
UN 1966 [Hydrogen, refrigerated liquid (cryogenic liquid)]

1333-82-0
CrO_3
100.00
Chromium(VI) oxide (1:3)
Anhydride chromique (French)
Anidride cromica (Italian)
Chrome (trioxyde de) (French)
Chromic acid
Chromic acid, Solid [NA 1463]
Chromic acid, Solution [UN 1755]
Chromic anhydride (DOT)
Chromic trioxide (DOT)
Chromic (VI) acid
Chromium oxide
Chromium trioxide
Chromium(6+) trioxide
Chromium trioxide, Anhydrous [UN 1463]
Chromium (VI) oxide
Chromsaeureanhydrid (German)
Chromtrioxid (German)
Chroomtrioxyde (Dutch)
Chroomzuuranhydride (Dutch)
Cromo(triossido di) (Italian)
Monochromium oxide
Monochromium trioxide
NA 1463 [Chromic acid, solid]
Puratronic chromium trioxide
UN 1463 [Chromium trioxide, anhydrous]
UN 1755 [Chromic acid solution]

1333-83-1
F_2HNa
62.00
Sodium-fluoride
Na-0101 T 1/8''
Sodium bifluoride, Solid (DOT)
Sodium bifluoride, Solution (DOT)
Sodium hydrogen difluoride
Sodium hydrogen fluoride [UN 2439]
UN 2439 [Sodium hydrogen fluoride]

1333-86-4
Unknown
Unknown
Carbon-Black
Acetylene Black
Aro
Aroflow
Arogen
Aromex
Arotone
Arovel
Arrow
Atlantic
Black Pearls
Cancarb
Carbodis
Carbolac
Carbolac 1
Carbomet
Carbon Black, Acetylene
Carbon Black (ACGIH,OSHA)
Carbon Black BV and V
Carbon Black, Channel
Carbon Black, Furnace
Carbon Black, Lamp
Carbon Black, Thermal
Channel Black
C.I. 77266
C.I. Pigment Black 6
C.I. Pigment Black 7
CK3
Collocarb
Columbia carbon
Conductex
Continental
Continex
Corax
Corax P
Croflex
Crolac
Degussa
Delussa Black FW
Dixie
Dixiecell
Dixiedensed
Dixitherm
Durex
Eagle Germantown
Elf
Elftex
Essex
Excelsior
Explosion Black
Explosion Acetylene Black
Farbruss
Fecto
Flamruss
Furnal
Furnex
Furnex N 765
Gas-Furnace Black
Gastex
Huber
Humenegro
Impingement Black
Ketjenblack EC
Kosmink
Kosmobil
Kosmolak
Kosmos
Kosmotherm
Kosmovar
Metanex
Magecol
Micronex
Miike 20
Modulex
Mogul

Mogul L
Molacco
Monarch
Neo-Spectra
Neo Spectra II
Neotex
Oil-Furnace Black
P-33
P68
P1250
Peerless
Pelletex
Philblack
Philblack N 550
Philblack N 765
Philblack O
Pigment Black 7
Printex
Printex 60
Raven
Raven 30
Raven 420
Raven 500
Raven 8000
Rebonex
Regal
Regal 99
Regal 300
Regal 330
Regal 600
Regal 400R
Regal SRF
Regent
Royal Spectra
Sevacarb
Seval
Shawinigan Acetylene Black
Shell Carbon
Special Black 1V & V
Special Schwarz
Spheron
Spheron 6
Statex
Statex N 550
Sterling
Sterling N 765
Sterling NS
Sterling SO 1
Superba
Super-Carbovar
Super-Spectra
Texas
Therma-Atomic Black
Thermal Acetylene Black
Thermatomic
Thermax
Thermblack
Tinolite
TM 30
Torch brand
Triangle
UCET
Ukarb
United
Velvetex
Vulcan
Witco
Witcoblak No. 100
Wyex

1334-74-3
$C_3H_9O_6P.2Na$
218.06
1,2,3-Propanetriol, mono(dihydrogen phosphate), disodium salt (9CI)

1334-76-5
$C_6H_6O_3$
126.11
Furancarboxylic acid, methyl ester (9CI)
Methyl furancarboxylate

1334-78-7
C_8H_8O
120.16
Tolualdehyde
Benzaldehyde, methyl-
Toluenecarboxaldehyde
Tolyl aldehyde
VHR X

1334-99-2
$C_{17}H_{24}O_3$
276.38
Oxiranecarboxylic acid, (2,4-bis(1-methylethyl) phenyl)-, ethyl ester (9CI)
2,4-Diisopropyl phenyl glycidic acid ethyl ester
Glycidic acid, (2,4-diisopropylphenyl)-, ethyl ester (8CI)

1335-08-6
$C_8H_8O_3$
152.15
Benzoic acid, methoxy- (9CI)
Anisic acid (8CI)

1335-09-7
$C_8H_{16}O$
128.24
CC(O)CCCC(C)=C
6-Hepten-2-ol, 6-methyl
Methylheptenol
6-Methyl-6-hepten-2-ol

1335-25-7
Unknown
Unknown
Lead oxide (VAN)(9CI)

1335-30-4
Unknown
Unknown
Silicic acid, aluminum salt (9CI)
Aluminosilicic acid
Aluminum silicate

1335-31-5
$C_2Hg_2N_2O$
469.20
Mercuric oxycyanide, Solid (Desensitized) [UN 1642]
Mercuric oxycyanide
Mercury cyanide oxide (9CI)
Mercury(II) oxide cyanide
Mercury oxycyanide
UN 1642 [Mercury oxycyanide, Desensitized]

1335-32-6
$C_4H_{10}O_8Pb_3$
807.71
CC(=O)O[Pb]OC(C)=O.O[Pb]O.O[Pb]O
Lead, bis(acetato)tetrahydroxytri
Basic Lead Acetate
Bis(aceto)dihydroxytrilead
Bis(acetato)tetrahydroxytrilead
BLA
Lead acetate, Basic
Lead, bis(acetato-O)tetrahydroxytri-
Lead monosubacetate
Lead subacetate
Monobasic lead acetate
RCRA waste number U146
Subacetate lead

1335-34-8
$C_3H_9O_6P.xK$
Unknown
Potassium glycerophosphate
Glycerol, mono(dihydrogen phosphate), potassium salt
Glycerophosphoric acid, potassium salt
1,2,3-Propanetriol, mono(dihydrogen phosphate), potassium salt

1335-39-3
$C_6H_{10}O$
98.14
CCCC=CC=O
Hexenal (VAN)(9CI)

1335-40-6
$C_7H_8O_3$
140.14
Furancarboxylic acid, ethyl ester (9CI)
Furoic acid, ethyl ester (8CI)

1335-42-8
$C_{10}H_{12}O$
148.20
Ethanone, 1-(dimethylphenyl)- (9CI)
1-(Dimethylphenyl)ethanone

1335-46-2
$C_{14}H_{22}O$
206.33

Methylionone
Ionone, methyl- (9CI)

1335-48-4
$C_{15}H_{28}O$
224.39
Dodecadien-1-ol, 3,7,11-trimethyl- (8CI,9CI)
Dihydrofarnesol

1335-85-9
$C_7H_6N_2O_5$
198.15
o-Cresol, dinitro
Dinitro-o-cresol
Dinitro-o-cresol, Liquid (DOT)
Dinitro-o-cresol, Solid [UN 1598]
Phenol, 2-methyldinitro- (9CI)
UN 1598 [Dinitro-o-cresol, solid]

1335-86-0
C_7H_{12}
96.17
Cyclohexene, methyl- (9CI)

1335-87-1
$C_{10}H_2Cl_6$
334.82
Naphthalene, hexachloro
Halowax 1014
Hexachlornaftalen (Czech)
Hexachloronaphthalene (ACGIH,OSHA)

1335-88-2
$C_{10}H_4Cl_4$
265.94
Naphthalene, tetrachloro
Halowax
Tetrachloronaphthalene (ACGIH,OSHA)

1336-21-6
$H_4N.HO$
35.06
[NH4+].[OH-]
Ammonium hydroxide
Ammonia aqueous
Ammonia solution, Containing 44% or less ammonia
Ammonia Water 29%
Ammonium hydroxide, Containing less than 12% ammonia
Ammonium hydroxide, Containing not less than 12% but not more than 44% ammonia
Aqua ammonia
Aqua ammonia, Solution
NA 2672 (DOT)

1336-36-3
$C_{12}H_6Cl_4$
291.99
Polychlorinated-biphenyls
Aroclor
Aroclor 1221
Aroclor 1232
Aroclor 1242
Aroclor 1248
Aroclor 1254
Aroclor 1260
Aroclor 1262
Aroclor 1268
Aroclor 2565
Aroclor 4465
Aroclor 5442
Biphenyl, polychloro-
Chlophen
Chlorextol
Chlorinated biphenyl
Chlorinated diphenyl
Chlorinated diphenylene
Chloro biphenyl
Chloro 1,1-biphenyl
Clophen
Dykanol
Fenclor
Inerteen
Kanechlor
Kanechlor 300
Kanechlor 400
Montar
Noflamol
PCB
PCBs
Phenochlor
Phenoclor
Polychlorinated biphenyls [UN 2315]
Polychlorobiphenyl
Pyralene
Pyranol
Santotherm
Santotherm FR
Sovol
Therminol FR-1
UN 2315 [Polychlorinated biphenyls]

1338-02-9
$C_{13}H_{25}CuO_2$
276.89
Naphthenic acid, copper salt
Chapco Cu-NAP
CNC
Copper naphthenate
Copper Uversol
Cunapsol
Cuprinol
Troysan Copper 8%
Wiltz-65
Wittox C

1338-16-5
$C_6H_{14}O_6.C_6H_8O_7.xFe$
765.29
Iron-sorbitol-citrate
Eisen-sorbitol-zitrat (German)
ESZ
Iron sorbitex
Iron-sorbitol-citric acid

1338-23-4
$C_8H_{18}O_2$
146.23
CC(=O)CCOOCCC(C)=O
2-Butanone, peroxide
2-Butanone peroxide, Containing more than 10% available oxygen
2-Butanone peroxide, Maximum concentration 60%
2-Butanone peroxide, With not more than 9% by weight active oxygen
Butanox LPT
Butanox M 50
Butanox M 105
Cadox
Chaloxyd MEKP-HA 1
Chaloxyd MEKP-LA 1
Esperfoam FR
Ethyl methyl ketone peroxide
Ethyl methyl ketone peroxide, Max. conc. 50% containing more than 10% available oxygen (UN 2550 DOT)
Ethyl methyl ketone peroxide, Maximum concentration 50% (DOT)
Ethyl methyl ketone peroxide, Maximum concentration 60%
FR 222
Hi-Point 90
Hi-Point 180
Hi-Point PD-1
Ketonox
Lucidol Deltax
Lupersol
Lupersol DDA 30
Lupersol DDM
Lupersol Delta-X
Lupersol DNF
Lupersol DSW
MEK peroxide
MEKP (OSHA)
Methyl ethyl ketone hydroperoxide
Methyl ethyl ketone peroxide (ACGIH,OSHA)
Methyl ethyl ketone peroxide, In solution with not more than 9% by wt. active oxygen (DOT)
Methyl ethyl ketone peroxide, Maximum concentration 60%
Permek N
Methylethylketonhydroperoxide
NCI-C55447
Quickset Extra
Quickset Super
RCRA waste number U160
Sprayset MEKP
Thermacure
UN 2127 (DOT)
UN 2550 (DOT)

1338-24-5
Unknown
Unknown
Naphthenic-acid
Agenap

Naphid
Sunaptic Acid B
Sunaptic Acid C

1338-32-5
Unknown
Unknown
Polychlorobenzoic acid, dimethylamine salt
Benzac
Benzac 354
Dimethylamine salts of mixed polychlorobenzoic acids
PBA, dimethylamine salt
Zobar

1338-39-2
$C_{18}H_{34}O_6$
346.52
Sorbitan, monolaurate
Emsorb 2515
Radiasurf 7125
Sorbitan, monododecanoate
Span 20

1338-41-6
$C_{24}H_{46}O_6$
430.70
Sorbitan, monostearate
Anhydrosorbitol stearate
Arlacel 60
Armotan MS
Crill 3
Crill K 3
Drewsorb 60
Durtan 60
Emsorb 2505
D-Glucitol, Anhydro-, monooctadecanoate
Glycomul S
Hodag SMS
Ionet S 60
Liposorb S
Liposorb S-20
Montane 60
MS 33
MS 33F
Newcol 60
Nikkol SS 30
Nissan Nonion SP 60
Nonion SP 60
Nonion SP 60R
Rikemal S 250
Sorbitan C
Sorbitan, monooctadecanoate (9CI)
Sorbitan stearate
Sorbon S 60
Sorgen 50
Span 55
Span 60

1338-43-8
$C_{24}H_{44}O_6$
428.68
Sorbitan, monooleate
Arlacel 80
Armotan MO
Emsorb 2500
Glycomul O
Ionet S-80
Liposorb O
Liposorb O-20
ML 33F
ML 55F
Monodehydrosorbitol monooleate
Montan 80
Nikkol SO 10
Nikkol SO-15
Nikkol SO-30
Nonion OP80R
O 250
Sorbester P 17
Sorbitan monooleic acid ester
Radiasurf 7155
Sorbitan O
Sorbitan oleate
Sorgen 40
Span 80

1341-24-8
C_8H_7ClO
154.60
Chloroacetophenone
Chloracetophenone
2-Chloroacetophenone
Mace

1341-49-7
F_2H_5N
57.06
Ammonium-hydrogen-fluoride
Acid ammonium fluoride
Ammonium bifluoride
Ammonium bifluoride, Solid (DOT)
Ammonium difluoride
Ammonium bifluoride, Solution (DOT)
Ammonium fluoride comp. with hydrogen fluoride (1:1)
Ammonium hydrofluoride
Ammonium hydrogen fluoride, Solid [UN 1727]
Ammonium hydrogen fluoride, Solution [UN 2817]
Ammonium hydrogen bifluoride
Ammonium hydrogen difluoride
UN 1727 [Ammonium hydrogen fluoride, solid]
UN 2817 [Ammonium hydrogen fluoride, solution]

1343-88-0
Unknown
Unknown
Florisil
Avibest
Bitesorb 40
Britesorb
Britesorb No. 40
Britesorb 90
Caswell No. 533
Celkate T 21
Chooz
EPA Pesticide Chemical Code 2601
Gastomag
KW 600S
Magmasil
Magnesium silicate
Magnesol
Magsorbent
Novasorb
Salisil
Silicic acid, magnesium salt CI)
Silicic acid, magnesium salt :1)
Trimax
Trinesium
Tri-Sil
Trisomin

1343-98-2
Unknown
Unknown
Hydrated silica
Caswell No. 734
EPA Pesticide Chemical Code 2602
Precipitated Silica
Silica gel
Silica hydrate
Silicic acid (VAN) (9CI)

1344-00-9
Unknown
Unknown
Sasil
Aluminosilicic acid, sodium salt
Aluminum sodium silicate
Alusil ET
AMSR 3
Decalso
Decalso F
Degussa P820
NCI-C55505
Silicic acid, aluminum sodium salt (9CI)
Sodium aluminosilicate
Sodium aluminum silicate
Sodium silicoaluminate
Vulkasil A 1
Zeolex
Zeolex 100
Zeolex 23A
Zeolex 23P
Zeolex 25
Zeolex 35
23P

1344-01-0
Unknown
Unknown
Aluminum calcium sodium silicate
Aluminosilicic acid, calcium ium salt
Aluminosilicic acid, calcium dium salt, hydrate
Aluminosilicic acid (Unspecified, calcium sodium salt, hydrate
Calcium sodium aluminosilicate
Calcium sodium aluminosilicate

hydrate
Silicic acid, aluminum calcium sodium salt (9CI)
Sodium calcium aluminosilicate
Sodium calcium aluminosilicate, hydrated
Sodium calcium silicoaluminate

1344-06-5
Unknown
Unknown
Sodium phosphoaluminate
Aluminum sodium oxide phosphate

1344-08-7
Unknown
Unknown
Sodium sulfide ($Na_2(S_x)$) (9CI)
Caswell No. 789
EPA Pesticide Chemical Code 006902
Sodium polysulfide
Sodium sulfide
Sulfure de sodium (French)
Sulfuro sodico (Spanish)

1344-09-8
Unknown
Unknown
Sodium silicate
Caswell No. 792
EPA Pesticide Chemical Code 072603
Silicic acid, sodium salt (9CI)
Sodium β-silicate
Sodium Silicate Glass
Sodium Silicate Solution
Water Glass

1344-13-4
Unknown
Unknown
Tin chloride (VAN)(9CI)

1344-28-1
Al_2O_3
101.96
Aluminum oxide (2:3)
A 1 (Sorbent)
Abramant
Abramax
Abrarex
Abrasit
Activated aluminum oxide
A1-3945 E 1/16''
A1-4126 E 1/16''
Alcoa F 1
Almite
Alon
Alon C
Aloxite
Alumina
α-Alumina (ACGIH,OSHA)
γ-Alumina
β-Alumina
Aluminite 37
Aluminium oxide
Aluminum oxide
α-Aluminum oxide
β-Aluminum oxide
γ-Aluminum oxide
Aluminum sesquioxide
Alumite
Alumite (Oxide)
Alundum
Alundum 600
A1-0109 P
A1-3916 P
A1-3970 P
A1-1401 P(MS)
A1-0104 T 3/16''
A1-1404 T 3/16''
A1-3438 T 1/8''
A1-3980 T 5/32''
A1-4028 T 3/16''
Brockmann, Aluminum oxide
Cab-O-Grip
Catapal S
Catapal SB Alumina
Compalox
Conopal
Diadur
Dialuminum trioxide
Dispal
Dispal Alumina
Dispal M
Dotment 324
Dotment 358
Dural
Exolon
Exolon XW 60
F 360 (Alumina)
Faserton
Fasertonerde
G 0 (Oxide)
G 2 (Oxide)
GK (Oxide)
Hypalox II
Jubenon R
KA 101
Ketjen B
KHP 2
LA 6
Lucalox
Ludox CL
Martoxin
Microgrit WCA
Poraminar
PS 1
PS 1 (Alumina)
Q-Loid A 30
RC 172DBM

1344-32-7
$C_7H_4Cl_4$
229.91
Toluene, α,ar,ar,ar-tetrachloro
TCBC
Trichlorobenzyl chloride

1344-37-2
Unknown
Unknown
C.I. Pigment Yellow 34 (9CI)
C.I. 77600
C.I. 77603
C.P. Chrome Yellow Light 1066
C.P. Chrome Yellow Light 1074
C.P. Chrome Yellow Medium 1074
C.P. Chrome Yellow Medium 1085
C.P. Chrome Yellow Medium 1298
Chromastral Green HM
Chromastral Green M
Chromastral Green Y
Chrome Fast Green CP
Chrome Orange
Chrome Yellow
Chrome Yellow A-241
Chrome Yellow GL Medium
Chrome Yellow Lemon
Chrome Yellow LF AA
Chrome Yellow Light
Chrome Yellow Light Y 434D
Chrome Yellow Medium
Chrome Yellow Medium Y 469D
Chrome Yellow Middle
Chrome Yellow Primrose
Chrome Yellow 4G
Chrome Yellow 10G
Chrome Yellow 5GF
Chrome Yellow 4GL Light
Chrome Yellow 6GL Primrose
Chrome Yellow 62E
Dainichi Chrome Yellow 5G
Dainichi Chrome Yellow 10G
Krolor Yellow KY 788D
Lemon Chrome A 3G
Lemon Chrome C 4G
Middle Chrome
Middle Chrome BHG
Primrose Chrome
Pure Lemon Chrome HL 3G
Pure Lemon Chrome L 3G
Pure Lemon Chrome L 3GS
Pure Lemon Chrome 3GN
Pure Lemon Chrome 24882
Pure Middle Chrome LG
Pure Middle Chrome 24883
Pure Primrose Chrome L 6G
Pure Primrose Chrome L 10G
Pure Primrose Chrome 24880
Pure Primrose Chrome 24881
Renol Chrome Yellow Y 2G
Renol Chrome Yellow Y 2RS

1344-38-3
Unknown
Unknown
Lead chromate oxide
C.I. Pigment Orange 21 (9CI)

1344-43-0
MnO
70.94
Manganese(II) oxide
Cassel Green
C.I. 77726
Manganese Green
Manganese monooxide

Manganese monoxide
Manganous oxide
Nu-Manese
Rosensthiel

1344-67-8
Unknown
Unknown
Copper chloride [UN 2802]
Kirticopper
UN 2802 [Copper chloride]

1344-73-6
Unknown
Unknown
Basic Copper Sulfate
Copper sulfate basic
Sulfuric acid, copper salt, basic (9CI)

1344-81-6
CaS.x
Polymer
Calcium polysulphide
Calcium polysulfide (9CI)
Caswell No. 150
Eau Grison
EPA Pesticide Chemical Code 076702
Lime sulfur
Lime sulphur
Orthorix
Polysulfure de calcium (French)
Sulka

1344-95-2
Unknown
Unknown
Silicic acid, calcium salt
Calcium hydrosilicate
Calcium monosilicate
Calcium polysilicate
Calcium silicate (OSHA)
Calflo E
Calsil
CS Lafarge
Florite R
Marimet 45
Microcal 160
Microcal ET
Micro-Cel
Micro-Cel A
Micro-Cel B
Micro-Cel C
Micro-Cel E
Micro-Cel T
Micro-Cel T26
Micro-Cel T38
Micro-Cel T41
Promaxon P60
Silene EF
Silmos T
Solex
Stabinex NW 7PS
Starlex L
SW 400
Toyofine A

1345-04-6
S_3Sb_2
339.68
Antimony-trisulfide
Antimonous sulfide
Antimony Glance
Antimony Orange
Antimony sulfide
Antimony sulfide, Solid (DOT)
Antimony trisulfide colloid
C.I. 77060
C.I. Pigment Red 107
Crimson Antimony
Lymphoscan
NA 1325
Needle Antimony

1345-09-1
Unknown
Unknown
Cadmium mercury sulfide (9CI)
Cadmium vermilion A
C.I. Pigment Red 113
C.I. 77201
Mercury Cadmium Reds

1345-13-7
Ce_2O_3
328.24
Cerium-trioxide
Dicerium trioxide

1345-25-1
FeO
71.85
Ferrous oxide
CI 77489
Iron monooxide
Iron monoxide
Iron oxide (9CI)
Iron(II) oxide
Natural Wuestite

1365-19-1
$C_{10}H_{18}O_2$
170.25
Linalool oxide (VAN)(9CI)
AI3-36113-X
Linalool oxide (α-ethenyl-α,3,3-trimethyloxirane-propanol + α-methyl-α-(4-methyl-3-pentenyl)oxir-anemethanol)

1390-65-4
$C_{22}H_{20}O_{13}$
492.40
Carmine
B Rose Liquid

1393-03-9
Unknown
Unknown
Quillaja saponin
β-D-Glucopyranosiduronic acid, (3β,4α,16α)-17-carboxy-16-hydroxy-23-oxo-28-norolean-12-en-3-yl (9CI)
Quillajasaponin

1393-88-0
Unknown
Unknown
Gramicidin D (8CI,9CI)
Gramicidin Dubos

1395-18-2
Unknown
Unknown
Azolitmin (8CI,9CI)

1397-89-3
$C_{47}H_{73}NO_{17}$
924.21
Amphotericin-B
AMB
Amphomoronal
Amphoteracin B
Amphotericin β
Amphotericine B
Amphozone
Fungilin
Fungisone
Fungizone
IAB
Iodoacetamide
Mysteclin-F
NSC-527017
Tegopen

1397-94-0
$C_{28}H_{40}N_2O_9$
548.62
Antimycin A
Antipiricullin
Caswell No. 052B
EPA Pesticide Chemical Code 006314
Verosin

1398-61-4
$C_{30}H_{50}N_4O_{19}$
770.84
Chitin
β-Chitin
Chitina (Italian)

1399-80-0
$C_{50}H_{94}Cl_3N_3$
843.69
Ammonium, N-alkyl(C9-15)-tolyl methyltrimethyl-, chloride
Alkyl(C9-15)tolyl methyltri-methyl ammonium chloride
Dodecyl p-tolyl trimethyl ammonium chloride
Hyamine 2389
Methyl dodecyl benzyl ammon-ium chloride

1400-61-9
$C_{46}H_{83}NO_{18}$
938.30
Nystatin
Biofanal
Candex
Candio-Hermal
Diastatin
Moronal
Mycostatin
Mycostatin 20
Nilstat
Nystan
Nystatine
Nystatyna (Polish)
Nystavescent
O-V Statin

1401-55-4
$C_{76}H_{52}O_{46}$
1701.28
Tannic-acid
Acacia Mollissima Tannin
Acid, Tannic
Chestnut-Tannin
Castanea Sativa Mill Tannin
Caswell No. 819
Catechins
D'acide tannique (French)
EPA Pesticide Chemical Code 078502
Gallotannic acid
Gallotannic acids
Gallotannin
Gallotannins
Glycerite
Hifix SL
Liquidambar Styraciflua
Mimosa-Tannin
Quebracho wood extract
Quebracho-Tannin
Schinopsis Lorentzii Tannin
Sunlife TN
Sweet-Gum
Tanaphen P 500
Tannic acids
Tannin
Tannin from Chestnut
Tannin from Mimosa
Tannin from Quebracho
Tannin from Sweet GUM
Tannins

1401-69-0
$C_{45}H_{77}NO_{17}$
904.23
Tylosin
Tylan

1402-68-2
$C_{51}H_{46}O_{20}$
978.92
Aflatoxin

1403-66-3
$C_{19}H_{39}N_5O_7$
449.55
Gentamicin
Garamycin
Gentamycin
Gentamycin-Creme (German)
Uromycine

1404-04-2
$C_{23}H_{46}N_6O_{13}$
614.66
Neomycin
Myacyne
Mycifradin
Neomcin
Nivemycin
Vonamycin Powder V

1404-90-6
$C_{66}H_{75}Cl_2N_9O_{24}$
1449.40
Vancomycin
Vancocin

1405-10-3
H_2O_4S.xUnknown
Unknown
Neomycin sulfate
Biosol
Biosol veterinary
Caswell No. 595A
EPA Pesticide Chemical Code 006313
Fradiomycin sulfate
Lidamycin creme
Mycaifradin sulfate
Mycifradin
Mycifradin-N
Mycigient
Myciguent
Neobiotic
Neo-Mantle Creme
Neomycin, sulfate (Salt) (9CI)
Neomycine sulfate
Neomycins sulfate
Neomix
Neotizol
Neovet
Otobiotic
Quintess-N
USAF CB-19

1405-41-0
$C_{19}H_{41}N_5O_{11}S$
547.63
Gentamicin, sulfate
Garamycin
Genoptic
Genoptic S.O.P.
Gentalline
Gentamycin sulfate
GM sulfate
NSC-82261
Refobacin
SCH 9724

1405-86-3
$C_{42}H_{62}O_{16}$
823.04
Glycyrrhizinic-acid
Liquorice

1405-87-4
Unknown
Unknown
Bacitracin
Ayfivin
Baciguent
Baci-Jel
Baciliquin
Bacitek Ointment
Fortracin
Parentracin
Penitracin
Topitracin
Topitrasin
USAF CB-7
Zutracin

1406-05-9
Unknown
Unknown
Penicillin (9CI)
Mykoin BF 510
Penizillin (German)

1406-16-2
Unknown
Unknown
Vitamin-D

1406-66-2
$C_{28}H_{48}O_2$
416.69
Tocopherols

1415-73-2
$C_{21}H_{22}O_9$
418.43
10-β-D-Glucopyranosyl-1,8-dihydroxy-3-(hydroxymethyl)-, 9(10H)-anthracenone-, (+)
ALOIN
10-(1',5'-Anhydroglucosyl)aloe-emodin-9-anthrone
Barbaloin
1,8-Dihydroxy-3-hydroxy-methyl-10-(6-hydroxymethyl-3,4,5-trihydroxy-2-pyranyl)-anthrone
10-Glucopyranosyl-1,8-di-hydroxy-3-(hydroxymethyl)-9(10H)-anthracenone

1415-93-6
Unknown
Unknown
Humic acids

1420-04-8
$C_{13}H_8Cl_2N_2O_4.C_2H_7NO$
388.23
Oc1ccc(Cl)cc1C(=O)Nc2ccc(cc2Cl)N(=O)=O
Salicylanilide, 2',5-dichloro-4'-nitro-, Compd. with

2-aminoethanol (1:1)
Bayer 73
Bayer 25648
Bayluscid
Bayluscide
5-Chloro-N-(2-chloro-4-nitrophenyl)-2-hydroxybenzamide Compd. with 2-aminoethanol (1:1)
Clonitralid
2',5-Dichloro-4'-nitrosalicylanilide, 2-aminoethanol salt
5,2'-Dichloro-4'-nitrosalicylanilide, ethanolamine salt
5,2-Dichloro-4-nitrosalicylic anilide 2-aminoethanol salt
2',5-Dichloro-4'-nitrosalicyloylanilide ethanolamine salt
Ethanolamine salt of 5,2'-dichloro-4'-nitrosalicyclicanilide
M 73
NCI-C00431
Molluscicide Bayer 73
Niclosamide
Salicylanilide, 2',5-dichloro-4-nitro-, ethanolamine salt
SR 73

1420-06-0
$C_{23}H_{23}NO$
329.47
C1CN(CCO1)C(c2ccccc2)(c3ccccc3)c4ccccc4
Morpholine, N-trityl
Frescon
Morpholine, 4-trityl-
Trifenmorph
Triphenmorphe
N-Tritylmorpholine
4-(Triphenylmethyl)morpholine
WL 8008

1420-07-1
$C_{10}H_{12}N_2O_5$
240.24
Phenol, o-t-butyl-4,6-dinitro
2-tert-Butyl-4,6-dinitrophenol
2-(1,1-Dimethylethyl)-4,6-dinitrophenol
2,4-Dinitro-6-tert-butylphenol
Dinoterb
Dinoterbe
DNTBP
Herbogil

1420-53-7
$C_{36}H_{42}N_2O_6.O_4S$
694.86
COc1ccc2CC5C3C=CC(O)C4Oc1c2C34CCN5C.OS(O)(=O)=O
Morphinan-6-α-ol, 7,8-didehydro-4,5-α-epoxy-3-methoxy-17-methyl-, sulfate (2:1) (Salt)
Codeine sulfate

1420-55-9
$C_{22}H_{29}N_3S_2$
399.66
CCSc4ccc3Sc1ccccc1N(CCCN2CCN(C)CC2)c3c4
Phenothiazine, 2-(ethylthio)-10-(3-(4-methyl-1-piperazinyl)propyl)
Ethylthioperazine
10H-Phenothiazine, 2-(ethylthio)-10-(3-(4-methyl-1-piperazinyl)propyl)-
Theithylperazine
Thiethylperazine
Torecan

1421-49-4
$C_{15}H_{22}O_3$
250.37
O=C(O)c(cc(c(O)c1C(C)(C)C)C(C)(C)C)c1
Benzoic acid, 3,5-di-tert-butyl-4-hydroxy
3,5-Di-tert-butyl-4-hydroxybenzoic acid

1421-63-2
$C_{10}H_{12}O_4$
196.22
O=C(c(c(O)cc(O)c1O)c1)CCC
Butyrophenone, 2',4',5'-trihydroxy
THBP
2',4',5'-Trihydroxybutyrophenone
USAF EK

1422-07-7
$C_{18}H_{21}NO_3.ClH$
335.86
Cl.COc1ccc2CC5C3C=CC(O)C4Oc1c2C34CCN5C
Morphinan-6-α-ol, 7,8-didehydro-4,5-α-epoxy-3-methoxy-17-methyl-, hydrochloride
Codeine hydrochloride

1422-26-0
$C_{10}H_{18}O_3$
186.25
Decanoic acid, 9-oxo- (8CI,9CI)

1423-46-7
$C_{10}H_{16}O$
152.24
O=CC(C(CC(=C1)C)C)C1C
3-Cyclohexene-1-carboxaldehyde, 2,4,6-trimethyl- (8CI)
AI3-07328
2,4,6-Trimethyl-3-cyclohexene-1-carboxaldehyde
2,4,6-Trimethyl-4-cyclohexene-1-carboxaldehyde

1429-50-1
$C_6H_{20}N_2O_{12}P_4$
436.11
O=P(O)(O)CN(CCN(CP(=O)(O)O)CP(=O)(O)O)CP(=O)(O)O
(Ethylenedinitrilo)-tetramethylenephosphonic acid
Dequest 2041
Editempa
Editempa acid
EDTPO
(1,2-Ethanediylbis(nitrilobis-(methylene)))tetrakisphosphonic acid
Ethylenediaminetetra(methylenephosphonic)acid
N,N,N',N' Ethylenediamine tetra(methylenephosphonic acid)
Phosphonic acid, (1,2-ethanediylbis(nitrilobis(methylene)))-tetrakis- (9CI)

1430-97-3
$C_{14}H_{12}$
180.25
9H-Fluorene, 2-methyl- (9CI)

1432-14-0
$C_{10}H_{11}ClO_3.C_4H_{11}NO_2$
319.78
Diethanolamine 2-(2-methyl-4-chlorophenoxy)propionate
Ethanol, 2,2-iminodi-, Compd. with 2-((4-chloro-o-tolyl)-oxy)propionic acid (1:1)
Mecoprop diethanolamine salt
2-(2-Methyl-4-chlorophenoxy)propionic acid, diethanolamine salt
Propionic acid, 2-((4-chloro-o-tolyl)oxy)-, Compd. with 2,2'-iminodiethanol (1:1)

1435-50-3
$C_6H_3BrCl_2$
225.90
Benzene, 2-bromo-1,4-dichloro- (8CI,9CI)

1435-53-6
$C_6H_3Br_2F$
253.90
Benzene, 2,4-dibromo-1-fluoro- (8CI,9CI)
NSC-88308

1435-60-5
$C_{12}H_9N_3O_3$
243.20
Oc2ccc(N=Nc1ccc(cc1)N(=O)=O)cc2
Phenol, 4-((4-nitrophenyl)azo)- (9CI)
4-Hydroxy-4'-nitroazobenzene
4-((4-Nitrophenyl)azo)phenol

1435-71-8
$C_{13}H_{11}N_3O_3$
257.23
O=N(=O)c(c(N=Nc(c(O)ccc1C)

c1)ccc2)c2
Phenol, 4-methyl-2-((2-nitrophenyl)azo)- (9CI)
2-Hydroxy-5-methyl-2'-nitroazobenzene
4-Methyl-2-((2-nitrophenyl)azo)phenol

1438-91-1
C_6H_8OS
128.19
O(C(=CC=1)CSC)C1
Furan, 2-((methylthio)methyl)- (9CI)
2-((Methylthio)methyl)furan

1438-94-4
C_9H_9NO
147.19
O(C(=CC=1)CN(C=CC=2)C2)C1
Pyrrole, 1-furfuryl
N-Furfuryl pyrrole
N-(2-Furfuryl)pyrrole
1-Furfurylpyrrole

1441-02-7
$C_8H_3Cl_5O_2$
308.36
Phenol, pentachloro-, acetate
Pentachlorophenyl acetate
Rabcon

1444-64-0
$C_{12}H_{16}O$
176.28
OC1CCCCC1c2ccccc2
Cyclohexanol, 2-phenyl
ESNN
2-Fenylcyklohexanol (Czech)
Insect Repellent 448
2-Phenyl cyclohexanol

1445-19-8
$C_{16}H_{13}NO_3S.Na$
322.33
1-Naphthalenesulfonic acid, 8-(phenylamino)-, monosodium salt (9CI)
1-Naphthalenesulfonic acid, 8-anilino-, monosodium salt

1445-45-0
$C_5H_{12}O_3$
120.15
O(C(OC)(OC)C)C
Ethane, 1,1,1-trimethoxy-
AI3-24332
1,1,1-Trimethoxyethane

1445-75-6
$C_7H_{17}O_3P$
180.21
Phosphonic acid, methyl-, diisopropyl ester
Diisopropyl methanephosphonate
Diisopropyl methylphosphonate
DIMP
Phosphonic acid, methyl-, bis-(1-methylethyl) ester

1445-79-0
C_3H_9Ga
114.83
Gallium, trimethyl- (9CI)
Trimethylgallium

1446-61-3
$C_{20}H_{31}N$
285.47
c(c(C(C(C(C(CC1)(CN)C)C2)(C1)C)ccc3C(C)C)(c3)C2
Dehydroabietylamine
Amine D
Caswell No. 276
EPA Pesticide Chemical Code 004206
1-Phenanthrenemethanamine, 1,2,3,4,4a,9,10,10a-octahydro-1,4a-dimethyl-7-(1-methylethyl)-, (1R-(1α,4aβ,10aα))- (9CI)
Podocarpa-8,11,13-trien-15-amine, 13-isopropyl- (8CI)

1449-49-6
$C_7H_{12}O$
112.17
Cyclobutanone, 2,2,3-trimethyl- (7CI,8CI,9CI)
2,2,3-Trimethylcyclobutanone

1450-72-2
$C_9H_{10}O_2$
150.18
O=C(c(c(O)ccc1C)c1)C
Ethanone, 1-(2-hydroxy-5-methylphenyl)- (9CI)
1-Hydroxy-2-acetyl-4-methylbenzene
1-(2-Hydroxy-5-methylphenyl)ethanone

1452-15-9
$C_4H_2N_2S$
110.13
N#CC(N=CS1)=C1
4-Cyanothiazole
4-Thiazolecarbonitrile (9CI)

1452-77-3
$C_6H_6N_2O$
122.11
O=C(N)c(nccc1)c1
Picolinamide (8CI)
2-Pyridinecarboxamide (9CI)

1453-06-1
$C_{12}H_{16}O_2$
192.26
Benzenebutanoic acid, 2,5-dimethyl- (9CI)
Butyric acid, 4-(2,5-xylyl)- (8CI)
NSC-63112

1453-82-3
$C_6H_6N_2O$
122.14
O=C(N)c(ccnc1)c1
Isonicotinamide
4-Carbamoylpyridine
Isonicotinic acid amide
γ-Pyridinecarboxamide
4-Pyridinecarboxamide (9CI)

1454-85-9
$C_{17}H_{36}O$
256.47
OCCCCCCCCCCCCCCCCC
1-Heptadecanol (9CI)
AI3-01234

1455-21-6
$C_9H_{20}S$
160.32
SCCCCCCCCC
1-Nonanethiol (9CI)

1455-77-2
$C_2H_5N_5$
99.12
N=C(NC(=N)N1)N1
s-Triazole, 3,5-diamino
Guanazole
NCI-C04819
NSC-1895

1456-28-6
$C_6H_{12}N_2O_2$
144.20
CC1CN(CC(C)O1)N=O
Morpholine, 2,6-dimethyl-N-nitroso
Dimethylnitrosomorpholine
2,6-Dimethylnitrosomorpholine
DMNM
Nitroso-2,6-dimethylmorpholine
Me2NMor
N-Nitroso-2,6-dimethylmorpholine

1459-09-2
$C_{22}H_{38}$
302.54
c(cccc1)(c1)CCCCCCCCCCCCCCC
Benzene, hexadecyl- (9CI)
Hexadecylbenzene
1-Phenylhexadecane

1459-10-5
$C_{20}H_{34}$
274.49
c(cccc1)(c1)CCCCCCCCCC

CCCC
Tetradecylbenzene
Benzene, tetradecyl- (9CI)
1-Phenyltetradecane
Tetradecane, 1-phenyl-

1459-11-6
$C_{18}H_{30}$
246.44
Benzene, 1,2,4-tris(1,1-dimethylethyl)- (9CI)
Benzene, 1,2,4-tri-tert-butyl- (7CI,8CI)
1,2,4-Tri-tert-butylbenzene

1459-93-4
$C_{10}H_{10}O_4$
194.20
O=C(OC)c(cccc1C(=O)OC)c1
Isophthalic acid, dimethyl ester
1,3-Benzenedicarboxylic acid, dimethyl ester
Dimethylester kyseliny isoftalove (Czech)
Dimethyl isophthalate

1460-02-2
$C_{18}H_{30}$
246.44
Benzene, 1,3,5-tris(1,1-dimethylethyl)- (9CI)
Benzene, 1,3,5-tri-tert-butyl- (8CI)

1460-06-6
$C_{14}H_{11}Cl$
214.70
Benzene, 1,1'-(1-chloro-1,2-ethenediyl)bis- (9CI)
1-Chloro-1,2-diphenylethene
α-Chlorostilbene
Stilbene, α-chloro- (6CI, 7CI,8CI)

1460-16-8
$C_8H_{14}O_2$
142.20
Cycloheptanecarboxylic acid (8CI,9CI)
NSC-18964

1460-34-0
$C_6H_{10}O_3$
130.14
Pentanoic acid, 3-methyl-2-oxo- (9CI)
α-Keto-β-methylvaleric acid
α-Keto-β-methyl-n-valeric acid
2-Keto-3-methylvaleric acid
KMVA
3-Methyl-2-oxopentanoic acid
2-Oxo-3-methylvaleric acid
Valeric acid, 3-methyl-2-oxo- (8CI)

1460-97-5
$C_{15}H_{24}$
204.36
Naphthalene, 1,2,3,4,4a,5,6,8a-octahydro-7-methyl-4-methylene-1-(1-methylethyl)-, (1R-(1α,4aβ,8aα))- (9CI)
1β,6α,7βH-Cadina-4,10(15)-diene (8CI)
γ-Cadinene, (-)-

1461-03-6
$C_{15}H_{24}$
204.36
1H-Benzocycloheptene, 2,4a,5,6,7,8-hexahydro-3,5,5,9-tetramethyl-, (R)- (9CI)
1H-Benzocycloheptene, 2,4aβ,5,6,7,8-hexahydro-3,5,5,9-te tramethyl-, (+)- (8CI)
β-Himachalene

1461-22-9
$C_{12}H_{27}ClSn$
325.53
CCCC[Sn](Cl)(CCCC)CCCC
Stannane, chlorotributyl
Chlorid tri-n-butylcinicity (Czech)
Chlorotributylstannane
Chlorotributyltin
Monochlorotributyltin
Stannane, tributylchloro-
Tin, tri-n-butyl-, chloride
Tributylchlorotin
Tributylstannium chloride
Tributylstannyl chloride
Tributyltin chloride
Tri-n-butylzinn-chlorid (German)
WR 3396

1461-25-2
$C_{16}H_{36}Sn$
347.21
CCCC[Sn](CCCC)(CCCC)CCCC
Stannane, tetrabutyl
Tetra-n-butylcin (Czech)
Tetrabutylstannane
Tetrabutyltin
Tin, tetrabutyl-

1462-54-0
$C_{15}H_{31}NO_2$
257.41
O=C(O)CCNCCCCCCCCCCCC
Lauraminopropionic acid
β-Alanine, N-dodecyl- (9CI)
N-Dodecyl-β-alanine
N-Lauryl, myristyl β-aminopropionic acid
Unitex 710-L

1462-84-6
$C_8H_{11}N$
121.18
n(c(ccc1C)C)c1C
Pyridine, 2,3,6-trimethyl- (9CI)
2,3,6-Trimethylpyridine

1463-17-8
$C_{11}H_{11}N$
157.21
Cc2ccc1cccc(C)c1n2
Quinoline, 2,8-dimethyl- (8CI, 9CI)
NSC-62133
o-Toluquinaldine

1464-42-2
$C_5H_{11}NO_2Se$
196.13
Butyric acid, 2-amino-4-(methylselenyl)
2-Amino-4-(methylselenyl)-butyric acid
Methionine, seleno
Selenomethionine

1464-53-5
$C_4H_6O_2$
86.10
O(C1C(O2)C2)C1
Butane, 1,2:3,4-diepoxy
1,1'-Bi(ethylene oxide)
Bioxiran
Bioxirane
2,2'-Bioxirane
Butadiendioxyd (German)
Butadiene diepoxide
1,3-Butadiene diepoxide
Butadiene dioxide
Butane diepoxide
DEB
Diepoxybutane
1,2:3,4-Diepoxybutane
2,4-Diepoxybutane
Dioxybutadiene
ENT 26,592
Erythritol anhydride
RCRA waste number U085
Threitol, 1,2:3,4-dianhydro-

1465-25-4
$C_{12}H_{14}N_2.2ClH$
259.20
Cl.Cl.NCCNc1cccc2ccccc12
Ethylenediamine, N-(1-naphthyl)-, dihydrochloride
N-1-Naphthalenyl-1,2-ethanediamine dihydrochloride
N-(1-Naphthyl)ethylenediamine dihydrochloride
NCI-C03281

1466-76-8
$C_9H_{10}O_4$
182.18
O=C(O)c(c(OC)ccc1)c1OC

Benzoic acid, 2,6-dimethoxy- (9CI)
2,6-Dimethoxybenzoic acid

1467-72-7
$C_9H_9N_5O$
203.20
1,3,5-Triazine-2,4-diamine, 6-phenoxy- (9CI)
Phenoxydiaminotriazine
Phenoxydiamino-s-triazine
s-Triazine, 2,4-diamino-6-phenoxy- (6CI,7CI,8CI)

1467-79-4
$C_3H_6N_2$
70.11
N#CN(C)C
Cyanamide, dimethyl
Dimethylcyanamide
Dimethylkyanamid (Czech)

1468-26-4
$C_4H_3N_5O_2$
153.12
O=c2[nH]c(=O)c1[nH]nnc1[nH]2
1H-v-Triazolo(4,5-d)pyrimidine-5,7(4H,6H)-dione
8-Azaxanthine
2,6-Dioxy-8-azapurine
NSC-756
v-Triazolo(4,5-d)pyrimidine-5,7-diol
USAF CB-26

1468-83-3
C_6H_6OS
126.18
CC(=O)c1ccsc1
Ethanone, 1-(3-thienyl)- (9CI)
Ketone, methyl 3-thienyl (8CI)

1470-94-6
$C_9H_{10}O$
134.19
Oc(ccc(c1CC2)C2)c1
5-Indanol
5-Hydroxyhydrindene
Indan-5-ol

1471-03-0
$C_6H_{12}O$
100.16
O(CCC)CC=C
Ether, allyl propyl (8CI)
1-Propene, 3-propoxy- (9CI)
3-Propoxy-1-propene

1471-17-6
$C_{14}H_{24}O_4$
256.34
1-Propanol, 3-(2-propenyloxy)-2,2-bis((2-propenyloxy)-methyl)-
AI3-09536

1471-18-7
$C_{17}H_{28}O_4$
296.41
O(CC(COCC=C)(COCC=C)COCC=C)CC=C
1-Propene, 3,3'-((2,2-bis((2-propenyloxy)methyl)-1,3-propanediyl)bis(oxy))bis-
AI3-15645
Allyl pentareythritol

1472-87-3
$C_{15}H_{28}O_4$
272.38
O=C(OC)CCCCCCCCCCCC(=O)OC
Dimethyl brassylate
Dimethyl tridecanedioate
Tridecanedioic acid, dimethyl ester (9CI)

1476-11-5
$C_4H_6Cl_2$
125.00
C(=CCCl)CCl
1,4-Dichloro-cis-2-butene
2-Butene, 1,4-dichloro-, cis-
2-Butene, 1,4-dichloro-, (Z)-(9CI)
cis-1,4-Dichloro-2-butene
(Z)-1,4-Dichloro-2-butene

1477-39-0
Unknown
Unknown
Noracymethadol
DEA No. 9633
α-Ethyl-β- (2-(methylamino)-propyl)-β-phenylbenzene-ethanol, acetate (ester)
Noracimetadol (Spanish)
Noracimetadolo
Noracymethadolum (Latin)

1477-42-5
$C_8H_8N_2S$
164.24
N=C(Nc(c1ccc2)c2C)S1
Benzothiazole, 2-amino-4-methyl
2-Amino-4-methylbenzothiazole
4-Methyl-2-aminobenzothiazole

1477-55-0
$C_8H_{12}N_2$
136.22
NCc(cccc1CN)c1
Methylamine, m-phenylenebis
1,3-Bis-aminomethylbenzen (Czech)
1,3-Bis(aminomethyl)benzene
MXDA
m-Phenylenebis(methylamine)
m-Xylene α,α'-diamine (ACGIH,OSHA)
m-Xylylendiamin (Czech)
m-Xylylenediamine

1478-61-1
$C_{15}H_{10}F_6O_2$
336.23
FC(F)(F)C(c(ccc(O)c1)c1)(c(ccc(O)c2)c2)C(F)(F)F
Phenol, 4,4'-(2,2,2-trifluoro-1-(trifluoromethyl)ethylidene)bis- (9CI)
Phenol, 4,4'-(bis(trifluoromethyl)methylene)di-
4,4'-(2,2,2-Trifluoro-1-(trifluoromethyl)ethylidene)bisphenol

1483-60-9
$C_{12}H_{18}$
162.27
Benzene, 2,4-dimethyl-1-(1-methylpropyl)- (9CI)
m-Xylene, 4-sec-butyl-

1484-12-4
$C_{13}H_{11}N$
181.23
N-Methylcarbazole
Carbazole, 9-methyl- (8CI)
9H-Carbazole, 9-methyl- (9CI)
9-Methylcarbazole

1484-13-5
$C_{14}H_{11}N$
193.26
N(c(c(c1cccc2)ccc3)c3)(c12)C=C
Carbazole, 9-vinyl
9H-Carbazole, 9-ethenyl- (9CI)
Vinylcarbazole
N-Vinylcarbazole
9-Vinylcarbazole
N-Vinylkarbazol (Czech)

1484-84-0
$C_7H_{15}NO$
129.20
OCCC(NCCC1)C1
2-Piperidineethanol (9CI)
AI3-36432

1487-15-6
C_5H_8O
84.12
Furan, 2,3-dihydro-5-methyl- (8CI,9CI)

1487-18-9
C_6H_6O
94.11
Furan, 2-ethenyl- (9CI)
Furan, 2-vinyl- (8CI)

1488-42-2
$C_{13}H_9IO_2$
324.12

O=C(O)c(c(ccc1)Ic(cccc2)c2)c1
Iodonium, (2-carboxyphenyl)-phenyl-, hydroxide, inner salt (9CI)
AI3-62156

1489-53-8
$C_6H_3F_3$
132.09
Benzene, 1,2,3-trifluoro- (6CI,7CI,8CI,9CI)
1,2,3-Trifluorobenzene

1490-04-6
$C_{10}H_{20}O$
156.27
OC(C(CCC1C)C(C)C)C1
Menthol
AI3-08161
Caswell No. 540
Cyclohexanol, 5-methyl-2-(1-methylethyl)-
EPA Pesticide Chemical Code 051601
5-Methyl-2-(1-methylethyl)-cyclohexanol

1491-59-4
$C_{16}H_{24}N_2O$
260.42
Cc1cc(c(O)c(C)c1CC2=NCCN2)C(C)(C)C
Phenol, 6-t-butyl-3-(2-imidazolin-2-ylmethyl)-2,4-dimethyl
6-t-Butyl-3-(2-imidazolin-2-yl-methyl)-2,4-dimethylphenol
Nasivine
Oxymetazoline
Oxymethazoline
Oxymetozoline

1493-02-3
CHFO
48.02
Formyl fluoride (8CI,9CI)

1493-13-6
CHF_3O_3S
150.08
O=S(=O)(O)C(F)(F)F
Trifluoromethanesulfonic acid
AI3-62912
Methanesulfonic acid, trifluoro-

1497-49-0
$C_{17}H_{16}N_2.ClH$
284.78
Benzenamine, N-(5-(phenylamino)-2,4-pentadienylidene)-, monohydrochloride (9CI)
Glutaconaldehydedianil hydrochloride
Glutaconaldehyde dianil monohydrochloride

1497-68-3
$C_4H_{10}ClOPS$
172.62
O(P(=S)(CC)Cl)CC
O-Ethyl ethylthiophosphonyl chloride
Ethanephosphonochloridothioic acid, O-ethyl ester
O-Ethyl ethanephosphonochloridothioate
O-Ethyl ethanephosphonothionochloridate
O-Ethyl ethylphosphonochloridothioate
Ethylthionophosphonic acid O-ethyl ester chloride
Phosphonochloridothioic acid, ethyl-, O-ethyl ester (9CI)

1498-51-7
$C_2H_5Cl_2O_2P$
162.94
O=P(OCC)(Cl)Cl
Phosphorodichloridic acid, ethyl ester
Dichlorophosphoric acid, ethyl ester
Ethyl phosphorodichloridate (DOT)
NA 1760 (DOT)

1499-10-1
$C_{26}H_{18}$
330.43
c(c(c(c(c1ccc2)c2)c(cccc3)c3)c cc4)(c1c(cccc5)c5)c4
9,10-Diphenylanthracene
Anthracene, 9,10-diphenyl- (9CI)

1501-27-5
$C_6H_{10}O_4$
146.14
O=C(OC)CCCC(=O)O
Glutaric acid, monomethyl ester (8CI)
Methyl hydrogen glutarate
Monomethyl pentanedioate
Pentanedioic acid, monomethyl ester (9CI)

1501-82-2
$C_{12}H_{22}$
166.31
C1CCCCCC=CCCCC1
Cyclododecene (8CI,9CI)

1502-05-2
$C_{10}H_{20}O$
156.27
OC(CCCCCCCC1)C1
Cyclodecanol (9CI)

1502-38-1
C_9H_{18}
126.24
Cyclooctane, methyl- (6CI, 7CI,8CI,9CI)
Methylcyclooctane

1502-95-0
$C_{21}H_{23}NO_5.ClH$
405.91
Morphinan-3,6-α-diol, 7,8-didehydro-4,5-α-epoxy-17-methyl-, diacetate (ester), hydrochloride
Diacetylmorphine hydrochloride
Diamorphine hydrochloride
Heroin hydrochloride
Heroine hydrochloride

1503-48-6
$C_{20}H_{10}N_2O_4$
342.30
O=C(c(c(nc(c1ccc2)c2)C(=O)c3c(O)c(c(n4)ccc5)c5)c1O)c34
Quino(2,3-b)acridine-6,7,13, 14(5H,12H)-tetrone (9CI)

1504-74-1
$C_{10}H_{10}O_2$
162.20
O=CC=Cc(c(OC)ccc1)c1
Cinnamaldehyde, o-methoxy
o-Methoxy cinnamaldehyde
2-Methoxycinnamaldehyde
o-Methoxycinnamic aldehyde
β-(o-Methoxyphenyl)acrolein
2-Propenal, 3-(2-methoxy-phenyl)-

1506-02-1
$C_{18}H_{26}O$
258.40
O=C(c(c(cc(c1C(CC2C)(C)C)C2(C)C)C)c1)C
Acetyl hexamethyl tetralin
2'-Acetonaphthone, 5',6',7',8'-tetrahydro-3',5',5',6',8',8'-hexamethyl- (VAN) (8CI)
6-Acetyl-1,1,2,4,4,7-hexa-methyl-1,2,3,4-tetrahydro-naphthalene
7-Acetyl-1,1,3,4,4,6-hexa-methyltetrahydronaphthalene
Ethanone, 1-(5,6,7,8-tetrahydro-3,5,5,6,8,8-hexamethyl-2-naphthalenyl)- (VAN) (9CI)
1-(5,6,7,8-Tetrahydro-3,5,5,6,8,8-hexamethyl-2-naphthalen-yl)ethanone

1510-16-3
$C_{12}H_{26}O_3S$
250.40
Dodecylsulfonic acid

1515-72-6
$C_{12}H_{13}NO_2$
203.23

O=C(N(C(=O)c1cccc2)CCCC)c12
Phthalimide, N-butyl- (8CI)
AI3-02418
2-Butyl-1H-isoindole-1,3(2H)-dione
N-Butylphthalimide
1H-Isoindole-1,3(2H)-dione, 2-butyl- (9CI)

1516-08-1
C_2D_6O
40.02
Ethanol-d6 (9CI)
Ethyl-d5 alcohol-d (8CI)

1516-27-4
$C_6H_{14}N.Cl$
135.66
Ammonium, allyltrimethyl-, chloride
Allyltrimethylammonium chloride
Homoneurine chloride

1517-15-3
C_5H_8O
84.12
Cyclobutanone, 2-methyl- (6CI,7CI,8CI,9CI)
2-Methylcyclobutanone

1518-54-3
$C_6H_{12}O_6$
180.16
D-Erythro-pentonic acid, 3-deoxy-2-C-(hydroxymethyl)- (8CI,9CI)
α-D-Glucoisosaccharinic acid
α-D-Isosaccharinic acid

1518-56-5
$C_6H_{12}O_6$
180.16
D-threo-Pentonic acid, 3-deoxy-2-C-(hydroxymethyl)- (8CI, 9CI)
β-D-Glucoisosaccharinic acid
β-D-Isosaccharinic acid

1518-59-8
$C_6H_{12}O_6$
180.16
D-Arabino-hexonic acid, 3-deoxy- (8CI,9CI)
β-D-Glucometasaccharinic acid

1518-83-8
$C_{11}H_{14}O$
162.23
Phenol, 4-cyclopentyl-
AI3-22766

1520-44-1
$C_{16}H_{18}$
210.32
c(cccc1)(c1)CCC(c(cccc2)c2)C
1,3-Diphenylbutane
Benzene, 1,1'-(1-methyl-1,3-propanediyl)bis- (9CI)
Butane, 1,3-diphenyl-
1,1'-(1-Methyl-1,3-propanediyl)bisbenzene

1520-77-0
$C_2H_6Cl_2Pb$
308.18
Plumbane, dichlorodimethyl- (8CI,9CI)

1520-78-1
C_3H_9ClPb
287.76
Plumbane, chlorotrimethyl
Chlorotrimethylplumbane
Lead, trimethyl-, chloride
TriML
Trimethyl lead chloride
Trimethylplumbane chloride

1521-38-6
$C_9H_{10}O_4$
182.18
COc1cccc(C(O)=O)c1OC
Benzoic acid, 2,3-dimethoxy- (9CI)
AI3-01432

1522-00-5
$C_8H_{15}NO_3$
173.24
Succinamic acid, N,N-diethyl
Butanoic acid, 4-(diethylamino)-4-oxo- (9CI)
N,N-Diethylsuccinamic acid
N,N-Diethylsuccinic acid monoamide
Mg 164
Monodietilamide dell'acido succinico (Italian)

1522-92-5
$C_5H_9Br_3O$
324.87
1-Propanol, 3-bromo-2,2-bis-(bromoethyl)
Pentaerythritol tribromide
Pentaerythritol tribromohydrin
Tribomoneopentyl alcohol
2,2,2-Tris(bromomethyl)ethanol

1523-06-4
$C_8H_7NO_4$
181.14
CC(=O)Oc1cccc(c1)N(=O)=O
Acetic acid, 3-nitrophenyl ester (9CI)
Acetic acid, m-nitrophenyl ester (8CI)
NSC-33906
Phenol, m-nitro-, acetate

1524-88-5
$C_{24}H_{33}FO_6$
436.57
Pregn-4-ene-3,20-dione, 6-α-fluoro-11-β,16-α,17, 21-tetrahydroxy-, cyclic 16,17- acetal with acetone
Alondra-F
Cordan
Cordran
Drenison
Drocort
Fluorandrenolone
Fluorandrenolone acetonide
Fluoroandrenolone acetonide
Flurandrenolide
Flurandrenolone
Flurandrenolone acetonide
Glucocorticoid
Haelan
Haldrone-F
L 33379
Sermaka

1528-00-3
$C_{10}H_{24}Sn$
263.01
Stannane, dibutyldimethyl- (8CI,9CI)

1528-01-4
$C_{13}H_{30}Sn$
305.09
Stannane, tributylmethyl- (8CI,9CI)

1528-49-0
$C_{27}H_{42}O_6$
462.63
O=C(OCCCCCC)c(ccc(c1C(=O)OCCCCCC)C(=O)OCCCCCC)c1
1,2,4-Benzenetricarboxylic acid, trihexyl ester (9CI)

1528-74-1
$C_{12}H_8N_2O_4$
244.22
O=N(=O)c(ccc(c(ccc(N(=O)=O)c1)c1)c2)c2
Biphenyl, 4,4'-dinitro
4,4'-Dinitrobifenyl (Czech)
4,4'-Dinitrobiphenyl

1528-82-1
$C_{13}H_{10}ClNO_2$
247.67
Benzamide, N-(4-chlorophenyl)-N-hydroxy- (9CI)
Benzohydroxamic acid, N-(p-chlorophenyl)- (8CI)

1529-57-3
$C_{60}H_{123}Al$
871.62

Trieicosylaluminum
Aluminum, trieicosyl- (9CI)
Tri(eicosyl) aluminum

1529-58-4
$C_{42}H_{87}Al$
619.13
Tris(tetradecyl)aluminum
Aluminum, tris(tetradecyl)-
Aluminum, tritetradecyl- (9CI)
Tritetradecylaluminum
Tri(tetradecyl) aluminum

1529-59-5
$C_{36}H_{75}Al$
534.97
Tris(dodecyl)aluminum
Aluminum, tridodecyl- (9CI)
Tridodecylaluminum
Tri(dodecyl) aluminum

1529-68-6
$C_4H_6Br_4$
373.74
BrCC(Br)C(Br)CBr
Butane, 1,2,3,4-tetrabromo
Fire Guard 5000
Tetrabromobutane
1,2,3,4-Tetrabromobutane

1531-12-0
$C_{16}H_{21}NO$
243.34
Norlevorphanol
DEA No. 9634
Norlevorfanol (Spanish)
Norlevorphanolum (Latin)

1533-45-5
$C_{28}H_{18}N_2O_2$
414.45
O(c(c(N=1)ccc2)c2)C1c(ccc(c3)C=Cc(ccc(C(Oc(c4ccc5)c5)=N4)c6)c6)c3
2,2'-(Vinylenedi-4-phenylene)-bis(benzoxazole)
Benzoxazole, 2,2'-(1,2-ethenediyldi-4,1-phenylene)bis- (9CI)
2,2'-(1,2-Ethenediyldi-4,1-phenylene)bisbenzoxazole

1533-76-2
$C_{23}H_{27}N_5O_7$
485.47
O=C(OCCN(c(ccc(N=Nc(ccc(N(=O)=O)c1)c1)c2NC(=O)CC)c2)CCOC(=O)C)C
Propanamide, N-(5-(bis-(2-(acetyloxy)ethyl)amino)-2-((4-nitrophenyl)azo)-phenyl)- (9CI)

1533-77-3
$C_{23}H_{27}N_5O_8$
501.47
O=C(OCCN(c(ccc(N=Nc(c(OC)cc(N(=O)=O)c1)c1)c2NC(=O)C)c2)CCOC(=O)C)C
Acetanilide, 5'-(bis(2-hydroxy-ethyl)amino)-2'-((2-methoxy-4-nitrophenyl)azo)-, di-acetate (ester) (8CI)

1533-78-4
$C_{22}H_{24}ClN_5O_7$
505.88
O=C(OCCN(c(ccc(N=Nc(c(cc(N(=O)=O)c1)Cl)c1)c2NC(=O)C)c2)CCOC(=O)C)C
Acetamide, N-(5-(bis(2-(acetyl-oxy)ethyl)amino)-2-((2-chloro-4-nitrophenyl)azo)-phenyl)- (9CI)
5'-(Bis(2-hydroxyethyl)amino)-2'-((2-chloro-4-nitrophenyl)-azo)acetanilide, diacetate (ester)
3-((N,N-Diacetyloxyethyl)-amino)-6-((2'-chloro-4'-nitro-phenyl)azo)acetanilide

1534-08-3
C_3H_6OS
90.14
Methyl thioacetate
Ethanethioic acid, S-methyl ester

1534-27-6
$C_{12}H_{24}O$
184.32
3-Dodecanone (8CI,9CI)
NSC-158522

1538-09-6
$C_{32}H_{36}N_4O_8S_2.C_{16}H_{20}N_2$
909.22
4-Thia-1-azabicyclo(3.2.0)-heptane-2-carboxylic acid, 3,3-dimethyl-7-oxo-6-(2-phenylacetamido)-, Compd. with N,N'-dibenzyl-ethylenediamine (2:1)
Beacillin
BEN-P
Benzacillin
Benzathine benzylpenicillin
Benzathine penicillin
Benzathine penicillin G
Benzethacil
Benzylpenicillin benzathine
Benzylpenicillin dibenzylethyl-enediamine salt
Bica-Penicillin
Bicillin
Cepacilina
Cepacillina
Cillenta
Dbed Dipenicillin G
Dbed Penicillin
Debecillin
Debecylina
Diamine Penicillin
Diaminocillina
Dibencil
Dibencillin
N,N'-Dibenzylethylenediamine bis(benzyl penicillin)
Dibenzylethylenediamine-di-penicillin G
N,N'-Dibenzylethylenediamine, Compounded with penicillin G (1:2)
N,N'-Dibenzylethylenediamine salt of benzylpenicillin
Dipo-Saft
Durabiotic
Dura-penita
Duropenin
Ethylenediamine, N,N'-di-benzyl-, Compd. with penicil-lin G (1:2)
Extencilline
Extenicilline
Lentocillin
Lentopenil
Leomypen
Longacilina
Longicil
LPG
Megacillin Suspension
Moldamin
NCI-C56100
Neolin
Penadur
Penadur L-A
Pendepon
Pen-Di-Ben
Penditan
Penduran
Penicillin G, Benzathine
Penicillin G, Compd. with N,N'-dibenzylethylenediamine (2:1)
Penicillin G salt of N,N'-di-benzylethylenediamine
Penidural
Penidure
Penilente
Permapen
Retarpen
Tardocillin
Vetarcillin
Vicin
Wycillina

1538-74-5
$C_{11}H_{15}NO_2$
193.24
CCCCOC(=O)Nc1ccccc1
Carbamic acid, phenyl-, butyl ester (9CI)
Carbanilic acid, butyl ester (8CI)
NSC-29705

1538-75-6
$C_{10}H_{18}O_3$
186.25
O=C(OC(=O)C(C)(C)C)C(C)(C)C
Pivalic anhydride
2,2-Dimethylpropanoic acid

anhydride
Propanoic acid, 2,2-dimethyl-, anhydride (9CI)

1539-04-4
$C_{20}H_{14}O_4$
318.33
O=C(Oc(cccc1)c1)c(ccc(c2)C(=O)Oc(cccc3)c3)c2
Terephthalic acid, diphenyl ester (8CI)
1,4-Benzenedicarboxylic acid, diphenyl ester (9CI)
Diphenyl 1,4-benzenedicarboxylate
Diphenyl terephthalate

1540-38-1
$C_8H_{14}O_2$
142.20
2,4-Pentanedione, 3-(1-methylethyl)- (9CI)
NSC-8341
2,4-Pentanedione, 3-isopropyl- (8CI)

1541-66-8
$C_{18}H_{39}NO_2$
301.51
OC(C)CN(CCCCCCCCCCCC)CC(O)C
2-Propanol, 1,1'-(dodecylimino)bis- (9CI)
Dodecylbis(2-hydroxypropyl)amine
1,1'-(Dodecylimino)bis-2-propanol

1541-67-9
$C_{16}H_{35}NO_2$
273.45
OCCN(CCCCCCCCCCCC)CCO
Ethanol, 2,2'-(dodecylimino)bis- (9CI)
AI3-16725
2,2'-(Dodecylimino)bisethanol
Lauryldiethanolamine

1541-81-7
$C_{16}H_{33}NO$
255.44
O(CCN(C1)CCCCCCCCCCCC)C1
N-Dodecylmorpholine
AI3-04702
N-Laurylmorpholine
Morpholine, 4-dodecyl-

1551-21-9
$C_4H_{10}S$
90.19
Propane, 2-(methylthio)- (9CI)
Sulfide, isopropyl methyl (8CI)

1551-31-1
$C_6H_{12}S$
116.23
Thiophene, tetrahydro-2,5-dimethyl- (8CI,9CI)

1552-12-1
C_8H_{12}
108.18
COD
1,5-Cyclooctadiene
1,5-Cyclooctadiene (Z,Z)
cis,cis-Cycloocta-1,5-diene

1552-42-7
$C_{26}H_{29}N_3O_2$
415.52
O=C(OC(c1ccc(N(C)C)c2)(c(ccc(N(C)C)c3)c3)c(ccc(N(C)C)c4)c4)c12
Crystal Violet Lactone
AI3-17349
3,3-Bis(p-dimethylaminophenyl)-6-dimethylaminophthalate
3,3-Bis(p-dimethylaminophenyl)-6-dimethylaminophthalide
3,3-Bis(4-dimethylaminophenyl)-6-dimethylaminophthalide
6-(Dimethylamino)-3,3-bis-(4-(dimethylamino)phenyl)-phthalide
1(3H)-Isobenzofuranone, 6-(dimethylamino)-3,3-bis(4-(dimethylamino)phenyl)- (9CI)

1555-53-9
$C_{18}H_{34}O_2.1/2Mg$
294.62
9-Octadecenoic acid (Z)-, magnesium salt (9CI)
AI3-19805
9-Octadecenoic acid, (Z)-, magnesium salt (2:1)

1555-66-4
$C_{12}H_{13}N_3$
199.24
N#CCCN(c(cccc1)c1)CCC#N
Propanenitrile, 3,3'-(phenylimino)bis- (9CI)
Aniline, N,N-dicyanoethyl-
3,3'-(Phenylimino)bispropanenitrile
Propionitrile, 3,3'-(phenylimino)di- (8CI)

1555-68-6
$C_{10}H_{25}N_3$
187.31
1,3-Propanediamine, N-(3-aminopropyl)-N-butyl- (9CI)
1-Butanamine, N,N-bis(3-aminopropyl)-
Butylamine, N,N-bis(3-aminopropyl)- (8CI)

1556-99-6
$C_{14}H_{12}$
180.25
9H-Fluorene, 4-methyl- (9CI)
Fluorene, 4-methyl- (8CI)

1558-17-4
$C_6H_8N_2$
108.13
n(c(cc(n1)C)C)c1
Pyrimidine, 4,6-dimethyl- (9CI)
4,6-Dimethylpyrimidine

1558-25-4
CH_2Cl_4Si
183.92
Trichloro(chloromethyl)silane
(Chloromethyl)trichlorosilane
Silane, chloromethyl(trichloro)-
Silane, trichloro(chloromethyl)- (8CI,9CI)
Trichloro(chloromethyl)-silane

1558-33-4
$C_2H_5Cl_3Si$
163.51
Silane, dichloro(chloromethyl)methyl- (9CI)
Dichloro(chloromethyl)methylsilane

1559-34-8
$C_{12}H_{26}O_5$
250.34
O(CCOCCOCCOCCO)CCCC
3,6,9,12-Tetraoxahexadecan-1-ol (9CI)
Tetraethylene glycol, monobutyl ether

1559-81-5
$C_{11}H_{14}$
146.23
Naphthalene, 1,2,3,4-tetrahydro-1-methyl- (8CI,9CI)

1560-06-1
$C_{10}H_{12}$
132.22
2-Butene, 1-phenyl
Benzene, 2-butenyl-
1-Phenylbutene-2
1-Phenyl-2-butene

1560-72-1
$C_{31}H_{64}$
436.85
Triacontane, 2-methyl- (8CI, 9CI)

1560-88-9

$C_{19}H_{40}$
268.53
Octadecane, 2-methyl-
AI3-35195

1560-89-0
$C_{18}H_{38}$
254.50
Heptadecane, 2-methyl- (9CI)
AI3-35565

1560-92-5
$C_{17}H_{36}$
240.47
Hexadecane, 2-methyl- (8CI, 9CI)

1560-93-6
$C_{16}H_{34}$
226.45
Pentadecane, 2-methyl- (9CI)

1560-95-8
$C_{15}H_{32}$
212.42
Tetradecane, 2-methyl- (8CI, 9CI)

1560-96-9
$C_{14}H_{30}$
198.39
Tridecane, 2-methyl- (8CI, 9CI)

1560-97-0
$C_{13}H_{28}$
184.37
2-Methyldodecane

1561-11-1
$C_7H_{14}O_2$
130.19
O=C(O)CCC(CC)C
Hexanoic acid, 4-methyl- (9CI)
4-Methylhexanoic acid

1561-20-2
$C_3H_2Cl_2O_2$
140.95
2-Propenoic acid, 3,3-dichloro- (9CI)

1561-49-5
$C_{14}H_{22}O_6$
286.33
O=C(OC(CCCC1)C1)OOC(=O)OC(CCCC2)C2
Dicyclohexyl peroxydicarbonate (Not more than 91% with water)
Dicyclohexyl peroxide carbonate
Dicyclohexyl peroxydicarbonate
Dicyclohexyl peroxydicarbonate, Not more than 91% with water
Dicyclohexyl peroxydicarbonate, Technically pure
Peroxidicarbonato de diciclohexilo (Spanish)
Peroxydicarbonate de dicyclohexyle (French)
Peroxydicarbonic acid, dicyclohexyl ester (8CI,9CI)
Peroxydicarbonic acid, dicyclohexyl ester, Not more than 91% with water
UN 2152
UN 2153

1561-86-0
$C_6H_{11}ClO$
134.61
Cyclohexanol, 2-chloro-
AI3-61848

1562-00-1
$C_2H_6O_4S.Na$
149.12
Sodium isethionate
Ethanesulfonic acid, 2-hydroxy-, monosodium salt (9CI)
Ethanesulfonic acid, 2-hydroxy-, sodium salt
2-Hydroxyethanesulfonic acid, sodium salt
Isethionic acid sodium salt
Sodium β-hydroxyethanesulfonate
Sodium 2-hydroxyethanesulfonate
Sodium 2-hydroxy-1-ethanesulfonate
Sodium 2-hydroxyethanesulfonic acid
Sodium hydroxyethylsulfonate
Sodium 2-hydroxyethylsulfonate

1562-93-2
$C_{13}H_{10}N_2O_2$
226.22
O=C(O)c(ccc(N=Nc(cccc1)c1)c2)c2
Benzoic acid, p-(phenylazo)- (8CI)
Benzoic acid, 4-(phenylazo)- (9CI)
4-(Phenylazo)benzoic acid

1562-94-3
$C_{14}H_{14}N_2O_3$
258.26
O=N(=Nc(ccc(OC)c1)c1)c(ccc(OC)c2)c2
Azoxybenzene, 4,4'-dimethoxy- (8CI)
AI3-52539
Bis(4-methoxyphenyl)diazene 1-oxide
Diazene, bis(4-methoxyphenyl)-, 1-oxide (9CI)

1563-38-8
$C_{10}H_{12}O_2$
164.20
O(c(c(ccc1)C2)c1O)C2(C)C
2,3-Dihydro-2,2-dimethyl-7-benzofuranol
AI3-27488
7-Benzofuranol, 2,3-dihydro-2,2-dimethyl-
Carbofuran phenol
Carbofuran 7-phenol
2,3-Dihydro-2,2-dimethyl-7-hydroxybenzofuran
NIA 10272

1563-66-2
$C_{12}H_{15}NO_3$
221.28
O=C(Oc(c(OC(C1)(C)C)c1cc2)c2)NC
Carbamic acid, methyl-, 2,3-dihydro-2,2-dimethyl-7-benzofuranyl ester
Bay 70143
Carbamic acid, methyl-, 2,2-dimethyl-2,3-dihydrobenzofuran-7-yl ester
Carbofuran (ACGIH,DOT, OSHA)
Carbofurane
Carbofuran mixture, Liquid (DOT)
Crisfuran
Curaterr
D 1221
2,3-Dihydro-2,2-dimethylbenzofuranyl-7-N-methylcarbamate
2,3-Dihydro-2,2-dimethylbenzofuran-7-yl methylcarbamate
2,3-Dihydro-2,2-dimethyl-7-benzofuranyl methylcarbamate
2,2-Dimethyl-7-coumaranyl N-methylcarbamate
2,2-Dimethyl-2,3-dihydro-7-benzofuranyl N-methylcarbamate
ENT 27,164
FMC 10242
Furadan (OSHA)
Furadane
Furodan
Karbofuranu (Polish)
Methyl carbamic acid 2,3-dihydro-2,2-dimethyl-7-benzofuranyl ester
NA 2757 (DOT)
NIA 10242
Niagara 10242
Niagara NIA-10242
OMS-864
Yaltox

1563-84-4
$C_7H_{13}NO_2$
143.18
Acetamide, N-acetyl-N-propyl- (9CI)

Diacetamide, N-propyl- (8CI)

1565-94-2
$C_{29}H_{36}O_8$
512.65
O=C(OCC(O)COc(ccc(c1)C(c(ccc(OCC(O)COC(=O)C(=C)C)c2)c2)(C)C)c1)C(=C)C
2-Propenoic acid, 2-methyl-, (1-methylethylidene)bis-(4,1-phenyleneoxy(2-hydroxy-3,1- propanediyl)) ester
Bis-GMA
Bisphenol A glycidylmethacrylate
Nupol 1629
Nupol 46-4005

1568-70-3
$C_7H_7NO_4$
169.13
Phenol, 4-methoxy-2-nitro- (8CI,9CI)

1569-01-3
$C_6H_{14}O_2$
118.20
O(CC(O)C)CCC
2-Propanol, 1-propoxy
Propasol solvent P
1-Propoxy-2-propanol
Propylene glycol n-propyl ether

1569-02-4
$C_5H_{12}O_2$
104.15
Propylene glycol ethyl ether
1-Ethoxy-2-propanol
2-Propanol, 1-ethoxy- (8CI,9CI)

1569-50-2
$C_5H_{10}O$
86.13
OC(C=CC)C
3-Penten-2-ol
AI3-28607

1569-69-3
$C_6H_{12}S$
116.24
SC(CCCC1)C1
Cyclohexanethiol
Cyklohexanthiol (Czech)
Cyklohexylmerkaptan (Czech)

1570-45-2
$C_8H_9NO_2$
151.18
O=C(OCC)c(ccnc1)c1
Isonicotinic acid, ethyl ester
Ethyl isonicotinate
4-Pyridinecarboxylic acid, ethyl ester

1570-64-5
C_7H_7ClO
142.59
Oc(c(cc(c1)Cl)C)c1
o-Cresol, 4-chloro
4-Chloro-o-cresol
4-Chloro-2-methylphenol

1570-65-6
$C_7H_6Cl_2O$
177.03
Cc1cc(Cl)cc(Cl)c1O
2,4-Dichloro-6-methylphenol
AI3-19024
o-Cresol, 4,6-dichloro- (8CI)
Phenol, 2,4-dichloro-6-methyl- (9CI)

1571-08-0
$C_9H_8O_3$
164.16
O=Cc(ccc(c1)C(=O)OC)c1
Methyl 4-formylbenzoate
Benzoic acid, 4-formyl-, methyl ester (9CI)
p-Carbomethoxybenzaldehyde
4-Carbomethoxybenzaldehyde
4-Carboxybenzaldehyde methyl ester
p-Formylbenzoic acid methyl ester
2-Methoxycarbonylbenzaldehyde
4-(Methoxycarbonyl)benzaldehyde
Methyl benzaldehyde-4-carboxylate
Methyl p-formylbenzoate
Methyl terephthalaldehydate
Methyl terephthaldehydate
Terephthalaldehydic acid, methyl ester

1571-33-1
$C_6H_7O_3P$
158.10
O=P(O)(O)c(cccc1)c1
Phosphonic acid, phenyl
Benzenephosphonic acid
Phenylphosphonic acid

1572-46-9
$C_{20}H_{16}$
256.35
9H-Fluorene, 9-(phenylmethyl)- (9CI)
9-Benzylfluorene
Fluorene, 9-benzyl- (6CI, 7CI,8CI)

1573-28-0
$C_9H_{20}O$
144.26
3-Heptanol, 3,6-dimethyl- (9CI)

1573-57-5
$C_4H_4Cl_6$
264.79
Butane, 1,2,2,3,3,4-hexachloro- (6CI,7CI,8CI, 9CI)
1,2,2,3,3,4-Hexachlorobutane

1573-58-6
$C_4H_3Cl_3$
157.43
1,3-Butadiene, 1,2,3-trichloro- (8CI,9CI)

1574-41-0
C_5H_8
68.12
C(=CC=C)C
1,3-Pentadiene, (Z)- (9CI)
cis-1,3-Pentadiene
(Z)-1,3-Pentadiene

1576-35-8
$C_7H_{10}N_2O_2S$
186.25
O=S(=O)(NN)c(ccc(c1)C)c1
Hydrazine, p-tolylsulfonyl
p-Toluenesulfonic acid, hydrazide
N-Toluolsulphonyl hydrazine
p-Tolylsulfonylhydrazine

1576-67-6
$C_{16}H_{14}$
206.29
Cc3ccc2ccc1ccc(C)cc1c2c3
Phenanthrene, 3,6-dimethyl- (9CI)

1576-69-8
$C_{16}H_{14}$
206.29
Phenanthrene, 2,7-dimethyl- (9CI)

1576-95-0
$C_5H_{10}O$
86.13
OCC=CCC
2-Penten-1-ol, (Z)- (9CI)
(Z)-2-Penten-1-ol

1577-19-1
$C_8H_{14}O_2$
142.20
O=C(O)CC=CCCCC
3-Octenoic acid (9CI)

1577-22-6
$C_6H_{10}O_2$
114.14
OC(=O)CCCC=C

5-Hexenoic acid (8CI,9CI)

1582-09-8
$C_{13}H_{16}F_3N_3O_4$
335.32
CCCN(CCC)c1c(cc(cc1N(=O)=O)C(F)(F)F)N(=O)=O
p-Toluidine, α,α,α-trifluoro-2,6-dinitro-N,N-dipropyl
Agreflan
Agriflan 24
Benzenamine, 2,6-dinitro-N,N-dipropyl-4-(trifluoromethyl)- (9CI)
Crisalin
Digermin
2,6-Dinitro-N,N-dipropyl-4-(trifluoromethyl)benzenamine
2,6-Dinitro-N,N-di-n-propyl-α,α,α-trifluoro-p-toluidine
4-(Di-n-propylamino)-3,5-dinitro-1-trifluoromethylbenzene
N,N-Dipropyl-2,6-dinitro-4-trifluormethylanilin (German)
N,N-Di-n-propyl-2,6-dinitro-4-trifluoromethylaniline
2,6-Dinitro-4-trifluormethyl-N,N-dipropylanilin (German)
N,N-Dipropyl-2,6-dinitro-4-trifluoromethylaniline
N,N-Dipropyl-4-trifluoromethyl-2,6-dinitroaniline
Elancolan
L-36352
Lilly 36,352
NCI-C00442
Nitran
Olitref
Su seguro carpidor
Trefanocide
Treficon
Treflam
Treflan
Treflanocide elancolan
Trifluoralin
Trifluralin
Trifluraline
α,α,α-Trifluoro-2,6-dinitro-N,N-dipropyl-p-toluidine
Trifurex
Trikepin
Trim

1583-67-1
$C_8H_5FO_4$
184.12
1,2-Benzenedicarboxylic acid, 3-fluoro- (9CI)
NSC-402999
Phthalic acid, 3-fluoro- (8CI)

1585-07-5
C_8H_9Br
185.06
c(ccc(c1)Br)(c1)CC
Benzene, 1-bromo-4-ethyl- (9CI)
1-Bromo-4-ethylbenzene

1585-40-6
$C_{11}H_6O_{10}$
298.16
OC(=O)c1cc(C(O)=O)c(C(O)=O)c(C(O)=O)c1C(O)=O
Benzenepentacarboxylate
Benzenepentacarboxylic acid (9CI)

1586-92-1
$C_6H_{15}AlO$
130.17
CCO[Al](CC)CC
Aluminum, ethoxydiethyl- (9CI)
Diethylaluminum ethoxide
Ethoxydiethylaluminum

1587-04-8
$C_{10}H_{12}$
132.21
Benzene, 1-methyl-2-(2-propenyl)- (9CI)
NSC-73972
Toluene, o-allyl- (8CI)

1589-47-5
$C_4H_{10}O_2$
90.12
1-Propanol, 2-methoxy- (8CI, 9CI)
2-Methoxy-1-propanol

1591-31-7
$C_{12}H_9I$
280.11
c(ccc(c(cccc1)c1)c2)(c2)I
4-Iodobiphenyl
AI3-15372
Biphenyl, 4-iodo- (8CI)
1,1'-Biphenyl, 4-iodo- (9CI)
4-Iodo-1,1'-biphenyl

1592-23-0
$C_{36}H_{70}O_4.Ca$
607.14
Stearic acid, calcium salt
Aquacal
Calcium distearate
Calcium stearate (ACGIH)
Calstar
Flexichem
Flexichem CS
G 339 S
Nopcote C 104
Octadecanoic acid, calcium salt
Stavinor 30
Synpro stearate
Witco G 339S

1593-77-7
Unknown
Unknown
4-Cyclododecyl-2,6-dimethylmorpholine

1594-08-7
$C_{21}H_{15}NO_6S$
409.41
O=C(c(c(C(=O)c1c(O)ccc2Nc(ccc(OS(=O)(=O)C)c3)c3)ccc4)c4)c12
9,10-Anthracenedione, 1-hydroxy-4-((4-((methylsulfonyl)oxy)phenyl)amino)- (9CI)
Anthraquinone, 1-hydroxy-4-(p-hydroxyanilino)-, 4-methanesulfonate (Ester)

1594-56-5
$C_7H_3N_3O_4S$
225.19
O=N(=O)c1ccc(SC#N)c(c1)N(=O)=O
Thiocyanic acid, 2,4-dinitrophenyl ester
2,4-Dinitrofenylthiokyanat (Czech)
2,4-Dinitrophenyl thiocyanate
2,4-Dinitro-rhodanbenzol (German)
2,4-Dinitrothiocyanatobenzene
2,4-Dinitrothiocyanobenzene
2,4-Dinitro-1-thiocyanobenzene
DNRB
DNTB
DRB
Gryzbol
Grzybol
NBT
Nirit
Nitrite
Rhodandinitrobenzol
Rodatox 60
Trirodazeen
Tri-Rodazene
2317-W

1596-84-5
$C_6H_{12}N_2O_3$
160.20
CN(C)NC(=O)CCC(O)=O
Succinic acid, mono(2,2-dimethylhydrazide)
Alar
Alar-85
Aminozide
B-9
B 995
Bernsteinsaeure-2,2-dimethylhydrazid (German)
B-Nine
Butanedioic acid mono(2,2-dimethylhydrazide)
Daminozide
Dimas
N-Dimethyl amino-β-carbamyl propionic acid
Dimethylaminosuccinamic acid
N-Dimethylamino-succinamidsaeure (German)
N-(Dimethylamino)succinamic acid
2,2-Dimethylhydrazid kyseliny jantarove (Czech)

DMASA
DMSA
Kylar
NCI-C03827
SADH
Succinic acid 2,2-dimethyl-hydrazide
Succinic 1,1-dimethyl hydrazide

1598-99-8
$C_{11}H_{20}FN_5$
241.36
s-Triazine, 2,4-bis(diethyl-amino)-6-fluoro
2,4-Bis(diethylamino)-6-fluoro-s-triazine

1599-67-3
$C_{22}H_{44}$
308.59
C(=C)CCCCCCCCCCCCCCCCCCCC
1-Docosene (9CI)
AI3-36497

1600-27-7
$C_4H_6O_4.Hg$
318.69
[Hg+2].CC([O-])=O.CC([O-])=O
Acetic acid, mercury(2+) salt
Bis(acetyloxy)mercury
Diacetoxymercury
Mercuriacetate
Mercuric acetate (DOT)
Mercuric diacetate
Mercury(2+) acetate
Mercury(II) acetate
Mercury acetate [UN 1629]
Mercury diacetate
Mercuryl acetate
UN 1629 [Mercury acetate]

1600-37-9
C_3HCl_5
214.29
C(=C(Cl)Cl)(C(Cl)Cl)Cl
Propene, 1,1,2,3,3-pentachloro
1,1,2,3,3-Pentachloropropene
1,1,2,3,3-Pentachloropropylene
1-Propene, 1,1,2,3,3-penta-chloro-

1602-00-2
$C_{12}H_8Cl_2N_2$
251.12
Azobenzene, 4,4'-dichloro
DCAB
Diazene, bis(4-chlorophenyl)-
p,p'-Dichloroazobenzene
4,4'-Dichloroazobenzene

1603-01-6
C_8H_{14}
110.20
1,4-Heptadiene, 3-methyl-(8CI,9CI)

1603-41-4
$C_6H_8N_2$
108.16
n(c(N)ccc1C)c1
3-Picoline, 6-amino
2-Amino-5-methylpyridine
6-Amino-3-picoline
2-Pyridinamine, 5-methyl-

1603-79-8
$C_{10}H_{10}O_3$
178.19
O=C(OCC)C(=O)c(cccc1)c1
Ethyl phenylglyoxylate
AI3-10033
Benzeneacetic acid, α-oxo-, ethyl ester (9CI)
Ethyl α-oxobenzeneacetate
Glyoxylic acid, phenyl-, ethyl ester (8CI)

1604-11-1
$C_7H_{12}O_4$
160.17
O=C(OC)C(CC(=O)OC)C
Butanedioic acid, methyl-, dimethyl ester
AI3-28480
Dimethyl methylbutanedioate
Dimethyl methylsuccinate

1604-32-6
$C_{18}H_{32}O$
264.45
3,5-Pentadecadien-2-one, 6,10,14-trimethyl- (8CI, 9CI)

1604-34-8
$C_{13}H_{26}O$
198.35
O=C(CCCC(CCCC(C)C)C)C
2-Undecanone, 6,10-dimethyl-(9CI)
AI3-15989
6,10-Dimethyl-2-undecanone

1604-35-9
$C_{15}H_{28}O$
224.39
1-Dodecyn-3-ol, 3,7,11-tri-methyl- (8CI,9CI)

1606-67-3
$C_{16}H_{11}N$
217.28
c(c(c(cc1)ccc2)c2cc3)(c3ccc4N)c14
1-Pyrenamine
1-Aminopyrene
3-Aminopyrene

1606-85-5
$C_8H_{14}O_4$
174.20
O(CC#CCOCCO)CCO
Ethanol, 2,2'-(2-butyne-1,4-diylbis(oxy))bis- (9CI)
2,2'-(2-Butyne-1,4-diylbis-(oxy))bisethanol
Ethanol, 2,2'-(2-butynylene-dioxy)di- (8CI)

1609-07-0
$C_3H_6N_4$
98.11
CNC(N)=NC#N
3-Methylcyanoguanidine
Guanidine, N-cyano-N'-methyl-,

1609-19-4
$C_4H_{11}ClSi$
122.67
Chlorodiethylsilane
Diethylchlorosilane
Silane, chlorodiethyl- (8CI,9CI)

1609-21-8
$C_{15}H_{26}N.Cl$
255.83
Pyridinium, 1-decyl-, chloride
AI3-15067
Decylpyridinium chloride
1-Decylpyridinium chloride

1609-93-4
$C_3H_3ClO_2$
106.51
OC(=O)C=CCl
Acrylic acid, 3-chloro-, (Z)-(8CI)
2-Propenoic acid, 3-chloro-, (Z)- (9CI)

1610-17-9
$C_9H_{17}N_5O$
211.31
CCNc1nc(NC(C)C)nc(OC)n1
s-Triazine, 2-ethylamino-4-isopropylamino-6-methoxy
Atraton
Atratone
Atroton
2-Ethylamino-4-isopropylamino-6-methoxy-s-triazine
4-Ethylamino-6-isopropylamino-2-methoxy-s-triazine
6-Ethylamino-4-isopropylamino-2-methoxy-1,3,5-triazine
N-Ethyl-6-methoxy-N'-(1-methylethyl)-1,3,5-triazine-2,4-diamine
G-32293
Geigy 32,293
Gesatamin
2-Methoxy-4-ethylamino-6-iso-propylamino-s-triazine
2-Methoxy-4-isopropylamino-6-ethylamino-s-triazine
Primatol

s-Triazine, 4-ethylamino-6-isopropylamino-2-methoxy-

1610-18-0
$C_{10}H_{19}N_5O$
225.34
O(c(nc(nc1NC(C)C)NC(C)C)n1)C
s-Triazine, 2,4-bis(isopropylamino)-6-methoxy
2,4-Bis(isopropylamino)-6-methoxy-s-triazine
2,6-Diisopropylamino-4-methoxytriazine
N,N'-Diisopropyl-6-methoxy-1,3,5-triazine-2,4-diyldiamine
G-31435
Gesafram
Gesafram 50
Gesagram
2-Methoxy-4,6-bis(isopropylamino)-s-triazine
2-Methoxy-4,6-bis(isopropylamino)-1,3,5-triazine
Methoxypropazine
Ontracic 800
Ontrack
Ontrack-WE-2
Pramitol
Primatol
Primatol 25E
Prometon
Prometone

1611-92-3
$C_7H_6Br_2$
249.94
Benzene, 1,3-dibromo-5-methyl- (9CI)
Toluene, 3,5-dibromo- (6CI, 7CI,8CI)

1613-51-0
$C_5H_{10}S$
102.20
2H-Thiopyran, tetrahydro- (8CI,9CI)
NSC-9459
Pentamethylene sulfide
Penthiophane
Thiacyclohexane
Thiane
Thiopyran, tetrahydro-

1613-52-1
$C_7H_{14}S$
130.25
2H-Thiopyran, 2-ethyltetrahydro- (8CI,9CI)
Thiopyran, 2-ethyltetrahydro-

1614-30-8
$C_{14}H_{14}N_4S_2$
302.44
Urea, 1,1'-(4,4'-biphenylyene)bis(2-thio
p,p'-Biphenylene-bis-1,1'-(2-thiourea)
4,4'-Biphenylene bis(thiourea)

1615-02-7
$C_9H_7ClO_2$
182.61
OC(=O)C=Cc1ccc(Cl)cc1
Cinnamic acid, p-chloro
p-Chlorocinnamic acid
4-Chlorocinnamic acid
3-(p-Chlorophenyl)acrylic acid
2-Propenoic acid, 3-(4-chlorophenyl)- (9CI)

1615-06-1
$C_6H_6N_2O_2S$
170.19
O=S2(=O)Nc1ccccc1N2
1H,3H-2,1,3-Benzothiadiazole 2,2-dioxide
2,1,3-Benzothiadiazole, 1,3-dihydro-, 2,2-dioxide
1,3-BTDZD

1615-80-1
$C_4H_{12}N_2$
88.18
CCNNCC
Hydrazine, 1,2-diethyl
1,2-Diaethylhydrazin (German)
N-N'-Diethylhydrazine
sym-Diethylhydrazine
1,2-Diethylhydrazine
Hydrazoethane
Hydroazoethane
RCRA waste number U086
SDEH

1615-91-4
$C_{30}H_{50}$
410.73
Diploptene
22(29)-Hopene

1616-88-2
$C_4H_9NO_3$
119.14
O=C(OCCOC)N
Ethanol, 2-methoxy-, carbamate
Carbamic acid, 2-methoxyethyl ester
2-Methoxyethyl carbamate
N-Methoxyurethane

1617-32-9
$C_5H_8O_2$
100.12
3-Pentenoic acid, (E)- (8CI,9CI)

1618-22-0
$C_{12}H_{22}$
166.31
Naphthalene, decahydro-2,6-dimethyl- (8CI,9CI)

1618-26-4
$C_3H_8S_2$
108.23
S(CSC)C
Methane, bis(methylthio)- (9CI)
Bis(methylthio)methane

1619-57-4
Unknown
Unknown
Diethyl acetoacetate

1619-62-1
$C_9H_{16}O_4$
188.22
Propanedioic acid, dimethyl-, diethyl ester (9CI)
Malonic acid, dimethyl-, diethyl ester (8CI)
NSC-28462

1622-32-8
$C_2H_4Cl_2O_2S$
163.02
O=S(=O)(CCCl)Cl
Ethanesulfonyl chloride, 2-chloro- (8CI,9CI)
2-Chloroethane sulfochloride
β-Chloroethanesulfonyl chloride
2-Chloroethanesulfonyl chloride
2-Chloroethylsulfonyl chloride

1622-61-3
$C_{15}H_{10}ClN_3O_3$
315.73
Clc1ccccc1C2=NCC(=O)Nc3ccc(cc23)N(=O)=O
2-H-1,4-Benzodiazepin-2-one, 5-(o-chlorophenyl)-1,3-dihydro-7-nitro
5-(o-Chlorophenyl)-1,3-dihydro-7-nitro-2H-1,4-benzodiazepin-2-one
5-(2-Chlorophenyl)-1,3-dihydro-7-nitro-2H-1,4-benzodiazepin-2-one
5-(o-Chlorophenyl)-7-nitro-1H-1,4-benzodiazepin-2(3H)-one
Cloazepam
Clonazepam
1,3-Dihydro-7-nitro-5-(2-chlorophenyl)-2H-1,4.benzodiazepin-2-one
Rivotril
Ro 4-8180
Ro 5-4023

1622-62-4
$C_{16}H_{12}FN_3O_3$
313.27
Flunitrazepam
2H-1,4-Benzodiazepin-2-one, 1,3-dihydro-5-(2-fluoro-

phenyl)-1-methyl-7-nitro-
2H-1,4-Benzodiazepin-2-one, 5-(2-fluorophenyl)-1,3-dihydro-1-methyl-7-nitro-
5-(o-Fluorophenyl)-1,3-dihydro-1-methyl-7-nitro-2H-1,4-benzodiazepin-2-one
DEA No. 2763
1,3-Dihydro-5-(o-fluorophenyl)-1-methyl-7-nitro-2H-1,4-benzodiazepin-2-one
Flunitrazepamum (Latin)
1-Methyl-7-nitro-5-(2-fluorophenyl)-3H-1,4-benzodiazepin-2(1H)-one
Narcozep
Ro 5-4200
Rohypnol

1623-05-8
$C_5F_{10}O$
266.04
FC(F)(OC(F)=C(F)F)C(F)(F)C(F)(F)F
Propane, 1,1,1,2,2,3,3-heptafluoro-3-((trifluoroethenyl)-oxy)- (9CI)
Perfluoro(propyl vinyl ether)

1623-06-9
$C_3H_9O_4P$
140.08
O=P(OCCC)(O)O
Phosphoric acid, monopropyl ester (9CI)
Monopropyl phosphate

1623-14-9
$C_2H_7O_4P$
126.05
O=P(OCC)(O)O
Phosphoric acid, monoethyl ester
AI3-15046
Monoethyl phosphate

1623-15-0
$C_4H_{11}O_4P$
154.10
O=P(OCCCC)(O)O
Phosphoric acid, monobutyl ester (9CI)
AI3-15021
Monobutyl phosphate

1623-19-4
$C_9H_{15}O_4P$
218.21
O=P(OCC=C)(OCC=C)OCC=C
Phosphoric acid, tri-2-propenyl ester
Allyl phosphate
Phosphoric acid, triallyl ester
Triallylfosfat (Czech)
Triallyl phosphate

1623-24-1
$C_3H_9O_4P$
140.09
O=P(OC(C)C)(O)O
Phosphoric acid, isopropyl ester
Isopropyl acid phosphate [UN 1793]
Isopropyl acid phosphate, Solid (DOT)
Isopropyl phosphoric acid, Solid (DOT)
UN 1793 [Isopropyl acid phosphate]

1624-02-8
$C_{36}H_{30}CrO_4Si_2$
634.84
Chromic acid, bis(triphenylsilyl) ester
Bis(triphenylsilyl)chromate

1628-58-6
$C_{17}H_{16}N_2S$
280.41
N(c(c(S1)ccc2)c2)=C1C=Cc(ccc(N(C)C)c3)c3
Benzothiazole, 2-(p-(dimethylamino)styryl)
2-(p-(Dimethylamino)styryl)-benzothiazole
2-(4-Dimethylaminostyryl)-benzothiazole

1629-58-9
C_5H_8O
84.13
O=C(C=C)CC
1-Penten-3-one
Ethyl vinyl ketone
Ketone, ethyl vinyl

1629-60-3
$C_6H_{10}O$
98.14
1-Hexen-3-one (8CI,9CI)

1630-08-6
$C_6H_5N_5O_6$
243.11
O=N(=O)c(c(N)c(N(=O)=O)c(N)c1N(=O)=O)c1
m-Phenylenediamine, 2,4,6-trinitro- (8CI)
1,3-Benzenediamine, 2,4,6-trinitro- (9CI)
DATB
2,4,6-Trinitro-1,3-benzenediamine

1630-17-7
$C_{14}H_{13}NO_3$
243.26
Benzene, 1,3-dimethyl-5-(4-nitrophenoxy)- (9CI)
DMNP
Ether, p-nitrophenyl 3,5-xylyl (8CI)
Farmaid

1630-77-9
$C_2H_2F_2$
64.03
Ethene, 1,2-difluoro-, (Z)- (9CI)
Ethylene, 1,2-difluoro-, (Z)- (8CI)

1630-78-0
$C_2H_2F_2$
64.03
Ethene, 1,2-difluoro-, (E)- (9CI)
Ethylene, 1,2-difluoro-, (E)- (8CI)

1631-73-8
$C_3H_{10}Sn$
164.82
Trimethyltin
Stannane, trimethyl- (9CI)
Trimethylstannane

1632-16-2
C_8H_{16}
112.24
C(=C)(CCCC)CC
1-Hexene, 2-ethyl
2-Ethyl hexene-1
2-Ethyl-1-hexene
Heptane, 3-methylene-
USAF DO-21

1632-26-4
$C_6H_{10}N_2O$
126.16
1,2-Diazabicyclo (2.2.2)octan-3-one-(7CI,8CI,9CI)

1632-73-1
$C_{10}H_{18}O$
154.25
OC(C(CC1C2)(C2)C)C1(C)C
Fenchol
AI3-00733
Bicyclo(2.2.1)heptan-2-ol, 1,3,3-trimethyl-
1,3,3-Trimethylbicyclo(2.2.1)-heptan-2-ol

1633-05-2
$CH_2O_3.Sr$
149.65
Strontium carbonate
Carbonic acid, strontium salt (1:1) (9CI)
C.I. 77837

1633-83-6
$C_4H_8O_3S$
136.18
O=S(=O)(OCCC1)C1
1,2-Oxathiane, 2,2-dioxide

Butanesulfone
Butane sultone
δ-Butane sultone
1,4-Butanesultone
1,4-Butylene sulfone
δ-Valerosultone

1634-02-2
$C_{18}H_{36}N_2S_4$
408.80
N(C(=S)SSC(N(CCCC)CCCC)=S)(CCCC)CCCC
Disulfide, bis(dibutylthiocarbamoyl)
Tetrabutylthiuram disulphide
Thiuram disulfide tetrabutyl
Thiuram, tetrabutyl-, disulfide

1634-04-4
$C_5H_{12}O$
88.17
O(C(C)(C)C)C
Ether, tert-butyl methyl
tert-Butyl methyl ether
Methyl-tert-butylether [UN 2398]
Methyl 1,1-dimethylethyl ether
Propane, 2-methoxy-2-methyl- (9CI)
UN 2398 [Methyl-tert-butylether]

1634-09-9
$C_{14}H_{16}$
184.28
Naphthalene, 1-butyl- (6CI,7CI,8CI,9CI)
α-Butylnaphthalene
1-Butylnaphthalene
1-Naphthyl-1-butane

1634-78-2
$C_{10}H_{19}O_7PS$
314.32
CCOC(=O)CC(SP(=O)(OC)OC)C(=O)OCC
Succinic acid, mercapto-, diethyl ester, S-ester with O,O-dimethylphosphorothioate
Butanedioic acid, ((dimethoxyphosphinyl)thio)-, diethyl ester (9CI)
Carbethoxy malaoxon
S-(1,2-Diethoxycarbonyl)ethyl O,O-dimethyl phosphorothioate
O,O-Dimethyl-S-1,2-bis(ethoxycarbonyl)ethyl phosphorothioate
O,O-Dimethyl-S-(1,2-dicarbethoxy)ethyl phosphorothioate
Liromat
Malaoxon
Malaoxone
Malathion-O-Analog
NCI-C08628
Oxycarbophos
Phosphorothioic acid, O,O-dimethyl ester, S-ester with 1,2-bis(methoxycarbonyl) ethanethiol

1635-61-6
$C_6H_5ClN_2O_2$
172.56
Nc1cc(Cl)ccc1N(=O)=O
Aniline, 5-chloro-2-nitro- (8CI)
Benzenamine, 5-chloro-2-nitro- (9CI)

1636-39-1
$C_{10}H_{18}$
138.25
1,1'-Bicyclopentyl (9CI)
Bicyclopentyl (8CI)
Cyclopentane, cyclopentyl-
NSC-38865

1637-31-6
$C_4H_2Cl_4$
191.87
1,3-Butadiene, 1,2,3,4-tetrachloro- (8CI,9CI)

1638-22-8
$C_{10}H_{14}O$
150.24
Oc(ccc(c1)CCCC)c1
Phenol, p-butyl
4-n-Butylphenol
p-Butylphenol, Liquid [UN 2228]
p-Butylphenol, Solid [UN 2229]
UN 2228 [Butylphenols, liquid]
UN 2229 [Butylphenols, solid]

1638-26-2
C_7H_{14}
98.19
Cyclopentane, 1,1-dimethyl- (9CI)

1639-09-4
$C_7H_{16}S$
132.29
SCCCCCCC
1-Heptanethiol
Heptyl mercaptan
n-Heptylmercaptan
USAF EK-2122

1639-39-0
$C_{10}H_{11}N_5O$
217.23
s-Triazine, 2,4-diamino-6-(p-tolyloxy)- (7CI,8CI)

1639-66-3
$C_{20}H_{37}O_7S.Na$
444.62
Succinic acid, sulfo-, 1,4-dioctyl ester, sodium salt
Butanedioic acid, sulfo-, 1,4-dioctyl ester, sodium salt (9CI)
Di-n-octyl sodium sulfosuccinate
Dioktylester sulfojantaranu sodneho (Czech)
Sulfosuccinic acid 1,4-dioctyl ester sodium salt

1640-39-7
$C_{11}H_{13}N$
159.22
N(c(c(ccc1)C2(C)C)c1)=C2C
3H-Indole, 2,3,3-trimethyl- (9CI)
AI3-51456
2,3,3-Trimethyl-3H-indole
2,3,3-Trimethylindolenine

1640-89-7
C_7H_{14}
98.21
Cyclopentane, ethyl
Ethylcyclopentane

1642-54-2
$C_{10}H_{21}N_3O.C_6H_8O_7$
391.48
1-Piperazinecarboxamide, N,N-diethyl-4-methyl-, citrate (1:1)
Banocide
Caricide
Caritrol
Dicarocide
Diethylcarbamazane citrate
Diethylcarbamazine acid citrate
Diethylcarbamazine citrate
Diethylcarbamazine hydrogen citrate
1-Diethylcarbamoyl-4-methylpiperazine dihydrogen citrate
N,N-Diethyl-4-methyl-1-piperazine carboxamide citrate
N,N-Diethyl-4-methyl-1-piperazinecarboxamide dihydrogen citrate
Ditrazin
Ditrazine
Ditrazin citrate
Ditrazine citrate
Ethodryl citrate
Francocide
Franozan
Hetrazan
Loxuran
1-Methyl-4-diethylcarbamoylpiperazine citrate
1-Piperazinecarboxamide, N,N-diethyl-4-methyl-, 2-hydroxy-1,2,3-propanetricarboxylate

1643-19-2
$C_{16}H_{36}N.Br$
322.37

[Br-].CCCC[N+](CCCC)(CCCC)CCCC
1-Butanaminium, N,N,N-tributyl-, bromide (9CI)
N,N,N-Tributyl-1-butanaminium bromide

1643-20-5
$C_{14}H_{31}NO$
229.46
O=N(CCCCCCCCCCCC)(C)C
Dodecylamine, N,N-dimethyl-, N-oxide
Ammonyx Lo
Amonyx AO
Aromox DMMC-W
Conco XAL
DDNO
Dimethyldodecylamine N-oxide
N,N-Dimethyldodecylamine oxide
N,N-Dimethyl-dodecylaminoxid (Czech)
Dodecyldimethylamine oxide
n-Dodecyldimethylamine oxide
Lauryldimethylamine oxide
NCI-C55129

1646-75-9
$C_5H_{11}NOS$
133.21
N(O)=CC(SC)(C)C
Propionaldehyde, 2-methyl-2-(methylthio)-, oxime
Aldicarb oxime
2-Methyl-2-(methylthio)-propanal oxime
Propanal, 2-methyl-2-(methylthio)-, oxime
Temik Oxime

1646-87-3
$C_7H_{14}N_2O_3S$
206.29
CNC(=O)ON=CC(C)(C)S(C)=O
Propionaldehyde, 2-methyl-2-(methylsulfinyl)-, O-(methylcarbamoyl)oxime
Aldicarb sulfoxide
2-Methyl-2-(methylsulfinyl)-propanal O-((methylamino)-carbonyl)oxime
2-Methyl-2-(methylsulfinyl)-propionaldehyde O-(methylcarbamoyl)oxime
Propanal, 2-methyl-2-(methylsulfinyl)-, O-((methylamino)-carbonyl)oxime
Temik Sulfoxide

1646-88-4
$C_7H_{14}N_2O_4S$
222.29
CNC(=O)ON=CC(C)(C)S(C)(=O)=O
Propionaldehyde, 2-methyl-2-(methylsulfonyl)-, O-(methylcarbamoyl)oxime
Aldicarb sulfone
Aldoxycarb
Aldoxycarbe (French)
ENT AI3-29,261
ENT 4.9
2-Mesyl-2-methylpropionaldehyde O-methylcarbamoyloxime
2-Methyl-2-(methylsulfonyl)propanal O-((methylamino)carbonyl)oxime
2-Methyl-2-(methylsulfonyl)propionaldehyde O-(methylcarbamoyl)oxime
Standak
Sulfocarb
UC-21865

1647-16-1
$C_{10}H_{18}$
138.25
C(=C)CCCCCCC=C
1,9-Decadiene (9CI)

1647-26-3
$C_8H_{15}Br$
191.11
Cyclohexane, (2-bromoethyl)-(9CI)
NSC-6078
NSC-46808

1649-08-7
$C_2H_2Cl_2F_2$
134.94
FC(F)(Cl)CCl
Ethane, 1,2-dichloro-1,1-difluoro
1,2-Dichloro-1,1-difluoroethane
HCFC-132b

1650-76-6
$C_{15}H_{12}FN_5$
281.29
s-Triazine, 2,4-dianilino-6-fluoro- (6CI,8CI)

1652-36-4
$C_3H_2F_2N_4$
132.07
1,3,5-Triazin-2-amine, 4,6-difluoro- (9CI)
2-Amino-4,6-difluoro triazine
2-Amino-4,6-difluoro-1,3,5-triazine
s-Triazine, 2-amino-4,6-difluoro- (7CI,8CI)

1652-63-7
$C_{14}H_{16}F_{17}N_2O_2S.I$
726.22
1-Propanaminium, 3-(((heptadecafluorooctyl)sulfonyl)amino)-N,N,N-trimethyl-, iodide (9CI)
3-(((Heptadecafluorooctyl)-sulfonyl)amino)-N,N,N-trimethyl-1-propanaminium iodide

1653-19-6
$C_4H_4Cl_2$
122.98
C(C(=C)Cl)(=C)Cl
1,3-Butadiene, 2,3-dichloro
2,3-Dichlor-1,3-butadien (Czech)
2,3-Dichloro-1,3-butadiene

1653-30-1
$C_{11}H_{24}O$
172.31
OC(CCCCCCCCC)C
2-Undecanol
AI3-35680

1653-31-2
$C_{13}H_{28}O$
200.36
2-Tridecanol (8CI,9CI)
NSC-9499

1653-40-3
$C_8H_{18}O$
130.23
1-Heptanol, 6-methyl- (8CI, 9CI)

1654-86-0
$C_{20}H_{40}O_2$
312.54
Decanoic acid, decyl ester (9CI)
Decyl decanoate

1655-35-2
$C_{10}H_8O_6S_2.2Na$
334.28
2,7-Naphthalenedisulfonic acid, disodium salt (9CI)

1656-48-0
$C_6H_8N_2O$
124.16
N#CCCOCCC#N
Propionitrile, 3,3'-oxydi
Ether, bis(2-cyanoethyl)
β,β'-Dicyanodiethyl ether
2,2'-Oxydiethankarbonitril (Czech)
β,β'-Oxydipropionitrile
3,3'-Oxydipropionitrile
Propanenitrile, 3,3'-oxybis-

1656-63-9
$C_{36}H_{75}PS_3$
635.16
S(P(SCCCCCCCCCCCC)SCCCCCCCCCCCC)CCCCCCCCC

CCC
Phosphorotrithious acid, tridodecyl ester (9CI)

1658-56-6
$C_{20}H_{14}N_2O_4S.Na$
401.41
1-Naphthalenesulfonic acid, 4-((2-hydroxy-1-naphthalenyl)azo)-, monosodium salt
Acid Cardinal G
Acid Leather Red ROC
Acid Red 88
Acid Red AV
Acid Red G
Acid Rose AV
Airedale Red A
Amacid Fast Red A
Anthrosin BRX
Atul Acid Fast Red A
Azo Acid Red GS
Benzyl Red ROC
Benzyl Red S
Brasilan Red S
Bucacid Fast Red A
Calcocid Fast Red A
C.I. 15620
C.I. Acid Red 88
Colacid Red AV
Dai-Ei Roccelline
Diacid Red A
Drimarene Turquoise X-2G
Eniacid Fast Red A
Eriosin Roccelline
Eriosin Roccelline SS
Ext. D & C Red No. 8
Fast Acid Red G
Fast Red A
Fast Red AE
Fast Red AG
Fast Red ALS
Fast Red AV
Fast Red MA
Fast Red S
Fenazo Red M
Hidacid Fast Red A
Hispacid Fast Red A
4-((2-Hydroxy-1-naphthalenyl)-azo)1-naphthalenesulfonic acid sodium salt
Kayaku Roccelline
Lurazol Red E
Naphthalene Red J (6CI)
Naphthalene Red JS
2-Naphthol Red J
Naphtocard Fast Red C
Neklacid Fast Red A
Neutral Red R
New Red WO
Nitto Roccelline
Peony
Plastoresin Red RC
Pontacyl Fast Red AS
11391 Red
Red No. 506
Red J
Rocceline
Roccelline
Roccelline A
Roccelline G
Roccelline K
Roccelline KG
Roccelline L
Roccelline NS
Roccelline S
Shikiso Roccelline
Solid Red A
Tertracid Red RO
Toyo Roccelline
Vondacid Red GN

1663-35-0
$C_5H_{10}O_2$
102.15
COCCOC=C
Ethane, 1-methoxy-2-(vinyloxy)
Ether, 2-methoxyethyl vinyl
Ethylene glykol methyl vinyl ether
2-Methoxyethyl vinyl ether
Methylvinylether ethylenglykolu (Czech)
Vinyl 2-methoxyethyl ether

1663-39-4
$C_7H_{12}O_2$
128.17
O=C(OC(C)(C)C)C=C
2-Propenoic acid, 1,1-dimethylethyl ester (9CI)
Acrylic acid, tert-butyl ester (8CI)
tert-Butyl acrylate
tert-Butyl propenoate
1,1-Dimethylethyl 2-propenoate

1665-48-1
$C_{12}H_{15}NO_3$
221.28
Cc2cc(C)cc(OCC1CNC(=O)O1)c2
2-Oxazolidinone, 5-((3,5-xylyloxy)methyl)
AHR-438
CL 39,148
5-((3,5-Dimethylphenoxy)-methyl)-2-oxazolidinone
Metaxalone
Methaxalonum
Methoxolone
Metazalone
Metazolone
Skelaxin
Zorane
5-(3,5-Xyloloxymethyl)oxazolidin-2-one
5-((3,5-Xylyloxy)methyl)-2-oxazolidinone

1667-01-2
$C_{11}H_{14}O$
162.23
O=C(c(c(cc(c1)C)C)c1C)C
Acetophenone, 2',4',6'-trimethyl- (8CI)
AI3-11164
Ethanone, 1-(2,4,6-trimethylphenyl)- (9CI)
1-(2,4,6-Trimethylphenyl)ethanone

1667-04-5
$C_9H_{11}Cl$
154.64
Benzene, 2-chloro-1,3,5-trimethyl- (9CI)
Mesitylene, 2-chloro- (8CI)
NSC-139128

1668-19-5
$C_{19}H_{21}NO$
279.41
CN(C)CCC=C2c1ccccc1COc3ccccc23
Dibenz(b,e)oxepin-$\delta^{11(6H)}$,γ-propylamine, N,N-dimethyl
11-(3-Dimethylaminopropylidene)-6,11-dihydrodibenz-(b,e)oxipin
Doxepin

1668-54-8
$C_5H_8N_4O$
140.17
s-Triazine, 2-amino-4-methoxy-6-methyl
CV 399
2-Methyl-4-amino-6-methoxy-s-triazine
1,3,5-Triazin-2-amine, 4-methoxy-6-methyl- (9CI)

1668-99-1
$C_{20}H_{24}N_2O_2.ClH$
360.92
Quinidine, monohydrochloride
Chinidin hydrochlorid (German)
Cinchonan-9-ol, 6'-methoxy-, monohydrochloride, (9S)-(9CI)
Quinidine hydrochloride

1669-44-9
$C_8H_{14}O$
126.20
O=C(C=CCCCC)C
3-Octen-2-one (9CI)

1671-82-5
$C_4H_5N_3O_2$
127.12
Cn1ccnc1N(=O)=O
Imidazole, 1-methyl-2-nitro
1-Methyl-2-nitroimidazole

1674-10-8
C_8H_{14}
110.20
Cyclohexene, 1,2-dimethyl- (9CI)

1674-33-5
$C_5H_{10}Cl_2$
141.04
Pentane, 1,2-dichloro- (7CI,8CI,9CI)
1,2-Dichloropentane

1675-54-3
$C_{21}H_{24}O_4$
340.45
O(C1COc(ccc(c2)C(c(ccc(OCC(O3)C3)c4)c4)(C)C)c2)C1
Propane, 2,2-bis(p-(2,3-epoxypropoxy)phenyl)
2,2-Bis(4-(2,3-epoxypropyloxy)-phenyl)propane
Bis(4-glycidyloxyphenyl)dimethyamethane
2,2-Bis(p-glycidyloxyphenyl)-propane
2,2-Bis(4-hydroxyphenyl)propane, diglycidyl ether
Bis(4-hydroxyphenyl)dimethylmethane diglycidyl ether
2,2-Bis(p-hydroxyphenyl)propane, diglycidyl ether
Bisphenol A diglycidyl ether
D.E.R. 332
Dian-bis-glycidylether (Czech)
Diglycidyl bisphenol A ether
Diglycidyl ether of 2,2-bis-(p-hydroxyphenyl)propane
Diglycidyl ether of 2,2-bis-(4-hydroxyphenyl)propane
Diglycidyl ether of bisphenol A
Diglycidyl ether of 4,4'-isopropylidenediphenol
4,4'-Dihydroxydiphenyldimethylmethane diglycidyl ether
p,p'-Dihydroxydiphenyldimethylmethane diglycidyl ether
Epi-Rez 508
Epi-Rez 510
Epon 828
Epoxide A
ERL-2774
4,4'-Isopropylidenediphenol diglycidyl ether
2,2'-((1-Methylethylidene)bis-(4,1-phenyleneoxymethylene))bisoxirane

1678-82-6
$C_{10}H_{20}$
140.27
C(CCC(C1)C(C)C)(C1)C
Cyclohexane, 1-methyl-4-(1-methylethyl)-, trans- (9CI)
p-Menthane, trans- (8CI)
trans-1-Methyl-4-(1-methylethyl)cyclohexane

1678-91-7
C_8H_{16}
112.22
Ethylcyclohexane
AI3-15348
Cyclohexane, ethyl- (9CI)
Ethyl cyclohexane

1678-92-8
C_9H_{18}
126.24
Cyclohexane, propyl- (9CI)

1678-93-9
$C_{10}H_{20}$
140.27
C(CCCC1)(C1)CCCC
Cyclohexane, butyl- (9CI)
Butylcyclohexane
1-Cyclohexylbutane

1678-97-3
C_9H_{18}
126.24
Cyclohexane, 1,2,3-trimethyl- (8CI,9CI)

1678-98-4
$C_{10}H_{20}$
140.27
Cyclohexane, (2-methylpropyl)- (9CI)
Cyclohexane, isobutyl- (8CI)
NSC-74187

1679-02-3
$C_{24}H_{18}$
306.41
Picene, 2,9-dimethyl- (8CI,9CI)
NSC-409492

1679-07-8
$C_5H_{10}S$
102.21
SC(CCC1)C1
Cyclopentanethiol
Cyclopentyl mercaptan
Mercaptocyclopentane

1679-09-0
$C_5H_{12}S$
104.22
SC(CC)(C)C
2-Butanethiol, 2-methyl- (9CI)
2-Methyl-2-butanethiol

1679-51-2
$C_7H_{12}O$
112.17
OCC(CCC=C1)C1
3-Cyclohexene-1-methanol (9CI)
AI3-24787

1679-64-7
$C_9H_8O_4$
180.16
O=C(OC)c(ccc(c1)C(=O)O)c1
Hydrogen methyl terephthalate
Acetic acid, (4-carboxyphenyl)-
1,4-Benzenedicarboxylic acid, monomethyl ester (9CI)
4-(Carbomethoxy)benzoic acid
p-Carboxy-α-toluic acid
Homoterephthalic acid
Methyl hydrogen terephthalate
4-Methoxycarbonylbenzoic acid
Methyl terephthalate
Monomethyl 1,4-benzenedicarboxylate
Monomethyl terephthalate
Terephthalic acid, monomethyl ester (8CI)

1680-21-3
$C_{12}H_{18}O_6$
258.30
O=C(OCCOCCOCCOC(=O)C=C)C=C
Acrylic acid, diester with triethylene glycol
2-Propenoic acid, 1,2-ethanediylbis(oxy-2,1-ethanediyl) ester (9CI)
Triethylene glycol diacrylate

1680-51-9
$C_{11}H_{14}$
146.23
Naphthalene, 1,2,3,4-tetra hydro-6-methyl- (6CI, 7CI,8CI,9CI)
2-Methyl-5,6,7,8-tetrahydronaphthalene
6-Methyl-1,2,3,4-tetrahydronaphthalene
6-Methyltetralin
1,2,3,4-Tetrahydro-6-methylnaphthalene

1685-82-1
$C_{11}H_{14}$
146.25
Indan, 4,6-dimethyl
4,6-Dimethylindan

1686-14-2
$C_{10}H_{16}O$
152.24
O(C1(C(CC2C3)C2(C)C)C)C13
2-Pinene oxide
2,3-Epoxypinane
3-Oxatricyclo(4.1.1.02,4)octane, 2,7,7-trimethyl- (9CI)
α-Pinene epoxide
Pinane, 2,3-epoxy- (8CI)
α-Pinene oxide
2,7,7-Trimethyl-3-oxatricyclo-(4.1.1.02,4)octane

1686-62-0
$C_{21}H_{32}O_2$
316.48
1-Phenanthrenecarboxylic acid,

7-ethenyl-1,2,3,4,4a,4b,5,6,7,8,10,10a-dodecahydro-1,4a,7-trimethyl-, methyl ester, (1R-(1α,4aβ,4bα,7α,10aα))- (9CI)
Podocarp-7-en-15-oic acid, 13β-methyl-13-vinyl-, methyl ester (8CI)

1687-30-5
$C_8H_{12}O_4$
172.18
O=C(O)C(C(C(=O)O)CCC1)C1
1,2-Cyclohexanedicarboxylic acid (9CI)
Hexahydrophthalic acid

1689-78-7
$C_8H_{12}S$
140.25
Thiophene, 2-(1,1-dimethylethyl)-
AI3-15885

1689-82-3
$C_{12}H_{10}N_2O$
198.24
Oc(ccc(N=Nc(cccc1)c1)c2)c2
Phenol, p-(phenylazo)
Atul Brilliant Oil Yellow G
p-Benzeneazophenol
Brasilazina Oil Yellow O
C.I. 11800
C.I. Solvent Yellow 7
Fast Oil Yellow 2G
p-Hydroxyazobenzene
4-Hydroxyazobenzene
Organol Yellow AP
p-Phenylazophenol
4-Phenylazophenol
Pirocard Green 491
Zlut Rozpoustedlova 7 (Czech)

1689-83-4
$C_7H_3I_2NO$
370.91
Oc1c(I)cc(C#N)cc1I
Benzonitrile, 3,5-diiodo-4-hydroxy
ACP 63303
Actril
Bantrol
Bentrol
Benzonitrile, 4-hydroxy-3,5-diiodo-
Certol
Certrol
4-Cyano-2,6-diiodophenol
4-Cyano-2,6-dijodphenol (German)
3,5-Diiodo-4-hydroxybenzonitrile
3,5-Dijod-4-hydroxy-benzonitril (German)
4-Hydroxy-3,5-diiodobenzonitrile
Iotox
Ioxynil
Loxynil (German)
M&B 8873
Mylone
Oxytril
Totril

1689-84-5
$C_7H_3Br_2NO$
276.93
Oc1c(Br)cc(C#N)cc1Br
Benzonitrile, 3,5-dibromo-4-hydroxy
Brittox
Brominal
Brominex
Brominil
Bromoxynil
Broxynil
Bucril
Buctril
Buctril Industrial
Butilchlorofos
Chipco Buctril
Chipco Crab-Kleen
2,6-Dibromo-4-cyanophenol
3,5-Dibromo-4-hydroxybenzonitrile
3,5-Dibromo-4-hydroxyphenylcyanide
ENT 20,852
Harness
4-Hydroxy-3,5-dibromobenzonitrile
MB 10064
M&B 10,064
M&B 10731
ME4 Brominal
Nu-Lawn Weeder
Oxytril M
Torch

1689-86-7
$C_7H_3BrClNO$
232.46
Benzonitrile, 3-bromo-5-chloro-4-hydroxy- (7CI,8CI,9CI)

1689-99-2
$C_{15}H_{17}Br_2NO_2$
403.15
O=C(Oc(c(cc(C#N)c1)Br)c1Br)CCCCCCC
Benzonitrile, 3,5-dibromo-4-octanoyloxy
Bromoxynil octanoate
Bronate
Buctril
2,6-Dibromo-4-cyanophenyl octanoate
3,5-Dibromo-4-octanoyloxybenzonitrile
M&B 10731
RP-16272

1691-13-0
$C_2H_2F_2$
64.03
Ethene, 1,2-difluoro- (9CI)
Ethylene, 1,2-difluoro- (8CI)

1691-99-2
$C_{12}H_{10}F_{17}NO_3S$
571.25
O=S(=O)(N(CCO)CC)C(F)(F)C(F)(F)C(F)(F)C(F)(F)C(F)(F)C(F)(F)C(F)(F)C(F)(F)F
1-Octanesulfonamide, N-ethyl-1,1,2,2,3,3,4,4,5,5,6,6,7,7,8,8,8-heptadecafluoro-N-(2-hydroxyethyl)- (9CI)
N-Ethyl-1,1,2,2,3,3,4,4,5,5,6,6,7,7,8,8,8-heptadecafluoro-N-(2-hydroxyethyl)-1-octanesulfonamide

1694-09-3
$C_{39}H_{41}N_3O_6S_2.Na$
734.94
[Na+].CCN(Cc1cccc(c1)S([O-])(=O)=O)c2ccc(cc2)C(=C3C=CC(C=C3)=[N+](CC)Cc4cccc(c4)S([O-])(=O)=O)c5ccc(cc5)N(C)C
Ammonium, (4-(p-(dimethylamino)-α-(p-(ethyl(m-sulfobenzyl)amino)phenyl)benzylidene)- 2,5-cyclohexadien-1-ylidene)ethyl(m-sulfobenzyl)-, hydroxide, inner salt, sodium salt
Acid Fast Violet 5BN
Acid Violet
Acid Violet 49
Acid Violet 4BNS
Acid Violet 5B
Acid Violet 6B
Acid Violet 5BN
Acid Violet S
Acilan Violet S4BN
A.F. Violet No. 1
Aizen Acid Violet 5BH
Aizen Food Violet No. 1
Benzyl Violet
Benzyl Violet 3B
Benzyl Violet 4B
Calcocid Violet 4BNS
C.I. 42640
C.I. Acid Violet 49
C.I. Acid Violet 49 (Sodium salt)
C.I. Food Violet 2
Cogilor Violet 411.12
Coomassie Violet
D & C Violet No. 1
Dispersed Violet 12197
Eriosin Violet 3B
Fast Acid Violet 5BN
FD & C Violet 1
FD & C Violet No. 1
Food Violet 2
Formyl Violet S4BN
Hidacid Wool Violet 5B
Intracid Violet 4BNS
Kiton Violet 4BNS
Pergacid Violet 2B
Pergacid Violet 3B
Polaxal Violet 6B
Solar Violet 5BN

Tetracid Brilliant Violet 6B
Violet 2
11386 Violet
Violet 6B
Violet 5B
Violet 5BN
Violet Kysela 49 (Czech)
Violet No. 1
Violet Potravinarska 2 (Czech)
Wool Violet
Wool Violet 4BN
Wool Violet 5BN

1694-31-1
$C_8H_{14}O_3$
158.20
Butanoic acid, 3-oxo-, 1,1-dimethylethyl ester (9CI)
Acetoacetic acid, tert-butyl ester (8CI)
NSC-42869

1694-92-4
$C_6H_4ClNO_4S$
221.61
O=S(=O)(c(c(N(=O)=O)ccc1)c1)Cl
Benzenesulfonyl chloride, o-nitro- (8CI)
Benzenesulfonyl chloride, 2-nitro- (9CI)
2-Nitrobenzenesulfonyl chloride

1696-17-9
$C_{11}H_{15}NO$
177.27
O=C(N(CC)CC)c(cccc1)c1
Benzamide, N,N-diethyl
Benzoic acid diethylamide
Benzoic acid N,N-diethylamide
Benzoyldiethylamine
N,N-Diethylbenzamide
R 2
Rebemid
REP

1696-20-4
$C_6H_{11}NO_2$
129.18
O=C(N(CCOC1)C1)C
Morpholine, 4-acetyl
N-Acetylmorfolin (Czech)
N-Acetylmorpholine
4-Acetylmorpholine

1698-60-8
$C_{10}H_8ClN_3O$
221.66
O=C(N(N=CC=1N)c(cccc2)c2)C1Cl
3(2H)-Pyridazinone, 5-amino-4-chloro-2-phenyl
5-Amino-4-chloro-2,3-dihydro-3-oxo-2-phenylpyridazine
5-Amino-4-chloro-2-phenyl-pyridazin-3(2H)-one
5-Amino-4-chloro-2-phenyl-3(2H)-pyridazinone
Burex (Czech)
Chloridazon
Chloridazone
1-Fenyl-4-amino-5-chlor-6-pyridazinon (Czech)
H 119
HS-119-1
PCA
Phenazon
Phenosane
1-Phenyl-4-amino-5-chloropyridazin-6-one
1-Phenyl-4-amino-5-chloropyridaz-6-one
1-Phenyl-4-amino-5-chloro-6-pyridazone
1-Phenyl-4-amino-5-chloropyridazone-6
1-Phenyl-4-amino-5-chlorpyridazon-(6) (German)
1-Phenyl-4-amino-5-chlorpyridaz-6-one
Pyramin
Pyramine
Pyramin RB
Pyrazon
Pyrazone
Pyrazonl

1700-02-3
$C_9H_5Cl_2N_3$
226.05
1,3,5-Triazine, 2,4-dichloro-6-phenyl- (9CI)
NSC-51871
s-Triazine, 2,4-dichloro-6-phenyl- (8CI)

1700-10-3
C_8H_{12}
108.18
C(=CC=CCCC1)C1
1,3-Cyclooctadiene (9CI)
AI3-26696

1701-68-4
$C_{10}H_{15}N$
149.23
Benzenamine, 3,5-diethyl- (9CI)
Aniline, 3,5-diethyl- (8CI)

1702-17-6
$C_6H_3Cl_2NO_2$
192.00
OC(=O)c1nc(Cl)ccc1Cl
Picolinic acid, 3,6-dichloro
Clopyralid
3,6-Dichloropicolinic acid
3,6-Dichloro-2-pyridinecarboxylic acid
DOWCO 290
Kyselina 3,6-dichlorpikolinova (Czech)
Lontrel
Lontrel 3
Matrigon
2-Pyridinecarboxylic acid, 3,6-dichloro- (9CI)
XRM 3972

1703-51-1
$C_7H_{12}O_2$
128.17
2,5-Heptanedione

1703-58-8
$C_8H_{10}O_8$
234.18
O=C(O)C(C(C(=O)O)CC(=O)O)CC(=O)O
1,2,3,4-Butanetetracarboxylic acid
Butanetetracarboxylic acid

1704-62-7
$C_6H_{15}NO_2$
133.22
O(CCN(C)C)CCO
Ethanol, 2-(2-dimethylaminoethoxy)
2-(2-Dimethylaminoethoxy)-ethanol

1705-85-7
$C_{19}H_{14}$
242.33
Cc3cc2c1ccccc1ccc2c4ccccc34
Chrysene, 6-methyl
6-Methylchrysene

1706-01-0
$C_{17}H_{12}$
216.29
Fluoranthene, 3-methyl
3-Methylfluoranthene

1707-67-1
$C_{21}H_{15}ClN_2$
330.80
N(C(c(cccc1)c1)=C(N2)c(cccc3)c3)=C2c(c(ccc4)Cl)c4
1H-Imidazole, 2-(2-chlorophenyl)-4,5-diphenyl- (9CI)
2-(o-Chlorophenyl)-4,5-diphenylimidazole
2-(2-Chlorophenyl)-4,5-diphenyl-1H-imidazole

1707-68-2
$C_{42}H_{28}Cl_2N_4$
659.59
N(C(c(cccc1)c1)=C(N2N(C(c(cccc3)c3)=C(N=4)c(cccc5)c5)C4c(c(ccc6)Cl)c6)c(cccc7)c7)=C2c(c(ccc8)Cl)c8
1,1'-Bi-1H-imidazole, 2,2'-bis-(2-chlorophenyl)-4,4',5,5'-tetraphenyl- (9CI)
1,1'-Biimidazole, 2,2'-bis-(o-chlorophenyl)-4,4',5,5'-tetraphenyl-

1708-29-8
C_4H_6O
70.09
O(CC=C1)C1
Furan, 2,5-dihydro- (9CI)
2,5-Dihydrofuran

1708-82-3
$C_8H_{14}O_2$
142.20
O=C(OCCC=CCC)C
3-Hexen-1-ol, acetate (9CI)

1709-70-2
$C_{54}H_{78}O_3$
775.32
Benzene, 1,3,5-trimethyl-2,4,6-tris(3,5-di-t-butyl-4-hydroxybenzyl)
Ahydol (Russian)
Antioxidant 330
AO-40
Santoquin emulsion
Santoquin mixture 6
Ethanox 330
1,3,5-Trimethyl-2,4,6-tris-(3,5-di-tert-butyl-4-hydroxy-benzyl)benzene

1712-64-7
$C_3H_7NO_3$
105.11
O=N(=O)OC(C)C
Nitric acid, isopropyl ester
Isopropylester kyseliny dusicne (Czech)
Isopropyl nitrate [UN 1222]
UN 1222 [Isopropyl nitrate]

1713-11-7
$C_{13}H_{17}ClO_3$
256.73
Isobutyl 2-methyl-4-chloro-phenoxyacetate

1713-12-8
$C_{13}H_{17}ClO_3$
256.75
CCCCOC(=O)COc1ccc(Cl)cc1C
Acetic acid, ((4-chloro-o-tolyl)oxy)-, butyl ester
Acetic acid, (4-chloro-2-methyl-phenoxy)-, butyl ester
Butyl 2-methyl-4-chlorophen-oxyacetate
MCPA Butyl
MCP Butyl ester
2-Methyl-4-chlorophenoxy-acetic acid n-butyl ester
Yamaclean M

1713-14-0
$C_{14}H_{19}ClO_3$
270.76
Propanoic acid, 2-(4-chloro-2-methylphenoxy)-, butyl ester (9CI)
Propionic acid, 2-((4-chloro-o-tolyl)oxy)-, butyl ester (8CI)

1713-15-1
$C_{12}H_{14}Cl_2O_3$
277.16
Acetic acid, (2,4-dichloro-phenoxy)-, isobutyl ester
Acetic acid, (2,4-dichloro-phenoxy)-, 2-methylpropyl ester (9CI)
(2,4-Dichlorophenoxy)acetic acid isobutyl ester
2,4-D Isobutyl ester
Isobutyl 2,4-D

1719-53-5
$C_4H_{10}Cl_2Si$
157.13
Silane, dichlorodiethyl
Diethyldichlorosilane [UN 1767]
UN 1767 [Diethyldichloro-silane]

1719-58-0
C_4H_9ClSi
94.65
Silane, chloroethenyldimethyl- (9CI)
AI3-25204
Chlorodimethylvinylsilane
Chloroethenyldimethylsilane
Dimethylchlorovinylsilane
Dimethylvinylchlorosilane
Dimethylvinylsilyl chloride
Silane, chlorodimethylvinyl-
Vinylchlorodimethylsilane
Vinyldimethylchlorosilane

1720-11-2
$C_{33}H_{68}$
464.90
Dotriacontane, 2-methyl- (8CI, 9CI)

1721-89-7
$C_{11}H_{11}N$
157.21
Quinoline, 2,3-dimethyl- (8CI, 9CI)

1721-93-3
$C_{10}H_9N$
143.18
Isoquinoline, 1-methyl- (9CI)

1722-15-2
$C_5H_7N_3$
109.13
1,3,5-Triazine, 2,4-di methyl- (9CI)
Dimethyl-s-triazine
2,4-Dimethyl-1,3,5-triazine
s-Triazine, 2,4-dimethyl- (6CI,7CI,8CI)

1722-18-5
$C_9H_7N_3$
157.18
c1ccc(cc1)c2ncncn2
1,3,5-Triazine, 2-phenyl- (9CI)
Phenyl-1,3,5-triazine
2-Phenyl-s-triazine
2-Phenyl-1,3,5-triazine
s-Triazine, 2-phenyl- (6CI, 7CI,8CI)

1724-39-6
$C_{12}H_{24}O$
184.32
OC(CCCCCCCCCC1)C1
Cyclododecanol (9CI)

1725-04-8
$C_{13}H_{24}O_2$
212.33
Oxacyclotetradecan-2-one (9CI)

1725-74-2
$C_6H_6Cl_6$
290.83
3-Hexene, 1,2,3,4,5,6-hexa-chloro- (6CI,7CI,9CI)
1,2,3,4,5,6-Hexachloro-3-hexene

1726-65-4
$C_{48}H_{99}Al$
703.30
Aluminum, trihexadecyl- (8CI,9CI)
Tri(hexadecyl) aluminum
Trihexadecylaluminum

1726-66-5
$C_{30}H_{63}Al$
450.81
Tri-n-decylaluminum
Aluminum, tris(decyl)- (9CI)
Tridecylaluminum
Tri-1-decylaluminum
Tris(decyl)aluminum
Tris(n-decyl)aluminum

1730-37-6
$C_{14}H_{12}$
180.25
1-Methylfluorene
Fluorene, 1-methyl- (8CI)
9H-Fluorene, 1-methyl- (9CI)

1731-81-3
$C_{13}H_{26}O_2$

214.35
1-Undecanol, acetate (9CI)
NSC-23056
Undecyl alcohol, acetate (8CI)

1731-84-6
$C_{10}H_{20}O_2$
172.27
O=C(OC)CCCCCCCC
Methyl pelargonate
AI3-28570
Methyl nonanoate
Nonanoic acid, methyl ester

1731-86-8
$C_{12}H_{24}O_2$
200.32
O=C(OC)CCCCCCCCCC
Undecanoic acid, methyl ester (9CI)
Methyl undecanoate

1731-88-0
$C_{14}H_{28}O_2$
228.38
O=C(OC)CCCCCCCCCCCC
Tridecanoic acid, methyl ester (9CI)
Methyl tridecanoate

1731-92-6
$C_{18}H_{36}O_2$
284.48
O=C(OC)CCCCCCCCCCCC CCCC
Heptadecanoic acid, methyl ester (9CI)
AI3-36453
Methyl heptadecanoate

1731-94-8
$C_{20}H_{40}O_2$
312.54
Nonadecanoic acid, methyl ester
AI3-36454

1732-09-8
$C_{10}H_{18}O_4$
202.25
O=C(OC)CCCCCCC(=O)OC
Octanedioic acid, dimethyl ester (9CI)
Dimethyl octanedioate
Dimethyl suberate

1732-10-1
$C_{11}H_{20}O_4$
216.28
O=C(OC)CCCCCCCC(=O)OC
Azelaic acid, dimethyl ester (8CI)
AI3-06080
Dimethyl nonanedioate
Nonanedioic acid, dimethyl ester (9CI)

1733-96-6
$C_{25}H_{40}N.Cl$
390.05
Alkyl dimethyl 1-napthyl-methyl ammonium chloride
N-Dodecyl-N,N-dimethyl-N-(1-naphthylmethyl)ammonium chloride
1-Naphthalenemethanaminium, N-dodecyl-N,N-dimethyl-, chloride (9CI)

1738-25-6
$C_5H_{10}N_2$
98.17
N#CCCN(C)C
Propionitrile, 3-(dimethyl-amino)
3-Dimethylaminopropannitril (Czech)
β-Dimethylaminopropionitrile
3-(Dimethylamino)propionitrile

1739-84-0
$C_5H_8N_2$
96.12
N(C=CN1C)=C1C
1,2-Dimethylimidazole
1,2-Dimethyl-1H-imidazole
Imidazole, 1,2-dimethyl- (8CI)
1H-Imidazole, 1,2-dimethyl- (9CI)

1740-19-8
$C_{21}H_{30}O_2$
314.51
O=C(O)C(C(C(c(c(cc(c1)C(C)C) C2)c1)(CC3)C)C2)(C3)C
Podocarpa-8,11,13-trien-15-oic acid, 13-isopropyl
Dehydroabietic acid
DHA
13-Isopropylpodocarpa-8,11,13-trien-15-oic acid

1741-41-9
$C_8H_{18}O_2$
146.23
O(C(OCC)C(C)C)CC
Propane, 1,1-diethoxy-2-methyl-
AI3-28228
1,1-Diethoxy-2-methylpropane
Isobutanal diethyl acetal

1741-83-9
$C_6H_{14}S$
118.24
Pentane, 1-(methylthio)-(9CI)
Amyl methyl sulfide
Methyl pentyl sulfide
Pentyl methyl sulfide
Sulfide, methyl pentyl (6CI, 7CI,8CI)
2-Thiaheptane

1745-81-9
$C_9H_{10}O$
134.19
Oc(c(ccc1)CC=C)c1
Phenol, o-allyl
2-Allylphenol

1746-01-6
$C_{12}H_4Cl_4O_2$
321.96
Clc3cc2Oc1cc(Cl)c(Cl)cc1Oc2 cc3Cl
Dibenzo-p-dioxin, 2,3,7,8-tetrachloro
2,3,7,8-Czterochlorodwubenzo-p-dwuoksyny (Polish)
Dibenzo(b,e)(1,4)dioxin, 2,3,7,8-tetrachloro-
Dioksyny (Polish)
Dioxine
Dioxin (herbicide contaminant)
NCI-C03714
TCDBD
TCDD
2,3,7,8-TCDD
2,3,7,8-Tetrachlorodibenzo(b,e)-(1,4)dioxan
2,3,6,7-Tetrachlorodibenzo-p-dioxin
2,3,7,8-Tetrachlorodibenzo-p-dioxin
2,3,7,8-Tetrachlorodibenzo-1,4-dioxin
Tetradioxin

1746-03-8
$C_2H_5O_3P$
108.03
O=P(O)(O)C=C
Phosphonic acid, ethenyl-(9CI)
Ethenylphosphonic acid

1746-23-2
$C_{12}H_{16}$
160.26
c(ccc(c1)C=C)(c1)C(C)(C)C
Benzene, 1-(1,1-dimethyl-ethyl)-4-ethenyl- (9CI)
1-(1,1-Dimethylethyl)-4-ethenyl-benzene

1746-77-6
$C_4H_9NO_2$
103.14
O=C(OC(C)C)N
Carbamic acid, isopropyl ester
Carbamic acid, 1-methylethyl ester
Isopropyl carbamate
Isopropylester kyseliny karbam-inove (Czech)

1746-81-2
$C_9H_{11}ClN_2O_2$
214.67
O=C(N(OC)C)Nc(ccc(c1)Cl)c1
Urea, 3-(p-chlorophenyl)-1-methoxy-1-methyl
Afesin
Aresin
Arezin
Arezine
Arresin
3-(4-Chlorphenyl)-1-methoxy-1-methylharnstoff (German)
N-(4-Chlorophenyl)-N'-methoxy-N-methylurea
3-(4-Chlorophenyl)-1-methoxy-1-methylurea
HOE 2747
Monolinuron
Premalin
Urea, N'-(4-chlorophenyl)-N-methoxy-N-methyl-

1747-60-0
$C_8H_8N_2OS$
180.24
O(c(ccc(NC(=N)S1)c12)c2)C
Benzothiazole, 2-amino-6-methoxy
6-Methoxy-2-aminobenzothiazole

1747-80-4
$C_{13}H_{10}FNO$
215.22
Benzamide, 2-fluoro-N-phenyl- (9CI)
Benzanilide, 2-fluoro- (8CI)

1752-30-3
$C_4H_9N_3S$
131.22
N(NC(=N)S)=C(C)C
Acetone, thiosemicarbazone
Acetonthiosemikarbazon (Czech)
Hydrazinecarbothioamide, 2-(1-methylethylidene)- (9CI)
Thiosemicarbazone acetone

1754-47-8
$C_{22}H_{39}O_3P$
382.58
O=P(OCCCCCCCC)(OCCCCCCCC)c(cccc1)c1
Phosphonic acid, phenyl-, dioctyl ester
Benzenephosphonic acid, dioctyl ester
Phenylphosphonic acid dioctyl ester

1754-58-1
$C_8H_{13}N_2O_2P$
200.20
Phosphorodiamidic acid, N,N'-dimethyl-, phenyl ester
Diamidafos
Diamidfos
DOWCO 169
Nellite
o-Phenyl-N,N'-dimethyl phosphorodiamidate

1757-18-2
$C_{12}H_{14}Cl_3O_3PS$
375.64
CCOP(=S)(OCC)OC(=CCl)c1cc(Cl)ccc1Cl
Phosphorothioic acid, O-(2-chloro-1-(2,5-dichlorophenyl)vinyl) O,O-diethyl ester
Akton
Axiom
O-2-Chlor-1-(2,5-dichlorfenyl)-vinyl-O,O-diethylthiofosfat (Czech)
O-(2-Chloro-1-(2,5-dichlorophenyl)vinyl) O,O-diethyl phosphorothioate
O,O-Diethyl O-(2-chloro-1-2,5-dichlorophenylvinyl) phosphorothioate
ENT 27,102
OMS-1344
Phosphorothioic acid, O-(2-chloro-1-(2,5-dichlorophenyl)ethenyl) O,O-diethyl ester
SD 9098
Shell SD-9098

1757-42-2
$C_6H_{10}O$
98.14
CC1CCC(=O)C1
Cyclopentanone, 3-methyl- (9CI)

1758-33-4
C_4H_8O
72.11
Oxirane, 2,3-dimethyl-, cis- (9CI)
Butane, 2,3-epoxy-, cis- (8CI)

1758-73-2
$CH_4N_2O_2S$
108.11
O=S(O)C(=N)N
Formamidine sulfinic acid
Aminoiminomethanesulfinic acid
Methanesulfinic acid, aminoimino- (9CI)

1758-88-9
$C_{10}H_{14}$
134.22
c(ccc(c1CC)C)(c1)C
Benzene, 2-ethyl-1,4-dimethyl- (9CI)
2-Ethyl-1,4-dimethylbenzene
p-Xylene, 2-ethyl- (8CI)

1759-28-0
C_6H_7NS
125.20
N(C(=C(S1)C=C)C)=C1
Thiazole, 4-methyl-5-vinyl
4-Methyl-5-vinyl thiazole

1759-53-1
$C_3H_6O_2$
74.09
O=C(O)C(C1)C1
Cyclopropanecarboxylic-acid

1759-58-6
C_7H_{14}
98.19
Cyclopentane, 1,3-dimethyl-, trans- (9CI)

1759-81-5
C_6H_{10}
82.15
Cyclopentene, 4-methyl- (8CI, 9CI)

1760-24-3
$C_8H_{22}N_2O_3Si$
222.41
COSi(CCCNCCN)(OC)OC
Ethylenediamine, N-(3-(trimethoxysilyl)propyl)
Silane, (3-(2-aminoethyl)aminopropyl)trimethoxy-
Silicone A-1120
N-(3-Trimethoxysilylpropyl)-ethylenediamine

1761-71-3
$C_{13}H_{26}N_2$
210.41
NC(CCC(C1)CC(CCC(N)C2)C2)C1
Cyclohexylamine, 4,4'-methylenebis
Bis(p-aminocyclohexyl)methane
Bis(4-aminocyclohexyl)methane
Cyclohexanamine, 4,4'-methylenebis- (9CI)
Di(p-aminocyclohexyl)methane
p,p'-Diaminodicyclohexylmethane
4,4'-Diaminodicyclohexylmethane
HLR 4219
HLR 4448
Methylenebis(4-aminocyclohexane)
4,4'-Methylenebis(cyclohexanamine)
4,4'-Methylenebis(cyclohexylamine)
4,4'-Methylenedicyclohexanamine
4,4'-Methylenedicyclohexane-

amine
4,4'-Methylenedicyclohexyl-amine
PACM 20
Wandamin HM

1762-26-1
$C_5H_{14}Pb$
281.37
CC[Pb](C)(C)C
Ethyltrimethyllead
Ethyltrimethylplumbane
Lead, ethyltrimethyl-
Plumbane, ethyltrimethyl- (9CI)
Trimethylethyllead

1762-27-2
$C_6H_{16}Pb$
295.39
CC[Pb](C)(C)CC
Diethyldimethyllead
Diethyldimethylplumbane
Dimethyldiethyllead
Lead, diethyldimethyl-
Plumbane, diethyldimethyl- (9CI)

1762-28-3
$C_7H_{18}Pb$
309.42
CC[Pb](C)(CC)CC
Methyltriethyllead
Lead, triethylmethyl-
Methyltriethylplumbane
Plumbane, triethylmethyl- (9CI)
Triethylmethyllead
Triethylmethylplumbane

1762-95-4
$CNS.H_4N$
76.13
Thiocyanic acid, ammonium salt
Ammonium rhodanate
Ammonium rhodanide
Ammonium sulfocyanate
Ammonium sulfocyanide
Ammoniumthiocyanate
Amthio
Rhodanid
Rhodanide
Trans-Aid
USAF EK-P-433
Weedazol TL

1768-24-7
C_5H_9NO
99.13
Cyanic acid, butyl ester (7CI,8CI,9CI)
Butyl cyanate

1768-31-6
C_3HCl_5O
230.29
FC(F)C(=O)C(F)(F)F
2-Propanone, 1,1,1,3,3-pentachloro
Pentachloroacetone
1,1,1,3,3-Pentachloropropanone

1771-07-9
$C_{12}H_{23}N_5O_3$
285.32
1,3,5-Triazine-2,4-diamine, 6-methoxy-N,N'-bis(3-methoxypropyl)- (9CI)
Caswell No. 549DD
EPA Pesticide Chemical Code 549300
Methometon
2-Methoxy-4,6-bis((3-methoxypropyl)amino)-s-triazine
s-Triazine, 2-methoxy-4,6-bis-((3-methoxypropyl)amino)- (8CI)

1772-25-4
$C_9H_{11}N_3$
161.19
N#CC(CCC#N)CCCC#N
1,3,6-Tricyanohexane
4-Cyanosuberonitrile
1,3,6-Hexanetricarbonitrile (9CI)

1773-37-1
$C_{10}H_{12}N_2O_6S$
288.28
Acetylasulam
Carbasulam
Caswell No. 573I
EPA Pesticide Chemical Code 573800

1777-33-9
$C_6H_{10}O$
98.15
4-Pentenal, 3-methyl- (6CI, 7CI,8CI,9CI)
3-Methyl-4-pentenal

1777-82-8
$C_7H_6Cl_2O$
177.03
OCc1ccc(Cl)cc1Cl
Benzyl alcohol, 2,4-dichloro
2,4-Dichlorobenzyl alcohol
Dybenal
Myacide SP

1777-84-0
$C_{10}H_{12}N_2O_4$
224.24
O=C(Nc(ccc(OCC)c1N(=O)=O)c1)C
p-Acetophenetidide, 3-nitro
4-Acetamino-2-nitrophenetole
p-Acetophenetide, 3'-nitro-
4-Ethoxy-3-nitroanilid kyseliny octove (Czech)
N-(4-Ethoxy-3-nitro)phenylacetamide
N-(4-Ethoxyphenyl)-3'-nitro-acetamide
NCI-C01978
2-Nitro-4-acetaminofenetol (Czech)
3-Nitro-p-acetophenetide
3'-Nitro-p-acetophenetide
5-Nitro-p-acetophenetidide
3'-Nitro-p-acetophenetidin

1779-19-7
$C_5H_{10}O_3$
118.13
1,3,6-Trioxocane (8CI,9CI)
Diethylene glycol formal
Diglycol formal
1,3,6-Trioxocin, tetrahydro-

1779-25-5
$C_8H_{18}AlCl$
176.69
Aluminum, chlorodiisobutyl
Alluminio diisobutil-mono-cloruro (Italian)
Aluminum, chlorobis(2-methyl-propyl)- (9CI)
Bis(isobutyl)aluminum chloride
Chlorodiisobutylaluminum
Diisobutylaluminum chloride
Diisobutylaluminum mono-chloride
Diisobutylchloroaluminum

1779-48-2
$C_6H_7O_2P$
142.09
O=P(O)c(cccc1)c1
Phosphinic acid, phenyl- (9CI)
AI3-15040
Phenylphosphinic acid

1779-81-3
$C_3H_6N_2S$
102.17
N=C(NCC1)S1
2-Thiazoline, 2-amino
2-Amino-2-thiazoline
2-Iminothiazolidine
2-Thiazolamine, 4,5-dihydro- (9CI)
2-Thiazolidinimine
USAF PD-57

1782-00-9
$C_6H_6Cl_4$
219.92
Cyclohexene, 3,4,5,6-tetra-chloro- (8CI,9CI)
Benzene tetrachloride

1783-84-2
$C_{20}H_{34}O_2$
306.49
8,11,14-Eicosatrienoic acid,

(Z,Z,Z)- (8CI,9CI)
Dihomo-γ-linolenic acid
Ro 12-1989

1785-02-0
$C_{21}H_{15}N_3O_3$
357.39
s-Triazine-2,4,6(1H,3H,5H)-trione, 1,3,5-triphenyl
Triphenyl isocyanurate
1,3,5-Triphenyl-s-triazine-2,4,6(1H,3H,5H)-trione

1785-51-9
$C_{16}H_8O_2$
232.24
O=C(c(c(c(c(c1)C=CC2=O)c2cc3)c3C=4)c1)C4
1,6-Pyrenedione
3,8-Pyrenedione
1,6-Pyrenequinone
3,8-Pyrenequinone

1786-12-5
$C_{20}H_{40}$
280.54
Cembrane
Cembrane I

1786-86-3
$C_8H_4Cl_4N_2O_2$
301.93
1,3-Benzenedicarboxamide, 2,4,5,6-tetrachloro- (9CI)
Isophthalamide, 2,4,5,6-tetrachloro-

1787-61-7
$C_{20}H_{13}N_3O_7S.Na$
462.38
Eriochrome Black T
C.I. Mordant Black 11
C.I. Mordant Black 11, Monosodium salt (8CI)
3-Hydroxy-4-((1-hydroxy-2-naphthalenyl)azo)-7-nitro-1-naphthalenesulfonic acid sodium salt
1-Naphthalenesulfonic acid, 3-hydroxy-4-((1-hydroxy-2-naphthalenyl)azo)-7-nitro-, monosodium salt (9CI)

1789-58-8
$C_2H_6Cl_2Si$
129.07
Silane, dichloroethyl
Ethyl dichlorosilane
Ethyldichlorosilane [UN 1183]
UN 1183 [Ethyldichlorosilane]

1790-22-3
$C_4H_7Cl_3$
161.46
Butane, 1,2,4-trichloro- (7CI,9CI)
1,2,4-Trichlorobutane

1795-04-6
$C_7H_{10}S$
126.22
Thiophene, 2,3,4-trimethyl- (8CI,9CI)

1795-09-1
$C_5H_{10}S$
102.20
Thiophene, tetrahydro-2-methyl- (8CI,9CI)

1795-15-9
$C_{14}H_{28}$
196.38
C(CCCC1)(C1)CCCCCCCC
Cyclohexane, octyl- (9CI)
Octane, 1-cyclohexyl- (8CI)
Octylcyclohexane

1795-16-0
$C_{16}H_{32}$
224.43
C(CCCC1)(C1)CCCCCCCCCC
Cyclohexane, decyl- (9CI)
1-Cyclohexyldecane
Decane, 1-cyclohexyl- (8CI)
Decylcyclohexane

1795-17-1
$C_{18}H_{36}$
252.48
C(CCCC1)(C1)CCCCCCCCCCCC
Dodecylcyclohexane
Cyclohexane, dodecyl- (9CI)
1-Cyclohexyldodecane

1795-18-2
$C_{20}H_{40}$
280.54
C(CCCC1)(C1)CCCCCCCCCCCCCC
Cyclohexane, tetradecyl- (9CI)
1-Cyclohexyltetradecane
Tetradecylcyclohexane

1795-27-3
C_9H_{18}
126.24
Cyclohexane, 1,3,5-trimethyl-, (1α,3α,5α)- (9CI)
Cyclohexane, 1,3,5-trimethyl-, cis- (8CI)

1797-33-7
$C_{20}H_{34}O_3S.Na$
377.54
Benzenesulfonic acid, 4-tetradecyl-, sodium salt (9CI)
Sodium 4-tetradecylbenzenesulfonate
4-Tetradecylbenzenesulfonic acid, sodium salt

1798-04-5
$C_{12}H_{16}O_3$
208.26
Acetic acid, (4-(1,1-dimethylethyl)phenoxy)- (9CI)
Acetic acid, (p-tert-butylphenoxy)- (8CI)
NSC-8481

1799-84-4
$C_{10}H_9F_9O_2$
332.17
O=C(OCCC(F)(F)C(F)(F)C(F)(F)C(F)(F)F)C(=C)C
2-Propenoic acid, 2-methyl-, 3,3,4,4,5,5,6,6,6-nonafluorohexyl ester (9CI)
Methacrylic acid, 3,3,4,4,5,5,6,6,6-nonafluorohexyl ester

1804-87-1
$C_6H_{10}O_5S.Na$
217.20
2-Propenoic acid, 2-methyl-, 2-sulfoethyl ester, sodium salt (9CI)

1804-93-9
$C_6H_{15}O_4P$
182.16
O=P(OCCC)(OCCC)O
Phosphoric acid, dipropyl ester
AI3-16970
Dipropyl phosphate

1805-32-9
$C_7H_6Cl_2O$
177.03
Benzenemethanol, 3,4-dichloro- (9CI)
Benzyl alcohol, 3,4-dichloro- (8CI)
NSC-26136
NSC-407829

1806-23-1
$C_{13}H_8ClFO$
234.66
O=C(c(ccc(F)c1)c1)c(c(ccc2)Cl)c2
Methanone, (2-chlorophenyl)-(4-fluorophenyl)- (9CI)
(2-Chlorophenyl)(4-fluorophenyl)methanone

1806-26-4
$C_{14}H_{22}O$
206.33
Oc(ccc(c1)CCCCCCCC)c1
4-Octylphenol
1-(p-Hydroxyphenyl)octane

p-Octylphenol
Phenol, p-octyl-
Phenol, 4-octyl- (9CI)

1806-29-7
$C_{12}H_{10}O_2$
186.22
Oc(c(c(c(O)ccc1)c1)ccc2)c2
2,2'-Biphenyldiol
o,o'-Biphenol
2,2'-Biphenol
2,2'-Dihydroxybiphenyl

1806-34-4
$C_{24}H_{16}N_2O_2$
364.39
O(C(c(cccc1)c1)=CN=2)C2c(ccc(c3)C(OC(c(cccc4)c4)=C5)=N5)c3
POPOP
Oxazole, 2,2'-p-phenylenebis-(5-phenyl- (8CI)
Oxazole, 2,2'-(1,4-phenylene)-bis(5-phenyl- (9CI)
2,2'-(1,4-Phenylene)bis(5-phenyloxazole)

1806-54-8
$C_{24}H_{51}O_4P$
434.64
Phosphoric acid, trioctyl ester
AI3-05904

1807-55-2
$C_{15}H_{18}N_2$
226.35
N(c(ccc(c1)Cc(ccc(NC)c2)c2)c1)C
Aniline, 4,4'-methylenebis-(N-methyl
Aniline, N,N'-dimethyl-4,4'-methylenedi-
Benzenamine, 4,4'-methylenebis(N-methyl- (9CI)
Bis(N-methylaniline)methane
Bis(N-methylanilino)methan (German)
Dimethyldiaminodiphenylmethane
N,N'-Dimethyl-4,4'-methylenedianiline
4,4'-Methylenebis(N-methylaniline)

1808-26-0
$C_{22}H_{36}O_2$
332.53
Ethyl arachidonate
5,8,11,14-Eicosatetraenoic acid, ethyl ester

1809-10-5
$C_5H_{11}Br$
151.05
BrC(CC)CC
Pentane, 3-bromo- (9CI)
3-Bromopentane

1809-19-4
$C_8H_{19}O_3P$
194.24
O=P(OCCCC)OCCCC
Dibutyl-phosphite
Butyl alcohol, hydrogen phosphite
Dibutylfosfit (Czech)
Dibutyl hydrogen phosphite
Mobil DBHP

1809-20-7
$C_6H_{15}O_3P$
166.18
O=P(OC(C)C)OC(C)C
Phosphonic acid, diisopropyl ester
Diisopropyl hydrogen phosphite
Diisopropyl phosphite
Diisopropylphosphonate
O,O-Diisopropyl phosphonate
Isopropyl phosphonate
Phosphonic acid, bis(1-methylethyl) ester

1809-21-8
$C_6H_{15}O_3P$
166.16
CCCOP(=O)OCCC
Phosphonic acid, dipropyl ester (8CI,9CI)
NSC-79865

1811-28-5
$C_{13}H_{11}N_3.1/2H_2O_4S$
258.29
Acridine, 3,6-diamino-, sulfate (2:1)
3,6-Diaminoacridine hemisulfate
Proflavine hemisulphate

1812-30-2
$C_{14}H_{10}BrN_3O$
316.14
Bromazepam
2H-1,4-Benzodiazepin-2-one, 7-bromo-1,3-dihydro-5-(2-pyridinyl)- (9CI)
2H-1,4-Benzodiazepin-2-one, 7-bromo-1,3-dihydro-5-(2-pyridyl)- (8CI)
2H-1,4-Benzodiazepin-2-one, 1,3-dihydro-7-bromo-5-(2-pyridyl)-
Bromazepamum (Latin)
7-Bromo-1,3-dihydro-5-(2-pyridyl)-2H-1,4-benzdiazepin-2-one
7-Bromo-1,3-dihydro-5-(2-pyridyl)-2H-1,4-benzodiazepin-2-one
7-Bromo-5-(2-pyridyl)-3H-1,4-benzodiazepin-2(1H)-one
Compendium
DEA No. 2748
1,3-Dihydro-7-bromo-5-(2-pyridyl)-2H-1,4-benzodiazepin-2-one
KL-001
LA XVII
Lectopam
Lexomil
Lexotan
Lexotanil
Ro 4-9253
Ro 5-3350

1812-51-7
$C_{14}H_{14}$
182.27
Biphenyl, 2-ethyl- (8CI)
1,1'-Biphenyl, 2-ethyl- (9CI)

1812-53-9
$C_{34}H_{72}N.Cl$
530.52
Ammonium dihexadecyldimethyl-, chloride
Aliquat 206
Dicetyldimethylammonium chloride
Dimethyldicetylammonium chloride
1-Hexadecanaminium, N-hexadecyl-N,N-dimethyl-, chloride (9CI)

1814-88-6
$C_3H_3F_5$
134.05
Propane, 1,1,1,2,2-pentafluoro- (8CI,9CI)

1817-13-6
$C_6H_2Cl_5N$
265.35
n(c(c(cc1)Cl)C(Cl)(Cl)Cl)c1Cl
Pyridine, 3,6-dichloro-2-(trichloromethyl)- (9CI)
2,5-Dichloro-6-(trichloromethyl)pyridine
3,6-Dichloro-2-(trichloro methyl)pyridine

1817-47-6
$C_9H_{11}NO_2$
165.19
O=N(=O)c(ccc(c1)C(C)C)c1
Benzene, 1-(1-methylethyl)-4-nitro- (9CI)
Cumene, p-nitro- (8CI)
1-(1-Methylethyl)-4-nitrobenzene
p-Nitroisopropylbenzene

1817-66-9
$C_7H_6N_2O_5$
198.12
Phenol, 3-methyl-2,4-dinitro- (9CI)

1817-68-1
$C_{23}H_{24}O$
316.47
Oc(c(cc(c1)C)C(c(cccc2)c2)C)c1C(c(cccc3)c3)C
p-Cresol, 2,6-bis(α-methylbenzyl)
Alkofen MBP
2,6-Bis(1-phenylethyl)-4-methylphenol
Ionol 6
Phenol, 4-methyl-2,6-bis-(1-phenylethyl)- (9CI)

1817-73-8
$C_6H_4BrN_3O_4$
262.04
O=N(=O)c(cc(N(=O)=O)c(N)c1Br)c1
Aniline, 2-bromo-4,6-dinitro
2-Bromo-4,6-dinitroaniline
Bromo DNA
2,4-Dinitro-6-bromanilin (Czech)
NCI-C60844

1817-74-9
$C_{13}H_{10}N_2O_4$
258.22
O=N(=O)c(ccc(c1)Cc(ccc(N(=O)=O)c2)c2)c1
Benzene, 1,1'-methylenebis-(4-nitro- (9CI)
AI3-03273
Methane, bis(p-nitrophenyl)-(8CI)
1,1'-Methylenebis(4-nitrobenzene)

1820-81-1
$C_4H_3ClN_2O_2$
146.54
Clc1c[nH]c(=O)[nH]c1=O
Uracil, 5-chloro
5-Chlorouracil
2,4(1H,3H)-Pyrimidinedione, 5-chloro-

1821-12-1
$C_{10}H_{12}O_2$
164.20
O=C(O)CCCc(cccc1)c1
Benzenebutanoic acid (9CI)
AI3-12065
Butyric acid, 4-phenyl- (8CI)

1823-91-2
C_9H_9N
131.17
N#CC(c(cccc1)c1)C
Benzeneacetonitrile, α-methyl-(9CI)
AI3-07025
α-Methylbenzeneacetonitrile

1824-09-5
$C_{10}H_{18}ClN_5O$
259.71
1,3,5-Triazine-2,4-diamine, 6-chloro-N-(3-methoxypropyl)-N'-(1-methylethyl)-(9CI)
G 34698
s-Triazine, 2-chloro-4-(isopropylamino)-6-((3-methoxypropyl)-amino)- (8CI)

1824-81-3
$C_6H_8N_2$
108.16
n(c(ccc1)C)c1N
Pyridine, 2-amino-6-methyl
2-Amino-6-methylpyridine

1825-19-0
$C_7H_3Cl_5S$
296.41
Sulfide, methyl pentachlorophenyl
Benzene, pentachloro(methylthio)- (9CI)
Methyl pentachlorophenyl sulfide
Methylthiopentachlorobenzene
PCTAS
Pentachloro(methylthio)benzene
Pentachlorophenyl methyl sulfide
Pentachlorothioanisole

1825-21-4
$C_7H_3Cl_5O$
280.35
COc1c(Cl)c(Cl)c(Cl)c(Cl)c1Cl
Anisole, 2,3,4,5,6-pentachloro
Benzene, pentachloromethoxy-(9CI)
Methyl pentachlorophenate
Methyl pentachlorophenyl ester
NCI-C56520
Pentachloroanisole
2,3,4,5,6-Pentachloroanisole
Pentachloromethoxybenzene
Pentachlorophenyl methyl ether

1825-30-5
$C_{10}H_6Cl_2$
197.06
c(c(c(cc1)Cl)ccc2)(c2Cl)c1
Naphthalene, 1,5-dichloro-(9CI)
1,5-Dichloronaphthalene

1825-31-6
$C_{10}H_6Cl_2$
197.06
c(c(c(cc1)Cl)ccc2)(c1Cl)c2
Naphthalene, 1,4-dichloro-(9CI)
1,4-Dichloronaphthalene

1825-61-2
$C_4H_{12}OSi$
104.22
COSi(C)(C)C
Silane, methoxytrimethyl-(9CI)
Methoxytrimethylsilane
Trimethylmethoxysilane

1825-62-3
$C_5H_{14}OSi$
118.28
Silane, ethoxytrimethyl
Ethoxytrimethylsilane
Silane, trimethylethoxy-

1832-53-7
$C_3H_9O_3P$
124.08
Phosphonic acid, methyl-, monoethyl ester (8CI,9CI)

1832-54-8
$C_4H_{11}O_3P$
138.10
Isopropyl methylphosphonic acid
IMPA

1832-68-4
$C_{17}H_{37}O_3P$
320.45
Phosphonic acid, methyl-, dioctyl ester (8CI,9CI)

1835-04-7
$C_{11}H_{14}O_3$
194.25
1-Propanone, 1-(3,4-dimethoxyphenyl)
3,4-Dimethoxyphenyl ethyl ketone
1-(3,4-Dimethoxyphenyl)-1-propanone
3,4-Dimethoxypropiophenone
3',4'-Dimethoxypropiophenone
Propiophenone, 3',4'-dimethoxy-
Propioveratrone

1835-14-9
$C_{10}H_{12}O_3$
180.20
1-Propanone, 1-(4-hydroxy-3-methoxyphenyl)- (9CI)

1836-75-5
$C_{12}H_7Cl_2NO_3$
284.10
O=N(=O)c(ccc(Oc(c(cc(c1)Cl)Cl)c1)c2)c2
Ether, 2,4-dichlorophenyl p-nitrophenyl
Benzene, 2,4-dichloro-1-(4-nitrophenoxy)-

2,4-Dechlorophenyl p-nitrophenyl ether
2',4'-Dichloro-4-nitrobiphenyl ether
2,4-Dichloro-4'-nitrodiphenyl ether
2,4-Dichloro-1-(4-nitrophenoxy)benzene
4-(2,4-Dichlorophenoxy)nitrobenzene
2,4-Dichlorophenyl p-nitrophenyl ether
2,4-Dichlorophenyl 4-nitrophenyl ether
2,4-Dichlorphenyl-4-nitrophenylaether (German)
FW 925
Mezotox
NCI-C00420
Niclofen
NIP
Nitofen
Nitrafen
Nitraphen
Nitrochlor
4'-Nitro-2,4-dichlorodiphenyl ether
Nitrofen
Nitrofene (French)
Nitrophen
Nitrophene
Preparation 125
TOK
TOK-2
TOK E
TOK E-25
TOK E 40
Tokkorn
TOK WP-50
Trizilin
Trizilin 25

1836-77-7
$C_{12}H_6Cl_3NO_3$
318.54
Clc2cc(Cl)c(Oc1ccc(cc1)N(=O)=O)c(Cl)c2
Ether, p-nitrophenyl 2,4,6-trichlorophenyl
Benzene, 1,3,5-trichloro-2-(4-nitrophenoxy)-
Chlornitrofen
CNP
CNP 1032
MC 338
MC 1478
MO
MO 338
p-Nitrophenyl 2,4,6-trichlorophenyl ether
2',4',6'-Trichloro-4-nitrobiphenyl ether
2,4,6-Trichloro-4'-nitrodiphenyl ether
2,4,6-Trichlorophenyl 4-nitrophenyl ether

1837-91-8
$C_6H_6Br_6$
557.54
BrC(C(Br)C(Br)C(Br)C1Br)C1Br
Cyclohexane, 1,2,3,4,5,6-hexabromo- (8CI)
1,2,3,4,5,6-Hexabromocyclohexane

1838-08-0
$C_{18}H_{39}N.ClH$
306.04
Octadecylamine-hydrochloride

1838-19-3
$C_{18}H_{37}N$
267.49
NCCCCCCCCC=CCCCCCCCC
Oleylamine
AI3-26645
9-Octadecen-1-amine (9CI)
9-Octadecenylamine (8CI)

1843-03-4
$C_{37}H_{52}O_3$
544.82
Oc(c(cc(c1C)C(c(c(cc(O)c2C(C)(C)C)C)c2)CC(c(c(cc(O)c3C(C)(C)C)C)c3)C)C(C)(C)C)c1
Tributylcresylbutane
4,4',4''-(1-Methyl-1-propanyl-3-ylidene)tris(2-(1,1-dimethylethyl)-5-methylphenol
Phenol, 4,4',4''-(1-methyl-1-propanyl-3-ylidene)tris(2-(1,1-dimethylethyl)-5-methyl-(9CI)
Stabilizer Mark 328
Tri(butylcresyl)butane
1,1,3-Tris(2-methyl-4-hydroxy-5-tert-butylphenyl)butane

1843-05-6
$C_{21}H_{26}O_3$
326.47
O=C(c(cccc1)c1)c(c(O)cc(OCCCCCCC)c2)c2
Benzophenone, 2-hydroxy-4-(octyloxy)
2-Hydroxy-4-(octyloxy)benzophenone
2-Hydroxy-4-oktyloxybenzofenon (Czech)

1843-48-7
$C_9H_{12}N_6O_6$
300.23
s-Triazine-1,3,5(2H,4H,6H)-triacetamide, 2,4,6-trioxo- (7CI,8CI)

1847-55-8
$C_{18}H_{36}O_4S.Na$
371.54
9-Octadecen-1-ol, hydrogen sulfate, sodium salt, (Z)-(9CI)

1847-58-1
$C_{14}H_{27}O_5S.Na$
330.46
Acetic acid, sulfo-, 1-dodecyl ester, sodium salt
Acetic acid, sulfo-, dodecyl ester, s-sodium salt (7CI)
Dodecyl sodium sulfoacetate
Lathanol
Lathanol LAL
Lathanol-LAL 70
Nacconol LAL
Sodium lauryl sulfoacetate
Sulfoacetic acid 1-dodecyl ester, sodium salt
Sulfoacetic acid dodecyl ester S-sodium salt

1849-18-9
$C_{14}H_{22}O$
206.36
Phenol, 2,4-di-sec-butyl
2,4-Di-sec-butylphenol

1849-36-1
$C_6H_5NO_2S$
155.18
Sc1ccc(cc1)N(=O)=O
Benzenethiol, p-nitro
Benzenethiol, 4-nitro-
p-Nitrobenzenethiol
4-Nitrobenzenethiol
p-Nitrophenyl mercaptan
p-Nitrothiophenol
4-Nitrothiophenol

1852-04-6
$C_{11}H_{20}O_4$
216.28
O=C(O)CCCCCCCCCC(=O)O
Undecanedioic acid (9CI)

1852-14-8
$C_4H_{10}N_4O_2$
146.18
O=C(NCCNC(=O)N)N
Urea, 1,1'-ethylenedi
Ethanediurea
1,1'-Ethylenebisurea
Ethylenediurea
1,1'-Ethylenediurea
Monoethylenediurea
Urea, N,N''-1,2-ethanediylbis-(9CI)

1852-16-0
$C_8H_{15}NO_2$
157.24
O=C(NCOCCCC)C=C
Acrylamide, N-butoxymethyl
N-(Butoxymethyl)acrylamide
N-Butoxymethylakrylamid (Czech)
2-Propenamide, N-(butoxymethyl)- (9CI)

1852-21-7

$C_7H_{15}N_3O_4$
205.20
O=C(N(CN(C1)CCO)CO)N1CO
1,3,5-Triazin-2(1H)-one, tetrahydro-5-(2-hydroxyethyl)-1,3-bis(hydroxymethyl)-(9CI)
1,3-Dimethylol-5-β-hydroxyethylhexahydrotriazinone-2
Dimethylol(hydroxyethyl)triazone
5-Hydroxyethyl-1,3-bis-(hydroxymethyl)hexahydro-s-triazin-2-one
Tetrahydro-5-(2-hydroxyethyl)-1,3-bis(hydroxymethyl)-1,3,5-triazin-2(1H)-one

1853-88-9
$C_{10}H_{11}N_5$
201.20
1,3,5-Triazine-2,4-diamine, 6-(phenylmethyl)- (9CI)
Acetoguanamine, phenyl-
NSC-4406
s-Triazine, 2,4-diamino-6-benzyl-(8CI)

1853-90-3
$C_5H_8N_4$
124.12
Cc1nc(C)nc(N)n1
1,3,5-Triazin-2-amine, 4,6-dimethyl- (9CI)
s-Triazine, 2-amino-4,6-dimethyl-(8CI)

1853-91-4
$C_{10}H_{10}N_4$
186.19
1,3,5-Triazin-2-amine, 4-methyl-6-phenyl- (9CI)
s-Triazine, 2-amino-4-methyl-6-phenyl- (8CI)

1853-95-8
$C_9H_8N_4$
172.19
1,3,5-Triazin-2-amine, 4-phenyl- (9CI)
s-Triazine, 2-amino-4-phenyl- (7CI,8CI)

1853-96-9
$C_{10}H_9ClN_4$
220.66
s-Triazine, 2-amino-4-(chloromethyl)-6-phenyl-(7CI,8CI)

1854-23-5
$C_{13}H_{25}N_3O_4$
287.36
4,4'-(2-Ethyl-2-nitrotrimethylene)dimorpholine

1854-26-8
$C_5H_{10}N_2O_5$
178.17
O=C(N(C(O)C1O)CO)N1CO
2-Imidazolidinone, 1,3-bis-(hydroxymethyl)-4,5-dihydroxy
Arkofix NG
Cassurit LR
Depremol G
4,5-Dihydroxy-1,3-bis(hydroxymethyl)-2-imidazolidinone
Dimethyloldihydroxyethyleneurea
Dimethylolglyoxalurea
Dmdheu
Firmatex RK
Fixapret CP
Fixapret CP 40
Fixapret CPK
Fixapret CPN
Fixapret CPNS
Hylite LF
Knittex LE
NCI-C60322
Neuperm GFN
NS 11
Permafresh 183
Permafresh 113B
Permafresh LF
Permafresh LH
Permafresh LKS
Protocol C
Prox DW
Readpret KPN
Sarcoset GM
Sumitex FSK
Sumitex NS
Sumitex NS 1SPE
Verapret DH
Verapret DKH
WNM

1855-09-0
$C_9H_{12}O_2$
152.19
OC(c(cccc1)c1)C(O)C
1,2-Dihydroxy-1-phenylpropane
1-Phenyl-1,2-propanediol
1,2-Propanediol, 1-phenyl-(9CI)

1860-39-5
$C_7H_{14}O$
114.19
Hexanal, 5-methyl- (8CI,9CI)

1861-32-1
$C_{10}H_6Cl_4O_4$
331.96
COC(=O)c1c(Cl)c(Cl)c(C(=O)OC)c(Cl)c1Cl
Terephthalic acid, tetra chloro-, dimethyl ester
1,4-Benzenedicarboxylic acid, 2,3,5,6-tetrachloro-, dimethyl ester
Chlorothal
Chlorthal-dimethyl
Chlorthal-methyl
DAC 893
Dacthal
Dacthalor
Daktal (Czech)
DCPA
Dimethylester kyseliny tetrachlortereftalove (Czech)
Dimethyl tetrachloroterephthalate
Dimethyl 2,3,5,6-tetrachloroterephthalate
Fatal
2,3,5,6-Tetrachlorphthalsauredimethylester (German)
Tetrachloroterephthalic acid dimethyl ester
2,3,5,6-Tetrachloroterephthalic acid, dimethyl ether

1861-40-1
$C_{13}H_{16}F_3N_3O_4$
335.32
CCCCN(CC)c1c(cc(cc1N(=O)=O)C(F)(F)F)N(=O)=O
p-Toluidine, N-butyl-N-ethyl-α,α,α-trifluoro-2,6-dinitro
Balan
Balfin
Banafine
Benalan
Benefex
Benefin
Benfluralin
Bethrodine
Binnell
Blulan
Bonalan
n-Butyl-2,6-dinitro-N-ethyl-4-trifluoromethylaniline
n-Butyl-N-ethyl-2,6-dinitro-4-(trifluoromethyl)benzenamine
n-Butyl-N-ethyl-α,α,α-trifluoro-2,6-dinitro-p-toluidine
Carpidor
EL-110
Emblem
L 54521
Quilan
α,α,α-Trifluoro-2,6-dinitro-N,N-ethylbutyl-p-toluidine

1861-44-5
$C_{10}H_{11}Cl_3O_2$
269.56
2-Propanol, 1-((2,3,6-trichlorophenyl)methoxy)-(9CI)
HRS 587
2-Propanol, 1-((2,3,6-tri chlorobenzyl)oxy)-(7CI,8CI)
1-(2,3,6-Trichlorobenzyl oxy)propan-2-ol
Tritac

1863-63-4
$C_7H_5O_2.H_4N$
139.17
Benzoic acid, ammonium salt
Ammonium benzoate
Vulnoc AB

1864-92-2
$C_{12}H_{19}NO$
193.28
O(c(cccc1N(CC)CC)c1)CC
Benzenamine, 3-ethoxy-N,N-diethyl- (9CI)
3-Ethoxy-N,N-diethylbenzenamine
m-Phenetidine, N,N-diethyl- (8CI)

1866-39-3
$C_{10}H_{10}O_2$
162.19
2-Propenoic acid, 3-(4-methylphenyl)- (9CI)
Cinnamic acid, p-methyl- (6CI, 7CI,8CI)
p-Methylcinnamic acid
4-Methylcinnamic acid

1866-39-3
$C_{10}H_{10}O_2$
162.19
2-Propenoic acid, 3-(4-methylphenyl)- (9CI)
Cinnamic acid, p-methyl- (6CI,7CI,8CI)
p-Methylcinnamic acid
4-Methylcinnamic acid

1867-66-9
$C_{13}H_{16}ClNO.ClH$
274.21
Cl.CNC1(CCCCC1=O)c2ccccc2Cl
Cyclohexanone, 2-(o-chlorophenyl)-2-(methylamino)-, hydrochloride
2-(o-Chlorophenyl)-2-(methylamino)cyclohexanone hydrochloride
C.I. 581
C.I. 369
CN-52,372-2
Ketaject
Ketalar
Ketamine
Ketamine hydrochloride
(+-)-Ketamine hydrochloride
Ketanest
Ketaset
Ketavet
Ketolar
Vetalar

1869-67-6
$C_{13}H_8F_3N_3O_4$
327.22
Benzenamine, 2,4-dinitro-N-(3-(trifluoromethyl)phenyl)- (9CI)
m-Toluidine, N-(2,4-dinitro phenyl)-α,α,α-trifluoro- (7CI,8CI)

1869-77-8
$C_{14}H_{12}F_{17}NO_4S$
613.29
O=C(OCC)CN(S(=O)(=O)C(F)(F)C(F)(F)C(F)(F)C(F)(F)C(F)(F)C(F)(F)C(F)(F)C(F)(F)F)CC
Glycine, N-ethyl-N-((heptadecafluorooctyl)sulfonyl)-, ethyl ester (8CI)
Ethyl N-ethyl-N-((heptadecafluorooctyl)sulfonyl) glycinate

1871-57-4
$C_4H_6Cl_2$
125.00
Propene, 3-chloro-2-(chloromethyl)
3-Chloro-2-(chloromethyl)-propene

1871-96-1
$C_{10}H_{16}N_2$
164.28
N#CCCCCCCCCC#N
Sebaconitrile
Decanedinitrile
1,8-Dicyanooctane
Octamethylene dicyanide
Oktamethylendikyanid (Czech)

1873-88-7
$C_7H_{22}O_2Si_3$
222.51
Trisiloxane, 1,1,1,3,5,5,5-heptamethyl- (9CI)
Bis(trimethylsiloxy)methylsilane
1,1,1,3,5,5,5-Heptamethyltrisiloxane

1875-92-9
$C_9H_{13}N.ClH$
171.69
Benzylamine, dimethyl-, hydrochloride
Dimethylbenzylamine hydrochloride
Dimethylbenzylammonium chloride
USAF EL-78

1877-72-1
$C_8H_5NO_2$
147.13
OC(=O)c1cccc(C#N)c1
Benzoic acid, 3-cyano- (9CI)
Benzoic acid, m-cyano- (8CI)

1878-18-8
$C_5H_{12}S$
104.22
SCC(CC)C
1-Butanethiol, 2-methyl- (9CI)
2-Methyl-1-butanethiol

1878-65-5
$C_8H_7ClO_2$
170.60
OC(=O)Cc1cccc(Cl)c1
Acetic acid, (m-chlorophenyl)- (8CI)
Benzeneacetic acid, 3-chloro- (9CI)

1878-66-6
$C_8H_7ClO_2$
170.60
O=C(O)Cc(ccc(c1)Cl)c1
Acetic acid, (p-chlorophenyl)
Benzeneacetic acid, 4-chloro- (9CI)
4-Chlorobenzeneacetic acid
(p-Chlorophenyl)acetic acid
2-(p-Chlorophenyl)acetic acid
(4-Chlorophenyl)acetic acid

1878-84-8
$C_8H_8O_4$
168.15
O=C(O)COc(ccc(O)c1)c1
4-Hydroxyphenoxyacetic acid
Acetic acid, (4-hydroxyphenoxy)- (9CI)
(4-Hydroxyphenoxy)acetic acid

1878-85-9
$C_9H_{10}O_4$
182.19
COc1ccccc1OCC(O)=O
Acetic acid, (o-methoxyphenoxy)
Acide o-methoxyphenoxy-acetique (French)
(o-Methoxyphenoxy)acetic acid

1878-87-1
$C_8H_7NO_5$
197.14
OC(=O)COc1ccccc1N(=O)=O
Acetic acid, (2-nitrophenoxy)- (9CI)
Acetic acid, (o-nitrophenoxy)- (8CI)
NSC-37409

1879-09-0
$C_{12}H_{18}O$
178.30
Oc(c(cc(c1)C)C(C)(C)C)c1C
2,4-Xylenol, 6-tert-butyl
2-tert-Butyl-4,6-dimethylphenol
6-t-Butyl-2,4-dimethylphenol
6-t-Butyl-2,4-xylenol
Prodox 340

1879-16-9
$C_8H_{10}OS$
154.23
Anisole, p-(methylthio)- (8CI)
Benzene, 1-methoxy-4-(methylthio)- (9CI)

1882-26-4
$C_{11}H_{15}N_3O_4$
253.29
CNC(=O)OCc1cccc(COC(=O)NC)n1
Carbamic acid, methyl-, 2,6-pyridinediyldimethylene ester
Anginin
Anginine
Methylcarbamic acid 2,6-pyridinediyldimethylene ester
Parmidin
Parmidine
Parmidine R
Piridinol carbamato (Spanish)
Prodectine
2,6-Pyridinedimethanol, bis-(methylcarbamate) (ester) (9CI)
Pyridinol carbamate
Sospitan

1885-14-9
$C_7H_5ClO_2$
156.57
O=C(Oc(cccc1)c1)Cl
Carbonochloridic acid, phenyl ester
Chloroformic acid phenyl ester
Fenylester kyseliny chlormravenci (Czech)
Formic acid, chloro-, phenyl ester
Phenyl chlorocarbonate
Phenyl chloroformate
Phenylchloroformate [UN 2746]
UN 2746 [Phenylchloroformate]

1885-29-6
$C_7H_6N_2$
118.15
N#Cc(c(N)ccc1)c1
Anthranilonitrile
o-Aminobenzonitrile
2-Aminobenzonitrile
Benzonitrile, 2-amino- (9CI)
o-Cyanoaniline
2-Cyanoaniline

1886-75-5
$C_8H_{18}O_2S$
178.30
Propane, 2-((1,1-dimethylethyl)-sulfonyl)-2-methyl- (9CI)
tert-Butyl sulfone (8CI)
NSC-179032

1888-57-9
$C_8H_{16}O$
128.21
3-Hexanone, 2,5-dimethyl- (9CI)

1888-71-7
C_3Cl_6
248.73
C(=C(Cl)Cl)(C(Cl)(Cl)Cl)Cl
Propene, hexachloro
Hexachloropropene
Hexachloropropylene
RCRA waste number U243

1888-89-7
$C_6H_{10}O_2$
114.16
Hexane, 1,2:5,6-diepoxy
1,2:5,6-Diepoxyhexane

1897-45-6
$C_8Cl_4N_2$
265.90
N#Cc(c(c(c(c1C#N)Cl)Cl)Cl)c1Cl
Isophthalonitrile, tetrachloro
1,3-Benzenedicarbonitrile, 2,4,5,6-tetrachloro-
Bravo
Bravo 6F
Bravo-W-75
Chloroalonil
Chlorothalonil
Chlorthalonil (German)
DAC 2787
Daconil
Daconil 2787
Daconil 2787 Flowable Fungicide
Dacosoil
1,3-Dicyanotetrachlorobenzene
Exotherm
Exotherm Termil
Forturf
Isophthalonitrile, 2,4,5,6-tetrachloro-
NCI-C00102
Nopcocide
Nopcocide N-96
Nopcocide N40D & N96
Sweep
TCIN
Termil
Tetrachlorisoftalonitril (Czech)
2,4,5,6-Tetrachloro-3-cyanobenzonitrile
Tetrachloroisophthalonitrile
meta-Tetrachlorophthalodinitrile
m-Tetrachlorophthalonitrile
m-TCPN
TPN (Pesticide)

1898-66-4
$C_{18}H_{12}N_5O_6$
394.35
O=N(=O)c(cc(N(=O)=O)c(NN(c(cccc1)c1)c(cccc2)c2)c3N(=O)=O)c3
Hydrazyl, 2,2-diphenyl-1-(2,4,6-trinitrophenyl)
Diphenylpicrylhydrazyl free radical
2,2-Diphenyl-1-(2,4,6-trinitrophenyl)hydrazyl

1898-74-4
$C_{15}H_{11}N_3$
233.28
1,3,5-Triazine, 2,4-di phenyl- (9CI)
s-Triazine, 2,4-diphenyl- (6CI,7CI,8CI)

1899-94-1
$C_7H_9NO_2S$
171.21
Cc1cccc(c1)S(N)(=O)=O
Benzenesulfonamide, 3-methyl- (9CI)
m-Toluenesulfonamide (8CI)

1906-79-2
$C_7H_{10}N.Br$
188.07
Pyridinium, 1-ethyl-, bromide
AI3-52378
1-Ethylpyridinium bromide

1908-87-8
$C_4H_7NS_2$
133.23
N(C(SC1)=S)(C1)C
2-Thiazolidinethione, 3-methyl- (9CI)
3-Methylimidazolidine-2-thione
3-Methyl-2-thiazolidinethione

1910-42-5
$C_{12}H_{14}N_2.2Cl$
257.18
[Cl-].[Cl-].C[n+]1ccc(cc1)c2cc[n+](C)cc2
4,4'-Bipyridinium, 1,1'-dimethyl-, dichloride
AH 501
Bipyridinium, 1,1'-dimethyl-4,4'-, dichloride
Cekuquat
Crisquat
Dextrone
Dextrone-X
Dexuron
N,N'-Dimethyl-4,4'-bipyridinium dichloride
N,N'-Dimethyl-4,4'-bipyridylium dichloride
1,1'-Dimethyl-4,4'-bipyridynium dichloride
1,1'-Dimethyl-4,4'-dipyridinium-dichlorid (German)
4,4'-Dimethyldipyridyl dichloride
1,1'-Dimethyl-4,4'-dipyridylium chloride
N,N'-Dimethyl-4,4'-dipyridyl-

ium dichloride
Dimethyl viologen chloride
Esgram
Goldquat 276
Gramixel
Gramonol
Gramoxon
Gramoxone
Gramoxone D
Gramoxone dichloride
Gramoxone S
Gramoxone W
Gramuron
Herbaxon
Herboxone
Methylviologen
Methyl viologen dichloride
Methyl viologen (Reduced)
OK 622
Ortho Paraquat Cl
Para-Col
Paraquat (ACGIH)
Paraquat chloride
Paraquat Cl
Paraquat, dichloride
Pathclear
Pillarquat
Pillarxone
PP148
Sweep
Terraklene
Totacol
Toxer Total
Viologen, methyl-
Weedol

1910-89-0
$C_4H_9N_3O$
115.14
1,3,5-Triazin-2(1H)-one, tetrahydro-5-methyl- (9CI)
s-Triazin-2(1H)-one, tetrahydro-5-methyl- (6CI,7CI, 8CI)

1912-24-9
$C_8H_{14}ClN_5$
215.72
n(c(nc(n1)NC(C)C)NCC)c1Cl
s-Triazine, 2-chloro-4-ethylamino-6-isopropylamino
A 361
Aatrex
Aatrex 4L
Aatrex Nine-O
Aatrex 80W
2-Aethylamino-4-chlor-6-isopropylamino-1,3,5-triazin (German)
2-Aethylamino-4-isopropylamino-6-chlor-1,3,5-triazin (German)
Aktikon
Aktikon PK
Aktinit A
Aktinit PK
Argezin
Atazinax
Atranex
Atrasine
Atratol A
Atrazin
Atrazine (ACGIH,OSHA)
Atred
Atrex
Candex
Cekuzina-T
2-Chloro-4-ethylamineisopropylamine-s-triazine
1-Chloro-3-ethylamino-5-isopropylamino-s-triazine
1-Chloro-3-ethylamino-5-isopropylamino-2,4,6-triazine
2-Chloro-4-ethylamino-6-isopropylamino-s-triazine
2-Chloro-4-ethylamino-6-isopropylamino-1,3,5-triazine
6-Chloro-N-ethyl-N'-(1-methylethyl)-1,3,5-triazine-2,4-diamine
2-Chloro-4-(2-propylamino)-6-ethylamino-s-triazine
Crisatrina
Crisazine
Cyazin
Farmco atrazine
Fenamin
Fenamine
Fenatrol
G 30027
Geigy 30,027
Gesaprim
Gesoprim
Griffex
Hungazin
Hungazin PK
Inakor
Oleogesaprim
Primatol
Primatol A
Primaze
Radazin
Radizine
Shell Atrazine Herbicide
Strazine
Triazine A 1294
1,3,5-Triazine-2,4-diamine, 6-chloro-N-ethyl-N'-(1-methylethyl)- (9CI)
Vectal
Vectal SC
Weedex A
Wonuk
Zeaphos
Zeazin
Zeazine

1912-25-0
$C_{10}H_{18}ClN_5$
243.78
CCN(CC)c1nc(Cl)nc(NC(C)C)n1
s-Triazine, 2-chloro-4-(diethylamino)-6-(isopropylamino)
2-Chloro-4-(diethylamino)-6-(isopropylamino)-s-triazine
G 30031
Geigy
Gesabal
Heptazine
Ipazine
Isodiazine

1912-26-1
$C_9H_{16}ClN_5$
229.75
n(c(nc(n1)N(CC)CC)NCC)c1Cl
s-Triazine, 2-chloro-4-(diethylamino)-6-(ethylamino)
Aventox
Bronox
2-Chloro-4-(diethylamino)-6-(ethylamino)-s-triazine
6-Chloro-N,N,N'-triethyl-1,3,5-triazine-2,4-diamine
2-Ethylamino-4-diethylamino-6-chloro-s-triazine
G 27901
Gesafloc
NC 1667
Remtal
1,3,5-Triazine-2,4-diamine, 6-chloro-n,n,N'-triethyl-
Trietazine
Triethazine

1912-30-7
$C_4H_{10}O_3S$
138.20
Ethanesulfonic acid, ethyl ester
Diethylsulfonate
EES
Ethyl ethane sulfonate

1917-34-6
$C_{11}H_9N_3O_2$
215.21
s-Triazine-2,4(1H,3H)-dione, 6-styryl- (8CI)
s-Triazine-2,4-diol, 6-styryl- (7CI)

1917-38-0
$C_{15}H_{10}ClN_3O$
283.72
1,3,5-Triazin-2(1H)-one, 4-(4-chlorophenyl)-6-phenyl- (9CI)
s-Triazin-2-ol, 4-(p-chlorophenyl)-6-phenyl- (7CI,8CI)

1917-40-4
$C_{11}H_{11}N_3O$
201.23
1,3,5-Triazin-2(1H)-one, 4-ethyl-6-phenyl- (9CI)
s-Triazin-2(1H)-one, 6-ethyl-4-phenyl- (8CI)
s-Triazin-2-ol, 4-ethyl-6-phenyl- (7CI)

1917-41-5
$C_{10}H_6Cl_3N_3O$
290.54

s-Triazin-2-ol, 4-phenyl-6-(trichloromethyl)- (7CI, 8CI)

1917-43-7
$C_{10}H_9N_3OS$
219.27
1,3,5-Triazin-2(1H)-one, 4-(methylthio)-6-phenyl- (9CI)
s-Triazin-2-ol, 4-(methylthio)-6-phenyl- (7CI, 8CI)

1917-44-8
$C_{15}H_{11}N_3O$
249.25
1,3,5-Triazin-2(1H)-one, 4,6-diphenyl- (9CI)
NSC-288740
s-Triazin-2(1H)-one, 4,6-diphenyl- (8CI)

1917-95-9
$C_{14}H_{18}Cl_2O_3$
305.20
Acetic acid, (2,4-dichlorophenoxy)-, hexyl ester (8CI, 9CI)

1917-97-1
Unknown
Unknown
2-Octyl 2,4-dichlorophenoxyacetate

1918-00-9
$C_8H_6Cl_2O_3$
221.04
COc1c(Cl)ccc(Cl)c1C(O)=O
Benzoic acid, 3,6-dichloro-2-methoxy
Acido (3,6-dicloro-2-metossi)-benzoico (Italian)
o-Anisic acid, 3,6-dichloro-
Banex
Banlen
Banvel
Banvel CST
Banvel D
Banvel Herbicide
Banvel II Herbicide
Banvel 4S
Banvel 4WS
Brush Buster
Compound B Dicamba
Dianat (Russian)
Dianate
Dicamba
Dicambe
3,6-Dichloor-2-methoxy-benzoeizuur (Dutch)
3,6-Dichlor-3-methoxy-benzoesaeure (German)
3,6-Dichloro-o-anisic acid
2,5-Dichloro-6-methoxybenzoic acid
3,6-Dichloro-2-methoxybenzoic acid
Kyselina 3,6-dichlor-2-methoxybenzoova (Czech)
MDBA
Mediben
2-Methoxy-3,6-dichlorobenzoic acid
Velsicol Compound "R"
Velsicol 58-CS-11

1918-02-1
$C_6H_3Cl_3N_2O_2$
241.46
Nc1c(Cl)c(Cl)nc(C(O)=O)c1Cl
Picolinic acid, 4-amino-3,5,6-trichloro
Amdon
Amdon Grazon
4-Amino-3,5,6-trichloropicolinic acid
4-Amino-3,5,6-trichloro-2-picolinic acid
4-Amino-3,5,6-trichlorpicolinsaeure (German)
ATCP
Borolin
Chloramp (Russian)
K-Pin
NCI-C00237
Picloram (ACGIH,OSHA)
Tordon
Tordon 10K
Tordon 22K
Tordon 101 Mixture
3,5,6-Trichloro-4-aminopicolinic acid

1918-08-7
$C_{13}H_{19}N_3O_4$
281.35
CCCN(CCC)c1c(cc(C)cc1N(=O)=O)N(=O)=O
p-Toluidine, 2,6-dinitro-N,N-dipropyl
2,6-Dinitro-N,N-dipropyl-p-toluidine
2,6-Dinitro-N,N-di-n-propyl-p-toluidine
Dipropalin
N,N-Di-n-propyl-2,6-dinitro-4-methylaniline
N,N-Dipropyl-2,6-dinitro-p-toluidine

1918-11-2
$C_{17}H_{27}NO_2$
277.41
Azak
2,6-Di-t-butyl-4-methylphenyl-N-methylcarbamate
2,6-Di-tert-butyl-p-tolyl methylcarbamate
Caswell No. 294
EPA Pesticide Chemical Code 038801
Terbucarb
Terbutol

1918-13-4
$C_7H_5Cl_2NS$
206.09
NC(=S)c1c(Cl)cccc1Cl
Benzamide, 2,6-dichlorothio
Benzenecarbothioamide, 2,6-dichloro- (9CI)
Chlorothiamide
Chlorthiamid
Chlorthiamide
Chlorthioamide
Chlortiamid
DCBN
2,6-Dichlorobenzenecarbothioamide
2,6-Dichlorothiobenzamide
2,6-Dichlor-thiobenzamid
(German)
Prefix
SD 7961
WL-5792

1918-16-7
$C_{11}H_{14}ClNO$
211.71
CC(C)N(C(=O)CCl)c1ccccc1
Acetanilide, 2-chloro-N-isopropyl
Acetamide, 2-chloro-N-(1-methylethyl)-N-phenyl-
Bexton
Bexton 4L
Chloressigsaeure-N-isopropylanilid (German)
α-Chloro-N-isopropylacetanilide
2-Chloro-N-isopropylacetanilide
2-Chloro-N-isopropyl-N-phenylacetamide
CP 31393
N-Isopropyl-α-chloroacetanilide
N-Isopropyl-2-chloroacetanilide
Niticid
Propachlor
Propachlore
Ramrod
Ramrod 65
Satecid

1918-18-9
$C_8H_7Cl_2NO_2$
220.06
COC(=O)Nc1ccc(Cl)c(Cl)c1
Carbanilic acid, 3,4-dichloro-, methyl ester
Carbamic acid, (3,4-dichlorophenyl)-, methyl ester
3,4-Dichlorocarbanilic acid methyl ester
FMC 2995
MCC
Methyl 3,4-dichlorocarbanilate
Methyl-N-(3,4-dichlorophenyl) carbamate
Methylester kyseliny 3,4-dichlorkarbanilove (Czech)
NIA 2,995
NIA 2995J
Swep

1918-77-0
$C_6H_6O_2S$
142.18
O=C(O)CC(SC=C1)=C1
2-Thienylacetic acid
2-Thiopheneacetic acid (9CI)

1918-79-2
$C_6H_6O_2S$
142.18
2-Thiophenecarboxylic acid, 5-methyl- (8CI,9CI)
NSC-89698

1919-45-5
$C_{21}H_{12}Cl_3N_3O_3$
460.71
1,3,5-Triazine, 2,4,6-tris-(4-chlorophenoxy)- (9CI)
s-Triazine, 2,4,6-tris-(p-chlorophenoxy)- (7CI, 8CI)
2,4,6-Tris(p-chloro-phenoxy)-s-triazine
2,4,6-Tris(4-chloro-phenoxy)-1,3,5-triazine

1919-46-6
$C_{24}H_{21}N_3O_3$
399.45
1,3,5-Triazine, 2,4,6-tris-(3-methylphenoxy)- (9CI)
s-Triazine, 2,4,6-tris(m-tolyloxy)- (7CI,8CI)

1919-48-8
$C_{21}H_{15}N_3O_3$
357.35
O(c(cccc1)c1)c(nc(Oc(cccc2)c2)nc3Oc(cccc4)c4)n3
1,3,5-Triazine, 2,4,6-tri-phenoxy- (9CI)
AI3-52873
2,4,6-Triphenoxy-s-triazine
2,4,6-Triphenoxy-1,3,5-tria-zine

1920-21-4
$C_8H_{12}O_2$
140.20
CC1=COC(C)(CC1)C=O
1,4-Pyran, 2,3-dihydro-2,5-di-methyl-2-formyl
2,3-Dihydro-2,5-dimethyl-2-formyl-1,4-pyran
Methacrolein dimer
Methacrylaldehyde dimer
2H-Pyran-2-carboxaldehyde, 3,4-dihydro-2,5-dimethyl- (8CI)

1920-90-7
$C_{16}H_{36}Pb$
435.66
CCCC[Pb](CCCC)(CCCC)CCCC
Plumbane, tetrabutyl- (9CI)
Tetrabutylplumbane

1921-70-6
$C_{19}H_{40}$
268.59
C(CCCC(CCCC(CCCC(C)C)C)C)(C)C
Pentadecane, 2,6,10,14-tetra-methyl
Pristane
2,6,10,14-Tetramethylpenta-decane

1928-01-4
$C_{10}H_6Cl_2O_4S_2$
325.19
1,5-Naphthalenedisulfonyl dichloride (9CI)
1,5-Naphthalenedisulfo-chloride
1,5-Naphthalenedisulfonyl chloride (6CI,7CI,8CI)

1928-37-6
$C_9H_7Cl_3O_3$
269.51
Acetic acid, (2,4,5-trichloro-phenoxy)-, methyl ester (9CI)

1928-38-7
$C_9H_8Cl_2O_3$
235.07
Acetic acid, (2,4-dichloro-phenoxy)-, methyl ester
2,4-D methyl ester
Methyl 2,4-D ester

1928-39-8
$C_{10}H_9Cl_3O_3$
283.54
Acetic acid, (2,4,5-trichloro-phenoxy)-, ethyl ester (8CI, 9CI)
NSC-190452

1928-43-4
$C_{16}H_{22}Cl_2O_3$
333.25
2-Ethylhexyl 2,4-dichloro-phenoxyacetate
Acetic acid, (2,4-dichloro-phenoxy)-, 2-ethylhexyl ester
Caswell No. 315AS
2,4-D 2-ethylhexyl ester
(2,4-Dichlorophenoxy)acetic acid 2-ethylhexyl ester
EPA Pesticide Chemical Code 030063
2-Ethylhexyl (2,4-dichloro-phenoxy)acetate

1928-44-5
$C_{16}H_{22}Cl_2O_3$
333.28
CCCCCCCCOC(=O)COc1ccc(Cl)cc1Cl
Acetic acid, 2,4-dichlorophen-oxy-, octyl ester
2,4-Dichlorphenoxyacetic acid octyl ester
2,4-D-octyl ester
Octyl 2,4-dichlorophenoxy-acetate
Octyl ester of 2,4-D
Oktylester 2,4-dichlorfenoxy-octove (Czech)

1928-45-6
$C_{15}H_{20}Cl_2O_4$
335.25
Acetic acid, 2,4-dichloro-phenoxy-, butoxypropyl ester
2,4-Dichlorophenoxyacetic acid, propylene glycol butyl ether ester
2,4-D PGBE
2,4-D propylene glycol butyl ether ester

1928-47-8
$C_{16}H_{21}Cl_3O_3$
367.72
Acetic acid, (2,4,5-tri-chlorophenoxy)-, 2-ethyl-hexyl ester
2,4,5-T ethylhexyl ester
2,4,5-T 2-ethylhexyl ester
(2,4,5-Trichlorophenoxy)acetic acid 2-ethylhexyl ester

1928-48-9
$C_{15}H_{19}Cl_3O_4$
369.67
Acetic acid, (2,4,5-trichloro-phenoxy)-, 3-butoxypropyl ester (8CI,9CI)
Caswell No. 881O
EPA Pesticide Chemical Code 082055
2,4,5-Trichlorophenoxyacetic acid 3-butoxypropyl ester

1928-57-0
Unknown
Unknown
Butoxyethoxypropyl 2,4-di-chlorophenoxyacetate

1928-58-1
$C_{17}H_{23}Cl_3O_5$
413.73
Acetic acid, (2,4,5-tri chlorophenoxy)-, 3-(2-butoxyethoxy)propyl ester (7CI,8CI)

1928-61-6

$C_{11}H_{12}Cl_2O_3$
263.12
Acetic acid, (2,4-dichlorophenoxy)-, propyl ester (8CI,9CI)

1929-73-3
$C_{14}H_{18}Cl_2O_4$
321.22
CCCCOCCOC(=O)COc1ccc(Cl)cc1Cl
Acetic acid, (2,4-dichlorophenoxy)-, butoxyethyl ester
Acetic acid, (2,4-dichlorophenoxy)-, 2-butoxyethyl ester (8CI,9CI)
2,4-D-Bee
Bladex-B
Brush Killer 64
Butoxyethyl 2,4-dichlorophenoxyacetate
2,4-D butoxyethanol ester
2,4-D butoxyethyl ester
2,4-D 2-butoxyethyl ester
2,4-Dichlorophenoxyacetic acid butoxyethanol ester
(2,4-Dichlorophenoxy)acetic acid butoxyethyl ester
Planotox
Weedone LV 4

1929-77-7
$C_{10}H_{21}NOS$
203.38
CCCSC(=O)N(CCC)CCC
Carbamic acid, dipropylthio-, S-propyl ester
Dipropylthiocarbamic acid S-propyl ester
Perbulate
PPTC
S-Propyl dipropylthiocarbamate
Propyl N,N-dipropylthiolcarbamate
n-Propyl-di-n-propylthiolcarbamate
R-1607
Surpass
Vanalate
Vernam
Vernam 7E
Vernolate

1929-82-4
$C_6H_3Cl_4N$
230.90
n(c(ccc1)C(Cl)(Cl)Cl)c1Cl
Pyridine, 2-chloro-6-(trichloromethyl)
2-Chloro-6-(trichloromethyl)-pyridine
2-Chloro-6-trichloromethyl pyridine (OSHA)
DOWCO-163
Nitrapyrin
N-Serve
N-Serve Nitrogen Stabilizer

1929-86-8
$C_{10}H_{10}ClO_3.K$
168.10
Propionic acid, 2-((4-chloro-o-tolyl)oxy)-, potassium salt
2-((4-Chloro-o-tolyl)oxy)propionic acid potassium salt
Gordon's Mecomec
Hedonal MCPP
MCPP potassium salt
Mecopex
Mecoprop potassium salt
Methoxone M
Propanoic acid, 2-(4-chloro-2-methylphenoxy)-, potassium salt (9CI)
Sys 67MPROP
Vi-Pex

1929-88-0
$C_9H_9N_3OS$
207.27
CNC(=O)Nc2nc1ccccc1s2
Urea, 1-(2-benzothiazolyl)-3-methyl
Bay 60618
Benzothiazole, 2-(1-methylureido)-
N-(2-Benzothiazolyl)-N'-methylurea
Benzthiazuron
N-(2-Benzthiazolyl)-N'-methyl-harnstoff (German)
3-(2-Benzthiazolyl)-1-methyl-harnstoff (German)
Gatinon
Gatnon
1-Methyl-3-(2-benzthiazolyl)-urea
S 22012
Urea, 3-(2-benzothiazolyl)-1-methyl-

1931-62-0
$C_8H_{12}O_5$
188.20
O=C(OOC(C)(C)C)C=CC(=O)O
Maleic monoperoxy acid, 1-tert-butyl ester
tert-Butyl monopermaleate
tert-Butyl monoperoxymaleate
tert-Butyl monoperoxymaleate, Maximum concentration 55% as a paste (DOT)
tert-Butyl monoperoxymaleate, Maximum concentration 55% in solution (DOT)
tert-Butyl peroxymaleate, Not more than 55% as a paste (DOT)
tert-Butyl peroxymaleate, Not more than 55% in solution (DOT)
tert-Butyl monoperoxymaleate, Technically pure (DOT)
tert-Butyl peroxymaleate, Technically pure (DOT)
tert-Butyl peroxymaleic acid
Esperox 25
Esperox 41-40
Maleic monoperoxy acid, 1-tert-butyl ester, Not more than 55% in solution
UN 2099 (DOT)
UN 2100 (DOT)
UN 2101 (DOT)

1932-50-9
$C_2H_4O_3.K$
115.15
Acetic acid, hydroxy-, monopotassium salt (9CI)
Hydroxyacetic acid, potassium salt
Monopotassium hydroxyacetate
Potassium hydroxyacetate

1934-20-9
$C_{16}H_{12}N_2O_4S.Na$
351.35
2-Naphthalenesulfonic acid, 6-hydroxy-5-(phenylazo)-, monosodium salt
Acid Bright Orange Zh
Acid Brilliant Orange Zh
Acidine Orange GN
Acid Orange 12
Acilan Orange G
Acilan Ponceau 4GBL
Amacid Brilliant Orange
Brilliant Orange (6CI)
Brilliant Orange G
Brilliant Orange GN
Brilliant Orange GN Type 8019
C.I. 15970
C.I. Acid Orange 12 (7CI)
C.I. Acid Orange 12, Monosodium salt (8CI)
C.I. Food Orange 1
Croceine Orange
Croceine Orange EN
Crocein Orange G
Croceine Orange 2G
Croceine Orange Y
Crocein Orange
Food Orange No. 1
Helio Orange CAG
Hexacol Orange RN
Hispacid Orange CG
Kiton Brilliant Orange G
Kiton Ponceau 4G
Lutetia Orange 2JR
Monolite Orange C
1008 Orange
Orange G
Orange LZS
Orange RN
Ponceau 4G
Segnale Light Orange GR
Siloton Orange GR
Tertracid Brilliant Orange P4G

1934-21-0
$C_{16}H_9N_4O_9S_2.3Na$
534.38
[Na+].[Na+].[Na+].Oc2c(N=Nc1ccc(cc1)S([O-])(=O)=O)c(nn2c3ccc(cc3)S([O-])(=O)=O)C([O-])=O
Pyrazole-3-carboxylic acid,

5-hydroxy-1-(p-sulfophenyl)-4-(p-sulfophenyl)azo-, trisodium salt
Acid Leather Yellow T
Acid Yellow 23
Acid Yellow T
Acilan Yellow GG
A.F. Yellow No. 4
Airedale Yellow T
Aizen Tartrazine
Amacid Yellow T
Atul Tartrazine
Bucacid Tartrazine
Calcocid Yellow MCG
Calcocid Yellow XX
Canacert Tartrazine
3-Carboxy-5-hydroxy-1-p-sulfophenyl-4-p-sulfophenylazopyrazole trisodium salt
Certicol Tartrazol Yellow S
Cilefa Yellow T
C.I. 640
C.I. 19140
C.I. Acid Yellow 23
C.I. Acid Yellow 23, Trisodium salt
C.I. Food Yellow 4
Curon Fast Yellow 5G
D and C Yellow No. 5
Dolkwal Tartrazine
Dye FD & C Yellow No. 5
E 102
Edicol Supra Tartrazine N
Egg Yellow A
Erio Tartrazine
Eurocert Tartrazine
FD & C Yellow 5
FD & C Yellow No. 5 Tartrazine
Fenazo Yellow T
Food Yellow 4
Food Yellow 5
Food Yellow No. 4
HD Tartrazine
HD Tartrazine Supra
Hexacert Yellow No. 5
Hexacol Tartrazine
Hidazid Tartrazine
Hispacid Fast Yellow T
Hydrazine Yellow
Hydroxine Yellow L
Kako Tartrazine
Kayaku Food Colour Yellow No. 4
Kayaku Tartrazine
KCA Foodcol Tartrazine PF
KCA Tartrazine PF
Kiton Yellow T
Lake Yellow
Lemon Yellow A
Lemon Yellow A Geigy
L-Gelb 2 (German)
Maple Tartrazol Yellow
Mitsui Tartrazine
Naphtocard Yellow O
Neklacid Yellow T
Oxanal Yellow T
San-Ei Tartrazine
Schultz No. 737
Sugai Tartrazine
Tartar Yellow N
Tartar Yellow S
Tartar Yellow FS
Tartar Yellow PF
Tartran Yellow
Tartraphenine
Tartrazine
Tartrazine A Expo T
Tartrazine B
Tartrazine B.P.C.
Tartrazine Extra Pure A
Tartrazine FD & C Yellow #5
Tartrazine FQ
Tartrazine G
Tartrazine Lake
Tartrazine Lake Yellow N
Tartrazine M
Tartrazine MCGL
Tartrazine N
Tartrazine NS
Tartrazine O
Tartrazine T
Tartrazine XX
Tartrazine XXX
Tartrazine Yellow
Tartrazol BPC
Tartrazol Yellow
Tartrine Yellow O
Trisodium 3-carboxy-5-hydroxy-1-p-sulfophenyl-4-p-sulfophenylazopyrazole
Trisodium salt of 3-carboxy-5-hydroxy-1-sulfophenylazo-pyrazole
Unitertracid Yellow TE
Usacert Yellow No. 5
Vondacid Tartrazine
Wool Yellow
Xylene Fast Yellow GT
Y-4
1310 Yellow
1409 Yellow
Yellow Lake 69
Yellow No. 5
Yellow No. 5 FDC
Zlut Kysela 23 (Czech)
Zlut Pigment 100 (Czech)
Zlut Potravinarska 4 (Czech)

1936-15-8
$C_{16}H_{10}N_2O_7S_2.2Na$
452.38
[Na+].[Na+].Oc2ccc1cc(cc(c1c2N=Nc3ccccc3)S([O-])(=O)=O)S([O-])(=O)=O
1,3-Naphthalenedisulfonic acid, 7-hydroxy-8-(phenylazo)-, disodium salt
Acidal Fast Orange
Acid Fast Orange Egg
Acid Fast Orange G
Acid Leather Orange KG
Acid Leather Orange PGW
Acid Light Orange G
Acid Light Orange J
Acid Light Orange JA Export
Acid Light Orange SX
Acid Orange 10
Acid Orange G
Acid Orange 2G
Acid Orange GG
Acilan Orange GX
Apocid Orange 2G
Atul Acid Crystal Orange G
Brasilan Orange 2G
Bucacid Fast Orange G
Calcocid Fast Light Orange 2G
Certicol Orange GS
Cetil Light Orange GG
C.I. 27
C.I. 16230
C.I. Acid Orange 10
C.I. Acid Orange 10, Disodium salt
C.I. Food Orange 4
Colacid Orange G
Crystal Orange 2G
D & C Orange No. 3
Eniacid Light Orange G
Erio Fast Orange AS
Fast Light Orange GA
Fast Light Orange GA-CF
Fast Orange G
Hexacol Orange GG Crystals
Hidacid Fast Orange G
Hidacid Fast Orange 2G
Hispacid Fast Orange 2G
7-Hydroxy-8-(phenylazo)-1,3-naphthalenedisulfonic acid, disodium salt
7-Hydroxy-8-(phenylazo)-1,3-naphthalenedisulphonic acid, disodium salt
Java Orange 2G
Kiton Fast Orange G
Ink Orange JSN
Intracid Fast Orange G
Naphthalene Fast Orange 2G
Naphthalene Fast Orange 2GS
Naphthalene Orange Solide GG
NCI-C53838
Neklacid Fast Light Orange GG
Neklacid Fast Orange 2G
Orange #10
1370 Orange
Orange BPC
Orange G
Orange 2G
Orange G (Biological stain)
Orange G BPC
Orange G Dye
Orange GG
Orange G (Indicator)
Orange GMP
Oranz G (Polish)
Oranz GG (Czech)
Oranz Kysela 10 (Czech)
Oranz Potravinarska 4 (Czech)
1-Phenylazo-2-naphthol-6,8-disulfonic acid, disodium salt
1-Phenylazo-2-naphthol-6,8-disulphonic acid, disodium salt
Schultz No. 39
Solar Light Orange GX
Standacol Orange G
Straight Orange G
Sulfacid Light Orange J
Tertracid Light Orange G
Unitertracid Light Orange G
Vendacid Light Orange 2G
Wool Orange G
Wool Orange 2G
Xylene Fast Orange G

1937-19-5
$CH_6N_4.ClH$
110.57
Guanidine, amino-, hydrochloride
Aminoguanidine hydrochloride
Guanylhydrazine hydrochloride
Hydrazinecarboximidamide hydrochloride

1937-37-7
$C_{34}H_{25}N_9O_7S_2.2Na$
781.78
Nc6ccc(N=Nc1ccc(cc1)c5ccc(N=Nc4c(N)c3c(O)c(N=Nc2ccccc2)c(cc3cc4S(O)(=O)=O)S(O)(=O)=O)cc5)c(N)c6
2,7-Naphthalenedisulfonic acid, 4-amino-3-((4'-((2,4-diaminophenyl)azo)(1,1'-biphenyl)-4-yl) azo)-5-hydroxy-6-(phenylazo)-, disodium salt
Ahco Direct Black GX
Airedale Black ED
Aizen Direct Deep Black EH
Aizen Direct Deep Black GH
Aizen Direct Deep Black RH
Amanil Black GL
Amanil Black WD
Apomine Black GX
Atlantic Black BD
Atlantic Black C
Atlantic Black E
Atlantic Black EA
Atlantic Black GAC
Atlantic Black GG
Atlantic Black GXCW
Atlantic Black GXOO
Atlantic Black SD
Atul Direct Black E
Azine Deep Black EW
Azocard Black EW
Azomine Black EWO
Belamine Black GX
Bencidal Black E
Benzamil Black E
Benzo Deep Black E
Benzo Leather Black E
Benzoform Black BCN-CF
Black 2EMBL
Black 4EMBL
Brasilamina Black GN
Brilliant Chrome Leather Black H
Calcomine Black
Calcomine Black Exl
Carbide Black E
Cern Prima 38 (Czech)
Chloramine Black C
Chloramine Black EC
Chloramine Black ERT
Chloramine Black EX
Chloramine Black EXR
Chloramine Black XO
Chloramine Carbon Black S
Chloramine Carbon Black SJ
Chloramine Carbon Black SN
Chlorazol Black E
Chlorazol Black E (Biological stain)
Chlorazol Black EA
Chlorazol Black EN
Chlorazol Burl Black E
Chlorazol Leather Black ENP
Chlorazol Silk Black G
Chrome Leather Black E
Chrome Leather Black EC
Chrome Leather Black EM
Chrome Leather Black G
Chrome Leather Brilliant Black ER
C.I. 30235
C.I. Direct Black 38
C.I. Direct Black 38, Disodium salt
Coir Deep Black C
Columbia Black EP
Diacotton Deep Black
Diacotton Deep Black RX
Diamine Deep Black EC
Diamine Direct Black E
Diaphtamine Black V
Diazine Black E
Diazine Direct Black E
Diazine Direct Black G
Diazol Black 2V
Diphenyl Deep Black G
Direct Black A
Direct Black BRN
Direct Black CX
Direct Black CXR
Direct Black E
Direct Black EW
Direct Black EX
Direct Black FR
Direct Black GAC
Direct Black GW
Direct Black GX
Direct Black GXR
Direct Black JET
Direct Black Meta
Direct Black Methyl
Direct Black N
Direct Black RX
Direct Black SD
Direct Black WS
Direct Black Z
Direct Black 3
Direct Black 38
Direct Deep Black E
Direct Deep Black E Extra
Direct Deep Black EA-CF
Direct Deep Black EAC
Direct Deep Black EW
Direct Deep Black EX
Enianil Black CN
Erie Black B
Erie Black BF
Erie Black GAC
Erie Black GXOO
Erie Black Jet
Erie Black Nug
Erie Black RXOO
Erie Brilliant Black S
Erie Fibre Black VP
Fenamin Black E
Fibre Black VF
Fixanol Black E
Formaline Black C
Formic Black C
Formic Black CW
Formic Black BA
Formic Black MTG
Formic Black TG
Hispamin Black EF
Interchem Direct Black Z
Kayaku Direct Deep Black EX
Kayaku Direct Deep Black GX
Kayaku Direct Deep Black S
Kayaku Direct Leather Black EX
Kayaku Direct Special Black AAX
Lurazol Black BA
Meta Black
Mitsui Direct Black EX
Mitsui Direct Black GX
NCI-C54557
Nippon Deep Black
Nippon Deep Black GX
Paper Black BA
Paper Black T
Paper Deep Black C
Paramine Black B
Paramine Black E
Peeramine Black E
Peeramine Black GXOO
Phenamine Black BCN-CF
Phenamine Black CL
Phenamine Black E
Phenamine Black E 200
Pheno Black EP
Pheno Black SGN
Pontamine Black E
Pontamine Black EBN
Sandopel Black EX
Seristan Black B
Telon Fast Black E
Tetrazo Deep Black G
Tertrodirect Black E
Tetrodirect Black EFD
Union Black EM
Vondacel Black N

1937-62-8
$C_{19}H_{36}O_2$
296.49
O=C(OC)CCCCCCCC=CCCCCCCC
9-Octadecenoic acid, methyl ester, (E)-
AI3-36449
(E)-Methyl 9-octadecenoate

1940-18-7
$C_8H_{16}O$
128.24
CCC1(O)CCCCC1
Cyclohexanol, 1-ethyl
1-Aethyl-cyclohexanol-(1) (German)
1-Ethylcyclohexanol

1940-29-0
$C_6H_4BrCl_2N$
240.91
Benzenamine, 4-bromo-3,5-dichloro- (9CI)
Aniline, 4-bromo-3,5-dichloro- (8CI)
4-Bromo-3,5-dichloro

aniline

1940-42-7
$C_6H_3BrCl_2O$
241.90
Oc(c(cc(c1Cl)Br)Cl)c1
Phenol, 4-bromo-2,5-dichloro (8CI,9CI)
4-Bromo-2,5-dichlorophenol
Leptophos phenol
Phosvel phenol

1942-45-6
C_8H_{14}
110.20
4-Octyne (8CI,9CI)

1942-46-7
$C_{10}H_{18}$
138.25
5-Decyne (9CI)
NSC-135002

1942-71-8
$C_{16}H_{24}O_2$
248.37
O(c(ccc(c1)C(C)(C)C)c1)C(C(O)CCC2)C2
2-(p-t-Butylphenoxy)cyclohexanol
Cyclohexanol, 2-(p-tert-butylphenoxy)-
Cyclohexanol, 2-(4-(1,1-dimethylethyl)phenoxy)- (9CI)
2-(4-(1,1-Dimethylethyl)phenoxy)cyclohexanol

1943-16-4
$CClN_3O_6$
185.49
Methane, chlorotrinitro
Chlorotrinitromethane

1943-79-9
$C_8H_9NO_2$
151.18
CNC(=O)Oc1ccccc1
Carbamic acid, methyl-, phenyl ester
Methylcarbamic acid phenyl ester
Phenyl monomethylcarbamate

1943-95-9
$C_{12}H_{16}O$
176.26
Oc(cccc1C(CCCC2)C2)c1
Phenol, 3-cyclohexyl- (9CI)
m-Cyclohexylphenol
3-Cyclohexylphenol

1945-53-5
$C_{20}H_{30}O_2$
302.46
Podocarpa-8,13-dien-15-oic acid, 13-isopropyl- (8CI)
1-Phenanthrenecarboxylic acid, 1,2,3,4,4a,5,6,9,10,10a-decahydro-1,4a-dimethyl-7-(1-methylethyl)-, (1R-(1α, 4aβ,10aα))- (9CI)

1948-33-0
$C_{10}H_{14}O_2$
166.24
Oc(c(cc(O)c1)C(C)(C)C)c1
Hydroquinone, tert-butyl
1,4-Benzenediol, 2-(1,1-dimethylethyl)- (9CI)
tert-Butylhydroquinone
MTBHQ
Sustane
TBHQ
Tenox TBHQ

1950-85-2
$CH_4O_2S_2.Na$
135.16
Methanesulfonothioic acid, sodium salt (9CI)
Sodium methanesulfonothioate

1951-12-8
$C_4H_7ClO_2$
122.56
O=C(O)CC(Cl)C
Butyric acid, 3-chloro
Butanoic acid, 3-chloro-
3-Chlorobutanoic acid
β-Chlorobutyric acid
3-Chlorobutyric acid

1953-89-5
$C_8H_7Cl_2NOS$
236.12
OCNC(=S)c1c(Cl)cccc1Cl
Benzamide, 2,6-dichloro-N-(hydroxymethyl)thio
N-Hydroxy-methyl-2,6-dichlorothiobenzamide
TH 073-H

1954-28-5
$C_{12}H_{22}O_6$
262.34
2,2'-(2,5,8,11-Tetraoxa-1,12-dodecane diyl)bis-oxirane
Ayerst 62013
1,2-Bis(2-(2,3-epoxypropoxy)-ethoxy)ethane
1,2:15,16-Diepoxy-4,7,10,13-tetraoxahexadecane
Diglycidyltriethylene glycol
Epodyl
Ethoglucid
Ethoglucide
Etoglucid
Etoglucide
ICI-32865
Oxirane, 2,2'-(2,5,8,11-tetraoxadodecane-1,12-diyl)bis- (9CI)
2,2'-(2,5,8,11-Tetraoxa-1,2-dodecanediyl)bisoxirane
4,7,10,13-Tetraoxahexadecane, 1,2:15,16-diepoxy- (8CI)
TDE
Triethylene glycol diglycidyl ether

1954-81-0
Unknown
Unknown
Sodium chloramben

1955-45-9
$C_5H_8O_2$
100.13
O=C(OC1)C1(C)C
2-Oxetanone, 3,3-dimethyl
3,3-Dimethyl-2-oxetanone
3,3-Dimethyl-2-oxethanone
Dimethyl propiolactone
3,3-Dimethyl-β-propiolactone
NCI-C04126
Pivalic acid lactone
Pivalolactone

1961-72-4
$C_{14}H_{28}O_3$
244.42
Tetradecanoic acid, 3-hydroxy
β-Hydroxymyristic acid
3-Hydroxymyristic acid
β-Hydroxytetradecanoic acid
3-Hydroxytetradecanoic acid

1962-75-0
$C_{16}H_{22}O_4$
278.38
CCCCOC(=O)c1ccc(cc1)C(=O)OCCCC
Terephthalic acid, dibutyl ester
1,4-Benzenedicarboxylic acid, dibutyl ester
Dibutyl terephthalate

1965-38-4
$C_8H_{14}O$
126.20
Bicyclo(3.2.1)octan-2-ol, exo- (8CI,9CI)
2-exo-Bicyclo(3.2.1)octanol
exo-Bicyclo(3.2.1)octan-2-ol
exo-2-Hydroxybicyclo-(3.2.1)octane

1966-58-1
$C_9H_9Cl_2NO_2$
234.09
CNC(=O)OCc1ccc(Cl)c(Cl)c1
Carbamic acid, N-methyl-, 3,4-dichlorobenzyl ester
Benzenemethanol, 3,4-dichloro-, methylcarbamate

3,4-Dichlorbenzylester kyseliny methylkarbaminove (Czech)
3,4-Dichlorobenzyl methylcarbamate
Methylcarbamate 3,4-dichlorobenzyl ester
Rowmate
Sirmate
UC22463

1967-16-4
$C_{11}H_{10}ClNO_2$
223.67
CC(OC(=O)Nc1cccc(Cl)c1)C#C
Carbanilic acid, m-chloro-, 1-methyl-2-propynyl ester
BICP
BIPC
BIPC (The herbicide)
3-Butyn-2-ol, m-chlorocarbanilate
Butyn-1-ol-3-ester of m-chlorophenylcarbamic acid
3-Butynyl-m-chlorocarbanilate
1-Butyn-3-yl-m-chlorophenylcarbamate
Carbamic acid, (3-chlorophenyl)-, 1-methyl-2-propynyl ester (9CI)
Chlorbufam
Chlorbufame
Chlorbupham
Chlorobufam
3-Chlorophenylcarbamic acid 1-methylpropynyl ester
3-Chlorphenyl-carbamidsaure-butin-(1)-yl(3)-ester (German)
Grisemin
Grisin
IEM-1-15
Isobutinyl-N-(3-chlorphenyl)-carbamat (German)
1-Methyl-2-propynyl m-chlorocarbanilate
1-Methyl-2-propynyl m-chlorophenylcarbamate
1-Methylpropynyl 3-chlorophenylcarbamate
1-Methylpropynyl ester of 3-chlorophenylcarbamic acid

1967-25-5
$C_7H_7BrN_2O$
215.04
NC(=O)Nc1ccc(Br)cc1
Urea, (4-bromophenyl)-
AI3-61301

1967-27-7
$C_7H_7ClN_2O$
170.59
NC(=O)Nc1cccc(Cl)c1
3-Chlorophenylurea
meta-Chlorophenylurea

1970-40-7
$C_5H_2Cl_3NO$
198.43
Oc1c(Cl)cnc(Cl)c1Cl
4-Pyridinol, 2,3,5-trichloro
Daxtron
Pyriclor
2,3,5-Trichloro-4-hydroxypyridine
2,3,5-Trichloro-4-pyridinol

1972-08-3
$C_{21}H_{30}O_2$
314.51
CCCCCc3cc(O)c2C1C=C(C)CC C1C(C)(C)Oc2c3
6H-Dibenzo(b,d)pyran-1-ol, 6a,7,8,10a-tetrahydro-6,6,9-trimethyl-3-pentyl
Abbott 40566
Cannabinol, 1-trans-δ^9-tetrahydro-
3-Pentyl-6,6,9-trimethyl-6a,7,8,10a-tetrahydro-6H-dibenzo(b,d)pyran-1-ol
QCD 84924
SP 104
(l)-δ^1-Tetrahydrocannabinol
δ^1-Tetrahydrocannabinol
(-)-δ^1-3,4-trans-Tetrahydrocannabinol
δ^9-Tetrahydrocannabinol
(-)-δ^9-trans-Tetrahydrocannabinol
trans-δ^9-Tetrahydrocannabinol
1-trans-δ^9-Tetrahydrocannabinol
THC
δ^1-THC
δ^9-THC
6,6,9-Trimethyl-3-pentyl-7,8,9,10-tetrahydro-6H-dibenzo(b,d)pyran-1-ol

1972-98-1
$C_{13}H_{17}N_5$
243.31
1,3,5-Triazine-2,4-diamine, N,N'-diethyl-6-phenyl- (9CI)
s-Triazine, 2,4-bis (ethylamino)-6-phenyl- (7CI,8CI)

1973-03-1
$C_{10}H_9ClN_4$
220.66
s-Triazine, 2-chloro-4-(methylamino)-6-phenyl- (7CI,8CI)

1973-04-2
$C_4H_3Cl_2N_3$
163.98
1,3,5-Triazine, 2,4-dichloro-6-methyl- (9CI)
s-Triazine, 2,4-dichloro-6-methyl- (8CI)

1973-05-3
$C_{21}H_{18}N_6$
354.38
n(c(nc(n1)Nc(cccc2)c2)Nc(cccc3) c3)c1Nc(cccc4)c4
1,3,5-Triazine-2,4,6-triamine, N,N',N''-triphenyl- (9CI)
AI3-51094
Triphenylmelamine

1973-06-4
$C_6H_{11}N_5$
153.19
1,3,5-Triazine-2,4-diamine, N,N',6-trimethyl- (9CI)
s-Triazine, 2-methyl-4,6-bis-(methylamino)- (6CI,7CI, 8CI)

1973-07-5
$C_8H_{15}N_5$
181.21
1,3,5-Triazine-2,4-diamine, N,N'-diethyl-6-methyl- (9CI)
s-Triazine, 2,4-bis(ethylamino)-6-methyl- (8CI)

1973-08-6
$C_{21}H_{16}N_4O_2$
356.39
1,3,5-Triazin-2-amine, 4,6-diphenoxy-N-phenyl- (9CI)
6-Anilino-2,4-diphenoxy-s-triazine
s-Triazine, 2-anilino-4,6-diphenoxy- (6CI,7CI,8CI)

1973-09-7
$C_{15}H_{12}ClN_5$
297.72
1,3,5-Triazine-2,4-diamine, 6-chloro-N,N'-diphenyl- (9CI)
NSC-76482
s-Triazine, 2,4-dianilino-6-chloro- (8CI)

1973-22-4
C_8H_9Br
185.06
CCc1ccccc1Br
Benzene, 1-bromo-2-ethyl- (8CI,9CI)

1974-04-5
$C_7H_{15}Br$
179.10
BrC(CCCCC)C
Heptane, 2-bromo- (9CI)
2-Bromoheptane

1974-05-6
$C_7H_{15}Br$
179.10
BrC(CCCC)CC
Heptane, 3-bromo- (9CI)
3-Bromoheptane

1975-50-4
$C_8H_7NO_4$
181.14
O=C(O)c(c(c(N(=O)=O)cc1)C)c1
Benzoic acid, 2-methyl-3-nitro- (9CI)
2-Methyl-3-nitrobenzoic acid
3-Nitro-o-toluic acid

1975-52-6
$C_8H_7NO_4$
181.14
Benzoic acid, 2-methyl-5-nitro- (9CI)
o-Toluic acid, 5-nitro- (8CI)

1975-78-6
$C_{10}H_{19}N$
153.26
N#CCCCCCCCCC
Decanenitrile (9CI)
AI3-11101

1977-10-2
$C_{18}H_{18}ClN_3O$
327.84
CN1CCN(CC1)C3=Nc2ccccc2Oc4ccc(Cl)cc34
Dibenz(b,f)(1,4)oxazepine, 2-chloro-11-(4-methyl-1-piperazinyl)
2-Chloro-11-(4-methyl-1-piperazinyl)-dibenzo(b,f)(1,4)oxazepine
2-Chloro-11-(4-methyl-1-piperazinyl)-dibenzo(b,f)(1,4)oxoazepine
CL-62362
CL-71563
Cloxazepine
Dibenzacepin
Dibenzoazepine
HF 3170
Loxapine
LW 3170
Oxilapine
S-805
SUM 3170

1982-37-2
$C_{18}H_{20}N_2S$
296.46
CN1CCC(C1)CN3c2ccccc2Sc4cccc34
Phenothiazine, 10-((1-methyl-3-pyrrolidinyl)methyl)
Dilosyn
Disyncram
Disyncran
Methdilazine
10-((1-Methyl-3-pyrrolidinyl)-methyl)-phenothiazine
MJ 5022
NCI-C60720
Product 5022
Tacaryl
Tacazyl
Tacryl

1982-47-4
$C_{15}H_{15}ClN_2O_2$
290.77
CN(C)C(=O)Nc2ccc(Oc1ccc(Cl)cc1)cc2
Urea, 3-(p-(p-chlorophenoxy)-phenyl)-1,1-dimethyl
C 1983
3-(4-(4-Chloor-fenoxy)-fenyl)-1,1-dimethylureum (Dutch)
3-(4-(4-Chloro-fenossil)fenil)-1,1-dimetil-urea (Italian)
N'-4-(4-Chlorophenoxy)phenyl-N,N-dimethylurea
3-(p-(p-Chlorophenoxy)phenyl)-1,1-dimethylurea
1-(4-(4-Chloro-phenoxy)-phenyl)-3,3-d'methyluree (French)
Chloroxifenidim
Chloroxuron
3-(4-(4-Chlor-phenoxy)-phenyl)-1,1-dimethylharnstoff (German)
Ciba 1983
Norex
Tenoran

1982-49-6
$C_{14}H_{20}N_2O$
232.36
CC1CCCCC1NC(=O)Nc2ccccc2
Urea, 1-(2-methylcyclohexyl)-3-phenyl
Du Pont 1318
Du Pont Herbicide 1,318
H 1318
1-(2-Methylcyclohexyl)-3-phenylurea
N-Phenyl-N'-(2-methylcyclohexyl)urea
Siduron
Tupersan

1982-55-4
$C_6HCl_3N_2S$
239.50
Clc2cc(Cl)c1nsnc1c2Cl
2,1,3-Benzothiadiazole, 4,5,7-trichloro
052 H
PH 40-21
TH 052
TH 052-H
4,5,7-Trichloro-2,1,3-benzothiadiazole
4,5,7-Trichlorobenzthiadiazole-2,1,3

1982-67-8
$C_5H_{12}N_2O_3S$
180.25
Sulfoximine, S-(3-amino-3-carboxypropyl)-S-methyl-, DL
Butanoic acid, 2-amino-4-(S-methylsulfonimidoyl)-(9CI)
Methionine sulfoximine
Methionine-DL-sulfoximine, DL-

1982-69-0
$C_8H_6Cl_2O_3.Na$
244.03
Sodium 2-methoxy-3,6-dichlorobenzoate
o-Anisic acid, 3,6-dichloro-, sodium salt
Benzoic acid, 3,6-dichloro-2-methoxy-, sodium salt (9CI)
Dicamba sodium salt
Dicamba-sodium
3,6-Dichloro-o-anisic acid, sodium salt
3,6-Dichloro-2-methoxybenzoic acid, sodium salt
2-Methoxy-3,6-dichlorobenzoic acid sodium salt
Sodium dicamba
Sodium 3,6-dichloro-2-methoxybenzoate

1983-10-4
$C_{12}H_{27}FSn$
309.08
Stannane, fluorotributyl
Fluorotributylstannane
Tin, tributyl-, fluoride
Tributylstannane fluoride
Tributyltin fluoride

1984-04-9
$C_{11}H_7N$
153.19
Naphthalene, 1-isocyano- (9CI)
1-Naphthyl isocyanide (7CI, 8CI)

1984-06-1
$C_8H_{16}O_2.Na$
167.23
Octanoic acid, sodium salt
Caprylic acid sodium salt
Sodium caprylate
Sodium octanoate
Sodium n-octanoate

1984-58-3
$C_7H_6Cl_2O$
177.03
O(c(c(ccc1Cl)Cl)c1)C
Benzene, 1,4-dichloro-2-methoxy- (9CI)
2,5-Dichloroanisole
1,4-Dichloro-2-methoxybenzene

1984-59-4
$C_7H_6Cl_2O$
177.03
O(c(c(c(cc1)Cl)Cl)c1)C

Benzene, 1,2-dichloro-3-methoxy- (9CI)
1,2-Dichloro-3-methoxybenzene

1984-65-2
$C_7H_6Cl_2O$
177.03
O(c(c(ccc1)Cl)c1Cl)C
Benzene, 1,3-dichloro-2-methoxy- (9CI)
2,6-Dichloroanisole
1,3-Dichloro-2-methoxybenzene

1984-77-6
$C_{15}H_{28}O_3$
256.43
Lauric acid, 2,3-epoxypropyl ester
Glycidyl laurate

1985-46-2
$C_5H_{10}N_6$
154.21
Melamine, N^2,N^2-dimethyl

1985-51-9
$C_{13}H_{20}O_4$
240.30
O=C(OCC(C)(C)COC(=O)C(=C)C)C(=C)C
2-Propenoic acid, 2-methyl-, 2,2-dimethyl-1,3-propanediyl ester (9CI)
2,2-Dimethylpropane dimethacrylate

1985-57-5
$C_{12}H_{18}$
162.28
Benzene, (1,1-dimethylbutyl)- (6CI,7CI,8CI,9CI)
(1,1-Dimethylbutyl)benzene
tert-Hexylbenzene
2-Methyl-2-phenylpentane

1985-59-7
$C_{12}H_{16}$
160.26
Naphthalene, 1,2,3,4-tetrahydro-1,1-dimethyl- (6CI, 7CI,8CI,9CI)
1,1-Dimethyltetralin

1986-70-5
$C_{29}H_{40}O_9$
532.69
Calotropin
Pecilocerin A
Pokilocerin A

1987-50-4
$C_{13}H_{20}O$
192.30
Oc(ccc(c1)CCCCCCC)c1
Phenol, 4-heptyl- (9CI)
4-Heptylphenol

1996-88-9
$C_{14}H_9F_{17}O_2$
532.20
O=C(OCCC(F)(F)C(F)(F)C(F)(F)C(F)(F)C(F)(F)C(F)(F)C(F)(F)C(F)(F)F)C(=C)C
2-Propenoic acid, 2-methyl-, 3,3,4,4,5,5,6,6,7,7,8,8,9,9, 10,10,10-heptadecafluorodecyl ester (9CI)

2000-43-3
$C_8H_7Cl_3O$
225.50
OC(c(cccc1)c1)C(Cl)(Cl)Cl
Benzyl alcohol, α-(trichloromethyl)
Benzenemethanol, α-(trichloromethyl)- (9CI)
Efiran 99
Phenyl(trichloromethyl)carbinol
α-(Trichloromethyl)benzenemethanol
α-(Trichloromethyl)benzyl alcohol
Trichloromethylphenyl carbinol

2001-95-8
$C_{54}H_{90}N_6O_{18}$
1111.50
Valinomycin
Antibiotic N-329 B
NSC-122023
Valinomicin

2002-24-6
$C_2H_7NO.ClH$
97.56
Ethanol, 2-amino-, hydrochloride
β-Aminoethanol hydrochloride
2-Aminoethanol hydrochloride
Colamine hydrochloride
Ethanolamine chloride
Ethanolamine hydrochloride
MEA Hydrochloride
Monoethanolamine hydrochloride

2002-59-7
$C_5H_4N_4OS$
168.15
6-Thioxanthine
AI3-52244
2H-Purin-2-one, 1,3,6,7-tetrahydro-6-thioxo-

2002-60-0
$C_5H_4N_4O_2S$
184.15
6-Thiouric acid
AI3-51991
1H-Purine-2,8(3H,6H)-dione, 7,9-dihydro-6-thioxo- (9CI)
Uric acid, 6-thio- (VAN) (8CI)

2004-03-7
$C_6H_6N_4$
134.16
Cc1ncnc2nc[nH]c12
Purine, 6-methyl
6-Methylpurine
1H-Purine, 6-methyl- (9CI)

2004-70-8
C_5H_8
68.13
C(=CC=C)C
1,3-Pentadiene, (E)
trans-1,3-Pentadiene
trans-Piperylene

2005-98-3
$C_{17}H_{23}N_7O_5$
405.42
Cytomycin (9CI)
Saitomycin

2008-39-1
$C_{10}H_{11}Cl_2NO_3$
264.12
CN(C)OC(=O)COc1ccc(Cl)cc1Cl
Acetic acid, (2,4-dichlorophenoxy)-, Compd. with dimethylamine (1:1)
Acetic acid, (2,4'dichlorophenoxy)-, Compd. with N-methylmethanamine (1:1) (9CI)
Bladex G
2,4-D acetate
2,4-D amine
2,4-D amine salt
2,4-D dimethylamine salt
Defy
Demise
(2,4-Dichlorophenoxy)acetic acid dimethylamine
Dimethylamine, (2,4-dichlorophenoxy)acetate
Dimethylamine salt of 2,4-D
Dimethylammonium 2,4-dichlorophenoxyacetate
Formula 40
Hormin
Phordene
Reed Amine 400

2008-41-5
$C_{11}H_{23}NOS$
217.41
O=C(N(CC(C)C)CC(C)C)SCC
Carbamic acid, diisobutylthio-, S-ethyl ester
Bis(2-methylpropyl)carbamothioic acid S-ethyl ester
Butilate

Butylate
Carbamothioic acid, bis(2-methylpropyl)-, S-ethyl ester
Diisobutylthiocarbamic acid S-ethyl ester
Diisocarb
S-Ethyl bis(2-methylpropyl)carbamothioate
Ethyl N,N-diisobutylthiocarbamate
S-Ethyldiisobutyl thiocarbamate
S-Ethyl N,N-diisobutylthiocarbamate
Ethyl-N,N-diisobutyl thiolcarbamate
R-1910
Stauffer R-1910
Sutan

2008-46-0
$C_8H_5Cl_3O_3.C_6H_{15}N$
308.62
2,4,5-Trichlorophenoxyacetic acid triethylamine salt
Acetic acid, (2,4,5-trichlorophenoxy)-, Compd. with N,N-diethylethanamine (1:1) (9CI)
Acetic acid, (2,4,5-trichlorophenoxy)-, Compd. with Et_3N
Acetic acid, (2,4,5-trichlorophenoxy)-, Compd. with triethylamine (1:1) (8CI)
Caswell No. 881K
EPA Pesticide Chemical Code 082034

2008-58-4
$C_7H_5Cl_2NO$
190.02
NC(=O)c1c(Cl)cccc1Cl
2,6-Dichlorobenzamide
Benzamide, 2,6-dichloro- (9CI)

2011-67-8
$C_{16}H_{13}N_3O_3$
295.28
Nimetazepam
2H-1,4-Benzodiazepin-2-one, 1,3-dihydro-1-methyl-7-nitro-5-phenyl-
DEA No. 2837
1,3-Dihydro-1-methyl-7-nitro-5-phenyl-2H-1,4-benzodiazepin-2-one
Elimin
Hypnon
1-Methylnitrazepam
1-Methyl-7-nitro-5-phenyl-1,3-dihydro-2H-1,4-benzodiazepin-2-one
1-Methyl-5-phenyl-7-nitro-1,3-dihydro-2H-1,4-benzodiazepin-2-one
Nimetazepamum (Latin)
S 1530

2012-74-0
$C_{11}H_{13}ClO_2$
212.69
O=C(O)C(c(ccc(c1)Cl)c1)C(C)C
Butyric acid, 2-(p-chlorophenyl)-3-methyl
2-(p-Chlorophenyl)-3-methylbutyric acid

2012-81-9
$C_{11}H_{12}ClN$
193.67
N#CC(c(ccc(c1)Cl)c1)C(C)C
Benzeneacetonitrile, 4-chloro-α-(1-methylethyl)- (9CI)
α-Isopropyl-p-chlorophenylacetonitrile

2016-42-4
$C_{14}H_{31}N$
213.40
NCCCCCCCCCCCCCC
Tetradecylamine (8CI)
AI3-15076
1-Tetradecanamine (9CI)

2016-48-0
$C_{14}H_{31}N.ClH$
249.92
[Cl-].CCCCCCCCCCCCN(C)C
Dodecylamine, N,N-dimethyl-, hydrochloride
Dimethyldodecylamine chlorhydrate (French)
N,N-Dimethyldodecylamine hydrochloride

2016-52-6
$C_{16}H_{35}N.C_2H_4O_2$
301.51
1-Hexadecanamine, acetate (9CI)
Armac 16D
Hexadecylamine, acetate (8CI)

2016-56-0
$C_{12}H_{27}N.C_2H_4O_2$
245.46
CCCCCCCCCCCCN.CC(O)=O
Dodecylamine, acetate
Dodecanamine acetate
1-Dodecylamine acetate

2016-57-1
$C_{10}H_{23}N$
157.34
NCCCCCCCCCC
1-Decanamine
1-Aminodecane
Decylamine

2016-88-8
$C_6H_8ClN_7O.ClH$
266.05
NC(=N)NC(=O)c1nc(Cl)c(N)nc1N
Pyrazinecarboxamide, 3,5-diamino-N-(aminoiminomethyl)-6-chloro-, monohydrochloride (9CI)
Amiloride hydrochloride
Amipramizide
Nilurid
Pyrazinecarboxamide, N-amidino-3,5-diamino-6-chloro-, monohydrochloride (8CI)

2021-20-7
$C_{11}H_{15}NO_2$
193.24
1H-Isoindole-1,3(2H)-dione, 3a,4,7,7a-tetrahydro-2-propyl- (9CI)
4-Cyclohexene-1,2-dicarboximide, N-propyl- (8CI)

2021-28-5
$C_{11}H_{14}O_2$
178.23
O=C(OCC)CCc(cccc1)c1
Benzenepropanoic acid, ethyl ester (9CI)
AI3-11591
Ethyl benzenepropanoate
Hydrocinnamic acid, ethyl ester (8CI)

2022-85-7
$C_4H_4FN_3O$
129.11
Nc1nc(=O)[nH]cc1F
Cytosine, 5-fluoro
Alcobon
4-Amino-5-fluoro-2(1H)-pyrimidinone
Ancobon
Ancotil
5-FC
Flucytosine
5-Fluorocystosine
5-Fluorocytosine
2-Hydroxy-4-amino-5-fluoropyrimidine
Ro 2-9915

2026-24-6
$C_{20}H_{31}N.C_2H_4O_2$
345.52
Dehydroabietylamine acetate
Caswell No. 276A
EPA Pesticide Chemical Code 004201
1-Phenanthrenemethanamine, 1,2,3,4,4a,9,10,10a-octahydro-1,4a-dimethyl-7-(1-methylethyl)-, acetate, (1R-(1α,4aβ,10aα))- (9CI)
Podocarpa-8,11,13-trien-15-amine, 13-isopropyl-, acetate (8CI)

2027-17-0
$C_{13}H_{14}$

170.27
c(c(ccc1C(C)C)ccc2)(c2)c1
Naphthalene, 2-isopropyl
2-Isopropylnaphthalene

2027-47-6
$C_{18}H_{34}O_2$
282.47
O=C(O)CCCCCCCC=CCCCC CCCC
9-Octadecenoic acid (VAN) (9CI)

2031-67-6
$C_7H_{18}O_3Si$
178.34
CCOSi(C)(OCC)OCC
Silane, methyltriethoxy
Methyltriethoxysilane
Silane, triethoxymethyl-
Triethoxy-methylsilane
Union Carbide A-162

2032-59-9
$C_{11}H_{16}N_2O_2$
208.29
CNC(=O)Oc1ccc(N(C)C)c(C)c1
Carbamic acid, methyl-, 4-dimethylamino-m-tolyl ester
A 363
Aminocarb
Aminocarbe (French)
Bay 44646
Bayer 5080
Bayer 44646
Carbamic acid, methyl-, 4-(dimethylamino)-3-methylphenyl ester
m-Cresol, 4-(dimethylamino)-, methylcarbamate (ester)
4-Dimethylamine m-cresyl methylcarbamate
4-Dimethylamino-3-cresyl methylcarbamate
4-(Dimethylamino)-3-methylphenol methyl carbamate (ester)
(4-Dimethylamino-3-methylphenyl)N-methyl-carbamaat (Dutch)
(4-Dimethylamino-3-methylphenyl)N-methyl-carbamat (German)
(4-Dimethylamino-3-methylphenyl)N-methyl-carbamate
4-(Dimethylamino)-m-tolyl methylcarbamate
(4-Dimetilamino-3-metil-fenil)-N-metil-carbammato (Italian)
ENT 25,784
Matacil
N-Methylcarbamate de 4-dimethylamino 3-methyl phenyle (French)
Mitacil

2032-65-7
$C_{11}H_{15}NO_2S$
225.33
CNC(=O)Oc1cc(C)c(SC)c(C)c1
Carbamic acid, N-methyl-, 4-(methylthio)-3,5-xylyl ester
B 37344
Bay 9026
Bay 37344
Bayer 37344
Carbamic acid, methyl-, 3,5-dimethyl-4-(methylthio)phenyl ester
3,5-Dimethyl-4-(methylthio)-phenol methylcarbamate
3,5-Dimethyl-4-methyl-thio-phenyl-N-carbamat (German)
3,5-Dimethyl-4-methylthio-phenyl N-methylcarbamate
Draza
ENT 25,726
H 321
Mercaptodimethur
Mesurol
Methiocarb
Methiocarbe
Methyl carbamic acid 4-(methylthio)-3,5-xylyl ester
4-Methylmercapto-3,5-dimethylphenyl N-methylcarbamate
4-Methylmercapto-3,5-xylyl methylcarbamate
4-Methylthio-3,5-dimethylphenyl methylcarbamate
4-(Methylthio)-3,5-xylyl methylcarbamate
Metmercapturon
OMS-93
3,5-Xylenol, 4-(methylthio)-, methylcarbamate

2033-89-8
$C_8H_{10}O_3$
154.17
O(c(c(OC)cc(O)c1)c1)C
Phenol, 3,4-dimethoxy- (9CI)
3,4-Dimethoxyphenol

2034-22-2
$C_3HBr_3N_2$
304.79
Brc1nc(Br)c(Br)[nH]1
Imidazole, 2,4,5-tribromo
2,4,5-Tribromoimidazole

2035-99-6
$C_{13}H_{26}O_2$
214.39
O=C(OCCC(C)C)CCCCCCC
Octanoic acid, isopentyl ester
Isoamyl caprylate
Isoamyl octanoate
Isopentyl octanoate

2036-15-9
Unknown
Unknown
Dipropylaluminum hydride

2037-26-5
C_7D_8
84.08
Benzene-d5, methyl-d3- (9CI)
Toluene-d8 (8CI)

2037-31-2
C_6H_5ClS
144.62
Sc(cccc1Cl)c1
Benzenethiol, m-chloro- (8CI)
Benzenethiol, 3-chloro- (9CI)
3-Chlorobenzenethiol

2038-03-1
$C_6H_{14}N_2O$
130.18
O(CCN(C1)CCN)C1
4-(2-Aminoethyl)-morpholine
AI3-52273
β-Aminoaethyl-morpholin (German)
N-2-Aminoethylmorfolin (Czech)
N-Aminoethylmorpholine
N-2-Aminoethylmorpholine
Morpholine, 4-(2-aminoethyl)-
4-Morpholineethanamine

2039-34-1
$C_{16}H_{15}N_5$
277.33
1,3,5-Triazine-2,4-diamine, 6-methyl-N,N'-diphenyl- (9CI)
s-Triazine, 2,4-dianilino-6-methyl- (7CI,8CI)

2039-37-4
$C_{16}H_{14}N_4$
262.32
s-Triazine, 2-(methyl amino)-4,6-diphenyl- (7CI,8CI)

2039-46-5
Unknown
Unknown
Dimethylamine 2-methyl-4-chlorophenoxyacetate
Caswell No. 557G
EPA Pesticide Chemical Code 030516
2-Methyl-4-chlorophenoxyacetic acid, dimethylamine salt

2039-82-9
C_8H_7Br
183.05
Brc1ccc(C=C)cc1
Benzene, 1-bromo-4-ethenyl- (9CI)
Styrene, p-bromo- (8CI)

2039-85-2
C_8H_7Cl
138.60
c(cccc1C=C)(c1)Cl
m-Chlorostyrene
Benzene, 1-chloro-3-ethenyl- (9CI)
1-Chloro-3-ethenylbenzene
3-Chlorostyrene
Styrene, m-chloro- (8CI)
Styrene, 3-chloro-

2039-87-4
C_8H_7Cl
138.60
Clc1ccccc1C=C
Styrene, o-chloro
o-Chlorostyrene (ACGIH, OSHA)
2-Chlorostyrene

2039-88-5
C_8H_7Br
183.05
Benzene, 1-bromo-2-ethenyl- (9CI)
Styrene, o-bromo- (8CI)

2039-89-6
$C_{10}H_{12}$
132.22
Styrene, 2,5-dimethyl
2,5-Dimethylstyrene

2039-90-9
$C_{10}H_{12}$
132.21
Benzene, 2-ethenyl-1,3-dimethyl- (9CI)
Styrene, 2,6-dimethyl- (8CI)

2040-00-8
$C_4H_{10}AlI$
212.01
Aluminum, diethyliodo- (9CI)
Diethyliodoaluminum

2040-04-2
$C_{10}H_{12}O_3$
180.22
Acetophenone, 2',6'-dimethoxy
2,6-Dimethoxyacetophenone
Ethanone, 1-(2,6-dimethoxyphenyl)- (9CI)
USAF K-2801

2040-88-2
C_6H_4BrClO
207.45
Phenol, 2-bromo-6-chloro- (8CI,9CI)

2040-95-1
C_9H_{18}
126.24
Cyclopentane, butyl- (8CI,9CI)
NSC-74179

2040-96-2
C_8H_{16}
112.24
Cyclopentane, propyl
Propylcyclopentane

2042-14-0
$C_7H_7NO_3$
153.13
Phenol, 4-methyl-3-nitro- (9CI)
p-Cresol, 3-nitro- (8CI)
NSC-41205

2042-37-7
C_7H_4BrN
182.01
Benzonitrile, o-bromo- (8CI)
Benzonitrile, 2-bromo- (9CI)

2043-47-2
$C_6H_5F_9O$
264.09
FC(F)(F)C(F)(F)C(F)(F)C(F)(F)CCO
1-Hexanol, 3,3,4,4,5,5,6,6,6-nonafluoro- (9CI)
3,3,4,4,5,5,6,6,6-Nonafluoro-1-hexanol
1,1,2,2-Tetrahydroperfluoro-1-hexanol

2043-55-2
$C_6H_4F_9I$
373.99
FC(F)(F)C(F)(F)C(F)(F)C(F)(F)CCI
Hexane, 1,1,1,2,2,3,3,4,4-nonafluoro-6-iodo- (9CI)
1,1,1,2,2,3,3,4,4-Nonafluoro-6-iodohexane
1,1,2,2-Tetrahydroperfluorohexyl iodide
1,1,2,2-Tetrahydroperfluorohexyliodide

2043-57-4
$C_8H_4F_{13}I$
474.00
FC(F)(F)C(F)(F)C(F)(F)C(F)(F)C(F)(F)C(F)(F)CCI
Octane, 1,1,1,2,2,3,3,4,4,5,5,6,6-tridecafluoro-8-iodo- (9CI)
1,1,2,2-Tetrahydroperfluorooctyl iodide
1,1,2,2-Tetrahydroperfluorooctyliodide

2043-61-0
$C_7H_{12}O$
112.17
Cyclohexanaecarboxaldehyde
AI3-05667
Cyclohexanecarboxaldehyde (9CI)

2044-00-0
$C_6H_8O_5$
160.13
threo-Hexaric acid, 2,5-anhydro-3,4-dideoxy- (9CI)
2,5-Furandicarboxylic acid, tetrahydro-, trans- (8CI)

2044-64-6
$C_6H_{11}NO_2$
129.18
O=C(N(C)C)CC(=O)C
Acetoacetamide, N,N-dimethyl
Butanamide, N,N-dimethyl-3-oxo- (9CI)
N,N-Dimethylacetoacetamide
Dimethylamid kyseliny acetoctove (Czech)

2045-52-5
$C_{17}H_{22}N_2.ClH$
290.87
[Cl-].CN(C)CCN(Cc1ccccc1)c2ccccc2
Ethylenediamine, N-benzyl-N',N'-dimethyl-N-phenyl-, hydrochloride
Aniline, N-benzyl-N-dimethylaminoethyl-, hydrochloride
Antergan hydrochloride
Compound 2339 RP
Corps 2339 R P (French)
N-Dimethylaminoethylbenzylaniline hydrochloride
Phenbenzamine hydrochloride
N-Phenyl-N-benzyl-N',N'-dimethylethylenediamine hydrochloride
2339 R.P. Hydrochloride

2049-70-9
$C_{10}H_{18}O_4$
202.25
Propanedioic acid, ethylmethyl-, diethyl ester (9CI)
Malonic acid, ethylmethyl-, diethyl ester (8CI)

2049-92-5
$C_{11}H_{17}N$
163.26
p-tert-Amylaniline
Benzenamine, 4-(1,1-dimethylpropyl)- (9CI)
NSC-4701
NSC-7128

2049-94-7
$C_{11}H_{16}$

148.25
Benzene, (3-methylbutyl)- (9CI)
Benzene, isopentyl- (8CI)
NSC-62142

2049-95-8
$C_{11}H_{16}$
148.27
c(cccc1)(c1)C(CC)(C)C
Benzene, tert-pentyl
tert-Amylbenzene
tert-Pentylbenzene

2049-96-9
$C_{12}H_{16}O_2$
192.28
O=C(OCCCCC)c(cccc1)c1
Benzoic acid, pentyl ester
Amyl benzoate
Pentyl benzoate
n-Pentyl benzoate

2050-08-0
$C_{12}H_{16}O_3$
208.28
O=C(OCCCCC)c(c(O)ccc1)c1
Salicylic acid, pentyl ester
Amylester kyseliny salicylove (Czech)
Amyl salicylate
Pentyl salicylate

2050-20-6
$C_{11}H_{20}O_4$
216.28
Heptanedioic acid, diethyl ester (9CI)
NSC-17503
Pimelic acid, diethyl ester (8CI)

2050-23-9
$C_{12}H_{22}O_4$
230.30
Octanedioic acid, diethyl ester (9CI)
NSC-62701
Suberic acid, diethyl ester (8CI)

2050-24-0
$C_{11}H_{16}$
148.27
c(cc(cc1CC)CC)(c1)C
Toluene, 3,5-diethyl
Benzene, 1,3-diethyl-5-methyl-
1,3-Diethyl-5-methylbenzene
3,5-Diethyltoluene

2050-43-3
$C_{10}H_{13}NO$
163.24
O=C(Nc(c(cc(c1)C)C)c1)C
2',4'-Acetoxylidide
Acetamide, N-(2,4-dimethyl-phenyl)- (9CI)
Acetanilide, 2',4'-dimethyl-
2',4'-Acetoxylidine
2,4-Dimethylacetanilide
2',4'-Dimethylacetanilide

2050-47-7
$C_{12}H_8Br_2O$
328.02
O(c(ccc(c1)Br)c1)c(ccc(c2)Br)c2
Ether, bis(4-bromophenyl)
Bis(p-bromophenyl) ether
USAF DO-61

2050-60-4
$C_{10}H_{18}O_4$
202.25
O=C(OCCCC)C(=O)OCCCC
Ethanedioic acid, dibutyl ester (9CI)
AI3-06011
Dibutyl ethanedioate
Oxalic acid, dibutyl ester (8CI)

2050-67-1
$C_{12}H_8Cl_2$
223.10
3,3'-Dichlorobiphenyl
3,3'-Dichloro-1,1'-biphenyl

2050-68-2
$C_{12}H_8Cl_2$
223.10
Clc1ccc(cc1)c2ccc(Cl)cc2
4,4'-Dichlorobiphenyl
AI3-09066
Biphenyl, 4,4'-dichloro- (8CI)
1,1'-Biphenyl, 4,4'-dichloro- (9CI)

2050-69-3
$C_{10}H_6Cl_2$
197.06
c(c(c(c(c1)Cl)Cl)ccc2)(c1)c2
Naphthalene, 1,2-dichloro- (9CI)
1,2-Dichloronaphthalene

2050-72-8
$C_{10}H_6Cl_2$
197.06
c(c(c(cc1)Cl)ccc2Cl)(c1)c2
Naphthalene, 1,6-dichloro- (9CI)
1,6-Dichloronaphthalene

2050-73-9
$C_{10}H_6Cl_2$
197.06
c(c(c(cc1)Cl)cc(c2)Cl)(c2)c1
Naphthalene, 1,7-dichloro- (9CI)
1,7-Dichloronaphthalene

2050-74-0
$C_{10}H_6Cl_2$
197.06
c(c(c(cc1)Cl)c(cc2)Cl)(c1)c2
Naphthalene, 1,8-dichloro- (9CI)
1,8-Dichloronaphthalene

2050-75-1
$C_{10}H_6Cl_2$
197.06
c(c(ccc1)cc(c2Cl)Cl)(c1)c2
Naphthalene, 2,3-dichloro- (9CI)
2,3-Dichloronaphthalene

2050-76-2
$C_{10}H_6Cl_2O$
213.06
Oc(c(c(c(c1)Cl)ccc2)c2)c1Cl
1-Naphthalenol, 2,4-dichloro- (9CI)
AI3-00183
2,4-Dichloro-1-naphthalenol
1-Naphthol, 2,4-dichloro- (8CI)

2050-77-3
$C_{10}F_{21}I$
646.00
C(CCCCCCCCC)I
Decane, 1-iodo
Decyl iodide
1-Decyl iodide
1-Iododecane

2050-92-2
$C_{10}H_{23}N$
157.34
N(CCCCC)CCCCC
1-Pentanamine, n-pentyl
Diamyl amine
Di-n-amylamine [UN 2841]
Dipentylamine
Pentylamine, pentyl-
UN 2841 [Di-n-amylamine]

2051-24-3
$C_{12}Cl_{10}$
498.66
Clc1c(Cl)c(Cl)c(c(Cl)c1Cl)c2c(Cl)c(Cl)c(Cl)c(Cl)c2Cl
Decachlorobiphenyl
AI3-16253
1,1'-Biphenyl, 2,2',3,3',4,4',5,5',6,6'-decachloro-

2051-30-1
$C_{10}H_{22}$
142.28
Octane, 2,6-dimethyl- (9CI)

2051-31-2
$C_{10}H_{22}O$
158.28
4-Decanol (8CI,9CI)
NSC-2637

2051-60-7
$C_{12}H_9Cl$
188.66
Clc1ccccc1c2ccccc2
Biphenyl, 2-chloro
2-Chlorobiphenyl
2-Chloro-1,1'-biphenyl
o-Chlorodiphenyl
2-Chlorodiphenyl

2051-61-8
$C_{12}H_9Cl$
188.66
Clc1cccc(c1)c2ccccc2
3-Chlorobiphenyl

2051-62-9
$C_{12}H_9Cl$
188.66
Clc1ccc(cc1)c2ccccc2
Biphenyl, 4-chloro
4-Chlorobiphenyl
4-Chloro-1,1'-biphenyl
p-Chlorodiphenyl
4-Chlorodiphenyl

2051-79-8
$C_{11}H_{18}N_2$.ClH
214.77
p-Phenylenediamine, N,N-diethyl-3-methyl-,hydrochloride
4-Amino-3-methyl-N,N-diethylaniline hydrochloride

2051-85-6
$C_{12}H_{10}N_2O_2$
214.24
Oc(ccc(N=Nc(cccc1)c1)c2O)c2
1,3-Benzenediol, 4-(phenylazo)
Benzeneazoresorcinol
Ceres Orange G
Ceres Orange GN
Cerisol Yellow GR
C.I. 11920
C.I. Food Orange 3
C.I. Solvent Orange 1 (8CI)
2,4-Dibenzeneazoresorcinol
2,4-Dihydroxyazobenzene
Fast Oil Orange T
Fast Oil Yellow G
Fast Oil Yellow 2G
Fat Orange A
Fat Orange G
Fat Orange GS
Fat Orange RG
Fat Victoria Yellow D
Grasol Yellow RSF
Hexacol Oil Yellow GG
Lacquer Orange V 3G
Oil Orange G
Oil Orange 4G
Oil Orange MO
Oil Orange MON
Oil Orange MON Extra
Oil-Sol. Yellow ZH
Oil Yellow G Extra
Oil Yellow GG
Oranz Potravinarska 3 (Czech)
Oranz Rozpoustedlova 1 (Czech)
Organol Orange 2J
4-(Phenylazo)resorcinol
Plastoresin Orange F 3A
Resinol Orange G
Solvent Orange 1
Sudan G
Sudan Orange G
Sudan Yellow AR
Tertrogras Orange SG
1504 Yellow
Yellow M Soluble in Grease

2051-90-3
$C_{13}H_{10}Cl_2$
237.13
c(cccc1)(c1)C(c(cccc2)c2)(Cl)Cl
Benzene, 1,1'-(dichloromethylene)bis- (9CI)
1,1'-(Dichloromethylene)bisbenzene
Methane, dichlorodiphenyl- (8CI)

2051-95-8
$C_{10}H_{10}O_3$
178.19
O=C(c(cccc1)c1)CCC(=O)O
Benzenebutanoic acid, γ-oxo- (9CI)
AI3-03666
γ-Oxobenzenebutanoic acid
Propionic acid, 3-benzoyl- (8CI)

2052-06-4
$C_{10}H_{14}BrN$
228.13
N(c(ccc(c1)Br)c1)(CC)CC
Aniline, p-bromo-N,N-diethyl- (8CI)
Benzenamine, 4-bromo-N,N-diethyl- (9CI)
p-Bromo-N,N-diethylaniline
4-Bromo-N,N-diethylbenzenamine

2052-07-5
$C_{12}H_9Br$
233.11
c(c(c(cccc1)c1)ccc2)(c2)Br
1,1'-Biphenyl, 2-bromo- (9CI)
AI3-11170
2-Bromobiphenyl
2-Bromo-1,1'-biphenyl

2052-15-5
$C_9H_{16}O_3$
172.25
O=C(OCCCC)CCC(=O)C
Levulinic acid, butyl ester
Butyl laevulinate
n-Butyl laevulinate
Butyl levulinate
n-Butyl levulinate
Butyl 4-oxopentanoate
4-Ketopentanoic acid butyl ester
Pentanoic acid, 4-oxo-, butyl ester (9CI)

2052-25-7
$C_{16}H_{12}N_2O_6S$.Na
383.32
Benzenesulfonic acid, 3-((1,5-dihydroxy-2-naphthalenyl)azo)-4-hydroxy-, monosodium salt (9CI)

2057-49-0
$C_{14}H_{15}N$
197.27
n(ccc(c1)CCCc(cccc2)c2)c1
Pyridine, 4-(3-phenylpropyl)- (9CI)
4-(3-Phenylpropyl)pyridine

2058-46-0
$C_{22}H_{24}N_2O_9$.ClH
496.94
[Cl-].CN(C)C2C1C(O)C4C(=C(O)C1(O)C(=O)C(=C2O)C(N)=O)C(=O)c3c(O)cccc3C4(C)O
2-Naphthacenecarboxamide, 4-(dimethylamino)-1,4,4a,5,5a,6,11,12a-octahydro-3,5,6,10,12,12a- hexahydroxy-6-methyl-1,11-dioxomonohydrochloride
Aquacycline
Biosolvomycin
Engemycin
Hydrocyclin
5-Hydroxytetracycline hydrochloride
Liquamycin injectable
Mepatar
NSC-9169
Otetryn
Oxlopar
Oxyject 100
Oxytetracycline hydrochloride
Terramycin hydrochloride
Tetramine
Tetran hydrochloride
TM 5

2062-98-8
$C_6F_{12}O_2$
332.05
O=C(F)C(F)(OC(F)(F)C(F)(F)C(F)(F)F)C(F)(F)F
Propanoyl fluoride, 2,3,3,3-tetrafluoro-2-(heptafluoropropoxy)- (9CI)
Propionyl fluoride, tetrafluoro-2-(heptafluoropropoxy)-

2064-28-0
$C_{16}H_{16}N_2$
236.32

5H-Dibenz(b,f)azepine-5-ethanamine (9CI)
5H-Dibenz(b,f)azepine, 5-(2-aminoethyl)- (7CI, 8CI)

2065-23-8
$C_9H_{10}O_3$
166.18
O=C(OC)COc(cccc1)c1
Acetic acid, phenoxy-, methyl ester (9CI)
AI3-04318
Methyl phenoxyacetate

2065-70-5
$C_{10}H_6Cl_2$
197.06
c(c(ccc1Cl)cc(c2)Cl)(c2)c1
Naphthalene, 2,6-dichloro- (9CI)
2,6-Dichloronaphthalene

2067-33-6
$C_5H_9BrO_2$
181.03
O=C(O)CCCCBr
Pentanoic acid, 5-bromo- (9CI)
AI3-11743
5-Bromopentanoic acid
5-Bromovaleric acid

2068-78-2
$C_{46}H_{56}N_4O_{10}.H_2O_4S$
923.14
CCC9(O)CC3CN(CCc1c([nH]c2ccccc12)C(C3)(C(=O)OC)c4cc5c(cc4OC)N(C=O)C6C(O)(C(OC(C)=O)C7(CC)C=CCN8CCC56C78)C(=O)OC)C9.OS(O)
Leurocristine sulfate (1:1)
Kyocristine
Leurocristine sulfate (1:1) (Salt)
Lilly 37231
NSC-67574
Oncovin
Onkovin
VCR sulfate
Vincaleukoblastine, 22-oxo-, sulfate (1:1) (Salt)
Vincristine sulfate
Vincristinsulfat (German)
Vincrisul

2074-50-2
$C_{12}H_{14}N_2.2CH_3O_4S$
408.48
COS([O-])(=O)=O.C[n+]1ccc(cc1)c2cc[n+](C)cc2
4,4'-Bipyridinium, 1,1'-dimethyl-, bis(methyl sulfate)
1,1'-Dimethyl-4,4'-bipyridynium dimethylsulfate
1',1'-Dimethyl-4,4'-dipyridinium di(methyl sulfate)
Gramoxone methyl sulfate
Paraquat (ACGIH)
Paraquat bis(methyl sulfate)
Paraquat dimethyl sulfate
Paraquat dimethyl sulphate
Paraquat I
PP 910

2077-13-6
$C_9H_{11}Cl$
154.64
c(c(ccc1)Cl)(c1)C(C)C
Benzene, 1-chloro-2-(1-methylethyl)- (9CI)
o-Chlorocumene
1-Chloro-2-(1-methylethyl)-benzene

2077-46-5
$C_7H_5Cl_3$
195.47
c(c(c(c(c1)Cl)C)Cl)(c1)Cl
Toluene, 2,3,6-trichloro
Benzene, 1,2,4-trichloro-3-methyl-
2,3,6-Trichlorotoluene

2077-99-8
$C_{10}H_{10}F_3N_3O_4$
293.18
CCCNc1c(cc(cc1N(=O)=O)C(F)(F)F)N(=O)=O
Benzenamine, 2,6-dinitro-N-propyl-4-(trifluoromethyl)- (9CI)
p-Toluidine, α,α,α-trifluoro-2,6-dinitro-N-propyl- (8CI)

2078-42-4
Unknown
Unknown
2,3,6-Trichlorobenzoic acid, sodium salt
2,3,6-TBA-sodium

2078-54-8
$C_{12}H_{18}O$
178.30
Oc(c(ccc1)C(C)C)c1C(C)C
Phenol, 2,6-diisopropyl
2,6-Bis(1-methylethyl)phenol
Diprivan
2,6-Diisopropylphenol
ICI 35868
Phenol, 2,6-bis(1-methylethyl)- (9CI)
Propofol

2079-00-7
$C_{17}H_{26}N_8O_5$
422.51
CN(CCC(N)CC(=O)NC1C=CC(OC1C(O)=O)n2ccc(N)nc2=O)C(N)=N
Blasticidin-S
BAB
BABS
BCS-3
BLA-S
Blasticidin
Cytovirin
TOA BLA-S

2080-89-9
C_8H_{14}
110.20
1,4-Hexadiene, 3-ethyl- (8CI, 9CI)

2082-79-3
$C_{35}H_{62}O_3$
530.88
O=C(OCCCCCCCCCCCCCCCCCC)CCc(cc(c(O)c1C(C)(C)C)C(C)(C)C)c1
Hydrocinnamic acid, 3,5-di-t-butyl-4-hydroxy-, octadecyl ester
Antioxidant 1076
AO 4
Benzenepropanoic acid, 3,5-bis-(1,1-dimethylethyl)-4-hydroxy-, octadecyl ester (9CI)
3,5-Bis(1,1-dimethylethyl)-4-hydroxybenzenepropanoic acid octadecyl ester
Hydrocinnamic acid, 3,5-di-tert-butyl-4-hydroxy-, octadecyl ester
Irganox 1076
Irganox 1906
Octadecyl 3,5-di-tert-butyl-4-hydroxyhydrocinnamate
Octadecyl 3-(3,5-di-tert-butyl-4-hydroxyphenyl)propionate

2082-81-7
$C_{12}H_{18}O_4$
226.27
O=C(OCCCCOC(=O)C(=C)C)C(=C)C
1,4-Butanediol dimethacrylate
Butanediol dimethacrylate
1,4-Butanediyl 2-methyl-2-propenoate
Butylene dimethacrylate
1,4-Butylene dimethacrylate
1,4-Butylene glycol dimethacrylate
Methacrylic acid, tetramethylene ester
Oligotetramethylene glycol dimethacrylate
2-Propenoic acid, 2-methyl-, 1,4-butanediyl ester (9CI)
Tetramethylene dimethacrylate
Tetramethylene glycol dimethacrylate
Tetramethylene methacrylate
X 970

2082-84-0
$C_{13}H_{30}N.Br$

280.35
[Br-].CCCCCCCCCC[N+](C)(C)C
Ammonium, decyltrimethyl-, bromide
Decyltrimethylammonium bromide

2084-18-6
$C_5H_{12}S$
104.22
3-Methyl-2-butanethiol
2-Butanethiol, 3-methyl- (8CI, 9CI)

2085-88-3
$C_9H_{10}O$
134.19
Oxirane, 2-methyl-2-phenyl
Cumene, α,β-epoxy- (6CI,7CI, 8CI)
2-Methyl-2-phenyloxirane
α-Methylstyrene epoxide
α-Methylstyrene oxide
2-Phenylpropene oxide

2088-07-5
$C_6H_{12}O$
100.18
1-Penten-3-ol, 2-methyl
2-Methyl-1-penten-3-ol
Propanol, 1-ethyl-2-methylene-

2088-72-4
$C_6H_{13}O_5PS$
228.22
CCOC(=O)CSP(=O)(OC)OC
Acetic acid, mercapto-, ethyl ester, S-ester with O,O-dimethyl phosphorothioate
Acetic acid, ((dimethoxyphosphinyl)thio)-, ethyl ester (9CI)
O,O-Dimethyl S-(carbethoxy)-methyl phosphorothiolate
O,O-Dimethyl S-carboethoxy-methyl thiophosphate
Methylacetaphos
Methyl acetophos
Methyl acetoxon
Phosphorothioic acid, O,O-dimethyl ester, S-ester with ethyl mercaptoacetate

2090-05-3
$C_7H_6O_2.1/2Ca$
142.15
Calcium benzoate
Benzoic acid, calcium salt (9CI)

2090-15-5
$C_{19}H_{36}$
264.50
Cyclohexane, 1,1'-heptylidenebis- (9CI)
Heptane, 1,1-dicyclohexyl- (6CI,7CI,8CI)

2090-64-4
$CH_2O_3.1/2Mg$
86.34
Carbonic acid, magnesium salt (2:1) (9CI)

2091-29-4
$C_{16}H_{30}O_2$
254.41
O=C(O)CCCCCCCC=CCCCCCC
Palmitoleic acid
9-Hexadecenoic acid (9CI)

2092-55-9
$C_{16}H_{12}N_2O_5S.Na$
367.35
Oc1ccc(cc1N=Nc2c(O)ccc3cc ccc23)S(O)(=O)=O
Benzenesulfonic acid, 4-hydroxy-3-((2-hydroxy-1-naphthalenyl)azo)-, monosodium salt
Acid Alizarine Violet
Acid Alizarine Violet B
Acid Alizarine Violet N
Acid Chrome Violet K
Acid Chrome Violet N
Aizen Chrome Violet BH
Alizarine Violet N
Alphacroic Violet B
Atlantichrome Violet B
Brasilan Chrome Violet B
Chromacid Violet R
Chromal Violet B
Chromaven Violet B
Chrome Fast Violet B
Chrome Violet B
Chrome Violet K
Chrome Violet R
C.I. 15670
C.I. Mordant Violet 5
C.I. Mordant Violet 5, Monosodium salt
Diacromo Violet N
Diamond Corinth N
Durochrome Violet B
Eriochrome Violet B
Erio Chrome Violet BA
Erio Chrome Violet BR
Hispacrom Violet B
Java Chrome Violet B
Magacrom Violet N
Mitsui Chrome Violet bc
Monchrome Violet B
Omega Chrome Dark Violet D
Pontachrome Violet SW
Solochrome Violet R
Solochrome Violet RS
Solocrom Violet RS
Sunchromine Violet B
Superchrome Violet B
Symulon Chrome Violet B
Tertrochrome Violet N
Yodochrome Violet B

2094-98-6
$C_{14}H_{20}N_4$
244.38
Cyclohexanecarbonitrile, 1,1'-azobis
Azodi-(1,1'-hexahydrobenzonitrile) [UN 2954]
UN 2954 [1,1'-Azodi-(hexahydrobenzonitrile)]

2094-99-7
$C_{13}H_{15}NO$
201.26
3-Isopropenyl-α,α-dimethylbenzyl isocyanate
Benzene, 1-(1-isocyanato-1-methylethyl)-3-(1-methylethenyl)- (9CI)
Isocyanic acid meta-isopropenyl-α,α-dimethylbenzyl ester
3-TMI

2095-03-6
$C_{19}H_{20}O_4$
312.37
Oxirane, 2,2'-(methylenebis-(4,1-phenyleneoxymethylene))-bis- (9CI)
Methane, bis(p-(2,3-epoxypropoxy)phenyl)- (8CI)

2095-06-9
$C_{12}H_{15}NO_2$
205.28
O(C1CN(c(cccc2)c2)CC(O3)C3)C1
Aniline, N,N-bis(2,3-epoxypropyl)
Bis(2,3-epoxypropyl)aniline
N,N-Bis(2,3-epoxypropyl)-aniline
Bis(epoxypropyl)phenylamine
N,N-Diglycidylanilin (Czech)
N-N-Diglycidylaniline
N-N-Diglycidylphenylamine
Oxiranemethanamine, N-(oxiranylmethyl)-N-phenyl- (9CI)

2097-19-0
$C_{12}H_{17}NO_3Si$
251.39
2,8,9-Trioxa-5-aza-1-silabicyclo(3.3.3)undecane, 1-phenyl
1-Fenylosilatranu (Polish)
Fenylsilatran (Czech)
Phenylsilatrane
1-Phenylsilatrane
Silatrane, phenyl-

2100-17-6
C_5H_8O
84.13
O=CCCC=C
4-Pentenal

2100-42-7
$C_8H_9ClO_2$
172.61
O(c(c(cc(OC)c1)Cl)c1)C
2,5-Dimethoxychlorobenzene
Benzene, 2-chloro-1,4-dimethoxy- (8CI,9CI)
1-Chloro-2,5-dimethoxybenzene
2-Chloro-1,4-dimethoxybenzene
Chlorohydroquinone dimethyl ether

2104-64-5
$C_{14}H_{14}NO_4PS$
323.32
CCOP(=S)(Oc1ccc(cc1)N(=O)=O)c2ccccc2
Phosphonothioic acid, phenyl-, O-ethyl O-(p-nitrophenyl)ester
O-Aethyl-O-(4-nitro-phenyl)-phenyl-monothiophosphonat (German)
Benzenephosphonic acid, thiono-, ethyl-p-nitrophenyl ester
ENT 17,798
EPN (ACGIH,OSHA)
EPN 300
Ethoxy-4-nitrophenoxyphenylphosphine sulfide
O-Ethyl-O-p-nitrofenylester kyseliny fenylthiofosfonove (Czech)
O-Ethyl-O-((4-nitro-fenyl)-fenyl)-monothiofosfonaat (Dutch)
O-Ethyl O-(4-nitrophenyl)benzenethionophosphonate
Ethyl p-nitrophenyl benzenethionophosphonate
Ethyl p-nitrophenyl benzenethiophosphate
Ethyl p-nitrophenyl benzenethiophosphonate
Ethyl p-nitrophenyl phenylphosphonothioate
O-Ethyl O-p-nitrophenyl phenylphosphonothiolate
O-Ethyl O-(4-nitrophenyl) phenylphosphonothioate
O-Ethyl O-p-nitrophenyl phenylphosphorothioate
Ethyl p-nitrophenyl thionobenzenephosphate
Ethyl p-nitrophenyl thionobenzenephosphonate
O-Ethyl phenyl p-nitrophenyl thiophosphonate
O-Etil-O-((4-nitro-fenil)-fenil)-monotiofosfonato (Italian)
O-(4-Nitrophenyl) O-ethyl phenyl thiophosphonate
OMS 219
Phenol, p-nitro-, O-ester with O-ethyl phenyl phosphonothioate
Phenylphosphonothioate, O-ethyl-O-p-nitrophenyl-
Phenylphosphonothioic acid O-ethyl O-p-nitrophenyl ester
Phenylthiophosphonate de O-ethyle et O-4-nitrophenyle (French)
PIN
Santox

2104-96-3
$C_8H_8BrCl_2O_3PS$
366.00
COP(=S)(OC)Oc1cc(Cl)c(Br)cc1Cl
Phosphorothioic acid, O-(4-bromo-2,5-dichlorophenyl) O,O-dimethyl ester
Brofene
O-(4-Brom-2,5-dichlor-phenyl)-O,O-dimethyl-monothiophosphat (German)
O-(4-Bromo-2,5-dicloro-fenil)-O,O-dimetil-monotiofosfato (Italian)
O-(4-Bromo-2,5-dichlorophenyl) O,O-dimethyl phosphorothioate
4-Bromo-2,5-dichlorophenyl dimethyl phosphorothionate
Bromofos
Bromofos-Methyl
O-(4-Broom-2,5-dichloor-fenyl)-O,O-dimethyl-monothiofosfaat (Dutch)
Bromophos
Bromovur
Bruomophos (Russian)
Cela S 1942
O,O-Dimethyl O-(4-bromo-2,5-dichlorophenyl) phosphorothioate
O,O-Dimethyl-O-(2,5-dichlor-4-bromphenyl)-thionophosphat (German)
O,O-Dimethyl-O-(2,5-dichloro-4-bromophenyl)phosphorothioate
O,O-Dimethyl O-(2,5-dichloro-4-bromophenyl) thiophosphate
Drillzid
EL 400
ENT 27,162
Metabrom
Monsanto CP 51969
Netal
Nexagan
Nexion
Nexion 40
Nexion 5g
Nexion LC40
OMS-658
Phenol, 4-bromo-2,5-dichloro-, O-ester with O,O-dimethyl phosphorothioate
S 1942
Sovinexion
Thiophosphate de O,O-dimethyle et de O-4-bromo-2,5-dichlorophenyle (French)

2105-40-0
$C_8H_{16}O$
128.21
Cyclohexanemethanol, 2-methyl- (8CI,9CI)
NSC-19178

2107-76-8
$C_{10}H_8O_4$
192.17
O=C(Oc(c(C=1C)c(O)cc2O)c2)C1
5,7-Dihydroxy-4-methylcoumarin
AI3-23192
2H-1-Benzopyran-2-one, 5,7-dihydroxy-4-methyl- (9CI)
Coumarin, 5,7-dihydroxy-4-methyl-
5,7-Dihydroxy-4-methyl-2H-1-benzopyran-2-one
4-Methyl-5,7-dihydroxycoumarin

2107-77-9
$C_{10}H_8O_4$
192.18
Coumarin, 7,8-dihydroxy-4-methyl
2H-1-Benzopyran-2-one, 7,8-dihydroxy-4-methyl- (9CI)
7,8-Dihydroxy-4-methylcoumarin
4-Methyldaphnetin

2108-92-1
$C_6H_{10}Cl_2$
153.05
Cyclohexane, 1,1-dichloro- (8CI,9CI)
NSC-9463

2109-64-0
$C_{11}H_{25}NO$
187.32
Dibutylisopropanolamine
N,N-Dibutyl(2-hydroxypropyl)amine
2-Propanol, 1-dibutylamino-
2-Propanol, 1-(dibutylamino)- (8CI,9CI)

2111-75-3
$C_{10}H_{14}O$
150.24
O=CC(=CCC(C(=C)C)C1)C1
1-Cyclohexene-1-carboxaldehyde, 4-isopropenyl
1-Cyclohexene-1-carboxaldehyde, 4-(1-methylethenyl)- (9CI)
Dihydrocuminyl aldehyde
4-Isopropenyl-1-cyclohexene-1-carboxaldehyde
para-Mentha-1,8-dien-7-al
Perilla aldehyde
Perillal
Perillaldehyde
Perillyl aldehyde

2113-00-0
$C_4Cl_5N_3$
267.33
1,3,5-Triazine, 2,4-dichloro-6-(trichloromethyl)- (9CI)
s-Triazine, 2,4-dichloro-6-(trichloromethyl)- (6CI, 7CI,8CI)

2113-57-7
$C_{12}H_9Br$
233.12
c(cccc1c(cccc2)c2)(c1)Br
Biphenyl, 3-bromo
3-Bromobiphenyl

2113-58-8
$C_{12}H_9NO_2$
199.22
Biphenyl, 3-nitro
1,1'-Biphenyl, 3-nitro- (9CI)
m-Nitrobiphenyl
3-Nitrobiphenyl

2114-11-6
$C_4H_7NO_2$
101.12
O=C(OCC=C)N
Carbamic acid, allyl ester
Allyl carbamate

2116-65-6
$C_{12}H_{11}N$
169.24
n(ccc(c1)Cc(cccc2)c2)c1
Pyridine, 4-benzyl
4-Benzylpyridine

2116-84-9
$C_{15}H_{32}O_3Si_4$
372.76
Phenyl trimethicone
Methyl phenyl polysiloxane
Phenyltris(trimethylsiloxyl)-silane
Polyphenylmethyl siloxane
Rhodorsil Oils 70641 V 200
Silicone Fluid PD 5
Silicone Fluid PK 20
Trisiloxane, 1,1,1,5,5,5-hexamethyl-3-phenyl-3-((trimethylsilyl)oxy)- (9CI)

2122-19-2
$C_4H_8N_2S$
116.20
2-Imidazolidinethione, 4-methyl
4-Methylethylenethiourea
4-Methyl-2-imidazolidinethione
PLTU
Propilentiourea (Italian)
Propylene thiourea
Propylenthioharnstoff (German)

2122-70-5
$C_{14}H_{14}O_2$
214.26
O=C(OCC)Cc(c(c(ccc1)cc2)c1)c2
Ethyl 1-naphthaleneacetate
AI3-02254
Caswell No. 589AA
EPA Pesticide Chemical Code 056008
Ethyl 1-naphthylacetate
1-Naphthaleneacetic acid, ethyl ester (8CI,9CI)
1-Naphththaleneacetic acid, ethyl ester
2-(1-Naphthyl)acetic acid ethyl ester

2122-77-2
$C_8H_7Cl_3O_2$
241.50
OCCOc1cc(Cl)c(Cl)cc1Cl
Ethanol, 2-(2,4,5-trichlorophenoxy)
Klorinol
TCPE
2-(2,4,5-Trichlorophenoxy)-ethanol

2123-27-5
$C_8H_6Cl_2$
173.04
Benzene, 2,4-dichloro-1-ethenyl- (9CI)
Styrene, 2,4-dichloro- (8CI)

2123-28-6
$C_8H_6Cl_2$
173.04
Benzene, 1,2-dichloro-3-ethenyl- (9CI)
2,3-Dichlorostyrene
Styrene, 2,3-dichloro- (7CI, 8CI)

2124-57-4
$C_{46}H_{64}O_2$
649.01
Vitamin K2(35)
Menaquinone K7
Menaquinone 7
MK7
1,4-Naphthalenedione, 2-(3,7,11,15,19,23,27-heptamethyl-2,6,10,14,18,22,26-octacosahept-aenyl)-3-methyl-, (All-E)-
1,4-Naphthoquinone, 2-(3,7,11,15,19,23,27-heptamethyl-2,6,10,14,18,22,26-octacosahept-aenyl)-3-methyl-, (All-E)-
Vitamin MK 7

2130-56-5
$C_{14}H_{12}N_2O_4$
272.28
O=C(O)c(c(N)ccc1c(ccc(N)c2C(=O)O)c2)c1
3,3'-Biphenyldicarboxylic acid, 4,4'-diamino
3,3'-Benzidinedicarboxylic acid
5,5'-Bianthranilic acid
4,4'-Diamino-3,3'-biphenyl-dicarboxylic acid
4,4'-Diaminobiphenyl-3,3'-di-carboxylic acid
3,3'-Dicarboxybenzidine
Kwas benzydynodwukaroks-ylowy (Polish)

2131-18-2
$C_{21}H_{36}$
288.52
c(cccc1)(c1)CCCCCCCCCCCCCCC
n-Pentadecylbenzene
Benzene, pentadecyl- (9CI)
Pentadecane, 1-phenyl- (8CI)
Pentadecylbenzene
1-Phenylpentadecane

2131-38-6
$C_{13}H_{14}$
170.25
Naphthalene, 1,3,7-trimethyl- (8CI,9CI)

2131-41-1
$C_{13}H_{14}$
170.25
Naphthalene, 1,4,5-trimethyl- (8CI,9CI)

2131-42-2
$C_{13}H_{14}$
170.25
Naphthalene, 1,4,6-trimethyl- (8CI,9CI)
NSC-91460

2131-55-7
C_7H_4ClNS
169.63
Isothiocyanic acid, p-chlorophenyl ester
4-Chlor-phenyl-isothiocyanat (German)

2132-80-1
$C_{14}H_{14}O_2$
214.26
1,1'-Biphenyl, 4,4'-dimethoxy- (9CI)
Biphenyl, 4,4'-dimethoxy- (8CI)
NSC-17524

2132-86-7
$C_{13}H_{20}$
176.30
Benzene, (1-propylbutyl)- (9CI)
Heptane, 4-phenyl- (6CI,

7CI,8CI)
(4-Heptyl)benzene
4-Phenylheptane

2133-34-8
$C_4H_7NO_2$
101.12
OC(=O)C1CCN1
Azetidine-2-carboxylic acid, L
Acide L-azetidine-2-carboxylic (French)
L-Azetidine-2-carboxylic acid
L-2-Azetidinecarboxylic acid
2-Azetidinecarboxylic acid, L-
2-Azetidinecarboxylic acid, (S)- (9CI)

2134-90-9
$C_{16}H_{14}O_8$
334.28
(1,1'-Biphenyl)-3,3'-dicarboxylic acid, 6,6'-dihydroxy-5,5'-dimethoxy- (9CI)
3,3'-Biphenyldicarboxylic acid, 6,6'-dihydroxy-5,5'-dimethoxy- (8CI)
5,5'-Bivanillic acid
Dehydrodivanillic acid

2134-91-0
$C_9H_8O_6$
212.16
1,3-Benzenedicarboxylic acid, 4-hydroxy-5-methoxy- (9CI)
Isophthalic acid, 4-hydroxy-5-methoxy- (8CI)

2136-70-1
$C_{16}H_{34}O_2$
258.44
Ethanol, 2-(tetradecyloxy)- (8CI,9CI)

2136-71-2
$C_{18}H_{38}O_2$
286.50
Ethanol, 2-(hexadecyloxy)- (8CI,9CI)

2136-79-0
$C_8H_2Cl_4O_4$
303.91
Chlorthal
AI3-33410
1,4-Benzenedicarboxylic acid, 2,3,5,6-tetrachloro- (9CI)
Caswell No. 833A
Terephthalic acid, tetrachloro- (8CI)
Tetrachloroterephthalic acid

2136-89-2
$C_7H_4Cl_4$
229.92
c(c(ccc1)Cl)(c1)C(Cl)(Cl)Cl
Benzene, 1-chloro-2-(trichloromethyl)- (9CI)
AI3-02820
1-Chloro-2-(trichloromethyl)-benzene
Toluene, o,α,α,α-tetrachloro- (8CI)

2136-99-4
$C_{12}H_2Cl_8$
429.77
Clc1cc(Cl)c(Cl)c(c1Cl)c2c(Cl)c(Cl)cc(Cl)c2Cl
1,1'-Biphenyl, 2,2',3,3',5,5',6,6'-octachloro- (9CI)
Biphenyl, 2,2',3,3',5,5',6,6'-octachloro- (8CI)
2,2',3,3',5,5',6,6'-Octachlorobiphenyl
2,2',3,3',5,5',6,6'-PCB

2138-22-9
$C_6H_5ClO_2$
144.56
Pyrocatechol, 4-chloro
1,2-Benzenediol, 4-chloro-
4-Chlorocatechol
4-Chloropyrocatechol

2138-43-4
$C_9H_{12}O_2$
152.21
Pyrocatechol, 4-isopropyl
1,2-Benzenediol, 4-(1-methylethyl)-
4-Isopropylcatechol

2138-48-9
$C_9H_{12}O_2$
152.21
Pyrocatechol, 3-isopropyl
1,2-Benzenediol, 3-(1-methylethyl)- (9CI)
3-Isopropylcatechol
3-Isopropylpyrocatechol

2141-62-0
C_5H_9NO
99.13
N#CCCOCC
Propanenitrile, 3-ethoxy- (9CI)
AI3-25450
3-Ethoxypropanenitrile
Propionitrile, 3-ethoxy- (8CI)

2142-63-4
C_8H_7BrO
199.05
O=C(c(cccc1Br)c1)C
Acetophenone, 3'-bromo- (8CI)
m-Bromoacetophenone
1-(3-Bromophenyl)ethanone
Ethanone, 1-(3-bromophenyl)- (9CI)

2142-64-5
$C_{10}H_{12}O$
148.20
Ethanone, 1-(2-ethylphenyl)- (9CI)
Acetophenone, 2'-ethyl- (8CI)

2142-65-6
$C_{11}H_{14}O$
162.23
Ethanone, 1-(2-(1-methylethyl)phenyl)- (9CI)
Acetophenone, 2'-isopropyl- (6CI,7CI,8CI)
2'-Isopropylacetophenone
o-Isopropylphenyl methyl ketone

2142-71-4
$C_{10}H_{12}O$
148.21
Ethanone, 1-(2,3-dimethylphenyl)- (9CI)
Acetophenone, 2',3'-dimethyl- (6CI,7CI,8CI)
2,3-Dimethylacetophenone

2142-76-9
$C_{10}H_{12}O$
148.20
Ethanone, 1-(2,6-dimethylphenyl)- (9CI)
Acetophenone, 2',6'-dimethyl- (8CI)

2144-41-4
$C_6H_{12}O$
100.16
Furan, tetrahydro-2,5-dimethyl-, cis- (8CI,9CI)

2144-45-8
$C_{16}H_{14}O_6$
302.28
Dibenzyl peroxydicarbonate (More than 87% with water)
Dibenzyl peroxydicarbonate
Dibenzyl peroxydicarbonate, Maximum concentration 87% with water
Dibenzyl peroxydicarbonate, More than 87% with water
Dibenzyl peroxydicarbonate (Not more than 87% with water)
Peroxidicarbonato de dibencilo (Spanish)
Peroxydicarbonate de dibenzyle (French)
Peroxydicarbonic acid, dibenzyl ester
Peroxydicarbonic acid, dibenzyl ester, More than 87% with water

UN 2149

2144-53-8
$C_{12}H_9F_{13}O_2$
432.18
O=C(OCCC(F)(F)C(F)(F)C(F)(F)C(F)(F)C(F)(F)C(F)(F)F)C(=C)C
2-Propenoic acid, 2-methyl-, 3,3,4,4,5,5,6,6,7,7,8,8,8-tridecafluorooctyl ester (9CI)

2144-54-9
$C_{16}H_9F_{21}O_2$
632.21
O=C(OCCC(F)(F)C(F)(F)C(F)(F)C(F)(F)C(F)(F)C(F)(F)C(F)(F)C(F)(F)C(F)(F)F)C(=C)C
2-Propenoic acid, 2-methyl-, 3,3,4,4,5,5,6,6,7,7,8,8,9,9,10,10,11,11,12,12,12-heneicosafluorododecyl ester (9CI)

2149-36-2
$C_9H_7NO_4S$
225.22
O=S(=O)(Oc(c(nccc1)c1cc2)c2)O
8-Quinolinol, hydrogen sulfate (ester) (9CI)

2150-25-6
$C_{10}H_{11}Cl_2NO_2$
248.11
Carbamic acid, (2,4-di chlorophenyl)-, 1-methylethyl ester (9CI)
Carbanilic acid, 2,4-dichloro-, isopropyl ester (7CI,8CI)
Isopropyl N-(2,4-dichlorophenyl)carbamate

2150-28-9
$C_{10}H_{11}Cl_2NO_2$
248.10
Carbamic acid, (3,4-dichlorophenyl)-, 1-methylethyl ester (9CI)
Carbanilic acid, 3,4-dichloro-, isopropyl ester (8CI)
3,4-DCIPC
H 22949

2150-32-5
$C_{10}H_{11}Cl_2NO_2$
248.10
Carbamic acid, (3-chlorophenyl)-, 2-chloro-1-methylethyl ester (9CI)
Carbanilic acid, m-chloro-, 2-chloro-1-methylethyl ester (8CI)
CPPC

2150-38-1
$C_{10}H_{12}O_4$
196.20
Benzoic acid, 3,4-dimethoxy-, methyl ester (9CI)
AI3-20957
Methyl 3,4-dimethoxybenzoate
Veratric acid, methyl ester (8CI)

2150-47-2
$C_8H_8O_4$
168.15
O=C(OC)c(c(O)cc(O)c1)c1
Benzoic acid, 2,4-dihydroxy-, methyl ester
AI3-31503
Methyl 2,4-dihydroxybenzoate
β-Resorcylic acid, methyl ester

2150-48-3
$C_{21}H_{27}N_2O.Cl$
358.95
Ammonium, (6-(diethylamino)-3H-xanthen-3-ylidene)diethyl-, chloride
Ammonium, diethyl(6-(diethylamino)-3H-xanthen-3-ylidene)-, chloride
C.I. 45010
(6-(Diethylamino)-3H-xanthen-3-ylidine)diethylammonium chloride
N-(6-(Diethylamino)-3H-xanthen-3-ylidine)-N-ethylethanaminium chloride
E Tetraethylpyronin
Ethanaminium, N-(6-(diethylamino)-3H-xanthen-3-ylidene)-N-ethyl-, chloride (9CI)
NSC-44690
Pyronin B
Pyronine B
Pyronine B (BY)

2150-54-1
$C_{34}H_{26}N_4O_{16}S_4.4Na$
966.80
C.I. Direct Blue 25, Tetrasodium salt (8CI)
2,7-Naphthalenedisulfonic acid, 3,3'-((3,3'-dimethyl(1,1'-biphenyl)-4,4'-diyl)bis(azo))bis-(4,5-dihydroxy-, tetrasodium salt (9CI)

2150-55-2
$C_4H_6N_2O_2S$
146.18
2-Thiazoline-4-carboxylic acid, 2-amino
2-Amino-4-carboxythiazoline
2-Amino-2-thiazoline-4-carboxylic acid
4-Thiazolecarboxylic acid, 2-amino-4,5-dihydro-

2150-88-1
$C_8H_8ClNO_2$
185.60
COC(=O)Nc1cccc(Cl)c1
Carbamic acid, (3-chlorophenyl)-, methyl ester (9CI)
Carbanilic acid, m-chloro-, methyl ester (8CI)

2150-93-8
$C_8H_7Cl_2NO$
204.05
CC(=O)Nc1ccc(Cl)c(Cl)c1
Acetamide, N-(3,4-dichlorophenyl)-
Acetanilide (8CI), 3',4'-dichloro-
AI3-31362

2152-34-3
$C_9H_8N_2O_2$
176.19
N=C1NC(=O)C(O1)c2ccccc2
4-Oxazolidinone, 2-imino-5-phenyl
A 13397
2-Amino-5-phenyl-4(5H)-oxazolone
Azoksodon
Azoxodone
Centramin
Constimol
CS 293
Cylert
Dantromin
Deltamine
Endolin
Fenoxazol
FIO
FWH-352
H 3104
Hyton
2-Imino-5-phenyl-4-oxazolidinone
Juston-Wirkstoff
Kethamed
LA 956
Myamin
Nitan
Notair
NPL 1
Okodon
P 10
Pemolina (Italian)
Pemoline
Phenalone
Phenilone
Phenoxazole
5-Phenyl-2-imino-4-oxazolidinone
5-Phenyl-2-imino-4-oxo-oxazolidine
Phenylisohydantoin
Phenylpseudohydantoin
Pio
Pioxol
Pomoline
Pondex

PT 360
PN/135
Ronyl
Sigmadyn
Sistral
Sofro
Stimul
Tradon
Tradone
Volital
Volitol
YH 1

2152-64-9
$C_{37}H_{29}N_3$.ClH
552.10
Spirit Blue
Benzenamine, N-phenyl-4-((4-(phenylamino)phenyl)-(4-(phenylimino)-2,5-cyclohexadien-1-y-lidene)methyl)-, monohydrochloride (9CI)
C.I. Solvent Blue 23
C.I. Solvent Blue 23, Monohydrochloride
C.I. 42760
Opal Blue SS

2155-70-6
$C_{16}H_{32}O_2Sn$
375.14
Tri-n-butylstannylmethacrylate
Caswell No. 867EF
EPA Pesticide Chemical Code 083120
Stannane, tributyl(methacryloyloxy)-
Stannane, tributyl((2-methyl-1-oxo-2-propenyl)oxy)- (9CI)
Tin tributylmethacrylate
Tributyl(methacryloxy)stannane
Tributyl(methacryloyloxy)-stannane
Tributyl((2-methyl-1-oxo-2-propenyl)oxy)stannane
Tributylstannyl methacrylate
Tributyltin methacrylate (8CI)

2155-71-7
$C_{16}H_{22}O_6$
310.35
O=C(OOC(C)(C)C)c(c(ccc1)C(=O)OOC(C)(C)C)c1
Di-(tert-butylperoxy) phthalate (More than 55% in solution)
1,2-Benzenedicarboperoxoic acid, bis(1,1-dimethylethyl) ester (9CI)
Di-(terc-butilperoxi)ftalato (Spanish)
Di-(tert-butylperoxy) phthalate
Di-(tert-butylperoxy) phthalate (Not more than 55% as a paste)
Di-(tert-butylperoxy) phthalate (Not more than 55% in solution)
Di-(tert-butylperoxy) phthalate (Technically pure)
Diperoxyphtalate de tert-butyle (French)
Peroxyphthalic acid, di-tert-butyl ester (8CI)
UN 2106
UN 2107
UN 2108

2156-96-9
$C_{13}H_{24}O_2$
212.37
O=C(OCCCCCCCCCC)C=C
Acrylic acid, decyl ester
Decyl acrylate
n-Decyl acrylate
2-Propenoic acid, decyl ester (9CI)

2156-97-0
$C_{15}H_{28}O_2$
240.39
O=C(OCCCCCCCCCCCC)C=C
2-Propenoic acid, dodecyl ester (9CI)
AI3-03198
Dodecyl 2-propenoate
Lauryl acrylate
n-Lauryl acrylate

2157-01-9
$C_{12}H_{22}O_2$
198.31
O=C(OCCCCCCCC)C(=C)C
2-Propenoic acid, 2-methyl-, octyl ester
AI3-08767
Octyl methacrylate
Octyl 2-methyl-2-propenoate

2157-42-8
$C_{12}H_{30}O_7Si_2$
342.54
Silicic acid, hexaethyl ester (9CI)
Hexaethoxydisiloxane

2158-09-0
$C_{13}H_{28}N_2O$
228.37
O=C(NCCCCCCCCCCCC)N
Urea, dodecyl- (9CI)
Dodecylurea

2160-93-2
$C_8H_{19}NO_2$
161.24
OCCN(C(C)(C)C)CCO
tert-Butyldiethanolamine
2,2'-((1,1-Dimethylethyl)imino)-bisethanol
Ethanol, 2,2'-((1,1-dimethylethyl)imino)bis- (9CI)

2162-74-5
$C_{25}H_{34}N_2$
362.55
N(=C=Nc(c(ccc1)C(C)C)c1C(C)C)c(c(ccc2)C(C)C)c2C(C)C
Benzenamine, N,N'-methanetetraylbis(2,6-bis(1-methylethyl)- (9CI)
2,2',6,6'-Tetraisopropyldiphenylcarbodiimide

2162-92-7
$C_6H_{12}Cl_2$
155.07
ClCC(Cl)CCCC
Hexane, 1,2-dichloro- (9CI)
1,2-Dichlorohexane

2163-00-0
$C_6H_{12}Cl_2$
155.07
ClCCCCCCCl
Hexane, 1,6-dichloro- (9CI)
1,6-Dichlorohexane

2163-48-6
$C_{10}H_{18}O_4$
202.25
Propanedioic acid, propyl-, diethyl ester (9CI)
Malonic acid, propyl-, diethyl ester (8CI)
NSC-53659

2163-68-0
$C_8H_{15}N_5O$
197.24
CCNc1nc(O)nc(NC(C)C)n1
2-Hydroxyatrazine
2-Hydroxy-4-(ethylamino)-6-(isopropylamino)-s-triazine

2163-69-1
$C_{11}H_{22}N_2O$
198.35
CN(C)C(=O)NC1CCCCCCC1
Urea, 3-cyclooctyl-1,1-dimethyl
Alipur
Alipur-O
3-Cyclooctyl-1,1-dimethylharnstoff (German)
N-Cyclooctyl-N',N'-dimethylurea
3-Cyclooctyl-1,1-dimethylurea
Cyclouron
Cycluron
HS 61
OMU

2163-79-3
$C_{13}H_{22}N_2O$
222.32
CN(C)C(=O)NC1CC2CC1C3CCCC23
Norea
AI3-28205

1,1-Dimethyl-3-tetrahydrodicyclopentadienylurea
3-(Hexahydro-4,7-methanoindan-5-yl)-1,1-dimethylurea
1-(Tetrahydrodicyclopentadienyl)-3,3-dimethylurea
Urea, N,N-dimethyl-N'-(octahydro-4,7-methano-1H-inden-5-yl)-
Urea, 3-(hexahydro-4,7-methanoindan-5-yl)-1,1-dimethyl- (8CI)

2163-80-6
$CH_4AsO_3.Na$
161.96
[Na+].C[As](O)([O-])=O
Methanearsonic acid, monosodium salt
Ansar 170
Ansar 170 H.C.
Ansar 1701
Ansar 529
Arsonic acid, methyl-, monosodium salt
Ansar 529 H.C.
Arsonate Liquid
Asazol
Bueno
Bueno 6
Daconate
Daconate 6
Dal-E-Rad
Dal-E-Rad 120
Gepiron
Herb-All
Herban M
Merge
Merge 823
Mesamate
Mesamate-400
Mesamate-600
Mesamate Concentrate
Mesamate H.C.
Methylarsenic acid, sodium salt
Methylarsonat monosodny (Czech)
Monate
Monosodium acid methanearsonate
Monosodium acid metharsonate
Monosodium methanearsonate
Monosodium methanearsonic acid
Monosodium methylarsonate
MSMA
NCI-C60071
Phyban
Phyban H.C.
Silvisar 550
Sodium methanearsonate
Target MSMA
Trans-Vert
Weed-E-Rad
Weed-HOE
Sodium acid methanearsonate
Weed 108

2163-81-7
$C_{15}H_{22}BrNO_2$
328.24
Acetamide, 2-bromo-N-(2-(1,1-dimethylethyl)-6-methylphenyl)-N-(methoxymethyl)- (9CI)
o-Acetotoluidide, 2-bromo-6'-tert-butyl-N-(methoxymethyl)- (8CI)
CP 45592

2164-07-0
$C_8H_{10}O_5.2K$
264.36
Dipotassium endothall
Caswell No. 625B
Endothall dipotassium salt
7-Endothall, dipotassium salt
EPA Pesticide Chemical Code 038904
7-Oxabicyclo(2.2.1)heptane-2,3-dicarboxylic acid, dipotassium salt (8CI,9CI)

2164-08-1
$C_{13}H_{18}N_2O_2$
234.33
O=c2[nH]c1CCCc1c(=O)n2C3CCCCC3
1H-Cyclopentapyrimidine-2,4(3H,5H)-dione, 6,7-dihydro-3-cyclohexyl
Ban-HOE
3-Cyclohexyl-5,6-trimethyleneuracil
3-Cyclohexyl-5,6-trimethylenuracil (German)
Du Pont 634
Lenacil
Venzar

2164-09-2
$C_{10}H_9Cl_2NO$
230.10
CC(=C)C(=O)Nc1ccc(Cl)c(Cl)c1
Acrylanilide, 3',4'-dichloro-2-methyl
Chloranocryl
DCM
DCMA
3,4-Dichloranilid kyseliny methakrylove (Czech)
3',4'-Dichloro-2-methacrylanilide
3',4'-Dichloro-2-methylacrylanilide
N-(3,4-Dichlorophenyl)methacrylamide
N-(3,4-Dichlorophenyl)-2-methyl-2-propenamide
Dicryl
FMC 4556
Methacrylic acid 3,4-dichloroanilide
α-Methylacrylic acid, 3,4-dichloroanilide
N 4,556
NIA 4556
Niagara 4556
2-Propenamide, N-(3,4-dichlorophenyl)-2-methyl- (9CI)

2164-13-8
$C_{11}H_{14}ClNO_2$
227.68
Carbamic acid, (3-chlorophenyl)-, 1-methylpropyl ester (9CI)
BCPC
Carbanilic acid, m-chloro-, sec-butyl ester (8CI)
NSC-2475
NSC-74793

2164-17-2
$C_{10}H_{11}F_3N_2O$
232.23
O=C(N(C)C)Nc(cccc1C(F)(F)F)c1
Urea, 1,1-dimethyl-3-(α,α,α-trifluoro-m-tolyl)
C 2059
Ciba 2059
Cottonex
Cotoran
Cotoran Multi 50WP
1,1-Dimethyl-3-(3-trifluoromethylphenyl)urea
1,1-Dimethyl-3-(α,α,α-trifluoro-m-tolyl) urea
Fluometuron
Herbicide C-2059
Higalcoton
Lanex
NCI-C08695
Pakhtaran
3-(5-Trifluormethylphenyl)-1,1-dimethylharnstoff (German)
N-(m-Trifluoromethylphenyl)-N',N'-dimethylurea
N-(3-Trifluoromethylphenyl)-N'-N'-dimethylurea
3-(m-Trifluoromethylphenyl)-1,1-dimethylurea
Urea, N,N-dimethyl-N'-(3-(trifluoromethyl)phenyl)-

2167-23-9
$C_{12}H_{26}O_4$
234.34
O(OC(C)(C)C)C(OOC(C)(C)C)(CC)C
2,2-Di-(tert-butylperoxy)-butane (More than 55% in solution)
Bis(tert-butylperoxy)-2,2 butane (French)
2,2-Bis(tert-butylperoxy)butane
2,2-Di-(terc-butilperoxi) butano (Spanish)
2,2-Di-(tert-butylperoxy)butane
2,2-Di-(tert-butylperoxy)butane (Not more than 55% in solution)
Peroxide, sec-butylidenebis(tert-butyl (8CI)
Peroxide, (1-methylpropylidene)bis((1,1-dimethylethyl)

(9CI)
UN 2111

2167-39-7
C_4H_8O
72.12
Butane, 1,3-epoxy
2-Methyloxetan
2-Methyloxetane
1-Methyltrimethylene oxide
Oxetane, 2-methyl- (9CI)

2168-68-5
$C_8H_{16}N_3OPS$
233.30
Phosphine sulfide, bis(1-aziridinyl)morpholino
Bis(1-aziridinyl)morpholinophosphine sulphide
N,N'-Diethylenemorpholinophosphinothioic diamide
N,N'-Diethylene-N'-(3-oxapentamethylene)phosphorothioic triamide
Diethylene oxapentamethylenethiophosphoramide
Morpholine, 4-(bis(1-aziridinyl)-phosphinothioyl)- (9CI)
Morzid
OPSPA
N-(3-Oxapentamethylene)-N',N''-diethylenethiophosphoramide
Thiomorpholidophosphoric diethylenimide

2170-44-7
$C_{12}Br_6Cl_2O_2$
726.46
Dibenzo(b,e)(1,4)dioxin, 1,2,4,6,7,9-hexabromo-3,8-dichloro- (9CI)
Dibenzo-p-dioxin, 1,2,4,6,7,9-hexabromo-3,8-dichloro- (8CI)
1,2,4,6,7,9-Hexabromo-3,8-dichlorodibenzo-p-dioxin

2170-45-8
$C_{12}Br_8O_2$
815.36
Dibenzo(b,e)(1,4)dioxin, octabromo- (9CI)
Dibenzo-p-dioxin, octabromo- (8CI)
Octabromodibenzo-p-dioxin

2172-33-0
$C_{42}H_{18}N_2O_6$
646.60
Dinaphtho(2,3-i:2',3'-i')-benzo(1,2-a:4,5-a')dicarbazole-5,7,12,17,19,24-hexone, 6,18-dihydro- (9CI)

2173-56-0
$C_{10}H_{20}O_2$
172.27
O=C(OCCCCC)CCCC
n-Pentyl valerate
AI3-01269
Pentanoic acid, pentyl ester (9CI)
Pentyl pentanoate
Valeric acid, pentyl ester (8CI)

2176-62-7
C_5Cl_5N
251.31
n(c(c(c(c1Cl)Cl)Cl)Cl)c1Cl
Pyridine, 2,3,4,5,6-pentachloro
Pentachloropyridine
2,3,4,5,6-Pentachloropyridine
Perchloropyridine

2177-47-1
$C_{10}H_{10}$
130.19
1H-Indene, 2-methyl- (9CI)
Indene, 2-methyl- (8CI)

2177-86-8
$C_{10}H_{20}O_2$
172.27
Octanoic acid, 2-methyl-, methyl ester (6CI,7CI, 8CI,9CI)
Methyl 2-methyloctanoate
2-Methyloctanoic acid methyl ester

2179-57-9
$C_6H_{10}S_2$
146.28
S(SCC=C)CC=C
Allyl-disulfide
Allyl disulphide
Diallyl disulfide
Diallyl disulphide
Disulfide, di-2-propenyl (9CI)
4,5-Dithia-1,7-octadiene
2-Propenyl disulphide

2179-59-1
$C_6H_{12}S_2$
148.30
CCCSSCC=C
Disulfide, allyl propyl
Allyl propyl disulfide (ACGIH, OSHA)

2179-60-4
$C_4H_{10}S_2$
122.26
S(SC)CCC
Disulfide, methyl propyl
AI3-38157
Methyl propyl disulfide

2183-56-4
$C_{17}H_{21}NO_4$
303.35
Hydromorphinol
DEA No. 9301
7,8-Dihydro-14-hydroxymorphine
Hidromorfinol (Spanish)
Hydromorphinolum (Latin)
Idromorfinolo
Morphinan-3,6-α,14-triol, 4,5-α-epoxy-17-methyl-
Morphine, 7,8-dihydro-14-hydroxy-
Ram-320

2185-86-6
$C_{29}H_{32}ClN_3$
458.09
Methanaminium, N-(4-((4-(dimethylamino)phenyl)(4-(ethylamino)-1-naphthalenyl)methylene)- 2,5-cyclohexadien-1-ylidene)-N-methyl-, chloride
Aizen Victoria Blue BOH
Basic Bluek
C.I. 44040
Hidaco Victoria Blue R
N,N'-(N,N'-Tetramethyl)-1-diaminodiphenylnaphthylaminomethane hydrochloride
Victoria Blue R
Victoria Blue RS
Victoria Lake Blue R

2185-92-4
$C_{12}H_{11}N.ClH$
205.70
Cl.Nc1ccccc1c2ccccc2
2-Biphenylamine, hydrochloride
NCI-C50282

2186-24-5
$C_{10}H_{12}O_2$
164.22
Propane, 1,2-epoxy-3-(p-tolyloxy)
p-Cresol glycidyl ether
p-Cresyl glycidyl ether
1,2-Epoxy-3-(p-tolyloxy)propane
Glycidyl 4-methylphenyl ether
Glycidyl p-tolyl ether
Methylphenyl glycidyl ether
Oxirane, ((4-methylphenoxy)-methyl)- (9CI)

2186-92-7
$C_{10}H_{14}O_3$
182.22
O(C(OC)c(ccc(OC)c1)c1)C
Benzene, 1-(dimethoxymethyl)-4-methoxy- (9CI)
Anisicaldehyde dimethylacetal
1-(Dimethoxymethyl)-4-methoxybenzene

2189-60-8
$C_{14}H_{22}$
190.33
c(cccc1)(c1)CCCCCCCC
n-Octylbenzene
AI3-16044
Benzene, octyl- (9CI)
Octane, 1-phenyl- (8CI)
Octylbenzene

2190-04-7
$C_{18}H_{39}N.C_2H_4O_2$
329.64
Octadecylamine, acetate
Armac 18D
Armac OD
Octadecanamine acetate
1-Octadedecylamine acetate

2191-10-8
$C_{16}H_{30}O_4.Cd$
398.86
Octanoic acid, cadmium salt (2:1)
Cadmium caprylate

2197-52-6
$C_{18}H_{32}O_2$
280.45
8,11-Octadecadienoic acid (8CI,9CI)

2197-63-9
$C_{32}H_{67}O_4P$
546.86
O=P(OCCCCCCCCCCCCCCCC)(OCCCCCCCCCCCCCCCC)O
Dicetylphosphate
Dicetyl phosphate
Dihexadecyl phosphate
1-Hexadecanol, hydrogen phosphate (9CI)

2198-20-1
$C_{10}H_{18}$
138.25
Cyclodecene, (E)- (8CI,9CI)
NSC-155648

2198-58-5
$C_{10}H_{16}N_2.ClH$
200.74
p-Phenylenediamine, N,N-diethyl-, hydrochloride
p-Amino diethylaniline hydrochloride
N,N-Diethyl-p-phenylenediamine hydrochloride

2198-61-0
$C_{11}H_{22}O_2$
186.33
O=C(OCCC(C)C)CCCCC
Hexanoic acid, isopentyl ester
Isoamyl caproate
Isoamyl hexanoate
Isopentyl hexanoate
Isopentyl-n-hexanoate

2198-75-6
$C_{10}H_6Cl_2$
197.06
c(c(c(cc1Cl)Cl)ccc2)(c2)c1
Naphthalene, 1,3-dichloro- (9CI)
1,3-Dichloronaphthalene

2198-77-8
$C_{10}H_6Cl_2$
197.06
c(c(ccc1Cl)ccc2Cl)(c1)c2
Naphthalene, 2,7-dichloro- (9CI)
2,7-Dichloronaphthalene

2199-41-9
$C_7H_{11}N$
109.17
Cc1cc(C)c(C)[nH]1
1H-Pyrrole, 2,3,5-trimethyl- (9CI)
Pyrrole, 2,3,5-trimethyl- (6CI,7CI,8CI)
2,3,5-Trimethylpyrrole
2,4,5-Trimethylpyrrole

2201-15-2
$C_{14}H_{21}N$
203.36
N(C(c(cccc1)c1)(CCCC2)C2)CC
Cyclohexanamine, N-ethyl-1-phenyl
Cyclohexamine
Cyclohexylamine, N-ethyl-1-phenyl- (8CI)
Ethylphencyclidine
N-Ethyl-1-phenylcyclohexylamine
PCE

2201-24-3
$C_{12}H_{17}N$
175.27
1-Phenylcyclohexylamine
AI3-05775
Cyclohexanamine, 1-phenyl-
DEA No. 7460

2201-39-0
$C_{16}H_{23}N$
229.36
Rolicyclidine
DEA No. 7458
PCBy
PCPY
1-(1-Phenylcyclohexyl)pyrrolidine
1-(1-Phenylcyclohexyl)-pyrrolidine
PHP
Pyrrolidine analog of phencyclidine
Pyrrolidine, 1-(1-phenylcyclohexyl)- (8CI,9CI)
Roliciclidina (Spanish)
Rolicyclidinum (Latin)

2206-27-1
C_2D_6OS
72.09
Methane-D3, sulfinylbis- (9CI)
(Methyl sulfoxide)-D6 (8CI)

2206-89-5
$C_5H_7ClO_2$
134.57
ClCCOC(=O)C=C
Acrylic acid, 2-chloroethyl ester
Acrylic acid β-chloroethyl ester
Chloroethyl acrylate
β-Chloroethyl acrylate
2-Chloroethyl acrylate
2-Chlorethylester kyseliny akrylove (Czech)
Ethanol, 2-chloro-, acrylate
2-Propenoic acid, 2-chloroethyl ester (9CI)

2207-01-4
C_8H_{16}
112.22
Cyclohexane, 1,2-dimethyl-, cis- (9CI)
AI3-28849
cis-1,2-Dimethylcyclohexane

2207-03-6
C_8H_{16}
112.22
C(CCCC1C)(C1)C
Cyclohexane, 1,3-dimethyl-, trans- (8CI,9CI)
NSC-74161

2207-04-7
C_8H_{16}
112.22
1,4-Dimethylcyclohexane-trans
Cyclohexane, 1,4-dimethyl-, trans- (9CI)

2210-25-5
$C_6H_{11}NO$
113.18
O=C(NC(C)C)C=C
Acrylamide, N-isopropyl
Isopropyl acrylamide
N-Isopropylacrylamide
Isopropylamid kyseliny akrylove (Czech)
Nipam
2-Propenamide, N-(1-methylethyl)- (9CI)

2210-28-8
$C_7H_{12}O_2$
128.19
O=C(OCCC)C(=C)C
Methacrylic acid, propyl ester
Propyl methacrylate
n-Propyl methacrylate

2210-79-9
$C_{10}H_{12}O_2$
164.22
O(C1COc(c(ccc2)C)c2)C1
Propane, 1,2-epoxy-3-(o-tolyoxy)
1,2-Epoxy-3-(o-tolyloxy)propane
Glycidyl o-methylphenyl ether
Glycidyl 2-methylphenyl ether
Glycidyl o-tolyl ether
o-Kresol-glycidaether (German)
2-((2-Methylphenoxy)methyl)-oxirane
2-Methylphenyl glycidyl ether
Oxirane, ((2-methylphenoxy)-methyl)- (9CI)

2211-94-1
$C_{10}H_{12}O_3$
180.22
Benzene, 1-(2,3-epoxypropoxy)-4-methoxy
Anisole, p-(2,3-epoxypropoxy)-
2,3-Epoxypropyl-4-methoxy-phenyl ether
Glycidyl p-methoxyphenyl ether
Glycidyl 4-methoxyphenyl ether
2-(p-Methoxyphenoxymethyl)-oxirane
Methoxyphenyl glycidyl ether
p-Methoxyphenyl glycidyl ether
Oxirane, ((4-methoxyphenoxy)-methyl)- (9CI)

2211-98-5
$C_{18}H_{30}O_3S.Na$
349.49
Benzenesulfonic acid, 4-dodecyl-, sodium salt (9CI)
Sodium 4-dodecylbenzene-sulfonate
p-Undecylbenzenesulfonic acid, sodium salt

2211-99-6
$C_{18}H_{30}O_3S.Na$
349.49
Benzenesulfonic acid, 4-(1-methylundecyl)-, sodium salt (9CI)
Benzenesulfonic acid, p-(1-methylundecyl)-, sodium salt (8CI)

2212-32-0
$C_7H_{18}N_2O$
146.22
OCCN(CCN(C)C)C
Ethanol, 2-((2-(dimethylamino)ethyl)methylamino)- (9CI)
2-((2-(Dimethylamino)ethyl)-methylamino)-ethanol
N-(2-(Dimethylamino)ethyl)-N-methylethanolamine
N,N,N'-Trimethyl-N'-(2-hydroxyethyl)-1,2-ethane-diamine

2212-52-4
$C_{18}H_{30}O_3S.Na$
349.49
Benzenesulfonic acid, 4-(1-pentylheptyl)-, sodium salt (9CI)
Benzenesulfonic acid, p-(1-pentylheptyl)-, sodium salt (8CI)

2212-53-5
Unknown
Unknown
Octylamine 2,4-dichlorophenoxyacetate

2212-54-6
Unknown
Unknown
Alkyl* amine 2,4-dichlorophenoxyacetate *(100% C12)

2212-59-1
Unknown
Unknown
N-Oleyl-1,3-propylenediamine 2,4-dichlorophenoxyacetate

2212-67-1
$C_9H_{17}NOS$
187.33
CCSC(=O)N1CCCCCC1
1H-Azepine-1-carbothioic acid, hexahydro-, S-ethyl ester
S-Aethyl-N-hexahydro-1H-azepinthiolcarbamat (German)
S-Ethyl hexahydro-1H-azepine-1-carbothioate
Ethyl 1-hexamethyleneiminecarbothiolate
S-Ethyl 1-hexamethyleneimino-thiocarbamate
Felan
Hydram
Jalan
Molinate
Molmate
Ordram
R-4572
Stauffer R-4,572
S-Ethyl-N-hexamethylenethio-carbamate
Yalan
Yulan

2212-81-9
$C_{20}H_{34}O_4$
338.49
Peroxide, (1,3-phenylenebis-(1-methylethylidene))bis-((1,1-dimethylethyl) (9CI)
Perkadox 14
Perkadox 14C
Perkadox 14/40
Perkadox 14/96
Peroxide, (m-phenylenedi-isopropylidene)bis(tert-butyl (8CI)

2213-23-2
C_9H_{20}
128.26
Heptane, 2,4-dimethyl- (8CI, 9CI)

2213-32-3
C_7H_{14}
98.19
1-Pentene, 2,4-dimethyl- (9CI)

2213-63-0
$C_8H_4Cl_2N_2$
199.04
n(c(c(nc1Cl)ccc2)c2)c1Cl
Quinoxaline, 2,3-dichloro
2,3-Dichloroquinoxaline

2215-35-2
$C_{24}H_{52}O_4P_2S_4Zn$
660.28
Zinc, bis(O,O-bis(1,3-dimethylbutyl) phosphorodithioato-S,S')-, (T-4)- (9CI)
O,O-Bis(1,3-dimethylbutyl)di-thiophosphate zinc salt

2216-30-0
C_9H_{20}
128.26
Heptane, 2,5-dimethyl- (8CI, 9CI)

2216-32-2
C_9H_{20}
128.26
Heptane, 4-ethyl- (8CI,9CI)

2216-33-3
C_9H_{20}
128.26
3-Methyloctane
Octane, 3-methyl- (9CI)

2216-34-4
C_9H_{20}
128.26
4-Methyloctane
Octane, 4-methyl- (9CI)

2216-51-5
$C_{10}H_{20}O$
156.30
OC(C(CCC1C)C(C)C)C1
L-Menthol
Cyclohexanol, 5-methyl-2-(1-methylethyl)-, (1R-(1α,2β,5α))- (9CI)
Menthol, (1R,3R,4S)-(-)- (8CI)
(L)-Menthol
(R)-(-)-Menthol
(-)-Menthol
1-Menthol
(1R,3R,4S)-(-)-Menthol
(-)-Menthyl alcohol
(1R-(1-α,2-β,5-α))-5-Methyl-2-(1-methylethyl)cyclohexanol
NSC-62788
U.S.P. Menthol

2216-69-5
$C_{11}H_{10}O$
158.20
O(c(c(c(ccc1)cc2)c1)c2)C
Naphthalene, 1-methoxy- (9CI)
AI3-02144
1-Methoxynaphthalene

2216-81-1
$C_{10}H_{20}O_2$
172.27
Propanoic acid, heptyl ester (9CI)
Propionic acid, heptyl ester (8CI)

2218-52-2
$C_2H_2F_2O_2.Na$
119.02
Acetic acid, difluoro-, sodium salt (9CI)
Sodium difluoroacetate

2218-68-0
$C_7H_{14}Cl_3NO_4$
282.54
Chloral betaine
β-Chlor
Betainchloralum
Cloral betaina (Spanish)
Cloral betaine
Cloralum betainum (Latin)
DEA No. 2460
Methanaminium, 1-carboxy-N,N,N-trimethyl-, hydroxide, inner salt, Compd. with 2,2,2-trichloro-1,1-ethanediol (1:1); (2) chloral hydrate betaine (1:1) compound
5107

2219-31-0
$C_5H_{12}N_4O_3.H_2O_4S$
274.23
L-Homoserine, O-((amino-iminomethyl)amino)-, sulfate (1:1)
AI3-52581
L-Canavanine sulfate

2219-72-9
$C_{11}H_{16}O$
164.25
Oc(ccc(c1C)C(C)(C)C)c1
Phenol, 4-(1,1-dimethylethyl)-3-methyl- (9CI)
4-(1,1-Dimethylethyl)-3-methyl-phenol

2219-82-1
$C_{11}H_{16}O$
164.27
Oc(c(ccc1)C(C)(C)C)c1C
Phenol, 2-tert-butyl-6-methyl
2-tert-Butyl-6-methylphenol

2222-33-5
$C_{15}H_{10}O$
206.25
O=C(c(c(C=Cc1cccc2)ccc3)c3)c12
5H-Dibenzo(a,d)cyclohepten-5-one

2223-82-7
$C_{11}H_{16}O_4$
212.27
O=C(OCC(C)(C)COC(=O)C=C)C=C
Acrylic acid, 2,2-dimethyl-trimethylene ester
Acrylic acid, 2,2-dimethyl-1,3-propanediol diester
Dimethylolpropane diacrylate
2,2-Dimethyltrimethylene acrylate
Neopentyl glycol diacrylate
1,3-Propanediol, 2,2-dimethyl-, diacrylate
2-Propenoic acid, 2,2-dimethyl-1,3-propanediyl ester (9CI)
SR 247

2223-93-0
$C_{36}H_{72}O_4.Cd$
681.48
[Cd+2].CCCCCCCCCCCCCCCCCC([O-])=O.CCCCCCCCCCCCCCCCCC([O-])=O
Octadecanoic acid, cadmium salt
Cadmium stearate
Kadmiumstearat (German)
Stearic acid, cadmium salt

2224-00-2
$C_{13}H_{12}O_3$
216.24
O=C(O)c(c(c(ccc1)cc2)c1)c2OCC
1-Naphthalenecarboxylic acid, 2-ethoxy- (9CI)
2-Ethoxy-1-naphthalenecarboxylic acid
2-Ethoxy-1-naphthoic acid

2224-15-9
$C_8H_{14}O_4$
174.22
O(C1COCCOCC(O2)C2)C1
Ethane, 1,2-bis(2,3-epoxy-propoxy)
1,2-Bis(glycidyloxy)ethane
Diglycidylethylene glycol
1,2-Diglycidyloxyethane
1,2-Ethanediol diglycidyl ether
Ethylene diglycidyl ether
Ethylene glycol diglycidyl ether
Ethylenglykoldiglycidylether (Czech)
Glycol diglycidyl ether
Oxirane, 2,2'-(1,2-ethanediyl-bis(oxymethylene))bis- (9CI)

2224-44-4
$C_8H_{16}N_2O_3$
188.23
4-(2-Nitrobutyl)morpholine
Caswell No. 601A
EPA Pesticide Chemical Code 100801
Morpholine, 4-(2-nitrobutyl)-
N-(2-Nitrobutyl)morpholine
Vancide F 5386
Vancide 40

2224-49-9
$C_{12}H_{24}O_2.C_6H_{15}NO_3$
349.51
Triethanolamine laurate
Caswell No. 887A
Dodecanoic acid, Compd. with 2,2',2''-nitrilotris(ethanol) (1:1) (9CI)
EPA Pesticide Chemical Code 079043

2225-40-3
$C_8H_8N_2O_3S$
212.22
1H-2,1,3-Benzothiadiazin-4(3H)-one, 3-methyl-, 2,2-dioxide (8CI,9CI)

2227-13-6
$C_{12}H_6Cl_4S$
324.04
Clc2ccc(Sc1cc(Cl)c(Cl)cc1Cl)cc2
Sulfide, p-chlorophenyl 2,4,5-trichlorophenyl
Animert
Animert V-10
Animert V-101
Animert V-10K
Benzene, 1,2,4-trichloro-5-((4-chlorophenyl)thio)-
p-Chlorophenyl 2,4,5-trichlorophenyl sulfide

4-Chlorophenyl 2,4,5-trichlorophenyl sulfide
ENT 27,115
V 101
Philips-Duphar V-101
3,4,6,4'-Tetrachlor-diphenylsulfid (German)
2,4,4',5-Tetrachlorodiphenyl sulfide
2,4,5,4'-Tetrachlorodiphenyl sulfide
Tetrasul
V-101

2227-17-0
$C_{10}Cl_{10}$
474.60
ClC1=C(Cl)C(Cl)(C(=C1Cl)Cl)C2(Cl)C(=C(Cl)C(=C2Cl)Cl)Cl
Bi-2,4-cyclopentadien-1-yl, decachloro
Bis(pentachlor-2,4-cyclopentadien-1-yl)
Bis(pentachlorocyclopentadienyl)
Bis(pentachloro-2,4-cyclopentadien-1-yl)
Decachlor
Decachlorobi-2,4-cyclopentadien-1-yl
1,1',2,2',3,3',4,4',5,5'-Decachlorobi-2,4-cyclopentadien-1-yl
Dienochlor
Dienochlore
ENT 25,718
Hooker HRS-16
Hooker HRS 1654
HRS-16
HRS 1654
HRS 16A
Pentac
Pentac WP

2227-79-4
C_7H_7NS
137.21
N=C(S)c(cccc1)c1
Benzamide, thio
Benzenecarbothioamide
Benzothiamide
Benzothioamide
Thiobenzamide
Tiobenzamide (Italian)

2228-98-0
$C_{10}H_{18}$
138.25
Cyclohexene, 4-(1,1-dimethylethyl)- (9CI)
Cyclohexene, 4-tert-butyl- (8CI)

2231-57-4
CH_6N_4S
106.17
N(N)=C(S)NN
Carbohydrazide, thio
Carbonothioic dihydrazide (9CI)
Hydrazinecarbohydrazonothioic acid
TCH
Thiocarbazide
Thiocarbohydrazide
Thiocarbonic dihydrazide
Thiocarbonohydrazide
USAF EK-7372

2233-00-3
$C_3H_3Cl_3$
145.41
C(C(Cl)(Cl)Cl)=C
1-Propene, 3,3,3-trichloro- (9CI)
3,3,3-Trichloropropene
3,3,3-Trichloro-1-propene

2233-29-6
$C_8H_{11}N$
121.18
Pyridine, 2,3,4-trimethyl- (8CI,9CI)

2234-13-1
$C_{10}Cl_8$
403.70
c(c(c(c(c1Cl)Cl)Cl)c(c(c2Cl)Cl)Cl)(c1Cl)c2Cl
Naphthalene, octachloro
Octachloronaphthalene (ACGIH,OSHA)

2234-16-4
$C_8H_6Cl_2O$
189.04
O=C(c(c(cc(c1)Cl)Cl)c1)C
2',4'-Dichloroacetophenone
Acetophenone, 2',4'-dichloro- (8CI)
1-(2,4-Dichlorophenyl)ethanone
Ethanone, 1-(2,4-dichlorophenyl)- (9CI)

2234-20-0
$C_{10}H_{12}$
132.22
Styrene, 2,4-dimethyl
2,4-Dimethylstyrene

2235-12-3
C_6H_8
80.14
C(C=CC=C)=C
1,3,5-Hexatriene
Divinylethylene

2235-25-8
$C_2H_7HgO_4P$
326.65
CC[Hg]OP(O)(O)=O
Mercury, ethyl(phosphato(1-))
EMP
Ethylmercuric phosphate
Ethylmercury phosphate
Ethylmerkuridihydrogenfosfat (Czech)
Granosan M
Lignasan
N. I. Ceresan
Ruberon
Soilsin

2235-43-0
$C_3H_{12}NO_9P_3.5Na$
413.99
Pentasodium aminotrimethylene phosphonate
Aminotri(methylenephosphonic acid), pentasodium salt
Aminotris(methylphosphonic acid), pentasodium salt
Dequest 2006
Nitrilo(methylenephosphonic acid), pentasodium salt
Nitrilotri(methylenephosphonic acid), pentasodium salt
Nitrilotris(methylenephosphonic acid) pentasodium salt
Pentasodium aminotris(methylphosphonic acid)
Pentasodium nitrilotris(methylenephosphonate)
Pentasodium (nitrilotris(methylene))trisphosphonate
Pentasodium(nitrilotris(methylene))triphosphonate
Phosphonic acid, (nitrilotris(methylene))tri-, pentasodium salt
Phosphonic acid, (nitrilotris(methylene))tris-, pentasodium salt (9CI)

2235-46-3
$C_8H_{15}NO_2$
157.24
O=C(N(CC)CC)CC(=O)C
Acetoacetamide, N,N-diethyl
Butanamide, N,N-diethyl-3-oxo- (9CI)
Diethylacetoacetamide
N,N-Diethylacetoacetamide
Diethylamid kyseliny acetoctové (Czech)

2235-54-3
$C_{12}H_{25}O_4S.H_4N$
283.48
N.CCCCCCCCCCCCOS(O)(=O)=O
Sulfuric acid, monododecyl ester, ammonium salt
Akyposal als 33
Ammonium dodecyl sulfate
Ammonium N-dodecyl sulfate
Ammonium lauryl sulfate
Conco Sulfate A
Dodecyl ammonium sulfate
Lauryl ammonium sulfate
Lauryl sulfate ammonium salt
Maprofix NH
Montopol La 20
Neopon Lam

Presulin
Richonol AM
Sinopon
Sipon LA 30
Siprol 122
Sterling AM
Sulfuric acid, lauryl ester, ammonium salt
Texa Pon A 400
Texapon Special

2235-83-8
$C_{11}H_{14}O$
162.23
CC(=O)CCCc1ccccc1
2-Pentanone, 5-phenyl- (8CI, 9CI)
NSC-167086

2237-30-1
$C_7H_6N_2$
118.15
Nc1cccc(C#N)c1
Benzonitrile, m-amino
m-Aminobenzonitrile
3-Aminobenzonitrile
m-Anthranilonitrile
Benzonitrile, 3-amino- (9CI)
m-Cyanoaniline
3-Cyanoaniline

2237-36-7
$C_8H_8O_4$
168.15
Benzoic acid, 2-hydroxy-4-methoxy- (9CI)
p-Anisic acid, 2-hydroxy- (8CI)
NSC-94304

2238-07-5
$C_6H_{10}O_3$
130.16
O(C1COCC(O2)C2)C1
Ether, bis(2,3-epoxypropyl)
DGE (OSHA)
Di(2,3-epoxy)propyl ether
Diglycidyl ether (ACGIH, OSHA)
Ether, diglycidyl

2239-78-3
$C_{19}H_{38}O_2$
298.51
Hexadecanoic acid, propyl ester
AI3-31575

2239-92-1
$C_{10}H_{12}ClNO_2$
213.68
O=C(OC(C)C)Nc(ccc(c1)Cl)c1
Carbanilic acid, p-chloro-, isopropyl ester

2243-27-8
$C_9H_{17}N$
139.27
N#CCCCCCCCC
Nonanenitrile
1-Cyanooctane
n-Nonanenitrile
Nonanonitrile
n-Nonanonitrile
Octyl cyanide
n-Octyl cyanide
1-Octyl cyanide
Pelargonitrile
Pelargononitrile

2243-35-8
$C_{10}H_{10}O_3$
178.19
O=C(OCC(=O)c(cccc1)c1)C
Acetophenone, 2-hydroxy-, acetate (8CI)
2-(Acetyloxy)-1-phenylethanone
AI3-10570
Ethanone, 2-(acetyloxy)-1-phenyl- (9CI)

2243-52-9
$C_{10}H_{10}O_2$
162.19
2-Butenoic acid, 4-phenyl- (9CI)
Crotonic acid, 4-phenyl- (6CI,8CI)
4-Phenylbut-2-enoic acid
4-Phenylcrotonic acid

2243-61-0
Unknown
Unknown
Nc1ccc(N)c2ccccc12
1,4-Naphthylenediamine

2243-62-1
$C_{10}H_{10}N_2$
158.22
Nc1cccc2c(N)cccc12
1,5-Naphthalenediamine
1,5-Diaminonaphthalene
1,5-Naphthylenediamine
NCI-C03021

2244-07-7
$C_{11}H_{21}N$
167.29
N#CCCCCCCCCCC
Undecanenitrile (9CI)
AI3-28302

2244-11-3
$C_4H_2N_2O_4.H_2O$
160.10
2,4,5,6(1H,3H)-Pyrimidinetetrone, monohydrate
Alloxan monohydrate
Mesoxalylcarbamide monohydrate
Mesoxalylurea monohydrate
2,4,5,6(1H,3H)-Pyrimidinetetrone hydrate
2,4,5,6-Tetraoxohexahydropyrimidine hydrate

2244-16-8
$C_{10}H_{14}O$
150.24
O=C(C(=CCC1C(=C)C)C)C1
p-Mentha-6,8-dien-2-one, (S)-(+)
Carvone, (+)-
(+)-Carvone
d-Carvone
D(+)-Carvone
(S)-Carvone
(S)-(+)-Carvone
2-Cyclohexen-1-one, 2-methyl-5-(1-methylethenyl)-, (S)-
d-p-Mentha-6,8,(9)-dien-2-one
d-1-Methyl-4-isopropenyl-6-cyclohexen-2-one

2244-21-5
$C_3HCl_2N_3O_3.K$
237.07
s-Triazine-2,4,6(1H,3H,5H)-trione, 1,3-dichloro-, potassium salt
ACL-59
1,3-Dichloro-s-triazine-2,4,6-(1H,3H,5H)trione potassium salt
Dichloroisocyanuric acid potassium salt [UN 2465]
Dichloro-s-triazine-2,4,6-(1H,3H,5H)-trione potassium deriv
Dichlor-s-triazin-2,4,6-(1H,3H,5H)trione potassium
Isocyanuric acid, dichloro-, potassium salt
NA 2465 (DOT)
Potassium dichloroisocyanurate
Potassium dichloro isocyanurate [UN 2465]
Potassium dichloro-s-triazinetrione
Potassium dichloro-s-triazinetrione, Dry, Containing more than 39% available chlorine [UN 2465]
1,3,5-Triazine-2,4,6(1H,3H,5H)-trione, 1,3-dichloro-, potassium salt
Potassium troclosene
s-Triazine-2,4,6(1H,3H,5H)-trione, dichloro-, potassium deriv
Troclosene potassium
UN 2465 [Dichloroisocyanuric acid, Dry or dichloroisocyanuric acid salts]

2245-38-7
$C_{13}H_{14}$
170.25
Naphthalene, 1,6,7-trimethyl- (9CI)
AI3-28792

2253-43-2
$C_6H_{15}O_2PS_2$
214.29
O(P(OCCC)(S)=S)CCC
Di-n-propylphosphorodithioic acid
SY:Dipropyl dithiophosphate
O,O-Dipropyl dithiophosphate
O,O-Dipropyldithiophosphoric acid
Dipropyl phosphorodithioate
O,O-Dipropyl phosphorodithioate
O,O-Dipropyl phosphorodithiotic acid
Phosphorodithioic acid, O,O-dipropyl ester (9CI)

2253-52-3
$C_8H_{19}O_2PS_2$
242.34
O(P(OCC(C)C)(S)=S)CC(C)C
Phosphorodithioic acid, O,O-bis(2-methylpropyl) ester (9CI)
O,O-Bis(2-methylpropyl) phosphorodithioate

2253-57-8
$C_{16}H_{35}O_2PS_2$
354.56
O(P(OCCCCCCCC)(S)=S)CCCCCCCC
Phosphorodithioic acid, O,O-dioctyl ester (9CI)
O,O-Dioctyl phosphorodithioate

2255-17-6
$C_9H_{12}NO_6P$
261.19
COP(=O)(OC)Oc1ccc(N(=O)=O)c(C)c1
Phosphoric acid, dimethyl 4-nitro-m-tolyl ester
Accothion o-analog
Bay 42247
m-Cresol, 4-nitro-, dimethyl phosphate
O,O-Dimethyl O-(3-methyl-4-nitrophenyl)phosphorate
Fenitrooxon
Fenitrooxone
Fenitrothion oxon
Fenitroxon
Metaoxon
3-Methyl-4-nitrophenyl dimethyl phosphate
Oxosumithion
Sumioxon
Sumioxone

2257-35-4
$C_3HCl_3O_2$
175.40
OC(=O)C(Cl)=C(Cl)Cl
2-Propenoic acid, 2,3,3-trichloro- (9CI)
Acrylic acid, trichloro- (8CI)

2259-96-3
$C_{14}H_{16}ClN_3O_4S_2$
389.90
NS(=O)(=O)c4cc1c(NC(NS1(=O)=O)C2CC3CC2C=C3)cc4Cl
2H-1,2,4-Benzothiadiazine-7-sulfonamide, 3,4-dihydro-6-chloro-3-(5-norbornen-2-yl)-, 1,1-dioxide
Anhydron
Aquirel
Cyclothiazide
Doburil
MDI 193
Renazide
Valmiran

2263-09-4
$C_{14}H_{14}F_{17}NO_3S$
599.30
O=S(=O)(N(CCCC)CCO)C(F)(F)C(F)(F)C(F)(F)C(F)(F)C(F)(F)C(F)(F)C(F)(F)C(F)(F)F
1-Octanesulfonamide, N-butyl-1,1,2,2,3,3,4,4,5,5,6,6,7,7,8,8,8-heptadecafluoro-N-(2-hydroxyethyl)- (9CI)

2269-21-8
$C_{22}H_{47}NO_2$
357.61
OC(C)CN(CCCCCCCCCCCCCCCC)CC(O)C
2-Propanol, 1,1'-(hexadecylimino)bis- (9CI)
Hexadecylbis(2-hydroxypropyl) amine
1,1'-(Hexadecylimino)bis-2-propanol

2269-22-9
$C_4H_{10}O.1/3Al$
83.12
2-Butanol, aluminum salt (9CI)
Aluminum sec-butoxide

2270-20-4
$C_{11}H_{14}O_2$
178.25
OC(=O)CCCCc1ccccc1
Valeric acid, 5-phenyl
Benzenepentanoic acid (9CI)
Phenylpentanoic acid
5-Phenylpentanoic acid
Phenylvaleric acid
γ-Phenylvaleric acid
5-Phenylvaleric acid

2272-23-3
$C_{10}H_8Cl_2N_4$
255.08
1,3,5-Triazin-2-amine, 4,6-dichloro-N-(2-methylphenyl)- (9CI)
s-Triazine, 2,4-dichloro-6-o-toluidino- (8CI)

2272-24-4
$C_{10}H_8Cl_2N_4$
255.08
1,3,5-Triazin-2-amine, 4,6-dichloro-N-(4-methylphenyl)- (9CI)
s-Triazine, 2,4-dichloro-6-p-toluidino- (8CI)

2272-28-8
$C_{21}H_{15}Cl_3N_6$
457.71
Melamine, N(2),N(4),N(6)-tris-(o-chlorophenyl)- (8CI)
NSC-57582

2272-29-9
$C_9H_5Cl_3N_4$
275.50
1,3,5-Triazin-2-amine, 4,6-dichloro-N-(4-chlorophenyl)- (9CI)
s-Triazine, 2,4-dichloro-6-(p-chloroanilino)- (8CI)

2272-33-5
$C_9H_4Cl_4N_4$
309.97
1,3,5-Triazin-2-amine, 4,6-dichloro-N-(2,5-dichlorophenyl)- (9CI)
s-Triazine, 2,4-dichloro-6-(2,5-dichloroanilino)-(6CI,7CI,8CI)

2272-39-1
$C_9H_3Cl_5N_4$
344.42
s-Triazine, 2,4-dichloro-6-(2,4,5-trichloroanilino)-(7CI,8CI)

2272-40-4
$C_9H_6Cl_2N_4$
241.05
1,3,5-Triazin-2-amine, 4,6-dichloro-N-phenyl- (9CI)
Anex
s-Triazine, 2-anilino-4,6-dichloro-(8CI)

2273-43-0
$C_4H_{10}O_2Sn$
208.83
Stannane, butylhydroxyoxo
1-Butanestannonic acid
Butylhydroxytin oxide
Butylstannoic acid
Butyltin hydroxide oxide

2274-67-1
$C_{10}H_{10}Cl_3O_4P$
331.52
COP(=O)(OC)OC(=CCl)c1ccc(Cl)cc1Cl
Phosphoric acid, 2-chloro-1-(2,4-dichlorophenyl)vinyl dimethyl ester
2-Chloro-1-(2,4-dichlorophenyl)ethenyl dimethyl phosphate
ENT 25,818
OMS-712
Phosphoric acid, 2-chloro-1-(2,4-dichlorophenyl)ethenyl dimethyl ester
SD 8280
Shell SD-8280

2275-14-1
$C_{11}H_{13}Cl_2O_2PS_3$
375.29
CCOP(=S)(OCC)SCSc1cc(Cl)ccc1Cl
Phosphorodithioic acid, S-(((2,5-dichlorophenyl)thio)methyl) O,O-diethyl ester
O,O-Diaethyl-S((2,5-dichlorphenyl-thio)-methyl)-dithiophosphat (German)
S-(2,5-Dichlorophenylthiomethyl) O,O-diethyl phosphorodithioate
2,5-Dichlorophenylthiomethyl O,O-diethyl phosphorodithioate
S-(2,5-Dichlorophenylthiomethyl) diethyl phosphorothiolothionate
O,O-Diethyl S-(2,5-dichlorophenylthiomethyl) dithiophosphate
O,O-Diethyl-S-(2,5-dichlorophenylthiomethyl) dithiophosphoran
O,O-Diethyl S-(2,5-dichlorophenylthiomethyl) phosphorothiolothionate
O,O-Diethyl S-(2,5-dichlorophenylthiomethyl) phosphorodithioate
Dithiophosphate de O,O-diethyle et de S(2,5-dichlorophenyl) thiomethyle (French)
ENT 25,585
Fenkapton (Dutch)
G 28029
Geigy 28029
Geigy G-28029
Methanethiol, ((2,5-dichlorophenyl)thio)-, S-ester with O,O-diethyl phosphorodithioate
NA 2783 [Organophosphorus pesticides, solid, toxic, N.O.S.]
Phenatol
Phencapton [NA 2783]
Phenkapton
Phenkaptone
Phenudin
Przedziorkofos (Polish)

2275-18-5
$C_9H_{20}NO_3PS_2$
285.39
CCOP(=S)(OCC)SCC(=O)NC(C)C
Phosphorodithioic acid, O,O-diethyl ester, S-ester with N-isopropyl-2-mercaptoacetamide
AC 18682
Acetamide, N-isopropyl-2-mercapto-, S-ester with O,O-diethyl phosphorodithioate
American Cyanamid 18682
O,O-Diethyl S-(N-isopropylcarbamoylmethyl) dithiophosphate
O,O-Diethyldithiophosphorylacetic acid, N-monoisopropylamide
O,O-Diethyl S-isopropylcarbamoylmethyl phosphorodithioate
O,O-Diethyl S-(N-isopropylcarbamoylmethyl) phosphorodithioate
EI 18682
ENT 24,652
FAC
FAC 20
Fostion
Isopropyl diethyldithiophosphorylacetamide
L 343
N-Monoisopropylamide of O,O-diethyldithiophosphorylacetic acid
Oleofac
Phosphorodithioic acid, O,O-diethyl S-(2-((1-methylethyl)amino)-2-oxoethyl) ester (9CI)
Prothoate
Protoat (Hungarian)
Telefos
Trimethoate

2275-23-2
$C_8H_{18}NO_4PS_2$
287.36
CNC(=O)C(C)SCCSP(=O)(OC)OC
Phosphorothioic acid, O,O-dimethyl ester, S-ester with 2-((2-mercaptoethyl)thio)-N- methylpropionamide
American Cyanamid 43073
O,O-Dimethyl S-2-(1-N-methylcarbamoylethylmercapto)ethyl thiophosphate
O,O-Dimethyl S-(2-(1-methylcarbamoylethylthio)ethyl) phosphorothioate
Dimethyl S-(2-(1-methylcarbamoylethylthio)ethyl) phosphorothiolate
ENT 26,613
Kilval
N-Methyl O,O-dimethylthiolophosphoryl-5-thia-3-methyl-2-valeramide
N-Methyl-3-thia-2-methyl-valeramid der O,O-dimethylthiolphosphorsaeure (German)
NPH 83
RP-9895
R.P. 10,465
Trucidor
Vamidoate
Vamidothion

2277-16-9
$C_9H_{16}O$
140.23
4-Nonenal, (E)- (8CI,9CI)

2278-22-0
$C_2H_3NO_5$
121.06
CC(=O)OON(=O)=O
Peroxyacetyl-nitrate
Nitric acid, anhydride with peroxyacetic acid
PAN

2279-64-3
$C_7H_8HgN_2O$
336.76
Mercury, phenylureido
Abavit
Leutosan
Leytosan
Phenylmercuric urea
Phenylmercuriurea
Phenyl mercury urea
Urea (phenylmercuri)-

2279-76-7
$C_9H_{21}ClSn$
283.44
CCC[Sn](Cl)(CCC)CCC
Stannane, chlorotripropyl
Chlorotripropylstannane
Tin, tripropyl-, chloride
Tripropylstannium chloride
Tri-n-propyltin chloride

2280-49-1
$C_{13}H_{10}Cl_3NO_2S_2$
382.71
O=S(=O)(N(SC(Cl)(Cl)Cl)c(cc cc1)c1)c(cccc2)c2
Benzenesulfonamide, N-phenyl-N-((trichloromethyl)thio)- (9CI)
N-Phenyl-N-(trichloromethylsulfenyl)benzene sulfonamide

2280-93-5
$C_7H_6Cl_2N_2O$
205.03
Benzamide, 3-amino-2,5-dichloro- (8CI,9CI)
NSC-153180

2282-34-0
$C_{13}H_{19}NO_2$
221.33
CCCC(C)c1cccc(OC(=O)NC)c1
Carbamic acid, methyl-, m-(1-methylbutyl)phenyl ester
3-sec-Amylphenyl N-methylcarbamate
Bufencarb
BUX
BUX Ten
Chevron RE 5353
ENT 27,127
m-(1-Methylbutyl)phenyl methylcarbamate
OMS 227
Ortho 5,353
Ortho RE-5353
Phenol, m-(1-methylbutyl)-, methylcarbamate
Phenol, m-sec-pentyl-, meth ylcarbamate
RE-5353
RE 9659

2283-08-1
$C_{11}H_8O_3$
188.19
O=C(O)c(c(c(ccc1)cc2)c1)c2O
1-Naphthoic acid, 2-hydroxy

2294-47-5
$C_6H_4N_6$
160.10
p-Diazidobenzene
Benzene, p-diazido-
Benzene, 1,4-diazido-
p-Diazidobenceno (Spanish)
1,4-Diazidobenzene
p-Phenylene diazide

2294-76-0
$C_{10}H_{15}N$
149.23
n(c(ccc1)CCCCC)c1
Pyridine, 2-pentyl- (9CI)
2-Pentylpyridine

2294-82-8
$C_{15}H_{14}$
194.28
9H-Fluorene, 9-ethyl- (9CI)
Fluorene, 9-ethyl- (8CI)

2300-66-5
$C_8H_6Cl_2O_3.C_2H_7N$
266.14
Benzoic acid, 3,6-dichloro-2-methoxy-, Compd. with N-methylmethanamine (1:1)
o-Anisic acid, 3,6-dichloro-, Compd. with dimethylamine (1:1)
Banex
Banvel D
Banvel 4S
Dianate
Dicamba amine
Dicamba dimethylamine salt
3,6-Dichloro-2-methoxybenzoic acid Compd. with N-methylmethanamine (1:1)
Dimethylamine salt of dicamba

2302-17-2
$C_8H_{10}N_2O_4S.Na$
253.22
Sodium asulam
ARD 13/02
Asulam sodium
Asulam-sodium
Asulam, sodium salt
Asulox
Asulox, sodium salt
Carbamic acid, ((4-aminophenyl)sulfonyl)-, methyl ester, monosodium salt (9CI)
Carbamic acid, sulfanilyl-, methyl ester, monosodium salt
Caswell No. 062B
Methyl sulfanilylcarbamate, sodium salt

2302-96-7
$C_{20}H_{14}N_2O_7S_2.2Na$
504.46
1-Naphthalenesulfonic acid, 4-((2-hydroxy-6-sulfo-1-naphthalenyl)azo)-, disodium salt
Acidal Red E
Acid Leather Red IBW
Acid Red 13
Acidol Red E
Acilan Red E
Brilliant Ponceau 3RF Extra
Certicol Fast Red E
Cherry Red A Geigy
C.I. 16045
C.I. Acid Red 13 (7CI)
C.I. Acid Red 13, Disodium salt (8CI)
C.I. Food Red 4
Cilefa Red E
Crispin Fast Red E
Dalf Fast Red
Fast Red E (6CI)
Fast Red Specially Pure
HD Fast Red E
HD Fast Red E Supra
Hexacol Fast Red E
Hexalan Fast Red E
Hispacid Fast Red E
Naphthalene Red EA
Naphthionic Red A
Neklacid Fast Red E
Neklacid Red E
1869 Red
Red E for Food
Tertracid Red E

2303-16-4
$C_{10}H_{17}Cl_2NOS$
270.24
CC(C)N(C(C)C)C(=O)SCC(Cl)=CCl
Carbamic acid, diisopropylthio-, S-(2,3-dichloroallyl) ester
Avadex
Bis(1-methylethyl) carbamothioic acid, S-(2,3-dichloro-2-propenyl)ester
Carbamothioic acid, bis(1-methylethyl)-, S-(2,3-dichloro-2-propenyl) ester
CP 15,336
DATC
2,3-DCDT
Diallaat (Dutch)
Diallat (German)
Diallate
Di-Allate
S-(2,3-Dichlor-allyl)-N,N-diisopropyl-monothiocarbamaat (Dutch)
2,3-Dichlorallyl-N,N-(diisopropyl)-thiocarbamat (German)
S-(2,3-Dichloro-allil)-N,N-diisopropil-monotiocarbammato (Italian)
Dichloroallyl diisopropylthiocarbamate
S-2,3-Dichloroallyl diisopropylthiocarbamate
2,3-Dichloroallyl N,N-diisopropylthiolcarbamate
2,3-Dichloro-2-propene-1-thiol, diisopropylcarbamate
Diisopropylthiocarbamic acid, S-(2,3-dichloroallyl) ester
Di-Isopropylthiolocarbamate de S-(2,3-dichloro allyle) (French)
2-Propene-1-thiol, 2,3-di chloro-, diisopropylcarbamate
Pyradex
RCRA waste number U062

2303-17-5
$C_{10}H_{16}Cl_3NOS$
304.68
CC(C)N(C(C)C)C(=O)SCC(Cl)=C(Cl)Cl
Carbamic acid, diisopropylthio-, S-(2,3,3-trichloroallyl) ester
Avadex BW
CP 23426
N-Diisopropylthiocarbamic acid S-2,3,3-trichloro-2-propenyl ester
N,N-Diisopropyl-2,3,3-trichlorallyl-thiolcarbamat (German)
Diisopropyltrichloroallylthiocarbamate
Dipthal
Far-Go
2-Propene-1-thiol, 2,3,3-trichloro-, diisopropylcarbamate
Thiocarbamic acid, N-diisopropyl-, S-2,3,3-trichloroallyl ester

Triallat (German)
Triallate
2,3,3-Trichlorallyl-N,N-(diisopropyl)-thiocarbamat (German)
S-2,3,3-Trichloroallyl N,N-diisopropylthiocarbamate
2,3,3-Trichloroallyl diisopropylthiocarbamate

2303-25-5
$C_{13}H_{11}NO_3$
229.23
Benzene, 1-methyl-3-(4-nitrophenoxy)- (9CI)
Ether, p-nitrophenyl m-tolyl (8CI)
HE 314
TOPE

2303-35-7
$C_9H_{20}NO_2.Cl$
209.75
Muscarine-chloride
(+)-Muscarine chloride
L-(+)-Muscarine chloride

2304-30-5
$C_{16}H_{36}P.Cl$
294.94
[Cl-].CCCC[P+](CCCC)(CCCC)CCCC
Phosphonium, tetrabutyl-, chloride
Tetra-n-butylphosphonium chloride

2304-85-0
$C_{16}H_8O_2$
232.24
O=C(c(c(c(c(cc1)C=CC2=O)c2c3)c1C=4)c3)C4
1,8-Pyrenedione
3,10-Pyrenedione
1,8-Pyrenequinone
3,10-Pyrenequinone

2305-05-7
$C_{12}H_{22}O_2$
198.34
O=C(OC(C1)CCCCCCCC)C1
2(3H)-Furanone, dihydro-5-octyl
γ-Dodecalactone
Dodecanolide-1,4
γ-n-Octyl-γ-n-butyrolactone

2305-13-7
$C_{10}H_{14}O_3$
182.22
Benzenepropanol, 4-hydroxy-3-methoxy- (9CI)

2305-21-7
$C_6H_{12}O$
100.16
OCC=CCCC
2-Hexen-1-ol (9CI)

2305-26-2
$C_8H_{10}O_4$
170.17
O=C(O)C(C(C(=O)O)CC=C1)C1
4-Cyclohexene-1,2-dicarboxylic acid, cis- (9CI)
AI3-23867
cis-4-Cyclohexene-1,2-dicarboxylic acid

2306-88-9
$C_{16}H_{32}O_2$
256.43
O=C(OCCCCCCCC)CCCCCCC
Octanoic acid, octyl ester (9CI)
AI3-31017
Octyl octanoate

2306-89-0
$C_{18}H_{36}O_2$
284.48
Octanoic acid, decyl ester (9CI)
Decyl octanoate

2306-92-5
$C_{18}H_{36}O_2$
284.48
Decanoic acid, octyl ester (9CI)
Octyl decanoate

2307-49-5
$C_8H_5Cl_3O_3$
255.48
COc1c(Cl)cc(Cl)c(Cl)c1C(O)=O
o-Anisic acid, 3,5,6-trichloro
Banvel T
Benzoic acid, 2-methoxy-3,5,6-trichloro-
Benzoic acid, 2,3,5-trichloro-6-methoxy- (9CI)
Kyselina 3,5,6-trichlor-2-methoxybenzoova (Czech)
2-Methoxy-3,5,6-trichlorobenzoic acid
Metriben
Tricamba
3,5,6-Trichloro-o-anisic acid
3,5,6-Trichloro-2-methoxybenzoic acid
Velsicol
Velsicol C
Velsicol Compound C
Velsicol 58-CS-25

2307-55-3
$C_8H_6Cl_2O_3.H_3N$
238.08
Acetic acid, (2,4-dichlorophenoxy)-, ammonium salt
Ammonium 2,4-D
Ammonium 2,4-dichlorophenoxyacetate
2,4-D ammonium salt
(2,4-Dichlorophenoxy)acetic acid ammonium salt

2307-68-8
$C_{13}H_{18}ClNO$
239.77
p-Valerotoluidide, 3'-chloro-2-methyl
N-(3-Chlor-methylphenyl)-2-methylpentanamid (German)
N-(3-Chloro-4-methylphenyl)-2-methylpentanamide
3'-Chloro-2-methyl-p-valerotoluidide
CMMP
Dakuron
Dutom
FMC 4512
Niagara 4512
Pentanochlor
Pentachlore
Solan

2310-17-0
$C_{12}H_{15}ClNO_4PS_2$
367.82
CCOP(=S)(OCC)SCn1c(=O)oc2cc(Cl)ccc12
Phosphorodithioic acid, S-((6-chloro-2-oxo-3(2H)-benzoxazolyl)methyl) O,O-diethyl ester
Azofene
S-((3-Benzoxazolinyl-6-chloro-2-oxo)methyl) O,O-diethyl phosphorodithioate
Benzphos
Benzophosphate
Chipman 11974
S-(6-Chloro-3-(mercaptomethyl)-2-benzoxazolinone) O,O-diethyl phosphorodithioate
3-(6-Chloro-2-oxobenzoxazolin-3-yl)methyl O,O-diethyl phosphorothiolothionate
S-((6-Chloro-2-oxo-3(2H)-benzoxazolyl)methyl) O,O-diethyl phosphorodithioate
O,O-Diaethyl-S-(6-chlor-benzoxazolon-2-on-3-yl)-dithiophosphat (German)
O,O-Diaethyl-S-(6-chlor-2-oxo-ben(b)-1,3-oxalin-3-yl)-methyl-dithiophosphat (German)
O,O-Diethyl-S-((6-chloor-2-oxo-benzoxazolin-3-yl)-methyl)-dithiofosfaat (Dutch)
O,O-Diethyl S-(6-chlorobenzoxazolinyl-3-methyl) dithiophosphate
O,O-Diethyl S-((6-chloro-2-oxo-benzoxazolin-3-yl)methyl) phosphorodithioate
O,O-Diethyl-S-(6-chloro-2-oxo

-benzoxazolin-3-yl)methyl-phosphorothiolothionate
3-Diethyldithiophosphoryl-methyl-6-chlorobenzoxazol-one-2
O,O-Diethyl phosphorodithioate, S-ester with 6-chloro-3-(mercaptomethyl)-2-benzoxazolinone
O,O-Dietil-S-((6-cloro-2-oxo-benzossazolin-3-il)-metil)-ditiofosfato (Italian)
ENT 27,163
Fosalon
Fozalon
NIA-9241
Niagara 9241
NPH-1091
P-974
Phasolon
Phosalon
Phozalon
Phosalone
Phosphorodithioic acid-S-ester of 6-chloro-3-mercapto-methylbenzoxazyol-2-one
Rhodia RP 11974
RP 11,974
Rubitox
Zolon
Zolone PM
Zoolon
Zolone

2311-46-8
$C_9H_{18}O_2$
158.24
Hexanoic acid, 1-methylethyl ester (9CI)
Hexanoic acid, isopropyl ester (8CI)

2311-59-3
$C_{13}H_{26}O_2$
214.35
Decanoic acid, 1-methylethyl ester (9CI)
Decanoic acid, isopropyl ester (8CI)

2312-35-8
$C_{19}H_{26}O_4S$
350.51
CC(C)(C)c2ccc(OC1CCCCC1OS(=O)OCC#C)cc2
Sulfurous acid, 2-(p-tert-butylphenoxy)cyclohexyl-2-propynyl ester
BPPS
2-(p-t-Butylphenoxy)cyclohexyl propargyl sulfite
2-(p-tert-Butylphenoxy)cyclohexyl 2-propynyl sulfite
Comite
2-(4-(1,1-Dimethylethyl)phenoxy)cyclohexyl 2-propynyl sulfite
DO 14
ENT 27,226
Naugatuck D-014
Omait
Omite
Propargite
Sulfurous acid, 2-(4-(1,1-dimethylethyl)phenoxy)cyclohexyl 2-propynyl ester
Uniroyal D014
U.S. Rubber D-014

2312-73-4
$C_{17}H_{19}N_5O$
309.41
C1CCC(OC1)n3cnc4c(NCc2ccccc2)ncnc34
9H-Purine, 6-benzylamino-9-tetrahydropyran-2-yl
Accel
Adenine, N-benzyl-9-(tetrahydro-2H-pyran-2-yl)- (8CI)
6-Benzylamino-9-tetrahydropyran-2-yl-9H-purine
BPA
PBA
PBA (Growth stimulant)
SD 8339

2312-76-7
$C_7H_5N_2O_5.Na$
220.12
Sodium 4,6-dinitro-o-cresylate
Caswell No. 390A
Corodinoc
o-Cresol, 4,6-dinitro-, sodium salt
Cresotol
Dinitro-o-cresate de sodium (French)
Dinitro-o-cresol sodium salt
3,5-Dinitro-o-cresol sodium salt
4,6-Dinitro-o-cresol sodium salt
Dinitro-o-cresolato sodico (Spanish)
2,4-Dinitro-6-methylphenol sodium salt
DINOC
DNOC
DNOC sodium salt
Dynosol
EK 54
Elgetol
EPA Pesticide Chemical Code 037508
Krenite (Obs.)
Krezonite
Krezotol
Krezotol DNOC
2-Methyl-4,6-dinitrophenol sodium salt
2-Methyl-4,6-dinitro-phenol sodium salt
Phenol, 2-methyl-4,6-dinitro-, sodium salt (9CI)
Sinox
Sodium dinitro-o-cresolate
Sodium 4,6-dinitro-o-cresoxide
Sodium salt of 4,6-dinitro-o-cresol

2313-65-7
$C_7H_{16}O$
116.20
2-Hexanol, 3-methyl- (8CI,9CI)
NSC-93810

2314-36-5
$C_7H_4Cl_2O_2$
191.01
Benzaldehyde, 3,5-dichloro-4-hydroxy- (8CI,9CI)
NSC-31590

2314-78-5
$C_6H_9NO_2$
127.14
2,5-Pyrrolidinedione, 1-ethyl- (9CI)
NSC-38693
Succinimide, N-ethyl- (8CI)

2314-97-8
CF_3I
195.91
FC(F)(F)I
Methane, trifluoroiodo- (9CI)
Trifluoroiodomethane
Trifluoromethyl iodide

2315-36-8
$C_6H_{12}ClNO$
149.64
O=C(N(CC)CC)CCl
Acetamide, 2-chloro-N,N-diethyl
CDEA
N-Chloroacetyldiethylamine
Diethylamid kyseliny chloroctove (Czech)
N,N-Diethylchloroacetamide
TL 83

2315-66-4
$C_{34}H_{62}O_{11}$
646.86
3,6,9,12,15,18,21,24,27-Nonaoxanonacosan-1-ol, 29-(4-(1,1,3,3-tetramethylbutyl)phenoxy)- (9CI)
3,6,9,12,15,18,21,24,27-Nonaoxanonacosan-1-ol, 29-(p-(1,1,3,3-tetramethylbutyl)-phenoxy)- (8CI)

2315-68-6
$C_{10}H_{12}O_2$
164.20
O=C(OCCC)c(cccc1)c1
Benzoic acid, propyl ester (9CI)
AI3-01973
Propyl benzoate

2316-10-1
$C_{18}H_{16}O_6$

328.32
Naphtho(2,3-c)furan-1(3H)-one, 4-(3,4-dihydroxyphenyl)-3a,4,9,9a-tetrahydro-6,7-di hydroxy-, (3aR-(3aα,4α,9aβ))- (9CI)
α-Conidendrol
Naphtho(2,3-c)furan-1(3H)-one, 4-(3,4-dihydroxy phenyl)-3a,4,9,9a-tetra-hydro-6,7-dihydroxy-(8CI)
α-Norconidendrin

2316-26-9
$C_{11}H_{12}O_4$
208.21
COc1ccc(C=CC(O)=O)cc1OC
Cinnamic acid, 3,4-dimethoxy-(8CI)
AI3-24427
2-Propenoic acid, 3-(3,4-dimethoxyphenyl)- (9CI)

2318-18-5
$C_{19}H_{28}NO_6$
366.48
CC=C1CC(C)C(C)(O)C(=O)OCC2=CCN(C)CCC(OC1=O)C2=O
Senecionanium, 8,12-dihydroxy-4-methyl-11,16-dioxo
2,12-Dihydroxy-4-methyl-11,16-dioxosenecionanium
trans-15-Ethylidene-12-β-hydroxy-4,12-α,13-β-trimethyl-8-oxo-4,8 secosenec-1-enine
NSC-89945
Renardin
Renardine
4,8-Secosenecionan-8,11,16-trione, 12-hydroxy-4-methyl-
Senkirkin
Senkirkine
Senkirkine (Neutral)

2318-25-4
$C_7H_{13}NO_3.ClH$
195.64
Propanoic acid, 3-ethoxy-3-imino-, ethyl ester, hydrochloride (9CI)

2321-07-5
$C_{20}H_{12}O_5$
332.31
O=C(OC(c(c(Oc1cc(O)cc2)cc(O)c3)c3)(c12)c4cccc5)c45
Spiro(isobenzofuran-1(3H),9'-(9H)xanthen)-3-one, 3',6'-dihydroxy- (9CI)
Acid Yellow 73
Benzoic acid, 2-(6-hydroxy-3-oxo-3H-xanthen-9-yl)- (VAN)
C-Ext. Gelb 16 (Germany)
C.I. 45350:1
9-(o-Carboxyphenyl)-6-hydroxy-3-isoxanthenone
9-(o-Carboxyphenyl)-6-hydroxy-3h-xanthen-3-one
D & C Yellow No. 7
3',6'-Dihydroxyfluoran
3',6'-Dihydroxyspiro(isobenzofuran-1(3h),9'-(9h)xanthen)-3-one
3,6-Dihydroxyspiro(xanthene-9,3'-phthalide)
3',6'-Fluorandiol
Fluorescein (8CI)
Fluoresceine
Germany: C-Ext. Gelb 16
Hidacid Fluorescein
Japan Yellow 201
Japan: Yellow No. 201
Resorcinolphthalein
Soap Yellow F
Spiro(isobenzofuran-1(3H),9'-(9H)xanthen)-3-one, 3',6'-dihydroxy-
Yellow No. 201 (Japan)
Zlut Kysela 73 (Czech)
11712 Yellow

2327-02-8
$C_7H_6Cl_2N_2O$
205.03
NC(=O)Nc1ccc(Cl)c(Cl)c1
Urea, (3,4-dichlorophenyl)-
AI3-61325

2338-25-2
$C_8H_3Cl_2F_3N_2$
255.03
FC(F)(F)c2nc1cc(Cl)c(Cl)cc1[nH]2
1H-Benzimidazole, 5,6-dichloro-2-(trifluoromethyl)
5,6-Dichloro-2-trifluoromethyl-benzimidazole
NC 2983

2338-37-6
$C_{22}H_{29}NO_2$
339.47
(L)-Propoxyphene
Benzeneethanol, α-(2-(dimethylamino)-1-methylethyl)-α-phenyl-, propanoate (Ester), (R-(R*,S*))-
2-Butanol, 4-(dimethylamino)-3-methyl-1,2-diphenyl-, propionate (Ester), (-)-
α-L-4-Dimethylamino-3-methyl-1,2-diphenyl-2-butanol propionate
(R)-α-(2-(Dimethylamino)-1-methylethyl)-α-phenyl-benzeneethanol propanoate
Leropropoxyphene
Levopropossifene
Levopropoxifeno (Spanish)
Levopropoxiphenum
Levopropoxyphenum (Latin)

2343-60-4
$C_9H_4F_2N_4O_3$
254.15
s-Triazine, 2,4-difluoro-6-(p-nitrophenoxy)- (6CI, 8CI)

2345-34-8
$C_9H_8O_4$
180.16
O=C(Oc(ccc(c1)C(=O)O)c1)C
Benzoic acid, 4-(acetyloxy)-(9CI)
4-(Acetyloxy)benzoic acid
AI3-20213

2345-61-1
$C_3H_3ClO_2$
106.51
Acrylic acid, 3-chloro-, (E)
(E)-3-Chloroacrylic acid
trans-β-Chloroacrylic acid
trans-3-Chloroacrylic acid
2-Propenoic acid, 3-chloro-, (E)- (9CI)

2347-72-0
$C_{16}H_{10}N_2Na_2O_7S_2$
452.37
E 111
C.I. Food Orange 2
C.I. 15980
E-111
6-Hydroxy-5-((3-sulfophenyl)-azo)-2-naphthalenesulfonic acid, disodium salt
Orange GGN
L-Orange 1
Orange No.1

2349-07-7
$C_{10}H_{20}O_2$
172.30
O=C(OCCCCCC)C(C)C
Isobutyric acid, hexyl ester
Hexyl isobutanoate
n-Hexyl isobutanoate
Hexyl isobutyrate
n-Hexyl isobutyrate
1-Hexyl isobutyrate
Propanoic acid, 2-methyl-, hexyl ester (9CI)

2349-67-9
$C_2H_3N_3S_2$
133.20
N(NC(=N)S1)=C1S
1,3,4-Thiadiazole-2-thiol, 5-amino
2-Amino-5-mercapto-1,3,4-thiadiazole
5-Amino-2-mercapto-1,3,4-thiadiazole
2-Amino-1,3,4-thiadiazole-5-thiol
5-Amino-1,3,4-thiadiazole-2-thiol

2-Amino-δ(2)-1,3,4-thiadiazoline-5-thione
5-Amino-1,3,4-thiadiazoline-2-thione
NSC-21402
2-Mercapto-5-amino-1,3,4-thiadiazole
1,3,4-Thiadiazole-2(3H)-thione, 5-amino- (9CI)
2-Thiol-5-amino-1,3,4-thiadiazole
USAF PD-25

2349-85-1
$C_{16}H_{26}O_2$
250.38
1,4-Benzenediol, 2,6-bis (1,1-dimethylpropyl)- (9CI)
2,6-Di-tert-amylhydroquinone
Hydroquinone, 2,6-di-tert -pentyl- (6CI,7CI,8CI)

2350-11-0
$C_{12}H_{25}Cl$
204.78
C(CCCCCCCCCC)(Cl)C
Dodecane, 2-chloro- (9CI)
2-Chlorododecane

2350-89-2
$C_{14}H_{12}$
180.25
1,1'-Biphenyl, 4-ethenyl- (9CI)

2352-36-5
$C_9H_5Cl_2N_5O_2$
286.05
1,3,5-Triazin-2-amine, 4,6-dichloro-N-(4-nitrophenyl)- (9CI)
s-Triazine, 2,4-dichloro-6-(p-nitroanilino)- (8CI)

2352-37-6
$C_{15}H_{10}ClN_7O_4$
387.74
1,3,5-Triazine-2,4-diamine, 6-chloro-N,N'-bis(4-nitrophenyl)- (9CI)
s-Triazine, 2-chloro-4,6-bis-(p-nitroanilino)-(6CI,7CI,8CI)

2353-33-5
$C_8H_{12}N_4O_4$
228.24
s-Triazin-2(1H)-one, 4-amino-1-(2-deoxy-β-d-erythro-pentofuranosyl)
4-Amino-1-(2-deoxy-β-d-erythro-pentofuranosyl)-s-triazin-2(1H)-one
5-Azadeoxycytidine
5-Aza-2'-deoxycytidine
1,3,5-Triazin-2(1H)-one, 4-amino-1-(2-deoxy-β-d-erythro-pentofuranosyl)- (9CI)

2353-45-9
$C_{37}H_{36}N_2O_{10}S_3.2Na$
810.91
[Na+].[Na+].CCN(Cc1cccc(c1)S([O-])(=O)=O)c2ccc(cc2)C(=C3C=CC(C=C3)=[N+](CC)Cc4cccc(c4)S([O-])(=O)=O)c5ccc(O)cc5S([O-])(=
Ammonium, ethyl(4-(α-(p-(ethyl(m-sulfobenzyl)-amino)phenyl)-4-hydroxy-3-sulfobenzylidene)- 2,5-cyclohexadien-1-ylidene)-(m-sulfobenzyl)-, hydroxide, inner salt, disodium salt
Aizen Food Green No. 3
C.I. 42053
C.I. Food Green 3
Fast Green FCF
FD & C Green No. 3
Food Green 3
1724 Green
Solid Green FCF
Zelen Potravinarska 3 (Czech)
Zelen Stala FCF (Czech)

2358-84-1
$C_{12}H_{18}O_5$
242.27
O=C(OCCOCCOC(=O)C(=C)C)C(=C)C
Diethylene glycol bis(methacrylate)
DGM 2
Diethylene glycol dimethacrylate
Diethylene glycol dimethacrylate monomer
Methacrylic acid, oxydiethylene ester
Oxydiethylene methacrylate
2-Propenoic acid, 2-methyl-, oxydi-2,1-ethanediyl ester (9CI)
TGM 2

2362-57-4
$C_{12}H_{10}N_2O$
198.21
Phenol, 2-(phenylazo)- (9CI)
Phenol, o-(phenylazo)- (8CI)

2363-71-5
$C_{21}H_{42}O_2$
326.56
Heneicosanoic acid (8CI,9CI)

2363-88-4
$C_{10}H_{16}O$
152.24
O=CC=CC=CCCCCC
2,4-Decadienal (9CI)

2363-89-5
$C_8H_{14}O$
126.20
O=CC=CCCCCC
2-Octenal
AI3-36269

2365-48-2
$C_3H_6O_2S$
106.15
O=C(OC)CS
Acetic acid, mercapto-, methyl ester
Methylmercaptoacetate
Methyl 2-mercaptoacetate
Methylthioglycolate
Thioglycolic acid methyl ester
Thioglykolsaeure-methylester (German)
USAF EK-7119

2366-52-1
C_4H_9F
76.11
FCCCC
Butane, 1-fluoro- (9CI)
1-Fluorobutane

2367-82-0
$C_6H_2F_4$
150.08
Benzene, 1,2,3,5-tetrafluoro- (8CI,9CI)

2370-12-9
$C_7H_{16}O$
116.20
1-Pentanol, 2,2-dimethyl- (9CI)

2370-63-0
$C_8H_{14}O_3$
158.20
O=C(OCCOCC)C(=C)C
2-Ethoxy ethyl methacrylate
2-Ethoxyethyl methacrylate
2-Ethoxyethyl 2-methyl-2-propenoate
2-Propenoic acid, 2-methyl-, 2-ethoxyethyl ester (9CI)

2371-42-8
$C_{11}H_{20}O$
168.28
Bicyclo(2.2.1)heptan-2-ol, 1,2,7,7-tetramethyl-, exo- (9CI)
2-Methylisoborneol
2-Norbornanol, 1,2,7,7-tetramethyl-
2-Norbornanol, 1,2,7,7-tetramethyl-, exo- (8CI)

2372-21-6
$C_8H_{16}O_4$
176.24
O=C(OC(C)C)OOC(C)(C)C
Peroxycarbonic acid, OO-tert-butyl O-isopropyl ester
tert-Butyl peroxy isopropyl carbonate, Technically pure (DOT)
Kayacarbon BIC
Lupersol TBIC
Lupersol TBIC-M75
UN 2103 (DOT)

2372-45-4
$C_4H_{10}O.Na$
97.11
Sodium n-butylate
Butanol, sodium salt
1-Butanol, sodium salt (9CI)
Butyl alcohol, sodium salt
Sodium butoxide
Sodium 1-butoxide
Sodium butylate

2373-23-1
$C_{20}H_{38}O_7S$
422.64
Succinic acid, sulfo-, 1,4-dioctyl ester
Butanedioic acid, sulfo-, 1,4-dioctyl ester (9CI)
Dioctyl sulfosuccinate
Empimin OT

2373-38-8
$C_{16}H_{30}O_7S.Na$
389.47
Butanedioic acid, sulfo-, 1,4-bis(1,3-dimethylbutyl) ester, sodium salt (9CI)

2373-80-0
$C_{10}H_8O_4$
192.17
OC(=O)C=Cc2ccc1OCOc1c2
2-Propenoic acid, 3-(1,3-benzodioxol-5-yl)- (9CI)
Cinnamic acid, 3,4-(methylenebis(oxy))-
Cinnamic acid, 3,4-(methylenedioxy)- (8CI)
NSC-5953

2373-98-0
$C_{12}H_{12}N_2O_2$
216.26
3,3'-Biphenyldiol, 4,4'-diamino
Benzidine, 3,3'-dihydroxy-
m,m'-Biphenol, 6,6'-diamino-
4,4'-Diamino-3,3'-biphenyldiol
3,3'-Dihydroxybenzidine
3,3'-Dioxybenzidine
3,3'-Dwuoksybenzydyna (Polish)

2374-14-3
$C_{12}H_{21}F_9O_3Si_3$
468.54
1,3,5-Tris(trifluoropropyl)trimethylcyclotrisiloxane
Cyclotrisiloxane, 2,4,6-trimethyl-2,4,6-tris(3,3,3-trifluoropropyl)- (9CI)
Fluorosilicone trimer
2,4,6-Trimethyl-2,4,6-tris(3,3,3-trifluoropropyl)cyclotrisiloxane

2379-74-0
$C_{18}H_{10}Cl_2O_2S_2$
393.30
O=C(c(c(S1)cc(c2)Cl)c2C)C1=C(Sc(c3c(cc4Cl)C)c4)C3=O
Benzo(b)thiophen-3(2H)-one, 6-chloro-2-(6-chloro-4-methyl-3-oxobenzo(b)thien-2(3H)-ylidene)-4- ethyl
Ahcovat Pink FFD
Ahcovat Printing Pink FF
Amanthrene Pink FF
Amanthrene Pink FFD
Amanthrene Pink FFWP
($\delta^{2,2'}$(3H,3'H)-Bibenzo(b)thiophene)-3,3'-dione, 6,6'-dichloro-4,4'-dimethyl- (8CI)
Calcoloid Pink FFC
Calcoloid Pink FFD
Calcoloid Pink FFRP
Calcoloid Printing Pink FFE
Calcophyl Red FF
Calophyl Pink ZFF
Chemithrene Brilliant Pink R
C.I. 73360
Ciba Brilliant Pink FR
Ciba Pink FF
Daltolite Pink FF
D & C Red No. 30
5,5'-Dichloro-3,3'-dimethylthioindigo
Durindone Pink FF
Durindone Pink FF-FA
Durindone Printing Pink FF
Fast Pink Y
Fenanthren Brilliant Pink R
Fenanthren Pink R Spura
Fenidon Pink R
Helanthrene Brilliant Pink R
Helanthrene Pink R
Helindone Pink CN
Helindon Pink R
Hostavat Brilliant Pink R
Indanthren Brilliant Pink R
Indanthren Brilliant Pink RB
Indanthren Brilliant Pink RP
Indanthren Brilliant Pink RS
Lithosol Fast Pink SVP
Mikethrene Brilliant Pink R
Nihonthrene Brilliant Pink R
Nyanthrene Brilliant Pink R
Oralith Brilliant Pink R
Palanthrene Brilliant Pink R
Paradone Brilliant PNK R
Permanent Pink
11484 Red
Romantrene Brilliant Pink FR
Sanyothrene Brillant Pink IR
Sandothrene Brilliant Pink R
Solanthrene Brilliant Pink R
Solanthrene Brilliant Pink RF
Vat Printing Pink FF
Vat Red 1
Thioindigo Brilliant Pink ZH
Thioindigo Brilliant Pink ZHP
Tina Brilliant Pink R
Tyrian Brilliant Pink I-R
Vat Pink FF

2379-79-5
$C_{29}H_{14}N_2O_5$
470.43
O=C(c(c(C(=O)c1cccc2)cc(OC(=N3)c(ccc(c4C(=O)c(c5ccc6)c6)C5=O)c4N)c37)c7)c12
Anthra(2,3-d)oxazole-5,10-dione, 2-(1-amino-2-anthraquinonyl)- (8CI)
Anthra(2,3-d)oxazole-5,10-dione, 2-(1-amino-9,10-dihydro-9,10-dioxo-2-anthracenyl)- (9CI)

2379-81-9
$C_{42}H_{23}N_3O_6$
665.68
16H-Dinaphtho(2,3-a:2',3'-i)-carbazole-5,10,15,17-tetraone, 6,9-dibenzamido
Ahcovat Olive ARN
Ahcovat Olive R
Amanthrene Olive R
Atic Vat Olive R
Benzadone Olive R
Calcoloid Olive R
Calcoloid Olive RC
Calcoloid Olive RL
Caledone Olive RP
Caledon Olive R
Carbanthrene Olive R
Cern Kypova 27 (Czech)
C.I. 69005
Cibanone Olive F2R
Cibanone Olive 2R
C.I. Vat Black 27
Fenanthren Olive R
Indanthrene Olive R
Indanthren Olive R
Mayvat Olive AR
Mikethrene Olive R
Nihonthrene Olive R
Nyanthrene Olive R
Oliv Ostanthrenovy R (Czech)
Ostanthren Olive R
Palanthrene Olive R
Paradone Olive R
Pernithrene Olive R
Ponsol Olive AR
Ponsol Olive ARD
Aromantrene Olive FR
Sandothrene Olive N2R
Solanthrene Olive R
Tinon Olive 2R
Tyrian Olive I-R

2380-94-1
C_8H_7NO

133.14
1H-Indol-4-ol (9CI)
Indol-4-ol (8CI)

2381-15-9
$C_{19}H_{14}$
242.33
Cc4ccc3cc2ccc1ccccc1c2cc3c4
Benz(a)anthracene, 10-methyl
10-Methylbenz(a)anthracene
7-Methyl-1,2-benzanthracene

2381-16-0
$C_{19}H_{14}$
242.33
Cc4ccc3cc1c(ccc2ccccc12)cc3c4
Benz(a)anthracene, 9-methyl
9-Methylbenz(a)anthracene
6-Methyl-1,2-benzanthracene

2381-18-2
$C_{18}H_{13}N$
243.32
Benz(a)anthracen-7-amine
10-Amino-1,2-benzanthracene

2381-21-7
$C_{17}H_{12}$
216.29
Pyrene, 1-methyl
1-Methylpyrene
3-Methylpyrene

2381-31-9
$C_{19}H_{14}$
242.33
Cc3cccc4cc1c(ccc2ccccc12)cc34
Benz(a)anthracene, 8-methyl
8-Methylbenz(a)anthracene
5-Methyl-1,2-benzanthracene

2381-39-7
$C_{21}H_{14}$
266.35
Cc2c1ccccc1c3ccc4cccc5ccc2c3c45
Benzo(a)pyrene, 6-methyl
6-Methylbenzo(a)pyrene
5-Methyl-3,4-benzopyrene
5-Methyl-3,4-benzpyrene

2381-40-0
$C_{19}H_{15}N$
257.35
Cc4ccc3c(C)c2ccc1ccccc1c2nc3c4
Benz(c)acridine, 7,10-dimethyl
2,10-Dimethyl-7,8-benzacridine (French)
6,9-Dimethyl-1,2-benzacridine
7,10-Dimethylbenz(c)acridine

2382-40-3
$C_{12}H_{22}O_2$
198.31
9-Dodecenoic acid (8CI,9CI)

2382-76-5
$C_5H_{13}O_4P$
168.13
O=P(OCCCCC)(O)O
Phosphoric acid, monopentyl ester (9CI)
Monopentyl phosphate

2384-86-3
$C_{10}H_{18}$
138.25
4-Decyne (8CI,9CI)

2385-81-1
$C_{21}H_{31}NO_4$
361.48
Furethidine
DEA No. 9626
Ethyl 4-phenyl-1-(2-tetrahydrofurfuryloxyethyl)piperidine-4-carboxylate
Furethidinum (Latin)
Furetidina
Isonipecotic acid, 4-phenyl-1-(2-((tetrahydrofurfuryl)oxy)-ethyl)-, ethyl ester

2385-85-5
$C_{10}Cl_{12}$
545.50
ClC2(Cl)C4(Cl)C1(Cl)C5(Cl)C(Cl)(Cl)C3(Cl)C1(Cl)C2(Cl)C3(Cl)C45Cl
1,3,4-Metheno-1H-cyclobuta-(cd)pentalene, 1,1a,2,2,3,3a,4,5,5,5a,5b,6- dodecachloro-octahydro
Bichlorendo
CG-1283
Cyclopentadiene, hexachloro-, dimer
Decane,perchloropentacyclo-
Dechlorane
Dechlorane 4070
Dodecachlorooctahydro-1,3,4-metheno-2H-cyclobuta(c,d)pentalene
1,1a,2,2,3,3a,4,5,5,5a,5b,6-Dodecachlorooctahydro-1,3,4-metheno-1H-cyclobuta(cd)pentalene
Dodecachloropentacyclodecane
Dodecachloropentacyclo-$(3.2.2.0^{2,6},0^{3,9},0^{5,10})$decane
ENT 25,719
Ferriamicide
GC 1283
Hexachlorocyclopentadiene dimer
1,2,3,4,5,5-Hexachloro-1,3-cyclopentadiene dimer
HRS 1276
1,3,4-Metheno-1H-cyclobuta-(cd)pentalene, dodecachloro-octahydro-
Mirex
NCI-C06428
Perchlorodihomocubane
Perchloropentacyclodecane
Perchloropentacyclo-$(5.2.1.0^{2,6}.0^{3,9}.0^{5,8})$decane

2386-53-0
$C_{12}H_{26}O_3S.Na$
273.39
CCCCCCCCCCCCS(O)(=O)=O
1-Dodecanesulfonic acid, sodium salt (9CI)
Sodium 1-dodecanesulfonate

2386-60-9
$C_4H_9ClO_2S$
156.63
O=S(=O)(CCCC)Cl
1-Butanesulfonyl chloride (9CI)

2386-64-3
C_2H_5ClMg
88.82
Magnesium, chloroethyl- (9CI)
Chloroethylmagnesium

2386-87-0
$C_{14}H_{20}O_4$
252.34
O=C(OCC(CC(O1)C1C2)C2)C(CC(O3)C3C4)C4
7-Oxabicyclo(4.1.0)heptane-3-carboxylic acid, 7-oxa-bicyclo(4.1.0)hept-3-yl-methyl ester
3,4-Epoxycyclohexylmethyl 3,4-epoxycyclohexane carboxylate
ERL-4221

2386-90-5
$C_{10}H_{14}O_3$
182.22
O(C1C(OC(C(O2)C2C3)C3)CC4)C14
Bis(2,3-epoxycyclopentyl)ether
Bis(2,3-epoxycyclopentyl) ether
EP-205
ERR 4205
Ether, bis(2,3-epoxycyclopentyl)
6-Oxabicyclo(3.1.0)hexane, 2,2'-oxybis- (8CI,9CI)
2,2'-Oxybis-6-oxabicyclo-(3.1.0)hexane

2388-12-7
$C_{12}H_{24}O_3$
216.32
Dodecaneperoxoic acid (9CI)
NSC-74803
Peroxylauric acid (8CI)

2388-14-9
$C_{12}H_{16}$
160.26
Benzene, 1-(1-methylethenyl)-4-(1-methylethyl)- (9CI)
Cumene, p-isopropenyl-
NSC-93801
Styrene, p-isopropyl-α-methyl-(8CI)

2390-54-7
$C_{17}H_{19}N_2S.Cl$
318.86
Thioflavin T
Benzothiazolium, 2-(p-(dimethylamino)phenyl)-3,6-dimethyl-, chloride (8CI)
Benzothiazolium, 2-(4-(dimethylamino)phenyl)-3,6-dimethyl-, chloride (9CI)
2-(p-(Dimethylamino)phenyl)-3,6-dimethylbenzothiazolium chloride
Thioflavin S

2390-59-2
$C_{31}H_{42}N_3.Cl$
492.21
Ethanaminium, N-(4-(bis-(4-(diethylamino)phenyl)-methylene)-2,5-cyclohexadien-1-ylidene)-N- ethyl-, chloride
C.I. 42600
C.I. Basic Violet 4 (8CI)
Ethyl Violet
Ethyl Violet AX
Ethyl Violet GGA
Shikiso Acid Brilliant Blue 6B

2390-60-5
$C_{33}H_{40}N_3.Cl$
514.14
C.I. Basic Blue 7 (8CI)
Ethanaminium, N-(4-((4-(diethylamino)phenyl)(4-(ethylamino)-1-naphthalenyl)-methylene)-2,5-cyclohexadien-1-ylidene)-N-ethyl-, chloride (9CI)

2390-68-3
$C_{22}H_{48}N.Br$
406.53
1-Decanaminium, N-decyl-N,N-dimethyl-, bromide (9CI)
Ammonium, didecyldimethyl-, bromide (8CI)
DDAB
Deciquam
Deciquam 222

2393-53-5
$C_{25}H_{26}O_3$
374.51
Estrone, benzoate
Benzoate d'oestrone (French)
3-(Benzoyloxy)estra-1,3,5(10)-trien-17-one
3-Hydroxyestra-1,3,5(10)-trien-17-one benzoate
Ketohydroxyestrin benzoate
Ketohydroxyoestrin benzoate
Oestronbenzoat (German)

2395-96-2
$C_{15}H_{12}O$
208.26
Anthracene, 9-methoxy-(8CI,9CI)
NSC-122699

2395-99-5
Unknown
Unknown
4-Nitrophenyl
para-Nitrophenyl

2396-03-4
C_7H_7O
107.13
Phenyl, 4-methoxy- (9CI)
Phenyl, p-methoxy- (8CI)

2396-60-3
$C_{13}H_{12}N_2O$
212.24
Diazene, (4-methoxyphenyl)-phenyl- (9CI)
Anisole, p-(phenylazo)-
Azobenzene, 4-methoxy- (8CI)
NSC-16044

2396-63-6
C_7H_8
92.15
C(#C)CCCC#C
1,6-Heptadiyne (8CI,(CI)
NSC-353895

2396-65-8
C_9H_{12}
120.19
C(#C)CCCCCC#C
1,8-Nonadiyne (9CI)
AI3-37714

2396-84-1
$C_8H_{12}O_2$
140.18
O=C(OCC)C=CC=CC
Sorbic acid, ethyl ester (8CI)
(E,E)-Ethyl 2,4-hexadienoate
2,4-Hexadienoic acid, ethyl ester
2,4-Hexadienoic acid, ethyl ester, (E,E)-

2398-37-0
C_7H_7BrO
187.04
O(c(cccc1Br)c1)C
Anisole, m-bromo- (8CI)
Benzene, 1-bromo-3-methoxy-(9CI)
m-Bromoanisole
1-Bromo-3-methoxybenzene

2398-64-3
$C_{26}H_{46}$
358.66
Benzene, (1-hexyltetradecyl)- (9CI)
Eicosane, 7-phenyl- (7CI, 8CI)

2398-65-4
$C_{26}H_{46}$
358.66
Benzene, (1-octyldodecyl)-(9CI)
Eicosane, 9-phenyl- (6CI, 7CI,8CI)

2398-66-5
$C_{26}H_{46}$
358.66
Benzene, (1-methylnonadecyl)- (9CI)
Eicosane, 2-phenyl- (7CI, 8CI)

2398-68-7
$C_{26}H_{46}$
358.66
Benzene, eicosyl- (9CI)
Eicosane, 1-phenyl- (6CI, 7CI,8CI)
Eicosylbenzene
1-Phenyleicosane

2398-81-4
$C_6H_5NO_3$
139.10
O=C(O)c(cccn1=O)c1
Oxiniacic acid
Acide oxiniacique (French)
Acido oxiniacico (Spanish)
Acidum oxiniacicum (Latin)
Nicotinic acid, 1-oxide (8CI)
3-Pyridinecarboxylic acid, 1-oxide (9CI)

2398-96-1
$C_{19}H_{17}NOS$
307.43
O(c(ccc(c1ccc2)c2)c1)C(N(c(cccc3C)c3)C)=S
Carbanilic acid, m,n-dimethylthio-, o-2-naphthyl ester
Carbamothioic acid, methyl-(3-methylphenyl)-, o-2-naphthalenyl ester (9CI)
Dermoxin
m,N-Dimethylthiocarbanilic acid o-2-naphthyl ester
Focusan

Hi-Alazin
Methyl (3-methylphenyl)carbamothioic acid o-2-naphthalenyl ester
Naphthiomate T
o-2-Naphthyl m,n-dimethylthiocarbanilate
2-Naphthyl N-methyl-N-(3-tolyl)thionocarbamate
Pitrex
Sporiline
Tinactin
Tinaderm
Tolnaftate
Tolnaphthate
Tolsanil
Tonoftal

2399-48-6
$C_8H_{12}O_3$
156.18
O=C(OCC(OCC1)C1)C=C
2-Propenoic acid, (tetrahydro-2-furanyl)methyl ester (9CI)
AI3-03203
2-Propenoic acid, tetrahydrofurfuryl ester
Tetrahydrofurfuryl acrylate

2399-81-7
Unknown
Unknown
Nitrilotriacetic acid and its salts
NTA, beryllium salt (1:1)

2399-83-9
Unknown
Unknown
Nitrilotriacetic acid and its salts
NTA, barium salt (1:1)

2399-85-1
$C_6H_6NO_6.3K$
305.41
Acetic acid, nitrilotri-, tripotassium salt
Glycine, N,N-bis(carboxymethyl)-, tripotassium salt (9CI)
NTA, K3
NTA, tripotassium salt
Tripotassium nitrilotriacetate

2399-86-2
Unknown
Unknown
Nitrilotriacetic acid, dipotassium salt
NTA, dipotassium salt

2399-88-4
Unknown
Unknown
Nitrilotriacetic acid, potassium magnesium salt (1:1:1)
NTA, potassium magnesium salt (1:1:1)

2399-89-5
Unknown
Unknown
Nitrilotriacetic acid and its salts
NTA, potassium strontium salt (1:1:1)

2399-94-2
$C_6H_{11}CaNO_6$
233.24
Nitrilotriacetic acid and its salts
NTA, calcium salt (1:1)

2400-00-2
$C_{18}H_{30}$
246.44
c(cccc1)(c1)C(CCCCCCCCC)CC
Benzene, (1-ethyldecyl)- (9CI)
(1-Ethyldecyl)benzene

2400-01-3
$C_{19}H_{32}$
260.46
Benzene, (1-hexylheptyl)- (9CI)
Tridecane, 7-phenyl- (8CI)

2400-02-4
$C_{26}H_{46}$
358.65
Benzene, (1-ethyloctadecyl)- (9CI)
Eicosane, 3-phenyl- (8CI)
NSC-219884

2400-03-5
$C_{26}H_{46}$
358.66
Benzene, (1-propylheptadecyl)- (9CI)
Eicosane, 4-phenyl- (7CI, 8CI)

2400-04-6
$C_{26}H_{46}$
358.66
Benzene, (1-butylhexadecyl)- (9CI)
Eicosane, 5-phenyl- (7CI, 8CI)

2401-64-1
$C_5H_6Cl_2N_4$
193.05
s-Triazine, 2,4-dichloro-6-dimethylamino
2,4-Dichloro-6-dimethylamino-s-triazine

2401-85-6
$C_{10}H_5ClN_2O_4$
252.62
O=N(=O)c(c(c(c(c1N(=O)=O)ccc2)c2)Cl)c1
Naphthalene, 1-chloro-2,4-dinitro
1-Chloro-2,4-dinitronaphthalene
2,4-Dinitro-1-chloro-naphthalene

2402-77-9
$C_5H_3Cl_2N$
147.99
Clc1cccnc1Cl
Pyridine, 2,3-dichloro

2402-78-0
$C_5H_3Cl_2N$
147.99
n(c(ccc1)Cl)c1Cl
Pyridine, 2,6-dichloro

2402-79-1
C_5HCl_4N
216.87
n(c(c(cc1Cl)Cl)Cl)c1Cl
Pyridine, 2,3,5,6-tetrachloro
2,3,5,6-Tetrachloropyridine

2403-88-5
$C_9H_{19}NO$
157.25
OC(CC(NC1(C)C)(C)C)C1
Lastar A
4-Piperidinol, 2,2,6,6-tetramethyl- (9CI)
2,2,6,6-Tetramethyl-4-piperidinol

2404-05-9
$C_5H_{13}O_3PS$
184.21
CCOP(=O)(OCC)SC
Phosphorothioic acid, O,O-diethyl S-methyl ester
O,O-Diethyl S-methyl phosphorothioate

2404-35-5
$C_7H_{13}Br$
177.08
BrC(CCCCC1)C1
Cycloheptane, bromo- (9CI)
Bromocycloheptane
Cycloheptyl bromide

2404-44-6
$C_{10}H_{20}O$
156.27
O(C1CCCCCCCC)C1
Oxirane, octyl- (9CI)
AI3-14198
1,2-Epoxydecane
Octyloxirane

2404-73-1
$C_9H_{21}O_3P$
208.24
Phosphonic acid, methyl-, dibutyl ester (8CI,9CI)

2406-65-7
$C_4H_{12}Sn$
178.85
Stannane, butyl- (8CI,9CI)

2407-13-8
$C_{25}H_{15}Cl_2N_5O_{11}S_3.3Na$
797.46
2,7-Naphthalenedisulfonic acid, 5-(((2,3-dichloro-6-quinoxalinyl)carbonyl)amino)-4-hydroxy-3-((2-sulfophenyl)azo)-, trisodium salt (9CI)
5-(((2,3-Dichloroquinoxalin-6-yl)carbonyl)amino)-4-hydroxy-3-((2-sulfophenyl)azo)-2,7-naphthalenedisulfonic acid, trisodium salt
4-(((2,3-Dichloro-6-quinoxalinyl)carbonyl)amino)-5-hydroxy-6-((2-sulfophenyl)azo)-2,7-naphthalenedisulfonic acid, trisodium salt

2407-94-5
$C_{12}H_{22}O_4$
230.30
O(OC(O)(CCCC1)C1)C(O)(CCCC2)C2
Di-(1-hydroxycyclohexyl) peroxide (Technically pure)
Bis(1-hydroxycyclohexyl)-peroxide
Cyclohexanol, 1,1'-dioxybis- (9CI)
Cyclohexanol, 1,1'-dioxydi-
Dihydroxycyclohexylperoxide
Di-(1-hydroxycyclohexyl) peroxide
Di-(1-hydroxycyclohexyl)-peroxide, Technically pure
1,1'-Dioxybiscyclohexanol
1,1'-Dioxybis-cyclohexanol
Peroxyde de bis (hydroxy-1 cyclohexyle) (French)
Peroxido de di-(1-hidroxiciclohexilo) (Spanish)
1,1'-Peroxydicyclohexanol
UN 2148

2408-37-9
$C_9H_{16}O$
140.23
O=C(C(CCC1)(C)C)C1C
Cyclohexanone, 2,2,6-trimethyl- (9CI)
2,2,6-Trimethylcyclohexanone

2409-55-4
$C_{11}H_{16}O$
164.27
Oc(c(cc(c1)C)C(C)(C)C)c1
p-Cresol, 2-tert-butyl
2-t-Butyl-p-cresol
2-terc.Butyl-p-kresol (Czech)
2-t-Butyl-4-methylphenol

2412-80-8
$C_7H_{14}O_2$
130.19
O=C(OC)CCC(C)C
Pentanoic acid, 4-methyl-, methyl ester (9CI)
Methyl 4-methylpentanoate
Methyl 4-methylvalerate

2414-98-4
$C_2H_6O.1/2Mg$
58.23
Ethanol, magnesium salt (9CI)
Magnesium ethoxide
Magnesium ethylate

2415-72-7
C_6H_{12}
84.16
Cyclopropane, propyl- (9CI)
Propane, 1-cyclopropyl- (6CI,7CI,8CI)
Propylcyclopropane

2415-85-2
$C_{11}H_{13}NO_2$
191.22
O=C(Nc(ccc(c1)C)c1)CC(=O)C
Butanamide, N-(4-methylphenyl)-3-oxo- (9CI)
p-Acetoacetotoluidide (8CI)
N-(4-Methylphenyl)-3-oxobutanamide

2416-94-6
$C_9H_{12}O$
136.19
Oc(c(ccc1C)C)c1C
2,3,6-Trimethylphenol
3-Hydroxypseudocumene
Phenol, 2,3,6-trimethyl- (9CI)

2419-74-1
$C_4H_8Cl_2O$
143.01
2-Butanol, 1,4-dichloro- (8CI, 9CI)

2420-16-8
$C_7H_5ClO_2$
156.57
Benzaldehyde, 3-chloro-4-hydroxy- (9CI)

2420-98-6
$C_8H_{16}O_2.1/2Cd$
200.41
Cadmium 2-ethylhexanoate
Cadmium di-2-ethylhexylate
Cadmium ethylhexanoate
Cadmium 2-ethylhexoate
Hexanoic acid, 2-ethyl-, cadmium salt (9CI)

2421-02-5
$C_{12}H_{27}NO_3$
233.35
2-Butanol, 1,1',1''-nitrilotris- (9CI)
2-Butanol, 1,1',1''-nitrilotri- (7CI,8CI)
Tris(sec-butanolamine)

2421-28-5
$C_{17}H_6O_7$
322.23
O=C(OC(=O)c1ccc(C(=O)c(ccc(c2C(=O)O3)C3=O)c2)c4)c14
Phthalic anhydride, 4,4'-carbonyldi- (8CI)
3,3',4,4'-Benzophenonetetracarboxylic acid dianhydride
5,5'-Carbonylbis-1,3-isobenzofurandione
1,3-Isobenzofurandione, 5,5'-carbonylbis- (9CI)

2422-79-9
$C_{19}H_{14}$
242.33
Cc3c1ccccc1cc4ccc2ccccc2c34
Benz(a)anthracene, 12-methyl
12-Methylbenz(a)anthracene
9-Methyl-1,2-benzanthracene

2422-85-7
C_8H_{12}
108.18
Cyclobutane, 1,2-di ethenyl- (9CI)
Cyclobutane, 1,2-divinyl- (6CI,7CI,8CI)
1,2-Divinylcyclobutane

2422-91-5
$C_{22}H_{13}N_3O_3$
367.35
O=C=Nc(ccc(c1)C(c(ccc(N=C=O)c2)c2)c(ccc(N=C=O)c3)c3)c1
Benzene, 1,1',1''-methylidynetris(4-isocyanato- (9CI)
1,1',1''-Methylidynetris(4-isocyanatobenzene)

2423-10-1
$C_{18}H_{34}O$
266.47
9-Octadecenal, (Z)- (9CI)
Olealdehyde (8CI)

2423-71-4

$C_8H_9NO_3$
167.18
Cc1cc(cc(C)c1O)N(=O)=O
2,6-Xylenol, 4-nitro
4-Nitro-2,6-xylenol

2423-72-5
$C_7H_6Cl_2O_2$
193.03
Phenol, 2,6-dichloro-4-methoxy- (7CI,8CI, 9CI)
2,6-Dichloro-4-methoxyphenol

2425-01-6
$C_{12}H_{14}O_4$
222.24
O(C1COc(ccc(OCC(O2)C2)c3)c3)C1
Oxirane, 2,2'-(1,4-phenylenebis(oxymethylene))bis- (9CI)
AI3-22178
Hydroquinone 1,4-diglycidyl ether
Oxirane, 2,2''-(1,4-phenylenebis(oxymethylene))bis-

2425-06-1
$C_{10}H_9Cl_4NO_2S$
349.06
ClC(Cl)C(Cl)(Cl)SN2C(=O)C1CC=CCC1C2=O
4-Cyclohexene-1,2-dicarboximide, N-((1,1,2,2-tetrachloroethyl)thio)
Captafol
Captatol
Captafol (OSHA)
Difolatan (OSHA)
Difosan
Folcid
Ortho 5865
Sanspor
Sulfonimide
Sulpheimide
N-(1,1,2,2-Tetrachloraethylthio)-cyclohex-4-en-1,4-diacarboximid (German)
N-(1,1,2,2-Tetrachloraethylthio)-tetrahydrophthalamid (German)
N-1,1,2,2-Tetrachloroethylmercapto-4-cyclohexene-1,2-carboximide
N-((1,1,2,2-Tetrachloroethyl)-sulfenyl)-cis-4-cyclohexene-1,2-dicarboximide
N-(1,1,2,2-Tetrachloroethylthio)-4-cyclohexene-1,2-dicarboximide

2425-10-7
$C_{10}H_{13}NO_2$
179.24
O=C(Oc(ccc(c1C)C)c1)NC
Carbamic acid, methyl-, 3,4-xylyl ester
Carbamic acid, N-methyl-, (3,4-dimethylphenyl) ester
3,4-Dimethylphenyl-N-methylcarbamate
Methylcarbamic acid 3,4-xylyl ester
Meobal
MPMC
Phenol, 3,4-dimethyl-, methylcarbamate (9CI)
S-1046
V 17004
Xylylcarb
Xylecarb
3,4-Xylylester kyseliny methylkarbaminove (Czech)
3,4-Xylyl methylcarbamate

2425-25-4
$C_8H_{17}O_5PS$
256.28
CCOC(=O)CSP(=O)(OCC)OCC
Acetic acid, mercapto-, ethyl ester, S-ester with O,O-diethyl phosphorothioate
Acetaphos
Acetic acid, ((diethoxyphosphinyl)thio)-, ethyl ester (9CI)
Acetofos
Acetophos
Acetoxon
O,O-Diethyl S-(carbethoxy)-methyl phosphorothiolate
O,O-Diethyl S-carboethoxymethyl phosphorothioate
O,O-Diethyl S-carboethoxymethyl thiophosphate
Phosphorothioic acid, O,O-diethyl ester, S-ester with ethyl mercaptoacetate

2425-54-9
$C_{14}H_{29}Cl$
232.84
ClCCCCCCCCCCCCCC
Tetradecane, 1-chloro- (9CI)
1-Chlorotetradecane

2425-66-3
$C_3H_6ClNO_2$
123.55
CC(CCl)N(=O)=O
Propane, 1-chloro-2-nitro
Chloronitropropan (Polish)
Chloronitropropane
1-Chloro-2-nitropropane
Korax
Lanstan
Lastan
NIA 5961
Niagara 5961

2425-74-3
$C_5H_{11}NO$
101.17
O=CNC(C)(C)C
Formamide, N-(tert-butyl)
tert-Butylformamide

2425-79-8
$C_{10}H_{18}O_4$
202.28
O(C1COCCCCOCC(O2)C2)C1
Butane, 1,4-bis(2,3-epoxypropoxy)
Araldit DY 026
1,4-Bis(2,3-epoxypropoxy)-butane
1,4-Bis(glycidyloxy)butane
1,4-Butane diglycidyl ether
Butanediol diglycidyl ether
Butane-1:4-diol diglycidyl ether
1,4-Butanediol diglycidyl ether
CD 15006 A
ChS-RR2
1,4-Diglycidloxybutane
Grilonit RV 1806
Oxirane, 2,2'-(1,4-butanediylbis(oxymethylene))bis- (9CI)
TK 10352

2425-85-6
$C_{17}H_{13}N_3O_3$
307.33
O=N(=O)c(c(N=Nc(c(c(ccc1)cc2)c1)c2O)ccc3C)c3
2-Naphthalenol, 1-((4-methyl-2-nitrophenyl)azo)
Accosperse Toluidine Red XL
ADC Toluidine Red B
Calcotone Toluidine Red YP
Carnelio Helio Red
Cerven Pigment 3 (Czech)
Chromatex Red J
C.I.12120
C.I. Pigment Red 3
C.P.Toluidine Toner A-2989
C.P.Toluidine Toner A-2990
C.P.Toluidine Toner Dark RS-3340
C.P.Toluidine Toner Deep X-1865
C.P.Toluidine Toner Light RS-3140
C.P.Toluidine Toner RT-6101
C.P.Toluidine Toner RT-6104
Dainichi Permanent Red 4 R
D & C Red No. 35
Deep Fastona Red
Duplex Toluidine Red L 20-3140
Eljon Fast Scarlet PV Extra
Eljon Fast Scarlet RN
Enialit Light Red RL
Fastona Red B
Fastona Scarlet RL
Fastona Scarlet YS
Fast Red A
Fast Red A (Pigment)
Fast Red AB
Fast Red J
Fast Red JE
Fast Red R
Graphtol Red A-4RL
Hansa Red B
Hansa Red G
Hansa Scarlet RB
Hansa Scarlet RN

Hansa Scarlet RNC
Helio Fast Red BN
Helio Fast Red RL
Helio Fast Red RN
Helio Red RL
Helio Red Toner
Hispalit Fast Scarlet RN
Independence Red
Irgalite Fast Red P4R
Irgalite Fast Scarlet RND
Irgalite Red PV2
Irgalite Red RNPX
Irgalite Scarlet RB
Isol Fast Red HB
Isol Fast Red RNB
Isol Fast Red RN2B
Isol Fast Red RNG
Isol Fast Red RN2G
Isol Toluidine Red HB
Isol Toluidine Red RNB
Isol Toluidine Red RN2B
Isol Toluidine Red RNG
Isol Toluidine Red RN2G
Kromon Helio Fast Red
Kromon Helio Fast Red YS
Lake Red 4R
Lake Red 4RII
Lithol Fast Scarlet RN
Lutetia Fast Red 3R
Lutetia Fast Scarlet RF
Lutetia Fast Scarlet RJN
Monolite Fast Scarlet CA
Monolite Fast Scarlet GSA
Monolite Fast Scarlet RB
Monolite Fast Scarlet RBA
Monolite Fast Scarlet RN
Monolite Fast Scarlet RNA
Monolite Fast Scarlet RNV
Monolite Fast Scarlet RT
NCI-C60366
No. 2 Forthfast Scarlet
Oralith Red P4R
Permanent Red 4R
Pigment Red 3
Pigment Red RL
Pigment Ruby
Pigment Scarlet
Pigment Scarlet (Russian)
Pigment Scarlet B
Pigment Scarlet N
Pigment Scarlet R
Polymo Red FGN
Recolite Fast Red RBL
Recolite Fast Red RL
Recolite Fast Red RYL
Sanyo Scarlet Pure
Sanyo Scarlet Pure No. 1000
Scarlet Pigment RN
Segnale Light Red B
Segnale Light Red 2B
Segnale Light Red BR
Segnale Light Red C4R
Segnale Light Red RL
Siegle Red 1
Siegle Red B
Siegle Red BB
Silogomma Red RLL
Silosol Red RBN
Silosol Red RN
Siloton Red BRLL
Siloton Red RLL
Symuler Fast Scarlet 4R
Syton Fast Scarlet RB
Syton Fast Scarlet RD
Syton Fast Scarlet RN
Tertropigment Red HAB
Tertropigment Scarlet LRN
Toluidine Red
Toluidine Red 10451
Toluidine Red 3B
Toluidine Red BFB
Toluidine Red BFGG
Toluidine Red D 28-3930
Toluidine Red Light
Toluidine Red M 20-3785
Toluidine Red R
Toluidine Red 4R
Toluidine Red RT-115
Toluidine Red Toner
Toluidine Red XL 20-3050
Toluidine Toner
Toluidine Toner Dark 5040
Toluidine Toner 4R X-2700
Toluidine Toner HR X-2741
Toluidine Toner Keep HR X-2742
Toluidine Toner L 20-3300
Toluidine Toner RT-252
Versal Scarlet PRNL
Versal Scarlet RNL
Vulcafor Scarlet A

2426-07-5
$C_8H_{14}O_2$
142.22
C(CCC1CO1)CC2CO2
Octane, 1,2:7,8-diepoxy
1,2:7,8-Diepoxyoctane
1,2-Epoxy-7,8-epoxyoctane
Oxirane, 2,2'-(1,4-butanediyl)bis- (9CI)

2426-08-6
$C_7H_{14}O_2$
130.21
O(C1COCCCC)C1
Propane, 1-butoxy-2,3-epoxy
Ageflex BGE
BGE (OSHA)
Butyl glycidyl ether
n-Butyl glycidyl ether (ACGIH, OSHA)
2,3-Epoxypropyl butyl ether
Ether, butyl 2,3-epoxypropyl
Ether, butyl glycidyl
Glycidyl butyl ether

2426-54-2
$C_9H_{17}NO_2$
171.27
O=C(OCCN(CC)CC)C=C
Acrylic acid, 2-(diethylamino)ethyl ester
Acrylic acid, N,N-diethylaminoethyl ester
Ageflex FA-2
Diethylaminoethyl acrylate
N,N-Diethylaminoethyl acrylate
β-Diethylaminoethyl acrylate
2-(Diethylamino)ethyl acrylate
2-Diethylaminoethylester kyseliny akrylove (Czech)
2-Propenoic acid, 2-(diethylamino)ethyl ester (9CI)

2428-04-8
$C_3Cl_6N_6$
332.75
1,3,5-Triazine-2,4,6-triamine, N,N,N',N',N'',N''-hexachloro-
AI3-61053
Hexachloromelamine

2429-71-2
$C_{34}H_{26}N_4O_{10}S_2.2Na$
760.69
Benzo Azurine G
Benzoazurine G
C.I. Direct Blue 8
C.I. Direct Blue 8, Disodium salt (8CI)
1-Naphthalenesulfonic acid, 3,3'-((3,3'-dimethoxy-(1,1'-biphenyl)-4,4'-diyl)bis-(azo))bis(4-hydroxy-, disodium salt (9CI)

2429-73-4
$C_{32}H_{24}N_6O_{11}S_3.3Na$
833.77
2,7-Naphthalenedisulfonic acid, 5-amino-3-((4'-((7-amino-1-hydroxy-3-sulfo-2-naphthalenyl) azo)(1,1'-biphenyl)-4-yl)azo)-4-hy droxy-, trisodium salt
Airedale Black BHD
Aizen Direct Black BH
Altazine Black BH
Amanil Developed Black BHSW
Atul Developed Black BT
Azine Diazo Black BHK
Azocard Blue BH
Azomine Black BH
Belamine Diazo Black BH
Bencidal Navy Blue BH
Benzanil Black BH
Benzo Black Blue BH
Benzo Black Blue FBH
Blue BH
Brasilazol Black BH
Calcoloid Diazo Black BHL
Calcomine Diazo Black BHD
Calcomine Diazo Black BTCW
Chloramine Black BH
Chlorazol Black BH
Chlorazol Leather Black BH
Chrome Leather Black BH
Chrome Leather Black CR
Chrome Leather Black DS
Chrome Leather Dark Blue BHM
C.I. 22590
C.I. Direct Blue 2
C.I. Direct Blue 2, Trisodium salt
Cutamin Dark Blue CB
Diacotton Black BH

Diamine Black BH
Diamine Black BHM
Diaminogene Velour Black B
Dianil Dark Blue H
Diaphtamine Black BH
Diazine Black BHC
Diazine Black H
Diazine Black HDW
Diazine Black HNJ
Diazo Black BH
Diazo Black BHN-CF
Diazo Black BHSW
Diazo Black BHSWK
Diazo Black CR
Diazo Direct Black N
Diazo Fast Black BH
Diazo Fast Black MBH
Diazol Black BH
Diazo Navy Blue BH
Diazophenyl Black BH
Diphenyl Blue Black GHS
Diphenyl Blue Black MBH
Direct Black BH
Direct Blue 2
Direct Blue Black BH
Direct Dark Blue BH
Direct Diazo Black
Direct Diazo Black C
Direct Diazo Black N
Direct Diazo Black S
Direct Navy Blue BH
Eniazol Blue Black BHN
Fenamin Navy Blue H
Fixanol Blue BH
Indoxine KL
Japanol Black BHK
Kayaku Direct Black BH
Melantherine BH
Melantherine BHX
Mitsui Direct Black BH
Navy Blue EMBL
NCI-C61110
Neklamin Black BH
Paramine Black BH
Phenazo Black BH
Pheno Navy Blue
Pontamine Deep Blue BH
Pontamine Diazo Black BHSW
Symulon Direct Black BH
Tertrodirect Black BH
Tertrodirect Black BHS
Union Fast Navy Blue DS
Vondacel Dark Blue BH
Zambesi Dark Blue BH

2429-74-5
$C_{34}H_{28}N_6O_{16}S_4.4Na$
996.88
2,7-Naphthalenedisulfonic acid, 3,3'-((3,3'-dimethoxy-4,4'-biphenylylene)bis(azo))-bis(5-amino-4-hydroxy-, tetrasodium salt
Airedale Blue D
Aizen Direct Sky Blue 5BH
Amanil Sky Blue
Atlantic Sky Blue A
Atul Direct Sky Blue
Azine Sky Blue 5B
Belamine Sky Blue A
Benzanil Sky Blue
Benzo Sky Blue S
Benzo Sky Blue A-CF
Chloramine Sky Blue A
Chloramine Sky Blue 4B
Chrome Leather Pure Blue
C.I. 24400
C.I. Direct Blue 15
C.I. Direct Blue 15, Tetrasodium salt
Cresotine Pure Blue
Diacotton Sky Blue 5B
Diamine Sky Blue CI
Diaphtamine Pure Blue
Diazol Pure Blue 4B
Diphenyl Brilliant Blue
Diphenyl Sky Blue 6B
Direct Blue 10G
Direct Blue 15
Direct Blue HH
Direct Pure Blue
Direct Pure Blue M
Direct Sky Blue A
Enianil Pure Blue AN
Fenamin Sky Blue
Hispamin Sky Blue 3B
Kayaku Direct Sky Blue 5B
Mitsui Direct Sky Blue 5B
Modr Prima 15 (Czech)
Naphtamine Blue 10G
NCI-C61290
Niagara Blue 4B
Niagara Sky Blue
Nippon Direct Sky Blue
Nitto Direct Sky Blue 5B
Phenamine Sky Blue A
Pontamine Sky Blue 5BX
Pontacyl Sky Blue 4BX
Shikiso Direct Sky Blue 5B
Sky Blue 4B
Sky Blue 5B
Tertrodirect Blue F
Vondacel Blue HH

2429-79-0
$C_{29}H_{21}N_5O_6S.2Na$
613.53
CI Direct Orange 8
Benzoic acid, 5-((4'-((1-amino-4-sulfo-2-naphthalenyl)azo)-(1,1'-biphenyl)-4-yl)azo)-2-hydroxy-, disodium salt (9CI)
C.I. Direct Orange 8, Disodium salt (8CI)

2429-81-4
$C_{46}H_{30}N_{10}O_{13}S_3.4Na$
1119.00
Benzoic acid, 5-((4'-((2,6-diamino-3-((8-hydroxy-3,6-disulfo-7-((4-sulfo-1-naphthalenyl)azo)- 2-naphthalenyl)-azo)-5-methylphenyl)azo)-(1,1'-biphenyl)-4-yl)azo)-2-hydroxy-, tetrasodium salt
Airedale Brown BSD
Amanil Fast Brown HP
Amanil Rayon Brown B
Atlantic Brown BCW
Atlantic Brown BP
Belamine Fast Brown BP
Benzanil Brown BS
Benzo Deep Brown NZ
Calcomine Brown B
Calcomine Catechu 2B
Chlorazol Brown LF
Chocolate EMBL
Chrome Leather Brown BS
C.I.35660
C.I. Direct Brown 31
C.I.Direct Brown 31, Tetrasodium salt
Cupranil Brown BCW
Cupranil Brown BCWR
Diaphtamine Fast Brown TB
Diazol Cutch F
Diazol Cutch FB
Diphenyl Brown BS
Diphenyl Brown TB
Diphenyl Fast Brown F
Direct Brown 31
Direct Brown B
Direct Brown 3B
Direct Brown BS
Direct Brown BSB
Direct Brown FS
Direct Brown TRB
Direct Fast Brown BP
Direct Fast Brown TSN
Direct Fast Brown TWC
Erie Fast Brown B
Fenamin Brown PBL
Fixanol Brown LF
Hispamin Fast Brown NZ
Phenamine Fast Brown T
Phenamine Fast Brown TWC
Pontamine Brown BCW
Pontamine Brown BT
Tertrodirect Brown TB
Triazol Brown B
Trisulphone Brown B
Vegentine Fast Brown B
Vondacel Brown S
Vondacel Brown SP

2429-82-5
$C_{29}H_{21}N_5O_7S.2Na$
629.53
Direct Brown KX
Airedale Brown MD
Aizen Direct Brown MH
Amanil Brown MR
5-((4'-((7-Amino-1-hydroxy-3-sulfo-2-naphthyl)azo)-4-biphenyl)azo)-salicylic acid
Atlantic Brown M
Atul Direct Brown MR
Azine Brown M
Azocard Brown M
Azomine Brown M
Belamine Fast Brown M
Bencidal Fast Brown M
Benzanil Brown M
Benzanol Brown M
Benzo Brown M
Benzoic acid, 5-((4'-((7-amino-1-hydroxy-3-sulfo-2-naphthalenyl)azo)(1,1'-biphenyl)-4-yl)-azo)-2-hydroxy-, disodium salt (9CI)
Brasilamina Fast Brown 3RA
Brown M

Calcomine Brown MCW
Chloramine Brown M
Chloramine Brown 2ME
Chlorazol Brown M
Chrome Leather Brown M
C.I. Direct Brown 2
C.I. Direct Brown 2, Disodium salt (8CI)
C.I. 22311
Columbia Brown M
Cresotine Brown RC
Cutamine Brown CM
Diacotton Brown M
Diamine Brown M
Diamine Brown MBA-CF
Diaphtamine Brown M
Diazine Brown M
Diazo Brown MC
Diazol Brown M
Diphenyl Brown V
Diphenyl Fast Brown MD
Direct Brown 2
Direct Brown 3RB
Direct Fast Brown M
Enianil Fast Brown M
Erie Fast Brown 3RB
Fenamin Brown M
Hispamin Fast Brown 3R2B
Japanol Brown M
Kayaku Direct Brown M
Mahogany EMBL
Mitsui Direct Brown M
Naphtamine Brown DC
Paramine Fast Brown M

2429-83-6
$C_{35}H_{29}N_9O_7S_2.2Na$
797.73
C.I. Direct Black 4
Ahco Direct Black RW
Airedale Black RWD
Atlantic Black RW
Azine Deep Black 3RL
Azocard Black RW
Bencidal Black RW
Benzanil Black RW
Benzo Deep Black RW
Black 3EMBL
Carbide Black ER
Chloramine Black E2B
Chlorazol Black LF
Chrome Leather Black ER
C.I. Direct Black 4, Disodium salt (8CI)
C.I. 30245
Coir Deep Black R
Cotton Black MT
Diamine Deep Black RW
Diaphtamine Black MT
Diazine Direct Black R
Diazo Black RW
Diazol Black ER
Diphenyl Deep Black VN
Direct Black K
Direct Black R
Direct Deep Black RW
Direct Diazo Black RW
Enianil Black RCN
Erie Black RB
Erie Black RF
Erie Black RRAC
Erie Black RW
Erie Black RX
Fenamin Black RW
Formic Black MTR
Hispamin Black 3RX
2,7-Naphthalenedisulfonic acid, 4-amino-3-((4'-((2,4-diamino-5-methylphenyl)azo)(1,1'-biphenyl)-4-yl)azo)-5-hydroxy-6-(phenylazo)-, disodium salt (9CI)
Nippon Deep Black RL
Nippon Deep Black RL Extra
Nippon Deep Black 3RL
Paper Black RW
Paper Deep Black R
Paraldehyde Black RW
Phenamine Black RW
Pontamine Black RRX
Tertrodirect Black RW
Tetrazo Deep Black R
Vondacel Black RW

2429-84-7
$C_{29}H_{21}N_5O_7S.2Na$
6291.53
Benzoic acid, 5-((4'-((2-amino-8-hydroxy-6-sulfo-1-naphthalenyl)azo)(1,1'-biphenyl)-4-yl)azo)-2-hydroxy-, disodium salt (9CI)

2430-22-0
$C_9H_{20}O$
144.29
CC(C)CCCCCCO
1-Octanol, 7-methyl
Isononyl alcohol
7-Methyl-1-octanol

2431-50-7
$C_4H_5Cl_3$
159.44
C(C(Cl)CCl)(=C)Cl
1-Butene, 2,3,4-trichloro
2,3,4-Trichlorobutene-1

2431-54-1
$C_4H_5Cl_3$
159.44
ClCC(Cl)C(Cl)=C
2-Butene, 1,2,4-trichloro
1,2,4-Trichlorobutene-2

2431-55-2
$C_4H_4Cl_6$
264.79
Butane, 1,1,2,2,3,4-hexachloro- (6CI,7CI,9CI)
1,1,2,2,3,4-Hexachlorobutane

2432-12-4
$C_7H_6Cl_2O$
177.03
Oc(c(cc(c1)C)Cl)c1Cl
p-Cresol, 2,6-dichloro- (8CI)
AI3-24009
2,6-Dichloro-4-methylphenol
Phenol, 2,6-dichloro-4-methyl- (9CI)

2432-74-8
$C_6H_{12}N_2$
112.16
N#CCCCCCN
6-Aminohexanenitrile
ω-Aminocapronitrile
6-Aminocapronitrile
Hexanenitrile, 6-amino- (9CI)

2432-79-3
$C_{10}H_{20}OS$
188.33
Hexanethioic acid, S-butyl ester (7CI,9CI)
Hexanoic acid, thio-, S-butyl ester (8CI)

2432-89-5
$C_{30}H_{58}O_4$
482.79
O=C(OCCCCCCCCCC)CCCCCCCC(=O)OCCCCCCCCCC
Decanedioic acid, didecyl ester (9CI)
Didecyl decanedioate

2432-90-8
$C_{32}H_{54}O_4$
502.86
O=C(OCCCCCCCCCCCC)c(c(ccc1)C(=O)OCCCCCCCCCCCC)c1
Phthalic acid, didodecyl ester
1,2-Benzenedicarboxylic acid, didodecyl ester (9CI)
Didodecyl phthalate
Di-n-dodecyl phthalate
Dilauryl phthalate

2432-99-7
$C_{11}H_{23}NO_2$
201.35
O=C(O)CCCCCCCCCCN
Undecanoic acid, 11-amino
Aminoundecanoic acid
11-Aminoundecanoic acid
11-Aminoundecylic acid
NCI-C50613

2433-96-7
$C_{23}H_{46}O_2$
354.62
Tricosanoic acid (8CI,9CI)

2435-53-2
$C_6Cl_4O_2$
245.88
O=C(C(=C(C(=C1Cl)Cl)Cl)Cl)C1=O

2-Chloranil
o-Benzoquinone, 3,4,5,6-tetrachloro- (8CI)
o-Chloranil
3,5-Cyclohexadiene-1,2-dione, 3,4,5,6-tetrachloro- (9CI)
3,4,5,6-Tetrachloro-3,5-cyclohexadiene-1,2-dione

2436-90-0
$C_{10}H_{18}$
138.28
C(=CCCC(C=C)C)(C)C
1,6-Octadiene, 3,7-dimethyl
Citronellene
Dihydromyrcene
3,7-Dimethyl-1,6-octadiene

2436-92-2
$C_{11}H_{11}N$
157.22
Quinoline, 3,4-dimethyl- (6CI,7CI,8CI,9CI)
3,4-Dimethylquinoline

2437-23-2
$C_{12}H_{24}O_2.H_3N$
217.35
Dodecanoic acid, ammonium salt
AI3-00287
Ammonium dodecanoate
Ammonium laurate
Lauric acid, ammonium salt

2437-25-4
$C_{12}H_{23}N$
181.36
N#CCCCCCCCCCCC
Dodecanenitrile
Lauronitrile

2437-29-8
$C_{46}H_{50}N_4.C_2H_2O_4.2C_2HO_4$
927.10
Ammonium, (4-(p-(dimethylamino)-α-phenylbenzylidene)-2,5-cyclohexadien-1-ylidene)- dimethyl-, oxalate (2:1), oxalate (1:1)
Malachite Green Oxalate

2437-56-1
$C_{13}H_{26}$
182.39
C(=C)CCCCCCCCCCC
1-Tridecene

2437-72-1
$C_{12}H_{13}N$
171.24
Quinoline, 2,3,4-trimethyl- (7CI,8CI,9CI)
2,3,4-Trimethylquinoline

2437-79-8
$C_{12}H_6Cl_4$
291.99
Clc1ccc(c(Cl)c1)c2ccc(Cl)cc2Cl
2,4,2',4'-Tetrachlorobiphenyl
1,1'-Biphenyl, 2,2',4,4'-tetrachloro-
2,4,2',4'-TCB

2437-92-5
$C_{20}H_{38}$
278.52
1,2-Hexadecadiene, 3,7,11,15-tetramethyl- (9CI)

2438-04-2
$C_{10}H_{12}O_2$
164.20
CC(C)c1ccccc1C(O)=O
Benzoic acid, 2-(1-methylethyl)- (9CI)
Benzoic acid, o-isopropyl- (8CI)

2438-53-1
$C_{19}H_{31}N_2O_2.Cl$
354.97
Ammonium, (5-hydroxycarvacryl)trimethyl-, chloride, 1-piperidinecarboxylate
Ammonium, (5-(piperidinocarbonyloxy)carvacryl)trimethyl-, chloride
AMO 1618
Carbamic acid, N,N-pentamethylene-, 4-dimethylaminothymyl ester, methochloride
1-Piperidinecarboxylic acid, 6-(trimethylammonio)thymyl ester, chloride
TL-1049

2438-80-4
$C_6H_{12}O_5$
164.16
O=CC(O)C(O)C(O)C(O)C
L-Galactose, 6-deoxy- (VAN) (9CI)
6-Deoxy-L-galactose

2438-88-2
$C_7H_3Cl_4NO_3$
290.91
COc1c(Cl)c(Cl)c(N(=O)=O)c(Cl)c1Cl
Anisole, 2,3,5,6-tetrachloro-4-nitro
Benzene, 1,2,4,5-tetrachloro-3-methoxy-6-nitro- (9CI)
ENT 22,335
NCI-C03032
4-Nitro-2,3,5,6-tetrachloranisole
TCNA
Tetrachloronitroanisole
2,3,5,6-Tetrachloro-4-nitroanisole

2439-00-1
$C_8H_5Cl_3O_2.Na$
262.47
Sodium 2,3,6-trichlorophenylacetate
Benzeneacetic acid, 2,3,6-trichloro-, sodium salt (9CI)
Caswell No. 882C
Chlorfenac-sodium
EPA Pesticide Chemical Code 082602
Sodium 2,3,6-trichlorobenzeneacetate
2,3,6-Trichlorophenylacetic acid sodium salt

2439-01-2
$C_{10}H_6N_2OS_2$
234.30
Cc3ccc2nc1sc(=O)sc1nc2c3
Carbonic acid, dithio-, cyclic S,S-(6-methyl-2,3-quinoxalinediyl)ester
Bay 36205
Bayer 4964
Bayer 36205
Bayer SS2074
Chinomethionat
Chinomethionate
Cyclic S,S-(6-methyl-2,3-quinoxalinediyl) dithiocarbonate
Dithiolo(4,5-b)quinoxalin-2-one, 6-methyl- (9CI)
Dithioquinox
ENT 25,606
Erade
Erazidon
Forstan
6-Methyl-chinoxalin-2,3-dithiol-cyclo-carbonat (German)
6-Methyl-1,3-dithiolo(4,5-b)-quinoxalin-2-one
6-Methyl-2-oxo-1,3-dithiolo(4,5-b)quinoxaline
6-Methyl-2,3-quinoxaline dithiocarbonate
6-Methyl-2,3-quinoxalinedithiol cyclic carbonate
6-Methyl-2,3-quinoxalinedithiol cyclic dithiocarbonate
6-Methyl-2,3-quinoxalinedithiol cyclic S,S-dithiocarbonate
6-Methyl-quinoxaline-2,3-dithiolcyclocarbonate
Morestan
Morestane
Oxythioquinox
Quinomethionate
2,3-Quinoxalinedithiol, 6-methyl-, cyclic carbonate
2,3-Quinoxalinedithiol, 6-methyl-, cyclic dithiocarbonate (ester)
SS 2074

2439-10-3
$C_{13}H_{29}N_3.C_2H_4O_2$
287.51
CCCCCCCCCCCCNC(N)=N.

CC(O)=O
Guanidine, dodecyl-, acetate
AC 5223
American Cyanamid 5223
Apadodine
Carpene
Curitan
Cyprex
Cyprex 65W
n-Dodecylguanidinacetat (German)
Dodecylguanidine acetate
n-Dodecylguanidine acetate
Dodin
Dodine
Dodine acetate
Dodine, Mixture with glyodin
Doguadine
Doquadine
ENT 16,436
Experimental fungicide 5223
Kyselina 3-dodecylguanidino-octova (Czech)
Laurylguanidine acetate
Melprex
Milprex
Syllit
Tsitrex
Venturol
Vondodine

2439-35-2
$C_7H_{13}NO_2$
143.18
O=C(OCCN(C)C)C=C
2-Propenoic acid, 2-(dimethyl-amino)ethyl ester (9CI)
AI3-08751
Dimethylaminoethyl acrylate
2-(Dimethylamino)ethyl 2-propenoate

2439-99-8
$C_4H_{11}NO_8P_2$
263.10
O=P(O)(O)CN(CP(=O)(O)O)CC(=O)O
Glycine, N,N-bis(phosphono-methyl)
N,N-Bis(phosphonomethyl)-glycine
CP 41845
Glyphosine
Polaris

2440-22-4
$C_{13}H_{11}N_3O$
225.27
Oc(c(N(N=C(C=1C=CC=2)C2)N1)cc(c3)C)c3
p-Cresol, 2-(2H-benzotriazol-2-yl)
Benazol P
Drometrizole
2-(2-Hydroxy-5-methylphenyl)-benzotriazole
Phenol, 2-(2H-benzotriazol-2-yl)-4-methyl- (9CI)
Porex P
Tin P
Tinuvin P
UV Absorber-1

2440-40-6
C_3H_7ClHg
279.14
CCC[Hg]Cl
Mercury, chloropropyl
Chloropropylmercury
Propylmercuric chloride
Propylmercury chloride

2441-97-6
C_6H_9Cl
116.60
Cyclohexene, 3-chloro
3-Chloro-1-cyclohexene
3-Cyclohexenyl chloride

2443-39-2
$C_{18}H_{34}O_3$
298.52
Octadecanoic acid, 9,10-epoxy
cis-9,10-Epoxyoctadecanoate
cis-9,10-Epoxyoctadecanoic acid
Epoxyoleic acid
9,10-Epoxystearic acid
cis-9,10-Epoxystearic acid
cis-3-Octyl-oxiraneoctanoic acid

2444-29-3
$C_{13}H_{11}NO_3$
229.23
p-Nitrophenyl o-tolyl ether
Benzene, 1-methyl-2-(4-nitro-phenoxy)-
Ether, p-nitrophenyl o-tolyl
2-Methylphenyl 4-nitrophenyl ether

2444-36-2
$C_8H_7ClO_2$
170.60
OC(=O)Cc1ccccc1Cl
Acetic acid, (o-chlorophenyl)-(8CI)
AI3-20877
Benzeneacetic acid, 2-chloro-(9CI)

2444-68-0
$C_{16}H_{12}$
204.27
Anthracene, 9-ethenyl- (9CI)

2444-89-5
$C_{12}H_8Cl_2O$
239.10
Benzene, 1,1'-oxybis(4-chloro-(9CI)
Ether, bis(p-chlorophenyl) (8CI)

2444-90-8
$C_{15}H_{16}O_2.2Na$
274.27
Bisphenol A disodium salt
Bisphenol A, disodium salt
4,4'-Isopropylidinebisphenol, disodium salt
Phenol, 4,4'-isopropylidenedi-, disodium salt
Phenol, 4,4'-(1-methylethylidene)bis-, disodium salt (9CI)

2445-07-0
$C_7H_{15}AsN_2S_4$
330.40
CN(C)C(=S)S[As](C)SC(=S)N(C)C
Carbamic acid, dimethyldithio-, bis(anhydrosulfide) with dithiomethanearsonous acid
Bis(dimethylthiocarbamoyl-thio)methyl-arsine
Methanearsonous acid, dithio-, bis(anhydrosulfide) with dimethyldithiocarbamic acid
Methylarsenic dimethyl dithiocarbamate
Methyl arsine-bis(dimethyl-dithiocarbamate)
Methylbis(dimethylthiocarbamoylthio)arsine
Monzet
Tuzet
Urbacid
Urbacide
Urbazid

2445-76-3
$C_9H_{18}O_2$
158.24
O=C(OCCCCCC)CC
Propanoic acid, hexyl ester
AI3-33593
Hexyl propanoate

2445-78-5
$C_{10}H_{20}O_2$
172.27
O=C(OCC(CC)C)C(CC)C
Butanoic acid, 2-methyl-, 2-methylbutyl ester (9CI)
2-Methylbutanoic acid, 2-methylbutyl ester
2-Methylbutyl 2-methylbutanoate
2-Methylbutyl 2-methylbutyrate

2446-69-7
$C_{12}H_{18}O$
178.27
Oc(ccc(c1)CCCCCC)c1
Phenol, 4-hexyl- (9CI)
4-Hexylphenol

2447-54-3

$C_{20}H_{14}NO_4$
332.35
Sanguinarine
(1,3)-Benzodioxolo(5,6-c)-1,3-dioxolo(4,5-i)phenanthridinium, 13-methyl- (9CI)
Dimethylenedioxy benzphenanthridine
Pseudochelerythrine
Sanguinarin
Sanguiritrin

2449-05-0
$C_{16}H_{14}N_2O_4$
298.29
O=C(OCc(cccc1)c1)N=NC(=O)OCc(cccc2)c2
Diazenedicarboxylic acid, bis-(phenylmethyl) ester (9CI)
Bis(phenylmethyl) diazenedicarboxylate
Dibenzyl azodicarboxylate

2449-49-2
$C_{10}H_{15}N$
149.23
α-Methylbenzyl dimethyl amine
Benzenemethanamine, N,N,α-trimethyl- (9CI)
Benzylamine, N,N,α-trimethyl-(8CI)
1-Dimethylamino-1-phenylethane
N,N-Dimethyl-α-methylbenzylamine

2450-71-7
C_3H_5N
55.09
C(#C)CN
2-Propynylamine
2-Propyn-1-amine

2451-01-6
$C_{10}H_{20}O_2.H_2O$
190.28
Terpin hydrate
AI3-01762
Cyclohexanemethanol, 4-hydroxy-α,α,4-trimethyl-, monohydrate
Cyclohexanemethanol, 4 -hydroxy-α,α,4-trimethyl-, monohydrate, cis-
p-Menthane-1,8-diol monohydrate
Terpinol

2451-62-9
$C_{12}H_{15}N_3O_6$
297.25
O=C(N(C(=O)N(C1=O)CC(O2)C2)CC(O3)C3)N1CC(O4)C4
1,3,5-Triglycidyl-s-triazinetrione
s-Triazine-2,4,6(1H,3H,5H)-trione, 1,3,5-tris(2,3-epoxypropyl)- (8CI)
1,3,5-Triazine-2,4,6(1H,3H,5H)-trione, 1,3,5-tris(oxiranylmethyl)- (9CI)
Tri(epoxypropyl)isocyanurate
Triglycidylisocyanurate
Triglycidyl isocyanurate

2452-01-9
$C_{12}H_{24}O_2.1/2Zn$
233.01
Zinc laurate
Dodecanoic acid, zinc salt (9CI)
Witco Zinc Soap #26
Zinc dodecanoate

2452-84-8
$C_{16}H_{19}N_3O_2$
285.33
C.I. Solvent Yellow 58 (8CI)
Ethanol, 2,2'-((4-(phenylazo)-phenyl)imino)bis- (9CI)

2452-99-5
C_7H_{14}
98.19
Cyclopentane, 1,2-dimethyl-(8CI,9CI)

2453-00-1
C_7H_{14}
98.19
Cyclopentane, 1,3-dimethyl-(8CI,9CI)
1,3-Dimethylcyclopentane

2454-37-7
$C_8H_{11}NO$
137.18
(m-Aminophenyl)methyl carbinol
m-Amino-α-methylbenzyl alcohol
3-Amino-α-methylbenzyl alcohol
1-(3-Aminophenyl)ethanol
Benzenemethanol, 3-amino-α-methyl- (9CI)
Benzyl alcohol, m-amino-α-methyl- (8CI)

2455-08-5
Unknown
Unknown
Nitrilotriacetic acid and its salts
NTA, calcium potassium salt (1:1:1)

2455-24-5
$C_9H_{14}O_3$
170.21
O=C(OCC(OCC1)C1)C(=C)C
Tetrahydrofurfuryl methacrylate
AI3-08497
Methacrylic acid, tetrahydrofurfuryl ester (8CI)
2-Propenoic acid, 2-methyl-, (tetrahydro-2-furanyl)methyl ester (9CI)
Sartomer SR 203
SR 203

2456-27-1
$C_{18}H_{38}O$
270.50
Nonane, 1,1'-oxybis- (9CI)
Nonyl ether (8CI)

2456-28-2
$C_{20}H_{42}O$
298.55
O(CCCCCCCCCC)CCCCCCCCCC
Decane, 1,1'-oxybis- (9CI)
Decyl ether (8CI)
1,1'-Oxybisdecane

2457-01-4
$C_8H_{16}O_2.1/2Ba$
212.88
Hexanoic acid, 2-ethyl-, barium salt (9CI)
Barium 2-ethylhexanoate

2457-47-8
$C_5H_3Cl_2N$
147.99
Clc1cncc(Cl)c1
Pyridine, 3,5-dichloro

2457-76-3
$C_7H_6ClNO_2$
171.59
O=C(O)c(c(cc(N)c1)Cl)c1
Benzoic acid, 4-amino-2-chloro
2-Chloro-4-aminobenzoic acid
USAF NB-1

2459-09-8
$C_7H_7NO_2$
137.13
O=C(OC)c(ccnc1)c1
Isonicotinic acid, methyl ester (8CI)
Methyl 4-pyridinecarboxylate
4-Pyridinecarboxylic acid, methyl ester (9CI)

2459-10-1
$C_{12}H_{12}O_6$
252.22
O=C(OC)c(ccc(c1C(=O)OC)C(=O)OC)c1
Trimethyl 1,2,4-benzenetricarboxylate
AI3-08219

1,2,4-Benzenetricarboxylic acid, trimethyl ester
Methyl trimellitate
Trimellitic acid trimethyl ester
Trimethyl trimellitate

2460-49-3
$C_7H_6Cl_2O_2$
193.03
COc1cc(Cl)c(Cl)cc1O
Phenol, 4,5-dichloro-2-methoxy
4,5-Dichloroguaiacol
4,5-Dichloro-2-methoxyphenol

2460-77-7
$C_{14}H_{20}O_2$
220.31
O=C(C(=CC(=O)C=1C(C)(C)C)C(C)(C)C)C1
2,5-Di-t-butyl-p-benzoquinone
AI3-16635
p-Benzoquinone, 2,5-di-tert-butyl- (8CI)
2,5-Cyclohexadien-1,4-dione, 2,5-bis(1,1-dimethylethyl)-
2,5-Cyclohexadiene-1,4-dione, 2,5-bis(1,1-dimethylethyl)- (9CI)
2,5-Di-tert-butylbenzoquinone
2,5-Di-tert-butyl-1,4-benzoquinone

2461-15-6
$C_{11}H_{22}O_2$
186.33
O(C1COCC(CCCC)CC)C1
Propane, 1,2-epoxy-3-((2-ethylhexyl)oxy)
2-Ethylhexyl glycidyl ether
(((2-Ethylhexyl)oxy)methyl)-oxirane
Glycidyl 2-ethylhexyl ether
Oxirane, (((2-ethylhexyl)-oxy)methyl)- (9CI)

2461-18-9
$C_{15}H_{30}O_2$
242.40
O(C1COCCCCCCCCCCCC)C1
Dodecyl glycidyl ether
n-Dodecyl glycidyl ether
((Dodecyloxy)methyl)oxirane
Ether, dodecyl 2,3-epoxypropyl
Lauryl glycidyl ether
Oxirane, ((dodecyloxy)methyl)- (9CI)
Propane, 1-(dodecyloxy)-2,3-epoxy-

2463-02-7
$C_{21}H_{38}O_2$
322.53
11,14-Eicosadienoic acid, methyl ester (8CI,9CI)

2463-53-8
$C_9H_{16}O$
140.25
O=CC=CCCCCCC
2-Nonenal
β-Hexylacrolein
2-Nonen-1-al
α-Nonenyl aldehyde

2463-63-0
$C_7H_{12}O$
112.17
O=CC=CCCCC
2-Heptenal
AI3-36270

2463-84-5
$C_8H_9ClNO_5PS$
297.66
COP(=S)(OC)Oc1ccc(cc1Cl)N(=O)=O
Phosphorothioic acid, O-(2-chloro-4-nitrophenyl) O,O-dimethyl ester
AC 4124
American Cyanamid 4,124
Bayer 22/190
Captec
O-(4-Chloor-3-nitro-fenyl)-O,O-dimethylmonothiofosfaat (Dutch)
O-(4-Chlor-3-nitro-phenyl)-O,O-dimethyl-monothio-phosphat (German)
O-(4-Cloro-3-nitro-fenil)-O,O-dimetil-monotiofosfato (Italian)
O-(2-Chloro-4-nitrophenyl) O,O-dimethyl phosphorothioate
Chlorthion
Dicaptan
Dicapthon
Dicapton
O,O-Dimethyl O-2-chloro-4-nitrophenyl phosphorothioate
O,O-Dimethyl-O-(2-chloro-4-nitrophenyl)thionophosphate
Dimethyl 2-chloronitrophenyl thiophosphate
ENT 17,035
Experimental Insecticide 4124
Insecticide ACC 4124
Isochloorthion (Dutch)
Isochlorthion
Isomeric chlorthion
p-Nitro-o-chlorophenyl dimethyl thionophosphate
OMS-214
Phenol, 2-chloro-4-nitro-, O-ester with O,O-dimethyl phosphorothioate
Thiophosphate de O,O-dimethyle et de O-4-chloro-3-nitrophenyle (French)

2464-37-1
$C_{14}H_9ClO_3$
260.68
Chlorflurecol
Caswell No. 192A
Chlorflurenol
Chloroflurenol (French)
2-Chloro-9-hydroxy-9H-fluorene-9-carboxylic acid
2-Chloro-9-hydroxyfluorene-9-carboxylic acid
EPA Pesticide Chemical Code 292200
Fluorene-9-carboxylic acid, 2-chloro-9-hydroxy- (8CI)
9H-Fluorene-9-carboxylic acid, 2-chloro-9-hydroxy- (9CI)
Methyl 2-chloro-9-hydroxy-fluorene-9-carboxylate

2465-27-2
$C_{17}H_{21}N_3.ClH.H_2O$
321.89
Cl.CN(C)c1ccc(cc1)C(=N)c2ccc(cc2)N(C)C
Aniline, 4,4'-(imidocarbonyl)-bis(N,N-dimethyl-, hydrochloride
ADC Auramine O
Aizen Auramine
Aizen Auramine Conc. SFA
Aizen Auramine OH
Auramine A1
Auramine Conc. Specially soluble in spirit
Auramine Extra
Auramine Extra Conc. A
Auramine FA
Auramine Hydrochloride
Auramine Lake Yellow O
Auramine N
Auramine O
Auramine O (Biological stain)
Auramine O Extra Conc. A export
Auramine ON
Auramine OO
Auramine OOO
Auramine OS
Auramine Pure
Auramine SP
Auramine Yellow
Benzenamine, 4,4'-carbonimidoylbis(N,N-dimethyl-, monohydrochloride (9CI)
4,4'-Bis(dimethylamino)-benzhydrylidenimine hydrochloride
4:4'-Bis(dimethylamino)benzophenone-imine hydrochloride
1,1-Bis(p-dimethylamino-phenyl)methylenimine hydrochloride
Calcozine Yellow OX
C.I. 41000
C.I. Basic Yellow 2
C.I. Basic Yellow 2, Monohydrochloride
4,4'-(Imidocarbonyl)bis(N,N-dimethylamine), monohydrochloride
Mitsui Auramine O
Zlut Zasadita 2

2465-29-4
$C_{15}H_{14}N_2O.ClH$
274.77
3H-Xanthen-6-amine, N-methyl-3-(methylimino)-, hydrochloride
Acridine Red
Acridine Red 3B
Acridine Red, Hydrochloride
C.I. 45000
Dimethyldiaminoxanthenyl chloride

2465-59-0
$C_5H_4N_4O_2$
152.09
Oc2nc(O)c1cn[nH]c1n2
Oxipurinol
AI3-50432
NSC-76239
Ossipurinolo
Oxipurinolum (Latin)
Oxypurinol
1H-Pyrazolo(3,4-d)pyrimidin-4,6(5H,7H)-dione
1H-Pyrazolo(3,4-d)pyrimidine-4,6-diol
1H-Pyrazolo(3,4-d)pyrimidine-4,6(5H,7H)-dione (9CI)

2465-65-8
$C_4H_{11}O_3PS$
170.17
O,O-Diethyl phosphorothionate
DETP

2471-83-2
$C_{11}H_{10}$
142.20
1H-Indene, 1-ethylidene- (9CI)
Indene, 1-ethylidene- (8CI)

2473-01-0
$C_9H_{19}Cl$
162.70
ClCCCCCCCCC
Nonane, 1-chloro- (9CI)
1-Chlorononane

2473-03-2
$C_{11}H_{23}Cl$
190.76
ClCCCCCCCCCCC
Undecane, 1-chloro- (9CI)
AI3-08977
1-Chloroundecane

2475-31-2
$C_{16}H_6Br_4N_2O_2$
577.84
O=C(c(c(N1)c(cc2Br)Br)c2)C1=C(Nc(c3cc(c4)Br)c4Br)C3=O
(δ2,2'-Biindoline)-3,3'-dione, 5,5',7,7'-tetrabromo- (8CI)
3H-Indol-3-one, 5,7-dibromo-2-(5,7-dibromo-1,3-dihydro-3-oxo-2H-indol-2-ylidene)-1,2-dihydro- (9CI)

2475-33-4
$C_{42}H_{18}N_2O_6$
646.62
Dinaphtho(2,3-a:2',3'-i)naphth(2',3':6,7)indolo(2,3-c)carbazole-5,10,15,17,22,24-hexaone, 16,23-dihydro
Ahcovat Brown BR
Amanthrene Brown BR
Benzadone Brown BR
Brown SK
Calcoloid Brown BR
Caledon Dark Brown 3R
Carbanthrene Brown BR
Chemithrene Brown BR
C.I. 70800
C.I. 70802
Cibanone Brown BR
Cibanone Brown FBR
C.I. Vat Brown 1
C.I. Vat Brown 44
Fenanthren Brown BR
Helanthrene Brown GR
HNED Kypova 1 (Czech)
Indanthren Bronze BR
Indanthren Brown BR
Indanthren Brown GR
Indanthrene Brown BR
Hned Ostanthrenova BR (Czech)
Mayvat Brown BR
Mikethrene Brown BR
Mikethrene Brown GR
Naphth(2',3':6,7)indolo(2,3-c)-dinaphtho(2,3-a:2',3'-i)carbazole-5,10,15,17,22,24-hexone
Nihonthrene Brown BR
Nihonthrene Brown GR
Nyanthrene Brown RB
Ostanthren Brown BR
Ostanthrene Brown BR
Palanthrene Brown BR
Paradone Red Brown 2RD
Ponsol Brown RBT
Romantrene Brown FBR
Romantrene Brown FGR
Sandothrene Brown NBR
Solanthrene Brown BR
Solanthrene Brown JR
Tinon Brown BR
Tyrian Brown I-BR

2475-44-7
$C_{16}H_{14}N_2O_2$
266.32
O=C(c(c(C(=O)c1c(NC)ccc2NC)ccc3)c3)c12
Anthraquinone, 1,4-bis-(methylamino)
Acetate Blue B
9,10-Anthracenedione, 1,4-bis-(methylamino)- (9CI)
Artisil Blue BRP
1,4-Bis(methylamino)anthraquinone
Celliton Fast Blue B
Cibacet Blue BR
C.I. 61500
C.I. Disperse Blue 14
C.I. Disperse Blue 110
Cilla Fast Blue B
Disperse Blue 14
Duranol Brilliant Blue G
Resiren Blue TB
Serisol Brilliant Blue G
Setacyl Blue BS
Supracet Fast Blue 2G

2475-45-8
$C_{14}H_{12}N_4O_2$
268.30
O=C(c(c(c(N)cc1)C(=O)c2c(N)ccc3N)c1N)c23
9,10-Anthracenedione, 1,4,5,8-tetraamino
Acetate Blue G
Acetoquinone Blue L
Acetoquinone Blue R
Acetylon Fast Blue G
Amacel Blue GG
Amacel Pure Blue B
Anthraquinone, 1,4,5,8-tetramino-
Artisil Blue SAP
Artisil Blue SAP Conc
Brasilazet Blue GR
Celanthrene Pure Blue BRS
Celliton Blue BB-CF
Celliton Blue Extra
Celliton Blue G
Celliton Blue GA-CF
Cibacet Sapphire Blue G
C.I. 64500
C.I. Disperse Blue 1
C.I. Solvent Blue 18
Cilla Blue Extra
Diacelliton Fast Blue R
Disperse Blue 1
Disperse Blue No. 1
Duranol Brilliant Blue CB
Fenacet Blue G
Grasol Blue 2GS
Kayalon Fast Blue BR
Microsetile Blue EB
Miketon Fast Blue
Miketon Fast Blue B
Nacelan Blue G
NCI-C54900
Neosetile Blue EB
Nyloquinone Blue 2J
Oracet Sapphire Blue G
Perliton Blue B
Serinyl Blue 2G
Serinyl Blue 3G
Serinyl Blue 3GN
Setacyl Blue 2GS
Setacyl Blue 2GS II
Supracet Brilliant Blue 2GN
Supracet Deep Blue R
1,4,5,8-Tetraaminoanthraquinone
1,4,5,8-Tetraminoanthraquinone

2475-46-9
$C_{17}H_{16}N_2O_3$

296.35
O=C(c(c(C(=O)c1c(NC)ccc2NC CO)ccc3)c3)c12
Anthraquinone, 1-((2-hydroxyethyl)amino)-4-(methylamino)
Acetate Brilliant Blue 4B
Acetoquinone Light Pure Blue R
Altocyl Brilliant Blue B
Amacel Blue BNN
Amacel Brilliant Blue B
9,10-Anthracenedione, 1-((2-hydroxyethyl)amino)-4-(methylamino)- (9CI)
Artisil Blue BSG
Artisil Blue BSQ
Calcosyn Sapphire Blue 2GS
Calcosyn Sapphire Blue R
Celanthrene Brilliant Blue
Celanthrene Brilliant Blue FFS
Celliton Blue FFR
Celliton Fast Blue FBBN
Celliton Fast Blue FFR
Celliton Fast Blue FFRN
Celliton Fast Blue FFRS
Celutate Blue BLT
Celutate Blue RNH
Celutate Brilliant Blue B
C.I. 61505
Cibacet Blue BNG
Cibacet Blue F3R
Cibacet Brilliant Blue BG New
Cibacete Brilliant Blue BG New
C.I. Disperse Blue 3
Cilla Fast Blue FFR
Diacelliton Fast Brilliant Blue B
Diacelliton Fast Brilliant Blue BF
Disperse Blue 3
Disperse Blue K
Dispersive Blue K
Duranol Brilliant Blue B
Duranol Brilliant Blue BN
Duranol Printing Blue B
Eastman Blue BNN
Eastman Blue GBN
Fenacet Fast Blue FF
Fenacet Fast Blue FFN
4-Hydroxyethylamino-1-methylaminoanthraquinone
Interchem Acetate Blue B
Interchem Acetate Blue NBN
Interchem Acetate Blue RBN
Interchem Acetate Blue WNBN
Kayalon Fast Blue FN
Lurafix Blue FFR
1-MA-4OEAA (Russian)
1-Methylamino-4-ethanolaminoanthraquinone
1-Methylamino-4-(β-hydroxyethylamino)anthraquinone
1-Methylamino-4-oxyethylaminoanthraquinone (Russian)
Microsetile Blue FF
Microsetile Blue FFR
Miketon Brilliant Blue B
Mireton Brilliant Blue B
Modr Disperzni 3 (Czech)
Modr Ostacetova P3R (Czech)
Nacelan Blue KLT
Nyloquinone Pure Blue
Nyloquinone Pure Blue R
Perliton Blue FFR
Serinyl Hosiery Blue
Serinyl Hosiery Blue BG
Serisol Brilliant Blue BG
Serisol Brilliant Blue BP
Serisol Brilliant Blue FF
Setacyl Blue BN
Setacyl Blue FG
Setacyl Blue RF
Setacyl Brilliant Blue
Setacyl Brilliant Blue BG
Supracet Brilliant Blue BG
Transetile Blue P-FER

2478-20-8
$C_{20}H_{16}N_2O_2$
316.35
O=C(N(C(=O)c(c1c(c(N)cc2)cc3)c3)c(c(cc(c4)C)C)c4)c12
Solvent Yellow 44
6-Amino-2-(2,4-dimethylphenyl)-1h-benz(de)isoquinoline-1,3(2h)-dione
4-Amino-N-(2,4-dimethylphenyl)naphthalene-1,8-dicarboximide
4-Amino-N-(2',4'-xylyl)-1,8-naphthalimide
1H-Benz(de)isoquinoline-1,3(2H)-dione, 6-amino-2-(2,4-dimethylphenyl)- (9CI)
C.I. Solvent Yellow 44
C.I. 56200

2478-38-8
$C_{10}H_{12}O_4$
196.22
COc1cc(cc(OC)c1O)C(C)=O
Acetophenone, 3,5-dimethoxy-4-hydroxy
Acetophenone, 4'-hydroxy-3',5'-dimethoxy- (8CI)
Acetosyringone
3,5-Dimethoxy-4-hydroxyacetophenone
Ethanone, 1-(4-hydroxy-3,5-dimethoxyphenyl)- (9CI)
4-Hydroksy-3',5'-dwumetoksyacetofenon (Polish)

2478-67-3
$C_{14}H_8ClNO_3$
273.67
O=C(c(c(C(=O)c1c(O)cc(c2N)Cl)ccc3)c3)c12
9,10-Anthracenedione, 1-amino-2-chloro-4-hydroxy- (9CI)
1-Amino-2-chloro-4-hydroxy-9,10-anthracenedione
1-Amino-2-chloro-4-hydroxyanthraquinone

2480-86-6
$C_9H_{12}O_3$
168.19
Benzenemethanol, 4-hydroxy-3-methoxy-α-methyl- (9CI)
Apocynol
NSC-47035
Vanillyl alcohol, α-methyl- (8CI)

2481-94-9
$C_{16}H_{19}N_3$
253.38
N(=Nc(cccc1)c1)c(ccc(N(CC)CC)c2)c2
Benzenamine, N,N-diethyl-4-(phenylazo)
Aniline, N,N-diethyl-p-(phenylazo)-
C-299
Ceres Yellow GGN
C.I. 11021
C.I. Solvent Yellow 56
p-(Diethylamino)azobenzene
4-(Diethylamino)azobenzene
N,N-Diethyl-4-aminoazobenzene
N,N-Diethyl-p-(phenylazo)-aniline
N,N-Diethyl-4-(phenylazo)-benzenamine
Diethyl Yellow
Fast Oil Yellow 64403
Fat Yellow GGN
Oil Yellow 2635
Oil Yellow DE
Oil Yellow DEA
Oil Yellow E190
Oil Yellow ENC
Oil Yellow GA
Oil Yellow NB
Orient Oil Yellow GGS
Sico Fat Yellow P
Sudan Yellow GGN
Waxoline Yellow ED

2482-68-0
C_4Cl_8
331.67
1-Butene, 1,1,2,3,3,4,4,4-octachloro- (8CI,9CI)
1-Butene, octachloro-

2484-88-0
$C_{12}H_{10}N_2O_3S$
262.28
O=S(=O)(O)c(ccc(N=Nc(cccc1)c1)c2)c2
Benzenesulfonic acid, 4-(phenylazo)- (9CI)
4-(Phenylazo)benzenesulfonic acid

2486-70-6
$C_8H_9NO_2$
151.16
Cc1cc(ccc1N)C(O)=O
Benzoic acid, 4-amino-3-methyl- (9CI)
NSC-227945
m-Toluic acid, 4-amino- (8CI)

2486-71-7
$C_7H_6ClNO_2$
171.58
Nc1ccc(cc1Cl)C(O)=O
Benzoic acid, 4-amino-3-chloro- (8CI,9CI)
NSC-212132

2487-40-3
$C_5H_4N_4OS$
168.19
n(c(O)c(N=CN1)c1n2)c2S
6H-Purin-6-one, 1,2,3,7-tetrahydro-2-thioxo
1,2,3,7-Tetrahydro-2-thioxo-6H-purin-6-one
2-Thio-6-oxypurine
2-Thioxanthine
Xanthine, 2-thio- (8CI)

2487-90-3
$C_3H_{10}O_3Si$
122.22
CO[SiH](OC)OC
Silane, trimethoxy
Trimethoxy silane

2489-05-6
$C_{22}H_{44}O_2.Ag$
448.46
Docosanoic acid, silver(1+) salt (9CI)
Docosanoic acid, silver salt
Silver(1+) docosanoate

2489-77-2
$C_4H_{10}N_2S$
118.22
N(=C(N(C)C)S)C
Urea, 1,1,3-trimethyl-2-thio
NCI-C02186
Thiate E
Trimethylthiourea
N,N,N'-Trimethylthiourea
1,1,3-Trimethyl-2-thiourea

2489-86-3
$C_{13}H_{12}$
168.24
Naphthalene, 1-(2-propenyl)- (9CI)
Naphthalene, 1-allyl- (8CI)

2490-48-4
$C_{17}H_{36}O$
256.47
1-Hexadecanol, 2-methyl- (8CI, 9CI)

2490-49-5
$C_{18}H_{36}O_2$
284.48
Hexadecanoic acid, 14-methyl-, methyl ester (8CI,9CI)

2491-38-5
$C_8H_7BrO_2$
215.05
1-(4-Hydroxyphenyl)-2-bromoethanone
2-Bromo-4-hydroxyacetophenone
2-Bromo-4'-hydroxyacetophenone
Busan 90
Caswell No. 115
EPA Pesticide Chemical Code 008707

2491-52-3
$C_{12}H_9N_3O_2$
227.24
O=N(=O)c2ccc(N=Nc1ccccc1)cc2
Azobenzene, 4-nitro
Diazene, (4-nitrophenyl)phenyl- (9CI)
p-Nitroazobenzene
4-Nitroazobenzene

2491-71-6
$C_{12}H_{11}N_3O_3S.Na$
300.28
Benzenesulfonic acid, 4-((4-aminophenyl)azo)-, monosodium salt (9CI)
Benzenesulfonic acid, 4-((4-aminophenyl)azo)-, sodium salt
C.I. Food Yellow 6, Monosodium salt (8CI)

2491-72-7
$C_{12}H_9N_3O_5S.Na$
330.26
Benzenesulfonic acid, 4-((4-nitrophenyl)azo)-, sodium salt (9CI)

2491-74-9
$C_{14}H_{14}N_4O_2$
270.27
O=N(=O)c(ccc(N=Nc(ccc(N(C)C)c1)c1)c2)c2
Benzenamine, N,N-dimethyl-4-((4-nitrophenyl)azo)- (9CI)
N,N-Dimethyl-4-((4-nitrophenyl)azo)benzenamine

2492-26-4
$C_7H_4NS_2.Na$
189.23
[Na+].Sc2nc1ccccc1s2
Sodium 2-mercaptobenzothiazole
AI3-17229
Benzothiazolethiol, sodium salt
2-Benzothiazolethiol, sodium deriv
2-Benzothiazolethiol, sodium salt
2(3H)-Benzothiazolethione, sodium salt (9CI)
Caswell No. 541C
Duodex
EPA Pesticide Chemical Code 051704
Mercaptobenzothiazol, sodium salt solution
Mercaptobenzothiazole sodium salt
2-Mercaptobenzothiazole sodium deriv
2-Mercaptobenzothiazole, sodium salt
2-Mercapto-benzothiazole, sodium
NACAP
NaMBT
Sodium 2-benzothiazolethioate
Sodium benzothiazolethiolate
Sodium 2-benzothiazolethiolate
Sodium 2(3H)-benzothiazolethionate
Sodium, (2-benzothiazolylthio)-
Sodium MBT
Sodium 2-mercaptobenzothiazol solution
Sodium mercaptobenzothiazolate
Sodium 2-mercaptobenzothiazolate
Sodium mercaptobenzothiazole

2493-84-7
$C_{15}H_{22}O_3$
250.34
O=C(O)c(ccc(OCCCCCCCC)c1)c1
Benzoic acid, p-(octyloxy)- (8CI)
Benzoic acid, 4-(octyloxy)- (9CI)
4-(Octyloxy)benzoic acid

2494-89-5
$C_8H_{11}NO_6S_2$
281.30
O=S(=O)(OCCS(=O)(=O)c(ccc(N)c1)c1)O
Ethanol, 2-((4-aminophenyl)-sulfonyl)-, hydrogen sulfate (ester) (9CI)

2494-93-1
$C_{17}H_{15}N_3O_6S_2$
421.44
O=S(=O)(O)c(cc(c(c1ccc2N=Nc(c(cc(N)c3)C)c3)c2)S(=O)(=O)O)c1
1,3-Naphthalenedisulfonic acid, 7-((4-amino-o-tolyl)-azo)- (8CI)
1,3-Naphthalenedisulfonic acid, 7-((4-amino-2-methylphenyl)-azo)- (9CI)

2495-25-2

$C_{17}H_{32}O_2$
268.44
O=C(OCCCCCCCCCCCCC)
C(=C)C
Tridecyl methacrylate
Methacrylic acid, tridecyl ester
2-Propenoic acid, 2-methyl-, tridecyl ester (9CI)
Tridecyl 2-methyl-2-propenoate

2495-27-4
$C_{20}H_{38}O_2$
310.52
O=C(OCCCCCCCCCCCCCCC
CC)C(=C)C
Hexadecyl methacrylate
Cetyl methacrylate
Hexadecyl 2-methyl-2-propenoate
Methacrylic acid, hexadecyl ester
2-Propenoic acid, 2-methyl-, hexadecyl ester (9CI)

2495-37-6
$C_{11}H_{12}O_2$
176.22
O=C(OCc(cccc1)c1)C(=C)C
2-Propenoic acid, 2-methyl-, phenylmethyl ester (9CI)
Benzyl methacrylate
Phenylmethyl 2-methyl-2-propenoate

2495-39-8
$C_3H_6O_3S.Na$
145.13
Sodium 2-propene-1-sulfonate
2-Propene-1-sulfonic acid, sodium salt (9CI)
Sodium allylsulfonate

2496-91-5
$C_8H_{19}O_5PS_2$
290.36
CCOP(=O)(OCC)SCCS(=O)(=O)
CC
Phosphorothioic acid, O,O-diethyl S-(2-(ethylsulfonyl)ethyl) ester
O,O-Diethyl S-(2-ethsulfonylethyl)phosphorothioate
Diethyl S-(2-ethsulfonylethyl)-thiophosphate
O,O-Diethyl S-ethyl 2-ethylmercaptophosphorothiolate sulfone
O,O-Diethyl 2-ethylmercaptoethyl thiophosphate, thiolo isomer
Isodemeton-Sulfone
Iso-Systox Sulfone
PO Systox Sulfone
Thiol Systox Sulfone

2497-06-5
$C_8H_{19}O_4PS_3$
306.42
CCOP(=S)(OCC)SCCS(=O)(=O)
CC
Phosphorodithioic acid, O,O-diethyl S-(2-(ethylsulfonyl)ethyl)ester
O,O-Diethyl S-(2-ethsulfonylethyl) phosphorodithioate
O,O-Diethyl S-(2-ethsulfonylethyl) thiothionophosphate
O,O-Diethyl-S-(2-ethylsulfonylethyl)phosphorodithioate
O,O-Diethyl S-(2-ethylsulfonylethyl)thionophosphate
Disyston sulfone
Ethanethiol, 2-(ethylsulfonyl)-, S-ester with O,O-diethyl phosphorodithioate

2497-07-6
$C_8H_{19}O_3PS_3$
290.42
CCOP(=S)(OCC)SCCS(=O)CC
Phosphorodithioic acid, O,O-diethyl S-((ethylsulfinyl)ethyl) ester
Bay 23323
DEPD
O,O-Diethyl-S-((ethylsulfinyl)-ethyl)phosphorodithioate
O,O-Diethyl S-(2-(ethylsulfinyl)ethyl) phosphorodithioate
Disulfoton disulide
Disulfoton sulfoxide
Disyston S
Disyston sulfoxide
Disyston sulphoxide
Ethylthiometon sulfoxide
L 16/184
Oxydisulfoton
S 309

2497-18-9
$C_8H_{14}O_2$
142.22
O=C(OCC=CCCC)C
2-Hexen-1-ol acetate
Hex-2-enyl acetate
2-Hexenyl acetate
2-Hexen-1-yl-acetate
(E)-2-Hexenyl acetate
trans-2-Hexenyl acetate

2497-21-4
$C_6H_{10}O$
98.16
O=C(C=CC)CC
2-Hexen-4-one
2-Hexene-4-one

2497-54-3
$C_{16}H_{14}$
206.29
Phenanthrene, 2,10-dimethyl- (8CI,9CI)

2497-58-7
$C_{26}H_{46}O_7$
470.65
3,6,9,12,15-Pentaoxaheptadecan-1-ol, 17-(4-(1,1,3,3-tetramethylbutyl)phenoxy)- (9CI)
3,6,9,12,15-Pentaoxaheptadecan-1-ol, 17-(p-(1,1,3,3-tetramethylbutyl)phenoxy)- (8CI)

2498-20-6
$C_{20}H_{44}N.I$
425.47
[I-].CCCCC[N+](CCCCC)(C
CCCC)CCCCC
1-Pentanaminium, N,N,N-tripentyl-, iodide (9CI)
N,N,N-Tripentyl-1-pentanaminium iodide

2498-63-7
$C_{19}H_{13}F$
260.31
Benz(a)anthracene, 5-fluoro-7-methyl- (9CI)

2498-66-0
$C_{18}H_{10}O_2$
258.28
O=C(c(c(c(c(c1)ccc2)c2)C(=O)
c3cccc4)c1)c34
Benz(a)anthracene-7,12-dione (9CI)

2498-75-1
$C_{19}H_{14}$
242.33
Cc4ccc2c(ccc3cc1ccccc1cc23)c4
Benz(a)anthracene, 3-methyl
3-Methylbenz(a)anthracene

2498-76-2
$C_{19}H_{14}$
242.33
Cc4ccc3ccc2cc1ccccc1cc2c3c4
Benz(a)anthracene, 2-methyl
2-Methylbenz(a)anthracene
2'-Methyl-1,2-benzanthracene

2498-77-3
$C_{19}H_{14}$
242.33
Cc3cccc4ccc2cc1ccccc1cc2c34
Benz(a)anthracene, 1-methyl
1-Methylbenz(a)anthracene
1'-Methyl-1,2-benzanthracene

2498-95-5
$C_{27}H_{24}N_4O_7S_3.2Na$
658.66
7-Benzothiazolesulfonic acid, 2-(4-((2-cyano-3-(4-(methyl-(2-sulfoethyl)amino)phenyl)-1-oxo-2-propenyl)amino)-phenyl)-6-methyl-, disodium

salt (9CI)
7-Benzothiazolesulfonic acid, 2-(p-(α-cyano-p-(methyl-(2-sulfoethyl)amino)cinnamamido)phenyl)-6-methyl-, disodium salt

2499-59-4
$C_{11}H_{20}O_2$
184.28
O=C(OCCCCCCCC)C=C
Acrylic acid, octyl ester (8CI)
AI3-03827
Octyl acrylate
Octyl 2-propenoate
2-Propenoic acid, octyl ester (9CI)

2499-95-8
$C_9H_{16}O_2$
156.25
O=C(OCCCCCC)C=C
Acrylic acid, hexyl ester
Ageflex n-HA
Hexyl acrylate
n-Hexyl acrylate
Hexyl 2-propenoate
2-Propenoic acid, hexyl ester (9CI)

2500-83-6
$C_{12}H_{16}O_2$
192.26
O=C(OC(C(C(C(C12)CC=3)C3)C1)C2)C
4,7-Methano-1H-inden-5-ol, 3a,4,5,6,7,7a-hexahydro-, acetate (9CI)
3a,4,5,6,7,7a-Hexahydro-4,7-methano-1H-inden-5-yl acetate
Tricyclo(5.2.1.02,6)dec-3-en-9-yl acetate

2502-15-0
$C_{27}H_{33}O_4P$
452.53
Phenol, 4-(1-methylethyl)-, phosphate (3:1) (9CI)
p-Cumenyl phosphate
Phenol, p-isopropyl-, phosphate (3:1) (8CI)

2503-46-0
$C_{10}H_{12}O_3$
180.20
2-Propanone, 1-(4-hydroxy-3-methoxyphenyl)- (9CI)

2503-56-2
$C_6H_6N_4O$
150.12
N(=C(N=C(C=C1O)C)N1N=2)C2
(1,2,4)Triazolo(1,5-a)pyrimidin-7-ol, 5-methyl- (9CI)
5-Methyl-7-hydroxy-1,3,4-triazaindolizine
5-Methyl-(1,2,4)triazolo(1,5-a)-pyrimidin-7-ol

2503-73-3
$C_{42}H_{29}N_7O_{13}S_4.4Na$
1059.91
1,4-Benzenedisulfonic acid, 2-((4-((4-((1-hydroxy-6-(phenylamino)-3-sulfo-2-naphthalenyl)azo)-1-naphthalenyl)azo)-6-sulfo-1-naphthalenyl)azo)-, tetrasodium salt (9CI)
Ismafast Blue 4GL

2507-55-3
$C_{14}H_{28}O_3$
244.37
α-Hydroxymyristic acid
2-Hydroxytetradecanoic acid
Tetradecanoic acid, 2-hydroxy-(9CI)

2508-20-5
$C_{13}H_9NO$
195.23
O=Nc3ccc1c(Cc2ccccc12)c3
Fluorene, 2-nitroso
Nitrosofluorene
2-Nitrosofluorene

2510-55-6
$C_{15}H_9N$
203.24
9-Phenanthrenecarbonitrile (9CI)
AI3-22124

2510-86-3
$C_{10}H_{15}O_4P$
230.20
CCOP(=O)(OCC)Oc1ccccc1
Phosphoric acid, diethyl phenyl ester (9CI)
AI3-19536

2511-10-6
$C_5H_{13}O_2PS$
168.21
Phosphonothioic acid, methyl-, O,S-diethyl ester
O,S-Diethyl methylphosphonothioate
O,S-Diethyl methylthiophosphonate
LG 61
Methylphosphonothioic acid O,S-diethyl ester
OSDMP

2512-29-0
$C_{17}H_{16}N_4O_4$
340.31
O=C(Nc(cccc1)c1)C(N=Nc(c(N(=O)=O)cc(c2)C)c2)C(=O)C
C.I. Pigment Yellow 1 (VAN) (8CI)
Accosperse Hansa Yellow G
Acrylamide Yellow G
Adc Pigment Yellow G
AI3-30763
Arylamide Yellow G
Brazil Yellow X 2866
Burma Yellow X 1622
Butanamide, 2-((4-methyl-2-nitrophenyl)azo)-3-oxo-N-phenyl- (9CI)
Calcotone Hansa Yellow
Carnelio Yellow G
Chromatex Yellow J
C.I. 11680
Dainichi Fast Yellow G
Eljon Fast Yellow GN-GX
Eljon Fast Yellow PV Extra
Ext. D & C Yellow No. 5
Fanchon Yellow G-YH 1
Fanchon Yellow WD 259
Fast Yellow J
Fast Yellow JT
Fastona Yellow G
Fastona Yellow G Transparent
Graphtol Yellow GL
Graphtol Yellow 4813-0
Hancock Yellow 1008
Hansa Yellow
Hansa Yellow G
Hansa Yellow G Extra
Hansa Yellow G Toner
Hansa Yellow G 45-4050
Hansa Yellow GAD
Hansa Yellow GT
Hansa Yellow S 3155
Hansa Yellow Toner G-YA 8365
Helio Fast Yellow GN
Helio Fast Yellow GNS
Helio Fast Yellow GT
Irgalite Fast Yellow PG
Irgalite Fast Yellow PG Transparent
Irgalite Yellow G
Irgalite Yellow GNS
Irgalite Yellow GTN
Irgalite Yellow PV 2
Isol Aryl Yellow G
Isol Aryl Yellow G Transp
Isol Aryl Yellow GX
Isol Fast Yellow G
Isol Fast Yellow G Transp
Kromon Yellow G
Light Yellow
Lightfast Yellow
2-((4-Methyl-2-nitrophenyl)-azo)-3-oxo-N-phenylbutanamide
Pigment Yellow 1

2516-96-3
$C_7H_4ClNO_4$
201.56
O=C(O)c(c(ccc1N(=O)=O)Cl)c1
Benzoic acid, 2-chloro-5-nitro-(9CI)
AI3-16578

2-Chloro-5-nitrobenzoic acid

2517-16-0
$C_4H_6Cl_2O_2.Na$
178.98
Butanoic acid, 2,2-dichloro-, sodium salt (9CI)
Alterungsschutz HS
Basinex
Butyric acid, 2,2-dichloro-, sodium salt (7CI,8CI)
2,2-Dichlorobutyric acid, sodium salt
HS
Sodium 2,2-dichlorobutyrate

2517-43-3
$C_5H_{12}O_2$
104.15
O(C(CCO)C)C
3-Methoxy-1-butanol
AI3-24920
1-Butanol, 3-methoxy- (9CI)

2518-72-1
$C_6H_8N_2O_3$
156.13
CCC1C(=O)NC(=O)NC1=O
2,4,6(1H,3H,5H)-Pyrimidinetrione, 5-ethyl- (9CI)
Barbituric acid, 5-ethyl- (8CI)
NSC-27274
NSC-66908

2519-30-4
$C_{28}H_{21}N_5O_{14}S_4.4Na$
871.74
[Na+].[Na+].[Na+].[Na+].CC(=O)Nc4ccc(c5cc(c(N=Nc2ccc(N=Nc1ccc(cc1)S([O-])(=O)=O)c3ccc(cc23)S([O-])(=O)=O)c(O)c45)S([O-])(=O)=O
1,7-Naphthalenedisulfonic acid, 4-acetamido-5-hydroxy-6-((7-sulfo-4-((p-sulfophenyl)azo)- 1-naphthyl)azo)-, tetrasodium salt
1743 Black
Black PN
Blue Black BN
Brilliant Acid Black BNA Export
Brilliant Acid Black BN Extra Pure A
Brilliant Black
Brilliant Black A
Brilliant Black BN
Brilliant Black NAF
Brilliant Black N.FQ
Brilliantschwarz BN (German)
Cern Brilantni PN (Czech)
Cern Potravinarska 1 (Czech)
Certicol Black PNW
Cilefa Black B
C.I. 28440
C.I. Food Black 1, Tetrasodium salt
E 151
Edicol Supra Black BN
Hexacol Black PN
L-Schwarz 1
Melan Black
Noir Brillant BN (French)
Xylene Black F

2519-50-8
$C_{10}H_9N_3O$
187.20
s-Triazin-2-ol, 4-methyl-6-phenyl- (7CI,8CI)

2519-77-9
$C_{19}H_{26}N_2O_2$
314.43
153C51
1,7-Bis(p-aminophenoxy)-heptane

2523-37-7
$C_{14}H_{12}$
180.26
Fluorene, 9-methyl
9H-Fluorene, 9-methyl-
9-Methylfluorene

2523-44-6
$C_{13}H_9Cl$
200.67
Fluorene, 2-chloro- (8CI)
AI3-01713
9H-Fluorene, 2-chloro- (9CI)

2523-46-8
$C_{14}H_{12}O$
196.25
9H-Fluorene, 2-methoxy- (9CI)
Fluorene, 2-methoxy- (6CI, 7CI,8CI)
Fluoren-2-yl methyl ether
2-Methoxyfluorene

2523-48-0
$C_{14}H_9N$
191.23
Fluorene-2-carbonitrile (8CI)
9H-Fluorene-2-carbonitrile (9CI)

2524-03-0
$C_2H_6ClO_2PS$
160.56
O(P(OC)(=S)Cl)C
Phosphorochloridothioic acid, O,O-dimethyl ester
Dimethyl chlorothiophosphate (DOT)
Dimethylchlorthiofosfat (Czech)
O,O-Dimethylester kyseliny chlorthiofosforecne (Czech)
O,O-Dimethylphosphorochloridothioate
Dimethyl phosphorochloridothioate (DOT)
Methyl PCT
NA 2267 (DOT)
Phosphonothioic acid, chloro-, O,O-dimethyl ester

2524-04-1
$C_4H_{10}ClO_2PS$
188.62
O(P(OCC)(=S)Cl)CC
Phosphorochloridothioic acid, O,O-diethyl ester
O,O-Diethylphosphorochloridothioate
Diethylchlorthiofosfat (Czech)
Diethylchlorothiophosphate
Diethylthiophosphoryl chloride [UN 2751]
Phosphonothioic acid, chloro-, O,O-diethyl ester
UN 2751 [Diethylthiophosphoryl chloride]

2524-05-2
$C_6H_{14}ClO_2PS$
216.67
O(P(OCCC)(=S)Cl)CCC
Phosphorochloridothioic acid, O,O-dipropyl ester (9CI)

2524-06-3
$C_6H_{14}ClO_2PS$
216.68
Phosphorochloridothioic acid, O,O-diisopropyl ester
Phosphorochloridothionic acid, diisopropyl ester

2524-09-6
$C_6H_{15}O_2PS_2$
214.29
Phosphorodithioic acid, O,O,S-triethyl ester (8CI,9CI)

2524-52-9
$C_8H_9NO_2$
151.16
O=C(OCC)c(nccc1)c1
Ethyl picolinate
Ethyl 2-pyridinecarboxylate
Picolinic acid, ethyl ester
2-Pyridinecarboxylic acid, ethyl ester (9CI)

2528-16-7
$C_{15}H_{12}O_4$
256.26
O=C(OCc(cccc1)c1)c(c(ccc2)C(=O)O)c2
1,2-Benzenedicarboxylic acid, mono(phenylmethyl) ester (9CI)

2528-36-1
$C_{14}H_{23}O_4P$
286.34
O=P(OCCCC)(OCCCC)Oc(ccc c1)c1
Phosphoric acid, dibutyl phenyl ester
Dibutyl phenyl phosphate (ACGIH)

2528-61-2
$C_7H_{13}ClO$
148.63
O=C(CCCCCC)Cl
Heptanoyl chloride (9CI)

2529-36-4
$C_{10}H_{12}O_2$
164.20
Benzoic acid, 2,3,6-trimethyl- (8CI,9CI)

2530-83-8
$C_9H_{20}O_5Si$
236.38
COSi(CCCOCC1CO1)(OC)OC
Silane, 3-(2,3-epoxypropoxy)-propyltrimethoxy
3-(2,3-Epoxypropoxy)propyltrimethoxysilane
γ-Glycidoxypropyltrimethoxysilane
Silane-Y-4087
Silicone A-187
Union Carbide A-187

2530-85-0
$C_{10}H_{20}O_5Si$
248.39
COSi(CCCOC(=O)C(C)=C)(OC) OC
1-Propanol, 3-(trimethoxysilyl)-, methacrylate
Methacrylic acid, 3-(trimethoxysilyl)propyl ester
γ-Methacryloxypropyltrimethoxysilane
2-Propenoic acid, 2-methyl-, 3-(trimethoxysilyl)propyl ester
Silane, (3-hydroxypropyl)trimethoxy-, methacrylate
Silicone A-174
3-(Trimethoxysilyl)-1-propanol methacrylate
Trimethoxysilyl-3-propylester kyseliny methakrylove (Czech)
Union Carbide A-174

2530-87-2
$C_6H_{15}ClO_3Si$
198.72
COSi(CCCCl)(OC)OC
Silane, (3-chloropropyl)trimethoxy- (9CI)
(3-Chloropropyl)trimethoxysilane

2531-04-6
$C_{17}H_{23}N_3O$
285.37
3H-Pyrazol-3-one, 4-ethyl-1,2-dihydro-2-(1-methyl-4-piperidinyl)-5-phenyl- (9CI)
Piperilona (Spanish)
Piperylone
Piperylonum (Latin)
3-Pyrazolin-5-one, 4-ethyl-1-(1-methyl-4-piperidyl)-3-phenyl- (8CI)

2531-84-2
$C_{15}H_{12}$
192.27
Phenanthrene, 2-methyl
2-Methylphenanthrene

2532-58-3
C_7H_{14}
98.19
C(CCC1C)(C1)C
Cyclopentane, 1,3-dimethyl-, cis- (9CI)
cis-1,3-Dimethylcyclopentane

2533-20-2
$C_{20}H_{39}N_5$
349.53
1,3,5-Triazine-2,4-diamine, 6-heptadecyl- (9CI)
NSC-7780
Stearoguanamine
s-Triazine, 2,4-diamino-6-heptadecyl- (8CI)

2533-82-6
CH_3AsS
122.02
C[As]=S
Arsine, methylthioxo
Asozin
Bay 4934
MAS
Methane, (thioarsenoso)-
Methylarsenic sulfide
Methylarsine sulfide
Methylarsinic sulfide
Methylarsinic sulphide
Monkil WP
Rhizoctol
Urbasulf

2533-89-3
$C_{12}H_{15}Cl_2NO$
260.18
Valeranilide, 3',4'-dichloro-2-methyl
3,4-Dichloranilid kyseliny α-methylvalerove (Czech)
N-(3,4-Dichlorophenyl)-2-methylpentamide
N-(2,4-Dichlorophenyl)-2-methylpentanamide
Karsil
α-Methylvaleric acid, 3,4-dichloroanilide
NCA
Niagara 4562

2536-31-4
$C_{15}H_{11}ClO_3$
274.71
O=C(OC)C(O)(c(c(c1cccc2)ccc3 Cl)c3)c12
Fluorene-9-carboxylic acid, 2-chloro-9-hydroxy-, methyl ester
Break-Thru
CF 125
Chloflurecol-methyl
Chloflurecol-methyl ester
Chlorflurecol
Chlorflurenol
Chlorflurenol-methyl ester
2-Chlor-9-hydroxyfluoren-carbonsaeure-(9)-methylester (German)
Chloroflurenol-methyl ester
2-Chloro-9-hydroxy-9-methylcarboxylatefluorene
Curbiset
IT 3456
Maintain A
Maintain CF125
Methyl-2-chloro-9-hydroxyfluorene-9-carboxylate
Morphactin
Multiprop

2538-85-4
$C_{20}H_{13}N_2O_5S.Na$
416.40
1-Naphthalenesulfonic acid, 3-hydroxy-4-(2-hydroxy-1-naphthylazo)-, sodium salt
Calcon
3-Hydroxy-4-(2-hydroxy-1-naphthylazo)-1-naphthalenesulfonic acid, sodium salt

2539-13-1
$C_9H_9Cl_3O_3$
271.53
Benzene, 1,2,3-trichloro-4,5,6-trimethoxy- (8CI,9CI)

2539-17-5
$C_7H_4Cl_4O_2$
261.91
COc1c(O)c(Cl)c(Cl)c(Cl)c1Cl
Phenol, 2-methoxy-3,4,5,6-tetrachloro
2-Methoxytetrachlorophenol
2-Methoxy-3,4,5,6-tetrachlorophenol
Tetrachloroguaiacol

2539-26-6
$C_8H_7Cl_3O_3$

257.50
3,4,5-Trichloro-2,6-dimethoxy-phenol
3,4,5-Trichlorosyringol

2540-82-1
$C_6H_{12}NO_4PS_2$
257.28
COP(=S)(OC)SCC(=O)N(C)C=O
Phosphorodithioic acid, O,O-dimethyl ester, S-ester with N-formyl-2-mercapto-N-methyl- acetamide
Acetamide, N-formyl-2-mercap-to-N-methyl-, S-ester with O,O-dimethyl phosphorodi-thioate
Aflix
Anthio
Antio
CP 53926
O,O-Dimethyl dithiophosphoryl-acetic acid N-methyl-N-formylamide
O,O-Dimethyl S-(N-formyl-N-methylcarbamoylmethyl) phosphorodithioate
O,O-Dimethyl-S-(3-methyl-2,4-dioxo-3-aza-butyl)-di-thiofosfaat (Dutch)
O,O-Dimethyl-S-(3-methyl-2,4-dioxo-3-aza-butyl)-di-thiophosphat (German)
O,O-Dimethyl-s-(N-methyl-N-formyl-carbamoylmethyl)-dithiophosphat (German)
O,O-Dimethyl S-(N-methyl-N-formylcarbamoylmethyl)-phosphorodithioate
O,O-Dimethyl phosphorodi-thioate N-formyl-2-mercapto-N-methylacetamide s-ester
O,O-Dimetil-S-(N-formil-N-metil-carbamoil-metil)-di-tiofosfato (Italian)
ENT 27,257
Formothion
S-(2-(Formylmethylamino)-2-oxoethyl) O,O-dimethyl-phosphorodithioate
N-Formyl-N-methylcarbamoyl-methyl O,O-dimethyl phos-phorodithioate
S-(N-Formyl-N-methylcar-bamoylmethyl) O,O-dimethyl phosphorodithioate
S-(N-Formyl-N-methylcarbamo-ylmethyl) dimethyl phos-phorothiolothionate
J-38
OMS-968
Phosphorodithioic acid, O,O-di-methyl ester, N-formyl-2-mer-capto-N-methylacetamide S-ester
Phosphorodithioic acid, S-(2-(formylmethylamino)-2-oxoethyl) O,O-dimethyl ester
S 6900
Sandoz S-6900
San 244 I
San 6913 I
San 7107 I
Spencer S-6900
Vel 4284

2540-99-0
$C_{16}H_6O_8S$
358.28
1,3-Isobenzofurandione, 5,5'-sulfonylbis- (9CI)
Phthalic anhydride, 4,4'-sulf-onyldi- (8CI)

2541-68-6
$C_{19}H_{13}F$
260.32
Benz(a)anthracene, 6-fluoro-7-methyl
6-Fluoro-7-methylbenz(a)anthra-cene

2541-69-7
$C_{19}H_{14}$
242.33
Cc2c1ccccc1cc3c2ccc4ccccc34
Benz(a)anthracene, 7-methyl
7-MBA
10-Methyl-1,2-benzanthracen (German)
7-Methylbenz(a)anthracene
10-Methyl-1,2-benzanthracene

2542-29-2
$C_{16}H_{14}ClN_3O_7$
326.44
Carbanilic acid, p-chloro-, α-ethoxy-4,6-dinitro-o-tolyl ester (7CI,8CI)

2545-59-7
$C_{14}H_{17}Cl_3O_4$
355.66
Acetic acid, (2,4,5-trichloro-phenoxy)-, 2-butoxyethyl ester
Bladex H
Butoxyethyl 2,4,5-T
Hormoslyr 500t
2,4,5-T butoxyethanol ester
2,4,5-T butoxyethyl ester
(2,4,5-Trichlorophenoxy)acetic acid 2-butoxyethyl ester
Trinoxol

2545-60-0
$C_6H_2Cl_3N_2O_2.K$
279.55
Picolinic acid, 4-amino-3,5,6-trichloro-, monopotas-sium salt
4-Amino-3,5,6-trichloropicolin-ic acid potassium salt
Chloramp
Pichloram K
Pichloram potassium salt
Picloram potassium salt
Potassium picloram
2-Pyridinecarboxylic acid, 4-amino-3,5,6-trichloro-, monopotassium salt (9CI)
Tordon K
Tordon 10K
Tordon 22K

2547-45-7
$C_7H_6O_6$
186.12
1,3-Butadiene-1,2,4-tri-carboxylic acid (8CI,9CI)
2,4-Hexadienedioic acid, 3-carboxy-

2549-51-1
$C_4H_5ClO_2$
120.54
O=C(OC=C)CCl
Acetic acid, chloro-, vinyl ester
UN 2589 [Vinyl chloroacetate]
Vinyl chloroacetate [UN 2589]

2549-53-3
$C_{18}H_{34}O_2$
282.47
O=C(OCCCCCCCCCCCCCC)C(=C)C
Tetradecyl methacrylate
Methacrylic acid, tetradecyl ester
Myristyl methacrylate
2-Propenoic acid, 2-methyl-, tetradecyl ester (9CI)
Tetradecyl 2-methyl-2-propen-oate

2549-67-9
C_4H_9N
71.14
N(C1CC)C1
Aziridine, 2-ethyl
Butyleneimine
1,2-Butylenimine
2-Ethylaziridine
Ethylethylenimine
2-Ethylethylenimine
N-Methylcaprolactam
1-Methylcaprolactam

2550-06-3
$C_3H_6Cl_4Si$
211.98
Silane, trichloro(3-chloro-propyl)- (9CI)
Chloropropyltrichlorosilane
(3-Chloropropyl)trichlorosilane
Trichloro(3-chloropropyl)silane

2550-26-7
$C_{10}H_{12}O$
148.22
O=C(CCc(cccc1)c1)C
2-Butanone, 4-phenyl

Benzylacetone
Methyl phenethyl ketone
Methyl phenylethyl ketone
Methyl 2-phenylethyl ketone
Phenethyl methyl ketone
4-Phenylbutan-2-one
4-Phenyl-2-butanone
β-Phenylethyl methyl ketone

2550-40-5
$C_{12}H_{22}S_2$
230.44
S(SC(CCCC1)C1)C(CCCC2)C2
Dicyclohexyl disulfide
Bis(cyclohexyl)disulfide
Cyclohexyl disulfide
Disulfide, dicyclohexyl (9CI)

2551-62-4
F_6S
146.06
FS(F)(F)(F)(F)F
Sulfur-fluoride
Hexafluorure de soufre (French)
Sulfur hexafluoride (ACGIH, OSHA) [UN 1080]
UN 1080 [Sulfur hexafluoride]

2553-08-4
$C_{21}H_{26}O_3$
326.44
O=C(Oc(ccc(c1)C(CC(C)(C)C)(C)C)c1)c(c(O)ccc2)c2
Benzoic acid, 2-hydroxy-, 4-(1,1,3,3-tetramethylbutyl)-phenyl ester (9CI)
4-(1,1,3,3-Tetramethylbutyl)-phenyl salicylate

2553-96-0
$C_{11}H_{22}O_2$
172.27
Octanoic acid, 4,6-dimethyl-, methyl ester, (4S,6S)-(+)- (8CI)
Octanoic acid, 4,6-dimethyl-, methyl ester (7CI)

2554-06-5
$C_{12}H_{24}O_4Si_4$
344.66
Cyclotetrasiloxane, 2,4,6,8-tetraethenyl-2,4,6,8-tetramethyl- (9CI)
Cyclotetrasiloxane, 2,4,6,8-tetramethyl-2,4,6,8-tetravinyl-
1,3,5,7-Tetramethyl-1,3,5,7-tetravinylcyclotetrasiloxane
Tetravinyltetramethylcyclotetrasiloxane

2555-49-9
$C_{10}H_{12}O_3$
180.20
O=C(OCC)COc(cccc1)c1
Acetic acid, phenoxy-, ethyl ester
AI3-04110
Ethyl phenoxyacetate

2555-99-9
$C_{16}H_{14}O_8$
334.28
Benzoic acid, 3-(4-carboxy-2-methoxyphenoxy)-4-hydroxy-5-methoxy- (9CI)
m-Anisic acid, 4'-hydroxy-4,5'-oxydi- (8CI)

2556-36-7
$C_8H_{10}N_2O_2$
166.17
O=C=NC(CCC(N=C=O)C1)C1
Cyclohexane diisocyanate
Cyclohexane, 1,4-diisocyanato- (9CI)
1,4-Diisocyanatocyclohexane

2562-52-9
$C_5HCl_3N_4$
223.42
Clc2nc(Cl)c1[nH]c(Cl)nc1n2
2,6,8-Trichloropurine
1H-Purine, 2,6,8-trichloro- (9CI)

2567-01-3
C_4H_7NO
85.10
Propanenitrile, 3-hydroxy-2-methyl- (9CI)
Hydracrylonitrile, 2-methyl- (8CI)

2567-14-8
$C_3H_3Cl_3$
145.41
1-Propene, 1,1,3-trichloro- (9CI)
Propene, 1,1,3-trichloro- (8CI)

2567-83-1
$C_8H_{20}N.ClO_4$
229.70
Tetraethylammonium perchlorate (Dry)
Ammonium, tetraethyl-, perchlorate, Dry
Ethanaminium, N,N,N-triethyl-, perchlorate (9CI)
Perclorato de tetraetilamonio (Spanish)
Tetraethylammonium perchlorate
Tetraethylammonium perchlorate, Dry
N,N,N-Triethylethanaminium perchlorate
UN 1325 [Flammable solids, N.O.S.]

2568-30-1
$C_4H_7ClO_2$
122.55
O(CCO1)C1CCl
1,3-Dioxolane, 2-(chloromethyl)- (9CI)
AI3-08042
Chloroacetaldehyde ethylene acetal
2-(Chloromethyl)-1,3-dioxolane

2568-90-3
$C_9H_{20}O_2$
160.26
O(CCCC)COCCCC
Butane, 1,1'-(methylenebis(oxy))bis- (9CI)
AI3-05675
Methane, dibutoxy- (8CI)
1,1'-(Methylenebis(oxy))bis-butane

2568-92-5
$C_9H_{20}O_2$
160.26
Butane, 2,2'-(methylenebis(oxy))bis- (9CI)
Methane, di-sec-butoxy- (7CI,8CI)

2568-96-9
$C_5H_{10}O_2$
102.13
1,3-Dioxolane, 2-ethyl- (8CI, 9CI)
Propanal, cyclic 1,2-ethanediyl acetal

2569-01-9
$C_{14}H_{21}Cl_2NO_6$
370.23
Triethanolamine 2,4-dichlorophenoxyacetate
Caswell No. 315AC
2,4-Dichlorophenoxyacetic acid triethanolamine salt
2,4-D-tris(2-hydroxyethyl)-ammonium
EPA Pesticide Chemical Code 030033

2570-26-5
$C_{15}H_{33}N$
227.43
NCCCCCCCCCCCCCCC
Pentadecylamine
1-Pentadecanamine (9CI)
n-Pentadecylamine
1-Pentadecylamine

2571-88-2
$C_{20}H_{43}NO$
313.56
O=N(CCCCCCCCCCCCCCCCCC)(C)C

Stearamine oxide
N,N-Dimethyl-1-octadecan-amine N-oxide
Incromine Oxide S
Jordamox SDA
1-Octadecanamine, N,N-di-methyl-, N-oxide (9CI)
Octadecyldimethylamine oxide
Stearyl dimethylamine oxide

2572-44-3
$C_{15}H_{10}Cl_3N_5$
366.64
1,3,5-Triazine-2,4-diamine, 6-chloro-N,N'-bis-(4-chlorophenyl)- (9CI)
s-Triazine, 2-chloro-4,6-bis-(p-chloroanilino)- (6CI, 7CI,8CI)

2575-07-7
$C_{11}H_{20}O_3$
200.28
COC(=O)CCCCCCCC(C)=O
Decanoic acid, 9-oxo-, methyl ester (6CI, 7CI,9CI)
Methyl 9-oxodecanoate

2577-72-2
$C_{13}H_9Br_2NO_2$
371.02
Metabromsalan
Benzamide, 3,5-dibromo-2-hydroxy-N-phenyl- (9CI)
Caswell No. 289
3,5-Dibromosalicylanilide
EPA Pesticide Chemical Code 077405
Metabromsalanum (Latin)
NSC-526280
Salicylanilide, 3,5-dibromo-(8CI)

2580-56-5
$C_{33}H_{32}N_3.Cl$
506.07
Victoria Blue B
4-(α-(4-Anilino-1-naphthyl)-p-dimethylamino)benzylidene-2,5-cyclohexadien-1-ylidene-dimethyl ammonium chloride
C.I. Basic Blue 26 (8CI)
N-(4-((4-(Dimethylamino)-phenyl)(4-(phenylamino)-1-naphthalenyl)methylene)-2,5-cyclohexadien-1-ylidene)-N-methylmethanaminium, chloride
Methanaminium, N-(4-((4-(di-methylamino)phenyl)(4-(phenylamino)-1-naphthalen-yl)methylene)-2,5-cyclohexa-dien-1-ylidene)-N-methyl-, chloride (9CI)

2580-77-0
$C_4H_{10}O_4S$
154.19
O=S(=O)(CCO)CCO
Ethanol, 2,2'-sulfonylbis-(9CI)
AI3-16183
Diethanol sulfone
Ethanol, 2,2'-sulfonyldi- (8CI)
2,2'-Sulfonylbisethanol

2580-78-1
$C_{22}H_{16}N_2O_{11}S_3.2Na$
626.56
2-Anthracenesulfonic acid, 1-amino-9,10-dihydro-4-(m-((2-hydroxyethyl)-sulfonyl)anilino- 9,10-dioxo-, hydrogen sulfate (ester), disodium salt
Cavalite Brilliant Blue R
C.I. 61200
C.I. Reactive Blue 19
C.I. Reactive Blue 19, Disod-ium salt
Reactive Blue 19
Remalan Brilliant Blue R
Remazol Brilliant Blue R

2581-34-2
$C_7H_7NO_3$
153.15
O=N(=O)c(c(cc(O)c1)C)c1
m-Cresol, 4-nitro
3-Methyl-4-nitrophenol
4-Nitro-m-cresol
4-Nitro-3-cresol

2581-69-3
$C_{18}H_{14}N_4O_2$
318.36
O=N(=O)c(ccc(N=Nc(ccc(Nc(cccc1)c1)c2)c2)c3)c3
Diphenylamine, 4-((p-nitro-phenyl)azo)
4-((p-Nitrophenyl)azo)diphenyl-amine

2586-57-4
$C_{34}H_{27}N_5O_{10}S_2.2Na$
775.70
C.I. Direct Blue 22, Disodium salt (8CI)
1,3-Naphthalenedisulfonic acid, 4-amino-5-hydroxy-6-((4'-((2-hydroxy-1-naphthalenyl)-azo)-3,3'-dimethoxy(1,1'-bi-phenyl)-4-yl)azo)-, disodium salt (9CI)

2586-58-5
$C_{32}H_{24}N_8O_6S.2Na$
694.68
Benzoic acid, 5-((4'-((2,6-di-amino-3-methyl-5-((4-sulfo-phenyl)azo)phenyl)azo)-(1,1'-biphenyl)- 4-yl)azo)-2-hydroxy-, disodium salt
Atlantic Brown D 3Y
Atul Direct Brown CN
Benzo Brown D 3GA-CF
Chlorazol Orange Brown X
C.I. 30110
C.I. Direct Brown 1:2
C.I. Direct Brown 1A, Disod-ium salt
Diphenyl Brown PT
Direct Brown 1:2
Direct Brown 1A
Direct Brown 5C
Direct Brown CGN
Direct Brown 5G
Direct Brown 2GS
Enianil Brown 2GS
Fixanol Orange Brown X
Honey Yellow 3GNT
Oxydiamine Brown 3GN
Phenamine Brown D 3G
Pontamine Brown D 3GN
Pontamine Brown NCR

2586-60-9
$C_{32}H_{22}N_6O_8S_2.2Na$
728.70
2-Naphthalenesulfonic acid, 5,5'-(4,4'-biphenylylenebis-(azo))bis(6-amino-4-hydr-oxy-, disodium salt
Airedale Violet ND
Amanil Fast Violet N
Atlantic Violet N
Atul Direct Violet N
Azocard Violet N
Bencidal Fast Violet N
Benzanil Violet N
Benzo Violet N
Brasilamina Violet 3R
Calcomine Violet N
Chlorazol Violet N
C.I. 22570
C.I. Direct Violet 1
C.I. Direct Violet 1, Disodium salt
Cotton Violet R
Diamine Violet N
Diaphtamine Violet N
Diazine Violet N
Diazol Violet N
Direct Brilliant Violet 2R
Direct Fast Violet N
Direct Fast Violet MN
Direct Violet C
Direct Violet N
Direct Violet R
Direct Violet FR
Erie Violet 3R
Fixanol Violet N
Hispamin Violet 3R
Japanol Violet J
Naphtamine Violet N
Paramine Fast Violet N
Pheno Violet N
Pontamine Violet N
Tertrodirect Violet N
Trisulfon Violet N
Violet Prima 1 (Czech)

2587-90-8
$C_5H_{13}O_3PS_2$
216.27
COP(=O)(OC)SCCSC
Phosphorothioic acid, O,O-dimethyl S-(2-(methylthio)-ethyl) ester
Cebetox
Cymetox
Demephion
Demephion-S
O,O-Dimethyl S-(2-(methylthio)ethyl) phosphorothioate
Ethanethiol, 2-(methylthio)-, O,O-dimethyl phosphorothioate
Ethanethiol, 2-(methylthio)-, S-ester with O,O-dimethyl phosphorothioate
Isonitox
Methyl demeton methyl
2-Methylthioethyl O,O-dimethyl phosphorothioate
Tinox

2588-03-6
$C_7H_{17}O_3PS_3$
276.39
Phosphorodithioic acid, O,O-diethyl S-((ethylsulfinyl)methyl) ester
O,O-Diethyl-S-ethylthionylmethylphosphorodithioate
O,O-Diethyl-S-ethylthionylmethyl thionophosphate
O,O-Diethyl S-eththionylmethyl thiothionophosphate
Thimet sulfoxide

2588-04-7
$C_7H_{17}O_4PS_3$
292.39
Phosphorodithioic acid, O,O-diethyl S-((ethylsulfonyl)methyl) ester
O,O-Diethyl S-ethsulfonylmethyl thiothionophosphate
O,O-Diethyl-S-ethylsulfonyl methylphosphorodithioate
O,O-Diethyl S-ethylsulfonylmethyl thionophosphate
Methanethiol, (ethylsulfonyl)-, S-ester with O,O-diethyl phosphorodithioate
Phorate sulfone
Thimet sulfone

2588-05-8
$C_7H_{17}O_4PS_2$
260.33
CCOP(=O)(OCC)SCS(=O)CC
Phosphorothioic acid, O,O-diethyl S-((ethylsulfinyl)-methyl) ester
O,O-Diethyl S-eththionylmethyl phosphorothioate
Diethyl S-eththionylmethyl thiophosphate
Phorate oxon sulfoxide
Phorate sulfoxide
Phoratoxon sulfoxide

2589-78-8
$C_{22}H_{38}O$
318.54
Oc(ccc(c1)CCCCCCCCCCCCCCCC)c1
Phenol, 4-hexadecyl- (9CI)
4-Hexadecylphenol

2591-86-8
$C_6H_{11}NO$
113.18
O=CN(CCCC1)C1
Piperidine, 1-formyl
n-Formylpiperidin (German)
1-Formylpiperidine

2593-15-9
$C_5H_5Cl_3N_2OS$
247.53
CCOc1nc(ns1)C(Cl)(Cl)Cl
1,2,4-Thiadiazole, 5-ethoxy-3-(trichloromethyl)
Aaterra
5-Aethoxy-3-trichlormethyl-1,2,4-thiadiazol (German)
Dwell
Echlomezol
Echlomezole
Etcmtb
Ethazol
Ethazole (Fungicide)
5-Ethoxy-3-trichloromethyl-1,2,4-thiadiazole
ETMT
Etridiazol
Etridiazole
Koban
MF-344
Olin Mathieson 2,424
OM 2424
Pansoil
TCMTB
Terrachlor-Super X
Terracoat
Terracoat 121
Terraflo
Terrazole
2-(Thiocyanomethylthio)benzothiazole
3-(Trichloromethyl)-5-ethoxy-1,2,4-thiadiazole
Truban

2595-54-2
$C_{10}H_{20}NO_5PS_2$
329.40
CCOC(=O)N(C)C(=O)CSP(=S)(OCC)OCC
Carbamic acid, (mercaptoacetyl)methyl-, ethyl ester, S-ester with O,O- diethyl phosphorodithioate
Afos
O,O-Diaethyl-S-(3-methyl-2,4-dioxo-5-oxa-3-aza-heptyl)-dithiophosphat (German)
O,O-Diethyl S-(N-ethoxycarbonyl-N-methylcarbamoylmethyl) phosphorodithioate
O,O-Diethyl S-(N-ethoxycarbonyl-N-methylcarbamoylmethyl) phosphorothiolothionate
O,O-Diethyl S-(N-methyl-N-carboethoxycarbamoylmethyl) dithiophosphate
O,O-Diethyl-S-(3-methyl-2,4-dioxo-5-oxa-3-aza-heptyl)-dithiofosfaat (Dutch)
O,O-Dietil-S-(N-etossi-carbonil-N-metil-carbamoil-metil)-ditiofosfato (Italian)
Dithiophosphate de O,O-diethyle et de S-N-methyl N-carboethoxy carbamoylmethyle (French)
N-Ethoxycarbonyl-N-methylcarbamoylmethyl O,O-diethyl phosphorodithioate
S-((Ethoxycarbonyl)methylcarbamoyl)methyl O,O-diethyl phosphorodithioate
S-(N-Ethoxycarbonyl-N-methylcarbamoylmethyl)-diethyl phosphorodithioate
Marfotoks
MC 474
Mecarbam
Mecarbame
MS 1053
MS 1143
Muratox
Murfotox
Murotox
Murphotox
Murutox
P 474
Pennsalt TD-72
Pestan

2597-03-7
$C_{12}H_{17}O_4PS_2$
320.38
CCOC(=O)C(SP(=S)(OC)OC)c1ccccc1
Acetic acid, mercaptophenyl-, ethyl ester, S-ester with O,O-dimethyl phosphorodi thioate
Acetic acid, (O,O-dimethyldithiophosphorylphenyl)-, ethyl ester
Aimsan
Bay 33051
Bayer 18510
Benzeneacetic acid, α-((dimethoxyphosphinothioyl)thio)-, ethyl ester
Cidemul
Cidial
Cydeal
Dimefenthoat
Dimephenthioate
Dimephenthoate
Dimethenthoate
O,O-Dimethyl S-(1-carboeth-

oxybenzyl) dithiophosphate
O,O-Dimethyl S-α-ethoxycarbonylbenzyl phosphorodithioate
O,O-Dimethyl-S-(phenylacetic acid ethyl ester) phosphorodithioate
O,O-Dimethyl S-(phenyl)(carboethoxy)methyl phosphorodithioate
(Dimethyl S-(phenylethoxycarbonylmethyl)phosphorothiolothionate)
Elsan
ENT 23,438
ENT 27,386
ENT 27,386GC
S-α-Ethoxycarbonylbenzyl O,O-dimethyl phosphorodithioate
S-α-Ethoxycarbonylbenzyl dimethyl phosphorothiolothionate
Ethyl α-((dimethoxyphosphenothioyl)thio)benzeneacetate
Ethyl O,O-dimethyl phosphorodithioylphenyl acetate
Ethyl ester of O,O-dimethyldithiophosphoryl α-phenyl acetate acid
Ethyl mercaptophenylacetate-O,O-dimethylphosphorodithioate
Fenthoate
L-561
Montecatini L-561
NSC-190978
OMS 1075
PAP
Papthion
Phendal
Phenthoate
Rogodial
S 2940
Tanone
TH 346-1
Tsidial

2597-93-5
$C_{11}H_7Cl_6HgNO_2$
598.48
CC[Hg]N1C(=O)C2C(C1=O)C3(Cl)C(=C(Cl)C2(Cl)C3(Cl)Cl)Cl
Mercury, ethyl(1,4,5,6,7,7-hexachloro-5-norbornene-2,3-dicarboximidato)
50-CS-46
EMMI
Ethylmercurichlorendimide
N-(Ethylmercuri)-1,4,5,6,7,7-hexachlorobicyclo(2.2.1)hept-5-ene-2,3-dicarboximide
N-Ethylmercuri-3,4,5,6,7,7-hexachloro-3,6-endomethylene-1,2,3,6-tetrahydrophthalimide
N-Ethylmercuri-1,2,3,6-tetrahydro-3,6-endomethano-3,4,5,6,7,7-hexachlorophthalimide
5-Norbornene-2,3-dicarboximide, 1,4,5,6,7,7-hexachloro-N-(ethylmercuri)-

2597-97-9
C_2H_3HgN
241.65
Mercury cyanomethyl
Chipcote
Chipcote 75
Methylmercuric cyanide
Methylmercury cyanide
Methylmercury nitrile

2599-01-1
$C_{30}H_{60}O_2$
452.81
O=C(OCCCCCCCCCCCCCCCC)CCCCCCCCCCCCC
Cetyl myristate
Hexadecyl tetradecanoate
Tetradecanoic acid, hexadecyl ester (9CI)

2599-11-3
$C_7H_{13}N_5O$
183.18
n(c(nc(n1)NCC)NCC)c1O
s-Triazin-2-ol, 4,6-bis(ethylamino)- (8CI)
4,6-Bis(ethylamino)-1,3,5-triazin-2(1H)-one
1,3,5-Triazin-2(1H)-one, 4,6-bis(ethylamino)- (9CI)

2602-46-2
$C_{32}H_{20}N_6O_{14}S_4.4Na$
932.78
[Na+].[Na+].[Na+].[Na+].Nc5cc(cc6cc(c(N=Nc1ccc(cc1)c4ccc(N=Nc3c(O)c2c(N)cc(cc2cc3S([O-])(=O)=O)S([O-])(=O)=O)cc4)c(O)c56
2,7-Naphthalenedisulfonic acid, 3,3'-((4,4'-biphenylylene)bis(azo))bis(5-amino-4-hydroxy-, tetrasodium salt
Airedale Blue 2BD
Aizen Direct Blue 2BH
Amanil Blue 2BX
Atlantic Blue 2B
Atul Direct Blue 2B
Azocard Blue 2B
Azomine Blue 2B
Belamine Blue 2B
Bencidal Blue 2B
Benzanil Blue 2B
Benzo Blue BBA-CF
Benzo Blue BBN-CF
Benzo Blue GS
Blue 2B
Blue 2B Salt
Brasilamina Blue 2B
Calcomine Blue 2B
Chloramine Blue 2B
Chlorazol Blue B
Chlorazol Blue BP
Chrome Leather Blue 2B
C.I. 22610
C.I. Direct Blue 6
C.I. Direct Blue 6, tetrasodium salt
Cresotine Blue 2B
Diacotton Blue BB
Diamine Blue 2B
Diamine Blue BB
Diaphtamine Blue BB
Diazine Blue 2B
Diazol Blue 2B
Diphenyl Blue 2B
Diphenyl Blue KF
Diphenyl Blue M2B
Direct Blue 6
Direct Blue A
Direct Blue 2B
Direct Blue BB
Direct Blue GS
Direct Blue K
Enianil Blue 2BN
Fenamin Blue 2B
Fixanol Blue 2B
Hispamin Blue 2B
Indigo Blue 2B
Kayaku Direct
Mitsui Direct Blue 2BN
Modr Prima 6 (Czech)
Naphtamine Blue 2B
NB2B
NCI-C54579
Niagara Blue 2B
Nippon Blue BB
Paramine Blue 2B
Phenamine Blue BB
Pheno Blue 2B
Pontamine Blue BB
Sodium diphenyl-4,4'-bis-azo-2''-8''-amino-1''-naphthol-3'',6''-disulphonate
Tertrodirect Blue 2B
Vondacel Blue 2B

2603-10-3
$C_8H_9NO_2$
151.16
COC(=O)Nc1ccccc1
Carbamic acid, phenyl-, methyl ester (9CI)
AI3-15077
Carbanilic acid, methyl ester (8CI)

2605-44-9
$C_{24}H_{48}O_4.Cd$
513.12
Lauric acid, cadmium salt (2:1)
Cadmium dilaurate
Cadmium dodecanoate
Cadmium laurate
Dodecanoic acid, cadmium salt (9CI)

2608-48-2
$C_{11}H_9NO_3$
203.20
5-(4-Nitrophenyl)-2,4-pentadienal

2609-46-3
$C_6H_7ClN_6O$
214.62
NC(=N)NC(=O)c1nc(Cl)c(N)nc1N
Amiloride
Amilorida (Spanish)
Amiloridum (Latin)

2610-05-1
$C_{34}H_{28}N_6O_{16}S_4.4Na$
996.88
6,8-Naphthalenedisulfonic acid, 3,3'-((3,3'-dimethoxy-4,4'-biphenylene)bis(azo))bis-(5-amino-4-hydroxy-, tetrasodium salt
Airedale Blue FFD
Amanil Sky Blue 6B
Amanil Sky Blue FF
Atlantic Resin Fast Blue
Atlantic Sky Blue 6B
Atlantic Sky Blue FF
Atul Direct Sky Blue FB
Azine Brilliant Blue 6B
Azocard Blue 6B
Belamine Sky Blue FF
Benzanil Sky Blue FF
Benzanil Supra Blue 2GN
Benzo Brilliant Blue 6BS
Brasilamina Sky Blue 6B
Brilliant Benzo Blue 6BA-CF
Calcodur Blue 6GFL
Calcodur Resin Fast Blue
Calcomine Sky Blue FF
Chicago Blue 6B
Chicago Sky Blue 6B
Chloramine Sky Blue FF
Chlorantine Fast Blue B5GL
Chlorazol Sky Blue FF
Chrome Leather Sky Blue
C.I. 24410
C.I. Direct Blue 1
C.I. Direct Blue 1, Tetrasodium salt
Cresotine Blue 6B
Diacotton Sky Blue 6B
Diaphtamine Blue BS
Diazine Sky Blue FF
Diazol Pure Blue 6B
Diphenyl Brilliant Blue FF
Direct Blue 1
Direct Blue 6B
Direct Blue 6BS
Direct Blue FF
Direct Blue FFN
Direct Bright Blue
Direct Brilliant Blue FF
Direct Brilliant Blue MFF
Direct Brilliant Sky Blue 6B
Direct Pure Blue 6B
Direct Pure Blue FF
Direct Sky Blue 6B
Direct Sky Blue 6BS
Direct Sky Blue FF
Direct Sky Blue Green Shade
Direct Sky Blue GS
Enianil Brilliant Blue FF
Fenamin Sky Blue 3F
Fixanol Sky Blue FF
Hispamin Sky Blue 6B
Ink Blue 6B
Japanol Brilliant Blue 6BKX
Kayaku Direct Sky Blue 6B
Lumicrease Blue 4GL
Lumicrease Sky Blue 6GUL
Mitsui Direct Brilliant Blue 6B
Modr Prima 1 (Czech)
Naphtamine Sky Blue DD
NCI-C61109
Niagara Sky Blue 6B
NSB6B
Nyanza Sky Blue 6B
Paper Blue 6B
Paramine Sky Blue FF
Phenamine Brilliant Blue 6B
Pheno Sky Blue 6BX
Pontamine Sky Blue
Pontamine Sky Blue 6BX
Pontamine Sky Blue 6BX Greenish
Pontamine Sky Blue 6x
Pure Sky Blue 6B
Pyrazol Fast Brilliant Blue VP
Shikiso Direct Sky Blue 6B
Sirius Supra Blue 4G
Sky Blue 6B
Solar Blue 4GL
Tertrodirect Blue FF
Vegentine Blue CSW
Vondacel Blue FF

2610-10-8
$C_{45}H_{32}N_{10}O_{21}S_6.6Na$
1379.08
C.I. Direct Red 80
CI 35780
Direct Red 80
Durazol Brillant Red BS
2-Naphthalenesulfonic acid, 7,7'-(carbonyldiimino)bis-(4-hydroxy-3-((2-sulfo-4-((4-sulfophenyl)azo)phenyl)-azo)-, hexasodium salt (9CI)
Picrosirius red
Saturn Red F 3B
Sirius Red F 3B
Sirius Red F 3BA

2610-11-9
$C_{29}H_{19}N_5O_8S_2.2Na$
675.63
2-Naphthalenesulfonic acid, 7-(benzoylamino)-4-hydroxy-3-((4-((4-sulfophenyl)azo)-phenyl)azo)-, disodium salt
Airedale Red KD
Aizen Primula Red 4BH
Amanil Fast Red 8BL
Amanil Fast Red 8BLW
Belamine Fast Red 8 BL
Benzanil Fast Red K
Benzo Fast Red 8BL
Bordeaux EMBL
Calcodur Red 8BL
Chloramine Fast Red 5BL
Chloramine Fast Red K
Chlorantine Fast Red
Chlorantine Fast Red 5B (6CI)
Chrome Leather Red 5B
C.I. 28160
C.I. Direct Red 81 (7CI)
Dialuminous Red 4B
Diaphtamine Light Red 4B
Diazine Fast Red 8BK
Diazo Light Red 8BD
Diazo Light Red 8B
Diphenyl Fast 5BL Supra I Red
Diphenyl Fast Red 5BL
Diphenyl Fast Red 5BLN
Direct Fast Red 5B
Direct Fast Red 8BL
Direct Fast Red 2S
Direct Lightfast Red 2S
Direct Light Red 4B
Direct Light Red 8B
Direct Light Red M 8BL
Direct Red 81
Durazol Red 2B
Durazol Red 2BP
Eliamina Red 8BL
Fastolite Red 8BL
Fastusol Red 4BA-CF
Fenaluz Red 4B
Helion Red 8B
Hispaluz Red 8BL
Japan Red No. 505
Paranol Fast Red 8BL
Pheno Fast Scarlet 9B
Pontamine Fast Red 8BLX
Pyrazol Fast Red 5BL
Pyrazol Fast Red 8BL
Pyrazoline Red 8BL
Saturn Red B
Sirius Red 4B
Sirius Red 4BA
Solantine Red 8BL
Solar Red B
Solius Red 4B
Sumilight Red 4B
Suprazo Red 4B
Suprexcel Red 8BL
Tertrodirect Fast Red 5B
Tetramine Fast Red 8B
Triantine Fast Red 4BN
Triantine Light Red 4BN

2611-00-9
$C_{14}H_{20}O_2$
220.31
O=C(OCC(CCC=C1)C1)C(CCC=C2)C2
3-Cyclohexenylmethyl 3-cyclohexenecarboxylate
3-Cyclohexene-1-carboxylic acid, 3-cyclohexen-1-yl-methyl ester (9CI)
3-Cyclohexene-1-carboxylic acid, 3-cyclohexen-1-yl methyl ester
3-Cyclohexenyl 3-cyclohexene 1-carboxylate
Diene 221

2611-82-7
$C_{20}H_{14}N_2O_{10}S_3.3Na$
607.51
[Na+].[Na+].[Na+].Oc2ccc1cc(cc(c1c2N=Nc3ccc(c4ccccc34)S([O-])(=O)=O)S([O-])(=O)

=O)S([O-])(=O)=O
1,3-Naphthalenedisulfonic acid, 7-hydroxy-8-((4-sulfo-1-naphthyl)azo)-, trisodium salt
Acidal Bright Ponceau 3R
Acid Brilliant Scarlet 3R
Acid Ponceau 4R
Acid Red 18
Acid Scarlet 3R
Acid Scarlet 4R
Acid Scarlet 3RZ
Acilan Scarlet V3R
Aizen Brilliant Scarlet 3RH
Atul Acid Scarlet 3R
Atul Scarlet F
Brilliant Ponceau 3R
Brilliant Ponceau 4R
Brilliant Ponceau 5R
Brilliant Ponceau 4RC
Brilliant Ponceau 4RC Specially Pure
Brilliant Ponceau 3RF
Brilliant Scarlet
Brilliant Scarlet 3R
Brilliant Scarlet 4R
Bucacid Brilliant Scarlet 3R
Calcocid Brilliant Scarlet 3RN
Certicol Ponceau 4RS
Cerven kosenilova A (Czech)
Cerven Kysela 18 (Czech)
Cerven Potravinarska 7 (Czech)
Cilefa Ponceau 4R
Coccine
Coccine Nouvelle (French)
Coccin Red
Cochenillerot A (German)
Cochineal Red A
Cochineal Red A Specially Pure
Cochineal Red 4R
Colacid Ponceau 4R
C.I. 185
C.I. 16255
C.I. Acid Red 18
C.I. Acid Red 18, Trisodium salt
C.I. Food Red 7
Crimson SX
Curol Bright Red 4R
Daishiki Brilliant Scarlet 3R
E 124
Edicol Supra Ponceau 4R
Eurocert Cochineal Red A
Fenazo Scarlet 3R
Food Red 6
Food Red 7
Food Red No. 102
HD Ponceau 4R
HD Ponceau 4R Supra
Hexacol Ponceau 4R
Hidacid Fast Scarlet 3R
Hispacid Brilliant Scarlet 3RF
Java Scarlet 3R
Kayaku Acid Brilliant Scarlet 3R
Kayaku Food Colour Red No. 102
Kiton Scarlet 4R
Kochineal Red A for Food
L-Rot 4 (German)
Naphthalene Ink Scarlet 4R
Naphthalene Scarlet 4R
Naphthalene Scarlet 4RS
Neklacid Red 3R
Neklacid Red 4R
Neucoccin (German)
New Coccin
New Coccine
New Coccine Extra Conc. A Export
New Coccine Extra Pure A
Ponceau 4R
Ponceau 4R Aluminum Lake
Ponceau 4RE
Ponceau 4RE.FQ
Ponceau 4RF
Ponceau 4RT
Pontacyl Scarlet RR
Purple Red
Purple SX
Rakuto Brilliant Scarlet 3R
1578 Red
Rouge de Cochenille A (French)
San-Ei Brilliant Scarlet 3R
Schultz Nr. 213 (German)
Strawberry Red A Geigy
Sugai Brilliant Scarlet 3R
1-(4-Sulpho-1-naphthylazo)-2-naphthol-6,8-disulphonic acid, trisodium salt
SX Purple
Symulon Acid Brilliant Scarlet 3R
Takaoka Brilliant Scarlet 3R
Victoria Scarlet 3R
Victoria Scarlet Red
Victoria Scharlach 4 R Extra (German)

2612-35-3
$C_4H_6Cl_2O_2$
157.00
1,3-Dioxolane, 2-(dichloromethyl)- (8CI,9CI)

2613-76-5
$C_{12}H_{16}$
160.26
1H-Indene, 2,3-dihydro-1,1,3-trimethyl- (9CI)
Indan, 1,1,3-trimethyl- (8CI)
NSC-16797

2614-76-8
$C_3H_8O_4$
108.09
2,2-Dihydroperoxy propane (Not more than 25% with inert organic solid)
Hydroperoxide, (1-methylethylidene)bis-
UN 2178

2615-15-8
$C_{12}H_{26}O_7$
282.38
OCCOCCOCCOCCOCCOCCO
Hexaethylene-glycol
Hexaoxyethylene glycol
3,6,9,12,15-Pentaoxaheptadecane-1,17-diol

2617-47-2
CH_5O_4P
112.02
O=P(O)(O)CO
Phosphonic acid, (hydroxymethyl)- (9CI)
(Hydroxymethyl)phosphonic acid

2617-79-0
C_7H_6ClNO
155.59
Clc1ccc(NC=O)cc1
Formanilide, 4'-chloro
p-Chlorfenylamid kyseliny mravenci (Czech)
p-Chloroformanilide
4'-Chloroformanilide
N-(p-Chlorophenyl)formamide
N-(4-Chlorophenyl)formamide
Formamide, N-(4-chlorophenyl)- (9CI)
Formic acid p-chlorophenylamide

2618-77-1
$C_{22}H_{26}O_6$
386.48
O=C(OOC(CCC(OOC(=O)c(cccc1)c1)(C)C)(C)C)c(cccc2)c2
Peroxybenzoic acid, 1,1,4,4-tetramethyltetramethylene ester
Benzenecarboperoxoic acid, 1,1,4,4-tetramethyl-1,4-butanediyl ester (9CI)
Dimethyl-2,5 bis(benzoylperoxy)-2,5 hexane (French)
2,5-Dimethyl-2,5-di-(benzoylperoxy) hexane
2,5-Dimethyl-2,5-di-(benzoylperoxy)hexane, Not more than 82% with inert solid (DOT)
2,5-Dimethyl-2,5-di-(benzoylperoxy)hexane, Not more than 82% with water (DOT)
2,5-Dimethyl-2,5-di-(benzoylperoxy)hexane, Technically pure (DOT)
2,5-Dimetil-2,5-di-(benzoilperoxi)hexano (Spanish)
Luperox 118
Lupersol 118
Peroxybenzoic acid, 1,1,4,4-tetramethyltetramethylene ester (8CI)
Peroxybenzoic acid, 1,1,4,4-tetramethyltetramethylene ester, Not more than 82% with water
UN 2172 (DOT)
UN 2173 (DOT)
UN 2959 (DOT)
USP 711

2619-00-3
$C_{13}H_{13}NO_2$
215.25
Carbamic acid, dimethyl-, 1-naphthalenyl ester
AI3-27522

2620-53-3
$C_8H_8ClNO_2$
185.61
CNC(=O)Oc1ccc(Cl)cc1
4-Chlorophenyl-N-methylcarbamate
p-Chlorophenyl-N-methylcarbamate

2621-46-7
$C_9H_{11}Cl$
154.64
c(ccc(c1)Cl)(c1)C(C)C
Benzene, 1-chloro-4-(1-methylethyl)- (9CI)
1-Chloro-4-(1-methylethyl)-benzene

2622-21-1
C_8H_{12}
108.20
C(=CCCC1)(C=C)C1
Cyclohexene, 1-vinyl
1-Ethenylcyclohexene
1-Vinylcyclohexene

2622-89-1
$C_{12}H_{11}BO$
182.03
Diphenylborinic acid
Borinic acid, diphenyl-

2623-33-8
$C_{10}H_{11}NO_3$
193.20
CC(=O)Nc1ccc(OC(C)=O)cc1
Diacetamate
Acetamide, N-(4-(acetyloxy)-phenyl)- (9CI)
Acetanilide, 4'-hydroxy-, acetate (ester) (8CI)
AI3-17250
Diacetamato (Spanish)
Diacetamatum (Latin)

2623-36-1
$C_{12}H_{10}N_2O_4S.Na$
301.27
Benzenesulfonic acid, 4-((4-hydroxyphenyl)azo)-, monosodium salt (9CI)
Benzenesulfonic acid, p-((p-hydroxyphenyl)azo)-, monosodium salt (8CI)
Benzenesulfonic acid, p-((p-hydroxyphenyl)azo)-, sodium salt

2623-50-9
$C_{11}H_{11}N$
157.21
Quinoline, 5,8-dimethyl- (8CI,9CI)

2623-82-7
$C_8H_{15}BrO_2$
223.12
2-Bromooctanoate

2623-86-1
$C_4H_7BrO_2$
167.00
Butanoic acid, 3-bromo- (9CI)
Butyric acid, 3-bromo- (8CI)

2624-01-3
$C_7H_{13}BrO_2$
209.08
Heptanoic acid, 2-bromo- (8CI,9CI)

2624-17-1
$C_3H_3N_3O_3.Na$
152.08
s-Triazine-2,4,6(1H,3H,5H)-trione, monosodium salt
Acovenoside B
Monosodium cyanurate
NCI-C56542
Sodium cyanurate
Sodium isocyanurate
1,3,5-Triazine-2,4,6(1H,3H,5H)-trione, monosodium salt (9CI)

2627-06-7
$C_{16}H_{26}O_3S.Na$
321.44
Sodium p-decylbenzenesulfonate
Benzenesulfonic acid, p-decyl-, sodium salt
Benzenesulfonic acid, 4-decyl-, sodium salt (9CI)
4-Decylbenzenesulfonic acid, sodium salt
Na-C10LAS.
Sodium 4-decylbenzenesulfonate

2627-35-2
$C_{12}H_{27}O_4P$
266.32
O=P(OCCCCCCCCCCCC)(O)O
Phosphoric acid, monododecyl ester (9CI)
Lauryl dihydrogen phosphate
Monododecyl phosphate

2627-86-3
$C_8H_{11}N$
121.18
NC(c(cccc1)c1)C
Benzenemethanamine, α-methyl-, (S)- (9CI)
(S)-α-Methylbenzenemethanamine

2627-95-4
$C_8H_{18}OSi_2$
186.40
Disiloxane, 1,3-diethenyl-1,1,3,3-tetramethyl- (9CI)
Bisvinyltetramethyldisiloxane
1,3-Diethenyl-1,1,3,3-tetramethyl disiloxane
1,3-Diethenyl-1,1,3,3-tetramethyldisiloxane
Divinyltetramethyldisiloxane
1,3-Divinyltetramethyldisiloxane
1,1,3,3-Tetramethyl-1,3-divinyldisiloxane

2628-17-3
C_8H_8O
120.16
Oc(ccc(c1)C=C)c1
Phenol, p-vinyl
p-Hydroxystyrene
4-Hydroxystyrene
Phenol, 4-ethenyl- (9CI)
p-Vinylphenol
4-Vinylphenol

2630-10-6
$C_5H_8N_4O_2$
156.12
1,3,5-Triazine-2,4(1H,3H)-dione, 6-(ethylamino)- (9CI)
G 30888
s-Triazine-2,4-diol, 6-(ethylamino)- (8CI)

2631-37-0
$C_{12}H_{17}NO_2$
207.30
CNC(=O)Oc1cc(C)cc(c1)C(C)C
Carbamic acid, methyl-, m-cym-5-yl ester
Carbamic acid, N-methyl-, 3-methyl-5-isopropylphenyl ester
Carbamic acid, methyl-, 3-methyl-5-(1-methylethyl)-phenyl ester
Carbamult
m-Cym-5-yl methylcarbamate
ENT 27,300
ENT 27,300-A
EP 316
Methylcarbamic acid m-cym-5-yl ester
3-Methyl-5-isopropyl N-methylcarbamate
(3-Methyl-5-isopropylphenyl)-N-methylcarbamat (German)
3-Methyl-5-isopropylphenyl-N-methylcarbamate
3-Methyl-5-(1-methylethyl)-phenolmethylcarbamate

Minacide
Morton EP-316
OMS 716
Promecarb
Promecarbe
Schering 34615
SN 34615
T-32
UC 9880
Union Carbide UC-9880

2631-40-5
$C_{11}H_{15}NO_2$
193.27
O=C(Oc(c(ccc1)C(C)C)c1)NC
Carbamic acid, methyl-, o-cumenyl ester
Bay 39731
Bay 105807
Bayer 39731
Carbamic acid, methyl-, o-isopropylphenyl ester
Carbamic acid, methyl-, 2-(1-methylethyl)phenyl ester
ENT 25,670
Etrofolan
Hytox
Isoprocarb
Isoprocarbe
Isopropylphenol methylcarbamate
o-Isopropylphenyl N-methylcarbamate
2-Isopropyl-phenyl-N-methylcarbamate
KHE 0145
2-(1-Methylethyl)phenyl methylcarbamate
Mipc
Mipcin
Mipsin
OMS-32
Phenol, o-isopropyl-, methylcarbamate
PPC 3

2631-68-7
$C_6Cl_3N_3O_6$
316.44
O=N(=O)c(c(c(N(=O)=O)c(c1N(=O)=O)Cl)Cl)c1Cl
Benzene, 1,3,5-trichloro-2,4,6-trinitro
Bulbosan
TCTNB
Trichloro-1,3,5-trinitrobenzene
sym-Trichlorotrinitrobenzene
1,3,5-Trichloro-2,4,6-trinitrobenzene

2633-54-7
$C_9H_{10}Cl_3O_3PS$
335.57
CCOP(=S)(OC)Oc1cc(Cl)c(Cl)cc1Cl
Phosphorothioic acid, O-ethyl O-methyl O-(2,4,5-trichlorophenyl) ester
O,O-Dimethyl 2-ethylmercaptoethyl thiophosphate
O-Methyl O-ethyl O-2,4,5-trichlorophenyl thiophosphate
Pantozol 2
TCM 3
Trichlormetafos-3
Trichlorol
Trichlorometaphos-3
Trichloro-3-methaphos

2634-33-5
C_7H_5NOS
151.18
O=C(NSc1cccc2)c12
1,2-Benzisothiazol-3(2H)-one (9CI)
1,2-Benzisothiazolin-3-one (8CI)
Caswell No. 079A
Caswell No. 513A
EPA Pesticide Chemical Code 098901
IPX
Proxan
Proxel
Proxel PL
Proxel Press Paste

2636-26-2
$C_9H_{10}NO_3PS$
243.23
COP(=S)(OC)Oc1ccc(C#N)cc1
Phosphorothioic acid, O,O-dimethyl ester, O-ester with p-hydroxybenzonitrile
Bay 34727
Bayer 34727
Ciafos
O-p-Cyanophenyl O,O-dimethyl phosphorothioate
O-(4-Cyanophenyl) O,O-dimethyl phosphorothioate
Cyanophos
Cyanox
Cyap
O,O-Dimethyl-O-(4-cyanophenyl)-monothiophosphat (German)
O,O-Dimethyl-O-p-cyanophenyl phosphorothioate
O,O-Dimethyl O-4-cyanophenyl phosphorothioate
O,O-Dimethyl O-4-cyanophenyl thiophosphate
ENT 25,675
May & Baker S-4084
OMS 226
OMS 869
Phosphorothioic acid O-(4-cyanophenyl)O,O-dimethyl ester
S 4084
Sumitomo S 4084
Sunitomo S 4084

2638-94-0
$C_{12}H_{16}N_4O_4$
280.32
O=C(O)CCC(N=NC(C#N)(CCC(=O)O)C)(C#N)C
Valeric acid, 4,4'-azobis-(4-cyano
Azobis(cyanovaleric acid)
4,4'-Azobis(4-cyanovaleric acid)
Kyselina 4,4'-azo-bis-(4-kyanvalerova) (Czech)
Pentanoic acid, 4,4'-azobis-(4-cyano- (9CI)

2639-63-6
$C_{10}H_{20}O_2$
172.30
O=C(OCCCCCC)CCC
Butyric acid, hexyl ester
Hexyl butanoate
n-Hexyl butanoate
n-Hexyl n-butanoate
Hexyl butyrate
n-Hexyl butyrate
1-Hexyl butyrate

2642-71-9
$C_{12}H_{16}N_3O_3PS_2$
345.40
CCOP(=S)(OCC)SCn2nnc1ccccc1c2=O
Phosphorodithioic acid, O,O-diethyl ester, S-ester with 3-(mercaptomethyl)-1,2,3-benzotriazin-4(3H)-one
Athyl-Gusathion
Azinfos-Ethyl (Dutch)
Azinophos-Ethyl
Azinos
Azinphos-Aethyl (German)
Azinphos Ethyl
Azinphos-Etile (Italian)
Bay 16255
Bayer 16259
Benzotriazine derivative of an ethyl dithiophosphate
Bionex
Cotnion-Ethyl
Crysthion
O,O-Diaethyl-S-(4-oxobenzotriazin-3-methyl)-dithiophosphat (German)
O,O-Diaethyl-S-((4-oxo-3H-1,2,3-benzotriazin-3-yl)-methyl)-dithiophosphat (German)
O,O-Diethyl S-(4-oxobenzotriazino-3-methyl)phosphorodithioate
O,O-Diethyl-S-(4-oxo-3H-1,2,3-benzotriazine-3-yl)-methyl-dithiophosphate
O,O-Diethyl-S-((4-oxo-3H-1,2,3-benzotriazin-3-yl)-methyl)-dithiofosfaat (Dutch)
O,O-Diethyl phosphorodithioate s-ester with 3-(mercaptomethyl)-1,2,3-benzotriazin-4(3H)-one
O,O-Dietil-S-((4-oxo-3H-1,2,3-benzotriazin-3-il)-metil)-ditiofosfato (Italian)
3,4-Dihydro-4-oxo-3-benzotriazinylmethyl O,O-diethyl phosphorodithioate

S-(3,4-Dihydro-4-oxo-1,2,3-benzotriazin-3-ylmethyl) O,O-diethyl phosphorodithioate
ENT 22,014
Ethyl gusathion
Ethyl guthion
Gusathion A
Gusathion H
Gusathion K
Guthion (Ethyl)
R 1513
Triazotion (Russian)

2642-80-0
$C_{14}H_{11}Cl_3$
285.60
Ethane, 2,2-bis(p-chlorophenyl)-1-chloro
2,2-Bis(p-chlorophenyl)-1-monochloroethane

2642-82-2
$C_{14}H_{12}Cl_2O$
267.16
Benzeneethanol, 4-chloro-β-(4-chlorophenyl)
2,2-Bis(p-chlorophenyl)ethanol
2,2-Bis(4-chlorophenyl)ethanol
2,2-Bis(4-chlorophenyl)-1-hydroxyethane
DDOH
p,p'-DDOH
DDOM
Ethanol, 2,2-bis(p-chlorophenyl)- (8CI)

2642-98-0
$C_{18}H_{13}N$
243.32
Nc3cc2c1ccccc1ccc2c4ccccc34
6-Chrysenamine
6-AMC
6-Aminochrysene
Chrysenex
Chrysonex

2644-70-4
$H_4N_2.ClH$
68.52
Hydrazine, hydrochloride
Hydrazine monochloride
Hydrazinium chloride
Hydrazinium monochloride

2645-07-0
$C_9H_8N_2O_5$
224.16
4-Nitrohippuric acid
para-Nitrohippuric acid

2646-17-5
$C_{17}H_{14}N_2O$
262.33
Oc(ccc(c1ccc2)c2)c1N=Nc(c(ccc3)C)c3
2-Naphthol, 1-(o-tolylazo)
A.F. Orange No. 2
Aizen Food Orange No. 2
Atul Oil Orange T
C.I. 12100
C.I. Solvent Orange 2
D & C Orange No. 2
Dolkwal Orange SS
Ext. D & C Orange No. 4
Fat Orange II
Fat Orange RR
FD & C Orange 2
FD & C Orange No. 2
Hexacol Oil Orange SS
1-((2-Methylphenyl)azo)-2-naphthalenol
Lacquer Orange V
Oil Orange Opel
Oil Orange O'pel
Oil Orange SS
Oil Orange TX
Oil Orange XO
Oleal Orange SS
Orange OT
Orange OT*
Orange 3R Soluble in Grease
Orange SS
Oranz Rozpoustedlova 2 (Czech)
Oranz SS (Czech)
Organol Orange 2R
Toluene-2-azonaphthol-2
o-Tolueno-azo-β-naphthol
1-(o-Tolylazo)-β-naphthol
1-o-Tolylazo-2-naphthol

2646-78-8
Unknown
Unknown
Triethylamine 2,4-dichlorophenoxyacetate

2648-59-1
$C_6H_{10}Cl_2O$
169.05
2-Hexanone, 1,1-dichloro- (7CI,8CI,9CI)

2648-60-4
$C_6H_{10}Cl_2O$
169.05
3-Hexanone, 4,4-dichloro- (7CI,8CI,9CI)
4,4-Dichloro-3-hexanone

2650-18-2
$C_{37}H_{36}N_2O_9S_3.2H_3N$
783.01
[NH4+].[NH4+].CCN(Cc1cccc(c1)S([O-])(=O)=O)c2ccc(cc2)C(=C3C=CC(C=C3)=[N+](CC)Cc4cccc(c4)S([O-])(=O)=O)c5ccccc5S([O-])(=O
Ammonium, ethyl(4-(p-(ethyl-(m-sulfobenzyl)amino)-α-(o-sulfophenyl)benzylidene)- 2,5-cyclohexadien-1-ylidene)(m-sulfobenzyl)-, hydroxide, inner salt, diammonium salt
Acid Blue 9
Acilan Turquoise Blue AE
A.F. Blue No. 1
Aizen Brilliant Blue FCF
Alphazurine
Alphazurine FG
Alphazurine FGND
Alphazurine (Indicator)
Amacid Blue FG Conc
Bleu Brilliant FCF
11388 Blue
Brilliant Blue
Brilliant Blue FCF
Bucacid Azure Blue
Calcocid Blue EG
Calcocid Blue 2 G
C.I. 671
C.I. 42090
C.I. Acid Blue 9
C.I. Acid Blue 9, Diammonium salt
C.I. Direct Brown 78
C.I. Direct Brown 78, Diammonium salt
C.I. Food Blue 2
D & C Blue No. 4
Disulphine Lake Blue EG
Edicol Supra Blue E6
Erioglaucine
Erioglaucine A
Erioglaucine E
Eriosky Blue
Fenazo Blue XR
Food Blue 1
Hidacid Azure Blue
H.K. Formula No. K. 7117
Kiton Blue AR
Kiton Pure Blue L
Maple Brilliant Blue FCF
Neptune Blue BRA
Neptune Blue BRA Concentration
Patent Blue AE
Patent Blue 2Y
Peacock Blue X-1756
Schultz No. 770
Triantine Light Brown 3RN
Xylene Blue VSG

2652-13-3
$C_{20}H_{60}O_8Si_9$
681.46
Nonasiloxane, eicosamethyl- (9CI)
Eicosamethylnonasiloxane

2652-25-7
$C_{26}H_{24}O_6$
432.47
6,12-Methano-12H-dibenzo-(d,g)(1,3)dioxocin-3-ol, 8-methoxy-6-(p-methoxyphenyl)-13-methyl-, acetate (8CI)

2654-57-1
$C_{10}H_{12}N_2O$

176.21
O=C(NN(c(cccc1)c1)C2)C2C
3-Pyrazolidinone, 4-methyl-1-phenyl- (9CI)
AI3-60150
4-Methyl-1-phenyl-3-pyrazolidinone
1-Phenyl-4-methyl-3-pyrazolidone

2654-58-2
$C_{11}H_{14}N_2O$
190.23
O=C(NN(c(cccc1)c1)C2)C2(C)C
3-Pyrazolidinone, 4,4-dimethyl-1-phenyl- (9CI)
4,4-Dimethyl-1-phenyl-3-pyrazolidinone

2655-14-3
$C_{10}H_{13}NO_2$
179.24
CNC(=O)Oc1cc(C)cc(C)c1
Carbamic acid, methyl-, 3,5-xylyl ester
Cosban
3,5-Dimethylphenyl N-methylcarbamate
DRC 3340
H-69
Macbal
Maqbarl
Methylcarbamic acid 3,5-xylyl ester
XMC
3,5-XMC
3,5-Xylenol, methylcarbamate
3,5-Xylenyl N-methylcarbamate
3,5-Xylylester kyseliny methylkarbaminove (Czech)
3,5-Xylyl-N-methylcarbamate

2655-15-4
$C_{11}H_{15}NO_2$
193.24
2,3,5-Trimethylphenyl methylcarbamate
AI3-27096
Carbamic acid, methyl-, 2,3,5-trimethylphenyl ester
2,3,5-Landrin
Landrin B
Phenol, 2,3,5-trimethyl-, methylcarbamate
SD 8786
2,3,5-Trimethyl-phenol methylcarbamate

2655-19-8
$C_{16}H_{25}NO_2$
263.42
CNC(=O)Oc1cc(cc(c1)C(C)(C)C)C(C)(C)C
Carbamic acid, methyl-, 3,5-di-t-butylphenyl ester
BTS 14639
Butacarb
Butacarbe (French)
3,5-Di-t-butylphenylmethylcarbamate
RD 14639

2657-00-3
$C_{10}H_6N_2O_4S.Na$
273.21
1-Naphthalenesulfonic acid, 6-diazo-5,6-dihydro-5-oxo-, sodium salt
AI3-62933
2-Diazo-1-naphthol-5-sulfonic acid sodium salt
2-Diazo-1-naphthone-5-sulfonic acid, sodium salt

2657-89-8
$C_{16}H_{12}N_2O_{10}S_3$
488.47
1,3-Naphthalenedisulfonic acid, 7-hydroxy-8-((4-sulfophenyl)azo)- (9CI)
1,3-Naphthalenedisulfonic acid, 7-hydroxy-8-((p-sulfophenyl)azo)- (7CI,8CI)
Orange AB

2658-24-4
C_4H_9N
71.14
Aziridine, 2,2-dimethyl
2,2-Dimethylaziridine
2,2-Dimethylethylenimine

2664-42-8
$C_{20}H_{39}NO$
309.53
O=C(N(C)C)CCCCCCCC=CCCCCCCCC
Oleamide, N,N-dimethyl- (8CI)
AI3-09125
AI3-26663-X
(Z)-N,N-Dimethyl-9-octadecenamide
N,N-Dimethyloleamide
9-Octadecenamide, N,N-dimethyl-, (Z)- (9CI)
9-Octadecenamide, N,N-dimethyl-, (Z)-, (80%) and related amides

2664-63-3
$C_{12}H_{10}O_2S$
218.28
Oc(ccc(Sc(ccc(O)c1)c1)c2)c2
Phenol, 4,4'-thiodi
DFS
4,4'-Dihydroxydiphenyl sulfide
4,4'-Dioxydiphenylsulfide
Sulfide, bis(4-hydroxyphenyl)
4,4'-Thiodiphenol

2665-13-6
$C_{12}H_{24}B_2O_6$
285.94
Biobor JF
1,3-Butanediol, cyclic ester with boric acid (H3BO3), 1-methyltrimethylene ester (8CI)
1,3,2-Dioxaborinane, 2,2'-((1-methyl-1,3-propanediyl)bis(oxy))bis(4-methyl- (9CI)
2,2'-(1-Methyltrimethylenedioxy)bis(4-methyl-1,3,2-dioxaborinane)

2665-30-7
$C_{13}H_{12}NO_4PS$
309.28
Phosphonothioic acid, methyl-, O-(4-nitrophenyl) O-phenyl ester (9CI)
AI3-25787
Colep
CP 40294
ENT 25,787
O-Fenyl-O-p-nitrofenylester kyseliny methylthiofosfonove (Czech)
Monsanto CP-40294
O-(4-Nitrophenyl) O-phenylmethylphosphonothioate
Phosphonothioic acid, methyl-, O-(p-nitrophenyl) O-phenyl ester (8CI)

2666-14-0
$C_2H_5O_7P_2.3Na$
271.98
[Na+].OC(COP(O)O)OP(O)O.[Na][Na]
Diphosphonic acid, (1-hydroxyethylidene)-, trisodium salt
Ethane-1-hydroxy-1,1-diphosphonic acid, trisodium salt
(1-Hydroxyethylidene)diphosphonic acid, trisodium salt
Trisodium etidronate

2666-17-3
$C_{23}H_{18}N_3O_6S.Na$
487.49
2-Anthracenesulfonic acid, 1-amino-9,10-dihydro-4-(p-(N-methylacetamido)-anilino)-9,10-dioxo-, monosodium salt
Acid Blue 41
Alizarine Blue A
Alizarine Blue AR
Alizarine Direct Blue AR
Alizarine Direct Blue ARA
Alizarine Sapphire AR
Anthralan Blue B
C.I. 62130
C.I. Acid Blue 41
C.I. Acid Blue 71
Diacid Light Blue BR
Erio Fast Blue BRL
Fenalan Blue B

Kayacyl Blue BR
Lanaperl Blue B
Lissamine Blue AR
Lissamine Ultra Blue AR
Modr Kysela 41 (Czech)
Nylomine Acid Blue B-B
Superan Blue AR
Unitertracid Light Blue AB

2668-24-8
$C_7H_5Cl_3O_2$
227.47
COc1cc(Cl)c(Cl)c(Cl)c1O
Phenol, 6-methoxy-2,3,4-trichloro
4,5,6-Trichloroguaiacol

2668-47-5
$C_{20}H_{26}O$
282.43
Oc(c(cc(c(cccc1)c1)c2)C(C)(C)C)c2C(C)(C)C
(1,1'-Biphenyl)-4-ol, 3,5-bis-(1,1-dimethylethyl)- (9CI)
3,5-Bis(1,1-dimethylethyl)-(1,1'-biphenyl)-4-ol

2669-32-1
$C_7H_{13}N_2O_4PS_3$
316.37
CCOc1nn(CSP(=S)(OC)OC)c(=O)s1
Phosphorodithioic acid, O,O-dimethyl ester, S-ester with 2-ethoxy- 4-(mercaptomethyl)-δ^2-1,3,4-thiadiazolin-5-one
O,O-Dimethyl-S-(5-ethoxy-1,3,4-thiadiazolinyl-3-methyl)-dithiophosphate
O,O-Dimethyl S-(5-ethoxy-1,3,4-thiadiazol-2(3H)-onyl-(3)-methyl)dithiophosphate
O,O-Dimethyl-S-(5-ethoxy-1,3,4-thiadiazol-2(3H)-onyl-(3)-methyl)phosphorodithioate
ENT 27,238
S-((5-Ethoxy-2-oxo-1,3,4-thiadiazol-3(2H)-yl)methyl) O,O-dimethyl phosphorodithioate
Geigy 12968
Geigy GS-12968
GS-12968
Lythidathion
NC 2962

2673-22-5
$C_{30}H_{58}O_7S.Na$
585.84
Ditridecyl sodium sulfosuccinate
Bis(tridecyl) sulfosuccinate, sodium salt
Butanedioic acid, sulfo-, 1,4-ditridecyl ester, sodium salt (9CI)
Succinic acid, sulfo-, 1,4-ditridecyl ester, sodium salt (8CI)

2675-77-6
$C_8H_8Cl_2O_2$
207.06
COc1cc(Cl)c(OC)cc1Cl
Benzene, 1,4-dichloro-2,5-dimethoxy
Chloroneb
Chloronebe (French)
Demasan
Demosan
1,4-Dichloro-2,5-dimethoxybenzene
Soil Fungicide 1823
Tersan-SP

2675-80-1
$C_9H_{11}ClO_3$
202.64
Benzene, 5-chloro-1,2,3-trimethoxy- (8CI,9CI)

2678-21-9
$C_6HCl_3N_2O_4$
271.43
Benzene, 1,2,4-trichloro-3,5-dinitro- (9CI)
AI3-23491

2679-87-0
$C_6H_{14}O$
102.18
O(C(CC)C)CC
Butane, 2-ethoxy- (9CI)
Ether, sec-butyl ethyl
2-Ethoxybutane

2680-03-7
C_5H_9NO
99.15
O=C(N(C)C)C=C
Acrylamide, N,N-dimethyl
N,N-Dimethylacrylamide
Dimethylamid kyseliny akrylove (Czech)
2-Propenamide, N,N-dimethyl-(9CI)

2682-20-4
C_4H_5NOS
115.15
Methylisothiazolinone
Caswell No. 572A
EPA Pesticide Chemical Code 107104
3(2H)-Isothiazolone, 2-methyl-(9CI)
4-Isothiazolin-3-one, 2-methyl-(8CI)
Methylisothiazolinone
2-Methyl-4-isothiazolin-3-one
2-Methyl-3(2H)-isothiazolone

2683-43-4
$C_6H_4Cl_2N_2O_2$
207.00
Nc1c(Cl)cc(Cl)cc1N(=O)=O
Benzenamine, 2,4-dichloro-6-nitro- (9CI)
Aniline, 2,4-dichloro-6-nitro-(8CI)

2686-99-9
$C_{11}H_{15}NO_2$
193.27
CNC(=O)Oc1cc(C)c(C)c(C)c1
Carbamic acid, methyl-, 3,4,5-trimethylphenyl ester
ENT 25,843
Landrin
Methylcarbamic acid 3,4,5-trimethylphenyl ester
OMS-597
SD 8530
Shell SD-8530
3,4,5-Trimethylfenylester kyseliny methylkarbaminove (Czech)
3,4,5-Trimethylphenyl methylcarbamate

2687-25-4
$C_7H_{10}N_2$
122.19
Nc(c(ccc1)C)c1N
Toluene-2,3-diamine
1,2-Benzenediamine, 3-methyl-(9CI)
2,3-Diaminotoluene
2,3-Toluylenediamine
2,3-Tolylenediamine

2687-91-4
$C_6H_{11}NO$
113.18
O=C(N(CC1)CC)C1
2-Pyrrolidinone, 1-ethyl
N-Ethylpyrrolidinone
1-Ethyl-2-pyrrolidinone
N-Ethylpyrrolidone

2691-41-0
$C_4H_8N_8O_8$
296.20
O=N(=O)N(CN(N(=O)=O)CN(N(=O)=O)CN1N(=O)=O)C1
1,3,5,7-Tetrazocine, octahydro-1,3,5,7-tetranitro
Cyclotetramethylene tetranitramine, Dry (DOT)
Cyclotetramethylenetetranitramine
Cyclotetramethylenetetranitramine, Containing at least 10%-25% water (DOT)
Cyclotetramethylenetetranitramine, Desensitized with not less than 10% phlegmatizer (DOT)
Cyclotetramethylenetetranitramine (HMX; Octogen), Wetted with not less than 15 per

cent water, by mass [UN 0226]
HMX (DOT)
β HMY
HW 4
LX 14-0
Octogen
Oktogen
Tetramethylenetetranitramine
UN 0226 [Cyclotetramethylenetetranitramine (HMX; Octogen), Wetted with not less than 15 per cent water, by mass]

2694-54-4
$C_{18}H_{18}O_6$
330.34
O=C(OCC=C)c(ccc(c1C(=O)OC C=C)C(=O)OCC=C)c1
1,2,4-Benzenetricarboxylic acid, tri-2-propenyl ester (9CI)
Triallyl trimellitate
Tri-2-propenyl 1,2,4-benzenetricarboxylate

2695-37-6
$C_8H_8O_3S.Na$
207.21
Benzenesulfonic acid, 4-ethenyl-, sodium salt (9CI)
Sodium 4-ethenylbenzenesulfonate

2696-43-7
Unknown
Unknown
Nonyl methacrylate

2696-84-6
$C_9H_{13}N$
135.23
Nc(ccc(c1)CCC)c1
Aniline, 4-propyl
1-Amino-4-propylbenzene
Aniline, p-propyl- (8CI)
Benzenamine, 4-propyl- (9CI)
p-Propylaniline
p-n-Propylaniline
4-Propylaniline
4-n-Propylaniline

2696-92-6
ClNO
65.46
Nitrosyl chloride [UN 1069]
Nitrogen oxychloride
UN 1069 [Nitrosyl chloride]

2698-40-0
Unknown
Unknown
Isopropyl 2-methyl-4-chlorophenoxyacetate

2698-41-1
$C_{10}H_5ClN_2$
188.62
N#CC(C#N)=Cc(c(ccc1)Cl)c1
Malononitrile, o-chlorobenzylidene
(o-Chlorobenzal)malononitrile
2-Chlorobenzalmalononitrile
o-Chlorobenzylidene malonitrile
o-Chlorobenzylidene malononitrile (ACGIH,OSHA)
2-Chlorobenzylidene malononitrile
2-Chlorobmn
CS
β,β-Dicyano-o-chlorostyrene
NCI-C55118
Propanedinitrile, ((2-chlorophenyl)methylene)
USAF KF-11

2699-79-8
F_2O_2S
102.06
FS(F)(=O)=O
Sulfuryl-fluoride
Fluorure de sulfuryle (French)
Sulfuric oxyfluoride
Sulfuryl fluoride (ACGIH, OSHA) [UN 2191]
Sulphuryl fluoride (DOT)
UN 2191 [Sulfuryl fluoride]
Vikane
Vikane Fumigant

2702-72-9
$C_8H_5Cl_2O_3.Na$
243.02
[Na+].[O-]C(=O)COc1ccc(Cl) cc1Cl
Acetic acid, (2,4-dichlorophenoxy)-, sodium salt
Agrion
2,4-Dichlorophenoxyacetic acid, sodium salt
Diconirt
Diconirt D
Dikonirt
Dikonirt D
2,4-D sodium salt
Fernoxene
Hormit
Pielika (Polish)
Pielik E
Sodium 2,4-D
Sodium 2,4-dichlorophenoxyacetate
Spray-Hormite
Spritz-Hormit

2703-13-1
$C_{10}H_{15}O_2PS_2$
262.33
Phosphonothioic acid, meth yl-, O-ethyl O-(4-(methylthio)phenyl) ester
AI3-25612
Bayer 29952
ENT 25,612

2704-78-1
$C_{10}H_{16}O_2$
168.24
Cyclobutaneacetaldehyde, 3-acetyl-2,2-dimethyl- (8CI, 9CI)
NSC-46246

2705-87-5
$C_{12}H_{20}O_2$
196.32
O=C(OCC=C)CCC(CCCC1)C1
Cyclohexanepropionic acid, allyl ester
Allyl cyclohexanepropionate
3-Allylcyclohexyl propionate
Allyl hexahydrophenylpropionate
Cyclohexanol, 3-allyl-, propionate

2712-83-6
$C_{10}H_5F_7N_2O_4$
350.14
O=C(Nc(c(O)cc(N(=O)=O)c1)c1) C(F)(F)C(F)(F)C(F)(F)F
Butanamide, 2,2,3,3,4,4,4-heptafluoro-N-(2-hydroxy-4-nitrophenyl)- (9CI)
Butyranilide, 2,2,3,3,4,4,4-heptafluoro-2'-hydroxy-4'-nitro-
2,2,3,3,4,4,4-Heptafluoro-2'-hydroxy-4'-nitrobutyranilide

2717-15-9
$C_{18}H_{34}O_2.C_6H_{15}NO_3$
431.65
TEA-oleate
Belloid FR
Caswell No. 887D
EPA Pesticide Chemical Code 079025
Ethanol, 2,2',2''-nitrilotri-, Mixed with oleic acid (1:1)
Ethanol, 2,2',2''-nitrilotri-, oleate (Salt)
Lavon
9-Octadecenoic acid, Compd. with 2,2',2''-nitrilotris-(ethanol) (1:1)
9-Octadecenoic acid (Z)-, Compd. with 2,2',2''-nitrilotris(ethanol) (1:1) (9CI)
Oleic acid, Compd with 2,2', 2''-nitrilotriethanol (1:1)
Oleic acid, triethanolamine salt
Oleic acid, triethanolamine soap
TEA-oleate
Triethanolamine oleate
Triethanolamine oleic acid salt
Triethanolamine salt of oleic acid
Triethanolammonium oleate
Trietol

2719-08-6
$C_{10}H_{11}NO_3$
193.20
O=C(Nc(c(ccc1)C(=O)OC)c1)C
Anthranilic acid, N-acetyl-, methyl ester (8CI)
Benzoic acid, 2-(acetylamino)-, methyl ester (9CI)
Methyl 2-(acetylamino)benzoate
Methyl N-acetylanthranilate

2719-52-0
$C_{11}H_{16}$
148.25
Benzene, (1-methylbutyl)- (8CI, 9CI)

2719-61-1
$C_{18}H_{30}$
246.44
Benzene, (1-methylundecyl)- (9CI)
Dodecane, 2-phenyl- (8CI)

2719-62-2
$C_{18}H_{30}$
246.44
Benzene, (1-pentylheptyl)- (9CI)
Dodecane, 6-phenyl- (8CI)

2719-63-3
$C_{18}H_{30}$
246.44
Benzene, (1-butyloctyl)- (9CI)
Dodecane, 5-phenyl- (8CI)

2719-64-4
$C_{18}H_{30}$
246.44
Benzene, (1-propylnonyl)- (9CI)
Dodecane, 4-phenyl- (8CI)

2720-73-2
$C_6H_{11}OS_2.K$
202.39
Carbonic acid, dithio-, O-pentyl ester, potassium salt
Aeroxanthate
Amyl potassium xanthate
Carbonodithioic acid, O-pentyl ester, potassium salt (9CI)
Dithiocarbonic acid O-pentyl ester potassium salt
Potassium amylxanthate
Potassium amylxanthogenate
Potassium n-amylxanthogenate
Potassium pentylxanthate
Potassium pentyl xanthogenate
Xanthic acid, pentyl-, potassium salt

2724-69-8
$C_8H_{10}N_2S$
166.26
N(c(cccc1)c1)=C(S)NC
Urea, 1-methyl-3-phenyl-2-thio
N-Methyl-N'-phenyl thiourea

2734-52-3
$C_{16}H_{18}N_4O_4$
330.32
O=N(=O)c(ccc(N=Nc(ccc(N(CCO)CCO)c1)c1)c2)c2
Ethanol, 2,2'-((4-((4-nitrophenyl)azo)phenyl)imino)bis- (9CI)

2735-04-8
$C_8H_{11}NO_2$
153.20
O(c(c(N)ccc1OC)c1)C
Aniline, 2,4-dimethoxy
2,4-Dimethoxyaniline

2738-19-4
C_7H_{14}
98.19
2-Hexene, 2-methyl- (8CI,9CI)
NSC-73928

2744-49-2
$C_{15}H_{14}ClN_3O_2S$
335.80
O=S(=O)(N)c(ccc(N(N=C(c(ccc(c1)Cl)c1)C2)C2)c3)c3
Benzenesulfonamide, 4-(3-(4-chlorophenyl)-4,5-dihydro-1H-pyrazol-1-yl)- (9CI)
4-(3-(4-Chlorophenyl)-4,5-dihydro-1H-pyrazol-1-yl)benzenesulfonamide

2747-17-3
$C_6H_3Br_3O_2$
346.80
1,2-Benzenediol, 3,4,5-tribromo- (9CI)
Pyrocatechol, 3,4,5-tribromo- (7CI)
3,4,5-Tribromocatechol
3,4,5-Tribromopyrocatechol

2748-40-5
$C_{21}H_{15}Cl_3N_6$
457.75
Melamine, N2,N4,N6-tris-(p-chlorophenyl)- (7CI, 8CI)

2749-59-9
$C_5H_8N_2O$
112.12
3H-Pyrazol-3-one, 2,4-dihydro-2,5-dimethyl- (9CI)
NSC-304
2-Pyrazolin-5-one, 1,3-dimethyl- (8CI)

2752-95-6
$C_{16}H_{19}O_4P$
306.30
O=P(OCCCC)(Oc(cccc1)c1)Oc(cccc2)c2
Phosphoric acid, butyl diphenyl ester
AI3-07856
Butyl diphenyl phosphate

2754-17-8
$C_{12}H_{18}O_6$
258.27
Hexanedioic acid, bis(oxiranylmethyl) ester (9CI)
Adipic acid, bis(2,3-epoxypropyl) ester (8CI)

2756-56-1
$C_{13}H_{22}O_2$
210.32
O=C(OC(C(C(C1C2)(C)C)(C2)C)C1)CC
Bicyclo(2.2.1)heptan-2-ol, 1,7,7-trimethyl-, propanoate, exo- (9CI)
AI3-04363
Isoborneol, propionate (8CI)

2757-18-8
$C_3H_2O_4.2Tl$
510.81
Thallous malonate
Dithallium propanedioate
Formomalenic thallium
Malonic acid, thallium salt (1:2)
Propanedioic acid, dithallium salt (9CI)
Thallium malonate
Thallium(I) malonate

2757-90-6
$C_{12}H_{17}N_3O_4$
267.32
NC(CCC(=O)NNc1ccc(CO)cc1)C(O)=O
Glutamic acid, 5-(2-(α-hydroxy-p-tolyl)hydrazide), L
Agaritine
L-Glutamic acid, 5-(2-(4-(hydroxymethyl)phenyl)hydrazide) (9CI)
L-Glutamic acid, 5-(2-(α-hydroxy-para-tolyl)hydrazide)
β-N-(γ-L(+)Glutamyl)4-hydroxymethylphenylhydrazine
NCI-C08899

2758-18-1
C_6H_8O
96.13
O=C(C=C(C1)C)C1

2-Cyclopenten-1-one, 3-methyl- (9CI)
1-Methyl-1-cyclopenten-3-one
3-Methyl-2-cyclopenten-1-one

2758-42-1
Unknown
Unknown
Dimethylamine 4-(2,4-dichlorophenoxy)butyrate
Caswell No. 316C
2,4-DB-dimethylammonium
4-(2,4-Dichlorophenoxy)butyric acid dimethylamine salt
EPA Pesticide Chemical Code 030819

2759-54-8
$C_9H_{10}ClNO$
183.63
4-Chloropropionanilide
AI3-17673
Propanamide, N-(4-chlorophenyl)-

2759-71-9
$C_{10}H_9Cl_2NO$
230.10
Clc2ccc(NC(=O)C1CC1)cc2Cl
Cyclopropanecarboxanilide, 3',4'-dichloro
Cipromid
Clobber
Cypromid
Cypromide
3,4-Dichloranilid kyseliny cyklopropankarboxylove (Czech)
3',4'-Dichlorocyclopropanecarboxanilide
N-(3,4-Dichlorophenyl)cyclopropanecarboxamide
S-6000

2760-98-7
$C_8H_{10}N_4O_2$
194.17
O=C(NN)c(cccc1C(=O)NN)c1
Isophthalic acid, dihydrazide (8CI)
1,3-Benzenedicarboxylic acid, dihydrazide (9CI)

2763-96-4
$C_4H_6N_2O_2$
114.12
O=C(NOC=1CN)C1
3-Isoxazolol, 5-(aminomethyl)
Agarin
5-Aminomethyl-3-hydroxyisoxazole
5-(Aminomethyl)-3-isoxazolol
5-(Aminomethyl)-3(2H)-isoxazolone
5-Aminomethyl-3-isoxyzole
3-Hydroxy-5-aminomethylisoxazole
3-Hydroxy-5-aminomethylisoxazole-agarin
Muscimol
Pantherin
RCRA waste number P007

2764-13-8
$C_{25}H_{53}N_2O_2.NO_3$
475.69
1-Propanaminium, N-(2-hydroxyethyl)-N,N-dimethyl-3-((1-oxooctadecyl)amino)-, nitrate (Salt) (9CI)

2764-72-9
$C_{12}H_{12}N_2$
184.26
[Br].[Br].C2C[n+]1ccccc1c3cccc[n+]23
Dipyrido(1,2-a:2',1'-c)pyrazinediium, 6,7-dihydro
Diquat
1,1'-Ethylene-2,2'-bipyridylium ion

2765-04-0
$C_6H_{12}S_3$
180.36
1,3,5-Trithiane, 2,4,6-trimethyl (9CI)
NSC-227897
Thioacetaldehyde
s-Trithiane, 2,4,6-trimethyl- (8CI)
Trithioacetaldehyde

2765-11-9
$C_{15}H_{30}O$
226.40
O=CCCCCCCCCCCCCCC
Pentadecanal
Pentadecanal (9CI)

2768-02-7
$C_5H_{12}O_3Si$
148.26
COSi(OC)(OC)C=C
Silane, trimethoxyvinyl
Trimethoxyvinylsilane
Vinyl trimethoxy silane

2768-16-3
$C_{10}H_{22}O$
158.28
1-Nonanol, 5-methyl- (8CI,9CI)

2770-75-4
$C_3H_4N_4S_2$
160.20
n(c(nc(n1)N)S)c1S
s-Triazine-2,4-dithiol, 6-amino- (8CI)
6-Amino-1,3,5-triazine-2,4(1H,3H)-dithione
1,3,5-Triazine-2,4(1H,3H)-dithione, 6-amino- (9CI)

2772-45-4
$C_{24}H_{26}O$
330.47
Phenol, 2,4-bis(1-methyl-1-phenylethyl)- (9CI)
2,4-Bis(1-methyl-1-phenylethyl)phenol
2,4-Di(α-methylstyryl)phenol

2772-46-5
$C_8H_8Cl_2O_2$
207.06
Benzene, 1,2-dichloro-4,5-dimethoxy- (8CI,9CI)

2777-05-1
$C_{12}H_9ClN_2O_3S.Na$
319.71
Benzenesulfonic acid, 4-((4-chlorophenyl)azo)-, sodium salt (9CI)
Benzenesulfonic acid, p-((p-chlorophenyl)azo)-, sodium salt

2778-04-3
$C_9H_{13}O_6PS$
280.25
COc1coc(CSP(=O)(OC)OC)cc1=O
Phosphorothioic acid, O,O-dimethyl ester, S-ester with 2-(mercaptomethyl)-5-methoxy- 4H-pyran-4-one
AC-18,737
O,O-Dimethyl S-(5-methoxy-4-oxo-4H-pyran-2-yl)phosphorothioate
O,O-Dimethyl-S-((5-methoxypyron-2-yl)-methyl)-thiolphosphat (German)
O,O-Dimethyl S-(5-methoxypyronyl-2-methyl) thiophosphate
Endocid
Endocide
Endothion
ENT 24,653
Exothion
FMC 5767
5-Methoxy-2-(dimethoxyphosphinylthiomethyl)pyrone-4
S-5-Methoxy-4-oxopyran-2-ylmethyl dimethyl phosphorothioate
S-((5-Methoxy-4H-pyron-2-yl)-methyl)-O,O-dimethyl-monothiofosfaat (Dutch)
S-((5-Methoxy-4H-pyron-2-yl)-methyl)-O,O-dimethyl-monothiophosphat (German)
S-(5-Methoxy-4-pyron-2-ylmethyl) dimethyl phosphorothiolate
S-((5-Metossi-4H-piron-2-il)-metil)-O,O-dimetil-monotiofosfato (Italian)
NIA-5767
Niagara 5767
Phosphate 100

Phosphopyron
Phosphopyrone
7175 RP
Thiophosphate de O,O-dimethyle et de S-((5-methoxy-4-pyronyl)-methyle) (French)

2778-41-8
$C_{14}H_{16}N_2O_2$
244.28
Benzene, 1,4-bis(1-isocyanato-1-methylethyl)- (9CI)
Isocyanic acid, p-phenylenediisopropylidene ester
Isocyanic acid, α,α,α',α'-tetramethyl-p-xylylene ester (8CI)

2778-42-9
$C_{14}H_{16}N_2O_2$
244.28
Benzene, 1,3-bis(1-isocyanato-1-methylethyl)- (9CI)
1,3-Bis(1-isocyanato-1-methylethyl)benzene

2778-96-3
$C_{36}H_{72}O_2$
536.97
O=C(OCCCCCCCCCCCCCCCCCC)CCCCCCCCCCCCCCCCC
Octadecyl stearate
Octadecanoic acid, octadecyl ester (9CI)
Stearic acid, stearyl ester
Stearyl stearate

2781-00-2
$C_{20}H_{34}O_4$
338.49
O(OC(C)(C)C)C(c(ccc(c1)C(OOC(C)(C)C)(C)C)c1)(C)C
1,4-Bis(α-(t-butyldioxy)isopropyl)benzene
α,α'-Bis(tert-butyldioxy)-p-diisopropylbenzene
1,4-Bis(α-(tert-butyldioxy)isopropyl)benzene
α,α'-Bis(tert-butylperoxy)-p-diisopropylbenzene
α,α'-Bis(tert-butylperoxy)-1,4-diisopropylbenzene
Perkadox U 14/40
Perkadox 14/40C
Peroxide, (1,4-phenylenebis-(1-methylethylidene))bis(-(1,1-dimethylethyl) (9CI)
Peroxide, (p-phenylenediisopropylidene)bis(tert-butyl
Peroximon F 40
Peroximon F 100

2781-10-4
$C_{24}H_{48}O_4Sn$
519.41
Stannane, bis(2-ethylhexanoyloxy)dibutyl
Dibutyltin bis(α-ethylhexanoate)
Dibutyltin bis(2-ethylhexanoate)
Dibutyltin di(2-ethylhexanoate)
Di-n-butyltin di-2-ethylhexanoate
Dibutyltin di(2-ethylhexoate)
Stannane, dibutylbis((2-ethylhexanoyl)oxy)-
Stannane, dibutylbis((2-ethyl-1-oxohexyl)oxy)- (9CI)

2781-11-5
$C_9H_{22}NO_5P$
255.24
O=P(OCC)(OCC)CN(CCO)CCO
Diethyl ((diethanolamino)-methyl)phosphonate
Diethyl (N,N-bis(2-hydroxyethyl)amino)methanephosphonate
Diethyl ((bis(2-hydroxyethyl)-amino)methyl)phosphonate
Diethyl ((N,N-bis(2-hydroxyethyl)amino)methyl)phosphonate
Fyrol 6
Phosphonic acid, ((bis(2-hydroxyethyl)amino)methyl)-, diethyl ester (9CI)

2782-57-2
$C_3H_2Cl_2N_3O_3$
198.98
O=C(N(C(=O)NC1=O)Cl)N1Cl
s-Triazine-2,4,6(1H,3H,5H)-trione, 1,3-dichloro
ACL 70
CDB 60
Dichlorocyanuric acid
Dichloroisocyanurate
Dichloroisocyanuric acid
Dichloroisocyanuric acid, Dry [UN 2465]
FI Clor 71
Hilite 60
Isocyanuric acid, dichloro-
Isocyanuric dichloride
Kyselina dichlorisokyanurova (Czech)
Orced
Troclosene
UN 2465 [Dichloroisocyanuric acid, Dry or dichloroisocyanuric acid salts]

2782-91-4
$C_5H_{12}N_2S$
132.25
N(C(N(C)C)=S)(C)C
Urea, 1,1,3,3-tetramethyl-2-thio
NA 101
Tetramethylthiourea
1,1,3,3-Tetramethylthiourea
TMTU

2783-17-7
$C_{12}H_{28}N_2$
200.42
NCCCCCCCCCCCCN
1,12-Dodecanediamine
1,12'-Diaminodecane
Dodecamethylenediamine
1,12'-Dodecamethylenediamine
1,12'-Dodecanediamine
Dodecyldiamine
1,12'-Dodecylenediamine

2783-94-0
$C_{16}H_{10}N_2O_7S_2.2Na$
452.38
[Na+].[Na+].Oc2ccc1cc(ccc1c2N=Nc3ccc(cc3)S([O-])(=O)=O)S([O-])(=O)=O
2-Naphthalenesulfonic acid, 6-hydroxy-5-((p-sulfophenyl)azo)-, disodium salt
Acid Yellow TRA
A.F. Yellow No. 5
Aizen Food Yellow No. 5
Alabaster No. 3
Atul Sunset Yellow FCF
Canacert Sunset Yellow FCF
Certicol Sunset Yellow CFS
Certolake Sunset Yellow
C.I. 15985
C.I. Food Yellow 3
C.I. Food Yellow 3, Disodium salt
Cilefa Orange S
Disodium salt of 1-p-sulphophenylazo-2-naphthol-6-sulphonic acid
Dispersed Orange 11348
Dispersed Yellow 12116
Dolkwal Sunset Yellow
Dye FD & C Yellow Lake 6
Dye FD & C Yellow No. 6
Dye FD & C Yellow Lake 6
Dye FD & C Yellow No. 6
Dye Sunset Yellow
E 110
Edicol Supra Yellow FC
Eniacid Sunset Yellow
Eurocert Orange FCF
FD & C No. 6
FD & C Yellow 6
FD & C Yellow Lake No. 6
FD & C Yellow No. 6
FD & C Yellow No. 6 Aluminium Lake
FDC Yellow No. 6
Foodcol Sunset Yellow FCF
Food Yellow 3
Food Yellow 6
Gelborange-S (German)
HD Sunset Yellow FCF
HD Sunset Yellow FCF Supra
Hexacol Sunset Yellow F & F Supra
Hexacol Sunset Yellow FCF
Hexacol Sunset Yellow FCF Supra
Hexacol Sunset Yellow FCP
6-Hydroxy-5-((p-sulfophenyl)azo)-2-naphthalenesulfonic acid, disodium salt
6-Hydroxy-5-((4-sulfophenyl)-azo)-2-naphthalenesulfonic

acid, disodium salt
6-Hydroxy-5-((p-sulphophenyl)-azo)-2-naphthalenesulphonic acid, disodium salt
6-Hydroxy-5-((4-sulphophenyl)-azo)-2-naphthalenesulphonic acid, disodium salt
KCA Foodcol Sunset Yellow FCF
Jaune Orange S
Jaune Soleil
L-Orange 2
L. Orange Z2010
Maple Sunset Yellow FCF
NCI-C53907
Orange Pal
Orange II R
Orange RGL Conc. Specially Pure
Orange Yellow S
Orange Yellow S.AF
Orange Yellow S.FQ
Para Orange
Standacol Sunset Yellow FCF
1-p-Sulfophenylazo-2-hydroxynaphthalene-6-sulfonate, disodium salt
1-p-Sulfophenylazo-2-naphthol-6-sulfonic acid, disodium salt
1-p-Sulphophenylazo-2-naphthol-6-sulphonic acid, disodium salt
Sun Orange A Geigy
Sunset Yellow
Sunset Yellow BSS
Sunset Yellow FCF
Sunset Yellow FCF Supra
Sunset Yellow FU
Sunset Yellow FU Supra
Sunset Yellow Lake
Sun Yellow
Sun Yellow A-CE
Sun Yellow A-FDC
Sun Yellow Extra Conc. A Export
Sun Yellow Extra Pure A
Sun Yellow FCF
Usacert Yellow No. 6
Usacert FD & C Yellow No. 6
Usalake FD & C Yellow No. 6 Lake
1351 Yellow
1899 Yellow
Yellow No. 6
Yellow Orange S
Yellow Orange S Specially Pure
Yellow Orange Specially Pure 85
Yellow SF for Food
Yellow Sun
Yellow SY for Food
Zlut Potravinarska 3 (Czech)

2784-94-3
$C_{11}H_{17}N_3O_4$
255.31
O=N(=O)c(c(NC)ccc1N(CCO)CCO)c1
Ethanol, 2,2'-((4-(methylamino)-3-nitrophenyl)imino)di
N',N'-Bis(2-hydroxyethyl)-N-methyl-2-nitro-p-phenylenediamine
HC Blue 1
N-Methyl-amino-2-nitro-4-N',N'-bis-(2-hydroxyethyl)-aminobenzene
2,2'-((4-(Methylamino)-3-nitrophenyl)imino)diethanol
NCI-C04159
p-Phenylenediamine, N,N-bis-(2-hydroxyethyl)-N'-methyl-2-nitro-

2785-54-8
$C_{19}H_{34}N.Cl$
311.93
Pyridinium, 1-tetradecyl-, chloride (9CI)
Myristylpyridinium chloride
1-Tetradecylpyridinium chloride

2785-75-3
$C_8H_{10}O_2$
138.17
1,2-Benzenediol, 3,5-dimethyl- (9CI)
3,5-Dimethyl-1,2-benzenediol
3,5-Dimethylcatechol
3,5-Dimethylpyrocatechol
Pyrocatechol, 3,5-dimethyl-(6CI,7CI,8CI)

2785-76-4
$C_8H_{10}O_2$
138.17
3,4-Dimethylcatechol
1,2-Benzenediol, 3,4-dimethyl-

2785-87-7
$C_{10}H_{14}O_2$
166.24
O(c(c(O)ccc1CCC)c1)C
Phenol, 2-methoxy-4-propyl
Cerulignol
Coerulignol
Dihydroeugenol
Guaiacylpropane
4-Hydroxy-3-methoxypropylbenzene
2-Methoxy-4-propylphenol
p-Propylguaiacol
p-n-Propylguaiacol
4-Propylguaiacol
1-Propyl-3-methoxy-4-hydroxybenzene

2785-89-9
$C_9H_{12}O_2$
152.19
O(c(c(O)ccc1CC)c1)C
Phenol, 4-ethyl-2-methoxy- (9CI)
4-Ethyl-2-methoxyphenol

2786-71-2
$C_{20}H_{14}N_2O_5S$
394.40
O=C(c(c(C(=O)c1c(Nc(cccc2)c2)cc(S(=O)(=O)O)c3N)ccc4)c4)c13
2-Anthracenesulfonic acid, 1-amino-9,10-dihydro-9,10-dioxo-4-(phenylamino)- (9CI)
1-Amino-4-anilino-2-anthraquinonesulfonic acid
1-Amino-9,10-dihydro-9,10-dioxo-4-(phenylamino)-2-anthracenesulfonic acid

2786-76-7
$C_{26}H_{22}N_4O_4$
454.46
O=C(Nc(c(OCC)ccc1)c1)c(c(O)c(N=Nc(ccc(C(=O)N)c2)c2)c(c3ccc4)c4)c3
CI Pigment Red 120
4-((4-(Aminocarbonyl)phenyl)-azo)-N-(2-ethoxyphenyl)-3-hydroxy-2-naphthalenecarboxamide
C.I. Pigment Red 170
2-Naphthalenecarboxamide, 4-((4-(aminocarbonyl)phenyl)-azo)-N-(2-ethoxyphenyl)-3-hydroxy- (9CI)
2-Naphtho-o-phenetidide, 4-((p-carbamoylphenyl)azo)-3-hydroxy-
Permanent Red F 3RK70
Permanent Red F 5RK
Spent pulping liquor, Furfural polymer

2795-39-3
$C_8HF_{17}O_3S.K$
539.23
1-Octanesulfonic acid, 1,1,2,2,3,3,4,4,5,5,6,6,7,7,8,8,8-heptadecafluoro-, potassium salt (9CI)
AI3-50950
1,1,2,2,3,3,4,4,5,5,6,6,7,7,8,8,8-Heptadecafluoro-1-octanesulfonic acid, potassium salt
Perfluorooctanesulfonic acid, potassium salt

2797-51-5
$C_{10}H_6ClNO_2$
207.62
NC2=C(Cl)C(=O)c1ccccc1C2=O
1,4-Naphthoquinone, 2-amino-3-chloro
ACN
ACNQ
2-Amino-3-chloro-1,4-naphthoquinone
2-Chloro-3-amino-1,4-naphthoquinone
06K
Mogeton G
Mogeton Granule
1,4-Naphthalenedione, 2-amino-

3-chloro-
06K-Quinone
06K-50W

2797-59-3
$C_7H_8Cl_6N_6O_2$
420.86
Ethanol, 1,1'-((6-amino-1,3,5-triazine-2,4-diyl)diimino)-bis(2,2,2-trichloro- (9CI)
Ethanol, 1,1'-((6-amino-s-triazine-2,4-diyl)diimino)bis-(2,2,2-trichloro- (8CI)

2801-68-5
$C_{11}H_{17}NO_3.ClH$
247.75
O(c(c(cc(OC)c1)CC(N)C)c1)C
Isopropylamine, 2-(2,5-dimethoxyphenyl)-, hydrochloride
C 1739
2-(2,5-Dimethoxyphenyl)isopropylamine hydrochloride

2801-87-8
$C_{16}H_{34}$
226.45
Pentadecane, 4-methyl- (6CI,7CI,8CI,9CI)
4-Methylpentadecane

2806-45-3
C_9H_{12}
120.20
Cyclohexane, 2-propyn ylidene- (7CI,8CI,9CI)
Propargylidenecyclohexane
1-Propyne, 3-cyclohexylidene-
Propynylidenecyclohexane

2806-85-1
$C_5H_{10}O_2$
102.15
Propionaldehyde, 3-ethoxy
3-Ethoxypropanal
β-Ethoxypropionaldehyde
3-Ethoxypropionaldehyde
Propanal, 3-ethoxy- (9CI)

2807-30-9
$C_5H_{12}O_2$
104.17
O(CCO)CCC
Ethanol, 2-propoxy
Ektasolve EP
Ethylene glycol mono propyl ether
Ethylene glycol mono-n-propyl ether
Monopropyl ether of ethylene glycol
2-Propoxyethanol
Propyl cellosolve

2808-86-8
C_5HCl_4N
216.87
n(c(c(c(c1Cl)Cl)Cl)Cl)c1
Pyridine, 2,3,4,5-tetrachloro
2,3,4,5-Tetrachloropyridine

2809-21-4
$C_2H_8O_7P_2$
206.04
O=P(O)(O)C(O)(P(=O)(O)O)C
Phosphonic acid, (1-hydroxyethylidene)di
Dequest 2010
Dequest 2015
Dequest Z 010
EHDP
Ethane-1-hydroxy-1,1-diphosphonate
1,1,1-Ethanetriol diphosphonate
Etidronic acid
Ferrofos 510
HEDP
1-Hydroxy-1,1-diphosphonoethane
Hydroxyethanediphosphonic acid
1-Hydroxyethanediphosphonic acid
Oxyethylidenediphosphonic acid
Phosphonic acid, 1-hydroxy-1,1-ethanediyl ester
Phosphonic acid, (1-hydroxyethylidene)bis-
1000SL
Turpinal SL

2809-64-5
$C_{11}H_{14}$
146.23
Naphthalene, 1,2,3,4-tetra hydro-5-methyl- (6CI,7CI,8CI,9CI)
1-Methyl-5,6,7,8-tetrahydronaphthalene
5-Methyl-1,2,3,4-tetrahydronaphthalene
5-Methyltetralin
1,2,3,4-Tetrahydro-5-methylnaphthalene

2809-67-8
C_8H_{14}
110.20
2-Octyne (8CI,9CI)

2810-69-7
$C_8H_6Br_4$
421.77
Benzene, 1,2,3,4-tetrabromo-5,6-dimethyl- (9CI)
1,2-Dimethyltetrabromobenzene
o-Xylene, 3,4,5,6-tetrabromo- (6CI,7CI)
Tetrabromo-o-xylene

2812-72-8
C_2H_3ClOS
110.56
Carbonochloridothioic acid, O-methyl ester (9CI)
Formic acid, chlorothio-, O-methyl ester (6CI, 7CI)
Methoxythiocarbonyl chloride
O-Methyl chlorothioformate
O-Methyl thiochloroformate

2812-73-9
C_3H_5ClOS
124.59
Formic acid, chlorothio-, ethyl ester
Ethyl chlorothioformate [UN 2826]
UN 2826 [Ethyl chlorothioformate]

2813-95-8
$C_{12}H_{14}N_2O_6$
282.28
CCC(C)c1cc(cc(N(=O)=O)c1OC(C)=O)N(=O)=O
Acetic acid, 2-(sec-butyl)-4,6-dinitrophenyl ester
Acetic acid, (2,4-dinitro-6-s-butylphenyl) ester
Acetic acid, (4,6-dinitro-2-s-butylphenyl) ester
o-Acetyl-2-sec-butyl-4,6-dinitrophenol
Aretit
Aretit (The phenol)
2-sek.Butyl-4,6-dinitrofenylester kyseliny octove (Czech)
2-sec-Butyl-4,6-dinitrophenylacetate
6-sec-Butyl-2,4-dinitrophenylacetate
2,4-Dinitro-6-s-butylfenylester kyseliny octove (Czech)
2,4-Dinitro-6-sek.butyl-phenylacetat (German)
4,6-Dinitro-2-s-butylphenyl acetate
Dinoseb-acetate
Dinosebe acetate
HOE 2904
β-(2-Hydroxy-3,5-dinitrophenyl)butane acetate
Ivosit
2-(1-Methylpropyl)-4,6-dinitrophenyl acetate
Phenol, 2-sec-butyl-4,6-dinitro-, acetate (ester) (8CI)
Phenol, 2-(1-methylpropyl)-4,6-dinitro-, acetate (ester) (9CI)
Phenotan

2814-20-2

$C_8H_{12}N_2O$
152.22
n(c(O)cc(n1)C)c1C(C)C
4(3H)-Pyrimidinone, 2-isopropyl-6-methyl
G 27550
2-Isopropyl-4-methyl-6-hydroxypyrimidine
6-Pyrimidinol, 2-isopropyl-4-methyl

2814-77-9
$C_{16}H_{10}ClN_3O_3$
327.74
O=N(=O)c(ccc(N=Nc(c(c(ccc1)cc2)c1)c2O)c3Cl)c3
2-Naphthol, 1-((2-chloro-4-nitrophenyl)azo)
ADC Permanent Red Toner R
American Vermilion
Blazing Red
Carnelio Red R
1-((2-Chloro-4-nitrophenyl)azo)-2-naphthol
Chlorparanitraniline Red
C.I. 12085
C.I. Pigment Red 4
Dainichi Permanent Red RX
D & C Red No. 36
Duplex Permaton Red L 20-7022
Fastona Red R
Fast Orange 3R
Fast Orange 3RJ
Flame Tones
Flaming Red
Graphtal Red RL
Graphtol Red RL
Irgalite Red PRR
Isol Fast Red R
Isol Fast Red RG
Kromon Red R
Latexol Red J
Lincoln Red 1002
Lutetia Fast Orange 3R
Monolite Fast Red G
Monolite Fast Red GA
Monolite Fast Red GF
2-Naphtalenol, 1-((2-chloro-4-nitrophenyl)azo)-
No. 1 Forthfast Red R
Oralith Red
Permanent Red BFR
Permanent Red F
Permanent Red R
Permanent Red R Extra
Permanent Red RG Extra
Permanent Red Toner R
Permansa Red
Permaton Red XL 20-7015
Pigment Red 4
Pigment Ruby ZH
Pigment Scarlet ZH
Pyrotone Red Toner RA-5520
12094 Red
Red Extract R
Red No. 228
Rubber Red R Extra
Segnale Light Red PRG
Silopol Red G
Silosol Red GN
Siloton Red 2g
Syton Fast Red R
Tanager Red X-761
Tiger Orange
Versal Fast Red R
Vulcafix Red J
Vulcafor Orange R
Vulcan Red R

2815-58-9
C_8H_{16}
112.22
Cyclopentane, 1,2,4-trimethyl-(8CI,9CI)

2816-57-1
$C_8H_{14}O$
126.20
Cyclohexanone, 2,6-dimethyl-(9CI)

2818-16-8
$C_9H_7Cl_3O_3.K$
308.61
Propanoic acid, 2-(2,4,5-trichlorophenoxy)-, potassium salt (9CI)
Caswell No. 739L
EPA Pesticide Chemical Code 082503
Kurosal SL
Propionic acid, 2-(2,4,5-trichlorophenoxy)-, potassium salt (8CI)
Silvex, potassium salt
2-(2,4,5-Trichlorophenoxy)propionic acid potassium salt

2819-86-5
$C_{11}H_{16}O$
164.25
Oc(c(c(c(c1C)C)C)C)c1C
Pentamethylphenol
Phenol, pentamethyl- (9CI)

2820-51-1
$C_{10}H_{14}N_2.xClH$
417.48
Nicotine hydrochloride (d,l)
Chlorhydrate de nicotine (French)
Nicotine hydrochloride [UN 1656]
Nicotine hydrochloride, Solution [UN 1656]
UN 1656 [Nicotine hydrochloride or Nicotine hydrochloride solution]

2825-15-2
$C_3H_3N_3O_4$
145.06
1-Nitro hydantoin
Hydantoin, 1-nitro-
2,4-Imidazolidinedione, 1-nitro-
1-Nitrohydantoin

2825-82-3
$C_{10}H_{16}$
136.26
C(C(C(C1C2)CC3)C3)(C1)C2
4,7-Methanoindan, hexahydro-, exo
exo-Hexahydro-4,7-methanoindan
JP-10
4,7-Methano-1H-indene, octahydro-, (3a-α,4-β,7-β,7a-α)-(9CI)
exo-Tetrahydrobicyclopentadiene
exo-Tetrahydrodi(cyclopentadiene)
exo-Tricyclo(5.2.1.0^{2,6})decane
exo-Trimethylenenorbornane
exo-5,6-Trimethylenenorbornane

2827-44-3
$C_5H_9N_5O$
155.19
s-Triazine, 4,6-diamino-2-ethoxy

2827-45-4
$C_4H_7N_5O$
141.16
s-Triazine, 4,6-diamino-2-methoxy

2827-46-5
$C_6H_{12}N_6$
168.24
Melamine, N^2,N^4,N^6-trimethyl
N^2,N^4,N^6-Trimethylmelamine
2,4,6-Tris(methylamino)-s-triazine
2,4,6-Tris-methylamino-1,3,5-triazine
1,3,5-Triazine-2,4,6-triamine, N,N',N''-trimethyl- (9CI)
s-Triazine, 2,4,6-tris(methylamino)-

2827-49-8
$C_{15}H_{30}N_6$
294.41
1,3,5-Triazine-2,4,6-triamine, N,N,N',N',N'',N''-hexaethyl-(9CI)
Melamine, hexaethyl- (8CI)
NSC-37756

2828-42-4
$C_{10}H_{12}N_2O_2$
192.24
CC(C)=NOC(=O)Nc1ccccc1
Acetone, oxime, o-(phenylcarbamate)
Acetone o-carbaniloyloxime
Acetone oxime N-phenylcarbamate
Acetone oxime phenylurethane

Acetone, O-(phenylcarbamoyl)-oxime (8CI)
o-(N-Fenylkarbamoyl)propanon-oxim (Czech)
o-(N-Phenylcarbamoyl)-propan-onoxim (German)
o-(N-Phenylcarbamoyl)propan-onoxime
2-Propanone, o-((phenylamino)-carbonyl)oxime (9CI)
Proximpham (German)
Proxypham

2831-60-9
$C_8H_8N_2O_6$
228.15
O=N(=O)c(ccc(OCCO)c1N(=O)=O)c1
Ethanol, 2-(2,4-dinitro-phenoxy)- (9CI)
2,4-Dinitrophenoxyethanol
2-(2,4-Dinitrophenoxy)ethanol

2831-66-5
$C_3HCl_2N_3$
149.95
1,3,5-Triazine, 2,4-dichloro- (9CI)
s-Triazine, 2,4-dichloro- (8CI)

2832-19-1
$C_3H_6ClNO_2$
123.53
O=C(NCO)CCl
N-Methylol-chloracetamide
Acetamide, 2-chloro-N-(hydroxymethyl)- (8CI,9CI)
AI3-62444
Caswell No. 192BB
Chloracetamide-N-metholol
2-Chloro-N-hydroxymethyl-acetamide
2-Chloro-N-(hydroxymethyl)-acetamide
EPA Pesticide Chemical Code 109501
Grota
Grotan DF-35

2832-40-8
$C_{15}H_{15}N_3O_2$
269.33
O=C(Nc(ccc(N=Nc(c(O)ccc1C)c1)c2)c2)C
Acetamide, N-(4-((2-hydroxy-5-methylphenyl)azo)phenyl)
Acetamine Yellow CG
Acetate Fast Yellow G
Acetoquinone Light Yellow
Acetoquinone Light Yellow 4JLZ
Altco Sperse Fast Yellow GFN New
Amacel Yellow G
Artisil Direct Yellow G
Artisil Yellow G
Artisil Yellow 2GN
Calcosyn Yellow GC
Calcosyn Yellow GCN
Celliton Discharge Yellow GL
Celliton Fast Yellow G
Celliton Fast Yellow GA
Celliton Fast Yellow GA-CF
Celliton Yellow G
Celutate Yellow GH
C.I. 11855
C.I. 3/11855
Cibacete Yellow GBA
Cibacet Yellow GBA
Cibacet Yellow 2GC
C.I. Disperse Yellow 3
Cilla Fast Yellow G
C.I. Solvent Yellow 77
Diacelliton Fast Yellow G
Disperse Fast Yellow G
Disperse Yellow 3
Disperse Yellow G
Disperse Yellow Z
Dispersol Fast Yellow G
Dispersol Printing Yellow G
Dispersol Yellow A-G
Durgacet Yellow G
Durosperse Yellow G
Eastone Yellow GN
Esteroquinone Light Yellow 4JL
Estone Yellow GN
Fenacet Fast Yellow G
Fenacet Yellow G
Genacron Yellow G
Hispacet Fast Yellow G
Hisperse Yellow G
N-(4-((2-Hydroxy-5-methyl-phenyl)azo)phenyl)acetamide
4'-((6-Hydroxy-m-tolyl)azo)-acetanilide
Interchem Acetate Yellow G
Interchem Disperse Yellow GH
Intraperse Yellow GBA
Intrasperse Yellow GBA Extra
Kayalon Fast Yellow G
Kayaset Yellow G
KCA acetate Fast Yellow G
Microsetile Yellow GR
Miketon Fast Yellow G
Nacelan Fast Yellow CG
NCI-C53781
Novalon Yellow 2GN
Nyloquinone Light Yellow 4JL
Nyloquinone Yellow 4J
Ostacet Yellow P2G
Palacet Yellow GN
Palanil Yellow G
Pamacel Yellow G-3
Perliton Yellow G
Reliton Yellow C
Resiren Yellow TG
Safaritone Yellow G
Samaron Yellow PA3
Serinyl Hosiery Yellow GD
Seriplas Yellow GD
Serisol Fast Yellow GD
Setacyl Yellow G
Setacyl Yellow 2GN
Setacyl Yellow P-2GL
Silotras Yellow TSG
Supracet Fast Yellow G
Synten Yellow 2G
Synton Yellow 2G
Terasil Yellow GBA Extra
Terasil Yellow 2GC
Tertranese Yellow N-2GL
Tuladisperse Fast Yellow 2G
Vonteryl Yellow G
Vonteryl Yellow R
Yellow Reliton G
Yellow Z
Zlut Disperzni 3 (Czech)
Zlut Rozpoustedlova 77 (Czech)

2834-90-4
$C_{10}H_9NO$
159.18
1-Naphthalenol, 4-amino- (9CI)
4-Amino-1-naphthol
1-Naphthol, 4-amino- (8CI)

2834-92-6
$C_{10}H_9NO$
159.18
Oc(c(c(c(ccc1)c2)c1)N)c2
2-Naphthalenol, 1-amino- (9CI)
1-Amino-2-naphthalenol
1-Amino-2-naphthol

2835-06-5
$C_8H_9NO_2$
151.16
O=C(O)C(N)c(cccc1)c1
Benzeneacetic acid, α-amino-, (+-)- (9CI)
(+-)-α-Aminobenzeneacetic acid
Glycine, 2-phenyl-, DL- (8CI)

2835-39-4
$C_8H_{14}O_2$
142.22
O=C(OCC=C)CC(C)C
Isovaleric acid, allyl ester
Allyl isovalerate
Allyl isovalerianate
Allyl 3-methylbutyrate
Butanoic acid, 3-methyl-, 2-propenyl ester
Butyric acid, 3-methyl-, allyl ester
Fema No. 2045
Isovaleric acid, allyl ester
3-Methylbutanoic acid, 2-propenyl ester
3-Methylbutyric acid, allyl ester
NCI-C54717
2-Propenyl isovalerate
2-Propenyl 3-methylbutanoate

2835-68-9
$C_7H_8N_2O$
136.14
O=C(N)c(ccc(N)c1)c1
4-Aminobenzamide
p-Aminobenzoic acid amide
Benzamide, p-amino- (8CI)
Benzamide, 4-amino- (9CI)

2835-81-6

$C_4H_9NO_2$
103.11
O=C(O)C(N)CC
Butanoic acid, 2-amino-, (+-)- (9CI)
AI3-15284
(+-)-2-Aminobutanoic acid
Butyric acid, 2-amino-, DL- (8CI)

2835-95-2
C_7H_9NO
123.17
Oc(c(ccc1N)C)c1
Phenol, 5-amino-2-methyl
5-Amino-o-cresol
4-Amino-2-hydroxytoluene
5-Amino-2-methylphenol

2836-00-2
C_7H_9NO
123.15
Phenol, 3-amino-4-methyl- (9CI)
p-Cresol, 3-amino- (8CI)

2836-32-0
$C_2H_3O_3.Na$
98.04
[Na+].OCC([O-])=O
Glycolic acid, monosodium salt
Acetic acid, hydroxy-, monosodium salt
Glykokolan sodny (Czech)
Monosodium glycolate
Sodium hydroxyacetate
Sodium α-hydroxyacetate

2837-89-0
C_2HClF_4
136.48
Ethane, 2-chloro-1,1,1,2-tetrafluoro- (8CI,9CI)
Freon 124
R 124

2840-51-9
$C_{14}H_{10}O$
194.23
9H-Fluoren-9-one, 2-methyl- (9CI)
Fluoren-9-one, 2-methyl- (8CI)

2842-38-8
$C_8H_{17}NO$
143.26
OCCNC1CCCCC1
Ethanol, 2-(cyclohexylamino)
Abromeen E-25
2-(Cyclohexylamino)ethanol
2-(Cyclohexylamino)ethanol Aerosol
Ethanol, 2-(cyclohexylamino)-, (Aerosol)
N-(2-Hydroxyethyl)cyclohexylamine

2845-89-8
C_7H_7ClO
142.58
O(c(cccc1Cl)c1)C
Benzene, 1-chloro-3-methoxy- (9CI)
1-Chloro-3-methoxybenzene

2847-16-7
$C_{10}H_{20}O_2.1/2Cd$
228.47
Decanoic acid, cadmium salt (9CI)
Cadmium decanoate

2847-72-5
$C_{11}H_{24}$
156.31
Decane, 4-methyl- (8CI,9CI)

2851-52-7
$Cl_{12}N_6P_6$
695.28
1,3,5,7,9,11-Hexaaza-2,4,6,8,10,12-hexaphosphacyclododeca-1,3,5,7,9,11-hexaene, 2,2,4,4,6,6,8,8,10,10,12,12-dodecachloro-2,2,4,4,6,6,8,8,10,10,12,12-dodecahydro- (8CI,9CI)
1,3,5,7,9,11-Hexaaza-2,4,6,8,10,12-hexaphosphacyclododecahexaene, 2,2,4,4,6,6,8,8,10,10,12,12-dodecachloride

2852-07-5
$C_4H_3Cl_3$
157.42
ClC(Cl)=C(Cl)C=C
1,3-Butadiene, 1,1,2-trichloro
TCBD
1,1,2-Trichlorobutadiene

2852-68-8
$C_{15}H_{14}O$
210.28
Methanone, bis(3-methylphenyl)- (9CI)
Benzophenone, 3,3'-dimethyl- (7CI,8CI)
3,3'-Dimethylbenzophenone
Di-m-tolyl ketone

2854-70-8
$C_9H_{16}N_8$
236.33
s-Triazine, 2-azido-4-(tert-butylamino)-6-(ethylamino)
2-Azido-4-ethylamino-4-tert-butylamino-s-triazine
WL 9385

2854-94-6
$C_8H_{14}N_8$
222.25
2-Azidoatrazine
2-Azido-4-(ethylamino)-6-(isopropylamino)-s-triazine

2854-95-7
$C_7H_{12}N_8$
208.18
1,3,5-Triazine-2,4-diamine, 6-azido-N,N'-diethyl- (9CI)
s-Triazine, 2-azido-4,6-bis-(ethylamino)- (8CI)

2854-96-8
$C_9H_{16}N_8$
236.23
1,3,5-Triazine-2,4-diamine, 6-azido-N,N,N'-triethyl- (9CI)
s-Triazine, 2-azido-4-(diethylamino)-6-(ethylamino)- (8CI)

2855-13-2
$C_{10}H_{22}N_2$
170.34
NCC(CC(N)CC1(C)C)(C1)C
Cyclohexanemethylamine, 5-amino-1,3,3-trimethyl
Cyclohexanemethanamine, 5-amino-1,3,3-trimethyl- (9CI)
Isophoronediamine [UN 2289]
UN 2289 [Isophoronediamine]

2855-19-8
$C_{12}H_{24}O$
184.36
O(C1CCCCCCCCCC)C1
Dodecane, 1,2-epoxy
Decyl oxirane
Dodecene epoxide
1,2-Epoxydodecane
1,2-Epoxydodekan (Czech)

2859-67-8
$C_8H_{11}NO$
137.20
n(cccc1CCCO)c1
3-Pyridinepropanol
3-Propanolpyridine

2860-64-2
$C_{11}H_{11}Cl_2NO$
244.11
2-Pyrrolidinone, 1-(3,4-dichlorophenyl)-3-methyl- (8CI,9CI)
BV 201

2866-43-5
$C_{18}H_{10}N_2O_2S$
318.34
O(c(c(N=1)ccc2)c2)C1C(SC(=C3)

C(Oc(c4ccc5)c5)=N4)=C3
FBA 185
Benzoxazole, 2,2'-(2,5-thiophenediyl)bis- (9CI)
Thiophene, 2,5-di(benzoxazol-2-yl)-
2,2'-(2,5-Thiophenediyl)bis-benzoxazole

2867-05-2
$C_{10}H_{16}$
136.24
C(C1C(=C2)C)(C2)(C1)C(C)C
Bicyclo(3.1.0)hex-2-ene, 2-methyl-5-(1-methylethyl)-(9CI)
α-Thujene

2867-47-2
$C_8H_{15}NO_2$
157.24
O=C(OCCN(C)C)C(=C)C
Methacrylic acid, 2-(dimethylamino)ethyl ester
Ageflex FM-1
2-Dimethylaminoethylester kyseliny methakrylove (Czech)
Dimethylaminoethyl methacrylate
β-Dimethylaminoethyl methacrylate
N,N-Dimethylaminoethyl methacrylate
2-Dimethylaminoethyl methacrylate
2-(Dimethylamino)ethyl methacrylate
Dimethylaminoethyl methacrylate [UN 2522]
Ethanol, 2-(dimethylamino)-, methacrylate
UN 2522 [Dimethylaminoethyl methacrylate]
USAF RH-3

2868-75-9
$C_{34}H_{28}N_6O_8S_2.2Na$
758.71
C.I. Direct Red 7, Disodium salt (8CI)
1-Naphthalenesulfonic acid, 3,3'-((3,3'-dimethoxy-(1,1'-biphenyl)-4,4'-diyl)bis-(azo))bis(4-amino-, disodium salt (9CI)

2869-34-3
$C_{13}H_{29}N$
199.37
NCCCCCCCCCCCCC
1-Tridecanamine (9CI)

2870-04-4
$C_{10}H_{14}$
134.22
c(c(c(cc1)C)CC)(c1)C
Benzene, 2-ethyl-1,3-dimethyl-(9CI)
2-Ethyl-1,3-dimethylbenzene

2870-32-8
$C_{30}H_{28}N_4O_8S_2.2Na$
682.66
Chrysophenine
Benzenesulfonic acid, 2,2'-(1,2-ethenediyl)bis(5-((4-ethoxyphenyl)azo)-, disodium salt (9CI)
4,4'-Bis((p-ethoxyphenyl)azo)-2,2'-stilbenesulfonic acid disodium salt
C.I. Direct Yellow 12 (8CI)
C.I. 24895
Direct Yellow 12
2,2'-(1,2-Ethenediyl)bis(5-((4-ethoxyphenyl)azo)benzenesulfonic acid), disodium salt

2870-71-5
$C_{18}H_{26}NO_3.Br$
384.36
1-α-H,5-α-H-Tropanium, 3-α-hydroxy-8-methyl-, bromide, (+-)-tropate (ester)
Atropine methobromide
Atropine methylbromide
3-α-Hydroxy-8-methyl-1-α,5-α-H-tropanium bromide (+-)tropate
Hyoscyamine methylbromide
Methylatropine
Methylatropine bromide
N-Methylatropine bromide
Methylatropinium bromide
8-Methylatropinium bromide
Mintussin
Mydriasin
Tropin

2871-01-4
$C_8H_{11}N_3O_3$
197.22
O=N(=O)c(c(NCCO)ccc1N)c1
Ethanol, 2-(4-amino-2-nitroanilino)
2-((4-Amino-2-nitrophenyl)-amino)ethanol
Ethanol, 2-((4-amino-2-nitrophenyl)amino)-
HC Red No. 3
NCI-C54922

2872-48-2
$C_{15}H_{12}N_2O_3$
268.29
O=C(c(c(C(=O)c1c(N)cc(OC)c2N)ccc3)c3)c12
Anthraquinone, 1,4-diamino-2-methoxy
Acetate Fast Pink 3B
Amacel Cerise B
9,10-Anthracenedione, 1,4-diamino-2-methoxy- (9CI)
Artisil Brilliant Rose 5BP
Celanthrene Fast Pink 3B
Celliton Fast Pink FF3B
Celliton Fast Pink FF3BA-CF
Celliton Rose FF3B
Cerven disperzni 11 (Czech)
Cibacet Brilliant Pink 4BN
Cibacete Brilliant Pink 4BN
C.I. 62015
C.I. Disperse Red 11
Cilla Fast Pink FF3B
C.I. Solvent Violet 26
1,4-DA-2-MOA (Russian)
1,4-Diamino-2-methoxyanthraquinone
Disperse Brilliant Pink
Disperse Brilliant Rose
Disperse Red 11
Dispersol Red B 3B
Duranol Red X3B
Fenacet Fast Pink 3BE
Interchem Acetate Pink 3B
Miketon Fast Pink FF 3B
Nacelan Pink 3B
Palanil Violet 6R
Perliton Red Violet FFB
Samaron Red Violet F3B
Seriplas Red X3B
Serisol Blilliant Red X 3B
Setacyl Red P-3B
Solvent Violet 26
Supracet Fast Pink 3B
Terasil Brilliant Pink 4BN
Violet Rozpoustedlova 26 (Czech)

2872-52-8
$C_{16}H_{18}N_4O_3$
314.32
O=N(=O)c(ccc(N=Nc(ccc(N(CCO)CC)c1)c1)c2)c2
C.I. Disperse Red 1 (8CI)
Ethanol, 2-(ethyl(4-((4-nitrophenyl)azo)phenyl)amino)-(9CI)

2873-74-7
$C_5H_6Cl_2O_2$
169.01
O=C(CCCC(=O)Cl)Cl
Pentanedioyl dichloride (9CI)
Glutaryl chloride
Pentanedioic acid, dichloride

2873-97-4
$C_9H_{15}NO_2$
169.25
O=C(NC(CC(=O)C)(C)C)C=C
Acrylamide, N-(1,1-dimethyl-3-oxobutyl)
Diacetone acrylamide
N-(1,1-Dimethyl-3-oxobutyl)-acrylamide
N-(1,1-Dimethyl-3-oxobutyl)-2-propenamide
N-(2-(2-Methyl-4-oxopentyl))-acrylamide
2-Propenamide, N-(1,1-dimethyl-3-oxobutyl)-

2876-13-3
$C_{12}H_{12}N.Cl$
205.68
Pyridinium, 1-(phenylmethyl)-, chloride (9CI)
1-(Phenylmethyl)pyridinium chloride

2876-35-9
$C_{14}H_{16}$
184.28
Naphthalene, 2-(1,1-dimethylethyl)- (9CI)
β-tert-Butylnaphthalene
2-tert-Butylnaphthalene
Naphthalene, 2-tert-butyl- (6CI,7CI,8CI)

2877-14-7
$C_8H_3Cl_5O_3$
324.37
Acetic acid, (pentachlorophenoxy)- (8CI,9CI)
NSC-78926

2882-21-5
$C_6H_8N_2O$
124.13
O(c(nc(cn1)C)c1)C
Pyrazine, 2-methoxy-6-methyl- (9CI)
2-Methoxy-6-methylpyrazine

2882-96-4
$C_{16}H_{34}$
226.45
Pentadecane, 3-methyl- (8CI, 9CI)

2883-02-5
$C_{15}H_{30}$
210.40
C(CCCC1)(C1)CCCCCCCCC
Cyclohexane, nonyl- (9CI)
1-Cyclohexylnonane
Nonylcyclohexane

2883-05-8
$C_{14}H_{28}$
196.38
Octane, 2-cyclohexyl- (7CI, 8CI)

2883-98-9
$C_{12}H_{16}O_3$
208.28
COc1cc(OC)c(C=CC)cc1OC
Benzene, 1,2,4-trimethoxy-5-propenyl-, (E)
Asaron
Asarone
Asarone, trans-
α-Asarone
trans-Asarone
Asarum Camphor
Benzene, 1,2,4-trimethoxy-5-propenyl-, trans-
Etherophenol
trans-Isoasarone

2884-69-7
$C_{12}H_{14}ClNO$
223.70
2-Pyrrolidinone, 1-(3-chloro-4-methylphenyl)-3-methyl- (9CI)
BV 207
2-Pyrrolidinone, 1-(3-chloro-p-tolyl)-3-methyl- (8CI)

2885-00-9
$C_{18}H_{38}S$
286.57
SCCCCCCCCCCCCCCCCCC
n-Octadecyl mercaptan
1-Octadecanethiol (9CI)

2887-91-4
$C_{18}H_{19}NO.ClH$
301.81
1-Propanamine, 3-dibenz(b,e)oxepin-11(6H)-ylidene-N-methyl-, hydrochloride (9CI)
Dibenz(b,e)oxepin-δ(11(6H),γ)-propylamine, N-methyl-, hydrochloride (8CI)

2888-06-4
$C_6H_4Cl_2O_2S$
211.07
Benzenesulfonyl chloride, 3-chloro- (9CI)
Benzenesulfonyl chloride, m-chloro- (8CI)

2889-58-9
$C_{13}H_{15}NO$
201.27
Benzene, 1-(1-isocyanato-1-methylethyl)-4-(1-methylethenyl)- (9CI)
Isocyanic acid, p-isopropenyl-α,α-dimethylbenzyl ester (7CI,8CI)
p-Isopropenyldimethylbenzylisocyanate
p-Isopropenyl-α,α-dimethylbenzyl isocyanate
4-Isopropenyl-α,α-dimethylbenzyl isocyanate

2893-78-9
$C_3HCl_2N_3O_3.Na$
220.96
[Na+].ClN1C(=O)[N-]C(=O)N(Cl)C1=O
s-Triazine-2,4,6(1H,3H,5H)-trione, dichloro-, sodium salt
ACL 60
CDB 63
Dichloroisocyanuric acid sodium salt [UN 2465]
Dikonit
Dimanin C
FI Clor 60S
Isocyanuric acid, dichloro-, sodium salt
OCI 56
SDIC
Simpla
Sodium dichlorocyanurate
Sodium dichlorisocyanurate
Sodium dichloroisocyanurate (DOT)
1-Sodium-3,5-dichloro-s-triazine-2,4,6-trione
1-Sodium-3,5-dichloro-1,3,5-triazine-2,4,6-trione
Sodium dichloro-s-triazinetrione, Dry, Containing more than 39% available chlorine [UN 2465]
Sodium salt of dichloro-s-triazinetrione
1,3,5-Triazine-2,4,6(1H,3H,5H)-trione, 1,3-dichloro-, sodium salt
UN 2465 [Dichloroisocyanuric acid, Dry or dichloroisocyanuric acid salts]

2893-80-3
$C_{31}H_{22}N_6O_8S.2Na$
684.56
Benzoic acid, 5-((4'-((2,4-dihydroxy-3-((4-sulfophenyl)azo)phenyl)azo)(1,1'-biphenyl)-4-yl)azo)-2-hydroxy-, disodium salt (9CI)
C.I. Direct Brown 6, Disodium salt (8CI)

2894-67-9
$C_{15}H_{10}Cl_2N_2O$
305.15
Chlordesmethyldiazepam
2H-1,4-Benzodiazepin-2-one, 7-chloro-5-(o-chlorophenyl)-1,3-dihydro- (8CI)
2H-1,4-Benzodiazepin-2-one, 7-chloro-5-(2-chlorophenyl)-1,3-dihydro- (9CI)
2H-1,4-Benzodiazepin-2-one, 1,3-dihydro-7-chloro-5-(2-chlorophenyl)-
Clordesmetildiazepam (Spanish)
DEA No. 2754
Delorazepam
Delorazepamum (Latin)
1,3-Dihydro-7-chloro-5-(o-chlorophenyl)-2H-1,4-benzodiazepin-2-one
EN
RV-12165

2896-60-8
$C_8H_{10}O_2$
138.17
Oc(c(ccc1O)CC)c1
Resorcinol, 4-ethyl- (8CI)

1,3-Benzenediol, 4-ethyl- (9CI)
2,4-Dihydroxy-1-ethylbenzene
4-Ethyl-1,3-benzenediol
4-Ethylresorcinol

2897-60-1
$C_{11}H_{24}O_4Si$
248.39
Silane, diethoxymethyl(3-(oxiranylmethoxy)propyl)- (9CI)
(γ-Glycidoxypropyl)methyldiethoxysilane

2898-12-6
$C_{16}H_{15}ClN_2$
270.75
Medazepam
Ansilan
1H-1,4-Benzodiazepine, 7-chloro-2,3-dihydro-1-methyl-5-phenyl-
1H-1,4-Benzodiazepine, 2,3-dihydro-7-chloro-1-methyl-5-phenyl-
7-Chloro-2,3-dihydro-1-methyl-5-phenyl-1H-1,4-benzodiazepine
DEA No. 2836
Diepin
2,3-Dihydro-7-chloro-1-methyl-5-phenyl-1H-1,4-benzodiazepine
Elbrus
Esmail
Medazepamum (Latin)
Medazepol
Megasedan
Mezepan
Narsis
Nobrium
Pazital
Psiquium
Resmit
Rudotel
Serenium
Siman
Tranquilax

2903-23-3
$C_{10}H_{18}O$
154.25
2-Nonen-4-one, 2-methyl- (8CI,9CI)

2905-24-0
$C_6H_4BrClO_2S$
255.52
Benzenesulfonyl chloride, 3-bromo- (9CI)
Benzenesulfonyl chloride, m-bromo- (8CI)
m-Bromobenzenesulfonyl chloride
3-Bromobenzenesulfonyl chloride

2905-60-4
$C_7H_3Cl_3O$
209.46
O=C(c(c(c(cc1)Cl)Cl)c1)Cl
2,3-Dichlorobenzoyl chloride
Benzoyl chloride, 2,3-dichloro- (9CI)

2905-61-5
$C_7H_3Cl_3O$
209.46
O=C(c(c(ccc1Cl)Cl)c1)Cl
2,5-Dichlorobenzoyl chloride
Benzoyl chloride, 2,5-dichloro- (9CI)

2905-62-6
$C_7H_3Cl_3O$
209.46
O=C(c(cc(cc1Cl)Cl)c1)Cl
Benzoyl chloride, 3,5-dichloro- (9CI)
3,5-Dichlorobenzoyl chloride

2905-65-9
$C_8H_7ClO_2$
170.60
O=C(OC)c(cccc1Cl)c1
Methyl 3-chlorobenzoate
Benzoic acid, m-chloro-, methyl ester
Benzoic acid, 3-chloro-, methyl ester (9CI)
m-Chlorobenzoic acid methyl ester
Methyl m-chlorobenzoate

2905-69-3
$C_8H_6Cl_2O_2$
205.04
O=C(OC)c(c(ccc1Cl)Cl)c1
Methyl 2,5-dichlorobenzoate
Benzoic acid, 2,5-dichloro-, methyl ester (9CI)

2909-38-8
C_7H_4ClNO
153.57
O=C=Nc(cccc1Cl)c1
Isocyanic acid, m-chlorophenyl ester
m-Chlorfenylisokyanat (Czech)
m-Chlorophenyl isocyanate

2911-36-6
$C_5H_8ClN_5$
173.63
s-Triazine, 4,6-bis(methylamino)-2-chloro
4,6-Bis(methylamino)-2-chloro-s-triazine

2915-52-8
$C_{28}H_{52}O_4$
452.72
O=C(OCCCCCCCCCCCC)C=CC(=O)OCCCCCCCCCCCC
Maleic acid, didodecyl ester (8CI)
AI3-07871-X
2-Butenedioic acid (Z)-, didodecyl ester (9CI)
2-Butenedioic acid, (Z)-, didodecyl ester (C10-C18)

2915-53-9
$C_{20}H_{36}O_4$
340.56
O=C(OCCCCCCCC)C=CC(=O)OCCCCCCCC
Maleic acid, dioctyl ester
Dioctyl maleate
Di-n-octyl maleate

2917-26-2
$C_{16}H_{34}S$
258.51
SCCCCCCCCCCCCCCCC
1-Hexadecanethiol (9CI)
AI3-07600-X
AI3-15914
1-Hexadecanethiol (50% concentrate)

2917-73-9
$C_{17}H_{32}O_4$
300.44
O=C(OCCCC)CCCCCCCC(=O)OCCCC
Dibutyl azelate
AI3-01982
Azelaic acid, dibutyl ester (8CI)
Dibutyl nonanedioate
Ergoplast AZDB
Nonanedioic acid, dibutyl ester (9CI)

2917-94-4
$C_{20}H_{34}O_6S.Na$
425.54
Ethanesulfonic acid, 2-(2-(2-(4-(1,1,3,3-tetramethylbutyl)phenoxy)ethoxy)ethoxy)-, sodium salt (9CI)
Entsufon sodium
Ethanesulfonic acid, 2-(2-(2-(p-(1,1,3,3-tetramethylbutyl)phenoxy)ethoxy)ethoxy)-, sodium salt (8CI)
Sodium octoxynol-2 ethane sulfonate

2920-38-9
$C_{13}H_9N$
179.22
N#Cc(ccc(c(cccc1)c1)c2)c2
(1,1'-Biphenyl)-4-carbonitrile (9CI)

2921-31-5
$C_8H_{14}ClO_2PS$
240.70
3-Phosphabicyclo(4.4.0)dec-

ane, p-chloro-2,4-dioxa-5-methyl-p-thiono
p-Chloro-2,4-dioxa-5-methyl-p-thiono-3-phosphabicyclo-(4.4.0)decane
ENT 23,970
UC 8305
Union Carbide UC-8305

2921-88-2
$C_9H_{11}Cl_3NO_3PS$
350.59
CCOP(=S)(OCC)Oc1nc(Cl)c(Cl)cc1Cl
Phosphorothioic acid, O,O-diethyl O-(3,5,6-trichloro-2-pyridyl) ester
Brodan
Chlorpyrifos (ACGIH,DOT, OSHA)
Chlorpyrifos-ethyl
Chlorpyriphos
Chlorpyriphos-ethyl
Detmol U.A.
Ethion, Dry
O,O-Diethyl O-3,5,6-trichloro-2-pyridyl phosphorothioate
O,O-Diaethyl-O-3,5,6-trichlor-2-pyridylmonothiophosphat (German)
DOWCO 179
Dursban
Dursban F
ENT 27,311
Eradex
Killmaster
Lorsban
NA 2783 (DOT)
OMS-0971
2-Pyridinol, 3,5,6-trichloro-, O-ester with O,O-diethyl phosphorothioate
Pyrinex
Stipend

2922-51-2
$C_{17}H_{34}O$
254.46
2-Heptadecanone (8CI,9CI)

2923-93-5
$C_{32}H_{41}F_7N_2O_4$
650.67
O=C(Nc(c(O)cc(NC(=O)C(Oc(c(cc(c1)C(CC)(C)C)C(CC)(C)C)c1)CCCC)c2)c2)C(F)(F)C(F)(F)C(F)(F)F
Hexanamide, 2-(2,4-bis(1,1-dimethylpropyl)phenoxy)-N-(4-((2,2,3,3,4,4,4-heptafluoro-1-oxobutyl)amino)-3-hydroxyphenyl)- (9CI)

2925-56-6
$C_9H_5ClF_2N_4$
242.62
s-Triazine, 2-(o-chloroanilino)-4,6-difluoro- (6CI,8CI)

2927-83-5
$C_4F_8O_2$
232.03
O=C(F)C(F)(OC(F)(F)F)C(F)(F)F
Propanoyl fluoride, 2,3,3,3-tetrafluoro-2-(trifluoromethoxy)- (9CI)

2929-86-4
$C_{30}H_{63}O_3P$
502.80
O(P(OCCCCCCCCCC)OCCCCCCCCCC)CCCCCCCCCC
Phosphorous acid, tris(decyl) ester (9CI)

2934-05-6
$C_{12}H_{18}O$
178.30
Phenol, 2,4-diisopropyl
2,4-Diisopropylphenol

2935-44-6
$C_6H_{14}O_2$
118.20
OC(CCC(O)C)C
2,5-Hexanediol

2935-90-2
$C_4H_8O_2S$
120.17
O=C(OC)CCS
Methyl 3-mercaptopropionate
3-Mercaptopropionic acid methyl ester
Methyl 3-mercaptopropanoate
Methyl β-mercaptopropionate
Propanoic acid, 3-mercapto-, methyl ester (9CI)
Propionic acid, 3-mercapto-, methyl ester (8CI)

2937-50-0
$C_4H_5ClO_2$
120.54
O=C(OCC=C)Cl
Formic acid, chloro-, allyl ester
Allyl chlorocarbonate
Allyl chlorocarbonate (DOT)
Allyl chloroformate [UN 1722]
Allylester kyseliny chlormravenci (Czech)
Chloroformic acid allyl ester
UN 1722 [Allyl chloroformate]

2939-80-2
$C_{10}H_9Cl_4NO_2S$
349.06
4-Cyclohexene-1,2-dicarboximide, N-((1,1,2,2-tetrachloroethyl)thio)-, cis
Captafol (ACGIH)
Crisfolatan
Difolatan
Folcid
Haipen
1H-Isoindole-1,3(2H)-dione, 3a,4,7,7a-tetrahydro-2-((1,1,2,2-tetrachloroethyl)thio)-, cis-
Merpafol
Mycodifol
Ortho-5865
Pillartan
Sanspor
cis-N-((1,1,2,2-Tetrachloroethyl)thio)-4-cyclohexene-1,2-dicarboxymide
cis-3a,4,7,7a-Tetrahydro-2-((1,1,2,2-tetrachloroethyl)-thio)-1H-isoindole-1,3(2H)-dione
TN 80

2941-55-1
$C_7H_{15}NOS$
161.29
CCSC(=O)N(CC)CC
Carbamic acid, diethylthio-, S-ethyl ester
Ethiolate
S-Ethyl diethylcarbamothioate
S-Ethyl diethylthiocarbamate
Prefox
S6176
S-15076

2941-64-2
C_3H_5ClOS
132.56
O=C(SCC)Cl
S-Ethyl chlorothiolformate
Carbonochloridothioic acid, S-ethyl ester (9CI)
S-Ethyl carbonochloridothioate
Ethyl chlorothioformate
S-Ethyl chlorothioformate
Ethyl chlorothiolformate [UN 2826]
Ethyl chlorothioloformate
Ethyl thiochloroformate
S-Ethyl thiochloroformate
Formic acid, chlorothio-, S-ethyl ester
UN 2826 [Ethyl chlorothioformate]

2943-75-1
$C_{14}H_{32}O_3Si$
276.49
Silane, triethoxyoctyl- (9CI)
Octyl(triethoxy)silane
Triethoxyoctylsilane

2944-67-4
$C_2H_2O_4.1/3Fe.H_3N$
MW:114.33
Ferric ammonium oxalate
Ethanedioic acid, ammonium iron(3+) salt (3:3:1) (9CI)

2948-46-1
$C_{12}H_{18}O_2$
194.27
OC(c(ccc(c1)C(O)(C)C)c1)(C)C
p-Bis(2-hydroxyisopropyl)-benzene
1,4-Benzenedimethanol, α,α,α',α'-tetramethyl- (9CI)
p-Bis(α-hydroxyisopropyl)-benzene
1,4-Bis(2-hydroxy-2-propyl)-benzene
α,α'-Dihydroxy-p-diisopropyl-benzene
α,α,α',α'-Tetramethyl-p-benz-enedimethanol
α,α,α',α'-Tetramethyl-1,4-benzenedimethanol
α,α,α',α'-Tetramethyl-p-xylyl-enediol
p-Xylene-α,α'-diol, α,α,α',α'-tetramethyl-

2950-45-0
$Cl_8N_4P_4$
463.52
Octachlorocyclotetraphos-phazene
Phosphonitrilic chloride
1,3,5,7,2,4,6,8-Tetrazatetra-phosphocine, 2,2,4,4,6,6,8,8-octachloro-2,2,4,4,6,6,8,8-octahydro- (9CI)

2953-29-9
$C_3H_9O_2PS_2$
172.21
COP(=S)(OC)SC
Phosphorodithioic acid, O,O,S-trimethyl ester
O,O,S-Trimethylphosphorodi-thioate

2955-38-6
$C_{19}H_{17}ClN_2O$
324.80
Prazepam
2H-1,4-Benzodiazepin-2-one, 7-chloro-1-(cyclopropyl-methyl)-1,3-dihydro-5-phenyl- (8CI,9CI)
2H-1,4-Benzodiazepin-2-one, 1,3-dihydro-7-chloro-1-(cy-clopropylmethyl)-5-phenyl-
Centrax
7-Chloro-1-(cyclopropyl-methyl)-1,3-dihydro-5-phenyl-2H-1,4-benzodiazepin-2-one
7-Chloro-1-cyclopropylmethyl-5-phenyl-1H-1,4-benzodi-azepin-2(3H)-one
DEA No. 2764
Demetrin
K-373
Prazepamum (Latin)
Verstran
W 4020

2955-56-8
$C_{20}H_{40}O$
296.54
3-Eicosanone (7CI,8CI, 9CI)
Ethyl heptadecyl ketone

2956-12-9
$C_9H_{16}OS_2$
204.36
O(C(=S)SCC=C)CCCCC
Carbonodithioic acid, O-pentyl S-2-propenyl ester (9CI)
O-Pentyl S-2-propenyl carbono-dithioate

2958-09-0
$C_{18}H_{39}O_4P$
350.48
O=P(OCCCCCCCCCCCCCCCC CC)(O)O
Phosphoric acid, monoocta-decyl ester (9CI)
Monooctadecyl phosphate
Monostearyl acid phosphate

2963-66-8
$C_{13}H_{10}N_2O$
210.22
Oc(c(ccc1)C(=Nc(c2ccc3)c3) N2)c1
Phenol, 2-(1H-benzimidazol-2-yl)- (9CI)
2-(1H-Benzimidazol-2-yl)-phenol

2966-50-9
$C_2HF_3O_2.Ag$
221.89
[Ag+].[O-]C(=O)C(F)(F)F
Acetic acid, trifluoro-, silver(1+) salt (9CI)
Silver trifluoroacetate
Silver(1+) trifluoroacetate

2971-22-4
$C_{14}H_{11}Cl_3$
285.60
c(cccc1)(c1)C(c(cccc2)c2)C(Cl)(Cl)Cl
Ethane, 2,2-diphenyl-1,1,1-tri-chloro
DT
Diphenyltrichloroethane
2,2-Diphenyl-1,1,1-trichloro-ethane
1,1,1-Trichloro-2,2-diphenyl-ethane

2971-38-2
$C_{12}H_{11}Cl_3O_3$
309.58
ClCC=CCOC(=O)COc1ccc(Cl) cc1Cl
Acetic acid, 2,4-dichloro-phenoxy-, 4-chloro-2-butenyl ester
Chlorocrotyl ester of 2,4-d
Crotilin
Crotiline
Crotylin
2,4-D, α-chlorocrotyl ester
2,4-Dichlorophenoxyacetic acid, 4-chlorocrotonyl ester
Krotilin
Krotiline

2971-90-6
$C_7H_7Cl_2NO$
192.05
4-Pyridinol, 3,5-dichloro-2,6-dimethyl
Clopidol (ACGIH,OSHA)
Coccidiostat C
Coyden
Coyden 25
3,5-Dichloro-2,6-dimethyl-4-pyridinol
Lerbek
Methylchloropindol
Methylchlorpindol
Meticlorpindol

2972-65-8
$C_{15}H_{10}ClN_3O_2$
299.72
1,3,5-Triazine, 2-chloro-4,6-diphenoxy- (9CI)
6-Chloro-2,4-diphenoxy-s-triazine
Diphenoxychloro-s-triazine
s-Triazine, 2-chloro-4,6-di-phenoxy- (6CI,7CI,8CI)

2973-10-6
$C_6H_{14}O_4S$
182.26
CC(C)OS(=O)(=O)OC(C)C
Sulfuric acid, diisopropyl ester
DIPS
Di-isopropylsulfat (German)
Di-isopropylsulfate
Isopropyl sulfate

2973-44-6
C_6H_4Br
255.52
Phenyl, 3-bromo- (9CI)
3-Bromophenyl radical
Phenyl, m-bromo- (8CI)

2973-76-4
$C_8H_7BrO_3$
231.05
O=Cc(cc(OC)c(O)c1Br)c1
Benzaldehyde, 3-bromo-4-hydroxy-5-methoxy- (9CI)
3-Bromo-4-hydroxy-5-methoxy-benzaldehyde

2973-77-5
$C_7H_4Br_2O_2$
279.92
Benzaldehyde, 3,5-dibromo-4-hydroxy- (8CI,9CI)
NSC-72944

2974-90-5
$C_{12}H_8Cl_2$
223.10
1,1'-Biphenyl, 3,4'-dichloro- (9CI)
Biphenyl, 3,4'-dichloro- (8CI)
3,4'-Dichlorobiphenyl
3,4'-PCB

2974-92-7
$C_{12}H_8Cl_2$
223.10
Biphenyl, 3,4-dichloro- (8CI)
1,1'-Biphenyl, 3,4-dichloro- (9CI)

2976-74-1
$C_8H_6Cl_2O_3$
221.04
Acetic acid, (2,3-dichlorophenoxy)- (9CI)

2978-58-7
C_5H_9N
83.13
C(#C)C(N)(C)C
3-Butyn-2-amine, 2-methyl- (9CI)
3-Amino-3-methyl-1-butyne
2-Methyl-3-butyn-2-amine

2984-42-1
$C_4H_4Cl_2$
122.98
1,3-Butadiene, 1,4-dichloro- (6CI,7CI,8CI,9CI)
1,4-Dichloro-1,3-butadiene

2984-50-1
$C_8H_{16}O$
128.24
Octane, 1,2-epoxy
1,2-Epoxyoctane
1,2-Epoxyoktan (Czech)
Octylene epoxide
Oktylenoxid (Czech)

2984-64-7
$C_{10}H_{14}ClOPS_2$
280.78
Phosphonodithioic acid, ethyl-, S-(p-chlorophenyl) O-ethyl ester
Bay 36743
Bayer 36743
S-(p-Chlorophenyl) O-ethyl ethanephosphonodithioate
S-(4-Chlorophenyl) O-ethyl ethylphosphonodithioate
ENT 25,723
O-Ethyl S-4-chlorophenyl ethylphosphonodithioate
N-2596
Stauffer N-2596

2985-59-3
$C_{25}H_{34}O_3$
382.54
O=C(c(c(O)cc(OCCCCCCCCCCCC)c1)c1)c(cccc2)c2
Methanone, (4-(dodecyloxy)-2-hydroxyphenyl)phenyl- (9CI)

2987-87-3
$C_4H_5NO_3$
115.08
2-Butenoic acid, 4-amino-4-oxo-, (E)- (9CI)
Fumaramic acid (8CI)
NSC-126766

2989-98-2
$C_7H_7BrN_2O$
215.04
NC(=O)Nc1cccc(Br)c1
Urea, (3-bromophenyl)-
AI3-61351

2991-51-7
$C_{12}H_7F_{17}NO_4S.K$
623.36
Glycine, N-ethyl-N-((heptadecafluorooctyl)sulfonyl)-, potassium salt
FC-128
Fluorad FC 128

2996-92-1
$C_9H_{14}O_3Si$
198.32
Silane, phenyltrimethoxy
Phenyltrimethoxysilane

2997-45-7
$C_{24}H_{14}$
302.38
Dibenz(a,e)acephanthrylene
Dibenz(b,e)fluoranthene

2997-92-4
$C_8H_{18}N_6.2ClH$
271.24
Propionamidine, 2,2'-azobis-(2-methyl-, dihydrochloride
2,2'-Azobis(2-methylpropionamidine) dihydrochloride
MS 1
MS 1 (Catalyst)
Propanimidamide, 2,2'-azobis-(2-methyl-, dihydrochloride (9CI)
V 50

2998-04-1
$C_{12}H_{18}O_4$
226.30
O=C(OCC=C)CCCCC(=O)OCC=C
Adipic acid, diallyl ester
Allyl adipate
Diallyl adipate
Diallylester kyseliny adipove (Czech)
Hexanedioic acid, di-2-propenyl ester (9CI)

2999-74-8
C_2H_6Mg
54.39
Magnesium, dimethyl
Dimethylmagnesium (DOT)
UN 1368 (DOT)

3001-15-8
$C_{12}H_8I_2$
406.00
c(ccc(c(ccc(c1)I)c1)c2)(c2)I
1,1'-Biphenyl, 4,4'-diiodo- (9CI)
4,4'-Diiodo-1,1'-biphenyl

3001-61-4
$C_7H_{14}N_2O_5$
206.19
O=C(N(C(O)C1O)COC)N1COC
1,3-Bis(methoxymethyl)-4,5-dihydroxy cyclic ethyleneurea
4,5-Dihydroxy-1,3-bis(methoxymethyl)-2-imidazolidinone
Dihydroxydi(methoxymethyl)-ethyleneurea
2-Imidazolidinone, 4,5-dihydroxy-1,3-bis(methoxymethyl)- (9CI)

3001-66-9
$C_8H_{18}S$
146.30
SC(CCCCCC)C
2-Octanethiol (9CI)

3002-18-4
$C_{12}H_{21}NO_6$
275.30
O=C(OCCN(CCOC(=O)C)CCOC(=O)C)C
Ethanol, 2,2',2''-nitrilotris-, triacetate (ester) (9CI)
AI3-18283
Triethanolamine triacetate

3002-22-0
$C_{60}H_{117}NO_6$
948.59
Tris(β-stearatoethyl)amine

Octadecanoic acid, nitrilotri-2,1-ethanediyl ester (9CI)
Stearic acid, triester with 2,2',2''-nitrilotriethanol
Triethanolamine tristearate

3002-48-0
$C_{16}H_{36}N.C_6H_4NO_3$
380.56
1-Butanaminium, N,N,N-tributyl-, Salt with 4-nitrophenol (1:1) (9CI)
Ammonium, tetrabutyl-, Salt with p-nitrophenol (1:1) (8CI)
Tetrabutylammonium p-nitrophenoxide

3003-84-7
C_5H_9ClO
120.59
O(C(CC1)CCl)C1
Furan, 2-(chloromethyl)-tetrahydro
2-Chloromethyltetrahydrofuran

3004-70-4
$C_{11}H_{21}N_5O$
239.37
CCN(CC)c1nc(NC(C)C)nc(OC)n1
s-Triazine, 2-(diethylamino)-4-(isopropylamino)-6-methoxy
D 31,717
2-Diethylamino-4-isopropylamino-6-methoxy-s-triazine
G-31,717
2-Methoxy-4-isopropylamino-6-diethylamino-s-triazine

3004-71-5
$C_7H_{12}ClN_5$
201.69
CNc1nc(Cl)nc(NC(C)C)n1
s-Triazine, 2-chloro-4-(isopropylamino)-6-(methylamino)
2-Chloro-4-isopropylamino-6-methylamino-s-triazine
G-30,026
Geigy 30026
2-Methylamino-4-isopropylamino-6-chloro-s-triazine
Norazine

3004-74-8
$C_9H_8Cl_2O_3$
235.07
Benzeneacetic acid, 3,6-dichloro-2-methoxy- (9CI)
Acetic acid, (3,6-dichloro-2-methoxyphenyl)- (8CI)
Methoxyfenac
Vel 59CS52

3006-15-3
$C_{16}H_{30}O_7S.Na$
389.47
Dihexyl sodium sulfosuccinate
AI3-18858
Bis(1-methylamyl) sodium sulfosuccinate
Butanedioic acid, sulfo-, 1,4-dihexyl ester, sodium salt
Sodium 1,4-dihexyl sulfobutanedioate

3006-82-4
$C_{12}H_{24}O_3$
216.32
O=C(OOC(C)(C)C)C(CCCC)CC
Hexaneperoxoic acid, 2-ethyl-, 1,1-dimethylethyl ester (9CI)

3006-86-8
$C_{14}H_{28}O_4$
260.37
O(OC(C)(C)C)C(OOC(C)(C)C)(CCCC1)C1
1,1-Di-(tert-butylperoxy)cyclohexane (Not more than 77% in solution)
Bis(tert-butylperoxy)-1,1 cyclohexane (French)
1,1-Di-(terc-butilperoxi) ciclohexano (Spanish)
1,1-Di-(tert-butylperoxy)cyclohexane
1,1-Di-(tert-butylperoxy)cyclohexane (Not more than 40% with inert inorganic solid with less than 13% in phlegmatizer)
1,1-Di-(tert-butylperoxy)cyclohexane (Not more than 50% with phlegmatizer)
1,1-Di-(tert-butylperoxy)cyclohexane (Technically pure)
Peroxide, cyclohexylidenebis-((1,1-dimethylethyl) (9CI)
UN 2179
UN 2180
UN 2885
UN 2897

3006-93-7
$C_{14}H_8N_2O_4$
268.24
O=C(N(c(cccc1N(C(=O)C=C2)C2=O)c1)C(=O)C=3)C3
Maleimide, N,N'-(m-phenylene)di
1,3-Bismaleimidobenzene
1,3-Dimaleimidobenzene
HVA 2
HVA-2 Curing Agent
M-PHDM
N,N'-(m-Phenylene)bismaleimide
N,N'-(m-Phenylenedimaleimide)
1,1'-(m-Phenylene)bis-1H-pyrrole-2,5-dione (9CI)
1H-Pyrrole-2,5-dione, 1,1'-(phenylene)bis-

3007-31-6
$C_{24}H_{51}N$
353.67
N(CCCCCCCCCCCC)CCCCCCCCCCCC
Didodecylamine (8CI)
AI3-16576
1-Dodecanamine, N-dodecyl- (9CI)
N-Dodecyl-1-dodecanamine

3007-43-0
$C_{10}H_9N.ClH$
179.66
Lepidine, hydrochloride
Lepidin hydrochlorid (German)
4-Methylchinolin hydrochlorid (German)
4-Methylquinoline hydrochloride
Quinoline, 4-methyl-, hydrochloride (9CI)

3008-50-2
$C_{37}H_{68}O_8$
640.94
O=C(OCC(COC(=O)CCCCCCC)(COC(=O)CCCCCCC)COC(=O)CCCCCCC)CCCCCCC
Octanoic acid, 2,2-bis(((1-oxooctyl)oxy)methyl)-1,3-propanediyl ester (9CI)
Pentaerythritol tetracaprylate

3010-38-6
$C_{14}H_{14}N_4O_2$
270.32
Aniline, N,N-dimethyl-p-((o-nitrophenyl)azo)
N,N-Dimethyl-p-((o-nitrophenyl)azo)aniline
2'-Nitro-4-dimethylaminoazobenzene

3011-34-5
$C_7H_5NO_4$
167.11
4-Hydroxy-3-nitrobenzaldehyde

3012-37-1
C_8H_7NS
149.22
N#CSCc(cccc1)c1
Thiocyanic acid, benzyl ester
Benzyl-thiocyanate
Solvat 14
α-Thiocyanatotoluene
Thiocyanic acid, phenylmethyl ester (9CI)
Tropeolin

3012-65-5
$C_6H_8O_7.2H_3N$
226.19
[NH4+].[NH4+].OC(=O)C(O)

(CC([O-])=O)CC([O-])=O
Diammonium citrate
AI3-26173
Ammonium citrate, dibasic
Ammonium monohydrogen citrate
Citric acid, diammonium salt
Diammonium citrate (Secondary)
Diammonium hydrogen citrate
Dibasic ammonium citrate
2-Hydroxy-1,2,3-propanetricarboxylic acid, diammonium salt
1,2,3-Propanetricarboxylic acid, 2-hydroxy-, diammonium salt

3017-60-5
CHNS.1/2Co
88.55
Cobalt thiocyanate
Cobalt(2+) thiocyanate
Thiocyanic acid, cobalt(2+) salt (9CI)

3017-95-6
C_3H_6BrCl
157.44
BrC(CCl)C
Propane, 2-bromo-1-chloro- (9CI)
2-Bromo-1-chloropropane

3017-96-7
C_3H_6BrCl
157.44
Propane, 1-bromo-2-chloro- (8CI,9CI)

3018-12-0
C_2HCl_2N
109.94
ClC(Cl)C#N
Acetonitrile, dichloro
Dichloroacetonitrile
Dichloromethyl cyanide

3018-20-0
$C_{16}H_{16}$
208.31
Naphthalene, 1,2,3,4-tetra hydro-1-phenyl- (6CI, 7CI,8CI,9CI)
1-Phenyl-1,2,3,4-tetra hydronaphthalene
1-Phenyltetralin
1,2,3,4-Tetrahydro-1-phenylnaphthalene

3019-16-7
$C_9H_5Cl_2N_3S$
258.11
1,3,5-Triazine, 2,4-dichloro-6-(phenylthio)- (9CI)
s-Triazine, 2,4-dichloro-6-(phenylthio)- (8CI)

3021-31-6
$C_{16}H_{23}ClO_3S$
330.88
O=S(OC(CCCC1)C1Oc(ccc(c2)C(C)(C)C)c2)Cl
Chlorosulfurous acid, 2-(p-t-butylphenoxy)cyclohexyl ester
Chlorosulfurous acid, 2-(p-tert-butylphenoxy)cyclohexyl ester (8CI)

3021-63-4
$C_{10}F_{22}$
538.07
Octane, 1,1,1,2,3,3,4,4,5,5,6,6,7,8,8,8-hexadecafluoro-2,7-bis(trifluoromethyl)- (9CI)
Octane, hexadecafluoro-2,7-bis-(trifluoromethyl)- (8CI)

3021-73-6
$C_{11}H_{18}O_2$
182.26
4a(2H)-Naphthalenecar boxylic acid, octahydro-, cis- (8CI,9CI)
cis-Decalin-9-carboxylic acid

3025-52-3
$C_{16}H_{18}N_4O_2$
298.32
Benzenamine, N,N-diethyl-4-((4-nitrophenyl)azo)- (9CI)
Aniline, N,N-diethyl-p-((p-nitrophenyl)azo)- (8CI)

3025-77-2
$C_{16}H_{12}N_4O_2$
292.27
O=N(=O)c(ccc(N=Nc(c(c(ccc1)cc2)c1)c2N)c3)c3
2-Naphthalenamine, 1-((4-nitrophenyl)azo)- (9CI)
1-((4-Nitrophenyl)azo)-2-naphthalenamine

3025-88-5
Unknown
Unknown
O(O)C(CCC(OO)(C)C)(C)C
Hexane, dimethyl-, dihydroperoxide
Dimethylhexane dihydroperoxide, Dry (DOT)
Dimethylhexane dihydroperoxide, (With 18% or more water) (DOT)
UN 2174 (DOT)

3026-22-0
$C_7H_6O_2.1/2Cd$
178.32
Benzoic acid, cadmium salt (9CI)
Cadmium benzoate

3026-63-9
$C_{13}H_{28}O_4S.Na$
303.42
Sodium tridecyl sulfate
1-Tridecanol, hydrogen sulfate, sodium salt (9CI)
Tridecyl sodium sulfate

3026-66-2
$C_{17}H_{30}N.I$
375.38
[I-].CCCCCCCCCCCC[n+]1cccc1
Pyridinium, 1-dodecyl-, iodide
Dodecylpyridinium iodide
n-Dodecylpyridinium iodide
1-Dodecylpyridinium iodide
N-(n-Dodecyl)-pyridinium-jodid (German)
Laurylpyridinium iodide

3026-81-1
$C_{12}H_{28}O_2S_2Sn$
387.20
Ethanol, 2,2'-((dibutylstannylene)bis(thio))bis- (9CI)
Dibutylbis((β-hydroxyethyl)-thio)tin
2,2'-((Dibutylstannylene)bis-(thio))bisethanol
Dibutyltin bis(2-hydroxyethylmercaptide)

3028-00-0
$C_8H_{11}Cl_4N_5S$
351.09
Methanesulfenamide, 1,1,1-trichloro-N-(4-chloro-6-(ethylamino)-s-triazin-2-yl)-N-ethyl- (7CI,8CI)

3030-47-5
$C_9H_{23}N_3$
173.35
N(CCN(C)C)(CCN(C)C)C
Diethylenetriamine, 1,1,4,7,7-pentamethyl
1,2-Ethanediamine, N-(2-(dimethylamino)ethyl)-N,N',N'-trimethyl- (9CI)
Pentamethyldiethylenetriamine
N,N,N',N',N''-Pentamethyldiethylenetriamine
1,1,4,7,7-Pentamethyldiethylenetriamine
PMDT
2,5,8-Trimethyl-2,5,8-triazanonane

3031-08-1
$C_{13}H_{14}$

170.25
Naphthalene, 1,3,6-trimethyl- (8CI,9CI)

3031-51-4
$C_{13}H_{16}N_4O_6.ClH$
360.73
5-(Morpholinomethyl)-3-((5-nitrofurfurylidene)-amino)-2-oxazolidinone
Furaltadone
Furmethonol
levo-5-(Morpholinomethyl)-3-((5-nitrofurfurylidene)-amino)-2-oxazolidinone hydrochloride
NF-260
2-Oxazolidinone, 5-(morpholinomethyl)-3-((5-nitrofurfurylidene)amino)-, L-, monohydrochloride

3031-66-1
$C_6H_{10}O_2$
114.16
OC(C#CC(O)C)C
3-Hexyne-2,5-diol
Hexyne-3-diol-2,5

3031-73-0
CH_4O_2
48.04
Hidroperoxido de p-mentilo (Spanish)
Hydroperoxyde de p-menthyle (French)
p-Menthyl hydroperoxide

3031-74-1
$C_2H_6O_2$
62.07
Ethyl hydroperoxide
Ethyl hydrogen peroxide
Hidroperoxido de etilo (Spanish)
Hydroperoxyde d'ethyle (French)
Hydroperoxide, ethyl

3031-75-2
$C_3H_8O_2$
76.10
Hydroperoxide, 1-methylethyl (9CI)
Isopropyl hydroperoxide (8CI)

3032-55-1
$C_5H_9N_3O_9$
255.12
O=N(=O)OCC(C)(CON(=O)=O)CON(=O)=O
1,3-Propanediol, 2-methyl-2-((nitrooxy)methyl)-, dinitrate (ester) (9CI)
2-Methyl-2-hydroxymethyl-1,3-propanediol trinitrate
Metriol trinitrate
1,1,1-Trimethylolethane trinitrate

3033-29-2
$C_{19}H_{38}O_2SSn$
449.32
6H-1,3,2-Oxathiastannin-6-one, dihydro-2,2-dioctyl
Di-n-octyltin β-mercaptopropionate
Di-n-octyl-zinn β-mercapto-propionat (German)
Tin, dioctyl-, β-mercaptopropionate

3033-62-3
$C_8H_{20}N_2O$
160.30
O(CCN(C)C)CCN(C)C
Ethylamine, 2,2'-oxybis-(N,N-dimethyl
Bis(2-dimethylaminoethyl)ether
Niax Catalyst Al

3033-77-0
$C_6H_{14}NO.Cl$
151.66
Ammonium, (2,3-epoxy-propyl)trimethyl-, chloride
Glycidyl-trimethyl-ammonium chloride

3034-38-6
$C_3H_3N_3O_2$
113.09
O=N(=O)C(N=CN1)=C1
Imidazole, 4-nitro
4-Nitro-1H-imidazole (9CI)

3034-41-1
$C_4H_5N_3O_2$
127.12
Cn1cnc(c1)N(=O)=O
Imidazole, 1-methyl-4-nitro
1-Methyl-4-nitro-1H-imidazole (9CI)

3034-42-2
$C_4H_5N_3O_2$
127.12
Cn1cncc1N(=O)=O
Imidazole, 1-methyl-5-nitro
1-Methyl-5-nitro-1H-imidazole (9CI)

3035-45-8
$C_8H_{15}N_5O$
197.28
CNc1nc(NC(C)C)nc(OC)n1
s-Triazine, 2-isopropylamino-4-methoxy-6-methylamino
Aratone
G 32292
2-Isopropylamino-4-methoxy-6-methylamino-s-triazine
2-Methoxy-4-methylamino-6-isopropylamino-s-triazine
Noretone
1,3,5-Triazine-2,4-diamine, 6-methoxy-N-methyl-N'-(1-methylethyl)- (9CI)

3037-72-7
$C_9H_{23}NO_2Si$
205.42
CCOSi(C)(CCCCN)OCC
Butylamine, 4-(diethoxy-methylsilyl)
(4-Aminobutyl)diethoxymethyl-silane
δ-Aminobutylmethyldiethoxy-silane
Silane, (4-aminobutyl)diethoxy-methyl-

3037-89-6
$C_{36}H_{75}O_4P$
602.96
O=P(OCCCCCCCCCCCCCCCCCC)(OCCCCCCCCCCCCCCCCCC)O
Phosphoric acid, dioctadecyl ester (9CI)
Dioctadecyl phosphate
Distearyl acid phosphate
Phosphated stearyl alcohol

3039-13-2
$C_2H_2Br_2O$
201.85
Acetaldehyde, dibromo- (8CI, 9CI)

3039-83-6
$C_2H_4O_3S.Na$
131.11
Ethenesulfonic acid, sodium salt (9CI)
Sodium ethenesulfonate

3041-23-4
$C_{54}H_{111}Al$
787.46
Aluminum, trioctadecyl- (9CI)
Trioctadecylaluminum
Tri(octadecyl) aluminum

3046-94-4
$C_{12}H_{19}NO$
193.28
OCCN(c(cccc1)c1)CCCC
Ethanol, 2-(butylphenyl-amino)- (9CI)
2-(Butylphenylamino)ethanol

3047-33-4
$C_3H_3N_3O_3.3Na$
198.03
1,3,5-Triazine-2,4,6(1H,3H,5H)-trione, trisodium salt

(9CI)
Trisodium cyanurate
Trisodium 1,3,5-triazine-2,4,6 (1H,3H,5H)-trione

3048-48-4
C_8H_7NS
149.21
Benzothiazole, 4-methyl- (8CI, 9CI)

3048-64-4
C_9H_{12}
120.21
C(C=CC1C2C=C)(C1)C2
2-Norbornene, 5-vinyl
Bicyclo(2.2.1)hept-2-ene, 5-ethenyl-
Vinylnorbornene
2-Vinylnorbornene
5-Vinylnorbornene
5-Vinyl-2-norbornene

3048-65-5
C_9H_{12}
120.21
C(C(C=C1)CC=C2)(C2)C1
Indene, 3a,4,7,7a-tetrahydro
Bicyclo(4,3,0)nona-3,7-diene
1H-Indene, 3a,4,7,7a-tetrahydro-
3a,4,7,7a-Tetrahydroindene
3a,4,7,7a-Tetrahydro-1H-indene
4,7,8,9-Tetrahydroindene
Tetrahydroindene (Russian)

3049-71-6
$C_{48}H_{26}N_6O_4$
750.74
Anthra(2,1,9-def:6,5,10-d'e'f')-diisoquinoline-1,3,8,10 (2H,9H)-tetrone, 2,9-bis-(4-(phenylazo)phenyl)- (9CI)

3050-42-8
C_4Cl_8
663.34
2-Butene, octachloro- (6CI, 7CI,8CI)

3051-11-4
$C_{26}H_{20}N_4O_8S_2.2Na$
626.56
Benzenesulfonic acid, 2,2'-(1,2-ethenediyl)bis(5-((4-hydroxyphenyl)azo)-, disodium salt (9CI)
C.I. Direct Yellow 4, Disodium salt (8CI)

3054-88-4
$C_{12}H_{24}O_5S$
280.39
O=C(O)C(S(=O)(=O)O)CCCCCCCCCC
Dodecanoic acid, 2-sulfo- (6CI,7CI,8CI,9CI)
2-Sulfododecanoic acid
α-Sulfolauric acid

3054-92-0
$C_8H_{18}O$
130.23
3-Pentanol, 2,3,4-trimethyl- (6CI,7CI,8CI, 9CI)
1,1-Diisopropylethanol
Diisopropylmethylcarbinol
2,3,4-Trimethyl-3-pentanol

3055-93-4
$C_{16}H_{34}O_3$
274.44
O(CCCCCCCCCCCC)CCOCCO
Laureth-2
2-(2-(Dodecyloxy)ethoxy)-ethanol
Ethanol, 2-(2-(dodecyloxy)-ethoxy)- (9CI)
Incropol L2
Lauryl alcohol mono(oxyethylene) ethanol
PEG-2 Lauryl Ether
Polyethylene glycol 100 lauryl ether
Polyoxyethylene (2) lauryl ether
Unihydol LS-2

3055-94-5
$C_{18}H_{38}O_4$
318.56
O(CCCCCCCCCCCC)CCOCCOCCO
Ethanol, 2-(2-(2-(dodecyloxy)-ethoxy)ethoxy)
2-(2-(2-(Dodecyloxy)ethoxy)-ethoxy)ethanol
Dodecyl triethylene glycol ether
Lauryl alcohol triglycol ether
Lauryl triethoxylate
Lauryltriglycol ether
LEA
Triethylene glycol dodecyl ether

3055-95-6
$C_{22}H_{46}O_6$
406.60
O(CCOCCOCCOCCOCCO)CCCCCCCCCCCC
Laureth-5
Lauryl alcohol tetra(oxyethylene) ethanol
PEG-5 Lauryl Ether
3,6,9,12,15-Pentaoxaheptacosan-1-ol (9CI)
Polyethylene glycol (5) lauryl ether
Polyoxyethylene (5) lauryl ether

3055-96-7
$C_{24}H_{50}O_7$
450.74
3,6,9,12,15,18-Hexaoxatriacontan-1-ol
Dodecyl hexaethylene glycol
Dodecyl hexaoxyethylene monoether
Hexaethylene glycol dodecyl ether
Hexaethylene glycol lauryl ether
Hexa(oxydiethanol) monodecyl ether
Hexaoxyethylene dodecyl monoether
Lauryl hexaethoxylate

3055-97-8
$C_{26}H_{54}O_8$
494.71
O(CCOCCOCCOCCOCCOCCO)CCCCCCCCCCCC
Laureth-7
3,6,9,12,15,18,21-Heptaoxatritriacontan-1-ol (9CI)
Incropol L-7
Incropol L7-90
Lauryl alcohol hexa(oxyethylene) ethanol
PEG-7 Lauryl Ether
Polyethylene glycol (7) lauryl ether
Polyoxyethylene (7) lauryl ether

3056-93-7
$C_{22}H_{22}N_2.ClH$
350.92
3H-Indolium, 2-(2-(2-methylindol-3-yl)vinyl)-1,3,3-trimethyl-, chloride
Aizen Cathilon Orange GL
Aizen Cathilon Orange GLH
Astrazon Orange
Astrazon Orange G
Basic Orange 21
Cationic Orange ZH
C.I. 48035
C.I. Basic Orange 21
Genacryl Orange G
Nabor Orange G
Orange ZH
Sandocryl Orange B-G
Sevron Orange G
Sumiacryl Orange G

3058-38-6
$C_6H_6N_6O_6$
258.18
O=N(=O)c(c(N)c(N(=O)=O)c(N)c1N(=O)=O)c1N
1,3,5-Benzenetriamine, 2,4,6-trinitro
TATB
s-Triaminotrinitrobenzene
1,3,5-Triamino-2,4,6-trinitrobenzene
2,4,6-Trinitro-1,3,5-benzenetriamine

3060-89-7
$C_9H_{11}BrN_2O_2$
259.13

CON(C)C(=O)Nc1ccc(Br)cc1
Urea, 3-(p-bromophenyl)-1-methoxy-1-methyl
3-(4-Bromphenyl)-1-methoxy-harnstoff (German)
N'-(4-Bromophenyl)-n-methoxy-n-methylurea
3-(p-Bromophenyl)-1-methoxy-1-methylurea
3-(p-Bromophenyl)-1-methyl-1-methoxyurea
C-3126
Ciba-3126
Metobromuron
Patoran
Pattonex

3061-75-4
$C_{22}H_{45}NO$
339.60
O=C(N)CCCCCCCCCCCCCCCCCCCCC
Behenamide
Behenic acid amide
Docosanamide (9CI)
Kenamide B

3064-70-8
$C_2Cl_6O_2S$
300.80
O=S(=O)(C(Cl)(Cl)Cl)C(Cl)(Cl)Cl
Bis(trichloromethyl)sulfone
Bis(trichloromethyl) sulfone
Caswell No. 104
Chlorosulfona
EPA Pesticide Chemical Code 035601
Methane, sulfonylbis(trichloro- (9CI)
N 1386
N-1386 Biocide
Slimicide E
Sulfone, bis(trichloromethyl) (8CI)
Sulfonylbis(trichloromethane)

3066-70-4
$C_7H_{10}Br_2O_2$
285.99
Methacrylic acid, 2,3-dibromopropyl ester
2,3-Dibromopropyl methacrylate
2-Propenoic acid, 2-methyl-, 2,3-dibromopropyl ester

3067-12-7
$C_{20}H_{10}O_2$
282.30
O=C2C=C4c1ccccc1C(=O)c5ccc3cccc2c3c45
Benzo(a)pyrene-6,12-dione
6,12-Benzo(a)pyrenedione
Benzo(a)pyrene 6,12-quinone
6,12-Benzopyrene quinone
BP-6,12-quinone

3067-13-8
$C_{20}H_{10}O_2$
282.30
O=C2C=Cc3ccc4C(=O)c1ccccc1c5ccc2c3c45
Benzo(a)pyrene-1,6-dione
1,6-Benzo(a)pyrenedione
Benzo(a)pyrene-1,6-quinone
BP-1,6-quinone

3067-14-9
$C_{20}H_{10}O_2$
282.30
O=C2C=Cc3ccc4c1ccccc1C(=O)c5ccc2c3c45
Benzo(a)pyrene-3,6-dione
3,6-Benzo(a)pyrenedione
Benzo(a)pyrene-3,6-quinone
BP-3,6-quinone

3068-00-6
$C_4H_{10}O_3$
106.14
OCCC(O)CO
1,2,4-Butanetriol
1,3,4-Butanetriol
1,2,4-Butantriol (German)
2-Deoxyerythritol
1,2,4-Trihydroxybutane
Triol 124

3068-39-1
$C_{27}H_{29}N_2O_3.Cl$
464.98
Xanthylium, 3,6-bis(ethylamino)-9-(2-(methoxycarbonyl)phenyl)-2,7-dimethyl-, chloride (9CI)

3068-88-0
$C_4H_6O_2$
86.10
O=C(OC1C)C1
2-Oxetanone, 4-methyl
β-Butyrolactone
β-Butyrolakton (Czech)
3-Hydroxybutanoic acid, β-lactone
Hydroxybutyric acid lactone
3-Hydroxybutyric acid lactone
3-Hydroxybutyric acid, β-lactone
4-Methyl-2-oxetanone

3070-15-3
$C_{11}H_{17}O_3PS_2$
292.37
O(P(OCC)(Oc(ccc(SC)c1)c1)=S)CC
Phosphorothioic acid, O,O-diethyl O-(p-methylthio)-phenyl ester

3070-53-9
C_7H_{12}
96.17
C(=C)CCCC=C
1,6-Heptadiene (9CI)

3071-32-7
$C_8H_{10}O_2$
138.17
O(O)C(c(cccc1)c1)C
Ethylbenzene hydroperoxide
EBHP
Hydroperoxide, 1-phenylethyl (9CI)

3071-66-7
$C_9H_{16}ClN_5$
229.68
1,3,5-Triazine-2,4-diamine, 6-chloro-N,N'-dipropyl- (9CI)
s-Triazine, 2-chloro-4,6-bis-(propylamino)- (8CI)

3071-73-6
$C_{36}H_{25}N_5O_6S_2.2Na$
733.71
C.I. Acid Black 24, Disodium salt (8CI)
1-Naphthalenesulfonic acid, 8-(phenylamino)-5-((4-((5-sulfo-1-naphthalenyl)azo)-1-naphthalenyl)azo)-, disodium salt (9CI)

3073-66-3
C_9H_{18}
126.27
C(CCCC1C)(C1)(C)C
Cyclohexane, 1,1,3-trimethyl
1,1,3-Trimethylcyclohexane

3074-00-8
$C_{19}H_{10}O$
254.29
Naphthanthrone
6H-Benzo[cd]pyren-6-one (6CI, 7CI, 8CI, 9CI)

3074-03-1
$C_{17}H_{10}O$
230.27
11H-Benzo(b)fluoren-11-one
Benzo(b)fluoren-11-one

3074-75-7
C_9H_{20}
128.26
2-Methyl-4-ethylhexane

3074-76-8
C_9H_{20}
128.26
Hexane, 3-ethyl-3-methyl- (8CI, 9CI)

3074-77-9

C_9H_{20}
128.26
3-Methyl-4-ethylhexane

3076-04-8
$C_{16}H_{30}O_2$
254.46
O=C(OCCCCCCCCCCCCC)C=C
Acrylic acid, tridecyl ester
2-Propenoic acid, tridecyl ester (9CI)
Tridecyl acrylate

3076-05-9
$C_8H_{17}NO_3S.Na$
230.28
Ethanesulfonic acid, 2-(cyclohexylamino)-, monosodium salt (9CI)
(2-Cyclohexylamino)ethanesulfonic acid, sodium salt

3076-26-4
$C_{19}H_{38}O_5S$
378.57
O=C(OC)C(S(=O)(=O)O)CCCCCCCCCCCCCCCC
Octadecanoic acid, 2-sulfo-, 1-methyl ester (9CI)
Methyl stearate α-sulfonic acid
1-Methyl 2-sulfooctadecanoate

3076-63-9
$C_{36}H_{75}O_3P$
586.96
O(P(OCCCCCCCCCCCC)OCCCCCCCCCCCC)CCCCCCCCCCCC
Phosphorous acid, tridodecyl ester (9CI)
AI3-51074

3077-30-3
$C_{12}H_{25}NO_3$
231.33
O=C(N(CCO)CCO)CCCCCCC
Octanamide, N,N-bis(2-hydroxyethyl)- (9CI)
N,N-Bis(2-hydroxyethyl)octanamide
Caprylic acid, diethanolamide
Caprylic diethanolamide

3077-37-0
$C_{13}H_{27}NO_3$
245.36
O=C(N(CCO)CCO)CCCCCCCC
Nonanamide, N,N-bis(2-hydroxyethyl)- (9CI)
N,N-Bis(2-hydroxyethyl)nonanamide
Pelargonic acid, diethanolamide
Pelargonic acid, diethanolamine amide
Pelargonic acid, diethanolamine soap
Pelargonic diethanolamide

3081-01-4
$C_{19}H_{26}N_2$
282.42
N(c(ccc(Nc(cccc1)c1)c2)c2)C(CCC(C)C)C
1,4-Benzenediamine, N-(1,4-dimethylpentyl)-N'-phenyl- (9CI)
p-Phenylenediamine, N-(1,4-dimethylpentyl)-N'-phenyl-

3081-14-9
$C_{20}H_{36}N_2$
304.58
N(c(ccc(NC(CCC(C)C)C)c1)c1)C(CCC(C)C)C
p-Phenylenediamine, N,N'-bis-(1,4-dimethylpentyl)
N,N'-Bis(1,4-dimethylpentyl)-p-phenylenediamine
N,N-Di(1,4-dimethylpentyl)-p-phenylenediamine
Eastozone
Eastozone 33
Elastozone 33
NCI-C56337
Santoflex 77
Tenamene

3083-23-6
$C_3H_3Cl_3O$
161.41
Propane, 1,2-epoxy-3,3,3-trichloro
1,2-Epoxy-3,3,3-trichloropropane
Oxirane, (trichloromethyl)- (9CI)
Propane, 1,1,1-trichloro-2,3-epoxy-
TCPO
Trichloropropane oxide
1,1,1-Trichloropropane-2,3-oxide
Trichloropropene oxide
1,1,1-Trichloropropene oxide
1,1,1-Trichloropropene-2,3-oxide
3,3,3-Trichloropropene oxide
1,1,1-Trichloropropylene oxide
3,3,3-Trichloropropylene oxide

3083-25-8
$C_4H_5Cl_3O$
175.44
O(C1CC(Cl)(Cl)Cl)C1
Butane, 1,2-epoxy-4,4,4-trichloro
1,2-Epoxy-4,4,4-trichlorobutane
Oxirene, 2,2,2-trichloroethyl-
Trichlorobutylene oxide
4,4,4-Trichloro-1,2-epoxybutane

3084-92-2
$C_6H_{10}ClN_5$
187.60
1,3,5-Triazine-2,4-diamine, 6-chloro-N-ethyl-N'-methyl- (9CI)
G 28509
s-Triazine, 2-chloro-4-(ethylamino)-6-(methylamino)- (8CI)

3084-94-4
$C_7H_{12}BrN_5$
246.08
1,3,5-Triazine-2,4-diamine, 6-bromo-N,N'-diethyl- (9CI)
s-Triazine, 2-bromo-4,6-bis-(ethylamino)- (8CI)

3085-30-1
$C_4H_{10}O.1/3Al$
83.12
1-Butanol, aluminum salt (9CI)
Aluminum n-butoxide

3085-35-6
$C_5H_{12}O_2$
104.15
O(CCCC)CO
Methanol, butoxy- (9CI)
Butoxymethanol

3085-82-3
$C_{15}H_{18}N_2$
226.31
N(c(ccc(N)c1)c1)(c(cccc2)c2)C(C)C
1,4-Benzenediamine, N-(1-methylethyl)-N-phenyl- (9CI)
N-(1-Methylethyl)-N-phenyl-1,4-benzenediamine

3087-36-3
$C_2H_6O.1/4Ti$
58.04
CCO[Ti](OCC)(OCC)OCC
Ethanol, titanium(4+) salt (9CI)
Ethyl titanate
Tetraethyl titanate

3088-31-1
$C_{16}H_{34}O_6S.Na$
377.50
Diethylene glycol monolauryl ether sodium sulfate
Diethylene glycol monododecyl ether sodium sulfate
Diethylene glycol monododecyl ether sulfate sodium salt
Diethylene glycol monolauryl ether sulfate sodium salt
2-(2-Dodecyloxyethoxy)ethyl sodium sulfate
Ethanol, 2-(2-(dodecyloxy)-ethoxy)-, hydrogen sulfate, sodium salt (9CI)

Lauristyl diglycol ether sulfate sodium salt
Lauryl diethylene glycol ether sulfonate sodium
Sodium diethylene glycol dodecyl ether sulfate
Sodium dioxyethylenedodecyl ether sulfate
Sodium lauryl alcohol diglycol ether sulfate
Sodium lauryl di(oxyethyl) sulfate
Sodiumlaurylglycolether sulfate
Sodium lauryloxyethoxyethyl sulfate
Sulfuric acid mono(2-(2-(dodecyloxy)ethoxy)ethyl) ether sodium salt
Tergentol

3089-11-0
$C_{15}H_{30}N_6O_6$
390.51
O(CN(c(nc(nc1N(COC)COC)N(COC)COC)n1)COC)C
1,3,5-Triazine-2,4,6-triamine, N,N,N',N',N'',N''-hexakis-(methoxymethyl)
Cymel 303
Cyrez 963 Resin
Hexakis-methoxymethyl-melamin (Czech)
Hexakis(methoxymethyl)-melamine
Hexakis(methoxymethyl)-s-triazine-2,4,6-triamine
Hexa(methoxymethyl)melamine
Hexamethylol-melamin-hexa-methylaether (German)
Hexamethyl methylolmelamine
LK 36
Malamine, hexakis(methoxy-methyl)
Metazin
Metazine

3089-16-5
$C_{20}H_{10}Cl_2N_2O_2$
381.20
Oc(c(c(nc1cc(c(nc(c2ccc3)c3Cl)c4)c2O)c(cc5)Cl)c5)c14
Quino(2,3-b)acridine-7,14-dione, 4,11-dichloro-5,12-dihydro- (9CI)
Quinacridone, 4,11-dichloro-

3089-17-6
$C_{20}H_{10}Cl_2N_2O_2$
381.20
Oc(c(c(nc1cc(c(nc(c2cc(c3)Cl)c3)c4)c2O)ccc5Cl)c5)c14
Quino(2,3-b)acridine-7,14-dione, 2,9-dichloro-5,12-dihydro- (9CI)

3091-25-6
$C_8H_{17}Cl_3Sn$
338.29
Stannane, octyltrichloro
Octyltrichlorostannane
Mono-n-octyltin trichloride
Mono-n-octyl-zinn-trichlorid (German)
Stannane, trichlorooctyl-
Tin, octyl-, trichloride

3095-95-2
$C_6H_{13}O_5P$
196.14
Acetic acid, (diethoxyphosphinyl)- (9CI)
Acetic acid, phosphono-, P,P-diethyl ester (8CI)
NSC-272281

3096-47-7
$C_{13}H_7ClO$
214.65
9H-Fluoren-9-one, 2-chloro- (9CI)
Fluoren-9-one, 2-chloro- (8CI)
NSC-107566

3097-08-3
$C_{12}H_{26}O_4S.1/2Mg$
278.56
Magnesium lauryl sulfate
Caswell No. 532A
EPA Pesticide Chemical Code 079017
Magnesium monododecyl sulfate
Sulfuric acid, monododecyl ester, magnesium salt (9CI)

3101-60-8
$C_{13}H_{18}O_2$
206.31
O(C1COc(ccc(c2)C(C)(C)C)c2)C1
Propane, 1-(p-tert-butylphenoxy)-2,3-epoxy
tert-Butylphenol glycidyl ether
3-(p-tert-Butylphenoxy)-1,2-epoxypropane
tert-Butylphenyl glycidyl ether
Oxirane, ((4-(1,1-dimethylethyl)phenoxy)methyl)- (9CI)
R 1007

3102-33-8
C_5H_8O
84.12
3-Penten-2-one, (E)- (8CI, 9CI)
trans-3-Penten-2-one
(E)-3-Penten-2-one

3105-97-3
$C_{20}H_{24}N_2O_2S$
356.52
CCN(CC)CCNc2ccc(CO)c3sc1ccccc1c(=O)c23
9H-Thioxanthen-9-one, 1-((2-(diethylamino)ethyl)-amino)-4-(hydroxymethyl)
1-((2-(Diethylamino)ethyl)-amino)-4-(hydroxymethyl)-thioxanthen-9-one
1-((2-(Diethylamino)ethyl)-amino)-4-(hydroxymethyl)-9H-thioxanthen-9-one
Hycanthon
Hycanthone
Lucanthone Metabolite
NSC-134434
WIN 24933

3112-85-4
$C_7H_8O_2S$
156.21
CS(=O)(=O)c1ccccc1
Sulfone, methyl phenyl
Benzene, (methylsulfonyl)-(9CI)
Methyl phenyl sulfone
Phenyl methyl sulfone
(Phenylsulfonyl)methane

3113-71-1
$C_8H_7NO_4$
181.14
O=C(O)c(ccc(N(=O)=O)c1C)c1
Benzoic acid, 3-methyl-4-nitro- (9CI)
3-Methyl-4-nitrobenzoic acid

3113-72-2
$C_8H_7NO_4$
181.14
O=C(O)c(c(N(=O)=O)ccc1C)c1
Benzoic acid, 5-methyl-2-nitro- (9CI)
3-Methyl-6-nitrobenzoic acid
5-Methyl-2-nitrobenzoic acid

3114-52-1
$C_{21}H_{14}ClN_3$
343.82
1,3,5-Triazine, 2-(4-chlorophenyl)-4,6-diphenyl- (9CI)
s-Triazine, 2-(p-chlorophenyl)-4,6-diphenyl- (7CI,8CI)

3114-54-3
$C_{21}H_{12}Cl_3N_3$
412.71
1,3,5-Triazine, 2,4,6-tris-(4-chlorophenyl)- (9CI)
s-Triazine, 2,4,6-tris-(p-chlorophenyl)- (6CI, 7CI,8CI)

3114-70-3
$C_6H_{14}N_2$
114.18
1,4-Cyclohexanediamine (8CI,9CI)

3115-39-7
$C_{16}H_{35}O_4P$
322.43
O=P(OCCCCCCCC)(OCCCCCC
CC)O
Phosphoric acid, dioctyl ester (9CI)
AI3-15045
Di(octyl) hydrogen phosphate
Dioctyl phosphate

3115-49-9
$C_{17}H_{26}O_3$
278.39
O=C(O)COc(ccc(c1)CCCCCCC
CC)c1
Acetic acid, (4-nonylphenoxy)- (9CI)
(4-Nonylphenoxy)acetic acid

3115-68-2
$C_{16}H_{36}P.Br$
339.40
[Br-].CCCC[P+](CCCC)(CCCC)
CCCC
Phosphonium, tetrabutyl-, bromide
Tetra-n-butylphosphonium bromide

3116-76-5
$C_{19}H_{17}Cl_2N_3O_5S$
470.31
Cc1onc(c1C(=O)NC3C2SC(C)
(C)C(N2C3=O)C(O)=O)c4c
(Cl)cccc4Cl
Dicloxacillin
BRL 1702
6-(3-(2,6-Dichlorophenyl)-5-methyl-4-isoxazolecarboxamido)-3,3-dimethyl-7-oxo-4-thia-1-azabicyclo(3.2.0)-heptane-2-carboxylic acid
6-(3-(2,6-Dichlorophenyl)-5-methyl-4-isoxazolecarboxamido)penicillanic acid
3-(2,6-Dichlorophenyl)-5-methyl-4-isoxazolylpenicillin
Diclossacillina
Dicloxacilin
Dicloxacilina (Spanish)
Dicloxacilline (French)
Dicloxacillinum (Latin)
Dicloxacycline
Maclicine
Methyldichlorophenylisoxazolylpenicillin
R-13423
4-Thia-1-azabicyclo(3.2.0)heptane-2-carboxylic acid, 6-(((3-(2,6-dichlorophenyl)-5-methyl-4-isoxazolyl)carbonyl)amino)-3,3-dimethyl-7-oxo-, (2S-(2α,5α,6β))-

3118-97-6
$C_{18}H_{16}N_2O$
276.36
Oc(ccc(c1ccc2)c2)c1N=Nc(c(cc
(c3)C)C)c3
2-Naphthol, 1-(2,4-xylylazo)
A.F. Red No. 5
Aizen Food Red No. 5
Brasilazina Oil Scarlet 6G
Brilliant Oil Scarlet B
Calco Oil Scarlet BL
Calco Oil Scarlet ZBL
Ceres Oranges RR
Cerisol Scarlet G
Cerotinscharlach G
C.I. 12140
C.I. Solvent Orange 7
1-((2,4-Dimethylphenyl)azo)-2-naphthalenol
Ext. D & C. Red. No. 14
Fast Oil Orange II
Fat red (Yellowish)
Fat Scarlet 2G
Fettorange B
Grasan Orange 3R
Lacquer Orange VR
Motirot G
Oil Orange KB
Oil Orange N Extra
Oil Orange R
Oil Orange 2R
Oil Orange X
Oil Orange XO
Oil Red GRO
Oil Red O
Oil Red RO
Oil Red XO
Oil Scarlet
Oil Scarlet 371
Oil Scarlet APYO
Oil Scarlet BL
Oil Scarlet 6G
Oil Scarlet L
Oil Scarlet Y
Oil Scarlet YS
Orange Insoluble OLG
Orange Insoluble RR
Orange Oil KB
Oranz Rozpoustedlova 7 (Czech)
Ponceau Insoluble OLG
Pyronalrot R
Red B
Red No. 5
Resin Scarlet 2R
Resoform Orange R
Rot B
Rot GG Fettloeslich
Somalia Orange A2R
Somalia Orange 2R
Soudan II
Sudan 2
Sudan II
Sudan Orange RPA
Sudan Orange RRA
Sudan Red
Sudan Scarlet 6G
Sudan X
Sudan AX
Sudan Orange
Waxakol Vermilion L
1-Xylylazo-2-naphthol
1-(o-Xylylazo)-2-naphthol
1-(2,4-Xylylazo)-2-naphthol

3120-74-9
$C_8H_{10}OS$
154.24
Oc(ccc(SC)c1C)c1
m-Cresol, 4-(methylthio)
3-Methyl-4-methylthiophenol
4-(Methylthio)-m-cresol
MMTP
Phenol, 3-methyl-4-(methylthio)- (9CI)
USAF MA-17

3121-61-7
$C_6H_{10}O_3$
130.16
O=C(OCCOC)C=C
Ethanol, 2-methoxy-, acrylate
Acrylic acid, 2-methoxyethyl ester
Ethylene glycol monomethyl ether acrylate
Glycol monomethyl ether acrylate
2-Methoxyethanol, acrylate
Methoxyethyl acrylate
2-Methoxyethyl acrylate
Methyl cellosolve acrylate
2-Propenoic acid, 2-methoxyethyl ester

3121-71-9
$C_{13}H_{12}O_2$
200.24
α-Naphthyl propionate
AI3-18247
1-Naphthalenol, propanoate

3121-79-7
$C_7H_{16}O$
116.20
1-Pentanol, 4,4-dimethyl- (9CI)

3123-97-5
$C_6H_{10}O_2$
114.14
2(3H)-Furanone, dihydro-5,5-dimethyl- (8CI,9CI)
γ-Isocaprolactone
NSC-128078
NSC-221122
Valeric acid, 4-hydroxy-4-methyl-, lactone
Valeric acid, 4-hydroxy-4-methyl-, γ-lactone

3126-95-2
$C_6H_{12}O_2$
116.18
Propane, 1,2-epoxy-3-propoxy
1,2-Epoxy-3-propoxypropane
Ether, 2,3-epoxypropyl propyl
Glycidyl propyl ether

(Propoxy methyl)oxirane
Propyl glycidyl ether

3128-06-1
$C_6H_{10}O_3$
130.14
O=C(O)CCCC(=O)C
Hexanoic acid, 5-oxo- (9CI)
5-Oxohexanoic acid

3129-90-6
CHNS
59.09
Isothiocyanic acid
Acide isothiocyanique (French)
Acido isotiocianico (Spanish)
Hydrogen isothiocyanate

3129-91-7
$C_{12}H_{23}N.HNO_2$
228.38
C1CCC(CC1)NC2CCCCC2
Dicyclohexylamine, nitrite
Cyclohexanamine, N-cyclohexyl-, nitrite (9CI)
DECHAN
DIANA
DICHAN (Czech)
Dicyclohexylaminonitrite
Dicyclohexylammonium nitrite
Dicyklohexylamin nitrit (Czech)
Dicynit (Czech)
Dodecahydrophenylamine nitrite
Dusitan dicyklohexylaminu (Czech)
Leukorrosin C
NDA

3129-92-8
$C_7H_6O_2.C_6H_{13}N$
221.29
Benzoic acid, Compd. with cyclohexanamine (1:1) (9CI)
Cyclohexylammonium benzoate

3130-19-6
$C_{20}H_{30}O_6$
366.50
O=C(OCC(CC(O1)C1C2)C2)CCCCC(=O)OCC(CC(O3)C3C4)C4
Hexanedioic acid, bis(7-oxabicyclo(4.1.0)hept-3-ylmethyl ester
Bis((3,4-epoxycyclohexyl)-methyl)adipate

3130-29-8
Unknown
Unknown
Potassium 4-tert-butylphenate

3130-95-8
Unknown
Unknown
Nitrilotriacetic acid and its salts
NTA, scandium(3+) salt (1:1)

3131-60-0
$C_8H_{12}N_4O_5$
244.24
as-Triazin-3(2H)-one, 5-amino-2-β-D-ribofuranosyl
5-Amino-2-β-D-ribofuranosyl-as-triazin-3(2H)-one
6-Azacytidine

3132-64-7
C_3H_5BrO
136.99
O(C1CBr)C1
Propane, 3-bromo-1,2-epoxy
3-Bromo-1,2-epoxypropane
Epibromhydrin
Epibromohydrin [UN 2558]
Epibromohydrine
UN 2558 [Epibromohydrin]

3132-99-8
C_7H_5BrO
185.02
O=Cc(cccc1Br)c1
3-Bromobenzaldehyde
Benzaldehyde, 3-bromo- (9CI)
m-Bromobenzaldehyde

3133-01-5
$C_{23}H_{48}O$
340.63
1-Tricosanol (8CI,9CI)

3134-12-1
$C_{16}H_{14}Cl_2N_2O_2$
337.22
CN(C)C(=O)N(C(=O)c1ccccc1)c2ccc(Cl)c(Cl)c2
Benzamide, N-(3,4-dichlorophenyl)-N-((dimethylamino)-carbonyl)
Benzomarc
Benzuride
Benzuron
N-Benzyl-N-(dichloro-3,4-phenyl)-N',N'-dimethylurea
N-(3,4-Dichlorophenyl)-N-((dimethylamino)carbonyl)benzamide
Phenobenzoron
Phenobenzuron
PP-65-25
Urea, 1-benzoyl-1-(3,4-dichlorophenyl)-3,3-dimethyl- (8CI)

3134-66-5
$C_{10}H_{19}NOS$
201.33
1H-Azepine-1-carbothioic acid, hexahydro-, S-propyl ester (8CI,9CI)

3134-71-2
$C_{11}H_{21}NOS$
215.35
1H-Azepine-1-carbothioic acid, hexahydro-, S-(2-methylpropyl) ester (9CI)
1H-Azepine-1-carbothioic acid, hexahydro-, S-isobutyl ester (8CI)

3138-42-9
$C_{10}H_{23}O_4P$
238.26
O=P(OCCCCC)(OCCCCC)O
Phosphoric acid, dipentyl ester
AI3-16972
Diamyl acid phosphate
Dipentyl phosphate
Phosphoric acid, diphenyl ester

3140-73-6
$C_5H_6ClN_3O_2$
175.56
1,3,5-Triazine, 2-chloro-4,6-dimethoxy- (9CI)
NSC-46520
s-Triazine, 2-chloro-4,6-dimethoxy- (8CI)

3141-11-5
$C_{10}H_{15}AsO_2$
242.15
Arsonous acid, phenyl-, diethyl ester (9CI)
Benzenearsonous acid, diethyl ester (6CI,7CI,8CI)
Diethoxyphenylarsine
Diethoxyphenylarsine oxide

3141-26-2
$C_4H_2Br_2S$
241.93
Thiophene, 3,4-dibromo- (9CI)

3141-27-3
$C_4H_2Br_2S$
241.93
S(C(=CC=1)Br)C1Br
Thiophene, 2,5-dibromo- (9CI)
AI3-08106
2,5-Dibromothiophene

3142-72-1
$C_6H_{10}O_2$
114.14
O=C(O)C(=CCC)C
2-Pentenoic acid, 2-methyl- (9CI)
2-Methyl-2-pentenoic acid

3144-16-9
$C_{10}H_{16}O_4S$
232.32

O=C(C(C(C1C2)(C)C)(C2)CS
(=O)(=O)O)C1
10-Bornanesulfonic acid, 2-oxo-, (1S,4R)-(+)
Bicyclo(2.2.1)heptane-1-methanesulfonic acid, 7,7-dimethyl-2-oxo-, (1S)- (9CI)
Camphersulfosaeure (German)
Camphorsulfonic acid
(+)-Camphorsulfonic acid
(+)-β-Camphorsulfonic acid
d-Camphorsulfonic acid
d-10-Camphorsulfonic acid
Reychler's Acid

3146-39-2
$C_7H_{10}O$
110.17
Norbornane, 2,3-epoxy-, exo
exo-2,3-Epoxynorbornane
Norbornane oxide, exo-2,3-
exo-Norbornene oxide
3-Oxatricyclo(3.2.1.0^{2,4})octane, (1-α,2-β,4-β,5-α)- (9CI)
exo-2,3-Oxidonorbornane

3147-75-9
$C_{20}H_{25}N_3O$
323.42
Oc(c(N(N=C(C=1C=CC=2)C2)
N1)cc(c3)C(CC(C)(C)C)(
C)C)c3
Octrizole
2-(2H-Benzotriazol-2-yl)-4-(1,1,3,3-tetramethylbutyl)-phenol
Cyasorb 5411
2-(2'-Hydroxy-5'-t-octyl-phenyl)benzotriazole
Octrizol (Spanish)
Octrizolum (Latin)
Phenol, 2-(2H-benzotriazol-2-yl)-4-(1,1,3,3-tetramethyl-butyl)- (9CI)
Spectra-Sorb UV 5411
UV Absorber-5

3148-72-9
$C_{11}H_{18}N_2O_9$
322.31
O=C(O)CN(CC(O)CN(CC(=O)O)
CC(=O)O)CC(=O)O
Acetic acid, ((2-hydroxy-1,3-trimethylene)dinitrilo)-tetra
Acetic acid, diaminopropanol-tetra-
Acetic acid, ((2-hydroxytri-methylene)dinitrilo)tetra- (8CI)
DHPTA
Diaminopropanol tetra acetic acid
DPTA
DTA
Glycine, N,N'-(2-hydroxy-1,3-propanediyl)bis(N-(car-boxymethyl)- (9CI)

3149-28-8
$C_5H_6N_2O$
110.10
O(c(nccn1)c1)C
Pyrazine, methoxy- (9CI)
Methoxypyrazine

3149-65-3
$C_{11}H_{22}O_6$
250.29
β-D-Glucopyranoside, methyl 2,3,4,6-tetra-O-methyl- (9CI)
Glucopyranoside, methyl tetra-O-methyl-, β-D- (8CI)
NSC-226830

3153-36-4
$C_6H_{11}ClO_2$
150.60
Butanoic acid, 4-chloro-, ethyl ester (9CI)
Butyric acid, 4-chloro-, ethyl ester (8CI)

3159-62-4
$C_{18}H_{36}O_3.1/2Ca$
320.52
Octadecanoic acid, 12-hydroxy-, calcium salt (2:1) (9CI)

3160-37-0
$C_{11}H_{12}O_3$
192.23
O=C(C=Cc(ccc(OCO1)c12)c2)C
2-Butanone, 4-(3,4-methylene-dioxyphenyl)
Heliotropyl acetone
3,4-Methylenedioxybenzyl acetone
4-(3,4-Methylenedioxyphenyl)-2-butanone
Piperonyl acetone

3161-99-7
C_6H_2
74.08
1,3,5-Hexatriyne (9CI)
Hexatriyne (6CI,7CI,8CI)
Triacetylene

3164-29-2
$C_4H_4O_6.2H_4N$
184.18
Tartaric acid, diammonium salt
Ammonium tartrate
Ammonium d-tartrate
Butanedioic acid, 2,3-di-hydroxy-, (R-(R*,R*))-, di-ammonium salt
Diammonium tartrate
2,3-Dihydroxy-butanedioic acid, diammonium salt (9CI)
L-Tartaric acid, ammonium salt

3164-60-1
$C_{22}H_{39}O_3P$
382.52
O(P(OCC(CCCC)CC)Oc(cccc1)
c1)CC(CCCC)CC
Phosphorous acid, bis(2-ethyl-hexyl) phenyl ester (9CI)

3164-85-0
$C_8H_{16}O_2.K$
183.31
Hexanoic acid, 2-ethyl-, potassium salt (9CI)
2-Ethylhexanoic acid, potassium salt
Potassium 2-ethylhexanoate

3165-93-3
$C_7H_8ClN.ClH$
178.07
Cl.Cc1cc(Cl)ccc1N
o-Toluidine, 4-chloro-, hydro-chloride
Amarthol Fast Red TR Base
Amarthol Fast Red TR Salt
2-Amino-5-chlorotoluene hydro-chloride
Azanil Red Salt TRD
Azoene Fast Red TR Salt
Azoic Diazo Component 11 Base
Azogene Fast Red TR
Benzenamine, 4-chloro-2-methyl-, hydrochloride
Brentamine Fast Red TR Salt
Chlorhydrate de 4-chloro-orthotoluidine (French)
5-Chloro-2-aminotoluene hydro-chloride
4-Chloro-2-methylbenzenamine hydrochloride
4-Chloro-2-methylaniline hydrochloride
4-Chloro-6-methylaniline hydro-chloride
4-Chloro-2-toluidine hydro-chloride
4-Chloro-o-toluidine hydro-chloride [UN 1579]
C.I. 37085
C.I. Azoic Diazo Component 11
Daito Red Salt TR
Devol Red K
Devol Red TA Salt
Devol Red TR
Diazo Fast Red TR
Diazo Fast Red TRA
Fast Red 5CT Salt
Fast Red Salt TR
Fast Red Salt TRA
Fast Red Salt TRN
Fast Red TR Salt
Hindasol Red TR Salt
Kromon Green B
2-Methyl-4-chloroaniline hydro-chloride
Natasol Fast Red TR Salt
NCI-C02368

Neutrosel Red TRVA
OFNA-Perl Salt RRA
RCRA waste number U049
Red Base Ciba IX
Red Base IRGA IX
Red Salt Ciba IX
Red Salt IRGA IX
Red TRS Salt
Sanyo Fast Red Salt TR
UN 1579 [4-Chloro-o-toluidine hydrochloride]

3167-49-5
$C_6H_6N_2O_2$
138.11
O=C(O)c(ccc(n1)N)c1
6-Aminonicotinic acid
6-Amino-3-pyridinecarboxylic acid
Nicotinic acid, 6-amino- (8CI)
3-Pyridinecarboxylic acid, 6-amino- (9CI)

3170-83-0
HO_2
33.01
Perhydroxyl radical
Peroxyl radical

3172-52-9
$C_4H_2Cl_2S$
153.03
S(C(=CC=1)Cl)C1Cl
Thiophene, 2,5-dichloro- (9CI)
2,5-Dichlorothiophene

3173-53-3
$C_7H_{11}NO$
125.19
O=C=NC(CCCC1)C1
Isocyanic acid, cyclohexyl ester
Cyclohexane, isocyanato- (9CI)
Cyclohexyl isocyanate [UN 2488]
Isocyanatocyclohexane
NSC-87419
UN 2488 [Cyclohexyl isocyanate]

3173-72-6
$C_{12}H_6N_2O_2$
210.20
O=C=Nc(c(c(c(N=C=O)cc1)cc2)c1)c2
Isocyanic acid, 1,5-naphthylene ester
1,5-Naphthalene diisocyanate
Naphthalene, 1,5-diisocyanato- (9CI)

3175-23-3
$C_3H_5Cl_3$
147.43
C(CCl)(Cl)(Cl)C
Propane, 1,2,2-trichloro (8CI, 9CI)
1,2,2-Trichloropropane

3176-03-2
$C_{19}H_{27}NO_4$
333.42
Drotebanol
DEA No. 9335
Dihydro-14-hydroxy-4-o-methyl-6-β-thebainol
Dihydro-14-hydroxy-6-β-thebainol 4 methyl ester
6-β,14-Dihydroxy-3,4-dimethoxy-N-methylmorphinan
3,4-Dimethoxy-17-methylmorphinan-6 β,14-diol
Drotebanolum (Latin)
14-Hydroxydihydro-6-β-thebainol 4-methyl ether
14-Hydroxy-6-β-thebainol 4-methyl ether
Metebanyl
Morphinan, 6-β,14-dihydroxy-3,4-dimethoxy-N-methyl-
Morphinan-6-β,14-diol, 3,4-dimethoxy-17-methyl-
Oxymetebanol
Oxymethebanol
Ram-326

3176-88-3
$C_{20}H_{17}N_3O_4$
363.35
1H-Naphth(2,3-f)isoindole-1,3,5,10(2H)-tetrone, 4,11-diamino-2-butyl- (9CI)
2,3-Anthracenedicarboximide, 1,4-diamino-N-butyl-9,10-dihydro-9,10-dioxo- (8CI)

3178-22-1
$C_{10}H_{20}$
140.27
tert-Butylcyclohexane
t-Butylcyclohexane
Cyclohexane, tert-butyl- (8CI)
Cyclohexane, (1,1-dimethylethyl)- (9CI)

3179-31-5
$C_2H_3N_3S$
101.14
N(=C(S)NN=1)C1
1H-1,2,4-Triazole-3-thiol
1H-1,2,4-Triazole, 3-mercapto-
1,2,4-Triazol-thiol-(3) (German)

3179-47-3
$C_{14}H_{26}O_2$
226.36
O=C(OCCCCCCCCCC)C(=C)C
2-Propenoic acid, 2-methyl-, decyl ester (9CI)
n-Decyl methacrylate
Decyl 2-methyl-2-propenoate
Methacrylic acid, decyl ester (8CI)
NSC-20975

3179-56-4
$C_8H_{14}O_5S$
222.26
O=C(OOS(=O)(=O)C(CCCC1)C1)C
Acetyl cyclohexanesulfonyl peroxide
Acetyl cyclohexanepersulfonate
Acetyl cyclohexanesulfonyl peroxide, More than 82% wetted with less than 12% water
Acetyl cyclohexanesulfonyl peroxide, Not more than 32% in solution
Acetyl cyclohexanesulfonyl peroxide, Not more than 82% wet with not less than 12% water
Acetyl cyclohexylsulfonyl peroxide
Lupersol 228Z
Peroxide, acetyl cyclohexylsulfonyl (8CI,9CI)
Peroxide, acetyl cyclohexylsulfonyl, More than 82% wetted with less than 12% water
Peroxide, acetyl cyclohexylsulfonyl, Not more than 32% in solution
Peroxide, acetyl cyclohexylsulfonyl, Not more than 82% wetted with not less than 12% water
Peroxido de acetil ciclohexano sulfonilo (Spanish)
Peroxyde d'acetyle et de cyclohexane sulfonyle (French)
UN 2082
UN 2083

3179-76-8
$C_8H_{21}NO_2Si$
191.39
CCOSi(C)(CCCN)OCC
Propylamine, 3-(diethoxymethylsilyl)
(3-Aminopropyl)diethoxymethylsilane
Silane, (3-aminopropyl)diethoxymethyl-

3179-80-4
$C_{17}H_{36}N_2O$
284.47
O=C(NCCCN(C)C)CCCCCCCCCCC
Lauramidopropyl dimethylamine
N-(3-Dimethylaminopropyl)dodecanamide
N-(3-(Dimethylamino)propyl)-dodecanamide
Dimethylaminopropyl lauramide
N-(3-Dimethylaminopropyl)-lauramide
N,N-Dimethyl-N-(3-dodecan-

amidopropyl)amine
N,N-Dimethyl-(3-dodecylamidopropyl)amine
Dodecanamide, N-(3-(dimethylamino)propyl)- (9CI)
Dodecanoylamidopropyldimethylamine
Jordamide DAPLM
Jordamide LAA
Lauramidopropyldimethylamine
Lauric 3-dimethylaminopropylamide
Lauroyl(dimethyl amino propyl)amine
N-Lauroyl-3-(dimethylamino)-propylamine

3179-89-3
$C_{17}H_{20}N_4O_4$
344.35
O=N(=O)c(ccc(N=Nc(c(cc(N(CCO)CCO)c1)C)c1)c2)c2
Ethanol, 2,2'-((3-methyl-4-((4-nitrophenyl)azo)phenyl)-imino)bis- (9CI)

3179-90-6
$C_{18}H_{18}N_2O_6$
358.38
O=C(c(c(c(NCCO)cc1)C(=O)c2c(O)ccc3O)c1NCCO)c23
Anthraquinone, 1,4-bis((2-hydroxyethyl)amino)-5,8-dihydroxy
5,8-Dihydroxy-1,4-dihydroxyethylaminoanthraquinone
1,4-Dioxyethylamino-5,8-dioxyanthraquinone (Russian)
1,4-DOEA-5,8-DAPFA (Russian)

3180-09-4
$C_{10}H_{14}O$
150.24
Phenol, o-butyl
2-n-Butylphenol
o-Butylphenol, Liquid [UN 2228]
o-Butylphenol, Solid [UN 2229]
UN 2228 [Butylphenols, liquid]
UN 2229 [Butylphenols, solid]

3180-81-2
$C_{16}H_{17}ClN_4O_3$
348.76
O=N(=O)c(ccc(N=Nc(ccc(N(CCO)CC)c1)c1)c2Cl)c2
Ethanol, 2-((4-((2-chloro-4-nitrophenyl)azo)phenyl)-ethylamino)- (9CI)

3181-86-0
$C_{18}H_{14}N_2O_3S_2$
370.44
7-Benzothiazolesulfonic acid, 6-methyl-2-(2-methyl-6-quinolinyl)- (9CI)
7-Benzothiazolesulfonic acid, 6-methyl-2-(2-methyl-6-quinolyl)-

3184-65-4
$C_{13}H_{11}ClO.Na$
241.67
Sodium 2-benzyl-4-chlorophenate
o-Benzyl-p-chlorophenol sodium salt
2-Benzyl-4-chlorophenol, sodium salt
Caswell No. 083B
4-Chloro-2-(phenylmethyl)-phenol sodium salt
EPA Pesticide Chemical Code 062203
Phenol, 4-chloro-2-(phenylmethyl)-, sodium salt (9CI)
Sodium o-benzyl-p-chlorophenate
Sodium o-benzyl-p-chlorophenolate

3185-99-7
$C_8H_{10}O_2S$
170.23
Benzene, 1-methyl-4-(methylsulfonyl)- (9CI)
Sulfone, methyl p-tolyl (8CI)

3188-00-9
$C_5H_8O_2$
100.13
O=C(CCO1)C1C
3(2H)-Furanone, dihydro-2-methyl
2-Methyltetrahydrofuran-3-one

3188-13-4
C_3H_7ClO
94.55
O(CC)CCl
Ether, chloromethyl ethyl
Chloromethoxyethane
Chloromethyl ethyl ether [UN 2354]
Ethane, (chloromethoxy)- (9CI)
Ethoxychloromethane
Ethoxymethyl chloride
Ethyl chloromethyl ether
UN 2354 [Chloromethyl ethyl ether]

3188-83-8
$C_{22}H_{26}O_5$
370.45
O(C1COc(ccc(c2)C(c(ccc(OCC(O3)C3)c4CO)c4)(C)C)c2)C1
Benzenemethanol, 5-(1-methyl-1-(4-(oxiranylmethoxy)phenyl)ethyl)-2-(oxiranylmethoxy)- (9CI)

3194-55-6
$C_{12}H_{18}Br_6$
641.70
BrC(C(Br)CCC(Br)C(Br)CCC(Br)C(Br)C1)C1
1,2,5,6,9,10-Hexabromocyclododecane
Cyclododecane, 1,2,5,6,9,10-hexabromo- (9CI)
Hexabromocyclododecane

3197-06-6
Unknown
Unknown
Aminoethyldiethanolamine, aminoethylethanolamine solution

3202-86-6
$C_{21}H_{15}N_3O$
325.37
Phenol, 2-(4,6-diphenyl-1,3,5-triazin-2-yl)- (9CI)
2,4-Diphenyl-6-(2-hydroxyphenyl)-s-triazine
Phenol, o-(4,6-diphenyl-s-triazin-2-yl)- (7CI,8CI)

3207-12-3
$C_4H_{11}NO.ClH$
125.59
1-Propanol, 2-amino-2-methyl-, hydrochloride (9CI)
2-Amino-2-methyl-1-propanol hydrochloride

3208-16-0
C_6H_8O
96.13
O(C(=CC=1)CC)C1
Furan, 2-ethyl- (9CI)
2-Ethylfuran

3209-22-1
$C_6H_3Cl_2NO_2$
192.00
O=N(=O)c(c(c(cc1)Cl)Cl)c1
Benzene, 1,2-dichloro-3-nitro
2,3-Dichloronitrobenzene

3214-47-9
$C_{35}H_{28}N_6O_{13}S_4.4Na$
960.83
C.I. Direct Yellow 50, Tetrasodium salt (8CI)
1,5-Naphthalenedisulfonic acid, 3,3'-(carbonylbis(imino-(2-methyl-4,1-phenylene)-azo))bis-, tetrasodium salt (9CI)

3217-00-3
$C_8H_{14}N_4$
166.20
N#CCCNCCNCCC#N
Propanenitrile, 3,3'-(1,2-ethanediyldiimino)bis- (9CI)

N,N'-Bis(2-cyanoethyl)-1,2-ethanediamine
4,7-Diazadecane-1,10-dinitrile
N,N'-Di(2-cyanoethyl)-1,2-ethylenediamine
3,3'-(1,2-Ethanediyldiimino)bispropanenitrile
Propionitrile, 3,3'-(ethylenediimino)di- (8CI)

3217-15-0
$C_6H_3BrCl_2O$
241.90
Phenol, 4-bromo-2,6-dichloro- (9CI)

3221-61-2
C_9H_{20}
128.26
2-Methyloctane
Octane, 2-methyl- (9CI)

3223-70-9
C_3H_3Cl
74.51
1,2-Propadiene, 1-chloro- (9CI)
Chloroallene
Chloropropadiene
Propadiene, chloro- (6CI, 7CI,8CI)

3224-15-5
$C_{18}H_{17}NO_5$
327.33
O=C(c(c(C(=O)c1c(N)c(OCCC(O)C)cc2O)ccc3)c3)c12
9,10-Anthracenedione, 1-amino-4-hydroxy-2-(3-hydroxybutoxy)- (9CI)
Anthraquinone, 1-amino-4-hydroxy-2-(3-hydroxybutoxy)-

3225-97-6
$C_{15}H_{10}ClNO_2$
271.70
9,10-Anthracenedione, 1-amino-4-chloro-2-methyl- (9CI)
Anthraquinone, 1-amino-4-chloro-2-methyl- (8CI)
NSC-82289

3228-04-4
$C_{10}H_{14}O$
150.24
Phenol, 2-isopropyl-6-methyl
2-Isopropyl-6-methylphenol

3232-26-6
$C_7H_{13}N_5$
167.18
1,3,5-Triazine-2,4-diamine, 6-butyl-
AI3-60157

3232-84-6
$C_2H_3N_3O_2$
101.05
O=C(NC(=O)N1)N1
Bicarbamimide (VAN) (8CI)
AI3-61104
1,2,4-Triazolidine-3,5-dione (9CI)

3234-02-4
$C_4H_6Br_2O_2$
245.92
OCC(=C(Br)CO)Br
2-Butene-1,4-diol, 2,3-dibromo

3234-28-4
$C_{14}H_{28}O$
212.38
O(C1CCCCCCCCCCCCC)C1
Oxirane, dodecyl-
AI3-14200
Dodecyloxirane
1,2-Epoxytetradecane
1,2-Tetradecaneoxide

3234-54-6
$C_6H_8O_2$
112.13
O=C(OC=C)C=CC
2-Butenoic acid, ethenyl ester, (E)- (9CI)
(E)-Ethenyl 2-butenoate

3234-85-3
$C_{28}H_{56}O_2$
424.75
O=C(OCCCCCCCCCCCCCC)CCCCCCCCCCCCC
Myristyl myristate
Hefti MYM-33
Tetradecanoic acid, tetradecyl ester (9CI)
Tetradecyl tetradecanoate

3237-50-1
$C_4H_4N_2O_5$
160.10
OC1(O)C(=O)NC(=O)NC1=O
Barbituric acid, 5,5-dihydroxy
Alloxan hydrate
Alloxan monohydrate
2,4,6(1H,3H,5H)-Pyrimidinetrione, 5,5-dihydroxy- (9CI)

3240-10-6
$C_9H_5ClO_2$
180.59
2-(4-Chlorophenyl)-2-propynoic acid
4-Chlorophenylpropiolic acid
Propiolic acid, (p-chlorophenyl)- (8CI)
Propynoic acid, (4-chlorophenyl)-
2-Propynoic acid, 2-(4-chlorophenyl)- (9CI)

3244-88-0
$C_{20}H_{17}N_3O_9S_3.2Na$
585.56
[Na+].[Na+].Cc1cc(cc(c1N)S([O-])(=O)=O)C(=C2C=CC(=N)C(=C2)S([O-])(=O)=O)c3ccc(N)c(c3)S(O)(=O)=O
Benzenesulfonic acid, 2-amino-5-((4-amino-3-sulfophenyl)(4-imino-3-sulfo-2,5-cyclohexadien- 1-ylidene)methyl)-3-methyl-, disodium salt
Acidal Fuchsine
Acidal Magenta
Acid Fuchsine
Acid Fuchsine FB
Acid Fuchsine N
Acid Fuchsine O
Acid Fuchsine S
Acid Leather Magenta A
Acid Magenta
Acid Magenta O
Acid Rosein
Acid Rubin
C.I. Acid Violet 19 (7CI)
C.I. Acid Violet 19, Disodium salt (8CI)
Fuchsin(E) Acid
Fuchsine Acid
p-Fuchsine Acid
Kiton Magenta A
Rubine S (6CI)

3244-90-4
$C_{12}H_{28}O_5P_2S_2$
378.46
CCCOP(=S)(OCCC)OP(=S)(OCCC)OCCC
Thiopyrophosphoric acid, tetrapropyl ester
A 42
ASP 51
Aspon
Bis-O,O-di-n-propylphosphorothionic anhydride
E 8573
ENT 16,894
NPD
Propyl thiopyrophosphate
Stauffer ASP-51
Tetrapropyldithiodifosfat (Czech)
Tetra-n-propyl dithionopyrophosphate
O,O,O,O-Tetrapropyl dithiopyrophosphate
Tetra-n-propyl dithiopyrophosphate

3247-75-4
$C_9H_{10}O_4$
182.18
3-(3-Hydroxyphenyl)-3-hydroxypropanoic acid

mHPHA
β-(meta-Hydroxyphenyl)-hydracrylic acid

3248-28-0
$C_6H_{10}O_4$
146.14
Propionyl peroxide (8CI)
Dipropionyl peroxide
Dipropionyl peroxide, Maximum conc. 28% in solution
Peroxide, bis(1-oxopropyl)
Peroxido de dipropionilo (Spanish)
Peroyxde de dipropionyle (French)
Propionyl peroxide, More than 28% in solution
Propionyl peroxide, Not more than 28% in solution
UN 2132

3248-91-7
$C_{22}H_{23}N_3.ClH$
365.89
[Cl-].CC1=CC(C=CC1=[NH2+])=C(c2ccc(N)c(C)c2)c3ccc(N)c(C)c3
New Fuchsin
Benzenamine, 4-((4-amino-3-methylphenyl)(4-imino-3-methyl-2,5-cyclohexadien-1-ylidene)methyl)-2-methyl-, monohydrochloride (9CI)
C.I. Basic Violet 2, Monohydrochloride (8CI)
Fuchsin NB

3248-93-9
$C_{20}H_{19}N_3$
301.42
N=C(C=CC(C=1)=C(c(ccc(N)c2C)c2)c(ccc(N)c3)c3)C1
2,4-Xylidine, α⁴-(p-aminophenyl)-α⁴-(4-imino-2,5-cyclohexadien- 1-ylidene)
4-((4-Aminophenyl)(4-imino-2,5-cyclohexadien-1-ylidene)methyl)-2-methylbenzenamine
Benzenamine, 4-((4-aminophenyl)(4-imino-2,5-cyclohexadien-1-ylidene)methyl)-2-methyl-
C.I. Basic Violet 14, Free base
C.I. Solvent Red 41
Fuchsin (Basic)
Fuchsine Base
Fuchsine HF Base
Magenta Base
para-Rosaniline
Rosaniline Base
o-Toluidine, 4-((p-aminophenyl)(4-imino-2,5-cyclohexadien-1-ylidene)methyl)-
Waxoline Red A

3251-23-8
$N_2O_6.Cu$
187.56
Nitric acid, copper(2+)salt
Copper dinitrate
Copper (II) nitrate
Copper(2+) nitrate
Cupric dinitrate
Cupric nitrate (DOT)
NA 1479

3252-43-5
C_2HBr_2N
198.86
N#CC(Br)Br
Acetonitrile, dibromo
Dibromoacetonitrile

3254-63-5
$C_9H_{13}O_4PS$
248.25
COP(=O)(OC)Oc1ccc(SC)cc1
Phosphoric acid, dimethyl p-(methylthio)phenyl ester
Allied GC-6506
O,O-Dimethyl O-(4-methylmercaptophenyl) phosphate
Dimethyl-p-(methylthiofenyl)-fosfat (Czech)
Dimethyl p-(methylthio)phenyl phosphate
ENT 25,734
GC 6506
HA-1200
4-Methylthiophenyldimethyl phosphate
Phenol, p-(methylthio)-, dimethyl phosphate

3257-28-1
$C_{18}H_{14}N_2O_7S_2.2Na$
480.44
C.I. Food Red 2, Disodium salt
Acid Scarlet GNA
Acid Scarlet JN Extra Pure A
Cerven Potravinarska 2 (Czech)
C.I. 14815
C.I. Food Red 2
Cilefa Red G
E 125
Ecarlate GN (French)
Eurocert Scarlet GN
L Red Z 3000
1-Naphthalenesulfonic acid, 6-((2,4-dimethyl-6-sulfophenyl)azo)-5-hydroxy-, disodium salt
Salmon Red Geigy
Sarlach GN (Czech)
Scarlet GN
Scarlet GN Specially Pure

3258-87-5
$C_{29}H_{50}$
398.72
A'-Neo-30-norgammacerane, (21β)- (9CI)

3263-31-8
$C_{20}H_{16}O_4S_2$
384.48
O=C(c(c(S1)cc(OCC)c2)c2)C1=C(Sc(c3ccc4OCC)c4)C3=O
($\delta^{2,2'}$(3H,3'H))-Bibenzo(b)thiophene)-3,3'-dione, 6,6'-diethoxy
Ahcovat Printing Orange R
Algol Orange RF
Amanthrene Orange R
Calcoloid Printing Orange RE
Calcaloid Printing Orange RYW
C.I. 73335
Ciba Orange R
Ciba Orange RDL
Ciba Orange RP
C.I. Vat Orange 5
6,6'-Diethoxythioindigo
Durindone Orange R
Durindone Orange RP
Durindone Printing Orange R
Heliane Orange RF
Helindon Orange R
Hostavat Orange R
Mikethrene Orange R
Nihonthrene Fast Orange R
Oranz Kypova 5 (Czech)
Palanthrene Orange R
Sandothrene Orange R
Solindene Orange R
Sulfanthrene Orange R
Sulfanthrene Orange RS
Thioindigo Orange KKh
Tina Orange R
Tyrian Orange A-RF
Vat Orange R
Vat Orange RF
Vat Printing Orange R

3266-23-7
C_4H_8O
72.12
O(C1C)C1C
Butane, 2,3-epoxy
2-Butene expoxide
2-Butene oxide
β-Butylene oxide
2,3-Butylene oxide
2,3-Dimethyloxirane
2,3-Epoxybutane
Oxirane, 2,3-dimethyl- (9CI)
β-Oxybutene

3267-76-3
$C_4H_6O_4.Co$
177.02
Cobaltous succinate
Butanedioic acid, cobalt(2+) salt (1:1)
Cobalt succinate
Succinic acid, cobalt(2+) salt (1:1)

3267-78-5
$C_{11}H_{24}O_2Sn$

307.04
CCC[Sn](CCC)(CCC)OC(C)=O
Stannane, acetoxytripropyl
Tripropylstannium acetate
Tripropyltin acetate

3268-49-3
C_4H_8OS
104.18
O=CCCSC
Propionaldehyde, 3-(methylthio)
Methional
β-(Methylmercapto)propionaldehyde
3-(Methylmercapto)propionaldehyde
Methylmercaptopropionic aldehyde
3-(Methylthio)propanal
β-(Methylthio)propionaldehyde
3-(Methylthio)propionaldehyde
Propanal, 3-(methylthio)- (9CI)
Thia-4-pentanal [UN 2785]
4-Thiapentanal
UN 2785 [Thia-4-pentanal]

3268-87-9
$C_{12}Cl_8O_2$
459.72
Clc3c(Cl)c(Cl)c2Oc1c(Cl)c(Cl)c(Cl)c(Cl)c1Oc2c3Cl
Dibenzo-p-dioxin, 1,2,3,4,6,7,8,9-octachloro
NCI-C03678
OCDD
Octachlorodibenzodioxin
Octachlorodibenzo(b,e)(1,4)dioxin
Octachlorodibenzo-p-dioxin
1,2,3,4,6,7,8,9-Octachlorodibenzodioxin

3269-10-1
$C_6O_{12}Pr_2$
545.87
Praseodymium, tris(ethanedioato(2-))di- (9CI)
Praseodymium oxalate
Tris(ethanedioato(2-))dipraseodymium

3270-86-8
$C_{10}H_{14}NO_5PS$
291.28
CCOP(=O)(OCC)Sc1ccc(cc1)N(=O)=O
Phosphorothioic acid, O,O-diethyl S-(p-nitrophenyl) ester
O,O-Diaethyl-S-(p-nitrophenyl)-phosphat (German)
O,O-Diethyl-S-p-nitrofenylester kyseliny thiofosforecne (Czech)
O,O-Diethyl-S-p-nitrofenylthiofosfat (Czech)
O,O-Diethyl S-(4-nitrophenyl) phosphorothioate
O,O-Diethyl S-(4-nitrophenyl)-thiophosphate
Parathion S
S-Phenyl parathion
Phosphorothioic acid, O,O-diethyl-S-(4-nitrophenyl) ester (9CI)

3270-98-2
$C_6H_6O_4$
142.11
Hydroxymuconic semialdehyde
2-Hydroxy-6-oxohexa-2,4-dienoate semialdehyde
2-Hydroxy-6-oxo-2,4-hexadienoic acid

3271-05-4
$C_{14}H_{11}NO_3$
241.26
O=C(N(C(=O)c(c1c(c(OC)cc2)cc3)c3)C)c12
Naphthalimide, 4-methoxy-N-methyl
4-Methoxy-N-methylnaphthalimide
N-Methyl-4-methoxynaphthalimide

3271-76-9
$C_{31}H_{15}NO_3$
449.47
O=C(c(c(c(Nc(c(c(c(c(c(C1=O)ccc2)c2)c3)c1cc4)c45)c3)c5c6)C(=O)c7cccc8)c6)c78
Anthra(2,1,9-mna)naphth-(2,3-h)acridine-5,10,15-trione
C.I. 69500
C.I. Vat Green 3
Vat Green 3
Zelen Kypova 3 (Czech)
Zelen Olivova Ostanthrenova B (Czech)

3274-20-2
$C_{15}H_{10}O_3$
238.24
9,10-Anthracenedione, 2-methoxy- (9CI)
Anthraquinone, 2-methoxy- (8CI)

3274-29-1
$C_9H_{18}O_2$
158.24
O=C(O)C(CCCCC)CC
Heptanoic acid, 2-ethyl- (9CI)
2-Ethylheptanoic acid

3277-26-7
$C_4H_{14}OSi_2$
134.36
C[SiH](C)O[SiH](C)(C)C
Disiloxane, 1,1,3,3-tetramethyl
Bis(dimethylsilyl) ether
Bis(dimethylsilyl) oxide
1,3-Dihydrotetramethyldisiloxane
Dimethylsilyl ether
sym-Tetramethyldisiloxane
1,1,3,3-Tetramethyldisiloxane

3278-22-6
$C_5H_8O_4S_2$
196.25
O=S(=O)(C=C)CS(=O)(=O)C=C
Ethene, 1,1'-(methylenebis(sulfonyl))bis- (9CI)
1,1'-(Methylenebis(sulfonyl))-bisethene

3278-35-1
$C_6H_{10}O_3S_2$
194.28
O=C(OCC)SC(OCC)=S
Carbonic acid, dithio-, anhydrosulfide with O-ethyl thiocarbonate, O-ethyl ester (8CI)
AI3-19742
Ethyl xanthogen ethyl formate
Thiodicarbonic acid, diethyl ester (9CI)

3278-46-4
$C_3H_3Cl_3O_2$
177.41
OC(=O)C(Cl)(Cl)CCl
Propionic acid, 2,2,3-trichloro
Chloropon
Kyselina 2,2,3-trichlorpropionova (Czech)
Latka 6249 (Czech)
2,2,3-Trichloropropionic acid

3279-07-0
$C_{10}H_{13}NO_3$
195.21
Phenol, 4-(1,1-dimethylethyl)-2-nitro- (9CI)
NSC-36629
Phenol, 4-tert-butyl-2-nitro- (8CI)

3279-27-4
$C_{11}H_{16}O$
164.25
Phenol, 2-(1,1-dimethylpropyl)- (9CI)
o-tert-Amylphenol
2-(1,1-Dimethylpropyl)-phenol

3280-08-8
$C_{11}H_{16}O$
164.25
Benzeneethanol, α,α,β-trimethyl- (9CI)
Phenethyl alcohol, α,α,β-trimethyl- (6CI,7CI,8CI)

3282-30-2
C_5H_9ClO
120.59
O=C(C(C)(C)C)Cl
Acetyl chloride, trimethyl
2,2-Dimethylpropanoyl chloride
2,2-Dimethylpropionyl chloride
Neopantanoyl chloride
Pivalic acid chloride
Pivalolyl chloride
Pivaloyl chloride (DOT)
Pivalyl chloride
Propanoyl chloride, 2,2-dimethyl- (9CI)
Trimethylacetyl chloride [UN 2438]
UN 2438 [Trimethylacetyl chloride]

3282-73-3
$C_{26}H_{56}N.Br$
462.64
Ammonium, didodecyldimethyl-, bromide (8CI)
1-Dodecanaminium, N-dodecyl-N,N-dimethyl-, bromide (9CI)

3282-85-7
$C_{17}H_{35}O_4S.Na$
358.57
6-Tridecanol, 3,9-diethyl-, hydrogen sulfate, sodium salt
3,9-Diethyl-tridecyl-6-sulfate
4-Ethyl-1-(3-ethylpentyl)oktylsiran sodny (Czech)
Sodium 4-ethyl-1-(3-ethylpentyl)-1-octyl sulfate
Tergitol 7
Tergitol No. 7

3283-07-6
$C_{30}H_{28}N_6$
472.56
N(c(ccc(N(c(ccc(N)c1)c1)c(ccc(N)c2)c2)c3)c3)(c(ccc(N)c4)c4)c(ccc(N)c5)c5
1,4-Benzenediamine, N,N,N',N'-tetrakis(4-aminophenyl)-(9CI)
N,N,N',N'-Tetrakis(4-aminophenyl)-1,4-benzenediamine

3283-17-8
$C_8H_{15}N_5$
181.21
1,3,5-Triazine-2,4-diamine, 6-pentyl- (9CI)
s-Triazine, 2,4-diamino-6-pentyl- (8CI)

3287-06-7
$C_{22}H_{31}O_3P$
374.46
O(P(Oc(cccc1)c1)Oc(cccc2)c2)CCCCCCCCCC
Phosphorous acid, decyl diphenyl ester
AI3-28072

3287-87-4
$C_{12}H_{27}O_2PS_2.K$
337.55
Potassium O,O-dihexyl dithiophosphate
Phosphorodithioic acid, O,O-dihexyl ester, potassium salt

3287-99-8
$C_7H_9N.ClH$
143.63
NCc1ccccc1
Benzylamine, hydrochloride
Benzenemethanamine, hydrochloride
Benzylammonium chloride
USAF EL-82

3288-58-2
$C_5H_{13}O_2PS_2$
200.27
Phosphorodithioic acid, O,O-diethyl S-methyl ester
O,O-Diethyl-S-methyl-dithiophosphate
RCRA waste number U087

3289-50-7
$C_6H_9N_3O_2$
155.14
O(c(nc(OC)cc1N)n1)C
4-Pyrimidinamine, 2,6-dimethoxy- (9CI)
6-Amino-2,4-dimethoxypyrimidine
2,6-Dimethoxy-4-pyrimidinamine

3290-24-2
$C_{14}H_{17}N$
199.29
n(c(c(ccc1)cc2CC)c1)c2CCC
Quinoline, 3-ethyl-2-propyl- (9CI)
3-Ethyl-2-propylquinoline

3290-53-7
$C_{10}H_{12}$
132.21
Benzene, (2-methyl-2-propenyl)- (9CI)
Benzene, (2-methylallyl)- (8CI)
1-Propene, 2-methyl-3-phenyl-
1-Propene, 2-(phenylmethyl)-

3290-70-8
$C_4H_6Cl_4O$
211.90
OCC(Cl)CC(Cl)(Cl)Cl
2,4,4,4-Tetrachloro-1-butanol
Butanol, 2,4,4,4-tetrachloro-
1-Butanol, 2,4,4,4-tetrachloro- (9CI)
TCBA
2,4,4,4-Tetrachlorobutanol

3290-92-4
$C_{18}H_{26}O_6$
338.44
O=C(OCC(CC)(COC(=O)C(=C)C)COC(=O)C(=C)C)C(=C)C
1,3-Propanediol, 2-ethyl-2-hydroxymethyl-, trimethacrylate
Trimethylolpropane trimethacrylate
Trimethylolpropane trimethan crylate

3294-03-9
$C_{28}H_{42}O_2S$
442.71
Oc(c(Sc(c(O)ccc1C(CC(C)(C)C)(C)C)c1)cc(c2)C(CC(C)(C)C)(C)C)c2
Phenol, 2,2'-thiobis(4-(1,1,3,3-tetramethylbutyl)- (9CI)

3296-90-0
$C_5H_{10}Br_2O_2$
261.97
OCC(CBr)(CBr)CO
1,3-Propanediol, 2,2-bis(bromomethyl)
2,2-Bis(bromomethyl)-1,3-propanediol
Dibromoneopentyl glycol
Dibromopentaerythritol
FR 1138
NCI-C55516
Pentaerythritol dibromide
Pentaerythritol dibromohydrin

3299-05-6
$C_{10}H_{14}O$
150.22
Benzene, (1-ethoxyethyl)- (9CI)
Ether, ethyl α-methylbenzyl (8CI)
NSC-104

3299-99-8
$C_{16}H_{16}$
208.30
9H-Fluorene, 9-(1-methylethyl)- (9CI)
Fluorene, 9-isopropyl- (8CI)

3301-94-8
$C_9H_{16}O_2$
156.22
O=C(OC(CC1)CCCC)C1
2H-Pyran-2-one, 6-butyltetrahydro-
AI3-36027
6-Butyltetrahydro-2H-pyran-2-one

3302-12-3
$C_9H_{18}O_2$
158.24
Pentanoic acid, 2,2,4,4-tetramethyl- (9CI)
2,2,4,4-Tetramethylpentanoic acid
2,2,4,4-Tetramethylvaleric acid
Valeric acid, 2,2,4,4-tetramethyl- (6CI,7CI,8CI)

3307-39-9
$C_9H_9ClO_3$
200.62
CC(Oc1ccc(Cl)cc1)C(O)=O
2-(4-Chlorophenoxy)propionic acid
2-(p-Chlorophenoxy)propionic acid
2-p-CPP
Propanoic acid, 2-(4-chlorophenoxy)- (9CI)
Propionic acid, 2-(p-chlorophenoxy)- (8CI)

3307-41-3
$C_9H_8Cl_2O_3$
235.07
Propanoic acid, 2-(3,4-dichlorophenoxy)- (9CI)
3,4-DP
Propionic acid, 2-(3,4-dichlorophenoxy)- (8CI)

3310-97-2
$C_{21}H_{32}O_2$
316.48
1-Phenanthrenecarboxylic acid, 1,2,3,4,4a,4b,5,6,7,9,10,10a-dodecahydro-1,4a-dimethyl-7-(1-methylethylidene)-, methyl ester, (1R-(1α,4aβ,4bα,10aα))-(9CI)
Podocarp-8(14)-en-15-oic acid, 13-isopropylidene-, methyl ester (8CI)

3312-58-1
$C_6H_{14}N_2O_2$
146.18
OCN(CCN(C1)CO)C1
1,4-Piperazinedimethanol (9CI)
N,N'-Dimethylolpiperazine

3312-60-5
$C_9H_{20}N_2$
156.26
N(C(CCCC1)C1)CCCN
N-(3-Aminopropyl) cyclohexylamine
Cyclohexylamine, N-(3-aminopropyl)-
N-Cyclohexyl-1,3-propanediamine
N-Cyclohexylpropylene 1,3-diamine
1,3-Propanediamine, N-cyclohexyl- (9CI)

3316-09-4
$C_6H_5NO_4$
155.10
4-Nitrocatechol

3317-67-7
$C_{32}H_{16}CoN_8$
571.42
Cobalt, (phthalocyaninato-(2-))- (8CI)
Cobalt, (29H,31H-phthalocyaninato(2-)-N29,N30,N31,N32)-, (SP-4-1)- (9CI)

3318-61-4
$C_{11}H_{12}O_2$
176.22
O=C(CC(=O)Cc(cccc1)c1)C
2,4-Pentanedione, 1-phenyl- (9CI)
1-Phenyl-2,4-pentanedione

3319-31-1
$C_{33}H_{54}O_6$
546.79
O=C(OCC(CCCC)CC)c(ccc(c1C(=O)OCC(CCCC)CC)C(=O)OCC(CCCC)CC)c1
Tri-(2-ethylhexyl)trimellitate
1,2,4-Benzenetricarboxylic acid, tris(2-ethylhexyl) ester (9CI)
Kodaflex TOTM
Morflex 510
Staflex TOTM
Tri-2-ethylhexyl trimellitate
Trimex T 08
Tris(2-ethylhexyl) trimellitate

3321-64-0
$C_4H_{11}O_3P$
138.10
CCCCP(O)(O)=O
Phosphonic acid, butyl- (8CI, 9CI)
1-Butanephosphonic acid
NSC-38367

3321-80-0
$C_{20}H_{23}NO_3$
325.40
Benzilic acid, 1-methyl-3-piperidyl ester
DEA No. 7484
α-Hydroxy-α-phenylbenzeneacetic acid, 1-methyl-3-piperidinyl ester
JB 336
N-Methyl-3-piperidyl benzilate

3322-93-8
$C_8H_{12}Br_4$
427.84
BrCC(Br)C(CCC(Br)C1Br)C1
Cyclohexane, 1,2-dibromo-4-(1,2-dibromoethyl)
Citex BCL 462
1,2-Dibromo-4-(1,2-dibromoethyl)cyclohexane
1-(1,2-Dibromoethyl)-3,4-dibromocyclohexane
Saytex BCL 462

3323-53-3
$C_6H_{16}N_2.C_6H_{10}O_4$
262.34
Hexamethylenediamine adipate
Adipan hexamethylendiaminu (Czech)
Adipic acid, Compd. with 1,6-hexanediamine (1:1)
Hexamethylenediamine adipate (1:1)
Hexamethylenediamine monoadipate
Hexamethylenediammonium adipate
Hexanedioic acid, Compd. with 1,6-hexanediamine (1:1) (9CI)
Nylon 66 salt

3324-71-8
$C_3H_9N_3$
87.15
Guanidine, dimethyl
Dimethyl-guanidin (German)
Dimethylguanidine

3327-22-8
$C_6H_{15}ClNO.Cl$
188.12
Ammonium, (3-chloro-2-hydroxypropyl)trimethyl-, chloride
(3-Chloro-2-hydroxypropyl)trimethylammonium chloride

3329-91-7
$C_{13}H_{19}NO_2$
221.33
Dioscorine (8CI)
Spiro(2-azabicyclo(2.2.2)octane-5,2'-(2H)pyran)-6'(3'H)-one, 2,4'-dimethyl-, (1R-(1α,4α,5α))- (9CI)

3332-27-2
$C_{16}H_{35}NO$
257.45
O=N(CCCCCCCCCCCCCC)(C)C
Myristamine oxide
N,N-Dimethyl-1-tetradecanamine N-oxide
Myristyl dimethyl amine oxide

Myristyldimethylamine oxide
Tetradecanamine, N,N-dimethyl-, N-oxide
1-Tetradecanamine, N,N-dimethyl-, N-oxide (9CI)

3333-15-1
$C_{15}H_{14}O_2$
226.27
CC(Cc1ccccc1)...

Benzenepropanoic acid, α-phenyl- (9CI)
Benzeneacetic acid, α-(phenylmethyl)-
NSC-49
NSC-11135
Propanoic acid, 2,3-diphenyl-
Propionic acid, 2,3-diphenyl- (8CI)

3333-52-6
$C_8H_{12}N_2$
136.22
CC(C)(C#N)C(C)(C)C#N
Succinonitrile, tetramethyl
Butanedinitrile, tetramethyl- (9CI)
Tetramethyl succinonitrile (ACGIH,OSHA)
Tetramethylsuccinic acid dinitrile
Tetramethylsuccinodinitrile
Tetramethylsuccinonitrile
Tetramethylsukcinonitril (Czech)
TMSN

3333-62-8
$C_{25}H_{15}N_3O_2$
389.39
O=C(Oc(c(ccc1N(N=C(c(c(C=C2)ccc3)c3)C2=4)N4)C=5)c1)C5c(cccc6)c6
2H-1-Benzopyran-2-one, 7-(2H-naphtho(1,2-d)triazol-2-yl)-3-phenyl- (9CI)
3-Phenyl-7-(1,2-2H-naphthotriazolyl)coumarin

3333-67-3
$CNiO_3$
118.72
Nickel(II) carbonate (1:1)
Carbonic acid, nickel salt (1:1)
C.I. 77779
Nickelous carbonate

3336-41-2
$C_7H_4Cl_2O_3$
207.01
Benzoic acid, 3,5-dichloro-4-hydroxy
3,5-Dichloro-4-hydroxybenzoic acid

3337-70-0
$C_8H_8N_2O_6S$
260.24
Carbamic acid, ((p-nitrophenyl)sulfonyl)-, methyl ester
MB 8882
Methyl N-(4-nitrobenzenesulphonyl)carbamate

3337-71-1
$C_8H_{10}N_2O_4S$
230.26
COC(=O)NS(=O)(=O)c1ccc(N)cc1
Carbamic acid, sulfanilyl-, methyl ester
4-Amino-benzolsulfonyl-methylcarbamat (German)
Asilan
Asulam
Asulox 40
Asulfox F
Jonnix
MB 9057
Methyl N-(4-aminobenzenesulfonyl)carbamate
Methyl 4-aminobenzenesulphonyl carbamate
Methyl ((4-aminophenyl)sulfonyl)carbamate
Methyl 4-aminophenylsulphonyl carbamate
Methyl sulfanilyl carbamate

3338-24-7
$C_4H_{10}O_2PS_2.Na$
208.22
[Na+].CCOP([S-])(=S)OCC
Phosphorodithioic acid, O,O-diethyl ester, sodium salt
O,O-Diethyldithiofosforecnan sodny (Czech)

3338-55-4
$C_{10}H_{16}$
136.24
C(=CCC=C(C=C)C)(C)C
1,3,6-Octatriene, 3,7-dimethyl-, (Z)- (9CI)
(Z)-3,7-Dimethyl-1,3,6-octatriene
cis-β-Ocimene

3342-98-1
$C_{21}H_{16}O$
284.37
Cc3ccc4cc1c(ccc2ccccc12)c5C(O)Cc3c45
1-Cholanthrenol, 3-methyl
1-Hydroxy-3-methylcholanthrene
15-Hydroxy-20-methylcholanthrene
3-Methylcholanthren-1-ol

3344-12-5
$C_{10}H_{19}O_6PS_2$
330.38
Succinic acid, mercapto-, diethyl ester, S-ester with O,S-dimethyl phosphorodithioate
Butanedioic acid, ((methoxy(methylthio)phosphinothioyl)thio)diethyl ester (9CI)
8063HC
Isomalathion
Mercaptosuccinic acid diethyl ester, S-ester with O,S-dimethylphosphorodithioate
(Methoxy(methylthio)phosphinothioyl)butanedioic acid diethyl ester
S-Methylmalathion

3344-14-7
$C_9H_{12}NO_5PS$
277.25
COP(=O)(Oc1ccc(N(=O)=O)c(C)c1)SC
Phosphorothioic acid, O,S-dimethyl O-(4-nitro-m-tolyl) ester
O,S-Dimethyl-O-(3-methyl-4-nitrofenyl)thiofosfat (Czech)
8062 HC
Isosumithion
Metathion, S-methyl isomer
S-Methyl fenitrooxon
S-Methyl fenitrothion
Phosphorothioic acid, O,S-dimethyl O-(3-methyl-4-nitrophenyl) ester (9CI)
Sumithion S-isomer
Thiophosphate de O,S-dimethyl et de O-(3-methyl-4-nitrophenyle) (French)

3347-22-6
$C_{14}H_4N_2O_2S_2$
296.32
O=c2c1ccccc1c(=O)c3sc(C#N)c(C#N)sc23
Naphtho(2,3-b)-p-dithiin-2,3-dicarbonitrile, 5,10-dihydro-5,10-dioxo
Delan
Delan-Col
2,3-Dicarbonitrilo-1,4-diathia-anthrachinon (German)
2,3-Dicyano-1,4-dithia-anthraquinone
5,10-Dihydro-5,10-dioxonaphtho(2,3-b)-p-dithiin-2,3-dicarbonitrile
2,3-Dinitrilo-1,4-dithioanthrachinon (German)
2,3-Dinitrilo-1,4-dithia-anthraquinone
1,4-Dithiaanthraquinone-2,3-dicarbonitrile
1,4-Dithiaanthraquinone-2,3-dinitrile
Dithianon
Dithianone
DTA
IT 931
MV 119A

Stauffer MV-119A
Thynon

3347-60-2
$C_{27}H_{48}O_2$
404.68
Cholestane-3,5-diol, (3β,5α)- (9CI)
5α-Cholestane-3β,5-diol (8CI)

3351-05-1
$C_{32}H_{23}N_5O_6S_2.2Na$
683.64
C.I. Acid Blue 113, Disodium salt (8CI)
1-Naphthalenesulfonic acid, 8-(phenylamino)-5-((4-((3-sulfophenyl)azo)-1-naphthalenyl)azo)-, disodium salt (9CI)

3351-28-8
$C_{19}H_{14}$
242.33
Cc1cccc3c1ccc4c2ccccc2ccc34
Chrysene, 1-methyl
1-Methylchrysene

3351-30-2
$C_{19}H_{14}$
242.33
Cc3cccc4ccc2c1ccccc1ccc2c34
Chrysene, 4-methyl
4-Methyl chrysene

3351-31-3
$C_{19}H_{14}$
242.33
Cc4ccc3ccc2c1ccccc1ccc2c3c4
Chrysene, 3-methyl
3-Methylchrysene

3351-32-4
$C_{19}H_{14}$
242.33
Cc4ccc2c(ccc3c1ccccc1ccc23)c4
Chrysene, 2-methyl
2-Methylchrysene

3352-57-6
HO
17.01
Hydroxyl radical
Hydrogen oxide
Hydroxyl

3352-87-2
$C_{16}H_{33}NO$
255.50
O=C(N(CC)CC)CCCCCCCCCCC
Nopcogen 14-l
N,N-Diethyldodecanamide
Diethyllauramide
N,N-Diethyllauramide
N,N-Diethyllaurylamide
Dodecanamide, N,N-diethyl-

3353-12-6
$C_{17}H_{12}$
216.28
Pyrene, 4-methyl- (8CI,9CI)

3366-61-8
$C_{14}H_{14}N_2O$
226.30
CC(=O)Nc1ccc(cc1)c2ccc(N)cc2
Acetanilide, 4'-(p-aminophenyl)
Acetamide, N-(4'-amino(1,1'-biphenyl)-4-yl)-
4'-Acetamidobenzidine
N-Acetylbenzidine
4'-(p-Aminophenyl)acetanilide
Monoacetylbenzidine

3370-35-2
$C_{19}H_{39}NO_2$
313.52
O=C(NCO)CCCCCCCCCCCCCCCCC
Octadecanamide, N-(hydroxymethyl)- (8CI,9CI)
AI3-18377
N-(Hydroxymethyl)octadecanamide
Nalan RF
NSC-3148

3375-22-2
$C_4H_6Cl_2$
125.00
CC(CCl)=CCl
Propene, 1,3-dichloro-2-methyl
1,3-Dichloroisobutylene
1,3-Dichloro-2-methylpropene
1,3-Dichloro-2-methyl-1-propene
1-Propene, 1,3-dichloro-2-methyl- (9CI)

3375-25-5
$C_{28}H_{30}N_2O_3$
442.55
O=C(O)c(c(C(c(c(OC1=CC(=N(CC)CC)C=C2)cc(N(CC)CC)c3)c3)=C12)ccc4)c4
Xanthylium, 9-(2-carboxyphenyl)-3,6-bis(diethylamino)-, hydroxide, inner salt (9CI)
Ethaminium, N-(9-(2-carboxyphenyl)-6-(diethylamino)-3H-xanthen-3-ylidene)-N-ethyl)-, hydroxide, inner salt

3375-84-6
$C_7H_{13}NO_3$
159.19
2-Oxazolidinone, 3-(2-hydroxypropyl)-5-methyl- (6CI,7CI,8CI, 9CI)
5-Methyl-3-(2-hydroxypropyl)-2-oxazolidinone

3377-86-4
$C_6H_{13}Br$
165.07
BrC(CCCC)C
Hexane, 2-bromo- (9CI)
2-Bromohexane

3377-87-5
$C_6H_{13}Br$
165.07
BrC(CCC)CC
Hexane, 3-bromo- (9CI)
3-Bromohexane

3380-34-5
$C_{12}H_7Cl_3O_2$
289.54
O(c(c(O)cc(c1)Cl)c1)c(c(cc(c2)Cl)Cl)c2
Ether, 2'-hydroxy-2,4,4'-trichlorodiphenyl
CH 3565
Irgasan
Irgasan DP300
Phenol, 5-chloro-2-(2,4-dichlorophenoxy)-
Phenyl ether, 2'-hydroxy-2,4,4'-trichloro-
TCC
2,4,4'-Trichloro-2'-hydroxydiphenyl ether
Triclosan

3381-52-0
$C_{34}H_{54}O_3S$
542.87
Cholestan-3-ol, 4-methylbenzenesulfonate, (3β, 5α)- (9CI)
5α-Cholestan-3β-ol, p-toluenesulfonate (6CI,7CI, 8CI)
5α-Cholestan-3β-ol tosylate
Cholestan-3β-yl toluene-p-sulfonate
3β-(Tosyloxy)-5α-cholestane

3383-96-8
$C_{16}H_{20}O_6P_2S_3$
466.48
O(P(OC)(Oc(ccc(Sc(ccc(OP(OC)(OC)=S)c1)c1)c2)c2)=S)C
Phosphorothioic acid, O,O-dimethyl ester, O,O-diester with 4,4'-thiodiphenol
Abat
Abate
Abathion
AC 52160
American Cyanamid AC 52,160

American Cyanamid CL-52160
American Cyanamid E.I. 52,160
Biothion
Bithion
CL 52160
Difenthos
O,O-Dimethyl phosphorothioate O,O-diester with 4,4'-thiodiphenol
Ecopro
EI 52160
ENT 27,165
Experimental Insecticide 52160
Nimitex
Nimitox
Phenol, 4,4'-thiodi-, O,O-diester with O,O-dimethyl phosphorothioate
Phosphorothioic acid, O,O'-(thiodi-4,1-phenylene) O,O,O',O'-tetramethyl ester
Swebate
Temefos
Temephos (ACGIH,OSHA)
Temophos
Tetramethyl-O,O'-thiodi-p-phenylene phosphorothioate
O,O,O',O'-Tetramethyl O,O'-thiodi-p-phenylene phosphorothioate
O,O'-(Thiodi-4,1-phenylene)bis-(O,O-dimethyl phosphorothioate)
O,O'-(Thiodi-p-phenylene) O,O,O',O'-tetramethyl bis-(phosphorothioate)

3384-24-5
$C_{18}H_{36}O_3$
300.48
Octadecanoic acid, 9-hydroxy- (8CI,9CI)

3385-66-8
$C_{11}H_{22}O_2$
186.29
O(C1COCCCCCCCC)C1
Oxirane, ((octyloxy)methyl)- (9CI)
NSC-86982
Propane, 1,2-epoxy-3-(octyloxy)- (8CI)

3386-18-3
$C_{18}H_{38}O_{10}$
414.56
Nonaethylene-glycol
3,6,9,12,15,18,21,24-Octaoxahexacosane-1,26-diol

3386-33-2
$C_{18}H_{37}Cl$
288.94
ClCCCCCCCCCCCCCCCCCC
Octadecyl chloride
AI3-28591
Octadecane, 1-chloro- (9CI)

3387-41-5
$C_{10}H_{16}$
136.24
Sabinene
Bicyclo(3.1.0)hexane, 4-methylene-1-(1-methylethyl)- (9CI)
4(10)-Thujene (8CI)

3388-03-2
$C_{15}H_{22}O_4$
266.34
O(CC(CC(O1)C1C2)(C2)CO3)C3C(CC(O4)C4C5)C5
Spiro(1,3-dioxane-5,3'-(7)oxabicyclo(4.1.0)heptane), 2-(7-oxabicyclo(4.1.0)hept-3-yl)- (9CI)

3388-04-3
$C_{11}H_{22}O_4Si$
246.42
COSi(CCC2CCC1OC1C2)(OC)OC
Silane, β-(3,4-epoxycyclohexyl)ethyltrimethoxy
Epoxycyclohexylethyl trimethoxy silane
β-(3,4-Epoxycyclohexyl)ethyltrimethoxysilane
Silane Y-4086
Silicone A-186
Union Carbide A-186

3388-72-5
$C_{22}H_{42}N_2O.C_2H_4O_2$
410.63
1H-Imidazole-1-ethanol, 2-(8-heptadecenyl)-4,5-dihydro-, monoacetate (Salt) (9CI)
1-(Hydroxyethyl)-2-(8-heptadecenyl)imidazoline acetate

3389-71-7
$C_7H_2Cl_6$
298.81
C(=C(C(C1(Cl)Cl)(C=C2)Cl)Cl)(C12Cl)Cl
1,2,3,4,7,7-Hexachloronorbornadiene
AI3-51332
Bicyclo(2.2.1)hepta-2,5-diene, 1,2,3,4,7,7-hexachloro- (9CI)
1,2,3,4,7,7-Hexachlorobicyclo-(2.2.1)hepta-2,5-diene
Hexachloronorbornadiene
1,2,3,4,7,7-Hexachloro-2,5-norbornadiene
2,5-Norbornadiene, 1,2,3,4,7,7-hexachloro- (8CI)

3390-61-2
$C_{33}H_{34}O_2Si_3$
546.89
Trisiloxane, 1,3,5-trimethyl-1,1,3,5,5-pentaphenyl- (9CI)
1,3,5-Trimethyl-1,1,3,5,5-pentaphenyltrisiloxane

3391-10-4
C_8H_9ClO
156.61
OC(c(ccc(c1)Cl)c1)C
Benzenemethanol, 4-chloro-α-methyl- (9CI)
AI3-02463
Benzyl alcohol, p-chloro-α-methyl- (8CI)
4-Chloro-α-methylbenzenemethanol

3391-86-4
$C_8H_{16}O$
128.24
OC(C=C)CCCCC
1-Octen-3-ol
Amylvinylcarbinol
Matsutake alcohol (Japanese)
1-Okten-3-ol (Czech)

3393-64-4
$C_5H_{10}O_2$
102.13
2-Butanone, 4-hydroxy-3-methyl- (8CI,9CI)
NSC-62078

3397-62-4
$C_3H_4ClN_5$
145.52
n(c(nc(n1)Cl)N)c1N
s-Triazine, 2,4-diamino-6-chloro- (8CI)
AI3-50982
2-Chloro-4,6-diamino-1,3,5-triazine
6-Chloro-1,3,5-triazine-2,4-diamine
2,4-Diamino-6-chloro-1,3,5-triazine
1,3,5-Triazine-2,4-diamine, 6-chloro- (9CI)

3398-69-4
$C_6H_{14}N_2O$
130.22
tert-Butylamine, N-ethyl-N-nitroso
Aethyl-t-butyl-nitrosoamin (German)
t-Butanamine, N-ethyl-N-nitroso-
EBNA
Ethyl-tert-butylnitrosamine
Ethyl-t-butylnitrosoamine
N-Nitroso-tert-butylethylamine
N-Nitrosoethyl-tert-butylamine

3399-73-3
$C_8H_{15}N$
125.21
NCCC(=CCCC1)C1
1-Cyclohexene-1-ethanamine

(9CI)
Cyclohexenylethylamine

3400-09-7
Cl_2HN
85.91
Chlorimide (8CI,9CI)
Dichloramine

3400-33-7
$C_{12}H_{11}NO$
185.23
1-Naphthalenecarboxamide, N-methyl- (9CI)
N-Methyl-α-naphthamide
N-Methyl-1-naphthamide
1-Naphthamide, N-methyl- (7CI,8CI)

3400-45-1
$C_6H_{10}O_2$
114.14
OC(=O)C1CCCC1
Cyclopentanecarboxylic acid (8CI,9CI)
NSC-59714

3401-73-8
$C_{26}H_{47}NO_{10}S.4Na$
657.68
Tetrasodium dicarboxyethyl stearyl sulfosuccinamate
L-Aspartic acid, N-(3-carboxy-1-oxosulfopropyl)-N-octadecyl-, tetrasodium salt
L-Aspartic acid, N-(3-carboxy-1-oxo-2-sulfopropyl)-N-octadecyl-, tetrasodium salt (9CI)
N-(3-Carboxy-1-oxosulfopropyl)-N-octadecyl-L-aspartic acid, tetrasodium salt
N-Octadecyl-N-(sulfosuccinyl)-aspartic acid, tetrasodium salt
Tetrasodium N-(1,2-dicarboxyethyl)-N-octadecyl sulfosuccinate

3401-74-9
$C_{26}H_{56}N.Cl$
418.18
Dilauryldimonium chloride
Dilauryl dimethyl ammonium chloride
1-Dodecanaminium, N-dodecyl-N,N-dimethyl-, chloride (9CI)
N-Dodecyl-N,N-dimethyl-1-dodecanaminium chloride
Quaternium-47

3401-80-7
$C_7H_4Cl_2O_3$
207.01
O=C(O)c(c(ccc1Cl)Cl)c1O
Salicylic acid, 3,6-dichloro
Benzoic acid, 3,6-dichloro-2-hydroxy- (9CI)
3,6-Dichlorosalicylic acid

3404-61-3
C_7H_{14}
98.19
C(C(CCC)C)=C
1-Hexene, 3-methyl- (9CI)
3-Methyl-1-hexene

3404-62-4
C_7H_{14}
98.19
C(=CC)CC(C)C
2-Hexene, 5-methyl- (9CI)
5-Methyl-2-hexene

3404-65-7
C_7H_{14}
98.19
3-Hexene, 3-methyl- (6CI, 7CI,8CI,9CI)
3-Methyl-3-hexene

3404-67-9
C_8H_{16}
112.22
C(C(CC)C)(=C)CC
Hexane, 3-methyl-4-methylene- (9CI)
2-Ethyl-3-methylpentene-1
2-Ethyl-3-methyl-1-pentene
3-Methyl-4-methylenehexane

3404-72-6
C_7H_{14}
98.19
1-Pentene, 2,3-dimethyl- (8CI,9CI)
NSC-74134

3404-73-7
C_7H_{14}
98.19
1-Pentene, 3,3-dimethyl- (8CI,9CI)
NSC-74136

3404-78-2
C_8H_{16}
112.22
2-Hexene, 2,5-dimethyl- (8CI,9CI)
NSC-74164

3405-32-1
$C_4H_6Cl_4$
195.90
ClCC(Cl)C(Cl)CCl
Butane, 1,2,3,4-tetrachloro
1,2,3,4-Tetrachlorobutane

3408-97-7
$C_9H_{11}BrN_2O$
243.09
Urea, N'-(4-bromophenyl)-N,N-dimethyl- (9CI)
Bromfenuron
p-Bromofenuron
N-(4-Bromophenyl)-N',N'-dimethylurea
N'-(4-Bromophenyl)-N,N-dimethylurea
1-(p-Bromophenyl)-3,3-dimethylurea
1-(4-Bromophenyl)-3,3-dimethylurea
Bromuron
Urea, 3-(p-bromophenyl)-1,1-dimethyl- (7CI, 8CI)

3411-95-8
$C_{13}H_9NOS$
227.28
Oc(c(ccc1)C(=Nc(c2ccc3)c3)S2)c1
Phenol, o-2-benzothiazolyl- (8CI)
o-(2-Benzothiazolyl)phenol
2-(2-Benzothiazolyl)phenol
2-(2-Hydroxyphenyl)benzothiazole
Phenol, 2-(2-benzothiazolyl)- (9CI)

3416-24-8
$C_6H_{13}NO_5$
179.20
O=CC(N)C(O)C(O)C(O)CO
D-Glucose, 2-amino-2-deoxy
2-Amino-2-deoxy-d-glucose
Chitosamine
Glucosamine
D-Glucosamine

3424-57-5
$C_{21}H_{24}O_3Si_3$
408.72
Cyclotrisiloxane, 2,4,6-trimethyl-2,4,6-triphenyl-, (Z)
cis-2,4,6-Trimethyl-2,4,6-triphenylcyclotrisiloxane

3424-82-6
$C_{14}H_8Cl_4$
318.02
Ethylene, 1,1-dichloro-2-(o-chlorophenyl)-2-(p-chlorophenyl)
Benzene, 1-chloro-2-(2,2-dichloro-1-(4-chlorophenyl)-ethenyl)- (9CI)
o,p'-DDE
Ethylene, 1-(o-chlorophenyl)-1-(p-chlorophenyl)-2,2-dichloro-

3425-61-4
$C_5H_{12}O_2$
104.17
O(O)C(CC)(C)C
tert-Pentyl hydroperoxide

t-Amyl hydroperoxide
tert-Amyl hydroperoxide
1,1-Dimethylpropyl hydroperoxide
Hydroperoxide, 1,1-dimethylpropyl (9CI)

3426-62-8
$C_7H_3Cl_3O_2.C_2H_7N$
270.55
Benzoic acid, 2,3,6-trichloro-, Compd. with dimethylamine (1:1)
Benzac 1281
Dimethylamine, 2,3,6-trichlorobenzoate
Kyselina 2,3,6-trichlorbenzoova dimethylamonna sul (Czech)
2,3,6-Trichlorobenzoic acid, dimethylamine salt
Trysben 200

3426-63-9
$C_{11}H_{20}ClN_5O_2$
289.77
s-Triazine, 2-chloro-4,6-bis((3-methoxypropyl)amino)- (6CI,7CI, 8CI)

3428-24-8
$C_6H_4Cl_2O_2$
179.00
Pyrocatechol, 4,5-dichloro
1,2-Benzenediol, 4,5-dichloro-
4,5-Dichlorocatechol
4,5-Dichloropyrocatechol

3433-16-7
$C_{11}H_{20}O_3$
200.28
Nonanoic acid, 9-oxo-, ethyl ester (9CI)
Azelaaldehydic acid, ethyl ester (7CI,8CI)
Azelaic semialdehyde monoethyl ester
Ethyl azelaaldehydate
Ethyl 9-oxononanoate

3437-33-0
$C_9H_{20}N_2$
156.26
1H-Azepine-1-propanamine, hexahydro- (9CI)
1H-Azepine, 1-(3-aminopropyl)-hexahydro- (8CI)
Hexamethylenimine, 1-(3-aminopropyl)-

3437-84-1
$C_8H_{14}O_4$
174.20
O=C(OOC(=O)C(C)C)C(C)C
Diisobutyryl peroxide
Diisobutyryl peroxide, Not more than 52% in solution
Isobutyroyl peroxide
Isobutyryl peroxide
Isobutyryl peroxide, Maximum concentration 52% in solution
Peroxide, bis(2-methyl-1-oxopropyl) (9CI)
Peroxido de diisobutirilo (Spanish)
Peroxyde de diisobutyryle (French)
UN 2182

3437-95-4
C_4H_3IS
210.04
S(C(=CC=1)I)C1
Thiophene, 2-iodo- (9CI)
2-Iodothiophene

3438-46-8
$C_5H_6N_2$
94.10
n(ccc(n1)C)c1
Pyrimidine, 4-methyl- (9CI)
4-Methylpyrimidine

3440-02-6
$C_{14}H_{16}O_2Si$
244.37
Silane, dimethyldiphenoxy- (8CI,9CI)

3440-19-5
$C_5H_6Cl_2N_4$
193.01
1,3,5-Triazin-2-amine, 4,6-dichloro-N-ethyl- (9CI)
Etatryn
s-Triazine, 2,4-dichloro-6-(ethylamino)- (8CI)

3440-75-3
$C_{12}H_{28}Pb$
379.59
CCC[Pb](CCC)(CCC)CCC
Plumbane, tetrapropyl
Lead, tetrapropyl-
Tetrapropyl lead
Tetrapropylplumbane

3441-14-3
$C_{35}H_{27}N_7O_{10}S_2.2Na$
815.71
Direct Red 23
C.I. Direct Red 23, Disodium salt (8CI)
C.I. 29160
Fast Scarlet 4BSA
2-Naphthalenesulfonic acid, 3-((4-(acetylamino)phenyl)-azo)-4-hydroxy-7-((((5-hydroxy-6-(phenylazo)-7-sulfo-2-naphthalenyl)amino)-carbonyl)amino)-, disodium salt (9CI)

3442-78-2
$C_{17}H_{12}$
216.28
2-Methylpyrene
Pyrene, 2-methyl-

3444-14-2
$CH_2O_3.2Cu$
189.12
Carbonic acid, copper(1+) salt (9CI)

3445-11-2
$C_6H_{11}NO_2$
129.18
O=C(N(CC1)CCO)C1
2-Pyrrolidinone, 1-(2-hydroxyethyl)
1-(2-Hydroxyethyl)-2-pyrrolidinone
n-(2-Hydroxyethyl)-2-pyrrolidone

3452-07-1
$C_{20}H_{40}$
280.54
C(=C)CCCCCCCCCCCCCCCCCC
1-Eicosene (9CI)
AI3-36496

3452-09-3
C_9H_{16}
124.23
C(#C)CCCCCCC
1-Nonyne (9CI)

3452-97-9
$C_9H_{20}O$
144.26
OCCC(CC(C)(C)C)C
3,5,5-Trimethylhexanol
AI3-22142
Caswell No. 892A
EPA Pesticide Chemical Code 492200
1-Hexanol, 3,5,5-trimethyl- (9CI)
3,5,5-Trimethyl-1-hexanol

3453-64-3
$C_9H_8O_2$
148.16
CC1OC(=O)c2ccccc12
1(3H)-Isobenzofuranone, 3-methyl- (9CI)
Benzoic acid, 2-(1-hydroxyethyl)-, γ-lactone
Phthalide, 3-methyl- (8CI)

3454-07-7
$C_{10}H_{12}$
132.21
c(ccc(c1)C=C)(c1)CC

Benzene, 1-ethenyl-4-ethyl- (9CI)
Styrene, p-ethyl- (8CI)

3454-66-8
$C_4H_{11}O_2PS_2.K$
225.33
Phosphorodithioic acid, O,O-diethyl ester, potassium salt (8CI,9CI)

3457-46-3
$C_{14}H_8Cl_2O_2$
279.12
O=C(c(ccc(c1)Cl)c1)C(=O)c(cc c(c2)Cl)c2
Ethanedione, bis(4-chloro-phenyl)- (9CI)
AI3-15871
Bis(4-chlorophenyl)ethanedione
4,4'-Dichlorobenzil

3457-61-2
$C_{13}H_{20}O_2$
208.33
O(OC(C)(C)C)C(c(cccc1)c1)(C)C
Peroxide, tert-butyl α,α-dimethylbenzyl
tert-Butyl cumene peroxide, Technically pure (DOT)
tert-Butyl cumyl peroxide
tert-Butyl cumyl peroxide, Technically pure (DOT)
Cumyl tert-butyl peroxide
Kayabutyl C
Peroxide, 1,1-dimethylethyl 1-methyl-1-phenylethyl (9CI)
Trigonox T
UN 2091 (DOT)

3458-22-8
$C_8H_{19}NO_6S_2.ClH$
325.86
Cl.CS(=O)(=O)OCCCNCCCOS (C)(=O)=O
1-Propanol, 3,3'-iminodi-, dimethanesulfonate (ester), hydrochloride
N,N-Bis(methylsulfonepropoxy)amine hydrochloride
Compound 864
3,3'-Imidodi-1-propanol, dimethanesulfonate (ester), hydrochloride
IPD
NCI-C01547
NSC-102627
Sakurai No. 864
Yoshi 864

3458-28-4
$C_6H_{12}O_6$
180.16
O=CC(O)C(O)C(O)C(O)CO
D-Mannose (9CI)
AI3-18442
Mannose, D- (8CI)

3467-59-2
$C_{10}H_{13}NO_3$
195.21
O=C(Nc(c(OC)ccc1OC)c1)C
Acetamide, N-(2,5-dimethoxyphenyl)- (9CI)
Acetanilide, 2',5'-dimethoxy- (8CI)
AI3-17947
N-(2,5-Dimethoxyphenyl)acetamide

3468-11-9
$C_8H_7N_3$
145.18
N=C(NC(=N)c1cccc2)c12
Isoindoline, 1,3-diimino
Afastogen Blue 5040
1,3-Diiminoisoindolin (Czech)
1,3-Diiminoisoindoline
Fastogen Blue FP-3100
Fastogen Blue SH-100
Modr Ftalostanova 3G (Czech)
Phthalimidimide
Phthalocyanine Blue 01206
Phthalogen

3468-63-1
$C_{16}H_{10}N_4O_5$
338.30
O=N(=O)c(c(N=Nc(c(c(ccc1)cc2) c1)c2O)ccc3N(=O)=O)c3
2-Naphthol, 1-((2,4-dinitrophenyl)azo)
Brilliant Tangerine 13030
Calcotone Orange 2R
Carnelio Red 2G
Chromatex Orange R
C.I. 12075
C.I. Pigment Orange 5
Dainichi Permanent Red GG
D & C Orange No. 17
Dinitraniline Orange
Dinitroaniline Orange ND-204
Dinitroaniline Red
Fastona Red 2G
Graphtol Red 2GL
Hansa Orange RN
Helio Fast Orange RN
Helio Fast Orange 3RN
Helio Fast Orange RT
Helio Fast Orange 3RT
Irgalite Fast Red 2GL
Irgalite Red 2G
Irgalite Red 2GW
Irgalite Red PV8
Isol Fast Red 2G
Lake Red 2GL
Light Orange R
Lutetia Fast Orange R
Monolite Fast Orange 2R
Monolite Fast Paper Orange 2R
Monolite Fast Red 2G
Nippon Orange X-881
Oralith Red 2GL
11048 Orange
Orange No. 203
Orange Pigment X
Permanent Orange
Permanent Orange DN Toner
Permanent Orange GG
Permanent Orange HD
Permanent Orange Toner RA-5650
Permanent Red GG
Permansa Orange
Permatone Orange
Permaton Orange XL 45-3015
Pigment Fast Orange
Segnale Light Orange RN
Segnale Light Orange RNG
Siegle Orange 2S
Signal Orange Orange Y-17
Silopol Orange R
Silosol Orange RN
Siloton Orange RL
Syton Fast Red 2G
Tertropigment Orange LRN
Tertropigment Red P2G
Versal Orange RNL

3470-97-1
$C_7H_{11}NO_2$
141.16
2,5-Pyrrolidinedione, 1-propyl- (9CI)
NSC-50330
Succinimide, N-propyl- (8CI)

3475-63-6
$C_4H_9N_3O_2$
131.16
CN(C)C(=O)N(C)N=O
Urea, 1,1,3-trimethyl-3-nitroso
N-Nitroso-trimethylharnstoff (German)
Nitrosotrimethylurea
N-Nitrosotrimethylurea
Trimethylnitrosoharnstoff (German)
Trimethylnitrosomocovina (Czech)
N-Trimethyl-N-nitrosourea
1,1,3-Trimethyl-3-nitrosourea

3476-90-2
$C_{35}H_{25}N_5O_7S.2Na$
705.62
Benzoic acid, 2-hydroxy-5-((4'-((1-hydroxy-7-(phenylamino)-3-sulfo-2-naphthalenyl)azo)(1,1'-biphenyl)-4-yl)azo)-, disodium salt (9CI)
C.I. Direct Brown 59 (VAN)
C.I. Direct Brown 59, Disodium salt (8CI)
Direct Fast Brown B (Polish)
NSC-47760

3478-94-2
$C_{16}H_{21}Cl_2NO_2$
330.25
3-(2-Methylpiperidino)propyl 3,4-dichlorobenzoate
Benzoic acid, 3,4-dichloro-, 3-(2-methylpiperidino)propyl

ester
Caswell No. 575
EL 211
EPA Pesticide Chemical Code 097003
γ-(2-Methylpiperidino)propyl 3,4-dichlorobenzoate
Piperalin
1-Piperidinepropanol, 2-methyl-, 3,4-dichlorobenzoate (Ester)
Pipron

3480-87-3
$C_9H_{10}O_3$
166.18
OC(CC(O)=O)c1ccccc1
β-Hydroxyphenylpropionic acid
3-Hydroxy-3-phenylpropionic acid

3481-09-2
$C_8H_4ClNO_2$
181.57
1H-Isoindole-1,3(2H)-dione, 2-chloro- (9CI)
NSC-76078
Phthalimide, N-chloro- (8CI)

3481-20-7
$C_6H_3Cl_4N$
230.90
Nc1c(Cl)c(Cl)cc(Cl)c1Cl
Aniline, 2,3,5,6-tetrachloro- (8CI)
Benzenamine, 2,3,5,6-tetrachloro- (9CI)

3486-35-9
$CO_3.Zn$
125.38
Carbonic acid, zinc salt (1:1)
Zinc carbonate (1:1)

3491-27-8
$C_{10}H_{16}O_2$
168.26
2,7-Octadien-1-ol, acetate
Acetic acid, 1-octa-2,7-dienyl ester
1-Acetoxyoctadiene
2,7-Octadienyl acetate

3497-00-5
$C_6H_5Cl_2PS$
211.04
S=P(c(cccc1)c1)(Cl)Cl
Phosphonothioic dichloride, phenyl
Benzene phosphorus thiodichloride (DOT)
Dichlorophenylphosphine sulfide
Phenyl phosphorus thiodichloride [UN 2799]
Phenylthionophosphonic dichloride
UN 2799 [Phenyl phosphorus thiodichloride]

3507-99-1
$C_{18}H_{36}O_2.Ag$
392.35
Octadecanoic acid, silver(1+) salt (9CI)
Silver(1+) octadecanoate

3508-00-7
$C_{17}H_{35}Br$
319.37
BrCCCCCCCCCCCCCCCCC
Heptadecane, 1-bromo- (9CI)
1-Bromoheptadecane
Heptadecyl bromide

3508-78-9
$C_8H_{12}O_2$
140.18
2,4-Pentanedione, 3-(2-propenyl)- (9CI)
2,4-Pentanedione, 3-allyl- (8CI)

3511-19-1
$C_4H_6Cl_2O$
141.00
Furan, 2,3-dichlorotetrahydro
2,3-Dichlorotetrahydrofuran

3518-05-6
$C_{19}H_{15}N$
257.35
Benz(a)acridine, 8,12-dimethyl
1,10-Dimethyl-5,6-benzacridine
8,12-Dimethylbenz(a)acridine

3519-87-7
$C_{18}H_{13}N$
243.30
Benz(c)acridine, 5-methyl- (8CI,9CI)

3520-72-7
$C_{32}H_{24}Cl_2N_8O_2$
623.54
O=C(N(N=C1C)c(cccc2)c2)C1 N=Nc(c(cc(c(ccc(N=NC(C(=O)N(N=3)c(cccc4)c4)C3C)c5Cl)c5)c6)Cl)c6
C.I. Pigment Orange 13
Atul Vulcan Fast Pigment Orange G
Benzidine Orange
Benzidine Orange 45-2850
Benzidine Orange 45-2880
Benzidine Orange Toner
Benzidine Orange wd 265
Calcotone Orange R
Carnelio Orange G
C.I. 21110
Dainichi Fast Orange RR
Daltolite Fast Orange G
Diarylide Orange
Eljon Fast Orange G
Fast Benzidene Orange YB 3
Fastona Orange G
Fast Orange G
Graphtol Orange GP
Irgalite Orange P
Irgalite Orange PG
Irgalite Orange PX
Irgaplast Orange G
Kromon Orange G
Latexol Fast Orange J
Lutetia Orange J
Monolite Fast Orange G
Monolite Fast Orange GA
No. 56 Conc. Permanent Orange G
No. 59 Forthfast Benzidine Yellow
Oralith Orange PG
Orange G
Oswego Orange X 2065
Permanent Orange G
Permanent Orange G Extra
Pigment Fast Orange G
Pigment Orange ERH
Pigment Orange G
Pigment Orange ZH
Plastol Orange G
Polymo Orange GR
Ponolith Orange Y
Pv-Orange G
Pyrazalone Orange NP 215
Pyrazolone Orange YB 3
Recolite Orange G
Resamine Fast Orange G
Sanyo Benzidine Orange
Segnale Light Orange G
Segnale Light Orange PG
Siegle Orange S
Silogomma Orange G
Silotermo Orange G
Siloton Orange GT
Symuler Fast Pyrazolone Orange G
Syton Fast Orange G
Tertropigment Orange PG
Vulcafix Orange J
Vulcafix Orange JV
Vulcafor Fast Orange G
Vulcafor Fast Orange GA
Vulcan Fast Orange G
Vulcan Fast Orange GA
Vulcan Fast Orange GN
Vulcol Fast Orange G
Vynamon Orange G

3521-06-0
$C_{23}H_{24}ClN_2.Cl$
399.35
Setoglaucine
AI3-22668
C.I. Basic Blue 1 (8CI)
Methanaminium, N-(4-((2-chlorophenyl)(4-(dimethylamino)phenyl)methylene)-2,5-cyclohexadien-1-ylidene)-N-methyl-, chloride (9CI)

3521-91-3
$C_7H_{14}O$
114.19
OC(CCC)CC=C
1-Hepten-4-ol (9CI)
AI3-28622

3522-94-9
C_9H_{20}
128.26
2,2,5-Trimethylhexane
Hexane, 2,2,5-trimethyl- (9CI)

3524-68-3
$C_{14}H_{18}O_7$
298.32
O=C(OCC(CO)(COC(=O)C=C)COC(=O)C=C)C=C
2-Propenoic acid, 2-(hydroxymethyl)-2-(((1-oxo-2-propenyl)oxy)methyl)-1,3-propanediyl ester
Acrylic acid, pentaerithritol triester
Pentaerythritol triacrylate
PETA

3528-17-4
C_9H_8OS
164.23
4H-1-Benzothiopyran-4-one, 2,3-dihydro- (9CI)

3528-63-0
$C_{22}H_{43}N_3$
349.59
N(=C(N(C1)CCN)CCCCCCCC=CCCCCCCCC)C1
1H-Imidazole-1-ethanamine, 2-(8-heptadecenyl)-4,5-dihydro- (9CI)

3529-08-6
$C_8H_{18}N_2$
142.23
1-Piperidinepropanamine (9CI)
Piperidine, 1-(3-aminopropyl)-(8CI)

3530-19-6
$C_{30}H_{24}N_4O_8S_2.2Na$
678.63
C.I. Direct Red 37, Disodium salt (8CI)
1,3-Naphthalenedisulfonic acid, 8-((4'-((4-ethoxyphenyl)azo)-(1,1'-biphenyl)-4-yl)azo)-7-hydroxy-, disodium salt (9CI)

3531-19-9
$C_6H_4ClN_3O_4$
217.58
O=N(=O)c(cc(N(=O)=O)c(N)c1Cl)c1
Aniline, 6-chloro-2,4-dinitro
Benzenamine, 2-chloro-4,6-dinitro-
6-Chloro-2,4-dinitroaniline

3539-43-3
$C_{16}H_{35}O_4P$
322.43
O=P(OCCCCCCCCCCCCCCCC)(O)O
Cetyl phosphate
1-Hexadecanol, dihydrogen phosphate (9CI)
Hexadecyl phosphate
Hexadecyl phosphate (1:1)

3542-36-7
$C_{16}H_{34}Cl_2Sn$
416.09
CCCCCCCC[Sn](Cl)(Cl)CCCCCCCC
Stannane, dichlorodioctyl
Dichlorodioctylstannane
Dioctylstannium dichloride
Dioctyltin dichloride
Di-n-octyltindichloride
DOTC
Di-n-octyl-zinn dichlorid (German)
Stannane, dioctyldichloro-
Tin, dioctyl-, dichloride

3546-10-9
$C_{39}H_{59}Cl_2NO_2$
644.89
CC(C)CCCC(C)C3CCC4C2CC=C1CC(CCC1(C)C2CCC34C)OC(=O)Cc5ccc(cc5)N(CCCl)CCCl
Acetic acid, (4-(bis(2-chloroethyl)amino)phenyl)-, cholesteryl ester
(p-(Bis(2-chloroethyl)amino)-phenyl)acetic acid cholesterol ester
Cholest-5-en-3-ol, (3-β)-, 4-(bis(2-chloroethyl)amino)-benzeneacetate
5-Cholesten-3-β-ol 3-(p-(bis-(2-chloroethyl)amino)phenyl)-acetate
Cholesterol, (p-(bis(2-chloroethyl)amino)phenyl) acetate
Cholesteryl p-bis(2-chloroethyl)amino phenylacetate
Fenesterin
Fenestrin
NCI-C01558
NSC-104469
Phenesterin
Phenesterine
Phenestrin

3547-07-7
$C_{10}H_{11}ClO_3$
214.65
Butanoic acid, 4-(4-chlorophenoxy)- (9CI)
Butyric acid, 4-(p-chlorophenoxy)- (8CI)
NSC-190562

3547-33-9
$C_{10}H_{22}OS$
190.38
OCCSCCCCCCCC
Ethanol, 2-(octylthio)
2-Hydroxyethyl-n-octyl sulfide
MGK Repellent 874
MGK Repellent R-874
2-(Octylthio)ethanol
2-(Oktylthio)ethanol (Czech)
R-874
R-874 Phillips

3555-18-8
$C_{10}H_{13}NO_3$
195.21
Phenol, 4-(1-methylpropyl)-2-nitro- (9CI)
Phenol, 4-sec-butyl-2-nitro-(8CI)

3555-47-3
$C_{12}H_{36}O_4Si_5$
384.84
Trisiloxane, 1,1,1,5,5,5-hexamethyl-3,3-bis((trimethylsilyl)oxy)- (9CI)
Tetrakis(trimethylsiloxy)silane
Trisiloxane, 1,1,1,5,5,5-hexamethyl-3,3-bis(trimethylsiloxy)-

3558-24-5
$C_{15}H_{13}N$
207.27
N(c(c(C=1)ccc2)c2)(C1c(cccc3)c3)C
Indole, 1-methyl-2-phenyl- (8CI)
1H-Indole, 1-methyl-2-phenyl-(9CI)
1-Methyl-2-phenyl-1H-indole

3562-63-8
$C_{22}H_{30}O_3$
342.48
Megestrol
Megestrolo
Megestrolum (Latin)
Pregna-4,6-diene-3,20-dione, 17-hydroxy-6-methyl-

3563-27-7
$C_{18}H_{22}O_2$
270.37
Estra-1,3,5(10),7-tetraene-3,17-diol, (17β)- (9CI)
β-Dihydroequilin
17β-Dihydroequilin
Estra-1,3,5(10),7-tetraene-3,17β-diol (6CI, 7CI, 8CI)

3563-34-6
$C_8H_{18}O_4S_2$
242.36
Propane, 1,1'-(1,2-ethanediyl-bis(sulfonyl))bis- (9CI)
Ethane, 1,2-bis(propylsulfonyl)-, trans- (8CI)

3563-36-8
$C_6H_{12}Cl_2S_2$
219.20
ClCCSCCSCCCl
Ethane, 1,2-bis(2-chloroethyl-thio)
Bis(2-chloroethylthio)ethane
1,2-Bis(β-chloroethylthio)ethane
1,2-Bis(2-chloroethylthio)ethane
Ethane, 1,2-bis(2-chloroethyl-mercapto)-
Sesquimustard
Sesquimustard Q
TL 86

3563-49-3
Unknown
Unknown
Pyrovalerone
DEA No. 1485
Pirovalerona (Spanish)
Pyrovaleronum (Latin)

3563-58-4
Unknown
Unknown
Chlorhexadol
DEA No. 2510

3564-09-8
$C_{19}H_{16}N_2O_7S_2.2Na$
494.47
[Na+].[Na+].Cc3cc(C)c(N=Nc1c(O)c(cc2cc(ccc12)S([O-])(=O)=O)S([O-])(=O)=O)cc3C
2,7-Naphthalenedisulfonic acid, 3-hydroxy-4-((2,4,5-tri-methylphenyl)azo)-, disod-ium salt
A.F. Red No. 1
Cerven kumidinova (Czech)
Cerven potravinarska 6 (Czech)
C.I. 16155
C.I. Food Red 6
C.I. Food Red 6, Disodium salt
Disodium 3-hydroxy-4-((2,4,5-trimethylphenyl)azo)-2,7-naphthalenedisulfonate
Disodium 3-hydroxy-4-((2,4,5-trimethylphenyl)azo)-2,7-naphthalenedisulfonic acid
Disodium 3-hydroxy-4-((2,4,5-trimethylphenyl)azo)-2,7-naphthalenedisulphonate
Disodium 3-hydroxy-4-((2,4,5-trimethylphenyl)azo)-2,7-naphthalenedisulphonic acid
Dolkwal Ponceau 3R
Ext. D & C Red No. 15
FD & C Red No. 1
3-Hydroxy-4-((2,4,5-trimethyl-phenyl)azo)-2,7-naphthalene-disulphonic acid, disodium salt
3-Hydroxy-4((2,4,5-trimethyl-phenyl)azo)-2,7-naphthalene-disulfonic acid, disodium salt
Maple Ponceau 3R
Ponceau 3R
Ponceau RN
Ponceau 3R Lake
Ponceau 3RN
Ponceau 3R Sodium salt
Sodium cumeneazo-β-naphthol disulfonate
Sodium cumeneazo-β-naphthol disulphonate
USAcert Red No. 1

3566-00-5
$C_{10}H_{13}NO_2S$
211.30
CNC(=O)Oc1ccc(SC)c(C)c1
Carbamic acid, methyl-, 4-methylthio-m-tolyl ester
Bay 32651
Carbamic acid, methyl-, 3-methyl-4-(methylthio)phenyl ester
Bayer 32651
Bay S 2758
ENT 25,777
3-Methyl-4-(methylthio)phenyl methylcarbamate

3566-10-7
$C_4H_8N_2S_4.2H_3N$
246.46
N.CC.SC(=S)NCCNC(S)=S
Amobam
Ambam
Carbamic acid, ethylenebis(di-thio-, diammonium salt (8CI)
Carbamodithioic acid, 1,2-eth-anediylbis-, diammonium salt (9CI)
Dithane Stainless

3567-12-2
$C_{21}H_{25}NO_3$
339.43
N-Ethyl-3-piperidyl benzilate
Benzeneacetic acid, α-hydroxy-α-phenyl-, 1-ethyl-3-piperidin-yl ester (9CI)
Benzilic acid, 1-ethyl-3-piperid-yl ester (8CI)
DEA No. 7482
JB 318

3567-25-7
$C_{19}H_{11}Cl_4N_2O_5S.Na$
544.17
Benzenesulfonic acid, 5-chloro-2-(4-chloro-2-(3-(3,4-dichlorophenyl)ureido)-phenoxy)-, sodium salt
N-3,4-Dichlorophenyl N-5-chloro-2-(2-sodium sulfonyl-4-chlorophenoxy)phenyl urea
N-(3,4-Dichlorophenyl)-N'-2-(2-sulfo-4-chlorophenoxy)-5-chlorophenyl urea sodium salt
Mitin
Mitin FF
Sodium 5-chloro-2-(4-chloro-2-(3-(3,4-dichlorophenyl)-ure-ido)phenoxy)benzenesulfonate
Sodium salt of N-(3,4-dichloro-phenyl)-N'-2(2-sulfo-4-chloro-phenoxy)-5-chlorophenyl urea
Sulcofuron

3567-62-2
$C_8H_8Cl_2N_2O$
219.06
CNC(=O)Nc1ccc(Cl)c(Cl)c1
Urea, N-(3,4-dichlorophenyl)-N'-methyl- (9CI)
DCPMU
Urea, 1-(3,4-dichlorophenyl)-3-methyl- (8CI)

3567-65-5
$C_{35}H_{26}N_4O_{10}S_3.2Na$
804.81
1,3-Naphthalenedisulfonic acid, 7-hydroxy-8-((4'-((4-(((4-methylphenyl)sulfonyl)-oxy)phenyl) azo)(1,1'-bi-phenyl)-4-yl)azo)-, disodium salt
Acidine Scarlet GD
Acid Leather Red GR
Acid Leather Scarlet G
Acid Red 85
Acid Red PG
Airedale Red PGM
Altochrome Milling Scarlet G
Amacid Milling Red PGS
Apocid Milling Red G
Benzyl Fast Red GRG
Benzyl Red GR
C.I. 22245
C.I. Acid Red 85
C.I. Acid Red 85, Disodium salt
Coomassie Red PG
Coomassie Red PGP
Cutamin Brilliant Red CG
Elite Fast Red G
Elite Fast Red GRS
Erionyl Red G
Fast Scarlet S
Fenafor Red PG
Folan Yellow G
Kayanol Milling Red PG
Kayanol Red PG
KCA Silk Red G
Lanaperl Fast Red 3G
Lanaperl Red G
Midlon Red PG
Milling Brilliant Scarlet GN
Milling Fast Red G
Milling Fast Red GL
Milling Fast Red PG
Milling Red J

Milling Red SWG
Milling Scarlet G
Mitsui Milling Scarlet G
Neonyl Scarlet R
Neutral Red PG
Nitto Acid Red PG
Nylomine Acid Scarlet C-R
Nylomine Acid Scarlet P-R
Optanol Fast Scarlet GN
Pharmanil Scarlet Y
Polar Red G
Polar Red G Supra
Shikiso Acid Red PG
Sulfonine Red G
Sulfonine Red GN
Sulfonine Red GS
Sulfonine Red SG
Sulphonol Red PG
Suminol Red PG
Supranol Fast Scarlet GN
Supranol Red PG-CF
Supranol Scarlet BN
Symulon Acid Red PG
Telon Fast Scarlet N
Tertracid Milling Red G
Vondamol Fast Red G

3567-66-6
$C_{16}H_{13}N_3O_7S_2.2Na$
469.42
2,7-Naphthalenedisulfonic acid, 4-amino-5-hydroxy-6-phenylazo-, disodium salt
Acetyl Red B
Acid Fuchsine D
Acid Fuchsin Fast B
Acid Red 33
Acid Red 2A
Acid Red B
Amacid Fuchsine 4B
Azo Fuchsine
Azo Grenadine
Azo Magenta G
Brasilan Fuchsine D
Certicol Red B
C.I. Acid Red 33
C.I. Food Red 12
C.I. Red 33, Disodium salt
C.I. Reducing Agent 6
Colacid Red 2A
D & C Red No. 33
Edicol Supra 10B
Edicol Supra 10bS
Eniacid Fuchsine BN
Fast Acid Magenta
Fast Acid Magenta B
Hexacol Red 10B
Hexalan Red B
Naphthalene Red B
1424 Red
11427 Red
Red 10B
Red No. 227

3567-69-9
$C_{20}H_{12}N_2O_7S_2.2Na$
502.44
[Na+].[Na+].Oc3c(N=Nc1ccc(c2ccccc12)S([O-])(=O)=O)cc(c4ccccc34)S([O-])(=O)=O
1-Naphthalenesulfonic acid, 4-hydroxy-3,4'-azodi-, disodium salt
Acetacid Red B
Acid Brilliant Rubine A2G Conc.
Acid Brilliant Rubine 2G
Acid Brilliant Rubine 2GT
Acid Chrome Blue BA
Acid Chrome Blue BA-CF
Acid Chrome Blue FBS
Acid Chrome Blue 2R
Acid Fast Red FB
Acid Red 14
Acid Red 2C
Acid Rubine
Acid Rubine Extra
Airedale Carmoisine
Amacid Carmoisine B
Amacid Chrome Blue R
Atul Acid Crystal Red
Atul Crystal Red F
Azorubin
Azo Rubin Extra
Azo Rubine
Azo Rubine AF
Azo Rubine (Biological stain)
Azo Rubine Extra LC
Azo Rubine for Food
Azo Rubine LZ
Azo Rubine S
Azo Rubine S Specially Pure
Azo Rubine XX
Azo Rubin S
Azo Rubin XX
Brasilan Azo Rubine 2NS
Brilliant Acid Rubine M
Brilliant Carmoisine
Brilliant Crimson Red
Brilliant Crimson 2R.FQ
Bucacid Azo Rubine
Calcocid Rubine XX
Carmoisin (German)
Carmoisine
Carmoisine Aluminum Lake
Carmoisine BA
Carmoisine BA-CF
Carmoisine BSS
Carmoisine FU
Carmoisine GRN
Carmoisine LAS
Carmoisine S
Carmoisine Supra
Carmoisine W
Carmoisine WS
Certicol Carmoisine S
Cerven Kysela 14 (Czech)
Cerven Potravinarska 3 (Czech)
Chrome Fast Blue 2R
Chromotrop FB
Chromotrope FB
C.I. 14720
C.I. Acid Red 14, Disodium salt
C.I. Food Red 3
Cilefa Rubine R
Crimson EMBL
Crimson 2embl
Diadem Chrome Blue G
Diadem Chrome Blue R
Disodium salt of 2-(4-sulpho-1-naphthylazo)-1-naphthol-4-sulphonic acid
Disodium 2-(4-sulfo-1-naphthylazo)-1-naphthol-4-sulfonate
Disodium 2-(4-sulpho-1-naphthylazo)-1-naphthol-4-sulphonate
E122
Edicol Supra Carmoisine W
Edicol Supra Carmoisine WS
Eniacid Brilliant Rubine 3B
Erio Rubine B
Eurocert Azorubine
Ext. D & C Red No. 10
Fenazo Red C
Food Red 5
Fruit Red A Extra Yellowish Geigy
HD Carmoisine
HD Carmoisine Supra
Hexacol Carmoisine
Hexacol Carmoisine Conc.
Hidacid Azo Rubine
Hispacid Rubine F
4-Hydroxy-3,4'-azodi-1-naphthalenesulfonic acid, disodium salt
4-Hydroxy-3,4'-azodi-1-naphthalenesulphonic acid, disodium salt
4-Hydroxy-3-((4-sulfo-1-naphthalenyl)azo)-1-naphthalenesulfonic acid, disodium salt
4-Hydroxy-3-((4-sulpho-1-naphthalenyl)azo)-1-naphthalenesulphonic acid, disodium salt
Java Rubine N
Karmesin
Kenachrome Blue 2R
Kiton Crimson 2R
Kiton Rubine R
Lighthouse Chrome Blue 2R
L. Red Z 3040
Nacarat
Nacarat A Export
Nacarat Extra Pure A
NCI-C53849
Neklacid Azorubine W
Neklacid Rubine W
Nylomine Acid Red P4B
Omega Chrome Blue FB
Pontacyl Rubine R
Poloxal Red 2B
Red #14
11959 Red
Schultz Nr. 208 (German)
Solar Rubine
Solochrome Blue FB
Standacol Carmoisine
2-(4-Sulfo-1-naphthylazo)-1-naphthol-4-sulfonic acid, disodium salt
2-(4-Sulpho-1-naphthylazo)-1-naphthol-4-sulphonic acid, disodium salt
Tertracid Red CA
Tertrochrome Blue FB

3568-29-4
$C_9H_{16}O_5$
204.22
O(C1COCC(O)COCC(O2)C2)C1
Glycerol 1,3-diglycidyl ether

1,3-Bis(oxiranylmethoxy)-2-propanol
Glycerine, 1,3-diglycidyl ether
2-Propanol, 1,3-bis(oxiranylmethoxy)- (9CI)

3569-57-1
$C_{11}H_{23}ClOS$
238.82
3-Chloropropyl n-octylsulfoxide
3-Chloropropyl-n-octylsulfoxide
MGK Repellent 1,207
Sulfoxide, 3-chloropropyl octyl
Sulfoxide, 3-chloropropyl n-octyl-

3570-55-6
$C_4H_{10}S_3$
154.32
SCCSCCS
Ethanethiol, 2,2'-thiobis- (9CI)
AI3-62012
Ethanethiol, 2,2'-thiodi- (8CI)
2,2'-Thiobisethanethiol

3570-61-4
$C_8H_6Cl_3O_5S.Na$
343.54
[Na+].OS(=O)(=O)OCCOc1cc(Cl)c(Cl)cc1Cl
Ethanol, 2-(2,4,5-trichlorophenoxy)-, hydrogen sulfate, sodium salt
Natrin
Natrin Herbicide
Sodium 2,4,5-trichlorophenoxyethyl sulfate
2,4,5-TES
2-(2,4,5-Trichlorophenoxy)ethyl sulfate, sodium salt

3570-75-0
$C_8H_6N_4O_4S$
254.24
O=CNNc1nc(cs1)c2ccc(o2)N(=O)=O
Formic acid, 2-(4-(5-nitro-2-furyl)-2-thiazolyl)hydrazide
AS-17665
FNT
Formic 2-(4-(5-nitrofuryl)-2-thiazolyl)hydrazide
2-(2-Formylhydrazino)-4-(5-nitro-2-furyl)thiazole
Hydrazinecarboxaldehyde, 2-(4-(5-nitro-2-furanyl)-2-thiazolyl)-
Nefurthiazole
Nifurthiazol
Nifurthiazole
NSC-525334

3570-80-7
$C_{24}H_{16}Hg_2O_9$
849.58
Mercury, bis(acetato)(mu-(3',6'-dihydroxy-2',7'-fluorandiyl))di
Fluorescein mercuriacetate
Fluorescein mercuric acetate
Fluorescein mercury acetate
FMA
FMA (Analytical reagent)

3572-43-8
$C_{14}H_{20}Br_2N_2$
376.15
CN(Cc1cc(Br)cc(Br)c1N)C2CCCCC2
Bromhexine
Bromhexina (Spanish)
Bromhexinum (Latin)

3577-01-3
$C_{18}H_{19}N_3O_6S$
405.46
CC(=O)OCC1=C(N3C(SC1)C(NC(=O)C(N)c2ccccc2)C3=O)C(O)=O
5-Thia-1-azabicyclo(4.2.0)oct-2-ene-2-carboxylic acid, 7-(2-amino-2-phenylacetamido)- 3-(hydroxymethyl)-8-oxo-, acetate (ester), D
7-(D-α-Aminophenyl-acetamido)cephalosporanic acid
Cefaloglycin
Cephaloglycin
Cephaloglycine
D-Cephaloglycine
Cephaoglycin acid
Kafocin
Kefglycin
Lilly 39435

3577-63-7
$C_7H_7NO_5S$
217.20
O=C(O)c(c(N)ccc1S(=O)(=O)O)c1
Benzoic acid, 2-amino-5-sulfo- (9CI)
2-Amino-5-sulfobenzoic acid

3582-17-0
$C_2H_6F_2Sn$
186.77
Stannane, difluorodimethyl
Difluorodimethylstannane
Dimethyltin fluoride
Dimethyltin difluoride
Tin, dimethyl-, fluoride

3583-47-9
$C_4H_6Cl_2O$
141.00
ClCC1OC1CCl
Butane, 1,4-dichloro-2,3-epoxy
1,4-Dichlorobutene-2,3-epoxide
1,4-Dichloro-2,3-epoxybutane

3584-22-3
$C_{12}H_7Cl_6N_3$
405.93
1,3,5-Triazine, 2-(4-methyl-phenyl)-4,6-bis-(trichloromethyl)- (9CI)
s-Triazine, 2-p-tolyl-4,6-bis-(trichloromethyl)- (7CI, 8CI)

3584-23-4
$C_{12}H_7Cl_6N_3O$
421.91
1,3,5-Triazine, 2-(4-methoxyphenyl)-4,6-bis(trichloromethyl)- (9CI)
s-Triazine, 2-(p-methoxyphenyl)-4,6-bis(trichloromethyl)-(8CI)

3584-24-5
$C_{12}H_7Cl_6N_3$
405.93
s-Triazine, 2-benzyl-4,6-bis(trichloromethyl)-(7CI,8CI)

3586-14-9
$C_{13}H_{12}O$
184.25
O(c(cccc1C)c1)c(cccc2)c2
Benzene, 1-methyl-3-phenoxy
Ether, phenyl m-tolyl-
3-Methyldiphenyl ether
1-Methyl-3-phenoxybenzene
m-Methylphenyl phenyl ether
3-Methylphenyl phenyl ether
m-Phenoxytoluene
3-Phenoxytoluene
Phenyl m-tolyl ether

3586-58-1
$C_5H_8O_2$
100.12
O=C(O)C(=C)CC
Butanoic acid, 2-methylene-(9CI)
Butyric acid, 2-methylene- (8CI)
Ethacrylic acid

3588-31-6
$C_6H_{10}O_2$
114.15
1,3-Butadiene, 2,3-dimethoxy- (6CI,7CI, 8CI,9CI)
2,3-Dimethoxybutadiene
2,3-Dimethoxybuta-1,3-diene

3590-84-9
$C_{32}H_{68}Sn$
571.69
Stannane, tetraoctyl
Tetraoctylstannane
Tetra-n-octylstannane

Tetraoctyltin
Tetra-n-octyltin
Tin, tetraoctyl-

3592-81-2
$C_9H_{19}N$
141.25
Cyclohexanamine, N-propyl- (9CI)
Cyclohexylamine, N-propyl- (8CI)

3597-91-9
$C_{13}H_{12}O$
184.25
4-Biphenylmethanol
(1,1'-Biphenyl)-4-methanol
4HMB
4-(Hydroxymethyl)biphenyl
p-Phenylbenzyl alcohol

3598-16-1
$C_8H_7O_3.Na$
174.14
Acetic acid, phenoxy-, sodium salt
Phenoxyacetate sodium
Phenoxyessigsaure natrium salz (German)
Sodium phenoxyacetate

3599-32-4
$C_{43}H_{48}N_2O_6S_2.Na$
776.04
1H-Benz(e)indolium, 2-(7(1,1-dimethyl-3-(4-sulfobutyl)benz(E)indolin-2-ylidene)- 1,3,5-heptatrienyl)-1,1-dimethyl-3-(4-sulfobutyl)-, hydroxide, inner salt, sodium salt
Cardio-Green
ICG
Indocyanine Green

3599-58-4
$C_{10}H_{13}Cl_2NO_4$
282.13
Ethanolamine 2,4-dichlorophenoxyacetate

3599-59-5
$C_5H_7N_3$
109.13
1,3,5-Triazine, 2-ethyl- (9CI)
s-Triazine, 2-ethyl- (6CI, 7CI,8CI)

3599-60-8
$C_7H_{11}N_3$
137.19
s-Triazine, 2,4-diethyl- (6CI,7CI,8CI)

3599-61-9
$C_{11}H_{11}N_3$
185.23
1,3,5-Triazine, 2,4-dimethyl-6-phenyl- (9CI)
s-Triazine, 2,4-dimethyl-6-phenyl- (7CI,8CI)

3599-62-0
$C_{16}H_{13}N_3$
247.30
1,3,5-Triazine, 2-methyl-4,6-diphenyl- (9CI)
2-Methyl-4,6-diphenyl-1,3,5-triazine
s-Triazine, 2-methyl-4,6-diphenyl- (6CI,7CI,8CI)

3599-66-4
$C_6H_9N_3O$
139.16
1,3,5-Triazine, 2-methoxy -4,6-dimethyl- (9CI)
s-Triazine, 2-methoxy-4,6-dimethyl- (7CI,8CI)

3599-71-1
$C_7H_5Cl_6N_3$
343.83
1,3,5-Triazine, 2-ethyl-4,6-bis-(trichloromethyl)- (9CI)
s-Triazine, 2-ethyl-4,6-bis (trichloromethyl)- (8CI)

3599-72-2
$C_{11}H_4Cl_6N_4O_2$
436.90
s-Triazine, 2-(m-nitrophenyl)-4,6-bis(trichloromethyl)- (7CI,8CI)

3599-73-3
$C_{10}H_4Cl_6N_4$
392.89
s-Triazine, 2-(4-pyridyl) -4,6-bis(trichloromethyl)- (7CI,8CI)

3599-74-4
$C_5HCl_6N_3$
315.80
1,3,5-Triazine, 2,4-bis-(trichloromethyl)- (9CI)
2,4-Bis(trichloromethyl)-s-triazine
4,6-Bis(trichloromethyl)-s-triazine
s-Triazine, 2,4-bis(trichloromethyl)- (7CI,8CI)

3599-75-5
$C_{11}H_5Cl_7N_4$
441.36
1,3,5-Triazin-2-amine, N-(4-chlorophenyl)-4,6-bis(trichloromethyl)- (9CI)
s-Triazine, 2-(p-chloroanilino)-4,6-bis(trichloromethyl)- (7CI, 8CI)

3599-76-6
$C_6H_3Cl_6N_3S$
361.89
1,3,5-Triazine, 2-(methylthio)-4,6-bis(trichloromethyl)- (9CI)
s-Triazine, 2-(methylthio)-4,6-bis(trichloromethyl)- (7CI,8CI)

3599-79-9
$C_{17}H_{12}Cl_3N_3$
364.66
1,3,5-Triazine, 2-(4-methylphenyl)-4-phenyl-6-(trichloromethyl)- (9CI)
s-Triazine, 2-phenyl-4-p-tolyl-6-(trichloromethyl)- (7CI,8CI)

3599-80-2
$C_{16}H_9Cl_4N_3$
385.08
s-Triazine, 2-(p-chlorophenyl)-4-phenyl-6-(trichloromethyl)- (7CI, 8CI)

3599-81-3
$C_{16}H_9Cl_3N_4O_2$
395.63
s-Triazine, 2-(m-nitrophenyl)-4-phenyl-6-(trichloromethyl)- (7CI, 8CI)

3599-82-4
$C_{17}H_{12}Cl_3N_3O$
380.66
s-Triazine, 2-(p-methoxyphenyl)-4-phenyl-6-(trichloromethyl)- (7CI, 8CI)

3599-87-9
$C_4H_5N_3$
95.10
1,3,5-Triazine, 2-methyl- (9CI)
Methyl-s-triazine
2-Methyl-1,3,5-triazine
C-Methyl-1,3,5-triazine
s-Triazine, 2-methyl- (6CI, 7CI,8CI)

3607-78-1
$C_3H_2Cl_6$

250.77
Propane, 1,1,1,3,3,3-hexa-chloro- (7CI,8CI,9CI)
1,1,1,3,3,3-Hexachloro-propane

3610-02-4
C_8H_6OS
150.20
Benzo(b)thiophene-4-ol
1-Benzothiophene-4-ol

3613-30-7
$C_{11}H_{22}O_2$
186.33
O=CCC(CCCC(OC)(C)C)C
1-Octanal, 3,7-dimethyl-7-methoxy
Hydroxycitronellal methyl ether
Methoxycitronellal methyl ether

3615-21-2
$C_8H_3Cl_2F_3N_2$
255.03
FC(F)(F)c2nc1c(Cl)c(Cl)ccc1[nH]2
Benzimidazole, 4,5-dichloro-2-(trifluoromethyl)
Chlorflurazole
Chloroflurazole
4,5-Dichloro-2-trifluoromethyl-benzimidazole
NC 3363

3615-41-6
$C_6H_{12}O_5$
164.16
O=CC(O)C(O)C(O)C(O)C
L-Mannose, 6-deoxy- (9CI)
6-Deoxy-L-mannose
Rhamnopyranose, L- (8CI)

3618-58-4
$C_{20}H_{13}N_3O_7S.Na$
462.38
C.I. Mordant Black 1 (VAN) (8CI)
1-Naphthalenesulfonic acid, 3-hydroxy-4-((2-hydroxy-1-naphthalenyl)azo)-7-nitro-, monosodium salt (9CI)

3618-72-2
$C_{23}H_{25}BrN_6O_{10}$
625.35
O=C(OCCN(c(c(OC)cc(N=Nc(c(cc(N(=O)=O)c1)Br)c1N(=O)=O)c2NC(=O)C)c2)CCOC(=O)C)C
Acetamide, N-(5-(bis(2-(acetyl-oxy)ethyl)amino)-2-((2-bromo-4,6-dinitrophenyl)-azo)-4-methoxyphenyl)-(9CI)
Benzeneamine, 5-acetamino-4-((2,4-dinitro-6-bromo-phenyl)azo)-2-methoxy-N,N-bis((2-acetoxy)ethyl)-
4-(6-Bromo-2,4-dinitrophenyl-azo)-3-acetylamino-6-meth-oxy-N-bis(acetoxyethyl)aniline
3-(2,4-Dinitro-6-bromophenyl-azo)-4-acetamido-6-(N,N-bis-(acetoxyethyl)amino)anisole

3618-73-3
$C_{23}H_{25}ClN_6O_{10}$
580.90
Acetamide, N-(5-(bis(2-(acetyl-oxy)ethyl)amino)-2-((2-chloro-4,6-dinitrophenyl)-azo)-4-methoxyphenyl)-(9CI)

3622-84-2
$C_{20}H_{19}NO_2S$
337.46
O=S(=O)(NCCCC)c(cccc1)c1
Benzenesulfonamide, n-butyl
Benzenesulfonic acid butyl amide
n-Butylbenzenesulfonamide

3623-05-0
$C_7H_{13}NO$
127.21
2H-Azepin-2-one, hexahydro-4-methyl
β-Methylcaprolactam
4-Methylcaprolactam
α-Methylhexanone isoxime
α-Methylhexanonisoxim (German)

3624-68-8
$C_{16}H_{11}ClN_2O_8S_2.2Na$
504.83
1-Naphthalenesulfonic acid, 6-((5-chloro-2-hydroxy-3-sulfophenyl)azo)-5-hydroxy-, disodium salt (9CI)

3624-77-9
$C_{21}H_{39}NO_3.Na$
376.53
Glycine, N-methyl-N-(1-oxo-9-octadecenyl)-, sodium salt (9CI)

3625-52-3
$C_{32}H_{64}O_2$
480.87
Dotriacontanoic acid
Lacceric acid

3626-28-6
$C_{34}H_{25}N_7O_8S_2.2Na$
769.76
2,6-Naphthalenedisulfonic acid, 4-amino-5-hydroxy-3-((4'-((4-hydroxyphenyl)-azo)(1,1'-biphenyl)-4-yl)azo)-6-(phenylazo)-, disodium salt
Airedale Green BWD
Aizen Direct Dark Green BH
Amanil Green LT
Atlantic Dark Green
Atlantic Green WT
Atul Direct Dark Green P
Azine Dark Green BH/C
Azocard Dark Green B
Bencidal Dark Green B
Benzanil Dark Green BW
Benzo Dark Green B
Benzo Dark Green BA-CF
Brasilamina Green B
Calcomine Dark Green BG
Chlorazol Dark Green PL
Chrome Leather Dark Green N
Chrome Leather Dark Green S
Chrome Leather Green B
C.I. 30280
C.I. Direct Green 1
Cresotine Dark Green B
Dark Green EMBL
Diacotton Dark Green
Diamine Dark Green B
Diamine Dark Green N
Diaphthamine Fast Black FE
Diazine Dark Green BO
Diazine Dark Green P
Diazol Green Black N
Diphenyl Dark Green B
Diphenyl Dark Green BN
Direct Black Green
Direct Dark Green A
Direct Dark Green B
Direct Dark Green BF
Direct Dark Green BG
Direct Dark Green MB
Direct Dark Green S
Direct Dark Green Supra
Direct Dark Green WS
Direct Deep Green A
Direct Green WAC
Enianil Dark Green BG
Erie Green WT
Fenamin Green M
Hispamin Green WT
Kayaku Direct Dark Green B
Mitsui Direct Dark Green BX
Naphthamine Dark Green B
Nippon Dark Green B
Phenamine Dark Green B
Polycor Dark Green S
Pontamine Green S
Sandopel Dark Green B
Shikiso Direct Dark Green B
Tertrodirect Green BG
Union Dark Green B
Vondacel Green DB

3638-04-8
$C_4H_3Cl_2N_3O$
179.97
1,3,5-Triazine, 2,4-dichloro-6-methoxy- (9CI)
NSC-50574
s-Triazine, 2,4-dichloro-6-methoxy- (8CI)

3648-18-8
$C_{40}H_{80}O_4Sn$
743.89
Stannane, dioctyldi(lauroyloxy)
Di-n-octyltin dilaurate
Dioctyltin dilaurate
Di-n-octyl-zinn dilaurat (German)
Stannane, bis(dodecanoyloxy)-dioctyl-
Stannane, dioctylbis(lauroyloxy)-
Stannane, bis(lauroyloxy)di-octyl-
Stannane, didodecanoyloxydi-octyl-
Stannane, dioctyldidodecanoyl-oxy-
Tin, dioctyl-, dilaurate

3648-20-2
$C_{30}H_{50}O_2$
442.80
O=C(OCCCCCCCCCCC)c(c(ccc1)C(=O)OCCCCCCCCCCC)c1
Phthalic acid, diundecyl ester
Diundecyl phthalate
Santicizer 711

3648-21-3
$C_{22}H_{34}O_4$
362.56
O=C(OCCCCCCC)c(c(ccc1)C(=O)OCCCCCCC)c1
Phthalic acid, diheptyl ester
1,2-Benzenedicarboxylic acid, diheptyl ester (9CI)
Diheptyl phthalate
Di-n-heptyl phthalate
Heptyl phthalate

3650-04-2
$C_{18}H_{24}O_2$
272.39
1-Phenanthrenecarboxylic acid, 1,2,3,4,4a,9,10,10a-octahydro-1,4a-dimethyl-, methyl ester, (1R-(1α,4aβ,10aα))- (9CI)
Podocarpa-8,11,13-trien-15-oic acid, methyl ester (8CI)

3651-23-8
$C_6H_{10}Cl_2Si$
181.14
Silane, dichlorodi-2-propenyl- (9CI)
Diallyldichlorosilane
Dichlorodi-2-propenylsilane

3652-91-3
$C_{13}H_{11}N$
181.23
Carbazole, 2-methyl- (8CI)
9H-Carbazole, 2-methyl- (9CI)

3653-48-3
$C_9H_8ClO_3.Na$
222.61
[Na+].Cc1cc(Cl)ccc1OCC([O-])=O
Acetic acid, ((4-chloro-o-tolyl)oxy)-, sodium salt
Acetic acid, (4-chloro-2-methylphenoxy)-, sodium salt
Agroxone 3
4-Chloro-2-methylphenoxy-acetic acid sodium salt
(p-Chloro-o-tolyloxy)acetic acid sodium salt
Chwastoks
Chwastox
Chwastox 80
Diamet
Dicotex
Dicotex 80
Dikoteks
Dikotex
Dikotex 30
MCPA Na salt
MCPA sodium salt
Metaxone
Methoxon
Methoxone
(2-Methyl-4-chlorophenoxy)-acetic acid, sodium salt
2M-4KH Sodium salt
2M-4X
Na MCPA
Phenoxylene
Sodium (4-chloro-2-methyl-phenoxy)acetate
Sodium MCPA
Sodium (2-methyl-4-chloro-phenoxy)acetate
Sys 67ME

3655-00-3
$C_{18}H_{35}NO_4.2Na$
375.45
Disodium lauriminodipropionate
β-Alanine, N-(2-carboxyethyl)-N-dodecyl-, disodium salt (9CI)
N-(2-Carboxyethyl)-N-dodecyl-β-alanine, disodium salt
Deriphat 160
Disodium 3,3'-(dodecylimino)-bis(propionate)
Disodium β,β'-(laurylimino)di-propionate
Disodium N-lauryl-β-iminodi-propionate
Disodium N-lauryl-β,β'-imino-dipropionate
N-Lauryl-β-iminodipropionate sodium salt
Propionic acid, 3,3'-(dodecyl-imino)di-, disodium salt
Sodium N-dodecyliminodipro-pionate
Sodium N-lauryl-β-iminodipro-pionate

3655-88-7
$C_6H_{10}N_2O_2S_2$
206.30
CN1CN(CSC1=S)CC(O)=O
4H,6H-1,3,5-Thiadiazineacetic acid, 5-methyl-6-thioxo
5-Carboxymethyl-3-methyl-2H-1,3,5-thiadiazine-2-thione
D 3520
N-Methyl-tetrahydrothiamidin-thione acetic acid
5-Methyl-6-thioxotetrahydro-3-thiadiazineacetic acid
Terracur
Thiadiazinthion

3658-48-8
$C_{16}H_{35}O_3P$
306.48
O=P(OCC(CCCC)CC)OCC(CCCC)CC
Phosphonic acid, bis(2-ethylhexyl) ester
Bis(2-ethylhexyl) hydrogen phosphite

3658-77-3
$C_6H_8O_3$
128.14
O=C(C(O)=C(O1)C)C1C
3(2H)-Furanone, 2,5-dimethyl-4-hydroxy
Alletone
Coe 536
2,5-Dimethyl-4-hydroxy-3(2H)-furanone
Fema 3174
Furaneol
4-Hydroxy-2,5-dimethyl-3(2H)furanone
Pineapple ketone

3658-80-8
$C_2H_6S_3$
126.27
S(SSC)C
Dimethyl trisulfide
AI3-26172
Methyl trisulfide (8CI)
Trisulfide, dimethyl (9CI)

3659-66-3
$C_3H_3Cl_4NO$
210.87
Formamide, N-(1,2,2,2-tetrachloroethyl)- (8CI,9CI)

3663-23-8
$C_{10}H_{16}N_2$
164.24
Nc(c(N)cc(c1)CCCC)c1
4-Butyl-1,2-benzenediamine
1,2-Benzenediamine, 4-butyl- (9CI)
4-n-Butyl-o-phenylenediamine
o-Phenylenediamine, 4-butyl-

3664-60-6
$C_8H_{14}O$
126.20
7-Octen-2-one (8CI,9CI)

3664-64-0
$C_{11}H_{20}O$
168.28
7-Nonen-2-one, 4,8-dimethyl- (7CI,8CI,9CI)

3665-80-3
$C_8H_{10}N_2O_2$
166.17
Benzenamine, N-ethyl-4-nitro- (9CI)
Aniline, N-ethyl-p-nitro- (8CI)

3674-65-5
$C_{16}H_{14}$
206.29
Phenanthrene, 2,3-dimethyl- (8CI,9CI)

3674-73-5
$C_{17}H_{16}$
220.31
Phenanthrene, 2,3,5-trimethyl- (8CI,9CI)

3674-74-6
$C_{16}H_{14}$
206.29
Phenanthrene, 2-ethyl- (8CI, 9CI)

3674-75-7
$C_{16}H_{14}$
206.29
Phenanthrene, 9-ethyl- (9CI)

3675-13-6
$C_4H_4O_2$
84.07
Maleic anhydride (8CI)
2-Butenedial, (Z)- (9CI)
NSC-250971

3681-71-8
$C_8H_{14}O_2$
142.20
O=C(OCCC=CCC)C
3-Hexen-1-ol, acetate, (Z)-
AI3-34392

3682-19-7
$C_8H_5N_3O_4$
207.16
O=N(=O)c2ccc1c(=O)[nH][nH]c(=O)c1c2
1,4-Phthalazinedione, 2,3-dihydro-6-nitro
6-Nitrophthalhydrazide

3682-56-2
$C_{13}H_{10}N_2O_2$
226.22
Benzoic acid, 2-(phenylazo)- (9CI)
Benzoic acid, o-(phenylazo)- (8CI)

3687-22-7
$C_{14}H_{20}N_2O_5$
296.31
Phenol, 2-(1-methylheptyl)-4,6-dinitro- (8CI,9CI)

3687-31-8
$As_2O_8.3Pb$
899.41
Arsenic acid, lead(2+) salt (2:3)
Arsinette
Gypsine
Lead arsenate (ACGIH) [UN 1617]
Nu Rexform
Ortho L10 Dust
Soprabel
Talbot
UN 1617 [Lead arsenates]

3687-45-4
$C_{36}H_{68}O_2$
532.93
O=C(OCCCCCCCCC=CCCCCCCC)CCCCCCCC=CCCCCCCCC
Oleyl oleate
9-Octadecenoic acid, 9-octadecenyl ester
9-Octadecenoic acid (Z)-, 9-octadecenyl ester, (Z)- (9CI)
Unitolate

3687-46-5
$C_{28}H_{54}O_2$
422.74
O=C(OCCCCCCCCCC)CCCCCCC=CCCCCCCCC
Decyl oleate
Decyl 9-octadecenoate
9-Octadecenoic acid, decyl ester
9-Octadecenoic acid (Z)-, decyl ester (9CI)
Unimul-CTV
Unitolate V

3688-53-7
$C_{11}H_8N_2O_5$
248.21
NC(=O)C(=Cc1ccc(o1)N(=O)=O)c2ccco2
Acrylamide, 2-(2-furyl)-3-(5-nitro-2-furyl)
AF-2
AF 2 (Preservative)
FF
2-Furanacetamide, α-((5-nitro-2-furanyl)methylene)- (9CI)
2-Furanacrylamide, α-2-furyl-5-nitro- (8CI)
Furylamide
Furylfuramide
α-2-Furyl-5-nitro-2-furanacrylamide
2-(2-Furyl)-3-(5-nitro-2-furyl)-acrylamide
2-(2-Furyl)-3-(5-nitro-2-furyl)-acrylic acid amide
α-(Furyl)-β-(5-nitro-2-furyl)-acrylic amide
Tofuron

3688-65-1
$C_{18}H_{21}NO_4$
315.36
Codeine-N-oxide
Codeigene
Codeine N-oxide
DEA No. 9053
Genocodein
Genocodeine
Morphinan-6-ol, 7,8-didehydro-4,5-epoxy-3-methoxy-17-methyl-, 17-oxide, (5-α,6-α)-
Morphinan-6-α-ol, 7,8-didehydro-4,5-α-epoxy-3-methoxy-17-methyl-, 17-oxide

3688-66-2
$C_{24}H_{24}N_2O_4$
404.45
Nicocodine
Codeine, nicotinate (ester)
DEA No. 9309
(5 α,6 α)-7,8-Didehydro-4,5-epoxy-3-methoxy-17-methylmorphinan-6-ol, 3-pyridinecarboxylate (Ester)
Lyopect
Morphinan-6-α-ol, 7,8-didehydro-4,5-α-epoxy-3-methoxy-17-methyl-, nicotinate (Ester)
Nicocodeine
Nicocodina (Spanish)
Nicocodinum (Latin)
Nicotinic acid, 7,8-didehydro-4,5-α-epoxy-3-methoxy-17-methylmorphinan-6-α-yl ester
Nicotinic acid, ester with codeine
RC 146

3689-24-5
$C_8H_{20}O_5P_2S_2$
322.34
CCOP(=S)(OCC)OP(=S)(OCC)OCC
Thiopyrophosphoric acid, tetraethyl ester
ASP 47
Bay-E-393
Bayer-E 393

Bis-O,O-diethylphosphoro-thionic anhydride
Bladafum
Bladafume
Bladafun
O,O,O,O-Tetraaethyl-dithiono-pyrophosphat (German)
Dithio
Dithiodiphosphoric acid, tetra-ethyl ester
Dithiofos
Dithion
Dithione
Dithiophos
Di(thiophosphoric) acid, tetra-ethyl ester
Dithiopyrophosphate de tetra-ethyle (French)
Dithiotep
E393
ENT 16,273
Ethyl thiopyrophosphate
Lethalaire G-57
Pirofos
Plant Dithio Aerosol
Plantfume 103 Smoke Generator
Pyrophosphorodithioic acid, tetraethyl ester
Pyrophosphorodithioic acid, O,O,O,O-tetraethyl ester
RCRA waste number P109
Sulfatep
Sulfotep (ACGIH,OSHA)
Sulfotepp
TEDP (OSHA)
TEDTP
O,O,O,O-Tetraethyl-dithio-di-fosfaat (Dutch)
Tetraethyldithiodifosfat (Czech)
Tetraethyl dithionopyrophos-phate
Tetraethyl dithiopyrophosphate [UN 1704]
O,O,O,O-Tetraethyl dithiopyro-phosphate
Tetraethyl dithiopyrophosphate, Liquid (DOT)
Tetraethyl dithiopyrophosphate Mixture, dry (DOT)
Tetraethyl dithiopyrophosphate Mixture, Liquid (DOT)
O,O,O,O-Tetraetil-ditio-piro-fosfato (Italian)
Thiodiphosphoric acid tetraethyl ester
Thiotepp
UN 1704 [Tetraethyl dithio-pyrophosphate]

3691-35-8
$C_{23}H_{15}ClO_3$
374.82
Chlorophacinone
CAID
Caswell No. 211C
Chloorfacinon (Dutch)
2(2-(4-Chloor-fenyl-2-fenyl)-acetyl)-indaan-1,3-dion (Dutch)
Chlorfacinon (German)
2-(α-p-Chlorophenylacetyl)-indane-1,3-dione
2-(α-p-Chlorophenyl-α-phenyl-acetyl)indane-1,3-dione
2-((p-Chlorophenyl)phenyl-acetyl)-1,3-indandione (8CI)
2(2-(4-Chlorophenyl)-2-phenyl-acetyl)indan-1,3-dione
2-(2-(4-Chlorophenyl)-2-phenyl-acetyl)indan-1,3-dione
2-(2-(4-Chlorophenyl)-2-phenyl-acetyl)indane-1,3-dione
2-((4-Chlorophenyl)phenyl-acetyl)-1H-indene-1,3(2H)-dione (9CI)
Chlorphacinon (Italian)
1-(4-Chlorphenyl)-1-phenyl-acetyl-indan-1,3-dion (German)
2(2-(4-Chlor-phenyl-2-phenyl)-acetyl)indan-1,3-dion (German)
((4- Chlorphenyl)-1-phenyl)-acetyl-1,3-indandion (German)
2(2-(4- Cloro-fenil-2fenil)-acet-il)indan-1,3-dione (Italian)
Delta
Drat
EPA Pesticide Chemical Code 067707
1,3-Indandione, 2-((p-chloro-phenyl)phenylacetyl)-
Liphadione
LM 91
Microzul
Muriol
Partox
2-(2-Phenyl-2-(4-chlorophenyl)-acetyl)-1,3-indandione
Quick
Ramucide
Ranac
Ratomet
Raviac
Razol
Saviac
Topitox

3691-78-9
$C_{23}H_{29}NO_3$
367.48
Benzethidine
Benzethidin
Benzethidinum (Latin)
Benzetidina (Spanish)
DEA No. 9606
Ethyl 1-(2-benzyloxyethyl)-4-phenylpiperidine-4-carboxylate
Isonipecotic acid, 1-(2-(benzyl-oxy)ethyl)-4-phenyl-, ethyl ester
NIH 7574
4-Piperidinecarboxylic acid, 4-phenyl-1-(2-(phenylmeth-oxy)ethyl)-, ethyl ester

3692-90-8
$C_{11}H_{11}NO_3$
205.23
CNC(=O)Oc1cccc(OCC#C)c1
Carbamic acid, methyl-, m-(2-propynyloxy)phenyl ester
ENT 25,732
H 8717
Hercules 8717
3-Propargyloxyphenyl-N-methyl-carbamate
3-(2-Propynyloxy)phenyl-N-methylcarbamate

3694-74-4
$C_{16}H_{34}O_5S.Na$
361.50
Ethanol, 2-(tetradecyloxy)-, hydrogen sulfate, sodium salt (9CI)
2-Tetradecyloxyethyl sodium sulfate

3696-28-4
$C_{10}H_8N_2O_2S_2$
252.32
O=n(c(SSc(n(=O)ccc1)c1)ccc2)c2
Pyridine, 2,2'-dithiodi-, 1,1'-dioxide
2,2'-Dithiodipyridine-1,1'-di-oxide
DS
Olin

3697-24-3
$C_{19}H_{14}$
242.33
Cc3cc1ccccc1c4ccc2ccccc2c34
Chrysene, 5-methyl
5-Methylchrysene

3697-27-6
$C_{20}H_{16}$
256.36
Chrysene, 5,6-dimethyl
5,6-Dimethylchrysene

3697-30-1
$C_{20}H_{16}$
256.36
Benz(a)anthracene, 7-ethyl
7-Ethylbenz(a)anthracene
10-Ethyl-1,2-benzanthracene

3697-42-5
$C_{22}H_{30}Cl_2N_{10}.2ClH$
578.31
Chlorhexidine dihydro-chloride
AY-5312
N,N'-Bis(4-chlorophenyl)-3,12-diimino-2,4,11,13-tetra-azatetradecanediimidamide, dihydrochloride
Caswell No. 481F
Chlorhexidine hydrochloride
EPA Pesticide Chemical Code

481700
1,1'-Hexamethylene bis(5-(p-chlorophenyl)biguanide), dihydrochloride
1,1'-Hexamethylenebis(5-(p-chlorophenyl)biguanide) dihydrochloride
2,4,11,13-Tetraazatetradecanediimidamide, N,N'-bis(4-chlorophenyl)-3,12-diimino-, dihydrochloride
2,4,11,13-Tetraazatetradecanediimidamide, N,N''-bis-(4-chlorophenyl)-3,12-diimino-, dihydrochloride

3698-54-2
$C_6H_3N_5O_8$
273.14
Aniline, 2,3,4,6-tetranitro
Tetranitraniline (French)
Tetranitroaniline
2,3,4,6-Tetranitroaniline
TNA

3699-54-5
$C_5H_{10}N_2O_2$
130.13
O=C(N(CC1)CCO)N1
2-Imidazolidinone, 1-(2-hydroxyethyl)- (9CI)
AI3-24563
Hydroxyethylethyleneurea
1-(2-Hydroxyethyl)-2-imidazolidinone

3703-10-4
$C_6H_8Cl_2N_4$
207.04
1,3,5-Triazin-2-amine, 4,6-dichloro-N-(1-methylethyl)-(9CI)
NSC-344238
s-Triazine, 2,4-dichloro-6-(isopropylamino)- (8CI)

3709-43-1
$C_{14}H_{10}N_2O_{10}S_2.2Na$
476.34
Benzenesulfonic acid, 2,2'-(1,2-ethenediyl)bis(5-nitro-, disodium salt (9CI)
4,4'-Dinitrostilbene-2,2'-disulfonic acid, disodium salt
Disodium 2,2'-(1,2-ethenediyl)-bis(5-nitrobenzenesulfonate)
2,2'-(1,2-Ethenediyl)bis(5-nitrobenzenesulfonic acid), disodium salt
2,2'-Stilbenedisulfonic acid, 4,4'-dinitro-, disodium salt

3710-30-3
C_8H_{14}
110.22
C(=C)CCCCC=C
1,7-Octadiene

3710-41-6
C_8H_{14}
110.20
1,6-Octadiene (6CI,7CI, 8CI,9CI)
1,6-n-Octadiene
2,7-Octadiene

3710-43-8
C_6H_8O
96.13
Furan, 2,4-dimethyl- (8CI,9CI)

3710-84-7
$C_4H_{11}NO$
89.16
ON(CC)CC
Hydroxylamine, N,N-diethyl
DEHA
N,N-Diethylhydroxyamine
Diethylhydroxylamine
N,N-Diethylhydroxylamine
N-Hydroxydiethylamine

3712-60-5
$C_{11}H_4Cl_7N_3$
426.35
1,3,5-Triazine, 2-(4-chlorophenyl)-4,6-bis(trichloromethyl)- (9CI)
s-Triazine, 2-(p-chlorophenyl)-4,6-bis(trichloromethyl)- (7CI,8CI)

3714-62-3
$C_6HCl_4NO_2$
260.88
Benzene, 1,2,3,5-tetrachloro-4-nitro- (8CI,9CI)

3717-15-5
C_8H_9NO
135.16
Benzaldehyde, 4-methyl-, oxime, (E)- (9CI)
p-Tolualdehyde, oxime, (E)-(8CI)

3718-65-8
C_7H_9NO
123.15
Pyridine, 3,5-dimethyl-, 1-oxide (9CI)
3,5-Lutidine, 1-oxide (8CI)
NSC-272271

3720-22-7
$C_7H_{12}O_2$
128.17
2H-Pyran-2-one, tetrahydro-3,6-dimethyl- (8CI, 9CI)
Hexanoic acid, 5-hydroxy-2-methyl-, δ-lactone (7CI)

3720-97-6
$C_3H_6N_2O_3$
118.08
O=C(NC(O)C1O)N1
2-Imidazolidinone, 4,5-dihydroxy- (9CI)
4,5-Dihydroxyimidazolidone-2
4,5-Dihydroxy-2-imidazolidinone
4,5-Dihydroxytetrahydroimidazol-2-one

3721-95-7
$C_5H_8O_2$
100.13
O=C(O)C(CC1)C1
Cyclobutanecarboxylic-acid
Cyclobutylcarboxylic acid

3724-65-0
$C_4H_6O_2$
86.10
O=C(O)C=CC
Crotonic acid, Solid [UN 2823]
Acrylic acid, 3-methyl-
α-Butenoic acid
2-Butenoic acid (9CI)
α-Crotonic acid
Kyselina krotonova (Czech)
β-Methylacrylic acid
3-Methylacrylic acid
Solid Crotonic Acid
UN 2823 [Crotonic acid, solid]

3726-47-4
C_8H_{16}
112.22
Cyclopentane, 1-ethyl-3-methyl-(8CI,9CI)
NSC-73952

3728-56-1
C_9H_{18}
126.24
Cyclohexane, 1-ethyl-4-methyl-(8CI,9CI)

3730-56-1
$C_{21}H_{32}O_2$
316.48
1-Phenanthrenecarboxylic acid, 7-ethenyl-1,2,3,4,4a,4b,5,6,7,9,10,10a-dodecahydro-1,4a,7-trimethyl-, methyl ester, (1R-(1α,4aβ,4bα,7β,10aα))- (9CI)
Podocarp-8(14)-en-15-oic acid, 13α-methyl-13-vinyl-, methyl ester (8CI)

3730-60-7
$C_8H_{18}O$

130.23
2-Hexanol, 2,5-dimethyl- (8CI,9CI)
NSC-5594

3731-38-2
$C_7H_{11}NO$
125.16
O=C(C(CCN1C2)C2)C1
1-Azabicyclo(2.2.2)octan-3-one (9CI)
3-Quinuclidinone (8CI)

3731-39-3
$C_{15}H_{17}N_3$
239.35
Aniline, N,N-dimethyl-4-(o-tolylazo)
Aniline, N,N-dimethyl-p-(2'-methylphenylazo)-
Benzenamine, N,N-dimethyl-4-((2-methylphenyl)azo)-
N,N-Dimethyl-p-((o-tolyl)-azo)aniline
o'-Methyl-p-dimethylaminoazo-benzene
2'-Methyl-4-dimethylaminoazo-benzene
2-Methyl-N,N-dimethyl-4-am-inoazobenzene

3734-33-6
$C_{21}H_{29}N_2O.C_7H_5O_2$
446.58
Denatonium benzoate
Ammonium, benzyldiethyl-((2,6-xylylcarbamoyl)-methyl)-, benzoate (8CI)
Benzenemethanaminium, N-(2-((2,6-dimethylphenyl)-amino)-2-oxoethyl)-N,N-di-ethyl-, benzoate (9CI)
Benzoate de denatonium (French)
Benzoato de denatonio (Spanish)
Benzyldiethyl((2,6-xylylcar-bamoyl)methyl)ammonium benzoate
Bitrex
Caswell No. 083BB
Denatonii benzoas (Latin)
Denatonium benzoate
N-(2-((2,6-Dimethylphenyl)-amino)-2-oxoethyl)-N,N-di-ethylbenzenemethanaminium benzoate
EPA Pesticide Chemical Code 009106
THS-839
WIN 16568
((2,6-Xylylcarbamoyl)methyl diethyl benzyl ammonium benzoate

3734-48-3
$C_{10}H_6Cl_6$
338.86
C(=C(C(C1(Cl)Cl)(C(C2CC=3)C3)Cl)Cl)(C12Cl)Cl
4,7-Methanoindene, 4,5,6,7,8,8-hexachloro-$\delta^{1,5}$-tetrahydro
Addukt hexachlorcyklopenta-dienu s cyklopentadienem (Czech)
Chlordene
4,5,6,7,8,8-Hexachlor-$\delta^{1,5}$-tetra-hydro-4,7-methanoinden
4,5,6,7,8,8-Hexachloro-3a,4,7,7a-tetrahydro-4,7-methano-indene
4,7-Methanoindene, 4,5,6,7,8,8-hexachloro-3a,4,7,7a-te-trahydro-

3734-49-4
$C_{10}H_5Cl_9$
444.22
Nonachlor
AI3-27005
4,7-Methano-1H-indene, 1,2,3,4,5,6,7,8,8-nonachloro-2,3,3a,4,7,7a-hexahydro-

3734-52-9
$C_{15}H_{21}NO$
231.33
Metazocine
DEA No. 9240
Metazocina
Metazocinum (Latin)

3734-67-6
$C_{18}H_{13}N_3O_8S.2Na$
477.38
2,7-Naphthalenedisulfonic acid, 5-(acetylamino)-4-hydroxy-3-(phenylazo)-, disodium salt
Acetyl Red G
Acetyl Red J
Acetyl Rose 2GL
Acidal Brilliant Red 2G
Acid Bright Red
Acid Brilliant Red
Acid Fast Red Egg
Acid Fast Red 3G
Acidine Red G
Acid Leather Red KG
Acid Naftol Red G
Acid Phloxine GA
Acid Red 1
Acid Red 2G
Acid Red GA
Acid Rose 2GL
Acilan Naphthol Red G
Acilan Naphtol Red G
Ahcocid Carmine 2G
Amacid Phloxine
Amido Naphthol Red G
Amido Naphthol Red 2G
Amido Naphthol Red GA
Amido Red 2G
Ariavit Red 2G
Atul Acid Geranine G
Azo Geranine 2G
Azo Geranine 2GA
Azonaphthol Red J
Azophloxin
Azophloxine
Azo Phloxine GA
Azo Phloxine GA-CF
Azo Rhodine 2G
Belacid Phloxine G
Brilliant Acid Red G
Brilliant Acid Rosamine 2G
Brilliant Colacid Red G
Bucacid Fast Crimson
Calcocid Phloxine 2G
Cetil Light Red GG
Cerven 2G (Czech)
Cerven Kysela 1 (Czech)
Cerven Potravinarska 10 (Czech)
C.I. 18050
C.I. Acid Red 1
C.I. Acid Red 1, Disodium salt
C.I. Food Red 10
Edicol Supra Geranine 2G
Edicol Supra Geranine 2GS
Egacid Red G
Eniacid Light Red 3G
Erio Floxine 2G
Erio Floxine 2GN
Ext. D & C Red No. 11
Fast Crimson GR
Fast Drimson GR
Fenazo Red B
Geranine 2GS
Hastings Carmine 2G
Hexacol Red 2G
Hexalan Red 2G
Hidacid Fast Crimson
Hispacid Fast Carmoisine G
Ink Red JSN
Java Naphtol Red G
Kiton Red G
Kiton Red 2G
Leather Red G
Lissamine Red 2G
Naphthazine Rose 2G
Naphtocard Red 2G
Phloxine G
Phloxine 2G
Pontacyl Carmine 2G
1379 Red
Red 2G
Solar Fast Red 3g
Unitertracid Red 2G
Vondacid Light Red NG

3734-97-2
$C_{10}H_{24}NO_3PS.C_2H_2O_4$
359.42
CCOP(=O)(OCC)SCCN(CC)CC.[O-]C(=O)C([O-])=O
Phosphorothioic acid, S-(2-(diethylamino)ethyl) O,O-diethyl ester, oxalate (1:1)
Amiton oxalate
Chipman 6199
Chipman R-6,199
Citram
S-(2-Diethylaminoethyl) O,O-diethylphosphorothioate hydrogenoxalate
O,O-Diethyl-S-(2-diethyl-

amino)ethylphosphorothioate hydrogen oxalate
O,O-Diethyl S-(β-diethylamino)ethyl phosphorothiolate hydrogen oxalate
O,O-Diethyl S-(2-ethyl-N,N-diethylamino)phosphorothioate hydrogen oxalate
ENT 20,993
Hydrogen Oxalate of Amiton
Tetram 75
Tetram, acid oxalate
Tetram monooxalate

3735-01-1
$C_{10}H_{16}NO_3PS$
261.30
CCOP(=S)(OCC)Oc1ccc(N)cc1
Phosphorothioic acid, O-(4-aminophenyl) O,O-diethyl ester
Aminoparathion
p-Aminoparathion
O-(4-Aminophenyl) O,O-diethyl phosphorothioate
O,O-Diethyl O-(4-aminophenyl) phosphorothioate
E 605 Reduced

3735-23-7
$C_9H_{11}Cl_2O_2PS_3$
349.26
Methyl phenkapton
Caswell No. 362A
S-(((2,5-Dichlorophenyl)thio)methyl) O,O-dimethyl phosphorodithioate
O,O-Dimethyl S-(2,5-dichlorophenylthio)methyl phosphorodithioate
O,O-Dimethyl S-(((2,5-dichlorophenyl)thio)methyl) phosphorodithioate
ENT 25,554-X
EPA Pesticide Chemical Code 362200
G 30494
Geigy G-30494
Geigy 30494
Methanethiol, ((2,5-dichlorophenyl)thio)-, S-ester with O,O-dimethyl phosphorodithioate
Methyl phencapton
Phosphorodithioic acid, S-(((2,5-dichlorophenyl)thio)methyl) O,O-dimethyl ester

3736-08-1
Unknown
Unknown
Fenethylline
DEA No. 1503
Fenetilina (Spanish)
Fenetillina
Fenetylinum
Fenetyllinum (Latin)
7-(2-((α-Methylphenethyl)amino)ethyl)theophylline

3737-00-6
C_3H_4BrCl
155.42
ClC=CCBr
1-Propene, 3-bromo-1-chloro- (9CI)
Propene, 3-bromo-1-chloro- (8CI)

3739-38-6
$C_{13}H_{10}O_3$
214.23
OC(=O)c2cccc(Oc1ccccc1)c2
Benzoic acid, m-phenoxy
Benzoic acid, 3-phenoxy-
m-Phenoxybenzoic acid
3-Phenoxybenzoic acid

3739-67-1
$C_{21}H_{24}O_2$
308.42
O(c(ccc(c1)C(c(ccc(OCC=C)c2)c2)(C)C)c1)CC=C
Benzene, 1,1'-(1-methylethylidene)bis(4-(2-propenyloxy)- (9CI)
AI3-02489

3741-00-2
$C_{10}H_{20}$
140.27
Cyclopentane, pentyl- (9CI)
NSC-174063
Pentane, 1-cyclopentyl- (8CI)

3744-02-3
$C_6H_{10}O$
98.14
4-Penten-2-one, 4-methyl-
AI3-24338

3745-18-4
$C_6H_9N_3S$
155.22
1,3,5-Triazine, 2,4-dimethyl-6-(methylthio)- (9CI)
s-Triazine, 2,4-dimethyl-6-(methylthio)- (7CI,8CI)

3746-39-2
$C_{14}H_{28}O_2S$
260.44
O=C(OCCCCCCCCCCCC)CS
Acetic acid, mercapto-, dodecyl ester (9CI)
Dodecyl mercaptoacetate
Dodecyl thioglycolate
Lauryl thioglycolate

3747-48-6
$C_9H_{10}ClNO_2$
199.63
Carbanilic acid, 2-chloroethyl ester
AI3-17233
N-(2-Chloroethoxycarbonyl)aniline
2-Chloroethyl phenylcarbamate
2-Chloroethyl N-phenylcarbamate
Ethanol, 2-chloro-, carbanilate (8CI)
Ethanol, 2-chloro-, phenylcarbamate (9CI)

3748-84-3
$C_9H_{13}N$
135.20
Cc1cc(C)c(C)nc1C
Pyridine, 2,3,5,6-tetramethyl- (8CI,9CI)

3757-76-4
$C_6H_4Cl_2O.Na$
185.99
Phenol, 2,4-dichloro-, sodium salt (9CI)
2,4-Dichlorophenol sodium salt

3760-20-1
$C_8H_{16}O$
128.21
Cyclohexanol, 2-ethyl- (8CI, 9CI)
NSC-62035

3761-41-9
$C_{10}H_{15}O_4PS_2$
294.34
COP(=S)(OC)Oc1ccc(S(C)=O)c(C)c1
Phosphorothioic acid, O,O-dimethyl O-(4-(methylsulfinyl)-m-tolyl) ester
O,O-Dimethyl O-(4-(methylsulfinyl)-m-tolyl) phosphorothioate
O,O-Dimethyl O-((4-methylthio)-m-tolyl)phosphorothioate sulfoxide
Fensulfoxide
Fenthion sulfoxide
Mesulfenos
Phosphorothioic acid, O,O-dimethyl O-(3-methyl-4-(methylsulfinyl)phenyl) ester (9CI)

3761-42-0
$C_{10}H_{15}O_5PS_2$
310.34
COP(=S)(OC)Oc1ccc(c(C)c1)S(C)(=O)=O
Phosphorothioic acid, O,O-dimethyl O-(4-(methylsulfonyl)-m-tolyl) ester
O,O-Dimethyl O-(4-(methylsulfonyl)-m-tolyl) phosphorothioate

O,O-Dimethyl O-((4-methyl-thio)-m-tolyl)phosphorothioate sulfone
Fenthione sulfone
Phosphorothioic acid, O,O-dimethyl O-(3-methyl-4-(methylsulfonyl)phenyl) ester (9CI)

3761-53-3
$C_{18}H_{14}N_2O_7S_2.2Na$
480.44
[Na+].[Na+].Cc3ccc(N=Nc1c(O)c(cc2cc(ccc12)S([O-])(=O)=O)S([O-])(=O)=O)c(C)c3
2,7-Naphthalenedisulfonic acid, 3-hydroxy-4-(2,4-xylylazo)-, disodium salt
Acetacid Red J
Acidal Ponceau G
Acid Leather Red KPR
Acid Leather Red P2R
Acid Leather Scarlet IRW
Acid Ponceau R
Acid Ponceau 2RL
Acid Ponceau Special
Acid Red 26
Acid Scarlet
Acid Scarlet 2B
Acid Scarlet 2BN
Acid Scarlet 2R
Acid Scarlet 2R for Lakes
Acid Scarlet 2R for Lakes Bluish
Acid Scarlet 2RL
Acid Scarlet 2RN
Acilan Ponceau RRL
Ahcocid Fast Scarlet R
Aizen Ponceau RH
Amacid Lake Scarlet 2R
Amacid Scarlet 2R
Brilliant Ponceau G
Calcocid 2RIL
Calcocid Scarlet 2R
Calcocid Scarlet 2RIL
Calcolake Scarlet 2R
Certicol Ponceau MXS
Cerven Kysela 26 (Czech)
Cerven Potravinarska 5 (Czech)
C.I. 79
C.I. 16150
C.I. Acid Red 26
C.I. Acid Red 26, Disodium salt
C.I. Food Red 5
Colacid Ponceau Special
Comacid Scarlet 2R
D & C Red No. 5
4-((2,4-Dimethylphenyl)azo)-3-hydroxy-2,7-naphthalene-disulfonic acid, disodium salt
4-((2,4-Dimethylphenyl)azo)-3-hydroxy-2,7-naphthalenedi-sulphonic acid, disodium salt
Disodium (2,4-dimethylphenyl-azo)-2-hydroxynaphthalene-3,6-disulfonate
Disodium (2,4-dimethylphenyl-azo)-2-hydroxynaphthalene-3,6-disulphonate
Disodium salt of 1-(2,4-xylyl-azo)-2-naphthol-3,6-disulfonic acid
Disodium salt of 1-(2,4-xylyl-azo)-2-naphthol-3,6-di-sulphonic acid
Edicol Ponceau RS
Edicol Supra Ponceau R
Fenazo Scarlet 2R
Food Red No. 101
Hexacol Ponceau MX
Hexacol Ponceau 2R
Hidacid Scarlet 2R
Hispacid Ponceau R
3-Hydroxy-4-(2,4-xylylazo)-3,7-naphthalenedisulfonic acid, disodium salt
3-Hydroxy-4-(2,4-xylylazo)-3,7-naphthalenedisulphonic acid, disodium salt
Java Ponceau 2R
Kiton Ponceau R
Kiton Ponceau 2R
Kiton Scarlet 2RC
Lake Ponceau
Lake Scarlet R
Lake Scarlet 2RBN
Naphthalene Lake Scarlet R
Naphthalene Scarlet R
Naphthazine Scarlet 2r
Neklacid Red RR
New Ponceau 4R
Paper Red HRR
Pigment Ponceau R
Ponceau BNA
Ponceau De Xylidine
Ponceau FR
Ponceau G
Ponceau GR
Ponceau J
Ponceau MX
Ponceau NR
Ponceau PXM
Ponceau R
Ponceau 2R
Ponceau R (Biological stain)
Ponceau 2R (Biological stain)
Ponceau 2RE
Ponceau Red
Ponceau Red R
Ponceau 2R Extra A Export
Ponceau RG
Ponceau 2RL
Ponceau RS
Ponceau RR
Ponceau RR Type 8019
Ponceau 2RX
Ponceau Xylidine
Ponceau Xylidine (Biological stain)
1695 Red
Red for Lakes J
Red R
Scarlet R
Scarlet 2R
Scarlet 2RB
Scarlet 2RL
Scarlet 2RL Bluish
Scarlet RRA
Schultz No. 95
Tertracid Ponceau 2R
Xylidine Ponceau
Xylidine Ponceau 3RS
Xylidine Red
1-Xylylazo-2-naphthol-3,6-disulfonic acid, disodium salt
1-(2,4-Xylylazo)-2-naphthol-3,6-disulphonic acid, disodium salt
1-(2,4-Xylylazo)-2-naphthol-3,6-disulphonic acid, disodium salt
1-Xylylazo-2-naphthol-3,6-disulphonic acid, disodium salt

3761-60-2
$C_{18}H_{29}ClO_5S$
392.94
Smite
PPPS
2-Propanol, 1-(2-(p-tert-butylphenoxy)-1-methylethoxy)-, 2-chloroethylsulfite
Sulfurous acid, 2-(2-(p-tert-butylphenoxy)-1-methylethoxy)-1-methylethyl-, 2-chloroethyl ester (8CI)
Sulfurous acid, 2-chloroethyl-, 2-(2-(4-(1,1-dimethylethyl)-phenoxy)-1-methylethoxy)-1-methylethyl ester (9CI)

3765-57-9
$C_{10}H_6Cl_4O_3S$
348.02
COC(=O)c1c(Cl)c(Cl)c(C(=O)SC)c(Cl)c1Cl
Thioterephthalic acid, tetrachloro-, O,S-dimethyl ester
O,S-Dimethylester kyseliny tetrachlorthiotereftalove (Czech)
O,S-Dimethyltetrachlorothio-terephthalate
Glenbar
OCS-21,944
Terephthalic acid, tetrachlorothio-, O,S-dimethyl ester (8CI)

3766-27-6
$C_8H_5Cl_2O_3.Li$
226.97
Lithium 2,4-dichlorophenoxyacetate
Acetic acid, 2,4-dichlorophenoxy-, lithium salt
Acetic acid, (2,4-dichlorophenoxy)-, lithium salt
AI3-25317
Caswell No. 315B
2,4-Dichlorophenoxyacetic acid lithium salt
EPA Pesticide Chemical Code 030002
Lithate 2,4-D
Lithium 2,4-D

3766-60-7
$C_{12}H_{13}ClN_2O$
236.72
CC(C#C)N(C)C(=O)Nc1ccc(Cl)cc1

Urea, 3-(p-chlorophenyl)-1-methyl-1-(1-methyl-2-propynyl)
Arisan
Buturon
Butyron
N'-(4-Chlorophenyl)-N-isobutinyl-N-methylurea
N'-(4-Chlorophenyl)-N-methyl-N-(1-methyl-2-propynyl)-urea
3-(p-Chlorophenyl)-1-methyl-1-(1-methyl-2-propynyl)urea
N-(4-Chlorphenyl)-N'-methyl-N'-isobutinylharnstoff (German)
3-(4-Chlorphenyl)-1-methyl-1-isobutinylharnstoff (German)
3-(4-Chlorophenyl)-1-methyl-1-(1-methylprop-2-ynyl)urea
Eptapur
H 95
H 95-1
HS 95

3766-81-2
$C_{12}H_{17}NO_2$
207.30
O=C(Oc(c(ccc1)C(CC)C)c1)NC
Carbamic acid, methyl-, o-sec-butylphenyl ester
Barizon
Bassa
Baycarb
Bayer 41367C
Bayer 41637
BPMC
2-sek.Butylfenylester kyseliny methylkarbaminove (Czech)
o-sec-Butylphenyl methylcarbamate
2-sec-Butylphenyl N-methylcarbamate
Carvil
Fenobcarb
Fenobucarb
Hopcin
Methylcarbamic acid o-sec-butylphenyl ester
2-(1-Methylpropyl)phenyl methylcarbamate
OSBAC
Phenol, 2-(1-methylpropyl)-, methylcarbamate

3769-23-1
C_7H_{14}
98.19
1-Hexene, 4-methyl- (8CI,9CI)
NSC-73926

3769-57-1
$C_{17}H_{19}ClN_4O_4$
378.79
O=N(=O)c(ccc(N=Nc(c(cc(N(CCO)CCO)c1)C)c1)c2Cl)c2
Ethanol, 2,2'-((4-((2-chloro-4-nitrophenyl)azo)-3-methylphenyl)imino)bis- (9CI)
4-(2-Chloro-4-nitrophenylazo)-3-methyl-N,N-bis(2-hydroxyethyl)aniline

3770-48-7
$C_{13}H_{11}N$
181.23
9H-Carbazole, 4-methyl- (9CI)
Carbazole, 4-methyl- (8CI)

3770-97-6
$C_{10}H_5ClN_2O_3S$
268.67
O=C(c(c(c(S(=O)(=O)Cl)cc1)C=C2)c1)C2=N=N
1-Naphthalenesulfonyl chloride, 6-diazo-5,6-dihydro-5-oxo- (9CI)
2-Diazo-1,2-dihydro-1-oxo-5-naphthalenesulfonyl chloride
6-Diazo-5,6-dihydro-5-oxo-1-naphthalenesulfonyl chloride
2-Diazo-1-naphthol-5-sulfonyl chloride
2-Diazo-1-naphthone-5-sulfonic acid chloride
2-Diazo-1-naphthone-5-sulfonyl chloride
1-Hydroxy-5-chlorosulfonyl-2-naphthalenediazonium hydroxide, inner salt

3771-19-5
$C_{20}H_{22}O_3$
310.42
CC(C)(Oc1ccc(cc1)C2CCCc3ccccc23)C(O)=O
Propionic acid, 2-methyl-2-(p-(1,2,3,4-tetrahydro-1-naphthyl)phenoxy)
CH 13-437
Ciba 13437 SU
C 13437 SU
2-Methyl-2-(4-(1,2,3,4-tetrahydro-1-naphthalenyl)phenoxy)propanoic acid
α-Methyl-α-(p-1,2,3,4-tetrahydronaphth-1-ylphenoxy)-propionic acid
2-Methyl-2-(p-(1,2,3,4-tetrahydro-1-naphthyl)phenoxy)-propionic acid
2-Methyl-2-(4-(1,2,3,4-tetrahydro-1-naphthyl)phenoxy)-propanoic acid
Melipan
Nafenoic acid
Nafenopin
SU-13437
TPIA

3771-38-8
$C_{17}H_{21}N_3O_2$
299.35
Ethanol, 2,2'-((3-methyl-4-(phenylazo)phenyl)imino)-bis- (9CI)
Ethanol, 2,2'-((4-(phenylazo)-m-tolyl)imino)di- (8CI)

3772-94-9
$C_{18}H_{23}Cl_5O_2$
448.64
O=C(Oc(c(c(c(c1Cl)Cl)Cl)Cl)c1Cl)CCCCCCCCCCC
Pentachlorophenol laurate
AI3-17004
Caswell No. 521A
Dodecanoic acid, pentachlorophenyl ester (9CI)
EPA Pesticide Chemical Code 063010
Lauryl pentachlorophenate
Pentachlorophenyl dodecanoate

3773-14-6
$C_6H_3Cl_3S$
213.50
Benzenethiol, 2,4,5-trichloro
Renacit II
2,4,5-Trichlorobenzenethiol
2,4,5-Trichlorothiophenol

3774-52-5
C_7H_5NOS
151.18
Thiocyanic acid, p-hydroxyphenyl ester (8CI)
AI3-19626
Thiocyanic acid, 4-hydroxyphenyl ester (9CI)

3775-85-7
$C_{10}H_{18}O_4$
202.28
CC2(COCCOCC1(C)CO1)CO2
Ethane, 1,2-bis(2,3-epoxy-2-methylpropoxy)
Ethylene glycol bis(2,3-epoxy-2-methylpropyl) ether
Ethylene glycol di(2,3-epoxy-2-methylpropyl) ether
Ethylene glycolide, (2,3-epoxy-2-methylpropyl)ether

3775-90-4
$C_{10}H_{19}NO_2$
185.30
O=C(OCCNC(C)(C)C)C(=C)C
Methacrylic acid, 2-(tert-butylamino)ethyl ester
Ageflex FM-4
tert-Butylaminoethyl methacrylate
2-(tert-Butylamino)ethyl methacrylate
Ethanol, 2-(tert-butylamino)-, methacrylate (ester)

3777-69-3
$C_9H_{14}O$
138.23
CCCCCc1ccco1
Furan, 2-pentyl

2-Amylfuran
2-Pentylfuran
2-n-Pentylfuran

3777-71-7
$C_{11}H_{18}O$
166.26
O(C(=CC=1)CCCCCCC)C1
Furan, 2-heptyl- (9CI)
2-Heptylfuran

3778-73-2
$C_7H_{15}Cl_2N_2O_2P$
261.11
ClCCNP1(=O)OCCCN1CCCl
1,3,2-Oxazaphosphorine, 3-(2-chloroethyl)-2-((2-chloroethyl)amino)tetrahydro-, 2-oxide
A 4942
Asta Z 4942
N,N-Bis(β-chloroethyl)-amino-N',O-propylene-phosphoric acid ester diamide
2,3-(N,N[1]-Bis(2-chloroethyl)-diamido)-1,3,2-oxazaphosphoridinoxyd
N,3-Bis(2-chloroethyl)tetrahydro-2H-1,3,2-oxazaphosphorin-2-amine 2-oxide
N-(2-Chloraethyl)-N'-(2-chloraethyl)-N',O-propylen-phosphorsaureester-diamid (German)
3-(2-Chloroethyl)-2-((2-chloroethyl)amino)perhydro-2H-1,3,2-oxazaphosphorine oxide
3-(2-Chloroethyl)-2-((2-chloroethyl)amino)tetrahydro-2H-1,3,2-oxazaphosphorine 2-oxide
N-(2-Chloroethyl)-N'-(2-chloroethyl)-N',O-propylenephosphoric acid diamide
N-(2-Chloroethyl)-N'-(2-chloroethyl)-N',O-propylenephosphoric acid ester diamide
Cyfos
Holoxan
Holoxan 1000
Ifosfamid
Ifosfamide
Iphosphamid
Iphosphamide
Isoendoxan
Isofosfamide
Isophosphamide
Mitoxana
MJF 9325
Naxamide
NCI-C01638
NSC-109724
2H-1,3,2-Oxazaphosphorin-2-amine, N,3-bis(2-chloroethyl)tetrahydro-, 2-oxide (9CI)
Z 4942

3779-61-1
$C_{10}H_{16}$
136.24
1,3,6-Octatriene, 3,7-dimethyl-, (E)- (8CI,9CI)

3779-63-3
$C_{24}H_{36}N_6O_6$
504.55
1,3,5-Triazine-2,4,6(1H,3H,5H)-trione, 1,3,5-tris(6-isocyanatohexyl)- (9CI)

3782-00-1
$C_{10}H_{10}O$
146.19
O(c(c(C=1C)ccc2)c2)C1C
Benzofuran, 2,3-dimethyl- (9CI)
2,3-Dimethylbenzofuran

3782-80-7
$C_{22}H_{19}NO_2$
329.40
Benzoic acid, 4-(((1,1'-biphenyl)-4-ylmethylene)amino)-, ethyl ester (9CI)
Benzoic acid, p-((p-phenylbenzylidene)amino)-, ethyl ester (7CI,8CI)
Ethyl 4-(4-phenylbenzalamino)benzoate

3785-20-4
$C_{13}H_{18}ClNO$
239.77
o-Acetotoluidide, 6'-tert-butyl-2-chloro
Acetamide, 2-chloro-N-(2-(1,1-dimethylethyl)-6-methylphenyl)- (9CI)
2-tert-Butyl-6-methylchloroacetanilide
2-Chloro-N-(2-methyl-6-tert-butylphenyl)acetamide
CP 31675
Monsanto 31675

3786-23-0
$C_{21}H_{21}N_9$
399.46
Propionitrile, 3,3',3'',3'''-((6-phenyl-s-triazine-2,4-diyl)dinitrilo)tetra- (7CI,8CI)

3786-76-3
$C_{41}H_{66}O_{13}$
767.07
5-β-Cardanolide, 3-β-((O-2,6-dideoxy-β-D-ribo-hexopyranosyl-(1-4)-O-2,6-dideoxy-β-D-hexopyranosyl-(1-4)-2,6-dideoxy-β-d-ribo-hexopyranosyl)oxy)-14-hydroxy
Dihydrodigitoxin
20,22-Dihydrodigitoxin

3786-91-2
$C_9H_{10}O_8$
246.17
OC(=O)C1CC(C(C1C(O)=O)C(O)=O)C(O)=O
1,2,3,4-Cyclopentanetetracarboxylic acid, (1α,2α,3α,4α)- (9CI)
1,2,3,4-Cyclopentanetetracarboxylic acid, all-cis- (8CI)
NSC-73712

3790-71-4
$C_{15}H_{26}O$
222.37
2,6,10-Dodecatrien-1-ol, 3,7,11-trimethyl-, (Z,E)- (8CI,9CI)

3794-83-0
$C_2H_4O_7P_2.4Na$
293.96
[Na+].[Na+].OC(COP(O)O)OP(O)O.[Na][Na]
Diphosphonic acid, (1-hydroxyethylidene)-, tetrasodium salt
Ethane-1-hydroxy-1,1-diphosphonic acid, tetrasodium salt
(1-Hydroxyethylidene)diphosphonic acid, tetrasodium salt
Tetrasodium etidronate

3795-88-8
$C_{13}H_{16}N_4O_6$
324.27
Levofuraltadone
Levofuraltadona (Spanish)
Levofuraltadonum (Latin)
5-(Morpholinomethyl)-3-((5-nitrofurfurylidene)amino)-2-oxazolidinone
(-)-5-(Morpholinomethyl)-3-((5-nitrofurfurylidene)amino)-2-oxazolidinone
NF-602
NF-902
NSC-527986
2-Oxazolidinone, 5-(morpholinomethyl)-3-((5-nitrofurfurylidene)amino)-, (-)-, (8CI)
2-Oxazolidinone, 5-(4-morpholinylmethyl)-3-(((5-nitro-2-furanyl)methylene) amino)-, (S)-, (9CI)
2-Oxazolidinone, 5-(4-morpholinylmethyl)-3-(((5-nitro-2-furanyl)methylene)amino)-, (-)-

3796-70-1
$C_{13}H_{22}O$
194.32
O=C(CCC=C(CCC=C(C)C)C)C
5,9-Undecadien-2-one, 6,10-dimethyl-, (E)- (9CI)
(E)-6,10-Dimethyl-5,9-undeca-

dien-2-one

3797-36-2
$C_9H_{16}FN_5$
213.26
s-Triazine, 2-(diethyl-amino)-4-(ethylamino)-6-fluoro- (7CI,8CI)

3806-34-6
$C_{11}H_{16}N.C_6H_3N_3O_9S$
455.46
O(P(OCC1(COP(O2)OCCCCCC CCCCCCCCCCCC)C2)OCC CCCCCCCCCCCCCCCC)C1
Aziridinium, 1-benzyl-1-ethyl-, picrylsulfonate
1-Benzyl-1-ethylaziridinium 2,4,6-trinitrobenzenesulfonate
P 1

3810-74-0
$C_{42}H_{78}N_{14}O_{24}.H_6O_{12}S_3$
1457.58
Streptomycin, sulfate (2:3) (Salt)
Agri-Mycin
Agristrep
AS-15
Phytomycin
Plantomycin
Strepcin
Strep-Gran
Strepsulfat
Streptomycin sesquisulfate
Streptomycin sulfate
Streptomycin sulphate B.P
Streptorex
Strepvet
Vetstrep

3811-04-9
$ClO_3.K$
122.55
Chloric acid, potassium salt
Berthollet Salt
Berthollet's Salt
Chlorate de potassium (French)
Chlorate of potash (DOT)
Fekabit
Kaliumchloraat (Dutch)
Kaliumchlorat (German)
Oxymuriate of Potash
Pearl Ash
Potash chlorate (DOT)
Potassio (chlorato di) (Italian)
Potassium chlorate [UN 1485]
Potassium chlorate, Aqueous solution [UN 2427]
Potassium (chlorate de) (French)
Potassium oxymuriate
Potcrate
Salt of Tarter
UN 1485 [Potassium chlorate]
UN 2427 [Potassium chlorate, solution]

3811-49-2
$C_8H_9O_3PS$
216.20
COP2(=S)OCc1ccccc1O2
4H-(1.3.2)Benzodioxaphos-phorine, 2-methoxy-, 2-sulfide
Dioxabenzofos
K-9
2-Methoxy-4H-1,2,3-benzodi-oxaphosphorine-2-sulfide
Phosphorothioic acid, cyclic O,O-(methylene-o-phenylene) O-methyl ester
Salithion
Salithion-Sumitomo

3811-68-5
$C_{18}H_{37}N.C_2H_4O_2$
327.55
9-Octadecen-1-amine, acetate (9CI)

3811-73-2
$C_5H_5NOS.Na$
150.16
2-Pyridinethiol, N-oxide, sod-ium salt
(1-Hydroxy-2-pyridinethione), sodium salt, tech.
2-Mercaptopyridine-N-oxide sodium salt
2-Pyridinethiol, 1-oxide, sodium salt
Sodium pyrithione
Thione (Reagent)

3812-32-6
CO_3
60.01
O=C(O)O
Carbonate (8CI,9CI)
Carbonic acid, ion(2-)

3813-05-6
$C_9H_6ClNO_3S$
243.67
OC(=O)Cn1c(=O)sc2cccc(Cl)c12
Acetic acid, (4-chloro-2-oxo-benzothiazol-3-yl)
Ben-30
Benazalox
Benazolin
Benazoline
Ben-Cornox
Benopan
Bensecal
Benzar
3(2H)-Benzothiazoleacetic acid, 4-chloro-2-oxo- (9CI)
3-Benzothiazolineacetic acid, 4-chloro-2-oxo- (8CI)
4-Chloro-2-oxobenzothiazol-3-ylacetic acid
Cornox CWK
Cresopur
Eunasin
EX 10781
Galipan
Grassland Weedkiller
Herbazolin
Herbitox
Keropur
Legumex Extra
Ley-Cornox
Leymin
Metizolin
RD 7693
Tri-Cornox
Tri-Cornox Special

3813-14-7
$C_8H_5Cl_3O_3.C_6H_{15}NO_3$
404.67
Acetic acid, (2,4,5-trichloro-phenoxy)-, Compd. with 2,2',2''-nitrilotris(ethanol) (1:1) (9CI)
Acetic acid, (2,4,5-trichloro-phenoxy)-, Compd. with 2,2',2''-nitrilotriethanol (1:1) (8CI)
Caswell No. 881J
EPA Pesticide Chemical Code 082033
2,4,5-T-Tris(2-hydroxyethyl)-ammonium
2,4,5-Trichlorophenoxyacetic acid triethanolamine salt

3817-11-6
$C_8H_{18}N_2O_2$
174.28
CCCCN(CCCCO)N=O
1-Butanol, 4-(butylnitroso-amino)
BBN
BBNOH
BHBN
Butanol (4)-butyl-nitrosamine
Butyl-butanol(4)-nitrosamin (German)
Butyl-butanol-nitrosamine
n-Butyl-N-(4-hydroxybutyl)-nitrosamine
n-Butyl-(4-hydroxybutyl)nitros-amine
4-(Butylnitrosamino)-1-butanol
4-(n-Butylnitrosamino)-1-but-anol
Dibutylamine, 4-hydroxy-N-ni-troso-
HBBN
4-Hydroxybutylbutylnitros-amine
NBHA
N-Nitroso-n-butyl-(4-hydroxy-butyl)amine
OH-BBN

3818-54-0
$C_{22}H_{30}O_2S$
358.55
Oc(cc(c(Sc(c(cc(O)c1)C(C)(C)C) c1C)c2C(C)(C)C)C)c2
Phenol, 4,4'-thiobis(3-(1,1-di-

methylethyl)-5-methyl- (9CI)

3819-18-9
C_9H_7NOS
177.23
8-Quinolinol, sulfate (Salt) (8CI,9CI) (VAN)

3825-26-1
$C_8F_{15}O_2.H_4N$
431.13
OC(=O)C(F)(F)C(F)(F)C(F)(F)C(F)(F)C(F)(F)C(F)(F)C(F)(F)F
Octanoic acid, pentadecafluoro-, ammonium salt
Ammonium pentadecafluoro-octanate
Ammonium perfluorocaprilate
Ammonium perfluorocaprylate
Ammonium perfluorooctanoate
APFO
FC-143
Perfluoroammonium octanoate

3841-15-4
$C_{37}H_{25}N_7O_{10}S_2.3Na$
860.71
Benzoic acid, 5-((4-((4-((6-amino-1-hydroxy-3-sulfo-2-naphthalenyl)azo)-1-naphthalenyl)-azo)-6-sulfo-1-naphthalenyl)-azo)-2-hydroxy-, trisodium salt (9CI)
C.I. Direct Blue 148 (VAN)
C.I. Direct Blue 148, Trisodium salt (8CI)

3842-03-3
$C_9H_{20}O_2$
160.26
O(C(OCC)CC(C)C)CC
Butane, 1,1-diethoxy-3-methyl- (9CI)
1,1-Diethoxy-3-methylbutane
Isovaleraldehyde, diethyl acetal
3-Methylbutanal, diethyl acetal

3842-52-2
$C_{15}H_{11}ClN_4$
282.73
1,3,5-Triazin-2-amine, 4-chloro-N,6-diphenyl- (9CI)
s-Triazine, 2-anilino-4-chloro-6-phenyl- (6CI,7CI,8CI)

3842-53-3
$C_9H_7ClN_4$
206.64
1,3,5-Triazin-2-amine, 4-chloro-6-phenyl- (9CI)
s-Triazine, 2-amino-4-chloro-6-phenyl- (6CI, 7CI,8CI)

3842-55-5
$C_{15}H_{10}ClN_3$
267.72
1,3,5-Triazine, 2-chloro-4,6-diphenyl- (9CI)
2-Chloro-4,6-diphenyl-s-triazine
6-Chloro-2,4-diphenyl-s-triazine
s-Triazine, 2-chloro-4,6-diphenyl- (6CI,7CI,8CI)

3844-45-9
$C_{37}H_{36}N_2O_9S_3.2Na$
794.91
[Na+].[Na+].CCN(Cc1cccc(c1)S([O-])(=O)=O)c2ccc(cc2)C(=C3C=CC(C=C3)=[N+](CC)Cc4cccc(c4)S([O-])(=O)=O)c5ccccc5S([O])(=O)=
Ammonium, ethyl(4-(p-(ethyl-(m-sulfobenzyl)amino)-α-(o-sulfophenyl)benzylidene)- 2,5-cyclohexadien-1-ylidene)(m-sulfobenzyl)-, hydroxide, inner salt, disodium salt
Acid Sky Blue A
Aizen Food Blue No. 2
1206 Blue
Blue 1206
Brilliant Blue
Brilliant Blue FCD No. 1
Brilliant Blue FCF
Canacert Brilliant Blue FCF
C.I. 671
C.I. 42090
C.I. Acid Blue 9, Disodium salt
C.I. Food Blue 2
Cogilor Blue 512.12
Cosmetic Blue Lake
D & C Blue No. 4
Dispersed Blue 12195
Dolkwal Brilliant Blue
Edicol Blue CL 2
Erioglaucine G
FD & C Blue 1
FD & C Blue No. 1
Fenazo Blue XI
Food Blue 2
Food Blue Dye No. 1
Hexacol Brilliant Blue A
Intracid Pure Blue L
Merantine Blue EG
Modr Brilantni FCF (Czech)
Modr Kysela 9 (Czech)
Modr Potravinarska 2 (Czech)
Schultz No. 770
Usacert Blue No. 1

3844-60-8
$C_4H_8N_6O_4$
204.18
Biurea, 1,6-dimethyl-1,6-dinitroso
1,6-Dimethyl-1,6-dinitrosobiurea
1,2-Hydrazinedicarboxamide, N,N'-dimethyl-N,N'-dinitroso- (9CI)
Hydrazodicarbonsaeureabis-(methylnitrosamid) (German)
Hydrazodicarboxylic acid bis-(methylnitrosamide)
Hydroazodicarboxybis(methylnitrosamide)
NSC-409425
SRI 1666

3846-74-0
$C_8H_{13}N_3S$
183.28
s-Triazine, 2,4-diethyl-6-(methylthio)- (7CI, 8CI)

3850-30-4
$C_6H_{15}N$
101.19
NC(C(C)(C)C)C
2-Butanamine, 3,3-dimethyl- (9CI)
2-Amino-3,3-dimethylbutane
3,3-Dimethyl-2-butanamine
1,2,2-Trimethylpropylamine

3852-09-3
$C_5H_{10}O_3$
118.15
O=C(OC)CCOC
Propionic acid, 3-methoxy-, methyl ester
β-Methoxypropionic acid, methyl ester
3-Methoxypropionic acid methyl ester
Methylester kyseliny 3-methoxypropionove (Czech)

3855-32-1
$C_{11}H_{27}N_3$
201.34
N(CCCN(CCCN(C)C)C)(C)C
1,3-Propanediamine, N-(3-(dimethylamino)propyl)-N,N',N'-trimethyl- (9CI)
N,N,N'-Trimethyl-N'-(3-(dimethylamino)propyl)-1,3-propanediamine

3856-25-5
$C_{15}H_{24}$
204.36
Tricyclo(4.4.0.0(2,7))dec-3-ene, 1,3-dimethyl-8-(1-methylethyl)-, stereoisomer (9CI) (VAN)
Aglaiene
Copaene
α-Copaene
Tricyclo(4.4.0.0(2,7))dec-3-ene, 8-isopropyl-1,3-dimethyl-
Tricyclo(4.4.0.0(2,7))dec-3-ene, 8-isopropyl-1,3-dimethyl-, (1R,2S,6S,7S,8S)-(-)- (8CI)

3858-78-4
$C_4H_{11}N.ClH$
109.59
1-Butanamine, hydrochloride (9CI)

3860-63-7
$C_{16}H_{14}N_2O_4$
298.29
O=C(c(c(c(NC)cc1)C(=O)c2c(O)ccc3NC)c1O)c23
9,10-Anthracenedione, 1,5-dihydroxy-4,8-bis(methylamino)- (9CI)

3861-41-4
Unknown
Unknown
Bromoxynil butyrate

3861-47-0
$C_{15}H_{17}I_2NO_2$
497.13
CCCCCCCC(=O)Oc1c(I)cc(C#N)cc1I
Octanoic acid, (4-cyano-2,6-diiodo)phenyl ester
Benzonitrile, 3,5-diiodo-4-hydroxy-, octanoate
Benzonitrile, 3,5-diiodo-4-octanoyloxy-
4-Cyano-2,6-dijodphenol caprysaeureester (German)
3,5-Diiodo-4-hydroxybenzonitrile octanoate
3,5-Diiodo-4-octanoyloxybenzonitrile
3,5-Dijod-4-hydroxy-benzonitril caprysaeureester (German)
Ioxynil octanoate
M&B 11,461
RIP-15830
Totril

3861-72-1
$C_{20}H_{25}NO_4$
343.42
Acetyldihydrocodeine
Acetyldihydrokodein (Czech)
Codeine, acetyldihydro-
DEA No. 9051
Morphinan-6-α-ol, 4,5-α-epoxy-3-methoxy-17-methyl-, acetate

3861-73-2
$C_{26}H_{19}N_3O_{10}S_3.3Na$
698.63
2,7-Naphthalenedisulfonic acid, 4-((4-anilino-5-sulfo-1-naphthyl)azo)-5-hydroxy-, trisodium salt
Acid Blue 92
Acid Blue A
Acid Leather Blue R
Acid Wool Blue RL
Acilan Fast Navy Blue R
Airedale Blue RL
Amacid Fast Blue R
Anazolene, Sodium
4-((4-Anilino-5-sulfo-1-naphthyl)azo)-5-hydroxy-2,7-naphthalenedisulfonic acid trisodium
Benzyl Blue R
Benzyl Fast Blue R
Bucacid Fast Wool Blue R
Calcocid Fast Blue SR
C.I. 13390
C.I. Acid Blue 92
C.I. Acid Blue 92, Trisodium salt
Cirene Brilliant Blue R
Colacid Blue A
Coomassie Blue
Coomassie Blue Medicinal
Coomassie Blue RL
Cyanine Acid Blue R
Cyanine Acid Blue R New
Fast Acid Blue RL
Fast Wool Blue R
Fenazo Blue SR
Hispacid Fast Blue R
Medium Blue EMBL
Modr Kysela 92 (Czech)
Pontacyl Fast Blue R
Sodium Amazolene
Sodium Anazolene
Sulfonine Acid Blue R
Sulphon Acid Blue R
Sulphon Acid Blue RA
Tertracid Fast Blue SR
Trisodium 4'-anilino-8-hydroxy-1,1'-azonaphthalene-3,6,5'-trisulfonate
Vondamol Fast Blue R
Wool Blue RL
Wool Fast Blue R

3862-11-1
$C_{24}H_{27}O_4P$
410.45
Tris(3,4-xylenyl)phosphate
Phenol, 3,4-dimethyl-, phosphate (3:1)
Phosphoric acid, tris(3,4-dimethylphenyl)ester
Tri-3,4-xylenyl phosphate
3,4-Xylenol, phosphate (3:1)

3862-12-2
$C_{24}H_{27}O_4P$
410.45
Tris(2,4-xylenyl)phosphate
2,4-Dimethylphenol phosphate (3:1)
Phenol, 2,4-dimethyl-, phosphate (3:1) (9CI)
Phosphoric acid, tris(2,4-dimethylphenyl)ester
Tri(2,4-dimethylphenyl) phosphate
Tri(2,4-xylenyl)phosphate
2,4-Xylenol, phosphate (3:1)

3864-99-1
$C_{20}H_{24}ClN_3O$
357.86
Oc(c(cc(c1)C(C)(C)C)C(C)(C)C)c1N(N=C(C=2C=C(C=3)Cl)C3)N2
Phenol, 2-(5-chloro-2H-benzotriazol-2-yl)-4,6-bis(1,1-dimethylethyl)- (9CI)
2-(3',5'-Di-tert-butyl-2'-hydroxyphenyl)-5-chloro-2H-benzotriazole

3867-15-0
$C_{12}H_{20}N_2$
192.29
1-Piperidinocyclohexanecarbonitrile
1-(1-Cyanocyclohexyl)piperidine
Cyclohexanecarbonitrile, 1-piperidino-
Cyclohexanecarbonitrile, 1-(1-piperidinyl)- (9CI)
DEA No. 8603
PCC
Piperidinocyclohexanecarbonitrile
1-(1-Piperidinyl)-cyclohexanecarbonitrile

3871-50-9
$C_{12}H_8F_{17}NO_4S.Na$
608.22
Glycine, N-ethyl-N-((heptadecafluorooctyl)sulfonyl)-, sodium salt (9CI)
N-Ethyl-N-((heptadecafluorooctyl)sulfonyl)glycine, sodium salt

3875-51-2
C_8H_{16}
112.22
Cyclopentane, (1-methylethyl)- (9CI)
Cyclopentane, isopropyl- (8CI)

3876-97-9
$C_{13}H_{14}$
170.25
Naphthalene, 1,2,8-trimethyl- (8CI,9CI)

3877-15-4
$C_4H_{10}S$
90.19
Propane, 1-(methylthio)- (9CI)
Sulfide, methyl propyl (8CI)

3877-19-8
$C_{11}H_{14}$
146.23
Naphthalene, 1,2,3,4-tetrahydro-2-methyl- (8CI,9CI)
NSC-66993

3878-19-1
$C_{11}H_8N_2O$
184.21
c1coc(c1)c3nc2ccccc2[nH]3
Benzimidazole, 2-(2-furyl)
B-33172
Bay 33172
Bayer 33172
1H-Benzimidazole, 2-(2-furanyl)-
Fuberidatol
Fuberidazole
Fuberisazol
Fubridazole
Furidazol
Furidazole
2-(2-Furyl)benzimidazole
2-(2'-Furyl)-benzimidazole
Voronit
Voronite
W VII/117

3878-55-5
$C_5H_8O_4$
132.12
O=C(OC)CCC(=O)O
Monomethyl succinate
AI3-03389
Butanedioic acid, monomethyl ester (9CI)
Methyl hydrogen succinate
Monomethyl butanedioate
Succinic acid, monomethyl ester (8CI)

3880-99-7
$C_7H_6O_2$
122.12
Benzoic-carboxy-13C acid (6CI,7CI,8CI,9CI)
Benzoic-1'-13C acid

3882-98-2
$C_3H_6N_2S.ClH$
138.61
2-Thiazolamine, 4,5-dihydro-, monohydrochloride (9CI)
Thiazolidine, 2-imino-, hydrochloride (8CI)

3883-43-0
$C_4H_6Cl_2O_2$
157.00
ClC1OCCOC1Cl
p-Dioxane, 2,3-dichloro-, trans
trans-2,3-Dichloro-p-dioxane
trans-2,3-Dichloro-1,4-dioxane
1,4-Dioxane, trans-2,3-dichloro-

3883-58-7
$C_7H_{10}O_2$
126.16
1,3-Cyclopentanedione, 2,2-dimethyl- (7CI,8CI, 9CI)
2,2-Dimethyl-1,3-cyclopentanedione

3884-95-5
$C_{14}H_{22}O$
206.33
Oc(c(ccc1)C(CC(C)(C)C)(C)C)c1
Phenol, 2-(1,1,3,3-tetramethylbutyl)- (9CI)
o-tert-Octylphenol
2-(1,1,3,3-Tetramethylbutyl)phenol

3886-69-9
$C_8H_{11}N$
121.18
NC(c(cccc1)c1)C
Benzenemethanamine, α-methyl-, (R)- (9CI)
(R)-α-Methylbenzenemethanamine

3891-98-3
$C_{15}H_{32}$
212.42
Dodecane, 2,6,10-trimethyl- (8CI,9CI)
Farnesane

3891-99-4
$C_{16}H_{34}$
226.45
Tridecane, 2,6,10-trimethyl- (8CI,9CI)

3892-00-0
$C_{18}H_{38}$
254.50
Pentadecane, 2,6,10-trimethyl- (9CI)
Norpristane

3895-17-8
$C_{10}H_{22}O_3$
190.29
Butane, 1-(2-(2-ethoxyethoxy)ethoxy)- (9CI)
Ethane, 1-butoxy-2-(2-ethoxyethoxy)- (8CI)

3896-11-5
$C_{17}H_{18}ClN_3O$
315.78
Oc(c(cc(c1)C)C(C)(C)C)c1N(N=C(C=2C=C(C=3)Cl)C3)N2
Bumetrizole
Bumetrizol (Spanish)
Bumetrizolum (Latin)
2-t-Butyl-6-(5-chloro-2H-benzotriazol-2-yl)-p-cresol
2-tert-Butyl-6-(5-chloro-2H-benzotriazol-2-yl)-p-cresol
2-(5-Chloro-2H-benzotriazol-2-yl)-6-(1,1-dimethylethyl)-4-methylphenol
2-(2'-Hydroxy-3'-t-butyl-5'-methylphenyl)-5-chlorobenzotriazole
2-(2'-Hydroxy-3'-tert-butyl-5'-methylphenyl)-5-chlorobenzotriazole
Phenol, 2-(5-chloro-2H-benzotriazol-2-yl)-6-(1,1-dimethylethyl)-4-methyl- (9CI)
Tinuvin
UV Absorber-6

3900-04-7
$C_6H_{15}O_4P$
182.16
O=P(OCCCCCC)(O)O
Hexyl phosphate
Monohexyl dihydrogen phosphate
Monohexyl phosphate
Phosphoric acid, monohexyl ester (9CI)

3900-31-0
$C_{16}H_{12}ClFN_2O$
302.75
CN2C(=O)CN=C(c1ccccc1F)c3cc(Cl)ccc23
2H-1,4-Benzodiazepin-2-one, 1,3-dihydro-7-chloro-5-(o-fluorophenyl)-1-methyl
2H-1,4-Benzodiazepin-2-one, 7-chloro-5-(o-fluorophenyl)-1,3-dihydro-1-methyl-
2H-1,4-Benzodiazepin-2-one, 7-chloro-5-(2-fluorophenyl)-1,3-dihydro-1-methyl- (9CI)
7-Chloro-5-(2-fluorophenyl)-1-methyl-1H-1,4-benzodiazepin-2(3H)-one
Erispan
Fludiazepam
ID 540
Ro 5-3438

3902-71-4
$C_{14}H_{12}O_3$
228.26
O=C(Oc(c(C=1C)cc(c2OC=3C)C3)c2C)C1
7H-Furo(3,2-g)(1)benzopyran-7-one, 2,5,9-trimethyl
6-Hydroxy-β,2,7-trimethyl-5-benzofuranacrylic acid γ-lactone
NSC-71047
2,5,9-Trimethyl-7H-furo-(3,2-g)(1)benzopyran-7-one
2',4,8-Trimethylpsoralen
4,5',8-Trimethylpsoralen
Trioxalen
Trioxsalen
Trioxysalen
Trisoralen

3905-19-9
$C_{40}H_{24}Cl_4N_6O_4$
794.44
2-Naphthalenecarboxamide, N,N'-1,4-phenylenebis(4-((2,5-dichlorophenyl)azo)-

3-hydroxy- (9CI)
N,N'-p-Phenylenebis(4-((2,5-dichlorophenyl)azo)-3-hydroxy-2-naphthalenecarboxamide)

3905-64-4
$C_{18}H_{24}$
240.39
Naphthalene, 2,6-bis(1,1-dimethylethyl)- (9CI)
Naphthalene, 2,6-di-tert-butyl- (8CI)
NSC-91463

3910-35-8
$C_{18}H_{20}$
236.36
c(c(ccc1)C(C2)(C)C)(c1)C2(c(cccc3)c3)C
Indan, 1,1,3-trimethyl-3-phenyl- (8CI)
2,3-Dihydro-1,1,3-trimethyl-3-phenyl-1H-indene
1H-Indene, 2,3-dihydro-1,1,3-trimethyl-3-phenyl- (9CI)
1,3,3-Trimethyl-1-phenylindane

3913-02-8
$C_{12}H_{26}O$
186.38
OCC(CCCCCC)CCCC
1-Octanol, 2-butyl
2-Butyl-1-octanol
2-Butyloctyl alcohol

3913-71-1
$C_{10}H_{18}O$
154.28
O=CC=CCCCCCCC
2-Decenal
2-Decen-1-al
Decenaldehyde

3913-81-3
$C_{10}H_{18}O$
154.25
2-Decenal, (E)- (8CI,9CI)

3917-15-5
C_5H_8O
84.12
Vinyl allyl ether
AI3-25058
Allyl vinyl ether
Ether, allyl vinyl (8CI)
1-Propene, 3-(ethenyloxy)- (9CI)

3921-30-0
$C_{10}H_{23}O_4P$
238.26
O=P(OCCCCCCCCCC)(O)O
n-Decyl phosphoric acid
AI3-17201
Monodecyl phosphate
Phosphoric acid, monodecyl ester

3926-62-3
$C_2H_2ClO_2.Na$
116.48
[Na+].OC(=O)CCl
Acetic acid, chloro-, sodium salt
Chloroacetic acid sodium salt
Chloroctan sodny (Czech)
Dow Defoliant
Monoxone
SMA
SMCA
Sodium chloroacetate [UN 2659]
Sodium monochloracetate
UN 2659 [Sodium chloroacetate]

3929-89-3
$C_8H_8O_4$
168.15
Benzoic acid, 2,3-dihydroxy-4-methyl- (9CI)
o-Pyrocatechuic acid, 4-methyl- (8CI)

3930-20-9
$C_{12}H_{20}N_2O_3S$
272.40
CC(C)NCC(O)c1ccc(NS(C)(=O)=O)cc1
Methanesulfonanilide, 4'-(1-hydroxy-2-(isopropylamino)ethyl)
β-Cardone
Methanesulfonamide, N-(4-(1-hydroxy-2-((1-methylethyl)amino)ethyl)phenyl)- (9CI)
Sotalol

3934-84-7
$C_8H_8O_5$
184.15
3-O-Methylgallic acid
Benzoic acid, 3,4-dihydroxy-5-methoxy-
3-O-Methylgallate

3937-56-2
$C_9H_{20}O_2$
160.26
OCCCCCCCCCO
1,9-Nonanediol (9CI)
AI3-06325

3938-16-7
$C_6H_4Cl_2O_2$
179.00
1,2-Benzenediol, 3,6-dichloro- (9CI)
Pyrocatechol, 3,6-dichloro- (8CI)

3938-95-2
$C_7H_{14}O_2$
130.19
O=C(OCC)C(C)(C)C
Ethyl pivalate
Ethyl 2,2-dimethylpropanoate
Pivalic acid, ethyl ester (8CI)
Propanoic acid, 2,2-dimethyl-, ethyl ester (9CI)

3942-54-9
$C_8H_8ClNO_2$
185.62
O=C(Oc(c(ccc1)Cl)c1)NC
Carbamic acid, methyl-, o-chlorophenyl ester
o-Chlorophenyl methylcarbamate
2-Chlorophenyl-N-methylcarbamate
CPMC
Etrofol
Hopcide

3944-87-4
$C_6H_{12}Cl_2O_4S_2$
283.20
Ethane, 1,2-bis((2-chloroethyl)sulfonyl)
1,2-Bis((2-chloroethyl)sulfonyl)ethane

3949-34-6
$C_{33}H_{21}N_3O_3$
507.55
s-Triazine, 2,4,6-tris-(1-naphthyloxy)- (7CI, 8CI)

3953-10-4
$C_9H_{16}O_2$
156.25
O=C(OCC(CC)CC)C=C
Acrylic acid, 2-ethylbutyl ester
2-Ethylbutylacrylate
2-Ethylbutylester kyseliny akrylove (Czech)
2-Propenoic acid, 2-ethylbutyl ester (9CI)

3956-73-8
$C_{46}H_{60}N_6$
696.99
1,4-Benzenediamine, N,N,N',N'-tetrakis(4-(diethylamino)phenyl)- (9CI)

3959-13-5
$C_7H_{10}O_2Si$
154.24
Silanediol, methylphenyl- (8CI,9CI)

3964-56-5
C_6H_4BrClO
207.45
Phenol, 4-bromo-2-chloro- (8CI,9CI)

3964-58-7
$C_7H_5ClO_3$
172.57
OC(=O)c1ccc(O)c(Cl)c1
3-Chloro-4-hydroxybenzoate
Benzoic acid, 3-chloro-4-hydroxy- (9CI)

3965-55-7
$C_{10}H_9O_7S.Na$
296.24
COC(=O)c1cc(cc(c1)S(=O)(=O)[O-][Na+])C(=O)OC
Benzenesulfonic acid, 3,5-bis-(methoxycarbonyl)-, sodium salt
3,5-Bis(methoxycarbonyl)benzenesulfonic acid sodium salt
3,5-Bis-(methoxykarbonyl)benzensulfonan sodny (Czech)
3,5-Bis-methylkarboxy-benzensulfonan sodny (Czech)

3966-11-8
$C_{15}H_{20}Cl_2O_4$
335.23
Acetic acid, (2,4-dichlorophenoxy)-, 2-butoxy-1-methylethyl ester (8CI,9CI)
Esteron 99

3970-35-2
$C_7H_4ClNO_4$
201.56
Benzoic acid, 2-chloro-3-nitro- (8CI,9CI)
NSC-92742

3970-62-5
$C_7H_{16}O$
116.20
3-Pentanol, 2,2-dimethyl- (9CI)

3971-33-3
$C_{10}H_{18}O_4$
202.25
Octanedioic acid, 2-ethyl- (8CI,9CI)

3973-18-0
$C_5H_8O_2$
100.12
O(CCO)CC#C
Ethanol, 2-(2-propynyloxy)- (9CI)
2-(2-Propynyloxy)ethanol

3976-35-0
$C_{15}H_{16}$
196.29
1,1'-Biphenyl, 2,4,6-trimethyl- (9CI)
Biphenyl, 2,4,6-trimethyl- (6CI,7CI,8CI)
Mesitylbenzene
2-Phenylmesitylene
2,4,6-Trimethylbiphenyl

3978-67-4
$C_6H_4Cl_2O_2$
179.00
Oc1ccc(Cl)c(Cl)c1O
3,4-Dichlorocatechol
DCBZ
3,4-Dichloro-1,2-benzenediol

3978-81-2
$C_9H_{13}N$
135.20
CC(C)(C)c1ccncc1
Pyridine, 4-(1,1-dimethylethyl)- (9CI)
AI3-36657
Pyridine, 4-tert-butyl- (8CI)

3982-82-9
$C_{28}H_{32}O_2Si_3$
484.82
Trisiloxane, 1,3,3,5-tetramethyl-1,1,5,5-tetraphenyl- (9CI)
1,3,3,5-Tetramethyl-1,1,5,5-tetraphenyltrisiloxane
1,1,5,5-Tetraphenyl-1,3,3,5-tetramethyltrisiloxane

3982-91-0
Cl_3PS
169.38
ClP(Cl)(Cl)=S
Thiophosphoryl chloride [UN 1837]
Fosforthiochlorid (Czech)
Phosphorothioic trichloride
Phosphorothionic trichloride
Phosphorous sulfochloride
Phosphorous thiochloride
Phosphorous trichloride sulfide
Sulfidotrichlorid fosforecny (Czech)
Thiochlorid fosforecny (Czech)
Thiophosphoric trichloride
Thiophosphoryl trichloride
Trichlorophosphine sulfide
TL 262
UN 1837 [Thiophosphoryl chloride]

3985-81-7
$C_{18}H_{38}O.1/3Al$
279.49
1-Octadecanol, aluminum salt (8CI)
Aluminum n-octadecoxide

3988-77-0
$C_{13}H_{10}OS$
214.29
2-Propen-1-one, 3-phenyl-1-(2-thienyl)- (8CI,9CI)
NSC-96359

3990-03-2
$C_6H_8O_4$
144.13
O=C(OCC)C=CC(=O)O
Maleic acid, monoethyl ester (8CI)
2-Butenedioic acid (Z)-, monoethyl ester (9CI)

3991-73-9
$C_8H_{19}O_4P$
210.21
O=P(OCCCCCCCC)(O)O
Monooctyl phosphate
AI3-15053
Octyl dihydrogen phosphate
Phosphoric acid, monooctyl ester

3995-42-4
$C_{10}H_8Cl_2N_4$
255.11
1,3,5-Triazin-2-amine, 4,6-dichloro-N-methyl-N-phenyl- (9CI)
s-Triazine, 2,4-dichloro-6-(N-methylanilino)- (6CI, 8CI)

3995-43-5
$C_{17}H_{16}ClN_5$
325.80
1,3,5-Triazine-2,4-diamine, 6-chloro-N,N'-dimethyl-N,N'-diphenyl- (9CI)
s-Triazine, 2-chloro-4,6-bis-(N-methylanilino)- (6CI, 7CI,8CI)

3996-59-6
$C_{11}H_{14}N_2O_5$
254.23
Medinoterb
AI3-19045
2-tert-Butyl-4,6-dinitro-m-cresol
6-tert-Butyl-3-methyl-2,4-dinitrophenol
Caswell No. 128BB
EPA Pesticide Chemical Code 228300
Phenol, 6-(1,1-dimethylethyl)-3-methyl-2,4-dinitro-

3999-70-0
$C_4H_{10}O.K$
113.22
1-Butanol, potassium salt (9CI)

3999-78-8
$C_8H_{11}N$
121.18
CCc1cncc(C)c1
Pyridine, 3-ethyl-5-methyl- (9CI)
3-Picoline, 5-ethyl- (8CI)

4000-78-6
$C_6H_9N_3O_2$
155.16
1,3,5-Triazine, 2,4-dimethoxy-6-methyl- (9CI)
s-Triazine, 2,4-dimethoxy-6-methyl- (6CI,7CI,8CI)

4003-94-5
$C_{14}H_{11}NO_2$
225.26
Stilbene, 4-nitro
4-Nitrostilben (German)
4-Nitrostilbene

4005-68-9
$C_{18}H_{15}N_3O_3S$
353.38
Metanil Yellow
Benzenesulfonic acid, 3-((4-(phenylamino)phenyl)azo)- (9CI)

4013-34-7
$C_9H_{12}O$
136.19
Benzene, (1-methoxyethyl)- (9CI)
(1-Methoxyethyl)benzene

4016-11-9
$C_5H_{10}O_2$
102.15
O(C1COCC)C1
Propane, 1,2-epoxy-3-ethoxy
1,2-Epoxy-3-ethoxypropane
1,2-Epoxy-3-ethyloxy propane [UN 2752]
(Ethoxymethyl)oxirane
Ethyl glycidyl ether
Oxirane, (ethoxymethyl)- (9CI)
UN 2752 [1,2-Epoxy-3-ethoxypropane]

4016-14-2
$C_6H_{12}O_2$
116.18
O(C1COC(C)C)C1
Propane, 1,2-epoxy-3-isopropoxy
Glycidyl isopropyl ether
IGE (OSHA)
(Isopropoxymethyl)oxirane
Isopropyl glycidyl ether (ACGIH,OSHA)
3-Isopropyloxypropylene oxide
NCI-C56439
Oxirane, ((1-methylethoxy)-methyl)- (9CI)

4016-21-1
$C_{13}H_{26}O_5S.Na$
317.40
Dodecanoic acid, 2-sulfo-, 1-methyl ester, sodium salt (9CI)
Methyl laurate α-sulfonic acid, sodium salt

4016-22-2
$C_{15}H_{30}O_5S.Na$
345.46
Tetradecanoic acid, 2-sulfo-, 1-methyl ester, sodium salt (9CI)
Methyl α-sulfomyristate, sodium salt

4016-24-4
$C_{17}H_{34}O_5S.Na$
373.51
Hexadecanoic acid, 2-sulfo-, 1-methyl ester, sodium salt (9CI)

4018-65-9
$C_6H_5ClO_2$
144.56
3-Chlorocatechol

4022-55-3
$C_{11}H_{21}N_5S$
255.39
1,3,5-Triazine-2(1H)-thione, 4,6-bis(diethylamino)- (9CI)
2,4-Bis(diethylamino)-6-mercapto-s-triazine
4,6-Bis(diethylamino)-s-triazine-2(1H)-thione
s-Triazine-2-thiol, 4,6-bis-(diethylamino)- (6CI,7CI, 8CI)

4022-58-6
$C_{11}H_{15}N_5$
217.24
CC1(C)N=C(N)N=C(N)N1c2cc ccc2
1,3,5-Triazine-2,4-diamine, 1,6-dihydro-6,6-dimethyl-1-phenyl- (9CI)
s-Triazine, 4,6-diamino-1,2-dihydro-2,2-dimethyl-1-phenyl- (8CI)

4024-81-1
C_6Cl_8O
371.69
2-Cyclohexen-1-one, 2,3, 4,4,5,5,6,6-octachloro- (8CI,9CI)
2-Cyclohexen-1-one, octa-chloro- (6CI,7CI)
OCH
Octone

4026-20-4
$C_6H_{12}O_3$
132.16
O=C(O)C(O)C(C)(C)C
Butanoic acid, 2-hydroxy-3,3-dimethyl- (9CI)
3,3-Dimethyl-2-hydroxybutyric acid
2-Hydroxy-3,3-dimethylbutanoic acid

4030-02-8
$C_4H_7N_5O$
141.13
s-Triazine, 2,4-diamino-6-methyl-, 3-oxide (7CI, 8CI)

4032-26-2
$C_{12}H_{12}N_2.2Cl$
255.16
[Cl-].[Cl-].C2C[n+]1ccccc1c3c ccc[n+]23
Dipyrido(1,2-a;2',1'-c)pyrazinediium, 6,7-dihydro-, dichloride
Diquat dichloride
1,1'-Ethylene-2,2'-dipyridinium dichloride

4032-86-4
Unknown
Unknown
3,3-Dimethylheptane

4032-93-3
$C_{10}H_{22}$
142.28
Heptane, 2,3,6-trimethyl- (8CI,9CI)

4035-89-6
$C_{23}H_{38}N_6O_5$
478.67
O=C(N(C(=O)NCCCCCCN=C=O)CCCCCCN=C=O)NCCC CCCN=C=O
Isocyanic acid, triester with 1,3,5-tris(6-hydroxyhexyl)-biuret
DES-N
Imidodicarbonic diamide, N,N',2-tris(6-isocyanato-hexyl)-
Tris(isocyanatohexyl)biuret

4038-04-4
C_7H_{14}
98.19
1-Pentene, 3-ethyl- (9CI)

4039-99-0
$C_4H_6N_4$
110.09
1,3,5-Triazin-2-amine, N-methyl- (9CI)
s-Triazine, 2-(methylamino)- (8CI)

4040-00-0
$C_5H_8N_4$
124.15
1,3,5-Triazin-2-amine, N,N-dimethyl- (9CI)
N,N-Dimethyl-1,3,5-triazin-2-amine
s-Triazine, 2-(dimethylamino)- (6CI,7CI,8CI)

4040-01-1
$C_{10}H_{10}N_4$
186.22
s-Triazine, 2-(benzylamino)- (6CI,7CI,8CI)

4040-07-7
$C_9H_8N_4$
172.19
1,3,5-Triazin-2-amine, N-phenyl- (9CI)
s-Triazine, 2-anilino- (6CI, 7CI,8CI)
N-(s-Triazin-2-yl)aniline

4041-11-6
$C_7H_{10}O$
110.16
2-Cyclopenten-1-one, 2,5-dimethyl- (8CI,9CI)

4044-65-9
$C_8H_4N_2S_2$
192.25
N(c(ccc(N=C=S)c1)c1)=C=S
Bitoscanate
AI3-28258
Benzene, 1,4-diisothiocyanato-
Biscomate
Bitoscanato (Spanish)
Bitoscanatum (Latin)
1,4-Diisothiocyanatobenzene
Isothiocyanic acid p-phenylene ester
Isothiocyanic acid, 1,4-phenylenedi-
Jonit
Phenylene thiocyanate
Phenylene-1,4-diisothiocyanate
16842

4049-81-4
C_7H_{12}
96.17
1,5-Hexadiene, 2-methyl- (8CI,9CI)
NSC-66540

4050-45-7
C_6H_{12}
84.16
2-Hexene, (E)- (9CI)

4051-63-2
$C_{28}H_{16}N_2O_4$
444.43
O=C(c(c(C(=O)c1c(N)ccc2c(c(c(C(=O)c(c3ccc4)c4)c(N)c5)C3=O)c5)ccc6)c6)c12
(1,1'-Bianthracene)-9,9',10,10'-tetrone, 4,4'-diamino- (9CI)

4054-38-0
C_7H_{10}
94.16
1,3-Cycloheptadiene (8CI,9CI)

4055-39-4
$C_{16}H_{19}N_3O_6$
349.38
COC4=C(C)C(=O)C2=C(C(COC(N)=O)C3(OC)C1NC1CN23)C4=O
Azirino(2',3':3,4)pyrrolo-(1,2-a)indole-4,7-dione, 1,1a,2,8,8a,8b-hexahydro-8-(hydroxy methyl)-6,8a-dimethoxy-5-methyl-, carbamate (ester)
Mitomycin A

4055-40-7
$C_{16}H_{19}N_3O_6$
349.38
COC4=C(C)C(=O)C1=C(C(COC(N)=O)C2(O)C3C(CN12)N3C)C4=O
Azirino(2',3':3,4)pyrrolo-(1,2-a)indole-4,7-dione, 1,1a,2,8,8a,8b-hexahydro-8a-hydroxy-8- (hydroxymethyl)-6-methoxy-1,5-dimethyl-, 8-carbamate
Mitomycin B

4058-30-4
$C_{18}H_{15}ClN_6O_2$
382.77
O=N(=O)c(ccc(N=Nc(ccc(N(CCC#N)CCC#N)c1)c1)c2Cl)c2
Propanenitrile, 3,3'-((4-((2-chloro-4-nitrophenyl)-azo)phenyl)imino)bis- (9CI)
4-((2-Chloro-4-nitrophenyl)azo)-N,N-biscyanoethylaniline
3,3'-((p-((2-Chloro-4-nitrophenyl)azo)phenylimino)bispropionitrile
3,3'-((4-((2-Chloro-4-nitrophenyl)azo)phenyl)imino)bispropanenitrile

4062-60-6
$C_{10}H_{24}N_2$
172.30
N(C(C)(C)C)CCNC(C)(C)C
1,2-Ethanediamine, N,N'-bis-(1,1-dimethylethyl)- (9CI)
N,N'-Bis(1,1-dimethylethyl)-1,2-ethanediamine
N,N'-Di-tert-butylethylenediamine

4062-78-6
$C_{19}H_{38}O_5S.Na$
401.56
Octadecanoic acid, 2-sulfo-, 1-methyl ester, sodium salt (9CI)

4063-41-6
$C_{13}H_{10}O_3$
214.23
2H-Furo(2,3-h)(1)benzopyran-2-one, 4,8-dimethyl
5-Benzofuranacrylic acid, 4-hydroxy-β,2-dimethyl-, δ-lactone
4',4-Dimethylangelicin
4,5'-Dimethyl angelicin
4,8-Dimethyl-2H-furo(2,3-h)-1-benzopyran-2-one
4,8-Dimethylisopsoralen

4064-06-6
$C_{12}H_{20}O_6$
260.29
α-D-Galactopyranose, 1,2:3,4-bis-O-(1-methylethylidene)- (9CI)
Galactopyranose, 1,2:3,4-di-O-isopropylidene-, α-D- (8CI)
NSC-89756

4065-24-1
$C_{11}H_{16}ClN_5O_2$
285.74
Acetamide, N,N'-(6-chloro-s-triazine-2,4-diyl)bis-(N-ethyl- (8CI)
s-Triazine, 2-chloro-4,6-bis-(N-ethylacetamido)- (7CI)

4065-45-6
$C_{14}H_{12}O_6S$
308.32
O=C(c(cccc1)c1)c(c(O)cc(OC)c2S(=O)(=O)O)c2
Benzenesulfonic acid, 5-benzoyl-4-hydroxy-2-methoxy
Benzophenone 4
2-Hydroxy-4-methoxybenzophenone-5-sulfonic acid
Sulisobenzone

4067-16-7
$C_1OH_{28}N_6$
124.35

N(CCNCCNCCNCCN)CCN
Pentaethylenehexamine
PEHA
3,6,9,12-Tetraazatetradecane-1,14-diamine

4074-88-8
$C_{10}H_{14}O_5$
214.24
O=C(OCCOCCOC(=O)C=C)C=C
Acrylic acid, 2-ethoxyethanol diester
Acrylic acid, oxydiethylene ester (8CI)
Diethylene glycol diacrylate
Oxydiethylene acrylate
Oxydiethylene diacrylate
2-Propenoic acid, oxydi-2,1-ethanediyl ester (9CI)
TGA 2

4075-79-0
$C_{14}H_{13}NO$
211.28
O=C(Nc(ccc(c(cccc1)c1)c2)c2)C
Acetanilide, 4'-phenyl
Acetamide, N-(1,1'-biphenyl)-4-yl- (9CI)
4-Acetamidobiphenyl
4-Acetylaminobiphenyl
N-Acetylxenylamin (Czech)
4-Biphenylacetamide
N-4-Biphenylacetamide
N-(4-Biphenylyl)acetamide
4'-Fenylacetanilid (Czech)
p-Phenylacetanilide
4'-Phenylacetanilide

4075-81-4
$C_6H_{10}O_4.Ca$
186.24
Propionic acid, calcium salt
Bioban-C
Calcium dipropionate
Calcium propanoate
Calcium propionate
Propanoic acid, calcium salt (9CI)

4080-31-3
$C_9H_{16}ClN_4.Cl$
251.19
3,5,7-Triaza-1-azoniaadamantane, 1-(3-chloroallyl)-, chloride
1-(3-Chloroallyl)-3,5,7-triaza-1-azoniaadamantane chloride
DOWCO 184
Dowicide Q
Dowicil 75
Dowicil 100
Quaternium-15

4082-55-7
$C_{30}H_{60}O_2$
452.81
Nonacosanoic acid, methyl ester (8CI,9CI)

4083-64-1
$C_8H_7NO_3S$
197.21
O=C=NS(=O)(=O)c(ccc(c1)C)c1
Benzenesulfonyl isocyanate, 4-methyl- (9CI)
4-Methylbenzenesulfonyl isocyanate

4088-22-6
$C_{37}H_{77}N$
536.02
N(CCCCCCCCCCCCCCCCCC)(CCCCCCCCCCCCCCCCCC)C
1-Octadecanamine, N-methyl-N-octadecyl- (9CI)
N-Methyl-N-octadecyl-1-octadecanamine

4088-60-2
C_4H_8O
72.11
OCC=CC
2-Buten-1-ol, (Z)- (9CI)
(Z)-2-Buten-1-ol

4089-58-1
$C_8F_{16}O_5S$
512.13
O=C(F)C(F)(OC(F)(F)C(F)(OC(F)(F)C(F)(F)S(=O)(=O)F)C(F)(F)F)C(F)(F)F
Propanoyl fluoride, 2,3,3,3-tetrafluoro-2-(1,1,2,3,3,3-hexafluoro-2-(1,1,2,2-tetrafluoro-2-(fluorosulfonyl)ethoxy)propoxy)- (9CI)
Propionyl fluoride, tetrafluoro-2-(hexafluoro-2-(tetrafluoro-2-(fluorosulfonyl)ethoxy)propoxy)-

4091-39-8
C_4H_7ClO
106.55
O=C(C(Cl)C)C
2-Butanone, 3-chloro- (9CI)
3-Chloro-2-butanone

4096-20-2
$C_{11}H_{15}N$
161.24
N(c(cccc1)c1)(CCCC2)C2
Piperidine, 1-phenyl- (9CI)
1-Phenylpiperidine

4097-33-0
$C_{14}H_{20}N_2O_5$
296.31
2,6-Dinitro-4-octylphenol
Phenol, 2,6-dinitro-4-octyl-

4097-36-3
$C_{11}H_{14}N_2O_5$
254.27
Phenol, 2-(1-methylbutyl)-4,6-dinitro
2-sec-Amyl-4,6-dinitrophenol
Chemox General
4,6-Dinitro-o-sec-amylphenol
Dinosam
Dinosame (French)
DNAP
DNOSAP
2-(1-Methylbutyl)-4,6-dinitrofenol (Dutch)
2-(1-Methyl-n-butyl)-4,6-dinitrophenol
2-(1-Metil-butil)-4,6-dinitrofenolo (Italian)

4097-47-6
$C_9H_{10}N_2O_5$
226.21
O=N(=O)c(c(O)c(N(=O)=O)cc1C(C)C)c1
Phenol, 4-isopropyl-2,6-dinitro
2,6-Dinitro-4-isopropylphenol
4-Isopropyl-2,6-dinitrophenol
Phenol, 2,6-dinitro-4-isopropyl-

4097-49-8
$C_{10}H_{12}N_2O_5$
240.20
Phenol, 4-(1,1-dimethylethyl)-2,6-dinitro- (9CI)
NSC-21491
Phenol, 4-tert-butyl-2,6-dinitro- (8CI)

4097-58-9
$C_{12}H_{14}N_2O_5$
266.24
Phenol, 4-cyclohexyl-2,6-dinitro- (8CI,9CI)

4097-89-6
$C_6H_{18}N_4$
146.21
N(CCN)(CCN)CCN
1,2-Ethanediamine, N,N-bis-(2-aminoethyl)- (9CI)
4-(2-Aminoethyl)diethylenetriamine
N,N-Bis(2-aminoethyl)-1,2-ethanediamine

4098-71-9
$C_{12}H_{18}N_2O_2$
222.32
O=C=NCC(CC(N=C=O)CC1(C)C)(C1)C
Isocyanic acid, methylene-(3,5,5-trimethyl-3,1-cyclohexylene) ester
Cyclohexane, 5-isocyanato-1-(isocyanatomethyl)-

1,3,3-trimethyl- (9CI)
IPDI
3-Isocyanatomethyl-3,5,5-trimethylcyclohexylisocyanate
Isophorone diamine diisocyanate
Isophoronediisocyanate (ACGIH,OSHA) [UN 2290]
UN 2290 [Isophoronediisocyanate]

4099-07-4
$C_{10}H_{18}O$
154.25
Cyclopentanol, 1,2-dimethyl-3-(1-methylethenyl)-, (1R-(1α,2β,3α))- (9CI)
Cyclopentanol, 3-isopropenyl-1,2-dimethyl-, (1R,2R,3S)-(-)- (8CI)
Plinol B

4099-65-4
$C_{12}H_{16}N_2O_5$
268.26
Phenol, 2-hexyl-4,6-dinitro- (8CI,9CI)

4100-80-5
$C_5H_6O_3$
114.10
O=C(OC(=O)C1)C1C
Succinic anhydride, methyl- (8CI)
AI3-11208
Dihydro-3-methyl-2,5-furandione
2,5-Furandione, dihydro-3-methyl- (9CI)
Methylsuccinic anhydride

4101-68-2
$C_{10}H_{20}Br_2$
300.08
BrCCCCCCCCCCBr
Decane, 1,10-dibromo- (9CI)
AI3-11007
Decamethylene dibromide
1,10-Dibromodecane

4104-14-7
$C_{14}H_{13}Cl_2N_2O_2PS$
375.20
Gophacide
AI3-27854
Bay 38819
Bayer 38819
O,O-Bis(p-chlorophenyl)acetimidophosphoroamidothioate
O,O-Bis(p-chlorophenyl) acetimidoylamidothiophosphat
O,O-Bis(p-chlorophenyl) acetimidoylphosphoramidothioate
O,O-Bis(p-chlorophenyl)acetimidoylphosphoramidothioate
O,O-Bis(4-chlorophenyl) (1-iminoethyl)phosphoramidothioate
Caswell No. 091A
DRC 714
Phosacetim
Phosacetime
Phosazetim
Phosphoramidothioic acid, acetimidoyl-, O,O-bis(p-chlorophenyl) ester (8CI)
Phosphoramidothioic acid, (1-iminoethyl)-, O,O-bis-(4-chlorophenyl) ester

4104-56-7
$C_{10}H_{18}$
138.25
Cyclohexene, 3-(2-methylpropyl)- (9CI)
Cyclohexene, 3-isobutyl- (6CI,7CI,8CI)
3-Isobutylcyclohexene
3-Isobutyl-1-cyclohexene

4109-96-0
H_2Cl_2Si
101.01
Silane, dichloro
Dichlorosilane [UN 2189]
UN 2189 [Dichlorosilane]

4110-50-3
$C_5H_{12}S$
104.22
Propane, 1-(ethylthio)- (9CI)
NSC-163319
Sulfide, ethyl propyl (8CI)

4116-10-3
$C_5H_8ClNO_2$
149.57
O=C(NC)C(C(=O)C)Cl
Butanamide, 2-chloro-N-methyl-3-oxo- (9CI)
2-Chloro-N-methylacetoacetamide
2-Chloro-N-methyl-3-oxobutanamide
N-Methyl-2-chloroacetoacetamide

4116-93-2
$C_{10}H_{14}$
134.22
2,8-Decadiyne (7CI,8CI,9CI)

4121-16-8
$C_{28}H_{54}O_4$
454.73
O=C(OCCCCCCCCC)CCCCCCCCC(=O)OCCCCCCCCC
Decanedioic acid, dinonyl ester (9CI)
Dinonyl decanedioate

4121-67-9
$C_{26}H_{16}N_2O_5S_2.Na$
523.53
7-Benzothiazolesulfonic acid, 2-(2-(2,3-dihydro-1,3-dioxo-1H-inden-2-yl)-6-quinolinyl)-6-methyl-, sodium salt (9CI)
2-(2-(1,3-Dioxo-2-indanyl)-6-quinolyl)-6-methyl-7-benzothiazolesulfonic acid, sodium salt

4122-04-7
$C_3H_4N_4$
96.07
1,3,5-Triazin-2-amine (9CI)
NSC-54656
s-Triazine, 2-amino- (8CI)

4122-05-8
$C_5H_8N_4$
124.12
1,3,5-Triazin-2-amine, N-ethyl- (9CI)
s-Triazine, 2-(ethylamino)- (8CI)

4124-31-6
$C_4Cl_6O_3$
308.76
O=C(OC(=O)C(Cl)(Cl)Cl)C(Cl)(Cl)Cl
Trichloroacetic anhydride
Acetic acid, trichloro-, anhydride (9CI)
Trichloroacetic acid anhydride

4126-78-7
C_8H_{16}
112.22
Cycloheptane, methyl- (8CI, 9CI)

4128-31-8
$C_8H_{18}O$
130.23
2-Octanol, (+-)- (8CI,9CI)

4128-37-4
$C_7H_{16}N_2O$
144.20
O=C(NC(C)C)NC(C)C
Urea, N,N'-bis(1-methylethyl)- (9CI)
N,N'-Bis(1-methylethyl)urea
Urea, 1,3-diisopropyl- (8CI)

4128-71-6
$C_{14}H_{13}N_3O$
239.30
Acetanilide, 4'-phenylazo
Acetamide, N-(4-(phenylazo)-phenyl)- (9CI)
p-Acetamidoazobenzene
4-Acetylaminoazobenzene
p-Phenylazoacetanilide
4'-Phenylazoacetanilide

4129-84-4
$C_{41}H_{45}N_3O_6S_2$.Na
762.93
Violet BNP
Acid Violet 17
Benzenemethanaminium, N-(4-((4-(diethylamino)phenyl)-(4-(ethyl((3-sulfophenyl)-methyl)amino)phenyl)methyl-ene)-2,5-cyclohexadien-1-yl-idene)-N-ethyl-3-sulfo-, hydroxide, inner salt, sodium salt (9CI)
C.I. Acid Violet 17
C.I. Acid Violet 17, Sodium salt (8CI)
C.I. Food Violet 3
C.I.42581

4130-08-9
$C_8H_{12}O_6Si$
232.27
Silanetriol, ethenyl-, triacetate (9CI)
Ethenylsilanetriol triacetate
Triacetoxyvinylsilane
Vinyltriacetoxysilane

4130-42-1
$C_{16}H_{26}O$
234.38
Oc(c(cc(c1)CC)C(C)(C)C)c1C(C)(C)C
Phenol, 2,6-bis(1,1-dimethyl-ethyl)-4-ethyl- (9CI)
2,6-Bis(1,1-dimethylethyl)-4-ethylphenol
Phenol, 2,6-di-tert-butyl-4-ethyl-(8CI)

4131-74-2
$C_8H_{14}O_4S$
206.26
O=C(OC)CCSCCC(=O)OC
Propanoic acid, 3,3'-thiobis-, dimethyl ester (9CI)
Dimethyl 3,3'-thiobispropano-ate

4138-38-9
$C_{10}H_{14}N_2O_2$
194.24
Benzenamine, N-(1,1-di-methylethyl)-4-nitro-(9CI)
Aniline, N-tert-butyl-p-nitro-(7CI,8CI)
p-(N-tert-Butylamino)-nitrobenzene
N-tert-Butyl-4'-nitroaniline
N-tert-Butyl-4-nitro-aniline

4147-51-7
$C_{11}H_{21}N_5S$
255.43
CCSc1nc(NC(C)C)nc(NC(C)C)n1
s-Triazine, 2,4-bis(isopropyl-amino)-6-ethylthio
Cotofor
Dipropetryn
Dipropetryne
2-Ethylthio-4,6-bis(isopropyl-amino)-s-triazine
6-(Ethylthio)-N,N'-bis(1-methylethyl)-1,3,5-triazine-2,4-diamine
GS-16068
Sancap

4147-62-0
$C_7H_3Cl_6N_3$
341.84
s-Triazine, 2,4-bis(tri-chloromethyl)-6-vinyl-(7CI,8CI)

4150-59-8
$C_7H_{13}N_5$
167.18
CCNc1ncnc(NCC)n1
1,3,5-Triazine-2,4-diamine, N,N'-diethyl- (9CI)
s-Triazine, 2,4-bis(ethylamino)-(8CI)

4150-79-2
$C_{17}H_{16}N_4O_2$
308.34
s-Triazine, 2-amino-4,6-bis(p-tolyloxy)- (7CI, 8CI)

4151-51-3
$C_{21}H_{12}N_3O_6PS$
465.36
O=C=Nc(ccc(OP(Oc(ccc(N=C=O)c1)c1)(Oc(ccc(N=C=O)c2)c2)=S)c3)c3
Phenol, 4-isocyanato-, phos-phorothioate (3:1) (ester) (VAN)(9CI)
Tris(4-isocyanatophenyl)thio-phosphate

4161-60-8
$C_5H_{10}O_2$
102.13
2-Pentanone, 4-hydroxy-(8CI,9CI)
NSC-263780

4162-45-2
$C_{19}H_{20}Br_4O_4$
631.98
O(c(c(cc(c1)C(c(cc(c(OCCO)c2Br)Br)c2)(C)C)Br)c1Br)CCO
Ethanol, 2,2'-((1-methylethyl-idene)bis((2,6-dibromo-4,1-phenylene)oxy))bis-(9CI)
2,2-Bis(3,5-dibromo-4-(2-hydroxyethoxy)phenyl)pro-pane
Ethoxylated tetrabromobis-phenol A

4163-80-8
$C_7H_5Cl_3O_2S$
259.54
Benzene, 1,2,4-trichloro-5-(methylsulfonyl)- (9CI)
Sulfone, methyl 2,4,5-tri-chlorophenyl (8CI)

4164-28-7
$C_2H_6N_2O_2$
90.10
CN(C)N(=O)=O
Dimethylamine, N-nitro
Dimethylnitramin (German)
Dimethylnitramine
DMNM
DMNO
Dimethylnitroamine
N-Nitrodimethylamine
N-Nitro-DMA

4164-32-3
$C_4H_8N_2O_3$
132.14
Morpholine, 4-nitro
N-Nitromorpholine
4-Nitromorpholine

4165-78-0
$C_{11}H_{14}$
149.26
Benzene, (1-ethyl-1-pro-penyl)-, (Z)- (9CI)
2-Pentene, 3-phenyl-, (Z)-(8CI)
cis-3-Phenyl-2-pentene
(Z)-3-Phenyl-2-pentene

4165-86-0
$C_{11}H_{14}$
149.26
Benzene, (1-ethyl-1-pro-penyl)-, (E)- (9CI)
2-Pentene, 3-phenyl-, (E)-(8CI)
trans-3-Phenyl-2-pentene
(E)-3-Phenyl-2-pentene

4169-04-4
$C_9H_{12}O_2$
152.21
O(c(cccc1)c1)C(CO)C
1-Propanol, 2-phenoxy
Dowanol PPH Glycol Ether
2-Phenoxypropanol
2-Phenoxypropyl alcohol
Propylene glycol phenyl ether

4170-07-4

$C_{24}H_{16}N_2O_4$
176.30
Benzene, 2-nitro-5-(p-nitrophenyl)-1,3-diphenyl- (7CI,8CI)

4170-24-5
$C_4H_7ClO_2$
122.55
Butanoic acid, 2-chloro- (9CI)
Butyric acid, 2-chloro- (8CI)

4170-30-3
C_4H_6O
70.10
O=CC=CC
Crotonaldehyde (DOT,OSHA)
2-Butenal (9CI)
Crotonaldehyde, Stabilized [UN 1143]
Crotonic aldehyde
Krotonaldehyd (Czech)
β-Methyl acrolein (DOT)
β-Methylacrolein
RCRA waste number U053
UN 1143 [Crotonaldehyde, stabilized]

4170-90-5
$C_{10}H_{14}O$
150.24
Cc1cc(C)c(CO)c(C)c1
Benzyl alcohol, 2,4,6-trimethyl
Mesitylcarbinol
2,4,6-Trimethyl-benzenemethanol (9CI)
2,4,6-Trimethylbenzyl alcohol

4175-37-5
$C_{20}H_{27}N$
281.44
N(c(ccc(c1)CCCCCCCC)c1)c(cccc2)c2
Benzenamine, 4-octyl-N-phenyl- (9CI)
4-Octyldiphenylamine
4-Octyl-N-phenylbenzenamine

4175-38-6
$C_{18}H_{28}N_2$
272.48
N(c(ccc(NC(CCCC1)C1)c2)c2)C(CCCC3)C3
p-Phenylenediamine, N,N'-dicyclohexyl
1,4-Benzenediamine, N,N'-dicyclohexyl-
N,N'-Dicyclohexyl-p-phenylenediamine
UOP 26

4175-53-5
$C_{11}H_{14}$
146.23
Indan, 1,3-dimethyl- (8CI)
1H-Indene, 2,3-dihydro-1,3-dimethyl- (9CI)

4178-93-2
$C_{12}H_{17}N_3O_3$
251.27
O=C(Nc(ccc(N(=O)=O)c1)c1)C(N)CC(C)C
Pentanamide, 2-amino-4-methyl-N-(4-nitrophenyl)-, (S)- (9CI)
L-Leucine-p-nitroanilide

4180-23-8
$C_{10}H_{12}O$
148.22
O(c(ccc(c1)C=CC)c1)C
Anisole, p-propenyl-, trans
trans-Anethol
trans-Anethole
Anisole, p-propenyl-, (E)- (8CI)
Benzene, 1-methoxy-4-(1-propenyl)-, (E)- (9CI)
trans-p-Propenylanisole

4181-95-7
$C_{40}H_{82}$
563.09
C(CCCCCCCCCCCCCCCCCCCCCCCCCCCCCCCCCCCCCCC)C
Tetracontane
AI3-36490

4190-05-0
$C_{12}H_{10}O_3$
202.21
Oc(c(c(cccc1O)c1)ccc2O)c2
(1,1'-Biphenyl)-2,3',4-triol (9CI)
2,3',4-Biphenyltriol
2,4,3'-Trihydroxybiphenyl
2,3',4-Trihydroxydiphenyl

4193-55-9
$C_{40}H_{40}N_{12}O_{10}S_2.2Na$
959.02
2,2'-Stilbenedisulfonic acid, 4,4'-bis((4-anilino-6-bis-(2-hydroxyethyl)amino-s-triazin-2-yl) amino)-, disodium salt

4194-69-8
$C_{18}H_{15}O_{21}.In$
682.15
Indium-citrate

4196-86-5
$C_{33}H_{28}O_8$
552.61
O=C(OCC(COC(=O)c(cccc1)c1)(COC(=O)c(cccc2)c2)COC(=O)c(cccc3)c3)c(cccc4)c4
Pentaerythritol, tetrabenzoate
Benzoflex S-552
Benzoic acid, tetraester with pentaerythritol
1,3-Propanediol, 2,2-bis((benzoyloxy)methyl)-, dibenzoate

4196-87-6
$C_{26}H_{24}O_6$
432.47
O=C(OCC(C)(COC(=O)c(cccc1)c1)COC(=O)c(cccc2)c2)c(cccc3)c3
1,3-Propanediol, 2-((benzoyloxy)methyl)-2-methyl-, dibenzoate (9CI)
1,3-Propanediol, 2-(hydroxymethyl)-2-methyl-, tribenzoate
Trimethylolethane tribenzoate

4196-89-8
$C_{19}H_{20}O_4$
312.37
O=C(OCC(C)(C)COC(=O)c(cccc1)c1)c(cccc2)c2
1,3-Propanediol, 2,2-dimethyl-, dibenzoate (9CI)
2,2-Dimethyl-1,3-propanediol dibenzoate
Neopentyl glycol dibenzoate

4196-99-0
$C_{22}H_{16}N_4O_7S_2.2Na$
558.48
[Na+].[Na+].Oc2ccc1ccccc1c2N=Nc4ccc(N=Nc3ccc(cc3)S([O-])(=O)=O)cc4S([O-])(=O)=O
Benzenesulfonic acid, 2-((2-hydroxy-1-naphthalenyl)-azo)-5-((4-sulfophenyl)azo)-, disodium salt (9CI)
Biebrich Scarlet
C.I. Acid Red 66, Disodium salt (8CI)

4197-07-3
$C_{16}H_{12}N_2O_8S_2.2Na$
470.40
2,7-Naphthalenedisulfonic acid, 4,5-dihydroxy-3-(phenylazo)-, disodium salt
Acilan Chromotrope RR
Brasilan Fast Fuchsine G
Chromotrope 2R (6CI)
Chromotrope Red 2R
C.I. 16570
C.I. Acid Red 29 (7CI)
C.I. Acid Red 29, Disodium salt (8CI)
Fast Fuchsine G
Fast Fuchsin G
Hispacid Fuchsin G

4197-25-5
$C_{29}H_{24}N_6$
456.59
N(=Nc(cccc1)c1)c(c(c(c(N=Nc(c(c(c(NC(N2)(C)C)cc3)c2c4)c3)c4)c5)ccc6)c6)c5
1H-Perimidine, 2,3-dihydro-

2,2-dimethyl-6-((4-(phenylazo)-1-naphthyl)azo)**
C.I. Solvent Black 3
2,3-Dihydro-2,2-dimethyl-6-((4-(phenylazo)-1-naphthyl)azo)perimidine
Sudan Black B

4198-19-0
$C_{34}H_{26}N_4O_{18}S_4.4Na$
998.80
2,7-Naphthalenedisulfonic acid, 3,3'-((3,3'-dimethoxy-(1,1'-biphenyl)-4,4'-diyl)bis(azo))bis(4,5-dihydroxy-, tetrasodium salt (9CI)
3,3'-((3,3'-Dimethoxy(1,1'-biphenyl)-4,4'-diyl)bis(azo))bis-(4,5-dihydroxy-2,7-naphthalenedisulfonic acid), tetrasodium salt

4200-55-9
$C_{30}H_{63}O_4P$
518.80
O=P(OCCCCCCCCCC)(OCCCCCCCCCC)OCCCCCCCCCC
Phosphoric acid, tris(decyl) ester (9CI)
Tris(decyl) phosphate

4202-74-8
$C_4H_8N_2O_3$
132.11
O=C(OC(=O)CN)CN
Glycine, anhydride (9CI)

4203-77-4
$C_{32}H_{22}N_4O_2$
494.58
O=C(c(c(C(N(N=C1C(C(C(=NN(C=2c(c(C(=O)C3=C4)ccc5)c5)CC)C23)=C4)=CC=6)CC)=C17)ccc8)c8)C67
(3,3'-Bianthra(1,9-cd)pyrazole)-6,6'(1H,1'H)-dione, 1,1'-diethyl
Ahcovat Rubine R
Carbanthrene Red G 2B
Carbanthrene Red G 2BP
Cerven Kypova 13 (Czech)
C.I. 70320
Cibanone Red 6B
Cibanone Red F 6B
C.I. Pigment Red 195
C.I. Vat Red 13
Fenanthren Rubine R
Indanthrene Rubine R
Indanthren Rubine R
Indanthren Rubine RS
Indo Maroon Lake RV 6666
Nihonthrene Red BB
Nyanthrene Red G 2B
Palanthrene Red G 2B
Ponsol Red 2B
Ponsol Red 2BD
Sandothrene Red N 6B
Tinon Red 6B
Vat Red 13

4205-90-7
$C_9H_9Cl_2N_3$
230.11
Clc1cccc(Cl)c1N=C2NCCN2
2-Imidazoline, 2-(2,6-dichloroanilino)
734571A
Benzenamine, 2,6-dichloro-N-2-imidazolidinylidene-(9CI)
Clonidin
Clonidine
2-(2,6-Dichlorophenylamino)-2-imidazoline

4206-61-5
$C_{10}H_{18}O_5$
218.28
Ether, bis(2-(2,3-epoxypropoxy)ethyl)
Diethylene glycol diglycidyl ether
Oxirane, 2,2'-(oxybis(2,1-ethanediyloxymethylene))bis-(9CI)

4208-80-4
$C_{21}H_{25}N_2O_2.Cl$
372.93
3H-Pseudoindolium, 2-(2-(2,4-dimethoxyanilino)vinyl)-1,3,3-trimethyl- chloride

4209-91-0
$C_8H_{18}O$
130.23
OC(CC(C)C)(CC)C
3-Hexanol, 3,5-dimethyl- (9CI)
AI3-24973
3,5-Dimethyl-3-hexanol

4213-45-0
$C_{23}H_{28}Cl_3N_3O.2ClH$
541.81
COc3ccc2nc1cc(Cl)ccc1c(NC(C)CCCN(CCCl)CCCl)c2c3
Acridine, 9-(4-bis(2-chloroethyl)amino-1-methylbutylamino)-6-chloro-2-methoxy-, dihydrochloride
9-(4-Bis(2-chloroethyl)amino-1-methylbutylamino)-6-chloro-2-methoxyacridine dihydrochloride
ICR 10
2-Methoxy-6-chloro-9-(4-bis-(2-chloroethyl)amino-1-methylbutylamino)acridine dihydrochloride
2-Methoxy-6-chloro-9-(3-(ethyl-2-chloroethyl)aminopropylamino)acridine dihydrochloride
Quinacrine mustard dihydrochloride

4216-02-8
$C_{26}H_{12}N_4O_2$
412.38
O=C(N(c(c(N=1)ccc2)c2)C1c(c3c(c(cc4)C(=O)N(c(c(N=5)ccc6)c6)C57)c7c8)c8)c34
Bisbenzimidazo(2,1-b:1',2'-j)benzo(lmn)(3,8)phenanthroline-6,9-dione (9CI)

4218-48-8
$C_{11}H_{16}$
148.25
Benzene, 1-ethyl-4-(1-methylethyl)- (9CI)
Cumene, p-ethyl- (8CI)

4219-55-0
$C_{10}H_{10}O_3$
178.19
Benzoic acid, 4-(1-oxopropyl)- (9CI)
Benzoic acid, p-propionyl-(7CI,8CI)

4221-03-8
$C_5H_{10}O_2$
102.13
5-Hydroxypentanal
AI3-19935
Pentanal, 5-hydroxy-

4221-80-1
$C_{29}H_{42}O_3$
438.65
O=C(Oc(c(cc(c1)C(C)(C)C)C(C)(C)C)c1)c(cc(c(O)c2C(C)(C)C)C(C)(C)C)c2
3,5-Ditert-butyl-4-hydroxy-(2,4-ditert-butylphenyl)-benzoate
Benzoic acid, 3,5-bis(1,1-dimethylethyl)-4-hydroxy-, 2,4-bis(1,1-dimethylethyl)-phenyl ester (9CI)
3,5-Di-tert-butyl-4-hydroxybenzoic acid, (2,4-di-tert-butylphenyl) ester

4221-98-1
$C_{10}H_{16}$
136.24
C(=CCC(C=1)C(C)C)(C1)C
1,3-Cyclohexadiene, 2-methyl-5-(1-methylethyl)-, (R)- (9CI)

4223-03-4
$C_{11}H_{21}NO$
183.33
O=C(NC(CC(C)(C)C)(C)C)C=C
Acrylamide, N-(5,5-dimethylhexyl)
N-(5,5-Dimethylhexyl)acrylamide

N-tert-Octylacrylamide
2-Propenamide, N-(1,1,3,3-tetramethylbutyl)- (9CI)

4223-11-4
$C_8H_{16}O_2$
144.24
CCCCOCCOC=C
Ethane, 1-butoxy-2-(vinyloxy)
Butane, 1-(2-(ethenyloxy)-ethoxy)-
2-Butoxyethyl vinyl ether
Butylvinylether ethylenglykolu (Czech)
Ethylene glycol butyl vinyl ether
Vinyl 2-(butoxyethyl) ether

4223-84-1
$C_{10}H_{12}O_3$
180.21
Ethanone, 1-(2-hydroxy-5-methoxy-4-methyl-phenyl)- (9CI)
Acetophenone, 2'-hydroxy-5'-methoxy-4'-methyl- (6CI,7CI,8CI)

4224-70-8
$C_6H_{11}BrO_2$
195.06
O=C(O)CCCCCBr
Hexanoic acid, 6-bromo- (9CI)
6-Bromohexanoic acid

4229-35-0
$C_{28}H_{54}O_7S.Na$
557.79
Butanedioic acid, sulfo-, 1,4-didodecyl ester, sodium salt (9CI)

4230-32-4
$C_{10}H_{14}O_2$
166.22
Bicyclo(2.2.1)heptane-2,5-dione, 1,7,7-trimethyl- (9CI)
2,5-Bornanedione (8CI)

4232-27-3
$C_8H_6N_2O_6$
226.13
O=C(Oc(c(N(=O)=O)cc(N(=O)=O)c1)c1)C
Phenol, 2,4-dinitro-, acetate (ester) (9CI)
AI3-00482
2,4-Dinitrophenol acetate (ester)

4234-79-1
$C_{17}H_{12}Cl_{10}O_4$
634.79
1,3,4-Metheno-1H-cyclobuta-(c,d)-pentalene-2-levulinic acid, 1,1a,3,3a,4,5,5a,5b,6-decachlorooctahydro-2-hydroxy-, ethyl ester
GC-9160
Allied GC 9160
Despirol
Kelevan
General Chemical 9160
General Chemicals 9160

4237-25-6
$C_{12}H_{18}O$
178.27
Oc(c(ccc1CC)C(C)(C)C)c1
Phenol, 2-(1,1-dimethylethyl)-5-ethyl- (9CI)
2-tert-Butyl-5-ethylphenol
2-(1,1-Dimethylethyl)-5-ethylphenol

4246-51-9
$C_{10}H_{24}N_2O_3$
220.36
O(CCOCCOCCCN)CCCN
Diethylene glycol, di(3-aminopropyl) ether
Di(3-aminopropyl) ether of diethylene glycol
1-Propanamine, 3,3'-(oxybis-(2,1-ethanediyloxy))bis-

4248-77-5
$C_{11}H_{24}O_6S_2$
316.47
CS(=O)(=O)OCCCCCCCCCOS(C)(=O)=O
1,9-Nonanediol, dimethanesulfonate
Nonamethylene dimethanesulfonate
Nonane-1,9-dimethanesulfonate

4250-38-8
$C_{29}H_{58}O_2$
438.78
Nonacosanoic acid (8CI,9CI)

4251-01-8
$C_{12}H_{20}N_2$
192.29
N(c(ccc(NC(C)C)c1)c1)C(C)C
1,4-Benzenediamine, N,N'-bis-(1-methylethyl)- (9CI)
N,N'-Bis(1-methylethyl)-1,4-benzenediamine

4252-78-2
$C_8H_5Cl_3O$
223.49
2,2',4'-Trichloroacetophenone

4252-85-1
$C_{42}H_{74}O_6$
675.05
O=C(OC(C(OCC1OC(=O)CCCCCCC=CCCCCCCCC)C1O2)C2)CCCCCCCC=CCCCCCCCC
D-Glucitol, 1,4:3,6-dianhydro-, di-9-octadecenoate, (Z,Z)- (9CI)
1,4:3,6-Dianhydro-D-glucitol, di(Z)-9-octadecenoate

4253-22-9
$C_8H_{18}SSn$
265.01
Stannane, dibutylthioxo
Dibutyltin sulfide
Tin dibutyl mercaptide

4253-34-3
$C_7H_{12}O_6Si$
220.28
CC(=O)OSi(C)(OC(C)=O)OC(C)=O
Silane, methyltriacetoxy
Methyltriacetoxysilane
Triacetoxy-methylsilane

4259-15-8
$C_{32}H_{68}O_4P_2S_4Zn$
772.49
Zinc, bis(O,O-bis(2-ethylhexyl) phosphorodithioato-S,S')-, (T-4)- (9CI)

4265-25-2
C_9H_8O
132.16
Cc2cc1ccccc1o2
Benzofuran, 2-methyl-
AI3-11240

4267-15-6
$C_4Cl_2N_4S$
207.04
1,3,5-Triazine, 2,4-dichloro-6-isothiocyanato- (9CI)
Isothiocyanic acid, 4,6-dichloro-s-triazin-2-yl ester (6CI,7CI,8CI)

4268-36-4
$C_{13}H_{26}N_2O_4$
274.41
CCCCNC(=O)OCC(C)(CCC)COC(N)=O
Carbamic acid, butyl-, ester with 2-(hydroxymethyl)-2-methylpentyl carbamate
Benvil
N-n-Butyl-2-methyl-2-propyl-1,3-propanediol dicarbamate
N-Butyl-2-methyl-2-propyl-1,3-propanediol dicarbamate
Carbamic acid, butyl-, 2-(((aminocarbonyl)oxy)methyl)-2-methylpentyl ester
Carbamic acid, butyl-, 2-

(hydroxymethyl)-2-methylpentyl ester, carbamate
Carbamic acid, ester with 2-(hydroxymethyl)-2-methylpentyl butylcarbamate
Carbamic acid, ester with 2-methyl-2-propyl-1,3-propanediol butylcarbamate
Effisax
2-(Hydroxymethyl)-2-(methylpentyl) butylcarbamate carbamate
Idalene
2-Methyl-2-propyl-1,3-propanediol butylcarbamate carbamate
2-Methyl-2-propyltrimethylene butylcarbamate carbamate
Nospan
1,3-Propanediol, 2-methyl-2-propyl-, butylcarbamate carbamate
Solacen
Solacin
Tibamato
Tybamate
Tybatran
W 713

4276-49-7
$C_{20}H_{41}Br$
361.45
BrCCCCCCCCCCCCCCCCCCCC
Eicosane, 1-bromo- (9CI)
1-Bromoeicosane

4279-22-5
$C_4H_8Cl_2$
127.01
C(CC)(Cl)(Cl)C
Butane, 2,2-dichloro- (9CI)
2,2-Dichlorobutane

4282-31-9
$C_6H_4O_4S$
172.16
O=C(O)C(SC(=C1)C(=O)O)=C1
2,5-Thiophenedicarboxylic acid (9CI)

4282-40-0
$C_7H_{15}I$
226.10
C(CCCCCC)I
Heptane, 1-iodo- (9CI)
Heptyl iodide
1-Iodoheptane

4282-42-2
$C_9H_{19}I$
254.15
C(CCCCCCCC)I
Nonane, 1-iodo- (9CI)
1-Iodononane
n-Nonyl iodide

4282-44-4
$C_{11}H_{23}I$
282.21
C(CCCCCCCCCC)I
Undecane, 1-iodo-
AI3-23975
1-Iodoundecane

4292-10-8
$C_{19}H_{38}N_2O_3$
342.51
O=C(NCCCN(CC(=O)O)(C)C)CCCCCCCCCCC
Lauramidopropyl betaine
N-(Carboxymethyl)-N,N-dimethyl-3-((1-oxododecyl)amino)-1-propanaminium hydroxide, inner salt
Jortaine LMAB
3-Lauroylamidopropyl betaine
N-Laurylamidopropyl-N,N-dimethylbetaine
Lauroylaminopropyldimethylaminoacetate
(3-Laurylaminopropyl)dimethylaminoacetic acid, hydroxide, inner salt
1-Propanaminium, N-(carboxymethyl)-N,N-dimethyl-3-((1-oxododecyl)amino)-, hydroxide, inner salt (9CI)

4292-19-7
$C_{12}H_{25}I$
296.23
C(CCCCCCCCCCC)I
Dodecane, 1-iodo- (9CI)
n-Dodecyl iodide
1-Iodododecane

4292-75-5
$C_{12}H_{24}$
168.32
C(CCCC1)(C1)CCCCCC
Cyclohexane, hexyl- (9CI)
1-Cyclohexylhexane
Hexane, 1-cyclohexyl- (8CI)
Hexylcyclohexane

4292-92-6
$C_{11}H_{22}$
154.30
C(CCCC1)(C1)CCCCC
Cyclohexane, pentyl- (9CI)
1-Cyclohexylpentane
Pentylcyclohexane

4297-95-4
$C_6H_7O_2P.Na$
165.08
Phosphinic acid, phenyl-, sodium salt (9CI)
Sodium phenylphosphinate

4300-97-4
$C_5H_8Cl_2O$
155.02
Propanoyl chloride, 3-chloro-2,2-dimethyl- (9CI)
Propionyl chloride, 3-chloro-2,2-dimethyl- (8CI)

4301-50-2
$C_{16}H_{15}FO_2$
258.29
Fluenethyl
AI3-27477
4-Biphenylacetic acid 2-fluoroethyl ester
4-Biphenylacetic acid, 2-fluoroethyl ester
(1,1'-Biphenyl)-4-acetic acid, 2-fluoroethyl ester
Caswell No. 462A
EPA Pesticide Chemical Code 462200
Fluenetil
Fluenthyl
Fluenyl
2-Fluorethylester kyseliny xenyloctove (Czech)
Labrol EC
Lambrol
M 2060
Mytrol
TH 367-1

4305-26-4
$C_8H_{14}O_3$
158.20
2-Hexanone, 6-(acetyloxy)- (9CI)
6-Acetoxy-2-hexanone
2-Hexanone, 6-hydroxy-, acetate (7CI,8CI)
5-Oxohexyl acetate

4306-88-1
$C_{23}H_{40}O$
332.57
Oc(c(cc(c1)CCCCCCCCC)C(C)(C)C)c1C(C)(C)C
Phenol, 2,6-bis(1,1-dimethylethyl)-4-nonyl- (9CI)
2,6-Bis(1,1-dimethylethyl)-4-nonylphenol

4313-03-5
$C_7H_{10}O$
110.16
O=CC=CC=CCC
2,4-Heptadienal, (E,E)- (9CI)
(E,E)-2,4-Heptadienal
(E,E)-2,4-Heptadien-1-al
trans-2-trans-4-Heptadienal
trans,trans-2,4-Heptadienal

4313-13-7
$C_{15}H_{16}N_4O_2$
284.29
Benzenamine, N,N,2-trimethyl-4-((4-nitrophenyl)azo)- (9CI)

NSC-45530
o-Toluidine, N,N-dimethyl-4-((p-nitrophenyl)azo)- (8CI)

4313-14-8
$C_{15}H_{16}N_4O_2$
284.29
Benzenamine, N,N,2-trimethyl-4-((3-nitrophenyl)azo)- (9CI)
o-Toluidine, N,N-dimethyl-4-((m-nitrophenyl)azo)- (8CI)

4316-23-8
$C_9H_8O_4$
180.16
O=C(O)c(c(ccc1C)C(=O)O)c1
Phthalic acid, 4-methyl- (8CI)
1,2-Benzenedicarboxylic acid, 4-methyl- (9CI)
4-Methyl-1,2-benzenedicarboxylic acid

4316-33-0
$C_{11}H_{16}O_4$
212.25
Benzene, 1-methoxy-4-(trimethoxymethyl)- (9CI)
ortho-p-Anisic acid, trimethyl ester (6CI,7CI,8CI)

4316-42-1
$C_7H_{12}N_2$
124.21
Imidazole, 1-butyl
Butylimidazole
n-Butylimidazole
N-n-Butyl imidazole [UN 2690]
N-(n-Butyl)imidazole
1-Butylimidazole
1H-Imidazole, 1-butyl- (9CI)
UN 2690 [N-n-Butyl imidazole]

4316-74-9
$C_3H_9NO_3S.Na$
162.16
Ethanesulfonic acid, 2-(methylamino)-, monosodium salt (9CI)

4318-76-7
$C_5H_7N_3$
109.15
Nc1ccc(N)nc1
Pyridine, 2,5-diamino
2,5-Diaminopyridine

4320-30-3
$C_6H_{14}N_4O_2.C_5H_9NO_4$
321.30
Arginine Glutamate
L-Arginine L-glutamate (1:1)
Arginylglutamate
Glutamic acid, L-, Compd. with L-arginine (1:1) (8CI)
L-Glutamic acid, Compd. with L-arginine (1:1) (9CI)
Modumate

4323-21-1
$C_6H_6O_2$
110.11
2,5-Cyclohexadien-1-one, 4-hydroxy- (8CI,9CI)
Benzoquinol (VAN)
Quinol (VAN)

4325-76-2
$C_{20}H_{14}$
254.33
Phenanthrene, 1-phenyl- (8CI,9CI)

4329-03-7
$C_{18}H_{19}Cl_3O_2$
373.72
Ethane, 2,2-bis(p-ethoxyphenyl)-1,1,1-trichloro
2,2-Bis(p-ethoxyphenyl)-1,1,1-trichloroethane
2,2-Bis(p-phenetyl)-1,1,1-trichloroethane

4331-54-8
$C_8H_{14}O_2$
142.20
Cyclohexanecarboxylic acid, 4-methyl- (8CI,9CI)

4335-09-5
$C_{34}H_{24}N_8O_{10}S_2.2Na$
814.68
C.I. Direct Green 6, Disodium salt (8CI)
2,7-Naphthalenedisulfonic acid, 4-amino-5-hydroxy-6-((4'-((4-hydroxyphenyl)azo)(1,1'-biphenyl)-4-yl)azo)-3-((4-nitrophenyl)azo)-, disodium salt (9CI)

4337-65-9
$C_{14}H_{26}O_4$
258.36
CCCCC(CC)COC(=O)CCCCC(O)=O
Hexanedioic acid, mono(2-ethylhexyl) ester (9CI)
Adipic acid, 2-ethylhexyl ester
Adipic acid, mono(2-ethylhexyl) ester (8CI)

4337-75-1
$C_{15}H_{31}NO_4S.Na$
344.47
Ethanesulfonic acid, 2-(methyl(1-oxododecyl)-amino)-, sodium salt (9CI)
Taurine, N-lauroyl-N-methyl-, sodium salt (8CI)

4340-77-6
$C_{12}H_9ClN_2$
216.66
Azobenzene, 4-chloro- (8CI)
Diazene, (4-chlorophenyl)-phenyl- (9CI)

4341-85-9
$C_4H_4N_2O$
96.08
Malononitrile (8CI)
Butanedinitrile, hydroxy- (9CI)

4342-03-4
$C_6H_{10}N_6O$
182.22
CN(C)N=Nc1[nH]cnc1C(N)=O
Imidazole-4-carboxamide, 5-(3,3-dimethyl-1-triazeno)
Dacarbazine
Deticene
DIC
(Dimethyltriazeno)imidazolecarboxamide
4-(Dimethyltriazeno)imidazole-5-carboxamide
4-(3,3-Dimethyl-1-triazeno)-imidazole-5-carboxamide
4-(5)-(3,3-Dimethyl-1-triazeno)imidazole-5(4)-carboxamide
5-(Dimethyltriazeno)imidazole-4-carboxamide
5-(3,3-Dimethyltriazeno)imidazole-4-carboxamide
5-(3,3-Dimethyl-1-triazeno)-imidazole-4-carboxamide
DTIC
DTIC-Dome
1H-Imidazole-4-carboxamide, 5-(3,3-dimethyl-1-triazenyl)-(9CI)
NCI-C04717
NSC-45388

4342-36-3
$C_{19}H_{32}O_2Sn$
411.20
Stannane, benzoyloxytributyl
Benzoyloxytributylstannane
Tin, tributyl-, benzoate
Tributyltin benzoate
Tri-n-butyl-zinn benzoate (German)

4342-60-3
$C_8H_{12}O_2$
140.18
3-Cyclohexene-1-carboxylic acid, 4-methyl- (8CI,9CI)
NSC-99286

4342-61-4
$C_4H_{12}Cl_2Si_2$
187.22
Disilane, 1,2-dichloro-1,1,2,2-tetramethyl- (9CI)
1,2-Dichloro-1,1,2,2-tetra-

methyldisilane
Tetramethyldichlorodisilane
1,1,2,2-Tetramethyldichlorodisilane

4345-03-3
$C_{33}H_{54}O_5$
530.79
O=C(Oc(c(c(c(OC(CC1)(CCCC(CCCC(CCCC(C)C)C)C)C)c12)C)C)c2C)CCC(=O)O
Tocopheryl succinate
Butanedioic acid, mono(3,4-dihydro-2,5,7,8-tetramethyl-2-(4,8,12-trimethyltridecyl)-2H-1-benzopyran-6-yl) ester
Butanedioic acid, mono(3,4-dihydro-2,5,7,8-tetramethyl-2-(4,8,12-trimethyltridecyl)-2H-1-benzopyran-6-yl) ester, (2R-(2R*(4R*,8R*)))- (9CI)
Succinic acid, mono(2,5,7,8-tetramethyl-2-(4,8,12-trimethyltridecyl)-6-chromanyl) ester, (+)- (8CI)
D-α Tocopheryl succinate
DL-α Tocopheryl succinate
Vitamin E Hemisuccinate
Vitamin E Succinate
Vitamine E Succinate

4346-18-3
$C_{10}H_{12}O_2$
164.20
Butanoic acid, phenyl ester (9CI)
Butyric acid, phenyl ester (8CI)

4346-51-4
$C_6H_6N_2O_5S.H_3N$
235.24
Benzenesulfonic acid, 2-amino-5-nitro-, ammonium salt
2-Amino-5-nitrobenzenesulfonic acid ammonium salt

4350-09-8
$C_{11}H_{12}N_2O_3$
220.25
O=C(O)C(N)CC(c(c(N1)ccc2O)c2)=C1
Tryptophan, 5-hydroxy-, L
L-5-HTP
5-Hydroxy-l-tryptophan
l-5-Hydroxytryptophan
l-Tryptophan, 5-hydroxy-, (9CI)

4351-54-6
$C_7H_{12}O_2$
128.17
Formic acid, cyclohexyl ester (9CI)
AI3-30436

4352-98-1
$C_7H_{14}O_2$
130.19
1,3-Dioxolane, 2-methyl-2-propyl- (7CI,8CI,9CI)
2-Pentanone, cyclic 1,2-ethanediyl acetal

4353-28-0
$C_{12}H_{26}O_5$
250.38
O(CCOCCOCCOCC)CCOCC
3,6,9,12,15-Pentaoxaheptadecane
Diethoxytetraethylene glycol
Diethylether tetraethylenglykolu (Czech)
Ether, bis(2-(2-ethoxyethoxy)ethyl)
Glycol, diethoxytetraethylene
Tetraethylene glycol diethyl ether

4354-58-9
$C_{12}H_{24}O$
184.32
Cyclohexanehexanol (6CI, 8CI,9CI)

4356-60-9
$C_9H_{18}N_2O_5$
234.24
O=C(N(C(OC)C1OC)COC)N1COC
2-Imidazolidinone, 4,5-dimethoxy-1,3-bis(methoxymethyl)- (9CI)
4,5-Dimethoxy-1,3-bis(methoxymethyl)-2-imidazolidinone

4368-28-9
$C_{11}H_{17}N_3O_8$
319.31
5,9:7,10a-Dimethano-10ah-(1,3)dioxocino(6,5-d)pyrimidine-4,7,10,11,12-pentol, octahydro- 12-(hydroxymethyl)-2-imino
Fugu Poison
Maculotoxin
Spheroidine
Tarichatoxin
Tetrodontoxin
Tetrodotoxin
Tetrodoxin
TTX

4368-56-3
$C_{20}H_{19}N_2O_5S.Na$
422.46
2-Anthracenesulfonic acid, 9,10-dihydro-1-amino-4-(cyclohexylamino)-9,10-dioxo-, monosodium salt
Acid Alizarine Pure Blue R
Acid Blue 62
Alizarine Brilliant Sapphire R
Alizarine Brilliant Sky Blue R
Alizarine Direct Pure Blue R
Alizarine Fast Blue RFE
Alizarine Sky Blue R
Alizarine Supra Blue R
Alizarine Supra Sky RA
Brilliant Alizarine Cyanine R
Brilliant Alizarine Light Blue 3FR
C.I. Acid Blue 62 (8CI)
C.I. 62045
9,10-Dihydro-1-amino-4-(cyclohexylamino)-9,10-dioxo-2-anthracenesulfonic acid sodium salt
Erio Anthracene Brilliant Blue RFF
Erionyl Blue E-RFF
Eriosin Fast Blue RFF
Fenazo Light Blue RA
Kayacyl Sky Blue R
Kayaku Alizarine Sky Blue R
Polan Navy Blue E 2R
Sol. Sulfur Blue 10
Suminol Levelling Sky Blue R
Telon Blue RRL
Tertracid Brilliant Light Blue R

4371-31-7
$C_{12}H_{10}O_4$
218.21
Oc(c(c(c(O)cc(O)c1)c1)ccc2O)c2
(1,1'-Biphenyl)-2,2',4,4'-tetrol (9CI)
2,2',4,4'-Biphenyltetrol (8CI)

4372-29-6
$C_{13}H_{24}O_4$
244.33
Propanedioic acid, decyl- (9CI)
Malonic acid, decyl- (8CI)
1,1-Undecanedicarboxylic acid

4376-18-5
$C_9H_8O_4$
180.16
Phthalic acid, monomethyl ester (8CI)
AI3-02332
1,2-Benzenedicarboxylic acid, monomethyl ester (9CI)

4376-20-9
$C_{16}H_{22}O_4$
278.38
CCCCC(CC)COC(=O)c1ccccc1C(O)=O
Phthalic acid, mono-(2-ethylhexyl) ester
MEHP
Monoethylhexyl phthalate
Mono(2-ethylhexyl)phthalate

4377-41-7
$C_{10}H_8ClN$
177.63
Quinoline, 2-(chloromethyl)-

(8CI,9CI)
NSC-158442

4378-61-4
$C_{22}H_8Br_2O_2$
464.11
O=C(c(c(c(c(c1Br)ccc2)c2C(=O)c3cc(c(c45)ccc6)Br)c34)c1)c56
Dibenzo(def,mno)chrysene-6,12-dione, 4,10-dibromo-(9CI)

4381-07-1
$C_2H_7N_5O_2$
133.14
Biurea, 1-amino
1-Aminobiurea
Carbohydrazide-N-carboxamide

4384-81-0
$CH_2NS_2.Na$
115.15
Carbamic acid, dithio-, sodium salt
Sodium dithiocarbamate

4385-05-1
$C_8H_{18}N_2O$
158.23
O(CCN(C1)CCN(C)C)C1
4-Morpholineethanamine, N,N-dimethyl- (9CI)
N,N-Dimethyl-4-morpholine-ethanamine

4390-04-9
$C_{16}H_{34}$
226.45
C(CC(CC(CC(C)(C)C)C)(C)C)(C)(C)C
2,2,4,4,6,8,8-Heptamethylnonane
Nonane, 2,2,4,4,6,8,8-heptamethyl- (9CI)

4394-85-8
$C_5H_9NO_2$
115.15
O=CN1CCOCC1
4-Morpholinecarboxaldehyde
N-Formylmorfolin (Czech)
N-Formylmorpholine
4-Formylmorpholine

4395-53-3
$C_{45}H_{22}N_2O_5$
670.67
Anthra(2,1,9-mna)naphth-(2,3-h)acridine-5,10,15(16H)-trione, 3-((9,10-dihydro-9,10-dioxo-1-anthracenyl)-amino)- (9CI)

4395-65-7
$C_{20}H_{14}N_2O_2$
314.33
O=C(c(c(C(=O)c1c(N)ccc2Nc(ccc3)c3)ccc4)c4)c12
9,10-Anthracenedione, 1-amino-4-(phenylamino)-(9CI)
1-Amino-4-(phenylamino)-9,10-anthracenedione
1-Amino-4-p-toluidinoanthraquinone

4395-79-3
$C_{10}H_{13}Cl$
168.67
Benzene, 2-chloro-1-methyl-4-(1-methylethyl)- (9CI)
Carvacryl chloride
p-Cymene, 2-chloro- (8CI)
NSC-409886

4396-27-4
$C_9H_{19}N$
141.26
Azecine, decahydro- (7CI, 8CI,9CI)
Azacyclodecane
Decahydroazecine
Nonamethylenimine

4399-55-7
$C_{40}H_{27}N_7O_{13}S_4.4Na$
1033.88
1,5-Naphthalenedisulfonic acid, 3-((4-((4-((6-amino-1-hydroxy-3-sulfo-2-naphthalenyl)azo)-6-sulfo-1-naphthalenyl)azo)-1-naphthalenyl)azo)-, tetrasodium salt (9CI)

4401-11-0
$C_4H_6O_2$
86.09
Ethanone, 1-oxiranyl-(9CI)
Acetyloxirane
2-Butanone, 3,4-epoxy-(6CI,7CI,8CI)
3,4-Epoxybutanone
1,2-Epoxy-3-butanone

4402-46-4
$C_9H_{15}ClN_4O$
230.70
s-Triazine, 2-butoxy-4-chloro-6-(ethylamino)-(7CI,8CI)

4403-61-6
C_5H_7N
81.11
2-Butenenitrile, 2-methyl- (9CI)
Crotononitrile, 2-methyl- (8CI)
NSC-44233

4403-90-1
$C_{28}H_{20}N_2O_8S_2.2Na$
622.60
Benzenesulfonic acid, 2,2'-(1,4-anthraquinonylenediimino)bis(5-methyl-, disodium salt
Acid Green 25
2,2'-(1,4-Anthraquinonylenediimino)bis(5-methylbenzene-sulfonic acid) disodium salt
C.I. 61570
C.I. Acid Green 25
Zelen Alizarinova Brilantni G-Extra (Czech)
Zelen Kysela 25 (Czech)

4404-43-7
$C_{40}H_{44}N_{12}O_{10}S_2$
916.92
C.I. Fluorescent Brightening Agent 28
Benzenesulfonic acid, 2,2'-(1,2-ethenediyl)bis(5-((4-(bis-(2-hydroxyethyl)amino)-6-(phenylamino)-1,3,5-triazin-2-yl)amino)- (9CI)
4,4'-Bis((4-anilino-6-(bis(2-hydroxyethyl)amino)-s-triazin-2-yl)amino)-2,2'-stilbene disulfonic acid
4,4'-Bis((4-anilino-6-(bis(2-hydroxyethyl)amino)-1,3,5-triazin-2-yl)amino)stilbene-2,2'-disulfonic acid

4405-16-7
$C_7H_{14}O_2$
130.19
1,3-Dioxolane, 2-methyl-2-(1-methylethyl)- (9CI)
1,3-Dioxolane, 2-isopropyl-2-methyl- (7CI,8CI)
2-Isopropyl-2-methyl-1,3-dioxolane
Isopropyl methyl ketone ethylene ketal

4406-22-8
Unknown
Unknown
Cyprenorphine
Ciprenorfina (Spanish)
N-Cyclopropylmethyl-6,14-endoetheno-7 α- (1-hydroxy-1-methylethyl)-6,7,8,14-tetrahydronororipavine
Cyprenorphinum (Latin)
DEA No. 9054

4407-40-3
$C_5H_6ClN_3S_2$
207.71
s-Triazine, 2,4-bis(methylthio)-6-chloro
2,4-Bis(methylthio)-6-chloro-

s-triazine
2-Chloro-4,6-bis(methylthio)-
s-triazine

4407-41-4
$C_5H_6N_6S_2$
214.27
s-Triazine, 2-azido-4,6-bis-(methylthio)- (7CI,8CI)

4407-44-7
$C_5H_6ClN_3OS$
191.64
s-Triazine, 2-chloro-4-methoxy-6-(methylthio)- (7CI,8CI)

4407-72-1
$C_7H_{11}ClN_4O$
202.64
1,3,5-Triazin-2-amine, 4-chloro-6-methoxy-N-propyl- (9CI)
s-Triazine, 2-chloro-4-methoxy-6-(propylamino)-(7CI,8CI)

4408-64-4
$C_5H_9NO_4$
147.12
Strombine
Acetic acid, (methylimino)di-(8CI)
Glycine, N-(carboxymethyl)-N-methyl- (9CI)
2-Methyliminodiacetic acid

4408-78-0
$C_2H_5O_5P$
140.04
Acetic acid, phosphono
Carboxymethanephosphonic acid
Fosfonet
Phosphonacetic acid
Phosphonoacetic acid

4411-89-6
$C_{10}H_{10}O$
146.19
O=CC(c(cccc1)c1)=CC
Benzeneacetaldehyde, α-ethylidene- (9CI)
α-Ethylidenebenzeneacetaldehyde

4412-91-3
$C_5H_6O_2$
98.10
OCc1ccoc1
3-Furanmethanol (8CI,9CI)

4413-31-4
$C_{16}H_{15}Cl_3$
313.66
c(ccc(c1)C)(c1)C(c(ccc(c2)C)c2)C(Cl)(Cl)Cl
Ethane, 1,1-bis(p-tolyl)-2,2,2-trichloro
Ethane, 1,1,1-trichloro-2,2-bis-(p-tolyl)-

4414-15-7
$C_{16}H_{30}Cl_3O_5P$
439.74
Dodecanoic acid, 2,2,2-trichloro-1-(dimethoxyphosphinyl)ethyl ester (9CI)
Lauric acid, ester with dimethyl (2,2,2-trichloro-1-hydroxyethyl)phosphonate (8CI)

4417-64-5
$C_8H_{13}ClN_4O$
216.67
1,3,5-Triazin-2-amine, 4-chloro-N-ethyl-6-propoxy- (9CI)
s-Triazine, 2-chloro-4-(ethylamino)-6-propoxy-(7CI,8CI)

4417-65-6
$C_9H_{15}ClN_4O$
230.70
s-Triazine, 2-chloro-4-propoxy-6-(propylamino)-(7CI,8CI)

4417-68-9
$C_9H_{15}ClN_4O$
230.70
s-Triazine, 2-chloro-4-isopropoxy-6-(isopropylamino)- (7CI,8CI)

4417-70-3
$C_8H_{11}ClN_4O$
214.66
s-Triazine, 2-(allyloxy)-4-chloro-6-(ethylamino)-(7CI,8CI)

4417-72-5
$C_9H_{11}ClN_4O$
226.67
s-Triazine, 2-(allylamino)-4-(allyloxy)-6-chloro-(7CI,8CI)

4418-26-2
$C_8H_7O_4.Na$
190.14
[Na+].CC1=CC(=O)C(C(C)=O)C(=O)O1
2H-Pyran-2,4(3H)-dione, 3-acetyl-6-methyl-, sodium salt
Dehydroacetic acid, sodium salt
DHA-S
DHA-Sodium
DHN
Harven
4-Hexenoic acid, 2-acetyl-5-hydroxy-3-oxo, δ-lactone, sodium deriv.
3-(1-Hydroxyethylidene)-6-methyl-2H-pyran-2,4(3H)-dione, sodium salt
Kyselina dehydroacetova sodna sul (Czech)
Sodium dehydroacetate
Sodium dehydroacetic acid

4418-61-5
CH_3N_5
85.09
N(NNN1)C1N
1H-Tetrazole, 5-amino
Aminotetrazole
5-Aminotetrazole
5-Amino-1H-tetrazole
1H-Tetrazol-5-amine

4418-66-0
$C_{14}H_{12}Cl_2O_2S$
315.22
Phenol, 2,2'-thiobis(4-chloro-6-methyl)-
Caswell No. 850
Chlorbisan
o-Cresol, 6,6'-thiobis(4-chloro-
2,2'-Dihydroxy-3,3'-dimethyl-5,5'-dichlorodiphenyl sulfide
EPA Pesticide Chemical Code 064208
Orbisan
Phenol, 2,2'-thiobis(4-chloro-6-methyl- (9CI)
6,6'-Thiobis(4-chloro-o-cresol)
2,2'-Thiobis(4-chloro-6-methylphenol)
2,2'-Thiobis(4-chloro-6-methylphenol)

4419-11-8
$C_{14}H_{24}N_4$
248.42
N#CC(N=NC(C#N)(CC(C)C)C)(CC(C)C)C
Valeronitrile, 2,2'-azobis-(2,4-dimethyl
2,2'-Azodi-(2,4-dimethylvaleronitrile) [UN 2953]
Pentanenitrile, 2,2'-azobis-(2,4-dimethyl-
UN 2953 [2,2'-Azodi-(2,4 dimethylvaleronitrile)]

4419-22-1
Unknown
Unknown
Bis(tributyltin) sulfosalicylate

4419-57-2

$C_9H_{17}NO$
155.23
Pyrrolidine, 1-(1-oxopentyl)- (9CI)
NSC-191020
Pyrrolidine, 1-valeryl- (8CI)

4420-74-0
$C_6H_{16}O_3SSi$
196.37
COSi(CCCS)(OC)OC
Propanethiol, trimethoxysilyl
γ-Mercaptopropyltrimethoxysilane
3-Mercaptopropyltrimethoxysilane
Silane, 3-mercaptopropyltrimethoxy-
Silicone A-189
Union Carbide A-189

4422-95-1
$C_9H_3Cl_3O_3$
265.48
O=C(c(cc(C(=O)Cl)cc1C(=O)Cl)c1)Cl
1,3,5-Benzenetricarbonyl trichloride (9CI)
1,3,5-Benzenetricarbonyl chloride
Trimesoyl chloride

4423-94-3
$C_8H_{14}O$
126.20
Cyclohexanone, 2-ethyl- (8CI, 9CI)
NSC-163961

4424-06-0
$C_{26}H_{12}N_4O_2$
412.38
O=C(N(c(c(N=1)ccc2)c2)C1c(c3c(c(c4)C(=O)N(c(c(N=5)ccc6)c6)C57)c7cc8)c4)c38
Bisbenzimidazo(2,1-b:2',1'-i)benzo(lmn)(3,8)phenanthroline-8,17-dione (9CI)
Bordeaux RRN
Brilliant Orange GR
C.I. Pigment Orange 43
C.I. Vat Orange 7
C.I. 71105
Cibanone Brilliant Orange GR
Fenanthren Brilliant Orange GR
Hostaperm Orange GR
Hostaperm Vat Orange GR
Hostavat Brilliant Orange GR
Indanthren Brilliant Orange GR
Indanthrene Brilliant Orange GR
Indanthrene Brilliant Orange GRP
Indofast Orange OV 5983
Mikethren Brilliant Orange GR
Mikethrene Orange GR
Ostanthren Orange GR
Ostanthrene Orange GR
Palanthrene Brilliant Orange GR
Paradone Brilliant Orange GR
Paradone Brilliant Orange GR New
trans-Perinone
PV Fast Orange GRL
Sanyo Permanent Orange D 213
Sanyo Permanent Orange D 616
Solanthrene Brilliant Orange JR
Symuler Fast Orange GRD
Threne Brilliant Orange GR
Tinon Brilliant Orange GR
Vat Brilliant Orange
Vat Scarlet 2Zh

4424-87-7
$C_{28}H_{14}N_2O_4$
442.42
O=C(c(c(C(=O)c1c(nc(c(c2O)ccc3)c3)c2cc4)ccc5c(O)c(c(n6)ccc7)c7)c56)c14
Benzo(1,2-c:4,5-c')diacridine-6,9,15,18(5H,14H)-tetrone (9CI)

4426-47-5
$C_4H_{11}BO_2$
101.94
n-Butylboronic acid
Butaneboronic acid
n-Butylboronate

4430-20-0
$C_{19}H_{12}Cl_2O_5S$
423.27
O=S(=O)(OC(c1cccc2)(c(ccc(O)c3Cl)c3)c(ccc(O)c4Cl)c4)c12
Chlorophenol Red
3',3''-Dichlorophenolsulfonphthalein
Phenol, 4,4'-(3H-2,1-benzoxathiol-3-ylidene)bis(2-chloro-, S,S-dioxide (9CI)

4431-89-4
$C_{12}H_{22}$
166.31
Cyclohexane, (cyclopentylmethyl)- (9CI)
Cyclopentylcyclohexylmethane
Methane, cyclohexylcyclopentyl- (8CI)

4433-11-8
$C_{14}H_{14}$
182.27
1,1'-Biphenyl, 3,4-dimethyl- (9CI)
Biphenyl, 3,4-dimethyl- (8CI)
3,4-Dimethylbiphenyl

4433-79-8
$C_{12}H_{14}ClNO_4$
271.69
O=C(Nc(c(OC)cc(c1OC)Cl)c1)CC(=O)C
Acetoacetanilide, 4'-chloro-2',5'-dimethoxy- (8CI)
Butanamide, N-(4-chloro-2,5-dimethoxyphenyl)-3-oxo- (9CI)
2,5-Dimethoxy-4-chloroacetoacetanilide

4433-85-6
$C_6H_{10}O_3$
130.14
Butanoic acid, 2-ethyl-3-oxo- (9CI)
Acetoacetic acid, 2-ethyl- (8CI)
Butyric acid, 2-acetyl-
α-Ethylacetoacetic acid
2-Ethyl-3-oxobutanoic acid

4434-66-6
$C_{19}H_{39}Br$
347.42
BrCCCCCCCCCCCCCCCCCCC
Nonadecane, 1-bromo- (9CI)
1-Bromononadecane

4435-50-1
$C_4H_{10}O_3$
106.12
1,2,3-Butanetriol (8CI,9CI)

4435-53-4
$C_7H_{14}O_3$
146.21
O=C(OCCC(OC)C)C
1-Butanol, 3-methoxy-, acetate
Acetic acid, 3-methoxybutyl ester
Butoxyl [UN 2708]
3-Methoxybutyl acetate
3-Methoxybutylester kyseliny octove (Czech)
Methyl-1,3-butylene glycol acetate
UN 2708 [Butoxyl]

4436-75-3
$C_6H_8O_2$
112.13
O=C(C=CC(=O)C)C
3-Hexene-2,5-dione (9CI)

4437-20-1
$C_{10}H_{10}O_2S_2$
226.32
O(C(=CC=1)CSSCC(OC=C2)=C2)C1
Furan, 2,2'-(dithiobis(methylene))bis- (9CI)
2,2'-(Dithiobis(methylene))bisfuran
2,2'-(Dithiodimethylene)difuran

4439-20-7

$C_6H_{16}N_2O_2$
148.19
OCCNCCNCCO
Ethanol, 2,2'-(ethylenediimino)di- (8CI)
2,2'-(1,2-Ethanediyldiimino)bisethanol
Ethanol, 2,2'-(1,2-ethanediyldiimino)bis- (9CI)

4439-24-1
$C_6H_{14}O_2$
118.20
O(CC(C)C)CCO
Ethanol, 2-isobutoxy
Ektasolve EIB
Ethylene glycol monoisobutyl ether
2-Isobutoxyethanol
Isobutyl cellosolve

4440-33-9
$C_{13}H_9N$
179.22
9H-Fluoren-9-imine (9CI)
Fluoren-9-imine (8CI)
NSC-12352

4442-79-9
$C_8H_{16}O$
128.24
OCCC(CCCC1)C1
Ethanol, 2-cyclohexyl
2-Cyclohexylethanol
Cyclohexylethyl alcohol
Hexahydrophenylethyl alcohol

4443-55-4
$C_{26}H_{52}$
364.70
Cyclohexane, eicosyl- (9CI)
Eicosane, 1-cyclohexyl- (8CI)
NSC-163587

4443-61-2
$C_{26}H_{52}$
364.70
Cyclohexane, (1-octyldodecyl)- (9CI)
Eicosane, 9-cyclohexyl- (6CI,8CI)

4444-68-2
$C_8H_{19}N$
157.30
1-Butanamine, N,N-diethyl- (9CI)
Butylamine, N,N-diethyl- (6CI,7CI,8CI)
N,N-Diethylbutylamine
Ethanamine, N-butyl-N-ethyl-

4445-06-1
$C_{24}H_{48}$
336.65
C(CCCC1)(C1)CCCCCCCCCCCCCCCCCC
Cyclohexane, octadecyl- (9CI)
1-Cyclohexyloctadecane
Octadecylcyclohexane

4445-07-2
$C_{24}H_{42}$
330.60
c(cccc1)(c1)CCCCCCCCCCCCCCCCCC
Benzene, octadecyl- (9CI)
Octadecane, 1-phenyl- (8CI)
Octadecylbenzene
1-Phenyloctadecane

4446-75-7
$C_7H_{11}ClN_4O$
202.62
1,3,5-Triazin-2-amine, 4-chloro-6-methoxy-N-(1-methylethyl)- (9CI)
s-Triazine, 2-chloro-4-(isopropylamino)-6-methoxy- (8CI)

4446-76-8
$C_7H_9ClN_4O$
200.63
1,3,5-Triazin-2-amine, 4-chloro-6-methoxy-N-2-propenyl- (9CI)
s-Triazine, 2-(allylamino)-4-chloro-6-methoxy- (7CI, 8CI)

4449-51-8
$C_{27}H_{41}NO_2$
411.69
CC1CNC6C(C1)OC5(CCC4C3CC=C2CC(O)CCC2(C)C3CC4=C5C)C6C
Cyclopamine
Alkaloid V
11-Deoxojervine

4454-05-1
$C_6H_{10}O_2$
114.14
O(C=CCC1)C1OC
2-Methoxy-3,4-dihydro-2H-pyran
3,4-Dihydro-2-methoxy-2H-pyran
2H-Pyran, 3,4-dihydro-2-methoxy- (9CI)

4454-16-4
$C_8H_{16}O_2.1/2Ni$
173.57
Hexanoic acid, 2-ethyl-, nickel(2+) salt (9CI)
Nickel(2+) 2-ethylhexanoate

4455-09-8
$C_{15}H_{16}O_3S$
276.36
Benzenesulfonic acid, 4-methyl-, 2-phenylethyl ester (9CI)
Phenethyl alcohol, p-toluenesulfonate (6CI)
2-Phenylethyl p-methylbenzenesulfonate
Phenethyl tosylate
2-Phenylethyl tosylate
β-Phenylethyl tosylate
Phenethyl p-toluenesulfonate
Phenethyl 4-toluenesulfonate
p-Toluenesulfonic acid, phenethyl ester (7CI,8CI)
2-Tosyloxyethylbenzene

4455-26-9
$C_{17}H_{37}N$
255.48
N(CCCCCCCC)(CCCCCCCC)C
1-Octanamine, N-methyl-N-octyl- (9CI)
Di(octyl)methylamine
N-Methyl-N-octyl-1-octanamine

4457-71-0
$C_6H_{14}O_2$
118.20
OCCC(CCO)C
1,5-Pentanediol, 3-methyl
3-Methyl-1,5-pentanediol

4460-86-0
$C_{10}H_{12}O_4$
196.22
O=Cc(c(OC)cc(OC)c1OC)c1
Benzaldehyde, 2,4,5-trimethoxy
Asaraldehyde
Asaronaldehyde
Azarylaldehyde
Gazarin
NCI-C61632
2,4,5-Trimethoxybenzaldehyde
3,4,6-Trimethoxybenzaldehyde

4461-41-0
C_4H_7Cl
90.55
2-Chlorobutene-2
2-Butene, 2-chloro-
2-Chloro-2-butene

4461-42-1
C_4H_7Cl
90.56
C(=CCl)CC
1-Butene, 1-chloro
1-Chloro-1-butene

4461-48-7
C_6H_{12}
84.16
C(=CC)C(C)C
4-Methyl-2-pentene

4461-52-3
$C_2H_6O_2$
62.07
O(CO)C
Methanol, methoxy- (9CI)
Methoxymethanol

4465-58-1
$C_{16}H_{13}NO_3$
267.28
O=C(c(c(c(NCCO)cc1)C(=O)c2cccc3)c1)c23
9,10-Anthracenedione, 1-((2-hydroxyethyl)amino)- (9CI)
1-((2-Hydroxyethyl)amino)-9,10-anthracenedione
1-(N-(2-Hydroxyethyl)amino)-anthraquinone

4465-94-5
$C_7H_{17}Cl_2N_2O_3P.C_6H_{13}N$
378.33
Phosphorodiamidic acid, N,N-bis(2-chloroethyl)-N'-(3-hydroxypropyl)-, cyclohexylamine salt
N,N-Bis(2-chloroethyl)-N'-(3-hydroxypropyl)phosphorodiamidate cyclohexylammonium salt
Cytoxal alcohol
Cytoxyl alcohol cyclohexylammonium salt
NCI-C04922
NSC-52695

4466-24-4
$C_8H_{12}O$
124.18
Furan, 2-butyl- (8CI,9CI)

4466-77-7
$C_{18}H_{18}$
234.36
Phenanthrene, 1,2,3,4-tetramethyl
1:2:3:4-Tetramethylphenanthrene

4468-02-4
$C_{12}H_{22}O_4Zn$
295.71
Zinc-gluconate
Zinc, bis(D-gluconato-O^1,O^2)-(9CI)

4468-42-2
$C_{12}H_{18}$
162.28
Benzene, (1-ethylbutyl)-(9CI)
Hexane, 3-phenyl- (6CI, 7CI,8CI)
(3-Hexyl)benzene
3-Phenylhexane

4471-41-4
$C_{18}H_{18}N_2O_4$
326.34
O=C(c(c(C(=O)c1c(NCCO)ccc2NCCO)ccc3)c3)c12
Anthraquinone, 1,4-bis((2-hydroxyethyl)amino)- (8CI)
9,10-Anthracenedione, 1,4-bis-((2-hydroxyethyl)amino)-(9CI)
1,4-Bis((2-hydroxyethyl)amino)-9,10-anthracenedione

4472-06-4
CHN_3S_2
119.15
Azidodithiocarbonic acid
Acide azidodithiocarbonique (French)
Acido azidoditiocarbonico (Spanish)
Azidodithioformic acid
Azidothiocarbonic acid
Carbonazidodithioic acid
Formic acid, azidodithio-

4474-24-2
$C_{32}H_{30}N_2O_8S_2.2Na$
680.70
Benzenesulfonic acid, 3,3'-((9,10-dihydro-9,10-dioxo-1,4-anthracenediyl)diimino)-bis(2,4,6-trimethyl-, disodium salt (9CI)
Alizarine Blue BL
2-Mesitylenesulfonic acid, 4,4'-(1,4-anthraquinonylenediimino)di-, disodium salt (8CI)

4477-28-5
$C_{12}H_{10}N_3.HO_4S$
293.28
Benzenediazonium, 4-(phenylamino)-, sulfate (1:1) (9CI)
4-(Phenylamino)benzenediazonium sulfate (1:1)

4478-10-8
$C_{10}H_{13}Br$
213.12
Benzene, 2-bromo-4-methyl-1-(1-methylethyl)- (9CI)
p-Cymene, 3-bromo- (8CI)
Thymyl bromide

4478-63-1
$C_6H_{10}O_2$
114.16
CC(=O)C1OC1(C)C
2-Pentanone, 3,4-epoxy-4-methyl
4-Methyl-3,4-epoxypentan-2-one

4480-45-9
$C_9H_{11}N_7O$
233.23
s-Triazine, 2-(allylamino)-4-(allyloxy)-6-azido-(8CI)

4482-55-7
$C_9H_{12}N_2O.C_2HCl_3O_2$
327.61
CN(C)C(=O)Nc1ccccc1.[O-]C(=O)C(Cl)(Cl)Cl
Acetic acid, trichloro-, Compd. with 1,1-dimethyl-3-phenylurea (1:1)
Acetic acid, trichloro-, Compd. with N,N-dimethyl-N'-phenylurea (1:1) (9CI)
Acetic acid, trichloro-, Mixed with urea, 1,1-dimethyl-3-phenyl- (1:1)
1,1-Dimethyl-3-phenylurea trichloroacetate
N,N-Dimethyl-N'-phenyluronium trichloracetate
Dozer
Fenuron TCA
Fenuron TCA Salt
Fenuron trichloroacetate
GC-2603
3-Phenyl-1,1-dimethylurea, trichloroacetate
URAB
Urea, 1,1-dimethyl-3-phenyl-, trichloroacetate

4482-70-6
$C_{34}H_{49}N_3$
499.77
N(c(ccc(c1C)C(c(c(cc(N(CC)CC)c2)C)c2)c(c(cc(N(CC)CC)c3)C)c3)c1)(CC)CC
Benzenamine, 4,4',4''-methylidynetris(N,N-diethyl-3-methyl- (9CI)

4483-62-9
$C_{18}H_{22}O_2$
270.37
Octanoic acid, 1-naphthalenyl ester (9CI)
Octanoic acid, 1-naphthyl ester (8CI)

4484-72-4
$C_{12}H_{25}Cl_3Si$
303.81
Silane, dodecyltrichloro
Dodecyl trichlorosilane
Dodecyltrichlorosilane [UN 1771]
Silane, trichlorododecyl-
Trichlorododecylsilane
UN 1771 [Dodecyltrichlorosilane]

4485-12-5

$C_{18}H_{35}O_2.Li$
290.47
Stearic acid, lithium salt
Lithalure
Lithium octadecanoate
Lithium stearate (ACGIH)
Litholite
Octadecanoic acid, lithium salt
Stavinor

4489-84-3
$C_{11}H_{14}$
146.23
Benzene, (3-methyl-2-butenyl)- (9CI)
2-Butene, 3-methyl-1-phenyl- (8CI)

4492-78-8
$C_{14}H_{30}O_4S.C_6H_{15}NO_3$
443.64
1-Tetradecanol, hydrogen sulfate, Compd. with 2,2',2''-nitrilotris(ethanol) (1:1) (9CI)
Sulfuric acid, monotetradecyl ester, Compd. with 2,2',2''-nitrilotris(ethanol) (1:1)
Tetradecyl sulfate (1:1), triethanolamine salt
Triethanolamine myristyl sulfate
Triethanolamine tetradecyl sulfate

4499-01-8
$C_{23}H_{12}Cl_2N_6O_8S_2.2Na$
681.41
Nc4c(cc(Nc2ccc(c(Nc1nc(Cl)nc(Cl)n1)c2)S(=O)(=O)[O-][Na+])c5C(=O)c3ccccc3C(=O)c45)S(=O)(=O)[O-][Na+]
Benzenesulfonic acid, 2-(4,6-dichloro-s-triazin-2-yl-amino)-4-(4-amino-3-sulfo-1- anthraquinonylamino)-, disodium salt
Modr Brilantni Ostazinova S-R (Czech)

4499-91-6
$C_{22}H_{44}O_2.Li$
347.53
Docosanoic acid, lithium salt (9CI)
Behenic acid, lithium salt
Lithium docosanoate

4500-29-2
$C_{10}H_{21}NO_2$
187.32
OCCN(C(CCCC1)C1)CCO
Ethanol, 2,2'-cyclohexyliminodi
Abbomeen E-2
Abbomeen E-2 Aerosol
2,2'-Cyclohexyliminodiethanol
2,2'-Cyclohexyliminodiethanol aerosol
N,N-Di(2-hydroxyethyl)cyclohexylamine
Ethanol, 2,2'-cyclohexyliminodi-, (Aerosol)

4501-58-0
$C_{10}H_{16}O$
152.24
O=CCC(C(C(=C1)C)(C)C)C1
3-Cyclopentene-1-acetaldehyde, 2,2,3-trimethyl-
AI3-23129
3-Cyclopentene-1-acetaldehyde, 2,2,3-trimethyl-, (R)- (9CI)
2,2,3-Trimethyl-3-cyclopent-acetaldehyde
2,2,3-Trimethyl-3-cyclopenten-1-acetaldehyde

4505-54-8
$C_6H_6O_3$
126.11
1,2,4-Cyclopentanetrione, 3-methyl- (8CI,9CI)
NSC-150953
NSC-403804

4506-36-9
$C_{13}H_{16}$
172.27
Naphthalene, 1,2-dihydro-1,5,8-trimethyl- (7CI, 8CI,9CI)
1,2-Dihydro-1,5,8-trimethylnaphthalene

4511-39-1
$C_{12}H_{16}O_3$
208.26
O=C(OOC(CC)(C)C)c(cccc1)c1
Benzenecarboperoxoic acid, 1,1-dimethylpropyl ester (9CI)
tert-Amyl peroxybenzoate

4514-19-6
$C_{21}H_{14}$
266.35
Benzo(a)pyrene, 12-methyl
12-Methylbenzo(a)pyrene

4514-53-8
$C_9H_8ClN_5$
221.62
Nc1nc(N)nc(n1)c2ccc(Cl)cc2
1,3,5-Triazine-2,4-diamine, 6-(4-chlorophenyl)- (9CI)
NSC-211975
s-Triazine, 2,4-diamino-6-(p-chlorophenyl)- (8CI)

4514-62-9
$C_3H_2Br_2N_4$
253.89
s-Triazine, 2-amino-4,6-dibromo- (7CI,8CI)

4516-69-2
C_8H_{16}
112.22
Cyclopentane, 1,1,3-trimethyl- (9CI)

4518-98-3
$C_2H_6Cl_4Si_2$
228.05
Disilane, 1,1,2,2-tetrachloro-1,2-dimethyl- (9CI)
1,2-Dimethyltetrachlorodisilane
1,1,2,2-Tetrachloro-1,2-dimethyldisilane

4523-49-3
$C_{12}H_9N$
167.20
5-Acenaphthylenamine (8CI, 9CI)

4524-77-0
$C_6H_3BrCl_2O$
241.90
Phenol, 2-bromo-4,6-dichloro- (6CI, 7CI, 8CI, 9CI)
2-Bromo-4,6-dichlorophenol
6-Bromo-2,4-dichlorophenol
2,4-Dichloro-6-bromophenol

4525-46-6
$C_{10}H_{16}N.I$
277.17
Ammonium, benzyltrimethyl-, iodide
Ammonium, trimethylbenzyl-, iodide
Benzenemethanaminium, N,N,N-trimethyl-, iodide (9CI)
Benzyldimethylamine methiodide
Benzyl-trimethylammonium iodide
Phenmethyl-trimethylammonium iodide

4526-56-1
$C_6H_3Br_2ClO$
286.35
Phenol, 2,4-dibromo-6-chloro- (8CI,9CI)

4528-26-1
$C_6H_{10}O_2$
114.14
1,3-Dioxolane, 2-(1-propenyl)- (9CI)
1,3-Dioxolane, 2-propenyl- (8CI)

4528-34-1
$C_2H_4N_2O_5$
136.05
Nitroethyl nitrate
Ethanol, 2-nitro-, nitrate (Ester)

4531-49-1
$C_{34}H_{30}Cl_2N_6O_6$
689.52
O=C(Nc(c(OC)ccc1)c1)C(N=Nc(c(cc(c(ccc(N=NC(C(=O)C)C(=O)Nc(c(OC)ccc2)c2)c3Cl)c3)c4)Cl)c4)C(=O)C
Butanamide, 2,2'-((3,3'-dichloro(1,1'-biphenyl)-4,4'-diyl)bis(azo))bis(N-(2-methoxyphenyl)-3-oxo- (9CI)

4531-54-8
$C_4H_6N_4O_2$
142.12
1H-Imidazol-5-amine, 1-methyl-4-nitro- (9CI)
Imidazole, 5-amino-1-methyl-4-nitro- (7CI,8CI)

4531-71-9
$C_{16}H_{11}ClN_2O_4S$
362.79
2-Naphthalenesulfonic acid, 5-((4-chlorophenyl)azo)-6-hydroxy- (9CI)
2-Naphthalenesulfonic acid, 5-((p-chlorophenyl)azo)-6-hydroxy- (7CI,8CI)

4531-79-7
$C_{12}H_{10}N_2O_2$
214.21
Benzenamine, 3-nitro-N-phenyl- (9CI)
Diphenylamine, 3-nitro- (8CI)

4533-95-3
$C_6H_3Cl_2NO_4S$
256.06
O=S(=O)(c(c(ccc1N(=O)=O)Cl)c1)Cl
Benzenesulfonyl chloride, 2-chloro-5-nitro- (9CI)
2-Chloro-5-nitrobenzenesulfonyl chloride

4533-96-4
$C_6H_3Cl_2NO_4S$
256.06
Benzenesulfonyl chloride, 4-chloro-2-nitro- (8CI,9CI)
NSC-81212

4534-49-0
$C_{19}H_{32}$
260.46
Benzene, (1-pentyloctyl)- (9CI)
Tridecane, 6-phenyl- (8CI)

4534-50-3
$C_{19}H_{32}$
260.46
Benzene, (1-butylnonyl)- (9CI)
Tridecane, 5-phenyl- (8CI)

4534-51-4
$C_{19}H_{32}$
260.46
Benzene, (1-propyldecyl)- (9CI)
Tridecane, 4-phenyl- (8CI)

4534-52-5
$C_{19}H_{32}$
260.46
Benzene, (1-ethylundecyl)- (9CI)
Tridecane, 3-phenyl- (8CI)

4534-53-6
$C_{19}H_{32}$
260.46
Benzene, (1-methyldodecyl)- (9CI)
Tridecane, 2-phenyl- (8CI)

4534-54-7
$C_{20}H_{34}$
274.49
Benzene, (1-hexyloctyl)- (9CI)
Tetradecane, 7-phenyl- (8CI)

4534-55-8
$C_{20}H_{34}$
274.49
Benzene, (1-pentylnonyl)- (9CI)
Tetradecane, 6-phenyl- (8CI)

4534-56-9
$C_{20}H_{34}$
274.49
Benzene, (1-butyldecyl)- (9CI)
Tetradecane, 5-phenyl- (8CI)

4534-57-0
$C_{20}H_{34}$
274.49
Benzene, (1-propylundecyl)- (9CI)
Tetradecane, 4-phenyl- (8CI)

4534-58-1
$C_{20}H_{34}$
274.49
Benzene, (1-ethyldodecyl)- (9CI)
Tetradecane, 3-phenyl- (8CI)

4534-59-2
$C_{20}H_{34}$
274.49
Benzene, (1-methyltridecyl)- (9CI)
Tetradecane, 2-phenyl- (8CI)

4536-23-6
$C_7H_{14}O_2$
130.19
O=C(O)C(CCCC)C
Hexanoic acid, 2-methyl- (9CI)
2-Methylhexanoic acid

4536-30-5
$C_{14}H_{30}O_2$
230.44
O(CCCCCCCCCCCC)CCO
Ethanol, 2-(dodecyloxy)
2-(Dodecyloxy)ethanol
Ethylene glycol monododecyl ether
Ethylene glycol monolauryl ether
Laureth-1
Lauryl ethoxylate
Lauryl monoethoxylate
Lipocol L-1

4536-86-1
$C_{17}H_{28}$
232.41
Benzene, (1-propyloctyl)- (9CI)
Undecane, 4-phenyl- (8CI)

4536-87-2
$C_{17}H_{28}$
232.41
Benzene, (1-ethylnonyl)- (9CI)
Undecane, 3-phenyl- (8CI)

4536-88-3
$C_{17}H_{28}$
232.41
Benzene, (1-methyldecyl)- (9CI)
Undecane, 2-phenyl- (8CI)

4537-11-5
$C_{16}H_{26}$
218.38
Benzene, (1-butylhexyl)- (9CI)
Decane, 5-phenyl- (8CI)

4537-12-6
$C_{16}H_{26}$
218.38
Benzene, (1-propylheptyl)- (9CI)
Decane, 4-phenyl- (8CI)

4537-13-7
$C_{16}H_{26}$
218.38
Benzene, (1-methylnonyl)- (9CI)

Decane, 2-phenyl- (8CI)

4537-14-8
$C_{17}H_{28}$
232.41
Benzene, (1-pentylhexyl)- (9CI)
Undecane, 6-phenyl- (8CI)

4537-15-9
$C_{17}H_{28}$
232.41
Benzene, (1-butylheptyl)- (9CI)
Undecane, 5-phenyl- (8CI)

4540-00-5
$C_{16}H_{17}ClN_4O_4$
364.76
O=N(=O)c(ccc(N=Nc(c(cc(N(CC O)CCO)c1)Cl)c1)c2)c2
Ethanol, 2,2'-((3-chloro-4-((4-nitrophenyl)azo)phenyl)-imino)bis- (9CI)

4540-44-7
C_3H_6BrClO
173.44
2-Propanol, 1-bromo-3-chloro- (9CI)

4541-13-3
$C_{10}H_{22}O_2$
174.29
1-Butanol, 4-(hexyloxy)- (7CI,8CI,9CI)

4542-47-6
$C_7H_{12}N_2O$
140.17
N#CCCN(CCOC1)C1
4-Morpholinepropanenitrile (9CI)
AI3-13188
Cyanoethylmorpholine

4542-57-8
$C_{24}H_{50}O$
354.66
O(CCCCCCCCCCCC)CCCCCC CCCCCC
Dodecane, 1,1'-oxybis- (9CI)
AI3-09529
Dodecyl ether (8CI)
1,1'-Oxybisdodecane

4548-53-2
$C_{18}H_{14}N_2O_7S_2.2Na$
480.44
[Na+].[Na+].Cc1cc(C)c(cc1N=N c3cc(c2ccccc2c3O)S([O-])(=O)=O)S([O-])(=O)=O
1-Naphthalenesulfonic acid, 4-hydroxy-3-((6-sulfo-2,4-xylyl)azo)-, disodium salt
Certicol Ponceau SXS
Cerven potravinarska 1 (Czech)
C.I. 14700
C.I. Food Red 1
C.I. Food Red 1, Disodium salt
Crimson 4R
3-((2,4-Dimethyl-5-sulfo-phenyl)azo)-4-hydroxy-1-naphthalenesulfonic acid, disodium salt
3-((2,4-Dimethyl-5-sulpho-phenyl)azo)-4-hydroxy-1-naphthalenesulphonic acid, disodium salt
Dye FD & C Red No. 4
Edicol Supra Ponceau SX
FD & C Red No. 4
FD & C Red No. 4 - Aluminium Lake
Food Red 4
Hexacol Ponceau SX
4-Hydroxy-3-((5-sulpho-2,4-xylyl)azo)-1-naphthalenesulphonic acid, disodium salt
Maple Ponceau SX
4-Hydroxy-3-((5-sulfo-2,4-xylyl)azo)-1-naphthalene-sulfonic acid, disodium salt
Ponceau SX
Ponceau SX Lake
Purple 4R
1306 Red
12101 Red
Red Lake 89865N
Red No. 1
Red No. 4
4R Purple
2-(6-Sulfo-2,4-xylylazo)-1-naphthol-4-sulfonic acid, disodium salt
2-(6-Sulpho-2,4-xylylazo)-1-naphthol-4-sulphonic acid, disodium salt
Usacert FD & C Red No. 4
Usacert Red No. 4

4549-33-1
$C_9H_{18}Br_2$
286.05
BrCCCCCCCCCBr
Nonane, 1,9-dibromo- (9CI)
1,9-Dibromononane

4549-40-0
$C_3H_6N_2O$
86.11
CN(C=C)N=O
Vinylamine, N-methyl-N-nitroso
Ethenylamine, N-methyl-N-nitroso-
N-Methyl-N-nitroso-ethenyl-amine
N-Methyl-N-nitrosovinylamine
Methylvinylnitrosamin (German)
Methylvinylnitrosamine
MVNA
N-Nitrosomethylvinylamine
NMVA
RCRA waste number P084

4549-43-3
$C_4H_8N_2O$
100.14
CN(CC=C)N=O
Allylamine, N-methyl-N-nitroso
Allylmethylnitrosamine
Methylallylnitrosamin (German)
Methylallylnitrosamine
N-Methyl-N-nitrosoallylamine
N-Nitrosoallylmethylamine
Nitrosomethylallylamine
N-Nitrosomethylallylamine
2-Propen-1-amine, N-methyl-N-nitroso- (9CI)

4549-44-4
$C_6H_{14}N_2O$
130.22
CCCCN(CC)N=O
Butylamine, N-ethyl-N-nitroso
Aethyl-n-butyl-nitrosoamin (German)
Butanamine, N-ethyl-N-nitroso-
Butylethylnitrosamin (Czech)
Ethyl-n-butylnitrosamine
N-Ethyl-N-nitrosobutylamine
N-Nitroso-n-butylethylamine
N-Nitrosoethyl-n-butylamine

4549-74-0
C_6H_{10}
82.15
CC=C(C)C=C
1,3-Pentadiene, 3-methyl- (7CI,8CI,9CI)
3,4-Dimethylbutadiene
1,2-Dimethyl-1,3-butadiene
3-Methylpentadiene
3-Methyl-1,3-pentadiene

4553-62-2
$C_6H_7N_2$
107.15
N#CC(CCC#N)C
1,5-Valerodinitrile, 2-methyl
Methyl glutaronitrile

4560-87-6
$C_9H_{12}N_6$
204.27
s-Triazine-1,3,5(2H,4H,6H)tri-acetonitrile
s-Triazine, hexahydro-1,3,5-tris-(cyanomethyl)-
1,3,5-Tris(cyanomethyl)hexa-hydro-s-triazine

4562-27-0
$C_4H_4N_2O$
96.10
n(ccc(n1)O)c1
4(1H)-Pyrimidinone
Deaminoisocytosine
4-Hydroxypyrimidine

6-Hydroxypyrimidine
4-Oxypyrimidine
4-Pyrimidinol

4568-28-9
$C_{18}H_{36}O_2.C_6H_{15}NO_3$
433.67
TEA-stearate
Octadecanoic acid, Compd. with 2,2',2''-nitrilotris-(ethanol) (1:1) (9CI)
Octadecanoic acid, triethanolamine salt
Stearic acid, triethanolamine soap
Triethanolamine stearate
Triethanolamine, stearic acid salt

4573-50-6
$C_{10}H_{16}O$
152.24
O=C(C=C(CC1)C)C1C(C)C
2-Cyclohexen-1-one, 3-methyl-6-(1-methylethyl)-, (R)- (9CI)

4574-04-3
$C_{17}H_{38}N.Cl$
291.94
[Cl-].CCCCCCCCCCCCCC[N+](C)(C)C
Ammonium, trimethyltetradecyl-, chloride (8CI)
1-Tetradecanaminium, N,N,N-trimethyl-, chloride (9CI)
N,N,N-Trimethyl-1-tetradecanaminium chloride

4576-40-3
$C_4H_6BrN_5$
218.06
1,3,5-Triazine-2,4-diamine, 6-(bromomethyl)- (9CI)
2-Bromomethyl-4,6-diaminotriazine
s-Triazine, 2,4-diamino-6-(bromomethyl)- (7CI, 8CI)

4579-31-1
$C_8H_{14}O$
126.20
Cyclobutanone, 2-(1,1-dimethylethyl)- (9CI)
Cyclobutanone, 2-tert-butyl- (8CI)

4584-46-7
$C_4H_{10}ClN.ClH$
144.06
Ethylamine, 2-chloro-N,N-dimethyl-, hydrochloride
Bis(methyl)-2-chloroethylamine hydrochloride
2-Chloro-N,N-dimethylethylamine hydrochloride
Dimethyl-β-chloroethylamine hydrochloride
Dimethyl(2-chloroethyl)amine hydrochloride

4584-49-0
$C_5H_{12}ClN.ClH$
158.09
Propane, 1-(N,N-dimethylamino)-2-chloro-, hydrochloride
2-Chloropropyl-dimethylamine hydrochloride

4587-03-5
$C_6H_9N_7O$
195.18
s-Triazine, 2-azido-4-(ethylamino)-6-methoxy-(7CI,8CI)

4591-46-2
$C_6H_3N_5O_8$
273.09
2,4,6-Trinitrophenyl nitramine
Aniline, N,2,4,6-tetranitro-
Benzenamine, N,2,4,6-tetranitro-
N,2,4,6-Tetranitroaniline
2,4,6-Trinitrophenylnitramine

4593-90-2
$C_{10}H_{12}O_2$
164.20
CC(CC(O)=O)c1ccccc1
3-Phenylbutyric acid
AI3-11112
Benzenepropanoic acid, β-methyl-

4595-59-9
$C_4H_3BrN_2$
158.97
n(cc(Br)cn1)c1
Pyrimidine, 5-bromo- (9CI)
5-Bromopyrimidine

4599-92-2
$C_{21}H_{12}O$
280.33
11H-Indeno(2,1-a)phenanthren-11-one (8CI,9CI)

4599-94-4
$C_{21}H_{12}O$
280.33
13H-Dibenzo(a,h)fluoren-13-one (8CI,9CI)

4602-84-0
$C_{15}H_{26}O$
222.41
OCC=C(CCC=C(CCC=C(C)C)C)C
2,6,10-Dodecatrien-1-ol, 3,7,11-trimethyl
Farnesyl alcohol
Farnesol
3,7,11-Trimethyl-2,6,10-dodecatrien-1-ol

4612-63-9
$C_{15}H_{14}$
194.28
9H-Fluorene, 2,3-dimethyl-(9CI)
Fluorene, 2,3-dimethyl- (8CI)

4613-11-0
$C_{16}H_{18}$
210.32
Benzene, 1,1'-(1,2-dimethyl-1,2-ethanediyl)-bis-, (R*,S*)- (9CI)
Bibenzyl, α,α'-dimethyl-, erythro- (8CI)
erythro-2,3-Diphenylbutane
meso-2,3-Diphenylbutane

4617-17-8
Unknown
Unknown
β,β'-Dithiocyano diethyl ether

4620-47-7
$C_{12}H_{12}O_4$
220.22
2,4-Pentanedione, 3-(benzoyloxy)- (9CI)
2,4-Pentanedione, 3-hydroxy-, benzoate (8CI)

4620-70-6
$C_6H_{15}NO$
117.19
OCCNC(C)(C)C
tert-Butylaminoethanol
AI3-28035
2-((1,1-Dimethylethyl)amino)-ethanol
Ethanol, 2-((1,1-dimethylethyl)-amino)- (9CI)

4621-04-9
$C_9H_{18}O$
142.27
OC(CCC(C1)C(C)C)C1
Cyclohexanol, p-isopropyl
p-Isopropylcyclohexanol
4-Isopropylcyclohexanol

4621-36-7
$C_{16}H_{26}$
218.38
Benzene, (1-ethyloctyl)- (9CI)
Decane, 3-phenyl- (8CI)

4628-08-4

$C_{15}H_{25}Cl_2N_3O$
334.29
s-Triazine, 2,4-dichloro-6-(dodecyloxy)- (6CI, 7CI,8CI)

4629-58-7
$C_{14}H_{12}N_2O_2$
240.25
Benzenamine, 4-(2-(4-nitrophenyl)ethenyl)- (9CI)
NSC-52232
4-Stilbenamine, 4'-nitro- (8CI)

4630-06-2
$C_8H_{16}O$
128.21
5-Hepten-2-ol, 6-methyl-, (+-)- (8CI,9CI)
Sulcatol

4630-20-0
$C_{13}H_{11}N$
181.23
9H-Carbazole, 3-methyl- (9CI)
Carbazole, 3-methyl- (8CI)
NSC-10154

4635-59-0
$C_4H_6Cl_2O$
141.00
O=C(CCCCl)Cl
Butanoyl chloride, 4-chloro- (9CI)
4-Chlorobutanoyl chloride
γ-Chlorobutyryl chloride

4635-87-4
C_5H_7N
81.11
N#CCC=CC
3-Pentenenitrile (9CI)

4637-56-3
$C_9H_8N_2O_2$
176.19
ONc1ccn(=O)c2ccccc12
Quinoline, 4-(hydroxyamino)-, 1-oxide
4HAQO
Hydroxylamine, N-(4-quinolyl)-, 1'-oxide
4-(Hydroxyamino)quinoline 1-oxide

4638-48-6
Unknown
Unknown
5-Chlorosalicyclanilide

4643-25-8
$C_7H_{12}O$
112.17
O=C(C=CC)CCC
2-Hepten-4-one (9CI)

4645-15-2
$C_{12}H_{22}O$
182.31
Cyclohexane, 1,1'-oxybis- (9CI)
Cyclohexyl ether (8CI)

4647-42-1
$C_9H_{10}O_2$
150.18
1H-Indene-1,2-diol, 2,3-dihydro-, cis- (9CI)
1,2-Indandiol, cis- (8CI)

4647-43-2
$C_9H_{10}O_2$
150.18
1H-Indene-1,2-diol, 2,3-dihydro-, trans- (9CI)
1,2-Indandiol, trans- (8CI)

4649-27-8
$C_{15}H_9NO_4$
267.23
1H-Isoindole-5-carboxylic acid, 2,3-dihydro-1,3-dioxo-2-phenyl- (9CI)
5-Isoindolinecarboxylic acid, 1,3-dioxo-2-phenyl- (8CI)
NSC-180808

4649-67-6
$C_3H_4BrN_5$
190.01
1,3,5-Triazine-2,4-diamine, 6-bromo- (9CI)
s-Triazine, 2,4-diamino-6-bromo- (7CI,8CI)

4653-94-5
$C_6H_9ClN_4O$
188.62
1,3,5-Triazin-2-amine, 4-chloro-N-ethyl-6-methoxy- (9CI)
s-Triazine, 2-chloro-4-(ethylamino)-6-methoxy- (7CI,8CI)

4655-34-9
$C_7H_{12}O_2$
128.19
O=C(OC(C)C)C(=C)C
Methacrylic acid, isopropyl ester
Isopropyl methacrylate
2-Propenoic acid, 2-methyl-, 1-methylethyl ester (9CI)

4657-00-5
$C_{28}H_{27}N_2.Cl$
426.98
3H-Indolium, 1,3,3-trimethyl-2-(2-(1-methyl-2-phenyl-1H-indol-3-yl)ethenyl)-, chloride (9CI)
AI3-22671
3H-Indolium, 1,3,3-trimethyl-2-(2-(1-methyl-2-phenylindol-3-yl)vinyl)-, chloride (8CI)
2-(2-(1-Methyl-2-phenyl-1H-indol-3-yl)ethenyl)-1,3,3-trimethyl-3H-indolium chloride

4657-20-9
$C_{18}H_{28}O_4Si_4$
420.82
Cyclotetrasiloxane, 2,6-diphenyl-2,4,4,6,8,8-hexamethyl
2,6-Diphenyl-2,4,6,6,8,8-hexamethylcyclotetrasiloxane

4657-93-6
$C_{12}H_{11}N$
169.24
5-Acenaphthenamine
5-Acenaphtheneamine
5-Acenaphthylenamine, 1,2-dihydro-
5-Aminoacenaphthene
1,2-Dihydro-5-acenaphthylenamine

4658-20-2
$C_7H_{10}Br_2N_4$
310.00
s-Triazine, 2,4-dibromo-6-(isobutylamino)- (7CI, 8CI)

4658-25-7
$C_6H_9N_7S$
211.25
s-Triazine, 2-azido-4-(ethylamino)-6-(methylthio)- (7CI,8CI)

4658-28-0
$C_7H_{11}N_7S$
225.31
CSc1nc(NC(C)C)nc(N=N=N)n1
s-Triazine, 2-azido-4-(isopropylamino)-6-(methylthio)
Azeprotyrne
2-Azido-4-isopropylamino-6-methylthio-s-triazine
2-Azido-4-isopropylamino-6-methylthio-1,3,5-triazine
4-Azido-N-(1-methylethyl)-6-(methylthio)-1,3,5-triazin-2-amine
Aziprotryn
Aziprotryne
Azirpotryne
Brasoran
Brassoron
C 7019
Ciba C 7019
Isopropylamino-4-azido-

6-methylthio-1,3,5-triazin (German)
Mesoranil
Mezaronil
Mezuron
1,3,5-Triazin-2-amine, 4-azido-N-(1-methylethyl)-6-(methylthio)-

4658-30-4
$C_8H_{13}ClN_4S$
232.74
1,3,5-Triazin-2-amine, 4-chloro-N-(2-methylpropyl)-6-(methylthio)- (9CI)
s-Triazine, 2-chloro-4-(isobutylamino)-6-(methylthio)- (7CI,8CI)

4659-47-6
$C_7H_3Cl_3O$
209.45
Benzaldehyde, 2,3,6-trichloro
2,3,6-Trichlorobenzaldehyde

4660-80-4
$C_5H_8O_3$
116.12
Glycidic acid ethyl ester
Ethyl glycidate

4672-26-8
$C_5H_{13}O_3P$
152.13
Phosphonic acid, pentyl- (8CI,9CI)
NSC-187676
1-Pentanephosphonic acid

4672-38-2
$C_3H_9O_3P$
124.09
CCCP(O)(O)=O
Phosphonic acid, propyl
NIA 10656
1-Propylphosphonic acid

4674-50-4
$C_{15}H_{22}O$
218.34
O=C(C=C(C(C1C)(CC(C(=C)C)C2)C)C2)C1
2(3H)-Naphthalenone, 4,4a,5,6,7,8-hexahydro-4,4a-dimethyl-6-(1-methylethenyl)-, (4R-(4α,4aα,6β))- (9CI)
4βH,5α-Eremorphila-1(10)11-dien-2-one
4,4a,5,6,7,8-Hexahydro-4,4a-dimethyl-6-(1-methylethenyl)-2(3H)-naphthalenone, (4R-(4α,4aα,6β))-
Nootkatone

4675-87-0
$C_5H_{10}O$
86.13
OCC(=CC)C
2-Methyl-2-buten-1-ol
2-Buten-1-ol, 2-methyl- (9CI)

4680-78-8
$C_{37}H_{36}N_2O_6S_2.Na$
691.86
[Na+].CCN(Cc1cccc(c1)S([O-])(=O)=O)c2ccc(cc2)C(=C3C=CC(C=C3)=[N+](CC)Cc4cccc(c4)S([O-])(=O)=O)c5ccccc5
Ammonium, ethyl(4-(p-(ethyl-(m-sulfobenzyl)amino)-α-phenylbenzylidene)-2,5-cyclohexadien-1-ylidene)(m-sulfobenzyl)-, hydroxide, inner salt, sodium salt
Acidal Green G
Acid Green 3
Acid Green
Acid Green B
Acid Green 2G
Acid Green S
Acid Green G
Acid Green L
Acid Leather Green F
Acid Leather Green 3G
Acilan Green B
A.F. Green No. 1
Amacid Green B
Brilliant Green 3EMBL
Bucacid Guinea Green BA
Calcocid Green G
C.I. 42085
C.I. Acid Green 3
C.I. Acid Green 3, Monosodium salt
C.I. Acid Green 3, Sodium salt
C.I. Food Green 1
FD & C Green 1
FD & C Green No. 1
Fenazo Green L
Food Green 1
Guinea Green
Guinea Green B
Guinea Green BA
Hidacid Emerald Green
Hispacid Green GB
Intracid Green F
Kiton Green F
Kiton Green FC
Leather Green B
Lissamine Green G
Merantine Green G
Naphthalene Green G
Naphthalene Lake Green G
Naphthalene Leather Green G
Neran Brilliant Green G
Pontacyl Green BL
Sulfacid Brilliant Green 1B
Sulpho Green 2B
Vondacid Green L
Zelen Kysela 3 (Czech)
Zelen Potravinarska 1 (Czech)

4682-03-5
$C_6H_2N_4O_5$
210.08
N#Nc(c(O)c(N(=O)=O)cc1N(=O)=O)c1
Diazodinitrophenol (Dry)
2,4-Cyclohexadien-1-one, 6-diazo-2,4-dinitro- (9CI)
6-Diazo-2,4-dinitro-2,4-cyclohexadien-1-one
Diazodinitrofenol (Spanish)

4682-78-4
$C_9H_5Cl_2N_3O$
242.05
1,3,5-Triazine, 2,4-dichloro-6-phenoxy- (9CI)
s-Triazine, 2,4-dichloro-6-phenoxy- (8CI)

4684-94-0
$C_6H_4ClNO_2$
157.56
OC(=O)c1cccc(Cl)n1
Picolinic acid, 6-chloro
6-Chloropicolinic acid
6-CPA
2-Pyridinecarboxylic acid, 6-chloro-

4685-14-7
$C_{12}H_{14}N_2$
186.28
[Cl-].[Cl-].C[n+]1ccc(cc1)c2cc[n+](C)cc2
4,4'-Bipyridinium, 1,1'-dimethyl
Dimethyl viologen
Gramoxone S
Methyl Viologen (2+)
Paraquat (ACGIH,OSHA)
Paraquat dication
Paraquat ion

4685-18-1
$C_7H_7N_5O$
177.19
s-Triazine, 2,4-diamino-6-(2-furyl)
2,4-Diamino-6-(2-furyl)-s-triazine

4693-19-0
$C_{14}H_{20}O_3$
236.31
O=C(c(c(O)ccc1O)c1)CCCCCCC
1-Octanone, 1-(2,5-dihydroxyphenyl)- (9CI)
1-(2,5-Dihydroxyphenyl)-1-octanone

4695-62-9
$C_{10}H_{16}O$
152.26
O=C(C(CC1C2)(C2)C)C1(C)C
2-Norbornanone, 1,3,3-tri-

methyl-, (1R,4S)-(+)
Bicyclo(2.2.1)heptan-2-one, 1,3,3-trimethyl-, (1R)- (9CI)
(+)-Fenchone
D(+)-Fenchone
d-Fenchone
1,3,3-Trimethylbicyclo(2.2.1)-heptan-2-one
d-1,3,3-Trimethyl-2-norbornan-one
d-1,3,3-Trimethyl-2-nor-camphanone

4696-46-2
$C_{18}H_{38}O_4S.H_3N$
367.59
Sulfuric acid, monooctadecyl ester, ammonium salt (9CI)

4696-47-3
$C_{16}H_{34}O_4S.H_3N$
339.53
1-Hexadecanol, hydrogen sulfate, ammonium salt (9CI)
Ammonium hexadecyl sulfate
Cetyl sulfate ammonium salt

4696-57-5
$C_{12}H_{24}O_2.1/2Ba$
268.98
Dodecanoic acid, barium salt (9CI)
Barium dodecanoate

4697-36-3
$C_{17}H_{18}N_2O_6S$
378.43
CC3(C)SC2C(NC(=O)C(C(O)=O)c1ccccc1)C(=O)N2C3C(O)=O
Malonamic acid, N-(2-carboxy-3,3-dimethyl-7-oxo-4-thia-1-azabicyclo(3.2.0)-hept-6-yl)-2- phenyl
Carbenicillin

4702-64-1
$C_{21}H_{16}N_2O_5$
376.39
O=C(c(c(c(N)cc1c(ccc(OC)c2)c2)C(=O)c3c(O)ccc4N)c1O)c34
Anthraquinone, 1,5-diamino-4,8-dihydroxy-3-(p-methoxyphenyl)
1,5-Diamino-4,8-dihydroxy-3-(p-methoxyphenyl)anthraquinone
Modr Ostacetova SE-LB (Czech)

4702-65-2
$C_{21}H_{16}N_2O_5$
376.36
O=C(c(c(c(N)cc1c(ccc(O)c2C)c2)C(=O)c3c(O)ccc4N)c1O)c34
9,10-Anthracenedione, 4,8-diamino-1,5-dihydroxy-2-(4-hydroxy-3-methylphenyl)-(9CI)
1,5-Diamino-4,8-dihydroxy-3-(4-hydroxy-3-methylphenyl)anthraquinone

4704-77-2
$C_3H_7BrO_2$
155.01
1,2-Propanediol, 3-bromo
Bromodeoxyglycerol
α-Bromohydrin
3-Bromo-1,2-propanediol
Monobromoglycerol

4705-34-4
$C_{16}H_{16}O_2$
240.30
Benzene, 1,1'-(1,2-ethenediyl)bis(4-methoxy- (9CI)
AI3-14664
Stilbene, 4,4'-dimethoxy- (8CI)

4706-81-4
$C_{14}H_{30}O$
214.39
2-Tetradecanol (9CI)
AI3-35271

4706-89-2
$C_{11}H_{16}$
148.25
Benzene, 2,4-dimethyl-1-(1-methylethyl)- (9CI)
Cumene, 2,4-dimethyl-(6CI,7CI,8CI)
1,3-Dimethyl-4-isopropylbenzene
1-Isopropyl-2,4-dimethylbenzene
4-Isopropyl-1,3-dimethylbenzene
4-Isopropyl-m-xylene

4706-90-5
$C_{11}H_{16}$
148.25
Benzene, 1,3-dimethyl-5-(1-methylethyl)- (9CI)
Cumene, 3,5-dimethyl- (8CI)

4707-47-5
$C_{10}H_{12}O_4$
196.20
O=C(OC)c(c(cc(O)c1C)C)c1O
Benzoic acid, 2,4-dihydroxy-3,6-dimethyl-, methyl ester (9CI)
2,4-Dihydroxy-3,6-dimethylbenzoic acid, methyl ester
Methyl 3,6-dimethylresorcylate
Methyl β-orcinolcarboxylate
β-Resorcylic acid, 3,6-dimethyl-, methyl ester

4707-50-0
$C_{10}H_{12}O_4$
196.20
Benzoic acid, 2,4-dihydroxy-6-propyl- (9CI)
β-Resorcylic acid, 6-propyl-(7CI,8CI)
Divaric acid

4712-38-3
C_4H_9DO
73.11
1-Butanol-d (9CI)
Butyl alcohol-d (8CI)

4712-39-4
C_4H_9DO
74.12
2-Butanol-d (9CI)
2-Butanol-d1
sec-Butyl alcohol-d (6CI, 7CI,8CI)

4712-55-4
$C_{12}H_{11}O_3P$
234.19
O=P(Oc(cccc1)c1)Oc(cccc2)c2
Phosphonic acid, diphenyl ester (9CI)
Diphenyl phosphonate

4714-14-1
$C_{14}H_{13}Cl$
216.71
Benzene, 1,1'-(1-chloro-1,2-ethanediyl)bis- (9CI)
Bibenzyl, α-chloro- (6CI, 7CI,8CI)
α-Chlorobibenzyl
1-Chloro-1,2-diphenylethane
1,2-Diphenylethyl chloride

4719-04-4
$C_9H_{21}N_3O_3$
219.33
OCCN(CN(CN1CCO)CCO)C1
s-Triazine-1,3,5(2H,4H,6H)-triethanol
Grotan B
Grotan BK
Hexahydro-1,3,5-tris(hydroxyethyl)triazine
Kalpur TE
KM 200
Onyxide 200
1,3,5-Triazine-1,3,5(2H,4H,6H)-triethanol (9CI)

4720-09-6
$C_{22}H_{36}O_7$
412.58
7,9a-Methano-9aH-cyclopenta-(b)heptalene-2,4,8,11,11a,12(1H)-hexol,dodecahydro-1,1,4,8- tetramethyl-,

12-acetate (2S,3aS,4R,4aR, 7R,8R,9aS,11R,11aR,12R)
Acetyllandromedol
Andromedotoxin
Asebotoxin
G-I
Grayanotoxane-3,5,6,10,14,16-hexol 14-acetate
Grayanotoxane-3,5,6,10,14,16-hexol, 14-acetate, (3-β,6-β,14R)-
Grayanotoxin I
Rhodotoxin

4721-24-8
$C_6H_{15}O_3P$
166.16
CCCCCCP(O)(O)=O
Phosphonic acid, hexyl- (8CI, 9CI)
1-Hexanephosphonic acid
NSC-222656

4726-14-1
$C_{13}H_{19}N_3O_6S$
345.36
CCCN(CCC)c1c(cc(cc1N(=O)=O)S(C)(=O)=O)N(=O)=O
Nitralin
AI3-61434
Aniline, 2,6-dinitro-N,N-dipropyl-4-(methylsulfonyl)-
Aniline, 4-(methylsulfonyl)-2,6-dinitro-N,N-dipropyl-
Benzenamine, 4-(methylsulfonyl)-2,6-dinitro-N,N-dipropyl-
Caswell No. 578
2,6-Dinitro-4-methylsulfonyl-N,N-dipropylaniline
EPA Pesticide Chemical Code 037601
4-Methylsulfonyl-2,6-dinitro-N,N-dipropylaniline
4-(Methylsulfonyl)-2,6-dinitro-N,N-dipropylaniline
4-(Methylsulfonyl)-2,6-dinitro-N,N-dipropylbenzenamine
4-(Methylsulfonyl)-2,6-dinitro-N,N-dipropylbenzeneamine
4-Methylsulphonyl-2,6-dinitro-N,N-dipropylaniline
Nitraline
Planavin
Planavin 75
Planuin
SD 11831

4726-22-1
$C_8H_{11}NO_7S_2$
297.30
Phenol, 2-amino-4-((2-(sulfooxy)ethyl)sulfonyl)- (9CI)
2-Amino-4-((2-(sulfooxy)ethyl)-sulfonyl)phenol

4730-22-7
$C_8H_{18}O$
130.23
CC(C)CCCC(C)O
2-Heptanol, 6-methyl- (8CI, 9CI)
NSC-75858

4731-53-7
$C_{24}H_{51}P$
370.64
P(CCCCCCCC)(CCCCCCCC)CCCCCCCC
Phosphine, trioctyl- (9CI)
Trioctylphosphine

4736-60-1
$C_{20}H_{20}P.I$
418.27
[I-].CC[P+](c1ccccc1)(c2ccccc2)c3ccccc3
Phosphonium, ethyltriphenyl-, iodide
Ethyltriphenylphosphonium iodide

4741-74-6
$C_{19}H_{24}O_2$
284.43
Propane, 1,1-bis(p-methoxyphenyl)-2,2-dimethyl
Benzene, 1,1'-(2,2-dimethylpropylidene)bis(4-methoxy-(9CI)
1,1-Bis(p-methoxyphenyl)-2,2-dimethylpropane
Dianisylneopentane

4743-13-9
$C_7H_{11}ClN_4O$
202.64
1,3,5-Triazin-2-amine, 4-chloro-6-ethoxy-N-ethyl- (9CI)
s-Triazine, 2-chloro-4-ethoxy-6-(ethylamino)- (6CI, 7CI,8CI)

4744-08-5
$C_7H_{16}O_2$
132.20
O(C(OCC)CC)CC
Propane, 1,1-diethoxy- (9CI)
1,1-Diethoxypropane
Propionaldehyde, diethyl acetal (8CI)

4744-10-9
$C_5H_{12}O_2$
104.15
Propane, 1,1-dimethoxy- (9CI)
Propionaldehyde, dimethyl acetal (8CI)

4746-61-6
$C_8H_9NO_2$
151.18
OCC(=O)Nc1ccccc1
Acetamide, 2-hydroxy-N-phenyl
Glycolanilide
2-Hydroxy-N-phenylacetamide

4748-78-1
$C_9H_{10}O$
134.18
O=Cc(ccc(c1)CC)c1
Benzaldehyde, 4-ethyl- (9CI)
Benzaldehyde, p-ethyl- (8CI)
4-Ethylbenzaldehyde

4749-27-3
$C_4H_5Cl_3$
159.44
1-Propene, 3,3,3-trichloro-2-methyl- (9CI)
Propene, 3,3,3-trichloro-2-methyl- (8CI)

4750-28-1
$C_{12}H_{10}ClNO_2S$
267.73
Benzenesulfonamide, N-(4-chlorophenyl)- (9CI)
Benzenesulfonanilide, 4'-chloro-(8CI)
NSC-62066

4751-43-3
$C_{20}H_{16}N_4O$
328.35
O(C(=CC=1)C(=Nc(c2ccc3)c3)N2C)C1C(=Nc(c4ccc5)c5)N4C
1H-Benzimidazole, 2,2'-(2,5-furandiyl)bis(1-methyl-(9CI)

4754-44-3
$C_{14}H_{30}O_4S$
294.46
O=S(=O)(OCCCCCCCCCCCCCC)O
1-Tetradecanol, hydrogen sulfate (9CI)
Tetradecyl sulfuric acid

4755-77-5
$C_4H_5ClO_3$
136.53
O=C(OCC)C(=O)Cl
Acetic acid, chlorooxo-, ethyl ester (9CI)
Ethyl chlorooxoacetate
Oxalyl chloride, ethyl ester

4755-81-1
$C_5H_7ClO_3$
150.56
O=C(OC)C(C(=O)C)Cl
Butanoic acid, 2-chloro-3-oxo-, methyl ester (9CI)

Methyl 2-chloroacetoacetate
Methyl 2-chloro-3-oxobutanoate

4759-48-2
$C_{20}H_{28}O_2$
300.48
O=C(O)C=C(C=CC=C(C=CC(=C(CCC1)C)C1(C)C)C)C
Retinoic acid, 13-cis
Accutane
Isotretinoin
Neovitamin A acid
13-RA
13-cis-RA
13-cis-Retinoic acid
Ro-4-3780
Roaccutane
13-cis-Vitamin A Acid

4764-17-4
$C_{10}H_{13}NO_2$
179.24
Phenethylamine, α-methyl-3,4-(methylenedioxy)
1,3-Benzodioxole-5-ethanamine, α-methyl- (9CI)
MDA
Methylenedioxyamphetamine
3,4-Methylenedioxy-amphetamine
α-Methyl-3,4-(methylenedioxy)-phenethylamine

4767-03-7
$C_5H_{10}O_4$
134.13
O=C(O)C(CO)(CO)C
Propanoic acid, 3-hydroxy-2-(hydroxymethyl)-2-methyl-(9CI)

4771-80-6
$C_7H_{10}O_2$
126.17
OC(=O)C1CCC=CC1
3-Cyclohexene-1-carboxylic acid
Kyselina 1,2,5,6-tetrahydrobenzoova (Czech)

4775-09-1
$C_6H_{15}O_2P$
150.16
Phosphinic acid, diethyl-, ethyl ester (6CI,7CI, 8CI,9CI)
Ethoxydiethylphosphine oxide
Ethyl diethylphosphinate

4776-06-1
$C_{14}H_8Br_2F_3NO_2$
439.02
O=C(Nc(cccc1C(F)(F)F)c1)c(c(O)c(cc2Br)Br)c2
Fluorophene
Benzamide, 3,5-dibromo-2-hydroxy-N-(3-(trifluoromethyl)phenyl)- (9CI)
Caswell No. 290
3,5-Dibromo-2-hydroxy-N-(3-(trifluoromethyl)phenyl)-benzamide
3,5-Dibromo-3'-trifluoromethyl-salicylanilide
3,5-Dibromo-3'-(trifluoromethyl)salicylanilide
3,5-Dibromo-α,α,α-trifluoro-m-salicylotoluidide
EPA Pesticide Chemical Code 027601
Fluorosalan
Flusalan
Flusalanum (Latin)
m-Salicylotoluidide, 3,5-dibromo-α,α,α-trifluoro- (8CI)

4780-79-4
$C_{11}H_{10}O$
158.20
1-Naphthalenemethanol
AI3-05977

4784-14-9
$C_6H_8N_6$
164.17
1,3,5-Triazine-2-propanenitrile, 4,6-diamino-(9CI)
Propioguanamine, β-cyano- s-Triazine-2-propionitrile, 4,6-diamino- (7CI,8CI)

4784-77-4
C_4H_7Br
135.00
BrCC=CC
1-Crotyl bromide
AI3-25728-X
1-Bromo-2-butene
2-Butene, 1-bromo- (9CI)
2-Butene, 1-bromo- (86%)
Crotyl bromide

4786-19-0
C_5H_7N
81.13
CC(=C)CC#N
3-Butenenitrile, 3-methyl
Methallyl cyanide
Methallylkyanid (Czech)
3-Methyl-3-butenonitrile
3-Methyl-3-butennitril (Czech)

4786-20-3
C_4H_5N
67.10
N#CC=CC
Crotononitrile
2-Butenenitrile
Crotonic nitrile
Crotonique nitrile (French)
Crotonitrile
1-Cyanopropene
β-Methylacrylonitrile
1-Propenyl cyanide

4789-76-8
$C_{16}H_{13}N$
219.28
Quinoline, 4-methyl-2-phenyl-(9CI)
Lepidine, 2-phenyl- (8CI)
NSC-42810

4792-15-8
$C_{10}H_{22}O_6$
238.32
O(CCOCCOCCOCCO)CCO
Pentaethylene-glycol
3,6,9,12-Tetraoxatetradocane-1,14-diol

4794-05-2
$C_{12}H_{12}$
156.23
Benzene, 2,5-cyclohexadien-1-yl- (6CI,7CI, 8CI,9CI)
1,4-Dihydrobiphenyl
3-Phenyl-1,4-cyclohexadiene

4798-44-1
$C_6H_{12}O$
100.16
OC(C=C)CCC
1-Hexen-3-ol (9CI)
AI3-28612

4798-45-2
$C_6H_{12}O$
100.16
1-Penten-3-ol, 4-methyl- (8CI, 9CI)
NSC-95412

4798-58-7
$C_6H_{12}O$
100.16
4-Hexen-3-ol (8CI,9CI)
NSC-93799

4799-62-6
$C_6H_{14}O_2$
118.18
1-Pentanol, 5-methoxy-(6CI,7CI,8CI,9CI)
5-Methoxy-1-pentanol

4801-39-2
$C_8H_{10}N_2O.ClH$
186.66
Cl.NCC(=O)Nc1ccccc1
Acetanilide, 2-amino-, monohydrochloride
Acetamide, 2-amino-N-phenyl-, monohydrochloride (9CI)
2-Aminoacetanilide hydro-

chloride
Glycinanilide, hydrochloride

4802-20-4
$C_{10}H_{20}S_2$
204.40
SCC(C(CCC(C1S)C)C1)C
Limonene dimercaptan
Cyclohexaneethanethiol, 3-mercapto-β,4-dimethyl- (9CI)
p-Menthane-2,9-dithiol (8CI)

4803-05-8
$C_4HCl_2N_5$
189.99
s-Triazine, 2,4-dichloro-6-(diazomethyl)- (6CI, 7CI,8CI)

4803-06-9
$C_6H_7N_5O_2$
181.15
1,3,5-Triazine, 2-(diazomethyl)-4,6-dimethoxy- (9CI)
s-Triazine, 2-(diazomethyl)-4,6-dimethoxy- (6CI,7CI, 8CI)

4803-17-2
$C_{15}H_{27}N_3$
249.40
s-Triazine, 2,4,6-tributyl- (7CI,8CI)

4806-61-5
Unknown
Unknown
Ethylcyclobutane

4810-09-7
C_8H_{16}
112.22
1-Heptene, 3-methyl- (8CI,9CI)

4812-20-8
$C_9H_{12}O_2$
152.19
O(c(c(O)ccc1)c1)C(C)C
2-Isopropoxyphenol
Caswell No. 508A
EPA Pesticide Chemical Code 205400
o-Isopropoxyphenol
2-(1-Methylethoxy)phenol
Phenol, 2-(1-methylethoxy)- (9CI)

4812-29-7
$C_{10}H_{20}O_2$
172.27
Octanoic acid, 3,6-dimethyl- (8CI,9CI)

4813-57-4
$C_{21}H_{40}O_2$
324.55
O=C(OCCCCCCCCCCCCCCCCCC)C=C
2-Propenoic acid, octadecyl ester
Acrylic acid, octadecyl ester
AI3-15687
Octadecyl acrylate
n-Octadecyl acrylate
Octadecyl 2-propenoate

4815-57-0
$C_{12}H_{18}$
162.27
Benzene, 1,4-dipropyl- (9CI)
Benzene, p-dipropyl- (8CI)

4819-67-4
$C_{10}H_{18}O$
154.25
O=C(C(CC1)CCCCC)C1
Cyclopentanone, 2-pentyl- (9CI)
2-Pentylcyclopentanone

4821-04-9
$C_{12}H_{20}O_2$
196.29
O=C(OC(CCC(=C1)C)(C1)C(C)C)C
3-Cyclohexen-1-ol, 4-methyl-1-(1-methylethyl)-, acetate (9CI)

4823-47-6
$C_5H_7BrO_2$
179.03
O=C(OCCBr)C=C
Acrylic acid, 2-bromoethyl ester
2-Bromoethyl acrylate
2-Propenoic acid, 2-bromoethyl ester (9CI)

4824-72-0
$C_6HCl_4NO_3$
276.88
Phenol, 2,3,5,6-tetrachloro-4-nitro- (8CI,9CI)
NSC-407822
NSC-57905

4824-78-6
$C_{10}H_{12}BrCl_2O_3PS$
394.06
CCOP(=S)(OCC)Oc1cc(Cl)c(Br)cc1Cl
Phosphorothioic acid, O-(4-bromo-2,5-dichlorophenyl) O,O-diethyl ester
O-(4-Bromo-2,5-dichlorophenyl) O,O-diethyl phosphorothioate
O-(4-Bromo-2,5 dichlorophenyl) O,O-diethylphosphorothionate
Bromofos-Ethyl
Bromophos-Ethyl
Cela S-2225
O,O-Diaethyl-O-(4-brom-2,5-dichlor)-phenyl-monothiophosphat (German)
O,O-Diaethyl-O-(2,5-dichlor-4-bromphenyl)-thionophosphat (German)
O,O-Diethyl-O-(4-broom-2,5-dichloor-fenyl)-monothiofosfaat (Dutch)
O,O-Diethyl O-(2,5-dichloro-4-bromophenyl) thiophosphate
O,O-Diethyl O-2,5-dichloro-4-bromophenyl-phosphorothioate
O,O-Dietil-O-(4-bromo-2,5 dicloro-fenil)-monotiofosfato (Italian)
ENT 27,258
Ethyl bromophos
Filariol
Nexagan
OMS-659
Phenol, 4-bromo-2,5-dichloro-, O-ester with O,O-diethyl phosphorothioate
S 2225
Thiophosphate de O,O-diethyle et de O-(2,5-dichloro-4-bromo) phenyle (French)

4825-86-9
$C_{20}H_{19}NO_6$
369.40
Alanine, N-((8-hydroxy-3-methyl-1-oxo-7-isochromanyl)carbonyl)-3-phenyl-, (-)
Ochratoxin B

4826-62-4
$C_{12}H_{22}O$
182.34
O=CC=CCCCCCCCCC
2-Dodecenal
β-Octyl acrolein

4827-55-8
$C_{25}H_{36}Cl_6O_4$
613.27
O=C(OCC(CCCC)CC)C(C(C(=C(C12Cl)Cl)Cl)(C1(Cl)Cl)Cl)C2C(=O)OCC(CCCC)CC
Bicyclo(2.2.1)hept-5-ene-2,3-dicarboxylic acid, 1,4,5,6,7,7-hexachloro-, bis(2-ethylhexyl) ester (9CI)
Di-2-ethylhexyl chlorendate

4830-99-3
$C_{11}H_{14}$
146.23
1H-Indene, 1-ethyl-2,3-dihydro- (9CI)

Indan, 1-ethyl- (8CI)
NSC-38861

4835-11-4
$C_{14}H_{32}N_2$
228.48
N(CCCCCCNCCCC)CCCC
1,6-Hexanediamine, N,N'-dibutyl
DBHMD
Dibutylhexamethylenediamine
N,N'-Dibutylhexamethylenediamine
N,N'-Dibutyl-1,6-hexanediamine

4835-90-9
$C_5H_{10}O_3$
118.13
Propanoic acid, 3-hydroxy-2,2-dimethyl- (9CI)
3-Hydroxy-2,2-dimethylpropanoic acid
Hydroxypivalic acid

4839-46-7
$C_7H_{12}O_4$
160.17
O=C(O)CC(CC(=O)O)(C)C
3,3-Dimethylglutarate
AI3-62519
3,3-Dimethylpentanedioic acid
Glutaric acid, 3,3-dimethyl- (8CI)
Pentanedioic acid, 3,3-dimethyl- (9CI)

4840-76-0
$C_9H_{14}O_2$
154.21
Cyclohexanecarboxylic acid, ethenyl ester (9CI)
Cyclohexanecarboxylic acid, vinyl ester (8CI)

4844-10-4
$C_{36}H_{42}N_2$
502.73
Hexafluorenium
Ammonium, hexamethylenebis-(fluoren-9-yldimethyl-
1,6-Hexanediaminium,N,N'-di-9H-fluoren-9-yl-N,N,N',N'-tetramethyl

4844-11-5
$C_9H_{12}O$
136.19
Bicyclo(3.3.1)non-2-en-9-one (8CI,9CI)

4849-32-5
$C_{14}H_{21}N_3O_3$
279.38
CN(C)C(=O)Nc1cccc(OC(=O)NC(C)(C)C)c1
Carbamic acid, tert-butyl-, (m-(3,3-dimethyureido)-phenyl) ester
tert-Butylcarbamic acid ester with 3-(m-hydroxyphenyl)-1,1-dimethylurea
CGA 61837
3-(((Dimethylamino)carbonyl)-amino)phenyl 1,1-dimethylethylcarbamate
1,1-Dimethyl-3-(3-(N-tert-butylcarbamyloxy)phenyl)urea
m-(3,3-Dimethylharnstoff)-phenyl-tert-butylcarbamat (German)
m-(3,3-Dimethylureido)phenyl-tert-butyl carbamate
FMC 11092
Karbutilate
NIA 11092
Tandex
Tanzene

4860-03-1
$C_{16}H_{33}Cl$
260.89
ClCCCCCCCCCCCCCCCC
1-Chlorohexadecane
Hexadecane, 1-chloro- (9CI)

4861-19-2
$CH_4N_2O.H_3O_4P$
158.04
Urea, phosphate (1:1)
AI3-17199

4861-58-9
$C_9H_{14}S$
154.28
S(C(=CC=1)CCCCC)C1
Thiophene, 2-pentyl- (9CI)
2-Pentylthiophene

4862-03-7
$C_{15}H_{31}Cl$
246.86
ClCCCCCCCCCCCCCCC
Pentadecane, 1-chloro- (9CI)
1-Chloropentadecane

4865-85-4
$C_{22}H_{22}ClNO_6$
431.87
Ochratoxin C
Alanine, N-((5-chloro-8-hydroxy-3-methyl-1-oxo-7-isochromanyl)carbonyl)-3-phenyl-, ethyl ester, L-
Ochratoxin A ethyl ester
L-Phenylalanine, N-((5-chloro-3,4-dihydro-8-hydroxy-3-methyl-1-oxo-1H-2-benzopyran-7-yl)carbonyl)-, ethyl ester, (R)-

4875-10-9
$C_6H_5NO_2S$
155.17
Benzenethiol, 2-nitro- (9CI)
Benzenethiol, o-nitro- (8CI)

4884-24-6
$C_{10}H_{16}O$
152.24
O=C(C(C(CCC1)C1)CC2)C2
(1,1'-Bicyclopentyl)-2-one (9CI)
2-Cyclopentylcyclopentanone

4885-02-3
$C_2H_4Cl_2O$
114.96
O(C(Cl)Cl)C
Ether, dichloromethyl methyl
α,α-Dichloromethyl ether
α,α-Dichloromethyl methyl ether

4887-30-3
$C_{14}H_{28}O_2$
228.38
Hexanoic acid, octyl ester (8CI,9CI)
NSC-53816

4891-15-0
$C_{23}H_{32}Cl_2NO_6P$
520.43
Estradiol, 3-(bis(2-chloroethyl)carbamate) dihydrogen phosphate
Estracyt
Estramustine phosphate
LEO 299
LS 299

4891-54-7
$C_8H_{19}O_5PS_2$
290.36
CCOP(=S)(OCC)OCCS(=O)(=O)CC
Phosphorothioic acid, O,O-diethyl O-(2-(ethylsulfonyl)ethyl) ester
Demeton-Sulfone
O,O-Diethyl 2-ethylmercaptoethyl thiophosphate, thiono isomer
O,O-Diethyl-O-(2-ethylsulfonylethyl)phosphorothioate
Diethyl 2-ethylsulfonylethyl thionophosphate
Ethanol, 2-(ethylsulfonyl)-, O-ester with O,O-diethyl phosphorothioate
Systox Sulfone
Thionodemeton Sulfone

4897-25-0
$C_5H_4ClN_3O_2$
173.57

Imidazole, 5-chloro-1-methyl-4-nitro
A(S50154-9)
5-Chloro-1-methyl-4-nitroimidazole

4897-31-8
$C_4H_4ClN_3O_2$
161.56
Imidazole, 4-chloro-1-methyl-5-nitro
Chlomizole
4-Chloro-1-methyl-5-nitroimidazole

4901-51-3
$C_6H_2Cl_4O$
231.88
Oc1cc(Cl)c(Cl)c(Cl)c1Cl
Phenol, 2,3,4,5-tetrachloro
2,3,4,5-Tetrachlorophenol

4904-55-6
Unknown
Unknown
Cyclohexanone peroxide and di-(1-hydroxycyclohexyl)-peroxide mixture
Cyclohexanone peroxide (As a paste with not more than 9% by weight active oxygen)
Cyclohexanone peroxide (In solution with not more than 9% by weight active oxygen)
Cyclohexanone peroxide (Not over 50% peroxide)
Cyclohexanone peroxide (50 to 85% peroxide)
UN 2118
UN 2119
UN 2896

4904-61-4
$C_{12}H_{18}$
162.30
C(=CCCC=CCCC=CC1)C1
1,5,9-Cyclododecatriene [UN 2518]
CDT
UN 2518 [1,5,9-Cyclododecatriene]

4906-91-6
$C_{18}H_{30}O_2$
278.44
8,11,14-Octadecatrienoic acid, (Z,Z,Z)- (8CI,9CI)

4911-70-0
$C_7H_{16}O$
116.20
2-Pentanol, 2,3-dimethyl- (9CI)

4914-30-1
$C_{29}H_{38}N_2O_4$
478.69
Emetine, 2,3-didehydro
Dehydroemetine
2-Dehydroemetine
2,3-Didehydroemetine
Emetan, 2,3-didehydro-6',7',10,11-tetramethoxy-(9CI)
Mebadin
Ro 1-9334

4916-63-6
C_8H_{12}
108.18
1,3,5-Hexatriene, 2,5-dimethyl- (6CI,7CI,8CI, 9CI)

4920-77-8
$C_7H_7NO_3$
153.13
Phenol, 3-methyl-2-nitro- (9CI)

4920-92-7
$C_{11}H_{14}O_2$
178.23
Propanoic acid, 2,2-dimethyl-, phenyl ester (9CI)
NSC-74550
Pivalic acid, phenyl ester (8CI)

4920-95-0
$C_{16}H_{18}$
210.32
1,1'-Biphenyl, 3,3',4,4'-tetramethyl- (9CI)
Biphenyl, 3,3',4,4'-tetramethyl-(8CI)
3,3',4,4'-Tetramethylbiphenyl

4920-99-4
$C_{11}H_{16}$
148.25
Benzene, 1-ethyl-3-(1-methylethyl)- (9CI)
Cumene, m-ethyl- (8CI)

4923-85-7
$C_{10}H_{14}ClOP$
216.65
Phosphinic chloride, (1,1-dimethylethyl)-phenyl- (9CI)
Phosphinic chloride, tert-butylphenyl- (7CI,8CI)

4938-52-7
$C_7H_{14}O$
114.19
OC(C=C)CCCC
1-Hepten-3-ol (9CI)
AI3-28621

4938-72-1
$C_{12}H_{13}Cl_3O_3$
311.59
2,4,5-T, isobutyl ester
Acetic acid, (2,4,5-trichlorophenoxy)-, isobutyl ester
Acetic acid, (2,4,5-trichlorophenoxy)-, 2-methylpropyl ester
Caswell No. 881S
EPA Pesticide Chemical Code 082062
Isobutyl 2,4,5-T
2,4,5-Trichlorophenoxyacetic acid isobutyl ester

4946-22-9
$C_8H_{11}NS$
153.26
Benzenethiol, p-(dimethylamino)
p-Dimethylaminobenzenethiol
4-Dimethylaminobenzenethiol
USAF PD-101

4948-15-6
$C_{40}H_{26}N_2O_4$
598.65
O=C(N(C(=O)c(c1c(c(c(c(c(c(C(=O)N(C2=O)c(cc(cc3C)C)c3)c4)c2cc5)c56)c4)cc7)c6c8)c8)c(cc(cc9C)C)c9)c17
Anthra(2,1,9-def:6,5,10-d'e'f')-diisoquinoline-1,3,8,10(2H, 9H)-tetrone, 2,9-bis(3,5-dimethylphenyl)- (9CI)
2,9-Bis(3,5-dimethylphenyl)-anthra(2,1,9-def:6,5,10-d',e',f')diisoquinoline-1,3,8,10-(2H,9H)-tetrone

4948-28-1
$C_{10}H_{18}O$
154.25
OC(C(CC1C2)C1(C)C)(C2)C
Bicyclo(3.1.1)heptan-2-ol, 2,6,6-trimethyl-, (1α,2α,5α)-(9CI)
Caswell No. 663L
cis-2-Pinanol
2,6,6-Trimethylbicyclo(3.1.1)-heptan-2-ol, (1α,2α,5α)-

4955-32-2
$C_{15}H_{22}O$
218.34
O=CC(=CCCC(=CCC=C(C=C)C)C)C
2,6,9,11-Dodecatetraenal, 2,6,10-trimethyl- (9CI)
2,6,10-Trimethyl-2,6,9,11-dodecatetraenal

4979-32-2
$C_{19}H_{26}N_2S_2$
346.59
N(c(c(S1)ccc2)c2)=C1SN(C(CCC

C3)C3)C(CCCC4)C4
2-Benzothiazolesulfenamide, N,N-dicyclohexyl
N,N-Dicyklohexylbenzthiazol-sulfenamid (Czech)
Sulfenamid DC

4981-47-9
$C_{11}H_{24}N_2O_3$
232.31
O=C(NCOCCCC)NCOCCCC
Urea, N,N'-bis(butoxymethyl)- (9CI)
N,N'-Bis(butoxymethyl)urea
Dimethylol urea dibutyl ether

4981-66-2
$C_{14}H_{10}O_2$
210.23
9,10-Anthracenediol (9CI)

4982-20-1
C_8H_{12}
108.18
Cyclohexane, 1,4-bis-(methylene)- (9CI)
Cyclohexane, 1,4-di-methylene- (7CI,8CI)
1,4-Dimethylenecyclo-hexane

4984-01-4
$C_{10}H_{20}$
140.27
1-Octene, 3,7-dimethyl- (8CI, 9CI)
NSC-157589

4985-85-7
$C_7H_{18}N_2O_2$
162.27
OCCN(CCCN)CCO
Ethanol, 2,2'-(aminopropyl-imino)
Aminopropyldiethanolamine (DOT)
NA 1760 (DOT)

4986-89-4
$C_{17}H_{20}O_8$
352.34
O=C(OCC(COC(=O)C=C)(COC(=O)C=C)COC(=O)C=C)C=C
2-Propenoic acid, 2,2-bis-(((1-oxo-2-propenyl)oxy)-methyl)-1,3-propanediyl ester (9CI)
Pentaerythritol tetraacrylate

4987-75-1
$C_6H_{13}NO_4$
163.17
O=C(OC(C)C)N(CO)CO
Carbamic acid, bis(hydroxy-methyl)-, 1-methylethyl ester (9CI)
Dimethylol isopropyl carbamate

4991-47-3
$C_{16}H_{32}O_2.1/2Zn$
289.12
Hexadecanoic acid, zinc salt (9CI)
Zinc hexadecanoate

4994-16-5
$C_{12}H_{14}$
158.24
Benzene, 3-cyclohexen-1-yl- (8CI,9CI)
Benzene, (3-cyclohexen-1-yl)-

4998-57-6
$C_6H_9N_3O_2$
155.14
O=C(O)C(N)CC(N=CN1)=C1
DL-Histidine (9CI)

4998-82-7
$C_{16}H_{10}N_4O_6$
354.25
O=N(=O)c(cc(N=Nc(c(c(ccc1)cc2)c1)c2O)c(O)c3N(=O)=O)c3
2-Naphthalenol, 1-((2-hydroxy-3,5-dinitrophenyl)-azo)- (9CI)

5004-48-8
$C_9H_8N_2O$
160.16
1(2H)-Phthalazinone, 4-methyl- (8CI,9CI)
NSC-116342

5009-32-5
$C_9H_{16}O$
140.23
8-Nonen-2-one (8CI,9CI)

5026-74-4
$C_{15}H_{19}NO_4$
277.31
O(C1CN(c(ccc(OCC(O2)C2)c3)c3)CC(O4)C4)C1
Oxiranemethanamine, N-(4-(oxiranylmethoxy)phenyl)-N-(oxiranylmethyl)- (9CI)
Aniline, p-(2,3-epoxypropoxy)-N,N-bis(2,3-epoxypropyl)

5026-76-6
C_8H_{16}
112.22
C(=C)CCCC(C)C
1-Heptene, 6-methyl-
AI3-37712
6-Methyl-1-heptene

5031-74-3
$C_9H_9NO_5S$
243.23
O=C(Oc(c1ccc2S(=O)(=O)CCO)c2)N1
2(3H)-Benzoxazolone, 6-((2-hydroxyethyl)sulfonyl)- (9CI)

5034-77-5
$C_8H_{12}Cl_2N_6O$
279.16
Imidazole-4-carboxamide, 5-(3,3-bis(2-chloroethyl)-1-triazeno)
Bic
5-(3,3-Bis(2-chloroethyl)-1-tri-azeno)imidazole-4-carbox-amide
1H-Imidazole-4-carboxamide, 5-(3,3-bis(2-chloroethyl)-1-triazenyl)- (9CI)
Imidazole Mustard
NCI-C01616
NSC-82196
SRI 2489
TIC Mustard

5035-58-5
Unknown
Unknown
Diphenylstibene 2-ethyl-hexanoate
Biomet 14
Caswell No. 399C
Diphenylstibine 2-ethyl-hexanoate
EPA Pesticide Chemical Code 006202

5039-78-1
$C_9H_{18}NO_2.Cl$
207.69
Ethanaminium, N,N,N-tri-methyl-2-((2-methyl-1-oxo-2-propenyl)oxy)-, chloride (9CI)

5041-09-8
$C_4H_{11}N.ClH$
109.59
1-Propanamine, 2-methyl-, hydrochloride (9CI)
Isobutylamine, hydrochloride (8CI)

5042-54-6
$C_{13}H_{14}N_4$
226.26
N(=Nc(cccc1)c1)c(c(N)cc(N)c2C)c2
1,3-Benzenediamine, 4-methyl-6-(phenylazo)- (9CI)
4-Methyl-6-(phenylazo)-1,3-benzenediamine

5042-55-7

$C_6H_7N_3O_2$
153.12
O=N(=O)c(cc(N)cc1N)c1
5-Nitro-1,3-benzenediamine
1,3-Benzenediamine, 5-nitro- (9CI)
m-Phenylenediamine, 5-nitro-

5045-23-8
$C_{20}H_{15}N_3O_{10}S_3.3Na$
622.50
2,7-Naphthalenedisulfonic acid, 5-amino-4-hydroxy-3-((1-sulfo-2-naphthalenyl)-azo)-, trisodium salt (9CI)

5045-40-9
$C_{23}H_8Cl_8N_4O_2$
655.94
O=C(NC(=Nc(c(c(N=C(NC(=O)c1c(c(c(c2Cl)Cl)Cl)Cl)c12)cc3)C)c3)c4c(c(c(c5Cl)Cl)Cl)Cl)c45
1H-Isoindol-1-one, 3,3'-((2-methyl-1,3-phenylene)di-imino)bis(4,5,6,7-tetrachloro- (9CI)
1-Methyl-2,6-phenylenediamine-bis(4,5,6,7-tetrachloroisoindo-lin-1-one-3-ylidene)

5051-62-7
$C_8H_8Cl_2N_4$
231.06
NC(=N)NN=Cc1c(Cl)cccc1Cl
Guanabenz
((2,6-Dichlorobenzylidene)-amino)guanidine
Guanabenzo (Spanish)
Guanabenzum (Latin)
Guanidine, ((2,6-dichlorobenz-ylidene)amino)- (8CI)
Hydrazinecarboximidamide, 2-((2,6-dichlorophenyl)methyl-ene)- (9CI)
NSC-68982
Wy-8678

5057-96-5
$C_6H_{10}O_6$
178.14
Butanedioic acid, 2,3-di hydroxy-, dimethyl ester, (R*,S*)- (9CI)
Dimethyl meso-tartrate
Tartaric acid, dimethyl ester, meso- (8CI)

5062-67-9
$C_{17}H_{16}N_2$
248.35
Cyclohexylamine, 1-amino-methyl
Amino-1 aminomethyl-1 cyclo-hexane
1-Aminomethylcyclohexylamine

5064-31-3
$C_6H_6NO_6.3Na$
257.10
[Na+].[Na+].[Na+].[O-]C(=O)CN(CC([O-])=O)CC([O-])=O
Glycine, N,N-bis(carboxy-methyl)-, trisodium salt
Acetic acid, nitrilotri-, trisodium salt
Cheelox NTA-14, -Na3
Hampshire NTA
Nitrilotriacetic acid, trisodium salt
NTA
Sodium nitrilotriacetate
Trisodium nitrilotriacetate
Trisodium nitrilotriacetic acid

5075-13-8
$C_8H_{10}ClOPS$
220.66
O(P(=S)(c(cccc1)c1)Cl)CC
Phosphonochloridothioic acid, phenyl-, O-ethyl ester
Ethyl p-phenylphosphonochlor-idothioate
Phenylphosphonochloridothioic acid O-ethyl ester

5076-19-7
$C_5H_{10}O$
86.15
CC1OC1(C)C
Butane, 2,3-epoxy-2-methyl
β-Isoamylene oxide
Oxirane, trimethyl- (9CI)
Trimethylethylene oxide
Trimethyloxacyclopropane
Trimethyloxirane
2,2,3-Trimethyloxirane

5076-20-0
$C_6H_{12}O$
100.16
O(C1(C)C)C1(C)C
Butane, 2,3-epoxy-2,3-di-methyl- (8CI)
2,3-Epoxy-2,3-dimethylbutane
Oxirane, tetramethyl- (9CI)
Tetramethyloxirane

5089-76-9
$C_{16}H_{36}O_4Si$
320.54
Silicic acid, tetrakis(1-methyl-propyl) ester (9CI)

5089-96-3
C_3H_9Sn
163.81
Stannylium, trimethyl- (9CI)
Tin(1+), trimethyl-, ion (8CI)

5090-41-5
$C_{18}H_{34}O$
266.47
9-Octadecenal (8CI,9CI)

5097-51-8
$C_{10}H_7Cl_2N_5O$
284.11
Urea, 1-(4,6-dichloro-s-tri-azin-2-yl)-3-phenyl- (7CI, 8CI)

5097-52-9
$C_6H_7Cl_2N_5O$
236.06
Urea, 3-(4,6-dichloro-s-tri-azin-2-yl)-1,1-dimethyl- (7CI,8CI)

5097-54-1
$C_7H_9Cl_2N_5O$
250.09
Urea, 1-(4,6-dichloro-s-tri-azin-2-yl)-3-isopropyl- (7CI,8CI)

5097-56-3
$C_7H_9Cl_2N_5S$
266.15
Urea, 1-(4,6-dichloro-s-tri-azin-2-yl)-3-isopropyl-2-thio- (7CI,8CI)

5097-58-5
$C_5H_4Cl_2N_4OS$
239.08
Carbamothioic acid, (4,6-dichloro-1,3,5-tri-azin-2-yl)-, O-methyl ester (9CI)
s-Triazine-2-carbamic acid, 4,6-dichlorothio-, O-methyl ester (7CI,8CI)

5099-06-9
$C_{13}H_9ClN_2O_4$
292.67
O=C(Nc(c(O)cc(N(=O)=O)c1Cl)c1)c(cccc2)c2
Benzamide, N-(5-chloro-2-hydroxy-4-nitrophenyl)- (9CI)
2-Benzamido-4-chloro-5-nitro-phenol
Benzanilide, 5'-chloro-2'-hydroxy-4'-nitro-
N-(5-Chloro-2-hydroxy-4-nitro-phenyl)benzamide

5102-83-0
$C_{36}H_{34}Cl_2N_6O_4$
685.57
O=C(Nc(c(cc(c1)C)C)c1)C(N=Nc(c(cc(c(ccc(N=NC(C(=O)C)C(=O)Nc(c(cc(c2)C)C)c2)c3Cl)c3)c4)Cl)c4)C(=O)C
C.I. Pigment Yellow 13 (VAN) (8CI)

Butanamide, 2,2'-((3,3'-dichloro(1,1'-biphenyl)-4,4'-diyl)bis(azo))bis(N-(2,4-dimethylphenyl)-3-oxo- (9CI)
C.I. 21100
2,2'-((3,3'-Dichloro(1,1'-biphenyl)-4,4'-diyl)bis(azo))-bis(N-(2,4-dimethylphenyl)-3-oxobutanamide)

5103-71-9
$C_{10}H_6Cl_8$
409.76
4,7-Methanoindan, 1-α,2-α, 4-β,5,6,7-β,8,8-octachloro-3a-α,4,7,7a-α- tetrahydro
α-Chlordan
cis-Chlordan
α-Chlordane
α(cis)-Chlordane
cis-Chlordane

5103-73-1
$C_{10}H_5Cl_9$
444.22
4,7-Methano-1H-indene, 1,2,3,4,5,6,7,8,8-nonachloro-2,3,3a,4,7,7a-hexahydro-, (1α,2α,3α,3aα,4β,7β,7aα)-(9CI)
4,7-Methanoindan, 1α,2α,3α,4β, 5,6,7β,8,8-nonachloro-3aα,4,7,7aα-tetrahydro- (8CI)
Nonachlor, cis-

5103-74-2
$C_{10}H_6Cl_8$
409.76
4,7-Methanoindan, 3a-β,4, 7,7a-β-tetrahydro-1-β,2-α, 4-α,5,6,7-α,8,8- octachloro
β-Chlordan
trans-Chlordan
β-Chlordane
trans-Chlordane

5116-94-9
$C_{13}H_{29}O_4P$
280.34
O=P(OCCCCCCCCCCCCC)(O)O
1-Tridecanol, dihydrogen phosphate (9CI)
Monotridecyl phosphate
Tridecyl acid phosphate

5116-95-0
$C_{26}H_{55}O_4P$
462.69
O=P(OCCCCCCCCCCCCC)(OCCCCCCCCCCCCC)O
1-Tridecanol, hydrogen phosphate (9CI)
Ditridecyl acid phosphate

5122-28-1
$C_{11}H_{11}ClN_4$
234.69
s-Triazine, 2-(benzylamino)-4-chloro-6-methyl- (7CI,8CI)

5123-63-7
$C_{10}H_{14}NO_3S.Na$
251.30
[Na+].CCN(CC)c1cccc(c1)S([O-])(=O)=O
Benzenesulfonic acid, 3-(diethylamino)-, sodium salt
3-(Diethylamino)benzenesulfonic acid sodium salt
N,N-Diethylmetanilan sodny (Czech)

5124-25-4
$C_{18}H_{15}N_3O_4S$
369.38
O=S(=O)(Nc(cccc1)c1)c(ccc(Nc(cccc2)c2)c3N(=O)=O)c3
Benzenesulfonamide, 3-nitro-N-phenyl-4-(phenylamino)-(9CI)
AI3-28269
Sulfanilanilide, 3-nitro-N4-phenyl- (8CI)

5124-30-1
$C_{15}H_{22}N_2O_2$
262.39
O=C=NC(CCC(C1)CC(CCC(N=C=O)C2)C2)C1
Isocyanic acid, methylenedi-4,1-cyclohexylene ester
Bis(4-isocyanatocyclohexyl)-methane
Methylene bis(4-cyclohexylisocyanate) (ACGIH,OSHA)
Methylene bis-(4-cyclohexyl-isocyanate)
Nacconate H 12

5128-10-9
$C_7H_{14}O_2.1/2Cu$
161.96
Heptanoic acid, copper(2+) salt (9CI)
Copper heptanoate
Copper(2+) heptanoate
Heptanoic acid, cupric salt

5129-58-8
$C_{15}H_{30}O_2$
242.40
Tridecanoic acid, 12-methyl-, methyl ester (8CI,9CI)

5129-60-2
$C_{17}H_{34}O_2$
270.46
Pentadecanoic acid, 14-methyl-, methyl ester (8CI,9CI)

5129-66-8
$C_{16}H_{32}O_2$
256.43
Tetradecanoic acid, 12-methyl-, methyl ester (8CI, 9CI)

5131-24-8
$C_{12}H_{14}NO_4PS$
299.30
CCOP(=S)(OCC)N2C(=O)c1ccccc1C2=O
Phosphonothioic acid, phthalimido-, O,O-diethyl ester
O,O-Diaethyl-N-phtalimido-thiophosphat (German)
O,O-Diethyl (1,2-dihydro-1,3-dioxo-2H-isoindol-2-yl)phosphonothioate
O,O-Diethyl phthalimido-phosphonothioate
O,O-Diethyl phthalimidothiophosphate
Ditalimfos
Ditalimphos
DOWCO-199
Laptran
M-2452
Phosphonothioic acid, (1,2-dihydro-1,3-dioxo-2H-isoindol-2-yl)-, O,O-diethyl ester
Plondrel

5131-58-8
$C_6H_7N_3O_2$
153.16
O=N(=O)c(c(N)cc(N)c1)c1
m-Phenylenediamine, 4-nitro
1,3-Benzenediamine, 4-nitro-
C.I. 76030
2,4-Diaminonitrobenzene
4-Nitro-1,3-fenylendiamin (Czech)
4-Nitro-1,3-phenylenediamine

5131-60-2
$C_6H_7ClN_2$
142.60
Nc(c(ccc1N)Cl)c1
m-Phenylenediamine, 4-chloro
1,3-Benzenediamine, 4-chloro-(9CI)
C.I. 76027
p-Chlor-m-fenylendiamin (Czech)
4-Chloro-1,3-benzenediamine
1-Chloro-2,4-diaminobenzene
4-Chlorophene-1,3-diamine
p-Chloro-m-phenylenediamine
4-Chloro-m-phenylenediamine
4-Chlorophenylene-1,3-diamine
4-Chloro-1,3-phenylenediamine
4-Cl-m-PD
NCI-C03305

5131-66-8
$C_7H_{16}O_2$

132.23
O(CC(O)C)CCCC
2-Propanol, 1-butoxy
1-Butoxy-2-propanol
Propasol Solvent B
Propylene glycol n-butyl ether

5133-47-1
$C_9H_{17}N_5S$
227.30
1,3,5-Triazine-2(1H)-thione, 4,6-bis((1-methylethyl)-amino)- (9CI)
s-Triazine-2-thiol, 4,6-bis-(isopropylamino)- (8CI)

5136-51-6
Unknown
Unknown
Disodium 2,2'-oxybis(4-dodecylbenzenesulfonate)

5137-55-3
$C_{25}H_{54}N.Cl$
404.25
[Cl-].CCCCCCCC[N+](C)(CCCCCCCC)CCCCCCCC
Ammonium, methyltrioctyl-, chloride
Aliquat 336
Aliquat 336N
Aliquat 336-PTC
Methyltricaprylylammonium chloride
Methyltrioctylammonium chloride
1-Octanaminium, N-methyl-N,N-dioctyl-, chloride (9CI)
Tricaprylmethylammonium chloride
Tricaprylylmethylammonium chloride
Trioctylmethylammonium chloride

5137-70-2
$C_{12}H_{27}O_3P$
250.36
O=P(O)(O)CCCCCCCCCCCC
Phosphonic acid, dodecyl
n-Dodecanephosphonic acid
1-Dodecanephosphonic acid

5138-18-1
$C_4H_6O_7S$
198.15
O=C(O)C(S(=O)(=O)O)CC(=O)O
Butanedioic acid, sulfo- (9CI)
Succinic acid, sulfo- (8CI)

5138-90-9
$C_6H_4ClO_3S.Na$
214.60
[Na+].[O-]S(=O)(=O)c1ccc(Cl)cc1
Benzenesulfonic acid, p-chloro-, sodium salt
p-Chlorbenzensulfonan sodny (Czech)

5141-20-8
$C_{37}H_{36}N_2O_9S_3.2Na$
794.91
[Na+].[Na+].CCN(Cc1cccc(c1)S([O-])(=O)=O)c2ccc(cc2)C(=C3C=CC(C=C3)=[N-](CC)Cc4cccc(c4)S([O-])(=O)=O)c5ccc(cc5)S([O-])(=O
Ammonium, ethyl(4-(p-(ethyl-(m-sulfobenzyl)amino)-α-(p-sulfophenyl)benzylidene)-2,5-cyclohexadien-1-ylidene)-(m-sulfobenzyl)-, hydroxide, inner salt, disodium salt
Acidal Light Green SF
Acid Brilliant Green SF
Acid Green 5
Acid Green A
Acilan Green SFG
A.F. Green No. 2
Amacid Green G
C.I. 670
C.I. 42095
C.I. Acid Green 5
C.I. Acid Green 5, Disodium salt
C.I. Food Green 2
D & C Green No. 4
Fast Acid Green N
FD & C Green No. 2
FD & C Green No. 2-Aluminum Lake
Fenazo Green 7G
Food Green 2
Green No. 203
Leather Green SF
Lichtgruen (German)
Light Green CF
Light Green FCF Yellowish
Light Green FS
Light Green G
Light Green 2GN
Light Green Lake
Light Green S
Light Green SF
Light Green SFA
Light Green SFD
Light Green SF Yellowish
Light Green Yellowish
Light SF Yellowish (Biological stain)
Lissamine Green SF
Lissamine Lake Green SF
Merantine Green SF
MY/68
Pencil Green SF
Sulfo Green J
Sumitomo Light Green SF Yellowish
Wool Brilliant Green SF
Zelen Kysela 5 (Czech)
Zelen Kysela F (Czech)
Zelen Potravinarska 2 (Czech)
Zelen Svetla SF (Czech)

5145-01-7
$C_6H_{10}O_2$
114.14
2(3H)-Furanone, dihydro-3,5-dimethyl- (8CI,9CI)
NSC-250665
Valeric acid, 4-hydroxy-2-methyl-, lactone
Valeric acid, 4-hydroxy-2-methyl-, γ-lactone

5145-99-3
$C_5H_{12}S$
104.22
Propane, 2-(ethylthio)- (9CI)
NSC-163314
Sulfide, ethyl isopropyl (8CI)

5146-66-7
$C_{10}H_{15}N$
149.23
N#CC=C(CCC=C(C)C)C
2,6-Octadienenitrile, 3,7-dimethyl- (9CI)
3,7-Dimethyl-2,6-octadienenitrile

5149-85-9
$C_{10}H_{14}N_3.1/2Cl_4Zn$
279.83
Benzenediazonium, 4-(diethylamino)-, (T-4)-tetrachlorozincate(2-) (2:1) (9CI)
p-Diazo-N,N'-diethylaniline chloride zinc chloride
p-Diazonium N,N-diethylaniline chloride, zinc chloride salt (2:1)
p-Diazonium-N,N-diethylaniline, zinc chloride (2:1)
4-Diethylaminobenzenediazonium chloride zinc chloride (2:1)

5155-70-4
$C_{20}H_{28}O_2$
300.44
1-Phenanthrenecarboxylic acid, 1,2,3,4,4a,9,10,10a-octahydro-1,4a-dimethyl-7-(1-methylethyl)-, (1S-(1α,4aα,10aβ))-(9CI)
Callitrisic acid
Dehydro-4-epiabietic acid
4-Epiabietic acid, dehydro-
Podocarpa-8,11,13-trien-16-oic acid, 13-isopropyl- (8CI)

5160-02-1
$C_{17}H_{12}ClN_2O_4S.1/2Ba$
444.49
[Ba+2].Cc3cc(N=Nc1c([O-])ccc2ccccc12)c(cc3Cl)S([O-])(=O)=O
Benzenesulfonic acid, 5-chloro-2-((2-hydroxy-1-naphthalenyl)azo)-4-methyl-, barium salt
Bright Red

Brilliant Red
Brilliant Scarlet
Brilliant Toner Z
Bronze Red RO
Bronze Red 16913 Yellowish
Bronze Scarlet
Bronze Scarlet C
Bronze Scarlet CA
Bronze Scarlet CBA
Bronze Scarlet CT
Bronze Scarlet CTA
Bronze Scarlet Toner
5-Chloro-2-((2-hydroxy-1-naphthalenyl)azo)-4-methylbenzenesulfonic acid, barium salt (2:1)
5-Chloro-2-((2-hydroxy-1-naphthalenyl)azo)-4-methylbenzenesulphonic acid, barium salt
5-Chloro-2-((2-hydroxy-1-naphthyl)azo)-p-toluenesulfonic acid, barium salt
1-(4-Chloro-o-sulfo-5-tolylazo)-2-naphthol, barium salt
1-(4-Chloro-o-sulpho-5-tolylazo)-2-naphthol, barium salt
C.I. Pigment Red
C.I. Pigment Red 53:1
C.I. Pigment Red 53, Barium salt
Cosmetic Coral Red KO Bluish
Cosmetic DVR
Cosmetic Pigment Yellow Red DVR
Dainichi Lake Red C
D & C Red No. 9
Desert Red
Duplex Red Lake C20-5925
Eljon Lake Red C
Hamilton Red
Helio Red Toner LCLL
Irgalite Red CBN
Irgalite Red CBR
Irgalite Red CBT
Irgalite Red MBC
Isol Lake Red LCS 12527
Isol Red LCR 2517
Lake Red 1520
Lake Red C
Lake Red C 18287
Lake Red C 18958
Lake Red C 21245
Lake Red C 27200
Lake Red C 27217
Lake Red C 27218
Lake Red C Barium Toner
Lake Red CC
Lake Red CCT
Lake Red CR
Lake Red CRLC-232 (Barium)
Lake Red C Toner 8195
Lake Red C Toner 8366
Lake Red GB Barium salt
Lake Red RRG
Lake Red Toner C
Lake Red Toner LCLL
Latexol Scarlet R
Latexol Scarlet R Solupowder
Ld Rubber Red 16913
Lutetia Red CLN
Lutetia Red CLN-ST
Microtex Lake Red CR
Mohican Red A-8008
NCI-C53792
No.3 Conc. Bronze Scarlet
No.3 Conc. Scarlet
Paridine Red LCL
Pigment Lake Red BFC
Pigment Lake Red CD
Pigment Lake Red LC
Pigment Red CD
Potomac Red
Recolite Red Lake C
Recolite Red Lake CR
1860 Red
Red 1860
Red 11938
Red for Lake C
Red for Lake C Toner RA-5190
Red for Lake Toner RA-5190
Red 16913H
Red Lake CM 20-5650
Red Lake CR-1
Red Lake C Toner
Red Lake C Toner 20-5650
Red Lake C Toner RA-5190
Red Lake R-91
Red Scarlet
Red Toner Z
Rubber Red 16913R
Sanyo Lake Red C
Scarlet Toner Y
Segnale Red LC
Segnale Red LCG
Segnale Red LCL
Sico Lake Red 2L
Superol Red C RT-265
Symuler Lake Red C
Termosolido Red LCG
Texan Red Toner D
Toner Lake Red C
Transparent Bronze Scarlet
Vulcafix Scarlet R
Vulcafix Scarlet R-D Masse
Vulcafor Red 2R
Vulcan Red LC
Vulcol Fast Red L
Wayne Red X-2486

5161-04-6
$C_{11}H_{16}$
148.25
Benzene, 1-methyl-4-(2-methylpropyl)- (9CI)
Toluene, p-isobutyl- (8CI)

5161-17-1
$C_6H_{12}S$
116.23
2H-Thiopyran, tetrahydro-4-methyl- (9CI)
Thiopyran, tetrahydro-4-methyl- (8CI)

5162-44-7
C_4H_7Br
135.00
BrCCC=C
1-Butene, 4-bromo- (9CI)
4-Bromo-1-butene

5162-48-1
$C_8H_{18}O$
130.23
3-Pentanol, 2,2,4-trimethyl- (8CI,9CI)

5165-97-9
$C_7H_{13}NO_4S.Na$
230.23
1-Propanesulfonic acid, 2-methyl-2-((1-oxo-2-propenyl)amino)-, monosodium salt (9CI)
2-Acrylamido-2-methylpropanesulfonic acid sodium salt
2-Methyl-2-((1-oxo-2-propenyl)amino)-1-propanesulfonic acid, sodium salt

5166-35-8
C_5H_9Cl
104.59
1-Butene, 3-chloro-2-methyl
3-Chloro-2-methyl-1-butene

5166-53-0
$C_7H_{12}O$
112.19
O=C(C=CC(C)C)C
3-Hexen-2-one, 5-methyl
5-Methyl-3-hexen-2-one

5169-51-7
$C_8H_{14}O_2$
142.20
3-Octenoic acid, (Z)- (9CI)
(Z)-3-Octenoic acid
cis-Oct-3-enoic acid

5180-59-6
$C_{14}H_{22}N_2O_2$
250.34
Benzenamine, 2,6-bis-(1,1-dimethylethyl)-4-nitro- (9CI)
Aniline, 2,6-di-tert-butyl-4-nitro- (6CI,7CI,8CI)
2,6-Di-tert-butyl-4-nitroaniline

5185-97-7
$C_7H_{12}O_3$
144.19
2-Pentanone, 5-hydroxy-, acetate
Acetopropyl acetate
γ-Acetylpropyl acetate
3-Acetylpropyl acetate
5-Hydroxy-2-pentanone acetate
4-Ketovaleryl acetate
4-Oxopentyl acetate
2-Pentanone, 5-(acetyloxy)- (9CI)

5186-68-5
$C_{14}H_{22}$
190.33
Benzene, 1,5-dimethyl-2,4-bis(1-methylethyl)-(9CI)
4,6-Diisopropyl-1,3-dimethylbenzene
4,6-Diisopropyl-m-xylene
1,5-Dimethyl-2,4-bis-(1-methylethyl)benzene
1,3-Dimethyl-4,6-diisopropylbenzene
m-Xylene, 4,6-diisopropyl-(6CI,7CI,8CI)

5188-07-8
$CH_4S.Na$
71.10
Methanethiol, sodium salt (9CI)
Methyl mercaptan sodium salt
Sodium thiomethylate

5194-50-3
C_6H_{10}
82.15
2,4-Hexadiene, (E,Z)- (8CI,9CI)

5194-51-4
C_6H_{10}
82.15
2,4-Hexadiene, (E,E)- (8CI,9CI)

5197-80-8
$C_{11}H_{18}N.Cl$
199.72
[Cl-].CC[N+](C)(C)Cc1ccccc1
Benzenemethanaminium, N-ethyl-N,N-dimethyl-, chloride (9CI)

5204-64-8
$C_5H_8O_2$
100.12
CC=CCC(O)=O
3-Pentenoic acid (8CI,9CI)

5204-80-8
$C_7H_{12}O$
112.17
4-Pentenal, 2-ethyl- (6CI, 7CI,8CI,9CI)
2-Ethyl-4-pentenal

5205-93-6
$C_9H_{18}N_2O$
170.24
O=C(NCCCN(C)C)C(=C)C
2-Propenamide, N-(3-(dimethylamino)propyl)-2-methyl- (9CI)

5208-87-7
$C_{10}H_{10}O_3$
178.20
OC(C=C)c2ccc1OCOc1c2
1,3-Benzodioxole-5-methanol, α-ethenyl
1,3-Benzodioxole-5-(2-propen-1-ol)
1'-Hydroxysafrole
1,2-Methylenedioxy-4-(1-hydroxyallyl)benzene
Piperonyl alcohol, α-vinyl-

5210-79-7
$C_9H_{13}N_5S$
223.30
1,3,5-Triazine-2(1H)-thione, 4,6-bis(2-propenylamino)- (9CI)
s-Triazine-2-thiol, 4,6-bis-(allylamino)- (7CI,8CI)

5213-49-0
$C_3H_2N_4O_2$
126.09
O=N(=O)c1cnc([nH]1)N(=O)=O
Imidazole, 2,4-dinitro
2,4-Dinitroimidazole
2,5-Dinitroimidazole
Imidazole, 2,5-dinitro-

5213-50-3
$C_4H_4N_4O_4$
172.08
1H-Imidazole, 1-methyl-2,4-dinitro- (9CI)
Imidazole, 1-methyl-2,4-dinitro-
NSC-342705

5216-25-1
$C_7H_4Cl_4$
229.91
c(ccc(c1)Cl)(c1)C(Cl)(Cl)Cl
Toluene, α,α,α,p-tetrachloro
Benzene, 1-chloro-4-(trichloromethyl)- (9CI)
p-Chlorobenzotrichloride
4-Chlorobenzotrichloride
p-Chlorophenyltrichloromethane
p,α,α,α-Tetrachlorotoluene
α,α,α,4-Tetrachlorotoluene
p-Trichloromethylchlorobenzene

5217-85-6
$C_5H_8ClN_5O_2$
205.57
1,3,5-Triazine-2,4-diamine, 6-chloro-N,N'-dimethoxy-(9CI)
s-Triazine, 2-chloro-4,6-bis-(methoxyamino)- (8CI)

5221-49-8
$C_{11}H_{20}N_3O_3PS$
305.37
CCOP(=S)(OCC)Oc1cc(C)nc(n1)N(C)C
Phosphorothioic acid, O-(2-(dimethylamino)-6-methyl-4-pyrimidinyl) O,O-diethyl ester
Diothyl
ICI 29661
Pyrimital
Pyrimithate

5224-23-7
$C_6H_{16}Pb$
295.99
CC[PbH](CC)CC
Triethyllead
Lead, triethylhydro- (7CI)
Plumbane, triethyl- (8CI,9CI)
Triethyl lead
Triethyllead hydride (6CI)
Triethylplumbane

5234-68-4
$C_{12}H_{13}NO_2S$
235.32
CC1=C(SCCO1)C(=O)Nc2ccccc2
1,4-Oxathiin-3-carboxamide, 5,6-dihydro-2-methyl-N-phenyl
D 735
Carbathiin
5-Carboxanilido-2,3-dihydro-6-methyl-1,4-oxathiin
Carboxin
Carboxine
DCMO
2,3-Dihydro-5-carboxanilido-6-methyl-1,4-oxathiin
5,6-Dihydro-2-methyl-3-carboxanilido-1,4-oxathiin (German)
2,3-Dihydro-6-methyl-1,4-oxathiin-5-carboxanilide
5,6-Dihydro-2-methyl-1,4-oxathiin-3-carboxanilide
5,6-Dihydro-2-methyl-N-phenyl-1,4-oxathiin-3-carboxamide
DMOC
F 735
Flo Pro V Seed Protectant
1,4-Oxathiin-3-carboxanilide, 5,6-dihydro-2-methyl-
1,4-Oxathiin, 2,3-dihydro-5-carboxanilido-6-methyl-
Vitaflo 250
Vitavax
Vitavax 100

5239-06-5
$C_3H_4N_4$
96.07
1,3-Diazopropane
Propane, 1,3-bis(diazo)-

5244-34-8
$C_6H_{14}O_2S_2$
182.31
OCCSCCSCCO
Ethanol, 2,2'-(1,2-ethanediylbis(thio))bis- (9CI)

3,6-Dithia-1,8-octanediol
2,2'-(1,2-Ethanediylbis(thio))-bisethanol
Ethylenedithioethanol

5246-57-1
$C_8H_{11}NO_3S$
201.24
O=S(=O)(c(cccc1N)c1)CCO
Ethanol, 2-((3-aminophenyl)-sulfonyl)- (9CI)
2-((3-Aminophenyl)sulfonyl)-ethanol

5248-41-9
$C_8H_{14}N_4O$
182.23
1,3,5-Triazin-2-amine, 4-ethoxy-N-ethyl-6-methyl- (9CI)
s-Triazine, 2-ethoxy-4-(ethylamino)-6-methyl- (7CI,8CI)

5248-42-0
$C_{10}H_{18}N_4O$
210.28
1,3,5-Triazin-2-amine, 4-butoxy-N-ethyl-6-methyl- (9CI)
s-Triazine, 2-butoxy-4-(ethylamino)-6-methyl- (7CI,8CI)

5248-48-6
$C_8H_{14}N_4O$
182.26
s-Triazine, 2-ethyl-4-ethylamino-6-methoxy
EEM
2-Ethyl-4-ethylamino-6-methoxy-s-triazine

5248-65-7
$C_4H_4ClN_7$
185.58
s-Triazine, 2-azido-4-chloro-6-(methylamino)- (7CI,8CI)

5248-69-1
$C_5H_6ClN_7$
199.60
s-Triazine, 2-azido-4-chloro-6-(dimethylamino)- (7CI,8CI)

5248-70-4
$C_3H_2ClN_7$
171.55
s-Triazine, 2-amino-4-azido-6-chloro- (7CI, 8CI)

5248-73-7
$C_4H_5N_7O$
167.13
s-Triazine, 2-amino-4-azido-6-methoxy- (7CI, 8CI)

5250-39-5
$C_{19}H_{17}ClFN_3O_5S$
453.86
Cc1onc(c1C(=O)NC3C2SC(C)(C)C(N2C3=O)C(O)=O)c4c(F)cccc4Cl
Floxacillin
BRL 2039
6-(3-(2-Chloro-6-fluorophenyl)-5-methyl-4-isoxazolecarboxamido)-3,3-dimethyl-7-oxo-4-thia-1-azabicyclo(3.2.0)-heptane-2-carboxylic acid
3-(2-Chloro-6-fluorophenyl)-5-methyl-4-isoxazolylpenicillin
Floxapen
Flucloxacilina (Spanish)
Flucloxacillin
Flucloxacilline (French)
Flucloxacillinum (Latin)
4-Thia-1-azabicyclo(3.2.0)-heptane-2-carboxylic acid, 6-(((3-(2-chloro-6-fluorophenyl)-5-methyl-4-isoxazolyl)carbonyl)amino)-3,3-dimethyl-7-oxo-, (2S(2α,5α,6β))

5251-34-3
$C_{21}H_{25}ClO_5$
392.88
Cloprednol
6-Chloro-11β,17,21-trihydroxy-pregna-1,4,6-triene-3,20-dione
6-Chloro-11 β, 17 α, 21-trihydroxypregn-1,4,6-triene-3,20-dione
Cloprednolum (Latin)
Pregna-1,4,6-triene-3,20-dione, 6-chloro-11,17,21-trihydroxy-, (11β)-
RS-4691

5251-93-4
$C_9H_9NO_4$
195.17
OC(=O)CONC(=O)c1ccccc1
Benzadox
Acetic acid, (benzamidooxy)- (8CI)
Acetic acid, ((benzoylamino)-oxy)- (9CI)
Benzamidooxyacetic acid
(Benzamidooxy)acetic acid
Caswell No. 075C
EPA Pesticide Chemical Code 275400
S 7173
Topcide

5254-41-1
$C_{21}H_{20}N_2O_3$
348.43
O=C(Nc(c(ccc1)CCNC(=O)c(ccc(c2ccc3)c3)c2O)c1)C
2-Naphthalenecarboxamide, N-(o-(acetylamino)phenethyl)-1-hydroxy
1-Hydroxy-2-(β-(2'-acetamidophenyl)-ethyl)-naphthamide

5255-75-4
$C_9H_9NO_4$
195.19
O=N(=O)c(ccc(OCC(O1)C1)c2)c2
Propane, 1,2-epoxy-3-(p-nitrophenoxy)
1,2-Epoxy-3(p-nitrophenoxy)-propane
Glycidyl p-nitrophenyl ether
Glycidyl 4-nitrophenyl ether
Nitrophenyl glycidyl ether
p-Nitrophenyl glycidyl ether
Oxirane, ((4-nitrophenoxy)-methyl)- (9CI)

5258-64-0
$C_7H_7NO_5S.Na$
240.19
Benzenesulfonic acid, 2-methyl-5-nitro-, sodium salt (9CI)

5259-88-1
$C_{12}H_{13}NO_4S$
267.32
1,4-Oxathiin-3-carboxanilide, 5,6-dihydro-2-methyl-, 4,4-dioxide
Carboxin sulfone
DCMOD
2,3-Dihydro-5-carboxanilido-6-methyl-1,4-oxathiin, 4,4-dioxide
5,6-Dihydro-2-methyl-3-carboxanilido-1,4-oxathiin-4,4-dioxid (German)
5,6-Dihydro-2-methyl-1,4-oxathiin-3-carboxanilide 4,4-dioxide
5,6-Dihydro-2-methyl-N-phenyl-1,4-oxathiin-3-carboxamide 4,4-dioxide
Dioxide of Vitavax
F 461
F 461 (Pesticide)
1,4-Oxathiin, 2,3-dihydro-5-carboxanilido-6-methyl-, 4,4-dioxide
Oxicarboxin
Oxycarboxin
Oxycarboxine
Plantvax
Plantvax 20
Vitavax sulfone

5261-31-4
$C_{19}H_{17}Cl_2N_5O_4$
450.25

O=C(OCCN(c(ccc(N=Nc(c(cc(N(=O)=O)c1)Cl)c1Cl)c2)c2)CCC#N)C
Propanenitrile, 3-((2-(acetyloxy)ethyl)(4-((2,6-dichloro-4-nitrophenyl)azo)phenyl)amino)- (9CI)
Benzenamine, N-(2-acetoxy)ethyl-N-(2-cyano)ethyl-4-(((2,6-dichloro-4-nitro)phenyl)azo)-
4-(2,6-Dichloro-4-nitrophenylazo)-N-(β-acetoxyethyl)-N-(β-cyanoethyl)aniline
4-((2,6-Dichloro-4-nitrophenyl)azo)-N-(cyanoethyl)-N-(acetoxyethyl)aniline
3-(4-((2,6-Dichloro-4-nitrophenyl)azo)-N-(2-hydroxyethyl)anilino)propionitrile, acetate (ester)

5263-87-6
$C_{10}H_9NO$
159.18
O(c(ccc(nccc1)c12)c2)C
6-Methoxyquinoline
AI3-16316
Quinoline, 6-methoxy- (9CI)

5264-47-1
$C_{16}H_{15}N_5O_4S$
373.36
O=C(N(N=C1C)c(cccc2)c2)C1N=Nc(c(O)ccc3S(=O)(=O)N)c3
Benzenesulfonamide, 3-((4,5-dihydro-3-methyl-5-oxo-1-phenyl-1H-pyrazol-4-yl)azo)-4-hydroxy- (9CI)

5267-27-6
$C_7H_7N_3O_4$
197.17
Cc1cc(N)c(cc1N(=O)=O)N(=O)=O
m-Toluidine, 4,6-dinitro
5-Amino-2,4-dinitrotoluene
4,6-Dinitro-m-toluidine

5274-68-0
$C_{20}H_{42}O_5$
362.55
O(CCOCCOCCOCCO)CCCCCCCCCCCC
Dodecyltetraethylene glycol monoether
Chemal LA-4
Hefti LA-55-4
Laureth-4
Lauryl alcohol tri(oxyethylene) ethanol
PEG-4 Lauryl Ether
Polyethylene Glycol 200 Lauryl Ether
Polyoxyethylene (4) Lauryl Ether
3,6,9,12-Tetraoxatetracosan-1-ol (9CI)
Unicol LA-4
Unihydol LS-4

5279-14-1
$C_9H_{11}NO_2$
165.19
Benzene, (1-nitropropyl)- (7CI,8CI,9CI)
1-Nitro-1-phenylpropane

5280-66-0
Unknown
Unknown
C.I. Pigment Red 48:4 (9CI)
2-Naphthalenecarboxylic acid, 4-((5-chloro-4-methyl-2-sulfophenyl)azo)-3-hydroxy-, manganese(2+) salt (1:1)

5280-68-2
$C_{33}H_{27}ClN_4O_6$
611.03
O=C(Nc(c(OC)cc(c1OC)Cl)c1)c(c(O)c(N=Nc(c(OC)ccc2C(=O)Nc(cccc3)c3)c2)c(c4ccc5)c5)c4
2-Naphthalenecarboxamide, N-(4-chloro-2,5-dimethoxyphenyl)-3-hydroxy-4-((2-methoxy-5-((phenylamino)carbonyl)phenyl)azo)- (9CI)

5280-74-0
$C_{46}H_{28}Cl_4N_6O_4$
870.54
2-Naphthalenecarboxamide, N,N'-(3,3'-dichloro(1,1'-biphenyl)-4,4'-diyl)bis(4-((2-chlorophenyl) azo)-3-hydroxy- (9CI)
4',4'''-Bi-2-naphthanilide, 2',2'''-dichloro-4,4''-bis-((o-chlorophenyl)azo)-3,3''-dihydroxy- (6CI,8CI)
C.I. Pigment Orange 31
Cromophtal Orange 4R
Pigment Orange 31

5280-78-4
$C_{40}H_{23}Cl_5N_6O_4$
828.89
2-Naphthalenecarboxamide, N,N'-(2-chloro-1,4-phenylene)bis(4-((2,5-dichlorophenyl)azo)-3-hydroxy- (9CI)

5280-80-8
$C_{44}H_{38}Cl_4N_8O_6$
916.60
Benzamide, 3,3'-((2,5-dimethyl-1,4-phenylene)bis(imino(1-acetyl-2-oxo-2,1-ethanediyl)azo))bis(4-chloro-N-(5-chloro-2-methylphenyl)- (9CI)

5281-04-9
$C_{18}H_{14}N_2O_6S.Ca$
426.45
D & C Red No. 7
CI 15850:1 (Ca Salt)
3-Hydroxy-4-((4-methyl-2-sulfophenyl)azo)-2-naphthalenecarboxylic acid, calcium salt
3-Hydroxy-4-((2-sulfo-p-tolyl)azo)-2-naphthalenecarboxylic acid, calcium salt (1:1)
Japan: Red No. 202
Lithol Rubin B Ca
2-Naphthalenecarboxylic acid, 3-hydroxy-4-((4-methyl-2-sulfophenyl)azo)-, calcium salt
2-Naphthalenecarboxylic acid, 3-hydroxy-4-((4-methyl-2-sulfophenyl)azo)-, calcium salt (1:1) (9CI)
Pigment Red 57:1
Red No. 202 (Japan)

5281-13-0
$C_{24}H_{40}O_8$
456.64
CCCCOCCOCCOC(OCCOCCOCCCC)c2ccc1OCOc1c2
1,3-Benzodioxole, 6-(bis-(2-(2-butoxyethoxy)ethoxy))-methyl
1-Bis(2-(2-butoxyethoxy)-ethoxy)methyl-3,4-methylene-dioxybenzene
ENT 28,344
Heliotropin acetal
Piperonal bis(2-(2-butoxyethoxy)ethyl)acetal
Piprotal
Tropital

5283-66-9
$C_8H_{17}Cl_3Si$
247.69
Silane, octyltrichloro
Octyl trichlorosilane
Octyltrichlorosilane [UN 1801]
Silane, trichlorooctyl-
UN 1801 [Octyltrichlorosilane]

5283-67-0
$C_9H_{19}Cl_3Si$
261.72
Silane, nonyltrichloro
Nonyl trichlorosilane
Nonyltrichlorosilane [UN 1799]
Silane, trichlorononyl-
UN 1799 [Nonyltrichlorosilane]

5285-60-9
$C_{21}H_{30}N_2$
310.47
N(c(ccc(c1)Cc(ccc(NC(CC)C)c2)c2)c1)C(CC)C
Benzenamine, 4,4'-methylene-

bis(N-(1-methylpropyl)- (9CI)
4,4'-Bis(sec-butylamino)diphenylmethane

5285-87-0
C_7H_5NS
135.19
N#CSc1ccccc1
Thiocyanic acid, phenyl ester
Phenylrhodanid (German)
Phenyl thiocyanate

5292-21-7
$C_8H_{14}O_2$
142.22
O=C(O)CC(CCCC1)C1
Cyclohexaneacetic-acid
Cyclohexylacetic acid

5292-45-5
$C_{10}H_9NO_6$
239.18
O=C(OC)c(ccc(c1N(=O)=O)C(=O)OC)c1
1,4-Benzenedicarboxylic acid, 2-nitro-, dimethyl ester (9CI)

5301-73-5
$C_7H_{17}O_3PS$
212.25
Phosphorothioic acid, S-methyl O,O-dipropyl ester (8CI,9CI)
S-Methyl O,O-dipropylphosphorothioate
Methyl propyl phosphorothioate ((MeS)(PrO)2PO) (7CI)

5302-39-6
$C_{14}H_{13}N_3O_2$
255.26
Acetamide, N-(4-((4-hydroxyphenyl)azo)phenyl)- (9CI)
Acetanilide, 4'-((p-hydroxyphenyl)azo)- (8CI)

5307-02-8
$C_7H_{10}N_2O$
138.19
O(c(c(N)ccc1N)c1)C
p-Phenylenediamine, 2-methoxy
1,4-Benzenediamine, 2-methoxy- (9CI)
2,5-Diaminoanisole
2-Methoxy-p-phenylenediamine

5307-14-2
$C_6H_7N_3O_2$
153.16
O=N(=O)c(c(N)ccc1N)c1
p-Phenylenediamine, 2-nitro
4-Amino-2-nitroaniline
C.I. 76070
C.I. Oxidation Base 22
1,4-Diamino-2-nitrobenzene
Durafur Brown
Durafur Brown 2R
Dye GS
Fouramine 2R
Fourrine 36
Fourrine Brown 2R
NCI-C02222
2NDB
Nitro-p-phenylenediamine
2-Nitro-1,4-benzenediamine
2-Nitro-1,4-diaminobenzene
o-Nitro-p-phenylenediamine
2-Nitro-1,4-phenylenediamine
2-Nitro-p-phenylenediamine
2-NP
2-NPPD
Oxidation Base 22
2-N-p-PDA
Ursol Brown RR
Zoba Brown RR

5309-52-4
$C_8H_{14}O_2$
142.20
2-Ethylhexenoic acid
2-Ethyl-2-hexenoic acid
2-Ethyl-3-propylacrylic acid
2-Hexenoic acid, 2-ethyl-
Kyselina 3-hepten-3-karboxylova (Czech)

5311-21-7
$C_6H_3Cl_6N_3$
329.81
1,3,5-Triazine, 2,4,6-tris-(dichloromethyl)- (9CI)
s-Triazine, 2,4,6-tris(dichloromethyl)- (8CI)

5311-23-9
$C_6H_5Cl_4N_3$
260.94
s-Triazine, 2,4-bis(dichloromethyl)-6-methyl- (6CI,7CI,8CI)

5311-24-0
$C_6H_7Cl_2N_3O_2$
224.05
s-Triazine, 2-(dichloromethyl)-4,6-dimethoxy- (6CI,7CI,8CI)

5311-25-1
$C_6H_6Cl_3N_3O_2$
258.49
s-Triazine, 2,4-dimethoxy-6-(trichloromethyl)- (6CI, 7CI,8CI)

5315-79-7
$C_{16}H_{10}O$
218.25
1-Hydroxypyrene
1-Pyrenol (9CI)

5323-95-5
$C_{18}H_{33}O_3.Na$
320.50
Ricinoleic acid, sodium salt
Sodium ricinoleate

5324-12-9
$C_3H_7Br_2O_4P$
297.89
1-Propanol, 2,3-dibromo-, dihydrogen phosphate
2,3-Dibromo-1-propanol dihydrogen phosphate
2,3-Dibromopropylphosphate
Mono(2,3-dibromopropyl)phosphate
Phosphoric acid, mono(2,3-dibromopropyl) ester
1-Propanol, 2,3-dibromo-, phosphate (1:1)

5324-13-0
$C_6H_3Br_2ClO$
286.35
Phenol, 2,6-dibromo-4-chloro- (9CI)

5324-84-5
$C_8H_{18}O_3S.Na$
217.28
[Na+].CCCCCCCCS([O-])(=O)=O
1-Octanesulfonic acid, sodium salt (9CI)
Sodium 1-octanesulfonate

5325-20-2
C_8H_7NOS
165.21
2H-1,4-Benzothiazin-3(4H)-one (8CI,9CI)
NSC-130

5325-88-2
$C_{10}H_8S_2$
192.30
1,5-Naphthalenedithiol (6CI,7CI,9CI)
1,5-Dimercaptonaphthalene

5325-97-3
$C_{14}H_{18}$
186.30
Phenanthrene, 1,2,3,4,5,6,7,8-octahydro- (8CI,9CI)
NSC-240

5328-01-8
$C_{12}H_{12}O$
172.23

Naphthalene, 1-ethoxy- (9CI)
AI3-00179

5328-04-1
$C_{26}H_{18}O_6.Cu$
489.98
[Cu].OC(=O)c2cccc(c1ccccc1)c2O
Copper(II), bis(3-phenylsalicylato)
(1,1'-Biphenyl)-3-carboxylic acid, 2-hydroxy-, copper(2+) salt (2:1) (9CI)
Copper 3-phenylsalicylate

5329-12-4
$C_6H_5Cl_3N_2$
211.47
N(N)c(c(cc(c1)Cl)Cl)c1Cl
Hydrazine, (2,4,6-trichlorophenyl)- (9CI)
(2,4,6-Trichlorophenyl)hydrazine

5329-14-6
H_3NO_3S
97.10
NS(O)(=O)=O
Sulfamic-acid [UN 2967]
Amidosulfonic acid
Amidosulfuric acid
Aminosulfonic acid
Kyselina amidosulfonova (Czech)
Kyselina sulfaminova (Czech)
Sulfamidic acid
Sulphamic acid
UN 2967 [Sulfamic acid]

5329-79-3
$C_6H_{15}N$
101.19
NC(CCCC)C
Pentylamine, 1-methyl- (8CI)
AI3-16553
2-Hexanamine (9CI)
1-Methylpentylamine

5330-17-6
$C_8H_{15}ClO_3$
194.66
O=C(OCCOCCCC)CCl
Acetic acid, chloro-, 2-butoxyethyl ester (9CI)
AI3-07400
2-Butoxyethyl chloroacetate

5331-08-8
$C_7H_{12}O_2$
128.17
Cyclohexanone, 2-(hydroxymethyl)- (6CI, 7CI,8CI,9CI)
2-(Hydroxymethyl)cyclohexanone
1-Hydroxymethyl-2-cyclohexanone

5331-48-6
$C_5H_{11}NO$
101.14
Acetamide, N-propyl- (8CI,9CI)
NSC-2292

5332-24-1
C_9H_6BrN
208.05
n(c(c(ccc1)cc2Br)c1)c2
Quinoline, 3-bromo- (9CI)
AI3-16560
3-Bromoquinoline

5332-52-5
$C_{11}H_{24}S$
188.38
1-Undecanethiol (8CI,9CI)

5332-73-0
$C_4H_{11}NO$
89.16
O(CCCN)C
Propylamine, 3-methoxy
3-Methoxypropylamine
3-MPA

5333-42-6
$C_{20}H_{42}O$
298.62
OCC(CCCCCCCCCC)CCCCCCCC
1-Dodecanol, 2-octyl
Eutanol G
Octyldodecanol
2-Octyldodecanol
2-Octyldodecyl alcohol

5333-84-6
$C_9H_{10}O_3$
166.18
O=C(OC(=O)C1C(C=CC2)C)C12
1,3-Isobenzofurandione, 3a,4,7,7a-tetrahydro-4-methyl- (9CI)
Maleic anhydride and 1,3-pentadiene adduct

5335-05-7
$C_8H_7ClO_2$
170.60
Methanol, chloro-, benzoate
AI3-17422

5335-24-0
$C_{12}H_8Cl_2O$
239.10
4,6-Dichloro-2-phenylphenol
AI3-18904
2-Biphenylol, 3,5-dichloro- (8CI)
(1,1'-Biphenyl)-2-ol, 3,5-dichloro- (9CI)
Caswell No. 323E
3.5-Dichloro-2-biphenylol
3.5-Dichloro-(1,1'-biphenyl)-2-ol
EPA Pesticide Chemical Code 064216

5336-24-3
$C_9H_{20}N_2O$
172.26
Urea, N,N'-bis(1,1-dimethylethyl)- (9CI)
NSC-1052
Urea, 1,3-di-tert-butyl- (8CI)

5336-94-7
$C_6H_9N_3S_2$
187.29
1,3,5-Triazine, 2-methyl-4,6-bis(methylthio)- (9CI)
s-Triazine, 2-methyl-4,6-bis-(methylthio)- (6CI,8CI)

5337-72-4
$C_8H_{16}O$
128.21
Cyclohexanol, 2,6-dimethyl- (8CI,9CI)
NSC-821

5337-93-9
$C_{10}H_{12}O$
148.20
1-Propanone, 1-(4-methylphenyl)- (9CI)
NSC-852
Propiophenone, 4'-methyl- (8CI)

5339-85-5
$C_8H_{11}NO$
137.18
OCCc(c(N)ccc1)c1
Benzeneethanol, 2-amino- (9CI)
AI3-18009
2-Aminobenzeneethanol

5340-36-3
$C_9H_{20}O$
144.29
OC(CCCCC)(CC)C
3-Octanol, 3-methyl
Amylethylmethylcarbinol
Aprol 161
3-Methyloctan-3-ol
3-Methyl-3-octanol

5341-95-7
$C_4H_{10}O_2$
90.12
2,3-Butanediol, (R*,S*)- (9CI)
AI3-00959
2,3-Butanediol, meso- (8CI)

5343-99-7
$C_{12}H_6O_3$
198.18
Naphtho(1,2-c)furan-1,3-dione (9CI)
1,3-αβ-Isonaphthofurandione
1,2-Naphthalenedicarboxylic anhydride (8CI)
NSC-521

5344-27-4
C_7H_9NO
123.17
n(ccc(c1)CCO)c1
4-Pyridineethanol
4-Ethanolpyridine

5344-44-5
$C_6H_5ClN_2O_2$
172.56
Benzenamine, 3-chloro-5-nitro- (9CI)
Aniline, 3-chloro-5-nitro- (8CI)
NSC-1117

5344-82-1
$C_7H_7ClN_2S$
186.67
N=C(S)Nc(c(ccc1)Cl)c1
Urea, 1-(o-chlorophenyl)-2-thio
2-Chlorophenyl thiourea
RCRA waste number P026

5345-54-0
C_7H_8ClNO
157.61
COc1ccc(N)cc1Cl
p-Anisidine, 3-chloro
Benzenamine, 3-chloro-4-methoxy- (9CI)
3-Chloroanisidine
Orthochloroparanisidine
OCPA

5350-41-4
$C_{10}H_{16}N.Br$
230.18
Ammonium, benzyltrimethyl-, bromide
Benzenemethanaminium, N,N,N-trimethyl-, bromide
Benzyltrimethylammonium bromide(btm)
N-Benzyl-N,N,N-trimethylammonium bromide
Trimethylbenzylammonium bromide
WV 562 (German)

5351-69-9
$C_7H_9N_3S$
167.25
N(c(cccc1)c1)=C(S)NN
Semicarbazide, 4-phenyl-3-thio
4-Phenyl-3-thiosemicarbazide
USAF EK-5426
USAF EL-45

5353-25-3
$C_{20}H_{40}O_2$
312.54
Ethanol, 2-(9-octadecenyloxy)-, (Z)- (8CI,9CI)

5362-50-5
$C_6H_{10}O$
98.14
3-Pentenal, 4-methyl- (8CI,9CI)

5362-56-1
$C_6H_{10}O$
98.14
O=CC=CC(C)C
2-Pentenal, 4-methyl- (9CI)
4-Methyl-2-pentenal
4-Methyl-2-pentene-1-al

5363-63-3
$C_7H_{14}O$
114.19
Pentane, 1-(ethenyloxy)- (9CI)
Ether, pentyl vinyl (6CI, 7CI,8CI)
Pentyl vinyl ether

5363-64-4
$C_8H_{16}O$
128.21
Hexane, 1-(ethenyloxy)- (9CI)
Ether, hexyl vinyl (8CI)
NSC-174082

5379-19-1
$C_{10}H_{14}O$
150.22
Benzenemethanol, α,2,4-trimethyl- (9CI)
Benzyl alcohol, α,2,4-trimethyl- (8CI)
NSC-78938

5379-20-4
$C_{10}H_{12}$
132.22
Styrene, 3,5-dimethyl
Benzene, 1-ethenyl-3,5-dimethyl- (9CI)
3,5-Dimethylstyrene

5385-75-1
$C_{24}H_{14}$
302.38
Dibenz(a,e)aceanthrylene
Dibenzo(a,e)fluoranthene
2,3,5,6-Dibenzofluoranthene

5388-62-5
$C_6H_4ClN_3O_4$
217.55
O=N(=O)c(c(N)c(N(=O)=O)cc1 Cl)c1
2,6-Dinitro-4-chloroaniline
AI3-62729
Aniline, 4-chloro-2,6-dinitro- (8CI)
Benzenamine, 4-chloro-2,6-dinitro- (9CI)
4-Chloro-2,6-dinitroaniline
4-Chloro-2,6-dinitrobenzenamine

5390-61-4
$C_3H_7Cl_2PS_2$
209.10
S=P(SCCC)(Cl)Cl
Phosphorodichloridodithioic acid, propyl ester (9CI)
Propyl phosphorodichloridodithioate

5390-94-3
$C_2H_7NO_4S.Na$
164.13
Sulfamic acid, methoxymethyl-, sodium salt (8CI)
Sodium N,O-dimethylhydroxylamine-N-sulfonate
Sodium methoxymethylsulfamate

5392-40-5
$C_{10}H_{16}O$
152.26
O=CC=C(CCC=C(C)C)C
2,6-Octadienal, 3,7-dimethyl
Citral
3,7-Dimethyl-2,6-octadienal
NCI-C56348

5393-81-7
$C_{10}H_{20}O_3$
188.27
2-Hydroxydecanoate
Decanoic acid, 2-hydroxy- (9CI)

5394-36-5
$C_6H_{10}N_2O_2$
142.18
O=C(NC(C1=O)(CC)C)N1
Hydantoin, 5-ethyl-5-methyl
5-Ethyl-5-methylhydantoin
T11

5394-83-2
$C_{10}H_{16}O_4$
200.23
CC1(C)C(CCC1(C)C(O)=O) C(O)=O
Camphoric acid (8CI)

1,3-Cyclopentanedicarboxylic acid, 1,2,2-trimethyl-, cis- (9CI)

5395-01-7
$C_3H_7NO_3$
105.11
Carbamic acid, 2-hydroxyethyl ester
β-Hydroxyethylcarbamate
2-Hydroxyethylcarbamate

5396-91-8
$C_{17}H_{18}O$
238.33
3-Pentanone, 1,5-diphenyl- (8CI,9CI)
NSC-4397
NSC-10969

5397-01-3
$C_4H_7N_5S$
157.17
1,3,5-Triazine-2,4-diamine, 6-(methylthio)-
AI3-22166

5397-04-6
$C_6H_{11}N_5$
153.22
s-Triazine, 2,4-diamino-6-isopropyl
s-Triazine, 2,4-diamino-6-(2-propyl)

5397-31-9
$C_{11}H_{25}NO$
187.37
O(CC(CCCC)CC)CCCN
Propylamine, 3-((2-ethylhexyl)oxy)
2-Ethylhexyl 3-aminopropyl ether
2-Ethylhexyloxypropylamine
3-(2-Ethylhexyloxy)propylamine

5399-02-0
$C_{17}H_{33}N$
251.45
N#CCCCCCCCCCCCCCCCC
Heptadecanedinitrile
AI3-07618
Heptadecanenitrile (9CI)

5401-94-5
$C_7H_5N_3O_2$
163.15
O=N(=O)c(ccc(NN=C1)c12)c2
1H-Indazole, 5-nitro
5-Nitroindazole
5-Nitro-1H-indazole

5405-53-8
$C_{13}H_8N_2O_4$
256.23
O=N(=O)c3ccc1c(Cc2cc(ccc12)N(=O)=O)c3
Fluorene, 2,7-dinitro
2,7-Dinitrofluorene

5405-58-3
$C_{14}H_{30}O_2$
230.39
O(CCCCCC)C(OCCCCCC)C
Acetaldehyde, dihexyl acetal (8CI)
1,1-Di(hexyloxy)ethane
1,1'-(Ethylidenebis(oxy))bishexane
Hexane, 1,1'-(ethylidenebis(oxy))bis- (9CI)

5407-87-4
$C_7H_{10}N_2$
122.19
n(c(cc(c1)C)C)c1N
Pyridine, 2-amino-4,6-dimethyl
2-Amino-4,6-dimethylpyridine

5408-74-2
$C_9H_{11}N$
133.19
2-Vinyl-5-ethylpyridine
3-Ethyl-6-vinylpyridine
5-Ethyl-2-vinylpyridine
Pyridine, 2-ethenyl-5-ethyl- (9CI)
Pyridine, 5-ethyl-2-vinyl- (8CI)

5408-86-6
$C_4H_8Br_2$
215.92
BrC(C(Br)C)C
2,3-Dibromobutane
Butane, 2,3-dibromo- (9CI)

5409-83-6
$C_{12}H_6Cl_2O$
237.08
Clc1ccc2c(c1)oc3c(Cl)cccc23
2,8-Dichlorodibenzofuran
Dibenzofuran, 2,8-dichloro- (9CI)

5411-22-3
$C_{17}H_{21}N.ClH$
275.82
Benzfetamine
Benzeneethanamine, N,α-dimethyl-N-(phenylmethyl)-, hydrochloride, (+)- (9CI)
Benzphetamine chloride
Benzphetamine hydrochloride
Didrex
Phenethylamine, N-benzyl-N,α-dimethyl-, hydrochloride, (+)- (8CI)

5413-60-5
$C_{12}H_{16}O_2$
192.28
O=C(OC(C(C(C(C12)C=C3)C3)C1)C2)C
4,7-Methanoinden-6-ol, 3a,4,5,6,7,7a-hexahydro-, acetate
Dihydro-nordicyclopentadienyl acetate
4,7-Methano-1H-inden-6-ol, 3a,4,5,6,7,7a-hexahydro-, acetate (9CI)
Tricyclodecen-4-yl 8-acetate
Verdyl acetate

5413-75-2
$C_{22}H_{16}N_4O_7S_2.2Na$
558.48
C.I. Acid Red 73, Disodium salt (8CI)
1,3-Naphthalenedisulfonic acid, 7-hydroxy-8-((4-(phenylazo)-phenyl)azo)-, disodium salt (9CI)

5414-19-7
$C_4H_8Br_2O$
231.91
O(CCBr)CCBr
Ethane, 1,1'-oxybis(2-bromo- (9CI)
AI3-17837
β,β'-Dibromodiethyl ether
Ether, bis(2-bromoethyl) (8CI)
1,1'-Oxybis(2-bromoethane)

5417-42-5
$C_5H_{11}NO_2$
117.15
Acetamide, N-(2-methoxyethyl)- (7CI,8CI,9CI)
N-(2-Methoxyethyl)-acetamide

5417-82-3
$C_7H_8N_2O$
136.14
Propanedinitrile, (1-ethoxyethylidene)- (9CI)
(1-Ethoxyethylidene)propanedinitrile

5418-07-5
$C_{15}H_{12}N_4$
248.29
1,3,5-Triazin-2-amine, 4,6-diphenyl- (9CI)
s-Triazine, 2-amino-4,6-diphenyl- (6CI,7CI, 8CI)

5419-55-6
$C_9H_{21}BO_3$
188.11
CC(C)OB(OC(C)C)OC(C)C

Boric acid, tris(1-methylethyl) ester
Boric acid, triisopropyl ester
Isopropyl borate
Triisopropyl borate [UN 2616]
UN 2616 [Triisopropyl borate]

5421-46-5
$C_2H_3O_2S.H_3N$
108.15
Acetic acid, mercapto-, monoammonium salt
Ammonium mercaptoacetate
Ammonium thioglycolate
Ammonium thioglycollate
Thioglycollic acid, ammonium salt
USAF MO-2

5421-66-9
$C_{21}H_{24}N_8.2ClH$
461.40
C.I. Basic Brown 4, Dihydrochloride (8CI)
AI3-52759
1,3-Benzenediamine, 4,4'-((4-methyl-1,3-phenylene)bis-(azo))bis(6-methyl-, dihydrochloride (9CI)

5422-17-3
$C_{35}H_{24}N_8O_{12}S_2.3Na$
881.68
Benzoic acid, 5-((4'-((8-amino-1-hydroxy-7-((4-nitrophenyl)azo)-3,6-disulfo-2-naphthalenyl)azo)(1,1'-biphenyl)-4-yl)azo)-2-hydroxy-, trisodium salt (9CI)
C.I. Direct Green 8, Trisodium salt (8CI)

5426-78-8
$C_{10}H_{14}O_2$
166.22
O(C(Oc(cccc1)c1)C)CC
Acetaldehyde, ethyl phenyl acetal (8CI)
AI3-21892
Benzene, (1-ethoxyethoxy)-(9CI)
(1-Ethoxyethoxy)benzene

5429-41-4
$C_{14}H_{19}N_3O.C_4H_4O_4$
361.40
1,2-Ethanediamine, N-(2-furanylmethyl)-N',N'-dimethyl-N-2-pyridinyl-, (E)-2-butenedioate (1:1) (9CI)
Forlamin
Methafurylene Fumarate
Pyridine, 2-((2-(dimethylamino)ethyl)furfurylamino)-, fumarate (1:1) (8CI)

5431-33-4
$C_{21}H_{38}O_3$
338.59
CCCCCCCCC=CCCCCCCCC(=O)OCC1CO1
Oleic acid, 2,3-epoxypropyl ester
2,3-Epoxy-1-propanol oleate
2,3-Epoxypropyl ester of oleic acid
2,3-Epoxypropyl oleate
Glycidol oleate
Glycidylester kyseliny olejove (Czech)
Glycidyl oleate
Glycidyl octadecenoate
Oleic acid glycidyl ester
Oxiranylmethyl ester of 9-octadecenoic acid

5432-61-1
$C_{14}H_{29}N$
211.38
N-2-(Ethylhexyl)-cyclohexylamine
Cyclohexanamine, N-(2-ethylhexyl)- (9CI)
Cyclohexylamine, N-(2-ethylhexyl)-
N-(2-Ethylhexyl)cyclohexylamine
Hexylamine, N-cyclohexyl-2-ethyl- (8CI)

5434-27-5
$C_5H_{10}Br_2$
229.94
BrCC(CBr)(C)C
Propane, 1,3-dibromo-2,2-dimethyl- (9CI)
1,3-Dibromo-2,2-dimethylpropane

5434-82-2
$C_{24}H_{21}N_3O_3$
399.43
1,3,5-Triazine, 1,3,5-tribenzoylhexahydro- (9CI)
NSC-15117
s-Triazine, 1,3,5-tribenzoylhexahydro- (8CI)

5437-38-7
$C_8H_7NO_4$
181.14
O=C(O)c(c(N(=O)=O)c(cc1)C)c1
Benzoic acid, 3-methyl-2-nitro- (9CI)
3-Methyl-2-nitrobenzoic acid

5437-45-6
Unknown
Unknown
Benzyl bromoacetate

5438-40-4
$C_{10}H_9ClO_4$
228.63
2-Propenoic acid, 3-(3-chloro-4-hydroxy-5-methoxyphenyl)-(9CI)
Cinnamic acid, 3-chloro-4-hydroxy-5-methoxy- (8CI)
NSC-16684

5441-52-1
$C_8H_{16}O$
128.21
Cyclohexanol, 3,5-dimethyl- (8CI,9CI)
NSC-21130

5444-75-7
$C_{15}H_{22}O_2$
234.34
O=C(OCC(CCCC)CC)c(cccc1)c1
Benzoic acid, 2-ethylhexyl ester (9CI)
2-Ethylhexyl benzoate

5448-22-6
$C_{10}H_{20}O$
156.27
OC(C(C(C)(C)C)CCC1)C1
Cyclohexanol, 2-(1,1-dimethylethyl)-, trans- (9CI)
trans-2-(1,1-Dimethylethyl)-cyclohexanol

5451-63-8
Unknown
Unknown
n-Tetradecylformate

5451-76-3
$C_{13}H_{18}O_3$
222.28
O=C(OCCOCCCC)c(cccc1)c1
Ethanol, 2-butoxy-, benzoate (9CI)
AI3-30512
2-Butoxyethanol benzoate
2-Butoxyethyl benzoate

5452-35-7
$C_7H_{15}N$
113.20
NC(CCCCC1)C1
Cycloheptylamine
AI3-52207
Cycloheptanamine (9CI)

5454-28-4
$C_{11}H_{22}O_2$
186.29
Heptanoic acid, butyl ester (9CI)
AI3-30738

Butyl heptanoate

5454-79-5
$C_7H_{14}O$
114.19
Cyclohexanol, 3-methyl-, cis- (9CI)

5455-34-5
$C_7H_{12}N_2O_2$
156.17
5,5-Diethylhydantoin
Hydantoin, 5,5-diethyl- (8CI)
2,4-Imidazolidinedione, 5,5-diethyl- (9CI)

5455-98-1
$C_{11}H_9NO_3$
203.21
O=C(N(C(=O)c1cccc2)CC(O3)C3)c12
Phthalimide, N-(2,3-epoxypropyl)
N-(2,3-Epoxypropyl)-phthalimide
N-Glycidylphthalimide
1H-Isoindole-1,3(2H)-dione, 2-(oxiranylmethyl)- (9CI)

5456-28-0
$C_{20}H_{40}N_4S_8.Se$
672.08
CCN(CC)C(=S)S[Se](SC(=S)N(CC)CC)(SC(=S)N(CC)CC)SC(=S)N(CC)CC
Selenium, tetrakis(diethyldithiocarbamato)
Ethyl selenac
Ethyl seleram
Selenium diethyldithiocarbamate
Tetrakis(diethylcarbamodithioato-S,S')selenium
Tetrakis(diethyldithiocarbamato)selenium

5458-59-3
$C_{11}H_{22}O_2$
186.29
Octanoic acid, 1-methylethyl ester (9CI)
NSC-23737
Octanoic acid, isopropyl ester (8CI)

5459-85-8
$C_7H_{10}N_2.ClH$
158.62
Toluene-2,4-diamine, monohydrochloride (8CI)
1,3-Benzenediamine, 4-methyl-, monohydrochloride (9CI)
4-Methyl-1,3-benzenediamine monohydrochloride

5459-93-8
$C_8H_{17}N$
127.26
N(C(CCCC1)C1)CC
Cyclohexylamine, N-ethyl
N-Ethyl(cyclohexyl)amine

5460-09-3
$C_{10}H_9NO_7S_2.Na$
342.30
2,7-Naphthalenedisulfonic acid, 4-amino-5-hydroxy-, monosodium salt (9CI)

5460-29-7
$C_{11}H_{10}BrNO_2$
268.10
Phthalimide, N-(3-bromopropyl)- (8CI)
1H-Isoindole-1,3(2H)-dione, 2-(3-bromopropyl)- (9CI)

5462-29-3
$C_{18}H_{10}Cl_2O_2S_2$
393.31
O=C(c(c(S1)c(cc2Cl)C)c2)C1=C(Sc(c3cc(c4)Cl)c4C)C3=O
(δ2,2'(3H,3'H)-Bibenzo(b)thiophene)-3,3'-dione, 5,5'-dichloro-7,7'-dimethyl- (8CI)
Benzo(b)thiophen-3(2H)-one, 5-chloro-2-(5-chloro-7-methyl-3-oxobenzo(b)thien-2(3H)-ylidene)-7-methyl- (9CI)

5463-50-3
$C_{10}H_8O_3$
176.17
1,3-Isobenzofurandione, 4,7-dimethyl- (9CI)
NSC-16057
Phthalic anhydride, 3,6-dimethyl- (8CI)

5464-79-9
$C_8H_8N_2OS$
180.24
COc1cccc2sc(N)nc12
Benzothiazole, 2-amino-4-methoxy
2-Amino-4-methoxybenzothiazole
4-Methoxy-2-aminobenzothiazole

5465-03-2
$C_{12}H_{24}N_6$
252.33
Melamine, N(2),N(4),N(6)-triisopropyl- (8CI)
NSC-26770

5466-06-8
$C_5H_{10}O_2S$
134.20
O=C(OCC)CCS
Propanoic acid, 3-mercapto-, ethyl ester (9CI)
Ethyl 3-mercaptopropanoate
Propionic acid, 3-mercapto-, ethyl ester (8CI)

5466-77-3
$C_{18}H_{26}O_3$
290.40
O=C(OCC(CCCC)CC)C=Cc(ccc(OC)c1)c1
Octyl methoxycinnamate
AI3-05710
2-Ethylhexyl methoxycinnamate
2-Ethylhexyl-4-methoxycinnamate
3-(4-Methoxyphenyl)-2-propenoic acid, 2-ethylhexyl ester
2-Propenoic acid, 3-(4-methoxyphenyl)-, 2-ethylhexyl ester (9CI)

5466-84-2
$C_8H_3NO_5$
193.12
O=C(OC(=O)c1ccc(N(=O)=O)c2)c12
Phthalic anhydride, 4-nitro
1,3-Isobenzofurandione, 5-nitro- (9CI)
4-Nitrophthalic acid anhydride
4-Nitrophthalic anhydride

5468-75-7
$C_{34}H_{30}Cl_2N_6O_4$
657.52
O=C(Nc(c(ccc1)C)c1)C(N=Nc(c(cc(c(ccc(N=NC(C(=O)C)C(=O)Nc(c(ccc2)C)c2)c3Cl)c3)c4)Cl)c4)C(=O)C
Butanamide, 2,2'-((3,3'-dichloro(1,1'-biphenyl)-4,4'-diyl)bis(azo))bis(N-(2-methylphenyl)-3-oxo- (9CI)
C.I. Pigment Yellow 14 (VAN) (8CI)

5470-02-0
$C_8H_{17}N$
127.26
Piperidine, 1-propyl
Propylpiperidine

5470-11-1
$H_3NO.ClH$
69.50
[Cl-].NO
Hydroxylamine, hydrochloride
Hydroxyamine hydrochloride
Hydroxyammonium chloride
Hydroxylamine chloride
Hydroxylamine chloride (1:1)
Hydroxylammonium chloride
Oxammonium hydrochloride

5470-65-5
$C_6H_4BrNO_3$
218.00
Phenol, 3-bromo-4-nitro- (9CI)

5471-08-9
$C_8H_{11}N.1/2H_2O_4S$
170.21
Benzeneethanamine, sulfate (2:1) (9CI)
β-Phenylethylamine sulfate

5471-63-6
$C_{20}H_{14}O$
270.33
1,3-Diphenylisobenzofuran

5473-16-5
$C_7H_4N_2O_6.C_6H_{13}N$
311.28
Benzoic acid, 3,5-dinitro-, Compd. with cyclohexanamine (1:1) (9CI)
Benzoic acid, 3,5-dinitro-, Compd. with cyclohexylamine (1:1) (8CI)
NSC-29737

5490-27-7
$(C_{21}H_{41}N_7O_{12})_2.3H_2O_4S$
1461.33
Dihydrostreptomycin sulfate
AI3-50133
Azimycin
Diathal
Didromycin
Dihydrostreptomycin sulfate
Fermi Strept
Panstreptin
Penicillin-Strep
D-Streptamine, O-2-deoxy-2-(methylamino)-α-L-glucopyranosyl-(1-2)-O-5-deoxy-3-C-(hydroxymethyl)-α-L-lyxofuranosyl-(1-4)-N,N'-bis-(aminoiminomethyl)-, sulfate (2:3) (Salt) (9CI)
Streptomycin, dihydro-, sulfate (2:3) (Salt) (8CI)

5493-45-8
$C_{14}H_{20}O_6$
284.34
O=C(OCC(O1)C1)C(C(C(=O)OCC(O2)C2)CCC3)C3
1,2-Cyclohexanedicarboxylic acid, bis(2,3-epoxypropyl) ester
1,2-Cyclohexanedicarboxylic acid, bis(oxiranylmethyl)ester (9CI)
Diglycidylester kyseliny hexahydroftalove (Czech)
Diglycidyl hexahydrophthalate
Lekutherm 2159
Lekutherm X 100
Phthalic acid, hexahydro-, bis-(2,3-epoxypropyl) ester
Phthalic acid, hexahydro-, diglycidyl ester

5495-84-1
$C_{16}H_{14}OS$
254.35
Oc(c(c(sc1cccc2)ccc3C(C)C)c3)c12
9H-Thioxanthen-9-one, 2-(1-methylethyl)- (9CI)
2-Isopropylthioxanthone
2-(1-Methylethyl)-9H-thioxanthen-9-one

5500-21-0
C_4H_5N
67.08
N#CC(C1)C1
Cyclopropanecarbonitrile (9CI)
AI3-07023
Cyclopropyl cyanide

5502-88-5
$C_{10}H_{18}$
138.25
C(=CCC(C1)C(C)C)(C1)C
Cyclohexene, 1-methyl-4-(1-methylethyl)- (9CI)
AI3-26469
1-Methyl-4-(1-methylethyl)-cyclohexene

5502-94-3
$C_{15}H_{30}O_2$
242.40
Tetradecanoic acid, 12-methyl- (8CI,9CI)
C(15)-Anteiso acid
Anteisopentadecanoic acid

5516-46-1
$C_7H_5Cl_6N_3S$
375.92
1,3,5-Triazine, 2-(ethylthio)-4,6-bis(trichloromethyl)- (9CI)
s-Triazine, 2-(ethylthio)-4,6-bis(trichloromethyl)- (7CI,8CI)

5516-47-2
$C_{11}H_5Cl_6N_3S$
423.96
1,3,5-Triazine, 2-(phenylthio)-4,6-bis(trichloromethyl)- (9CI)
s-Triazine, 2-(phenylthio)-4,6-bis(trichloromethyl)- (7CI,8CI)

5516-48-3
$C_8H_7Cl_6N_3S$
389.95
1,3,5-Triazine, 2-(methylthio)-4,6-bis(1,1,2-trichloroethyl)- (9CI)
s-Triazine, 2-(methylthio)-4,6-bis(1,1,2-trichloroethyl)- (7CI,8CI)

5516-50-7
$C_6H_5Cl_4N_3S$
293.00
s-Triazine, 2,4-bis(dichloromethyl)-6-(methylthio)- (7CI,8CI)

5516-51-8
$C_{12}H_7Cl_6N_3S$
437.99
1,3,5-Triazine, 2-((phenylmethyl)thio)-4,6-bis(trichloromethyl)- (9CI)
s-Triazine, 2-(benzylthio)-4,6-bis(trichloromethyl)- (7CI,8CI)

5521-31-3
$C_{26}H_{14}N_2O_4$
418.40
O=C(N(C(=O)c(c1c(c(c(c(c(C(=O)N(C2=O)C)c3)c2cc4)c45)c3)cc6)c5c7)c7)C)c16
Anthra(2,1,9-def:6,5,10-d'e'f')-diisoquinoline-1,3,8,10(2H,9H)-tetrone, 2,9-dimethyl- (9CI)

5522-43-0
$C_{16}H_9NO_2$
247.26
O=N(=O)c(c(c(c(c(c1)ccc2)c2cc3)c3c4)c1)c4
Pyrene, 1-nitro
1-Nitropyrene
3-Nitropyrene

5524-05-0
$C_{10}H_{16}O$
152.24
Cyclohexanone, 2-methyl-5-(1-methylethenyl)-, (2R-trans)- (9CI)
d-Dihydrocarvone

5532-90-1
$C_{10}H_{13}NO_2$
179.21
CCCOC(=O)Nc1ccccc1
Carbamic acid, phenyl-, propyl ester (9CI)
Carbanilic acid, propyl ester (8CI)
NSC-28524

5534-95-2
$C_{37}H_{49}N_7O_9S$
767.99
Alaninamide, N-carboxy-

β-alanyl-l-tryptophyl-l-methionyl-l-aspartylphenyl-, N-tert-butyl ester, L
AY 6608
ICI 50123
Pentagastrin
Peptavlon
Petavlon
Petogasrin

5535-49-9
C_8H_9ClS
172.68
Benzene, ((2-chloroethyl)thio)- (9CI)

5536-61-8
$C_4H_6O_2.Na$
109.08
Sodium methacrylate
AI3-52402
Methacrylic acid, sodium salt
2-Methyl-2-propenoic acid, sodium salt
2-Propenoic acid, 2-methyl-, sodium salt
Sodium 2-methyl-2-propenoate

5538-94-3
$C_{18}H_{40}N.Cl$
305.97
Dioctyl dimethyl ammonium chloride
Caswell No. 392H
N,N-Dimethyl-N-octyl-1-octanaminium chloride
Dioctyldimethylammonium chloride
EPA Pesticide Chemical Code 069166
1-Octanaminium, N,N-dimethyl-N-octyl-, chloride (9CI)

5556-16-1
$C_6H_6O_2S$
142.18
Ethanone, 1-(4-hydroxy-3-thienyl)- (9CI)
Ketone, 4-hydroxy-3-thienyl methyl (7CI,8CI)

5560-72-5
$C_{19}H_{28}N_2$
284.49
CN(C)CCCn2c1CCCCCCc1c3cccc23
5H-Cyclooct(b)indole, 5-(3-(dimethylamino)propyl)-6,7,8,9,10,11-hexahydro
5-(3-(Dimethylamino)propyl)-6,7,8,9,10,11-hexahydro-5H-cyclooct(b)indole
Galatur
Iprindole
Pramindole
Prondol
WY-3263

5566-34-7
$C_{10}H_6Cl_8$
409.76
4,7-Methanoindan, 2,2,4,5,6,7,8,8-octachloro-3a,4,7,7a-tetrahydro
γ-Chlordan
trans-Chlordan
γ(trans)-Chlordane
4,7-Methano-1H-indene, 2,2,4,5,6,7,8,8-octachloro-2,3,3a,4,7,7a-hexahydro- (9CI)
2,2,4,5,6,7,8,8-Octachloro-3a,4,7,7a-tetrahydro-4,7-methanoindan

5567-15-7
$C_{36}H_{32}Cl_4N_6O_8$
818.46
COc4cc(NC(=O)C(N=Nc1ccc(cc1Cl)c3ccc(N=NC(C(C)=O)C(=O)Nc2cc(OC)c(Cl)cc2OC)c(Cl)c3)C(C)=O)c(OC)cc4Cl
Butanamide, 2,2'-((3,3'-dichloro(1,1'-biphenyl)-4,4'-diyl)bis(azo))bis(N-(4-chloro-2,5-dimethoxyphenyl)-3-oxo- (9CI)

5574-34-5
$C_{15}H_{16}O_2$
228.29
1,2-Naphthalenedione, 3,8-dimethyl-5-(1-methylethyl)- (9CI)
Mansonone C
1,2-Naphthoquinone, 5-isopropyl-3,8-dimethyl- (8CI)

5577-35-5
$C_{10}H_7Cl_2N_5S$
300.17
Urea, 1-(4,6-dichloro-s-triazin-2-yl)-3-phenyl-2-thio- (7CI,8CI)

5578-88-1
$C_{12}H_9ClO.Na$
227.65
4-Biphenylol, 2-chloro-, sodium salt (8CI)

5580-57-4
$C_{43}H_{35}Cl_5N_8O_6$
937.02
Benzamide, 3,3'-((2-chloro-5-methyl-1,4-phenylene)bis(imino(1-acetyl-2-oxo-2,1-ethanediyl)azo))bis(4-chloro-N-(3-chloro-2-methylphenyl)- (9CI)

5580-80-3
$C_6H_3F_4N$
165.08
Aniline, 2,3,4,5-tetrafluoro- (8CI)
Benzenamine, 2,3,4,5-tetrafluoro- (9CI)

5582-57-0
$C_9H_{21}AsS_3$
300.38
Arsenotrithious acid, tripropyl ester (9CI)
Propyl thioarsenite (7CI)
Thioarsenious acid (H3AsS3), tripropyl ester (8CI)
Tris(propylthio)arsine

5582-58-1
$C_{10}H_{15}AsS_2$
447.15
Arsonodithious acid, phenyl-, diethyl ester (9CI)
Benzenearsonous acid, dithio-, diethyl ester (7CI, 8CI)
Bis(ethylthio)phenylarsine

5582-82-1
$C_8H_{18}O$
130.23
3-Heptanol, 3-methyl- (9CI)

5585-39-7
$C_{10}H_{15}N$
149.26
N#CC=C(CCC=C(C)C)C
2,6-Octadienenitrile, 3,7-dimethyl-, (E)
(E)-3,7-Dimethyl-2,6-octadienenitrile
Geranonitrile
Geranyl nitrile

5587-89-3
$C_{12}H_{13}I_3N_2O_2$
597.97
CN(C)C=Nc1c(I)cc(I)c(CCC(O)=O)c1I
Hydrocinnamic acid, 3-(((dimethylamino)methylene)amino)-2,4,6-triiodo
Benzenepropanoic acid, 3-(((dimethylamino)methylene)amino)-2,4,6-triiodo- (9CI)
Bilopten
Iopodic acid
Ipodate

5588-33-0
$C_{21}H_{26}N_2OS_2$
386.61
CN1CCCCC1CCN3c2ccccc2Sc4ccc(cc34)S(C)=O
Phenothiazine, 10-(2-(1-methyl-2-piperidyl)ethyl)-2-(methylsulfinyl)

Lidanar
Lidanil
Mesoridazine
10-(2-(1-Methyl-2-piperidyl)-ethyl)-2-methylsulfinyl phenothiazine
NC 123
Serentil
Thioridazien thiomethyl sulfoxide
TPS23

5590-18-1
$C_{22}H_6Cl_8N_4O_2$
641.91
O=C(NC(=Nc(ccc(N=C(NC(=O)c1c(c(c(c2Cl)Cl)Cl)Cl)c12)c3)c3)c4c(c(c(c5Cl)Cl)Cl)Cl)c45
1H-Isoindol-1-one, 3,3'-(1,4-phenylenediimino)bis(4,5,6,7-tetrachloro- (9CI)
1,4-Phenylenediaminebis-(4,5,6,7-tetrachloroisoindolin-1-one-3-ylidene)

5591-45-7
$C_{23}H_{29}N_3O_2S_2$
443.67
CN4CCN(CCC=C2c1ccccc1Sc3cc(cc23)S(=O)(=O)N(C)C)CC4
Thioxanthene-2-sulfonamide, N,N-dimethyl-9-(3-(4-methyl-1-piperazinyl)propylidene)
CP-12,252-1
N,N-Dimethyl-9-(3-(4-methyl-1-piperazinyl)propylidene)thiaxanthene-2-sulfonamide
N,N-Dimethyl-9-(3-(4-methyl-1-piperazinyl)propylidene)-thioxanthene-2-sulfonamide
2-(Dimethylsulfamoyl)-(9-(4-methyl-1-piperazinyl)-propylidene)thioxanthene
Navane
Navaron
Orbinamon
P-4657-B
Thiothixene
Thiothixine
Tiotixene
9H-Thioxanthene-2-sulfonamide, N,N-dimethyl-9-(3-(4-methyl-1-piperanizyl)-propylidene)-, (Z)-

5593-70-4
$C_{16}H_{36}O_4.Ti$
340.42
Titanic acid, tetrabutyl ester
Butyl titanate
Tetrabutyltitanate (Czech)
Tyzor TBT

5598-13-0
$C_7H_7Cl_3NO_3PS$
322.53
COP(=S)(OC)Oc1nc(Cl)c(Cl)cc1Cl
Phosphorothioic acid, O,O-dimethyl O-(3,5,6-trichloro-2-pyridyl) ester
Chloropyriphos-methyl
Chlorpyrifos-methyl
O,O-Dimethyl O-(3,5,6-trichloro-2-pyridyl)phosphorothioate
DOWCO 214
Dursban Methyl
ENT 27,520
M 3196
Methyl chlorpyrifos
Methyl chlorpyriphos
Methyl dursban
Noltran
NSC-60380
OMS-1155
Reldan
Zertell

5598-52-7
$C_7H_7Cl_3NO_4P$
306.47
Phosphoric acid, dimethyl 3,5,6-trichloro-2-pyridyl ester
Chlorpyrifos-methyl oxon
Dimethyl 3,5,6-trichloro-2-pyridyl phosphate
Dimethyl-3,5,6-trichlor-2-pyridylfosfat (Czech)
DOWCO 217
ENT 27,521
Fospirat
Fospirate
Fospirate methyl
NSC-195058
OMS 1168
Torelle

5599-20-2
$C_7H_{12}N_4$
152.20
s-Triazine, 2-amino-4,6-diethyl- (7CI,8CI)

5599-21-3
$C_{17}H_{16}N_4$
276.34
s-Triazine, 2-amino-4,6-di-p-tolyl- (7CI,8CI)

5599-24-6
$C_9Cl_7N_3O$
414.27
1,3,5-Triazine, 2,4-dichloro-6-(pentachlorophenoxy)- (9CI)
s-Triazine, 2,4-dichloro-6-(pentachlorophenoxy)- (8CI)

5600-21-5
$C_5H_6ClN_3$
143.59
Pyrimidine, 2-amino-4-chloro-6-methyl
2-Amino-4-chloro-6-methyl-pyrimidine

5601-60-5
$C_{11}H_{22}O_2$
186.29
Decanoic acid, 8-methyl- (8CI,9CI)

5606-16-6
$C_7H_{14}N_6$
182.19
CCNc1nc(N)nc(NCC)n1
1,3,5-Triazine-2,4,6-triamine, N,N'-diethyl-
AI3-60350

5606-18-8
$C_{15}H_{14}N_6$
278.32
1,3,5-Triazine-2,4,6-triamine, N,N'-diphenyl- (9CI)
Melamine, N2,N4-diphenyl- (6CI,7CI,8CI)

5606-20-2
$C_{11}H_{22}N_6$
238.34
1,3,5-Triazine-2,4,6-triamine, N,N,N',N'-tetraethyl- (9CI)
Melamine, N2,N2,N4,N4-tetraethyl- (6CI,7CI,8CI)

5606-23-5
$C_5H_{10}N_6$
154.14
CCNc1nc(N)nc(N)n1
1,3,5-Triazine-2,4,6-triamine, N-ethyl- (9CI)
Melamine, ethyl- (8CI)
NSC-298103

5606-27-9
$C_9H_{10}N_6$
202.18
Nc2nc(N)nc(Nc1ccccc1)n2
1,3,5-Triazine-2,4,6-triamine, N-phenyl- (9CI)
Melamine, phenyl- (8CI)
NSC-43644

5610-59-3
CNO.Ag
149.88
Silver fulminate
Fulminate d'argent (French)
Fulminato de plata (Spanish)
Fulminic acid, silver(1+) salt
Silver fulminate, Dry

5610-64-0
$C_{40}H_{20}CrN_6O_{14}S_2.H.2Na$
948.72
Acid Black 52
C.I. Acid Black 52 (9CI)
Chromate(3-), bis(3-hydroxy-4-((2-hydroxy-1-naphthalenyl)azo)-7-nitro-1-naphthalenesulfonato(3-))-, disodium hydrogen
3-Hydroxy-4-((2-hydroxy-1-naphthalenyl)azo)-7-nitro-1-naphthalenesulfonic acid, monosodium salt, chromium complex

5610-94-6
$C_{43}H_{22}N_6O_{13}S_3$
926.84
1-Naphthalenesulfonic acid, 6-diazo-5,6-dihydro-5-oxo-, 4-benzoyl-1,2,3-benzenetriyl ester (9CI)

5614-38-0
$C_5H_{10}O$
86.13
CCOC1CC1
Cyclopropane, ethoxy- (9CI)
Ether, cyclopropyl ethyl (8CI)

5617-28-7
$C_{20}H_{20}N_2O_5S$
400.44
2-Anthracenesulfonic acid, 1-amino-4-(cyclohexylamino)-9,10-dihydro-9,10-dioxo- (9CI)
2-Anthraquinonesulfonic acid, 1-amino-4-(cyclohexylamino)-

5617-41-4
$C_{13}H_{26}$
182.35
C(CCCC1)(C1)CCCCCCC
Cyclohexane, heptyl- (9CI)
1-Cyclohexylheptane
Heptane, 1-cyclohexyl- (8CI)
Heptylcyclohexane

5623-32-5
$C_{21}H_{12}O$
280.33
7H-Benzo(de)naphthacen-7-one (8CI,9CI)
9',10'-Benzo-meso-benzanthrone

5623-78-9
C_8H_{14}
110.20
Cyclopentane, 1-propenyl- (9CI)
Cyclopentane, propenyl- (8CI)

5633-20-5
$C_{22}H_{31}NO_3$
357.49
Oxybutynin
Benzeneacetic acid, α-cyclohexyl-α-hydroxy-, 4-(diethylamino)-2-butynyl ester
Cyclohexaneglycolic acid, α-phenyl-,4-(diethylamino)-2-butynyl ester
4-Diethylamino-2-butynyl α-phenylcyclohexaneglycolate
Oxibutinina (Spanish)
Oxybutynine (French)
Oxybutyninum (Latin)

5634-39-9
$C_6H_{11}IO_3$
258.07
1,3-Dioxolane-4-methanol, 2-(1-iodoethyl)
Iodinated glycerol
2-(1-Iodoethyl)-1,3-dioxolane-4-methanol
Iodopropylidene glycerol
NCI-C55469
Organidin

5637-82-1
$C_9H_{12}N_6O_3$
252.20
Acetamide, N,N',N''-1,3,5-triazine-2,4,6-triyltris- (9CI)
Acetamide, N,N',N''-s-triazine-2,4,6-triyltris- (8CI)
s-Triazine, 2,4,6-triacetamido-
1,3,5-Triazine-2,4,6-triamine, N,N',N''-triacetyl-

5637-83-2
C_3N_{12}
204.15
s-Triazine, 2,4,6-triazido
Cyanuric triazide (DOT)
1,3,5-Triazine, 2,4,6-triazido-

5637-84-3
$C_{24}H_{18}N_6O_3$
438.45
Benzamide, N,N',N''-1,3,5-triazine-2,4,6-triyltris- (9CI)
Benzamide, N,N',N''-s-triazine-2,4,6-triyltris- (8CI)
Melamine, N2,N4,N6-tribenzoyl- (7CI)
1,3,5-Triazine-2,4,6-triamine, N,N',N''-tribenzoyl-
s-Triazine, 2,4,6-tribenzamido- (6CI)
2,4,6-Tribenzamido-s-triazine
N2,N4,N6-Tribenzoylmelamine

5638-12-0
$C_{18}H_{36}O_2$
284.48
O=C(O)C(CCCCCCCCCCCCCCC)C
Isostearic acid
Heptadecanoic acid, 2-methyl- (9CI)
2-Methylheptadecanoic acid

5650-10-2
$C_{15}H_{17}N$
211.30
Benzenamine, 4-(1-methylethyl)-N-phenyl- (9CI)
Cumidine, N-phenyl- (8CI)

5650-20-4
$C_{10}H_{22}O_5$
222.28
O(CCOCCOCCOCC)CCO
3,6,9,12-Tetraoxatetradecan-1-ol (9CI)
Tetraethylene glycol, monoethyl ether

5653-21-4
Unknown
Unknown
Methazonic acid
Acide methazoique (French)
Acido metazoico (Spanish)

5656-90-6
$C_{12}H_8O_3$
200.19
O=C(OC(O)c(c1c(ccc2)cc3)c3)c12
1H,3H-Naphtho(1,8-cd)pyran-1-one, 3-hydroxy- (9CI)
3-Hydroxy-1H,3H-naphtho(1,8-cd)pyran-1-one

5659-41-6
$C_5HCl_5O_2$
270.33
2,4-Pentadienoic acid, 2,3,4,5,5-pentachloro- (8CI,9CI)
NSC-41927
NSC-61406
NSC-61938

5660-60-6
$C_{11}H_{12}O_2$
176.22
O(C(OC1)C=Cc(cccc2)c2)C1
1,3-Dioxolane, 2-(2-phenylethenyl)- (9CI)
AI3-22613
1,3-Dioxolane, 2-styryl- (8CI)
2-(2-Phenylethenyl)-1,3-dioxolane
2-Styryl-1,3-dioxolane

5664-17-5
C_9H_{14}

122.21
Cyclohexane, 1,2-propa-dienyl- (9CI)
Cyclohexane, propadienyl- (8CI)
Cyclohexylallene
1-Cyclohexyl-1,2-propa-diene

5666-11-5
$C_{25}H_{32}N_2O_2$
392.53
Levomoramide
DEA No. 9629
(-)-2,2-Diphenyl-3-methyl-4-morpholinobutyrylpyrroli-dine
Levomoramida (Spanish)
Levomoramidum (Latin)
Levoramide
Pyrrolidine, 1-(2,2-diphenyl-3-methyl-4-morpholinobutyr-yl)-, (-)-
Pyrrolidine, 1-(3-methyl-4-(4-morpholinyl)-1-oxo-2,2-di-phenylbutyl)-, (R)- (9CI)
R 898

5669-15-8
$C_{10}H_{10}O_3$
178.19
Oxiranecarboxylic acid, 3-methyl-3-phenyl- (9CI)
Butyric acid, 2,3-epoxy-3-phenyl-
Glycidic acid, 3-methyl-3-phenyl-
Hydrocinnamic acid, α,β-epoxy-β-methyl- (6CI, 7CI,8CI)
3-Methyl-3-phenylglycidic acid
β-Methyl-β-phenylglycidic acid

5669-17-0
$C_{11}H_{14}O_2$
178.25
Hydrocinnamic acid, β-ethyl
Benzenepropanoic acid, β-ethyl- (9CI)
β-Ethylhydrocinnamic acid
3-Phenylvaleric acid

5673-07-4
$C_9H_{12}O_2$
152.19
Benzene, 1,3-dimethoxy-2-methyl- (9CI)
NSC-62674
Toluene, 2,6-dimethoxy- (8CI)

5689-12-3
$C_{13}H_{16}O$
188.27
2H-Inden-2-one, 1,3-di-hydro-1,1,3,3-tetra-methyl- (9CI)
2-Indanone, 1,1,3,3-tetra-methyl- (6CI,7CI,8CI)
1,1,3,3-Tetramethyl-2-indan-one

5692-66-0
$C_{18}H_{22}N_2$
266.39
Diazene, bis(2,4,6-tri-methylphenyl)- (9CI)
Azobenzene, 2,2',4,4', 6,6'-hexamethyl- (7CI, 8CI)
Azomesitylene
2,2',4,4',6,6'-Hexamethyl-azobenzene
Mesitylene, 2,2'-azodi-

5697-00-7
$C_{14}H_{12}O_4$
244.25
O=C(Oc(c(c(c(OC(=O)C)c1)ccc2)c2)c1)C
1,4-Naphthalenediol, diacetate (9CI)

5697-56-3
$C_{34}H_{50}O_7$
570.84
O=C(OC(C(C(C(C(C(C(=O)C=C(C1(CCC2(CCC(C(=O)O)(C3)C)C)C)C23)C1(C4)C)(C5)C)C4)(C)C)C5)CCC(=O)O
Olean-12-en-30-oic acid, 3-β-hydroxy-11-oxo-, hydrogen succinate
Biogastrone
Bioral
Carbenoxolone
3-β-(3-Carboxypropionyloxy)-11-oxo-olean-12-en-30-oic acid

5698-98-6
$C_3H_4O_2.1/2Mg$
84.22
2-Propenoic acid, magnesium salt (9CI)
Magnesium 2-propenoate

5699-58-1
$C_5H_6O_4$
130.10
CC(=O)CC(=O)C(O)=O
Acetylpyruvic acid
Acetopyruvate
2,4-Dioxovalerate

5700-49-2
$C_2H_8N_2.2HI$
315.91
Ethylenediamine dihydro-iodide
EDDHI
EDDI
1,2-Ethanediamine, dihydriodide (9CI)
Ethylenediamine, dihydriodide (8CI)
Hi-Boot
Hi-O-Dide
Hiamine
Hoffman Bonded E-D-D Iodine Compound
Hy-Odide
Hydriodide-Boot
Hydriodide-O-Dide
Hydrodine
Iod-Ethamine
Iomide
Jodethamine
Orgadine
Whitmoyer Ethylene Diamine Dihydriodide

5700-53-8
$C_8H_{18}N_2$
142.23
1,2-Ethanediamine, N-cyclo-hexyl- (9CI)
Cyclohexanamine, N-(2-amino-ethyl)-
Ethylenediamine, N-cyclohexyl- (8CI)

5707-04-0
C_6H_5ClSe
191.52
Benzeneselenenyl chloride (8CI,9CI)

5707-69-7
$C_{10}H_8ClN_3O_2$
237.66
CC1=NOC(=O)C1=NNc2ccccc2Cl
4,5-Isoxazoledione, 3-methyl-, 4-((o-chlorophenyl)hydra-zone)
4-(2-Chlorophenylhydrazone)-3-methyl-5-isoxazolone
4-(2-Chlorophenylhydrazono)-3-methyl-5(4H)-isoxazolone
Drazoxolon
Drazoxolone
Ganocide
3-Methyl-4-((o-chlorophenyl)-hydrazone)-4,5-isoxazoledione
3-Methyl-4-(o-chlorophenyl-hydrazono)-5-isoxazolone
3-Methyl-4,5-isoxazoledione 4-((2-chlorophenyl)hydrazone)
Mil-Col
PP781
Saisan
Sopracol
Sopracol 781

5707-73-3
$C_{10}H_8ClN_3O_2$
237.63
4,5-Isoxazoledione, 3-methyl-,

4-((3-chlorophenyl)hydrazone) (9CI)
4,5-Isoxazoledione, 3-methyl-, 4-((m-chlorophenyl)hydrazone) (8CI)
Metazoxolone

5714-22-7
$F_{10}S_2$
254.12
FS(F)(F)(F)(F)S(F)(F)(F)(F)F
Sulfur-decafluoride
Disulfur decafluoride
Sulfur pentafluoride (ACGIH, OSHA)
TL 70

5716-15-4
$C_4H_{11}NO_2.C_4H_4N_2O_2$
217.26
OCCNCCO.O=c1ccc(=O)[nH][nH]1
3,6-Pyridazinedione, 1,2-dihydro-, Compd. with 2,2'-iminodiethanol (1:1)
Diethanolammonium maleic hydrazide
Ethanol, 2,2'-iminodi-, Compd. with 1,2-dihydro-3,6-pyridazinedione (1:1)
6-Hydroxy-3-(2H)-pyridazinone diethanolamine
Maleic hydrazide diethanolamine salt
Mazide 30
NCI-C54660
Slo-Gro

5729-47-5
$C_5H_6O_2$
98.11
2-Pentenal, 4-oxo
4-Ketopentenal

5736-15-2
$C_{13}H_5Cl_6O_2.Na$
428.89
Monosodium 2,2'-methylenebis(3,4,6-trichlorophenate)
Caswell No. 566A
EPA Pesticide Chemical Code 044902
Hexachlorophene
Isobac 20
2,2'-Methylenebis(3,4,6-trichlorophenol) sodium salt
Monosodium salt of 2,2'-methylene bis(3,4,6-trichlorophenol)
Phenol, 2,2'-methylenebis-(3,4,6-trichloro-, sodium salt
Seribak

5737-13-3
$C_{15}H_8O$
204.23
4H-Cyclopenta(def)phenanthren-4-one (8CI,9CI)
NSC-132541

5739-83-3
$C_7H_{12}O_3$
144.17
O=C(OC)C=CC(O)(C)C
2-Pentenoic acid, 4-hydroxy-4-methyl-, methyl ester, (E)-(9CI)

5742-17-6
$C_{11}H_{15}Cl_2NO_3$
280.15
Isopropylamine 2,4-dichlorophenoxyacetate

5742-19-8
Unknown
Unknown
Diethanolamine 2,4-dichlorophenoxyacetate
Caswell No. 315K
2,4-D-bis(2-hydroxyethyl)-ammonium
2,4-Dichlorophenoxyacetic acid diethanolamine salt
EPA Pesticide Chemical Code 030016

5743-27-1
$C_6H_8O_6.1/2Ca$
196.17
Calcium ascorbate
Ascorbic acid calcium salt
L-Ascorbic acid, calcium salt (2:1) (9CI)
Calci-C

5743-97-5
$C_{14}H_{24}$
192.34
C(C(C(C(C(C1)CCC2)C2)CCC3)(C1)C3
Phenanthrene, tetradecahydro- (9CI)
Tetradecahydrophenanthrene

5744-03-6
$C_{13}H_{22}$
178.32
C1CCC2C(C1)CC3CCCCC23
Fluorene, dodecahydro- (8CI)
1H-Fluorene, dodecahydro-(9CI)

5749-44-0
$C_{20}H_{34}O$
290.49
Kauran-13-ol (8CI,9CI)
1H-2,10a-Ethanophenanthrene, kauran-13-ol deriv.
13-Hydroxystevane

5759-58-0
$C_6H_9N_3S_3$
219.36
s-Triazine, 2,4,6-tris(methylthio)
2,4,6-Tris(methylthio)-s-triazine

5761-97-7
C_7H_8ClOP
174.57
Phosphinic chloride, methylphenyl- (7CI,8CI, 9CI)
Methylphenylphosphinic chloride
Methylphenylphosphinyl chloride

5770-08-1
$C_{24}H_{27}O_4P$
410.45
Phosphoric acid, triphenethyl ester (8CI)

5779-72-6
$C_{10}H_{12}O$
148.20
Benzaldehyde, 2,4,5-trimethyl-(8CI,9CI)
Durylaldehyde

5779-79-3
$C_{20}H_{14}$
254.34
Benz(k)acephenanthrylene, 4,5-dihydro
Acenaphthanthracene
Benz(k)acephenanthrene
4,5-Dihydrobenz(k)acephenanthrylene
3:4-Dimethylene-1:2-benzanthracene

5779-94-2
$C_9H_{10}O$
134.18
Benzaldehyde, 2,5-dimethyl-(8CI,9CI)

5780-07-4
$C_9H_8O_4$
180.16
O=Cc(cc(OCO1)c1c2OC)c2
1,3-Benzodioxole-5-carboxaldehyde, 7-methoxy-
AI3-24290
7-Methoxy-1,3-benzodioxole-5-carboxaldehyde

5787-50-8
$C_{10}H_{14}O.Na$
173.21
Sodium 4-tert-butylphenate
p-tert-Butylphenol sodium salt

Caswell No. 130G
4-(1,1-Dimethylethyl)phenol sodium salt
EPA Pesticide Chemical Code 064115
Phenol, 4-(1,1-dimethylethyl)-, sodium salt (9CI)

5796-89-4
$C_3H_5NO_5$
135.08
Peroxypropionyl nitrate
POPN

5798-43-6
$C_{18}H_{27}N_3O_{18}.C_6H_8BiNO_7.7Na$
1149.44
Nitrilotriacetic acid and its salts
Bismuth, oxo(dihydrogen nitrilotriacetato)-, sodium salt, Compd. with disodium nitrilotriacetate (1:3)
Bismuth sodium triglycollamate
Bistrimate
Nitriloacetic acid bismuth complex sodium salt
NTA, disodium salt, Compd. with oxo(dihydrogen nitriloacetato)bismuth

5798-75-4
$C_9H_9BrO_2$
229.07
O=C(OCC)c(ccc(c1)Br)c1
Benzoic acid, p-bromo-, ethyl ester (8CI)
AI3-11086
Benzoic acid, 4-bromo-, ethyl ester (9CI)
Ethyl 4-bromobenzoate

5798-79-8
C_8H_6BrN
196.06
BrC(C#N)c1ccccc1
Acetonitrile, bromophenyl
Acetic acid, bromophenyl-, nitrile
BBC
BBN
Benzeneacetonitrile, α-bromo- (9CI)
Brombenzyl cyanide
α-Brombenzylkyanid (Czech)
Bromobenzyl cyanide (DOT)
Bromobenzyl cyanide, Liquid [UN 1694]
α-Bromobenzyl cyanide
Bromobenzylnitrile
α-Bromobenzylnitrile
α-Bromophenylacetonitrile
α-Bromo-α-tolunitrile
CA
Camite
UN 1694 [Bromobenzyl cyanides, liquid]

5798-94-7
$C_7H_6BrNO_3$
232.03
Benzamide, 5-bromo-N,2-dihydroxy- (9CI)
5-Bromosalicylhydroxamic acid
Salicylohydroxamic acid, 5-bromo- (8CI)

5802-82-4
$C_{27}H_{54}O_2$
410.72
Hexacosanoic acid, methyl ester (8CI,9CI)

5807-02-3
$C_6H_{10}N_2O$
126.18
N#CCN1CCOCC1
4-Morpholineacetonitrile
Acetonitrile, morpholino-
N-Cyanomethylmorpholine

5809-23-4
$C_{18}H_{19}NO_4$
313.35
O=C(O)c(c(ccc1)C(=O)c(c(O)cc(N(CC)CC)c2)c2)c1
Benzoic acid, 2-(4-(diethylamino)-2-hydroxybenzoyl)- (9CI)

5809-41-6
$C_8H_{15}NO$
141.21
1H-Azepine, 1-acetylhexahydro- (8CI,9CI)
Hexamethylenimine, 1-acetyl-
NSC-54157

5810-11-7
$C_6H_{10}ClNO_2$
163.60
O=C(N(C)C)C(C(=O)C)Cl
Butanamide, 2-chloro-N,N-dimethyl-3-oxo- (9CI)
2-Chloro-N,N-dimethylacetoacetamide
2-Chloro-N,N-dimethyl-3-oxobutanamide

5810-88-8
$C_{16}H_{35}O_2PS_2$
354.60
O(P(OCC(CCCC)CC)(S)=S)CC(CCCC)CC
Phosphorodithioic acid, O,O-bis(2-ethylhexyl) ester
O,O'-Di(2-ethylhexyl) dithiophosphoric acid

5814-85-7
$C_{15}H_{16}$
196.29
Benzene, 1,1'-(1-methyl-1,2-ethanediyl)bis- (9CI)
Bibenzyl, α-methyl- (8CI)
NSC-403881
Propane, 1,2-diphenyl-

5815-11-2
$C_6H_{15}NO_3$
149.18
O(CCN(O)CCOC)C
Ethanamine, N-hydroxy-2-methoxy-N-(2-methoxyethyl)- (9CI)

5819-01-2
$C_{24}H_{50}Se$
417.62
Dodecane, 1,1'-selenobis- (9CI)
Dodecyl selenide
1,1'-Selenobisdodecane

5819-08-9
$C_4H_8Cl_2OS$
175.08
Sulfoxide, bis(2-chloroethyl)
Bis(β-chloroethyl)sulfoxide
H Sulfoxide

5822-97-9
$C_{10}H_{15}HgNO_3$
397.83
Caswell No. 657F
EPA Pesticide Chemical Code 066013
Phenylmercuric monoethanol ammonium acetate

5825-64-9
$C_7H_5Cl_5$
266.38
Bicyclo(2.2.1)hept-2-ene, 1,2,3,4,7-pentachloro- (9CI)
2-Norbornene, 1,2,3,4,7-pentachloro-

5825-87-6
$C_9H_{10}ClNO_2$
199.63
O=C(N)C(Oc(cccc1Cl)c1)C
2-(m-Chlorophenoxy)propionamide
Caswell No. 205A
2-(3-Chlorophenoxy)propanamide
EPA Pesticide Chemical Code 021203
Propanamide, 2-(3-chlorophenoxy)- (9CI)

5826-91-5
$C_{12}H_{21}N_2O_3PS$
304.38
CCCc1nc(C)cc(OP(=S)(OCC)OCC)n1
Phosphorothioic acid,

O,O-diethyl O-(2-propyl-4-methyl-6-pyrimidyl) ester
O,O-Diethyl-O-(2-propyl-4-methylpyrimidinyl-6) phosphorothioate
O,O-Diethyl-O-(2-n-propyl-4-methyl-pyrimidyl-6)phosphorothioate
Diethyl propylmethylpyrimidyl thiophosphate
G-24622
Pirazinon
Pyrazinon
RCRA waste number P040

5827-05-4
$C_9H_{21}O_3PS_3$
304.45
CCS(=O)CSP(=S)(OC(C)C)OC(C)C
Phosphorodithioic acid, S-((ethylsulfinyl)methyl) O,O-diisopropyl ester
Aphidan
O,O-Diisopropyl-S-ethylsulfinylmethyldithiophosphate
O,O-Diisopropyl-S-ethylsulfinylmethyl phosphorodithioate
S-Ethylsulfinylmethyl-O,O-diisopropyldithiofosfat (Czech)
S-(Ethylsulfinyl)methyl O,O-diisopropyl phosphorodithioate
IPSP
Methanethiol, (ethylsulfinyl)-, S-ester with O,O-diisopropylphosphorodithioate
PSP
PSP 204

5835-26-7
$C_{20}H_{30}O_2$
302.46
CC3(CCC1C(=CCC2C1(C)CCCC2(C)C(O)=O)C3)C=C
1-Phenanthrenecarboxylic acid, 7-ethenyl-1,2,3,4,4a,4b,5,6,7,8,10,10a-dodecahydro-1,4a,7-trimethyl-, (1R-(1α,4aβ,4bα,7α,10aα))- (9CI)
Isopimaric acid
Podocarp-7-en-15-oic acid, 13β-methyl-13-vinyl-, (-)-(8CI)

5836-10-2
$C_{17}H_{16}Cl_2O_3$
339.23
CC(C)OC(=O)C(O)(c1ccc(Cl)cc1)c2ccc(Cl)cc2
Benzilic acid, 4,4'-dichloro-, isopropyl ester
Acaralate
Chlormite
Chloropropylate
Chlorpropylat (Czech)
ENT 26,999
G 24,163
Geigy 24163
Gesakur
Isopropyl-4,4'-dichlorobenzilate
Isopropylester kyseliny 4,4'-dichlorbenzilove (Czech)
1-Methylethyl 4-chloro-α-(4-chlorophenyl)-α-hydroxybenzeneacetate
Propyl p,p'-dichlorobenzilate
Rospan
Rospin

5836-29-3
$C_{19}H_{16}O_3$
292.35
Oc3c(C1CCCc2ccccc12)c(=O)oc4ccccc34
Coumarin, 4-hydroxy-3-(1,2,3,4-tetrahydro-1-naphthyl)
Bay 25634
Bay ENE 11183 B
Bayer 25 634
2H-1-Benzopyran-2-one, 4-hydroxy-3-(1,2,3,4-tetrahydro-1-naphthalenyl)- (9CI)
Coumatetralyl
Cumatetralyl (German, Dutch)
Endox
Endrocid
Endrocide
ENE 11183 B
4-Hydroxy-3-(1,2,3,4-tetrahydro-1-naftyl)-cumarine (Dutch)
4-Hydroxy-3-(1,2,3,4-tetrahydro-1-naphthyl)-cumarin
4-Idrossi-3-(1,2,3,4-tetraidro-1-naftil)-cumarina (Italian)
Racumin
Raucumin 57
Rodentin
3-(1,2,3,4-Tetrahydro-1-naphthyl)-4-hydroxycumarin (German)
3-(1,2,3,4-Tetrahydro-1-naphtyl)-4-hydroxycoumarine (French)
3-(α-Tetral)-4-oxycoumarin
3-(α-Tetralyl)-4-hydroxycoumarin
3-(d-Tetralyl)-4-hydroxycoumarin

5847-55-2
$C_{44}H_{88}O_4Sn$
800.01
Stannane, dibutylbis(stearoyloxy)
Dibutylbis(stearoyloxy)stannane
Dibutyltin distearate
Dibutyltin stearate
Stannane, dibutylbis((1-oxooctadecyl)oxy)- (9CI)

5847-57-4
$C_6H_3Cl_2NO_3$
207.99
Phenol, 2,5-dichloro-4-nitro-(8CI,9CI)

5847-59-6
$C_6H_4BrNO_3$
218.00
Phenol, 2-bromo-4-nitro-(9CI)

5848-93-1
$C_3Cl_4N_2S$
237.91
N(C(=NS1)C(Cl)(Cl)Cl)=C1Cl
1,2,4-Thiadiazole, 5-chloro-3-(trichloromethyl)- (9CI)

5850-16-8
$C_{26}H_{18}N_4O_8S_2.2Na$
624.54
C.I. Acid Brown 14, Disodium salt (8CI)
1-Naphthalenesulfonic acid, 4,4'-((2,4-dihydroxy-1,3-phenylene)bis(azo))bis-, disodium salt (9CI)

5850-44-2
$C_{20}H_{14}N_2O_{13}S_4.4Na$
710.56
1,3,6-Naphthalenetrisulfonic acid, 7-hydroxy-8-((4-sulfo-1-naphthalenyl)azo)-, tetrasodium salt
Acid Red 41
Acilan Ponceau 6R
Cerven Kysela 41 (Czech)
Cerven Potravinarska 8 (Czech)
C.I. 16290
C.I. Acid Red 41
C.I. Acid Red 41, Tetrasodium salt (8CI)
C.I. Food Red 8
E 126
Eurocert Ponceau 6R
Hispacid Brilliant Scarlet
l-Red 5
Neklacid Red 6r
Ponceau 6R
Ponceau 6RA
Ponceau Red 6R
Ponceau 6RPA
Ponceau 6R Specially Pure
Scarlet 6R

5850-80-6
$C_{17}H_{11}ClN_2O_6S.1/2Ca.Na$
426.83
Benzoic acid, 2-chloro-5-((2-hydroxy-1-naphthalenyl)-azo)-4-sulfo-, calcium sodium salt (2:1:2) (9CI)

5850-81-7
$C_{16}H_{10}Cl_2N_2O_4S.Na$
420.22
Benzenesulfonic acid, 4,5-dichloro-2-((2-hydroxy-1-naphtha-lenyl)azo)-, monosodium salt (9CI)

C.I. Pigment Orange 7, Monosodium salt (8CI)

5850-86-2
$C_{17}H_{14}N_2O_4S.Na$
365.35
Benzenesulfonic acid, 4-((2-hydroxy-1-naphthalenyl)azo)-3-methyl-, monosodium salt (9CI)
C.I. Acid Orange 8, Monosodium salt (8CI)

5850-90-8
$C_{17}H_{13}ClN_2O_4S.Na$
398.80
Benzenesulfonic acid, 4-chloro-2-((2-hydroxy-1-naphthalenyl)azo)-5-methyl-, monosodium salt (9CI)
C.I. 15595
C.I. Pigment Red 69
C.I. Pigment Red 69, monosodium salt (8CI)
Lithol Red GGS
Monaco Red
Pigment Red 69

5851-14-9
$C_6H_{11}O_2PS_2$
210.26
Phosphorodithioic acid, O,O-di-2-propenyl ester (9CI)
Allyl phosphorodithioate (6CI,7CI)
O,O-Diallyl dithiophosphate
O,O-Diallyl phosphorodithioate

5851-43-4
$C_{10}H_{12}N_2$
160.24
CC(C)c2nc1ccccc1[nH]2
Benzimidazole, 2-isopropyl
2-Isopropylbenzimidazole

5853-29-2
$C_{28}H_{38}N_2O_4.2ClH$
539.60
Cephaeline-hydrochloride
(-)-Cephaeline dihydrochloride
Emetan-6'-ol, 7',10,11-trimethoxy-, dihydrochloride (9CI)

5856-62-2
$C_4H_{11}NO$
89.13
1-Butanol, 2-amino-, (S)- (9CI)
1-Butanol, 2-amino-, (S)-(+)- (8CI)

5858-39-9
$C_{17}H_{14}N_2O_5S.Na$
381.35
C.I. Acid Red 4, Monosodium salt (8CI)
1-Naphthalenesulfonic acid, 4-hydroxy-3-((2-methoxyphenyl)azo)-, monosodium salt (9CI)

5858-81-1
$C_{18}H_{14}N_2O_6S.2Na$
432.38
[Na+].[Na+].Cc3ccc(N=Nc1c(O)c(cc2ccccc12)C([O-])=O)c(c3)S([O-])(=O)=O
2-Naphthalenecarboxylic acid, 3-hydroxy-4-((4-methyl-2-sulfophenyl)azo)-, disodium salt
Cerven Pigment 57 (Czech)
C.I. 15850
C.I. Pigment Red 57 (7CI)
C.I. Pigment Red 57, Disodium salt (8CI)
D & C Red No. 6
Irgalite Red 4B
Irgalite Rubine PB
Isol Bona Rubine BK
Isol Bona Rubine BKS
Isol Bona Rubine KBK
Isol Ruby BK
Isol Ruby BKS
Japan Red 201
Kromon Permanent Red 4B
Lithol Rubine
Lithol Rubin B
Lithol Rubine BNA
Permanent Red 4B
Permanent Red F 6R
Pigment Red 57
Pigment Rubine B
Pigment Rubine BCL
Plastol Rubine BC
11070 Red
Resamine Rubine BC
Rubine Red RR 1253
Segnale Red 3R
Silotermo Carmine G
Siloton Rubine B
Siloton Rubine 2B
Vynamon Claret Y

5858-93-5
$C_{20}H_{14}N_2O_7S_2.2Na$
481.39
Acid Red 25
CI Acid Red 25
C.I. Acid Red 25, Disodium salt
C.I. 16050
Croceine Scarlet 3BX
Croceine 3BX
Cyanine Fast Yellow R New
7-Hydroxy-8-((4-sulfo-1-naphthalenyl)azo)-, disodium salt
Kiton Red A
1-Naphthalenesulfonic acid,

5859-00-7
$C_{16}H_{12}N_2O_7S_2.2Na$
454/38
C.I. Acid Orange 14, Disodium salt (8CI)
2,7-Naphthalenedisulfonic acid, 3-hydroxy-4-(phenylazo)-, disodium salt (9CI)

5859-04-1
$C_{16}H_{11}N_3O_6S$
373.35
2-Naphthalenesulfonic acid, 6-hydroxy-5-((4-nitrophenyl)azo)- (9CI)
2-Naphthalenesulfonic acid, 6-hydroxy-5-((p-nitrophenyl)azo)- (7CI,8CI)
2-Naphthol-6-sulfonic acid, 1-(p-nitrophenylazo)- (6CI)

5859-07-4
$C_{17}H_{14}N_2O_4S$
344.39
2-Naphthalenesulfonic acid, 6-hydroxy-5-((4-methylphenyl)azo)- (9CI)
1-(4'-Methylphenylazo)-2-naphthol-6-sulfonic acid
2-Naphthalenesulfonic acid, 6-hydroxy-5-(p-tolylazo)- (8CI)

5859-11-0
$C_{16}H_{12}N_2O_7S_2$
408.40
2-Naphthalenesulfonic acid, 6-hydroxy-5-((4-sulfophenyl)azo)- (9CI)

5862-38-4
$C_{20}H_{14}N_2O_2$
314.33
Oc(c(c(nc1Cc(c(nc(c2ccc3)c3)C4)c2O)ccc5)c5)c14
Quino(2,3-b)acridine-7,14-dione, 5,6,12,13-tetrahydro- (9CI)
6,13-Dihydroquinacridone
5,6,12,13-Tetrahydro-quino-(2,3-b)acridine-7,14-dione

5863-46-7
$C_{42}H_{47}N_3O_6S_2.Na$
776.96
Benzenemethanaminium, N-(4-((4-(diethylamino)-2-methylphenyl)(4-(ethyl-((3-sulfophenyl)methyl)-amino)phenyl)methylene)-2,5-cyclohexadien-1-ylidene)-N-ethyl-3-sulfo-, hydroxide, inner salt, sodium salt (9CI)

5871-17-0
$C_4H_{11}O_3PS.K$
209.27
Phosphorothioic acid, O,O-diethyl ester, potassium salt (8CI,9CI)

5873-54-1
$C_{15}H_{10}N_2O_2$
250.24
O=C=Nc(c(ccc1)Cc(ccc(N=C=O)c2)c2)c1
Benzene, 1-isocyanato-2-((4-isocyanatophenyl)methyl)- (9CI)

5875-45-6
$C_{14}H_{22}O$
206.33
Oc(c(ccc1C(C)(C)C)C(C)(C)C)c1
Phenol, 2,5-bis(1,1-dimethylethyl)- (9CI)
2,5-Bis(1,1-dimethylethyl)-phenol
Phenol, 2,5-di-tert-butyl- (8CI)

5876-87-9
$C_{12}H_{22}$
166.31
1,11-Dodecadiene (8CI,9CI)

5877-42-9
$C_{10}H_{18}O$
154.25
OC(C#C)C(CCCC)CC
1-Octyn-3-ol, 4-ethyl- (9CI)
4-Ethyl-1-octyn-3-ol

5878-19-3
$C_4H_8O_2$
88.12
O=C(C)COC
2-Propanone, 1-methoxy
Methoxyacetone
1-Methoxyacetone
Methoxymethyl methyl ketone
Methoxy-2-propanone
1-Methoxy-2-propanone

5888-33-5
$C_{13}H_{20}O_2$
208.30
O=C(OC(C(C(C1C2)(C)C)(C2)C)C1)C=C
2-Propenoic acid, 1,7,7-trimethylbicyclo(2.2.1)hept-2-yl ester, exo- (9CI)
Isobornyl acrylate

5892-10-4
CBi_2O_5
509.97
Bismuth subcarbonate
2,4-Dioxa-1,5-dibismapentane, 1,3,5-trioxo- (9CI)
1,3,5-Trioxo-2,4-dioxa-1,5-dibismapentane

5893-66-3
Unknown
Unknown
Cupric oxalate

5894-60-0
$C_{16}H_{33}Cl_3Si$
359.93
Silane, hexadecyltrichloro
Hexadecyltrichlorosilane [UN 1781]
Silane, trichlorohexadecyl-
UN 1781 [Hexadecyltrichlorosilane]

5894-79-1
$C_{10}H_{10}O_3$
178.19
O=CC(c(cccc1)c1)C(=O)OC
Benzeneacetic acid, α-formyl-, methyl ester (9CI)
Benzeneacetic acid, 2-formyl-, methyl ester
α-Formyl methyl phenyl acetate
Malonaldehydic acid, phenyl-, methyl ester
Methyl α-formylbenzeneacetate

5895-45-4
$CH_2O_3.2/3Pr$
155.96
Carbonic acid, praseodymium(3+) salt (3:2) (9CI)
Praseodymium carbonate

5895-46-5
$CH_2O_3.2/3Nd$
158.19
Carbonic acid, neodymium(3+) salt (3:2) (9CI)

5902-51-2
$C_9H_{13}ClN_2O_2$
216.69
Cc1[nH]c(=O)n(c(=O)c1Cl)C(C)(C)C
Uracil, 3-tert-butyl-5-chloro-6-methyl
3-tert.Butyl-5-chlor-6-methyl-uracil (German)
3-tert-Butyl-5-chloro-6-methyl-uracil
5-Chloro-3-tert-butyl-6-methyl-uracil
5-Chloro-3-(1,1-dimethylethyl)-6-methyl-2,4(1H,3H)-pyrimidinedione
Compound 732
Du Pont 732
Du Pont Herbicide 732
Experimental Herbicide 732
2,4(1H,3H)-Pyrimidinedione, 5-chloro-3-(1,1-dimethylethyl)-6-methyl-
Sinbar
Terbacil
Turbacil

5902-95-4
$C_2H_8As_2O_6.Ca$
318.02
Methanearsonic acid, calcium salt (2:1)
Calar
Calcium acid methanearsonate
Calcium acid methyl arsonate
Calcium hydrogen methanearsonate
Calcium methanearsonate
CAMA
Super Dal-E-Rad
Super Crab-E-Rad-Calar
Super Dal-E-Rad-Calar
USAF AN-11

5905-52-2
$C_6H_{10}O_6.Fe$
234.01
[Fe].CC(O)C(O)=O
Lactic acid, iron(2+) salt (2:1)
Ferrous lactate
Iron(2+) lactate

5908-87-2
$C_{24}H_{48}O_2$
368.64
Docosanoic acid, ethyl ester (8CI,9CI)

5910-75-8
$C_{26}H_{55}N$
381.82
N(CCCCCCCCCCCCC)CCCCCCCCCCCCC
Ditridecylamine

5910-89-4
$C_6H_8N_2$
108.16
n(c(c(nc1)C)C)c1
Pyrazine, 2,3-dimethyl
2,3-Dimethylpyrazine

5911-04-6
$C_{10}H_{22}$
142.28
Nonane, 3-methyl- (8CI,9CI)

5912-86-7
$C_{10}H_{12}O_2$
164.22
Phenol, 2-methoxy-4-propenyl-, (Z)
cis-Isoeugenol
cis-2-Methoxy-4-propenylphenol

5915-41-3
$C_9H_{16}ClN_5$
229.75
n(c(nc(n1)NC(C)(C)C)NCC)c1Cl
s-Triazine, 2-(tert-butylamino)-4-chloro-6-(ethylamino)
2-tert.Butylamino-4-aethylamino-6-chlor-1,3,5-triazin (German)
2-tert-Butylamino-4-chloro-6-ethylamino-s-triazine
Gardoprim
GS 13529
Primatol M
Primatol-M80
Sorgoprim
Terbuthylazine
Turbulethylazin (German)
Terbutylethylazine

5918-29-6
$C_{17}H_{34}O_2$
270.46
14-Methylhexadecanoic acid
14-Methylpalmitic acid

5918-93-4
$C_3H_4N_2O$
84.09
4-Imidazolin-2-one
ENT 27,439
2-Hydroxyimidazole
2H-Imidazol-2-one, 1,3-dihydro- (9CI)
SD 8591
Shell SD-8591

5921-65-3
$C_{12}H_{23}N_5$
237.32
1,3,5-Triazine-2,4-diamine, 6-nonyl- (9CI)
s-Triazine, 2,4-diamino-6-nonyl- (8CI)

5921-80-2
$C_{12}H_{26}O_2$
202.34
Butane, 1,1-dibutoxy- (9CI)
Butyraldehyde, dibutyl acetal (6CI,7CI,8CI)
1,1-Dibutoxybutane
Lageracetal

5926-90-9
$C_9H_{18}O_2$
158.27
O(C1COCCCCCC)C1
Propane, 1,2-epoxy-3-(hexyloxy)
2,3-Epoxypropylhexyl ether
Glycidyl hexyl ether
Hexyl glycidyl ether
Oxirane, ((hexyloxy)methyl)- (9CI)

5930-71-2
$C_4H_{12}O_4P_2S_4$
314.34
O(P(OC)(=S)SSP(OC)(OC)=S)C
Thioperoxydiphosphoric acid (((HO)$_2$P(S))$_2$S$_2$), tetramethyl ester (9CI)
Bis(O,O-dimethylthiophosphoryl)disulfide
Methyl thioperoxydiphosphate (7CI)
Thioperoxydiphosphoric acid, O,O,O,O-tetramethyl-
Thioperoxydiphosphoric acid (((HO)2PS)2S2), tetramethyl ester (8CI)

5932-68-3
$C_{10}H_{12}O_2$
164.22
Phenol, 2-methoxy-4-propenyl-, (E)
trans-Isoeugenol
trans-2-Methoxy-4-propenyl phenol

5932-79-6
$C_9H_{20}O$
144.26
OC(CCCCC)CCC
4-Nonanol
AI3-37212

5932-91-2
$C_8H_{16}O$
128.22
Hexanal, 4,4-dimethyl- (7CI,8CI,9CI)
4,4-Dimethylhexanal

5943-83-9
$C_{10}H_{19}N_5O_2$
241.30
1-Propanol, 3,3'-((6-methyl-s-triazine-2,4-diyl)diimino)di- (7CI, 8CI)

5944-07-0
$C_8H_{15}N_5O_2$
213.24
Ethanol, 2,2'-((6-methyl-1,3,5-triazine-2,4-diyl)diimino)bis- (9CI)
Ethanol, 2,2'-((6-methyl-s-triazine-2,4-diyl)diimino)di- (7CI,8CI)

5949-05-3
$C_{10}H_{18}O$
154.25
O=CCC(CCC=C(C)C)C
6-Octenal, 3,7-dimethyl-, (S)- (9CI)
(S)-3,7-Dimethyl-6-octenal

5952-26-1
$C_6H_{12}N_2O_5$
192.20
O=C(OCCOCCOC(=O)N)N
Ethanol, 2,2'-oxydi-, dicarbamate
Diethylene glycol, dicarbamate (8CI)
Diglycolurethane
2,2'-Oxydiethanol dicarbamate

5954-72-3
C_4H_8S
88.17
2-Butene-1-thiol (8CI,9CI)

5959-52-4
$C_{11}H_9NO_2$
187.21
O=C(O)c(c(N)cc(c1ccc2)c2)c1
2-Naphthoic acid, 3-amino
3-Aminoisonaphthoic acid
3-Amino-2-naphthalenecarboxylic acid
3-Amino-2-naphthoic acid

5959-89-7
$C_{18}H_{34}O_6$
346.46
O=C(OCC(O)C(OCC1O)C1O)CCCCCCCCCCC
D-Glucitol, 1,4-anhydro-, 6-dodecanoate (9CI)
1,4-Anhydro-D-glucitol 6-dodecanoate

5961-59-1
$C_8H_{11}NO$
137.18
CNc1ccc(OC)cc1
Benzenamine, 4-methoxy-N-methyl- (9CI)
p-Anisidine, N-methyl- (8CI)

5961-99-9
$C_{15}H_{14}S$
226.34
2H-1-Benzothiopyran, 3,4-dihydro-2-phenyl- (9CI)
2-Phenylthiochroman
Thiaflavan
Thioflavan (8CI)

5962-23-2
$C_6H_{11}N_5$
153.16
Butyroguanamine
AI3-51439
1,3,5-Triazine-2,4-diamine, 6-propyl-

5963-49-5
$C_{14}H_{12}Cl_2$
251.16
Benzene, 1,1'-(1,2-dichloro-1,2-ethanediyl)bis- (9CI)
Bibenzyl, α,α'-dichloro- (6CI,7CI,8CI)
Stilbene dichloride
1,2-Dichloro-1,2-diphenylethane

5964-24-9
$C_8H_9HgO_3S_2.Na$
440.87
Mercury, ethyl((p-sulfophenyl)thio)-, sodium salt
Ethyl(hydrogen p-mercaptobenzenesulfonato)mercury sodium salt
Ethyl ((p-sulfophenyl)thio)-mercury, sodium salt
Mercury, ethyl(4-mercaptobenzenesulfonato-S-[4])-, sodium salt
Sodium thimerfonate
Sodium p-((ethylmercuri)thio)-benzenesulfonate
Sulfo-Merthiolate
Thimerfonate sodium

5964-35-2
$C_{10}H_{16}N_2O_8.4K$
448.63
Tetrapotassium ethylenediaminetetraacetate
Glycine, N,N'-1,2-ethanediylbis(N-(carboxymethyl)-, tetrapotassium salt (9CI)

5967-77-1
$C_{19}H_{21}NO_3.C_6H_3N_3O_7$
540.49
Morphinan, 6,7,8,14-tetradehydro-4,5-epoxy-3,6-dimethoxy-17-methyl-, (5α)-, Compd. with 2,4,6-trinitrophenol (1:1) (9CI)
Thebaine, picrate

5968-84-3
$C_2H_7AsO_2.1/3Fe$
156.61
Iron cacodylate
Arsinic acid, dimethyl-, iron-(3+) salt
Cacodilato de hierro
Ferric cacodylate
Ferric dimethylarsinate
Iron(III) cacodylate
Iron dimethylarsonate

5968-88-7
$C_6H_8O_7.Mn$
247.06
1,2,3-Propanetricarboxylic acid, 2-hydroxy-, manganese(3+) salt (1:1) (9CI)
Citric acid, manganese(3+) salt (1:1) (8CI)
Manganese citrate

5970-32-1
$C_7H_4HgO_3$
336.70
Mercuric salicylate, Solid
Mercuric salicylate
Mercurisalicylic acid
Mercury salicylate
Mercury, (salicylato(2-))-
Mercury subsalicylate
Salicilato de mercurio (Spanish)
Salicylate de mercure (French)
UN 1644 [Mercury salicylate]

5970-45-6
$C_4H_6O_4.Zn.2H_2O$
219.51
Zinc acetate, dihydrate
Acetic acid, zinc salt, dihydrate
Octan zinecnaty (Czech)
Zinc diacetate, dihydrate
Zinc acetate

5970-61-6
$C_2O_{11}Zn_5.4H$
531.00
Zincate(4-), bis(carbonato-(2-))pentaoxopenta-, tetrahydrogen (9CI)

5970-62-7
$CH_2O_2.H_2O.1/2Zn$
96.73
Formic acid, zinc salt, dihydrate (8CI,9CI)
Diformatozinc dihydrate
Zinc diformate dihydrate
Zinc formate, dihydrate (7CI)
Zinc formate ($Zn(HCO_2)_2$) dihydrate

5972-73-6
$C_2H_7NO_5$
125.08
Ammonium oxalate

5973-71-7
$C_9H_{10}O$
134.18
Benzaldehyde, 3,4-dimethyl- (8CI,9CI)

5976-47-6
C_3H_5ClO
92.53
2-Propen-1-ol, 2-chloro
β-Chloro allyl alcohol
2-Chloro-2-propen-1-ol

5979-28-2
$C_{34}H_{28}Cl_4N_6O_4$
726.41
O=C(Nc(c(cc(c(ccc(NC(=O)C(N=Nc(c(cc(c1)Cl)Cl)c1)C(=O)C)c2C)c2)c3)C)c3)C(N=Nc(c(cc(c4)Cl)Cl)c4)C(=O)C
Butanamide, N,N'-(3,3'-dimethyl(1,1'-biphenyl)-4,4'-diyl)bis(2-((2,4-dichlorophenyl)azo)-3-oxo- (9CI)
C.I. Pigment Yellow 16 (VAN) (8CI)

5980-23-4
$C_7H_5Cl_2NO$
190.02
Benzamide, 3,5-dichloro- (8CI,9CI)

5980-26-7
$C_7H_5Cl_2NO$
190.02
Benzamide, 2,5-dichloro- (8CI,9CI)

5981-06-6
$C_4H_8O_3$
104.11
1,3,5-Trioxepane (8CI,9CI)
NSC-511991
1,3,5-Trioxacycloheptane

5981-39-5
$C_8H_8O_4$
168.15
Benzoic acid, 2,5-dihydroxy-3-methyl- (9CI)
Gentisic acid, 3-methyl- (6CI,8CI)
2,5-Dihydroxy-3-methylbenzoic acid
2,5-Dihydroxy-m-toluic acid
5-Hydroxy-2,3-cresotic acid
3-Methyl-2,5-dihydroxybenzoic acid
3-Methylgentisic acid

5988-76-1
$C_{32}H_{48}O_8$
560.80
19-Nor-9-β,10-α-lanosta-5,23-diene-11,22-dione, 3-β,16-α,20,25-tetrahydroxy-9-(hydroxymethyl)-, 25-acetate
Cucurbitacine (C)

5988-91-0
$C_{10}H_{20}O$
156.27
O=CCC(CCCC(C)C)C
Octanal, 3,7-dimethyl- (VAN) (9CI)
3,7-Dimethyloctanal

5989-27-5
$C_{10}H_{16}$
136.26
C(=CCC(C(=C)C)C1)(C1)C
Cyclohexene, 1-methyl-4-(1-methylethenyl)-, (R)
(+)-4-Isopropenyl-1-methylcyclohexene
d-Limonene
D-(+)-Limonene
(+)-R-Limonene
d-Limoneno (Spanish)
d-p-Mentha-1,8-diene
p-Mentha-1,8-diene
NCI-C55572
Refchole

5989-33-3
$C_{10}H_{18}O_2$
170.25
O(C(C=C)(CC1)C)C1C(O)(C)C
2-Furanmethanol, 5-ethenyltetrahydro-α,α,5-trimethyl-, cis- (VAN)(9CI)
cis-5-Ethenyltetrahydro-α,α,5-trimethyl-2-furanmethanol
Linalool oxide

5989-54-8
$C_{10}H_{16}$
136.26
C(=CCC(C(=C)C)C1)(C1)C
p-Mentha-1,8-diene, (S)-(-)
Cyclohexene, 1-methyl-4-(1-methylethenyl)-, (S)-
l-Limonene
(-)-Limonene

5994-45-6
$C_{26}H_{43}NO_3$.Na
440.62
Benzoic acid, 4-((octadecylamino)carbonyl)-, monosodium salt (9CI)
N-Octadecylterephthalamate, monosodium salt
N-Octadecylterephthalamic acid, monosodium salt
Sodium N-octadecylterephthalamate
Terephthalamic acid, N-octadecyl-, monosodium salt

5994-61-6
$C_5H_{10}NO_7P$
227.10
Glycine, N-(carboxymethyl)-N-(phosphonomethyl)- (9CI)
Acetic acid, ((phosphonomethyl)imino)di- (8CI)
MON 820

5995-42-6
$C_4H_{13}NO_7P_2$
249.09
O=P(O)(O)CN(CP(=O)(O)O)CCO
Phosphonic acid, (((2-hydroxyethyl)imino)bis-(methylene))bis- (9CI)
2-Hydroxyethyliminobis-(methylene phosphonic acid)

6000-43-7
$C_2H_5NO_2$.ClH
111.54
[Cl-].NCC(O)=O
Glycine, hydrochloride
Glycocoll hydrochloride
Glyco-Hydrochloride

6001-97-4
$C_{16}H_{30}O_7$S.Na
389.47
Butanedioic acid, sulfo-, 1,4-bis(1-methylpentyl) ester, sodium salt (9CI)

6004-38-2
$C_{10}H_{16}$
136.24
C(C(C(C1C2)CC3)C3)(C1)C2
4,7-Methano-1H-indene, octahydro- (9CI)
AI3-51352
Octahydro-4,7-methano-1H-indene
Tricyclo(5.2.1.02,6)decane

6004-60-0
$C_7H_{12}O$
112.17
Ethanone, 1-cyclopentyl- (9CI)
Ketone, cyclopentyl methyl (8CI)

6006-33-3
$C_{19}H_{38}$
266.51
C(CCCC1)(C1)CCCCCCCCCCCCC
Cyclohexane, tridecyl- (9CI)
1-Cyclohexyltridecane
Tridecylcyclohexane

6006-95-7
$C_{21}H_{42}$
294.56
C(CCCC1)(C1)CCCCCCCCCCCCCCC
Cyclohexane, pentadecyl- (9CI)
1-Cyclohexylpentadecane
Pentadecylcyclohexane

6008-36-2
$C_{15}H_{22}O$
218.34
1-Nonanone, 1-phenyl- (9CI)
Nonanophenone (8CI)

6009-70-7
$C_2H_2O_4$.2H3N.H2O
142.10
Ethanedioic acid, diammonium salt, monohydrate (9CI)
Ammonium oxalate monohydrate
Ammonium oxalate ($(NH_4)_2C_2O_4$) monohydrate
Diammonium oxalate monohydrate
Oxalic acid, diammonium salt, monohydrate (8CI)

6012-83-5
$C_{16}H_{16}Cl_2O_2$
311.22
CCOCC(O)(c1ccc(Cl)cc1)c2ccc(Cl)cc2
Benzhydrol, 4,4'-dichloro-α-(ethoxymethyl)
1,1-Bis(4-chlorophenyl)-2-ethoxyethanol
4,4'-Dichloro-α-(ethoxymethyl)benzhydrol
Ethoxymethylbis(p-chlorophenyl)carbinol
Ethoxymethyldichlorophenylcarbinol
Ethoxymethyl-di-(p-chlorophenyl)carbinol
1-Ethoxymethyl-1,1-di-(p-chlorophenyl)carbinol
Etoxinol
G-23645
Geigy 337

6012-92-6
$C_{10}H_8ClNOS_2$
257.76
Rhodanine, 3-(p-chlorophenyl)-5-methyl
3-(p-Chlorophenyl)-5-methyl rhodanine
N-244

6012-97-1
C_4Cl_4S
221.90
Clc1sc(Cl)c(Cl)c1Cl
Thiophene, 2,3,4,5-tetrachloro
2,3,4,5-Chlorothiophene
ENT 25,764
IF
IF (Fumigant)
Penn Salt TD-183
Penphene
Perchlorothiophene
TCTP
TD-183
Tetrachlorothiophene
Tetrachlorothiofene
Tetrachlorthiofen (Czech)
Thiophene, tetrachloro-

6014-75-1

$C_{18}H_9F_{25}O_2$
732.23
O=C(OCCC(F)(F)C(F)(F)C(F)(F)C(F)(F)C(F)(F)C(F)(F)C(F)(F)C(F)(F)C(F)(F)C(F)(F)C(F)(F)C(F)(F)F)C(=C)C
2-Propenoic acid, 2-methyl-, 3,3,4,4,5,5,6,6,7,7,8,8,9,9,10,10,11,11,12,12,13,13,14,14,14-pentacosafluorotetra decyl ester (9CI)

6018-89-9
$C_4H_6O_4.Ni.4H_2O$
248.89
Nickel-acetate-tetrahydrate
Nickel(II) acetate tetrahydrate
Nickel diacetate tetrahydrate
Nickelous acetate tetrahydrate

6023-29-6
$C_{14}H_{20}N_3O_3.1/2Cl_4Zn$
381.92
Benzenediazonium, 2,5-diethoxy-4-(4-morpholinyl)-, (T-4)-tetrachlorozincate(2-) (2:1) (9CI)
N-(4-Diazo-2,5-diethoxyphenyl)morpholine, chloride, zinc chloride (2:1)
2,5-Diethoxy-4-morpholinobenzenediazonium chloride, zinc chloride double salt

6023-57-0
$C_{10}H_9ClN_4$
220.66
s-Triazine, 2-anilino-4-chloro-6-methyl- (7CI, 8CI)

6028-47-3
$C_{12}H_{27}O_2PS_2$
298.45
O(P(OC(CC(C)C)C)(S)=S)C(CC(C)C)C
2-Pentanol, 4-methyl-, hydrogen phosphorodithioate (9CI)
O,O-Bis(1,3-dimethylbutyl) phosphorodithioate

6028-57-5
$C_8H_{16}O_2.1/3Al$
153.21
Aluminum caprylate
Aluminum octanoate
Octanoic acid, aluminum salt (9CI)

6029-84-1
$C_{15}H_{25}NO_5$
299.41
Rinderine

6031-02-3
$C_{12}H_{18}$
162.28
Benzene, (1-methylpentyl)- (9CI)
Hexane, 2-phenyl- (6CI,7CI, 8CI)
(2-Hexyl)benzene
2-Phenylhexane

6032-29-7
$C_5H_{12}O$
88.17
OC(CCC)C
2-Pentanol
sec-Amyl alcohol [UN 1105]
Isoamyl alcohol, secondary (OSHA)
Methyl propyl carbinol
Pentanol-2
sec-Pentyl alcohol
UN 1105 [Amyl alcohols]

6033-05-2
$C_{18}H_{34}O_2.C_{17}H_{19}NO_3$
567.80
Morphine oleate
Morphinan-3,6-diol, 7,8-didehydro-4,5-epoxy-17-methyl-(5α,6α)-, (Z)-9-octadecenoate (Salt) (9CI)
Morphinan-3,6α-diol, 7,8-didehydro-4,5α-epoxy-17-methyl-, oleate (Salt)
Morphine oleate, 20%

6035-40-1
$C_{22}H_{23}NO_7$
413.46
Narcotine
Coscopin
Gnoscopine
Noscapine
NSC-5366
Opian

6041-94-7
$C_{23}H_{15}Cl_2N_3O_2$
436.28
O=C(Nc(cccc1)c1)c(c(O)c(N=Nc(c(ccc2Cl)Cl)c2)c(c3ccc4)c4)c3
2-Naphthalenecarboxamide, 4-((2,5-dichlorophenyl)azo)-3-hydroxy-N-phenyl- (9CI)

6048-82-4
$C_{16}H_{24}O$
232.37
O=C(c(cccc1)c1)CCCCCCCCC
1-Decanone, 1-phenyl- (9CI)
Decanophenone
1-Phenyl-1-decanone

6051-87-2
$C_{19}H_{12}O_2$
272.31
O=c2cc(oc3ccc1ccccc1c23)c4ccccc4
1H-Naphtho(2,1-b)pyran-1-one, 3-phenyl
β-Naphthoflavone
β-NF
3-Phenyl-1H-naphtho(2,1-b)-pyran-1-one

6051-98-5
$C_{17}H_{10}O$
230.27
7H-Benzo(c)fluoren-7-one (8CI,9CI)
Allochrysoketone
allo-Chrysoketone

6054-48-4
$C_{16}H_{14}N_4$
262.29
N(=Nc(ccc(N)c1)c1)c(c(c(c(N)c2)ccc3)c3)c2
1-Naphthalenamine, 4-((4-aminophenyl)azo)- (9CI)
4-((4-Aminophenyl)azo)-1-naphthalenamine

6054-58-6
$C_{18}H_{19}N_5O_3$
353.35
Propanenitrile, 3-((2-hydroxyethyl)(3-methyl-4-((4-nitrophenyl)azo)phenyl)amino)- (9CI)
Propionitrile, 3-(N-(2-hydroxyethyl)-4-((p-nitrophenyl)azo)-m-toluidino)- (8CI)

6055-19-2
$C_7H_{15}Cl_2N_2O_2P.H_2O$
279.13
2H-1,3,2-Oxazaphosphorine, tetrahydro-2-(bis(2-chloroethyl)amino)-, 2-oxide, monohydrate
B 518
N,N-Bis(β-cloraethyl) N'-O-propylenphosphorildiamid monohydratum (Romanian)
2-(Bis(2-chloroethyl)amino)-1-oxa-3-aza-2-phosphocyclohexane 2-oxide monohydrate
1-Bis(2-chloroethyl)amino-1-oxo-2-aza-5-oxaphosphoridine monohydrate
(Bis(chloro-2-ethyl)amino)-2-tetrahydro-3,4,5,6-oxazaphosphorine-1,3,2-oxide-2 monohydrate
Bis(2-chloroethyl)phosphoramide cyclic propanolamide ester monohydrate
N,N-Bis(β-chloroethyl)-N',O-propylenephosphoric acid ester amide monohydrate
N,N-Bis(2-chloroethyl)tetrahydro-2H-1,3,2-oxaphosphor-

in-2-amine, 2-oxide monohydrate
N,N-Bis(β-chloroethyl)-N',O-trimethylenephosphoric acid ester diamide monohydrate
CB-4564
Clafen
Cyclic N',O-propylene ester of N,N-bis(2-chloroethyl)phosphorodiamidic acid monohydrate
Cyclophosphamide hydrate
Cyclophosphamide monohydrate
Cyclophosphamidum
Cyclophosphan
Cyclophosphane
Cyclophosphanum
Cytophosphan
Cytoxan
2-(Di(2-chloroethyl)amino)-1-oxa-3-aza-2-phosphacyclohexane-2-oxide monohydrate
N,N-Di(2-chloroethyl)amino-N,O-propylene phosphoric acid ester diamide monohydrate
Endoxan monohydrate
Endoxana
Endoxan-ASTA
Endoxan R
Enduxan
Genoxal
Mitoxan
NSC-26271
Procytox
Semdoxan
Sendoxan
Senduxan

6055-52-3
$C_6H_{16}N_2.2ClH$
189.16
1,6-Hexanediamine, dihydrochloride
1,6-Diaminohexane dihydrochloride
Hexamethylenediamine dihydrochloride
1,6-Hexamethylenediamine dihydrochloride
Hexanediamine dihydrochloride

6061-06-9
$C_4H_4Cl_2$
122.98
1,3-Butadiene, 1,1-dichloro- (8CI,9CI)

6062-02-8
$C_7H_{14}N_2O_2$
158.19
Propanal, 2,2-dimethyl-, O-((methylamino)carbonyl)-oxime (9CI)
Pivalaldehyde, O-(methylcarbamoyl)oxime (8CI)

6062-26-6
$C_{11}H_{12}ClO_3.Na$
250.66
Sodium 4-(2-methyl-4-chlorophenoxy)butyrate
Butanoic acid, 4-(4-chloro-2-methylphenoxy)-, sodium salt
Butyric acid, 4-(4-chloro-2-methylphenoxy)-, sodium salt
Butyric acid, 4-((4-chloro-o-tolyl)oxy)-, sodium salt
Cantrol
Caswell No. 558A
4-(4-Chlor-2-methyl-phenoxy)-buttersaeure natriumsalz (German)
Chloromethylphenoxybutyric acid sodium salt
4-(4-Chloro-2-methylphenoxy)-butyric acid sodium salt
(4-Chloro-o-tolyloxy)butyric acid sodium salt
EPA Pesticide Chemical Code 019202
M&B 3046
MCPB
4-(MCPD)
4-(2-Methyl-4-chlorophenoxy)-butyric acid, sodium salt
Thistrol
Tropotox

6064-52-4
$C_9H_{16}O_3$
172.22
Nonanoic acid, 4-oxo- (8CI, 9CI)

6064-90-0
$C_{22}H_{44}O_2$
340.59
Heneicosanoic acid, methyl ester (8CI,9CI)

6065-01-6
$C_9H_{17}NO_2$
171.27
4-Nonene, 5-nitro
5-Nitro-4-nonene

6065-04-9
$C_9H_{17}NO_2$
171.27
CCCCCC=C(CC)N(=O)=O
3-Nonene, 3-nitro
3-Nitro-3-nonene

6065-09-4
$C_8H_{15}NO_2$
157.24
CCCCC=C(CC)N(=O)=O
3-Octene, 3-nitro
3-Nitro-3-octene

6065-10-7
$C_8H_{15}NO_2$
157.24
CCCCCC(=CC)N(=O)=O
2-Octene, 3-nitro
3-Nitro-2-octene

6065-11-8
$C_8H_{15}NO_2$
157.24
CCCCCC=C(C)N(=O)=O
2-Octene, 2-nitro
2-Nitro-2-octene

6065-13-0
$C_7H_{13}NO_2$
143.21
CCCCC(=CC)N(=O)=O
2-Heptene, 3-nitro
3-Nitro-2-heptene

6065-14-1
$C_7H_{13}NO_2$
143.21
CCCCC=C(C)N(=O)=O
2-Heptene, 2-nitro
2-Nitro-2-heptene

6065-17-4
$C_6H_{11}NO_2$
129.18
CCCC=C(C)N(=O)=O
2-Hexene, 2-nitro
2-Nitro-2-hexene

6065-18-5
$C_5H_9NO_2$
115.15
2-Pentene, 3-nitro
3-Nitro-2-pentene
3-Nitro-3-pentene
3-Pentene, 3-nitro-

6065-19-6
$C_5H_9NO_2$
115.15
CCC=C(C)N(=O)=O
2-Pentene, 2-nitro
2-Nitro-2-pentene

6069-98-3
$C_{10}H_{20}$
140.27
C(CCC(C1)C(C)C)(C1)C
Cyclohexane, 1-methyl-4-(1-methylethyl)-, cis- (9CI)
p-Menthane, cis- (8CI)
cis-p-Menthane
cis-1-Methyl-4-(1-methylethyl)-cyclohexane

6075-11-2
$C_{13}H_{14}O_3$
218.27

O=C(CCC(OC=C1)=C1)CCC
(OC=C2)=C2
3-Pentanone, 1,5-di-2-furyl
1,5-Di-2-furyl-3-pentanone

6080-56-4
$C_4H_6O_4.Pb.3H_2O$
379.35
Lead acetate (II), trihydrate
Acetic acid, lead(+2) salt trihydrate
Bis(acetato)trihydroxytrilead
Bleiazetat (German)
Lead acetate trihydrate
Lead diacetate trihydrate
Plumbous acetate

6086-21-1
$C_3H_5N_3$
83.07
1H-1,2,4-Triazole, 1-methyl- (8CI,9CI)

6086-22-2
$C_6H_{11}N_3$
125.16
1H-1,2,4-Triazole, 1-butyl- (8CI,9CI)

6088-51-3
$C_{20}H_{14}O_2S_2$
350.46
Oc(ccc(c1ccc2SSc(ccc(c3ccc4O)
c4)c3)c2)c1
2,2'-Dihydroxy-6,6'-dinaphthyldisulfide
AI3-61075
6,6'-Dithiobis-2-naphthalenol
2-Naphthalenol, 6,6'-dithiobis- (9CI)
2-Naphthol, 6,6'-dithiodi- (8CI)

6091-44-7
$C_5H_{11}N.ClH$
121.63
[Cl-].C1CC[N]CC1
Piperidine, hydrochloride
Hexahydropyridine hydrochloride

6092-54-2
$C_7H_{13}ClO_2$
164.63
O=C(OCCCCCC)Cl
n-Hexyl chloroformate
Carbonochloridic acid, hexyl ester (9CI)
Chloroformic acid n-hexyl ester
Formic acid, chloro-, hexyl ester
Hexyl carbonochloridate
Hexyl chlorocarbonate
Hexyl chloroformate

6094-02-6
C_7H_{14}
98.19
1-Hexene, 2-methyl- (8CI,9CI)
NSC-73924

6104-30-9
$C_6H_{14}N_4O_2$
174.24
O=C(NC(NC(=O)N)C(C)C)N
Urea, 1,1'-isobutylidenedi
Diureidoisobutane
1,1-Diureidisobutane
IBDU
Isobutyldiurea
Isobutylenediurea
Isobutylidenediurea
1,1'-Isobutylidenebisurea
Isodur
Urea, N,N''-(2-methylpropylidene)bis-, (9CI)

6104-58-1
$C_{47}H_{49}N_3O_7S_2.Na$
855.03
Coomassie Brilliant Blue
Acid Blue 90
Benzenemethanaminium, N-(4-((4-((4-ethoxyphenyl)-amino)phenyl)(4-(ethyl((3-sulfophenyl)methyl)amino)-2-methylphenyl)methylene)-3-methyl-2,5-cyclohexadien-1-ylidene)-N-ethyl-3-sulfo-, hydroxide, inner salt, monosodium salt (9CI)
C.I. Acid Blue 90, Monosodium salt (8CI)
C.I. 42655
Coomassie Blue G
Coomassie Brilliant Blue G250
Serva Blue G
Supranol Cyanine G
Xylene Brilliant Cyanine G

6106-41-8
$C_5H_{10}O_2.Na$
125.12
Pentanoic acid, sodium salt (9CI)
Valeric acid, sodium salt (8CI)

6106-46-3
$C_{17}H_{21}NO_4.CH_3NO_3$
380.44
Scopolamine, Compd. with methyl nitrate (1:1)
Epoxytropine tropate methylnitrate
Mescomine
Methyl nitrate, Compd. with scopolamine (1:1)
Methyl scopolamine nitrate
Scopolamine methyl nitrate
Skopolate
Skopyl
Viscope

6106-81-6
$C_{17}H_{21}NO_5.BrH$
400.26
Aminoxyscopolamine hydrobromide
Benzeneacetic acid, α-(hydroxymethyl)-, 9-methyl-3-oxa-9-azatricyclo-(3.3.1.02,4)non-7-yl ester, N-oxide, hydrobromide, (7(S)-(1α,2β,4β,5α,7β))- (9CI)
Benzeneacetic acid, α-(hydroxymethyl)-, 9-methyl-3-oxa-9-azatricyclo(3.3.1.0-(2,4))non-7-yl ester, N-oxide, hydrobromide, (7(S)-(1α,2β,4β,5α,7β))-
Scopolamine aminoxide hydrobromide
Scopolamine, N-oxide, hydrobromide
1αH,5αH-Tropan-3α-ol, 6β,7β-epoxy-, (-)-tropate (ester), 8-oxide, hydrobromide (8CI)

6108-10-7
$C_6H_6Cl_6$
290.83
C(C(C(C(C1Cl)Cl)Cl)Cl)
(C1Cl)Cl
Cyclohexane, 1,2,3,4,5,6-hexachloro-, (1α,2α,3α,4β,5β,6β)
AI3-15109
E-HCH

6108-11-8
$C_6H_6Cl_6$
290.83
Cyclohexane, 1,2,3,4,5,6-hexachloro-, (1α,2α,3α,4α,5α,6α) (9CI)

6108-12-9
$C_6H_6Cl_6$
290.83
Cyclohexane, 1,2,3,4,5,6-hexachloro-, (1α,2α,3α,4α,5β,6β) (8CI,9CI)

6108-13-0
$C_6H_6Cl_6$
290.83
Cyclohexane, 1,2,3,4,5,6-hexachloro-, (1α,2α,3α,4α,5α,6β) (9CI)

6109-97-3
$C_{14}H_{14}N_2.ClH$
246.76
Cl.CCn2c1ccccc1c3cc(N)ccc23
Carbazole, 3-amino-9-ethyl-, hydrochloride
3-Amino-9-ethylcarbazole hydrochloride
NCI-C03043

6111-78-0

$C_{19}H_{14}$
242.33
Cc3cccc4cc2ccc1ccccc1c2cc34
Benz(a)anthracene, 11-methyl
11-Methylbenz(a)anthracene
8-Methyl-1:2-benzanthracene

6112-76-1
$C_5H_4N_4S.H_2O$
170.21
S=c1[nH]cnc2nc[nH]c12
Purine-6-thiol, monohydrate
1,7-Dihydro-6H-purin-6-thion, monohydrat (Czech)
6-Mercaptopurine monohydrate
6-Merkaptopurin, monohydrat (Czech)
6H-Purine-6-thione, 1,7-dihydro-, monohydrate (9CI)
Purin-6-thiol, monohydrat (Czech)
6H-Purin-6-thion, monohydrat (Czech)

6117-80-2
$C_4H_8O_2$
88.11
2-Butene-1,4-diol, (Z)- (8CI, 9CI)

6117-91-5
C_4H_8O
72.12
OCC=CC
2-Buten-1-ol
2-Butene-1-ol
2-Butenol
2-Butenyl alcohol
Crotonyl alcohol
Crotyl alcohol
Krotylalkohol (Czech)

6119-53-5
$C_{20}H_{24}N_2O_2.H_3O_2P$
390.41
Cinchonan-9-ol, 6'-methoxy-, (8α,9R)-, monophosphinate (Salt) (9CI)
Quinine, monophosphinate (Salt) (8CI)

6124-90-9
$C_6H_{12}O$
100.16
Oxirane, 2-methyl-3-propyl-, cis- (9CI)
cis-2,3-Epoxyhexane
Hexane, 2,3-epoxy-, cis- (8CI)

6130-64-9
$C_{16}H_{18}N_2O_4S.C_{13}H_{20}N_2O_2.H_2O$
588.79
4-Thia-1-azabicyclo(3.2.0)heptane-2-carboxilic acid, 3,3-dimethyl-7-oxo-6-(2-phenyl acetamido)-, Compd. with 2-(diethylamino)ethyl p-aminobenzoate (1:1), monohydrate
Abbocillin-DC
Afsillin
Ampin-Penicillin
Aquacillin
Aquasuspen
Avloprocil
Benzylpenicillinic acid procaine salt
Benzylpenicillin procaine monohydrate
Benzylpenicillin procaine salt
Cilicaine
Crysticillin
Depo-Penicillin
Despacilina
Distaquaine
Diurnal-Penicillin
Dorsallin A.R.
Duracillin
Eskacillin
Flo-Cillin
Flo-Cillin Aqueous
Hydracillin
Hypercillin
Ilcocillin P
Kabipenin
Ledercillin
Lenticillin
Lentopen
Megapen
Mylipen
Neoproc
Novocaine Penicillin
Parencillin
Penaquacaine G
Pen-Fifty
Penlator
Penicillin G compd. with 2-(diethylamino)ethyl p-aminobenzoate
Penicillin G Procaine
Pfizerpen-AS
Premocillin
Procaine benzylpenicillinate
Procaine penicillin
Procaine penicillin G
Procanodia
Pro-Pen
Sharcillin
Wycillin

6130-72-9
$C_{30}H_{32}O_6$
488.58
Oxirane, 2,2',2''-(1-propanyl-3-ylidenetris-(4,1-phenyleneoxymethylene))tris- (9CI)
Propane, 1,1,3-tris(p-(2,3-epoxypropoxy)-phenyl)- (7CI,8CI)

6130-75-2
$C_7H_5Cl_3O$
211.47
O(c(c(cc(c1Cl)Cl)Cl)c1)C
Anisole, 2,4,5-trichloro- (8CI)
AI3-09172
Benzene, 1,2,4-trichloro-5-methoxy- (9CI)
1,2,4-Trichloro-5-methoxybenzene

6130-82-1
C_4Cl_6
260.76
Cyclobutene, hexachloro- (8CI,9CI)
NSC-136548

6135-54-2
$C_8H_{16}O_2$
144.22
1,3-Dioxolane, 2-(1,1-dimethylethyl)-2-methyl- (9CI)
2-tert-Butyl-2-methyl-1,3-dioxolane
1,3-Dioxolane, 2-tert-butyl-2-methyl- (7CI,8CI)

6136-37-4
$C_6H_6N_4O_2$
166.16
Cn2c(=O)[nH]c1nc[nH]c1c2=O
Xanthine, 1-methyl
1-Methylxanthine

6137-08-2
$C_9H_{18}O$
142.27
2-Octanone, 3-methyl
3-Methyl-2-octanone

6137-11-7
$C_8H_{16}O$
128.21
O=C(C(CCC)C)CC
4-Methyl-3-heptanone
3-Heptanone, 4-methyl-

6138-53-0
$C_{21}H_{24}O_3Si_3$
408.68
Cyclotrisiloxane, 2,4,6-trimethyl-2,4,6-triphenyl-, (2α,4α,6β)- (9CI)
Cyclotrisiloxane, 2,4,6-trimethyl-2,4,6-triphenyl-, cis,trans- (8CI)

6138-79-0
$C_{19}H_{22}N_2.ClH.H_2O$
332.86
Triprolidine hydrochloride
Actidil
Alleract
Pyridine, 2-(1-(4-methylphenyl)-3-(1-pyrrolidinyl)-1-propenyl)-, monohydrochloride, monohydrate, (E)-
(E)-2-(3-(1-Pyrrolidinyl)-1-p-tolylpropenyl)pyridine

monohydrochloride mono-hydrate

6140-74-5
$C_{19}H_{36}O_2$
296.49
O=C(OCCCCCCCCCCCCCCC)C(=C)C
2-Propenoic acid, 2-methyl-, pentadecyl ester (9CI)
Pentadecyl methacrylate
Pentadecyl 2-methyl-2-propenoate

6143-52-8
$C_6H_{15}N.HNO_3$
164.19
2-Propanamine, N-(1-methylethyl)-, nitrate (9CI)
Diisopropylamine, nitrate (8CI)

6144-04-3
$(C_9H_{10})_2$
236.35
α-Methylstyrene dimer
Benzene, (1-methylethenyl)-, dimer (9CI)
(1-Methylethenyl)benzene dimer
Styrene, α-methyl-, dimer

6145-73-9
$C_9H_{18}Cl_3O_4P$
327.57
O=P(OCC(C)Cl)(OCC(C)Cl)OCC(C)Cl
Fyrol PCF
2-Chloro-1-propanol phosphate (3:1)
1-Propanol, 2-chloro-, phosphate (3:1) (9CI)

6147-53-1
$C_4H_6O_4.Co.4H_2O$
249.11
CC(=O)[O-][Co+2][O-]C(C)=O
Acetic acid, cobalt(2+) salt, tetrahydrate
Cobalt acetate tetrahydrate
Cobalt(II) acetate tetrahydrate
Cobalt diacetate tetrahydrate
Cobaltous acetate tetrahydrate
Octan kobaltnaty (Czech)

6149-03-7
$C_{14}H_{22}O_3S.Na$
293.38
Benzenesulfonic acid, 4-octyl-, sodium salt (9CI)
Benzenesulfonic acid, p-octyl-, sodium salt (8CI)

6149-34-4
$C_{12}H_{11}N_3O_2$
229.22
1,4-Benzenediamine, N-(4-nitrophenyl)- (9CI)
NSC-159352
p-Phenylenediamine, N-(p-nitrophenyl)- (8CI)

6155-96-0
$C_5H_9ClO_2$
136.58
Pentanoic acid, 2-chloro- (9CI)
Valeric acid, 2-chloro- (8CI)

6156-18-9
$C_5H_{12}S_2$
136.28
Propane, 2,2-bis(methylthio)- (9CI)
Acetone, dimethyl mercaptole (8CI)

6156-78-1
$C_4H_6O_4.Mn.4H_2O$
245.12
O.O.O.O.[Mg+].CC(O)=O
Acetic acid, manganese(2+) salt, tetrahydrate
Manganese acetate tetrahydrate
Manganese(II) acetate tetrahydrate
Manganese diacetate, tetrahydrate
Manganous acetate tetrahydrate

6158-45-8
$C_{13}H_{14}$
170.25
c(c(c(cc1)C(C)C)ccc2)(c2)c1
1-Isopropylnaphthalene
1-(1-Methylethyl)naphthalene
Naphthalene, 1-isopropyl- (8CI)
Naphthalene, 1-(1-methylethyl)- (9CI)

6163-66-2
$C_8H_{18}O$
130.26
tert-Butyl ether (DOT)
Di-tert-butyl ether [UN 1149]
Propane, 2,2'-oxybis(2-methyl- (9CI)
UN 1149 [Dibutyl ethers]

6164-47-2
$C_{20}H_{19}NO_5.ClH$
389.86
Protopine, hydrochloride
Bis(1,3)benzodioxolo(4,5-c:5',6'-g)azecin-13(5H)-one, 4,6,7,14-tetrahydro-5-methyl-, hydrochloride (9CI)
NSC-11440
7,13a-Secoberbin-13a-one, 7-methyl-2,3:9,10-bis(methylenedioxy)-, hydrochloride (8CI)

6164-98-3
$C_{10}H_{13}ClN_2$
196.70
CN(C)C=Nc1ccc(Cl)cc1C
Formamidine, N'-(4-chloro-o-tolyl)-N,N-dimethyl
Acaron
Bermat
C 8514
Carzol
CDM
Chlordimeform
Chlorfenamidine
N'-(4-Chloro-2-methylphenyl)-N,N-dimethylmethanimidamide
Chlorophenamidin
Chlorophenamidine
N'-(4-Chloro-o-tolyl)-N,N-dimethylformamidine
N^2-(4-Chloro-o-tolyl)-N^1,N^1-dimethylformamidine
Chlorphenamidine
N'-(4-Chlor-o-tolyl)-N,N-dimethylformamidin (German)
Ciba 8514
Ciba-C8514
N,N-Dimethyl-N'-(2-methyl-4-chlorophenyl)formamidine
N,N-Dimethyl-N'-(2-methyl-4-chlorphenyl)-formadin (German)
ENT 27,335
ENT 27,567
EP-333
Fundal
Fundal 500
Fundex
Galecron
Methanimidamide, N'-(4-chloro-2-methylphenyl)-N,N-dimethyl-
N'-(2-Methyl-4-chlorophenyl)-N,N-dimethylformamidine
N'-(2-Methyl-4-chlorphenyl)-formamidin-hydrochlorid (German)
NSC-190935
Ovatoxion
RS 141
Schering 36268
SN 36268
Spanon
Spanone

6165-40-8
$C_{16}H_{34}$
226.45
Pentadecane, 7-methyl- (8CI, 9CI)

6165-44-2
$C_{16}H_{30}$
222.42
Cyclohexane, 1,1'-(1,4-butanediyl)bis- (9CI)
Butane, 1,4-dicyclohexyl- (6CI,7CI,8CI)
1,4-Dicyclohexylbutane

6165-96-4
C_8H_2
98.10
1,3,5,7-Octatetrayne (9CI)
Octatetrayne (8CI)

6169-78-4
$C_{10}H_{12}O$
148.20
1-Benzoxepin, 2,3,4,5-tetrahydro- (8CI,9CI)
Homochroman

6175-45-7
$C_{12}H_{16}O_3$
208.28
O=C(c(cccc1)c1)C(OCC)OCC
Glyoxal, phenyl-, 2-(diethyl acetal)
α,α-Diethoxyacetophenone
2,2-Diethoxyacetophenone
Ethanone, 2,2-diethoxy-1-phenyl-
Phenylglyoxal diethyl acetal

6175-49-1
$C_{12}H_{24}O$
184.32
O=C(CCCCCCCCCC)C
2-Dodecanone
AI3-28136
Methyl decyl ketone

6178-32-1
$C_{18}H_{28}O_2$
276.42
O(C1COc(ccc(c2)CCCCCCCCC)c2)C1
Oxirane, ((4-nonylphenoxy)-methyl)- (9CI)
((4-Nonylphenoxy)methyl)-oxirane
p-Nonylphenyl glycidyl ether

6187-24-2
$C_7H_{13}NO_2$
143.21
3-Heptene, 3-nitro
3-Nitro-3-heptene

6190-65-4
$C_6H_{10}ClN_5$
152.15
Desethyl atrazine
Deethylatrazine
4-Deethylatrazine
G 30033
s-Triazine, 2-amino-4-chloro-6-(isopropylamino)-
1,3,5-Triazine-2,4-diamine, 6-chloro-n-(1-methylethyl)-

6191-90-8
C_9H_{16}
124.23
Spiropentane, butyl- (7CI, 8CI,9CI)
Butylspiropentane

6196-54-9
$C_{15}H_{15}N$
209.28
Dibenz(c,e)azocine, 5,6,7,8-tetrahydro- (8CI,9CI)

6196-95-8
$C_{16}H_{18}$
210.32
c(cccc1)(c1)C(c(ccc(c2C)C)c2)C
Benzene, 1,2-dimethyl-4-(1-phenylethyl)- (9CI)
1,2-Dimethyl-4-(1-phenylethyl)-benzene
1-(3,4-Dimethylphenyl)-1-phenylethane
1-Phenyl-1-(3,4-dimethyl-phenyl)ethane

6197-30-4
$C_{24}H_{27}NO_2$
361.48
O=C(OCC(CCCC)CC)C(C#N)=C(c(cccc1)c1)c(cccc2)c2
Octocrilene
2-Ethylhexyl 2-cyano-3,3-diphenylacrylate
2-Ethylhexyl 2-cyano-3,3-diphenyl-2-propenoate
Octocrileno (Spanish)
Octocrilenum (Latin)
Octocrylene
2-Propenoic acid, 2-cyano-3,3-diphenyl-, 2-ethylhexyl ester (9CI)
UV Absorber-3

6199-67-3
$C_{32}H_{46}O_8$
558.78
19-Nor-9-β,10-α-lanosta-5,23-diene-3,11,22-trione, 2-β,16-α,20,25- tetrahydroxy-9-methyl-, 25-acetate
Amarine
Cucurbitacin B
Cucurbitacine b

6201-64-5
$C_{18}H_{14}N_2O$
274.31
O=C(C=CC(=Nc(ccc(Nc(cccc1)c1)c2)c2)C=3)C3
2,5-Cyclohexadien-1-one, 4-((4-(phenylamino)phenyl)-imino)- (9CI)

6201-86-1
$C_7H_7NO_6S$
233.20
Benzoic acid, 3-amino-2-hydroxy-5-sulfo- (9CI)
Salicylic acid, 3-amino-5-sulfo-(8CI)

6201-87-2
$C_7H_7NO_6S$
233.20
Benzoic acid, 5-amino-2-hydroxy-3-sulfo- (9CI)
Salicylic acid, 5-amino-3-sulfo-(8CI)

6203-88-9
$C_6H_{10}O_2$
114.14
2-Buten-2-ol, acetate (8CI,9CI)

6206-10-6
$C_{16}H_{18}N_2O_8S$
398.39
Uridine, 2'-(4-methylbenzenesulfonate) (9CI)
2'-O-(p-Toluenesulfonyl)-uridine
Uridine, 2'-p-toluene-sulfonate (7CI,8CI)

6214-20-6
$C_7H_7NO_5S$
217.20
O=S(=O)(OC)c(ccc(N(=O)=O)c1)c1
Methyl-4-nitrobenzenesulfonate
Benzenesulfonic acid, 4-nitro-, methyl ester (9CI)

6219-66-5
$C_{20}H_{14}O_{10}$
414.33
O=C(c(c(c(O)c(c1C(=O)c(c2cc(O)c3C(=O)O)c3C(=O)O)C2=O)CC)c1O)C
1,2-Anthracenedicarboxylic acid, 7-acetyl-6-ethyl-9,10-dihydro-3,5,8-trihydroxy-9,10-dioxo- (9CI)

6219-71-2
$C_6H_5ClN_2.H_2O_4S$
238.66
p-Phenylenediamine, 2-chloro-, sulfate
3-Chloro-4-aminoaniline sulfate
2-Chloro-1,4-benzenediamine sulfate
2-Chloro-p-phenylenediamine sulfate
C.I. 76066
C.I. Oxidation Base 13A
2-Cl-p-PD
Fourrine 81
Fourrine SO
NCI-C03316
Renal SO

6219-77-8
$C_6H_7N_3O_2.2ClH$
225.36
4-Nitro-o-phenylenediamine HCl
1,2-Benzenediamine, 4-nitro-, dihydrochloride (9CI)
Fourrine 4G
4-Nitro-1,2-benzenediamine dihydrochloride
o-Phenylenediamine, 4-nitro-, dihydrochloride

6219-89-2
$C_{13}H_{14}N_2O$
214.25
Oc(ccc(Nc(ccc(N)c1C)c1)c2)c2
Phenol, p-(4-amino-m-toluidino)- (8CI)
4-((4-Amino-3-methylphenyl)-amino)phenol
Phenol, 4-((4-amino-3-methylphenyl)amino)- (9CI)

6221-88-1
$C_{15}H_{34}OSi$
258.52
Silane, (dodecyloxy)trimethyl- (8CI,9CI)

6221-90-5
$C_{19}H_{42}OSi$
314.63
Silane, (hexadecyloxy)-trimethyl- (7CI,8CI,9CI)

6221-95-0
$C_{17}H_{34}O_2$
270.46
O=C(OCCCCCCCCCCCCCC)CC
Myristyl propionate
1-Tetradecanol, propanoate (9CI)

6222-63-5
$C_{20}H_{18}N_4O_6S.Na$
465.42
C.I. Acid Red 137, Monosodium salt (8CI)
2-Naphthalenesulfonic acid, 7-(acetylamino)-3-((4-(acetylamino)phenyl)azo)-4-hydroxy-, monosodium salt (9CI)

6225-10-1
$C_6H_{13}NO$
115.17
Pentanamide, N-methyl- (9CI)
Valeramide, N-methyl- (8CI)

6227-02-7
$C_{26}H_{19}N_5O_8S_2.2Na$
639.55
2-Naphthalenesulfonic acid, 7-amino-4-hydroxy-3-((5-hydroxy-6-(phenylazo)-7-sulfo-2-naphthalenyl)azo)-, disodium salt (9CI)

6227-14-1
$C_{30}H_{25}N_5O_8S_2.2Na$
693.64
2-Naphthalenesulfonic acid, 4-hydroxy-3-((2-methoxy-5-methyl-4-((4-sulfophenyl)-azo)phenyl)azo)-7-(phenylamino)-, disodium salt (9CI)

6228-73-5
C_4H_7NO
85.10
O=C(N)C(C1)C1
Cyclopropanecarboxamide (9CI)
AI3-62011
Cyclopropylcarboxamide

6232-56-0
$C_{15}H_{14}Cl_2N_4O_3$
369.18
O=N(=O)c(cc(c(N=Nc(ccc(N(CCO)C)c1)c1)c2Cl)Cl)c2
Ethanol, 2-((4-((2,6-dichloro-4-nitrophenyl)azo)phenyl)-methylamino)- (9CI)
4-(Methyl(β-hydroxyethyl)-amino)-2',6'-dichloro-4'-nitroazobenzene

6232-88-8
$C_8H_7BrO_2$
215.05
O=C(O)c(ccc(c1)CBr)c1
Benzoic acid, 4-(bromomethyl)- (9CI)
4-(Bromomethyl)benzoic acid
α-Bromo-p-toluic acid

6237-24-7
$C_{18}H_{20}N_4O_2$
324.42
Acridine, 9-((3-(dimethylamino)propyl)amino)-3-nitro
C-257
3-Nitro-9-(3'-dimethylaminopropylamino)acridine
1,3-Propanediamine, N,N-dimethyl-N'-(3-nitro-9-acridinyl)- (9CI)

6237-86-1
$C_4H_5N_3S_2$
159.22
1,3,5-Triazine-2,4(1H,3H)-dithione, 6-methyl- (9CI)
s-Triazine-2,4-dithiol, 6-methyl- (8CI)
s-Triazine-2,4(1H,3H)-dithione, 6-methyl-

6247-34-3
$C_{22}H_{17}N_3O_6S$
451.44
O=C(Nc(ccc(Nc(c(c(C(=O)c(c1ccc2)c2)c(N)c3S(=O)(=O)O)C1=O)c3)c4)c4)C
2-Anthracenesulfonic acid, 4-((4-(acetylamino)phenyl)-amino)-1-amino-9,10-dihydro-9,10-dioxo- (9CI)

6250-23-3
$C_{18}H_{14}N_4O$
302.31
Oc(ccc(N=Nc(ccc(N=Nc(cccc1)c1)c2)c2)c3)c3
C.I. Disperse Yellow 23 (8CI)
Phenol, 4-((4-(phenylazo)-phenyl)azo)- (9CI)
4-((4-(Phenylazo)phenyl)azo)-phenol

6253-10-7
$C_{22}H_{16}N_4O$
352.37
Oc(ccc(N=Nc(c(c(c(N=Nc(cccc1)c1)c2)ccc3)c3)c2)c4)c4
Phenol, 4-((4-(phenylazo)-1-naphthalenyl)azo)- (9CI)
4-((4-(Phenylazo)-1-naphthalenyl)azo)phenol

6259-42-3
$C_7H_8ClN.ClH$
178.05
Benzenamine, 5-chloro-2-methyl-, hydrochloride (9CI)
5-Chloro-2-methylbenzenamine hydrochloride

6259-76-3
$C_{13}H_{18}O_3$
222.28
O=C(OCCCCCC)c(c(O)ccc1)c1
Benzoic acid, 2-hydroxy-, hexyl ester
AI3-07842
Hexyl 2-hydroxybenzoate

6260-97-5
$C_{17}H_{14}ClNO_2$
299.74
1-(p-Chlorobenzoyl)-5-methoxy-2-methylindole
Indole, 1-(p-chlorobenzoyl)-5-methoxy-2-methyl-
1H-Indole, 1-(4-chlorobenzoyl)-5-methoxy-2-methyl-

6262-21-1
$C_{20}H_8Cl_4O_5$
470.09

O=C(OC(c(c(Oc1cc(O)cc2)cc(O)c3)c3)(c12)c4c(c(c(c5Cl)Cl)Cl)Cl)c45
Spiro(isobenzofuran-1(3H),9'-(9H)xanthen)-3-one, 4,5,6,7-tetrachloro-3',6'-dihydroxy-(9CI)

6262-51-7
C_3HCl_5
214.29
Cyclopropane, pentachloro
Pentachlorocyclopropane

6263-38-3
$C_8H_{10}N_2O_2$
166.20
Urea, 1-hydroxy-3-methyl-1-phenyl
N-Phenyl-N-hydroxy-N'-methylurea

6265-01-6
$C_{14}H_{13}NO_3S$
275.32
O=S(=O)(O)c(c(ccc1N)C=Cc(ccc2)c2)c1
Benzenesulfonic acid, 5-amino-2-(2-phenylethenyl)-(9CI)

6265-05-0
$C_7H_8N_2O_3$
168.15
Phenol, 2-amino-4-methyl-3-nitro- (9CI)
p-Cresol, 2-amino-3-nitro-(8CI)

6265-06-1
$C_7H_8N_2O_3$
168.15
Phenol, 2-amino-4-methyl-5-nitro- (9CI)
p-Cresol, 2-amino-5-nitro-(8CI)

6265-09-4
$C_{11}H_{16}N_2O_2$
208.26
Phenol, 5-(diethylamino)-4-methyl-2-nitroso- (9CI)
p-Cresol, 5-(diethylamino)-2-nitroso- (8CI)

6265-11-8
$C_9H_{12}N_2O_2$
180.21
Phenol, 5-(dimethylamino)-4-methyl-2-nitroso- (9CI)
p-Cresol, 5-(dimethylamino)-2-nitroso- (8CI)

6265-13-0
$C_8H_{11}NO$
137.18
Phenol, 4-methyl-3-(methylamino)- (9CI)
p-Cresol, 3-(methylamino)-(8CI)
4-Methyl-3-methylamino-phenol

6265-15-2
$C_{15}H_{14}N_2O_4$
286.29
Benzoic acid, 5-((4-aminobenzoyl)amino)-2-hydroxy-3-methyl- (9CI)
2,3-Cresotic acid, 5-(p-aminobenzamido)-(6CI,8CI)

6267-02-3
$C_{15}H_{15}N$
209.28
N(c(c(ccc1)C(c2cccc3)(C)C)c1)c23
9,9-Dimethylcarbazine
Acridan, 9,9-dimethyl- (8CI)
Acridine, 9,10-dihydro-9,9-dimethyl- (9CI)
AI3-00193
9,10-Dihydro-9,9-dimethylacridine

6268-49-1
$C_{15}H_{15}N_3O_2$
269.28
CN(C)c2ccc(N=Nc1ccc(cc1)C(O)=O)cc2
Benzoic acid, p-((p-(dimethylamino)phenyl)azo)- (8CI)
Benzoic acid, 4-((4-(dimethylamino)phenyl)azo)- (9CI)

6272-74-8
$C_{21}H_{23}N_2O_3.Cl$
386.91
Pyridinium, 1-(2-hydroxyethylcarbamoylmethyl)-, chloride, dodecanoate
Emcol E-607
1-(((2-Hydroxyethyl)carbamoyl)methyl)pyridinium chloride laurate (ester)
Lapyrium chloride
N-(Lauroylcolamenoformylmethyl)pyridinium chloride
NSC-33659
Pyridinium, 1-(2-oxo-2-((2-((1-oxododecyl)oxy)ethyl)amino)ethyl)-, chloride

6274-12-0
$C_4H_{11}N.BrH$
154.04
Ethanamine, N-ethyl-, hydrobromide (9CI)
Diethylamine, hydrobromide (8CI)
NSC-35762

6274-27-7
$C_8H_6Cl_2O_2S$
237.11
O=C(O)CSc(c(ccc1Cl)Cl)c1
Acetic acid, ((2,5-dichlorophenyl)thio)- (9CI)
((2,5-Dichlorophenyl)thio)acetic acid

6275-02-1
$C_{14}H_{10}N_2O_4$
270.23
2,2-Dinitrostilbene
Benzene, 1,1'-(1,2-ethenediyl)-bis(2-nitro-
2,2-Dinitroestilbeno (Spanish)
Dinitro-2,2 stilbene (French)
2,2'-Dinitrostilbene

6281-42-1
$C_5H_{11}N_3O$
129.14
O=C(N(CC1)CCN)N1
2-Imidazolidinone, 1-(2-aminoethyl)- (9CI)
AI3-24564
1-(2-Aminoethyl)-2-imidazolidinone
1-(2-Aminoethyl)-2-imidazolidone

6283-25-6
$C_6H_5ClN_2O_2$
172.58
O=N(=O)c(ccc(c1N)Cl)c1
Aniline, 2-chloro-5-nitro
2-Chloro-5-nitroaniline

6284-40-8
$C_7H_{17}NO_5$
195.21
OCC(O)C(O)C(O)C(O)CNC
Meglumine
1-Deoxy-1-(methylamino)-D-glucitol
Glucitol, 1-deoxy-1-(methylamino)-, D- (8CI)
D-Glucitol, 1-deoxy-1-(methylamino)- (9CI)
Meglumina (Spanish)
Megluminum (Latin)

6284-84-0
Unknown
Unknown
Dimethyl piperazine-cis

6285-57-0
$C_7H_5N_3O_2S$
195.21
O=N(=O)c(ccc(NC(=N)S1)c12)c2
2-Benzothiazolamine,

6-nitro

6287-38-3
$C_7H_4Cl_2O$
175.01
O=Cc(ccc(c1Cl)Cl)c1
Benzaldehyde, 3,4-dichloro- (9CI)
3,4-Dichlorobenzaldehyde

6289-46-9
$C_{10}H_{12}O_6$
228.20
O=C(OC)C(C(=O)CC(C1=O)C(=O)OC)C1
1,4-Cyclohexanedicarboxylic acid, 2,5-dioxo-, dimethyl ester (9CI)
AI3-14663
Dimethyl 2,5-dioxo-1,4-cyclohexanedicarboxylate
Dimethyl succinylsuccinate

6291-84-5
$C_4H_{12}N_2$
88.14
N(CCCN)C
1,3-Propanediamine, N-methyl- (9CI)
AI3-25443
(3-Aminopropyl)methylamine
N-Methyl-1,3-propanediamine

6291-87-8
$C_6H_9N_5O$
167.14
Nc1nc(N)nc(OCC=C)n1
s-Triazine, 2-(allyloxy)-4,6-diamino- (8CI)
AI3-60313
1,3,5-Triazine-2,4-diamine, 6-(2-propenyloxy)- (9CI)

6294-34-4
$C_6H_{12}Cl_3O_3P$
269.49
O=P(OCCCl)(OCCCl)CCCl
Phosphonic acid, (2-chloroethyl)-, bis(2-chloroethyl) ester (9CI)
AI3-25413

6294-40-2
$C_7H_{13}Br$
177.08
BrC(CCC(C1)C)C1
Cyclohexane, 1-bromo-4-methyl- (9CI)
1-Bromo-4-methylcyclohexane
p-Methylcyclohexyl bromide

6294-89-9
$C_2H_6N_2O_2$
90.07
O=C(OC)NN
Methyl carbazate
AI3-62053
Carbazic acid, methyl ester (8CI)
Hydrazinecarboxylic acid, methyl ester (9CI)
Methyl hydrazinecarboxylate

6295-15-4
$C_6H_{11}N_5O$
169.22
s-Triazine, 4,6-diamino-2-propyl (8CI)
NSC-11809
NSC-13876
1,3,5-Triazine-2,4-diamine, 6-propoxy- (9CI)

6299-66-7
$C_8H_{16}O_2$
144.21
Hexanoic acid, 4-ethyl- (8CI, 9CI)
NSC-44869

6300-07-8
$C_{13}H_{13}N_3O_4S.Na$
330.30
Benzenesulfonic acid, 3-((4-amino-3-methoxyphenyl)azo)-, monosodium salt (9CI)
3-(4-Amino-3-methoxyphenylazo)benzenesulfonic acid, sodium salt
3-((4-Amino-3-methoxyphenyl)azo)benzenesulfonic acid, sodium salt

6300-37-4
$C_{19}H_{16}N_4O$
316.39
Oc(c(cc(N=Nc(ccc(N=Nc(cccc1)c1)c2)c2)c3)C)c3
Phenol, 2-methyl-4-((4-(phenylazo)phenyl)azo)
Acetate Fast Yellow 5RL
Azofenol 4K
Benzene-1-azobenzene-4-azo-o-cresol
Celliton Discharge Yellow 5RL
Celliton Fast Yellow 5R
Celliton Yellow 5R
C.I. 26090
C.I. Disperse Yellow 7 (7CI, 8CI)
Cilla Fast Yellow 5R
Dianix Fast Yellow 5R
Dianix Yellow 5R-E
Disperse Dye Fast Yellow 4K
Disperse Fast Yellow 4K
Disperse Yellow 7
DZhp-4K
Kayalon Fast Yellow 4R
Kayalon Polyester Yellow 4R-E
Kayalon Polyester Yellow RF
Miketon Polyester Yellow 5R
Palanil Yellow 5R
Palanil Yellow 5RX
Resolin Yellow 5R
Samaron Yellow 5RL
Serisol Fast Yellow N 5RD
Supracet Fast Yellow 4R
Tersetile Yellow 5R
Tersetile Yellow 5RL
Tulasteron Fast Yellow 5R-B
Vonteryl Yellow 3R
Yellow Fast Dye 4K
Yellow Stable 4K

6300-50-1
$C_{22}H_{17}N_5O_7S_2.2Na$
573.49
2-Naphthalenesulfonic acid, 7-amino-4-hydroxy-3-((4-((4-sulfophenyl)azo)phenyl)azo)-, disodium salt (9CI)
7-Amino-4-hydroxy-3-(4-(4-(sulfophenyl)azo)phenylazo)-2-naphthalenesulfonic acid, disodium salt

6303-21-5
H_3O_2P
66.00
Phosphinic acid (9CI)
Hypophosphorous acid

6303-58-8
$C_{10}H_{12}O_3$
180.22
Butyric acid, 4-phenoxy
4-Phenoxybutyric acid

6305-71-1
$C_7H_{16}O$
116.20
1-Pentanol, 2,4-dimethyl- (9CI)
AI3-38565

6306-07-6
$C_{12}H_{10}O$
170.21
Acenaphthene-1-ol
1-Acenaphthenol (8CI)
1-Acenaphthylenol, 1,2-dihydro- (9CI)

6311-44-0
$C_{16}H_{18}N_2$
238.32
Diazene, bis(2,5-dimethylphenyl)- (9CI)
Azobenzene, 2,2',5,5'-tetramethyl- (8CI)
NSC-43206

6311-92-8
$C_{11}H_{17}N$
163.26
Pyridine, 3-hexyl- (9CI)
NSC-42647

6317-18-6
$C_3H_2N_2S_2$
130.19
N#CSCSC#N
Thiocyanic acid, methylene ester
Methylendithiokyanat (Czech)
Methylendirhodanid (Czech)
Methylenedirhodanid (German)
Methylene dithiocyanate

6319-21-7
$C_{13}H_{12}N_2O$
212.25
Diazene, (2-methoxy-phenyl)phenyl- (9CI)
Anisole, o-(phenylazo)- (6CI,7CI)
Azobenzene, 2-methoxy- (8CI)
o-Methoxyazobenzene
2-Methoxyazobenzene

6319-23-9
$C_{14}H_{14}N_2O_2$
242.28
Diazene, bis(3-methoxy-phenyl)- (9CI)
Anisole, 3,3'-azodi- (7CI)
Azobenzene, 3,3'-di-methoxy- (8CI)
3,3'-Dimethoxyazobenzene

6319-26-2
$C_{15}H_{16}N_2$
224.31
Diazene, (2,6-dimethyl-phenyl)(2-methylphenyl)- (9CI)
Azobenzene, 2,2',6-tri-methyl- (7CI,8CI)

6324-11-4
$C_8H_8O_4$
168.15
OC(=O)COc1ccccc1O
Acetic acid, (2-hydroxy-phenoxy)- (9CI)
Acetic acid, (o-hydroxyphenoxy)- (8CI)
NSC-30092

6325-93-5
$C_6H_6N_2O_4S$
202.20
O=S(=O)(N)c(ccc(N(=O)=O)c1)c1
Benzenesulfonamide, p-nitro

6328-48-9
$C_{10}H_{11}NO$
161.20
N#CCCOCc(cccc1)c1
Propanenitrile, 3-(phenyl-methoxy)- (9CI)
3-(Benzyloxy)propionitrile
3-(Phenylmethoxy)propane-nitrile

6333-15-9
$C_{13}H_9N_3O_5$
287.21
O=C(Nc(ccc(N(=O)=O)c1)c1)c(ccc(N(=O)=O)c2)c2
Benzamide, 4-nitro-N-(4-nitro-phenyl)- (9CI)
4,4'-Dinitrobenzanilide
4-Nitro-N-(4-nitrophenyl)-benzamide

6334-11-8
$C_9H_{13}N.ClH$
171.69
Cl.Cc1cc(C)c(N)c(C)c1
Aniline, 2,4,6-trimethyl, hydrochloride
Aminomesitylene hydrochloride
2-Aminomesitylene hydro-chloride
2-Amino-1,3,5-trimethyl-benzene hydrochloride
Benzenamine, 2,4,6-trimethyl-, hydrochloride (9CI)
Mesidin hydrochloride
Mesidine hydrochloride
Mesitylamine hydrochloride
2,4,6-Trimethylaniline hydro-chloride

6334-96-9
$C_8H_{16}Cl_2O$
199.14
O(CCCCCl)CCCCCl
Ether, bis(4-chlorobutyl)
Bis-(4-chlorbut-1-yl)-ether (German)
Bis(4-chlorobutyl) ether
Butane, 1,1'-oxybis(4-chloro- (9CI)
4,4'-Dichlorodibutyl ether
1,1'-Oxydi-4-chlorobutane

6334-97-0
$C_{12}H_{11}NO_5S$
281.28
O=C(Nc(ccc(c1cc(S(=O)(=O)O)c2)c2O)c1)C
2-Naphthalenesulfonic acid, 7-(acetylamino)-4-hydroxy- (9CI)
2-Naphthalenesulfonic acid, 7-acetamido-4-hydroxy- (8CI)

6335-02-0
$C_{10}H_{18}O_5$
218.25
Propanedioic acid, (2-methoxyethyl)-, diethyl ester (9CI)
Malonic acid, (2-methoxy-ethyl)-, diethyl ester (7CI, 8CI)

6342-56-9
$C_5H_{10}O_3$
118.13
O=C(C)C(OC)OC
2-Propanone, 1,1-dimethoxy- (9CI)
AI3-37790
1,1-Dimethoxy-2-propanone

6351-10-6
$C_9H_{10}O$
134.18
1-Indanol (8CI)
AI3-05996
2,3-Dihydro-1H-inden-1-ol
1H-Inden-1-ol, 2,3-dihydro- (9CI)

6358-07-2
$C_6H_5ClN_2O_3$
188.58
O=N(=O)c(c(cc(N)c1O)Cl)c1
Phenol, 2-amino-4-chloro-5-nitro
2-Amino-4-chloro-5-nitrophenol

6358-08-3
$C_6H_5ClN_2O_3$
188.56
Phenol, 2-amino-4-chloro-6-nitro- (8CI,9CI)

6358-09-4
$C_6H_5ClN_2O_3$
188.56
O=N(=O)c(cc(N)c(O)c1Cl)c1
2-Amino-6-chloro-4-nitro-phenol
6-Chloro-4-nitro-2-aminophenol
Phenol, 2-amino-6-chloro-4-nitro- (9CI)

6358-15-2
$C_6H_4Cl_3NO$
212.46
Phenol, 2-amino-3,4,6-trichloro- (8CI,9CI)
NSC-7536

6358-18-5
$C_{12}H_8ClN_3O_5$
309.67
Phenol, 4-chloro-2-((2,4-di-nitrophenyl)amino)- (9CI)
Phenol, 4-chloro-2-(2,4-di-nitroanilino)- (7CI,8CI)

6358-20-9
$C_{10}H_{14}N_2O_2$
194.22
Phenol, 5-(diethylamino)-2-nitroso- (8CI,9CI)

6358-23-2
$C_{12}H_9N_3O_5$
275.24
Oc1ccccc1Nc2ccc(cc2N(=O)=O)N(=O)=O
Phenol, o-(2,4-dinitroanilino)
2-(2,4-Dinitroanilino)phenol

6358-29-8
$C_{32}H_{28}N_4O_8S_2.2Na$
706.74
1,3-Naphthalenedisulfonic acid, 8-((4'-((4-ethoxyphenyl)azo)-3,3'-dimethyl-(1,1'-biphenyl)-4-yl)azo)-7-hydroxy-, disodium salt
Airedale Scarlet 3BD
Amanil Fast Scarlet 3B
Atlantic Scarlet 3B
Benzanil Scarlet 3B
Benzanol Brilliant Scarlet 3B
Benzo Red 3B
Benzyl Scarlet 3BS
Calcomine Scarlet 3B
Chloramine Red 3B
Chrome Leather Scarlet 3BS
C.I. 23630
C.I. Direct Red 39
Diamine Scarlet 3BA-CF
Diaphtamine Fast Scarlet
Diazol Scarlet 3B
Diphenyl Red 3BS
Diphenyl Scarlet 3BS
Direct Fast Scarlet 3B
Direct Red 39
Direct Scarlet 3BS
Erie Scarlet 3B
Fenamin Scarlet 3B
Kayaku Direct Scarlet 3B
Mitsui Direct Scarlet 3BX
Paper Scarlet 3BX
Paramine Fast Scarlet 3B
Phenamine Scarlet 3B
Pheno Fast Scarlet 4B
Pontamine Scarlet 3B
Shikiso Direct Scarlet 3B
Triazol Fast Scarlet 3B
Union Fast Scarlet 3B

6358-30-1
$C_{34}H_{22}Cl_2N_4O_2$
589.46
O(c(c(N=C1C(=C(Oc(c(N=2)cc(c(c(N3CC)ccc4)c4)c35)c5)C2C=6Cl)Cl)cc(c(c(N7CC)ccc8)c8)c79)c9)C16
Diindolo(3,2-b:3',2'-m)triphenodioxazine, 8,18-dichloro-5,15-diethyl-5,15-dihydro- (9CI)
C.I. Pigment Violet 23

6358-31-2
$C_{18}H_{18}N_4O_6$
386.34
O=C(Nc(c(OC)ccc1)c1)C(N=Nc(c(OC)cc(N(=O)=O)c2)c2)C(=O)C
CI Pigment Yellow 74
Butanamide, 2-((2-methoxy-4-nitrophenyl)azo)-N-(2-methoxyphenyl)-3-oxo- (9CI)
C.I. 11741
Dalamar Yellow
Hansa Brilliant Yellow 5GX
Luna Yellow
2-((2-Methoxy-4-nitrophenyl)-azo)-o-acetoacetanisidide
Permanent Yellow, Lead Free
Pigment Yellow 74
Ponolith Yellow Y

6358-36-7
$C_{21}H_{29}N_3.ClH$
359.92
Aniline, 4,4'-imidocarbonylbis(N,N-diethyl-, monohydrochloride (8CI)
Benzenamine, 4,4'-carbonimidoylbis(N,N-diethyl-, monohydrochloride (9CI)

6358-37-8
$C_{34}H_{30}Cl_2N_6O_4$
657.52
O=C(Nc(ccc(c1)C)c1)C(N=Nc(c(cc(c(ccc(N=NC(C(=O)C)C(=O)Nc(ccc(c2)C)c2)c3Cl)c3)c4)Cl)c4)C(=O)C
Butanamide, 2,2'-((3,3'-dichloro(1,1'-biphenyl)-4,4'-diyl)bis(azo))bis(N-(4-methylphenyl)-3-oxo- (9CI)

6358-53-8
$C_{18}H_{16}N_2O_3$
308.36
COc3ccc(OC)c(N=Nc1c(O)ccc2ccccc12)c3
2-Naphthol, 1-((2,5-dimethoxyphenyl)azo)
Cerven Rozpoustedlova 80 (Czech)
C.I. 12156
C.I. Solvent Red 80
Citrus Red 2
Citrus Red No. 2
2,5-Dimethoxybenzeneazo-β-naphthol
1-((2,5-Dimethoxyphenyl)azo)-2-naphthalenol
1-(2,5-Dimethoxyphenylazo)-2-naphthol
2,5-Dimethoxy-1-(phenylazo)-2-naphthol
1-(1-(2,5-Dimethoxyphenyl)-azo)-2-naphthol
1-(2,5-Dimethyloxyphenylazo)-2-naphthol

6358-57-2
$C_{37}H_{30}N_4O_{10}S_3.2Na$
832.82
2,7-Naphthalenedisulfonic acid, 3-((2,2'-dimethyl-4'-((4-(((4-methylphenyl)sulfonyl)oxy)phenyl)azo)(1,1'-biphenyl)-4-yl)azo)-4-hydroxy-, disodium salt (9CI)

6358-64-1
$C_8H_{10}ClNO_2$
187.64
O(c(c(cc(OC)c1N)Cl)c1)C
Aniline, 4-chloro-2,5-dimethoxy

6358-69-6
$C_{16}H_7O_{10}S_3.3Na$
524.38
1,3,6-Pyrenetrisulfonic acid, 8-hydroxy-, trisodium salt
C.I. 59040
C.I. Solvent Green 7
C.I. Solvent Green 9
D & C Green 8
D & C Green No. 8
11389 Green
Green No. 204
8-Hydroxy-1,3,6-pyrenetrisulfonic acid trisodium salt
8-Hydroxypyrene-1,3,6-trisulfonic acid sodium salt
Pyranine
Pyranine Concentrated
Solvent Green 7
Trisodium 1-hydroxy-3,6,8-pyrenetrisulfonate

6358-83-4
$C_{19}H_{20}N_8O$
376.37
O(c(c(N=Nc(c(N)cc(N)c1)c1)cc(N=Nc(c(N)cc(N)c2)c2)c3)c3)C
1,3-Benzenediamine, 4,4'-((4-methoxy-1,3-phenylene)bis(azo))bis- (9CI)

6358-85-6
$C_{32}H_{26}Cl_2N_6O_4$
629.54
O=C(Nc(cccc1)c1)C(N=Nc(c(cc(c(ccc(N=NC(C(=O)C)C(=O)Nc(cccc2)c2)c3Cl)c3)c4)Cl)c4)C(=O)C
Acetoacetanilide, 2,2'-((3,3'-dichloro-4,4'-biphenylylene)-diazo)bis
Amazon Yellow X2485
Benzidine Lacquer Yellow G
Benzidene Yellow ABZ-245
Benzidene Yellow WD-266 (Water dispersible)
Benzidene Yellow YB-1
Benzidine Yellow
Benzidine Yellow 45-2650
Benzidine Yellow 45-2680
Benzidine Yellow 45-2685
Benzidine Yellow E
Benzidine Yellow G
Benzidine Yellow GF
Benzidine Yellow GR
Benzidine Yellow GT
Benzidine Yellow GTR

Benzidine Yellow HG
Benzidine Yellow HG PLV
Benzidine Yellow Toner
Benzidine Yellow Toner YA-8081
Benzidine Yellow Toner YT-378
Bis(acetyl-N-phenylcarbamyl-methyl)-4,4'-disazo-3,3'-dichlorobiphenyl
Brilliant Yellow Slurry
Butanamide, 2,2'-((3,3'-dichloro(1,1'-diphenyl)-4,4'-diyl)bis(azo))bis(3-oxo-N-phenyl- (9CI)
Carnelio Yellow GX
C.I. 21090
C.I. Pigment Yellow 12
Dainichi Benzidine Yellow G
Dainichi Benzidine Yellow GRT
Dainichi Benzidine Yellow GT
Dainichi Benzidine Yellow GY
Dainichi Benzidine Yellow GYT
Diarylide Yellow AAA
Dairylide Yellow YT 553D
Daltolite Fast Yellow GT
Diarylanilide Yellow
2,2'-((3,3'-Dichloro(1,1'-biphenyl)-4,4'-diyl)-bis(azo))bis-(3-oxo-N-phenyl)-butanamide
Eljon Yellow BG
Graphtol Yellow A-HG
Hancock Yellow 10010
Helic Yellow GW
Helioyellow GW
Helio Yellow GWN
Irgalite Yellow BO
Irgalite Yellow BST
Irgalite Yellow BTR
Isol Benzidine Yellow G
Isol Benzidine Yellow G 2537
Isol Benzidine Yellow Gapropyl
Isol Benzidine Yellow Gbpropyl
Isol Benzidine Yellow G Special
Kromon Yellow GXT Conc
Kromon Yellow MTB
Light Yellow JB
Light Yellow JBO
Light Yellow JBT
Lodestone Yellow YB-57
Monolite Yellow GRA
Monolite Yellow 2GRA
Monolite Yellow GT
Monolite Yellow GTA
Monolite Yellow GTN
Monolite Yellow GTNA
Monolite Yellow GTS
NCI-C03269
No. 49 Conc. Benzidine Yellow
Permanent Yellow DHG
Permanent Yellow GHG
Pigment Yellow 12
Pigment Yellow GT
Rangoon Yellow
Recolite Yellow BG
Recolite Yellow BGT
Sanyo Benzidine Yellow-B
Segnale Light Yellow 2GR
Segnale Light Yellow 2GRT
Siloton Yellow GTX
Siloton Yellow 3GX
Symuler Fast Yellow GF
Verona Yellow X-1791
Vulcafor Fast Yellow GT
Vulcafor Fast Yellow GTA
Vulcol Fast Yellow GR
Zlut Pigment 12 (Czech)

6358-87-8
$C_{36}H_{28}Cl_2N_8O_6$
739.53
C.I. Pigment Red 38 (VAN) (8CI)
1H-Pyrazole-3-carboxylic acid, 4,4'-((3,3'-dichloro(1,1'-biphenyl)-4,4'-diyl)bis(azo))bis-(4,5-dihydro-5-oxo-1-phenyl-, diethyl ester (9CI)

6359-45-1
$C_{23}H_{29}N_2.Cl$
368.94
3H-Indolium, 2-(2-(4-(diethylamino)phenyl)ethenyl)-1,3,3-trimethyl-, chloride (9CI)

6359-85-9
$C_{23}H_{21}N_5O_6S_2.Na$
550.54
Benzenesulfonic acid, 4-(4,5-dihydro-3-methyl-4-((4-methyl-3-((phenylamino)-sulfonyl)phenyl)azo)-5-oxo-1H-pyrazol-1-yl)-, monosodium salt (9CI)
C.I. Acid Yellow 25, Monosodium salt (8CI)
1H-Pyrazole, 5-oxo-4,5-dihydro-3-methyl-4-((3-((phenylamino)sulfonyl)-4-methylphenyl)azo)-1-(4-sulfophenyl)-, sodium salt

6359-90-6
$C_{16}H_{13}ClN_4O_4S.Na$
415.79
Benzenesulfonic acid, 4-chloro-3-(4,5-dihydro-3-methyl-5-oxo-4-(phenylazo)-1H-pyrazol-1-yl)-, monosodium salt (9CI)
C.I. Acid Yellow 34, Monosodium salt (8CI)

6359-97-3
$C_{16}H_{12}Cl_2N_4O_4S.Na$
450.23
Benzenesulfonic acid, 2,5-dichloro-4-(4,5-dihydro-3-methyl-5-oxo-4-(phenylazo)-1H-pyrazol-1-yl)-, sodium salt (9CI)

6359-98-4
$C_{16}H_{12}Cl_2N_4O_7S_2.2Na$
553.29
Benzenesulfonic acid, 2,5-dichloro-4-(4,5-dihydro-3-methyl-5-oxo-4-((4-sulfophenyl)azo)-1H-pyrazol-1-yl)-, disodium salt (9CI)
C.I. Acid Yellow 17, Disodium salt (VAN) (8CI)

6360-54-9
$C_{33}H_{28}N_8O_6S.2Na$
710.63
Benzoic acid, 5-((4'-((2,6-diamino-3-methyl-5-((4-sulfophenyl)azo)phenyl)azo)(1,1'-biphenyl)-4-yl)azo)-2-hydroxy-3-methyl-, disodium salt (9CI)

6362-79-4
$C_8H_5O_7S.Na$
268.18
[Na+].OC(=O)c1cc(cc(c1)S([O-])(=O)=O)C(O)=O
Benzenesulfonic acid, 3,5-dicarboxy-, sodium salt
Benzene-1,3-dicarboxylic acid, 5-sulfo-, monosodium salt
3,5-Dicarboxybenzenesulfonic acid, sodium salt
3,5-Dikarboxybenzensulfonan sodny (Czech)
Kyselina 3,5-dikarboxybenzensulfonova sodny (Czech)

6363-87-7
$C_{15}H_9NO_3$
251.24
2-Anthracenecarboxaldehyde, 1-amino-9,10-dihydro-9,10-dioxo- (9CI)
1-Aminoanthraquinone-2-carboxaldehyde
2-Anthraldehyde, 1-amino-9,10-dihydro-9,10-dioxo- (8CI)
2-Anthraquinonecarboxaldehyde, 1-amino-

6364-34-7
$C_{12}H_{13}N_5$
227.24
1,3-Benzenediamine, 4-((4-aminophenyl)azo)- (9CI)
m-Phenylenediamine, 4-((p-aminophenyl)azo)- (8CI)

6365-50-0
$C_{18}H_{11}NO_2$
273.28
O=C(OC(c1cccc2)=Cc(nc(c(ccc3)c4)c3)c4)c12
1(3H)-Isobenzofuranone, 3-(2-quinolinylmethylene)- (9CI)

6365-72-6
Unknown
Unknown
Isopropanolamine 2,4-dichlorophenoxyacetate

6365-73-7
Unknown
Unknown
Morpholine 2,4-dichlorophenoxyacetate

6365-83-9
$C_{10}H_{12}N_2O_5.H_3N$
257.28
[NH4+].CCC(C)c1cc(cc(N(=O)=O)c1[O-])N(=O)=O
Phenol, 2-sec-butyl-4,6-dinitro-, ammonium salt
Butophen
2-sec-Butyl-4,6-dinitrophenol, ammonium salt
Chemox Selective
4,6-Dinitro-2-sec.butylfenolate ammony (Czech)
4,6-Dinitro-o-sec-butylphenol ammonium salt
4,6-Dinitro-2-sec-butylphenol ammonium salt
Dinoseb (Amine)
DNBP Ammonium salt
Dow Selective
2-(1-Methyl-n-propyl) 4,6-dinitrophenol, ammonium salt
Phenol, 2-sec-butyl-4,6-dinitro-, amine deriv.
Selective
Sinox W

6368-72-5
$C_{24}H_{21}N_5$
379.50
N(=Nc(cccc1)c1)c(ccc(N=Nc(c(c(ccc2)cc3)c2)c3NCC)c4)c4
2-Naphthylamine, N-ethyl-1-((p-(phenylazo)phenyl)azo)
Ceres Red 7B
Cerven Rozpoustedlova 19 (Czech)
C.I. 26050
C.I. Solvent Red 19
N-Ethyl-1-((p-(phenylazo)-phenyl)azo)-2-naphthalenamine
N-Ethyl-1-((4-(phenylazo)-phenyl)azo)-2-naphthylamine
Fat Red 7B
Hexatype Carmine B
Lacquer Red V3B
2-Naphthalenamine, N-ethyl-1-((4-(phenylazo)phenyl)azo)-(9CI)
Oil Violet
Organol Bordeaux B
(Phenylazo-4-phenylazo)-1-ethylamino-2-naphthalene
1-(4-Phenylazo-phenylazo)-2-ethylaminonaphthalene
Solvent Red 19
Special Blue X 2137
Sudan 7B
Sudan Red 7B
Sudanrot 7B
Typogen Carmine

6369-59-1
$C_7H_{10}N_2.xH_2O_4S$
808.75
Cc1cc(N)ccc1N.OS(O)(=O)=O
Toluene-2,5-diamine, sulfate
1,4-Benzenediamine, 2-methyl-, sulfate (9CI)
C.I. 76043
C.I. Oxidation Base 4
Fouramine STD
NCI-C01832
2,5-TDS
p-Toluenediamine sulfate
2,5-Toluenediamine sulfate

6369-96-6
$C_8H_5Cl_3O_3.C_3H_9N$
314.49
Acetic acid, (2,4,5-trichlorophenoxy)-, Compd. with N,N-dimethylmethanamine (1:1) (9CI)
Acetic acid, (2,4,5-trichlorophenoxy)-, Compd. with trimethylamine (1:1) (8CI)
2,4,5-T amines

6369-97-7
$C_8H_5Cl_3O_3.C_2H_7N$
300.56
Acetic acid, (2,4,5-trichlorophenoxy)-, Compd. with N-methylmethanamine (1:1) (9CI)
Acetic acid, (2,4,5-trichlorophenoxy)-, Compd. with dimethylamine (1:1) (8CI)
Caswell No. 881E
EPA Pesticide Chemical Code 082019
TA 20 (VAN)
2,4,5-Trichlorophenoxyacetic acid dimethylamine salt

6370-08-7
$C_{20}H_{11}CrN_2O_9S_2.2Na$
585.41
Chromate(2-), hydroxy(3-hydroxy-4-((1-hydroxy-8-sulfo-2-naphthalenyl)azo)-1-naphthalenesulfonato(4-))-, disodium (9CI)

6370-93-0
$C_{16}H_{14}N_2O_{16}S_4.4Na$
710.50
2,6-Anthracenedisulfonic acid, 9,10-dihydro-1,5-dihydroxy-9,10-dioxo-4,8-bis((sulfomethyl)amino)-, tetrasodium salt (9CI)

6373-20-2
$C_{34}H_{12}Cl_4O_2$
594.28
O=C(c(c(c(c1c(c(c(c(c(c(c(C2=O)cc(c3)Cl)c3)cc4Cl)c2c5)c46)c5)cc7)c6c8Cl)c8)ccc9Cl)c9)c17
Anthra(9,1,2-cde)benzo(rst)pentaphene-5,10-dione, 3,12,16,17-tetrachloro-(9CI)

6373-73-5
$C_{12}H_{10}N_4O_4$
274.21
O=N(=O)c(ccc(Nc(ccc(N)c1)c1)c2N(=O)=O)c2
p-Phenylenediamine, N-(2,4-dinitrophenyl)- (8CI)
1,4-Benzenediamine, N-(2,4-dinitrophenyl)- (9CI)
N-(2,4-Dinitrophenyl)-1,4-benzenediamine

6373-74-6
$C_{18}H_{13}N_4O_7S.Na$
452.40
OS(=O)(=O)c2cc(Nc1ccc(cc1N(=O)=O)N(=O)=O)ccc2Nc3ccccc3
Benzenesulfonic acid, 2-anilino-5-(2,4-dinitroanilino)-, monosodium salt
Acid Fast Yellow AG
Acid Fast Yellow E5R
Acid Leather Light Brown G
Acid Orange No. 3
Acid Yellow E
Airedale Yellow E
Amido Yellow E
Amido Yellow EA
Amido Yellow EA-CF
Anthralan Yellow RRT
Benzenesulfonic acid, 5-((2,4-dinitrophenyl)amino)-2-(phenylamino)-, monosodium salt
C.I. 10385
C.I. Acid Orange 3
Derma Yellow P
Erio Fast Yellow AE
Erio Fast Yellow AEN
Erionyl Yellow E-AEN
Fast Light Yellow E
Fenalan Yellow E
Kiton Fast Yellow A
Light Fast Yellow ES
Lissamine Fast Yellow AE
NCI-C54911
Nylomine Acid Yellow B-RD
Sodium 4-(2,4-dinitroanilino)-diphenylamine-2-sulfonate
Superian Yellow R
Tectilon Orange 3GT
Tertracid Light Yellow 2R
Vondacid Fast Yellow AE
Xylene Fast Yellow ES

6375-17-3
$C_9H_{11}NO_2$
165.19
O=C(Nc(c(O)ccc1C)c1)C
Acetamide, N-(2-hydroxy-5-methylphenyl)- (9CI)
m-Acetotoluidide, 6'-hydroxy- (8CI)
N-(2-Hydroxy-5-methylphenyl)-acetamide

6375-27-5
$C_{12}H_{15}NO_4$
237.25
O=C(Nc(c(OC)ccc1OC)c1)CC(=O)C
Acetoacetanilide, 2',5'-dimethoxy- (8CI)
Acetoacet-2,5-dimethoxyanilide
Acetoacetyl-2,5-dimethoxy-anilide
Butanamide, N-(2,5-dimethoxy-phenyl)-3-oxo- (9CI)
N-(2,5-Dimethoxyphenyl)-3-oxobutanamide

6375-46-8
$C_{12}H_{18}N_2O$
206.28
O=C(Nc(cccc1N(CC)CC)c1)C
Acetamide, N-(3-(diethyl-amino)phenyl)- (9CI)
N-(3-(Diethylamino)phenyl)-acetamide

6375-47-9
$C_9H_{12}N_2O_2$
180.19
O=C(Nc(ccc(OC)c1N)c1)C
Acetamide, N-(3-amino-4-methoxyphenyl)- (9CI)
3-Amino-4-methoxyacetanilide
N-(3-Amino-4-methoxyphenyl)-acetamide

6375-55-9
$C_{32}H_{24}N_8O_8S_2.2Na$
758.74
(1,1'-Biphenyl)-2,2'-disulfonic acid, 4,4'-bis((4,5-dihydro-3-methyl-5-oxo-1-phenyl-1H-pyrazol-4-yl)azo)-, disodium salt
Acid Anthracene Yellow GR
Acid Fast Yellow MR
Acid Leather Yellow CRS
Acid Yellow 42
Acid Yellow K
Airedale Yellow 3GM
Amacid Fast Yellow RS
Belacid Milling Yellow R
Benzyl Fast Yellow RS
Calcocid Milling Yellow R
C.I. 22910
C.I. Acid Yellow 42
C.I. Acid Yellow 42 Disodium salt
Coomassie Yellow R
Coomassie Yellow RP
Cyanine Fast Yellow M
Erio Fast Yellow RL
Fast Silk Yellow SH
KCA Acid Milling Yellow M
Midlon Yellow Propyl
Milling Yellow 3G
Milling Yellow 3J
Milling Yellow RX
Naphthalene Leather Yellow 2G
Optanol Yellow R
Pharmacine Yellow R
Pharmatex Yellow G
Shikiso Acid Fast Yellow MR
Sulfonine Yellow CSR
Sulphon Yellow RS-CF
Suminol Milling Yellow MR
Supranol Yellow R
Symulon Acid Fast Yellow MR
Tertracid Milling Yellow R
Xylene Milling Yellow SH

6378-65-0
$C_{12}H_{24}O_2$
200.36
O=C(OCCCCCC)CCCCC
Hexanoic acid, hexyl ester
Hexyl caproate
n-Hexyl hexanoate
Hexyl hexoate

6379-37-9
$C_8H_{19}N.CH_5AsO_3$
269.21
Octylammonium methane-arsonate
Caswell No. 612
EPA Pesticide Chemical Code 013804
Methanearsonous acid, Compd. with octylamine (1:1)
Octylamine, methanearsonate (1:1)
Octylammonium methylarsonate

6379-46-0
$C_6HCl_3N_2O_4$
271.44
Clc1c(Cl)c(cc(N(=O)=O)c1Cl)N(=O)=O
Benzene, 4,6-dinitro-1,2,3-trichloro
4,6-Dinitro-1,2,3-tri-chlorobenzene
1,2,3-Trichloro-4,6-dinitro-benzene
Vancide PB

6381-61-9
$C_7H_5NO_3S.H_3N$
200.23
1,2-Benzisothiazolin-3-one, 1,1-dioxide, ammonium salt
Ammonium saccharin
1,2-Benzisothiazol-3(2H)-one, 1,1-dioxide, ammonium salt (9CI)
Daramin
Saccharin ammonium
Saccharinate ammonium (French)

6381-77-7
$C_6H_8O_6.Na$
199.12
[Na+].OCC(O)C1OC(=O)C(=C1O)[O-]
Sodium erythorbate
neo-Cebitate
D-Erythro-hex-2-enonic acid, γ-lactone, monosodium salt (9CI)
Isona
Mercate 20
Sodium isoascorbate
Uantox sebate

6386-38-5
$C_{18}H_{28}O_3$
292.42
O=C(OC)CCc(cc(c(O)c1C(C)(C)C)C(C)(C)C)c1
Benzenepropanoic acid, 3,5-bis(1,1-dimethylethyl)-4-hydroxy-, methyl ester (9CI)
Hydrocinnamic acid, 3,5-di-tert-butyl-4-hydroxy-, methyl ester

6387-89-9
$C_5H_8O_3$
116.12
Oxiranemethanol, acetate (9CI)
Glycidol acetate
1-Propanol, 2,3-epoxy-, acetate (8CI)

6388-26-7
$C_{50}H_{38}N_{12}O_{18}S_4.5Na$
1338.06
Chlorantine Fast Green BLL
Benzoic acid, 2-hydroxy-5-((4-((4-((8-hydroxy-7-((4-((8-hydroxy-3,6-disulfo-1-naph-thalenyl)azo)-2-methoxy-5-methylphenyl)azo)-3,6-di-sulfo-1-naphthalenyl)amino)-6-(phenylamino)-1,3,5-triazin-2-yl)amino)phenyl)azo)-, pentasodium salt (9CI)

6393-42-6
$C_7H_7N_3O_4$
197.17
Cc1cc(N(=O)=O)c(N)c(c1)N(=O)=O
p-Toluidine, 2,6-dinitro
4-Amino-3,5-dinitrotoluene
Aniline, 2,6-dinitro-4-methyl-
Benzenamine, 4-methyl-2,6-di-nitro-
2,6-Dinitro-4-methylaniline
2,6-Dinitro-p-toluidine

4-Methyl-2,6-dinitroaniline

6406-45-7
$C_{32}H_{23}N_5O_7S_2$.2Na
697.66
1-Naphthalenesulfonic acid, 4-hydroxy-3-((4-((4-(phenylamino)-3-sulfophenyl)azo)-1-naphthalenyl) azo)-, disodium salt (9CI)
C.I. 26690
C.I. Acid Black 26B (8CI)
C.I. Acid Black 26:2
Nerol 2B

6406-56-0
$C_{22}H_{16}N_4O_4$S.Na
455.43
Benzenesulfonic acid, 4-((4-((2-hydroxy-1-naphthalenyl)-azo)phenyl)azo)-, monosodium salt (9CI)
C.I. Acid Red 151, Monosodium salt (8CI)

6408-22-6
$C_{34}H_{18}CrN_8O_{26}S_6$.H.6Na
1222.90
Chromate(7-), bis(4-hydroxy-3-((2-hydroxy-5-nitro-3-sulfophenyl)azo)-6-((sulfomethyl)amino)-2-naphthalenesulfonato(5-))-, hexasodium hydrogen (9CI)
C.I. Acid Black 84

6408-31-7
$C_{16}H_{10}ClCrN_4O_9S_2$.2Na
599.81
C.I. Acid Red 183 (8CI)
Acid Red 4ZhM
Benzenesulfonic acid, 5-chloro-3-[[4,5-dihydro-3-methyl-5-oxo-1-(3-sulfophenyl)-1H-pyrazol-4-yl]azo]-2-hydroxy-, chromium complex
Bucolan Red GRE
C.I. 18800
Chromate(2-), (5-chloro-3-((4,5-dihydro-3-methyl-5-oxo-1-(3-sulfophenyl)-1H-pyrazol-4-yl)azo)-2-hydroxybenzenesulfonato(4-))hydroxy-, disodium
Chrome Intra Red GRE
Chromolan Red GRE
Efdolan Red GRN
Gycolan Red GRL
Inochrome Red 3J
Neolan Red GRE
Nyasol Red GGS
Palatin Fast Red GREN
Palatine Fast Red GREN
Pilate Fast Red GREN
Super Fast Red GRE
Tertroxane Red G
Vitrolan Red GRE

6408-33-9
Unknown
Unknown
C.I. Acid Orange 61 (9CI)

6408-78-2
$C_{20}H_{14}N_2O_5$S.Na
417.39
2-Anthracenesulfonic acid, 1-amino-9,10-dihydro-9,10-dioxo-4-(phenylamino)-, monosodium salt (9CI)

6410-10-2
$C_{16}H_{11}N_3O_3$
293.30
O=N(=O)c(ccc(N=Nc(c(c(ccc1)cc2)c1)c2O)c3)c3
2-Naphthol, 1-((4-nitrophenyl)azo)
2-Naphthalenol, 1-((4-nitrophenyl)azo)-
1-((4-Nitrophenyl)azo)-2-naphthol
Para Red

6410-20-4
$C_{18}H_{16}N_2O_2$
292.33
2-Naphthalenol, 1-((2-methoxy-5-methylphenyl)azo)- (9CI)
C.I. Solvent Red 17
2-Naphthol, 1-((6-methoxy-m-tolyl)azo)- (8CI)

6410-30-6
$C_{24}H_{17}ClN_4O_4$
460.85
O=C(Nc(ccc(c1)Cl)c1)c(c(O)c(N=Nc(c(ccc2N(=O)=O)C)c2)c(c3ccc4)c4)c3
2-Naphthalenecarboxamide, N-(4-chlorophenyl)-3-hydroxy-4-((2-methyl-5-nitrophenyl)azo)- (9CI)

6410-38-4
$C_{24}H_{17}Cl_2N_3O_3$
466.30
O=C(Nc(c(OC)ccc1)c1)c(c(O)c(N=Nc(c(ccc2Cl)Cl)c2)c(c3ccc4)c4)c3
2-Naphthalenecarboxamide, 4-((2,5-dichlorophenyl)azo)-3-hydroxy-N-(2-methoxyphenyl)- (9CI)

6416-39-3
$C_{16}H_{16}$
208.30
1H-Indene, 2,3-dihydro-1-methyl-3-phenyl- (9CI)
Indan, 1-methyl-3-phenyl-(8CI)

6416-57-5
$C_{16}H_{14}N_4$
262.34
N(=Nc(c(N)cc(N)c1)c1)c(c(c(ccc2)cc3)c2)c3
m-Phenylenediamine, 4-(1-naphthylazo)
C.I. 11285
C.I. Solvent Brown PR
C.I. Solvent Brown 1
Fat Brown 2G
Fat Brown 2R
Fat Brown RR
Grasan Brown DT New
Hexatype Brown N
HNED Rozpoustedlova 1 (Czech)
HNED Sudan RR (Czech)
Lithofor Brown A
1-Naphthalenazo-2',4'-diaminobenzene
4-(1-Naphthalenylazo)-1,3-phenylenediamine
4-(1-Naphthylazo)-m-phenylenediamine
Organol Brown 2R
Organol 2R
Resinol Brown RRN
Resinol RRN
Silotras Brown TRN
Sudan Brown RR
Sudan Brown YR
Typogen Brown N

6416-68-8
$C_{24}H_{16}N_3O_3$S.Na
449.48
OS(=O)(=O)c1cc(ccc1C=Cc2ccccc2)n5nc4ccc3ccccc3c4n5
Benzenesulfonic acid, 5-(2H-naphtho(1,2-d)triazol-2-yl)-2-(2-phenylethenyl)-, sodium salt
C.I. 40645
C.I. Fluorescent Brightener 46
C.I. Fluorescent Brightening Agent 46, Sodium salt (8CI)
Fluorescent Brightener 46
2H-Naphtho(1,2-d)triazole, 2-(4-(2-phenylethenyl)-3-sulfophenyl)-, sodium salt
Sodium 2-(4-styryl-3-sulfophenyl)-2H-naphtho-(1,2-d)-triazole
2-Stilbenesulfonic acid, 4-(2H-naphtho(1,2-d)triazol-2-yl)-, sodium salt
Tinopal RBS
Tinopal RBS 200

6417-46-5
$C_{40}H_{35}N_3O_3$S
637.78
O=S(=O)(O)c(c(cc(Nc(ccc(c1)C(=C(C=CC(=Nc(cccc2C)c2)C=3)C3)c(ccc(Nc(cccc4C)c4)c5)c5)c1)c6)C)c6
Benzenesulfonic acid, 2-

methyl-4-((4-((4-((3-methylphenyl)amino)phenyl)(4-((3-methylphenyl)imino)-2,5-cyclohexadien-1-ylidene)methyl)phenyl)amino)-(9CI)

6417-83-0
$C_{21}H_{14}N_2O_6S.Ca$
462.48
D & C Red No. 34
CI 15880:1 (Ca Salt)
C-Rot 14 (Germany)
Germany: C-Rot 14
3-Hydroxy-4-((1-sulfo-2-naphthalenyl)azo)-2-naphthalenecarboxylic acid, calcium salt(1:1)
Japan: Red No. 220
2-Naphthalenecarboxylic acid, 3-hydroxy-4-((1-sulfo-2-naphthalenyl)azo)-, calcium salt (1:1) (9CI)
Pigment Red 63:1
Red No. 220 (Japan)

6419-19-8
$C_3H_{12}NO_9P_3$
299.07
O=P(O)(O)CN(CP(=O)(O)O)CP(=O)(O)O
Phosphonic acid, (nitrilotris(methylene))tri
Aminotri(methylenephosphonic acid)
Aminotri(methylphosphonic acid)
Aminotris(methanephosphonic acid)
Aminotris(methylphosphonic acid)
Dequest 2000
Dowell L 37
Ferrofos 509
Nitrilotrimethanephosphonic acid
Nitrilotrimethylenephosphonic acid
Nitrilotrimethylphosphonic acid
(Nitrilotris(methylene))trisphosphonic acid
Nitrilotris(methylphosphonic acid)
Tris(phosphonomethyl)amine

6420-33-3
$C_{37}H_{32}N_6O_{15}S_4.4Na$
1020.88
1,5-Naphthalenedisulfonic acid, 3,3'-(carbonylbis(imino(5-methoxy-2-methyl-4,1-phenylene)azo))bis-, tetrasodium salt (9CI)

6420-40-2
$C_{34}H_{24}N_6O_{11}S_2.3Na$
825.66
Benzoic acid, 3-((1-hydroxy-6-((((5-hydroxy-6-(phenylazo)-7-sulfo-2-naphthalenyl)amino)carbonyl)amino)-3-sulfo-2-naphthalenyl)azo)-, trisodium salt (9CI)

6420-44-6
$C_{35}H_{28}N_6O_{13}S_3.3Na$
905.77
C.I. Direct Red 24, Trisodium salt (8CI)
2-Naphthalenesulfonic acid, 4-hydroxy-7-((((5-hydroxy-6-((2-methoxyphenyl)azo)-7-sulfo-2-naphthalenyl)amino)carbonyl)amino)-3-((2-methyl-4-sulfophenyl)azo)-, trisodium salt (9CI)

6420-47-9
$C_{16}H_{27}N_3O_8$
389.46
Phenol, 2-sec-butyl-4,6-dinitro-, 2,2',2''-nitrilotriethanol salt
2-sec-Butyl-4,6-dinitrophenol 2,2',2''-nitrilotriethanol salt
o-sec-Butyl-4,6-dinitrophenol triethanolamine salt
Dinitrobutylphenol 2,2',2''-nitrilotriethanol salt
2-(1-Methyl-n-propyl)-4,6-dinitrophenol triethanolamine salt

6422-86-2
$C_{24}H_{38}O_4$
390.62
O=C(OCC(CCCC)CC)c(ccc(c1)C(=O)OCC(CCCC)CC)c1
Terephthalic acid, bis(2-ethylhexyl)ester
1,4-Benzenedicarboxylic acid, bis(2-ethylhexyl)ester (9CI)
Bis(2-ethylhexyl) terephthalate
Kodaflex DOTP

6422-99-7
$C_{10}H_{18}O_4.C_6H_{16}N_2$
318.52
Sebacic acid, Compd. with 1,6-hexanediamine (1:1)
Decanedioic acid, Compd. with 1,6-hexanediamine (1:1) (9CI)
Hexamethylenediamine sebacate
Hexamethylenediammonium sebacate
Nylon 610 salt
Sebakan hexamethylendiaminu (Czech)

6423-43-4
$C_3H_6N_2O_6$
166.11
O=N(=O)OCC(ON(=O)=O)C
1,2-Propanediol, dinitrate
PGDN
Propylene dinitrate
Propylene glycol dinitrate (ACGIH,OSHA)
Propylene glycol 1,2-dinitrate
1,2-Propylene glycol dinitrate

6424-76-6
$C_{36}H_{18}O_4$
514.54
Dinaphtho(1',2',3':3,4;3'',2'',1'':9,10)perylo(1,12-efg)-(1,4)dioxocin-5,10-dione, 17,18- dihydro
Calcoloid Navy Blue 2GC
Caledon Dark Blue G
Caledon Printing Navy G
Carbanthrene Navy Blue G
C.I. 71200
C.I. Vat Blue 16
16,17-Ethylenedioxyviolanthrone
Indanthrene Navy Blue G
Indanthren Navy Blue G
Mikethrene Marine Blue G
Modr Kypova 16 (Czech)
Modr Namornicka Ostanthrenova G (Czech)
Nihonthrene Navy Blue G
Palanthrene Navy Blue G
Paradone Navy Blue G
Ponsol Navy Blue
Ponsol Navy Blue D
Romantrene Navy Blue FG
Romantrene Navy Blue G

6424-77-7
$C_{38}H_{22}N_2O_6$
602.59
O=C(N(C(=O)c(c1c(c(c(c(c(c(C(=O)N(C2=O)c(ccc(OC)c3)c3)c4)c2cc5)c56)c4)cc7)c6c8)c8)c(ccc(OC)c9)c9)c17
Anthra(2,1,9-def:6,5,10-d'e'f')-diisoquinoline-1,3,8,10(2H,9H)-tetrone, 2,9-bis(4-methoxyphenyl)- (9CI)

6424-85-7
$C_{22}H_{17}N_3O_6S.Na$
474.43
2-Anthracenesulfonic acid, 4-((4-(acetylamino)phenyl)amino)-1-amino-9,10-dihydro-9,10-dioxo-, monosodium salt (9CI)

6425-39-4
$C_{12}H_{24}N_2O_3$
244.32
O(CCN(C1)CCOCCN(CCOC2)C2)C1
Morpholine, 4,4'-(oxydi-2,1-ethanediyl)bis- (9CI)
Dimorpholinodiethyl ether
4,4'-(Oxydi-2,1-ethanediyl)bismorpholine

6426-67-1

$C_{32}H_{22}N_4O_{11}S_3.3Na$
803.69
1,3,6-Naphthalenetrisulfonic acid, 8-hydroxy-7-((4'-((2-hydroxy-1-naphthalenyl)azo)-(1,1'-biphenyl)-4-yl)-azo)-, trisodium salt (9CI)
Aizen Direct Violet LNH
Benzanil Violet BXN
C.I. 22480
C.I. Direct Violet 22
C.I. Direct Violet 22, trisodium salt (8CI)
Chlorazol Violet WB
Chlorazol Violet WBS
Chrome Leather Violet BS
Diphenyl Violet TS
Direct Violet BS
Direct Violet 22
Erie Violet BW
Kayaku Direct Violet LN
Mitsui Direct Violet LN
Nippon Violet LN
Pontamine Brilliant Violet RN
Triazol Violet B
Trisulfon Violet B

6428-31-5
$C_{34}H_{29}N_{13}O_7S_2.2Na$
841.86
C.I. Direct Black 19, Disodium salt
Artificial Silk Black G
Artificial Silk Black GN
Artificial Silk Black GR
Atlantic Artificial Silk Black G
Bali Viscose Black G
Bali Viscose Black N
Bencidal Fast Black G
Benzanil Fast Black G
Benzo Fast Black G
Benzoform Black RRA-CF
Cern Prima 19 (Czech)
Chlorazol Viscose Black B
Chrome Leather Black GNA
C.I. 35255
C.I. Direct Black 19
Columbia Fast Black G
Columbia Fast Black GB
Coranil Direct Black B
Cutamin Black CG
Diamine Fast Black B
Diaphtamine Fast Black KG
Direct Artificial Silk Black G
Direct Black 19
Direct Fast Black G
Direct Fast Black GU
Direct Fast Black SA
Direct Rayon Black KSG
Fenamin Black GR
Formal Fast Black 2B
Hispamin Fast Black CG
Indoblack GR
Kayarus Black G
Kayarus Black G Conc.
Phenamine Viscose Black RR
Pyrazol Fast Black GS
Rayon Black G
Rayon Black GSN
Rayon Black M
Rayon Fast Black B
Solar Black G
Sumilight Black G
Tetrodirect Black V
Tetrazo Deep Black GC Extra
Visco Black N
Viscose Black G
Viscose Black GNA
Viscose Black J
Viscose Black N
Viscose Black NG
Vondacel Black VG
Vondacel Black VN
Water Black 100

6428-60-0
$C_{42}H_{29}N_7O_{13}S_4.4Na$
1059.91
2-Naphthalenesulfonic acid, 5-((1-hydroxy-6-(phenylamino)-3-sulfo-2-naphthalenyl)azo)-8-((6-sulfo-4-(-(3-sulfophenyl)azo)-1-naphthalenyl)azo)-, tetrasodium salt (9CI)

6428-94-0
$C_{34}H_{27}N_5O_9S_2.2Na$
759.70
Cc1cc(N)ccc1N.OS(O)(=O)=O
1-Naphthalenesulfonic acid, 4-amino-3-((4'-((1-hydroxy-4-sulfo-2-naphthalenyl)azo)-3,3'-dimethoxy(1,1'-biphenyl)-4-yl)azo)-, disodium salt (9CI)
C.I. Direct Violet 32
C.I. Direct Violet 32, Disodium salt (8CI)

6440-58-0
$C_7H_{12}N_2O_4$
188.21
O=C(N(C(C1=O)(C)C)CO)N1CO
Hydantoin, 1,3-bis(hydroxymethyl)-5,5-dimethyl
1,3-Bis(hydroxymethyl)-5,5-dimethylhydantoin
Dantoin DMDMH 55
Dimethylol-5,5-dimethylhydantoin
DMDMH
DMDMH 55
DMDM Hydantoin
Glydant
2,4-Imidazolidinedione, 1,3-bis-(hydroxymethyl)-5,5-dimethyl-(9CI)

6441-77-6
$C_{20}H_6Br_4Cl_2O_5.2K$
795.00
Fluorescein, 4,7-dichloro-2',4',5',7'-tetrabromo-, dipotassium salt
C.I. 45405
C.I. Acid Red 98
4,6-Dichloro-2',4',5',7'-tetrabromofluorescein dipotassium salt
Fluorescein, 2',4',5',7'-tetrabromo-4,7-dichloro-, dipotassium salt
Food Dye Red No. 104
Phloxin
Phloxine
Phloxine K
2,4,5,7-Tetrabromo-12,15-dichlorofluorescein, dipotassium salt
Toyo acid phloxine

6443-92-1
C_7H_{14}
98.19
2-Heptene, (Z)- (8CI,9CI)

6448-90-4
$C_{16}H_{12}O_4$
268.28
O=C(c(c(c(OC)cc1)C(=O)c2ccc c3OC)c1)c23
Anthraquinone, 1,5-dimethoxy
9,10-Anthracenedione, 1,5-dimethoxy-
1,5-Dimethoxyanthrachinon (Czech)
1,5-Dimethoxyanthraquinone

6448-95-9
$C_{24}H_{18}N_4O_4$
426.41
O=C(Nc(cccc1)c1)c(c(O)c(N=Nc (c(ccc2N(=O)=O)C)c2)c(c3c cc4)c4)c3
2-Naphthalenecarboxamide, 3 -hydroxy-4-((2-methyl-5-nitrophenyl)azo)-N-phenyl-(9CI)
3-Hydroxy-4-((5-nitro-o-tolyl)-azo)-2-naphthanilide

6448-96-0
$C_{31}H_{23}N_5O_6$
561.52
O=C(Nc(cccc1N(=O)=O)c1)c(c (O)c(N=Nc(c(OC)ccc2C(=O) Nc(cccc3)c3)c2)c(c4ccc5) c5)c4
2-Naphthalenecarboxamide, 3-hydroxy-4-((2-methoxy-5-((phenylamino)carbonyl)-phenyl)azo)-N-(3-nitrophenyl)- (9CI)
1-(2-Methoxy-5-phenylcarbamoylphenylazo)-2-hydroxy-3-(3-nitrophenylcarbamoyl)naphthalene

6449-35-0
$C_{34}H_{27}N_5O_{10}S_2.2Na$
775.70
C.I. Direct Blue 151, Disodium salt (8CI)

1-Naphthalenesulfonic acid, 3-((4'-((6-amino-1-hydroxy-3-sulfo-2-naphthalenyl)azo)-3,3'-dimethoxy(1,1'-biphenyl)-4-yl)azo)-4-hydroxy-, disodium salt (9CI)

6452-71-7
$C_{15}H_{23}NO_3$
265.39
CC(C)NCC(O)COc1ccccc1O
CC=C
2-Propanol, 1-(2-allyloxyphenoxy)-3-(isopropylamino)
1-(o-(Allyloxy)phenoxy)-3-(isopropylamino)-2-propanol
Coretal
1-(Isopropylamino)-2-hydroxy-3-(o-(allyloxy)phenoxy)-propane
1-((1-Methylethyl)amino)-3-(2-(2-propenyloxy)-phenoxy)-2-propanol
Oxprenolol
2-Propanol, 1-((1-methylethyl)-amino)-3-(2-(2-propenyloxy)-phenoxy)-

6459-94-5
$C_{37}H_{30}N_4O_{10}S_3.2Na$
832.87
1,3-Naphthalenedisulfonic acid, 8-((3,3'-dimethyl-4'-((4-(((4-methylphenyl)sulfonyl)oxy) phenyl)azo)(1,1'-biphenyl)-4-yl)azo)-7-hydroxy-, disodium salt
Acid Leather Red BG
Acid Red 114
Benzyl Fast Red BG
Benzyl Red BR
Cerven Kysela 114 (Czech)
C.I. 23635
C.I. Acid Red 114
C.I. Acid Red 114, Disodium salt
Erionyl Red RS
Fenafor Red PB
Folan Red B
Kayanol Milling Red RS
Leather Fast Red B
Levanol Red GG
Midlon Red PRS
Milling Fast Red B
Milling Red B
Milling Red BB
Milling Red SWB
NCI-C61096
Polar Red RS
Sella Fast Red RS
Sulphonol Red R
Suminol Milling Red RS
Supranol Fast Red GG
Supranol Red PBX-CF
Supranol Red R
Telon Fast Red GG
Tetracid Milling Red G
Vondamol Fast Red RS

6460-01-1
$C_{38}H_{28}N_6O_{12}S_3.3Na$
925.81
C.I. Direct Red 73, Trisodium salt (8CI)
2-Naphthalenesulfonic acid, 4-hydroxy-7-((((5-hydroxy-6-((2-methylphenyl)azo)-7-sulfo-2-naphthalenyl)amino)carbonyl)-amino)-3-((6-sulfo-2-naphthalenyl)azo)-, trisodium salt (9CI)

6465-02-7
$C_{21}H_{19}N_5O_3$
389.38
O=C(OC)Nc(ccc(N=Nc(c(cc(N=N
c(ccc(O)c1)c1)c2)C)c2)c3)c3
Carbamic acid, (4-((4-((4-hydroxyphenyl)azo)-2-methylphenyl)azo)phenyl)-, methyl ester (9CI)
4-((4-((4-((Methoxycarbonyl)-amino)phenyl)azo)-3-methyl-phenyl)azo)phenol

6470-31-1
$C_{32}H_{22}Cl_2N_6O_6S_2.2Na$
767.55
1-Naphthalenesulfonic acid, 3,3'-((3,3'-dichloro(1,1'-biphenyl)-4,4'-diyl)bis(azo))-bis(4-amino-, disodium salt (9CI)
C.I. Direct Red 61 (VAN)
C.I. Direct Red 61, Disodium salt (8CI)

6470-98-0
$C_{13}H_{11}N_3O_3.Na$
280.22
Benzoic acid, 5-((4-aminophenyl)azo)-2-hydroxy-, monosodium salt (9CI)

6471-49-4
$C_{24}H_{17}N_5O_7$
487.46
O=C(Nc(cccc1N(=O)=O)c1)c(c
(O)c(N=Nc(c(OC)ccc2N(=O)
=O)c2)c(c3ccc4)c4)c3
2-Naphthalenecarboxamide, 3-hydroxy-4-((2-methoxy-5-nitrophenyl)azo)-N-(3-nitrophenyl)
Alkali Resistant Red Dark
Calcotone Red 3B
Carnation Red Toner B
C.I. 12355
C.I. Pigment Red 23
Congo Red R-138
Fenalac Red FKB Extra
Malta Red X 2284
Naphthol Red B
Naphthol Red B 20-7575
Naphthol Red Deep 10459
Naphthol Red D Toner 35-6001
NCI-C60377
Pigment Red 23
Pigment Red BH
Rubescence Red MT-21
Sanyo Fast Red 10B
Sapona Red Lake RL-6280
Segnale Light Rubine RG
Textile Red WD-263

6471-51-8
$C_{25}H_{19}Cl_2N_3O_2$
464.33
O=C(Nc(c(cc(c1)Cl)C)c1)c(c(O)
c(N=Nc(c(cc(c2)Cl)C)c2)c(c3
ccc4)c4)c3
2-Naphthalenecarboxamide, N-(4-chloro-2-methylphenyl)-4-((4-chloro-2-methylphenyl)azo)-3-hydroxy- (9CI)

6471-78-9
$C_8H_{11}NO_4S$
217.24
O=S(=O)(O)c(c(cc(N)c1OC)C)c1
Benzenesulfonic acid, 4-amino-5-methoxy-2-methyl-(9CI)
4-Amino-5-methoxy-2-methyl-benzenesulfonic acid

6473-13-8
$C_{44}H_{35}N_{13}O_{11}S_3.3Na$
1086.93
C.I. Direct Black 22
2-Naphthalenesulfonic acid, 6-((2,4-diaminophenyl)azo)-3-((4-((4-((7-((2,4-diaminophenyl)azo)-1-hydroxy-3-sulfo-2-naphthalenyl)azo)-phenyl)amino)-3-sulfophenyl)-azo)-4-hydroxy-, trisodium salt (9CI)

6474-16-4
C_4H_9NSe
150.08
Selenazolidine, 4-methyl-(7CI,8CI,9CI)

6483-64-3
$C_{41}H_{32}N_4O_4$
644.77
O(c(c(N=Nc(c(c(ccc1)cc2)c1)
c2O)ccc3C(c(cccc4)c4)c(ccc
(N=Nc(c(c(ccc5)cc6)c5)c6O)
c7OC)c7)c3)C
Di-2-naphthol, 1,1'-(benzylidene)bis((2-methoxy-p-phenylene)(azo))
1,1'-(Benzylidenebis((2-methoxy-p-phenylene)(azo))di-2-naphthol)
3,3'-Dimethoxytriphenylmethane-4,4'-bis(1''-azo-2''-naphthol)

6484-52-2
$HNO_3.H_3N$
80.06
Ammonium(I) nitrate (1:1)
Ammonium nitrate
Ammonium nitrate, No organic coating (DOT)
Ammonium nitrate, Organic coating (DOT)
Ammonium nitrate, Solution (containing not less than 15% water) (DOT)
Ammonium nitrate, With more than 0.2% combustible substances [UN 0222]
Ammonium nitrate, With not more than 0.2% combustible substances [UN 1942]
Ammonium saltpeter
Herco Prills
NA 1942 [Ammonium nitrate fertilizers with not more than 0.2 percent carbon; which meet the definition in the Fertilizer Institute publication "Definition and Test Procedures for Ammonium Nitrate Fertilizer", dated May 8, 1971]
Nitric acid, ammonium salt
UN 0222 [Ammonium nitrate, with more than 0.2 per cent combustible substances, including any organic substance calculated as carbon, to the exclusion of any other added substance]
UN 1942 [Ammonium nitrate, with not more than 0.2 per cent of combustible substances, including any organic substance calculated as carbon, to the exclusion of any other added substance]
UN 2426 [Ammonium nitrate, liquid (hot concentrated solution)]
Varioform I

6485-34-3
$C_{14}H_{10}N_2O_6S_2.Ca$
406.46
1,2-Benzisothiazolin-3-one, 1,1-dioxide, calcium salt
1,2-Benzisothiazol-3(2H)-one, 1,1-dioxide, calcium salt (9CI)
Calcium o-benzosulfimide
Calcium o-benzosulphimide
Calcium 2-benzosulphimide
Calcium saccharin
Calcium saccharina
Calcium saccharinate
Daramin
Saccharin calcium
Sulphobenzoic imide calcium salt

6485-39-8
$C_6H_{12}O_7.1/2Mn$
223.63
Manganese gluconate
Gluconic acid, manganese salt (2:1)
D-Gluconic acid, manganese salt (2:1)

6485-40-1
$C_{10}H_{14}O$
150.24
O=C(C(=CCC1C(=C)C)C)C1
p-Mentha-6,8-dien-2-one, (R)-(-)
(-)-Carvone
l-Carvone
L(-)-Carvone
(r)-Carvone
2-Cyclohexen-1-one, 2-methyl-5-(1-methylethenyl)-, (R)- (9CI)
p-Mentha-6,8-dien-2-one, (-)-
l-6,8(9)-p-Menthadien-2-one
l-1-Methyl-4-isopropenyl-6-cyclohexen-2-one

6485-91-2
$C_4H_{11}ClO_2Si$
154.69
CCO[SiH](Cl)OCC
Silane, chlorodiethoxy
Chlorodiethoxysilane
Diethoxychlorosilane

6486-23-3
$C_{16}H_{12}Cl_2N_4O_4$
395.18
O=C(Nc(c(ccc1)Cl)c1)C(N=Nc(c(N(=O)=O)cc(c2)Cl)c2)C(=O)C
Pigment Yellow 3
Butanamide, 2-((4-chloro-2-nitrophenyl)azo)-N-(2-chlorophenyl)-3-oxo- (9CI)
2-((4-Chloro-2-nitrophenyl)azo)-N-(2-chlorophenyl)-3-oxo-butanamide
C.I. Pigment Yellow 3 (VAN) (8CI)
C.I. 11710

6492-73-5
$C_{16}H_9BrN_2O_2$
341.15
3H-Indol-3-one, 5-bromo-2-(1,3-dihydro-3-oxo-2H-indol-2-ylidene)-1,2-dihydro- (9CI)
(δ(2,2')-Biindoline)-3,3'-dione, 5-bromo- (8CI)
C.I. Vat Blue 3
C.I. Vat Blue 34
Mitsui Tsuya Indigo RN

6494-90-2
$C_{13}H_{16}N_6O_3$
304.31
s-Triazine, 2,4-bis(ethylamino)-6-(p-nitrophenoxy)- (7CI,8CI)

6494-91-3
$C_{13}H_{17}N_5O$
259.28
1,3,5-Triazine-2,4-diamine, N,N'-diethyl-6-phenoxy- (9CI)
s-Triazine, 2,4-bis(ethylamino)-6-phenoxy- (8CI)

6494-92-4
$C_{14}H_{19}N_5S$
289.41
s-Triazine, 2,4-bis(ethylamino)-6-(p-tolylthio)- (8CI)
s-Triazine, 3,4-bis(ethylamino)-6-(p-tolylthio)- (7CI)

6494-99-1
$C_{14}H_{19}N_5O$
273.31
1,3,5-Triazine-2,4-diamine, N-ethyl-N'-(1-methylethyl)-6-phenoxy- (9CI)
s-Triazine, 2-(ethylamino)-4-(isopropylamino)-6-phenoxy- (8CI)

6495-03-0
$C_{13}H_{16}ClN_5O$
293.76
s-Triazine, 2-(o-chlorophenoxy)-4,6-bis(ethylamino)- (7CI,8CI)

6495-07-4
$C_{13}H_{17}N_5S$
275.35
1,3,5-Triazine-2,4-diamine, N,N'-diethyl-6-(phenylthio)- (9CI)
s-Triazine, 2,4-bis(ethylamino)-6-(phenylthio)- (8CI)

6502-20-1
$C_{13}H_{18}O$
190.29
O=CC(C)Cc(c(ccc1)C(C)C)c1
Benzenepropanal, α-methyl-2-(1-methylethyl)- (9CI)
α-Methyl-2-(1-methylethyl)-benzenepropanal

6505-28-8
$C_{34}H_{32}N_6O_6$
620.63
O=C(Nc(cccc1)c1)C(N=Nc(c(OC)cc(c(ccc(N=NC(C(=O)C)C(=O)Nc(cccc2)c2)c3OC)c3)c4)c4)C(=O)C
Butanamide, 2,2'-((3,3'-di-

methoxy(1,1'-biphenyl)-4,4'-diyl)bis(azo))bis(3-oxo-N-phenyl- (9CI)

2,9-Triphenodioxazinedisulfonic acid, 6,13-dichloro-3,10-bis(phenylamino)-, disodium salt (9CI)

6505-30-2
$C_{43}H_{49}N_3O_6S_2.Na$
790.98
Benzenemethanaminium, N-(4-((4-(diethylamino)-phenyl)(4-(ethyl((3-sulfo-phenyl)methyl)amino)-2-methylphenyl)methylene)-3-methyl-2,5-cyclohexadien-1-ylidene)-N-ethyl-3-sulfo-, hydroxide, inner salt, sodium salt (9CI)

6510-65-2
$C_{13}H_{11}N$
181.23
9H-Carbazole, 1-methyl- (9CI)
Carbazole, 1-methyl- (8CI)

6515-09-9
$C_5H_2Cl_3N$
182.43
Clc1ccc(Cl)c(Cl)n1
Pyridine, 2,3,6-trichloro
2,3,6-Trichloropyridine

6515-38-4
$C_5H_2Cl_3NO$
198.44
Oc1nc(Cl)c(Cl)cc1Cl
3,5,6-Trichloro-2-pyridinol
Caswell No. 821AA
EPA Pesticide Chemical Code 206900
TCP
3,5,6-Trichloro-2(1H)-pyridinone

6527-70-4
$C_{30}H_{18}Cl_2N_4O_8S_2.2Na$
743.49
Sirius Supra Blue FGL-CF stain
C.I. Direct Blue 106
Solophenyl Brilliant Blue BL

6528-34-3
$C_{18}H_{18}N_4O_6$
386.34
O=C(Nc(c(OC)ccc1)c1)C(N=Nc(c(N(=O)=O)cc(OC)c2)c2)C(=O)C
Butanamide, 2-((4-methoxy-2-nitrophenyl)azo)-N-(2-methoxyphenyl)-3-oxo- (9CI)
2-((4-Methoxy-2-nitrophenyl)-azo)-o-acetoacetanisidide

6528-46-7
$C_{10}H_9NO_3S$
223.25
2-Naphthalenesulfonamide, 6-hydroxy- (8CI,9CI)
2-Naphthol-6-sulfonamide

6528-53-6
$C_{16}H_{10}O_{12}S_4$
522.51
O=S(=O)(O)c(c(c(c(c(c(S(=O)(=O)O)cc1S(=O)(=O)O)c2)c1cc3)c3c4S(=O)(=O)O)c2)c4
1,3,6,8-Pyrene tetrasulfonate
PTSA
1,3,6,8-Pyrenetetrasulfonic acid (9CI)

6529-53-9
C_8H_8BrCl
219.51
Benzene, 1-(2-bromoethyl)-4-chloro- (6CI,7CI,8CI,9CI)
1-Bromo-2-(4-chlorophenyl)ethane
1-(2-Bromoethyl)-4-chlorobenzene
1-Chloro-4-(2-bromoethyl)-benzene
4-Chlorophenethyl bromide
p-Chlorophenethyl bromide
2-(p-Chlorophenyl)ethyl bromide
p-Chloro-β-phenethyl bromide

6531-38-0
$C_8H_{20}N_4$
172.25
N(CCN(C1)CCN)(C1)CCN
1,4-Piperazinediethanamine (9CI)

6531-86-8
$C_{12}H_{22}O$
182.31
OC(C(C(CCCC1)C1)CCC2)C2
(Bicyclohexyl)-2-ol (8CI)
AI3-01023
(1,1'-Bicyclohexyl)-2-ol (9CI)
Caswell No. 271
2-Cyclohexylcyclohexanol
EPA Pesticide Chemical Code 065001

6533-00-2
$C_{21}H_{28}O_2$
312.49
18,19-Dinor-17-α-pregn-4-en-20-yn-3-one, 13-ethyl-17-hydroxy-, (+-)
18,19-Dinorpregn-4-en-20-yn-3-one, 13-ethyl-17-hydroxy-, (17-α)-(+-)- (9CI)
dl-13-β-Ethyl-17-α-ethynyl-17-β-hydroxygon-4-en-3-one
dl-13-β-Ethyl-17-α-ethynyl-19-nortestosterone
(+-)-13-Ethyl-17-hydroxy-18,19-dinor-17-α-pregn-4-en-20-yn-3-one
FH 122-A
LD Norgestrel (French)
Monovar
Norgestrel
(+-)-Norgestrel
α-Norgestrel
dl-Norgestrel
SH 850
SH 70850
WY 3707

6533-73-9
$CO_3.2Tl$
468.75
Thallium(I) carbonate (2:1)
Carbonic acid, dithallium(1+) salt
Dithallium carbonate
RCRA waste number U215
Thallous carbonate
Thiochroman-4-one, oxime

6535-46-2
$C_{24}H_{16}Cl_3N_3O_2$
484.75
O=C(Nc(c(ccc1)C)c1)c(c(O)c(N=Nc(c(cc(c2Cl)Cl)Cl)c2)c(c3ccc4)c4)c3
Pigment Red 112
CI 12370
3-Hydroxy-N-(2-methylphenyl)-4-((2,4,5-trichlorophenyl)azo)-2-naphtha enecarboxamide
2-Naphthalenecarboxamide, 3-hydroxy-N-(2-methylphenyl)-4-((2,4,5-trichlorophenyl)azo)- (9CI)

6537-68-4
$C_{16}H_{12}N_2O_2$
264.27
(2,2'-Bi-1H-indole)-3,3'-diol (9CI)
(2,2'-Biindole)-3,3'-diol (8CI)
Leucoindigo
NSC-8648

6542-37-6
$C_6H_{11}NO_3$
145.15
O(CC(N1CO2)(CO)C2)C1
5-Hydroxymethyl-1-aza-3,7-dioxabicyclo(3.3.0)octane
Caswell No. 495AB
EPA Pesticide Chemical Code 107002
5-(Hydroxymethyl)-1-aza-3,7-dioxabicyclo(3.3.0)octane
Nuosept 95
1H,3H,5H-Oxazolo(3,4-c)oxazole-7a(7H)-methanol (9CI)

6542-67-2
$C_6Cl_9N_3$
433.14
1,3,5-Triazine, 2,4,6-tris(trichloromethyl)- (9CI)
Acetonitrile, trichloro-, trimer
NSC-26769
s-Triazine, 2,4,6-tris(trichloromethyl)- (8CI)

6548-29-4
$C_{32}H_{22}Cl_2N_6O_{12}S_4.4Na$
973.70
C.I. Direct Red 46, Tetrasodium salt
Acetopurpurine 8B
Amanil Chloramine Red 8BS
Benzo Brilliant Red 8BS
Chloramine Brilliant Red 8B
Chloramine Red 8B
Chlorazol Brilliant Purpurine 8B
C.I. 23050
C.I. Direct Red 46
Diaphtamine Fast Purpurine 8B
Diazol Fast Purpurine 8B
Diphenyl Red 8B
Direct Fast Purpurine 8B
Direct Red 8BS
Fast Purpurine
Mitsui Direct Brilliant Scarlet 8B
Nippon Purpurine 8B
Triazol Fast Red 8B

6555-95-9
$C_{10}H_{18}O$
154.25
Bicyclo(2.2.2)octane, 1-methoxy-4-methyl- (7CI,8CI,9CI)

6560-09-4
$C_{13}H_{16}ClN_5S$
309.82
s-Triazine, 2-((p-chlorophenyl)thio)-4,6-bis-(ethylamino)- (7CI,8CI)

6575-09-3
C_8H_6ClN
151.59
Benzonitrile, 2-chloro-6-methyl- (9CI)
NSC-80657
o-Tolunitrile, 6-chloro- (8CI)

6582-52-1
$C_{13}H_{14}N_2$
198.26
Nc(c(ccc1)Cc(c(N)ccc2)c2)c1
Benzenamine, 2,2'-methylenebis- (9CI)
Aniline, 2,2'-methylenedi-
2,2'-Methylenebisbenzenamine
2,2'-Methylenebis(benzeneamine)

6585-53-1
$CH_{13}AsFeN_3O_2$
229.94
Ferric-methanearsonate
Ferric monomethylarsonate
Neo-Asozin
Neo So Sin Gin
Neo So Sin Gin-S

6585-96-2
$C_{13}H_{16}ClN_5O$
293.76
s-Triazine, 2-(p-chlorophenoxy)-4,6-bis(ethylamino)- (7CI,8CI)

6596-35-6
$C_{14}H_{24}$
192.34
C(C(CCC1)CC(C2CCC3)C3)(C1)C2
Perhydroanthracene
Anthracene, tetradecahydro- (9CI)

6600-31-3
$C_{19}H_{28}O_4$
320.47
O(CC(CO1)(COC(O2)C(CCC=C3)C3)C2)C1C(CCC=C4)C4
2,4,8,10-Tetraoxaspiro(5,5)undecane, 3,9-di(3-cyclohexenyl)
3,9-Di-(3-cyclohexenyl)-2,4,8,10-tetraoxaspiro(5,5)-undecane

6606-59-3
$C_{14}H_{22}O_4$
254.33
O=C(OCCCCCCOC(=O)C(=C)C)C(=C)C
2-Propenoic acid, 2-methyl-, 1,6-hexanediyl ester (9CI)
1,6-Hexanediol dimethacrylate
1,6-Hexanediyl 2-methyl-2-propenoate

6607-45-0
$C_8H_6Cl_2$
173.04
ClC=C(Cl)c1ccccc1
Styrene, α,β-dichloro
Dichlorostyrene
Dwuchlorostyren (Polish)

6610-29-3
$C_2H_7N_3S$
105.15
N(=C(S)NN)C
Hydrazinecarbothioamide, N-methyl- (9CI)
N-Methylhydrazinecarbothioamide
Methylthiosemicarbazide
4-Methylthiosemicarbazide
4-Methyl-3-thiosemicarbazide
Semicarbazide, 4-methyl-3-thio- (8CI)

6616-62-2
$C_{16}H_{12}N_2O_4S$
312.35
Benzenesulfonic acid, 2-((2-hydroxy-1-naphthalenyl)azo)- (9CI)
Benzenesulfonic acid, o-((2-hydroxy-1-naphthyl)-azo)- (8CI)
1-(2'-Sulfophenylazo)-2-naphthol

6617-49-8
$C_{15}H_{22}$
202.34
Naphthalene, 1,2,3,4-tetrahydro-1,6-dimethyl-4-(1-methylethyl)- (9CI)
Naphthalene, 1,2,3,4-tetrahydro-4-isopropyl-1,6-dimethyl- (8CI)

6618-03-7
$C_9H_{22}Pb$
337.50
CCC[PbH](CCC)CCC
Plumbane, tripropyl
Lead tripropyl
Tripropyl lead

6622-76-0
$C_6H_{10}O_2$
114.16
O=C(OC)C(=CC)C
2-Butenoic acid, 2-methyl-, methyl ester, (E)
Crotonic acid, 2-methyl-, methyl ester, (E)- (8CI)
Methyl α-methylcrotonate
Methyl (E)-2-methylcrotonate
Methyl trans-2-methylcrotonate
Methyl tiglate
Tiglic acid, methyl ester (6CI, 7CI)

6625-46-3
$C_{19}H_{17}N_3O_9S_2.2Na$
541.45
C.I. Acid Violet 12, Disodium salt (8CI)
2,7-Naphthalenedisulfonic acid, 5-(acetylamino)-4-hydroxy-3-((2-methoxyphenyl)azo)-, disodium salt (9CI)

6627-34-5
$C_6H_4Cl_2N_2O_2$
207.02
O=N(=O)c(c(cc(N)c1Cl)Cl)c1
Aniline, 2,5-dichloro-4-nitro

2,5-Dichloro-4-nitroaniline

6628-18-8
$C_4H_{10}S_2$
122.26
Ethane, 1,2-bis(methylthio)- (9CI)

6629-10-3
$C_{16}H_{14}N_4O_2$
294.29
O=C(NN=Cc(cccc1)c1)C(=O)NN =Cc(cccc2)c2
Ethanedioic acid, bis((phenylmethylene)hydrazide) (9CI)

6630-01-9
$C_{13}H_{20}$
176.30
Benzene, 1-(1,1-dimethylethyl)-3-ethyl-5-methyl- (9CI)
NSC-60003
Toluene, 3-tert-butyl-5-ethyl-

6634-82-8
$C_{14}H_{12}N_2O_8S_2.2Na$
446.36
Benzenesulfonic acid, 5-amino-2-(2-(4-nitro-2-sulfophenyl)ethenyl)-, disodium salt (9CI)
4-Amino-4'-nitro-2,2'-stilbenedisulfonic acid, disodium salt
5-Amino-2-(2-(4-nitro-2-sulfophenyl)ethenyl)benzenesulfonic acid, disodium salt
Nitroaminostilbene Disa

6637-87-2
$C_{28}H_{17}N_9O_{16}S_2.2Na$
845.55
2,7-Naphthalenedisulfonic acid, 4-((2,4-dihydroxy-5-((2-hydroxy-3,5-dinitrophenyl)azo)-3-((4-nitrophenyl)azo)phenyl)azo)-5-hydroxy-, disodium salt (9CI)
4-((2,4-Dihydroxy-3-((4-nitrophenyl)azo)-5-((3,5-dinitro-2-hydroxyphenyl)azo)phenyl)azo)-5-hydroxy-2,7-naphthalenedisulfonic acid, disodium salt

6637-88-3
$C_{28}H_{26}N_6O_6S.2Na$
620.56
Benzoic acid, 5-((4'-((2,6-diamino-3-methyl-5-sulfophenyl)azo)-3,3'-dimethyl-(1,1'-biphenyl)-4-yl)azo)-2-hydroxy-, disodium salt (9CI)
C.I. Direct Orange 6, Disodium salt (8CI)

6639-30-1
$C_7H_5Cl_3$
195.47
c(c(cc(c1Cl)Cl)Cl)(c1)C
Benzene, 1,2,4-trichloro-5-methyl- (9CI)
Toluene, 2,4,5-trichloro- (8CI)
1,2,4-Trichloro-5-methylbenzene

6639-99-2
$C_{18}H_{19}O_2$
267.37
α-Estra-1,3,5,7,9-pentaen-3,17-diol
α-Dihydroequilenin
alfa-Dihydroequilenina (Spanish)

6640-27-3
C_7H_7ClO
142.58
Oc(c(cc(c1)C)Cl)c1
p-Cresol, 2-chloro- (8CI)
2-Chloro-4-methylphenol
Phenol, 2-chloro-4-methyl- (9CI)

6641-64-1
$C_6H_4Cl_2N_2O_2$
207.00
Nc1cc(Cl)c(Cl)cc1N(=O)=O
Benzenamine, 4,5-dichloro-2-nitro- (9CI)
Aniline, 4,5-dichloro-2-nitro- (8CI)
NSC-17012

6651-25-8
$C_{66}H_{135}Al$
955.78
Aluminum, tridocosyl- (9CI)
Tridocosylaluminum
Tri(docosyl) aluminum

6651-26-9
$C_{72}H_{147}Al$
1039.94
Aluminum, tritetracosyl- (9CI)
Tritetracosylaluminum
Tri(tetracosyl) aluminum

6651-27-0
$C_{84}H_{171}Al$
1208.26
Aluminum, trioctacosyl- (8CI)
Trioctacosylaluminum
Tri(octacosyl) aluminum

6655-84-1
$C_{25}H_{20}N_4O_4$
440.43
O=C(Nc(c(ccc1)C)c1)c(c(O)c (N=Nc(c(ccc2N(=O)=O)C) c2)c(c3ccc4)c4)c3
2-Naphthalenecarboxamide, 3-hydroxy-4-((2-methyl-5-nitrophenyl)azo)-N-(2-methylphenyl)- (9CI)

6656-02-6
$C_{15}H_8CrN_4O_9S.Na$
495.28
Chromate(1-), hydroxy(4-hydroxy-3-nitro-5-((1,2,3,4-tetrahydro-2,4-dioxo-3-quinolinyl)azo)benzenesulfonato(3-))-, sodium (9CI)

6656-03-7
$C_{38}H_{20}Cu_2N_5O_{13}S_3.3Na$
1046.83
Cuprate(3-), (mu-(7-((3,3'-dihydroxy-4'-((1-hydroxy-6-(phenylamino)-3-sulfo-2-naphthalenyl)azo)(1,1'-biphenyl)-4-yl)azo)-8-hydroxy-1,6-naphthalenedisulfonato(7-)))di-, trisodium (9CI)

6657-05-2
$C_{12}H_9ClN_2O$
232.68
Phenol, 2-chloro-4-(phenylazo)
2-Chloro-4-(phenylazo)phenol

6657-33-6
$C_{17}H_{16}ClN_5O_3$
373.77
O=N(=O)c(ccc(N=Nc(ccc(N(CC C#N)CCO)c1)c1)c2Cl)c2
Propanenitrile, 3-((4-((2-chloro-4-nitrophenyl)azo)phenyl)(2-hydroxyethyl)amino)- (9CI)
Aniline, 4-((4-nitro-2-chlorophenyl)azo)-N-hydroxyethyl-N-cyanoethyl
Benzenamine, 4-(((2-chloro-4-nitro)phenyl)azo)-N-(2-cyanoethyl)-N-(2-hydroxyethyl)-

6659-45-6
$C_{23}H_{24}O_6$
396.44
COc5cc4OCC3Oc2c1CC(Oc1ccc 2C(=O)C3c4cc5OC)C(C)C
(1)Benzopyrano(3,4-b)furo-(2,3-h)(1)benzopyran-6(6aH)-one, 1,2,12,12a-tetrahydro-8,9-dimethoxy-2-(1-methylethyl)-, (2R-(2α,6aα,12aα))- (9CI)
(1)Benzopyrano(3,4-b)furo-(2,3-h)(1)benzopyran-6(6aαH)-one, 1,2,12,12aα-tetrahydro-2α-isopropyl-8,9-dimethoxy- (8CI)
Caswell No. 353

Dihydroretenone
Dihydrorotenone
EPA Pesticide Chemical Code 071002
NSC-351138
NSC-53866
Rotenone, dihydro-

6659-60-5
$C_4H_7N_3O_9$
241.14
O=N(=O)OCCC(ON(=O)=O)CON(=O)=O
1,2,4-Butanetriol trinitrate (DOT)

6673-35-4
$C_{14}H_{22}N_2O_3$
266.38
CC(C)NCC(O)COc1ccc(NC(C)=O)cc1
Acetanilide, 4'-(2-hydroxy-3-(isopropylamino)propoxy)
Acetamide, N-(4-(2-hydroxy-3-((1-methylethyl)amino)propoxy)phenyl)- (9CI)
1-(4-Acetamidophenoxy)-3-isopropylamino-2-propanol
AY 21011
Dalzic
Eraldin
4'-(2-Hydroxy-3-(isopropylamino)propoxy)acetanilide
N-(4-(2-Hydroxy-3-((1-methylethyl)amino)propoxy)phenyl)acetamide
ICI 50172
Practalol
Practolol
Praktololu (Polish)
Teranol

6674-22-2
$C_9H_{16}N_2$
152.23
N(=C(N(CCCC1)CC2)C1)C2
1,8-Diazabicyclo(5.4.0)undec-7-ene
1,5-Diazabicyclo(5.4.0)undec-5-ene
Pyrimido(1,2-a)azepine, 2,3,4,6,7,8,9,10-octahydro- (9CI)

6675-28-1
$C_{12}H_{19}NS$
209.35
N(=C(C(S1)CCC2)C2)C1(CCCC3)C3
Spiro(benzothiazole-2(4H), 1'-cyclohexane), 5,6,7,7a-tetrahydro- (9CI)
5,6,7,7a-Tetrahydrospiro(benzothiazole-2(4H),1'-cyclohexane)

6676-90-0
$C_{13}H_{12}N_2$
196.25
Diazene, (2-methylphenyl)-phenyl- (9CI)
Azobenzene, 2-methyl- (6CI,7CI,8CI)
2-Methylazobenzene

6682-06-0
$C_{12}H_{16}$
160.26
1H-Indene, 2,3-dihydro-4,5,7-trimethyl- (9CI)
Indan, 4,5,7-trimethyl- (8CI)

6682-71-9
$C_{11}H_{14}$
146.23
1H-Indene, 2,3-dihydro-4,7-dimethyl- (9CI)
Indan, 4,7-dimethyl- (8CI)
NSC-81389

6683-19-8
$C_{73}H_{108}O_{12}$
1177.65
Irganox 1010
Benzenepropanoic acid, 3,5-bis-(1,1-dimethylethyl)-4-hydroxy-, 2,2-bis((3-(3,5-bis-(1,1-dimethylethyl)-4-hydroxyphenyl)-1-oxopropoxy)-methyl)-1,3-propanediyl ester (9CI)
Tetrakis(methylene(3,5-di-tert-butyl-4-hydroxyhydrocinnamate)methane

6684-27-1
$C_6H_7Cl_2N_3O$
208.03
1,3,5-Triazine, 2,4-dichloro-6-(1-methylethoxy)- (9CI)
s-Triazine, 2,4-dichloro-6-isopropoxy- (8CI)

6693-29-4
$C_{13}H_{26}N_2$
210.35
NC(CCC(C1)CC(CCC(N)C2)C2)C1
Cyclohexanamine, 4,4'-methylenebis-, (trans(trans))-(9CI)

6693-30-7
$C_{13}H_{26}N_2$
210.35
NC(CCC(C1)CC(CCC(N)C2)C2)C1
Cyclohexanamine, 4,4'-methylenebis-, (trans(cis))-(9CI)

6693-31-8
$C_{13}H_{26}N_2$
210.35
NC(CCC(C1)CC(CCC(N)C2)C2)C1
Cyclohexanamine, 4,4'-methylenebis-, (cis(cis))-(9CI)
(cis(cis))-4,4'-Methylenebiscyclohexanamine

6703-81-7
$C_{26}H_{52}$
364.70
Cyclopentane, (1-decylundecyl)- (9CI)
Heneicosane, 11-cyclopentyl- (8CI)

6703-82-8
$C_{26}H_{52}$
364.70
Cyclopentane, heneicosyl-(9CI)
1-Cyclopentylheneicosane
Heneicosane, 1-cyclopentyl-(8CI)

6705-89-1
$C_6H_{10}O_4$
146.14
2,2'-Bi-1,3-dioxolane (8CI,9CI)
Ethanedial, cyclic 1,1:2,2-bis-(1,2-ethanediyl acetal)

6711-48-4
$C_{10}H_{25}N_3$
187.38
N(CCCNCCCN(C)C)(C)C
Dipropylenetriamine, N,N,N',N'-tetramethyl
Bis(3-dimethylaminopropyl)-amine
N,N,N',N'-Tetramethyldipropylenetriamine

6714-00-7
$C_7H_{12}O$
112.17
O=C(CCC=CC)C
5-Hepten-2-one (9CI)

6726-45-0
$C_{24}H_{21}N_3$
351.46
1,3,5-Triazine, 2,4,6-tris-(4-methylphenyl)- (9CI)
s-Triazine, 2,4,6-tri-p-tolyl-(6CI,7CI,8CI)

6728-26-3
$C_6H_{10}O$
98.16
O=CC=CCCC
2-Hexenal, (E)
2-trans-Hexenal
trans-2-Hexenal

β-Propyl acrolein

6731-36-8
$C_{17}H_{34}O_4$
302.51
O(OC(C)(C)C)C(OOC(C)(C)C)(CC(CC1(C)C)C)C1
Peroxide, (3,3,5-trimethylcyclohexylidene)bis(tert-butyl (8CI)
Bis(tert-butylperoxy)-1,1 tri-methyl-3,3,5 cyclohexane (French)
1,1-Di-(terc-butilperoxi)-3,3,5-trimetilciclohexano (Spanish)
1,1-Di-(tert-butylperoxy)-3,3,5-trimethylcyclohexane
1,1-Di-(tert-butylperoxy)-3,3,5-trimethylcyclohexane, Maximum 57% in solution (DOT)
1,1-Di-(tert-butylperoxy)-3,3,5-trimethylcyclohexane, Not > 58% inert solid (DOT)
1,1-Di-(tert-butylperoxy)-3,3,5-trimethylcyclohexane, Technically pure (DOT)
Luperco 231G
Luperco 231XL
Luperco 231XLP
Luperox 231
Lupersol 231
Perhexa 3M
Peroxide, (3,3,5-trimethylcyclohexylidene)bis((1,1-dimethylethyl)- (9CI)
Peroxide, (3,3,5-trimethylcyclohexylidene)bis(tert-butyl-, Not more than 57% in solution
Peroxide, (3,3,5-trimethylcyclohexylidene)bis(tert-butyl-, Not more than 58% with inert solid
Trigonox 29
Trigonox 29/40
Trigonox 29B50
Trigonox 29B75
UN 2145 (DOT)
UN 2146 (DOT)
UN 2147 (DOT)

6734-80-1
$C_2H_4NS_2.Na.2H_2O$
165.22
O.O.[Na+].CNC(S)=S
Carbamic acid, methyldithio-, sodium salt, dihydrate
Herbatim
Karbation
Maposol
Metam
Metham dihydrate
Methyldithiokarbaman sodny dihydrat (Czech)
Monam
SMDC
Sodium N-methyldithiocarbamate dihydrate
Trimaton
Vapam
Vaporooter
VPM

6737-62-8
$C_{17}H_{16}ClN_5O_2$
357.80
1,3,5-Triazine-2,4-diamine, 6-chloro-N,N'-bis-(4-methoxyphenyl)- (9CI)
s-Triazine, 2,4-di-p-anisidino-6-chloro- (6CI,7CI, 8CI)

6739-62-4
$C_{38}H_{28}N_8O_{13}S.3Na$
905.67
Benzoic acid, 2-((2-amino-6-((4'-((3-carboxy-4-hydroxyphenyl)azo)-3,3'-dimethoxy(1,1'-biphenyl)-4-yl)azo)-5-hydroxy-7-sulfo-1-naphthalenyl)azo)-5-nitro-, trisodium salt (9CI)

6742-54-7
$C_{17}H_{28}$
232.41
c(cccc1)(c1)CCCCCCCCCCC
Benzene, undecyl- (9CI)
NSC-251008
1-Phenylundecane
Undecane, 1-phenyl- (8CI)
Undecylbenzene
n-Undecylbenzene

6746-27-6
$C_4H_9N_3S$
131.18
1,3,5-Triazine-2(1H)-thione, tetrahydro-5-methyl- (9CI)
NSC-246414
s-Triazine-2-thiol, 1,4,5,6-tetrahydro-5-methyl- (8CI)
s-Triazine-2(1H)-thione, tetrahydro-5-methyl-

6746-48-1
$C_8H_7O_4$
167.14
Benzoic acid, 4-hydroxy-3-methoxy-, ion(1-) (9CI)

6746-59-4
$C_{19}H_{23}NO_3.ClH.2H_2O$
385.93
Morphinan-6-α-ol, 7,8-didehydro-4,5-α-epoxy-3-ethoxy-17-methyl-, hydrochloride, dihydrate
7,8-Didehydro-4,5-α-epoxy-3-ethoxy-17-methylmorphinan-6-α-ol hydrochloride dihydrate
Dionin
Ethylmorphine hydrochloride
Ethylmorphine hydrochloride dihydrate

6750-03-4
$C_9H_{14}O$
138.21
O=CC=CC=CCCCC
2,4-Nonadienal
AI3-37786

6753-24-8
$C_{14}H_{18}Cl_2O_3$
305.20
Butyl 4-(2,4-dichlorophenoxy)butyrate
Butanoic acid, 4-(2,4-dichlorophenoxy)-, butyl ester (9CI)
Butyric acid, 4-(2,4-dichlorophenoxy)-, butyl ester
Caswell No. 316B
2,4-DB butyl ester
4-(2,4-Dichlorophenoxy)butyric acid butyl ester
EPA Pesticide Chemical Code 030856

6753-47-5
Unknown
Unknown
Triisopropanolamine picloram

6753-98-6
$C_{15}H_{24}$
204.36
C(=CCCC(=CCC(C=C1)(C)C)C)(C1)C
Humulene
1,4,8-Cycloundecatriene, 2,6,6,9-tetramethyl-, (E,E,E)-(9CI)
α-Humulene

6754-13-8
$C_{15}H_{18}O_4$
262.33
CC2CC1OC(=O)C(=C)C1C(O)C3(C)C2C=CC3=O
Ambrosa-2,11(13)-dien-12-oic acid, 6-α,8-β-dihydroxy-4-oxo-, 12,8-lactone
Helenalin

6765-39-5
$C_{17}H_{34}$
238.46
C(=C)CCCCCCCCCCCCCCC
1-Heptadecene (9CI)
AI3-36483

6779-09-5
$C_2H_7O_3P$
110.05
CCP(O)(O)=O
Phosphonic acid, ethyl- (8CI,

9CI)
Ethanephosphonic acid

6781-42-6
$C_{10}H_{10}O_2$
162.19
CC(=O)c1cccc(c1)C(C)=O
Ethanone, 1,1'-(1,3-phenylene)-bis- (9CI)
Benzene, m-diacetyl- (8CI)

6786-83-0
$C_{33}H_{33}N_3O$
487.63
N(c(ccc(c1)C(c(ccc(N(C)C)c2)c2)c(c(c(c(Nc(cccc3)c3)c4)ccc5)c5)c4)c1)(C)C
1-Naphthalenemethanol, α,α-bis(4-(dimethylamino)-phenyl)-4-(phenylamino)-(9CI)

6789-80-6
$C_6H_{10}O$
98.16
O=CCC=CCC
3-Hexenal, (Z)
cis-3-Hexenal
3-(Z)-Hexenal
cis-β,γ-Hexylenic aldehyde

6789-88-4
$C_{13}H_{18}O_2$
206.31
O=C(OCCCCCC)c(cccc1)c1
Benzoic acid, hexyl ester
Hexyl benzoate
n-Hexylbenzoate
Hexylester kyseliny benzoove (Czech)

6795-23-9
$C_{17}H_{12}O_7$
328.29
COc4cc2OC1OC=CC1(O)c2c5oc(=O)c3C(=O)CCc3c45
Cyclopenta(c)furo(3',2':4,5)-furo(2,3-h)(1)benzopyran-1,11-dione, 2,3,6a,9a-tetra-hydro- 9a-hydroxy-4-meth-oxy
Aflatoxin M1
4-Hydroxyaflatoxin B1

6795-87-5
$C_5H_{12}O$
88.15
Butane, 2-methoxy- (9CI)
Ether, sec-butyl methyl (8CI)

6804-07-5
$C_{11}H_{10}N_4O_4$
262.25
COC(=O)NN=Cc2c[n+]([O-])c1ccccc1[n+]2[O-]
Carbazic acid, 3-(2-quin-oxalinylmethylene)-, methyl ester, N^1,N^4-dioxide
Carbadox
3-(2-Chinoxalinylmethylen-1,4-dioxid)methylkarbazat (Czech)
2-Formylquinoxaline 1,4-di-oxide carbomethoxyhydrazone
Fortigro
GS 6244
Hydrazinecarboxylic acid, (2-quinoxalinylmethylene)-, methyl ester, N,N'-dioxide (9CI)
Karbadox (Czech)
Mecadox

6806-86-6
CH_2Cl
49.48
Methyl, chloro- (8CI,9CI)

6812-38-0
$C_{22}H_{44}$
308.60
Cyclohexane, hexadecyl-(9CI)
1-Cyclohexylhexadecane
Hexadecane, 1-cyclohexyl-(6CI,8CI)
Hexadecylcyclohexane

6819-19-8
$C_{13}H_{20}O_2$
208.30
O=C(OCCC(C(CCC(=C1)C)C1)=C)C
3-Cyclohexene-1-propanol, 4-methyl-γ-methylene-, acet-ate (9CI)
Homolinalyl acetate

6820-74-2
C_4Cl_{10}
402.57
Butane, decachloro- (6CI, 7CI,8CI,9CI)
Decachlorobutane

6828-41-7
$C_{10}H_{12}O_3$
180.20
Benzenebutanoic acid, β-hydroxy- (9CI)
Butyric acid, 3-hydroxy-4-phenyl- (6CI,7CI,8CI)

6830-82-6
$C_9H_{11}NO$
149.19
O=C(NC)Cc(cccc1)c1
Benzeneacetamide, N-methyl-(9CI)
AI3-03603
N-Methylbenzeneacetamide

6833-13-2
$C_{13}H_{10}ClNO$
231.68
Benzamide, 2-chloro-N-phenyl- (9CI)
AI3-14148
Benzanilide, 2-chloro- (8CI)

6834-92-0
$O_3Si.2Na$
122.07
Silicic acid (H_2SiO_3), di-sodium salt
B-W
Crystamet
Disodium metasilicate
Disodium monosilicate
Metso 20
Metso Beads 2048
Metso Beads, Drymet
Metso Pentabead 20
Orthosil
Sodium metasilicate
Sodium metasilicate, Anhydrous
Sodium silicate
Water Glass

6836-38-0
$C_{12}H_{26}O$
186.34
6-Dodecanol (8CI,9CI)

6838-85-3
$C_8H_6O_4.Pb$
373.33
Lead phthalate
1,2-Benzenedicarboxylic acid, lead(2+) salt (1:1) (9CI)

6841-96-9
$(C_4H_9NO)_2$
348.46
Propane, 2-methyl-2-nitroso-, dimer (8CI,9CI)

6842-15-5
$C_{12}H_{24}$
168.36
Propene, tetramer
Amsco tetramer
Dodecene
Dodecylene
1-Propene, tetramer (9CI)
Propylene tetramer [UN 2850]
Tetrapropylene
UN 2850 [Propylene tetramer]

6843-66-9
$C_{14}H_{16}O_2Si$
244.37
Silane, dimethoxydiphenyl-(9CI)
Dimethoxydiphenylsilane
Diphenyldimethoxysilane

6843-97-6
$C_{18}H_{39}N_3O_2$
329.51
O=C(O)CNCCNCCNCCCCCCC CCCCC
Dodicin
Caswell No. 413A
N-(2-((-(Dodecylamino)ethyl)-amino)ethyl)glycine
N-(2-((2-(Dodecylamino)ethyl)-amino)ethyl)glycine
Dodecylbis(aminoethyl)glycine
8-Dodecyl-2,5,8-triazaoctane-1-carboxylic acid
EPA Pesticide Chemical Code 413300
Glycine, N-(2-((2-(dodecyl-amino)ethyl)amino)ethyl)-(9CI)
Lauryl diethylenediamino-glycine

6846-50-0
$C_{16}H_{30}O_4$
286.46
O=C(OCC(C)(C)C(OC(=O)C(C)C)C(C)C)C(C)C
1,3-Pentanediol, 2,2,4-tri-methyl-, diisobutyrate (ester)
Kodaflex TXIB
Isobutyric acid, 1-isopropyl-2,2-dimethyltrimethylene ester
2,2,4-Trimethylpentanediol-1,3-diisobutyrate
2,2,4-Trimethyl-1,3-pentanediol diisobutyrate
TXIB

6848-13-1
$C_8H_{10}ClN$
155.62
N(c(cccc1Cl)c1)(C)C
Benzenamine, 3-chloro-N,N-dimethyl- (9CI)
3-Chloro-N,N-dimethylaniline
3-Chloro-N,N-dimethylbenzen-amine
N,N-Dimethyl-3-chloroaniline

6850-38-0
$C_6H_{13}NO$
115.17
Cyclohexanol, 2-amino- (8CI, 9CI)

6853-57-2
$C_{12}H_{16}O$
176.26
O=Cc(ccc(c1)CCCCC)c1
Benzaldehyde, 4-pentyl- (9CI)
4-Pentylbenzaldehyde

6854-81-5
$C_{53}H_{44}N_{14}O_{13}S_4.4Na$
1305.16
Benzenesulfonic acid, 2,2'-(car-bonylbis(imino(2-methyl-4,1-phenylene)azo(2-methyl-4,1-phenylene)azo))bis(5-((4-sulfophenyl)azo)-, tetra-sodium salt (9CI)
C.I. Direct Brown 106 (8CI)
Sirius Supra Brown G

6859-99-0
$C_5H_{11}NO$
101.14
3-Piperidinol (9CI)

6863-58-7
$C_8H_{18}O$
130.26
O(C(CC)C)C(CC)C
sec-Butyl ether (DOT)
Bis(2-butyl)ether
Butane, 2,2'-oxybis- (9CI)
Di-sec-butyl ether
Di-sec-butyl ether [UN 1149]
UN 1149 [Dibutyl ethers]

6864-23-9
$C_9H_{16}N_4O$
196.25
1,3,5-Triazin-2-amine, 4-ethoxy-N,6-diethyl-(9CI)
s-Triazine, 2-ethoxy-4-ethyl-6-(ethylamino)- (7CI,8CI)

6864-53-5
$C_{14}H_{30}$
198.39
Undecane, 2,6,10-trimethyl-(8CI,9CI)
Norfarnesane

6865-35-6
$C_{36}H_{70}O_4.Ba$
704.40
Stearic acid, barium salt
Barium distearate
Barium stearate
Octadecanoic acid, barium salt (9CI)
Stavinor 40

6865-97-0
$(C_4H_8ClNO)_2$
243.12
2-Chloro-3-nitrosobutane dimer
Butane, 2-chloro-3-nitroso-, dimer (8CI)

6870-67-3
$C_{18}H_{25}NO_6$
351.44
Jacobine
15,20-Epoxy-15,20-dihydro-12-hydroxysenecionan-11,16-dione
NSC-89936
Senecionan-11,16-dione, 15,20-epoxy-15,20-dihydro-12-hydroxy-, (15-α,20s)-(9CI)

6872-06-6
$C_9H_{11}N$
133.21
N(c(c(ccc1)C2)c1)C2C
Indoline, 2-methyl
2-Methylindoline
PE-11

6876-13-7
$C_{10}H_{18}$
138.25
C(CC1C(C2)C)(C1(C)C)C2
Bicyclo(3.1.1)heptane, 2,6,6-tri-methyl-, (1α,2β,5α)- (9CI)
Pinane, stereoisomer (8CI)
cis-Pinane

6876-23-9
C_8H_{16}
112.22
CC1CCCCC1C
Cyclohexane, 1,2-dimethyl-, trans- (9CI)
AI3-28850
trans-1,2-Dimethylcyclohexane

6881-94-3
$C_7H_{16}O_3$
148.20
O(CCOCCO)CCC
Ethanol, 2-(2-propoxyethoxy)-(9CI)
2-(2-Propoxyethoxy)ethanol

6885-57-0
$C_{17}H_{14}O_7$
330.31
Cyclopenta(c)furo(3',2':4,5)-furo(2,3-h)(1)benzopyran-1,11-dione, 2,3,6a,8,9,9a-hexahydro- 9a-hydroxy-4-methoxy
Aflatoxin M2
4-Hydroxyaflatoxin B2

6891-44-7
$C_9H_{18}NO_2.CH_3O_4S$
283.34
Ethanaminium, N,N,N-tri-methyl-2-((2-methyl-1-oxo-2-propenyl)oxy)-, methyl sulfate (9CI)
N,N,N-Trimethyl-2-(1-oxo-2-methyl-2-propenyloxy)-ethanaminium methyl sulfate

6893-02-3
$C_{15}H_{12}I_3NO_4$
650.98

NC(Cc2cc(I)c(Oc1ccc(O)c(I)c1)c(I)c2)C(O)=O
Alanine, 3-(4-(4-hydroxy-3-iodophenoxy)-3,5-diiodophenyl)-, L
L-3-(4-(4-Hydroxy-3-iodophenoxy)-3,5-diiodophenyl)alanine
o-(4-Hydroxy-3-iodophenyl)-3,5-diiodo-l-tyrosine
4-(3-Iodo-4-hydroxyphenoxy)-3,5-diiodophenylalanine
Liothyronin
Liothyronine
L-Liothyronine
T3
T_3
L-T3
Tresitope
Triiodothyronine
Triiodo-l-thyronine
L-Triiodothyronine
3,3',5-Triiodothyronine
3,5,3'-Triiodothyronine
Triothyrone
L-Tyrosine, o-(4-hydroxy-3-iodophenyl)-3,5-diiodo-(9CI)

6899-10-1
$C_{19}H_{42}N$
284.62
N(CCCCCCCCCCCCCCCC)(C)(C)C
Ammonium, hexadecyltrimethyl
Cetrimonium
Cetyltrimethylammonium
Cetyltrimethylammonium cation
Cetyltrimethylammonium ion
1-Hexadecanaminium, N,N,N-trimethyl- (9CI)
Hexadecyltrimethylammonium
Hexadecyltrimethylammonium ion
Trimethylhexadecylammonium

6900-35-2
$C_4H_6O_2.K$
125.19
2-Propenoic acid, 2-methyl-, potassium salt (9CI)
Potassium methacrylate
Potassium 2-methyl-2-propenoate

6903-65-7
$C_{12}H_8Cl_2O$
239.10
Benzene, 1-chloro-2-(4-chlorophenoxy)- (9CI)
Ether, o-chlorophenyl p-chlorophenyl (8CI)

6907-59-1
$C_{16}H_{13}N$
219.28
Isoquinoline, 1-(phenylmethyl)-(9CI)
Isoquinoline, 1-benzyl- (8CI)

6912-86-3
$C_{11}H_{12}N_2O_2$
204.23
Tryptophan
Trofan
Tryptacin
L-Tryptophan
levoTryptophan

6915-15-7
$C_4H_6O_5$
134.10
O=C(O)CC(O)C(=O)O
Malic-acid
Butanedioic acid, hydroxy-(9CI)
Deoxytetraric acid
Hydroxysuccinic acid
α-Hydroxysuccinic acid
Kyselina hydroxybutandiova (Czech)
Kyselina jablecna (Czech)
Pomalus acid
Succinic acid, hydroxy-

6915-18-0
$C_4H_4O_4$
116.07
O=C(O)C=CC(=O)O
2-Butenedioic acid (9CI)
Butenedioic acid (8CI)
1,2-Ethenedicarboxylic acid

6921-35-3
$C_5H_{10}O$
86.15
Propane, 1,3-epoxy-2,2-dimethyl
3,3-Dimethyloxetane
β,β-Dimethyltrimethylene oxide
3,3-Dimethyltrimethylene oxide
1,3-Epoxy-2,2-dimethylpropane
Oxetane, 3,3-dimethyl- (9CI)

6923-22-4
$C_7H_{14}NO_5P$
223.19
CNC(=O)C=C(C)OP(=O)(OC)OC
Phosphoric acid, dimethyl ester, ester with (E)-3-hydroxy-N-methylcrotonamide
Apadrin
Azodrin (OSHA)
Azodrin Insecticide
Azodrin-71
Bilobran
Biloborn
C 1414
Crisodrin
Ciba 1414
Crisodin
Crotonamide, 3-hydroxy-N-methyl-, dimethylphosphate, cis-
Crotonamide, 3-hydroxy-N-methyl-, dimethylphosphate, (E)-
3-(Dimethoxyphosphinyloxy)n-methyl-cis-crotonamide
(E)-Dimethyl 1-methyl-3-(methylamino)-3-oxo-1-propenylphosphate
O,O-Dimethyl-O-(2-N-methyl-carbamoyl-1-methyl-vinyl)-fosfaat (Dutch)
O,O-Dimethyl-O-(2-N-methyl-carbamoyl-1-methyl)-vinyl-phosphat (German)
O,O-Dimethyl-O-(2-N-methyl-carbamoyl-1-methyl-vinyl) phosphate
O,O-Dimethyl-O-(1-methyl-2-N-methyl-carbamoyl)-vinyl-phosphat (German)
Dimethyl 1-methyl-2-(methyl-carbamoyl)vinyl phosphate, cis
Dimethyl phosphate ester with (E)-3-hydroxy-N-methylcrotonamide
Dimethyl phosphate ester of 3-hydroxy-N-methyl-cis-crotonamide
Dimethyl phosphate of 3-hydroxy-N-methyl-cis-crotonamine
O,O-Dimetil-O-(2-N-metilcarbamoil-1-metil-vinil)-fosfato (Italian)
E-Monocrotophos
ENT 27,129
Glore Phos 36
Hazodrin
3-Hydroxy-N-methylcrotonamide dimethyl phosphate
3-Hydroxy-N-methyl-cis-crotonamide dimethyl phosphate
Monokrotofosz (Hungarian)
cis-1-Methyl-2-methyl carbamoyl vinyl phosphate
Monocil 40
Monocron
Monocrotophos (ACGIH, OSHA)
Monodrin
Nuvacron
Nuvacron 20
OMS 834
Phosphate de dimethyle et de 2-methylcarbamoyl 1-methyl vinyle (French)
Phosphoric acid, dimethyl ester, ester with cis-3-hydroxy-N-methylcrotonamide
Phosphoric acid, dimethyl 1-methyl-3-(methylamino)-3-oxo-1-propenyl ester, (E)-
Plantdrin
Pillardrin
SD 9129
Shell SD 9129
Susvin
Ulvair

6925-69-5
$C_{18}H_{10}N_2O$
270.28
O=C(N(c(c(c(ccc1)cc2)c1N=3)c2)C3c4cccc5)c45
12H-Phthaloperin-12-one (9CI)
12-Phthaloperinone

6933-10-4
C_7H_8BrN
186.05
Cc1cc(N)ccc1Br
Benzenamine, 4-bromo-3-methyl- (9CI)
m-Toluidine, 4-bromo- (8CI)

6936-40-9
$C_7H_4Cl_4$
229.92
Anisole, 2,3,5,6-tetrachloro- (8CI)
AI3-22330
Benzene, 1,2,4,5-tetrachloro-3-methoxy- (9CI)

6938-45-0
$C_{13}H_{18}O_2$
206.28
Hexanoic acid, phenylmethyl ester (9CI)
Hexanoic acid, benzyl ester (8CI)
NSC-53964

6938-94-9
$C_{12}H_{22}O_4$
230.34
O=C(OC(C)C)CCCCC(=O)OC(C)C
Adipic acid, diisopropyl ester
Ceraphyl 230
Crodamol DA
Diisopropyl adipate
Hexanedioic acid, bis-(1-methylethyl) ester (9CI)
Iso-Adipate 2/043700
Isopropyl adipate
Prodipate
Schercemol DIA
Standamul DIPA
Tegester 504-D
Wickenol 116

6945-35-3
$C_7H_{12}O_2$
128.17
2-Pentenoic acid, 4,4-dimethyl- (6CI,7CI,9CI)

6949-77-5
$C_8H_{15}NO_2$
157.21
1-Aminocycloheptanecarboxylic acid
Cycloheptanecarboxylic acid, 1-amino- (9CI)

6952-59-6
C_7H_4BrN
182.03
Benzonitrile, m-bromo
m-Bromobenzonitrile
3-Bromobenzonitrile

6959-47-3
$C_6H_6ClN.ClH$
164.04
ClCc1ccccn1
Pyridine, 2-chloromethyl-, hydrochloride
2-(Chloromethyl) pyridine, hydrochloride
NCI-C03907
2-Picolyl chloride hydrochloride
2-Pyridylmethylchloride hydrochloride

6959-48-4
$C_6H_6ClN.ClH$
164.04
Cl.ClCc1cccnc1
Pyridine, 3-chloromethyl-, hydrochloride
3-(Chloromethyl) pyridine, hydrochloride
NCI-C03838

6961-73-5
$C_8H_{15}NOS$
173.27
1-Piperidinecarbothioic acid, S-ethyl ester (8CI,9CI)
NSC-62748

6962-44-3
$C_{10}H_{13}NO_2$
179.21
Acetamide, N-(2-methoxy-5-methylphenyl)- (9CI)
N-(2-Methoxy-5-methylphenyl)-acetamide
5'-Methyl-o-acetanisidide

6963-65-1
$C_3H_2BrN_3O_2$
191.95
1H-Imidazole, 4-bromo-5-nitro- (9CI)
Imidazole, 4-bromo-5-nitro- (8CI)
NSC-54255

6964-19-8
$C_{21}H_{35}ClO$
338.96
Phenol, 4-chloro-3-pentadecyl- (8CI,9CI)
NSC-66318

6964-21-2
$C_6H_6O_2S$
142.18
3-Thiopheneacetic acid
3-Thienylacetic acid

6966-10-5
$C_9H_{12}O$
136.19
Benzenemethanol, 3,4-dimethyl- (9CI)
Benzyl alcohol, 3,4-dimethyl- (8CI)
NSC-18728

6967-70-0
$C_{10}H_{15}NO_2$
181.24
2(1H)-Pyridinone, 3-butyl-4-hydroxy-6-methyl- (9CI)
3-Butyl-6-methyl-2,4-pyridinediol
2,4-Pyridinediol, 3-butyl-6-methyl- (6CI,8CI)

6973-13-3
$C_7H_6ClNO_5S$
251.64
O=S(=O)(O)c(c(N(=O)=O)cc(c1Cl)C)c1
Benzenesulfonic acid, 5-chloro-4-methyl-2-nitro- (9CI)
5-Chloro-4-methyl-2-nitrobenzenesulfonic acid
2-Chloro-5-nitro-1-toluene-4-sulfonic acid

6975-71-9
$C_8H_{11}N$
121.18
N#CCC(=CCCC1)C1
1-Cyclohexene-1-acetonitrile (9CI)
AI3-04979
Cyclohexenylacetonitrile

6975-98-0
$C_{11}H_{24}$
156.31
Decane, 2-methyl- (9CI)

6982-25-8
$C_4H_{10}O_2$
90.12
2,3-Butanediol, (R*,R*)-(+-)- (9CI)
2,3-Butanediol, (+-)- (8CI)

6985-92-8
$C_{26}H_{19}N_5O_5$
481.44
O=C(Nc(ccc(NC(=O)N1)c12)c2)c(c(O)c(N=Nc(c(C(=O)OC)ccc3)c3)c(c4ccc5)c5)c4

Benzoic acid, 2-((3-(((2,3-dihydro-2-oxo-1H-benzimidazol-5-yl)amino)carbonyl)-2-hydroxy-1-naphthalenyl)-azo)-, methyl ester (9CI)

6986-48-7
$C_4H_8Cl_2O$
143.02
Ether, bis(1-chloroethyl)
Bis(α-chloroethyl) ether
1,1'-Oxybis(1-chloroethane)

6988-21-2
$C_{11}H_{13}NO_4$
223.25
CNC(=O)Oc1ccccc1C2OCCO2
Carbamic acid, methyl-, o-(1,3-dioxolan-2-yl)phenyl ester
C-8353
Ciba 8353
Dioxacarb
Dioxacarbe
Dioxocarb
2-(1,3-Dioxolane-2-yl)phenyl N-methylcarbamate
2-(1,3-Dioxolan-2-yl)phenyl-N-methylcarbamat (German)
o-(1,3-Dioxolan-2-yl)phenyl methylcarbamate
Du Pont 1519
Du Pont Insecticide 1519
Elcron
Elecron 50
Elocron
Elocron 8353
Elocron 50WP
ENT 27,389
Famid
NSC-190981
Rovlinka

6990-43-8
$C_{16}H_{36}O_4P_2S_4Zn$
548.06
Zinc, bis(O,O-dibutyl phosphorodithioato-S,S')-, (T-4)-(9CI)

6994-46-3
$C_{18}H_{18}N_2O_2$
294.34
O=C(c(c(C(=O)c1c(NCC)ccc2NCC)ccc3)c3)c12
Sudan Blue
9,10-Anthracenedione, 1,4-bis-(ethylamino)- (9CI)
1,4-Bis(ethylamino)-9,10-anthracenedione

6994-95-2
$C_9H_{14}O_3$
170.21
2-Hexenoic acid, 3,4,4-trimethyl-5-oxo-, (E)- (8CI)

6996-81-2
$C_5H_{13}O_2PS$
168.20
Phosphonothioic acid, methyl-, O,O-diethyl ester (8CI,9CI)

6996-88-9
$C_{14}H_{19}NOS$
249.37
1H-Azepine-1-carbothioic acid, hexahydro-, S-(phenylmethyl) ester (9CI)
1H-Azepine-1-carbothioic acid, hexahydro-, S-benzyl ester (8CI)

6996-92-5
$C_6H_6O_2Se$
189.07
Benzeneseleninic acid (8CI,9CI)

7003-32-9
$C_7H_{15}N$
113.20
NC(C(CCC1)C)C1
Cyclohexanamine, 2-methyl-(9CI)
2-Methylcyclohexanamine
2-Methylcyclohexylamine

7003-48-7
$C_{10}H_{16}O_2$
168.24
8-Nonynoic acid, methyl ester (7CI,8CI,9CI)
Methyl 8-nonynoate

7005-47-2
$C_6H_{15}NO$
117.22
OCC(N(C)C)(C)C
1-Propanol, 2-dimethylamino-2-methyl
2-Dimethylamino-2-methyl-1-propanol
DMAMP
USAF CS-1

7005-72-3
$C_{12}H_9ClO$
204.66
O(c(ccc(c1)Cl)c1)c(cccc2)c2
p-Chlorophenyl phenyl ether
AI3-32895
Benzene, 1-chloro-4-phenoxy-(9CI)
4-Chlorodiphenyl ether
p-Chlorodiphenyl oxide
1-Chloro-4-phenoxybenzene
4-Chlorophenyl phenyl ether
Ether, p-chlorophenyl phenyl (8CI)

7008-42-6
$C_{20}H_{19}NO_3$
321.40
COc3cc1OC(C)(C)C=Cc1c4n(C)c2ccccc2c(=O)c34
7H-Pyrano(2,3-c)acridin-7-one, 3,12-dihydro-6-methoxy-3,3,12-trimethyl
Acromycine
Acronine
Acronycine
Compound 42339
3,12-Dihydro-6-methoxy-3,3,12-trimethyl-7H-pyrano(2,3-c)acridin-7-one
NCI-C01536
NSC-403169

7012-37-5
$C_{12}H_7Cl_3$
257.55
1,1'-Biphenyl, 2,4,4'-trichloro-(9CI)
Biphenyl, 2,4,4'-trichloro-
2,4,4'-Trichlorobiphenyl

7023-61-2
$C_{18}H_{13}ClN_2O_6S.Ca$
460.90
2-Naphthalenecarboxylic acid, 4-((5-chloro-4-methyl-2-sulfophenyl)azo)-3-hydroxy-, calcium salt (1:1) (9CI)
4-((5-Chloro-2-sulfo-p-tolyl)-azo)-3-hydroxy-2-naphthalene-carboxylic acid, calcium salt (1:1)

7030-18-4
$C_{13}H_{11}N_3O_2$
241.25
Diazene, (2-methylphenyl)-(4-nitrophenyl)- (9CI)
Azobenzene, 2-methyl-4'-nitro- (7CI,8CI)

7040-58-6
$C_{13}H_{29}O_3P$
264.35
Phosphonic acid, methyl-, bis-(1,2,2-trimethylpropyl) ester (8CI,9CI)

7045-71-8
$C_{12}H_{26}$
170.34
Undecane, 2-methyl- (8CI,9CI)

7045-83-2
$C_{10}H_6O_5S.Na$
261.21
2-Naphthalenesulfonic acid, 1,4-dihydro-1,4-dioxo-, sodium salt (9CI)
2-Sulfo-1,4-naphthoquinone, sodium salt

7047-84-9
$C_{18}H_{37}AlO_4$
344.53
Aluminum, dihydroxy-(stearato)
Aluminum stearate (ACGIH)
Stearic acid, aluminium salt

7055-03-0
$C_{14}H_{13}NO$
211.28
Cc1ccccc1C(=O)Nc2ccccc2
Benzamide, 2-methyl-N-phenyl
BAS 305
BAS-3050
BAS 3050F
Bebenil
F 368
Mebenil
o-Methylbenzanilide
2-Methylbenzanilide
2-Methylbenzoanilide
2-Methylbenzoic acid anilide
2-Methyl-N-phenylbenzamide
o-Toluanilide

7057-92-3
$C_{24}H_{51}O_4P$
434.64
O=P(OCCCCCCCCCCCC)(OCCCCCCCCCCCC)O
Di-n-dodecyl phosphate
AI3-23305
Didodecyl hydrogen phosphate
Didodecyl phosphate
Dilauryl acid phosphate
Dilauryl phosphate
ELA
Ortholeum 162
Phosphoric acid, didodecyl ester (9CI)

7058-01-7
$C_{10}H_{20}$
140.27
Cyclohexane, (1-methylpropyl)-(9CI)
Cyclohexane, sec-butyl-
NSC-73718

7059-16-7
$C_{32}H_{68}O_4P_2S_4Zn$
772.49
Zinc, bis(O,O-dioctyl phosphorodithioato-S,S')-, (T-4)-(9CI)
Phosphorodithioic acid, O,O-dioctyl ester, zinc salt

7068-83-9
$C_5H_{12}N_2O$
116.19
CCCCN(C)N=O
Butylamine, N-methyl-N-nitroso
Butanamine, N-methyl-N-nitroso-
Butylmethylnitrosamine
MBNA
Methyl-butyl-nitrosamin (German)
Methylbutylnitrosamine
Methyl-n-butylnitrosamine
N-Methyl-N-nitrosobutylamine
N-Nitroso-n-butylmethylamine
Nitrosomethyl-n-butylamine
N-Nitrosomethyl-n-butylamine
N-Nitroso-N-methyl-N-n-butylamine
NMBA

7070-15-7
$C_{12}H_{22}O_2$
198.31
O(C(C(C(C1C2)(C)C)(C2)C)C1)CCO
Ethanol, 2-((1,7,7-trimethylbicyclo(2.2.1)hept-2-yl)oxy)-, exo- (9CI)
exo-2-Camphanyl β-hydroxyethyl ether
β-Hydroxyethyl isobornyl ether

7073-42-9
$C_9H_{10}ClNO_2$
199.63
COc1ccc(NC(C)=O)cc1Cl
Acetamide, N-(3-chloro-4-methoxyphenyl)- (9CI)
p-Acetanisidide, 3'-chloro-(8CI)

7073-94-1
$C_9H_{11}Br$
199.09
CC(C)c1ccccc1Br
Benzene, 1-bromo-2-(1-methylethyl)- (9CI)
Cumene, o-bromo- (8CI)

7081-78-9
C_4H_9ClO
108.57
O(C(Cl)C)CC
Ethane, 1-chloro-1-ethoxy-(9CI)
1-Chloro-1-ethoxyethane

7082-31-7
$C_{44}H_{32}N_6O_{14}S_3.3Na$
1033.90
1-Naphthalenesulfonic acid, 3-((7-((4'-((1,8-dihydroxy-4-sulfo-2-naphthalenyl)azo)-3,3'-dimethyl (1,1'-biphenyl)-4-yl)azo)-8-hydroxy-6-sulfo-2-naphthalenyl)-azo)-4,5- dihydroxy-, trisodium salt (9CI)
Amanil Chrome Navy Blue B
Benzo Chrome Black Blue BS
C.I. 31930
C.I. Direct Blue 26
C.I. Direct Blue 26, trisodium salt (8CI)
Calcomine Dark Blue BN
Chlorazol Steel Blue 6B
Diacotton Navy Blue BS
Diphenyl Chrome Black Blue 2B
Diphenyl Chrome Blue Black 2B
Direct Blue 26
Direct Chrome Black Blue B
Direct Chrome Black Blue 2B
Direct Chrome Dark Blue 2B
Direct Fast Blue Black MB
Dirochrome Dark Blue B
Niagara Chrome Blue Black B
Pontamine Navy Blue BFN
Trisulfon Fast Blue B
Viscoform Navy Blue 2GB
Viscoform Navy Blue BG

7085-19-0
$C_{10}H_{11}ClO_3$
214.66
Propionic acid, 2-((4-chloro-o-tolyl)oxy)-, (+-)
Anicon P
BH Mecoprop
CMPP
Compitox
Compitox Plus
Isocarnox
Iso-Cornox 57
Kilprop
MCPP
MCPP-D-4
MCPP-K-4
Mechlorprop
Mecomec
Mecoprop
Methoxone
Mepro
Okultin MP
Propanoic acid, 2-(4-chloro-2-methylphenoxy)-, (+-)-
Proponex-Plus
Rankotex
RD 4593

7085-85-0
$C_6H_7NO_2$
125.12
O=C(OCC)C(C#N)=C
Ethyl 2-cyanoacrylate
Ethyl cyanoacrylate
Ethyl 2-cyano-2-propenoate
2-Propenoic acid, 2-cyano-, ethyl ester (9CI)

7087-68-5

$C_8H_{19}N$
129.24
N(C(C)C)(C(C)C)CC
N,N-Diisopropylethylamine
N-Ethyl-N-(1-methylethyl)-2-propanamine
2-Propanamine, N-ethyl-N-(1-methylethyl)- (9CI)
Triethylamine, 1,1'-dimethyl-(8CI)

7091-57-8
Unknown
Unknown
8-Quinolinol benzoate

7097-60-1
Unknown
Unknown
Tetraglycine hydroperiodide
Globalin

7098-08-0
$C_{20}H_{14}N_2O_5$
362.33
O=C(c(c(c(N)cc1c(ccc(O)c2)c2)C(=O)c3c(O)ccc4N)c1O)c34
9,10-Anthracenedione, 4,8-diamino-1,5-dihydroxy-2-(4-hydroxyphenyl)- (9CI)
4,8-Diamino-1,5-dihydroxy-2-(p-hydroxyphenyl)anthracen-9,10-dione
4,8-Diamino-1,5-dihydroxy-2-(4-hydroxyphenyl)-9,10-anthracenedione
1,5-Diamino-4,8-dihydroxy-3-(4-hydroxyphenyl)anthraquinone

7098-22-8
$C_{44}H_{90}$
619.20
C(CCCCCCCCCCCCCCCCCCCCCCCCCCCCCCCCCCCCCCCCC)C
n-Tetratetracontane
AI3-36493
Tetratetracontane (9CI)

7101-31-7
$C_2H_6Se_2$
187.99
Dimethyldiselenide

7116-86-1
C_8H_{16}
112.22
C(=C)CCC(C)(C)C
1-Hexene, 5,5-dimethyl- (9CI)
5,5-Dimethylhexene-1
5,5-Dimethyl-1-hexene

7119-94-0
$C_5H_{10}N_2O_2$
130.13
Piperidine, 1-nitro- (8CI,9CI)

7128-64-5
$C_{26}H_{26}N_2O_2S$
430.56
O(c(c(N=1)cc(c2)C(C)(C)C)c2)C1C(SC(=C3)C(Oc(c4cc(c5)C(C)(C)C)c5)=N4)=C3
Benzoxazole, 2,2'-(2,5-thiophenediyl)bis(5-(1,1-dimethylethyl)- (9CI)

7128-91-8
$C_{18}H_{39}NO$
285.51
O=N(CCCCCCCCCCCCCCCC)(C)C
Palmitamine oxide
AI3-61416-X
Cetamine oxide
Cetyl dimethyl amine oxide
N,N-Dimethyl-1-hexadecanamine N-oxide
1-Hexadecanamine, N,N-dimethyl-, N-oxide (9CI)
1-Hexadecanamine, N,N-dimethyl-, N-oxide (40% solution)
Jordamox CDA
Jordamox CDA-40
Palmityl dimethylamine oxide

7132-64-1
$C_{16}H_{32}O_2$
256.43
O=C(OC)CCCCCCCCCCCCCC
Pentadecanoic acid, methyl ester
AI3-36452
Methyl pentadecanoate

7133-46-2
$C_{12}H_{22}S$
198.37
Cyclohexane, 1,1'-thiobis- (9CI)
Cyclohexyl sulfide (8CI)

7138-40-1
$C_{27}H_{54}O_2$
410.72
Heptacosanoic acid (8CI,9CI)

7144-05-0
$C_6H_{14}N_2$
114.18
N(CCC(C1)CN)C1
4-Piperidinemethanamine (9CI)

7144-37-8
$C_7H_8O_3S.1/2Cu$
203.97
Benzenesulfonic acid, 4-methyl-, copper(2+) salt (9CI)
Copper(2+) 4-methylbenzenesulfonate

7144-65-2
$C_{15}H_{14}O_2$
226.27
Oxirane, (((1,1'-biphenyl)-2-yloxy)methyl)- (9CI)
NSC-52971
Propane, 1-(2-biphenylyloxy)-2,3-epoxy- (8CI)

7145-20-2
C_8H_{16}
112.22
2-Hexene, 2,3-dimethyl- (8CI,9CI)
NSC-74163

7146-60-3
$C_{10}H_{22}$
142.28
Octane, 2,3-dimethyl- (9CI)

7147-42-4
$C_{36}H_{36}N_6O_6$
648.68
O=C(Nc(c(ccc1)C)c1)C(N=Nc(c(OC)cc(c(ccc(N=NC(C(=O)C)C(=O)Nc(c(ccc2)C)c2)c3OC)c3)c4)c4)C(=O)C
Butanamide, 2,2'-((3,3'-dimethoxy(1,1'-biphenyl)-4,4'-diyl)bis(azo))bis(N-(2-methylphenyl)-3-oxo- (9CI)

7147-89-9
$C_7H_6ClNO_3$
187.58
O=N(=O)c(c(O)cc(c1Cl)C)c1
Phenol, 4-chloro-5-methyl-2-nitro- (9CI)
4-Chloro-5-methyl-2-nitrophenol

7148-92-7
$C_{15}H_9N$
203.24
Indeno(1,2,3-de)isoquinoline (9CI)

7149-26-0
$C_{17}H_{23}NO_2$
273.41
O=C(OC(C=C)(CCC=C(C)C)C)c(c(N)ccc1)c1
Anthranilic acid, linalyl ester
Anthranilic acid, 1,5-dimethyl-1-vinyl-4-hexenyl ester
3,7-Dimethyl-1,6-octadien-3-yl o-aminobenzoate
1,5-Dimethyl-1-vinyl-4-hexen-1-yl o-aminobenzoate
4-Hexen-1-ol, 1,5-dimethyl-1-vinyl-, o-aminobenzoate

Linalyl o-aminobenzoate
Linalyl anthranilate
1,6-Octadien-3-ol, 3,7-dimethyl-, o-aminobenzoate
1,6-Octadien-3-ol, 3,7-dimethyl-, 2-aminobenzoate (9CI)

7149-75-9
C_7H_8ClN
141.59
Benzenamine, 4-chloro-3-methyl- (9CI)
NSC-72329
m-Toluidine, 4-chloro- (8CI)

7154-79-2
C_9H_{20}
128.26
2,2,3,3-Tetramethylpentane
Pentane, 2,2,3,3-tetramethyl- (9CI)

7154-80-5
$C_{10}H_{22}$
142.28
Heptane, 3,3,5-trimethyl- (9CI)

7159-34-4
$C_7H_5Cl_4NO$
260.93
COc1cc(cc(Cl)n1)C(Cl)(Cl)Cl
4-Picoline, 2-chloro-6-methoxy-α,α,α-trichloro
2-Chloro-6-methoxy-4-(trichloromethyl)pyridine
DOWCO 269
Lorvec
Lorvek
Nurelle
Pyroxychlor
Pyroxychlore

7160-01-2
$C_{10}H_{14}N_2O$
178.22
CN(C)C(=O)Nc1ccc(C)cc1
Urea, N,N-dimethyl-N'-(4-methylphenyl)-
AI3-61362

7160-02-3
$C_{10}H_{14}N_2O_2$
194.22
COc1ccc(NC(=O)N(C)C)cc1
Urea, N'-(4-methoxyphenyl)-N,N-dimethyl- (9CI)
Urea, 3-(p-methoxyphenyl)-1,1-dimethyl- (8CI)

7166-19-0
$C_8H_6BrNO_2$
228.04
O=N(=O)C(Br)=Cc(cccc1)c1
β-Bromo-β-nitrostyrene
Benzene, (2-bromo-2-nitroethenyl)- (9CI)
(2-Bromo-2-nitroethenyl)-benzene
Caswell No. 116B
EPA Pesticide Chemical Code 101401
Giv-Cide BNS
Giv-Gard
Slime-Trol

7169-34-8
$C_8H_6O_2$
134.13
3(2H)-Benzofuranone (8CI,9CI)
Coumaranone
NSC-512726

7170-77-6
$C_8H_5N_3S_2$
207.26
N#CSc(ccc(NC(=N)S1)c12)c2
Thiocyanic acid, 2-amino-6-benzothiazolyl ester (9CI)

7173-51-5
$C_{22}H_{48}N.Cl$
362.16
[OH].[Cl-].CCCCCCCCCC[N+](C)(C)CCCCCCCCCC
Ammonium, didecyldimethyl-, chloride
Aliquat 203
Bardac 22
Bio-Dac 50-22
BTC 1010
BTCO 1010
1-Decanaminium, N-decyl-N,N-dimethyl-, chloride (9CI)
Didecyl dimethyl ammonium chloride
Dimethyldidecylammonium chloride
Quaternium-12

7173-62-8
$C_{21}H_{44}N_2$
324.58
N(CCCCCCCCC=CCCCCCCCC)CCCN
1,3-Propanediamine, N-9-octadecenyl-, (Z)- (9CI)
(Z)-N-9-Octadecenyl-1,3-propanediamine

7175-47-5
C_8H_7N
117.15
Benzene, 1-isocyano-4-methyl- (9CI)
1-Isocyano-p-toluene
p-Methylphenyl isocyanide
4-Methylphenyl isocyanide
p-Tolyl isocyanide (6CI,7CI,8CI)
p-Tolyl isonitrile

7177-48-2
$C_{16}H_{19}N_3O_4S.3H_2O$
403.50
O.O.O.CC3(C)SC2C(NC(=O)C(N)c1ccccc1)C(=O)N2C3C(O)=O
4-Thia-1-azabicyclo(3.2.0)-heptane-2-carboxylic acid, 6-(2-amino-2-phenylacetamido)- 3,3-dimethyl-7-oxo-, trihydrate, D- (-)
Acillin
Amcap
Amcill
Aminobenzylpenicillin trihydrate
α-Aminobenzylpenicillin trihydrate
Amperil
Ampichel
Ampicillin trihydrate
Ampikel
Ampinova
Amplin
Cymbi
Divercillin
Lifeampil
Morepen
NCI-C56086
Pen A
Pensyn
Polycillin
Princillin
Principen
Ro-Ampen
Trafarbiot
Ukopen
Vidopen

7179-50-2
Unknown
Unknown
Calcium oxytetracycline

7180-62-3
$C_7H_8O_3$
140.14
4-Cyclopentene-1,3-dione, 4-methoxy-5-methyl- (7CI,8CI,9CI)

7181-73-9
$C_{17}H_{22}NO$
256.40
Ammonium, benzyldimethyl-(2-phenoxyethyl)
Bephenium

7187-62-4
$C_{26}H_{28}N_3$
382.51
Pyrvinium
6-(Dimethylamino)-2-(2-(2,5-dimethyl-1-phenyl-1H-pyrrol-3-yl)ethenyl)-1-methylquinolinium

Pyrvinum
Quinolinium, 6-(dimethyl-amino)-2-(2-(2,5-dimethyl-1-phenyl-1H-pyrrol-3-yl)-ethenyl)-1-methyl-
Quinolinium, 6-(dimethyl-amino)-2-(2-(2,5-dimethyl-1-phenylpyrrol-3-yl)vinyl)-1-methyl-

7194-84-5
$C_{37}H_{76}$
521.01
C(CCCCCCCCCCCCCCCCCCCC CCCCCCCCCCCCCCCCC)C
Heptatriacontane (9CI)

7194-85-6
$C_{38}H_{78}$
535.04
C(CCCCCCCCCCCCCCCCCCCC CCCCCCCCCCCCCCCCCC)C
Octatriacontane
AI3-36495

7195-43-9
$Cl_4H_{14}O_6$
251.94
Isophthalic acid, bis(2,3-epoxypropyl)ester
1,3-Benzenedicarboxylic acid, bis(oxiranylmethyl) ester (9CI)
Bis(2,3-epoxypropyl)iso-phthalate
Diglycidyl isophthalate

7195-45-1
$C_{14}H_{14}O_6$
278.28
O=C(OCC(O1)C1)c(c(ccc2)C (=O)OCC(O3)C3)c2
Phthalic acid, bis(2,3-epoxy-propyl) ester
Diglycidylester kyseliny ftalove (Czech)
Diglycidyl phthalate
Phthalic acid, diglycidyl ester

7199-02-2
$C_{19}H_{24}N_3O.xCl_2Zn.Cl$
482.15
Phenoxazin-5-ium, 3-(di-ethylamino)-7-(dimethyl-amino)-2-methyl-, chlor-ide, Compd. with zinc chloride ($ZnCl_2$) (9CI)
C.I. 51015
Capri Blue GON
Phenoxazin-5-ium, 3-(di-ethylamino)-7-(dimethyl-amino)-2-methyl-, chlor-ide, Compd. with zinc chloride (8CI)
Zinc chloride, Compd. with 3,7-bis(diethylamino)-2-methylphenazoxonium chloride
Zinc chloride ($ZnCl_2$), Compd. with 3,7-bis(di-ethylamino)-2-methylphen-azoxonium chloride (9CI)

7203-90-9
$C_8H_{10}ClN_3$
183.66
CN(C)N=Nc1ccc(Cl)cc1
Triazene, 3,3-dimethyl-1-(p-chlorophenyl)
1-p-Chlorfenyl-3,3-dimethyl-triazen (Czech)
Chloro-PDMT
1-(p-Chlorophenyl)-3,3-di-methyl-triazene
1-(4-Chlorophenyl)-3,3-di-methyltriazene
1-(p-Chlor-phenyl)-3,3-di-methyl-triazen (German)
Triazene, 1-(p-chlorophenyl)-3,3-dimethyl-

7204-16-2
$C_{12}H_{18}O_4$
226.27
Ethanol, 2-(2-(2-phenoxy-ethoxy)ethoxy)- (8CI,9CI)

7206-76-0
$C_{11}H_{14}N_2O_2$
206.27
CCC(C(N)=O)(C(N)=O)c1ccccc1
Propanediamide, 2-ethyl-2-phenyl
2-Ethyl-2-phenylpropane-diamide
Malonamide, 2-ethyl-2-phenyl-(8CI)
PEMA
PEMA (Amide)
Phenylethylmalonamide
Phenylethylmalondiamide

7208-47-1
$C_{18}H_{26}O_{12}$
434.40
O=C(OCC(OC(=O)C)C(OC(=O) C)C(OC(=O)C)C(OC(=O)C) COC(=O)C)C
D-Glucitol, hexaacetate (9CI)
AI3-19577
Glucitol, hexaacetate, D- (8CI)

7212-44-4
$C_{15}H_{26}O$
222.37
OC(C=C)(CCC=C(CCC=C(C)C) C)C
Nerolidol
AI3-10519
1,6,10-Dodecatrien-3-ol, 3,7,11-trimethyl-
3,7,11-Trimethyldodeca-1,6,10-trien-3-ol
3,7,11-Trimethyl-1,6,10-do-decatrien-3-ol

7214-61-1
$C_8H_9NO_2$
151.17
Benzene, (1-nitroethyl)-(6CI,7CI,8CI,9CI)
1-Nitro-1-phenylethane
1-Phenyl-1-nitroethane

7220-81-7
$C_{17}H_{14}O_6$
314.31
COc4cc2OC1OCCC1c2c5oc(=O) c3C(=O)CCc3c45
Cyclopenta(c)furo(3',2':4,5)-furo(2,3-h)(1)benzopyran-1,11-dione, 2,3,6a-α,8,9,9a-α-hexahydro-4-methoxy
Aflatoxin B2
Dihydroaflatoxin B1

7225-67-4
$C_{14}H_{30}$
198.39
Heptane, 2,2,3,3,5,6,6-hepta-methyl- (8CI,9CI)
NSC-109494

7227-91-0
$C_8H_{11}N_3$
149.22
CN(C)N=Nc1ccccc1
Triazene, 3,3-dimethyl-1-phenyl
3,3-Dimethyl-1-phenyltriazene
DMPT
1-Fenyl-3,3-dimethyltriazin
NSC-3094
PDMT
PDT
Phenyl-dimethyl-triazine
1-Phenyl-3,3-dimethyltriazene
1-Phenyl-3,3-dimethyl-triazine
1-Triazene, 3,3-dimethyl-1-phenyl- (9CI)
X 119

7235-40-7
$C_{40}H_{56}$
536.96
C(=C(CCC1)C)(C=CC(=CC=CC (=CC=CC=C(C=CC=C(C=CC (=C(CCC2)C)C2(C)C)C)C) C)C)C1(C)C
β-Carotene, all-trans (8CI)
Betacarotene
Betacaroteno (Spanish)
Betacarotenum (INN-Latin)
C.I. 75130
C-Orange 11 (Germany)
all-trans-β-Carotene
β-Carotene
β,β-Carotene (9CI)
Cyclohexene, 1,1'-(3,7,12,16-te-tramethyl-1,3,5,7,9,11,13,15, 17-octadecanonaene-1,18-

diyl)bis(2,6,6-trimethyl-, (all-E)-
Germany: C-Orange 11
Karotin (Czech,Sweden)
KPMK
Natural Yellow 26
NSC-62794
Serlabo
Solatene
Sweden: Karotin
Zlut Prirodni 26 (Czech)

7241-98-7
$C_{17}H_{14}O_7$
330.31
COc4cc2OC1OCCC1c2c5oc(=O)c3C(=O)OCCc3c45
1H,12H-Furo(3',2':4,5)furo-(2,3-h)pyrano(3,4-c)(1)benzo-pyran-1,12-dione, 3,4,7a-α, 9, 10,10a-α-hexahydro-5-methoxy
Aflatoxin G2

7244-67-9
$C_{11}H_{13}NO_3$
207.22
L-Alanine, N-benzoyl-, methyl ester (9CI)
Alanine, N-benzoyl-, methyl ester
Alanine, N-benzoyl-, methyl ester, L- (8CI)

7248-45-5
$C_{27}H_{22}N_6O_8S.2Na$
636.52
Benzoic acid, 2-hydroxy-5-((4-((((2-methoxy-4-((3-sulfo-phenyl)azo)phenyl)amino)-carbonyl)amino)phenyl) azo)-, disodium salt (9CI)

7250-85-3
$C_8H_{16}O_2$
144.22
2-Butene, 1,4-diethoxy- (7CI,8CI,9CI)
1,4-Diethoxy-2-butene

7251-61-8
$C_9H_8N_2$
144.19
Cc2cnc1ccccc1n2
Quinoxaline, 2-methyl
2-Methylquinoxaline

7254-11-7
$C_4H_5ClN_4O$
160.54
1,3,5-Triazin-2-amine, 4-chloro-6-methoxy- (9CI)
NSC-72983
s-Triazine, 2-amino-4-chloro-6-methoxy- (8CI)

7261-97-4
$C_{14}H_{10}N_4O_5$
314.28
O=C3CN(N=Cc1ccc(o1)c2ccc(cc2)N(=O)=O)C(=O)N3
Hydantoin, 1-((5-(p-nitro-phenyl)furfurylidene)amino)
Dantrolene

7267-90-5
$C_{23}H_{12}O$
304.35
11H-Indeno(2,1-a)pyren-11-one (8CI,9CI)

7274-88-6
$C_6H_{14}N_2O_2.ClH$
182.68
Lysine, hydrochloride, D
D-Lysine hydrochloride

7277-87-4
$C_{14}H_{12}N_2O_7S_2$
384.38
O=C(N(N=C1C)c(ccc(c2cc(S(=O)(=O)O)c3)c3S(=O)(=O)O)c2)C1
1,3-Naphthalenedisulfonic acid, 6-(4,5-dihydro-3-methyl-5-oxo-1H-pyrazol-1-yl)- (9CI)

7278-65-1
$C_{15}H_{32}O$
228.42
OC(CCCC(CCCC(C)C)C)(CC)C
3-Dodecanol, 3,7,11-trimethyl- (VAN)(9CI)
3,7,11-Trimethyldodecan-3-ol
3,7,11-Trimethyl-3-dodecanol

7280-37-7
$C_{18}H_{22}O_5S.C_4H_{10}N_2$
436.56
Estropipate
Conjugated estrogen
Estra-1,3,5(10)-trien-17-one, 3-(sulfooxy)-, Compd. with piperazine (1:1)
Estrone hydrogen sulfate compound with piperazine (1:1)
Ogen
Piperazine estrone sulfate
Sulestrex

7281-04-1
$C_{21}H_{38}N.Br$
384.51
[Br-].CCCCCCCCCCCC[N+](C)(C)Cc1ccccc1
Ammonium, benzyldodecyl-dimethyl-, bromide
Ammonyl BR 1244
Bacfor BL
Benzalkonium bromide
Benzenemethanaminium, N-do-decyl-N,N-dimethyl-, bromide (9CI)
Benzododecinium bromide
Benzyldimethyldodecyl-ammonium bromide
Benzyldodecyldimethyl-ammonium bromide
Bromek dwumetylolaurylo-benzyloamoniowy (Polish)
Dimethyl laurylbenzene ammonium bromide
Sinnoquat BL 80
Sinnoquat BL 95
Sterinol
Sterinolu (Polish)

$C_{24}H_{52}O_4P_2S_4Zn$
660.28
Zinc O,O-dihexyl dithiophos-phate
Di-n-hexyl phosphorodithioate, zinc salt
Phosphorodithioic acid, O,O-di-hexyl ester, zinc salt
Zinc, bis(O,O-dihexyl phos-phorodithioato-S,S')-, (T-4)-(9CI)
Zinc dihexyl dithiophosphate

7286-69-3
$C_9H_{16}ClN_5$
229.75
n(c(nc(n1)NC(CC)C)NCC)c1Cl
s-Triazine, 2-(sec-butylamino)-4-chloro-6-(ethylamino)
2-Aethylamino-4-sek.butyl-amino-6-chlor-1,3,5-triazin (German)
GS 13528
Sebuthylazine
1,3,5-Triazine-2,4-diamine, 6-chloro-N-ethyl-N'-(1-methylpropyl)- (9CI)

7286-76-2
$C_9H_{13}BrN_2O_2$
261.15
Uracil, 5-bromo-3-tert-butyl-6-methyl
5-Bromo-3-tert-butyl-6-methyl-uracil
Compound 733

7286-84-2
$C_8H_7Cl_2NO_2$
220.05
O=C(OC)c(c(c(N)cc1Cl)Cl)c1
Chloramben, methyl ester
3-Amino-2,5-dichlorobenzoic acid methyl ester
Benzoic acid, 3-amino-2,5-di-chloro-, methyl ester (9CI)
Caswell No. 168E
EPA Pesticide Chemical Code 029905

7287-19-6
$C_{10}H_{19}N_5S$
241.40
CSc1nc(NC(C)C)nc(NC(C)C)n1
s-Triazine, 2,4-bis(isopropylamino)-6-(methylthio)
2,4-Bis(isopropylamino)-6-methyl mercapto-s-triazine
2,4-Bis(isopropylamino)-6-methylthio-s-triazine
2,4-Bis(isopropylamino)-6-methylthio-1,3,5-triazine
N,N'-Bis(1-methylethyl)-6-methyl-thio-1,3,5-triazine-2,4-diamine
2,4-Bis(propylamino)-6-methylthio-1,3,5-triazin (German)
Caparol
N,N'-Diisopropyl-6-methylthio-1,3,5-triazine-2,4-diyldiamine
G 34161
Gesagard
Merkazin
2-Methylmercapto-4,6-bis(isopropylamino)-s-triazine
2-Methylthio-4,6-bis(isopropylamino)-s-triazine
Polisin
Primatol Q
Promethryn
Prometrex
Prometrin
Prometryn
Prometryne
Selektin
Sesagard
s-Triazine, 4,6-bis(isopropylamino)-2-methylmercapto-

7287-36-7
$C_{13}H_{18}ClNO$
239.77
CCCC(C)(C)C(=O)Nc1ccc(Cl)cc1
Valeranilide, 4'-chloro-2,2-dimethyl
4-Chloranilid kyseliny 2,2-dimethylvalerove (Czech)
N-(4-Chlorophenyl)-2,2-dimethylpentanamide
N-(4-Chlorophenyl)-2,2-dimethylvaleroamide
N-(4-Chlorphenyl)-2,2-dimethylpentamid (German)
N-(4-Chlor-phenyl)-2,2-dimethyl-valeriansaeureamid (German)
D-90-A
Monalide
Potablan
Schering-35830
SN 35830

7292-16-2
$C_{13}H_{21}O_4PS$
304.37
Phosphoric acid, p-(methylthio)phenyl dipropyl ester
O,O-Di-n-propyl-O-(4-methylthiophenyl)phosphate
Kayaphos
Kayphosnac
4-Methylthiophenyl dipropyl phosphate
NK-1158
Phosphoric acid, dipropyl 4-methylthiophenyl ester
Phosphoric acid, 4-(methylthio)-phenyl dipropyl ester (9CI)
Propaphos

7294-05-5
$C_8H_{18}O$
130.23
3-Pentanol, 2,2,3-trimethyl- (8CI,9CI)

7295-76-3
C_6H_7NO
109.12
COc1cccnc1
Pyridine, 3-methoxy- (8CI,9CI)

7296-12-0
$C_{15}H_{10}Cl_2N_4$
317.18
s-Triazine, 2-anilino-4-chloro-6-(p-chlorophenyl)- (7CI,8CI)

7299-91-4
$C_8H_{14}O_2$
142.22
Crotonic acid, butyl ester
2-Butenoic acid, butyl ester (9CI)
Butyl 2-butenoate
Butyl crotonate

7304-99-6
$C_9H_{10}N_2O_4$
210.18
Carbamic acid, dimethyl-, 3-nitrophenyl ester
AI3-33125

7305-61-5
$C_4H_{11}O_3P$
138.12
Phosphonic acid, ethyl-, monoethyl ester
O-Ethyl ethanephosphonic acid
O-Ethyl ethylphosphonate
O-Ethyl hydrogen ethylphosphonate
Ethylphosphonic acid ethyl ester
Phosphonic acid, ethyl-, ethyl ester

7307-02-0
$C_7H_{12}O_2$
128.17
CCCC(=O)CC(C)=O
2,4-Heptanedione (8CI,9CI)

7311-27-5
$C_{23}H_{40}O_5$
396.57
O(CCOc(ccc(c1)CCCCCCCCC)c1)CCOCCOCCO
Nonoxynol-4
Chemax NP-4
Ethanol, 2-(2-(2-(2-(4-nonylphenoxy)ethoxy)ethoxy)-ethoxy)- (9CI)
Hefti NP-55-40
PEG-4 Nonyl Phenyl Ether
Polyethylene Glycol 200 Nonyl Phenyl Ether
Polyoxyethylene (4) nonyl phenyl ether
Unicol NP-4
Uniterge NP-4

7313-54-4
$C_5H_9N_5O$
155.13
s-Triazin-2-ol, 4-amino-6-(ethylamino)- (8CI)
1,3,5-Triazin-2(1H)-one, 4-amino-6-(ethylamino)- (9CI)

7314-30-9
$C_5H_{10}O_2S$
134.20
Sulfonium, (2-carboxyethyl)-dimethyl-, hydroxide, inner salt (8CI,9CI)
Dimethylpropiothetin
Propiothetin, dimethyl-

7315-68-6
$C_{11}H_{14}O_2$
178.25
Butyric acid, 3-methyl-4-phenyl
3-Methyl-4-phenyl-butyric acid

7319-23-5
$C_6H_{12}O_2$
116.16
3-Hexene-2,5-diol (8CI,9CI)
NSC-60702

7320-34-5
$H_4O_7P_2.4K$
334.37
Tetrapotassium pyrophosphate
Diphosphoric acid, tetrapotassium salt (9CI)
Potassium pyrophosphate
Tetrapotassium diphosphorate
TKPP

7320-37-8
$C_{16}H_{32}O$
240.48
O(C1CCCCCCCCCCCCCC)C1
Hexadecane, 1,2-epoxy

1,2-Epoxyhexadecane
Hexadecene epoxide
NCI-C55538

7321-55-3
$C_5H_6N_2$
94.12
Propanedinitrile, dimethyl- (9CI)
2,2-Dicyanopropane
Dimethylmalononitrile
Malononitrile, dimethyl- (6CI,7CI,8CI)

7326-46-7
$C_5H_{10}O_2$
102.13
2-Furanol, tetrahydro-2-methyl- (8CI,9CI)

7327-60-8
$C_6H_6N_4$
134.12
N#CCN(CC#N)CC#N
Acetonitrile, nitrilotri- (8CI)
Acetonitrile, 2,2',2''-nitrilotris- (9CI)
2,2',2''-Nitrilotrisacetonitrile

7327-69-7
$C_5H_{10}N_2O_4$
162.13
O=C(N(CO)COC1)N1CO
4H-1,3,5-Oxadiazin-4-one, tetrahydro-3,5-bis(hydroxymethyl)- (9CI)
3,5-Dimethyloluron

7328-17-8
$C_9H_{16}O_4$
188.22
O=C(OCCOCCOCC)C=C
2-Propenoic acid, 2-(2-ethoxyethoxy)ethyl ester (9CI)
Carbitol acrylate
2-(2-Ethoxyethoxy)ethyl 2-propenoate

7328-91-8
$C_5H_{14}N_2$
102.17
NCC(CN)(C)C
1,3-Propanediamine, 2,2-dimethyl- (9CI)
AI3-28052
2,2-Dimethyl-1,3-propanediamine
Neopentyldiamine

7328-97-4
$C_{38}H_{38}O_8$
622.71
O(C1COc(ccc(c2)C(c(ccc(OCC(O3)C3)c4)c4)C(c(ccc(OCC(O5)C5)c6)c6)c(ccc(OCC(O7)C7)c8)c8)c2)C1
Oxirane, 2,2',2'',2'''-(1,2-ethanediylidenetetrakis(4,1-phenyleneoxymethylene))tetrakis- (9CI)
Tetraphenylolethane, epichlorohydrin epoxy resin

7332-46-9
$C_{24}H_{51}O_{10}P$
530.64
O=P(OCCOCCOCCCC)(OCCOCCOCCCC)OCCOCCOCCCC
Ethanol, 2-(2-butoxyethoxy)-, phosphate (3:1) (9CI)
2-(2-Butoxyethoxy)ethanol phosphate (3:1)

7333-86-0
$C_{27}H_{45}NO_3$
431.65
O=C(NCCCCCCCCCCCCCCCCCC)c(ccc(c1)C(=O)OC)c1
Benzoic acid, 4-((octadecylamino)carbonyl)-, methyl ester (9CI)

7334-33-0
$C_{12}H_8Cl_2N_2$
251.10
N(=Nc(c(ccc1)Cl)c1)c(c(ccc2)Cl)c2
Diazene, bis(2-chlorophenyl)- (9CI)
Bis(2-chlorophenyl)diazene
2,2'-Dichloroazobenzene

7335-65-1
$C_2H_4O_2.H_4N_2$
92.09
Hydrazine, monoacetate (9CI)

7336-20-1
$C_{14}H_{14}N_2O_6S_2.2Na$
416.38
Benzenesulfonic acid, 2,2'-(1,2-ethenediyl)bis(5-amino-, disodium salt (9CI)
2,2'-(1,2-Ethenediyl)bis-(5-aminobenzenesulfonic acid) disodium salt
2,2'-Stilbenedisulfonic acid, 4,4'-diamino-, disodium salt

7337-45-3
$C_4H_{11}N.BH_3$
87.00
tert-Butylamine, Compd. with borane (1:1)
tert-Butylamine borane
2-Propanamine, 2-methyl-, Compd. with borane (1:1) (9CI)

7339-53-9
$CH_6N_2.ClH$
82.55
[Cl-].CNN
Hydrazine, methyl-, hydrochloride
Methylhydrazine hydrochloride

7339-87-9
$C_8H_8O_2$
136.15
Benzeneacetaldehyde, 4-hydroxy- (9CI)
Acetaldehyde, (p-hydroxyphenyl)- (8CI)

7342-47-4
$C_{12}H_{27}ISn$
416.98
Stannane, iodotributyl
Iodotributylstannane
Stannane, tributyliodo-
Tin, tri-n-butyl-, iodide
Tri-n-butyl tin iodide

7343-06-8
$C_{18}H_{18}$
234.34
Phenanthrene, 3,4,5,6-tetramethyl- (8CI,9CI)

7346-80-7
$C_{21}H_{43}NO_3S.Na$
412.63
Ethanesulfonic acid, 2-(methyl-9-octadecenylamino)-, sodium salt, (Z)- (9CI)
Caswell No. 780AA
EPA Pesticide Chemical Code 079068
Sodium N-methyl-N-oleoyltaurate

7351-61-3
$C_7H_{11}Cl_3O_2Si$
261.61
2-Propenoic acid, 2-methyl-, 3-(trichlorosilyl)propyl ester (9CI)

7359-72-0
$C_7H_5Cl_3$
195.48
Benzene, 1,2,3-trichloro-4-methyl- (9CI)
Toluene, 2,3,4-trichloro- (7CI,8CI)
2,3,4-Trichlorotoluene

7360-53-4
$CH_2O_2.1/3Al$
55.02
Formic acid, aluminum salt (9CI)
Aluminum formate

7364-20-7
$C_{10}H_{12}O_2$
164.20
Benzoic acid, 4-ethyl-, methyl ester (9CI)
Benzoic acid, p-ethyl-, methyl ester (8CI)

7367-82-0
$C_{10}H_{18}O_2$
170.25
2-Octenoic acid, ethyl ester, (E)- (9CI)

7367-83-1
$C_{11}H_{20}O_2$
184.28
4-Decenoic acid, methyl ester, (Z)- (8CI,9CI)

7367-84-2
$C_{12}H_{22}O_2$
198.31
4-Decenoic acid, ethyl ester, (Z)- (9CI)
4-Decenoic acid, ethyl ester, cis- (8CI)

7372-87-4
$C_{16}H_{14}$
206.29
Phenanthrene, 1,8-dimethyl- (8CI,9CI)

7372-88-5
$C_{13}H_{10}S$
198.29
Dibenzothiophene, 4-methyl- (8CI,9CI)

7374-47-2
$C_9H_7Cl_3O_3.C_2H_7NO$
330.60
Propionic acid, 2-(2,4,5-trichlorophenoxy)-, Compd. with 2-aminoethanol (1:1) (8CI)
Ethanol, 2-amino-, 2-(2,4,5-trichlorophenoxy)-propionate (Salt) (8CI)

7374-53-0
$C_9H_{17}N_5O$
211.24
CC(C)Nc1nc(O)nc(NC(C)C)n1
1,3,5-Triazin-2(1H)-one, 4,6-bis-((1-methylethyl)amino)- (9CI)
GS 11526
Hydroxypropazine
s-Triazin-2-ol, 4,6-bis-(isopropylamino)- (8CI)

7376-31-0
$C_6H_{15}NO_3.xH_2O_4S$
Unknown
TEA-sulfate
Ethanol, 2,2',2''-nitrilotris-, sulfate (Salt) (9CI)
2,2',2''-Nitrilotrisethanol sulfate (Salt)
2,2',2''-Nitrilotris(ethanol) sulfate (Salt)
Triethanolamine sulfate

7378-23-6
$C_6H_9NaO_6$
200.12
D-Erythro-hex-2-enonic acid, γ-lactone, sodium salt (9CI)

7378-99-6
$C_{10}H_{23}N$
157.29
N(CCCCCCCC)(C)C
Octylamine, N,N-dimethyl- (8CI)
N,N-Dimethyl-1-octanamine
1-Octanamine, N,N-dimethyl- (9CI)

7379-12-6
$C_7H_{11}O$
111.18
CCCC(=O)C(C)C
3-Hexanone, 2-methyl
2-Methyl-3-hexanone

7379-26-2
$C_{10}H_{16}N_2O_8.xH_3N$
Unknown
Ammonium ethylenediamine-tetraacetate
Caswell No. 438A
EPA Pesticide Chemical Code 039117
Glycine, N,N'-1,2-ethanediyl-bis(N-(carboxymethyl)-, ammonium salt (9CI)

7379-27-3
$C_{10}H_{16}N_2O_8.xK$
Unknown
Glycine, N,N'-1,2-ethanediyl-bis(N-(carboxymethyl)-, potassium salt (9CI)

7379-28-4
$C_{10}H_{16}N_2NaO_8$
315.24
Glycine, N,N'-1,2-ethanediyl-bis(N-(carboxymethyl)-, sodium salt (9CI)

7379-35-3
$C_5H_4ClN.ClH$
150.00
4-Chloropyridine
4-Chloropyridine hydrochloride
Pyridine, 4-chloro-, hydrochloride (9CI)

7379-51-3
$C_9H_{12}OS$
168.26
Oc(cc(c(SC)c1C)C)c1
Phenol, 3,5-dimethyl-4-(methylthio)- (9CI)
3,5-Dimethyl-4-(methylthio)-phenol
4-Methylthio-3,5-xylenol

7382-59-4
$C_{17}H_{20}O_6$
320.35
Guaiacylglycerol-β-guaiacyl ether
GGBGE
Guaiacylglycerol-β-O-4-guaiacyl ether
1,3-Propanediol, 1-(4-hydroxy-3-methoxyphenyl)-2-(2-methoxyphenoxy)-

7383-90-6
$C_{14}H_{14}$
182.27
1,1'-Biphenyl, 3,4'-dimethyl- (9CI)
m,p'-Bitolyl (8CI)
3,4'-Dimethylbiphenyl

7385-78-6
C_7H_{14}
98.19
1-Pentene, 3,4-dimethyl- (9CI)

7388-22-9
$C_{14}H_{22}O$
206.33
O=C(C(=CC(C(CCC1)=C)C1(C)C)C)C
3-Buten-2-one, 4-(2,2-dimethyl-6-methylenecyclohexyl)-3-methyl- (9CI)

7388-32-1
$C_{18}H_{20}Cl_2O_2$
339.28
Benzene, 1,1'-(2,2-dichloroethylidene)bis(4-ethoxy
1,1'-(2,2-Dichloroethylidene)bis(4-ethoxybenzene)

7388-44-5
$C_7H_{14}N_2O_4$
190.19
O=C(N(COC)COC1)N1COC
4H-1,3,5-Oxadiazin-4-one, tetrahydro-3,5-bis-(methoxymethyl)- (9CI)
AI3-09491
Dimethoxymethyluron

7390-81-0
$C_{18}H_{36}O$
268.48
O(C1CCCCCCCCCCCCCC
CC)C1
Oxirane, hexadecyl- (9CI)
AI3-26311
1,2-Epoxyoctadecane
Hexadecyloxirane

7391-61-9
$C_8H_{12}N_2O_3$
184.22
Barbituric acid, 1,3-dimethyl-5-ethyl
1,3-Dimethyl-5-ethylbarbituric acid

7396-28-3
$C_9H_5ClO_2$
180.59
3-(3-Chlorophenyl)-2-propynoic acid
Propiolic acid, (m-chlorophenyl)- (8CI)
Propynoic acid, (3-chlorophenyl)-
2-Propynoic acid, 3-(3-chlorophenyl)- (9CI)

7396-38-5
$C_{18}H_{18}$
234.34
Phenanthrene, 2,4,5,7-tetramethyl- (9CI)

7396-58-9
$C_{21}H_{45}N$
311.59
N(CCCCCCCCCC)(CCCCCCCC
CC)C
1-Decanamine, N-decyl-N-methyl- (9CI)
N-Decyl-N-methyl-1-decanamine
Di(decyl)methylamine

7397-62-8
$C_6H_{12}O_3$
132.16
O=C(OCCCC)CO
Acetic acid, hydroxy-, butyl ester
AI3-07975
Butyl hydroxyacetate

7398-69-8
$C_8H_{16}N.Cl$
161.67
2-Propen-1-aminium, N,N-dimethyl-N-2-propenyl-, chloride (9CI)
Diallyldimethylammonium chloride
Dimethyldiallylammonium chloride

7400-08-0
$C_9H_8O_3$
164.17
O=C(O)C=Cc(ccc(O)c1)c1
Cinnamic acid, p-hydroxy
p-Coumaric acid
4-Coumaric acid
p-Cumaric acid
p-Hydroxycinnamic acid
4-Hydroxycinnamic acid
4'-Hydroxycinnamic acid
p-Hydroxyphenylacrylic acid
β-(4-Hydroxyphenyl)acrylic acid
3-(4-Hydroxyphenyl)-2-propenoic acid
2-Propenoic acid, 3-(4-hydroxyphenyl)- (9CI)

7400-27-3
$C_4H_{12}N_2.ClH$
124.60
Hydrazine, (1,1-dimethylethyl)-, monohydrochloride (9CI)
Hydrazine, tert-butyl-, monohydrochloride (8CI)

7408-18-6
$C_8H_{10}O_9$
250.16
O=C(O)CC(OC(C(=O)O)CC(=O)
O)C(=O)O
Butanedioic acid, 2,2'-oxybis- (9CI)
Malomalic ether
2,2'-Oxybisbutanedioic acid

7411-16-7
$C_4H_7N_3$
97.10
1H-1,2,4-Triazole, 3-ethyl- (9CI)
s-Triazole, 3-ethyl- (8CI)

7411-49-6
$C_{12}H_{14}N_4.4ClH$
360.14
Cl.Cl.Cl.Cl.Nc1ccc(cc1N)c2ccc
(N)c(N)c2
3,3',4,4'-Biphenyltetramine, tetrahydrochloride
3,3'-Diaminobenzidine tetrahydrochloride
3,3',4,4'-Tetraaminobiphenyl tetrahydrochloride

7414-83-7
$C_2H_6O_7P_2.2Na$
250.00
[Na+].[Na+].CC(O)(P(O)
([O-])=O)P(O)([O-])=O
Phosphonic acid, (1-hydroxyethylidene)di-, disodium salt
Didronel R
Disodium dihydrogen (1-hydroxyethylidene)diphosphonate
Disodium ethanol-1,1-diphosphonate
Disodium ethydronate
Disodium etidronate
Disodium 1-hydroxyethylidene phosphonate
Ethane-1-hydroxy-1,1-diphosphonic acid, disodium salt
Etidronate disodium
(1-Hydroxyethane-1,1-diyl)-diphosphonic acid disodium salt
(1-Hydroxyethylidene)diphosphonic acid, disodium salt
1-Hydroxyethylidene-1,1-diphosphonic acid disodium salt
Phosphonic acid, (1-hydroxyethane-1,1-diyl)di-, disodium salt
Sodium ethidronate
Sodium ethydronate
Sodium etidronate

7415-31-8
$C_4H_6Cl_2$
125.00
1,3-Dichloro-2-butene
2-Butene, 1,3-dichloro-, (E)-
trans-1,3-Dichlorobutene-2

7416-34-4
$C_{16}H_{24}N_2O_2$
276.42
CCc2c(C)[nH]c3CCC(CN1CCO
CC1)C(=O)c23
Indol-4(5H)-one, 3-ethyl-6,7-dihydro-2-methyl-5-(morpholinomethyl)
EN-1733A
Molindone

7417-67-6
$C_3H_6N_2O_2$
102.11
CN(N=O)C(C)=O
Acetamide, N-methyl-N-nitroso
Methylnitrosoacetamid (German)
Methylnitrosoacetamide
N-Methyl-N-nitrosoacetamide
N-Nitroso-N-methylacetamide

7421-93-4
$C_{12}H_8Cl_6O$
380.91
Endrin aldehyde
1,2,4-Methenecyclopenta(c,d)-pentalene-r-carboxaldehyde, 2,2a,3,3,4,7-hexachlorodecahydro
1,2,4-Methenocyclopenta(cd)-

pentalene-5-carboxaldehyde, 2,2a,3,3,4,7-hexachloro-decahydro-, (1α,2β,2aβ,4β, 4aβ,5β,6aβ,6bβ,7R*)-
SD 7442

7422-52-8
$C_{13}H_{32}O_4Si_3$
336.65
Trisiloxane, 1,1,1,3,5,5,5-heptamethyl-3-(3-(oxiranyl-methoxy)propyl)- (9CI)
(3-Glycidoxypropyl)bis(tri-methylsiloxy)methylsilane

7422-92-6
$C_5H_{12}N_2$
100.16
Propionaldehyde, ethyl-hydrazone (7CI,8CI)

7423-42-9
$C_{12}H_{20}O_4$
228.29
O=C(OCC(CCCC)CC)C=CC(=O)O
2-Butenedioic acid (Z)-, mono-(2-ethylhexyl) ester (9CI)
Mono(2-ethylhexyl)maleate

7423-53-2
$C_{12}H_{18}Cl_3N_3O_6$
406.65
1,3,5-Triazine-2,4,6-(1H,3H,5H)-trione, 1,3,5-tris(3-chloro-2-hydroxypropyl)- (9CI)
s-Triazine-2,4,6(1H,3H,5H)-trione, tris(3-chloro-2-hydroxypropyl)- (7CI,8CI)
Tris(3-chloro-2-hydroxy-propyl) isocyanurate

7425-14-1
$C_{16}H_{32}O_2$
256.48
O=C(OCC(CCCC)CC)C(CCCC)CC
Hexanoic acid, 2-ethyl-, 2-ethylhexyl ester
2-Ethylhexanoic acid, 2-ethyl-hexyl ester
2-Ethylhexylester kyseliny 2-ethylkapronove (Czech)
2-Ethylhexyl-2-ethylhexanoate

7426-35-9
$C_{10}H_{11}N_5$
201.26
s-Triazine, 2-amino-4-anilino-6-methyl
2-Amino-4-anilino-6-methyl-s-triazine
2-Amino-4-methyl-6-phenyl-amino-1,3,5-triazine
CB 2487

7428-48-0
$C_{36}H_{70}O_4.Pb$
774.25
Stearic acid, lead salt
Bleistearat (German)
Hal-Lub-N
Lead stearate

7429-90-5
Al
26.98
Aluminum (ACGIH)
A 00
A 95
A 99
A 995
A 999
AA 1099
AA1199
AD 1
AD1M
ADO
AE
Alaun (German)
Allbri Aluminum Paste and Powder
Alumina Fibre
Aluminium
Aluminium Bronze
Aluminium Flake
Aluminum 27
Aluminum A00
Aluminum dehydrated
Aluminum, Metallic, Powder (DOT)
Aluminum metal (OSHA)
Aluminum Powder
Aluminum Powder, Coated [UN 1309]
Aluminum, Powder, Pyrophoric [UN 1383]
Aluminum, Powder, Uncoated, Non-pyrophoric [UN 1396]
Aluminum Pyro Powders (OSHA)
Aluminum Welding Fumes (OSHA)
AO A1
AR2
AV00
AV000
C.I. 77000
Emanay Atomized Aluminum Powder
JISC 3108
JISC 3110
L16
Metana
Metana Aluminum Paste
Noral Aluminum
Noral Extra Fine Lining Grade
Noral Ink Grade Aluminum
Noral Non-Leafing Grade
PAP-1
UN 1309 [Aluminum powder, coated]
UN 1383 [Pyrophoric metals, N.O.S., or Pyrophoric alloys, N.O.S.]
UN 1396 [Aluminum powder, uncoated]

7429-91-6
Dy
162.5
[Dy]
Dysprosium (8CI,9CI)

7437-35-6
Unknown
Unknown
Bis(tributyltin) adipate

7439-89-6
Fe
55.85
Iron
Ancor EN 80/150
Armco iron
Carbonyl iron
EFV 250/400
EO 5A
Ferrovac E
GS 6
Loha
NC 100
PZh2M
PZhO
Remko
Suy-B 2
3ZhP

7439-90-9
Kr
83.80
[Kr]
Krypton
Krypton, Compressed [UN 1056]
Krypton, Refrigerated liquid [UN 1970]
UN 1056 [Krypton, compressed]
UN 1970 [Krypton, refrigerated liquid (cryogenic liquid)]

7439-91-0
La
138.91
[La]
Lanthanum (9CI)

7439-92-1
Pb
207.19
Lead (ACGIH)
C.I. Pigment Metal 4
C.I. 77575
Glover
KS-4
Lead Flake
Lead inorganic (OSHA)
Lead S2
Olow (Polish)
Omaha

Omaha & Grant
SI
SO

7439-93-2
Li
6.94
Lithium [UN 1415]
Lithium metal (DOT)
Lithium metal, In cartridges (DOT)
UN 1415 [Lithium]

7439-94-3
Lu
174.97
[Lu]
Lutetium (9CI)

7439-95-4
Mg
24.31
Magnesium
Magnesio (Italian)
Magnesium Borings [UN 1869]
Magnesium Clippings [UN 1869]
Magnesium Metal [UN 1869]
Magnesium Pellets [UN 1869]
Magnesium Powdered [UN 1418]
Magnesium Ribbons [UN 1869]
Magnesium Turnings [UN 1869]
NA 1869 (DOT)
RMC
UN 1418 [Magnesium, powder or Magnesium alloys, powder]
UN 1869 [Magnesium or Magnesium alloys with more than 50 per cent magnesium in pellets, turnings or ribbons]
UN 2950 [Magnesium granules, coated particle size not less than 149 microns]

7439-96-5
Mn
54.94
Manganese
Colloidal Manganese
Magnacat
Mangan (Polish)
Manganese
Mangan nitridovany (Czech)
Tronamang

7439-97-6
Hg
200.59
Mercury (ACGIH,OSHA)
Colloidal Mercury
Kwik (Dutch)
Mercure (French)
Mercurio (Italian)
Mercury, Metallic (DOT)
Metallic mercury
NA 2809 (DOT)
NCI-C60399
Quecksilber (German)
Quick Silver
RCRA waste number U151
RTEC (Polish)
UN 2809 (DOT)

7439-98-7
Mo
95.94
Molybdenum (ACGIH)
Molybdate

7440-01-9
Ne
20.18
[Ne]
Neon (DOT)
Neon, Compressed [UN 1065]
Neon, Refrigerated liquid [UN 1913]
UN 1065 [Neon, compressed]
UN 1913 [Neon, refrigerated liquid (cryogenic liquid)]

7440-02-0
Ni
58.71
[Ni]
Nickel (ACGIH)
Carbonyl nickel Powder
C.I. 77775
Ni 270
Nickel 270
Nickel (Dust)
Nichel (Italian)
Nickel, Metal (OSHA)
Nickel Particles
Nickel Sponge
Ni 0901-S
Ni 4303T
NP 2
Raney Alloy
Raney Nickel

7440-03-1
Nb
92.91
[Nb]
Niobium (9CI)

7440-04-2
Os
190.20
Osmium
Metallic osmium

7440-05-3
Pd
106.42
[Pd]
Palladium (9CI)

7440-06-4
Pt
195.09
Platinum (ACGIH)
C.I. 77795
Liquid Bright Platinum
Platin (German)
Platinum Black
Platinum metal (OSHA)
Platinum Sponge

7440-07-5
Pu
244
Plutonium

7440-08-6
Po
209
Polonium

7440-09-7
K
39.10
Potassium [UN 2257]
Potassium, (Liquid alloy)
Potassium, Metal (DOT)
Potassium, Metal alloys [UN 1420]
Potassium, Metal liquid alloy (DOT)
Potassium, Metallic (DOT)
UN 1420 [Potassium, metal alloys]
UN 2257 [Potassium]

7440-10-0
Pr
140.91
[Pr]
Praseodymium (9CI)

7440-14-4
Ra
226.03
Radium (9CI)

7440-15-5
Re
186.21
[Re]
Rhenium (9CI)

7440-16-6
Rh
102.91
Rhodium (ACGIH)
Rh
Rhodium metal (OSHA)

7440-17-7
Rb
85.47
Rubidium [UN 1423]

Rubidium metal (DOT)
Rubidium metal, In cartridges (DOT)
UN 1423 [Rubidium]

7440-18-8
Ru
101.07
[Ru]
Ruthenium (9CI)

7440-19-9
Sm
150.36
[Sm]
Samarium (9CI)

7440-20-2
Sc
44.96
[Sc]
Scandium (9CI)

7440-21-3
Si
28.09
Silicon (ACGIH,DOT,OSHA)
Defoamer S-10
Silicon powder, Amorphous [UN 1346]
UN 1346 [Silicon powder, amorphous]

7440-22-4
Ag
107.87
Silver (ACGIH,OSHA)
Argentum
C.I. 77820
L-3
Shell Silver
Silber (German)
Silver Atom

7440-23-5
Na
22.99
Sodium [UN 1428]
Natrium
Sodium metal (DOT)
UN 1428 [Sodium]
UN 1429 (DOT)

7440-24-6
Sr
87.62
Strontium (9CI)

7440-25-7
Ta
180.95
Tantalum (ACGIH)
Tantalum-181
Tantalum, Metal (OSHA)

7440-26-8
Tc
98
Technetium

7440-27-9
Tb
158.93
[Tb]
Terbium (9CI)

7440-28-0
Tl
204.37
Thallium (ACGIH,OSHA)
Ramor

7440-29-1
Th
232.00
Thorium
Thorium-232
Thorium metal, Pyrophoric [UN 2975]
UN 2975 [Thorium metal, pyrophoric]

7440-30-4
Tm
168.93
[Tm]
Thulium (9CI)

7440-31-5
Sn
118.69
Tin (ACGIH,OSHA)
Silver Matt Powder
Tin (α)
Tin Flake
Tin Powder
Zinn (German)

7440-32-6
Ti
47.90
Titanium
Contimet 30
C.P. Titanium
IMI 115
NCI-C04251
Oremet
T 40
Titanate
Titanium, Powder wetted with 20% or more water
Titanium, Powder wetted with not less than 25 per cent water (a visible excess of water must be present) (a) mechanically produced, particle size less than 53 microns; (b) chemically produced, particle size less than 840 microns [UN 1352]
Titanium, Wet, with less than 20% water (DOT)
Titanium 50A
Titanium alloy
Titanium (DOT)
Titanium metal, Powder, Dry [UN 2546]
UN 2546 [Titanium powder, dry]
UN 2878 [Titanium sponge granules or Titanium sponge powders]
VT 1

7440-33-7
W
183.85
Tungsten (ACGIH)
Wolfram

7440-34-8
Ac
227.03
Actinium

7440-35-9
Am
24
Americium (9CI)

7440-36-0
Sb
121.75
[B]=S
Antimony (ACGIH,OSHA)
Antimony Black
Antimony, Powder [UN 2871]
Antimony, Regulus
Antymon (Polish)
C.I. 77050
Stibium
UN 2871 [Antimony powder]

7440-37-1
Ar
39.95
[Ar]
Argon (DOT)
Argon-40
Argon, Compressed [UN 1006]
Argon, Refrigerated liquid [UN 1951]
UN 1006 [Argon, compressed]
UN 1951 [Argon, refrigerated liquid (cryogenic liquid)]

7440-38-2
As
74.92
Arsenic (ACGIH,OSHA) [UN 1558]
Arsenicals
Arsen (German,Polish)
Arsenic Black
Arsenic, Metallic (DOT)

Arsenic-75
Arsenic, Solid (DOT)
Colloidal Arsenic
Grey arsenic
Metallic arsenic
UN 1558 [Arsenic]

7440-39-3
Ba
137.34
Barium (ACGIH,OSHA) [UN 1400]
Barium, Alloys, Non-pyrophoric (DOT)
Barium, Alloys, Pyrophoric [UN 1854]
Barium, Metal, Non-pyrophoric (DOT)
UN 1399 (DOT)
UN 1400 [Barium]
UN 1854 [Barium alloys, pyrophoric]

7440-41-7
Be
9.01
Beryllium (ACGIH,OSHA)
Beryllium-9
Beryllium, Metal powder [UN 1567]
Glucinium
Glucinum
RCRA waste number P015
UN 1567 [Beryllium, powder]

7440-42-8
B
10.81
Boron

7440-43-9
Cd
112.40
Cadmium (ACGIH,OSHA)
C.I. 77180
Colloidal Cadmium
Kadmium (German)

7440-44-0
C
12.01
Carbon
Acticarbone
Activated Carbon (DOT)
AG 3
AG 3 (Adsorbent)
AG 5
AG 5 (Adsorbent)
AK (Adsorbent)
Anthrasorb
AR 3
ART 2
AU 3
BAU
BG 6080
Black Lead
Carbon-12
Carbon, Activated [UN 1362]
Carbopol Extra
Carbopol M
Carbopol Z 4
Carbopol Z Extra
Carbosieve
Carbosorbit R
Cecarbon
CF 8
CF 8 (Carbon)
C.I. 77265
C.I. Pigment Black 10
CLF II
CMB 50
CMB 200
Coke Powder
Columbia LCK
Conductex
CUZ 3
CWN 2
Darco
Filtrasorb
Filtrasorb 200
Filtrasorb 400
Graphite Synthetic (ACGIH, OSHA)
Grosafe
Hydrodarco
Irgalite 1104
Jado
K 257
MA 100 (Carbon)
Norit
Nuchar
OU-B
Pelikan C 11/1431a
Plumbago
SKG
SKT
SKT (Adsorbent)
SU 2000
Suchar 681
Supersorbon IV
Supersorbon S 1
U 02
UN 1362 [Carbon, activated]
Watercarb
Witcarb 940
XE 340
XF 4175L

7440-45-1
Ce
140.12
Cerium
Cerium, Crude, Powder (DOT)
Cerium, Crude, Slabs or ingots [UN 1333]
UN 1333 [Cerium, slabs, ingots, or rods]

7440-46-2
Cs
132.91
Cesium [UN 1407]
Caesium [UN 1407]
Caesium, Powdered (DOT)
Caesium, Metal (DOT)
Cesium-133
Cesium, Powdered (DOT)
Cesium metal (DOT)
UN 1407 [Cesium or Caesium]

7440-47-3
Cr
52.00
Chromium (ACGIH)
Chrome
Chromium metal (OSHA)

7440-48-4
Co
58.93
Cobalt (ACGIH,OSHA)
Aquacat
C.I. 77320
Cobalt-59
Kobalt (German, Polish)
NCI-C60311
Super Cobalt

7440-50-8
Cu
63.54
Copper (ACGIH,OSHA)
Allbri Natural Copper
Anac 110
Arwood Copper
Bronze Powder
CDA 101
CDA 102
CDA 110
CDA 122
C.I. 77400
C.I. Pigment Metal 2
Copper-Airborne
Copper Bronze
Copper-Milled
Copper Slag-Airborne
Copper Slag-Milled
1721 Gold
Gold Bronze
Kafar Copper
M 1
M 3
M 4
M1 (Copper)
M2 (Copper)
M3 (Copper)
M4 (Copper)
M3R
M3S
OFHC Cu
Raney Copper

7440-52-0
Er
167.26
[Er]
Erbium (9CI)

7440-53-1
Eu
151.97
[Eu]
Europium (9CI)

7440-54-2
Gd
157.25
Gadolinium

7440-55-3
Ga
69.72
Gallium [UN 2803]
Gallium, metal (DOT)
Gallium metal, Liquid (DOT)
Gallium metal, Solid (DOT)
UN 2803 [Gallium]

7440-56-4
Ge
72.61
[Ge]
Germanium (9CI)

7440-57-5
Au
196.97
Gold
Burnish Gold
C.I. 77480
C.I. Pigment Metal 3
Colloidal Gold
Gold Flake
Gold Leaf
Gold Powder
Magnesium Gold Purple
Shell Gold

7440-58-6
Hf
178.49
Hafnium (ACGIH,OSHA)
Hafnium metal (DOT)
Hafnium metal, Wet (DOT)
UN 1326 [Hafnium powder, Wetted with not less than 25 per cent water (a visible excess of water must be present) (a) mechanically produced, particle size less than 53 microns; (b) chemically produced, particle size less than 840 microns]
UN 2545 [Hafnium powder, dry]

7440-59-7
He
4.00
[He]
Helium (DOT)
Helium, Compressed [UN 1046]
Helium, Refrigerated liquid [UN 1963]
UN 1046 [Helium, compressed]
UN 1963 [Helium, refrigerated liquid (cryogenic liquid)]

7440-60-0
Ho
164.93
[Ho]
Holmium (9CI)

7440-61-1
U
238.00
Uranium (ACGIH,OSHA)
UN 2979 [Uranium metal, pyrophoric]
Uranium metal, Pyrophoric [UN 2979]

7440-62-2
V
50.94
Vanadium

7440-63-3
Xe
131.30
[Xe]
UN]
SY: UN 2036 [Xenon]
UN 2591 [Xenon, refrigerated liquid (cryogenic liquids)]
Xenon, Refrigerated liquid [UN 2591]

7440-64-4
Yb
173.04
Ytterbium

7440-65-5
Y
88.91
Yttrium (ACGIH,OSHA)
Yttrium-89

7440-66-6
Zn
65.37
Zinc
Blue Powder
C.I. 77945
C.I. Pigment Black 16
C.I. Pigment Metal 6
Emanay Zinc Dust
Granular zinc
Jasad
Merrillite
Pasco
UN 1383 [Pyrophoric metals, N.O.S.]
UN 1436 [Zinc powder or Zinc dust]
Zinc, Ashes (DOT)
Zinc, Powder or dust, Non-pyrophoric (DOT)
Zinc, Powder or dust, Pyrophoric [UN 1383]
Zinc Dust
Zinc Powder

7440-67-7
Zr
91.22
Zirconium (ACGIH,OSHA)
UN 1308 [Zirconium suspended in a liquid]
UN 1358 [Zirconium powder, Wetted with not less than 25 per cent water (a visible excess of water must be present) (a) mechanically produced, particle size less than 53 microns; (b) chemically produced, particle size less than 840 microns]
UN 1932 [Zirconium scrap]
UN 2008 [Zirconium powder, dry]
UN 2009 [Zirconium, dry, finished sheets, strip or coiled wire]
UN 2858 [Zirconium, Dry, coiled wire, finished metal sheets, strip (thinner than 254 microns but not thinner than 18 microns)]
Zircat
Zirconium, Borings (DOT)
Zirconium, Clippings (DOT)
Zirconium metal (DOT)
Zirconium, Scrap [UN 1932]
Zirconium, Suspended in fammable liquid (DOT)
Zirconium, Turnings (DOT)

7440-69-9
Bi
208.98
[B]I
Bismuth
Bismuth-209

7440-70-2
Ca
40.08
Calcium [UN 1401]
Calcicat
Calcium, Non-pyrophoric (DOT)
Calcium, Pyrophoric [UN 1855]
Calcium, Metal, Crystalline (DOT)
Calcium, Metal (DOT)
NA 1401 (DOT)
UN 1401 [Calcium]
UN 1855 [Calcium, pyrophoric or Calcium alloys, pyrophoric]

7440-74-6
In
114.82
Indium (ACGIH,OSHA)

7442-13-9
$C_3H_{10}Pb$
253.32

C[PbH](C)C
Plumbane, trimethyl
Lead, trimethyl-
Trimethyl lead
Trimethylplumbane

7443-52-9
$C_7H_{14}O$
114.19
Cyclohexanol, 2-methyl-, trans-(8CI,9CI)
NSC-244887
NSC-245854

7443-55-2
$C_7H_{14}O$
114.19
Cyclohexanol, 3-methyl-, trans-(8CI,9CI)

7443-70-1
$C_7H_{14}O$
114.19
Cyclohexanol, 2-methyl-, cis-(8CI,9CI)
NSC-100902

7446-08-4
O_2Se
110.96
Selenium(IV) dioxide (1:2)
Selenious anhydride
RCRA waste number U204
Selenium dioxide
Selenium oxide

7446-09-5
O_2S
64.06
O=S=O
Sulfur-dioxide
Bisulfite
Fermenicide Liquid
Fermenicide Powder
Schwefeldioxyd (German)
Siarki dwutlenek (Polish)
Sulfur dioxide (ACGIH,DOT, OSHA)
Sulfurous acid anhydride
Sulfurous anhydride
Sulfurous oxide
Sulfur oxide
Sulphur dioxide
Sulphur dioxide, Liquefied [UN 1079]
UN 1079 [Sulfur dioxide, liquefied]

7446-11-9
O_3S
80.06
Sulfur trioxide (DOT)
Sulfan
Sulfuric anhydride (DOT)
Sulfuric oxide
Sulfur trioxide, Stabilized [UN 1829]
UN 1829 [Sulfur trioxide, inhibited]

7446-14-2
$O_4S.Pb$
303.25
Lead(II) sulfate (1:1)
Anglislite
Bleisulfat (German)
C.I. 77630
C.I. Pigment White 3
Fast White
Freemans White Lead
Lead Bottoms
Lead Dross (DOT)
Lead sulfate, Solid, Containing more than 3% free acid [UN 1794]
Lead sulphate, With more than 3% free acid (DOT)
Milk White
Mulhouse White
Sulfate de plomb (French)
Sulfuric acid, lead(2+) salt (1:1)
UN 1794 [Lead sulfate with more than 3 per cent free acid]

7446-15-3
$H_2O_4Se.Pb$
352.17
Lead selenate
Selenic acid, lead(2+) salt (1:1) (9CI)

7446-18-6
$O_4S.2Tl$
504.80
Thallium(I) sulfate (2:1)
C.F.S.
CSF-Giftweizen
Dithallium sulfate
Dithallium(1+) sulfate
Eccothal
M7-Giftkoerner
Rattengiftkonserve
RCRA waste number P115
Sulfuric acid, dithallium(1+) salt (8ci,9CI)
Sulfuric acid, thallium(1+) salt (1:2)
Thallium sulfate
Thallium(1) sulfate
Thallous sulfate

7446-19-7
H_4O_5SZn
181.46
Zinc sulfate
Sulfuric acid, zinc salt (1:1)

7446-20-0
$O_4SZn.7H_2O$
287.57
Zinc sulfate heptahydrate (1:1:7)
NUZ
Sulfuric acid, zinc salt (1:1), heptahydrate
White Vitriol
Zinc sulfate
Zinc sulphate
Zinc vitriol
Zinc sulfate (1:1) heptahydrate

7446-27-7
$O_8P_2.3Pb$
811.51
Lead(II) phosphate (3:2)
Bleiphosphat (German)
C.I. 77622
Collodial Lead Phosphate
Lead orthophosphate
Lead phosphate
Lead phosphate (3:2)
Lead(2+) phosphate
Normal Lead Orthophosphate
Perlex Paste 500
Perlex Paste 600A
Phosphoric acid, lead(2+) salt (2:3)
Plumbous phosphate
RCRA waste number U145
Trilead phosphate

7446-34-6
SSe
111.02
S=[Se]
Selenium-sulfide
NCI-C50033
RCRA waste number U205
Selenium monosulfide
Selenium sulphide
Selensulfid (German)
Sulfur selenide

7446-70-0
$AlCl_3$
133.33
Aluminum-chloride
Alluminio(cloruro di) (Italian)
Aluminiumchlorid (German)
Aluminum chloride (1:3)
Aluminum chloride, Anhydrous [UN 1726]
Aluminum chloride, Solution [UN 2581]
Aluminum trichloride
Chlorure d'aluminium (French)
Pearsall
Trichloroaluminum
UN 1726 [Aluminum chloride, anhydrous]
UN 2581 [Aluminum chloride, solution]

7446-81-3
$C_3H_4O_2.Na$
95.05
Sodium acrylate
Acrylic acid, sodium salt
2-Propenoic acid, sodium salt (9CI)

Sodium 2-propenoate

7447-39-4
Cl_2Cu
134.44
Copper(II) chloride (1:2)
Copper bichloride
Copper(2+) chloride
Copper(II) chloride
Cupric chloride
Cupric dichloride

7447-40-7
ClK
74.55
[Cl-].[K+]
Potassium-chloride
Chlorid draselny (Czech)
Chloropotassuril
Dipotassium dichloride
Emplets potassium chloride
Enseal
Kalitabs
Kaochlor
Kaon-Cl
Kaon-Cl 10
Kaon-Cl Tabs
Kay Ciel
K-Lor
Klotrix
K-Lyte/Cl
K-Predne-Dome
Pfiklor
Potassium monochloride
Potavescent
Rekawan
Slow-K
Tripotassium trichloride

7447-41-8
ClLi
42.39
[Li+].[Cl-]
Lithium-chloride
Chlorku litu (Polish)
Chlorure de lithium (French)

7447-67-8
Unknown
Unknown
Diethylene glycol phthalate

7450-69-3
$C_6H_9N_2O_2P$
172.11
Phosphoric phenyl ester diamide
AI3-51284
Phosphorodiamidic acid, phenyl ester

7452-01-9
$C_{10}H_{14}O_2$
166.22
Benzenemethanol, α-ethyl-2-methoxy- (9CI)
Benzyl alcohol, α-ethyl-o-methoxy- (8CI)

7452-79-1
$C_7H_{14}O_2$
130.19
O=C(OCC)C(CC)C
Butanoic acid, 2-methyl-, ethyl ester (9CI)
AI3-35155
Butyric acid, 2-methyl-, ethyl ester (8CI)
Ethyl 2-methylbutanoate

7459-63-4
$C_9H_7N_3O_2$
189.16
1,3,5-Triazine-2,4(1H,3H)-dione, 6-phenyl- (9CI)
s-Triazine-2,4-diol, 6-phenyl-
s-Triazine-2,4(1H,3H)-dione, 6-phenyl- (8CI)

7460-84-6
$C_{21}H_{40}O_3$
340.61
Stearic acid, 2,3-epoxypropyl ester
2,3-Epoxy-1-propanol stearate
2,3-Epoxypropyl ester of stearic acid
2,3-Epoxypropyl stearate
Glycidol stearate
Glycidyl octadecanoate
Glycidyl stearate
Oxiranylmethyl ester of octadecanoic acid
Stearic acid, glycidyl ester

7463-22-1
$C_8H_{10}ClNO_2S$
219.69
O=S(=O)(N(C)C)c(ccc(c1)Cl)c1
Benzenesulfonamide, 4-chloro-N,N-dimethyl- (9CI)
AI3-17741
4-Chloro-N,N-dimethylbenzenesulfonamide

7464-68-8
$C_5H_7N_3O_2$
141.15
Imidazole, 1,5-dimethyl-4-nitro
1,5-Dimethyl-4-nitro-1H-imidazole (9CI)

7466-42-4
$C_{18}H_{14}N_2$
258.33
Diazene, (1,1'-biphenyl)-4-ylphenyl- (9CI)
Azobenzene, 4-phenyl- (6CI,8CI)
4-(Phenylazo)biphenyl
4-Phenylazodiphenyl

7469-40-1
$C_{16}H_{14}$
206.29
Naphthalene, 1,2-dihydro-4-phenyl- (7CI,8CI,9CI)
1,2-Dihydro-4-phenylnaphthalene
3,4-Dihydro-1-phenylnaphthalene
1-Phenyl-3,4-dihydronaphthalene

7469-77-4
$C_{11}H_{10}O$
158.20
1-Naphthalenol, 2-methyl- (9CI)
1-Naphthol, 2-methyl- (8CI)

7473-98-5
$C_{10}H_{12}O_2$
164.20
O=C(c(cccc1)c1)C(O)(C)C
1-Propanone, 2-hydroxy-2-methyl-1-phenyl- (9CI)
2-Hydroxy-2-methyl-1-phenyl-1-propanone
2-Hydroxy-2-methylpropiophenone
1-Phenyl-2-hydroxy-2-methylpropan-1-one

7473-99-6
$C_{10}H_{11}ClO$
182.65
O=C(c(cccc1)c1)C(Cl)(C)C
1-Propanone, 2-chloro-2-methyl-1-phenyl- (9CI)
2-Chloro-2-methyl-1-phenyl-1-propanone
1-Phenyl-2-chloro-2-methyl-1-propanone

7481-89-2
$C_9H_{13}N_3O_3$
211.25
Cytidine, 2',3'-dideoxy
2',3'-Dideoxycytidine

7486-38-6
$C_6H_8O_4.2Na$
190.12
Adipic acid, disodium salt
1,4-Butanedicarboxylic acid disodium salt
Disodium adipate
Hexanedioic acid, disodium salt (9CI)
Sodium adipate

7487-28-7
$C_8H_{14}O_3$
158.22
CC2(COCC1(C)CO1)CO2

Ether, bis(2-methylglycidyl)
Bis(2,3-epoxy-2-methylpropyl)-ether
Bis(2-methylglycidyl) ether
EP-2017
Ether, bis(2,3-epoxy-2-methyl-propyl)
Oxirane, 2,2'-oxybis(methyl-ene)bis-2-methyl-

7487-79-8
$C_{12}H_{24}O_2.C_4H_{11}NO_2$
305.45
Lauric acid diethanolamine salt
Diethanolamine monolaurate
Dodecanoic acid, Compd. with 2,2'-iminobis(ethanol) (1:1) (9CI)
Lauric acid, Compd. with 2,2'-iminodiethanol (1:1)
Lauric diethanolamine

7487-88-9
$O_4S.Mg$
120.37
Magnesium sulfate (1:1)
Epsom Salts
Magnesium sulphate

7487-94-7
Cl_2Hg
271.49
Cl[Hg]Cl
Mercury(II) chloride
Bichloride of mercury
Bichlorure de mercure (French)
Calochlor
Chlorid rtutnaty (Czech)
Chlorure mercurique (French)
Cloruro di mercurio (Italian)
Corrosive mercury chloride
Corrosive sublimate
Dichloromercury
Emisan 6
Fungchex
MC
Mercuric bichloride
Mercuric chloride [UN 1624]
Mercuric chloride, Solid (DOT)
Mercury bichloride
Mercury perchloride
NCI-C60173
Quecksilber chlorid (German)
Perchloride of mercury
Sulema (Russian)
Sublimat (Czech)
TL 898
UN 1624 [Mercuric chloride]

7488-51-9
$H_2O_3Se.Pb$
336.17
Lead selenite
Selenious acid, lead(2+) salt (1:1) (9CI)

7488-56-4
S_2Se
143.08
S=[Se]=S
Selenium(IV) disulfide (1:2)
Exsel
RCRA waste number U205
Selenium disulphide [UN 2657]
Selenium sulfide
Selsun Blue
Selsun
UN 2657 [Selenium disulfide]

7492-30-0
$C_{18}H_{34}O_3.K$
337.56
Potassium ricinoleate
Caswell No. 701A
EPA Pesticide Chemical Code 079023
12-Hydroxy-9-octadecenoic acid, monopotassium salt
9-Octadecenoic acid, 12-hydroxy-, monopotassium salt
9-Octadecenoic acid, 12-hydroxy-, monopotassium salt, (R-(Z))- (9CI)
Ricinoleic acid, monopotassium salt

7492-55-9
$C_6H_8O_2.1/2Ca$
132.17
Calcium sorbate
(E,E)-Calcium 2,4-hexadienoate
2,4-Hexadienoic acid, calcium salt
2,4-Hexadienoic acid, calcium salt, (E,E)- (9CI)
Sorbic acid, calcium salt

7492-66-2
$C_{14}H_{26}O_2$
226.40
O(C(OCC)C=C(CCC=C(C)C)C)CC
2,6-Octadienal, 3,7-dimethyl-, diethyl acetal
Citral diethyl acetal
1,1-Diethoxy-3,7-dimethyl-2,6-octadiene
3,7-Dimethyl-2,6-octadienal diethyl acetal
2,6-Octadiene, 1,1-diethoxy-3,7-dimethyl- (9CI)

7492-68-4
CH_3CuO_3
126.58
Carbonic acid, copper salt (8CI,9CI)

7493-57-4
$C_{13}H_{20}O_2$
208.30
O(C(OCCc(cccc1)c1)C)CCC
Benzene, (2-(1-propoxy-ethoxy)ethyl)- (9CI)
Acetaldehyde propyl phenyl-ethyl acetal
(2-(1-Propoxyethoxy)ethyl)-benzene

7493-58-5
$C_6H_{10}O_2$
114.14
CC(C)C(=O)C(C)=O
2,3-Pentanedione, 4-methyl- (8CI,9CI)
NSC-31666

7493-63-2
$C_{10}H_{11}NO_2$
177.20
Anthranilic acid, allyl ester (8CI)
AI3-36005
Benzoic acid, 2-amino-, 2-propenyl ester (9CI)

7495-84-3
$C_{10}H_{12}O_2$
164.20
Propanoic acid, 4-methyl-phenyl ester
AI3-22019
4-Methylphenyl propanoate

7496-02-8
$C_{18}H_{11}NO_2$
273.30
Chrysene, 6-nitro
6-Nitrochrysene

7500-53-0
$C_{10}H_{10}O_4$
194.19
1,2-Benzenediacetic acid (9CI)
o-Benzenediacetic acid (8CI)
NSC-401681

7501-27-1
$C_5H_7N_3O_2$
141.13
s-Triazine-2,4-diol, 6-ethyl- (8CI)
s-Triazine-2,4(1H,3H)-dione, 6-ethyl- (7CI)

7501-79-3
$C_8H_{17}NO$
143.23
Acetamide, N-hexyl- (6CI, 7CI,8CI,9CI)
N-Hexylacetamide

7507-35-9
$C_8H_6Br_2O_3$
309.94
(3,5-Dibromophenoxy) acetic acid

Acetic acid, (3,5-dibromophen-oxy)- (9CI)

7507-89-3
$C_9H_{10}O_4$
182.18
Acetophenone, 2',6'-di-hydroxy-4'-methoxy- (8CI)
Ethanone, 1-(2,6-dihydroxy-4-methoxyphenyl)- (9CI)

7508-73-8
$C_{14}H_{13}N_2O_7PS$
384.32
CCOP(=S)(Oc1ccc(cc1)N(=O)=O)Oc2ccc(cc2)N(=O)=O
Phosphorothioic acid, O,O-bis-p-(nitrophenyl) O-ethyl ester
O,O-Bis-p-nitrofenyl-O-ethyl-ester kyseliny thiofosforecne (Czech)
Bis-parathion (Czech)
O-Ethyl-O,O-bis-(p-nitrofenyl)-thiofosfat (Czech)

7519-36-0
$C_5H_8N_2O_3$
144.15
OC(=O)C1CCCN1N=O
Proline, N-nitroso-, L
1-Nitroso-l-proline
N-Nitroso-l-proline
No-Pro
NPRO

7521-80-4
$C_4H_9Cl_3Si$
191.57
Silane, butyltrichloro
Butyltrichlorosilane [UN 1747]
Trichlorobutylsilane
UN 1747 [Butyltrichlorosilane]

7525-62-4
$C_{10}H_{12}$
132.22
c(cccc1C=C)(c1)CC
Styrene, m-ethyl
m-Ethylstyrene
m-Ethyl vinylbenzen (Czech)
m-Ethylvinylbenzene

7534-94-3
$C_{14}H_{22}O_2$
222.33
O=C(OC(C(C(C1C2)(C)C)(C2)C)C1)C(=C)C
Isobornyl methacrylate
Methacrylic acid, isobornyl ester
2-Propenoic acid, 2-methyl-, 1,7,7-trimethylbicyclo(2.2.1)-hept-2-yl ester, exo- (9CI)

7535-00-4
$C_6H_{13}NO_5$
179.20
Galactose, 2-amino-2-deoxy-, D
D-2-Amino-2-deoxygalactose
Chondrosamine
Galactosamine
D-Galactosamine
D-(+)-Galactosamine

7542-37-2
$C_{23}H_{45}N_5O_{14}$
615.73
Streptamine, O-2,6-diamino-2,6-dideoxy-β-l-idopyranosyl-(1-3)-O-β-D-ribofuranosyl-(1-5)-O-(2-amino-2-deoxy-α-D-glucopyranosyl-(1-4))-2-deoxy
Aminosidin
Aminosidine
Aminosidine I
Amminosidin
Antibiotic 2230D
Antibiotic SF 767B
Antibiotic 503-3
Catenulin
Crestomycin
Estomycin
Gabbromycin
Humatin
Humycin
Hydroxymycin
Monomycin A
Neomycin E
Pargonyl
Paromomycin
Paromomycine
Paromomycin I
Paucimycin
Quintomycin C
R 400
Zygomycin A1

7549-37-3
$C_{12}H_{22}O_2$
198.34
O(C(OC)C=C(CCC=C(C)C)C)C
2,6-Octadiene, 1,1-dimethoxy-3,7-dimethyl-, (cis and trans)
Citral dimethyl acetal
1,1-Dimethoxy-3,7-dimethyl-2,6-octadiene

7550-35-8
BrLi
86.85
[Li+].[Br-]
Lithium-bromide

7550-45-0
Cl_4Ti
189.70
Cl[Ti](Cl)(Cl)Cl
Titanium chloride ($TiCl_4$)
Tetrachlorure de titane (French)
Titaantetrachloride (Dutch)
Titane (tetrachlorure de) (French)
Titanio (tetracloruro di) (Italian)
Titanium chloride
Titanium tetrachloride [UN 1838]
Titantetrachlorid (German)
UN 1838 [Titanium tetra-chloride]

7553-56-2
I_2
253.80
II
Iodine (ACGIH,OSHA)
Iode (French)
Iodine Crystals
Iodine Sublimed
Iodio (Italian)
Jod (German, Polish)
Jood (Dutch)

7554-12-3
$C_8H_{14}O_5$
190.20
O=C(OCC)C(O)CC(=O)OCC
Diethyl malate
AI3-00932
Butanedioic acid, hydroxy-, di-ethyl ester
Diethyl hydroxybutanedioate

7558-63-6
$C_5H_8NO_4.H_4N$
164.19
Glutamic acid, mono-ammonium salt, L
Ammoniumglutaminat (German)
MAG
Monoammonium glutamate

7558-79-4
$HO_4P.2Na$
141.96
Sodium monohydrogen phos-phate (2:1:1)
Dibasic sodium phosphate
Disodium hydrogen phosphate
Disodium monohydrogen phos-phate
Disodium orthophosphate
Disodium phosphate
Disodium phosphoric acid
DSP
Exsiccated sodium phosphate
Natriumphosphat (German)
Soda phosphate
Sodium hydrogen phosphate
Sodium phosphate, dibasic
Phosphoric acid, disodium salt

7558-80-7
$H_2O_4P.Na$
119.98
Sodium dihydrogen phosphate (1:2:1)

Monosodium dihydrogen phosphate
Monosodium phosphate
Monosorb XP-4
Primary sodium phosphate
Sodium acid phosphate
Sodium biphosphate
Sodium biphosphate anhydrous
Sodium phosphate, monobasic

7559-34-4
$C_8H_{13}N_3$
151.21
s-Triazine, 2,4-dimethyl-6-propyl- (7CI,8CI)

7560-83-0
$C_{13}H_{25}N$
195.39
N(C(CCCC1)C1)(C(CCCC2)C2)C
Dicyclohexylamine, N-methyl
N-Methyldicyclohexylamine

7564-63-8
$C_{10}H_{12}$
132.21
Benzene, 1-ethenyl-2-ethyl- (9CI)
Styrene, o-ethyl- (8CI)

7564-64-9
$C_6H_{14}O_3$
134.18
OCCC(O)(CCO)C
1,3,5-Pentanetriol, 3-methyl- (9CI)
3-Methyl-1,3,5-pentanetriol

7568-93-6
$C_8H_{11}NO$
137.20
OC(c(cccc1)c1)CN
Benzyl alcohol, α-(aminomethyl)
2-Amino-1-phenyl-1-ethanol
Ethanol, 2-amino-1-phenyl-
Ethylamine, β-hydroxy-β-phenyl-
β-Hydroxyphenethylamine
β-Hydroxy-β-phenylethylamine
Phenethylamine, β-hydroxy-

7570-26-5
$C_2H_4N_2O_4$
120.05
1,2-Dinitroethane
1,2'-Dinitroetano (Spanish)
Dinitro-1,2 ethane (French)
Ethane, 1,2-dinitro-

7572-29-4
C_2Cl_2
94.92
ClC#CCl
Acetylene, dichloro
Dichloroacetylene (ACGIH, DOT,OSHA)
Dichloroethyne
Ethyne, dichloro- (9CI)

7575-23-7
$C_{17}H_{28}O_8S_4$
488.67
O=C(OCC(COC(=O)CCS)(COC(=O)CCS)COC(=O)CCS)CCS
Propanoic acid, 3-mercapto-, 2,2-bis((3-mercapto-1-oxopropoxy)methyl)-1,3-propanediyl ester (9CI)

7575-62-4
Unknown
Unknown
Disodium 4-dodecyl-2,4'-oxydibenzenesulfonate

7576-65-0
$C_{18}H_{11}NO_3$
289.28
O=C(c(c(C1=O)ccc2)c2)C1c(nc(c(ccc3)c4)c3)c4O
1H-Indene-1,3(2H)-dione, 2-(3-hydroxy-2-quinolinyl)- (9CI)
3-Hydroxy-2-(1,3-indandione-2-yl)quinoline
2-(3-Hydroxy-2-quinolyl)-1,3-indanedione

7578-36-1
$C_2H_3BF_3O_2.H$
127.86
Boron trifluoride acetic acid complex
Borate(1-), (acetato-O)trifluoro-, hydrogen, (T-4)- (9CI)
Boron fluoride, Compound with acetic acid
Boron trifluoride-acetic acid complex [UN 1742]
Trifluorure de bore et d'acide acetique, Complexe de (French)
Trifluoruro de boro y acido acetico, Complejo de (Spanish)
UN 1742 [Boron trifluoride acetic acid complex]

7580-31-6
$C_8H_{18}NiO_2$
204.94
Hexanoic acid, 2-ethyl-, nickel salt (9CI)
Nickel 2-ethylhexanoate

7580-37-2
$C_4H_{12}O_4P.C_2H_3O_2$
214.18
Phosphonium, tetrakis-(hydroxymethyl)-, acetate
Tetrakis(hydroxymethyl)phosphonium acetate

7580-67-8
HLi
7.95
Lithium-hydride
Hydrure de lithium (French)
Lithium hydride (ACGIH, OSHA) [UN 1414]
Lithium hydride, In fused solid form [UN 2805]
UN 1414 [Lithium hydride]
UN 2805 [Lithium hydride, fused solid]

7580-85-0
$C_6H_{14}O_2$
118.20
Ethanol, 2-tert-butoxy
2-tert-Butoxyethanol
Ethanol, 2-(1,1-dimethylethoxy)- (9CI)
Ethylene glycol, mono-tert-butyl ether
Swasolve ETB

7581-97-7
$C_4H_8Cl_2$
127.01
C(C(Cl)C)(Cl)C
2,3-Dichlorobutane
Butane, 2,3-dichloro- (9CI)

7585-20-8
$C_2H_7O_2Zr$
154.30
Zirconium acetate
Acetic acid, zirconium salt (9CI)

7585-41-3
C18$H_{13}ClN_2O_6S$.Ba
558.14
2-Naphthalenecarboxylic acid, 4-((5-chloro-4-methyl-2-sulfophenyl)azo)-3-hydroxy-, barium salt (1:1) (9CI)
4-((5-Chloro-2-sulfo-p-tolyl)-azo)-3-hydroxy-2-naphthalenecarboxylic acid barium salt (1:1)

7597-97-9
$C_{13}H_{20}O$
192.33
Phenol, 2-tert-butyl-4-isopropyl
2-tert-Butyl-4-isopropylphenol

7601-54-9
$O_4P.3Na$
163.94
Phosphoric acid, trisodium

salt
Dri-Tri
Emulsiphos 440/660
Nutrifos STP
Sodium phosphate
Sodium phosphate, Anhydrous
Sodium phosphate, tribasic
Tribasic sodium phosphate
Trinatriumphosphat (German)
Trisodium orthophosphate
Trisodium phosphate
Tromete
TSP

7601-89-0
$ClO_4.Na$
122.44
Perchloric acid, sodium salt
Natriumperchloraat (Dutch)
Natriumperchlorat (German)
Perchlorate de sodium (French)
Sodio (perclorato di) (Italian)
Sodium perchlorate [UN 1502]
Sodium (perchlorate de) (French)
UN 1502 [Sodium perchlorate]

7601-90-3
$ClHO_4$
100.46
Perchloric-acid
Perchloric acid, More than 72% strength (DOT)
Perchloric acid, More than 50% but not more than 72% strength [UN 1873]
Perchloric acid, Not over 50% acid [UN 1802]
UN 1802 [Perchloric acid not more than 50 per cent acid by mass]
UN 1873 [Perchloric acid more than 50 per cent but not more than 72 per cent acid, by mass]

7613-16-3
$C_8H_{20}N_2.2ClH$
217.22
1,8-Octanediamine, dihydrochloride
1,8-Diaminooktan hydrochlorid (Czech)
MW 217
Octane-1,8-diamine dihydrochloride
Oktamethylendiamin hydrochlorid (Czech)

7614-93-9
$C_{16}H_{16}$
208.31
Benzene, 1,1'-(3-methyl-1-propene-1,3-diyl)bis- (9CI)
1-Butene, 1,3-diphenyl- (7CI,8CI)
1,3-Diphenyl-1-butene

7615-57-8
C_6N_6
156.11
1,3,5-Triazine-2,4,6-tricarbonitrile (9CI)
s-Triazinetricarbonitrile
s-Triazine-2,4,6-tricarbonitrile (7CI,8CI)

7616-94-6
$ClFO_3$
102.45
Perchloryl-fluoride
Chlorine fluoride oxide
Chlorine oxyfluoride
Perchloryl fluoride (ACGIH, OSHA)

7617-57-4
$C_6H_4N_2O_2$
136.12
Benzene, o-dinitroso

7617-78-9
$C_{13}H_{29}NO$
215.37
O(CCCCCCCCCC)CCCN
1-Propanamine, 3-(decyloxy)- (9CI)
3-Decoxypropane-1-amine
3-(Decyloxy)-1-propanamine

7620-77-1
$C_{18}H_{36}O_3.Li$
307.42
Octadecanoic acid, 12-hydroxy-, monolithium salt
AI3-19768

7621-86-5
$C_{13}H_{12}N_4$
224.24
N(c(c(N1)cc(N)c2)c2)=C1c(ccc(N)c3)c3
1H-Benzimidazol-5-amine, 2-(4-aminophenyl)- (9CI)
2-(4-Aminophenyl)-5-aminobenzimidazole
2-(4-Aminophenyl)-1H-benzimidazol-5-amine

7623-09-8
$C_3H_4Cl_2O$
126.97
O=C(C(Cl)C)Cl
Propanoyl chloride, 2-chloro- (9CI)
2-Chloropropanoyl chloride
2-Chloropropionyl chloride

7631-86-9
O_2Si
60.09
Silica, Amorphous fumed
Acticel
Aerosil
Amorphous Silica Dust
Aquafil
Cab-O-Grip II
Cab-O-Sil
Cab-O-Sperse
Cataloid
Colloidal Silica
Colloidal Silicon Dioxide
Davison SG-67
Dicalite
Dri-Die Insecticide 67
ENT 25,550
Fossil Flour
Fumed Silica
Fumed Silicon Dioxide
Ludox
Nalcoag
Nyacol
Nyacol 830
Nyacol 1430
Santocel
SG-67
Silica, Amorphous
Silicic anhydride
Silikill
Vulkasil

7631-89-2
$AsH_3O_4.xNa$
302.88
[Na+].[Na+].O[As]([O-])([O-])=O
Arsenic acid, sodium salt
Fatsco Ant Poison
Sodium arsenate [UN 1685]
Sodium metaarsenate
Sodium orthoarsenate
Sweeney's Ant-Go
UN 1685 [Sodium arsenate]

7631-90-5
$HO_3S.Na$
104.06
Sodium bisulfite (1:1)
Bisulfite de sodium (French)
Hydrogen sulfite sodium
NA 2693 (DOT)
Sodium acid sulfite
Sodium bisulfite (ACGIH, OSHA)
Sodium bisulfite, Solid
Sodium bisulfite, Solution [UN 2693]
Sodium bisulphite
Sodium hydrogen sulfite
Sodium hydrogen sulfite, Solid (DOT)
Sodium hydrogen sulfite, Solution (DOT)
Sodium sulhydrate
Sulfurous acid, monosodium salt
UN 2693 [Bisulfites, inorganic, aqueous solutions, N.O.S.]

7631-95-0
$MoO_4.2Na$
205.92

Molybdic acid, disodium salt
Disodium molybdate
Natriummolybdat (German)
Sodium molybdate
Sodium molybdate(VI)

7631-99-4
$NO_3.Na$
85.00
[Na+].[O-]N(=O)=O
Sodium(I) nitrate (1:1)
Chile Saltpeter
Cubic Niter
Nitratine
Nitrate de sodium (French)
Nitric acid, sodium salt
Soda Niter
Sodium nitrate [UN 1498]
Sodium Saltpeter
UN 1498 [Sodium nitrate]

7632-00-0
$NO_2.Na$
69.00
[Na+].[O-]N=O
Nitrous acid, sodium salt
Anti-Rust
Diazotizing salts
Dusitan sodny (Czech)
Erinitrit
Filmerine
Natrium nitrit (German)
NCI-C02084
Nitrite de sodium (French)
Sodium nitrite [UN 1500]
UN 1500 [Sodium nitrite]

7632-04-4
$BHO_3.Na$
82.81
Perboric acid, sodium salt
Sodium borate
Sodium perborate
Sodium peroxoborate

7632-05-5
$H_3O_4P.xNa$
Unknown
Sodium phosphate
Phosphoric acid, sodium salt (9CI)
Sodium salt of phosphoric acid

7632-50-0
$C_6H_8O_7.xH_3N$
Unknown
Ammonium citrate
1,2,3-Propanetricarboxylic acid, 2-hydroxy-, ammonium salt (9CI)

7632-51-1
Cl_4V
192.74
Vanadium tetrachloride [UN 2444]
UN 2444 [Vanadium tetrachloride]
Vanadium chloride

7637-07-2
BF_3
67.81
FB(F)F
Borane, trifluoro
Boron fluoride
Boron trifluoride (ACGIH, OSHA) [UN 1008]
Fluorure de bore (French)
UN 1008 [Boron trifluoride]

7637-13-0
$C_{18}H_{36}O_2.xSn$
Unknown
Octadecanoic acid, tin salt (9CI)
Tin octadecanoate

7642-04-8
C_8H_{16}
112.22
2-Octene, (Z)- (8CI,9CI)
NSC-244460

7642-09-3
C_6H_{12}
84.16
3-Hexene, (Z)- (9CI)

7642-10-6
C_7H_{14}
98.19
3-Heptene, (Z)- (8CI,9CI)
cis-3-Heptene
(Z)-3-Heptene

7645-25-2
$AsH_3O_4.xPb$
1592.28
Arsenic acid, lead salt
Arseniate de plomb (French)
Lead arsenate [UN 1617]
Lead arsenate, Solid (DOT)
UN 1617 [Lead arsenates]

7646-69-7
HNa
24.00
Sodium hydride [UN 1427]
NaH 80
UN 1427 [Sodium hydride]

7646-78-8
Cl_4Sn
260.49
Tin(IV) chloride (1:4)
Etain (tetrachlorure d') (French)
Libavius fuming spirit
Stagno (tetracloruro di) (Italian)
Stannic chloride
Stannic chloride, Anhydrous [UN 1827]
Tin chloride (DOT)
Tin perchloride (DOT)
Tin tetrachloride
Tin tetrachloride, Anhydrous (DOT)
Tintetrachloride (Dutch)
UN 1827 [Stannic chloride, anhydrous]
Zinntetrachlorid (German)

7646-79-9
Cl_2Co
129.83
Cobalt(II) chloride
Cobalt chloride
Cobalt dichloride
Cobalt muriate
Cobaltous chloride
Cobaltous dichloride
Kobalt chlorid (German)

7646-85-7
Cl_2Zn
136.27
Zinc-chloride
Butter of Zinc
Chlorure de zinc (French)
Tinning Flux (DOT)
UN 1840 [Zinc chloride, solution]
UN 2331 [Zinc chloride, anhydrous]
Zinc Butter
Zinc chloride (ACGIH,OSHA)
Zinc chloride, Anhydrous [UN 2331]
Zinc chloride, Solution [UN 1840]
Zinc (chlorure de) (French)
Zinc dichloride
Zinc muriate, Solution (DOT)
Zinco (cloruro di) (Italian)
Zinkchlorid (German)
Zinkchloride (Dutch)

7646-88-0
$C_2HCl_3O_2.H_3N$
180.41
Acetic acid, trichloro-, ammonium salt (8CI,9CI)
Ammonium trichloroacetate
Caswell No. 049
EPA Pesticide Chemical Code 076801

7646-93-7
$HO_4S.K$
136.17
Potassium-bisulfate
Acid potassium sulfate
Monopotassium sulfate
Potassium acid sulfate
Potassium bisulphate
Potassium hydrogen sulfate, Solid (DOT)
Potassium hydrogen sulphate [UN 2509]

Potassium sulfate
Sal Enixum
Sulfuric acid, monopotassium salt
UN 2509 [Potassium hydrogen sulfate]

7647-01-0
ClH
36.46
Cl
Hydrochloric acid (DOT)
Acide chlorhydrique (French)
Acido cloridrico (Italian)
Chloorwaterstof (Dutch)
Chlorohydric acid
Chlorowodor (Polish)
Chlorwasserstoff (German)
Hydrochloric acid, Anhydrous [UN 1050]
Hydrochloric acid, Solution [UN 1789]
Hydrochloric acid, Solution, Inhibited (DOT)
Hydrochloride
Hydrogen chloride (ACGIH, DOT,OSHA)
Hydrogen chloride, Anhydrous (DOT)
Hydrogen chloride (Aerosol)
Hydrogen chloride, Refrigerated liquid [UN 2186]
Muriatic acid (DOT)
Spirits of Salt (DOT)
UN 1050 [Hydrogen chloride, anhydrous]
UN 1789 [Hydrochloric acid, solution]
UN 2186 [Hydrogen chloride, refrigerated liquid (cryogenic liquid)]

7647-10-1
Cl_2Pd
177.30
Palladium(2+) chloride
NCI-C60184
Palladium chloride
Palladous chloride

7647-14-5
ClNa
58.44
[Na+].[Cl-]
Sodium-chloride
Common Salt
Dendritis
Extra Fine 200 Salt
Extra Fine 325 Salt
Halite
H.G. Blending
Natriumchlorid (German)
Purex
Rock Salt
Saline
Salt
Sea Salt
Sterling
Table Salt
Top Flake
USP Sodium Chloride
White Crystal

7647-15-6
BrNa
102.90
Sodium-bromide
Bromide salt of sodium
Bromnatrium (German)
Sedoneural

7647-17-8
ClCs
168.36
[Cl-].[Cs+]
Cesium-chloride
Cesium monochloride
Dicesium dichloride
Tricesium trichloride

7647-18-9
Cl_5Sb
299.00
Cl[Sb](Cl)(Cl)(Cl)Cl
Antimony (V) chloride
Antimoine (pentachlorure d') (French)
Antimonio (pentacloruro di) (Italian)
Antimonpentachlorid (German)
Antimony pentachloride (DOT)
Antimony pentachloride, Liquid [UN 1730]
Antimoonpentachloride (Dutch)
Antimony pentachloride, Solution [UN 1731]
Antimony perchloride
Butter of Antimony
Pentachloroantimony
Pentachlorure d'antimoine (French)
Perchlorure d'antimoine (French)
Tin chloride
UN 1730 [Antimony pentachloride, liquid]
UN 1731 [Antimony pentachloride, solutions]

7647-19-0
F_5P
125.97
Phosphorus pentafluoride [UN 2198]
UN 2198 [Phosphorus pentafluoride]

7650-84-2
$C_{15}H_{17}P$
228.27
Phosphine, diphenylpropyl- (8CI,9CI)

7651-02-7
$C_{23}H_{48}N_2O$
368.63
O=C(NCCCN(C)C)CCCCCCCC CCCCCCCCC
Stearamidopropyl dimethylamine
N-(3-Dimethylaminopropyl)-octadecamide
N-(3-(Dimethylamino)propyl)-octadecanamide
Dimethylaminopropyl stearamide
N-Dimethylaminopropylstearamide
N-(3-Dimethylamidopropyl)-stearamide
N,N-Dimethyl-N-(3-stearamidopropyl)amine
Incromine SB
Jordamide DAPSA
Octadecanamide, N-(3-(dimethylamino)propyl)- (9CI)
Octadecanoylamidopropyldimethylamine
Stearamidopropyldimethylamine
Stearic acid, 3-dimethylaminopropylamide
Stearic 3-dimethylaminopropylamide
Unizeen SA

7652-64-4
$C_{14}H_{16}N_2O_2$
244.32
O=C(N(C1C)C1)c(cccc2C(=O)N (C3C)C3)c2
Aziridine, 1,1'-isophthaloylbis(2-methyl
Aziridine, 1,1'-(1,3-phenylenedicarbonyl)bis(2-methyl- (9CI)
N,N'-Bispropyleneisophthalamide
HX 752

7659-86-1
$C_{10}H_{20}O_2S$
204.36
O=C(OCC(CCCC)CC)CS
Acetic acid, mercapto-, 2-ethylhexyl ester
2-Ethylhexyl mercaptoacetate
2-Ethylhexyl thioglycolate
Mercaptoacetic acid 2-ethylhexyl ester
Thioglycolic acid 2-ethylhexyl ester
Thioglykolsaeure-2-aethylhexyl ester (German)

7660-25-5
$C_6H_{12}O_6$
180.18
Fructopyranose, β-D
Fructose
Fruit Sugar
Frutabs
Laevoral
Laevosan
Levugen
Levulose

7661-47-4
$C_{11}H_{11}N$
157.22
Quinoline, 7-ethyl- (6CI, 8CI,9CI)
7-Ethylquinoline

7664-38-2
H_3O_4P
98.00
OP(O)(O)=O
Phosphoric acid (ACGIH, OSHA) [UN 1805]
Acide phosphorique (French)
Acido fosforico (Italian)
Fosforzuuroplossingen (Dutch)
Orthophosphoric acid
Phosphoric acid, Liquid (DOT)
Phosphoric acid, Solid (DOT)
Phosphorsaeureloesungen (German)
UN 1805 [Phosphoric acid]

7664-39-3
FH
20.01
[H+].[F-]
Hydrofluoric-acid
Acide fluorhydrique (French)
Acido fluoridrico (Italian)
Anhydrous hydrofluoric acid (DOT)
Fluoric acid (DOT)
Fluorowodor (Polish)
Fluorwasserstoff (German)
Fluorwaterstof (Dutch)
Hydrofluoric acid, Anhydrous (DOT)
Hydrofluoric acid, Solution (DOT)
Hydrofluoride
Hydrogen fluoride (ACGIH, OSHA) [UN 1052]
RCRA waste number U134
Rubigine
UN 1052 [Hydrogen fluoride, anhydrous]
UN 1790 [Hydrofluoric acid, solution, more than 60 per cent strength]

7664-41-7
H_3N
17.04
N
Ammonia (ACGIH,OSHA) (8CI,9CI)
Am-Fol
Ammonia, Anhydrous [UN 1005]
Ammonia Solution, Strong
Ammoniac (French)
Ammoniaca (Italian)
Ammonia Gas
Ammoniak (German)
Ammonia solution, Containing more than 44% ammonia [UN 1005]
Ammonia solution, Containing more than 50% ammonia (DOT)
Ammonia solution, More than 10% and not more than 35% ammonia [UN 2672]
Ammonia solution, More than 35% and not more than 50% ammonia [UN 2073]
Amoniaco (Spanish)
Amoniak (Polish)
Anhydrous ammonia (DOT)
Aromatic Ammonia, Vaporole
Caswell No. 041
EPA Pesticide Chemical Code 005302
Nitro-Sil
R 717
Spirit of Hartshorn
UN 1005 [Ammonia, anhydrous, liquefied or Ammonia solutions with more than 50 per cent ammonia]
UN 2073 [Ammonia solutions, with more than 35 per cent but not more than 50 per cent ammonia]
UN 2672 [Ammonia solutions, With more than 10 per cent but not more than 35 percent ammonia by mass]

7664-93-9
H_2O_4S
98.08
OS(O)(=O)=O
Sulfuric acid (ACGIH,OSHA) [UN 1830]
Acide sulfurique (French)
Acido solforico (Italian)
Bov
Dipping Acid
Hydrogen sulfate (DOT)
Matting Acid (DOT)
Nordhausen Acid (DOT)
Oil of Vitriol (DOT)
Spent sulfuric acid (DOT)
Sulfuric acid, Spent [UN 1832]
Sulphuric acid
Schwefelsaeureloesungen (German)
UN 1830 [Sulfuric acid]
UN 1832 [Sulfuric acid, spent]
Vitriol Brown Oil
Vitriol, Oil of (DOT)
Zwavelzuuroplossingen (Dutch)

7665-72-7
$C_7H_{14}O_2$
130.21
O(C1COC(C)(C)C)C1
Oxirane, ((1,1-dimethylethoxy)methyl)
t-BGE
t-Butyl glycidyl ether
1,1-Dimethylethyl glycidyl ether
Propane, 1-tert-butoxy-2,3-epoxy-

7667-55-2
C_9H_{18}
126.24
Cyclohexane, 1,2,3-trimethyl-, (1α,2α,3β)-(9CI)
Cyclohexane, 1,2,3-trimethyl-, cis-1,2,trans-1,3-(8CI)

7667-80-3
$C_{60}H_{122}$
843.63
Hexacontane (8CI,9CI)

7669-54-7
$C_6H_4ClNO_2S$
189.62
O=N(=O)c(c(SCl)ccc1)c1
2-Nitrophenylsulphenyl chloride
Benzenesulfenyl chloride, o-nitro- (8CI)
Benzenesulfenyl chloride, 2-nitro- (9CI)
2-Nitrobenzenesulfenyl chloride

7673-07-6
$C_{21}H_{21}N.ClH$
323.86
Benzenemethanamine, N,N-bis(phenylmethyl)-, hydrochloride (9CI)
Tribenzylamine, hydrochloride

7673-09-8
$C_3H_3Cl_3N_6$
229.47
n(c(nc(n1)NCl)NCl)c1NCl
1,3,5-Triazine, N,N',N''-trichloro-2,4,6-triamino
Chloromelamine
Decco Salt No. 5
TCM
Trichloromelamine
N,N',N''-Trichloro-2,4,6-triamine-1,3,5-triazine
2,4,6-Tris(chloroamine)triazine

7680-73-1
$C_6H_8N.Cl$
129.60
Pyridinium, 1-methyl-, chloride
Methyl pyridinium chloride
n-Methylpyridinium chloride
1-Methylpyridinium chloride
Pyridine methochloride

7681-11-0
IK
166.00
[K+].[I-]
Potassium-iodide
K1-N

Knollide
Potide

7681-38-1
$HO_4S.Na$
120.06
Sodium-bisulfate
GBS
Nitre Cake
Sodium acid sulfate
Sodium acid sulfate, Solid (DOT)
Sodium acid sulfate, Solution (DOT)
Sodium bisulfate, Fused
Sodium bisulfate, Solid (DOT)
Sodium bisulfate, Solution (DOT)
Sodium hydrogen sulfate, Solid [UN 1821]
Sodium hydrogen sulfate, Solution [UN 2837]
Sodium pyrosulfate
Sulfuric acid, monosodium salt
UN 1821 [Sodium hydrogen sulfate, solid]
UN 2837 [Sodium hydrogen sulfate, solution]

7681-49-4]
FNa
41.99
F[Na]
Sodium fluoride [UN 1690]
Alcoa Sodium Fluoride
Antibulit
Cavi-Trol
Chemifluor
Credo
Disodium difluoride
FDA 0101
F1-Tabs
Floridine
Florocid
Flozenges
Fluoral
Fluorident
Fluoride, sodium
Fluorid sodny (Czech)
Fluorigard
Fluorineed
Fluorinse
Fluoritab
Fluorocid
Fluor-O-Kote
Fluorol
Fluorure de sodium (French)
Flura
Flura Drops
Flura-Gel
Flura-Loz
Flurcare
Flursol
Fungol B
Gel II
Gelution
Gleem
Iradicav
Karidium
Karigel
Kari-Rinse
Lea-Cov
Lemoflur
Luride
Luride Lozi-Tabs
Luride-SF
Nafeen
Nafpak
Na Frinse
Natrium fluoride
NCI-C55221
Nufluor
Ossalin
Ossin
Pediaflor
Pedident
Pennwhite
Pergantene
Phos-Flur
Point Two
Predent
Rafluor
Rescue Squad
Roach Salt
Sodium fluoride cyclic dimer
Sodium fluoride, Solid (DOT)
Sodium fluoride, Solution (DOT)
Sodium fluorure (French)
Sodium hydrofluoride
Sodium monofluoride
So-Flo
Stay-Flo
Studafluor
Super-Dent
T-Fluoride
Thera-Flur
Thera-Flur-N
Trisodium trifluoride
UN 1690 [Sodium fluoride]
Villiaumite

7681-52-9
ClO.Na
74.44
Hypochlorous acid, sodium salt
Antiformin
B-K Liquid
Carrel-Dakin Solution
Chloros
Chlorox
Clorox
Dakins Solution
Hyclorite
Milton
NA 1791 (DOT)
Sodium hypochlorite (DOT)
Surchlor

7681-53-0
$H_2O_2P.Na$
87.98
Sodium-hypophosphite
Natriumhypophosphit (German)

7681-55-2
$IO_3.Na$
197.89
Iodic acid, sodium salt
Sodium iodate
Natriumjodat (German)

7681-57-4
$O_5S_2.2Na$
190.10
Pyrosulfurous acid, disodium salt
Disodium pyrosulfite
NA 2693 (DOT)
Sodium metabisulfite (ACGIH, DOT,OSHA)
Sodium metabisulphite
Sodium pyrosulfite

7681-65-4
CuI
190.45
Copper(I) iodide
Copper iodide (9CI)
Copper monoiodide
Cuprous iodide
Hydro-Giene
Natural Marshite

7681-82-5
INa
149.89
[Na+].[I-]
Sodium-iodide
Anayodin
Ioduril
Jodid sodny (Czech)
Natriumjodid (German)
Sodium iodine
Sodium monoiodide

7682-38-4
$C_3HCl_3N_2$
171.40
Clc1nc(Cl)c(Cl)[nH]1
2,4,5-Trichloroimidazole

7683-64-9
$C_{30}H_{50}$
410.73
CC(C)=CCCC(C)=CCCC(C)=CC CC=C(C)CCC=C(C)CCC=C (C)C
Squalene

7688-21-3
C_6H_{12}
84.16
2-Hexene-cis
2-Hexene, (Z)- (9CI)

7693-13-2
$C_{12}H_{10}Ca_3O_{14}$
498.44
Calcium citrate
Citric acid, calcium salt
1,2,3-Propanetricarboxylic acid, 2-hydroxy-, calcium salt

(9CI)

7693-52-9
$C_6H_4BrNO_3$
218.00
Phenol, 4-bromo-2-nitro- (9CI)

7696-12-0
$C_{19}H_{25}NO_4$
331.45
Cyclopropanecarboxylic acid, 2,2-dimethyl-3-(2-methyl-1-propenyl)-, (1,3,4,5,6,7-hexahydro- 1,3-dioxo-2H-isoindol-2-yl)methyl ester
Bioneopynamin
(1-Cyclohexane-1,2-dicarboximido)methyl chrysanthemumate
ENT 27,339
FMC-9260
Insectol
Multicide
Neo-Pynamin
Neopynamin forte
NIA-9260
Niagara NIA-9260
NSC-190939
Phthalthrin
d-Phthalthrin
(+-)-cis/trans-Phthalthrin
SP-1103
Sumitomo SP-1103
2,3,4,5-Tetrahydrophthalimidomethylchrysanthemate
N-(3,4,5,6-Tetrahydrophthalimido)-methyl dl-cis-trans-chrysanthemate
3,4,5,6-Tetrahydrophthalimidomethyl cis and trans dl chrysanthemummonocarboxylic acid
3,4,5,6-Tetrahydro-phthalimidomethylester der dl-cis-trans-chrysanthemumsaeure (German)
Tetramethrin
Tetramethrine (French)

7697-37-2
HNO_3
63.02
ON(=O)=O
Nitric-acid
Acide nitrique (French)
Acido nitrico (Italian)
Aqua Fortis
Azotic acid
Azotowy kwas (Polish)
Hydrogen nitrate
Kyselina dusicne (Czech)
Nitric acid (ACGIH,DOT, OSHA)
Nitric acid, Red fuming [UN 2032]
Nitric acid, Fuming (DOT)
Nitric acid, Over 40% (DOT)
Nitrous Fumes
Red Fuming Nitric Acid
RFNA
Salpetersaure (German)
Salpeterzuuroplossingen (Dutch)
UN 2031 (DOT)
UN 2032 [Nitric acid, red fuming]

7699-31-2
$C_4H_{12}N_2.2ClH$
161.10
Hydrazine, 1,2-diethyl-, dihydrochloride
1,2-Diethylhydrazine dihydrochloride

7699-43-6
Cl_2OZr
178.12
Zirconium, dichlorooxo
Basic Zirconium Chloride
Chlorozirconyl
Dichlorooxozirconium
NCI-C60811
Zirconium oxychloride
Zirconyl chloride

7699-45-8
Br_2Zn
225.19
Zinc-bromide
Zinc dibromide

7700-17-6
$C_{14}H_{19}O_6P$
314.30
COP(=O)(OC)OC(C)=CC(=O)OC(C)c1ccccc1
Crotonic acid, 3-hydroxy-, α-methylbenzyl ester, dimethyl phosphate, (E)
2-Butenoic acid, 3-((dimethoxyphosphinyl)oxy)-, 1-phenylethyl ester, (E)- (9CI)
Ciodrin
Ciovap
Crotoxyfos
Crotoxyphos
Cyodrin
Cypona E.C.
O,O-Dimethyl O-(1-methyl-2-carboxy-α-phenylethyl)-vinyl phosphate
Decrotox
Dimethyl-cis-1-methyl-2-(1-phenylethoxycarbonyl)-vinyl phosphate
Dimethyl phosphate of α-methylbenzyl 3-hydroxy-cis-crotonate
Duo-Kill
Duravos
ENT 24,717
1-Methylbenzyl-3-(dimethoxyphosphinyloxo)isocrotonate
α-Methyl benzyl-3-(dimethoxyphosphinyloxy)-cis-crotonate
α-Methylbenzyl 3-hydroxycrotonate dimethyl phosphate
OMS 239
Pantozol 1
cis-2-(1-Phenylethoxy)carbonyl-1-methylvinyl dimethylphosphate
1-Phenylethyl 3-(dimethoxyphosphinoyloxy)isocrotonate
SD 4294
Shell SD 4294
Volfazol

7704-34-9
S
32.06
Sulfur [UN 1350]
Bensulfoid
Brimstone
Collokit
Colloidal-S
Colloidal Sulfur
Colsul
Corosul D and S
Cosan
Cosan 80
Crystex
Elosal
Flour Sulphur
Flowers of Sulfur (DOT)
Flowers of Sulphur
Ground Vocle Sulphur
Hexasul
Kocide
Kolofog
Kolospray
Kumulus
Kumulus S
Magnetic 70, 90, and 95
Microflotox
Precipitated Sulfur
Sofril
Sperlox-S
Spersul
Spersul Thiovit
Sublimed Sulfur
Sulfidal
Sulforon
Sulfur Flower (DOT)
Sulfur, Solid (DOT)
Sulkol
Super Cosan
Sulphur (DOT)
Sulphur, Lump or powder (DOT)
Sulphur, Molten [UN 2448]
Sulsol
Technetium Tc 99M Sulfur Colloid
Tesuloid
Thiolux
Thiovit
UN 1350 [Sulfur]
UN 2448 [Sulfur, molten]

7704-99-6
H_2Zr
93.24
Zirconium hydride (8CI,9CI)
UN 1437

7705-07-9
Cl_3Ti
154.25
Titanium-chloride
TAC 121
TAC 131
Titanium(III) chloride
Titanium trichloride
Titanium trichloride, Pyrophoric [UN 2441]
Titanous chloride
Trichlorotitanium
UN 2441 [Titanium trichloride, pyrophoric or Titanium trichloride mixtures, pyrophoric]

7705-08-0
Cl_3Fe
162.20
Ferric chloride [UN 1773]
Chlorure perrique (French)
Ferric chloride, Anhydrous (DOT)
Ferric chloride, Solid (DOT)
Ferric chloride, Solid, Anhydrous (DOT)
Ferric chloride, Solution [UN 2582]
Flores martis
Iron chloride
Iron(III) chloride
Iron chloride, Solid (DOT)
Iron sesquichloride, Solid (DOT)
Iron trichloride
Perchlorure de fer (French)
UN 1773 [Ferric chloride]
UN 2582 [Ferric chloride, solution]

7705-14-8
$C_{10}H_{16}$
136.24
C(=CCC(C(=C)C)C1)(C1)C
Cyclohexene, 1-methyl-4-(1-methylethenyl)-, (+-)- (9CI)
p-Mentha-1,8-diene, (+-)- (8CI)
(+-)-1-Methyl-4-(1-methylethenyl)cyclohexene

7709-13-9
$C_3H_3ClN_4$
130.51
1,3,5-Triazin-2-amine, 4-chloro- (9CI)
s-Triazine, 2-amino-4-chloro- (8CI)

7718-54-9
Cl_2Ni
129.60
Nickel chloride (9CI)
Nickel(II) chloride (1:2)
Nickel(2+) chloride
Nickel dichloride
Nickelous chloride

7718-98-1
Cl_3V
157.29
Vanadium trichloride [UN 2475]
Vanadium(III) chloride
UN 2475 [Vanadium trichloride]

7719-09-7
Cl_2OS
118.96
Thionyl chloride (ACGIH,-OSHA) [UN 1836]
Sulfinyl chloride
Sulfur chloride oxide
Sulfurous dichloride
Sulfurous oxychloride
Thionyl dichloride
UN 1836 [Thionyl chloride]

7719-12-2
Cl_3P
137.32
Phosphorus-chloride
Chloride of phosphorus (DOT)
Fosforo(tricloruro di) (Italian)
Fosfortrichloride (Dutch)
Phosphore(trichlorure de) (French)
Phosphorous chloride
Phosphortrichlorid (German)
Phosphorus chloride (DOT)
Phosphorus trichloride (ACGIH, OSHA) [UN 1809]
Trojchlorek fosforu (Polish)
UN 1809 [Phosphorus trichloride]

7720-78-7
$O_4S.Fe$
151.91
Iron(II) sulfate (1:1)
Copperas
Duretter
Duroferon
Exsiccated ferrous sulfate
Exsiccated ferrous sulphate
Feosol
Feospan
Fer-In-Sol
Fero-Gradumet
Ferro-Gradumet
Ferralyn
Ferrosulfat (German)
Ferrosulfate
Ferro-Theron
Ferrous sulfate
Ferrous sulfate (1:1)
Ferrous sulphate
Fersolate
Green Vitriol
Iron monosulfate
Iron protosulfate
Iron sulfate (1:1)
Iron(II) sulfate
Iron(2+) sulfate
Iron(2+) sulfate (1:1)
Iron Vitriol
Irospan
Irosul
Slow-Fe
Sulferrous
Sulfuric acid iron salt (1:1)
Sulfuric acid, iron(2+) salt (1:1)

7721-01-9
Cl_5Ta
358.20
Tantalum-chloride
Tantalum pentachloride

7721-15-5
$C_{18}H_{32}O_2.xCu$
Unknown
(Z,Z)-9,12-Octadecadienoic acid, copper salt
Copper linoleate
9,12-Octadecadienoic acid (Z,Z)-, copper salt (9CI)

7722-64-7
$MnO_4.K$
158.04
Permanganic acid, potassium salt
Cairox
Chameleon mineral
C.I. 77755
Condy's Crystals
Kaliumpermanganaat (Dutch)
Kaliumpermanganat (German)
Permanganate de potassium (French)
Permanganate of potash (DOT)
Potassio (permanganato di) (Italian)
Potassium permanganate [UN 1490]
Potassium (permanganate de) (French)
UN 1490 [Potassium permanganate]

7722-71-6
$C_{18}H_{37}O_4P$
348.46
O=P(OCCCCCCCCC=CCCCCC CCC)(O)O
9-Octadecen-1-ol, dihydrogen phosphate, (Z)- (9CI)

7722-73-8
$C_{21}H_{27}N_3O_3$
369.51
O=C(N(C1CC)C1)c(cc(C(=O)N (C2CC)C2)cc3C(=O)N(C4CC) C4)c3
Benzene, 1,3,5-tris((2-ethylaziridinyl)-carbonyl)
HX-868
1,3,5-Tris(carbonyl-2-ethyl-1-azidine)benzene

7722-76-1
$H_3O_4P.H_3N$
115.03
Ammonium phosphate
AI3-26062
Ammonium acid phosphate
Ammonium biphosphate
Ammonium diacid phosphate
Ammonium dihydrogen orthophosphate
Ammonium dihydrogen phosphate
Ammonium dihydrophosphate
Ammonium monobasic phosphate
Ammonium orthophosphate dihydrogen
Ammonium phosphate, monobasic
Ammonium primary phosphate
Dihydrogen ammonium phosphate
Monoammonium acid phosphate
Monoammonium dihydrogen orthophosphate
Monoammonium dihydrogen phosphate
Monoammonium hydrogen phosphate
Monoammonium orthophosphate
Monoammonium phosphate
Monobasic ammonium phosphate
Phosphoric acid, monoammonium salt (9CI)
Primary ammonium phosphate

7722-84-1
H_2O_2
34.02
OO
Hydrogen peroxide (ACGIH, OSHA)
Albone
Albone 35
Albone 50
Albone 70
Albone 35CG
Albone 50CG
Albone 70CG
Dihydrogen dioxide
Hioxyl
Hydrogen dioxide
Hydrogen peroxide, Solution (8% to 40% Peroxide) (DOT)
Hydrogen peroxide, Solution (over 52% peroxide) (DOT)
Hydrogen peroxide, Stabilized (over 60% peroxide) (DOT)
Hydrogen peroxide, 30%
Hydrogen peroxide, 90%
Hydroperoxide
Inhibine
Interox
Kastone
Oxydol
Perhydrol
Perone
Perone 30
Perone 35
Perone 50
Perossido di idrogeno (Italian)
Peroxan
Peroxide
Peroxyde d'hydrogene (French)
Superoxol
T-Stuff
UN 2014 [Hydrogen peroxide, aqueous solutions with more than 40 per cent but not more than 60 per cent hydrogen peroxide (stabilized as necessary)]
UN 2015 [Hydrogen peroxide, stabilized or Hydrogen peroxide aqueous solutions, stabilized with more than 60 per cent hydrogen peroxide]
Wasserstoffperoxid (German)
Waterstofperoxyde (Dutch)

7722-88-5
$O_7P_2.4Na$
265.90
Pyrophosphoric acid, tetrasodium salt
Anhydrous tetrasodium pyrophosphate
Natrium pyrophosphat (German)
Phosphotex
Pyrophosphate
Sodium pyrophosphate
Tetranatriumpyrophosphat (German)
Tetrasodium diphosphate
Tetrasodium pyrophosphate (ACGIH,OSHA)
Tetrasodium pyrophosphate, Anhydrous
TSPP
Victor TSPP

7723-14-0
P
30.97
Phosphorus (8CI,9CI)
Bonide Blue Death Rat Killer
Caswell No. 663
Common sense cockroach and rat preparations
EPA Pesticide Chemical Code 066502
Exolit LPKN 275
Exolit VPK-n 361
Fosforo (Spanish)
Fosforo bianco (Italian)
Fosforo blanco (Spanish)
Gelber phosphor (German)
Phosphore (French)
Phosphore blanc (French)
Phosphorous (white)
Phosphorus
Phosphorus, red (DOT)
Phosphorus, white (DOT)
Phosphorus-31
Rat-Nip
Red phosphorus
Tetrafosfor (Dutch)
Tetraphosphor (German)
Weiss phosphor (German)
White phosphorus (DOT)
Yellow phosphorus (DOT)

7725-93-1
$C_6H_7NOS_2$
173.25
O=C(C(SC(=N1)S)=C1C)C
Ethanone, 1-(2,3-dihydro-4-methyl-2-thioxo-5-thiazolyl)- (9CI)

7726-95-6
Br_2
159.82
BrBr
Bromine (ACGIH,OSHA) [UN 1744]
Brom (German)
Brome (French)
Bromine, Solution (DOT)
Bromo (Italian)
Broom (Dutch)
UN 1744 [Bromine]

7727-15-3
$AlBr_3$
266.71
Aluminum bromide
Aluminum bromide, Anhydrous [UN 1725]
Aluminum bromide, Solution [UN 2580]
Aluminum tribromide
Tribromoaluminum
UN 1725 [Aluminum bromide, anhydrous]
UN 2580 [Aluminum bromide, solution]

7727-18-6
Cl_3OV
173.29
Vanadium, trichlorooxo
UN 2443 [Vanadium oxytrichloride]
Vanadium oxytrichloride [UN 2443]
Vanadyl trichloride

7727-21-1
$H_2O_8S_2.2K$
272.34
Peroxydisulfuric acid, dipotassium salt
Anthion
Dipotassium persulfate
Potassium peroxydisulfate
Potassium peroxydisulphate
Potassium persulfate (ACGIH) [UN 1492]
Potassium persulphate (DOT)
UN 1492 [Potassium persulfate]

7727-37-9

N_2
28.02
[N][N]
Nitrogen (DOT)
Nitrogen Gas
Nitrogen, Compressed [UN 1066]
Nitrogen, Refrigerated liquid [UN 1977]
UN 1066 [Nitrogen, compressed]
UN 1977 [Nitrogen, refrigerated liquid (cryogenic liquid)]

7727-43-7
$O_4S.Ba$
233.40
Barium-sulfate
Actybaryte
Artificial Barite
Artificial Heavy Spar
Bakontal
Baridol
Barite
Baritop
Barium sulfate (1:1)
Barium sulfate (ACGIH,OSHA)
Barium sulphate
Barosperse
Barotrast
Baryta White
Barytes
Barytes 22
Bayrites
Blanc Fixe
C.I. 77120
C.I. Pigment White 21
Citobaryum
Colonatrast
Enamel White
Esophotrast
Eweiss
E-Z-Paque
Finemeal
Lactobaryt
Liquibarine
Macropaque
Neobar
Oratrast
Permanent White
Precipitated Barium Sulphate
Raybar
Redi-Flow
Solbar
Sulfuric acid, barium salt (1:1)
Supramike
Travad
Unibaryt

7727-54-0
$O_8S_2.2H_4N$
228.22
Peroxydisulfuric acid, diammonium salt
Ammonium peroxydisulfate
Ammonium persulfate (ACGIH, dot)
Ammonium persulphate [UN 1444]
Diammonium peroxydisulfate
Diammonium peroxydisulphate
Diammonium persulfate
Persulfate d'ammonium (French)
UN 1444 [Ammonium persulfate]

7728-97-4
$C_6H_{14}NO_2Se$
211.14
Selenonium, (3-amino-3-carboxypropyl)dimethyl- (9CI)
Se-Methylselenomethionine

7728-98-5
$C_4H_9NO_2S$
135.20
Alanine, 3-(methylthio)-, L
Cysteine, S-methyl-
S-Methylcysteine
S-Methyl-L-cysteine
USAF CB-24

7731-28-4
$C_7H_{14}O$
114.19
Cyclohexanol, 4-methyl-, cis-(8CI,9CI)

7731-29-5
$C_7H_{14}O$
114.19
Cyclohexanol, 4-methyl-, trans-(8CI,9CI)

7732-18-5
H_2O
18.02
O
Water
Dihydrogen oxide

7732-92-5
$C_{18}H_{23}NO_3$
301.38
Methyldihydromorphine
DEA No. 9304
Dihydroheterocodeine
Morphinan-3-ol, 4,5-α-epoxy-6-α-methoxy-17-methyl-

7733-02-0
$O_4S.Zn$
161.43
Zinc sulfate (1:1)
Bonazen
Bufopto zinc sulfate
Op-Thal-Zin
Sulfate de zinc (French)
Sulfuric acid, zinc salt (1:1)
White Copperas
White vitriol
Verazinc
Zinc sulfate
Zinc sulphate
Zinc vitriol
Zinkosite

7738-94-5
CrH_2O_4
118.01
Chromic acid (OSHA)
Acide chromique (French)
Acido cromico (Spanish)
AI3-51760
Caswell No. 221
Chromic acid (H2CrO4) (9CI)
Chromic acid (OSHA)
Chromic acid, Solid
Chromic(VI) acid
Chromium oxide
EPA Pesticide Chemical Code 021101

7740-69-4
$C_7H_{12}O$
112.17
1-Butyne, 3-ethoxy-3-methyl- (9CI)
Ether, 1,1-dimethyl-2-propynyl ethyl (6CI,7CI, 8CI)

7745-89-3
$C_7H_8ClN.ClH$
178.05
3-Chloro-p-toluidine hydrochloride
Caswell No. 216A
CTH
DRC-1339
EPA Pesticide Chemical Code 009901
4-Methyl, 3-chloroaniline hydrochloride
Starlicide
p-Toluidine, 3-chloro-, hydrochloride

7747-35-5
$C_7H_{13}NO_2$
143.18
O(CC(N1CO2)(CC)C2)C1
1H,3H,5H-Oxazolo(3,4-c)oxazole, 7a-ethyldihydro- (9CI)
1-Aza-5-ethyl-3,7-dioxabicyclo-(3.3.0)octane

7747-84-4
C_3H_3Cl
74.51
1-Propyne, 1-chloro- (9CI)
Chloromethylacetylene
1-Chloro-2-methylacetylene
1-Chloropropyne
1-Chloro-1-propyne
Methylchloroacetylene
Propyne, 1-chloro- (6CI, 7CI,8CI)

7753-01-7
$C_{19}H_{13}N_3O$
299.33
s-Triazine, 2-(2-furyl)-4,6-diphenyl- (7CI,8CI)

7753-05-1
$C_{21}H_{14}N_4O_2$
354.37
s-Triazine, 2-(m-nitrophenyl)-4,6-diphenyl- (6CI,7CI,8CI)

7753-12-0
$C_{24}H_{21}N_3O_3$
399.45
1,3,5-Triazine, 2,4,6-tris-(4-methoxyphenyl)- (9CI)
s-Triazine, 2,4,6-tris(p-methoxyphenyl)- (6CI,7CI, 8CI)

7756-94-7
$C_{12}H_{24}$
168.36
Propene, 2-methyl-, trimer
Isobutene trimer
1-Propene, 2-methyl-, trimer (9CI)
Triisobutylene [UN 2324]
UN 2324 [Triisobutylene]

7756-96-9
$C_{11}H_{15}NO_2$
193.24
O=C(OCCCC)c(c(N)ccc1)c1
Anthranilic acid, butyl ester (8CI)
AI3-03442
Benzoic acid, 2-amino-, butyl ester (9CI)
Butyl 2-aminobenzoate

7757-79-1
KNO_3
140.21
Potassium nitrate [UN 1486]
Kaliumnitrat (German)
Niter
Nitre
Nitric acid, potassium salt
Saltpeter
UN 1486 [Potassium nitrate]
Vicknite

7757-81-5
$C_6H_7O_2.Na$
134.12
Sorbic acid, sodium salt
Sodium sorbate
Sorban sodny (Czech)

7757-82-6
$O_4S.2Na$
142.04
[Na+].[Na+].[O-]S([O-])(=O)=O
Sodium sulfate (2:1)
Disodium sulfate
Natriumsulfat (German)
Salt Cake
Sodium sulfate anhydrous
Sodium sulphate
Sulfuric acid, disodium salt
Thenardite
Trona

7757-83-7
$O_3S.2Na$
126.04
Sodium sulfite (2:1)
Anhydrous sodium sulfite
Disodium sulfite
Exsiccated sodium sulfite
Sulftech
Natriumsulfit (German)
Sodium sulfite
Sodium sulfite anhydrous
Sodium sulphite
Sulfurous acid, sodium salt (1:2)

7757-87-1
$H_3O_4P.3/2Mg$
134.46
Trimagnesium phosphate
Magnesium orthophosphate
Magnesium phosphate
Magnesium phosphate, neutral
Magnesium phosphate, tribasic
"Neutral" Magnesium Phosphate
Phosphoric acid, magnesium salt (2:3) (9CI)
Tertiary magnesium phosphate
Tribasic magnesium phosphate
Trimagnesium diorthophosphate
Trimagnesium diphosphate

7757-93-9
$Ca.H_3O_4P$
138.07
[Ca+2].OP([O-])([O-])=O
Calcium hydrogen phosphate
Bicalcium phosphate
Calcium acid phosphate
Calcium dibasic phosphate
Calcium hydrogen orthophosphate
Calcium hydrogenphosphate
Calcium monohydrogen phosphate
Calcium monohydrogen phosphate anhydrous
Calcium phosphate (1:1)
Calcium phosphate dibasic anhydrous
Calcium secondary phosphate
DCP 340
Dicalcium orthophosphate
Dicalcium phosphate
Dicalcium-O-phosphate
Dicalcium phosphate anhydrous
Monocalcium acid phosphate
Monocalcium orthophosphate
Phosphoric acid, calcium salt (1:1) (9CI)
Secondary calcium phosphate

7757-96-2
$C_{12}H_{18}O_3$
210.27
1-Octenyl succinic anhydride
2,5-Furandione, dihydro-3-(1-octenyl)- (9CI)
1-Octenylsuccinic anhydride
Succinic anhydride, (1-octenyl)- (8CI)

7758-01-2
$BrO_3.K$
167.01
[K+].[O-]Br(=O)=O
Bromic acid, potassium salt
EEC No. E924
Potassium bromate [UN 1484]
UN 1484 [Potassium bromate]

7758-02-3
BrK
119.01
[K+].[Br-]
Potassium-bromide
Bromide salt of potassium
Tripotassium tribromide

7758-04-5
$C_6H_{13}NO_3S.K$
218.33
Potassium cyclamate
Cyclohexanesulfamic acid, monopotassium salt
Potassium cyclohexyl sulfamate
Potassium N-cyclohexylsulfamate
Sulfamic acid, cyclohexyl-, monopotassium salt

7758-05-6
$IO_3.K$
214.00
Iodic acid, potassium salt
Potassium iodate

7758-09-0
$NO_2.K$
85.11
[K+].O=[N-]=O
Potassium nitrite [UN 1488]
Nitrous acid, potassium salt
Potassium nitrite (1:1)
UN 1488 [Potassium nitrite]

7758-11-4
$H_3O_4P.2K$
176.20
Dipotassium phosphate
Dibasic potassium phosphate
Dikalium phosphate
Dipotassium hydrogen phos-

phate
Dipotassium monohydrogen phosphate
Dipotassium monophosphate
Dipotassium orthophosphate
Dipotassium phosphate, dibasic
Dipotassium-O-phosphate
DKP
Hydrogen dipotassium phosphate
Phosphoric acid, dipotassium salt (9CI)
Potassium biphosphate
Potassium dibasic phosphate
Potassium hydrogen phosphate
Potassium monohydrogen phosphate
Potassium monophosphate
Potassium orthophosphate, mono-H
Potassium phosphate, dibasic
Potassium phosphate NF XII

7758-16-9
$H_2O_7P_2.2Na$
221.94
Pyrophosphoric acid, disodium salt
Dinatriumpyrophosphat (German)
Diphosphoric acid, disodium salt
Disodium dihydrogen pyrophosphate
Disodium diphosphate
Disodium pyrophosphate
Sodium acid pyrophosphate
Sodium pyrophosphate

7758-19-2
$ClNaO_2$
90.44
Sodium chlorite [UN 1496]
Caswell No. 755
Chlorite de sodium (French)
Clorito sodico (Spanish)
Chlorous acid, sodium salt (8CI,9CI)
EPA Pesticide Chemical Code 020502
Sodium chlorite, Solution containing more than 5% available chlorine [UN 1908]
Sodium chlorite, Solution not exceeding 42% sodium chlorite (DOT)
Textile
Textone
UN 1496 [Sodium chlorite]
UN 1908 [Sodium chlorite solution with more than 5 per cent available chlorine]

7758-23-8
$Ca.2H_3O_4P.H_2O$
223.13
Phosphoric acid, calcium salt, Hydrate (2:1:1)
Calcium phosphate, monobasic

7758-29-4
$O_{10}P_3.5Na$
367.86
Triphosphoric acid, pentasodium salt
Armofos
Empiphos STP-D
Natriumtripolyphosphat (German)
Pentasodium triphosphate
Pentasodium tripolyphosphate
Polygon
Poly
Sodium triphosphate
Sodium tripolyphosphate
S 400
Sodium phosphate ($Na_5P_3O_{10}$)
Sodium triphosphate ($Na_5P_3O_{10}$)
Sodium tripolyphosphate
STPP
Tripoly
Thermphos
Thermphos L 50
Thermphos N
Thermphos SPR

7758-87-4
$Ca.2/_3H_3O_4P$
89.06
Calcium phosphate
Bonarka
Calcigenol Simple
Calcium orthophosphate
Calcium orthophosphate, tri-(tert)
Calcium phosphate (3:2)
Calcium phosphate, tribasic
Calcium tertiary phosphate
Caswell No. 148
EPA Pesticide Chemical Code 076401
Natural Whitlockite
Phosphoric acid, calcium salt (2:3) (9CI)
Phosphoric acid calcium(2+) salt (2:3)
Synthos
Tertiary calcium phosphate
Tricalcium diphosphate
Tricalcium orthophosphate
Tricalcium phosphate

7758-88-5
CeF_3
197.12
Cerium-fluoride
Cerium fluorure (French)
Cerium trifluoride
Cerous fluoride

7758-89-6
ClCu
98.99
Copper(I) chloride
Chlorid medny (Czech)
Copper chloride
Copper(1+) chloride
Copper monochloride
Cuprous chloride
Cuprous dichloride
Dicopper dichloride

7758-94-3
Cl_2Fe
126.75
Iron(II) chloride (1:2)
Ferrous chloride
Ferrous chloride, Solid [NA 1759]
Ferrous chloride, Solution (DOT)
Iron dichloride
Iron protochloride
NA 1759 [Ferrous chloride, solid]
NA 1760

7758-95-4
Cl_2Pb
278.09
Lead chloride (DOT)
Lead(2+) chloride
Lead (II) chloride
Lead dichloride
NA 2291 (DOT)
Plumbous chloride

7758-97-6
$CrO_4.Pb$
323.19
Chromic acid, lead(2+) salt (1:1)
Canary Chrome Yellow 40-2250
Chromate de plomb (French)
Chrome Green
Chrome Green UC61
Chrome Green UC74
Chrome Green UC76
Chrome Lemon
Chrome Yellow
Chrome Yellow G
Chrome Yellow 5G
Chrome Yellow GF
Chrome Yellow LF
Chrome Yellow Light 1066
Chrome Yellow Light 1075
Chrome Yellow Medium 1074
Chrome Yellow Medium 1085
Chrome Yellow Medium 1295
Chrome Yellow Medium 1298
Chrome Yellow Primrose 1010
Chrome Yellow Primrose 1015
Chromium Yellow
C.I. 77600
C.I. Pigment Yellow 34
Cologne Yellow
C.P. Chrome Yellow Light
C.P. Chrome Yellow Medium
C.P. Chrome Yellow Primose
Crocoite
Dainichi Chrome Yellow G
Giallo Cromo (Italian)
King's Yellow
Lead Chromate (ACGIH)

Lead Chromate(VI)
Leipzig Yellow
Lemon Yellow
Paris Yellow
Pigment Green 15
Plumbous Chromate
Pure Lemon Chrome L3GS

7758-98-7
$O_4S.Cu$
159.60
Copper (II) sulfate (1:1)
Basic Copper Sulfate
BCS Copper Fungicide
Blue Copper
Blue Stone
Blue Vitriol
Copper monosulfate
Copper sulfate
Copper sulfate (1:1)
Copper(II) sulfate
Copper(2+) sulfate
Copper(2+) sulfate (1:1)
Copper sulfate basic
CP Basic Sulfate
Cupric sulfate
Cupric sulfate anhydrous
Cupric sulphate
Griffin super Cu
Incracide 10A
Incracide E 51
Kilcop 53
Kobasic
Kupfersulfat (German)
Phyto-Bordeaux
Roman Vitriol
Sulfate de cuivre (French)
Sulfuric acid, copper(2+) salt (1:1)
TNCS 53
Trinagle

7758-99-8
$O_4S.Cu.5H_2O$
249.70
Copper(II) sulfate, pentahydrate (1:1:5)
Blue Copper AS
Bluestone
Blue Vitriol
Copperfine-Zinc
Copper sulfate
Copper(II) sulfate pentahydrate
Copper(2+) sulfate pentahydrate
Copper sulphate
CSP
Cupric sulfate pentahydrate
Kupfersulfat (German)
Kupfersulfat-pentahydrat (German)
Kupfervitriol (German)
Roman Vitriol
Salzburg Vitriol
Sulfuric acid, copper(2+) salt, pentahydrate
Sulfuric acid copper(2+) salt (1:1), pentahydrate (8CI,9CI)
Triangle

7759-01-5
$H_2O_4W.Pb$
457.06
Lead tungstate
Lead tungstate (IV)
Lead tungsten oxide (9CI)
Lead tungsten tetraoxide
Lead Wolframate
Natural Raspite
Natural Stolzite
Scheelite
Tungstic acid, lead(2+) salt (1:1)

7759-02-6
$H_2O_4S.Sr$
185.70
Sulfuric acid, strontium salt (1:1) (9CI)

7760-50-1
$C_4H_2Mg_5O_{14}$
395.58
Magnesium carbonate hydroxide
Basic Magnesium Carbonate
Magnesite
Magnesium Basic Carbonate
Magnesium, tetrakis(carbonato(2-))dihydroxypenta- (9CI)
Tetrakis(carbonato(2-))dihydroxypentamagnesium

7761-88-8
$NO_3.Ag$
169.88
Silver(I) nitrate (1:1)
Lunar Caustic
Nitrate d'argent (French)
Nitric acid, silver(1+) salt
Silver(1+) nitrate
Silbernitrat
Silver nitrate [UN 1493]
UN 1493 [Silver nitrate]

7763-77-1
C_2H_3ClO
78.50
ClC1CO1
Ethane, chloroepoxy
Chloroethylene oxide
Chlorooxirane
Monochloroethylene oxide
Oxirane, chloro- (9CI)

7764-50-3
$C_{10}H_{16}O$
152.26
O=C(C(CCC1C(=C)C)C)C1
p-Menth-8-en-2-one
Dihydrocarvone
2-Methyl-5-(1-methylethenyl)-cyclohexanone

7772-98-7
$O_3S_2.2Na$
158.10
Thiosulfuric acid, disodium salt
Chlorine control
Chlorine cure
Declor-It
Disodium thiosulfate
S-Hydril
Hypo
Sodium hyposulfite
Sodium oxide sulfide
Sodium thiosulfate
Sodium thiosulfate anhydrous
Sodium thiosulphate
Sodothiol

7772-99-8
Cl_2Sn
189.59
Cl[Sn]Cl
Tin(II) chloride (1:2)
C.I. 77864
NA 1759 (DOT)
NCI-C02722
Stannochlor
Stannous chloride
Stannous chloride, Solid (DOT)
Tin dichloride
Tin protochloride

7773-01-5
Cl_2Mn
125.84
Manganese(II) chloride (1:2)
Manganese chloride
Manganese dichloride
Manganous chloride

7773-03-7
$H_2O_3S.K$
121.18
Potassium bisulfite
Potassium acid sulfite
Potassium hydrogen sulfite
Potassium sulfite, hydrogen
Sulfurous acid, monopotassium salt (9CI)

7773-06-0
$H_2NO_3S.H_4N$
114.14
Sulfamic acid, monoammonium salt
Amcide
Amicide
Ammat
Ammate
Ammate X
Ammonium amidosulfonate
Ammonium amidosulphate
Ammonium sulfamate
Ammonium sulphamate
Ammonium sulphamidate
Ammoniumsalz der amidosulfonsaure (German)
AMS
Ikurin
Monoammonium sulfamate

Sulfamate
Sulfamate dammonium
Sulfaminsaure (German)

7774-29-0
HgI_2
454.39
[I-].[I-].[Hg+2]
Mercury(II) iodide [UN 1638]
Hydrargyrum bijodatum (German)
Mercuric iodide
Mercuric iodide, Solid (DOT)
Mercuric iodide, Solution (DOT)
Mercuric iodide, Red
Mercury biniodide
Red Mercuric Iodide
UN 1638 [Mercury iodide]

7775-09-9
$ClO_3.Na$
106.44
Chloric acid, sodium salt
Asex
Atlacide
Atratol
B-Herbatox
Chlorate of soda (DOT)
Chlorate salt of sodium
Chlorax
Chlorsaure (German)
De-Fol-Ate
Desolet
Drexel Defol
Drop Leaf
Evau-Super
Fall
Grain Sorghum Harvest-Aid
Granex O
Harvest-Aid
Klorex
Kusa-Tohru
Kusatol
Natriumchloraat (Dutch)
Natriumchlorat (German)
Ortho C-1 Defoliant & Weed Killer
Oxycil
Rasikal
Shed-A-Leaf
Shed-A-Leaf "L"
Soda chlorate (DOT)
Sodio (clorato di) (Italian)
Sodium chlorate [UN 1495]
Sodium(chlorate de) (French)
Sodium chlorate, Aqueous solution [UN 2428]
Travex
Tumbleaf
UN 1495 [Sodium chlorate]
UN 2428 [Sodium chlorate, solution]
United Chemical Defoliant No. 1
Val-Drop

7775-11-3
$CrO_4.2Na$
161.98
Chromic acid, disodium salt
Chromate of soda
Chromium disodium oxide
Chromium sodium oxide
Disodium chromate
Neutral Sodium Chromate
Sodium chromate
Sodium chromate(VI)

7775-14-6
$O_4S_2.2Na$
174.10
Dithionous acid, disodium salt
D-OX
Hydrolin
K-Brite
Reductone
Sodium dithionite [UN 1384]
Sodium hydrosulfite [UN 1384]
Sodium hydrosulphite
Sodium sulfoxylate
UN 1384 [Sodium dithionite or sodium hydrosulfite]
Vatrolite
V-Brite
Virchem
Virtex CC
Virtex D
Virtex L
Virtex RD

7775-19-1
$BHO_2.Na$
66.81
Boric acid, monosodium salt
Sodium metaborate

7775-27-1
$O_8S_2.2Na$
238.10
Peroxydisulfuric acid, disodium salt
Persulfate de sodium (French)
Sodium peroxydisulfate
Sodium persulfate (ACGIH) [UN 1505]
Sodium persulphate (DOT)
UN 1505 [Sodium persulfate]

7775-41-9
AgF
126.87
Silver fluoride (8CI,9CI)
Caswell No. 736
EPA Pesticide Chemical Code 072502

7775-50-0
$C_{60}H_{116}O_7$
949.58
Tristearyl citrate
Citric acid, trioctadecyl ester
2-Hydroxy-1,2,3-propanetricarboxylic acid, trioctadecyl ester
Lubrol TSC 5110
1,2,3-Propanetricarboxylic acid, 2-hydroxy-, trioctadecyl ester (9CI)
Trioctadecyl citrate

7776-05-8
$C_{21}H_{26}N_2OS_2$
386.59
10H-Phenothiazine, 10-[2-(1-methyl-2-piperidinyl)ethyl]-2-(methylthio)-, 5-oxide (9CI)
Phenothiazine, 10-[2-(1-methyl-2-piperidyl)ethyl]-2-(methylthio)-, 5-oxide (7CI, 8CI)
Thioridazine sulfoxide
Thioridazine 5-sulfoxide
Thioridazine 5-oxide

7776-28-5
$C_6H_{18}O_{24}P_6.6Ca$
900.51
Calcium phytate
Hexacalcium phytate
Inositol, hexakis(dihydrogen phosphate) calcium salt(1:6), myo-
Inositol hexaphosphate calcium salt
myo-Inositol, hexakis(dihydrogen phosphate), calcium salt (1:6)

7778-18-9
$O_4S.Ca$
136.14
Sulfuric acid, calcium salt (1:1)
Calcium sulfate (ACGIH, OSHA)
Crysalba
Drierite
Gibs
Thiolite

7778-39-4
AsH_3O_4
141.94
Arsenic acid (9CI)
Acide arsenique liquide (French)
Arsenate
Arsenic acid, Liquid [UN 1553]
Arsenic acid, Solid [UN 1554]
Arsenic acid, Solution (DOT)
Caswell No. 056
Desiccant L-10
Dessicant L-10
EPA Pesticide Chemical Code 006801
Hi-Yield Desiccant H-10
Orthoarsenic acid
RCRA waste No. P010
Scorch
UN 1553 [Arsenic acid, Liquid]
UN 1554 [Arsenic acid, Solid]
Zotox
Zotox Crab Grass Killer

7778-43-0
$AsHO_4.2Na$
185.91
Arsenic acid, disodium salt
Disodium arsenate
Disodium arsenic acid
Disodium hydrogen arsenate
Disodium hydrogen orthoarsenate
Disodium monohydrogen arsenate
Sodium acid arsenate
Sodium arsenate
Sodium arsenate, dibasic
Sodium arsenate dibasic, Anhydrous

7778-44-1
$As_2O_8.3Ca$
398.08
Arsenic acid, calcium salt(2:3)
Arseniate de calcium (French)
Calciumarsenat
Calcium arsenate [UN 1573]
Calcium arsenate, Solid (DOT)
Calcium orthoarsenate
Chip-Cal
Chip-Cal Granular
Cucumber Dust
Fencal
Flac
Kalo
Kalziumarseniat (German)
Kilmag
Pencal
Security
Spracal
Tricalciumarsenat (German)
Tricalcium arsenate
Turf-Cal
UN 1573 [Calcium arsenate]

7778-50-9
$Cr_2O_7.2K$
294.20
Dichromic acid, dipotassium salt
Bichromate of Potash
Chromic acid, dipotassium salt
Dipotassium dichromate
Iopezite
Kaliumdichromat (German)
NA 1479 (DOT)
Potassium bichromate
Potassium dichromate (DOT)
Potassium dichromate (VI)

7778-53-2
$H_3O_4P.3K$
215.30
Potassium phosphate, tribasic
Caswell No. 700A
EPA Pesticide Chemical Code 076407
Phosphoric acid, tripotassium salt (9CI)
Potassium orthophosphate
Potassium phosphate
Tripotassium phosphate

7778-54-3
$Cl_2O_2.Ca$
142.98
Hypochlorous acid, calcium salt
B-K Powder
Bleaching Powder
Bleaching Powder, Containing 39% or less chlorine (DOT)
Calcium chlorohydrochlorite
Calcium hypochloride
Calcium hypochlorite
Calcium hypochlorite, Dry [UN 1748]
Calcium hypochlorite mixture, Dry (containing more than 39% available chlorine) (DOT)
Calcium hypochlorite mixtures, Dry, with 10% to 39% available chlorine [UN 2208]
Calcium oxychloride
Caporit
CCH
Chloride of Lime (DOT)
Chlorinated Lime (DOT)
HTH
Hy-Chlor
Hypochlorous acid, calcium salt, (Dry mixture)
Hypochlorous acid, calcium salt, Dry mixture with 10% to 39% available chlorine
Lime chloride
Lo-Bax
Losantin
Perchloron
Pittchlor
Pittcide
Pittclor
Sentry
UN 1748 [Calcium hypochlorite, Dry or calcium hypochlorite mixtures dry with more than 39 per cent available chlorine (8.8 per cent available oxygen)]
UN 2208 [Calcium hypochlorite mixtures, Dry, with more than 10 per cent but not more than 39 per cent available chlorine]

7778-70-3
$C_7H_5NS_2.K$
206.35
2(3H)-Benzothiazolethione, potassium salt (9CI)
2-Benzothiazolethiol, potassium salt (8CI)
Caswell No. 541B
EPA Pesticide Chemical Code 051707
2-Mercaptobenzothiazole potassium salt

7778-73-6
$C_6HCl_5O.K$
305.44
Potassium pentachlorophenate
Caswell No. 641A
EPA Pesticide Chemical Code 063002
Pentachlorophenol potassium salt
Phenol, pentachloro-, potassium salt (8CI,9CI)

7778-74-7
$ClO_4.K$
138.55
Perchloric acid, potassium salt (1:1)
Astrumal
Irenal
Irenat
Periodin
Potassium hyperchloride
Potassium perchlorate [UN 1489]
UN 1489 [Potassium perchlorate, solid or solution]

7778-77-0
$H_3O_4P.K$
137.10
Monopotassium phosphate
KDP
MKP
Monobasic potassium phosphate
Monopotassium dihydrogen phosphate
Monopotassium orthophosphate
Phosphoric acid, monopotassium salt (9CI)
Potassium acid phosphate
Potassium biphosphate
Potassium dihydrogen orthophosphate
Potassium dihydrogen phosphate
Potassium diphosphate
Potassium hydrogen phosphate
Potassium orthophosphate, dihydrogen
Potassium phosphate
Potassium phosphate, monobasic
Sorensen's Potassium Phosphate

7778-80-5
$O_4S.2K$
174.26
Potassium sulfate (2:1)
Sulfuric acid, dipotassium salt

7778-83-8
$C_{12}H_{14}O_2$
190.26
O=C(OCCC)C=Cc(cccc1)c1
Cinnamic acid, propyl ester
2-Propenoic acid, 3-phenyl-, propyl ester (9CI)

n-Propyl cinnamate
Propylester kyseliny skoricove (Czech)

7778-85-0
$C_5H_{12}O_2$
104.15
Propane, 1,2-dimethoxy- (8CI, 9CI)

7779-16-0
$C_{13}H_{17}NO_2$
219.28
Benzoic acid, 2-amino-, cyclohexyl ester (9CI)
Anthranilic acid, cyclohexyl ester (8CI)

7779-27-3
$C_9H_{21}N_3$
171.33
N(CN(CN1CC)CC)(C1)CC
s-Triazine, hexahydro-1,3,5-triethyl
Hexahydro-1,3,5-triethyl-s-triazine
Vancide TH

7779-60-4
$C_{11}H_{22}O_3$
202.30
Undecanoic acid, 4-hydroxy- (6CI,8CI,9CI)
4-Hydroxyundecanoic acid

7779-77-3
$C_{11}H_{15}NO_2$
193.24
O=C(OCC(C)C)c(c(N)ccc1)c1
Benzoic acid, 2-amino-, 2-methylpropyl ester
AI3-03443
2-Methylpropyl 2-aminobenzoate

7779-86-4
$O_4S_2.Zn$
193.49
Dithionous acid, zinc salt (1:1)
UN 1931 [Zinc dithionite or Zinc hydrosulfite]
Zinc dithionite
Zinc hydrosulfite [UN 1931]

7779-88-6
$N_2O_6.Zn$
189.39
Zinc nitrate [UN 1514]
Nitrate de zinc (French)
Nitric acid, zinc salt
UN 1514 [Zinc nitrate]

7779-90-0
$O_8P_2.3Zn$
386.05
Phosphoric acid, zinc salt (2:3)
Neutral Zinc Phosphate
Trizinc diphosphate
Zinc orthophosphate
Zinc phosphate

7779-95-5
$C_{12}H_{24}O_3$
216.32
Dodecanoic acid, 5-hydroxy- (8CI,9CI)

7781-98-8
$C_9H_{10}O_3$
166.18
Benzoic acid, 3-hydroxy-, ethyl ester (9CI)
Benzoic acid, m-hydroxy-, ethyl ester (8CI)
NSC-32427

7782-39-0
D_2
4.02
Deuterium [UN 1957]
UN 1957 [Deuterium]

7782-40-3
C
12.01
Diamond (9CI)

7782-41-4
F_2
38.00
FF
Fluorine (ACGIH,DOT, OSHA)
Bifluoriden (Dutch)
Fluor (Dutch, French, German, Polish)
Fluorine, Compressed [UN 1045]
Fluoro (Italian)
Fluorures acide (French)
Fluoruri acidi (Italian)
RCRA waste number P056
Saeure fluoride (German)
UN 1045 [Fluorine, compressed]

7782-42-5
C
12.01
Graphite
Aerodag G
Ag 1500
Aquadag
AS 1
AT 20
ATJ-S
ATJ-S Graphite
Black Lead
Canlub
CB 50
Ceylon Black Lead
C.I. 77265
C.I. Pigment Black 10
CPB 5000
DC 2
EG 0
Electrographite
Exp-F
Fortafil 5Y
GK 2
GK 3
GP 60
GP 63
GP 60S
Grafoil
Grafoil GTA
Graphite, Natural (ACGIH, OSHA)
Graphnol N 3M
GS 2
GY 70
H 451
Hitco HMG 50
IG 11
Korobon
MG 1
Mineral Carbon
MPG 6
Papyex
PG 50
Plumbago
Plumbago (Graphite)
Pyro-Carb 406
Rocol X 7119
S 1
S 1 (Graphite)
Schungite
Shungite
Silver Graphite
SKLN 1
Stove Black
Swedish Black Lead
UCAR 38
VVP 66-95

7782-44-7
O_2
32.00
O=O
Oxygen (DOT)
Oxygen, Compressed [UN 1072]
Oxygen, Refrigerated liquid [UN 1073]
UN 1072 [Oxygen, compressed]
UN 1073 [Oxygen, refrigerated liquid (cryogenic liquid)]

7782-49-2
Se
78.96
[Se]
Selenium (ACGIH)
C.I. 77805
Colloidal Selenium
Elemental Selenium
Selen (Polish)
Selenate

Selenium alloy
Selenium Base
Selenium (Colloidal)
Selenium Dust
Selenium Elemental
Selenium Homopolymer
Selenium metal powder [UN 2658]
Selenium metal powder, Non-pyrophoric (DOT)
UN 2658 [Selenium powder]
Vandex

7782-50-5
Cl_2
70.90
ClCl
Chlorine (ACGIH,OSHA) [UN 1017]
Bertholite
Chloor (Dutch)
Chlor (German)
Chlore (French)
Chlorine Mol.
Cloro (Italian)
Molecular Chlorine
UN 1017 [Chlorine]

7782-63-0
$O_4S.Fe.7H_2O$
278.05
Iron(II) sulfate (1:1), heptahydrate
Copperas
Feosol
Fer-In-Sol
Fero-Gradumet
Ferrous sulfate heptahydrate
Fesofor
Fesotyme
Green vitrol
Haemofort
Ironate
Iron vitrol
Irosul
Mol-Iron
Presfersul
Sulferrous

7782-64-1
F_2Mn
92.94
Manganese(II) fluoride
Manganese fluorure (French)

7782-65-2
GeH_4
76.63
Germane [UN 2192]
Germanium hydride
Germanium tetrahydride (ACGIH,OSHA)
Monogermane
UN 2192 [Germane]

7782-77-6
HNO_2
47.02
Nitrous-acid
Kyselina dusite (Czech)
Nitrosyl hydroxide

7782-79-8
HN_3
43.04
[H+].[N-]=[N+]=[N-]
Hydrazoic-acid
Azoimide
Diazoimide
Hydrazoic acid (ACGIH)
Hydrogen azide
Hydronitric acid
Stickstoffwasserstoffsaeure (German)
Triazoic acid

7782-82-3
$HO_3Se.Na$
150.96
Selenious acid, monosodium salt
Hydrogen sodium selenite
Hydrogen sodium selenium oxide
Sodium hydrogen trioxoselenite
Sodium hydroselenite
Sodium selenite (7CI)

7782-86-7
H_3HgNO_4
281.62
Mercurous nitrate

7782-87-8
$H_3O_2P.K$
105.09
Potassium hypophosphite
Phosphinic acid, potassium salt (9CI)
Potassium hypophosphite, monobasic
Potassium phosphinate

7782-89-0
H_2LiN
22.97
Lithium amide (DOT)
Lithamide
Lithium amide, Powdered (DOT)
UN 1412 (DOT)

7782-92-5
H_2NNa
39.02
Sodium amide (DOT)
Sodamide
UN 1425 (DOT)

7782-94-7
$H_2N_2O_2$
62.02
Nitramide (8CI,9CI)
Nitramine

7782-95-8
$H_4O_6P_2.2Na$
207.96
Disodium dihydrogen hypophosphate
Disodium dihydrogen subphosphate
Disodium hypophosphate
Disodium hypophosphorate
Hypophosphoric acid, disodium salt (9CI)
Sodium acid hypophosphate
Sodium dihydrogen hypophosphate

7782-99-2
H_2O_3S
82.08
Sulfurous acid [UN 1833]
Schweflige saure (German)
Sulfur dioxide Solution
Sulphurous acid (DOT)
UN 1833 [Sulfurous acid]

7783-00-8
H_2O_3Se
128.98
Selenious-acid
RCRA waste number U204
Selenium dioxide

7783-06-4
H_2S
34.08
S
Hydrogen sulfide (ACGIH, DOT,OSHA)
Acide sulfhydrique (French)
Hydrogen sulfide, Liquefied [UN 1053]
Hydrogen sulfuric acid
Hydrogen sulphide (DOT)
Hydrogene sulfure (French)
Idrogeno solforato (Italian)
RCRA waste number U135
Schwefelwasserstoff (German)
Siarkowodor (Polish)
Stink Damp
Sulfureted hydrogen
Sulfur hydride
Zwavelwaterstof (Dutch)

7783-07-5
H_2Se
80.98
Hydrogen-selenide
Electronic E-2
Hydrogen selenide (ACGIH, DOT,OSHA)
Hydrogen selenide, Anhydrous [UN 2202]
Selenium hydride
UN 2202 [Hydrogen selenide, anhydrous]

7783-08-6
H_2O_4Se
144.98
Selenic acid [UN 1905]
Selenic acid, Liquid (DOT)
UN 1905 [Selenic acid]

7783-18-8
$O_3S_2.2H_4N$
148.22
Thiosulfuric acid, diammonium salt
Ammonium hyposulfite
Ammonium thiosulfate
Diammonium thiosulfate

7783-20-2
$O_4S.2H_4N$
132.16
Ammonium sulfate (2:1)
Ammonium sulphate
Diammonium sulfate
Diammonium sulphate
Dolamin
Mascagnite
Sulfatom ammoniya (Russian)
Sulfuric acid, diammonium salt

7783-24-6
$H_3N.H_2O_4S.3H_2O.1/2Zn$
201.84
Zinc ammonium sulfate
Ammonium zinc sulfate
Ammonium zinc sulfate hexahydrate
Diammonium zinc disulfate hexahydrate
Sulfuric acid, ammonium zinc salt (2:2:1), hexahydrate
Zinc diammonium disulfate hexahydrate

7783-28-0
$(NH_4)_2HPO_4$
132.04
Ammonium phosphate
AI3-25349
Ammonium hydrogen phosphate
Ammonium hydrogen phosphate solution
Ammonium monohydrogen orthophosphate
Ammonium phosphate, dibasic
Caswell No. 286C
DAP
Diammonium acid phosphate
Diammonium hydrogen orthophosphate
Diammonium hydrogen phosphate
Diammonium monohydrogen phosphate
Diammonium orthophosphate
Diammonium phosphate
Dibasic ammonium phosphate
Fyrex
Pelor
Phosphoric acid, diammonium salt
Secondary ammonium phosphate

7783-33-7
$HgI_4.2K$
786.39
Mercurate (2-), tetraiodo-, dipotassium
Channing's Solution
Mercuric potassium iodide
Mercury(II) potassium iodide
Mercuric potassium iodide, Solid (DOT)
Mercury potassium iodide [UN 1643]
Nessler Reagent
Potassium mercuric iodide
Potassium tetraiodomercurate (II)
Solution potassium iodohydragyrate
UN 1643 [Mercury potassium iodide]

7783-35-9
$O_4S.Hg$
296.65
Mercury(II) sulfate (1:1)
Mercuric sulfate
Mercuric sulphate (DOT)
Mercuric sulfate, Solid (DOT)
Mercury bisulfate
Mercury bisulphate (DOT)
Mercury persulfate
Mercury sulfates [UN 1645]
Sulfate mercurique (French)
Sulfuric acid, mercury(2+) salt (1:1)
UN 1633 (DOT)
UN 1645 [Mercury sulfates]

7783-36-0
$O_4S.2Hg$
497.24
Mercury(I) sulfate
Mercurous sulfate
Mercurous sulfate, Solid (DOT)
Mercurous sulphate (DOT)
UN 1628 (DOT)

7783-41-7
F_2O
54.00
FOF
Oxygen-difluoride
Fluorine monoxide
Fluorine oxide
Oxygen difluoride (ACGIH, OSHA) [UN 2190]
Oxygen fluoride
UN 2190 [Oxygen difluoride]

7783-42-8
F_2OS
86.06
Thionyl-fluoride
Fluorure de thionyle (French)
Sulfur difluoride monoxide
Sulfur difluoride oxide
Sulfurous oxyfluoride
Thionyl difluoride

7783-46-2
F_2Pb
245.19
Lead(II) fluoride
Lead difluoride
Lead fluoride (DOT)
NA 2811 (DOT)
Plomb fluorure (French)
Plumbous fluoride

7783-47-3
F_2Sn
156.69
Tin-fluoride
Fluoristan
Stannous fluoride
Tin bifluoride
Tin difluoride

7783-48-4
F_2Sr
125.62
Strontium-fluoride

7783-49-5
F_2Zn
103.37
Zinc-fluoride
Zinc fluorure (French)

7783-50-8
F_3Fe
112.85
Iron-fluoride
Ferric fluoride
Iron trifluoride

7783-54-2
F_3N
71.01
FN(F)F
Nitrogen trifluoride (ACGIH, OSHA) [UN 2451]
Nitrogen fluoride
UN 2451 [Nitrogen trifluoride]

7783-56-4
F_3Sb
178.75
Antimony(III) fluoride (1:3)
Antimoine fluorure (French)
Antimonous fluoride
Antimony trifluoride
Antimony trifluoride, Solid [NA 1549]
Antimony trifluoride, Solution (DOT)
NA 1549 [Antimony trifluoride,

solid]
Stibine, trifluoro- (9CI)
Trifluoroantimony

7783-60-0
F_4S
108.06
Sulfur tetrafluoride (ACGIH, OSHA) [UN 2418]
Tetrafluorosulfurane
UN 2418 [Sulfur tetrafluoride]

7783-61-1
F_4Si
104.09
Silicon-fluoride
Silane, tetrafluoro-
Silicon tetrafluoride [UN 1859]
UN 1859 [Silicon tetrafluoride]

7783-66-6
F_5I
221.90
Iodine-fluoride
Iodine pentafluoride [UN 2495]
Pentafluoroiodine
UN 2495 [Iodine pentafluoride]

7783-70-2
F_5Sb
216.75
F[Sb](F)(F)(F)F
Antimony(V) pentafluoride
Antimony fluoride
Antimony (V) fluoride
Antimony pentafluoride [UN 1732]
Pentafluoroantimony
UN 1732 [Antimony pentafluoride]

7783-71-3
F_5Ta
275.95
Tantalum-fluoride

7783-79-1
F_6Se
192.96
Selenium-fluoride
Selenium hexafluoride (ACGIH, OSHA) [UN 2194]
UN 2194 [Selenium hexafluoride]

7783-80-4
F_6Te
241.60
Tellurium hexafluoride (ACGIH,OSHA) [UN 2195]
UN 2195 [Tellurium hexafluoride]

7783-81-5
F_6U
352.00
Uranium fluoride (Fissle)
UN 2977 [Uranium hexafluoride, fissile (Containing more than 1% U-235)]
UN 2978 [Uranium hexafluoride, fissile excepted or non-fissile]
Uranium hexafluoride, Fissile (containing more than 1% U-235) [UN 2977]
Uranium hexafluoride, Fissile excepted or non-fissile [UN 2978]
Uranium hexafluoride, Low specific activity (DOT)

7783-82-6
F_6W
297.84
Tungsten hexafluoride
Hexafluorure de tungstene (French)
Hexafluoruro de tungsteno (Spanish)
Tungsten fluoride, (OC-6-11)- (9CI)
(OC-6-11)Tungsten fluoride
UN 2196 [Tungsten hexafluoride]

7783-90-6
AgCl
143.32
Silver chloride (9CI)
Caswell No. 735A
EPA Pesticide Chemical Code 072506
Silver(I) chloride
Silver monochloride

7783-91-7
$Ag.ClHO_2$
176.33
Silver chlorite (Dry)
Chlorite d'argent (French)
Chlorous acid, silver(1+) salt
Clorito de plata (Spanish)
Silver chlorite
Silver chlorite, Dry

7783-93-9
$Ag.ClHO_4$
208.33
Perchloric acid, silver(1+) salt (9CI)
Silver(1+) perchlorate

7783-95-1
AgF_2
145.87
Silver(II) fluoride
Argent fluorure (French)
Argentic fluoride
Silver difluoride

7783-96-2
AgI
234.77
Silver iodide (9CI)
Colloidal Silver Iodide
Neosiluol
Neosilvol
Silver(1+) iodide
Silver monoiodide

7784-01-2
$Ag.1/2CrH_2O_4$
225.88
Silver chromate
Chromic acid, disilver(1+) salt (8CI,9CI)

7784-13-6
$AlCl_3.6H_2O$
241.45
Aluminum chloride, hexahydrate
Aluminum(III) chloride, hexahydrate
Aluminum trichloride hexahydrate

7784-18-1
AlF_3
83.98
Aluminum-fluoride
Aluminium fluorure (French)
Aluminum trifluoride
Fluorid hlinity (Czech)

7784-21-6
AlH_3
30.01
Aluminum hydride [UN 2463]
Alane
Aluminum trihydride
α-Aluminum trihydride
UN 2463 [Aluminum hydride]

7784-25-0
$Al.H_3N.2H_2O_4S$
162.24
Aluminum ammonium sulfate
Alum
Alum ammonium
Aluminum ammonium alum
Aluminum ammonium disulfate
Ammonium alum
Ammonium aluminum alum
Ammonium aluminum sulfate
Burnt Ammonium Alum
Caswell No. 041B
Curb
EPA Pesticide Chemical Code 098501
Exsiccated Ammonium Alum
Monoammonium monoaluminum sulfate

Sulfuric acid, aluminum ammonium salt (2:1:1) (9CI)
Sulfuric acid, aluminum ammonium salt (2:1:1), dodecahydrate
Unichem AMAL

7784-26-1
$AlNH_4(O_4S)_2.12H_2O$
453.32
Alum, ammonium
Aluminum ammonium sulfate (1:1:2) dodecahydrate
Sulfuric acid, aluminum ammonium salt (2:1:1), dodecahydrate

7784-28-3
$AlH_{24}NaO_{20}S_2$
408.31
Sodium alum
Alum
Aluminum sodium disulfate dodecahydrate
Monoaluminum monosodium disulfate dodecahydrate
Porous
SAS
Soda alum
Sodium aluminum sulfate
Sodium aluminum sulfate dodecahydrate

7784-30-7
$O_4P.Al$
121.95
Phosphoric acid, aluminum salt (1:1)
Aluminophosphoric acid
Aluminum acid phosphate
Aluminum monophosphate
Aluminum phosphate
Aluminum phosphate (1:1)
Aluphos
FFB 32
Monoaluminum phosphate
Phosphalugel

7784-33-0
$AsBr_3$
314.65
Arsenic(II) bromide
Arsenic bromide [UN 1555]
Arsenic bromide, Solid (DOT)
Arsenic tribromide
Arsenous bromide
Arsenous tribromide
Tribromoarsine
UN 1555 [Arsenic bromide]

7784-34-1
$AsCl_3$
181.27
Arsenic chloride (DOT)
Arsenic Butter
Arsenic(III) chloride
Arsenic chloride, Liquid (DOT)
Arsenic trichloride [UN 1560]
Arsenic trichloride, Liquid (DOT)
Arsenious chloride
Arsenous chloride
Arsenous trichloride (9CI)
Butter of Arsenic
Chlorure arsenieux (French)
Chlorure d'arsenic (French)
Fuming Liquid Arsenic
Trichloroarsine
Trichlorure d'arsenic (French)
UN 1560 [Arsenic trichloride]

7784-35-2
AsF_3
131.92
Arsenous-trifluoride
Arsenic fluoride
Arsenic trifluoride
Arsenous fluoride
TL 156
Trifluoroarsine

7784-40-9
$AsHO_4.Pb$
347.12
Arsenic acid, lead(2+) salt(1:1)
Acid lead arsenate
Acid lead orthoarsenate
Arsenate of lead
Arsinette
Dibasic lead arsenate
Gypsine
Lead acid arsenate
Lead arsenate [UN 1617]
Lead arsenate (Standard)
Ortho L10 Dust
Ortho L40 Dust
Schultenite
Security
Soprabel
Standard Lead Arsenate
Talbot
UN 1617 [Lead arsenates]

7784-41-0
$AsH_2O_4.K$
180.04
Arsenic acid, monopotassium salt
Macquer's Salt
Monopotassium arsenate
Monopotassium dihydrogen arsenate
Potassium acid arsenate
Potassium arsenate [UN 1677]
Potassium arsenate, monobasic
Potassium arsenate, Solid (DOT)
Potassium dihydrogen arsenate
Potassium hydrogen arsenate
UN 1677 [Potassium arsenate]

7784-42-1
AsH_3
77.95
[AsH3]
Arsine (ACGIH,OSHA) [UN 2188]
Arsenic hydrid
Arsenic hydride
Arsenic trihydride
Arseniuretted hydrogen
Arsenous hydride
Arsenowodor (Polish)
Arsenwasserstoff (German)
Hydrogen arsenide
UN 2188 [Arsine]

7784-44-3
$AsHO_4.2H_4N$
176.03
Arsenic acid, diammonium salt
Ammonium acid arsenate
Ammonium arsenate [UN 1546]
Ammonium arsenate, Solid (DOT)
Diammonium arsenate
Diammonium hydrogen arsenate
Diammonium monohydrogen arsenate
Dibasic ammonium arsenate
Secondary ammonium arsenate
UN 1546 [Ammonium arsenate]

7784-45-4
AsI_3
455.62
Arsenic iodide, Solid [NA 1557]
Arsenic triiodide
Arsenous iodide
Arsenous triiodide (9CI)
NA 1557 [Arsenic compounds, solid, N.O.S.]
Triiodoarsine

7784-46-5
$AsO_2.Na$
129.91
[Na+].[O-][As]=O
Arsenious acid, monosodium salt
Arsenenous acid, sodium salt (9CI)
Arsenious acid, sodium salt
Arsenite de sodium (French)
Atlas "A"
Chem Pels C
Chem-Sen 56
Kill-All
Penite
Prodalumnol
Prodalumnol double
Sodanit
Sodium arsenite
Sodium arsenite, Aqueous solutions [UN 1686]
Sodium arsenite, Liquid (solution) (DOT)
Sodium arsenite, Solid [UN 2027]
Sodium metaarsenite
UN 1686 [Sodium arsenite, Aqueous solutions]

UN 2027 [Sodium arsenite, solid]

7785-23-1
AgBr
187.77
Silver bromide (9CI)

7785-26-4
$C_{10}H_{16}$
136.24
C(C(CC1C2)C1(C)C)(=C2)C
Bicyclo(3.1.1)hept-2-ene, 2,6,6-trimethyl-, (1S)- (9CI)

7785-53-7
$C_{10}H_{18}O$
154.25
OC(C(CCC(=C1)C)C1)(C)C
3-Cyclohexene-1-methanol, α,α,4-trimethyl-, (R)- (9CI)

7785-84-4
$H_3O_9P_3.3Na$
308.91
Metaphosphoric acid, trisodium salt
Sodium trimetaphosphate
Trimetaphosphate

7785-87-7
$O_4S.Mn$
151.00
Manganese(II) sulfate (1:1)
Manganese sulfate (1:1)
Manganese(2+) sulfate (1:1)
Manganese sulphate
Manganous sulfate
Man-Gro
NCI-C61143
Sorba-Spray Mn
Sulfuric acid, manganese (II) salt (1:1) (8CI)

7785-88-8
H_7AlNaO_4P
152.00
Aluminum sodium phosphate
Phosphoric acid, aluminum sodium salt (9CI)
Sodium aluminum phosphate

7786-17-6
$C_{33}H_{52}O_2$
480.85
Oc(c(cc(c1)C)Cc(c(O)c(cc2C)CCCCCCCCC)c2)c1CCCCCCCCC
p-Cresol, 2,2'-methylenebis-(6-nonyl
2,2'-Methylene-bis(4-methyl-6-nonylphenol)
Nauga White
Phenol, 2,2'-methylenebis-(4-methyl-6-nonyl- (9CI)

7786-30-3
Cl_2Mg
95.21
Magnesium-chloride
Dus-Top

7786-34-7
$C_7H_{13}O_6P$
224.17
COC(=O)C=C(C)OP(=O)(OC)OC
Crotonic acid, 3-hydroxy-, methyl ester, dimethyl phosphate
Apavinphos
2-Butenoic acid, 3-((dimethoxyphosphinyl)oxy)-, methyl ester (9CI)
2-Carbomethoxy-1-methylvinyl dimethyl phosphate
α-2-Carbomethoxy-1-methylvinyl dimethyl phosphate
2-Carbomethoxy-1-propen-2-yl dimethyl phosphate
CMDP
Compound 2046
3-((Dimethoxyphosphinyl)oxy)-2-butenoic acid methyl ester
O,O-Dimethyl-O-(2-carbomethoxy-1-methylvinyl) phosphate
Dimethyl-1-carbomethoxy-1-propen-2-yl phosphate
O,O-Dimethyl-O-2-methoxycarbonyl-1-methyl-vinyl-phosphat (German)
Dimethyl 2-methoxycarbonyl-1-methylvinyl phosphate
Dimethyl methoxycarbonylpropenyl phosphate
Dimethyl (1-methoxycarboxypropen-2-yl)phosphate
O,O-Dimethyl O-(1-methyl-2-carboxyvinyl) phosphate
Duraphos
ENT 22,374
Fosdrin
Gesfid
Gestid
3-Hydroxycrotonic acid methyl ester dimethyl phosphate
Meniphos
Menite
(2-Methoxycarbonyl-1-methylvinyl)-dimethyl-fosfaat (Dutch)
(2-Methoxycarbonyl-1-methylvinyl)-dimethyl-phosphat (German)
2-Methoxycarbonyl-1-methylvinyl dimethyl phosphate
1-Methoxycarbonyl-1-propen-2-yl dimethyl phosphate
Methyl 3-(dimethoxyphosphinyloxy)crotonate
(2-Metossicarbonil-1-metilvinil)-dimetil-fosfato (Italian)
Mevinfos (Dutch)
Mevinox
Mevinphos (ACGIH,DOT, OSHA)
Mevinphos Mixture, Dry (DOT)
Mevinphos Mixture, Wet (DOT)
NA 2783 (DOT)
OS 2046
PD 5
Phosdrin (OSHA)
Phosfene
Phosphate de dimethyle et de 2-methoxycarbonyl-1 methylvinyle (French)
Phosphene (French)
Phosphoric acid, dimethyl ester, ester with methyl 3-hydroxycrotonate
Phosphoric acid, (1-methoxycarboxypropen-2-yl) dimethyl ester

7786-67-6
$C_{10}H_{18}O$
154.28
OC(C(C(=C)C)CCC1C)C1
p-Menth-8-en-3-ol
Isopulegol
8(9)-p-Menthen-3-ol
1-Methyl-4-isopropenylcyclohexan-3-ol

7786-81-4
$O_4S.Ni$
154.77
Nickel(II) sulfate (1:1)
NCI-C60344
Nickelous sulfate
Nickel sulfate
Nickel sulfate(1:1)
Nickel(II) sulfate
Nickel(2+)sulfate(1:1)
Sulfuric acid, nickel(2+)salt
Sulfuric acid, nickel(2+) salt (1:1)

7787-32-8
BaF_2
175.34
Barium-fluoride
Baryum fluorure (French)

7787-36-2
$Mn_2O_8.Ba$
375.22
Permanganic acid, barium salt
Barium permanganate [UN 1448]
UN 1448 [Barium permanganate]

7787-47-5
$BeCl_2$
79.91
Beryllium chloride (DOT)
Beryllium dichloride
NA 1566 (DOT)

7787-49-7

BeF_2
47.01
Beryllium fluoride (DOT)
Beryllium difluoride
NA 1566 (DOT)

7787-55-5
$BeNO_3$
71.02
Beryllium nitrate

7787-56-6
$O_4S.Be.4H_2O$
177.15
Beryllium sulfate, tetrahydrate (1:1:4)
Beryllium sulphate tetrahydrate
Sulfuric acid, beryllium salt (1:1), tetrahydrate

7787-69-1
BrCs
212.82
Cesium-bromide
Tricesium tribromide

7787-70-4
BrCu
143.45
[Cu+].[Br-]
Copper(I) bromide
Copper bromide (9CI)
Copper(1+) bromide
Copper monobromide
Cuprous bromide

7787-71-5
BrF_3
136.91
Bromine trifluoride [UN 1746]
UN 1746 [Bromine trifluoride]

7788-97-8
CrF_3
108.99
Chromic fluoride, Solid [UN 1756]
Chrome fluorure (French)
Chromic fluoride
Chromic fluoride solution [UN 1757]
Chromic trifluoride
Chromium fluoride (8CI,9CI)
Chromium(III) fluoride
Chromium trifluoride
Fluorure de chrome III (French)
Fluoruro cromico (Spanish)
UN 1756 [Chromic fluoride, Solid]
UN 1757 [Chromic fluoride, Solution]

7788-98-9
$CrO_4.2H_4N$
152.10
Chromic acid, diammonium salt
Ammonium chromate
Ammonium chromate(VI)
Diammonium chromate
Neutral Ammonium Chromate

7788-99-0
$O_8S_2.Cr.K.12H_2O$
499.41
Chromium(III) potassium sulfate (1:1:2), dodecahydrate
Chrome Alum
Chrome Alum (Dodecahydrate)
Potassium Chromium Alum
Sulfuric acid, chromium(3+)-potassium salt(2:1:1), dodecahydrate

7789-00-6
$CrO_4.2K$
194.20
Chromic acid, dipotassium salt
Bipotassium chromate
Chromate of Potass
Dipotassium chromate
Dipotassium monochromate
Neutral Potassium Chromate
Potassium chromate
Potassium chromate (VI)
Tarapacaite

7789-02-8
$N_3O_9.Cr.9H_2O$
400.21
Chromium(III) nitrate, nonahydrate (1:3:9)
Chromic nitrate nonahydrate
Chromium nitrate nonahydrate
Chromium trinitrate nonahydrate
Nitric acid, chromium(3+) salt, nonahydrate

7789-04-0
$Cr.H_3O_4P$
150.00
Chromium-phosphate
Arnaudon's Green
Arnaudon's Green (Hemiheptahydrate)
Chromic phosphate
Chromium orthophosphate
Phosphoric acid chromium (III) salt
Phosphoric acid, chromium(3+) salt (1:1)
Plessy's Green (Hemiheptahydrate)

7789-06-2
$CrO_4.Sr$
203.62
Chromic acid, strontium salt (1:1)
C.I. Pigment Yellow 32
Deep Lemon Yellow
Strontium chromate
Strontium chromate (1:1)
Strontium chromate 12170
Strontium chromate A
Strontium chromate (VI)
Strontium Chromate X-2396
Strontium Yellow

7789-09-5
$Cr_2O_7.2H_4N$
252.10
Dichromic acid, diammonium salt
Ammonio (bicromato di) (Italian)
Ammonio (dicromato di) (Italian)
Ammoniumbichromaat (Dutch)
Ammoniumdichromat (German)
Ammonium bichromate
Ammoniumdichromaat (Dutch)
Ammonium dichromate [UN 1439]
Ammonium (dichromate d') (French)
Ammonium dichromate (VI)
Bichromate d'ammonium (French)
UN 1439 [Ammonium dichromate]

7789-17-5
CsI
259.81
[I-].[Cs+]
Cesium-iodide
Cesium monoiodide
Dicesium diiodide
Tricesium triiodide

7789-20-0
D_2O
20.02
Water, Heavy (D2-O)
Deuterium oxide
Dideuterium oxide
Heavy Water
Heavy Water-D2
Water-D2 (9CI)
$Water^2$-H2

7789-21-1
FHO_3S
100.07
Fluorosulfonic acid [UN 1777]
Fluorosulphonic acid (DOT)
Fluosulfonic acid (DOT)
UN 1777 [Fluorosulfonic acid]

7789-23-3
FK
58.10
Potassium fluoride [UN 1812]
Fluorure de potassium (French)
Potassium fluoride, Solution (DOT)

Potassium fluorure (French)
UN 1812 [Potassium fluoride]

7789-24-4
FLi
25.94
Lithium-fluoride
TLD 100
Lithium fluorure (French)

7789-29-9
FK.FH
78.10
Potassium fluoride (8CI,9CI)
Bifluorure de potassium (French)
Bifluoruro potasico (Spanish)
Fluorure acide de potassium (French)
Hydrogen potassium fluoride
Potassium acid fluoride
Potassium bifluoride
Potassium bifluoride, Solid
Potassium bifluoride, Solution [UN 1811]
Potassium hydrogen difluoride
Potassium hydrogen fluoride
Potassium hydrogen fluoride, Solution
Potassium monohydrogen difluoride
UN 1811 [Potassium bifluoride, Solution]

7789-30-2
BrF_5
174.91
Bromine pentafluoride (ACGIH,OSHA) [UN 1745]
UN 1745 [Bromine pentafluoride]

7789-38-0
BrO_3.Na
150.90
Bromic acid, sodium salt
Bromate de sodium (French)
Dyetone
Sodium bromate [UN 1494]
UN 1494 [Sodium bromate]

7789-41-5
Br_2Ca
199.90
Calcium-bromide
Calcium dibromide

7789-42-6
Br_2Cd
272.22
Cadmium-bromide
Cadmium dibromide

7789-43-7
Br_2Co
218.75
Cobalt(II) bromide
Cobalt dibromide
Cobaltous bromide

7789-45-9
Br_2Cu
223.35
[Cu+2].[Br-].[Br-]
Copper(II) bromide
Copper bromide (9CI)
Copper dibromide
Cupric bromide

7789-47-1
Br_2Hg
360.41
Mercury bromide [UN 1634]
Mercuric bromide, Solid (DOT)
Mercury(II) bromide (1:2)
UN 1634 [Mercury bromides]

7789-59-5
Br_3OP
286.70
Phosphoryl-bromide
Phosphorous oxybromide
Phosphorus oxybromide [UN 1939]
Phosphorus oxybromide, Molten [UN 2576]
Phosphorus oxybromide, Solid (DOT)
Phosphoryl tribromide
UN 1939 [Phosphorus oxybromide]
UN 2576 [Phosphorus oxybromide, molten]

7789-60-8
Br_3P
270.70
Phosphorus tribromide [UN 1808]
Extrema
Phosphorous bromide (DOT)
Phosphorus bromide (DOT)
Tribromophosphine
UN 1808 [Phosphorus tribromide]

7789-61-9
Br_3Sb
361.48
Antimony bromide
Antimony tribromide
Antimony tribromide, Solid (DOT)
Antimony tribromide, Solution [NA 1549]
NA 1549 [Antimony tribromide, solution]
Tribromostibine

7789-75-5
CaF_2
78.08
Calcium-fluoride
Acid-Spar
Calcium difluoride
Fluorite (9CI)
Fluorspar
Irtran 3
Liparite
Met-Spar

7789-78-8
CaH_2
42.09
[CaH2]
Calcium hydride (9CI)

7789-79-9
Ca.$2H_3O_2P$
89.06
Calcium hypophosphite
Calcium phosphinate
Phosphinic acid, calcium salt (9CI)

7789-80-2
Ca.$2HIO_3$
391.90
Calcium iodate
Autarite
Iodic acid, calcium salt (9CI)
Lautarite

7789-89-1
$C_3H_5Cl_3$
147.43
CCC(Cl)(Cl)Cl
Propane, 1,1,1-trichloro
1,1,1-Trichloropropane

7789-90-4
$C_3H_3Cl_3O$
161.41
ClCC(Cl)(Cl)C=O
Propionaldehyde, 2,2,3-trichloro
2,2,3-Trichloropropanal
2,2,3-Trichloropropionaldehyde

7790-07-0
$C_{22}H_{42}O_6$
402.64
CCC(CC)COCCOC(=O)CCCCC(=O)OCCOCC(CC)CC
Adipic acid, bis(2-(2-ethylbutoxy)ethyl) ester
Adipic acid, di(2-(2-ethylbutoxy)ethyl) ester
Di-2-(2-ethylbutoxy)ethyl adipate
Di-(2-(2-ethylbutoxy)ethyl)ester kyseliny adipove (Czech)

7790-44-5
I_3Sb
502.46

Antimony triiodide
Antimony iodide
Stibine, triiodo- (9CI)
Triiodostibine

7790-47-8
I_4Sn
626.29
Tin(IV) iodide (1:4)
Stannic iodide
Tin tetraiodide

7790-69-4
$HNO_3.Li$
69.96
Nitric acid, lithium salt
Lithium nitrate [UN 2722]
UN 2722 [Lithium nitrate]

7790-76-3
$Ca.1/2H_4O_7P_2$
129.97
Calcium pyrophosphate
Calcium diphosphate
Calcium phosphate
Dicalcium pyrophosphate
Diphosphoric acid, calcium salt (1:2) (9CI)
Pyrophosphoric acid, calcium salt (1:2)

7790-79-6
CdF_2
150.40
Cadmium-fluoride
Cadmium fluorure (French)

7790-91-2
ClF_3
92.45
Chlorine fluoride
Chlorine trifluoride (ACGIH, OSHA) [UN 1749]
Chlorotrifluoride
Trifluorure de chlore (French)
UN 1749 [Chlorine trifluoride]

7790-93-4
$ClHO_3$
84.46
Chloric acid (DOT)
Chloric acid solution, Containing not more than 10% acid [UN 2626]
NA 2626 (DOT)
UN 2626 [Chloric acid solution, with not more than 10 per cent chloric acid]

7790-94-5
$ClHO_3S$
116.52
Chlorosulfuric-acid
Chlorosulfonic acid (DOT)
Chlorosulfonic acid (with or without sulfur trioxide) [UN 1754]
Chlorosulphonic acid (DOT)
Monochlorosulfuric acid
Sulfonic acid, monochloride
Sulfuric chlorohydrin
UN 1754 [Chlorosulfonic acid (with or without sulfur trioxide)]

7790-98-9
$ClO_4.H_4N$
117.50
Perchloric acid, ammonium salt
Ammonium perchlorate
Ammonium perchlorate, Average particle size less than 45 microns (DOT)
UN 0402 [Ammonium perchlorate, explosive]
UN 1442 [Ammonium perchlorate, oxidizer]

7790-99-0
ClI
162.35
[Cl-].[I+]
Iodine-chloride
Iodine monochloride [UN 1792]
Protochlorure d'iode (French)
UN 1792 [Iodine monochloride]
Wijs' Chloride

7791-10-8
Cl_2O_6Sr
254.52
Strontium chlorate, Solid [UN 1506]
Chloric acid, strontium salt
Strontium chlorate, Wet [UN 1506]
UN 1506 [Strontium chlorate, solid or solution]

7791-12-0
ClTl
239.82
Thallium(I) chloride
RCRA waste number U216
Thallium chloride
Thallium(1+) chloride
Thallium monochloride
Thallous chloride

7791-18-6
$Cl_2Mg.6H_2O$
203.33
Magnesium chloride, hexahydrate
Chlorure de magnesium hydrate (French)
CMH
Magnesium dichloride hexahydrate

7791-20-0
$Cl_2Ni.6H_2O$
237.73
Nickel(II) chloride, hexahydrate (1:2:6)

7791-21-1
Cl_2O
86.90
Chlorine monoxide
Chlorine oxide (9CI)

7791-23-3
Cl_2OSe
165.86
Seleninyl-chloride
Selenium chloride oxide
Selenium oxychloride [UN 2879]
UN 2879 [Selenium oxychloride]

7791-25-5
Cl_2O_2S
134.96
Sulfuryl chloride [UN 1834]
Sulfuric oxychloride
Sulphuryl chloride (DOT)
UN 1834 [Sulfuryl chloride]

7791-27-7
$Cl_2O_5S_2$
215.02
Pyrosulfuryl chloride [UN 1817]
Chlorosulfonic anhydride
Disulfur pentoxydichloride
Disulfuryl chloride
Pyro sulfuryl chloride
UN 1817 [Pyrosulfuryl chloride]

7795-91-7
$C_6H_{14}O_2$
118.20
CCC(C)OCCO
Ethanol, 2-sec-butoxy
2-sec-Butoxyethanol
Ethylene glycol mono-sec-butyl ether

7795-95-1
$C_8H_{17}ClO_2S$
212.74
O=S(=O)(CCCCCCCC)Cl
1-Octanesulfonyl chloride (9CI)

7798-23-4
$Cu.2/_3H_3O_4P$
134.28
Copper(II) phosphate
Copper phosphate
Copper phosphate (3:2)
Cupric phosphate
Phosphoric acid, copper(2+) salt

Phosphoric acid, copper(2+) salt (2:3) (9CI)

7803-49-8
H_3NO
33.04
NO
Hydroxylamine
Oxammonium

7803-51-2
H_3P
34.00
C.CCC
Phosphine (ACGIH,OSHA) [UN 2199]
Celphos
Delicia
Detia
Detia Gas EX-B
Fosforowodor (Polish)
Gas-EX-B
Hydrogen phosphide
Phosphorus trihydride
Phosphorwasserstoff (German)
RCRA waste number P096
UN 2199 [Phosphine]

7803-52-3
H_3Sb
124.78
Stibine (ACGIH,OSHA) [UN 2676]
Antimonwasserstoffes (German)
Antimony hydride
Antimony trihydride
Antymonowodor (Polish)
Hydrogen antimonide
UN 2676 [Stibine]

7803-55-6
$O_3V.H_4N$
116.99
Vanadic acid, ammonium salt
Ammonium metavanadate [UN 2859]
Ammonium vanadate
RCRA waste number P119
UN 2859 [Ammonium metavanadate]

7803-57-8
$H_4N_2.H_2O$
50.08
Hydrazine, monohydrate
Hydrazine hydrate
Idrazina idrata (Italian)

7803-62-5
H_4Si
32.13
Silane [UN 2203]
Monosilane
Silicane
Silicon tetrahydride (ACGIH, OSHA)
UN 2203 [Silane]

7803-63-6
$HO_4S.H_4N$
115.12
Sulfuric acid, monoammonium salt
Acid ammonium sulfate
Ammonium acid sulfate
Ammonium bisulfate
Ammonium hydrogen sulfate [UN 2506]
Ammonium monohydrogen sulfate
Monoammonium hydrogen sulfate
Monoammonium sulfate
UN 2506 [Ammonium hydrogen sulfate]

7803-65-8
$H_3N.H_3O_2P$
83.02
Phosphinic acid, ammonium salt (8CI,9CI)
Ammonium hypophosphite
Ammonium phosphinate

8000-21-3
$C_3H_3Br.CH_3Br.CCl_3NO_2$
378.27
1-Propyne, 3-bromo-, Mixt. with bromomethane and trichloronitromethane (9CI)
Trizone

8000-27-9
Unknown
Unknown
Cedarwood-Oil
Cedar Wood Oil
Cedrus Atlantica Oil
Oil Cedar
Oil of Cedar Wood
Oils, Cedarwood
Red Cedarwood Oil

8000-29-1
Unknown
Unknown
Oil of Citronella
Caswell No. 618
Citronella oil
EPA Pesticide Chemical Code 021901
Essential Oil of Cymbopogon Nardus
Oils, Citronella
Zitronell Oel (German)

8000-41-7
$C_{10}H_{18}O$
154.25
Terpineol (9CI)
Caswell No. 823
EPA Pesticide Chemical Code 067005
α-Terpineol

8000-42-8
Unknown
Unknown
Caraway-Oil
Kuemmel Oil (German)
Oil of Caraway

8000-46-2
Unknown
Unknown
Oil of Geranium
Caswell No. 618B
EPA Pesticide Chemical Code 597500
Geranium Absolute
Geranium Concrete
Geranium Oil
Geranium Oil, Algerian
Geranium Oil, Grasse
Geranium Oil, Morocco
Geranium Sur Roses
Hyperabsolute Geranium
Oils, Geranium
Oil of Pelargonium
Oil of Rose Geranium
Oil Rose Geranium Algerian
Pelargonium Oil
Rose Geranium Oil Algerian

8000-48-4
Unknown
Unknown
Eucalyptus Oil
Caswell No. 618A
Dinkum Oil
EPA Pesticide Chemical Code 040503
Eucalyptus Citriodora
Eucalyptus Citriodora Distillate
Eucalyptus Citriodora Oil
Eucalyptus Globulus Distillate
Eucalyptus Oil Citriodora
Eukalyptus Oel (German)
Globulus
Ingalipt
Oil Eucalyptus
Oil Eucalyptus Globulus or Macarthuri
Oil of Eucalyptus
Oils, Eucalyptus
Red Gum
Yacca Gum

8000-78-0
Unknown
Unknown
Allium sativum
Garlic Oil
Oil of Garlic
Oils, Garlic

8001-20-5
Unknown
Unknown
Tung Oil

Abrasin Oil
China Wood Oil
Japanese Wood Oil

8001-21-6
Unknown
Unknown
Sunflower Seed Oil
Anise LS
Basil LS
Bearberyy LS
Bitter Almond LS
Blackcurrant LS
Buckwheat LS
Coltsfoot LS
Common Bean LS
Common Fumitory LS
Cornflower LS
Cumin LS
Fig Tree LS
Gentian LS
Ginseng LS
Great Burdock LS
Incense LS
Ivy LS
Line LS
Lupine LS
Maritime Pine LS
Mate LS
Melilot LS
Mullein LS
Oat LS
Oils, Sunflower Seed
Olive Tree LS
Peppermint LS
Plantain LS
Roman Chamomile LS
Silver Birch LS
Solvent Extracted Sunflower Oil
Solvent Sunflower Oil
Stinging Nettle LS
Sunflower LS
Sunflower Oil
Sweet Almond LS
Tea LS
Traveler's Joy LS
Tropical Resins LS
Wheat LS

8001-22-7
Unknown
Unknown
Soybean-Oil
Degummed Soybean Oil
Soya-Bean Oil
Soybean Oil Degummed

8001-23-8
Unknown
Unknown
Safflower-Oil

8001-25-0
Unknown
Unknown
Olive-Oil

8001-26-1
Unknown
Unknown
Linseed-Oil
Groco
L-310

8001-28-3
Unknown
Unknown
Croton-Oil
Crotonoel (German)
Croton Resin
Croton Tiglium L. Oil
Oils, Croton
Oleum Tiglii
Olio di Croton (Italian)

8001-29-4
Unknown
Unknown
Cottonseed Oil (Deodorized winterized)
Deodorized Winterized Cottonseed Oil
NCI-C50168

8001-30-7
Unknown
Unknown
Corn-oil
Maise Oil
Maydol
Mazola Oil

8001-31-8
Unknown
Unknown
Coconut-Oil
Cocoanut Oil
Coconut Butter
Coconut Meal Pellets, Containing 6% to 13% moisture and no more than 10% residual fat (DOT)
Coconut Palm Oil
Copra [UN 1363]
Copra Oil
Copra Pellets (DOT)
Free Coconut Oil
UN 1363 [Copra]

8001-35-2
$C_{10}H_{10}Cl_8$
413.80
Toxaphene (DOT,OSHA)
Agricide Maggot Killer (F)
Alltex
Alltox
Attac 6
Attac 8
Attac 4-2
Attac 4-4
Attac 6-3
Camphechlor
Camphochlor
Camphoclor
Camphofene Huileux
Chem-Phene
Chlorinated camphene (OSHA)
Chlorinated camphene 60% (ACGIH)
Chlorocamphene
Clor Chem T-590
Compound 3956
Crestoxo
Cristoxo
Cristoxo 90
ENT 9,735
Estonox
Fasco-Terpene
Geniphene
Gy-Phene
Hercules 3956
Hercules toxaphene
Kamfochlor
Latka 3956 (Czech)
M 5055
Melipax
Motox
NA 2761 (DOT)
NCI-C00259
Octachlorocamphene
PCC
Penphene
Phenacide
Phenatox
Polychlorcamphene
Polychlorinated camphenes
Polychlorocamphene
RCRA waste number P123
Strobane-T
Strobane T-90
Synthetic 3956
Toxadust
Toxafeen (Dutch)
Toxakil
Toxaphen (German)
Toxon 63
Toxyphen
Vertac 90%
Vertac Toxaphene 90

8001-41-0
Unknown
Unknown
Hyoscyamus Oil
Hyoscyamus Oil-Expressed

8001-50-1
Unknown
Unknown
Terpene-polychlorinates
Chlorten (Czech)
Compound 3961
Dichloricide aerosol
Dichloricide mothproofer
ENT 19,442
Insecticide 3960-X14
Latka 3960x14 (Czech)
Strobane
3960-X14

8001-54-5
Unknown

Unknown
Ammonium, alkyldimethylbenzyl-, chloride
Alkyl dimethylbenzyl ammonium chloride
Alkyldimethyl(phenylmethyl)-quaternary ammonium chlorides
Ammonyx
Arquad DMMCB-75
Barquat MB-50
Barquat MB-80
Bayclean
Benirol
Benzalkonium chloride
Bionol
Bio-Quat 50-24
Bio-Quat 50-25
Bio-Quat 50-30
Bio-Quat 50-40
Bio-Quat 50-42
Bio-Quat 50-60
Bio-Quat 50-65
Bio-Quat 80-24
Bio-Quat 80-28
Bio-Quat 80-40
Bio-Quat 80-42
BTC
BTC 50
BTC 65
BTC 100
BTC 824
BTC 835
BTC 2565
BTC 8248
BTC 8249
BTC E-8358
BTC 50 USP
BTC 65 USP
Catamin AB
Catamine AB
Cequartyl
Dimanin A
Dodigen 226
Drapolene
Drapolex
Enuclen
Gardiquat 1450
Gardiquart SV480
Germicin
Germinol
Germitol
Hyamine 3500
Intexan LB-50
Katamin AB
Katamine AB
Marinol
Neo Germ-I-Tol
Onyx BTC (Onyx Oil & Chem Co)
Osvan
Paralkan
Pheneene Germicidal Solution and Tincture
Quaternary Ammonium Compounds, Alkylbenzyldimethyl, chlorides
Quaternium-1
Triton K-60
Vikrol RQ
Zephiral
Zephiran chloride

8001-58-9
Unknown
Unknown
Coal-Tar-Creosote
AWPA #1
Brick Oil
Coal Tar Oil (DOT)
Creosote
Creosote, From coal tar
Creosote Oil
Creosote P1
Creosotum
Cresylic creosote
Heavy Oil
Liquid Pitch Oil
Naphthalene Oil
RCRA waste number U051
Preserv-O-Sote
Tar Oil
UN 1136 [Coal tar distillates, flammable]
Wash Oil

8001-78-3
Unknown
Unknown
Hydrogenated Castor Oil
Castor Oil, Hydrogenated
Castorwax
Castorwax MP-70
Castorwax MP-80
Castorwax NF
Caswell No. 486A
EPA Pesticide Chemical Code 031604
Rice Syn Wax
Unitina HR

8001-79-4
Unknown
Unknown
Castor-Oil
Aromatic Castor Oil
Castor Oil Aromatic
Cosmetol
Crystal O
Gold Bond
NCI-C55163
Neoloid
Oil of Palma Christi
Phorbyol
Ricinus Oil
Ricirus Oil
Tangantangan Oil

8001-85-2
Unknown
Unknown
Bone Oil
Animal Oil
Caswell No. 107
Dippel's Oil
EPA Pesticide Chemical Code 010801
Oil of Hartshorn

8001-86-3
Unknown
Unknown
Isano oil

8001-88-5
Unknown
Unknown
Birch-Tar-Oil

8002-03-7
Unknown
Unknown
Peanut-Oil
Arachis Oil
Earthnut Oil
Groundnut Oil
Indigenous Peanut Oil
Katchung Oil
Pecan Shell Powder

8002-05-9
Unknown
Unknown
Petroleum
Base Oil
Coal Liquid
Coal Oil
Crude Oil
Crude Oil, Petroleum (DOT)
Crude Petroleum
NA 1270 (DOT)
Petrol
Petroleum Crude, Crude Oil [UN 1267]
Petroleum Oil [UN 1270]
Rock Oil
Seneca Oil
UN 1267 [Petroleum crude oil]
UN 1268 [Petroleum distillates, N.O.S.]
UN 1270 [Petroleum oil]

8002-09-3
Unknown
Unknown
Pine Oil [UN 1272]
Arizole
Oil of Pine
Oils, Pine
Oleum Abietis
Terpentinoel (German)
UN 1272 [Pine oil]
Unipine
Yarmor
Yarmor Pine Oil

8002-11-7
Unknown
Unknown
Poppy Seed Oil
Oil, Edible
Poppy Oil

8002-13-9
Unknown

Unknown
Rapeseed Oil
Brassica Campestris Oil
Oil, Edible
Oil of Rapeseed
Rape Oil

8002-16-2
Unknown
Unknown
Oil, Misc.
Aceite de colofonia (Spanish)
Huile de colophane (French)
Rosin Oil
UN 1286 [Rosin Oil]

8002-23-1
Unknown
Unknown
Spermaceti
Spermaceti Wax
Spermaceti Wax, Refined

8002-24-2
Unknown
Unknown
Oil, Misc.
Caswell No. 801C
EPA Pesticide Chemical Code 099701
Oils, Sperm
Sperm Oil
Sperm Oil, Acidulated

8002-26-4
Unknown
Unknown
Tall Oil
Acintol C
Lignin Liquor
Liquid Rosin
Oil, Misc.
Tall Oil Rosin
Tall Oil Rosin and Fatty Acids
Talleol
Tallol

8002-29-7
Unknown
Unknown
Coal Tar Neutral Oils
Caswell No. 225A
Coal Tar Hydrocarbons
EPA Pesticide Chemical Code 025001
Tar Oils

8002-31-1
Unknown
Unknown
Cocoa Butter
Cocoa Absolute
Cocoa Bean Extract
Cocao Bean Oil
Cocoa Beans Absolute, Colourless MD
Cocoa Beans, Methanol Extract
Cocoa Essence, Dark
Cocoa Essence, White
Cocoa Oil
Cocoa Oil Absolute
Cocoa Shell Extract
Dark Cocoa Essence
Theobroma Oil
White Cocoa Essence

8002-33-3
Unknown
Unknown
Sulfated Castor Oil
Castor Oil, Sulfated
Caswell No. 899
EPA Pesticide Chemical Code 079014
Sulfonated Castor Oil
Turkey Red Oil
Turkey-Red Oil
Unipol SCO

8002-43-5
Unknown
Unknown
Lecithin
Acti-Flow 68-SB
Alcolec S
Emulthin M-35
Gliddex
Granulestin
Kelecin
Lecithin, Soybean
Lecithins
Lecithol
Oleo-Coll LP
Ovovitellin
Phosphatidylcholine
Phospholutein
Soybean Lecithin
Unilex
Unilex DS
Unilex S
Unilex SH
Vitellin

8002-48-0
Unknown
Unknown
Malt Extract
Extract of Malt
Malt, Ext
Malt Extract, Powder
Malted Barley Extract
Maltine

8002-64-0
Unknown
Unknown
Neatsfoot Oil
Oils, Neat's-foot

8002-68-4
Unknown
Unknown
Juniper Oil
Juniper Berry Oil
Juniper Berry Oil, Terpenes
Juniperberry Oil
Oil of Juniper
Oil of Juniper Berry
Oils, Juniper
Oils, Juniperus Communis
Wacholderbeer Oel

8002-68-4
Unknown
Unknown
Juniper-Berry-Oil
Juniper Oil
Oil of Juniper Berry
Oils, Juniper
Wacholderbeer Oel (German)

8002-74-2
Unknown
Unknown
Paraffin
Paraffin wax fume (ACGIH, OSHA)

8002-75-3
Unknown
Unknown
Oils, Palm
Palm Butter
Palm Oil

8003-03-0
$C_{10}H_{13}NO_2.C_9H_8O_4.C_8H_{10}N_4O_2$
553.63
CCOc1ccc(NC(C)=O)cc1.CC(=O)Oc1ccccc1C(O)=O.Cn1cnc2n(C)c(=O)n(C)c(=O)c12
Benzoic acid, 2-(acetyloxy)-, Mixed with 3,7-dihydro-1,3,7-trimethyl-1H-purine-2,6-dione, and N-(4-ethoxyphenyl)acetamide
APC
APC (Pharmaceutical)
Ascophen
Aspirin, Phenacetin and caffeine
Citramon
Empirin Compound
NCI-C02697
Oscophen
Thomapyrin

8003-19-8
$C_3H_6Cl_2.C_3H_4Cl_2$
223.96
Propane, dichloro- mixed with propene, dichloro
D-D
DD Mixture
DD Soil Fumigant
Dichlorpropan-dichlorpropengemisch (German)
Dichloropropane-dichloropropene mixture
1,3-Dichloropropene and 1,2-dichloropropane mixture
Dowfume N

ENT 8,420
Nemafene
Telone
Vidden D

8003-22-3
Unknown
Unknown
C.I. Solvent Yellow 33
Arlosol Yellow S
Chinoline Yellow D Sol. In spirits
Chinoline Yellow ZSS
C.I. 47000
D & C Yellow No. 11
Nitro Fast Yellow SL
Oil Yellow SIS
Petrol Yellow C
Quinoline Yellow A Spirit Soluble
Quinoline Yellow Base
Quinoline Yellow Spirit Soluble
Quinoline Yellow SS
Solvant Yellow 33
Waxoline Yellow T

8003-34-7
Unknown
Unknown
Pyrethrum (ACGIH,OSHA)
Buhach
Chrysanthemum cinerareae-folium
Cinerin I or II
Dalmation Insect Flowers
Firmotox
Insect Powder
Jasmolin I or II
Pyrethrin I or II
Pyrethrins
Pyrethrum (Insecticide)
Trieste Flowers

8003-46-1
$C_6HCl_3N_2O_4$
271.44
Clc1cc(Cl)c(N(=O)=O)c(Cl)c1N(=O)=O
Benzene, dinitrotrichloro
Dinitrotrichlorobenzene

8003-69-8
$C_{36}H_{26}N_8O_{11}S_3.3Na$
911.77
2-Naphthalenesulfonic acid, 6-((7-amino-1-hydroxy-3-sulfo-2-naphthalenyl)azo)-3-((4-((4-amino-6(or 7)-sulfo-1-naphthalenyl)azo)phenyl)-azo)-4-hydroxy-, trisodium salt (9CI)

8004-09-9
$CH_3Br.CCl_3NO_2$
259.32
Methane, trichloronitro-, Mixt. with Bromomethane
Agel TG 37
Agel TG 67
Chloropicrin-methyl bromide mixt.
Chloropicrin and methyl bromide, Mixture [UN 1581]
Dowfume MC 2
Dowfume MC 33
MBC 33
M-B-C Fumigant
Methyl bromide-chloropicrin mixt.
Methyl bromide and more than 2% chloropicrin mixture, Liquid (DOT)
NA 1581 (DOT)
Terr-O-Gel
UN 1581 [Chloropicrin and methyl bromide mixtures]
Vertafume

8004-13-5
$C_{12}H_{10}.C_{12}H_{10}O$
324.44
Biphenyl, Mixed with biphenyl oxide (3:7)
Biphenyl-diphenyl ether mixture
1,1'-Biphenyl, Mixt. with 1,1'-oxybis(benzene)
Diphenyl mixed with diphenyl oxide
Dinil
Dinyl
Diphyl
Dowtherm
Dowtherm A
Phenyl ether-biphenyl Mixture (OSHA)

8004-87-3
Unknown
Unknown
C.I. Basic Violet
Basic Violet K
C.I. 42535
C.I. Basic Violet 1
Methyl Violet
Methyl Violet 2B
Methyl Violet BB
Methyl Violet FN
Methyl Violet N
Methyl-Violett (German)
Paris Violet R
Violet Methylova (Czech)
Violet Powder H 2503
Violet Zasadita 1 (Czech)

8004-92-0
$C_{18}H_9NO_8S_2.2Na$
477.38
C.I. Acid Yellow 3
Acid Yellow 3
Chinogelb (German)
Chinogelb Extra (German)
Chinogelb wasserloeslich (German)
C.I. 801
C.I. 918
C.I. 47005
C.I. Food Yellow 3
C.I. Food Yellow 13
D & C Yellow No. 10
Dye Quinoline Yellow
E 104
FD & C Yellow No. 10
Japan Yellow 203
Jaune de Quinoleine (French)
Lemon Yellow ZN 3
L-Gelb 3 (German)
Quinidine Yellow KT
Quinoline Yellow
2-(2-Quinolyl)-1,3-indandione disulfonic acid disodium salt
Schultz No. 918
Zlut Chinolonova (Czech)
Zlut Kysela 3 (Czech)
Zlut Potravinarska 13 (Czech)

8005-02-5
Unknown
Unknown
C.I. Solvent Black 7 (9CI)

8005-03-6
Unknown
Unknown
Nigrosin
Calco Nigrosine O 2P
C.I. Acid Black 2 (9CI)
C.I. 50420
Lurazol Deep Blue EB
Nigrosine
Nigrosine B
Nigrosine WL Water Soluble
Nigrosine WSB

8005-33-2
Unknown
Unknown
C.I. Natural Black 1 (9CI)

8005-40-1
$C_{28}H_{22}N_2O_2$
418.48
9,10-Anthracenedione, 1,5(or 1,8)-bis((4-methylphenyl)-amino)- (9CI)

8005-43-4
$C_7H_4Cl_4$
229.91
Randox-T
CDAA-T
CDAA + TCBC

8005-49-0
$C_{24}H_{21}Cl_6O_6P.C_{10}H_{12}N_2O_5$
889.36
Phosphorous acid, bis(2,4-dichlorophenoxyethyl) ester, Mixed with tris(2,4-dichloro phenoxyethyl)phosphite
Bis(2,4-dichlorophenoxyethyl)-phosphite mixed with tris-(2,4-dichlorophenoxyethyl)-phosphite

2,4-DEP
Falodin
Falone
Tris(2,4-dichlorophenoxyethyl)-phosphite mixed with bis-(2,4-dichlorophenoxyethyl)-phosphite

8005-56-9
$C_{26}H_{12}N_4O_2$
412.38
C.I. Vat Red 14 (9CI)

8005-72-9
Unknown
Unknown
C.I. Direct Yellow 28 (9CI)

8006-14-2
Unknown
Unknown
Gas, Natural
Natural Gas
Synthetic Natural Gas

8006-20-0
Unknown
Unknown
Gas, Producer
Fuel Gases, Low and medium B.T.U.

8006-24-4
Unknown
Unknown
Named reagents and solutions, Goulard's ext.
Goulard's extract
Lead subacetate soln.
Named solutions, Goulard's extract

8006-28-8
Unknown
Unknown
Soda lime
Soda lime, Solid
Sodasorb
UN 1907 [Soda lime with more than 4 per cent sodium hydroxide]

8006-39-1
$C_{10}H_{18}O$
154.28
Terpineol
Mixture of p-Methenols
Terpineols

8006-44-8
Unknown
Unknown
Candelilla Wax
Candelilla
Koster Keunen Candelilla Wax
Waxes, Candelilla

8006-54-0
Unknown
Unknown
Lanolin
Adeps Lane
Agnin
Agnolin
Agnolin No. 1
Alapurin
Aloe Extract #103
Amber Lanolin
Anhydrous lanolin
Anhydrous lanum
Caswell No. 518
Cosmelan
Cosmetic Lanolin
Dewaxed Lanolin
EPA Pesticide Chemical Code 031601
Ivarlan 3000
Ivarlan 3001
Ivarlan 3100
Lanain
Lanalin
Lanesin
Lanichol
Laniol
Lanolin anhydrous USP
Lanolin oil
Lantrol
Lanum
Liquid Lanolin Oil
Oesipos
Oils, Lanolin
Processed Lanolin
Wool Fat
Wool Grease
Wool Wax
Wool Wax, Refined

8006-61-9
Unknown
Unknown
Gasoline (ACGIH) [UN 1203]
Casing Head Gasoline (DOT)
Motor Fuel (DOT)
Motor Spirit (DOT)
Natural Gasoline [UN 1257]
Petrol (DOT)
UN 1203 [Gasoline]
UN 1257 [Natural gasoline]

8006-64-2
Unknown
Unknown
Turpentine (ACGIH,OSHA) [UN 1299]
Oil of Turpentine
Oil of Turpentine, Rectified
Spirit of Turpentine
Spirits of Turpentine
Terebenthine (French)
Terpentin Oel (German)
Turpentine Oil
Turpentine Oil, Rectifier
Turpentine Steam Distilled
UN 1299 [Turpentine]

8006-82-4
Unknown
Unknown
Black-Pepper-Oil

8006-84-6
Unknown
Unknown
Fennel-Oil
Bitter Fennel Oil
Fenchel Oel (German)
Oil of Fennel

8006-87-9
Unknown
Unknown
CC(CO)=CCCC2(C)C1CCC(C1)C2=C
Oils, Sandalwood
Arheol
East Indian Sandalwood Oil
Oil of Santal
Sandalwood Oil
Santal Oil

8006-90-4
Unknown
Unknown
Peppermint-Oil
Pfefferminz Oel (German)

8007-02-1
Unknown
Unknown
Lemongrass Oil
Caswell No. 618D
Citral terpenes
EPA Pesticide Chemical Code 040502
Lemongrass Oil West Indian
Oil of Lemon Grass
Oil of Lemongrass
Oil of Lemongrass, West Indian
Oils, Lemongrass
West Indian Lemongrass Oil

8007-12-3
Unknown
Unknown
Oil of Nutmeg, Expressed
Mace Oil
NCI-C56484
Nutmeg Butter
Oil of Mace
Oils, Mace

8007-18-9
Unknown
Unknown
Nickel Rutile Yellow
C.I. Pigment Yellow 53 (9CI)
Nickel Antimony Titanium Yellow Rutile

8007-24-7
Unknown
Unknown
Cashew Nut Shell Oil (Untreated)
Cashew, Nutshell Liq.

8007-27-0
Unknown
Unknown
Erigeron Oil
Fleabane Oil
Oil of Canada Fleabane
Oil of Erigeron
Oil of Fleabane
Oils, Fleabane
Oils, Fleabane, Erigeron Canadensis

8007-32-7
Unknown
Unknown
Perchloron (8CI)
Clori-Clean

8007-35-0
$C_{12}H_{20}O_2$
196.29
Terpinyl acetate
Terpineol, acetate (9CI)

8007-43-0
Unknown
Unknown
Sorbitan, Sesquioleate
Anhydrohexitol Sesquioleate
Arlacel 83
Arlacel C
Crill 16
Crill K 16
Emasol 41S
Emsorb 2502
Emulgator 8972
Glycolmul SOC
Glycomul SOC
Glycomul SOC Special
Hodag SSO
Liposorb SQ0
Montane 83
Nikkol SO 15
Nissan Nonion OP 83
Nissan Nonion OP 83RAT
Protachem SOC
SO 15
Sorbitan, (Z)-9-octadecenoate (2:3)
Sorgen 30
Span 83

8007-44-1
Unknown
Unknown
Pennyroyal Oil
Caswell No. 638B
EPA Pesticide Chemical Code 040509
Oil of Pennyroyal
Oils, Pennyroyal, Hedeoma Pulegioides
Pennyroyal
Pennyroyal Oil, European

8007-45-2
Unknown
Unknown
Coal-Tar
Coal Tar, Aerosol
Coal-Tar-Solution-USP
Carbo-Cort
Crude Coal Tar
Estar
Impervotar
Lav
Lavatar
Pixalbol
Pix Carbonis
Pix Lithanthracis
Polytar Bath
Supertah
Syntar
Tar
Tar, Coal
Tar, Liquid [UN 1999]
UN 1999 [Tars, liquid including road asphalt and oils, bitumen and cut backs]
Zetar

8007-47-4
Unknown
Unknown
Balsam Canada
Balsam of Fir
Balsams, Canada
Canadian Balsam
Fir Balsam Absolute

8007-56-5
$ClH.HNO_3$
99.47
Aqua Regia
Acide chlorhydrique et acide nitrique, melange d' (French)
Hydrochloric acid, Mixed with nitric acid (3:1)
Mezclas de acido clorhidricoy acido nitrico (Spanish)
Nitrohydrochloric acid
Nitrohydrochloric acid, Diluted
Nitromuriatic acid
UN 1798 [Nitrohydrochloric acid]

8007-70-3
Unknown
Unknown
Anise-Oil
Aniseed Oil
Anis Oel (German)
Oil of Anise
Oils, Anise
Star Anise Oil

8007-75-8
Unknown
Unknown
Bergamot-Oil-Rectified
Bergamotte Oel (German)
Oil of Bergamot, Rectified

8007-80-5
Unknown
Unknown
Cassia-Oil
Artificial Cinnamon Oil
Cinnamon Bark Oil
Cinnamon Oil
Kassia Oel (German)
Oil of Cassia
Oil of Chinese Cinnamon
Oil of Cinnamon
Oil of Cinnamon, Ceylon
Oils, Cinnamon

8008-20-6
Unknown
Unknown
Kerosene [UN 1223]
Coal Oil
Coal Oil (Export shipment only) (DOT)
Deobase
Kerosine
Kerosine (Petroleum)
Straight-Run Kerosene
UN 1223 [Kerosene]

8008-45-5
Unknown
Unknown
Oils, Nutmeg
Myristica Oil
Nutmeg Oil
Oil of Myristica
Oil of Nutmeg

8008-51-3
Unknown
Unknown
Camphor Oil
Aceite de alcanfor (Spanish)
Camphor Oil Brown
Camphor Oil (Light)
Camphor Oil, Rectified
Camphor Oil White
Camphor Oil Yellow
Camphor White Oil
Caswell No. 156
Formosa Camphor Oil
Formose Oil of Camphor
Huile de camphre (French)
Japanese Camphor Oil
Japanese, Oil of Camphor
Light Camphor Oil
Light Oil of Camphor
Liquid Camphor
Oils, camphor
Oil Camphor Sassafrassy
Oil of Camphor Rectified
Oil of Camphor White

UN 1130 [Camphor oil]
White Camphor Oil
White Oil of Camphor

8008-56-8
Unknown
Unknown
Lemon-Oil
Cedro Oil
Lemon-Petitgrain-Oil
Oil of Lemon
Zitronen Oel (German)

8008-57-9
Unknown
Unknown
Oil-of-Orange
Neat Oil of Sweet Orange
Oil of Sweet Orange
Orange Oil

8008-60-4
Unknown
Unknown
Opium
Crude Opium
DEA No. 9639
Opium, Crude
Powdered opium

8008-74-0
Unknown
Unknown
Sesame Oil
Caswell No. 733
EPA Pesticide Chemical Code 072401
Gingilli Oil
Oils, Sesame
Sextra
Uniderm SSME

8008-79-5
Unknown
Unknown
Spearmint-Oil
Oil of Spearmint

8009-03-8
Unknown
Unknown
Petrolatum
Caswell No. 645A
Cosmoline
Cream White
EPA Pesticide Chemical Code 598400
Extra Amber
Fonoline, White
Fonoline, Yellow
Fybrene
Mineral Fat
Mineral Grease (Petrolatum)
Mineral Jelly
Mineral Jelly No. 14
Mineral Jelly No. 17
Mineral Wax
Molo-Jel
Paraffin Jelly
Pennsoline Soft Yellow
Penreco White
Perfecta
Perlatum 310
Perlatum 315
Perlatum 320
Perlatum 325
Petrolatum Amber
Petrolatum USP
Petrolatum White
Petroleum Jelly
Protopet, Alba
Protopet, White 1S
Protopet, White 2L
Protopet, White 3C
Protopet, Yellow 1E
Protopet, Yellow 2A
Saxoline
Snow White
Sonojell #4
Sonojell #9
Ultima White
Vaseline
Vasoliment
White Petrolatum USP
White Petroleum Jelly
White Protopet
White Vaseline
Yellow Petrolatum

8011-48-1
Unknown
Unknown
Pine Tar
Caswell No. 666
EPA Pesticide Chemical Code 067204
Tar, Pine

8012-69-9
H_6Cu_6
387.32
Copper oxychloride sulfate

8012-74-6
Unknown
Unknown
London Purple [UN 1621]
London Purple, Solid (DOT)
UN 1621 [London Purple]

8012-89-3
Unknown
Unknown
Beeswax
Beeswax Absolute
Beeswax Oil, Absolute
Beeswax White
Beeswax (White)
Beeswax Yellow
Wax, Yellow
Yellow Beeswax

8012-95-1
Unknown
Unknown
Thermia-C
Adepsine Oil
Alboline
Bayol F
Blandlube
Crystol 325
Crystosol
Drakeol
Fonoline
Glymol
Bayol 55
Kaydol
Kondremul
Liquid paraffin
Mineral-Oil
Molol
Neo-Cultol
Nujol
Oil Mist (ACGIH)
Oil Mist, Mineral (OSHA)
Paraffin oil
Parol
Paroleine
Peneteck
Penreco
Perfecta
Petrogalar
Petrolatum, Liquid
Primol 355
Primol D
Protopet
Saxol
Tech Pet F

8013-00-1
$C_{10}H_{16}$
136.24
CC(C)C1=CC=C(C)CC1
Terpinene (9CI)

8013-07-8
Unknown
Unknown
Fatty acid, Soybean Oil, Epoxidized
Epoxidized Soybean Oil
Flexol-Epo
PX-800

8013-10-3
Unknown
Unknown
Juniper-Tar
Cade Oil Ractified

8013-17-0
Unknown
Unknown
Invert sugar
Calorose
Insubeta
Inverdex
Invertogen
Invertose
Nulomoline
Sugar, Invert (9CI)

Travert

8013-65-8
Unknown
Unknown
Methyl acetone
Methylacetone (French)
Metilacetona (Spanish)
UN 1232

8013-75-0
Unknown
Unknown
Fusel Oil
Aceite de fusel (Spanish)
Fusel Oil, Sugar Beet
Fuseloel (German)
Huile de fusel (French)
UN 1201 [Fusel oil]

8014-13-9
Unknown
Unknown
Cumin-Oil
Oils, Cumin

8014-17-3
Unknown
Unknown
Essential Oils
Mandarin Petitgrain Oil
Oil Citrus Reticulata
Oil Mandarin
Oils, Petitgrain
Orange Leaf Oil, Bitter
Orange Leaf Water, Absolute
Petitgrain Oil
Petitgrain Oil Saponified
Pettitgrain Oil

8014-91-3
$C_{80}H_{52}N_{16}O_{20}S_4.6Na$
1823.50
Benzoic acid, 3,3'-((3,7-disulfo-1,5-naphthalenediyl)-bis(azo(6-hydroxy-3,1-phenylene)azo(6(or 7)-sulfo-4,1-naphthalenediyl)azo(1,1'-biphenyl)-4,4'-diylazo))bis(6-hydroxy-, hexasodium salt
Benzoic acid, 3,3'-((3,7-disulfo-1,5-naphthalenediyl)bis(azo(6-hydroxy-3,1-phenylene)azo(6 (or-biphenyl)-4,4'-diylazo))-bis(6-hydroxy-, hexasodium salt
C.I. Direct Brown 74 (8CI)

8014-95-7
$H_2O_4S.O_3S$
178.14
Sulfuric acid, Fuming [UN 1831]
Disulphuric acid
Dithionic acid
Fuming Sulfuric acid (DOT)
NA 1831 (DOT)
Oleum (DOT)
Pyrosulphuric acid
Sulphuric acid, Fuming (DOT)
UN 1831 [Sulfuric acid, fuming]

8015-35-8
$C_8H_6Cl_2O_3.C_8H_5Cl_3O_3$
476.52
Acetic acid, (2,4-dichlorophenoxy)- Mixed with (2,4,5-trichlorophenoxy)-acetic acid (2:1)
Acetic acid, (2,4-dichlorophenoxy)- Mixed with acetic acid, (2,4,5-trichlorophenoxy)-, (2:1)
2,4-D and 2,4,5-T (2:1)
Hormoslyr 64

8015-55-2
$C_{11}H_{22}N_2O.C_{11}H_{10}ClNO_2$
422.02
Urea, 3-cyclooctyl-1,1-dimethyl- mixed with butynyl-3N-3-chlorophenylcarbamate
Alipur
Butynyl-3N-3-chlorophenylcarbamate mixed with 3-cyclooctyl-1,1-dimethyl urea
Chlorbufan mixed with cyceuron
Cyceuron plus chlorbufan
3-Cyclooctyl-1,1-dimethyl urea mixed with butynyl-3n-3-chlorophenylcarbamate
Cycluron-chlorbufam mixture
Cycluron-chlorbuyam mixture
HS 55

8015-62-1
Unknown
Unknown
Ambrette Seed Oil
Abelmosco, Semillas (Spanish)
Ambretta, Semi (Italian)
Ambrette
Ambrette Graines (French)
Ambrette Oil
Ambrette Seed Absolute
Ambrette Seed Liquid
Moschus Korner (German)
Oils, Ambrette

8015-64-3
Unknown
Unknown
Angelica-Root-Oil
Angelica-Seed-Oil
Angelika Oel (German)
Oil of Angelica Seed
Oils, Angelica

8015-65-4
Unknown
Unknown
Amyris-Oil
Oils, Amyris
West Indian Sandalwood Oil

8015-86-9
Unknown
Unknown
Carnauba
Brazil Wax
Carnauba Wax
Carnauba Waxes
Koster Keunen Carnauba Wax
Waxes, Carnauba

8016-13-5
Unknown
Unknown
Fish Oil
Oils, Fish

8016-20-4
Unknown
Unknown
Oil-of-Grapefruit
Grapefruit Oil

8016-23-7
Unknown
Unknown
Guaiac-Wood-Oil

8016-28-2
Unknown
Unknown
Lard Oil
Oils, Lard

8016-35-1
Unknown
Unknown
Oil, Misc.
Oils, Glyceridic, Oiticica
Oiticica Oil

8016-68-0
Unknown
Unknown
Savory Oil (Summer variety)

8016-85-1
Unknown
Unknown
Tangerine-Oil

8017-16-1
Unknown
Unknown
Polyphosphoric acid
Condensed Phosphoric Acid
Phospholeum
Polyphosphoric acids
Superphosphoric acid
Tetraphosphoric acid

8017-34-3
$C_{14}H_9Cl_5$
354.49
DDT-Technical

8018-01-7
$C_4H_6MnN_2S_4.C_4H_6N_2S_4Zn$
541.03
Vondozeb
Carbamic acid, ethylenebis-(dithio-, manganese zinc complex (8CI)
Carmazine
Dithane M 45
Dithane S 60
Dithane SPC
Dithane Ultra
F 2966
Fore
Green-Daisen M
Karamate
Mancofol
Mancozeb
Maneb-Zinc
Maneb-Zineb-Komplex (German)
Maneb-Zineb-Mischkomplex (German)
Mangan-zink-aethylendiamin-bis-dithio-carbamat (German)
Manoseb
Manzate 200
Manzeb
Marzin
Manzin 80
Nemispor
Policar MZ
Policar S
Triziman
Triziman D
Zimanat
Zimaneb
Ziman-Dithane

8018-07-3
$C_{14}H_{14}N_3.C_{13}H_{11}N_3.2ClH.Cl$
540.87
Acridinium, 3,6-diamino-10-methyl-, chloride, mono-hydrochloride, Mixt. with 3,6-acridinediamine mono-hydrochloride (9CI)
Acriflavinii chloridum (Latin)
Acriflavinio cloruro
Acriflavinium chloride (8CI)
Chlorure d'acriflavinium (French)
Cloruro de acriflavinio (Spanish)

8020-83-5
Unknown
Unknown
Deobase
Deodorized Kerosene
Deodorized Kerosine

8021-28-1
Unknown
Unknown
Canadian Fir Needle Oil
Abies Alba Oil
Abies Balsamea, Pinaceae
Abies Excelsa Oil
Abies Oil
Abies Picea Oil
Balsam Fir Oil
Fir Balsam
Fir Balsam Resinoid
Fir Needle Oil, Balsam
Fir Needle Oil, Siberian
Hemlock Oil
Oils, Fir

8021-39-4
Unknown
Unknown
Cresoate, Wood
Beechwood Cresoate
RCRA waste number U051

8022-00-2
$C_6H_{15}O_3PS_2$
230.30
CCSCCOP(=S)(OC)OC.CCSCCS P(=O)(OC)OC
Phosphorothioic acid, O,O-dimethyl O-(2-(ethyl-thio)ethyl) ester, Mixed with O,O- dimethyl s-(2-(ethyl-thio)ethyl) ester (7:3)
Bay 15203
Bayer 21/116
Demeton methyl
Duratox
ENT 18,862
S(and O)-2-(Ethylthio)ethyl O,O-dimethyl phosphoro-thioate
Metasystox
Methyl demeton (ACGIH, OSHA)
Methyl-mercaptophos
Methylmerkaptofos (Czech)
Methyl Systox

8022-19-3
Unknown
Unknown
Saffron Oil
Oils, Crocus Sativus
Oils, Saffron

8022-22-8
Unknown
Unknown
Spikenard Oil
Oils, Spikenard

8023-53-8
Unknown
Unknown
Alkyl dimethyl 3,4-dichloro-benzyl ammonium chloride
Alkyl(C8H17 to $C_{18}H_{37}$) di-methyl 3,4-dichlorobenzyl ammonium chloride
Ammonium, alkyl(C8-C18)di-methyl 3,4-dichlorobenzyl-, chloride
Dichlorobenzalkonium chloride
Tetrosan

8023-79-8
Unknown
Unknown
Palm Kernel Oil
Oil, Edible
Oils, Glyceridic, Palm kernel
Oils, Palm Kernel
Palm Nut Oil
Palm Oil
Palm Oil (From seed)

8023-94-7
Unknown
Unknown
Hyacinth-Absolute

8024-09-7
Unknown
Unknown
Walnut Oil
Oils, Walnut
Walnut Hull Extract

8024-14-4
Unknown
Unknown
Deertongue-Incolore
Deer's Tongue
Liatris
Liatrix Oleoresin
Vanilla Plant

8024-37-1
Unknown
Unknown
Curcumin
Curcuma Oil
Curcumine
NCI-C60015
Oils, Curcuma
Oil of Turmeric
Turmeric Oil
Turmeric Oleoresin

8027-00-7
$C_{16}H_{15}Cl_2NO_2.C_{15}H_{13}Cl_2NO_2$
634.41
Propane, 1,1-bis(p-chloro-phenyl)-2-nitro- mixed with 1,1-bis(p-chlorophenyl)-2-nitro butane (1:2)
1,1-Bis(p-chlorophenyl)-2-nitro-propane mixed with 1,1-bis-(p-chlorophenyl)-2-nitro-butane(1:2)
Dilan

ENT 18,066
Mixture of Bulan and Prolan (2:1)

8028-36-2
Unknown
Unknown
Thyroid
Dessicated Thyroid
Thyradin
Thyrocrine
Thyroid Extract
Thyroid Gland
Thyroid Tablets
Tiroidina

8028-73-7
Unknown
Unknown
Arsenical Dust
Arsenical Flue Dust
Flue Dust, Arsenic-contg.
Polvo arsenical (Spanish)
Poussiere arsenicale (French)
UN 1562 [Arsenical dust]

8028-89-5
Unknown
Unknown
Caramel (Color)
Caramel

8029-43-4
Unknown
Unknown
Corn Syrup
Corn Sugar Syrup
Glucose Syrup
Syrups, Corn
Syrups, Hydrolyzed starch

8029-99-0
Unknown
Unknown
Paregoric
Camphorated Opium Tincture USP XVI
Compound Tincture of Camphor
Paregoric Elixir
Parepectolin
Tinctura Opii Benzoica
Tinctura Thebaica Benzoica

8030-30-6
Unknown
Unknown
Benzin
Amsco H-J
Amsco H-SB
Benzin B70
Coal Tar Naphtha (DOT)
Hi-Flash Naphtha
Naphtha (OSHA) [UN 2553]
Naphtha
Naphtha Coal Tar (OSHA)
Naphtha Distillate (DOT)
Naphtha Petroleum [UN 1255]
Naphtha, Solvent [UN 1256]
Petroleum Benzin
Petroleum-Derived Naphtha
Petroleum Distillates (Naphtha)
Petroleum Naphtha (DOT)
Super VMP
UN 1255 [Naphtha, petroleum]
UN 1256 [Naphtha, solvent]
UN 2553 [Naphtha]

8030-55-5
Unknown
Unknown
Gurjun-Balsam
Balsam Gurjun
East Indian Copaiba
Wood Oil

8030-78-2
Unknown
Unknown
Ammonium, trimethyltallow alkyl-, chlorides
Adogen 471
Arquad T
Arquad T-50
Quaternary ammonium compounds, trimethyltallow alkyl-, chloride
(Tallow)trimethylammonium chloride
N-Tallow-trimethylammonium surfactant

8030-89-5
Unknown
Unknown
Ethyl-oxyhydrate
Pyroligneous acids, Reaction products with et alc., distillates
Rum Ether
ZV8-253

8032-32-4
Unknown
Unknown
Ligroine
Benzine (Light petroleum distillate)
Benzoline
Canadol
Ligroin
Painters Naphtha
Petroleum ether (DOT)
Petroleum Spirit [UN 1271]
Refined Solvent Naphtha
Skellysolve F
Skellysolve G
UN 1271 [Petroleum spirit]
Varnish Marker's Naphtha
VM and P Naphtha
VM & P Naphtha (ACGIH, OSHA)

8042-47-5
Unknown
Unknown
Mineral Oil, Slab Oil
Slab Oil (9CI)
White Mineral Oil

8047-42-5
$C_{21}H_{41}N_7O_{12}.Al_2H_4O_9Si_2.xAlH_3O_3$
868.70
D-Streptamine, O-2-deoxy-2-(methylamino)-α-L-glucopyranosyl-(1.fwdarw.2)-O-5- deoxy-3-C-(hydroxymethyl)-α-L-lyxofuranosyl-(1.fwdarw.4)-N,N'- bis(aminoiminomethyl)-, Mixt. with aluminum hydroxide ($Al(OH)_3$), kaolinite ($H_4Al_2SiO_9$) and pectin (9CI)
Aluminum hydroxide ($Al(OH)_3$), Mixt. contg. (9CI)
Dihydrostreptomycin-kaolinite-pectin-aluminum hydroxide mixture
Kaolinite ($H_4A_{12}Si_2O_9$), Mixt. contg. (9CI)
Pectin, Mixt. contg. (9CI)

8047-67-4
Unknown
Unknown
Iron oxide, saccharated
Colliron I.V.
Feojectin
Ferric oxide, saccharated
Ferric saccharate ...Iron oxide (Mix.)
Ferrivenin
Iron saccharate
Iron Sugar
Iviron
Neo-Ferrum
Proferrin
Saccharated ferric oxide
Saccharated iron
Sucrofer

8047-99-2
$C_9H_{13}NO_2S$
199.27
Benzenesulfonamide, N-ethyl-2(or 4)-methyl-
AI3-04487
AI3-08014

8048-52-0
$C_{14}H_{14}N_3.Cl.C_{13}H_{11}N_3$
469.03
Acridinium, 3,6-diamino-10-methyl-, chloride mixed with 3,6-acridinediamine
Acriflavin

Acriflavine
Acriflavine mixture with proflavine
Acriflavinii chloridum
Acriflavinium chloride
Acriflavon
Angiflan
Assiflavine
Avlon
Bialflavina
Bioacridin
Bovoflavin
Burnol
Buroflavin
Choliflavin
Chromoflavine
Diacrid
3,6-Diaminoacridine mixture with 3,6-diamino-10-methylacridinium chloride
3,6-Diamino-10-methylacridinium chloride mixture with 3,6-acridinediamine
2,8-Diamino-10-methylacridinium chloride mixture with 2,8-diaminoacridine
Euflavin
Euflavine
Flavacridinum hydrochloricum
Flavine
Flavinetten
Flavioform
Flavipin
Flavisept
Glyco-Flavine
Gonacin
Gonacrine
Isravin
Mediflavin
Neutral Acriflavine
Neutroflavin
Neutroflavine
Panflavin
Pantonsiletten
Tolivalin
Trachosept
Tripla-Etilo
Trypaflavin
Trypaflavine
Trypaflavine Neutral
Trypaflavinum
Vetaflavin
Xanthacridinum
Zoriflavin

8049-17-0
Unknown
Unknown
Ferrosilicon
Ferrosilicon, Containing more than 30% but less than 90% silicon [UN 1408]
UN 1408 [Ferrosilicon, with 30 percent or more but less than 90 percent silicon]

8049-19-2
Unknown
Unknown
Ferrophosphorus
Iron alloy, Base, Fe,P (Ferrophosphorus)

8049-47-6
Unknown
Unknown
Pancreatin

8049-99-8
Unknown
Unknown
Milorganite (8CI,9CI)

8050-81-5
Unknown
Unknown
Simethicone
Antifoam A
DC Antifoam A

8051-02-3
$C_{32}H_{49}NO_9$
591.75
Sabadilla alkaloids
Asagraea Officinalis
Caswell No. 728
Caustic Barley
Cevadilla
Cevadine
ENT 123
EPA Pesticide Chemical Code 002201
Sabacide
Sabadilla
Sabane Dust
Veratridine
Veratrin (German)
Veratrine

8052-41-3
Unknown
Unknown
Stoddard-Solvent
Stoddard Solvent (ACGIH, OSHA)

8052-42-4
Unknown
Unknown
Asphalt
Asphalt, At or above its flashpoint [NA 1999]
Asphalt, Cut back (DOT)
Asphalt, Fumes (ACGIH)
Asphalt, Petroleum
Asphaltum
Bitumen
Cut-Back, Asphalt or Bitumen (DOT)
Judean Pitch
Mineral Pitch
NA 1999 [Asphalt, at or above its flashpoint]
Petroleum Asphalt
Petroleum Bitumen
Petroleum Pitch
Petroleum Roofing Tar
Road Asphalt [UN 1999]
Road Asphalt, Liquid, Tars or oil (DOT)
Road Tar [UN 1999]
Road Tar, Liquid [UN 1999]
UN 1999 [Tars, liquid including road asphalt and oils, bitumen and cut backs]

8057-49-6
Unknown
Unknown
Valeriana Officinalis l., Root extract
V-103
Valerian Root

8057-62-3
Unknown
Unknown
Passionflower-Extract
Passiflorae Incarnatae Extractum
Passiflora Extract
Passiflora Incarnata, Extract

8061-51-6
Unknown
Unknown
Lignosulfonic acid, sodium salt (9CI)
AHR 2438B
Banirex N
BETZ 402
Desulfonated Spent Pulping Liquor
Lignosite 458
Lignosite 854
Lignosol D 10
Lignosol XD
Maracell C
Maracell E
Marasperse B
Marasperse CBS
Marasperse N
Marasperse N 22
Marasperse N 22 Dispersant
Orzan S
Peritan NA
Polignate Sodium
Polyfon
Polyfon F
Polyfon H
Polyfon HUN
Polyfon O
Polyfon T
Raylig 260LR
Reax 45A
Reax 80C
Reax 85A
Reveal NM
Reveal SM
Reveal SM 5
Reveal WM
Sanekis P 213
Sanekis P 552
Sodium Base Spent Sulfite Liquor
Sodium ligninsulfonate
Sodium lignosulfite

Sodium lignosulfonate
Sodium lignosulfonic acid
Sodium polignate
Sulfonated lignin sodium salt
UF 10000A
Urzan S
Vanicell
Vanirex HW
Vanirex N
Vanisperse
Wanin S

8061-51-6
Unknown
Unknown
Lignosulfonic acid, sodium salt
Marasperse B
Marasperse CBS
Marasperse N
Marasperse N 22
Marasperse N 22 Dispersant
Sodium ligninsulfonate

8061-52-7
$C_{20}H_{24}CaO_{10}S_2$
528.62
Calcium lignosulfonate
Caswell No. 146
EPA Pesticide Chemical Code 115101
Lignosulfonic acid, calcium salt (9CI)
Lime Fractionated, Spent Pulping Liquor, Precipitate

8062-15-5
Unknown
Unknown
Lignosulfonic acid (9CI)
AHR 2438B
Lignosulfonates

8063-14-7
Unknown
Unknown
Marihuana
Bhang
Cannabis
Cannabis Resin
Charas
CME
DEA No. 7360
Ganja
Hasach
Hashish
Indian Cannabis
Indian Hemp
Marijuana
SSC-85213

8063-77-2
$CO_2.O_2$
76.01
Carbogen (8CI)
Carbon dioxide mixed with oxygen
Carbon dioxide-oxygen, Mixture
Dioxyde de carbone et oxygene en melange (French)
Mezclas de dioxido de carbonoy oxigeno (Spanish)
Oxygen-carbon dioxide, Mixture
UN 1014 [Carbon dioxide and oxygen mixtures]

8064-42-4
$C_{26}H_{40}MnN_6O_6S_8Zn$
909.47
Dikar (8CI)
Mancokar
Manganese, ((1,2-ethanediylbis-(carbamodithioato))(2-))-, Mixt. with ((1,2-ethanediyl-bis(carbamodithioato))(2-))-zinc and 2(or 4)-isooctyl-4,6(or 2,6)-dinitrophenyl 2-butenoate (9CI)

8064-77-5
$C_{19}H_{35}NO_2.C_{17}H_{22}N_2O.C_8H_{11}NO_3.C_4H_6O_4.2ClH$
940.18
Butanedioic acid, Compd. with N,N-dimethyl-2-(1-phenyl-1-(2-pyridinyl)-ethoxy)ethanamine (1:1), Mixt. with 2-(diethylamino)-ethyl(1,1'-bicyclohexyl)-1-carboxylate hydrochloride and 5- hydroxy-6-methyl-3,4-pyridinedimethanol hydrochloride
Bendectin
Debendox
Lenotan

8064-90-2
$C_{14}H_{18}N_4O_3.C_{10}H_{11}N_3O_3S$
543.66
Pyrimidine, 2,4-diamino-5-(3,4,5-trimethoxybenzyl)- and N'-(5-methyl-3-isoxazol-yl) sulfanilamide
A 033
Abacin
Abactrim
Aposulfatrim
Bactramin
Bactrim
Bactrim DS
Bactrin
Bactromin
Baktar
Biseptol
Chemitrim
Co-Trimoxazole
Drylin
Eltrianyl
Eusaprim
Fectrim
Gantaprin
Gantrim
Kepinol
Linaris
Microtrim
Momentol
Nopil
Omsat
Oxaprim
Pantoprim
Septra
Septran
Septrim
Septrin
Sigaprin
Sulfamethoxazol-trimethoprim
Sulfotrim
Sulfotrimin
Sulprim
Sumetrolim
Suprin
Tacumil
Teleprin
TMS 480
Trigonyl
Trimesulf
Trimethoprimsulfa
Trimethoprim-sulfamethoxazole
Trimethoprim and sulphameth-oxazole
Trimforte
Trimosulfa
Uro-Septra

8065-36-9
$C_{13}H_{19}NO_2.C_{13}H_{19}NO_2$
442.66
Carbamic acid, methyl-, m-(1-methylbutyl)phenyl ester mixed with carbamic acid, methyl-, m-(1-ethyl-propyl)phenyl ester
Bufencarb
BUX
BUX-Ten
Metalkamate
3-(1-Methylbutyl)phenyl methylcarbamate mixed with 3-(1-ethylpropyl)phenyl methylcarbamate
Ortho 5353

8065-41-6
Unknown
Unknown
Aureofungin
D-Mannose, 3-amino-3,6-di-deoxy-, Mixt. with 1-(4-(methylamino)phenyl)ethanone

8065-48-3
$C_8H_{19}O_3PS_2.C_8H_{19}O_3PS_2$
516.72
CCOP(=S)(OCC)OCCSCC.CCOP(=O)(OCC)SCCSCC
Phosphorothioic acid, O,O-diethyl O-(2-(ethyl-thio)ethyl) ester, Mixed with O,O- diethyl S-(2-(ethyl-thio)ethyl) phosphorothioate (7:3)
Bay 10756

Bayer 8169
Demeton (ACGIH,OSHA)
Demeton-O + Demeton-S
Demox
Diethoxy thiophosphoric acid ester of 2-ethylmercapto-ethanol
O,O-Diethyl 2-ethylmercapto-ethyl thiophosphate
O,O-Diethyl O(and S)-2-(ethyl-thio)ethyl phosphorothioate Mixture
E 1059
ENT 17,295
Mercaptophos
Systemox
Systox (OSHA)
ULV

8065-62-1
$C_5H_{13}O_3PS_2.C_5H_{13}O_3PS_2$
432.52
Phosphorothioic acid, O,O-dimethyl O-(2-(methylthio)-ethyl) ester, Mixt. with O,O-dimethyl S-(2-(methyl-thio)ethyl) phosphorothioate (9CI)
Atlasetox
Demephion
Methyl demeton-methyl
Tinox
Tonix

8065-83-6
$Cl_2Hg_2.Cl_2Hg$
743.57
Calo-Clor

8066-01-1
$C_3H_6Cl_2.C_3H_4Cl_2.C_2H_3NS$
297.08
Propane, 1,2-dichloro-, Mixt. with 1,3-Dichloropropene and isothiocyanatomethane
D-D Methylisothiocyanate
DD-Mencs
DD-Methyl isothiocyanate mixt.
Di-Trapex
EP-162
Forlex
1-Propene, 1,3-dichloro-, Mixt. with 1,2-Dichloropropane and isothiocyanatomethane
Vorlex

8066-11-3
$C_{10}H_{19}N_5S.C_9H_{16}ClN_5$
471.01
1,3,5-Triazine-2,4-diamine, 6-chloro-N-(1,1-dimethyl-ethyl)-N'-ethyl-, Mixt. with N-(1,1-dimethylethyl)-N'-ethyl-6-(methylthio)-1,3,5-tri-azine-2,4-diamine (9CI)
A 3620
Camparol 3303
Opogard
Solanex
Topogard
Topogard 3623
Vuagt 179

8066-69-1
$C_{13}H_{11}N_3O_2.CH_3AsS$
363.25
Benzoic acid, (4-(hydroxy-imino)-2,5-cyclohexadien-1-ylidene)hydrazide, Mixt. with methylthioxoarsine (9CI)
Rhizoctol combi
Rhizoctol slurry

8067-55-8
$C_9H_{21}NO_3.C_9H_{21}NO_3.C_8H_6Cl_2O_3.C_6H_3Cl_3N_2O_2$
845.02
2-Pyridinecarboxylic acid, 4-amino-3,5,6-trichloro-, Compd. with 1,1',1''-nitrilo-tris(2-propanol) (1:1), Mixt. with 1,1',1''-nitrilotris(2-pro-panol) (2,4-dichlorophenoxy)-acetate (Salt) (9CI)
Tordon 50D
Tordon 101
Tordon 101 Mixture

8067-69-4
Unknown
Unknown
Halquinol
5,7-Dichloro-8-quinolinol, 5-chloro-8-quinolinol, and 7-chloro-8-quinolinol in proportions resulting natural-ly from chlorination of 8-quinolinol
5,7-Dichloro-8-quinolinol mixt. with 5-chloro-8-quinolinol and 7-chloro-8-quinolinol
Halquinols
8-Quinolinol, 5,7-dichloro-, Mixt. with 5-chloro-8-quin-olinol and 7-chloro-8-quin-olinol
Quinolor
SQ 16,401
Tarquinor

8068-02-8
Unknown
Unknown
Lignin, chlorinated
Chlorolignin

8068-05-1
Unknown
Unknown
Lignin, alkali (9CI)

8068-44-8
Unknown
Unknown
Clophen A 50
A 50

8068-77-7
$C_8H_6Cl_2O_3.C_8H_6Cl_2O_3$
442.08
COc1c(Cl)ccc(Cl)c1C(O)=O.OC(=O)COc1ccc(Cl)cc1Cl
Benzoic acid, 3,6-dichloro-2-methoxy-, Mixt. with (2,4-dichlorophenoxy)acetic acid
Aminopielik D
Banvel 72D
Banvel 720
Banvel-2,4-D mixture
2,4-D-Dicamba mixt.
Dicamba-2,4-D mixt.
Super D

8072-81-9
$C_{19}H_{35}ClN_{10}S$
471.07
Caragard
Carragard
1,3,5-Triazine-2,4-diamine, 6-chloro-N-(1,1-dimethyl-ethyl)-N'-ethyl-, Mixt. with N-(1,1-dimethylethyl)-N'-ethyl-6-methoxy-1,3,5-tri-azine-2,4-diamine

8073-77-6
$C_{10}H_{19}N_5S.C_8H_{14}ClN_5$
457.12
1,3,5-Triazine-2,4-diamine, N,N'-bis(1-methylethyl)-6-(methylthio)-, Mixt. with 6-Chloro- N-ethyl-N'-(1-methylethyl)-1,3,5-triazine-2,4-diamine
A 1798
Agelon
Agelon 1798
Apropin
Atrazine-Gesagard Mixt.
Atrazine-Prometryne Mixt.
Atrazine-Prometryn Mixt.
Gesaprim 1798
Gesaprim Multy
Inakor
Inakor T
Prozin
Prozin 50
Zeaprim

8074-35-9
$C_{14}H_{20}ClNO_2.C_9H_{19}NOS$
459.08
Carbamothioic acid, dipropyl-, S-ethyl ester, Mixt. with 2-chloro-N-(2,6-diethyl-phenyl)-N-(methoxymethyl) acetamide (9CI)

8075-74-9
Unknown

Unknown
Lignosulfonic acid, chromium iron salt (9CI)
Chromium iron lignosulfonate
Ferrochrome lignosulfonate
Lignosulfonic acid, ferro chromium salt
Lignosulfonic acid, iron chromium salt

8075-80-7
$C_{12}H_7Cl_2NO_3.C_9H_{10}Cl_2N_2O_2$
533.18
Urea, N'-(3,4-dichlorophenyl)-N-methoxy-N-methyl-, Mixt. with 2,4-dichloro-1-(4-nitrophenoxy)benzene (9CI)
FE3
Multitok
Rofen 240
Tok-Ultra
Tok Ultra B

9000-01-5
Unknown
Unknown
Arabic-Gum
Acacia
Acacia Dealbata Gum
Acacia Gum
Acacia Senegal
Acacia Syrup
Australian Gum
Gum Acacia
Gum Arabic
Gum Ovaline
Gum Senegal
Indian Gum
NCI-C50748
Senegal Gum
Starsol No. 1
Wattle Gum

9000-05-9
Unknown
Unknown
Gum Benzoin
Benjamin Gum
Benzoin
Benzoin Gum
Benzoin Gum, Sumatra
Benzoin Malasia
Benzoin Resin
Benzoin Resinoid
Benzoin Resinoid, Siam
Benzoin Siam
Benzoin Sumatra
Gum Benjamin
Gum Benzoin Siam
Gum Sumatra
Hyperabsolute Benzoin, Siam
Resin Benjamin
Resin Benzoin
Siam Benzoin
Styrax benzoin
Sumatra Benzoin

9000-07-1
Unknown
Unknown
Carrageen
3,6-Anhydro-D-galactan
Aubygel GS
Aubygum DM
Burtonite V-40-E
Carastay
Carastay C
Carrageenan
Carrageenan Gum
Carrageenin
Carragheanin
Carragheen
Carragheenan
Chondrus
Chondrus Extract
Colloid 775
Coreine
Eucheuma Spinosum Gum
Flanogen Ela
Galozone
Gelcarin
Gelcarin HMR
Gelozone
Genu
Genugel
Genugel CJ
Genugol RLV
Genuvisco J
Gum Carrageenan
Gum Chon 2
Gum Chond
Gum Chrond
Irish Gum
Irish Moss Extract
Irish Moss Gelose
Killeen
Lygomme CDS
Pearlpuss
Pellugel
Pellugel ID
Pencogel
Pig-Wrack
Satiagel GS350
Satiagum 3
Satiagum Standard
Seakem Carrageenin
Seakem 3 & LCM
Seaspen PF
Seatrem
Self Rock Moss
Viscarin
Viscarin 402 & TP-4

9000-11-7
Unknown
Unknown
Cellulose, carboxymethyl ether
Almelose
Apergel
Apeyel
Carbose
Carboxymethylated Cellulose Pulp
Carboxymethylcellulose
Carboxymethyl cellulose ether
Cellulose carboxymethylate
Celluloseglycolic acid
Cellulose Gum 7H
CMC
CM-Cellulose
CMC-4LF
Colloresine
Glycocel TA
Glycolic acid cellulose ether
7H
KMTs
Thylose

9000-16-2
Unknown
Unknown
Damar
Damar Gum
Damar Resin
Dammar
Gum Damar

9000-29-7
Unknown
Unknown
Guaiac-Resin
Gum Guaiac

9000-30-0
Unknown
Unknown
Guar-Gum
1212A
A-20D
Burtonite V-7-E
Cyamopsis Gum
Dealca TP1
Dealca TP2
Decorpa
Galactasol
Gendriv 162
Guar
Guaran
Guar Flour
Gum Cyamopsis
Gum Guar
Indalca AG
Indalca AG-BV
Indalca AG-HV
J 2FP
Jaguar
Jaguar 6000
Jaguar A 20 B
Jaguar A 20D
Jaguar A 40F
Jaguar Gum A-20-D
Jaguar No.124
Jaguar Plus
Lycoid DR
NCI-C50395
Regonol
Rein Guarin
Supercol G.F.
Supercol U Powder
Syngum D 46D
Uni-Guar

9000-36-6
Unknown
Unknown
Sterculia-Gum

Gum Sterculia

9000-40-2
Unknown
Unknown
Locust-Bean-Gum
Algaroba
Carob Bean Gum
Carob Flour
NCI-C50419
St. John's Bread

Supercol

9000-65-1
Unknown
Unknown
Tragacanth-Gum
Gum Tragacanth

9000-69-5
Unknown
Unknown
Pectin
Colyer Pectin
Genu
Mexpectin
Methoxypectin
Methyl pectin
Methyl pectinate
Pectinate
Pectinic acid
Pectins

9000-70-8
Unknown
Unknown
Gelatins
Absorbable Gelatin Sponge
Gelatin
Gelatine
Gelatin Foam
Gelfoam
GT
Pharmagel A
Pharmagel AdB
Pharmagel B
Puragel
Spongiofort
Vee Gee Gelatin

9000-71-9
Unknown
Unknown
Casein
Alacid
Alanate
Alaren
Alatate
Caseins
Milk Protein
Proteins, Milk
TMP
Unipro CAL-CASE
Unipro CO-CASE

9001-08-5
Unknown
Unknown
Esterase, choline (9CI)

9001-63-2
Unknown
Unknown
Lysozyme
N,O-Diacetylmuramidase
E.C. 3.2.1.17
Globulin G
Globulin G1
Leftose
Lysozyme G
Mucopeptide glucohydrolase
Muramidase

9001-90-5
Unknown
Unknown
Fibrinolysin (Human)
Actase
E.C. 3.4.4.14
E.C. 3.4.21.7
Fibrinase
Fibrinolysin
Plasmin (9CI)
Serum Tryptase
Thrombolysin

9001-98-3
Unknown
Unknown
Rennin
Abomasal Enzyme
Chymase
Chymosin
EC 3443
EC 34234
Lab
Lab Ferment (German)
Rennase

9002-13-5
Unknown
Unknown
Urease
Jack Bean Urease

9002-18-0
Unknown
Unknown
Agar
Agar-Agar
Agar Agar Flake
Agar-Agar Gum
Bengal
Bengal Gelatin
Bengal Isinglass
Ceylon
Ceylon Isinglass
Chinese Isinglass
Digenea Simplex Mucilage
Gelose
Japan Agar
Japan Isinglass
Layor Carang
NCI-C50475

9002-61-3
Unknown
Unknown
Gonadotropin, chorionic
Ambinon
Antuitrin S
APL
APL (Hormone)
Apoidina
Chorigon
Choriogonadotropin
Choriogonin
Chorionic Gonadotrophin
Chorionic Gonadotropic Hormone
Chorionic Gonadotropin
Chorulon
Coriantin
Follutein
Gonabion
Gonadex
HCG
Human Chorionic Gonadotropin
Korotrin
Physex
Praedyn
Pregnyl
Primogonyl
Randonos
Synaphorin

9002-62-4
Unknown
Unknown
Prolactin
Adenohypophyseal luteotropin
Anterior Pituitary Luteotropin
Bovine Lactogenic Hormone
Bovine Prolactin
Galctin
Lactogen
Lactogenic Hormone
Lactosomatotropic Hormone
Luteotrophin
Luteotropic Hormone
Luteotropic Hormone LTH
Luteotropin
Mammotropin
Paralctin
Pituitary Lactogenic Hormone

9002-64-6
Unknown
Unknown
Parathormone
Kakerbin
Parathyrin
Parathyroid Hormone
PTH

9002-68-0
Unknown
Unknown
Follicle-Stimulating Hormone
Anthrogon
Follitropin

FSH
FSH-P
Gonad Stimulating Factor
Hebin
Luteoantine
Menotrophin
Menotropins
Prolan B
Thylakentrin
Urinary Hebin

9002-69-1
Unknown
Unknown
Relaxin
Cervilaxin
Releasin
W 1164-3

9002-72-6
Unknown
Unknown
Pituitary-Growth-Hormone
Adenohypophyseal Growth Hormone
Anterior Pituitary Growth Hormone
GH
Growth Hormone
Hormone Somatotrope (French)
Hypophyseal Growth Hormone
Phyol
Phyone
Somacton
Somatotrophic Hormone
Somatotrophin
Somatotropic Hormone
Somatotropin
STH

9002-81-7
$(CH_2O)x$
Polymer
Poly(oxymethylene) (9CI)

9002-88-4
$(C_2H_4)n$
Unknown
Polyethylene
AC 8
AC 394
AC 680
AC 1220
AC GA
ACP 6
AC 8 (Polymer)
Acroart
Agilene
Alathon
Alathon 14
Alathon 15
Alathon 1560
Alathon 6600
Alathon 7026
Alathon 7040
Alathon 7050
Alathon 7140
Alathon 7511
Alathon 5B
Alathon 71XHN
Alcowax 6
Aldyl A
Alithon 7050
Alkathene
Alkathene 17/04/00
Alkathene 22 300
Alkathene 200
Alkathene ARN 60
Alkathene WJG 11
Alkathene WNG 14
Alkathene XDG 33
Alkathene XJK 25
Allied PE 617
Alphex Fit 221
Ambythene
Amoco 610A4
A 60-20R
A 60-70R
Bakelite DFD 330
Bakelite DHDA 4080
Bakelite DYNH
Bareco Polywax 2000
Bareco Wax C 7500
Bicolene C
BPE-I
Bralen KB 2-11
Bralen RB 03-23
Bulen A
Bulen A 30
Carlona 58-030
Carlona 900
Carlona 18020 FA
Carlona PXB
Chemcor
Chemplex 3006
Cipe
Coathylene HA 1671
Courlene-X3
CPE
CPE 16
CPE 25
Cryopolythene
Cry-O-Vac L
Daisolac
Daplen
Daplen 1810 H
DFD 0173
DFD 0188
DFD 2005
DFD 6005
DFD 6032
DFD 6040
DFDJ 5505
DGNB 3825
Diothene
Dixopak
DMDJ 4309
DMDJ 5140
DMDJ 7008
Dowlex Film
DQDA 1868
DQWA 0355
DXM 100
Dyall
Dylan
Dylan Super
Dylan WPD 205
Dynh
Dynk 2
Eltex
Eltex 6037
Eltex A 1050
Epolene C
Epolene C 10
Epolene C 11
Epolene E
Epolene E 10
Epolene E 12
Epolene N
Ethene Polymer
Etherin
Etherol E
Ethylene Homopolymer
Ethylene Polymer
Ethylene Polymers (8CI)
23F203
Fabritone PE
FB 217
Fertene
Flamolin MF 15711
Flothene
FM 510
Fortiflex 6015
Fortiflex A 60/500
FP 4
2100 GP
G-Resins
Grex
Grex PP 60-002
Grisolen
HFDb 4201
Hi-Fax
Hi-Fax 1900
Hi-Fax 4401
Hi-Fax 4601
Hizex
Hizex 5000
Hizex 5100
Hizex 3000B
Hizex 3300F
Hizex 7000F
Hizex 7300F
Hizex 1091J
Hizex 1291J
Hizex 1300J
Hizex 2100J
Hizex 2200J
Hizex 2100LP
Hizex 5100LP
Hizex 6100P
Hizex 3000S
Hizex 3300S
Hizex 5000S
Hoechst PA 190
Hoechst Wax PA 520
Hostalen
Hostalen GD 620
Hostalen GD 6250
Hostalen GF 4760
Hostalen GF 5750
Hostalen GM 5010
Hostalen GUR
Hostalen HDPE
Interflo
Irax
Irrathene R
Lacqten 1020
LD 400
LD 600
LDPE 4
Lupolen 4261A
Lupolen 6042D

Lupolen 1010H
Lupolen 1800H
Lupolen 1810H
Lupolen 6011H
Lupolen KR 1032
Lupolen KR 1051
Lupolen KR 1257
Lupolen 6011L
Lupolen L 6041D
Lupolen N
Lupolen 1800S
Manolene 6050
Marlex 9
Marlex 50
Marlex 60
Marlex 960
Marlex 6003
Marlex 6009
Marlex 6015
Marlex 6050
Marlex 6060
Marlex EHM 6001
Marlex M 309
Marlex TR 704
Marlex TR 880
Marlex TR 885
Marlex TR 906
Microthene
Microthene 510
Microthene 704
Microthene 710
Mecrothene F
Microthene FN 500
Microthene FN 510
Microthene MN 754-18
Mikrolour
Mirason 9
Mirason 16
Mirason M 15
Mirason M 50
Mirason M 68
Mirason Neo 23H
Mirathen
Mirathen 1313
Mirathen 1350
Moplen RO-QG 6015
Neopolen
Neopolen 30N
Neozex 45150
Neozex 4010B
Nopol (Polymer)
Novatec JUO 80
Novatec JVO 80
NVC 9025
Okiten G 23
Orizon
Orizon 805
6020P
PA 130
PA 190
PA 520
PA 560
PAD 522
P 2010B
PE 512
PE 617
PEN 100
PEP 211
PES 100
PES 200
Petrothene
Petrothene LB 861
Petrothene LC 731
Petrothene LC 941
Petrothene NA 219
Petrothene NA 227
Petrothene XL 6301
P 4007EU
P 4070L
Planium
Plaskon PP 60-002
Plastazote X 1016
Plastronga
Plastylene MA 2003
Plastylene MA 7007
Politen
Politen I 020
Polyaethylen (German)
Poly-Em 12
Poly-Em 40
Poly-Em 41
Polyethylene AS
Polyethylene Resins
Polymist A12
Polymul CS 81
Polysion N 22
Polythene
Polywax 1000
Porolen
P 2070P
PPE 2
Procene UF 1.5
Profax A 60-008
P 2020T
P 2050T
P 4007T
PTS 2
PVP 8T
PY 100
RCH 1000
Repoc
Rigidex
Rigidex 35
Rigidex 50
Rigidex Type 2
Ropol
Ropothene OB.03-110
Sanwax 161P
Sclair 59
Sclair 2911
Sclair 19A
Sclair 96A
Sclair 59C
Sclair 79D
Sclair 11K
Sclair 19X6
SDP 640
Sholex 5003
Sholex 5100
Sholex 6000
Sholex 6002
Sholex F 171
Sholex F 6050C
Sholex F 6080C
Sholex 4250HM
Sholex L 131
Sholex S 6008
Sholex Super
Sholex XMO 314
Socarex
SRM 1475
SRM 1476
Staflen E 650
Stamylan 900
Stamylan 1000
Stamylan 1700
Stamylan 8200
Stamylan 8400
Sumikathene
Sumikathene F 101-1
Sumikathene F 210-3
Sumikathene F 702
Sumikathene G 201
Sumikathene G 202
Sumikathene G 701
Sumikathene G 801
Sumikathene G 806
Sumikathene Hard 2052
Sunwax 151
Super Dylan
Suprathen
Suprathen C 100
Takathene
Takathene P 3
Takathene P 12
Telcotene
Telecothene
Tenaplas
Tenite 800
Tenite 1811
Tenite 2910
Tenite 2918
Tenite 3300
Tenite 3340
Trovidur PE
Tyrin
Tyvek
Unifos Dyob S
Unifos EFD 0118
Valeron
Valspex 155-53
Velustral KPA
Vestolen
Vestolen A 616
Vestolen A 6016
Wax LE
WJG 11
WNF 15
WVG 23
XL 335-1
XL 1246
XNM 68
XO 440
Yukalon EH 30
Yukalon HE 60
Yukalon K 3212
Yukalon LK 30
Yukalon MS 30
Yukalon PS 30
Yukalon YK 30
ZF 36
Zinpol

9002-89-5
$(C_2H_4O)x$
Polymer
Polyvinyl-alcohol
Alcotex 88/05
Alcotex 88/10
Alkotex
Alvyl
Aracet APV
Cipoviol W 72
Covol
Covol 971

Elvanol
Elvanol 50-42
Elvanol 52-22
Elvanol 70-05
Elvanol 71-30
Elvanol 90-50
Elvanol 522-22
Elvanol 73125G
EP 160
Ethenol homopolymer (9CI)
Galvatol 1-60
Gelvatol
Gelvatol 1-30
Gelvatol 1-90
Gelvatol 3-91
Gelvatol 20-30
Gelvatol 2090
GH 20
GL 02
GL 03
GLO 5
GM 14
Gohsenol
Gohsenol AH 22
Gohsenol GH
Gohsenol GH 17
Gohsenol GH 20
Gohsenol GH 23
Gohsenol GL 03
Gohsenol GL 05
Gohsenol GL 08
Gohsenol GM 14
Gohsenol GM 94
Gohsenol GM 14l
Gohsenol KH 17
Gohsenol NH 05
Gohsenol NH 17
Gohsenol NH 18
Gohsenol NH 20
Gohsenol NH 26
Gohsenol NK 114
Gohsenol NL 05
Gohsenol NM 14
Ivalon
Kuralon VP
Kurare Poval 1700
Kurare PVA 205
Kurate Poval 120
Lemol
Lemol 5-88
Lemol 5-98
Lemol 12-88
Lemol 16-98
Lemol 24-98
Lemol 30-98
Lemol 51-98
Lemol 60-98
Lemol 75-98
Lemol GF-60
M 13/20
Mowiol
Mowiol N 30-88
Mowiol N 50-98
Mowiol N 70-98
NH 18
NM 11
NM 14
Polydesis
Polysizer 173
Polyvinol
Polyviol
Polyviol M 13/140
Polyviol MO 5/140
Polyviol W 25/140
Polyviol W 40/140
Poval 117
Poval 120
Poval 203
Poval 205
Poval 217
Poval 1700
Poval C 17
PVA
PVA 008
PVS 4
Resistoflex
Rhodoviol
Rhodoviol 4/125
Rhodoviol 16/200
Rhodoviol 4-125P
Rhodoviol R 16/20
Solvar
Sumitex H 10
Vibatex S
Vinacol MH
Vinalak
Vinarol
Vinarol DT
Vinarole
Vinarol ST
Vinavilol 2-98
Vinnarol
Vinol
Vinol 125
Vinol 205
Vinol 351
Vinol 523
Vinol Unisize
Vinyl Alcohol Polymer
Vinylon Film 2000

9002-91-9
$(C_2H_4O)x$
Polymer
Metaldehyde
Acetaldehyde, Homopolymer (9CI)
Acetaldehyde, Polymers
Antimilace
Limax
Limovet
META
Namekil
Polyacetaldehyde
Schneckokorn
Schnex-Schneckentod
Slugit
Terrasan-Schneckentod Gekoernt
2,4,6,8-Tetramethyl-1,3,5,7-tetraoxacyclooctane or acetaldehyde homopolymer

9002-92-0
$(C_2H_4O)x.C_{12}H_{26}O$
Polymer
Glycols, polyethylene, monododecyl ether
Aldosperse L 9
Atlas G-2133
Atlas G-3705
Base LP 12
BL 9
BL 9EX
BRIJ 23
BRIJ 30
BRIJ 35
BRIJ 30SP
Chimipal AE 3
Cimagel
Dehydol LS 4
DEPEG
DO 9
Dodecanol, ethoxylate
Dodecanol-ethylene oxide (9.5 moles) condensate
Dodecanol, polyethoxylated
Dodecyl alcohol, ethoxylated
α-Dodecyl-ω-hydroxy-polyoxyethylene
Dodecyl-polyaethylenoxydaether (German)
Dodecyl poly(oxyethylene)ether
Du Pont WK
Emal 108
Emalex 715
Emulgen 100
Emulgen 105
Emulgen 108
Emulgen 109
Emulgen 120
Emulgen 147
Emulmin L 500
Ethal LA-X
Ethosperse LA-4
Ethosperse LA 12
Ethosperse LA 23
Ethoxylated lauryl alcohol
G 3707
G 3711
G-2130A
Hydroxypolyethoxydodecane
LA
LA 7
LA (Alcohol)
Lanettes
Laureth
Laureth 9
Laureth 12
Laureth 23
Lauromacrogol
Lauromacrogol 400
Lauryl alcohol, ethoxylated
Lauryl polyethylene glycol ether
Lauryl poly(oxyethylene) ether
Lipal 4LA
Lipocol L-4
Lipocol L 12
Lipocol L-23
Lubrol 12A9
Lubrol PX
Marlipal 1217
Mergital LM 11
NCI-C54875
Newcol 1203
Nikkol BL
Nikkol BL 25
Nikkol BL 42
Nikkol BL 9EX
Nissan Nonion K 220
Noigen 160
Noigen 170
Noigen ET 83

Noigen ET 102
Noigen ET 143
Noigen ET 160
Noigen ET 170
Noigen ET 190
Noigen P
Noigen YX 400
Noigen YX 500
Noniolite AL 20
Oxyethylenated dodecyl alcohol
PEG n-Dodecyl Ether
Pegnol L 12
Pegnol L 20
Peregal O 20
Plurafac RA 43
Polidocanol
Polyethoxylated dodecanol
Polyethylene glycol dodecyl ether
Polyethylene glycol lauryl ether
Poly(ethylene oxide) dodecyl ether
Poly(oxy-1,2-ethanediyl), α-dodecyl-ω-hydroxy-
Polyoxyethylene lauric alcohol
Polyoxyethylene lauryl alcohol
Poly(oxyethylene) monolauryl ether
Rokanol L
Rokanol L 2
Rokanol L 4
Rokanol L 6
Rokanol L 10
Rokanol L 10/80
Romopal LN
Simulol 330 M
Simulsol P 4
Simulsol P 23
Siponic L
Siponic L 3
Siponic L-4
Siponic L 10
Siponic L 15
Siponic L 150
Siponic L 7-90
Slovasol S
Slovasol SF
Standamul LA 2
Surfactant WK
Teric 12A
Texofor B 9
Texofor B 30
Thesat
Thesit
Trycol LAL 12
Trycol LAL Series
Value 2205
YX 500

9002-93-1
$(C_2H_4O)x.C_{14}H_{22}O$
Polymer
CC(C)(C)CC(C)(C)c1ccc(OCCO)cc1
Glycols, polyethylene, mono-(p-(1,1,3,3-tetramethylbutyl)-phenyl) ether
Alfenol 3
Alfenol 9
Antarox A-200
Conco NIX-100
3,6,9,12,15,18,21,24,27,30-Decaoxatriacontan-1-ol, 30-(p-(1,1,3,3-tetramethylbutyl)-phenyl)-
Hydrol SW
Hyonic PE-250
Igepal CA-630
Marlophen 820
Neutronyx 605
Octoxinol
Octoxynol
Octoxynol 3
Octoxynol 9
Octyl phenol condensed with 12-13 moles ethylene oxide
p-tert-Octylphenoxypolyethoxyethanol
OPE 30
PEG-9 Octyl Phenyl Ether
Polyethylene glycol monoether with p-tert-octylphenyl
Polyethylene glycol mono-(4-octylphenyl) ether
Polyethylene glycol mono-(4-tert-octylphenyl) ether
Polyethylene glycol mono-(p-tert-octylphenyl) ether
Polyethylene glycol mono-(p-(1,1,3,3-tetramethylbutyl)phenyl) ether
Polyethylene glycol octylphenol ether
Polyethylene glycol 450 octyl phenyl ether
Polyethylene glycol p-octylphenyl ether
Polyethylene glycol p-tert-octylphenyl ether
Polyethylene glycol p-1,1,3,3-tetramethylbutylphenyl ether
Poly(oxy-1,2-ethanediyl), α-(4-(1,1,3,3-tetramethylbutyl)phenyl)-ω-hydroxy-(9CI)
Polyoxyethylene mono(octylphenyl) ether
Polyoxyethylene (9) octylphenyl ether
Polyoxyethylene (13) octylphenyl ether
Poly(oxyethylene)p-tert-octylphenyl ether
Preceptin
Triton X
Triton X 35
Triton X 45
Triton X 100
Triton X 102
Triton X 165
Triton X 305
Triton X 405
Triton X 705
TX 100

9002-98-6
$(C_2H_5N)x$
Polymer
Aziridine, Homopolymer (9CI)
CF 218 (Polymer)
Corcat P 18
Corcat P 100
Corcat P 145
Corcat P 200
Corcat P 600
EL 402
EL 420
Emerlube 6717
Emery 6717
Epomine D 3000
Epomine P 500
Epomine P 1000
Epomine P 1500
Epomine SP 003
Epomine SP 006
Epomine SP 012
Epomine SP 018
Epomine SP 103
Epomine SP 110
Epomine SP 200
Epomine 1000
Epomine 150T
Ethyleneimine, Homopolymer
Ethylenimine, Polymers (8CI)
Everamine
Everamine 50T
Everamine 210T
Montrek 6
NSC-124034
NSC-134422
NSC-196335
PAZ 33
PEI
PEI-7
PEI-15
PEI-30
PEI-45
PEI-1000
PEI-1500
PEI-2500
Polyaziridine
Polyethyleneimine
Polyethylenimine
Polyethylenimine 7
Polyethylenimine 15
Polyethylenimine 30
Polyethylenimine 45
Polyethylenimine 1000
Polyethylenimine 1500
Polyethylenimine 2500
Polyethylenimine (10,000)
Polyethylenimine (20,000)
Polyethylenimine (35,000)
Polyethylenimine (40,000)
Polymin FL
Polymin G 35
Polymin HS
Polymin Water Free
Tydex 12
15T
2MB

9003-01-4
$(C_3H_4O_2)_4$
168.06
Acrylic acid, Polymers
Acrylic acid homopolymer
Acrylic acid polymer
Acrylic acid resin
Acrylic polymer
Acrylic Resin
Acrysol A 1

Acrysol A 3
Acrysol A 5
Acrysol AC 5
Acrysol ASE-75
Acrysol WS-24
Alcogum
Antiprex A
Antiprex 461
Arolon
Aron
Aron A 10H
Atactic Poly(acrylic acid)
Carbomer 940
Carbomer 934P
Carbopol 934
Carbopol 940
Carbopol 941
Carbopol 960
Carbopol 961
Carbopol 934P
Carboset
Carboset 515
Carboset Resin No. 515
Carpolene
Dispex C40
G-Cure
Good-Rite K 37
Good-Rite K-700
Good-Rite K 702
Good-Rite K727
Good-Rite WS 801
Haloflex 202
Haloflex 208
Junlon 110
Jurimer AC 10H
Jurimer AC 10P
Nalfloc 636
Neocryl A-1038
Old 01
PAA-25
PA 11M
P 11H
Polyacrylate
Poly(acrylic acid)
Polytex 973
Primal ASE 60
2-Propenoic acid homopolymer (9CI)
R968
Racryl
76 Res
Revacryl A 191
Rohagit SD 15
Synthemul 90-588
Tecpol
Texcryl
Versicol E 7
Versicol E9
Versicol E15
Versicol S 25
Viscalex HV 30
Viscon 103
WS 24
WS 801
XPA
Zinpol

9003-04-7
$(C_3H_4O_2)x.xNa$
Polymer
Sodium-poly-acrylate
Polyco
Rhotex GS

9003-05-8
$(C_3H_5NO)x$
Polymer
Acrylamide, Polymers
Polyacrylamide
2-Propenamide, Homopolymer

9003-07-0
$(C_3H_6)x$
Polymer
Propene-polymers
Admer PB 02
A-Fax
Amco
Amerfil
Amoco 1010
Ampol C 60
AT 36
Atactic Polypropylene
Avisun
Avisun 101
Avisun 12-270A
Avisun 12-407A
Azdel
Beamette
Bicolene P
Carlona K 571
Carlona KM 61
Carlona P
Carlona PM 61 Naturel
Carlona PPLZ 074
CD 419
Celgard 2400W
Celgard KKX 2
Celgard 2500
Celgard 3501
Chisso 507b
Chisso Polypro 1014
Clysar
Coathylene PF 0548
Courlene PY
CPP 25S
D 151
Daplen AD
Daplen APP
Daplen AS 50
Daplen AT 10
Daplen ATK 92
Daplen DM 55U
Dexon E 117
DLP
DS 8620
Eastobond L 8080-270A
Eastobond M 3
Eastobond M 5
Eastobond M 5H
Elpon
El Rexene PP 115
EM 490
Enjay CD 392
Enjay CD 460
Enjay CD 490
Enjay E 11S
Enjay E 117
Epolene M 5H
Epolene M 5K
Epolene M 5W
Escon 622
Escon CD 44A
Escon EX 375
F 080PP
Gerfil
Gpcd 398
Hercoflat 135
Hercotuf 110A
Hercotuf PB 681
Hercules 6523
Herculon
HF 20
HO 50
Hostalen N 1060
Hostalen PP
Hostalen PPH 1050
Hostalen PPN
Hostalen PPN 1060
Hostalen PPN 1075 F
Hostalen PPN 1076 F
Hostalen PPR 1042
Hostalen PPT VP 7090A
Hostalen PP-U
Huls P 6500
ICI 543
Isotactic Polypropylene
J 400
J 700
JGD 1800
JMD 4500
K 300
Lambeth
Lanco Wax PP 1362D
Lupareen
Lym 42
Marlex 9400
Marlex HGH 050-01
Maurylene
Meraklon
MFR 4
MH 4
Mikrolour
Mitsui Polypro B 220
MM2A
Moplen
Moplen AD 50N
Moplen AS 50
Moplen Q 51C
Moplen T 30G
Mosten
Noblen
Noblen BC 8
Noblen D 101
Noblen D 501
Noblen EBG
Noblen FA 3
Noblen FL
Noblen FL 4
Noplen FL 6314
Noblen FP
Noblen FS 101
Noblen FS 2011
Noblen H
Noblen H 101
Noblen H 501
Noblen HS
Noblen JHHG
Noblen JK-M
Noblen MA 4
Noblen MH 6
Noblen MM 2A
Nablen S 50

Noblen S 101
Noblen SHG
Noblen 2VH501
Noblen W 101
Noblen W 501
Noblen W 502
Noblen WF 464
Novamont 2030
Novolen
Novolen KR 1300P
Novolen 1300ZX
Oletac 100
P 6500
Paisley 750
Paisley Polymer
Pellon 2505
Pellon 2506
Pellon FT 2140
Pistac CC
Pistac L
Polypro 1014
Polypro B 220
Polypro G 400P
Polypro J 600
Polypro J 400P
Polypropene
Polypropylene
Polytac
Poprolin
PP 1
PP 2
PP 4
PP 1151
PP 1 (Polymer)
PPSD 30
PR 144
Profax
Profax 6301
Profax 6401
Profax 6423
Profax 6501
Profax 6523
Profax 6601
Profax 6723
Profax 6823
Profax 6523F
Profax PCO 72
Propathene
Propathene 22/44
Propathene 101/24
Propathene 112/00/Grey 9897
Propathene GSE 108
Propathene GSE 180
Propathene GWE 21
Propathene GW 522 M
Propathene GW 601M
Propathene GY 702M
Propathene HF 20
Propathene HW 70GR
Propathene HWM 25
Propathene LWF 31
Propafilm
Propathene LY 542M
Propathene O
Propathene PXC 3830
Propathene PXC 4515
Propathene PXC 8639
Propathene PXC 9617
1-Propene homopolymer (9CI)
Propolin
Propophane
Propylene polymer
PS 2011
PXC 3391
PXC 8639
Rexall 413S
Rexene
413S
SD 5220
Shell 5520
Shoallomer
Shoallomer FA 120
Shoallomer FA 530
Shoallomer MA 210
Syndiotactic Polypropylene
TA 3
Tatren 141
Tatren EB 111
Tenite 423
Tenite 4231
Tenite 423DF
Tenite P 7673-079F
Trespaphan
Trespaphan CEA
Trespaphan N 12
Tuff-Lite
Ulstron
USI 11-4-0047
Vestolen 5200
Vestolen P 5500
Viscol 350P
Viscol 550P
Viscol 660 P
W 101
WEX 1242

9003-08-1
$(C_3H_6N_6.CH_2O)x$
Polymer
Melamine, Polymer with formaldehyde
Accobond 3524
Solapret
1,3,5-Triazine-2,4,6-triamine, Polymer with formaldehyde (9CI)

9003-11-6
$(C_3H_6O.C_2H_4O)x$
Polymer
Oxirane, methyl-, Polymer with oxirane (9CI)
Actinol P 3035
Adeka Carpol GH 10
Adeka Carpol MH 500
Adeka Carpol PH 2000
Adeka CM 294
Balab 615
Berol 370
Berol TVM 370
Bloatguard
Breox 75W270
BSP 5000
Carpol 2040
Carpol 2050
CE
Desmophen 7100
Dissolvan 4411
Emkalyx EP 64
Emkalyx L 101
Emulgen PP
Emulgen PP 150
Emulgen PP 250
Emulgen PP 290
Epan 420
Epan 450
Epan 485
Epan 710
Epan 720
Epan 740
Epan 742
Epan 750
Epan U 102
Epan U 103
Epan U 105
Epan U 108
Ethylene glycol polyethylene-polypropylene glycol ether (1:2)
Ethylene glycol-propylene glycol copolymer
Ethylene glycol-propylene glycol polymer
Ethylene oxide-propylene oxide copolymer
Ethylene oxide-propylene oxide copolymer ethylene glycol ether
FT 257
Genapol PF
Genapol PF 10
Genapol PF 20
Genapol PF 40
Genapol PF 80
Glycols, polyethylene-poly-propylene (8CI)
Highflex 443
Industrol N 3
Jeffox FF 200
Jeffox PPG 2000
Laprol 1502
Laprol 1502-2-70
Laprol 1601
Laprol 5003-2B10
M 90/20
Magcyl
Methyloxirane-oxirane co-polymer
Methyloxirane-oxirane polymer
Monolan 12000E80
Monolan 8000E80
Monolan PB
N 480
Nalco dispersant SPF-WTB 33
Nalco SPF-WTB 33
Nissan Unilube 25DE
Nissan Unilube 50MB168X
Nissan Unilube 50MB26X
Nissan Unilube 70DP950B
Nissan Unilube 750DE2620
Nissan Unilube 75DE3800
Nissan Unilube DE 60
Nixolen
Nixolen NS 4
Nixolen SL 19
Nixolen SL 8
Nixolen VS 13
Nixolen VS 2600
Nutek 7C
Oligoether L 1502-2-30
Oxalgon
Oxilube 50/150
Oxilube 50000
Oxirane, Polymer with methyl-

oxirane
Oxirane-methyloxirane copolymer
Oxirane-methyloxirane polymer
Oxyethylene-oxypropylene polymer
PAG
PAG 1
PAG 1 (Polyglycol)
PAG 2
PF 80
Pluracol 686
Pluracol V
Pluriol L 64
Pluriol SC 9361
Poloxalcol
Poloxalene
Poloxalene 2930
Poloxalene L 64
Poloxalkol
Poly(oxyethylene) poly(oxypropylene) glycol
Poly(oxyethylene)-poly(oxypropylene) polymer
Poly(propylene oxide-ethylene oxide)
Poly-G WT 90000
Poly-G WT 9150
Polyethylene oxide-polypropylene oxide
Polyethylene oxide-polypropylene oxide copolymer
Polyethylene-polypropylene glycol
Polylon 13-5
Polyoxyethylenated poly(oxypropylene)
Polyoxyethylene oxypropylene
Polyoxyethylene-polyoxypropylene
Polyoxyethylene-polyoxypropylene copolymer
Polyoxyethylene-polyoxypropylene polymer
Polyoxyethylenepropylene glycol ether
Polyoxypropylene-polyoxyethylene copolymer
Polypropylene glycol-ethylene oxide copolymer
PPG Diol 3000EO
Pr 168
Prevocell EO
PRO 21
Proksanol
Propylene oxide-ethylene oxide copolymer
Propylene oxide-ethylene oxide polymer
Proxanol
Proxanol 158
Proxanol 168
Proxanol 186
Proxanol 224
Proxanol 228
Proxanol P 168
Proxanol P 268
Proxanol TSL 3
PS 072
RC 102
Regulaid
Rokopol 30P9
Sannix PL 910
Separol 29
Separol WF 34
Separol WF 41
SKF 18667
Slovanik
Slovanik 610
Slovanik 630
Slovanik 660
Slovanik M
Slovanik PV 670
Slovanik T 310
Slovanik T 320
Slovanik T 630
Supronic B 50
Supronic B 75
Supronic E 400
Supronic E 800
Surflo HS 1
Teric PE
Teric PE 40
Teric PE 60
Teric PE 61
Teric PE 62
Teric PE 68
Teric PE 70
TsL 431
TVM 370
Ucon 25H
Ucon 75H
Ucon 75H1400
Ucon 75H90000
Unilube 50MB168X
Unilube 50MB26X
Unilube 70DP950B
Velvetol OE 2NT1
Voranol P 2001
WS 661
Wyandotte 7135
X 423
X 427
XD 8379
75H380000
75H90000

9003-13-8

$(C_3H_6O)x.C_4H_{10}O$

Polymer

Poly(oxy(methyl-1,2-ethanediyl)), α-butyl-ω-hydroxy

Butoxypolypropylene glycol
Butoxypropanediol polymer
Caswell No. 122
Crag Fly Repellent
ENT 8,286
EPA Pesticide Chemical Code 011901
Exp. Miticide No. 7
Fluid-AP
Newpol LB3000
OPSB
Poly(oxy(methyl-1,2-ethanediyl)), α-butyl-ω-hydroxy- (9CI)
Poly(oxypropylene) butyl ether
Polyoxypropylene (4) butyl ether
Polyoxypropylene (5) butyl ether
Polyoxypropylene (9) butyl ether
Polyoxypropylene (14) butyl ether
Polyoxypropylene (15) butyl ether
Polyoxypropylene (16) butyl ether
Polyoxypropylene (18) butyl ether
Polyoxypropylene (22) butyl ether
Polyoxypropylene (24) butyl ether
Polyoxypropylene (30) butyl ether
Polyoxypropylene (33) butyl ether
Polyoxypropylene (40) butyl ether
Polyoxypropylene (53) butyl ether
Polyoxypropylene glycol butyl monoether
Polyoxypropylene monobutyl ether
Polypropylene glycol butyl ether
Polypropylene glycol (4) butyl ether
Polypropylene glycol (5) butyl ether
Polypropylene glycol (9) butyl ether
Polypropylene glycol (14) butyl ether
Polypropylene glycol (15) butyl ether
Polypropylene glycol (16) butyl ether
Polypropylene glycol (18) butyl ether
Polypropylene glycol (22) butyl ether
Polypropylene glycol (24) butyl ether
Polypropylene glycol (30) butyl ether
Polypropylene glycol (33) butyl ether
Polypropylene glycol (40) butyl ether
Polypropylene glycol (53) butyl ether
Polypropylene glycol monobutyl ether
Poly(propylene oxide), monobutyl ether
PPG-4 Butyl Ether
PPG-5 Butyl Ether
PPG-9 Butyl Ether
PPG-14 Butyl Ether
PPG-15 Butyl Ether
PPG-16 Butyl Ether
PPG-18 Butyl Ether
PPG-22 Butyl Ether
PPG-24 Butyl Ether
PPG-30 Butyl Ether
PPG-33 Butyl Ether
PPG-40 Butyl Ether
PPG-53 Butyl Ether
Stabilene
Stabilene Fly Repellent
Ucon LB-250

Ucon LB 1145
Ucon LB 1800X
Witconol APB

9003-17-2
$(C_4H_6)x$
Polymer
Polybutadiene
Alfine
Atactic butadiene polymer
B 7
B 11
B 3000
Budium RK 622
Butadiene homopolymer
1,3-Butadiene, Homopolymer (9CI)
Butadiene oligomer
Butadiene polymer
1,3-Butadiene, Polymers
Butadiene resin
Butarez 15
CB 10
Diene 35 NF
Dienite X 555
Dienite X 644
Dienite 556
Dienite 643
FCR 126
FCR 1261
FCR 1261PD
Hystl
Hystl B 300
Hystl B 1000
Hystl B 2000
Hystl B 3000
LCB 150
Nisso BN 1000
Nisso PB 100
Nisso PB 3000
Nisso PB 4000
Nisso PB-B 4000
Nisso PB-GQ 3000
Nisso PR 2000
PBC200
Poly-1,3-butadiene
Polyoil 110
Polyoil 130
Quintol B 1000
S 820
XPDR-A 288
XRDR-A-288

9003-20-7
$(C_4H_6O_2)x$
Polymer
Acetic acid, vinyl ester, Polymer
Acetic acid ethenyl ester homopolymer (9CI)
Asahisol 1527
ASB 516
AYAA
AYAF
AYJV
Bakelite AYAA
Bakelite AYAF
Bakelite AYAT
Bakelite LP 90
Bond CH 3
Bond CH 18
Bond CH 1200
Booksaver
Borden 2123
Cascorez
Cemedine 196
Cevian 380
Cevian A 678
D 50
Danfirm
Daratak
DCA 70
D 50 M
D 50 (Polymer)
Duvilax
Duvilax BD 20
Duvilax HN
Duvilax LM 52
Elmer's Glue All
Elvacet 81-900
Emultex F
EN-COR
EP 1208
EP 1436
EP 1437
EP 1463
Esnil P 18
Everflex B
Formvar 1285
Gelva
Gelva 25
Gelva CSV 16
Gelva GP 702
Gelva S 55H
Gelva TS 22
Gelva TS 23
Gelva TS 30
Gelva TS 85
Gelva V 15
Gelva V 25
Gelva V 100
Gelva V 800
Gohensil E 50Y
Gohsenyl E 50 Y
Kurare OM 100
Lemac
Lemac 1000
Meikatex 5000NG60
Merckogel OR
Merckogen 6000
Mokotex D 2602
Movinyl
Movinyl 114
Movinyl 801
Movinyl 50M
Mowilith 30
Mowilith 50
Mowilith 70
Mowilith 90
Mowilith D
Mowilith DV
Mowilith M70
National 120-1207
National Starch 1014
NS 2842
OM 100
OR 1500
P-170
Pioloform F
Plyamul 40-155
Plyamul 40-350
Polisol S-3
Polyco 953
Polyco 2116
Polyco 2134
Polyco 117FR
Polyfox P 20
Polyfox PO
Polysol 1000
Polysol 1200
Polysol 1000AX
Polysol PS 10
Polysol S 5
Polysol S 6
Poly(vinylacetate)
Protex
Protex (Polymer)
PS 3H
PVAE
R 10688
Raviflex 43
76 Res
Resyn 25-1014
Resyn 25-1025
Rhodopas
Rhodopas 010
Rhodopas 5425
Rhodopas A 10
Rhodopas AM 041
Rhodopas B
Rhodopas BB
Rhodopas HV 2
Rhodopas M
Rhodopas 5000SMR
RV225-5B
Sakunol SN 08
S-Nyl-P 42
Soloid
Soviol
SP 60
SP 60 (Ester)
Toabond 2
Toabond 6
Toabond 40H
TS2
UCAR 15
UCAR 130
UK 131
V 501
VA 0112
Vinac
Vinac ASB 10
Vinac B 7
Vinac RP251
Vinacet D
Vinalite D 50N
Vinalite DS 41/11
Vinamul 9300
Vinapol A 16
Vinipaint 555
Vinnapas B
Vinnapas B 17
Vinnapas B 100
Vinnapas UW 50
Vinyl acetate homopolymer
Vinyl acetate polymer
Vinyl acetate resin
Vinylite AYAF
Vinylite AYAT
Vinyl Products R 10688
Wikol
Winacet D

9003-22-9

9003-27-4

$(C_4H_6O_2.C_2H_3Cl)x$
Polymer
Acetic acid, vinyl ester, Polymer with chloroethylene
A15
A15-0
Acetic acid ethenyl ester polymer with chlorethene (9CI)
A 15 (Polymer)
A 15S
Bakelite LP 70
Bakelite VLFV
Bakelite VMCC
Bakelite VSJD 10
Bakelite VYHD
Bakelite VYHH
Bakelite VYHN
Bakelite VYNS
Bakelite VYNW
Breon 351
Breon 425
Breon AS 60/41
Chloroethylenevinyl acetate polymer
Corvic 51/83
Corvic 236581
Corvic R 46/88
Denka AC 50
Denkalac 61
Denkalac 41M
Denka Vinyl MM 90
Diamond Shamrock 744
Diamond Shamrock 7401
Eslec C
Exon 450
Exon 454
Exon 470
Exon 481
Exon 760
Flovic
Geon 135
Geon 351
Geon 421
Geon 427
Geon 434
Geon 103EP-J
Geon 440L2
Geon 1032X
Geon 130X10
Geon 400X47
Geon 100X150
Geon 150XML
Geon 450X150PN
Geon 103 ZX
Hostaflex VP 150
Hostalit PVP
Leucovyl PA 1302
Lucovyl GA 8502
Lucovyl MA 6028
Lucovyl PA 1208
Marvinol VP 56
50ME
Norvinyl P 6
Opalon 400
Pevikon C 870
Pliovac AO
Pliovic AO
Polyvinyl chloride-polyvinyl acetate
PVC Cordo
Resin 4301
Rhodopas 6000
Rhodopas AX
Rhodopas AX 30/10
Rhodopas AX 85/15
S 2803
Sarpifan HP 1
Sconatex
Solvic 523KC
Solvic PA 513
Solvic 513PB
Solvinc PA 513
Sumilit PCX
Tennus 0565
Tygon
VA 3 (Copolymer)
VAGD
VH 10/60
Vilit 40
Vinnol H 10/60
Vinnol H 15/45
Vinnol H 40/60
Vinyl acetate-vinyl chloride copolymer
Vinyl acetate-vinyl chloride polymer
Vinyl chloride-vinyl acetate polymer
Vinylite VGHH
Vinylite VYDR
Vinylite VYDR 21
Vinylite VYFS
Vinylite VYHD
Vinylite VYHH
Vinylite VYHH-1
Vinylite VYNS
Vinylite VYNW
Vinyon
VLVF
VMCC
VYGEN 220
VYHH
VYNS
VYNW

9003-27-4
$(C_4H_8)x$
Polymer
Polyisobutene
AMOCO 600
Caswell No. 676A
EPA Pesticide Chemical Code 011403
Hyvis 30
Hyvis 200
Hyvis 2000
Indopol H 1900
Isobutene Homopolymer
Isobutene Polymer
Isobutylene Homopolymer
Isobutylene Polymer
Isobutylene Resin
KP 5
Maxvis 2000
2-Methyl-1-propene homopolymer
2-Methylpropene polymer
Napvis 30
Oktol
Oppanol B
Oppanol B 3
Oppanol B 15
Oppanol B 100
Oppanol B 150
Oppanol B 200
P 20
P 85
P 118
P 200
Paratac
Paxon 3204
PB 150
PIB 100
Polybutene
Polybutylene
Polyisobutylene
Polyisobutylene PSG
Poly(2-methylpropene)
Polyvis 06SH
Polyvis 30SH
Polyvis 150SH
Polyvis 200SH
1-Propene, 2-methyl-, Homopolymer (9CI)
Propene, 2-methyl-, Polymers
PSB
PSB (Aliphatic polymer)
USB

9003-29-6
$(C_4H_8)x$
Polymer
Polybutene
AI3-26247
AI3-26248
AI3-26249
AI3-26250
AI3-26251
AI3-26252
AI3-26253
AI3-26254
AI3-26255
AI3-26256
AI3-29162-X
Amoco H 300
Amoco 15H
Butene, Homopolymer (9CI)
Butene, Homopolymer (n=ca. 5)
Butene, Homopolymer (n=ca. 8)
Butene, Homopolymer (n=ca. 9)
Butene, Homopolymer (n=ca. 12)
Butene, Homopolymer (n=ca. 13)
Butene, Homopolymer (n=ca. 14)
Butene, Homopolymer (n=ca. 15)
Butene, Homopolymer (n=ca. 20)
Butene, Homopolymer (n=ca. 32)
Butene, Homopolymer (n=ca. 34)
Butene, Homopolymer (44% Formulation)
Butene, Polymers
Chevron 6
Chevron 12
Chevron 16
Chevron 18
H 100

H 300
H 1500
H 1900
HE 375
HE 1975
HV 1900
HV-1900
Hyvis 10
Hyvis 7000/45
Indopol
Indopol H 50
Indopol H 100
Indopol H 300
Indopol L 10
Indopol L 14
Indopol L 100
L 14
L 14 (Polymer)
L 100
LV 50
Oktol 600
Oronite 6
PB 2110
Petrofin 100
Polybuden 300H
Polybutene SH 015
Polybutylene
Polyisobutene
Polyisobutylene
Polyvis OO
Polyvis 015SH
Polyvis 2000CH
SH 015
SV 7000
Witron 131
15H
300H
2000H

9003-31-0
$(C_5H_8)x$
Polymer
Polyisoprene
Betaprene H
1,3-Butadiene, 2-methyl-, Homopolymer (9CI)
Isoprene D
Isoprene oligomer
Isoprene polymer
Isoprene, Polymers
2-Methyl-1,3-butadiene, Homopolymer
cis-1,4-Polyisoprene
trans-Polyisoprene
trans-1,4-Polyisoprene
Poly(2-methyl-1,3-butadiene)
Poly-1-methylbutenylene

9003-35-4
$(C_6H_6O.(CH_2O)x)x$
Polymer
Phenol-formaldehyde resin
Formaldehyde, phenol polymer
Formaldehyde, polymer with paraformaldehyde and phenol
Paraformaldehyde, formaldehyde, phenol polymer
Paraformaldehyde, phenol polymer
Phenol-formaldehyde copolymer
Phenol, formaldehyde polymer
Phenol, Polymer with formaldehyde
Phenol, polymer with paraformaldehyde

9003-39-8
$(C_6H_9NO)x$
Polymer
Poly(1-vinyl-2-pyrrolidinone) homopolymer
Agent AT 717
Albigen A
Aldacol Q
AT 717
Bolinan
1-Ethenyl-2-pyrrolidinone polymers
Ganex P 804
Hemodesis
Hemodez
K15
K25
K25 (Polymer)
K30
K30 (Polymer)
K60
K60 (Polymer)
K90
K115
K115 (Polyamide)
Kollidon
Kollidon 17
Kollidon 25
Kollidon 30
Luviskol
Luviskol K30
Luviskol k90
MPK 90
NCI-C60582
Neocompensan
Peragal ST
Peregal ST
Periston
Periston-N
Peviston
Plasdone
Plasdone K-26/28
Plasdone K 29-32
Plasdone XL
Plasmosan
Polyclar AT
Polyclar H
Polyclar L
Poly(1-(2-oxo-1-pyrrolidinyl)-ethylene)
Polyvidone
Poly(n-vinylbutyrolactam)
Polyvinylpyrrolidone
Poly(vinylpyrrolidinone)
Poly(n-vinylpyrrolidinone)
Poly(1-vinylpyrrolidinone)
Poly(1-vinyl-2-pyrrolidinone) Hueper's polymer No.1
Poly(1-vinyl-2-pyrrolidinone) Hueper's polymer No.2
Poly(1-vinyl-2-pyrrolidinone) Hueper's polymer No.3
Poly(1-vinyl-2-pyrrolidinone) Hueper's polymer No.4
Poly(1-vinyl-2-pyrrolidinone) Hueper's polymer No.5
Poly(1-vinyl-2-pyrrolidinone) Hueper's polymer No.6
Poly(1-vinyl-2-pyrrolidinone) Hueper's polymer No.7
Povidone
Povidone (USP XIX)
Protagent
PVP
PVP 1
PVP 2
PVP 3
PVP 5
PVP 4
PVP 6
PVP 7
PVP 40
PVP-K 15
PVP-K 30
PVP-K 60
PVP-K 90
PVPP
2-Pyrrolidinone, 1-ethenyl, Homopolymer
2-Pyrrolidinone, 1-vinyl-, Polymers
143 RP
Subtosan
Vinisil
n-Vinylbutyrolactam polymer
Vinylpyrrolidinone polymer
n-Vinylpyrrolidinone polymer
n-Vinylpyrrolidone polymer
Vinylpyrrolidone polymer

9003-53-6
$(C_8H_8)x$
Polymer
Styrene-polymer
3A
A 3-80
Afcolene
Afcolene 666
Afcolene S 100
Atactic Polystyrene
Bactolatex
Bakelite SMD 3500
BASF III
BDH 29-790
Benzene, ethenyl-, Homopolymer (9CI)
Bextrene XL 750
Bicolastic A 75
Bicolene H
Bio-Beads S-S 2
BP-KLP
BSB-S 40
BSB-S-E
Bustren
Bustren K 500
Bustren K 525-19
Bustren U 825
Bustren U 825E11
Bustren Y 825
Bustren Y 3532
Cadco 0115
Carinex GP
Carinex HR
Carinex HRM
Carinex SB 59
Carinex SB 61

9003-53-6 - continued

Carinex SL 273
Carinex TGX/MF
Copal Z
Cosden 550
Cosden 945E
Denka QP3
Diarex 43G
Diarex HF 55
Diarex HF 77
Diarex HF 55-247
Diarex HS 77
Diarex HT 88
Diarex HT 90
Diarex HT 190
Diarex HT 500
Diarex HT 88A
Diarex YH 476
Dorvon
Dorvon FR 100
Dow 360
Dow 456
Dow 665
Dow 860
Dow 1683
Dow MX 5514
Dow MX 5516
Dylark 250
Dylene
Dylene 8
Dylene 9
Dylene 8G
Dylite F 40
Dylite F 40L
686E
Edistir RB
Esbrite
Esbrite 2
Esbrite 4
Esbrite 4-62
Esbrite 8
Esbrite G 10
Esbrite G-P 2
Esbrite 500HM
Esbrite LBL
Escorez 7404
Estyrene 4-62
Estyrene G 15
Estyrene G 20
Estyrene G-P 4
Estyrene H 61
Estyrene 500SH
Ethenylbenzene homopolymer
FC-MY 5450
FG 834
Foster Grant 834
Gedex
454H
HF 10
HF 55
HF 77
HH 102
HHI 11
Hi-Styrol
Hostyren N
Hostyren N 4000
Hostyren N 7001
Hostyren N 4000V
Hostyren S
HT 88
HT 91-1
HT 88A
HT-F 76
IT 40
K 525
KB (Polymer)
KM
KM (Polymer)
Koplen 2
KR 2537
Krasten 1.4
Krasten 052
Krasten SB
Lacqren 506
Lacqren 550
LS 061A
LS 1028E
Lustrex
Lustrex H 77
Lustrex HP 77
Lustrex HH 101
Lustrex HT 88
MX 4500
MX 5514
MX 5516
MX 5517-02
168N15
Napst
NBS 706
N 4000v
Poligostyrene
Owispol GF
Pelaspan 333
Pelaspan ESP 109s
Piccolastic
Piccolastic A
Piccolastic A 5
Piccolastic A 25
Piccolastic A 50
Piccolastic A 75
Piccolastic C 125
Piccolastic D
Piccolastic D-100
Piccolastic D 125
Piccolastic D 150
Piccolastic E 75
Piccolastic E 100
Piccolastic E 200
Polyco 220NS
Polyflex
Polystrol D
Polystyrene
Polystyrene BW
Polystyrene Beads, Expandable, evolving flammable vapor. [UN 2211]
Polystyrene Latex
Polystyrol
Printel's
PRX 1195
PS 1
PS 2
PS 200
PS 209
PS-B
PSB-C
PSB-S
PSB-S 40
PSB-S-E
PS 454H
PS 2 (Polymer)
PS 5 (Polymer)
PSV-L
PSV-L 1
PSV-L 2
PSV-L 1S
PY2763
R 3
R 3612
Rexolite 1422
Rhodolne
S 173
SB 475K
SD 188
Shell 300
SMD 3500
SPS 600
SRM 705
SRM 706
ST 90
Sternite 30
Sternite ST 30VL
ST 30UL
Styrafoil
Styragel
Styrene Polymers
Styrex C
Styrocell PM
Styrofan 2D
Styroflex
Styrofoam
Styrolux
Styron
Styron 475
Styron 492
Styron 666
Styron 678
Styron 679
Styron 683
Styron 685
Stryon 686
Styron 690
Styron 69021
Styron 440A
Styron 470A
Styron 475D
Styron GP
Styron 666K27
Styron PS 3
Styron T 679
Styron 666U
Styron 666V
Styropian
Styropian FH 105
Styropol HT 500
Styropol IBE
Styropol JQ 300
Styropol KA
Styropor
TC 3-30
TGD 5161
TMDE 6500
Toporex 500
Toporex 830
Toporex 550-02
Toporex 850-51
Toporex 855-51
Trolitul
Trycite 1000
825TV
825TV-PS
475U
U625
666U
Ubatol U 2001
UCC 6863
UN 2211 [Polystyrene beads,

expandable, evolving flammable vapor.]
UP 1
UP 2
UP 27
UPM
UPM703
UPM508L
Vestolen P 5232g
Vestyron
Vestyron 512
Vestyron 114-12
Vestyron MB
Vestyron N
Vinamul N 710
Vinamul N 7700
Vinylbenzene polymer
Vinyl Products R 3612
X 600

9003-54-7
$(C_8H_8.C_3H_3N)x$
Polymer
Acrylonitrile, Polymer with styrene
Acrilafil
Acrylonitrile-styrene copolymer
Acrylonitrile-styrene polymer
Acrylonitrile-styrene Resin
ACS
Afcolene
AS 61CL
Bakelite RMD 4511
Cevian HL
Cevian N
Cevian NF
Denka AS-CY
Dialux
Dikaril
Estyrene AS
FN 20
FN 25
Kostil
Kostil 235
Kostil AN (ATX) 2010
Kralac 1155
Litac
Litac C 100P
Lna 21-1000
Lorkaril
Luran
Luran 378P
Luran 368R
Lustran
Lustran 28
Lustran A
Lustran A 21
Lustran A 2121
Lustran LNA 21
Lustran SAN
Piccoflex
Polysan
Polystyrene-acrylonitrile
2-Propenenitrile, Polymer with ethenylbenzene (9CI)
Rexene 106
RMD 4500
Ronfalin S
Sahrekkusu
SAN-C
Sanrex
Sanrex C
Sanrex Sanc
Sanrex SAN-H
Sepian N 050
Sepian N 6000HL
SN 20
SN 25
SN 20A
SNP 2
SN 20P
Styren-acrylonitrile polymer
Styrene-acrylonitrile copolymer
Terulan KP 2540
Terulan KR 2540
Thermocomp BF
Trolitul AN
Trolitul EN
Tyril
Tyril 760
Tyril 767
Tyril 780
Tyril 783
Tyril 790
Tyril 860
Tyril 867
Tyrilfoam

9003-55-8
$(C_8H_8.C_4H_6)x$
Polymer
Styrene polymer with 1,3-butadiene
8000A
Afcolac B 101
Andrez
Base 661
Benzene, ethenyl-, Polymer with 1,3-butadiene
1,3-Butadiene-styrene copolymer
Butadiene-styrene polymer
1,3-Butadiene-styrene polymer
Butadiene-styrene Resin
Butakon 85-71
Diarex 600
Dienol S
Dow 209
Dow 234
Dow 460
Dow 620
Dow 680
Dow 816
Dow Latex 612
Dowtex TL 612
DST 50
DST 75
Duranit
Duranit 40
Edistir RB 268
Ethenylbenzene polymer with 1,3-butadiene
Goodrite 1800X73
Histyrene S 6F
Hycar LX 407
K 55E
Kopolymer butadien styrenovy (Czech)
KRO 1
KRO 2
Litex CA
Lytron 5202
Marbon 9200
Marbon 8000a
Nipol 407
Pharos 100.1
Plioflex
Pliolite 151
Pliolite 160
Pliolite 491
Pliolite 55B
Pliolite S5
Pliolite S 50
Pliolite S 5A
Pliolite S-5B
Pliolite S-5C
Pliolite S 5D
Pliolite S-5E
Polybutadiene-polystyrene copolymer
Polyco 2410
Polyco 2415
PS-SU2
R-104
Ricon 100
Ricon 109
Ricon 181
Ricon 182
Ricon 183
Ricon A-84
SBS
SBS (Block polymer)
SBS Copolymer
SD 354
SD 345 (Polymer)
S6F Histyrene Resin
SKS 85
Soil Stabilizer 661
Solprene 300
Solprene 303
Styrene-butadiene copolymer
Styrene-1,3-butadiene copolymer
Styrene-butadiene polymer
Synpol 1500
Thermoplastic 125
TR 201
UP 1E
Vestyron HI

9003-56-9
$(C_8H_8.C_4H_6.C_3H_3N)x$
Polymer
Acrylonitrile, Polymer with 1,3-butadiene and styrene
Acrylonitrile-butadiene-styrene copolymer
2-Propenenitrile, Polymer with 1,3-butadiene and ethenylbenzene (9CI)

9003-73-0
$(C_{10}H_{16})x$
Polymer
Polylimonene
Cyclohexene, 1-methyl-4-(1-methylethenyl)-, Homopolymer (9CI)
Dipentene polymer
Limonene polymer
p-Mentha-1,8-diene, Polymers
Polydipentene

Polymerized dipentene
Zonarez 7085
Zonarez 7115

9003-74-1
Unknown
Unknown
Terpene resin (9CI)

9004-07-3
Unknown
Unknown
Chymotrypsin
α Chymar
α-Chymar Ophth
Avazyme
Chymar
Chymotest
α-Chymotrypsin
Chymotrypsin A
Chymotrypsin B
E.C. 3.4.4.5
E.C. 3.4.4.6
E.C. 3.4.21.1
Enzeon
Quimar
Quimotrase

9004-09-5
Unknown
Unknown
Plasmin (Human) (9CI)
Fibrinolysin (Human)

9004-10-8
Unknown
Unknown
Insulin
Actrapid
Decurvon
Endopancrine
Iletin
Insular
Insulin Injection
Insulyl
Iszilin

9004-30-2
Unknown
Unknown
Carboxymethyl hydroxyethyl cellulose
Carboxymethyl hydroxyethyl-cellulose
Cellulose, carboxymethyl hydroxyethyl ether
Cellulose, carboxymethyl hydroxyethyl mixed ether
Cellulose, carboxymethyl 2 -hydroxyethyl ether
Cellulose, carboxymethyl-2-hydroxyethyl ether, sodium salt
CMHEC
Sodium carboxymethyl hydr-oxyethylcellulose
SPX 5338
Tylose CH 50

9004-32-4
Unknown
Unknown
Cellulose, carboxymethyl ether, sodium salt
AC-Di-Sol. NF
AKU-W 515
Aquaplast
B 10
Blanose BS 190
Blanose BWM
B 10 (Polysaccharide)
Carbose 1M
Carboxymethyl cellulose
Carboxymethylcellulose sodium
Carboxymethylcellulose, sodium salt
Carmethose
Cellofas
Cellofas B
Cellofas B5
Cellofas B6
Cellofas B50
Cellofas C
Cellogel C
Cellogen 3H
Cellogen PR
Cellogen WS-C
Cellpro
Cellufix FF 100
Cellugel
Cellulose glycolic acid, sodium salt
Cellulose sodium glycolate
CMC
CMC 2
CMC 41A
CM-Cellulose Na Salt
CM-Cellulose Sodium Salt
CMC 7H
CMC 4H1
CMC 7H3SF
CMC 7L1
CMC 7M
CMC 4M6
CMC 7MT
CMC 3M5T
CMC Sodium Salt
Collowel
Copagel PB 25
Courlose A 590
Courlose A 610
Courlose A 650
Courlose F 4
Courlose F 8
Courlose F 20
Courlose F 370
Courlose F 1000G
Daicel 1150
Daicel 1180
Edifas B
Ethocel
Ethoxose
Fine Gum HES
Glikocel TA
7H3SF
KMTs 212
KMTs 300
KMTs 500
KMTs 600
Lovosa
Lovosa 20alk.
Lovosa TN
Lucel
Lucel (Polysaccharide)
Majol PLX
Modocoll 1200
Nacm-Cellulose Salt
Nymcel S
Nymcel SLC-T
Nymcel ZSB 10
Nymcel ZSB 16
Polyfibron 120
Sanlose SN 20A
Sarcell TEL
S 75M
Sodium carboxmethylcellulose
Sodium carboxyl methyl cellu-lose
Sodium carboxymethyl cellu-lose
Sodium cellulose glycolate
Sodium CMC
Sodium CM-Cellulose
Sodium glycolate cellulose
Sodium salt of carboxymethyl-cellulose
Tylose 666
Tylose C
Tylose C 30
Tylose C 300
Tylose C 600
Tylose CB 200
Tylose CBR 400
Tylose CBR Series
Tylose CBS 30
Tylose CBS 70
Tylose CB Series
Tylose C 1000P
Tylose CR
Tylose CR 50
Tylose DKL
Unisol RH

9004-34-6
Unknown
Unknown
Cellulose (ACGIH,OSHA)
Abicel
β-Amylose
Arbocel
Arbocel BC 200
Arbocell B 600/30
Avicel
Avicel 101
Avicel 102
Avicel PH 101
Avicel PH 105
Cellex MX
α-Cellulose
Cellulose 248
Cellulose Crystalline
Celufi
Cepo
Cepo CFM
Cepo S 20
Cepo S 40
Chromedia CC 31
Chromedia CF 11
Cupricellulose

Elcema F 150
Elcema G 250
Elcema P 050
Elcema P 100
Fresenius D 6
Heweten 10
Hydroxycellulose
Kingcot
LA 01
MN-Cellulose
Onozuka P 500
Pyrocellulose
Rayophane
Rayweb Q
Rexcel
Sigmacell
Solka-Fil
Solka-Floc
Solka-Floc BW
Solka-Floc BW 20
Solka-Floc BW 100
Solka-Floc BW 200
Solka-Floc BW 2030
Spartose OM-22
Sulfite Cellulose
Tomofan
Tunicin
Whatman CC-31

9004-35-7
Unknown
Unknown
Acetylcellulose
A 432-130B
Acetate cotton
Acetate ester of cellulose
Acetic acid, cellulose ester
Acetose
Acetyl 35
Allogel
Ampacet C/A
Bioden
Cellidor
Cellidor A
Cellit K 700
Cellit L 700
Cellulose, acetate (9CI)
Cellulose 2,5-acetate
Cellulose, diacetate
Cellulose, 2,5-diacetate
Cellulose monoacetate
Cellulose, triacetate
Crellate
DP 02
DP 06
E 376-40
E 383-40
E 394-30
E 394-40
E 394-45
E 394-60
E 398-10
E 400-25
E-400-25
Eastman 298-10
Etrol OEM
Monoacetylcellulose
Nicollembal
Nixon C/A
Plastacele
PP 612
PP 613
PP 628
Stripmix
Strux
T-Cellit
Tenite I
Vladipor

9004-36-8
Unknown
Unknown
Cabufocon A
Cabufocon B
Celaburato (Spanish)
Cellaburato
Cellaburatum (Latin)
Cellulose, acetate butanoate (9CI)
Cellulose acetate butyrate
Cellulose acetate-butyrate
Paracab II
Tenite Butyrate Formula 264 H4

9004-51-7
Unknown
Unknown
Iron-dextrin Complex
Astrafer
Dextriferron
Dextriferron Injection
Ferrigen
Iron Carbohydrate Complex
Iron Dextrin Injection

9004-53-9
$(C_6H_{10}O_5)x.xH_2O$
Unknown
Dextrins

9004-57-3
Unknown
Unknown
Cellulose, ethyl ether
Ampacet E/C
Cellulose ethyl
Cellulose ethylate
ETS
ETS (Polysaccharide)
Ethocel
Ethocel 150
Ethocel 890
Ethocel E7
Ethocel E50
Ethocel MED
Ethocel N7
Ethocel N10
Ethocel N200
Ethocel STD
Ethylcellulose
G 50
G 200
G 50 (Polysaccharide)
N 5
Nixon E/C
SPT 50 CPS
T 100
T 100 (Polysaccharide)

9004-59-5
Unknown
Unknown
Methyl ethyl cellulose
Celacol EM
Cellocol EM
Cellofas A
Cellofas WLD
Cellulose, ethyl methyl ether
Edifas A
Edifas Grade "A"
Ethyl methyl cellulose
Methylethylcellulose

9004-61-9
Unknown
Unknown
Hyaluronic acid (9CI)
Acid, hyaluronic
Amvisc
Etamucine
Healon
Hyaluronan
Hyaluronate, sodium
Hyvisc
Luronit
Sodium hyaluronate

9004-62-0
Unknown
Unknown
Cellulose, 2-hydroxyethyl ether
AW 15 (Polysaccharide)
BL 15
Cellosize 4400H16
Cellosize QP
Cellosize QP3
Cellosize QP 1500
Cellosize QP 4400
Cellosize QP 30000
Cellosize UT 40
Cellosize WP
Cellosize WP 300
Cellosize WP 4400
Cellosize WP 300H
Cellosize WP 400H
Cellosize WPO 9H17
Cellulose hydroxyethylate
Cellulose hydroxyethyl ether
Fuji HEC-BL 20
Glutofix 600
HEC
HEC-AL 5000
Hercules N 100
Hespan
Hetastarch
Hydroxyethyl cellulose
2-Hydroxyethyl cellulose
Hydroxyethyl cellulose ether
2-Hydroxyethyl cellulose ether
Hydroxyethyl ether cellulose
Hydroxyethyl starch
J 164
Natrosol
Natrosol 250
Natrosol 250G
Natrosol 250H
Natrosol 300H

Natrosol 250HHP
Natrosol 250HHR
Natrosol 250HR
Natrosol 250H4R
Natrosol 250HX
Natrosol 240JR
Natrosol 150L
Natrosol 180L
Natrosol 250L
Natrosol LR
Natrosol 250M
Natrosol 250MH
Oets
Tylose H 20
Tylose H 300
Tylose H Series
Tylose MB
Tylose MH
Tylose MHB
Tylose MHB-Y
Tylose MHB-YP
Tylose MH-K
Tylose MH-XP
Tylose P
Tylose PS-X
Tylose P-X
Tylose P-Z Series

9004-64-2
Unknown
Unknown
Hydroxypropyl-cellulose
Hydroxypropyl ether of cellulose
Klucel

9004-66-4
Unknown
Unknown
Imferon
A 100
A 100 (Pharmaceutical)
B 75
Chinofer
Dextran Iron Complex
Dextrofer 75
Eisendextran (German)
Fe-Dextran
Fenate
Ferdex 100
Ferric dextran
Ferridextran
Ferrodextran
Ferroglucin
Ferroglukin 75
Imposil
Iro-Jex
Iron dextran
Iron-Dextran Complex
Iron Dextran Injection
Iron hydrogenated dextran
Ironorm Injection
Myofer 100
Polyfer
Prolongal
RCRA waste number U139
Ursoferran

9004-67-5
Unknown
Unknown
Cellulose, methyl ether (1/2%)
Adulsin
Bagolax
Bufapto Methalose
Bulkaloid
Celacol M
Celacol M20
Celacol M450
Celacol MM
Celacol MM 10P
Celacol M 20P
Cellapret
Cellogran
Cellothyl
Cellulose methyl
Cellulose methylate
Cellumeth
Cethylose
Cethytin
Culminal K 42
Edisol M
Hydrolose
Mapolose M25
Mapolose 60SH50
MCO 8000
MC 4000 cP
MC 20000S
Mellose
Methocel 10
Methocel 15
Methocel 181
Methocel 400
Methocel 4000
Methocel A
Methocel CHG
Methocel 400CPS
Methocel 4000CPS
Methocel MC
Methocel MC 25
Methocel MC4000
Methocel MC 8000
Methocel SM 100
Methulose
Methylcellulose
Methylcellulose (1/2%)
Methyl Cellulose-A
Methyl cellulose ether
Metolose MC 8000
Metolose 60SH
Metolose 60SH400
Metolose SM 15
Metolose SM 100
Metolose SM 4000
MMTs-BTR
MTs
Napolone
Nicel
Rhomellose
Syncelose
Tylose 444
Tylose A4S
Tylose MF
Tylose MH
Tylose MH20
Tylose MH50
Tylose MH300
Tylose MH1000
Tylose MH2000
Tylose MH4000
Tylose MH300P
Tylose SAP
Tylose SL
Tylose SL 100
Tylose SL 400
Tylose SL 600
Tylose TWA
USP Methylcellulose
Viscol
Viscontran L52
Viscosol
Walsroder MC 20000S

9004-70-0
Unknown
Unknown
Nitrocellulose (DOT)
AS
C 2018
CA 80-15
Celex
Celloidin
Cellulose nitrate
Cellulose, nitrate (9CI)
Cellulose tetranitrate
Collodion (DOT)
Collodion Cotton
Collodion Wool
Colloxylin
Corial EM Finish F
E 1440
Film (DOT)
Flexible Collodion
FM-NTS
Guncotton (DOT)
HX 3/5
Kodak LR 115
LR 115
NA 1324 (DOT)
NA 1325 (DOT)
NA 2059 (DOT)
NA 2555 (DOT)
NA 2556 (DOT)
Nitrocellulose, Block, Wet with not less than 25% alcohol [UN 0342]
Nitrocellulose, Colloided, Granular or flake, Wet with not less than 20% alcohol or solvent (DOT)
Nitrocellulose, Colloided, Granular or flake, Wet with not less than 20% water (DOT)
Nitrocellulose, Containing at least 25% alcohol and not exceeding 12.6% nitrogen (DOT)
Nitrocellulose, Containing at least 25%, by weight, water (DOT)
Nitrocellulose, Dry (DOT)
Nitrocellulose, Dry or wetted with less than 25% alcohol [UN 0340]
Nitrocellulose, Dry or wetted with less than 25% water [UN 0340]
Nitrocellulose, In solution in flammable liquids (DOT)
Nitrocellulose, Wet with less than 25% alcohol or 25%

water [UN 0340]
Nitrocellulose, Wet with not less than 30% alcohol or solvent (DOT)
Nitrocellulose, Wet with not less than 20% water
Nitrocellulose, Wetted with more than 40% flammable liquids (DOT)
Nitrocellulose E950
Nitrocotton
Nitron
Nitron (Nitrocellulose)
Nixon N/C
NTs 62
NTs 218
NTs 222
NTs 539
NTs 542
Parlodion
Pyralin
Pyroxylin
Pyroxylin Plastic (DOT)
Pyroxylin Rods (DOT)
Pyroxylin Rolls (DOT)
Pyroxylin Scrap (DOT)
Pyroxylin Sheets (DOT)
Pyroxylin Tubes (DOT)
RF 10
RS
R.S.Nitrocellulose
Soluble Gun Cotton
SS
Synpor
Tsapolak 964
UN 0340 [Nitrocellulose, Dry or wetted with less than 25 per cent water (or alcohol), by mass]
UN 0342 [Nitrocellulose, Wetted with not less than 25 percent alcohol, by mass]
UN 1324 [Films, nitrocellulose base, gelatine coated (except scrap)]
UN 2555 [Nitrocellulose with water not less than 25 per cent water, by mass]
UN 2556 [Nitrocellulose with alcohol not less than 25 per cent alcohol by mass, and not more than 12.6 per cent nitrogen, by dry mass]
UN 2059 [Nitrocellulose, solution, flammable with not more than 12.6 per cent nitrogen, by mass, and not more than 55 per cent nitrocellulose]
UN 2060 (DOT)
Xyloidin

9004-74-4
$(C_2H_4O)x.CH_4O$
Polymer
Poly(oxy-1,2-ethanediyl), α-methyl-.ω.-hydroxy- (9CI)
Breox MPEG 550
Carbowax 350
Carbowax 550
Carbowax 750
Carbowax 2000
Carbowax 5000
CP 2000
CP 2000 (polyoxyalkylene)
Ethylene oxide-methanol adduct
Glycols, polyethylene, monomethyl ether (8CI)
GN 8384
Hymol PM
O-Methoxypolyethylene glycol
Methoxypoly(ethylene glycol)
Methoxy polyethylene glycol 350
Methoxy polyethylene glycol 550
Methoxy polyethylene glycol 750
Methyl polyglycol
Monomethoxy poly(ethylene oxide)
Monomethoxypolyethylene glycol
Monomethoxypolyoxyethylene
MPEG
MPEG 5000
MPG 025
MPG 081
Nissan Uniox M 400
Nissan Uniox M 550
Nissan Uniox M 2000
Polyethylene glycol methyl ether
Polyethylene glycol monomethyl ether

9004-77-7
$(C_2H_4O)xC_4H_{10}O$
Polymer
Poly(oxy-1,2-ethanediyl), α-butyl-ω-hydroxy- (9CI)
Polyethylene glycol butyl ether
Polyethylene glycol, monobutyl ester

9004-78-8
$(C_2H_4O)xC_6H_6O$
Polymer
Glycols, polyethylene, monophenyl ether
Ethylan HB 4
FR 214
Marlophen P
Marlophen P 7
Phenol-ethylene oxide adduct
Polyethylene glycol phenyl ether
Poly(oxy-1,2-ethanediyl), α-phenyl-ω-hydroxy- (9CI)
Polyoxyethylene phenyl ether
Spermicide 741
Tritonyl 45

9004-80-2
$(C_2H_4O)xC_{10}H_{23}O_4P$
Polymer
Deceth-4 phosphate
Decyl alcohol, ethoxylated, dihydrogen phosphate
Jordaphos 236
PEG-4 Decyl Ether Phosphate
Polyethylene glycol 200 decyl ether phosphate
Poly(oxy-1,2-ethanediyl), α-phosphono-ω-(decyloxy)- (9CI)
Polyoxyethylene (4) decyl ether phosphate

9004-81-3
$(C_2H_4O)xC_{12}H_{24}O_2$
Polymer
Polyglycol-laurate
Nopalcol 6-L

9004-82-4
$(C_2H_4O)x.C_{12}H_{26}O_4S.Na$
Polymer
Sodium-lauryl-ether-sulfate
Alkanate 3SL3
Avirol 100E
Calfoam ES 30
Conco Sulfate WE
Cycloryl NA
Elfan 242
Elfan NS 242
Elfan NS 243
Elfan NS 243S
Empicol ESB 3
Empicol ESB 30
Empimin KSN
Empimin KSN 27
Empimin KSN 60
Empimin KSN 70
Genapol LRO
Glycols, polyethylene, mono-(hydrogen sulfate), dodecyl ether, sodium salt (8CI)
Maprofix ES
Maprofix 60S
Polyethylene glycol sulfate monododecyl ether sodium salt
Poly(oxy-1,2-ethanediyl), α-sulfo-ω-(dodecyloxy)-, sodium salt (9CI)
Retzolate 60
Rewopol NL-2
Sipon ES
Sipon LES 25
Sodium dodecylpoly(oxyethylene) sulfate
Sodium Laureth-5 Sulfate
Sodium Laureth-7 Sulfate
Sodium Laureth-12 Sulfate
Sodium (lauryloxypolyethoxy)-ethyl sulfate
Sodium laurylpoly(oxyethylene) sulfate
Sodium polyethylene glycol (5) lauryl ether sulfate
Sodium polyethylene glycol (7) lauryl ether sulfate
Sodium polyethylene glycol 600 lauryl ether sulfate
Sodium polyoxyethylene (12) lauryl ether sulfate
Sodium polyoxyethylene (5) lauryl ether sulfate
Sodium polyoxyethylene (7)

lauryl ether sulfate
Sodium poly(oxyethylene) lauryl ether sulfate
Standapol ES 2
Standapol ES 3
Steinapol NL 3
Texapon N 25
Texapon N 40
Texapon NSO
Ultrasulfate SE 5
Unipol 125-E
Unipol 130-E
Zetesol LES 2
Zetesol NL

9004-83-5
$(C_2H_4O)xC_{14}H_{30}OS$
Polymer
Poly(oxy-1,2-ethanediyl), α-(2-(tert-dodecylthio)ethyl)-ω-hydroxy- (9CI)

9004-86-8
$(C_2H_4O)xC_{14}H_{10}O_3$
Polymer
Glycols, polyethylene, dibenzoate
Benzoflex P 200
Benzoflex P-600
Benzoic acid diester with polyethylene glycol 600
Polyethylene glycol dibenzoate
Polyethylene glycol 220 dibenzoate
Polyethylene 600 dibenzoate
Poly(oxy-1,2-ethanediyl), α-benzoyl-ω-(benzoyloxy)- (9CI)
Polyoxyethylene dibenzoate

9004-94-8
$(C_2H_4O)xC_{16}H_{32}O_2$
Polymer
PEG-18 Palmitate
α-(1-Oxohexadecyl)-ω-hydroxy-poly(oxy-1,2-ethanediyl)
PEG-6 Palmitate
PEG-20 Palmitate
Polyethylene glycol (18) monopalmitate
Polyethylene glycol 300 monopalmitate
Polyethylene glycol 1000 monopalmitate
Polyethyleneglycol palmitate
Poly(oxy-1,2-ethanediyl), α-(1-oxohexadecyl)-ω-hydroxy- (9CI)
Polyoxyethylene (6) monopalmitate
Polyoxyethylene (18) monopalmitate
Polyoxyethylene (20) monopalmitate

9004-95-9
$(C_2H_4O)xC_{16}H_{34}O$
Polymer
Ceteth-1
Ceteth-2
Ceteth-4
Ceteth-5
Ceteth-6
Ceteth-10
Ceteth-12
Ceteth-15
Ceteth-16
Ceteth-20
Ceteth-24
Ceteth-25
Ceteth-30
Ceteth-45
Cetomacrogol
Cetomacrogol 1000
Cetomacrogol 1000 BPC
Cetyl alcohol, ethoxylated
Ethoxylated cetyl alcohol
Ethylene glycol cetyl ether
Ethylene glycol monocetyl ether
Hefti CE-55-2
Hefti CE-55-20
PEG-1 Cetyl Ether
PEG-2 Cetyl Ether
PEG-4 Cetyl Ether
PEG-5 Cetyl Ether
PEG-6 Cetyl Ether
PEG-10 Cetyl Ether
PEG-12 Cetyl Ether
PEG-15 Cetyl Ether
PEG-16 Cetyl Ether
PEG-20 Cetyl Ether
PEG-24 Cetyl Ether
PEG-25 Cetyl Ether
PEG-30 Cetyl Ether
PEG-45 Cetyl Ether
Polyethylene glycol (5) cetyl ether
Polyethylene glycol (15) cetyl ether
Polyethylene glycol (16) cetyl ether
Polyethylene glycol (24) cetyl ether
Polyethylene glycol (25) cetyl ether
Polyethylene glycol (30) cetyl ether
Polyethylene glycol (45) cetyl ether
Polyethylene glycol 100 cetyl ether
Polyethylene glycol 200 cetyl ether
Polyethylene glycol 300 cetyl ether
Polyethylene glycol 500 cetyl ether
Polyethylene glycol 600 cetyl ether
Polyethylene glycol 1000 cetyl ether
Polyethylene glycol monocetyl ether
Poly(oxy-1,2-ethanediyl), α-hexadecyl-ω-hydroxy- (9CI)
Polyoxyethylated cetyl alcohol
Polyoxyethylene (2) cetyl ether
Polyoxyethylene (4) cetyl ether
Polyoxyethylene (5) cetyl ether
Polyoxyethylene (6) cetyl ether
Polyoxyethylene (20) cetyl ether
Polyoxyethylene (24) cetyl ether
Polyoxyethylene (25) cetyl ether
Polyoxyethylene (30) cetyl ether
Polyoxyethylene (45) cetyl ether
Unicol CA-2
Unicol CA-4
Unicol CA-10
Unicol CA-20

9004-96-0
$(C_2H_4O)x.C_{18}H_{34}O_2$
Polymer
Polyglycol-oleate
Akyporox O 50
Atlas G-2142
Atlas G-2144
Cemulsol 1050
Cemulsol A
Cemulsol C 105
Cemulsol D-8
Chemester 300-OC
Cithrol PO
Crodet O 6
E2
Emanon 4115
Emcol H-2A
Emcol H 31A
Emerest 2646
Emerest 2660
Empilan BP 100
Empilan BQ 100
Emulphor A
Emulphor UN-430
Emulphor VN 430
Ethofat O
Ethofat O 15
Ethylan A3
Ethylan A6
Extrex P 60
Ionet MO-400
Lannagol LF
Lipal 400-OL
Lipal 30W
Macrogol Oleate 600
Nikkol MYO 2
Nikkol MYO 10
Noigen ES 160
Nonex 25
Nonex 30
Nonex 52
Nonex 64
Nonion O2
Nonion O4
Nonion 06
Nonisol 200
Nopalcol 1-0
Nopalcol 4-O
Nopalcol 6-0
OK 7
Oleic acid poly(oxyethylene) ester
Oleox 5
Olepal I
Olepal III

PEG 200MO
PEG 600MO
PEG 1000MO
PEG-6 Oleate
PEG-20 Oleate
PEG-32 Oleate
Pegosperse 400MO
Polyethylene glycol monooleate
Polyethylene glycol oleate
Polyethylene oxide monooleate
Poly(ethylene oxide) oleate
Polyglycol monooleate
Polyglycol oleate
Poly(oxy-1,2-ethanediyl), α-(1-oxo-9-octadecenyl)-ω-hydroxy-, (Z)- (9CI)
Poly(oxyethylene) monooleate
Poly(oxyethylene) oleate
Poly(oxyethylene) oleic acid ester
POOA
Prodhyphore B
Rokacet
Rokacet O 7
S 1006
S 1132
Slovasol A
Trydet OS Series
Unisol 4-O
Witco 31
X-539-R

9004-98-2
$(C_2H_4O)xC_{18}H_{36}O$
Polymer
Oleylpolyoxethylene-glycol-ether
Ameroxol OE 2
Ameroxol OE 10
Ameroxol OE 20
BRIJ 92
BRIJ 92((2)-Oleyl)
BRIJ 93
BRIJ 96
BRIJ 96((10) Oleyl)
BRIJ 97
BRIJ 98
BRIJ 99
Brox OL-4
Brox OL-5
Brox OL-10
Brox OL-20
Chemal OA-2
Chemal OA-4
Chemal OA-5
Chemal OA-10
Chemal OA-20
Chemal OA-23
Decaethoxy oleyl ether
Dehydol 100
EL-620
EL-719
Emalex 515
Emulgen 404
Emulgen 408
Emulgen 409P
Emulgen 420
Emulgen 430
Emulgin O 5
Emulgin O 10
Emulphor
Emulphor ON 870
Emulphor Surfactants
Emulsogen MS 12
G 3910
G 3920
Genapol O
Glycols, polyethylene, mono-9-octadecenyl ether, (Z)- (8CI)
Hefti OL-55-F-2
Hefti OL-55-F-10
Hefti OL-55-F-20
Lipal OA
Merpol HC
Merpoxen OLF 80
Nikkol BO
Noigen ET 60
Noigen ET 77
Noigen ET 80
Noigen ET 120
Noigen ET 180
Noniolite AO 5
Noniolite AO 20
OA 8
OA 20
OL 55F2
OL 55F10
Oleinol 7
Oleol 18
Oleth
Oleth-2
Oleth-3
Oleth-4
Oleth-5
Oleth-6
Oleth-7
Oleth-8
Oleth-9
Oleth-10
Oleth-12
Oleth-15
Oleth-16
Oleth-20
Oleth-23
Oleth-25
Oleth-44
Oleth-50
Oleyl alcohol condensed with 2 moles ethylene oxide
Oleyl alcohol condensed with 20 moles ethylene oxide
Oleyl Alcohol EO (2)
Oleyl Alcohol EO (10)
Oxanol O 18
PEG-2 Oleyl Ether
PEG-3 Oleyl Ether
PEG-4 Oleyl Ether
PEG-5 Oleyl Ether
PEG-6 Oleyl Ether
PEG-7 Oleyl Ether
PEG-8 Oleyl Ether
PEG-9 Oleyl Ether
PEG-10 Oleyl Ether
PEG-12 Oleyl Ether
PEG-15 Oleyl Ether
PEG-16 Oleyl Ether
PEG-20 Oleyl Ether
PEG-23 Oleyl Ether
PEG-25 Oleyl Ether
PEG-44 Oleyl Ether
PEG-50 Oleyl Ether
Polyethylene glycol oleyl ether
Polyethylene glycol monoleyl ether
Polyethylene glycol (3) oleyl ether
Polyethylene glycol (5) oleyl ether
Polyethylene glycol (7) oleyl ether
Polyethylene glycol (15) oleyl ether
Polyethylene glycol (16) oleyl ether
Polyethylene glycol (23) oleyl ether
Polyethylene glycol (25) oleyl ether
Polyethylene glycol (44) oleyl ether
Polyethylene glycol (50) oleyl ether
Polyethylene glycol 100 oleyl ether
Polyethylene glycol 200 oleyl ether
Polyethylene glycol 400 oleyl ether
Polyethylene glycol 450 oleyl ether
Polyethylene glycol 500 oleyl ether
Polyethylene glycol 600 oleyl ether
Polyethylene glycol 1000 oleyl ether
Polyethylene glycol (300) oleyl ether
Poly(oxy-1,2-ethanediyl), α-9-octadecenyl-ω-hydroxy-, (Z)- (9CI)
Polyoxy-1,2-ethanediyl, α-((Z)-9-octadecenyl-ω-hydroxy-
Polyoxyethylated-vegetable-oil
Polyoxyethylene monooleyl ether
Polyoxyethylene (2) oleyl ether
Polyoxyethylene (3) oleyl ether
Polyoxyethylene (4) oleyl ether
Polyoxyethylene (5) oleyl ether
Polyoxyethylene (6) oleyl ether
Polyoxyethylene (7) oleyl ether
Polyoxyethylene (8) oleyl ether
Polyoxyethylene (9) oleyl ether
Polyoxyethylene (10) oleyl ether
Polyoxyethylene (12) oleyl ether
Polyoxyethylene (15) oleyl ether
Polyoxyethylene (16) oleyl ether
Polyoxyethylene (20) oleyl ether
Polyoxyethylene (23) oleyl ether
Polyoxyethylene (25) oleyl ether
Polyoxyethylene (44) oleyl ether
Polyoxyethylene (50) oleyl ether
Polyoxyl 10 Oleyl Ether
Procol OA-25
Rokanol O

Simulsol 92
Simulsol 96
Simulsol 98
Siponic Y 500
Slovaton U
Trycol OAL 23
Unicol OA-10
Unicol OA-2
Unicol OA-4
Unicol OA-20
Unimul-05
Unimul-10
Volpo 20

9004-99-3
$(C_2H_4O)nC_{18}H_{36}O_2$
Unknown
Poly(oxy-1,2-ethanediyl), α-(1-oxooctadecyl)-ω-hydroxy-(9CI)
Akyporox S 100
Arosurf 1855E40
Carbowax 1000 monostearate
Cerasynt M
Cerasynt MN
Chemax E-400-MS
Chemax E-600-MS
Chemax E-1000-MS
Cithrol PS
Cithrol 10MS
Cithrol 4MS
Clearate G
Cremophor A
CRILL 20,21,22,23
Emanon 3113
Emanon 3115
Emanon 3199
Emerest 2640
Emery 15393
Emulphor VT-650
Ethofat 60/15
Ethofat 60/20
Ethofat 60/25
Ethoxylated stearic acid
Glycol polyethylene mono-stearate #200
Glycol, polyethylene mono-stearate #6000
Glycols, polyethylene, mono-stearate (8CI)
Hefti PGE-1000-MS
Hefti PGE-400-MS
Hefti RS-55-100
Hefti RS-55-40
Hefti RS-55-9
Lactine
Lamacit CA
LIPAL 15S
LX 3
Macrogol Ester 400
Macrogol Ester 2000
Macrogol Stearate 400
Macrogol Stearate 2000
Myrj
Myrj 45
Myrj 49
Myrj 51
Myrj 52
Myrj 53
Nikkol MYS
Nissan Nonion S 15
NONEX 28
NONEX 29
NONEX 53
NONEX 54
NONEX 63
Noniolite S 100
Nonion S 2
Nonion S 4
Nonion S 15
NSC-31811
PEG 1000MS
PEG 42
PEG 600MS
PEG-5 Stearate
PEG-6 Stearate
PEG-7 Stearate
PEG-8 Stearate
PEG-9 Stearate
PEG-10 Stearate
PEG-12 Stearate
PEG-14 Stearate
PEG-18 Stearate
PEG-20 Stearate
PEG-25 Stearate
PEG-30 Stearate
PEG-32 Stearate
PEG-35 Stearate
PEG-36 Stearate
PEG-40 Stearate
PEG-45 Stearate
PEG-50 Stearate
PEG-75 Stearate
PEG-90 Stearate
PEG-100 Stearate
PEG-120 Stearate
PEG-150 Stearate
PEG-6-32 Stearate
Pegosperse S 9
Perphinol 45/100
PMS No. 1
PMS No. 2
Polyethylene-glycol-mono-stearate
Polyethylene glycol mono-stearate #200
Polyethylene glycol mono-stearate #400
Polyethylene glycol mono-stearate #1000
Polyethylene glycol mono-stearate #6000
Polyethylene glycol 300 mono-stearate
Polyethylene glycol 400 mono-stearate
Polyethylene glycol 450 mono-stearate
Polyethylene glycol 500 mono-stearate
Polyethylene glycol 600 mono-stearate
Polyethylene glycol 1000 monostearate
Polyethylene glycol 1500 monostearate
Polyethylene glycol 1540 monostearate
Polyethylene glycol 1800 monostearate
Polyethylene glycol 2000 monostearate
Polyethylene glycol 4000 monostearate
Polyethylene glycol 6000 monostearate
Polyethylene glycol (5) mono-stearate
Polyethylene glycol (7) mono-stearate
Polyethylene glycol (14) mono-stearate
Polyethylene glycol (18) mono-stearate
Polyethylene glycol (25) mono-stearate
Polyethylene glycol (30) mono-stearate
Polyethylene glycol (35) mono-stearate
Polyethylene glycol (45) mono-stearate
Polyethylene glycol (50) mono-stearate
Polyethylene glycol (90) mono-stearate
Polyethylene glycol (100) monostearate
Polyethylene glycol (120) monostearate
Polyethylene glycol stearate
Polyethyleneglycols mono-stearate
Poly(oxy-1,2-ethanediyl), α-1-(oxooctadecyl)-ω-hydroxy- (9CI)
Polyoxyethylene monostearate
Polyoxyethylene 1500 mono-stearate
Polyoxyethylene (5) mono-stearate
Polyoxyethylene (6) mono-stearate
Polyoxyethylene (7) mono-stearate
Polyoxyethylene (8) mono-stearate
Polyoxyethylene (9) mono-stearate
Polyoxyethylene (10) mono-stearate
Polyoxyethylene (12) mono-stearate
Polyoxyethylene (14) mono-stearate
Polyoxyethylene (18) mono-stearate
Polyoxyethylene (20) mono-stearate
Polyoxyethylene (25) mono-stearate
Polyoxyethylene (30) mono-stearate
Polyoxyethylene (32) mono-stearate
Polyoxyethylene (35) mono-stearate
Polyoxyethylene (36) mono-stearate
Polyoxyethylene (40) mono-stearate
Polyoxyethylene (45) mono-stearate
Polyoxyethylene (50) mono-stearate

Polyoxyethylene (75) monostearate
Polyoxyethylene (90) monostearate
Polyoxyethylene (100) monostearate
Polyoxyethylene (120) monostearate
Polyoxyethylene (150) monostearate
Polyoxyethylene-8-monostearate
Polyoxyethylene 50 stearate
Polyoxyethylene(8)stearate
Poly(oxyethylene) stearate
Polyoxyl 8 stearate
Polyoxyl 40 stearate
Polyoxyl 50 stearate
Polystate
Polystate B
Rokacet S 10
Ropol 24
S 541
S 1004
S 1012
S 1054
S 1116
S 1042
Simulsol M
Slovasol MKS 16
Stearethate 40
Stearox 6
Stearox 920
Teric SF
Trydet SA Series
Trydet SA 40
Unipeg-S-40
Unipeg-400 MS
Unipeg-600 MS
Unipeg-1000 MS
Unipeg-1540 MS
Unipeg-4000 MS
Unipeg-6000 MS
USAF KE-9
USAF KE-12
USAF KE-14
Witconol H35A
60S

9005-00-9
$(C_2H_4O)xC_{18}H_{38}O$
Polymer
Poly(oxy-1,2-ethanediyl), α-octadecyl-ω-hydroxy- (9CI)
Avivan SO 6
Berol 043
Berol 08
BRIJ 72
BRIJ 76
BRIJ 78
BRIJ 721
Brox HLB-13
Brox S-2
Brox S-20
Brox S-30
Cetalox AT
Ekaline G 80
Emulgen 306P
Emulgen 320P
ESK 1 (demulsifier)(9005-00-9)
G 3710
G 3720
Genapol S
Glycols, polyethylene, mono-octadecyl ether (8CI)
Hefti ST-55-2
Hefti ST-55-20
Levenol PW
Marlipal 1850
Nonion S 220
OS 20A
1-Octadecanol, monoether with polyethylene glycol
Octadecyl polyoxyethylene ether
PEG-2 Stearyl Ether
PEG-4 Stearyl Ether
PEG-7 Stearyl Ether
PEG-10 Stearyl Ether
PEG-11 Stearyl Ether
PEG-13 Stearyl Ether
PEG-15 Stearyl Ether
PEG-16 Stearyl Ether
PEG-20 Stearyl Ether
PEG-25 Stearyl Ether
PEG-27 Stearyl Ether
PEG-30 Stearyl Ether
PEG-40 Stearyl Ether
PEG-50 Stearyl Ether
PEG-100 Stearyl Ether
Polyethylene glycol (7) stearyl ether
Polyethylene glycol (11) stearyl ether
Polyethylene glycol (13) stearyl ether
Polyethylene glycol (15) stearyl ether
Polyethylene glycol (16) stearyl ether
Polyethylene glycol (25) stearyl ether
Polyethylene glycol (27) stearyl ether
Polyethylene glycol (30) stearyl ether
Polyethylene glycol (50) stearyl ether
Polyethylene glycol (100) stearyl ether
Polyethylene glycol 100 stearyl ether
Polyethylene glycol 200 stearyl ether
Polyethylene glycol 500 stearyl ether
Polyethylene glycol 2000 stearyl ether
Polyethylene glycol 1000 stearyl ether
Polyoxyethylated stearyl alcohol
Polyoxyethylene (2) stearyl ether
Polyoxyethylene (4) stearyl ether
Polyoxyethylene (7) stearyl ether
Polyoxyethylene (10) stearyl ether
Polyoxyethylene (11) stearyl ether
Polyoxyethylene (13) stearyl ether
Polyoxyethylene (15) stearyl ether
Polyoxyethylene (16) stearyl ether
Polyoxyethylene (20) stearyl ether
Polyoxyethylene (25) stearyl ether
Polyoxyethylene (27) stearyl ether
Polyoxyethylene (30) stearyl ether
Polyoxyethylene (40) stearyl ether
Polyoxyethylene (50) stearyl ether
Polyoxyethylene (100) stearyl ether
Simulsol 72
Simulsol 76
Steareth
Steareth-2
Steareth-4
Steareth-7
Steareth-10
Steareth-11
Steareth-13
Steareth-15
Steareth-16
Steareth-20
Steareth-25
Steareth-27
Steareth-30
Steareth-40
Steareth-50
Steareth-100
Stearyl alcohol, ethoxylated
Sunmorl S 1500
Unicol SA-2
Unicol SA-10
Unicol SA-13
Unicol SA-15
Unicol SA-20
Unicol SA-40

9005-07-6
$(C_2H_4O)xC_{36}H_{66}O_3$
Polymer
Polyethylene glycol dioleate
PEG-4 Dioleate
PEG-6 Dioleate
PEG-6-32 Dioleate
PEG-8 Dioleate
PEG-10 Dioleate
PEG-12 Dioleate
PEG-20 Dioleate
PEG-32 Dioleate
PEG-75 Dioleate
PEG-150 Dioleate
Hefti PGE-400-DO
Polyethylene glycol 200 dioleate
Polyethylene glycol 300 dioleate
Polyethylene glycol 400 dioleate
Polyethylene glycol 500 dioleate
Polyethylene glycol 600 dioleate

Polyethylene glycol 1000 dioleate
Polyethylene glycol 1500 dioleate
Polyethylene glycol 1540 dioleate
Polyethylene glycol 4000 dioleate
Polyethylene glycol 6000 dioleate
Poly(oxy-1,2-ethanediyl), α-(1-oxo-9-octadecenyl)-ω-((1-oxo-9-octadecenyl)oxy)-, (Z,Z)- (9CI)
Polyoxyethylene (4) dioleate
Polyoxyethylene (6) dioleate
Polyoxyethylene (8) dioleate
Polyoxyethylene (10) dioleate
Polyoxyethylene (12) dioleate
Polyoxyethylene (20) dioleate
Polyoxyethylene (32) dioleate
Polyoxyethylene (75) dioleate
Polyoxyethylene (150) dioleate
Polyoxyethylene 1500 dioleate
Unipeg-400 DO
Unipeg-600 DO

9005-08-7
$(C_2H_4O)xC_{36}H_{70}O_3$
Polymer
Poly(oxy-1,2-ethanediyl), α-(1-oxooctadecyl)-ω-((1-oxo-octadecyl)oxy)- (9CI)
Carbowax 1000 distearate
Chemax PEG-1000-DS
Chemax PEG-400-DS
Cithrol 10DS
Cithrol 60DS
Emerest 2642
Glycols, polyethylene, distearate (8CI)
Hefti PGE-400-DS
Hefti PGE-600-DS
Lipal 15-DS
Nonex 80
PEG-3 Distearate
PEG-4 Distearate
PEG-6 Distearate
PEG-8 Distearate
PEG-9 Distearate
PEG-12 Distearate
PEG-20 Distearate
PEG-32 Distearate
PEG-75 Distearate
PEG-150 Distearate
PEG-175 Distearate
PEG 1540DS
PEG 6000DS
Polyethylene glycol distearate #1000
Polyethylene glycol distearate
Polyethylene glycol 3 distearate
Polyethylene glycol 175 distearate
Polyethylene glycol 200 distearate
Polyethylene glycol 300 distearate
Polyethylene glycol 400 distearate
Polyethylene glycol 450 distearate
Polyethylene glycol 600 distearate
Polyethylene glycol 1000 distearate
Polyethylene glycol 1540 distearate
Polyethylene glycol 4000 distearate
Polyethylene glycol 6000 distearate
Polyoxyethylene (3) distearate
Polyoxyethylene (4) distearate
Polyoxyethylene (6) distearate
Polyoxyethylene (8) distearate
Polyoxyethylene (9) distearate
Polyoxyethylene (20) distearate
Polyoxyethylene (32) distearate
Polyoxyethylene (75) distearate
Polyoxyethylene (12) distearate
Polyoxyethylene (150) distearate
Polyoxyethylene (175) distearate
S 1009
S 1013
Stabogel
Stearic acid, polyethylene glycol diester
Triethyleneglycol distearate
Triglycol distearate
Unipeg-400 DS
Unipeg-600 DS
Unipeg-4000 DS
Unipeg-6000 DS
Witconol L32-45
62S

9005-12-3
C_7H_8OSi
136.23
Phenyl dimethicone
Phenyl simethicone

9005-25-8
Unknown
Unknown
Corn-Starch
Amaizo W 13
Amylomaize VII
Amylum
Aquapel (Polysaccharide)
Argo Brand Corn Starch
Arrowroot Starch
Claro 5591
Clearjel
Clearjrel
Corn Products
CPC 3005
CPC 6448
Farinex 100
Galactasol A
Genvis
HRW 13
Keestar
Maizena
Maranta
Melogel
Meluna
OK Pre-Gel
Penford Gum 380
Remyline Ac
Rice Starch
Sorghum Gum
Staramic 747
Starch (OSHA)
α-Starch
Starch, Corn
Sta-RX 1500
Tapioca Starch
Tapon
Trogum
W-Gum
W-13 Stabilizer

9005-32-7
Unknown
Unknown
Alginic-acid
Kelacid
Landalgine
Norgine
Polymannuronic acid
Sazzio

9005-34-9
Unknown
Unknown
Ammonium alginate
Alginic acid, ammonium salt (9CI)
Ammonium polymannurate
Analgine
Callatex
Collatex Arm Extra
Digamon
Protomon
Superloid

9005-35-0
Unknown
Unknown
Alginic acid, calcium salt
Ca 33
Calcium alginate
Calginate
Combinace

9005-36-1
Unknown
Unknown
Potassium alginate
Alginic acid, potassium salt (9CI)
Kelmar
Kelmar Improved
Potassium polymannuronate
Stercofuge

9005-37-2
Unknown
Unknown
Propylene-glycol-alginate
Kelcoloid

9005-38-3

Unknown
Unknown
Alginic acid, sodium salt
Algin
Algin (Polysaccharide)
Alginate KMF
Algipon L-1168
Amnucol
Antimigrant C 45
Cecalgine TBV
Cohasal-IH
Darid QH
Dariloid QH
Duckalgin
Halltex
Kelco Gel LV
Kelcosol
Kelgin
Kelgin F
Kelgin HV
Kelgin LV
Kelgin XL
Kelgum
Kelset
Kelsize
Keltex
Keltone
L'-Algiline
Lamitex
Manucol
Manucol DM
Manucol KMF
Manucol SS/LD2
Manugel F 331
Manutex
Manutex F
Manutex RS
Manutex RS 1
Manutex RS-5
Manutex SA/KP
Manutex SH/LH
Meypralgin R/LV
Minus
Mosanon
Nouralgine
OG 1
Pectalgine
Proctin
Protacell 8
Protanal
Protatek
Snow Algin H
Snow Algin L
Snow Algin M
Sodium alginate
Sodium polymannuronate
Stipine
Tagat
Tragaya

9005-49-6
Unknown
Unknown
Heparin
α-Heparin
Heparinate
Heparinic acid
Heparin sulfate
Lipo-Hepin
Liquaemin
Liquemin
Novoheparin
Sublingula
Thromboliquine
Vetren
Vitamin AB
Vitrum AB

9005-53-2
Unknown
Unknown
Lignin (9CI)

9005-59-8
Unknown
Unknown
Sodium pectinate
Pectin, sodium salt

9005-64-5
$C_{58}H_{114}O_{26}$
1227.72
Polysorbate 20
Armotan PML-20
Capmul
Dxewmulse POE-SML
Emsorb 6915
Glycosperse L-20
Glycosperse L-20X
Hodag PSML-20
Liposorb L-20
POE 20 Sorbitan monolaurate
Polyoxyethylene sorbitan monolaurate
Polyoxyethylene (20) sorbitan monolaurate
Protasorb L-20
PSML
Sorbimacrogol laurate 300
Sorbitan, monolaurate polyoxyethylene deriv.
Sorbitan, monododecanote, poly(oxy-1,2-ethanediyl) derivatives
Tween 20

9005-65-6
Unknown
Unknown
Sorbitan, monooleate polyoxyethylene deriv.
Armotan PMO-20
Atlox 1087
Atlox 8916TF
Capmul POE-O
Crill 10
Crill 11
Crillet 4
Crill S 10
Drewmulse POE-SMO
Durfax 80
Emsorb 6900
Ethoxylated sorbitan monooleate
Glycosperse O-20
Glycosperse O-20 VEG
Glycosperse O-20X
Hodag SVO 9
Liposorb O-20
MO 55F
Monitan
Montanox 80
NCI-C60286
Nikkol TO
Nikkol TO 10
Olothorb
Polyoxyethylene sorbitan monooleate
Polyoxyethylene sorbitan oleate
Polysorban 80
Polysorbate 80
Polysorbate 81
Polysorbate 80 B.P.C.
Polysorbate 80, U.S.P.
Protasorb O-20
Romulgin O
Sorbimacrogol Oleate
Sorbimacrogol Oleate 300
Sorbital 0 20
Sorbon T 80
Sorethytan (20) Monooleate
Sorlate
SVO 9
TO 10
Tween 80
Tween 81
Tween 80 A

9005-66-7
$C_{62}H_{122}O_{26}$
1283.84
Sorbitan, monopalmitate polyoxyethylene deriv.
Polyoxyethylene sorbitan monopalmitate
Tween 40
Polyoxyethylene 20 sorbitan monopalmitate
Polysorbate 40
Sorbitan, monohexadecanoate, poly(oxy-1,2-ethanediyl) derivs.
Radiasurf 7145

9005-67-8
$C_{64}H_{126}O_{26}$
1311.90
Sorbitan, monostearate polyoxyethylene deriv.
Capmul
Glycosperse S-20
Liposorb S-20
Polyoxyethylene sorbitan monostearate
Polyoxyethylene 20 sorbitan monostearate
Polysorbate 60
Sorbitan, monooctadecanoate, poly(oxy-1,2-ethanediyl) derivs.
Tween 60

9005-79-2
Unknown
Unknown
Glycogen (9CI)

9005-81-6
$(C_6H_{10}O_5)x$
Unknown
Cellophane
Visking cellophane

9005-84-9
Unknown
Unknown
Amylodextrin (9CI)
Amylodextrins

9005-90-7
Unknown
Unknown
Turpentine [UN 1299]
Essence de terebenthine (French)
Essence de terebenthine, succedane d' (French)
Gum Turpentine
Sucedaneo de trementina (Spanish)
Trementina (Spanish)
Turpentine Substitute
UN 1299 [Turpentine]

9006-03-5
Unknown
Unknown
Rubber, chlorinated
Adbond 1000 Clear
Adeka CR 5
Adeka CR 10
Adeka CR 20
Adeka CR 40
Adeka CR 150
Alloprene
C-Raban 1HB
Chlorinated natural rubber
Chlorinated rubber
Hardlen 15L
KCH 749
KCH 770
KCH 771
KCH 1116
Natural rubber, chlorinated
NB Coat CR 503
Parlon
Parlon S 5
Parlon S 10
Parlon S 20
Parlon S 125
Parlon S 300
Parlon 300CP
Pergut S 10
Pergut S 20
Pergut S 40
Pergut S 90
Rubber, chlorinated
Rubber, natural, chlorinated
Superchlone CR 5
Superchlone CR 10
Superchlone CR 20

9006-42-2
Unknown
Unknown
Polyram
Amarex
Carbatene
FMC-9102
Metiram (9CI)
Metirame zinc (French)
NIA 9102
Polikarbatsin (Russian)
Polycarbacin
Polycarbacine
Polycarbazin
Polycarbazine
Polymarcin
Polymarcine
Polymarsin
Polymarzin
Polymarzine
Polymat
Polyram 80
Polyram Combi
Polyram 80WP
Zinc metiram
Zineb-ethylene thiuram disulfide adduct

9006-65-9
Unknown
Unknown
Dimethicone
Dimethicones
Dimethylpolysiloxane
Dimethyl silicone
Dimeticona (Spanish)
Dimeticone
Dimeticonum (Latin)
Poly(oxy(dimethylsilylene)), α-(trimethylsilyl)-ω-methyl-
Polysilane
Rhodorsil Oils 70047
Sentry Dimethicone
Silbione Oils 70047
Silicone Fluids M
α-(Trimethylsilyl)-ω-methyl-poly(oxy(dimethylsilylene))
360 Medical Fluid

9006-97-7
Unknown
Unknown
Bagasse

9007-12-9
Unknown
Unknown
Thyrocalcitonin
TCT

9007-13-0
Unknown
Unknown
Calcium resinate [UN 1313]
Calcium resinate, Technically pure (DOT)
Calcium resinate, Fused [UN 1314]
Limed Rosin
UN 1313 [Calcium resinate]
UN 1314 [Calcium resinate, fused]

9007-39-0
Unknown
Unknown
Copper salts of fatty and rosin acids
Caswell No. 254A
Copper resinate
EPA Pesticide Chemical Code 023104
Resin acids and rosin acids, copper salts
Rosin, copper hydroxide reaction product

9007-48-1
Unknown
Polymer
1,2,3-Propanetriol, Homopolymer, (Z)-9-octadecenoate
Demal 14
Emcol 12-14-18
Oleic acid polyglyceride
Plurol oleique
Polyglycerol monooleate
Polyglycerol oleate
Polyglyceryl oleate

9007-57-2
Unknown
Unknown
Edestin

9007-73-2
Unknown
Unknown
Ferritin

9008-02-0
Unknown
Unknown
Deoxyhemoglobin
Hemoglobins
Hb(T)
T-State Hemoglobin

9009-54-5
Unknown
Unknown
Polyurethane-Foam
Andur
Curene
Etheron
Etheron Sponge
Isourethane
NCI-C56451
Pliogrip
Polyfoam Plastic Sponge
Polyfoam Sponge
Polyurethane Ester Foam
Polyurethane Ether Foam
Polyurethane Sponge
Spenkel
Spenlite
Urethane polymers

9009-86-3
Unknown
Unknown
Ricin
Ricin, Reconstituted
Ricine

9010-34-8
Unknown
Unknown
Thyroglobulin
Proloid
Thyractin
Thyroglobuline (French)
Thyroglobulins
Thyroglobulinum (Latin)
Thyroid globulin
Thyroprotein
Tiroglobulina (Spanish)

9010-69-9
Unknown
Unknown
Resin acid, zinc salt
Resin acids and Rosin acids, Zinc salts
UN 2714 [Zinc resinate]
Zinc resinate [UN 2714]

9010-85-9
$(C_5H_8.C_4H_8)x$
Polymer
Butyl Rubber
1,3-Butadiene, 2-methyl-, Polymer with 2-methyl-1-propene (9CI)
Isobutylene/isoprene copolymer
2-Methyl-1,3-butadiene polymer with 2-methyl-1-propene

9010-88-2
$(C_5H_8O_2.C_5H_8O_2)x$
Polymer
2-Propenoic acid, 2-methyl-, methyl ester, Polymer with ethyl 2-propenoate (9CI)
Ethyl acrylate, methyl methacrylate polymer
Ethyl acrylate, Polymer with methyl methacrylate
Methyl methacrylate, ethyl acrylate polymer
Methyl methacrylate, Polymer with ethyl acrylate

9010-98-4
$(C_4H_5Cl)x$
Polymer
1,3-Butadiene, 2-chloro-, Polymers
2-Chloro-1,3-butadiene homopolymer (9CI)
Chlorobutadiene polymer
2-Chloro-1,3-butadiene polymer
Chloroprene Polymer
Duprene
GR-M
Nairit
Neoprene
Perbunan C
Plastifix PC
Poly(2-chlorobutadiene)
Poly(2-chloro-1,3-butadiene)
Polychloroprene
Sovprene
Svitpren

9011-05-6
$CH_4N_2O.CH_2O$
90.10
Urea, Polymer with formaldehyde
Acrisin FS 017
Aerolite 300
Aerolite A 300
Aerolite FFD
Agroform
Amikol 65
Anaflex
BASF
BC 20
BC 40
BC 77
BC 20 (Polymer)
Beckamine 21-511
Beckamine NF 5
Beckamine P 136
Beckamine P 138
Beckamine P 138-60
Beckamine P 196M
Beetle 55
Beetle 60
Beetle 65
Beetle 80
Beetle 212-9
Beetle BE 685
Beetle BU 700
Beetle XB 1050
BU 700
Carbamol
Cascamite
Casco 5H
Casco PR 335
Casco Resin
Casco UL 30
Casco WS 114-79
Casco WS 138-43
Casco WS 138-44
Cyrez 933
Depremol M
Diaform UR
Diakol DM
Diakol F
Diakol M
Dynomin UI 16
Dynomin UM 15
Epok U 9048
Fibraset TC
Formaldehyde copolymer with urea
Formaldehyde-urea condensate
Formaldehyde-urea copolymer
Formaldehyde-urea polymer
Formaldehyde-urea precondensate
Formaldehyde-urea prepolymer
Formaldehyde-urea resin
Formalin-urea copolymer
Gabrite
Hygromull
K 0
K 17
K 17(Polymer)
K385
K 8870
K 411-02
Karbamol
Karbamol B/M
Kauresin K244
Kaurit 420
Kaurit 285 FL
KFS
KM 2
KM 2 (Polymer)
Knittex TC
Knittex TS
Koprez 87-110
KS 11
KS 35
KS 68M
KS-M 0.3P
L 195
Larex
M 2
Poly(methibis(hydroxymethyl)-ureylene)amer
M 2 (Polymer)
M 60
M 60 (Formaldehyde polymer)
M 70
M 70 (Polymer)
MCH 52
Melan 11
Methylolurea resin
MF
MF 1
MF 17
MF 27
MFPS 1
MF Resin
Mirbane SU 118K
MKH 52
Mouldrite A256
MPF 2
N 50
Noxylin
Paraformaldehyde-urea polymer
Paraformaldehyde-urea resin
Pianizol
Piatherm
Piatherm D
Plastopal BT
Plyamine HD 1129A
Plyamine P 364BL
Polynoxylin
Ponoxylan
PR 703-78
Resamin 155F
Resamin HW 505
Resimene X 970
Resimene X 975
Resimene X 980
Resimine 975
Resina X
SFK 70
SK 75
SK 75V
S-Resin AER 20

Sumirez 614
Sumitekkusu Rejin 810
Sumitex 260
Sumitex 810
Sumitex NF 113
Sumitex Resin 810
T 101
U 963
UF 33
UF 240
Uformite 700
Uformite F 240N
UKS 72
UKS 73
Uloid 22
Uloid 100
Uloid 301
UL 52R
Umalur
UM-G
Uralite
Uralite (Polymer)
Uramine T101
Uramine T105
Uramine TSL 58
Uramite
Urea-formaldehyde adduct
Urea-formaldehyde condensate
Urea-formaldehyde copolymer
Urea-formaldehyde oligomer
Urea-formaldehyde polymer
Urea-formaldehyde precondensate
Urea-formaldehyde prepolymer
Urea-formaldehyde resin
Ureapap W
Urecoll K
Urecoll KL
Urecoli S
Urelit C
Urelit HM
Urelit R
Urepret
Urofix
UST
W 70
Yuban 10HV
Yuban 10S

9011-06-7
$(C_2H_3Cl.C_2H_2Cl_2)x$
Polymer
Ethylene, 1,1-dichloro-, Polymer with chloroethylene
Breon 202
Breon CS 100/30
Chloroethylene-1,1-dichloroethylene polymer
Daran
Daran CR 6795H
1,1-Dichloroethene polymer with chloroethene
1,1-Dichloroethylene-monochloroethylene polymer
1,1-Dichloroethylene polymer with chloroethylene
Dow 874
Dow Latex 874
ET 67
Ethene, 1,1-dichloro-, Polymer with chloroethene (9CI)
Geon 222
Geon 652
IKhS 1
KhS 596
Kurehalon A0
Laplen
Latex SVKh
Polyco 2611
QX 2168
Saran 683
Saran 746
Saran Resin 683
SP 489
SVKh 1
SVKh 40
UP 925
Velon
Vikh 65
Viniden 60
Vinyl chloride copolymer with vinylidene chloride
Vinyl chloride-1,1-dichloroethylene copolymer
Vinyl chloride-vinylidene chloride copolymer
Vinyl chloride-vinylidene chloride polymer
Vinylidene chloride-vinyl chloride polymer
VKhVD 40
Winiden 60

9011-14-7
$(C_5H_8O_2)x$
Polymer
Polymethylmethacrylate
Acronal S 320 D
Acrylite
Acryloid A-15
Acrypet
Acrypet M 001
Acrypet V
Acrypet VH
Acrysol ASE
Akuripetto VH
Altulor M 70
A 21LV
AO 10
CMW Bone Cement
Crinothene
Degalan LP 59/03
Degalan S 85
Delpet 50M
Delpet 60N
Delpet 80N
Diakon
Diakon LO 951
Diakon MG
Diakon MG 101
Disapol M
DV 400
Elvacite
Elvacite 2008
Elvacite 2009
Elvacite 2010
Elvacite 2021
Elvacite 2041
Elvacite 6011
Elvacite 6012
Kallocryl K
Kallodent 222
Kallodent Clear
Kaneace PA 20
K 120 N
Korad
LPT
LPT 1
LSO-M
LSO-M 4B
Lucite
Lucite 30
Lucite 47
Lucite 120
Lucite 129
Lucite 130
Lucite 140
Lucite 147
Lucite 180
Metaplex NO
Metaplex 4002t
Methacrylic acid methyl ester polymers
Methyl methacrylate homopolymer
Methyl methacrylate polymer
Methyl methacrylate resin
2-Methyl-2-propenoic acid methyl ester homopolymer (9CI)
MH 101-2
50N
50N (Polymer)
Organic Glass E 2
Osteobond
Osteobond Surgical Bone Cement
Palacos
Palacos R
Paraglas
Parapet 60N
Paraplex P 543
Paraplex P 681
Perspex
Plex 8572-F
Plexiglas
Plexigum M 920
PMMA
PMMA-A
Pontalite
Repairsin
Resarit 4000
Rhoplex B 85
Riston
Romacryl
Shinkolite
SO 95
SO 120
SO 140
Sol
Sol 90
Sol 95
ST 1
Stellon Pink
ST 1 (Polymer)
Sumipex B-LG
Sumipex B-MH
Sumipex-B MHD
Sumipex LG
Sumipex LO
Sumipex MHO
Superacryl AE
Superacryl O

Surgical Simplex
Surgical Simplex P
Tensol 7
Torex G
Vedril
Vedril 5
Vedril 8

9011-19-2
Unknown
Unknown
Polysiloxane

9011-70-5
Unknown
Unknown
Polybor-chlorate

9012-00-4
Unknown
Unknown
Protamine
Protamines

9014-01-1
Unknown
Unknown
Bacillus-Subtilis-Carlsberg
Alcalase
Alk-Enzyme
Bacillopeptidase A
Bacillopeptidase B
Bioprase
Colistinase
E.C. 3.4.4.16
E.C. 3.4.21.14
Maxatase
Nagarse
Subtilisin (ACGIH) (9CI)
Subtilisin BPN'
Subtilisin Carlsburg
Subtilisin Novo
Subtilopeptidase A
Subtilopeptidase B
Subtilopeptidase BPN'
Subtilopeptidase C
Thermoase PC-10

9014-76-0
Unknown
Unknown
Sephadex
Sephadex G-25
Sephadex G-75
Sephadex G-100
Sephadex G-200
Sephadex Gels
Sephadex LH-20

9014-90-8
$(C_2H_4O)xC_{15}H_{24}O_4S.Na$
Polymer
Poly(oxy-1,2-ethanediyl), α-sulfo-ω-(nonylphenoxy)-, sodium salt
AF 5
Alipal CO 430
Alipal CO 433
Celanol 252
CO 433
Etoxon AF5
Fenopon CO 433N
Levenol WZ
Newcol 560SN
Nissan Trax N300
Perlankrol RN
Retzolate 1075
Solumin FP 85SD
Steinapol NOS 25
Sunsolt WA

9014-92-0
$(C_2H_4O)xC_{18}H_{30}O$
Unknown
Tergitol 12-P-9

9014-93-1
$(C_2H_4O)xC_{24}H_{42}O$
Polymer
Nonyl Nonoxynol-10
Chemax DNP-150
Nonyl Nonoxynol-5
Nonyl Nonoxynol-49
Nonyl Nonoxynol-100
Nonyl Nonoxynol-150
PEG-5 Dinonyl Phenyl Ether
PEG-10 Dinonyl Phenyl Ether
PEG-49 Dinonyl Phenyl Ether
PEG-100 Dinonyl Phenyl Ether
PEG-150 Dinonyl Phenyl Ether
Polyethylene glycol (5) dinonyl phenyl ether
Polyethylene glycol (49) dinonyl phenyl ether
Polyethylene glycol (100) dinonyl phenyl ether
Polyethylene glycol (150) dinonyl phenyl ether
Polyethylene glycol 500 dinonyl phenyl ether
Poly(oxy-1,2-ethanediyl), α-(dinonylphenyl)-ω-hydroxy- (9CI)
Polyoxyethylene (5) dinonyl phenyl ether
Polyoxyethylene (10) dinonyl phenyl ether
Polyoxyethylene (49) dinonyl phenyl ether
Polyoxyethylene (100) dinonyl phenyl ether
Polyoxyethylene (150) dinonyl phenyl ether

9016-00-6
$(C_2H_6OSi)x$
Polymer
Polydimethyl-siloxane
AF 72
AF 75
Dimethicone 350
Dow Corning 346
Geon
Good-Rite
Gum
Hycar
Latex
Methyl silicone
Poly(oxy(dimethylsilylene))

9016-45-9
$(C_2H_4O)x.C_{15}H_{24}O$
Polymer
Tergitol NP-33 (Nonionic)
Agral 90
Arkopal N-090
Carsonon N-9
Chemax NP Series
Conco NI-90
Glycols, polyethylene, mono-(nonylphenyl) ether
Igepal CO-630
Lissapol NX
Neutronyx 600
Nonoxynol
Nonylphenol, polyoxyethylene ether
Nonyl phenyl polyethylene glycol
Nonyl phenyl polyethylene glycol ether
PEG-9 Nonyl Phenyl Ether
Polyethylene glycol 450 nonyl phenyl ether
Polyoxyethylene nonylphenol
Polyoxyethylene (9) nonyl phenyl ether
Protachem 630
Rewopol HV-9
Synperonic NX
Tergitol TP-9
Tergitol NP-14
Tergitol NP-27
Tergitol NP-35
Tergitol NP-40
Tergitol NPX
Triton N-100
Trycol NP-1

9016-87-9
Unknown
Polymer
Polymethylenepolyphenyl-isocyanate

9036-19-5
$(C_2H_4O)x.C_{14}H_{22}O$
Polymer
Triton X100
Antarox CA 620
Caswell No. 614
Cemulsol OP 16
Charger E
Emulgen 808
Emulgen 810
EPA Pesticide Chemical Code 079100
Ethoxylated octyl phenol
Ethylan CP
Ethylan CPX
Glycols, polyethylene, mono-((1,1,3,3-tetramethylbutyl)-phenyl) ether (8CI)
Hyonic PE 260

Igepal CA
Igepal CA 520
Nekanil LN
Newcol 808
Newcol 865
Neutronyx 622
Neutronyx 675
Nikkol OP
Nissan Nonion HS 206
Nissan Nonion HS 208
Nissan Nonion HS 210
Noigen EA 102
Noigen EA 110
Noigen EA 120
Noigen EA 140
Noigen EA 142
Noigen EA 160
Noigen EA 170
Nonidet P 40
Nonidet P40
Nonion HS 206
Nonion HS 208
Octapol 100
Octyl phenol condensed with 3 moles ethylene oxide
Octyl phenol condensed with 8-10 moles ethylene oxide
Octyl phenol condensed with 16 moles ethylene oxide
Octyl phenol condensed with 20 moles ethylene oxide
Octylphenol EO (3)
Octylphenol EO (10)
Octyl Phenol EO (16)
Octyl phenol EO (20)
Octylphenoxypolyethoxyethanol
Octylphenoxypoly(ethoxyethanol)
tert-Octylphenoxypoly(ethoxyethanol)
Octylphenoxypoly(ethyleneoxy)ethanol
tert-Octylphenoxy poly(oxyethylene)ethanol
OF 7
Polyethylene glycol mono-(octylphenyl) ether
Polyethylene glycol octylphenyl ether
Poly(ethylene oxide)octylphenyl ether
Poly(oxy-1,2-ethanediyl), α-((1,1,3,3-tetramethylbutyl)phenyl)-ω-hydroxy- (9CI)
Polyoxyethylene monooctylphenyl ether
Poly(oxyethylene)octylphenol ether
Poly(oxyethylene)octylphenyl ether
OP (Chinese surfactant)
OP 13.2
OP 17.7
OP 30
OP 115
OP 1062
OPE-3
Rokafenol O
Secopal OP 20
Sinnopal OP 8
Siponic F 300
Siponic F 400
Synperonic OP
Synperonic OP 10
T45
T 45 (Polyglycol)
Teric GX 13
Triton AG 98
Triton X 15
Triton X 114
Triton X 207
Value 3608

9036-63-9
$(C_{11}H_{20}O_2)x$
Polymer
2-Propenoic acid, isooctyl ester, Homopolymer (9CI)
Isooctyl 2-propenoate homopolymer

9037-22-3
Unknown
Unknown
Amylopectin (9CI)
Kosol

9038-95-3
Unknown
Polymer
Glycols, polyethylenepolypropylene, monobutyl ether (Nonionic)
Butoxyslovanik BK 61
Glycols, polyethylene-polypropylene, monobutyl ether (8CI)
Methyloxirane, Polymer with oxirane, monobutyl ether
Newpol 50HB5100
Newpol 75H90000
Oxirane, methyl-, Polymer with oxirane, monobutyl ester (9CI)
Pluracol W 3520N
Polyoxyethylene (3) polyoxypropylene (2) monobutyl ether
Polyoxyethylene (5) polyoxypropylene (3) monobutyl ether
Polyoxyethylene (7) polyoxypropylene (5) monobutyl ether
Polyoxyethylene (10) polyoxypropylene (7) monobutyl ether
Polyoxyethylene (12) polyoxypropylene (9) monobutyl ether
Polyoxyethylene (16) polyoxypropylene (12) monobutyl ether
Polyoxyethylene (20) polyoxypropylene (15) monobutyl ether
Polyoxyethylene (27) polyoxypropylene (24) monobutyl ether
Polyoxyethylene (30) polyoxypropylene (20) monobutyl ether
Polyoxyethylene (35) polyoxypropylene (28) monobutyl ether
Polyoxyethylene (45) polyoxypropylene (33) monobutyl ether
Polyoxypropylene (2) polyoxyethylene (3) monobutyl ether
Polyoxypropylene (3) polyoxyethylene (5) monobutyl ether
Polyoxypropylene (5) polyoxyethylene (7) monobutyl ether
Polyoxypropylene (7) polyoxyethylene (10) monobutyl ether
Polyoxypropylene (9) polyoxyethylene (12) monobutyl ether
Polyoxypropylene (12) polyoxyethylene (16) monobutyl ether
Polyoxypropylene (15) polyoxyethylene (20) monobutyl ether
Polyoxypropylene (20) polyoxyethylene (30) monobutyl ether
Polyoxypropylene (24) polyoxyethylene (27) monobutyl ether
Polyoxypropylene (28) polyoxyethylene (35) monobutyl ether
Polyoxypropylene (33) polyoxyethylene (45) monobutyl ether
PPG-2-Buteth-3
PPG-3-Buteth-5
PPG-5-Buteth-7
PPG-7-Buteth-10
PPG-9-Buteth-12
PPG-12-Buteth-16
PPG-15-Buteth-20
PPG-20-Buteth-30
PPG-24-Buteth-27
PPG-28-Buteth-35
PPG-33-Buteth-45
Tergitol Nonionic XD
Tergitol XD (Nonionic)
U-660
U-2000
U-5100
Ucon Fluid 50-HB-260
Ucon LB-285 (Fluid)
Ucon LB 165
Ucon 50-HB-55
Ucon 50-HB-100
Ucon 50-HB-260
Ucon 50-HB-400
Ucon 50-HB-660
Ucon 50-HB-2000
Ucon 50-HB-3520
Ucon 50-HB-5100
Ucon 50-HB-280-X

9039-76-3
Unknown
Unknown
Butyl phenol, formaldehyde

resin in xylene

9041-07-0
$C_{30}H_{62}O_{21}$
758.81
Decaglycerol (9CI)

9041-93-4
Unknown
Unknown
Bleomycin, sulfate
Blenoxane
Bleomycin, sulfate (Salt) (9CI)
Blexane

9043-30-5
$(C_2H_4O)xC_{13}H_{28}O$
Polymer
Poly(oxy-1,2-ethanediyl), α-isotridecyl-ω-hydroxy- (9CI)
Polyethylene glycol, isotridecyl ether

9046-01-9
Unknown
Unknown
Trideceth-6 phosphate
Jordaphos JS-61
PEG-6 Tridecyl Ether Phosphate
Phosphoric acid, (ethoxylated tridecyl alcohol) esters
Polyethylene glycol 300 tridecyl ether phosphate
Poly(oxy-1,2-ethanediyl), α-tridecyl-ω-hydroxy-, phosphate (9CI)
Polyoxyethylene (6) tridecyl ether phosphate

9046-10-0
$(C_3H_6O)xC_6H_{16}N_2O$
Polymer
Poly(oxy(methyl-1,2-ethanediyl)), α-(2-aminomethylethyl)-ω-(2-aminomethylethoxy)- (9CI)

9046-40-6
Unknown
Unknown
Pectic acid (9CI)

9048-46-8
Unknown
Unknown
Albumin, Iodinated I 125 serum
Albumin, Blood serum, Labeled with iodine-125
Albumin, Blood serum, Labeled with iodine-131
Albumin, Iodinated I 131 serum
Albumins, Blood serum
Albumotope I-125
Albumotope I-131
Bovine serum albumin
Calf serum ultralyzate
IHSA I-125
IHSA I-131
Moisturizing Extract SV
RISA-125
RISA-131
Seroalbumina humana iodada (125 I) (Spanish)
Seroalbumina humana iodada (131 I) (Spanish)
Seroalbuminum humanum iodinatum (125 I) (Latin)
Seroalbuminum humanum iodinatum (131 I) (Latin)
Seroalbuminum humanum jodinatum (125 I)
Serum albumin, iodinated (125I) human
Serum albumin, iodinated (131I) human
Serum-albumine humaine iodee (125 I) (French)
Serum-albumine humaine iodee (131 I) (French)
Serum albuminum radio-iodatum (131 I)
Serum proteins

9050-36-6
Unknown
Unknown
Maltodextrin (9CI)

9051-57-4
$(C_2H_4O)xC_{15}H_{24}O_4S.H_3N$
Polymer
Poly(oxy-1,2-ethanediyl), α-sulfo-ω-(nonylphenoxy)-, ammonium salt
Alipal CO 436
Alipal EP
Alipal EP 110
Alipal EP 120
Ammonium nonoxynol-4-sulfate
CO 436
Fenopon CO 436
Fenopon EP 110
Fenopon EP 120
Hitenol N 093
Newcol 560SF
Nikkol SNP

9056-38-6
Unknown
Unknown
Nitrostarch
Nitroalmidon (Spanish)
Nitroamidon (French)
Nitrostarch, Containing less than 20% water [UN 0146]
Nitrostarch, Dry [UN 0146]
Nitrostarch, Wet with not less than 30% alcohol or solvent
Nitrostarch, Wet with not less than 20% water [UN 1337]
Starch, nitrate (9CI)
UN 0146 [Nitrostarch, Dry or wetted with less than 20 per cent water, by mass]
UN 1337 [Nitrostarch, Wetted with not less than 20 per cent water, by mass]

9057-06-1
Unknown
Unknown
Carboxymethyl starch
Starch, carboxymethyl ether (9CI)

9061-29-4
Unknown
Unknown
Carboxyhemoglobin
Carbomonoxyhemoglobin
Carbonmonoxyhemoglobin
Carboxyhemoglobin A
Carboxyhemoglobin C

9062-07-1
Unknown
Unknown
Iota-carrageenan (9CI)
Aubygel X 52
Eucheuma Spinosum Gum
Gelcarin SI
Pellugel ID

9064-57-7
Unknown
Unknown
lambda-Carrageenan
lambda-Carrageenin

9066-50-6
Unknown
Unknown
Lignosulfonic acid, chromium salt (9CI)
Chromium lignosulfonate
Chromium salt of oxidized lignosulfonate
Chromium salt of spent sulfite liquor

9080-17-5
Unknown
Unknown
Ammonium polysulfide solution [UN 2818]
Ammonium polysulfide
Ammonium sulfide
Caswell No. 045A
EPA Pesticide Chemical Code 076701
Sulfure d'ammonium (French)
Sulfuro amonico (Spanish)
UN 2818 [Ammonium polysulfide, Solution]

9080-49-3
Unknown
Unknown

Polysulfide
Permlastic

9080-79-9
$(C_8H_8O_3S)x.xNa$
Polymer
Benzenesulfonic acid, ethenyl-, Homopolymer, sodium salt
AEP 1
Flexan 500
Oligo Z
Poly(styrenesulfonate) sodium salt
Sodium carbonate stabilized sulfonated polystyrene sodium salt
Sodium polystyrenesulfonate
Versa TL 71
Versa TL 400
Versa TL 500

9081-17-8
$(C_2H_4O)xC_{15}H_{24}O_4S$
Polymer
Poly(oxy-1,2-ethanediyl), α-sulfo-ω-(nonylphenoxy)- (9CI)
Nonylphenol polyoxyethylene sulfuric acid
α-Sulfo-ω-(nonylphenoxy)-poly(oxy-1,2-ethanediyl)

9082-00-2
Unknown
Unknown
Glycerol, ethylene oxide, propylene oxide polymer
Adeka GH-200
Glycerol poly(oxyethylene, oxypropylene) ether
Glycerol, propylene oxide, ethylene oxide polymer
Oxirane, methyl-, Polymer with oxirane, ether with 1,2,3-propanetriol (3:1) (9CI)
PEG/PPG-24/24 Glycerine
Polyoxyethylene (12) polyoxypropylene (66) glyceryl ether
Polyoxyethylene (24) polyoxypropylene (24) glyceryl ether
Polyoxyethylene (30) polyoxypropylene (20) glyceryl ether
Polyoxypropylene (20) polyoxyethylene (30) glyceryl ether
Polyoxypropylene (24) polyoxyethylene (24) glyceryl ether
Polyoxypropylene (66) polyoxyethylene (12) glyceryl ether
PPG-20-Glycereth-30
PPG-24-Glycereth-24
PPG-66-Glycereth-12
1,2,3-Propanetriol, Polymer with methyloxirane and oxirane
Propylene oxide, ethylene oxide, glycerol adduct

9084-06-4
$(C_{10}H_8O_3S.CH_2O)x.xNa$
Unknown
[O-]S(=O)(=O)c2cc(NC(=O)Nc1ccccc1)ccc2C=Cc4ccc(NC(=O)Nc3ccccc3)cc4.[Na][Na].O=S(=O)=O
Blancol
Blancol Dispersant
Sodium salt of sulfonated naphthaleneformaldehyde condensate

9085-22-7
Unknown
Unknown
Cellulose, octadecanoate (9CI)

10004-44-1
$C_4H_5NO_2$
99.10
Cc1cc(O)no1
Isoxazole, 3-hydroxy-5-methyl
F-319
Hydroxyisoxazole
3-Hydroxy-5-methylisoxazole
Hymexazol
3-Isoxazolol, 5-methyl-
Itachigarden
5-Methyl-3(2H)-isoxazolone
SF-6505
Tachigaren

10017-56-8
$C_6H_{15}NO_3.xH_3O_4P$
Unknown
Ethanol, 2,2',2''-nitrilotris-, phosphate (Salt) (9CI)
2,2',2''-Nitrilotrisethanol phosphate (Salt)
Phosphoric acid, triethanolamine salt

10018-87-8
$C_8H_{15}NO_2.C_2H_4O_2$
217.26
2-Propenoic acid, 2-methyl-, 2-(dimethylamino)ethyl ester, acetate (9CI)

10021-55-3
$C_6H_6O_5S.Na$
213.17
Benzenesulfonic acid, 2,5-dihydroxy-, monosodium salt (9CI)
2,5-Dihydroxybenzenesulfonic acid, monosodium salt

10022-31-8
$N_2O_6.Ba$
261.36
Barium(II) nitrate (1:2)
Barium dinitrate
Barium nitrate [UN 1446]
Dusicnan barnaty (Czech)
Nitrate de baryum (French)
Nitric acid, barium salt
UN 1446 [Barium nitrate]

10022-60-3
$C_{18}H_{34}O_4$
314.52
CCC(CC)COC(=O)CCCCC(=O)OCC(CC)CC
Adipic acid, bis(2-ethylbutyl) ester
Adipic acid, di(2-ethylbutyl) ester
Bis-(2-ethylbutyl)ester kyseliny adipove (Czech)
Di(2-ethylbutyl) adipate

10022-70-5
$ClO.Na.5H_2O$
164.54
Hypochlorous acid, sodium salt, pentahydrate
Sodium hypochlorite

10024-47-2
$C_{22}H_{42}O_2$
338.57
O=C(OCC(C)C)CCCCCCCC=CCCCCCCCC
9-Octadecenoic acid (Z)-, 2-methylpropyl ester (9CI)

10024-74-5
$C_{16}H_{19}N$
225.36
CC(NC(C)c1ccccc1)c2ccccc2
Dibenzylamine, α,α'-dimethyl
Bis(α-methylbenzyl)amine

10024-89-2
$C_4H_9NO.ClH$
123.60
[Cl-].C1COCCN1
Morpholine, hydrochloride
Morpholine hydrochloride

10024-97-2
N_2O
44.02
[N-]=[N+]=O
Nitrogen-oxide
Dinitrogen monoxide
Factitious Air
Hyponitrous acid anhydride
Laughing Gas
Nitrous oxide (DOT)
Nitrous oxide, Compressed [UN 1070]
Nitrous oxide, Refrigerated liquid [UN 2201]
UN 1070 [Nitrous oxide, compressed]
UN 2201 [Nitrous oxide, refrigerated liquid (cryogenic liquid)]

10025-65-7
Cl_2Pt
265.99
Platinum-chloride
Muriate of platinum
Platinous chloride

10025-67-9
Cl_2S_2
135.02
Sulfur-chloride [UN 1828]
Chloride of sulfur [UN 1828]
Disulfur dichloride
Siarki chlorek (Polish)
Sulfur chloride(di) (DOT)
Sulfur monochloride (ACGIH, OSHA)
Sulfur subchloride
Thiosulfurous dichloride
UN 1828 [Sulfur chlorides]

10025-73-7
Cl_3Cr
158.35
Cl[Cr](Cl)Cl
Chromium(III) chloride (1:3)
Chromic chloride
Chromium chloride
Chromium chloride, Anhydrous
Chromium trichloride
C.I. 77295
Puratronic chromium chloride
Trichlorochromium

10025-77-1
$Cl_3Fe.6H_2O$
270.32
Iron(3+) chloride, hexahydrate
Ferric chloride, hexahydrate
Ferric trichloride hexahydrate
Iron (III), chloride, hexahydrate
Iron trichloride hexahydrate

10025-78-2
Cl_3HSi
135.45
ClSi(Cl)Cl
Silane, trichloro
Silici-chloroforme (French)
Siliciumchloroform (German)
Silicochloroform
Trichloorsilaan (Dutch)
Trichloromonosilane
Trichlorosilane [UN 1295]
Trichlorsilan (German)
Triclorosilano (Italian)
UN 1295 [Trichlorosilane]

10025-82-8
Cl_3In
221.17
Indium-chloride
Indium trichloride

10025-85-1
Cl_3N
120.36
Nitrogen chloride
Agene
Chlorine nitride
Nitrogen trichloride
Trichloramine
Trichlorine nitride
Trichlorure d'azote (French)
Tricloruro de nitrogeno (Spanish)

10025-87-3
Cl_3OP
153.32
Phosphoryl chloride (DOT)
Fosforoxychlorid (Czech)
Oxychlorid fosforecny (Czech)
Phosphorus oxychloride (ACGIH,OSHA) [UN 1810]
Phosphorus oxytrichloride
UN 1810 [Phosphorus oxychloride]

10025-91-9
Cl_3Sb
228.10
Antimony (III) chloride
Antimoine (trichlorure d') (French)
Antimonio (tricloruro di) (Italian)
Antimonous chloride (DOT)
Antimontrichlorid (German)
Antimony Butter
Antimony chloride (DOT)
Antimony trichloride
Antimony trichloride, Liquid [UN 1733]
Antimony trichloride, Solid (DOT)
Antimony trichloride, Solution (DOT)
Antimoontrichlride (Dutch)
Butter of Antimony
Chlorid antimonity (Czech)
Chlorure antimonieux (French)
C.I. 77056
Stibine, trichloro-
Trichlorostibine
Trichlorure d'antimoine (French)
UN 1733 [Antimony trichloride, Liquid]

10025-97-5
Cl_4Ir
334.00
Iridium(IV) chloride
Iridium tetrachloride

10026-04-7
Cl_4Si
169.89
Silicon chloride (DOT)
Chlorid kremicity (Czech)
Extrema
Silane, tetrachloro-
Silicio(tetracloruro di)
Siliciumtetrachlorid (German)
Siliciumtetrachloride (Dutch)
Silicium(tetrachlorure de) (French)
Silicon tetrachloride [UN 1818]
Tetrachlorosilane
Tetrachlorure de silicium (French)
UN 1818 [Silicon tetrachloride]

10026-07-0
Cl_4Te
269.40
Tellurium-chloride
Telluric chloride
Tellurium chloride, (T-4)-
Tellurium tetrachloride
Tetrachlorotellurium

10026-08-1
Cl_4Th
373.80
Thorium-chloride
Tetrachlorothorium
Thorium tetrachloride

10026-10-5
Cl_4U
379.80
Uranium(IV) chloride
Uranium tetrachloride

10026-11-6
Cl_4Zr
233.02
Zirconium(IV) chloride (1:4)
UN 2503 [Zirconium tetrachloride]
Zirconium chloride
Zirconium tetrachloride [UN 2503]
Zirconium tetrachloride, Solid (DOT)

10026-12-7
Cl_5Nb
270.16
Niobium-chloride
Columbium pentachloride
Niobium pentachloride

10026-13-8
Cl_5P
208.22
Phosphorane, pentachloro
Fosforo(pentacloruro di) (Italian)
Fosforpentachloride (Dutch)
Pieciochlorek fosforu (Polish)
Phosphore(pentachlorure de) (French)
Phosphoric chloride
Phosphorous pentachloride
Phosphorpentachlorid (German)
Phosphorus pentachloride

(ACGIH,OSHA) [UN 1806]
Phosphorus pentachloride, Solid (DOT)
Phosphorus perchloride
UN 1806 [Phosphorus pentachloride]

10026-17-2
CoF_2
96.93
Cobalt(II) fluoride
Cobalt difluoride
Cobaltous fluoride

10026-24-1
$CoO_4S.7H_2O$
281.13
Cobalt(II) sulfate (1:1), heptahydrate
Cobalt monosulfate heptahydrate
Cobaltous sulfate heptahydrate
Cobalt(II) sulphate heptahydrate
Sulfuric acid, cobalt(2+) salt (1:1), heptahydrate (8CI,9CI)

10028-14-5
No
259
Nobelium

10028-15-6
O_3
48.00
Ozone (ACGIH,OSHA)
Ozon (Polish)
Triatomic oxygen

10028-17-8
T_2
3.00
Tritium (9CI)

10028-18-9
F_2Ni
96.71
Nickel(II) fluoride (1:2)
Nickel difluoride
Nickelous fluoride

10028-22-5
$Fe_2O_{12}S_3$
399.88
Iron(III) sulfate
Diiron trisulfate
Ferric sulfate
Iron persulfate
Iron sesquisulfate
Iron sulfate (2:3)
Iron(3+) sulfate
Iron tersulfate
Sulfuric acid, iron(3+) salt (3:2)

10029-31-9
$C_{16}H_{20}N_2$
240.34
N(c(ccc(c1)Cc(ccc(N)c2)c2)c1)C(C)C
Aniline, N-isopropyl-4,4'-methylenedi- (8CI)
N-Isopropylmethylenedianiline
N-Isopropyl-4,4'-methylenedianiline

10030-73-6
$C_{16}H_{30}O_2$
254.41
O=C(O)CCCCCCCC=CCCCCCC
9-Hexadecenoic acid, (E)-
AI3-36444
(E)-9-Hexadecenoic acid

10030-78-1
$C_6H_7N_5O$
165.12
6H-Purin-6-one, 1,7-dihydro-2-(methylamino)- (9CI)

10031-16-0
Unknown
Unknown
Barium dichromate

10031-18-2
BrHg
280.50
Mercury(I) bromide (1:1)
Mercurous bromide (DOT)
Mercurous bromide, Solid (DOT)
Mercury bromides [UN 1634]
UN 1634 [Mercury bromides]

10031-22-8
Br_2Pb
367.01
Br[Pb]Br
Lead bromide (9CI)

10031-59-1
$O_4S.xTl$
1526.65
Thallium-sulfate
NA 1707 [Thallium sulfate, solid]
Ratox
RCRA waste number P115
Sulfuric acid, thallium salt
Thallium sulfate, Solid [NA 1707]
Zelio

10031-82-0
$C_9H_{10}O_2$
150.19
O=Cc(ccc(OCC)c1)c1
Benzaldehyde, p-ethoxy
Ethoxybenzaldehyde
p-Ethoxybenzaldehyde
4-Ethoxybenzaldehyde

10031-96-6
$C_{11}H_{12}O_3$
192.23
COc1cc(CC=C)ccc1OC=O
Phenol, 4-allyl-2-methoxy-, formate (Ester)
4-Allyl-2-methoxyphenol formate
Eugenol formate
Eugenyl formate
4-(2-Propenyl)-2-methoxyphenyl formate

10032-15-2
$C_{11}H_{22}O_2$
186.33
O=C(OCCCCCC)C(CC)C
Butyric acid, 2-methyl-, hexyl ester
Hexyl 2-methylbutyrate
2-Methylbutanoic acid, n-hexyl ester

10034-81-8
$Cl_2O_8.Mg$
223.21
Perchloric acid, magnesium salt
Anhydrone
Dehydrite
Magnesium perchlorate [UN 1475]
Perchlorate de magnesium (French)
UN 1475 [Magnesium perchlorate]

10034-82-9
$CrO_4.2Na.4H_2O$
303.03
Chromic acid, disodium salt, tetrahydrate

10034-85-2
HI
127.91
Hydriodic acid (DOT)
Anhydrous hydriodic acid
Hydriodic acid, Solution [UN 1787]
Hydrogen iodide
Hydrogen iodide, Anhydrous [UN 2197]
Hydrogen iodide, Solution (DOT)
UN 1787 [Hydriodic acid, solution]
UN 2197 [Hydrogen iodide, anhydrous]

10034-93-2
$H_4N_2.H_2O_4S$
130.14
NN.OS(O)(=O)=O

Hydrazine, sulfate (1:1)
HS
Hydrazine monosulfate
Hydrazine sulphate
Hydrazinium sulfate
Hydrazonium sulfate
Idrazina solfato (Italian)
NSC-150014
Siran hydrazinu (Czech)

10034-96-5
$MnO_4S.H_2O$
169.02
Manganese, monosulfate, monohydrate
Manganese sulfate monohydrate
Manganese(2+) sulfate monohydrate
Manganous sulfate monohydrate
Sulfuric acid, manganese(2+) salt (1:1), monohydrate

10035-10-6
BrH
80.92
Br
Hydrobromic-acid
Acide bromhydrique (French)
Acido bromidrico (Italian)
Anhydrous hydrobromic acid
Bromowodor (Polish)
Bromwasserstoff (German)
Broomwaterstof (Dutch)
Hydrobromic acid, Anhydrous [UN 1048]
Hydrobromic acid, More than 49% strength (DOT)
Hydrobromic acid, Not more than 49% strength (DOT)
Hydrobromic acid, Solution [UN 1788]
Hydrogen bromide (ACGIH, DOT,OSHA)
Hydrogen bromide, Anhydrous (DOT)
UN 1048 [Hydrogen bromide, anhydrous]
UN 1788 [Hydrobromic acid solution]

10039-32-4
$HO_4P.2Na.12H_2O$
358.20
Phosphoric acid, disodium salt, dodecahydrate
Sodium monohydrogen phosphate dodecahydrate (2:1:1:12)

10039-54-0
$H_6N_2O_2.H_2O_4S$
164.16
Hydroxylamine, sulfate [UN 2865]
Bis(hydroxylamine) sulfate
Hydroxylamine neutral sulfate
Hydroxylamine, sulfate (2:1)
Hydroxylamine sulphate (DOT)
Hydroxylammonium sulfate
Oxammonium sulfate
UN 2865 [Hydroxylamine sulfate]

10042-59-8
$C_{10}H_{22}O$
158.32
OCC(CCCCC)CCC
1-Heptanol, 2-propyl
2-Propylheptanol

10042-76-9
$N_2O_6.Sr$
211.64
Strontium(II) nitrate (1:2)
Nitrate de strontium (French)
Nitric acid, strontium salt
Strontium nitrate [UN 1507]
UN 1507 [Strontium nitrate]

10042-84-9
$C_6H_9NO_6.xNa$
352.09
Acetic acid, nitrilotri-, sodium salt
Glycine, N,N-bis(carboxymethyl)-, sodium salt (9CI)
Nitrilotriacetic acid sodium salt
Sodium aminotriacetate
Sodium nitriloacetate
Sodium nitrilotriacetate

10043-01-3
$O_{12}S_3.2Al$
342.14
Aluminum sulfate (2:3)
Alum
Aluminum alum
Aluminum sulfate
Aluminum sulfate, Solution (DOT)
Aluminum sulphate
Aluminum trisulfate
Cake Alum
Dialuminum sulphate
Dialuminum trisulfate
NA 1760
Sulfuric acid, aluminum salt (3:2)

10043-09-1
$C_{10}H_{16}O_6$
232.26
C2COC(OCC1CO1)C(O2)OC C3CO3
p-Dioxane, 2,3-bis(2,3-epoxypropoxy)
2,3-Bis(2,3-epoxypropoxy)-p-dioxan
2,3-Bis(2,3-epoxypropoxy)-1,4-dioxane
2,3-Bis(glycidyloxy)-1,4-dioxane
p-Dioxane, 2,3-bis(glycidyloxy)-

10043-35-3
BH_3O_3
61.84
OB(O)O
Boric-acid
Boracic acid
Borofax
Borsaure (German)
NCI-C56417
Orthoboric acid
Three Elephant

10043-52-4
$CaCl_2$
110.98
Calcium-chloride
Calplus
Caltac
Dowflake
Liquidow
Peladow
Snomelt
Superflake Anhydrous

10043-66-0
I
131
[I]
Iodine, Isotope of mass 131 (8CI,9CI)

10043-67-1
$Al.2H_2O_4S.K$
301.33
Alum, potassium
Alum
Alum, N.F.
Aluminum potassium alum
Aluminum potassium disulfate
Aluminum potassium sulfate
Aluminum potassium sulfate alum
Burnt Alum
Dialuminum dipotassium sulfate
Exsiccated alum
Potash alum
Potassium alum
Potassium aluminum sulfate (1:1:2)
Sulfuric acid, aluminum potassium salt (2:1:1) (9CI)

10043-92-2
Rn
222
[Rn]
Radon
Actinon
Alphatron
Niton
Radium Emanation
Radon and Compounds

10045-86-0
$Fe.H_3O_4P$
153.84
Ferric phosphate

Ferric orthophosphate
Iron phosphate
Phosphoric acid, iron(3+) salt (1:1) (9CI)

10045-87-1
$Fe.H_4O_7P_2.Na$
256.80
Iron sodium pyrophosphate
Diphosphoric acid, iron(3+) sodium salt (1:1:1)
Ferric sodium pyrophosphate
Pyrophosphoric acid, iron(3+) sodium salt (1:1:1)

10045-89-3
$Fe.2H_3N.2H_2O_4S$
286.06
Ferrous ammonium sulfate
Ammonium ferrous sulfate
Ammonium iron sulfate
Ammonium iron sulfate (2:2:1)
Caswell No. 459B
EPA Pesticide Chemical Code 050506
Ferrous ammonium sulfate hexahydrate
Ferrous diammonium disulfate
Mohr's Salt
Sulfuric acid, ammonium iron-(2+) salt (2:2:1) (9CI)

10045-94-0
$N_2O_6.Hg$
324.61
Mercury(II) nitrate (1:2)
Mercuric nitrate [UN 1625]
Mercury nitrate
Mercury pernitrate
Nitrate mercurique (French)
Nitric acid, mercury(II) salt
UN 1625 [Mercuric nitrate]

10045-97-3
Cs
137
Cesium, Isotope of mass 137 (8CI,9CI)

10046-00-1
$H_3NO.H_2O_4S$
131.10
Hydroxylamine, sulfate (1:1) (Salt) (9CI)
Hydroxylammonium acid sulfate

10048-13-2
$C_{18}H_{16}O_6$
328.34
COc4cc2OC1OC=CC1c2c5oc3c ccc(O)c3c(=O)c45
7H-Furo(3',2':4,5)furo(2,3-c)-xanthen-7-one, 3a,12c-di-hydro-8-hydroxy-6-methoxy
3a,12c-Dihydro-8-hydroxy-6-methoxy-7H-furo(3',2':4,5)-furo(2,3-c)xanthen-7-one
Sterigmatocystin

10048-32-5
$C_6H_8O_2$
112.14
2H-Pyran-2-one, 5,6-dihydro-6-methyl
(S)-(+)-5,6-Dihydro-6-methyl-2H-pyran-2-one
2-Hexenoic acid, 5-hydroxy-, δ-lactone
γ-Hexenolactone
2-Hexen-5,1-olide
D''-Hexenollactone
5-Hydroxy-2-hexenoic acid lactone
Kyselina paraskorbova (Czech)
Parascorbic acid
Parasorbic acid
(+)-Parasorbinsaeure (German)
Sorbic Oil

10048-95-0
$AsHO_4.2Na.7H_2O$
427.00
Arsenic acid, disodium salt, heptahydrate
Dibasic sodium arsenate
Disodium arsenate, heptahydrate
Sodium acid arsenate, hepta-hydrate
Sodium arsenate, dibasic, hepta-hydrate
Sodium arsenate heptahydrate
Sodium arseniate

10049-04-4
ClO_2
67.45
Chlorine-oxide
Alcide
Anthium dioxcide
Chlorine dioxide
Chlorine dioxide (ACGIH, OSHA)
Chlorine dioxide, Not hydrated (DOT)
Chlorine dioxide hydrate, Frozen [NA 9191]
Chlorine(IV) oxide
Chlorine peroxide
Chloroperoxyl
Chloryl Radical
Doxcide 50
NA 9191 [Chlorine dioxide, hydrate, frozen]

10049-05-5
Cl_2Cr
122.90
Chromium(II) chloride (1:2)
Chromium chloride
Chromium dichloride
Chromium(II) chloride
Chromous chloride

10049-07-7
Cl_3Rh
209.26
[Cl-].[Cl-].[Cl-].[Rh+3]
Rhodium(III) chloride (1:3)
Rhodium chloride
Rhodium trichloride

10051-44-2
$C_6H_{12}O_2.Na$
139.15
Hexanoic acid, sodium salt (9CI)
Sodium hexanoate

10054-29-2
$C_{14}H_{31}O_4P$
294.37
O=P(OCCCCCCCCCCCCCC) (O)O
1-Tetradecanol, dihydrogen phosphate (9CI)

10058-23-8
$H_2O_5S.K$
153.18
Potassium peroxymonosulfuric acid
Caswell No. 699A
EPA Pesticide Chemical Code 063604
Monopotassium peroxymono-sulfurate
Peroxymonosulfuric acid, monopotassium salt (8CI,9CI)
Potassium peroxymonosulfate

10058-44-3
$Fe.3/_4H_4O_7P_2$
189.33
Ferric pyrophosphate
Diphosphoric acid, iron(3+) salt (3:4) (9CI)
Iron pyrophosphate
Pyrophosphoric acid, iron(3+) salt (3:4)

10061-01-5
$C_3H_4Cl_2$
110.97
ClCC=CCl
Propene, 1,3-dichloro-, (Z)
cis-1,3-Dichloropropene
(Z)-1,3-Dichloropropene
cis-1,3-Dichloropropylene
1-Propene, 1,3-dichloro-, (Z)-(9CI)

10061-02-6
$C_3H_4Cl_2$
110.97
ClCC=CCl
Propene, 1,3-dichloro-, (E)
(E)-1,3-Dichloropropene
trans-1,3-Dichloropropene

trans-1,3-Dichloropropylene
1-Propene, 1,3-dichloro-, (E)-(9CI)

10061-32-2
Unknown
Unknown
DEA No. 9631
Levophenacylmorphan

10075-50-0
C_8H_6BrN
196.04
Brc2ccc1[nH]ccc1c2
1H-Indole, 5-bromo- (9CI)

10075-74-8
$C_{12}H_{12}S_2$
220.36
Naphthalene, 1,5-bis-(methylthio)- (8CI,9CI)
1,5-Bis(methylthio)naphthalene

10081-67-1
$C_{30}H_{31}N$
405.58
N(c(ccc(c1)C(c(cccc2)c2)(C)C)c1)c(ccc(c3)C(c(cccc4)c4)(C)C)c3
Benzenamine, 4-(1-methyl-1-phenylethyl)-N-(4-(1-methyl-1-phenylethyl)-phenyl)- (9CI)

10088-45-6
$C_7H_9O_3P$
172.12
Methyl phenylphosphonate
Phenylphosphonic acid, monomethyl ester

10088-95-6
$C_{12}H_{12}O_5$
236.24
4H,5H-Pyrano(4,3-b)pyran-4,5-dione, 2,3-dihydro-3-α-hydroxy-2-β-methyl-7-propenyl
2,3-Dihydro-3-α-hydroxy-2-β-methyl-7-propenyl-4H,5H-pyrano(4,3-b)pyran-4,5-dione
Radicinin
Stemphylone

10094-45-8
$C_{40}H_{79}NO$
590.07
O=C(NCCCCCCCCCCCCCCCCCC)CCCCCCCCCCCC=CCCCCCCCC
Stearyl erucamide
13-Docosenamide, N-octadecyl-
13-Docosenamide, N-octadecyl-, (Z)- (9CI)
Kenamide E-180
N-Octadecyl-13-docosenamide
(Z)-N-Octadecyl-13-docosenamide

10098-89-2
$C_6H_{15}ClN_2O_2$
182.65
L-Lysine, hydrochloride (VAN) (9CI)

10098-97-2
Sr
90
Strontium, Isotope of mass 90 (8CI,9CI)

10099-58-8
Cl_3La
245.26
Lanthanum-chloride

10099-59-9
$N_3O_9.La$
324.94
Lanthanum-nitrate

10099-66-8
Cl_3Lu
281.32
Lutetium-chloride

10099-71-5
$C_{14}H_{24}O_4$
256.38
O=C(OCCCCC)C=CC(=O)OCCCCC
Maleic acid, dipentyl ester
Diamylester kyseliny maleinove (Czech)
Diamyl maleate
Dipentyl maleate

10099-74-8
$N_2O_6.Pb$
331.21
Lead(II) nitrate (1:2)
Lead dinitrate
Lead nitrate
Lead(2+) nitrate
Lead(II) nitrate
Lead nitrate [UN 1469]
Nitric acid, lead(2+) salt
Nitrate de plomb (French)
Plumbous Nitrate
UN 1469 [Lead nitrate]

10099-79-3
O_6PbV_2
405.08
Lead vanadate
Lead vanadium oxide (9CI)

10101-21-0
$C_6H_{12}O_7.1/2Sr$
239.97
D-Gluconic acid, strontium salt (2:1) (9CI)
Gluconic acid, strontium salt (2:1) (8CI)

10101-39-0
$Ca.H_2O_3Si$
118.18
Calcium silicate
Calcium metasilicate
Silicic acid, calcium salt
Silicic acid, calcium salt (1:1) (8CI,9CI)

10101-50-5
$MnO_4.Na$
141.93
Permanganic acid, sodium salt
Permanganate de sodium (French)
Sodium permanganate [UN 1503]
UN 1503 [Sodium permanganate]

10101-53-8
$O_{12}S_3.2Cr$
392.18
Chromium (III) sulfate (2:3)
Baychrom A
Baychrom F
Chromic sulfate
Chromic sulphate
Chromitan B
Chromitan MS
Chromitan NA
Chromium III sulfate
Chromium sulfate (2:3)
Chromium sulphate
Chromium sulphate (2:3)
C.I. 77305
Dichromium sulfate
Dichromium sulphate
Dichromium trisulfate
Dichromium trisulphate
Sulfuric acid, chromium(3+) salt (3:2)

10101-63-0
I_2Pb
461.01
Lead iodide (9CI)
Lead diiodide
Lead(II) iodide
Plumbous iodide

10101-75-4
Unknown
Unknown
Tin(IV) chromate

10101-89-0
$O_4P.3Na.12H_2O$

380.18
Phosphoric acid, trisodium salt, dodecahydrate
Phosphoric acid, trisodium salt, dodeahydrate
Sodium phosphate tribasic dodecahydrate
Sodium phosphate dodecahydrate

10101-97-0
$NiO_4S.6H_2O$
262.89
Nickel(II) sulfate hexahydrate (1:1:6)
Nickel monosulfate hexahydrate
Nickel sulfate hexahydrate
Nickel(II) sulfate hexahydrate
Nickel(2+) sulfate hexahydrate
Nickel sulphate hexahydrate
Sulfuric acid, nickel(2+) salt, hexahydrate

10102-06-4
N_2O_8U
394.02
Uranium, bis(nitrato-O,O')-dioxo- (Solid)
UN 2981 [Uranyl nitrate, solid]
Uranyl nitrate, Solid [UN 2981]

10102-17-7
$O_3S_2.2Na.5H_2O$
317.17
Sodium thiosulfate, pentahydrate
Ametox
Antichlor
Hypo
NSC-45624
Sodium hyposulfite
Sodothiol
Sulfothiorine
Thiosulfuric acid, disodium salt, pentahydrate

10102-18-8
$O_3Se.2Na$
172.94
Selenious acid, disodium salt
Disodium selenite
Natriumselenit (German)
Sodium selenite [UN 2630]
UN 2630 [Selenates or Selenites]

10102-20-2
$H_2O_3Te.2Na$
223.59
Sodium tellurate(IV)
Sodium tellurate
Sodium tellurite
Telluric acid, disodium salt (8CI,9CI)
Tellurous acid, disodium salt

10102-24-6
$H_2O_3Si.2Li$
91.98
Silicic acid, dilithium salt (9CI)
Lithium silicate

10102-43-9
NO
30.01
[N]=O
Nitrogen-monoxide
Bioxyde d'azote (French)
Nitric oxide (ACGIH,OSHA) [UN 1660]
Oxyde nitrique (French)
RCRA waste number P076
Stickmonoxyd (German)
UN 1660 [Nitric oxide]

10102-44-0
NO_2
46.01
[O-]N=O
Nitrogen-dioxide
Azote (French)
Azoto (Italian)
NA 1067 (DOT)
Nitrito
Nitrogen dioxide (ACGIH, OSHA)
Nitrogen dioxide, Liquid (DOT)
Nitrogen peroxide
Nitrogen peroxide, Liquid [UN 1067]
RCRA waste number P078
Stickstoffdioxid (German)
Stikstofdioxyde (Dutch)
UN 1067 [Nitrogen dioxide, Liquefied]

10102-45-1
$NO_3.Tl$
266.38
Thallium(I) nitrate (1:1)
Nitric acid, thallium(1+) salt
RCRA waste number U217
Thallium mononitrate
Thallium nitrate [UN 2727]
Thallous nitrate
UN 2727 [Thallium nitrate]

10102-48-4
H_2AsO_4Pb
348.13
Lead arsenate
Lead(IV) arsenate

10102-49-5
$AsO_4.Fe$
194.77
Iron(III) arsenate (1:1)
Arsenate of iron, ferric
Ferric arsenate [UN 1606]
Ferric arsenate, Solid (DOT)
UN 1606 [Ferric arsenate]

10102-50-8
$As_2O_8.3Fe$
445.39
Iron(II) arsenate (3:2)
Arsenate of iron, ferrous
Ferrous arsenate [UN 1608]
Ferrous arsenate, Solid (DOT)
UN 1608 [Ferrous arsenate]

10102-71-3
$Al.2H_2O_4S.Na$
267.10
Aluminum sodium sulfate
Soda alum
Sodium alum
Sodium aluminum sulfate
Sulfuric acid, aluminum sodium salt (2:1:1) (9CI)

10102-90-6
$H_5CuO_7P_2$
242.53
Copper (II) pyrophosphate
Caswell No. 252
Copper diphosphorate
Copper pyrophosphate
Diphosphoric acid, copper salt (9CI)
EPA Pesticide Chemical Code 069701
Pyrophosphoric acid, copper salt

10103-43-2
$H_3N.H_2O_3S_2$
131.17
Ammonium thiosulfate solution
Thiosulfuric acid ($H_2S_2O_3$), monoammonium salt (8CI, 9CI)

10103-48-7
H_4CuO_4P
162.55
Copper orthophosphate
Copper phosphate
Phosphoric acid, copper salt (9CI)
Primary copper phosphate

10103-50-1
$AsH_3O_4.xMg$
312.12
Arsenic acid, magnesium salt
Arseniate de magnesium (French)
Magnesium arsenate [UN 1622]
Magnesium arsenate, Solid (DOT)
Magnesium arsenate phosphor
UN 1622 [Magnesium arsenate]

10103-62-5

Unknown
Unknown
Arsenic compound
Calcium arsenate

10103-89-6
$C_{14}H_{14}N_2O_6$
306.30
2,3-Quinoxalinedimethanol, diacetate, 1,4-dioxide
2,3-Bis(acetoxymethyl)quinoxaline di-N-oxide
Chinoxidin
Quinoxidine

10105-38-1
$C_{16}H_{34}$
226.45
Pentadecane, 6-methyl- (8CI, 9CI)
NSC-158676

10108-56-2
$C_{10}H_{21}N$
155.32
CCCCNC1CCCCC1
Cyclohexylamine, N-butyl
N-Butylcyclohexylamine

10108-64-2
$CdCl_2$
183.30
Cl[Cd]Cl
Cadmium-chloride
Caddy
Cadmium dichloride
Kadmiumchlorid (German)
VI-Cad

10108-73-3
$N_3O_9.Ce$
326.15
Cerium(III) nitrate
Cerium nitrate
Cerium(3+) nitrate
Cerium trinitrate
Cerous nitrate
Dusicnan cerity (Czech)
Nitric acid, cerium(3+) salt (8CI,9CI)

10108-86-8
$C_{11}H_{26}N.Cl$
207.78
1-Octanaminium, N,N,N-trimethyl-, chloride (9CI)
Ammonium, trimethyloctyl-, chloride (8CI)

10108-87-9
$C_{13}H_{30}N.Cl$
235.83
[Cl-].CCCCCCCCCC[N+](C)(C)C
1-Decanaminium, N,N,N-trimethyl-, chloride (9CI)
Ammonium, decyltrimethyl-, chloride (8CI)

10108-91-5
$C_{30}H_{64}N.Cl$
474.29
1-Tetradecanaminium, N,N-dimethyl-N-tetradecyl-, chloride (9CI)
Ammonium, dimethylditetradecyl-, chloride (8CI)
NSC-61373

10109-95-2
$C_{20}H_{16}N_2O_4$
348.35
O=C(O)c(c(Nc(cccc1)c1)cc(c2Nc(cccc3)c3)C(=O)O)c2
1,4-Benzenedicarboxylic acid, 2,5-bis(phenylamino)- (9CI)
2,5-Dianilinoterephthalic acid

10111-08-7
$C_4H_4N_2O$
96.09
1H-Imidazole-2-carboxaldehyde (9CI)
2-Formylimidazole
2-Imidazolecarbaldehyde
Imidazole-2-carboxaldehyde (7CI,8CI)

10112-91-1
Cl_2Hg_2
472.08
Mercury-chloride
Calogreen
Calomel
Calotab
Chlorure mercureux (French)
Cyclosan
Dimercury dichloride
Mercurous chloride
Mercury subchloride

10114-24-6
$C_{38}H_{28}N_6O_{13}S_3.3Na$
941.81
2-Naphthalenesulfonic acid, 4-hydroxy-7-((((5-hydroxy-6-((2-methoxyphenyl)azo)-7-sulfo-2-naphthalenyl)-amino)carbonyl)amino)-3-((6-sulfo-2-naphthalenyl)-azo)-, trisodium salt (9CI)

10114-86-0
$C_{27}H_{24}N_6O_9S_2.2Na$
MW:686.60
Benzenesulfonic acid, 3,3'-(carbonylbis(imino(3-methoxy-4,1-phenylene)azo))bis-, disodium salt (9CI)
3,3'-(Carbonylbis(imino(3-methoxy-4,1-phenylene)azo))bis-(benzenesulfonic acid), disodium salt
3,3'-(Ureylenebis((3-methoxy-p-phenylene)azo))dibenzenesulfonic acid, disodium salt

10114-96-2
$C_{17}H_{12}N_2O_6S.2Na$
418.33
Benzoic acid, 2-hydroxy-5-((5-sulfo-2-naphthalenyl)azo)-, disodium salt (9CI)
2-Hydroxy-5-((5-sulfonaphth-2-yl)azo)benzoic acid, disodium salt

10114-97-3
$C_{17}H_{12}N_2O_6S.2Na$
418.33
Benzoic acid, 2-hydroxy-5-((8-sulfo-2-naphthalenyl)-azo)-, disodium salt (9CI)
2-Hydroxy-5-((8-sulfonaphth-2-yl)azo)benzoic acid, disodium salt

10117-38-1
$H_2O_3S.2K$
160.28
[K+].[K+].[O-]S([O-])=O
Potassium sulfite
Dipotassium sulfite
Sulfurous acid, dipotassium salt (9CI)
Sulfurous acid, potassium salt

10118-76-0
$Mn_2O_8.Ca$
277.96
Calcium permanganate [UN 1456]
Kaliumpermanganat (German)
UN 1456 [Calcium permanganate]

10118-90-8
$C_{23}H_{27}N_3O_7$
457.53
CN(C)C4C3CC2Cc1c(ccc(O)c1C(=O)C2=C(O)C3(O)C(=O)C(=C4O)C(N)=O)N(C)C
2-Naphthacenecarboxamide, 4,7-bis(dimethylamino)-1,4,4a,5,5a,6,11,12a-octahydro-3,10,12,12a- tetrahydroxy-1,11-dioxo
CL 59806
7-Dimethylamino-6-demethyl-6-deoxytetracycline
Minocyclin
Minocycline

10119-53-6
$C_{18}H_{36}O_2.xCe$
Unknown
Octadecanoic acid, cerium salt (9CI)

Cerium octadecanoate

10124-36-4
$O_4S.Cd$
208.46
[Cd].CC.CCCC[O-]
Cadmium sulfate (1:1)
Cadmium sulfate
Cadmium sulphate
Sulfuric acid, cadmium(2+) salt
Sulphuric acid, cadmium salt (1:1)

10124-37-5
$N_2O_6.Ca$
164.10
Calcium(II) nitrate (1:2)
Calcium nitrate [UN 1454]
Calcium saltpeter
UN 1454 [Calcium nitrate]

10124-41-1
$O_3S_2.Ca$
152.20
Thiosulfuric acid, calcium salt (1:1)
Calcium thiosulfate

10124-43-3
$O_4S.Co$
154.99
Cobalt(II) sulfate (1:1)
Cobaltous sulfate
Cobalt sulfate
Cobalt sulfate (1:1)
Cobalt(II) sulfate
Cobalt (2+) sulfate
Cobalt(II) sulphate
Sulfuric acid, cobalt(2+) salt (1:1)

10124-48-8
ClH_2HgN
252.07
N[Hg]Cl
Mercury-amide-chloride
Aminomercuric chloride
Ammoniated mercury
Mercuric ammonium chloride, Solid (DOT)
Mercuric chloride, ammoniated
Mercury amine chloride
Mercury ammoniated
Mercury ammonium chloride [UN 1630]
UN 1630 [Mercury ammonium chloride]
White Mercury Precipitated
White Precipitate

10124-50-2
$AsH_3O_3.xK$
399.65
Arsenious acid, potassium salt
Arsenenous acid, potassium salt
Arsenite de potassium (French)
Arsonic acid, potassium salt
Kaliumarsenit (German)
NSC-3060
Potassium arsenite [UN 1678]
Potassium arsenite, Solid (DOT)
Potassium metaarsenite
UN 1678 [Potassium arsenite]

10124-54-6
$H_3O_4P.xMn$
Unknown
Phosphoric acid, manganese salt (9CI)
Manganese phosphate

10124-56-8
$O_{18}P_6.6Na$
611.76
Metaphosphoric acid, hexasodium salt
Calgon
Chemi-Charl
Hexametaphosphate, sodium salt
HMP
Medi-Calgon
Natrium hexametaphosphat (German)
Polyphos
SHMP
Sodium hexametaphosphate

10124-65-9
$C_{12}H_{23}O_2.K$
238.45
Dodecanoic acid, potassium salt
Potassium dodecanoate
Potassium laurate

10125-13-0
$CuCl_2.2H_2O$
170.48
Cupric chloride
Caswell No. 235A
Copper chloride dihydrate
Copper(2+) chloride dihydrate
Coppertrace
EPA Pesticide Chemical Code 023701

10127-05-6
$C_{18}H_{13}CrN_4O_{10}S_2.2Na$
607.41
C.I. Acid Yellow 54 (8CI)
Chromate(5-), bis(2-((4,5-dihydro-3-methyl-1-(2-methyl-4-sulfophenyl)-5-oxo-1H-pyrazol-4-yl)azo)-4-sulfobenzoato(4-))-, tetrasodium hydrogen (9CI)

10137-69-6
$C_6H_9Cl_3Si$
215.58
Cyclohexenyltrichlorosilane
Ciclohexenitriclorosilano (Spanish)
Cyclohexene, 4-(trichlorosilyl)-
Cyclohexenyl trichlorosilane
3-Cyclohexenyltrichlorosilane
Silane, (3-cyclohexenyl)trichloro-
Silane, trichloro-3-cyclohexen-1-yl- (9CI)
Trichloro-3-cyclohexenylsilane
Trichloro-3-cyclohexen-1-ylsilane
UN 1762 [Cyclohexenyltrichlorosilane]

10137-73-2
$C_{10}H_{18}O$
154.28
C1CCC(C1)OC2CCCC2
Ether, dicyclopentyl
Cyclopentyl ether

10137-74-3
$Cl_2O_6.Ca$
206.98
Chloric acid, calcium salt
Calcium chlorate
Calcium chlorate [UN 1452]
Calcium chlorate, Aqueous solution [UN 2429]
Chlorate de calcium (French)
UN 1452 [Calcium chlorate]
UN 2429 [Calcium chlorate solution]

10137-80-1
$C_{14}H_{23}N$
205.34
N-2-(Ethylhexyl) aniline
Aniline, N-(2-ethylhexyl)-
Benzenamine, N-(2-ethylhexyl)- (9CI)
N-(2-Ethylhexyl)aniline
N-(2-Ethylhexyl)-benzenamine
Hexylamine, 2-ethyl-n-phenyl-

10138-21-3
$C_4H_8Cl_2Si$
155.11
CCSi(Cl)(Cl)C=C
Silane, dichloroethylvinyl
Dichloroethylvinylsilane
Ethylvinyldichlorosilane

10138-34-8
$C_8H_{14}O_3$
158.22
CCCCOC(=O)C1OC1C
Butyric acid, 2,3-epoxy-, butyl ester
2,3-Epoxybutyric acid, butyl ester

10138-47-3
$C_9H_{20}O_2$
160.29

CCCCC(CC)OCCO
Ethanol, 2-((1-ethylpentyl)oxy)
2-(1-Ethylamyloxy)ethanol
2-((1-Ethylpentyl)oxy)ethanol
2-(3-Heptyloxy)ethanol

10138-74-6
$C_5H_{14}N_2O$
118.17
N-(2-Hydroxyethyl) propylene diamine
Ethanol, 2-((2-amino-1-methylethyl)amino)-
Hydroxyethylpropylenediamine

10138-93-9
$C_6H_{15}N.xH_3O_4P$
Unknown
Ethanamine, N,N-diethyl-, phosphate (9CI)
N,N-Diethylethanamine phosphate

10139-47-6
I_2Zn
319.20
[Zn+2].[I-].[I-]
Zinc iodide (9CI)
Diiodozinc
NSC-39113
Zinc diiodide

10140-65-5
$H_3Na_2O_5P$
159.97
Sodium phosphate, dibasic

10140-87-1
$C_4H_6Cl_2O_2$
157.00
Ethanol, 1,2-dichloro-, acetate
Acetic acid 1,2-dichloroethyl ester
1,2-Dichloroethyl acetate
1,2-Dichlorethylester kyseliny octove (Czech)

10141-00-1
Cr.2H$_2$O$_4$S.K
255.20
Chromium potassium sulfate (1:1:2)
0% Basicity Chrome Alum
Chrome Alum
Chrome Potash Alum
Chromic potassium sulfate
Chromic potassium sulphate
Chromium potassium sulfate
Chromium potassium sulphate
Crystal Chrome Alum
Potassium chromic sulfate
Potassium chromic sulphate
Potassium Chromium Alum
Potassium disulphatochromate (III)
Sulfuric aicd, chromium(3+) potassium salt (2:1:1)

10141-05-6
N_2O_6.Co
182.95
Cobalt(II) nitrate (1:2)
Cobalt dinitrate
Cobalt nitrate
Cobalt(2+) nitrate
Cobalt(II) nitrate
Cobaltous nitrate
Nitric acid, cobalt(2+) salt

10141-22-7
$C_4H_6Cl_2O$
141.00
CC(Cl)(CCl)C=O
Propionaldehyde, 2,3-dichloro-2-methyl
2,3-Dichloro-2-methylpropionaldehyde

10143-22-3
$C_6H_{13}NO_5$
179.20
O=C(OCCOC)N(CO)CO
Carbamic acid, bis(hydroxymethyl)-, (2-methoxy)ethyl ester
N,N-Dimethylol-2-methoxyethyl carbamate

10143-23-4
$C_7H_{16}O$
116.23
CCC(C)C(C)CO
1-Pentanol, 2,3-dimethyl
2,3-Dimethylpentanol
2,3-Dimethyl-1-pentanol

10143-60-9
$C_{16}H_{34}O$
242.50
O(CC(CCCC)CC)CC(CCCC)CC
Ether, bis(2-ethylhexyl)
Di-(2-ethylhexyl) ether
Hexane, 1,1'-oxybis(2-ethyl-

10152-76-8
C_4H_8S
88.17
S(CC=C)C
1-Propene, 3-(methylthio)- (9CI)
Allyl methyl sulfide
3-(Methylthio)-1-propene

10152-77-9
C_4H_8S
88.17
1-Propene, 1-(methylthio)- (9CI)
Sulfide, methyl propenyl (8CI)

10155-47-2
$C_{18}H_{22}N_2O_3$
314.37
O=C(NCCCN(CCOC1)C1)c(c(O)cc(c2ccc3)c3)c2
2-Naphthalenecarboxamide, 3-hydroxy-N-(3-(4-morpholinyl)propyl)- (9CI)
N-(3-Morpholinopropyl)-3-hydroxy-2-naphthoamide

10163-15-2
FO_3P.2Na
143.95
Phosphorofluoridic acid, disodium salt
Disodium fluorophosphate
Disodium monofluorophosphate
Disodium phosphorofluoridate
Sodium fluorophosphate (Na_2PO_3F)
Sodium monofluorophosphate
Sodium phosphorofluoridate
Sodium phosphorofluridate

10169-02-5
$C_{32}H_{20}N_4O_8S_2$.2Na
698.66
(1,1'-Biphenyl)-2,2'-disulfonic acid, 4,4'-bis((2-hydroxy-1-naphthalenyl)azo)-, disodium salt
Acid Anthracene Red G
Acid Anthracene Red GA-CF
Acid Red 97
Airedale Scarlet GM
Alizarine Chrome Red G
Altochrome Scarlet G
Amacid Milling Scarlet G
Anthra Red G
Azo Milling Red G
Belacid Milling Red G
Benzyl Reg GS
Benzyl Red MG
C.I. Acid Red 97, Disodium salt (8CI)
Calcocid Milling Red G
C.I. 22890
C.I. Acid Red 97
Coomassie Milling Scarlet G
Coomassie Milling Scarlet GP
Coriacid Scarlet R
Crispin Red GM
Cyanine Fast Scarlet G
Fenazo Red FG
Hexaderm Red MRG
Korostan Red G
Milling Red A
Milling Scarlet DH
Milling Scarlet 2G
Milling Scarlet R
Naphthalene Leather Scarlet G
Optanol Scarlet GS
Pharmaglo Red G
Polycor Red GS
Shikiso Acid Anthracene Red G
Suminol Brilliant Scarlet DH
Sumitomo Fast Scarlet G
Supranol Scarlet GS
Tertracid Milling Red AGE

Vondamol Brilliant Red G
Xylene Milling Red G

10173-38-3
$C_4H_9BO_2$
99.93
1,3,2-Dioxaborolane, 2-ethyl- (9CI)
Cyclic ethylene ethylboronate ((C_2H_4O_2)EtB)
Ethaneboronic acid, cyclic ethylene ester (7CI,8CI)
Ethylene glycol, cyclic ethaneboronate

10182-91-9
$C_{15}H_{34}N$
228.44
N(CCCCCCCCCCCC)(C)(C)C
Dodecyltrimethylammonium
Dodecyltrimethylammonium bromide
Lauryltrimethylammonium chloride
N,N,N-Trimethyl-1-dodecanaminium

10190-55-3
MoO_4Pb
367.14
Lead molybdate
Lead molybdenum oxide (9CI)

10190-66-6
$C_{12}H_{11}N_4O_3S.Na$
314.32
[Na+].Nc2ccc(N=Nc1ccc(cc1)S([O-])(=O)=O)c(N)c2
Benzenesulfonic acid, 4-((2,4-diaminophenyl)azo)-, monosodium salt
Benzenesulfonic acid, p-((2,4-diaminophenyl)azo)-, monosodium salt (8CI)
C.I. 13220
C.I. Direct Brown 191
Padding Brown J
Padding Brown N

10190-68-8
$C_{25}H_{22}N_4O_9S_3.2Na$
664.63
C.I. Direct Yellow 27, Disodium salt (8CI)
7-Benzothiazolesulfonic acid, 2-(4-((1-(((2-methoxyphenyl)-amino)carbonyl)-2-oxopropyl)-azo)-3-sulfophenyl)-6-methyl-, disodium salt (9CI)
Benzo Viscose Yellow 5GL
Chlorantine Fast Yellow 7GL
C.I. Direct Yellow 27
C.I. 13950
Diazol Light Yellow 7JL
Diphenyl Fast Brilliant Yellow 8GL
Direct Yellow 27
Fastusol Yellow L5GA
Fenaluz Yellow 4G
Helion Yellow 5G
Hispaluz Yellow 5G
NSC-326243
Pyrazol Fast Flavine 5G
Sirius Supra Yellow 5G
Solamine Fast Yellow 5G
Solantine Yellow 8GL
Solar Flavine 5G
Solex Canary Yellow 5G
Solius Light Yellow 5G
Solophenyl Yellow 7GL
Tertrodirect Fast Yellow 8G
Tetramine Fast Yellow Extra-Greenish

10191-18-1
$C_6H_{15}NO_5S$
213.25
O=S(=O)(O)CCN(CCO)CCO
BES
AI3-62516
N,N-Bis(hydroxyethyl)-2-aminoethanesulfonic acid
N,N-Bis(2-hydroxyethyl)aminoethanesulfonic acid
N,N-Bis(2-hydroxyethyl)-2-aminoethanesulfonic acid
2-(Bis(2-hydroxyethyl)amino)-ethanesulfonic acid
Ethanesulfonic acid, 2-(bis-(2-hydroxyethyl)amino)- (9CI)
NSC-166667
Taurine, N,N-bis(2-hydroxyethyl)-

10192-29-7
$ClO_3.H_3N$
100.48
Ammonium chlorate
Chlorate d'ammonium (French)
Chloric acid, ammonium salt
Clorato amonico (Spanish)

10192-30-0
$HO_3S.H_4N$
99.12
Sulfurous acid, monoammonium salt
Ammonium bisulfite
Ammonium bisulfite, Solid (DOT)
Ammonium bisulfite, Solution (DOT)
Ammonium hydrogen sulfite
Ammonium monosulfite
Monoammonium sulfite
NA 2693 (DOT)

10192-85-5
$C_3H_4O_2.K$
111.16
2-Propenoic acid, potassium salt (9CI)
Potassium acrylate
Potassium 2-propenoate

10194-00-0
$C_{10}H_{18}O_4S_2$
266.38
O=C(OCCS)CCCCC(=O)OCCS
Hexanedioic acid, bis(2-mercaptoethyl) ester (9CI)
Bis(2-mercaptoethyl)hexanedioate
Hexanedioic acid, bis(2-mercaptoethyl)ester)
2-Mercaptoethanol adipate

10196-04-0
$H_3N.1/2H_2O_3S$
58.06
Ammonium sulfite
Diammonium sulfite
Sulfurous acid, diammonium salt (9CI)

10196-18-6
$N_2O_6.Zn.6H_2O$
297.51
Zinc(II) nitrate, hexahydrate (1:2:6)
Dusicnan zinecnaty (Czech)
Nitric acid, zinc salt, hexahydrate

10196-66-4
$C_{14}H_{28}O_2.1/2Ba$
297.05
Tetradecanoic acid, barium salt (9CI)
Barium myristate
Barium tetradecanoate

10196-67-5
$C_{14}H_{28}O_2.1/2Cd$
284.58
Tetradecanoic acid, cadmium salt (9CI)
Cadmium myristate
Cadmium tetradecanoate

10196-69-7
$C_{18}H_{36}O_2.1/2Sr$
328.29
Octadecanoic acid, strontium salt (9CI)
Strontium octadecanoate

10198-40-0
Co
60
Cobalt, Isotope of mass 60 (8CI,9CI)

10202-46-7
$C_9H_4Cl_3N_3$
260.49
1,3,5-Triazine, 2,4-dichloro-6-(4-chlorophenyl)- (9CI)
s-Triazine, 2,4-dichloro-

6-(p-chlorophenyl)- (8CI)

10203-28-8
$C_{12}H_{26}O$
186.34
2-Dodecanol (8CI,9CI)
NSC-86142

10203-33-5
$C_{12}H_{26}O$
186.34
5-Dodecanol (8CI,9CI)
NSC-158520

10203-58-4
$C_{11}H_{20}O_4$
216.28
O=C(OCC)C(C(=O)OCC)CC(C)C
Propanedioic acid, (2-methylpropyl)-, diethyl ester (9CI)
AI3-05627
Diethyl isobutylmalonate
Malonic acid, isobutyl-, diethyl ester
NSC-68522

10210-68-1
$C_8Co_2O_8$
341.94
Cobalt, di-mu-carbonylhexacarbonyldi-, (Co-Co)
Cobalt carbonyl (ACGIH, OSHA)
Cobalt octacarbonyl
Cobalt tetracarbonyl
Cobalt tetracarbonyl dimer
Di-mu-carbonylhexacarbonyldicobalt
Dicobalt carbonyl
Dicobalt octacarbonyl
Octacarbonyldicobalt

10212-58-5
$C_{22}H_{44}N_2O$
352.59
O=C(N(C1C)C1)NCCCCCCCCCCCCCCCCC
1-Aziridinecarboxamide, 2-methyl-N-octadecyl- (9CI)

10213-74-8
$C_9H_{18}O_3$
174.24
3-(2-Ethylbutoxy) propionic acid
3-(2-Ethylbutoxy)propionic acid
Kyselina 3-(2-ethylbutoxy)propionova (Czech)
Propionic acid, 3-(2-ethylbutoxy)-

10213-75-9
$C_{11}H_{21}NO$
183.33
N#CCCOCC(CCCC)CC
Propionitrile, 3-(2-ethylhexyloxy)
3-(2-Ethylhexyloxy)propannitril (Czech)
3-(2-Ethylhexyloxy)propionitrile
Propanenitrile, 3-((2-ethylhexyl)oxy)-

10213-77-1
$C_{10}H_{22}O_4$
206.32
O(CC(OCC(OC)C)C)C(CO)C
1-Propanol, 2-(2-(2-methoxypropoxy)propoxy)
Tripropylene glycol methyl ether
Dowanol
Dowanol 62B
Glycol Ether TPM
Poly-Solv TPM

10213-78-2
$C_{22}H_{47}NO_2$
357.61
OCCN(CCCCCCCCCCCCCCCCCC)CCO
Ethanol, 2,2'-(octadecylimino)bis- (9CI)
2,2'-(Octadecylimino)bisethanol

10215-30-2
$C_6H_{14}O_2$
118.18
1-Propanol, 2-propoxy- (8CI, 9CI)

10220-23-2
$C_7H_{15}NO_2$
145.20
O(CCN(C1)CCOC)C1
Morpholine, 4-(2-methoxyethyl)- (9CI)
4-(2-Methoxyethyl)morpholine

10222-01-2
$C_3H_2Br_2N_2O$
241.89
O=C(N)C(C#N)(Br)Br
Acetamide, 2-cyano-2,2-dibromo
Acetamide, 2,2-dibromo-2-cyano- (8CI,9CI)
DBNPA
Dibromocyanoacetamide
α,α-Dibromo-α-cyanoacetamide
2,2-Dibromo-3-nitrilopropionamide

10224-91-6
$C_{18}H_{22}$
238.37
Benzene, 1,1'-ethylidenebis-(4-ethyl- (9CI)
Ethane, 1,1-bis(p-ethylphenyl)-(8CI)

10226-29-6
$C_6H_{11}BrO$
179.06
2-Hexanone, 6-bromo- (8CI, 9CI)

10233-00-8
Unknown
Unknown
Potassium thiosulfate

10233-13-3
$C_{15}H_{30}O_2$
242.40
O=C(OC(C)C)CCCCCCCCCCC
Isopropyl laurate
AI3-01094-X
Dodecanoic acid, 1-methylethyl ester (9CI)
Dodecanoic acid, 1-methylethyl ester (Crude)
1-Methylethyl dodecanoate

10236-47-2
$C_{27}H_{32}O_{14}$
580.59
Naringin
Naringenin-7-β-neohesperidoside
Naringoside

10241-05-1
Cl_5Mo
273.19
Molybdenum pentachloride [UN 2508]
UN 2508 [Molybdenum pentachloride]

10241-21-1
$C_{16}H_{10}CrN_4O_7S.Na$
477.31
Chromate(1-), (6-amino-5-((2-hydroxy-4-nitrophenyl)-azo)-1-naphthalenesulfonato-(3-))hydroxy-, sodium (9CI)

10243-82-0
$C_5Cl_6N_4$
328.80
Acetimidoyl chloride, 2,2,2-trichloro-N-(4,6-dichloro-s-triazin-2-yl)-(7CI,8CI)

10243-83-1
$C_7Cl_9N_5$
473.19
Acetimidoyl chloride, N,N'-(6-chloro-s-triazine-2,4-diyl)bis(2,2,2-tri-

chloro- (7CI,8CI)

10248-74-5
$C_{24}H_{49}NO_4$
415.65
O=C(OCCN(CCO)CCO)CCCCC CCCCCCCCCCCC
Octadecanoic acid, 2-(bis-(2-hydroxyethyl)amino)ethyl ester
AI3-03103

10250-71-2
$C_6H_4ClN_3O_4$
217.55
Benzenamine, 3-chloro-2,6-dinitro- (9CI)
Aniline, 3-chloro-2,6-dinitro- (7CI,8CI)

10254-48-5
Unknown
Unknown
Sodium 2,2'-methylenebis-(4-chlorophenate)

10254-57-6
$C_{19}H_{38}N_2S_4$
MW 422.78
N(C(=S)SCSC(N(CCCC)CCC C)=S)(CCCC)CCCC
Carbamodithioic acid, dibutyl-, methylene ester (9CI)
Bis(di-n-butylthiocarbamoylthio)methane
Methylenebis(di-n-butylthiocarbamate)
Methylene dibutylcarbamodithioate

10257-54-2
Unknown
Unknown
Copper sulfate, monohydrate
Caswell No. 257
Copper sulfate monohydrate
EPA Pesticide Chemical Code 024402

10262-69-8
$C_{20}H_{23}N$
277.44
CNCCCC13CCC(c2ccccc12) c4ccccc34
9,10-Ethanoanthracene-9(10h)-propylamine, N-methyl
276-Ba
3-(9,10-Dihydro-9,10-ethanoanthracen-9-yl)propylmethylamine
Ludiomil
Maprotiline
Maprotylina (Polish)

10264-17-2
$C_{10}H_{21}NO$
171.28
Butanamide, N-hexyl- (9CI)
Butyramide, N-hexyl- (8CI)
NSC-8225
NSC-405015

10264-29-6
$C_{12}H_{25}NO$
199.34
Hexanamide, N-hexyl- (7CI,8CI,9CI)
N-Hexylhexanamide

10265-69-7
$C_8H_9NO_2.Na$
174.15
Glycine, N-phenyl-, monosodium salt (9CI)
N-Phenylglycine monosodium salt

10265-92-6
$C_2H_8NO_2PS$
141.14
COP(N)(=O)SC
Phosphoramidothioic acid, O,S-dimethyl ester
Acephate-Met
Bay 71628
Bayer 71628
Chevron 9006
Chevron Ortho 9006
O,S-Dimethyl ester amide of amidothioate
O,S-Dimethyl phosphoramidothioate
ENT 27,396
Hamidop
Metamidofos estrella
Methamidophos
Monitor
MTD
NSC-190987
Ortho 9006
Pillaron
SRA 5172
Tahmabon
Tamaron
Thiophosphorsaeure-O,S-dimethylesteramid (German)

10271-57-5
$C_{14}H_9Cl$
212.68
Benzene, 1-chloro-2-(phenylethynyl)- (9CI)
Acetylene, (o-chlorophenyl)phenyl- (6CI,7CI,8CI)
(o-Chlorophenyl)phenylacetylene

10275-58-8
$C_{18}H_{24}$
240.39
Naphthalene, 2,7-bis-(1,1-dimethylethyl)- (9CI)
2,7-Di-tert-butylnaphthalene
Naphthalene, 2,7-di-tert-butyl- (6CI,7CI,8CI)

10276-21-8
$C_9H_{14}O_2$
154.21
7-Oxabicyclo(4.1.0)heptan-2-one, 4,4,6-trimethyl-(6CI,7CI,8CI,9CI)
2,3-Epoxy-3,5,5-trimethylcyclohexanone
2,3-Epoxy-3,5,5-trimethyl-1-cyclohexanone
Isophorone epoxide
Isophorone oxide
3,5,5-Trimethyl-2,3-epoxycyclohexanone

10277-04-0
$C_{24}H_{47}NO_4$
413.64
O=C(OCCN(CCO)CCO)CCCCC CCC=CCCCCCCCC
9-Octadecenoic acid (Z)-, 2-(bis(2-hydroxyethyl)-amino)ethyl ester (9CI)
AI3-02282
Triethanolamine, oleic acid monoester

10277-43-7
$N_3O_9.La.6H_2O$
433.06
Lanthanum(III) nitrate, hexahydrate (1:3:6)
Nitric acid, lanthanum(3+) salt, hexahydrate

10278-71-4
$C_{10}H_{14}N_2$
162.26
N(c(c(ccc1)C)c1)=CN(C)C
Formamidine, N,N-dimethyl-N'-(o-tolyl)
N,N-Dimethyl-N'-(o-tolyl)formamidine

10282-57-2
$C_{13}H_{10}BrNO$
276.13
Benzamide, 2-bromo-N-phenyl-
AI3-01074

10286-75-6
$C_{13}H_9Cl_2NO$
266.12
Clc2ccc(NC(=O)c1ccccc1)cc2Cl
Benzamide, N-(3,4-dichlorophenyl)- (9CI)
Benzanilide, 3',4'-dichloro-(8CI)

10287-53-3
$C_{11}H_{15}NO_2$
193.24
O=C(OCC)c(ccc(N(C)C)c1)c1
Parbenate
Benzoic acid, 4-(dimethyl-amino)-, ethyl ester (9CI)
Ethyl 4-(dimethylamino)benzo-ate

10290-12-7
$AsHO_3.Cu$
187.47
Arsenious acid, copper(II) salt (1:1)
Acid copper arsenite
Air-Flo Green
Arsonic acid, copper(2+) salt (1:1) (9CI)
Copper arsenite [UN 1586]
Copper arsenite, Solid (DOT)
Copper orthoarsenite
Cupric arsenite
Cupric Green
Scheeles Green
Scheele's Mineral
Swedish Green
UN 1586 [Copper arsenite]

10291-28-8
$C_{22}H_{20}N_2O_4$
376.40
O=C(O)c(c(Nc(ccc(c1)C)c1)cc(c2 Nc(ccc(c3)C)c3)C(=O)O)c2
Terephthalic acid, 2,5-di-p-toluidino- (8CI)
1,4-Benzenedicarboxylic acid, 2,5-bis((4-methylphenyl)-amino)- (9CI)
NSC-319997
2,5-Bis(p-toluidino)terephthalic acid
2,5-Bis(4-toluidino)terephthalic acid
2,5-Di-p-toluidinoterephthalic acid

10292-98-5
$C_{10}H_{16}O$
152.24
Bicyclo(2.2.1)heptan-2-one, 4,7,7-trimethyl-, (1S)- (9CI)
3-Bornanone, (1S,4S)-(-)- (8CI)
β-Camphor
epi-Camphor
Epicamphor

10294-33-4
BBr_3
250.54
Boron tribromide (ACGIH, OSHA) [UN 2692]
Boron-bromide
Trona
UN 2692 [Boron tribromide]

10294-34-5
BCl_3
117.16
Borane, trichloro
Boron chloride
Boron trichloride [UN 1741]
Chlorure de bore (French)
UN 1741 [Boron trichloride]

10294-40-3
$Ba.CrH_2O_4$
255.36
Barium chromate(VI)
Barium chromate
Barium chromate (1:1)
Barium chromate oxide
Baryta Yellow
Chromic acid, barium salt (1:1)
C.I. 77103
C.I. Pigment Yellow 31
Lemon Chrome
Lemon Yellow
Permanent Yellow
Steinbuhl Yellow
Ultramarine Yellow

10294-53-8
Unknown
Unknown
Iron (III) dichromate

10294-56-1
H_3O_3P
82.00
Phosphorous acid (VAN)(9CI)
Caswell No. 663I
EPA Pesticide Chemical Code 076002
Phosphorous acid

10302-15-5
$C_8H_9NO_2S$
183.22
O=S(=O)(N(C1)C1)c(cccc2)c2
1-(Phenylsulfonyl)aziridine
AI3-50705
Aziridine, 1-(phenylsulfonyl)-
n-Benzenesulfonylaziridine

10309-79-2
$C_8H_{12}N_2$
136.22
Hydrazine, 2-benzyl-1-methyl
1-Benzyl-2-methylhydrazine
Hydrazine, 1-methyl-2-(phenyl-methyl)-
1-Methyl-2-benzyl-hydrazine

10310-21-1
$C_5H_4ClN_5$
169.59
Purine, 2-amino-6-chloro
6-Chloroguanine
1H-Purin-2-amine, 6-chloro- (9CI)

10310-38-0
$C_{20}H_{28}N_6P_2.2Br$
574.30
Phosphonium, ethylenebis-(tris(2-cyanoethyl)-, di-bromide
Cyagard RF-1
Ethylenebis(tris(2-cyanoethyl)-phosphonium bromide)
Phosphonium, 1,2-ethanediyl-bis(tris(2-cyanoethyl)-, di-bromide

10311-84-9
$C_{14}H_{17}ClNO_4PS_2$
393.86
Phosphorodithioic acid, S-(2-chloro-1-(1,3-dihydro-1,3-dioxo-2H-isoindol-2-yl)-ethyl) O,O- diethyl ester
S-(2-Chloro-1-(1,3-dihydro-1,3-dioxo-2H-isoindol-2-yl)-ethyl)O,O-diethyl phosphoro-dithioate
S-(2-Chloro-1-phthalimido-ethyl) O,O-diethyl phosphoro-dithioate
O,O-Diaethyl-S-2-chlor-1-(phthalimido)-aethyl-di-thiophosphat (German)
Dialifor
Dialifos
O,O-Diethyl S-(2-chloro-1-phthalimidoethyl)phosphoro-dithioate
ENT 27,320
Hercules 14503
Phosphorodithioic acid, S-(2-chloro-1-phthalimido-ethyl) O,O-diethyl ester
Torak

10318-26-0
$C_6H_{12}Br_2O_4$
308.00
OC(CBr)C(O)C(O)C(O)CBr
Galactitol, 1,6-dibromo-1,6-dideoxy
DBD
Dibromdulcit
Dibromdulcitol
Dibromodulcitol
1,6-Dibromodideoxydulcitol
1,6-Dibromo-1,6-dideoxy-dulcitol
1,6-Dibromo-1,6-dideoxygal-actitol
1,6-Dibromo-1,6-dideoxy-D-galactitol
1,6-Dibromodulcitol
Elobromol
Galacticol
Mitolac
Mitolactol
NCI-C04795
NSC-104800

10319-14-9

$C_{18}H_{10}BrNO_3$
368.18
O=C(c(c(C1=O)ccc2)c2)C1c(nc
(c(c3Br)ccc4)c4)c3O
1H-Indene-1,3(2H)-dione, 2-(4-bromo-3-hydroxy-2-quinolinyl)- (9CI)
4-Bromo-3-hydroxy-2-(1,3-indandion-2-yl)quinoline
C.I. Disperse Yellow 64

10325-94-7
CdN_2O_6
236.42
[Cd+2].[O-]N(=O)=O.
[O-]N(=O)=O
Cadmium-nitrate
Cadmium dinitrate
Nitric acid, cadmium salt

10326-21-3
$Cl_2O_6.Mg$
191.21
Chloric acid, magnesium salt
Chlorate salt of magnesium
De-Fol-Ate
E-Z-Off
KRMD 58
Magnesium chlorate [UN 2723]
Magnesium dichlorate
Magron
MC Defoliant
Ortho MC
UN 2723 [Magnesium chlorate]

10326-27-9
$BaCl_2.2H_2O$
244.28
Barium chloride, dihydrate
Barium dichloride dihydrate
NCI-C61074

10332-32-8
$C_{23}H_{44}O_5$
400.60
O=C(OCC(CO)(CO)CO)CCCCC
CCC=CCCCCCCCC
9-Octadecenoic acid (Z)-, 3-hydroxy-2,2-bis(hydroxymethyl)propyl ester (9CI)
AI3-07501
Pentaerythritol monooleate

10332-40-8
$C_{10}H_{20}O_2$
158.24
1-Butanol, 2,2-diethyl-, acetate (6CI,8CI,9CI)
Di-2-ethyl-butyl-acetate
Plastolein 9050 DHZ

10344-93-1
$C_3H_3O_2$
71.06
O=C(O)C=C
2-Propenoic acid, ion(1-) (9CI)
Acrylic acid, ion(1-) (8CI)

10350-33-1
$C_{21}H_{14}$
266.35
Perylene, 1-methyl- (6CI, 7CI,9CI)
1-Methylperylene

10356-76-0
$C_9H_{12}FN_3O_4$
245.24
Nc1nc(=O)n(cc1F)C2CC(O)
C(CO)O2
Cytidine, 2'-deoxy-5-fluoro
FCdR
FCDR
5-Fluor-desoxycytidin (German)
5-Fluorodeoxycytidine
5-Fluoro-2'-deoxycytidine

10359-36-1
$C_{18}H_{14}NO_6P$
371.28
Phosphoric acid, 4-nitrophenyl diphenyl ester (9CI)
Phosphoric acid, p-nitrophenyl diphenyl ester (8CI)

10359-69-0
$C_{14}H_{11}NO_5S.Na$
328.29
Benzenesulfonic acid, 5-nitro-2-(2-phenylethenyl)-, sodium salt (9CI)
5-Nitro-2-(2-phenylethenyl)-benzenesulfonic acid, sodium salt

10361-03-2
$O_3P.Na$
101.96
Metaphosphoric acid, sodium salt
Graham's Salt
Metafos
Sodium metaphosphate

10361-16-7
Unknown
Unknown
Octyl dodecyl dimethyl ammonium chloride
BTC 812
Caswell No. 613B
EPA Pesticide Chemical Code 069190
Octyldodecyldimethylammonium chloride

10361-31-6
$C_6H_{15}NO_7$
213.19
Ammonium gluconate
Ammonium D-gluconate
Gluconic acid, ammonium salt, D-
D-Gluconic acid, ammonium salt (9CI)

10361-37-2
$BaCl_2$
208.24
Barium-chloride
Barium dichloride
NCI-C61074
SBA 0108E

10361-89-4
$H_{20}Na_3O_{14}P$
344.09
Sodium phosphate, tribasic

10361-95-2
$Cl_2O_6.Zn$
232.27
Zinc chlorate [UN 5113]
UN 1513 [Zinc chlorate]

10373-78-1
$C_{10}H_{14}O_2$
166.22
Bicyclo(2.2.1)heptane-2,3-dione, 1,7,7-trimethyl-, (+-)- (9CI)
2,3-Bornanedione, (+-)- (8CI)

10374-74-0
$C_{14}H_{28}$
196.38
7-Tetradecene (8CI,9CI)

10375-96-9
$C_{14}H_{22}$
190.33
Benzene, 1,4-dimethyl-2,5-bis-(1-methylethyl)- (9CI)
NSC-84199
p-Xylene, 2,5-diisopropyl-(8CI)

10377-48-7
$O_4S.2Li$
109.94
Lithium sulfate (2:1)
Lithium sulphate
Sulfuric acid, dilithium salt
Sulfuric acid, lithium salt (1:2)

10377-51-2
ILi
133.85
[Li+].[I-]
Lithium iodide (9CI)

10377-60-3
$N_2O_6.Mg$
148.33

Magnesium(II) nitrate (1:2)
Magnesium nitrate [UN 1474]
Nitric acid, magnesium salt (2:1)
UN 1474 [Magnesium nitrate]

10377-98-7
$C_{12}H_{20}O_{12}.4Na.Zr$
813.16
Lactic acid, sodium zirconium salt (4:4:1)
Sodium zirconium lactate
Zirconium sodium lactate

10378-01-5
$C_{16}H_{32}O$
240.43
OCCCCCCCCC=CCCCCCC
9-Hexadecen-1-ol, (Z)- (9CI)
(Z)-9-Hexadecen-1-ol

10378-48-0
$AsHO_2.2H_2O.1/2SR$
112.85
Strontium arsenite tetrahydrate
Arsenenous acid, strontium salt, tetrahydrate
Arsenious acid, strontium salt
Arsenious acid, strontium salt, tetrahydrate
Strontium arsenite

10379-14-3
$C_{16}H_{17}ClN_2O$
288.76
Tetrazepam
2H-1,4-Benzodiazepin-2-one, 7-chloro-5-(1-cyclohexen-1-yl)-1,3-dihydro-1-methyl-
CB 4261
4261 CB
4361 CB
7-Chloro-5-(cyclohexen-1-yl)-1,3-dihydro-1-methyl-2H-1,4-benzodiazepin-2-one
7-Chloro-5-(1-cyclohexen-1-yl)-1,3-dihydro-1-methyl-2H-1,4-benzodiazepin-2-one
7-Chloro-5-(1-cyclohexenyl)-1-methyl-2-oxo-2,3-dihydro-1H-(1,4)-benzo(f)diazepine
Clinoxan
DEA No. 2886
Musaril
Myolastan
Tetrazepamum (Latin)

10380-28-6
$C_{18}H_{12}CuN_2O_2$
351.86
O([Cu]Oc1cccc2cccnc12)c3cccc4cccnc34
8-Quinolinol, copper (II) chelate
Bioquin
Bioquin 1
Bis(8-oxyquinoline)copper
Bis(8-quinolinato)copper
Bis(8-quinolinolato)copper
Cellu-Quin
Copper-8
Copper, bis(8-quinolinolato-N^1,O^8)-
Copper 8-hydroxyquinolate
Copper hydroxyquinolinate
Copper 8-hydroxyquinolinate
Copper 8-hydroxyquinoline
Copper oxinate
Copper (2+) oxinate
Copper oxine
Copper oxyquinolate
Copper oxyquinoline
Copper quinolate
Copper 8-quinolate
Copper 8-quinolinolate
Copper 8-quinolinol
Copper quinolinolate
Cunilate
Cunilate 2472
Cupric 8-hydroxyquinolate
Cupric 8-quinolinolate
Dokirin
Fruitdo
8-Hydroxyquinoline copper complex
Milmer
Oxime copper
Oxine copper
Oxine cuivre
Oxyquinolinoleate de cuivre (French)
Quinondo

10380-29-7
$H_{14}CuN_4O_4S$
229.75
Cupric sulfate, ammoniated

10381-75-6
$C_7H_7BrN_4O_2$
259.04
O=C(N(C(=O)C(N=C(N1)Br)=C12)C)N2C
1H-Purine-2,6-dione, 8-bromo-3,7-dihydro-1,3-dimethyl- (9CI)
Bromotheophylline
8-Bromotheophylline
NSC-164940
Theophylline, 8-bromo-

10389-73-8
$C_{10}H_{14}ClN$
184.67
Clortermine
Benzeneethanamine, 2-chloro-α,α-dimethyl-
o-Chloro-α,α-dimethylphenethylamine
Clortermina (Spanish)
Clorterminum (Latin)
1-(o-Chlorophenyl)-2-methyl-2-aminopropane
1-(o-Chlorophenyl)-2-methyl-2-propylamine
DEA No. 1647
Phenethylamine, o-chloro-α,α-dimethyl-

10393-51-8
$C_{11}H_{11}N_5.xClH$
UNKNOWN
2,6-Pyridinediamine, 3-(phenylazo)-, hydrochloride (9CI)
Pyridine, 2,6-diamino-3-(phenylazo)-, hydrochloride (6CI,7CI,8CI)

10396-10-8
$C_8H_{11}N_3O_3S$
229.24
O=C(NNS(=O)(=O)c(ccc(c1)C)c1)N
4-Toluenesulfonyl semicarbazide
Benzenesulfonic acid, 4-methyl-, 2-(aminocarbonyl)hydrazide (9CI)
p-Toluenesulfonyl semicarbazide

10402-15-0
Unknown
Unknown
Chelates of copper citrate
Caswell No. 235B
Copper citrate
EPA Pesticide Chemical Code 044005

10402-16-1
$C_{18}H_{35}CuO_2$
347.02
(Z)-9-Octadecenoic acid, copper salt
9-Octadecenoic acid (Z)-, copper salt (9CI)

10405-27-3
F_2HN
53.01
Fluorimide (8CI,9CI)
Difluoramine
NHF2

10409-78-6
$C_7H_{15}N_7$
197.20
1,3,5-Triazin-2(1H)-one, 4,6-bis(dimethylamino)-, hydrazone (9CI)
NSC-99856
s-Triazine, 2,4-bis(dimethylamino)-6-hydrazino- (8CI)

10411-52-6
$C_4H_6Cl_2O_2$
157.00
Propanoic acid, 2,3-dichloro-2-methyl- (9CI)

2,3-Dichloroisobutyrate
2,3-Dichloroisobutyric acid
NSC-85847

10413-71-5
Unknown
Unknown
Nitrilotriacetic acid and its salts
NTA, erbium(3+) salt (3:1)

10415-75-5
$NO_3.Hg$
262.60
Mercury(I) nitrate (1:1)
Mercurous nitrate [UN 1627]
Mercurous nitrate, Solid (DOT)
Nitrate mercureux (French)
Nitric acid, mercury(I) salt
UN 1627 [Mercurous nitrate]

10418-03-8
$C_{21}H_{32}N_2O$
328.49
Stanozolol
Anabol
Androstanazol (VAN)
Androstanazole (VAN)
2'H-Androst-2-eno(3,2-c)pyrazol-17-ol, 17-methyl-, (5α,17β)- (9CI)
2'H-5α-Androst-2-eno(3,2-c)-pyrazol-17β-ol, 17-methyl-
Estanozolol (Spanish)
Estazol
1,2,3,3a,3b,4,5,5a,6,8,10,10a,10b,11,12,12a-Hexadecahydro-1,10a,12a-trimethylcyclopenta(7,8)-phenanthro-(2,3-c)pyrazol-1-ol
17β-Hydroxy-17α-methylandrostano(3,2-c)pyrazole
17-Methyl-2H-5α-androst-2-eno(3,2-c)pyrazol-17β-ol
17-Methyl-2'H-5α-androst-2-eno(3,2-c)pyrazol-17β-ol
NSC-43193
NSC-233046
Stanazol
Stanazolol
Stanozol
Stanozolo
Stanozololum (Latin)
Stromba
Strombaject
Tevabolin
Win 14833 (VAN)
Winstroid
Winstrol (VAN)
Winstrol V (VAN)

10420-33-4
$C_8H_{12}O_5$
188.18
O=C(OC)C(C(=O)C)CC(=O)OC
Butanedioic acid, acetyl-, dimethyl ester (9CI)
Dimethyl acetylbutanedioate

10421-48-4
FeN_3O_9
241.88
Nitric acid, iron(3+) salt
Ferric nitrate [UN 1466]
Iron nitrate
Iron (III) nitrate, Anhydrous
Iron trinitrate
UN 1466 [Ferric nitrate]

10421-98-4
$C_7H_{15}N_7$
197.24
1,3,5-Triazin-2(1H)-one, 4,6-bis(ethylamino)-, hydrazone (9CI)
s-Triazine, 2,4-bis(ethylamino)-6-hydrazino- (7CI, 8CI)

10422-01-2
$C_{12}H_{16}N_8O_3$
320.26
2-Furaldehyde, 5-nitro-, (4,6-bis(ethylamino)-s-triazin-2-yl)hydrazone (8CI)

10422-35-2
$C_5H_{11}Br$
151.05
BrCC(CC)C
Butane, 1-bromo-2-methyl- (9CI)
1-Bromo-2-methylbutane

10431-98-8
C_5H_9NO
99.13
O(C(=NC1)CC)C1
Oxazole, 2-ethyl-4,5-dihydro- (9CI)
2-Ethyl-4,5-dihydrooxazole
2-Ethyloxazoline
2-Ethyl-2-oxazoline
NSC-136557
2-Oxazoline, 2-ethyl-

10433-59-7
$C_{10}H_9Cl_2O_3.Na$
271.07
Sodium 4-(2,4-dichlorophenoxy)butyrate
Butoxone SB
Butyrac
Butyric acid, 4-(2,4-dichlorophenoxy)-, sodium salt
Caswell No. 316E
2,4-DB sodium salt
2,4-DB-sodium
γ-(2,4-Dichlorophenoxy)butyric acid, sodium salt
2,4-Dichlorophenoxybutyric acid, sodium salt
4-(2,4-Dichlorophenoxy)butyric acid sodium salt
Embutox
EPA Pesticide Chemical Code 030804
MB 2878

10436-39-2
$C_3H_2Cl_4$
179.85
C(=C(Cl)Cl)(CCl)Cl
Propene, 1,1,2,3-tetrachloro
1,1,2,3-Tetrachloropropene

10439-77-7
$C_8H_{10}N_2O_2$
166.17
Benzenamine, N,2-dimethyl-4-nitro- (9CI)
o-Toluidine, N-methyl-4-nitro- (8CI)

10441-36-8
$C_{12}H_{10}O_4$
218.21
Acetic acid, ((6-hydroxy-2-naphthalenyl)oxy)- (9CI)
Acetic acid, ((6-hydroxy-2-naphthyl)oxy)- (8CI)

10443-70-6
$C_{13}H_{17}ClO_3$
256.73
CCOC(=O)CCCOc1ccc(Cl)cc1C
Butyric acid, 4-((4-chloro-o-tolyl)oxy)-, ethyl ester
Ethyl 4-(4-chloro-2-methylphenoxy)butylate
Ethyl 4-(4-chloro-2-methylphenoxy)butyrate
MCPB-ethyl

10448-09-6
$C_{13}H_{26}O_4Si_4$
358.75
Cyclotetrasiloxane, heptamethylphenyl
Heptamethylphenylcyclotetrasiloxane
Monophenylheptamethylcyclotetrasiloxane
PM_1MM^3

10448-10-9
$C_{23}H_{30}O_4Si_4$
482.84
Cyclotetrasiloxane, 2,2,4,6,8-pentamethyl-4,6,8-triphenyl- (6CI,7CI,8CI,9CI)
2,2,4,6,8-Pentamethyl-4,6,8-triphenylcyclotetrasiloxane

10449-71-5
$C_{78}H_{159}Al$
1124.10

Aluminum, trihexacosyl- (8CI)
Trihexacosylaluminum
Tri(hexacosyl) aluminum

10453-86-8
$C_{22}H_{26}O_3$
338.48
CC(C)=CC3C(C(=O)OCc2coc(Cc1ccccc1)c2)C3(C)C
Cyclopropanecarboxylic acid, 2,2-dimethyl-3-(2-methylpropenyl)-, (5-benzyl-3-furyl) methyl ester
Benzofuroline
Benzyfuroline
5-Benzyl-3-furylmethyl(+-)-cis,trans-chrysanthemate
(5-Benzyl-3-furyl) methyl-2,2-dimethyl-3-(2-methylpropenyl)-cyclopropanecarboxylate
Chryson
Chrysron
Dimethyl 3-(2-methyl-1-propenyl)cyclopropanecarboxylate
ENT 27,474
FMC 17370
For-Syn
NIA 17370
NRDC 104
NSC-195022
OMS-1206
Premgard
Pynosect
Pyretherm
Resmethrin
Resmetrina (Portuguese)
SBP-1382
S.B. Penick 1382
Synthrin

10453-89-1
$C_{10}H_{16}O_2$
168.26
O=C(O)C(C1(C)C)C1C=C(C)C
Cyclopropanecarboxylic acid, 2,2-dimethyl-3-(2-methylpropenyl)
Chrysanthemic acid
Chrysanthemumic acid
Chrysanthemummonocarboxylic acid

10458-14-7
$C_{10}H_{18}O$
154.25
Cyclohexanone, 5-methyl-2-(1-methylethyl)- (9CI)
2-Isopropyl-5-methylcyclohexanone
p-Menthanone
p-Menthan-3-one (8CI)
Menthone
5-Methyl-2-(isopropylcyclo)-hexanone
5-Methyl-2-(1-methylethyl)-cyclohexanone
NSC-113134

10460-00-1
$C_{18}H_{37}N.C_2H_4O_2$
327.55
9-Octadecen-1-amine, (Z)-, acetate (9CI)
(Z)-9-Octadecen-1-amine acetate
cis-9-Octadecenylamine, acetate
Oleyl amine, acetate
Oleylamine acetate
Oleylamine, acetic acid salt

10468-30-1
$C_{18}H_{34}O_2.1/2Cd$
338.67
9-Octadecenoic acid (Z)-, cadmium salt (9CI)
Cadmium oleate

10469-09-7
$C_6HCl_4NO_2$
260.88
O=C(O)c(nc(c(c1Cl)Cl)Cl)c1Cl
2-Pyridinecarboxylic acid, 3,4,5,6-tetrachloro- (9CI)
Tetrachloropicolinic acid
3,4,5,6-Tetrachloro-2-pyridinecarboxylic acid

10471-14-4
$C_5H_{12}O_2$
104.15
O(C(OCC)C)C
Ethane, 1-ethoxy-1-methoxy- (9CI)
Acetaldehyde methyl ethyl acetyl
1-Ethoxy-1-methoxyethane

10471-78-0
C_6H_9NO
111.14
O(C(=NC1)C(=C)C)C1
Oxazole, 4,5-dihydro-2-(1-methylethenyl)- (9CI)
4,5-Dihydro-2-(1-methylethenyl)oxazole
2-(1-Methylethenyl)-4,5-dihydrooxazole

10474-14-3
$C_4H_7Br_2Cl$
250.38
CC(Br)(CCl)CBr
Propane, 3-chloro-1,2-dibromo-2-methyl
1,2-Dibromo-3-chloro-2-methylpropane
MDBCP
Methyl-DBCP
2-Methyl-1,2-dibromo-3-chloropropane

10476-82-1
$C_{21}H_{22}N_2O_2.AsH_3O_4$
476.35
Strychnidin-10-one, Compd. with arsenic acid (H_3AsO_4) (1:1) (9CI)
Strychnine, Compd. with arsenic acid (H_3AsO_4) (1:1) (8CI)

10476-85-4
Cl_2Sr
158.52
Strontium-chloride

10476-95-6
$C_8H_{12}O_4$
172.20
O=C(OC(OC(=O)C)C(=C)C)C
2-Propene-1,1-diol, 2-methyl-, diacetate
Acetic acid, 2-methyl-2-propene-1,1-diol diester
2-Methyl-2-propene-1,1-diol diacetate

10482-43-6
$C_{24}H_{19}N_5O_7S.2Na$
567.46
Benzoic acid, 2-(((2-carboxyphenyl)amino)sulfonyl)-5-((4,5-dihydro-3-methyl-5-oxo-1-phenyl-1H-pyrazol-4-yl)azo)-, disodium salt (9CI)

10482-56-1
$C_{10}H_{18}O$
154.25
OC(C(CCC(=C1)C)C1)(C)C
(L)-α-Terpineol
3-Cyclohexene-1-methanol, α,α,4-trimethyl-, (S)- (9CI)
p-Menth-1-en-8-ol, (S)-(-)-α-Terpineol
L-α-Terpineol
(-)-α-Terpineol

10486-19-8
$C_{13}H_{26}O$
198.35
O=CCCCCCCCCCCCC
Tridecanal (9CI)

10487-92-0
$C_{11}H_{12}O_4$
208.22
Benzeneacetic acid, α-(acetyloxy)-α-methyl-, (S)- (9CI)
Mandelic acid, α-methyl-, acetate, (+)- (8CI)

10520-38-4
$C_6H_{12}N_2O_2$
144.16
Propanal, 2-methyl-, O-((methylamino)carbonyl)-oxime (9CI)

Isobutyraldehyde, O-(methyl-carbamoyl)oxime (8CI)

10533-67-2
C_3H_7NOS
105.15
N(O)=CCSC
Acetaldehyde, (methylthio)-, oxime (9CI)
(Methylthio)acetaldehyde oxime
(Methylthio)acetaldoxime

10535-50-9
$C_{32}H_{60}O_6$
540.82
O=C(OCC(C)(COC(=O)CCCCCCCC)COC(=O)CCCCCCCC)CCCCCCCC
Nonanoic acid, 2-methyl-2-(((1-oxononyl)oxy)methyl)-1,3-propanediyl ester (9CI)
Trimethylolethane trinonanoate
Trimethylolethane tripelargonate

10544-63-5
$C_6H_{10}O_2$
114.16
O=C(OCC)C=CC
Crotonic acid, ethyl ester
2-Butenoic acid, ethyl ester
Ethyl crotonate
Ethylester kyseliny krotonove (Czech)

10544-72-6
N_2O_4
92.02
Nitrogen tetroxide (DOT)
Dinitrogen tetroxide (DOT)
Nitrogen dioxide, di-
NA 1067 (DOT)
Nitrogen tetroxide, Liquid [UN 1067]
UN 1067 [Dinitrogen tetroxide; (Nitrogen dioxide), Liquefied]

10544-73-7
N_2O_3
76.02
Nitrogen trioxide [UN 2421]
Dinitrogen trioxide
Nitrogen sesquioxide
Nitrous anhydride
UN 2421 [Nitrogen trioxide]

10545-99-0
Cl_2S
102.96
Sulfur-dichloride
Chloride of sulfur (DOT)
Chlorine sulfide
Dichlorosulfane
Monosulfur dichloride
Sulfur chloride
Sulfur chloride (8CI,9CI)
Sulfur chloride(mono) [UN 1828]
UN 1828 [Sulfur chlorides]

10548-10-4
$C_9H_{21}O_3PS_3$
304.43
CCOP(=S)(OCC)SCS(=O)C(C)(C)C
Terbufos sulfoxide

10551-21-0
$C_{14}H_{16}N.Br$
278.19
Pyridinium, 2-methyl-1-(2-phenylethyl)-, bromide (9CI)
2-Methyl-1-(2-phenylethyl)-pyridinium bromide
NSC-35717
N-Phenethyl-α-picolinium bromide
1-Phenethyl-2-picolinium bromide
α-Picolinium-β-phenylethyl bromide
2-Picolinium, 1-phenethyl-, bromide (8CI)

10556-98-6
$C_{12}H_{27}N_3$
213.35
1,3,5-Triazine, hexahydro-1,3,5-tris(1-methylethyl)- (9CI)
NSC-166334
s-Triazine, hexahydro-1,3,5-triisopropyl- (8CI)

10563-26-5
$C_8H_{22}N_4$
174.27
N(CCNCCCN)CCCN
N,N'-Bis(γ-aminopropyl)diaminoethane
N,N'-Bis(3-aminopropyl)-1,2-diaminoethane
N,N'-Bis(3-aminopropyl)ethylenediamine
N,N'-Di(3-aminopropyl)-1,2-ethylenediamine
N,N''-1,2-Ethanediylbis-1,3-propanediamine
NSC-180823
1,3-Propanediamine, N,N''-1,2-ethanediylbis- (9CI)
1,3-Propanediamine, N,N''-ethylenebis-
1,5,8,12-Tetraazadodecane

10574-01-3
$C_{21}H_{41}NO$
323.56
O=C(NC(C)C)CCCCCCCC=CCCCCCCCC
9-Octadecenamide, N-(1-methylethyl)-, (Z)- (9CI)
(Z)-N-(1-Methylethyl)-9-octadecenamide
Oleic acid, isopropylamide

10574-36-4
C_7H_{14}
98.19
2-Hexene, 3-methyl-, (Z)-(9CI)
3-Methyl-cis-2-hexene
(Z)-3-Methyl-2-hexene
cis-3-Methyl-2-hexene
NSC-73929

10574-37-5
C_7H_{14}
98.19
2-Pentene, 2,3-dimethyl-(8CI,9CI)

10580-24-2
$C_{15}H_{30}O_2$
242.40
Undecanoic acid, butyl ester (8CI,9CI)

10580-52-6
Cl_2V
121.84
Vanadium-dichloride

10581-62-1
$C_4H_6ClN_5$
159.58
1,3,5-Triazine-2,4-diamine, 6-(chloromethyl)- (9CI)
Chloroacetoguanamine
s-Triazine, 2,4-diamino-6-(chloromethyl)- (6CI,7CI,8CI)

10588-01-9
$Cr_2O_7.2Na$
261.98
Dichromic acid, disodium salt
Bichromate de sodium (French)
Bichromate of soda
Chromic acid, disodium salt
Chromium sodium oxide
Disodium dichromate
NA 1479 (DOT)
Natriumbichromaat (Dutch)
Natriumdichromaat (Dutch)
Natriumdichromat (German)
Sodio (dicromato di) (Italian)
Sodium bichromate
Sodium chromate
Sodium dichromate(VI)
Sodium dichromate (DOT)
Sodium(dichromate de) (French)

10592-27-5
$C_7H_8N_2$
118.14
1H-Pyrrolo(2,3-b)pyridine,

2,3-dihydro- (6CI,8CI, 9CI)
7-Azaindoline
2,3-Dihydro-1H-pyrrolo-(2,3-b)pyridine

10594-03-3
$C_{18}H_{27}N$
257.41
n(c(c(c(c1CCC2)C2)CCC3)C3)c1CCCCC
Phenanthridine, 1,2,3,4,7,8,9,10-octahydro-6-pentyl- (9CI)

10595-49-0
$C_{18}H_{39}N_2O.CH_3O_4S$
410.61
1-Propanaminium, N,N,N-trimethyl-3-((1-oxododecyl)-amino)-, methyl sulfate (9CI)

10595-60-5
$C_{16}H_{33}N_3$
267.44
N(CCNCCN=C(CC(C)C)C)=C(CC(C)C)C
1,2-Ethanediamine, N-(1,3-dimethylbutylidene)-N'-(2-((1,3-dimethylbutylidene)-amino)ethyl)- (9CI)
1,7-Bis(1,3-dimethylbutylidene) diethylenetriamine
1,7-Bis(1,3-dimethylbutylidene)-diethylenetriamine

10595-72-9
$C_{32}H_{62}O_4S$
542.91
O=C(OCCCCCCCCCCCCC)CCSCCC(=O)OCCCCCCCCCCCCC
Ditridecyl thiodipropionate
Ditridecyl 3,3'-thiobispropanoate
Di(tridecyl) thiodipropionate
Propanoic acid, 3,3'-thiobis-, ditridecyl ester (9CI)
3,3'-Thiobispropanoic acid, ditridecyl ester

10595-80-9
$C_6H_{10}O_5S$
194.21
O=C(OCCS(=O)(=O)O)C(=C)C
2-Propenoic acid, 2-methyl-, 2-sulfoethyl ester (9CI)
Sulfoethyl methacrylate
2-Sulfoethyl 2-methyl-2-propenoate

10595-95-6
$C_3H_8N_2O$
88.13
CCN(C)N=O
Ethylamine, N-methyl-N-nitroso
Ethylmethylnitrosamine
Methylaethylnitrosamin (German)
Methylethylnitrosamine
N,N-Methylethylnitrosamine
N-Methyl-N-nitroso-ethamine
N-Methyl-N-nitrosoethylamine
NEMA
N-Nitrosoethylmethylamine
N-Nitrosomethylethylamine
NMEA

10599-57-2
$C_7H_{10}O$
110.16
Furan, 2,3,4-trimethyl- (7CI,8CI,9CI)
2,3,4-Trimethylfuran

10599-90-3
ClH_2N
51.48
Chloramide
Chloramine
Chloramine (Inorganic compound)
Chloroamine
Monochloramide
Monochloramine
Monochloroamine
Monochloroammonia

10601-69-1
$C_7H_{13}NO_2$
143.19
Pentanamide, N-acetyl- (9CI)
Valeramide, N-acetyl- (7CI, 8CI)

10601-70-4
$C_8H_{15}NO_2$
157.21
Hexanamide, N-acetyl- (7CI,8CI,9CI)

10604-69-0
$C_3H_4O_2.H_3N$
89.09
2-Propenoic acid, ammonium salt (9CI)
Ammonium 2-propenoate

10605-10-4
$C_{12}H_9ClO.Na$
227.65
Sodium 4-chloro-2-phenylphenate
2-Biphenylol, 5-chloro-, sodium salt
Caswell No. 210B
5-Chloro-2-biphenylol sodium salt
5-Chloro-(1,1'-biphenyl)-2-ol sodium salt
4-Chloro-2-phenylphenol sodium salt
Dowicide 31
EPA Pesticide Chemical Code 062212
Sodium-2-chloro-4-phenyl phenate

10605-11-5
Unknown
Unknown
Sodium 6-chloro-2-phenylphenate

10605-21-7
$C_9H_9N_3O_2$
191.21
O=C(OC)N=C(Nc(c1ccc2)c2)N1
2-Benzimidazolecarbamic acid, methyl ester
BAS-3460
BAS 67054
Bavistin
BCM
Benzimidazole-2-carbamic acid, methyl ester
1H-Benzimidazol-2-ylcarbamic acid methyl ester
N-2-(Benzimidazolyl) carbamate
BMC
Carbendazim
Carbendazime
Carbendazole
Carbendazym
CTR 6669
Delsene
Derosal
HOE 17411
Kemdazin
MBC
2-(Methoxy-carbonylamino)-benzimidazol (German)
2-(Methoxycarbonylamino)-benzimidazole
Methyl 1H-benzemedazol-2-ylcarbamate
Methyl 2-benzimidazole-carbamate
Methyl benzimidazole-2-yl carbamate
Methyl benzimidazol-2-yl carbamate
U-32.104

10606-46-9
$C_{14}H_9Cl_5O$
370.49
Benzenemethanol, 2-chloro-α-(4-chlorophenyl)-α-(trichloromethyl)- (9CI)
Benzhydrol, 2,4'-dichloro-α-(trichloromethyl)-

11000-17-2
Unknown
Unknown
Vasopressin
ADH

Antidiuretic hormone
β-Hypophamine
Leiormone
Pitressin
Tonephin
Vasophysin
Vasopressina
Vasopressine
Vasopressinum

11003-38-6
Unknown
Unknown
Capreomycin
Capastat
Capostatin
Capromycin
Kapreomycin

11006-55-6
Unknown
Unknown
Tauroglycocholic acid
Sodium tauroglycocholate

11006-96-5
Unknown
Unknown
Peyote
DEA No. 7415

11024-24-1
$C_{56}H_{92}O_{29}$
1229.48
Digitonin
Digitin

11024-67-2
$C_{19}H_{22}O_7$
362.38
Oxiranecarboxylic acid, 2-methyl-, decahydro-8-hydroxy-3,6-bis(methylene)-2-oxospiro(azuleno(4,5-b)furan-9(2H),2'-oxiran)-4-yl ester, (3aR-(3aα,4α(S*),6aα,8β,9α,9aα,9bβ))- (9CI)
Repin (8CI)

11041-12-6
Unknown
Unknown
Cholestyramine
Cholestyramine chloride
Cholestyramine resin
Colestyramin
Cuemid
Quantalan
Questran

11043-90-6
Unknown
Unknown
Tipol

11056-06-7
Unknown
Unknown
Bleomycin
Blenoxane
Bleo
Bleocin
BLM
NDC 0015-3010
NSC-125066

11057-45-7
Unknown
Unknown
Benzoperylene

11062-77-4
O_2
32.00
Superoxide
Anion, Superoxide
Ion, Oxygen
Oxygen Ion
Radical, Superoxide
Superoxide Anion
Superoxide Radical

11063-25-5
$C_{10}H_4O_4$
188.14
Naphthalenetetrone (8CI,9CI)

11066-21-0
$C_{14}H_{28}O_3S.Na$
299.43
Tetradecenesulfonic acid, sodium salt (9CI)
Sodium tetradecenesulfonate

11067-81-5
$C_{18}H_{30}O_3S$
326.54
Benzenesulfonic acid, tetrapropylene
ABS
Tetrapropylene benzenesulfonate

11067-82-6
$C_{18}H_{29}O_3S.Na$
348.52
[Na+].CCCCCCCCCCCCOS(=O)(=O)c1ccccc1
Benzenesulfonic acid, tetrapropylene-, sodium salt
ABS
Sodium tetrapropylbenzene sulfonate
Tetrapropylenebenenesulphonate, sodium salt

11068-27-2
$C_{20}H_{14}$
254.33
Binaphthalene (9CI)
Binaphthyl (8CI)

11069-19-5
$C_4H_6Cl_2$
125.00
Dichlorobutene
Butene, dichloro-
Dichlorobutylene
NA 2924

11070-44-3
$C_9H_{10}O_3$
166.18
1,3-Isobenzofurandione, tetrahydromethyl- (9CI)
Methyl tetrahydrophthalic anhydride
Tetrahydromethyl-1,3-isobenzofurandione

11070-68-1
$C_5H_8NO_4$
146.12
L-Glutamic acid, ion(1-) (9CI)
Glutamic acid, ion(1-), L- (8CI)

11070-82-9
$Al_2O_{18}S_3.6Ca.xH_2O$
Unknown
Aluminate(12-), hexaoxotris(sulfato(2-))di-, calcium (1:6), hydrate (9CI)
Aluminum calcium oxide sulfate, hydrate (8CI)
C.I. Pigment White 33
Satin white

11071-15-1
$C_4H_5O_7Sb.K$
325.94
L-Antimony potassium tartrate
Potassium antimonyl L-tartrate
L-Tartaric acid, antimony potassium salt

11071-47-9
C_8H_{16}
112.22
Isooctene
Isoocteno (Spanish)
1-Pentene, 2,2,4-trimethyl-
UN 1216 [Isooctenes]

11071-85-5
Br_2H
160.82
Hydrogen bromide (9CI)
Hydrogen dibromide (8CI)

11072-43-8
$C_{16}H_{36}N.C_6H_4NO_4$
396.58

Ammonium, tetrabutyl-, salt with 4-nitropyrocatechol (1:1) (8CI)

11084-85-8
$ClNa_{13}O_{17}P_4$
730.20
Chlorinated-trisodium-phosphate
NCI-C55754
Sodium hypochlorite phosphate ($Na_{13}(ClO)(PO_4)_4$) (9CI)

11095-43-5
C_8H_6S
134.20
Benzothiophene (9CI) (VAN)

11096-42-7
$(C_2H_4O)xC_{15}H_{24}O.xI_2$
Polymer
Ethanol, nonylphenoxypolyethyleneoxy-, iodine complex
Biopal CVL-10
Biopal NR-20
Biopal VRO 10
Biopal VRO-20
Nonylphenoxypolyethyleneoxy ethanol-iodine complex

11096-82-5
Unknown
Unknown
Polychlorinated biphenyl (Aroclor 1260)
Arochlor 1260
Aroclor 1260
Chlorodiphenyl (60% Cl)
Clophen A60
Phenoclor DP6

11097-59-9
$CH_{16}Al_2Mg_6O_{19}$
531.92
Magnesium, (carbonato(2-))-hexadecahydroxybis(aluminum)hexa- (9CI)
Magnesium aluminum hydroxide carbonate

11097-69-1
Unknown
Unknown
Polychlorinated biphenyl (Aroclor 1254)
Arochlor 1254
Aroclor 1254
Chlorierte biphenyle, Chlorgehalt 54% (German)
Chlorodiphenyl (54% Chlorine) (ACGIH,OSHA)
Chlorodiphenyl (54% Cl)
Clorodifenili, Cloro 54% (Italian)
Diphenyle chlore, 54% de chlore (French)
NCI-C02664
PCB

11099-02-8
Unknown
Unknown
Nickel oxide (9CI) (VAN)

11099-03-9
Unknown
Unknown
C.I. Solvent Black 5
C.I. 50415
Lake Black Extra
Nigrosine SB
Nigrosine Spirit Soluble
Nigrosine SSB
Nigrosine SSBZ 14
Nigrosine SSBZ 30
Nigrosine SSJ
Nigrosine SSJJ
Nigrosine WN
Orient Spirit Black AB
Orient Spirit Black SB
Solvent Black 5
Spirit Black SB
Spirit Nigrosine SSB

11099-06-2
Unknown
Unknown
Silicic acid, ethyl ester (9CI)
Ethyl silicate
Polysilicic acid, ethyl ester
Silicic acid, tetraethyl ester, Homopolymer
Silicic acid, tetraethylester polymer
Tetraethyl orthosilicate polymer

11100-14-4
Unknown
Unknown
Polychlorinated biphenyl (Aroclor 1268)
Aroclor 1268
Chlorodiphenyl (68% Cl)

11102-90-2
Cu.Ni
122.24
Nickel alloy, base, Ni,Cu (9CI)
Copper alloy, nonbase, Ni, Cu (9CI)
Copper, nickel base
Nickel base, copper

11103-57-4
Unknown
Unknown
Vitamin A (VAN)(9CI)
Retinol
11 cis Retinol
Vitamin A
Vitamin A1

11103-86-9
$Cr_2HO_9Zn_2.K$
418.85
Chromate(1-), hydroxyoctaoxodizincatedi-, potassium
Buttercup Yellow
Chromic acid, potassium zinc salt (2:2:1)
Citron Yellow
Potassium zinc chromate
Potassium zinc chromate hydroxide
Zinc chrome
Zinc Yellow

11104-28-2
Unknown
Unknown
Polychlorinated biphenyl (Aroclor 1221)
Aroclor 1221
Chlorodiphenyl (21% Cl)

11104-93-1
Unknown
Unknown
Nitrogen oxide (9CI) (VAN)

11106-35-7
Unknown
Unknown
Nonidet A 50 (9CI)

11111-34-5
Unknown
Unknown
Poloxamine 1101
Oxirane, methyl-, Polymer with oxirane, ether with (1,2-ethanediyldinitrilo)tetrakis-(propanol) (4:1) (9CI)
Poloxamine 304
Poloxamine 504
Poloxamine 701
Poloxamine 702
Poloxamine 704
Poloxamine 707
Poloxamine 901
Poloxamine 904
Poloxamine 908
Poloxamine 1102
Poloxamine 1104
Poloxamine 1301
Poloxamine 1302
Poloxamine 1304
Poloxamine 1307
Poloxamine 1501
Poloxamine 1502
Poloxamine 1504
Poloxamine 1508
Tetronic 701

11113-50-1
Unknown
Unknown
Boric acid (VAN)(9CI)

11113-70-5
Unknown
Unknown
Lead chromate silicate
Chromium lead silicate
Silicic acid, chromium lead salt (9CI)

11113-74-9
H_2NiO_2
92.72
Nickel hydroxide

11113-75-0
NiS
90.77
Nickel-sulfide
Nickel monosulfide
Nickelous sulfide
Nickel(II) sulfide
α-Nickel sulfide (1:1) crystalline

11113-81-8
Unknown
Unknown
Praseodymium oxide (9CI)

11114-20-8
Unknown
Unknown
kappa-Carrageenan
kappa-Carrageen
kappa-Carrageenin
Satiagel GS 350

11114-46-8
Unknown
Unknown
Ferrochrome (Exothermic)
Carbon ferrochromium
Chrome Ferroalloy
Chromium Alloy, Cr,C,Fe,N,Si (9CI)
Chromium Alloy, Base, Cr,C,Fe,N,Si (Ferrochromium)
Ferrochrome
Ferrochrome, Exothermic (DOT)
Ferrochromium

11114-92-4
Unknown
Unknown
Cobalt Alloy, Co,Cr
Chromium-cobalt alloy
Cobalt-chromium alloy
DIN 2.4602
DIN 2.4964
Hastelloy C
Haynes 25
Haynes Alloy Number 25
Haynes Stellite 21
HEV-4
HS 21
HS 25
Kh15N55M16
Kh15N55M16V
S 816
Stellite 8
Stellite 21
Stellite 23
Stellite 25
Stellite 27
Stellite 30
Stellite 31
Stellite 36
Stellite 8A
Stellite C
Vitallium
Zimalloy

11115-67-6
Unknown
Unknown
Ammonium-vanadium-oxide
Ammonium vanadate
Vanadic acid, ammonium salt

11118-57-3
Unknown
Unknown
Chromium oxide (VAN)(9CI)
Chromium chromate

11120-22-2
Unknown
Unknown
Lead silicate
Silicic acid, lead salt (9CI)

11120-29-9
Unknown
Unknown
Polychlorinated biphenyl (Aroclor 4465)
Aroclor 4465

11121-16-7
Unknown
Unknown
Boric acid, aluminum salt (VAN)(9CI)
Aluminum borate

11126-42-4
Unknown
Unknown
Polychlorinated triphenyl (Aroclor 5460)
Arochlor 5460
Aroclor 5460

11130-12-4
$B_4O_7.2Na.5H_2O$
291.30
Borates, tetra, sodium salt, pentahydrate (ACGIH, OSHA)
Boric acid, sodium salt, pentahydrate (9CI)
Sodium borate, pentahydrate

11130-73-7
Unknown
Unknown
Tungsten carbide

11132-73-3
Unknown
Unknown
Lignocellulose (9CI)

11133-98-5
Unknown
Unknown
Copper alloy, Cu,Be
Beryllium-copper alloy

11135-81-2
Unknown
Unknown
Sodium potassium alloys
Potassium sodium alloy
Potassium-sodium, alloy
Sodium-potassium alloy
Sodium potassium alloy, Liquid
Sodium potassium alloy, Solid
UN 1422 [Potassium sodium alloys]

11138-66-2
Unknown
Unknown
Xanthan Gum (9CI)
Corn Sugar Gum
WT-5100
WT-6500
Xanthan

11141-16-5
Unknown
Unknown
Polychlorinated biphenyl (Aroclor 1232)
Aroclor 1232
Chlorodiphenyl (32% Cl)

11148-32-6
Fe.Ni
114.54
Iron alloy, Fe,Ni (9CI)

12001-26-2
Unknown
Unknown
Silicate, Mica
Mica (ACGIH,OSHA)
Suzorite mica

12001-28-4

$ONa_2Fe_2O_{33}FeO_8SiO_2H_2O$
765.98
Asbestos, crocidolite
Amorphous crocidolite asbestos
Asbestos (ACGIH)
Blue asbestos [UN 2212]
Crocidolite [UN 2212]
Crocidolite asbestos
Crociodolite
Fibrous crocidolite asbestos
Krokydolith (German)
NCI-C09007
UN 2212 [Blue Asbestos (Crocidolite) or Brown asbestos (amosite, mysorite)]

12001-29-5
Unknown
Unknown
Asbestos, chrysotile
7-45 Asbestos
Asbestos (ACGIH)
Asbestos, White (DOT)
Avibest C
Calidria RG 100
Calidria RG 144
Calidria RG 600
Cassiar AK
Chrysotile [UN 2590]
Chrysotile asbestos
Hooker No. 1 Chrysotile Asbestos
K6-30
Metaxite
NCI-C61223A
Plastibest 20
5R04
RG 600
Serpentine
Serpentine chrysotile
Sylodex
UN 2590 [White asbestos, (chrysotile, actinolite, anthophyllite, tremolite)]
White Asbestos [UN 2590]

12001-31-9
Unknown
Unknown
Bentone 38 (8CI,9CI)

12001-40-0
$C_{10}H_{16}O_2$
168.24
Bicyclo(2.2.1)heptan-2-one, hydroxy-1,7,7-trimethyl- (9CI)
2-Bornanone, hydroxy- (8CI)
Camphorol

12001-85-3
Unknown
Unknown
Naphthenic acid, zinc salt
Zinc naphthenate
Zinc Uversol

12002-03-8
$C_4H_6As_6Cu_4O_{16}$
1013.78
Copper, bis(acetato)hexameta-arsenitotetra
(Acetato-o)(trimetaarsenito)-dicopper
Acetoarsenite de cuivre (French)
Basle Green
C.I. 77410
C.I. Pigment Green 21 (9CI)
Copper acetate arsenite
Copper aceto-arsenite
Copper acetoarsenite [UN 1585]
Copper acetoarsenite, Solid (DOT)
Cupric acetoarsenite
Emerald Green
ENT 884
French Green
Genuine Paris Green
Imperial Green
King's Green
Meadow Green
Mineral Green
Mitis Green
Moss Green
Mountain Green
Neuwied Green
New Green
Ortho P-G Bait
Paris Green
Paris Green, Solid (DOT)
Parrot Green
Patent Green
Powder Green
Schweinfurtergrun
Schweinfurt Green
Schweinfurth Green
Sowbug & Cutworm Bait
Sowbug Cutworm Control
Swedish Green
UN 1585 [Copper acetoarsenite]
Vienna Green
Wuerzberg Green
Zwickau Green

12002-19-6
Unknown
Unknown
Mercurol (DOT)
Mercury nucleate [UN 1639]
Mercury nucleate, Solid (DOT)
UN 1639 [Mercury nucleate]

12002-22-1
$C_{15}H_{26}O_{11}$
382.36
α-D-Glucopyranoside, β-D-fructofuranosyl, mono-2-propenyl ether (9CI)
Allyl sucrose

12002-27-6
Unknown
Unknown
Benzoic acid, chlorinated
Polychlorobenzoic acid

12002-43-6
Unknown
Unknown
Gilsonite
NCI-C55185
Uintahite
Uintaite

12002-48-1
$C_6H_3Cl_3$
181.44
Benzene, trichloro
Trichlorobenzene
Trichlorobenzene, Liquid [UN 2321]
UN 2321 [Trichlorobenzenes, liquid]

12002-51-6
$C_7H_8O.K$
147.24
Phenol, methyl-, potassium salt (9CI)
Cresylic acid, potassium salt
Methylphenol potassium salt
Potassium cresylate

12003-38-2
$AlF_2O_{10}Si_3.K.3Mg$
372.63
Fluorophlogopite
Fluorphlogopite (9CI)
Mica
Synthetic Mica

12004-14-7
$Al_2O_{18}S_3.6Ca$
678.59
Aluminate(12-), hexaoxotris-(sulfato(2-))di-, calcium (1:6) (9CI)
Satin White Pigment

12004-83-0
Al.Zr
118.21
Aluminum, Compd. with zirconium (3:1) (9CI)
Zirconium aluminide
Zirconium aluminum

12005-48-0
$Al_{11}O_{17}.Na$
591.78
Aluminate, sodium (9CI)
β-Alumina

12007-60-2
$B_4Li_2O_7$
169.12
Lithium borate
Boric acid, dilithium salt
Boron lithium oxide (9CI)

Lithium tetraborate

12007-89-5
$B_5HO_8.H_3N$
200.08
Ammonium pentaborate
AI3-36531
Ammonium borate
Ammonium boron oxide (9CI)
Ammonium decaborate
Boric acid, ammonium salt
Boric acid, ammonium salt, octahydrate

12014-56-1
CeH_4O_4
208.14
Cerium hydroxide, (T-4)- (9CI)
Ceric hydroxide
Cerium hydrate
Cerium hydroxide
Cerium (IV) hydroxide
(T-4)Cerium hydroxide

12017-38-8
Co_2O_4Ti
229.74
Cobalt titanium oxide (9CI)

12018-01-8
CrO_2
83.99
Chromium dioxide
Chromium oxide (9CI)
Chromium (IV) oxide

12018-19-8
Cr_2O_4Zn
233.37
Chromium-zinc-oxide
Zinc chromate (ACGIH,OSHA)
Zinc chromite
Zinc chromium oxide
Zn-0312 T 1/4''

12023-91-5
$Fe_{12}O_{19}.Sr$
1061.77
Ferrate, strontium (1:1) (9CI)
Strontium ferrite
Strontium hexaferrite
Strontium iron oxide

12024-21-4
Ga_2O_3
187.44
Gallium-oxide
Digallium trioxide
Gallia
Gallium sesquioxide
Gallium trioxide

12027-06-4
H_4IN
144.94
[NH4+].[I-]
Ammonium iodide (9CI)

12027-67-7
$Mo_7O_{24}.6H_4N$
1163.88
Molybdic acid, hexaammonium salt
Ammonium heptamolybdate
Ammonium molybdate
Ammonium paramolybdate
Molybdate, hexaammonium (9CI)

12030-88-5
KO_2
71.10
Potassium superoxide ($K(O_2)$) (9CI)
Potassium dioxide
Superoxido potasico (Spanish)
Superoxyde de potassium (French)
UN 2466 [Potassium superoxide]

12030-97-6
$K.1/2O_3Ti$
87.04
Potassium titanate
Titanate, dipotassium (9CI)

12031-80-0
Li_2O_2
45.88
Lithium peroxide [UN 1472]
UN 1472 [Lithium peroxide]

12033-59-9
NSe
92.96
Selenium nitride
Nitrogen selenide
Nitrure de selenium (French)
Nitruro de selenio (Spanish)
Selenium mononitride

12034-12-7
NaO_2
54.99
Sodium superoxide
Superoxido sodico (Spanish)
Superoxyde de sodium (French)
UN 2547 [Sodium superoxide]

12034-88-7
Nb_2O_6Pb
489.01
Lead neobate
Lead niobate
Lead niobium oxide (8CI,9CI)

12035-36-8
NiO_2
90.69
Nickel oxide (9CI)

12035-72-2
Ni_3S_2
240.25
Nickel sulfide (3:2)
Heazlewoodite
Nickel subsulfide
Nickel subsulphide
α-Nickel sulfide (3:2) crystalline
Nickel sulphide
Nickel tritadisulphide
Trinickel disulphide

12036-71-4
O_4U
302.03
Uranium peroxide
Uranium oxide
Uranium oxide peroxide

12036-76-9
O_5Pb_2S
526.46
Lead sulfate
Lead oxide sulfate (8CI,9CI)
Lead sulfate (With more than 3% free acid) [UN 1794]
UN 1794 [Lead sulfate with more than 3 per cent free acid]

12037-82-0
P_4S_7
348.36
Phosphorus heptasulfide
AI3-15015
Phosphorus heptasulphide, Free from yellow or white phosphorus [UN 1339]
Phosphorus sulfide
UN 1339 [Phosphorus heptasulfide, Free from yellow or white phosphorus]

12039-52-0
Unknown
Unknown
Thallium-selenide
RCRA waste number P114
Thallium monoselenide

12040-58-3
Unknown
Unknown
Boric acid, calcium salt (9CI)
Calcium borate

12042-47-6
Unknown
Unknown
Sodium aluminate, Solid [UN 2812]

Sodium aluminate, Solution [UN 1819]
UN 1819 [Sodium aluminate, Solution]
UN 2812 [Sodium aluminate, Solid]

12042-91-0
$Al_2ClH_5O_5$
174.46
Aluminum-chloride-hydroxide
Aluminum chlorhydrate
Aluminum chlorhydrol
Aluminum chlorhydroxide
Aluminum chlorohydroxide
Aluminum hydroxide chloride
Aluminum hydroxychloride
Astringen
Basic Aluminum Chlorate
Chlorhydrol
Chlorhydrol, Granular
Chlorhydrol, Impalpable
Chlorohydrol
Chloropentahydroxydialuminum
Locron Extra
Locron Flakes
Locron Powder
Locron Solution
Macrospherical 95
Micro Dry
Wickenol 303
Wickenol 321
Wickenol 323
Wickenol 324

12054-48-7
H_2NiO_2
92.73
Nickel(II) hydroxide
Nickelous hydroxide

12056-53-0
KO_8Ti_4
358.70
Potassium-titanium-oxide
Potassium octatitanate

12057-24-8
Li_2O
29.88
Lithium oxide (9CI)

12057-74-8
Mg_3P_2
134.86
Magnesium phosphide (8CI, 9CI)
AI3-29275-X
Caswell No. 532B
EPA Pesticide Chemical Code 218100
Fosfuri di magnesio (Italian)
Fosfuro magnesico (Spanish)
Magnesiumfosfide (Dutch)
Phosphure de magnesium (French)
UN 2011 [Magnesium phosphide]

12060-00-3
$O_3Ti.Pb$
459.56
Lead titanium trioxide
Lead titanate
Lead titanium oxide ($PbTiO_3$) (8CI,9CI)
Titanic acid, lead salt

12060-01-4
O_3PbZr
346.42
Lead zirconate
Lead zirconium oxide (8CI,9CI)

12060-58-1
O_3Sm_2
348.72
Samarium oxide (9CI)

12060-59-2
$O_3Ti.Sr$
183.50
Titanate, strontium (1:1) (9CI)
Strontium titanate

12064-62-9
Gd_2O_3
362.50
Gadolinium oxide (9CI)
Digadolinium trioxide
Gadolinia

12065-68-8
O_6PbTa_2
665.09
Lead tantalate
Lead tantalum oxide ($PbTa_2O_6$) (8CI,9CI)

12068-03-0
$C_7H_8O_3S.Na$
195.19
Sodium toluenesulfonate
Benzenesulfonic acid, methyl-, sodium salt (9CI)
Methylbenzenesulfonic acid, sodium salt
Sodium methylbenzenesulfonate
Witconate STS

12068-56-3
$Al_6O_{13}Si_2$
426.05
Aluminum oxide silicate (9CI)
Aluminum silicate
Mullite

12069-00-0
PbSe
286.2
Lead selenide (8CI,9CI)

12069-32-8
CB_4
55.26
Boron carbide (9CI)

12069-69-1
$CO_3.H_2O_2.2Cu$
221.11
Copper(II) carbonate hydroxide (2:1:2)
Basic copper carbonate
Basic Copper(II) Carbonate
Basic Cupric Carbonate
(Carbonato)dihydroxydicopper
Cheshunt Compound
Copper carbonate, Basic
Copper carbonate hydroxide
Copper, (carbonato)dihydroxydi-
Copper, (carbonato(2-))dihydroxydi- (9CI)
Cupric carbonate
Cupric carbonate, Basic
Dicopper dihydroxycarbonate
Kop Karb
Kupfercarbonat (German)
Malachite

12070-12-1
CW
195.86
Tungsten-carbide
NCI-C61198
Tungsten monocarbide

12071-83-9
$(C_5H_8N_2S_4Zn)x$
Polymer
Zinc, (N,N'-propylene-1,2-bis-(dithiocarbamate))
Airone
Antracol
Bay 46131
Bayer 46131
LH 3012
LH 30/Z
(((1-Methyl-1,2-ethanediyl)bis-(carbamodithioato))(2-))zinc homopolymer
Methylzineb
Mezineb
Propineb
Propinebe
Propylenebis(dithiocarbamato)-zinc
Taifen
Tsipromat (Russian)
Zinc 1,2-propylene bisdithiocarbamate
Zink-(N,N'-propylen-1,2-bis(dithiocarbamat)) (German)
Zipromat

12075-68-2

$C_6H_{15}Al_2Cl_3$
247.52
Aluminum, trichlorotriethyldi
Ethylaluminum sesquichloride
Ethyl aluminum sesquichloride (DOT)
Sesquiethylaluminum chloride
Trichlorotriethyldialuminium
Trichlorotriethyldialuminum
Triethylaluminum sesquichloride
Triethyldialuminum trichloride
Triethyltrichlorodialuminum
UN 1925 (DOT)

12079-65-1
$C_8H_5MnO_3$
204.07
Manganese, cyclopentadienyltricarbonyl
Cyclopentadienylmanganese tricarbonyl
Cyklopentadientrikarbonylmanganium (Czech)
Cymantrene
Manganese cyclopentadienyl tricarbonyl (ACGIH,OSHA)
Manganese, tricarbonyl-pi-cyclopentadienyl-
MCT

12083-48-6
$C_{10}H_{10}Cl_2V$
252.04
Vanadium, dichlorodi-pi-cyclopentadienyl
Dichlorovanadocene

12108-13-3
$C_9H_7MnO_3$
218.10
Manganese, tricarbonyl methylcyclopentadienyl
AK-33X
Antiknock-33
CI-2
Combustion Improver -2
Manganese, (methylcyclopentadienyl)tricarbonyl-
Methylcyclopentadienyl manganese tricarbonyl (OSHA)
2-Methylcyclopentadienyl manganese tricarbonyl (ACGIH)
2-Methylcyclopentadienyl manganesetricarbonyl
Methylcyklopentadientrikarbonylmanganium (Czech)
MMT

12111-24-9
$C_{14}H_{18}N_3O_{10}.Ca.3Na$
497.40
Glycine, N,N-bis(2-(bis(carboxymethyl)amino)ethyl)-, calcium trisodium salt
Acetic acid, ((carboxymethylimino)bis(ethylenenitrilo))-tetra-, calcium trisodium salt
Ba 2797
N,N-Bis(2-(bis(carboxymethyl)-amino)ethyl)glycine calcium trisodium salt
Calciate(3)-, (N,N-bis(2-(bis-(carboxymethyl)amino)ethyl)-glycinato(5-))-, trisodium (9CI)
Calcium Chel-330
Calcium-DTPA
Calcium Trisodium CHEL 330
Calcium trisodium diethylenetriaminepentaacetate
Calcium Trisodium DTPA
Calcium trisodium pentetate
Calcium trisodium salt of diethylenetriaminepentaacetic acid
Diethylenetriamine pentaacetic acid, calcium trisodium salt
Ditripentat
DTPA Calcium trisodium salt
Pentacin
Pentacine
Pentetate trisodium calcium
Penthamil
Trisodium calcium diethylenetriaminepentaacetate

12113-07-4
$C_{18}H_{16}BO.Na$
282.13
Borate(1-), hydroxytriphenyl-, sodium, (T-4)- (9CI)
(T-4)-Hydroxytriphenylborate-(1-) sodium
Sodium hydroxytriphenylborate

12122-67-7
$C_4H_6N_2S_4.Zn$
275.73
S=C1NCCNC(=S)S[Zn]S1
Zinc, (ethylenebis(dithiocarbamato))
Aspor
Asporum
Bercema
Blightox
Blitex
Blizene
Carbadine
Carbamodithioic acid, 1,2-ethanediylbis-, zinc salt
Caswell No. 930
Chem zineb
Cineb
Crittox
Cynkotox
Daisen
Dipher
Discon
Dithane 65
Dithane Z
Dithane Z-78
Dithiane Z-78
Ditiamina
ENT 14,874
EPA Pesticide Chemical Code 014506
((1,2-Ethanediylbis(carbamodithioato))(2-))zinc
1,2-Ethanediylbis(carbamodithioato) (2-)-S,S'-zinc
1,2-Ethanediylbiscarbamodithioic acid, zinc complex
1,2-Ethanediylbiscarbamothioic acid, zinc salt
Ethylenebis(dithiocarbamato)-zinc
(Ethylenebis(dithiocarbamato))-zinc (8CI)
Ethylenebis(dithiocarbamic acid), zinc salt
Ethyl zimate
Fungo-Pulvit
Funjeb
Hexathane
Kupratsin
Kypzin
Lirotan
Lonacol
Micide
Miltox
Miltox Special
Novosir N
Novozin N 50
Novozir
Novozir N
Novozir N 50
NSC-49513
Pamosol 2 Forte
Parzate
Parzate C
Parzate Zineb
Perosin
Perosin 75B
Perozin
Perozine
Phytox
Polyram Z
Sperlox-Z
Thiodow
Tiezene
Tritoftorol
Tsineb (Russian)
Unizeb
Z-78
Zebenide
Zebtox
Zidan
Zimate
Zinc ethylene bisdithiocarbamate
Zinc ethylenebisdithiocarbamate
Zinc ethylenebis(dithiocarbamate)
Zinc, (ethylenebis(dithiocarbamato))-
Zinc ethylene-1,2-bisdithiocarbamate
Zineb
Zineb 75
Zineb 80
Zineb 75 WP
Zinebe (French)
Zink-(N,N'-aethylen-bis(dithiocarbamat)) (German)
Zinosan

12124-97-9
BrH_4N
97.94
[NH4+].[Br-]
Ammonium bromide (9CI)
Hydrobromic acid mono-ammoniate

12124-99-1
$H_4N.HS$
51.12
Ammonium-sulfide
Ammonium bisulfide
Ammonium hydrogen sulfide
Ammonium hydrosulfide
Ammonium hydrosulfide, Solution (DOT)
Ammonium mercaptan
Ammonium sulfhydrate
Monoammonium sulfide
NA 2683 (DOT)
Sirnik amonny (Czech)
True Ammonium Sulfide

12125-01-8
$H_4N.F$
37.05
Ammonium fluoride [UN 2505]
Ammonium fluorure (French)
Neutral Ammonium Fluoride
UN 2505 [Ammonium fluoride]

12125-02-9
$H_4N.Cl$
53.50
[NH4+].[Cl-]
Ammonium chloride (OSHA)
Amchlor
Ammoneric
Ammoniumchlorid (German)
Ammonium muriate
Chlorid amonny (Czech)
Darammon
Sal Ammonia
Sal Ammoniac
Salammonite
Salmiac

12125-56-3
H_3NiO_3
109.74
Nickel(III) hydroxide
Nickel Black
Nickelic hydroxide

12126-57-7
Unknown
Unknown
Ergot alkaloid
Alkaloids, ergot

12135-76-1
H_8N_2S
68.16
Ammonium sulfide (Solution)
Ammonium monosulfide
Ammonium sulfide, Solution [UN 2683]
Ammonium sulphide, Solution (DOT)
Diammonium sulfide
True Ammonium Sulfide
UN 2683 [Ammonium sulfide solution]

12136-15-1
Hg_3N_2
629.77
Mercury nitride
Nitrure de mercure (French)
Nitruro de mercurio (Spanish)

12136-93-5
Na.Tl
227.37
Sodium, Compd. with thallium (1:1) (8CI,9CI)

12161-82-9
$H_{10}O_9Si_2.H_2O.Be_4$
264.34
Bertrandite
Beryllium silicate hydrate

12164-94-2
H_4N_4
60.03
Ammonium azide
Azida amonica (Spanish)
Azoture d'ammonium (French)

12165-69-4
P_2S_3
158.15
Phosphorus trisulfide
Phosphorus trisulphide, Free from yellow or white phosphorus [UN 1343]
UN 1343 [Phosphorus trisulfide, Free from yellow or white phosphorus]

12167-20-3
$C_7H_7NO_3$
153.15
Cresol, nitro
Mononitrotoluol
Nitrocresol
Nitrocresols [UN 2446]
UN 2446 [Nitrocresols]

12168-20-6
Unknown
Unknown
Cupric ferric subsulfate complex

12168-85-3
Ca_3O_5Si
224.32
Tricalcium silicate
Calcium oxide silicate (9CI)
Calcium silicate
Calcium silicon oxide
C3S (Cement component)
Natural Alite
Silicic acid, calcium salt (1:3)
Silicic acid, tricalcium salt

12172-67-7
Unknown
Unknown
Actinolite (8CI,9CI) (VAN)
Ferroactinolite
Ferroactinolyte
Nephrite

12172-73-5
Unknown
Unknown
Asbestos, amosite
Amosite asbestos
Asbestos (ACGIH)
Mysorite
NCI-C60253A

12174-11-7
Unknown
Unknown
Palygorskite (8CI,9CI) (VAN)
Activated Attapulgite
Attaclay
Attacote
Attagel
Attagel 40
Attagel 50
Attagel 150
Attapulgite (VAN)
Attasorb
Diluex
Min-U-Gel FG
Min-U-Gel 200
Palygorskit (German)
Permagel
Polygorskite
RVM-FG
X 250
Zeogel
200U/P-RVM

12174-11-7
Unknown
Unknown
Palygorskite
Activated Attapulgite
Attaclay
Attaclay X 250
Attacote
Attagel
Attagel 40
Attagel 50
Attagel 150
Attasorb
Attapulgite
Diluex
Min-U-Gel 200
Min-U-Gel 400
Min-U-Gel FG

Palygorskit (German)
Permagel
Pharmasorb-Colloidal
RVM-FG
200U/P-RVM
X 250
Zeogel

12179-04-3
Unknown
Unknown
Borates, tetra, sodium salts, pentahydrate

12192-57-3
$C_6H_{11}O_5S.Au$
392.20
OCC1OC(S[Au])C(O)C(O)C1O
Gold, (1-thio-D-glucopyranosato)
Aureotan
Auromyose
Aurotan
1-Aurothio-D-glucopyranose
Aurothioglucose
Aurumine
Authron
Brenol
(D-Glucopyranosylthio)gold
(1-D-Glucosylthio)gold
Glysanol B
Gold, (D-glucopyranosylthio)-
Gold thioglucose
GTG
Oronol
Romosol
SKF 10056
Solganal
Solganal B
(1-Thio-D-glucopyranosato)gold
1-Thio-D-glucopyranose, gold complex
1-Thio-glucopyranose, monogold(1+) salt
1-Thio-D-glucopyranose, monogold(1+) salt
Thioglucose D'or (French)

12196-43-9
$I_2O_5Sb_4$
820.81
Antimony iodide oxide (8CI, 9CI)

12200-88-3
$Na.1/6O_{28}V_{10}$
182.55
Vanadate, hexasodium (9CI)
Sodium decavanadate

12202-48-1
P_3Rh
195.83
Rhodium phosphide (RhP_3) (6CI,7CI,9CI)
Rhodium triphosphide

12208-13-8
$H_6O_6Sb.K$
262.89
Antimonate, potassium, (OC-6-11)- (9CI)
Antimonic acid, monopotassium salt (8CI)

12217-38-8
$C_{24}H_{21}Cl_2N_5O_6S_2.Na$
633.46
Benzenesulfonic acid, 2,5-dichloro-4-(4-((2-((ethylphenylamino)sulfonyl)phenyl)azo)-4,5-dihydro-3-methyl-5-oxo-1H-pyrazol-1-yl)-, sodium salt (9CI)
C.I. Acid Yellow 61 (8CI)
Midlon Fast Yellow E
Suminol Fast Yellow G

12217-41-3
Unknown
Unknown
C.I. Basic Blue 22 (8CI,9CI)
Astrazon Blue FGL
Deorlene Fast Blue BL
Sevron Blue 2G
Synacril Blue R
Yoracryl Blue G

12217-79-7
$C_{14}H_9ClN_2O_4$
304.70
9,10-Anthracenedione, 1,5-diaminochloro-4,8-dihydroxy
C.I. 63285
C.I. Disperse Blue 56 (8CI)
C.I. Disperse Blue 59
C.I. Disperse Blue 71
Disperse Blue 56
Dispersol Blue B-R
Duranol Blue TR
Kayalon Polyester Blue EBL-E
Latyl Blue BCN
Miketon Polyester Blue FBL
Modr Disperzni 56 (Czech)
Palanil Blue R
Resolin Blue FBL
Samaron Blue FBL
Sumikaron Blue E-BL
Sumikaron Blue E-FBL
Sumikaron Blue R
Terasil Brilliant Blue 3RL
Tersetile Blue RBL

12217-80-0
$C_{20}H_{17}N_3O_5$
379.35
1H-Naphth(2,3-f)isoindole-1,3,5,10(2H)-tetrone, 4,11-diamino-2-(3-methoxypropyl)- (9CI)
1,4-Diaminoanthraquinon-N-γ-methoxypropyl-2,3-dicarboximide
1,4-Diamino-9,10-dihydro-N-(3-methoxypropyl)-9,10-dioxo-2,3-anthracenedicarboximide
1,4-Diamino-N-(3-methoxypropyl)anthraquinone-2,3-dicarboximide

12217-83-3
Unknown
Unknown
C.I. Disperse Orange 21 (8CI,9CI)
Foron Orange E-RFL
Ostacet Orange E-R

12218-95-0
$C_{38}H_{32}CrN_8O_{10}S_2.H$
877.81
Chromate(1-), bis(N-(7-hydroxy-8-((2-hydroxy-5-((methylamino)sulfonyl)-phenyl)azo)-1-naphthalenyl)-acetamidato(2-))-, hydrogen (9CI)

12218-96-1
Unknown
Unknown
C.I. Acid Black 107 (9CI)
Chromate(3-)(1-(3,5-dinitro-2-hydroxyphenyl)azo)-2-naphthalenolato(2-))(3-hydroxy-4-((2-hydroxy-1-naphthyl)azo)-7-nitro-1-naphthalenesulfonato(3-)), sodium dihydrogen

12220-06-3
$C_{26}H_{22}N_4O_8S_2.Na$
605.58
Benzenesulfonic acid, 4-((4-((2-methyl-4-(((4-methylphenyl)sulfonyl)oxy)phenyl)azo)phenyl)amino)-3-nitro-, monosodium salt (9CI)

12221-52-2
Unknown
Unknown
N(N(C(N=Nc(ccc(N(C)C)c1)c1)=N2C)C)=C2
Basic Red 22
C.I. Basic Red 22 (9CI)
5-((4-(Dimethylamino)phenyl)-azo)-1,4-dimethyl-1H-1,2,4-triazolium
1H-1,2,4-Triazolium, 5-((4-(dimethylamino)phenyl)azo)-1,4-dimethyl-

12221-69-1
Unknown
Unknown
Basic Red 46

12222-00-3
$C_{32}H_{14}Cu_2N_4O_{16}S_4.4Na$

1057.77
Cuprate(4-), (mu-((4,4'-((3,3'-dihydroxy(1,1'-biphenyl)-4,4'-diyl)bis(azo))bis(3-hydroxy-2,7-naphthalenedisulfonato))(8-)))di-, tetrasodium (9CI)

12222-60-5
$C_{48}H_{32}N_8O_{18}S_6.6Na$
1339.11
2H-Naphtho(1,2-d)triazole-5-sulfonic acid, 2,2'-(azobis((2-sulfo-4,1-phenylene)-2,1-ethenediyl(3-sulfo-4,1-phenylene)))bis-, hexasodium salt (9CI)

12222-78-5
Unknown
Unknown
C.I. Disperse Blue 73
Artisil Blue BGL
C.I. 60756
C.I. Disperse Blue 113
Dianix Blue BG-FS
Dianix Fast Blue BG-FS
Disperse Blue 73
Foron Blue SBGL
Modr Disperzni 73 (Czech)
Palanil Blue GL
Resolin Blue BSL
Sumikaron Blue S-BG
Terasil Blue R
Tersetile Blue 2BL

12222-79-6
Unknown
Unknown
C.I. Disperse Blue 81 (8CI,9CI)
Palanil Blue RT
Resolin Blue GRL
Sumikaron Blue E-GRL

12223-35-7
Unknown
Unknown
C.I. Disperse Red 50 (8CI,9CI)
Foron Scarlet E 2GFL
Tersetile Red 5GL
Transetile Red P 5GL

12223-97-1
Unknown
Unknown
C.I. Disperse Yellow 86 (8CI,9CI)
Eastman Fast Yellow 2R-GLF
Eastman Polyester Yellow 2R
Serisol Fast Yellow 2RGL

12224-98-5
$C_{28}H_{31}N_2O_3$.xUnknown
Unknown
Xanthylium, 9-(2-(ethoxycarbonyl)phenyl)-3,6-bis(ethylamino)-2,7-dimethyl-, molybdatetungstatephosphate (9CI)

12225-06-8
$C_{32}H_{24}N_6O_5$
572.54
O=C(Nc(ccc(NC(=O)N1)c12)c2)c(c(O)c(N=Nc(c(OC)ccc3C(=O)Nc(cccc4)c4)c3)c(c5ccc6)c6)c5
2-Naphthalenecarboxamide, N-(2,3-dihydro-2-oxo-1H-benzimidazol-5-yl)-3-hydroxy-4-((2-methoxy-5-((phenylamino)carbonyl)phenyl)azo)- (9CI)

12225-08-0
$C_{27}H_{24}N_6O_7S$
576.55
O=C(Nc(ccc(NC(=O)N1)c12)c2)c(c(O)c(N=Nc(c(OC)cc(S(=O)(=O)NC)c3OC)c3)c(c4ccc5)c5)c4
2-Naphthalenecarboxamide, N-(2,3-dihydro-2-oxo-1H-benzimidazol-5-yl)-4-((2,5-dimethoxy-4-((methylamino)sulfonyl)phenyl)azo)-3-hydroxy- (9CI)

12225-18-2
$C_{26}H_{27}ClN_4O_8S$
591.02
O=C(Nc(c(OC)cc(c1OC)Cl)c1)C(N=Nc(c(OC)cc(S(=O)(=O)Nc(cccc2)c2)c3OC)c3)C(=O)C
Butanamide, N-(4-chloro-2,5-dimethoxyphenyl)-2-((2,5-dimethoxy-4-((phenylamino)sulfonyl)phenyl)azo)-3-oxo- (9CI)

12225-21-7
Unknown
Unknown
FD & C Yellow No. 5 Aluminum Lake
C.I. Pigment Yellow 100 (9CI)
C.I. 19140:1

12225-26-2
Unknown
Unknown
C.I. Reactive Black 8
Cern Reaktivni 8 (Czech)
Cibacron Black B-D
Helaktyn Black DN
Ostazin Black H-N
Procion Black H-N
Reactive Black 8

12225-80-8
Unknown
Unknown
C.I. Reactive Green 12 (8CI,9CI)
Cibacron Brilliant Green T 3G-E
Drimarene Brilliant Green X 3G

12225-84-2
Unknown
Unknown
C.I. Reactive Orange 12 (9CI)
Procion Golden Yellow H-R
Procion Golden Yellow HRS
Procion Yellow H 3R
Xiron Golden Yellow 2R-HD

12226-12-9
Unknown
Unknown
C.I. Reactive Red 35 (8CI,9CI)
Remazol Brilliant Red 5B

12226-38-9
Unknown
Unknown
C.I. Reactive Violet 5 (9CI)

12226-50-5
Unknown
Unknown
C.I. Reactive Yellow 23 (8CI,9CI)
Remazol Yellow GNL

12226-63-0
Unknown
Unknown
C.I. Reactive Yellow 42 (8CI,9CI)
Remazol Yellow FG

12227-78-0
$C_{20}H_8I_4O_5$.xAl
Unknown
FD & C Red No. 3 Aluminum Lake
C.I. 45430:1
Spiro(isobenzofuran-1(3H),9'-(9H)xanthen)-3-one, 3',6'-dihydroxy-2',4',5',7'-tetraiodo-, aluminum salt (9CI)

12234-86-5
Unknown
Unknown
C.I. Acid Brown 311 (8CI,9CI)

12235-47-1
Unknown
Unknown
C.I. Basic Blue 69 (9CI)
Astrazon Blue FRR

12236-62-3

$C_{17}H_{13}ClN_6O_5$
416.74
O=C(Nc(c1cc(NC(=O)C(N=Nc(c(N(=O)=O)cc(c2)Cl)c2)C(=O)C)c3)c3)N1
CI Pigment Orange
Butanamide, 2-((4-chloro-2-nitrophenyl)azo)-N-(2,3-dihydro-2-oxo-1H-benzimidazol-5-yl)-3-oxo- (9CI)
C.I. Pigment Orange 36
Permanent Orange HL
PV Orange HL

12236-64-5
$C_{26}H_{20}ClN_5O_4$
501.90
O=C(Nc(ccc(NC(=O)c(c(O)c(N=Nc(c(ccc1C(=O)N)Cl)c1)c(c2ccc3)c3)c2)c4)c4)C
2-Naphthalenecarboxamide, N-(4-(acetylamino)phenyl)-4-((5-(aminocarbonyl)-2-chlorophenyl)azo)-3-hydroxy- (9CI)

12236-86-1
Unknown
Unknown
C.I. Reactive Blue 21 (8CI,9CI)
Remazol Turquoise Blue B
Remazol Turquoise Blue G

12236-92-9
Unknown
Unknown
C.I. Reactive Blue 49 (9CI)
Kayacion Blue P 3R
Procion Brilliant Blue H 3R

12237-00-2
Unknown
Unknown
C.I. Reactive Red 31 (9CI)

12237-16-0
Unknown
Unknown
C.I. Reactive Yellow 37 (8CI,9CI)
Remazol Brilliant Yellow GL

12237-63-7
Unknown
Unknown
C.I. Pigment Red 169 (9CI)

12238-00-5
Unknown
Unknown
C.I. Reactive Red 24 (8CI,9CI)
Cibacron Brilliant Red B-D
Ostazin Brilliant Red H-B

12238-07-2
Unknown
Unknown
C.I. Reactive Red 16 (8CI,9CI)
Cibacron Scarlet 4G-P
Intracron Scarlet 4G-P
Reactive Scarlet 2SKh

12238-31-2
Unknown
Unknown
C.I. Pigment Red 52:2 (9CI)

12239-00-8
$C_{24}H_{21}N_7O_{11}S_3$
679.63
Benzenesulfonic acid, 5-((4-(aminosulfonyl)-2-nitrophenyl)amino)-2-((4-((4-(amino-sulfonyl)-2-nitrophenyl)amino)phenyl)amino)- (9CI)

12239-15-5
$C_{16}H_{13}Cl_2N_5O_3S$
426.25
Benzenesulfonic acid, 4-((5-amino-3-methyl-1-phenyl-1H-pyrazol-4-yl)azo)-2,5-dichloro- (9CI)
5-Amino-1-phenyl-4-((2,5-dichloro-4-sulfophenyl)azo)-3-methyl-pyrazole
2,5-Dichloro-4-((5-amino-3-methyl-1-phenyl-1H-pyrazol-4-yl)azo)benzenesulfonic acid

12239-34-8
$C_{24}H_{27}BrN_6O_{10}$
639.38
Acetamide, N-(5-(bis(2-(acetyloxy)ethyl)amino)-2-((2-bromo-4,6-dinitrophenyl)azo)-4-ethoxyphenyl)- (9CI)
4-(2-Bromo-4,6-dinitrophenylazo)-5-acetylamino-2-ethoxy-N,N-bis(β-acetoxyethyl)-aniline

12239-87-1
$C_{32}H_{15}ClCuN_8$
610.48
Copper, (chloro-29H,31H-phthalocyaninato(2-)-N29,N30,N31,N32)- (9CI)
(Chloro-29H,31H-phthalocyaninato(2-)-N29,N30,N31,N32)copper

12262-20-3
Unknown
Unknown
C.I. Direct Orange 60 (8CI,9CI)

12262-26-9
Unknown
Unknown
C.I. Leuco Sulphur Blue 13 (9CI)

12262-32-7
Unknown
Unknown
C.I. Leuco Sulphur Green 2 (9CI)

12263-85-3
$C_3H_9Al_2Br_3$
338.81
Aluminum, tribromotrimethyldi
Methyl aluminium sesquibromide (DOT)
Methyl aluminum sesquibromide (DOT)
Tribromotrimethyldialuminum
UN 1926 (DOT)

12266-38-5
Pb.Sb
329.00
Lead antimonide
Antimony, Compd. with lead (1:1) (9CI)

12270-08-5
Unknown
Unknown
C.I. Acid Yellow 176 (8CI,9CI)

12270-13-2
$C_{19}H_{23}N_4O_2S.CH_3O_4S$
482.56
Basic Blue 41
Astrazon Blue FGGL
Basacryl Blue X 3GL
Benzothiazolium, 2-((4-((2,3-dimethoxypropyl)methylamino)phenyl)azo)-6-methoxy-3-methyl-, chloride
Benzothiazolium, 2-((4-(ethyl-(2-hydroxyethyl)amino)phenyl)azo)-6-methoxy-3-methyl-, methyl sulfate (Salt) (9CI)
C.I. Basic Blue 41 (8CI)
C.I. 11154
Deorlene Fast Blue RL
2-((4-((2,3,-Dimethoxypropyl)methylamino)phenyl)azo)-6-methoxy-3-methylbenzothiazolium chloride
Ethanol, 2-(ethyl(4-((6-methoxy-3-methyl-2-benzothiazolium)azo)phenyl)amino)-, methyl sulfate
Maxilon Blue GRL
Sandocryl Blue B-RLE
Synacril Blue G

12270-19-8
Unknown
Unknown
C.I. Basic Orange 40 (8CI,9CI)

12270-25-6
Unknown
Unknown
C.I. Basic Red 51 (8CI,9CI)
Basacryl Red X-BL

12280-03-4
Unknown
Unknown
Disodium octaborate tetrahydrate

12286-66-7
$C_{17}H_{16}N_4O_7S.1/2Ca$
440.44
Benzenesulfonic acid, 4-((1-(((2-methylphenyl)-amino)carbonyl)-2-oxopropyl)azo)-3-nitro-, calcium salt (2:1) (9CI)

12323-32-9
$Br_2O_5Sb_4$
726.81
Antimony bromide oxide (8CI,9CI)

12336-95-7
$CrHO_5S$
165.07
Chromium-hydroxide-sulfate
Cromo solfato basificato (Spanish)

12389-75-2
$C_{14}H_{18}FeN_3O_{10}.H.Na$
468.14
Ferrate(2-), (N,N-bis(2-(bis-(carboxymethyl)amino)-ethyl)glycinato(5-))-, sodium hydrogen, (PB-7-13-12564)-(9CI)

12392-64-2
$C_{40}H_{22}CrN_4O_{10}S_2.H.2Na$
858.74
Chromate(3-), bis(3-hydroxy-4-((2-hydroxy-1-naphthalenyl)azo)-1-naphthalenesulfonato(3-))-, disodium hydrogen (9CI)

12397-35-2
CKO
67.11
Potassium carbonyl
Potasiocarbonilo (Spanish)
Potassium carbonyle (French)

12400-75-8
CuO_4S
159.61
Cuprate(1-), (sulfato-(2-)-O)- (9CI)
Sulfuric acid, copper complex (9CI)

12401-86-4
Na_2O
61.98
Sodium monoxide
Calcined Soda
Disodium monoxide
Disodium oxide
Monoxido sodico (Spanish)
Monoxyde de sodium (French)
Sodium monoxide, Solid
Sodium oxide
UN 1825 [Sodium monoxide]

12403-82-6
$C_6H_3N_3O_{10}Pb_2$
691.49
Lead, dihydroxy(2,4,6-trinitro-1,3-benzenediolato(2-))di-(9CI)
Dihydroxy(2,4,6-trinitro-1,3-benzenediolato(2-)dilead

12407-86-2
$C_{11}H_{15}NO_2$
193.27
Carbamic acid, methyl-, 2,3,5-(or 3,4,5)-trimethylphenyl ester
Landrin
SD 8530
Trimethylphenyl methylcarbamate
Trimethacarb
UC 27867

12408-10-5
$C_6H_2Cl_4$
215.89
Benzene, tetrachloro- (9CI)

12408-11-6
$C_9H_6N_2O_2$
174.17
Quinoline, nitro
Nitroquinoline

12408-14-9
$C_{10}H_6Cl_6O$
354.87
4,7-Methano-1H-indenol, 4,5,6,7,8,8-hexachloro-3a,4,7,7a-tetrahydro- (9CI)
4,7-Methanoindenol, 4,5,6,7,8,8-hexachloro-3a,4,7,7a-tetrahydro- (8CI)

12412-52-1
Unknown
Unknown
Senarmontite

12427-27-9
Unknown
Unknown
Perlite
Perlite, Containing < 1% quartz

12427-38-2
$C_4H_6N_2S_4.Mn$
265.30
S=C1NCCNC(=S)S[Mn]S1
Manganese, (ethylenebis(dithiocarbamato))
Aamangan
Akzo Chemie Maneb
BASF-Maneb Spritzpulver
Carbamic acid, ethylenebis-(dithio-, manganese salt
Chem NEB
Chloroble M
CR 3029
Curzate M
Delsene M
Dithane M 22
Dithane M 22 Special
ENT 14,875
1,2-Ethanediylbis(carbamodithioato)(2-)-manganese
1,2-Ethanediylbiscarbamodithioic acid, manganese complex
1,2-Ethanediylbiscarbamodithioic acid, manganese(2+) salt (1:1)
1,2-Ethanediylbismaneb, manganese (2+) salt (1:1)
Ethylenebisdithiocarbamate manganese
N,N'-Ethylene bis(dithiocarbamate manganeux) (French)
Ethylenebis(dithiocarbamato), manganese
Ethylenebis(dithiocarbamic acid), manganese salt
Ethylenebis(dithiocarbamic acid) manganous salt
1,2-Ethylenediylbis(carbamodithioato)manganese
N,N'-Etilen-bis(ditiocarbammato) di manganese (Italian)
F 10 (Pesticide)
Griffin Manex
Kypman 80
Lonocol M
Manam
Maneb
Maneb 80
Maneba
Maneb, Stabilized against self-heating [UN 2968]
Maneb, With not less than 60% Maneb [UN 2210]
Maneb ZL4
Manebe (French)

Manebe 80
Manebgan
Manesan
Manex
Mangaan (II)-(N,N'-ethyleen -bis(dithiocarbamaat)) (Dutch)
Mangan(II)-(N,N-aethylen-bis-(dithiocarbamat)) (German)
Mangan (II)-(N,N'-aethylen-bis-(dithiocarbamate)) (German)
Manganese ethylene-1,2-bisdithiocarbamate
Manganese ethylene bis-dithiocarbamate (DOT)
Manganese (II) ethylene di(dithiocarbamate)
Manganous ethylenebis(dithiocarbamate)
Manzate
Manzate 200
Manzate D
Manzate Maneb Fungicide
Manzeb
Manzin
M-Diphar
MEB
MnEBD
Nereb
Nespor
Plantifog 160M
Polyram M
Remasan Chloroble M
Rhodianebe
Sopranebe
Superman Maneb F
Sup 'R Flo
Tersan-LSR
Trimangol
Trimangol 80
Tubothane
Unicrop Maneb
Vancide Maneb 80
UN 2210 [Maneb or Maneb preparations with not less than 60 per cent Maneb]
UN 2968 [Maneb stabilized or Maneb preparations, stabilized against self-heating]

12433-50-0
$Cr_4O_{17}Zn_4.2K$
819.73
Chromate(2-), heptadecaoxotetrazincatetetra-, dipotassium (9CI)
Potassium zinc chromate oxide

12436-94-1
$H_{12}N_4O_7S$
212.16
Ammonium sulfate nitrate
Ammonium nitrate sulfate (8CI,9CI)
NA 1477

12441-09-7
$C_6H_{12}O_5$
164.16
Sorbitan (9CI)

12510-42-8
Unknown
Unknown
Erionite

12514-32-8
BrFe
135.76
Iron bromide (FeBr) (6CI, 7CI,8CI,9CI)

12540-13-5
C_2Cu
87.57
Copper acetylide
Copper carbide

12542-85-7
$C_3H_9Al_2Cl_3$
205.43
Methylaluminum sesquichloride
Aluminum, trichlorotrimethyldi- (8CI,9CI)
Methyl aluminium sesquichloride
Methyl aluminum sesquichloride
Trichlorotrimethyldialuminum
UN 1927

12593-60-1
$H_{11}N_2O_8PS$
230.12
Ammonium phosphate sulfate (9CI)

12597-68-1
Unknown
Unknown
Stainless Steel
Steel, Stainless

12602-23-2
$C_2Co_2O_6.Co_3H_6O_6$
516.73
Cobalt carbonate, cobalt dihydroxide (2:3)

12604-53-4
Unknown
Unknown
Manganese Alloy, Base, Mn 74-82, Fe 8-19, C 6.9-C7.5, Si O-1.2, P O-O.4 (ASTM A99), Exothermic
Ferromanganese, Exothermic (DOT)

12604-58-9
Unknown
Unknown
Ferrovanadium (ACGIH, OSHA)
Ferrovanadium-Dust

12607-70-4
$CH_4Ni_3O_7$
304.11
Basic Nickel(II) Carbonate
(Carbonato(2-))tetrahydroxytrinickel
Nickel carbonate hydroxide
Nickel, (carbonato(2-))tetrahydroxytri- (9CI)

12607-93-1
$C_{37}H_{51}NO_{10}$
669.89
Taxine
Taxin (German)

12612-55-4
Unknown
Unknown
Nickel carbonyl
UN 1259

12616-35-2
Unknown
Unknown
Halowax 1013 (9CI)

12616-36-3
Unknown
Unknown
Halowax 1014

12623-78-8
Unknown
Unknown
Actinomycin-S
Actinomycin 1048A

12626-25-4
Unknown
Unknown
Amberlite 200 (9CI)

12627-13-3
H_2O_3Si
78.10
Silicate (9CI)

12627-52-0
Unknown
Unknown
Antimony sulfide, Solid
NA 1325

12642-23-8
Unknown
Unknown
Polychlorinated triphenyl

(Aroclor 5442)
Aroclor 5442

12645-31-7
Unknown
Unknown
Phosphoric acid, 2-ethylhexyl ester (9CI)
2-Ethylhexanol phosphate
2-Ethylhexanol, phosphate
2-Ethylhexyl phosphate
Phosphated 2-ethyl hexanol
Phosphoric acid, esters with 2-ethylhexanol
Phosphoric acid, 2-ethylhexyl esters

12645-53-3
$C_8H_{19}O_4P$
210.21
Isooctyl acid phosphate
Isooctyl phosphate
Phosphoric acid, isooctyl ester (9CI)

12653-76-8
Unknown
Unknown
Nickel titanium oxide (9CI)

12656-43-8
Unknown
Unknown
Aluminum carbide [UN 1394]
UN 1394 [Aluminum carbide]

12656-69-8
$C_4H_8N_2S_4$
212.38
SC(=S)NCCNC(S)=S
Dithane
1,2-Ethanedicarbamic acid, tetrathio-

12656-85-8
Unknown
Unknown
Molybdate-Orange
Chrome vermilion
C.I. 77605
C.I. Pigment Red 104
Mineral Fire Red 5DDS
Mineral Fire Red 5GS
Molybdate Red
Molybden Red
Molybdenum Red
NCI-C54626

12663-46-6
$C_{24}H_{30}Cl_2N_5O_7$
571.49
CCC2NC(=O)C1C(Cl)C(Cl)CN1C(=O)C(CO)NC(=O)CC(NC(=O)C(CO)NC2=O)c3ccccc3
Cyclochlorotine
Islanditoxin

12672-29-6
Unknown
Unknown
Polychlorinated biphenyl (Aroclor 1248)
Aroclor 1248
Chlorodiphenyl (48% Cl)

12674-11-2
Unknown
Unknown
Clc1ccc(cc1)c2ccc(Cl)cc2Cl
Polychlorinated biphenyl (Aroclor 1016)
Aroclor 1016
Chlorodiphenyl (41% Cl)

12676-97-0
Unknown
Unknown
Shellsol (9CI)

12679-43-5
$C_{10}H_6O_2$
158.16
Naphthalenedione (9CI)

12680-10-3
Unknown
Unknown
MC 3761 (9CI)

12680-12-5
Unknown
Unknown
MC 5127 (9CI)

12687-85-3
$C_7H_{18}NO_3$.xUnknown
Unknown
Ethanaminium, 2-hydroxy-N,N-bis(2-hydroxyethyl)-N-methyl-, salt with silicic acid (9CI)
Methyl triethanol ammonium silicate
Methyltriethanolammonium silicate
Tris(2-hydroxyethyl)methylammonium silicate

12688-94-7
$C_{18}H_{11}ClMnN_2O_6S$
473.73
Manganese, (4-((5-chloro-4-methyl-2-sulfophenyl)azo)-3-hydroxy-2-naphthalenecarboxylato(2-))- (9CI)
Manganese, (4-((5-chloro-2-sulfo-p-tolyl)azo)-3-hydroxy-2-naphthoato(2-))-
2-Naphthalenecarboxylic acid, 4-((5-chloro-4-methyl-2-sulfophenyl)azo)-3-hydroxy-, manganese salt (1:1)

12710-10-0
Unknown
Unknown
MA 40 (9CI)

12715-61-6
$C_{32}H_{28}CoN_8O_{10}S_2$.2H
809.65
C.I. Acid Yellow 151
Cobaltate(2-), bis(2-((5-(aminosulfonyl)-2-hydroxyphenyl)azo)-3-oxo-N-phenylbutanamidato- (2-))-, dihydrogen (9CI)

12737-18-7
Unknown
Unknown
Calcium silicon (Powder)
Calciumsilicide [UN 1405]
Calcium silicon, Powder
UN 1405 [Calcium silicide]
UN 1406

12737-87-0
Unknown
Unknown
Polychlorinated biphenyl (Kanechlor 400)
Kanechlor 400
KC-400

12738-64-6
Unknown
Unknown
Sucrose benzoate
β-D-Fructofuranosyl-α-D-glucopyranoside benzoate
α-D-Glucopyranoside, β-D-fructofuranosyl, benzoate (9CI)

12743-20-3
Al_2O_3.Al.Ni.Si
224.81
Aluminum alloy, base, Al 60-66,Si 25-30,Ni 5-7,Al_2O_3 3-4 (SAS 1) (9CI)
Aluminum oxide, alloy, Al $_{60-66}$,Si $_{25-30}$,Ni $_{5-7}$,Al_2O_3 3-4 (SAS 1)
Aluminum oxide (Al_2O_3), alloy, Al $_{60-66}$,Si $_{25-30}$, Ni $_{5-7}$,Al_2O_3 3-4 (SAS 1) (9CI)
Silicon alloy, nonbase, Al 60-66,Si 25-30,Ni 5-7,Al_2O_3 3-4 (SAS 1) (9CI)
Nickel alloy, nonbase, Al 60-66,Si 25-30,Ni

5-7,Al_2O_3 3-4 (SAS 1) (9CI)

12750-71-9
Unknown
Unknown
Advastab TM 180 (9CI)

12751-23-4
Unknown
Unknown
Phosphoric acid, dodecyl ester (9CI)
Dodecyl alcohol, phosphate ester
Dodecyl phosphate
Lauryl phosphate

12765-56-9
Unknown
Unknown
Santicizer 140 (9CI)

12767-79-2
Unknown
Unknown
Aroclor

12770-50-2
Unknown
Unknown
Beryllium-Aluminum-Alloy
Aluminium alloy, Al,Be
Aluminum alloy, Al,Be (9CI)
Aluminum beryllium alloy
Beryllium-aluminium alloy

12771-08-3
Unknown
Unknown
Sulfur monochloride

12771-68-5
$C_{15}H_{16}N_2O_2$
256.33
COc1ccc(cc1)C(O)(C2CC2)c3cncnc3
5-Pyrimidinemethanol, α-cyclopropyl-α-(p-methoxyphenyl)
Ancymidol
Ancymidole
Arest
α-Cyclopropyl-4-methoxy-α-(pyrimidin-5-yl)-benzyl-alkohol (German)
α-Cyclopropyl-α-(p-methoxyphenyl)-5-pyrimidinemethanol
α-Cyclopropyl-α-(4-methoxyphenyl)-5-pyrimidinemethanol
α-Cyclopropyl-4-methoxy-α-(pyrimidin-5-yl)benzyl alcohol
EL-531
Quel
Reducymol

12771-91-4
Unknown
Unknown
FC 310 (9CI)

12778-12-0
Unknown
Unknown
Phosphoric acid, 4-nitrophenyl ester (9CI)

12788-93-1
Unknown
Unknown
Acid butyl phosphate
n-Butyl acid phosphate
Butyl phosphoric acid
Phosphoric acid, butyl ester
UN 1718 [Butyl acid phosphate]

12789-03-6
Unknown
Unknown
ClC1CC2C(C1Cl)C3(Cl)C(=C(Cl)C2(Cl)C3(Cl)Cl)Cl
Chlordane
Chlordane, Technical
4,7-Methanoinden, 1,2,4,5,6,7,8,9-octachloro-3a,4,7,7a-tetrahydro-
RCRA waste number U036

12789-46-7
$C_5H_{13}O_4P$
168.13
Amyl acid phosphate
Phosphoric acid, pentyl ester
UN 2819 [Amyl acid phosphate]

12794-10-4
$C_9H_8N_2$
144.19
Benzodiazepine

13005-36-2
$C_8H_8O_2.K$
175.25
Benzeneacetic acid, potassium salt (9CI)
Potassium benzeneacetate

13007-85-7
$C_7H_{14}O_8.Na$
249.17
Gluceptate sodium
D-Glycero-D-gulo-heptonic acid, monosodium salt (9CI)
Monosodium D-glycero-D-gulo-heptonate

13007-92-6
C_6CrO_6
220.06
Chromium-carbonyl
Chromium carbonyl (OC-6-11) (9CI)
Chromium hexacarbonyl
Hexacarbonyl chromium

13010-47-4
$C_9H_{16}ClN_3O_2$
233.73
ClCCN(N=O)C(=O)NC1CCCCC1
Urea, 1-(2-chloroethyl)-3-cyclohexyl-1-nitroso
Belustine
CCNU
Cecenu
Ceenu
Chloroethylcyclohexylnitrosourea
N-(2-Chloroethyl)-N'-cyclohexyl-N-nitrosourea
1-(2-Chloroethyl)-3-cyclohexyl-1-nitrosourea
(Chloro-2-ethyl)-1-cyclohexyl-3-nitrosourea
Cinu
(Cloro-2-etil)-1-cicloesil-3-nitrosourea (Italian)
ICIG 1109
Lomustine
NCI-C04740
1-Nitrosourea, 1-(2-chloroethyl)-3-cyclohexyl-
NSC-79037
RB 1509
SRI 2200
Urea, N-(2-chloroethyl)-N'-cyclohexyl-N-nitroso- (9CI)

13014-18-1
$C_7H_3Cl_5$
264.36
c(ccc(c1Cl)C(Cl)(Cl)Cl)(c1)Cl
Benzene, 2,4-dichloro-1-(trichloromethyl)- (9CI)
AI3-02376
2,4-Dichlorobenzotrichloride
2,4-Dichlorophenyltrichloromethane
1,3-Dichloro-4-(trichloromethyl)benzene
2,4-Dichloro-1-(trichloromethyl)benzene
NSC-403840
2,4,α,α,α-Pentachlorotoluene
Toluene, α,α,α,2,4-pentachloro- (8CI)

13014-24-9
$C_7H_3Cl_5$
264.36
c(ccc(c1Cl)Cl)(c1)C(Cl)(Cl)Cl
Benzene, 1,2-dichloro-4-(trichloromethyl)- (9CI)
AI3-02582

3,4-Dichlorobenzotrichloride
3,4-Dichlorophenyltrichloromethane
1,2-Dichloro-4-trichloromethylbenzene
1,2-Dichloro-4-(trichloromethyl)benzene
NSC-163901
α,α,α,3,4-Pentachlorotoluene
Toluene, α,α,α,3,4-pentachloro- (8CI)

13018-37-6
$C_{27}H_{57}O_4P$
476.72
Phosphoric acid, trinonyl ester (8CI,9CI)

13019-16-4
$C_{12}H_{22}O$
182.31
2-Octenal, 2-butyl- (8CI,9CI)

13019-20-0
$C_8H_{16}O$
128.21
3-Heptanone, 2-methyl- (9CI)
2-Methyl-3-heptanone
NSC-21978

13023-00-2
$C_4H_6Cl_2O_2$
157.00
Butanoic acid, 2,2-dichloro- (9CI)
Butyric acid, 2,2-dichloro- (8CI)

13029-08-8
$C_{12}H_8Cl_2$
223.10
Clc1ccccc1c2ccccc2Cl
Biphenyl, 2,2'-dichloro
2,2'-Dichlorbiphenyl (German)
2,2'-Dichlorobiphenyl
2,2'-Dichloro-1,1'-biphenyl

13037-71-3
$C_{10}H_{12}O$
148.22
Phenol, p-2-butenyl

13037-86-0
$C_{13}H_{20}O_2$
208.30
O(c(ccc(O)c1)c1)CCCCCCC
Phenol, 4-(heptyloxy)- (9CI)
4-(Heptyloxy)phenol

13038-45-4
$C_{19}H_{36}O_2$
296.50
10-Octadecenoic acid, methyl ester, (E)- (8CI, 9CI)
Methyl trans-10-octadecenoate

13038-47-6
$C_{19}H_{34}O_2$
294.48
9,11-Octadecadienoic acid, methyl ester, (E,E)- (8CI,9CI)
Methyl trans-9,trans-11-octadecadienoate
Methyl trans,trans-9,11-octadecadienoate

13040-18-1
$C_{10}H_{20}O_2.K$
211.37
Decanoic acid, potassium salt (9CI)
Capric acid, potassium salt
Potassium decanoate

13042-02-9
$C_6H_6N_2$
106.14
N#CC=CCCC#N
2-Hexenedinitrile
Dihydromucodinitrile

13042-18-7
$C_{23}H_{25}N$
315.45
[Cl-].CC(NCCC(c1ccccc1)c2ccccc2)c3ccccc3
Fendiline
Fendilina (Spanish)
Fendilinum (Latin)

13045-94-8
$C_{13}H_{18}Cl_2N_2O_2$
305.23
Alanine, 3-(p-(bis(2-chloroethyl)amino)phenyl)-, D
4-(Bis(2-chloroethyl)amino)-D-phenylalanine
(+)-3-(p-(Bis(2-chloroethyl)-amino)phenyl)alanine
D-3-(p-(Bis(2-chloroethyl)-amino)phenyl)alanine
3026 C.B.
CB-3026
p-Di-(2-chloroethyl)-amino-D-phenylalanine
p-Di(2-chloroethyl)amino-d-phenylalanine
Medfalan
Medphalan
NSC-35051
D-Phenylalanine, 4-(bis-(2-chloroethyl)amino)- (9CI)
D-Phenylalanine mustard
D-Sarcolysine

13047-13-7
$C_{11}H_{14}N_2O_2$
206.23
O=C(NN(c(cccc1)c1)C2)C2(CO)C
3-Pyrazolidinone, 4-(hydroxymethyl)-4-methyl-1-phenyl- (9CI)
4-Hydroxymethyl-4-methyl-1-phenyl-3-pyrazolidone

13048-33-4
$C_{12}H_{18}O_4$
226.30
O=C(OCCCCCCOC(=O)C=C)C=C
Acrylic acid, hexamethylene ester
1,6-Hexanediol diacrylate
2-Propenoic acid, 1,6-hexanediyl ester

13052-09-0
$C_{24}H_{46}O_6$
430.70
O=C(OOC(CCC(OOC(=O)C(CCCC)CC)(C)C)(C)C)C(CCCC)CC
Peroxyhexanoic acid, 2-ethyl-, 1,1,4,4-tetramethyltetramethylene ester
2,5-Dimethyl-2,5-di-(2-ethylhexanoylperoxy)hexane, Technically pure (DOT)
Hexaneperoxoic acid, 2-ethyl-, 1,1,4,4-tetramethyl-1,4-butanediyl ester
UN 2157 (DOT)

13052-19-2
$C_2H_5NO_2$
75.06
O=CNCO
N-Hydroxymethylformamide
Formamide, N-(hydroxymethyl)- (9CI)
(Hydroxymethyl)formamide
N-(Hydroxymethyl)formamide
N-Methylolformamide
NSC-348403

13057-78-8
Unknown
Unknown
Chloroisocyanuric acid
Monochloro-s-triazinetrione

13065-07-1
$C_{12}H_{16}$
160.26
Naphthalene, 1,2,3,4-tetrahydro-2,7-dimethyl- (8CI, 9CI)
2,7-Dimethyltetralin

13067-93-1
$C_{15}H_{14}NO_2PS$
303.33

CCOP(=S)(Oc1ccc(C#N)cc1)
c2ccccc2
Phosphonothioic acid, phenyl-, O-ethyl ester, O-ester with p-hydroxybenzonitrile
B-10094
Benzonitrile, p-hydroxy-, O-ester with O-ethyl phenylphosphonothioate
CP 19699
Cyanofenphos
Cyanophenphos
O-p-Cyanophenyl O-ethyl phenyl-phosphonothioate
O-(4-Cyanophenyl) O-ethyl phenylphosphonothioate
CYP
ENT 25,832
ENT 25,832-A
O-Ethyl O-4-cyanophenyl phenylphosphorothioate
O-Ethyl phenylphosphonothioate, O-ester with p-hydroxybenzonitrile
Experimental Insecticide S-4087
Monsanto CP-19699
Phenylphosphonothioic acid O-ethyl ester O-ester with p-hydroxybenzonitrile
Phosphorothioic acid, O-(4-cyanophenyl) O-ethyl phenyl ester
Phosphonothioic acid, phenyl-, O-(4-cyanophenyl) O-ethyl ester
S 4087
Stauffer B-10094
Surecide
UpJohn U-32714

13071-79-9
$C_9H_{21}O_2PS_3$
288.45
Phosphorodithioic acid, O,O-diethyl S-(((1,1-dimethylethyl)thio)methyl) ester
AC 92100
S-((tert-Butylthio)methyl)-O,O-diethylphosphorodithioate
Counter
Counter 15G Soil Insecticide
Counter 15G Soil Insecticide -Nematicide
S-(((1,1-Dimethylethyl)thio)-methyl)-O,O-diethyl phosphorodithioate
ENT 27,920
Phosphorodithioic acid S-((tert-butylthio)methyl) O,O-diethyl ester
Phosphorodithioic acid S-(((1,1-dimethylethyl)thio)-methyl) O,O-diethyl ester
Terbufos

13073-35-3
$C_6H_{13}NO_2S$
163.26
O=C(O)C(N)CCSCC
Butyric acid, 2-amino-4-(ethylthio)-, L
L-2-Amino-4-(ethylthio)butyric acid
Ethionine
L-Ethionine
L-Homocysteine, S-ethyl-

13075-01-9
$C_6Br_3Cl_3$
418.14
1,2,4-Tribromo-3,5,6-trichlorobenzene
Benzene, 1,2,4-tribromo-3,5,6-trichloro-

13081-97-5
$C_{41}H_{80}O_6$
669.08
O=C(OCC(CO)(CO)COC(=O)CC
CCCCCCCCCCCCCCC)CC
CCCCCCCCCCCCCCC
Octadecanoic acid, 2,2-bis-(hydroxymethyl)-1,3-propanediyl ester (9CI)

13092-75-6
C_2HAg
132.90
Silver acetylide
Acetiluro de plata (Spanish)
Acetylure d'argent (French)
Silver acetylide, Dry

13093-12-4
$C_{17}H_{26}O_4Si_4$
406.74
Spiro(cyclotetrasiloxane-2,1'(2'H)-(1)sila-acenaphthylene), 4,4,6,6,8,8-hexamethyl- (9CI)
Hexamethyl(silaacenaphthenyl)cyclotetrasiloxane
Spiro(cyclotetrasiloxane-2,1'-(1)silaacenaphthene), 4,4,6,6,8,8-hexamethyl-(8CI)

13098-39-0
$C_3F_6O.3/2H_2O$
193.06
O.O.O.FC(F)(F)C(=O)C(F)(F)F.
FC(F)(F)C(=O)C(F)(F)F
2-Propanone, hexafluoro-, sesquihydrate
Acetone, hexafluoro-, sesquihydrate
Hexafluoroacetone sesquihydrate
Hexafluoro-2-propanone sesquihydrate

13098-41-4
$C_{10}H_{20}O_2.1/2Ba$
240.94
Decanoic acid, barium salt (9CI)
Barium decanoate

13101-26-3
$C_3H_6N_4O$
114.08
N#CNC(=N)NCO
Guanidine, N-cyano-N'-(hydroxymethyl)- (9CI)
N-Cyano-N'-(hydroxymethyl)-guanidine

13103-52-1
$C_{36}H_{70}O_4S$
599.02
O=C(OCCCCCCCCCCCCCCCCC
CC)CCSCCC(=O)OCCCCCC
CCCCCCC
Propanoic acid, 3-((3-(dodecyloxy)-3-oxopropyl)-thio)-, octadecyl ester (9CI)

13104-21-7
$C_{10}H_{10}BrCl_2O_4P$
375.98
Phosphoric acid, 2-bromo-1-(2,4-dichlorophenyl)vinyl dimethyl ester
Bromfenvinphos-methyl
O-1-(2,4-Dichlorophenyl)-2-bromovinyl-O,O-dimethyl phosphate
O,O-Dwumetylo-O-1-(2,4-dwuchlorofenylo)-2-bromowinylofosforan (Polish)
Enolofos
ENT 27,043
IPO 63
Methylbromfenvinphos
Phosphoric acid, 2-bromo-1-(2,4-dichlorophenyl)ethenyl dimethyl ester (9CI)
Polfos
Polphos
SD 8988
SD 8988 (Shell)
Shell SD-8988

13106-47-3
$CO_3.Be$
69.02
Beryllium carbonate (1:1)
Carbonic acid beryllium salt (1:1)

13106-76-8
$MoO_4.2H_4N$
196.04
Molybdic acid, diammonium salt
Ammonium molybdate
Ammonium paramolybdate
Diammonium molybdate

13107-54-5
$C_{15}H_{13}N_5$
263.30
1,3,5-Triazine-2,4-diamine, N,N'-diphenyl- (9CI)
2,4-Dianilino-s-triazine
s-Triazine, 2,4-dianilino- (7CI,8CI)

13108-52-6
Unknown
Unknown
2,3,5,6-Tetrachloro-4-(methylsulfonyl)pyridine

13110-37-7
$C_{12}H_{17}NO_2$
207.27
CCCCCOC(=O)c1ccc(N)cc1
Benzoic acid, 4-amino-, pentyl ester (9CI)
Benzoic acid, p-amino-, pentyl ester (8CI)
NSC-69110
Sun Block Gel
Sun Screen Gel
Weatherproofer

13114-72-2
$C_{14}H_{14}N_2O$
226.27
O=C(N(c(cccc1)c1)c(cccc2)c2)NC
Urea, N'-methyl-N,N-diphenyl- (9CI)
N,N-Diphenyl-N'-methylurea
N'-Methyl-N,N-diphenylurea

13114-87-9
$C_8H_7F_3N_2O$
204.14
O=C(Nc(cccc1C(F)(F)F)c1)N
Urea, (3-(trifluoromethyl)phenyl)- (9CI)
NSC-136288
N-(3-Trifluoromethylphenyl)-urea
3-Trifluoromethylphenylurea
(3-(Trifluoromethyl)phenyl)urea
3-(α,α,α-Trifluoro-m-tolyl)urea
Urea, (α,α,α-trifluoro-m-tolyl)-

13116-53-5
$C_3H_4Cl_4$
181.88
C(CCl)(CCl)(Cl)Cl
1,2,2,3-Tetrachloropropane
Propane, 1,2,2,3-tetrachloro- (9CI)

13116-60-4
$C_3H_4Cl_4$
181.88
Propane, 1,1,2,2-tetrachloro- (8CI,9CI)

13121-70-5
$C_{18}H_{34}OSn$
385.21
O[Sn](C1CCCCC1)(C2CCCCC2)C3CCCCC3
Stannane, tricyclohexylhydroxy
Cyhexatin (ACGIH,OSHA)
DOWCO-213
ENT 27,395
ENT 27,395-X
M 3180
Plictran
Plyctran
TCTH
Tin, tricyclohexylhydroxy-
Tricyclohexylhydroxystannane
Tricyclohexylhydroxytin
Tricyclohexylstannanol
Tricyclohexylstannium hydroxide
Tricyclohexyltin hydroxide
Tricyclohexylzinnhydroxid (German)

13132-25-7
$C_9H_{14}OSi$
166.30
CSi(C)(C)c1ccc(O)cc1
Phenol, 4-(trimethylsilyl)- (9CI)
NSC-83941
p-Trimethylsilylphenol
4-(Trimethylsilyl)phenol

13133-29-4
$C_{10}H_{22}O_4$
206.28
1-Propanol, 3-(3-(3-methoxypropoxy)propoxy)- (9CI)

13137-43-4
$C_6H_{10}Br_2O_2$
273.95
Hexanoic acid, 2,6-dibromo- (7CI,8CI,9CI)
2,6-Dibromocaproic acid
2,6-Dibromohexanoic acid

13138-45-9
$N_2O_6.Ni$
182.73
Nickel(II) nitrate (1:2)
Nickel nitrate [UN 2725]
Nitric acid, nickel(II) salt
UN 2725 [Nickel nitrate]

13138-51-7
$C_4H_6Cl_4$
195.90
Butane, 1,2,3,3-tetrachloro- (7CI,8CI,9CI)
1,2,3,3-Tetrachlorobutane

13140-86-8
$C_{10}H_{12}N_2O$
176.21
Urea, N-cyclopropyl-N'-phenyl- (9CI)
Urea, 1-cyclopropyl-3-phenyl- (8CI)

13140-89-1
$C_{12}H_{16}N_2O$
204.26
O=C(NC1CCCC1)Nc2ccccc2
Urea, N-cyclopentyl-N'-phenyl- (9CI)
NSC-131956
Urea, 1-cyclopentyl-3-phenyl- (8CI)

13142-64-8
$C_8H_9ClN_2O$
184.61
Urea, (3-chloro-4-methylphenyl)- (9CI)
CGA 16340
NSC-211456
Urea, (3-chloro-p-tolyl)- (8CI)

13146-28-6
$C_{13}H_{17}ClN_4O_6$
360.76
2-Oxazolidinone, 5-(4-morpholinylmethyl)-3-(((5-nitro-2-furanyl)methylene)amino), hydrochloride, (S)- (9CI)
2-Oxazolidinone, 5-(morpholinomethyl)-3-((5-nitrofurfurylidene)amino), hydrochloride, L- (8CI)

13147-09-6
Unknown
Unknown
4-Propionyloxy-4-phenyl-N-methylpiperidine
DEA No. 9661
1-Methyl-4-phenyl-4-propionoxypiperidine
MPPP
4-Piperidinol, 1-methyl-4-phenyl-, propanoate (Ester)
PPMP

13147-25-6
$C_4H_{10}N_2O_2$
118.16
CCN(CCO)N=O
Ethanol, 2-(ethylnitrosamino)
Aethyl-aethanol-nitrosoamin (German)
EENA
EHEN
Ethyl-2-hydroxyethylnitrosamine
N-Ethyl-N-hydroxyethylnitrosamine
2-(Ethylnitrosamino)ethanol
N-Nitrosoaethylaethanolamin (German)
N-Nitrosoethylethanolamine

N-Nitrosoethyl-2-hydroxyethyl-amine
N-Nitroso-N-ethyl-N-(2-hydroxyethyl)amine

13149-87-6
$C_{34}H_{70}O_9$
622.92
3,6,9,12,15,18,21,24-Octaoxa-dotetracontan-1-ol (8CI,9CI)
Ethanol, 2-(2-(2-(2-(2-(2-(2-(2-(octadecyloxy)ethoxy) ethoxy)ethoxy)ethoxy)ethoxy) ethoxy)ethoxy)-

13150-00-0
$C_{18}H_{37}O_7S.Na$
420.60
[Na+].CCCCCCCCCCCCOCCO CCOCC[O-].O=S(=O)=O
Ethanol, 2-(2-(2-(dodecyloxy)-ethoxy)ethoxy)-, hydrogen sulfate sodium salt
2-(2-(2-(Dodecyloxy)ethoxy)-ethoxy)ethanol hydrogen sulfate sodium salt
Ethanesulfonic acid, 2-(2-(2-(dodecyloxy)ethoxy)-ethoxy)-, sodium salt
Ethanol, 2-(2-(2-(dodecyloxy)-ethoxy)ethoxy)-, sulfate ester, monosodium salt
Sodium lauryl trioxyethylene sulfate
Sulfuric acid, mono(2-(2-(2-(do-decyloxy)ethoxy)ethoxy)ethyl) ester, sodium salt

13152-02-8
$C_{10}H_{20}$
140.27
Cyclooctane, ethyl- (6CI, 7CI,8CI,9CI)
Ethylcyclooctane

13164-93-7
$C_{29}H_{22}N_6O_7S.2Na$
644.54
Benzoic acid, 5-((4'-((4,5-di-hydro-3-methyl-5-oxo-1-(4-sulfophenyl)-1H-pyrazol-4-yl)azo)(1,1'-biphenyl)-4-yl)azo)-2-hydroxy-, disodium salt (9CI)
Salicylic acid, 5-((4'-((3-methyl-5-oxo-1-(p-sulfophenyl)-2-pyra-zolin-4-yl)azo)-4-biphenyl-yl)azo)-, disodium salt (8CI)

13171-00-1
$C_{17}H_{24}O$
244.38
O=C(c(c(c(cc1C(C)(C)C(C2)(C)C)C2)c1)C
Celestolide
AI3-28573
Ethanone, 1-(6-(1,1-dimethyl-ethyl)-2,3-dihydro-1,1-di-methyl-1H-inden-4-yl)-

13171-21-6
$C_{10}H_{19}ClNO_5P$
299.72
CCN(CC)C(=O)C(Cl)=C(C)OP(=O)(OC)OC
Phosphoric acid, dimethyl ester, ester with 2-chloro-N,N-diethyl-3-hydroxy-crotonamide
Apamidon
C 570
(2-Chloor-3-diethylamino-1-methyl-3-oxo-prop-1-en-yl)-dimethyl-fosfaat (Dutch)
(2-Chlor-3-diaethylamino-1-methyl-3-oxo-prop-1-en-yl)-dimethyl-phosphat (German)
2-Chloro-2-diethylcarbamoyl-1-methylvinyl dimethylphos-phate
1-Chloro-diethylcarbamoyl-1-propen-2-yl dimethyl phos-phate
(2-Cloro-3-dietilamino-1-metil-3-oxo-prop-1-en-il)-dimetil-fosfato (Italian)
Ciba 570
Crotonamide, 2-chloro-N,N-di-ethyl-3-hydroxy-, dimethyl phosphate
Dimecron
Dimecron 100
Dimethyl 2-chloro-2-diethyl-carbamoyl-1-methylvinyl phosphate
O,O-Dimethyl O-(2-chloro-2-(N,N-diethylcarbamoyl)-1-methylvinyl) phosphate
Dimethyl diethylamido-1-chlorocrotonyl (2) phosphate
O,O-Dimethyl-O-(1-methyl-2-chlor-2-N,N-diaethyl-car-bamoyl)-vinyl-phosphat (German)
(O,O-Dimethyl-O-(1-methyl-2-chloro-2-diethylcarbamoyl-vinyl) phosphate)
Dimethyl phosphate of 2-chloro-N,N-diethyl-3-hydroxycrotonamide
Dixon
ENT 25,515
Famfos
Fosfamidon (Dutch)
Fosfamidone (Italian)
ML 97
NCI-C00588
OMS 1325
OR 1191
Foszfamidon (Hungarian)
Phosphamidon
Phosphate de dimethyle et de (2-chloro-2-diethylcarbamoyl-1-methyl-vinyle) (French)

13171-61-4
$C_{10}H_{14}N_2O_2$
194.24
Cc1c(C)c(N(=O)=O)c(C)c(C)c1N
Benzenamine, 2,3,5,6-tetra-methyl-4-nitro- (9CI)
Aminonitrodurene
Aniline, 2,3,5,6-tetramethyl-4-nitro- (6CI,7CI,8CI)
Tetramethyl-p-nitroanil-ine

13177-28-1
$C_{16}H_9NO_2$
247.26
Fluoranthene, 1-nitro
1-Nitrofluoranthene

13177-29-2
$C_{16}H_9NO_2$
247.26
Fluoranthene, 2-nitro
2-Nitrofluoranthene

13181-17-4
$C_{13}H_7Br_2N_3O_6$
461.05
Oc2c(Br)cc(C=NOc1ccc(cc1N(=O)=O)N(=O)=O)cc2Br
Bromofenoxim
Benzaldehyde, 3,5-dibromo-4-hydroxy-, (2,4-dinitro-phenyl)oxime
Bromophenoxim
C9122
3,5-Dibrom-4-hydroxyl-benzaldoxim-O-(2',4'-di-nitrophenyl)-aether (German)
3,5-Dibromo-4-hydroxybenz-aldehyde 2,4-dinitrophenyl oxime
3,5-Dibromo-4-hydroxybenz-aldehyde (2',4'-dinitro-phenyl)oxime
3,5-Dibromo-4-hydroxybenz-aldehyde-O-(2',4'-dinitro-phenyl)oxime
Faneron

13183-09-0
$C_8H_5BrN_2O_2$
241.03
Sydnone, 4-bromo-3-phenyl-(8CI,9CI)
NSC-41404

13183-79-4
$C_2H_4N_4S$
116.16
N(N=NN1C)=C1S
1H-Tetrazole-5-thiol, 1-methyl
1-Methyl-5-mercaptotetrazole
1-Methyl-5-mercapto-1,2,3,4-te-trazole
N-Methyltetrazolethiol
1-Methyl-1H-tetrazole-5-thiol

13188-60-8

$C_{12}H_{22}O_4.C_6H_{16}N_2$
346.50
Dodecanedioic acid, Compd. with 1,6-hexanediamine (1:1) (9CI)
Nylon 6-12 Salt

13190-97-1
$C_{45}H_{74}O$
631.08
OCC=C(CCC=C(CCC=C(CCC=C(CCC=C(CCC=C(CCC=C(CCC=C(CCC=C(C)C)C)C)C)C)C)C)C)C
Solanesol
2,6,10,14,18,22,26,30,34-Hexatriacontanonaen-1-ol, 3,7,11,15,19,23,27,31,35-nonamethyl-, (Z,Z,Z,Z,Z,Z,E,E)-(9CI)

13192-04-6
$C_7H_{10}O_5$
174.15
Pentanedioic acid, 2-oxo-, dimethyl ester (9CI)
Dimethyl α-ketoglutarate
Dimethyl 2-oxoglutarate
Dimethyl α-oxoglutarate
Glutaric acid, 2-oxo-, dimethyl ester (6CI,7CI,8CI)
2-Oxoglutaric acid dimethyl ester

13194-48-4
$C_8H_{19}O_2PS_2$
242.36
CCCSP(=O)(OCC)SCCC
Phosphorodithioic acid, O-ethyl S,S-dipropyl ester
ENT 27,318
Ethoprop
Ethoprophos
O-Ethyl S,S-dipropylphosphorodithioate
Jolt
Mobil V-C 9-104
Mocap
Prophos
Rovokil
V-C 9-104
V-C Chemical V-C 9-104
Virginia-Carolina VC 9-104

13195-76-1
$C_{12}H_{27}BO_3$
230.16
Triisobutyl borate
AI3-09515
Boric acid, triisobutyl ester (8CI)
Boric acid, tris(2-methylpropyl) ester (9CI)
NSC-65329

13225-10-0
$C_7H_{10}N_4O_{14}$
374.15
α-Methylglucoside tetranitrate
α-D-Glucopyranoside, methyl-, tetranitrate

13230-04-1
$C_5H_7N_3O_2$
141.15
Imidazole, 1,2-dimethyl-4-nitro
1,2-Dimethyl-4-nitro-1H-imidazole (9CI)

13231-81-7
$C_7H_{16}O$
116.20
1-Hexanol, 3-methyl- (8CI,9CI)

13231-90-8
$C_4H_{10}Cl_2Pb$
336.23
Plumbane dichlorodiethyl
Dichlorodiethylplumbane
Diethyllead dichloride
Diethylplumbium dichloride
Lead, dichlorodiethyl-

13236-02-7
$C_{12}H_{20}O_6$
260.29
O(C1COC(COCC(O2)C2)COCC(O3)C3)C1
Triglycidylglycerol
Glycerine triglycidyl ether
Glycerol triglycidyl ether
Glycerol tris(2,3-epoxypropyl) ether
Oxirane, 2,2',2''-(1,2,3-propanetriyltris(oxymethylene))-tris- (9CI)
Propane, 1,2,3-tris(2,3-epoxypropoxy)-
2,2',2''-(1,2,3-Propanetriyltris-(oxymethylene))trisoxirane

13236-84-5
$C_4H_6N_6O$
154.10
Formamide, N-(4,6-diamino-1,3,5-triazin-2-yl)- (9CI)
Formamide, N-(4,6-diamino-s-triazin-2-yl)- (8CI)

13238-84-1
$C_8H_{12}N_2$
136.18
Pyrazine, 2,5-diethyl- (8CI, 9CI)

13244-33-2
$C_7H_9NO_4S$
203.23
O=S(=O)(O)c(c(N)ccc1OC)c1
Benzenesulfonic acid, 2-amino-5-methoxy
4-Aminoanisole-3-sulfonic acid
2-Amino-5-methoxy benzenesulfonic acid
Kyselina 4-aminoanisol-3-sulfonova (Czech)
4-Methoxy-2-sulfoaniline

13245-65-3
$C_4H_5Cl_3$
159.44
1-Propene, 1,3-dichloro-2-(chloromethyl)- (9CI)
1,3-Dichloro-2-chloromethyl-1-propene
Propene, 1,3-dichloro-2-(chloromethyl)- (7CI, 8CI)

13253-44-6
$C_6H_{10}N_2O_2S$
174.21
O=C(NC(C1=O)CCSC)N1
2,4-Imidazolidinedione, 5-(2-(methylthio)ethyl)- (9CI)
5-(2-(Methylthio)ethyl)-2,4-imidazolidinedione

13254-34-7
$C_9H_{20}O$
144.26
OC(CCCC(C)C)(C)C
2-Heptanol, 2,6-dimethyl-(9CI)
2,6-Dimethylheptan-2-ol
2,6-Dimethyl-2-heptanol

13255-27-1
$C_6H_3N_3O_8.Mg$
269.39
1,3-Benzenediol, 2,4,6-trinitro-, magnesium salt (1:1) (9CI)
Magnesium styphnate
2,4,6-Trinitro-1,3-benzenediol, magnesium salt (1:1)

13256-06-9
$C_{10}H_{22}N_2O$
186.34
CCCCCN(CCCCC)N=O
Dipentylamine, N-nitroso
Diamylnitrosamin (German)
Di-n-amylnitrosamine
Dipentylnitrosamine
Di-n-pentylnitrosamine
N-Nitrosodi-n-pentylamine

13256-07-0
$C_6H_{14}N_2O$
130.22
CCCCCN(C)N=O
Pentylamine, N-methyl-N-nitroso
AMN
N-Amyl-N-methylnitrosamine
Methylamylnitrosamin (German)

Methylamylnitrosamine
Methyl-N-amylnitrosamine
N-Methyl-N-nitrosopentylamine
Methyl-N-pentylnitrosamine
N-Nitroso-N-methyl-n-amyl-amine
Nitrosomethyl-n-amylamine
Nitrosomethyl-n-pentylamine

13256-13-8
$C_4H_8N_2O$
100.14
CCN(C=C)N=O
Vinylamine, N-ethyl-N-nitroso
Aethyl-vinyl-nitrosoamin (German)
Ethenamine, N-ethyl-N-nitroso- (9CI)
Ethenylamine, N-ethyl-N-nitroso-
N-Ethyl-N-nitrosovinylamine
Ethylvinylnitrosamine
N-Nitrosoethylvinylamine
N-Nitroso-N-ethylvinylamine
Vinylethylnitrosamin (German)
Vinylethylnitrosamine

13256-22-9
$C_3H_6N_2O_3$
118.11
CN(CC(O)=O)N=O
Sarcosine, N-nitroso
Glycine, N-methyl-N-nitroso-
N-Methyl-N-nitrosoglycine
N-Nitrosomethylglycine
N-Nitrososarcosine
Nitroso sarkosin (German)
NSAR

13265-60-6
$C_6H_{14}NO_3PS_2$
243.30
COP(=S)(OC)SCCNC(C)=O
Phosphorodithioic acid, O,O-dimethyl S-(2-acetamidoethyl) ester
S-(2-(Acetylamino)ethyl) O,O-dimethyl phosphorodithioate
Amiphos
CP 49674
DAEP
Phosphorodithioic acid, S-(2-(acetylamino)ethyl) O,O-dimethyl ester
O,O-Dimethyl-S-(2-(acetylamino)ethyl) dithiophosphate
O,O-Dimethyl S-(2-acetylaminoethyl) phosphorodithioate
N-((O,O-Dimethylphosphorodithioyl)ethyl)acetamide
ENT 27,346
Monsanto CP-49674
NSC-190945
Phosphorodithioic acid, O,O-dimethyl ester, S-ester with N-(2-mercaptoethyl)-acetamide

13269-35-7
$C_{17}H_{23}NO_3$
289.41
Hyoscyamine, (+)
(+)-Hyoscyamine

13269-52-8
C_6H_{12}
84.16
CCC=CCC
3-Hexene, (E)- (9CI)
(E)-3-Hexene
trans-3-Hexene
NSC-74125

13269-74-4
C_2H_6SSn
180.85
Stannane, dimethylthioxo- (9CI)
Dimethylthioxostannane
Dimethyltin sulfide

13270-03-6
$C_{21}H_{12}Cl_3N_3S_3$
508.90
s-Triazine, 2,4,6-tris-((p-chlorophenyl)thio)- (8CI)

13270-05-8
$C_{24}H_{21}N_3S_3$
447.65
s-Triazine, 2,4,6-tris(p-tolylthio)- (8CI)

13270-97-8
$C_{20}H_{34}O_3Si_4$
434.84
Tetrasiloxane, 1,1,1,3,5,7,7,7-octamethyl-3,5-diphenyl- (6CI,7CI,8CI,9CI)
1,1,1,3,5,7,7,7-Octamethyl-3,5-diphenyltetrasiloxane

13271-58-4
$C_{34}H_{50}O_5Si_6$
707.29
Hexasiloxane, 1,1,1,3,5,7,9,11,11,11-decamethyl-3,5,7,9-tetraphenyl- (7CI,8CI,9CI)

13274-84-5
$C_4H_9Cl_2O_2P$
190.99
Phosphinic acid, bis-(chloromethyl)-, ethyl ester (7CI,8CI,9CI)

13275-18-8
$C_6H_{12}Cl_2$
155.07
Hexane, 2,5-dichloro- (6CI,7CI,8CI,9CI)
2,5-Dichlorohexane

13279-58-8
$C_{10}H_{12}ClNO_5S$
293.72
O=C(Nc(c(OC)cc(S(=O)(=O)Cl)c1OC)c1)C
Benzenesulfonyl chloride, 4-(acetylamino)-2,5-dimethoxy- (9CI)

13279-86-2
$C_4H_5Cl_3$
159.44
1-Butene, 1,1,3-trichloro- (8CI,9CI)

13284-42-9
C_5H_7N
81.11
2-Pentenenitrile (8CI,9CI)

13286-32-3
$C_{11}H_{17}O_3PS$
260.31
CCOP(=O)(OCC)SCc1ccccc1
Phosphorothioic acid, S-benzyl O,O-diethyl ester
S-Benzyl-O,O-diethylester kyseliny thiofosforecne (Czech)
O,O-Diethyl S-benzyl thiophosphate
IBP
Kitazin
Ricid

13287-21-3
$C_{14}H_{30}$
198.40
Tridecane, 6-methyl- (6CI,7CI,8CI,9CI)
6-Methyltridecane

13290-96-5
$C_{10}H_9NO_6$
239.18
O=C(OC)c(cc(N(=O)=O)cc1C(=O)OC)c1
Isophthalic acid, 5-nitro-, dimethyl ester (8CI)
1,3-Benzenedicarboxylic acid, 5-nitro-, dimethyl ester (9CI)
Dimethyl 5-nitroisophthalate
5-Nitroisophthalic acid, dimethyl ester
NSC-93786

13292-46-1
$C_{43}H_{58}N_4O_{12}$

823.05
CCN(CC)C(=S)SSC(=S)N(CC)CC
Rifomycin SV, 8-(N-(4-methyl-1-piperazinyl)formidoyl)
Archidyn
Arficin
Dione 21-Acetate
L-5103
3-(4-Methylpiperazinylimino-methyl)-rifamycin SV
8-(4-Methylpiperazinylimino-methyl) rifamycin SV
8-(((4-Methyl-1-piperazinyl)-imino)methyl)rifamycin SV
NSC-113926
R/AMP
Rifa
Rifadin
Rifadine
Rifagen
Rifaldazin
Rifaldazine
Rifaldin
Rifamate
Rifampicin
Rifampicine (French)
Rifampicin SV
Rifampicinum
Rifampin
Rifamycin AMP
Rifamycin, 3-(((4-methyl-1-piperazinyl)imino)methyl)-
Rifaprodin
Rifinah
Rifobac
Rifoldin
Rifoldine
Riforal
Rimactan
Rimactane
Rimactazid
Rimactizid
Tubocin

13292-87-0
$C_2H_6S.BH_3$
75.98
Methyl sulfide, Compd. with borane (1:1)
Borane, Compd. with di-methylsulfide
Di-methylsulfide borane

13294-71-8
C_4H_7Br
135.00
C(Br)(=CC)C
2-Butene, 2-bromo- (9CI)
2-Bromo-2-butene

13301-33-2
$C_{16}H_{12}N_4O_{10}S_2$
484.40
O=S(=O)(O)c(c(N=Nc(c(O)ccc1N(=O)=O)c1)c(O)c(c2cc(S(=O)(=O)O)c3)c3N)c2
2,7-Naphthalenedisulfonic acid, 5-amino-4-hydroxy-3-((2-hydroxy-5-nitrophenyl)-azo)- (9CI)
5-Amino-4-hydroxy-3-((2-hydroxy-5-nitrophenyl)azo)-2,7-naphthalenedisulfonic acid

13301-35-4
$C_5H_6N_6$
150.14
1,3,5-Triazine-2-acetonitrile, 4,6-diamino- (9CI)
2,4-Diamino-6-cyanomethyl-1,3,5-triazine
s-Triazine-2-acetonitrile, 4,6-diamino- (8CI)

13301-61-6
$C_{17}H_{15}Cl_2N_5O_2$
392.21
O=N(=O)c(cc(c(N=Nc(ccc(N(CCC#N)CC)c1)c1)c2Cl)Cl)c2
Propanenitrile, 3-((4-((2,6-dichloro-4-nitrophenyl)azo)-phenyl)ethylamino)- (9CI)
3-(p-((2,6-Dichloro-4-nitro-phenyl)azo)-N-ethylanilino)-propionitrile
3-((4-((2,6-Dichloro-4-nitro-phenyl)azo)phenyl)ethyl-amino)propanenitrile
Propanonitrile, 3-(ethyl(4-((2,6-dichloro-4-nitrophenyl)azo)-phenyl)amino)-

13302-14-2
$C_{12}H_{27}ClPb$
414.03
Plumbane, chlorotributyl
Lead, tributyl-, chloride
Plumbane, tributylchloro-
Tributyl lead chloride
Tributylplumbium chloride

13306-69-9
$C_9H_{14}NO_3PS$
247.27
Phosphorothioic acid, O-(4-amino-m-tolyl) O,O-dimethyl ester
Aminofenitrothion
Aminosumithion
O,O-Dimethyl-O-(3-methyl-4-aminophenyl) phosphorothioate
Fenitrothion Amino Analog
Phosphorothioic acid, O-(4-amino-3-methylphenyl) O,O-dimethyl ester (9CI)

13306-70-2
$C_8H_{12}NO_3PS$
233.22
Phosphorothioic acid, O-(4-amino-3-methylphenyl) O-methyl ester (9CI)
Phosphorothioic acid, O-(4-amino-m-tolyl) O-methyl ester (8CI)

13308-40-2
$C_{18}H_{35}CrO_2$
335.47
9-Octadecenoic acid (Z)-, chromium salt (9CI)
Oleic acid, chromium salt (8CI)

13316-70-6
$C_{22}H_{48}N.Br$
406.62
[Br-].CCCCCCCCCCCCCCCC[N+](CC)(CC)CC
Ammonium, cetyldiethylethyl-, bromide
Ammonium, triethylhexadecyl-, bromide (8CI)
CDEA Br
Cetyltriethylammonium bromide
1-Hexadecanaminium, N,N,N-triethyl-, bromide (9CI)
Hexadecyltriethylammonium bromide
Triethylhexadecylammonium bromide

13327-32-7
$H_2O_2.Be$
43.03
Beryllium-hydroxide
Beryllium dihydroxide
Beryllium hydrate

13329-71-0
$C_{21}H_{46}N$
312.68
N(CCCCCCCCCCCCCCCCCC)C(C)C
Octadecylamine, N-isopropyl
Isopropyloctadecylamine

13331-52-7
Unknown
Unknown
Tributyltin acrylate

13344-99-5
$C_{11}H_{13}Cl_2N_5$
286.13
1,3,5-Triazine-2,4-diamine, 1-(3,4-dichlorophenyl)-1,6-dihydro-6,6-dimethyl- (9CI)
NSC-159729
s-Triazine, 4,6-diamino-1-(3,4-dichlorophenyl)-1,2-dihydro-2,2-dimethyl- (8CI)

13347-42-7
$C_{11}H_{13}ClO$
196.68
4-Chloro-2-cyclopentylphenol
Caswell No. 186
4-Chloro-2-cyclopentyl phenol
2-Cyclopentyl-p-chlorophenol

Dowicide 9
EPA Pesticide Chemical Code 064202
Phenol, 4-chloro-2-cyclopentyl- (8CI,9CI)

13350-71-5
$C_{12}H_8Cl_6$
364.91
2,4,7-Metheno-1H-cyclopenta-(a)pentalene, 1,1,2,3,3a, 7a-hexachloro-2,3,3a,3b,4, 6a,7,7a-octahydro-, (2α,3α, 3aα,3bα,4β,6aα,7β,7aα)-(9CI)
2,4,7-Metheno-1H-cyclopenta-(a)pentalene, 1,1,2,3,3a, 7a-hexachloro-2,3,3a,3b, 4,6a,7,7a-octahydro-, stereoisomer (8CI)
Photoaldrin
SD 18303

13351-73-0
$C_7H_7N_3$
133.17
Cn1nnc2ccccc12
1H-Benzotriazole, 1-methyl
1-Methylbenzotriazole

13355-96-9
$C_4H_{11}ClO_2Sn$
245.29
Stannane, butylchlorodihydroxy- (9CI)
Butylchlorodihydroxystannane
Butylchlorotin dihydroxide
NSC-323990

13356-08-6
$C_{60}H_{78}OSn_2$
1052.76
Distannoxane, hexakis(β,β-dimethylphenethyl)
Bendex
Bis(tris(β,β-dimethylphenethyl)-tin)oxide
Bis(tris(2-methyl-2-phenyl-propyl)tin)oxide
Di(tri-(2,2-dimethyl-2-phenyl-ethyl)tin)oxide
ENT 27,738
Fenbutatin oxide
Fenbutatin-oxyde
Fenylbutatin oxide
Fenylbutylstannium oxide (Czech)
Hexakis(β,β-dimethylphen-ethyl)-distannoxane
Hexakis (2-methyl-2-phenyl-propyl)-distannoxane
Hexakis(2-methyl-2-phenyl-propyl)distannoxane
Osdaran
SD 14114
Shell SD-14114
Torque
Vendex

13360-45-7
$C_9H_{10}BrClN_2O_2$
293.57
CON(C)C(=O)Nc1ccc(Br)c(Cl)c1
Urea, 3-(4-bromo-3-chloro-phenyl)-1-methoxy-1-methyl
Bromex
N-(4-Bromo-3-chlorophenyl)-N'-methoxy-N'-methylurea
3-(4-Bromo-3-chlorophenyl)-1-methoxy-1-methylurea
C-6313
Chlorbromuron
1-(3-Chloro-4-bromophenyl)-3-methyl-3-methoxyurea
Chlorobromuron
Ciba 6313
Maloran
Urea, N'-(4-bromo-3-chloro-phenyl)-N-methoxy-N-methyl-

13360-61-7
$C_{15}H_{30}$
210.40
C(=C)CCCCCCCCCCCCC
1-Pentadecene (9CI)
NSC-77125
Pentadecene,1-

13360-63-9
$C_6H_{15}N$
101.22
N(CCCC)CC
Butylamine, N-ethyl
Butylethylamine
N-Ethylbutylamine

13360-64-0
$C_7H_{10}N_2$
122.19
n(c(cnc1CC)C)c1
Pyrazine, 2-ethyl-5-methyl
2-Ethyl-5-methyl pyrazine

13360-65-1
$C_8H_{12}N_2$
136.18
n(c(c(nc1C)CC)C)c1
Pyrazine, 3-ethyl-2,5-dimethyl- (9CI)
3-Ethyl-2,5-dimethylpyrazine

13361-34-7
$C_{11}H_{19}NO_2$
197.27
O=C(OCC(CCCC)CC)CC#N
Acetic acid, cyano-, 2-ethylhexyl ester (9CI)
AI3-07380
Cyanoacetic acid, 2-ethylhexyl ester
2-Ethylhexyl cyanoacetate
NSC-69963

13366-73-9
$C_{12}H_8Cl_6O$
380.90
ClC3C6(Cl)C4C2C1OC1C5C2C3(Cl)C(Cl)(C45)C6(Cl)Cl
2,4,7-Metheno-1H-cyclopenta-(a)pentalene, 1,1,2,3,3a, 7a-hexachloro-5,6-epoxy-decahydro-, stereoisomer
Dieldrin, photo-
NCI-C00599
Photodieldrin

13372-77-5
$C_{10}H_{17}NO$
167.25
N(O)=CC=C(CCC=C(C)C)C
2,6-Octadienal, 3,7-dimethyl-, oxime
AI3-12094
Citral oxime
3,7-Dimethyl-2,6-octadienal oxime

13382-47-3
$C_4H_8O_4$
120.11
β-Hydroxyethoxyacetic acid
(2-Hydroxyethoxy)acetic acid

13389-42-9
C_8H_{16}
112.22
2-Octene, (E)- (8CI,9CI)
NSC-97522

13392-69-3
$C_5H_{10}O_3$
118.13
Pentanoic acid, 5-hydroxy- (9CI)
Valeric acid, 5-hydroxy- (8CI)

13393-93-6
$C_{20}H_{36}O$
292.51
OCC(C(C(C(C(C1)CC(C2)C(C)C)C2)(CC3)C)C1)(C3)C
1-Phenanthrenemethanol, tetradecahydro-1,4a-dimethyl-7-(1-methylethyl)-
AI3-04505

13395-16-9
$C_{10}H_{14}O_4.Cu$
261.78
CC1=CC(=[O+][Cu-2]2(O1)OC(=CC(=[O+]2)C)C)C
Copper, bis(2,4-pentanedionato)
Bis(acetylacetonato)copper
Bis(acetylacetone)copper
Bis(2,4-pentanedionato)copper
CD 9
Copper acetylacetonate
Copper(II) acetylacetonate

Copper bis(acetylacetonate)
Copper bis(acetylacetone)
Copper bis(2,4-pentanedionate)
Copper, bis(2,4-pentanedionato-O,O')-
Copper, bis(2,4-pentanedionato-O,O')-, (SP-4-1)- (9CI)
Copper diacetylacetonate
Cupric acetylacetonate

13396-41-3
$O_{12}P_4.4Na$
407.84
Metaphosphoric acid, tetrasodium salt
Cyclophos
Sodium tetrametaphosphate

13396-80-0
$C_8H_{15}P$
142.18
P(C(CC1)CCCC2)C12
9-Phosphabicyclo(4.2.1)nonane (9CI)

13397-24-5
$O_4S.Ca.2H_2O$
172.18
Gypsum (OSHA)
Gips
Gypsite
Landplaster
Phosphogypsum

13397-26-7
$CH_2O_3.Ca$
102.10
Calcite (8CI,9CI)
Calcspar
Hydrocarb 90
Iceland Spar
Omyalen

13402-02-3
$C_{19}H_{36}O_2$
296.49
O=C(OCCCCCCCCCCCCCCCC)C=C
2-Propenoic acid, hexadecyl ester (9CI)
Acrylic acid, hexadecyl ester
AI3-15694
Cetyl acrylate
Hexadecyl acrylate
Hexadecyl 2-propenoate
NSC-72788
Palmityl acrylate

13402-96-5
$C_{18}H_{28}O_3$
292.42
O=C(O)COc(c(cc(c1)C(CC)(C)C)C(CC)(C)C)c1
Acetic acid, (2,4-bis(1,1-dimethylpropyl)phenoxy)- (9CI)
Acetic acid, (2,4-di-tert-pentylphenoxy)-

13403-01-5
Unknown
Unknown
O=C(O)C(Oc(c(cc(c1)C(CC)(C)C)C(CC)(C)C)c1)CC
Butyric acid, 2-(2,4-di-tert-pentylphenoxy)
2-(2,4-Di-tert-pentylphenoxy)-butyric acid

13403-37-7
C_3H_5ClO
92.52
Oxirane, (chloromethyl)-, (+-)- (9CI)
Propane, 1-chloro-2,3-epoxy-, (+-)- (8CI)

13405-83-9
$C_6H_{10}O_4$
146.14
(1,4)Dioxino(2,3-b)-1,4-dioxin, hexahydro-, cis- (9CI)
p-Dioxino(2,3-b)-p-dioxin, hexahydro-, cis- (8CI)
cis-1,4,5,8-Tetraoxadecalin

13408-63-4
$C_6H_2FeN_6$
213.97
Hexacyanoferrate II
Ferrocyanide
(OC-6-11)-Hexakis(cyano-C)-ferrate(4-)

13410-01-0
$O_4Se.2Na$
188.94
Selenic acid, disodium salt
Disodium selenate
Natriumseleniat (German)
P-40
Sel-Tox SSO2 and SS-20
Sodium selenate

13414-54-5
$C_{10}H_{11}NO_3$
193.20
O=N(=O)c(c(OCC(=C)C)ccc1)c1
Benzene, 1-((2-methyl-2-propenyl)oxy)-2-nitro- (9CI)
Methallyl 2-nitrophenyl ether
1-((2-Methyl-2-propenyl)oxy)-2-nitrobenzene

13414-55-6
$C_{10}H_{11}NO_3$
193.20
O=N(=O)c(c(OC(C1)(C)C)c1cc2)c2
Benzofuran, 2,3-dihydro-2,2-dimethyl-7-nitro- (9CI)
2,3-Dihydro-2,2-dimethyl-7-nitrobenzofuran

13414-58-9
$C_{10}H_{11}NO_3$
193.20
O=N(=O)c(c(O)c(cc1)CC(=C)C)c1
Phenol, 2-(2-methylallyl)-6-nitro- (8CI)
2-Methallyl-6-nitrophenol
2-(2-Methylallyl)-6-nitrophenol
2-(2-Methyl-2-propenyl)-6-nitrophenol

13417-01-1
$C_{13}H_{13}F_{17}N_2O_2S$
584.29
O=S(=O)(NCCCN(C)C)C(F)(F)C(F)(F)C(F)(F)C(F)(F)C(F)(F)C(F)(F)C(F)(F)C(F)(F)F
1-Octanesulfonamide, N-(3-(dimethylamino)propyl)-1,1,2,2,3,3,4,4,5,5,6,6,7,7,8,8,8-heptadecafluoro- (9CI)
N-(3-(Dimethylamino)propyl)-1,1,2,2,3,3,4,4,5,5,6,6,7,7,8,8,8-heptadecafluoro-1-octanesulfonamide

13417-43-1
C_5H_9Cl
104.59
2-Butene, 1-chloro-2-methyl
1-Chloro-2-methyl-2-butene

13418-49-0
$C_{20}H_{18}N_4O_4$
378.36
O=C(N(C(=N)c1c(c(c(C(=O)c(c2ccc3)c3)c4N)C2=O)N)CCCOC)c14
1H-Naphth(2,3-f)isoindole-1,5,10-trione, 4,11-diamino-2,3-dihydro-3-imino-2-(3-methoxypropyl)- (9CI)

13419-15-3
$C_{18}H_{35}AlO_3$
326.46
Aluminum, (octadecanoato-O)oxo- (9CI)
(Octadecanoato-O)oxoaluminum
Oxo aluminium stearate

13419-31-3
$C_{20}H_{34}O_3S.Na$
377.54
Benzenesulfonic acid, 4-(1-methyltridecyl)-, sodium salt (9CI)
Benzenesulfonic acid, p-(1-methyltridecyl)-, sodium salt (8CI)

13419-59-5
$C_4H_6O_7S.3Na$
267.12
Butanedioic acid, sulfo-, trisodium salt (9CI)
Trisodium sulfobutanedioate

13422-51-0
$C_{62}H_{89}CoN_{13}O_{15}P$
1346.30
Hydroxocobalamin
Axion
Axlon
Ciplamin H
Cobalamin, hydroxo-
Cobalex
Cobinamide dihydroxide dihydrogen phosphate (ester), mono(inner salt), 3'-ester with 5,6-dimethyl-1-α-D-ribofuranosylbenzimidazole
Cobinamide, dihydroxide, dihydrogen phosphate (ester), mono(inner salt), 3'-ester with 5,6-dimethyl-1-α-D-ribofuranosyl-1H-benzimidazole (9CI)
Cobinamide hydroxide phosphate 3'-ester with 5,6-dimethyl-1-α-D-ribofuranosylbenzimidazole inner salt
α Cobione
Codroxomin
Depogamma
α-(5,6-Dimethylbenzimidazolyl)hydroxocobamide
Docclan
Docevita
Ducobee-Hy
Duradoce
Duralta-12
Hidroxocobalamina (Spanish)
Hydrocobalamin
Hydrogrisevit
Hydrovit
Hydroxocobalamine
Hydroxocobalamine (French)
Hydroxocobalaminum (Latin)
Hydroxocobemine
Hydroxy Vitamin B12
Hydroxycobalamin
Hydroxycobalamine
Hyxobamine
Idrogrisevit
Idrossocobalamina
Lyovit-H
Neo-Betalin 12
Neo-Cytamen
Neo-Macrabin
Neo-Rojamin
OH-Duphar
OHB12
Oxobemin
Oxolamine (arcum)
Primabalt RP
αRedisol
Sytobex-H
Vibeden
Vitadurin
Vitamin B12A

13423-61-5
$CrH_2O_4.Mg$
142.31
Magnesium chromate
Chromic acid, magnesium salt (1:1) (9CI)

13424-46-9
N_6Pb
291.20
Lead azide (8CI,9CI)
Azida de plomo (Spanish)
Azoture de plomb (French)
Initiating Explosive Lead Azide, Dextrinated type only
Lead azide, Containing, by weight, at least 20% water or mixture of alcohol and water [UN 0129]
Lead azide (Dry)
UN 0129 [Lead azide, Wetted with not less than 20 per cent water or mixture of alcohol and water, by mass]

13426-91-0
$C_2H_{10}N_2.xCu$
506.92
Ethane, 1,2-diamino-, copper complex
Cupriethylene diamine
Cupriethylenediamine, Solution [UN 1761]
UN 1761 [Cupriethylene-diamine solution]

13427-42-4
CHO_3Pb
268.21
Carbonic acid, lead salt (8CI, 9CI)

13427-43-5
C_9H_{18}
126.24
1-Hexene, 3,3,5-trimethyl- (9CI)

13427-80-0
$C_8H_5KO_4$
204.23
1,4-Benzenedicarboxylic acid, potassium salt (9CI)
Terephthalic acid, potassium salt (8CI)

13429-07-7
$C_7H_{16}O_3$
148.20
2-Propanol, 1-(2-methoxypropoxy)- (8CI,9CI)
Methoxy dipropylene glycol
1-(2-Methoxypropoxy)-2-propanol
Polyoxypropylene (2) methyl ether
Polypropylene glycol (2) methyl ether

13429-24-8
$(C_3F_6)_2$
600.09
1-Propene, 1,1,2,3,3,3-hexafluoro-, dimer (9CI)
Propene, hexafluoro-, cyclic dimer

13429-27-1
$C_{14}H_{28}O_2.K$
267.47
Potassium myristate
Caswell No. 696A
EPA Pesticide Chemical Code 079022
Potassium tetradecanoate
Tetradecanoic acid, potassium salt (9CI)

13432-25-2
$C_8H_{18}O$
130.23
3-Hexanol, 2,4-dimethyl- (8CI,9CI)
NSC-93914

13432-51-4
$C_6H_{11}N_2O_3PS_4$
318.39
Phosphorodithioic acid, O,O-dimethyl S-((5-(methylthio)-2-oxo-1,3,4-thiadiazol-3(2H)-yl)methyl) ester (9CI)
Phosphorodithioic acid, O,O-dimethyl ester, S-ester with 4-(mercaptomethyl)-2-(methylthio)-δ(2)-1,3,4-thiadiazolin-5-one (8CI)

13433-11-9
$C_8H_{13}N_3O_6$
247.19
L-Asparagine, N(2)-L-α-aspartyl- (9CI)
Asparagine, N(2)-L-α-aspartyl-, L- (8CI)

13441-22-0
$C_{45}H_{91}N_2O_5.CH_3O_4S$
851.31
Ammonium, bis(2-hydroxyethyl)(2-(N-(2-hydroxyethyl)octadecanamido)ethyl)-methyl-, methyl sulfate, stearate (ester) (8CI)

13444-75-2
Unknown
Unknown
Mercury (II) chromate

13444-85-4
I_3N
394.71
Nitrogen iodide
Nitrogen triiodide
Triiodure d'azote (French)
Triyoduro de nitrogeno (Spanish)

13446-10-1
$MnO_4.H_4N$
136.96
Ammonium permanganate
NA 9190
Permanganate d'ammonium (French)
Permanganato amonico (Spanish)
Permanganic acid, ammonium salt

13446-34-9
$Cl_2Mn.4H_2O$
197.92
Manganese(II) chloride, tetrahydrate
Manganese dichloride tetrahydrate
Manganous chloride tetrahydrate

13446-48-5
$NO_2.H_4N$
64.06
Nitrous acid, ammonium salt
Ammonium nitrite (DOT)

13446-72-5
$CrH_2O_4.2Rb$
288.95
Rubidium chromate
Chromic acid, dirubidium salt (8CI,9CI)

13446-73-6
$Cr_2O_7Rb_2$
386.92
Rubidium dichromate

13450-90-3
Cl_3Ga
176.08
Gallium chloride (8CI,9CI)
Gallium(3+) chloride
Gallium trichloride
NSC-94002
Trichlorogallium

13450-92-5
Br_4Ge
392.23
Germanium-bromide
Germanium tetrabromide
Germanium, tetrabromo-

13450-97-0
$Cl_2O_8.Sr$
286.52
Perchloric acid, strontium salt
Strontium diperchlorate
Strontium perchlorate [UN 1508]
UN 1508 [Strontium perchlorate]

13453-07-1
$AuCl_3$
303.32
Gold-chloride
Auric chloride
Gold(III) chloride
Gold trichloride

13453-66-2
$H_4O_7P_2.2Pb$
592.37
Lead pyrophosphate
Diphosphoric acid, lead(2+) salt (1:2) (9CI)
Lead(2+) pyrophosphate

13453-69-5
$BHO_2.Li$
50.76
Boric acid, lithium salt (9CI)

13454-78-9
$CrH_2O_4.2Cs$
383.82
Cesium chromate
Chromic acid, dicesium salt (8CI,9CI)

13454-96-1
Cl_4Pt
336.89
Platinum (IV) chloride
Platinum(IV) tetrachloride
Platinum tetrachloride

13455-25-9
$Co.CrH_2O_4$
176.94
Cobalt chromate
Chromic acid, cobalt(2+) salt (1:1) (8CI,9CI)

13457-18-6
$C_{14}H_{20}N_3O_5PS$
373.40
CCOC(=O)C2=Cn1nc(OP(=S)(OCC)OCC)cc1N=C2C
Phosphorothioic acid, O,O-diethyl ester, O-ester with (6-ethoxycarbonyl-5-methyl) pyrazolo(1,5-a)pyrimidin-2-ol
Afugan
Curamil
O,O-Diethyl-O-(5-methyl-6-ethoxy-carbonyl-pyrazolo-(1.5-a)pyrimid-2-yl)-thionophosphate
2-(O,O-Diethyl-thionophosphoryl)-5-methyl-6-carbethoxy-pyrazolo-(1,5a)pyrimidine
Ethyl 2-((diethoxyphosphinothioyl)oxy)-5-methylpyrazolo(1,5-a)pyrimidine-6-carboxylate
HOE 2873
Missile
Pyrazophos
Pyrazolo-(1,5a)pyrimidine, 2-(O,O-diethyl-thionophosphoryl)-5-methyl-6-ethoxycarbonyl-

13462-88-9
Br_2Ni
218.50
Br[Ni]Br
Nickel bromide (9CI)
Nickel dibromide
Nickelous bromide
NSC-128153

13462-90-3
I_2Ni
312.50
Nickel iodide (9CI)
Nickel diiodide
Nickel(2+) iodide
Nickelous iodide

13463-39-3
C_4NiO_4
170.75
Nickel carbonyl (ACGIH, OSHA) [UN 1259]
Nickel carbonyle (French)
Nichel tetracarbonile (Italian)
Nickel tetracarbonyl
Nikkeltetracarbonyl (Dutch)
Nickel tetracarbonyle (French)
RCRA waste number P073
UN 1259 [Nickel carbonyl]

13463-40-6
C_5FeO_5
195.90
Iron carbonyl (DOT)
Fer pentacarbonyle (French)
Iron pentacarbonyl (ACGIH, OSHA) [UN 1994]
Pentacarbonyliron
UN 1994 [Iron pentacarbonyl]

13463-41-7
$C_{10}H_8N_2O_2S_2.Zn$
317.69
Zinc, bis(2-pyridylthio)-, N,N'-dioxide
Bis(1-hydroxy-2(1H)-pyridinethionato)zinc

OM-1563
Omadine zinc
2-Pyridinethiol-1-oxide, zinc salt
Pyrithione zinc
Vancide P
Zinc, bis(1-hydroxy-2(1H)-pyridinethionato)- (8CI)
Zinc, bis(1-hydroxy-2(1H)-pyridinethionato-O,S)-, (T-4)- (9CI)
Zinc, bis(2-pyridylthio)-, 1,1'-dioxide
Zinc omadine
Zincpolyanemine
Zinc PT
Zinc pyridine-2-thiol-1-oxide
Zinc 2-pyridinethiol-1-oxide
Zinc pyridinethione
Zinc pyrion
Zinc pyrithione
ZnPT
ZPT

13463-67-7
O_2Ti
79.90
O=[Ti]=O
Titanium-oxide
A-Fil Cream
Atlas White Titanium Dioxide
Austiox
Bayeritian
Bayertitan
Baytitan
Calcotone White T
C.I. 77891
C.I. Pigment White 6
Cosmetic White C47-5175
Cosmetic White C47-9623
C-Weiss 7 (German)
Flamenco
Hombitan
Horse Head A-410
Horse Head A-420
Horse Head R-710
KH 360
Kronos
Kronos Titanium Dioxide
Levanox White RKB
NCI-C04240
Rayox
Runa RH20
Rutile
Rutiox CR
Tiofine
Tiona T.D.
Tioxide
Tipaque
Ti-Pure
Titafrance
Titandioxid (Swedish)
Titanium dioxide (ACGIH, OSHA)
Titanium peroxide
Titanox
Titanox 2010
Titan White
Trioxide(S)
Tronox
Unitane
Unitane O-110
Unitane O-220
Unitane OR-150
Unitane OR-340
Unitane OR-342
Unitane OR-350
Unitane OR-540
Unitane OR-640
1700 White
Zopaque

13464-10-3
$C_7H_{14}N_2O_3$
174.19
O=C(N(C(OC)C1OC)C)N1C
2-Imidazolidinone, 4,5-dimethoxy-1,3-dimethyl-
AI3-51623

13464-35-2
$AsHO_2.K$
147.03
Arsenic compound
Arsenenous acid, potassium salt (9CI)
Arsenite de potassium (French)
Arsenito potasico (Spanish)
NSC-3060
Potassium arsenite
X 13

13464-37-4
$AsH_3O_3.3Na$
194.91
Sodium arsenite, Liquid (Solutions) [UN 1686]
AI3-01066
Arsenious acid sodium salt (Na_3AsO_3)
Arsenious acid (H_3AsO_3), trisodium salt (8CI)
Arsenous acid, trisodium salt (9CI)
Sodium arsenite (Na_3AsO_3) (6CI,7CI)
Trisodium arsenite
UN 1686 [Sodium arsenite, Aqueous solutions]

13464-38-5
$AsH_3O_4.3Na$
210.91
Sodium arsenate
Arsenic acid, trisodium salt (8CI,9CI)
Caswell No. 743
EPA Pesticide Chemical Code 013505
Sodium orthoarsenate
Trisodium arsenate

13464-42-1
Unknown
Unknown
Sodium pyroarsenate
UN 1685

13464-80-7
$H_4N_2.1/2H_2O_4S$
81.07
Hydrazine, sulfate (2:1) (9CI)

13464-82-9
$O_{12}S_3.In_2$
517.82
Indium-sulfate
Indisulfat (German)
Sulfuric acid, indium salt

13465-08-2
$H_3NO.HNO_3$
96.03
Hydroxylamine, nitrate (Salt) (9CI)
Hydroxylamine nitrate

13465-09-3
Br_3In
354.53
Indium bromide (9CI)

13465-34-4
Unknown
Unknown
Mercury(I) chromate

13465-73-1
BrH_3Si
111.01
Bromosilane
Bromosilano (Spanish)
Silane, bromo-
Silyl bromide

13465-95-7
$Ba.2ClHO_4$
338.24
Barium perchlorate
Perchlorate de baryum (French)
Perchloric acid, barium salt (8CI,9CI)
Perclorato barico (Spanish)
UN 1447 [Barium perchlorate]

13466-78-9
$C_{10}H_{16}$
136.26
C(=CCC(C12)C1(C)C)(C2)C
3-Carene
Bicyclo(4.1.0)hept-3-ene, 3,7,7-trimethyl- (9CI)
δ^3-Carene
S-3-Carene
Isodiprene
3-Karen (Czech)
3-Norcarene, 3,7,7-trimethyl-
3,7,7-Trimethylbicyclo(4.1.0)-3-heptene
4,7,7-Trimethyl-3-norcarene

13470-50-3
Unknown
Unknown
2-Heptadecyl-1-methyl-1-(2-(stearoylamido)ethyl)-2-imidazolinium methyl sulfate

13472-08-7
$C_{10}H_{16}N_4$
192.24
Butanenitrile, 2,2'-azobis-(2-methyl- (9CI)
2,2'-Azobis(2-methylbutanenitrile)

13472-30-5
$H_4O_4Si.4Na$
188.08
Silicic acid, tetrasodium salt (9CI)

13472-45-2
$O_4W.2Na$
293.83
[Na+].[Na+].[O-][W]([O-])(=O)=O
Tungstic acid, disodium salt
Sodium tungstate

13473-75-1
Unknown
Unknown
Thallium(I) chromate

13473-90-0
$N_3O_9.Al$
213.01
Aluminum(111) nitrate (1:3)
Aluminum nitrate [UN 1438]
Aluminum trinitrate
Nitrato de aluminio (Spanish)
Nitric acid, aluminum salt
Nitric acid, aluminum(3+) salt
UN 1438 [Aluminum nitrate]

13475-78-0
$C_{10}H_{22}$
142.28
Heptane, 5-ethyl-2-methyl- (8CI,9CI)

13475-81-5
$C_{10}H_{22}$
142.28
Hexane, 2,2,3,3-tetramethyl- (8CI,9CI)

13475-82-6
$C_{12}H_{26}$
170.34
C(CC(CC(C)(C)C)C)(C)(C)C
Heptane, 2,2,4,6,6-pentamethyl- (9CI)
2,2,4,6,6-Pentamethylheptane

13477-00-4
$Cl_2O_6.Ba$
304.24
Chloric acid, barium salt
Barium chlorate [UN 1445]
Barium chlorate, Wet (DOT)
Chloric acid, barium salt (Wet)
UN 1445 [Barium chlorate]

13478-00-7
$N_2O_6.Ni.6H_2O$
290.85
Nickel(II) nitrate, hexahydrate (1:2:6)
Nitric acid, nickel(2+) salt, hexahydrate
Nickel(2+) nitrate, hexahydrate

13478-10-9
$Cl_2Fe.4H_2O$
198.83
Iron(2+) chloride, tetrahydrate
Ferrous chloride, tetrahydrate
Iron chloride tetrahydrate
Iron (II) chloride, tetrahydrate
Iron dichloride tetrahydrate

13478-50-7
$H_2O_3S_2.Pb$
321.35
Lead thiosulfate
Thiosulfuric acid, lead(2+) salt (1:1) (8CI,9CI)

13483-18-6
$C_4H_8Cl_2O_2$
159.02
O(CCOCCl)CCl
Ethane, 1,2-bis(chloromethoxy)
Bis-1,2-(chloromethoxy)ethane
Ethylene glycol bis(chloromethyl)ether

13483-47-1
$C_9H_{18}O_3$
174.24
Oxirane, ((2-butoxyethoxy)-methyl)- (9CI)
Ethane, 1-butoxy-2-(2,3-epoxypropoxy)- (8CI)

13485-66-0
$C_{12}H_{18}O$
178.28
2(1H)-Naphthalenone, 4a,5,6,7,8,8a-hexahydro-4a,8a-dimethyl-, cis- (8CI,9CI)

13486-13-0
$C_{17}H_{18}N_6$
306.33
N(=C(N=Nc(ccc(N(Cc(cccc1)c1)CC)c2)c2)NN=3)C3
Benzenemethanamine, N-ethyl-N-(4-(1H-1,2,4-triazol-3-ylazo)phenyl)- (9CI)
3-((p-(Benzylethylamino)-phenyl)azo)-1H-1,2,4-triazole

13486-43-6
$C_{18}H_{20}N_4O_2S$
356.42
O(c(ccc(N=C(N=Nc(ccc(N(CCO)CC)c1)c1)S2)c23)c3)C
Ethanol, 2-(ethyl(4-((6-methoxy-2-benzothiazolyl)azo)-phenyl)amino)- (9CI)
2-(Ethyl(4-((6-methoxy-2-benzothiazolyl)azo)phenyl)-amino)ethanol

13491-79-7
$C_{10}H_{20}O$
156.27
OC(C(C(C)(C)C)CCC1)C1
Cyclohexanol, 2-(1,1-dimethyl-ethyl)- (9CI)
o-tert-Butylcyclohexanol
2-(1,1-Dimethylethyl)cyclohexanol

13494-80-9
Te
127.60
[Br]
Tellurium (ACGIH,OSHA)
NCI-C60117
Tellur (Polish)
Telloy

13495-09-5
Unknown
Unknown
DEA No. 9730
Piminodina (Spanish)
Piminodine
Piminodinum (Latin)

13506-76-8
$C_8H_7NO_4$
181.14
Benzoic acid, 2-methyl-6-nitro- (9CI)
o-Toluic acid, 6-nitro- (8CI)

13509-52-9
$C_7H_{10}N_2O_2$
154.17
1,3,6-Trimethyluracil

13510-49-1
$O_4S.Be$
105.07

[Be+2].[O-]S([O-])(=O)=O
Beryllium sulfate (1:1)
Beryllium sulphate
Sulfuric acid, beryllium salt (1:1)

13510-89-9
$O_8Pb_3Sb_2$
993.10
Lead antimonate
Antimony lead oxide (9CI)

13511-38-1
$C_5H_9ClO_2$
136.58
O=C(O)C(CCl)(C)C
Propanoic acid, 3-chloro-2,2-dimethyl- (9CI)
3-Chloro-2,2-dimethylpropanoic acid
β-Chloro-α,α-dimethylpropionic acid
3-Chloro-2,2-dimethylpropionic acid
Chloropivalic acid
β-Chloropivalic acid
3-Chloropivalic acid
Chlorotrimethylacetic acid
NSC-89696
Propionic acid, 3-chloro-2,2-dimethyl- (8CI)

13515-40-7
$C_{17}H_{15}ClN_4O_5$
390.76
O=C(Nc(c(OC)ccc1)c1)C(N=Nc(c(N(=O)=O)cc(c2)Cl)c2)C(=O)C
Pigment Yellow 73
Butanamide, 2-((4-chloro-2-nitrophenyl)azo)-N-(2-methoxyphenyl)-3-oxo- (9CI)
C.I. Pigment Yellow 73
C.I. 11738
2-((4-Chloro-2-nitrophenyl)azo)-N-(2-methoxyphenyl)-3-oxobutanamide

13516-27-3
$C_{18}H_{41}N_7$
355.57
Guazatine
Bis(8-guanidino-octyl)amine
Caswell No. 471D
EPA Pesticide Chemical Code 498200
Iminobis(octamethylene)diguanidine

13517-17-4
$CrO_4.2Na.10H_2O$
342.13
Sodium chromate decahydrate
Chromic acid, disodium salt, decahydrate

13519-20-5
$C_{12}H_{22}O_5$
246.30
Ether, bis(2-(2,3-epoxy-2-methylpropoxy)ethyl) (7CI,8CI)

13520-61-1
$Cl_2O_8.Ni.6H_2O$
365.73
Perchloric acid, nickel(2+) salt, hexahydrate
Nickel(2+) perchlorate, hexahydrate

13520-83-7
$N_2O_8U.6H_2O$
502.12
Uranium, bis(nitrato)dioxo-, hexahydrate
Uranium, dinitratodioxo-, hexahydrate
Uranyl nitrate hexahydrate solution
Uranylnitrate hexahydrate
UN 2980 [Uranyl nitrate hexahydrate solution]

13524-04-4
C_8H_9ClO
156.61
OC(c(c(ccc1)Cl)c1)C
Benzenemethanol, 2-chloro-α-methyl- (9CI)
2-Chloro-α-methylbenzenemethanol

13524-76-0
$C_{10}H_{10}O_2$
112.13
2(3H)-Benzofuranone, 3,3-dimethyl- (8CI,9CI)

13528-88-6
$C_3H_9Cl_3Si_2$
207.63
Disilane, 1,1,2-trichloro-1,2,2-trimethyl- (9CI)
1,1,2-Trichloro-1,2,2-trimethyldisilane
1,1,2-Trimethyltrichlorodisilane

13530-65-9
$CrH_2O_4.Zn$
183.39
Chromic acid, zinc salt
Basic Zinc Chromate
Basic Zinc Chromate X-2259
Buttercup Yellow
Chromium zinc oxide
C.I. 77955
C.I. Pigment Yellow 36
Citron Yellow
C.P. Zinc Yellow X-883
C.P. Zinc Yellow X-2127
Primrose Yellow
Pure Zinc Chrome
Pure Zinc Chrome A
Pure zinc Yellow
Zinc chromate (ACGIH,OSHA)
Zinc Chromate AM
Zinc Chromate C
Zinc Chromate O
Zinc Chromate T
Zinc chromate(VI) hydroxide
Zinc Chromate Z
Zinc chrome
Zinc chrome (Anti-corrosion)
Zinc Chrome Yellow
Zinc chromium oxide
Zinc hydroxychromate
Zinc tetraoxychromate
Zinc Tetraoxychromate 76A
Zinc Tetraoxychromate 780B
Zinc tetroxychromate
Zinc Yellow
Zinc Yellow 1
Zinc Yellow 1425
Zinc Yellow 40-9015
Zinc Yellow AZ-16
Zinc Yellow AZ-18
Zinc Yellow KSH
Zinc Yellow 386N
Zinc Yellows

13530-68-2
$Cr_2H_2O_7$
218.00
Chromic acid (9CI)
Chromic acid, Solid [NA 1463]
Chromic acid solution [UN 1755]
Dichromic acid
NA 1463 [Chromic acid, Solid]
UN 1755 [Chromic acid solution]

13531-52-7
$C_5H_{15}N_3$
117.18
N(CCCN)CCN
1,3-Propanediamine, N-(2-aminoethyl)- (9CI)
N-(2-Aminoethyl)-1,3-propanediamine
N-(3-Aminopropyl)ethylenediamine

13532-26-8
$C_{10}H_{19}N_5O$
225.26
CCNc1nc(OC)nc(n1)N(CC)CC
1,3,5-Triazine-2,4-diamine, N,N,N'-triethyl-6-methoxy- (9CI)
G 31432
s-Triazine, 2-(diethylamino)-4-(ethylamino)-6-methoxy- (8CI)
Trietatone

13534-15-1
$C_6H_{12}N_2O_2$

144.16
O(C(=N)C(OCC)=N)CC
Ethanediimidic acid, diethyl ester (9CI)
Diethyl ethanediimidate

13534-98-0
$C_5H_5BrN_2$
173.02
Nc1ccncc1Br
4-Pyridinamine, 3-bromo- (9CI)
4-Amino-3-bromopyridine
Pyridine, 4-amino-3-bromo- (6CI,7CI,8CI)

13536-84-0
F_2O_2U
308.00
Uranium, difluorodioxo
Uranium oxyfluoride
Uranium fluoride oxide
Uranyl fluoride

13537-32-1
FH_2O_3P
99.99
Phosphorofluoridic-acid
Fluorophosphoric acid, Anhydrous [UN 1776]
Monofluorophosphoric acid
Monofluorophosphoric acid, Anhydrous (DOT)
UN 1776 [Fluorophosphoric acid anhydrous]

13540-56-2
$C_{15}H_{16}$
196.29
Benzene, 1,2-dimethyl-4-(phenylmethyl)- (9CI)
Methane, phenyl-3,4-xylyl- (8CI)

13547-06-3
$C_8H_{10}Cl_2$
177.07
Cyclohexene, 1-chloro-4-(1-chloroethenyl)- (9CI)
Cyclohexene, 1-chloro-4-(1-chlorovinyl)- (8CI)

13547-07-4
$C_8H_{10}Cl_2$
177.07
Cyclohexene, 1-chloro-5-(1-chloroethenyl)- (9CI)
Cyclohexene, 1-chloro-5-(1-chlorovinyl)- (8CI)

13547-70-1
$C_6H_{11}ClO$
134.61
O=C(C(C)(C)C)CCl
2-Butanone, 1-chloro-3,3-dimethyl- (9CI)
1-Chloro-3,3-dimethyl-2-butanone

13548-42-0
$CrH_2O_4.Cu$
181.56
Copper chromate
Chromic acid, copper(2+) salt (1:1) (8CI,9CI)
Copper chromium oxide

13548-68-0
$C_5H_3ClN_4O_2$
186.53
Xanthine, 8-chloro- (8CI)
AI3-52416
8-Chloroxanthine
NSC-45762
1H-Purine-2,6-dione, 8-chloro-3,7-dihydro- (9CI)

13551-17-2
$C_{13}H_{15}NO_4$
249.27
Cyclohexanecarboxylic acid, 4-nitrophenyl ester (9CI)
Cyclohexanecarboxylic acid, p-nitrophenyl ester (7CI, 8CI)
p-Nitrophenyl cyclohexanecarboxylate
p-Nitrophenyl cyclohexylcarboxylate

13551-87-6
$C_7H_{11}N_3O_4$
201.21
COCC(O)Cn1ccnc1N(=O)=O
Imidazole-1-ethanol, α-(methoxymethyl)-2-nitro
1-(2-Hydroxy-3-methoxypropyl)-2-nitroimidazole
1H-Imidazole-1-ethanol, α-(methoxymethyl)-2-nitro- (9CI)
α-(Methoxymethyl)-2-nitroimidazole-1-ethanol
α-(Methoxymethyl)-2-nitro-1H-imidazole-1-ethanol
Misonidazole
NSC-261,037
1-(2-Nitro-1-imidazolyl)-3-methoxy-2-propanol
1-(2-Nitroimidazol-1-yl)-3-methoxypropan-2-ol
Ro 7-0582
SR 1354
SRI 1354

13551-92-3
$C_6H_9N_3O_4$
187.18
1,2-Propanediol, 3-(2-nitroimidazol-1-yl)
Demethylmisonidazole
Desmethylmisonidazole
1-(2-Nitro-1-imidazolyl)-3-hydroxy-2-propanol
3-(2-Nitroimidazol-1-yl)-1,2-propanediol
NSC-261036
1,2-Propanediol, 3-(2-nitro-1H-imidazol-1-yl)-, (9CI)
Ro 5-9963
SR 1530

13552-21-1
$C_4H_{11}NO$
89.13
CCC(O)CN
2-Butanol, 1-amino- (9CI)
NSC-17695

13552-44-8
$C_{13}H_{14}N_2.2ClH$
271.21
Cl.Cl.Nc2ccc(Cc1ccc(N)cc1)cc2
Aniline, 4,4'-methylenedi-, dihydrochloride
Benzenamine, 4,4'-methylenebis-, dihydrochloride
p,p'-Methylenedianiline dihydrochloride
4,4'-Methylenedianiline dihydrochloride
NCI-C54604

13560-89-9
$C_{18}H_{12}Cl_{12}$
653.70
C(=C(C(C1(Cl)Cl)(C(C2CCC(C(C(=C(C34Cl)Cl)Cl)Cl)(C3(Cl)Cl)Cl)C4C5)C5)Cl)Cl)(C12Cl)Cl
1,4:7,10-Dimethanodibenzo-(a,e)cyclooctane, 1,2,3,4,7,8,9,10,13,13,14,14-dodecachloro- 1,4,4a,5,6,6a,7,10,10a,11,12,12a-dodecahydro
Dechlorane 605
Dechlorane Plus
Dechlorane Plus 515
Dechlorane Plus 2520

13560-99-1
$C_8H_4Cl_3O_3.Na$
277.46
Acetic acid, (2,4,5-trichlorophenoxy)-, sodium salt
(2,4,5-Trichlorophenoxy)acetic acid sodium salt
2,4,5-T sodium

13561-08-5
$C_{15}H_{18}O_4$
262.33
O(C1Cc(c(OCC(O2)C2)c(cc3)CC(O4)C4)c3)C1
Ether, 2,6-bis(2,3-epoxypropyl)phenyl 2,3-epoxy-

propyl
Bis(2,6-(2,3-epoxypropyl))-phenyl glycidyl ether
2,6-Di(2,3-epoxypropyl)phenyl 2,3-epoxypropyl

13567-11-8
$C_{39}H_{54}H_{10}O_{12}S$
757.09
γ-Amanitine

13573-18-7
$H_3Na_2O_{10}P_3$
301.92
Triphosphoric acid, sodium salt (8CI,9CI)

13586-68-0
$C_{14}H_{18}O$
202.30
O=CC(=Cc(ccc(c1)C(C)(C)C)c1)C
2-Propenal, 3-(4-(1,1-dimethylethyl)phenyl)-2-methyl- (9CI)
Cinnamaldehyde, p-tert-butyl-α-methyl-

13586-82-8
$C_8H_{17}CoO_2$
204.16
Hexanoic acid, 2-ethyl-, cobalt salt (9CI)
Cobalt 2-ethylhexanoate

13586-84-0
$C_{18}H_{36}O_2.xCo$
Unknown
Octadecanoic acid, cobalt salt (9CI)
Cobalt octadecanoate

13588-28-8
$C_7H_{16}O_3$
148.20
1-Propanol, 2-(2-methoxypropoxy)- (8CI,9CI)

13590-71-1
CH_5O_3P
96.03
[H]P(O)(=O)OC
Phosphonic acid, monomethyl ester
Methylester kyseliny fosforite (Czech)
Methylfosfit (Czech)
Methylfosfonat (Czech)
Methyl phosphite
Methyl phosphonate
Monomethylfosfit (Czech)

13590-97-1
$C_{13}H_{29}N_3.ClH$
263.91
Guanidine, dodecyl-, monohydrochloride
Dodecylguanidine hydrochloride

13593-03-8
$C_{12}H_{15}N_2O_3PS$
298.32
CCOP(=S)(OCC)Oc2cnc1ccccc1n2
Phosphorothioic acid, O,O-diethyl O-(2-quinoxalinyl) ester
Bay 5821
Bay 77049
Bayrusil
Chinalphos
Chinoin
O,O-Diaethyl-O-(chinoxalyl-(2))-monothiophosphat (German)
Diethquinalphion
Diethquinalphione
O,O-Diethyl O-(2-chinoxalyl)-phosphorothioate
O,O-Diethyl O-quinoxalin-2-yl phosphorothioate
O,O-Diethyl-O-(2-quinoxalinyl) phosphorothioate
O,O-Diethyl-O-(2-quinoxalyl) phosphorothioate
O,O-Diethyl-O-2-quinoxalyl-thiophosphate
Ekalux
Ekalux 25EC
ENT 27,394
NSC-190986
Quinalphos
S-6538
San 6538 I
San 6626 I
Sandoz 6538
Spencer S-6538
SRA 7312
Wie Oben

13596-41-3
$Cl_{10}N_5P_5$
579.40
1,3,5,7,9,2,4,6,8,10-Pentazapentaphosphecine, 2,2,4,4,6,6,8,8,10,10-decachloro-2,2,4,4,6,6,8,8,10,10-decahydro- (8CI,9CI)
1,3,5,7,9,2,4,6,8,10-Pentazapentaphosphecine, 2,2,4,4,6,6,8,8,10,10-decachloride
Phosphonitrile chloride, cyclic pentamer

13596-46-8
$Co.2H_3N.2H_2O_4S$
289.15
Sulfuric acid, ammonium cobalt(2+) salt (2:2:1) (9CI)
Cobaltous ammonium sulfate

13597-44-9
$H_2O_3S.Zn$
147.47
Sulfurous acid, zinc salt (1:1) (9CI)

13597-99-4
BeN_2O_6
133.03
Beryllium nitrate [UN 2464]
Beryllium dinitrate
Nitric acid, beryllium salt
UN 2464 [Beryllium nitrate]

13598-00-0
$Be.1/2H_4O_4Si$
57.07
Phenakite (8CI,9CI)
Phenacite

13598-15-7
$BeHO_4P$
104.99
Beryllium hydrogen phosphate (1:1)
Beryllium phosphate ($BeHPO_4$)
Phosphoric acid, beryllium salt (1:1)
Phosphorous acid, beryllium salt

13598-36-2
H_3O_3P
82.00
Phosphonic-acid
Orthophosphorus acid
Phosphorous acid
Phosphorous acid, ortho [UN 2834]
Phosphorus trihydroxide
Trihydroxyphosphine
UN 2834 [Phosphorous acid, ortho]

13598-37-3
$H_3O_4P.1/2Zn$
130.68
Phosphoric acid, zinc salt (2:1) (9CI)
Zinc dihydrogen phosphate
Zinc orthophosphate, dihydrogen

13598-51-1
H_3O_3PS
114.06
Phosphorothioic-acid
Monothiophosphoric acid
Thiophosphoric acid

13599-69-4
$C_{14}H_{13}NO_2$
227.27
Carbamic acid, methylphenyl-, phenyl ester (9CI)

Carbanilic acid, N-methyl,-
phenyl ester (6CI,7CI,8CI)
Phenyl methylphenyl-
carbamate
Phenyl N-methyl-N-phenyl-
carbamate

13601-19-9
$C_6FeN_6.4Na$
303.88
Sodium ferrocyanide
AI3-28762
Ferrate(4-), hexacyano-, tetra-
sodium
Ferrate(4-), hexakis(cyano-C)-,
tetrasodium, (OC-6-11)-
(9CI)
Sodium hexacyanoferate(II)
Sodium hexacyanoferrate (II)
Sodium Prussiate Yellow
Tetrasodium ferrocyanide
Tetrasodium hexacyanoferrate
Tetrasodium hexacyanoferrate-
(4-)
Yellow Prussiate of Soda

13602-12-5
$C_6H_5NO_3$
139.10
O=C(O)c(ccn(=O)c1)c1
**4-Pyridinecarboxylic acid,
1-oxide (9CI)**
NSC-63044
4-Pyridinecarboxylic acid
1-oxide

13603-07-1
$C_6H_{12}N_2O$
128.20
CC1CCCN(C1)N=O
3-Pipecoline, 1-nitroso
3-Methylnitrosopiperidine
1-Nitroso-3-pipecoline

13608-87-2
$C_8H_5Cl_3O$
223.49
2,3,4-Trichloroacetophenone

13615-38-8
$C_{11}H_9NO_2$
187.19
**Naphthalene, 3-methyl-1-nitro-
(8CI,9CI)**

13618-93-4
$C_8H_{15}N$
125.21
**Indolizine, octahydro- (8CI,
9CI)**
1-Azabicyclo(4.3.0)nonane
δ-Coniceine
Indolizidine

13623-06-8
$C_{15}H_{34}N.CH_3O_4S$
339.53
**1-Dodecanaminium, N,N,N-
trimethyl-, methyl sulfate
(9CI)**
Ammonium, dodecyltrimethyl-,
methyl sulfate
Dodecyltrimethylammonium
methosulfate
Dodecyltrimethylammonium
methyl sulfate
NSC-176805
Trimethyldodecylammonium
methosulfate

13631-64-6
$C_{10}H_{10}ClN_3S$
239.73
**4-Thiazolecarboxamidine,
N-phenyl-, hydrochloride
(8CI)**
N-Phenyl-4-thiazolecarbox-
amidine hydrochloride

13637-71-3
$ClHO_4.1/2Ni$
129.81
Nickel perchlorate
Nickel diperchlorate
Nickel(2+) perchlorate
Perchloric acid, nickel(2+) salt
(9CI)

13637-76-8
$Cl_2O_8.Pb$
406.09
Perchloric acid, lead(2+) salt
Lead diperchlorate
Lead perchlorate [UN 1470]
Lead(2+) perchlorate
UN 1470 [Lead perchlorate,
solid or solution]

13647-35-3
$C_{20}H_{27}NO_3$
329.43
Trilostane
Androstane-2-carbonitrile,
4,5-epoxy-17-hydroxy-3-oxo-,
(2-α,4-α,5-α,17-β)-
5-α-Androstane-2-α-carbo-
nitrile, 4-α,5-epoxy-17-β-
hydroxy-3-oxo-
Androst-2-ene-2-carbonitrile,
4,5-epoxy-3,17-dihydroxy-,
(4α,5α,17β)-
4α,5-Epoxy-3,17β-dihydroxy-
5α-androst-2-ene-2-carbo-
nitrile
4-α-5-Epoxy-17-β-hydroxy-
3-oxo-5-α-androstane-
2-α-carbonitrile
(2-α,4-α,5-α,17-β)-4,5-Epoxy-
17-hydroxy-3-oxoandrostane-
2-carbonitrile
Modrenal
Trilostano (Spanish)
Trilostanum (Latin)
WIN 24450

13654-09-6
$C_{12}Br_{10}$
943.17
c(c(c(c(c1Br)Br)Br)Br)(c1c(c(c
(c(c2Br)Br)Br)Br)c2Br)Br
Adine 0102
Berkflam B 10
1,1'-Biphenyl, 2,2',3,3',4,4',
5,5',6,6'-decabromo- (9CI)
2,2',3,3',4,4',5,5',6,6'-Deca-
bromo-1,1'-biphenyl
Flammex B 10
Polybrominated biphenyls

13655-52-2
$C_{15}H_{23}NO_2$
249.39
CC(C)NCC(O)COc1ccccc1CC=C
**2-Propanol, 1-(o-allyl-
phenoxy)-3-(isopropylamino)**
1-(o-Allylphenoxy)-3-(isopropylamino)-2-propanol
Alfeprol (Russian)
Alprenolol
H 56/28
2-Propanol, 1-((1-methylethyl)-
amino)-3-(2-(2-propenyl)-
phenoxy)-

13669-70-0
$C_{17}H_{19}NO$
253.37
CN3CCOC(c1ccccc1)c2ccccc2C3
**1H-2,5-Benzoxazocine,
3,4,5,6,7-tetrahydro-
5-methyl-1-phenyl**
Fenazoxine
Nefopam

13673-92-2
$C_6H_4Cl_2O_2$
179.00
**1,2-Benzenediol, 3,5-dichloro-
(9CI)**
Pyrocatechol, 3,5-dichloro-
(8CI)

13674-05-0
$C_{11}H_{15}NO_3$
209.24
MMDA
DEA No. 7401
5-Methoxy-3,4-methylenedioxy-
amphetamine
β-Methoxy-α-methyl-4,5-
(methylenedioxy)phenethyl
amine
3-Methoxy-α-methyl-4,5-
methylenedioxyphenethyl-
amine
Phenethylamine, 3-methoxy-
α-methyl-4,5-(methylene-
dioxy)-
Phenethylamine, α-methyl-
3-methoxy-4,5-(methylene-

dioxy)-

13674-84-5
$C_9H_{18}Cl_3O_4P$
327.59
O=P(OC(CCl)C)(OC(CCl)C)OC(CCl)C
Phosphoric acid, tris(2-chloro-1-methylethyl) ester
Tris(2-chloroisopropyl)phosphate

13674-87-8
$C_9H_{15}Cl_6O_4P$
430.91
O=P(OC(CCl)CCl)(OC(CCl)CCl)OC(CCl)CCl
2-Propanol, 1,3-dichloro-, phosphate (3:1)
1,3-Dichloro-2-propanol phosphate (3:1)
Emulsion 212
Fosforan troj-(1,3-dwuchloro-izopropylowy) (Polish)
Fyrol FR 2
PF 38
Phosphoric acid tris(1,3-dichloro-2-propyl)ester
TCPP
TDCPP
Tri(β,β'-dichloroisopropyl)-phosphate
Tris(1-chloromethyl-2-chloroethyl)phosphate
Tris(1,3-dichloroisopropyl)-phosphate
Tris-(1,3-dichloro-2-propyl)-phosphate

13676-54-5
$C_{21}H_{14}N_2O_4$
358.34
O=C(N(c(ccc(c1)Cc(ccc(N(C(=O)C=C2)C2=O)c3)c3)c1)C(=O)C=4)C4
1H-Pyrrole-2,5-dione, 1,1'-(methylenedi-4,1-phenylene)-bis- (9CI)
4,4'-Biphenylmethanebismaleimide
Bismaleimide, 4,4'-diphenylmethane
N,N'-Bismaleimido-4,4'-diphenylmethane
4,4'-Bis(maleimido)diphenylmethane
4,4'-(N,N'-Bismaleimido)diphenylmethane
Bis(4-maleimidophenyl)methane
N,N'-4,4'-Diaminodiphenylmethanebismaleimide
p,p'-Dimaleimidodiphenylmethane
4,4'-Dimaleimidodiphenylmethane
Diphenylmethanebismaleimide
N,N'-p,p'-Diphenylmethanebismaleimide
N,N'-4,4'-Diphenylmethanebismaleimide
4,4'-Diphenylmethanebismaleimide
4,4'-Diphenylmethanedimaleimide
Maleimide, N,N'-(methylenedi-p-phenylene)di-
4,4'-Methylenebis(N-phenylenemaleimide)
4,4'-Methylenebis(phenylmaleimide)
p,p'-Methylenebis(N-phenylmaleimide)
4,4'-Methylenebis(N-phenylmaleimide)
4,4'-Methylenedianiline bismaleimide
Methylenedi-p-phenylene-N,N'-bismaleimide
N,N'-(Methylenedi-4,1-phenylene)bismaleimide
N,N'-(Methylenedi-p-phenylene)dimaleimide
NSC-44754

13679-75-9
C_7H_8OS
140.21
1-Propanone, 1-(2-thienyl)- (9CI)
NSC-76041
2-Propionylthiophene

13682-92-3
$C_2H_6AlNO_4$
135.05
Aluminum glycinate
Aluminum, (glycinato-N,O)dihydroxy-
Aluminum, (glycinato-N,O)dihydroxy-, (T-4)- (9CI)
Dihydroxyaluminum aminoacetate
(Glycinato)dihydroxyaluminum
(Glycinato-N,O)dihydroxyaluminum
(T-4)-(Glycinato-N,O)dihydroxyaluminum

13684-56-5
$C_{16}H_{16}N_2O_4$
300.34
CCOC(=O)Nc2cccc(OC(=O)Nc1ccccc1)c2
Carbamic acid, phenylcarbamoyloxyphenyl-, ethyl ester
3-(Aethoxycarbonylamino-phenyl)-N-phenyl-carbamat (German)
Bentanex
Betanal AM
Betanex
Carbanilic acid, m-carbaniloyloxy-, ethyl ester
Desmedipham
EP-475
3-Ethoxycarbonylaminophenyl-N-phenylcarbamate
Ethyl m-hydroxycarbanilate carbanilate (Ester)
Ethyl phenylcarbamoyloxyphenylcarbamate
Schering 38107
SN 38107

13684-63-4
$C_{16}H_{16}N_2O_4$
300.34
COC(=O)Nc2cccc(OC(=O)Nc1cccc(C)c1)c2
Carbanilic acid, m-hydroxy-, methyl ester, m-methylcarbanilate
Betanal
Carbamic acid, (3-methylphenyl)-3-((methoxycarbonyl)amino)phenyl ester (9CI)
EP-452
Fenmedifam
m-Hydroxycarbanilic acid methyl ester m-methylcarbanilate
3-Methoxycarbonylamino-phenyl N-3'-methylphenylcarbamate
3-Methoxycarbonyl-N-(3'-methylphenyl)-carbamat (German)
Methyl m-hydroxycarbanilate, m-methylcarbanilate
3-(Methylphenyl)carbamic acid 3-((methoxycarbonyl)amino)-phenyl ester
Methyl 3-(m-tolylcarbamoyloxy)phenylcarbamate
Phendipham
Phenmedipham
Phenmediphame
Schering-38584
SN 4075
SN-38584

13693-11-3
$H_2O_4S.1/2Ti$
122.03
Titanium disulfate
NA 1760
Sulfuric acid, titanium(4+) salt (2:1) (9CI)
Titaniium disulfate
Titanium sulfate
Titanium sulfate (1:2)
Titanium(IV) sulfate
Titanium(4+) sulfate
Titanium sulfate, basic
Titanium sulfate solution
Titanium sulfate solution (Containing not more than 45% sulfuric acid)

13698-16-3
$C_3H_5Cl_2NO_2$
157.98
Carbamic acid, dichloro-, ethyl ester (8CI,9CI)

13700-81-7
$C_{14}H_{10}Cl_4$
320.05
Benzene, 1,1'-(1,1,2,2-tetrachloro-1,2-ethanediyl)bis- (9CI)
Bibenzyl, α,α,α',α'-tetrachloro- (6CI,7CI,8CI)
1,2-Diphenyltetrachloroethane
Tetrachloro-1,2-diphenylethane
1,1,2,2-Tetrachloro-1,2-diphenylethane
Tolane tetrachloride

13701-59-2
BHO_2.1/2Ba
176.54
Barium metaborate
Boric acid, barium salt (8CI, 9CI)
Caswell No. 071
EPA Pesticide Chemical Code 011101

13703-52-1
$C_{12}H_{12}$
156.23
Benzene, 1,4-cyclohexadien-1-yl- (9CI)
Benzene, (1,4-cyclohexadien-1-yl)- (8CI)

13704-90-0
$C_{10}H_5Cl_4N_3$
308.98
1,3,5-Triazine, 2-chloro-4-phenyl-6-(trichloro methyl)- (9CI)
s-Triazine, 2-chloro-4-phenyl-6-(trichloromethyl)- (7CI,8CI)

13704-97-7
$C_{11}H_{10}ClN_3$
219.68
1,3,5-Triazine, 2-chloro-4-ethyl-6-phenyl- (9CI)
s-Triazine, 2-chloro-4-ethyl-6-phenyl- (7CI,8CI)

13705-05-0
$C_4H_3Cl_2N_3S$
196.06
n(c(nc(n1)SC)Cl)c1Cl
s-Triazine, 2,4-dichloro-6-(methylthio)
2,4-Dichloro-6-(methylthio)-s-triazine

13705-07-2
$C_{10}H_8ClN_3S$
237.71
1,3,5-Triazine, 2-chloro-4-(methylthio)-6-phenyl- (9CI)
s-Triazine, 2-chloro-4-(methylthio)-6-phenyl- (7CI,8CI)

13707-65-8
$C_{12}H_{10}O$.K
209.31
Potassium 2-phenylphenate
(1,1'-Biphenyl)-2-ol, potassium salt (9CI)
Caswell No. 658D
EPA Pesticide Chemical Code 064108
o-Phenylphenol potassium salt

13709-38-1
F_3La
195.90
F[La](F)F
Lanthanum fluoride (9CI)

13718-26-8
O_3V.Na
121.93
Vanadic acid, monosodium salt
Metawanadanem sodowym (Polish)
Sodium metavanadate
Sodium vanadate

13733-90-9
$C_5H_5Cl_2N_3S$
210.09
1,3,5-Triazine, 2,4-dichloro-6-(ethylthio)- (9CI)
s-Triazine, 2,4-dichloro-6-(ethylthio)- (6CI,8CI)

13733-91-0
$C_9H_8N_4S_2$
236.30
1,3,5-Triazine-2,4(1H,3H)-dithione, 6-(phenylamino)- (9CI)
AF (Accelerator)
s-Triazine-2,4-dithiol, 6-anilino- (8CI)
Zisnet AF
Zisnet AS

13733-93-2
$C_{10}H_7Cl_2N_3S$
272.14
1,3,5-Triazine, 2,4-dichloro-6-((4-methylphenyl)thio)- (9CI)
s-Triazine, 2,4-dichloro-6-(p-tolylthio)- (8CI)

13733-95-4
$C_8H_{10}Cl_2N_4S_2$
297.23
Carbamic acid, diethyldithio-, 4,6-dichloro-s-triazin-2-yl ester (7CI,8CI)
s-Triazine-2-thiol, 4,6-dichloro-, diethyldithiocarbamate (ester)

13733-96-5
$C_9H_{13}ClN_4OS_2$
292.81
Carbamic acid, diethyldithio-, 4-chloro-6-methoxy-s-triazin-2-yl ester (7CI,8CI)
s-Triazine-2-thiol, 4-chloro-6-methoxy-, diethyldithiocarbamate (ester)

13733-97-6
$C_{14}H_{15}ClN_4S_2$
338.88
Carbamic acid, diethyldithio-, 4-chloro-6-phenyl-s-triazin-2-yl ester (7CI,8CI)
s-Triazine-2-thiol, 4-chloro-6-phenyl-, diethyldithiocarbamate (ester)

13738-63-1
$C_{12}H_6Cl_2FNO_3$
302.09
Fc1cc(Cl)cc(Cl)c1Oc2ccc(cc2)N(=O)=O
Ether, 2,4-dichloro-6-fluorophenyl p-nitrophenyl
CFNP
2,4-Dichloro-6-fluorophenyl p-nitrophenyl ether
2,4-Dichloro-6-fluorophenyl-4'-nitrophenyl ether
MO-500

13744-79-1
$C_{13}H_8F_3N_3O_4$
327.22
Benzenamine, 2,4-dinitro-N-(4-(trifluoromethyl)-phenyl)- (9CI)
2,4-Dinitro-4'-(perfluoromethyl)diphenylamine
p-Toluidine, N-(2,4-dinitrophenyl)-α,α,α-trifluoro- (8CI)

13746-89-9
N_4O_{12}.Zr
339.26
Zirconium nitrate [UN 2728]
Dusicnan zirkonicity (Czech)
UN 2728 [Zirconium nitrate]

13747-73-4
$C_9H_{14}O$
138.21
Cyclohexanone, 2-(1-methylethylidene)- (9CI)
Cyclohexanone, 2-isopropyl-

idene- (6CI,7CI,8CI)
2-Isopropylidenecyclohexan-one

13749-94-5
C_3H_7NOS
105.15
N(O)=C(SC)C
Methomyl oxime
Ethanimidothioic acid, N-hydroxy-, methyl ester (9CI)
Methyl N-hydroxyethanimidothioate

13752-51-7
$C_9H_{16}N_2O_2S_2$
248.39
O(CCN(SC(N(CCOC1)C1)=S)C2)C2
Morpholine, 4-((morpholinothiocarbonyl)thio)
Accelerator OTOS
Cure-Rite 18
Morpholine, 4-((4-morpholinylthio)thioxomethyl)- (9CI)
4-((Morpholinothiocarbonyl)-thio)morpholine
OTOS
N-Oxydiethylene thiocarbamyl-N-oxydiethylene sulfenamide

13757-90-9
$C_{12}H_{22}O_2$
198.31
6,7-Dodecanedione (8CI,9CI)

13762-51-1
$BH_3.K$
52.94
Potassium borohydride [UN 1870]
Borate(1-), tetrahydro-, potassium (8CI,9CI)
Borohydrure de potassium (French)
UN 1870 [Potassium borohydride]

13764-49-3
$C_{25}H_{33}NO_4.ClH$
448.00
DEA No. 9059
6,14-endoEtheno-7-(2-hydroxy-2-pentyl)-tetrahydro-oripavine hydrochloride
6,14-endo-Ethenotetrahydro-oripavine, 7-α-(1-(R)-hydroxy-1-methylbutyl)-, hydrochloride
Etorphine hydrochloride
7-α-(1-(R)-Hydroxy-1-methylbutyl)-6,14-endoethenotetrahydro-oripavine hydrochloride
M 99 Reckitt
Propylorvinol hydrochloride

13765-19-0
$CrO_4.Ca$
156.08
Chromic acid, calcium salt (1:1)
Calcium chromate
Calcium chromate (VI)
Calcium Chrome Yellow
Calcium chromium oxide ($CaCrO_4$)
Calcium monochromate
C.I. 77223
C.I. Pigment Yellow 33
Gelbin
RCRA waste number U032
Yellow Ultramarine

13770-61-1
InN_3O_9
300.85
Indium-nitrate

13770-89-3
$H_4N_2NiO_6S_2$
250.89
Nickel (II) sulfamate

13770-96-2
$AlH_4.Na$
54.01
Aluminate (1-), tetrahydro-, sodium
Aluminate(1-), tetrahydro-, sodium, (T-4)- (9CI)
Aluminum sodium hydride
SAH 22
Sodium aluminum hydride [UN 2835]
Sodium aluminum tetrahydride
Sodium tetrahydroaluminate(1-)
UN 2835 [Sodium aluminum hydride]

13771-22-7
AlB_3H_{12}
71.51
Aluminum, tris(tetrahydroborato(1-)-H,H')-, (OC-6-11)- (9CI)
Borate(1-), tetrahydro-, aluminum (8CI) (VAN)

13778-36-4
BrClPb
322.56
Lead bromide chloride (8CI, 9CI)

13779-41-4
F_2HO_2P
101.98
Difluorophosphoric acid, Anhydrous [UN 1768]
Acide difluorophosphorique (French)
Acido difluofosforico (Spanish)
Difluorophosphoric acid
Phosphorodifluoridic acid (8CI,9CI)
UN 1768 [Difluorophosphoric acid, Anhydrous]

13780-03-5
$Ca.2H_2O_3S$
204.24
Calcium-bisulfite
Calcium bisulfite, Solution (DOT)
Calcium hydrogen sulfite, Solution (DOT)
Calcium hydrosulphite [UN 1923]
NA 2693 (DOT)
UN 1923 [Calcium dithionite (calcium hydrosulfite)]

13780-06-8
$Ca.2HNO_2$
134.10
Nitrous acid, calcium salt (9CI)
Calcium nitrite
Caswell No. 147
EPA Pesticide Chemical Code 076202

13796-22-0
$C_{22}H_{18}N_2O_2$
342.39
Oc(c(c(nc1Cc(c(nc(c2cc(c3)C)c3)C4)c2O)ccc5C)c5)c14
Quino(2,3-b)acridine-7,14-dione, 5,6,12,13-tetrahydro-2,9-dimethyl- (9CI)
Quinacridone, 6,13-dihydro-2,9-dimethyl-

13814-96-5
$B_2F_8.Pb$
380.81
Borate(1-), tetrafluoro-, lead (2+)
Lead fluoborate (DOT)
NA 2291 (DOT)

13822-56-5
$C_6H_{17}NO_3Si$
179.29
1-Propanamine, 3-(trimethoxysilyl)- (9CI)
(γ-Aminopropyl)trimethoxysilane
(3-Aminopropyl)trimethoxysilane
KBE 903
NSC-83845
Propylamine, 3-(trimethoxysilyl)-
SC 3900
Silane SC 3900
3-(Trimethoxysilyl)-1-propanamine
N-(Trimethoxysilylpropyl)amine

3-(Trimethoxysilyl)propylamine

13823-29-5
$H_4N_4O_{12}.Th$
484.08
Thorium(IV), nitrate
Nitric acid, thorium(4+) salt (8CI,9CI)
Thorium nitrate, Solid [UN 2976]
Thorium(4+) nitrate
Thorium tetranitrate
UN 2976 [Thorium nitrate, solid]

13825-86-0
Unknown
Unknown
Chromous(II) sulfate

13825-90-6
H_2Si
30.10
Silylene (9CI)

13826-35-2
$C_{13}H_{12}O_2$
200.25
O(c(cccc1CO)c1)c(cccc2)c2
Benzyl alcohol, m-phenoxy
Benzenemethanol, 3-phenoxy- (9CI)
3-(Hydroxymethyl)diphenyl ether
m-Phenoxybenzyl alcohol
3-Phenoxybenzyl alcohol
(3-Phenoxyphenyl)methanol

13826-65-8
$HNO_2.1/2Pb$
150.60
Lead nitrate
Nitrous acid, lead(2+) salt (8CI,9CI)

13826-83-0
BFH_4N
47.86
Ammonium fluoroborate (DOT)
Ammonium fluoroborate
Ammonium tetrafluoroborate
Ammonium tetrafluoroborate-(1-)
Borate(1-), tetrafluoro-, ammonium
NA 9088 (DOT)

13828-37-0
$C_{10}H_{20}O$
156.27
OCC(CCC(C1)C(C)C)C1
Cyclohexanemethanol, 4-(1-methylethyl)-, cis- (9CI)
p-Methan-7-ol, (cis)-
cis-4-(1-Methylethyl)cyclohexanemethanol

13837-71-3
$C_{10}H_{18}$
138.25
Cyclohexane, 1-methylene-3-(1-methylethyl)-, (R)-(9CI)
m-Menth-1(7)-ene, (R)-(-)-(8CI)

13837-95-1
$C_{10}H_{16}$
136.24
Cyclohexane, 1-methylene-3-(1-methylethenyl)-, (R)-(9CI)
m-Mentha-1(7),8-diene, (R)-(-)- (8CI)

13838-16-9
$C_3H_2ClF_5O$
184.50
FC(F)OC(F)(F)C(F)Cl
Ether, 2-chloro-1,1,2-trifluoroethyl difluoromethyl
Anesthetic Compound No. 347
2-Chloro-1-(difluoromethoxy)-1,1,2-trifluoroethane
2-Chloro-1,1,2-trifluoroethyl difluoromethyl ether
Compound 347
Enflurane (ACGIH)
Ethane, 2-chloro-1-(difluoromethoxy)-1,1,2-trifluoro-
Ethrane
Methylflurether
NSC-115944
Ohio 347

13838-29-4
$C_{15}H_{11}ClN_4O$
298.73
1,3,5-Triazin-2-amine, 4-chloro-6-phenoxy-N-phenyl- (9CI)
s-Triazine, 2-anilino-4-chloro-6-phenoxy-(8CI)

13838-32-9
$C_7H_9Cl_2N_3O$
222.06
1,3,5-Triazine, 2-butoxy-4,6-dichloro- (9CI)
s-Triazine, 2-butoxy-4,6-dichloro-(8CI)

13838-34-1
$C_{10}H_7Cl_2N_3O$
256.07
1,3,5-Triazine, 2,4-dichloro-6-(4-methylphenoxy)- (9CI)
s-Triazine, 2,4-dichloro-6-(p-tolyloxy)- (8CI)

13840-33-0
ClHO.Li
59.40
Lithium hypochlorite
Caswell No. 528A
EPA Pesticide Chemical Code 014702
Hipoclorito de litio (Spanish)
Hypochlorite de lithium (French)
Hypochlorous acid, lithium salt (8CI,9CI)
Hypochlorous acid, lithium salt, Mixture with sodium chloride [UN 1471]
Lithium hypochlorite compound, Dry, Containing more than 39% available chlorine [UN 1471]
Lithium hypochlorite, Dry [UN 1471]
UN 1471 [Lithium hypochlorite, Dry or Lithium hypochlorite mixtures, Dry]

13843-59-9
$BrHO_3.H_3N$
145.93
Ammonium bromate
Bromate d'ammonium (French)
Bromato amonico (Spanish)
Bromic acid, ammonium salt

13843-81-7
Unknown
Unknown
Lithium dichromate

13845-12-0
Al_2Cl_6
266.68
Aluminum hexachloride
Aluminum, di-mu-chlorotetrachlorodi- (9CI)
Di-mu-chlorotetrachlorodialuminum

13845-35-7
$H_2O_4Te.Pb$
400.81
Lead tellurite
Lead tellurate
Telluric acid, lead(2+) salt (1:1) (8CI,9CI)

13849-96-2
$C_{30}H_{52}$
412.74
A'-Neogammacerane, (17α)-(9CI)
A'-Neo-17α-gammacerane (8CI)

13861-97-7
$C_6H_{10}O_2$

114.16
2(3H)-Furanone, dihydro-4,4-dimethyl
Dihydro-4,4-dimethyl-2(3H)-furanone
β,β-Dimethylbutylrolacton (German)
4,4-Dimethylbutyrolactone

13861-99-9
$C_8H_{13}NO_2$
155.19
2,5-Pyrrolidinedione, 3-ethyl-1,3-dimethyl- (9CI)
Succinimide, 2-ethyl-N,2-dimethyl- (8CI)

13863-31-5
$C_{38}H_{36}N_{12}O_8S_2.2Na$
898.96
CN(CCO)c6nc(Nc1ccccc1)nc(Nc5ccc(C=Cc4ccc(Nc3nc(Nc2ccccc2)nc(n3)N(C)CCO)cc4S(O)(=O)=O)c(c5)S(O)(=O)=O)n6
2,2'-Stilbenedisulfonic acid, 4,4'-bis((4-anilino-6-((2-hydroxyethyl)methylamino)-s-triazin- 2-yl)-amino)-, disodium salt
Tinopal 5BM

13863-41-7
BrCl
115.36
Bromine chloride [UN 2901]
Bromine monochloride
Bromochloride
UN 2901 [Bromide chloride]

13863-88-2
AgN_3
149.90
Silver-azide
Silver azide, Dry (DOT)

13867-27-1
$C_{12}H_6Cl_2N_2O_5$
329.08
Benzene, 1,1'-oxybis(2-chloro-4-nitro- (9CI)
Ether, bis(2-chloro-4-nitrophenyl) (8CI)

13877-91-3
$C_{10}H_{16}$
136.24
C(=CCC=C(C=C)C)(C)C
1,3,6-Octatriene, 3,7-dimethyl- (9CI)
3,7-Dimethyl-1,3,6-octatriene

13877-93-5
$C_{15}H_{24}$
204.36
Caryophyllene

13878-54-1
$C_{12}H_{20}N_2S_4Zn$
385.95
Zinc, bis(1-piperidinecarbodithioato)
Vulkacit ZP
Zinc, bis(1-piperidinecarbodithioato-S,S')-, (T-4)- (9CI)
Zinc pentamethylenedithiocarbamate

13882-55-8
$C_6H_9ClN_4O$
188.62
1,3,5-Triazin-2-amine, 4-chloro-6-methoxy-N,N-dimethyl- (9CI)
s-Triazine, 2-chloro-4-(dimethylamino)-6-methoxy- (7CI,8CI)

13882-61-6
$C_5H_9N_5O_2$
171.16
1,3,5-Triazin-2(1H)-one, 4,6-dimethoxy-, hydrazone (9CI)
s-Triazine, 2-hydrazino-4,6-dimethoxy- (6CI,7CI, 8CI)

13886-99-2
$C_{28}H_{55}P$
422.72
P(C(CC1)CCCC2)(C12)CCCCCCCCCCCCCCCCCCCC
9-Phosphabicyclo(4.2.1)-nonane, 9-eicosyl- (9CI)
9-Eicosyl-9-phosphabicyclo-(4.2.1)nonane

13887-00-8
$C_{28}H_{55}P$
422.72
P(C(CCC1)CCC2)(C12)CCCCCCCCCCCCCCCCCCCC
9-Phosphabicyclo(3.3.1)nonane, 9-eicosyl- (9CI)
9-Eicosyl-9-phosphabicyclo-(3.3.1)nonane

13887-02-0
$C_8H_{15}P$
142.18
P(C(CCC1)CCC2)C12
9-Phosphabicyclo(3.3.1)nonane (9CI)

13889-92-4
C_4H_7ClOS
138.62
O=C(SCCC)Cl
Formic acid, chlorothio-, S-propyl ester
S-Propyl chlorothioformate

13893-53-3
$C_6H_{12}N_2$
112.16
Butanenitrile, 2-amino-2,3-dimethyl- (9CI)
Butyronitrile, 2-amino-2,3-dimethyl- (8CI)

13898-68-5
$C_{13}H_{19}NO_3$
237.33
Benzamide, N,N-diethyl-3-ethoxy-4-hydroxy
Anacardiol
DEHB
Diethylamide of 3-ethoxy-4-hydroxy-benzoic acid
3-Ethoxy-N,N-diethyl-4-hydroxybenzamide
3-Ethoxy-4-hydroxy-diethylbenzamide
4-Hydroxy-3-ethoxy-benzoic acid diethylamide

13907-45-4
CrO_4
115.99
Chromate (9CI)

13907-47-6
Cr_2O_7
215.99
Chromate (8CI,9CI)
Chromic acid, ion(2-)

13909-09-6
$C_{10}H_{18}ClN_3O_2$
247.76
CC1CCC(CC1)NC(=O)N(CCCl)N=O
Urea, 1-(2-chloroethyl)-3-(4-methylcyclohexyl)-1-nitroso-, (E)
1-(2-Chloroethyl)-3-(4-methylcyclohexyl)-1-nitrosourea
1-(2-Chloroethyl)-3-(trans-4-methyl-cyclohexyl)-1-nitrosourea
N-(2-Chloroethyl)-N'-(trans-4-methylcyclohexyl)-N-nitrosourea
ICIG 1110
Lomustine, methyl-
Me-CCNU
Methyl-CCNU
trans-Methyl-CCNU
NCI-C04955
NSC-95441
Semustine
Urea, N-(2-chloroethyl)-N'-(4-methylcyclohexyl)-N-nitroso-
Urea, 1-(2-chloroethyl)-3-(4-methylcyclohexyl)-1-nitroso-, trans-

13925-00-3
$C_6H_8N_2$
108.16
n(ccnc1CC)c1
Pyrazine, ethyl
Ethylpyrazine
2-Ethylpyrazine

13925-03-6
$C_7H_{10}N_2$
122.16
Pyrazine, 2-ethyl-6-methyl- (8CI,9CI)

13925-06-9
$C_9H_{14}N_2$
150.21
n(c(c(nc1)CC(C)C)C)c1
Pyrazine, 2-methyl-3-(2-methylpropyl)- (9CI)
2-Isobutyl-3-methylpyrazine
2-Methyl-3-(2-methylpropyl)-pyrazine

13925-07-0
$C_8H_{12}N_2$
136.22
n(c(c(nc1C)C)CC)c1
Pyrazine, 3,5-dimethyl-2-ethyl
2-Ethyl-3,5-dimethyl pyrazine
FEMA 3150
Pyrazine, 2,6-dimethyl-3-ethyl-

13925-08-1
$C_7H_8N_2$
120.14
n(c(C=C)cnc1C)c1
Pyrazine, 2-ethenyl-5-methyl- (9CI)
2-Ethenyl-5-methylpyrazine
2-Methyl-5-vinylpyrazine

13927-77-0
$C_{18}H_{36}N_2S_4.Ni$
467.51
CCCCN(CCCC)C(=S)S[Ni]SC(=S)N(CCCC)CCCC
Nickel, bis(dibutyldithiocarbamato)
Bis(dibutyldithiocarbamato)-nickel
Carbamic acid, dibutyldithio-, nickel salt
Dibutyldithiocarbamic acid, nickel salt
Nickel dibutyldithiocarbamate
UV Chek AM 104
Vanguard N

13931-79-8
ClSn
154.16
Tin chloride (8CI,9CI)

13932-13-3
$H_2O_6S_4.2K$
304.47
Potassium tetrathionate
Caswell No. 702A
EPA Pesticide Chemical Code 075903
Tetrathionic acid, dipotassium salt (8CI,9CI)

13936-21-5
$C_{19}H_{18}O_2$
278.35
O=C(c(c(C(=O)c1cccc2)ccc3CCCCC)c3)c12
9,10-Anthracenedione, 2-pentyl- (9CI)
2-Amylanthraquinone
2-Pentyl-9,10-anthracenedione

13952-84-6
$C_4H_{11}N$
73.16
NC(CC)C
sec-Butylamine
2-AB
2-Aminobutane
(RS)-2-Aminobutane
2-Aminobutane Base
Butafume
2-Butanamine
(RS)-sec-Butylamine
Deccotane
Frucote
Propylamine, 1-methyl
Tutane

13954-62-6
$C_{14}H_{10}N_4O_6S_2.2Cl$
465.29
Benzenediazonium, 4,4'-(1,2-ethenediyl)bis(3-sulfo-, dichloride (9CI)
2,2'-Disulfo-4,4'-stilbenetetrazonium dichloride
NSC-408471
4,4'-Stilbenebis(diazonium), 2,2'-disulfo-, dichloride

13956-29-1
$C_{21}H_{30}O_2$
314.51
Resorcinol, 2-p-mentha-1,8-dien-3-yl-5-pentyl-, (-)-(E)
1,3-Benzenediol, 2-(3-methyl-6-(1-methylethenyl)-2-cyclohexen-1-yl)-5-pentyl-, (1R-trans)-
Cannabidiol
(-)-Cannabidiol
(-)-trans-Cannabidiol
CBD
(-)-trans-2-p-Mentha-1,8-dien-3-yl-5-pentylresorcinol

13960-26-4
$C_{11}H_{10}Cl_2N_4O_2$
301.13
s-Triazine, 2-(3,4-dichloroanilino)-4,6-dimethoxy- (8CI)

13960-29-7
$C_{11}H_9Cl_3N_4O_2$
335.58
s-Triazine, 2,4-dimethoxy-6-(2,4,5-trichloroanilino)- (8CI)

13960-31-1
$C_{24}H_{21}N_3$
351.46
1,3,5-Triazine, 2,4,6-tris-(phenylmethyl)- (9CI)
s-Triazine, 2,4,6-tribenzyl- (6CI,7CI,8CI)
Tribenzyl-s-triazine

13960-33-3
$C_{21}H_{14}N_4O_2$
354.37
1,3,5-Triazine, 2-(4-nitrophenyl)-4,6-diphenyl- (9CI)
s-Triazine, 2-(p-nitrophenyl)-4,6-diphenyl- (7CI,8CI)

13961-86-9
$C_{18}H_{34}O_2.C_4H_{11}NO_2$
387.68
Oleic acid, Compd. with 2,2'-iminodiethanol (1:1)
Diethanolamine oleate
Diethanolammonium oleate
NCI-C55334
9-Octadecenoic acid (Z)-, Compd. with 2,2'-iminobis(ethanol) (1:1)
Oleic acid diethanolamine (1:1)

13963-57-0
$C_{15}H_{21}AlO_6$
324.34
Aluminum, tris(2,4-pentanedionato)
Aluminum acetylacetonate
Aluminum(III) acetylacetonate
Aluminum triacetylacetonate
Aluminum tris(acetylacetonate)
Aluminum, tris(2,4-pentanedionato-O,O')-, (OC-6-11)- (9CI)
Tris(acetylacetonato)aluminum
Tris(acetylacetone)aluminum
Tris(acetylacetonyl)aluminum
Tris(2,4-pentanedionato)-aluminum
Tris(2,4-pentanedione)-aluminum

13964-21-1
$C_{19}H_{26}O_8S$
414.48
α-D-Allofuranose, 1,2:5,6-bis-O-(1-methylethylidene)-, 4-methylbenzenesulfonate (9CI)
Allofuranose, 1,2:5,6-di-O-isopropylidene-, p-toluenesulfonate, α-D- (8CI)

13966-86-4
CdCl
147.86
Cadmium chloride (8CI,9CI)

13967-48-1
Ru
106
Ruthenium, Isotope of mass 106 (8CI,9CI)

13967-63-0
Sc
46
Scandium, Isotope of mass 46 (8CI,9CI)

13967-70-9
Cs
134
Cesium, Isotope of mass 134 (8CI,9CI)

13967-71-0
Zr
95
Zirconium, Isotope of mass 95 (8CI,9CI)

13967-74-3
Ce
141
Cerium, Isotope of mass 141 (8CI,9CI)

13968-50-8
Sb
127
Antimony, Isotope of mass 127 (8CI,9CI)

13972-68-4
$CdMoO_4$
272.35
Cadmium molybdenum oxide (9CI)

13973-87-0
Unknown
Unknown
Bromine azide (DOT)
Bromine nitride
Nitrogen bromide

13973-88-1
ClN_3
77.45
Chlorine azide
Azida de cloro (Spanish)
Azoture de chlore (French)
Nitrogen chloride

13980-00-2
$C_6H_{11}N_7O_7$
293.15
1,3,5,7-Tetrazocine, 1-acetyloctahydro-3,5,7-trinitro- (8CI,9CI)
QDX
SEX (Explosive)

13981-16-3
Pu
238
Plutonium, Isotope of mass 238 (8CI,9CI)

13981-28-7
La
140
Lanthanum, Isotope of mass 140 (8CI,9CI)

13981-52-7
Unknown
Unknown
Polonium-210
Radium F

13982-39-3
Zn
65
Zinc, Isotope of mass 65 (8CI,9CI)

13982-63-3
Ra
226
Radium, Isotope of mass 226 (8CI,9CI)

13983-17-0
CaH_2O_3Si
118.19
Wollastonite
Aedelforsite
Cab-O-Lite
Cab-O-Lite 100
Cab-O-Lite 130
Cab-O-Lite 160
Cab-O-Lite F 1
Dab-O-Lite P 4
Casiflux
Casiflux VP 413-004
F 1
FW 50
FW 325
FW 200 (Mineral)
Gillebachite
NCI-C55470
NYAD
NYAD 10
NYAD 325
NYAD G
NYCOR
NYCOR 200
NYCOR 300
Okenite
Rivaite
Schalstein
Tabular Spar
Tremin
Vansil
Vansil W 10
Vansil W 20
Vansil W 30
Vilnite
Wollastokup

13986-18-0
$F_2Zn.4H2O$
175.45
Zinc fluoride, tetrahydrate (8CI,9CI)
Zinc difluoride tetrahydrate

13987-01-4
C_9H_{18}
126.27
Propene, trimer
Propylene trimer
Tripropylene [UN 2057]
UN 2057 [Tripropylene]

13988-26-6
$C_{12}H_{12}O_5$
236.24
O=C(OCCOCCOC(=O)c1cccc2)c12
Phthalic acid, cyclic oxydiethylene ester
2,5,8-Benzotrioxacycloundecin-1,9-dione, 3,4,6,7-tetrahydro- (9CI)
Diethylene glycol bisphthalate
Howflex GBP

13991-37-2
$C_5H_8O_2$
100.12
2-Pentenoic acid, (E)- (8CI,9CI)

13998-73-7
$C_{42}H_{85}N_3O_2$
664.14
Octadecanamide, N,N'-(iminodi-3,1-propanediyl)bis- (9CI)
N,N'-Iminobis(dipropylenedistearamide)

14002-21-2

C_3H_9NO
75.10
OCN(C)C
Methanol, (dimethylamino)- (9CI)
(Dimethylamino)methanol

14003-66-8
$C_4H_5N_3O_2$
127.12
O=N(=O)C(N=CN1)=C1C
Imidazole, 4-methyl-5-nitro
Imidazole, 5-methyl-4-nitro-
4-Methyl-5-nitro-1H-imidazole (9CI)

14009-71-3
$C_{10}H_{18}O_2$
170.25
2H-Pyran-3-ol, 6-ethenyltetrahydro-2,2,6-trimethyl-, cis- (9CI)
Linalool oxide D
2H-Pyran-3-ol, tetrahydro-2,2,6-trimethyl-6-vinyl-, cis- (8CI)

14010-23-2
$C_{19}H_{38}O_2$
298.51
Heptadecanoic acid, ethyl ester (8CI,9CI)
NSC-137831

14017-41-5
$Co.2H_3NO_3S$
156.03
Cobaltous sulfamate
Sulfamic acid, cobalt(2+) salt (2:1) (8CI,9CI)

14018-95-2
$Cr_2O_7.Zn$
281.37
Dichromic acid, zinc salt (1:1)
Zinc bichromate
Zinc chromate (ACGIH,OSHA)
Zinc chromium oxide
Zinc dichromate
Zinc dichromate (VI)

14024-48-7
$C_{10}H_{14}CoO_4$
257.15
CC(=O)C=C(C)O[Co]OC(C)=CC(C)=O
Cobalt, bis(2,4-pentanedionato)- (8CI)
Acetylacetone cobalt(II)
Bis(acetylacetonato)cobalt(II)
Bis(acetylacetonyl)copper
Bis(2,4-pentanedionato)cobalt
(T-4)-Bis(2,4-pentanedionato-O,O')cobalt
Cobalt(II) acetylacetonate
Cobalt bis(acetylacetonate)
Cobalt(II) bis(acetylacetonate)
Cobalt, bis(2,4-pentanedionato-O,O')-, (T-4)- (9CI)
Cobalt diacetylacetonate
Cobaltous acetylacetonate
NSC-4652

14024-63-6
$C_{10}H_{14}O_4Zn$
263.61
Zinc, bis(2,4-pentanedionato-O,O')
Zinc acetoacetonate

14024-64-7
$C_{10}H_{14}O_5Ti$
262.14
Titanium, oxobis(2,4-pentanedionato)
Bis(acetylacetonato)titanium oxide
Bis(2,4-pentanedionato)titanium oxide
Titanium acetonyl acetonate
Titanium oxide bis(acetylacetonate)
Titanium, oxobis(2,4-pentanedionato-O,O')
Titanyl bis(acetylacetonate)

14025-15-1
$C_{10}H_{12}CuN_2O_8.2Na$
397.76
Cuprate(2-), ((ethylenedinitrilo)tetraacetato)-, disodium
Disodium cupric EDTA
EDTA disodium copper salt
((Ethylenedinitrilo)tetraacetato)-cuprate(2-) disodium

14025-21-9
$C_{10}H_{12}N_2O_8Zn.2Na$
399.57
Zincate(2-), ((N,N'-1,2-ethanediylbis(N-(carboxymethyl)-glycinato))(4-))-, disodium
AI3-19668
(Ethylenedinitrilo)tetraacetic acid, disodium zinc salt
Sodium zinc EDTA
Zincate(2-), ((N,N'-1,2-ethanediylbis(N-(carboxymethyl)-glycinato))(4-)-N,N',O,O',ON,ON')-, disodium, (OC-6-21)- (9CI)

14027-78-2
$C_{16}H_{30}O_4$
286.41
Hexanedioic acid, dipentyl ester (9CI)
Adipic acid, dipentyl ester (8CI)

14031-86-8
$C_{12}H_{24}$
168.32
C(=C)(CC(CC(C)(C)C)(C)C)C
1-Heptene, 2,4,4,6,6-pentamethyl- (9CI)
2,4,4,6,6-Pentamethylheptene-1
2,4,4,6,6-Pentamethyl-1-heptene

14038-43-8
$C_6FeN_6.4/3Fe$
286.38
Ferric ferrocyanide
C-Blau 17 (Germany)
C.I. 77510
C.I. 77520
Chinese Blue
Ferrate(4-), hexacyano-, iron(3+) (3:4) (8CI)
Ferrate(1-), hexakis(cyano-C)di-
Ferrate(4-), hexakis(cyano-C)-, iron(3+) (3:4), (OC-6-11)- (9CI)
Ferric hexacyanoferrate (II)
Germany: C-Blau 17
Iron Blue
Iron cyanide
Iron ferrocyanide
Iron(III) ferrocyanide
Iron(3+) ferrocyanide
Iron hexacyanoferrate
Milori Blue
NSC-8665
Paris Blue
Pigment Blue 27
Prussian Blue
Tetrairon tris(hexacyanoferrate)
Tetrairon(3+) tris(hexacyanoferrate(4-))

14043-38-0
$C_{21}H_{12}N_6O_6$
444.37
s-Triazine, 2,4,6-tris-(m-nitrophenyl)- (6CI, 8CI)

14047-09-7
$C_{12}H_6Cl_4N_2$
320.00
Clc2ccc(N=Nc1ccc(Cl)c(Cl)c1)cc2Cl
Azobenzene, 3,3',4,4'-tetrachloro
Diazene, bis(3,4-dichlorophenyl)- (9CI)
TCAB
3,3',4,4'-Tetrachloroazobenzene
3,4,3',4'-Tetrachloroazobenzene

14047-56-4
$C_4H_6O_4.xNa$
Unknown
Butanedioic acid, sodium salt (9CI)
Sodium butanedioate

14049-11-7

$C_{10}H_{18}O_2$
170.25
O(C(C=C)(CCC1O)C)C1(C)C
2H-Pyran-3-ol, 6-ethenyltetrahydro-2,2,6-trimethyl- (9CI)
3-Hydroxy-2,2,6-trimethyl-6-vinyltetrahydropyran
2H-Pyran-3-ol, 6-ethenyl-tetrahydro-2,2,6-trimethyl-
2H-Pyran-3-ol, tetrahydro-2,2,6-trimethyl-6-vinyl-

14055-02-8
$C_{32}H_{16}N_8Ni$
571.17
Nickel, (phthalocyaninato(2-))- (8CI)
Nickel, (29H,31H-phthalocyaninato(2-)-N29,N30,N31,N32)-, (SP-4-1)- (9CI)
Nickel phthalocyanine
Nickel(II) phthalocyanine
Nickel phthalocyanine blue
NSC-173214
(Phthalocyaninato(2-))nickel
2,1-Phthalocyaninato nickel

14064-03-0
$C_4H_{10}O_2.1/2Mg$
102.28
Ethanol, 2-ethoxy-, magnesium salt (9CI)
2-Ethoxyethanol magnesium salt
Magnesium ethoxyethoxide

14064-48-3
$C_{15}H_{14}$
194.28
Benzene, 1-methyl-3-(2-phenylethenyl)-, (E)- (9CI)
trans-o-Methylstilbene
Stilbene, 3-methyl-, (E)- (8CI)

14072-86-7
C_8H_{14}
110.20
Cyclohexene, 4,4-dimethyl- (8CI,9CI)
NSC-134990

14073-00-8
$C_{10}H_8N_2O_3$
204.20
Cc2cn(=O)c1ccccc1c2N(=O)=O
Quinoline, 3-methyl-4-nitro-, 1-oxide
3-Methyl-4-nitroquinoline 1-oxide

14090-88-1
$C_9H_{16}O_2$
156.22
O=C(CC(=O)CCC)CCC
4,6-Nonanedione (9CI)

14097-03-1
$C_{19}H_{25}ClN_5O_2$
390.94
O=N(=O)c(ccc(N=Nc(ccc(N(CCN(C)(C)C)CC)c1)c1)c2Cl)c2
Ethanaminium, 2-((4-((2-chloro-4-nitrophenyl)-azo)phenyl)ethylamino)-N,N,N-trimethyl
Aizen Cathilon Red GTLH
Astrazon Red GTL
Ammonium, (2-(p-((2-chloro-4-nitrophenyl)azo)phenethyl-amino)ethyl)trimethyl-
Basic Red 18
C.I. 11085
C.I. Basic Red 18
Diacryl Supra Red GTL
Novacryl Red 2G
Red GTL
Sevron Red GL
Sumiacryl Red G
Sumiacryl Red GT
Synacryl Fast Red 2G
Synacryl Red 2G

14104-85-9
Unknown
Unknown
Magnesium dichromate

14112-00-6
$C_8H_{10}Cl_2$
177.07
Cyclobutane, 1,2-dichloro-1,2-diethenyl- (9CI)
Cyclobutane, 1,2-dichloro-1,2-divinyl- (8CI)
1,2-Dichloro-1,2-divinyl-cyclobutane

14112-98-2
$C_8H_{14}O_3$
158.20
7-Ketooctanoic acid
7-Oxooctanoic acid

14117-96-5
$C_{44}H_{78}O_4$
671.10
O=C(OCCCCCCCCCCCCCCCCCC)c(c(ccc1)C(=O)OCCCCCCCCCCCCCCCCCC)c1
1,2-Benzenedicarboxylic acid, dioctadecyl ester (9CI)
Dioctadecyl 1,2-benzenedicarboxylate
Distearyl phthalate
Phthalic acid, distearyl ester

14119-15-4
Mo
99
Molybdenum, Isotope of mass 99 (8CI,9CI)

14124-67-5
O_3Se
126.96
Selenite (9CI,8CI)

14124-68-6
O_4Se
142.96
Selenate (9CI,8CI)

14128-61-1
$C_{13}H_{18}O$
190.29
2-Hexanone, 5-methyl-5-phenyl- (8CI,9CI)

14129-82-9
$C_4H_3Cl_5$
228.33
Propene, 1,3,3-trichloro-2-(dichloromethyl)- (8CI)

14130-06-4
$C_{22}H_{47}N$
325.62
NCCCCCCCCCCCCCCCCCCCCCC
1-Docosanamine (9CI)

14133-76-7
Tc
99
Technetium, Isotope of mass 99 (8CI,9CI)

14143-60-3
$C_6H_2Cl_3N_3$
222.44
N#Cc(nc(c(c1N)Cl)Cl)c1Cl
2-Pyridinecarbonitrile, 4-amino-3,5,6-trichloro- (9CI)
4-Amino-3,5,6-trichloro-2-pyridinecarbonitrile

14147-71-8
$CH_5N.CHN_3O_6$
182.07
Methylamine nitroform
Methanamine, Compd. with trinitromethane (1:1)
Methylamine, Compd. with trinitromethane

14148-99-3
$C_{16}H_{19}N.ClH$
261.79
DEA No. 1635
Benzeneethanamine, N,N-dimethyl-α-phenyl-, hydro-

chloride, (R)-, (9CI)
(-)-1-Dimethylamino-1,2-diphenyl-ethane
(-)-N,N-Dimethyl-1,2-diphenylethylamine hydrochloride
l-1,2-Diphenyl-1-dimethylaminoethane hydrochloride
Ethylamine, N,N-dimethyl-1,2-diphenyl-, hydrochloride, (R) (-)-
SPA

14149-99-6
$C_{24}H_{50}O_7$
450.66
3,6,9,12,15,18-Hexaoxatetracosan-1-ol, 21,23-dimethyl-19-(2-methylpropyl)- (9CI)
3,6,9,12,15,18-Hexaoxatetracosan-1-ol, 19-isobutyl-21,23-dimethyl- (8CI)

14151-45-2
$C_{12}H_{19}N_2O_4PS$
318.32
Phosphoramidothioic acid, (1-methylethyl)-, O-ethyl O-(5-methyl-2-nitrophenyl) ester (9CI)
Phosphoramidothioic acid, isopropyl-, O-ethyl O-(6-nitro-m-tolyl) ester (8CI)
S 2571

14158-27-1
Sr
89
Strontium, Isotope of mass 89 (8CI,9CI)

14167-18-1
$C_{16}H_{14}CoN_2O_2$
325.22
N,N-Ethylenebis(salicylideneiminato)cobalt(II)
AI3-30876
Cobalt, ((2,2'-(1,2-ethanediylbis(nitrilomethylidyne))bis(-phenolato))(2-)-N,N',O,O')-
Cobalt, ((2,2'-(1,2-ethanediylbis(nitrilomethylidyne))bis-(phenolato))(2-)-N,N',O,O')-, (SP-4-2)- (9CI)
Cobalt, N,N'-ethylenebis(salicylideneiminato)-
Cobalt, ((α,α'-(ethylenedinitrilo)di-o-cresolato)(2-))- (8CI)
(N,N'-Ethylenebis(salicylaldehyde iminato))cobalt(II)
N,N-Ethylenebis(salicylidene iminato)cobalt II
N,N'-Ethylenebis(salicylideneiminato)cobalt (II)
N,N'-Ethylenebis(salicylideniminato)cobalt(II)
(N,N'-Ethylenebis(salicylideniminato))cobalt
NSC-32965
Salcomin
Salcomine
Salcomine Powder
Salicylaldehyde ethylenediimine cobalt

14167-59-0
$C_{34}H_{70}$
478.93
Tetratriacontane (9CI)
AI3-36492
NSC-2998
n-Tetratriacontane

14167-66-9
$C_{28}H_{58}$
394.77
Heptacosane, 3-methyl- (8CI, 9CI)

14167-67-0
$C_{30}H_{62}$
422.82
Nonacosane, 3-methyl- (8CI, 9CI)

14168-01-5
$C_{10}H_7Cl_7$
375.32
ClC1C=CC2C1C3(Cl)C(Cl)C(Cl)C2(Cl)C3(Cl)Cl
4,7-Methanoindan, 3a-α,4,7,7a-α-tetrahydro-2-α,4-β,5,6,7-β,8,8-heptachloro
BL 2487
β-DHC
β-Dihydroheptachlor
Dilor
3a-α,4,7,7a-α-Tetrahydro-2-α,4-β,5,6,7-β,8,8-heptachloro-4,7-methanoindan

14168-42-4
$C_5H_9N_5O_5$
219.13
1,3,5-Triazine, 1-acetylhexahydro-3,5-dinitro- (9CI)
TAX
s-Triazine, 1-acetylhexahydro-3,5-dinitro- (8CI)

14171-89-2
$C_{12}H_{16}O$
176.26
2-Hexanone, 6-phenyl- (8CI, 9CI)
NSC-210804

14186-60-8
$C_{11}H_{12}O_4$
208.21
O=C(OC)c(ccc(c1C)C(=O)OC)c1
1,4-Benzenedicarboxylic acid, 2-methyl-, dimethyl ester (9CI)
Dimethyl 2-methyl-1,4-benzenedicarboxylate
Dimethyl 3-methylterephthalate

14191-95-8
C_8H_7NO
133.16
Acetonitrile, (p-hydroxyphenyl)
Benzeneacetonitrile, 4-hydroxy- (9CI)
p-Hydroxybenzyl cyanide
4-Hydroxybenzyl cyanide
p-Hydroxyphenylacetonitrile
(4-Hydroxyphenyl)acetonitrile

14199-15-6
$C_9H_{10}O_3$
166.18
O=C(OC)Cc(ccc(O)c1)c1
Benzeneacetic acid, 4-hydroxy-, methyl ester
AI3-36062
Methyl 4-hydroxybenzeneacetate

14202-62-1
$C_9H_{20}O$
144.26
1-Pentanol, 2,2-diethyl- (8CI,9CI)
2,2-Diethyl-1-pentanol

14206-62-3
$C_{11}H_8O_3.Na$
211.17
[Na+].Oc2cc1ccccc1cc2C([O-])=O
2-Naphthalenecarboxylic acid, 3-hydroxy-, monosodium salt (9CI)

14212-91-0
$C_3H_7Cl_2O_2P$
176.97
Phosphinic acid, bis-(chloromethyl)-, methyl ester (8CI,9CI)

14212-97-6
$C_5H_9Cl_2O_2P$
203.01
Phosphinic acid, bis-(chloromethyl)-, 2-propenyl ester (9CI)
Phosphinic acid, bis-(chloromethyl)-, allyl ester (8CI)

14212-98-7
$C_8H_9Cl_2O_2P$
239.04
Phosphinic acid, bis-(chloromethyl)-, phenyl

ester (8CI,9CI)
Phenyl bis(chloromethyl)-phosphinate

14213-97-9
BO_3
58.81
Borate (8CI,9CI)
Boric acid, ion(3-)
Orthoborate

14214-32-5
$C_{16}H_{18}N_2O_3$
286.36
COc2ccc(Oc1ccc(NC(=O)N(C)C)cc1)cc2
Urea, 1,1-dimethyl-3-(p-(p-methoxyphenoxy)phenyl)
C 3470
Difenoxuron
Lironion
N-(4-(4-Methoxyphenoxy)-phenyl)-N,N-dimethylurea
Pinoran

14216-75-2
H_2NNiO_3
122.73
Nickel nitrate

14221-47-7
$C_6FeO_{12}.3H_4N$
374.06
Ferrate(3-), tris(oxalato)-, triammonium
Ammonium ferric oxalate
Ammonium ferrioxalate
Ammonium trioxalatoferrate(III)
Ferrate(3-), tris(ethanedioato-(2-)-O,O')-, triammonium, (OC-6-11)- (9CI)
Ferric ammonium oxalate
Triammonium tris-(ethanedioato(2-)-O,O')ferrate(3-1)

14228-73-0
$C_{14}H_{24}O_4$
256.34
O(C1COCC(CCC(C2)COCC(O3)C3)C2)C1
Oxirane, 2,2'-(1,4-cyclohexanediylbis(methyleneoxymethylene))bis- (9CI)
1,4-Bis((2,3-epoxypropoxy)-ethyl)cyclohexane

14230-52-5
$C_{25}H_{42}O_3$
390.61
O=C(O)C(Oc(cccc1CCCCCCCCCCCCCCC)c1)CC
Butanoic acid, 2-(3-pentadecylphenoxy)- (9CI)
2-(3-Pentadecylphenoxy)-butanoic acid

14233-37-5
$C_{20}H_{22}N_2O_2$
322.39
O=C(c(c(C(=O)c1c(NC(C)C)ccc2NC(C)C)ccc3)c3)c12
1,4-Bis(isopropylamino)-anthraquinone
9,10-Anthracenedione, 1,4-bis-((1-methylethyl)amino)- (9CI)
Anthraquinone, 1,4-bis(iso-propylamino)- (8CI)
1,4-Bis(isopropylamino)-9,10-anthracenedione
1,4-Bis(N-isopropylamino)-anthraquinone
Brilliant Oil Blue BGS
C.I. Disperse Blue 134
C.I. Solvent Blue 36
Duranol Blue PP
NSC-58039
Oil Blue A
Solvent Blue 36
Waxoline Blue AP

14234-29-8
Cs
136
Cesium, Isotope of mass 136 (8CI,9CI)

14234-35-6
Sb
125
Antimony, Isotope of mass 125 (8CI,9CI)

14234-82-3
$C_{12}H_{20}O_4$
228.29
O=C(OCC(C)C)C=CC(=O)OCC(C)C
2-Butenedioic acid (Z)-, bis-(2-methylpropyl) ester (9CI)

14239-68-0
$C_{10}H_{20}CdN_2S_4$
408.96
CCN(CC)C(=S)S[Cd]SC(=S)N(CC)CC
Cadmium, bis(diethyldithiocarbamato)
Bis(diethyldithiocarbamato)-cadmium
Cadmium diethyl dithiocarbamate
Carbamic acid, diethyldithio-, cadmium salt
Ethyl Cadmate
Ethyl Tuads

14255-04-0
Pb
210
Lead-210
Radium D

14255-72-2
$C_{11}H_{17}O_5PS_2$
324.37
CCOP(=S)(OCC)Oc1ccc(cc1)S(C)(=O)=O
Phosphorothioic acid, O,O-diethyl O-(p-methylsulfonyl)phenyl ester
Dasanit sulfone
Dasanit sulphone
O,O-Diethyl O-(p-methylsulfonyl)phenyl phosphorothioate
Fensulfothion sulfone
Phosphorothioic acid, O,O-diethyl O-(4-(methylsulfonyl)phenyl) ester (9CI)

14255-87-9
$C_{13}H_{17}N_3O_2$
247.33
CCCCc2ccc1[nH]c(NC(=O)OC)nc1c2
2-Benzimidazolecarbamic acid, 5-butyl-, methyl ester
5-Butyl-2-benzimidazole-carbamic acid methyl ester
n-(Butyl-5, benzimidazolyl)-2, carbamate de methyle (French)
(4-Butyl-1H-benzimidazol-2-yl)-carbamic acid methyl ester
5-Butyl-2-(carbomethoxy-amino)benzimidazole
Carbamic acid, (5-butyl-1H-benzimidazol-2-yl)-, methyl ester
Helmatac
Methyl 5-butyl-2-benzimidazolecarbamate
Parbendazole
PBDZ
SKF 29044
SK&F 29044
Verminum
Worm Guard

14255-88-0
$C_{15}H_7Cl_2F_3N_2O_2$
375.14
FC(F)(F)c2nc1cc(Cl)c(Cl)cc1n2C(=O)Oc3ccccc3
1-Benzimidazolecarboxylic acid, 5,6-dichloro-2-(trifluoromethyl)-, phenyl ester
1H-Benzimidazole-1-carboxylic acid, 5,6-dichloro-2-(trifluoromethyl)-, phenyl ester
5,6-Dichloro-1-phenoxycarbonyl-2-trifluoromethylbenzimidazole
5,6-Dichloro-2-trifluoromethyl-benzimidazole-1-carboxylate
ENT 27,438
Fenazaflor
Fenoflurazole
Fenozaflor
Fenzaflor
Fisons NC 5016
Lovozal
NC 5016

NSC-191025
OMS 1243
Phenyl-5,6-dichloro-2-trifluoromethyl-benzimidazole-1-carboxylate
Tarzol

14258-49-2
$C_2H_5NO_4$
107.07
Ammonium oxalate
Ammonium ethanedioate
Ethanedioic acid, ammonium salt (9CI)

14263-73-1
$C_4CuN_4.3K$
284.89
Cuprate(3-), tetrakis(cyano-C)-, tripotassium, (T-4)- (9CI)

14263-89-9
$C_6H_3ClN_3O_2.1/2Cl_4Zn$
288.15
Benzenediazonium, 4-chloro-2-nitro-, tetrachlorozincate-(2-) (2:1) (9CI)
4-Chloro-2-nitrobenzenediazonium, chloride, zinc chloride

14264-31-4
$C_3CuN_3.2Na$
187.58
Cuprate(2-), tris(cyano-C)-, disodium
Copper sodium cyanide
Sodium cuprocyanide, Solid [UN 2316]
Sodium cuprocyanide, Solution [UN 2317]
UN 2316 [Sodium cuprocyanide, solid]
UN 2317 [Sodium cuprocyanide, solution]

14265-44-2
O_4P
94.97
Phosphate (9CI)

14265-45-3
O_3S
80.06
Sulfite (8CI,9CI)
Sulfurous acid, ion(2-)

14265-71-5
Se
75
Selenium, Isotope of mass 75 (8CI,9CI)

14273-76-8
$C_9H_{11}ClN_2O_2$
214.65
(6-Chloro-2-methyl-5-pyrimidyl)acetic acid, ethyl ester
Acetic acid, (6-chloro-2-methyl-5-pyrimidyl)-, ethyl ester
5-Pyrimidineacetic acid, 4-chloro-2-methyl-, ethyl ester

14275-57-1
Unknown
Unknown
Tributyltin maleate

14280-50-3
Pb
207.
Lead, ion (Pb(2+)) (8CI,9CI)

14284-89-0
$C_{15}H_{21}MnO_6$
352.26
Manganese, tris(2,4-pentanedionato-O,O')-, (OC-6-11)- (9CI)
Manganese, tris(2,4-pentanedionato)- (8CI)
NSC-82319

14285-59-7
$C_{32}H_{12}CoN_8O_{12}S_4.4H$
891.67
Cobaltate(4-), (29H,31H-phthalocyanine-2,9,16,23-tetrasulfonato(6-)-N29,N30,N31,N32)-, tetrahydrogen, (SP-4-1)- (9CI)
Cobalt 4,4',4'',4'''-phthalocyaninetetrasulfonate

14287-04-8
$C_4H_8O_2.H_3N$
105.13
Butanoic acid, ammonium salt (9CI)
Ammonium butanoate
Ammonium butyrate

14287-61-7
$C_6H_{12}O_2$
116.16
Butanoic acid, 2,3-dimethyl- (9CI)
Butyric acid, 2,3-dimethyl- (8CI)

14292-26-3
$C_{10}H_{20}O_3$
188.27
Decanoic acid, 3-hydroxy- (8CI,9CI) (VAN)

14292-27-4
$C_8H_{16}O_3$
160.21
Octanoic acid, 3-hydroxy- (8CI,9CI)

14295-43-3
$C_{16}H_4Cl_4O_2S_2$
434.15
O=C(c(c(S1)c(cc2)Cl)c2Cl)C1=C(Sc(c3c(cc4)Cl)c4Cl)C3=O
Benzo(b)thiophen-3(2H)-one, 4,7-dichloro-2-(4,7-dichloro-3-oxobenzo(b)thien-2(3H)-ylidene)- (9CI)

14297-87-1
$C_{24}H_{25}NO_3$
375.46
Benzylmorphine
DEA No. 9052
Morphinan-6-α-ol, 3-(benzyloxy)-7,8-didehydro-4,5-α-epoxy-17-methyl-
Morphine, benzyl-

14300-21-1
$C_{15}H_{11}NO$
221.26
3,1-Benzoxazepine, 2-phenyl- (8CI,9CI)

14302-13-7
$C_{32}Br_6Cl_{10}CuN_8$
1393.90
Copper, (1,3,8,16,18,24-hexabromo-2,4,9,10,11,15,17,22,23,25-decachlorophthalocyaninato(2-))
C.I. 74265
C.I. Pigment Green 36
C.I. Pigment Green 38
C.I. Pigment Green 41
Fastogen Green Y
Fastogen Green 2YK
Helio Fast Green GN
Helio Fast Green GT
Heliogen Green 9360
Heliogen Green 6G
Heliogen Green 6GA
Heliogen Green 8GA
Hostaperm Green 8G
Monastral Fast Green 3Y
Monastral Fast Green 6Y
Monastral Fast Green 3YA
Monastral Fast Green 6YA
Monastral Green Y-GT 805D
Phthalocyanine Green 6G
Pigment Green 38
Sandorin Green 8GLS
Vynamon Green 6Y

14302-87-5
Hg
200.60
Mercury, ion (Hg(2+)) (8CI, 9CI)

14307-33-6
$Ca.Cr_2O_7$
256.08
Dichromic acid, calcium salt (1:1)
Calcium chromate
Calcium dichromate(VI)
Chromic acid, calcium salt (1:1) (9CI)

14307-35-8
$CrH_2O_4.2Li$
131.89
Lithium chromate
Chromic acid, dilithium salt (9CI)
Chromium lithium oxide
Dilithium chromate
Lithium chromate(VI)

14307-43-8
$C_4H_9NO_6$
167.12
Ammonium tartrate
Butanedioic acid, 2,3-dihydroxy- (R-(R*,R*))-, ammonium salt (9CI)

14309-40-1
$C_{14}H_{21}NO_2$
235.32
Benzoic acid, 4-amino-, heptyl ester (9CI)
Benzoic acid, p-amino-, heptyl ester (8CI)

14309-41-2
$C_{15}H_{23}NO_2$
249.35
O=C(OCCCCCCCC)c(ccc(N)c1)c1
Benzoic acid, 4-amino-, octyl ester (9CI)
NSC-522884
Octyl 4-aminobenzoate

14309-42-3
$C_{12}H_{15}NO_4$
237.25
O=C(OCCCCC)c(ccc(N(=O)=O)c1)c1
Benzoic acid, 4-nitro-, pentyl ester (9CI)
Amyl p-nitrobenzoate
Pentyl 4-nitrobenzoate

14309-57-0
$C_9H_{16}O$
140.23
O=C(C=CCCCCC)C
3-Nonen-2-one (9CI)

14324-55-1
$C_{10}H_{22}N_2S_4.Zn$
363.95
CCN(CC)C(=S)S[Zn]SC(=S)N(CC)CC
Zinc, bis(diethyldithiocarbamato)
Bis(diethyldithiocarbamato)zinc
Diethyldithiocarbamic acid zinc salt
Ethazate
Ethyl Cymate
Ethyl zimate
Ethyl ziram
Vulcacure
Zimate, Ethyl
Zinc diethyldithiocarbamate
Zinc N,N-diethyldithiocarbamate
Vulkacit LDA

14331-79-4
Bi
210
Bismuth-210
Radium E

14336-68-6
Cd
115
Cadmium, Isotope of mass 115 (8CI,9CI)

14350-72-2
$C_{21}H_{42}O_6S.Na$
445.62
Octadecanoic acid, 9-hydroxy-, isopropyl ester, hydrogen sulfate, sodium salt (8CI)
Isopropyl 9-(sodiumsulfooxy)-octadecanoate
Oleic acid, isopropyl ester, sulfated, sodium salt
Sulfated isopropyl oleate, sodium salt

14351-50-9
$C_{20}H_{41}NO$
311.55
O=N(CCCCCCCCC=CCCCCCCCC)(C)C
Oleamine Oxide
N,N-Dimethyl-9-octadecen-1-amine-N-oxide
(Z)-N,N-Dimethyl-9-octadecen-1-amine N-oxide
Dimethyloleylamine oxide
Incromine Oxide OD 50
Jordamox ODA
9-Octadecen-1-amine, N,N-dimethyl-, N-oxide
9-Octadecen-1-amine, N,N-dimethyl-, N-oxide, (Z)- (9CI)
Oleyl dimethyl amine oxide
Unimox OL

14351-66-7
$C_{20}H_{30}O_2.Na$
325.45
1-Phenanthrenecarboxylic acid, 1,2,3,4,4a,4b,5,6,10,10a-decahydro-1,4a-dimethyl-7-(1-methylethyl)-, sodium salt, (1R-(1α,4aβ,4bα,10aα))- (9CI)

14357-78-9
$C_{26}H_{35}NO_4$
425.56
Diprenorphine
DEA No. 9058
Diprenorfina (Spanish)
Diprenorphinum (Latin)
M. 5050
M50-50 Injection

14362-31-3
$C_{18}H_{21}ClN_2.ClH$
337.28
Chlorcyclizine hydrochloride
AH-289 hydrochloride
Chlorcyclizinium chloride
Chlorcylizine
1-(p-Chlorobenzhydryl)-4-methylpiperazine hydrochloride
Chlorocyclizine hydrochloride
1-(p-Chloro-α-phenylbenzyl)-4-methylpiperazine hydrochloride
Di-paralene
Diparalene hydrochloride
Di-paralene monohydrochloride
Eramide
Histantin
NSC-169496
Perazil
Piperazine, 1-(p-chloro-α-phenylbenzyl)-4-methyl-, hydrochloride
Piperazine, 1-(p-chloro-α-phenylbenzyl)-4-methyl-, monohydrochloride (8CI)
Piperazine, 1-((4-chlorophenyl)phenylmethyl)-4-methyl-, monohydrochloride (9CI)

14371-10-9
C_9H_8O
132.17
Cinnamaldehyde, (E)
(E)-Cinnamaldehyde
trans-Cinnamaldehyde
trans-Cinnamic aldehyde
trans-Cinnamylaldehyde
(E)-3-Phenylpropenal
2-Propenal, 3-phenyl-, (E)- (9CI)

14376-82-0
C_5H_8BrCl
183.48
Cyclopentane, 1-bromo-2-chloro-, trans- (8CI, 9CI)
trans-1-Bromo-2-chloro-

cyclopentane
trans-2-Chlorocyclopentyl bromide

14377-11-8
$C_9H_{14}O$
138.21
Ethanone, 1-(1-cyclohepten-1-yl)- (9CI)
1-Acetyl-1-cycloheptene
1-Cyclohepten-1-yl methyl ketone
Ketone, 1-cyclohepten-1-yl methyl (6CI,7CI,8CI)

14379-28-3
$C_{17}H_{33}NO_3$
299.45
L-Valine, N-(1-oxododecyl)- (9CI)
Valine, N-lauroyl-, L- (8CI)

14381-51-2
$C_6H_5BrO_2$
189.01
1,2-Benzenediol, 3-bromo- (9CI)
Pyrocatechol, 3-bromo- (8CI)

14383-50-7
O_3S_2
112.13
Thiosulfate (8CI,9CI)
Thiosulfuric acid, ion(2-)

14391-76-5
Ag
110
Silver, Isotope of mass 110 (8CI,9CI)

14392-02-0
Cr
51
Chromium, Isotope of mass 51 (8CI,9CI)

14400-94-3
$C_6H_2Br_4O$
409.70
Phenol, 2,3,4,6-tetrabromo- (8CI,9CI)

14402-75-6
$C_4CdN_4.2K$
294.66
Cadmate(2-), tetrakis(cyano-C)-, dipotassium, (T-4)- (9CI)

14402-88-1
$C_{10}H_{12}MgN_2O_8.2Na$
358.49
Magnesate(2-), ((N,N'-1,2-ethanediylbis(N-(carboxymethyl)glycinato))(4-)-N,N',O,O',ON,ON')-, disodium, (OC-6-21)- (9CI)
Disodium magnesium ethylenediaminetetraacetate
N,N'-1,2-Ethanediylbis(N-carboxymethyl)glycine, magnesium disodium salt
Magnesium sodium ethylenediaminetetraacetate

14426-21-2
$C_4H_{11}NO_2.ClH$
141.62
Ethanol, 2,2'-iminodi-, hydrochloride
Diethanolamine hydrochloride
Diethanolammonium chloride
Ethanol, 2,2'-iminobis-, hydrochloride

14426-42-7
$C_{10}H_{11}ClO_3$
214.65
Acetic acid, (4-chlorophenoxy)-, ethyl ester (9CI)
Acetic acid, (p-chlorophenoxy)-, ethyl ester (8CI)
NSC-66307

14433-76-2
$C_{12}H_{25}NO$
199.38
O=C(N(C)C)CCCCCCCCC
Decanamide, N,N-dimethyl
N,N-Dimethylcapramide
N,N-Dimethyldecanamide

14434-22-1
$C_6FeN_6.10H_2O.4Na$
484.03
Ferrate(4-), hexakis(cyano-C)-, tetrasodium, decahydrate, (OC-6-11)- (9CI)
Ferrate(4-), hexacyano-, tetrasodium, decahydrate (8CI)

14437-17-3
$C_{10}H_{10}Cl_2O_2$
233.10
COC(=O)C(Cl)Cc1ccc(Cl)cc1
Propionic acid, 2-chloro-3-(4-chlorophenyl)-, methyl ester
Bay 70533
Bayer 70533
Bidisin
Chlorfenprop-methyl
2-Chloro-3-(4-chlorophenyl)-methylpropionate
2-Chloro-3-(4-chlorophenyl)-propionic acid methyl ester
Chlorphenprop-methyl
3-(4-Chlorphenyl)-2-chlorpropionsaeuremethylester (German)
Methachlorphenprop
Methyl 2-chloro-3-(4-chlorophenyl)propionate
Methyl α,4-dichlorophenylpropanoate
Methylester kyseliny 2-chlor-3-p-chlorfenylpropionove (Czech)
W5769

14437-20-8
$C_9H_8Cl_2O_2$
219.07
Benzenepropanoic acid, α,4-dichloro- (9CI)
Hydrocinnamic acid, p,α-dichloro- (8CI)

14448-38-5
$H_2N_2O_2$
62.02
Hyponitrite
Acide hyponitreux (French)
Acido hiponitroso (Spanish)
Hyponitrous acid
Hyponitrous acid, disodium salt
N-Nitrosohydroxylamine

14450-05-6
$C_{41}H_{76}O_8$
697.05
O=C(OCC(COC(=O)CCCCCCCC)(COC(=O)CCCCCCCC)COC(=O)CCCCCCCC)CCCCCCCC
Nonanoic acid, 2,2-bis(((1-oxononyl)oxy)methyl)-1,3-propanediyl ester
AI3-14798

14450-07-8
$C_{36}H_{71}O_4P$
598.93
O=P(OCCCCCCCCC=CCCCCCCC)(OCCCCCCCCC=CCCCCCCCC)O
Dioleoyl phosphate
Dioctadecenyl phosphate
Dioleyl phosphate
9-Octadecen-1-ol, hydrogen phosphate, (Z,Z)- (9CI)

14452-57-4
MgO_2
56.30
Magnesium peroxide (8CI, 9CI)
IXPER 25M
Magnesium peroxide, Solid
UN 1476 [Magnesium peroxide]

14464-46-1
O_2Si
60.09

Silica, Crystalline-cristobalite (ACGIH,OSHA)
Calcined diatomite
Cristobalite (ACGIH)

14477-61-3
$C_{12}H_6Cl_4O_2$
323.99
(1,1'-Biphenyl)-2,2'-diol, 4,4',6,6'-tetrachloro- (9CI)
2,2'-Biphenyldiol, 4,4', 6,6'-tetrachloro- (8CI)
3,3',5,5'-Tetrachloro-2,2'-biphenyldiol

14481-26-6
$C_4O_9Ti.2K$
318.12
Titanate(2-), bis(ethanedioato-(2-)-O,O')oxo-, dipotassium, (SP-5-21)- (9CI)
Titanium potassium oxalate

14481-29-9
$C_6FeN_6.4H_4N$
284.08
Ammonium ferrocyanide
AI3-28802
Ammonium hexacyanoferrate (II)
Ferrate(4-), hexacyano-, tetraammonium
Ferrate(4-), hexakis(cyano-C)-, tetraammonium, (OC-6-11)- (9CI)
Tetraammonium hexacyanoferrate
Tetraammonium hexacyanoferrate(4-)
Triammonium hexakis-(cyano-C)ferrate(4-)

14481-60-8
$C_{22}H_{43}NO_6S.2Na$
495.63
Disodium stearyl sulfosuccinamate
Butanedioic acid, 4-(octadecylamino)-4-oxo-2-sulfo-, disodium salt
Butanedioic acid, sulfo-, monooctadecyl ester, disodium salt
Butanoic acid, 4-(octadecylamino)-4-oxo-2-sulfo-, disodium salt (9CI)
4-(Octadecylamino)-4-oxo-2-sulfobutanedioic acid, disodium salt
Sulfobutanedioic acid, monooctadecyl ester, disodium salt

14484-64-1
$C_9H_{18}N_3S_6.Fe$
416.49
[Fe+3].CN(C)C([S-])=S.CN(C)C([S-])=S.CN(C)C([S-])=S
Ferbam
Aafertis
AI3-14689
Bercema Fertam 50
Carbamate
Carbamic acid, dimethyldithio-, iron salt
Caswell No. 458
Dimethylcarbamodithioic acid, iron complex
Dimethylcarbamodithioic acid, iron(3+) salt
Dimethyldithiocarbamic acid, iron salt
Dimethyldithiocarbamic acid iron(3+) salt
Eisendimethyldithiocarbamat
Eisen(III)-tris(N,N-dimethyldithiocarbamat)
ENT 14,689
EPA Pesticide Chemical Code 034801
Ferbam (ACGIH,OSHA)
Ferbam 50
Ferbam, Iron salt
Ferbame
Ferbame (French)
Ferbeck
Ferberk
Fermate
Fermate Ferbam Fungicide
Fermocide
Ferradow
Ferric dimethyl dithiocarbamate
Ferric dimethyldithiocarbamate
Fuklasin
Fuklasin Ultra
Fuklazin
Hexaferb
Hokmate
Iron dimethyldithiocarbamate
Iron(III) dimethyldithiocarbamate
Iron, tris(dimethylcarbamodithioato-S,S')-, (OC-6-11)-
Iron tris(dimethyldithiocarbamate)
Iron, tris(dimethyldithiocarbamato)-
Karbam Black
Knockmate
Liromate
Niacide
Stauffer Ferbam
Sup'r-Flo Ferbam Flowable
Trifungol
Tris(dimethylcarbamodithioato-S,S')iron
(OC-6-11)-Tris(dimethylcarbamodithioato-S,S')iron
Tris(dimethyldithiocarbamato)-iron
Tris(N,N-dimethyldithiocarbamato) iron(III)
Tris(N,N-dimethyldithiocarbamato)iron(III)
Vancide Fe95

14486-19-2
B_2CdF_8
286.02
Cadmium-fluoborate
Cadmium fluorborate
TL 1026

14488-53-0
$C_8H_{18}Sn$
232.94
Tin(2+), dibutyl- (9CI)
Tin(2+), dibutyl-, ion (8CI)

14491-59-9
$C_{11}H_{10}N_2O$
186.23
Pyridazine, 3-(2-methylphenoxy)
Credazine
H-722
Kusakira
3-(2-Methylphenoxy)pyridazine
NIA 20439
SW-6701
SW-6721

14499-87-7
$C_4H_6Cl_4$
195.90
Butane, 2,2,3,3-tetrachloro- (6CI,8CI,9CI)
2,2,3,3-Tetrachlorobutane

14516-71-3
$C_{32}H_{51}NNiO_2S$
572.51
Nickel, (1-butanamine)((2,2'-thiobis(4-(1,1,3,3-tetramethylbutyl)phenolato))(2-)-O,O',S)- (9CI)

14521-96-1
$C_{25}H_{33}NO_4$
411.54
Etorphine
DEA No. 9056
7,8-Dihydro-7-α-(1-(R)-hydroxy-1-methylbutyl)-O⁶-methyl-6,14-endo-ethenomorphine
6,14-endo-Ethenotetrahydrooripavine, 7-α-(1-hydroxy-1-methylbutyl)-
Etorfina
Etorphine (Except hydrochloride salt)
7-α-Etorphine
(-)-Etorphine
Etorphinum (Latin)
7-α-(1-(R)-Hydroxy-1-methylbutyl)-6,14-endo-ethenotetrahydrooripavine
M. 99
M99 Injection
Propylorvinol
19-Propylorvinol
Tetrahydro-7-α-(1-hydroxy-1-methylbutyl)-6,14-endo-ethenooripavine

Tetrahydro-7-α-(2-hydroxy-2-pentyl)-6,14-endo-etheno-oripavine

14542-23-5
CaF_2
78.07
Fluorite (9CI)

14542-93-9
$C_9H_{17}N$
139.23
N(#C)C(CC(C)(C)C)(C)C
Butyl isocyanide, 1,1,3,3-tetramethyl- (8CI)
2-Isocyano-2,4,4-trimethylpentane
NSC-141688
Pentane, 2-isocyano-2,4,4-trimethyl- (9CI)
1,1,3,3-Tetramethylbutyl isocyanide
1,1,3,3-Tetramethylbutylisonitrile

14546-44-2
H_3N_5
73.03
Hydrazine azide
Hydrazine, azido-

14548-01-7
$C_{13}H_9NO$
195.21
Phenanthridine, 5-oxide (9CI)
NSC-263827
Phenanthridine N-oxide

14548-46-0
$C_{12}H_9NO$
183.22
O=C(c(cccc1)c1)c(ccnc2)c2
Ketone, phenyl 4-pyridyl
4-Benzoylpyridine
Phenyl 4-pyridyl ketone
Pyridine, 4-benzoyl-

14548-60-8
$C_8H_{10}O_2$
138.17
O(Cc(cccc1)c1)CO
Methanol, (phenylmethoxy)- (9CI)
Benzyl hemiformal
(Phenylmethoxy)methanol

14567-73-8
Unknown
Unknown
Tremolite

14570-15-1
$C_6H_{15}Pb$
294.39
Plumbylium, triethyl- (9CI)
Triethyl lead

14570-16-2
C_3H_9Pb
252.30
Plumbylium, trimethyl- (9CI)
Trimethyl lead

14590-60-4
$C_6H_{13}Cl_2O_2P$
219.05
Phosphinic acid, bis-(chloromethyl)-, butyl ester (8CI,9CI)

14593-28-3
$C_{10}H_{12}ClNO_3$
229.66
Phenol, 2-chloro-4-(1,1-dimethylethyl)-6-nitro- (9CI)
Phenol, 4-tert-butyl-2-chloro-6-nitro- (8CI)

14596-10-2
Am
241
Americium, Isotope of mass 241 (8CI,9CI)

14596-12-4
Fe
59
Iron, Isotope of mass 59 (8CI,9CI)

14596-37-3
P
32
Phosphorus, Isotope of mass 32 (8CI,9CI)

14624-13-6
$C_8H_{18}O.1/3Al$
139.22
1-Octanol, aluminum salt (9CI)
Aluminum n-octoxide

14624-15-8
$C_{12}H_{26}O.1/3Al$
195.33
1-Dodecanol, aluminum salt (9CI)
Aluminum n-dodecoxide

14639-97-5
$Cl_4Zn.2H_4N$
243.28
Zinc ammonium chloride
(T-4)-Tetrachlorozincate(2-) diammonium
Zincate(2-), tetrachloro-, diammonium, (T-4)- (9CI)

14639-98-6
$Cl_5Zn.3H_4N$
296.77
Zincate(3-), pentachloro-, triammonium (9CI)
Pentachlorozincate(3-) triammonium
Zinc ammonium chloride

14643-87-9
$C_3H_4O_2.1/2Zn$
104.75
2-Propenoic acid, zinc salt (9CI)
Acrylic acid, zinc salt
Zinc diacrylate
Zinc dipropenoate
Zinc 2-propenoate

14644-61-2
$O_8S_2.Zr$
283.34
[Zr+4].[O-]S([O-])(=O)=O.[O-]S([O-])(=O)=O
Zirconium(IV) sulfate (1:2)
Disulfatozirconic acid
NA 9163 [Zirconium sulfate]
Sulfuric acid, zirconium(4+) salt (2:1)
Zirconium sulfate [NA 9163]
Zirconium sulphate
Zirconyl sulfate

14650-24-9
$C_{15}H_{12}F_{17}NO_4S$
625.30
O=C(OCCN(S(=O)(=O)C(F)(F)C(F)(F)C(F)(F)C(F)(F)C(F)(F)C(F)(F)C(F)(F)C(F)(F)F)C)C(=C)C
2-Propenoic acid, 2-methyl-, 2-(((heptadecafluorooctyl)-sulfonyl)methylamino)ethyl ester (9CI)

14654-05-8
$CrH_2O_4.Pb$
325.21
Crocoite (Pb(CrO4)) (9CI)
Crocoite (7CI,8CI)

14666-78-5
$C_6H_{10}O_6$
178.14
Diethyl peroxydicarbonate (More than 27% in solution)
Diethyl peroxydicarbonate
Diethyl peroxydicarbonate, Not more than 27% in solution
Diethyl peroxydiformate
Ethyl peroxycarbonate
Peroxidicarbonato de dietilo (Spanish)
Peroxydicarbonate d'ethyle

(French)
Peroxydicarbonic acid, diethyl ester, More than 27% in solution
Peroxydicarbonic acid, diethyl ester, Not more than 27% in solution
UN 2175

14666-94-5
$C_{18}H_{35}CoO_2$
342.41
9-Octadecenoic acid (Z)-, cobalt salt (9CI)

14667-55-1
$C_7H_{10}N_2$
122.19
n(c(c(nc1C)C)C)c1
Pyrazine, trimethyl
Trimethylpyrazine
2,3,5-Trimethylpyrazine

14674-72-7
$CaCl_2O_4$
174.98
Calcium chlorite
Chlorite de calcium (French)
Clorito calcico (Spanish)
UN 1453 [Calcium chlorite]

14681-59-5
Fe
55
Iron, Isotope of mass 55 (8CI,9CI)

14683-12-6
Te
131
Tellurium, Isotope of mass 131 (8CI,9CI)

14683-16-0
I
132
Iodine, Isotope of mass 132 (8CI,9CI)

14686-13-6
C_7H_{14}
98.19
CCCCC=CC
Heptylene-2-trans
2-Heptene, trans-
2-Heptene, (E)- (9CI)
trans-2-Heptene
(E)-2-Heptene
Heptylene-2, trans-
NSC-74131

14686-14-7
C_7H_{14}
98.19
CCCC=CCC
3-Heptene, (E)- (9CI)
trans-3-Heptene
(E)-3-Heptene
NSC-74133

14695-88-6
Unknown
Unknown
Nitrilotriacetic acid, Compound with iron chloride
NTA, Compound with iron chloride

14696-82-3
IN_3
168.91
Iodine azide
Azida de yodo (Spanish)
Azoture d'iode (French)
Iodine azide, Dry
Iodine(I) azide
Iodoazide
Nitrogen iodide

14697-48-4
$C_{20}H_{38}O_4$
342.52
O=C(OCCCCCCC)CCCCC(=O)OCCCCCCC
Hexanedioic acid, diheptyl ester (9CI)
Diheptyl hexanedioate

14697-50-8
$C_{12}H_{24}B_2O_5$
269.94
2,2'-Oxybis(4,4,6-trimethyl-1,3,2-dioxaborinane)
AI3-50502
Biobor
Caswell No. 627
1,3,2-Dioxaborinane, 2,2'-oxybis(4,4,6-trimethyl-
EPA Pesticide Chemical Code 012402
2,2'-(1-Methyltrimethylenedioxy)bis(4-methyl-1,3,2-dioxaborinane)

14698-29-4
$C_{13}H_{11}NO_5$
261.25
CCn2cc(C(O)=O)c(=O)c3cc1OCOc1cc23
1,3-Dioxolo(4,5-g)quinoline-7-carboxylic acid, 5-ethyl-5,8-dihydro-8-oxo
Emyrenil
1-Ethyl-1,4-dihydro-6,7-methylenedioxy-4-oxo-3-quinolinecarboxylic acid
5-Ethyl-5,8-dihydro-8-oxo-1,3-dioxolo(4,5-g)quinoline-7-carboxylic acid
1-Ethyl-6,7-methylenedioxy-4-quinolone-3-carboxylic acid
Nidantin
NSC-110364
Ossian
Oxoboi
Oxolinic acid
Pietil
Prodoxol
Uritrate
Uro-Alvar
Urotrate
Uroxol
Utibid
W 4565

14706-41-3
$C_{13}H_{13}NO_4S_2$
311.38
O=S(=O)(NS(=O)(=O)c(ccc(c1)C)c1)c(cccc2)c2
Benzenesulfonamide, 4-methyl-N-(phenylsulfonyl)-(9CI)
4-Methyl-N-(phenylsulfonyl)-benzenesulfonamide
p-Toluenesulfonamide, N-(phenylsulfonyl)-

14719-47-2
$C_{10}H_{15}N$
149.23
Benzenamine, 2,4-diethyl- (9CI)
Aniline, 2,4-diethyl- (8CI)

14720-53-7
$BHO_2.1/2Pb$
147.41
Lead borate
Boric acid, lead(2+) salt (8CI, 9CI)

14720-74-2
$C_{10}H_{22}$
142.28
Heptane, 2,2,4-trimethyl-(8CI,9CI)

14726-36-4
$C_{30}H_{28}N_2S_4Zn$
610.21
Zinc dibenzyldithiocarbamate
(T-4)-Bis(bis(phenylmethyl)carbamodithioato-S,S')zinc
Dibenzyldithiocarbamic acid, zinc salt
Zinc, bis(bis(phenylmethyl)-carbamodithioato-S,S')-, (T-4)- (9CI)

14728-39-3
Unknown
Unknown
Ammonium polyphosphate solution

14733-03-0
Unknown
Unknown

Bismuth-214
Radium C

14752-75-1
$C_{23}H_{40}$
316.57
c(cccc1)(c1)CCCCCCCCCCCCC CCCC
Benzene, heptadecyl- (9CI)
Heptadecylbenzene
1-Phenylheptadecane

14758-11-3
$AsH_3O_4.Pb$
349.14
Schultenite (Pb(HAsO4)) (9CI)
Schultenite (8CI)

14759-06-9
$C_{21}H_{26}N_2O_2S_2$
402.61
Phenothiazine, 10-(2-(1-methyl-2-piperidyl)ethyl)-2-methylsulfonyl-,
Imagotan
Inofal
10-(2-(1-Methyl-2-piperidyl)-ethyl)-2-methylsulfonylpheno-thiazine
Psychoson
Sulforidazine
TPN-12

14762-55-1
He
3
Helium, Isotope of mass 3 (8CI,9CI)

14762-75-5
C
14
Carbon, Isotope of mass 14 (8CI,9CI)

14762-78-8
Ce
144
Cerium, Isotope of mass 144 (8CI,9CI)

14763-77-0
C_2CuN_2
115.58
Copper(II) cyanide
Copper cyanide [UN 1587]
Copper cynanamide
Cupric cyanide (DOT)
Cyanure de cuivre (French)
UN 1587 [Copper cyanide]

14769-73-4
$C_{11}H_{12}N_2S$
204.31
N(=C(N(CC1)C2)S1)C2c(cccc3) c3
Imidazo(2,1-b)thiazole, 2,3,5,6-tetrahydro-6-phenyl-, (S)
Ketrax
Levamisol
Levamisole
Levomysol
L-2,3,5,6-Tetrahyro-6-phenyl-imidazo(2,1-b)thiazole
(-)-Tetramisole
l-Tetramisole

14779-78-3
$C_{14}H_{21}NO_2$
235.36
O=C(OCCCCC)c(ccc(N(C)C) c1)c1
Benzoic acid, p-dimethyl-amino-, pentyl ester
Amyl-p-dimethylaminobenzoate
Amyl dimethyl PABA
p-Dimethylaminobenzoic acid, pentyl ester

14782-75-3
$C_{12}H_{23}AlO_5$
274.29
Aluminum, (ethyl 3-oxobut-anoato-O1',O3)bis(2-propan-olato)-, (T-4)- (9CI)
Diisopropyl aluminum ethyl acetoacetate
Ethyl acetoacetate aluminum diidopropylate di isopropyl aluminium ethyl acetoacetate

14797-55-8
NO_3
62.00
Nitrate (8CI,9CI)
Nitric acid, ion(1-) (VAN)

14797-65-0
NO_2
46.01
Smooth Muscle Inhibitory Factor
SMIF

14798-03-9
H_4N
18.03
Ammonium ion (8CI,9CI)
Ammonium ion, ion $(NH_4(+1))$

14800-16-9
$C_{18}H_{30}$
246.44
Benzene, 1,2,4-tributyl- (8CI,9CI)
1,2,4-Tributylbenzene
1,3,4-Tributylbenzene

14806-72-5
$C_6H_{15}NO_3.C_2H_4O_2$
209.24
Ethanol, 2,2',2''-nitrilotris-, acetate (Salt) (9CI)
Acetic acid, triethanolamine salt
2,2',2''-Nitrilotrisethanol acetate
2,2',2''-Nitrilotrisethanol acetate (Salt)
Triethanolamine acetate

14807-96-6
$H_{203}Si.3/4Mg$
96.33
Talc (ACGIH,OSHA)
Agalite
Asbestine
B 9
B 13
Beaver White 200
B 13 (Mineral)
CP 10-40
CP 38-33
Crystalite CRS 6002
Desertalc 57
EMTAL 500
EMTAL 549
EMTAL 596
EMTAL 599
EX-IT
Fibrene C 400
Finntalc C10
Finntalc M05
Finntalc M15
Finntalc P40
Finntalc PF
FW-XO
IT Extra
LMR 100
Micro Ace K1
Micro Ace L1
Micron White 5000A
Micron White 5000P
Micron White 5000S
Microtalco IT Extra
Mistron 139
Mistron Frost P
Mistron RCS
Mistron 2SC
Mistron Star
Mistron Super Frost
Mistron Vapor
MP 12-50
MP 25-38
MP 40-27
MP 45-26
MST
Mussolinite
Nytal 200
Nytal 400
P 3
PK-C
PK-N
P 3 (Mineral)
Polytal 4641
Polytal 4725
Steawhite
Supreme

Supreme Dense
Talc, Containing asbestos fibers
Talc (Powder), Containing no asbestos fibers
Talcan PK-P
Talcron CP 44-31
Talcum
TY 80

14808-60-7
O_2Si
60.09
Silica, Crystalline quartz (OSHA)
Agate
Amethyst
Celite
Chalcedony
Cherts
D & D
DQ12
Flint
Flintshot
Gold Bond R
Imsil
Min-U-Sil
Novaculite
Onyx
α-Quartz
Quazo Puro (Italian)
Snowit
Rose Quartz
Sand
Sil-Co-Sil
Silica Flour (Powdered crystalline silica)
Silica, Quartz (ACGIH)
Silicic anhydride
Silicon oxide, di- (Sand)
Silver Bond B

14808-79-8
O_4S
96.06
Sulfate (8CI,9CI)
Sulfuric acid, ion(2-)

14814-89-2
$C_{17}H_{12}O_2S$
280.35
2-Naphthol, 1-mercapto-, 2-benzoate (8CI)

14816-18-3
$C_{12}H_{15}N_2O_3PS$
298.32
CCOP(=S)(OCC)ON=C(C#N)c1ccccc1
Glyoxylonitrile, phenyl-, oxime, O,O-diethyl phosphorothioate
B 77488
Bay 5621
Bay 77488
Bayer 77488
Bay SRA 7502
Baythion
Benzeneacetonitrile, α-(((diethoxyphosphinothioyl)oxy)imino)-
Benzoyl cyanide O-(diethoxyphosphinothioyl)oxime
O,O-Diaethyl-O-(α-cyanbenzyliden-amino)-thionphosphat (German)
O,O-Diaethyl-O-(α-cyanobenzylidenamino)-monothiophosphat (German)
α-(((Diethoxyphosphinothioyl)oxy)imino)benzeneacetonitrile
(Diethoxy-thiophosphoryloxyimino)-phenyl acetonitrile
O,O-Diethyl-α-cyanobenzylidineaminooxyphosphonothiate
O,O-Diethyl phosphorothioate, O-ester with phenylglyoxylonitrile oxime
ENT 27,488
4-Ethoxy-7-phenyl-3,5-dioxa-6-aza-4-phosphaoct-6-ene-8-nitrile 4-sulfide
OMS 1170
Phenylglyoxylonitrile oxime O,O-diethyl phosphorothioate
Phoxim
Phoxime
Phoxin
Sebacil
Valexon
Valexone
Volaton

14816-20-7
$C_{12}H_{14}ClN_2O_3PS$
332.74
Chlorphoxim
AI3-27449
2-Chloro-α-((diethoxyphosphinothioyloxy)imino)benzeneacetonitrile
2-(2-Chlorophenyl)-2-(diethoxyphosphinothioyloxyimino)-acetonitrile
7-(2-Chlorophenyl)-4-ethoxy-3,5-dioxa-6-aza-4-phosphaoct-6-ene-8-nitrile 4-sulfide
Chlorphoxime (French)
3,5-Dioxa-6-aza-4-phosphaoct-6-ene-8-nitrile, 7-(2-chlorophenyl)-4-ethoxy-, 4-sulfide
O,O-Diethyl 2-chloro-α-cyanobenzylideneamino-oxyphosphonothioate

14832-14-5
$C_{32}Cl_{16}CuN_8$
1127.15
Copper, (1,2,3,4,8,9,10,11,15,16,17,18,22,23,24,25-hexadecachloro-29H,31H-phthalocyaninato(2-)-N29,N30,N31,N32)-, (SP-4-1)-(9CI)
AI3-26191

14832-90-7
Unknown
Unknown
Selenium oxide
NA 2811 [Selenium oxide]

14834-67-4
I
133
Iodine, Isotope of mass 133 (8CI,9CI)

14838-15-4
$C_9H_{13}NO$
151.23
OC(c(cccc1)c1)C(N)C
dl-Norephedrine
dl-α-(1-Aminoethyl)benzyl alcohol
dl-2-Amino-1-hydroxy-1-phenylpropane
dl-α-Hydroxy-β-aminopropylbenzene
dl-1-Phenyl-2-aminopropanol-1
Propadrine
dl-Propadrine

14840-89-2
$C_{15}H_{24}O$
220.35
O(C1CC(C=CCC(C=CCC2C)C)C)C12
13-Oxabicyclo(10.1.0)trideca-4,8-diene, 2,6,10-trimethyl-(8CI)

14850-23-8
C_8H_{16}
112.22
4-Octene, (E)- (9CI)
NSC-73940
(E)-4-Octene
n-trans-4-Octene
trans-n-4-Octene
trans-4-Octene

14858-54-9
$C_{18}H_{38}O_5S.Na$
389.55
Ethanol, 2-(hexadecyloxy)-, hydrogen sulfate, sodium salt (9CI)
2-(Hexadecyloxy)ethyl sodium sulfate

14861-06-4
$C_6H_8O_2$
112.13
Vinyl crotonate
AI3-25333
2-Butenoic acid, ethenyl ester (9CI)
Crotonic acid, vinyl ester (8CI)
NSC-5215
Vinyl 2-butenoate
Vinylester kyseliny krotonove (Czech)

14882-18-9
$C_7H_5BiO_4$
362.10
Bismuth, oxo(salicylato)
Bismuth salicylate, Basic
Bismuth subsalicylate
Salicylic acid, bismuth basic salt
Stabisol

14901-07-6
$C_{13}H_{20}O$
192.33
O=C(C=CC(=C(CCC1)C)C1(C)C)C
3-Buten-2-one, 4-(2,6,6-trimethyl-1-cyclohexen-1-yl)
β-Cyclocitrylideneacetone
β-Ionone
4-(2,6,6-Trimethyl-1-cyclohexen-1-yl)-3-buten-2-one

14901-08-7
$C_8H_{16}N_2O_7$
252.26
CN(=O)=NCC1OC(CO)C(O)C(O)C1O
β-D-Glucopyranoside, (methyl-onn-azoxy)methyl
Cycasin
Cycas Revoluta Glucoside
Cykazine
β-D-Glucosyloxyazoxymethane
β-D-Glucosyloxyazoxymethase
Methylazoxymethanol glucoside
Methylazoxymethanol-β-D-glucoside
(Methyl-ONN-azoxy)-methyl-β-D-glucopyranoside

14906-97-9
$C_6H_{11}NaO_7$
218.14
D-Gluconic acid, sodium salt (9CI)
Sodium D-gluconate

14913-33-8
$Cl_2H_6N_2Pt$
300.07
Platinum (II), diamminedichloro-, trans
trans-Diamminedichloroplatinum (II)
trans-Dichlorodiammineplatinum (II)
trans-Platinum(II)diamminedichloride

14920-92-4
$C_{11}H_{20}OSi_2$
224.45
Disiloxane, pentamethylphenyl- (8CI,9CI)
NSC-147799

14920-93-5
$C_{21}H_{24}OSi_2$
348.60
Disiloxane, 1,1,3-trimethyl-1,3,3-triphenyl- (8CI,9CI)
1,1,3-Trimethyl-1,3,3-triphenyldisiloxane
1,3,3-Trimethyl-1,1,3-triphenyldisiloxane

14921-00-7
$C_3Br_3N_3$
317.78
1,3,5-Triazine, 2,4,6-tribromo- (9CI)
Cyanuric bromide (6CI)
s-Triazine, 2,4,6-tribromo- (8CI)
2,4,6-Tribromo-s-triazine
2,4,6-Tribromo-sym-triazine

14925-96-3
$C_7H_{14}O$
114.19
Oxirane, 2-butyl-3-methyl- (9CI)
2,3-Epoxyheptane
Heptane, 2,3-epoxy- (7CI, 8CI)
Hept-2-ene oxide

14930-96-2
$C_{29}H_{37}NO_5$
479.67
CC1CCCC(O)C=CC(=O)OC24C(C=CC1)C(O)C(=C)C(C)C2C(Cc3ccccc3)NC4=O
2H-Oxacyclotetradec(2,3-d)-isoindole-2,18(5H)-dione, 16-benzyl-6,7,8,9,10,12a,13,14,15,15a, 16,17-dodecahydro-5,13-dihydroxy-9,15-dimethyl-14-methylene-, (E)-(5S,9R,12aS,13S,15S,15aS, 16aS,18aS)
Cytochalasin B
Phomin

14933-08-5
$C_{17}H_{37}NO_3S$
335.55
Lauryl sultaine
1-Dodecanaminium, N,N-dimethyl-N-(3-sulfopropyl)-, hydroxide, inner salt (9CI)
Lauryl sulfobetaine
Zwittergent 3-12

14938-35-3
$C_{11}H_{16}O$
164.27
Oc(ccc(c1)CCCCC)c1
Phenol, p-pentyl
4-n-Amylphenol

14939-34-5
F_2HOP
85.98
Phosphonic difluoride
Phosphine oxide, difluoro-

14946-02-2
$C_5H_5Cl_3N_4S$
259.55
1,3,5-Triazin-2-amine, 4-(methylthio)-6-(trichloromethyl)- (9CI)
s-Triazine, 2-amino-4-(methylthio)-6-(trichloromethyl)- (8CI)

14946-18-0
$C_6H_6Cl_3N_3S_2$
290.62
1,3,5-Triazine, 2,4-bis-(methylthio)-6-(trichloromethyl)- (9CI)
s-Triazine, 2,4-bis-(methylthio)-6-(trichloromethyl)- (8CI)

14959-86-5
$C_{14}H_{26}O_2$
226.36
O=C(OCCCCCCC=CCCCC)C
cis-7-Dodecen-1-ol acetate
AI3-33266
7-Dodecen-1-ol, acetate, (Z)-
(Z)-7-Dodecen-1-ol acetate

14960-06-6
$C_{18}H_{35}NO_4.Na$
352.46
Sodium Lauriminodipropionate
β-Alanine, N-(2-carboxyethyl)-N-dodecyl-, monosodium salt (9CI)
N-(2-Carboxyethyl)-N-dodecyl-β-alanine, monosodium salt
N-Lauryl-β-iminodipropionic acid, sodium salt
Sodium N-lauryl-β-iminodipropionate
Unitex 610-L

14970-71-9
$H_2O_6S_2$
162.14
Sodium dithionate
Dithionic acid, disodium salt

14970-87-7
$C_6H_{14}O_2S_2$
182.31
O(CCOCCS)CCS
Ethanethiol, 2,2'-(1,2-ethanediylbis(oxy))bis- (9CI)
1,2-Bis(2-mercaptoethoxy)ethane
2,2'-(1,2-Ethanediylbis(oxy))-

bisethanethiol
Ethanethiol, 2,2'-(ethylenedioxy)di-
2,2'-(Ethylenedioxy)diethanethiol
NSC-94782
Triethyleneglycol dimercaptan
Triglycol dimercaptan

14977-61-8
Cl_2CrO_2
154.90
Chromium, dichlorodioxo
Chlorure de chromyle (French)
Chromic oxychloride
Chromium chloride oxide
Chromium dichloride dioxide
Chromium dioxide dichloride
Chromium (VI) dioxychloride
Chromium oxychloride
[UN 1758]
Chromoxychlorid (German)
Chromylchlorid (German)
Chromyl chloride (ACGIH, DOT)
Chroomoxychloride (Dutch)
Cromile, cloruro di (Italian)
Cromo, ossicloruro di (Italian)
Dichlorodioxochromium
Dioxodichlorochromium
Oxychlorure chromique (French)
UN 1758 [Chromium oxychloride]

14979-34-1
$C_{10}Cl_{12}$
545.54
Dodecachlorodicyclopentadiene
AI3-27056
Hexachlorocyclopentadiene cyclic dimer
4,7-Methanoindene, 1,1,2,3,3a,4,5,6,7,7a,8,8-dodecachloro-3a,4,7,7a-tetrahydro-
4,7-Methano-1H-indene, 1,1,2,3,3a,4,5,6,7,7a,8,8-dodecachloro-3a,4,7,7a-tetrahydro-

14979-39-6
$C_8H_{18}O$
130.26
3-Heptanol, 4-methyl
4-Methyl-3-heptanol

14981-08-9
Unknown
Unknown
Nitrilotriacetic acid, calcium salt
NTA, calcium salt

14987-04-3
$H_4O_8Si_3.2Mg$
264.91
Magnesium trisilicate
Dicarbocalm
Magnesium mesotrisilicate
Magnesium silicate
Magnesium silicon oxide (9CI)
Magnesium trisilicate USP
Magnosil
Meerschaum
Parasepiolite
Petimin
Sepiolite
Silicic acid, magnesium salt (1:2)
Trisilicalm

14989-32-3
Unknown
Unknown
Sulfur chloride
UN 1828

14991-93-6
$C_7H_{10}Cl_2N_3O_2PS_2$
334.18
Phosphorodithioic acid, S-(4,6-dichloro-s-triazin-2-yl) O,O-diethyl ester (6CI,7CI,8CI)
s-Triazine-2-thiol, 4,6-dichloro-, S-ester with O,O-diethyl phosphorodithioate

14995-38-1
$C_8H_{10}O_3S.Na$
209.22
Benzenesulfonic acid, 4-ethyl-, sodium salt (9CI)
Benzenesulfonic acid, p-ethyl-, sodium salt
p-Ethylbenzenesulfonic acid, sodium salt
NSC-249835
Sodium p-ethylbenzenesulfonate
Sodium 4-ethylbenzenesulfonate

15009-91-3
$C_5H_4N_2O_2$
124.09
O=N(=O)c1ccccn1
Pyridine, 2-nitro- (9CI)
2-Nitropyridine
NSC-159025

15012-36-9
$C_{10}H_{12}O_2$
164.20
Benzoic acid, 2,3-dimethyl-, methyl ester (8CI,9CI)

15022-08-9
$C_7H_{10}O_3$
142.15
O=C(OCC=C)OCC=C
Carbonic acid, di-2-propenyl ester (9CI)
Allyl carbonate
Carbonic acid, diallyl ester
Diallyl carbonate
Di-2-propenyl carbonate
NSC-19177

15045-43-9
$C_8H_{16}O$
128.21
Furan, tetrahydro-2,2,5,5-tetramethyl- (8CI,9CI)

15046-84-1
I
129
Iodine, Isotope of mass 129 (8CI,9CI)

15067-28-4
Unknown
Unknown
Lead-214
Radium B

15067-64-8
$C_{11}H_{11}N_3S_2$
249.36
1,3,5-Triazine, 2,4-bis-(methylthio)-6-phenyl-(9CI)
s-Triazine, 2,4-bis(-methylthio)-6-phenyl-(8CI)

15075-85-1
$C_{18}H_{36}O_3S.Na$
355.54
9-Octadecene-1-sulfonic acid, sodium salt, (Z)- (9CI)

15086-27-8
$C_6H_6O.1/3Al$
103.11
Phenol, aluminum salt (9CI)
Aluminum phenoxide

15086-94-9
$C_{20}H_8Br_4O_5$
647.92
O=C(OC(c(c(Oc1c(c(O)c(c2)Br)Br)c(c(O)c3Br)Br)c3)(c12)c4cccc5)c45
Fluorescein, 2',4',5',7'-tetrabromo
Bromeosin
Bromoeosin
Bromofluoresceic acid
C.I. 45380:2
C.I. Solvent Red 43
D & C Red No. 21
Eosin
Eosine
2,4,5,7-Tetrabromo-3,6-fluorandiol
Tetrabromofluorescein
2',4',5',7'-Tetrabromofluorescein

15087-24-8
$C_{17}H_{20}O$
240.35
O=C(C(C(C1(C)C)CC2)=Cc(cccc3)c3)C12C
3-Benzylidene camphor
3-Benzylidene-2-bornanone
Bicyclo(2.2.1)heptan-2-one, 1,7,7-trimethyl-3-(phenylmethylene)- (9CI)
Mexoryl SD
1,7,7-Trimethylbicyclo(2.2.1)-heptan-2-one-3-benzylidene
1,7,7-Trimethyl-3-(phenylmethylene)bicyclo(2.2.1)-heptan-2-one
Unisol S-22

15096-41-0
$C_4H_6Cl_4Cr_2O_3$
347.89
Chromium, tetrachloro-mu-hydroxy(mu-(2-methyl-2-propenoato-O:O'))di- (9CI)
AI3-17757
Chromium, tetrachloro-u-hydroxy(u-(2-methyl-2-propenoato)-O:O'))di-

15096-52-3
$AlF_6.3Na$
209.95
Sodium aluminum fluoride
Aluminum sodium fluoride
Cryolite
ENT 24,984
Koyoside
Kryocide
Kryolith (German)
Natriumaluminiumfluorid (German)
Natriumhexafluoroaluminate (German)
Sodium-fluoaluminate
Sodium aluminofluoride
Sodium hexafluoroaluminate
Villiaumite

15104-61-7
$C_3H_3Cl_5$
216.32
Propane, 1,1,2,3,3-pentachloro- (8CI,9CI)

15110-74-4
$C_{13}H_8N_2O_4$
256.23
O=N(=O)c3ccc1c(Cc2cccc(N(=O)=O)c12)c3
Fluorene, 2,5-dinitro
2,5-Dinitrofluorene
9H-Fluorene, 2,5-dinitro- (9CI)

15112-89-7
$C_6H_{19}N_3Si$
161.37
Silanetriamine, N,N,N',N',N'',N''-hexamethyl
N,N,N',N',N'',N''-Hexamethylsilanetriamine
Tris(dimethylamino)silane

15118-60-2
$C_{10}H_{13}NO_2$
179.21
Benzenebutanoic acid, 4-amino- (9CI)
Butyric acid, 4-(p-aminophenyl)- (8CI)
NSC-27531

15120-99-7
$C_7H_{12}O_2$
128.17
Ethanone, 1-(trimethyloxiranyl)- (9CI)
2-Pentanone, 3,4-epoxy-3,4-dimethyl- (8CI)

15121-84-3
$C_8H_9NO_3$
167.16
O=N(=O)c(c(ccc1)CCO)c1
Benzeneethanol, 2-nitro-
AI3-36271
2-Nitrobenzeneethanol

15142-96-8
$C_6H_{20}N_2O_{12}P_4.6Na$
574.05
Phosphonic acid, (1,2-ethanediylbis(nitrilobis(methylene)))tetrakis-, hexasodium salt (9CI)
Ethylene bis(nitrilodimethylene)tetraphosphonic acid, hexasodium salt

15158-11-9
Cu
63.55
Copper, ion (Cu(2+)) (8CI,9CI)

15165-79-4
Unknown
Unknown
Potassium 1-naphthaleneacetate

15175-04-9
$C_{12}H_{23}ClO_2$
234.77
Decanoic acid, 2-chloroethyl ester (8CI,9CI)

15187-71-0
$C_4H_7Cl_3$
161.46
Butane, 1,3,3-trichloro- (8CI,9CI)
1,3,3-Trichlorobutane

15191-85-2
$O_4Si.2Be$
110.11
Silicic acid, beryllium salt
Beryllium silicate
Beryllium silicic acid
Beryllium orthosilicate
Orthosilicate
Phenacite
Phenakite
Phenazite

15194-98-6
Unknown
Unknown
Arsenic compound
Calcium arsenite (2:1)

15194-99-7
$AsHO_3.1/2Ca$
143.97
Arsenenic acid, calcium salt (9CI)
Arsenic acid, calcium salt (8CI)
Calcium arsenate

15214-89-8
$C_7H_{13}NO_4S$
207.24
O=C(NC(CS(=O)(=O)O)(C)C)C=C
2-Acrylamido-2-methylpropanesulfonate
2-AMPS
1-Propanesulfonic acid, 2-methyl-2-((1-oxo-2-propenyl)amino)- (9CI)

15220-85-6
$(C_4H_8)_4$
224.43
1-Propene, 2-methyl-, tetramer (9CI)
2-Methyl-1-propene tetramer

15233-47-3
$C_{20}H_{28}N_2$
296.44
N(c(ccc(Nc(cccc1)c1)c2)c2)C(CCCCCC)C
1,4-Benzenediamine, N-(1-methylheptyl)-N'-phenyl- (9CI)
N-(1-Methylheptyl)-N'-phenyl-1,4-benzenediamine
N-(1-Methylheptyl)-N'-phenyl-p-phenylenediamine
UOP 688

15234-85-2
$C_{16}H_{24}O_3$
264.36

Acetic acid, (p-octylphenoxy)- (8CI)

15242-96-3
$C_{18}H_{36}Cl_4Cr_2O_3$
546.34
Chromium, tetrachloro-mu-hydroxy(mu-(octadecanoato-O:O'))di
Chromic chloride stearate
Chromium, tetrachloro-mu-hydroxy(mu-stearato)di-
Khromolan
NCI-C60800
Quilon S
Stearate chromic chloride
Stearato chromic chloride
Stearato-chromic chloride complex
Stearatochromium chloride

15245-12-2
$H_6CaN_2O_3$
122.14
Calcium ammonium nitrate
Ammonium calcium nitrate
Nitric acid, ammonium calcium salt (9CI)

15245-44-0
$C_6H_3N_3O_8.Pb$
452.29
Lead styphnate (Dry)
1,3-Benzenediol, 2,4,6-trinitro-, lead(2+) salt (1:1) (9CI)
Estifnato de plomo (Spanish)
Lead styphnate
Lead trinitroresorcinate
Styphnate de plomb (French)

15250-22-3
$C_{10}H_{22}O$
158.29
1-Octanol, 2,7-dimethyl- (8CI,9CI)

15258-73-8
$C_7H_6Cl_2O$
177.03
Benzyl alcohol, 2,6-dichloro
Benzenemethanol, 2,6-dichloro- (9CI)
2,6-Dichlorobenzyl alcohol

15262-20-1
Ra
228
Radium, Isotope of mass 228 (8CI,9CI)
Mesothorium 1

15263-52-2
$C_7H_{15}N_3O_2S_2.ClH$
273.83
[Cl-].CN(C)C(CSC(N)=O)CSC(N)=O
Carbamic acid, thio-, S,S'-(2-(dimethylamino)trimethylene) ester, hydrochloride
1,3-Bis(carbamoylthio)-2-(N,N-dimethylamino)propane hydrochloride
Cadan
Caldan
Carbamothioic acid, S,S'-(2-(dimethylamino)-1,3-propanediyl) ester, monohydrochloride (9CI)
Cartap
Cartap Hydrochloride
1,3-Dicarbamoylthio-2-(N,N-dimethylamino)propane hydrochloride
Dihydronereistoxin dicarbamate
S,S'-(2-(Dimethylamino)trimethylene)bis(thiocarbamate) hydrochloride
Nereistoxin dicarbamate hydrochloride
NTD 2
PADAN
PADAN 4 G
PATAP
Propane, 1,3-bis(carbamoylthio)-2-(N,N-dimethylamino)-, hydrochloride
1,3-Propanedithiol, 2-(dimethylamino)-, dicarbamate (ester), hydrochloride
Sanvex
Thiobel
Thiocarbamic acid-S,S-(2-(dimethylamino)trimethylene) ester hydrochloride
TI-1258
Vegetox

15267-77-3
$C_{10}H_{12}O_5S$
244.27
Acetic acid, (3-methyl-4-(methylsulfonyl)phenoxy)- (9CI)
Acetic acid, ((4-(methylsulfonyl)-m-tolyl)oxy)- (8CI)

15268-40-3
$C_{25}H_{54}N_2$
382.71
N(CCCCCCCCCCCCCCCCCCCCCC)CCCN
1,3-Propanediamine, N-docosyl- (9CI)
N-Docosyl-1,3-propanediamine

15271-41-7
$C_{10}H_{12}ClN_3O_2$
241.66
Bicyclo(2.2.1)heptane-2-carbonitrile, 5-chloro-6-((((methylamino)carbonyl)oxy)imino)-, (1S-(1-α,2-β,4-α,5-α,6E))-
AI3-25962-X
Bicyclo(2.2.1)heptane-2-carbonitrile, 5-chloro-6-((((methylamino)carbonyl)oxy)imino)-
endo-3-Chloro-exo-6-cyano-2-norbornanone O-(methylcarbamoyl)oxime
2-exo-Chloro-6-endo-cyano-2-norbornanone O-(methyl carbamoyl) oxime
2-exo-Chloro-6-endo-cyano-2-norbornanone-O-(methylcarbamoyl)oxime
3-Chloro-6-cyano-2-norbornanone-O-(methylcarbamoyl)-oxime
3-Chloro-6-cyanonorbornanone-2 oxime O,N-methylcarbamate
5-Chloro-6-((((methylamino)carbonyl)oxy)imino)bicyclo-(2.2.1)heptane-2-carbonitrile
exo-5-Chloro-6-oxo-endo-2-norbornanecarbonitrile O-(methylcarbamoyl)oxime
Compound UC-20047 A
ENT 25,962
2-Norbornanone, endo-3-chloro-exo-6-cyano-, O-(methylcarbamoyl)oxime
Tranid
UC 20,047A
UC 20047
UC 26089
Union Carbide UC 20047

15284-51-2
$C_{14}H_{28}O_2.1/2Ca$
248.42
Tetradecanoic acid, calcium salt (9CI)
Calcium tetradecanoate

15287-32-8
$C_{10}H_4Br_3ClN_2O$
443.31
1H-Imidazole, 2,4,5-tribromo-1-(4-chlorobenzoyl)- (9CI)
Imidazole, 2,4,5-tribromo-1-(p-chlorobenzoyl)- (8CI)

15288-12-7
$C_9H_{16}OS_2$
204.36
Carbonic acid, dithio-, S-methyl O-(2-methylcyclohexyl) ester, cis- (8CI)

15299-99-7
$C_{17}H_{21}NO$
255.39
CCN(CC)C(=O)C(C)Oc1cccc2ccccc12
Propionamide, N,N-diethyl-2-(1-naphthyloxy)
Devrinol
N,N-Diethyl-2-(1-naphthalenyloxy)propionamide
N,N-Diethyl-2-(1-naphthyloxy)-propionamide
2-(α-Naphthoxy)-N,N-diaethyl-

propionsaeureamid (German)
2-(α-Naphthoxy)-N,N-diethyl-propionamide
Napropamide
R-7465
R-7475

15301-48-1
$C_{31}H_{32}N_4O_2$
492.60
Bezitramide
2-Benzimidazolinone, 1-(1-(3-cyano-3,3-diphenylpropyl)-4-piperidyl)-3-propionyl-
Benzitramide
Bezitramida (Spanish)
Bezitramidum (Latin)
Burgodin
1-(3-Cyano-3,3-diphenylpropyl)-4-(2-oxo-3-propionyl-1-benzimidazolinyl)piperidine
1-(1-(3-Cyano-3,3-diphenylpropyl)-4-piperidyl)-3-propionyl-2-benzimidazolinone
DEA No. 9800
R 4845

15307-79-6
$C_{14}H_{10}Cl_2NO_2.Na$
318.14
[Na+].[O-]C(=O)Cc1ccccc1Nc2c(Cl)cccc2Cl
Acetic acid, o-(2,6-dichloroanilino)phenyl-, monosodium salt
Benzeneacetic acid, 2-((2,6-dichlorophenyl)amino)-, monosodium salt
(o-(2,6-Dichloroanilino)phenyl)-acetic acid monosodium salt
(o-(2,6-Dichloroanilino)phenyl)-acetic acid sodium salt
2-((2,6-Dichlorophenyl)amino)-benzeneacetic acid monosodium salt
Dichronic
Diclofenac sodium
Diclophenac sodium
GP 45840
Kriplex
Neriodin
Ortofen
Prophenatin
Sodium (o-(2,6-dichloroanilino)-phenyl)acetate
Sodium (o-((2,6-dichlorophenyl)amino)phenyl)acetate
Tsudohmin
Valetan
Voltaren
Voltarol

15310-01-7
$C_{13}H_{10}INO$
323.14
Ic1ccccc1C(=O)Nc2ccccc2
Benzanilide, 2-iodo
BAS-3170
BAS 3170F
Benodanil
Calirus
2-Iodobenzoic acid anilide
2-Iodo-N-phenylbenzamide

15333-24-1
$C_4N_4Zn.2Na$
215.42
Zincate(2-), tetrakis(cyano-C)-, disodium, (T-4)- (9CI)
Disodium tetracyanozincate
(T-4)-Disodium tetrakis(cyano-C)zincate(2-)
Sodium zinc cyanide
(T-4)-Tetrakis(cyano-C)zincate-(2-) disodium

15336-82-0
$C_{12}H_{18}N_2O_4$
254.32
O=C(N(C(C1=O)(CC)C)CC(O2)C2)N1CC(O3)C3
Hydantoin, bis(2,3-epoxypropyl)-5-ethyl-5-methyl
Bis(2,3-epoxypropyl)-5-ethyl-5-methylhydantoin
5-Ethyl-1,3-diglycidyl-5-methylhydantoin
2,4-Imidazolidinedione, 5-ethyl-5-methyl-1,3-bis(oxiranylmethyl)-

15337-18-5
$C_{22}H_{44}N_2S_4Zn$
530.25
Zinc, bis(dipentylcarbamodithioato-S,S')-, (T-4)- (9CI)
(T-4)-Bis(dipentylcarbamodithioato-S,S')zinc

15337-60-7
Unknown
Unknown
[Cd].[Ba+2].CCCCCCCCCCCC([O-])=O.CCCCCCCCCCCC([O-])=O.OC(=O)CCCCCCCCCCCCCCCCCCCCCCC([O-])=O
Lauric acid, barium cadmium salt
Barium cadmium laurate

15339-38-5
$C_6H_{12}FeN_2S_4$
296.27
Iron, bis(dimethylcarbamodithioato-S,S')- (9CI)
Carbamodithioic acid, dimethyl-, iron(2+) salt
Iron, bis(dimethyldithiocarbamato)- (8CI) (VAN)
Trifungol

15340-76-8
$C_{17}H_{24}O$
244.38
3-Phenanthrenol, 4b,5,6,7,8,8a,9,10-octahydro-4b,8,8-trimethyl-, (4bS-trans)- (9CI)
Podocarpa-8,11,13-trien-12-ol (8CI)

15347-57-6
$C_2H_4O_2.xPb$
1510.39
Acetic acid, lead salt
Lead acetate

15352-77-9
$C_{15}H_{26}O$
222.37
3-Cyclohexen-1-ol, 1-(1,5-dimethyl-4-hexenyl)-4-methyl- (8CI,9CI) (VAN)
β-Bisabolol

15356-70-4
$C_{10}H_{20}O$
156.30
OC(C(CCC1C)C(C)C)C1
dl-Menthol
Cyclohexanol, 5-methyl-2-(1-methylethyl)-, (1-α,2-β,5-α)
4-Isopropyl-1-methylcyclohexan-3-ol
Menthol, cis-1,3-trans-1,4-(+-)-
(+-)-Menthol
3-p-Menthol
dl-3-p-Menthanol
Menthol racemic
NCI-C50000
Menthol racemique (French)

15356-74-8
$C_{11}H_{16}O_2$
180.25
2-Hydroxy-2,6,6-trimethylcyclohexylideneacetic acid γ-lactone
2(4H)-Benzofuranone, 5,6,7,7a-tetrahydro-4,4,7a-trimethyl-(VAN)(9CI)

15359-99-6
$C_{11}H_{16}O_2$
180.25
Benzene, 1-(1,1-dimethylethoxy)-3-methoxy- (9CI)
Benzene, 1-tert-butoxy-3-methoxy- (8CI)
1-tert-Butoxy-3-methoxybenzene

15366-08-2
C_4H_9ClMg
116.87
Magnesium, chloro(1-methylpropyl)- (9CI)
Chloro(1-methylpropyl)magnesium

15375-84-5
$C_{10}H_{12}MnN_2O_8.2Na$
389.11
Manganate(2-), ((ethylenedinitrilo)tetraacetato)-, disodium (8CI)
Disodium ((ethylenedinitrilo)-tetraacetato)manganese
Disodium manganese EDTA
Disodium manganese ethylenediaminetetraacetate
EDTA disodium manganese salt
Ethylenediaminetetraacetic acid, disodium manganese salt
Glycine, N,N'-1,2-ethanediylbis(N-(carboxymethyl)-, manganese disodium salt
Manganate(2-), ((N,N'-1,2-ethanediylbis(N-(carboxymethyl)glycinato))(4-)-N,N',O,O',ON,ON')-, disodium, (OC-6-21)- (9CI)
Manganese disodium ethylene diamine tetraacetate
NSC-7345

15385-57-6
Hg_2I_2
654.99
Mercurous iodide
Dimercury diiodide
Iodure de mercure (French)
Mercurous iodide, Solid
Mercury diiodide
Mercury iodide [UN 1638]
Mercury protoiodide
UN 1638 [Mercury iodide]
Yellow Mercury Iodide
Yoduro de Mercurio (Spanish)

15402-94-5
$C_{10}H_{16}$
138.25
Cycloheptene, 5-ethylidene-1-methyl- (8CI, 9CI)

15404-57-6
$C_{10}H_{18}O$
166.27
2-Oxabicyclo(2.2.1)heptane, 1,3,3,7-tetramethyl-, (1R,4S,7S)-(+)- (8CI)

15404-63-4
$C_{15}H_{28}$
208.39
Naphthalene, decahydro-1,8a-dimethyl-7-(1-methylethyl)-, (1R-(1α,4aβ,7β,8aα))-(9CI)
4βH,5α-Eremophilane (8CI)
Nootkatane
Nootkatane, (+)-
(+)-Nootkatane

15414-25-2
Unknown
Unknown
Nitrilotriacetic acid, yttrium(3+) salt (1:1)
NTA, yttrium(3+) salt (1:1)

15418-16-3
$C_{33}H_{16}Cu_2N_6O_{17}S_4.4Na$
1115.80
Cuprate(4-), (mu-((7,7'-(carbonyldiimino)bis(4-hydroxy-3-((2-hydroxy-5-sulfophenyl)-azo)-2-naphthalenesulfonato))(8-)))di-, tetrasodium (9CI)

15422-74-9
Unknown
Unknown
Polonium-218
Radium A

15426-14-9
$C_{12}H_8Cl_2N_2$
251.10
Diazene, bis(3-chlorophenyl)-(9CI)
Azobenzene, 3,3'-dichloro-(8CI)

15427-93-7
$C_{19}H_{24}N_{10}O.2CH_4O_3S$
600.61
Carbanilide, 4,4'-diacetyl-, 4,4'-bis(amidinohydrazone), dimethanesulfonate (8CI)
DDUG
DDUG diMS
Hydrazinecarboximidamide, 2,2'-(carbonylbis(imino-4,1-phenyleneethylidyne))bis-, dimethanesulfonate (9CI)
NSC-109555

15438-71-8
$C_5H_9NO_2$
115.13
O=C(N(CC1)CO)C1
2-Pyrrolidinone, 1-(hydroxymethyl)- (9CI)
AI3-28014
N-(Hydroxymethyl)-2-pyrrolidinone
1-(Hydroxymethyl)-2-pyrrolidinone
1-Hydroxymethyl-2-pyrrolidinone
N-(Hydroxymethyl)-2-pyrrolidone
1-(Hydroxymethyl)-2-pyrrolidone
N-Methylolpyrrolidinone
N-Methylol-2-pyrrolidinone
N-Methylolpyrrolidone
N-Methylol-2-pyrrolidone
NSC-84227

15438-85-4
$C_{12}H_{23}N_5O$
253.32
1,3,5-Triazine-2,4-diamine, N,N,N',N'-tetraethyl-6-methoxy- (9CI)
Tetraetatone
s-Triazine, 2,4-bis(diethylamino)-6-methoxy- (8CI)

15441-06-2
$C_8H_{14}O_4S_2$
238.33
O=C(OC)CCSSCCC(=O)OC
Propanoic acid, 3,3'-dithiobis-, dimethyl ester (9CI)
Dimethyl 3,3'-dithiobispropanoate
Dimethyl dithiodipropionate
Dimethyl 3,3'-dithiodipropionate

15442-64-5
$C_{34}H_{30}N_4O_4Zn.2H$
626.02
Zinc protoporphyrin
(SP-4-2)-Dihydrogen (7,12-diethenyl-3,8,13,17-tetramethyl-21H,23H-porphine-2,18-dipropanoato(4-)-N(21),N(22), N(23),N(24))zincate(2-)
Protoporphyrin IX, zinc chelate
Zincate(2-), (7,12-diethenyl-3,8,13,17-tetramethyl-21H, 23H-porphine-2,18-dipropanoato(4-)-N21,N22,N23,N24)-, dihydrogen, (SP-4-2)- (9CI)

15446-39-6
$C_{16}H_{16}N_4O_2$
296.36
O=C(Nc(ccc(N=Nc(ccc(NC(=O)C)c1)c1)c2)c2)C
Acetanilide, 4',4'''-azobis
p,p'-Azodiacetanilide
4,4'-Bis(acetamido)azobenzene

15448-99-4
$C_8H_7NO_3S$
197.21
1,2-Benzisothiazol-3(2H)-one, 2-methyl-, 1,1-dioxide (9CI)
AI3-23674
1,2-Benzisothiazolin-3-one, 2-methyl-, 1,1-dioxide (8CI)
2-Methyl-1,2-benzisothiazole-3(2H)-one-1,1-dioxide
N-Methylsaccharin
NSC-39120

15457-05-3
$C_{13}H_7F_3N_2O_5$
328.22
O=N(=O)c(c(Oc(ccc(N(=O)=O)

c1)c1)ccc2C(F)(F)F)c2
Ether, (p-nitrophenyl) (α,α,α-trifluoro-2-nitro-p-tolyl)
C-6989
2,4'-Dinitro-4-trifluoromethyl-diphenyl ether
Fluorodifen
Fluorodifene
2-Nitro-1-(4-nitrophenoxy)-4-(trifluoromethyl)benzene
p-Nitrophenyl 2-nitro-4-(trifluoromethyl) phenyl ether
4-Nitrophenyl α,α,α-trifluoro-2-nitro-p-tolyl ether
p-Nitrophenyl α,α,α-trifluoro-2-nitro-p-tolyl ether
Preforan
Soyex

15459-85-5
$C_{27}H_{44}O$
384.65
Cholest-7-en-3-one, (5α)- (9CI)
5α-Cholest-7-en-3-one (8CI)

15467-15-9
$C_2H_9ClN_2$
96.56
1,2-Ethanediamine, hydrochloride (9CI)
Ethylenediamine, hydrochloride (8CI) (VAN)

15467-20-6
$C_6H_7NO_6.2Na$
235.10
Nitrilotriacetic acid and its salts
Acetic acid, nitrilotri-, disodium salt
N,N-Bis(carboxymethyl)glycine disodium salt
N,N-Bis(carboxymethyl)-glycine disodium salt
Disodium nitrilotriacetate
Glycine, N,N-bis(carboxymethyl)-, disodium salt (9CI)
Kiresuto NTB
Nitrilotriacetic acid, disodium salt

15468-32-3
O_2Si
60.09
Silica, Crystalline-tridymite (ACGIH,OSHA)
Tridimite (French)
Tridymite

15468-86-7
$C_9H_{12}ClN_5$
225.65
1,3,5-Triazine-2,4-diamine, 6-chloro-N,N'-di-2-propenyl- (9CI)
s-Triazine, 2,4-bis(allylamino)-6-chloro- (8CI)

15471-17-7
$C_8H_{11}NO_3S$
201.24
O=S(=O)(O)CCCn(cccc1)c1
Pyridinium, 1-(3-sulfopropyl)-, hydroxide, inner salt (9CI)
1-(1-Pyridyl)propane-3-sulfonate
N-3-Sulfopropylpyridinium betaine
1-(3-Sulfopropyl)pyridinium hydroxide, inner salt

15475-56-6
$C_{20}H_{21}N_8O_5.Na$
476.38
Methotrexate sodium
Glutamic acid, N-(p-(((2,4-diamino-6-pteridinyl)methyl)-methylamino)benzoyl)-, sodium salt
MTX sodium

15477-33-5
$Al.3ClHO_3$
841.07
Chloric acid, aluminum salt (9CI)
Aluminum chlorate

15481-70-6
$C_7H_{10}N_2.2ClH$
195.11
Cl.Cl.Cc1c(N)cccc1N
Toluene-2,6-diamine, dihydrochloride
2,6-Diaminotoluene dihydrochloride
NCI-C50317

15482-54-9
$C_{10}H_{10}O_3$
178.19
Benzoic acid, 4-(2-oxopropyl)- (9CI)
Benzoic acid, p-acetonyl- (8CI)
4-(2-Oxopropyl)benzoic acid

15500-66-0
$C_{35}H_{60}N_2O_4.2Br$
732.79
[Br-].[Br-].CC(=O)OC5CC4CCC1C(CCC2(C)C1CC(C2OC(C)=O)[N+]3(C)CCCCC3)C4(C)CC5[N+]6(C)CCCCC6
Piperidinium, 1,1'-(2-β,16-β-(3-α,17-β-dihydroxy-5-α-androstanylene)) bis-(1-methyl-, dibromide, diacetate
5α-Androstan-3α,17β-diol, 2β,16β-dipipecolinio-, dibromide, diacetate
1,1'-(3,17-Bis(acetyloxy)androstane-2,16-diyl)bis(1-methylpiperidinium) dibromide
3-α,17-β-Diacetoxy-2-β,16-β-dipiperidino-5-α-androstane dimethobromide
Mioblock
NA 97
ORG NA 97
Pancuronium bromide
Pavulon

15502-74-6
Unknown
Unknown
Arsenite

15503-86-3
$C_{18}H_{25}NO_7$
367.44
Retrorsine, N-oxide
Isatidine
cis-Retronecic acid ester of retronecine-N-oxide
Retrorsine oxide
Senecionan-11,16-dione, 12,18-dihydroxy-, 4-oxide (9CI)

15505-13-2
$C_8H_{19}O_4P.2Na$
256.19
Phosphoric acid, mono(2-ethylhexyl) ester, disodium salt (9CI)

15506-53-3
$C_4H_4O_2$
84.08
1,3-Cyclobutanedione (7CI,8CI,9CI)

15511-81-6
$C_{12}H_{26}N_2O_4$
262.35
Hexanedioic acid, Compd. with 1,6-hexanediamine (9CI)

15516-76-4
$C_7H_5ClO_2.1/2Hg$
256.86
Benzoic acid, 4-chloro-, mercury(2+) salt (9CI)
Benzoic acid, p-chloro-, mercury(2+) salt (8CI)

15519-38-7
$C_4H_4Cl_2O_4$
186.98
Butanedioic acid, 2,2-dichloro- (9CI)
Succinic acid, 2,2-dichloro- (8CI)

15520-10-2

$C_6H_{16}N_2$
116.19
NCC(CCCN)C
1,5-Pentanediamine, 2-methyl- (9CI)
2-Methyl-1,5-pentanediamine

15520-11-3
$C_{22}H_{38}O_6$
398.54
O=C(OC(CCC(C(C)(C)C)C1)C1)OOC(=O)OC(CCC(C(C)(C)C)C2)C2
Bis(4-tert-butylcyclohexyl)peroxydicarbonate
Di-(4-tert-butylcyclohexyl)peroxydicarbonate
Di-(4-tert-butylcyclohexyl)peroxydicarbonate (Not more than 42% stable dispersion, in water)
Di-(4-tert-butylcyclohexyl)peroxydicarbonate (Technically pure)
Peroxidicarbonato de di-(4-terc-butilciclohexilo) (Spanish)
Peroxydicarbonate de bis (tert-butyl-4 cyclohexyle) (French)
Peroxydicarbonic acid, bis(4-(1,1-dimethylethyl)cyclohexyl) ester (9CI)
UN 2154
UN 2894

15521-65-0
$C_6H_{12}N_2S_4.Ni$
299.15
Nickel, bis(dimethyldithiocarbamato)

15535-29-2
$C_2H_7NO.1/2H_2O_3S$
102.12
Ethanol, 2-amino-, sulfite (2:1) (Salt) (9CI)
2-Aminoethanol sulfite (2:1) (Salt)

15541-45-4
Br
79.91
Bromate
Bromato (Spanish)

15545-48-9
$C_{10}H_{13}ClN_2O$
212.70
CN(C)C(=O)Nc1ccc(C)c(Cl)c1
Urea, 3-(3-chloro-p-tolyl)-1,1-dimethyl
C 2242
3-(3-Chlor-4-methylphenyl)-1,1-dimethylharnstoff (German)
N-(3-Chloro-4-methylphenyl)-N',N'-dimethylurea
3-(3-Chloro-4-methylphenyl)-1,1-dimethyl-urea
Chlorotoluron
Chlortoluron
Clortokem
Dicuran

15546-11-9
$C_{18}H_{28}O_8Sn$
491.15
Stannane, bis(methoxymaleoyloxy)dibutyl
Dibutyltin bis(methyl maleate)
Dibutyltin bis(monomethyl maleate)
Dibutyltin methyl maleate
Di-n-butylzinn-dimonomethylmaleinat (German)
Stan-Guard 156
Stannane, dibutylbis((3-carboxyacryloyl)oxy)-, dimethyl ester (Z,Z)- (8CI)
5,7,12-Trioxa-6-stannatrideca-2,9-dienoic acid, 6,6-dibutyl-4,8,11-trioxo-, methyl ester

15546-16-4
$C_{24}H_{40}O_8Sn$
575.33
Stannane, bis(butoxymaleoyloxy)dibutyl
Di-n-butyltin di(monobutyl)-maleate
Di-n-butyl-zinn-di(monobutyl)-maleinat (German)
Stannane, bis(monobutoxymaleoyloxy)dibutyl-

15547-17-8
$C_{16}H_{16}O_2$
240.32
O=C(c(c(C(=O)C=1CCCC2)ccc3CC)c3)C12
Anthraquinone, 2-ethyl-5,6,7,8-tetrahydro
9,10-Anthracenedione, 6-ethyl-1,2,3,4-tetrahydro-
2-Ethyl-5,6,7,8-tetrahydroanthraquinone
USAF SO-2

15566-80-0
$C_{22}H_{45}N_3O$
367.60
O=C(NCCNCCN)CCCCCCCC=CCCCCCCCC
9-Octadecenamide, N-(2-((2-aminoethyl)amino)ethyl)-, (Z)- (9CI)
Oleic acid, diethylenetriamine amide

15571-58-1
$C_{36}H_{72}O_4S_2Sn$
751.89
8-Oxa-3,5-dithia-4-stannatetradecanoic acid, 10-ethyl-4,4-dioctyl-7-oxo-, 2-ethylhexyl ester
Di-n-octyltin bis(2-ethylhexyl mercaptoacetate)
Di-n-octyltin-dithioglycolic acid 2-ethylhexyl ester
Di-n-octyltin-2-ethylhexyl-dimercaptoethanoate
OTS 11
Tin, bis(mercaptoacetate)dioctyl-, bis(2-ethylhexyl) ester
Tin, dioctyl-, bis(2-ethylhexylthioglycolate)

15576-47-3
BrPb
367.01
Lead bromide (PbBr) (6CI,8CI,9CI)
Lead monobromide

15578-51-5
$H_3O_4P.3/_4Ti$
133.92
Phosphoric acid, titanium(4+) salt (4:3) (9CI)
Titanium phosphate

15584-04-0
AsO_4
138.92
Arsenate (8CI,9CI)
Arsenic acid, ion(3-)

15588-95-1
$C_{13}H_{21}NO_2$
223.31
DOM
DEA No. 7395
2,5-Dimethoxy-4-methylamphetamine
"DOM" and "STP"
4-Methyl-2,5-dimethoxyamphetamine
4-Methyl-2,5-dimethoxy-α-methylphenethyl-amine
Phenethylamine, 2,5-dimethoxy-4-ethyl-α-methyl-
STP

15592-74-2
$C_9H_{12}O_3S$
200.26
Benzenesulfonic acid, 4-propyl- (9CI)
Benzenesulfonic acid, p-propyl- (7CI,8CI)

15598-34-2
$C_5H_5N.ClHO_4$
179.55
Pyridine perchlorate
NSC-249268
Pyridine, perchlorate (8CI,9CI)
Pyridinium perchlorate

15619-48-4
$C_{16}H_{14}N.Cl$
255.75
Quinolinium, 1-(phenylmethyl)-, chloride (9CI)
AI3-51333
Benzylquinolinium chloride
N-Benzylquinolinium chloride
1-Benzylquinolinium chloride
NSC-190376
1-(Phenylmethyl)quinolinium chloride
Quinoline-N-benzyl chloride quaternary
Quinolinium, 1-benzyl-, chloride (8CI)

15625-88-4
$C_4H_5N_3O$
111.10
1H-1,2,4-Triazole, 1-acetyl- (6CI,7CI,8CI,9CI)
1-Acetyltriazole
1-Acetyl-1,2,4-triazole
1-Acetyl-1H-1,2,4-triazole
N-Acetyl-1,2,4-triazole

15625-89-5
$C_{15}H_{20}O_6$
296.35
O=C(OCC(CC)(COC(=O)C=C)C OC(=O)C=C)C=C
Acrylic acid, triester with 2-ethyl-2-(hydroxymethyl)-1,3-propanediol
Acrylic acid, 1,1,1-(trihydroxymethyl)propane triester
2-Ethyl-2-(hydroxymethyl)-1,3-propanediol triacrylate
MFM
NK Ester A-TMPT
1,3-Propanediol, 2-ethyl-2-(hydroxymethyl)-, triacrylate
2-Propenoic acid, 2-ethyl-2-(((1-oxo-2-propenyl)oxy)-methyl)-1,3-propanediyl ester (9CI)
Sartomer SR 351
SR 351
TMPTA
Trimethylolpropane triacrylate

15640-03-6
$C_6H_8ClN_3$
157.60
1,3,5-Triazine, 2-(chloromethyl)-4,6-dimethyl- (9CI)
s-Triazine, 2-(chloromethyl)-4,6-dimethyl- (7CI,8CI)

15640-05-8
$C_6H_6Cl_3N_3$
226.49
1,3,5-Triazine, 2,4-dimethyl-6-(trichloromethyl)- (9CI)
s-Triazine, 2,4-dimethyl-6-(trichloromethyl)- (7CI, 8CI)

15640-10-5
$C_6H_4Cl_5N_3$
295.38
s-Triazine, 2-(dichloromethyl)-4-methyl-6-(trichloromethyl)- (8CI)

15646-96-5
$C_{11}H_{18}N_2O_2$
210.26
O=C=NCC(CC(CCN=C=O)(C) C)C
Hexane, 1,6-diisocyanato-2,4,4-trimethyl- (9CI)
1,6-Diisocyanato-2,4,4-trimethylhexane
2,4,4-Trimethyl-1,6-diisocyanatohexane

15647-08-2
$C_{20}H_{27}O_3P$
346.41
O(P(Oc(cccc1)c1)Oc(cccc2)c2) CC(CCCC)CC
Phosphorous acid, 2-ethylhexyl diphenyl ester (9CI)
Diphenyl 2-ethylhexyl phosphite
2-Ethylhexyl diphenyl phosphite

15648-73-4
$C_{19}H_{26}N_8O_{14}P_2$
652.36
5'-Cytidylic acid, adenylyl-(5'-3')- (9CI)
Adenosine, 5'-O-phosphorylcytidylyl-(3'-5')- (8CI)

15659-56-0
$C_{14}H_{28}Cl_4Cr_2O_3$
490.18
Chromium, tetrachloro-mu-hydroxy(mu-(tetradecanoato-O:O'))di- (9CI)
Tetradecanoato chromic chloride hydroxide (1:2:4:1)
Myristo chromic chloride

15662-33-6
$C_{25}H_{35}NO_9$
493.55
Ryanodine
Bonide Ryatox
Ground Ryania Specisa(vahl) Stemwood (Alkoloid Ryanodine)
Ryanexel
Ryania
Ryania Powder
Ryania Speciosa
Ryania Speciosa, Powdered stems of
Ryanicide

15663-27-1
$Cl_2H_6N_2Pt$
300.07
Platinum(II), diamminedichloro-, cis
CACP
CDDP
CPDD
Cisplatin
Cisplatino (Spanish)
Cisplatyl
CPDC
DDP
cis-DDP
Diamminedichloroplatinum
cis-Diaminedichloroplatinum
cis-Diaminodichloroplatinum(II)
cis-Diamminedichloroplatinum
cis-Dichlorodiammineplatinum
cis-Dichlorodiammine platinum(II)
cis-Dichlorodiamminoplatinum(II)
NCI-C55776
Neoplatin
NSC-119875
Peyrone's Chloride
Platiblastin
cis-Platin
Platinex
Platinol
cis-Platinous diaminodichloride
cis-Platinous diammine dichloride
cis-Platinum
cis-Platinum(II)
cis-Platinum(II) diamminedichloride
Platinum, diamminedichloro-, (SP-4-2)- (9CI)
Pt-01
cisPT(II)

15666-29-2
$C_8H_{18}S_3Sn_2$
447.85
Distannathiane, dibutyldithioxo- (9CI)
Dibutyldithioxodistannathiane
1,3-Dibutyl-1,3-dithioxodistannthiane
Distannthiane, 1,3-dibutyl-1,3-dithioxo-
Monobutyltin sulfide

15676-16-1
$C_{15}H_{23}N_3O_4S$
341.47
CCN1CCCC1CNC(=O)c2cc(ccc2 OC)S(N)(=O)=O
o-Anisamide, N-((1-ethyl-2-pyrrolidinyl)methyl)-5-sulfamoyl
Abilit
Aiglonyl
5-(Aminosulfonyl)-N-((1-ethyl-2-pyrrolidinyl)methyl)-2-methoxybenzamide

Benzamide, 5-(aminosulfonyl)-N-((1-ethyl-2-pyrrolidinyl)-methyl)-2-methoxy- (9CI)
Coolspan
Dobren
Dogmatil
Dogmatyl
Eglonyl
N-((1-Ethyl-2-pyrrolidinyl)-methyl)-2-methoxy-5-sulfamoylbenzamide
N-((1-Ethyl-2-pyrrolidinyl)-methyl)-5-sulfamoyl-o-anisamide
Guastil
Miradol
Mirbanil
Misulvan
Omperan
Pyrkappl
R.D. 1403
Sernevin
Splotin
Sulpirid
Sulpiride
Sulpyrid
Sursumid
Trilan

15679-12-6
C_6H_9NS
127.20
N(C(=CS1)C)=C1CC
Thiazole, 2-ethyl-4-methyl- (9CI)
2-Ethyl-4-methylthiazole

15679-24-0
$C_{18}H_{14}$
230.31
2,7-Dimethylpyrene
Pyrene, 2,7-dimethyl-

15684-36-3
BF_4.3H2O.1/2Ni
170.19
Borate(1-), tetrafluoro-, nickel(2+) (2:1), hexahydrate (9CI)
Borate(1-), tetrafluoro-, nickel(2+), hexahydrate (8CI)

15686-61-0
$C_{12}H_{16}N_2$
188.26
Fenproporex
DEA No. 1575
Femproporex (Spanish)
Fenproporexum (Latin)
Propanenitrile, 3-((1-methyl-2-phenylethyl)amino)-, (+-)- (9CI)
Propionitrile, 3-((α-methyl-phenethyl)amino)-, (+-)-

15686-71-2
$C_{16}H_{17}N_3O_4S$
347.42
CC1=C(N3C(SC1)C(NC(=O)C(N)c2ccccc2)C3=O)C(O)=O
5-Thia-1-azabicyclo(4.2.0)oct-2-ene-2-carboxylic acid, 7-(2-amino-2-phenylacetamido)-3- methyl-8-oxo-, D
7-(D-α-Aminophenylacetamido)desacetoxycephalosporanic acid
7-(D-2-Amino-2-phenylacetamido)-3-methyl-δ^3-cephem-4- carboxylic acid
Cefa-Iskia
Cefalexin
Cefaloto
Cephalexin
Ceporex
Ceporexin
Ceporexine
CEX
Keflex
Keforal
Larixin
Lexibiotico
Madlexin
Neolexina
Oracef
Oroxin
Ortisporina
S 6437
Sartosona
Sencephalin
Syncl

15686-83-6
$C_{11}H_{14}N_2S$
206.33
Pyrimidine, 1,4,5,6-tetrahydro-1-methyl-2-(2-(2-thienyl)vinyl)-, (E)
Antiminth
Banminth
CP-10,423-16
Pyrequan
Pyrantel
E-1,4,5,6-Tetrahydro-1-methyl-2-(2-(2-thienyl)vinyl)pyrimidine

15686-91-6
$C_{16}H_{25}N_3O$
275.38
Propiram
DEA No. 9649
(+-)-N-(1-Methyl-2-piperidinoethyl)-N-2-pyridylpropionamide
N-(1-Methyl-2-(1-piperidinyl)ethyl)-N-2-pyridinylpropanamide
Propanamide, N-(1-methyl-2-(1-piperidinyl)ethyl)-N-2-pyridinyl-
Propionamide, N-(1-methyl-2-piperidinoethyl)-N-2-pyridyl-, (+-)-
Propiramo (Spanish)
Propiramum (Latin)

15687-27-1
$C_{13}H_{18}O_2$
206.31
O=C(O)C(c(ccc(c1)CC(C)C)c1)C
Hydratropic acid, p-isobutyl
Acide (isobutyl-4 phenyl)-2 propionique (French)
Adran
Anflagen
Artril 300
Benzeneacetic acid, α-methyl-4-(2-methylpropyl)-
Bluton
Brufanic
Brufen
Buburone
Butylenin
Dolgin
Emodin
Epobron
Ibufen
Ibuprocin
Ibuprofen
IP-82
p-Isobutylhydratropic acid
4-Isobutylhydratropic acid
2-(4-Isobutylphenyl)propanoic acid
α-p-Isobutylphenylpropionic acid
α-(4-Isobutylphenyl)propionic acid
2-(p-Isobutylphenyl)propionic acid
Lamidon
Liptan
Motrin
Mynosedin
Napacetin
Nobfelon
Nobfen
Nobgen
Nurofen
R.D. 13621
Rebugen
Roidenin

15694-70-9
$CH_2O_2.H_2O$.1/2Ni
93.39
Formic acid, nickel(2+) salt, dihydrate (8CI,9CI)
Nickel diformate dihydrate
Nickel formate, dihydrate (7CI)
Nickel formate ($Ni(HCO_2)_2$) dihydrate
Nickel(2+) formate dihydrate

15696-43-2
$C_8H_{16}O_2$.xPb
Unknown
Lead caprylate
Lead octanoate
Lead octoate
Octanoic acid, lead salt (8CI, 9CI)

15699-18-0
$O_8S_2.Ni.2H_4N$
286.93
Sulfuric acid, ammonium nickel(2+) salt (2:2:1)
Ammonium disulfatonickelate(II)
Ammonium nickel sulfate
Nickel ammonium sulfate

15706-37-3
Pu
243
Plutonium, Isotope of mass 243 (8CI,9CI)

15707-23-0
$C_7H_{10}N_2$
122.19
n(c(c(nc1)CC)C)c1
Pyrazine, 2-ethyl-3-methyl
2-Ethyl-3-methyl pyrazine

15707-24-1
$C_8H_{12}N_2$
136.18
n(c(c(nc1)CC)CC)c1
Pyrazine, 2,3-diethyl- (9CI)
2,3-Diethylpyrazine

15707-34-3
$C_8H_{12}N_2$
136.18
Pyrazine, 5-ethyl-2,3-dimethyl- (8CI,9CI)

15708-41-5
$C_{10}H_{12}N_2O_8.Na.Fe$
367.08
[Na+].[Fe].OC(=O)CN(CCN(CC(O)=O)CC(O)=O)CC(O)=O
Acetic acid, (ethylenedinitrilo)tetra-, sodium salt, iron complex
Calmosine
Edathamil monosodium ferric salt
Ferisan
Ferrate(1-), ((ethylenedinitrilo)-tetraacetato)-, sodium (8CI)
Ferric sodium EDETATE
Ferric sodium EDTA
Monosodium Ferric EDTA
Rexene
Sequestrene NaFe Iron Chelate
Sodium Feredetate
Sodium ferric EDTA
Sodium iron EDTA
Sytron

15715-19-2
$C_{20}H_{12}Cl_2N_2O_2$
383.22
Oc(c(c(nc1Cc(c(nc(c2ccc3)c3Cl)C4)c2O)c(cc5)Cl)c5)c14
Quino(2,3-b)acridine-7,14-dione, 4,11-dichloro-5,6,12,13-tetrahydro- (9CI)
Quinacridone, 4,11-dichloro-6,13-dihydro-

15721-02-5
$C_{12}H_8Cl_4N_2$
322.02
Nc(c(cc(c1Cl)c(c(cc(N)c2Cl)Cl)c2)Cl)c1
Benzidine, 2,2',5,5'-tetrachloro
(1,1'-Biphenyl)-4,4'-diamine, 2,2',5,5'-tetrachloro- (9CI)
Tetrachlorobenzidine
2,2',5,5'-Tetrachlorobenzidine
3,3',6,6'-Tetrachlorobenzidine
2,2',5,5'-Tetrachloro-4,4'-diaminodiphenyl

15726-15-5
$C_8H_{16}O$
128.21
4-Heptanone, 3-methyl- (8CI, 9CI)

15735-67-8
Unknown
Unknown
Polonium 214
Radium C

15739-80-7
HO_4PbS
304.26
Lead sulfate
Sulfuric acid, lead salt (9CI)

15763-06-1
$C_{11}H_{15}N_5O_4$
281.24
1-Methyladenosine
Adenosine, N,6-didehydro-1,6-dihydro-1-methyl- (9CI)
Adenosine, 1-methyl- (VAN) (8CI)
N1-Methyladenosine
NSC-92165

15763-57-2
$C_6H_6O_8S_2.2K$
348.44
p-Benzenedisulfonic acid, 2,5-dihydroxy-, dipotassium salt (8CI)
1,4-Benzenedisulfonic acid, 2,5-dihydroxy-, dipotassium salt (9CI)
2,5-Dihydroxy-p-benzenedisulfonic acid, dipotassium salt
2,5-Dihydroxy-para-benzenedisulfonic acid dipotassium salt
NSC-158150

15772-26-6
$C_{14}H_{17}NO_5$
279.29
O=C(OCCN(c(cccc1)c1)CCO)C=CC(=O)O
Maleic acid, mono(2-(N-(2-hydroxyethyl)anilino)-ethyl) ester (8CI)

15774-82-0
$C_{14}H_{10}OS$
226.30
Oc(c(c(sc1cccc2)ccc3C)c3)c12
9H-Thioxanthen-9-one, 2-methyl- (9CI)
2-Methyl-9H-thioxanthen-9-one
2-Methylthioxanthone
NSC-263518
Thioxanthen-9-one, 2-methyl-

15782-05-5
$C_{18}H_{13}ClN_2O_6S.Sr$
508.44
2-Naphthalenecarboxylic acid, 4-((5-chloro-4-methyl-2-sulfophenyl)azo)-3-hydroxy-, strontium salt (1:1) (9CI)
4-((5-Chloro-4-methyl-2-sulfophenyl)azo)-3-hydroxy-2-naphthalenecarboxylic acid, strontium salt (1:1)
4-((5-Chloro-2-sulfo-p-tolyl)-azo)-3-hydroxy-2-naphthalenecarboxylic acid, strontium salt

15782-06-6
$C_{17}H_{12}N_2O_9S_2.3/_2Ba$
658.42
Benzoic acid, 2-((2-hydroxy-3,6-disulfo-1-naphthalenyl)-azo)-, barium salt (2:3) (9CI)

15790-07-5
Unknown
Unknown
C.I. Pigment Yellow 104 (9CI)
Certolake Sunset Yellow
2-Naphthalenesulfonic acid, 6-hydroxy-2-((4-sulfophenyl)-azo)-, aluminum lake

15791-78-3
$C_{22}H_{16}N_2O_7$
420.40
O=C(c(c(c(O)cc1)C(=O)c2c(O)ccc3N(=O)=O)c1Nc(ccc(c4)CCO)c4)c23
Anthraquinone, 1,8-dihydroxy-4-(p-(2-hydroxyethyl)anilino)-5-nitro
9,10-Anthracenedione, 1,8-dihydroxy-4-((4-(2-hydroxyethyl)phenyl)amino)-5-nitro-
C.I. 60767
C.I.Disperse Blue 27
1,8-Dihydroxy-4-(p-(2-hydroxy-

ethyl)anilino)-5-nitro-anthraquinone
1,8-Dihydroxy-4-(4'-β-hydroxy-ethyl)anilino-6-nitro-anthroquinone
Eastman Fast Blue B-GLF
Serisol Fast Blue BGLW

15792-67-3
$C_{37}H_{36}N_2O_9S_3.2/_3Al$
766.87
Benzenemethanaminium, N-ethyl-N-(4-((4-(ethyl((3-sulfophenyl)methyl)amino)-phenyl)(2-sulfophenyl)-methylene)-2,5-cyclohexa-dien-1-ylidene)-3-sulfo-, hydroxide, inner salt, aluminum salt (3:2) (9CI)

15793-73-4
$C_{34}H_{28}Cl_2N_8O_2$
651.51
O=C(N(N=C1C)c(ccc(c2)C)c2)C1N=Nc(c(cc(c(ccc(N=NC(C(=O)N(N=3)c(ccc(c4)C)c4)C3C)c5Cl)c5)c6)Cl)c6
Irgalite Orange F2G
4,4'-(3,3'-Dichloro-4,4'-bi-phenylene)bis(azo))bis(1-(4-methylphenyl)-3-methyl-5-pyrazolone)
3H-Pyrazol-3-one, 4,4'-((3,3'-dichloro(1,1'-biphenyl)-4,4'-diyl)bis(azo))bis(2,4-dihydro-5-methyl-2-(4-methylphenyl)-(9CI)

15825-70-4
$C_6H_8N_6O_{18}$
452.12
O=N(=O)OCC(ON(=O)=O)C(ON(=O)=O)C(ON(=O)=O)C(ON(=O)=O)CON(=O)=O
Mannitol hexanitrate
Dilangil
Hexanitrate de mannitol (French)
Hexanitrato de manitol (Spanish)
Hexanitrol
Hypertenain
Initiating Explosive Nitro Mannite
Manexin
Manhexin
Manicole
Manite
Mannex
Mannitol hexanitrate
D-Mannitol, hexanitrate (9CI)
Mannitol hexanitrate, Containing, by weight, at least 40% water [UN 0133]
Mannitoli hexanitras (Latin)
Mannitolo esanitrato
Mannitrin
Mannityl nitrate
Mannityli nitras
Mannitylium hexanitricum
Maxitate
Medemanol
Nitranitol
Nitro mannite
Nitromannite
Nitromannite (Dry)
Nitromannitol
SDM No.5
UN 0133 [Mannitol hexanitrate (Nitromannite), Wetted with not less than 40 per cent water, or mixture of alcohol and water, by mass]

15826-16-1
$C_{14}H_{30}O_5S.Na$
333.49
Ethanol, 2-(dodecyloxy)-, hydrogen sulfate, sodium salt
2-(Dodecyloxy)ethanol hydrogen sulfate sodium salt
Dodecyl sodium ethoxysulfate
Sodium lauryl ethoxysulphate

15826-37-6
$C_{23}H_{14}O_{11}.2Na$
512.35
[Na+].[Na+].OC(COc1cccc2oc(cc(=O)c12)C([O-])=O)COc3cccc4oc(cc(=O)c34)C([O-])=O
4H-1-Benzopyran-2-carboxylic acid, 5,5'-((2-hydroxytri-methylene)dioxy)bis(4-oxo-, disodium salt
Aarane
Aararre
Cromoglycate
Cromoglycate disodium
Cromolyn sodium
Cromolyn sodium salt
Disodium chromoglycate
Disodium cromoglicate
Disodium cromoglycate
Disodium 5,5'-((2-hydroxy-trimethylene)dioxy)-bis(4-oxo-4H-1-benzopyran-2-carboxyl-ate)
FPL 670
Frenasma
5,5'-((2-Hydroxytrimethylene)-dioxy)bis(4-oxo-4H-1-benzo-pyran-2-carboxylic acid) di-sodium salt
Inostral
Intal
Lomudal
Lomudas
Nalcrom
Nasmil
Rynacrom
Sodium cromoglycate
Sodium cromolyn

15827-60-8
$C_9H_{28}N_3O_{15}P_5$
573.18
O=P(O)(O)CN(CCN(CP(=O)(O)O)CP(=O)(O)O)CCN(CP(=O)(O)O)CP(=O)(O)O
Diethylenetriaminepenta-(methylenephosphonic) acid
Diethylenetriamine, penta-methylenepentaphosphonic acid
Phosphonic acid, (((phosphono-methyl)imino)bis(2,1-ethane-diylnitrilobis(methylene)))-tetrakis- (9CI)

15829-53-5
Hg_2O
417.18
Mercury oxide [UN 1641]
Mercurous oxide, Black, Solid (DOT)
Mercury(I) oxide
Quecksilberoxid (German)
UN 1641 [Mercury oxide]

15834-05-6
$C_{23}H_{44}O_4$
384.60
O=C(OCC(C)(C)COC(=O)CCCCCCC)CCCCCCCC
Nonanoic acid, 2,2-dimethyl-1,3-propanediyl ester (9CI)
2,2-Dimethyl-1,3-propanediyl nonanoate
Neopentyl glycol dipelargonate

15843-02-4
CH_3NiO_2
105.74
Nickel formate
AI3-26109
Formic acid, nickel salt
Formic acid, nickel(2+) salt

15844-52-7
Unknown
Unknown
Cupric nitrilotriacetate
Cu-NTA
Nitrilotriacetic acid, copper(2+) complex
NTA, copper(2+) complex

15845-52-0
$H_3O_4P.Pb$
305.20
Lead hydrogen phosphate
Phosphoric acid, lead(2+) salt (1:1) (8CI,9CI)

15845-66-6
$C_2H_7O_3P$
110.05
Ethyl hydrogen phosphonate (9CI)

15849-14-6

$C_5H_4O_3$
112.09
1,2,4-Cyclopentanetrione (6CI,8CI,9CI)

15860-96-5
$C_{31}H_{40}O$
428.66
Oc(c(cc(c1)CCCCCCCCC)C(c(cccc2)c2)C)c1C(c(cccc3)c3)C
Phenol, 4-nonyl-2,6-bis(1-phenylethyl)- (9CI)
2,6-Bis(1-phenylethyl)-4-nonylphenol
4-Nonyl-2,6-bis(1-phenylethyl)phenol

15862-07-4
$C_{12}H_7Cl_3$
257.55
Clc1cc(Cl)c(cc1Cl)c2ccccc2
1,1'-Biphenyl, 2,4,5-trichloro- (9CI)
Biphenyl, 2,4,5-trichloro- (8CI)
2,4,5-PCB
2,4,5-Trichlorobiphenyl

15862-72-3
$C_8H_{15}NO_2$
157.21
2-Piperidinecarboxylic acid, ethyl ester (9CI)
NSC-49360
Pipecolic acid, ethyl ester (8CI)

15869-85-9
$C_{10}H_{22}$
142.28
Nonane, 5-methyl- (9CI)
5-Methylnonane

15870-10-7
C_8H_{16}
112.22
1-Heptene, 2-methyl- (9CI)
2-Methyl-1-heptene
NSC-73949

15872-73-8
$C_7H_{10}N_2O$
138.16
Oc(c(cc(N)c1)C)c1N
2,4-Diamino-6-methylphenol
o-Cresol, 4,6-diamino-
2,4-Diamino-o-cresol
2-Methyl-4,6-diaminophenol
Phenol, 2,4-diamino-6-methyl- (9CI)

15873-51-5
$C_6H_6N_2O_2$.ClH
174.57
Benzenamine, 4-nitro-, monohydrochloride (9CI)
4-Nitrobenzenamine monohydrochloride

15874-15-4
$C_{24}H_{52}O_4P_2S_4Zn$
660.28
Zinc, bis(O,O-bis(4-methylpentyl) phosphorodithioato-S,S')-, (T-4)- (9CI)
1-Pentanol, 4-methyl-, hydrogen phosphorodithioate, zinc salt
Phosphorodithioic acid, O,O-diisohexyl ester, zinc salt (8CI)

15874-48-3
$C_6H_{15}O_2PS_2$.1/3Sb
254.87
Phosphorodithioic acid, O,O-dipropyl ester, antimony(3+) salt (9CI)
Antimony O,O'-di-n-propyl phosphorodithioate

15874-52-9
$C_{16}H_{35}O_2PS_2$.1/3Sb
395.14
Phosphorodithioic acid, O,O-bis(2-ethylhexyl) ester, antimony(3+) salt (9CI)
Antimony O,O'-di-2-ethylhexylphosphorodithioate

15875-13-5
$C_{18}H_{42}N_6$
342.66
N(CN(CN1CCCN(C)C)CCCN(C)C)(C1)CCCN(C)C
s-Triazine, hexahydro-1,3,5-tris(dimethylaminopropyl)
N,N',N''-Tris(dimethylaminopropyl)-s-hexahydrotriazine

15877-57-3
$C_6H_{12}O$
100.16
Pentanal, 3-methyl- (9CI)
NSC-102764
Valeraldehyde, 3-methyl- (8CI)

15879-93-3
$C_8H_{11}Cl_3O_6$
309.54
O(C(OC(C1O)C(O)CO)C1O2)C2C(Cl)(Cl)Cl
Chloralose, α
AGC
Alfamat
Anhydroglucochloral
Aphosal
Chloralosane
Chloralose
α-Chloralose
Chloroalosane
Dulcidor
Glucochloral
Glucochloralose
α-D-Glucochloralose
α-D-Glucofuranose, 1,2-O-(2,2,2-trichloroethylidene)-, (R)- (9CI)
Kalmettumsomniferum
Monotrichlor-aethyliden-α-glucose (German)
Murex
Somio
1,2-O-(2,2,2-Trichloroethylidene)-α-D-glucofuranose
(R)-1,2-O-(2,2,2-Trichloroethylidene)-α-D-glucofuranose

15883-59-7
$C_{14}H_{11}N_2NaO_{10}S_2$
454.37
Benzenesulfonic acid, 2,2'-(1,2-ethenediyl)bis(5-nitro-, sodium salt (9CI)

15890-25-2
$C_{33}H_{66}N_3S_6Sb$
819.04
Antimony, tris(dipentylcarbamodithioato-S,S')-, (OC-6-11)- (9CI)
Antimony diamyldithiocarbamate

15890-40-1
C_8H_{16}
112.22
Cyclopentane, 1,2,3-trimethyl-, (1α,2α,3β)- (9CI)
Cyclopentane, 1,2,3-trimethyl-, cis-1,2,trans-1,3- (8CI)

15901-40-3
$C_{19}H_{39}N_3Si$
337.61
Silanetriamine, N,N',N''-tricyclohexyl-1-methyl- (9CI)
Methyltris(cyclohexylamino)silane

15905-32-5
$C_{20}H_8I_4O_5$
835.88
O=C(OC(c(c(Oc1c(c(O)c(c2)I)I)c(c(O)c3I)I)c3)(c12)c4cccc5)c45
Fluorescein, 2',4',5',7'-tetraiodo
Iodeosin
Spiro(isobenzofuran-1(3H),9'-(9H)xanthen)-3-one, 3',6'-dihydroxy-2',4',5',7'-tetraiodo-
Tetraiodofluorescein
2',4',5',7'-Tetraiodofluorescein

15910-22-2
C_8H_{16}
112.22
3-Hexene, 2,5-dimethyl- (6CI,7CI,8CI,9CI)

2,5-Dimethyl-3-hexene

15914-23-5
$C_{20}H_{16}$
256.36
Chrysene, 1,2-dimethyl
1:2-Dimethylchrysene

15922-78-8
$C_5H_5NOS.Na$
150.15
Omadine sodium
Caswell No. 790A
EPA Pesticide Chemical Code 088004
1-Hydroxy-2(1H)-pyridinethionato sodium
1-Hydroxy-2(1H)-pyridinethione, sodium salt
Omacide 24
Omadine-sodium
2-(1H)-Pyridinethione, 1-hydroxy-, sodium salt
Sodium 1-hydroxypyridine-2-thione
Sodium omadine
Sodium 2-pyridinethiol-1-oxide
Sodium pyridinethione
Sodium pyrithione
SQ 3277

15930-66-2
$C_{11}H_{25}NO$
187.32
O(CCCCCCCC)CCCN
1-Propanamine, 3-(octyloxy)- (9CI)
3-Octoxypropane-1-amine
3-Octyloxypropanamine
3-(Octyloxy)-1-propanamine

15930-94-6
$CrO_4.H_2O_2.Zn_2.H_2O$
298.78
Chromic acid, zinc hydroxide hydrate (1:2:2:1)
Buttercup Yellow
Chromic acid, zinc salt (1:2)
Chromium(6+) zinc oxide hydrate (1:2:6:1)
Zinc chromate hydroxide
Zinc chromate (VI) hydroxide
Zinc hydroxychromate
Zinc Yellow

15932-66-8
$C_7H_{16}N_2$
128.21
2-Piperidineethanamine (9CI)
NSC-143025
Piperidine, 2-(2-aminoethyl)- (8CI)

15934-02-8
Unknown
Unknown
Nitrilotriacetic acid, monoammonium salt
NTA, monoammonium salt

15945-07-0
$C_6H_2Cl_4O_2S$
279.96
O=S(=O)(c(c(cc(c1Cl)Cl)Cl)c1)Cl
Benzenesulfonyl chloride, 2,4,5-trichloro- (9CI)
NSC-26958
2,4,5-Trichlorobenzenesulfonyl chloride

15950-66-0
$C_6H_3Cl_3O$
197.45
Oc1ccc(Cl)c(Cl)c1Cl
2,3,4-Trichlorophenol

15956-58-8
$C_8H_{16}O_2.xMn$
Unknown
Hexanoic acid, 2-ethyl-, manganese salt (9CI)
Manganese 2-ethylhexanoate

15958-68-6
$C_{26}H_{30}N_2O_2$
402.52
O=C(c(c(c(NC(CCCC1)C1)cc2)C(=O)c3cccc4NC(CCCC5)C5)c2)c34
9,10-Anthracenedione, 1,5-bis-(cyclohexylamino)- (9CI)
1,5-Bis(cyclohexylamino)-9,10-anthracenedione
1,5-Di(cyclohexylamino)anthracene-9,10-dione

15965-99-8
$C_{19}H_{38}O_2$
298.51
O(C1COCCCCCCCCCCCCCCCC)C1
Oxirane, ((hexadecyloxy)-methyl)- (9CI)
(Cetyloxymethyl)oxirane
((Hexadecyloxy)methyl)oxirane

15968-05-5
$C_{12}H_6Cl_4$
291.99
c(c(c(cc1)Cl)c(c(ccc2)Cl)c2Cl)(c1)Cl
1,1'-Biphenyl, 2,2',6,6'-tetrachloro- (9CI)
2,2',6,6'-Tetrachloro-1,1'-biphenyl

15972-60-8
$C_{14}H_{20}ClNO_2$
269.80
CCc1cccc(CC)c1N(COC)C(=O)CCl
Acetanilide, 2-chloro-2',6'-diethyl-N-(methoxymethyl)
Acetamide, 2-chloro-N-(2,6-diethylphenyl)-N-(methoxymethyl)- (9CI)
Alachlor
Alachlore
Alanex
Alanox
Alatox 480
Alochlor
Chloressigsaeure-N-(methoxymethyl)-2,6-diaethylanilid (German)
2-Chloro-2',6'-diethyl-N-(methoxymethyl)acetanilide
2-Chloro-N-(2,6-diethyl)phenyl-N-methoxymethylacetamide
CP 50144
Lasso
Lasso Micro-Tech
Lazo
Metachlor
Methachlor
Pillarzo

15979-79-0
$C_{12}H_9Cl_2N$
238.12
Benzenamine, 3-chloro-N-(4-chlorophenyl)- (9CI)
3,4'-Dichlorodiphenylamine
Diphenylamine, 3,4'-dichloro- (8CI)

15979-81-4
$C_{13}H_{13}NO_2S$
247.32
Benzenamine, 4-(methylsulfonyl)-N-phenyl- (9CI)
Diphenylamine, 4-(methylsulfonyl)- (8CI)

15979-82-5
$C_{13}H_{12}N_2O_2$
228.25
Benzenamine, 3-methyl-N-(4-nitrophenyl)- (9CI)
3'-Methyl-4-nitrodiphenylamine
m-Toluidine, N-(p-nitrophenyl)- (8CI)

15979-85-8
$C_{12}H_9ClN_2O_2$
248.67
Benzenamine, 3-chloro-N-(4-nitrophenyl)- (9CI)
3'-Chloro-4-nitrodiphenylamine
Diphenylamine, 3-chloro-4'-nitro- (8CI)

15979-87-0

$C_{12}H_9N_3O_4$
259.22
Benzenamine, 3-nitro-N-(4-nitrophenyl)- (9CI)
Diphenylamine, 3,4'-dinitro- (7CI,8CI)

15980-15-1
C_4H_8OS
104.18
C1CSCCO1
1,4-Oxathiane
Oxathiane
p-Thioxane

15989-50-1
$C_{15}H_{13}N_5S$
295.34
1,3,5-Triazine-2(1H)-thione, 4,6-bis(phenylamino)- (9CI)
s-Triazine-2-thiol, 4,6-dianilino- (8CI)

15990-43-9
$C_{13}H_{11}NO_2$
213.23
O=C(O)c(cccn1Cc(cccc2)c2)c1
Pyridinium, 3-carboxy-1-(phenylmethyl)-, hydroxide, inner salt (9CI)

16007-72-0
$C_9H_8ClN_5$
221.65
1,3,5-Triazine-2,4-diamine, 6-chloro-N-phenyl- (9CI)
Triazine, 2-amino-4-anilino-6-chloro- (6CI)
s-Triazine, 2-amino-4-anilino-6-chloro- (7CI,8CI)

16008-36-9
Unknown
Unknown
Methyldesorphine
DEA No. 9302

16014-23-6
$C_{18}H_{14}N_2O_6S$
386.37
O=C(O)c(c(O)c(N=Nc(c(S(=O)(=O)O)cc(c1)C)c1)c(c2ccc3)c3)c2
2-Naphthalenecarboxylic acid, 3-hydroxy-4-((4-methyl-2-sulfophenyl)azo)- (9CI)

16015-11-5
$C_6H_{10}O$
98.15
2H-Pyran, 3,4-dihydro-6-methyl- (6CI,7CI,8CI,9CI)
2-Methyl-δ2-dihydropyran
6-Methyl-2,3-dihydro-4H-pyran

16022-69-8
$C_7H_3Cl_5O$
280.36
Benzenemethanol, 2,3,4,5,6-pentachloro- (9CI)
Benzyl alcohol, 2,3,4,5,6-pentachloro- (8CI)
Blastin
PCBA

16024-56-9
$C_5H_{10}O_4$
134.13
Acetic acid, (2-methoxyethoxy)- (8CI,9CI)
NSC-127858

16033-74-2
$C_{15}H_{10}Cl_2N_4$
317.18
1,3,5-Triazin-2-amine, 4,6-dichloro-N,N-diphenyl- (9CI)
(Diphenylamino)cyanuric chloride
2-Diphenylamino-4,6-dichloro-s-triazine
2-Diphenylamino-4,6-dichloro-1,3,5-triazine
Diphenylamino-4,6-dichloro-s-triazine
2,4-Dichloro-6-(diphenylamino)-s-triazine
2,4-Dichloro-6-diphenylamino-1,3,5-triazine
s-Triazine, 2,4-dichloro-6-(diphenylamino)- (6CI,7CI,8CI)

16034-77-8
$C_{12}H_{13}I_3N_2O_3$
613.97
Propionic acid, 3-(acetyl-(3-amino-2,4,6-triiodophenyl)amino)-2-methyl
N-Acetyl-N-(3-amino-2,4,6-triiodophenyl)-β-aminoisobutyric acid
3-(Acetyl-(3-amino-2,4,6-triiodophenyl)amino)-2-methylpropanoic acid
N-Acetyl-N-(3-amino-2,4,6-triiodophenyl)-2-methyl-β-alanine
N-Acetyl-N-(2,4,6-triiodo-3-aminophenyl)-β-aminoisobutyric acid
Cholebrine
Cholimil
DRC 1201
Iocetamic acid
MP 620

16045-92-4
$C_4H_5ClO_4$
152.53
Butanedioic acid, chloro- (9CI)
Succinic acid, chloro- (8CI)

16056-34-1
CH_3Hg
215.62
Mercury, methyl- (8CI,9CI)

16058-17-6
$C_{17}H_{34}N_2O$
282.46
N(=C(N(C1)CCO)CCCCCCCCCCCC)C1
1H-Imidazole-1-ethanol, 2-dodecyl-4,5-dihydro- (9CI)
2-Dodecyl-4,5-dihydro-1H-imidazole-1-ethanol

16063-69-7
$C_5H_2Cl_3N$
182.43
Clc1cc(Cl)nc(Cl)c1
Pyridine, 2,4,6-trichloro
2,4,6-Trichloropyridine

16063-70-0
$C_5H_2Cl_3N$
182.43
Clc1cnc(Cl)c(Cl)c1
Pyridine, 2,3,5-trichloro
2,3,5-Trichloropyridine

16065-83-1
Cr
52.00
Chromium, ion (Cr 3+)
Chromic ion
Chromium (3+)
Chromium (III)
Chromium (III) ion
Chromium ion (3+)

16066-35-6
$C_9H_{12}O_3S$
200.26
O=S(=O)(O)c(ccc(c1)C(C)C)c1
Benzenesulfonic acid, 4-(1-methylethyl)- (9CI)
4-(1-Methylethyl)benzenesulfonic acid

16066-38-9
$C_8H_{14}O_6$
206.22
O=C(OCCC)OOC(=O)OCCC
Peroxydicarbonic acid, dipropyl ester
Di-n-propyl peroxydicarbonate
Di-n-propyl peroxydicarbonate, Technically pure (DOT)
n-Propyl percarbonate
UN 2176 (DOT)

16067-01-9
$C_8H_{15}NO_3$
173.21
2-Octanone, 1-nitro- (8CI,9CI)

16071-86-6
$C_{31}H_{20}N_6O_9S.Cu.2Na$
762.15
[Na+].[Na+].OC(=O)c7cc(N=Nc1ccc(cc1)c6ccc(N=Nc4c(O)cc5N3=Nc2cc(ccc2O[Cu]3Oc45)S([O-])(=O)=O)cc6)ccc7O
Copper, (5-((4'-((2,5-dihydroxy-4-((2-hydroxy-5-sulfophenyl)azo)phenyl)azo)-(1,1'-biphenyl)- 4-yl)azo)-2-hydroxybenzoato(2-))-, disodium salt
Aizen Primula Brown BRLH
Aizen Primula Brown PLH
Amanil Fast Brown BRL
Amanil Supra Brown LBL
Atlantic Fast Brown BRL
Atlantic Resin Fast Brown BRL
Belamine Fast Brown BRLL
Benzamil Supra Brown BRLL
Benzanil Supra Brown BRLN
Brown 4EMBL
Calcodur Brown BRL
Chloramine Fast Brown BRL
Chloramine Fast Brown BRLL
Chloramine Fast Cutch Brown PL
Chrome Leather Brown BRLL
Chrome Leather Brown BRSL
C.I. 30145
C.I. Direct Brown
Cuprofix Brown GL
Derma Fast Brown W-GL
Dermafix Brown PL
Dialuminous Brown BRS
Diaphtamine Light Brown BRLL
Diazine Fast Brown RSL
Diazol Light Brown BRN
Dicorel Brown LMR
Diphenyl Fast Brown BRL
Direct Brown 95
Direct Brown BRL
Direct Fast Brown BRL
Direct Fast Brown LMR
Direct Light Brown BRS
Direct Supra Light Brown ML
Durazol Brown BR
Durofast Brown BRL
Eliamina Light Brown BRL
Enianil Light Brown BRL
Fastolite Brown BRL
Fastusol Brown LBRSA
Fastusol Brown LBRSN
Fenaluz Brown BRL
Helion Brown BRSL
Hispaluz Brown BRL
HNED Prima 95 (Czech)
Kayarus Supra Brown BRS
KCA Light Fast Brown BR
NCI-C54568
Paranol Fast Brown BRL
Peeramine Fast Brown BRL
Pontamine Fast Brown BRL
Pontamine Fast Brown NP
Pyrazol Fast Brown BRL
Pyrazoline Brown BRL
Saturn Brown LBR
Sirius Supra Brown BRL
Sirius Supra Brown BRS
Solantine Brown BRL
Solar Brown PL
Solex Brown R
Solius Light Brown BRLL
Solius Light Brown BRS
Sumilight Supra Brown BRS
Suprazo Brown BRL
Suprexcel Brown BRL
Tertrodirect Fast Brown BR
Tetramine Fast Brown BRDN Extra
Tetramine Fast Brown BRP
Tetramine Fast Brown BRS
Triantine Brown BRS
Triantine Fast Brown OG
Triantine Fast Brown OR
Triantine Light Brown BRS
Triantine Light Brown OG

16079-88-2
$C_5H_{10}BrClN_2O_2$
245.51
Di-Halo
1-Bromo-3-chloro-5,5-dimethyl-2,4-imidazolidinedione
Caswell No. 114A
EPA Pesticide Chemical Code 006315

16088-72-5
$C_4H_4Br_3N_5$
361.84
s-Triazine, 2,4-diamino-6-(tribromomethyl)-(8CI)

16088-73-6
$C_4H_4Cl_3N_5$
228.44
1,3,5-Triazine-2,4-diamine, 6-(trichloromethyl)- (9CI)
NSC-120281
s-Triazine, 2,4-diamino-6-(trichloromethyl)- (8CI)

16090-02-1
$C_{40}H_{38}N_{12}O_8S_2.2Na$
925.00
OS(=O)(=O)c4cc(Nc2nc(Nc1ccccc1)nc(n2)N3CCOCC3)ccc4C=Cc8ccc(Nc6nc(Nc5ccccc5)nc(n6)N7CCOCC7)cc8S(O)(=O)=O
2,2'-Stilbenedisulfonic acid, 4,4'-bis((4-anilino-6-morpholino-s-triazin-2-yl)amino)-, disodium salt
Blankophor BBH
Blankophor MBBH
C.I. Fluorescent Brightener 260
Disodium 4,4'-bis((4-anilino-6-morpholino-1,3,5-triazin-2-yl)amino)stilbene-2,2'-disulfonate
MBBH 766
Mikephor TB
Tinopal AMS
Tinopal EMS

16090-14-5
$C_7F_{14}O_4S$
446.12
O=S(=O)(F)C(F)(F)C(F)(F)OC(F)(C(F)(F)F)C(F)(F)OC(F)=C(F)F
Ethanesulfonyl fluoride, 2-(1-(difluoro((trifluoroethenyl)-oxy)methyl)-1,2,2,2-tetrafluoroethoxy)-1,1,2,2-tetrafluoro- (9CI)

16090-49-6
C_2H_6GeS
126.69
Dimethyl germanium sulfide
Germane, dimethylthioxo-
Germanium sulfide, dimethyl-

16091-18-2
$C_{20}H_{36}O_4Sn$
459.25
CCCCCCCC[Sn]1(CCCCCCCC)OC(=O)C=CC(=O)O1
1,3,2-Dioxastannepin-4,7-dione, 2,2-dioctyl
2,2-Dioctyl-1,3,2-dioxastannepin-4,7-dione
Dioctylstannylene maleate
Dioctyltin maleate
Di-n-octyltin maleate
Di-n-octylzinn maleinat (German)
Estabex U 18
LIV 1176
Mellite 825
Stann OMF
Thermolite 813
TVS 8105

16106-59-5
C_8H_{16}
112.22
1-Hexene, 4,5-dimethyl- (8CI, 9CI)
NSC-244854

16107-88-3
$C_{22}H_{17}N_3$
323.40
1,3,5-Triazine, 2-(4-methylphenyl)-4,6-diphenyl- (9CI)
s-Triazine, 2,4-diphenyl-6-p-tolyl- (6CI,8CI)

16110-09-1
$C_5H_3Cl_2N$
147.99
Clc1ccc(Cl)nc1
Pyridine, 2,5-dichloro

16110-51-3
$C_{23}H_{16}O_{11}$
468.37
OC(COc1cccc2oc(cc(=O)c12)C(O)=O)COc3cccc4oc(cc(=O)c34)C(O)=O
Cromolyn
Acide cromoglicique (French)
Acido cromoglicico (Spanish)
Acidum cromoglicicum (Latin)
4H-1-Benzopyran-2-carboxylic acid, 5,5'-((2-hydroxy-1,3-propanediyl)bis(oxy))bis-(4-oxo-
4H-1-Benzopyran-2-carboxylic acid, 5,5'-((2-hydroxytrimethylene)dioxy)bis(4-oxo-
1,3-Bis(2-carboxychromon-5-yloxy)-2-hydroxypropane
Cromoglicic acid
Cromoglycic acid
1,3-Di(2-carboxy-4-oxo-chromen-5-yloxy)propan-2-ol
5,5'-(2-Hydroxytrimethylenedioxy)bis(4-oxochromene-2-carboxylic acid)

16110-89-7
$C_9H_6Cl_2N_4O_3S$
321.12
O=S(=O)(O)c(ccc(Nc(nc(nc1Cl)Cl)n1)c2)c2
4-(2,4-Dichloro-1,3,5-triazinylamino)benzenesulfonic acid
Benzenesulfonic acid, 4-((4,6-dichloro-1,3,5-triazin-2-yl)amino)- (9CI)
4-((4,6-Dichloro-1,3,5-triazin-2-yl)amino)benzenesulfonic acid
Sulfanilic acid, N-(4,6-dichloro-s-triazin-2-yl)-

16111-27-6
$C_5H_{13}N_3S.2ClH$
220.19
Pseudourea, 2-(2-(dimethylamino)ethyl)-2-thio-, dihydrochloride
Carbamimidothioic acid, 2-(dimethylamino)ethyl ester, dihydrochloride
N,N-Dimethyl-β-aminoaethylisothiuronium dihydrochlorid (German)
2-Dimethylaminoethylisothiuronium chloride hydrochloride
S-(2-(Dimethylamino)ethyl)-pseudothiourea dihydrochloride
2-(2-(Dimethylamino)ethyl)-2-thiopseudourea dihydrochloride
Nordimaprit
SKF 91487

16111-62-9
$C_{18}H_{34}O_6$
346.52
O=C(OCC(CCCC)CC)OOC(=O)OCC(CCCC)CC
Peroxydicarbonic acid, di-(2-ethylhexyl) ester
Di(2-ethylhexyl)peroxydicarbonate
Di-(2-ethylhexyl)peroxydicarbonate, Maximum concentration 32% (DOT)
Di-(2-ethylhexyl)peroxydicarbonate, Technically pure (DOT)
Di-(2-ethylhexyl)peroxydicarbonate, 77% In solution (DOT)
Peroxydicarbonic acid, bis-(2-ethylhexyl) ester
Peroxydicarbonic acid, di-(2-ethylhexyl) ester, Not more than 77% in solution
UN 2122 (DOT)
UN 2123 (DOT)
UN 2960 (DOT)

16111-79-8
$C_7H_7N_7O_2$
221.18
Hydrazine, 1-((4-amino-pyrazolo(5,1-c)-as-triazin-3-yl)carbonyl)-2-formyl- (8CI)

16118-45-9
$C_{12}H_{16}N_2O_3$
236.26
Propanamide, N-ethyl-2-(((phenylamino)carbonyl)-oxy)- (9CI)
Lactamide, N-ethyl-, carbanilate
Lactamide, N-ethyl-, carbanilate (ester) (8CI)
Legurame

16118-49-3
$C_{12}H_{16}N_2O_3$
236.30
CCNC(=O)C(C)OC(=O)Nc1ccccc1
Carbanilic acid, (1-ethylcarbamoyl)ethyl ester, D-(-)
Carbetamex
Carbetamid (German)
Carbetamide
D-N-Ethylacetamide carbanilate
D-(-)-1-(Ethylcarbamoyl)ethyl phenylcarbamate
D-N-Ethyllactamide carbanilate (ester)
(R)-N-Ethyl-2-(((phenylamino)-carbonyl)oxy)propanamide
Legurame
2-Phenyl-carbamoyloxy-N-aethyl-propionamid (German)
(Phenylcarbamoyloxy)-2-N-ethylpropionamide
N-Phenyl-1-(ethylcarbamoyl-1)-ethylcarbamate, D isomer
11,561 RP

16143-79-6
$C_{32}H_{16}Cu_2N_6O_{16}S_4.4Na$
1087.79
Cuprate(4-), (mu-((6,6'-((3,3'-dihydroxy(1,1'-biphenyl)-4,4'-diyl)bis(azo))bis(4-amino-5-hydroxy-1,3-naphthalenedisulfonato))(8-)))di-, tetrasodium (9CI)

16197-90-3
$C_4H_5ClO_2$
120.54
2-Butenoic acid, 4-chloro-(9CI)
γ-Chlorocrotonic acid
Crotonic acid, 4-chloro-(6CI,7CI,8CI)

16202-79-2
$C_{15}H_{22}O_2$
234.34
1H-3a,6-Methanoazulene-3-carboxylic acid, 2,3,4,5,6,7-hexahydro-7,7,8-trimethyl-, (3S-(3α,3aα,6α))- (9CI)
Isokhusenic acid
Isozizanoic acid
1H-3aα,6-Methanoazulene-3-carboxylic acid, 2,3β,4,5,6β,7-hexahydro-7,7,8-trimethyl-(8CI)

16212-28-5
C_3Cl_3N
156.39
Acrylonitrile, 2,3,3-trichloro
Acrylonitrile, trichloro- (8CI)
2-Propenenitrile, 2,3,3-trichloro-(9CI)
TL 391
Trichloroacrylonitrile
α,β,β-Trichloroacrylonitrile

16214-98-5
$C_{13}H_{12}NO_2.Cl$
249.69
Pyridinium, 3-carboxy-1-(phenylmethyl)-, chloride (9CI)
N(1)-Benzyl-3-carboxylpyridinium chloride
N-Benzyl-3-carboxypyridinium chloride
1-Benzyl-3-carboxypyridinium chloride
N-Benzyl-3-carboxypyridinum chloride
3-Carboxy-1-benzylpyridinium chloride
3-Carboxy-1-(phenylmethyl)-pyridinium chloride
Nicotinic acid-benzyl chloride quat

NSC-139652
Pyridinium, 1-benzyl-3-carboxy-, chloride

16215-49-9
$C_{10}H_{18}O_6$
234.25
Butyl peroxydicarbonate
n-Butyl peroxydicarbonate, More than 27% to a maximum concentration of 52%
n-Butyl peroxydicarbonate, Not more than 27% in solution
n-Butyl peroxydicarbonate, More than 52% in solution
n-Butyl peroxydicarbonate, Not more than 52% in solution
Dibutyl peroxydicarbonate
Di-n-butyl peroxydicarbonate
Peroxidicarbonato de di-n-butilo (Spanish)
Peroxydicarbonate de di-n-butyle (French)
Peroxydicarbonic acid, dibutyl ester, More than 52% in solution
Peroxydicarbonic acid, dibutyl ester, Not more than 27% in solution
Peroxydicarbonic acid, dibutyl ester, Not more than 52% in solution
UN 2169
UN 2170

16219-25-3
$C_{14}H_{22}O_2$
222.33
1,3-Cyclopentadiene-1-carboxylic acid, 6-methylheptyl ester (8CI)

16219-75-3
C_9H_{12}
120.21
C(C(C=CC12)C1)(=CC)C2
2-Norbornene, 5-ethylidene
5-Ethylidenebicyclo(2.2.1)hept-2-ene
Ethylidene norbornene (ACGIH, OSHA)
5-Ethylidene-2-norbornene

16220-58-9
$C_{12}H_8ClN_3O_4$
293.65
Benzenamine, N-(3-chlorophenyl)-2,4-dinitro- (9CI)
Diphenylamine, 3'-chloro-2,4-dinitro- (8CI)
NSC-157494

16224-33-2
$C_7H_{15}ClO_2$
166.67
O(CC(O)CCl)CCCC
Ether, butyl (3-chloro-2-hydroxypropyl)
1-Butoxy-3-chloro-2-propanol
Butyl-chlorhydrinether (Czech)

16227-10-4
$C_6H_{11}N_3$
125.20
4H-1,2,4-Triazole, 4-butyl
BT
Butrizol
4-n-Butyl-4H-1,2,4-triazole
Dithane R-24
Indar
RH-124
Triazbutil
s-Triazole, 4-butyl-

16230-35-6
$C_{19}H_{24}N_2O_3Si$
356.48
Benzamide, N,N'-(ethoxymethylsilylene)bis(N-methyl- (9CI)
Bis(N-methylbenzamido)-methylethoxysilane

16238-56-5
$C_{20}H_{15}Br$
335.26
Cc3c1ccccc1c(CBr)c4ccc2ccccc2c34
Benz(a)anthracene, 7-bromomethyl-12-methyl
7-Bromomethyl-12-methylbenz(a)anthracene
ICR 502

16245-77-5
$C_6H_8N_2.H_2O_4S$
206.21
p-Phenylenediamine sulfate
1,4-Benzenediamine, sulfate (1:1) (9CI)

16245-79-7
$C_{14}H_{23}N$
205.34
Benzenamine, 4-octyl- (9CI)
Aniline, p-octyl- (8CI)

16245-97-9
$C_{21}H_{42}O_2$
326.56
O(C1COCCCCCCCCCCCCCCCCCC)C1
Oxirane, ((octadecyloxy)-methyl)- (9CI)
Octadecyl glycidyl ether
((Octadecyloxy)methyl)oxirane

16260-09-6
$C_{34}H_{67}NO$
505.91
O=C(NCCCCCCCCC=CCCCCCCCC)CCCCCCCCCCCCCCC
Oleyl palmitamide
Hexadecanamide, N-9-octadecenyl-
Hexadecanamide, N-9-octadecenyl-, (Z)- (9CI)
Kenamide P-181
N-9-Octadecenyl hexadecanamide
(Z)-N-9-Octadecenylhexadecanamide

16268-62-5
$C_8H_{16}N_6$
196.30
CNc1nc(nc(n1)N(C)C)N(C)C
Melamine, pentamethyl
Pentamethylmelamine
1,3,5-Triazine-2,4,6-triamine, N,N,N',N''-pentamethyl- (9CI)
UNT 51239

16268-92-1
$C_9H_{18}N_6$
210.25
CCNc1nc(NCC)nc(NCC)n1
1,3,5-Triazine-2,4,6-triamine, N,N',N''-triethyl- (9CI)
Melamine, N(2),N(4),N(6)-triethyl- (8CI)

16274-44-5
$C_9H_{18}N_6$
210.25
Melamine, N(2),N(4)-diisopropyl- (8CI)

16274-81-0
$C_6H_{12}N_6$
168.17
1,3,5-Triazine-2,4,6-triamine, N-(1-methylethyl)- (9CI)
Melamine, isopropyl- (8CI)
NSC-298102

16277-67-1
$C_{10}H_{12}O$
148.20
Benzene, (3-methoxy-1-propenyl)- (9CI)
Ether, cinnamyl methyl (8CI)

16291-96-6
Unknown
Unknown
Charcoal
Charcoal Briquettes [NA 1361]
Charcoal Screenings, Made from "pinon" wood (DOT)
NA 1361 [Charcoal briquettes, shell, screenings, wood, etc.]

16301-26-1
$C_4H_{10}N_2O$
102.16

Diazene, diethyl-, 1-oxide
Azoxyaethan (German)
Azoxyethane
Ethane, azoxy-

16308-92-2
$C_9H_{12}O$
136.19
Benzenemethanol, 2,4-dimethyl- (9CI)
Benzyl alcohol, 2,4-dimethyl- (8CI)

16323-13-0
$C_{14}H_{13}NO_2$
227.26
Carbamic acid, phenyl-, 4-methylphenyl ester (9CI)
Carbanilic acid, p-tolyl ester (8CI)

16336-83-7
$C_{12}H_{20}$
164.29
5-Dodecen-7-yne, (Z)- (8CI,9CI)
cis-5,7-Dodecenyne

16338-97-9
$C_6H_{10}N_2O$
126.18
C=CCN(CC=C)N=O
Diallylamine, N-nitroso
Diallylnitrosamin (German)
Diallylnitrosamine
N-Nitrosodiallyl amine

16339-04-1
$C_5H_{12}N_2O$
116.19
CCN(N=O)C(C)C
Diethylamine, 1-methyl-N-nitroso
Aethyl-isopropyl-nitrosoamin (German)
Ethylisopropylnitrosoamine
1-Methyl-N-nitrosodiethylamine
N-Nitrosoethylisopropylamine

16352-06-0
$C_4H_6N_4O$
126.09
1,3,5-Triazin-2(1H)-one, 4-amino-6-methyl- (9CI)
Acetoguanide
s-Triazin-2-ol, 4-amino-6-methyl- (8CI)

16352-07-1
$C_5H_8N_4O$
140.12
1,3,5-Triazin-2(1H)-one, 4-amino-6-ethyl- (9CI)
s-Triazin-2-ol, 4-amino-6-ethyl- (8CI)
s-Triazin-2(1H)-one, 4-amino-6-ethyl-

16354-52-2
$C_{22}H_{20}$
284.42
Benz(a)anthracene, 7,12-diethyl
9,10-Diethyl-1,2-benzanthracene

16354-95-3
$C_9H_{12}O_3$
168.19
1,2,3-Propanetriol, 1-phenyl-, (S-(R*,S*))- (9CI)
1,2,3-Propanetriol, 1-phenyl-, D-erythro- (8CI)

16365-27-8
$C_8H_9NO_4$
183.16
O=N(=O)c(ccc(OCCO)c1)c1
Ethanol, 2-(4-nitrophenoxy)- (9CI)
AI3-19441
Ethanol, 2-(p-nitrophenoxy)-
β-Hydroxyethyl p-nitrophenyl ether
p-Nitrophenoxyethanol
2-(p-Nitrophenoxy)ethanol
2-(4-Nitrophenoxy)ethanol
NSC-30512

16368-97-1
$C_{22}H_{39}O_4P$
398.58
CCCCC(CC)COP(=O)(OCC(CC)CCCC)Oc1ccccc1
Phosphoric acid, bis(2-ethylhexyl)phenyl ester
Bis(2-ethylhexyl)phenyl phosphate
DAFF
DEPP
Di(2-ethylhexyl)phenyl phosphate

16369-21-4
$C_5H_{13}NO$
103.16
OCCNCCC
Ethanol, 2-(propylamino)- (9CI)
2-(Propylamino)ethanol

16370-63-1
$C_5H_8N_4O_2$
156.12
1,3,5-Triazin-2-amine, 4,6-dimethoxy- (9CI)
NSC-100284
s-Triazine, 2-amino-4,6-dimethoxy- (8CI)

16376-89-9
$C_3H_4N_8$
152.07
1,3,5-Triazine-2,4-diamine, 6-azido- (9CI)
s-Triazine, 2,4-diamino-6-azido- (8CI)

16389-88-1
$CH_2O_3.1/2Ca.1/2Mg$
94.22
Dolomite (9CI)

16391-07-4
$C_3H_9O_2P$
108.08
Phosphinic acid, methyl-, ethyl ester (8CI,9CI)

16397-91-4
Mn
54.94
Manganese, ion (Mn2+)
Manganese(2+)
Manganese(II)
Manganese (Mn2+)
Manganese cation
Manganese(II) ion
Manganese ion(2+)
Manganese(2+) ion
Manganous ion
Mn2+

16399-10-3
$C_7H_{12}N_4O$
168.20
1,3,5-Triazin-2-amine, N-ethyl-4-methoxy-6-methyl- (9CI)
s-Triazine, 2-(ethylamino)-4-methoxy-6-methyl- (8CI)

16400-09-2
$C_{13}H_{10}ClNO_2$
247.68
Carbamic acid, phenyl-, 3-chlorophenyl ester (9CI)
Carbanilic acid, m-chlorophenyl ester (7CI,8CI)
3-Chlorophenyl carbanilate
3-Chlorophenyl phenylcarbamate
3-Chlorophenyl N-phenylcarbamate
m-Chlorophenyl N-phenylcarbamate
Phenol, m-chloro-, carbanilate (6CI)

16403-84-2
$C_{25}H_{20}N_4O_3$
424.43
O=C(Nc(cccc1)c1)c(c(O)c(N=Nc(c(ccc2C(=O)N)C)c2)c(c3ccc4)c4)c3
2-Naphthanilide, 4-((5-carbamoyl-o-tolyl)azo)-3-hydroxy- (8CI)

4-((5-(Aminocarbonyl)-2-methylphenyl)azo)-3-hydroxy-N-phenyl-2-naphthalenecarboxamide
4-((5-Carbamyl-2-methylphenyl)azo)-3-hydroxy-2-naphthanilide
1-(2-Methyl-5-benzamide)azo-2-hydroxy-3-naphthanalide
1-(2-Methyl-5-carbamylphenylazo)-2-hydroxy-3-phenylcarbamoylnaphthalene
4-((2-Methyl-5-carboxamidophenyl)azo)-3-hydroxy-N-phenyl-2-naphthalenecarboxamide
2-Naphthalenecarboxamide, 4-((5-aminocarbonyl-2-methylphenyl)azo)-3-hydroxy-N-phenyl-
2-Naphthalenecarboxamide, 3-hydroxy-4-((2-methyl-5-aminocarbonylphenyl)azo)-N-phenyl-
2-Naphthalenecarboxamide, 3-hydroxy-4-((2-methyl-5-phenylcarboxyamide)azo)-N-phenyl-

16408-14-3
C_2H_6Sn
148.78
Stannanediylium, dimethyl- (9CI)
Tin(2+), dimethyl-, ion (8CI)

16408-15-4
CH_3Sn
133.72
Methyltin
Tin(3+), methyl- (9CI)
Tin(3+), methyl-, ion (8CI)

16409-43-1
$C_{10}H_{18}O$
154.28
O(C(C=C(C)C)CC(C1)C)C1
Pyran, tetrahydro-2-(2-methyl-1-propenyl)-4-methyl
Pyran, 2-(2-methyl-1-propenyl)-4-methyltetrahydro-
Rose Oxide Levo

16409-45-3
$C_{12}H_{22}O_2$
198.34
CC1CCC(C(C)C)C(C1)OC(C)=O
Acetic acid, p-menth-3-yl ester, dl
Cyclohexanol, 5-methyl-2-(1-methylethyl)-, acetate
Menthol, acetate
Menthyl acetate
dl-Menthyl acetate
Menthyl acetate racemic

16411-33-9
$C_{13}H_{33}N_3Si$
259.49
Silanetriamine, N,N',N''-tributyl-1-methyl- (9CI)
1-Methyl-N,N',N''-tributylsilanetriamine
N,N',N''-Tributyl-1-methylsilanetriamine

16416-32-3
$C_{24}D_{50}$
388.97
Tetracosane-d50 (8CI,9CI)
Perdeutero-tetracosane

16420-13-6
C_3H_6ClNS
123.60
N(C(=S)Cl)(C)C
Carbamothioic chloride, dimethyl- (9CI)
Dimethylcarbamothioic chloride

16423-68-0
$C_{20}H_6I_4O_5.2Na$
879.84
[Na+].[Na+].[O-]C(=O)c1ccccc1c3c2cc(I)c([O-])c(I)c2oc4c(I)c(=O)c(I)cc34
Fluorescein, 2',4',5',7'-tetraiodo-, disodium salt
Aizen Erythrosine
Calcocid Erythrosine N
Canacert Erythrosine BS
9-(o-Carboxyphenyl)-6-hydroxy-2,4,5,7-tetraiodo-3-isoxanthone
Cerven Kysela 51 (Czech)
Cerven Potravinarska 14 (Czech)
C.I. 773
C.I. 45430
C.I. Acid Red 51
C.I. Food Red 14
Cilefa Pink B
D & C Red No. 3
Dolkwal Erythrosine
Dye FD & C Red No. 3
E 127
EBS
Edicol Supra Erythrosine A
Erythrosin
Erythrosin B
Erythrosine
Erythrosine B
Erythrosine 3B
Erythrosine B (Biological stain)
Erythrosine B-FO (Biological stain)
Erythrosine Bluish
Erythrosine Bluish (Biological stain)
Erythrosine BS
Erythrosine Extra Bluish
Erythrosine Extra Conc. A Export
Erythrosine Extra Pure A
Erythrosine (Indicator)
Erythrosine K-FO (Biological stain)
Erythrosine Lake
Erythrosine sodium
Erythrosine TB
Erythrosine TB Extra
FD & C Red 3
FD & C Red No. 3
Food Red 14
Hexacert Red No. 3
Hexacol Erythrosine BS
LB-Rot 1
Maple Erythrosine
New Pink Bluish Geigy
1427 Red
1671 Red
Schultz No. 887
Tetraiodofluorescein sodium salt
Usacert Red No. 3

16424-67-2
$C_{15}H_{30}O$
226.41
4-Heptanone, 5,5-diethyl-2,2,3,3-tetramethyl- (8CI, 9CI)

16426-62-3
$C_7H_6O_6$
186.12
2-Furanacetic acid, 3-carboxy-2,5-dihydro-5-oxo- (8CI,9CI)
1-Butene-1,2,4-tricarboxylic acid, 3-hydroxy-, γ-lactone

16432-45-4
$C_{16}H_{13}N_3O_4S$
343.34
O=S(=O)(N)c(ccc(O)c1N=Nc(c(c(ccc2)cc3)c2)c3O)c1
Benzenesulfonamide, 4-hydroxy-3-((2-hydroxy-1-naphthalenyl)azo)- (9CI)

16432-81-8
$C_{18}H_{16}O_5$
312.32
O=C(OCCOc(ccc(c1O)C(=O)c(cccc2)c2)c1)C=C
2-Propenoic acid, 2-(4-benzoyl-3-hydroxyphenoxy)ethyl ester (9CI)
Acrylic acid, 4-ester with 2-hydroxy-4-(2-hydroxyethoxy)benzophenone

16433-43-5
$C_{11}H_{14}O_2$
178.25
CC(CCC(O)=O)c1ccccc1
Valeric acid, 4-phenyl
Benzenebutanoic acid, γ-methyl- (9CI)
4-Phenylpentanoic acid
4-Phenylvaleric acid

16440-97-4
$C_{11}H_{12}O$
160.22
1H-Inden-1-one, 2,3-dihydro-5,6-dimethyl- (9CI)
1-Indanone, 5,6-dimethyl- (6CI,8CI)

16448-54-7
$C_6H_6FeNO_6$
243.98
Iron, (nitrilotriacetato)
Acetic acid, nitrilotri-, iron(III) chelate
Ferric nitrilotriacetate
Iron, (N,N-bis(carboxymethyl)-glycinato(3-)-N,O,O',O'')-, (T-4)- (9CI)
Iron nitrilotriacetate
Iron-nitrilotriacetate chelate
Iron(3+) NTA

16452-01-0
$C_8H_{11}NO$
137.20
COc1cc(N)ccc1C
Benzenamine, 3-methoxy-4-methyl
o-Cresidine

16470-24-9
$C_{40}H_{44}N_{12}O_{16}S_4.4Na$
1169.01
Benzenesulfonic acid, 2,2'-(1,2-ethenediyl)bis(5-((4-(bis(2-hydroxyethyl)amino)-6-((4-sulfophenyl)amino)-1,3,5-triazin-2-yl)amino)-, tetrasodium salt (9CI)
2,2'-Stilbenedisulfonic acid, 4,4'-bis((4-bis(2-hydroxyethyl)amino)-6-(p-sulfoanilino)-s-triazin-2-yl)amino)-, tetrasodium salt

16485-10-2
$C_9H_{19}NO_4$
205.25
O=C(NCCCO)C(O)C(CO)(C)C
Panthenol
Butanamide, 2,4-dihydroxy-N-(3-hydroxypropyl)-3,3-dimethyl-, (+-)- (9CI)
(+-)-2,4-Dihydroxy-N-(3-hydroxypropyl)-3,3-dimethylbutyramide
Pantenol (Spanish)
Pantenolo
Panthenolum (Latin)
(+-)-Pantothenyl alcohol

16485-47-5
$C_{10}H_{15}FeN_2O_7.Na$
354.06
Ferrate(1-), (N-(2-(bis(carboxymethyl)amino)ethyl)-N-(2-hydroxyethyl)glycinato-(3-))-, sodium (9CI)
Ferrous sodium HEDTA

16502-88-8
$C_4H_4Cl_2O_2$
154.98
2-Butenoic acid, 4,4-dichloro-, (E)- (9CI)
Crotonic acid, 4,4-dichloro-, (E)- (8CI)

16509-79-8
$C_{12}H_{25}N_3S_4Zn$
404.99
Ziram - cyclohexylamine complex
Caswell No. 931A
EPA Pesticide Chemical Code 034806
Zinc bis(dimethyldithiocarbamate)cyclohexylamine complex
Zinc, (cyclohexylamine)bis(dimethyldithiocarbamato)-
Zinc, dimethyldithiocarbamate cyclohexylamine complex
Zincdimethyl dithiocarbamate cyclohexylamine complex
ZIP
Ziram, cyclohexylamine complex

16516-78-2
$C_{10}H_{17}NOS$
199.31
1H-Azepine-1-carbothioic acid, hexahydro-, S-2-propenyl ester (9CI)
1H-Azepine-1-carbothioic acid, hexahydro-, S-allyl ester (8CI)

16519-24-7
$C_{10}H_{20}O$
156.27
Heptane, 1-(2-propenyloxy)- (9CI)
Ether, allyl heptyl (8CI)

16521-38-3
Unknown
Unknown
C.I. Pigment Blue 63 (9CI)

16523-06-1
$C_8H_{12}O_2$
140.18
Cyclopentanecarboxylic acid, ethenyl ester (9CI)
Cyclopentanecarboxylic acid, vinyl ester (8CI)

16525-05-6
$C_9H_{16}O$
140.23
2-Hepten-4-one, 2,6-dimethyl- (8CI,9CI)

16529-56-9
C_5H_7N
81.11
N#CC(C=C)C
3-Butenenitrile, 2-methyl-
AI3-30534
2-Methyl-3-butenenitrile

16532-79-9
C_8H_6BrN
196.06
N#CCc(ccc(c1)Br)c1
Acetonitrile, (p-bromophenyl)
Benzeneacetonitrile, 4-bromo- (9CI)
4-Bromobenzeneacetonitrile
p-Bromobenzyl cyanide [UN 1694]
4-Bromobenzylcyanide
p-Bromophenylacetonitrile
2-(4-Bromophenyl)acetonitrile
4-Bromophenylacetonitrile
UN 1694 [Bromobenzyl cyanides, Liquid]

16543-55-8
$C_9H_{11}N_3O$
177.19
O=NN1CCCC1c2cccnc2
Pyridine, 3-(1-nitroso-2-pyrrolidinyl)-, (S)- (9CI)
Nicotine, 1'-demethyl-1'-nitroso- (8CI)
Nicotine, 1'-nitroso-1'-demethyl-
1'-Nitroso-1'-demethylnicotine
Nitrosonornicotine
N-Nitrosonornicotine
N'-Nitrosonornicotine
1'-Nitrosonornicotine
1-Nitroso-2-(3-pyridyl)pyrrolidine
3-(1-Nitroso-2-pyrrolidinyl)-pyridine
Nornicotine, N-nitroso-

16545-54-3
$C_{34}H_{66}O_4S$
570.96
O=C(OCCCCCCCCCCCCCC)CCSCCC(=O)OCCCCCCCCCCCCCC
Dimyristyl thiodipropionate
Ditetradecyl 3,3'-thiobispropanoate
Propanoic acid, 3,3'-thiobis-, ditetradecyl ester (9CI)

16545-85-0
$C_{19}H_{39}NO_2$
313.52
O=C(O)C(N(C)(C)C)CCCCCCCCCCCCCC
1-Pentadecanaminium, 1-carboxy-N,N,N-trimethyl-, hydroxide, inner salt (9CI)

16554-83-9
C_7H_{10}
94.16
Bicyclo(4.1.0)hept-3-ene (9CI)
3-Norcarene (6CI,7CI,8CI)
δ3-Norcarene

16561-29-8
$C_{36}H_{56}O_8$
616.84
CCCCCCCCCCCCCC(=O)OC3C(C)C1(O)C(CC(CO)CC2(O)C1C=C(C)C2=O)C4C(C)(C)C34OC(C)=O
Tetradecanoic acid, 9a-(acetyloxy)-1a,1b,4,4a,5,7a,7b,8,9,9a-decahydro-4a,7b-dihydroxy-3-(hydroxymethyl)-1,1,6,8-tetramethyl-5-oxo-1H-cyclopropa(3,4)benz(1,2-e)azulen-9-yl ester, (1aR-(1aα,1bβ,4aβ,7aα,7bα,8α,9β,9aα))- (9CI)
13-O-Acetylphorbol 12-myristate
Factor A1 (VAN)
Factor A1 (Croton oil)
Myristic acid, 9-ester with 1,1aα,1bβ,4,4a,7aα,7b,8,9,9a-decahydro-4aβ,7bα,9β,9a-α-tetrahydroxy-3-(hydroxymethyl)-1,1,6,8α-tetramethyl-5H-cyclopropa(3,4)benz(1,2-e)-azulen-5-one 9a-acetate, (+)- (8CI)
Pentahydroxy-tigliadienone-monoacetate(c)monomyristate(b)
Phorbol acetate, myristate
Phorbol monoacetate monomyristate
Phorbol myristate acetate
Phorbol 12-myristate 13-acetate
Phorbol 12-tetradecanoate 13-acetate
PMA (Tumor promoter)
Tetradecanoylphorbol acetate
12-O-Tetradecanoyl phorbol acetate
12-Tetradecanoylphorbol 13-acetate
12-O-Tetradecanoylphorbol-13-acetate
12-O-Tetradekanoylphorbol-13-acetat (German)
TPA (VAN)

16568-02-8
$C_4H_8N_2O$
100.14
CC=NN(C)C=O
Formic acid, ethylidenemethylhydrazide
Acetaldehyde-N-formyl-N-methylhydrazone
Cetaldehyde methylformylhydrazone
Acetaldehyde-N-methyl-N-formylhydrazone
Ethylidene gyromitrin
Gyromitrin
Hydrazine carboxaldehyde, ethylidenemethyl-
N-Methyl-N-formyl hydrazone of acetaldehyde

16577-13-2
$C_{18}H_{30}O_3S$
326.50
O=S(=O)(O)c(cccc1CCCCCCCCCCCC)c1
Benzenesulfonic acid, 3-dodecyl- (9CI)
Benzenesulfonic acid, m-dodecyl- (8CI)

16580-24-8
$C_{10}H_{20}$
140.27
Cyclohexane, 1-methyl-3-(1-methylethyl)- (9CI)
m-Menthane (8CI)

16586-42-8
$C_{19}H_{18}N_6O_2S$
394.42
O=N(=O)c(ccc(N=C(N=Nc(c(cc(N(CCC#N)CC)c1)C)c1)S2)c23)c3
Propanenitrile, 3-(ethyl(3-methyl-4-((6-nitro-2-benzothiazolyl)azo)phenyl)amino)- (9CI)
3-(N-Ethyl-4-((6-nitro-2-benzothiazolyl)azo)-m-toluidino)-propionitrile

16586-43-9
$C_{18}H_{18}ClN_5O_2$
371.80
O=N(=O)c(ccc(N=Nc(c(cc(N(CCC#N)CC)c1)C)c1)c2Cl)c2
Propanenitrile, 3-((4-((2-chloro-4-nitrophenyl)azo)-3-methylphenyl)ethylamino)- (9CI)
Benzenamine, 4-(((2-chloro-4-nitro)phenyl)azo)-3-methyl-N-(2-cyanoethyl)-N-ethyl-
3-(4-((2-Chloro-4-nitrophenyl)-azo)-N-ethyl-m-toluidino)-propionitrile

16587-52-3
$C_{13}H_{10}S$
198.29
Dibenzothiophene, 3-methyl- (8CI,9CI)

16587-71-6
$C_{11}H_{20}O$
168.31
O=C(CCC(C(CC)(C)C)C1)C1
Cyclohexanone, 4-tert-pentyl
4-tert-Amylcyclohexanone
Cyclohexanone, 4-(1,1-dimethylpropyl)-
4-(1,1-Dimethylpropyl)cyclohexanone

16588-67-3
$C_{20}H_{21}N_5O_2S_2$
427.52
O=S(=O)(c(ccc(N=C(N=Nc(c(cc(N(CCC#N)CC)c1)C)c1)S2)c23)c3)C
Propanenitrile, 3-(ethyl(3-methyl-4-((6-(methylsulfonyl)-2-benzothiazolyl)azo)-phenyl)amino)- (9CI)

16589-43-8
$CH_6O_3Si.xNa$
Unknown
Silanetriol, methyl-, sodium salt (9CI)
Methylsilanetriol sodium salt
Sodium methylsiliconate

16604-76-5
$C_{14}H_{13}N_2O_7PS$
384.29
Phosphorothioic acid, S-ethyl O,O-bis(4-nitrophenyl) ester (9CI)
Phosphorothioic acid, S-ethyl O,O-bis(p-nitrophenyl) ester (8CI)

16605-91-7
$C_{12}H_8Cl_2$
223.10
Clc2cccc(c1ccccc1)c2Cl
2,3-Dichlorobiphenyl

16606-02-3
$C_{12}H_7Cl_3$
257.55
Clc1ccc(cc1)c2cc(Cl)ccc2Cl
1,1'-Biphenyl, 2,4',5-trichloro- (9CI)
Biphenyl, 2,4',5-trichloro- (8CI)
2,4',5-PCB
TCB
2,4',5-Trichlorobiphenyl
4,2',5'-Trichlorobiphenyl
2,4',5-Trichloro-1,1'-biphenyl

16608-68-7
$C_{10}H_{11}Cl$
166.65
Benzene, (3-chloro-2-butenyl)-, (Z)- (9CI)
2-Butene, 3-chloro-1-phenyl-, (Z)- (8CI)

16611-66-8
$C_{14}H_{15}O_3PS$
294.31
Phosphorothioic acid, S-ethyl

O,O-diphenyl ester (8CI,9CI)

16624-06-9
$C_{11}H_{22}O$
170.30
Cyclooctanemethanol, α,α-dimethyl- (8CI)

16635-54-4
$C_6H_{10}O$
98.14
2-Hexenal, (Z)- (8CI,9CI)

16645-06-0
$C_2H_7NO.ClH$
97.54
Methanamine, N-hydroxy-N-methyl-, hydrochloride (9CI)
NSC-45353

16647-04-4
$C_8H_{12}O$
124.18
3,5-Heptadien-2-one, 6-methyl-, (E)- (8CI,9CI)

16647-05-5
$C_{13}H_{20}O$
192.30
4,5,9-Undecatrien-2-one, 6,10-dimethyl- (8CI,9CI)

16650-10-5
C_2Cl_4O
181.82
Ethane, tetrachloroepoxy
Epoxyperchlorovinyl
Oxirane, tetrachloro- (9CI)
PCEO
Tetrachloroepoxyethane
Tetrachloroethylene oxide

16655-82-6
$C_{12}H_{15}NO_4$
237.28
Carbamic acid, methyl-, 2,3-dihydro-2,2-dimethyl-3-hydroxy-7-benzofuranyl ester
3,7-Benzofurandiol, 2,3-dihydro-2,2-dimethyl-, 7-(methylcarbamate) (9CI)
3-Hydroxycarbofuran

16664-45-2
$C_{16}H_{14}$
206.29
Phenanthrene, 1,3-dimethyl- (8CI,9CI)

16666-42-5
$C_5H_8O_2$
100.12
2-Pentenoic acid, (Z)- (8CI,9CI)
cis-3-Ethylacrylic acid
cis-2-Pentenoic acid

16669-59-3
$C_8H_{15}NO_2$
157.24
O=C(NCOCC(C)C)C=C
Acrylamide, N-(isobutoxymethyl)
N-Isobutoxymethylacrylamide
2-Propenamide, N-((2-methylpropoxy)methyl)- (9CI)

16672-87-0
$C_2H_6ClO_3P$
144.50
Phosphonic acid, (2-chloroethyl)
Amchem 68-250
Bromoflor
Camposan
CEP
CEPHA
2-CEPA
CEPHA 10LS
Cerone
2-Chloraethyl-phosphonsaeure (German)
Chlorethephon
2-Chloroethanephosphonic acid
2-Chloroethylphosphonic acid
Ethefon
Ethel
Ethephon
Ethepon
Etheverse
Ethrel
Flordimex
Florel
G 996
Gagro
Kamposan
Roll-Fruct
Tomathrel

16680-47-0
$C_{18}H_{19}NaO_5S$
370.40
Conjugated estrogen
Sodium Equilin Sulfate

16681-63-3
$C_4H_4BrN_3O_2$
205.98
1H-Imidazole, 2-bromo-1-methyl-4-nitro- (9CI)
Imidazole, 2-bromo-1-methyl-4-nitro- (8CI)
NSC-347484

16709-30-1
$C_{12}H_{13}NO_4$
235.26
Carbamic acid, methyl-, ester with 2,2-dimethyl-7-hydroxy-3(2H)-benzofuranone
3(2H)-Benzofuranone, 2,2-dimethyl-7-(((methylamino)carbonyl)oxy)- (9CI)
3-Ketocarbofuran

16712-64-4
$C_{11}H_8O_3$
188.18
O=C(O)c(ccc(c1ccc2O)c2)c1
2-Naphthalenecarboxylic acid, 6-hydroxy- (9CI)
6-Hydroxy-2-naphthalenecarboxylic acid
NSC-148862

16714-68-4
$C_3H_3Cl_5$
216.31
C(C(Cl)Cl)(CCl)(Cl)Cl
Propane, 1,1,2,2,3-pentachloro
1,1,2,2,3-Pentachloropropane

16715-77-8
$C_8H_8O_4$
168.16
1-Naphthoic acid, 2,3-dihydroxy
2,3-Dihydroxynaphthoic acid
1-Naphthalenecarboxylic acid, 2,3-dihydroxy-
β-Oxynaphtoic acid

16715-83-6
$C_{12}H_{23}NO_2$
213.31
O=C(OCCN(C(C)C)C(C)C)C(=C)C
2-(Diisopropylamino)ethyl methacrylate
Diisopropylaminoethyl methacrylate
N,N-Diisopropylaminoethyl methacrylate
Methacrylic acid, 2-(diisopropylamino)ethyl ester
2-Propenoic acid, 2-methyl-, 2-(bis(1-methylethyl)amino)ethyl ester (9CI)

16721-80-5
HNaS
56.06
Sodium-sulfide
NA 2922 [Sodium hydrosulfide, solution]
NA 2923 (DOT)
Sodium bisulfide
Sodium hydrogen sulfide
Sodium hydrosulfide
Sodium hydrosulfide, Solution [NA 2922]
Sodium hydrosulphide, With less than 25% water of crystallization [UN 2318]
Sodium hydrosulphide, Solid

(DOT)
Sodium hydrosulphide, Solid with not < 25% water of crystallization [UN 2949]
Sodium mercaptan
Sodium mercaptide
Sodium sulfhydrate
Sodium sulfide, Not less than 25% water of crystallization
UN 2318 [Sodium hydrosulfide, with less than 25 per cent water of crystallization]
UN 2949 [Sodium hydrosulfide with not less than 25 per cent water of crystallization]

16722-32-0
$C_{22}H_{38}O_3S$
382.61
O=S(=O)(O)c(ccc(c1)CCCCCCC CCCCCCCCC)c1
Benzenesulfonic acid, p-hexadecyl- (8CI)
p-Hexadecylbenzenesulfonic acid
4-Hexadecylbenzenesulfonic acid

16725-53-4
$C_{16}H_{30}O_2$
254.41
O=C(OCCCCCCCCC=CCCCC)C
9-Tetradecen-1-ol, acetate, (Z)-
AI3-33474
(Z)-9-Tetradecen-1-ol acetate
(Z)-9-Tetradecenyl acetate

16727-91-6
$C_{14}H_{16}$
184.28
Naphthalene, 1-(2-methylpropyl)- (9CI)
Naphthalene, 1-isobutyl- (6CI,8CI)
α-Isobutylnaphthalene
1-Isobutylnaphthalene

16728-49-7
$C_{11}H_{21}NO$
183.29
N#CCCOCCCCCCCC
Propanenitrile, 3-(octyloxy)- (9CI)
3-Octoxypropanenitrile
3-Octyloxypropanenitrile
3-(Octyloxy)propanenitrile

16728-51-1
$C_{13}H_{25}NO$
211.34
N#CCCOCCCCCCCCCC
Propanenitrile, 3-(decyloxy)- (9CI)
3-Decoxypropanenitrile
3-(Decyloxy)propanenitrile

16731-55-8
$O_5S_2.K$
183.22
Potassium-pyrosulfite
NA 2693 (DOT)
Potassium metabisulfite (DOT)
Pyrosulfurous acid, dipotassium salt

16747-25-4
C_9H_{20}
128.26
Hexane, 2,2,3-trimethyl- (8CI,9CI)

16747-26-5
C_9H_{20}
128.26
Hexane, 2,2,4-trimethyl-
AI3-28854

16747-28-7
C_9H_{20}
128.26
Hexane, 2,3,3-trimethyl- (8CI,9CI)

16747-30-1
C_9H_{20}
128.26
Hexane, 2,4,4-trimethyl- (8CI,9CI)

16747-31-2
C_9H_{20}
128.26
Hexane, 3,3,4-trimethyl- (9CI)
3,3,4-Trimethylhexane

16747-32-3
C_9H_{20}
128.26
Pentane, 3-ethyl-2,2-dimethyl- (8CI,9CI)

16747-33-4
C_9H_{20}
128.26
Pentane, 3-ethyl-2,3-dimethyl- (8CI,9CI)

16747-38-9
C_9H_{20}
128.26
Pentane, 2,3,3,4-tetramethyl- (7CI,8CI,9CI)
2,3,3,4-Tetramethylpentane

16747-42-5
$C_{10}H_{22}$
142.28
Hexane, 2,2,4,5-tetramethyl- (8CI,9CI)

16747-50-5
C_8H_{16}
112.22
Cyclopentane, 1-ethyl-1-methyl- (8CI,9CI)

16752-77-5
$C_5H_{10}N_2O_2S$
162.23
CNC(=O)ON=C(C)SC
Acetimidic acid, thio-N-((methylcarbamoyl)oxy)-, methyl ester
Acetimidic acid, N-((methylcarbamoyl)oxy)thio-, methyl ester (8CI)
Acetimidothioic acid, methyl-, N-(methylcarbamoyl) ester
DuPont 1179
Du Pont Insecticide 1179
ENT 27,341
Ethanimidothioic acid, N-(((methylamino)carbonyl)-oxy)-, methyl ester (9CI)
IN 1179
Insecticide 1,179
Lannate (OSHA)
Lannate L
Mesomile
Methomyl (ACGIH,OSHA)
N-(((Methylamino)carbonyl)-oxy)ethanimidothioic acid methyl ester
N-((Methylcarbamoyl)oxy)thio-acetimidic acid methyl ester
Methyl N-(((methylamino)carbonyl)oxy)ethanimidothioate
Methyl-N-((methylcarbamoyl)-oxy)thioacetimidate
S-Methyl N-(methylcarbamoyl-oxy)thioacetimidate
Methyl O-(methylcarbamoyl)-thiolacethohydroxamate
2-Methylthio-acetaldehyd-O-(methylcarbamoyl)-oxim (German)
2-Methylthio-propionaldehyd-O-(methylcarbamoyl)-oxim (German)
Metomil (Italian)
Nu-Bait II
Nudrin
RCRA waste number P066
SD 14999
3-Thiabutan-2-one, O-(methylcarbamoyl)oxime
WL 18236

16757-80-5
$C_{21}H_{14}$
266.35
Benzo(a)pyrene, 11-methyl
11-Methylbenzo(a)pyrene
6-Methyl-3,4-benzpyrene

16757-81-6
$C_{21}H_{14}$
266.35
Benzo(a)pyrene, 3-methyl
8-Methyl-3,4-benzpyrene
3-Methylbenzo(a)pyrene

16757-82-7
$C_{21}H_{14}$
266.35
Benzo(a)pyrene, 2-methyl
2-Methylbenzo(a)pyrene
9-Methyl-3,4-benzpyrene

16757-83-8
$C_{21}H_{14}$
266.35
Benzo(a)pyrene, 4-methyl
4-Methylbenzo(a)pyrene

16757-84-9
$C_{22}H_{16}$
280.38
Benzo(a)pyrene, 3,12-dimethyl
3,12-Dimethylbenzo(a)pyrene

16757-85-0
$C_{22}H_{16}$
280.38
Benzo(a)pyrene, 1,2-dimethyl
1,2-Dimethylbenzo(a)pyrene

16757-86-1
$C_{22}H_{16}$
280.38
Benzo(a)pyrene, 1,3-dimethyl
1,3-Dimethylbenzo(a)pyrene

16757-87-2
$C_{22}H_{16}$
280.38
Benzo(a)pyrene, 2,3-dimethyl
2,3-Dimethylbenzo(a)pyrene

16757-88-3
$C_{22}H_{16}$
280.38
Benzo(a)pyrene, 1,4-dimethyl
1,4-Dimethylbenzo(a)pyrene

16757-89-4
$C_{22}H_{16}$
280.38
Benzo(a)pyrene, 4,5-dimethyl
4,5-Dimethylbenzo(a)pyrene

16757-90-7
$C_{22}H_{16}$
280.38
Benzo(a)pyrene, 1,6-dimethyl
1,6-Dimethylbenzo(a)pyrene

16757-91-8
$C_{22}H_{16}$
280.38
Benzo(a)pyrene, 3,6-dimethyl
3,6-Dimethylbenzo(a)pyrene

16757-92-9
$C_{23}H_{18}$
294.41
Benzo(a)pyrene, 1,3,6-tri-methyl
1,3,6-Trimethylbenzo(a)pyrene

16761-12-9
$C_7H_{14}O_2.K$
169.29
Heptanoic acid, potassium salt (9CI)
Potassium heptanoate

16766-27-1
$C_8H_9ClO_2$
172.61
Benzene, 4-chloro-1,2-dimeth-oxy- (8CI,9CI)

16766-29-3
$C_8H_7Cl_3O_2$
241.50
Benzene, 1,2,3-trichloro-4,5-di-methoxy- (8CI,9CI)

16766-30-6
$C_7H_7ClO_2$
158.58
Phenol, 4-chloro-2-methoxy- (8CI,9CI)

16766-31-7
$C_7H_6Cl_2O_2$
193.03
Phenol, 2,4-dichloro-6-methoxy- (8CI,9CI)

16773-42-5
$C_7H_{10}ClN_3O_3$
219.65
Cc1ncc(N(=O)=O)n1CC(O)CCl
Imidazole-1-ethanol, α-(chloromethyl)-2-methyl-5-nitro
1-(3-Chloro-2-hydroxypropyl)-2-methyl-5-nitroimidazole
α-(Chlormethyl)-2-methyl-5-nitro-imidazol-1-aethanol (German)
α-(Chloromethyl)-2-methyl-5-nitro-1H-imidazole-1-ethan-ol
1H-Imidazole-1-ethanol, α-(chloromethyl)-2-methyl-5-nitro-
Ornidazole
Ro 7-0207
Tiberal

16782-30-2
$C_{11}H_{22}O$
170.30
Cyclooctanepropanol (8CI, 9CI)
1-Propanol, 3-cyclooctyl-

16812-54-7
NiS
90.76
Nickel sulfide (9CI)
Millerite

16813-72-2
Unknown
Unknown
Propylene dimer

16824-02-5
$C_{10}H_{10}O_3$
178.19
1(3H)-Isobenzofuranone, 3-ethoxy- (9CI)
NSC-60055
NSC-132303
Phthalide, 3-ethoxy- (8CI)

16825-16-4
$C_{18}H_{36}O$
268.48
2-Pentadecanone, 6,10,14-tri-methyl-, (R-(R*,R*))- (9CI)
2-Pentadecanone, 6,10,14-tri-methyl-, (6R,10R)- (8CI)
Phytol ketone
Phytone

16828-95-8
Unknown
Unknown
Copper in the form of an ammonia complex

16842-03-8
C_4HCoO_4
171.98
Cobalt hydrocarbonyl (ACGIH,OSHA)
Cobalt, tetracarbonylhydro- (8CI,9CI)
Hydridotetracarbonylcobalt
Hydrocobalt tetracarbonyl
Tetracarbonylhydridocobalt
Tetracarbonylhydrocobalt

16853-85-3
$AlH_4.Li$
37.96
Aluminate (1-), tetrahydro-, lithium
Aluminate(1-), tetrahydro-, lithium, (T-4)- (9CI)
Aluminum lithium hydride
Lithium alanate

Lithium aluminohydride
Lithium aluminum hydride [UN 1410]
Lithium aluminum hydride, ethereal [UN 1411]
Lithium aluminum tetrahydride
Lithium tetrahydroaluminate
Lithium tetrahydroaluminate(1-)
UN 1410 [Lithium aluminum hydride]
UN 1411 [Lithium aluminum hydride, ethereal]

16867-03-1
$C_5H_6N_2O$
110.10
3-Pyridinol, 2-amino- (9CI)
AI3-61061
2-Amino-3-hydroxypyridine
2-Amino-3-pyridinol
3-Hydroxy-2-pyridinamine
NSC-136806

16867-04-2
$C_5H_5NO_2$
111.11
2(1H)-Pyridone, 3-hydroxy
2,3-Dihydroxypyridine
2,3-Pyridinediol
2(1H)-Pyridinone, 3-hydroxy- (9CI)

16871-71-9
$F_6Si.Zn$
207.46
Silicate(2-), hexafluoro-, zinc
Fungol
Fungonit gf 2
Silicon zinc fluoride
UN 2855 [Zinc fluorosilicate]
Zinc fluorosilicate [UN 2855]
Zinc hexafluorosilicate
Zinc silicofluoride (DOT)

16872-11-0
$BF_4.H$
87.82
Borate(1-), tetrafluoro-, hydrogen
Borofluoric acid
Fluoboric acid [UN 1775]
Hydrofluoboric acid
Hydrofluoroboric acid (DOT)
Hydrogen tetrafluoroborate
Tetrafluoroboric acid
UN 1775 [Fluoboric acid]

16883-83-3
$C_{27}H_{34}O_6$
454.56
O=C(OCC(C)(C)C(OC(=O)c(c(cc c1)C(=O)OCc(cccc2)c2)c1) C(C)C)C(C)C
1,2-Benzenedicarboxylic acid, 2,2-dimethyl-1-(1-methylethyl)-3-(2-methyl-1-oxopropoxy)propyl phenylmethyl ester (9CI)

16887-00-6
Cl
35.45
Chloride (8CI,9CI)
Chlorine, ion (Cl(1-))
Hydrochloric acid, ion(1-)

16889-10-4
$C_{18}H_{16}N_6O_2$
348.40
N#Cc(c(N=Nc(ccc(N(CCC#N) CC)c1)c1)ccc2N(=O)=O)c2
Benzonitrile, 2-((4-((2-cyanoethyl)ethylamino)phenyl)azo)-5-nitro
Dispersive Rubin Polyether

16889-14-8
$C_{24}H_{50}N_2O$
382.66
O=C(NCCN(CC)CC)CCCCCCC CCCCCCCCCC
Stearamidoethyl diethylamine
N-(2-Diethylamino)ethyl)octadecanamide
N-(2-(Diethylamino)ethyl)octadecanamide
Diethylaminoethyl stearamide
N-(2-(Diethylamino)ethyl)stearamide
Lexamine 22
NSC-126195
Octadecanamide, N-(2-(diethylamino)ethyl)- (9CI)
Sapamine COB-ST (VAN)
Stearic acid-N,N-diethylethylenediamine condensate

16893-85-9
$F_6Si.2Na$
188.07
Silicate(2-), hexafluoro-, disodium
Destruxol Applex
Disodium hexafluorosilicate
Disodium hexafluorosilicate (2-)
Disodium silicofluoride
Ens-Zem Weevil Bait
ENT 1,501
Fluosilicate de sodium
Natriumsilicofluorid (German)
Ortho Earwig Bait
Ortho Weevil Bait
Prodan
PSC Co-Op Weevil Bait
Safsan
Salufer
Silicon sodium fluoride
Sodium fluorosilicate [UN 2674]
Sodium fluosilicate
Sodium hexafluorosilicate
Sodium hexafluosilicate
Sodium silicofluoride (DOT)
Sodium silicon fluoride
Super Prodan
UN 2674 [Sodium fluorosilicate]

16898-52-5
$C_{13}H_{26}N_2$
210.35
N(CCC(C1)CCCC(CCNC2) C2)C1
Piperidine, 4,4'-(1,3-propanediyl)bis- (9CI)
AI3-61833
1,3-Bis(4-piperidyl)propane
Di-PIP
1,3-Di-4-piperidylpropane
NSC-96364
Piperidine, 4,4'-trimethylenedi- (8CI)
4,4'-(1,3-Propanediyl)bispiperidine
4,4'-Trimethylenedipiperidine

16899-08-4
$C_{14}H_{28}O_3$
244.37
Tetradecanoic acid, 10-hydroxy- (8CI,9CI)

16909-78-7
$C_{15}H_{14}O_4$
258.27
O=C(c(cccc1)c1)c(c(O)cc(OC CO)c2)c2
Methanone, (2-hydroxy-4-(2-hydroxyethoxy)phenyl)-phenyl- (9CI)

16919-19-0
$F_6Si.2H_4N$
178.15
[NH4+].[NH4+].F[Si-2](F)(F)(F) (F)F
Ammonium silicofluoride
AI3-25550-X
Ammonium fluorosilicate [UN 2854]
Ammonium fluosilicate
Ammonium hexafluorosilicate
Ammonium silicon fluoride
Bye Bugs
Caswell No. 043
(Component of) Drianone
(Component of) Dri-Die
Diammonium fluorosilicate
Diammonium fluosilicate
Diammonium hexafluorosilicate
Diammonium hexafluorosilicate(2-)
Diammonium silicon hexafluoride
EPA Pesticide Chemical Code 075301
Fluorosilicic acid, ammonium salt
Hexafluorosilicate(2-) diammonium
Laidlaw U-San-O Moth Proofing Spray
NSC-310005

Silicate, hexafluoro-, diammonium
Silicate(2-), hexafluoro-, diammonium (9CI)
Silicate(2-), hexafluoro-, diammonium + silica
Superior Dri-Die
UN 2854 [Ammonium fluorosilicate]

16919-31-6
$F_6Zr.2H_4N$
241.29
Zirconium ammonium fluoride
Ammonium fluozirconate
Ammonium hexafluoro zirconate
Ammonium zirconium fluoride
Zirconate(2-), hexafluoro-, diammonium
Zirconate(2-), hexafluoro-, diammonium, (OC-6-11)- (9CI)
Zironate(2-), hexafluoro-, diammonium, (OC-6-11)-

16919-58-7
$Cl_6Pt.2H_4N$
443.89
Ammonium-chloroplatinate
Ammonium hexachloroplatinate(IV)
Ammonium platinic chloride
Diammonium hexachloroplatinate(2-)
1-Hexadecanaminium, N-ethyl-N,N-dimethyl-, bromide
Platinate(2-), hexachloro-, diammonium, (OC-6-11)-
Platinic ammonium chloride
Quaternium-17

16923-95-8
$F_6Zr.2K$
283.42
Zirconate(2-), hexafluoro-, dipotassium
Dipotassium hexafluorozirconate
Dipotassium zirconium hexafluoride
Potassium fluorozirconate
Potassium fluozirconate
Potassium hexafluorozirconate
Potassium hexafluorozirconate(IV)
Potassium zirconium fluoride
Potassium zirconium hexafluoride

16924-00-8
$F_7Ta.2K$
392.15
Potassium-heptafluorotantalate
Potassium fluotantalate
Potassium tantalum fluoride
Tantalum potassium fluoride

16938-22-0
$C_{11}H_{18}N_2O_2$
210.26
O=C=NCC(CC(CCN=C=O)C)(C)C
Hexane, 1,6-diisocyanato-2,2,4-trimethyl- (9CI)
1,6-Diisocyanato-2,2,4-trimethylhexane
2,2,4-Trimethyl-1,6-diisocyanatohexane

16940-66-2
$BH_4.Na$
37.84
Borate(1-), tetrahydro-, sodium
Borohydrure de sodium (French)
Sodium borohydride [UN 1426]
Sodium tetrahydroborate(1-)
UN 1426 [Sodium borohydride]

16940-81-1
F_6HP
145.98
Phosphate(1-) hexafluoro-, hydrogen
Hexafluorophosphoric acid [UN 1782]
Hydrogen hexafluorophosphate
UN 1782 [Hexafluorophosphoric acid]

16941-10-9
$AlH_4.1/2Ca$
51.05
Aluminate(1-), tetrahydro-, calcium (2:1), (T-4)- (9CI)
Aluminum calcium hydride (6CI)
Aluminate(1-), tetrahydro-, calcium (8CI)
Calcium bis(tetrahydroaluminate)
Calcium bis(tetrahydroaluminate(1-))
Calcium tetrahydroaluminate (7CI)

16941-12-1
$Cl_6Pt.2H$
409.81
Chloroplatinic acid
Acide chloroplatinique (French)
Acido cloroplatinico (Spanish)
Chloroplatinic (IV) acid
Chloroplatinic acid, Solid [UN 2507]
Dihydrogen hexachloroplatinate
Dihydrogen hexachloroplatinate(2-)
Hexachloroplatinic acid
Hexachloroplatinic (IV) acid
Hexachloroplatinic(4+) acid, hydrogen-
Hydrogen hexachloroplatinate(IV)
Hydrogen hexachloroplatinate(4+)
NSC-4958
Platinate(2-), hexachloro-, dihydrogen (8CI)
Platinate(2-), hexachloro-, dihydrogen, (OC-6-11)- (9CI)
Platinic chloride (VAN)
Platinum chloride
Platinum chloride (H_2PtCl_6)
UN 2507 [Chloroplatinic acid, Solid]

16941-32-5
$C_{153}H_{225}N_{43}O_{49}S$
3482.54
Glucagon
Glucagon (Human)
Glucagon (Pig)
Glucagon, Porcine, For bioassay
Glucagon, Porcine, For immunoassay
Glucagone
Glucagonum (Latin)
Glukagon novo

16947-63-0
$C_8H_{10}N_2O_2$
166.17
Cc1cc(cc(C)c1N)N(=O)=O
Benzenamine, 2,6-dimethyl-4-nitro- (9CI)
NSC-101580
2,6-Xylidine, 4-nitro- (8CI)

16949-15-8
$BH_4.Li$
21.78
Lithium borohydride
Borate(1-), tetrahydro-, lithium (8CI,9CI)
Borohidruro de litio (Spanish)
Borohydrure de lithium (French)
Lithium tetrahydroborate
Tetrahydroborate(1-) lithium
UN 1413 [Lithium borohydride]

16949-65-8
$F_6Si.Mg$
166.38
Magnesium fluorosilicate
Caswell No. 532
EPA Pesticide Chemical Code 075304
Hexafluorosilicate(2-) magnesium (1:1)
Magnesium fluosilicate
Silicate(2-), hexafluoro-, magnesium (1:1) (9CI)

16954-69-1
$C_8H_8N_2S$
164.24

Benzothiazole, 2-(methylamino)
2-Methylaminobenzothiazole

16958-92-2
$C_{32}H_{62}O_4$
510.84
O=C(OCCCCCCCCCCCCC)CCCCC(=O)OCCCCCCCCCCCCC
Ditridecyl dilinoleate
Dilinoleic acid, ditridecyl ester
Dimer acid, ditridecyl ester
Ditridecyl dimerate
Ditridecyl hexanedioate
Hexanedioic acid, ditridecyl ester (9CI)

16960-16-0
$C_{136}H_{210}N_{40}O_{31}S$
2933.92
α^{1-24}-Corticotropin
$ACTH^{1-24}$
ACTH-α^{1-24}
α^{1-24}-ACTH
β^{1-24}-ACTH
Actholain
Ba 30920
Ba 36716
Corticotropin-(1-24)
β-1,24-Corticotrophin
β^{1-24}-Corticotropin
Cortrophin S
Cortrosinta
Cortrosyn
Cosyntropin
Nuvacthen Depot
Synacthen
Tetracosactid
Tetracosactide
β^{1-24}-Tetracosactide
Tetracosactrin
Tetracosapeptide

16961-83-4
$F_6Si.2H$
144.11
Silicate(2-), hexafluoro-, dihydrogen
Acide fluorosilicique (French)
Acide fluosilicique (French)
Acido fluosilicico (Italian)
Dihydrogen hexafluorosilicate
Dihydrogen hexafluorosilicate (2-)
FKS
Fluorosilicic acid [UN 1778]
Fluosilicic acid (DOT)
Hexafluorokieselsaiure (German)
Hexafluorokiezelzuur (Dutch)
Hexafluosilicic acid
Hydrofluorosilicic acid (DOT)
Hydrofluosilicic acid (DOT)
Hydrogen hexafluorosilicate
Hydrosilicofluoric acid (DOT)
Kiezelfluorwaterstofzuur (Dutch)
NA 1778 (DOT)
Sand Acid (DOT)
Silicofluoric acid (DOT)
Silicon hexafluoride dihydride
UN 1778 [Fluorosilicic acid]

16965-90-5
$C_{10}H_{21}O_4P$
236.25
Phosphonic acid, (2-methyl-3-oxopentyl)-, diethyl ester (8CI,9CI)

16967-61-6
$C_8H_{13}N$
123.20
2-Azacyclopropa(cd)pentalene, octahydro-2-methyl- (8CI)

16967-79-6
C_2HCl_3O
147.38
Ethane, trichloroepoxy
Epoxy-1,1,2-trichloroethane
Oxirane, trichloro- (9CI)
TCEO
1,1,2-Trichloroepoxyethane
Trichloroethylene epoxide
Trichloroethylene oxide
Trichloro-oxirane

16971-82-7
$C_{15}H_{17}N_3.C_{12}H_8BO_4.H$
467.31
Borate(1-), bis(1,2-benzenediolato(2-)-O,O')-, (T-4)-, hydrogen, Compd. with N,N'-bis(2-methylphenyl)-guanidine (1:1) (9CI)

16974-11-1
$C_{14}H_{26}O_2$
226.36
O=C(OCCCCCCCCC=CCC)C
9-Dodecen-1-ol, acetate, (Z)-
AI3-33971
Caswell No. 411B
Disrupt
(Z)-9-Dodecen-1-ol acetate
(Z)-9-Dodecenyl acetate
EPA Pesticide Chemical Code 117701
Nomate Shootgard
Sonolure
WPSB Pheromone

16974-12-2
Unknown
Unknown
(Z)-9-Dodecenyl acetate

16977-58-5
$C_4H_5Cl_3N_4O$
231.44
Ethanol, 2,2,2-trichloro-1-(1H-1,2,4-triazol-3-ylamino)-(9CI)
Ethanol, 2,2,2-trichloro-1-(s-triazol-3-ylamino)- (8CI)
Sazol

16984-48-8
F
19.00
Fluoride
Fluoride(1-)
Fluoride ion
Fluoride ion(1-)
Perfluoride

16995-35-0
$C_3H_2Cl_4O$
195.85
2-Propanone, 1,1,1,3-tetrachloro
1,1,1,3-Tetrachloroacetone
1,1,1,3-Tetrachloro-2-propanone

16996-40-0
$C_8H_{16}O_2.xPb$
Unknown
Lead 2-ethylhexoate
Hexanoic acid, 2-ethyl-, lead salt (9CI)
Lead 2-ethylhexanoate

16996-51-3
$C_{18}H_{32}O_2.xPb$
Unknown
Lead linoleate
9,12-Octadecadienoic acid (Z,Z)-, lead salt (9CI)

17012-98-5
$C_{14}H_{22}O_3S$
270.39
Benzenesulfonic acid, 4-octyl-(9CI)
Benzenesulfonic acid, p-octyl-(8CI)

17014-71-0
K_2O_2
110.20
Potassium peroxide [UN 1491]
UN 1491 [Potassium peroxide]

17021-26-0
$C_{21}H_{32}O_2$
316.53
CC3CC1=CC(=O)CCC1(C)C4CCC2(C)C(CCC2(C)O)C34
Androst-4-en-3-one, 7-β,17-α-dimethyl-17-β-hydroxy
Androst-4-en-3-one, 17-hydroxy-7,17-dimethyl-, (7-β,17-β)- (9CI)
Androst-4-en-3-one, 17-β-hydroxy-7-β,17-dimethyl-
Calusterone

7-β,17-Dimethyltestosterone
7-β,17-α-Dimethyl testosterone
17-β-Hydroxy-7-β,17-α-dimethylandrost-4-ene-3-one
Methosarb
NSC-88536
U-22,550

17024-18-9
$C_{14}H_9NO_2$
223.24
Phenanthrene, 2-nitro
2-Nitrophenanthrene

17026-81-2
$C_{10}H_{14}N_2O_2$
194.26
O=C(Nc(ccc(OCC)c1N)c1)C
Acetanilide, 3'-amino-4'-ethoxy
Acetanilide, 3-amino-4-ethoxy-
2-Amino-4-acetaminifenetol (Czech)
3-Amino-4-ethoxyacetanilide
3-Amino-4-ethoxyanilid kyseliny octove (Czech)
NCI-C01887

17040-19-6
$C_6H_{15}O_5PS_2$
262.30
CCS(=O)(=O)CCSP(=O)(OC)OC
Phosphorothioic acid, O,O-dimethyl S-(2-(ethylsulfonyl)-ethyl) ester
Bayer 20315
Demeton-S-methylsulfon (German)
Demeton-S-methyl-sulphone
O,O-Dimethyl-S-(2-aethyl-sulfonyl-aethyl)-thiolphosphat (German)
O,O-Dimethyl S-(2-ethsulfonyl-ethyl)phosphorothioate
Dimethyl S-(2-ethsulfonyl-ethyl)thiophosphate
O,O-Dimethyl S-ethyl-2-sulfonylethyl phosphorothiolate
O,O-Dimethyl S-ethylsulphonyl-ethyl phosphorothiolate
Dioxydemeton-S-methyl
E 158
Isometasystox sulfone
Isomethylsystox sulfone
M 3/158
Metaisosystox-solfon 20 315
Sulfone, demeton-S-methyl-

17052-15-2
$C_4H_9ClIO_2P$
282.45
Phosphinic acid, (chloromethyl)(iodomethyl)-, ethyl ester (8CI)

17052-17-4
$C_4H_9I_2O_2P$
373.90
Phosphinic acid, bis(iodomethyl)-, ethyl ester (8CI)

17052-18-5
$C_6H_{13}I_2O_2P$
401.95
Phosphinic acid, bis(iodomethyl)-, butyl ester (8CI)

17057-82-8
$C_{11}H_{14}$
146.23
1H-Indene, 2,3-dihydro-1,2-dimethyl- (9CI)
1,2-Dimethylindan
1,2-Dimethylindane
Indan, 1,2-dimethyl-

17057-91-9
$C_{13}H_{14}$
170.25
Naphthalene, 1,3,8-trimethyl- (8CI,9CI)

17059-48-2
$C_{11}H_{14}$
146.23
1H-Indene, 2,3-dihydro-1,6-dimethyl- (9CI)
Indan, 1,6-dimethyl-

17066-67-0
$C_{15}H_{24}$
204.36
Naphthalene, decahydro-4a-methyl-1-methylene-7-(1-methylethenyl)-, (4aR-(4aα,7α,8aβ))- (9CI)
Eudesma-4(14),11-diene (8CI)
β-Selinene

17068-78-9
Unknown
Unknown
Anthophyllite
Azbllen asbestos
16 F

17080-02-3
$C_{21}H_{26}O_4$
342.47
Cyclopropanecarboxylic acid, 2,2-dimethyl-3-(2-methylpropenyl)-, ester with (+-)-2-furfuryl- 4-hydroxy-3-methyl-2-cyclopenten-1-one, (+-)-cis,trans
(+-)-Furethionyl (+-)-cis, trans-chrysanthemate
Furethrin
DL-Furfurylrethronyl DL-cis-trans-chrysanthemate

17084-02-5
$C_{10}H_{15}FeN_2O_7$
331.07
Iron, (N-(2-(bis(carboxymethyl)amino)ethyl)-N-(2-hydroxyethyl)glycinato(3-)) (9CI)
Ferric HEDTA
NSC-97346

17085-91-5
$C_{14}H_{16}$
184.28
Naphthalene, 1-(1,1-dimethylethyl)- (9CI)
Naphthalene, 1-tert-butyl- (8CI)
NSC-122456

17088-28-7
$C_{16}H_{18}O_4$
274.32
O=C(OCC)C=Cc(ccc(c1)C=CC(=O)OCC)c1
2-Propenoic acid, 3,3'-(1,4-phenylene)bis-, diethyl ester (9CI)

17090-79-8
$C_{36}H_{62}O_{11}$
670.98
1,6-Dioxaspiro(4.5)decane-7-butyric acid, 2-(5-ethyl-tetrahydro-5-(tetrahydro-3-methyl- 5-(tetrahydro-6-hydroxy-6-(hydroxymethyl)-3,5-dimethyl-2H-pyran-2-yl)-2-furyl)-2-furyl)- 9-hydroxy-β-methoxy-α,γ,2,8-tetramethyl
A 3823A
Elancoban
Monelan
Monensic acid
Monensin
Monensin A

17091-31-5
$C_6H_{10}BrMgNO$
216.35
Magnesium, bromo(hexahydro-2H-azepin-2-onato-N)-(9CI)
Bromo(hexahydro-2H-azepin-2-onato-N)magnesium

17092-92-1
$C_{11}H_{16}O_2$
180.25
Dihydroactinidiolide
Actinidiolide, dihydro-
2(4H)-Benzofuranone, 5,6,7,7a-tetrahydro-4,4,7a-trimethyl-(VAN) (8CI)
2(4H)-Benzofuranone, 5,6,7,7a-tetrahydro-4,4,7a-trimethyl-,

(R)- (9CI)
2(4H)-Benzofuranone, 5,6,7,7a-tetrahydro-4,4,7a-trimethyl-, (S)-
NSC-357087

17095-24-8
$C_{26}H_{25}N_5O_{19}S_6.4Na$
995.88
2,7-Naphthalenedisulfonic acid, 3,6-(bis(4-((2-hydroxyethyl)sulfonyl)phenyl)bis-(azo))- 5-amino-4-hydroxy-, di(hydrogen sulfate) ester, tetrasodium salt
Remazol Black B

17105-75-8
$C_{28}H_{48}O$
400.69
Ergost-7-en-3-ol, (3B,24XI)-(9CI)
24XI-Ergost-7-en-3β-ol (8CI)

17109-49-8
$C_{14}H_{15}O_2PS_2$
310.38
CCOP(=O)(Sc1ccccc1)Sc2ccccc2
Phosphorodithioic acid, O-ethyl-S,S-diphenyl ester
O-Aethyl-S,S-diphenyl-dithiophosphat (German)
Bay 78418
Bayer 78418
Dithiophosphorsaeure-O-aethyl-S,S-diphenylester (German)
EDDP
Edifenphos
Ediphenphos
O-Ethyl S,S-diphenyl dithiophosphate
O-Ethyl-S,S-diphenyl phosphorodithioate
Hinosan
Lutrol
SRA 7847

17114-78-2
$C_{17}H_{18}$
222.33
9H-Fluorene, 9-(1,1-dimethylethyl)- (9CI)
9-tert-Butylfluorene
Fluorene, 9-tert-butyl- (6CI, 8CI)

17121-34-5
$C_{19}H_{43}N_3O_5$
393.55
OC(C)CN(CCN(CC(O)C)CC(O)C)CCN(CC(O)C)CC(O)C
2-Propanol, 1,1',1'',1'''-(((2-hydroxypropyl)imino)bis-(2,1-ethanediylnitrilo))-tetrakis- (9CI)
N,N,N',N'',N''-Penta(2-hydroxypropyl)diethylenetriamine

17125-80-3
$F_6Si.Ba$
279.43
Barium-silicofluoride
Barium fluorosilicate
Barium fluosilicate
Barium hexafluorosilicate
Barium hexafluorosilicate(2-)
Bariumsilicofluorid (German)
Barium silicon fluoride
Silicate(2-), hexafluoro-, barium
Silicate(2-), hexafluoro-, barium (1:1) (9CI)
Silicon fluoride barium salt

17132-78-4
$C_7H_3N_3$
129.10
2,3-Pyridinedicarbonitrile (8CI,9CI)

17138-28-2
$C_{10}H_{12}O_3$
180.20
O=C(OCC)Cc(ccc(O)c1)c1
Benzeneacetic acid, 4-hydroxy-, ethyl ester (9CI)
Ethyl 4-hydroxybenzeneacetate
Ethyl 4-hydroxyphenylacetate

17165-86-5
$C_{25}H_{18}$
318.42
9H-Fluorene, 9-(1,1'-biphenyl)-4-yl- (9CI)
Fluorene, 9-(4-biphenylyl)-(8CI)

17181-54-3
$C_3H_9O_6P$
172.08
β-Glycerophosphoric acid
β-Glycerol phosphate
β-Glycerophosphate
2-Glycerophosphoric acid

17194-00-2
BaH_2O_2
171.34
Barium hydroxide (9CI)
Barium dihydroxide
Barium hydroxide lime
Caustic Baryta

17199-21-2
$C_{14}H_{20}ClNO_3$
285.76
Phenol, 2-chloro-6-nitro-4-(1,1,3,3-tetramethylbutyl)-(8CI,9CI)

17199-22-3
$C_{14}H_{20}BrNO_3$
330.22
Phenol, 2-bromo-6-nitro-4-(1,1,3,3-tetramethylbutyl)-(8CI,9CI)

17199-23-4
$C_{10}H_{12}BrNO_3$
274.11
Phenol, 2-bromo-4-(1,1-dimethylethyl)-6-nitro- (9CI)
Phenol, 2-bromo-4-tert-butyl-6-nitro- (8CI)

17199-24-5
$C_{14}H_{21}ClO$
240.77
Oc(c(cc(c1)C(CC(C)(C)C)(C)C)Cl)c1
Phenol, 2-chloro-4-(1,1,3,3-tetramethylbutyl)- (9CI)
2-Chloro-4-tert-octylphenol
2-Chloro-4-(1,1,3,3-tetramethylbutyl)phenol
NSC-9891

17199-54-1
Unknown
Unknown
αMethadol
DEA No. 9605

17199-55-2
$C_{21}H_{29}NO$
311.46
Betamethadol
DEA No. 9609
Betametadol (Spanish)
Betametadolo
Betamethadolum (Latin)
Dimepheptanol
6-Dimethylamino-4,4-diphenyl-3-heptanol
3-Heptanol, 6-(dimethylamino)-4,4-diphenyl-
β-Methadol

17199-58-5
Unknown
Unknown
αCetylmethadol
DEA No. 9603

17199-59-6
Unknown
Unknown
βCetylmethadol
DEA No. 9607

17219-21-5
$C_{10}H_{18}O_4$
202.25
Hexanedioic acid, 2,2-dimethyl-, dimethyl ester (6CI,7CI,8CI,9CI)

Dimethyl 2,2-dimethyl-
adipate

17219-94-2
$C_{14}H_8Cl_2$
247.12
Phenanthrene, 9,10-dichloro- (8CI,9CI)
NSC-97575

17230-88-5
$C_{22}H_{27}NO_2$
337.50
CC23Cc1cnoc1C=C2CCC4C3CCC5(C)C4CCC5(O)C#C
17-α-Pregna-2,4-dien-20-yno-(2,3-d)isoxazol-17-ol
Danazol
Danocrine
Danol
Pregna-2,4-dien-20-yno(2,3-d)-isoxazol-17-ol, (17-α)- (9CI)
17-α-2,4-Pregnadien-20-yno-(2,3-d)isoxazol-17-ol
17-α-Pregn-4-en-20-yno(2,3-d)-isoxazol-17-ol
WIN 17757

17233-65-7
$C_{18}H_{14}N_2O_2$
290.31
O(c(c(N=1)cc(c2)C)c2)C1C=CC(Oc(c3cc(c4)C)c4)=N3
Benzoxazole, 2,2'-(1,2-ethene-diyl)bis(5-methyl-, (E)- (9CI)
trans-2,2'-Ethylenebis(5-methyl-benzoxazole)

17243-57-1
$C_{12}H_{18}ClN$
211.73
Mefenorex
Benzeneethanamine, N-(3-chloropropyl)-α-methyl- (9CI)
N-(3-Chloropropyl)-α-methyl-phenethylamine
Mefenorexum (Latin)
Phenethylamine, N-(3-chloro-propyl)-α-methyl-

17311-31-8
$C_{10}H_{10}N_2O_4$
222.22
2,3-Quinoxalinedimethanol, 1,4-dioxide
2,3-Bis(hydroxymethyl)quinoxaline di-N-oxide
1,4-Di-N-oxide 2,3-bis(oxy-methyl)quinoxline
1,4-Di-N-oxide of dihydroxy-methylquinoxaline
Dioxidin
Dioxidine
Dioxydine

17341-40-1
$C_9H_{18}N_2O_2$
186.29
O=C(NN(CC(O)C)(C)C)C(=C)C
Hydrazinium, 1,1-dimethyl-1-(2-hydroxypropyl)-2-(2-methyl-1-oxo-2-propen-yl)-, hydroxide, inner salt
Ashland 10303
Hydrazinium, 1-(2-hydroxy-propyl)-2-methacryloyl-1,1-di-methyl-, hydroxide, inner salt (8CI)
YPH 103

17369-59-4
$C_{11}H_{10}O_2$
174.21
O=C(OC(c1cccc2)=CCC)c12
1(3H)-Isobenzofuranone, 3-propylidene
Propylidene phthalide
3-Propylidenephthalide

17372-87-1
$C_{20}H_8Br_4O_5.2Na$
693.90
[Na+].[Na+].[O-]C(=O)c1ccccc1c3c2cc(Br)c([O-])c(Br)c2oc4c(Br)c(=O)c(Br)cc34
Fluorescein, 2',4',5',7'-tetra-bromo-, disodium salt
Aizen Eosine GH
Bromo Acid
Bromo B
Bromo 4DC
Bromoeosine
Bromo FL
Bromofluoresceic acid
Bromo fluorescein
Bromo JPS
Bromo TS
Bromo X-100
Bromo XX
Bronze Bromo
Certiqual Eosine
Cerven Kysela 87 (Czech)
C.I. 45380
C.I. Acid Red 87
D & C 22
D & C Red No. 22
Disodium Eosin
Eosin
Eosine
Eosine B
Eosine BPC
Eosine BS
Eosine BS-SF
Eosine DA
Eosine DWC 73
Eosine Extra Conc. A. Export
Eosine FA
Eosine G
Eosine 3G
Eosine GF
Eosine J
Eosine OJ
Eosine S13 (Bluish)
Eosine Sodium Salt
Eosine W/S
Eosine Y
Eosine 3Y
Eosine YB
Eosine Yellowish
Eosine Yellowish-(YS)
Eosine YS
Eosin G
Eosin Gelblich (German)
Eosin Y
Eosin YS
Fenazo Eosine XG
Hidacid Boiling Bromo
Hidacid Bromo Acid Regular
Hidacid Dibromo Fluorescein
Hidacid Eosine Soda Salt
Hidacid White Bromo
Irgalite Bronze Red CL
Phloxine Red 20-7600
Phloxine Toner B
Phlox Red Toner X-1354
Pure Eosine YY
11445 Red
11731 Red
Sodium Eosinate
Sodium Eosine
Spiro(isobenzofuran-1(3H), 9'-(9H)xanthen)-3-one, 2',4',5',7'-tetrabromo-3',6'-dihydroxy-, 2Na
Symuler Eosin Toner
2,4,5,7-Tetrabromo-3,6-fluor-andiol
Tetrabromofluorescein
Tetrabromofluorescein D
2',4',5',7'-Tetrabromofluores-cein disodium salt
Tetrabromofluorescein S
Tetrabromofluorescein Soluble
Toyo Eosine G
1903 Yellow Pink

17375-41-6
Unknown
Unknown
Ferrous sulfate mono-hydrate

17401-48-8
$C_{19}H_{15}N$
257.35
Benz(a)acridine, 9,12-dimethyl
2,10-Dimethyl-5,6-benzacridine
9,12-Dimethylbenz(a)acridine

17406-45-0
$C_{50}H_{83}NO_{21}$
1034.34
Tomatine
A''-Tomatidine
Lycopersicin
Tomatidine, glycoside
Tomatin
α-Tomatine

17413-73-9
$C_{10}H_{11}ClO_3$
214.66

Propionic acid, 2-(m-chloro-phenoxy)-2-methyl
Acide (m-chlorophenoxy)-2 methyl-2 propionique (French)
2-(m-Chlorophenoxy)-2-methyl-propionic acid

17418-58-5
$C_{20}H_{13}NO_4$
331.34
O=C(c(c(C(=O)c1c(N)c(Oc(ccc c2)c2)cc3O)ccc4)c4)c13
Anthraquinone, 1-amino-4-hydroxy-2-phenoxy
1-Amino-4-hydroxy-2-phenoxy-anthraquinone
Cerven Brilantni Ostacetova F-LB (Czech)
Cerven Disperzni 60 (Czech)
C.I. Disperse Red 60 (8CI)
C.I. Disperse Red 71
C.I. Disperse Red 83
Disperse Polyester Pink 2S
Disperse Red 60
Dispersol Red B 2B
Duranol Brilliant Red T 2B
Foron Brilliant Red E 2BL
Hostatherm Pink FBL
Latyl Cerise N
Miketon Polyester Red FB
Ostacet Brilliant Red E-LB
Palanil Red BF
Resiren Red TB
Resolin Red FB
Resolin Red FBE
Resorin Red FBE
Samaron Pink FBL
Serilene Brilliant Red 2BL
Serilene Red 2BL
Sumikaron Red E-FBL
Teraprint
Tersetile Rubine FL
Transetile Rubine P-FL

17420-30-3
$C_7H_5N_2O_2$
149.14
N#Cc(c(N)ccc1N(=O)=O)c1
Aniline, 2-cyano-4-nitro
2-Cyano-4-nitroaniline
2-Kyan-4-nitroanilin (Czech)

17421-79-3
$C_{10}H_{16}N_2O_8.Na$
315.22
Sodium ethylenediaminetetra-acetate
Caswell No. 438C
EPA Pesticide Chemical Code 039103
Ethylenediaminetetraacetic acid, sodium salt
Glycine, N,N'-1,2-ethanediyl-bis(N-(carboxymethyl)-, monosodium salt (9CI)
Sodium edetate

17429-29-7
$C_9H_{14}O$
138.21
2-Cyclohexen-1-one, 4,4,5-tri-methyl- (8CI,9CI)

17433-31-7
$C_8H_9N_3O_2$
179.20
CC(=O)NNC(=O)c1ccccn1
Hydrazine, 1-acetyl-2-picol-inoyl
N-Acetyl-N'-isonicotinyl hydrazide
1-Acetyl-2-picolinolhydrazine
1-Acetyl-2-picolinoylhydrazine
Azapicyl
NCI-C04739
NSC-68626
P-2292
2-Pyridinecarboxylic acid, 2-acetylhydrazide (9CI)

17455-13-9
$C_{12}H_{24}O_6$
264.36
O(CCOCCOCCOCCOCCOC1)C1
1,4,7,10,13,16-Hexanoxacyclo-octadecane
18-Crown-6

17462-58-7
$C_5H_9ClO_2$
136.58
O=C(OC(CC)C)Cl
Carbonochloridic acid, 1-methylpropyl ester (9CI)
sec-Butyl chlorocarbonate
sec-Butyl chloroformate
sec-Butylchloroformate
1-Methylpropyl carbonochlor-idate
1-Methylpropyl chloroformate

17463-44-4
$C_4H_{10}N_2O_2$
118.16
Propionic acid, α-amino-β-methylamino-, DL
DL-α-Amino-β-methylamino-propionic acid
Propionic acid, α-amino-β-methylamino-

17464-91-4
$C_{16}H_{15}BrCl_2N_4O_4$
478.10
O=N(=O)c(cc(c(N=Nc(c(cc(N(C CO)CCO)c1)Cl)c1)c2Cl)Br)c2
Ethanol, 2,2'-((4-((2-bromo-6-chloro-4-nitrophenyl)azo)-3-chlorophenyl)imino)bis-(9CI)
2,2'-((4-((2-Bromo-6-chloro-4-nitrophenyl)azo)-3-chloro-phenyl)imino)bis(ethanol)

17465-58-6
$C_{18}H_{26}$
242.41
s-Indacene, 1,2,3,5,6,7-hex-ahydro-1,1,4,7,7,8-hexa-methyl- (9CI)
s-Hydrindacene, 1,1,4,7,7,8-hexamethyl- (8CI)

17465-59-7
$C_{18}H_{26}$
242.40
s-Indacene, 1,2,3,5,6,7-hexa-hydro-1,1,4,5,5,8-hexamethyl-(9CI)
s-Hydrindacene, 1,1,4,5,5,8-hex-amethyl- (8CI)

17466-45-4
$C_{35}H_{48}N_8O_{11}S$
788.97
Phalloidin
Phalloidine

17467-15-1
$C_8H_7N_3S$
177.24
N(=C(NS1)c(cccc2)c2)C1=N
1,2,4-Thiadiazol-5-amine, 3-phenyl
5-Amino-3-phenyl-1,2,4-thia-diazole
3-Phenyl-1,2,4-thiadiazol-5-amine

17494-99-4
$C_{11}H_{18}ClN_3S_2$
291.87
s-Triazine, 2,4-bis(butyl-thio)-6-chloro- (6CI,7CI, 8CI)

17505-11-2
$C_3H_4N_2Se$
147.04
1,2,5-Selenadiazole, 3-methyl- (8CI,9CI)
3-Methyl-1,2,5-selena-diazole

17511-60-3
$C_{13}H_{18}O_2$
206.31
O=C(OC(C(C(C(C12)C=C3)C3) C1)C2)CC
4,7-Methanoindene-6-carbox-ylic acid, 3a,4,5,6,7,7a-hexa-hydro-, ethyl ester
Cyclaprop
Tricyclodecenyl propionate

17526-74-8
$C_9H_{16}O_3$
172.25
Hexanoic acid, 2,3-epoxy-propyl ester

Glycidyl ester of hexanoic acid

17526-94-2
$C_{13}H_{20}N_4O_2$
264.30
O=C(N(C)C)Nc(c(ccc1NC(=O)N(C)C)C)c1
Urea, N,N''-(4-methyl-1,3-phenylene)bis(N',N'-dimethyl- (9CI)
NSC-375994

17540-75-9
$C_{18}H_{30}O$
262.44
Oc(c(cc(c1)C(CC)C)C(C)(C)C)c1C(C)(C)C
Phenol, 2,6-bis(1,1-dimethylethyl)-4-(1-methylpropyl)- (9CI)
2,6-Di-tert-butyl-4-sec-butylphenol
NSC-14460

17557-23-2
$C_{11}H_{20}O_4$
216.31
O(C1COCC(C)(C)COCC(O2)C2)C1
Propane, 1,3-bis(2,3-epoxypropoxy)-2,2-dimethyl
1,3-Bis(2,3-epoxypropoxy)-2,2-dimethylpropane
Diglycidyl ether of neopentyl gylcol
2,2'-((2,2-Dimethyl-1,3-propanediyl)bis(oxymethylene))-bisoxirane
Heloxy WC68
Neopentyl glycol diglycidyl ether
Oxirane, 2,2'-((2,2-dimethyl-1,3-propanediyl)bis(oxymethylene))bis- (9CI)

17557-67-4
$C_8H_8N_2O_2S_2$
228.28
O=S(=O)(c(ccc(NC(=N)S1)c12)c2)C
2-Benzothiazolamine, 6-(methylsulfonyl)- (9CI)
2-Aminobenzothiazole-6-methyl sulfone
2-Aminobenzothiazolyl-6 methyl sulfone
6-(Methylsulfonyl)-2-benzothiazolamine

17559-81-8
$C_6H_8O_2$
112.13
3-Hexene-2,5-dione, (Z)- (8CI,9CI)

17560-51-9
$C_{16}H_{16}ClN_3O_3S$
365.86
CC2Nc1cc(Cl)c(cc1C(=O)N2c3ccccc3C)S(N)(=O)=O
6-Quinazolinesulfonamide, 1,2,3,4-tetrahydro-7-chloro-2-methyl-4-oxo-3-o-tolyl
7-Chloro-1,2,3,4-tetrahydro-2-methyl-3-(2-methylphenyl)-4-oxo-6-quinazolinesulfonamide
7-Chloro-1,2,3,4-tetrahydro-2-methyl-4-oxo-3-o-tolyl-6-quinazolinesulfonamide
Diulo
2-Methyl-3-o-tolyl-6-sulfamyl-7-chloro-1,2,3,4-tetrahydro-4-quinazolinone
Metenix
Metolazone
6-Quinazolinesulfonamide, 7-chloro-1,2,3,4-tetrahydro-2-methyl-4-oxo-3-o-tolyl-
SR 720-22
Zaroxolyn

17564-64-6
$C_9H_6ClNO_2$
195.60
O=C(N(C(=O)c1cccc2)CCl)c12
1H-Isoindole-1,3(2H)-dione, 2-(chloromethyl)- (9CI)
2-(Chloromethyl)-1H-isoindole-1,3(2H)-dione
Chloromethylphthalimide
N-(Chloromethyl)phthalimide
N-Chloromethyltrimellitimide
NSC-29558
Phthalimide, N-chloromethyl-
Phthalimide, N-(chloromethyl)-

17572-97-3
$C_{10}H_{16}N_2O_8.3K$
409.53
Tripotassium EDTA
N,N'-1,2-Ethanediylbis(N-(carboxymethyl)glycine), tripotassium salt
Glycine, N,N'-1,2-ethanediylbis(N-(carboxymethyl)-, tripotassium salt (9CI)
Tripotassium ethylenediaminetetraacetate

17584-14-4
$C_9H_7ClN_4O$
222.64
s-Triazin-2-ol, 4-amino-6-(p-chlorophenyl)- (8CI)

17587-22-3
$C_{10}H_{11}F_7O_2$
296.18
O=C(C(F)(F)C(F)(F)C(F)(F)F)CC(=O)C(C)(C)C
6,6,7,7,8,8,8-Heptafluoro-2,2-dimethyl-3,5-octanedione
1,1,1,2,2,3,3-Heptafluoro-7,7-dimethyl-4,6-octanedione
3,5-Octanedione, 6,6,7,7,8,8,8-heptafluoro-2,2-dimethyl- (9CI)

17590-87-3
$C_{14}H_{14}N_2$
210.28
Diazene, (2,6-dimethylphenyl)phenyl- (9CI)
Azobenzene, 2,6-dimethyl- (6CI,7CI,8CI)
2,6-Dimethylazobenzene

17601-96-6
$C_8H_{11}NO_4S$
217.24
O=S(=O)(c(ccc(O)c1N)c1)CCO
Phenol, 2-amino-4-((2-hydroxyethyl)sulfonyl)- (9CI)
2-Amino-4-((2-hydroxyethyl)-sulfonyl)phenol

17611-82-4
$C_6H_8N_2$
108.13
N#CC(CC)CC#N
Butanedinitrile, ethyl- (9CI)
Ethylbutanedinitrile
Ethylsuccinonitrile

17612-35-0
$C_7H_{14}O$
114.19
Oxirane, 2,2-dimethyl-3-propyl- (9CI)
Hexane, 2,3-epoxy-2-methyl- (8CI)

17617-23-1
$C_{21}H_{23}ClFN_3O$
387.92
CCN(CC)CCN2C(=O)CN=C(c1ccccc1F)c3cc(Cl)ccc23
2H-1,4-Benzodiazepin-2-one, 1,3-dihydro-7-chloro-1-(2-(diethylamino)ethyl)-5-(o-fluorophenyl)
7-Chloro-1-(2-(diethylamino)-ethyl)-5-(2-fluorophenyl)-1H-1,4-benzodiazepin-2(3H)-one
Flurazepam
Ro-5-6901/3

17618-77-8
C_7H_{14}
98.19
2-Hexene, 3-methyl- (6CI, 7CI,8CI,9CI)
3-Methyl-2-hexene

17619-97-5
$C_{14}H_{12}O$
196.25

Stilbene oxide
Bibenzyl, α,α'-epoxy- (8CI)
NSC-155516
Oxirane, 2,3-diphenyl- (9CI)

17626-93-6
$C_{14}H_{26}O_7$
306.36
Oxirane, 2,2'-(2,5,8,11,14-pentaoxapentadecane-1,15-diyl)-bis- (9CI)
4,7,10,13,16-Pentaoxanonadecane, 1,2:18,19-diepoxy- (8CI)

17635-40-4
$C_4H_5N_3O$
111.10
1,3,5-Triazine, 2-methoxy- (9CI)
2-Methoxy-1,3,5-triazine
s-Triazine, 2-methoxy- (6CI, 8CI)

17639-93-9
$C_4H_7ClO_2$
122.56
O=C(OC)C(Cl)C
Propionic acid, 2-chloro-, methyl ester
Methyl 2-chloropropanoate
Methyl-2-chloropropionate [UN 2933]
UN 2933 [Methyl-2-chloropropionate]

17640-28-7
$C_6H_{12}O_4$
148.16
Acetic acid, (2-methoxyethoxy)-, methyl ester (8CI,9CI)

17648-03-2
$C_{14}H_{10}O_4$
242.23
O=C(c(c(C(=O)C1)c(O)c(c2ccc3)c3)c2O)C1
1,4-Anthracenedione, 2,3-dihydro-9,10-dihydroxy- (9CI)
2,3-Dihydro-9,10-dihydroxy-1,4-anthracenedione

17661-50-6
$C_{32}H_{64}O_2$
480.96
O=C(OCCCCCCCCCCCCCC)CCCCCCCCCCCCCCCCC
Stearic acid, tetradecyl ester
Myristyl stearate
Octadecanoic acid, tetradecyl ester (9CI)
Tetradecyl octadecanoate
Tetradecyl stearate

17663-27-3
$C_{12}H_{22}O_2$
198.31
2,7-Octanedione, 4,4,5,5-tetramethyl- (8CI,9CI)
4,4,5,5-Tetramethyl-2,7-octanedione

17670-75-6
$C_{17}H_{34}O_2$
270.46
Tetradecanoic acid, 7-methyl-, ethyl ester (8CI)
Ethyl 7-methylmyristates

17673-25-5
$C_{20}H_{27}O_6$
363.47
CC2C(O)C4(O)C(C3C=C(CO)CC1(O)C(C=C(C)C1=O)C23O)C4(C)C
5H-Cyclopropa(3,4)benz-(1,2-e)azulen-5-one, 1,1a-β,1b-α,4,4a,7a-β,7b,8,9,9a-decahydro-4a-α,7b-α,9-β,9a-α-tetrahydroxy-3-(hydroxymethyl)-1,1,6,8-α- tetramethyl
Phorbol

17675-60-4
$C_2H_9N_4O_5P$
200.09
Urea, (aminoiminomethyl)-, phosphate (9CI)
(Aminoiminomethyl)urea phosphate

17688-68-5
$C_{10}H_{13}NO_2S$
211.28
O=S(=O)(CCN(c(cccc1)c1)C2)C2
Thiomorpholine, 4-phenyl-, 1,1-dioxide (9CI)
NSC-261427
4-Phenylthiomorpholine-1,1-dioxide

17689-77-9
$C_8H_{14}O_6Si$
234.28
Silanetriol, ethyl-, triacetate (9CI)
Ethylsilanetriol triacetate
Ethyltriacetoxysilane

17696-11-6
$C_8H_{15}BrO_2$
223.11
Octanoic acid, 8-bromo- (8CI, 9CI)

17696-37-6
$C_{12}H_{18}O$
178.27
Oc(c(cc(c1C)C(C)(C)C)C)c1
Phenol, 4-(1,1-dimethylethyl)-2,5-dimethyl- (9CI)
4-(1,1-Dimethylethyl)-2,5-dimethylphenol

17700-09-3
$C_6H_2Cl_3NO_2$
226.44
Clc1ccc(N(=O)=O)c(Cl)c1Cl
Benzene, 4-nitro-1,2,3-trichloro
4-Nitro-1,2,3-trichlorobenzene
1,2,3-Trichloro-4-nitrobenzene

17702-41-9
$B_{10}H_{14}$
122.24
CC.CCCCCCCCCC.CCCCCCCCCCCC
Decaborane (ACGIH,OSHA) [UN 1868]
Decaborane(14)
UN 1868 [Decaborane]

17702-57-7
$C_{12}H_{17}N_3O_2$
235.27
Formparanate
AI3-27305
Carbamic acid, methyl-, 4-(((dimethylamino)methylene)amino)-m-tolyl ester
Caswell No. 359DD
4-(((Dimethylamino)methylene)amino)-m-tolylmethylcarbamate
N,N-Dimethyl-N'-(2-methyl-4-(((methylamino)carbonyl)oxy)phenyl)methanimidamide
ENT 27,305
EPA Pesticide Chemical Code 359700
Methanimidamide, N,N-dimethyl-N'-(2-methyl-4-(((methylamino)carbonyl)oxy)phenyl)-
N-(4-Methylcarbamoyloxy-O-tolyl)-N,N-dimethylformamidine
Schering 36103
UC-34096
UC-25074
Union Carbide UC-25074

17708-57-5
$C_{10}H_{17}Cl_2NOS$
270.24
Carbamic acid, diisopropylthio-, S-(2,3-dichloroallyl)-ester, (Z)
cis-Diallate

17741-62-7
$C_{16}H_{14}Cl_2N_4O_4S$
429.26

O=S(=O)(CCN(c(ccc(N=Nc(c(cc(N(=O)=O)c1)Cl)c1Cl)c2)c2)C3)C3
Thiomorpholine, 4-(4-((2,6-dichloro-4-nitrophenyl)azo)phenyl)-, 1,1-dioxide (9CI)

17752-10-2
$C_{11}H_{17}N_2O_2$
209.26
Benzenaminium, N,N,N-trimethyl-3-(((methylamino)carbonyl)oxy)- (9CI)
Ammonium, (m-hydroxyphenyl)trimethyl-, methylcarbamate (ester) (8CI)

17752-71-5
$C_9H_5Cl_3N_4O_2S$
339.56
Benzenesulfonamide, 4-chloro-N-(4,6-dichloro-1,3,5-triazin-2-yl)- (9CI)
Benzenesulfonamide, p-chloro-N-(4,6-dichloro-s-triazin-2-yl)-(8CI)

17752-85-1
$C_{19}H_{12}Cl_2N_6O_{10}S_3.3Na$
720.38
Procion Brilliant Red M-2BS
C.I. Reactive Red 1
C.I. Reactive Red 1, Trisodium salt (8CI)
Mikacion Brilliant Red 2BS
2,7-Naphthalenedisulfonic acid, 5-((4,6-dichloro-1,3,5-triazin-2-yl)amino)-4-hydroxy-3-((2-sulfophenyl)azo)-, trisodium salt (9CI)
NSC-240567
Procion Brilliant Red 2BS
Procion Brilliant Red M 2B
Procion Red 2BS

17754-90-4
$C_{11}H_{15}NO_2$
193.24
O=Cc(c(O)cc(N(CC)CC)c1)c1
Benzaldehyde, 4-(diethylamino)-2-hydroxy- (9CI)
4-(Diethylamino)-2-hydroxybenzaldehyde
4-N,N-Diethylaminosalicylic aldehyde
2-Hydroxy-4-diethylaminobenzaldehyde

17757-70-9
$C_{12}H_{13}NO_3S$
251.30
1,4-Oxathiin-3-carboxamide, 5,6-dihydro-2-methyl-N-phenyl-, 4-oxide (9CI)
F 831
1,4-Oxathiin-3-carboxanilide, 5,6-dihydro-2-methyl-, 4-oxide (8CI)

17759-88-5
$C_3H_5BrCl_2$
191.89
Propane, 2-bromo-1,2-dichloro- (7CI,8CI,9CI)

17760-93-9
Unknown
Unknown
Polychlorinated-triphenyl
PCT
Polychlorotriphenyl

17773-41-0
C_5H_9NOS
131.19
N#CC(O)CCSC
Butanenitrile, 2-hydroxy-4-(methylthio)- (9CI)
2-Hydroxy-4-(methylmercapto)-butyronitrile
2-Hydroxy-4-(methylthio)butanenitrile

17773-65-8
C_5H_9Cl
104.58
2-Butene, 2-chloro-3-methyl- (8CI,9CI)

17773-66-9
$C_5H_{10}Cl_2$
141.04
Butane, 2,2-dichloro-3-methyl- (8CI,9CI)
2,2-Dichloro-3-methylbutane

17781-31-6
$C_{18}H_{13}Cl_2NO$
330.22
3-Pyridinemethanol, α,α-bis-(p-chlorophenyl)
α,α-Bis(4-chlorophenyl)-3-pyridinemethanol
Bis(4-chlorophenyl)-3-pyridylmethanol
EL 241
Parinol
Parnon

17784-12-2
$C_{12}H_{14}N_4O_3S$
294.31
Sulfacitine
Benzenesulfonamide, 4-amino-N-(1-ethyl-1,2-dihydro-2-oxo-4-pyrimidinyl)- (9CI)
C.I. 636
N1-(1-Ethyl-1,2-dihydro-2-oxo-4-pyrimidinyl)sulfanilamide
N^1-(1-Ethyl-1,2-dihydro-2-oxo-4-pyrimidinyl)sulfanilamide
1-Ethyl-N-sulfanilylcytosine
NSC-356717
Renoquid
Solfacitina
Sulfacitina (Spanish)
Sulfacitinum (Latin)
Sulfacytine
Sulfanilamide, N1-(1-ethyl-1,2-dihydro-2-oxo-4-pyrimidinyl)-(8CI)
Sulfanilamide, N(1)-(1-ethyl-1,2-dihydro-2-oxo-4-pyrimidinyl)-
N-Sulfanilyl-1-ethylcytosine
NL-Sulfanilyl-1-ethylcytosine

17790-61-3
$C_{15}H_{15}Cl$
230.74
4-Chloro-4'-isopropylbiphenyl

17796-82-6
$C_{14}H_{15}NO_2S$
261.36
O=C(N(SC(CCCC1)C1)C(=O)c2cccc3)c23
Phthalimide, N-(cyclohexylthio)
N-Cyclohexylsulfenylphthalimide
N-(Cyclohexylthio)phthalimide
1H-Isoindole-1,3(2H)-dione, 2-(cyclohexylthio)-
Santogard PVI

17797-03-4
$C_6H_{11}ClS$
150.67
S(C(CCCC1)C1)Cl
Cyclohexanesulfenyl chloride (9CI)

17804-35-2
$C_{14}H_{18}N_4O_3$
290.36
O=C(OC)N=C(N(c(c1ccc2)c2)C(=O)NCCCC)N1
2-Benzimidazolecarbamic acid, 1-(butylcarbamoyl)-, methyl ester
Arilate
BBC
Benlat
Benlate
Benlate 50
Benlate 50 W
Benomyl (ACGIH,OSHA)
Benomyl 50W
BNM
1-(Butylcarbamoyl)-2-benzimidazolecarbamic acid, methyl ester
1-(Butylcarbamoyl)-2-benzimidazol-methylcarbamat (German)
1-(N-Butylcarbamoyl)-2-(methoxy-carboxamido)-benzimidazol (German)
Carbamic acid, methyl-,

1-(butylcarbamoyl)-2-benz-imidazole ester
D 1991
Du Pont 1991
F1991
Fundasol
Fundazol
Fungicide 1991
MBC
Methyl 1-(butylcarbamoyl)-2-benzimidazolylcarbamate
Tersan 1991

17804-49-8
$C_{19}H_{10}Cl_2N_6O_7S_2.2Na$
615.35
2,7-Naphthalenedisulfonic acid, 5-((4,6-dichloro-s-triazin-2-yl)amino)-4-hydroxy-3- (phenylazo)-, disodium salt
Brilliant Red 5SKH
Cerven reaktivni 2 (Czech)
Chemictive Brilliant Red 5b
C.I. Reactive Red 2
Mikacion Brilliant Red 5BS
Ostazin Brilliant Red S 5b
Procion Brilliant Red 5BS
Procion Brilliant Red M 5B
Procion Brilliant Red MX 5B
Procion Red MX 5B
Reactive Brilliant Red 5SKH

17811-28-8
$C_{15}H_{22}O_4$
266.34
Zinniol
NSC-125427

17824-83-8
$C_6Cl_4N_2$
241.88
N#Cc(nc(c(c1Cl)Cl)Cl)c1Cl
2-Pyridinecarbonitrile, 3,4,5,6-tetrachloro- (9CI)
Tetrachloro-2-cyanopyridine
3,4,5,6-Tetrachloro-2-pyridine-carbonitrile

17825-86-4
$C_6H_7NO_2$
125.12
1H-Pyrrole-2,5-dione, 3,4-dimethyl- (9CI)
Maleimide, 2,3-dimethyl- (8CI)

17831-71-9
$C_{14}H_{22}O_7$
302.36
O=C(OCCOCCOCCOCCOC(=O)C=C)C=C
Acrylic acid, diester with tetraethylene glycol
Acrylic acid, oxybis(ethylene-oxyethylene) ester
2-Propenoic acid, oxybis-(2,1-ethanediyloxy-2,1-ethane-diyl)ester
Tetraethylene glycol diacrylate

17844-07-4
$C_{15}H_{24}O_3$
252.35
1-Cyclohexene-1-carboxylic acid, 4-(1,5-dimethyl-3-oxo-hexyl)- (8CI,9CI) (VAN)

17849-38-6
C_7H_7ClO
142.58
Benzenemethanol, 2-chloro-
AI3-20627

17849-64-8
$C_9H_7Cl_2N$
200.06
Benzenepropanenitrile, α,4-dichloro- (9CI)
Hydrocinnamonitrile, p,α-dichloro- (8CI)

17851-27-3
$C_{11}H_{16}$
148.25
Benzene, 1-ethyl-2,4,5-trimethyl- (8CI,9CI)

17851-53-5
$C_{16}H_{22}O_4$
278.35
1,2-Benzenedicarboxylic acid, butyl 2-methylpropyl ester (9CI)
Phthalic acid, butyl isobutyl ester (8CI)

17852-98-1
$C_{18}H_{14}N_2O_6S.Ba$
523.70
2-Naphthalenecarboxylic acid, 3-hydroxy-4-((4-methyl-2-sulfophenyl)azo)-, barium salt (1:1) (9CI)
D & C Red No. 6, Barium Lake

17852-99-2
$C_{18}H_{13}ClN_2O_6S.Ca$
460.90
2-Naphthalenecarboxylic acid, 4-((4-chloro-5-methyl-2-sulfophenyl)azo)-3-hydroxy-, calcium salt (1:1) (9CI)

17869-76-0
$C_7H_{14}OSi$
142.28
Silane, trimethyl((1-methyl-2-propynyl)oxy)- (8CI,9CI)
3-(Trimethylsiloxy)-1-but-yne

17872-56-9
$C_{11}H_{16}N_2O_3$
224.25
O=C(NNc(cccc1)c1)C(CO)(CO)C
Propanoic acid, 3-hydroxy-2-(hydroxymethyl)-2-methyl-, 2-phenylhydrazide (9CI)

17901-16-5
$C_6H_6Cl_2N_2O_2S$
241.10
Pyrimidine, 4,5-dichloro-6-methyl-2-(methylsulfonyl)- (8CI,9CI)

17904-23-3
$C_{19}H_{32}O_3$
308.47
15-Nor-5β,8βH,8βH, 10α-labdan-14-al, 8,13-epoxy-19-hydroxy-, (13R)-(-)- (8CI)

17906-09-1
$C_{15}H_{32}O_3Si_4$
372.76
Tetrasiloxane, 1,1,1,3,3,5,7,7,7-nonamethyl-5-phenyl- (7CI,8CI,9CI)

17909-77-2
$C_{15}H_{22}O$
218.34
2,6,9,11-Dodecatetraenal, 2,6,10-trimethyl-, (E,E,E)- (9CI)
α-Sinensal

17924-92-4
$C_{18}H_{22}O_5$
318.40
O=C(OC(CCCC(=O)CCCC=Cc1cc(O)cc2O)C)c12
1H-2-Benzoxacyclotetradecin-1-one, 3,4,5,6,7,8,9,10-octahydro-14,16-dihydroxy-3-methyl- 7-oxo-, (E)
Benzoxacyclotetradec-11-en-1-one, 14,16-dihydroxy-3-methyl-7-oxo-, trans-
Compound F-2
F2
FES
F-2 Toxin
Fusarium Toxin
6-(10-Hydroxy-6-oxo-trans-1-undecenyl)-β-resorcylic acid-N-lactone
Mycotoxin F2
NCI-C50226
Resorcylic acid, 6-(10-hydroxy-6-oxo-1-undecenyl)-, mu-lactone, trans-

Toxin F2
Zearalenone
(-)-Zearalenone
(S)-Zearalenone
(10S)-Zearalenone
trans-Zearalenone

17925-97-2
$C_{15}H_{12}Cl_4$
334.07
Benzene, 1-chloro-4-(2,2,2-trichloro-1-(4-methylphenyl)-ethyl)- (9CI)
Ethane, 1,1,1-trichloro-2-(p-chlorophenyl)-2-p-tolyl-(8CI)

17927-72-9
$C_{16}H_{28}O_6Ti$
364.27
Titanium, bis(2,4-pentanedionato-O,O')bis(2-propanolato)- (9CI)

17928-28-8
$C_{10}H_{30}O_3Si_4$
310.69
Trisiloxane, 1,1,1,3,5,5,5-heptamethyl-3-((trimethylsilyl)oxy)- (9CI)
Methyltris(trimethylsiloxy)-silane

17943-83-8
$C_{12}H_{23}N$
181.32
Cyclopentanemethylamine, 2-isopropylidene-N,N,5-trimethyl-, (1R,5R)-(-)- (8CI)

17945-05-0
$C_{12}H_{27}NO_5Si$
293.43
Carbamic acid, (3-(triethoxysilyl)propyl)-, ethyl ester (8CI,9CI)
(3-Triethoxysilyl)propylcarbamic acid ethyl ester

17963-04-1
$C_{10}H_{22}O_3Si$
218.37
Silane, ethoxydimethyl(3-(oxiranylmethoxy)propyl)- (9CI)
(3-Glycidoxypropyl)dimethyl-ethoxysilane

17977-09-2
$C_6H_8N_2O_6$
204.13
O=C(OCC(N(=O)=O)(N(=O)=O)C)C=C
2,2-Dinitropropyl acrylate
Acrylic acid, 2,2-dinitropropyl ester (8CI)
2,2-Dinitropropyl 2-propenoate
NSC-166462
2-Propenoic acid, 2,2-dinitropropyl ester (9CI)

17980-09-5
$C_{16}H_{14}$
206.29
Phenanthrene, 2,9-dimethyl- (8CI,9CI)

17980-16-4
$C_{16}H_{14}$
206.29
Phenanthrene, 2,6-dimethyl- (8CI,9CI)

17983-22-1
$C_{10}H_{18}O$
154.25
Ketone, methyl 2,2,3-trimethylcyclopentyl (8CI)

18005-40-8
$C_3H_9O_2PS$
140.14
Phosphonothioic acid, methyl-, O-ethyl ester (8CI,9CI)

18017-73-7
$C_{16}H_{24}O_2$
248.37
Benzenedecanoic acid (9CI)
Decanoic acid, 10-phenyl- (8CI)
NSC-263840

18024-00-5
$C_2H_5ClO_3S$
144.58
Ethanesulfonic acid, 2-chloro- (8CI,9CI)
Hol 1302

18039-42-4
$C_7H_6N_4$
146.13
N(NNN1)C1c(cccc2)c2
1H-Tetrazole, 5-phenyl- (8CI,9CI)
Expandex OX 5PT
Expandex 5PT
Kempore 50XPT
MA 1623
NSC-11138
5-Phenyl tetrazole
5-Phenyltetrazole (VAN)
5-Phenyl-1H-tetrazole
5-Phenyl-2H-tetrazole
2H-Tetrazole, 5-phenyl-

18077-53-7
$C_{12}H_{34}O_2Si_4$
322.75
Trisiloxane, 1,1,1,3,5,5,5-heptamethyl-3-(2-(trimethylsilyl)ethyl)- (6CI, 8CI,9CI)

18105-03-8
Unknown
Unknown
Nitrilotriacetic acid, mercury(2+) salt (2:3)
NTA, mercury(2+) salt (2:3)

18113-03-6
$C_7H_7ClO_2$
158.58
Phenol, 2-chloro-4-methoxy- (8CI,9CI)

18113-14-9
$C_7H_6Cl_2O_2$
193.03
2,5-Dichloro-4-methoxyphenol

18122-77-5
$C_{14}H_{20}Br_6$
667.74
BrC(C(CCC(Br)C1Br)C1)C(Br)C(CCC(Br)C2Br)C2
Cyclohexane, 1,1'-(1,2-dibromo-1,2-ethanediyl)bis(3,4-dibromo- (9CI)
1,1'-(1,2-Dibromo-1,2-ethanediyl)bis(3,4-dibromocyclohexane)
Ethane, 1,2-dibromo-1,2-bis-(3,4-dibromocyclohexyl)- (8CI)

18128-16-0
Unknown
Unknown
Potassium 2-chloro-4-phenylphenate

18128-17-1
Unknown
Unknown
Potassium 6-chloro-2-phenylphenate

18138-04-0
$C_9H_{14}N_2$
150.21
n(c(c(nc1C)CC)CC)c1
Pyrazine, 2,3-diethyl-5-methyl- (9CI)
2,3-Diethyl-5-methylpyrazine

18138-05-1
$C_9H_{14}N_2$
150.21
Pyrazine, 3,5-diethyl-2-methyl- (8CI,9CI)

18138-18-6
C_8H_5NOS

163.20
1H-Isoindol-1-one, 2,3-dihydro-3-thioxo- (9CI)
Monothiophthalimide
Phthalimide, thio- (6CI,8CI)
Thiophthalimide

18153-42-9
$C_{20}H_{16}$
256.35
9H-Fluorene, 9-(3-methylphenyl)- (9CI)
Fluorene, 9-m-tolyl- (8CI)
9-m-Tolylfluorene

18153-43-0
$C_{20}H_{16}$
256.35
9H-Fluorene, 9-(4-methylphenyl)- (9CI)
Fluorene, 9-p-tolyl- (8CI)

18168-01-9
$C_{15}H_{16}N_2O$
240.29
Urea, N'-ethyl-N,N-diphenyl- (9CI)
Urea, 3-ethyl-1,1-diphenyl- (8CI)

18169-57-8
$C_4H_9Cl_3Si$
191.56
Silane, trichloro(2-methylpropyl)- (9CI)
Isobutyltrichlorosilane
Trichloro(2-methylpropyl)silane

18172-67-3
$C_{10}H_{16}$
136.24
C(C(CC1C2)C1(C)C)(C2)=C
Bicyclo(3.1.1)heptane, 6,6-dimethyl-2-methylene-, (1S)- (9CI)

18181-70-9
$C_8H_8Cl_2IO_3PS$
412.99
COP(=S)(OC)Oc1cc(Cl)c(I)cc1Cl
Phosphorothioic acid, O-(2,5-dichloro-4-iodophenyl) O,O-dimethyl ester
Alfacron
C-9491
Ciba C-9491
Ciba 9491
Ciba-Geigy C-9491
Compound C-9491
O-(2,5-Dichloro-4-iodophenyl) O,O-dimethyl phosphorothioate
O,O-Dimethyl-O-(2,5-dichlor-4-jodphenyl)-thionophosphat (German)
O,O-Dimethyl-O-(2,5-dichlor-4-jodphenyl)-monothiophosphat (German)
O,O-Dimethyl-O-2,5-dichloro-4-iodophenyl thiophosphate
ENT 27,408
Iodofenophos
Iodofenphos
Iodophos
Jodfenphos
NSC-190998
Nuvanol N
OMS-1211
Phenol, 3,4-dichloro-, O-ester with O-methyl methylphosphoramidothioate

18181-80-1
$C_{17}H_{16}Br_2O_3$
428.15
CC(C)OC(=O)C(O)(c1ccc(Br)cc1)c2ccc(Br)cc2
Benzilic acid, 4,4'-dibromo-, isopropyl ester
Acarol
Ascarol 2E
Benzeneacetic acid, 4-bromo-α-(4-bromophenyl)-α-hydroxy-, 1-methylethyl ester
4-Bromo-α-(4-bromophenyl)-α-hydroxybenzeneacetic acid 1-methyl ethyl ester
Bromopropylate
Ciba-Geigy GS 19851
4,4'-Dibromobenzilic acid isopropyl ester
ENT 27,552
Geigy GS-19851
GS-19851
Isopropyl 4,4'-dibromobenzilate
1-Methylethyl 4-bromo-α-(4-bromophenyl)-α-hydroxybenzeneacetate
Neoron
NSC-195087
Phenisobromolate

18185-81-4
$C_8H_{16}O$
128.21
OCCC=CCCCC
3-Octen-1-ol
AI3-07194

18202-24-9
$C_{19}H_{30}O_2$
290.45
10,13-Octadecadiynoic acid, methyl ester (8CI, 9CI)

18202-28-3
$C_{18}H_{32}O$
264.46
17-Octadecen-14-yn-1-ol (8CI)

18213-73-5
$C_{11}H_{11}N_3O_2$
217.23
1,3,5-Triazine, 2,4-dimethoxy-6-phenyl- (9CI)
s-Triazine, 2,4-dimethoxy-6-phenyl- (7CI,8CI)

18217-12-4
$C_8H_{16}O$
128.21
2-Heptanone, 5-methyl- (8CI, 9CI)

18220-90-1
$C_{15}H_{14}O$
210.28
Methanone, (4-ethylphenyl)-phenyl- (9CI)
Benzophenone, 4-ethyl- (8CI)

18226-46-5
$C_{12}H_{15}Cl_3$
265.61
Benzene, 1,3,5-tris-(2-chloroethyl)- (8CI)

18237-29-1
$C_9H_7Cl_2N_5O_2S$
320.16
Benzenesulfonic acid, 2-(4,6-dichloro-s-triazin-2-yl)hydrazide (6CI,8CI)

18247-77-3
$C_9H_3Cl_4N_3O$
310.94
1,3,5-Triazine, 2,4-dichloro-6-(2,4-dichlorophenoxy)- (9CI)
s-Triazine, 2,4-dichloro-6-(2,4-dichlorophenoxy)- (8CI)

18259-05-7
$C_{12}H_5Cl_5$
326.44
1,1'-Biphenyl, 2,3,4,5,6-pentachloro- (9CI)
Biphenyl, 2,3,4,5,6-pentachloro- (8CI)
2,3,4,5,6-PCB
2,3,4,5,6-Pentachlorobiphenyl

18263-25-7
$C_{16}H_{31}BrO_2$
335.32
O=C(O)C(Br)CCCCCCCCCCCCCC
2-Bromopalmitate
2-Bromohexadecanoic acid
α-Bromopalmitic acid
2-Bromopalmitic acid
Hexadecanoic acid, 2-bromo- (9CI)
NSC-58378

18266-52-9
$C_6H_7N_3O_2.2ClH$
226.06
2-Nitro-1,4-benzenediamine dihydrochloride
1,4-Benzenediamine, 2-nitro-, dihydrochloride (9CI)
p-Phenylenediamine, 2-nitro-, dihydrochloride

18266-55-2
$C_5H_{11}NO_2$
117.14
O=C(NCCO)CC
Propanamide, N-(2-hydroxyethyl)- (9CI)
AI3-15237
N-(2-Hydroxyethyl)propanamide
β-Hydroxyethylpropionamide
N-(β-Hydroxyethyl)propionamide
NSC-6001

18268-69-4
$C_8H_6Cl_2O_3$
221.04
Benzaldehyde, 2,3-dichloro-4-hydroxy-5-methoxy- (9CI)
Vanillin, 5,6-dichloro- (8CI)

18268-70-7
$C_{24}H_{46}O_7$
446.70
O=C(OCCOCCOCCOCCOC(=O)C(CCCC)CC)C(CCCC)CC
Hexanoic acid, 2-ethyl-, diester with tetraethylene glycol
Flexol 4GO
Hexanoic acid, 2-ethyl-, oxybis(2,1-ethyldiyloxy-2,1-ethanediyl)ester (9CI)
Plasticizer 4GO
Polyethylene glycol 200 di-(2-ethylhexoate)
Tetraethylene glycol di-(2-ethylhexoate)

18268-76-3
$C_8H_7ClO_3$
186.59
Benzaldehyde, 2-chloro-4-hydroxy-5-methoxy- (9CI)
NSC-95796
Vanillin, 6-chloro- (8CI)

18277-91-3
$C_{15}H_{14}N_2O_2$
254.29
Benzoic acid, 2-(phenylazo)-, ethyl ester (9CI)
Benzoic acid, o-(phenylazo)-, ethyl ester (7CI,8CI)
Ethyl azobenzene-2-carboxylate

18281-04-4
$C_{21}H_{42}O_2$
326.56
Nonadecanoic acid, ethyl ester (8CI,9CI)
NSC-136559

18281-05-5
$C_{22}H_{44}O_2$
340.59
Eicosanoic acid, ethyl ester (8CI,9CI)

18281-07-7
$C_{25}H_{50}O_2$
382.68
Tricosanoic acid, ethyl ester (6CI,8CI,9CI)
Ethyl tricosanoate

18282-10-5
O_2Sn
150.69
Tin-oxide

18282-59-2
$C_6H_3BrCl_2$
225.90
c(ccc(c1Cl)Cl)(c1)Br
Benzene, 4-bromo-1,2-dichloro- (9CI)
4-Bromo-1,2-dichlorobenzene

18292-29-0
$C_7H_{16}Si$
128.29
Silane, ethenyldiethylmethyl- (9CI)
Diethylmethylvinylsilane
Methyldiethylvinylsilane
Silane, diethylmethylvinyl- (6CI,7CI,8CI)
Vinyl(diethyl)methylsilane

18292-97-2
$C_7H_5N_3O_6$
227.15
Toluene, 2,3,6-trinitro
Benzene, 1-methyl-2,3,6-trinitro-
2,3,6-Trinitrotoluene

18294-89-8
$C_8H_{14}O_2$
142.20
4-Octenoic acid (6CI,7CI,8CI,9CI)

18307-23-8
$H_2O_3Si.5/6H_2O.3/4Mg$
111.35
Sepiolite
Meerschaum

18309-32-5
$C_{10}H_{14}O$
150.24
O=C(C(CC1C=2C)C1(C)C)C2
2-Pinen-4-one, (1R,5R)-(+)
(1R-cis)-4,6,6-Trimethylbicyclo-(3.1.1)hept-3-en-2-one
d-Verbenone

18312-66-8
$C_3H_7NO_2Se$
168.07
Alanine, 3-selenyl-, (+-)
Alanine, 3-selenyl-, DL (8CI)
DL-Alanine, 3-selenyl- (9CI)
Selenium-cystine
d,l-Selenocysteine
Seleno-DL-cysteine

18317-90-3
$C_7H_5Cl_5$
266.38
Bicyclo(2.2.1)hept-2-ene, 1,2,3,4,7-pentachloro-, syn- (9CI)
2-Norbornene, 1,2,3,4,7-pentachloro-, syn- (8CI)
NSC-143932

18323-44-9
$C_{18}H_{33}ClN_2O_5S$
425.04
CCCC1CC(N(C)C1)C(=O)NC(C(C)Cl)C2OC(SC)C(O)C(O)C2O
L-threo-D-Galacto-octopyranoside, methyl 7-chloro-6,7,8-trideoxy-6-(1-methyl-4-propyl-L-2- pyrrolidinecarboxamido)-1-thio-, trans-α
7(S)-Chloro-7-deoxylincomycin
Cleocin
Clindamycin
Clindamycine (French)
U-21,251

18328-11-5
$C_{10}H_{12}O$
148.20
Benzenebutanal (9CI)
Butyraldehyde, 4-phenyl- (8CI)

18328-90-0
$C_6H_{13}N$
99.17
N(CC(=C)C)CC
2-Propen-1-amine, N-ethyl-2-methyl- (9CI)
N-Ethyl-2-methylallylamine
N-Ethyl-2-methyl-2-propen-1-amine

18343-30-1

$C_5H_5Cl_2N_3O$
194.00
1,3,5-Triazine, 2,4-dichloro-6-ethoxy- (9CI)
s-Triazine, 2,4-dichloro-6-ethoxy- (8CI)

18344-37-1
$C_{21}H_{44}$
296.58
Heptadecane, 2,6,10,14-tetramethyl- (8CI,9CI)

18358-13-9
$C_4H_5O_2$
85.08
2-Propenoic acid, 2-methyl-, ion(1-) (9CI)
Methacrylic acid, ion(1-) (8CI)

18371-12-5
$C_{14}H_{15}N_5O$
269.28
Acetamide, N-(4-((2,4-diaminophenyl)azo)phenyl)- (9CI)
Acetanilide, 4'-((2,4-diaminophenyl)azo)- (8CI)

18378-89-7
$C_{52}H_{76}O_{24}$
1085.28
NC(=O)Nc1ccccc1
Mithramycin
A-2371
Antibiotic LA 7017
Aurlelic acid
Aureolic acid
Mithracin
Mithramycin A
Mitramycin
NSC-24559
PA 144

18395-30-7
$C_7H_{18}O_3Si$
178.30
Silane, trimethoxy(2-methylpropyl)- (9CI)
Isobutyltrimethoxysilane
Trimethoxy(2-methylpropyl)-silane

18407-16-4
$C_{13}H_{26}O_2Si_3$
298.60
Trisiloxane, 1,1,1,3,3,5,5-heptamethyl-5-phenyl- (8CI,9CI)

18409-17-1
$C_8H_{16}O$
128.21
OCC=CCCCCC
2-Octen-1-ol, (E)-
AI3-36043
(E)-2-Octen-1-ol

18414-36-3
$C_{14}H_{14}Cl_2Si$
281.26
Dibenzyldichlorosilane
Silane, dibenzyldichloro-
Silane, dichlorobis(phenylmethyl)-
UN 2434 [Dibenzyldichlorosilane]

18428-88-1
$ClHO_2Zr$
159.68
Zirconium, chlorohydroxyoxo-
AI3-29111-X
Chlorohydroxyoxozirconium

18429-70-4
$C_{20}H_{16}$
256.36
Benz(a)anthracene, 4,5-dimethyl
3,4'-Dimethyl-1,2-benzanthracene
4,5-Dimethylbenz(a)anthracene

18431-82-8
$C_{15}H_{24}$
204.36
Spiro(5.5)undec-2-ene, 3,7,7-trimethyl-11-methylene-, (R)- (8CI,9CI)
Chamigrene
β-Chamigrene

18432-54-7
Unknown
Unknown
Nitrilotriacetic acid, cadmium(2+) complex
NTA, cadmium(2+) complex

18433-97-1
$C_9H_{14}N_2$
150.21
n(c(c(nc1C)CCC)C)c1
Pyrazine, 2,5-dimethyl-3-propyl- (9CI)
2,5-Dimethyl-3-propylpyrazine
3,6-Dimethyl-2-propylpyrazine
2-Propyl-3,6-dimethylpyrazine

18435-45-5
$C_{19}H_{38}$
266.51
C(=C)CCCCCCCCCCCCCCCCC
1-Nonadecene (9CI)
AI3-36475
NSC-77135

18448-65-2
$C_{23}H_{48}NO_2.Cl$
406.09
9-Octadecen-1-aminium, N,N-bis(2-hydroxyethyl)-N-methyl-, chloride, (Z)- (9CI)

18454-12-1
$CrO_4Pb.OPb$
546.38
Lead chromate(VI) oxide
Arancio cromo (Italian)
Austrian cinnabar
Basic Lead Chromate
Chinese Red
Chrome Orange
Chrome Orange 54
Chrome Orange 56
Chrome Orange 57
Chrome Orange 58
Chrome Orange Dark
Chrome Orange Extra Light
Chrome Orange G
Chrome Orange Medium
Chrome Orange NC-22
Chrome Orange R
Chrome Orange 5R
Chrome Orange RF
Chrome Orange XL
Chrome Red
Chromium lead oxide
Chromium Orange
C.I. 77601
C.I. Pigment Oragne 21
C.I. Pigment Red
C.P. Chrome Orange Dark 2030
C.P. Chrome Orange Extra Dark 2040
C.P. Chrome Light 2010
C.P. Chrome Orange Medium 2020
Dainichi Chrome Orange R
Dainichi Chrome Orange 5R
Genuine Acetate Chrome Orange
Genuine Orange Chrome
Indian Red
International Orange 2221
Irgachrome Orange OS
Lead chromate, Basic
Lead chromate oxide
Lead chromate, Red
Light Orange Chrome
No. 156 Orange Chrome
Orange Chrome
Orange Nitrate Chrome
Pale Orange Chrome
Persian Red
Pure Orange Chrome M
Pure Orange Chrome Y
Red Lead Chromate
Vynamon Orange CR

18472-36-1
$C_{29}H_{48}O$
412.70
Stigmasta-5,24(28)-dien-3-ol, (3β)- (9CI) (VAN)
δ(5)-Avenasterol
Stigmasta-5,24(28)-dien-3β-ol (8CI) (VAN)

18472-51-0
$C_{22}H_{30}Cl_2N_{10}.2C_6H_{12}O_7$
897.88
D-Gluconic acid, Compd. with N,N''-bis(4-chlorophenyl)-3,12-diimino-2,4,11,13- tetraazatetradecane diimidamide (2:1)
Abacil
Arlacide G
Biguanide, 1,1'-hexamethylene-bis(5-(p-chlor phenyl)-, di-gluconate
1,6-Bis(5-(p-c lorophenyl)bi-guandino)he cane digluconate
Caswell No. 81G
Chlorhexidi digluconate
Chlorhexidi e gluconate
Chlorhexid glukonatu (Czech)
Corsodyl
Disteryl
Gluconic cid, Compd. with 1,1'-he amethylene bis(5-(p-chlo ophenyl)biguanide) (2:1), - (8CI)
1,1'-He methylenebis(5-(p-chlc ophenyl)biguanide)di-glucon te
Hibicler
Hibiscru
Hibitan
Septeal

18479-51 1
$C_{10}H_{20}O$
156.27
OC(CCC C(C)C)(CC)C
6-Octe -3-ol, 3,7-dimethyl-
AI3-24 06
Dihydr linalool
3,7-Dir ethyl-6-octen-3-ol

18479-5 -4
$C_{10}H_{18}O$
154.25
OC(C=C =C(C)C)(CC)C
4,6-Oc adien-3-ol, 3,7-di-meth l- (9CI)
3,7-Di ethyl-4,6-octadien-3-ol

18479-57-7
$C_{10}H_{22}O$
158.28
OC(CCCC(CC)C)(C)C
2-Octanol, 2,6-dimethyl-
AI3-24902
2,6-Dimethyl-2-octanol
Tetrahydromyrcenol

18479-58-8
$C_{10}H_{20}O$
156.30
OC(CCCC(C=C)C)(C)C
7-Octen-2-ol, 2,6-dimethyl
Dihydromyrcenol
2,6-Dimethyl-7-octen-2-ol

18480-07-4
H_2O_2Sr
121.63
Strontium hydroxide (9CI)

18487-39-3
$C_6H_4Cl_3N$
196.46
Benzenamine, 2,3,5-trichloro- (9CI)
Aniline, 2,3,5-trichloro- (8CI)

18495-30-2
$C_3H_4Cl_4$
181.88
ClCC(C(Cl)Cl)Cl
Propane, 1,1,2,3-tetrachloro- (9CI)
1,1,2,3-Tetrachloropropane

18496-25-8
S
32.06
Sulfide (8CI,9CI)
Sulfur, ion (S(2-))

18508-00-4
$C_{18}H_{10}O_2$
258.28
Benz(a)anthracene-5,6-dione
NSC-171732

18521-07-8
$C_9H_{18}O$
142.24
3-Octen-2-ol, 2-methyl-, (Z)- (8CI,9CI)

18521-59-0
$C_{10}H_{14}O_3S$
214.29
Benzenesulfonic acid, 4-butyl- (9CI)
Benzenesulfonic acid, p-butyl- (8CI)

18521-63-6
$C_6H_{12}O_6$
180.16
D-Xylo-hexonic acid, 3-deoxy- (8CI,9CI)
α-D-Galactometasaccharinic acid

18530-56-8
$C_{13}H_{22}N_2O$
222.37
Urea, 3-(hexahydro-4,7-meth-anoindan-5-yl)-1,1-dimethyl-, endo,exo-5
Asepta Herban
Herban
Hercules 7531
1-(3a,4,5,6,7,7a-Hexahydro-4,7-methano-5-indanyl)-3,3-dimethylurea
1-(5-(3a,4,5,6,7,7a-Hexahydro-4,7-methanoindanyl))-3,3-di-methylurea
3-(Hexahydro-4,7-methano-indan-5-yl)-1,1-dimethylurea
3-(5-(3a,4,5,6,7,7a-Hexahydro-4,6-methanoindanyl))-1,1-di-methylurea
Narea
Norea
Nores
Noruron
1-(Tetrahydrodicyclopenta-dienyl)-3,3-dimethylurea

18540-29-9
Cr
52.00
Chromium, ion (Cr 6+)
Chromium(6+)
Chromium (Cr +6)
Chromium (Cr^{6+})
Chromium hexavalent ion
Chromium(6+) ion
Chromium, ion (Cr^{6+}) (8CI,9CI)
Chromium(VI)
NCI-C04273

18585-38-1
$C_6H_8Cl_6$
292.85
Hexane, 1,2,3,4,5,6-hexa-chloro- (8CI)

18599-20-7
$C_4H_4Br_2F_4$
287.88
FC(F)(Br)C(F)(F)CCBr
Butane, 1,4-dibromo-1,1,2,2-tetrafluoro- (9CI)
1,4-Dibromo-1,1,2,2-tetrafluoro-butane

18604-02-9
$C_{18}H_{28}O_4Si_4$
420.82
Cyclotetrasiloxane, 2,4-di-phenyl-2,4,6,6,8,8-hexam-ethyl-, racemic mixture
racemic-2,4-Diphenyl-2,4,6,6,8,8-hexamethylcyclo-tetrasiloxane

18608-30-5
C_3Cl_4
177.84
1,2-Propadiene, 1,1,3,3-tetra-chloro- (9CI)
Propadiene, tetrachloro- (8CI)

18619-18-6
$C_8H_7ClO_2S$
202.66
o-Chlorophenyl thioacetic acid

Acetic acid, ((o-chlorophenyl)-thio)- (8CI)
Acetic acid, ((2-chlorophenyl)-thio)- (9CI)
o-Chlorophenylmercaptoacetic acid
NSC-513490

18623-80-8
ClZn
101.83
Zinc chloride (ZnCl) (6CI, 7CI,8CI,9CI)

18624-44-7
FeH_2O_2
89.86
Iron hydroxide (9CI)

18633-25-5
$C_{15}H_{30}O$
226.40
O(C1CCCCCCCCCCCCC)C1
Oxirane, tridecyl- (9CI)
1,2-Pentadecane oxide
Tridecyloxirane

18636-55-0
$C_{11}H_{12}$
144.22
1H-Indene, 1,1-dimethyl- (9CI)
Indene, 1,1-dimethyl- (7CI, 8CI)
1,1-Dimethylindene

18636-66-3
$C_9H_{20}O$
144.26
Pentane, 1-butoxy- (9CI)
Ether, butyl pentyl (6CI, 8CI)

18641-71-9
$C_9H_{18}O$
142.24
3-Heptanone, 2,4-dimethyl- (8CI,9CI)

18641-81-1
$C_7H_{16}O$
116.20
Butane, 2-(1-methylethoxy)- (9CI)
Ether, sec-butyl isopropyl (8CI)

18662-53-8
$C_6H_6NO_6.3Na.H_2O$
311.16
O.[Na+].[Na+].[Na+].[O-]C(=O)CN(CC([O-])=O)CC([O-])=O
Acetic acid, nitrilotri-, trisodium salt, monohydrate
N,N-Bis(carboxymethyl)glycine trisodium salt monohydrate
Glycine, N,N-bis(carboxymethyl)-, trisodium salt, monohydrate
NCI-C01445
Nitriloacetic acid trisodium salt monohydrate
Nitrilotriacetic acid trisodium salt monohydrate
NTA, sodium hydrate
Trisodium nitrilotriacetate monohydrate

18669-52-8
C_8H_{14}
110.20
1,4-Hexadiene, 2,3-dimethyl- (8CI,9CI)
2,3-Dimethyl-1,4-hexadiene

18684-55-4
$C_{20}H_{26}O_3$
314.46
Podocarpa-8,11,13-trien-15-oic acid, 13-isopropyl-7-oxo
13-Isopropyl-7-oxopodocarpa-8,11,13-trien-15-oic acid
7-Ketodehydroabietic acid
7-Oxodehydroabietic acid

18691-97-9
$C_{10}H_{11}N_3OS$
221.30
CNC(=O)N(C)c2nc1ccccc1s2
Urea, 1-(2-benzothiazolyl)-1,3-dimethyl
Bay 74283
1,3-Dimethyl-3-(2-benzthiazolyl)-harnstoff (German)
1,3-Dimethyl-3-(2-benzothiazolyl)urea
Methabenzthiazuron
Methibenzuron
N-Methyl-N'-methyl-N'-(2-benzothiazolyl)urea
Tribunil

18707-60-3
$C_5H_8O_2$
100.12
Crotonic acid, methyl ester (8CI)
AI3-06008
2-Butenoic acid, methyl ester (9CI)
Methyl 2-butenoate
Methyl crotonate
NSC-18745

18708-70-8
$C_6H_2Cl_3NO_2$
226.44
O=N(=O)c(c(cc(c1)Cl)Cl)c1Cl
Benzene, 1,3,5-trichloro-2-nitro- (9CI)
NSC-10244
2,4,6-Trichloronitrobenzene
1,3,5-Trichloro-2-nitrobenzene
2,4,6-Trichloro-1-nitrobenzene

18713-51-4
$C_9H_{15}Br_6O_4P$
697.67
2-Propanol, 1,3-dibromo-, phosphate (3:1)
Tris(1-bromomethyl-2-bromoethyl)phosphate

18717-72-1
$C_{16}H_{19}NO_4$
289.36
1-α-H,5-α-H-Nortropane-2-β-carboxylic acid, 3-β-hydroxy-, methyl ester, benzoate (ester)
Norcocaine

18760-44-6
$C_{14}H_{28}O_3S$
276.44
O=S(=O)(CC(OCCCCCCCCCC)C1)C1
Thiophene, 3-(decyloxy)tetrahydro-, 1,1-dioxide (8CI)
3-Decyloxysulfolane
3-(Decyloxy)tetrahydrothiophene 1,1-dioxide

18771-50-1
$C_9H_{16}N_2O_6$
248.22
Tetrahydrouridine
5,6-Dihydrouridine
NSC-112907
2(1H)-Pyrimidinone, tetrahydro-4-hydroxy-1-β-D-ribofuranosyl- (9CI)
3,4,5,6-Tetrahydrouridine
THU
U 23284

18777-54-3
$C_{18}H_{30}O_3S$
326.50
O=S(=O)(O)c(ccc(c1)C(CCCCCCCC)CC)c1
Benzenesulfonic acid, 4-(1-ethyldecyl)- (9CI)
4-(1-Ethyldecyl)benzenesulfonic acid

18791-19-0
$C_4H_2Cl_8$
333.68
Butane, 1,1,1,2,3,4,4,4-octachloro- (8CI,9CI)
1,1,1,2,3,4,4,4-Octachlorobutane

18801-52-0

$C_5H_{12}N_4S$
160.22
Carbonothioic dihydrazide, (1-methylpropylidene)- (9CI)
Carbohydrazide, 1-sec-butylidene-3-thio- (8CI)

18808-10-1
$C_{17}H_{16}N_4$
276.34
1,3,5-Triazin-2-amine, N,N-dimethyl-4,6-diphenyl- (9CI)
s-Triazine, 2-(dimethylamino)-4,6-diphenyl- (8CI)

18810-58-7
BaN_6
221.40
Barium-azide
Barium azide, Dry or containing less than 50% water [UN 0224]
Barium azide, Wet, 50% or more water [UN 1571]
UN 0224 [Barium azide, Dry or wetted with less than 50 per cent water, by mass]
UN 1571 [Barium azide, Wetted with not less than 50 per cent water, by mass]

18815-11-7
C_7H_7O
107.13
Phenyl, 3-methoxy- (9CI)
Phenyl, m-methoxy- (8CI)

18829-55-5
$C_7H_{12}O$
112.19
O=CC=CCCCC
2-Heptenal, (E)
β-Butylacrolein
3-Butylacrolein
(E)-2-Hepten-1-al

18832-83-2
C_3H_7BrHgO
339.59
CC(O)C[Hg]Br
1-Bromomercuri-2-hydroxypropane

18835-33-1
$C_{26}H_{52}$
364.70
C(=C)CCCCCCCCCCCCCCCCCCCCCCCC
1-Hexacosene
AI3-10513

18854-01-8
$C_{13}H_{16}NO_4PS$
313.33
CCOP(=S)(OCC)Oc1cc(on1)c2ccccc2
Phosphorothioic acid, O,O-diethyl O-(5-phenyl-3-isoxazolyl) ester
O,O-Diethyl O-(5-phenyl-3-isoxazolyl) phosphorothioate
O,O-Diethyl O-(3-(5-phenyl)-1,2-isoxazolyl)phosphorothionate
E-48
Isoxathion
Karphos
SI-6711

18855-13-5
$C_{12}H_{10}O_2$
186.21
Oc(ccc(c(cccc1O)c1)c2)c2
(1,1'-Biphenyl)-3,4'-diol (9CI)
3,4'-Dihydroxydiphenyl

18868-66-1
$C_{20}H_{16}$
256.36
Benz(a)anthracene, 12-ethyl
12-Ethylbenz(a)anthracene

18869-73-3
$C_{26}H_{21}NO_6$
443.48
O=C(Oc(ccc(c1)C(c(c(N2C(=O)C)ccc3)c3)(c(ccc(OC(=O)C)c4)c4)C2=O)c1)C
2H-Indol-2-one, 1,3-dihydro-1-acetyl-3,3-bis(4-(acetyloxy)phenyl)
1-Acetyl-3,3-bis(4-(acetyloxy)phenyl)-1,3-dihydro-2H-indol-2-one
1-Acetyl-3,3-bis(p-hydroxyphenyl)oxindole diacetate
17,17-Dimethyl-18-norandrost-4,13-dien-3-one
Isatex
Laxagen
Laxagetten
Phenisatin
TDI
Triacetyldiphenolisatin
Trisatin
Unilax

18871-14-2
$C_{12}H_{22}O_3$
214.30
O=C(OC(C(CCCCC)COC1)C1)C
2H-Pyran-4-ol, tetrahydro-3-pentyl-, acetate (9CI)
4-Acetoxy-3-pentyltetrahydropyran
3-Amyl-4-acetoxytetrahydropyran
Tetrahydro-3-pentyl-2H-pyran-4-ol acetate

18874-52-7
$C_4H_4BrN_3O_2$
206.00
1H-Imidazole, 4-bromo-2-methyl-5-nitro- (9CI)
4-Bromo-2-methyl-5-nitroimidazole
5-Bromo-2-methyl-4-nitroimidazole
Imidazole, 4-bromo-2-methyl-5-nitro- (8CI)
2-Methyl-5-bromo-4-nitroimidazole

18880-36-9
$C_6H_{13}NO_3S_3.Na$
266.36
1-Propanesulfonic acid, 3-(((dimethylamino)thioxomethyl)thio)-, sodium salt (9CI)
3-((Dimethylamino)thioxomethyl)thio)-1-propanesulfonic acid, sodium salt
3-Sulfopropyl N,N-dimethyldithiocarbamate, sodium salt

18883-66-4
$C_8H_{15}N_3O_7$
265.26
CN(N=O)C(=O)NC1C(O)OC(CO)C(O)C1O
Glucopyranose, 2-deoxy-2-(3-methyl-3-nitrosoureido)-, D
2-Deoxy-2-(((methylnitrosoamino)carbonyl)amino)-D-glucopyranose
2-Deoxy-2-(3-methyl-3-nitrosoureido)-D-glucopyranose
2-Deoxy-2-(3-methyl-3-nitrosoureido)-α(and β)-D-glucopyranose
D-Glucopyranose, 2-deoxy-2-(((methylnitrosoamino)carbonyl)amino)-
D-Glucose, 2-deoxy-2-(((methylnitrosoamino)carbonyl)amino)- (9CI)
D-Glucose, 2-deoxy-2-(3-methyl-3-nitrosoureido)-
N-d-Glucosyl(2)-N'-nitrosomethylharnstoff (German)
N-D-Glucosyl-(2)-N'-nitrosomethylurea
NCI-C03167
NSC-85598
NSC-85998
RCRA waste number U206
STR
Streptozocin
Streptozoticin
Streptozotocin
STRZ
STZ
U-9889
Zanosar

18886-16-3
$C_{19}H_{12}Cl_2N_6O_7S_2.2Na$

617.32
2-Naphthalenesulfonic acid, 7-((4,6-dichloro-1,3,5-triazin-2-yl)amino)-4-hydroxy-3-((2-sulfophenyl)azo)-, disodium salt (9CI)

18895-89-1
$C_{11}H_{20}ClN_3O_4P_2S_4$
483.96
Phosphorodithioic acid, S,S'-(6-chloro-s-triazine-2,4-diyl) O,O,O',O'-tetraethyl ester (6CI,8CI)
s-Triazine-2,4-dithiol, 6-chloro-, S,S'-diester with O,O-diethyl phosphorodithioate

18917-89-0
$C_{14}H_{10}MgO_6$
298.53
Magnesium, bis(2-hydroxybenzoato-O1,O2)-, (T-4)-(9CI)
(T-4)-Bis(2-hydroxybenzoato-O1,O2)magnesium

18917-91-4
$C_9H_{15}AlO_9$
294.22
CC(O)C(=O)O[Al](OC(=O)C(C)O)OC(=O)C(C)O
Aluminum, tris(lactato)
Aluminum lactate
Aluminum, tris(2-hydroxypropanoato-O^1,O^2)- (9CI)

18923-87-0
$(C_4H_8)_2$
224.43
1-Propene, 2-methyl-, dimer (9CI)
Propene, 2-methyl-, dimer (8CI)

18936-17-9
C_5H_9N
83.15
CCC(C)C#N
Butyronitrile, 2-methyl
Butanenitrile, 2-methyl- (9CI)
2-Methylbutyronitrile

18936-75-9
$C_{17}H_{12}N_2$
244.31
Benz(a)acridine, 10-amino
10-Aminobenz(a)acridine
Benz(a)acridin-10-amine (9CI)

18946-94-6
Unknown
Unknown
Nitrilotriacetic acid, neodymium(3+) salt (1)
NTA, neodymium(3+) salt (1)

18979-50-5
$C_9H_{12}O_2$
152.19
O(c(ccc(O)c1)c1)CCC
Phenol, p-propoxy- (8CI)
NSC-82358
Phenol, 4-propoxy- (9CI)
p-Propoxyphenol
4-Propoxyphenol

18979-53-8
$C_{11}H_{16}O_2$
180.25
O(c(ccc(O)c1)c1)CCCCC
Phenol, 4-(pentyloxy)- (9CI)
4-(Pentyloxy)phenol

18979-55-0
$C_{12}H_{18}O_2$
194.27
O(c(ccc(O)c1)c1)CCCCCC
Phenol, 4-(hexyloxy)- (9CI)
4-(Hexyloxy)phenol

18979-90-3
C_8H_9ClO
156.62
Phenol, 4-chloro-2-ethyl
4-Chlor-2-aethylphenol (German)
4-Chloro-2-ethylphenol

18979-96-9
$C_{13}H_{19}ClO$
226.75
Phenol, 4-chloro-2-heptyl- (8CI,9CI)

18980-00-2
C_8H_9ClO
156.61
Phenol, 2-chloro-4-ethyl- (8CI,9CI)

18980-02-4
$C_{10}H_{13}ClO$
184.67
Phenol, 4-butyl-2-chloro- (8CI,9CI)

18980-06-8
$C_{13}H_{19}ClO$
226.75
Phenol, 2-chloro-4-heptyl- (8CI,9CI)

18983-72-7
Unknown
Unknown
Nitrilotriacetic acid, beryllium potassium salt (1:1)
NTA, beryllium potassium salt (1:1)

18991-62-3
$C_6H_{10}D_2O_6$
178.14
D-Glucose-6,6-C-d2 (9CI)
D-Glucose-6,6-d2 (8CI)

18994-66-6
Unknown
Unknown
Nitrilotriacetic acid and its salts

18996-35-5
$C_6H_7O_7.Na$
214.12
Citric acid, sodium salt
Citric acid, monosodium salt (8CI)
Citrofluyl
Monosodium citrate
Monosodium dihydrogen citrate
1,2,3-Propanetricarboxylic acid, 2-hydroxy-, monosodium salt (9CI)
Sodium citrate
Sodium dihydrogen citrate

18996-86-6
$C_9H_4Cl_4N_4$
309.94
1,3,5-Triazin-2-amine, 4,6-dichloro-N-(3,4-dichlorophenyl)-(9CI)
s-Triazine, 2,4-dichloro-6-(3,4-dichloroanilino)- (8CI)

19010-45-8
$C_{10}H_{13}ClO$
184.67
Phenol, 2-butyl-4-chloro- (8CI, 9CI)

19010-66-3
$C_6H_{12}N_2S_4.Pb$
447.63
CN(C)C(=S)S[Pb]SC(=S)N(C)C
Lead, bis(dimethyldithiocarbamato)
Bis(dimethylcarbamodithioato-S,S')lead
Bis(dimethyldithiocarbamato)-lead
Carbamic acid, dimethyldithio-, lead salt
Lead dimethyldithiocarbamate
Methyl ledate
NCI-C02891

19010-73-2
Unknown
Unknown
Nitrilotriacetic acid, alumin-

ium(3+) complex
NTA, aluminium(3+) complex

19013-11-7
$C_8H_8N_2O_3$
180.15
O=C(N)c(ccc(c1N(=O)=O)C)c1
Benzamide, 4-methyl-3-nitro- (9CI)
4-Methyl-3-nitrobenzamide

19014-53-0
$C_{27}H_{24}N_2O_5$
456.49
O=C(N(CCCC1)Cc(ccc(Oc(c(c(c(C(=O)c(c2ccc3)c3)c4O)C2=O)N)c4)c5)c5)C1
Anthraquinone, 1-amino-2-(p-((hexahydro-2-oxo-1H-azepin-1-yl)methyl)phenoxy)-4-hydroxy- (8CI)
1-((4-((1-Amino-4-hydroxy-2-anthraquinonyl)oxy)phenyl)-methyl)hexahydro-2H-azepin-2-one

19044-88-3
$C_{12}H_{18}N_4O_6S$
346.40
CCCN(CCC)c1c(cc(cc1N(=O)=O)S(N)(=O)=O)N(=O)=O
Sulfanilamide, 3,5-dinitro-N^4,N^4-dipropyl
3,5-Dinitro-N^4,N^4-dipropyl-sulfanilamide
4-(Dipropylamino)-3,5-dinitro-benzenesulfonamide
Dirimal
EL-119
Oryzalin
Rycelan
Rycelon
Ryzelan
Surflan

19045-79-5
$C_8H_{17}O_4P.2K$
286.38
Octyl potassium phosphate
Phosphoric acid, monooctyl ester, dipotassium salt (9CI)

19049-40-2
$C_{12}H_{18}Be_4O_{13}$
406.34
Beryllium, hexakis(mu-acetato)-mu^4-oxotetra
Beryllium acetate, basic
Beryllium, hexakis(mu-(acetato-O:O'))-mu^4-oxotetra- (9CI)
Beryllium oxide acetate
Beryllium oxyacetate
Hexakis(mu-(acetato-O:O'))-mu^4-oxotetraberyllium
Hexakis(mu-acetato)-mu^4-oxo-tetraberyllium

19064-69-8
$C_8H_7N_3$
145.15
Nc1nncc2ccccc12
1-Phthalazinamine (9CI)
Phthalazine, 1-amino- (8CI)

19071-21-7
$C_4H_3ClO_4$
150.52
2-Butenedioic acid, 2-chloro- (9CI)

19074-59-0
$C_{24}H_{27}O_4P$
410.45
O=P(Oc(c(ccc1C)C)c1)(Oc(c(ccc2C)C)c2)Oc(c(ccc3C)C)c3
Tris(2,5-xylenyl)phosphate
NSC-66475
Phenol, 2,5-dimethyl-, phosphate (3:1) (9CI)
Phosphoric acid, tris(2,5-dimethylphenyl)ester
Tri(2,5-dimethylphenyl) phosphate
Tri(2,5-xylenyl)phosphate
2,5-Xylenol, phosphate (3:1) (8CI)

19077-97-5
$C_9H_{12}N_4O_3S$
256.26
O=C(Nc(ccc(S(=O)(=O)NC(=N)N)c1)c1)C
Acetamide, N-(4-(((amino-iminomethyl)amino)sulfonyl)phenyl)- (9CI)
N-(4-(((Aminoiminomethyl)-amino)sulfonyl)phenyl)acet-amide
NSC-28593

19078-35-4
$C_{15}H_{26}$
206.37
1H-3a,7-Methanoazulene, octahydro-1,4,9,9-tetramethyl- (8CI,9CI) (VAN)

19089-47-5
$C_5H_{12}O_2$
104.17
CCOC(C)CO
1-Propanol, 2-ethoxy
2-Ethoxy-1-propanol
Propylene glycol monoethyl ether, α

19095-79-5
$C_{14}H_{20}N_2O$
232.33
O=C(NC1CCCCCC1)Nc2ccccc2
Urea, N-cycloheptyl-N'-phenyl- (9CI)
1-Cycloheptyl-3-phenyl-urea
Urea, 1-cycloheptyl-3-phenyl- (8CI)

19102-74-0
$C_{16}H_{34}O_2$
258.44
Dioctyl peroxide
Caprylyl peroxide
Octane, 1,1'-dioxybis-
Octyl peroxide

19125-99-6
$C_{20}H_{24}N_2O_2$
324.41
O=C(N(C(=O)c(c1c(c(NCCCC)cc2)cc3)c3)CCCC)c12
1H-Benz(de)isoquinoline-1,3(2H)-dione, 2-butyl-6-(butylamino)- (9CI)
4-Butylamino-N-butyl-1,8-naphthalimide
4-(Butylamino)-N-butyl-1,8-naphthalimide
2-Butyl-6-(butylamino)-1H-benz(de)isoquinoline-1,3(2H)-dione

19132-06-0
$C_4H_{10}O_2$
90.12
2,3-Butanediol, (S-(R*,R*))- (9CI)
2,3-Butanediol, L- (8CI)

19141-82-3
$C_{16}H_{34}O.1/3Al$
251.44
1-Hexadecanol, aluminum salt (9CI)
Aluminum n-hexadecoxide

19159-68-3
$C_{12}H_4N_8O_{12}$
452.16
O=N(=O)c(cc(N(=O)=O)c(N=Nc(c(N(=O)=O)cc(N(=O)=O)c1)c1N(=O)=O)c2N(=O)=O)c2
Hexanitroazoxybenzene
Bis(2,4,6-trinitrophenyl)diazene
Diazene, bis(2,4,6-trinitrophenyl)- (9CI)
2,2',4,4',6,6'-Hexanitroazobenzene
Hexanitroazoxi benceno (Spanish)
Hexanitroazoxy benzene (French)
HNAB

19183-14-3
$C_3H_2N_4O_4$
158.05
1H-Imidazole, 4,5-dinitro-

(9CI)
Imidazole, 4,5-dinitro- (8CI)

19183-15-4
$C_4H_4N_4O_4$
172.10
1H-Imidazole, 1-methyl-4,5-dinitro- (9CI)
Imidazole, 1-methyl-4,5-dinitro- (8CI)
1-Methyl-4,5-dinitroimidazole

19183-16-5
$C_4H_4N_4O_4$
172.08
1H-Imidazole, 2-methyl-4,5-dinitro- (9CI)
Imidazole, 2-methyl-4,5-dinitro- (8CI)

19183-17-6
$C_5H_6N_4O_4$
186.13
1H-Imidazole, 1,2-dimethyl-4,5-dinitro- (9CI)
1,2-Dimethyl-4,5-dinitroimidazole
Imidazole, 1,2-dimethyl-4,5-dinitro- (8CI)

19184-65-7
$C_5H_{11}BrO_3$
199.04
OCC(CBr)(CO)CO
1,3-Propanediol, 2-(bromomethyl)-2-(hydroxymethyl)- (9CI)
NSC-151735
Tris(hydroxymethyl)(bromomethyl)methane

19216-56-9
$C_{19}H_{21}N_5O_4$
383.45
COc2cc1nc(nc(N)c1cc2OC)N3CCN(CC3)C(=O)c4ccco4
Quinazoline, 4-amino-6,7-dimethoxy-2-(4-(2-furoyl)piperazin-1-yl)
1-(4-Amino-6,7-dimethoxy-2-quinazolinyl)-4-(2-furanylcarbonyl)piperazine
Furazosin
2-(4-(2-Furoyl)piperazin-1-yl)-4-amino-6,7-dimethoxyquinazoline
Prazosin

19218-94-1
$C_{14}H_{29}I$
324.29
C(CCCCCCCCCCCCC)I
Tetradecane, 1-iodo- (9CI)
1-Iodotetradecane
Tetradecyl iodide

19219-48-8
$C_{14}H_{13}NO_3$
243.26
Phenol, 4-methoxy-, phenylcarbamate (9CI)
p-Anisyl phenylcarbamate
p-Methoxyphenylcarbanilate
4-Methoxyphenyl phenylcarbamate
4-Methoxyphenyl N-phenylcarbamate
Phenol, p-methoxy-, carbanilate (6CI,8CI)

19219-85-3
$C_{12}H_{18}$
162.28
Benzene, (2-ethylbutyl)- (6CI,7CI,8CI)

19219-99-9
C_8H_6ClNO
167.60
O(c(c(N=1)cc(c2)Cl)c2)C1C
Benzoxazole, 5-chloro-2-methyl
5-Chloro-2-methylbenzoxazole

19223-55-3
$C_{20}H_{42}N_2O_5S$
422.62
O=C(NCCCN(CC(O)CS(=O)(=O)O)(C)C)CCCCCCCCCCC
1-Propanaminium, 2-hydroxy-N,N-dimethyl-N-(3-((1-oxododecyl)amino)propyl)-3-sulfo-, hydroxide, inner salt (9CI)
Betaine
N-(3-Laurylamidopropyl)-N,N-dimethyl-N-(2-hydroxy-3-sulfopropyl)ammonium, inner salt

19224-26-1
$C_{17}H_{16}O_4$
284.31
O=C(OCC(OC(=O)c(cccc1)c1)C)c(cccc2)c2
1,2-Propanediol, dibenzoate (9CI)

19248-13-6
$C_{11}H_{18}N_2$
178.27
N(c(cccc1C)c1)(CCN)CC
Ethylenediamine, N-ethyl-N-m-tolyl- (8CI)
N-(2-Aminoethyl)-N-ethyl-m-toluidine
1,2-Ethanediamine, N-ethyl-N-(3-methylphenyl)- (9CI)
N-Ethyl-N-(β-aminoethyl)-m-toluidine
N-Ethyl-N-(3-methylphenyl)-1,2-ethanediamine
NSC-151043

19249-34-4
$C_{14}H_{19}NO_4$
265.34
O=C(OCCN(c(cccc1)c1)CCOC(=O)C)C
Ethanol, 2,2'-(phenylimino)bis-, diacetate (Ester)
N,N-Bis(2-acetoxyethyl)aniline
2,2'-Phenyliminodiethanol diacetate

19262-20-5
$C_{12}H_{18}$
162.28
Benzene, (1,2,2-trimethylpropyl)- (6CI,8CI,9CI)

19277-91-9
$C_8H_{10}O.Na$
145.16
Phenol, 4-ethyl-, sodium salt (9CI)
p-Ethylphenol, sodium salt
4-Ethylphenol sodium salt
Sodium p-ethylphenoxide

19287-45-7
B_2H_6
27.67
Diborane (ACGIH,OSHA) [UN 1911]
Boroethane
Boron hydride
Diborane mixture (DOT)
Diborane or diborane mixture
Diborane(6) (9CI)
Diborano (Spanish)
Diboron hexahydride
UN 1911 [Diborane]

19287-96-8
$C_6H_8O_7.H_2O.3Na$
276.09
Citric acid, trisodium salt, monohydrate (8CI)

19314-74-0
$C_{29}H_{30}O_{10}$
538.56
Benzoic acid, 4-((2,4-dimethoxy-6-methylbenzoyl)oxy)-2-methoxy-6-methyl-, 3-methoxy-4-(methoxycarbonyl)-5-methylphenyl ester (9CI)
4,2-Cresotic acid, 6-methoxy-, bimol. ester, methyl ester, 4,6-dimethoxy-o-toluate (8CI)
4,2-Cresotic acid, 6-methoxy-, methyl ester, ester with 6-methoxy-4,2-cres-

otic acid 4,6-dimethoxy-o-toluate (7CI)
o-Toluic acid, 4,6-dimethoxy-, ester with 6-methoxy-4,2-eresotic acid bimol. ester, methyl ester

19336-97-1
$C_{15}H_{13}NO_4$
271.27
Benzoic acid, 2-(((3-methoxyphenyl)amino)carbonyl)- (9CI)
Phthalanilic acid, 3'-methoxy- (8CI)

19338-12-6
$C_{10}H_{11}N_5$
201.20
1,3,5-Triazine-2,4-diamine, 6-(4-methylphenyl)- (9CI)
s-Triazine, 2,4-diamino-6-p-tolyl- (8CI)

19343-78-3
$C_{10}H_{13}N$
147.21
N(c(c(ccc1)C(C2)C)c1)C2
Quinoline, 1,2,3,4-tetrahydro-4-methyl-
AI3-22251
Lepidine, 1,2,3,4-tetrahydro-
4-Methyl-1,2,3,4-tetrahydroquinoline
1,2,3,4-Tetrahydrolepidine
1,2,3,4-Tetrahydro-4-methylquinoline

19355-69-2
$C_4H_8N_2$
84.11
N#CC(N)(C)C
Propanenitrile, 2-amino-2-methyl- (9CI)
α-Aminoisobutyronitrile
2-Amino-2-methylpropanenitrile

19356-17-3
$C_{28}H_{46}O_2$
414.68
CC(CCCC(C)(C)O)C1CCC2C(CCCC12C)=CC=C3CC(O)CCC3=C
Calcifediol
Calcifediolum (Latin)

19377-95-8
$C_{16}H_{28}O$
236.40
Bicyclo(3.1.0)hexan-2-one, 1,5-bis(1,1-dimethylethyl)-3,3-dimethyl- (9CI)
Bicyclo(3.1.0)hexan-2-one, 1,5-di-tert-butyl-3,3-dimethyl- (8CI)

19377-97-0
$C_{12}H_{20}O$
180.29
4,5-Octadien-3-one, 2,2,7,7-tetramethyl- (8CI, 9CI)

19393-92-1
$C_6H_3BrCl_2$
225.90
Benzene, 2-bromo-1,3-dichloro- (9CI)
2-Bromo-1,3-dichlorobenzene
2,6-Dichlorobromobenzene
1,3-Dichloro-2-bromobenzene
NSC-155332

19395-62-1
$C_{18}H_{22}N_4.2ClH$
367.30
1,9-Acridinediamine, N(9)-(3-(dimethylamino)propyl)-, dihydrochloride (9CI)
Acridine, 1-amino-9-((3-(dimethylamino)propyl)amino)-, dihydrochloride (8CI)
C 663

19398-53-9
$C_5H_{10}Br_2$
229.94
BrC(CC(Br)C)C
Pentane, 2,4-dibromo- (9CI)
2,4-Dibromopentane

19398-61-9
$C_7H_6Cl_2$
161.03
c(ccc(c1C)Cl)(c1)Cl
Benzene, 1,4-dichloro-2-methyl- (9CI)
1,4-Dichloro-2-methylbenzene
2,5-Dichlorotoluene
NSC-86117
Toluene, 2,5-dichloro- (8CI)

19406-51-0
$C_7H_7N_3O_4$
197.17
m-Toluidine, 2,4-dinitro
4-ADNT
3-Amino-2,6-dinitrotoluene
2,4-Dinitro-m-toluidine
4-Methyl-3,5-dinitrobenzenamine

19406-86-1
$C_8H_{10}N_2O$
150.17
O=C(N)c(ccc(c1N)C)c1
Benzamide, 3-amino-4-methyl- (9CI)
3-Amino-4-methylbenzamide

19408-74-3
$C_{12}H_2Cl_6O_2$
390.84
Dibenzo-p-dioxin, 1,2,3,7,8,9-hexachloro
1,2,3,7,8,9-Hexachlorodibenzo-p-dioxin

19420-61-2
$C_{10}H_{12}O_4$
196.20
Benzoic acid, 2,3-dihydroxy-4-(1-methylethyl)- (9CI)
o-Pyrocatechuic acid, 4-isopropyl- (8CI)

19424-29-4
$C_7H_{12}O_3$
144.17
1,3-Dioxolan-2-one, 4,4,5,5-tetramethyl- (9CI)
2,3-Butanediol, 2,3-dimethyl-, cyclic carbonate (8CI)
Carbonic acid, cyclic tetramethylethylene ester (6CI, 8CI)
Pinacolone cyclic carbonate

19424-34-1
$C_7H_{13}N$
111.19
Hexanenitrile, 5-methyl- (6CI,7CI,8CI,9CI)
5-Methylhexanenitrile

19430-93-4
$C_6H_3F_9$
246.08
FC(F)(F)C(F)(F)C(F)(F)C(F)(F)C=C
1-Hexene, 3,3,4,4,5,5,6,6,6-nonafluoro- (9CI)
3,3,4,4,5,5,6,6,6-Nonafluoro-1-hexene

19433-86-4
$C_7H_9NO_4S.Na$
226.20
Benzenesulfonic acid, 2-amino-5-methoxy-, monosodium salt (9CI)
2-Amino-5-methoxybenzenesulfonic acid, monosodium salt

19433-93-3
$C_{13}H_{20}N_2O_2$
236.30
O=C(Nc(ccc(OC)c1N(CC)CC)c1)C
Acetamide, N-(3-(diethylamino)-4-methoxyphenyl)- (9CI)
p-Acetanisidide, 3'-(diethyl-

amino)-
N-(3-(Diethylamino)-4-methoxyphenyl)acetamide

19433-94-4
$C_{14}H_{19}N_3O_2$
261.31
O=C(Nc(ccc(OC)c1N(CCC#N)CC)c1)C
Acetamide, N-(3-((2-cyanoethyl)ethylamino)-4-methoxyphenyl)- (9CI)
p-Acetanisidide, 3'-((2-cyanoethyl)ethylamino)-
3-(N-Cyanoethyl-N-ethylamino)-4-methoxyacetanilide
3-(N-Cyanoethyl-N-ethyl)amino-4-methoxyacetanilide
N-(3-((2-Cyanoethyl)ethylamino)-4-methoxyphenyl)acetamide

19434-42-5
$C_{12}H_{11}NO$
185.22
(1,1'-Biphenyl)-2-ol, 5-amino- (9CI)
2-Biphenylol, 5-amino- (8CI)
NSC-409777
Phenol, 4-amino-2-phenyl-

19434-65-2
C_3H_5ClO
92.52
Propanal, 3-chloro- (9CI)
Propionaldehyde, 3-chloro- (8CI)

19437-42-4
$C_{10}H_8O_9S_3.xNa$
Unknown
1,3,6-Naphthalenetrisulfonic acid, sodium salt (9CI)

19438-61-0
$C_9H_6O_3$
162.14
O=C(OC(=O)c1ccc(c2)C)c12
1,3-Isobenzofurandione, 5-methyl- (9CI)
5-Methyl-1,3-isobenzofurandione

19456-58-7
Unknown
Unknown
Nitrilotriacetic acid, indium(3+) complex
NTA, indium(3+) complex

19463-48-0
$C_8H_7ClO_3$
186.59
Benzaldehyde, 3-chloro-4-hydroxy-5-methoxy- (9CI)
5-Chlorovanillin
NSC-45929
Vanillin, 5-chloro- (8CI)

19464-92-7
$C_{12}H_{14}O_3$
206.24
Oxiranecarboxylic acid, 3-methyl-3-phenyl-, ethyl ester, trans- (9CI)
trans-Ethyl 3-methyl-3-phenylglycidate
Hydrocinnamic acid, α,β-epoxy-β-methyl-, ethyl ester, trans- (8CI)

19479-83-5
$C_{10}H_{22}N_4$
198.29
N(CCNC1)(C1)CCN(CCNC2)C2
Piperazine, 1,1'-(1,2-ethanediyl)bis- (9CI)
1,1'-(1,2-Ethanediyl)bispiperazine

19480-39-8
$C_9H_8ClO_3.H_4N$
217.67
N.Cc1cc(Cl)ccc1OCC(O)=O
Acetic acid, (4-chloro-2-methylphenoxy)-, ammonium salt
Acetic acid, ((4-chloro-o-tolyl)oxy)-, amine salt
Acetic acid, ((4-chloro-o-tolyl)oxy)-, ammonium salt (8CI)
Dikoteks 40
Dikoteks AM
Dikotex 40

19480-43-4
$C_{15}H_{21}ClO_4$
300.78
O=C(OCCOCCCC)COc(c(cc(c1)Cl)C)c1
2-Butoxyethyl 2-methyl-4-chlorophenoxyacetate
Acetic acid, (4-chloro-2-methylphenoxy)-, 2-butoxyethyl ester (9CI)
Caswell No. 557D
EPA Pesticide Chemical Code 030553
2-Methyl-4-chlorophenoxyacetic acid, butoxyethyl ester
2-Methyl-4-chlorophenoxyacetic acid, 2-butoxyethyl ester

19484-26-5
$C_{13}H_{28}S$
216.43
SCCCCCCCCCCCCC
1-Tridecanethiol (9CI)
NSC-851
Tridecyl mercaptan

19485-03-1
$C_{10}H_{14}O_4$
198.24
O=C(OCCC(OC(=O)C=C)C)C=C
Acrylic acid, 1,3-butylene glycol diester
Acrylic acid, 1-methyltrimethylene ester (8CI)
1,3-Butanediol diacrylate
1,3-Butylene diacrylate
1,3-Butylene glycol diacrylate
2-Propenoic acid, 1-methyl-1,3-propanediyl ester (9CI)

19489-10-2
C_9H_{18}
126.24
Cyclohexane, 1-ethyl-3-methyl-, cis- (8CI,9CI)

19519-45-0
$C_8H_{13}NO$
139.19
Oxazole, 4,5-dimethyl-2-(1-methylethyl)- (9CI)
Oxazole, 2-isopropyl-4,5-dimethyl- (8CI)

19524-06-2
$C_5H_4BrN.ClH$
194.45
Pyridine, 4-bromo-, hydrochloride (8CI,9CI)
NSC-76559

19525-59-8
$C_8H_9NO_2.K$
190.26
Glycine, N-phenyl-, monopotassium salt (9CI)
AI3-15398
NSC-405072
N-Phenylglycine monopotassium salt
N-Phenylglycine potassium salt

19532-03-7
$C_{10}H_9NO_6S_2.Na$
326.30
1,5-Naphthalenedisulfonic acid, 2-amino-, monosodium salt (8CI)

19549-73-6
$C_9H_{20}O$
144.26
3-Heptanol, 2,6-dimethyl- (8CI, 9CI)

19549-77-0
$C_9H_{20}O$
144.26
4-Heptanol, 2,4-dimethyl- (8CI,

9CI)

19549-79-2
$C_9H_{20}O$
144.26
4-Heptanol, 3,5-dimethyl- (8CI, 9CI)

19549-80-5
$C_9H_{18}O$
142.24
O=C(CC(CC(C)C)C)C
2-Heptanone, 4,6-dimethyl- (9CI)
4,6-Dimethyl-2-heptanone

19550-03-9
$C_8H_{18}O$
130.23
2-Hexanol, 2,3-dimethyl- (8CI, 9CI)

19550-30-2
$C_6H_{14}O$
102.18
CC(C)C(C)CO
1-Butanol, 2,3-dimethyl- (8CI, 9CI)

19561-31-0
$C_{17}H_{12}$
216.29
4H-Cyclohepta(def)phenanthrene (8CI,9CI)

19583-54-1
$C_8H_{16}O_2.xFe$
Unknown
Hexanoic acid, 2-ethyl-, iron salt (9CI)
2-Ethylhexanoic acid, iron salt
Iron 2-ethylhexanoate

19600-63-6
$C_8H_{14}O$
126.20
Oxirane, 5-hexenyl- (9CI)
1-Octene, 7,8-epoxy- (8CI)

19615-27-1
$C_7H_{11}NO$
125.17
Pyridine, 1-acetyl-1,2,3, 4-tetrahydro- (8CI,9CI)
1-Acetyl-1,2,3,4-tetrahydropyridine

19622-19-6
$C_8H_{18}N_2OS.ClH$
226.80
[Cl-].CCSC(=O)NCCCN(C)C
Carbamic acid, (3-dimethylaminopropyl)thio-, S-ethyl ester, hydrochloride
N-(3-Dimethylaminopropyl)-thiocarbaminsaeure-S-aethyl-ester-hydrochlorid (German)
Dynone
S-Ethyl N-(3-dimethylamino-propyl)thiol carbamate hydrochloride
Previcur
Prothiocarb
SN-41703

19624-22-7
B_5H_9
63.14
Pentaborane (ACGIH,OSHA) [UN 1380]
Pentaborane(9)
UN 1380 [Pentaborane]

19660-16-3
$C_6H_8Br_2O_2$
271.96
O=C(OCC(Br)CBr)C=C
Acrylic acid, 2,3-dibromopropyl ester
2,3-Dibromopropyl acrylate
2-Propenoic acid, 2,3-dibromopropyl ester (9CI)

19666-30-9
$C_{15}H_{18}Cl_2N_2O_3$
345.25
CC(C)Oc1cc(c(Cl)cc1Cl)n2nc(oc2=O)C(C)(C)C
δ^2-1,3,4-Oxadiazolin-5-one, 2-tert-butyl-4-(2,4-dichloro-5-isopropyloxyphenyl)
2-tert-Butyl-4-(2,4-dichloro-5-isopropyloxyphenyl)-1,3,4-oxadiazolin-5-one
3-(2,4-Dichloro-5-isopropyloxyphenyl)-δ^4-5-(tert-butyl)-1,3,4-oxadiazoline-2-one
Oxadiazon
1,3,4-Oxazol-2(3H)-one, 3-(2,4-dichloro-5-(1-methylethoxy)phenyl)-5-(1,1-dimethylethyl)-
Ronstar
RP 17623

19679-38-0
$C_{16}H_{15}Cl_3S_2$
377.78
Ethane, 2,2-bis(p-(methylthio)-phenyl)-1,1,1-trichloro
Benzene, 1,1'-(2,2,2-trichloroethylidene)bis(4-(methylthio)-(9CI)
2,2-Bis(p-(methylthio)phenyl)-1,1,1-trichloroethane
Methiochlor

19686-73-8
C_3H_7BrO
139.01
OC(CBr)C
2-Propanol, 1-bromo
1-Bromo-2-propanol
2-Hydroxypropyl bromide

19693-75-5
$C_4H_8O_3$
104.11
1,3-Dioxolane, 2-methoxy-(9CI)
Orthoformic acid, cyclic ethylene methyl ester (8CI)

19700-21-1
$C_{12}H_{22}O$
182.34
CC1CCCC2(C)CCCCC12O
4a(2H)-Naphthalenol, octahydro-4,8a-dimethyl-, (4S-(4-α,4a-α,8a-β))-
Geosmin; (-)-geosmin; 4a-α-(2H)-naphthol, octahydro-4-α,8a-β-dimethyl- (8CI)

19704-83-7
$C_{18}H_{32}O_2.1/2Ca$
300.49
9,12-Octadecadienoic acid (Z,Z)-, calcium salt (9CI)

19710-01-1
$C_{19}H_{42}N.BH_4$
299.47
Ammonium, hexadecyltrimethyl-, borohydride
Cetyltrimethylammonium borohydride
Hexadecyltrimethyl-ammonium borohydride

19727-17-4
Unknown
Unknown
N-Dodecylguanidine terephthalate

19730-04-2
$C_{11}H_{16}N_2$
176.25
Piperidine, 1-methyl-2-(3-pyridyl)- (8CI)
NSC-127744

19745-44-9
$C_{20}H_{18}N_6O_2S$
406.43
O=N(=O)C(SC(=N1)N=Nc(ccc(N(CCc(cccc2)c2)CCC#N)c3)c3)=C1
Propanenitrile, 3-(4-((5-nitro-2-thiazolyl)azo)(2-phenylethyl)amino)- (9CI)

19750-95-9

$C_{10}H_{13}ClN_2.ClH$
233.16
[Cl-].CN(C)C=Nc1ccc(Cl)cc1C
Formamidine, N'-(4-chloro-o-tolyl)-N,N-dimethyl-, hydrochloride
Chlordimeform hydrochloride
Chlorophenamidine hydrochloride
N'-(4-Chloro-o-tolyl)-N,N-dimethylformamidine, hydrochloride
N,N-Dimethyl-N'-(2-methyl-4-chlorophenyl)-formamidine, hydrochloride
ENT 27,567
EP 333
Fundal SP
Galecron SP
Methanimidamide, N'-(4-chloro-2-methylphenyl)-, monohydrochloride
Morton EP 333
Nor-Am EP 333
NSC-195102
Schering 36268

19752-55-7
$C_6H_3BrCl_2$
225.90
c(cc(cc1Cl)Br)(c1)Cl
Benzene, 1-bromo-3,5-dichloro- (9CI)
1-Bromo-3,5-dichlorobenzene

19757-13-2
$C_{12}H_{10}N_4O_6S$
338.28
Sulfamide, N,N'-bis(4-nitrophenyl)-
AI3-61640

19766-89-3
$C_8H_{16}O_2.Na$
167.20
Hexanoic acid, 2-ethyl-, sodium salt (9CI)
Sodium 2-ethylhexanoate

19780-10-0
$C_{12}H_{24}O$
184.32
Butyl heptyl ketone

19780-11-1
$C_{16}H_{26}O_3$
266.38
2,5-Furandione, 3-(2-dodecenyl)dihydro- (9CI)
3-(2-Dodecenyl)dihydro-2,5-furandione

19780-40-6
$C_9H_{20}O$
144.26
2-Heptanol, 5-ethyl- (8CI, 9CI)

19781-72-7
$C_{21}H_{42}O$
310.56
11-Heneicosanone (8CI,9CI)
NSC-93984

19781-73-8
$C_{23}H_{46}$
322.62
Cyclohexane, heptadecyl- (9CI)
Heptadecane, 1-cyclohexyl- (8CI)

19783-14-3
H_2O_2Pb
241.21
Lead hydroxide (8CI,9CI)

19800-42-1
$C_{19}H_{15}N_5O_4$
377.33
O=N(=O)c(ccc(N=Nc(ccc(N=Nc(ccc(O)c1)c1)c2OC)c2)c3)c3
Phenol, 4-((2-methoxy-4-((4-nitrophenyl)azo)phenyl)azo)- (9CI)
4-((2-Methoxy-4-((4-nitrophenyl)azo)phenyl)azo)phenol
4-((4-((p-Nitrophenyl)azo)-2-methoxyphenyl)azo)phenol

19833-78-4
$C_4H_{11}N.HI$
201.04
Ethanamine, N-ethyl-, hydriodide (9CI)
Diethylamine, hydriodide (8CI)

19837-00-4
$C_{15}H_{27}N_3O_3$
297.40
1,3,5-Triazine, 2,4,6-tributoxy- (9CI)
s-Triazine, 2,4,6-tributoxy- (8CI)
2,4,6-Tributoxytriazine
2,4,6-Tributyloxy-s-triazine

19855-61-9
$C_{20}H_{43}N.C_2H_4O_2$
357.61
1-Octadecanamine, N,N-dimethyl-, acetate (9CI)
N,N-Dimethyl-1-octadecanamine acetate

19860-71-0
$C_9H_{16}O$
140.23
2-Octen-4-one, 2-methyl- (8CI, 9CI)

19889-37-3
$C_7H_{14}O_2$
130.19
Butanoic acid, 2-ethyl-2-methyl- (9CI)
2-Ethyl-2-methylbutanoic acid

19895-44-4
$C_{10}H_{14}N_2O$
178.22
Urea, N-(1-methylethyl)-N'-phenyl- (9CI)
Urea, 1-isopropyl-3-phenyl- (8CI)

19900-46-0
$C_7H_{10}O_3$
142.16
2-Propenoic acid, (2-methyloxiranyl)methyl ester (9CI)
Acrylic acid, 2,3-epoxy-2-methylpropyl ester (6CI, 7CI,8CI)
2-Methyl-2,3-epoxypropyl acrylate
2-Methylglycidyl acrylate
β-Methylglycidyl acrylate

19910-65-7
$C_{10}H_{18}O_6$
234.28
O=C(OC(CC)C)OOC(=O)OC(CC)C
Peroxydicarbonic acid, di-sec-butyl ester
sec-Butyl peroxydicarbonate
Di-sec-butyl peroxydicarbonate
Di-sec-butyl peroxydicarbonate, Not more than 52% in solution (DOT)
Di-sec-butyl peroxydicarbonate, Technically pure (DOT)
UN 2150 (DOT)
UN 2151 (DOT)

19937-59-8
$C_{10}H_{13}ClN_2O_2$
228.70
COc1ccc(NC(=O)N(C)C)cc1Cl
Urea, N'-(3-chloro-4-methoxyphenyl)-N,N-dimethyl
N'-(3-Chlor-4-methoxy-phenyl)-N,N-dimethylharnstoff (German)
3-(3-Chlor-4-methoxyphenyl)-1,1-dimethylharnstoff (German)
N-(3-Chloro-4-methoxyphenyl)-N',N'-dimethylurea
Deftor
N,N-Dimethyl-N'-(4-methoxy-3-chlorophenyl)urea
Dosaflo
Dosagran
Dosanex
Dosanex FL
Dosanex MG

FL
Herbicide 6602
Metoxuron
Purivel
San 6915H
San 7102H
Sulerex
Urea, 3-(3-chloro-4-methoxy-phenyl)-1,1-dimethyl-

19947-22-9
$C_{11}H_{14}$
146.23
Benzene, (1-ethyl-2-propenyl)- (9CI)
Benzene, (1-ethylallyl)-
1-Pentene, 3-phenyl- (7CI, 8CI)
3-Phenyl-1-pentene

19952-47-7
$C_7H_5ClN_2S$
184.65
N=C(Nc(c1ccc2)c2Cl)S1
Benzothiazole, 2-amino-4-chloro
2-Amino-4-chlorobenzothiazole

19962-04-0
$C_9H_{12}N_2O_2$
180.19
Carbamic acid, dimethyl-, 3-aminophenyl ester (9CI)
Carbamic acid, dimethyl-, m-aminophenyl ester (8CI)

19962-25-5
$C_{13}H_{11}CLO$
183.23
p-Chlorobenzyl phenyl ether
Ether, p-chlorobenzyl phenyl

19981-17-0
CH_2N_2Na
65.03
Cyanamide, sodium salt (8CI, 9CI)

20007-87-8
$C_{17}H_{14}N_2O_3$
294.30
Cyclopenin
(-)-Cyclopenin
Cyclopenine (8CI)
(-)-Cyclopenine
4-Methyl-3'-phenylspiro(3H-1,4-benzodiazepine-3,2'-oxirane)-2,5(1H,4H)-dione
NSC-114538
Spiro(3H-1,4-benzodiazepine-3,2'-oxirane)-2,5(1H,4H)-dione, 4-methyl-3'-phenyl-, cis-(-)- (9CI)

20017-67-8
$C_{15}H_{16}O$
212.29
Benzenepropanol, γ-phenyl- (9CI)
NSC-74499
1-Propanol, 3,3-diphenyl- (8CI)

20018-09-1
$C_8H_8I_2O_2S$
422.03
O=S(=O)(c(ccc(c1)C)c1)C(I)I
Diiodomethyl p-tolyl sulfone
Amical 48
Benzene, 1-((diiodomethyl)-sulfonyl)-4-methyl- (9CI)
Caswell No. 353B
1-((Diiodomethyl)sulfonyl)-4-methylbenzene
EPA Pesticide Chemical Code 101002

20019-64-1
$C_6H_8O_2$
112.13
2(5H)-Furanone, 5,5-dimethyl- (8CI,9CI)
2-Pentenoic acid, 4-hydroxy-4-methyl-, lactone

20020-02-4
$C_{10}H_4Cl_4$
265.95
Clc2c(Cl)c(Cl)c1ccccc1c2Cl
1,2,3,4-Tetrachloronaphthalene
AI3-26504
Naphthalene, 1,2,3,4-tetrachloro-
Napthalene, 1,2,3,4-tetrachloro-
NSC-524443

20030-30-2
$C_9H_{14}O$
138.21
O=C(C(=CCC1C)C)C1C
2-Cyclohexen-1-one, 2,5,6-trimethyl- (9CI)
2,5,6-Trimethyl-2-cyclohexen-1-one

20039-91-2
$C_7H_6BrNO_3$
232.03
Phenol, 2-bromo-4-methyl-6-nitro- (9CI)
p-Cresol, 2-bromo-6-nitro- (8CI)

20053-88-7
$C_{10}H_{16}O$
152.24
1,5,7-Octatrien-3-ol, 3,7-dimethyl-, (R-(E))- (9CI)
Hotrienol
Hotrienol, trans-(-)-
trans-(-)-Hotrienol
1,5,7-Octatrien-3-ol, 3,7-dimethyl-, (E)-(R)-(-)- (8CI)

20056-92-2
$C_{12}H_{24}O$
184.36
7-Dodecen-1-ol, (Z)
Looplure Inhibitor

20068-02-4
C_5H_7N
81.11
N#CC(=CC)C
cis-2-Methyl-2-butenenitrile
Angelic acid nitrile
Angeliconitrile
trans-Angeliconitrile
2-Butenenitrile, 2-methyl-, (Z)- (9CI)
Crotononitrile, 2-methyl-, (Z)-
2-Methyl-cis-2-butenenitrile
(Z)-2-Methyl-2-butenenitrile
cis-2-Methyl-2-butennitrile
cis-2-Methyl-2-butenonitrile

20073-24-9
$C_{14}H_{10}O_5$
258.24
7H-Furo(3,2-g)(1)benzopyran-6-carboxylic acid, 7-oxo-, ethyl ester
3-Carbethoxypsoralen
3-CPs (20073-24-9)
3-Ethoxycarbonylpsoralen
Ethyl 7-oxo-7H-furo(3,2-g)-(1)benzopyran-6-carboxylate
Ethyl 3-psoralencarboxylate
((6-Hydroxy-5-benzofuranyl)-methylene)malonic acid, γ-lactone, ethyl ester
7-Oxo-7H-furo(3,2-g)(1)benzopyran-6-carboxylic acid ethyl ester

20073-50-1
$C_5H_{11}NO_2$
117.14
O(CCN1CCO)C1
3-Oxazolidineethanol (9CI)

20074-52-6
Fe
55.85
Iron, ion (Fe(3+)) (8CI,9CI)

20091-61-6
$C_{19}H_{40}BrNO$
378.45
Cyclohexanaminium, N-heptyl-2-hydroxy-N,N,2-trimethyl-5-(1-methylethyl)-
AZ 2088
Dimethylheptyl(1-hydroxy-p-menth-2-yl)ammonium.Br
Z 2008

20103-09-7
$C_6H_6Cl_2N_2$
177.04
Nc(c(cc(N)c1Cl)Cl)c1
p-Phenylenediamine, 2,5-dichloro
1,4-Diamino-3,6-dichlorobenzene
2,5-Dichlor-1,4-fenylendiamin (Czech)
2,5-Dichloro-p-phenylenediamine

20115-23-5
$C_{11}H_{14}O_2$
178.23
Phenyl valerate

20115-34-8
$C_{12}H_7ClN_2O_5$
294.64
Benzene, 4-chloro-2-nitro-1-(4-nitrophenoxy)- (9CI)
DNCDE
Ether, 4-chloro-2-nitrophenyl p-nitrophenyl (8CI)
KK 60

20116-65-8
$C_{11}H_{12}O_4$
208.21
O=C(OC)c(c(ccc1C)C(=O)OC)c1
Phthalic acid, 4-methyl-, dimethyl ester (8CI)
1,2-Benzenedicarboxylic acid, 4-methyl-, dimethyl ester (9CI)
Dimethyl 4-methyl-1,2-benzenedicarboxylate
NSC-165800

20120-32-5
$C_{12}H_{18}Cl_3N_3O_3$
358.65
1,3,5-Triazine, 1,3,5-tris-(3-chloro-1-oxopropyl)-hexahydro- (9CI)
s-Triazine, 1,3,5-tris-(3-chloropropionyl)hexahydro- (8CI)

20120-33-6
$C_6H_{14}NO_5P$
211.18
O=P(OC)(OC)CCC(=O)NCO
Phosphonic acid, (2-((hydroxymethyl)carbamoyl)ethyl)-, dimethyl ester
N-Methylol dimethylphosphonopropionamide
Phosphonic acid, (3-((hydroxymethyl)amino)-3-oxopropyl)-, dimethyl ester (9CI)
Pyrovatex 3805
Pyrovatex CP
Spolapret OS

20129-49-1
$C_{33}H_{68}$
464.90
Dotriacontane, 3-methyl- (8CI, 9CI)

20139-55-3
$C_{11}H_{12}ClNO_2$
225.69
O=C(Nc(c(cc(c1)Cl)C)c1)CC(=O)C
o-Acetoacetanisidide, 4'-chloro
Acetoacet-4-chloro-2-methylanilide
Butanamide, N-(4-chloro-2-methylphenyl)-3-oxo-
Butyranilide, 4'-chloro-2'-methyl-3-oxo-

20139-66-6
$C_{36}H_{34}Cl_2N_6O_6$
717.57
O=C(Nc(ccc(OCC)c1)c1)C(N=Nc(ccc(c2Cl)c(c(cc(N=NC(C(=O)C)C(=O)Nc(ccc(OCC)c3)c3)c4)Cl)c4)c2)C(=O)C
p-Acetoacetophenetidide, 2,2''-((2,2'-dichloro-4,4'-biphenylylene)bis(azo))-bis- (8CI)

20178-34-1
$C_{11}H_{24}O_4$
220.31
2-Propanol, 1-(2-(2-ethoxy-1-methylethoxy)-1-methylethoxy)- (8CI,9CI)

20189-42-8
$C_7H_9NO_2$
139.16
1H-Pyrrole-2,5-dione, 3-ethyl-4-methyl- (9CI)
Ethylmethylmaleimide
2-Ethyl-3-methylmaleimide
Maleimide, 2-ethyl-3-methyl- (6CI,7CI,8CI)
Methylethylmaleimide

20193-23-1
$C_9H_{21}N$
143.27
1-Hexanamine, N-propyl- (9CI)
Hexylamine, N-propyl- (8CI)

20195-08-8
$C_9H_{21}O_4P$
224.24
Phosphoric acid, diethyl pentyl ester (8CI,9CI)

20200-86-6
$C_{11}H_{13}NO$
175.22
2H-Indol-2-one, 1,3-dihydro-1,3,3-trimethyl- (9CI)
2-Indolinone, 1,3,3-trimethyl- (8CI)

20217-01-0
$C_9H_8Br_2O_2$
307.97
O(C1COc(c(cc(c2)Br)Br)c2)C1
Oxirane, ((2,4-dibromophenoxy)methyl)- (9CI)
((2,4-Dibromophenoxy)methyl)-oxirane
2,4-Dibromophenyl glycidyl ether

20222-29-1
$C_{25}H_{20}N_2O_2$
380.45
Benzenamine, 4-nitro-N-(triphenylmethyl)-(9CI)
Aniline, p-nitro-N-trityl-(7CI,8CI)
Aniline, 4-nitro-N-trityl-
N-Trityl-p-nitroaniline

20223-84-1
$C_{16}H_{25}HgNO_6S.2H$
562.09
Mercury, (3-(3-carboxy-2,2,3-trimethylcyclopentanecarboxamido)-2-methoxypropyl) (hydrogen mercaptoacetato)
Mercaptomerin
Thiomerin

20225-24-5
$C_7H_{14}O_2$
130.19
CCCC(CC)C(O)=O
Pentanoic acid, 2-ethyl- (9CI)
Valeric acid, 2-ethyl- (8CI)

20227-53-6
$C_{53}H_{77}O_4P$
809.17
Phosphorous acid, 2-(1,1-dimethylethyl)-4-(1-(3-(1,1-dimethylethyl)-4-hydroxyphenyl)-1-methylethyl)phenyl bis(4-nonylphenyl) ester (9CI)

20227-92-3
$C_7H_{15}NO_3$
161.20
Carbonic acid, Compd. with cyclohexanamine (9CI)
Carbonic acid, Compd. with cyclohexylamine (8CI)
KTsA
VPI 300

20237-34-7

C_6H_{10}
82.15
1,3-Hexadiene, (E)- (8CI,9CI)

20241-68-3
$C_{15}H_{16}N_2O_4S$
320.36
O=C(Nc(cccc1S(=O)(=O)CCO)c1)c(ccc(N)c2)c2
Benzamide, 4-amino-N-(3-((2-hydroxyethyl)sulfonyl)-phenyl)- (9CI)
4-Amino-N-(3-(2-hydroxy-ethyl)sulfonylphenyl)benz-amide
4-Amino-N-(3-((2-hydroxy-ethyl)sulfonyl)phenyl)benz-amide

20241-74-1
$C_{34}H_{34}N_2O_2$
502.64
O=C(c(c(C(=O)c1c(Nc(c(ccc2)CC)c2CC)ccc3Nc(c(ccc4)CC)c4CC)ccc5)c5)c13
9,10-Anthracenedione, 1,4-bis-((2,6-diethylphenyl)amino)-(9CI)

20241-76-3
$C_{20}H_{12}N_2O_6$
376.31
O=C(c(c(c(O)cc1)C(=O)c2c(O)ccc3N(=O)=O)c1Nc(cccc4)c4)c23
9,10-Anthracenedione, 1,8-di-hydroxy-4-nitro-5-(phenyl-amino)- (9CI)
4-Anilino-5-nitrochrysazin
Anthraquinone, 1-anilino-4,5-di-hydroxy-8-nitro-

20241-77-4
$C_{20}H_{14}N_2O_4$
346.33
O=C(c(c(c(O)cc1)C(=O)c2c(O)ccc3N)c1Nc(cccc4)c4)c23
9,10-Anthracenedione, 1-amino-4,5-dihydroxy-8-(phenylamino)- (9CI)
4-Amino-5-anilino-1,8-di-hydroxyanthraquinone
Anthraquinone, 1-amino-4,5-di-hydroxy-8-(phenylamino)-

20244-70-6
$C_2H_2I_2$
279.85
Ethene, 1,2-diiodo- (9CI)
1,2-Diiodoethylene
Ethylene, 1,2-diiodo- (6CI, 7CI,8CI)

20252-66-8
$C_{22}H_{40}O_4Si_5$
508.99
Pentasiloxane, 1,1,1,3,5,5,7,9,9,9-decamethyl-3,7-di-phenyl- (8CI,9CI)

20262-58-2
$C_{20}H_{19}N_3O_{11}S_3.2Na$
619.54
2-Naphthalenesulfonic acid, 6-(acetylamino)-4-hydroxy-3-((4-((2-(sulfooxy)ethyl)sulf-onyl)phenyl)azo)-, disodium salt (9CI)

20265-96-7
$C_6H_6ClN.ClH$
164.04
Aniline, p-chloro-, hydro-chloride
Benzenamine, 4-chloro-, hydro-chloride
p-Chloroaniline hydrochloride
4-Chloroaniline hydrochloride
p-Chloroanilinium chloride
p-Chlorophenylamine hydro-chloride

20265-97-8
$C_7H_9NO.ClH$
159.63
Cl.COc1ccc(N)cc1
p-Anisidine, hydrochloride
p-Aminoanisole hydrochloride
4-Aminoanisole hydrochloride
1-Amino-4-methoxybenzene hydrochloride
4-Anisidine hydrochloride
p-Anisidine monohydrochloride
p-Anisylamine hydrochloride
Benzenamine, 4-methoxy-, hydrochloride (9CI)
4-Methoxy-1-aminobenzene hydrochloride
p-Methoxyaniline hydrochloride
4-Methoxyaniline hydrochloride
4-Methoxybenzeneamine hydro-chloride
p-Methoxyphenylamine hydro-chloride
NCI-C03758

20268-51-3
$C_{18}H_{11}NO_2$
273.29
7-Nitrobenzanthracene
Benz(a)anthracene, 7-nitro-
7-Nitrobenz(a)anthracene

20273-24-9
$C_5H_{10}O$
86.13
2-Penten-1-ol (8CI,9CI)

20273-27-2
$C_{18}H_{26}$
242.40
Benzene, (1,1'-bicyclohexyl)-4-yl- (9CI)
Bicyclohexyl, 4-phenyl- (8CI)

20282-70-6
$C_{14}H_{12}N_4O_2$
268.25
N#Nc(c(OC)cc(c(ccc(N#N)c1OC)c1)c2)c2
Fast Blue B
(1,1'-Biphenyl)-4,4'-bis(diazon-ium), 3,3'-dimethoxy- (9CI)
3,3'-Dimethoxybiphenyl-4,4'-bisdiazonium
3,3'-Dimethoxy-(1,1'-biphenyl)-4,4'-bis(diazonium)

20291-74-1
$C_{16}H_{14}$
206.29
Phenanthrene, 1,6-dimethyl-(8CI,9CI)

20304-47-6
$C_{29}H_{40}O_9$
532.69
Calactin

20304-49-8
$C_{29}H_{40}O_{10}$
548.69
5-α-Card-20(22)-enolide, 3-β-((6-deoxy-β-d-hexopy-ranos-2-ulos-1-yl)oxy)-2-α,14-dihydroxy-19-oxo
Calotoxin

20306-75-6
$C_5H_9NO_2$
115.13
O=C(NC)CC(=O)C
Butanamide, N-methyl-3-oxo-(9CI)
Acetoacet-monomethylamide
N-Methyl-3-oxobutanamide
N-Monomethylacetoacetamide

20317-19-5
$C_{20}H_{22}N_4O_{12}S_3.2K$
684.78
Benzenesulfonic acid, 4-(4-((2,5-dimethoxy-4-((2-(sulfo-oxy)ethyl)sulfonyl)phenyl)-azo)-4,5-dihydro-3-methyl-5-oxo-1H-pyrazol-1-yl)-, di-potassium salt (9CI)
Benzenesulfonic acid, p-(4-((4-((2-hydroxyethyl)sulfonyl)-2,5-dimethoxyphenyl)azo)-3-methyl-5-oxo-2-pyrazolin-1-yl)-, hydrogen sulfate, di-potassium salt
C.I. Reactive Yellow 17
C.I. Reactive Yellow 17, Dipotassium salt (8CI)
Diamira Golden Yellow G
Remazol Golden Yellow G

Remazol Golden Yellow GGL

20324-32-7
$C_7H_{16}O_3$
148.20
O(CC(OCC(O)C)C)C
2-Propanol, 1-(2-methoxy-1-methylethoxy)- (9CI)
1-(2-Methoxy-1-methylethoxy)-2-propanol

20324-33-8
$C_{10}H_{22}O_4$
206.32
O(CC(OCC(OCC(O)C)C)C)C
2-Propanol, 1-(2-(2-methoxy-1-methylethoxy)-1-methylethoxy)
Dowanol 62B
Dowanol TPM glycol ether
Tripropylene glycol methyl ether
Tripropylenglykolmonomethylether (Czech)

20324-34-9
$C_{13}H_{28}O_5$
264.36
2,5,8,11-Tetraoxatetradecan-13-ol, 4,7,10-trimethyl- (9CI)
Tetrapropylene glycol, monomethyl ether (8CI) (VAN)

20324-87-2
$C_{21}H_{16}N_2O_9S_2.2Na$
550.47
2-Naphthalenesulfonic acid, 7,7'-(carbonyldiimino)bis-(4-hydroxy-, disodium salt (9CI)

20325-40-0
$C_{14}H_{16}N_2O_2.2ClH$
317.24
Benzidine, 3,3'-dimethoxy-, dihydrochloride
(1,1'-Biphenyl)-4,4'-diamine, 3,3-dimethoxy-, dihydrochloride (9CI)
o-Dianisidine dihydrochloride
C.I. Disperse Black 6 Dihydrochloride
3,3'-Dimethoxybenzidine dihydrochloride

20328-87-4
$C_{21}H_{28}N_4O_2S_2.2I$
686.39
Thiazolium, 2,2'-((2-carboxy-1,4-phenylene)bis(imino-2,1-ethenediyl))bis(3-ethyl-4,5-dihydro-, diiodide (9CI)

20338-26-5
$C_4H_2Cl_8$
333.68
Butane, 1,1,2,2,3,3,4,4-octachloro- (6CI,7CI,8CI)

20350-15-6
$C_{16}H_{24}O_4$
280.40
4H-Cyclopent(f)oxacyclotridecin-4-one, 1,6,7,8,9,11a-β,12,13,14,14a-α-decahydro-1- β-13-α-dihydroxy-6-β-methyl
Ascotoxin
Brefeldin A
Cyanaein
Decumbin
Nectrolide

20354-26-1
$C_9H_6Cl_2N_2O_3$
261.07
Cn2c(=O)on(c1ccc(Cl)c(Cl)c1)c2=O
1,2,4-Oxadiazolidine-3,5-dione, 2-(3,4-dichlorophenyl)-4-methyl
Bioxone
2-(3,4-Dichlorophenyl)-4-methyl-1,2,4-oxadiazolidine-3,5-dione
Methazole
Mezopur
Oxydiazol
Paxilon
Probe
Tunic
VCS 438

20376-31-2
$C_5H_2Cl_6N_4$
330.79
1,3,5-Triazin-2-amine, 4,6-bis-(trichloromethyl)- (9CI)
s-Triazine, 2-amino-4,6-bis-(trichloromethyl)- (8CI)

20376-34-5
$C_9H_5Cl_3N_4$
275.53
1,3,5-Triazin-2-amine, 4,6-dichloro-N-(3-chlorophenyl)- (9CI)
s-Triazine, 2,4-dichloro-6-(m-chloroanilino)- (8CI)

20376-36-7
$C_9H_5BrCl_2N_4$
319.98
1,3,5-Triazin-2-amine, N-(4-bromophenyl)-4,6-dichloro- (9CI)
s-Triazine, 2-(p-bromoanilino)-4,6-dichloro- (6CI, 8CI)

20380-58-9
Unknown
Unknown
Tilidine
DEA No. 9750
Tilidina
Tilidino (Spanish)
Tilidinum (Latin)

20395-24-8
$C_8H_{16}Cl_2$
183.12
Octane, 1,1-dichloro- (8CI, 9CI)
1,1-Dichlorooctane

20403-41-2
$C_{14}H_{28}O_2.xPb$
Unknown
Lead myristate
Lead tetradecanoate
Tetradecanoic acid, lead salt (9CI)

20404-02-8
$C_6H_2Cl_3NO_3$
242.44
Phenol, 2,3,6-trichloro-4-nitro- (8CI,9CI)

20405-19-0
$C_9H_9ClO_3.C_4H_{11}NO_2$
305.75
Diethanolamine 2-methyl-4-chlorophenoxyacetate
Acetic acid, ((4-chloro-o-tolyl)oxy)-, Compd. with 2,2'-iminodiethanol (1:1)
Caswell No. 557E
EPA Pesticide Chemical Code 030511
MCPA diethanolamine salt
2-Methyl-4-chlorophenoxy-acetic acid diethanolamine salt
(2-Methyl-4-chlorophenoxy)-acetic acid, diethanolamine salt

20427-59-2
$H_2O_2.Cu$
97.56
Copper-hydroxide
Comac
Copper dihydroxide
Copper(II) hydroxide
Copper(2+) hydroxide
Criscobre
Cuidrox
Cupravit Blau
Cupravit Blue
Cupric hydroxide
Kocide
Kocide 101
Kocide 220
Kocide 404
Kocide SD
Kuprablau

Parasol

20427-84-3
$C_{19}H_{32}O_3$
308.46
O(CCOc(ccc(c1)CCCCCCCCC)
c1)CCO
**Ethanol, 2-(2-(4-nonylphen-
oxy)ethoxy)- (9CI)**
2-(2-(4-Nonylphenoxy)ethoxy)-
ethanol

20442-06-2
$C_{10}H_{20}O_3$
188.27
O=C(OCCOCCCC)CCC
**Butanoic acid, 2-butoxyethyl
ester (9CI)**
2-Butoxyethyl butanoate
Butoxyethyl isobutyrate

20455-68-9
$C_{14}H_{15}N.ClH$
233.76
Dibenzylamine, hydrochloride
Benzenemethanamine, N-
(phenylmethyl)-, hydro-
chloride (9CI)
Dibenzylammonium chloride

20461-54-5
I
126.90
Iodide (8CI,9CI)
Hydriodic acid, ion(1-)
Iodine, ion (I(1-))
Polyiodide

20461-60-3
$C_8H_{12}N_2O_2$
168.18
**Pyrimidine, 2,4-diethoxy-
(8CI,9CI)**
2,4-Diethoxypyrimidine

20469-71-0
$CH_4N_2S_2.N_2H_4$
140.25
**Carbazic acid, dithio-, hydra-
zine (Salt)**
Diammonium dithiocarbazate
Dithiocarbazic acid hydrazine
(Salt)
Hydrazinecarbodithioic acid,
Compd. with hydrazine (1:1)

20479-71-4
C_7H_{12}
96.17
**Cyclopropane, (1-methyl-
1-propenyl)-, (E)- (9CI)**
2-Butene, 2-cyclopropyl-,
(E)- (8CI)

20479-72-5
C_7H_{12}
96.17
**Cyclopropane, (1-methyl
-1-propenyl)-, (Z)- (9CI)**
2-Butene, 2-cyclopropyl-,
(Z)- (8CI)

20521-42-0
C_6H_8O
96.13
2-Butenal, 2-ethenyl- (9CI)
Crotonaldehyde, 2-vinyl- (8CI)

20547-99-3
$C_9H_{14}O_2$
154.21
**1,4-Cyclohexanedione,
2,2,6-trimethyl- (6CI,
7CI,8CI,9CI)**
Dihydrooxophorone
2,6,6-Trimethyl-1,4-cyclo-
hexanedione
3,5,5-Trimethyl-1,4-cyclo-
hexanedione

20548-54-3
CaS
72.14
Calcium sulfide (9CI)
Calcium monosulfide

20562-02-1
$C_{45}H_{73}NO_{15}$
868.19
CC9CCC8C(C)C3C(CC4C2C
C=C1CC(CCC1(C)C2CCC34
C)OC6OC(CO)C(O)C(OC5O
C(CO)C(O)C(O)C5O)C6O
C7OC(C)C(O)C(O)C7O)N8C9
**Solanid-5-ene, 3-β-((O-6-de-
oxy-α-l-mannopyranosyl-
(1-2)-O-(β-D-glucopyranosyl-
(1-3))-β-d-galactopyranosyl)-
oxy)**
Solanine
α-Solanin
α-Solanine

20589-63-3
$C_{20}H_{11}NO_2$
297.31
3-Nitroperylene
Perylene, 3-nitro-

20589-85-9
$C_3H_2Cl_4$
179.86
1,2,3,3-Tetrachloro-1-propene

20591-89-3
$C_{12}H_{22}O_5$
246.31
**Propanedioic acid, ethyl
(2-methoxyethyl)-, di-
ethyl ester (9CI)**
Malonic acid, ethyl(2-meth-
oxyethyl)-, diethyl ester
(8CI)

20591-90-6
$C_{13}H_{24}O_5$
260.33
**Propanedioic acid,
(2-methoxyethyl)propyl-,
diethyl ester (9CI)**
Malonic acid, (2-methoxy-
ethyl)propyl-, diethyl ester
(7CI,8CI)

20591-91-7
$C_{14}H_{26}O_5$
274.36
**Propanedioic acid, butyl-
(2-methoxyethyl)-, di-
ethyl ester (9CI)**
Malonic acid, butyl(2-meth-
oxyethyl)-, diethyl ester
(8CI)

20592-85-2
$C_3H_{13}NNaO_9P_3$
323.05
**Phosphonic acid, (nitrilotris-
(methylene))tris-, sodium
salt (9CI)**
Aminotri(methylene phosphonic
acid), sodium salt
Aminotris(methylphosphonic
acid), sodium salts
Nitrilotris(methylene phosphonic acid), sodium salt

20600-96-8
$C_6H_{10}N_4O_{13}$
346.14
Tetranitro diglycerin
Diglycerol tetranitrate
1,2-Propanediol, 3,3'-oxydi-,
tetranitrate
UN 1510 [Tetranitromethane]

20624-25-3
$C_5H_{10}NS_2.Na.3H_2O$
225.33
O.O.O.[Na+].CC.CC#N.S=C=S
**Carbamic acid, diethyldithio-,
sodium salt, trihydrate**
Diethyldithiocarbamate sodium
trihydrate
Diethyldithiocarbamic acid
sodium salt trihydrate
Diethyldithiokarbaman sodny
trihydrat (Czech)
Dithiocarb
Sodium diethyldithiocarbamate
trihydrate

20624-96-8
$C_9H_8Cl_2O_5$
267.06

Benzoic acid, 2,6-dichloro-4-hydroxy-3,5-dimethoxy- (8CI,9CI)

20627-31-0
$C_{20}H_{16}$
256.36
Benz(a)anthracene, 8,12-dimethyl
5,9-Dimethyl-1,2-benzanthracene
8,12-Dimethylbenz(a)anthracene

20627-32-1
$C_{21}H_{18}$
270.39
Benz(a)anthracene, 6,7,8-trimethyl
6,7,8-Trimethylbenz(a)anthracene

20627-33-2
$C_{21}H_{18}$
270.39
Benz(a)anthracene, 6,7,12-trimethyl
4,9,10-Trimethyl-1,2-benzanthracene
6,7,12-Trimethylbenz(a)anthracene

20627-34-3
$C_{21}H_{18}$
270.39
Benz(a)anthracene, 6,8,12-trimethyl
6,8,12-Trimethylbenz(a)anthracene

20628-36-8
$C_8H_{14}O_2$
142.20
2H-Pyran-2-one, tetrahydro-4,6,6-trimethyl- (8CI,9CI)
Hexanoic acid, 5-hydroxy-3,5-dimethyl-, δ-lactone
NSC-134776

20632-35-3
$C_{13}H_6Cl_6N_2O$
418.91
Carbanilide, 2,2',4,4',6,6'-hexachloro
2,2',4,4',6,6'-Hexachloro-N,N'-diphenylurea

20633-11-8
$C_6H_{13}NO_3$
147.17
O=N(=O)OCCCCCC
Nitric acid, hexyl ester (9CI)
Hexyl nitrate

20633-12-9
$C_7H_{15}NO_3$
161.19
Nitric acid, heptyl ester (8CI, 9CI)

20642-93-7
$C_{10}H_{13}NO_2$
179.21
Carbamic acid, (2,6-dimethylphenyl)-, methyl ester (9CI)
Carbanilic acid, 2,6-dimethyl-, methyl ester (8CI)

20651-71-2
$C_{11}H_{14}O_2$
178.23
O=C(O)c(ccc(c1)CCCC)c1
Benzoic acid, 4-butyl- (9CI)
p-Butylbenzoic acid
4-Butylbenzoic acid

20662-83-3
C_5H_7NO
97.11
Oxazole, 4,5-dimethyl- (8CI, 9CI)

20662-84-4
C_6H_9NO
111.14
Oxazole, trimethyl- (8CI,9CI)
Oxazole, 2,4,5-trimethyl-

20668-13-7
$C_8H_8ClNO_2$
185.61
Carbanilic acid, o-chloro-, methyl ester
Carbamic acid, (o-chlorophenyl)-, methyl ester
Carbamic acid, (2-chlorophenyl)-, methyl ester

20679-58-7
$C_8H_{10}Br_2O_4$
329.97
1,4-Bis(bromoacetoxy)-2-butene
Acetic acid, bromo-, 2-butene-1,4-diyl ester (9CI)
Acetic acid, bromo-, 2-butenylene ester (8CI)
2-Butene-1,4-diol bis(bromoacetate)
Caswell No. 088A
NSC-23989
Slimacide V 10

20680-10-8
$C_7H_{16}O_2$
132.20
O(C(OCCC)C)CC
Propane, 1-(1-ethoxyethoxy)-
AI3-38086
1-(1-Ethoxyethoxy)propane
1-Ethoxy-1-propoxyethane

20685-45-4
$C_8H_{14}O$
126.20
3-Hexen-2-one, 3,4-dimethyl-, (Z)- (8CI)

20691-84-3
$C_{15}H_{15}N_3O_2$
269.33
CN(C)c2ccc(N=Nc1cccc(c1)C(O)=O)cc2
Benzoic acid, m-((p-(dimethylamino)phenyl)azo)
3'-Carboxy-4-dimethylaminoazobenzene
3-((p-(Dimethylamino)phenyl)azo)benzoic acid

20702-77-6
$C_{28}H_{36}O_{15}$
612.64
Glucopyranoside, 3,5-dihydroxy-4-(3-hydroxy-4-methoxyhydrocinnamoyl)-phenyl- 2-O-(6-deoxy-α-l-mannopyranosyl)-, β-d
NCI-C60764
Neohesperidin dihydrochalcone
Neohesporidin dihydrochalcone

20721-50-0
$C_{16}H_{20}N_4O_2$
300.34
OCCN(c(ccc(N=Nc(ccc(N)c1)c1)c2)c2)CCO
Ethanol, 2,2'-((4-((4-aminophenyl)azo)phenyl)imino)-bis- (9CI)
2,2'-((4-((4-Aminophenyl)azo)-phenyl)imino)bis(ethanol)
2,2'-((p-((p-Aminophenyl)azo)-phenyl)imino)diethanol

20721-76-0
$C_{11}H_{20}O_5$
232.28
Propanedioic acid, (2-methoxyethyl)methyl-, diethyl ester (9CI)
Malonic acid, (2-methoxyethyl)methyl-, diethyl ester (8CI)

20721-77-1
$C_{13}H_{24}O_5$
260.33
Propanedioic acid, (2-methoxyethyl)(1-methylethyl)-, diethyl ester (9CI)
Malonic acid, isopropyl-(2-methoxyethyl)-, diethyl ester (7CI,8CI)

20727-33-7
$C_{20}H_{38}O_7S.Na$
445.57
Butanedioic acid, sulfo-, 1,4-bis(1-methylheptyl) ester, sodium salt (9CI)

20736-64-5
$C_6H_{15}CrNO_4$
217.19
Chromic acid, Compd. with cyclohexanamine (9CI)
KhTsA

20748-72-5
$C_2H_7NO.HNO_3$
124.08
Ethanol, 2-amino-, nitrate (Salt) (9CI)
2-Aminoethanol mononitrate
2-Aminoethanol, nitrate
2-Aminoethanol nitrate (Salt)
Ethanolamine nitrate

20762-60-1
KN_3
81.13
Potassium-azide

20763-88-6
$C_{12}H_{27}Sn$
290.06
Stannyl, tributyl- (8CI,9CI)

20776-81-2
$C_8H_{12}N_6$
192.23
s-Triazine-2-carbonitrile, 4,6-bis(ethylamino)- (7CI,8CI)

20776-86-7
$C_9H_{14}N_6$
206.25
s-Triazine-2-carbonitrile, 4-(ethylamino)-6-(isopropylamino)- (7CI,8CI)

20782-57-4
$C_9H_{11}ClN_2O_2$
214.64
CNC(=O)Nc1ccc(OC)c(Cl)c1
Urea, N-(3-chloro-4-methoxyphenyl)-N'-methyl- (9CI)
Urea, 1-(3-chloro-4-methoxyphenyl)-3-methyl- (8CI)

20809-46-5
$C_9H_{18}O$
142.24
4-Octanone, 7-methyl- (8CI, 9CI)

20810-06-4
$C_8H_{19}N$
129.24
N(CCCC)CC(C)C
1-Butanamine, N-(2-methylpropyl)- (9CI)
N-(2-Methylpropyl)-1-butanamine

20816-12-0
O_4Os
254.20
O=[Os](=O)(=O)=O
Osmium tetroxide (ACGIH, OSHA) [UN 2471]
Osmic acid
RCRA waste number P087
UN 2471 [Osmium tetroxide]

20820-44-4
$C_4H_6N_4O_{11}$
286.09
O=N(=O)OCC(N(=O)=O)(CON(=O)=O)CON(=O)=O
Nitro isobutane triol trinitrate
Nitroisobutane triol trinitrate
Nitroisobutanetriol trinitrate
Nitroisobutyl glyceryl trinitrate
Nitroisobutyl glycol trinitrate
1,3-Propanediol, 2-(hydroxymethyl)-2-nitro-, trinitrate (ester)
1,3-Propanediol, 2-nitro-2-((nitrooxy)methyl)-, dinitrate (ester) (9CI)
Trimethylol nitromethane trinitrate
Trinitrate de nitroisobutane triol (French)
Trinitrato de nitro isobutanotriol (Spanish)

20822-30-4
$C_9H_{21}O_3PS$
240.30
Phosphorothioic acid, O,O-dibutyl S-methyl ester (8CI,9CI)

20824-56-0
$C_{10}H_{16}N_2O_8.2H_3N$
326.29
Glycine, N,N'-1,2-ethanediylbis(N-(carboxymethyl)-, diammonium salt (9CI)

20829-66-7
$C_2H_8N_2.2HNO_3$
186.11
1,2-Ethanediamine, dinitrate (9CI)
Ethylenediamine dinitrate

20830-75-5
$C_{41}H_{64}O_{14}$
781.05
CC1OC(CC(O)C1O)OC2C(O)CC(OC2C)OC8C(O)CC(OC7CCC3(C)C(CCC4C3CC(O)C5(C)C(CCC45O)C6=CC(=O)OC6)C7)OC8C
Digoxin
Chloroformic digitalin
Digacin
Digitalis glycoside
Digoksyna (Polish)
Digoxigenin-tridigitoxosid (German)
Digoxine
Homolle's Digitalin
Lanicor
Lanoxin
Rougoxin
SK-Digoxin

20830-81-3
$C_{27}H_{29}NO_{10}$
527.57
COc4cccc5C(=O)c3c(O)c2CC(O)(CC(OC1CC(N)C(O)C(C)O1)c2c(O)c3C(=O)c45)C(C)=O
Daunomycin
Acetyladriamycin
Cerubidin
Daunamycin
Daunoblastina
Daunorubicin
Daunorubicine
DM
FI6339
Leukaemomycin C
NCI-C04693
NSC-82151
RCRA waste number U059
RP 13057
13.057 R.P.
Rubidomycin
Rubidomycine
Rubomycin C
Rubomycin C 1
Streptomyces Peucetius

20837-86-9
$CH_2N_2.Pb$
249.23
Lead cyanamide
Cyanamide, lead(2+) salt (1:1) (8CI,9CI)

20856-57-9
$C_9H_7Cl_5N_2O$
336.43
Clc1ccc(NC(NC=O)C(Cl)(Cl)Cl)cc1Cl
Formamide, N-(1-(3,4-dichloroanilino)-2,2,2-trichloroethyl)
Bay 79770
Chloraniformethan
Chloraniformethane
1-(3,4-Dichloranilino)-1-formylamino-2,2,2-trichloraethan (German)
N-Formyl-N'-(3',4'-dichlorphenyl)-2,2,2-trichloracetaldehydam (German)

Imugan
Milfaron
N-(2,2,2-Trichloro-1-(3,4-dichloroanilino)ethyl)formamide

20859-73-8
AlP
57.95
Aluminum-phosphide
AIP
Al-Phos
Aluminium Fosfide (Dutch)
Aluminium phosphide
Aluminum monophosphide
Aluminum phosphide [UN 1397]
Celphide
Celphine
Celphos
Delicia
Delicia gastoxin
Detia
Detia-Ex-B
Detia Gas Ex-B
Fosfuri di alluminio (Italian)
Fumitoxin
Phosphures d'alumium (French)
Phostoxin
Quickphos
RCRA waste number P006
UN 1397 [Aluminum phosphide]

20880-93-7
$C_{12}H_{20}O_6$
260.29
Fructopyranose, 1,2:4,5-di O-isopropylidene-, α-D- (8CI)

20881-04-3
$C_{11}H_{18}O_5$
230.26
α-D-Xylofuranose, 1,2:3,5-bis-O-(1-methylethylidene)- (9CI)
Xylofuranose, 1,2:3,5-di-O-isopropylidene-
Xylofuranose, 1,2:3,5-di-O-isopropylidene-, α-D- (8CI)

20893-30-5
C_6H_5NS
123.17
N#CCC(SC=C1)=C1
2-Thiopheneacetonitrile
AI3-08540

20895-41-4
$C_9H_8O_2$
148.16
3(2H)-Benzofuranone, 6-methyl- (7CI,8CI,9CI)
6-Methylbenzo(b)furan-3(2H)-one

20895-45-8
$C_{10}H_{10}O_2$
162.19
3(2H)-Benzofuranone, 4,7-dimethyl- (8CI,9CI)

20925-85-3
C_7Cl_5N
275.34
Pentachlorobenzonitrile
AI3-23239
Benzonitrile, pentachloro-

20928-02-3
$C_{13}H_{10}S$
198.29
Dibenzothiophene, 2-methyl- (8CI,9CI)

20940-37-8
$C_8H_6Cl_2O_3.C_4H_{11}N$
294.17
2,4-Dichlorophenoxyacetic acid, diethanolamine salt solution
Acetic acid, (2,4-dichlorophenoxy)-, diethylamine salt
Caswell No. 315M
DEA
EPA Pesticide Chemical Code 030017
2,4-D diethylamine salt
2,4-Dichlorophenoxyacetic acid, diethylamine salt
Diethylamine 2,4-dichlorophenoxyacetate

20940-42-5
$C_8H_9ClN_2O$
184.61
CNC(=O)Nc1cccc(Cl)c1
Urea, N-(3-chlorophenyl)-N'-methyl- (9CI)
Urea, 1-(m-chlorophenyl)-3-methyl- (8CI)

20941-65-5
$C_20H_40N_4S_8.Te$
720.67
CCN(CC)C(=S)S[Te](SC(=S)N(CC)CC)(SC(=S)N(CC)CC)SC(=S)N(CC)CC
Ethyl tellurac
Carbamodithioic acid, diethyl-, tetrakis(anhydrosulfide) with thiotelluric acid (H_4TeS_4)
Diethyldithio carbamic acid tellurium salt
NCI-C02857
Tellurac
Tellurium bis(diethyldithiocarbamate)
Tellurium(IV) diethyl dithiocarbamate
Tellurium diethyldithiocarbamate
Tellurium, tetrakis(diethylcarbamodithioato-S,S')-
Tellurium, tetrakis(diethylcarbamodithioato-S,S')-, (DD-8-111''1''1'1'1'''1''')-
Tellurium, tetrakis(diethyldithiocarbamato)- (8CI)
Tetrakis(diethylcarbamodithioato-S,S')tellurium
Tetrakis(diethyldithiocarbamato)tellurium

20944-88-1
$C_{10}H_{12}O$
148.20
Phenol, 2-(2-methyl-2-propenyl)- (9CI)
Phenol, o-(2-methylallyl)- (8CI)

20959-33-5
$C_{18}H_{38}$
254.50
Heptadecane, 7-methyl- (8CI, 9CI)

20964-55-0
$C_9H_8N_4O_2$
204.19
1,3,5-Triazine-2,4(1H,3H)-dione, 6-(phenylamino)- (9CI)
s-Triazine-2,4-diol, 6-anilino- (6CI,7CI,8CI)

21020-24-6
C_4H_5Cl
88.54
C(#C)C(Cl)C
1-Butyne, 3-chloro- (9CI)
3-Chloro-1-butyne

21078-95-5
$C_{16}H_{26}O$
234.38
Benzenemethanol, α-nonyl- (9CI)
1-Decanol, 1-phenyl- (8CI)
α-Nonylbenzenemethanol
α-Nonylbenzyl alcohol
1-Phenyl-1-decanol

21085-19-8
$C_{10}H_{16}N_4OS$
240.31
CSc2nnc(C1CCCCC1)c(=O)n2N
1,2,4-Triazin-5(4H)-one, 4-amino-6-cyclohexyl-3-(methylthio)- (9CI)
BAY 86791
as-Triazin-5(4H)-one, 4-amino-6-cyclohexyl-3-(methylthio)- (8CI)

21087-61-6
$C_7H_{12}N_4OS$
200.24
CSc1nnc(C(C)C)c(=O)n1N

1,2,4-Triazin-5(4H)-one, 4-amino-6-(1-methylethyl)-3-(methylthio)-
AI3-52881

21087-63-8
$C_{10}H_{10}N_4OS$
234.28
CSC2N=NC(c1ccccc1)C(=O)N2N
Aglypt
4-Amino-3-methylthio-6-phenyl-1,2,4-triazine-5-one
Bay 79758

21087-64-9
$C_8H_{14}N_4OS$
214.32
CSc1nnc(c(=O)n1N)C(C)(C)C
as-Triazin-5(4H)-one, 4-amino-6-tert-butyl-3-(methylthio)
4-Amino-6-tert-butyl-3-methylthio-as-triazin-5-one
4-Amino-6-tert-butyl-3-(methylthio)-1,2,4-triazin-5-one
4-Amino-6-(1,1-dimethylethyl)-3-(methylthio)-1,2,4-triazin-5(4H)-one
Bay 61597
Bay DIC
Bayer 94337
Bayer 6159H
Bayer 6443H
DIC 1468
Lexone
Metribuzin (ACGIH,OSHA)
Sencor
Sencoral
Sencorer
Sencorex
1,2,4-Triazin-5-one, 4-amino-6-tert-butyl-3-(methylthio)-
1,2,4-Triazin-5(4H)-one, 4-amino-6-(1,19dimethylethyl)-3-(methylthio)-
Zenkor

21089-06-5
$C_2H_8O_7P_2.2K$
284.23
Phosphonic acid, (1-hydroxyethylidene)bis-, dipotassium salt (9CI)
1-Hydroxyethylidene-1,1'-diphosphonate, dipotassium salt
Phosphonic acid, (1-hydroxyethylidene)bis-, dipotassium salt

21113-55-3
$C_{18}H_{20}$
236.36
Benzene, 1,1'-cyclohexylidenebis- (9CI)
Cyclohexane, 1,1-diphenyl- (7CI,8CI)

21117-52-2
$C_5H_6BrN_3O_2$
220.01
1H-Imidazole, 5-bromo-1,2-dimethyl-4-nitro- (9CI)
Imidazole, 5-bromo-1,2-dimethyl-4-nitro- (8CI)
NSC-226232

21145-77-7
$C_{18}H_{26}O$
258.40
O=C(c(c(cc(c1C(CC2C)(C)C)C2(C)C)C)c1)C
Tonalid
AI3-28517
Ethanone, 1-(5,6,7,8-tetrahydro-3,5,5,6,8,8-hexamethyl-2-naphthalenyl)-

21150-01-6
$C_7H_{13}N_3$
139.18
1H-Imidazole-4-ethanamine, β,β-dimethyl- (9CI)
Imidazole, 4-(2-amino-1,1-dimethylethyl)- (8CI)

21150-21-0
$C_{39}H_{53}N_9O_{14}S$
904.07
9,18-(Iminoethaniminoethaniminoethaniminoethaniminomethano)-pyrrolo(1',2':8,9)(1,5,8,11,14)thiatetraaza cyclooctadecino(18,17-b)indole-6-acetic acid, 1,2,3,5,6,7,8,9,10,12,17,18,19,20,21,22,23,23a-octadecahydro-29-sec-butyl-21-(2,3-dihydroxy-1-methylpropyl)-2-hydroxy-5,8,20,23,24,27,30,33-octaoxo-, 11-oxide-5,8,20,23,24,27, 30,33-octaoxo-, 11-oxide
Amanin
Amanine
α-Amanitin, 1-l-aspartic acid-4-(2-mercapto-l-tryptophan)-

21150-22-1
$C_{39}H_{53}N_9O_{15}S$
920.07
9,18-(Iminoethaniminoethaniminoethaniminoethaniminomethano)-pyrrolo(1',2':8,9)(1,5,8,11,14)thiatetraaza cyclooctadecino(18,17-b)-indole-6-acetic acid, 1,2,3,5,6,7,8,9,10,12,17,18,19,20,21,22,23, 23a-octadecahydro-29-sec-butyl-2,14-dihydroxy-21-(2,3-dihydroxy-1-methylpropyl)-5,8,20,23,24,27,30,33-Octaoxo-, 11-oxide-5,8,20,23, 24,27,30,33-octaoxo-, 11-oxide
α-Amanitin, 1-l-aspartic acid-(9CI)
β-Amanitin

21161-58-0
$C_{10}H_5Cl_7$
373.32
1,2,4-Metheno-1H-cyclobuta(cd)pentalene, 1,3,4,5,5,5a,6-heptachlorooctahydro- (8CI,9CI)

21184-58-7
$C_{17}H_{12}N_2O_3$
292.28
Benzoic acid, 4-((4-hydroxy-1-naphthalenyl)azo)- (9CI)
Benzoic acid, p-((4-hydroxy-1-naphthyl)azo)- (8CI)
NSC-74770

21213-69-4
$C_{18}H_{24}O_4$
304.39
1,3-Benzodioxole, 5-((5-(3-ethyl-3-methyloxiranyl)-3-methyl-2-pentenyl)oxy)-
AI3-34601

21227-47-4
$C_5H_5Cl_3N_4$
227.46
1,3,5-Triazin-2-amine, 4-methyl-6-(trichloromethyl)-(9CI)
s-Triazine, 2-amino-4-methyl-6-(trichloromethyl)- (8CI)

21232-47-3
$C_{12}H_6Cl_4N_2O$
336.00
Clc2ccc(N=N(=O)c1ccc(Cl)c(Cl)c1)cc2Cl
Azoxybenzene, 3,3',4,4'-tetrachloro
Diazene, bis(3,4-dichlorophenyl)-, 1-oxide (9CI)
TCAOB
3,3',4,4'-Tetrachloroazoxybenzene
3,4,3',4'-Tetrachloroazoxybenzene

21245-02-3
$C_{17}H_{27}NO_2$
277.40
O=C(OCC(CCCC)CC)c(ccc(N(C)C)c1)c1
Octyl dimethyl PABA
Arlatone UVB
Benzoic acid, 4-(dimethylamino)-, 2-ethylhexyl ester (9CI)
4-(Dimethylamino)benzoic acid, 2-ethylhexyl ester
Escalol 507
2-Ethylhexyl p-dimethylaminobenzoate

2-Ethylhexyl p-(dimethyl-amino)benzoate
Octyl dimethyl p-aminobenzoate
Padimate O
Solarchem O
UVasorb DMO

21248-00-0
$C_{20}H_{11}Br$
331.22
Benzo(a)pyrene, 6-bromo
6-Bromobenzo(a)pyrene

21259-20-1
$C_{24}H_{34}O_9$
466.58
CC(C)CC(=O)OC4CC1(COC(C)=O)C(OC2C(O)C(OC(C)=O)C1(C)C23CO3)C=C4C
Trichothec-9-ene-3-α,4-β,8-α,15-tetrol, 12,13-epoxy-, 4,15-diacetate 8- isovalerate
4,15-Diacetoxy-8-(3-methyl-butyryloxy)-12,13-epoxy-δ-9-trichothecen-3-ol
4-β,15-Diacetoxy-8-α-(3-methylbutyryloxy)-3-α-hydroxy-12,13-epoxytri-chothec-9-ene
4,15-Diacetoxy-8-(3-methyl-butyryloxy)scirp-9-en-3-ol
Fusariotoxine T2
Fusariotoxin T 2
3-Hydroxy-4,15-diacetoxy-8-(3-methylbutyryloxy)-12,13-epoxy-δ^9-trichothecene
Insariotoxin
Isariotoxin
8-Isovalerate
8-(3-Methylbutyryloxy)-diacet-oxyscirpenol
Mycotoxin T-2
NSC-138780
Scirpenol, 8-(3-methylbutyryl-oxy)-diacetoxy-
T-2 Mycotoxin
Toxin T2
T-2 Toxin
T_2-Trichothecene

21260-46-8
$C_9H_{18}N_3S_6.Bi$
569.64
CN(C)C(=S)S[Bi](SC(=S)N(C)C)SC(=S)N(C)C
Bismuth, tris(dimethyldithio-carbamato)
Bismate
Bismuth dimethyldithioc-arbamate
Carbamic acid, dimethyldithio-, bismuth salt
Tris(dimethyldithiocarbamato)-bismuth

21265-50-9
$C_{10}H_{12}FeN_2O_8.H_4N$
362.08
Ferrate(1-), ((N,N'-1,2-ethane-diylbis(N-(carboxymethyl)-glycinato))(4-)-N,N',O,O',ON,ON')-, ammonium, (OC-6-21)- (9CI)
Ethylenediaminetetracetic acid, ferric ammonium salt
Ferrate(-1), ((N,N'-1,2-ethane-diylbis(N-(carboxymethyl)-glycinato))(4-)-N,N',O,O',ON,ON')-, ammonium
Ferric ammonium edetate

21267-72-1
$C_{12}H_{12}ClNO$
221.70
ClCC(=O)N(CCC#C)c1ccccc1
Acetanilide, 2-chloro-N-(3-butynyl)
Acetamide, 2-chloro-N-(1-methyl-2-propynyl)-N-phenyl- (9CI)
Acetanilide, 2-chloro-N-(1-methyl-2-propynyl)- (8CI)
N-Acetyl-N-3-butinyl-2-chlor-anilin (Czech)
Basamaize
BAS-290-H
BAS 2900H
BAS 2903H
Butisan
Butisane
Chloressigsaeure-N-isobutinyl-anilid (German)
Chloretin
2-Chloro-N-(1-methyl-2-propyn-yl)acetanilide
2903 H
Prynachlor
Prynachlore

21282-97-3
$C_{10}H_{14}O_5$
214.22
O=C(OCCOC(=O)CC(=O)C)C(=C)C
Butanoic acid, 3-oxo-, 2-((2-methyl-1-oxo-2-propenyl)-oxy)ethyl ester (9CI)

21293-02-7
$C_{12}H_{22}$
166.31
3,5-Octadiene, 4,5-di-ethyl-, (E,Z)- (8CI,9CI)
cis,trans-4,5-Diethyl-3,5-octadiene

21297-72-3
$C_{10}H_{25}NO_2Si_2$
247.48
1-Aza-2-silacyclopentane, 2,2-diethoxy-1-(trimethyl-silyl)- (9CI)

21302-09-0
$C_{24}H_{51}O_3P$
418.64
O=P(OCCCCCCCCCCCC)OCCCCCCCCCCCC
Phosphonic acid, didodecyl ester (9CI)
Didodecyl phosphite
Di-n-dodecyl phosphite
Didodecyl phosphonate
Dilauryl hydrogen phosphite
Dilauryl phosphite
NSC-41924

21306-32-1
$C_5H_{11}N_3S$
145.21
1,3,5-Triazine-2(1H)-thione, tetrahydro-4,6-dimethyl- (9CI)
s-Triazine-2(1H)-thione, tetra-hydro-4,6-dimethyl- (8CI)

21310-38-3
$C_6H_{11}ClNO_2P$
195.59
Phosphinic acid, (chloro-methyl)(2-cyanoethyl)-, ethyl ester (8CI)

21320-62-7
$C_4H_5ClN_4$
144.54
1,3,5-Triazin-2-amine, 4-chloro-6-methyl- (9CI)
s-Triazine, 2-amino-4-chloro-6-methyl- (8CI)

21320-64-9
$C_5H_9N_5$
139.13
s-Triazine, 2-amino-4-methyl-6-(methylamino)- (8CI)

21342-85-8
$C_{10}H_9Cl_2N$
214.10
Cc1c(Cl)cccc1CC(Cl)C#N
Propionitrile, 2-chloro-3-(3-chloro-o-tolyl)
Benzenepropanenitrile, α,3-di-chloro-2-methyl-
Bi-Act
PRB-8

21351-79-1
CsHO
149.92
Cesium hydroxide (ACGIH, OSHA)
Caesium hydroxide, Solid [UN 2682]
Caesium hydroxide, Solution [UN 2681]
Cesium hydrate
Cesium hydroxide dimer
Cesium hydroxide, Solid (DOT)

Cesium hydroxide, Solution (DOT)
UN 2681 [Caesium hydroxide solution]
UN 2682 [Caesium hydroxide, solid]

21361-93-3
$C_{10}H_{14}N_2$.ClH
198.72
Nicotine, hydrochloride, (-)
l-Nicotine hydrochloride

21384-33-8
$C_7H_4Cl_6N_4$
356.85
s-Triazine, 2-(1-aziridinyl)-4,6-bis(trichloromethyl)- (6CI,8CI)

21399-51-9
$C_{10}H_{16}O$
152.24
Naphth(2,3-b)oxirene, decahydro- (9CI)
Naphthalene, 2,3-epoxydecahydro- (8CI)

21400-25-9
$C_3H_3Cl_3$
145.41
1-Propene, 1,1,2-trichloro- (9CI)
Propene, 1,1,2-trichloro- (8CI)

21416-87-5
$C_{11}H_{16}N_4O_4$
268.31
CC(CN1CC(=O)NC(=O)C1)N2CC(=O)NC(=O)C2
2,6-Piperazinedione, 4,4'-propylenedi-, (+-)
(+,-)-1,2-Bis(3,5-dioxopiperazine-1-yl)propane
(+-)-1,2-Bis(3,5-dioxopiperazinyl)propane
ICRF-159
NCI-C01627
NSC-129943
2,6-Piperazinedione, 4,4'-(1-methyl-1,2-ethanediyl)bis-, (+-)- (9CI)
2,6-Piperazinedione-4,4'-propylene dioxopiperazine
Propane, (+-)-1,2-bis(3,5-dioxopiperazin-1-yl)-
Razoxin
(+-)-(3,5,3',5'-Tetraoxo)-1,2-dipiperazinopropane

21431-58-3
$C_5H_6BrN_3O_2$
220.01
1H-Imidazole, 4-bromo-1,2-dimethyl-5-nitro- (9CI)
Imidazole, 4-bromo-1,2-dimethyl-5-nitro- (8CI)

21436-96-4
$C_8H_{11}N$.ClH
157.66
Cl.Cc1ccc(N)c(C)c1
2,4-Xylidine, hydrochloride
1-Amino-2,4-dimethylbenzene hydrochloride
4-Amino-1,3-dimethylbenzene hydrochloride
4-Amino-3-methyltoluene hydrochloride
4-Amino-1,3-xylene hydrochloride
2,4-Dimethylaniline hydrochloride
2,4-Dimethylbenzenamine hydrochloride
4-Methyl-o-toluidine hydrochloride
2-Methyl-p-toluidine hydrochloride
m-Xylidine hydrochloride
m-4-Xylidine hydrochloride

21436-97-5
$C_9H_{13}N$.ClH
171.69
Cl.Cc1cc(C)c(N)cc1C
Aniline, 2,4,5-trimethyl, hydrochloride
1-Amino-2,4,5-trimethylbenzene hydrochloride
Benzenamine, 2,4,5-trimethyl-, hydrochloride (9CI)
psi-Cumidine hydrochloride
Pseudocumidine hydrochloride
1,2,4-Trimethyl-5-aminobenzene hydrochloride
2,4,5-Trimethylaniline hydrochloride

21450-13-5
C_5H_9Cl
104.59
CCCC=CCl
1-Pentene, 1-chloro
1-Chloro-1-pentene

21460-88-8
$C_8H_6Cl_2O_2$
205.04
Benzoic acid, 2,5-dichloro-4-methyl- (9CI)
p-Toluic acid, 2,5-dichloro- (6CI,8CI)

21465-51-0
$C_7H_5NO_3S_2$.Na
238.24
2-Benzothiazolesulfonic acid, sodium salt (9CI)
Sodium 2-benzothiazolesulfonate

21483-62-5
C_4HCl_9
368.13
Butane, 1,1,1,2,2,3,3,4,4-nonachloro- (8CI)
Butane, 1,1,2,2,3,3,4,4,4-nonachloro- (6CI)

21490-63-1
C_4H_8O
72.11
Oxirane, 2,3-dimethyl-, trans- (9CI)
Butane, 2,3-epoxy-, trans- (8CI)

21493-04-9
$C_{18}H_{15}N_5O_4S_2$.Na
452.44
7-Benzothiazolesulfonic acid, 2-(4-((4,5-dihydro-3-methyl-5-oxo-1H-pyrazol-4-yl)azo)phenyl)-6-methyl-, monosodium salt (9CI)
6-Methyl-2-(p-((3-methyl-5-oxo-2-pyrazolin-4-yl)azo)phenyl)-7-benzothiazolesulfonic acid, sodium salt

21500-98-1
$C_{15}H_{23}NS$
249.41
Tenocyclidine
DEA No. 7470
Piperidine, 1-(1-(2-thienyl)cyclohexyl)-
Tenociclidina (Spanish)
Tenocyclidinum (Latin)
((Thienyl-2)-1 cyclohexyle)-N piperidine (French)
1-(1-(2-Thienyl)cyclohexyl)piperidine
1-(1-(2-Thienyl)-cyclohexyl)piperidine
Thienylphencyclidine
Thiophene analog of phencyclidine

21504-45-0
$C_4H_{11}O_4P$.Hg
354.69
Phosphoric acid, diethyl ester, mercury(1+) salt (8CI)

21528-31-4
$C_{16}H_{15}NO_3$
269.29
O=C(O)c(c(ccc1)C(=O)c(ccc(N(C)C)c2)c2)c1
Benzoic acid, 2-(4-(dimethylamino)benzoyl)- (9CI)
2-(4-(Dimethylamino)benzoyl)benzoic acid

21544-02-5
$C_6H_{14}N_2$
114.18

NCC(C(N)CC1)C1
Cyclopentanemethanamine, 2-amino- (9CI)
1-Amino-2-(aminomethyl)cyclopentane
2-Aminocyclopentanemethanamine
2-Aminocyclopentanemethylamine
1-(Aminomethyl)-2-aminocyclopentane
2-(Aminomethyl)cyclopentylamine
Cyclopentanemethylamine, 2-amino- (8CI)
Cyclopentylamine, 2-(aminomethyl)-
NSC-141555

21548-32-3
$C_6H_{12}NO_3PS_2$
241.28
O=P(OCC)(OCC)N=C(SC1)S1
Imidocarbonic acid, phosphonodithio-, cyclic methylene p,p-diethyl ester
AC 64475
Acconem
CL 64475
(Diethoxyphosphinylimino)-1,3-dithietane
Diethoxyphosphinylimino-2 dithietanne-1,3 (French)
Fosthietan
Geofos
Nem-A-Tak
Phosphoramidic acid, 1,3-dithietan-2-ylidene-, diethyl ester

21548-73-2
Ag_2S
247.80
Silver sulfide (9CI)

21564-17-0
$C_9H_6N_2S_3$
238.35
Thiocyanic acid, 2-(benzothiazolylthio)methyl ester, 60%
Busan 72A
TCMTB
TCMB 80%
Thiocyanic acid, 2-(benzothiazolylthio)methyl ester, 80%
2-Thiocyanomethylthiobenzothiazole, 80%
2-(Thiocyanomethylthio)benzothiazole, 60%
2-Tiocianometiltiobenzotiazolo, 80% (Italian)

21564-92-1
$C_{12}H_{16}$
160.26
Naphthalene, 1,2,3,4-tetrahydro-2,3-dimethyl- (8CI,9CI)

21569-63-1
$C_5H_{10}O_5$
150.13
D-threo-Pentonic acid, 3-deoxy (8CI,9CI)

21571-58-4
$C_{13}H_9Cl_3O$
167.46
p-Chlorobenzyl 2,4-dichlorophenyl ether
Ether, p-chlorobenzyl 2,4-dichlorophenyl

21577-41-3
$C_{27}H_{21}N_3$
387.49
1,3,5-Triazine, 2,4,6-tris-(2-phenylethenyl)- (9CI)
s-Triazine, 2,4,6-tristyryl-(8CI)

21577-80-0
$C_{20}H_{30}O_4$
334.46
1,2-Benzenedicarboxylic acid, monododecyl ester (9CI)
Phthalic acid, monododecyl ester (8CI)

21593-23-7
$C_{17}H_{17}N_3O_6S_2$
423.49
CC(=O)OCC1=C(N3C(SC1)C(NC(=O)CSc2ccncc2)C3=O)C(O)=O
5-Thia-1-azabicyclo(4.2.0)oct-2-ene-2-carboxylic acid, 3-(hydroxymethyl)-8-oxo-7-(2-(4- pyridylthio)acetamido)-, acetate (ester)
Cefapirin (German)
Cephapirin

21609-90-5
$C_{13}H_{10}BrCl_2O_2PS$
412.07
COP(=S)(Oc1cc(Cl)c(Br)cc1Cl)c2ccccc2
Phosphonothioic acid, phenyl-, O-(4-bromo-2,5-dichlorophenyl) O-methyl ester
ABAR
O-(4-Bromo-2,5-dichlorophenyl) O-methyl phenylphosphonothioate
O-(2,5-Dichloro-4-bromophenyl) O-methyl phenylthiophosphonate
Fosvel
K62-105
Leptophos
MBCP
O-Methyl-O-(4-bromo-2,5-dichlorophenyl)phenyl thiophosphonate
O-Methyl O-2,5-dichloro-4-bromophenyl phenylthiophosphonate
NK 711
OMS 1438
Phenylphosphonothioic acid O-(4-bromo-2,5-dichlorophenyl) O-methyl ester
Phosvel
PSL
V.C.S.
VCS-506
Velsicol 506
Velsicol VCS 506

21615-36-1
$C_{15}H_{21}NO_4$
279.33
O=C(OCCN(c(cccc1C)c1)CCOC(=O)C)C
Ethanol, 2,2'-((3-methylphenyl)imino)bis-, diacetate (ester) (9CI)

21620-54-2
$C_{25}H_{19}N_3$
361.45
s-Triazine, 2-phenyl-4,6-distyryl- (8CI)

21635-69-8
$C_{10}H_{15}NO_7S_2$
325.36
O=S(=O)(OCCS(=O)(=O)c(c(cc(N)c1OC)C)c1)O
Ethanol, 2-((4-amino-5-methoxy-2-methylphenyl)sulfonyl)-, hydrogen sulfate (ester) (9CI)

21639-41-8
$C_7H_5ClO_3S$
204.63
Benzenesulfonyl chloride, 2-formyl- (9CI)
Benzenesulfonyl chloride, o-formyl- (8CI)

21641-70-3
$C_{10}H_3Cl_9$
442.21
4,7-Methano-1H-indene, 1,2,3,4,5,6,7,8,8-nonachloro-3a,4,7,7a-tetrahydro- (9CI)
4,7-Methanoindene, 1,2,3,4,5,6,7,8,8-nonachloro-3a,4,7,7a-tetrahydro-(8CI) (VAN)
Trichlordene

21645-51-2
AlH_3O_3
78.01
Aluminum-hydroxide

AF 260
Alcoa 331
Alcoa C 30BF
Alumigel
Alumina hydrate
Alumina hydrated
Alumina trihydrate
α-Alumina trihydrate
Aluminic acid
Aluminium hydroxide
Aluminum hydrate
Aluminum(III) hydroxide
Aluminum Hydroxide Gel
Aluminum oxide-3H2O
Aluminum oxide hydrate
Aluminum oxide trihydrate
Aluminum trihydrat
Aluminum trihydrate
Aluminum trihydroxide
Alusal
Amberol ST 140F
Amphojel
Baco AF 260
British Aluminum AF 260
C 31
C 33
C 31C
C 4D
C 31F
C-31-F
C.I. 77002
GHA 331
GHA 332
H 46
Higilite
Higilite H 32
Higilite H 42
Higilite H 31S
Hychol 705
Hydral 705
Hydral 710
Hydrated alumina
Liquigel
Martinal
P 30BF
PGA
Trihydrated alumina
Trihydroxyaluminum

21662-15-7
$C_{12}H_{20}O$
180.29
O=CC=CC=CCCCCCCC
2,4-Dodecadienal, (E,Z)- (9CI)
(E,E)-2,4-Dodecadienal
(E,Z)-2,4-Dodecadienal
(trans,cis)-2,4-Dodecadienal

21662-16-8
$C_{12}H_{20}O$
180.29
O=CC=CC=CCCCCCCC
2,4-Dodecadienal, (E,E)- (9CI)
(trans,trans)-2,4-Dodecadienal
(E,E)-2,4-Dodecadien-1-al

21677-57-6
$C_5H_8N_4O_2$
156.15
1H-Imidazol-5-amine, 1,2-dimethyl-4-nitro- (9CI)
5-Amino-1,2-dimethyl-4-nitroimidazole
Imidazole, 5-amino-1,2-dimethyl-4-nitro- (8CI)

21689-84-9
$C_{10}H_{16}N_6S$
252.34
Cyanatryn
2-(1-Cyano-1-methylethylamino)-4-ethylamino-6-methylthio-1,3,5-triazine

21693-51-6
$C_{13}H_{18}$
174.29
Naphthalene, 1,2,3,4-tetrahydro-1,5,8-trimethyl- (6CI,8CI,9CI)

21715-90-2
$C_9H_9NO_3$
179.17
O=C(N(O)C(=O)C1C(C=CC23)C2)C13
N-Hydroxy-5-norbornene-2,3-dicarboximide
4,7-Methano-1H-isoindole-1,3-(2H)-dione, 3a,4,7,7a-tetrahydro-2-hydroxy- (9CI)
5-Norbornene-2,3-dicarboximide, N-hydroxy-
NSC-100740

21722-85-0
$C_{15}H_{17}O_2PS$
292.35
Phosphonothioic acid, S-benzyl O-ethyl ester
ESBP
Inezin

21725-46-2
$C_9H_{13}ClN_6$
240.73
CCNc1nc(Cl)nc(NC(C)(C)C#N)n1
Propionitrile, 2-((4-chloro-6-(ethylamino)-s-triazin-2-yl)amino)-2-methyl
Bladex
Bladex 80WP
2-Chloro-4-(1-cyano-1-methylethylamino)-6-ethylamino-1,3,5-triazine
2-(4-Chloro-6-ethylamino-s-triazine-2-ylamino)-2-methyl-propionitrile
2-((4-Chloro-6-(ethylamino)-s-triazin-2-yl)amino)-2-methylpropionitrile
2-(4-Chloro-6-ethylamino-1,3,5-triazine-2-ylamino)-2-methylpropionitrile
Cyanazine
DW3418
Fortrol
Payze
Propanenitrile, 2-((4-chloro-6-(ethylamino)-1,3,5-triazin-2-yl)amino)-2-methyl-
SD 15418
s-Triazine, 2-chloro-4-ethylamino-6-(1-cyano-1-methyl)-ethylamino-
WL 19805

21729-98-6
$C_3H_4N_2O_2$
100.06
O=C(OC)NC#N
Carbamic acid, cyano-, methyl ester (9CI)
Methyl cyanocarbamate

21739-91-3
$C_{11}H_8BrO_4.Na$
307.09
[Na+].[O-]C(=O)C=C(Br)C(=O)c1ccccc1
Acrylic acid, 3-p-anisoyl-3-bromo-, sodium salt, (E)
(E)-3-p-Anisoyl-3-bromoacrylic acid sodium salt
2-Butenoic acid, 3-bromo-4-(4-methoxyphenyl)-4-oxo-, sodium salt, (E)- (9CI)
Cytembena
MBBA
NCI-C50737
NSC-104801
Sodna sul kyseliny cis-β-4-methoxybenzoyl-β-bromakrylove (Czech)

21766-50-7
$C_{10}H_{16}O_2$
168.24
1(2H)-Naphthalenone, octahydro-4-hydroxy- (8CI,9CI)

21770-86-5
$C_9H_{22}NO_2PS$
239.35
Phosphonothioic acid, methyl-, S-(2-(diethylamino)-ethyl) O-ethyl ester
S-2-Diethylaminoethyl-O-ethylester kyseliny methylthiofosfonove (Czech)
EDEMO
EDEMO 3

21774-13-0
C_2H_6Pb
237.27
Lead(2+), dimethyl- (9CI)
Lead(2+), dimethyl-, ion (8CI)

21777-84-4

$C_{13}H_{20}$
176.30
Benzene, (2-methyl-1-(1-methylethyl)propyl)-(9CI)
Benzene, (1-isopropyl-2-methylpropyl)- (8CI)
α,α-Diisopropyltoluene

21794-01-4
$C_{26}H_{30}O_{11}$
518.56
CCCCCCC(O)C1C(O)C4=C(CC(CC2=C1C(=O)OC2=O)C(O)C3CC=CC(=O)O3)C(=O)OC4=O
1H-Cyclonona(1,2-c:5,6-c')difuran-1,3,6,8(4H)-tetrone, 10-((3,6-dihydro-6-oxo-2H-pyran-2-yl) hydroxymethyl)-5,9,10,11-tetrahydro-4-hydroxy-5-(1-hydroxyheptyl)
Rubratoxin B

21799-87-1
$C_6H_6O_5S.K$
229.27
Benzenesulfonic acid, 2,5-dihydroxy-, monopotassium salt (9CI)
2,5-Dihydroxybenzenesulfonic acid, monopotassium salt

21810-39-9
$C_{10}H_{20}NO_2.CH_3O_4S$
297.37
Ethanaminium, N,N-diethyl-N-methyl-2-((1-oxo-2-propenyl)oxy)-, methyl sulfate (9CI)
N,N-Diethyl-N-methyl-2-(1-oxo-2-propenyloxy)ethanaminium methyl sulfate
N,N-Diethyl-N-methyl-2-((1-oxo-2-propenyl)oxy)ethanaminium, methyl sulfate

21811-92-7
$C_{16}H_{16}N_4O_5S$
376.37
O=C(Nc(cccc1)c1)C(N=Nc(c(O)ccc2S(=O)(=O)N)c2)C(=O)C
Butanamide, 2-((5-(aminosulfonyl)-2-hydroxyphenyl)azo)-3-oxo-N-phenyl- (9CI)
2-((2-Hydroxy-5-sulfonamidophenyl)azo)-N-phenyl-3-oxobutanamide

21829-25-4
$C_{17}H_{18}N_2O_6$
346.37
COC(=O)C1=C(C)NC(=C(C1c2ccccc2N(=O)=O)C(=O)OC)C
3,5-Pyridinedicarboxylic acid, 1,4-dihydro-2,6-dimethyl-4-(2-nitrophenyl)-, dimethyl ester
Adalat
Bay 1040
Bay A 1040
Citilat
Cordipin
1,4-Dihydro-2,6-dimethyl-4-(2-nitrophenyl)-3,5-pyridinedicarboxylic acid dimethyl ester
Dimethyl 1,4-dihydro-2,6-dimethyl-4-(2'-nitrophenyl)-3,5-pyridinedicarboxylate
Nifedin
Nifedipine
Nifelat
4-(2'-Nitrophenyl)-2,6-dimethyl-3,5-dicarbomethoxy-1,4-dihydropyridine
4-(2'-Nitrophenyl)-2,6-dimethyl-1,4-dihydropyridin-3,5-dicarbonsaeuredimethylester (German)
Oxcord
Procardia

21838-75-5
$C_8H_{19}NO_2$
161.24
2-Butanol, 1,1'-iminobis- (9CI)
2-Butanol, 1,1'-iminodi- (8CI)

21846-07-1
$C_{19}H_{13}Cl$
276.77
9H-Fluorene, 9-(4-chlorophenyl)- (9CI)
Fluorene, 9-(p-chlorophenyl)- (8CI)

21846-08-2
$C_{20}H_{16}O$
272.35
9H-Fluorene, 9-(4-methoxyphenyl)- (9CI)
Anisole, p-fluoren-9-yl- (6CI)
Fluorene, 9-(p-methoxyphenyl)- (8CI)
9-(p-Methoxyphenyl)fluorene

21850-44-2
$C_{21}H_{20}Br_8O_2$
943.62
O(c(c(cc(c1)C(c(cc(c(OCC(Br)CBr)c2Br)Br)c2)(C)C)Br)c1Br)CC(Br)CBr
Benzene, 1,1'-(1-methylethylidene)bis(3,5-dibromo-4-(2,3-dibromopropoxy)- (9CI)
1,1'-(1-Methylethylidene)bis-(3,5-dibromo-4-(2,3-dibromopropoxy))benzene
Tetrabromobisphenol A bis-(2,3-dibromopropyl) ether

21858-40-2
$C_{11}H_8Cl_6O$
368.90
4,7-Methanonaphth(1,2-b)oxirene, 4,5,6,7,8,8-hexachloro-1a,2,3,3a,4,7,7a,7b-octahydro-, (1aα,3aα,4α,7α,7aα,7bα)- (9CI)
1,4-Methanonaphthalene, 1,2,3,4,9,9-hexachloro-5,6-epoxy-1,4,4a,5,6,7,8,8a-octahydro-, stereoisomer (8CI)

21861-11-0
$C_7H_{13}NO_4S$
207.24
L-Cysteine, S-(2-carboxy-1-methylethyl)- (9CI)
Butyric acid, 3-((2-amino-2-carboxyethyl)thio)-, Stereoisomer (8CI)

21884-44-6
$C_{30}H_{22}O_{12}$
574.52
Cc8cc(O)c7C(=O)C5=C(O)C2C(O)C6C3C(O)C(C(=C4C(=O)c1c(O)cc(C)c(O)c1C(=O)C234)O)C56C(=O)c7c8O
5H,6H-6,5a,13a,14-(1,2,3,4)-Butanetetraylcycloocta-(1,2-b:5,6-b')dinaphthalene-5,8,13,16(14H)- tetrone, 1,4,7,9,12,15,17,20-octahydroxy-3,11-dimethyl
8,8'-Dihydroxy-rugulosin
Flavomycelin
Luteoskyrin
(-)-Luteoskyrin

21902-34-1
$C_{17}H_{14}ClN_3$
295.77
1,3,5-Triazine, 2-chloro-4,6-bis(4-methylphenyl)-(9CI)
s-Triazine, 2-chloro-4,6-di-p-tolyl- (8CI)

21908-53-2
HgO
216.59
Mercury(II) oxide
C.I. 77760
Mercuric oxide
Mercuric oxide, Red
Mercuric oxide, Solid (DOT)
Mercuric oxide, Yellow
Mercury oxide [UN 1641]
Oxyde de mercure (French)
Quecksilberoxid (German)
Red Oxide of Mercury
Red Precipitate
Santar
UN 1641 [Mercury oxide]
Yellow Mercuric Oxide
Yellow Oxide of Mercury

Yellow Precipitate

21921-96-0
$C_5H_{13}O_3P$
152.15
Phosphonic acid, propyl-, monoethyl ester
Ethyl hydrogen 1-propylphosphonate
NIA 10637

21923-23-9
$C_{11}H_{15}Cl_2O_3PS_2$
361.25
CCOP(=S)(OCC)Oc1cc(Cl)c(SC)cc1Cl
Phosphorothioic acid, O,O-diethyl O-(2,5-dichloro-4-(methylthio)phenyl) ester
Celamerck S-2957
Cela S-2957
Celathion
Chlorthiophos
CM S 2957
O-(Dichloro(methylthio)phenyl) O,O-diethyl phosphorothioate (3 isomers)
O,O-Diethyl-O-2,4,5-dichloro-(methylthio)phenyl thionophosphate
ENT 27,635
NSC-195164
OMS 1342
Phosphorothioic acid, O-(2,5-dichloro-4-(methylthio)phenyl) O,O-diethyl ester
S 2957

21951-32-6
$C_7H_6N_2O_3S_2$
230.26
O=S(=O)(O)c(ccc(NC(=N)S1)c12)c2
6-Benzothiazolesulfonic acid, 2-amino- (9CI)
2-Amino-6-benzothiazolesulfonic acid

21951-33-7
$C_{10}H_8N_2O_6S$
284.24
O=C(O)C(=NN(c(ccc(S(=O)(=O)O)c1)c1)C=2O)C2
1H-Pyrazole-3-carboxylic acid, 5-hydroxy-1-(4-sulfophenyl)- (9CI)

21964-48-7
$C_{13}H_{24}$
180.33
1,12-Tridecadiene (8CI,9CI)

21964-49-8
$C_{14}H_{26}$
194.36
1,13-Tetradecadiene
AI3-36491

21985-87-5
$C_6H_2N_6O_{10}$
318.08
Pentanitroaniline (Dry)
Aniline, 2,3,4,5,6-pentanitro-, Dry
Benzenamine, 2,3,4,5,6-pentanitro-
Pentanitroanilina (seca) (Spanish)
Pentanitroaniline, Dry
2,3,4,5,6-Pentanitroaniline

21993-11-3
$C_{16}H_{19}O_4P$
306.30
Phosphinic acid, bis(phenoxymethyl)-, ethyl ester (8CI,9CI)

22020-14-0
$C_{19}H_{41}N$
283.54
N(CCCCCCCCCC)(CCCCCCCC)C
1-Decanamine, N-methyl-N-octyl- (9CI)
N-Methyl-N-octyl-1-decanamine
Octyldecylmethylamine

22025-44-1
$C_{14}H_{15}N_3O_5S$
337.34
O=S(=O)(N)c(ccc(Nc(ccc(OCC)c1)c1)c2N(=O)=O)c2
Sulfanilamide, N4-(p-ethoxyphenyl)-3-nitro- (8CI)
4-(p-Ethoxyphenylamino)-3-nitrobenzenesulfonamide
N4-(p-Ethoxyphenyl)-3-nitrosulfanilamide

22031-17-0
$C_4H_{12}O_4P.1/3O_4P$
186.77
Phosphonium, tetrakis-(hydroxymethyl)-, phosphate (3:1) (Salt) (9CI)

22031-33-0
$C_{13}H_{16}N_2O_2$
232.27
O=C(OCCN(c(cccc1)c1)CCC#N)C
Propanenitrile, 3-((2-(acetyloxy)ethyl)phenylamino)- (9CI)
Aniline, N-acetoxyethyl-N-cyanoethyl-

22038-69-3
C_8H_{12}
108.18
1,3,6-Octatriene, (E,E)- (8CI, 9CI)

22039-38-9
$C_7H_4Cl_6$
300.82
NSC-143933

22042-96-2
$C_9H_{29}N_3NaO_{15}P_5$
597.20
Phosphonic acid, (((phosphonomethyl)imino)bis(2,1-ethanediylnitrilobis(methylene)))tetrakis-, sodium salt (9CI)

22047-25-2
$C_6H_6N_2O$
122.11
O=C(c(nccn1)c1)C
Ethanone, 1-pyrazinyl- (9CI)
Acetylpyrazine
2-Acetylpyrazine
AI3-34445
Ketone, methyl pyrazinyl
NSC-72374
1-Pyrazinylethanone

22047-27-4
$C_7H_8N_2O$
136.14
Ethanone, 1-(5-methylpyrazinyl)- (9CI)
Ketone, methyl 5-methylpyrazinyl (8CI)

22047-49-0
$C_{26}H_{52}O_2$
396.78
O=C(OCC(CCCC)CC)CCCCCCCCCCCCCCCC
Stearic acid, 2-ethylhexyl ester
2-Ethylhexyl octadecanoate
2-Ethylhexyl stearate
Octadecanoic acid, 2-ethylhexyl ester (9CI)
Octyl stearate
Wickenol 156

22071-15-4
$C_{16}H_{14}O_3$
254.30
O=C(c(cccc1C(C(=O)O)C)c1)c(cccc2)c2
Propionic acid, 2-(3-benzoylphenyl)
L'Acide (benzoyl-3-phenyl)-2-propionique (French)
Alrheumat
Alrheumun
m-Benzoylhydratropic acid
3-Benzoylhydratropic acid
3-Benzoyl-α-methylbenzeneacetic acid
2-(m-Benzoylphenyl)propionic acid

2-(3-Benzoylphenyl)propionic acid
Capisten
Fastum
Iso-K
Kefenid
Ketoprofen
Ketopron
Lertus
Meprofen
Orudis
Oruvail
Profenid
19583 RP

22080-08-6
$C_{16}H_{12}N_2O_4S$
328.34
Benzenesulfonic acid, 3-((2-hydroxy-1-naphthalenyl)azo)- (9CI)
Benzenesulfonic acid, m-((2-hydroxy-1-naphthyl)azo)- (8CI)

22083-74-5
$C_{10}H_{14}N_2$
162.26
Nicotine, (+-)
dl-β-Nicotine
DL-Nicotine
(R,S)-Nicotine
Pyridine, 3-(1-methyl-2-pyrrolidinyl)-, (+-)- (9CI)

22084-89-5
$C_{10}H_{12}O_2$
164.21
Cc1ccccc1CCC(O)=O
Benzenepropanoic acid, 2-methyl- (9CI)
Hydrocinnamic acid, o-methyl- (7CI,8CI)
ortho-Methylhydrocinnamic acid
3-(2-Methylphenyl)propanoic acid
2-Methyl-β-phenylpropionic acid

22092-38-2
$C_{17}H_{34}O$
254.46
O(C1CCCCCCCCCCCCCCC)C1
Oxirane, pentadecyl- (9CI)
1,2-Epoxyheptadecane
1,2-Heptadecane oxide
Pentadecyloxirane

22094-93-5
$C_{36}H_{32}Cl_4N_6O_4$
754.46
O=C(Nc(c(cc(c1)C)C)c1)C(N=Nc(c(cc(c2Cl)c(c(cc(N=NC(C(=O)C)C(=O)Nc(c(cc(c3)C)C)c3)c4Cl)Cl)c4)Cl)c2)C(=O)C
Butanamide, 2,2'-((2,2',5,5'-tetrachloro(1,1'-biphenyl)-4,4'-diyl)bis(azo))bis(N-(2,4-dimethylphenyl)-3-oxo- (9CI)
2',4'-Acetoacetoxylidide, 2,2''-((2,2',5,5'-tetrachloro-4,4'-biphenylene)bis(azo))bis-

22113-87-7
$CH_5N.HNO_3$
94.06
Methanamine, nitrate (9CI)
Methylamine nitrate

22117-06-2
$C_{15}H_{22}O_2$
234.34
7,10-Pentadecadiynoic acid (8CI)

22117-09-5
$C_{17}H_{30}O$
250.43
5,8,11-Heptadecatrien-1-ol (8CI,9CI)

22122-36-7
$C_5H_6O_2$
98.10
2(5H)-Furanone, 3-methyl- (8CI,9CI)

22144-77-0
$C_{30}H_{37}NO_6$
507.68
CC4CC=CC2C(O)C(=C)C(C)C3C(Cc1ccccc1)NC(=O)C23C(OC(C)=O)C=CC(C)(O)C4=O
1H-Cycloundec(d)isoindole-1,11(2H)-dione, 3-benzyl-3,3-α,4,5,6,6-α,9,10,12,15-decahydro-6,12,15-trihydroxy-4,10,12-trimethyl-5-methylene-, 15-acetate
Cytochalasin D
Zygosporin A

22175-22-0
$C_9H_{11}ClN_2O$
198.64
CNC(=O)Nc1ccc(C)c(Cl)c1
Urea, N-(3-chloro-4-methylphenyl)-N'-methyl- (9CI)
CGA 16339
NSC-164350
Urea, 1-(3-chloro-p-tolyl)-3-methyl- (8CI)

22204-53-1
$C_{14}H_{14}O_3$
230.28
COc2ccc1cc(ccc1c2)C(C)C(O)=O
Propionic acid, 2-(6-methoxy-2-naphthyl)-, (+)
CG 3117
Equiproxen
Floginax
(+)-6-Methoxy-α-methyl-2-naphthaleneacetic acid
(S)-6-Methoxy-α-methyl-2-naphthalene acetic acid
(+)-2-(Methoxy-2-naphthyl)-propionic acid
d-2-(6-Methoxy-2-naphthyl)propionic acid
D-2-(6'-Methoxy-2'-naphthyl)-propionsaeure (German)
(+)-2-(Methoxy-2-naphthyl)-propionsaeure (German)
MNPA
Naixan
2-Naphthaleneacetic acid, 6-methoxy-α-methyl-, (+)- (8CI)
2-Naphthaleneacetic acid, 6-methoxy-α-methyl-, (S)- (9CI)
Naprosine
Naprosyn
Naproxen
Naprux
Naxen
Naxyn
Proxen
RS-3540

22205-45-4
Cu_2S
159.14
Copper(I) sulfide
Copper sulfide
Cuprasulfide
Cuprous sulfide
Dicopper monosulfide
Dicopper sulfide

22212-55-1
$C_{18}H_{17}Cl_2NO_3$
366.26
Propionic acid, 2-(N-benzoyl-N-(3,4-dichlorophenyl))-amino-, ethyl ester
Alanine, N-benzoyl-N-(3,4-dichlorophenyl)-, ethyl ester, DL-
N-Benzoyl-N-(3,4-dichlorophenyl)-DL-alanine ethyl ester
Benzoylprop ethyl
Enaven
Ethyl N-benzoyl-N-(3,4-dichlorophenyl)-2-aminopropionate
SD 30,053
Suffix
Suffix 25

22221-10-9
$C_8H_{16}O_2.xCu$
Unknown
Copper 2-ethylhexanoate
Caswell No. 239
EPA Pesticide Chemical Code 041201
Hexanoic acid, 2-ethyl-, copper

salt (9CI)

22224-92-6
$C_{13}H_{22}NO_3PS$
303.39
CCOP(=O)(NC(C)C)Oc1ccc(SC)c(C)c1
Phosphoramidic acid, isopropyl-, 4-(methylthio)-m-tolyl ethyl ester
O-Aethyl-O-(3-methyl-4-methylthiophenyl)-isopropylamido-phosphorsaeureester (German)
Bay 68138
Bay SRA 3886
ENT 27,572
Ethyl 3-methyl-4-(methylthio)-phenyl(1-methylethyl)phosphoramidate
Ethyl 4-(methylthio)-m-tolyl isopropylphosphoramidate
Fenamiphos (ACGIH,OSHA)
Isopropylamino-O-ethyl-(4-methylmercapto-3-methylphenyl)phosphate
1-(Methylethyl)-ethyl 3-methyl-4-(methylthio)phenyl phosphoramidate
Nemacur
Nemacur P
NSC-195106
Phenamiphos
Phosphoramidic acid, (1-methylethyl)-, ethyl (3-methyl-4-(methylthio)phenyl) ester
SRA 3886

22228-82-6
$C_4H_8O_2.H_3N$
105.13
Propanoic acid, 2-methyl-, ammonium salt (9CI)
AIB
Ammonium isobutyrate
Ammonium 2-methylpropanoate
Caswell No. 044AA
EPA Pesticide Chemical Code 101501
GR-9234

22232-25-3
$C_{13}H_{10}Cl_2O_2.2Na$
315.12
Phenol, 2,2'-methylenebis-(4-chloro-, disodium salt (9CI)
Caswell No. 406
Disodium-2,2'-methylenebis-(4-chlorophenate)
EPA Pesticide Chemical Code 055003
2,2'-Methylenebis(4-chlorophenol), disodium salt

22232-54-8
$C_7H_{10}N_2O_2S$
186.25
4-Imidazoline-1-carboxylic acid, 3-methyl-2-thioxo-, ethyl ester
Athyromazole
Basolest
Carbethoxymethimazole
Carbimazol
Carbimazole
Carbinazole
CG1
1H-Imidazole-1-carboxylic acid, 2,3-dihydro-3-methyl-2-thioxo, ethyl ester
3-Methyl-2-thioxo-4-imidazoline-1-carboxylic acid ethyl ester
Neomercazole
Tyrazol

22232-71-9
$C_{16}H_{13}ClN_2O$
284.76
OC2(N1CCN=C1c3ccccc23)c4ccc(Cl)cc4
5H-Imidazo(2,1-a)isoindol-5-ol, 5-(4-chlorophenyl)-2,3-dihydro
5-p-Chlorophenyl-2,3-dihydro-5H-imidazo(2,1-a)isoindol-5-ol
Mazindol
SA 42-548
Sanorex
Teronac

22239-54-9
$C_{15}H_{16}O$
212.31
4-Biphenylol, 4'-isopropyl
4-Isopropyl-4'-hydroxy biphenyl

22248-79-9
$C_{10}H_9Cl_4O_4P$
365.96
Phosphoric acid, 2-chloro-1-(2,4,5-trichlorophenyl)-vinyl dimethyl ester, (Z)
(Z)-2-Chloro-1-(2,4,5-trichlorophenyl)vinyl dimethyl phosphate
Appex
CVMP
Debantic
Dietreen
Dust M
ENT 25,841
Gardcide
Gardona
Gordona
Phosphoric acid, 2-chloro-1-(2,4,5-trichlorophenyl)-ethenyl dimethyl ester, (Z)-
Rabon
Rabond
Rol
SD 8447
Stirofos
Stirophos
Tetrachlorvinphos
2,4,5-Trichloro-α-(chloromethylene)benzyl phosphate ester
Vinfos
Vinylphosphate

22257-44-9
$C_6H_5N_3.Ag$
226.98
1H-Benzotriazole, silver(1+) salt (9CI)

22259-30-9
$C_{11}H_{15}N_3O_2$
221.29
Carbamic acid, methyl-, ester with N'-(m-hydroxyphenyl)-N,N-dimethylformamidine
Carbamic acid, methyl-, m-(((dimethylamino)methylene)amino)phenyl ester
Dicarzol
Formetanat
Formetanate
Methanimidamide, N,N-dimethyl-, N'-(3-(((methylamino)carbonyl)oxy)phenyl)-(9CI)
Methylcarbamic acid ester with N'-(m-hydroxyphenyl)-N,N-dimethylformamidine

22268-16-2
$C_{16}H_{16}O_8$
336.30
Altersolanol A
NSC-173943

22287-11-2
$C_8H_{14}O$
126.20
3-Penten-2-one, 3-ethyl-4-methyl- (7CI,8CI,9CI)

22305-35-7
$C_{11}H_{21}N_5O$
239.32
1,3,5-Triazin-2(1H)-one, 4,6-bis(diethylamino)-(9CI)
s-Triazin-2-ol, 4,6-bis-(diethylamino)- (7CI,8CI)

22306-28-1
$C_{16}H_{34}$
226.45
Pentadecane, 8-methyl- (8CI, 9CI)

22306-37-2
$C_2H_4O_2.1/3Bi$
129.71
Bismuth acetate
Acetic acid, bismuth(3+) salt
Bismuth triacetate

NSC-370498

22311-25-7
$C_6H_2Br_4$
393.70
Benzene, 1,2,3,4-tetrabromo- (8CI,9CI)

22313-62-8
$C_{12}H_{16}O_3$
208.26
O=C(OOC(C)(C)C)c(c(ccc1)C)c1
Benzenecarboperoxoic acid, 2-methyl-, 1,1-dimethylethyl ester (9CI)
tert-Butyl peroxy-o-methylbenzoate
tert-Butyl peroxy-2-methylbenzoate

22316-47-8
$C_{16}H_{13}ClN_2O_2$
300.76
CN2C(=O)CC(=O)N(c1ccccc1)c3cc(Cl)ccc23
1H-1,5-Benzodiazepine-2,4(3H,5H)-dione, 7-chloro-1-methyl-5-phenyl
Chlorepin
7-Chloro-1-methyl-5-phenyl-1H-1,5-benzodiazepine-2,4(3H,5H)-dione
Clobazam
Clorepin
Frisium
H-4723
HR 376
LM-2717
1-Phenyl-5-methyl-8-chloro-1,2,4,5-tetrahydro-2,4-dioxo-3H-1,5-benzodiazepine
RU-4723
Urbanyl

22319-24-0
$C_8H_{14}O$
126.20
2-Hepten-4-one, 2-methyl- (8CI,9CI)

22319-25-1
$C_8H_{14}O$
126.20
3-Hepten-2-one, 4-methyl- (8CI,9CI)
4-Methyl-3-hepten-2-one

22319-31-9
$C_8H_{14}O$
126.20
4-Hepten-3-one, 4-methyl- (8CI,9CI)

22330-14-9
Unknown
Unknown
Bis(tributyltin) salicylate

22339-23-7
$C_{15}H_{22}$
202.34
Naphthalene, 1,2,3,4-tetrahydro-1,6-dimethyl-4-(1-methylethyl)-, (1R-cis)- (9CI)
7βH,10βH-Cadina-1,3,5-triene (8CI)
Calamenene, (+)-
(+)-Calamenene

22349-59-3
$C_{16}H_{14}$
206.30
Phenanthrene, 1,4-dimethyl
1,4-Dimethylphenanthrene

22367-76-6
$C_9H_{11}NO_2$
165.19
Benzenepropanamide, 2-hydroxy- (9CI)
Hydrocinnamamide, o-hydroxy- (6CI,7CI,8CI)
o-Hydroxyphenylpropionamide

22373-78-0
$C_{36}H_{61}O_{11}.Na$
692.96
CCC1(CCC(O1)C3(C)CCC2(CC(O)C(C)C(O2)C(C)C(OC)C(C)C(O)=O)O3)C4CCCO4
1,6-Dioxaspiro(4.5)decane-7-butyric acid, 2-(5-ethyltetrahydro-5-(tetrahydro-3-methyl-5- (tetrahydro-6-hydroxy-6-(hydroxymethyl)-3,5-dimethyl-2H-pyran-2-yl)-2-furyl)-2-furyl)-9- hydroxy-β-methoxy-α,γ,2,8-tetramethyl-, monosodium salt
Coban
Coban 45
Monensin, monosodium salt (9CI)
Monensin sodium
Monensin sodium salt
Rumensin
Sodium monensin

22397-33-7
$C_{11}H_{22}O_4$
218.29
O(OC(CCC(OO1)(C)C)(C)C)C1(C)C
1,2,4,5-Tetroxonane, 3,3,6,6,9,9-hexamethyl- (8CI,9CI)
3,3,6,6,9,9-Hexamethyl-1,2,4,5-tetraoxacyclononane (Not more than 52% in solution)
3,3,6,6,9,9-Hexamethyl-1,2,4,5-tetraoxacyclononane (Not more than 52% with inert solid)
3,3,6,6,9,9-Hexamethyl-1,2,4,5-tetraoxacyclononane (Technically pure)
Hexamethyl-3,3,6,6,9,9 tetraoxanonane-1,2,4,5 (French)
3,3,6,6,9.9-Hexamethyl-1,2,4,5-tetraoxocyclononane
3,3,6,6,9,9-Hexamethyl-1,2,4,5-tetroxonane
3,3,6,6,9,9-Hexametil-1,2,4,5-tetraoxaciclononano (Spanish)
UN 2165
UN 2166
UN 2167

22398-80-7
InP
145.79
Indium phosphide (9CI)

22413-02-1
$C_{38}H_{76}O_2$
565.02
O=C(OCCCCCCCCCCCCCCCCCCCC)CCCCCCCCCCCCCCCCC
Octadecanoic acid, eicosyl ester (9CI)
Eicosyl octadecanoate
Eicosyl stearate

22421-59-6
$C_{10}H_{10}Br_2O_2$
322.00
Oxirane, ((2,6-dibromo-4-methylphenoxy)methyl)- (9CI)
((2,6-Dibromo-4-methylphenoxy)methyl)oxirane

22431-62-5
$C_{24}H_{28}O_3$
364.52
CC2(C)C(C=C1CCCC1)C2C(=O)OCc4coc(Cc3ccccc3)c4
Cyclopropanecarboxylic acid, 3-(cyclopentylidenemethyl)-2,2-dimethyl-, (5-benzyl-3-furyl) methyl ester, (1R,3R)
AI 3-27985
5-Benzyl-3-furylmethyl (+)-trans-ethanochrysanthemate
Bioethanomethrin
Ethanochrysanthemate
Ethanomethrin
K-Othrine
NIA 24110
Niagara NIA-24110
RU 11679

22454-86-0
$C_5H_4N_2O_4.1/2Ca$
176.14

4-Pyrimidinecarboxylic acid, 1,2,3,6-tetrahydro-2,6-dioxo-, calcium salt (2:1) (9CI)
Calcium orotate

22464-99-9
$C_8H_{16}O_2$.xZr
Unknown
Hexanoic acid, 2-ethyl-, zirconium salt (9CI)
Zirconium 2-ethylhexanoate

22467-31-8
$C_{26}H_{32}O_{11}$
520.58
1H-Cyclonona(c)furan-6,7-dicarboxylic anhydride, 9-((3,6-dihydro-6-oxo-2H-pyran-2-yl) hydroxymethyl)-3,4,5,8,9,10-hexahydro-1,5-dihydroxy-4-(1-hydroxyheptyl)-3-oxo
Rubratoxin A

22473-78-5
$C_{10}H_{16}N_2O_8$.4H_3N
360.36
Glycine, N,N'-1,2-ethanediylbis(N-(carboxymethyl)-, tetraammonium salt (9CI)

22474-57-3
$C_{16}H_{33}F_9O_5Si_5$
616.85
Cyclopentasiloxane, 2,2,4,4,6,8,10-heptamethyl-6,8,10-tris(3,3,3-trifluoropropyl)- (8CI, 9CI)

22483-09-6
$C_4H_{11}NO_2$
105.13
O(C(OC)CN)C
Acetaldehyde, amino-, dimethyl acetal (8CI)
Aminoacetaldehyde dimethyl acetal
2,2-Dimethoxyethylamine
Ethanamine, 2,2-dimethoxy- (9CI)
NSC-73701

22499-12-3
$C_{18}H_{20}O_2$
268.36
O=C(c(cccc1)c1)C(OCC(C)C)c(cccc2)c2
Ethanone, 2-(2-methylpropoxy)-1,2-diphenyl- (9CI)
2-(2-Methylpropoxy)-1,2-diphenylethanone

22506-53-2
$C_{16}H_8N_2O_4$
292.24
3,9-Dinitrofluoranthene
4,12-Dinitrofluoranthene
Fluoranthene, 3,9-dinitro-

22509-74-6
$C_{11}H_9NO_4$
219.21
O=C(OCC)N(C(=O)c(c1ccc2)c2)C1=O
2-Isoindolinecarboxylic acid, 1,3-dioxo-, ethyl ester
N-Carbethoxyphthalimide
N-Carboethoxyphthalimide
N-(Ethoxycarbonyl)phthalimide
Ethyl N-phthaloylcarbamate
2H-Isoindole-2-carboxylic acid, 1,3-dihydro-1,3-dioxo-, ethyl ester (9CI)
N-Karbetoksi-ftalimid (Yugoslavian)

22527-63-5
$C_{19}H_{28}O_4$
320.43
O=C(OCC(C)(C)C(OC(=O)c(cccc1)c1)C(C)C)C(C)C
Isobutyric acid, 3-hydroxy-2,2,4-trimethylpentyl ester benzoate (8CI)

22531-20-0
$C_{12}H_{16}$
160.26
Naphthalene, 6-ethyl-1,2,3,4-tetrahydro- (6CI, 8CI,9CI)
6-Ethyltetralin

22534-71-0
$C_6H_{11}NO_2$
129.16
Butanamide, N-acetyl- (9CI)
Butyramide, N-acetyl- (7CI, 8CI)
N-Acetylbutyramide

22537-48-0
Cd
112.40
Cadmium, ion (Cd2+)
Cadmium(2+)
Cadmium ion

22571-95-5
$C_{20}H_{31}NO_6$
381.52
CC=C(C)C(=O)OC1CCN2CC=C(COC(=O)C(O)(C(C)C)C(C)O)C12
Butenoic acid, 2-methyl-, 7-((2,3-dihydroxy-2-(1-methylethyl)-1-oxobutoxy)-methyl)-2,3,5,7a- tetrahydro-1H-pyrrolizin-1-yl ester
Symphytine
7-Tiglylretronecine viridiflorate
7-Tiglyl-9-viridiflorylretronecine

22578-86-5
$C_{20}H_{20}BrN_7O_6$
534.28
O=C(Nc(c(N=Nc(c(cc(N(=O)=O)c1)Br)c1N(=O)=O)cc(OC)c2N(CCC#N)CC)c2)C
Acetamide, N-(2-((2-bromo-4,6-dinitrophenyl)azo)-5-((2-cyanoethyl)ethylamino)-4-methoxyphenyl)- (9CI)
p-Acetanisidide, 2'-((2-bromo-4,6-dinitrophenyl)azo)-5'-((2-cyanoethyl)ethylamino)-
p-Acetanisidide, 2'-((2-bromo-4,6-dinitrophenyl)azo)-5'-(N-(2-cyanoethyl)-N-ethylamino)-
5-((2-Bromo-4,6-dinitrophenyl)-azo)-4-acetamido-2-((2-cyanoethyl)ethylamino)anisole
2'-((2-Bromo-4,6-dinitrophenyl)azo)-5'-((2-cyanoethyl)-ethylamino)-p-acetanisidide
3-(N-Cyanoethyl-N-ethylamino)-4-methoxy-6-((2-bromo-4,6-dinitrophenyl)azo)-acetanilide
N-Ethyl-N-(2-cyanoethyl)-5-acetamido-4-((6-bromo-2,4-dinitrophenyl)azo)-2-methoxyaniline

22583-29-5
$C_3H_9N_3$.ClH
123.61
Guanidine, 1,1-dimethyl-, monohydrochloride
as-Dimethylguanidine hydrochloride

22588-78-9
$C_{14}H_{19}N_3O_3$
277.31
O=C(Nc(ccc(OC)c1N(CCC#N)CCO)c1)C
Acetamide, N-(3-((2-cyanoethyl)(2-hydroxyethyl)-amino)-4-methoxyphenyl)- (9CI)

22591-21-5
$C_6H_{10}Cl_2O$
169.06
O=C(C(C)(C)C)C(Cl)Cl
2-Butanone, 1,1-dichloro-3,3-dimethyl
1,1-Dichloro-3,3-dimethyl-2-butanone
Dichloromethyl tert-butyl ketone
Dichloropinacolin
α,α-Dichloropinacolin

Dichloropinakolin
ω,ω-Dichlorpinakolin (German)

22607-13-2
$C_{10}H_{14}O_2$
166.22
1,2-Butanediol, 1-phenyl- (6CI,8CI)

22608-53-3
$C_3H_9O_2PS_2$
172.21
COP(=O)(SC)SC
Phosphorodithioic acid, O,S,S-trimethyl ester
O,S,S-Trimethyl phosphorodithioate

22633-33-6
$C_8H_6N_2O_7$
242.13
O=C(OC)c(c(O)c(N(=O)=O)cc1N(=O)=O)c1
Benzoic acid, 2-hydroxy-3,5-dinitro-, methyl ester (9CI)
3,5-Dinitrosalicylic acid, methyl ester
Methyl 3,5-dinitrosalicylate
NSC-203306
Salicylic acid, 3,5-dinitro-, methyl ester

22676-00-2
$C_{18}H_{22}O_6$
334.37
1,3-Propanediol, 1-(3,4-dimethoxyphenyl)-2-(4-methoxyphenoxy)- (9CI)
Veratryl alcohol, α-(2-hydroxy-1-(p-methoxyphenoxy)ethyl)- (8CI)

22694-96-8
$C_6H_{11}NO$
113.18
3-Oxazoline, 2,4,5-trimethyl
2,4,5-Trimethyl-3-oxazoline

22699-70-3
$C_{10}H_{12}O$
148.20
Ethanone, 1-(3-ethylphenyl)- (9CI)
Acetophenone, 3'-ethyl- (8CI)

22707-35-3
$C_{22}H_{42}O_4$
370.57
O=C(OCCCCCCCCCC)CCCCC(=O)OCCCCCC
Hexanedioic acid, decyl hexyl ester (9CI)
Decyl hexyl hexanedioate

22722-98-1
$C_6H_{16}AlO_4.Na$
202.16
[Na+].COCCO[AlH2-]OCCOC
Dihydrobis(2-methoxyethoxy)-aluminate
Aluminate(1-), dihydrobis-(2-methoxyethanolato-O,O')-, sodium (9CI)

22750-93-2
$C_2H_5ClO_4$
128.51
Ethyl perchlorate
Perchlorate d'ethyle (French)
Perchloric acid, ethyl ester
Perclorato de etilo (Spanish)

22752-98-3
$C_{10}H_9ClN_2$
192.64
1,4'-Bipyridinium, chloride
AI3-61629

22771-17-1
$C_{12}H_{22}OSn$
301.03
Stannane, dicyclohexyloxo
Dicyclohexyltin oxide

22771-18-2
$C_6H_{12}O_2Sn$
234.87
Stannane, cyclohexylhydroxyoxo
Monocyclohexyltin acid

22775-37-7
$C_8H_6Cl_2O_3$
221.04
Benzoic acid, 3,5-dichloro-2-methoxy-
AI3-33373

22781-23-3
$C_{11}H_{13}NO_4$
223.25
CNC(=O)Oc1cccc2OC(C)(C)Oc12
Carbamic acid, methyl-, 2,3-(dimethylmethylenedioxy)phenyl ester
Bencarbate
Bendiocarb
Bendiocarbe
1,3-Benzodioxole, 2,2-dimethyl-4-(N-methylaminocarboxylato)-
1,3-Benzodioxole, 2,2-dimethyl-4-(N-methylcarbamato)-
1,3-Benzodioxol-4-ol, 2,2-dimethyl-, methylcarbamate
Carbamic acid, methyl-, 2,3-(isopropylidenedioxy)-phenyl ester
2,2-Dimethyl-1,3-benzdioxol-4-yl N-methylcarbamate
2,2-Dimethyl-1,3-benzodioxol-4-ol methylcarbamate
2,2-Dimethylbenzo-1,3-dioxol-4-yl methylcarbamate
Dycarb
Ficam
Ficam D
Ficam ULV
Ficam W
Ficam 80W
Garvox
2,3-Isopropylidene-dioxyphenyl methylcarbamate
Methylcarbamic acid 2,3-(isopropylidenedioxy)phenyl ester
Multamat
Multimet
NC 6897
Niomil
OMS-1394
Rotate
Seedox
Tattoo
Turcam

22788-18-7
$C_7H_{15}Cl_2N_2O_4P$
293.09
Hydracrylic acid, N,N-bis-(2-chloroethyl)phosphorodiamidate
Carboxycyclophosphamide
Carboxyphosphamide
Propanoic acid, 3-((amino(bis-(2-chloroethyl)amino)phosphinyl)oxy)-

22808-06-6
$C_{10}H_{20}$
140.27
3-Hexene, 2,2,5,5-tetramethyl- (8CI,9CI)

22818-40-2
$C_8H_9NO_3$
167.16
O=C(O)C(N)c(ccc(O)c1)c1
Benzeneacetic acid, α-amino-4-hydroxy-, (R)- (9CI)
(R)-α-Amino-4-hydroxybenzeneacetic acid

22822-99-7
$C_{11}H_{18}$
150.27
Cyclopropane, 1-ethenyl-2-hexenyl-, (1α,2β(E))-(.+-.)- (9CI)
(.+-.)-Dictyopterene A
1-Hexene, 1-(2-vinylcyclopropyl)-, (E)-trans-(.+-.)- (8CI)

22824-32-4
$C_{13}H_{18}$
174.29

Naphthalene, 1,2,3,4-tetrahydro-1,4,6-trimethyl- (7CI,8CI,9CI)
1,4,6-Trimethyl-1,2,3,4-tetrahydronaphthalene

22826-61-5
$C_6N_{12}O_2$
272.07
Tetraazido benzene quinone
Azidanil
p-Benzoquinone, 2,3,5,6-tetraazido-
2,5-Cyclohexadiene-1,4-dione, 2,3,5,6-tetraazido-
Tetraazido-p-benzoquinone
Tetrazido-1,4-benzoquinone

22839-47-0
$C_{14}H_{18}N_2O_5$
294.34
O=C(NC(C(=O)OC)Cc(cccc1)c1)C(N)CC(=O)O
Succinamic acid, 3-amino-N-(α-carboxyphenethyl)-, N-methyl ester, stereoisomer
3-Amino-N-(α-carboxyphenethyl)succinamic acid N-methyl ester, stereoisomer
Aspartame
Aspartylphenylalanine methyl ester
Canderel
Dipeptide sweetener
Equal
Methyl aspartylphenylalanate
1-Methyl N-l-α-aspartyl-l-phenylalanine
Nutrasweet
l-Phenylalanine, N-l-α-aspartyl-, 1-methyl ester (9CI)
Sweet Dipeptide

22862-76-6
$C_{14}H_{19}NO_4$
265.34
COc2ccc(CC1NCC(O)C1OC(C)=O)cc2
Anisomycin
Antibiotic PA-106
Flagecidin
2-(p-Methoxybenzyl)-3,4-pyrrolidinediol 3-acetate
2-p-Methoxyphenylmethyl-3-acetoxy-4-hydroxypyrrolidine
3,4-Pyrrolidinediol, 2-(p-methoxybenzyl)-, 3-acetate, (2S,3R,4R)-

22907-64-8
$C_7H_{17}O_3PS$
212.25
Phosphorothioic acid, S-methyl O,O-bis(1-methylethyl) ester (9CI)
Phosphorothioic acid, O,O-diisopropyl S-methyl ester (8CI)

22936-75-0
$C_{11}H_{21}N_5S$
255.43
CCNc1nc(NC(C)C(C)C)nc(SC)n1
s-Triazine, 2-((1,2-dimethylpropyl)amino)-4-ethylamino-6-methylthio
Avirosan
C 18898
Dimethametryn
Dimethametryne
4-(1,2-Dimethyl-N-propylamino)-2-ethylamino-6-methylthio-s-triazine

22936-86-3
$C_9H_{14}ClN_5$
227.73
s-Triazine, 2-chloro-4-cyclopropylamino-6-isopropylamino
2-Chloro-4-cyclopropylamino-6-isopropylamino-s-triazine
2-Chloro-4-cyclopropylamino-6-isopropylamino-1,3,5-triazine
Cyprazine
Outfox
S-6115
S-9115
1,3,5-Triazine-2,4-diamine, 6-chloro-N-cyclopropyl-N'-(1-methylethyl)-

22941-83-9
$C_{14}H_{21}N_2O_5PS$
360.36
Phosphorothioic acid, O,O-diethyl O-(4-(1-((((methylamino)carbonyl)oxy)imino)ethyl)phenyl) ester
AI3-27542

22948-02-3
C_6H_7NS
125.19
Benzenethiol, 3-amino- (9CI)
Benzenethiol, m-amino- (8CI)

22965-60-2
Unknown
Unknown
Nitrilotriacetic acid, nickel(3+) complex
NTA, nickel(3+) complex

22966-79-6
$C_{42}H_{50}Cl_4N_2O_4$
788.74
N(CCCl)(CCCl)c6ccc(CC(=O)Oc4ccc3C2CCC1(C)C(CCC1C2CCc3c4)OC(=O)Cc5ccc(cc5)N(CCCl)CCCl)cc6
Estra-1,3,5(10)-triene-3,17-β-diol, bis(p-(bis(2-chloroethylamino)phenyl)acetate)
Bis((4-(bis(2-chloroethyl)-amino)benzene)acetate)estra-1,3,5(10)-triene-3,17-diol(17β)
Bis((4-(bis(2-chloroethyl)-amino)benzene)acetate)oestra-1,3,5(10)-triene-3,17-diol(17β)
Bis((p-(bis(2-chloroethyl)-amino)phenyl)acetate)estradiol
Bis((p-(bis(2-chloroethyl)-amino)phenyl)acetate)estra-1,3,5(10)-triene-3,17-β-diol
Bis((p-(bis(2-chloroethyl)-amino)phenyl)acetate)oestradiol
Bis((p-bis(2-chloroethyl)amino-phenyl)acetate)oestra-1,3,5(10)-triene-3,17-β-diol
Estradiol Mustard
NCI-C01570
NSC-112259
Oestradiol Mustard

22967-92-6
CH_3Hg
215.63
C[Hg]
Mercury(1+), methyl-, ion
Mercury(1+), methyl- (9CI)
Methylmercury
Methylmercury(1+)
Methylmercury(II) cation
Methylmercury ion
Methylmercury ion(1+)

22975-58-2
$C_{12}H_{18}$
162.27
Benzene, 1-(1-ethylpropyl)-4-methyl- (9CI)
Toluene, p-(1-ethylpropyl)- (8CI)

22975-76-4
$C_{13}H_{10}O_3$
214.23
2H-Furo(2,3-h)(1)benzopyran-2-one, 4,9-dimethyl
4,4'-Dimethylangelicin plus ultraviolet a radiation
4,9-Dimethyl-2H-furo(2,3-h)-1-benzopyran-2-one plus ultraviolet a radiation
4,4'-Dimethylisopsoralen plus ultraviolet a radiation

22984-54-9
$C_{13}H_{27}N_3O_3Si$
301.44
2-Butanone, O,O',O''-(methylsilylidyne)trioxime (9CI)

22986-69-2
$C_{33}H_{66}O$
478.89
17-Tritriacontanone (8CI,9CI)
17-Triacontanone

23003-22-7
C_5H_5NOS
127.16
3-Hydroxypyridine-2-thiol
2-Mercapto-3-hydroxypyridine
NSC-283470
2(1H)-Pyridinethione, 3-hydroxy- (9CI)

23005-56-3
$C_{17}H_{16}O_4$
284.31
1,2-Benzenedicarboxylic acid, mono(1-(4-methylphenyl)-ethyl) ester (9CI)
Phthalic acid, mono(p,α-dimethylbenzyl) ester (8CI)

23010-07-3
$C_5H_{10}Cl_2$
141.04
Butane, 1,3-dichloro-2-methyl- (8CI,9CI)
1,3-Dichloro-2-methyl-butane

23031-25-6
$C_{12}H_{19}NO_3$
225.32
CC(C)(C)NCC(O)c1cc(O)cc(O)c1
Benzyl alcohol, α-((t-butyl-amino)methyl)-3,5-dihydroxy
1,3-Benzenediol, 5-(2-((1,1-dimethylethyl)amino)-1-hydroxyethyl)- (9CI)
Brican
Bricanyl
Bricar
Bricaril
Bricyn
5-(2-((1,1-Dimethylethyl)-amino)-1-hydroxyethyl)-1,3-benzenediol
KWD 2019
Terbutalin
Terbutaline

23031-38-1
$C_{18}H_{22}O_3$
286.40
Cyclopropanecarboxylic acid, 2,2-dimethyl-3-(2-methyl-propenyl)-, 5-(2-propynyl)-furfuryl ester
Furamethrin
5-Propargylfurfuryl chrys-anthemate
5-Propargyl-2-furylmethyl dl-cis,trans-chrysanthemate
Prothrin

23066-18-4
CH_3ClSn
169.20
Stannanediylium, chloromethyl- (9CI)
Tin(2+), chloromethyl-, ion (8CI)

23087-46-9
$C_2H_4O_2.xCu.xH_3N$
Unknown
Acetic acid, ammonium copper salt (9CI)
Ammonium copper acetate
Copper ammonium acetate

23092-17-3
$C_{17}H_{12}ClF_3N_2O$
352.73
Halazepam
2H-1,4-Benzodiazepin-2-one, 1,3-dihydro-7-chloro-5-phenyl-1-(2,2,2-trifluoroethyl)-
7-Chloro-1,3-dihydro-5-phenyl-1-(2,2,2-trifluoroethyl)-2H-1,4-benzodiazepin-2-one
DEA No. 2762
Halazepamum (Latin)
Paxipam
Sch 12041
Schering 12041

23097-98-5
C_9H_{16}
124.23
3-Heptyne, 5,5-dimethyl- (8CI,9CI)
5,5-Dimethyl-3-heptyne

23102-02-5
$C_8H_{15}NO_4$
161.16
Pentanoic acid, 4-methyl-4-nitro-, ethyl ester (9CI)
Valeric acid, 4-methyl-4-nitro-, ethyl ester (8CI)

23103-98-2
$C_{11}H_{18}N_4O_2$
238.33
CN(C)C(=O)Oc1nc(nc(C)c1C)N(C)C
Carbamic acid, dimethyl-, 2-(dimethylamino)-5,6-dimethyl-4-pyrimidyl ester
Abol
Aficida
Aphox
2-(Dimethylamino)-5,6-dimethyl-4-pyrimidinyldimethylcarbamate
Dimethylcarbamic acid 2-(dimethylamino)-5,6-dimethyl-4-pyrimidinyl ester
5,6-Dimethyl-2-dimethylamino-4-pyrimidinyldimethylcarbamate
ENT 27,766
Fernos
Pirimicarb
Primicarbe
Pirimor
PP 062
Pyrimor
Rapid

23109-05-9
$C_{39}H_{54}N_{10}O_{14}S$
919.09
α-Amanitine
α-Amanitin (8CI,9CI)

23120-99-2
C_2H_6Sn
148.76
Stannylene, dimethyl- (8CI, 9CI)
Dimethylstannylene
Dimethyltin
Tin, dimethyl- (6CI)

23128-51-0
$C_{17}H_{24}N_2O_6$
352.38
O=C(OCCN(c(c(OC)ccc1NC(=O)C)c1)CCOC(=O)C)C
Acetamide, N-(3-(bis(2-(acetyl-oxy)ethyl)amino)-4-methoxy-phenyl)- (9CI)
N-(3-(Bis(2-acetoxyethyl)-amino)-4-methoxyphenyl)-acetamide
3-Diacetoxyethylamino-4-methoxyacetanilide

23128-74-7
$C_{40}H_{64}N_2O_4$
636.95
O=C(NCCCCCCNC(=O)CCc(cc(c(O)c1C(C)(C)C)C(C)(C)C)c1)CCc(cc(c(O)c2C(C)(C)C)C(C)(C)C)c2
Benzenepropanamide, N,N'-1,6-hexanediylbis(3,5-bis-(1,1-dimethylethyl)-4-hydroxy- (9CI)

23135-22-0
$C_7H_{13}N_3O_3S$
219.29
CNC(=O)ON=C(SC)C(=O)N(C)C
Oxamimidic acid, N',N'-dimethyl-N-((methylcarbamoyl)oxy)-1-methylthio
D-1410
2-(Dimethylamino)-N-(((methyl-amino)carbonyl)oxy)-2-oxo-ethanimidothioic acid methyl ester
2-Dimethylamino-1-(methyl-thio)glyoxal O-methylcar-bamoylmonoxime
N,N-Dimethyl-α-methylcar-bamoyloxyimino-α-(methyl-thio)acetamide
N',N'-Dimethyl-N-((methylcar-bamoyl)oxy)-1-thiooxam-imidic acid methyl ester
DPX 1410

Insecticide-Nematicide 1410
Methyl 2-(dimethylamino)-N-(((methylamino)carbonyl)-oxy)-2-oxoethanimidothioate
Methyl 1-(dimethylcarbamoyl)-N-(methylcarbamoyloxy)thioformimidate
S-Methyl 1-(dimethylcarbamoyl)-N-((methylcarbamoyl)-oxy)thioformimidate
Methyl N',N'-dimethyl-N-((methylcarbamoyl)oxy)-1-thiooxamimidate
Oxamyl
Thioxamyl
Vydate
Vydate L Insecticide/Nematicide
Vydate L Oxamyl Insecticide/Nematocide

23143-01-3
$C_{21}H_{14}$
266.35
1H-Dibenzo(a,h)fluorene (8CI,9CI)

23159-07-1
$C_7H_{16}N_2$
128.21
N(CCC1)(C1)CCCN
1-Pyrrolidinepropanamine (9CI)
N-(3-Aminopropyl)pyrrolidine
1-(3-Aminopropyl)pyrrolidine
NSC-345670
Pyrrolidine, 1-(3-aminopropyl)-
3-Pyrrolidinopropylamine
3-(1-Pyrrolidinyl)propylamine

23184-66-9
$C_{17}H_{26}ClNO_2$
311.89
CCCCOCN(C(=O)CCl)c1c(CC)cccc1CC
Acetanilide, 2-chloro-2',6'-diethyl-N-(butoxymethyl)
Acetamide, N-(butoxymethyl)-2-chloro-N-(2,6-diethylphenyl)- (9CI)
Acetanilide, N-(butoxymethyl)-2-chloro-2',6'-diethyl- (8CI)
Butachlor
Butanex
N-Butoxymethyl-2-chloro-2',6'-diethylacetanilide
N-(Butoxymethyl)-2-chloro-N-(2,6-diethylphenyl)acetamide
2-Chloro-2',6'-diethyl-N-(butoxymethyl)acetanilide
CP 53619
Lambast
Machete
Machete (Herbicide)
Machette
Rasayanchlor

23185-94-6
$C_{33}H_{51}NO_7$
573.76
Cycloposine
17,23 β-Epoxyveratraman-3 β-yl-β-D-glucopyranoside
β-D-Glucopyranoside, (3β,23β)-17,23-epoxyveratraman-3-yl
3-Glucosyl-11-deoxojervine

23189-64-2
$C_{17}H_{16}$
220.32
Phenanthrene, 1,2,4-trimethyl- (6CI,7CI,8CI, 9CI)

23209-59-8
$HO_3P.Ca.Na$
143.05
Metaphosphoric acid, calcium sodium salt
Calcium sodium metaphosphate

23214-92-8
$C_{27}H_{29}NO_{11}$
543.57
COc4cccc5C(=O)c3c(O)c2CC(O)(CC(OC1CC(N)C(O)C(C)O1)c2c(O)c3C(=O)c45)C(=O)CO
Adriamycin
ADM
Adriamycin semiquinone
Adriblastin
Adriblastina
Doxorubicin
DX
F.I 106
14-Hydroxydaunomycin
14'-Hydroxydaunomycin
14-Hydroxydaunorubicine
KW-125
NCI-C01514
NDC 38242-874
NSC-123127

23239-32-9
$C_{10}H_{15}NO$
165.23
Benzeneethanamine, 4-methoxy-α-methyl-, (+-)- (9CI)
DEA No. 7411
p-Methoxyamphetamine
4-Methoxyamphetamine
(+-)-p-Methoxyamphetamine
(+-)-4-Methoxy-α-methyl-benzeneethanamine
4-Methoxy-α-methylphenethylamine
(+-)-p-Methoxy-α-methylphenethylamine
Paramethoxy-amphetamine
Phenethylamine, p-methoxy-α-methyl-, (+-)-
PMA

23239-41-0
$C_{13}H_{12}N_3O_6S.Na$
361.33
[Na+].CC(=O)OCC1=C(N2C(SC1)C(NC(=O)CC#N)C2=O)C([O-])=O
5-Thia-1-azabicyclo(4.2.0)oct-2-ene-2-carboxylic acid, 7-(2-cyanoacetamido)-3-(hydroxy methyl)-8-oxo-, acetate (ester), monosodium salt
Ba 36278
36278-Ba
Cefacetrile sodium
Celospor
Cephacetrile sodium
Ciba 36278-BA
Sodium cefacetril
Sodium cephacetrile
Sodium 7-(2-cyanoacetamido)-cephalosporanic acid

23246-96-0
$C_{18}H_{23}NO_6$
349.42
CC=C1CC(=C)C(O)(CO)C(=O)OCC2=CCN3CCC(OC1=O)C23
Riddelline
Riddelliin
Riddelliine
Senecionan-11,16-dione, 13,19-didehydro-12,18-dihydroxy- (9CI)

23248-23-9
$C_2H_6O_2.Li$
69.01
1,2-Ethanediol, monolithium salt (9CI)
Lithium ethylene glycoxide
Lithium 2-hydroxyethoxide

23251-72-1
$C_4H_{11}NO_2.C_2H_4O_2$
165.18
Ethanol, 2,2'-iminobis-, acetate (Salt) (9CI)
2,2'-Iminobisethanol acetate
2,2'-Iminobisethanol acetate (Salt)

23255-03-0
$C_6H_9NO_6.H_2O.2Na$
255.14
Glycine, N,N-bis(carboxymethyl)-, disodium salt, monohydrate (9CI)
Acetic acid, nitrilotri-, disodium salt, monohydrate (8CI)
Disodium nitrilotriacetic acid monohydrate
Nitrilotriacetic acid, disodium salt, monohydrate

23255-69-8

$C_{17}H_{22}O_8$
354.39
CC(=O)OC2C(O)C3OC1C=C(C)C(=O)C(O)C1(CO)C2(C)C34CO4
Trichothec-9-en-8-one, 4-(acetyloxy)-12,13-epoxy-3,7,15-trihydroxy-, (3-α,4-β,7-β)
4-Acetyloxy-12,13-epoxy-3,7,15-trihydroxy-(3-α,4-β,7-β)-trichothec-9-en-8-one
12,13-Epoxy-3-α,4-β,7-β,15-tetrahydroxytrichothec-9-en-8-one 4-acetate
Fusarenon
Fusarenon X
Fusarenone X
Nivalenol-4-O-acetate
3,7,15-Trihydroxy-4-acetoxy-8-oxo-12,13-epoxy-δ^9-trichothecene
3,7,15-Trihydroxyscirp-4-acetoxy-9-en-8-one

23255-93-8
$C_{20}H_{24}N_2O_2S.CH_4O_3S$
452.63
CCN(CC)CCNc2ccc(CO)c3sc1ccccc1c(=O)c23.CS(O)(=O)=O
9H-Thioxanthen-9-one, 1-((2-(diethylamino)ethyl)-amino)-4-(hydroxymethyl)-, monomethanesulfonate (Salt)
Etrenol
HCT
Hycanthone mesylate
Hycanthone methanesulfonate
Hycanthone methanesulphonate
Hycanthone monomethanesulphonate

23262-78-4
$C_{12}H_{20}O_6$
260.29
Talofuranose, 2,3:5,6-di-O-isopropylidene-, α-D-(8CI)

23262-79-5
$C_{12}H_{20}O_6$
260.29
β-D-Talofuranose, 1,2:5,6-bis-O-(1-methylethylidene)- (9CI)
Furo(2,3-d)-1,3-dioxole, β-D-talofuranose deriv. (9CI)
Talofuranose, 1,2:5,6-di-O-isopropylidene-, β-D-(8CI)

23275-26-5
$C_6H_{14}O.1/3Al$
111.17
1-Hexanol, aluminum salt (9CI)
Aluminum n-hexoxide

23282-20-4
$C_{15}H_{20}O_7$
312.35
Trichothec-9-en-8-one, 12,13-epoxy-3,4,7,15-tetrahydroxy-, (3-α,4-β,7-α)
12,13-Epoxy-3,4,7,15-tetrahydroxytrichothec-9-en-8-one
Nivalenol
3-α,4-β,7-α,15-Tetrahydroxyscirp-9-en-8-one

23287-26-5
$C_9H_{10}O_3$
166.19
O=C(OC)c(c(O)c(cc1)C)c1
Benzoic acid, 2-hydroxy-3-methyl-, methyl ester
2,3-Cresotic acid, methyl ester (8CI)
2-Hydroxy-3-methylbenzoic acid methyl ester
Levegal PT
Methyl 2-hydroxy-3-methylbenzoate

23290-55-3
$C_8H_{12}OS$
156.25
Thiophene, 2-(1,1-dimethylethoxy)- (9CI)
2-tert-Butoxythiophene
2-(1,1-Dimethylethoxy)thiophene
Thiophene, 2-tert-butoxy-(6CI,7CI,8CI)

23297-24-7
$C_6H_6N_6O_3$
210.15
s-Triazine-2,4,6-tricarboxamide (8CI)
2,4,6-Tricarbamoyl-s-triazine

23307-05-3
$C_2H_7N.H_2O_4S$
143.16
Methanamine, N-methyl-, sulfate (1:1) (9CI)
N-Methylmethanamine sulfate (1:1)

23311-84-4
$C_6H_{10}O_4.xNa$
Unknown
Hexanedioic acid, sodium salt (9CI)
Sodium hexanedioate

23315-05-1
$C_{13}H_{26}N_2O_3$
258.41
2-Butanol, 4-methoxy-3-(1-octenyl-ONN-azoxy)-, (E,Z)-(2S,3S)
Elaiomycin
(E,Z)-(2S,3S)-4-Methoxy-3-(1-octenyl-ONN-azoxy)-2-butanol

23319-51-9
Unknown
Unknown
Nitrilotriacetic acid, cobalt(3+) complex
NTA, cobalt(3+) complex

23319-66-6
$C_{12}H_{20}HgNO_3.C_3H_5O_3$
515.95
Phenylmercuric triethanolammonium lactate
Ammonium, tris(2-hydroxyethyl)(phenylmercurio)-, lactate
Caswell No. 657N
EPA Pesticide Chemical Code 066021
Fenylmerkuri-tris-(2-hydroxyethyl)ammoniumlaktat (Czech)
Lactic acid, tris(2-hydroxyethyl)(phenylmercuri)-ammonium deriv.
Lactic acid, ion(1-), tris(2-hydroxyethyl)(phenylmercurio)ammonium
Mercury(1+), (2,2',2''-nitrilotriethanol)phenyl-, lactate (Salt) (8CI)
Phenylmercuritriethanolammonium lactate
Phenylmercury triethanolamine lactate
PTAB
Puratized
Puratized Agricultural Spray
Puratized N5E
Puratizedat Agricultural Spray
Puraturf
Tris(2-hydroxyethyl)phenylmercuriammonium lactate

23328-69-0
C_4H_8ClNO
121.56
O(CCN(Cl)Cl)Cl
Morpholine, 4-chloro- (9CI)
4-Chloromorpholine

23342-25-8
$C_{14}H_{20}$
188.32
Naphthalene, 1,2,3,4-tetrahydro-2,2,5,7-tetramethyl- (8CI,9CI)
2,2,5,7-Tetramethyltetralin

23349-61-3
$C_6H_8O_7.xC_4H_{11}NO_2$
Unknown
Ethanol, 2,2'-iminobis-, 2-hydroxy-1,2,3-propanetricarboxylate (Salt) (9CI)
Citric acid, diethanolamine salt

23351-51-1
$C_7H_{14}O_8$
226.18
O=C(O)C(O)C(O)C(O)C(O)C(O)CO
D-Gluco-heptonic acid, (2.XI.)- (9CI)
(2.XI.)-D-Gluco-heptonic acid

23355-64-8
$C_{16}H_{15}Cl_3N_4O_4$
433.65
O=N(=O)c(cc(c(N=Nc(c(cc(N(CCO)CCO)c1)Cl)c1)c2Cl)Cl)c2
C.I. Disperse Brown 1
2,2'-((3-Chloro-4-((2,6-dichloro-4-nitrophenyl)azo)phenyl)-amino)bisethanol
Ethanol, 2,2'-((3-chloro-4-((2,6-dichloro-4-nitrophenyl)azo)-phenyl)imino)bis- (9CI)

23360-92-1
$C_{46}H_{56}N_4O_9$
809.06
Leurosine
Amotin (Russian)
NSC-528004
Vincaleukoblastine, 4'-deoxy-3',4'-epoxy-, (3'-α,4'-α)-(9CI)
Vinleurosine

23383-11-1
$C_6H_8O_7.xFe$
Unknown
Ferrous citrate
Citric acid, iron(2+) salt
Monoferrous acid citrate
1,2,3-Propanetricarboxylic acid, 2-hydroxy-, iron(2+) salt

23386-52-9
$C_{16}H_{26}O_7S.Na$
385.43
Dicyclohexyl sodium sulfosuccinate
Butanedioic acid, sulfo-, 1,4-dicyclohexyl ester, sodium salt (9CI)
Succinic acid, sulfo-, 1,4-dicyclohexyl ester, sodium salt
Sulfosuccinic acid, dicyclohexyl ester, sodium salt
Sulfosuccinic acid, 1,4-dicyclohexyl ester, sodium salt

23413-80-1
$C_{18}H_{15}AlO_9$
402.31
Aluminum, hydroxybis(salicylic acid acetato)
Aluminium acetylsalicylate
Aluminum Aspirin
Aluminum, bis(2-(acetyloxy)-benzoato-O[1])hydroxy- (9CI)
Aluminum bis(acetylsalicylate)
Aspirin aluminium
Hydroxybis(salicylic acid acetato)aluminum
Hypyrin

23414-72-4
$Mn_2O_8.Zn$
303.25
Zinc permangante [UN 1515]
UN 1515 [Zinc permanganate]

23422-53-9
$C_{11}H_{15}N_3O_2.ClH$
257.75
[Cl-].CNC(=O)Oc1cccc(N=CN(C)C)c1
Carbamic acid, methyl-, ester with N'-(m-hydroxyphenyl)-N,N-dimethylformamidine hydrochloride
Carzol
Carzol SP
Dicarzol
m-(((Dimethylamino)methylene)amino)phenyl methyl carbamate, hydrochloride
3-Dimethylaminomethyleneiminophenyl-N-methylcarbamate, hydrochloride
N,N-Dimethyl-N'-(((methylamino)carbonyl)oxy)phenylmethanimidamide monohydrochloride
ENT 27,566
EP-332
Formetanate hydrochloride
Methanimidamide, N,N-dimethyl-N'-(3-(((methylamino)-carbonyl)oxy)phenyl)-, monohydrochloride
Morton EP 332
Nor-Am EP 332
Optunal
Schering 36056
SN 36056

23436-19-3
$C_7H_{16}O_2$
132.23
O(CC(O)C)CC(C)C
2-Propanol, 1-isobutoxy
Dowanol PIB-T
1-Isobutoxy-2-propanol
2-Propanol, 1-(2-methylpropoxy)- (9CI)
Propylene glycol isobutyl ether

23451-24-3
$C_{25}H_{39}NO_6.ClH$
486.11
8-β-Podocarpan-16-oic acid, 13-(carboxymethylene)-3-β-hydroxy-14-methyl-7-oxo-, 13- (2-(dimethylamino)ethyl) 16-methyl ester, hydrochloride
Erythrophlamine hydrochloride

23474-91-1
$C_8H_{14}O_3$
158.20
O=C(OOC(C)(C)C)C=CC
tert-Butyl peroxycrotonate (Not more than 76% in solution)
2-Buteneperoxoic acid, 1,1-dimethylethyl ester (9CI)
tert-Butyl percrotonate
tert-Butyl peroxycrotonate
tert-Butyl peroxycrotonate, Not more than 76% in solution
1,1-Dimethylethyl 2-buteneperoxoate
Peroxicrotonato de terc-butilo (Spanish)
Peroxycrotonate de tert-butyle (French)
Peroxycrotonic acid, tert-butyl ester, Not more than 76% in solution
UN 2183

23481-33-6
$C_{16}H_{12}N_2O_4S$
328.34
2-Naphthalenesulfonic acid, 6-hydroxy-5-(phenylazo)-(9CI)

23486-02-4
$C_{11}H_{17}AsS$
256.24
Arsinothious acid, ethylphenyl-, 1-methylethyl ester (9CI)
Arsine, ethyl(isopropylthio)-phenyl- (8CI)

23488-38-2
$C_8H_6Br_4$
421.78
c(c(c(c(c1Br)C)Br)Br)(c1Br)C
p-Xylene, α,α,α',α'-tetrabromo
1,4-Bis(dibrommethyl)benzen (Czech)
p-TBX
Tetrabromo-p-xylen (Czech)
Tetrabromo-p-xylene

23491-45-4
$C_{25}H_{24}N_6O.3ClH$
533.93
Phenol, p-(5-(5-(4-methyl-1-piperazinyl)-2-benzimidazolyl)-2-benzimidazolyl)-,

trihydrochloride
Bisbenzimidazole
Bisbenzimide
H 33258
HOE 33058
HOE 33258
33258 Hoechst
Hoechst 33258
Hoechst Dye 33258
NSC-322921
Phenol, 4-(5-(4-methyl-1-piperazinyl)(2,5'-bi-1H-benzimidazol)-2'-yl)-, trihydrochloride

23500-79-0
$C_{13}H_{19}ClO$
226.75
Oc(c(cc(c1CCl)C)C(C)(C)C)c1C
Phenol, 3-(chloromethyl)-6-(1,1-dimethylethyl)-2,4-dimethyl- (9CI)
6-tert-Butyl-3-chloromethyl-2,4-xylenol
3-(Chloromethyl)-6-(1,1-dimethylethyl)-2,4-dimethylphenol
2,4-Dimethyl-3-(chloromethyl)-6-tert-butylphenol

23505-21-7
$C_7H_6Cl_2N_2O$
205.03
Benzenecarboximidamide, 2,6-dichloro-N-hydroxy- (9CI)
Benzamidoxime, 2,6-dichloro- (8CI)

23505-41-1
$C_{13}H_{24}N_3O_3PS$
333.43
CCOP(=S)(OCC)Oc1cc(C)nc(n1)N(CC)CC
Phosphorothioic acid, O,O-diethyl O-(2-(diethylamino)-6-methyl-4-pyrimidinyl) ester
O-(2-(Diethylamino)-6-methyl-4-pyrimidinyl)O,O-diethyl phosphorothioate
2-Diethylamino-6-methylpyrimidin-4-yl diethylphosphorothionate
O,O-Diethyl O-(2-diethylamino-6-methyl-4-pyrimidinyl)phosphorothioate
Ethyl pirimiphos
Fernex
Pirimifosethyl
Pirimiphos-ethyl
PP211
Primicid
Primotec
Prinicid
R 42211
Solgard

23509-16-2
$C_{31}H_{42}N_2O_6$
538.75
Batrachotoxin
Batrachotoxinin A, 20-(2,4-dimethyl-1H-pyrrole-3-carboxylate)
Betrachotoxinin A, 20-α-(2,4-dimethyl-1H-pyrrole-3-carboxylate) (9CI)
BTX

23519-77-9
$C_3H_8O.1/4Zr$
82.90
1-Propanol, zirconium(4+) salt (9CI)
Propyl alcohol, zirconium(4+) salt
Tetrapropyl zirconate

23537-16-8
$C_{30}H_{22}O_{10}$
542.52
Rugulosin
Radicalisin
(+)-Rugulosin

23552-76-3
$C_{21}H_{15}NO_4$
345.35
O=C(c(c(C(=O)c1c(O)ccc2Nc(ccc(OC)c3)c3)ccc4)c4)c12
9,10-Anthracenedione, 1-hydroxy-4-((4-methoxyphenyl)amino)- (9CI)
1-p-Anisidino-4-hydroxyanthraquinone

23555-96-6
Unknown
Unknown
Nitrilotriacetic acid, potassium strontium salt (2:4:1)
NTA, potassium strontium salt (2:4:1)

23555-98-8
Unknown
Unknown
Nitrilotriacetic acid, calcium potassium salt (2:1:4)
NTA, calcium potassium salt (2:1:4)

23564-05-8
$C_{12}H_{14}N_4O_4S_2$
342.42
COC(=O)NC(=S)Nc1ccccc1NC(=S)NC(=O)OC
Allophanic acid, 4,4'-o-phenylenebis(3-thio-, dimethyl ester
BAS 32500F
o-Bis(3-methoxycarbonyl-2-thioureido)benzene
1,2-Bis(methoxycarbonylthioureido)benzene
1,2-Bis(3-(methoxycarbonyl)-2-thioureido)benzene
Carbamic acid, (1,2-phenylenebis(iminocarbonothioyl))bis-, dimethyl ester
Cercobin M
Cercobin methyl
Cycosin
Dimethyl ((1,2-phenylene)bis-(iminocarbonothioyl))bis(carbamate)
Dimethyl-4,4'-o-phenylene-bis-(3-thioallophanate)
Ditek
Enovit M
Enovit Methyl
Enovit-Supper
Frumidor
Fungitox
Fungo
Fungo 50
Labilite
Methylthiofanate
Methyl thiophanate
Methyl Topsin
Mildothane
Neotopsin
NF 44
Pelt 14
Pelt-44
Sigma
Sipcaplant
Sipcasan
Sipcavit
TD 1771
Thiophanate M
Thiophanate methyl
Tiofanate metile (Italian)
Topsin M
Topsin NF-44
Topsin Turf and Ornamentals
Topsin WP Methyl
Trevin
Zyban

23564-06-9
$C_{14}H_{18}N_4O_4S_2$
370.48
CCOC(=O)NC(=S)Nc1ccccc1NC(=S)NC(=O)OCC
Allophanic acid, 4,4'-o-phenylenebis(3-thio-, diethyl ester
BAS 3220
1,2-Bis-(3-ethoxycarbonyl-thioureido)-benzene
1,2-Bis(3-ethoxycarbonyl-2-thioureido) benzene
Carbamic acid, (1,2-phenylenebis(iminocarbonothioyl))bis-, diethyl ester (9CI)
Cercobin
Cerobin
Cleary 3336
Diethyl 4,4'-o-phenylenebis-(3-thioallophanate)
Enovit
Ethyl thiophanate
Nemafax
NF 35
NF 35 (Fungicide)

Pelt
Pelt Sol
Thiofanate
Thiophanat (German)
Thiophanate
Thiophanate ethyl
Tiofanate etile (Italian)
Thiophenite
Topsin
Topsin E
Topsin NF 35
3336 Turf Fungicide
Vermadax

23576-23-0
$C_{13}H_{11}ClF_3N_3O$
317.70
SAN 6706
4-Chloro-5-(dimethylamino)-2-α,α,α- (trifluoro-m-tolyl)-3-(2H)-pyridazinone
SAN 6706-3197
Sandoz 6706

23586-64-3
$C_{18}H_{20}$
236.36
Benzene, 1,1'-(3,3-dimethyl-1-butenylidene)-bis- (9CI)
1-Butene, 3,3-dimethyl-1,1-diphenyl- (8CI)
β-tert-Butyl-1,1-diphenylethylene

23593-75-1
$C_{22}H_{17}ClN_2$
344.86
Clc1ccccc1C(c2ccccc2)(c3ccccc3)n4ccnc4
Imidazole, 1-(o-chloro-α,α-diphenylbenzyl)
Bay 5097
Bay B 5097
Bay-B 5097
Bisphenyl-(2-chlorphenyl)-1-imidazolyl-methan (German)
Canesten
1-(o-Chloro-α,α-diphenylbenzyl)imidazole
1-(α-(2-Chlorophenyl)benzhydryl)imidazole
1-((2-Chlorophenyl)diphenylmethyl)-1H-imidazole (9CI)
(Chlorotrityl)imidazole
1-(o-Chlorotrityl)imidazole
Chlotrimazole
Clotrimazol
Clotrimazole
Diphenyl-(2-chlorophenyl)-1-imidazolylmethane
Empecid
FB 5097
Gyne-Lotrimin
1H-Imidazole, 1-((2-chlorophenyl)diphenylmethyl)-
Lotrimin
Mycelax
Mycelex
Mycelex G
Mycosporin
Trimysten

23601-39-0
$C_{14}H_{30}O_6$
294.39
O(CCOCCOCCOCCOCC)CCOCC
3,6,9,12,15,18-Hexaoxaeicosane (9CI)
Pentaethylene glycol, monobutyl ether

23605-74-5
$C_{10}H_{28}N_2O_{12}P_4$
492.22
O=P(O)(O)CN(CCCCCCN(CP(=O)(O)O)CP(=O)(O)O)CP(=O)(O)O
Phosphonic acid, (1,6-hexanediylbis(nitrilobis(methylene)))tetrakis- (9CI)

23617-71-2
$C_{10}H_{12}O_2$
164.20
Benzoic acid, 2,4-dimethyl-, methyl ester (8CI,9CI)

23654-92-4
$C_4H_8S_3$
152.31
S(C(SS1)C)C1C
1,2,4-Trithiolane, 3,5-dimethyl- (9CI)
3,5-Dimethyl-1,2,4-trithiolane

23666-93-5
$C_{22}H_{23}O_4P$
382.40
O=P(Oc(cccc1)c1)(Oc(c(ccc2)C)c2C)Oc(c(ccc3)C)c3C
Phosphoric acid, bis(2,6-dimethylphenyl) phenyl ester (9CI)
Di(2,6-dimethylphenyl) phenyl phosphate

23666-94-6
$C_{20}H_{19}O_4P$
354.34
O=P(Oc(cccc1)c1)(Oc(cccc2)c2)Oc(c(ccc3)C)c3C
Phosphoric acid, 2,6-dimethylphenyl diphenyl ester (9CI)
Diphenyl 2,6-dimethylphenyl phosphate

23681-60-9
$C_{37}H_{31}N_3O$
533.65
OC(c(ccc(Nc(cccc1)c1)c2)c2)(c(ccc(Nc(cccc3)c3)c4)c4)c(ccc(Nc(cccc5)c5)c6)c6
Benzenemethanol, 4-(phenylamino)-α,α-bis(4-(phenylamino)phenyl)- (9CI)
Pararosaniline base, N,N',N''-triphenyl-

23708-56-7
$C_{11}H_{24}O$
172.31
6-Undecanol (8CI,9CI)
NSC-158434

23710-76-1
C_{14}-H_{13}-O_3.1/2Ca
249.30
Calcium 2-isovaleryl-1,3-indandione
1,3-Indandione, 2-isovaleryl-, ion(1-), calcium (8CI)
2-Isovaleryl-1,3-indandione calcium salt

23713-49-7
Zn
65.40
Zinc, ion (Zn(2+)) (8CI,9CI)

23746-34-1
$C_5H_{10}NO_2S_2$.K
219.38
[K+].OCCN(CCO)C([S-])=S
Carbamic acid, bis(2-hydroxyethyl)dithio-, monopotassium salt
Bis(2-hydroxyethyl)carbamodithioic acid, monopotassium salt
Bis(2-hydroxyethyl)dithiocarbamic acid, monopotassium salt
Bis(2-hydroxyethyl)dithiocarbamic acid, potassium salt
Potassium bis(2-hydroxyethyl)dithiocarbamate

23747-48-0
$C_8H_{10}N_2$
134.17
n(c(c(nc1)CC2)C2C)c1
5H-Cyclopentapyrazine, 6,7-dihydro-5-methyl- (9CI)
6,7-Dihydro-5-methyl-5H-cyclopentapyrazine
5H-5-Methyl-6,7-dihydrocyclopentapyrazine
5-Methyl-6,7-dihydro-5H-cyclopentapyrazine

23749-65-7
$C_7H_5Cl_3$
195.47
Benzene, 1,3,5-trichloro-2-methyl- (9CI)
Toluene, 2,4,6-trichloro- (8CI)

23758-27-2
$C_7H_{12}O$
112.17
2-Cyclohexen-1-ol, 1-methyl- (8CI,9CI)

23779-32-0
$C_{10}H_{24}N_2O_4Si$
264.39
Urea, (3-(triethoxysilyl)-propyl)- (9CI)
N-(Triethoxysilylpropyl)urea
(3-(Triethoxysilyl)propyl)urea

23779-99-9
$C_{20}H_{17}F_3N_2O_4$
406.39
OCC(O)COC(=O)c1ccccc1Nc2ccnc3c(cccc23)C(F)(F)F
Benzoic acid, 2-(8'-trifluoromethyl-4'-quinolylamino)-, 2,3-dihydroxypropyl ester
4-(o-(2',3'-Dihydroxypropyloxycarbonyl)phenyl)-amino-8-trifluoromethylquinoline
2,3-Dihydroxypropyl N-(8-(trifluoromethyl)-4-quinolyl)-anthranilate
Diralgan
Floctafenine
Idarac
Novodolan
R 4318
RU 15750
8-Trifluoromethyl-7-deschloroglafenine
2-(8'-Trifluoromethyl-4'-quinolylamino)benzoic acid 2,3-dihydroxy propyl ester

23783-42-8
$C_9H_{20}O_5$
208.25
O(CCOCCOCCOC)CCO
2,5,8,11-Tetraoxatridecan-13-ol (9CI)
NSC-345692
Tetraethylene glycol monomethyl ether
3,6,9,12-Tetraoxatridecan-1-ol

23783-98-4
$C_{10}H_{19}ClNO_5P$
299.69
(Z)-Phosphamidon
cis-Phosphamidon
Phosphoric acid, 2-chloro-3-(diethylamino)-1-methyl-3-oxo-1-propenyldimethyl ester, (Z)- (9CI)
Phosphoric acid, dimethyl ester, ester with 2-chloro-N,N-diethyl-3-hydroxycrotonamide, (Z)- (8CI)

23787-80-6
$C_7H_8N_2O$
136.14
O=C(c(nccn1)c1C)C
Ethanone, 1-(3-methylpyrazinyl)- (9CI)
2-Acetyl-3-methylpyrazine
2-Methyl-3-acetylpyrazine
1-(3-Methylpyrazinyl)ethanone

23787-97-5
$C_{14}H_{11}NO_4$
257.25
WHR 169
2-(p-Aminobenzoyloxy)benzoic acid
WHR-169

23797-84-4
$C_3H_3B_2D_9$
77.80
Diborane(6), tris(methyl-d3)- (8CI)

23799-25-9
C_9H_{16}
124.23
Cycloheptane, 1-methyl-4-methylene- (8CI,9CI)

23820-80-6
$C_{20}H_{18}O_6$
354.36
Viridiol

23850-94-4
$C_{28}H_{54}O_6Sn$
605.44
Stannane, butyltris((2-ethyl-1-oxohexyl)oxy)- (9CI)
Butyltin tris(2-ethylhexoate)
Butyltris((2-ethyl-1-oxohexyl)-oxy)stannane

23851-46-9
$C_{14}H_{10}Cl_2O_3$
297.14
4,4'-Dichlorobenzilic acid

23887-31-2
$C_{16}H_{11}ClN_2O_3$
314.76
Clorazepate
1H-1,4-Benzodiazepine-3-carboxylic acid, 7-chloro-2,3-dihydro-2-oxo-5-phenyl-
Clorazepic Acid
DEA No. 2768

23947-60-6
$C_{11}H_{19}N_3O$
209.33
CCCCc1c(C)nc(NCC)nc1O
4(1H)-Pyrimidinone, 5-butyl-2-(ethylamino)-6-methyl
5-n-Butyl-2-ethylamino-4-hydroxy-6-methylpyrimidine
5-Butyl-2-(ethylamino)-6-methyl-4(1H)-pyrimidinone
Ethirimol
2-Ethylamino-4-methyl-5-n-butyl-6-hydroxypyrimidine
Ethyrimol
Milcurb
Milcurb Super
Milgo
Milgo E
Milstem
Milstem Seed dressing
New Milstem
PP149

23950-58-5
$C_{12}H_{11}Cl_2NO$
256.14
CC(C)(NC(=O)c1cc(Cl)cc(Cl)c1)C#C
Benzamide, 3,5-dichloro-N-(1,1-dimethyl-2-propynyl)
3,5-Dichloro-N-(1,1-dimethyl-2-propynyl)benzamide
N-(1,1-Dimethylpropynyl)-3,5-dichlorobenzamide
Kerb
Kerb 50W
Promamide
Pronamide
Propyzamide
RCRA waste number U192
RH 315

23976-66-1
$C_8H_7Br_3O_2$
374.85
O(c(c(cc(c1)Br)Br)c1Br)CCO
Ethanol, 2-(2,4,6-tribromophenoxy)- (9CI)
2-(2,4,6-Tribromophenoxy)-ethanol

24009-05-0
$C_{10}H_6Cl_6O$
354.86
OC1C=CC2C1C3(Cl)C(=C(Cl)C2(Cl)C3(Cl)Cl)Cl
4,7-Methanoinden-1-ol, 4,5,6,7,8,8-hexachloro-3a,4,7,7a-tetrahydro-, endo,exo
Hydroxychlordene
1-exo-Hydroxychlordene

24017-47-8
$C_{12}H_{16}N_3O_3PS$
313.34
CCOP(=S)(OCC)Oc1ncn(n1)c2ccccc2
Phosphorothioic acid, O,O-diethyl O-(1-phenyl-1,2,4-triazolyl) ester
O,O-Diethyl O-(1-phenyl-1H-1,2,4-triazol-3-yl)phosphorothioate
HOE 2960
HOE 2960 OJ

Hostathion
1-Phenyl-3-(O,O-diethyl-thionophosphoryl)-1,2,4-triazole
1-Phenyl-1,2,4-triazolyl-3-(O,O-diethylthionophosphate)
Triazofosz (Hungarian)
Triazophos

24019-46-3
$C_{12}H_{14}O_9S.Na$
357.29
1,3-Benzenedicarboxylic acid, 5-sulfo-, 1,3-bis(2-hydroxyethyl) ester, monosodium salt (9CI)
5-Sulfoisophthalic acid, bis-(2-hydroxyethyl) ester, sodium salt

24019-80-5
$C_{42}H_{42}N_{12}O_8S_2.2Na$
953.06
Benzenesulfonic acid, 2,2'-(1,2-ethenediyl)bis(5-((4-((2-methylphenyl)amino)-6-(4- morpholinyl)-1,3,5-triazin-2-yl)amino)-, disodium salt
C.I. Fluorescent Brightener 225
Kayaphor LSK
2,2'-Stilbenedisulfonic acid, 4,4'-bis((4-morpholino-6-o-toluidino-s-triazin-2-yl)-amino)-, 2Na
Whitex SKC

24035-50-5
$C_{20}H_{28}O$
284.44
1-Phenanthrenecarboxaldehyde, 1,2,3,4,4a,9,10,10a-octahydro-1,4a-dimethyl-7-(1-methylethyl)-, (1S-(1α,4aα,10aβ))- (9CI)
Dehydro-4-epiabietal
4-Epiabietal, dehydro-
Podocarpa-8,11,13-trien-16-al, 13-isopropyl- (8CI)

24057-28-1
$C_7H_8O_3S.C_5H_5N$
251.30
Pyridine, 4-methylbenzenesulfonate (9CI)
p-Toluenesulfonic acid, pyridine adduct

24070-77-7
$C_6H_{12}O$
100.16
Cyclopentanol, 2-methyl- (8CI,9CI)

24072-75-1
$C_7H_4Cl_2N_2S$
219.09
N=C(Nc(c1cc(c2Cl)Cl)c2)S1
Benzothiazole, 2-amino-5,6-dichloro
5,6-Dichloro-2-benzothiazolamine

24094-44-8
$C_{15}H_{11}NO_3$
253.26
9,10-Anthracenedione, 1-amino-2-(hydroxymethyl)- (9CI)
1-Amino-2-(hydroxymethyl)anthraquinone
Anthraquinone, 1-amino-2-(hydroxymethyl)- (8CI)

24108-89-2
$C_{40}H_{26}N_2O_6$
630.64
O=C(N(C(=O)c(c1c(c(c(c(c(C(=O)N(C2=O)c(ccc(OCC)c3)c3)c4)c2cc5)c56)c4)cc7)c6c8)c8)c(ccc(OCC)c9)c9)c17
Anthra(2,1,9-def:6,5,10-d'e'f')-diisoquinoline-1,3,8,10(2H,9H)-tetrone, 2,9-bis(4-ethoxyphenyl)- (9CI)
C.I. Pigment Red 123

24124-25-2
$C_{30}H_{58}O_2Sn$
569.50
Tributyltin linoleate
Caswell No. 867E
EPA Pesticide Chemical Code 083109
Stannane, (linoleoyloxy)tributyl-
Stannane, tributyl(linoleoyloxy)-
Stannane, tributyl((1-oxo-9,12-octadecadienyl)oxy)-, (Z,Z)- (9CI)
Tin, tributyl-, linoleate

24126-20-3
$C_9H_{17}N_5O$
211.31
s-Triazine, 4,6-diamino-2-hexoxy

24126-22-5
$C_{15}H_{29}N_5O$
295.43
s-Triazine, 2,4-diamino-6-(dodecyloxy)- (6CI,8CI)

24140-30-5
$C_{22}H_{25}NO_3$
351.44
O=C(OCC(CC)C)C=Cc(ccc(N=Cc(ccc(OC)c1)c1)c2)c2
2-Propenoic acid, 3-(4-(((4-methoxyphenyl)methylene)-amino)phenyl)-, 2-methylbutyl ester, (S-(E,E))- (9CI)

24142-77-6
$C_{12}H_{18}O$
178.27
Benzene, (1-methyl-1-propoxyethyl)- (9CI)
α,α-Dimethylbenzyl propyl ether
(1-Methyl-1-propoxyethyl)benzene

24143-17-7
$C_{18}H_{17}ClN_2O_2$
328.79
Oxazolam
Benzo(6,7)-1,4-diazepino-(5,4-b)-oxazol-6-one, 10-chloro-2,3,5,6,7,11b-hexahydro-2-methyl-11b-phenyl-
DEA No. 2839
Oxazolamum (Latin)
Oxazolazepam
Serenal

24151-93-7
$C_{14}H_{28}NO_3PS_2$
353.52
CCCOP(=S)(OCCC)SCC(=O)N1CCCCC1C
Phosphorodithioic acid, O,O-dipropyl S-(2-pipecolinocarbonylmethyl) ester
Avirosan
C 19490
1-(Di-N-propoxyphosphinothioylthiomethylcarbonyl-2-methylpiperidine)
O,O-Dipropyl S-2-methyl-piperidinocarbonyl-methyl phosphorodithioate
Piperophos
Rilof

24156-95-4
$C_8H_{12}O$
124.18
2-Cyclopenten-1-one, 3,5,5-trimethyl- (6CI,8CI,9CI)
3,5,5-Trimethyl-2-cyclopentenone
3,5,5-Trimethyl-2-cyclopenten-1-one

24157-81-1
$C_{16}H_{20}$
212.33
c(c(ccc1C(C)C)cc(c2)C(C)C)(c2)c1
2,6-Diisopropylnaphthalene
2,6-Bis(1-methylethyl)naphthalene
Naphthalene, 2,6-bis(1-methylethyl)- (9CI)
Naphthalene, 2,6-diisopropyl-
NSC-166467

24166-13-0
$C_{17}H_{14}Cl_2N_2O_2$
349.20
Cloxazolam
10-Chloro-2,3,5,6,7,11b-hexahydro-11b-(o-chlorophenyl)-benzo(6,7)-1,4-diazepino-(5,4-b)-oxazol-6-one
Cloxazolamum (Latin)
Cloxazolazepam
CS 370
DEA No. 2753
Enadel
MT 14-411
Olcadil
Oxazolo(3,2-d)(1,4)benzodiazepin-6(5h)-one, 10-chloro-11b-(o-chlorophenyl)-2,3,7,11b-tetrahydro-
Sepazon

24168-70-5
$C_9H_{14}N_2O$
166.21
O(c(nccn1)c1C(CC)C)C
Pyrazine, 2-methoxy-3-(1-methylpropyl)- (9CI)
2-sec-Butyl-3-methoxypyrazine
2-Methoxy-3-sec-butylpyrazine
2-Methoxy-3-(1-methylpropyl)-pyrazine

24169-02-6
$C_{18}H_{15}Cl_3N_2O.HNO_3$
444.68
Imidazole, 1-(2,4-dichloro-β-((p-chlorobenzyl)oxy)phenethyl)-, mononitrate (8CI)
Econazole nitrate
Gyno-Pevaryl
Gyno-Pevaryl 150
Ifenec
1H-Imidazole, 1-(2-((4-chlorophenyl)methoxy)-2-(2,4-dichlorophenyl)ethyl)-, mononitrate (9CI)
NSC-243115
Palavale
Pevaryl
R 14827

24170-60-3
$C_{19}H_{19}N_7O_5$
425.36
O=C(Nc(c(N=Nc(c(C#N)cc(N(=O)=O)c1)c1N(=O)=O)cc c2N(CC)CC)c2)C
Acetamide, N-(2-((2-cyano-4,6-dinitrophenyl)azo)-5-(diethylamino)phenyl)- (9CI)
Acetanilide, 2'-((2-cyano-4,6-dinitrophenyl)azo)-5'-(diethylamino)-
Acetanilide, 2-((2,4-dinitro-6-cyanophenyl)azo)-5-diethylamino

24201-58-9
$C_{11}H_9Cl_2NO_3$
274.11
2,4-Oxazolidinedione, 3-(3,5-dichlorophenyl)-5,5-dimethyl
CS 8890
DDOD
3-(3,5-Dichlorophenyl)-5,5-dimethyl oxazoline-dione-2,4
Dichlozoline
Ortho 8890
Sclex

24222-05-7
$C_9H_{16}O_3$
172.22
Valeric acid, 2,3-epoxy-3,4-dimethyl-, ethyl ester, cis- (8CI)

24222-06-8
$C_{11}H_{20}O_3$
200.28
Valeric acid, 2,3-epoxy-3,4-dimethyl-, tert-butyl ester, cis- (8CI)

24255-23-0
$C_8H_5ClN_2S$
196.65
1,2,4-Thiadiazole, 5-chloro-3-phenyl- (8CI,9CI)
NSC-518113

24267-56-9
Unknown
Unknown
Iodine-131

24271-16-7
$C_{18}H_{30}O_3S$
326.50
Benzenesulfonic acid, p-(1,1-dimethyldecyl)- (8CI)

24271-17-8
$C_{18}H_{30}O_3S$
326.50
Benzenesulfonic acid, p-(2,2-dimethyldecyl)- (8CI)

24271-18-9
$C_{18}H_{30}O_3S$
326.50
Benzenesulfonic acid, p-(3,3-dimethyldecyl)- (8CI)

24271-19-0
$C_{18}H_{30}O_3S$
326.50
Benzenesulfonic acid, p-(9,9-dimethyldecyl)- (8CI)

24295-27-0
$C_9H_8Cl_2O_3$
235.07
Benzoic acid, 3,5-dichloro-4-methoxy-, methyl ester (9CI)
p-Anisic acid, 3,5-dichloro-, methyl ester (6CI,8CI)
Methyl 3,5-dichloro-4-methoxybenzoate

24298-49-5
$C_4H_8S_2$
120.24
Disulfide, ethenyl ethyl (9CI)
Disulfide, ethyl vinyl (8CI)

24305-97-3
$C_{36}H_{28}CrN_8O_{10}S_2.Na$
871.74
Chromate(1-), bis(N-(8-((5-(aminosulfonyl)-2-hydroxyphenyl)azo)-7-hydroxy-1-naphthalenyl)acetamidato-(2-))-, sodium (9CI)
Chromate(1-), bis(8-((2-hydroxy-5-sulfamoylphenyl)-azo)-1-acetylamino-7-naphthalenolato(2-))-, sodium

24306-23-8
$C_{12}H_{16}O_2$
192.26
Benzenebutanoic acid, 4-methyl-, methyl ester (9CI)
Butyric acid, 4-p-tolyl-, methyl ester (8CI)

24307-26-4
$C_7H_{16}N.Cl$
149.66
1,1-Dimethylpiperidinium chloride
BAS-083
BAS 083W
BAS-08300W
BAS 08301W
BAS 08305 W
BAS 08306 W
BAS 08307 W
BAS85559X
Caswell No. 380AB
N,N-Dimethylpiperidinium chloride
N,N-Dimethyl-piperidinium chloride
EPA Pesticide Chemical Code 109101
Mepiquat chloride
Methylpiperidine hydrochloride
Piperidinium, 1,1-dimethyl-, chloride
Pix
Terpal

24310-41-6
$C_8H_6ClN_3O$
195.59
O=C(N(N=Nc1cccc2)CCl)c12
1,2,3-Benzotriazin-4(3H)-one, 3-(chloromethyl)- (9CI)
3-Chloromethyl-4-keto-1,2,3-benzotriazine

24313-88-0
$C_9H_{13}NO_3$
183.20
O(c(c(OC)cc(N)c1)c1OC)C
Benzenamine, 3,4,5-trimethoxy- (9CI)
AI3-52691
NSC-37006
3,4,5-Trimethoxyaniline
3,4,5-Trimethoxybenzenamine

24327-56-8
$C_4H_9BrClO_2P$
235.45
Phosphinic acid, (bromomethyl)(chloromethyl)-, ethyl ester (8CI)

24327-58-0
$C_5H_{12}ClO_2P$
170.58
Phosphinic acid, (chloromethyl)ethyl-, ethyl ester (8CI,9CI)

24327-59-1
$C_{10}H_{23}O_2P$
206.27
Phosphinic acid, ethylhexyl-, ethyl ester (8CI)

24340-76-9
$C_4H_6N_2S$
114.16
5-Isothiazolamine, 3-methyl- (9CI)
Isothiazole, 5-amino-3-methyl- (8CI)

24345-02-6
$C_7H_8O_2S.1/2Zn$
188-89
Benzenesulfinic acid, 4-methyl-, zinc salt (9CI)
Zinc 4-methylbenzenesulfinate

24347-58-8
$C_4H_{10}O_2$
90.12
2,3-Butanediol, (R-(R*,R*))- (9CI)
2,3-Butanediol, (2R,3R)-(-)- (8CI)

24370-25-0
$C_8H_8N_4O$
176.20
Urea, (2-benzimidazolyl)
2-Benzimidazoleurea
Benzimidazol-2-ylurea
Urea, 1H-benzimidazol-2-yl- (9CI)
2-Ureidobenzimidazole

24382-04-5
$C_3H_3O_2.Na$
94.05
[Na+].O=C[CH-]C=O
Malonaldehyde, ion(1-), sodium
3-Hydroxy-2-propenal sodium salt
Malonaldehyde sodium salt
Propanedial, ion(1-), sodium (9CI)
Sodium malondialdehyde

24410-19-3
$C_7H_5N_3O$
147.12
Pyrido(2,3-d)pyrimidin-4(1H)-one (8CI,9CI)
NSC-112518
Pyrido(2,3-d)pyrimidin-4-ol

24423-11-8
$C_{14}H_9Cl$
212.68
Phenanthrene, 2-chloro- (8CI, 9CI)

24442-57-7
$C_4H_6Br_2O_2$
245.90
O=C(OC(Br)CBr)C
1,2-Dibromoethyl acetate
Acetic acid, 1,2-dibromoethyl ester
1,2-Dibromoethanol acetate
α,β-Dibromoethyl acetate
Ethanol, 1,2-dibromo-, acetate (9CI)

24448-09-7
$C_{11}H_8F_{17}NO_3S$
557.22
O=S(=O)(N(CCO)C)C(F)(F)C(F)(F)C(F)(F)C(F)(F)C(F)(F)C(F)(F)C(F)(F)C(F)(F)F
1-Octanesulfonamide, 1,1,2,2,3,3,4,4,5,5,6,6,7,7,8,8,8-heptadecafluoro-N-(2-hydroxyethyl)-N-methyl- (9CI)
1,1,2,2,3,3,4,4,5,5,6,6,7,7,8,8,8-Heptadecafluoro-N-(2-hydroxyethyl)-N-methyl-1-octanesulfonamide

24448-20-2
$C_{27}H_{32}O_6$
452.55
O=C(OCCOc(ccc(c1)C(c(ccc(OCCOC(=O)C(=C)C)c2)c2)(C)C)c1)C(=C)C
2,2-Bis-(4-(2-methacryloxyethoxy)phenyl)propane
2,2-Bis(4-(2-(methacryloxy)ethoxy)phenyl) propane
2,2'-(Isopropylidenebis(p-phenyleneoxy))diethylene ester of methacrylic acid
2-Methyl (1-methylethylidene)-bis(4,1-phenyleneoxy-2,1-ethanediyl) ester of 2-propenoic acid
2-Propenoic acid, 2-methyl-, (1-methylethylidene)bis(4,1-phenyleneoxy-2,1-ethanediyl) ester (9CI)

24448-89-3
$C_8H_{17}NO$
143.22
2-Piperidinepropanol (8CI,9CI)
NSC-143038

24458-48-8
$C_{10}H_{13}N_3O_3$
223.26
O=N(=O)C(C)(C)CNc(ccc(N=O)c1)c1
Aniline, N-(2-methyl-2-nitropropyl)-p-nitroso
Benzenamine, N-(2-methyl-2-nitropropyl)-p-nitroso- (9CI)
CP 25017
N-(2-Methyl-2-nitropropyl)-p-nitrosoaniline
N-(2-Methyl-2-nitropropyl)-4-nitrosobenzamine
Nitrol
Nitrol (Promoter)
N-(p-Nitrosoanilinomethyl)-2-nitropropane

24460-06-8
$C_{24}H_{22}N_2O_3$
386.44
O=C(OC(c(c(Oc1cc(N(CC)CC)cc2)ccc3N)c3)(c12)c4cccc5)c45
Spiro(isobenzofuran-1(3H),9'-(9H)xanthen)-3-one, 2'-amino-6'-(diethylamino)- (9CI)

24468-13-1
$C_9H_{17}ClO_2$
192.71
O=C(OCC(CCCC)CC)Cl
Carbonochloridic acid, 2-ethylhexyl ester
Ethylhexyl chloroformate
2-Ethylhexylchloroformate [UN 2748]
UN 2748 [2-Ethylhexylchloroformate]

24471-47-4
$C_{21}H_{14}$
266.34
Perylene, 3-methyl- (8CI,9CI)

24477-99-4
$C_{12}H_7Cl_6N_3S$
437.99
1,3,5-Triazine, 2-((2-methylphenyl)thio)-4,6-bis(trichloromethyl)- (9CI)
s-Triazine, 2-(o-tolylthio)-4,6-bis(trichloromethyl)- (8CI)

24478-00-0
$C_{12}H_7Cl_6N_3S$
437.99
1,3,5-Triazine, 2-((3-methylphenyl)thio)-4,6-bis(trichloromethyl)- (9CI)
s-Triazine, 2-(m-tolylthio)-4,6-bis(trichloromethyl)- (8CI)

24478-01-1
$C_{12}H_7Cl_6N_3S$
437.99
1,3,5-Triazine, 2-((4-methylphenyl)thio)-4,6-bis(trichloromethyl)- (9CI)
s-Triazine, 2-(p-tolylthio)-4,6-bis(trichloromethyl)- (8CI)

24478-02-2
$C_{12}H_7Cl_6N_3OS$
453.99
1,3,5-Triazine, 2-((2-methoxyphenyl)thio)-4,6-bis-(trichloromethyl)- (9CI)
s-Triazine, 2-((o-methoxyphenyl)thio)-4,6-bis(trichloromethyl)- (8CI)

24478-03-3
$C_{12}H_7Cl_6N_3OS$
453.99
1,3,5-Triazine, 2-((4-methoxyphenyl)thio)-4,6-bis-(trichloromethyl)- (9CI)
s-Triazine, 2-((p-methoxyphenyl)thio)-4,6-bis(trichloromethyl)- (8CI)

24478-04-4
$C_{13}H_9Cl_6N_3OS$
468.02
1,3,5-Triazine, 2-((2-ethoxyphenyl)thio)-4,6-bis-(trichloromethyl)- (9CI)
s-Triazine, 2-((o-ethoxyphenyl)thio)-4,6-bis(trichloromethyl)- (8CI)

24478-05-5
$C_{13}H_9Cl_6N_3OS$
468.02
1,3,5-Triazine, 2-((4-ethoxyphenyl)thio)-4,6-bis-(trichloromethyl)- (9CI)
s-Triazine, 2-((p-ethoxyphenyl)thio)-4,6-bis(trichloromethyl)- (8CI)

24478-06-6
$C_{11}H_4Cl_7N_3S$
458.41
1,3,5-Triazine, 2-((2-chlorophenyl)thio)-4,6-bis(trichloromethyl)- (9CI)
s-Triazine, 2-((o-chlorophenyl)thio)-4,6-bis(trichloromethyl)- (8CI)

24478-07-7
$C_{11}H_4Cl_7N_3S$
458.41
1,3,5-Triazine, 2-((3-chlorophenyl)thio)-4,6-bis(trichloromethyl)- (9CI)
s-Triazine, 2-((m-chlorophenyl)thio)-4,6-bis(trichloromethyl)- (8CI)

24478-08-8
$C_{11}H_4Cl_7N_3S$
458.41
1,3,5-Triazine, 2-((4-chlorophenyl)thio)-4,6-bis(trichloromethyl)- (9CI)
s-Triazine, 2-((p-chlorophenyl)thio)-4,6-bis(trichloromethyl)- (8CI)

24478-09-9
$C_{11}H_3Cl_8N_3S$
492.85
1,3,5-Triazine, 2-((2,4-dichlorophenyl)thio)-4,6-bis(trichloromethyl)- (9CI)
s-Triazine, 2-((2,4-di chlorophenyl)thio)-4,6-bis-(trichloromethyl)- (8CI)

24478-10-2
$C_{11}H_3Cl_8N_3S$
492.85
1,3,5-Triazine, 2-((3,4-dichlorophenyl)thio)-4,6-bis(trichloromethyl)- (9CI)
s-Triazine, 2-((3,4-dichlorophenyl)thio)-4,6-bis(trichloromethyl)- (8CI)

24478-11-3
$C_{11}H_2Cl_9N_3S$
527.30
1,3,5-Triazine, 2,4-bis(trichloromethyl)-6-((2,4,5-trichlorophenyl)thio)- (9CI)
s-Triazine, 2,4-bis(trichloromethyl)-6-((2,4,5-trichlorophenyl)thio)- (8CI)

24478-12-4
$C_{11}H_4BrCl_6N_3S$
502.87
1,3,5-Triazine, 2-((3-bromophenyl)thio)-4,6-bis-(trichloromethyl)- (9CI)
s-Triazine, 2-((m-bromophenyl)thio)-4,6-bis(trichloromethyl)- (8CI)

24478-13-5
$C_{12}H_6Cl_7N_3S$
472.44
1,3,5-Triazine, 2-((4-chloro-3-methylphenyl)thio)-4,6-bis(trichloromethyl)- (9CI)
s-Triazine, 2-((4-chloro-m-tolyl)thio)-4,6-bis(trichloromethyl)- (8CI)

24478-14-6
$C_{11}H_4Cl_6N_4O_2S$
468.96
1,3,5-Triazine, 2-((4-nitrophenyl)thio)-4,6-bis(trichloromethyl)- (9CI)
s-Triazine, 2-((p-nitrophenyl)thio)-4,6-bis-(trichloromethyl)- (8CI)

24478-15-7
$C_{11}H_3Cl_7N_4O_2S$
503.41
1,3,5-Triazine, 2-((4-chloro-3-nitrophenyl)thio)-4,6-bis(trichloromethyl)- (9CI)
s-Triazine, 2-((4-chloro-3-nitrophenyl)thio)-4,6-bis-(trichloromethyl)- (8CI)

24478-16-8
$C_{12}H_6Cl_6N_4O_3S$
498.99
1,3,5-Triazine, 2-((2-methoxy-4-nitrophenyl)thio)-4,6-bis(trichloromethyl)- (9CI)
s-Triazine, 2-((2-methoxy-4-nitrophenyl)thio)-4,6-bis-(trichloromethyl)- (8CI)

24478-18-0
$C_8H_9Cl_4N_3S_2$

353.12
s-Triazine-2-carboxaldehyde, 4,6-bis(dichloromethyl)-, dimethyl mercaptal (8CI)
s-Triazine, 2-(bis(methylthio)methyl)-4,6-bis(dichloromethyl)- (8CI)

24478-73-7
$C_{12}H_5Cl_3O$
271.53
Dibenzofuran, 1,2,4-trichloro- (8CI,9CI)

24478-74-8
$C_{12}H_6Cl_2O$
237.08
Dibenzofuran, 2,4-dichloro- (8CI,9CI)

24480-99-7
$C_8H_{12}O$
124.18
3-Cyclohexene-1-acetaldehyde (8CI,9CI)

24481-33-2
$C_7H_2Cl_9N_3$
447.19
1,3,5-Triazine, 2-(1,1,2-trichloroethyl)-4,6-bis(trichloromethyl)- (9CI)
s-Triazine, 2-(1,1,2-trichloroethyl)-4,6-bis(trichloromethyl)- (8CI)
2-(1,1,2-Trichloroethyl)-4,6-bis(trichloromethyl)-1,3,5-triazine

24481-35-4
$C_{22}H_{35}Cl_6N_3$
554.26
1,3,5-Triazine, 2-heptadecyl-4,6-bis(trichloromethyl)- (9CI)
s-Triazine, 2-heptadecyl-4,6-bis(trichloromethyl)- (8CI)

24481-36-5
$C_{14}H_{19}Cl_6N_3$
442.05
1,3,5-Triazine, 2-nonyl-4,6-bis(trichloromethyl)- (9CI)
s-Triazine, 2-nonyl-4,6-bis-(trichloromethyl)- (8CI)

24481-37-6
$C_{10}H_{11}Cl_6N_3$
385.94
1,3,5-Triazine, 2-pentyl-4,6-bis(trichloromethyl)- (9CI)
s-Triazine, 2-pentyl-4,6-bis-(trichloromethyl)- (8CI)

24481-40-1
$C_9H_9Cl_6N_3$
371.91
s-Triazine, 2-isobutyl-4,6-bis(trichloromethyl)- (8CI)

24481-41-2
$C_9H_9Cl_6N_3$
371.89
1,3,5-Triazine, 2-butyl-4,6-bis-(trichloromethyl)- (9CI)
s-Triazine, 2-butyl-4,6-bis-(trichloromethyl)- (8CI)

24481-42-3
$C_8H_7Cl_6N_3$
357.86
1,3,5-Triazine, 2-(1-methylethyl)-4,6-bis(trichloromethyl)- (9CI)
s-Triazine, 2-isopropyl-4,6-bis-(trichloromethyl)- (8CI)

24481-43-4
$C_8H_7Cl_6N_3$
357.86
1,3,5-Triazine, 2-propyl-4,6-bis-(trichloromethyl)- (9CI)
s-Triazine, 2-propyl-4,6-bis-(trichloromethyl)- (8CI)

24481-45-6
$C_{15}H_7Cl_6N_3$
441.96
1,3,5-Triazine, 2-(2-naphthalenyl)-4,6-bis(trichloromethyl)- (9CI)
s-Triazine, 2-(2-naphthyl)-4,6-bis(trichloromethyl)- (8CI)

24481-46-7
$C_{15}H_7Cl_6N_3$
441.96
1,3,5-Triazine, 2-(1-naphthalenyl)-4,6-bis(trichloromethyl)- (9CI)
s-Triazine, 2-(1-naphthyl)-4,6-bis(trichloromethyl)- (8CI)

24481-49-0
$C_{11}H_4BrCl_6N_3$
470.80
s-Triazine, 2-(p-bromophenyl)-4,6-bis(trichloromethyl)- (8CI)

24481-50-3
$C_{11}H_2Cl_9N_3$
495.24
s-Triazine, 2,4-bis(trichloromethyl)-6-(2,4,5-trichlorophenyl)- (8CI)

24481-51-4
$C_{11}H_3Cl_8N_3$
475.83
s-Triazine, 2-(3,4-dichlorophenyl)-4,6-bis-(trichloromethyl)- (8CI)

24481-52-5
$C_{11}H_3Cl_8N_3$
460.79
1,3,5-Triazine, 2-(2,4-dichlorophenyl)-4,6-bis-(trichloromethyl)- (9CI)
s-Triazine, 2-(2,4-dichlorophenyl)-4,6-bis(trichloromethyl)- (8CI)

24481-54-7
$C_{11}H_4Cl_7N_3$
426.35
s-Triazine, 2-(m-chlorophenyl)-4,6-bis(trichloromethyl)- (8CI)

24481-55-8
$C_{11}H_4Cl_7N_3$
426.35
s-Triazine, 2-(o-chlorophenyl)-4,6-bis(trichloromethyl)- (8CI)

24481-68-3
$C_8H_7Cl_6N_3S$
389.95
1,3,5-Triazine, 2-((1-methylethyl)thio)-4,6-bis-(trichloromethyl)- (9CI)
s-Triazine, 2-(isopropylthio)-4,6-bis(trichloromethyl)- (8CI)

24481-69-4
$C_9H_9Cl_6N_3S$
403.97
1,3,5-Triazine, 2-((2-methylpropyl)thio)-4,6-bis(trichloromethyl)- (9CI)
s-Triazine, 2-(isobutylthio)-4,6-bis(trichloromethyl)- (8CI)

24481-70-7
$C_{10}H_{11}Cl_6N_3S$
418.00
1,3,5-Triazine, 2-(pentylthio)-4,6-bis(trichloromethyl)- (9CI)
s-Triazine, 2-(pentylthio)-

4,6-bis(trichloromethyl)-
(8CI)

24481-71-8
$C_{11}H_{13}Cl_6N_3S$
432.03
1,3,5-Triazine, 2-(hexylthio)-4,6-bis(trichloromethyl)- (9CI)
s-Triazine, 2-(hexylthio)-4,6-bis(trichloromethyl)-(8CI)

24481-72-9
$C_{17}H_{25}Cl_6N_3S$
516.19
1,3,5-Triazine, 2-(dodecylthio)-4,6-bis-(trichloromethyl)- (9CI)
s-Triazine, 2-(dodecylthio)-4,6-bis(trichloromethyl)-(8CI)

24481-73-0
$C_{12}H_6Cl_7N_3S$
472.44
1,3,5-Triazine, 2-(((4-chlorophenyl)methyl)thio)-4,6-bis(trichloromethyl)-(9CI)
s-Triazine, 2-((p-chlorobenzyl)thio)-4,6-bis(trichloromethyl)- (8CI)

24481-74-1
$C_{12}H_6Cl_6N_4O_2S$
482.99
1,3,5-Triazine, 2-(((4-nitrophenyl)methyl)thio)-4,6-bis(trichloromethyl)-(9CI)
s-Triazine, 2-((p-nitrobenzyl)thio)-4,6-bis(trichloromethyl)- (8CI)

24504-17-4
$C_8H_7Cl_6N_3S$
389.95
1,3,5-Triazine, 2-(propylthio)-4,6-bis(trichloromethyl)- (9CI)
s-Triazine, 2-(propylthio)-4,6-bis(trichloromethyl)-(8CI)

24504-18-5
$C_9H_9Cl_6N_3S$
403.97
1,3,5-Triazine, 2-(butylthio)-4,6-bis(trichloromethyl)- (9CI)
s-Triazine, 2-(butylthio)-4,6-bis(trichloromethyl)-(8CI)

24504-22-1
$C_{11}H_5Cl_6N_3$
391.90
1,3,5-Triazine, 2-phenyl-4,6-bis(trichloromethyl)-(9CI)
s-Triazine, 2-phenyl-4,6-bis-(trichloromethyl)- (6CI, 7CI,8CI)

24524-52-5
$C_{10}H_{18}$
138.25
Cyclopropane, 1-methyl-2-(1-methylethyl)-3-(1-methylethylidene)-, cis- (9CI)
Propane, 2-(2-isopropylidene-3-methylcyclopropyl)-, cis- (8CI)

24524-53-6
$C_{12}H_{22}$
166.31
Hexane, 1-(isopropylidenecyclopropyl)- (8CI)

24526-64-5
$C_{16}H_{18}N_2$
238.36
CN3CC(c1ccccc1)c2cccc(N)c2C3
Isoquinoline, 1,2,3,4-tetrahydro-8-amino-2-methyl-4-phenyl
8-Amino-2-methyl-4-phenyl-1,2,3,4-tetrahydroisoquinoline
8-Isoquinolinamine, 1,2,3,4-tetrahydro-2-methyl-4-phenyl-(9CI)
Nomifensin
Nomifensine

24544-04-5
$C_{12}H_{19}N$
177.32
Nc(c(ccc1)C(C)C)c1C(C)C
Aniline, 2,6-diisopropyl
Benzenamine, 2,6-bis(1-methylethyl)-
2,6-Diisopropyl aniline

24549-06-2
$C_9H_{13}N$
135.23
Nc(c(ccc1)CC)c1C
Aniline, 2-methyl-6-ethyl
Benzenamine, 2-ethyl-6-methyl-
2-Ethyl-6-methylaniline
6-Ethyl-o-toluidine
2-Methyl-6-ethyl aniline
o-Toluidine, 6-ethyl-

24554-26-5
$C_8H_5N_3O_4S$
239.22
O=CNc1nc(cs1)c2ccc(o2)N(=O)=O
Formamide, N-(4-(5-nitro-2-furyl)-2-thiazolyl)
FANFT
2-Formylamino-4-(5-nitro-2-furyl)thiazole
N-(4-(5-Nitro-2-furyl)-2-thiazolyl)formamid (German)
N-(4-(5-Nitro-2-furyl)-2-thiazolyl)formamide

24560-98-3
$C_{18}H_{34}O_3$
298.47
Oxiraneoctanoic acid, 3-octyl-, cis- (9CI)
Octadecanoic acid, 9,10-epoxy-, cis- (8CI)

24579-73-5
$C_9H_{20}N_2O_2$
188.27
Propamocarb
Carbamic acid, (3-(dimethylamino)propyl)-, propyl ester
Propamocarbe (French)
Propyl 3-(dimethylamino)propylcarbamate
Propyl (3-(dimethylamino)propyl)carbamate

24582-52-3
$C_8H_{11}AsO_2$
214.10
Arsonous acid, phenyl-, dimethyl ester (9CI)
Benzenearsonous acid, dimethyl ester (8CI)
Dimethoxyphenylarsine
Dimethyl benzenearsonite
Dimethyl phenylarsenite

24582-54-5
$C_{13}H_{13}AsO$
260.17
Arsinous acid, diphenyl-, methyl ester (9CI)
Arsine, methoxydiphenyl-(8CI)
Diphenylmethoxyarsine
Methoxydiphenylarsine
Methyl diphenylarsinite

24582-55-6
$C_{14}H_{15}AsO$
274.20
Arsinous acid, diphenyl-, ethyl ester (9CI)
Arsine, ethoxydiphenyl-(8CI)
Diphenylethoxyarsine
Ethoxydiphenylarsine
Ethyl diphenylarsinite

24582-56-7
$C_9H_{13}AsO$

212.12
Arsinous acid, ethylphenyl-, methyl ester (9CI)
Arsine, ethylmethoxyphenyl- (8CI)
Methyl ethylphenylarsinite

24582-59-0
$C_{18}H_{15}AsS_2$
370.37
Arsonodithious acid, phenyl-, diphenyl ester (9CI)
Benzenearsonous acid, dithio-, diphenyl ester (7CI, 8CI)

24582-60-3
$C_{14}H_{15}AsS$
290.26
Arsine, (ethylthio)diphenyl- (8CI)

24582-61-4
$C_{16}H_{19}AsS$
318.32
Arsine, butylphenyl(phenylthio)- (8CI)

24588-61-2
$C_9H_{14}O_3$
170.21
2H-Pyran-2-carboxylic acid, 3,6-dihydro-4,5-dimethyl-, methyl ester (7CI,8CI,9CI)

24602-86-6
$C_{19}H_{39}NO$
297.59
CCCCCCCCCCCCCN1CC(C)OC(C)C1
Morpholine, 2,6-dimethyl-N-tridecyl
BAS 2205-F
Calixin
2,6-Dimethyl-4-tridecylmorpholine
E-236
N-Tridecyl-2,6-dimethylmorpholin (German)
N-Tridecyl-2,6-dimethylmorpholine
4-Tridecyl-2,6-dimethylmorpholine
Tridemorph

24607-12-3
$C_{10}H_{12}O_4$
196.20
Acetic acid, methoxyphenoxy-, methyl ester (6CI, 9CI)
Glyoxylic acid, methyl ester, 2-(methyl phenyl acetal) (8CI)
Methyl 2-methoxy-2-phenoxyacetate

24613-61-4
$C_{18}H_{37}O_4P$
348.46
O=P(OCCCCCCCCC=CCCCCCCCC)(O)O
9-Octadecen-1-ol, dihydrogen phosphate (9CI)

24613-89-6
$Cr_3O_{12}.2Cr$
452.00
Chromic acid, chromium (3+) salt (3:2)
Chromic chromate
Chromium chromate

24615-84-7
$C_6H_8O_4$
144.13
O=C(OCCC(=O)O)C=C
3-Acryloyloxypropionic acid
β-(Acryloyloxy)propanoic acid
β-Carboxyethyl acrylate
2-Carboxyethyl 2-propenoate
2-Propenoic acid, 2-carboxyethyl ester (9CI)

24634-61-5
$C_6H_7O_2.K$
150.23
Sorbic acid, potassium salt, (E,E)
2,4-Hexadienoic acid, potassium salt, (E,E)- (9CI)
Potassium sorbate
Sorbistat potassium

24650-10-0
$C_{13}H_{26}N_2$
210.35
NC(C(CCC1)CC(CCC(N)C2)C2)C1
Cyclohexanamine, 2-((4-aminocyclohexyl)methyl)- (9CI)
2-((4-Aminocyclohexyl)methyl)-cyclohexanamine
Cyclohexylamine, 2,4'-methylenebis-

24650-42-8
$C_{16}H_{16}O_3$
256.30
O=C(c(cccc1)c1)C(OC)(OC)c(ccc c2)c2
Ethanone, 2,2-dimethoxy-1,2-diphenyl- (9CI)
2,2-Dimethoxy-1,2-diphenylethanone
α,α-Dimethoxy-α-phenylacetophenone
1,2-Diphenylethane-1,2-dione, dimethyl ketal

24654-08-8
$C_9H_5ClO_2$
180.59
3-(2-Chlorophenyl)-2-propynoic acid
o-Chlorophenylpropiolic acid
2-Chlorophenylpropiolic acid
NSC-61873
Propiolic acid, (o-chlorophenyl)- (8CI)
Propynoic acid, (2-chlorophenyl)-
2-Propynoic acid, 3-(2-chlorophenyl)- (9CI)

24683-00-9
$C_9H_{14}N_2O$
166.21
O(c(nccn1)c1CC(C)C)C
2-Isobutyl-3-methoxypyrazine
3-Isobutyl-2-methoxypyrazine
2-Methoxy-3-isobutylpyrazine
2-Methoxy-3-(2-methylpropyl)-pyrazine
Pyrazine, 2-methoxy-3-(2-methylpropyl)- (9CI)

24683-32-7
C_5H_9Al
96.11
Aluminum, dihydro(2-methyl-1,3-butadienyl)- (9CI)
Dihydro(2-methyl-1,3-butadienyl)aluminum
Isoprenylaluminum

24687-55-6
$C_6Br_9N_3$
833.20
1,3,5-Triazine, 2,4,6-tris-(tribromomethyl)- (9CI)
s-Triazine, 2,4,6-tris-(tribromomethyl)- (8CI)

24687-57-8
$C_{20}H_{40}O_2.Ag$
420.40
Eicosanoic acid, silver salt (8CI)
Silver eicosanoate

24691-76-7
$C_{13}H_{15}NO_2$
217.29
CC1=C(CCCO1)C(=O)Nc2ccccc2
4H-Pyran, 5,6-dihydro-2-methyl-3-(phenylcarbamoyl)
5-6-Dihydro-2-methyl-3-(phenylcarbamoyl)-4H-pyrane
3,4-Dihydro-6-methyl-N-phenyl-2H-pyran-5-carboxamide
3,4-Dihydro-6-methyl-2H-pyran-5-carboxanilide

HOE 2989
HOE 6052
HOE 6053
HOE 13764
HOE 13764 OF
2-Methyl-5,6-dihydro-4,4-pyran-3-carbonsaeureanilid (German)
2-Methyl-5,6-dihydro-4-H-pyrane-3-carboxylic acid anilide
Pyracarbolid
Pyracarbolide
Sicarol

24719-19-5
$AsH_3O_4.3/_2Co$
181.23
Arsenic acid (H_3AsO_4), cobalt(2+) salt (2:3) (9CI)
Cobalt arsenate
NSC-309957

24740-88-3
$C_6H_6O_5$
158.11
Maleylacetate
Maleoylacetic acid

24767-66-6
$C_6H_{13}Cl_2O_2P$
219.05
Phosphinic acid, bis(chloromethyl)-, sec-butyl ester (8CI)

24772-51-8
$C_{14}H_{27}AlO_5$
302.35
Aluminum, bis(2-butanolato)(ethyl 3-oxobutanoato-O1',O3)-, (T-4)- (9CI)
Aluminum di-sec-butoxide acetoacetic ester chelate

24794-58-9
$C_6H_{15}NO_3.CH_2O_2$
195.21
Formic acid, Compd. with 2,2',2''-nitrilotris(ethanol) (1:1) (9CI)
2,2',2''-Nitrilotrisethanol formate

24800-44-0
$C_9H_{20}O_4$
192.29
Tripropylene-glycol
2-(2-(2-Hydroxypropoxy)-propoxy)-1-propanol
Propanol, ((1-methyl-1,2-ethanediyl)bis(oxy))bis- (9CI)

24802-82-2
$C_9H_{10}Cl_6N_4$
386.92
s-Triazine, 2-(butylamino)-4,6-bis(trichloromethyl)-(8CI)

24802-83-3
$C_9H_{10}Cl_6N_4$
386.92
s-Triazine, 2-(sec-butylamino)-4,6-bis(trichloromethyl)- (8CI)

24802-84-4
$C_9H_{10}Cl_6N_4$
386.92
s-Triazine, 2-(isobutylamino)-4,6-bis(trichloromethyl)- (8CI)

24802-85-5
$C_9H_{10}Cl_6N_4$
386.92
s-Triazine, 2-(tert-butylamino)-4,6-bis(trichloromethyl)- (8CI)

24802-86-6
$C_{10}H_{12}Cl_6N_4$
400.95
s-Triazine, 2-(pentylamino)-4,6-bis(trichloromethyl)- (8CI)

24802-87-7
$C_{11}H_{14}Cl_6N_4$
414.98
s-Triazine, 2-(hexylamino)-4,6-bis(trichloromethyl)- (8CI)

24802-88-8
$C_{13}H_{18}Cl_6N_4$
443.03
s-Triazine, 2-((1-ethylhexyl)amino)-4,6-bis(-trichloromethyl)-(8CI)

24802-89-9
$C_{10}H_{15}Cl_3N_4$
297.62
s-Triazine, 2-sec-butyl-4-(ethylamino)-6-(trichloromethyl)- (8CI)

24802-90-2
$C_{10}H_{15}Cl_3N_4$
297.62
s-Triazine, 2-(ethylamino)-4-isobutyl-6-(trichloromethyl)- (8CI)

24802-91-3
$C_{12}H_{19}Cl_3N_4$
325.67
s-Triazine, 2-pentyl-4-(propylamino)-6-(trichloromethyl)- (8CI)

24802-92-4
$C_{15}H_{25}Cl_3N_4$
367.75
s-Triazine, 2-(ethylamino)-4-nonyl-6-(trichloromethyl)- (8CI)

24802-93-5
$C_{23}H_{41}Cl_3N_4$
479.97
s-Triazine, 2-(ethylamino)-4-heptadecyl-6-(trichloromethyl)- (8CI)

24802-94-6
$C_8H_{10}Cl_4N_4$
304.01
s-Triazine, 2-(2-chloroethyl)-4-(ethylamino)-6-(trichloromethyl)-(8CI)

24802-95-7
$C_9H_{12}Cl_4N_4$
318.03
s-Triazine, 2-(2-chloroethyl)-4-(propylamino)-6-(trichloromethyl)-(8CI)

24802-96-8
$C_8H_{10}BrCl_3N_4$
348.46
s-Triazine, 2-(2-bromoethyl)-4-(ethylamino)-6-(trichloromethyl)- (8CI)

24802-97-9
$C_8H_8Cl_6N_4$
372.90
s-Triazine, 2-(ethylamino)-4-(1,1,2-trichloroethyl)-6-(trichloromethyl)-(8CI)

24802-98-0
$C_8H_9Cl_3N_4$
279.56
s-Triazine, 2-(ethylamino)-4-(trichloromethyl)-6-vinyl- (8CI)

24802-99-1
$C_{11}H_9Cl_3N_4$
303.58
s-Triazine, 2-(methylamino)-4-phenyl-6-(trichloromethyl)- (8CI)

24803-00-7
$C_{12}H_{11}Cl_3N_4$
317.61
s-Triazine, 2-(ethylamino)-4-phenyl-6-(trichloromethyl)- (8CI)

24803-01-8
$C_{13}H_{13}Cl_3N_4$
331.63
s-Triazine, 2-phenyl-4-(propylamino)-6-(trichloromethyl)- (8CI)

24803-02-9
$C_{13}H_{13}Cl_3N_4$
331.63
s-Triazine, 2-(isopropylamino)-4-phenyl-6-(trichloromethyl)- (8CI)

24803-04-1
$C_{17}H_{26}Cl_6N_4$
499.14
s-Triazine, 2-(dodecylamino)-4,6-bis(trichloromethyl)- (8CI)

24803-05-2
$C_7H_6Cl_6N_4$
358.87
1,3,5-Triazin-2-amine, N,N-dimethyl-4,6-bis(trichloromethyl)- (9CI)
s-Triazine, 2-(dimethylamino)-4,6-bis(trichloromethyl)- (6CI,8CI)

24803-06-3
$C_9H_{10}Cl_6N_4$
386.92
s-Triazine, 2-(diethylamino)-4,6-bis(trichloromethyl)- (8CI)

24803-07-4
$C_{11}H_{14}Cl_6N_4$
414.98
s-Triazine, 2-(dipropylamino)-4,6-bis(trichloromethyl)- (8CI)

24803-09-6
$C_{10}H_{10}Cl_6N_4$
398.94
s-Triazine, 2-piperidino-4,6-bis(trichloromethyl)- (8CI)

24803-10-9
$C_9H_8Cl_6N_4O$
400.91
1,3,5-Triazine, 2-(4-morpholinyl)-4,6-bis(trichloromethyl)- (9CI)
s-Triazine, 2-morpholino-4,6-bis(trichloromethyl)- (8CI)

24803-11-0
$C_7H_6Cl_6N_4O$
374.87
Ethanol, 2-((4,6-bis(trichloromethyl)-1,3,5-triazin-2-yl)amino)- (9CI)
Ethanol, 2-((4,6-bis(trichloromethyl)-s-triazin-2-yl)amino)- (8CI)

24803-12-1
$C_8H_8Cl_6N_4O$
388.90
1-Propanol, 3-((4,6-bis-(trichloromethyl)-s-triazin-2-yl)amino)- (8CI)

24803-13-2
$C_8H_8Cl_6N_4O$
388.90
s-Triazine, 2-((2-methoxyethyl)amino)-4,6-bis(trichloromethyl)- (8CI)

24803-14-3
$C_9H_{11}Cl_6N_5$
401.94
s-Triazine, 2-((2-(dimethylamino)ethyl)amino)-4,6-bis(trichloromethyl)- (6CI,8CI)

24803-15-4
$C_{11}H_{12}Cl_6N_4$
412.96
s-Triazine, 2-(cyclohexylamino)-4,6-bis(trichloromethyl)- (8CI)

24803-16-5
$C_{12}H_8Cl_6N_4$
420.94
1,3,5-Triazin-2-amine, N-(phenylmethyl)-4,6-bis-(trichloromethyl)- (9CI)
s-Triazine, 2-(benzylamino)-4,6-bis(trichloromethyl)- (8CI)

24803-17-6
$C_8H_6Cl_6N_4$
370.88
1,3,5-Triazin-2-amine, N-2-propenyl-4,6-bis(trichloromethyl)- (9CI)
s-Triazine, 2-(allylamino)-4,6-bis(trichloromethyl)- (8CI)

24803-18-7
$C_7H_9Cl_3N_4$
255.54
s-Triazine, 2-ethyl-4-(methylamino)-6-(trichloromethyl)- (8CI)

24803-19-8
$C_8H_{11}Cl_3N_4$
269.56
s-Triazine, 2-ethyl-4-(ethylamino)-6-(trichloromethyl)- (8CI)

24803-20-1
$C_9H_{13}Cl_3N_4$
283.59
s-Triazine, 2-ethyl-4-(propylamino)-6-(trichloromethyl)- (8CI)

24803-21-2
$C_9H_{13}Cl_3N_4$
283.59
s-Triazine, 2-ethyl-4-(isopropylamino)-6-(trichloromethyl)- (8CI)

24803-22-3
$C_{10}H_{15}Cl_3N_4$
297.62
s-Triazine, 2-(butylamino)-4-ethyl-6-(trichloromethyl)- (8CI)

24803-23-4
$C_{10}H_{15}Cl_3N_4$
297.62
s-Triazine, 2-ethyl-4-(isobutylamino)-6-(trichloromethyl)- (8CI)

24803-24-5
$C_8H_{11}Cl_3N_4$
269.56
s-Triazine, 2-(methylamino)-4-propyl-6-(trichloromethyl)- (8CI)

24803-25-6
$C_9H_{13}Cl_3N_4$
283.59
s-Triazine, 2-(ethylamino)-4-propyl-6-(trichloromethyl)- (8CI)

24803-26-7
$C_8H_{11}Cl_3N_4$
269.56
s-Triazine, 2-isopropyl-4-(methylamino)-6-(trichloromethyl)- (8CI)

24803-27-8
$C_9H_{13}Cl_3N_4$
283.59
s-Triazine, 2-(ethylamino)-4-isopropyl-6-(trichloromethyl)- (8CI)

24803-28-9
$C_{10}H_{15}Cl_3N_4$
297.62
s-Triazine, 2-butyl-4-(ethylamino)-6-(trichloromethyl)- (8CI)

24803-29-0
$C_7H_6Cl_6N_4$
358.84
1,3,5-Triazin-2-amine, N-ethyl-4,6-bis(trichloromethyl)- (9CI)
s-Triazine, 2-(ethylamino)-4,6-bis-(trichloromethyl)-(8CI)

24803-30-3
$C_8H_8Cl_6N_4$
372.90
s-Triazine, 2-(propylamino)-4,6-bis(trichloromethyl)- (8CI)

24803-31-4
$C_8H_8Cl_6N_4$
372.90
s-Triazine, 2-(isopropylamino)-4,6-bis(trichloromethyl)- (8CI)

24803-51-8
$C_7H_9Cl_3N_4$
255.54
s-Triazine, 2-(ethylamino)-4-methyl-6-(trichloromethyl)- (8CI)

24803-52-9
$C_8H_{11}Cl_3N_4$
269.56
1,3,5-Triazin-2-amine, 4-methyl-N-propyl-6-(trichloromethyl)- (9CI)
s-Triazine, 2-methyl-4-(propylamino)-6-(trichloromethyl)- (8CI)

24803-53-0
$C_9H_{13}Cl_3N_4$
283.59
1,3,5-Triazin-2-amine, 4-methyl-N-(2-methylpropyl)-6-(trichloromethyl)- (9CI)
s-Triazine, 2-(isobutylamino)-4-methyl-6-(trichloromethyl)- (8CI)

24803-54-1
$C_{10}H_{15}Cl_3N_4$
297.62
s-Triazine, 2-methyl-4-(pentylamino)-6-(trichloromethyl)- (8CI)

24803-55-2
$C_{11}H_{17}Cl_3N_4$
311.64
s-Triazine, 2-(hexylamino)-4-methyl-6-(trichloromethyl)- (8CI)

24803-56-3
$C_{17}H_{29}Cl_3N_4$
395.81
s-Triazine, 2-(dodecylamino)-4-methyl-6-(trichloromethyl)- (8CI)

24803-57-4
$C_7H_9Cl_3N_4$
255.54
1,3,5-Triazin-2-amine, N,N,4-trimethyl-6-(trichloromethyl)- (9CI)
s-Triazine, 2-(dimethylamino)-4-methyl-6-(trichloromethyl)- (8CI)

24803-58-5
$C_9H_{13}Cl_3N_4$
283.59
s-Triazine, 2-(diethylamino)-4-methyl-6-(trichloromethyl)- (8CI)

24803-59-6
$C_{10}H_{13}Cl_3N_4$
295.60
s-Triazine, 2-methyl-4-piperidino-6-(trichloromethyl)- (8CI)

24803-60-9
$C_9H_{11}Cl_3N_4O$
297.57
1,3,5-Triazine, 2-methyl-4-(4-morpholinyl)-6-(trichloromethyl)- (9CI)
s-Triazine, 2-methyl-4-morpholino-6-(trichloromethyl)- (8CI)

24803-61-0
$C_7H_9Cl_3N_4O$
271.53
Ethanol, 2-((4-methyl-6-(trichloromethyl)-s-triazin-2-yl)amino)- (8CI)

24803-62-1
$C_8H_{11}Cl_3N_4O$
285.56
1-Propanol, 3-((4-methyl-6-(trichloromethyl)-s-triazin-2-yl)amino)- (8CI)

24803-63-2
$C_8H_9Cl_3N_4$
267.55
s-Triazine, 2-(allylamino)-4-methyl-6-(trichloromethyl)- (8CI)

24803-64-3
$C_6H_4Cl_6N_4$
344.84
s-Triazine, 2-(methylamino)-4,6-bis(trichloromethyl)- (7CI,8CI)

24815-24-5
$C_{35}H_{42}N_2O_9$
634.79
COC5C(CC4CN3CCc1c([nH]c2c c(OC)ccc12)C3CC4C5C(=O) OC)OC(=O)C=Cc6cc(OC)c (OC)c(OC)c6
3-β,20-α-Yohimban-16-β-carboxylic acid, 18-β-hydroxy-11,17-α-dimethoxy-, methyl ester, 3,4,5-trimethoxycinnamate (ester)
Anapral
Anaprel
Cinamine
Cinatabs
Methyl reserpate 3,4,5-trimethoxycinnamic acid ester
Methyl 18-O-(3,4,5-trimethoxycinnamoyl)reserpate
Moderil
Normorescina
Raupyrol
Raurescine
Recinnamine
Recitensina
Rescaloid
Rescamin
Rescidan
Rescin
Rescinnamine
Rescinpal
Rescisan
Rescitens
Reserpinene
Reserpinin
Reserpinine
Resipal
Reskinnamin
Scinnamina
Tenamine
Trimethoxycinnamoyl methyl reserpate
3,4,5-Trimethylcinnamic acid, ester with methyl reserpate
3,4,5-Trimethylcinnamoyl methyl reserpate
Tuareg

24830-33-9
$C_6H_7Cl_3N_4$
241.48
1,3,5-Triazin-2-amine, N,4-dimethyl-6-(trichloromethyl)- (9CI)
s-Triazine, 2-methyl-4-(methylamino)-6-(trichloromethyl)- (8CI)

24830-34-0
$C_8H_{11}Cl_3N_4$
269.56
s-Triazine, 2-(isopropylamino)-4-methyl-6-(trichloromethyl)- (8CI)

24830-35-1
$C_9H_{13}Cl_3N_4$
283.59
s-Triazine, 2-(butylamino)-4-methyl-6-(trichloromethyl)- (8CI)

24848-37-1
$C_{14}H_{15}Cl_3N_4$
345.66
s-Triazine, 2-(butylamino)-4-phenyl-6-(trichloromethyl)- (8CI)

24848-38-2
$C_{12}H_{11}Cl_3N_4$
317.61
s-Triazine, 2-(dimethylamino)-4-phenyl-6-(trichloromethyl)- (8CI)

24848-39-3
$C_{12}H_{10}Cl_4N_4$
352.05
s-Triazine, 2-(p-chlorophenyl)-4-(ethylamino)-6-(trichloromethyl)- (8CI)

24848-40-6
$C_8H_5Cl_6N_5$
383.88
Propionitrile, 3-((4,6-bis(trichloromethyl)-s-triazin-2-yl)amino)- (8CI)

24848-41-7
$C_9H_7Cl_6N_5$
397.91
Propionitrile, 3-((4,6-bis(trichloromethyl)-s-triazin-2-yl)methylamino)- (8CI)

24848-42-8
$C_{10}H_9Cl_6N_5$
411.93
Propionitrile, 3-((4,6-bis(trichloromethyl)-s-triazin-2-yl)ethylamino)- (8CI)

24848-43-9
$C_{11}H_{11}Cl_6N_5$
425.96
Propionitrile, 3-((4,6-bis(trichloromethyl)-s-triazin-2-yl)propylamino)- (8CI)

24848-44-0
$C_8H_8Cl_3N_5$
280.55
Propionitrile, 3-((4-methyl-6-(trichloromethyl)-s-triazin-2-yl)amino)- (8CI)

24848-45-1
$C_9H_{10}Cl_3N_5$
294.57
Propionitrile, 3-(methyl-(4-methyl-6-(trichloromethyl)-s-triazin-2-yl)-amino)- (8CI)

24848-46-2
$C_{10}H_{12}Cl_3N_5$
308.60
Propionitrile, 3-(ethyl-(4-methyl-6-(trichloromethyl)-s-triazin-2-yl)-amino)- (8CI)

24848-47-3
$C_{12}H_{16}Cl_3N_5$
336.65
Propionitrile, 3-(butyl-(4-methyl-6-(trichloromethyl)-s-triazin-2-yl)-amino)- (8CI)

24848-48-4
$C_{12}H_{16}Cl_3N_5$
322.63
Propionitrile, 3-(sec-butyl(4-methyl-6-(trichloromethyl)-s-triazin-2-yl)amino)- (8CI)

24848-51-9
$C_{11}H_{14}Cl_3N_5$
322.63
Propionitrile, 3-(methyl-(4-propyl-6-(trichloromethyl)-s-triazin-2-yl)-amino)- (8CI)

24848-52-0
$C_{12}H_{16}Cl_3N_5$
336.65
Propionitrile, 3-(ethyl-(4-propyl-6-(trichloromethyl)-s-triazin-2-yl)-amino)- (8CI)

24848-53-1
$C_{13}H_{18}Cl_3N_5$
350.68
Propionitrile, 3-(propyl-(4-propyl-6-(trichloromethyl)-s-triazin-2-yl)-amino)- (8CI)

24848-54-2
$C_{14}H_{20}Cl_3N_5$
364.71
Propionitrile, 3-(butyl-(4-propyl-6-(trichloromethyl)-s-triazin-2-yl)-amino)- (8CI)

24848-59-7
$C_{13}H_{18}Cl_3N_5$
350.68
Propionitrile, 3-(methyl-(4-pentyl-6-(trichloromethyl)-s-triazin-2-yl)-amino)- (8CI)

24848-60-0
$C_{15}H_{22}Cl_3N_5$
378.74
Propionitrile, 3-((4-pentyl-6-(trichloromethyl)-s-triazin-2-yl)propylamino)- (8CI)

24848-61-1
$C_{16}H_{24}Cl_3N_5$
392.76
Propionitrile, 3-(butyl-(4-pentyl-6-(trichloromethyl)-s-triazin-2-yl)-amino)- (8CI)

24848-62-2
$C_{16}H_{24}Cl_3N_5$
392.76
Propionitrile, 3-(sec-butyl(4-pentyl-6-(trichloromethyl)-s-triazin-2-yl)amino)- (8CI)

24851-98-7
$C_{13}H_{22}O_3$
226.32
O=C(OC)CC(C(C(=O)C1)CCCC C)C1
Cyclopentaneacetic acid, 3-oxo-2-pentyl-, methyl ester (9CI)
Methyl dihydrojasmonate

24860-40-0

$C_6H_{11}N_5O$
169.22
s-Triazine, 4,6-diamino-2-isopropoxy

24863-51-2
$C_{12}H_{13}Cl_6N_5$
439.99
Propionitrile, 3-((4,6-bis-(trichloromethyl)-s-triazin-2-yl)butylamino)- (8CI)

24863-52-3
$C_{12}H_{13}Cl_6N_5$
439.99
Propionitrile, 3-((4,6-bis-(trichloromethyl)-s-triazin-2-yl)-sec-butylamino)- (8CI)

24863-53-4
$C_{11}H_{14}Cl_3N_5$
322.63
Propionitrile, 3-((4-methyl-6-(trichloromethyl)-s-triazin-2-yl)propylamino)- (8CI)

24863-54-5
$C_{14}H_{20}Cl_3N_5$
364.71
Propionitrile, 3-(sec-butyl(4-propyl-6-(trichloromethyl)-s-triazin-2-yl)amino)- (8CI)

24863-55-6
$C_{14}H_{20}Cl_3N_5$
364.71
Propionitrile, 3-(ethyl-(4-pentyl-6-(trichloromethyl)-s-triazin-2-yl)-amino)- (8CI)

24884-69-3
$C_4H_8N_2O_5$
164.11
2-Nitro-2-methylpropanol nitrate
2-Methyl-2-nitro-propanol nitrate
1-Propanol, 2-methyl-2-nitro-, nitrate

24887-06-7
$C_2H_6O_6S_2Zn$
255.59
Zinc formaldehyde sulfoxylate
Bis(hydroxymethanesulfinato-O,O')zinc
(T-4)-Bis(hydroxymethanesulfinato-OS,O1)zinc
Zinc, bis(hydroxymethanesulfinato-O,O')-
Zinc, bis(hydroxymethanesulfinato-OS,O1)-, (T-4)- (9CI)

24903-95-5
$C_9H_{14}O$
138.21
O=C(C(CC1C2)C1(C)C)C2
Bicyclo(3.1.1)heptan-2-one, 6,6-dimethyl- (9CI)
6,6-Dimethylbicyclo(3.1.1)-heptan-2-one
Nopinon
Nopinone
2-Norpinanone, 6,6-dimethyl- (8CI)
NSC-135004
β-Pinone

24910-63-2
C_7H_{14}
98.19
2-Pentene, 3,4-dimethyl- (9CI)

24911-15-7
$C_{12}H_{12}ClNO$
221.68
Benzamide, 4-chloro-N-(1,1-dimethyl-2-propynyl)- (9CI)
Benzamide, p-chloro-N-(1,1-dimethyl-2-propynyl)- (8CI)

24925-59-5
$C_{30}H_{47}N$
421.70
N(c(ccc(c1)CCCCCCCCC)c1)c(ccc(c2)CCCCCCCCC)c2
Benzenamine, 4-nonyl-N-(4-nonylphenyl)- (9CI)
Diphenylamine, 4,4'-dinonyl- (8CI)

24928-72-1
$C_{22}H_{40}N_2O_4S_2$
460.69
O=C(OCCN=C(S)C(=NCCOC(=O)CCCCCCC)S)CCCCCCC
Octanoic acid, (1,2-dithioxo-1,2-ethanediyl)bis(imino-2,1-ethanediyl) ester (9CI)
Ethanedithioamide-N,N'-bis(2-octanoyloxyethyl)-

24934-91-6
$C_5H_{12}ClO_2PS_2$
234.71
CCOP(=S)(OCC)SCCl
Phosphorodithioic acid, S-(chloromethyl) O,O-diethyl ester
Chlormefos
Chlormephos
S-(Chloromethyl) O,O-diethyl-phosphorodithioate
S-Chloromethyl-O,O-diethyl-phosphorothiolothionate
Dotan
MC 2188

24935-97-5
$C_5H_9NO_3$
131.13
O=C(OCC=C)NCO
Carbamic acid, (hydroxymethyl)-, 2-propenyl ester (9CI)
Allyl (hydroxymethyl)carbamate
Allyl N-methylolcarbamate
2-Propenyl (hydroxymethyl)-carbamate

24937-78-8
$(C_4H_6O_2.C_2H_4)x$
Polymer
Ethylene/Vinyl Acetate Copolymer
Acetic acid ethenyl ester, Polymer with ethene (9CI)
Ethylene-vinyl acetate copolymer emulsion
Ethylenevinylacetate copolymer
Ethylene-vinylacetate resin
Vinyl acetate, ethene polymer
Vinyl acetate, ethylene polymer

24937-79-9
$(C_2H_2F_2)x$
Polymer
Polyvinylidene fluoride
1,1-Difluoroethene homopolymer
Ethene, 1,1-difluoro-, Homopolymer (9CI)

24938-37-2
$(C_6H_{10}O_4.C_2H_6O_2)x$
Polymer
Hexanedioic acid, Polymer with 1,2-ethanediol (9CI)

24938-91-8
$(C_2H_4O)xC_{13}H_{28}O$
Polymer
Tridecanol Condensed with 6 moles ethylene oxide
Alkyl(C-13) polyethoxylates-(ethoxy-6)
Polyoxiethylene (6) alkyl (13) ether

24948-81-0
$C_6H_{14}ClN$
135.63
N(C(C)C)(C(C)C)Cl
2-Propanamine, N-chloro-N-(1-methylethyl)- (9CI)
N-Chloro-N-(1-methylethyl)-2-propanamine

24949-42-6
$C_{14}H_{28}$
196.38

6-Tridecene, 7-methyl- (6CI,8CI,9CI)
7-Methyl-6-tridecene

24952-65-6
$C_4H_{12}Pb$
267.33
Diethyllead

24955-63-3
$C_6H_{10}Cl_2$
153.05
Cyclohexane, 1,3-dichloro-, cis- (8CI,9CI)
cis-1,3-Dichlorocyclohexane

24959-67-9
HBr
80.92
Caswell No. 499D
EPA Pesticide Chemical Code 053200
Inorganic bromides

24966-39-0
$C_6H_4Cl_2S$
179.07
Benzenethiol, 2,6-dichloro- (8CI,9CI)

24968-12-5
$C_{12}H_{14}O_4$
222.24
Polytetramethylene terephthalate

24973-25-9
$C_{11}H_{17}NO_2.ClH$
231.72
Benzeneethanamine, 2,5-dimethoxy-α-methyl-, hydrochloride (9CI)
Amphetamine, 2,5-dimethoxy-, hydrochloride
DEA No. 7396
2,5-Dimethoxyamphetamine hydrochloride
2,5-Dimethoxy-α-methylphenethylamine hydrochloride
1-(2,5-Dimethoxyphenyl)-2-aminopropanehydrochloride
2,5-DMA hydrochloride
Isopropylamine, β-(2,5-dimethoxyphenyl)-, hydrochloride
Phenethylamine, 2,5-dimethoxy-α-methyl-, hydrochloride (6CI,8CI)

24979-70-2
$(C_8H_8O)x$
Polymer
Phenol, 4-ethenyl-, Homopolymer (9CI)
4-Ethenylphenol homopolymer
p-Hydroxystyrene polymer
Maruzen M
NSC-114470
Phenol, p-vinyl-, Polymers
Poly(p-hydroxystyrene)
Poly(4-hydroxystyrene)
Poly(p-vinylphenol)
Poly(4-vinylphenol)
p-Vinylphenol polymer
4-Vinylphenol polymer

24980-41-4
$(C_6H_{10}O_2)x$
Polymer
Aquaplast, Caprolactone
2-Oxepanone, Homopolymer (9CI)
PCL 700
PCL-700
Poly(ε-caprolactone)

24981-14-4
$(C_2H_3F)x$
Polymer
Ethene, fluoro-, Homopolymer (9CI)
Fluoroethene homopolymer

24985-48-6
$C_9H_{16}O_2$
156.23
2-Octen-4-one, 2-methoxy- (8CI)

24991-55-7
$(C_2H_4O)xC_2H_6O$
Polymer
Glycols, polyethylene, dimethyl ether
Dimethoxy polyethylene glycol
Glyme-23
α,ω-Methoxypoly(ethylene oxide)
Polyethylene glycol dimethyl ether
Poly(oxy-1,2-ethanediyl), α-methyl-ω-methoxy- (9CI)
Polyoxyethylene dimethyl ether
Selexol

25013-15-4
C_9H_{10}
118.19
Styrene, methyl
Methylstyrene
NCI-C56406
Toluene, vinyl- (Mixed isomers)
UN 2618 [Vinyl toluene, inhibited mixed isomers]
Vinyl toluene (ACGIH,OSHA)
Vinyltoluene
Vinyl toluenes (Mixed isomers), Inhibited [UN 2618]

25013-16-5
$C_{11}H_{16}O_2$
180.27
COc1ccc(O)c(c1)C(C)(C)C
Phenol, (1,1-dimethylethyl)-4-methoxy
Antioxyne B
Antrancine 12
BHA
Butylated hydroxyanisole
Butylhydroxyanisole
tert-Butylhydroxyanisole
tert-Butyl-4-hydroxyanisole
2(3)-tert-Butyl-4-hydroxyanisole
2-terc.Butyl-4-methoxyfenol (Czech)
tert-Butyl-4-methoxyphenol
2-tert-Butyl-4-methoxyphenol
Butylohydroksyanizol (Polish)
(1,1-Dimethylethyl)-4-methoxyphenol
EEC No. E320
Embanox
Nipantiox 1-F
Protex
Premerge Plus
Sustane
Sustane 1-F
Vertac
Tenox BHA

25014-41-9
$(C_3H_3N)x$
Polymer
Polyacrylonitrile
Acrylonitrile homopolymer
Acrylonitrile, Polymers (8CI)
Barex 210 Resin
Bulana
Dralon T
NSC-7763
PAN (VAN)
PAN (Polymer)
Poly(acrylonitrile), Fibers
2-Propenenitrile, Homopolymer (9CI)

25033-65-2
$C_{15}H_{23}NO$
233.36
Benzamide, N,N-dibutyl- (6CI,8CI,9CI)
N,N-Dibutylbenzamide
Di-N-n-butylbenzoic acid amide

25034-79-1
$(Cl_2NP)x$
Polymer
Phosphonitrile chloride, Homopolymer (9CI)
Cyclic PNCl2
Phosphonitrilic chlorides

25035-67-0
$(C_7H_{14}N_2O)n$
Polymer NA
Poly(iminocarbonylimino-1,6-hexanediyl) (9CI)
1,6-Hexamethylenediamine-1,6-hexamethylenediiso-

cyanate copolymer, SRU
Poly(iminocarbonyliminohex amethylene)
Polyurea 6
Poly(ureylenehexamethyl ene) (8CI)

25035-78-3
$(C_{14}H_{14}O_4)x$
Polymer
1,3-Benzenedicarboxylic acid, di-2-propenyl ester, Homopolymer (9CI)
Poly(diallyl isophthalate)
Poly(di-2-propenyl 1,3-benzene-dicarboxylate)

25036-25-3
$(C_{21}H_{24}O_4.C_{15}H_{16}O_2)x$
Polymer
Phenol, 4,4'-(1-methylethylidene)bis-, Polymer with 2,2'-((1-methylethylidene)-bis(4,1-phenyleneoxymethylene))bis(oxirane) (9CI)

25037-58-5
$(C_3H_4O_2)x$
Polymer NA
2-Oxetanone, Homopolymer (9CI)
2-Oxetanone, Polyesters (8CI)

25037-66-5
$(C_8H_4O_3.C_4H_2O_3.C_3H_8O_2)x$
Polymer
1,3-Isobenzofurandione, Polymer with 2,5-furandione and 1,2-propanediol (9CI)
Maleic anhydride, phthalic anhydride, propylene glycol terpolymer
Maleic anhydride, propylene glycol, 1,2-benzenedi-carboxylic anhydride polymer
Maleic anhydride, propylene glycol, phthalic anhydride polymer
Phthalic anhydride, maleic anhydride, and propylene glycol polymer
Phthalic anhydride, Polymer with maleic anhydride and propylene glycol
Polymer of propylene glycol, maleic anhydride, phthalic anhydride
1,2-Propanediol, maleic anhydride, phthalic anhydride polymer
Propylene glycol, maleic anhydride, phthalic anhydride polymer
Propylene glycol, phthalic anhydride, maleic anhydride polymer
Propylene glycol, phthalic anhydride, maleic anhydride resin

25038-04-4
$(C_3H_8O_3.C_3H_5ClO)x$
Polymer
1,2,3-Propanetriol, Polymer with (chloromethyl)oxirane (9CI)
Epichlorohydrin-glycerine co-polymer

25038-54-4
$(C_6H_{11}NO)x$
Polymer
Poly(iminocarbonylpentamethylene)
A 1030
A 1030N0
Akulon
Akulon M 2W
Alkamid
Amilan CM 1001
Amilan CM 1011
Amilan CM 1031
Amilan CM 1001C
Amilan CM 1001G
6-Aminohexanoic acid homo-polymer
ATM 2 (Nylon)
Aviamide-6
B-35
B-203
B-216
B-300
B-350
Bonamid
Capran 80
Capran 77C
Caproamide Polymer
Caprolactam Oligomer
Caprolactam Polymer
ε-Caprolactam polymer
ε-Caprolactam polymere (German)
Caprolon B
Caprolon V
Capron
Capron 8250
Capron 8252
Capron 8253
Capron 8256
Capron B
Capron GR 8256
Capron GR 8257
Capron GR 8258
Capron PK 4
Chemlon
CM 1001
CM 1011
CM 1031
CM 1041
Danamid
Dull 704
Durethan BK
Durethan BK 30S
Durethan BKV 30H
Durethan BKV 55H
Ertalon 6SA
Extron 6N
Grilon
Hexahydro-2H-azepin-2-one homopolymer
Itamid
Itamid 250
Itamide 25
Itamide 35
Itamide 250
Itamide 350
Itamide 250G
Itamide S
Kaprolit
Kaprolit B
Kaprolon
Kaprolon B
Kapromin
Kapron
Kapron A
Kapron B
KS 30P
Maranyl F 114
Maranyl F 124
Maranyl F 500
Metamid
Miramid H 2
Miramid WM 55
Nylon-6
Nylon A1035SF
Nylon CM 1031
Nylon X 1051
Orgamide
Orgamid RMNOCD
PA 6
PA 6 (Polymer)
PK 4
PKA
Plaskin 8200
Plaskon 201
Plaskon 8201
Plaskon 8205
Plaskon 8207
Plaskon 8252
Plaskon 8202C
PLASKON 8201HS
Plaskon XP 607
Policapran
Polyamide 6
Polyamide PK 4
Poly(ε-aminocaproic acid)
Polycaproamide
Poly(ε-caproamide)
Polycaprolactam
Poly(ε-caprolactam)
Poly(imino(1-oxo-1,6-hexane-diyl)) (9CI)
P 6 (Polyamide)
Relon P
Renyl MV
Sipas 60
Spencer 401
Spencer 601
Stee:PM
Stilon
Stylon
Tarlon X-A
Tarlon XB
Tarnamid T
Tarnamid T 2
Tarpamid T 27
TNK 2G5
Torayca N 6
UBE 1022B

Ultramid B 3
Ultramid B 4
Ultramid B 5
Ultramid BMK
Vidlon
Widlon
Zytel 211

25038-59-9
$(C_{10}H_8O_4)x$
Polymer
Poly(oxyethyleneoxytere-phthaloyl)
Alathon
Amilar
Arnite A
Arnite A 200
Arnite A-049000
Arnite FP 800
Arnite G
Arnite G 600
Cassappret SR
Celanar
Cleartuf
Clertuf
Crastin S 330
Crastin S 350
Crastin S 440
Daiya Foil
Dowlex
Estar
Estrofol
Estrofol B
Estrofol Ow
Ethylene terephthalate polymer
Fiber V
Hostadur
Hostadur A
Hostadur K
Hostadur K-VP 4022
Hostaphan
Hostaphan BNH
Hostaphan RN
Iambolen
KLT 40
Lavsan
Lawsonite
Lumilar 100
Lumirror
Lumirror 38S
Meliform
Melinex
Melinex O
Mylar
Mylar A
Mylar C
Mylar C-25
Mylar HS
Mylar T
Nitron Lavsan
Nitron (Polyester)
Pegoterate
Polyethylene terephthalate
Polyethylene terephthalate film
Poly(oxy-1,2-ethanediyloxy-carbonyl-1,4-phenylenecarbon-yl) (9CI)
Scotch PAR
Superfloc
Terephtahlic acid-ethylene glycol polyester
Terfan
Tergal
Terom
Terphan
VFR 3801
Vituf

25044-01-3
$C_6H_{10}O$
98.14
1-Penten-3-one, 2-methyl- (8CI,9CI)

25053-15-0
$(C_{14}H_{14}O_4)x$
Polymer
1,2-Benzenedicarboxylic acid, di-2-propenyl ester, Homo-polymer (9CI)
Allyl phthalate, Homopolymer

25057-89-0
$C_{10}H_{12}N_2O_3S$
240.30
O=C(N(S(=O)(=O)Nc1cccc2)C(C)C)c12
1H-2,1,3-Benzothiadiazin-4(3H)-one, 3-isopropyl-, 2,2-dioxide
BAS 3510
BAS 351-H
BAS 3510H
BAS 3512H
BAS 3517H
BAS 351-07H
Basagran
Bendioxide
Bentazon
Bentazone
3-Isopropyl-2,1,3-benzothia-diazinon-(4)-2,2-dioxid (German)
3-Isopropyl-1H-2,1,3-benzothia-diazin-4(3H)-one-2,2-dioxide
3-(1-Methylethyl)-1H-2,1,3-benzothiazain-4(3H)-one, 2,2-dioxide
Pentazone

25059-78-3
$C_8H_6Cl_2O_3.C_4H_{11}NO_2$
326.20
Diethanolamine dicamba
o-Anisic acid, 3,6-dichloro-, Compd. with 2,2'-iminodi-ethanol
o-Anisic acid, 3,6-dichloro-, Compd. with 2,2'-iminodi-ethanol (1:1)
Benzoic acid, 3,6-dichloro-2-methoxy-, Compd. with 2,2'-iminobis(ethanol) (1:1)
Caswell No. 295A
Dicamba diethanolamine salt
EPA Pesticide Chemical Code 029803
Ethanol, 2,2'-iminobis-, 3,6-di-chloro-2-methoxybenzoate (Salt)
Ethanol, 2,2'-iminodi-, 3,6-di-chloro-o-anisate (Salt)
2-Methoxy-3,6-dichlorobenzoic acid diethanolamine salt

25066-20-0
$C_{23}H_{48}N_2O_2$
384.63
O=C(NCCCN(=O)(C)C)CCCCCCCCCCCCCCCCC
Octadecanamide, N-(3-(di-methylamino)propyl)-, N-oxide (9CI)
Stearylamidopropyl-N,N-di-methylamine, oxide

25067-59-8
$(C_{14}H_{11}N)x$
Polymer
Poly-N-vinylcarbazole
9H-Carbazole, 9-ethenyl-, Homopolymer (9CI)
9-Ethenyl-9H-carbazole homo-polymer

25068-38-6
$(C_{15}H_{16}O_2.C_3H_5ClO)x$
Polymer
Phenol, 4,4'-isopropylidene-di-, Polymer with 1-chloro-2,3-epoxypropane
E 828
E 1001
E 1004
Epidian 5
Epikote 828
Epikote 1001
Epikote 1004
EPON 820
EPON 828
EPON 1001
EPON 1007
ERL-2795
Epoxy Resin ERL-2795
Phenol, 4,4'-isopropylidenedi-, Dimer with 1-chloro-2,3-epoxypropane
Phenol, 4,4'-isopropylidenedi-, Monomer with 1-chloro-2,3-epoxypropane
Phenol, 4,4'-isopropylidenedi-, tetramer with 1-chloro-2,3-epoxypropane

25074-67-3
$C_{12}H_7ClO$
202.64
3-Chlorodibenzofuran

25079-96-3
$C_{10}H_{13}NO$
163.22
Acetamide, N-((4-methyl-phenyl)methyl)- (9CI)
Acetamide, N-(p-methyl-benzyl)- (6CI,7CI,8CI)

25081-39-4
$C_{10}H_{12}O_2$
164.20
Benzoic acid, 3,5-dimethyl-, methyl ester (8CI,9CI)

25085-98-7
$(C_{14}H_{20}O_4)x$
Polymer
7-Oxabicyclo(4.1.0)heptane-3-carboxylic acid, 7-oxabicyclo(4.1.0)hept-3-ylmethyl ester, Homopolymer (9CI)

25085-99-8
$(C_{21}H_{24}O_4)x$
Polymer
Araldite B
Araldite F
2,2-Bis(p-(2,3-epoxypropoxy)-phenyl)propane polymers
Oxirane, 2,2'-((1-methylethylidene)bis(4,1-phenyleneoxymethylene))bis-, Homopolymer (9CI)

25086-25-3
$C_8H_{16}O_2$
144.22
ERL 4206
7-Oxabicyclo(4.1.0)heptane, 3-oxiranyl-, Homopolymer

25087-26-7
$(C_4H_6O_2)x$
Polymer
Polymethacrylic acid
2-Methyl-2-propenoic acid homopolymer
2-Propenoic acid, 2-methyl-, Homopolymer (9CI)

25088-57-7
$C_{36}H_{71}O_3P$
582.93
Dioleyl hydrogen phosphite
Dioleyl phosphite
Dioleyl phosphonate
Phosphonic acid, di-9-octadecenyl ester (Z,Z)-

25097-44-3
$C_{14}H_{22}ClOP$
272.76
Phosphinic chloride, (1,1-dimethylethyl)-(4-(1,1-dimethylethyl) phenyl)- (9CI)
Phosphinic chloride, tert-butyl(p-tert-butylphenyl)- (8CI)

25101-03-5
$(C_6H_{10}O_4.C_3H_8O_2)x$
Polymer
Hexanedioic acid, Polymer with 1,2-propanediol (9CI)
Poly(propylene glycol adipate)
Propylene glycol, Adipic acid resin

25103-09-7
$C_{10}H_{20}O_2S$
204.36
CC(C)CCCCCOC(=O)CS
Acetic acid, mercapto-, isooctyl ester
Isooctyl mercaptoacetate
Isooctyl thioglycolate

25103-12-2
$C_{24}H_{54}O_3P$
421.75
CC(C)CCCCCOP(=O)(CCCCCC(C)C)OCCCCCC(C)C
Phosphorous acid, triisooctyl ester
Triisooctyl phosphite

25103-52-0
$C_8H_{16}O_2$
144.21
Isooctanoic acid (9CI)

25103-54-2
$C_{20}H_{43}O_2PS_2.1/2Zn$
410.67
Zinc O,O-bisisodecyl dithiophosphate
ELCO 106
Isodecanol, hydrogen phosphorodithioate, zinc salt (9CI)
Isodecyl ZDDP
Phosphorodithioic acid, O,O-diisodecyl ester, zinc salt
ZDTP
Zinc O,O-diisodecyl dithiophosphate

25103-58-6
$C_{12}H_{26}S$
202.44
CCCCCCCCCCCCS
t-Dodecanethiol
terc.Dodecylmerkaptan (Czech)
tert-Dodecylmercaptan
tert-Dodecylthiol
2,3,3,4,4,5-Hexamethyl-2-hexanethiol

25103-87-1
$(C_6H_{10}O_4.C_4H_{10}O_2)x$
Polymer
Hexanedioic acid, Polymer with 1,4-butanediol (9CI)

25104-37-4
Unknown
Unknown
Barbituric acid, 5-ethyl-5-(3 or 6-oxo-1-cyclohexen-1-yl)
EHB-M
5-Ethyl-5-(3 or 6-oxo-1-cyclohexen-1-yl)barbituric acid
5-(3 or 6-Oxo-1-cyclohexen-1-yl)-5-ethylbarbituric acid

25111-05-1
$C_{42}H_{78}O_5$
663.08
O=C(OCC(CC)(CO)COC(=O)CCCCCCCC=CCCCCCCCC)CCCCCCCC=CCCCCCCCC
9-Octadecenoic acid (Z)-, 2-ethyl-2-(hydroxymethyl)-1,3-propanediyl ester (9CI)

25113-45-5
$C_6H_{11}N_3O_2$
157.17
s-Triazine-2,4(1H,3H)-dione, dihydro-6-isopropyl-(8CI)

25117-33-3
$C_{16}H_{34}$
226.45
Pentadecane, 5-methyl-(6CI,7CI,8CI,9CI)
5-Methylpentadecane

25134-01-4
$(C_8H_{10}O)x$
Polymer
Phenol, 2,6-dimethyl-, Homopolymer (9CI)
2,6-Dimethylphenol homopolymer

25134-08-1
$C_7H_3Cl_3O$
209.46
Benzoyl chloride, dichloro-(9CI)
Dichlorobenzoyl chloride

25134-21-8
$C_{10}H_{10}O_3$
178.20
5-Norbornene-2,3-dicarboxylic anhydride, methyl
Methylbicyclo(2.2.1)heptene-2,3-dicarboxylic anhydride isomers
Nadic methyl anhydride
NMA

25136-53-2
$C_6H_{12}O_3$
132.18
Propanediol, (allyloxy)
Glycerin monoallyl ether
Glycerol allyl ether

Glycerol monoallyl ether
Propanediol, (2-propenyloxy)-

25136-55-4
$C_6H_{12}O_2$
116.18
p-Dioxane, dimethyl
Dimethyl dioxane [UN 2707]
Dimethyl-p-dioxane [UN 2702]
UN 2707 [Dimethyldioxanes]

25144-04-1
$C_6H_{12}O$
100.16
Cyclopentanol, 2-methyl-, trans- (8CI,9CI)

25144-05-2
$C_6H_{12}O$
100.16
Cyclopentanol, 2-methyl-, cis- (8CI,9CI)
cis-2-Methylcyclopentanol

25152-84-5
$C_{10}H_{16}O$
152.24
O=CC=CC=CCCCCC
2,4-Decadienal, (E,E)- (9CI)
(E,E)-2,4-Decadienal
2,4-trans,trans-Decadienal

25154-38-5
$C_6H_{14}N_2O$
130.18
Monohydroxyethylpiperazine
Hydroxyethylpiperazine
Piperazineethanol (9CI)

25154-52-3
$C_{15}H_{24}O$
220.39
Phenol, nonyl
Hydroxyl No. 253
Nonyl phenol (Mixed isomers)

25154-54-5
$C_6H_4N_2O_4$
168.12
Benzene, dinitro
Dinitrobenzene (ACGIH)
Dinitrobenzene, Solid [UN 1597]
Dinitrobenzene, Solution (DOT)
Dinitrobenzol, Solid (DOT)
UN 1597 [Dinitrobenzenes, solid]

25154-55-6
$C_6H_5NO_3$
139.10
Nitrophenols
AI3-24342
Hydroxynitrobenzene
Mononitrophenol
Nitrophenol
Nitrophenol, Mixed
Phenol, nitro-

25154-86-3
$(C_8H_{15}NO_2)x$
Polymer
2-Propenoic acid, 2-methyl-, 2-(dimethylamino)ethyl ester, Homopolymer (9CI)
Dimethylaminoethyl methacrylate homopolymer
N,N-Dimethylaminoethyl methacrylate polymer

25155-15-1
$C_{10}H_{14}$
134.22
Cymene
Benzene, methyl(1-methylethyl)- (9CI)
Methyl(1-methylethyl)benzene

25155-18-4
$C_{10}H_8O_3S$
208.24
Methylbenzethonium
AI3-04543-X
Bactine
Benzenemethanaminium, N,N-dimethyl-N-(2-(2-(methyl-4-(1,1,3,3-tetraethylbutyl)phenoxy)ethoxy)ethyl)-, chloride
Benzenemethanaminium, N,N-dimethyl-N-(2-(2-(methyl-4-(1,1,3,3-tetramethylbutyl)phenoxy)ethoxy)ethyl)-, chloride (25% Aqueous)
Caswell No. 355
Chlorure de methylbenzethonium (French)
Cloruro de metilbenzetonio (Spanish)
Diisobutyl cresoxy ethoxy ethyl dimethyl benzyl ammonium chloride
Diisobutylcresoxyethoxyethyl dimethyl benzyl ammonium chloride
2-(2-(p-(Diisobutyl)cresoxy)ethoxy)ethyl dimethyl benzyl ammonium chloride
N,N-Dimethyl-N-(2-(2-(methyl-4-(1,1,3,3-tetramethylbutyl)phenoxy)ethoxy)ethyl)-benzenemethanaminium chloride
EPA Pesticide Chemical Code 069134
Hyamine 10
Methylbenzethonii chloridum (Latin)
Methylbenzethonium chloride
Metilbenzetonio cloruro
Octyl cresoxyethoxyethyl dimethyl benzyl ammonium chloride

25155-23-1
$C_{24}H_{27}O_4P$
410.48
Xylenol, phosphate (3:1)
Coalite NTP
Dimethylphenol phosphate (3:1)
Phenol, dimethyl-, phosphate (3:1) (9CI)
Reofos 95
Trixylenyl phosphate
Trixylyl phosphate
Xylyl phosphate

25155-25-3
$C_{20}H_{34}O_4$
338.54
Peroxide, (phenylenediisopropylidene)bis(tert-butyl
α,α'-Bis(tert-butylperoxy)diisopropylbenzene
Peroxide, (phenylenebis-(1-methylethylidene))bis-(1,1-dimethylethyl)-
(Phenylenediisopropylidene)-bis(tert-butylperoxide)
Vul-Cup
Vul-Cup 40KE
Vul-Cup R

25155-29-7
$C_3H_8N_2O_3$
120.10
Urea, bis(hydroxymethyl)-(9CI)
Bis(hydroxymethyl)urea

25155-30-0
$C_{18}H_{29}O_3S.Na$
348.52
CCCCC(C)(C)CCC(C)(C)c1ccc(cc1)S(=O)(=O)[O-][Na+]
Benzenesulfonic acid, dodecyl-, sodium salt
AA-9
AA-10
Abeson Nam
Bio-Soft D-40
Bio-Soft D-60
Bio-Soft D-62
Bio-Soft D-35X
Calsoft F-90
Calsoft L-40
Calsoft L-60
Conco AAS-35
Conco AAS-40
Conco AAS-65
Conco AAS-90
Conoco C-50
Conoco C-60
Conoco SD 40
Detergent HD-90
Dodecyl benzene sodium sulfonate
Dodecylbenzenesulfonic acid sodium salt
Dodecylbenzenesulphonate,

sodium salt
Dodecylbenzensulfonan sodny (Czech)
p-Dodecylbenzensulfonan sodny (Czech)
Mercol 25
Mercol 30
Naccanol NR
Naccanol SW
Nacconol 40F
Nacconol 90F
Nacconol 35SL
Neccanol SW
Pilot HD-90
Pilot SF-40
Pilot SF-60
Pilot SF-96
Pilot SF-40B
Pilot SF-40FG
Pilot SP-60
Richonate 1850
Richonate 45B
Richonate 60B
Santomerse 3
Santomerse No. 1
Santomerse No. 85
Sodium dodecylbenzene-sulfonate
Sodium dodecylbenzene-sulfonate, Dry
Sodium laurylbenzenesulfonate
Solar 40
Solar 90
Sol sodowa kwasu laurylo-benzenosulfonowego (Polish)
Sulfapol
Sulfapolu (Polish)
Sulframin 85
Sulframin 40 Flakes
Sulframin 90 Flakes
Sulframin 40 Granular
Sulframin 40RA
Sulframin 1238 Slurry
Sulframin 1250 Slurry
Ultrawet K
Ultrawet 60K
Ultrawet KX
Ultrawet SK

25167-32-2
$C_{36}H_{58}O_7S_2.2Na$
712.96
Benzenesulfonic acid, 2,2'(or 3,3')-oxybis(5(or 2)-dodecyl-, disodium salt (9CI)

25167-67-3
C_4H_8
56.12
Butene (DOT)
n-Butene
Butylene [UN 1012]
n-Butylene
Butylene (DOT)
UN 1012 [Butylene see also Petroleum gases, Liquefied]

25167-70-8
C_8H_{16}
112.24
Pentene, 2,4,4-trimethyl
Diisobutene
Diisobutylene (DOT)
Diisobutylene, isomeric compounds [UN 2050]
2,4,4-Trimethyl pentene
UN 2050 [Diisobutylene, isomeric compounds]

25167-80-0
C_6H_5ClO
128.56
Phenol, chloro- (8CI,9CI)

25167-81-1
$C_6H_4Cl_2O$
163.00
Phenol, dichloro-
AI3-15332
Dichlorophenol

25167-82-2
$C_6H_3Cl_3O$
197.44
Phenol, trichloro
NA 2020 (DOT)
Trichlorophenol (DOT)

25167-83-3
$C_6H_2Cl_4O$
231.88
Phenol, tetrachloro
Tetrachlorophenol

25167-84-4
$C_{12}H_6Cl_6O_4$
426.90
Trichlorocatechol
Trichloropyrocatechol

25167-85-5
$C_6H_4Cl_2O_2$
179.00
1,2-Benzenediol, dichloro- (9CI)
Pyrocatechol, dichloro- (8CI)

25167-93-5
$C_6H_4ClNO_2$
157.56
Benzene, chloronitro- (Mixed isomers)
Chloronitrobenzene
Mononitrochlorobenzene
Nitrochlorobenzene

25168-04-1
$C_8H_9NO_2$
151.18
Xylene, nitro
NA 1665 (DOT)
Nitrodimethylbenzene
Nitroxylene
Nitroxylol (DOT)

25168-05-2
C_7H_7Cl
126.59
Toluene, ar-chloro (8CI)
Benzene, chloromethyl- (9CI)
Chloromethylbenzene
Chlorotoluene
ar-Chlorotoluene

25168-06-3
$C_9H_{12}O$
136.19
Phenol, (1-methylethyl)- (9CI)
Isopropylphenol
(1-Methylethyl)phenol

25168-07-4
C_8H_{12}
108.18
Cyclohexene, ethenyl- (9CI)
Ethenylcyclohexene

25168-10-9
$C_{10}H_9N$
143.18
Naphthalenamine (9CI)
Naphthylamine (8CI)

25168-15-4
$C_{16}H_{21}Cl_3O_3$
367.72
CC(C)CCCCCOC(=O)Cc1cc(Cl)c(Cl)cc1Cl
Acetic acid, 2,4,5-trichlorophenoxy-, isooctyl ester
2,4,5-T isooctyl ester
U 46T

25168-21-2
$C_{32}H_{56}O_8Sn$
687.50
2-Butenoic acid, 4,4'-((dibutylstannylene)bis(oxy))bis(4-oxo-, diisooctyl ester, (Z,Z)-(9CI)

25168-24-5
$C_{28}H_{56}O_4S_2Sn$
639.65
Stannane, bis(isooctyloxycarbonylmethylthio)dibutyl
Bis(2-ethylhexyloxycarbonylmethylthio)dibutylstannane
Dibutylzinn-S,S'-bis(isooctylthioglycolat) (German)
Tin, dibutyl-, bis(isooctylthioglycollate)

25168-26-7
$C_{16}H_{22}Cl_2O_3$
333.28
CC(C)CCCCCOC(=O)Cc1ccc(Cl)cc1Cl
Acetic acid, (2,4-dichloro-

phenoxy)-, isooctyl ester
2,4-Dichlorophenoxyacetic acid, isooctyl ester
2,4-D isooctyl ester
Isooctyl alcohol, (2,4-dichlorophenoxy)acetate
Isooctyl 2,4-dichlorophenoxyacetate
Isooktylester kyseliny 2,4-dichlorfenoxyoctove (Czech)
Reed LV 2,4-D
Reed LV 400 2,4-D
Reed LV 600 2,4-D
Weedtrine-II

25168-73-4
$C_{30}H_{56}O_{12}$
608.77
Sucrose, monostearate (8CI)
β-D-Fructofuranosyl-α-D-glucopyranoside, monooctadecanoate
α-D-Glucopyranoside, β-D-fructofuranosyl, monooctadecanoate (9CI)
NSC-192745
Ryoto Sugar Ester S-1170
Ryoto Sugar Ester S-1570
Ryoto Sugar Ester S-1670
Saccharose monostearate
Saccharosemonostearate
Saccharose stearate
Sucrose monostearic acid ester
Sucrose stearate
Sucrose stearic acid ester

25172-06-9
$C_8H_{12}O$
124.18
3,7-Octadien-2-one, (E)-(8CI,9CI)

25176-37-8
$C_{12}H_{21}N_3$
207.32
1,3,5-Triazine, 2,4,6-tris-(1-methylethyl)- (9CI)
s-Triazine, 2,4,6-triisopropyl- (6CI,8CI)
2,4,6-Triisopropyl-1,3,5-triazine

25177-16-6
$C_{25}H_{20}N_2O_5$
428.43
O=C(N(CC1)Cc(ccc(Oc(c(c(c(c2O)C(=O)c(c3ccc4)c4)C3=O)N)c2)c5)c5)C1
Anthraquinone, 1-amino-4-hydroxy-2-((α-(2-oxo-1-pyrrolidinyl)-p-tolyl)oxy)- (8CI)
9,10-Anthracenedione, 1-amino-4-hydroxy-2-((α-(2-oxo-1-pyrrolidinyl)-p-tolyl)oxy)-

25182-84-7
$C_3H_7NO_3$
105.11
OCCCN(=O)=O
1-Propanol, 3-nitro
3-Nitropropanol
NPOH

25186-43-0
$C_9H_{12}N_2O_2$
180.19
O=N(=O)c(ccc(NC(C)C)c1)c1
Benzenamine, N-(1-methylethyl)-4-nitro- (9CI)
N-(1-Methylethyl)-4-nitrobenzenamine

25186-47-4
$C_7H_6Cl_2$
161.03
Benzene, 1,3-dichloro-5-methyl-(9CI)
Toluene, 3,5-dichloro- (8CI)

25190-01-6
$(C_2H_4O)xC_{14}H_{31}NO$
Polymer NA
Poly(oxy-1,2-ethanediyl), α-(2-(dodecylamino)ethyl)-ω-hydroxy- (9CI)
Glycols, polyethylene, mono-(2-(dodecylamino)ethyl) ether (8CI)

25190-06-1
$(C_4H_8O)xH_2O$
Polymer
Polytetramethylene glycol
Poly(oxy-1,4-butanediyl), α-hydro-ω-hydroxy- (9CI)

25191-48-4
$(C_8H_8.C_5H_8O)x$
Polymer NA
3-Buten-2-one, 3-methyl-, Polymer with ethenylbenzene (9CI)
Benzene, ethenyl-, Polymer-with 3-methyl-3-buten-2-one (9CI)
3-Buten-2-one, 3-methyl-, Polymer with styrene-(8CI)
Ecolyte PS 102
Ecolyte PS 108
Isopropenyl methyl ketone-styrene copolymer
3-Methyl-3-buten-2-one-styrene copolymer
3-Methyl-3-buten-2-one-styrene polymer
Methyl isopropenyl ketone -styrene copolymer
Methyl isopropenyl ketone-styrene polymer
Methyl 1-methylvinyl ketone-styrene copolymer
Styrene-methyl isopropenyl ketone polymer
Styrene, Polymer with 3-methyl-3-buten-2-one-(8CI)

25213-39-2
$(C_8H_{14}O_2.C_8H_8)x$
Polymer
2-Propenoic acid, 2-methyl-, butyl ester, Polymer with ethenylbenzene (9CI)
Benzene, ethenyl-, Polymer with butyl 2-methyl-2-propenoate
Ethenylbenzene, butyl methacrylate polymer
Ethenylbenzene, Polymer with butyl methacrylate
Styrene, butyl methacrylate polymer

25214-70-4
$(C_6H_7N.CH_2O)x$
Polymer
Formaldehyde, Polymer with benzenamine
AF 10
Aniline-formaldehyde Condensate
Aniline-formaldehyde Polymer
Aniline, Polymer with formaldehyde (8CI)
Formaldehyde-aniline Copolymer
Jeffamine AP22
Jeffamine AP27
MDA 150
MDA 220
Polyamine T

25215-10-5
$CH_5N_3.H$
60.06
Guanidine, Conjugate monoacid (9CI)

25238-43-1
$C_3H_9N.HNO_3$
122.11
Methanamine, N,N-dimethyl-, nitrate (9CI)
Trimethylamine, nitrate (8CI)

25238-98-6
$C_8H_{19}O_4P.K$
249.31
Phosphoric acid, dibutyl ester, potassium salt (9CI)

25238-99-7
$C_4H_{11}O_4P.K$
193.20
Phosphoric acid, monobutyl ester, monopotassium salt (8CI)
Butyl potassium phosphate (1:1:1)

25248-42-4
Unknown
Unknown
Polycaprolactone 700

25249-39-2
$C_{13}H_{10}Cl_2$
237.13
Benzene, 1,1'-methylenebis(chloro- (9CI)
Methane, bis(chlorophenyl)- (6CI,7CI,8CI)

25254-67-5
$C_7H_{13}N_5O$
183.25
s-Triazine, 2-butoxy-4,6-diamino

25264-93-1
C_6H_{12}
84.18
CCCC=CC
Hexene
Hexylene

25265-71-8
$C_6H_{14}O_3$
134.20
Propanol, oxybis
Dipropylene glycol

25265-75-2
$C_4H_{10}O_2$
90.12
Butylene glycol
Butanediol (8CI,9CI)

25265-76-3
$C_6H_8N_2$
108.13
Benzenediamine (9CI)
Bant
Benzenamine, ar-amino-
Benzene, diamino-
Phenylenediamine (8CI)

25265-77-4
$C_{12}H_{24}O_3$
216.36
CC(C)C(O)C(C)(C)COC(=O)C(C)C
Propionic acid, 2-methyl-, monoester with 2,2,4-trimethyl-1,3-pentanediol
1,3-Pentanediol, 2,2,4-trimethyl-, monoisobutyrate
Propanoic acid, methyl-, monoester with 2,2,4-trimethyl-1,3-pentanediol
Texanol
2,2,4-Trimethyl-1,3-pentanediol monoisobutyrate

25267-27-0
C_4H_9I
184.02
Butane, iodo- (8CI,9CI)

25267-55-4
$C_{12}H_6Cl_6O_2.Cu$
458.42
Phenol, trichloro-, copper(2+) salt
Copper trichlorophenolate
Copper 2,4,5-trichlorophenolate
CTCP
Trikhlorfenolyat medi (Russian)

25268-77-3
$C_{14}H_{10}F_{17}NO_4S$
611.27
O=C(OCCN(S(=O)(=O)C(F)(F)C(F)(F)C(F)(F)C(F)(F)C(F)(F)C(F)(F)C(F)(F)C(F)(F)F)C)C=C
2-Propenoic acid, 2-(((heptadecafluorooctyl)sulfonyl)methylamino)ethyl ester (9CI)

25277-05-8
$C_8H_9ClN_2O_2$
200.61
COc1ccc(NC(N)=O)cc1Cl
Urea, (3-chloro-4-methoxyphenyl)- (8CI,9CI)

25279-09-8
$C_{11}H_{20}O_2$
184.28
O=COC(CCCC(C=C)C)(C)C
7-Octen-2-ol, 2,6-dimethyl-, formate (9CI)
2,6-Dimethyl-7-octen-2-ol formate
2,6-Dimethyl-7-octen-2-yl formate

25306-75-6
$C_5H_{10}OS_2.Na$
173.26
Carbonodithioic acid, O-(2-methylpropyl) ester, sodium salt (9CI)

25307-17-9
$C_{22}H_{45}NO_2$
355.60
OCCN(CCCCCCCCC=CCCCCCCCC)CCO
Ethanol, 2,2'-(9-octadecenylimino)bis- (9CI)
2,2'-(9-Octadecenylimino)bisethanol

25311-71-1
$C_{15}H_{24}NO_4PS$
345.43
CCOP(=S)(NC(C)C)Oc1ccccc1C(=O)OC(C)C
Salicylic acid, isopropyl ester, O-ester with O-ethyl isopropylphosphoramidothioate
2-(O-Aethyl-N-isopropylamidothiophosphoryloxy)-benzosaeure-isopropylester (German)
Amaze
Bay-92114
Bay-SRA-12869
Benzoic acid, 2-((ethoxy((1-methylethyl)amino)phosphinothioyl)oxy)-, 1-methyl ester (9CI)
2-((Ethoxy((1-methylethyl)amino)phosphinothioyl)oxy)benzoic acid 1-methylethyl ester
O-Ethyl-O-(2-isopropoxy-carbonyl)-phenyl isopropylphosphoramidothioate
Isofenphos
Isophenphos
Isopropyl salicylate O-ester with O-ethylisopropylphosphoramidothioate
1-Methylethyl-2-((ethoxy((1-methylethyl)amino)phosphinothioyl)oxy) benzoate
Oftanol
Phosphoramidothioic acid, isopropyl-, O-ethyl O-(2-isopropoxycarbonylphenyl) ester
40 SD
SRA 12869

25316-40-9
$C_{27}H_{29}NO_{11}.ClH$
580.03
[Cl-].COc4cccc5C(=O)c3c(O)c2CC(O)(CC(OC1CC(N)C(O)C(C)O1)c2c(O)c3C(=O)c45)C(=O)CO
5,12-Naphthacenedione, 10-((3-amino-2,3,6-trideoxy-α-l-lyxo-hexopyranosyl)oxy)-7,8,9,10-tetrahydro-6,8,11-trihydroxy-8-(hydroxyacetyl)-1-methoxy-, hydrochloride, (8s-cis)
ADM hydrochloride
ADR
Adriacin
Adriamycin
Adriamycin, hydrochloride
Adriblastina
Adriblastin
Dox
Dox hydrochloride
Doxorubicin
Doxorubicin hydrochloride
FI 106
FI 6804
Hydroxydaunorubicin hydrochloride

25319-90-8
$C_{11}H_{13}ClO_2S$
244.75

CCSC(=O)COc1ccc(Cl)cc1C
Acetic acid, ((4-chloro-o-tolyl)-oxy)thio-, S-ethyl ester
Ethanethioic acid, (4-chloro-2-methylphenoxy)-, S-ethyl ester (9CI)
Ethylester kyseliny 4-chlor-2-tolyloxythiooctove (Czech)
Herbit
Hok 7501
MCPA-Thioethyl
2-Methyl-4-chlorophenoxythiol acetic acid S-ethyl ester
Phenothiol
Tripion CB
Zero One

25321-09-9
$C_{12}H_{18}$
162.30
Benzene, diisopropyl
Diisopropylbenzene

25321-14-6
$C_7H_6N_2O_4$
182.15
Cc1cccc(N(=O)=O)c1N(=O)=O
Toluene, dinitro
Benzene, methyldinitro-
Dinitrophenylmethane
Dinitrotoluene
Dinitrotoluene, Liquid [UN 2038]
Dinitrotoluene, Molten [UN 1600]
Dinitrotoluene, Solid (DOT)
Methyldinitrobenzene
Toluene, ar,ar-dinitro-
UN 1600 [Dinitrotoluenes, molten]
UN 2038 [Dinitrotoluenes, Liquid]

25321-22-6
$C_6H_4Cl_2$
147.00
Benzene, dichloro- (8CI,9CI)
Dichlorobenzene
Dilatin DBI

25321-41-9
$C_8H_{10}O_3S$
186.24
Xylenesulfonic-acid
XSA

25321-43-1
$C_{14}H_{22}O_3S$
270.39
Benzenesulfonic acid, octyl- (9CI)
Octylbenzenesulfonic acid

25322-01-4
$C_3H_7NO_2$
89.09
Nitropropane
Propane, nitro- (8CI,9CI)
UN 2608 [Nitropropanes]

25322-17-2
$C_{28}H_{44}O_3S$
460.72
Naphthalenesulfonic acid, dinonyl- (9CI)
Dinonylnaphthalenesulfonic acid

25322-20-7
$C_2H_2Cl_4$
167.84
Ethane, tetrachloro
Tetrachloroethane [UN 1702]
UN 1702 [Tetrachloroethane]

25322-68-3
$(C_6H_{11}NO)x$
Polymer
Poly(oxy-1,2-ethanediyl), α-hydro-ω-hydroxy- (9CI)
Alkapol PEG-200
Alkapol PEG-300
Alkapol PEG-600
Alkapol PEG-6000
Alkapol PEG-8000
Carbowax
Carbowax 1000
Carbowax 1500
Carbowax 1540
Carbowax 4000
Carbowax 6000
Carsonon PEG-4000
Jeffox
Jorchem 400 ML
Lutrol 9
Lutrol
Macrogol 1000
Macrogol 4000
P.E.G. 400
P.E.G. 1000
P.E.G. 1500
P.E.G. 4000
P.E.G. 6000
Pluracol E-200
Pluracol E-300
Pluracol E-400
Pluracol E-600
Pluracol E-1500
Pluracol E-4000
Pluracol E-6000
Pluracol P-410
Pluracol P-710
Pluracol P-1010
Pluracol P-2010
Pluracol P-3010
Pluracol P-4010
Polyaethylenglykole #200 (German)
Polyaethylenglykole #282 (German)
Polyaethylenglykole #238 (German)
Polyaethylenglykole #300 (German)
Polyaethylenglykole #400 (German)
Polyaethylenglykole #600 (German)
Polyaethylenglykole #810 (German)
Polyaethylenglykole #1000 (German)
Polyaethylenglykole #1250 (German)
Polyaethylenglykole #1500 (German)
Polyaethylenglykole #1540 (German)
Polyaethylenglykole #4000 (German)
Polyaethylenglykole #6000 (German)
Polyaethylenglykole #10000 (German)
Polyethylene glycol E 600
Polyethylene glycol 20M
Polyethylene glycol 425
Polyethylene glycol 1200
Polyethylene glycol 1500
Polyethylene glycol 2000
Polyethylene glycol 4000
Polyethylene glycol 6000
Polyethylene glycol 2000000
Polyethylene glycol 4000000
Polyethylene glycol 5000000
Polyethylene glycol #200
Polyethylene glycol #238
Polyethylene glycol #282
Polyethylene glycol #300
Polyethylene glycol #350
Polyethylene glycol #400
Polyethylene glycol #600
Polyethylene glycol #810
Polyethylene glycol #1000
Polyethylene glycol #1250
Polyethylene glycol #1540
Polyethylene glycol #10000
Poly(ethylene oxide)
Polyglycol E-300
Polyglycol E-1000
Polyglycol E-4000
Polyglycol E-4000 USP
Polyglycol P-425
Polyglycol P-1200
Polyglycol 4000
Polyglycol 1000
Poly-G Series
Poly G 400
Polyox
Poly(oxy-1,2-ethanediyl), α-hydro-ω-hydroxy-
Polyoxyethylene (75)
Polyoxyethylene 1500
Polyoxyethylene 2000000
Polyoxyethylene 5000000
WSR-301

25322-69-4
$(C_3H_6O)xH_2O$
Polymer
Poly(oxy(methyl-1,2-ethanediyl)), α-hydro-ω-hydroxy- (9CI)
Alkapol PPG-1200
Alkapol PPG-2000
Alkapol PPG-4000

Jeffox
Laprol 702
P.P.G. 150
P.P.G. 400
P.P.G. 425
P.P.G. 750
P.P.G. 1000
P.P.G. 1025
P.P.G. 1200
P.P.G. 1800
P.P.G. 2025
P.P.G. 3025
P.P.G. 4025
Polyglycol P-2000
Polypropylene-glycol
Polypropylene glycol 150
Polypropylene glycol 1025
Polypropylene glycol 2000
Polypropylene glycol 2025
Polypropylene glycol 3025
Polypropylene glycol 4025
Polypropylene glycol #400
Polypropylene glycol #425
Polypropylene glycol #750
Polypropylene glycol #1000
Polypropylene glycol #1200
Polypropylene glycol #1800
Polypropylenglykol (Czech)

25323-30-2
$C_2H_2Cl_2$
96.94
Ethylene, dichloro
Dichloroethylene [UN 1150]
UN 1150 [Dichloroethylene]

25323-41-5
C_8H_9Cl
140.61
Benzene, chlorodimethyl- (9CI)
Xylene, ar-chloro- (8CI)

25323-68-6
$C_{12}H_7Cl_3$
257.54
Biphenyl, trichloro
Apirolio 1431 C
1,1'-Biphenyl, trichloro- (9CI)
Pyranol 1499
Trichlorobiphenyl
Trichlorodiphenyl

25323-89-1
$C_2H_3Cl_3$
133.40
Trichloroethane
AI3-15422
Ethane, trichloro-

25324-56-5
PSn
149.68
Stannic phosphide (8CI,9CI)
Tin phosphide
Tin (IV) phosphide
UN 1433 [Stannic phosphide]

25327-89-3
$C_{21}H_{20}Br_4O_2$
624.00
O(c(c(cc(c1)C(c(cc(c(OCC=C)c2Br)Br)c2)(C)C)Br)c1Br)CC=C
Benzene, 1,1'-(1-methylethylidene)bis(3,5-dibromo-4-(2-propenyloxy)- (9CI)
2,2-Bis(4-allyloxy-3,5-dibromophenyl)propane
Tetrabromobisphenol A, bis-(allyl ether)

25329-35-5
C_5HCl_5
238.31
1,3-Cyclopentadiene, 1,2,3,4,5-pentachloro
Pentachlorocyclopentadiene
1,2,3,4,5-Pentachlorocyclopentadiene
Pentachloro-2,4-cyclopentadien-1-yl

25332-20-1
$C_5H_7N_3O_2.ClH$
177.57
1H-Imidazole, 1,2-dimethyl-5-nitro-, monohydrochloride (9CI)

25333-77-1
$C_{27}H_{35}NO_5$
453.58
Acetorphine
Acetorfina
Acetorphinum
Acetylpropylorvinol
DEA No. 9319
6,14-Ethenomorphinan-7-methanol, 3-(acetyloxy)-4,5-epoxy-6-methoxy-α,17-dimethyl-α-propyl-, (5α,7α(R))- (9CI)
6,14-endo-Ethenotetrahydrooripavine, 7α-(R)-1-hydroxy-1-methylbutyl)-, 3-acetate (8CI)

25338-55-0
$C_9H_{13}NO$
151.20
Phenol, ((dimethylamino)-methyl)- (9CI)
((Dimethylamino)methyl)phenol

25339-09-7
$C_{34}H_{68}O_2$
509.02
Stearic acid, isohexadecyl ester
Isocetyl stearate
Kessco ICS
Octadecanoic acid, isohexadecyl ester (9CI)
Standamul 7061

25339-17-7
$C_{10}H_{22}O$
158.32
CC(C)CCCCCCCO
Isodecyl-alcohol
Isodecanol

25339-53-1
$C_{10}H_{20}$
140.27
Decene

25339-56-4
C_7H_{14}
98.19
Heptylene
Heptene (VAN)(9CI)

25339-57-5
C_4H_6
54.09
Butadiene
Butadiene, Inhibited [UN 1010]
Pliolite
UN 1010 [Butadienes, Inhibited]

25339-99-5
$C_{24}H_{46}O_{13}$
542.63
Sucrose laurate
β-D-Fructofuranosyl-α-D-glucopyranoside, monododecanoate
α-D-Glucopyranoside, β-D-fructofuranosyl, monododecanoate
Sucrose monolaurate

25340-17-4
$C_{10}H_{14}$
134.24
Benzene, diethyl
Diethyl benzene
Diethylbenzene [UN 2049]
UN 2049 [Diethylbenzene]

25340-18-5
$C_{12}H_{18}$
162.30
Benzene, triethyl- (Mixed isomers)
Triethylbenzene

25354-39-6
$C_6H_8Cl_2N_4$
207.04
1,3,5-Triazin-2-amine, 4,6-dichloro-N-propyl- (9CI)
s-Triazine, 2,4-dichloro-6-(propylamino)- (8CI)

25360-10-5
$C_9H_{20}S$
160.32
tert-Nonyl mercaptan
tert-Nonanethiol (9CI)

25371-54-4
$C_{20}H_{43}O_3P$
362.53
O=P(OC)(OC)CCCCCCCCCC CCCCCCCC
Phosphonic acid, octadecyl-, dimethyl ester (9CI)
Dimethyl octadecylphosphonate
Dimethyloctadecylphosphonate
Octadecylphosphonic acid, dimethyl ester

25371-75-9
$C_7H_{17}O_2PS$
196.25
Phosphonothioic acid, methyl-, O,O-dipropyl ester (8CI,9CI)
NSC-202854

25376-38-9
$C_6H_3Br_3O$
330.80
Phenol, tribromo- (9CI)
Tribromophenol

25376-45-8
$C_7H_{10}N_2$
122.19
Toluenediamine (DOT)
Benzenediamine, ar-methyl-
Diaminotoluene
Methylphenylenediamine
NA 1709 (DOT)
RCRA waste number U221
Tolylenediamine

25377-72-4
C_5H_{10}
70.15
Pentene
Amylene
n-Amylene [UN 1108]
Amylene, normal (DOT)
Pentylene
UN 1108 [n-Amylene]

25377-73-5
$C_{16}H_{26}O_3$
266.42
Succinic anhydride, dodecenyl
DDS
DDS A
Dodecenylsuccinic anhydride
2,5-Furandione, 3-(dodecenyl)-dihydro-

25377-82-6
$C_{13}H_{26}$
182.35
Tridecene

25377-83-7
C_8H_{16}
112.22
Octene (9CI)
Octylene

25378-22-7
$C_{12}H_{24}$
168.32
Dodecene (9CI)
1-Dodecene
Dodecylene
α-Dodecylene
Propene, Polymers, Tetramer
Propene, Tetramer
1-Propene, Tetramer
Propylene tetramer
Tetrapropylene

25378-26-1
$C_{22}H_{42}O_2$
338.57
Docosenoic acid (8CI,9CI) (VAN)

25384-17-2
$C_{18}H_{25}NO_2$
287.40
4-Piperidinol, 1-methyl-4-phenyl-3-(2-propenyl)-, propanoate (ester) (9CI)
Allylprodine
Alperidine
Ro 2-7113
NIH 7440
4-Piperidinol, 3-allyl-1-methyl-4-phenyl-, propionate (ester) (8CI)
Propionic acid, 3-allyl-1-methyl-4-phenyl-4-piperidyl ester (6CI)

25394-13-2
$C_{14}H_{14}N_2O_6S_2.xNa$
Unknown
Benzenesulfonic acid, 2,2'-(1,2-ethenediyl)bis(5-amino-, sodium salt (9CI)
4,4'-Diaminostilbene-2,2'-disulfonic acid, sodium salt

25395-31-7
$C_7H_{12}O_5$
176.19
Acetin, di
Diacetin
Diacetylglycerol
Glycerin diacetate
Glycerine diacetate
Glycerol diacetate
Glyceryl diacetate
1,2,3-Propanetriol, diacetate (9CI)

25399-81-9
$Cl_2OZr.6H_2O$
286.22
Zirconium, dichlorooxo-, hexahydrate (8CI,9CI)

25401-86-9
$C_{22}H_{38}O$
318.54
Phenol, 2-hexadecyl- (9CI)
o-Hexadecylphenol
2-Hexadecylphenol

25415-71-8
$C_{11}H_{22}O_2$
186.29
Pentanoic acid, 4-methyl-, pentyl ester (9CI)
Valeric acid, 4-methyl-, pentyl ester (8CI)

25417-20-3
$C_{18}H_{24}O_3S.Na$
343.47
Naphthalenesulfonic acid, dibutyl-, sodium salt
Dibutyl-naphthalene sulfate, sodium salt
Naftalin-butil-solfonato (Italian)
Nekal
Sodium dibutylnaphthalene sulfate
Sodium dibutylnapthalenesulfonate
Sodium dibutylnaphthylsulfonate

25419-33-4
$C_{12}H_{16}$
160.26
Naphthalene, 1,2,3,4-tetrahydro-1,8-dimethyl- (8CI, 9CI)
1,8-Dimethyl-1,2,3,4-tetrahydronaphthalene
1,8-Dimethyltetralin

25429-23-6
$C_2H_2Br_2$
185.85
Ethene, dibromo- (9CI)
Dibromoethene

25429-29-2
$C_{12}H_5Cl_5$
326.42
Biphenyl, pentachloro
Apirolio 1476 C
1,1'-Biphenyl, pentachloro- (9CI)
Diphenyl pentachloride
Kanekrol 500
Pentachlorobiphenyl
Pentachlorodiphenyl
Pyralene 1476

Pyroclor 5

25429-37-2
$C_8H_{10}O$
122.17
Phenol, ethyl- (8CI,9CI)

25429-38-3
$C_9H_8O_3$
164.16
2-Propenoic acid, 3-(hydroxyphenyl)- (9CI)
Cinnamic acid, ar-hydroxy- (8CI)

25430-52-8
$C_{18}H_{34}O_2$
282.47
OC(C#CC(O)C(CCCC)CC)C(CCCC)CC
7-Tetradecyne-6,9-diol, 5,10-diethyl- (9CI)
5,10-Diethyl-7-tetradecyn-6,9-diol
5,10-Diethyl-7-tetradecyne-6,9-diol

25446-78-0
$C_{19}H_{40}O_7S.Na$
435.58
Sodium trideceth sulfate
Ethanol, 2-(2-(2-(tridecyloxy)ethoxy)ethoxy)-, hydrogen sulfate, sodium salt (9CI)
Sodium polyoxyethylene tridecyl sulfate
Sodium tridecyl ether sulfate
Sodium tridecyl tri(oxyethyl) sulfate

25446-80-4
$C_{20}H_{42}O_7S.Na$
449.61
Sodium myreth sulfate
Ethanol, 2-(2-(2-(tetradecyloxy)ethoxy)ethoxy)-, hydrogen sulfate, sodium salt (9CI)
Sodium myristyl ether sulfate
Unipol ES-40

25447-69-2
$C_8H_{14}O$
126.20
Octenal (8CI,9CI)

25448-24-2
$C_{13}H_{26}O_2$
214.35
Isotridecanoic acid (9CI)

25448-25-3
$C_{30}H_{63}O_3P$
502.80
Phosphorous acid, triisodecyl ester (9CI)

25457-47-0
$C_{10}H_{19}NO_2$
185.27
Acetamide, N-acetyl-N-hexyl- (9CI)
Diacetamide, N-hexyl- (7CI, 8CI)

25465-18-3
$C_6H_{10}O_2$
114.15
1,4-Dioxin, 2,3-dihydro-5,6-dimethyl- (9CI)
p-Dioxin, 2,3-dihydro-5,6-dimethyl- (8CI)

25482-47-7
$C_{30}H_{54}O$
430.76
Phenol, didodecyl- (9CI)
Didodecylphenol

25485-34-1
$C_{34}H_{48}O_2$
488.75
O=C(OC(C=C(C(C(C(C(C(C(C1)C(CCCC(C)C)C)C)(C2)C)C1)C=3)C2)(C4)C)C3)C4)c(cccc5)c5
Cholesta-4,6-dien-3-ol, benzoate, (3β)- (9CI)
4,6-Cholestadien-3β-ol, benzoate

25495-90-3
$C_6H_{13}Cl$
120.62
Hexane, chloro- (8CI,9CI)

25495-92-5
$C_6H_{13}I$
212.07
Hexane, iodo- (8CI,9CI)

25496-01-9
$C_{19}H_{32}O_3S$
340.53
Tridecylbenzenesulfonic acid
Benzenesulfonic acid, tridecyl- (9CI)
Tridecylbenzenesulfonic acid

25496-08-6
C_7H_7F
110.13
Benzene, fluoromethyl- (9CI)
Toluene, ar-fluoro- (8CI)

25496-72-4
$C_{21}H_{40}O_4$
356.61
Olein, mono
Adchem GMO
Ajax GMO
Aldo 40
Aldo MO-FG
Dur-EM 204
Emcol O
Emery Oleic Acid Ester 2221
Emrite 6009
Glycerine monooleate
Glycerin monooleate
Glycerol monooleate
Glycerol oleate
Glyceryl monooleate
Glyceryl oleate
GMO 8903
Harowax L 9
Loxiol G 10
Monoglyceryl oleate
Monoolein
Monooleoylglycerol
9-Octadecenoic acid (Z)-, monoester with 1,2,3-propanetriol (9CI)
Oleic acid glycerol monoester
Oleic acid monoglyceride
Oleoylglycerol
Oleylmonoglyceride
Olicine
Rikemal O 71D
Rikemal OL 100
S 1096
S 1097
Sinnoester OGC
S 1096R
Sunsoft O 30B
Supeol

25497-28-3
$C_2H_4F_2$
66.05
Difluoroethane
Difluoretano (Spanish)
Difluorethane (French)
Ethane, difluoro-
R-152a
UN 1030 [Difluoroethane]

25497-29-4
$C_2H_3ClF_2$
100.50
Chlorodifluoroethane
Chlorodifluorethane (French)
Chlorodifluoroethanes
Clorodifluoretano (Spanish)
Difluoromonochloroethane
Ethane, chlorodifluoro-
R 142
UN 2517 [Chlorodifluoroethanes or Difluorochloroethanes]

25498-49-1
$C_{10}H_{22}O_4$
206.32
COCCCOCCCOCCCO
Propanol, 3-(3-(3-methoxypropoxy)propoxy)
Dowanol TPM
Propanol, (2-(2-methoxymethyl-

ethoxy)methylethoxy)- (9CI)
Tripropylene glycol methyl ether

25502-52-7
$C_{20}H_{20}N_2O_4$
352.38
O=C(N(C(=O)C1C(C=CC23)C3)CCN(C(=O)C(C4C(C=C5)C6)C56)C4=O)C12
4,7-Methano-1H-isoindole-1,3(2H)-dione, 2,2'-(1,2-ethanediyl)bis(3a,4,7,7a-tetrahydro- (9CI)
N,N'-Ethylenebis(1,2,3,6-tetrahydro-3,6-endomethylene-phthalimide)

25510-81-0
$C_{18}H_{16}N_6O_2S$
380.39
O=N(=O)c(ccc(N=C(N=Nc(ccc(N(CCC#N)CC)c1)c1)S2)c23)c3
Propanenitrile, 3-(ethyl(4-((6-nitro-2-benzothiazolyl)azo)-phenyl)amino)- (9CI)
Benzothiazole, 2-((p-(N-(2-cyanoethyl)-N-ethylamino)-phenylazo)-6-nitro-

25512-11-2
$C_{32}H_{13}CuN_8O_9S_3.3H_4N$
867.31
Cuprate(3-), (29H,31H-phthalocyaninetrisulfonato-(5-)-N29,N30,N31,N32)-, tri-ammonium (9CI)
Copper, (29H,31H-phthalo-cyanine-ar,ar',ar''-trisulf-onato(2-)-N29,N30,N31,N32)-, triammonium salt

25512-42-9
$C_{12}H_8Cl_2$
223.10
Biphenyl, dichloro
1,1'-Biphenyl, dichloro- (9CI)
Dichlorobiphenyl
Dichlorodiphenyl

25523-14-2
$C_3H_2Cl_2$
108.96
1-Propyne, 3,3-dichloro- (9CI)
Propyne, 3,3-dichloro- (8CI)

25523-97-1
$C_{16}H_{19}ClN_2$
274.79
Dexchlorpheniramine
Chlo-Amine
D-2-(p-Chloro-α-(2-dimethyl-aminoethyl)benzyl)pyridine
D-Chlorpheniramine
(+)-Chlorpheniramine
γ-(4-Chlorophenyl)-N,N-di-methyl-2-pyridinepropanamine
Dexchlorpheniraminum (Latin)
Dexclorfeniramina (Spanish)
Fortamine
Isomerine
Phendextro
Polaramine
Pyridine, 2-(p-chloro-α-(2-(di-methylamino)ethyl)benzyl)-
2-Pyridinepropanamine, γ-(4-chlorophenyl)-N,N-dimethyl-, (S)-

25525-76-2
$C_{20}H_{35}ClO_2$
342.95
2-Chloroethyl linoleate
9,12-Octadecadienoic acid (Z,Z)-, 2-chloroethyl ester

25537-26-2
$C_{16}H_{21}Cl_3O_4$
383.70
Propionic acid, 2-(2,4,5-tri-chlorophenoxy)-, 3-but-oxypropyl ester (8CI)

25545-89-5
Unknown
Unknown
Ammonium 1-naphthalene-acetate

25549-16-0
$C_{24}H_{51}N$
353.67
Isooctanamine, N,N-diisooctyl- (9CI)
AI3-25359-X
N,N-Diisooctylisooctanamine
Isooctanamine, N,N-diisooctyl- (Mixed isomers)
Triisooctylamine

25550-13-4
$C_{11}H_{16}$
148.25
Benzene, diethylmethyl- (9CI)
Toluene, diethyl-
Toluene, ar,ar-diethyl- (8CI)

25550-14-5
C_9H_{12}
120.19
Ethyl toluene
Benzene, ethylmethyl- (9CI)
Ethylmethylbenzene
Ethyltoluene

25550-51-0
$C_9H_{12}O_3$
168.19
1,3-Isobenzofurandione, hexa-hydromethyl- (9CI)
Hexahydromethyl-1,3-isobenzo-furandione
Methylhexahydrophthalic anhydride

25550-52-1
$C_8H_{10}S$
138.23
Benzenethiol, dimethyl-
AI3-25192-X
Dimethylbenzenethiol

25550-58-7
$C_6H_4N_2O_5$
184.12
Phenol, dinitro
Dinitrophenol
Dinitrophenol, Dry or containing, by weight, less than 15% water [UN 0076]
Dinitrophenol, Solution in water or flammable liquid [UN 1599]
Dinitrophenol, Wetted with, by weight, at least 15% water [UN 1320]
Phenol, dinitro-, Wetted with at least 15% water
UN 0076 [Dinitrophenol, Dry or wetted with less than 15 per cent water, by mass]
UN 1320 [Dinitrophenol, Wetted with not less than 15 per cent water, by mass]
UN 1599 [Dinitrophenol solutions]

25550-98-5
$C_{26}H_{47}O_3P$
438.63
Phosphorous acid, diisodecyl phenyl ester (9CI)

25551-13-7
C_9H_{12}
120.21
Benzene, trimethyl- (Mixed isomers)
Trimethyl benzene (ACGIH, OSHA)

25551-28-4
Unknown
Unknown
Naphthalene diisocyanate

25551-49-9
$C_{12}H_{22}$
166.31
Naphthalene, ethyldecahydro- (8CI,9CI)

25560-00-3
$C_9H_{20}N_2$
156.26
1-Piperidinepropanamine,

2-methyl- (9CI)
2-Pipecoline, 1-(3-aminopropyl)- (8CI)

25567-10-6
$C_8H_8O_2$
136.15
Benzoic acid, methyl- (9CI)
Methylbenzoic acid

25567-11-7
$C_9H_{10}O_2$
150.18
Benzoic acid, methyl-, methyl ester (9CI)
Methyl methylbenzoate

25567-40-2
$C_{11}H_{16}O$
164.25
Phenol, (1,1-dimethylethyl)-4-methyl- (9CI)
(1,1-Dimethylethyl)-4-methyl-phenol

25567-55-9
$C_6H_2Cl_4O.Na$
254.89
Phenol, tetrachloro-, sodium salt (8CI,9CI)
Caswell No. 796
EPA Pesticide Chemical Code 063005
Sodium tetrachlorophenate
Sodium, (tetrachlorophenoxy)-

25567-67-3
$C_6H_3ClN_2O_4$
202.56
Benzene, chlorodinitro- (Mixed isomers)
Chlorodinitrobenzene [UN 1577]
Dinitrochlorobenzene (DOT)
UN 1577 [Chlorodinitro-benzenes]

25568-84-7
Unknown
Unknown
Cyclopentadiene polymers

25569-53-3
$(C_4H_6O_4.C_2H_6O_2)x$
Polymer
Butanedioic acid, Polymer with 1,2-ethanediol (9CI)

25569-80-6
$C_{12}H_8Cl_2$
223.10
Clc1cccc(c1)c2ccccc2Cl
1,1'-Biphenyl, 2,3'-dichloro- (9CI)
Biphenyl, 2,3'-dichloro- (8CI)
2,3'-Dichlorobiphenyl

25584-83-2
$C_6H_{10}O_3$
130.14
Hydroxypropyl acrylate
Acrylic acid, hydroxypropyl ester
Acrylic acid, monoester with 1,2-propanediol
1,2(or 3)-Propanediol, 1-acrylate
2-Propenoic acid, monoester with 1,2-propanediol (9CI)
Propylene glycol acrylate
Propylene glycol monoacrylate

25586-38-3
C_9H_8O
132.16
Benzofuran, methyl- (8CI,9CI)

25586-39-4
$C_{10}H_{10}O$
146.19
Benzofuran, dimethyl- (8CI, 9CI)

25586-42-9
$C_{21}H_{21}O_3P$
352.37
Phosphorous acid, tris(methyl-phenyl) ester (9CI)
Tritolyl phosphite

25586-43-0
$C_{10}H_7Cl$
162.62
Naphthalene, chloro- (8CI,9CI)
Chloronaphthalene
Halowax 1031

25608-33-7
$(C_8H_{14}O_2.C_5H_8O_2)x$
Polymer
2-Propenoic acid, 2-methyl-, butyl ester, Polymer with methyl 2-methyl-2-propenoate (9CI)
Acryloid B 66
Acryloid K 125
Butyl methacrylate, methyl methacrylate polymer
Elvacite 2013
Elvacite 6016
Glasure 40X
Lucite 2013
Metakril 40BM
Metakril 80BM
Methacrylic acid, butyl ester, Polymer with methyl meth-acrylate (8CI)
Plexigum PM 381
Plexisol PM 709
Solakryl BMX
Sulfix 6
80BM

25618-55-7
$(C_3H_8O_3)x$
Polymer
Glycerol, Polymers
Polyglycerin
1,2,3-Propanetriol, Homo-polymer (9CI)

25619-56-1
$C_{28}H_{44}O_3S.1/2Ba$
529.39
Naphthalenesulfonic acid, di-nonyl-, barium salt (9CI)
Barium dinonylnaphthalene-sulfonate
Dinonylnaphthalene sulfonic acid barium salt
NSC-49580

25619-60-7
$C_{10}H_{14}$
134.22
Benzene, tetramethyl- (8CI, 9CI)

25619-63-0
$C_{24}H_{34}O$
338.53
Benzene, dodecylphenoxy- (9CI)
Dodecylphenoxybenzene
(Dodecylphenoxy)benzene

25620-58-0
$C_9H_{22}N_2$
158.33
1,6-Hexanediamine, trimethyl
Trimethylhexamethylenediamine
Trimethylhexamethylene di-amine [UN 2327]
Trimethyl-1,6-hexanediamine
UN 2327 [Trimethylhexa-methylenediamines]

25620-59-1
$C_{14}H_9NO_2$
223.22
9,10-Anthracenedione, amino- (9CI)
Amino-9,10-anthracenedione

25620-62-6
$C_2H_4Br_2$
187.86
Ethane, dibromo- (8CI,9CI)

25637-84-7
$C_{39}H_{72}O_5$
621.00
Diolein
Glyceryl dioleate

9-Octadecenoic acid, diester with 1,2,3-propanetriol
9-Octadecenoic acid (Z)-, diester with 1,2,3-propanetriol (9CI)

25637-99-4
$C_{12}H_{18}Br_6$
641.70
Cyclododecane, hexabromo- (9CI)
Hexabromocyclododecane

25638-14-6
$C_7H_4BrClO_2$
235.46
Benzoic acid, bromochloro- (8CI,9CI)

25638-17-9
$C_{14}H_{16}O_3S.Na$
287.34
Naphthalenesulfonic acid, butyl-, sodium salt (9CI)
Butylnaphthalenesulfonate sodium salt

25639-25-2
$(C_6H_{10}O_2)x$
Polymer
Ether, poly(allyl-glycidyl)
Poly(allyl glycidyl ether)

25639-42-3
$C_7H_{14}O$
114.21
Cyclohexanol, methyl
Hexahydrocresol
Hexahydromethylphenol
Methyl cyclohexanols, Flash point not more than 60.5 degrees C [UN 2617]
Methylcyclohexanol (ACGIH, DOT,OSHA)
Metylocykloheksanol (Polish)
UN 2617 [Methyl cyclohexanols, flash point not more than 60.5 degrees C]

25640-78-2
$C_{15}H_{16}$
196.31
Biphenyl, isopropyl
1,1'-Biphenyl, (1-methylethyl)- (9CI)
Isopropylbiphenyl
Isopropyldiphenyl
Monoisopropylbiphenyl
Wemcol

25641-99-0
$C_8H_6Cl_4$
243.95
c(c(ccc1)C(Cl)Cl)(c1)C(Cl)Cl
Benzene, 1,2-bis(dichloromethyl)- (9CI)
1,2-Bis(dichloromethyl)benzene
o-Xylene, α,α,α',α'-tetrachloro-

25646-71-3
$C_{12}H_{21}N_3O_2S.3/2H_2O_4S$
418.49
Methanesulfonamide, N-(2-(4-amino-N-ethyl-m-toluidino)ethyl)-, sulfate (2:3)
N-(2-(4-Amino-N-ethyl-m-toluidino)ethyl)methanesulfonamide sulfate (2:3)
CD 3
CD III
Kodak CD-3
Methanesulfonamide, N-(2-((4-amino-3-methylphenyl)-ethylamino)ethyl)-, sulfate (2:3) (9CI)

25646-77-9
$C_{11}H_{18}N_2O.H_2O_4S$
292.39
Ethanol, 2-((4-amino-3-methylphenyl)ethylamino)-, sulfate (1:1) (Salt)
2-((4-Amino-3-methylphenyl)-ethylamino)ethanol sulfate
CD 4

25653-16-1
$C_{24}H_{27}O_4P$
410.45
Tris(3,5-xylenyl)phosphate
NSC-66515
Phenol, 3,5-dimethyl-, phosphate (3:1) (9CI)
Phosphoric acid, tris(3,5-dimethylphenyl)ester
Tri(3,5-xylenyl)phosphate
3,5-Xylenol, phosphate (3:1)

25655-41-8
$(C_6H_9NO)x.xI$
Polymer
II.C=CN1CCCC1=O
Poly(1-(2-oxo-1-pyrrolidinyl)-ethylene)iodine complex
Betadine
Efo-Dine
Isodine
Povidone-iodine
PVP-iodine
2-Pyrrolidinone, 1-ethenyl-, Homopolymer, Compd. with iodine
Ultradine
1-Vinyl-2-pyrrolidinone Polymer, Compd. with iodine

25659-22-7
$C_6H_{10}O$
98.16
4-Hexen-2-one
Acetone methylallyl
Methylallyl acetone

25659-31-8
$HIO_3.1/2Pb$
279.81
Lead iodate
Iodic acid, lead(2+) salt (8CI, 9CI)

25660-70-2
$C_4H_7N_3S_2$
161.23
N(NC(=N)S1)=C1SCC
1,3,4-Thiadiazol-2-amine, 5-(ethylthio)- (9CI)
5-(Ethylthio)-1,3,4-thiadiazol-2-amine
NSC-522480
1,3,4-Thiadiazole, 2-amino-5-(ethylthio)- (8CI)

25667-93-0
$(C_8H_{14}O_2.C_8H_8)x$
Polymer
2-Propenoic acid, 2-methyl-, 2-methylpropyl ester, Polymer with ethenylbenzene (9CI)
Ethenylbenzene, isobutyl methacrylate
Ethenylbenzene, isobutyl methacrylate polymer

25680-58-4
$C_7H_{10}N_2O$
138.16
O(c(nccn1)c1CC)C
Pyrazine, 2-ethyl-3-methoxy- (9CI)
2-Ethyl-3-methoxypyrazine
2-Methoxy-3-ethylpyrazine

25707-70-4
$C_{14}H_{28}N_2$
224.38
N(=C(CC(C)C)C)CCN=C(CC(C)C)C
1,2-Ethanediamine, N,N'-bis-(1,3-dimethylbutylidene)- (9CI)
Ethylenediamine, N,N'-bis(1,3-dimethylbutylidene)-

25712-08-7
$C_{27}H_{24}N_6O_9S_2$
640.62
O=C(Nc(c(OC)cc(N=Nc(cccc1S(=O)(=O)O)c1)c2)c2)Nc(c(OC)cc(N=Nc(cccc3S(=O)(=O)O)c3)c4)c4
Benzenesulfonic acid, 3,3'-(carbonylbis(imino(3-methoxy-4,1-phenylene)azo))bis- (9CI)
Benzenesulfonic acid, 3,3'-(ureylenebis((3-methoxy-p-phenylene)azo))di

25713-56-8
$C_6H_5Cl_2N_3S$
222.10
s-Triazine, 2-(allylthio)-4,6-dichloro- (6CI,7CI,8CI)

25713-57-9
$C_{10}H_7Cl_2N_3S$
272.16
s-Triazine, 2-(benzylthio)-4,6-dichloro- (8CI)

25721-38-4
$C_6H_3N_3O_7.xPb$
Unknown
Lead picrate
Lead picrate (Dry)
Picrate de plomb (French)
Picrato de plomo (Spanish)
Picric acid, lead salt (8CI)
Phenol, 2,4,6-trinitro-, lead salt (9CI)
2,4,6-Trinitrophenoate lead

25721-76-0
$(C_{10}H_{14}O_4)x$
Polymer
2-Propenoic acid, 2-methyl-, 1,2-ethanediyl ester, Homopolymer (9CI)
Poly(ethylene glycol dimethacrylate)

25723-16-4
$(C_3H_6O)x(C_3H_6O)x(C_3H_6O)x$ $C_6H_{14}O_3$
Polymer
Poly(oxy(methyl-1,2-ethanediyl)), α-hydro-ω-hydroxy-, ether with 2-ethyl-2-(hydroxymethyl)-1,3-propanediol (3:1) (9CI)

25724-58-7
$C_{22}H_{38}O_4$
366.60
O=C(OCCCCCCCCCC)c(c(ccc1)C(=O)OCCCCCC)c1
Phthalic acid, decyl hexyl ester

25735-29-9
$C_3H_5Cl_3$
147.43
Propane, trichloro- (6CI, 7CI,8CI,9CI)
AI3-08446
Trichloropropane

25773-40-4
$C_8H_{12}N_2O$
152.18
O(c(nccn1)c1C(C)C)C
2-Isopropyl-3-methoxypyrazine
2-Methoxy-3-isopropylpyrazine
2-Methoxy-3-(1-methylethyl)-pyrazine
Pyrazine, 2-methoxy-3-(1-methylethyl)- (9CI)

25790-28-7
$C_{15}H_{21}NO_6$
311.33
O=C(OC)OCCN(c(cccc1C)c1)CCOC(=O)OC
2,4,10-Trioxa-7-azaundecan-11-oic acid, 7-(3-methylphenyl)-3-oxo-, methyl ester (9CI)
N,N-(Bis(2-methoxycarbonyloxy)ethyl)-3-methylbenzenamine

25790-55-0
C_4H_5Cl
88.54
1,2-Butadiene, 4-chloro- (8CI, 9CI)

25791-96-2
$(C_3H_6O)x.(C_3H_6O)x.(C_3H_6O)x.$ $C_3H_8O_3$
Polymer
Niax Polyol L-56
Bypolet 34
Bypolet 36
Glycerol poly(oxypropylene)-triol
Glycerol-propylen oxide polymer
Glycols, polypropylene, 1,2,3-propanetriyl ether
GP 3000
L-3003 (Russian)
Laprol 263
Laprol 503
Laprol 1003
Laprol 3003
Laprol 5003
LG 56
LG 56 (Polymer)
Niax LG 56
Niax LG 240
Niax Polyol LG-168
Nisso TG 4400
Pluracol GP 430
Poly(oxy(methyl-1,2-ethanediyl)), α,α',α''-1,2,3-propanetriyltris(ω-hydroxy-
Poly(oxypropylene)glyceryl ether
Polyurax G 3000
Propylan 3
Propylene oxide-glycerol polymer
Thanol SF 1500
Voranol CP 260
Voranol CP 301
Voranol CP 450
Voranol CP 700
Voranol CP 1500
Voranol CP 3000
Voranol CP 3001

25797-78-8
$C_{10}H_{16}N_2O_3S$
244.30
O=S(=O)(NCCO)c(c(c(cc1N)C)C)c1
2,3-Xylenesulfonamide, 5-amino-N-(2-hydroxyethyl)-(8CI)
2,3-Dimethyl-5-aminobenzene-sulfethanolamide

25797-81-3
$C_{24}H_{23}N_3O_8S_2.Na$
568.57
2-Anthracenesulfonic acid, 1-amino-9,10-dihydro-4-((5-(((2-hydroxyethyl)amino)-sulfonyl)-3,4-dimethyl-phenyl)amino)-9,10-dioxo-, monosodium salt (9CI)

25808-74-6
$F_6Si.Pb$
349.28
Lead silicon fluoride
Hexafluorosilicate(2-) lead(2+) (1:1)
Lead(II) fluorosilicate
Silicate(2-), hexafluoro-, lead-(2+) (1:1) (9CI)
Silicate(2-), hexafluoro-, lead-(II) salt, dihydrate

25817-24-7
Unknown
Unknown
Nitrilotriacetic acid, potassium salt
NTA, potassium salt

25832-09-1
$C_{10}H_{10}O_3$
178.19
Benzenebutanoic acid, β-oxo- (9CI)
4-Phenylacetoacetic acid
4-Phenyl-3-oxobutyric acid

25834-80-4
$C_{20}H_{21}N_3$
303.39
Nc(c(cc(c1)Cc(ccc(N)c2)c2)Cc(ccc(N)c3)c3)c1
Benzenamine, 2,4-bis((4-aminophenyl)methyl)- (9CI)
2,4-Bis(p-aminobenzyl)aniline
2,4-Bis((4-aminophenyl)-methyl)benzenamine

25843-45-2
$C_2H_6N_2O$
74.10

CN=N(C)=O
Methane, azoxy
AOM
Azoxymethane

25852-47-5
$(C_2H_4O)xC_8H_{10}O_3$
Polymer
Hydrogel
Polyethylene glycol dimethacrylate
Poly(oxy-1,2-ethanediyl), α-(2-methyl-1-oxo-2-propenyl)-ω-((2-methyl-1-oxo-2-propenyl)oxy)- (9CI)

25852-70-4
$C_{34}H_{66}O_6S_3Sn$
785.87
Stannane, butyltris(isooctyloxycarbonylmethylthio)
Butyltris(2-ethylhexyloxycarbonylmethylthio)stannane
Tin, butyl-, tris(isooctylthioglycollate)

25854-16-4
$C_{10}H_8N_2O_2$
188.17
Benzene, bis(isocyanatomethyl)- (9CI)
Bis(isocyanatomethyl)benzene

25857-05-0
$C_{36}H_{40}N_6O_4$
620.72
O=C(OCCN(c(ccc(c1C)C=C(C#N)C#N)c1)CC)CCCCC(=O)OCCN(c(ccc(c2C)C=C(C#N)C#N)c2)CC
Hexanedioic acid, bis(2-((4-(2,2-dicyanoethenyl)-3-methylphenyl)ethylamino)-ethyl) ester (9CI)

25857-20-9
$C_5H_{13}NO_5S.Na$
222.21
Methanesulfonic acid, (bis-(2-hydroxyethyl)amino)-, monosodium salt (8CI)
Bis(2-hydroxyethyl)sulfomethylamine, sodium salt

25869-00-5
$C_6FeN_6.Fe.H_4N$
285.80
Ammonium iron (III) hexacyanoferrate
AFCF
Ammonium-ferric-cyanoferrate(II)
Ammonium ferric hexacyanoferrate
Ferrate(4-), hexakis(cyano-C)-, ammonium iron(3+) (1:1:1), (OC-6-11)- (9CI)

25870-62-6
$C_{12}H_{16}O$
176.26
2-Hexanone, 1-phenyl- (8CI, 9CI)
NSC-15336

25876-07-7
$C_{11}H_{16}O_5$
228.27
CCCCOC(=O)C=CC(=O)OCC1CO1
Fumaric acid, butyl 2,3-epoxypropyl ester
Butyl 2,3-epoxypropyl fumarate

25876-47-5
$C_9H_{12}O_5$
200.21
CCOC(=O)C=CC(=O)OCC1CO1
Fumaric acid, ethyl 2,3-epoxypropyl ester

25899-50-7
C_5H_7N
81.11
N#CC=CCC
cis-2-Pentenenitrile
2-Pentenenitrile, (Z)- (9CI)
(Z)-2-Pentenenitrile

25904-89-6
$C_2H_4O_3.xK$
Unknown
Acetic acid, hydroxy-, potassium salt (9CI)
Potassium hydroxyacetate

25911-51-7
$C_{16}H_9Cl$
236.70
Fluoranthene, 3-chloro- (8CI, 9CI)

25915-78-0
$C_2H_2Cl_2F_2$
134.94
Ethane, dichlorodifluoro- (8CI, 9CI)

25917-35-5
$C_6H_{14}O$
102.18
Hexanol (VAN)(9CI)

25939-05-3
$C_{13}H_8Cl_4N_2$
334.03
ClC(=NNc1c(Cl)cc(Cl)cc1Cl)c2ccccc2
Benzoyl chloride, α-(2,4,6-trichlorophenyl)hydrazono
Banamite
Benzoyl chloride, (2,4,6-trichlorophenyl)hydrazone
U-27,415

25951-54-6
$(C_2H_3Br)x$
Polymer
Ethylene, bromo-, Polymer
Polybromoethylene
Polyvinylbromide
PVBR

25953-06-4
$C_{10}H_{14}N_2O_2.ClH$
230.72
CCN(CC)c1ccc(N=O)c(O)c1
Phenol,5-(diethylamino)-2-nitroso-, monohydrochloride
5-(Diethylamino)-2-nitrosophenol hydrochloride

25953-19-9
$C_{14}H_{14}N_8O_4S_3$
454.54
5-Thia-1-azabicyclo(4.2.0)oct-2-ene-2-carboxylic acid, 3-(((5-methyl-1,3,4-thiadiazol-2-yl) thio)methyl)-8-oxo-7-(2-(1H-tetrazol-1-yl)acetamido)
Cefamezin
Cefazolin
Cefazoline
Cephamezine
Cephazolin
Cephazoline
CEZ
Elzogram

25954-13-6
$H_4N.C_3H_7NO_4P$
170.13
N.CCC(=O)NP(O)(O)=O
Ammonium, ethyl carbamoylphosphonate
Ammonium-aethyl-carbamoylphosphonat (German)
Ammonium ethyl carbamoylphosphonate solution
DPX 1108
Fosamine ammonium
Krenite
Krenite Brush Control Agent

25956-17-6
$C_{18}H_{16}N_2O_8S_2.2Na$
498.46
[Na+].[Na+].COc1cc(c(C)cc1N=Nc2c(O)ccc3cc(ccc23)S([O-])(=O)=O)S([O-])(=O)=O
2-Naphthalenesulfonic acid, 6-hydroxy-5-((6-methoxy-4-sulfo-m-tolyl)azo)-, disodium salt

Allura Red
Allura Red AC
C.I. 16035
C.I. Food Red 17
FD & C Red No. 40
Red No. 40

25959-70-0
$C_{10}H_{14}N_2O_5S$
274.29
O=S(=O)(NCCO)c(c(c(cc1N(=O)=O)C)C)c1
2,3-Xylenesulfonamide, N-(2-hydroxyethyl)- (8CI)
2,3-Dimethyl-5-nitrobenzene-sulfethanolamide
N-(2-Hydroxyethyl)-2,3-xylene-sulfonamide

25961-84-6
$C_{10}H_{20}O_3$
188.27
Ethanol, 2-(2-(cyclohexyl-oxy)ethoxy)- (8CI,9CI)
2-(2-(Cyclohexyloxy)-ethoxy)ethanol
Diethylene glycol mono-cyclohexyl ether

25962-77-0
$C_{11}H_{11}N_5O_4$
277.27
Formamidine, N,N-dimethyl-N'-(5-(2-(5-nitro-2-furyl)-vinyl)-1,3,4-oxadiazol-2-yl)
Methanimidamide, N,N-di-methyl-N'-(5-(2-(5-nitro-2-furanyl)ethenyl)-1,3,4-oxa-diazol-2-yl)-
1,3,4-Oxadiazole, 2-(((dimethyl-amino)methylene)amino)-5-(2-(5-nitro-2-furyl)vinyl)-

25973-55-1
$C_{22}H_{29}N_3O$
351.47
Oc(c(cc(c1)C(CC)(C)C)C(CC)(C)C)c1N(N=C(C=2C=CC=3)C3)N2
Phenol, 2-(2H-benzotriazol-2-yl)-4,6-bis(1,1-dimethyl-propyl)- (9CI)

26002-80-2
$C_{23}H_{26}O_3$
350.49
CC(C)=CC3C(C(=O)OCc2cccc(Oc1ccccc1)c2)C3(C)C
Cyclopropanecarboxylic acid, 2,2-dimethyl-3-(2-methyl-propenyl)-, m-phenoxybenzyl ester
Phenothrin
3-Phenoxybenzyl D-Z/E chrys-anthemate
Phenoxythrin
S 2539
Sumithrin
Sumitrin

26006-22-4
$(C_9H_{18}NO_2.C_3H_5NO.CH_3O_4S)x$
Polymer
Polyquaternium-5
Acrylamide, dimethylamino-ethyl methacrylate copolymer, dimethyl sulfate quaternized
Acrylamide, methacryloyloxy-ethyltrimethylammonium methylsulfate copolymer
Acrylamide, N,N,N-trimethyl-2-((2-methyl-1-oxo-2-pro-penyl)oxy)ethanaminium methyl sulfate polymer
Emcol Q
Ethanaminium, N,N,N-tri-methyl-2-((2-methyl-1-oxo-2-propenyl)oxy)-, methyl sulf-ate, Polymer with 2-propen-amide (9CI)
Ethanaminium, N,N,N-tri-methyl-2-((2-methyl-1-oxo-2-propenyl)oxy)-, methylsulf-ate, Polymer with 2-propen-amine
Nalquat P
Poly (acrylamide-dimethyl-aminoethyl methacrylate, di-methyl sulfate quat)
Quaternium-39

26007-63-6
$(C_{12}H_{10}O.CH_2O)x$
Polymer NA
Formaldehyde, Polymer with 1,1'-oxybis-(benzene) (9CI)
Benzene, 1,1'-oxybis-, Polymer with formalde-hyde (9CI)
DF
DF (Formaldehyde polymer)
Diphenyl ether-formalde-hyde copolymer
Diphenyl ether-formalde-hyde polymer
Diphenyl ether-formalde-hyde resin
Diphenyl oxide-formalde-hyde copolymer
Diphenyl oxide-formalde-hyde polymer
Doryl V 505-50
Phenyl ether, Polymer with formaldehyde (8CI)

26021-90-9
$C_7H_9NO_3S.Na$
210.20
Methanesulfonic acid, (phenylamino)-, mono-sodium salt (9CI)
Anilinomethanesulfonate sodium salt
Anilinomethanesulfonic acid, monosodium salt
Anilinomethanesulfonic acid, sodium salt
Methanesulfonic acid, anilino-, monosodium salt
NSC-57030
Phenylaminomethanesulfonic acid, monosodium salt
(Phenylamino)methanesulfonic acid, sodium salt
Sodium anilinomethanesulfonate

26027-37-2
$(C_2H_4O)xC_{20}H_{39}NO_2$
Polymer
Poly(oxy-1,2-ethanediyl), α-(2-((1-oxo-9-octadecenyl)-amino)ethyl)-ω-hydroxy-, (Z)- (9CI)
Ethoxylated oleic monoethanol-amide
Ethoxylated oleoamide

26027-38-3
$(C_2H_4O)xC_{15}H_{24}O$
Polymer
Glycols, polyethylene, mono-(p-nonylphenyl) ether
Nonylphenoxypoly(ethylene-oxy)ethanol
NP-9
Phenol, p-nonyl-, monoether with polyethylene glycol
N-9
Nonoxynol-9

26028-46-6
$C_9H_{15}N_3O_3$
213.22
1,3,5-Triazine, 1,3,5-tri-acetylhexahydro- (9CI)
NSC-194838
s-Triazine, 1,3,5-triacetyl-hexahydro- (8CI)

26062-94-2
$(C_8H_6O_4.C_4H_{10}O_2)x$
Polymer
Poly(1,4-butylene tereph thalate)
1,4-Benzenedicarboxylic acid, Polymer with 1,4-butanediol (9CI)
1,4-Butanediol, 1,4-benzene-dicarboxylic acid, α-hydro-ω-hydroxypoly(oxy-1,2-ethanediyl) polymer
Butanediol-terephthalic acid copolymer
1,4-Butanediol-terephthalic acid copolymer
1,4-Butanediol-terephthalic acid polymer
Butylene glycol-terephthalic acid copolymer
1,4-Butylene glycol-terephthalic acid copolymer
Butylene glycol-terephthalic acid polymer

1,4-Butylene glycol-terephthalic acid polymer
Poly(1,4-butanediol terephthalate)
Poly(butylene terephthalate)
Poly(tetramethylene terephthalate)
Terephthalic acid-1,4-butanediol polymer
Terephthalic acid, Polyester with 1,4-butanediol
Terephthalic acid-tetramethylene glycol copolymer
Terephthalic acid-tetramethylene glycol polymer
Tetramethylenediol-terephthalic acid polymer
Tetramethylene glycol-terephthalic acid polymer

26063-00-3
$C_4H_8O_2$
88.11
Poly-β-hydroxybutyrate
Poly(3-hydroxybutyrate)
Poly-β-hydroxybutyric acid

26087-47-8
$C_{13}H_{21}O_3PS$
288.37
CC(C)OP(=O)(OC(C)C)SCc1ccccc1
Phosphorothioic acid, S-benzyl O,O-diisopropyl ester
O,O-Bis(1-methylethyl) S-(phenylmethyl)phosphorothioate
O,O-Diisopropyl-S-benzylester kyseliny thiofosforecne (Czech)
O,O-Diisopropyl S-benzyl phosphorothiolate
O,O-Diisopropyl S-benzyl thiophosphate
IBP
Iprobenfos
Kitazin L
Kitazin P
Ricid II
Ricid P

26094-13-3
$C_{18}H_{34}O_2.C_4H_{11}N$
355.60
Butylamine oleate
Butylammonium oleate
9-Octadecenoic acid (Z)-, Compd. with 1-butanamine (1:1) (9CI)
Oleic acid, Compd. with butylamine (1:1)

26110-32-7
$C_{33}H_{47}ClN_2O_4$
571.19
O=C(Nc(c(ccc1NC(=O)CCCOc(c(cc(c2)C(CC)(C)C)C(CC)(C)C)c2)Cl)c1)CC(=O)C(C)(C)C
Pentanamide, N-(5-((4-(2,4-bis(1,1-dimethylpropyl)phenoxy)-1-oxobutyl)amino)-2-chlorophenyl)-4,4-dimethyl-3-oxo- (9CI)
N-(2-(Chloro-5-(4-(2,4-di-tert-pentylphenoxy)butyramido)-phenyl)-4,4-dimethyl-3-oxo-pentanamide

26113-25-7
$C_{15}H_{26}Cl_2N_4$
333.31
1,3,5-Triazin-2-amine, 4,6-dichloro-N-dodecyl- (9CI)
4,6-Dichloro-2-(dodecylamino)-s-triazine
s-Triazine, 2,4-dichloro-6-(dodecylamino)- (6CI, 7CI,8CI)

26115-70-8
$C_{21}H_{45}N_3O_{12}Si_3$
615.84
1,3,5-Triazine-2,4,6(1H,3H,5H)-trione, 1,3,5-tris(3-(trimethoxysilyl)propyl)- (9CI)
1,3,5-Tris(γ-trimethoxysilylpropyl)isocyanurate

26118-38-7
$C_8H_{16}O$
128.22
2-Hexanone, 3,3-dimethyl- (6CI,8CI,9CI)
3,3-Dimethyl-2-hexanone

26118-67-2
$C_3H_8ClNO_2S$
157.61
O=S(=O)(NC(C)C)Cl
Sulfamoyl chloride, (1-methylethyl)- (9CI)
(1-Methylethyl)sulfamoyl chloride

26124-68-5
$(C_2H_4O_3)x$
Polymer
Polyglycolic acid
Acetic acid, hydroxy-, Homopolymer (9CI)
Hydroxyacetic acid homopolymer
Polyglycollic acid

26130-84-7
$C_{16}H_{18}$
210.32
Naphthalene, 1,2,3-trimethyl-4-(1-propenyl)-, (Z)- (9CI)
Naphthalene, 1,2,3-trimethyl-4-propenyl-, (Z)- (8CI)

26134-62-3
Li_3N
34.82
Lithium nitride (8CI,9CI)
Nitrure de lithium (French)
Nitruro de litio (Spanish)
UN 2806 [Lithium nitride]

26137-53-1
$C_{16}H_{18}$
210.32
Naphthalene, 1,2,3-trimethyl-4-propenyl-, (E)- (8CI)

26140-60-3
$C_{18}H_{14}$
230.32
c1ccc(cc1)c2ccc(cc2)c3ccccc3
Terphenyl
Delowas S
Delowax OM
Diphenylbenzene
Gilotherm OM 2
Terbenzene
Triphenyl
Terphenyls (ACGIH,OSHA)

26148-68-5
$C_{11}H_9N_3$
183.23
c1ccc2c(c1)[nH]c3ncccc23
9H-Pyrido(2,3-b)indole, 2-amino
Amino-α-carboline
2-Amino-α-carboline
2-Amino-9H-pyrido(2,3-b)-indole
Glob-P-2
1H-Pyrido(2,3-b)indol-2-amine

26171-23-3
$C_{15}H_{15}NO_3$
257.31
Cc1ccc(cc1)C(=O)c2ccc(CC(O)=O)n2C
Pyrrole-2-acetic acid, 1-methyl-5-p-toluoyl
Acido 1-metil-5-(p-tolnil)-pirrol-2-acetico (Spanish)
MCN 2559
1-Methyl-5-(4-methylbenzoyl)-pyrrole-2-acetic acid
1-Methyl-5-p-toluoyl-pyrrole-2-acetic acid
1H-Pyrrole-2-acetic acid, 1-methyl-5-(4-methylbenzoyl)- (9CI)
Tolmetin
Tolmetine

26172-55-4
C_4H_4ClNOS
149.59
Methylchloroisothiazolinone
5-Chloro-2-methyl-4-isothia-

zolin-3-one
4-Isothiazolin-3-one, 5-chloro-2-methyl-
3(2H)-Isothiazolone, 5-chloro-2-methyl-

26183-52-8
$(C_2H_4O)xC_{10}H_{22}O$
Polymer
Decylpolyethyleneglycol 300
Decyl alcohol, ethoxylated
Poly(oxy-1,2-ethanediyl), α-decyl-ω-hydroxy- (9CI)

26225-79-6
$C_{13}H_{18}O_5S$
286.37
CCOC2Oc1ccc(OS(C)(=O)=O)cc1C2(C)C
5-Benzofuranol, 2-ethoxy-2,3-dihydro-3,3-dimethyl-, methanesulfonate, (+-)
CR 14658
Ethofumesate
(+-)-2-Ethoxy-2,3-dihydro-3,3-dimethyl-5-benzofuranol methanesulfonate
NC 8438
Nortron
Nortron (New)
Tramat

26227-73-6
$C_{18}H_{21}NO$
267.40
O(c(ccc(c1)C=Nc(ccc(c2)CCCC)c2)c1)C
Aniline, p-butyl-N-(p-methoxybenzylidene)
4-Methoxybenzylidene-4'-n-butylaniline
MBBA

26228-72-8
$C_{10}H_{23}NO$
173.29
Hydroxylamine, N-decyl- (8CI)

26234-33-3
$C_{10}H_{19}N_5$
209.30
s-Triazine, 2-methyl-4,6-bis(propylamino)- (7CI,8CI)

26234-34-4
$C_{12}H_{23}N_5$
237.35
s-Triazine, 2,4-bis(butylamino)-6-methyl- (8CI)

26234-35-5
$C_{12}H_{23}N_5$
237.32
s-Triazine, 2,4-bis(isobutylamino)-6-methyl- (8CI)

26234-36-6
$C_{12}H_{23}N_5$
237.35
s-Triazine, 2,4-bis(sec-butylamino)-6-methyl- (8CI)

26234-37-7
$C_{16}H_{31}N_5$
293.46
s-Triazine, 2,4-bis(hexylamino)-6-methyl- (8CI)

26234-38-8
$C_{28}H_{55}N_5$
461.78
s-Triazine, 2,4-bis(dodecylamino)-6-methyl- (8CI)

26234-39-9
$C_{10}H_{15}N_5$
205.26
s-Triazine, 2,4-bis-(allylamino)-6-methyl- (8CI)

26234-40-2
$C_{12}H_{23}N_5$
237.35
s-Triazine, 2,4-bis(diethylamino)-6-methyl- (7CI, 8CI)

26234-41-3
$C_{14}H_{23}N_5$
261.42
Triazine, 2,4-dipiperidino-6-methyl
2,4-Dipiperidino-6-methyltriazine

26234-42-4
$C_{12}H_{19}N_5O_2$
265.36
Triazine, 2,4-dimorpholino-6-methyl
2,4-Dimorpholino-6-methyltriazine

26234-95-7
$C_{10}H_{19}N_5O_2$
241.30
Ethanol, 2,2'-((6-methyl-1,3,5-triazine-2,4-diyl)-bis(methylimino))bis- (9CI)
Ethanol, 2,2'-((6-methyl-s-triazine-2,4-diyl)-bis(methylimino))di- (8CI)

26234-96-8
$C_6H_8Cl_3N_5$
256.52
1,3,5-Triazine-2,4-diamine, N,N'-dimethyl-6-(trichloromethyl)- (9CI)
s-Triazine, 2,4-bis(methylamino)-6-(trichloromethyl)- (7CI,8CI)

26234-97-9
$C_8H_{12}Cl_3N_5$
284.58
s-Triazine, 2,4-bis(ethylamino)-6-(trichloromethyl)- (7CI,8CI)

26234-98-0
$C_{10}H_{16}Cl_3N_5$
312.63
s-Triazine, 2,4-bis(propylamino)-6-(trichloromethyl)- (8CI)

26234-99-1
$C_{12}H_{20}Cl_3N_5$
340.69
s-Triazine, 2,4-bis(butylamino)-6-(trichloromethyl)- (8CI)

26235-00-7
$C_{16}H_{28}Cl_3N_5$
396.79
s-Triazine, 2,4-bis(hexylamino)-6-(trichloromethyl)- (8CI)

26235-01-8
$C_{28}H_{52}Cl_3N_5$
565.12
s-Triazine, 2,4-bis(dodecylamino)-6-(trichloromethyl)- (8CI)

26235-02-9
$C_{10}H_{12}Cl_3N_5$
308.60
s-Triazine, 2,4-bis(allylamino)-6-(trichloromethyl)- (7CI,8CI)

26235-03-0
$C_{12}H_{20}Cl_3N_5$
340.69
1,3,5-Triazine-2,4-diamine, N,N,N',N'-tetraethyl-6-(trichloromethyl)- (9CI)
s-Triazine, 2,4-bis(diethylamino)-6-(trichloromethyl)- (8CI)

26235-04-1

$C_{14}H_{20}Cl_3N_5$
364.71
s-Triazine, 2,4-dipiperid-ino-6-(trichloromethyl)- (8CI)

26235-06-3
$C_8H_{12}Cl_3N_5O_2$
316.58
Ethanol, 2,2'-((6-(tri-chloromethyl)-s-triazine-2,4-diyl)diimino)di- (8CI)

26235-07-4
$C_{10}H_{16}Cl_3N_5O_2$
344.63
1-Propanol, 3,3'-((6-(tri-chloromethyl)-s-triazine-2,4-diyl)diimino)di- (8CI)

26235-08-5
$C_{10}H_{16}Cl_3N_5O_2$
344.63
2-Propanol, 1,1'-((6-(tri-chloromethyl)-s-triazine-2,4-diyl)diimino)di- (8CI)

26235-09-6
$C_{10}H_{16}Cl_3N_5O_2$
344.63
Ethanol, 2,2'-((6-(tri-chloromethyl)-s-triazine--2,4-diyl)bis(methyl-imino))di- (8CI)

26235-11-0
$C_{10}H_{10}Cl_3N_7$
334.60
Propionitrile, 3,3'-((6-(tri-chloromethyl)-s-triazine-2,4-diyl)diimino)di- (8CI)

26235-12-1
$C_7H_{13}N_5$
167.22
s-Triazine, 2-ethyl-4,6-bis-(methylamino)- (8CI)

26235-13-2
$C_9H_{17}N_5$
195.27
s-Triazine, 2-ethyl-4,6-bis-(ethylamino)- (7CI,8CI)

26235-14-3
$C_{11}H_{21}N_5$
223.32
s-Triazine, 2-ethyl-4,6-bis-(isopropylamino)- (7CI,8CI)

26235-15-4
$C_{13}H_{25}N_5$
251.38
s-Triazine, 2-ethyl-4,6-bis-(isobutylamino)- (8CI)

26235-16-5
$C_{13}H_{25}N_5$
251.38
s-Triazine, 2,4-bis(diethyl-amino)-6-ethyl- (7CI,8CI)

26235-17-6
$C_8H_{15}N_5$
181.24
s-Triazine, 2,4-bis(methyl-amino)-6-propyl- (8CI)

26235-18-7
$C_{10}H_{19}N_5$
209.30
s-Triazine, 2,4-bis(ethyl-amino)-6-propyl- (7CI, 8CI)

26235-19-8
$C_{12}H_{23}N_5$
237.35
s-Triazine, 2-propyl-4,6-bis(propylamino)- (8CI)

26235-20-1
$C_{12}H_{23}N_5$
237.35
s-Triazine, 2,4-bis(iso-propylamino)-6-propyl- (7CI,8CI)

26235-21-2
$C_{14}H_{27}N_5$
265.40
s-Triazine, 2,4-bis(butyl-amino)-6-propyl- (8CI)

26235-22-3
$C_{14}H_{27}N_5$
265.40
s-Triazine, 2,4-bis(isobutyl-amino)-6-propyl- (8CI)

26235-23-4
$C_8H_{15}N_5$
181.24
s-Triazine, 2-isopropyl-4,6-bis(methylamino)- (8CI)

26235-24-5
$C_{10}H_{19}N_5$
209.30
s-Triazine, 2,4-bis(ethyl-amino)-6-isopropyl- (7CI, 8CI)

26235-25-6
$C_{12}H_{23}N_5$
237.35
s-Triazine, 2-isopropyl-4,6-bis(isopropylamino)- (7CI,8CI)

26235-26-7
$C_{14}H_{27}N_5$
265.40
s-Triazine, 2,4-bis(iso-butylamino)-6-isopropyl- (8CI)

26235-27-8
$C_{11}H_{21}N_5$
223.32
s-Triazine, 2-butyl-4,6-bis-(ethylamino)- (7CI,8CI)

26235-28-9
$C_{13}H_{25}N_5$
251.38
s-Triazine, 2-butyl-4,6-bis-(isopropylamino)- (8CI)

26235-29-0
$C_9H_{17}N_5$
195.27
s-Triazine, 2-isobutyl-4,6-bis(methylamino)- (8CI)

26235-30-3
$C_{11}H_{21}N_5$
223.32
s-Triazine, 2,4-bis(ethyl-amino)-6-isobutyl- (8CI)

26235-31-4
$C_{13}H_{25}N_5$
251.38
s-Triazine, 2-isobutyl-4,6-bis(propylamino)- (8CI)

26235-32-5
$C_{15}H_{29}N_5$
279.43
s-Triazine, 2,4-bis(butyl-amino)-6-isobutyl- (8CI)

26235-33-6
$C_{15}H_{29}N_5$
279.43
s-Triazine, 2-isobutyl-4,6-bis(isobutylamino)- (8CI)

26235-34-7
$C_{11}H_{21}N_5$
223.32
s-Triazine, 2-sec-butyl-4,6-bis(ethylamino)-(8CI)

26235-35-8
$C_{13}H_{25}N_5$
251.38
s-Triazine, 2-sec-butyl-4,6-bis(propylamino)-(8CI)

26235-36-9
$C_{10}H_{19}N_5$
209.30
s-Triazine, 2,4-bis(methylamino)-6-pentyl- (8CI)

26235-37-0
$C_{12}H_{23}N_5$
237.35
s-Triazine, 2,4-bis(ethylamino)-6-pentyl- (8CI)

26235-38-1
$C_{14}H_{27}N_5$
265.40
s-Triazine, 2-pentyl-4,6-bis(propylamino)-(8CI)

26235-39-2
$C_{16}H_{31}N_5$
293.46
s-Triazine, 2,4-bis(butylamino)-6-pentyl- (8CI)

26235-87-0
$C_{16}H_{31}N_5$
293.46
s-Triazine, 2,4-bis(isobutylamino)-6-pentyl- (8CI)

26235-88-1
$C_{16}H_{31}N_5$
293.46
s-Triazine, 2,4-bis(ethylamino)-6-nonyl- (8CI)

26235-90-5
$C_{24}H_{47}N_5$
405.68
s-Triazine, 2,4-bis(ethylamino)-6-heptadecyl-(8CI)

26235-91-6
$C_6H_8Br_3N_5$
389.89
s-Triazine, 2,4-bis-(methylamino)-6-(tribromomethyl)- (8CI)

26235-92-7
$C_9H_{16}ClN_5$
229.71
s-Triazine, 2-(2-chloroethyl)-4,6-bis(ethylamino)- (8CI)

26245-56-7
C_3H_8ClN
93.55
N(C(C)C)Cl
2-Propanamine, N-chloro-(9CI)
N-Chloro-2-propanamine

26248-24-8
$C_{19}H_{32}O_3S.Na$
363.52
Sodium tridecylbenzene sulfonate
Benzenesulfonic acid, tridecyl-, sodium salt (9CI)
Conoco C 650
Sodium tridecylbenzenesulfonate
Sodium n-tridecylbenzenesulfonate
Tridecylbenzene sulfonic acid sodium salt
Tridecylbenzenesulfonic acid, sodium salt
Witconate TDB

26248-42-0
$C_{13}H_{28}O$
200.41
Tridecanol

26248-87-3
$C_9H_{18}Cl_3O_4P$
327.59
CCCOP(=O)(OCC(C)Cl)OCC(Cl)CCC(Cl)COP(=O)(OCC(C)Cl)OCC(C)Cl.CC(Cl)COP(=O)(OCC(C)Cl)OCC(C)Cl
1-Propanol, 2-chloro-, phosphate (3:1), Mixed with 1-chloro-2-propanol phosphate (3:1)
Tris(chloropropyl)phosphate

26249-12-7
$C_6H_4Br_2$
235.92
Benzene, dibromo
Dibromobenzene [UN 2711]
UN 2711 [Dibromobenzene]

26249-20-7
Unknown
Unknown
UCAR Butylene Oxide 12 (Obs.)

26258-70-8
$C_{19}H_{23}NO_4$
329.43
Propane, 1,1-bis(p-ethoxyphenyl)-2-nitro
AI3-27990
Benzene, 1,1'-(2-nitropropylidene)bis(4-ethoxy- (9CI)
1,1-Bis(p-ethoxyphenyl)-2-nitropropane
GH 74
OMS 1356

26258-71-9
$C_{20}H_{25}NO_4$
343.46
Butane, 1,1-bis(p-ethoxyphenyl)-2-nitro
1,1-Bis(p-ethoxyphenyl)-2-nitrobutane

26259-45-0
$C_{10}H_{19}N_5O$
225.34
O(c(nc(nc1NC(CC)C)NCC)n1)C
s-Triazine, 2-(sec-butylamino)-4-(ethylamino)-6-methoxy
2-sec-Butylamino-4-ethylamino-6-methoxy-s-triazine
2-sec-Butylamino-4-ethylamino-6-methoxy-1,3,5-triazine
Etazine
Ezitan
Geigy G.S. 14254
GS 14254
2-Methoxy-4-sec-butylamino-6-aethylamino-s-triazin (German)
Secbumeton
Secumbeton
Sumitol
Sumitol 80W

26263-49-0
$C_{10}H_{19}N_5$
209.26
1,3,5-Triazine-2,4-diamine, 6-methyl-N,N'-bis(1-methylethyl)- (9CI)
s-Triazine, 2,4-bis(isopropylamino)-6-methyl- (8CI)

26263-50-3
$C_{10}H_{16}Cl_3N_5$
312.63
s-Triazine, 2,4-bis(isopropylamino)-6-(trichloromethyl)- (7CI,8CI)

26264-05-1
$C_{18}H_{30}O_3S.C_3H_9N$
385.61
Isopropylamine dodecylbenzenesulfonate

Benzenesulfonic acid, dodecyl-, Compd. with 2-propanamine (1:1) (9CI)
Dodecylbenzenesulfonic acid, Comp. with 2-Propanamine (1:1)
Dodecylbenzenesulfonic acid monoisopropanolamine salt

26264-06-2
$C_{18}H_{30}O_3S.1/2Ca$
346.54
Calcium dodecylbenzenesulfonate
Benzenesulfonic acid, dodecyl-, calcium salt (9CI)
Calcium alkylaromatic sulfonate
Calcium alkylbenzenesulfonate
Calcium dodecylbenzene sulfonate
Calcium n-dodecylbenzenesulfonate
Calcium dodecylbenzensulfonate
Casul 70HF
Dodecylbenzenesulfonic acid calcium salt
Dodecylbenzensulfonic acid calcium salt
Sinnozon NCX 70
Soprofor S 70
1371A

26264-09-5
$C_7H_5ClO_2$
156.57
Benzoic acid, chloro- (8CI,9CI)

26265-08-7
$(C_{15}H_{16}O_2.C_{15}H_{12}Br_4O_2.C_3H_5ClO)x$
Polymer
Phenol, 4,4'-(1-methylethylidene)bis(2,6-dibromo-, Polymer with (chloromethyl)oxirane and 4,4'-(1-methylethylidene)bis-(phenol) (9CI)
Bisphenol A, epichlorohydrin, tetrabromobisphenol A polymer
4,4'-(1-Methylethylidene)bisphenol, (chloromethyl)oxirane, 4,4'-(1-methylethylidene)bis-(2,6-dibromophenol polymer
Tetrabromobisphenol A, bisphenol A, epichlorohydrin polymer

26265-65-6
$H_2O_3S_2.1/2Pb$
217.75
Lead thiosulfate
Lead hyposulfite
Thiosulfuric acid, lead salt

26266-05-7
$C_{17}H_{34}$
238.46
Heptadecene (8CI,9CI)

26266-57-9
$C_{22}H_{42}O_6$
402.64
Sorbitan, monopalmitate
Arlacel 40
Crill 2
Emsorb 2510
Glycomul P
Liposorb P
Montane 40
Nikkol SP10
Nissan Nonion PP 40
Nissan Nonion PP 40R
Nonion PP40
Protachem SMP
Rheodol SP-P 10
Sorbitan Palmitate
Sorgen 70
Span 40

26266-58-0
$C_{60}H_{108}O_8$
957.68
Sorbitan, tris(9-octadecenoate), (Z)
Arlacel 85
Crill 5
Emasol 430
Emsorb 2503
Glycomul TO
Ionet S 85
Liposorb TO
Nissan Nonion OP 85
Nissan Nonion OP 85R
OP 85R
Protachem STO
Rheodol SP 030
Sorbitan, tri-9-octadecenoate, (Z,Z,Z)- (9CI)
Sorbitan trioleate
Span 85
TE 33

26266-68-2
$C_8H_{14}O$
126.22
CCCC=C(CC)C=O
Hexenal, 2-ethyl

26266-69-3
C_8H_{16}
112.22
Hexene, 2,5-dimethyl- (8CI, 9CI)

26266-76-2
$C_{24}H_{47}N_2O.C_2H_5O_4S$
504.76
1H-Imidazolium, 1-ethyl-2-(heptadecenyl)-4,5-dihydro-1-(2-hydroxyethyl)-, ethyl sulfate (Salt) (9CI)
1-Ethyl-2-(heptadecenyl)-1-(2-hydroxyethyl)-2-imidazolinium ethyl sulfate

26290-70-0
$C_4H_{11}O_4P.2K$
232.30
Phosphoric acid, monobutyl ester, dipotassium salt (9CI)
Butyl dipotassium phosphate

26292-91-1
$C_4H_5N_3S$
127.17
s-Triazine, 2-(methylthio)-(6CI,8CI)
Methylthio-s-triazine

26294-98-4
C_5H_7N
81.11
N#CC=CCC
2-Pentenenitrile, (E)- (9CI)
(E)-2-Pentenenitrile

26301-10-0
$(C_{12}H_{22}O_{11}.C_3H_6O.C_2H_4O)x$
Polymer
α-D-Glucopyranoside, β-D-fructofuranosyl, Polymer with methyloxirane and oxirane (9CI)
Sucrose, propylene oxide, ethylene oxide polymer

26303-54-8
$C_{10}H_{22}O.1/3Al$
167.28
1-Decanol, aluminum salt (9CI)
Aluminum n-decoxide

26322-14-5
$C_{34}H_{66}O_6$
570.89
O=C(OCCCCCCCCCCCCCCC C)OOC(=O)OCCCCCCCCC CCCCCCC
Dicetyl peroxydicarbonate (Not more than 42% stable dispersion, in water)
Dicetyl peroxydicarbonate
Dicetyl peroxydicarbonate, Not more than 42% in water
Dicetyl peroxydicarbonate, Technically pure
Dihexadecyl peroxydicarbonate
Perkadox 24W40
Peroxidicarbonato de dicetilo (Spanish)
Peroxydicarbonate de cetyle (French)
Peroxydicarbonic acid, dicetyl ester
Peroxydicarbonic acid, dihexadecyl ester (8CI,9CI)
Peroxydicarbonic acid, dihexadecyl ester, Not more than

42% in water
UN 2164
UN 2895

26322-44-1
$C_{10}H_{19}N_5O_2$
241.30
2-Propanol, 1,1'-((6-methyl-s-triazine-2,4-diyl)diimino)di- (8CI)

26322-45-2
$C_{12}H_{20}Cl_3N_5$
340.69
s-Triazine, 2,4-bis(isobutylamino)-6-(trichloromethyl)- (8CI)

26322-48-5
$C_{26}H_{51}N_5$
433.73
s-Triazine, 2-heptadecyl-4,6-bis(isopropylamino)-(8CI)

26351-32-6
$C_{21}H_{40}N_2$
320.55
N#CCCNCCCCCCCCC=CCCCCCCCC
Propanenitrile, 3-(9-octadecenylamino)-, (Z)- (9CI)
(Z)-3-(9-Octadecenylamino)-propanenitrile
N-(9-Octadecenyl)-3-aminopropionitrile

26377-29-7
$C_2H_6O_2PS_2.Na$
180.16
[Na+].COP([S-])(=S)OC
Phosphorodithioic acid, O,O-dimethyl ester, sodium salt
O,O-Dimethyldithiofosforecnan sodny (Czech)

26389-60-6
$C_7H_{15}N$
113.20
N(CC(C1)C1)CCC
Cyclopropanemethanamine, N-propyl- (9CI)
N-Propylcyclopropanemethanamine

26389-78-6
$C_{11}H_{13}Cl_2N_3O_4$
322.13
Benzenamine, N,N-bis(2-chloroethyl)-4-methyl-2,6-dinitro-
AI3-62692
N,N-Bis(2-chloroethyl)-2,6-dinitro-p-toluidine
Caswell No. 089A
Chlornidine
Torpedo

26392-63-2
$C_{17}H_{39}N_5O$
329.50
O=C(NCCNCCNCCNCCN)CCCCCCCC
Nonanamide, N-(2-((2-((2-((2-aminoethyl)amino)ethyl)amino)ethyl)amino)ethyl)-(9CI)
N-(2-((2-((2-((2-Aminoethyl)amino)ethyl)amino)ethyl)amino)ethyl)nonanamide
Tetraethylenetetramine pelargonamide

26396-34-9
$Cl_{11}H_{15}N_3O_2$
479.13
s-Triazine, 4,6-bis(pentachlorophenoxy)-2-chloro
4,6-Bis(pentachlorophenoxy)-2-chloro-s-triazine
2-Chloro-4,6-di(pentachlorophenoxy)-2-triazine

26399-36-0
$C_{14}H_{16}F_3N_3O_4$
347.33
CCCN(CC1CC1)c2c(cc(cc2N(=O)=O)C(F)(F)F)N(=O)=O
p-Toluidine, N-(cyclopropylmethyl)-2,6-dinitro-N-propyl-α,α,α-trifluoro
N-(Cyclopropylmethyl)-α,α,α-trifluoro-2,6-dinitro-N-propyl-p-toluidine
CGA 10832
ER5461
GA-10832
Pregard
Profluralin
Profluraline
Tolban

26400-24-8
$C_8H_{11}NO_2$
153.20
1H-Pyrrolizine-7-methanol, 2,3-dihydro-1-hydroxy-, (S)
Dehydroheliotridine
Heliotridine, 3,8-didehydro-

26401-27-4
$C_{20}H_{27}O_3P$
346.41
Phosphorous acid, isooctyl diphenyl ester (9CI)

26401-47-8
$(C_2H_4O)xC_{18}H_{30}O$
Polymer
Poly(oxy-1,2-ethanediyl), α-(4-dodecylphenyl)-ω-hydroxy- (9CI)

26401-86-5
$C_{38}H_{74}O_6S_3Sn$
841.91
Acetic acid, 2,2',2''-((octylstannylidyne)tris(thio))tris-, triisooctyl ester (9CI)
Monooctyltin tris(isooctylthioglycollate)

26401-97-8
$C_{36}H_{72}O_4S_2Sn$
751.89
Stannane, bis(isooctyloxycarbonylmethylthio)dioctyl
Acetic acid, 2,2'-((dioctylstannylene)bis(thio))bis-, diisooctyl ester
Advastab 17 MO
Diisooctyl ((dioctylstannylene)-dithio)diacetate
Dioctyltin bis(isooctyl mercaptoacetate)
Dioctyltin S,S'-bis(isooctyl mercaptoacetate)
Di-(n-octyl)tin-S,S'-bis(isooctylmercaptoacetate)
Dioctyltin bis(isooctyl thioglycolate)
Di-n-octyltin diisooctyl thioglycolate
Di-n-octyl-zinn-di-isooctylthioglykolat (German)
DOTG
Thermolite 831
Tin, dioctyl-, bis(isooctylthioglycollate)
Tin, bis(mercaptoacetate)-dioctyl-, bis(isooctyl) ester

26403-12-3
$C_8H_{19}O_4P$
210.21
Isooctyl alcohol, dihydrogen phosphate (8CI)
Isooctanol, dihydrogen phosphate
Monoisooctyl phosphate
NSC-41917
Phosphoric acid, monoisooctyl ester (9CI)

26403-14-5
$C_{13}H_{26}O_2$
214.35
Neotridecanoic acid (8CI,9CI)

26403-17-8
$C_{10}H_{20}O_2$
172.27
Isodecanoic acid (9CI)

26408-28-6
$C_{12}H_{15}N_3O_2$

233.25
O=C(Nc(ccc(OC)c1NCCC#N)c1)C
Acetamide, N-(3-((2-cyanoethyl)amino)-4-methoxyphenyl)- (9CI)
3-(N-Cyanoethyl)amino-4-methoxyacetanilide

26419-73-8
$C_8H_{14}N_2O_2S_2$
234.33
Tirpate
AI3-27696
Carbamic acid, methyl-, O-(((2,4-dimethyl-1,3-dithiolan-2-yl)methylene)amino)-
Carbamic acid, methyl-, O-(((2,4-dimethyl-1,3-dithiolan-2-yl)methylene)amino) deriv.
Caswell No. 364B
2,4-Dimethyl-1,3-dithiolane-2-carboxaldehyde O-(methylcarbamoyl)oxime
2,4-Dimethyl-2-formyl-1,3-dithiolane oxime methylcarbamate
2,4-Dimethyl-2-formyl-1,3-dithiolane-oxime-methylcarbamate
2,4-Dimethyl-2-formyl-1,3-dithiolanoximmethylkarbamat (Czech)
1,3-Dithiolane-2-carboxaldehyde, 2,4-dimethyl-, O-((methylamino)carbonyl)oxime
1,3-Dithiolane-2-carboxaldehyde, 2,4-dimethyl-, O-(methycarbamoyl)oxime
ENT 27,696
EPA Pesticide Chemical Code 364300
MBR 6168
3M MBR 6168

26428-41-1
$(C_7H_{12}O_2.C_4H_7NO_2.C_4H_6O_2)x$
Polymer
2-Propenoic acid, butyl ester, Polymer with ethenyl acetate and N-(hydroxymethyl)-2-propenamide (9CI)
Butyl acrylate, N-methylolacrylamide, vinyl acetate polymer
Butyl acrylate, N-methylolacrylamide, vinyl acetate polymer
Butyl acrylate, vinyl acetate, methylolacrylamide polymer
Butyl acrylate, vinyl acetate, N-methylolacrylamide polymer
Ethenyl acetate, Polymer with butyl 2-propenoate and N-(hydroxymethyl)-2-propenamide
Vinyl acetate, butyl acrylate, N-methylolacrylamide polymer

26444-19-9
$C_9H_{10}O$
134.18
Ethanone, 1-(methylphenyl)- (9CI)
1-(Methylphenyl)ethanone

26444-20-2
$C_{14}H_{14}N_2$
210.27
Diazene, bis(methylphenyl)- (9CI)
ar,ar'-Azotoluene (8CI)

26444-49-5
$C_{19}H_{17}O_4P$
340.33
Phosphoric acid, diphenyl tolyl ester
Cresol diphenyl phosphate
Cresyl diphenyl phosphate
Diphenyl cresol phosphate
Diphenyl cresyl phosphate
Diphenyl tolyl phosphate
Disflamoll DPK
Kronitex CDP
Methylphenyl diphenyl phosphate
Monocresyl diphenyl phosphate
Phosflex 112
Phosphoric acid, methylphenyl diphenyl ester (9CI)
Santicizer 140
Tolyl diphenyl phosphate

26445-01-2
$C_{12}H_{10}O_2$
186.21
Naphthaleneacetic acid (8CI, 9CI) (VAN)

26446-35-5
$C_5H_{10}O_4$
134.15
CC(=O)OCC(O)CO
Acetin, mono
Acetin

26446-38-8
$C_{28}H_{52}O_{12}$
580.71
Sucrose monopalmitate
α-D-Glucopyranoside, β-D-fructofuranosyl, monohexadecanoate (9CI)
Nitto Ester T-1570
NSC-192746
P 1570
Palmitic acid sucrose monoester
Palmitic sucrose ester
Sucrapan P
Sucrodet
Sucrose, monopalmitate (8CI)
Sucrose palmitate (VAN)
Sucrose palmitic acid ester

26446-73-1
$C_{20}H_{19}O_4P$
354.34
Phosphoric acid, bis(methylphenyl) phenyl ester (9CI)
Bis(methylphenyl) phenyl phosphate

26446-77-5
C_3H_7Br
122.99
Propane, bromo- (8CI,9CI)

26447-10-9
$C_8H_{10}O_3S.H_3N$
203.26
Ammonium xylenesulfonate
Ammonium dimethylbenzenesulfonate
Benzenesulfonic acid, dimethyl-, ammonium salt (9CI)
Witconate NXS

26447-14-3
$C_{10}H_{12}O_2$
164.22
Propane, 1,2-epoxy-3-(tolyloxy)
Cresol glycidyl ether
Cresylglycide ether
Cresyl glycidyl ether
1,2-Epoxy-3-(tolyloxy)propane
Glycidyl methylphenyl ether
((Methylphenoxy)methyl)-oxirane
Oxirane ((methylphenoxy)-methyl)- (9CI)
Tolyl glycidyl ether

26447-28-9
$C_5H_4O_3$
112.09
OC(=O)c1ccco1
Furoic-acid

26447-40-5
$C_{15}H_{10}N_2O_2$
250.24
Benzene, 1,1'-methylenebis-(isocyanato- (9CI)
1,1'-Methylenebis(isocyanatobenzene)

26447-45-0
$C_{13}H_{18}O_2$
206.28
Oxirane, (((1,1-dimethylethyl)phenoxy)methyl)- (9CI)
tert-Butylphenol glycidyl ether
tert-Butylphenyl 2,3-epoxypropyl ether

tert-Butylphenyl glycidyl ether
Propane, 1-(tert-butylphenoxy)-2,3-epoxy- (7CI,8CI)

26447-63-2
$C_{13}H_{18}$
174.29
Benzene, 2-heptenyl- (9CI)
2-Heptene, 1-phenyl- (8CI)

26452-80-2
$C_5H_3Cl_2N$
147.98
Pyridine, 2,4-dichloro- (8CI, 9CI)

26468-86-0
$(C_2H_4O)xC_8H_{18}O$
Polymer
Poly(oxy-1,2-ethanediyl), α-(2-ethylhexyl)-ω-hydroxy-(9CI)
Polyethylene glycol mono-(2-ethylhexyl) ether

26471-56-7
$C_6H_5N_3O_4$
183.14
Aniline, dinitro- (mixed isomers)
Benzenamine, ar,ar-dinitro-(9CI)
Dinitroanilines [UN 1596]
UN 1596 [Dinitroanilines]

26471-62-5
$C_9H_6N_2O_2$
174.17
Cc1ccc(N=C=O)cc1N=C=O.Cc1c(N=C=O)cccc1N=C=O
Isocyanic acid, methyl-m-phenylene ester
Benzene-, 1,3-diisocyanato-methyl-
Desmodur T100
Diisocyanatomethylbenzene
Diisocyanatotoluene
Hylene-T
Isocyanic acid, methylphenylene ester
Methyl-meta-phenylene diisocyanate
Methylphenylene isocyanate
Mondur-TD
Mondur-TD-80
Nacconate-100
Niax Isocyanate TDI
RCRA waste number U223
Rubinate TDI
Rubinate TDI 80/20
T 100
TDI
TDI-80
TDI 80-20
Toluene diisocyanate [UN 2078]
Tolylene diisocyanate
Tolylene isocyanate
UN 2078 [Toluene diisocyanate]

26472-00-4
$C_{12}H_{16}$
160.28
4,7-Methanoindene, 3a,4,7,7a-tetrahydrodimethyl
Bis(methylcyclopentadiene)
4,7-Methano-1H-indene, 3a,4,7,7a-tetrahydrodimethyl-(9CI)
Methylcyclopentadiene dimer
3a,4,7,7a-Tetrahydrodimethyl-4,7-methanoindene

26489-01-0
$C_{10}H_{20}O$
156.27
OCCC(CCC=C(C)C)C
6-Octen-1-ol, 3,7-dimethyl-, (+-)- (9CI)
(+-)-3,7-Dimethyl-6-octen-1-ol

26519-91-5
C_6H_8
80.14
1,3-Cyclopentadiene, methyl
Methylcyclopentadiene
Methyl-1,3-cyclopentadiene

26523-64-8
$C_2Cl_3F_3$
187.38
Ethane, trichlorotrifluoro- (8CI,9CI) (VAN)

26523-78-4
$C_{45}H_{69}O_3P$
689.01
Tris(nonylphenyl)phosphite
Nonylphenol phosphite (3:1)
Phenol, nonyl-, phosphite (3:1) (9CI)
Stabilizer Mark 1178

26530-20-1
$C_{11}H_{19}NOS$
213.37
O=C(N(SC=1)CCCCCCCC)C1
4-Isothiazolin-3-one, 2-octyl
3(2H)-Isothiazolone, 2-octyl-(9CI)
Kathon LP Preservative
Kathon SP 70
Micro-Chek 11
Micro-Chek 11D
Micro-Chek Skane
Octhilinone
2-Octyl-4-isothiazolin-3-one
Pancil
Pancil-T
RH 893
Skane M8

26532-24-1
$C_{10}H_{16}O$
152.24
O=CC=C(CCCC1(C)C)C1
Acetaldehyde, (3,3-dimethyl-cyclohexylidene)-, (Z)- (9CI)
cis-3,3-Dimethylcyclohexyl-ideneethanal

26535-50-2
$C_{21}H_{43}NO_5S.Na$
444.63
Octadecanamide, N-methyl-N-(2-(sulfooxy)ethyl)-, sodium salt (9CI)
Octadecanamide, N-(2-hydroxyethyl)-N-methyl-, hydrogen sulfate (ester), sodium salt (8CI)

26537-71-3
$C_{18}H_{30}O$
262.44
9,12,15-Octadecatrienal (8CI, 9CI)

26537-89-3
$C_{12}H_{27}O_2PS_2.Na$
298.45
Sodium dihexyl phosphorodithioate
Phosphorodithioic acid, O,O-dihexyl ester, sodium salt

26538-44-3
$C_{18}H_{26}O_5$
322.44
O=C(OC(CCCC(O)CCCCCc1cc(O)cc2O)C)c12
1H-2-Benzoxacyclotetradecin-1-one, 3,4,5,6,7,8,9,10,11,12-decahydro-7,14,16-trihydroxy-3- methyl-, (3s,7x)
6-(6,10-Dihydroxyundecyl)-β-resorcylic acid, mu-lactone
Frideron
MK-188
P1496
Ralabol
Ralgro
Ralone
Zearalanol
Zearanol
Zeranol

26543-36-2
$C_{19}H_{34}O_2$
294.48
10-Octadecynoic acid, methyl ester (8CI,9CI)
Methyl 10-octadecynoate

26544-17-2
$C_{25}H_{48}O_4$

412.65
Nonanedioic acid, diisooctyl ester (9CI)
Diisooctyl nonanedioate

26544-20-7
$C_{17}H_{25}ClO_3$
312.87
Acetic acid, ((4-chloro-o-tolyl)-oxy)-, isooctyl ester
Acetic acid, (4-chloro-2-methylphenoxy)-, isooctyl ester (9CI)
((4-Chloro-o-tolyl)oxy)acetic acid isooctyl ester
MCPA-Isooctyl

26544-23-0
$C_{22}H_{31}O_3P$
374.46
Phosphorous acid, isodecyl diphenyl ester (9CI)

26544-27-4
$C_{25}H_{50}O_6P_2$
508.62
2,4,8,10-Tetraoxa-3,9-diphosphaspiro(5.5)undecane, 3,9-bis(isodecyloxy)- (9CI)

26544-38-7
$C_{16}H_{26}O_3$
266.42
Succinic anhydride, (tetrapropenyl)
2,5-Furandione, dihydro-3-(tetrapropenyl)- (9CI)
RD 174
Tetrapropenylsuccinic anhydride

26545-53-9
$C_{18}H_{30}O_3S.C_4H_{11}NO_2$
431.63
DEA-Dodecylbenzenesulfonate
Benzenesulfonic acid, dodecyl-, Compd. with 2,2'-iminobis-(ethanol) (1:1) (9CI)
Caswell No. 413D
Diethanolamine dodecylbenzene sulfonate
Diethanolamine dodecylbenzenesulfonate
Dodecylbenzenesulfonic acid diethanolamine salt
EPA Pesticide Chemical Code 079015

26545-58-4
$C_{21}H_{16}O_6S_2.2Na$
474.47
Naphthalenesulfonic acid, methylenebis-, disodium salt (9CI)
Naphthalenesulfonic acid, methylenedi-, disodium salt

26545-73-3
$C_3H_6Cl_2O$
128.99
Propanol, dichloro- (9CI)
Dichloropropanol

26566-95-0
$C_{24}H_{52}O_4P_2S_4Zn$
660.28
Zinc, bis(O-(2-ethylhexyl) O-(2-methylpropyl) phosphorodithioato-S,S')-, (T-4)- (9CI)

26569-53-9
$C_{45}H_{69}O_4P$
705.01
Phenol, nonyl-, phosphate (3:1) (9CI)
Nonylphenol phosphate (3:1)

26570-85-4
$C_{12}H_{16}O$
176.26
Phenol, cyclohexyl- (9CI)
Cyclohexylphenol

26571-49-3
$(C_3H_6O)xC_{36}H_{66}O_3$
Polymer
PPG-17 Dioleate
Poly(oxy(methyl-1,2-ethanediyl)), α-(1-oxo-9-octadecenyl)-ω-((1-oxo-9-octadecenyl)oxy)-, (Z,Z)- (9CI)
Polyoxypropylene (17) dioleate
Polypropylene glycol dioleate
Polypropylene glycol (17) dioleate

26571-79-9
$C_6H_4Cl_4Si$
245.99
Chlorophenyltrichlorosilane
Clorofeniltriclorosilano (Spanish)
Silane, chlorophenyltrichloro-
UN 1753 [Chlorophenyltrichlorosilane]

26577-87-7
$C_{21}H_{43}NO_5S.Na$
444.63
Octadecanamide, N-(2-(sulfooxy)propyl)-, monosodium salt (9CI)
Octadecanamide, N-(2-hydroxypropyl)-, hydrogen sulfate (ester), monosodium salt (8CI)

26601-64-9
$C_{12}H_4Cl_6$
360.88
1,1'-Biphenyl, hexachloro- (9CI)
Biphenyl, hexachloro- (8CI)
Hexachlorobiphenyl

26603-40-7
$C_{27}H_{18}N_6O_6$
522.44
1,3,5-Triazine-2,4,6(1H,3H,5H)-trione, 1,3,5-tris(3-isocyanatomethylphenyl)- (9CI)

26604-41-1
$C_{13}H_{20}N_4O_2$
264.30
Urea, N,N''-(methyl-1,3-phenylene)bis(N',N'-dimethyl- (9CI)
Urea, 1,1'-(methyl-m-phenylene)bis(3,3-dimethyl-

26604-51-3
$C_9H_{15}Cl_6O_4P$
430.91
1-Propanol, dichloro-, phosphate (3:1) (8CI,9CI)

26616-35-3
$C_{12}H_{18}N.Cl$
211.73
Benzenemethanaminium, ar-ethenyl-N,N,N-trimethyl-, chloride (9CI)
Trimethyl(vinylbenzyl)ammonium chloride

26617-87-8
Unknown
Unknown
Prepodyne
Poloxamer-iodone
Polyethoxypolypropoxyethanol-iodine complex

26628-22-8
N_3Na
64.99
[Na+].[N-]=N#N
Sodium azide (ACGIH,OSHA) [UN 1687] (9CI)
AI3-50436
Azida sodica (Spanish)
Azide
Azium
Azoture de sodium (French)
Caswell No. 744A
EPA Pesticide Chemical Code 107701
Hydrazoic acid, sodium salt.
Kazoe
Natriumazid
Natriummazide (Dutch)
NCI-C06462
NSC-3072
RCRA waste number P105
Smite
Sodium, azoture de (French)

Sodium, azoturo di (Italian)
U-3886
UN 1687 [Sodium azide]

26635-64-3
C_8H_{18}
114.26
CCCCCC(C)C
Isooctane [UN 1261]
UN 1262 [Octanes]

26635-92-7
$(C_2H_4O)x(C_2H_4O)xC_{22}H_{47}NO_2$
Polymer
N-Polyoxyethylated-N-octadecylamine
Ethoxylated stearylamine
Poly(oxy-1,2-ethanediyl), α,α'-((octadecylimino)di-2,1-ethanediyl)bis(ω-hydroxy- (9CI)
Stearylamine, ethoxylated

26635-93-8
$(C_2H_4O)x(C_2H_4O)xC_{22}H_{45}NO_2$
Polymer
N-Polyoxyethylated-N-oleylamine hydrochloride
Ethoxylated oleylamine
α,α'-((9-Octadecenylimino)di-2,1-ethanediyl)bis(ω-hydroxy-poly- (oxy-1,2-ethanediyl)-, (Z)-
Oleylamine, ethoxylated
Poly(oxy-1,2-ethanediyl), α,α'-((9-octadecenylimino)di-2,1-ethanediyl)bis(ω-hydroxy-, (Z)- (9CI)

26636-01-1
$C_{22}H_{44}O_4S_2Sn$
555.47
Stannane, bis(isooctyloxycarbonylmethylthio)dimethyl
Dimethylzinn-S,S'-bis(isooctylthioglycolat) (German)
Tin, dimethyl-, bis(isooctylthioglycollate)

26637-46-7
$(C_{14}H_8O_4)x$
Polymer NA
Poly(oxy-1,3-phenyleneoxycarbonyl-1,3-phenylenecarbonyl) (9CI)
Isophthalic acid-resorcinol copolymer SRU
Isophthalic acid-resorcinol polymer, SRU
Isophthaloyl chloride-resorcinol polymer, SRU
Polyarylate R 1
Poly(oxy-m-phenyleneoxyisophthaloyl) (8CI)
Poly(1,3-phenylene isophthalate)
R 1
R 1 (Polyester)

26638-01-7
$C_{10}H_{10}O_4$
194.19
Benzenedicarboxylic acid, dimethyl ester (8CI,9CI)

26638-19-7
$C_3H_6Cl_2$
112.99
Propane, dichloro
Dichloropropane (DOT)
Dichlorpropan
Propylene dichloride [UN 1279]
UN 1279 [Propylene dichloride]

26638-28-8
$C_{19}H_{33}Cl_5O_2$
470.73
Octadecanoic acid, pentachloro-, methyl ester (9CI)

26638-43-7
$C_8H_7ClO_4S$
234.66
Benzoic acid, 2-(chlorosulfonyl)-, methyl ester (9CI)
Methyl 2-(chlorosulfonyl)benzoate

26644-46-2
$C_{10}H_{14}Cl_6N_4O_2$
434.98
ClC(Cl)(Cl)C(NC=O)N1CCN(CC1)C(NC=O)C(Cl)(Cl)Cl
Piperazine, 1,4-bis(1-formamido-2,2,2-trichloroethyl)
Biformychlorazin
Biformylchlorazin
N,N'-Bis(1-formamido-2,2,2-trichloroethyl)piperazine
1,4-Bis(1-formamido-2,2,2-trichloroethyl)piperazine
CA 70203
Cela 50
Cela W 524
CME 74770
Compound W
CW 524
Formamide, N,N'-(1,4-piperazinediylbis(2,2,2-trichloroethylidene))bis- (8CI,9CI)
Funginex
N,N'-(Piperazinediylbis(2,2,2-trichloroethylidene)) bis-(formamide)
Saprol
Triforine
W 524

26645-10-3
$C_4H_{10}AuBr$
334.99
Diethylgold bromide
Bromodiethylgold
Gold, bromodiethyl-

26650-75-9
$C_6H_7Cl_2N_3O$
208.03
1,3,5-Triazine, 2,4-dichloro-6-propoxy- (9CI)
s-Triazine, 2,4-dichloro-6-propoxy- (8CI)

26650-76-0
$C_6H_5Cl_2N_3O$
206.01
1,3,5-Triazine, 2,4-dichloro-6-(2-propenyloxy)- (9CI)
s-Triazine, 2-(allyloxy)-4,6-dichloro- (8CI)

26651-96-7
$C_{14}H_{22}O$
206.36
O=C(C=CC=C(CCC=C(C)C)C)CC
4,6,10-Dodecatrien-3-one, 7,11-dimethyl
2,6-Dimethyldodeca-2,6,8-trien-10-one
7,11-Dimethyl-4,6,10-dodecatrien-3-one
Pseudomethylionone

26655-49-2
$C_8H_{17}Cl$
446.03
Octane, chloro- (8CI,9CI)
sec-Octyl chloride

26658-09-3
$C_{16}H_{36}O_7P_2$
402.40
Diphosphoric acid, dioctyl ester
AI3-15051
Dioctyl acid pyrophosphate
Dioctyl diphosphorate
Dioctyl pyrophosphate

26658-19-5
$C_{60}H_{114}O_8$
963.56
Sorbitan tristearate
Anhydrosorbitol tristearate
Hefti TS-33-F
Sorbax STS
Sorbitan, esters, trioctadecanoate
Sorbitan, trioctadecanoate (9CI)
Sorbitan tristearate. (Compound usually contains also associated fatty acids.)
Sorbitani tristearas (Latin)
Span 65
Triestearato de sorbitano (Spanish)
Tristearate de sorbitan (French)

26672-24-2
$C_{10}H_{15}NO_8S_2$
341.36
O=S(=O)(OCCS(=O)(=O)c(c(OC)cc(N)c1OC)c1)O
Ethanol, 2-((4-amino-2,5-dimethoxyphenyl)sulfonyl)-, hydrogen sulfate (ester) (9CI)

26675-46-7
$C_3H_2ClF_5O$
184.50
FC(F)OC(Cl)C(F)(F)F
Ether, 1-chloro-2,2,2-trifluoroethyl difluoromethyl
Ethane, 2-chloro-2-(difluoromethoxy)-1,1,1-trifluoro- (9CI)
Forane
Isoflurane

26680-54-6
$C_{12}H_{18}O_3$
210.27
2,5-Furandione, dihydro-3-(octenyl)- (9CI)
Dihydro-3-(octenyl)-2,5-furandione

26719-40-4
$C_{35}H_{68}O_4$
552.92
O=C(OCCCCCCCCCCCCC)CCCCCCCC(=O)OCCCCCCCCCCCCC
Nonanedioic acid, ditridecyl ester (9CI)
Ditridecyl azelate
Ditridecyl nonanedioate

26720-21-8
$C_{38}H_{74}O_4$
595.00
O=C(OCCCCCCCCCCCCCCCCC)CCCCC(=O)OCCCCCCCCCCCCCCCC
Dicetyl adipate
Di-n-hexadecyl adipate
Dihexadecyl hexanedioate
Dipalmityl adipate
Hexanedioic acid, dihexadecyl ester (9CI)

26730-14-3
$C_{14}H_{30}$
198.40
Tridecane, 7-methyl- (6CI, 8CI,9CI)
7-Methyltridecane

26741-53-7
$C_{33}H_{50}O_6P_2$
604.77
O(P(OCC1(COP(O2)Oc(c(cc(c3)C(C)(C)C)C(C)(C)C)c3)C2)Oc(c(cc(c4)C(C)(C)C)C(C)(C)C)c4)C1
Phosphorous acid, cyclic neopentanetetrayl bis(2,4-di-tert-butylphenyl)ester
2,4,8,10-Tetraoxa-3,9-diphosphaspiro(5.5)undecane, 3,9-bis(2,4-bis(1,1-dimethylethyl)phenoxy)-
Ultranox 624
Ultranox 626
Weston 626
Weston MDW 626

26745-88-0
$(C_{10}H_{18}O_4.C_6H_{14}O_2)x$
Polymer
Decanedioic acid, Polymer with 1,6-hexanediol (9CI)
Sebacic acid, 1,6-hexanediol polymer

26746-29-2
$C_{10}H_{14}O_3S.Na$
237.28
Benzenesulfonic acid, butyl-, sodium salt (9CI)

26747-90-0
$C_{18}H_{12}N_4O_4$
348.29
1,3-Diazetidine-2,4-dione, 1,3-bis(3-isocyanatomethylphenyl)- (9CI)
Toluene diisocyanate dimer

26747-91-1
$C_{13}H_{14}N.Cl$
219.71
Pyridinium, methyl-1-(phenylmethyl)-, chloride (9CI)
N-Benzylpicolinonium chloride
Methyl-1-(phenylmethyl) pyridinium chloride

26748-41-4
$C_{14}H_{28}O_3$
244.42
Peroxyneodecanoic acid, tert-butyl ester
tert-Butyl perneodecanoate
tert-Butyl peroxyneodecanoate, Not more than 77% in solution (DOT)
tert-Butyl peroxyneodecanoate, Technically pure (DOT)
Esperox
Esperox 33M
Lupersol 10
Lupersol 1OM75
Peroxyneodecanoic acid, tert-butyl ester, Not more than 77% in solution
Trigonox 23-C75
UN 2177 (DOT)
UN 2594 (DOT)

26748-47-0
$C_{19}H_{30}O_3$
306.45
Neodecaneperoxoic acid, 1-methyl-1-phenylethyl ester (9CI)
α-Cumyl peroxyneodecanoate

26750-44-7
$C_4H_{10}S_2$
122.26
Ethanethiol, 2-(ethylthio)- (8CI,9CI)

26750-50-5
$C_6H_{10}O_5S_2$
226.27
O=S(=O)(C=C)COCS(=O)(=O)C=C
Ethene, 1,1'-(oxybis(methylenesulfonyl))bis- (9CI)
1,1'-(Oxybis(methylenesulfonyl))bisethene

26760-64-5
C_5H_{10}
70.15
Butene, 2-methyl
tert-Amylene
Isoamylene
Isopentene [UN 2371]
Methylbutene
2-Methylbutene
2-Methyl-2-butene [UN 2460]
UN 2371 [Isopentenes]
UN 2460 [2-Methyl-2-butene]

26761-40-0
$C_{28}H_{46}O_4$
446.74
Phthalic acid, diisodecyl ester
1,2-Benzenedicarboxylic acid, diisodecyl ester (9CI)
Bis(isodecyl)phthalate
DIDP (Plasticizer)
Diisodecyl phthalate
Palatinol Z
Sicol 184
Vestinol DZ

26761-45-5
$C_{13}H_{24}O_3$
228.37
Neodecanoic acid, 2,3-epoxypropyl ester
Glycidyl neodecanoate
Neodecanoic acid, oxiranylmethyl ester (9CI)

26761-50-2
$C_{26}H_{50}O_2$
394.68
9-Octadecenoic acid (Z)-, isooctyl ester (9CI)

Isooctyl 9-octadecenoate
$C_{18}H_{34}O_2$
282.47
Octadecenoic acid (VAN)(9CI)

26761-78-4
$C_{14}H_{16}O_3S$
264.35
Naphthalenesulfonic acid, butyl- (9CI)
Butylnaphthalenesulfonic acid

26762-44-7
$C_{18}H_{38}O$
270.50
Octadecanol (8CI,9CI)

26762-92-5
$C_{10}H_{20}O_2$
172.27
Cyclohexane, 1-methyl-4-(1-methylethyl)-, mono-hydroperoxy deriv. (9CI)

26762-93-6
$C_{12}H_{19}O_2$
195.31
Hydroperoxide, bis(1-methyl-ethyl)phenyl (9CI)
Diisopropylbenzene hydroper-oxide (DOT)
Hydroperoxide, diisopropyl-phenyl (8CI)
UN 2171 (DOT)

26763-63-3
$C_{13}H_{12}N_2O$
212.24
Urea, diphenyl- (8CI,9CI)

26763-69-9
$C_{34}H_{17}NO_2$
471.51
Anthra(9,1,2-cde)benzo(rst)-pentaphene-5,10-dione, amino- (9CI)
Aminodibenzanthrone
Violanthrone, amino-

26764-26-1

26766-27-8
$C_{18}H_{13}Cl_2N_2O$
344.23
OC(c1ccccc1)(c2cncnc2)c3ccc(Cl)cc3Cl
5-Pyrimidinemethanol, α-(2,4-dichlorophenyl)-α-phenyl
α-(2,4-Dichlorophenyl)-α-phenyl-5-pyrimidinemethanol
EL-273
Triarimol
Trimidal

26780-96-1
$(C_{11}H_{15}N)x$
Polymer
Poly(1,2-dihydro-2,2,4-tri-methylquinoline)
Acetonanil
Agerite MA
Antigene RDF
Antioxidant HS
Antioxidant HSL
Flectol H, Polymer
Nocrac 224
Nonflex RD
Permanax TQ
Permanax 45
Polnoks R
Quinoline, 1,2-dihydro-2,2,4-tri-methyl-, Homopolymer
Trimethyldihydroquinoline polymer
2,2,4-Trimethyl-1,2-dihydro-quinoline polymer

26782-43-4
$C_{19}H_{27}NO_7$
381.47
Senecionanium, 8,12,18-tri-hydroxy-4-methyl-11,16-dioxo
Hydroxysenkirkine
Senkirkine, hydroxy-

26787-78-0
$C_{16}H_{19}N_3O_5S$
365.44
4-Thia-1-azabicyclo(3.2.0)-heptane-2-carboxylic acid, 6-((amino(4-hydroxyphenyl)-acetyl) amino)-3,3-dimethyl-7-oxo
α-Amino-p-hydroxybenzylpeni-cillin
Amoxicillin
Amoxil
6-(D-(-)-α-Amino-p-hydroxy-phenylacetamido)penicillanic acid
AMPC
Amoxycillin
Amolin
Amopenixin
Amoxi
Amoxipen
Anemolin
BLP 1410
BRL 2333
Bristamox
Clamoxyl
Delacillin
Efpenix
Histocillin
6-(p-Hydroxy-α-aminophenyl-acetamido)penicillanic acid
p-Hydroxyampicillin
Ibiamox
Piramox
Sumox

26834-28-6
$C_{28}H_{44}O_3S.Na$
483.71
Naphthalenesulfonic acid, di-nonyl-, sodium salt (9CI)
Dinonylnaphthalenesulfonic acid, sodium salt

26836-07-7
$C_{18}H_{30}O_3S.C_2H_7NO$
387.58
Ethanolamine dodecyl-benzenesulfonate
Benzenesulfonic acid, dodecyl-, 2-aminoethanol salt (1:1)
Benzenesulfonic acid, dodecyl-, Compd. with 2-aminoethanol (1:1) (8CI,9CI)
Dodecylbenzene sulfonic acid, monoethanolamine salt
Dodecylbenzenesulfonic acid, 2-aminoethanol salt
Dodecylbenzenesulfonic acid, Compound with 2-amino-ethanol (1:1)
Dodecylbenzenesulfonic acid, ethanolamine salt
Dodecylbenzenesulfonic acid monoethanolamine salt
Dodecylbenzenesulfonic acid, monoethanolamine salt
Monoethanolamine dodecyl-benzenesulfonate
Monoethanolammonium do-decylbenzenesulfonate

26836-28-2
$C_{16}H_{36}O_7P_2$
402.40
Diphosphoric acid, bis(2-ethyl-hexyl) ester (9CI)
Bis(2-ethylhexyl) diphosphorate
Pyrophosphoric acid, bis(2-ethylhexyl) ester

26838-05-1
$C_{16}H_{30}O_7S.2Na$
412.45
Butanedioic acid, sulfo-, C-dodecyl ester, di-sodium salt (9CI)
Butanedioic acid, ((dodecyl-oxy)sulfonyl)-, disodium salt
Disodium lauryl sulfo-succinate
Dodecyl disodium sulfo-succinate
Lauryl disodium sulfo-succinate
Succinic acid, sulfo-, do-decyl ester, disodium salt (6CI)
Succinic acid, sulfo-, mono-dodecyl ester, disodium salt (8CI)
Sulfosuccinic acid mono-dodecyl ester disodium

salt
Sulfosuccinic acid monolauryl ester disodium salt
Steinapol SBF 12

26839-75-8
$C_{13}H_{24}N_4O_3S$
316.40
O(CCN(C(=NSN=1)C1OCC(O)CNC(C)(C)C)C2)C2
Timolol
2-Propanol, 1-((1,1-dimethylethyl)amino)-3-((4-(4-morpholinyl)-1,2,5-thiadiazol-3-yl)oxy)-, (S)- (9CI)
Timololum (Latin)

26850-24-8
$C_9H_{16}N_2O_4$
216.22
O=C(N(C(C1=O)(C)C)CCO)N1CCO
DEDM hydantoin
1,3-Bis(2-hydroxyethyl)-5,5-dimethyl-2,4-imidazolidinedione
Diethylol dimethyl hydantoin
1,3-Di(hydroxyethyl)-5,5-dimethylhydantoin
2,4-Imidazolidinedione, 1,3-bis-(2-hydroxyethyl)-5,5-dimethyl-(9CI)

26854-10-4
$C_{44}H_{46}CuN_{11}.3Cl$
898.82
Copper(3+), (N,N,N,N',N',N',N'',N'',N''-nonamethyl-29H,31H-phthalocyaninetrimethanaminiumato(2-)-N29,N30,N31,N32)-, trichloride (9CI)

26856-30-4
C_6H_{10}
82.15
Hexyne (9CI)

26856-61-1
Unknown
Sodium nonylbenzenesulfonate

26872-84-4
$C_9H_{17}NO_2$
171.23
β-Alanine, N-cyclohexyl- (8CI, 9CI)
NSC-62840

26886-05-5
$C_{12}H_{18}O$
178.27
Phenol, 3,5-bis(1-methylethyl)-(9CI)
Phenol, 3,5-diisopropyl- (8CI)

26896-18-4
$C_9H_{18}O_2$
158.24
Isononanoic acid (VAN)(9CI)

26896-20-8
$C_{10}H_{20}O_2$
172.30
Neodecanoic-acid
Wiltz-65

26896-48-0
$C_{12}H_{20}O_2$
196.29
Tricyclodecanedimethanol (9CI)

26897-24-5
$C_8H_{10}O$
122.17
Benzene, methoxymethyl- (9CI)
Anisole, methyl- (8CI)

26898-17-9
$C_{21}H_{20}$
272.39
Dibenzyltoluene
Benzene, methylbis(phenylmethyl)- (9CI)
Dibenzylmethylbenzene
Methylbis(phenylmethyl)-benzene

26903-07-1
Cl_3Sn
225.07
Tin chloride (9CI)

26914-02-3
C_3H_7I
170.00
Propane, iodo
Iodopropane
Iodopropanes [UN 2392]
UN 2392 [Iodopropanes]

26914-17-0
$C_{14}H_{12}$
180.25
9H-Fluorene, methyl- (9CI)
Fluorene, methyl-

26914-18-1
$C_{15}H_{12}$
192.26
Anthracene, methyl- (8CI,9CI)

26914-33-0
$C_{12}H_6Cl_4$
291.99
1,1'-Biphenyl, tetrachloro-(9CI)
Biphenyl, tetrachloro- (8CI)
Pyralene 1498

26914-40-9
$C_2H_6S_2$
94.20
Ethanedithiol (VAN)(9CI)

26915-12-8
C_7H_9N
107.15
Benzenamine, ar-methyl-(9CI)
ar-Methylbenzenamine
Toluidina (Spanish)
Toluidine

26919-50-6
$C_{13}H_{19}NO_4S.C_6H_{15}NO_3$
434.54
Hexanoic acid, 6-(methyl-(phenylsulfonyl)amino)-, Compd. with 2,2',2''-nitrilotris(ethanol) (1:1) (9CI)

26952-13-6
$C_{14}H_{28}$
196.38
Tetradecene
n-Tetradecene
Tetradecylene

26952-20-5
$C_{14}H_{19}Cl_3N_2O_2$
353.68
Isooctyl picloram
Picolinic acid, 4-amino-3,5,6-trichloro-, isooctyl ester (8CI)
2-Pyridinecarboxylic acid, 4-amino-3,5,6-trichloro-, isooctyl ester (9CI)
Tordon 3220

26952-21-6
$C_8H_{18}O$
130.26
CC(C)CCCCCO
Isooctyl-alcohol
Isooctanol
Isooctyl alcohol (ACGIH, OSHA)

26952-23-8
$C_3H_4Cl_2$
110.97
1-Propene, dichloro
Dichloropropene [UN 2047]
Dichloropropylene
UN 2047 [Dichloropropene]

26952-42-1
$C_6H_4N_4O_6$
228.14
Aniline, trinitro
Benzenamine, trinitro- (9CI)
Trinitroaniline [UN 0153]
UN 0153 [Trinitroaniline or Picramide]

26952-44-3
$C_3H_3ClO_2$
106.51
2-Propenoic acid, chloro- (9CI)
Acrylic acid, chloro- (8CI)

26967-65-7
$C_{10}H_{14}O$
150.22
Phenol, diethyl- (8CI,9CI)

26967-76-0
$C_{27}H_{33}O_4P$
452.53
Phenol, (1-methylethyl)-, phosphate (3:1) (9CI)

26968-58-1
$C_9H_{11}Cl$
154.64
Benzene, (chloromethyl)ethyl- (9CI)
(Chloromethyl)ethylbenzene

26978-65-4
$C_2H_5BrO_3S$
189.03
2-Bromoethanesulfonic acid

26983-51-7
$C_4H_{10}HgO_2$
290.71
Mercury, (ethoxyethyl)hydroxy- (8CI,9CI)

26983-52-8
$C_{12}H_{10}O_2$
186.21
(1,1'-Biphenyl)-ar,ar'-diol (9CI)
Biphenol
ar,ar'-Biphenyldiol (8CI)

26997-02-4
$C_{13}H_{20}O$
192.30
Phenol, heptyl- (8CI,9CI)

26998-80-1
$C_9H_{12}O$
136.19
Phenol, trimethyl- (9CI)
Trimethylphenol

26998-97-0
$C_{36}H_{58}O_2S.Ca$
595.00
Phenol, thiobis(dodecyl-, calcium salt (1:1) (9CI)
Calcium salt of thiobis(C12-alkylated phenol)
Thiobis(dodecylphenol) calcium salt (1:1)

26999-29-1
$C_{16}H_{35}O_2PS_2$
354.56
Phosphorodithioic acid, O,O-diisooctyl ester (9CI)
O,O-Diisooctyl phosphorodithioate
O,O'-Diisooctyl phosphorodithioic acid
O,O'-Diisoctylphosphorodithioic acid

27011-46-7
$C_{11}H_{17}Cl$
184.71
Tricyclo(4.3.1.13,8)undecane, 1-chloro- (8CI)

27020-65-1
$Cr_2H_2O_7.2K$
296.20
Lopezite ($K_2(Cr_2O_7)$) (9CI)
Lopezite (8CI)

27032-78-6
$C_3H_5N_3O_2$
115.07
Dioxohexahydrotriazine
Dihydro-1,3,5-triazine-2,4-(1H,3H)-dione
NSC-119749

27043-36-3
$C_9H_{16}O_5$
204.22
Propanol, bis(oxiranylmethoxy)- (9CI)
Propanol, bis(2,3-epoxypropoxy)- (8CI)

27043-37-4
$C_{38}H_{38}O_8$
622.71
Oxirane, 2,2',2'',2'''-(1,2-ethanediylidenetetrakis-(phenyleneoxymethylene))-tetrakis- (9CI)
Ethane, 1,1,2,2-tetrakis-((2,3-epoxypropoxy)phenyl)- (8CI)

27046-19-1
$C_7H_{13}Cl_2N_2O_3P$
275.09
4H-1,3,2-Oxazophosphorin-4-one, 2-(bis(2-chloroethyl)-amino)tetrahydro-, 2-oxide
4-Ketocyclophosphamide
4-Oxocyclophosphamide

27059-08-1
$C_{16}H_{22}N_2O_5$
322.35
O=C(OCCN(c(cccc1NC(=O)C)c1)CCOC(=O)C)C
Acetamide, N-(3-(bis(2-(acetyloxy)ethyl)amino)phenyl)- (9CI)
Acetanilide, 3-(N,N-di-β-acetoxyethyl)amino-
3-(N,N-Bisacetoxyethyl)aminoacetanilide

27070-58-2
$C_{18}H_{36}$
252.48
Octadecene (9CI)
Linear octadecene

27070-59-3
$C_{12}H_{18}$
162.27
Cyclododecatriene (9CI)

27073-01-4
$C_{26}H_{55}O_4P$
462.69
Phosphoric acid, diisotridecyl ester (9CI)
Diisotridecyl phosphate

27076-30-8
$C_5H_{12}N_2O_2$
132.15
O=C(N)CCNCCO
Propionamide, 3-((2-hydroxyethyl)amino)- (8CI)
3-((2-Hydroxyethyl)amino)propionamide

27090-63-7
$C_{22}H_{48}N_2$
340.62
N(CCCCCCN(CCCC)CCCC)(CCCC)CCCC
1,6-Hexanediamine, N,N,N',N'-tetrabutyl- (9CI)
N,N,N',N'-Tetrabutyl-1,6-hexanediamine

27096-29-3
$C_6H_{15}NO_3.HNO_3$
212.19
Ethanol, 2,2',2''-nitrilotris-, nitrate (Salt) (9CI)
Ethanol, 2,2',2''-nitrilotri-, nitrate (Salt) (8CI)

27103-90-8
$(C_9H_{18}NO_2.CH_3O_4S)x$
Polymer
Polyquaternium-14
Ethanaminium, N,N,N-trimethyl-2-((2-methyl-1-oxo-2-propenyl)oxy)-, methyl sulfate, Homopolymer (9CI)

27119-07-9
$(C_7H_{13}NO_4S)x$
Polymer
1-Propanesulfonic acid, 2-methyl-2-((1-oxo-2-propenyl)amino)-, Homopolymer (9CI)

27121-30-8
$C_{32}H_{13}Cl_3CuN_8O_6S_3$
871.56
Copper, (29H,31H-phthalocyaninetrisulfonyl trichloridato(2-)-N29,N30,N31,N32)-(9CI)
Phthalocyaninetrisulfonyl chloride copper(II) complex

27125-68-4
$C_{18}H_{10}Cl_5N_3$
445.57
4-(3,4-Dichloroanilino)-3,3',4'-trichloroazobenzene

27129-87-9
$C_9H_{12}O$
136.19
Benzenemethanol, 3,5-dimethyl-(9CI)
Benzyl alcohol, 3,5-dimethyl-(8CI)

27133-93-3
$C_{10}H_{12}$
132.21
1H-Indene, 2,3-dihydromethyl-(9CI)
Indan, methyl- (8CI)

27134-26-5
C_6H_6ClN
127.57
Benzenamine, chloro- (9CI)
NSC-174207

27134-27-6
$C_6H_5Cl_2N$
162.02
Aniline, dichloro- (mixed isomers)
Benzenamine, ar,ar-dichloro-(9CI)
o-Dichloroaniline
Dichloroanilines [UN 1590]
UN 1590 [Dichloroanilines, solid or liquid]

27136-73-8
$C_{22}H_{42}N_2O$
350.58
1-(2-Hydroxyethyl)-2-heptadecenyl-2-imidazoline
AI3-01744
1H-Imidazole-1-ethanol, 2-(heptadecenyl)-4,5-dihydro-

27137-85-5
$C_6H_3Cl_5Si$
280.43
Silane, (dichlorophenyl)trichloro
Dichlorophenyltrichlorosilane [UN 1766]
Trichloro(dichlorophenyl)silane
UN 1766 [Dichlorophenyltrichlorosilane]

27138-19-8
$C_{12}H_{12}$
156.23
Naphthalene, ethyl- (8CI,9CI)

27138-21-2
$C_{11}H_{16}$
148.25
Benzene, (1,1-dimethylethyl)-methyl- (9CI)
Toluene, tert-butyl- (8CI)

27138-31-4
$C_{20}H_{22}O_5$
342.39
Dipropylene glycol dibenzoate
Oxybispropanol dibenzoate
Propanol, oxybis-, dibenzoate (9CI)

27152-57-4
$AsH_3O_3.3/2Ca$
186.06
Arsenic compound
AI3-01165
Arsenious acid, calcium salt
Arsenous acid, calcium salt (2:3)
Calcium arsenite
Calcium arsenite, Solid [NA 1574]
Calcium arsenite (2:3)
Monocalcium arsenite
NA 1574 [Calcium arsenite, Solid]

27154-43-4
C_5H_9NO
99.13
Piperidinone (9CI)
Piperidone (8CI)

27154-44-5
$C_6H_6Cl_6$
290.83
Cyclohexane, hexachloro-(9CI)

27156-03-2
$C_2Cl_2F_2$
132.92
Dichlorodifluoroethylene
Ethylene, dichlorodifluoro-
NA 9018

27157-66-0
$C_{16}H_{26}O$
234.38
Phenol, decyl- (9CI)
Decylphenol

27157-94-4
$C_{14}H_{15}O_2PS_2$
310.38
O,O-Ditolyl phosphorodithioate
Cresyl aerofloat
Phosphorodithioic acid, O,O-bis(methylphenyl) ester (9CI)
Phosphorodithioic acid, O,O-ditolyl ester

27159-90-6
$C_{12}H_{20}N_2O_2S.Na$
279.35
Methanesulfonamide, N-(2-(ethyl(3-methylphenyl)-amino)ethyl)-, sodium salt (9CI)

27165-17-9
$C_7H_6N_3O_3$
180.12
N#Nc(c(OC)ccc1N(=O)=O)c1
Benzenediazonium, 2-methoxy-5-nitro- (9CI)
2-Methoxy-5-nitrobenzenediazonium

27165-22-6
$C_6H_3ClN_3O_2$
184.54
N#Nc(c(N(=O)=O)cc(c1)Cl)c1
Benzenediazonium, 4-chloro-2-nitro- (9CI)
4-Chloro-2-nitrobenzenediazonium

27175-64-0
C_7H_9N
107.15
Pyridine, dimethyl- (9CI)
Lutidine (8CI)

27176-87-0
$C_{18}H_{30}O_3S$
326.54
O=S(=O)(OCCCCCCCCCCCC)

c(cccc1)c1
Benzenesulfonic acid, dodecyl (8CI,9CI)
Bio-Soft S 100
Calsoft LAS 99
Caswell No. 413C
DBS
Dobanic Acid 83
Dobanic Acid JN
Dodecyl benzene sulfonic acid
Dodecylbenzene sulfonic acid
Dodecylbenzenesulfonic acid [NA 2584]
n-Dodecylbenzenesulfonic acid
Dodecylbenzenesulphonic acid
E 7256
Elfan WA Sulphonic Acid
EPA Pesticide Chemical Code 098002
LAS 99
Laurylbenzenesulfonic acid
Marlon AS 3
NA 2584 [Dodecylbenzene-sulfonic acid]
Nacconol 98SA
Nansa 1042P
Nansa SSA
P 3 Vetralat
Polystep A 13
Richonic acid B
Sulframin 1298
Sulframin Acid 1298
Witco 1298 Sulfonic Acid

27176-93-8
$C_{19}H_{32}O_3$
308.46
Ethanol, 2-(2-(nonylphenoxy)-ethoxy)- (9CI)
Nonylphenol mono(oxyethyl-ene) ethanol
2-(2-(Nonylphenoxy)ethoxy)-ethanol

27177-77-1
$C_{18}H_{30}O_3S.K$
365.60
Potassium dodecylbenzene-sulfonate
Benzenesulfonic acid, dodecyl-, potassium salt (9CI)
Caswell No. 691A
Dodecyl benzene sulfonic acid, potassium salt
Dodecylbenzenesulfonic acid, potassium salt
EPA Pesticide Chemical Code 079008

27178-16-1
$C_{26}H_{50}O_4$
426.76
Adipic acid, diisodecyl ester
Diisodecyl adipate
Hexanedioic acid, diisodecyl ester (9CI)

27178-34-3
$C_{10}H_{14}O$
150.22
Phenol, (1,1-dimethylethyl)- (9CI)
tert-Butylphenol
(1,1-Dimethylethyl)phenol

27179-64-2
$C_{12}H_{15}NO_2$
205.26
Acetamide, N-acetyl-N-(2-phenylethyl)- (9CI)
Diacetamide, N-phenethyl- (8CI)

27185-77-9
$C_{13}H_{18}O_2$
206.28
2-Cyclohexen-1-one, 2,4,4-tri-methyl-3-(3-oxo-1-butenyl)- (8CI,9CI)

27193-28-8
$C_{14}H_{22}O$
206.36
Phenol, octyl (8CI)
Caswell No. 613D
EPA Pesticide Chemical Code 064118
Octylphenol
Phenol, (1,1,3,3-tetramethyl-butyl)- (9CI)
(1,1,3,3-Tetramethylbutyl)-phenol
USAF RH-6

27193-86-8
$C_{18}H_{30}O$
262.48
Phenol, dodecyl-, Mixed isomers
Dodecylphenol (Mixed isomers)
T-DET

27195-67-1
$C_{24}H_{48}$
336.65
Dimethylcyclohexane
Dimetilciclohexano (Spanish)

27196-00-5
$C_{14}H_{30}O$
214.44
Tetradecanol, Mixed isomers
Myristyl alcohol (Mixed isomers)
Tetradecyl alcohol

27205-99-8
$C_6H_{15}O_2PS_2.Na$
237.28
Sodium diisopropyl phos-phorodithioate
Phosphorodithioic acid, O,O-bis(1-methylethyl) ester, sodium salt (9CI)
Phosphorodithioic acid, O,O-diisopropyl ester, sodium salt
Sodium diisopropyldithiophos-phate
Sodium O,O-diisopropyl dithio-phosphate

27206-35-5
$C_6H_{14}O_6S_4.2Na$
356.42
1-Propanesulfonic acid, 3,3'-dithiobis-, disodium salt (9CI)
3,3'-Dithiobis(1-propanesulfonic acid), disodium salt
Di(thiopropane sodium sulfon-ate)
γ,γ'-Sulfopropyldisulfide, di-sodium salt

27208-37-3
$C_{18}H_{10}$
226.28
C1=Cc2cc4cccc5ccc3ccc1c2 c3c45
Cyclopenta(cd)pyrene
Acepyrene
Acepyrylene
Cyclopenteno(c,d)pyrene

27213-78-1
$C_{10}H_{14}O_2$
166.22
tert-Butylcatechol
1,2-Benzenediol, (1,1-dimethyl-ethyl)- (9CI)
(1,1-Dimethylethyl)-1,2-benzenediol

27213-90-7
$C_{18}H_{24}O_3S.Na$
343.44
Naphthalenesulfonic acid, bis-(2-methylpropyl)-, sodium salt (9CI)
Bis(2-methylpropyl)naphthal-enesulfonic acid, sodium salt
Diisobutylnaphthalenesulfonic acid, sodium salt

27215-10-7
$C_{16}H_{35}O_4P$
322.48
Phosphoric acid, diisooctyl ester
Diisooctyl acid phosphate [UN 1902]
Diisooctyl phosphate
UN 1902 [Diisooctyl acid phosphate]

27215-95-8
C_9H_{18}
126.24

Nonene (9CI)
Nonene (Mixed isomers)
Nonene (Non-linear)
Propylene trimer
Tripropylene

27223-35-4
$C_{20}H_{17}ClN_2O_3$
368.81
Ketazolam
Ansieten
Anxon
11-Chloro-8,12b-dihydro-2,8-dimethyl-12b-phenyl-4H-(1,3)-oxazino(3,2-d)-(1,4)benzodiazepine-4,7(6H)dione
DEA No. 2772
Ketazolamum (Latin)
NSC-338158
4H-(1,3)Oxazino(3,2-d)(1,4)benzodiazepine-4,7(6H)-dione, 11-chloro-8,12b-dihydro-2,8-dimethyl-12b-phenyl-(8CI,9CI)
4H-(1,3)Oxazino(3,2-d)(1,4)benzodiazepine-4,7(6H)-dione, 8,12b-dihydro-11-chloro-2,8-dimethyl-12b-phenyl-
U 28774
Unakalm

27223-49-0
$C_{19}H_{24}O_3$
300.43
Cyclopropanecarboxylic acid, 2,2-dimethyl-3-(2-methylpropenyl)-, (2-methyl-5-(2-propynyl)- 3-furyl)methyl ester
Kikuthrin
2-Methyl-5-(2-propynyl)-3-furylmethyl-cis-trans-chrysanthemate
Proparthrin

27236-46-0
C_6H_{12}
84.16
2-Methyl-1-pentene
Isohexene
Isohexeno (Spanish)
2-Methylpentene
4-Methylpentene
2-Methyl-pentene
4-Methyl-1-pentene
Pentene, 2-methyl- (9CI)
UN 2288 [Isohexenes]

27236-65-3
$C_{28}H_{36}Hg_2O_4$
837.77
Di(phenylmercury) dodecenylsuccinate
Caswell No. 399A
EPA Pesticide Chemical Code 066001
Mercury, diphenyl(mu-((tetrapropenyl)butanedioato-(2-)-O:O'))di- (9CI)
Super Ad-IT

27241-31-2
$C_9H_6Cl_3N_3O$
278.51
O=C(N(NC1=N)c(c(cc(c2)Cl)Cl)c2Cl)C1
3H-Pyrazol-3-one, 5-amino-2,4-dihydro-2-(2,4,6-trichlorophenyl)- (9CI)
AI3-52586
3-Amino-1-(2,4,6-trichlorophenyl)-2-pyrazolin-5-one
3-Amino-1-(2,4,6-trichlorophenyl)-5-pyrazolone
NSC-113482
2-Pyrazolin-5-one, 3-amino-1-(2,4,6-trichlorophenyl)-
1-(2,4,6-Trichlorophenyl)-3-anilinopyrazolone

27247-96-7
$C_8H_{17}NO_3$
175.22
O=N(=O)OCC(CCCC)CC
2-Ethylhexyl nitrate
Nitric acid, 2-ethylhexyl ester (9CI)

27251-75-8
$C_{33}H_{54}O_6$
546.79
Triisooctyl trimellitate
1,2,4-Benzenetricarboxylic acid, triisooctyl ester (9CI)

27253-28-7
$C_{10}H_{20}O_2.xPb$
Unknown
Lead neodecanoate
Neodecanoic acid, lead salt (9CI)

27253-31-2
$C_{10}H_{20}O_2.xCo$
Unknown
Neodecanoic acid, cobalt salt (9CI)
Cobalt neodecanoate

27253-32-3
$C_{10}H_{20}O_2.xMn$
Unknown
Neodecanoic acid, manganese salt (9CI)
Manganese neodecanoate

27253-33-4
$C_{10}H_{20}O_2.1/2Ca$
192.31
Neodecanoic acid, calcium salt (9CI)
Calcium neodecanoate

27253-41-4
$C_9H_{18}O_2.xPb$
Unknown
Isononanoic acid, lead salt (9CI)
Lead isononanoate

27254-36-0
$C_{10}H_7NO_2$
173.18
Naphthalene, mononitro
Mononitronaphthalene
Naphthalene, nitro-
Nitronaphthalene [UN 2538]
UN 2538 [Nitronaphthalene]

27288-44-4
$C_{10}H_{20}O_2S.1/3Sb$
217.65
Acetic acid, mercapto-, isooctyl ester, antimony(3+) salt (9CI)
Antimony tris(isooctyloxycarbonylmethylmercaptide)
(Stibylidynetrithio)triacetic acid, triisooctyl ester

27304-13-8
$C_{10}H_4Cl_8O$
423.74
4,7-Methanoindan, 1,2,4,5,6,7,8,8-octachloro-2,3-epoxy-3a,4,7,7a-tetrahydro-, exo, endo
4,7-Methanoindan, 3a,4,7,7a-tetrahydro-2,3-epoxy-1,2,4,5,6,7,8,8-octachloro-, exo, endo-
Octachlor epoxide
Oxychlordan
Oxychlordane
3a,4,7,7a-Tetrahydro-1,2-epoxy-4,5,6,7,8,8-hexachloro-4,7-methanoindan

27312-17-0
$C_{14}H_9BrN_2O_4$
349.16
O=C(c(c(c(O)cc1Br)C(=O)c2c(N)ccc3O)c1N)c23
Anthraquinone, 2-bromo-1,5-diamino-4,8-dihydroxy
2-Bromo-1,5-diamino-4,8-dihydroxyanthraquinone
Modr Ostacetova LR (Czech)

27314-13-2
$C_{12}H_9ClF_3N_3O$
303.69
CNc2cnn(c1cccc(c1)C(F)(F)F)c(=O)c2Cl
3(2H)-Pyridazinone, 4-chloro-5-(methylamino)-2-(α,α,α-trifluoro-m-tolyl)
4-Chloro-5-(methylamino)-2-(α,α,α-trifluoro-m-tolyl)-3(2H)-pyridazinone

Evital
H 9789
H 52143
Monometflurazone
Norflurazon
Norflurazone
San 9789
San 9789 H
Solicam
Zorial

27315-26-0
$C_{11}H_{11}ClN_4O_2$
266.66
1,3,5-Triazin-2-amine, N-(2-chlorophenyl)-4,6-dimethoxy- (9CI)
s-Triazine, 2-(o-chloroanilino)-4,6-dimethoxy- (8CI)

27315-27-1
$C_{11}H_{11}ClN_4O_2$
266.69
1,3,5-Triazin-2-amine, N-(4-chlorophenyl)-4,6-dimethoxy- (9CI)
s-Triazine, 2-(p-chloroanilino)-4,6-dimethoxy- (6CI,8CI)

27317-59-5
$C_7H_{13}NO_3$
159.18
O=C(OCC)CC(OCC)=N
Propanoic acid, 3-ethoxy-3-imino-, ethyl ester (9CI)
Ethyl 3-ethoxy-3-iminopropanoate

27323-18-8
$C_{12}H_9Cl$
188.66
Biphenyl, chloro
Aroclor 1254
1,1'-Biphenyl, chloro-
Chlorobiphenyl
Chlorodiphenyl
Chlorodwufenol (Polish)
Diphenylchloride
Monochlorobiphenyl

27323-28-0
C_9H_9N
131.17
1H-Indole, methyl- (9CI)
Indole, methyl- (8CI)

27323-29-1
$C_{13}H_{11}N$
181.23
9H-Carbazole, methyl- (9CI)
Carbazole, methyl- (8CI)

27323-41-7
$C_{18}H_{30}O_3S.C_6H_{15}NO_3$
475.68
TEA-dodecylbenzenesulfonate
AI3-26730-X
Benzenesulfonic acid, dodecyl-, Compd. with 2,2',2''-nitrilotris(ethanol) (1:1) (9CI)
Caswell No. 887AA
Dodecylbenzenesulfonic acid, Compd. with 2,2',2''-nitrilotris(ethanol) (1:1)
Dodecylbenzenesulfonic acid triethanolamine salt
EPA Pesticide Chemical Code 079020
Triethanolamine dodecylbenzene sulfonate
Triethanolamine dodecylbenzenesulfonate
Witconate S-1280
Witconate TAB
Witconate 60L
Witconate 60T
Witconate 79S
Witconate 5725

27326-17-6
$C_{23}H_{27}N_2O.Cl$
382.92
Quinolinium, 1-((1,3-dihydro-1,3,3-trimethyl-2H-indol-2-ylidene)ethylidene)-1,2,3,4-tetrahydro-6-methoxy-, chloride (9CI)
1,3,3-Dimethyl-2-(2-(1,2,3,4-tetrahydro-6-methoxy-1-quinolyl)vinyl)-3H-indolium chloride

27341-33-9
$C_{21}H_{16}N_2O_3$
344.36
9,10-Anthracenedione, 1-amino-4-((methoxyphenyl)-amino)- (9CI)

27342-88-7
$C_{12}H_{26}O$
186.34
Dodecanol (VAN)(9CI)
Alcohol C-12
n-Dodecanol
Dodecyl alcohol
Lauryl alcohol

27344-06-5
$C_{42}H_{46}N_{14}O_{10}S_2.2Na$
1016.94
Benzenesulfonic acid, 2,2'-(1,2-ethenediyl)bis(5-((4-((3-amino-3-oxopropyl)-(2-hydroxyethyl)amino)-6-(phenylamino)-1,3,5-triazin-2-yl)amino)-, disodium salt (9CI)

27344-41-8
$C_{28}H_{20}O_6S_2.2Na$
562.58
[Na+].[Na+].[O-]S(=O)(=O)c1ccccc1C=Cc2ccc(cc2)c4ccc(C=Cc3ccccc3S([O-])(=O)=O)cc4
Benzenesulfonic acid, 2,2'-(4,4'-biphenylylenedivinylene)di-, disodium salt
Benzenesulfonic acid, 2,2'-((1,1'-biphenyl)-4,4'-diyldi-2,1-ethenediyl)bis-, disodium salt
Disodium 4,4'-bis(2-sulfostyryl)biphenyl
FBA 351
Stilbene 3
Tinopal CBS
Tinopal CBS-X

27355-22-2
$C_8H_2Cl_4O_2$
271.90
Clc2c(Cl)c(Cl)c1C(=O)OCc1c2Cl
Phthalide, 4,5,6,7-tetrachloro
Bayer 96610
Fthalide
KF-32
Phthalide
Rabcide
TCP
4,5,6,7-Tetrachlorophthalide

27356-46-3
$C_{16}H_{13}N$
219.28
Quinoline, 6-methyl-2-phenyl- (8CI,9CI)

27358-28-7
$C_{17}H_{16}$
220.31
Anthracene, trimethyl- (8CI, 9CI)

27359-10-0
$C_7H_5F_3$
146.11
Benzene, methyl-, trifluoro deriv. (9CI)
Toluene, trifluoro- (6CI,7CI, 8CI)
Trifluorotoluene

27375-52-6
$C_{10}H_{13}NO_4S$
243.30
O=C(Nc(ccc(S(=O)(=O)CCO)c1)c1)C
Acetanilide, 4'-(2-hydroxyethylsulfonyl)
p-Acetaminofenyl-β-hydroxyethylsulfon (Czech)
p-Acetaminofenyl-2-hydroxyethylsulfon (Czech)
4'-(2-Hydroxyethylsulfonyl)-acetanilide

27396-39-0
$C_8H_{11}NO_2$

153.18
2(5H)-Furanone, 5-(butylimino)- (9CI)
Acrylic acid, 3-(N-butyl-1-hydroxyformimidoyl)-, cyclic anhydride (7CI, 8CI)
Acrylic acid, 3-(N-butyl-1-hydroxyformimidoyl)-, γ-lactone (6CI)
N-Butylisomaleimide

27425-58-7
$C_{43}H_{27}CrN_6O_8S.2Na$
885.73
Chromate(2-), (3-hydroxy-4-((2-hydroxy-1-naphthalenyl)azo)-1-naphthalenesulfonato-(3-))(1-((2-hydroxy-5-((2-methoxyphenyl)azo)phenyl)azo)-2-naphthalenolato(2-))-, disodium (9CI)

27430-88-2
$C_7H_{13}N_5$
167.22
s-Triazine, 2-(ethylamino)-4-methyl-6-(methylamino)- (8CI)

27430-89-3
$C_8H_{15}N_5$
181.24
s-Triazine, 2-methyl-4-(methylamino)-6-(propylamino)- (8CI)

27430-90-6
$C_8H_{15}N_5$
181.24
s-Triazine, 2-(isopropylamino)-4-methyl-6-(methylamino)- (8CI)

27430-91-7
$C_9H_{17}N_5$
195.27
s-Triazine, 2-(butylamino)-4-methyl-6-(methylamino)- (8CI)

27430-92-8
$C_9H_{17}N_5$
195.27
s-Triazine, 2-(ethylamino)-4-methyl-6-(propylamino)- (8CI)

27430-93-9
$C_9H_{17}N_5$
195.24
1,3,5-Triazine-2,4-diamine, N-ethyl-6-methyl-N'-(1-methylethyl)- (9CI)
s-Triazine, 2-(ethylamino)-4-(isopropylamino)-6-methyl- (8CI)

27430-94-0
$C_{10}H_{19}N_5$
209.30
s-Triazine, 2-(butylamino)-4-(ethylamino)-6-methyl- (8CI)

27430-95-1
$C_{10}H_{19}N_5$
209.30
s-Triazine, 2-(sec-butylamino)-4-(ethylamino)-6-methyl- (8CI)

27430-96-2
$C_{10}H_{19}N_5$
209.30
1,3,5-Triazine-2,4-diamine, N-(1,1-dimethylethyl)-N'-ethyl-6-methyl- (9CI)
s-Triazine, 2-(tert-butylamino)-4-(ethylamino)-6-methyl- (8CI)

27430-97-3
$C_{11}H_{21}N_5$
223.32
s-Triazine, 2-(ethylamino)-4-methyl-6-(pentylamino)- (8CI)

27430-98-4
$C_{12}H_{23}N_5$
237.35
s-Triazine, 2-(ethylamino)-4-(hexylamino)-6-methyl- (8CI)

27430-99-5
$C_{14}H_{27}N_5$
265.40
s-Triazine, 2-(ethylamino)-4-((1-ethylhexyl)amino)-6-methyl- (8CI)

27431-00-1
$C_{18}H_{35}N_5$
321.51
s-Triazine, 2-(dodecylamino)-4-(ethylamino)-6-methyl- (8CI)

27431-01-2
$C_9H_{15}N_5$
193.25
s-Triazine, 2-(allylamino)-4-(ethylamino)-6-methyl- (8CI)

27431-02-3
$C_{10}H_{19}N_5$
209.30
s-Triazine, 2-(isopropylamino)-4-methyl-6-(propylamino)- (8CI)

27431-03-4
$C_{11}H_{21}N_5$
223.32
s-Triazine, 2-(butylamino)-4-(isopropylamino)-6-methyl- (8CI)

27431-04-5
$C_{13}H_{25}N_5$
251.38
s-Triazine, 2-(hexylamino)-4-(isopropylamino)-6-methyl- (8CI)

27431-05-6
$C_{19}H_{37}N_5$
335.54
s-Triazine, 2-(dodecylamino)-4-(isopropylamino)-6-methyl- (8CI)

27431-06-7
$C_{10}H_{17}N_5$
207.28
s-Triazine, 2-(allylamino)-4-(isopropylamino)-6-methyl- (8CI)

27431-07-8
$C_7H_{13}N_5$
167.22
s-Triazine, 2-(dimethylamino)-4-methyl-6-(methylamino)- (8CI)

27431-08-9
$C_9H_{17}N_5$
195.27
s-Triazine, 2-(dimethylamino)-4-methyl-6-(propylamino)- (8CI)

27431-09-0
$C_{10}H_{19}N_5$
209.30
s-Triazine, 2-(butylamino)-4-(dimethylamino)-6-methyl- (8CI)

27431-10-3
$C_{10}H_{19}N_5$
209.30
s-Triazine, 2-(diethylamino)-4-(ethylamino)-6-methyl- (8CI)

27431-11-4

$C_{12}H_{23}N_5$
237.35
s-Triazine, 2-(butylamino)-4-(diethylamino)-6-methyl- (8CI)

27431-12-5
$C_{11}H_{19}N_5$
221.31
s-Triazine, 2-(allylamino)-4-(diethylamino)-6-methyl- (8CI)

27431-14-7
$C_7H_{10}Cl_3N_5$
270.55
s-Triazine, 2-(ethylamino)-4-(methylamino)-6-(trichloromethyl)- (7CI,8CI)

27431-15-8
$C_8H_{12}Cl_3N_5$
284.58
s-Triazine, 2-(isopropylamino)-4-(methylamino)-6-(trichloromethyl)- (7CI, 8CI)

27431-16-9
$C_9H_{14}Cl_3N_5$
298.60
s-Triazine, 2-(ethylamino)-4-(propylamino)-6-(trichloromethyl)- (8CI)

27431-17-0
$C_8H_{12}Cl_3N_5$
284.58
s-Triazine, 2-(dimethylamino)-4-(ethylamino)-6-(trichloromethyl)- (7CI, 8CI)

27431-18-1
$C_{10}H_{19}N_5$
209.30
s-Triazine, 2-ethyl-4-(ethylamino)-6-(isopropylamino)- (8CI)

27431-19-2
$C_5H_6Cl_3N_5$
242.50
s-Triazine, 2-amino-4-(methylamino)-6-(trichloromethyl)- (8CI)

27457-18-7
$C_8H_{16}O$
128.21
Octanone (9CI)

27457-28-9
$C_8H_8O_3S.Na$
207.21
Benzenesulfonic acid, ethenyl-, sodium salt (9CI)
Sodium ethenylbenzenesulfonate

27458-06-6
$C_{14}H_{10}O_3$
226.23
Benzoic acid, benzoyl- (9CI)

27458-20-4
Unknown
Unknown
Butyl toluene

27458-90-8
$C_{24}H_{50}S_2$
402.79
Disulfide, di-tert-dodecyl (9CI)
Di-tert-dodecyl disulfide

27458-92-0
$C_{13}H_{28}O$
200.41
Isotridecyl-alcohol
Isotridecanol

27458-94-2
$C_9H_{20}O$
144.26
Isononanol (9CI)
Isononyl alcohol (8CI)

27470-63-9
$C_{10}H_{19}N_5$
209.30
s-Triazine, 2-(ethylamino)-4-(isobutylamino)-6-methyl- (8CI)

27470-64-0
$C_{11}H_{21}N_5$
223.32
s-Triazine, 2-(isobutylamino)-4-(isopropylamino)-6-methyl- (8CI)

27470-65-1
$C_{11}H_{21}N_5$
223.32
s-Triazine, 2-(sec-butylamino)-4-(isopropylamino)-6-methyl- (8CI)

27470-66-2
$C_8H_{15}N_5$
181.24
s-Triazine, 2-(dimethylamino)-4-(ethylamino)-6-methyl- (8CI)

27470-67-3
$C_{11}H_{21}N_5$
223.32
s-Triazine, 2-(diethylamino)-4-methyl-6-(propylamino)- (8CI)

27470-68-4
$C_8H_{12}Cl_3N_5$
284.58
s-Triazine, 2-(methylamino)-4-(propylamino)-6-(trichloromethyl)- (8CI)

27470-69-5
$C_9H_{14}Cl_3N_5$
298.60
s-Triazine, 2-(ethylamino)-4-(isopropylamino)-6-(trichloromethyl)- (7CI, 8CI)

27470-95-7
$C_7H_{10}Cl_3N_5$
270.55
s-Triazine, 2-(dimethylamino)-4-(methylamino)-6-(trichloromethyl)- (7CI, 8CI)

27470-96-8
$C_9H_{14}Cl_3N_5$
298.60
s-Triazine, 2-(diethylamino)-4-(methylamino)-6-(trichloromethyl)- (7CI, 8CI)

27470-97-9
$C_{10}H_{16}Cl_3N_5$
312.63
s-Triazine, 2-(diethylamino)-4-(ethylamino)-6-(trichloromethyl)- (8CI)

27470-98-0
$C_6H_8Cl_3N_5$
256.52
s-Triazine, 2-amino-4-(ethylamino)-6-(trichloromethyl)- (8CI)

27476-22-8
$C_7H_5Br_3$
986.53
Benzene, methyl-, tribromo deriv. (9CI)
Toluene, tribromo- (8CI)
Tribromotoluene

27476-93-3
$C_{22}H_{43}N_3$

349.59
1H-Imidazole-1-ethanamine, 2-(heptadecenyl)-2,3-dihydro- (9CI)
1-(β-Aminoethyl)-2-heptadecenylimidazoline

27478-24-6
$C_{14}H_{16}O_3S.H_3N$
281.37
Naphthalenesulfonic acid, butyl-, ammonium salt (9CI)
Butylnaphthalenesulfonic acid, ammonium salt

27478-34-8
$C_{10}H_6N_2O_4$
218.18
Naphthalene, dinitro
Dinitronaphthalene

27479-28-3
$C_{23}H_{42}N.Cl$
368.04
Quaternium-14
Benzenemethanaminium, N-dodecyl-ar-ethyl-N,N-dimethyl-, chloride (9CI)
Dodecyl dimethyl ethylbenzyl ammonium chloride
Dodecyldimethyl(ethylbenzyl)-ammonium chloride
N-Dodecyl-ar-ethyl-N,N-dimethylbenzenemethanaminium chloride

27479-29-4
$C_{25}H_{46}N.Cl$
396.09
Benzenemethanaminium, ar-ethyl-N,N-dimethyl-N-tetradecyl-, chloride (9CI)
Tetradecyldimethyl(ethylbenzyl)ammonium chloride

27496-82-8
$C_{15}H_{12}O_6$
288.27
Salicylic acid, methylenedi
Benzoic acid, methylenebis-(2-hydroxy- (9CI)
MDA
Methylenedisalicylic acid

27503-81-7
$C_{13}H_{10}N_2O_3S$
274.29
O=S(=O)(O)c(ccc(NC(=N1)c(cccc2)c2)c13)c3
2-Phenylbenzimidazole-5-sulfonic acid
1H-Benzimidazole-5-sulfonic acid, 2-phenyl- (9CI)
2-Phenyl-1H-benzimidazole-5-sulfonic acid

27515-66-8
$C_{14}H_{22}O$
206.33
Phenol, bis(2-methylpropyl)- (9CI)
Bis(2-methylpropyl)phenol
Phenol, diisobutyl-

27519-02-4
$C_{23}H_{46}$
322.62
C(=CCCCCCCCC)CCCCCCCCCCCCC
(Z)-9-Tricosene
AI3-35349
Caswell No. 883C
EPA Pesticide Chemical Code 103201
Muscalure
Muscamone
9-Tricosene, (Z)-

27529-92-6
$C_{10}H_{16}Cl_3N_5$
312.63
s-Triazine, 2-(ethylamino)-4-(isobutylamino)-6-(trichloromethyl)- (8CI)

27546-07-2
$H_4N.1/2Mo_2O_7$
169.97
Molybdate, diammonium (9CI)

27550-64-7
$C_{16}H_{21}N_3O_3$
303.34
O=C(Nc(cc(N(CCN(C(=O)CC1)C1=O)CC)cc2)c2)C
Acetamide, N-(3-(((2,5-dioxo-1-pyrrolidinyl)ethyl)ethylamino)phenyl)- (9CI)

27554-26-3
$C_{24}H_{38}O_4$
390.56
CC(C)(C)CC(C)(C)OC(=O)c1ccccc1C(=O)OC(C)(C)CC(C)(C)C
Diisooctyl phthalate
AI3-27697-X
1,2-Benzenedicarboxylic acid, diisooctyl ester
Corflex 880
Diisooctyl 1,2-benzenedicarboxylate
Diisooctyl phthalate
Di-iso-octyl phthalate
DIOP
Flexol Plasticizer DIOP
Hexaplas DIOP
Hexaplas M/O
Isooctyl phthalate
NSC-6381
Phthalic acid, bis(6-methylheptyl)ester (9CI)
Phthalic acid, diisooctyl ester
Unem 5005

27574-34-1
$C_{28}H_{40}NiO_2S$
499.38
Nickel, ((2,2'-thiobis(4-(1,1,3,3-tetramethylbutyl)-phenolato))(2-)-O,O',S)- (9CI)
Nickel bis(p-octylphenol)sulfide
Reaction product of bis(p-1,1,3,3-tetramethylbutyl-phenol)-2,2'-sulfide with nickel salts

27576-03-0
$C_{10}H_{12}$
132.21
Benzene, ethenyl-, dimethyl deriv. (9CI)
Styrene, dimethyl- (8CI)

27576-86-9
$C_{15}H_{16}O$
212.29
Phenol, (1-methyl-1-phenylethyl)- (9CI)
(1-Methyl-1-phenylethyl)-phenol

27577-90-8
$C_{17}H_{12}$
216.28
Pyrene, methyl- (8CI,9CI)

27577-96-4
$C_{10}H_{14}O$
450.67
Ethanol, 2-xylyl- (8CI)
Phenethyl alcohol, ar,ar-dimethyl- (6CI,7CI)

27578-60-5
$C_7H_{16}N_2$
128.21
NCCN1CCCCC1
1-Piperidineethanamine (9CI)
NSC-54993
Piperidine, 1-(2-aminoethyl)- (8CI)

27583-37-5
$C_6H_{12}O_2$
116.16
1,2-Cyclopentanediol, 3-methyl- (8CI,9CI)
NSC-403839

27593-23-3
$C_{10}H_{14}O_2$
166.22
O=C(OC(=CC=1)CCCCC)C1

2H-Pyran-2-one, 6-pentyl- (9CI)
6-Amyl-α-pyrone
6-Pentyl-2H-pyran-2-one

27598-81-8
$C_8H_{10}O_2$
138.17
Benzene, dimethoxy- (9CI)

27601-00-9
$C_{10}H_9N$
143.18
Quinoline, methyl- (8CI,9CI)

27622-91-9
$C_4H_6N_4$
110.12
1,3,5-Triazin-2-amine, 4-methyl- (9CI)
s-Triazine, 2-amino-4-methyl- (8CI)

27636-75-5
$C_{17}H_{28}O_3S.Na$
335.46
Benzenesulfonic acid, undecyl-, sodium salt (9CI)
Sodium undecylbenzenesulfonate

27636-85-7
C_3F_7I
295.93
Propane, heptafluoroiodo
Heptafluorjodpropan (Czech)
Heptafluoroiodopropane
Iodoheptafluoropropane

27668-52-6
$C_{26}H_{58}NO_3Si.Cl$
496.28
3-(Trimethoxysilyl)propyldimethyloctadecylammonium
Caswell No. 892B
EPA Pesticide Chemical Code 107401
1-Octadecanaminium, N,N-dimethyl-N-(3-(trimethoxysilyl)propyl)-, chloride (9CI)
Octadecyldimethyl(3-(trimethoxysilyl)propyl)ammonium chloride
Q9-5700
X9-5700
3-(Trimethoxysilyl)propyl dimethyl octadecyl ammonium chloride

27683-60-9
$C_8H_7Cl_3O$
225.51
2,2-Dichloro-1-(2-chlorophenyl) ethanol
Benzenemethanol, 2-chloro-α-(dichloromethyl)-
Benzyl alcohol, o-chloro-α-(dichloromethyl)-
o-Chloro-α-(dichloromethyl)-benzyl alcohol

27697-51-4
$C_5H_{14}N.Cl$
123.62
Ethanaminium, N,N,N-trimethyl-, chloride (9CI)
Ethyltrimethylammonium chloride
N,N,N-Trimethylethanaminium chloride

27725-17-3
$C_{29}H_{44}O_2$
424.67
Oc(c(cc(c1)C(CC(C)(C)C)(C)C)Cc(c(O)ccc2C(CC(C)(C)C)(C)C)c2)c1
Phenol, 2,2'-methylenebis(4-(1,1,3,3-tetramethylbutyl)- (9CI)

27731-61-9
$(C_2H_4O)xC_{14}H_{30}O_4S.H_3N$
Unknown
Ammonium myreth sulfate
Ammonium myristyl ether sulfate
Poly(oxy-1,2-ethanediyl), α-sulfo-ω-(tetradecyloxy)-, ammonium salt (9CI)
α-Sulfo-tetradecyloxypoly(oxy-1,2-ethanediyl), ammonium salt
Unipol EA-40

27733-08-0
$C_{14}H_9BrN_2O_4$
349.13
9,10-Anthracenedione, 1,8-diaminobromo-4,5-dihydroxy- (9CI)

27753-52-2
$C_{12}HBr_9$
864.32
Biphenyl, nonabromo
1,1-Biphenyl, nonabromo-
Bromkal 80-9D
Nonabromobiphenyl

27757-85-3
C_5H_7NS
113.18
NCc1cccs1
2-Thiophenemethanamine (9CI)
2-Thenylamine (8CI)

27774-13-6
O_5SV
163.00
OS(=O)(=O)O[V]=O
Vanadium, oxysulfato
C.I. 77940
NA 9152 (DOT)
UN 2931 [Vanadyl sulfate]
Vanadyl sulfate [UN 2931]

27794-93-0
$C_3H_{12}NO_9P_3.xK$
Unknown
Phosphonic acid, (nitrilotris-(methylene))tris-, potassium salt (9CI)
Amino tris(methylene phosphonic acid), potassium salt

27813-02-1
$C_7H_{12}O_3$
144.17
Hydroxypropyl methacrylate
Methacrylic acid, monoester with 1,2-propanediol
1,2-Propanediol, 2-methyl, monomethacrylate
2-Propenoic acid, 2-methyl-, 2-hydroxymethylethyl ester
2-Propenoic acid, 2-methyl-, monoester with 1,2-propanediol (9CI)

27831-13-6
$C_{10}H_{12}$
132.21
Benzene, 4-ethenyl-1,2-dimethyl- (9CI)
Styrene, 3,4-dimethyl- (8CI)

27858-07-7
$C_{12}H_2Br_8$
785.42
Biphenyl, octabromo
BB-8
Bromkal 80
OBB
Octabromobiphenyl
ar,ar,ar,ar,ar',ar',ar',ar'-Octabromo-1,1'-biphenyl
Octabromodiphenyl

27858-32-8
$C_{18}H_{32}O_8Ti$
424.33
Titanium, bis(ethyl 3-oxobutanoato-O1',O3)bis(2-propanolato)- (9CI)
Diisopropoxydi(ethoxyacetoacetyl)titanate

27859-58-1
$C_{16}H_{28}O_4$
284.40
Butanedioic acid, (tetrapropenyl)- (9CI)
(Tetrapropenyl)butanedioic acid

27883-12-1
$C_{22}H_{41}NO_3$
367.57
O=C(N(CCO)CCO)CCCCC=CC C=CCCCCCCCC
6,9-Octadecadienamide, N,N-bis(2-hydroxyethyl)-, (Z,Z)- (9CI)
Linoleamide DEA

27900-75-0
$C_6H_3Cl_2NO_2$
191.99
Benzene, dichloronitro-
AI3-15074

27938-76-7
$C_{14}H_8O_3$
224.22
9,10-Anthracenedione, hydroxy- (9CI)
Anthraquinone, hydroxy- (8CI)

27939-60-2
$C_9H_{14}O$
138.21
3-Cyclohexene-1-carboxaldehyde, dimethyl- (VAN)(8CI)
Dimethyl-3-cyclohexene-1-carboxaldehyde

27941-08-8
$(C_9H_{14}O_4)x$
Polymer NA
Poly(oxy(methyl-1,2-ethanediyl)oxy(1,6-dioxo-1,6-hexanediyl)) (9CI)
Adekacizer P 200
Harflex 321
Hexaplas PPA
Paraplex G 53
Plastolein 9765
Poly(oxy(methylethylene)oxyadipoyl) (8CI)
Polysizer W 2300
Polysizer W 2600
PPAL 6
Reoplex 400

27941-09-9
$(C_{10}H_{16}O_4)x$
Polymer NA
Poly(oxy(methyl-1,3-propanediyl)oxy(1,6-dioxo-1,6-hexanediyl)) (9CI)
Harflex 330
Hexaplas BUT
Poly(oxy(methyltrimethylene)oxyadipoyl) (8CI)

27944-79-2
$C_7H_{14}O$
114.19
Pentanal, 2,4-dimethyl- (9CI)
NSC-523741
Valeraldehyde, 2,4-dimethyl- (8CI)

27949-30-0
$C_{12}H_{10}O_3$
202.21
Oc(c(c(cccc1O)c1)c(O)cc2)c2
2,3',6-Biphenyltriol (8CI)
2,3',6-Trihydroxybiphenyl
2,3',6-Trihydroxydiphenyl

27949-36-6
$C_2H_3Br_2Cl$
222.32
Ethane, 1,1-dibromo-2-chloro- (8CI)

27955-87-9
$C_{21}H_{28}O_2$
312.49
Propane, 1,1-bis(p-ethoxyphenyl)-2,2-dimethyl
Benzene, 1,1'-(2,2-dimethylpropylidene)bis(4-ethoxy- (9CI)
1,1-Bis(p-ethoxyphenyl)-2,2-dimethylpropane
GH 44

27959-50-8
$C_{17}H_{14}N_2O_5S$
358.38
2-Naphthalenesulfonic acid, 6-hydroxy-5-((4-methoxyphenyl)azo)- (9CI)
2-Naphthalenesulfonic acid, 6-hydroxy-5-((p-methoxyphenyl)azo)- (8CI)

27963-33-3
$C_7H_{13}N_5O$
183.25
s-Triazine, 4,6-diamino-2-isobutoxy

27978-54-7
$ClHO_4.xH_4N_2$
Unknown
Hydrazine perchlorate

27985-70-2
$C_{14}H_{22}O$
206.33
Phenol, (1-methylheptyl)- (9CI)
(1-Methylheptyl)phenol

27986-36-3
$C_{17}H_{28}O_2$
264.41
Terics
Ethanol, 2-(nonylphenoxy)- (9CI)
2-(Nonylphenoxy)ethanol

27987-00-4
Unknown
Unknown
Sodium (1-methylundecyl)-benzenesulfonate

27987-10-6
$C_8H_{11}N$
121.18
Pyridine, ethylmethyl- (9CI)
Picoline, ethyl-

28005-74-5
$C_{11}H_{17}NO_2$
195.29
OCCN(c(c(ccc1)C)c1)CCO
o-Toluidine, N,N-bis(2-hydroxyethyl)
N,N-Bis(2-hydroxyethyl)-o-toluidine
Di-(hydroxyethyl)-o-tolylamine
Emery 5712
Ethanol, 2,2'-((2-methylphenyl)-imino)bis- (9CI)
Ethanol, 2,2'-(o-tolylimino)di- (8CI)
o-Tolyldiethanolamine

28013-11-8
$C_{14}H_{14}$
182.27
1,1'-Biphenyl, ar,ar'-dimethyl- (9CI)
Bitolyl
ar,ar'-Dimethylbiphenyl

28014-46-2
$(C_{18}H_{24}O_2.H_3O_4P)x$
Polymer
Estradiol, Polyester with phosphoric acid
Estradiol phosphate, Polymer
Estradurin
Estra-1,3,5(10)-triene-3,17 diol (17-β)-, Polymer with phoshporic acid
(17-β)-Estra-1,3,5(10)-triene-3,17-diol, Polymer with phosphoric acid
Oestradiol phosphate Polymer
Oestradiol polyester with phosphoric acid
PEP
Poly(estradiol phosphate)
Polyoestradiol phosphate

28016-00-4
$C_{28}H_{44}O_3S.1/2Zn$
493.41
Naphthalenesulfonic acid, dinonyl-, zinc salt (9CI)
Zinc dinonylnaphthalenesulfonate

28016-01-5
$C_6H_2F_4$
150.08
Benzene, tetrafluoro- (8CI,9CI)

28057-48-9
Unknown
Unknown
d-trans-Allethrin

28061-69-0
$C_{20}H_{41}N$
295.55
Octadecenylamine, N,N-dimethyl- (8CI)
N,N-Dimethyloctadecenylamine

28076-73-5
$C_{12}H_6Cl_4O$
307.99
Bis(2,4-dichlorophenyl)ether
Benzene, 1,1'-oxybis(2,4-dichloro-
2,2',4,4'-Tetrachlorodiphenyl ether
2,2',4,4'-Tetrachlorophenyl oxide

28080-86-6
$C_{12}H_{18}$
162.28
Benzene, (2,2-dimethylbutyl)- (6CI,7CI,8CI,9CI)
(2,2-Dimethylbutyl)benzene

28098-80-8
$C_7H_{10}O_4$
158.16
2-Butenedioic acid, 2-ethyl-3-methyl-, (E)- (9CI)
2-Ethyl-3-methylfumaric acid
Fumaric acid, ethylmethyl- (8CI)

28106-30-1
$C_{10}H_{12}$
132.21
Benzene, ethenylethyl- (9CI)
Ethenylethylbenzene

28108-99-8
$C_{21}H_{21}O_4P$
368.37
Phosphoric acid, (1-methylethyl)phenyl diphenyl ester (9CI)
Isopropylphenyl diphenyl phosphate

28109-00-4
$C_{24}H_{27}O_4P$
410.45
Phosphoric acid, bis((1-methylethyl)phenyl) phenyl ester (9CI)
Di(isopropylphenyl) phenyl phosphate

28109-02-6
$C_{22}H_{23}O_4P$
382.40
Phosphoric acid, sec-butylphenyl diphenyl ester (8CI)

28109-99-1
$C_{12}H_8$
152.20
Biphenyl, 2,2'-didehydro- (8CI)

28134-31-8
$C_9H_{10}O_2$
150.18
Benzoic acid, ethyl- (6CI, 7CI,8CI,9CI)
Ethylbenzoic acid

28137-64-6
$C_{24}H_{51}NO_2$
385.67
OC(C)CN(CCCCCCCCCCCCCCCCCC)CC(O)C
2-Propanol, 1,1'-(octadecylimino)bis- (9CI)
1,1'-(Octadecylimino)bis-2-propanol

28139-02-8
$C_6H_6N_4OS$
182.20
6H-Purin-6-one, 1,2,3,7-tetrahydro-3-methyl-2-thioxo- (9CI)
3-Methyl-2-thioxanthine
Xanthine, 3-methyl-2-thio- (6CI,7CI,8CI)

28141-13-1
$C_9H_{10}N_2O_2$
178.18
O=C(N(C(O)=CC=1C)CC)C1C#N
3-Pyridinecarbonitrile, 1-ethyl-1,2-dihydro-6-hydroxy-4-methyl-2-oxo- (9CI)
1-Ethyl-1,2-dihydro-6-hydroxy-4-methyl-2-oxo-3-pyridinecarbonitrile

28158-16-9
$(C_8H_{10}O_4)x$
Polymer
2-Propenoic acid, 1,2-ethanediyl ester, Homopolymer (9CI)

28165-52-8
$C_6H_4Br_2O$
251.91
Phenol, 2,5-dibromo- (8CI,9CI)

28169-46-2
$C_8H_6N_2O_6$
226.13
O=C(O)c(c(c(N(=O)=O)cc1N(=O)=O)C)c1
Benzoic acid, 2-methyl-3,5-dinitro- (9CI)
3,5-Dinitrotoluic acid
3,5-Dinitro-o-toluic acid
2-Methyl-3,5-dinitrobenzoic acid
NSC-168527
o-Toluic acid, 3,5-dinitro-

28170-54-9
$C_{10}H_{14}N_2O_2$
194.22
COc1cccc(NC(=O)N(C)C)c1
Urea, N'-(3-methoxyphenyl)-N,N-dimethyl- (9CI)
Urea, 3-(m-methoxyphenyl)-1,1-dimethyl- (8CI)

28175-98-6
$C_8H_{12}N_2O_2$
168.18
4(3H)-Pyrimidinone, 2-(2-hydroxy-1-methylethyl)-6-methyl- (8CI)

28178-42-9
$C_{13}H_{17}NO$
203.28
Benzene, 2-isocyanato-1,3-bis-(1-methylethyl)- (9CI)
2,6-Diisopropylphenyl isocyanate
2-Isocyanato-1,3-bis(1-methylethyl)benzene

28213-80-1
$(C_8H_8)_3$
312.45
Benzene, ethenyl-, trimer (9CI)
Ethenylbenzene trimer
Styrene trimer

28214-91-7
$C_{28}H_{44}O_3S.Li$
467.66
Naphthalenesulfonic acid, dinonyl-, lithium salt (9CI)
Dinonylnaphthalenesulfonic acid, lithium salt

28219-61-6
$C_{14}H_{24}O$
208.34

OCC(=CCC(C(C(=C1)C)(C)C)C1)CC
2-Buten-1-ol, 2-ethyl-4-(2,2,3-trimethyl-3-cyclopenten-1-yl)- (9CI)
2-Ethyl-4-(2,2,3-trimethyl-3-cyclopenten-1-yl)-2-buten-1-ol

28227-92-1
$C_{35}H_{48}N_8O_{10}S$
772.97
Phalloin

28249-77-6
$C_{12}H_{16}ClNOS$
257.80
CCN(CC)C(=O)SCc1ccc(Cl)cc1
Carbamic acid, diethylthio-, S-(p-chlorobenzyl) ester
B-3015
Benthiocarb
Bolero
Carbamothioic acid, diethyl-, S-((4-chlorophenyl)methyl) ester
S-(4-Chlorobenzyl) N,N-diethylthiocarbamate
S-((4-Chlorophenyl)methyl)diethylcarbamothioate
IMC 3950
Saturn
Saturno
Siacarb
Tamariz
Thiobencarb
Thiobencarbe

28258-64-2
Unknown
Unknown
Phenyl-β-naphthylamine

28260-61-9
$C_6H_2ClN_3O_6$
247.56
Benzene, chlorotrinitro
Trinitrochlorobenzene [UN 0155]
UN 0155 [Trinitrochlorobenzene; (Picryl chloride)]

28279-36-9
$C_{41}H_{43}Cl_3N_6O_5$
806.15
Benzamide, 4-(((2,4-bis(1,1-dimethylpropyl)phenoxy)acetyl)amino)-N-(4,5-dihydro-5-((4-methoxyphenyl)azo)-5-oxo-1-(2,4,6-trichlorophenyl)-1H-pyrazol-3-yl)- (9CI)

28291-69-2
$C_9H_{10}N_2S$
178.25
2-Benzothiazolamine, N-ethyl- (9CI)
Benzothiazole, 2-(ethylamino)- (8CI)

28291-83-0
$C_{11}H_{14}N_2O$
190.25
Benzoxazole, 2-(isobutylamino)- (8CI)

28299-41-4
$C_{14}H_{14}O$
198.26
Benzene, 1,1'-oxybis(methyl-
AI3-02478
1,1'-Oxybis(methylbenzene)

28300-74-5
$C_4H_4O_7Sb.K$
324.93
[K+].[O-]C(=O)C1O[Sb]2OC1C(=O)O2
Antimony potassium tartrate [UN 1551]
Antimonate(2)-, bis(mu-tartrato(4-))di-, dipotassium, trihydrate
Antimonyl potassium tartrate
Antimony potassium tartrate, Solid (DOT)
Emetique (French)
ENT 50,434
Potassium antimonyl tartrate
Potassium antimonyl d-tartrate
Potassium antimony tartrate
Tartar Emetic
Tartaric acid, antimony potassium salt
Tartarized Antimony
Tartox
Tartrate antimonio-potassique (French)
Tartrated antimony
UN 1551 [Antimony potassium tartrate]

28322-02-3
$C_{15}H_{13}NO$
223.29
CC(=O)NC2c1ccccc1c3ccccc23
Acetamide, N-fluoren-4-yl
Acetamide, N-9H-fluoren-4-yl- (9CI)
4-Acetylaminofluoren (German)
4-Acetylaminofluorene
N-Fluoren-4-ylacetamide
N-4-Fluorenylacetamide

28324-52-9
$C_{10}H_{18}O_2$
170.28
Pinanyl hydroperoxide
Hydroperoxide, 2,6,6-trimethylbicyclo(3.1.1)heptyl
Hydroperoxide, 2,6,6-trimethylbicyclo(3.1.1)heptyl-, Not over 45% peroxide
Pinane hydroperoxide, Technically pure (DOT)
Pinane hydroperoxide, Solution, Not over 45% peroxide (DOT)
Pinanyl hydroperoxide, Technically pure (DOT)
2,6,6-Trimethyl norpinanyl hydroperoxide, Technically pure (DOT)
UN 2162 (DOT)

28338-69-4
$C_{27}H_{46}$
370.67
Cholest-3-ene, (5α)- (9CI)
5α-Cholest-3-ene (8CI)

28343-61-5
$C_8HCl_3N_2O$
247.46
Oc1c(Cl)c(Cl)c(C#N)c(Cl)c1C#N
1,3-Benzenedicarbonitrile, 4-hydroxy-2,5,6-trichloro
4-Hydroxy-2,5,6-trichloro-1,3-benzenedicarbonitrile
4-Hydroxy-2,5,6-trichloroisophthalonitrile

28345-91-7
$C_{10}H_{20}O_2S$
204.33
Octane, 1-(ethenylsulfonyl)- (9CI)
Alvison 8
NSC-138831
Sulfone, octyl vinyl (8CI)

28347-13-9
$C_8H_8Cl_2$
175.06
Xylene, α,α'-dichloro
Benzene, bis(chloromethyl)- (9CI)
Bis(chloromethyl)benzene
Dichloroxylylene
α,α'-Dichloroxylene
Xylylene chloride
Xylylene dichloride

28348-53-0
$C_9H_{12}O_3S.Na$
223.25
Benzenesulfonic acid, (1-methylethyl)-, sodium salt (9CI)

28348-61-0
$C_{20}H_{34}O_3S.Na$
377.54
Benzenesulfonic acid, tetradecyl-, sodium salt (9CI)
Sodium tetradecylbenzenesulfonate

28351-04-4
$C_{11}H_{11}N$
157.21
Quinoline, dimethyl- (8CI,9CI)

28351-09-9
$C_9H_{10}O$
134.18
Benzaldehyde, dimethyl- (8CI,9CI)

28356-58-3
$C_7H_7NO_2$
137.13
O=C(O)Cc(ccnc1)c1
4-Pyridineacetic acid (9CI)

28382-15-2
$C_4H_4N_2O_2.K$
151.17
Potassium 1,2-dihydro-3,6-pyridazinedione
1,2-Dihydro-3,6-pyridazinedione monopotassium salt
Maleic hydrazide potassium salt
Potassium maleic hydrazide
3,6-Pyridazinedione, 1,2-di hydro-, monopotassium salt
Royal MH

28390-91-2
$(C_{13}H_{14}N_2.C_3H_5ClO)x$
Polymer
Benzenamine, 4,4'-methylenebis-, Polymer with (chloromethyl)oxirane (9CI)
4,4'-Methylenebis(benzenamine), (chloromethyl)oxirane polymer

28407-37-6
$C_{32}H_{16}Cu_2N_6O_{16}S_4.4Na$
1087.79
Direct Blue 218
Amanil Supra Blue 9GL
C.I. Direct Blue 218
C.I. 24401
Copper, (mu-((tetrahydrogen 3,3'-((3,3'-dihydroxy-4,4'-biphenylene)bis(azo))bis(5-amino-4-hydroxy-2,7-naphthalenedisulfonato))(4-)))di-, tetrasodium salt
Cuprate(4-), (mu-((3,3'-((3,3'-dihydroxy(1,1'-biphenyl)-4,4'-diyl)bis(azo))bis(5-amino-4-hydroxy- 2,7-naphthalenedisulfonato))(8-)))di-, tetrasodium (9CI)
(3,3'-((3,3'-Dihydroxy-1,1'-biphenyl-4,4'-diyl)bis(azo)bis(5-amino-2,7-naphthalenedisulfonato-(O4,O3)))dicopper, tetrasodium salt
Fastusol Blue 9GLP
2,7-Naphthalenedisulfonic acid, 3,3'((3,3'-dihydroxy(1,1'-biphenyl)-4,4'-diyl)bis(azo)bis-(5-amino-4-hydroxy-, sodium salt, copper complex
NCI-C60877
Pontamine Bond Blue B
Pontamine Fast Blue 7GLN

28411-49-6
$(C_{14}H_{14}O_4.C_5H_8O_2.C_4H_6O_2)x$
Polymer
Phthalic acid, diallyl ester, Polymer with ethyl acrylate and methacrylic acid (8CI)
Diallyl phthalate, ethyl acrylate, methacrylic acid polymer
Ethyl acrylate, methacrylic acid, 1,2-benzenedicarboxylic acid, di-2-propenyl ester polymer

28427-24-9
$C_5H_{14}NO.HO_3S$
185.24
Choline, sulfite (1:1) (Salt) (8CI)
Choline bisulfite
N,N,N-Trimethyl(2-hydroxyethyl)ammonium bisulfite

28434-00-6
$C_{19}H_{26}O_3$
302.45
CC(C)=CC2C(C(=O)OC1CC(=O)C(=C1C)CC=C)C2(C)C
Cyclopropanecarboxylic acid, 2,2-dimethyl-3-(2-methylpropenyl)-, (+)-(E)-, ester with (+)- 2-allyl-4-hydroxy-3-methyl-2-cyclopenten-1-one
AI 3-29024
d-t-Allethrin
trans-(+)-Allethrin
d-Allethrolone chrysanthemumate
(+)-Allethronyl (+)-trans-chrysanthemumate
s-Bioallethrin
s-trans-Bioallethrin
Esbiothrin
Esbioallethrin
Esdepallethrine
Esbiol
Esbiol Concentrate 90%
RU 16121

28434-01-7
$C_{22}H_{26}O_3$
338.48
CC(C)=CC3C(C(=O)OCc2ccc(Cc1ccccc1)o2)C3(C)C
Cyclopropanecarboxylic acid, 2,2-dimethyl-3-(2-methylpropenyl)-, (5-benzyl-3-furyl) methyl ester, d-trans
5-Benzyl-3-furylmethyl(+)-trans-chrysanthemate
Biobenzyfuroline
Bioresmethrin
Bioresmethrine
Bioresmetrina (Portuguese)
Combat White Fly Insecticide
FMC 18739
NIA-18739
NRDC 107
Resbuthrin
(+)-trans-Resmethrin
d-trans-Resmethrin
RU-11484
SBP-1390

28434-86-8
$C_{12}H_{10}Cl_2N_2O$
269.14
Nc2ccc(Oc1ccc(N)c(Cl)c1)cc2Cl
Ether, bis(4-amino-3-chlorophenyl)
Aniline, 4,4'-oxybis(2-chloro-
Bis(4-amino-3-chlorophenyl) ether
3,3'-Dichlor-4,4'-diamino-diphenylaether (German)
3,3'-Dichloro-4,4'-diaminodiphenyl ether
4,4'-Oxybis(2-chloroaniline)
4,4'-Oxybis(2-chloro-benzenamine)

28444-53-3
Unknown
Unknown
Nitrilotriacetic acid, monopotassium salt
NTA, monopotassium salt

28469-92-3
$C_8H_6Cl_2$
173.04
Benzene, 1,3-dichloro-2-ethenyl- (9CI)
NSC-89716
Styrene, 2,6-dichloro- (8CI)

28472-97-1
$C_{29}H_{56}O_4$
468.76
Nonanedioic acid, diisodecyl ester (9CI)
Azelaic acid, diisodecyl ester
Diisodecyl nonanedioate

28473-03-2
Unknown
Unknown
Isooctyl 2-(2-methyl-4-chlorophenoxy)propionate

28473-21-4
$C_9H_{20}O$
144.26
Nonanol (9CI)
AI3-28310-X

Nonanol (Mixed isomers)

28482-15-7
$C_{14}H_{31}NO_2$
245.40
OC(C)CN(CCCCCCCC)CC(O)C
2-Propanol, 1,1'-(octylimino)-bis- (9CI)
1,1'-(Octylimino)bis-2-propanol
2-Propanol, 1,1'-(octylimino)di-

28484-22-2
$C_{14}H_{28}O_2$
228.38
Undecanoic acid, 2,4,6-trimethyl- (8CI,9CI)
2,4,6-Trimethylhendecanoic acid
2,4,6-Trimethylundecanoic acid

28503-70-0
$C_7H_8O.1/4Ti$
120.11
Phenol, methyl-, titanium(4+) salt (9CI)
Methylphenol titanium(4+) salt
Tetracresyl titanate

28503-85-7
$C_{30}H_{46}O_2S$
470.76
Nonylphenol sulfide
Nonylphenyl sulfide
Phenol, thiobis(nonyl-

28514-45-6
$C_6H_4Br_2O$
251.91
Phenol, dibromo- (9CI)

28517-81-9
$C_{25}H_{23}NO_7S$
481.52
Benzenesulfonic acid, ((1-amino-9,10-dihydro-4-hydroxy-9,10-dioxo-2-anthracenyl)oxy)(1,1-dimethylpropyl)- (9CI)

28519-02-0
$C_{24}H_{34}O_7S_2.2Na$
544.64
Benzenesulfonic acid, dodecyl-(sulfophenoxy)-, disodium salt (9CI)
Benzenesulfonic acid, dodecyl (sulfophenoxy)-, disodium salt
Dodecyl(sulfophenoxy)benzenesulfonic acid, disodium salt

28519-06-4
$C_{10}H_{21}Cl$
176.76
CCCCCCCCCCCl
Decane, chloro
Decyl chloride (Mixed isomers)

28519-07-5
$C_{12}H_{25}Cl$
204.78
Dodecane, chloro- (8CI,9CI)

28520-00-5
$C_6H_2Cl_4O_2$
247.88
Resorcinol, tetrachloro

28523-79-7
$C_2H_7O_3PS.K$
181.21
Phosphorothioic acid, O,O-dimethyl ester, potassium salt (8CI,9CI)

28537-55-5
$C_9H_{11}NO_2$
165.19
Nitrous acid, 3-phenylpropyl ester (9CI)
3-Phenylpropyl nitrite
1-Propanol, 3-phenyl-, nitrite (7CI,8CI)

28540-82-1
$C_{22}H_{30}N_4O_4$
414.51
Tentoxin
Cycloleucyl-N-methylalanyl-glycyl-N-methyl dehydro-phenylalanine

28553-12-0
$C_{26}H_{42}O_4$
418.62
CC(C)CCCCCCOC(=O)c1cccc c1C(=O)OCCCCCCC(C)C
Diisononyl phthalate
1,2-Benzenedicarboxylic acid, diisononyl ester (9CI)
ENJ 2065
Diisononyl 1,2-benzenedicarboxylate
DINP
ENJ 2065
Palatinol DN
Palatinol N
Phthalic acid, diisononyl ester
Sansocizer DINP
Vestinol NN
Witamol 150

28554-00-9
$C_3H_5ClO_2$
108.52
Chloropropionic acid
α-Chloropropionic acid
Chloro-propanoic acid
Monochloropropionic acid
Propanoic acid, chloro- (9CI)
Propionic acid, chloro-
UN 2511 [α-Chloropropionic acid]

28558-32-9
$C_{10}H_7N_3S.H_3O_2P$
267.23
2-(4-Thiazolyl)benzimidazole, hypophosphite salt
Arbotect
Arbotect S
Arbotect 20-S
Benzimidazole, 2-(4-thiazolyl)-, monophosphinate
Phosphinic acid, Compd. with 2-(4-thiazolyl)-1H-benzimidazole (1:1)
Thiabendazole hypophosphite

28570-24-3
$C_{36}H_{72}O_4S_2Sn$
751.81
Dodecanoic acid, (dibutylstannylene)bis(thio-2,1-ethanediyl) ester (9CI)
Dibutyltin bis(2-mercaptoethyl dodecanoate)

28577-62-0
$C_4H_4Cl_2$
122.98
1,3-Butadiene, dichloro
DCBD
Dichlorobutadiene

28602-27-9
$C_9H_{11}NO$
149.19
Benzaldehyde, (dimethylamino)- (9CI)
(Dimethylamino)benzaldehyde

28605-74-5
$C_2Cl_4F_2$
203.83
Ethane, tetrachlorodifluoro- (9CI)
Tetrachlorodifluoroethane

28605-81-4
$C_{15}H_{22}N_2O_2$
262.34
Cyclohexane, 1,1'-methylenebis(isocyanato- (9CI)
Isocyanic acid, methylenedicyclohexylene ester

28609-66-7
$C_{21}H_{12}O$
280.33
8H-Dibenz(a,de)anthracen-8-one (8CI,9CI)

28623-46-3
$C_{19}H_{37}N$
279.50
N#CCCCCCCCCCCCCCCCCC
Nonadecanenitrile (9CI)
1-Cyanooctadecane
Nonadecanonitrile
NSC-148361
Octadecyl cyanide

28629-66-5
$C_{16}H_{35}O_2PS_2.1/2Zn$
387.25
Zinc O,O-diisooctyl dithiophosphate
Dithiophosphoric acid, O,O'-diisooctyl ester, zinc salt
Isooctyl ZDDP
Oronite
Phosphorodithioic acid, O,O-diisooctyl ester, zinc salt (9CI)
Zinc bis(diisooctyl dithiophosphate)
Zinc diisoctyl dithiophosphate
Zinc diisooctyl dithiophosphate

28631-35-8
Unknown
Unknown
Isooctyl 2-(2,4-dichlorophenoxy)propionate

28631-44-9
$C_{20}H_{43}O_2PS_2$
410.67
Phosphorodithioic acid, O,O-diisodecyl ester (9CI)
O,O-Diisodecyl phosphorodithioate

28632-15-7
C_6H_8S
112.20
Thiophene, dimethyl- (9CI)

28633-36-5
$C_{22}H_{40}N_2$
332.56
1,4-Benzenediamine, N,N'-di-sec-octyl- (9CI)
p-Phenylenediamine, N,N'-di-sec-octyl- (8CI)

28645-03-6
$C_{10}H_{16}O_2$
168.24
1-Pentalenecarboxylic acid, 1α,2,3,3aβ,4,5,6,6aβ-octahydro-3β-methyl- (8CI)

28652-04-2
$C_{16}H_{26}O$
234.42
Phenol, dipentyl
Diamylphenol
Phenol, diamyl-

28652-72-4
$C_{13}H_{12}$
168.24
Methylbiphenyl
Biphenyl, methyl-
1,1'-Biphenyl, methyl- (9CI)
Methyl-1,1'-biphenyl
Phenyltoluene

28652-77-9
$C_{13}H_{14}$
170.25
Naphthalene, trimethyl- (8CI,9CI)

28655-62-1
$C_{10}H_{12}O_2$
164.20
Benzenemethanol, (2-propenyloxy)- (9CI)
Allyloxy(hydroxymethyl)-benzene
Benzenemethanol, ar-(2-propenyloxy)-
(2-Propenyloxy)benzenemethanol

28655-71-2
$C_{12}H_3Cl_7$
395.32
1,1'-Biphenyl, heptachloro- (9CI)
Biphenyl, heptachloro- (8CI)
Heptachlorobiphenyl

28657-80-9
$C_{12}H_{10}N_2O_5$
262.24
CCn2nc(C(O)=O)c(=O)c3cc1OCOc1cc23
(1,3)Dioxolo(4,5-g)cinnoline-3-carboxylic acid, 1,4-dihydro-1-ethyl-4-oxo
Cinobac
Cinoxacin
Cinx
Compound 64716
1-Ethyl-1,4-dihydro-4-oxo(1,3)-dioxolo(4,5-g)cinnoline-3-carboxylic acid
1-Ethyl-6,7-methylenedioxy-4(1H)-oxocinnoline-3-carboxylic acid

28675-08-3
$C_{12}H_8Cl_2O$
239.10
Ether, dichlorophenyl
Dichloro diphenyl ether
Dichloro diphenyl oxide
Phenyl ether dichloro

28677-93-2
$C_4H_{10}O_2$
90.12
1-Propanol, methoxy- (9CI)
Methoxy-1-propanol

28679-13-2
$C_{17}H_{28}O$
248.41
Benzene, ethoxynonyl- (9CI)

28680-45-7
$C_7H_3Cl_7$
335.27
Heptachloronorbornene
Bicyclo(2.2.1)heptene, heptachloro-
Bicyclo(2.2.1)hept-2-ene, heptachloro- (9CI)
Heptachlorobicyclo(2.2.1)hept-2-ene

28685-18-9
$C_{14}H_{31}N.C_8H_6Cl_2O_3$
434.44
Acetic acid, (2,4-dichlorophenoxy)-, Compd. with 1-tetradecanamine (1:1) (9CI)
Acetic acid, (2,4-dichlorophenoxy)-, Compd. with tetradecylamine (1:1) (8CI)

28693-00-7
$C_{10}H_{13}ClO_2$
200.66
O=C(OCC(C(C=CC12)C1)C2)CCl
Acetic acid, chloro-, bicyclo(2.2.1)hept-5-en-2-ylmethyl ester (9CI)
Acetic acid, chloro-, 5-norbornen-2-ylmethyl ester
Bicyclo(2.2.1)hept-5-en-2-ylmethyl monochloroacetate
5-((Chloroacetoxy)methyl)-2-norbornene
NSC-44517

28699-88-9
$C_{10}H_6Cl_2$
197.06
Naphthalene, dichloro- (9CI)

28701-67-9
$C_{13}H_{29}NO.C_2H_4O_2$
275.43
1-Propanamine, 3-(isodecyloxy)-, acetate (9CI)
3-Isodecoxypropylamine, acetate
3-(Isodecyloxy)-1-propanamine acetate

28706-19-6

$C_{31}H_{26}N_6O_{11}S_3.3Na$
823.71
1,5-Naphthalenedisulfonic acid, 3-((4-((((2-methoxy-4-((3-sulfophenyl)azo) phenyl)-amino)carbonyl)amino)-2-methylphenyl)azo)-, trisodium salt (9CI)

28706-21-0
$C_{35}H_{28}N_6O_{13}S_4.4Na$
960.83
1,3-Naphthalenedisulfonic acid, 7,7'-(ureylenebis-((2-methyl-p-phenylene)-azo))di-, tetrasodium salt (8CI)

28706-22-1
$C_{35}H_{28}N_6O_{15}S_4.4Na$
992.83
1,5-Naphthalenedisulfonic acid, 3,3'-(carbonylbis-(imino(3-methoxy-4,1-phenylene)azo))bis-, tetrasodium salt (9CI)

28706-25-4
$C_{41}H_{28}N_6O_{15}S_4.4Na$
1064.89
2-Naphthalenesulfonic acid, 7,7'-(carbonyldiimino)bis-(4-hydroxy-3-((6-sulfo-2-naphthalenyl)azo)-, tetrasodium salt (9CI)
7,7'-(Carbonyldiimino)bis(4-hydroxy-3-((6-sulfo-2-naphthalenyl)azo)-2-naphthalenesulfonic acid), tetrasodium salt
7,7'-Ureylenebis(4-hydroxy-3-((6-sulfo-2-naphthyl)azo)-2-naphthalenesulfonic acid), tetrasodium salt

28715-26-6
$C_{10}H_{10}O$
146.19
Benzofuran, 4,7-dimethyl- (8CI,9CI)

28729-52-4
C_7H_{14}
98.19
Cyclopentane, dimethyl- (9CI)

28757-00-8
$C_{16}H_{20}O_3S$
292.40
Naphthalenesulfonic acid, bis-(1-methylethyl)- (9CI)
Bis(1-methylethyl)naphthalenesulfonic acid

28761-27-5
$C_{11}H_{22}$
154.30
Undecene (9CI)

28767-61-5
$C_{16}H_6N_4O_8$
382.26
Pyrene, 1,3,6,8-tetranitro
1,3,6,8-Tetranitropyrene

28768-32-3
$C_{25}H_{30}N_2O_4$
422.51
O(C1CN(c(ccc(c2)Cc(ccc(N(CC(O3)C3)CC(O4)C4)c5)c5)c2)CC(O6)C6)C1
Tetraglycidyl-4,4'-methylene dianiline
4,4'-Methylenebis(diglycidyl aniline)
Oxiranemethanamine, N,N'-(methylenedi-4,1-phenylene)-bis(N-(oxiranylmethyl)- (9CI)

28770-01-6
$C_8H_{17}NO_2$
159.22
O(CCN1CCO)C1C(C)C
3-Oxazolidineethanol, 2-(1-methylethyl)- (9CI)
2-Isopropyl-3-oxazolidine-ethanol
2-(1-Methylethyl)-3-oxazolidineethanol

28772-56-7
$C_{30}H_{23}BrO_4$
527.41
Bromadiolone
Boldo
Bromadialone
3-(3-(4'-Bromo(1,1-biphenyl)-4-yl)-3-hydroxy-1-phenyl-propyl)-4-hydroxy-2H-1-benzopyran-2-one (9CI)
3-(3-(4'-Bromo(1,1'-biphenyl)-4-yl)3-hydroxy-1-phenyl-propyl)-4-hydroxy-2H-1-benzopyran-2-one
3-(3-(4'-Bromo(1,1'-biphenyl)-4-yl)-3-hydroxy-1-phenyl-propyl)-4-hydroxy-2H-1-benzopyran-2-one
3-(3-(4'-Bromo-(1,1'-biphenyl)-4-yl)-3-hydroxy-1-phenyl-propyl)-4-hydroxy-2H-1-benzopyran-2-one
3-(3-(4'-Bromobiphenyl-4-yl)-3-hydroxy-1-phenylpropyl-4-hydroxycoumarin
3-(3-(4'-Bromo-(1,1'-biphenyl)-4-yl)-3-hydroxy-1-phenyl-propyl)-4-hydroxycoumarin
Bromone
3-(α-(p-(p-Bromophenyl)-β-hydroxyphenethyl)benzyl)-4-hydroxycoumarin
Broprodifacoum
Canadien 2000
Caswell No. 486AB
Contrac
Coumarin, 3-(3-(4'-bromo-1,1'-biphenyl-4-yl)-3-hydroxy-1-phenylpropyl)-4-hydroxy-
Coumarin, 3-(α-(p-(p-bromo-phenyl)-β-hydroxyphenethyl)-benzyl)-4-hydroxy-
EPA Pesticide Chemical Code 112001
(Hydroxy-4 coumarinyl 3)-3 phenyl-3 (bromo-4 bi-phenylyl-4)-1 propanol-1 (French)
LM 637
LM-637
Maki
Ratimus
Sup'Operats
Super-Caid
Super-Rozol
Temus

28777-60-8
C_8H_9Br
185.06
Benzene, (bromomethyl)methyl- (9CI)
Xylene, α-bromo- (8CI)

28777-67-5
C_8H_{18}
114.23
Hexane, dimethyl- (8CI,9CI)

28777-70-0
$C_{30}H_{39}O_4P$
494.66
Phosphoric acid, tris(tert-butylphenyl) ester
Mil-H-19457C
Phenol, (1,1-dimethylethyl)-, phosphate (3:1)
Tri(tert-butylphenyl) phosphate

28777-98-2
$C_{22}H_{38}O_3$
350.54
2,5-Furandione, dihydro-3-(octadecenyl)- (9CI)
Dihydro-3-(octadecenyl)-2,5-furandione
Octadecenylsuccinic anhydride

28779-08-0
$C_6H_3Br_3$
314.80
Benzene, tribromo- (8CI,9CI)

28779-32-0
$C_{16}H_{12}$
204.27
Pyrene, dihydro- (9CI)

28782-42-5
$C_{28}H_{28}N_2O_2$
424.53

Difenoxin
1-(3-Cyano-3,3-diphenylpropyl)-4-phenylisonipecotic acid
DEA No. 9168
Difenossina
Difenoxina (Spanish)
Difenoxine (French)
Difenoxinum (Latin)
Diphenoxin
Diphenoxylic acid
Isonipecotic acid, 1-(3-cyano-3,3-diphenylpropyl)-4-phenyl-
4-Piperidinecarboxylic acid, 1-(3-cyano-3,3-diphenyl propyl)-4-phenyl-
McN-JR-15,403-11
R 15,403

28789-80-2
$C_{14}H_{20}NO_5PS$
345.38
Carbamic acid, N-(O,O-dimethylphosphorothioyl)-N-methyl-, 2,2-dimethyl-2,3-dihydrobenzofuran- 7-yl ester
2,2-Dimethyl-2,3-dihydrobenzofuranyl-7-N-(O,O-dimethylphosphorothioyl)-N-methylcarbamate

28790-86-5
$C_8H_{12}O$
124.18
2-Cyclopenten-1-one, 2,3,4-trimethyl- (8CI,9CI)

28801-69-6
$C_{22}H_{46}O_2Sn$
461.32
Tributyltin neodecanoate
Carban T-10
Caswell No. 867G
EPA Pesticide Chemical Code 083111
Hydroxytributylstannane 4,4-dimethyloctanoate
Octanoic acid, 4,4-dimethyl-, tributylstannyl ester
Stannane, (4,4-dimethyloctanoyloxy)tributyl-
Stannane, (neodecanoyloxy)-tributyl-
Stannane, tributyl(neodecanoyloxy)-
Tin, tributyl-, neodecanoate

28802-49-5
C_6H_8O
96.14
Furan, dimethyl
Dimethyl furane

28804-67-3
$C_6HCl_4NO_2$
260.88
Tetrachloronitrobenzene
Benzene, tetrachloronitro- (9CI)
NSC-57595
Tetrachloronitrobenzene (VAN)

28804-85-5
$C_{12}H_7Cl_2N$
236.10
9H-Carbazole, dichloro- (9CI)
Carbazole, dichloro- (6CI, 8CI)

28804-88-8
$C_{12}H_{12}$
156.23
Naphthalene, dimethyl-
AI3-00957
AI3-24403-X
Dimethylnaphthalene

28804-96-8
$C_{13}H_9N$
179.22
(1,1'-Biphenyl)carbonitrile (9CI)
Biphenylcarbonitrile (8CI)

28805-75-6
$C_{15}H_9N$
203.24
Anthracenecarbonitrile (9CI)
Anthronitrile (8CI)

28805-90-5
$C_8H_8Br_2$
263.96
Benzene, dimethyl-, dibromo deriv. (9CI)
Xylene, dibromo- (8CI)

28807-97-8
C_7H_7Br
171.04
Benzene, bromomethyl- (9CI)
Bromomethylbenzene

28825-96-9
$(C_{12}H_{15}N_3O_6)x$
Polymer NA
1,3,5-Triazine-2,4,6(1H,3H,5H)-trione, 1,3,5-tris(oxiranylmethyl)-, Homopolymer (9CI)
Araldite PT 810
Araldite PT 816
Araldite TGIC
Epikote RXE 15
ETS (Cyanuric acid derivative)
Metallon E 5010
TEPIC-G
s-Triazine-2,4,6(1H,3H,5H)-trione, 1,3,5-tris(2,3-epoxypropyl)-, Polymers (8CI)
XB 2615

28832-11-3
$C_{22}H_{44}N_2O.C_2H_4O_2$
412.64
1H-Imidazole-1-ethanol, 2-heptadecyl-4,5-dihydro-, monoacetate (Salt) (9CI)

28836-03-5
$C_{16}H_{13}NO_3S.H_3N$
316.37
1-Naphthalenesulfonic acid, 8-(phenylamino)-, monoammonium salt
AI3-52565
Phenyl peri acid, ammonium salt

28865-36-3
$C_{20}H_{43}NO_4$
361.56
O=N(CC(O)CCCCCCCCCCCCCC)(CCO)CCO
2-Hexadecanol, 1-(bis(2-hydroxyethyl)amino)-, N-oxide (8CI)
N-(2-Hydroxyhexadecyl)diethanolamine oxide

28901-96-4
$C_{32}H_{18}CuN_8O_3S$
658.16
Cuprate(1-), (29H,31H-phthalocyaninesulfonato(3-)-N29,N30,N31,N32)-, hydrogen (9CI)

28903-26-6
$C_{16}H_{21}Cl_3O_4$
383.70
Propanoic acid, 2-(2,4,5-trichlorophenoxy)-, ester with butoxypropanol (9CI)
Propionic acid, 2-(2,4,5-trichlorophenoxy)-, butoxypropanol ester (8CI)
Propionic acid, 2-(2,4,5-trichlorophenoxy)-, ester with butoxypropanol

28906-38-9
C_6H_4BrCl
191.45
Benzene, bromochloro- (8CI, 9CI)

28906-96-9
Unknown
Polymer
Formaldehyde, Polymer with (chloromethyl)oxirane and 4,4'-(1-methylethylidene)bis-(phenol) (9CI)

28908-00-1
$C_8H_6ClNS_2$
215.72
N(c(c(S1)ccc2)c2)=C1SCCl
Benzothiazole, 2-((chloromethyl)thio)- (9CI)

28911-01-5
$C_{17}H_{12}Cl_2N_4$
343.19
Triazolam
8-Chloro-6-(o-chlorophenyl)-1-methyl-4H-s-triazolo(4,3-a)-(1,4)benzodiazepine
DEA No. 2887
Halcion
Triazolamum (Latin)
4H-s-Triazolo(4,3-a)(1,4)benzodiazepine, 8-chloro-6-(o-chlorophenyl)-1-methyl-
4H-(1,2,4)Triazolo(4,3-a)-(1,4)benzodiazepine, 8-chloro-6-(2-chlorophenyl)-1-methyl-
U-33,030

28927-38-0
Unknown
Unknown
Nitrilotriacetic acid, holmium salt
NTA, holmium salt

28928-97-4
$C_{13}H_{22}O_3$
226.32
2,5-Furandione, dihydro-3-(nonenyl)- (9CI)

28961-43-5
Unknown
Unknown
Ethanol, 2,2',2''-(propylidynetris(methyleneoxy))tri-, triacrylate
Trimethylolpropane ethoxytriacrylate

28981-97-7
$C_{17}H_{13}ClN_4$
308.75
Alprazolam
Alprazolamum (Latin)
8-Chloro-1-methyl-6-phenyl-4H-s-triazolo(4,3-a)(1,4)benzodiazepine
D 65MT
DEA No. 2882
Xanax
4H-s-Triazolo(4,3-a)(1,4)benzodiazepine, 8-chloro-1-methyl-6-phenyl-
4H-(1,2,4)Triazolo(4,3-a)-(1,4)benzodiazepine, 8-chloro-1-methyl-6-phenyl-
TUS-1
U 31889

28983-26-8
$C_{16}H_{26}O$
234.39
Phenol, 2-isononyl-4-methyl- (9CI)

28983-37-1
$C_{14}H_{30}S$
230.46
tert-Tetradecyl mercaptan
tert-Tetradecanethiol (9CI)

28984-69-2
$C_{22}H_{41}NO_3$
367.58
4,4(5H)-Oxazoledimethanol, 2-(heptadecenyl)-

28984-80-7
$C_4Cl_3F_7$
287.39
Butane, trichloroheptafluoro- (8CI,9CI)

28984-85-2
$C_{12}H_9NO_2$
199.21
Nitrobiphenyl

28984-89-6
$C_{18}H_{14}O$
246.31
1,1'-Biphenyl, phenoxy- (9CI)

28986-55-2
$C_{36}H_{58}S_2$
554.99
Disulfide, bis(dodecylphenyl) (8CI)

28987-17-9
$C_{15}H_{23}BaO$
356.69
Phenol, nonyl-, barium salt (9CI)

28994-41-4
$C_{13}H_{12}O$
184.24
Oc(c(ccc1)Cc(cccc2)c2)c1
2-Benzylphenol
Phenol, 2-(phenylmethyl)-

28995-89-3
$C_{10}H_4N_4O_8$
308.18
Naphthalene, 1,3,6,8-tetranitro
1,3,6,8-Tetranitronaphthalene

29006-00-6
$C_7H_{14}O_2$
130.19
2-Hexanone, 6-methoxy- (8CI,9CI)
4-Methoxybutyl methyl ketone
6-Methoxyhexan-2-one

29006-06-2
$C_9H_{18}O_3$
174.24
Butyric acid, 4-butoxy-, methyl ester (8CI)

29010-86-4
$C_8H_6O_4$
166.13
Benzenedicarboxylic acid (8CI,9CI)

29027-90-5
Unknown
Unknown
Nitrilotriacetic acid, cerium salt
NTA, cerium salt

29036-02-0
$C_{24}H_{18}$
306.41
Quaterphenyl (9CI)

29036-25-7
$C_{10}H_{10}$
130.19
1H-Indene, methyl- (9CI)
Indene, methyl-

29043-70-7
$C_{38}H_{80}O_2Si$
597.15
Silane, dimethylbis(octadecyloxy)- (9CI)

29062-98-4
$C_{16}H_{14}$
206.29
Phenanthrene, dimethyl- (9CI)

29063-00-1
$C_{16}H_{14}$
206.30
Anthracene, dimethyl
Dimethylanthracene

29063-28-3
$C_8H_{18}O$
130.23
Octanol (9CI) (VAN)

29071-93-0

$C_7H_{12}N_2O_3$
172.19
O=C(N(C(=O)C1(C)C)CCO)N1
Hydantoin, 3-(2-hydroxyethyl)-5,5-dimethyl- (8CI)
2,4-Imidazolidinedione, 3-(2-hydroxyethyl)-5,5-dimethyl- (9CI)

29082-74-4
C_8Cl_8
379.71
Octachlorostyrene

29086-38-2
C_8HCl_7
345.26
Benzene, pentachloro(1,2-dichloroethenyl)-, (E)- (9CI)
Styrene, α,β,2,3,4,5,6-heptachloro-, (E)- (8CI)

29086-39-3
C_8HCl_7
345.26
Benzene, pentachloro(1,2-dichloroethenyl)-, (Z)- (9CI)
Styrene, α,β,2,3,4,5,6-heptachloro-, (Z)- (8CI)

29091-05-2
$C_{11}H_{13}F_3N_4O_4$
322.28
CCN(CC)c1c(cc(c(N)c1N(=O)=O)C(F)(F)F)N(=O)=O
Toluene-2,4-diamine, N[4],N[4]-diethyl-3,5-dinitro-α,α,α-trifluoro
1,3-Benzenediamine, N[3],N[3]-diethyl-2,4-dinitro-6-(trifluoromethyl)-
Cobex
Cobex (Herbicide)
Cobexo
3-Diethylamino-2,4-dinitro-6-trifluoromethylaniline
N[4],N[4]-Diethyl-α,α,α-trifluoro-3,5-dinitrotoluene, 2,4-diamine
N[3],N[3]-Diethyl-2,4-dinitro-6-(trifluoromethyl)-1,3-benzenediamine
Dinitramine
Dinitroamine
Toluene-2,4-diamine, N[4],N[4]-diethyl-α,α,α-trifluoro-3,5-dinitro-
USB-3584

29091-09-6
$C_7HCl_2F_3N_2O_4$
305.00
O=N(=O)c(c(c(N(=O)=O)c(c1C(F)(F)F)Cl)Cl)c1
Benzene, 2,4-dichloro-1,3-dinitro-5-(trifluoromethyl)- (9CI)

29103-58-0
$C_{17}H_{20}N_2O$
268.36
O=C(Nc(cccc1N(Cc(cccc2)c2)CC)c1)C
Acetamide, N-(3-(ethyl-(phenylmethyl)amino)-phenyl)- (9CI)

29110-22-3
$C_3H_4Cl_2O_2.1/2Mg$
155.13
Magnesium dalapon
2,2-Dichloropropionic acid, magnesium salt
Magnesium 2,2-dichloropropanoate
Propanoic acid, 2,2-dichloro-, magnesium salt (9CI)

29122-68-7
$C_{14}H_{22}N_2O_3$
266.38
CC(C)NCC(O)COc1ccc(CC(N)=O)cc1
Acetamide, 2-(p-(2-hydroxy-3-(isopropylamino)propoxy)-phenyl)
Atenolol
Benzeneacetamide, 4-(2-hydroxy-3-((1-methylethyl)-amino)propoxy)- (9CI)
1-p-Carbamoylmethylphenoxy-3-isopropylamino-2-propanol
2-(p-(2-Hydroxy-3-(isopropyl-amino)propoxy)phenyl)acetamide
4-(2-Hydroxy-3-((1-methylethyl)amino)propoxy)benzeneacetamide
ICI 66082
Normiten
Tenormin

29128-56-1
$C_{17}H_{12}N_2O_3$
292.30
Benzoic acid, 2-((2-hydroxy-1-naphthalenyl)-azo)- (9CI)
1-(1'-Carboxy-2'-benzeneazo)-2-naphthol
1-((2-Carboxyphenyl)azo)-2-naphthol.
1-(2'-Carboxyphenylazo)-2-naphthol
2-(2-Hydroxy-1-naphthyl-azo)benzoic acid

29135-62-4
$C_{14}H_6N_8O_{14}$
510.20
Hexanitrooxanilide
N,N'-Bis(2,4,6-trinitrophenyl)ethanediamide
Dipicryloxamide
Ethanediamide, N,N'-bis-(2,4,6-trinitrophenyl)- (9CI)
Hexanitrooxanilida (Spanish)
Hexanitrooxanilide (French)
2,2',4,4',6,6'-Hexanitrooxanilide
Oxamide, N,N'-dipicryl-

29136-19-4
$C_{25}H_{44}$
344.63
c(cccc1)(c1)CCCCCCCCCCCCCCCCCCC
Benzene, nonadecyl- (9CI)

29154-12-9
C_5H_4ClN
113.54
Pyridine, chloro- (8CI,9CI)

29171-20-8
$C_{10}H_{16}O$
152.24
OC(C#C)(CCC=C(C)C)C
6-Octen-1-yn-3-ol, 3,7-dimethyl-

29171-23-1
$C_{20}H_{38}O$
294.52
1-Hexadecyn-3-ol, 3,7,11,15-tetramethyl- (8CI,9CI)
Dehydroisophytol
Isophytol, dehydro-

29173-31-7
$C_7H_{14}NO_4PS_2$
271.31
Phosphonodithioic acid, methyl-, S-((N-methoxycarbonyl)-N-methylcarbamoyl) methyl O-methyl ester
MC 2420
Mecarphon
Mecarphos
S-(N-Methoxycarbonyl-N-methylcarbamoylmethyl)dimethyl phosphonothiolothionate
Methyl (((methoxymethylphosphinothioyl)thio)acetyl)-methylcarbamate
2-Oxa-4-thia-7-aza-3-phosphaoctan-8-oic acid, 3,7-dimethyl-6-oxo-, methyl ester, 3-sulfide

29177-84-2
$C_{18}H_{14}ClFN_2O_3$
360.76
Ethyl loflazepate
1H-1,4-Benzodiazepine-3-carboxylic acid, 7-chloro-5-(2-fluorophenyl)-2,3-dihydro-2-oxo-, ethyl ester
CM 6912
DEA No. 2758
Ethyl 7-chloro-5-(o-fluoro-

phenyl)-2,3-dihydro-2-oxo-1H-1,4-benzodiazepine-3-carboxylate
Ethyl fluclozepate
Ethylis loflazepas (Latin)
Ethylloflazepate
Loflazepate d'ethyle (French)
Loflazepato de etilo (Spanish)

29181-67-7
$C_5H_2F_6N_4$
232.09
s-Triazine, 2-amino-4,6-bis(trifluoromethyl)-(8CI)

29181-68-8
$C_6H_4F_6N_4$
246.12
s-Triazine, 2-(methylamino)-4,6-bis(trifluoromethyl)- (8CI)

29188-28-1
$C_{32}H_{18}CuN_8O_6S_2$
738.22
Cuprate(2-), (29H,31H-phthalocyaninedisulfonato-(4-)-N29,N30,N31,N32)-, dihydrogen (9CI)

29191-52-4
C_7H_9NO
123.16
Aminoanisole
Anisidine
Anisidine, isomers (8CI)
Anisidine (o-,p-Isomers)
Benzenamine, ar-methoxy-(9CI)
Methoxyaniline

29204-84-0
$C_{20}H_{22}N_6NiO_6$
501.14
Nickel, bis(2,3-bis(hydroxyimino)-N-phenylbutanamidato-N2,N3)- (9CI)

29218-27-7
$C_{11}H_{13}NO_3$
207.25
2-Oxazolidinone, 5-hydroxymethyl-3-(m-tolyl)
Delalande 69276
5-(Hydroxymethyl)-3-(3-methylphenyl)-2-oxazolidinone
5-Hydroxymethyl-3-(m-tolyl)-2-oxazolidinone
69276 MD
Toloxatone

29222-39-7
$C_{12}H_{10}O_3$
202.22
1,2,4-Benzenetriol, phenyl

29222-48-8
C_8H_{18}
114.23
Pentane, trimethyl- (8CI,9CI)

29224-55-3
$C_{10}H_{14}$
134.22
Benzene, ethyldimethyl- (9CI)
Xylene, ethyl- (8CI)

29225-54-5
$C_{11}H_{16}O$
164.25
Benzene, butoxymethyl- (9CI)
Ether, butyl tolyl (8CI)

29232-93-7
$C_{11}H_{20}N_3O_3PS$
305.37
CCN(CC)c1nc(C)cc(OP(=S)(OC)OC)n1
Phosphorothioic acid, O-(2-(diethylamino)-6-methyl-4-pyrimidinyl) O,O-dimethyl ester
Actelic
Actellic
Actellifog
Blex
2-Diethylamino-6-methylpyrimidin-4-yl dimethyl phosphorothionate
O-(2-(Diethylamino)-6-methyl-4-pyrimidinyl)O,O-dimethyl phosphorothioate
O-(2-Diethylamino-6-methylpyrimidin-4-yl) O,O-dimethyl phosphorothioate
ENT 27,699GC
Methylpirimiphos
OMS 1424
Pirimifosmethyl
Pirimiphos-methyl
Plant Protection PP511
PP511
Pyridimine phosphate
Pyrimiphos methyl
Silosan
Sybol 2

29240-17-3
$C_{10}H_{20}O_3$
188.27
O=C(OOC(CC)(C)C)C(C)(C)C
Propaneperoxoic acid, 2,2-dimethyl-, 1,1-dimethylpropyl ester (9CI)

29253-36-9
$C_{26}H_{28}$
340.51
Naphthalene, (1-methylethyl)-

29256-79-9
$C_2Br_2ClF_3$
276.29
Ethane, chlorodibromotrifluoro
Chlorodibromotrifluoroethane
Dibromotrifluoromonochloroethane
Halon 2312
Monochlorodibromotrifluoroethane

29263-10-3
$C_{12}H_{21}N_3O_3$
255.30
1,3,5-Triazine, 2,4,6-tripropoxy-(9CI)
s-Triazine, 2,4,6-tripropoxy-(8CI)

29263-11-4
$C_{12}H_{21}N_3O_3$
255.32
1,3,5-Triazine, 2,4,6-tris-(1-methylethoxy)- (9CI)
s-Triazine, 2,4,6-triisopropoxy- (8CI)
2,4,6-Triisopropyloxy-s-triazine

29267-75-2
$C_6H_2N_4O_6.K$
265.18
Potassium dinitrobenzofuroxan
Benzofurazan, dinitro-, ion(1-), 1-oxide, potassium
Benzofurazan, 4,6-dinitro-, 1-oxide, potassium salt (8CI)
4-Benzofurazanol, 1,4-dihydro-5,7-dinitro-, 3-oxide, ion(1-), potassium (9CI)
KDNBF
Potassium dinitrobenzofuroxan

29268-77-7
$C_{14}H_{14}N_2O$
226.28
Azobenzene, 2-methoxy-6-methyl- (8CI)

29268-78-8
$C_{14}H_{14}N_2O$
226.28
Azobenzene, 2-methoxy-2'-methyl- (8CI)
2-Methoxy-2'-methylazobenzene
Anisole, o-(o-tolylazo)-(7CI)

29291-35-8
$C_{19}H_{18}N_8O_7$
470.45
Glutamic acid, N-nitroso-N-pteroyl-, L

Nitrosofolic acid

29298-03-1
$(C_{13}H_{18}O_2)x$
Polymer NA
Oxirane, ((4-(1,1-dimethylethyl)phenoxy)methyl)-, Homopolymer (9CI)
Epi-Rez 5014
Propane, 1-(p-tert-butylphenoxy)-2,3-epoxy-, Polymers (8CI)

29304-40-3
$C_9H_{16}O_3$
172.22
Hexanoic acid, 2-methyl-3-oxo-, ethyl ester (8CI,9CI)

29316-05-0
$C_{11}H_{16}$
148.27
Benzene, sec-pentyl
sec-Amylbenzene
sec-Pentylbenzene

29321-75-3
Unknown
Unknown
2-Naphthalenesulfonic acid, sodium salt, Polymer with formaldehyde (9CI)

29329-71-3
$C_2H_9NaO_7P_2$
230.03
Phosphonic acid, (1-hydroxyethylidene)bis-, sodium salt (9CI)
1-Hydroxyethanediphosphonic acid, sodium salt
Hydroxyethylidene diphosphonic acid, sodium salt

29350-73-0
$C_{15}H_{26}$
206.41
CC(C)C1CC=C(C)C2CC=C(C)CC12
Cadinene

29354-98-1
$C_{16}H_{34}O$
242.45
Hexadecanol (8CI,9CI) (VAN)

29366-72-1
$C_9H_8N_6O_2$
232.17
Nc1nc(N)nc(n1)c2cccc(c2)N(=O)=O
1,3,5-Triazine-2,4-diamine, 6-(3-nitrophenyl)- (9CI)
NSC-121172
s-Triazine, 2,4-diamino-6-(m-nitrophenyl)- (8CI)

29383-29-7
$C_{32}H_{14}CoN_8O_6S_2.2H$
731.55
Cobaltate(2-), (29H,31H-phthalocyaninedisulfonato-(4-)-N29,N30,N31,N32)-, dihydrogen (9CI)

29385-43-1
$C_7H_7N_3$
133.17
1H-Benzotriazole, methyl
Cobratec TT 100
Methyl-1H-benzotriazole

29387-86-8
$C_7H_{16}O_2$
132.23
CCCCOC(C)CO
1,2-Propanediol, monobutyl ether
Butoxypropanol
2-Butoxy-1-propanol
Propasol B
Propasol Solvent P
Propylene glycol butoxy ether
α-Propylene mono-n-butyl ether

29398-96-7
$C_{26}H_{20}N_6O_{10}$
576.44
O=N(=O)c(c(Nc(c(OC)cc(c(ccc(Nc(c(N(=O)=O)cc(N(=O)=O)c1)c1)c2OC)c2)c3)c3)ccc4N(=O)=O)c4
(1,1'-Biphenyl)-4,4'-diamine, N,N'-bis(2,4-dinitrophenyl)-3,3'-dimethoxy- (9CI)
4,4'-Bis((2,4-dinitrophenyl)-amino)-3,3'-dimethoxybiphenyl

29407-84-9
$(C_{15}H_{16}O_2.C_7H_{14}O_2.C_3H_5ClO)x$
Unknown
Epon 815

29414-47-9
$C_2H_6S_2$
94.20
Methanethiol, (methylthio)-(8CI,9CI)

29418-21-1
$C_{14}H_{14}N_2$
210.28
Azobenzene, 2,4-dimethyl-(8CI)

29418-22-2
$C_{14}H_{14}N_2$
210.28
Diazene, (2-methylphenyl)(4-methylphenyl)-(9CI)
o,p'-Azotoluene (7CI,8CI)

29418-23-3
$C_{15}H_{16}N_2$
224.31
Azobenzene, 2,2',4-trimethyl- (8CI)

29418-24-4
$C_{15}H_{16}N_2$
224.31
Azobenzene, 2,4,4'-trimethyl- (8CI)

29418-25-5
$C_{16}H_{18}N_2$
238.32
Diazene, bis(2,4-dimethylphenyl)- (9CI)
Azobenzene, 2,2',4,4'-tetramethyl- (8CI)

29418-26-6
$C_{15}H_{16}N_2$
224.31
Azobenzene, 2,4,6-trimethyl- (6CI,7CI,8CI)

29418-59-5
$C_{13}H_{11}N_3O_3$
257.27
Azobenzene, 4'-methoxy-4-nitro
4'-Methoxy-4-nitroazobenzene

29420-49-3
$C_4HF_9O_3S.K$
339.20
1-Butanesulfonic acid, 1,1,2,2,3,3,4,4,4-nonafluoro-, potassium salt (9CI)
1,1,2,2,3,3,4,4,4-Nonafluorobutane-1-sulfonic acid, potassium salt

29426-52-6
$C_{22}H_{26}N_4O_8S$
506.51
O=C(OCCN(c(ccc(N=Nc(c(S(=O)(=O)C)cc(N(=O)=O)c1)c1)c2C)c2)CCOC(=O)C)C
Ethanol, 2,2'-((3-methyl-4-((2-(methylsulfonyl)-4-nitrophenyl)azo)phenyl)imino)-bis-, diacetate (ester) (9CI)
Ethanol, 2,2'-((4-((2-(methylsulfonyl)-4-nitrophenyl)azo)-m-tolyl)imino)di-, diacetate (ester)

29446-15-9
$C_{12}H_6Cl_2O_2$

253.08
Dibenzo-p-dioxin, 2,3-dichloro
Dibenzo(b,e)(1,4)dioxin, 2,3-di-chloro-
2,3-Dichlorodibenzodioxin
2,3-Dichlorodibenzo-para-dioxin

29454-23-7
$C_{15}H_{30}O_5S$
322.47
O=C(OC)C(S(=O)(=O)O)CCCCCCCCCCCC
Tetradecanoic acid, 2-sulfo-, 1-methyl ester (9CI)
Methyl myristate α-sulfonic acid
1-Methyl 2-sulfotetradecanoate

29460-90-0
$C_7H_{10}N_2$
122.16
n(ccnc1C(C)C)c1
Pyrazine, (1-methylethyl)- (9CI)
(1-Methylethyl)pyrazine

29473-77-6
$C_{10}H_{18}O_4$.Pb
409.45
Lead sebacate
Decanedioic acid, lead(2+) salt (1:1) (9CI)

29507-58-2
Unknown
Unknown
Nitrilotriacetic acid, zinc(3+) complex sodium salt
NTA, zinc(3+) complex sodium salt

29512-49-0
$C_{31}H_{28}N_2O_3$
476.56
O=C(OC(c(c(Oc1cc(N(CC)CC)cc2)cc(c3Nc(cccc4)c4)C)c3)(c12)c5cccc6)c56
Spiro(isobenzofuran-1(3H),9'-(9H)xanthen)-3-one, 6'-(diethylamino)-3'-methyl-2'-(phenylamino)- (9CI)
7-Anilino-3-diethylamino-6-methylfluoran

29556-33-0
$C_{22}H_{24}N_3O_2$.Cl
397.89
1H-Benzimidazolium, 2-(7-(diethylamino)-2-oxo-2H-1-benzopyran-3-yl)-1,3-dimethyl-, chloride (9CI)

29560-84-7
C_3H_5ClO
92.53
OCC=CCl
2-Propen-1-ol, 3-chloro
pi-Chloro allyl alcohol

29590-42-9
$C_{11}H_{20}O_2$
184.28
2-Propenoic acid, isooctyl ester
AI3-28217
Isooctyl 2-propenoate

29598-76-3
$C_{65}H_{124}O_8S_4$
1161.96
Propanoic acid, 3-(dodecylthio)-, 2,2-bis((3-(dodecylthio)-1-oxopropoxy)methyl)-1,3-propanediyl ester (9CI)

29611-84-5
$C_8H_{11}N$
121.20
Pyridine, trimethyl
Collidine
Trimethylpyridine

29611-97-0
$(C_{10}H_{18}O_4)x$
Polymer
Oxirane, 2,2'-(1,4-butanediylbis(oxymethylene))bis-, Homopolymer (9CI)

29637-28-3
$C_{20}H_{14}N_2O_{14}S_4$.4Na
726.55
1,3-Naphthalenedisulfonic acid, 7-((1,8-dihydroxy-3,6-disulfo-2-naphthyl)azo)-, tetrasodium salt (8CI)

29637-52-3
$C_{42}H_{46}N_{14}O_{16}S_4$.4Na
1223.05
2,2'-Stilbenedisulfonic acid, 4,4'-bis((4-((2-carbamoylethyl)(2-hydroxyethyl)amino)-6-(p-sulfoanilino)-s-triazin-2-yl)amino)-, tetrasodium salt (8CI)

29649-47-6
$C_{22}H_{23}ClN_6O_5$
486.88
O=C(Nc(c(N=Nc(c(cc(N(=O)=O)c1)Cl)c1)ccc2N(CCN(C(=O)CC3)C3=O)CC)c2)C
Acetamide, N-(2-((2-chloro-4-nitrophenyl)azo)-5-((2-(2,5-dioxo-1-pyrrolidinyl)ethyl)ethylamino)phenyl)- (9CI)
4-(2-Chloro-4-nitrophenylazo)-N-ethyl-N-(β-succinimido-ethyl)-3-acetamidoaniline

29656-52-8
$C_{27}H_{50}N$.Cl
424.15
Ammonium, (ethylbenzyl)-hexadecyldimethyl-, chloride (VAN)(8CI)
Hexadecyldimethyl(ethylbenzyl)ammonium chloride

29658-97-7
$C_{16}H_{28}O_4$
284.40
Butanedioic acid, dodecenyl- (9CI)
Dodecenylbutanedioic acid
Dodecenylsuccinic acid

29662-90-6
Unknown
Unknown
O=C(O)C(CCCCCC)(C)C
Dimethyloctanoic acid

29689-14-3
CH_4CrO_3
116.04
Chromium carbonate
Carbonic acid, chromium salt (9CI)

29714-87-2
$C_{10}H_{16}$
136.26
Octatriene, dimethyl
Dimethyloctatriene (Mixed isomer)
Ocimene

29718-44-3
$C_{14}H_{30}O_2$
230.39
Ethoxy-1-dodecanol
1-Dodecanol, ethoxy-
Dodecyl alcohol, ethoxylated
Dodecyl alcohol (ethoxylated)
NCI-C54875

29733-18-4
$C_{25}H_{48}O_4$
412.65
Pentanedioic acid, diisodecyl ester (9CI)
Diisodecyl glutarate
Diisodecyl pentanedioate

29736-75-2
$C_6H_{12}O_6Sb_2$
423.66
2,5,7,10,11,14-Hexaoxa-1,6-distibabicyclo(4.4.4)tetradecane (9CI)
Antimony tris(ethylene glyc-

oxide)

29756-37-4
$C_7H_{15}Cl$
538.60
Heptane, chloro- (8CI,9CI)
Chloroheptane

29757-24-2
$C_{19}H_{22}N_6O_6$
430.42
Nitroaniline
UN 1661

29759-77-1
C_8H_{10}
318.51
Cyclooctatriene (6CI,8CI, 9CI)

29761-21-5
$C_{22}H_{31}O_4P$
390.46
Phosphoric acid, isodecyl diphenyl ester (9CI)
Isodecyl diphenyl phosphate

29765-00-2
$C_{27}H_{27}N_5O_7$
533.51
O=C(OCCN(c(ccc(N=Nc(ccc(N(=O)=O)c1)c1)c2NC(=O)c(cccc3)c3)c2)CCOC(=O)C)C
Benzamide, N-(5-(bis(2-(acetyloxy)ethyl)amino)-2-((4-nitrophenyl)azo)phenyl)-(9CI)
5'-(Bis(2-hydroxymethyl)-amino)-2'-((p-nitrophenyl)-azo)benzanilide, diacetate (ester)
2-(4-Nitrophenylazo)-5-(N,N-bis(acetoxyethyl)amino)benzanilide
N-(2-((4-Nitrophenyl)azo)-5-(N,N-bis(2-acetoxyethyl)-amino)phenyl)benzamide

29790-52-1
$C_{10}H_{14}N_2.C_7H_6O_3$
300.35
Nicotine salicylate
Nicotine, monosalicylate
Salicilato de nicotina (Spanish)
Salicylate de nicotine (French)
UN 1657 [Nicotine salicylate]

29797-09-9
C_6H_8
80.13
Cyclohexadiene (8CI,9CI)

29797-40-8
$C_7H_6Cl_2$
161.03
Benzene, dichloromethyl-(9CI)
Dichloromethylbenzene
Dichlorotoluene

29803-57-4
Unknown
Unknown
S-(2-Hydroxypropyl)thiomethanesulfonate

29804-22-6
$C_{19}H_{38}O$
282.51
O(C1CCCCCCCCCC)C1CCCCC(C)C
cis-7,8-Epoxy-2-methyloctadecane
AI3-34886
Caswell No. 424A
cis-2-Decyl-3-(5-methylhexyl)-oxirane
Disparlure
EPA Pesticide Chemical Code 114301
Oxirane, 2-decyl-3-(5-methylhexyl)-, cis-

29806-73-3
$C_{24}H_{48}O_2$
368.72
O=C(OCC(CCCC)CC)CCCCCCCCCCCCCCC
Palmitic acid, 2-ethylhexyl ester
Ceraphyl 368
2-Ethylhexyl palmitate
Hexadecanoic acid, 2-ethylhexyl ester (9CI)
Octyl palmitate
Wickenol 155

29808-66-0
$C_9H_6Cl_9N_3O_3$
523.24
s-Triazine, 2,4,6-tris-(2,2,2-trichloroethoxy)-(8CI)

29812-79-1
$C_{10}H_{23}NO$
173.29
Hydroxylamine, O-decyl-(8CI,9CI)

29813-38-5
$C_9H_{18}O_2.1/2Ca$
178.28
Nonanoic acid, calcium salt (9CI)
Calcium nonanoate
Calcium pelargonate

29828-28-2
$C_{10}H_{10}$
130.19
Naphthalene, dihydro- (8CI, 9CI)

29849-01-2
$C_6H_9ClN_2O$
160.60
Phenol, 2,4-diamino-, hydrochloride (8CI,9CI)

29857-13-4
$C_{24}H_{46}O_7S.Na$
501.68
Butanedioic acid, sulfo-, 1,4-diisodecyl ester, sodium salt (9CI)
Diisodecyl sulfosuccinate, sodium salt

29868-05-1
$C_2H_7NO.xH_3O_4P$
Unknown
Ethanol, 2-amino-, phosphate (Salt) (9CI)
2-Aminoethanol phosphate (Salt)
Monoethanolamine phosphate
Phosphoric acid, Compound with 2-aminoethanol

29868-16-4
Unknown
Unknown
(E)-9-Dodecenyl acetate

29870-99-3
$C_6H_{10}O_6.Sr$
265.78
Lactic acid, strontium salt (2:1)
2-Hydroxypropanoic acid strontium salt
Strontium lactate
Strontolac

29878-91-9
$C_{22}H_{14}O_4$
342.35
O=C(O)c(c(c(cc1)ccc2)c2c(c(cc c3)cc4)c3C(=O)O)c4)c1
(1,1'-Binaphthalene)-8,8'-dicarboxylic acid (9CI)
1,1'-Binaphthyl-8,8'-dicarboxylic acid
Dina acid
NSC-7810

29883-15-6
$C_{20}H_{27}NO_{11}$
457.48
OCC3OC(OCC2OC(OC(C#N)c1ccccc1)C(O)C(O)C2O)C(O)C(O)C3O
D(-)-Mandelonitrile-β-D-

gentiobioside
Amygdalin
D-Amygdalin
R-Amygdalin
Amygdaloside
Benzeneacetonitrile, α-((6-O-β-D-glucopyranosyl-β-D-glucopyranosyl)oxy)-, (R)-
(R)-α-((6-O-β-D-Glucopyranosyl-β-D-glucopyranosyl)oxy)-benzeneacetonitrile
D-Mandelonitrile-β-D-glucosido-6-β-D-glucoside
Mandelonitrile-β-gentiobioside
NSC-15780

29901-85-7
$C_7H_{12}O_2$
128.17
3-Heptenoic acid (8CI,9CI)

29911-28-2
$C_{10}H_{22}O_3$
190.32
O(CC(OCC(O)C)C)CCCC
2-Propanol, 1-(2-butoxy-1-methoxy)
Dipropylene glycol, butyl ether

29927-08-0
$C_9H_{10}N_2S$
178.25
2-Benzothiazolamine, 5,6-dimethyl- (9CI)
Benzothiazole, 2-amino-5,6-dimethyl- (8CI)
NSC-140729

29946-28-9
$C_{16}H_{32}O_2S$
288.49
O=C(OCCS)CCCCCCCCCCCCC
Tetradecanoic acid, 2-mercaptoethyl ester (9CI)
2-Mercaptoethyl tetradecanoate

29963-76-6
$(C_{18}H_{16}O_5)x$
Polymer
2-Propenoic acid, 2-(4-benzoyl-3-hydroxyphenoxy)-ethyl ester, Homopolymer (9CI)

29964-84-9
$C_{14}H_{26}O_2$
226.40
Methacrylic acid, isodecyl ester
Ageflex FM-10
Isodecyl methacrylate

29973-13-5
$C_{11}H_{15}NO_2S$
225.33
CCSCc1ccccc1OC(=O)NC
Carbamic acid, methyl-, 2-(ethylthiomethyl)phenyl ester
Bay-Hox-1901
Croneton
Ethiofencarb
Ethiophencarbe
Ethiophencarp
2-Ethyl-mercaptomethyl-phenyl-N-methylcarbamate
2-((Ethylthio)methyl)phenol methylcarbamate
2-((Ethylthio)methyl)phenyl methylcarbamate
(2-Ethylthiomethyl-phenyl)-N-methylcarbamate
α-Ethylthio-o-tolyl methylcarbamate
Hox 1901

29975-16-4
$C_{16}H_{11}ClN_4$
294.72
Estazolam
8-Chloro-6-phenyl-4H-s-triazolo(4,3-a)(1,4)benzodiazepine
8-Chloro-6-phenyl-4H-(1,2,4)triazolo(4,3-a)(1,4)benzodiazepine
8-Chloro-6-phenyl-4H-(1,2,4)triazolo-(4,3-a)(1,4)benzodiazepine
D-40TA
DEA No. 2756
Esilgan
Estazolamum (Latin)
Eurodin
Julodin
NSC-290818
Nuctalon
U 33737
4H-s-Triazolo(4,3-a)(1,4)benzodiazepine, 8-chloro-6-phenyl- (8CI)
4H-(1,2,4)Triazolo(4,3-a)-(1,4)benzodiazepine, 8-chloro 6-phenyl- (9CI)
4H-(1,2,4)Triazolo(4,3-a)(1,4)-benzodiazepine, 8-chloro-6-phenyl-

29988-16-7
$C_{22}H_{38}O$
318.54
Phenol, dioctyl- (9CI)
Dioctylphenol

29990-39-4
$C_9H_7Cl_3O_2$
253.51
Propionic acid, (2,4,5-trichlorophenoxy)
NA 2765 (DOT)
2,4,5-TP (DOT)
2,4,5-Trichlorophenoxypropionic acid (DOT)
(2,4,5-Trichlorophenoxy)propionic acid

29994-68-1
$(C_6H_{14}O_6.C_3H_5ClO)x$
Polymer
D-Glucitol, Polymer with (chloromethyl)oxirane (9CI)
D-Glucitol, 1-chloro-2,3-epoxy propane polymer

30025-33-3
$C_8H_6O_2$
134.13
Benzenedicarboxaldehyde (8CI, 9CI)

30026-85-8
$C_{18}H_{28}O_4Si_4$
420.76
Cyclotetrasiloxane, hexamethyldiphenyl- (8CI,9CI)

30026-92-7
$C_{13}H_{20}O_2$
208.30
tert-Butyl isopropyl benzene hydroperoxide
t-Butyl isopropyl benzene hydroperoxide
Hydroperoxide, t-butyl-α,α-dimethylbenzyl-
NA 2091

30027-44-2
$C_{10}H_{10}S$
162.26
Benzo(b)thiophene, dimethyl- (8CI,9CI)

30030-25-2
C_9H_9Cl
152.62
Vinylbenzyl chloride
Benzene, (chloromethyl)-ethenyl- (9CI)
(Chloromethyl)ethenylbenzene

30031-64-2
$C_4H_6O_2$
86.10
Butane, l-1,2:3,4-diepoxy
(S-(R*,R*))-2,2'-Bioxirane
l-Butadiene diepoxide
l-Diepoxybutane
l-1,2:3,4-Diepoxybutane
(2S,3S)-Diepoxybutane
(2S,3S)-1,2:3,4-Diepoxybutane
NSC-32606

30080-50-3
$C_7H_{14}O_8.Na$
249.17
D-Glycero-D-ido-heptonic acid, monosodium salt (9CI)

30084-25-4
$C_9H_{16}ClN_5O_2$
261.71
1-Propanol, 3,3'-((6-chloro-s-triazine-2,4-diyl)diimino)di- (8CI)

30086-02-3
$C_8H_{12}O$
124.18
3,5-Octadien-2-one, (E,E)- (8CI,9CI)

30113-45-2
$C_{13}H_{29}NO$
215.37
1-Propanamine, 3-(isodecyloxy)- (9CI)
3-(Isodecyloxy)-1-propanamine
Isodecyloxypropylamine
3-Isodecyloxypropylamine
3-(Isodecyloxy)propylamine
Propylamine, 3-isodecoxy-

30124-94-8
$C_{21}H_{21}N_5O_6$
439.40
O=C(OCCN(c(ccc(N=Nc(c(C#N)cc(N(=O)=O)c1)c1)c2)c2)CCOC(=O)C)C
Benzonitrile, 2-((4-(bis(2-(acetyloxy)ethyl)amino)phenyl)azo)-5-nitro- (9CI)
N,N-Bis(2-(acetyloxy)ethyl)-4-((2-cyano-4-nitrophenyl)azo)benzeneamine
2-((4-((Bis(2-hydroxyethyl))amino)phenyl)azo)-5-nitrobenzonitrile, diacetate (ester)
3-Nitro-6-((4-(N,N-diacetoxyethylamino)phenyl)azo)benzonitrile)

30125-47-4
$C_{26}H_6Cl_8N_2O_4$
693.95
1H-Isoindole-1,3(2H)-dione, 4,5,6,7-tetrachloro-2-(2-(4,5,6,7-tetrachloro-2,3-dihydro-1,3-dioxo-1H-inden-2-yl)-8-quinolinyl)- (9CI)

30136-13-1
$C_6H_{14}O_2$
118.18
n-Propoxypropanol
Propanol, n-propoxy-
n-Propoxypropanol (Mixed isomers)

30140-42-2
$C_9H_6O_3$
162.14
1,3-Isobenzofurandione, methyl- (9CI)

30140-46-6
$C_{12}H_{20}S$
196.36
Thiophene, bis(1,1-dimethylethyl)- (9CI)
Thiophene, di-tert-butyl-

30145-38-1
Unknown
Unknown
Ethyl tellurac

30145-51-8
$C_{16}H_{24}O_6$
312.36
O=C(OCC(C)(C)COC(=O)C(C)(C)COC(=O)C=C)C=C
2-Propenoic acid, 3-(2,2-dimethyl-1-oxo-3-((1-oxo-2-propenyl)oxy)propoxy)-2,2-dimethylpropyl ester (9CI)
Propanoic acid, 3-hydroxy-2,2-dimethyl-, 3-hydroxy-2,2-dimethyl propyl ester, diacrylate

30174-58-4
$C_{10}H_{22}S$
174.35
tert-Decylmercaptan
tert-Decanethiol (9CI)

30207-98-8
$C_{11}H_{24}O$
172.31
Undecanol (9CI)

30230-52-5
$C_9H_{12}O$
136.19
Phenol, ethylmethyl- (9CI)
Cresol, ar-ethyl- (8CI)

30232-26-9
$C_{17}H_{16}$
220.31
Phenanthrene, trimethyl- (8CI, 9CI)

30260-72-1
$C_{24}H_{34}O_7S_2$
498.66
Benzenesulfonic acid, dodecyl(sulfophenoxy)- (9CI)
Dodecyl(sulfophenoxy)benzenesulfonic acid

30260-73-2
$C_{36}H_{58}O_7S_2$
666.98
Benzenesulfonic acid, oxybis(dodecyl- (9CI)
Oxybis(dodecylbenzenesulfonic acid)

30270-60-1
$C_{16}H_{20}O_6$
308.33
5H-Furo(3,2-c)(2)benzopyran-5-one, 2,3,3a,9b-tetrahydro-6-hydroxy-7,8-dimethoxy-2-propyl-, (2S-(2α,3aβ,9bβ))- (9CI)
5H-Furo(3,2-c)(2)benzopyran-5-one, 2,3,3a,9b-tetrahydro-6-hydroxy-7,8-dimethoxy-2-propyl- (8CI) (VAN)
Monocerin

30275-76-4
$C_{13}H_{18}O_2$
206.29
Butyric acid, 3-methyl-4-(2,5-xylyl)- (8CI)

30283-95-5
$C_{23}H_{16}$
292.38
Picene, methyl- (8CI,9CI)

30286-23-8
C_9H_8O
132.16
Indenone, dihydro- (9CI)
Indanone (VAN)

30303-58-3
$C_{15}H_{13}N_5O$
279.27
1,3,5-Triazin-2(1H)-one, 4,6-bis(phenylamino)- (9CI)
2,4-Dianilino-6-hydroxy-s-triazine
s-Triazin-2-ol, 4,6-dianilino - (6CI,7CI,8CI)
s-Triazin-2(1H)-one, 4,6-dianilino-

30304-41-7
$C_{36}H_{59}O_2PS_2$
618.97
Phenol, dodecyl-, hydrogen phosphorodithioate (9CI)
Dodecylphenol hydrogen phosphorodithioate
Phosphorodithioic acid, O,O-bis(dodecylphenyl) ester

30310-80-6
$C_5H_8N_2O_4$
160.15
OC1CC(N(C1)N=O)C(O)=O
Proline, 4-hydroxy-1-nitroso-, L
trans-4-Hydroxy-1-nitroso-L-proline
N-Nitrosohydroxyproline

30316-14-4
$C_{13}H_{18}O_2$
206.29
Butyric acid, 2-methyl-4-(2,5-xylyl)- (8CI)

30316-19-9
$C_{13}H_{16}$
172.27
Naphthalene, 1,4-dihydro-2,5,8-trimethyl- (8CI)

30316-23-5
$C_{13}H_{16}$
172.27
Naphthalene, 1,2-dihydro-2,5,8-trimethyl- (8CI)

30339-33-4
$C_7H_8Br_4N_4$
467.80
s-Triazine, 2,4-bis(dibromomethyl)-6-(ethylamino)- (8CI)

30339-34-5
$C_5H_5Br_3N_4$
360.85
s-Triazine, 2-amino-4-methyl-6-(tribromomethyl)- (7CI,8CI)

30339-35-6
$C_7H_9Br_3N_4$
388.90
s-Triazine, 2-(ethylamino)-4-methyl-6-(tribromomethyl)- (8CI)

30339-36-7
$C_5H_2Br_6N_4$
597.55
s-Triazine, 2-amino-4,6-bis(tribromomethyl)- (7CI,8CI)

30339-37-8
$C_6H_4Br_6N_4$
611.58
s-Triazine, 2-(methylamino)-4,6-bis(tribromomethyl)- (8CI)

30339-38-9
$C_7H_6Br_6N_4$
625.61
s-Triazine, 2-(ethylamino)-4,6-bis(tribromomethyl)- (8CI)

30339-39-0
$C_{11}H_6Br_6N_4$
673.65
s-Triazine, 2-anilino-4,6-bis(tribromomethyl)- (8CI)

30339-40-3
$C_6H_7Cl_3N_4$
241.51
s-Triazine, 2-amino-4-ethyl-6-(trichloromethyl)- (8CI)

30339-47-0
$C_6H_6Cl_4N_4$
275.95
s-Triazine, 2-amino-4-(2-chloroethyl)-6-(trichloromethyl)- (8CI)

30339-50-5
$C_6H_4Cl_6N_4$
344.84
s-Triazine, 2-amino-4-(1,1,2-trichloroethyl)-6-(trichloromethyl)- (8CI)

30339-52-7
$C_6H_6BrCl_3N_4$
320.41
s-Triazine, 2-amino-4-(2-bromoethyl)-6-(trichloromethyl)- (8CI)

30339-54-9
$C_6H_5Cl_3N_4$
239.49
s-Triazine, 2-amino-4-(trichloromethyl)-6-vinyl- (8CI)

30339-56-1
$C_8H_{14}N_4$
166.23
s-Triazine, 2,4-diethyl-6-(methylamino)- (8CI)

30339-57-2
$C_7H_8Cl_4N_4$
289.98
s-Triazine, 2-amino-4,6-bis(1,1-dichloroethyl)- (6CI,8CI)

30339-58-3
$C_8H_{10}Cl_4N_4$
304.01
s-Triazine, 2,4-bis(1,1-dichloroethyl)-6-(methylamino)- (8CI)

30339-59-4
$C_9H_{12}Cl_4N_4$
318.03
s-Triazine, 2,4-bis(1,1-dichloroethyl)-6-(ethylamino)- (8CI)

30339-60-7
$C_7H_6Cl_6N_4$
358.87
s-Triazine, 2-amino-4,6-bis(1,1,2-trichloroethyl)- (8CI)

30339-67-4
$C_7H_9Cl_3N_4$
255.54
s-Triazine, 2-amino-4-isopropyl-6-(trichloromethyl)- (8CI)

30339-70-9
$C_8H_{11}Cl_3N_4$
269.56
s-Triazine, 2-amino-4-butyl-6-(trichloromethyl)- (8CI)

30339-73-2
$C_8H_{11}Cl_3N_4$
269.56
s-Triazine, 2-amino-4-sec-butyl-6-(trichloromethyl)- (8CI)

30339-74-3
$C_9H_{13}Cl_3N_4$
283.59
s-Triazine, 2-amino-4-pentyl-6-(trichloromethyl)- (8CI)

30339-78-7
$C_{11}H_5Br_6ClN_4$
708.10
s-Triazine, 2-(p-chloroanilino)-4,6-bis(tribromomethyl)- (8CI)

30339-79-8
$C_8H_{11}Cl_3N_4$
269.56
s-Triazine, 2-amino-4-isobutyl-6-(trichloromethyl)- (8CI)

30339-80-1
$C_7H_9Cl_3N_4$
255.54
s-Triazine, 2-amino-4-propyl-6-(trichloromethyl)- (8CI)

30345-27-8
$C_6H_7N_5O_2$
181.18

Guanine, 3-hydroxy-7-methyl
2-Amino-3-hydroxy-1,7-dihydro-7-methyl-6H-purin-6-one
3-Hydroxy-7-methylguanine

30345-28-9
$C_6H_7N_5O_2$
181.18
Guanine, 3-hydroxy-9-methyl
2-Amino-3-hydroxy-1,7-dihydro-8-methyl-6H-purin-6-one
3-Hydroxy-9-methylguanine

30346-73-7
$C_8H_{10}O_3S.K$
225.33
Potassium xylenesulfonate
Benzenesulfonic acid, dimethyl-, potassium salt (9CI)
Caswell No. 704A
EPA Pesticide Chemical Code 079024
Potassium dimethylbenzenesulfonate
Potassium xylene sulfonate

30354-65-5
$C_6H_{10}ClN_5$
187.63
s-Triazine, 2,4-diamino-6-(1-chloro-1-methylethyl)- (8CI)

30354-68-8
$C_7H_{13}N_5$
167.22
1,3,5-Triazine-2,4-diamine, 6-(2-methylpropyl)-(9CI)
s-Triazine, 2,4-diamino-6-isobutyl- (8CI)

30354-74-6
$C_7H_{13}N_5$
167.18
1,3,5-Triazine-2,4-diamine, 6-(1-methylpropyl)-
AI3-60104

30354-83-7
$C_{18}H_{35}N_5$
321.51
s-Triazine, 2,4-bis(isopropylamino)-6-nonyl-(8CI)

30354-86-0
$C_{20}H_{37}Cl_2N_5$
418.46
s-Triazine, 2,4-diamino-6-(1,1-dichloroheptadecyl)- (8CI)

30354-89-3
$C_9H_7Cl_2N_5$
256.10
1,3,5-Triazine-2,4-diamine, 6-(3,4-dichlorophenyl)-(9CI)
2,4-Diamino-6-(3,4-dichlorophenyl)-s-triazine
s-Triazine, 2,4-diamino-6-(3,4-dichlorophenyl)-(8CI)

30354-91-7
$C_{10}H_{11}N_5O$
217.20
1,3,5-Triazine-2,4-diamine, 6-(4-methoxyphenyl)- (9CI)
s-Triazine, 2,4-diamino-6-(p-methoxyphenyl)- (8CI)

30354-93-9
$C_{13}H_{11}N_5$
237.27
s-Triazine, 2,4-diamino-6-(1-naphthyl)- (8CI)

30354-97-3
$C_7H_{13}N_5O$
183.21
s-Triazine, 2,4-diamino-6-(2-ethoxyethyl)- (8CI)

30354-98-4
$C_6H_9N_5O$
167.17
2-Propanone, 1-(4,6-diamino-s-triazin-2-yl)-(8CI)

30354-99-5
$C_7H_{11}N_5O$
181.20
2-Butanone, 4-(4,6-diamino-s-triazin-2-yl)-(8CI)

30355-00-1
$C_9H_{14}FN_3$
213.26
s-Triazine, 2-fluoro-4,6-diisopropyl- (8CI)

30355-01-2
$C_6H_8ClN_5$
185.62
s-Triazine, 2-(allylamino)-4-amino-6-chloro- (6CI, 8CI)

30355-02-3
$C_{15}H_{28}ClN_5$
313.88
s-Triazine, 2-chloro-4,6-bis(hexylamino)-(8CI)

30355-03-4
$C_{27}H_{52}ClN_5$
482.20
1,3,5-Triazine-2,4-diamine, 6-chloro-N,N'-didodecyl-(9CI)
2,4-Bis(dodecylamino)-6-chloro-s-triazine
4-Chloro-2,6-bis(dodecylamino)-s-triazine
2-Chloro-4,6-bis(dodecylamino)-s-triazine
4-Chloro-2,6-bis(laurylamino)-s-triazine
2,4-Didodecylamino-6-chloro-1,3,5-triazine
s-Triazine, 2-chloro-4,6-bis-(dodecylamino)- (8CI)

30355-04-5
$C_{17}H_{16}ClN_5$
325.80
1,3,5-Triazine-2,4-diamine, 6-chloro-N,N'-bis-(phenylmethyl)- (9CI)
s-Triazine, 2,4-bis(benzylamino)-6-chloro- (6CI, 8CI)

30355-05-6
$C_{12}H_{13}Cl_2N_5$
298.18
s-Triazine, 2-chloro-4-(p-chloroanilino)-6-(isopropylamino)- (8CI)

30355-06-7
$C_{15}H_8Cl_5N_5$
435.53
s-Triazine, 2-chloro-4,6-bis(3,4-dichloroanilino)- (7CI,8CI)

30355-07-8
$C_{23}H_{16}ClN_5$
397.87
1,3,5-Triazine-2,4-diamine, 6-chloro-N,N'-di-1-naphthalenyl- (9CI)
2,6-Bis(α-naphthylamino)-4-chloro-triazine
s-Triazine, 2-chloro-4,6-bis-(1-naphthylamino)- (6CI, 8CI)

30355-52-3
$C_{16}H_{19}N_9$
337.39
Propionitrile, 3,3',3'',3'''-((6-methyl-s-triazine-2,4-diyl)dinitrilo)tetra-(8CI)

30355-53-4
$C_6H_9N_5O$
167.17
Acetamide, N-(4-amino-6-methyl-1,3,5-triazin-2-yl)- (9CI)
Acetamide, N-(4-amino-6-methyl-s-triazin-2-yl)- (8CI)

30355-54-5
$C_8H_{11}N_5O_2$
209.21
Acetamide, N,N'-(6-methyl-s-triazine-2,4-diyl)bis- (8CI)
s-Triazine, 2,4-diacetamido-6-methyl- (6CI)

30355-55-6
$C_{10}H_{11}N_7O_3$
277.24
2-Furaldehyde, 5-nitro-, (4-methyl-6-(methylamino)-s-triazin-2-yl)-hydrazone (8CI)

30355-56-7
$C_4H_5N_7$
151.13
s-Triazine, 2,4-diamino-6-(diazomethyl)- (6CI, 7CI,8CI)

30355-57-8
$C_8H_{13}N_7$
207.24
s-Triazine, 2-(diazomethyl)-4,6-bis(ethylamino)- (8CI)

30355-58-9
$C_6H_8F_3N_5$
207.16
s-Triazine, 2-amino-4-(ethylamino)-6-(trifluoromethyl)- (8CI)

30355-59-0
$C_6H_{10}ClN_5$
187.63
s-Triazine, 2-amino-4-(chloromethyl)-6-(ethylamino)- (8CI)

30355-60-3
$C_{10}H_{10}ClN_5$
235.68
1,3,5-Triazine-2,4-diamine, 6-(chloromethyl)-N-phenyl- (9CI)
s-Triazine, 2-amino-4-anilino-6-(chloromethyl)- (6CI,8CI)

30355-61-4
$C_{10}H_9Cl_2N_5$
270.12
s-Triazine, 2-amino-4-(p-chloroanilino)-6-(chloromethyl)- (6CI, 8CI)

30355-62-5
$C_8H_{14}ClN_5$
215.69
s-Triazine, 2-(chloromethyl)-4,6-bis(ethylamino)- (7CI,8CI)

30355-63-6
$C_4H_5Cl_2N_5$
194.02
s-Triazine, 2,4-diamino-6-(dichloromethyl)- (6CI, 8CI)

30355-64-7
$C_6H_9Cl_2N_5$
222.08
s-Triazine, 2-amino-4-(dichloromethyl)-6-(ethylamino)- (8CI)

30355-65-8
$C_{10}H_9Cl_2N_5$
270.12
s-Triazine, 2-amino-4-anilino-6-(dichloromethyl)- (6CI,8CI)

30355-66-9
$C_8H_{13}Cl_2N_5$
250.13
s-Triazine, 2-(dichloromethyl)-4,6-bis(ethylamino)- (7CI,8CI)

30355-69-2
$C_{10}H_8Cl_3N_5$
304.57
s-Triazine, 2-amino-4-anilino-6-(trichloromethyl)- (8CI)

30355-70-5
$C_6H_9Cl_3N_6$
271.54
s-Triazine, 2-(dimethylamino)-4-hydrazino-6-(trichloromethyl)- (8CI)

30355-71-6
$C_{10}H_7Cl_4N_5$
339.01
s-Triazine, 2-amino-4-(p-chloroanilino)-6-(trichloromethyl)- (8CI)

30355-91-0
$C_{16}H_{12}Cl_3N_5$
380.67
s-Triazine, 2,4-dianilino-6-(trichloromethyl)- (8CI)

30355-92-1
$C_{16}H_{10}Cl_5N_5$
449.56
s-Triazine, 2,4-bis(p-chloroanilino)-6-(trichloromethyl)- (8CI)

30355-96-5
$C_7H_8Cl_3N_5O$
284.53
Propionamide, N-(4-amino-6-(trichloromethyl)-s-triazin-2-yl)- (8CI)

30356-34-4
$C_{13}H_{18}Cl_6N_4$
443.03
s-Triazine, 2-(dibutylamino)-4,6-bis(trichloromethyl)- (8CI)

30356-35-5
$C_{13}H_{18}Cl_6N_4$
443.03
s-Triazine, 2-(diisobutylamino)-4,6-bis(trichloromethyl)- (8CI)

30356-50-4
$C_{11}H_6Cl_6N_4$
406.92
s-Triazine, 2-anilino-4,6-bis(trichloromethyl)- (8CI)

30356-51-5
$C_{11}H_5Cl_7N_4$
441.36
s-Triazine, 2-(o-chloroanilino)-4,6-bis(trichloromethyl)- (8CI)

30356-52-6
$C_{11}H_5Cl_7N_4$
441.36
s-Triazine, 2-(m-chloroanilino)-4,6-bis(trichloromethyl)- (8CI)

30356-54-8
$C_{11}H_4Cl_8N_4$
475.81
s-Triazine, 2-(2,5-dichloroanilino)-4,6-bis(trichloromethyl)- (8CI)

30356-55-9
$C_{11}H_4Cl_8N_4$
475.81
1,3,5-Triazin-2-amine, N-(3,4-dichlorophenyl)-4,6-bis(trichloromethyl)- (9CI)
s-Triazine, 2-(3,4-dichloroanilino)-4,6-bis(trichloromethyl)- (8CI)

30356-56-0
$C_{11}H_3Cl_9N_4$
510.25
s-Triazine, 2-(2,4,5-trichloroanilino)-4,6-bis(trichloromethyl)- (8CI)

30356-57-1
$C_{11}HCl_{11}N_4$
579.14
s-Triazine, 2-(2,3,4,5,6-pentachloroanilino)-4,6-bis(trichloromethyl)- (8CI)

30356-58-2
$C_{11}H_5BrCl_6N_4$
485.82
s-Triazine, 2-(m-bromoanilino)-4,6-bis(trichloromethyl)- (8CI)

30356-59-3
$C_{11}H_5BrCl_6N_4$
485.82
s-Triazine, 2-(p-bromoanilino)-4,6-bis(trichloromethyl)- (8CI)

30357-60-9
$C_5H_3Cl_5N_4$
296.37
s-Triazine, 2-amino-4-(dichloromethyl)-6-(trichloromethyl)- (8CI)

30357-61-0
$C_5H_2Cl_6N_4O$
346.82
s-Triazine, 2-(hydroxyamino)-4,6-bis(trichloromethyl)- (8CI)
Hydroxylamine, N-(4,6-bis(trichloromethyl)-s-triazin-2-yl)- (6CI)

30357-71-2
$C_{13}H_{18}Cl_6N_4$
443.03
s-Triazine, 2-((1,1,3,3-tetramethylbutyl)amino)-4,6-bis(trichloromethyl)- (8CI)

30357-74-5
$C_6H_4Cl_6N_4S$
376.91
Methanesulfenamide, 1,1,1-trichloro-N-(4-methyl-6-(trichloromethyl)-s-triazin-2-yl)- (8CI)

30357-76-7
$C_5H_4Cl_4N_4$
261.90
1,3,5-Triazin-2-amine, 4-(chloromethyl)-6-(trichloromethyl)- (9CI)
s-Triazine, 2-amino-4-(chloromethyl)-6-(trichloromethyl)- (8CI)

30357-78-9
$C_9H_6Cl_3N_5O_2S$
354.60
Benzenesulfonic acid, p-chloro-, 2-(4,6-dichloro-s-triazin-2-yl)hydrazide (8CI)

30357-79-0
$C_{10}H_9Cl_2N_5O_2S$
334.19
p-Toluenesulfonic acid, 2-(4,6-dichloro-s-triazin-2-yl)hydrazide (6CI,8CI)

30357-80-3
$C_5H_6Br_2N_4$
281.95
s-Triazine, 2,4-dibromo-6-(ethylamino)- (8CI)

30357-82-5
$C_9H_6Br_2N_4$
329.99
s-Triazine, 2-anilino-4,6-dibromo- (8CI)

30357-83-6
$C_9H_5Br_2ClN_4$
364.40
1,3,5-Triazin-2-amine, 4,6-dibromo-N-(4-chlorophenyl)- (9CI)
s-Triazine, 2,4-dibromo-6-(p-chloroanilino)- (8CI)

30357-84-7
$C_{10}H_{17}ClN_4O$
244.73
s-Triazine, 2-chloro-4-(hexylamino)-6-methoxy- (8CI)

30357-85-8
$C_{10}H_9ClN_4O$
236.66
1,3,5-Triazin-2-amine, 4-chloro-6-methoxy-N-phenyl- (9CI)
s-Triazine, 2-anilino-4-chloro-6-methoxy- (6CI,8CI)

30357-89-2
$C_9H_{15}ClN_4O$
230.70
s-Triazine, 2-chloro-4-(ethylamino)-6-isobutoxy- (8CI)

30357-90-5
$C_9H_7FN_4O$
206.18
s-Triazine, 2-amino-4-fluoro-6-phenoxy- (8CI)

30357-91-6
$C_9H_7ClN_4O$
222.64
1,3,5-Triazin-2-amine, 4-chloro-6-phenoxy- (9CI)
s-Triazine, 2-amino-4-chloro-6-phenoxy- (7CI,8CI)

30357-92-7
$C_{11}H_{11}ClN_4O$
250.69
s-Triazine, 2-chloro-4-(ethylamino)-6-phenoxy- (8CI)

30357-93-8
$C_9H_7BrN_4O$
267.09
s-Triazine, 2-amino-4-bromo-6-phenoxy- (8CI)

30357-94-9
$C_4H_5ClN_4S$
176.63
1,3,5-Triazin-2-amine, 4-chloro-6-(methylthio)- (9CI)
s-Triazine, 2-amino-4-chloro-6-(methylthio)- (8CI)

30357-95-0
$C_6H_9ClN_4S$
204.68
1,3,5-Triazin-2-amine, 4-chloro-N-ethyl-6-(methylthio)- (9CI)
s-Triazine, 2-chloro-4-(ethylamino)-6-(methylthio)- (8CI)

30357-96-1
$C_{10}H_9ClN_4S$
252.73
1,3,5-Triazin-2-amine, 4-chloro-6-(methylthio)-N-phenyl- (9CI)
s-Triazine, 2-anilino-4-chloro-6-(methylthio)-(8CI)

30357-97-2
$C_9H_7ClN_4S$
238.70
s-Triazine, 2-amino-4-chloro-6-(phenylthio)-(8CI)

30357-98-3
$C_6H_{10}N_4O_2$
170.17
1,3,5-Triazin-2-amine, 4,6-dimethoxy-N-methyl-(9CI)
2,4-Dimethoxy-6-(methylamino)-1,3,5-triazine
s-Triazine, 2,4-dimethoxy-6-(methylamino)- (8CI)

30357-99-4
$C_9H_{16}N_4O_2$
212.25
s-Triazine, 2-(butylamino)-4,6-dimethoxy-(8CI)

30358-00-0
$C_{12}H_{14}N_4O_2$
246.27
s-Triazine, 2-(benzylamino)-4,6-dimethoxy-(8CI)

30358-01-1
$C_{11}H_{12}N_4O_2$
232.24
1,3,5-Triazin-2-amine, 4,6-dimethoxy-N-phenyl-(9CI)
2-Anilino-4,6-dimethoxy-s-triazine
s-Triazine, 2-anilino-4,6-dimethoxy- (6CI,8CI)

30358-04-4
$C_{11}H_{11}BrN_4O_2$
311.14
1,3,5-Triazin-2-amine, N-(4-bromophenyl)-4,6-dimethoxy- (9CI)
s-Triazine, 2-(p-bromoanilino)-4,6-dimethoxy- (8CI)

30358-05-5
$C_{11}H_{11}N_5O_4$
277.24
s-Triazine, 2,4-dimethoxy-6-(p-nitroanilino)- (8CI)

30358-06-6
$C_{12}H_{14}N_4O_2$
246.27
1,3,5-Triazin-2-amine, 4,6-dimethoxy-N-methyl-N-phenyl- (9CI)
s-Triazine, 2,4-dimethoxy-6-(N-methylanilino)- (8CI)

30358-07-7
$C_9H_{16}N_4O_2$
212.25
s-Triazine, 2,4-diethoxy-6-(ethylamino)- (6CI,8CI)

30358-08-8
$C_{13}H_{16}N_4O_2$
260.30
s-Triazine, 2-anilino-4,6-diethoxy- (8CI)

30358-09-9
$C_{13}H_{15}ClN_4O_2$
294.74
s-Triazine, 2-(o-chloroanilino)-4,6-diethoxy- (8CI)

30358-10-2
$C_9H_{16}N_4O_2$
212.25
s-Triazine, 2-amino-4,6-dipropoxy- (8CI)

30358-11-3
$C_9H_{12}N_4O_2$
208.20
1,3,5-Triazin-2-amine, 4,6-bis-(2-propenyloxy)- (9CI)
NSC-8181
s-Triazine, 2,4-bis(allyloxy)-6-amino- (8CI)

30358-12-4
$C_{11}H_{16}N_4O_2$
236.28
1,3,5-Triazin-2-amine, N-ethyl-4,6-bis(2-propenyloxy)- (9CI)
s-Triazine, 2,4-bis(allyloxy)-6-(ethylamino)- (8CI)

30358-13-5
$C_9H_{16}N_4O_2$
212.25
s-Triazine, 2-amino-4,6-diisopropoxy- (8CI)

30358-14-6
$C_{15}H_{10}Cl_2N_4O_2$
349.18
s-Triazine, 2-amino-4,6-bis(o-chlorophenoxy)-(8CI)

30358-15-7
$C_{21}H_{14}Cl_2N_4O_2$
425.28
s-Triazine, 2-anilino-4,6-bis(o-chlorophenoxy)-(6CI,8CI)

30358-16-8
$C_{15}H_{10}Cl_2N_4O_2$
349.18
1,3,5-Triazin-2-amine, 4,6-bis(4-chlorophenoxy)-(9CI)
s-Triazine, 2-amino-4,6-bis-(p-chlorophenoxy)- (8CI)

30358-17-9
$C_{21}H_{14}Cl_2N_4O_2$
425.28
s-Triazine, 2-anilino-4,6-bis(p-chlorophenoxy)-(6CI,8CI)

30358-18-0
$C_5H_8N_4OS$
172.21
1,3,5-Triazin-2-amine, 4-methoxy-6-(methylthio)- (9CI)
s-Triazine, 2-amino-4-methoxy-6-(methylthio)- (7CI, 8CI)

30358-19-1
$C_5H_8N_4S_2$
188.25
1,3,5-Triazin-2-amine, 4,6-bis-(methylthio)- (9CI)
NSC-100281
s-Triazine, 2-amino-4,6-bis-(methylthio)- (8CI)

30359-60-5
$C_8H_8Cl_3N_5O_2$
312.54
Acetamide, N,N'-(6-(trichloromethyl)-s-triazine-2,4-diyl)bis- (8CI)

30359-61-6
$C_{12}H_{16}Cl_3N_5O_2$
368.65
Acetamide, N,N'-(6-(trichloromethyl)-s-triazine-2,4-diyl)bis(N-ethyl- (8CI)
s-Triazine, 2,4-bis(N-ethylacetamido)-6-(trichloromethyl)- (7CI)

30359-62-7
$C_{13}H_{13}Cl_3N_6$
359.65
Benzaldehyde, (4-(dimethylamino)-6-(trichloromethyl)-s-triazin-2-yl)hydrazone (8CI)

30359-63-8
$C_{13}H_8Cl_8N_6$
531.87
Benzaldehyde, pentachloro-, (4-(dimethylamino)-6-(trichloromethyl)-s-triazin-2-yl)-hydrazone (8CI)

30359-64-9
$C_{10}H_{10}BrN_5$
280.13
s-Triazine, 2-amino-4-anilino-6-(bromomethyl)-(8CI)

30359-65-0
$C_8H_{14}BrN_5$
260.14
s-Triazine, 2-(bromomethyl)-4,6-bis(ethylamino)- (7CI,8CI)

30359-66-1
$C_4H_5Br_2N_5$
282.94
s-Triazine, 2,4-diamino-6-(dibromomethyl)- (7CI, 8CI)

30359-67-2
$C_6H_9Br_2N_5$
310.99
s-Triazine, 2-amino-4-(dibromomethyl)-6-(ethylamino)- (8CI)

30359-68-3
$C_8H_{13}Br_2N_5$
339.04
s-Triazine, 2-(dibromomethyl)-4,6-bis-(ethylamino)- (7CI, 8CI)

30359-70-7
$C_6H_8Br_3N_5$
389.89
s-Triazine, 2-amino-4-(ethylamino)-6-(tribromomethyl)- (8CI)

30359-71-8
$C_{10}H_8Br_3N_5$
437.94
s-Triazine, 2-amino-4-anilino-6-(tribromomethyl)-(8CI)

30359-73-0
$C_{16}H_{12}Br_3N_5$
514.03
s-Triazine, 2,4-dianilino-6-(tribromomethyl)-(8CI)

30359-74-1
$C_4H_6IN_5$
251.03
s-Triazine, 2,4-diamino-6-(iodomethyl)- (7CI, 8CI)

30359-75-2
$C_8H_{14}IN_5$
307.14
s-Triazine, 2,4-bis(ethylamino)-6-(iodomethyl)-(7CI,8CI)

30359-76-3
$C_4H_5I_2N_5$
376.93
s-Triazine, 2,4-diamino-6-(diiodomethyl)- (8CI)

30359-77-4
$C_8H_{13}I_2N_5$
433.04
s-Triazine, 2-(diiodomethyl)-4,6-bis(ethylamino)- (8CI)

30359-79-6
$C_{11}H_{13}N_5$
215.26
1,3,5-Triazine-2,4-diamine, 6-ethyl-N-phenyl- (9CI)
s-Triazine, 2-amino-4-anilino-6-ethyl- (6CI,8CI)

30359-86-5
$C_5H_8ClN_5$
173.61
s-Triazine, 2,4-diamino-6-(1-chloroethyl)-(8CI)

30359-87-6
$C_5H_8ClN_5$
173.61
s-Triazine, 2,4-diamino-6-(2-chloroethyl)-(8CI)

30359-89-8
$C_5H_7Cl_2N_5$
208.05
s-Triazine, 2,4-diamino-6-(1,1-dichloroethyl)-(8CI)

30359-90-1
$C_5H_7Cl_2N_5$
208.05
s-Triazine, 2,4-diamino-6-(1,2-dichloroethyl)-(8CI)

30359-91-2
$C_5H_6Cl_3N_5$
242.50
s-Triazine, 2,4-diamino-6-(1,1,2-trichloroethyl)-(8CI)

30359-92-3
$C_5H_5Cl_4N_5$
276.94
s-Triazine, 2,4-diamino-6-(1,1,2,2-tetrachloroethyl)- (8CI)

30359-93-4
$C_5H_4Cl_5N_5$
311.39
s-Triazine, 2,4-diamino-6-(pentachloroethyl)-(8CI)

30359-94-5
$C_5H_7Br_2N_5$
296.96
s-Triazine, 2,4-diamino-6-(1,2-dibromoethyl)-(8CI)

30359-95-6
$C_{12}H_{15}N_5$
229.29
1,3,5-Triazine-2,4-diamine, N-phenyl-6-propyl-(9CI)
s-Triazine, 2-amino-4-anilino-6-propyl- (8CI)

30360-01-1
$C_6H_8Cl_3N_5$
256.52
s-Triazine, 2,4-diamino-6-(1,1,2-trichloropropyl)-(8CI)

30360-05-5
$C_{10}H_{12}N_6S$
248.31
Thiocyanic acid, 4,6-bis-(allylamino)-s-triazin-2-yl ester (8CI)

30360-06-6
$C_{16}H_{12}N_6S$
320.38

Thiocyanic acid, 4,6-dianilino-s-triazin-2-yl ester (8CI)

30360-07-7
$C_8H_{12}N_6S$
224.29
Isothiocyanic acid, 4,6-bis(ethylamino)-s-triazin-2-yl ester (8CI)

30360-08-8
$C_{10}H_{16}N_6S$
252.34
Isothiocyanic acid, 4,6-bis(isopropylamino)-s-triazin-2-yl ester (8CI)

30360-09-9
$C_{16}H_{12}N_6$
288.31
1,3,5-Triazine-2-carbonitrile, 4,6-bis(phenylamino)- (9CI)
s-Triazine-2-carbonitrile, 4,6-dianilino- (8CI)

30360-10-2
$C_{15}H_{12}N_8$
304.32
s-Triazine, 2,4-dianilino-6-azido- (8CI)

30360-11-3
$C_9H_9ClN_6$
236.67
Melamine, (o-chlorophenyl)- (6CI,8CI)

30360-12-4
$C_9H_9ClN_6$
236.67
Melamine, (p-chlorophenyl)- (8CI)

30360-13-5
$C_9H_9N_7O_2$
247.22
Melamine, (p-nitrophenyl)- (8CI)

30360-14-6
$C_{10}H_{12}N_6$
216.25
Melamine, N2-methyl-N2-phenyl- (6CI,7CI,8CI)

30360-15-7
$C_9H_{14}N_6$
206.25
1,3,5-Triazine-2,4,6-triamine, N,N'-di-2-propenyl- (9CI)
N,N'-Diallylmelamine
Melamine, N2,N4-diallyl- (6CI,7CI,8CI)

30360-17-9
$C_{12}H_{15}ClN_6$
278.75
1,3,5-Triazine-2,4,6-triamine, N-(4-chlorophenyl)-N'-(1-methylethyl)- (9CI)
Melamine, N2-(p-chlorophenyl)-N4-isopropyl- (8CI)

30360-18-0
$C_{15}H_{12}Cl_2N_6$
347.21
1,3,5-Triazine-2,4,6-triamine, N,N'-bis(4-chlorophenyl)- (9CI)
Melamine, N2,N4-bis-(p-chlorophenyl)- (7CI,8CI)

30360-19-1
$C_{10}H_{20}N_6$
224.27
1,3,5-Triazine-2,4,6-triamine, N,N'-diethyl-N''-(1-methylethyl)- (9CI)
Melamine, N(2),N(4)-diethyl-N(6)-isopropyl- (8CI)

30360-20-4
$C_{13}H_{17}ClN_6$
292.77
Melamine, N2-(p-chlorophenyl)-N4,N6-diethyl- (8CI)

30360-21-5
$C_{12}H_{18}N_6$
246.32
1,3,5-Triazine-2,4,6-triamine, N,N',N''-tri-2-propenyl- (9CI)
Melamine, N2,N4,N6-triallyl- (7CI,8CI)
N,N',N''-Triallylmelamine
2,4,6-Tris(allylamino)-s-triazine

30360-22-6
$C_{17}H_{16}Cl_2N_6$
375.26
Melamine, N2,N4-bis-(p-chlorophenyl)-N6-ethyl- (8CI)

30360-24-8
$C_{33}H_{24}N_6$
504.60
Melamine, N2,N4,N6-tri-1-naphthyl- (8CI)

30360-27-1
$C_9H_9Cl_3N_6O_3$
355.57
Acetamide, N,N',N''-s-triazine-2,4,6-triyltris-(2-chloro- (8CI)
s-Triazine, 2,4,6-tris-(2-chloroacetamido)- (6CI)

30360-28-2
$C_9H_6Cl_6N_6O_3$
458.90
Acetamide, N,N',N''-s-triazine-2,4,6-triyltris-(2,2-dichloro- (8CI)
s-Triazine, 2,4,6-tris(2,2-dichloroacetamido)- (6CI)

30360-29-3
$C_9H_3Cl_9N_6O_3$
562.24
Acetamide, N,N',N''-s-triazine-2,4,6-triyltris-(2,2,2-trichloro- (8CI)
s-Triazine, 2,4,6-tris-(2,2,2-trichloroacetamido)- (6CI)

30360-30-6
$C_9H_7ClN_4$
206.64
s-Triazine, 2-(p-chloroanilino)- (8CI)

30360-32-8
$C_6H_{10}N_4$
138.17
s-Triazine, 2-(ethylamino)-4-methyl- (8CI)

30360-33-9
$C_4H_3Cl_3N_4$
213.43
1,3,5-Triazin-2-amine, 4-(trichloromethyl)- (9CI)
s-Triazine, 2-amino-4-(trichloromethyl)- (8CI)

30360-34-0
$C_5H_8N_4$
124.15
s-Triazine, 2-amino-4-ethyl- (8CI)

30360-35-1
$C_{11}H_{12}N_4$
200.25
s-Triazine, 2-anilino-4,6-dimethyl- (8CI)

30360-36-2
$C_5H_6Cl_2N_4$
193.04
s-Triazine, 2-amino-4,6-bis(chloromethyl)- (6CI,8CI)

30360-37-3
$C_7H_{10}Cl_2N_4$
221.09
s-Triazine, 2,4-bis(chloromethyl)-6-(ethylamino)- (8CI)

30360-38-4
$C_7H_{10}Cl_2N_4$
221.09
s-Triazine, 2,4-bis(chloromethyl)-6-(dimethylamino)- (6CI,8CI)

30360-39-5
$C_{11}H_{10}Cl_2N_4$
269.14
s-Triazine, 2-anilino-4,6-bis(chloromethyl)- (8CI)

30360-40-8
$C_{11}H_9Cl_3N_4$
303.58
s-Triazine, 2-(p-chloroanilino)-4,6-bis(chloromethyl)- (8CI)

30360-41-9
$C_5H_4Cl_4N_4$
261.93
s-Triazine, 2-amino-4,6-bis(dichloromethyl)- (6CI,8CI)

30360-42-0
$C_7H_8Cl_4N_4$
289.98
s-Triazine, 2,4-bis(dichloromethyl)-6-(ethylamino)- (8CI)

30360-43-1
$C_7H_8Cl_4N_4$
289.98
s-Triazine, 2,4-bis(dichloromethyl)-6-(dimethylamino)- (6CI,8CI)

30360-44-2
$C_{11}H_8Cl_4N_4$
338.03
s-Triazine, 2-anilino-4,6-bis(dichloromethyl)- (8CI)

30360-47-5
$C_{11}H_{11}ClN_4$
234.69
s-Triazine, 2-(p-chloroanilino)-4,6-dimethyl- (8CI)

30360-48-6
$C_6H_{10}N_4$
138.15
1,3,5-Triazin-2-amine, N-(1-methylethyl)- (9CI)
s-Triazine, 2-(isopropylamino)- (8CI)

30360-51-1
$C_8H_{14}BrN_5$
260.14
s-Triazine, 2-bromo-4-(ethylamino)-6-(propylamino)- (7CI,8CI)

30360-52-2
$C_8H_{12}BrN_5$
258.13
s-Triazine, 2-(allylamino)-4-bromo-6-(ethylamino)- (7CI,8CI)

30360-53-3
$C_8H_{14}BrN_5$
260.11
1,3,5-Triazine-2,4-diamine, 6-bromo-N-ethyl-N'-(1-methylethyl)- (9CI)
s-Triazine, 2-bromo-4-(ethylamino)-6-(isopropylamino)- (8CI)

30360-54-4
$C_9H_{16}BrN_5$
274.17
s-Triazine, 2-bromo-4,6-bis(isopropylamino)- (6CI,8CI)

30360-55-5
$C_9H_{16}BrN_5$
274.13
1,3,5-Triazine-2,4-diamine, 6-bromo-N,N,N'-triethyl- (9CI)
s-Triazine, 2-bromo-4-(diethylamino)-6-(ethylamino)- (8CI)

30360-56-6
$C_6H_{11}N_5O$
169.16
1,3,5-Triazine-2,4-diamine, N-ethyl-6-methoxy- (9CI)
G 31709
s-Triazine, 2-amino-4-(ethylamino)-6-methoxy- (8CI)

30360-57-7
$C_6H_{11}N_5O$
169.16
1,3,5-Triazine-2,4-diamine, 6-methoxy-N,N'-dimethyl- (9CI)
s-Triazine, 2-methoxy-4,6-bis(methylamino)- (8CI)

30360-58-8
$C_{16}H_{15}N_5O$
293.33
1,3,5-Triazine-2,4-diamine, 6-methoxy-N,N'-diphenyl- (9CI)
s-Triazine, 2,4-dianilino-6-methoxy- (6CI,8CI)

30360-59-9
$C_9H_{17}N_5O$
211.24
1,3,5-Triazine-2,4-diamine, 6-ethoxy-N,N'-diethyl- (9CI)
s-Triazine, 2-ethoxy-4,6-bis(ethylamino)- (8CI)

30360-60-2
$C_{13}H_{25}N_5O$
267.38
s-Triazine, 2,4-bis(diethylamino)-6-ethoxy- (6CI,7CI,8CI)

30360-62-4
$C_{10}H_{19}N_5O$
225.30
s-Triazine, 2,4-bis(ethylamino)-6-propoxy- (7CI,8CI)

30360-64-6
$C_{12}H_{23}N_5O$
253.35
1,3,5-Triazine-2,4-diamine, 6-(1-methylethoxy)-N,N'-bis(1-methylethyl)- (9CI)
s-Triazine, 2-isopropoxy-4,6-bis(isopropylamino)- (7CI,8CI)

30360-65-7
$C_{14}H_{27}N_5O$
281.40
s-Triazine, 2,4-bis(diethylamino)-6-isopropoxy- (7CI,8CI)

30360-67-9
$C_{11}H_{21}N_5O$
239.32
s-Triazine, 2-butoxy-4,6-bis(ethylamino)- (7CI,8CI)

30360-68-0
$C_{10}H_{17}N_5O$
223.28
s-Triazine, 2-(allyloxy)-4,6-bis(ethylamino)- (7CI, 8CI)

30360-70-4
$C_8H_{15}N_5O$
197.28
s-Triazine, 4,6-diamino-2-pentoxy

30360-73-7
$C_{11}H_{21}N_5O_2$
255.32
1,3,5-Triazine-2,4-diamine, 6-(2-ethoxyethoxy)-N,N'-diethyl- (9CI)
s-Triazine, 2-(2-ethoxyethoxy)-4,6-bis(ethylamino)- (7CI,8CI)

30360-74-8
$C_{10}H_{11}N_5O$
217.23
s-Triazine, 2,4-diamino-6-(benzyloxy)- (8CI)

30360-76-0
$C_{21}H_{17}N_5O$
355.40
1,3,5-Triazine-2,4-diamine, 6-phenoxy-N,N'-diphenyl- (9CI)
s-Triazine, 2,4-dianilino-6-phenoxy- (8CI)

30360-77-1
$C_9H_8ClN_5O$
237.65
s-Triazine, 2,4-diamino-6-(o-chlorophenoxy)- (8CI)

30360-78-2
$C_9H_8ClN_5O$
237.65
s-Triazine, 2,4-diamino-6-(p-chlorophenoxy)- (8CI)

30360-79-3
$C_9H_8N_6O_3$
248.20
s-Triazine, 2,4-diamino-6-(p-nitrophenoxy)- (8CI)

30360-80-6
$C_{12}H_{23}N_5S$
269.38
1,3,5-Triazine-2,4-diamine, N,N,N',N'-tetraethyl-6-(methylthio)- (9CI)
s-Triazine, 2,4-bis(diethylamino)-6-(methylthio)- (8CI)

30360-81-7
$C_5H_9N_5S$
171.20
1,3,5-Triazine-2,4-diamine, 6-(ethylthio)- (9CI)
s-Triazine, 2,4-diamino-6-(ethylthio)- (8CI)

30360-82-8
$C_9H_{17}N_5S$
227.30
1,3,5-Triazine-2,4-diamine, N,N'-diethyl-6-(ethylthio)- (9CI)
s-Triazine, 2,4-bis(ethylamino)-6-(ethylthio)- (8CI)

30360-83-9
$C_7H_{13}N_5S$
199.28
s-Triazine, 2,4-diamino-6-(butylthio)- (8CI)

30360-84-0
$C_6H_9N_5S$
183.24
s-Triazine, 2-(allylthio)-4,6-diamino- (8CI)

30360-85-1
$C_{10}H_{11}N_5S$
233.30
1,3,5-Triazine-2,4-diamine, 6-((phenylmethyl)thio)- (9CI)
s-Triazine, 2,4-diamino-6-(benzylthio)- (6CI,8CI)

30360-86-2
$C_9H_9N_5S$
219.27
s-Triazine, 2,4-diamino-6-(phenylthio)- (8CI)

30360-87-3
$C_{21}H_{17}N_5S$
371.47
1,3,5-Triazine-2,4-diamine, N,N'-diphenyl-6-(phenylthio)- (9CI)
s-Triazine, 2,4-dianilino-6-(phenylthio)- (8CI)

30360-89-5
$C_9H_{14}N_6S$
238.32
Acetonitrile, ((4,6-bis(ethylamino)-s-triazin-2-yl)thio)- (7CI,8CI)

30360-90-8
$C_{10}H_{16}N_6S$
252.34
Acetonitrile, ((4-(ethylamino)-6-(isopropylamino)-s-triazin-2-yl)thio)- (7CI,8CI)

30360-92-0
$C_{10}H_{16}N_6S$
252.34
Thiocyanic acid, 4,6-bis(isopropylamino)-s-triazin-2-yl ester (7CI,8CI)

30360-93-1
$C_4H_4N_6S$
168.18
Thiocyanic acid, 4,6-diamino-s-triazin-2-yl ester (7CI,8CI)

30360-95-3
$C_8H_{12}N_6S$
224.29
Thiocyanic acid, 4,6-bis(ethylamino)-s-triazin-2-yl ester (7CI,8CI)

30361-82-1
$C_4H_4ClN_3$
129.55
s-Triazine, 2-(chloromethyl)- (7CI,8CI)

30361-83-2
$C_4H_2Cl_3N_3$
198.44
1,3,5-Triazine, 2-(trichloromethyl)- (9CI)
s-Triazine, 2-(trichloromethyl)- (6CI,7CI,8CI)
(Trichloromethyl)triazine

30361-84-3
$C_4H_4BrN_3$
174.01
s-Triazine, 2-(bromomethyl)- (8CI)

30361-85-4
$C_5H_6ClN_3$
143.58
s-Triazine, 2-(1-chloroethyl)- (7CI,8CI)

30361-86-5
$C_5H_5Cl_2N_3$
178.02
s-Triazine, 2-(1,1-dichloroethyl)- (7CI,8CI)

30361-87-6
$C_6H_9N_3$
123.16
1,3,5-Triazine, 2-(1-methylethyl)- (9CI)
s-Triazine, 2-isopropyl- (7CI,8CI)

30361-88-7
$C_6H_8ClN_3$
157.60
s-Triazine, 2-(1-chloro-1-methylethyl)- (7CI,8CI)

30361-89-8
$C_{10}H_9N_3$
171.20
s-Triazine, 2-benzyl- (6CI, 8CI)

30361-90-1
$C_9H_6N_4O_2$
202.17
1,3,5-Triazine, 2-(3-nitrophenyl)- (9CI)
s-Triazine, 2-(m-nitrophenyl)- (6CI,8CI)

30361-91-2
$C_5H_4Cl_3N_3$
212.47
s-Triazine, 2-methyl-4-(trichloromethyl)- (8CI)

30361-92-3
$C_{17}H_{15}N_3$
261.33
s-Triazine, 2,4-dibenzyl- (6CI,7CI,8CI)

30361-93-4
$C_6H_3F_6N_3$
231.10
s-Triazine, 2-methyl-4,6-bis(trifluoromethyl)- (8CI)

30361-94-5
$C_6H_5Cl_4N_3$
260.94
s-Triazine, 2,4-bis(chloromethyl)-6-(dichloromethyl)- (8CI)

30361-95-6
$C_8H_{11}Cl_2N_3S_2$
284.23
s-Triazine-2-carboxaldehyde, 4-(dichloromethyl)-6-methyl-, dimethyl mercaptal (6CI,8CI)
s-Triazine, 2-(bis(methylthio)methyl)-4-(dichloromethyl)-6-methyl- (8CI)

30361-97-8
$C_6H_2Cl_7N_3$
364.25
1,3,5-Triazine, 2-(chloromethyl)-4,6-bis(trichloromethyl)- (9CI)
s-Triazine, 2-(chloromethyl)-4,6-bis(trichloromethyl)- (8CI)

30361-98-9
$C_6H_8BrN_3$
202.06
s-Triazine, 2-(bromomethyl)-4,6-dimethyl- (7CI,8CI)

30361-99-0
$C_6H_6Br_3N_3$
359.86
s-Triazine, 2,4,6-tris(bromomethyl)- (8CI)

30362-00-6
$C_6H_5Br_4N_3$
438.76
1,3,5-Triazine, 2,4-bis(dibromomethyl)-6-methyl- (9CI)
s-Triazine, 2,4-bis(dibromomethyl)-6-methyl- (7CI, 8CI)

30362-01-7
$C_6H_3Br_6N_3$
596.52
1,3,5-Triazine, 2,4,6-tris(dibromomethyl)- (9CI)
s-Triazine, 2,4,6-tris(dibromomethyl)- (8CI)

30362-02-8
$C_6H_3Br_6N_3$
596.56
1,3,5-Triazine, 2-methyl-4,6-bis(tribromomethyl)- (9CI)
s-Triazine, 2-methyl-4,6-bis(tribromomethyl)- (7CI, 8CI)

30362-03-9
$C_7H_{11}N_3$
137.19
1,3,5-Triazine, 2-ethyl-4,6-dimethyl- (9CI)
s-Triazine, 2-ethyl-4,6-dimethyl- (7CI,8CI)

30362-04-0
$C_8H_{12}ClN_3$
185.66
s-Triazine, 2-(3-chloropropyl)-4,6-dimethyl- (8CI)

30362-05-1
$C_9H_{15}N_3$
165.24
s-Triazine, 2-butyl-4,6-dimethyl- (7CI,8CI)

30362-06-2
$C_9H_{15}N_3$
165.24
s-Triazine, 2-isobutyl-4,6-dimethyl- (8CI)

30362-07-3
$C_{13}H_{15}N_3$
213.28
s-Triazine, 2,4-dimethyl-6-phenethyl- (7CI,8CI)

30362-08-4
$C_{11}H_{10}ClN_3$
219.68
1,3,5-Triazine, 2-(4-chlorophenyl)-4,6-dimethyl- (9CI)
s-Triazine, 2-(p-chlorophenyl)-4,6-dimethyl- (7CI,8CI)

30362-09-5
$C_{11}H_{10}N_4O_2$
230.23
s-Triazine, 2,4-dimethyl-6-(m-nitrophenyl)- (8CI)

30362-10-8
$C_7H_5F_6N_3$
245.13
s-Triazine, 2-ethyl-4,6-bis(trifluoromethyl)- (8CI)

30362-11-9
$C_6H_{10}N_4S_2$
202.30
1,3,5-Triazin-2-amine, N-methyl-4,6-bis(methylthio)- (9CI)
s-Triazine, 2-(methylamino)-4,6-bis(methylthio)- (8CI)

30362-12-0
$C_7H_{12}N_4S_2$
216.33
1,3,5-Triazin-2-amine, N-ethyl-4,6-bis(methylthio)- (9CI)
s-Triazine, 2-(ethylamino)-4,6-bis(methylthio)- (8CI)

30362-13-1
$C_{11}H_{12}N_4S_2$
264.37
s-Triazine, 2-anilino-4,6-bis(methylthio)- (8CI)

30362-14-2
$C_7H_{12}N_4S_2$
216.31
1,3,5-Triazin-2-amine, 4,6-bis-(ethylthio)- (9CI)
s-Triazine, 2-amino-4,6-bis-(ethylthio)- (8CI)

30362-15-3
$C_{11}H_{20}N_4S_2$
272.41
1,3,5-Triazin-2-amine, 4,6-bis-(butylthio)- (9CI)
s-Triazine, 2-amino-4,6-bis-(butylthio)- (8CI)

30362-17-5
$C_9H_7ClN_4S_2$
270.76
s-Triazine-2,4-dithiol, 6-(o-chloroanilino)- (8CI)

30362-18-6
$C_9H_{12}N_4O_4$
240.22
s-Triazine-2,4-dicarboxylic acid, 6-amino-, diethyl ester (6CI,8CI)

30362-19-7
$C_7H_7Cl_3N_4O_2$
285.52
s-Triazine-2-carboxylic acid, 4-amino-6-(trichloromethyl)-, ethyl ester (6CI,8CI)

30362-20-0
$C_5H_4Cl_3N_5O$
255.49
s-Triazine-2-carboxamide, 4-amino-6-(trichloromethyl)- (6CI,8CI)

30362-21-1
$C_4H_5N_5O_2$
155.12
s-Triazine-2-carboxamide, 4-amino-6-hydroxy-(6CI,8CI)

30362-22-2
$C_{11}H_5ClN_6S_2$
320.78
Isothiocyanic acid, 6-(o-chloroanilino)-s-triazine-2,4-diyl ester (8CI)

30362-23-3
$C_{11}H_5ClN_6S_2$
320.78
Isothiocyanic acid, 6-(p-chloroanilino)-s-triazine-2,4-diyl ester (8CI)

30362-24-4
$C_{11}H_5ClN_6S_2$
320.78
Thiocyanic acid, 6-(o-chloroanilino)-s-triazine-2,4-diyl ester (7CI, 8CI)

30362-26-6
$C_5H_6BrN_7$
244.06
s-Triazine, 2-azido-4-bromo-6-(ethylamino)- (8CI)

30362-27-7
$C_{11}H_{11}N_7O$
257.26
s-Triazine, 2-azido-4-(ethylamino)-6-phenoxy-(8CI)

30362-29-9
$C_7H_{11}N_7S$
225.28
1,3,5-Triazin-2-amine, 4-azido-N-ethyl-6-(ethylthio)- (9CI)
s-Triazine, 2-azido-4-(ethylamino)-6-(ethylthio)- (8CI)

30362-30-2
$C_{10}H_9N_7O$
243.23
s-Triazine, 2-anilino-4-azido-6-methoxy- (8CI)

30362-31-3
$C_6HCl_8N_3$
398.72
1,3,5-Triazine, 2-(dichloromethyl)-4,6-bis(trichloromethyl)- (9CI)
s-Triazine, 2-(dichloromethyl)-4,6-bis(trichloromethyl)- (8CI)

30362-32-4
$C_{11}H_5F_6N_3$
293.17
s-Triazine, 2-phenyl-4,6-bis(trifluoromethyl)-(8CI)

30362-33-5
$C_{11}H_8Cl_3N_3$
288.57
s-Triazine, 2-methyl-4-phenyl-6-(trichloromethyl)- (7CI,8CI)

30362-35-7
$C_7Cl_{11}N_3$
516.08
s-Triazine, 2-(pentachloroethyl)-4,6-bis-(trichloromethyl)- (8CI)

30362-44-8
$C_8H_5Cl_6N_3O_2$
387.87
s-Triazine-2-methanol, 4,6-bis(trichloromethyl)-, acetate (ester) (8CI)
s-Triazine-2-methanol, 4,6-bis(trichloromethyl)-, acetate (7CI)

30362-45-9
$C_9H_7Cl_6N_3O_2$
401.89
s-Triazine-2-methanol, α-methyl-4,6-bis(trichloromethyl)-, acetate (ester) (8CI)
s-Triazine-2-methanol, α-methyl-4,6-bis(trichloromethyl)-, acetate (7CI)

30362-46-0
$C_9H_4Cl_9N_3O_2$
505.23
s-Triazine-2-methanol, α,4,6-tris(trichloromethyl)-, acetate (ester) (8CI)
s-Triazine-2-methanol, α,4,6-tris(trichloromethyl)-, acetate (7CI)

30362-47-1
$C_{10}H_9Cl_6N_3O_2$
415.92
s-Triazine-2-methanol, α,α-dimethyl-4,6-bis-(trichloromethyl)-, acetate (ester) (8CI)
s-Triazine-2-methanol, α,α-dimethyl-4,6-bis(trichloromethyl)-, acetate (7CI)

30362-48-2
$C_{11}H_{11}Cl_6N_3$
397.95
s-Triazine, 2-cyclohexyl-4,6-bis(trichloromethyl)-(8CI)

30362-55-1
$C_{11}Cl_{11}N_3$
564.13
s-Triazine, 2-(pentachlorophenyl)-4,6-bis(trichloromethyl)- (8CI)

30362-59-5
$C_{10}H_4Cl_6N_4$
392.89
s-Triazine, 2-(3-pyridyl)-4,6-bis(trichloromethyl)- (8CI)

30362-60-8
$C_8H_{13}N_3$
151.21
1,3,5-Triazine, 2,4-diethyl-6-methyl- (9CI)
s-Triazine, 2,4-diethyl-6-methyl- (7CI,8CI)

30362-61-9
$C_{16}H_{11}N_5O_4$
337.30
s-Triazine, 2-methyl-4,6-bis(m-nitrophenyl)- (8CI)

30362-62-0
$C_8H_{10}Cl_3N_3$
254.55
s-Triazine, 2,4-diethyl-6-(trichloromethyl)- (7CI, 8CI)

30362-63-1
$C_{16}H_{10}Cl_3N_3$
350.64
1,3,5-Triazine, 2,4-diphenyl-6-(trichloromethyl)- (9CI)
s-Triazine, 2,4-diphenyl-6-(trichloromethyl)- (7CI, 8CI)

30362-65-3
$C_{16}H_8Cl_5N_3$
419.53
s-Triazine, 2,4-bis(p-chlorophenyl)-6-(trichloromethyl)- (8CI)

30362-66-4
$C_{18}H_{14}Cl_3N_3$
378.69
s-Triazine, 2,4-di-p-tolyl-6-(trichloromethyl)- (7CI, 8CI)

30362-67-5
$C_9H_{14}BrN_3$
244.14
s-Triazine, 2-(1-bromoethyl)-4,6-diethyl- (7CI, 8CI)

30362-68-6
$C_9H_{13}N_3$
163.22
s-Triazine, 2,4-diethyl-6-vinyl- (7CI,8CI)

30362-69-7
$C_{13}H_{15}N_3$
213.28
1,3,5-Triazine, 2,4-diethyl-6-phenyl- (9CI)
s-Triazine, 2,4-diethyl-6-phenyl- (7CI,8CI)

30362-70-0
$C_{11}H_{17}N_3O_2$
223.28
s-Triazine-2-methanol, 4,6-diethyl-α-methyl-, acetate (ester) (8CI)
s-Triazine-2-methanol, 4,6-diethyl-α-methyl-, acetate (7CI)

30362-71-1
$C_{17}H_{15}N_3$
261.33
1,3,5-Triazine, 2-ethyl-4,6-diphenyl- (9CI)
s-Triazine, 2-ethyl-4,6-diphenyl- (6CI,7CI,8CI)

30362-72-2
$C_9H_{12}Cl_3N_3$
268.58
s-Triazine, 2,4,6-tris-(1-chloroethyl)- (6CI, 8CI)

30362-73-3
$C_{17}H_{14}ClN_3$
295.77
s-Triazine, 2-(1-chloroethyl)-4,6-diphenyl- (7CI, 8CI)

30362-74-4
$C_9H_9Cl_6N_3$
371.91
s-Triazine, 2,4,6-tris(1,1-dichloroethyl)- (7CI,8CI)

30362-75-5
$C_9H_6Cl_9N_3$
475.25
s-Triazine, 2,4,6-tris(1,1,2-trichloroethyl)- (7CI, 8CI)

30362-76-6
$C_9H_3Cl_{12}N_3$
578.58
s-Triazine, 2,4,6-tris-(1,1,2,2-tetrachloroethyl)- (8CI)

30362-77-7
$C_9Cl_{15}N_3$
681.92
s-Triazine, 2,4,6-tris(pentachloroethyl)- (8CI)

30362-78-8
$C_9H_{12}Br_3N_3$
401.94
s-Triazine, 2,4,6-tris-(1-bromoethyl)- (7CI, 8CI)

30362-79-9
$C_{17}H_{14}BrN_3$
340.23
s-Triazine, 2-(1-bromoethyl)-4,6-diphenyl- (8CI)

30362-94-8
$C_9H_9Br_6N_3$
638.65
s-Triazine, 2,4,6-tris(1,1-dibromoethyl)- (7CI,8CI)

30362-95-9
$C_{18}H_{17}N_3$
275.36
1,3,5-Triazine, 2,4-diphenyl-6-propyl- (9CI)
s-Triazine, 2,4-diphenyl-6-propyl- (8CI)

30362-98-2
$C_{30}H_{57}N_3$
459.81
s-Triazine, 2,4,6-trinonyl- (8CI)

30362-99-3
$C_{27}H_{18}Cl_3N_3$
490.82
s-Triazine, 2,4,6-tris-(p-chlorostyryl)- (8CI)

30363-00-9
$C_{21}H_{16}N_4$
324.39
s-Triazine, 2-(m-aminophenyl)-4,6-diphenyl- (8CI)

30363-01-0
$C_{21}H_{13}N_5O_4$
399.37

s-Triazine, 2,4-bis(m-nitrophenyl)-6-phenyl- (8CI)

30363-02-1
$C_{21}H_{13}N_5O_4$
399.37
1,3,5-Triazine, 2,4-bis(4-nitrophenyl)-6-phenyl- (9CI)
s-Triazine, 2,4-bis(p-nitrophenyl)-6-phenyl- (8CI)

30363-03-2
$C_{21}H_{12}Br_3N_3$
546.08
1,3,5-Triazine, 2,4,6-tris(4-bromophenyl)- (9CI)
s-Triazine, 2,4,6-tris(p-bromophenyl)- (8CI)

30363-04-3
$C_{27}H_{24}Cl_3N_3O_3$
544.87
s-Triazine-2,4,6-triethanol, α,α',α''-tris(o-chlorophenyl)- (8CI)

30363-05-4
$C_{27}H_{21}Cl_6N_3O_3$
648.20
s-Triazine-2,4,6-triethanol, α,α',α''-tris(2,4-dichlorophenyl)- (8CI)

30368-49-1
$C_5H_9N_5$
139.13
1,3,5-Triazine-2,4-diamine, N-ethyl- (9CI)
s-Triazine, 2-amino-4-(ethylamino)- (8CI)

30368-50-4
$C_5H_9N_5$
139.13
1,3,5-Triazine-2,4-diamine, N,N'-dimethyl- (9CI)
s-Triazine, 2,4-bis(methylamino)- (8CI)

30368-51-5
$C_6H_{11}N_5$
153.19
s-Triazine, 2-amino-4-(ethylamino)-6-methyl- (8CI)

30368-52-6
$C_5H_{10}N_6$
154.18
s-Triazine, 2-hydrazino-4-methyl-6-(methylamino)- (8CI)

30368-85-5
$C_{10}H_{19}N_5$
209.30
s-Triazine, 2-(diethylamino)-4-(dimethylamino)-6-methyl- (8CI)

30368-93-5
$C_{16}H_{13}Cl_2N_5$
346.22
1,3,5-Triazine-2,4-diamine, N,N'-bis(4-chlorophenyl)-6-methyl- (9CI)
s-Triazine, 2,4-bis(p-chloroanilino)-6-methyl- (8CI)

30368-96-8
$C_{10}H_{19}N_5O_2$
241.30
s-Triazine, 2,4-bis((2-methoxyethyl)amino)-6-methyl- (8CI)

30368-97-9
$C_{10}H_{13}N_7$
231.26
Propionitrile, 3,3'-((6-methyl-s-triazine-2,4-diyl)diimino)di- (8CI)

30369-01-8
$C_{13}H_{24}N_4$
236.36
s-Triazine, 2-amino-4,6-dipentyl- (8CI)

30369-02-9
$C_{13}H_{21}Cl_3N_4$
339.70
s-Triazine, 2-amino-4-nonyl-6-(trichloromethyl)- (8CI)

30369-04-1
$C_{21}H_{37}Cl_3N_4$
451.91
1,3,5-Triazin-2-amine, 4-heptadecyl-6-(trichloromethyl)- (9CI)
s-Triazine, 2-amino-4-heptadecyl-6-(trichloromethyl)- (8CI)

30369-07-4
$C_{10}H_7Cl_3N_4$
289.55
s-Triazine, 2-amino-4-phenyl-6-(trichloromethyl)- (6CI,8CI)

30369-15-4
$C_{10}H_5Cl_5N_4$
358.44
s-Triazine, 2-amino-4-(2,4-dichlorophenyl)-6-(trichloromethyl)- (8CI)

30369-16-5
$C_{11}H_9Cl_3N_4$
303.58
s-Triazine, 2-amino-4-p-tolyl-6-(trichloromethyl)- (8CI)

30369-17-6
$C_{11}H_9Cl_3N_4O$
319.58
s-Triazine, 2-amino-4-(p-methoxyphenyl)-6-(trichloromethyl)- (8CI)

30369-18-7
$C_{10}H_6Cl_3N_5O_2$
334.55
s-Triazine, 2-amino-4-(m-nitrophenyl)-6-(trichloromethyl)- (8CI)

30369-20-1
$C_{21}H_{16}N_4$
324.39
1,3,5-Triazin-2-amine, N,4,6-triphenyl- (9CI)
s-Triazine, 2-anilino-4,6-diphenyl- (6CI,8CI)

30369-21-2
$C_{15}H_{10}Cl_2N_4$
317.18
s-Triazine, 2-amino-4,6-bis(p-chlorophenyl)- (8CI)

30369-24-5
$C_6H_9ClN_4$
172.62
1,3,5-Triazin-2-amine, 4-chloro-N-ethyl-6-methyl- (9CI)
s-Triazine, 2-chloro-4-(ethylamino)-6-methyl- (8CI)

30369-27-8
$C_4H_3ClN_6$
170.56
s-Triazine, 2-amino-4-chloro-6-(diazomethyl)- (6CI,7CI,8CI)

30369-28-9
$C_5H_7ClN_4$
158.59
s-Triazine, 2-amino-4-chloro-6-ethyl- (8CI)

30369-29-0
$C_7H_{11}ClN_4$
186.65
1,3,5-Triazin-2-amine, 4-chloro-N,6-diethyl- (9CI)
s-Triazine, 2-chloro-4-ethyl-6-(ethylamino)- (8CI)

30369-30-3
$C_6H_9ClN_4$
172.59
1,3,5-Triazin-2-amine, 4-chloro-6-propyl- (9CI)
s-Triazine, 2-amino-4-chloro-6-propyl- (8CI)

30369-33-6
$C_9H_6Cl_2N_4$
241.05
1,3,5-Triazin-2-amine, 4-chloro-6-(4-chlorophenyl)- (9CI)
s-Triazine, 2-amino-4-chloro-6-(p-chlorophenyl)- (8CI)

30369-37-0
$C_{11}H_{12}N_4O$
216.24
s-Triazine, 2-anilino-4-methoxy-6-methyl- (8CI)

30369-38-1
$C_{10}H_{10}N_4O$
202.22
1,3,5-Triazin-2-amine, 4-methoxy-6-phenyl- (9CI)
s-Triazine, 2-amino-4-methoxy-6-phenyl- (8CI)

30369-39-2
$C_{12}H_{14}N_4O$
230.27
1,3,5-Triazin-2-amine, N-ethyl-4-methoxy-6-phenyl- (9CI)
s-Triazine, 2-(ethylamino)-4-methoxy-6-phenyl- (8CI)

30369-40-5
$C_{12}H_{14}N_4O$
230.27
s-Triazine, 2-(ethylamino)-4-methyl-6-phenoxy- (8CI)

30369-41-6
$C_7H_9Cl_3N_4S$
287.60
1,3,5-Triazin-2-amine, N-ethyl-4-(methylthio)-6-(trichloromethyl)- (9CI)
s-Triazine, 2-(ethylamino)-4-(methylthio)-6-(trichloromethyl)- (8CI)

30369-42-7
$C_8H_{11}Cl_3N_4S$
301.63
1,3,5-Triazin-2-amine, 4-(methylthio)-N-propyl-6-(trichloromethyl)- (9CI)
s-Triazine, 2-(methylthio)-4-(propylamino)-6-(trichloromethyl)- (8CI)

30369-43-8
$C_8H_9Cl_3N_4S$
299.61
1,3,5-Triazin-2-amine, 4-(methylthio)-N-2-propenyl-6-(trichloromethyl)- (9CI)
s-Triazine, 2-(allylamino)-4-(methylthio)-6-(trichloromethyl)- (8CI)

30369-44-9
$C_8H_{11}Cl_3N_4S$
301.63
1,3,5-Triazin-2-amine, N-(1-methylethyl)-4-(methylthio)-6-(trichloromethyl)- (9CI)
s-Triazine, 2-(isopropylamino)-4-(methylthio)-6-(trichloromethyl)- (8CI)

30369-45-0
$C_9H_{13}Cl_3N_4S$
315.65
1,3,5-Triazin-2-amine, N-butyl-4-(methylthio)-6-(trichloromethyl)- (9CI)
s-Triazine, 2-(butylamino)-4-(methylthio)-6-(trichloromethyl)- (8CI)

30369-46-1
$C_9H_{13}Cl_3N_4S$
315.65
1,3,5-Triazin-2-amine, N-(2-methylpropyl)-4-(methylthio)-6-(trichloromethyl)- (9CI)
s-Triazine, 2-(isobutylamino)-4-(methylthio)-6-(trichloromethyl)- (8CI)

30369-47-2
$C_{17}H_{29}Cl_3N_4S$
427.87
1,3,5-Triazin-2-amine, N-dodecyl-4-(methylthio)-6-(trichloromethyl)- (9CI)
s-Triazine, 2-(dodecylamino)-4-(methylthio)-6-(trichloromethyl)- (8CI)

30369-48-3
$C_9H_{13}Cl_3N_4S$
315.65
1,3,5-Triazin-2-amine, N,N-diethyl-4-(methylthio)-6-(trichloromethyl)- (9CI)
s-Triazine, 2-(diethylamino)-4-(methylthio)-6-(trichloromethyl)- (8CI)

30369-49-4
$C_6H_7Cl_3N_4S$
273.57
1,3,5-Triazin-2-amine, 4-(ethylthio)-6-(trichloromethyl)- (9CI)
s-Triazine, 2-amino-4-(ethylthio)-6-(trichloromethyl)- (8CI)

30369-50-7
$C_7H_9Cl_3N_4S$
287.60
1,3,5-Triazin-2-amine, 4-(ethylthio)-N-methyl-6-(trichloromethyl)- (9CI)
s-Triazine, 2-(ethylthio)-4-(methylamino)-6-(trichloromethyl)- (8CI)

30369-51-8
$C_8H_{11}Cl_3N_4S$
301.63
1,3,5-Triazin-2-amine, N-methyl-4-((1-methylethyl)thio)-6-(trichloromethyl)- (9CI)
s-Triazine, 2-(isopropylthio)-4-(methylamino)-6-(trichloromethyl)- (8CI)

30369-52-9
$C_8H_{11}Cl_3N_4S$
301.63
1,3,5-Triazin-2-amine, 4-(butylthio)-6-(trichloromethyl)- (9CI)
s-Triazine, 2-amino-4-(butylthio)-6-(trichloromethyl)- (8CI)

30369-53-0
$C_8H_{11}Cl_3N_4S$
301.63
1,3,5-Triazin-2-amine, 4-((2-methylpropyl)thio)-6-(trichloromethyl)- (9CI)
s-Triazine, 2-amino-4-(iso-

butylthio)-6-(trichloromethyl)- (8CI)

30369-54-1
$C_{11}H_{17}Cl_3N_4S$
343.71
1,3,5-Triazin-2-amine, 4-(hexylthio)-N-methyl-6-(trichloromethyl)- (9CI)
s-Triazine, 2-(hexylthio)-4-(methylamino)-6-(trichloromethyl)- (8CI)

30369-55-2
$C_{10}H_7Cl_3N_4S$
321.62
1,3,5-Triazin-2-amine, 4-(phenylthio)-6-(trichloromethyl)- (9CI)
s-Triazine, 2-amino-4-(phenylthio)-6-(trichloromethyl)- (8CI)

30369-56-3
$C_{11}H_9Cl_3N_4S$
335.64
1,3,5-Triazin-2-amine, N-methyl-4-(phenylthio)-6-(trichloromethyl)- (9CI)
s-Triazine, 2-(methylamino)-4-(phenylthio)-6-(trichloromethyl)- (8CI)

30369-57-4
$C_{10}H_6Cl_4N_4S$
356.06
1,3,5-Triazin-2-amine, 4-((4-chlorophenyl)thio)-6-(trichloromethyl)- (9CI)
s-Triazine, 2-amino-4-((p-chlorophenyl)thio)-6-(trichloromethyl)- (8CI)

30369-58-5
$C_{11}H_8Cl_4N_4S$
370.09
1,3,5-Triazin-2-amine, 4-((4-chlorophenyl)thio)-N-methyl-6-(trichloromethyl)- (9CI)
s-Triazine, 2-((p-chlorophenyl)thio)-4-(methylamino)-6-(trichloromethyl)- (8CI)

30369-59-6
$C_{12}H_{10}Cl_4N_4S$
384.12
1,3,5-Triazin-2-amine, 4-((4-chlorophenyl)thio)-N-ethyl-6-(trichloromethyl)- (9CI)
s-Triazine, 2-((p-chlorophenyl)thio)-4-(ethylamino)-6-(trichloromethyl)- (8CI)

30369-60-9
$C_{13}H_{12}Cl_4N_4S$
398.14
1,3,5-Triazin-2-amine, 4-((4-chlorophenyl)thio)-N-(1-methylethyl)-6-(trichloromethyl)- (9CI)
s-Triazine, 2-((p-chlorophenyl)thio)-4-(isopropylamino)-6-(trichloromethyl)- (8CI)

30369-61-0
$C_{11}H_7Cl_5N_4S$
404.53
1,3,5-Triazin-2-amine, 4-((2,4-dichlorophenyl)thio)-N-methyl-6-(trichloromethyl)- (9CI)
s-Triazine, 2-((2,4-dichlorophenyl)thio)-4-(methylamino)-6-(trichloromethyl)- (8CI)

30369-62-1
$C_{10}H_4Cl_6N_4S$
424.95
1,3,5-Triazin-2-amine, 4-(trichloromethyl)-6-((2,4,5-trichlorophenyl)thio)- (9CI)
s-Triazine, 2-amino-4-(trichloromethyl)-6-((2,4,5-trichlorophenyl)thio)- (8CI)

30369-63-2
$C_{11}H_6Cl_6N_4S$
438.98
1,3,5-Triazin-2-amine, N-methyl-4-(trichloromethyl)-6-((2,4,5-trichlorophenyl)thio)- (9CI)
s-Triazine, 2-(methylamino)-4-(trichloromethyl)-6-((2,4,5-trichlorophenyl)thio)- (8CI)

30369-64-3
$C_6H_{10}N_4O$
154.17
1,3,5-Triazin-2(1H)-one, 4-(ethylamino)-6-methyl- (9CI)
s-Triazin-2-ol, 4-(ethylamino)-6-methyl- (8CI)

30369-65-4
$C_4H_3Cl_3N_4O$
229.45
s-Triazin-2-ol, 4-amino-6-(trichloromethyl)- (8CI)

30369-67-6
$C_7H_{12}N_4O$
168.20
s-Triazin-2-ol, 4-amino-6-isobutyl- (8CI)

30369-68-7
$C_{10}H_{10}N_4O$
202.22
s-Triazin-2-ol, 4-amino-6-benzyl- (8CI)

30369-69-8
$C_9H_7N_5O_3$
233.19
s-Triazin-2-ol, 4-amino-6-(p-nitrophenyl)- (8CI)

30369-70-1
$C_4H_6N_4S$
142.18
1,3,5-Triazine-2(1H)-thione, 4-amino-6-methyl- (9CI)
s-Triazine-2-thiol, 4-amino-6-methyl- (8CI)

30369-71-2
$C_5H_8N_4S$
156.21
s-Triazine-2-thiol, 4-amino-6-ethyl- (8CI)

30369-72-3
$C_7H_{12}N_4S$
184.26
s-Triazine-2-thiol, 4-amino-6-isobutyl- (8CI)

30369-73-4
$C_{10}H_{10}N_4S$
218.28
s-Triazine-2-thiol, 4-amino-6-benzyl- (8CI)

30369-74-5
$C_9H_8N_4S$
204.25
1,3,5-Triazine-2(1H)-thione, 4-amino-6-phenyl- (9CI)
s-Triazine-2-thiol, 4-amino-6-phenyl- (8CI)

30369-75-6
$C_9H_7ClN_4S$
238.70
s-Triazine-2-thiol, 4-amino-6-(p-chlorophenyl)- (8CI)

30369-76-7
$C_5H_6F_2N_4$
160.13
1,3,5-Triazin-2-amine, N-ethyl-4,6-difluoro- (9CI)
2,4-Difluoro-6-(ethylamino)-s-triazine
s-Triazine, 2-(ethylamino)-4,6-difluoro- (8CI)

30369-78-9
$C_9H_5ClF_2N_4$
242.59
1,3,5-Triazin-2-amine, N-(4-chlorophenyl)-4,6-difluoro- (9CI)
s-Triazine, 2-(p-chloroanilino)-4,6-difluoro- (8CI)

30369-80-3
$C_6H_6Cl_2N_4$
205.02
1,3,5-Triazin-2-amine, 4,6-dichloro-N-2-propenyl- (9CI)
s-Triazine, 2-(allylamino)-4,6-di chloro- (8CI)

30369-82-5
$C_{10}H_8Cl_2N_4$
255.11
1,3,5-Triazin-2-amine, 4,6-dichloro-N-(phenylmethyl)- (9CI)
s-Triazine, 2-(benzylamino)-4,6-dichloro- (6CI, 7CI,8CI)

30369-84-7
$C_9H_5BrCl_2N_4$
319.98
s-Triazine, 2-(o-bromoanilino)-4,6-dichloro- (6CI,8CI)

30369-85-8
$C_9H_5BrCl_2N_4$
319.98
s-Triazine, 2-(m-bromoanilino)-4,6-dichloro- (6CI,8CI)

30369-86-9
$C_{15}H_{10}Cl_2N_4$
317.18
s-Triazine, 2-(4-biphenylylamino)-4,6-dichloro- (6CI,8CI)

30369-87-0
$C_{10}H_8Cl_2N_4O$
271.08
1,3,5-Triazin-2-amine, 4,6-di chloro-N-(2-methoxyphenyl)- (9CI)
s-Triazine, 2-o-anisidino-4,6-dichloro- (8CI)

30369-88-1
$C_{13}H_8Cl_2N_4$
291.15
s-Triazine, 2,4-dichloro-6-(1-naphthylamino)
2,4-Dichloro-6-(1-naphthylamino)-s-triazine

30369-89-2
$C_9H_6Cl_2N_4O_2S$
305.12
Benzenesulfonamide, N-(4,6-dichloro-1,3,5-triazin-2-yl)- (9CI)
Benzenesulfonamide, N-(4,6-dichloro-s-triazin-2-yl)- (8CI)

30377-13-0
$C_{18}H_{17}N_3$
275.36
1,3,5-Triazine, 2-methyl-4,6-bis(4-methylphenyl)- (9CI)
s-Triazine, 2-methyl-4,6-di-p-tolyl- (8CI)

30377-15-2
$C_9H_{11}Cl_6N_5S$
434.00
Methanesulfenamide, 1,1,1-trichloro-N-ethyl-N-(4-(ethylamino)-6-(trichloromethyl)-s-tria zin-2-yl)- (8CI)

30377-16-3
$C_6H_{10}BrN_5$
232.09
s-Triazine, 2-amino-4-(bromomethyl)-6-(ethylamino)- (8CI)

30377-17-4
$C_{16}H_{10}Br_3Cl_2N_5$
582.92
s-Triazine, 2,4-bis-(p-chloroanilino)-6-(tribromomethyl)- (8CI)

30377-19-6
$C_{11}H_{21}N_5O$
239.32
1,3,5-Triazine-2,4-diamine, 6-ethoxy-N,N'-bis-(1-methylethyl)- (9CI)
s-Triazine, 2-ethoxy-4,6-bis-(isopropylamino)- (7CI, 8CI)

30377-20-9
$C_{17}H_{18}N_6$
306.37
1,3,5-Triazine-2,4,6-triamine, N,N'-dimethyl -N,N'-diphenyl- (9CI)
Melamine, N2,N4-dimethyl-N2,N4-diphenyl- (8CI)

30377-22-1
$C_{11}H_{12}N_4$
200.25
s-Triazine, 2-(benzylamino)-4-methyl- (8CI)

30377-23-2
$C_7H_{12}N_4$
152.20
s-Triazine, 2-(ethylamino)-4,6-dimethyl- (8CI)

30377-24-3
$C_{10}H_6Cl_4N_4$
324.00
s-Triazine, 2-amino-4-(p-chlorophenyl)-6-(trichloromethyl)- (8CI)

30377-26-5
$C_6H_7Cl_3N_4S$
273.57
1,3,5-Triazin-2-amine, N-methyl-4-(methylthio)-6-(trichloromethyl)- (9CI)
s-Triazine, 2-(methylamino)-4-(methylthio)-6-(trichloromethyl)- (8CI)

30377-27-6
$C_{10}H_8Cl_2N_4O$
271.11
1,3,5-Triazin-2-amine, 4,6-dichloro-N-(4-methoxyphenyl)- (9CI)
s-Triazine, 2-p-anisidino-4,6-dichloro- (6CI,8CI)

30381-98-7
$C_{24}H_{19}F_{34}N_2O_8PS_2.H_3N$
1221.49
1-Octanesulfonamide, N,N'-(phosphinicobis(oxy-2,1-ethanediyl))bis(N-ethyl-1,1,2,2,3,3,4,4,5,5,6,6,7,7,8,8,8-heptadecafluoro-, ammonium salt (9CI)
Bis(N-ethyl-2-perfluorooctylsulfonaminoethyl)phosphate, ammonium salt
N,N'-((Hydroxyphosphinylidene)bis(oxy-2,1-ethanediyl))bis(N-ethylheptadecafluoro-1-octanesulfonamide), ammonium salt

30384-46-4
$C_{15}H_{16}N_4O_2$
284.32
s-Triazine, 2,4-bis(allyloxy)-6-anilino- (8CI)

30384-47-5
$C_{11}H_{20}N_4O_2$
240.31
s-Triazine, 2-amino-4,6-dibutoxy- (8CI)

30384-48-6
$C_{17}H_{16}N_4O_2$
308.34
s-Triazine, 2-(ethylamino)-4,6-diphenoxy- (8CI)

30385-25-2
$C_{10}H_{20}O$
156.27
OC(CCCC(=CC)C)(C)C
6-Octen-2-ol, 2,6-dimethyl- (9CI)
2,6-Dimethyl-6-octen-2-ol

30388-85-3
$C_{10}H_{16}Cl_3N_5O_2$
344.63
s-Triazine, 2,4-bis((2-methoxyethyl)amino)-6-(trichloromethyl)- (8CI)

30388-86-4
$C_9H_{16}FN_5O$
229.26
s-Triazine, 2-(ethylamino)-4-fluoro-6-((3-methoxypropyl)amino)- (8CI)

30388-90-0
$C_9H_7Cl_3N_6$
305.56
Acetonitrile, ((4-methyl-6-(trichloromethyl)-s-triazin-2-yl)imino)di- (8CI)

30388-91-1
$C_7H_4Cl_6N_4O$
372.85
Acetamide, 2,2,2-trichloro-N-(4-methyl-6-(trichloromethyl)-s-triazin-2-yl)- (8CI)

30388-94-4
$C_9H_5Br_2ClN_4$
364.44
s-Triazine, 2,4-dibromo-6-(o-chloroanilino)- (8CI)

30388-95-5
$C_7H_{11}ClN_4S$
218.71
s-Triazine, 2-chloro-4-(ethylamino)-6-(ethylthio)- (6CI,8CI)

30394-81-1
$(C_5H_8O_2.C_5H_8O_2.C_4H_7NO_2.C_3H_5NO)x$
Polymer
2-Propenoic acid, 2-methyl-, methyl ester, Polymer with ethyl 2-propenoate, N-(hydroxymethyl)-2-propenamide and 2-propenamide (9CI)
Acrylamide, ethyl acrylate, methylolacrylamide, methyl methacrylate polymer
Ethyl acrylate-acrylamide-N-methylolacrylamide-methylmethacrylate polymer
Ethyl acrylate, methyl methacrylate, acrylamide, methylolacrylamide polymer
Ethyl 2-propenoate, methyl 2-methyl-2-propenoate, N-(hydroxymethyl)-2-propenamide, 2-propenamide polymer

30399-84-9
$C_{18}H_{36}O_2$
284.48
Isooctadecanoic acid (VAN) (9CI)
Isostearic acid

30402-14-3
$C_{12}H_4Cl_4O$
305.97
Dibenzofuran, ar,ar,ar',ar'-tetrachloro- (8CI,9CI)

30402-15-4
$C_{12}H_3Cl_5O$
340.42
Dibenzofuran, pentachloro- (8CI,9CI)

30419-67-1
$C_4H_6O_2$
86.10
D-Threitol, 1,2:3,4-dianhydro
2,2'-Bioxirane, (R-(R*,R*))- (9CI)
Butane, D-1,2:3,4-diepoxy-
D-Diepoxybutane
2R:3R-Diepoxybutane

30431-53-9
$C_{12}H_{13}Cl_3O_3$
311.59
O=C(OCCCCC)c(c(c(cc1Cl)Cl)Cl)c1O
Benzoic acid, 2,3,5-trichloro-6-hydroxy-, pentyl ester (9CI)

30431-54-0
$C_{26}H_{24}Cl_6O_8$
677.19
O=C(Oc(c(cc(c1Cl)Cl)Cl)c1C(=O)OCCCCC)C(=O)Oc(c(cc(c2Cl)Cl)Cl)c2C(=O)OCCCCC
Bis(2,4,5-trichloro-6-carbopentoxyphenyl)oxalate
CPPO
Ethanedioic acid, bis(3,4,6-trichloro-2-((pentyloxy)carbonyl)phenyl) ester (9CI)

30453-31-7
$C_5H_{12}S_2$
136.28
Disulfide, ethyl propyl (8CI, 9CI)
3,4-Dithiaheptane

30496-78-7
$C_{12}H_{20}$
164.29
4,7-Methano-1H-indene, octahydrodimethyl- (9CI)
Methylcyclopentadiene, dimer, hydrogenated
Octahydrodimethyl-4,7-methano-1H-indene

30497-87-1
$C_9H_8O_4$
180.16
1,2-Benzenedicarboxylic acid, methyl- (9CI)
Phthalic acid, methyl-

30498-35-2
$C_7H_3Cl_2F_3$
215.00
Benzene, dichloro(trifluoromethyl)- (9CI)

30498-63-6
C_9H_{18}
126.24
Cyclohexane, trimethyl- (9CI)

30498-64-7
C_8H_{16}
112.22
Cyclopentane, trimethyl- (9CI)

30498-66-9
C_9H_{20}
128.26
Heptane, dimethyl- (9CI)

30499-70-8
$(C_6H_{14}O_3.C_3H_5ClO)x$
Polymer
1,3-Propanediol, 2-ethyl-2-(hydroxymethyl)-, Polymer with (chloromethyl)oxirane (9CI)
Trimethylolpropane, (chloromethyl)oxirane polymer

30516-87-1
$C_{10}H_{13}N_5O_4$
267.28
Thymidine, 3'-azido-3'-deoxy
3'-Azido-3'-deoxythymidine
Azidothymidine
AZT
BW-A 509U
Retrovir
Zidovudine

30525-89-4
$(CH_2O)x$
Unknown
Paraformaldehyde [UN 2213]
Flo-Mor
Formagene
Paraform
Triformol
Trioxymethylene
UN 2213 [Paraformaldehyde]

30526-22-8
$C_7H_8O_3S.K$
211.30
Benzenesulfonic acid, methyl-, potassium salt (9CI)
Caswell No. 703B
EPA Pesticide Chemical Code 079031
Potassium methylbenzenesulfonate
Potassium toluene sulfonate

30551-09-8
$C_{14}H_{14}$
182.27
Anthracene, tetrahydro- (8CI, 9CI)

30554-72-4
$C_6H_6Br_4Cl_2$
468.64
Cyclohexane, tetrabromodichloro- (9CI)
Tetrabromodichlorocyclohexane

30554-73-5
$C_6H_6Br_3Cl_3$
424.18
Cyclohexane, tribromotrichloro- (9CI)
Tribromotrichlorocyclohexane

30558-43-1
$C_5H_{10}N_2O_2S$
162.20
O=C(N(C)C)C(=NO)SC
Ethanimidothioic acid, 2-(dimethylamino)-N-hydroxy-2-oxo-, methyl ester (9CI)
2-(Hydroxyimino)-N,N-dimethyl-2-(methylmercapto)-acetamide

30560-19-1
$C_4H_{10}NO_3PS$
183.18
COP(=O)(NC(C)=O)SC
Phosphoramidothioic acid, N-acetyl-, O,S-dimethyl ester
Acephat (German)
Acephate
Acetylphosphoramidothioic acid O,S-dimethyl ester
Chevron RE 12,420
O,S-Dimethylacetylphosphoroamidothioate
ENT 27,822
N-(Methoxy(methylthio)phosphinoyl)acetamide
Orthene
Orthene-755
Ortho 124120
Ortran
Ortril
RE 12420
75 SP

30574-97-1
C_5H_7N
81.11
N#CC(=CC)C
2-Butenenitrile, 2-methyl-, (E)- (9CI)
(E)-2-Methyl-2-butenenitrile

30576-26-2
$C_{11}H_{10}Cl_3N_7O_3$
396.62
2-Furaldehyde, 5-nitro-, (4-(dimethylamino)-6-(trichloromethyl)-s-triazin-2-yl)hydrazone (8CI)

30576-27-3
$C_5H_6Cl_2N_4$
193.01
1,3,5-Triazin-2-amine, 4-(dichloromethyl)-6-methyl- (9CI)
s-Triazine, 2-amino-4-(dichloromethyl)-6-methyl- (8CI)

30576-30-8
$C_{11}H_{17}Cl_3N_4S$
343.71
1,3,5-Triazin-2-amine, N-hexyl-4-(methylthio)-6-(trichloromethyl)- (9CI)
s-Triazine, 2-(hexylamino)-4-(methylthio)-6-(trichloromethyl)- (8CI)

30576-31-9
$C_{10}H_5Cl_5N_4S$
390.51
1,3,5-Triazin-2-amine, 4-((2,4-dichlorophenyl)-thio)-6-(trichloromethyl)- (9CI)
s-Triazine, 2-amino-4-((2,4-dichlorophenyl)thio)-6-(trichloromethyl)- (8CI)

30576-32-0
$C_7H_{12}N_4O_2$
184.20
1,3,5-Triazin-2-amine, 4,6-diethoxy- (9CI)
s-Triazine, 2-amino-4,6-diethoxy- (8CI)

30583-33-6
$C_7H_5Cl_3$
195.47
Benzene, trichloromethyl- (9CI)
Methyltrichlorobenzene
Trichloromethylbenzene
Trichlorotoluene

30583-72-3
$(C_{15}H_{28}O_2.C_3H_5ClO)x$
Polymer
Cyclohexanol, 4,4'-(1-methylethylidene)bis-, Polymer with (chloromethyl)oxirane (9CI)
2,2-Bis(4-hydroxycyclohexyl)-propane, epichlorohydrin polymer
4,4'-(1-Methylethylidene)bis-cyclohexanol, Polymer with (chloromethyl)oxirane

30586-10-8
$C_5H_{10}Cl_2$
141.05
Pentane, dichloro
Dichloropentane [UN 1152]
UN 1152 [Dichloropentanes]

30587-19-0
$C_9H_{10}O_2$
150.18
Benzoic acid, dimethyl- (8CI, 9CI)

30622-37-8
Unknown
Unknown
Penta-s-triazinetrione

30638-08-5

$C_{32}H_{15}CoN_8O_3S.H$
651.48
Cobaltate(1-), (29H,31H-phthalocyaninesulfonato(3-)-N29,N30,N31,N32)-, hydrogen (9CI)
Cobalt, (hydrogen phthalocyaninesulfonato(2-))-

30638-09-6
$C_{32}H_{13}CuN_8O_9S_3.3H$
816.22
Cuprate(3-), (29H,31H-phthalocyaninetrisulfonato-(5-)-N29,N30,N31,N32)-, trihydrogen (9CI)
Copper,(29H,31H-phthalocyaninetrisulfonato(2-)-N29,N30,N31,N32)-
Phthalocyaninetrisulfonic acid, copper complex

30642-36-5
$C_{11}H_{13}N$
159.22
1H-Indole, trimethyl- (9CI)
Indole, trimethyl-

30667-99-3
$C_{16}H_{15}Cl_3O_2$
345.65
Benzene, 1-methoxy-2-(2,2,2-trichloro-1-(4-methoxyphenyl)ethyl)- (9CI)
AI3-16095
Ethane, 1,1,1-trichloro-2-(o-methoxyphenyl)-2-(p-methoxyphenyl)- (8CI)
o,p-Methoxychlor
o,p'-Methoxychlor
NSC-123014

30673-36-0
$C_{14}H_{28}O_2$
228.38
Decanoic acid, butyl ester
AI3-33573
Butyl decanoate

30674-80-7
$C_7H_9NO_3$
155.17
O=C(OCCN=C=O)C(=C)C
Metharcylic acid, 2-isocyanatoethyl ester
IEM
β-Isocyanatoethyl methacrylate
2-Isocyanatoethyl methacrylate
Methacryloyloxyethyl isocyanate
2-Propenoic acid, 2-methyl-, 2-isocyanatoethyl ester

30677-34-0
C_9H_{18}
126.24
Cyclohexane, ethylmethyl- (9CI)

30678-61-6
$C_{11}H_8O$
156.18
Naphthalenecarboxaldehyde (9CI)

30707-68-7
$C_{15}H_8Cl_4N_4O_3$
434.04
O=C(N(N=C1Nc(c(ccc2N(=O)=O)Cl)c2)c(c(cc(c3)Cl)Cl)c3Cl)C1
3H-Pyrazol-3-one, 5-((2-chloro-5-nitrophenyl)amino)-2,4-dihydro-2-(2,4,6-trichlorophenyl)- (9CI)

30714-78-4
Unknown
Unknown
Ethyl butyl carbonate

30714-88-6
$C_{10}H_{11}ClO_4$
230.65
Benzoic acid, 2-chloro-4,5-dimethoxy-, methyl ester (9CI)
Methyl 6-chloro-3,4-dimethoxybenzoate
Methyl 6-chloroveratrate
Veratric acid, 6-chloro-, methyl ester (8CI)

30744-85-5
$C_{52}H_{61}ClN_2O_8S$
909.57
Pentanamide, N-(5-((4-(2,4-bis(1,1-dimethylpropyl)phenoxy)-1-oxobutyl)amino)-2-chlorophenyl)-4,4-dimethyl-3-oxo-2-(4-((4-(phenylmethoxy)phenyl)sulfonyl)phenoxy)- (9CI)

30746-58-8
$C_{12}H_4Cl_4O_2$
321.96
Clc3c(Cl)c(Cl)c2Oc1ccccc1Oc2c3Cl
Dibenzo-p-dioxin, 1,2,3,4-tetrachloro
Dibenzo(b,e)(1,4)dioxin, 1,2,3,4-tetrachloro-
1,2,3,4-Tetrachlorodibenzodioxin
1,2,3,4-Tetrachlorodibenzo-para-dioxin

30773-71-8
$C_{10}H_{10}O_2$
162.19
Ethanone, 1,1'-(phenylene)bis-(9CI)
Benzene, diacetyl- (8CI)

30774-95-9
$C_{14}H_{10}O$
194.23
Phenanthrenol (9CI)
Phenanthrol

30776-59-1
$C_{20}H_{34}O_3S$
354.55
Isooctylphenoxypolyethoxyethanol
Benzenesulfonic acid, tetradecyl- (9CI)
Tetradecylbenzenesulfonic acid

30777-18-5
$C_{17}H_{12}$
216.28
Benzo(a)fluorene (9CI) (VAN)

30777-19-6
$C_{17}H_{12}$
216.29
Benzo(b)fluorene

30796-92-0
$C_{14}H_8S$
208.28
Phenanthro(4,5-bcd)thiophene (8CI,9CI)
Benzo(def)dibenzothiophene

30801-96-8
$C_7H_{14}O$
114.19
2-Hexen-1-ol, 3-methyl-, (E)- (8CI,9CI)

30804-91-2
$C_8H_{10}Cl_3N_5O$
298.56
Urea, 1,1-dimethyl-3-(4-methyl-6-(trichloromethyl)-s-triazin-2-yl)- (8CI)

30804-92-3
$C_{12}H_{10}Cl_3N_5O$
346.61
Urea, 1-(4-methyl-6-(trichloromethyl)-s-triazin-2-yl)-3-phenyl- (8CI)

30804-93-4
$C_{12}H_9Cl_4N_5O$
381.05
Urea, 1-(o-chlorophenyl)-3-(4-methyl-6-(trichloromethyl)-s-triazin-2-yl)-

(8CI)

30804-94-5
$C_{12}H_9Cl_4N_5O$
380.04
Urea, 1-(m-chlorophenyl)-3-(4-methyl-6-(trichloromethyl)-s-triazin-2-yl)-(8CI)

30804-95-6
$C_{12}H_9Cl_4N_5O$
381.05
Urea, 1-(p-chlorophenyl)-3-(4-methyl-6-(trichloromethyl)-s-triazin-2-yl)-(8CI)

30804-96-7
$C_{12}H_8Cl_5N_5O$
415.50
Urea, 1-(2,4-dichlorophenyl)-3-(4-methyl-6-(trichloromethyl)-s-triazin-2-y l)- (8CI)

30804-97-8
$C_{13}H_{12}Cl_3N_5O$
360.63
Urea, 1-(4-methyl-6-(trichloromethyl)-s-triazin-2-yl)-3-p-tolyl- (8CI)

30804-98-9
$C_{12}H_9Cl_3N_6O_3$
391.60
Urea, 1-(4-methyl-6-(trichloromethyl)-s-triazin-2-yl)-3-(m-nitrophenyl)-(8CI)

30804-99-0
$C_8H_7Cl_6N_5O$
401.90
Urea, 3-(4,6-bis(trichloromethyl)-s-triazin-2-yl)-1,1-dimethyl- (8CI)

30805-00-6
$C_{12}H_7Cl_6N_5O$
449.94
Urea, 1-(4,6-bis(trichloromethyl)-s-triazin-2-yl)-3-phenyl- (8CI)

30805-01-7
$C_{18}H_{17}N_7O_2$
363.38
Urea, 1,1'-(6-methyl-s-triazine-2,4-diyl)bis-(3-phenyl- (8CI)

30805-02-8
$C_{18}H_{14}Cl_3N_7O_2$
466.72
Urea, 1,1'-(6-(trichloromethyl)-s-triazine-2,4-diyl)bis(3-phenyl- (8CI)

30805-03-9
$C_{19}H_{19}N_7O_2$
377.41
Urea, 1,1'-(6-ethyl-s-triazine-2,4-diyl)bis(3-phenyl- (8CI)

30805-04-0
$C_{23}H_{19}N_7O_2$
425.45
Urea, N,N''-(6-phenyl-1,3,5-triazine-2,4-diyl)bis(N'-phenyl- (9CI)
Urea, 1,1'-(6-phenyl-s-triazine-2,4-diyl)bis(3-phenyl- (8CI)

30805-05-1
$C_{24}H_{21}N_9O_3$
483.49
Urea, 1,1',1''-(s-triazine-2,4,6-triyl)tris(3-phenyl-(8CI)

30805-06-2
$C_3Cl_2N_6$
190.99
s-Triazine, 2-azido-4,6-dichloro
2-Azido-4,6-dichloro-s-triazine

30805-07-3
$C_5H_6N_6O_2$
182.14
1,3,5-Triazine, 2-azido-4,6-dimethoxy- (9CI)
s-Triazine, 2-azido-4,6-dimethoxy- (8CI)

30805-08-4
$C_5H_6N_6$
150.14
s-Triazine, 2-azido-4,6-dimethyl- (8CI)

30805-09-5
$C_5Cl_6N_6$
356.81
s-Triazine, 2-azido-4,6-bis-(trichloromethyl)- (8CI)

30805-11-9
$C_4H_4Cl_3N_5O$
244.47
s-Triazine, 2,4-diamino-6-(trichloromethyl)-, 3-oxide (7CI,8CI)

30805-12-0
$C_5H_5Cl_3N_4O$
243.48
s-Triazine, 2-amino-4-methyl-6-(trichloromethyl)-, 3-oxide (7CI, 8CI)

30805-14-2
$C_{21}H_{18}Cl_3N_3$
418.73
1,3,5-Triazine, 1,3,5-tris-(4-chlorophenyl)hexahydro-(9CI)
s-Triazine, 1,3,5-tris(p-chlorophenyl)hexahydro- (8CI)

30805-18-6
$C_9H_6Cl_9N_3O_3$
523.24
s-Triazine, hexahydro-1,3,5-tris(trichloroacetyl)-(8CI)

30805-19-7
$C_{12}H_{21}N_3O_3$
255.32
1,3,5-Triazine, hexahydro-1,3,5-tris(1-oxopropyl)-(9CI)
s-Triazine, hexahydro-1,3,5-tripropionyl-(6CI,8CI)
1,3,5-Tripropionylhexahydro-s-triazine

30805-20-0
$C_{12}H_{12}Cl_9N_3O_3$
565.32
s-Triazine, hexahydro-1,3,5-tris(2,2,3-trichloropropionyl)- (8CI)

30805-22-2
$C_6H_6Cl_9N_3$
439.21
s-Triazine, hexahydro-1, 3,5-tris(trichloromethyl)-(8CI)

30805-23-3
$C_{27}H_{21}N_3$
387.49
1,3,5-Triazine, 1,2-dihydro-2,2,4,6-tetraphenyl- (9CI)
s-Triazine, 1,2-dihydro-2,2, 4,6-tetraphenyl- (6CI,8CI)

30805-24-4
$C_{10}H_{12}ClN_5$
237.66
1,3,5-Triazine-2,4-diamine, 5-(4-chlorophenyl)-5,6-dihydro-6-methyl- (9CI)
s-Triazine, 4,6-diamino-1-

(p-chlorophenyl)-1,2-dihydro-2-methyl- (8CI)

30805-25-5
$C_{12}H_{16}ClN_5$
265.75
s-Triazine, 4,6-diamino-1-(p-chlorophenyl)-1,2-dihydro-2-isopropyl- (8CI)

30805-26-6
$C_{12}H_{15}Cl_2N_5$
300.19
s-Triazine, 4,6-diamino-1-(3,4-dichlorophenyl)-1,2-dihydro-2-isopropyl- (8CI)

30805-27-7
$C_{11}H_{14}N_6O_2$
262.24
1,3,5-Triazine-2,4-diamine, 1,6-dihydro-6,6-dimethyl-1-(4-nitrophenyl)- (9CI)
s-Triazine, 4,6-diamino-1,2-dihydro-2,2-dimethyl-1-(p-nitrophenyl)- (8CI)

30805-28-8
$C_5H_7N_3O_3$
157.13
1,3,5-Triazine-2,4,6(1H,3H,5H)-trione, 1-ethyl- (9CI)
s-Triazine-2,4,6(1H,3H,5H)-trione, 1-ethyl- (8CI)

30805-30-2
$C_9H_9N_3O_2$
191.19
1,3,5-Triazine-2,4(1H,3H)-dione, dihydro-6-phenyl- (9CI)
s-Triazine-2,4(1H,3H)-dione, dihydro-6-phenyl- (6CI,8CI)

30805-32-4
$C_9H_9N_3S_2$
223.32
1,3,5-Triazine-2,4(1H,3H)-dithione, dihydro-6-phenyl- (9CI)
s-Triazine-2,4-dithiol, 5,6-dihydro-6-phenyl- (6CI)
s-Triazine-2,4(1H,3H)-dithione, dihydro-6-phenyl- (8CI)

30805-33-5
$C_5H_{11}N_3O$
129.16
1,3,5-Triazin-2(1H)-one, tetrahydro-4,6-dimethyl- (9CI)
s-Triazin-2(1H)-one, tetrahydro-4,6-dimethyl- (7CI, 8CI)

30805-34-6
$C_5H_{12}N_4$
128.18
s-Triazine, hexahydro-2-imino-4,6-dimethyl- (7CI,8CI)

30805-36-8
$C_5H_5Cl_6N_3O$
335.83
s-Triazin-2(1H)-one, tetrahydro-4,6-bis(trichloromethyl)- (8CI)

30805-37-9
$C_5H_5Cl_6N_3S$
351.90
s-Triazine-2(1H)-thione, tetrahydro-4,6-bis(trichloromethyl)- (8CI)

30805-38-0
$C_9H_{19}N_3O$
185.27
s-Triazin-2(1H)-one, tetrahydro-4,6-diisopropyl- (8CI)

30812-87-4
$C_8H_8Br_2$
263.98
Benzene, dibromoethyl
Alkazene 42
Diazene 42
Ethyl dibromobenzene
PRL-3191

30824-81-8
$C_{15}H_{28}$
208.39
Naphthalene, decahydro-1,4a-dimethyl-7-(1-methylethyl)-, (1S-(1α,4aα,7α,8aβ))- (9CI)
4αH-Eudesmane (8CI)
Selinane

30849-48-0
$C_{16}H_{22}O_4$
278.35
1,2-Benzenedicarboxylic acid, monoisooctyl ester (9CI)
Phthalic acid, isooctyl ester

30863-00-4
$C_8H_9Cl_4N_3S$
321.06
1,3,5-Triazine, 2,4-bis-(1,1-dichloroethyl)-6-(methylthio)- (9CI)
s-Triazine, 2,4-bis(1,1-dichloroethyl)-6-(methylthio)- (8CI)

30863-01-5
$C_{13}H_9Cl_6N_3S$
452.02
1,3,5-Triazine, 2-(phenylthio)-4,6-bis(1,1,2-trichloroethyl)- (9CI)
s-Triazine, 2-(phenylthio)-4,6-bis(1,1,2-trichloroethyl)- (8CI)

30863-02-6
$C_8H_3Cl_{10}N_3S$
527.73
1,3,5-Triazine, 2-(methylthio)-4,6-bis(pentachloroethyl)- (9CI)
s-Triazine, 2-(methylthio)-4,6-bis(pentachloroethyl)- (8CI)

30863-04-8
$C_{21}H_{15}N_3S$
341.44
s-Triazine, 2,4-diphenyl-6-(phenylthio)- (8CI)

30863-05-9
$C_9H_{12}ClN_5S_4$
353.94
Carbamodithioic acid, dimethyl-, 6-chloro-1,3,5-triazine-2,4-diyl ester (9CI)
Carbamic acid, dimethyldithio-, 6-chloro-s-triazine-2,4-diyl ester (6CI,8CI)

30863-06-0
$C_{13}H_{20}ClN_5S_4$
410.05
Carbamic acid, diethyldithio-, 6-chloro-s-triazine-2,4-diyl ester (6CI,8CI)
Carbamic acid, diethyldithio-, ester with 6-chloro-s-triazine-2,4-dithiol (7CI)

30863-07-1
$C_{11}H_{16}ClN_5S_4$
381.99
Carbamic acid, isopropyldithio-, 6-chloro-s-triazine-2,4-diyl ester (6CI, 8CI)

30863-08-2
$C_{17}H_{12}ClN_5S_4$
450.03
Carbanilic acid, dithio-, 6-chloro-s-triazine-2,4-diyl ester (6CI,8CI)

30863-09-3
$C_{17}H_{10}Cl_3N_5S_4$
518.92
Carbanilic acid, p-chloro-dithio-, 6-chloro-s-triazine-2,4-diyl ester (6CI, 8CI)

30863-10-6
$C_{11}H_{18}N_6S_4$
362.56
Carbamic acid, dimethyl-dithio-, 6-(dimethyl-amino)-s-triazine-2,4-diyl ester (6CI,8CI)

30863-11-7
$C_{17}H_{30}N_6S_4$
446.72
Carbamic acid, diethyldi-thio-, 6-(diethylamino)-s-triazine-2,4-diyl ester (6CI,8CI)

30863-12-8
$C_{18}H_{30}N_6S_6$
522.86
Carbamodithioic acid, di-ethyl-, 1,3,5-triazine-2,4,6-triyl ester (9CI)
Carbamic acid, diethyldi-thio-, s-triazine-2,4,6-triyl ester (8CI)
Carbamic acid, diethyldi-thio-, tris(anhydrosulfide) with trithiocyanuric acid (6CI)

30863-13-9
$C_9H_{16}N_3O_4PS_2$
325.35
Phosphorodithioic acid, S-(4,6-dimethoxy-s-tri-azin-2-yl) O,O-diethyl ester (7CI,8CI)

30863-14-0
$C_9H_{16}N_3O_2PS_2$
293.35
Phosphorodithioic acid, S-(4,6-dimethyl-s-triazin-2-yl) O,O-diethyl ester (8CI)

30863-15-1
$C_7H_6Cl_6N_4O$
374.87
Ethanol, 2,2,2-trichloro-1-((4-methyl-6-(trichloro-methyl)-s-triazin-2-yl)-amin o)- (8CI)

30863-16-2
$C_8H_9Cl_6N_5O_2$
419.91
Ethanol, 1,1'-((6-methyl-s-triazine-2,4-diyl)-diimino)bis(2,2,2-tri-chloro- (8CI)

30863-17-3
$C_8H_6Cl_9N_5O_2$
523.25
Ethanol, 1,1'-((6-(tri-chloromethyl)-s-triazine-2,4-diyl)diimino)bis-(2,2,2-tri chloro- (8CI)

30863-19-5
$C_{18}H_{27}N_3O_6$
381.43
s-Triazine-2,4,6-tricar-boxylic acid, triisobutyl ester (8CI)

30863-20-8
$C_9H_{12}N_6O_3$
252.23
s-Triazine-2,4,6-tricar-boxamide, N,N',N''-tri-methyl- (8CI)

30863-21-9
$C_{10}H_{10}Cl_3N_3O_4$
342.57
s-Triazine-2,4-dicarboxylic acid, 6-(trichloro-methyl)-, diethyl ester (8CI)

30863-22-0
$C_8H_5Cl_6N_3O_2$
387.87
s-Triazine-2-carboxylic acid, 4,6-bis(trichloro-methyl)-, ethyl ester (8CI)

30863-24-2
$C_6N_6S_3$
252.30
Thiocyanic acid, s-triazine-2,4,6-triyl ester (8CI)

30863-25-3
$C_6Cl_6N_4S$
372.88
Isothiocyanic acid, 4,6-bis-(trichloromethyl)-s-tri-azin-2-yl ester (8CI)

30863-26-4
$C_7H_8Cl_2N_4OS$
267.14
Carbamothioic acid, (4,6-dichloro-1,3,5-tri-azin-2-yl)-, O-(1-methyl-ethyl) ester (9CI)
s-Triazine-2-carbamic acid, 4,6-dichlorothio-, O-iso-propyl ester (7CI,8CI)

30863-28-6
$C_{12}H_{10}Cl_3N_5S$
362.67
Urea, 1-(4-methyl-6-(tri-chloromethyl)-s-triazin-2-yl)-3-phenyl-2-thio-(8CI)

30863-30-0
$C_5Cl_5N_5$
307.35
Imidocarbonyl chloride, (6-chloro-s-triazine-2,4-diyl)bis- (8CI)

30863-32-2
$C_9H_{12}Cl_3N_5O$
312.59
Urea, 1-isopropyl-3-(4-methyl-6-(trichloro-methyl)-s-triazin-2-yl)-(8CI)

30863-34-4
$C_9H_9N_3O_6$
255.19
s-Triazine-2,4,6-tricar-boxylic acid, trimethyl ester (6CI,8CI)

30863-35-5
$C_7H_{14}N_5O_2PS_2$
295.32
Phosphorodithioic acid, S-(4,6-diamino-s-triazin-2-yl) O,O-diethyl ester (8CI)

30863-36-6
$C_9H_{15}N_3S_3$
261.43
1,3,5-Triazine, 2,4,6-tris-(ethylthio)- (9CI)
s-Triazine, 2,4,6-tris(ethyl-thio)- (8CI)

30863-37-7
$C_{12}H_{15}N_3S_3$
297.47
s-Triazine, 2,4,6-tris-(allylthio)- (8CI)

30863-39-9
$C_6H_8ClN_3O_2$
189.60
s-Triazine, 2-(chloro-methyl)-4,6-dimethoxy-(6CI,8CI)

30863-40-2
$C_8H_{10}Cl_3N_3O_2$
286.55
s-Triazine, 2,4-diethoxy-6-(trichloromethyl)- (6CI, 7CI,8CI)

30863-41-3
$C_{10}H_{14}Cl_3N_3O_2$
314.60
s-Triazine, 2,4-diisopropoxy-6-(trichloromethyl)- (8CI)

30863-42-4
$C_{12}H_{18}Cl_3N_3O_2$
342.66
s-Triazine, 2,4-dibutoxy-6-(trichloromethyl)- (6CI, 7CI,8CI)

30863-43-5
$C_6H_8BrN_3O_2$
234.06
s-Triazine, 2-(bromomethyl)-4,6-dimethoxy- (8CI)

30863-44-6
$C_6H_7Br_2N_3O_2$
312.96
s-Triazine, 2-(dibromomethyl)-4,6-dimethoxy- (6CI,7CI,8CI)

30863-45-7
$C_6H_6Cl_3N_3O$
242.49
s-Triazine, 2-methoxy-4-methyl-6-(trichloromethyl)- (8CI)

30863-46-8
$C_7H_8Cl_3N_3O$
256.52
s-Triazine, 2-ethoxy-4-methyl-6-(trichloromethyl)- (6CI,8CI)

30863-48-0
$C_8H_{10}Cl_3N_3O$
270.55
s-Triazine, 2-isopropoxy-4-methyl-6-(trichloromethyl)- (8CI)

30863-49-1
$C_9H_{12}Cl_3N_3O$
284.57
s-Triazine, 2-butoxy-4-methyl-6-(trichloromethyl)- (6CI,8CI)

30863-50-4
$C_{11}H_{16}Cl_3N_3O$
312.63
s-Triazine, 2-(hexyloxy)-4-methyl-6-(trichloromethyl)- (8CI)

30863-51-5
$C_6H_3Cl_6N_3O$
345.83
s-Triazine, 2-methoxy-4,6-bis(trichloromethyl)- (6CI,7CI,8CI)

30863-52-6
$C_7H_5Cl_6N_3O$
359.86
s-Triazine, 2-ethoxy-4,6-bis(trichloromethyl)- (6CI,7CI,8CI)

30863-53-7
$C_8H_7Cl_6N_3O$
373.88
s-Triazine, 2-propoxy-4,6-bis(trichloromethyl)- (8CI)

30863-54-8
$C_8H_5Cl_6N_3O$
371.87
s-Triazine, 2-(allyloxy)-4,6-bis(trichloromethyl)- (8CI)

30863-55-9
$C_8H_7Cl_6N_3O$
373.88
s-Triazine, 2-isopropoxy-4,6-bis(trichloromethyl)- (6CI,7CI,8CI)

30863-56-0
$C_9H_9Cl_6N_3O$
387.91
s-Triazine, 2-butoxy-4,6-bis(trichloromethyl)- (6CI,7CI,8CI)

30863-57-1
$C_{10}H_{11}Cl_6N_3O$
401.94
s-Triazine, 2-(pentyloxy)-4,6-bis(trichloromethyl)- (6CI,8CI)

30863-58-2
$C_{11}H_{13}Cl_6N_3O$
415.96
s-Triazine, 2-(hexyloxy)-4,6-bis(trichloromethyl)- (8CI)

30863-59-3
$C_{17}H_{25}Cl_6N_3O$
500.13
s-Triazine, 2-(dodecyloxy)-4,6-bis(trichloromethyl)- (8CI)

30863-60-6
$C_{11}H_8Cl_3N_3O$
304.56
s-Triazine, 2-methoxy-4-phenyl-6-(trichloromethyl)- (8CI)

30863-61-7
$C_{12}H_{10}Cl_3N_3O$
318.59
s-Triazine, 2-ethoxy-4-phenyl-6-(trichloromethyl)- (6CI,8CI)

30863-62-8
$C_{11}H_7Cl_4N_3O$
339.01
s-Triazine, 2-(p-chlorophenyl)-4-methoxy-6-(trichloromethyl)- (8CI)

30863-63-9
$C_6H_3Br_6N_3O$
612.56
1,3,5-Triazine, 2-methoxy-4,6-bis(tribromomethyl)- (9CI)
s-Triazine, 2-methoxy-4,6-bis(tribromomethyl)- (7CI,8CI)

30863-65-1
$C_6H_7N_5S_2$
213.28
s-Triazine, 2-(diazomethyl)-4,6-bis(methylthio)- (6CI,7CI,8CI)

30863-82-2
$C_{21}H_{15}N_3S_3$
405.57
1,3,5-Triazine, 2,4,6-tris(phenylthio)- (9CI)
s-Triazine, 2,4,6-tris(phenylthio)- (7CI,8CI)
Triphenyl trithiocyanurate

30863-83-3
$C_{10}H_{14}Cl_3N_3O_2$
314.60
s-Triazine, 2,4-dipropoxy-6-(trichloromethyl)- (8CI)

30885-95-1
$C_4H_5N_3O_3$
143.10
1,3,5-Triazine-2,4(1H,3H)-dione, 6-methoxy- (9CI)

30885-97-3

s-Triazine-2,4-diol, 6-meth-
oxy- (8CI)

30885-97-3
$C_6H_9N_3O_2$
155.16
**s-Triazine-2,4-diol, 6-iso-
propyl- (8CI)**

30885-98-4
$C_9H_6ClN_3O_2$
223.62
**s-Triazine-2,4-diol, 6-
(p-chlorophenyl)- (8CI)**

30885-99-5
$C_5H_7N_3O$
125.13
**1,3,5-Triazin-2(1H)-one,
4,6-dimethyl- (9CI)**
s-Triazin-2-ol, 4,6-dimethyl-
(6CI,7CI,8CI)

30886-00-1
$C_5H_5Cl_2N_3O.C_2H_5ClN_2$
286.55
**Acetamidine, 2-chloro-,
Compd. with 4,6-bis-
(chloromethyl)-s-triazin-
2-ol (1:1) (8CI)**
Acetamidine, 2-chloro-,
Compd. with 4,6-bis-
(chloromethyl)-s-triazin-
2-ol (6CI)

30886-01-2
$C_5H_3Cl_4N_3O.C_2H_4Cl_2N_2$
389.88
**Acetamidine, 2,2-dichloro-,
Compd. with 4,6-bis(di-
chloromethyl)-s-triazin-
2-ol (1:1) (8CI)**
Acetamidine, 2,2-dichloro-,
Compd. with 4,6-bis(di-
chloromethyl)-s-triazin-
2-ol (6CI)

30886-02-3
$C_5HCl_6N_3O.C_2H_3Cl_3N_2$
493.22
**Acetamidine, 2,2,2-tri-
chloro-, Compd. with
4,6-bis(trichloromethyl)-
s-triazin-2-ol (1:1) (8CI)**
Acetamidine, 2,2,2-tri-
chloro-, Compd. with
4,6-bis(trichloromethyl)-
s-triazin-2-ol (6CI)

30886-03-4
$C_6H_{15}N.C_5HCl_6N_3O$
432.99
**s-Triazin-2-ol, 4,6-bis(tri-
chloromethyl)-, Compd.
with triethylamine (1:1)
(6CI,7CI,8CI)**

30886-04-5
$C_7H_7Cl_4N_3O.C_3H_6Cl_2N_2$
431.97
**Propionamidine, 2,2-di-
chloro-, Compd. with
4,6-bis(1,1-dichloro-
ethyl)-s-triazin-2-ol (1:1)
(8CI)**
Propionamidine, 2,2-di-
chloro-, Compd. with
4,6-bis(1,1-dichloroethyl)-
s-triazin-2-ol (6CI)

30886-05-6
$C_7H_5Cl_6N_3O.C_3H_5Cl_3N_2$
535.30
**Propionamidine, 2,2,3-tri-
chloro-, Compd. with
4,6-bis(1,1,2-trichloro-
ethyl)-s-triazin-2-ol (1:1)
(8CI)**

30886-09-0
$C_{15}H_9Cl_2N_3O$
318.16
**1,3,5-Triazin-2(1H)-one,
4,6-bis(4-chlorophenyl)-
(9CI)**
s-Triazin-2-ol, 4,6-bis-
(p-chlorophenyl)- (8CI)

30886-10-3
$C_{17}H_{15}N_3O$
277.33
**1,3,5-Triazin-2(1H)-one,
4,6-bis(4-methylphenyl)-
(9CI)**
s-Triazin-2(1H)-one, 4,6-di-
p-tolyl- (7CI,8CI)

30886-11-4
$C_{15}H_9N_5O_5$
339.27
**s-Triazin-2-ol, 4,6-bis-
(m-nitrophenyl)- (8CI)**

30886-13-6
$C_9H_7N_3S_2$
221.30
**1,3,5-Triazine-2,4(1H,
3H)-dithione, 6-phenyl-
(9CI)**
s-Triazine-2,4-dithiol,
6-phenyl- (6CI,8CI)
s-Triazine-2,4(1H,3H)-di-
thione, 6-phenyl- (7CI)

30886-14-7
$C_5H_7N_3O_2S$
173.19
**1,3,5-Triazine-2(1H)-thi-
one, 4,6-dimethoxy- (9CI)**
4,6-Dimethoxy-2-mercapto-
1,3,5-triazine
4,6-Dimethoxy-1,3,5-tri-
azine-2-thiol
s-Triazine-2-thiol, 4,6-di-
methoxy- (8CI)

30886-15-8
$C_5H_7N_3S$
141.20
**s-Triazine-2-thiol, 4,6-di-
methyl- (8CI)**

30886-16-9
$C_{15}H_{11}N_3S$
265.34
**1,3,5-Triazine-2(1H)-thi-
one, 4,6-diphenyl- (9CI)**
s-Triazine-2-thiol, 4,6-di-
phenyl- (8CI)
s-Triazine-2(1H)-thione,
4,6-diphenyl- (7CI)

30886-17-0
$C_{17}H_{15}N_3S$
293.39
**s-Triazine-2-thiol, 4,6-di-
p-tolyl- (7CI,8CI)**

30886-18-1
$C_9H_5F_2N_3O$
209.16
**1,3,5-Triazine, 2,4-di-
fluoro-6-phenoxy- (9CI)**
s-Triazine, 2,4-difluoro-
6-phenoxy- (8CI)

30886-23-8
$C_7H_9Cl_2N_3O$
222.08
**1,3,5-Triazine, 2,4-di-
chloro-6-(2-methylpro-
poxy)- (9CI)**
s-Triazine, 2,4-dichloro-
6-isobutoxy- (6CI,8CI)

30886-24-9
$C_{10}H_7Cl_2N_3O$
256.07
**1,3,5-Triazine, 2,4-dichloro-
6-(phenylmethoxy)- (9CI)**
s-Triazine, 2-(benzyloxy)-4,6-di-
chloro- (8CI)

30886-25-0
$C_9H_4Cl_3N_3O$
276.49
**1,3,5-Triazine, 2,4-dichloro-
6-(2-chlorophenoxy)- (9CI)**
s-Triazine, 2,4-dichloro-6-
(o-chlorophenoxy)- (8CI)

30886-26-1
$C_9H_4Cl_3N_3O$

276.49
1,3,5-Triazine, 2,4-dichloro-6-(4-chlorophenoxy)- (9CI)
s-Triazine, 2,4-dichloro-6-(p-chlorophenoxy)- (8CI)

30886-27-2
$C_9H_2Cl_5N_3O$
345.40
1,3,5-Triazine, 2,4-dichloro-6-(2,4,5-trichlorophenoxy)- (9CI)
s-Triazine, 2,4-dichloro-6-(2,4,5-trichlorophenoxy)- (6CI,7CI,8CI)

30886-29-4
$C_9H_4Cl_2N_4O_3$
287.04
1,3,5-Triazine, 2,4-dichloro-6-(4-nitrophenoxy)- (9CI)
s-Triazine, 2,4-dichloro-6-(p-nitrophenoxy)- (8CI)

30886-30-7
$C_{13}H_7Cl_2N_3O$
292.11
1,3,5-Triazine, 2,4-dichloro-6-(1-naphthalenyloxy)- (9CI)
s-Triazine, 2,4-dichloro-6-(1-naphthyloxy)- (8CI)

30894-58-7
$C_6H_7Cl_2N_3S$
224.11
1,3,5-Triazine, 2,4-dichloro-6-(propylthio)- (9CI)
s-Triazine, 2,4-dichloro-6-(propylthio)- (6CI,8CI)

30894-60-1
$C_6H_7Cl_2N_3S$
224.11
s-Triazine, 2,4-dichloro-6-(isopropylthio)- (6CI, 8CI)

30894-61-2
$C_7H_9Cl_2N_3S$
238.15
s-Triazine, 2-butylthio-4,6-dichloro
2-(n-Butylthio)-4,6-dichloro-s-triazine

30894-63-4
$C_9H_4Cl_3N_3S$
292.56
1,3,5-Triazine, 2,4-dichloro-6-((4-chlorophenyl)thio)- (9CI)
s-Triazine, 2,4-dichloro-6-((p-chlorophenyl)thio)- (8CI)

30894-64-5
$C_4H_2Cl_3N_3$
198.44
1,3,5-Triazine, 2,4-dichloro-6-(chloromethyl)- (9CI)
s-Triazine, 2,4-dichloro-6-(chloromethyl)- (6CI, 8CI)

30894-65-6
$C_4HCl_4N_3$
232.88
s-Triazine, 2,4-dichloro-6-(dichloromethyl)- (6CI, 7CI,8CI)

30894-67-8
$C_4H_2BrCl_2N_3$
242.90
s-Triazine, 2-(bromomethyl)-4,6-dichloro- (8CI)

30894-68-9
$C_4HBr_2Cl_2N_3$
321.80
s-Triazine, 2,4-dichloro-6-(dibromomethyl)- (8CI)

30894-69-0
$C_4H_2Cl_2IN_3$
289.89
s-Triazine, 2,4-dichloro-6-(iodomethyl)- (8CI)

30894-70-3
$C_4HCl_2I_2N_3$
415.79
s-Triazine, 2,4-dichloro-6-(diiodomethyl)- (6CI, 7CI,8CI)

30894-71-4
$C_5H_3Cl_4N_3$
246.91
s-Triazine, 2,4-dichloro-6-(1,1-dichloroethyl)- (8CI)

30894-72-5
$C_5H_2Cl_5N_3$
281.36
s-Triazine, 2,4-dichloro-6-(1,1,2-trichloroethyl)- (7CI,8CI)

30894-73-6
$C_6H_7Cl_2N_3$
192.05
1,3,5-Triazine, 2,4-dichloro-6-propyl- (9CI)
s-Triazine, 2,4-dichloro-6-propyl- (8CI)

30894-74-7
$C_6H_7Cl_2N_3$
192.05
s-Triazine, 2,4-dichloro-6-isopropyl- (7CI,8CI)

30894-75-8
$C_7H_{10}ClN_3O_2$
203.63
1,3,5-Triazine, 2-chloro-4,6-diethoxy- (9CI)
s-Triazine, 2-chloro-4,6-diethoxy- (6CI,8CI)

30894-76-9
$C_9H_{14}ClN_3O_2$
231.68
s-Triazine, 2-chloro-4,6-diisopropoxy- (8CI)

30894-77-0
$C_{15}H_8Cl_3N_3O_2$
368.61
1,3,5-Triazine, 2-chloro-4,6-bis(2-chlorophenoxy)- (9CI)
s-Triazine, 2-chloro-4,6-bis-(o-chlorophenoxy)- (6CI, 8CI)

30894-78-1
$C_{15}H_8Cl_3N_3O_2$
368.61
s-Triazine, 2-chloro-4,6-bis(p-chlorophenoxy)- (6CI,8CI)

30894-79-2
$C_{17}H_{14}ClN_3O_2$
327.77
s-Triazine, 2-chloro-4,6-bis(p-tolyloxy)- (6CI,8CI)

30894-80-5
$C_{15}H_8ClN_5O_6$
389.71
s-Triazine, 2-chloro-4,6-bis(p-nitrophenoxy)- (8CI)

30894-81-6
$C_7H_{10}ClN_3OS$
219.69
s-Triazine, 2-chloro-4-ethoxy-6-(ethylthio)- (6CI, 8CI)

30894-82-7

$C_{11}H_{18}ClN_3OS$
275.80
s-Triazine, 2-butoxy-4-(butylthio)-6-chloro- (6CI,7CI,8CI)

30894-83-8
$C_7H_{10}ClN_3S_2$
235.76
s-Triazine, 2-chloro-4,6-bis(ethylthio)- (6CI, 8CI)

30894-84-9
$C_5H_6ClN_3$
143.56
1,3,5-Triazine, 2-chloro-4,6-dimethyl- (9CI)
s-Triazine, 2-chloro-4,6-dimethyl- (8CI)

30894-85-0
$C_5H_3Cl_4N_3$
246.91
1,3,5-Triazine, 2-chloro-4-methyl-6-(trichloromethyl)- (9CI)
s-Triazine, 2-chloro-4-methyl-6-(trichloromethyl)- (8CI)

30894-86-1
$C_5H_4ClN_5$
169.57
s-Triazine, 2-chloro-4-(diazomethyl)-6-methyl- (6CI,7CI,8CI)

30894-87-2
$C_5H_4Cl_3N_3$
212.47
s-Triazine, 2-chloro-4,6-bis(chloromethyl)- (6CI,8CI)

30894-88-3
$C_5H_2Cl_5N_3$
281.36
s-Triazine, 2-chloro-4,6-bis(dichloromethyl)- (6CI,8CI)

30894-89-4
$C_5Cl_7N_3$
350.23
1,3,5-Triazine, 2-chloro-4,6-bis(trichloromethyl)- (9CI)
s-Triazine, 2-chloro-4,6-bis(trichloromethyl)- (8CI)

30894-90-7
$C_7H_6Cl_5N_3$
309.41
s-Triazine, 2-chloro-4,6-bis(1,1-dichloroethyl)- (6CI,8CI)

30894-91-8
$C_7H_4Cl_7N_3$
378.30
s-Triazine, 2-chloro-4,6-bis(1,1,2-trichloroethyl)- (7CI,8CI)

30894-92-9
$C_{10}H_6ClN_5$
231.65
s-Triazine, 2-chloro-4-(diazomethyl)-6-phenyl- (6CI,7CI,8CI)

30894-93-0
$C_{15}H_9Cl_2N_3$
302.17
s-Triazine, 2-chloro-4-(p-chlorophenyl)-6-phenyl- (7CI,8CI)

30894-94-1
$C_{15}H_8Cl_3N_3$
336.61
1,3,5-Triazine, 2-chloro-4,6-bis(4-chlorophenyl)- (9CI)
s-Triazine, 2-chloro-4,6-bis(p-chlorophenyl)- (8CI)

30894-95-2
$C_{15}H_8ClN_5O_4$
357.72
1,3,5-Triazine, 2-chloro-4,6-bis(3-nitrophenyl)- (9CI)
s-Triazine, 2-chloro-4,6-bis(m-nitrophenyl)- (8CI)

30894-96-3
$C_9H_5Br_2N_3O$
330.98
s-Triazine, 2,4-dibromo-6-phenoxy- (8CI)

30894-97-4
$C_9H_5Br_2N_3S$
347.04
s-Triazine, 2,4-dibromo-6-(phenylthio)- (8CI)

30894-98-5
$C_5H_2Br_2Cl_3N_3$
370.27
s-Triazine, 2,4-dibromo-6-(1,1,2-trichloroethyl)- (8CI)

30894-99-6
$C_7H_4BrCl_6N_3$
422.76
s-Triazine, 2-bromo-4,6-bis(1,1,2-trichloroethyl)- (8CI)

30895-02-4
$C_{12}H_6Cl_{15}N_3O_3$
771.99
s-Triazine, 2,4,6-tris-(2,2,3,3,3-pentachloropropoxy)- (8CI)

30895-05-7
$C_{15}H_{27}N_3O_3$
297.40
1,3,5-Triazine, 2,4,6-tris-(2-methylpropoxy)- (9CI)
s-Triazine, 2,4,6-triisobutoxy- (8CI)
Triisobutoxytriazine
2,4,6-Tri(2-methylpropyloxy)-s-triazine

30913-44-1
$C_6H_{13}N_3O$
143.19
s-Triazin-2(1H)-one, tetrahydro-5-isopropyl- (8CI)

30915-79-8
C_7H_7ClO
142.58
Phenol, (chloromethyl)- (9CI)

30937-70-3
$C_{10}H_8ClN_3$
205.65
1,3,5-Triazine, 2-chloro-4-methyl-6-phenyl- (9CI)
s-Triazine, 2-chloro-4-methyl-6-phenyl- (7CI, 8CI)

30947-30-9
$C_{17}H_{29}O_4P.1/2Ni$
357.74
Phosphonic acid, ((3,5-bis(1,1-dimethylethyl)-4-hydroxyphenyl)methyl)-, monoethyl ester, nickel(2+) salt (2:1) (9CI)

30951-95-2
C_8H_{16}
112.22
C(=C(CC)C)(CC)C
3-Hexene, 3,4-dimethyl- (9CI)
3,4-Dimethylhexene-3
3,4-Dimethyl-3-hexene

30964-01-3
$C_9H_{14}O_2$
154.21
8-Nonynoic acid (8CI,9CI)

30968-43-5
Unknown
Unknown
2-Heptadecenyl-2-imidazoline

30977-64-1
$C_{15}H_{24}O.1/2Ca$
240.39
Phenol, nonyl-, calcium salt (9CI)
Calcium, bis(nonylphenoxy)-

30979-48-7
$C_8H_{15}N_3O_2$
185.26
Imidazolin-2-one, 1-isobutylcarbamoyl
Azolamide
Bay 94871
1-Imidazolidinecarboxamide, N-isobutyl-2-oxo- (8CI)
1-Imidazolidinecarboxamide, N-(2-methylpropyl)-2-oxo- (9CI)
Imidazolidin-2-on-1-carbonsaeure-isobutylamid (German)
Imizolamid
1-Isobutylcarbamoyl-imidazolin-2-one
Isocarbamid
Isocarbamide
Isolamid
Merpelan AZ
MNF 166
Ozolamid

30995-64-3
$C_{13}H_{10}S$
198.29
Dibenzothiophene, methyl- (9CI)

30995-65-4
$C_8H_{10}O_3S.Na$
209.22
Benzenesulfonic acid, ethyl-, sodium salt (9CI)
NSC-234415
Sodium ethylbenzenesulfonate

30997-38-7
$C_{16}H_{14}$
206.29
Phenanthrene, ethyl- (9CI)

30997-39-8
$C_{17}H_{12}$
216.28
Fluoranthene, methyl- (9CI)

31005-02-4
$C_{11}H_{10}O_3$
190.20
7-Ethoxycoumarin

31007-95-1
$C_{11}Cl_{11}N_3S$
596.19
1,3,5-Triazine, 2-((pentachlorophenyl)thio)-4,6-bis(trichloromethyl)- (9CI)
s-Triazine, 2-((pentachlorophenyl)thio)-4,6-bis(trichloromethyl)- (8CI)

31017-40-0
$C_{12}H_{14}$
158.26
Benzene, cyclohexenyl
Phenylcyclohexene

31044-12-9
$C_8H_{20}NaO_4P$
234.21
2-Ethylhexyl sodium phosphate
Phosphoric acid, mono(2-ethylhexyl) ester, sodium salt (9CI)

31047-64-0
$C_{18}H_{26}O_4$
306.40
1,2-Benzenedicarboxylic acid, monoisodecyl ester (9CI)
Phthalic acid, isodecyl ester

31061-64-0
$C_{11}H_{18}O$
166.27
Tricyclo(4.3.1.13,8)undecan-1-ol (8CI,9CI)
1-Homoadamantanol
1-Hydroxyhomoadamantane

31062-69-8
$C_{35}H_{65}N_9O_9.C_2H_4O_2$
815.95
Valine, N-(N(2)-(N-(N-(N-(N-(N-glycyl-L-isoleucyl)glycyl)-L-alanyl)-L-valyl)-L-leucyl)-L-lysyl)-, monoacetate, L- (8CI)

31080-37-2
$C_{10}H_{20}O_2$
172.27
Hexanoic acid, 2-methyl-2-propyl- (8CI,9CI)
2-Methyl-2-butylpentanoic acid

31080-39-4
$C_{10}H_{20}O_2$
172.30
CCCCCC(CCC)C(O)=O
Heptanoic acid, 2-propyl
4-Nonanecarboxylic acid
2-Propylheptanic acid
2-Propylheptansaeure (German)

31095-87-1
$(C_8H_{10}O_3.C_3H_5ClO)x$
Polymer
1,3-Isobenzofurandione, hexahydro-, Polymer with (chloromethyl)oxirane (9CI)
Hexahydrophthalic anhydride, (chloromethyl)oxirane polymer

31107-44-5
$C_{14}H_4Cl_{12}O$
613.62
O(C(C(C(=C(C12Cl)Cl)Cl)(C2(Cl)Cl)Cl)C1C3C(C(=C(C45Cl)Cl)Cl)(C4(Cl)Cl)Cl)C35
1,4:6,9-Dimethanodibenzofuran, 1,2,3,4,6,7,8,9,10,10,11,11-dodecachloro-1,4,4a,5a,6,9,9a,9b-octahydro-
AI3-27887

31120-23-7
$C_7H_8Cl_3N_3O_2$
272.52
s-Triazine, 2,4-dimethoxy-6-(1,1,2-trichloroethyl)- (8CI)

31120-85-1
$C_{15}H_{24}NO_5P$
329.33
Benzoic acid, 2-((ethoxy-((1-methylethyl)amino)phosphinyl)oxy)-, 1-methylethyl ester (9CI)
Salicylic acid, isopropyl ester, ethyl isopropylphosphoramidate (8CI)

31121-12-7
$C_9H_{19}NO$
157.25
1-Propanol, 3-(cyclohexylamino)- (8CI,9CI)
NSC-44899

31135-63-4
$C_{18}H_{34}Cl_2O_2$
353.37
Octadecanoic acid, dichloro- (8CI,9CI)

31138-65-5
$C_7H_{14}O_8.Na$
249.17
D-Gluco-heptonic acid, monosodium salt, (2.XI.)- (9CI)
(2.XI.)-Monosodium D-glucoheptonate

31155-09-6
$C_7H_4Cl_2O$
175.01
Benzaldehyde, dichloro- (9CI)
Dichlorobenzaldehyde

31188-91-7
$C_{34}H_{37}Cl_3N_4O_4$
672.03
O=C(N=C(NN(C1=O)c(c(cc(c2)Cl)Cl)c2Cl)C1)c(cccc3NC(=O)COc(c(cc(c4)C(CC)(C)C)C(CC)(C)C)c4)c3
Benzamide, 3-(((2,4-bis(1,1-dimethylpropyl)phenoxy)acetyl)amino)-N-(4,5-dihydro-5-oxo-1-(2,4,6-trichlorophenyl)-1H-pyrazol-3-yl)- (9CI)

31198-76-2
$C_{10}H_{15}NO$
165.23
N(O)=C(C(=CCC1C(=C)C)C)C1
2-Cyclohexen-1-one, 2-methyl-5-(1-methylethenyl)-, oxime (9CI)
Carvone oxime
Carvoxime
p-Mentha-6,8-dien-2-one, oxime
1,8-p-Menthadienyl-6-oxime
NSC-97240

31212-28-9
$C_6H_5NO_5S$
203.17
Nitrobenzenesulfonic acid
Acide nitrobenzenesulfonique (French)
Acido nitrobencenosulfonico (Spanish)
Benzenesulfonic acid, nitro-
Nitrobenzenesulphonic acid
UN 2305 [Nitrobenzenesulfonic acid]

31215-04-0
$C_{17}H_{22}O$
242.36
2-Naphthalenol, heptyl- (9CI)
Heptyl-2-naphthalenol
x-Heptyl-2-naphthol

31218-83-4
$C_{10}H_{20}NO_4PS$
281.34
CCNP(=S)(OC)OC(C)=CC(=O)OC(C)C
Crotonic acid, 3-hydroxy-, isopropyl ester, O-ester with O-methyl ethylphosphoramidothioate, (E)
Blotic
2-Butenoic acid, 3-(((ethylamino)methoxyphosphinothioyl)oxy)-, isopropyl ester, (E)-
2-Butenoic acid, 3-(((ethylamino)methoxyphosphinothioyl)oxy)-, 1-methylethyl ester, (E)-
ENT 27,989
(E)-O-2-Isopropoxy-carbonyl-1-methylvinyl O-methyl ethylphosphoramidothioate
O-(1-Isopropoxycarbonyl-1-propen-2-yl)-O-methyl N-ethylphosphoramidothionate
(E)-1-Methylethyl 3-(((ethylamino)methoxyphosphinothioyl)oxy)-2-butenoate
1-Methylethyl (E)-3-(((ethylamino)methoxyphosphinothioyl)oxy)-2-butenoate
(E)-1-Metiletil-3-(((etilamino)-metoxifosfinotiol)oxi)-2-butenoato (Spanish)
OMS 1502
Ovidip
Propetamphos
Safrotin
San 52139
Sandoz 52139
San 322I
San 52 139 I
Seraphos
VEL 4283 (Obs.)

31230-13-4
$C_{15}H_{28}$
208.39
1H-Indene, octahydro-2,3a,4-trimethyl-2-(1-methylethyl)-, (2α,3aβ,4β,7aβ)-(+)- (9CI)
Fukinan
Fukinane
Indan, 3a,4,5,6,7,7aβ-hexahydro-2α-isopropyl-2,3aβ,4β- trimethyl-, (+)-(8CI)

31242-93-0
$C_{12}H_4Cl_6O$
376.86
Ether, trichlorophenyl
Benzene, 1,1'-oxybis-, hexachloro deriv. (9CI)
Chlorinated diphenyl oxide (ACGIH,OSHA)
Ether, hexachlorophenyl
Hexachlorodiphenyl ether
Hexachloro diphenyl oxide
Phenyl ether, hexachloro deriv. (8CI)
1,1'-Oxybisbenzene hexachloro deriv.
Trichloro diphenyl ether
Trichloro diphenyl oxide

31242-94-1
$C_{12}H_6Cl_4O$
307.98
Ether, tetrachlorophenyl
Phenyl ether tetrachloro
Tetrachloro diphenyl ether
Tetrachloro diphenyl oxide

31250-78-9
$C_6H_7Br_3N_2$
346.83
1H-Imidazole, 2,4,5-tribromo-1-propyl- (9CI)
Imidazole, 2,4,5-tribromo-1-propyl- (8CI)

31265-39-1
$C_{20}H_{17}ClO_5$
372.80
Oc(c(cc(c1)Cl)Cc(c(O)cc(O)c2)c2)c1Cc(c(O)cc(O)c3)c3
1,3-Benzenediol, 4,4'-((5-chloro-2-hydroxy-1,3-phenylene)bis(methylene))bis- (9CI)

31288-44-5
$C_{21}H_{16}N_2O_5$
376.36
9,10-Anthracenedione, 1,5-diamino-4,8-dihydroxy(4-methoxyphenyl)- (9CI)
1,5-Diamino-4,8-dihydroxy(p-methoxyphenyl)anthraquinone

31291-60-8
$C_{14}H_{22}O$
206.33
Phenol, bis(1-methylpropyl)- (9CI)
Bis(1-methylpropyl)phenol
Di-sec-butylphenol

31291-71-1
$C_{11}H_{14}$
146.25
Naphthalene, 1,2,3,4-tetrahydromethyl
Methyltetralin (Czech)
1,2,3,4-Tetrahydromethylnaphthalene

31295-46-2
$C_8H_{23}N_5$
189.27
N(CCNCCN)(CCN)CCN
1,2-Ethanediamine, N,N,N'-tris(2-aminoethyl)- (9CI)
N,N,N'-Tris(2-aminoethyl)-1,2-ethanediamine

31295-49-5
$C_{10}H_{25}N_5$
215.31
N(CCNC1)(C1)CCNCCNCCN
1,2-Ethanediamine, N-(2-aminoethyl)-N'-(2-(1-piperazinyl)ethyl)- (9CI)
N-(2-Aminoethyl)-N-(2-(1-piperazinyl)ethyl)-1,2-ethanediamine

31295-54-2
$C_{10}H_{25}N_5$
215.31
N(CCN(C1)CCN)(C1)CCNCCN
1,4-Piperazinediethanamine, N-(2-aminoethyl)- (9CI)
N-(2-Aminoethyl)-1,4-piperazinediethanamine

31295-56-4
$C_{15}H_{32}$
212.42
Dodecane, 2,6,11-trimethyl- (8CI,9CI)

31303-42-1
$C_{33}H_{37}CrN_2O_{11}$
689.65
Chromium, aqua(2-(3-(2,5-dihydroxyphenyl)-1-oxopropyl)cyclopentanonato)-(2-(((2-hydroxy-5-nitrophenyl)imino)methyl)-3,5-dipropoxyphenolato(2-))-(9CI)
Chrome(III) complex of 2-hydroxy-4,6-dipropoxybenzald(2'-oxy-5'-nitrophenyl)imine, 2-(3'-(2'',5''-dihydroxyphenyl)propionyl)-cyclopentanone, and water

31305-91-6
$(C_{12}H_{20}O_6)x$
Polymer NA
Oxirane, 2,2',2''-(1,2,3-propanetriyltris(oxymethylene))-tris-, Homopolymer (9CI)
Denacol EX 314
Epikote 812
Epon 812
Propane, 1,2,3-tris(2,3-epoxypropoxy)-, Polymers (8CI)

31317-07-4
$C_{13}H_{10}S$
198.29
Dibenzothiophene, 1-methyl- (8CI,9CI)

31328-15-1
$C_8H_8NO_3PS$
229.19
Phosphorothioic acid, O-(4-cyanophenyl) O-methyl ester (9CI)
Phosphorothioic acid, O-methyl ester, O-ester with p-hydroxybenzonitrile (8CI)

31328-16-2
$C_8H_8NO_4P$
213.12
Phosphoric acid, mono(4-cyanophenyl) monomethyl ester (9CI)
Phosphoric acid, monomethyl ester, monoester with p-hydroxybenzonitrile (8CI)

31361-57-6
$C_{32}H_{12}Cl_2CuN_8O_{10}S_4.2H$
933.18
Cuprate(2-), (bis(chlorosulfonyl)-29H,31H-phthalocyaninedisulfonato(4-)-N29,N30, N31,N32)-, dihydrogen (9CI)

31366-95-7
$C_{11}H_{16}O.Na$
187.24
Sodium 4-tert-amylphenate
p-tert-Amylphenol sodium salt
Caswell No. 050B
4-(1,1-Dimethylpropyl)phenol sodium salt
EPA Pesticide Chemical Code 064112
Phenol, 4-(1,1-dimethylpropyl)-, sodium salt (9CI)
Sodium p-tert-amylphenate
Sodium p-tert-amylphenolate

31366-97-9
Unknown
Unknown
Sodium 2-chloro-4-phenylphenate

31367-46-1
$C_8H_{18}O$
130.23
CCCCC(C)C(C)O
2-Heptanol, 3-methyl- (9CI)
3-Methyl-2-heptanol
NSC-92762

31389-11-4
$C_{15}H_{32}O$
228.42
Pentadecanol (8CI,9CI)

31391-49-8
$C_{12}H_{18}O$
178.27
Phenol, (1,1-dimethylethyl)-2,5-dimethyl- (9CI)
2,5-Xylenol, tert-butyl-

31393-23-4
C_9H_8S
148.23
Benzo(b)thiophene, methyl- (9CI)

31394-54-4
C_7H_{16}
100.23
Isoheptane [UN 1206]
UN 1206 [Heptanes]

31394-71-5
$(C_3H_6O)xC_{18}H_{34}O_2$
Polymer
PPG-26 Oleate
Poly(oxy(methyl-1,2-ethanediyl)), α-(1-oxo-9-octadecenyl)-ω-hydroxy-, (Z)-(9CI)
Polyoxypropylene (26) monooleate
Polyoxypropylene (36) monooleate
Polyoxypropylene oleate
Polypropylene glycol (26) monooleate
Polypropylene glycol (36) monooleate
PPG-36 Oleate
Witconol F26-46

31410-01-2
$C_6H_8N_2$
108.13
1H-Imidazole, 1-(2-propenyl)-(9CI)
Imidazole, 1-allyl- (8CI)

31423-92-4
$C_4H_6Cl_2$
125.00
Butene, 1,4-dichloro- (9CI)
1,4-Dichlorobutene

31430-18-9
$C_{14}H_{11}N_3O_3S$
301.34
2-Benzimidazolecarbamic acid, 5-(2-thenoyl)-, methyl ester
2-Benzimidazolecarbamic acid, 5-(2-thienoyl)-, methyl ester
2-Benzimidazolecarbamic acid, 5-(2-thienylcarbonyl)-, methyl ester
Carbamic acid, (5-(2-thienylcarbonyl)-1H-benzimidazol-2-yl)-, methyl ester
Methyl 5-(2-thenoyl)-2-benzimidazolecarbamate
Methyl (5-(2-thienylcarbonyl)-1H-benzimidazole-2-yl)carbamate
Nocodazole
NSC-238 159
Oncodazole
R 17934
N-(5-(2-Thienoyl)-2-benzimidazolyl)carbamic acid methyl ester

31431-39-7
$C_{16}H_{13}N_3O_3$
295.32
COC(=O)Nc2nc1cc(ccc1[nH]2)

C(=O)c3ccccc3
Carbamic acid, N-(5-benzoyl-benzimidazol-2-yl)-, methyl ester
2-Benzimidazolecarbamic acid, 5-benzoyl-, methyl ester
N-2 (5-Benzoyl-benzimidazole) carbamate de methyle (French)
5-Benzoyl-2-benzimidazole-carbamic acid methyl ester
N-(Benzoyl-5, benzimidazolyl)-2, carbamate de methyle (French)
(5-Benzoyl-1H-benzimidazol-2-yl)-carbamic acid methyl ester
MBDZ
Mebendazole
Methyl 5-benzoyl benzimidazole-2-carbamate
Methyl 5-benzoyl-2-benzimidazolecarbamate
Ovitelmin
Pantelmin
R 17635
Telmin
Vermirax
Vermox

31464-38-7
$C_{16}H_{15}N_5O_2$
309.30
O=N(=O)c(ccc(N=Nc(ccc(N(CCC#N)C)c1)c1)c2)c2
Propanenitrile, 3-(methyl(4-((4-nitrophenyl)azo)phenyl)-amino)- (9CI)
Aniline, N-(2-cyanoethyl)-N-methyl-4-((p-nitrophenyl)azo)-
3-(N-Methyl-p-((p-nitrophenyl)-azo)anilino)propionitrile

31472-83-0
$C_{12}H_2Cl_8$
429.77
1,1'-Biphenyl, ar,ar,ar,ar,ar', ar',ar',ar'-octachloro- (9CI)
ar,ar,ar,ar,ar',ar',ar',ar'-Octachlorobiphenyl

31474-57-4
$C_{24}H_{46}O_4$
398.63
Hexanedioic acid, isodecyl isooctyl ester (9CI)
Isodecyl isooctyl hexanedioate
Isooctyl isodecyl adipate

31482-56-1
$C_{17}H_{17}N_5O_2$
323.32
O=N(=O)c(ccc(N=Nc(ccc(N(CCC#N)CC)c1)c1)c2)c2
Propanenitrile, 3-(ethyl(4-((4-nitrophenyl)azo)phenyl)-amino)- (9CI)
3-(Ethyl(4-((4-nitrophenyl)azo)-phenyl)amino)propanenitrile
4'-Nitrophenylazo-4-(1-cyano-ethyl, (N-ethyl)phenylamine)

31502-23-5
$C_9H_{16}O_2$
156.22
6-Nonenoic acid, (E)- (8CI,9CI)

31506-32-8
$C_9H_4F_{17}NO_2S$
513.17
O=S(=O)(NC)C(F)(F)C(F)(F)C(F)(F)C(F)(F)C(F)(F)C(F)(F)C(F)(F)C(F)(F)F
1-Octanesulfonamide, 1,1,2,2,3,3,4,4,5,5,6,6,7,7,8,8,8-heptadecafluoro-N-methyl- (9CI)
Heptadecafluoro-N-methyl-1-octanesulfonamide

31508-00-6
$C_{12}H_5Cl_5$
326.44
Clc1ccc(cc1Cl)c2cc(Cl)c(Cl)cc2Cl
1,1'-Biphenyl, 2,3',4,4',5-pentachloro- (9CI)
Biphenyl, 2,3',4,4',5-pentachloro- (8CI)
2,3',4,4',5-PCB
2,3',4,4',5-Pentachlorobiphenyl

31512-74-0
$(C_{10}H_{24}N_2O.Cl_2)x$
Polymer
Poly(oxyethylene(dimethyliminio)ethylene(dimethylimino)ethylene dichloride)
Busan 77
Bualta
WSCP

31529-83-6
$C_{20}H_{14}N_2O_5$
362.33
9,10-Anthracenedione, 1,5-diamino-4,8-dihydroxy(4-hydroxyphenyl)- (9CI)
1,5-Diamino-4,8-dihydroxy(p-hydroxyphenyl)anthraquinone

31541-02-3
$C_{32}H_{18}$
402.50
Benzo(h)naphtho(1,2,3,4-rst)pentaphene (8CI, 9CI)

31541-03-4
$C_{32}H_{18}$
402.50
Benzo(fgh)trinaphthylene (8CI,9CI)

31541-07-8
$C_{32}H_{18}$
402.50
Anthra(1,2,3,4-rst)pentaphene (8CI,9CI)

31541-10-3
$C_{36}H_{20}$
452.56
Dinaphtho(3,2,1-fg:1',2',3'-st)pentacene (8CI,9CI)

31543-75-6
$C_7H_6Br_2$
249.93
Benzene, 2,4-dibromo-1-methyl- (9CI)
2,4-Dibromo-1-methylbenzene
NSC-139877

31551-28-7
C_6H_9N
95.14
N#CC(=CCC)C
cis-2-Methyl-2-pentenenitrile
trans-α-Methyl-β-ethylacrylonitrile
(E)-2-Methyl-2-pentenenitrile
2-Pentenenitrile, 2-methyl-, (E)- (9CI)

31551-45-8
$C_{13}H_6N_2O_5$
270.21
O=C(c(c(c1ccc(N(=O)=O)c2)ccc3N(=O)=O)c3)c12
9-Fluorenone, 2,7-dinitro
2,7-Dinitrofluorenone
2,7-Dinitro-9-fluorenone

31556-45-3
$C_{31}H_{62}O_2$
466.83
O=C(OCCCCCCCCCCCCC)CCCCCCCCCCCCCCCCC
Tridecyl stearate
Cirrasol LN-GS
NSC-152080
Octadecanoic acid, tridecyl ester (9CI)
Stearic acid, tridecyl ester
Tridecanol stearate
Tridecyl octadecanoate

31565-37-4
$C_{31}H_{62}O_2$
466.83
Octadecanoic acid, isotridecyl ester (9CI)
Isotridecyl octadecanoate
Isotridecyl stearate

31565-38-5
$C_{28}H_{56}O_2$
424.75

Octadecanoic acid, isodecyl ester (9CI)
Isodecyl octadecanoate
Isodecyl stearate

31566-31-1
$C_{21}H_{42}O_4$
358.63
CCCCCCCCCCCCCCCCCC(=O)OCC(O)CO.CCCCCCCCCCCCCCCC(=O)OCC(O)CO
Octadecanoic acid, monoester with 1,2,3-propanetriol
Abracol S.L.G.
Admul
Advawax 140
Aldo HMS
Aldo MS
Aldo MSA
Aldo MSLG
Aldo-28
Aldo-72
Arlacel 161
Arlacel 169
Armostat 801
Atmos 150
Atmul 67
Atmul 84
Atmul 124
Cefatin
Celinhol -A
Cerasynt 1000-D
Cerasynt S
Cerasynt SD
Cerasynt SE
Cerasynt WM
Citomulgan M
Cyclochem GMS
Dermagine
Distearin
Drewmulse TP
Drewmulse V
Drumulse AA
Emerest 2400
Emerest 2401
Emcol CA
Emcol MSK
Emul P.7
Estol 603
Glycerin monostearate
Glycerol monostearate
Glyceryl monostearate
Glyceryl stearate (ACGIH)
Grocor 5500
Grocor 6000
Hodag GMS
Imwitor 191
Imwitor 900K
Kessco 40
Lipo GMS 410
Lipo GMS 450
Lipo GMS 600
Monelgin
Monostearin
Ogeen 515
Ogeen GRB
Ogeen M
Ogeen MAV
Orbon
Protachem GMS
Sedetine
Starfol GMS 450
Starfol GMS 600
Starfol GMS 900
Stearic acid, monoester with glycerol
Stearic monoglyceride
Tegin
Tegin 503
Tegin 515
Unimate GMS
USAF KE-7
Witconol MS
Witconol MST

31570-04-4
$C_{42}H_{63}O_3P$
646.93
O(c(c(cc(c1)C(C)(C)C)C(C)(C)C)c1)P(Oc(c(cc(c2)C(C)(C)C)C(C)(C)C)c2)Oc(c(cc(c3)C(C)(C)C)C(C)(C)C)c3
Phenol, 2,4-bis(1,1-dimethylethyl)-, phosphite (3:1) (9CI)
Tris(2,4-di-tert-butylphenyl) phosphite

31587-78-7
$(C_2H_4O)x(C_2H_4O)xC_{16}H_{33}NO_3$
Polymer
Poly(oxy-1,2-ethanediyl), α,α'-(((1-oxododecyl)imino)di-2,1-ethanediyl)bis(ω-hydroxy- (9CI)
Polyoxyethylene lauramide

31603-77-7
$C_{14}H_{11}N$
193.24
(1,1'-Biphenyl)-4-acetonitrile (9CI)
Acetonitrile, (4-biphenylyl)-
NSC-114981

31621-91-7
$(C_2H_4O)xC_9H_{18}O_2$
Polymer
Poly(oxy-1,2-ethanediyl), α-(1-oxononyl)-ω-hydroxy- (9CI)

31624-59-6
$C_{15}H_{28}$
208.39
1,1'-Bicyclohexyl, (1-methylethyl)- (9CI)
Bicyclohexyl, isopropyl- (6CI,7CI,8CI)
Isopropylbicyclohexyl

31626-02-5
$C_9H_{14}N_2$
150.21
Benzenediamine, ar-(1-methylethyl)- (9CI)
Isopropylphenylenediamine
ar-(1-Methylethyl)benzenediamine

31627-33-5
$C_{19}H_{36}O_2$
296.49
Nonadecenoic acid, (Z)- (8CI, 9CI)

31642-67-8
$C_9H_{16}O_2$
156.23
8-Nonenoic acid (6CI,8CI, 9CI)

31671-77-9
$C_{15}H_{10}O$
206.24
Anthracenecarboxaldehyde (9CI)
Anthraldehyde

31691-97-1
$(C_2H_4O)xC_{15}H_{24}O_4S.H_3N$
Unknown
Ammonium nonoxynol-4 sulfate
p-Nonylphenol, ethoxylate, sulfate, ammonium salt
Poly(oxy-1,2-ethanediyl), α-sulfo-ω-(4-nonylphenoxy)-, ammonium salt (9CI)

31694-55-0
$(C_2H_4O)x(C_2H_4O)x(C_2H_4O)xC_3H_8O_3$
Polymer
Glycereth-12
Ethoxylated glycerin
Ethoxylated glycerine
Glycereth-26
Glycerin, ethylene oxide condensate
Glycerine ethoxylate
Glycerine, ethoxylated
Glycerol, ethoxylated
Glycerol poly(oxyethylene) ether
Glyceryl polyethylene glycol ether
PEG-12 Glyceryl Ether
PEG-26 Glyceryl Ether
Polyethylene glycol (26) glyceryl ether
Polyethylene glycol 600 glyceryl ether
Poly(oxy-1,2-ethanediyl), α,α',α''-1,2,3-propanetriyltris(ω-hydroxy- (9CI)
Polyoxyethylene glyceryl ether
Polyoxyethylene (12) glyceryl ether
Polyoxyethylene (26) glyceryl ether
Unipeg-ETG-12
Unipeg-ETG-26
1,2,3-Propanetriol, ethoxylated

1,2,3-Propanetriol, Polymer with oxirane

31702-33-7
$C_4H_5Cl_3$
159.44
1-Propene, 1,1,3-trichloro-2-methyl- (9CI)
Propene, 1,1,3-trichloro-2-methyl- (8CI)

31710-30-2
$C_{12}Cl_{10}O$
514.66
Benzene, 1,1'-oxybis(2,3,4,5,6-pentachloro- (9CI)
Ether, bis(pentachlorophenyl) (8CI)

31711-53-2
$C_{15}H_{12}$
192.26
Phenanthrene, methyl- (8CI, 9CI)

31714-55-3
$C_{32}H_{18}Cl_2CrN_4O_4.H$
646.40
Chromate(1-), bis(1-((5-chloro-2-hydroxyphenyl)-azo)-2-naphthalenolato(2-))-, hydrogen (9CI)
Chromate(1-), bis((1-(2-hydroxy-5-chlorophenyl)azo)-2-naphthalenolato)-, hydrogen

31717-87-0
$C_{18}H_{36}NO.C_2H_4O_2$
342.54
Dodemorph acetate
Acetate de dodemorphe (French)
Apadodine
BAS 238F
BAS 2382 F
BASF mehltaumittel
Caswell No. 268C
Caswell No. 268E
N-Cyclododecyl-2,6-dimethylmorpholinacetat (German)
Cyclododecyl-2,6-dimethylmorpholine acetate
4-Cyclododecyl-2,6-dimethylmorpholine acetate
N-Cyclododecyl-2,6-dimethylmorpholinium acetate
4-Cyclododecyl-2,6-dimethylmorpholinium acetate
Cyclomorph
Dodemorfe (French)
Dodemorph acetate
EPA Pesticide Chemical Code 110401
EPA Pesticide Chemical Code 213600
Mehltaumittel
Meltox
Milban
Morpholine, n-cyclododecyl-2,6-dimethyl-, acetate
Morpholine, 4-cyclododecyl-2,6-dimethyl-, acetate (8CI, 9CI)

31726-34-8
$(C_2H_4O)xC_6H_{14}O$
Polymer
Poly(oxy-1,2-ethanediyl), α-hexyl-ω-hydroxy- (9CI)
Hexyl alcohol, ethoxylated
α-Hexyl,ω-hydroxypoly(oxy-1,2-ethanediyl)
Hexyl poly(oxyethylene) ether

31774-90-0
$C_8H_{20}NO_3.C_2H_5O_4S$
303.37
Ethanaminium, N-ethyl-2-hydroxy-N,N-bis(2-hydroxyethyl)-, ethyl sulfate (Salt) (9CI)
Ethyltris(2-hydroxyethyl)-ammonium ethyl sulfate
Triethanolamine ethosulfate

31778-10-6
$C_{29}H_{25}N_5O_5$
523.52
O=C(Nc(ccc(NC(=O)N1)c12)c2)c(c(O)c(N=Nc(c(ccc3)C(=O)OCCCC)c3)c(c4ccc5)c5)c4
Benzoic acid, 2-((3-(((2,3-dihydro-2-oxo-1H-benzimidazol-5-yl)amino)carbonyl)-2-hydroxy-1-naphthalenyl)-azo)-, butyl ester (9CI)

31793-07-4
$C_{13}H_{14}ClNO_2$
251.73
CC(C(O)=O)c2ccc(N1CC=CC1)c(Cl)c2
Hydratropic acid, 3-chloro-4-(3-pyrrolin-1-yl)
3-Chloro-4-(3-pyrrolin-1-yl)-hydratropic acid
Pirprofen
Rangasil
Rengasil
SU 21524

31807-55-3
$C_{12}H_{26}$
170.34
Isododecane (9CI)

31835-45-7
$C_{12}H_{10}O_2$
186.21
Oc(c(c(cccc1O)c1)ccc2)c2
(1,1'-Biphenyl)-2,3'-diol (9CI)
2,3'-Dihydroxybiphenyl

31837-42-0
$C_{18}H_{15}N_5O_5$
381.32
Benzoic acid, 2-((1-(((2,3-dihydro-2-oxo-1H-benzimidazol-5-yl)-amino)carbonyl)-2-oxo-propyl)azo)- (9CI)
C.I. Pigment Yellow 151
Hostaperm Yellow H 4G

31844-98-1
C_4H_7Br
135.00
BrC=CCC
1-Butene, 1-bromo- (9CI)
1-Bromo-1-butene

31853-85-7
$(C_9H_{10}O_2)x$
Polymer NA
Phenol, 4-ethenyl-2-methoxy-, Homopolymer (9CI)
4-Hydroxy-3-methoxystyrene polymer
3-Methoxy-4-hydroxystyrene polymer
Phenol, 2-methoxy-4-vinyl-, Polymers (8CI)
Poly(4-hydroxy-3-methoxystyrene)
Poly(3-methoxy-4-hydroxystyrene)
Poly(vinylguaiacol)
Poly(4-vinylguaiacol)

31858-10-3
$C_9H_{14}ClN_3$
199.69
s-Triazine, 2-(1-chloroethyl)-4,6-diethyl- (7CI, 8CI)

31858-13-6
$C_{10}H_{19}N_5O_2$
241.26
1,3,5-Triazine-2,4-diamine, N,N'-diethyl-6-(2-methoxyethoxy)- (9CI)
s-Triazine, 2,4-bis(ethylamino)-6-(2-methoxyethoxy)- (8CI)

31872-14-7
$(C_{10}H_{12}O_3)x$
Polymer NA
Phenol, 4-ethenyl-2,6-dimethoxy-, Homopolymer (9CI)
Phenol, 2,6-dimethoxy-4-vinyl-, Polymers (8CI)
Poly(3,5-dimethoxy-4-hydroxystyrene)
Poly(4-hydroxy-3,5-dimethoxystyrene)

31879-05-7
$C_{15}H_{14}O_3$
242.29
CC(C(O)=O)c2cccc(Oc1ccccc1)c2
Hydratropic acid, m-phenoxy-, (+-)
Fenoprofen

31895-22-4
$C_5H_{11}NS_3.C_2H_2O_4$
271.39
v-Trithiane, 5-(dimethylamino)-, oxalate
5-Dimethylamino-1,2,3-trithiane hydrogeneoxalate
N,N-Dimethyl-1,2,3-trithian-5-amine, ethanedioate (1:1)
N,N-Dimethyl-1,2,3-trithian-5-amine hydrogenoxalate
N,N-Dimethyl-1,2,3-trithian-5-ylammonium hydrogen oxalate
Evisect
Evisekt
San 155
San 1551
San 155 I
Thiocyclam (ethanedioate 1:1)
Thiocyclam hydrogen oxalate
1,2,3-Trithian-5-amine, N,N-dimethyl-, ethanedioate (1:1)

31906-04-4
$C_{13}H_{22}O_2$
210.32
O=CC(CCC(=C1)CCCC(O)(C)C)C1
3-Cyclohexene-1-carboxaldehyde, 4-(4-hydroxy-4-methylpentyl)- (9CI)

31934-88-0
$C_{16}H_{15}ClO_4$
306.76
Cc1cc(Br)cc(O)c1O
Catechol, 5-chloro-3-methyl
5-Chloro-3-methyl-catechol

31949-57-2
$C_4H_3Br_3N_4O$
362.82
s-Triazin-2-ol, 4-amino-6-(tribromomethyl)- (8CI)

31972-43-7
$C_{13}H_{22}NO_4PS$
319.36
Phosphoramidic acid, (1-methylethyl)-, ethyl 3-methyl-4-(methylsulfinyl)phenyl ester (9CI)
Phosphoramidic acid, isopropyl-, ethyl 4-(ethylsulfinyl)-m-tolyl ester (8CI)

31972-44-8
$C_{13}H_{22}NO_5PS$
335.35
Phosphoramidic acid, (1-methylethyl)-, ethyl 3-methyl-4-(methylsulfonyl)phenyl ester (9CI)
Phosphoramidic acid, isopropyl-, ethyl 4-(methylsulfonyl)-m-tolyl ester (8CI)

31983-27-4
$C_{10}H_{15}N$
149.23
N#CC=C(CCC=C(C)C)C
2,6-Octadienenitrile, 3,7-dimethyl-, (Z)- (9CI)
cis-3,7-Dimethyl-2,6-octadienenitrile
(Z)-3,7-Dimethyl-2,6-octadienenitrile

31991-61-4
$C_{16}H_{20}O_2$
244.34
Benz(e)-as-indacene-7,10-dione, 1,2,3,4,5,6,6a,6b,8,9,10a,10b-dodecahydro- (8CI)

32064-70-3
$C_{10}H_{18}O$
154.25
2-Heptanone, 3-propylidene- (8CI,9CI)

32072-96-1
$C_{20}H_{34}O_3$
322.49
2,5-Furandione, 3-(hexadecenyl)dihydro- (9CI)
3-(Hexadecenyl)dihydro-2,5-furandione

32075-31-3
$C_6H_5NO_2$
123.11
Pyridinecarboxylic acid (8CI, 9CI)

32131-17-2
$(C_{12}H_{22}N_2O_2)x$
Polymer
Nylon
Nylon SI-N
Orgasol 1002D Natural Cos
Orgasol 1002D White 5 Cos
Orgasol 20030 White 5 Cos
Poly(imino(1,6-dioxo-1,6-hexanediyl)imino-1,6-hexanediyl) (9CI)
Rilsan BHV Nat Cos
SP-500 Nylon Powder
Upamid Resin UPC-1283

32139-72-3
$C_6H_3Cl_3O_2$
213.44
Oc1c(Cl)cc(Cl)c(Cl)c1O
Pyrocatechol, 3,4,6-trichloro
1,2-Benzenediol, 3,4,6-trichloro- (9CI)
3,4,6-Trichlorocatechol

32167-31-0
$(C_2H_4O)x(C_2H_4O)xC_4H_6O_2$
Polymer
Poly(oxy-1,2-ethanediyl), α,α'-2-butyne-1,4-diylbis(ω-hydroxy- (9CI)
1,4-Butynediol, ethoxylated

32180-75-9
$C_{31}H_{41}NO_3$
475.67
O=C(NCCCCOc(c(cc(c1)C(CC)(C)C)C(CC)(C)C)c1)c(ccc(c2ccc3)c3)c2O
2-Naphthalenecarboxamide, N-(4-(2,4-bis(1,1-dimethylpropyl)phenoxy)butyl)-1-hydroxy- (9CI)
N-(4-(2,4-Bis(1,1-dimethylpropyl)phenoxy)butyl)-1-hydroxy-2-naphthalenecarboxamide

32196-63-7
$(C_2H_4O)xC_4H_8O_2$
Polymer NA
Glycols, polyethylene, mono(2,3-epoxy-2-methylpropyl) ether (8CI)

32210-23-4
$C_{12}H_{22}O_2$
198.34
O=C(OC(CCC(C(C)(C)C)C1)C1)C
Acetic acid, (4-tert-butylcyclohexyl) ester
p-tert-Butylcyclohexyl acetate
4-tert-Butylcyclohexyl acetate
4-tert-Butylhexahydrophenyl acetate
Cyclohexanol, 4-tert-butyl-, acetate
Vertenex

32222-06-3
$C_{27}H_{44}O_3$
416.71
CC(CCCC(C)(C)O)C1CCC2C(CCC12C)=CC=C3CC(O)CC(O)C3=C
Cholecalciferol, 1a,25-dihydroxy
Calcitriol
1a,25-Dihydroxycholecalciferol

1-α,25-Dihydroxycholecalciferol
1,25-Dihydroxycholecalciferol
Dihydroxyvitamin D3
1-α,25-Dihydroxyvitamin D3
Ro 215535
Rocaltrol
(5Z,7E)-9,10-Secochesta-5,7,10(19)-triene-1-α,3-β,25-triol
9,10-Secocholesta-5,7,10(19)-triene-1,3,25-triol, (1-α,3-β,5Z,7E)- (9CI)

32224-61-6
$C_5H_8O_4.xNa$
Unknown
Pentanedioic acid, sodium salt (9CI)
Sodium pentanedioate

32241-08-0
$C_{10}HCl_7$
369.29
Naphthalene, heptachloro- (9CI)
Heptachloronaphthalene

32272-57-4
$C_7H_{11}NS$
141.23
Thiazole, 4-ethyl-2,5-dimethyl- (8CI,9CI)

32273-77-1
C_9H_{16}
124.23
Pentalene, octahydro-1-methyl- (8CI,9CI)

32276-75-8
$C_{10}H_{20}O_2.1/2Cu$
204.04
Octanoic acid, 2,2-dimethyl-, copper(2+) salt (9CI)
Copper neodecanoate
Cupric neodecanoate
Neodecanoic acid, copper salt

32280-46-9
Unknown
Unknown
N,N-Diethyl-1,3-butanediamine

32281-79-1
$C_{13}H_{18}O$
190.29
1-Naphthol, 1,2,3,4-tetrahydro-4,5,7-trimethyl- (6CI,8CI)

32288-17-8
$C_6H_{16}O_5P_2$
230.14
Diphosphonic acid, dimethyl-, diethyl ester (9CI)
Phosphonic acid, methyl-, bimol. monoanhydride, diethyl ester (8CI)

32341-80-3
Unknown
Unknown
Triisopropanolamine 2,4-dichlorophenoxyacetate

32345-29-2
$C_{10}H_{15}O_3PS$
246.28
CCOP(=S)(OCC)Oc1ccccc1
Phosphorothioic acid, O,O-diethyl O-phenyl ester
DEPPT
O,O-Diethyl-O-phenylphosphorothioate

32351-70-5
Unknown
Unknown
Dimethylamine 2-(2-methyl-4-chlorophenoxy)propionate
Caswell No. 559B
EPA Pesticide Chemical Code 031519
2-(2-Methyl-4-chlorophenoxy)-propionic acid, dimethylamine salt

32357-46-3
$C_{16}H_{22}Cl_2O_4$
349.25
O=C(OCCOCCCC)CCCOc(c(cc(c1)Cl)Cl)c1
Butoxyethanol 4-(2,4-dichlorophenoxy)butyrate
Butanoic acid, 4-(2,4-dichlorophenoxy)-, 2-butoxyethyl ester (9CI)
Caswell No. 316A
4-(2,4-Dichlorophenoxy)butyric acid butoxyethanol ester
2,4-Dichlorophenoxybutyric acid, butoxyethyl ester
EPA Pesticide Chemical Code 030853

32357-83-8
$C_{11}H_{24}O$
172.31
Hexane, 1-(pentyloxy)- (9CI)
1-(Pentyloxy)hexane

32360-05-7
$C_{22}H_{42}O_2$
338.57
O=C(OCCCCCCCCCCCCCCCCCC)C(=C)C
2-Propenoic acid, 2-methyl-, octadecyl ester
AI3-25418
Methacrylic acid, stearyl ester
Octadecyl 2-methyl-2-propenoate

32367-54-7
$C_{12}H_{16}$
160.26
Naphthalene, 2-ethyl-1,2,3,4-tetrahydro- (6CI, 8CI,9CI)
2-Ethyltetralin

32368-69-7
$C_7H_5NO_5$
183.12
Peroxybenzoyl nitrate

32377-09-6
$C_{20}H_{13}N$
267.33
9H-Fluorene-4-carbonitrile, 9-phenyl- (9CI)
Fluorene-4-carbonitrile, 9-phenyl- (8CI)

32377-10-9
$C_{23}H_{16}$
292.38
7H-Benzo(c)fluorene, 7-phenyl- (8CI,9CI)
9-Phenyl-3,4-benzfluorene
7-Phenyl-7H-benzo-(c)fluorene

32377-11-0
$C_{19}H_{13}Cl$
276.77
9H-Fluorene, 9-(3-chlorophenyl)- (9CI)
9-m-Chlorophenylfluorene
Fluorene, 9-(m-chlorophenyl)- (8CI)

32377-12-1
$C_{20}H_{13}F_3$
310.32
9H-Fluorene, 9-(3-(trifluoromethyl)phenyl)- (9CI)
Fluorene, 9-(α,α,α-trifluoro-m-tolyl)- (8CI)

32377-13-2
$C_{20}H_{16}O$
272.35
9H-Fluorene, 9-(3-methoxyphenyl)- (9CI)
Fluorene, 9-(m-methoxyphenyl)- (8CI)

32377-15-4
$C_{21}H_{19}N$
285.39
Benzenamine, 4-(9H-fluoren-9-yl)-N,N-dimethyl-

(9CI)
Aniline, p-fluoren-9-yl-N,N-dimethyl- (8CI)

32388-22-0
$C_6H_{11}BrO$
179.06
trans-4-Bromocyclohexanol
Cyclohexanol, 4-bromo-, trans-

32388-55-9
$C_{17}H_{26}O$
246.39
O=C(C(=C(C(CC1(C2CC3)C3C)C2(C)C)C)C1)C
Ethanone, 1-(2,3,4,7,8,8a-hexahydro-3,6,8,8-tetramethyl-1H-3a,7-methanoazulen-5-yl)-, (3R-(3α,3aβ,7β,8aα))-(9CI)
9-Acetylcedr-8-ene
9-Acetyl-8-cedrene

32399-56-7
$C_{14}H_{28}O_2$
228.38
Ethanol, 2-(cyclododecyloxy)- (8CI,9CI)
Cyclododecane, ethanol deriv. (9CI)
2-(Cyclododecyloxy)-ethanol
Ethylene glycol mono-cyclododecyl ether

32407-99-1
$C_9H_{13}HgNS_2$
399.93
Caswell No. 657
EPA Pesticide Chemical Code 066008
Phenylmercuric dimethyldithiocarbamate

32426-11-2
$C_{20}H_{44}N.Cl$
334.02
Quaternium-24
Ammonium, decyldimethyl-octyl, chloride
Caswell No. 613A
1-Decaminium, N-octyl-N,N-dimethyl-, chloride
1-Decanaminium, N,N-dimethyl-N-octyl-, chloride (9CI)
Decyl dimethyl octyl ammonium chloride
Decyloctyldimethylammonium chloride
N,N-Dimethyl-N-octyl-1-decanaminium chloride
EPA Pesticide Chemical Code 069165
Octyl decyl dimethyl ammonium chloride

32432-45-4
$C_{17}H_{14}Cl_2N_4O_4$
409.20
Butanamide, N-(4-chloro-2-methylphenyl)-2-((4-chloro-2-nitrophenyl)azo)-3-oxo-(9CI)

32503-27-8
$C_{16}H_{36}N.HO_4S$
339.53
CCCC[N+](CCCC)(CCCC)CCCC.OS([O-])(=O)=O
1-Butanaminium, N,N,N-tributyl-, sulfate (1:1) (9CI)
N,N,N-Tributyl-1-butanaminium sulfate (1:1)

32510-27-3
$C_7H_5NS_2.xCu$
Unknown
2(3H)-Benzothiazolethione, copper salt (9CI)

32511-06-1
$C_{10}H_{16}O_2$
168.24
Cyclopropanecarboxylic-14C acid, 2,2-dimethyl-3-(2-methyl-1-propenyl)-, (1R-trans)- (9CI)
(+)-trans-Chrysanthemic acid-carboxy-14C
Cyclopropanecarboxylic-carboxy-14C acid, 2,2-dimethyl-3-(2-methylpropenyl)-, trans-(+)- (8CI)

32517-36-5
$C_{32}H_{18}CrN_6O_8.H$
667.50
Chromate(1-), bis(1-((2-hydroxy-4-nitrophenyl)azo)-2-naphthalenolato(2-))-, hydrogen, (OC-6-22')- (9CI)

32527-15-4
$C_{20}H_{18}O_6S_2$
418.50
O=S(=O)(OCC#CC#CCOS(=O)(=O)c(ccc(c1)C)c1)c(ccc(c2)C)c2
2,4-Hexadiyne-1,6-diol, di-p-toluenesulfonate
2,4-Hexadiyn-1,6-bis-p-toluenesulfonate

32534-66-0
$C_2H_7O_2PS_2$
158.18
Phosphorodithioic acid, dimethyl ester (8CI,9CI)

32534-81-9
$C_{12}H_5Br_5O$
564.69
Benzene, 1,1'-oxybis-, pentabromo deriv. (9CI)
1,1'-Oxybisbenzene pentabromo deriv.
Pentabromodiphenyl ether
Pentabromodiphenyl oxide
Pentabromophenoxybenzene

32534-95-5
$C_{17}H_{23}Cl_3O_3$
381.75
Propionic acid, 2-(2,4,5-trichlorophenoxy)-, isooctyl ester
Propanoic acid, 2-(2,4,5-trichlorophenoxy)-, isooctyl ester (9CI)
2-(2,4,5-Trichlorophenoxy)propionic acid isooctyl ester
Silvex isooctyl ester

32536-52-0
$C_{12}H_2Br_8O$
801.38
Benzene, 1,1'-oxybis-, octabromo deriv. (9CI)
1,1'-Oxybisbenzene octabromo deriv.

32555-29-6
$C_6H_{12}O_4$
148.16
Propanediol, (oxiranylmethoxy)- (9CI)
Propanediol, (2,3-epoxypropoxy)-(8CI)

32568-89-1
$C_{14}H_{22}N_2O_5$
298.33
O=C(N(C(C1=O)(C)C)CC(O2)C2)N1CC(OCC(O3)C3)C
2,4-Imidazolidinedione, 5,5-dimethyl-3-(2-(oxiranylmethoxy)propyl)-1-(oxiranylmethyl)- (9CI)

32582-63-1
$C_{10}H_{23}NO$
173.29
OCCNCCCCCCCC
Ethanol, 2-(octylamino)- (9CI)
N-(2-Hydroxyethyl)octylamine
NSC-78728
2-(Octylamino)ethanol
N-Octylethanolamine

32588-74-2
$C_{29}H_{10}Br_8N_2O_4$
1089.63
O=C(N(c(ccc(c1)Cc(ccc(N(C(=O)c(c2c(c(c3Br)Br)Br)c3Br)C2=O)c4)c4)c1)C(=O)c5c(c(c(c6Br)Br)Br)Br)c56

1H-Isoindole-1,3(2H)-dione, 2,2'-(methylenedi-4,1-phenylene)bis(4,5,6,7-tetrabromo- (9CI)
Methylenebis(4-(3',4',5',6'-tetrabromophthalimido)benzene)

32588-76-4
$C_{18}H_4Br_8N_2O_4$
951.46
O=C(N(C(=O)c1c(c(c(c2Br)Br)Br)Br)CCN(C(=O)c(c3c(c(c4Br)Br)Br)c4Br)C3=O)c12
1H-Isoindole-1,3(2H)-dione, 2,2'-(1,2-ethanediyl)bis-(4,5,6,7-tetrabromo- (9CI)

32598-10-0
$C_{12}H_6Cl_4$
291.98
Clc1ccc(c(Cl)c1)c2ccc(Cl)c(Cl)c2
Biphenyl, 2,3',4,4'-tetrachloro
1,1'-Biphenyl, 2,3',4,4'-tetrachloro- (9CI)
2,4,3',4'-TCB
2,3',4,4'-Tetrachlorobiphenyl
2,3',4,4'-Tetrachloro-1,1'-biphenyl
2,4,3',4'-Tetrachlorobiphenyl
3,4,2',4'-Tetrachlorobiphenyl

32598-11-1
$C_{12}H_6Cl_4$
291.99
Clc1ccc(Cl)c(c1)c2ccc(Cl)c(Cl)c2
1,1'-Biphenyl, 2,3',4',5-tetrachloro- (9CI)
Biphenyl, 2,3',4',5-tetrachloro- (8CI)
2,3',4',5-PCB
2,3',4',5-Tetrachlorobiphenyl

32598-13-3
$C_{12}H_6Cl_4$
291.98
Clc1ccc(cc1Cl)c2ccc(Cl)c(Cl)c2
Biphenyl, 3,3',4,4'-tetrachloro
1,1'-Biphenyl, 3,3',4,4'-tetrachloro- (9CI)
4-CB
TCB
3,3',4,4'-Tetrachlorobiphenyl
3,4,3',4'-Tetrachlorobiphenyl

32598-14-4
$C_{12}H_5Cl_5$
326.42
Biphenyl, 2,3,3',4,4'-pentachloro
1,1'-Biphenyl, 2,3,3'4,4'-pentachloro- (9CI)
PenCB
2,3,3',4,4'-Pentachlorobiphenyl
2,3,4,3',4'-Pentachlorobiphenyl
3,4,2',3',4'-Pentachlorobiphenyl

32612-48-9
$C_{12}H_{25}(C_2H_4O)x_{(14)}.H_2O_4S.H_4N$
Polymer
Glycols, polyethylene, dodecyl ether, monosulfonate, ammonium salt
Ammonium laureth sulfate
Ammonium lauryl ether sulfate

32624-40-1
$C_{17}H_{12}N_2O_3$
292.28
Benzoic acid, 4-((2-hydroxy-1-naphthalenyl)azo)- (9CI)
Benzoic acid, p-((2-hydroxy-1-naphthyl)azo)- (8CI)

32624-41-2
$C_{17}H_{12}N_2O_3$
292.30
Benzoic acid, 3-((2-hydroxy-1-naphthalenyl)azo)- (9CI)
Benzoic acid, m-((2-hydroxy-1-naphthyl)azo)- (7CI, 8CI)
1-(3'-Carboxyphenylazo)-2-naphthol

32650-55-8
$C_{10}H_{23}O_2PS_2$
270.40
O(P(OCCC(C)C)(S)=S)CCC(C)C
1-Butanol, 3-methyl-, hydrogen phosphorodithioate (9CI)
3-Methyl-1-butanol hydrogen phosphorodithioate

32651-66-4
$C_{38}H_{28}N_8O_{14}S_4.4Na$
1040.86
Benzenesulfonic acid, 2,2'-(1,2-ethenediyl)bis(5-((4-((4-sulfophenyl)azo)phenyl)-NNO-azoxy)-, tetrasodium salt (9CI)

32685-17-9
Unknown
Unknown
Nitrilotriacetic acid, triammonium salt
NTA, triammonium salt

32687-77-7
$C_{17}H_{28}N_2O_2$
292.41
O=C(NN)CCc(cc(c(O)c1C(C)(C)C)C(C)(C)C)c1
Benzenepropanoic acid, 3,5-bis(1,1-dimethylethyl)-4-hydroxy-, hydrazide (9CI)
3,5-Di-tert-butyl-4-hydroxyhydrocinnamic acid, hydrazide

32687-78-8
$C_{34}H_{52}N_2O_4$
552.79
O=C(NNC(=O)CCc(cc(c(O)c1C(C)(C)C)C(C)(C)C)c1)CCc(cc(c(O)c2C(C)(C)C)C(C)(C)C)c2
Benzenepropanoic acid, 3,5-bis(1,1-dimethylethyl)-4-hydroxy-, 2-(3-(3,5-bis-(1,1-dimethylethyl)-4-hydroxyphenyl)-1-oxopropyl)hydrazide (9CI)
Bis(3,5-di-tert-butyl-4-hydroxyhydrocinnamoyl)hydrazine

32690-93-0
$C_{12}H_6Cl_4$
291.99
Clc1ccc(cc1)c2cc(Cl)c(Cl)cc2Cl
1,1'-Biphenyl, 2,4,4',5-tetrachloro- (9CI)
Biphenyl, 2,4,4',5-tetrachloro- (8CI)
2,4,4',5-PCB
2,4,4',5-Tetrachlorobiphenyl

32694-76-1
$C_4H_2Cl_8$
333.68
Butane, 1,1,1,2,3,3,4,4-octachloro- (8CI,9CI)

32694-95-4
$C_{38}H_{40}N_{12}O_8S_2.2Na$
902.85
Benzenesulfonic acid, 2,2'-(1,2-ethenediyl)bis(5-((4-((2-hydroxypropyl)amino)-6-(phenylamino)-1,3,5-triazin-2-yl)amino)-, disodium salt (9CI)
2,2'-Stilbenedisulfonic acid, 4,4'-bis((4-anilino-6-((2-hydroxypropyl)amino)-s-triazin-2-yl)amino-, disodium salt

32736-90-6
$C_8H_{10}N_2$
134.18
Pyrazine, 2-ethenyl-6-ethyl- (9CI)
2-Ethyl-6-vinylpyrazine
Pyrazine, 2-ethyl-6-vinyl- (8CI)

32736-91-7
$C_9H_{14}N_2$
150.21
Pyrazine, 2,5-diethyl-3-methyl- (8CI,9CI)

32740-01-5
$C_{19}H_{15}N$

257.35
Benz(c)acridine, 7,11-dimethyl
1,10-Dimethyl-7,8-benzacridine (French)
7,11-Dimethylbenz(c)acridine

32749-94-3
$C_7H_{14}O$
114.19
2,3-Dimethyl pentaldehyde
AI3-33228
2,3-Dimethylpentaldehyde
2,3-Dimethylpentanal
2,3-Dimethyl-pentanal
2,3-Dimethylvaleraldehyde
NSC-73707
Pentanal, 2,3-dimethyl- (9CI)
Valeraldehyde, 2,3-dimethyl- (8CI)

32761-96-9
$(C_{16}H_{18}O_3.C_{13}H_{12}N_3O.C_9H_{11}O_3S)x$
Polymer
Benzenediazonium, 2-methoxy-4-(phenylamino)-, Salt with 2,4,6-trimethylbenzenesulfonic acid (1:1) polymer with 1,1'-oxybis-(4-(methoxymethyl)benzene) (9CI)

32762-51-9
C_6H_5BrO
173.02
Phenol, bromo
Bromophenol

32768-54-0
$C_7H_6Cl_2$
161.03
c(c(c(cc1)Cl)Cl)(c1)C
Benzene, 1,2-dichloro-3-methyl- (9CI)
1,2-Dichloro-3-methylbenzene

32774-16-6
$C_{12}H_4Cl_6$
360.86
Clc1cc(cc(Cl)c1Cl)c2cc(Cl)c(Cl)c(Cl)c2
Biphenyl, 3,3',4,4',5,5'-hexachloro
1,1'-Biphenyl, 3,3',4,4',5,5'-hexachloro- (9CI)
3,3',4,4',5,5'-Hexachlorobiphenyl
3,4,5,3',4',5'-Hexachlorobiphenyl

32793-63-8
$C_{10}H_{17}N.xH_2O_4S$
Unknown
Tricyclo(3.3.1.13,7)decan-1-amine, sulfate (9CI)

32809-16-8
$C_{13}H_{11}Cl_2NO_2$
284.15
CC12CC1(C)C(=O)N(C2=O)c3cc(Cl)cc(Cl)c3
1,2-Cyclopropanedicarboximide, N-(3,5-dichlorophenyl)-1,2-dimethyl
3-Azabicyclo(3.1.0)hexane-2,4-dione, 3-(3,5-dichlorophenyl)-1,5-dimethyl-
N-(3',5'-Dichlorophenyl)-1,2-dimethylcyclopropane-1,2-dicarboximide
Procymidone
S 7131
Sumilex
Sumisclex

32844-67-0
$C_{16}H_{32}O_2$
256.43
Isohexadecanoic acid (VAN) (9CI)

32846-21-2
$C_{23}H_{26}N_4O_6S_2.Na$
541.58
2-Naphthalenesulfonic acid, 6-amino-5-((2-((cyclohexylmethylamino)sulfonyl)phenyl)azo)-4-hydroxy-, monosodium salt (9CI)

32861-85-1
$C_{13}H_9Cl_2NO_4$
314.13
COc2cc(Oc1ccc(Cl)cc1Cl)ccc2N(=O)=O
Ether, (2,4-dichlorophenyl) (3-methoxy-4-nitrophenyl)
Chlomethoxyfen
Chlomethoxynil
4-(2,4-Dichlorophenoxy)-2-methoxy-1-nitrobenzene
5-(2,4-Dichlorophenoxy)-2-nitroanisole
2,4-Dichlorophenyl 3'-methoxy-4'-nitrophenyl ether
Diphenex
Ekkusagoni
X-52

32862-97-8
$C_9H_7BrO_2$
227.07
O=C(O)C=Cc(cccc1Br)c1
Cinnamic acid, m-bromo
m-Bromocinnamic acid
2-Propenoic acid, 3-(3-bromophenyl)- (9CI)

32889-48-8
$C_{10}H_{13}ClN_6$
252.74
CC(CNc2nc(Cl)nc(NC1CC1)n2)C#N
Propionitrile, 2-(4-chloro-6-(cyclopropylamino)-s-triazin-2-ylamino)-2-methyl
2-(4-Chloro-6-(cyclopropylamino)-s-triazin-2-yl)amino-2-methylpropionitrile
2-((4-Chloro-6-(cyclopropylamino)-1,3,5-triazin-2-yl)amino)-2-methylpropanenitrile
CGA-18762
Cycle
Procyazine

32904-22-6
$C_{25}H_{38}O_2$
370.63
6H-Dibenzo(b,d)pyran-1-ol, 3-(1',2'-dimethylheptyl)-7,8,9,10-tetrahydro-6,6,9-trimethyl
Dimethylheptylpyran
3-(1,2-Dimethylheptyl)-7,8,9,10-tetrahydrocannabinol
DMHP
δ^3-THC

32953-89-2
$C_{12}H_{17}NO_3$
223.27
Rimiterol
Rimiterolum (Latin)

32954-58-8
$C_9H_{12}O_3$
168.21
Pentanone, 1-(3-furyl)-4-hydroxy
1-(β-Furyl)-4-hydroxypentanone
1-(3-Furyl)-4-hydroxypentanone
Ipomeanol
4-Ipomeanol

32970-45-9
$C_8H_{18}O$
130.23
Butane, 2-(1,1-dimethylethoxy)- (9CI)

32976-87-7
$C_{20}D_{16}$
272.36
Benz(a)anthracene-d16, 7,12-dimethyl
7,12-Dimethylbenz(a)anthracene-d16
7,12-Dimethylbenz(a)anthracene, deuterated

32976-88-8
$C_4H_8N_4O_3$
160.16
CCN(N=O)C(=O)NC(N)=O
Biuret, 1-ethyl-1-nitroso
ENBU
Ethylnitrosobiuret
N-Ethyl-N-nitrosobiuret

N-Nitroso-N-ethylbiuret

32986-56-4
$C_{18}H_{37}N_5O_9$
467.60
NCC3OC(OC2C(N)CC(N)C(OC1 OC(CO)C(O)C(N)C1O)C2O) C(N)CC3O
Streptamine, O-3-amino-3-deoxy-α-D-glucopyranosyl-(1-4)-O-(2,6-diamino-2,3,6-trideoxy-α-d-ribohexo-pyranosyl-(1-6))-2-deoxy-, D
Distobram
Gernebcin
Nebramycin Factor 6
NF 6
Obramycin
Tobradistin
Tobramycin
Tobrex

33007-83-9
$C_{15}H_{26}O_6S_3$
398.57
O=C(OCC(CC)(COC(=O)CCS) COC(=O)CCS)CCS
Propanoic acid, 3-mercapto-, 2-ethyl-2-((3-mercapto-1-oxopropoxy)methyl)-1,3-propanediyl ester (9CI)
3-Mercaptopropanoic acid, 2-ethyl-2-(hydroxymethyl)-1,3-propanediol triester
Trimethylolpropane tris(mercaptopropionate)

33025-41-1
$C_{12}H_6Cl_4$
291.99
Clc1ccc(cc1)c2ccc(Cl)c(Cl)c2Cl
1,1'-Biphenyl, 2,3,4,4'-tetrachloro- (9CI)
Biphenyl, 2,3,4,4'-tetrachloro-(8CI)
2,3,4,4'-PCB
2,3,4,4'-Tetrachlorobiphenyl

33032-17-6
$C_6H_7Cl_2N_3S$
224.11
1,3,5-Triazine, 2,4-bis-(chloromethyl)-6-(methylthio)- (9CI)
s-Triazine, 2,4-bis(chloromethyl)-6-(methylthio)-(8CI)

33037-07-9
$C_3H_5BrCl_2$
191.88
Propane, 1-bromo-2,3-dichloro-(8CI,9CI)

33039-81-5
$C_{12}H_8Cl_2$
223.10
1,1'-Biphenyl, ar,ar'-dichloro-(9CI)
Biphenyl, ar,ar'-dichloro- (8CI)
ar,ar'-Dichlorobiphenyl

33046-81-0
$C_9H_{16}O$
140.23
3-Octen-2-one, 7-methyl- (9CI)

33058-12-7
$C_{12}H_8Cl_6O$
380.91
1,5,2,4-Ethanediylidenecyclopenta(cd)pentalen-1(2H)-ol, 2,2a,3,3,4,8-hexachloro-octahydro- (8CI,9CI)
Endrin alcohol
NSC-59452
NSC-122237

33061-16-4
$C_7H_{12}O$
112.17
3-Cyclohexen-1-ol, 1-methyl- (7CI,9CI)
1-Methyl-3-cyclohexenol

33069-62-4
$C_{47}H_{51}NO_{14}$
853.99
Tax-11-en-9-one, 5-β,20-epoxy-1,2-α,4,7-β,10-β,13-α-hexahydroxy-, 4,10- diacetate 2-benzoate 13-ester with (2R,3S)-N-benzoyl-3-phenylisoserine
NSC-125973
Taxol

33079-08-2
Unknown
Unknown
tert-Butylamine 2-pyridinethiol-1-oxide

33086-18-9
$C_{14}H_9Cl_5$
354.49
Benzene, 1,1'-(2,2,2-trichloroethylidene)bis-(chloro- (9CI)
Ethane, 1,1,1-trichloro-2,2-bis(chlorophenyl)-(8CI)
1,1,1-Trichloro-2,2-bis-(chlorophenyl)ethane

33089-61-1
$C_{19}H_{23}N_3$
293.45
CN(C=Nc1ccc(C)cc1C)C=Nc2ccc (C)cc2C
2,4-Xylidine, N,N'-(methyliminodimethylidyne)bis
Acarac
Amitraz
Amitraze (French)
Amitraz estrella
Azadieno
Baam
N,N-Bis(2,4-xylyliminomethyl)-methylamine
Boots BTS 27419
BTS 27,419
1,5-Di(2,4-dimethylphenyl)-3-methyl-1,3,5-triazapenta-1,4-diene
N'-(2,4-Dimethylphenyl)-N-(((2,4-dimethylphenyl)imino)-methyl)-N-methylmethanimidamide
N,N-Di-(2,4-xylyliminomethyl)-methylamine
Ectodex
ENT 27,967
Methanimidamide, N'-(2,4-dimethylphenyl)-N-(((2,4-dimethylphenyl)imino)methyl)-N-methyl-
N-Methyl-bis(2,4-xylyliminomethyl)amine
2-Methyl-1,3-di(2,4-xylylimino)-2-azapropane
N,N'-((Methylimino)dimethylidyne)di-2,4-xylidine
N-Methyl-N'-2,4-xylyl-N-(N-2,4-xylylformimidoyl)-formamidine
Mitaban
Mitac
R.D. 27419
Taktic
Triatix
Triatox
U-36059
Upjohn U-36059

33091-17-7
$C_{12}H_2Cl_8$
429.77
2,2',3,3',4,4',6,6'-Octachlorobiphenyl

33113-08-5
$CH_2O_3.xCu.xH_3N$
Unknown
Copper ammonium carbonate
Ammonium copper carbonate
Carbonic acid, ammonium copper salt (9CI)
Caswell No. 229B
EPA Pesticide Chemical Code 022703

33125-86-9
$C_{10}H_{20}Cl_4O_8P_2$
472.02
O=P(OCCCl)(OCCCl)OCCOP (=O)(OCCCl)OCCCl
Phosphoric acid, 1,2-ethanediyl tetrakis(2-chloroethyl) ester (9CI)

33145-10-7
$C_{20}H_{26}O_2$
298.43
Oc(c(cc(c1)C)C(c(c(O)c(cc2C)C)
c2)C(C)C)c1C
Phenol, 2,2'-(2-methylpropyl-idene)bis(4,6-dimethyl- (9CI)
2,2'-Isobutylidenebis(4,6-di-methylphenol)

33146-45-1
$C_{12}H_8Cl_2$
223.10
Clc1cccc(Cl)c1c2ccccc2
2,6-Dichlorobiphenyl
2,6-Dichloro-1,1'-biphenyl

33156-92-2
$C_{11}H_{16}$
148.25
1,4-Cycloheptadiene, 6-(1-butenyl)-, (S-(Z))- (9CI)
1,4-Cycloheptadiene, 6-(1-butenyl)-, (Z)-(S)-(+)- (8CI)
Dictyopteren D'
Dictyopterene D'
S-(+)-Ectocarpene
(+)-(6S)-Ectocarpene
Sirenin
Sirenin (Etocarpus)

33184-55-3
$C_{14}H_{11}ClO$
230.69
Methanone, (5-chloro-2-methyl-phenyl)phenyl- (9CI)
Benzophenone, 5-chloro-2-methyl- (8CI)

33213-65-9
$C_9H_6Cl_6O_3S$
406.91
5-Norbornene-2,3-dimethanol, 1,4,5,6,7,7-hexachloro-, cyclic sulfite, exo
β-Benzoepin
Endosulfan B
b-Endosulfan-β
β-Endosulfan
Endosulfan 2
General Weed Killer
β-Thiodan
α-Thionex

33228-45-4
$C_{12}H_{19}N$
177.32
Aniline, (p-hexyl)
Benzenamine, 4-hexyl- (9CI)
p-Hexylaniline

33229-34-4
$C_{12}H_{19}N_3O_5$
285.34
OCCNc1ccc(cc1N(=O)=O)
N(CCO)CCO
Ethanol, 2,2'-((4-((2-hydroxy-ethyl)amino)-3-nitrophenyl)-imino)di
Ethanol, 2,2'-((4-((2-hydroxy-ethyl)amino)-3-nitrophenyl)-imino)bis- (9CI)
HC Blue No. 2
NCI-C54897
3-Nitro-N^1,N^1,N^4-tris(2-hydroxyethyl)-
p-Phenylenediamine, 3-nitro-N^1,N^1,N^4-tris(2-hydroxyethyl)-

33244-86-9
$C_{12}H_{13}NO$
187.24
Benzamide, N-(1,1-dimethyl-2-propynyl)- (8CI,9CI)

33245-39-5
$C_{12}H_{13}ClF_3N_3O_4$
355.73
CCCN(CCCl)c1c(cc(cc1N(=O)
=O)C(F)(F)F)N(=O)=O
p-Toluidine, N-(2-chloroethyl)-2,6-dinitro-N-propyl-α,α,α-trifluoro
BAS 3920
BAS 3922
BAS 392-H
Basalin
N-(2-Chloroethyl)-2,6-dinitro-N-propyl-4-(trifluoromethyl)-aniline
N-(2-Chloroethyl)-2,6-dinitro-N-propyl-4-(trifluoromethyl)-benzenamide
N-(2-Chloroethyl)-α,α,α-tri-fluoro-2,6-dinitro-N-propyl-p-toluidine
Fluchloralin
N-Propyl-N-(2-chloroethyl)-2,6-dinitro-4-trifluoromethyl aniline
N-Propyl-N-(2-chloroethyl)-α,α,α-trifluoro-2,6-dinitro-p-toluidine

33284-50-3
$C_{12}H_8Cl_2$
223.10
Clc1ccc(c(Cl)c1)c2ccccc2
1,1'-Biphenyl, 2,4-dichloro- (9CI)
Biphenyl, 2,4-dichloro- (8CI)
2,4-Dichlorobiphenyl

33284-52-5
$C_{12}H_6Cl_4$
291.99
Clc1cc(Cl)cc(c1)c2cc(Cl)cc(Cl)c2
3,5,3',5'-Tetrachlorobiphenyl
Biphenyl, 3,3',5,5'-tetrachloro- (8CI)
1,1'-Biphenyl, 3,3',5,5'-tetra-chloro- (9CI)
NSC-113288
3,3',5,5'-Tetrachlorodiphenyl

33284-53-6
$C_{12}H_6Cl_4$
291.99
Clc2cc(c1ccccc1)c(Cl)c(Cl)c2Cl
1,1'-Biphenyl, 2,3,4,5-tetra-chloro- (9CI)
Biphenyl, 2,3,4,5-tetrachloro- (8CI)
2,3,4,5-PCB
2,3,4,5-Tetrachlorobiphenyl

33284-54-7
$C_{12}H_6Cl_4$
291.99
Clc2cc(Cl)c(Cl)c(c1ccccc1)c2Cl
1,1'-Biphenyl, 2,3,5,6-tetra-chloro- (9CI)
Biphenyl, 2,3,5,6-tetrachloro- (8CI)
2,3,5,6-PCB
2,3,5,6-Tetrachlorobiphenyl

33312-01-5
$C_{22}H_{44}N_2O_2S_2$
432.72
O=C(NCCCCCCCC)CCSSCCC
(=O)NCCCCCCCC
Propionamide, 3,3'-dithio-bis(N-octyl- (8CI)
N,N'-Bis-n-octyl-3,3'-dithio-propionamide
3,3'-Dithiobis(N-octylpropion-amide)

33329-35-0
$C_{15}H_{36}N_4$
272.45
N(CCCN(CCCN(C)C)CCCN(C)
C)(C)C
1,3-Propanediamine, N,N-bis-(3-(dimethylamino)propyl)-N',N'-dimethyl- (9CI)
N,N-Bis(3-(dimethylamino)pro-pyl)-N',N'-dimethyl-1,3-pro-panediamine

33374-28-6
$C_{18}H_{26}O_5$
322.40
1,2-Benzenedicarboxylic acid, 2-butoxyethyl butyl ester (9CI)
Phthalic acid, 2-butoxyethyl butyl ester (8CI)

33397-79-4
$C_2H_6S_3Sn_2$
363.69
Distannathiane, dimethyldi-thioxo- (9CI)
Dimethyldithioxodistanna-

thiane

33401-49-9
$C_{17}H_{30}N_2O_3Si.ClH$
374.97
1,2-Ethanediamine, N-((4-ethenylphenyl)methyl)-N'-(3-(trimethoxysilyl)propyl)-, monohydrochloride (9CI)

33423-92-6
$C_{12}H_4Cl_4O_2$
321.96
Clc3cc(Cl)c2Oc1cc(Cl)cc(Cl)c1Oc2c3
Dibenzo-p-dioxin, 1,3,6,8-tetrachloro
Dibenzo(b,e)(1,4)dioxin, 1,3,6,8-tetrachloro-
1,3,6,8-TCDD
1,3,6,8-Tetrachlorodibenzo-dioxin
1,3,6,8-Tetrachlorodibenzo-para-dioxin

33433-95-3
$C_8H_5Cl_3O_3$
255.48
Acetic acid, (2,3,5-trichlorophenoxy)- (8CI,9CI)
2,3,5-T

33434-63-8
C_6HCl_7Si
349.33
Silane, trichloro(tetrachlorophenyl)- (9CI)
(Tetrachlorophenyl)trichlorosilane
Trichloro(tetrachlorophenyl)-silane

33439-45-1
$C_{10}H_{10}Cl_2F_2N_2OS$
315.18
Urea, 1-(3-chloro-4-(chlorodifluoromethylthio)phenyl)-3,3-dimethyl
Bay Kue 2079A
(3-Chloro-4-chlorodifluoro-methylthiophenyl)-1,1-dimethylurea
Clearcide
Fluothiuron
Kue 2079A
Thiochlormethyl

33442-83-0
$C_{10}H_5Cl_7$
373.32
Photoheptachlor

33455-24-2
$C_4H_6Cl_4$
195.90
Butane, 1,1,4,4-tetrachloro- (6CI,8CI)

33466-31-8
$C_{40}H_{72}O_8Sn$
799.72
5,7,12-Trioxa-6-stannatetracosa-2,9-dienoic acid, 6,6-dibutyl-4,8,11-trioxo-, dodecyl ester, (Z,Z)- (9CI)

33467-76-4
$C_7H_{14}O$
114.19
OCC=CCCCC
2-Hepten-1-ol, (E)- (9CI)
AI3-36042
trans-2-Hepten-1-ol
(E)-2-Hepten-1-ol
NSC-244909

33508-02-0
$C_{18}H_{20}$
236.36
1H-Indene, 2,3-dihydro-1,1,2-trimethyl-3-phenyl- (9CI)
Indan, 1,1,2-trimethyl-3-phenyl- (8CI)

33509-43-2
$C_7H_{12}N_4OS$
200.24
O=C(N(N)C(=NN=1)S)C1C(C)(C)C
1,2,4-Triazin-5(2H)-one, 4-amino-6-(1,1-dimethylethyl)-3,4-dihydro-3-thioxo- (9CI)
4-Amino-6-tert-butyl-3-mercapto-1,2,4-triazine-5(4H)-one

33533-53-8
$C_{16}H_{14}O_4$
270.28
1,2-Benzenedicarboxylic acid, mono(1-phenylethyl) ester (9CI)
NSC-108264
Phthalic acid, mono(α-methylbenzyl) ester (8CI)

33533-56-1
$C_{20}H_{22}O_4$
326.40
1,2-Benzenedicarboxylic acid, mono(1-(4-(1,1-dimethylethyl)phenyl)ethyl) ester (9CI)
Phthalic acid, mono(p-tert-butyl-α-methylbenzyl) ester (8CI)

33533-57-2
$C_{17}H_{16}O_5$
300.31
1,2-Benzenedicarboxylic acid, mono(1-(4-methoxyphenyl)ethyl) ester (9CI)
Phthalic acid, mono(p-methoxy-α-methylbenzyl) ester (8CI)

33543-31-6
$C_{17}H_{12}$
216.29
Fluoranthene, 2-methyl
2-Methylfluoranthene

33562-89-9
$C_8H_6O_7S.Na$
269.19
1,2-Benzenedicarboxylic acid, 4-sulfo-, monosodium salt (9CI)
Monosodium 4-sulfophthalate
NSC-66547
Phthalic acid, 4-sulfo-, monosodium salt
5-Sulfoisophthalic acid 5-sodium salt
4-Sulfophthalic acid monosodium salt

33583-02-7
$C_{13}H_{14}O$
186.26
1-Naphthalenol, 2,5,8-trimethyl- (9CI)
1-Naphthol, 2,5,8-trimethyl- (8CI)

33603-39-3
$C_{18}H_{20}$
236.36
Indan, 1,2,3-trimethyl-1-phenyl-, stereoisomer (8CI)

33611-16-4
$C_{18}H_{20}$
236.36
Indan, 1,2,3-trimethyl-1-phenyl-, stereoisomer (8CI)

33619-92-0
$C_8H_{19}O_2PS_2.Na$
265.33
Sodium di-sec-butyl phosphorodithioate
Aeroflat 238
Phosphorodithioic acid, O,O-bis(1-methylpropyl) ester, sodium salt (9CI)
Phosphorodithioic acid, O,O-di-sec-butyl ester, sodium salt

33629-47-9
$C_{14}H_{21}N_3O_4$

295.38
O=N(=O)c(c(NC(CC)C)c(N(=O)=O)cc1C(C)(C)C)c1
Aniline, N-sec-butyl-4-tert-butyl-2,6-dinitro
A 820
72-A34
Amchem 70-25
Amchem A-280
Amex
Amex 820
70-314B
Benzenamine, 4-(1,1-dimethylethyl)-N-(1-methylpropyl)-2,6-dinitro- (9CI)
Butalin
Butralin
Butraline
N-sec-Butyl-4-tert-butyl-2,6-dinitroaniline
Dibutalin
4-(1,1-Dimethylethyl)-N-(1-methylpropyl)-2,6-dinitrobenzenamine
Rutralin
Tamex

33637-20-6
$C_{14}H_{22}$
190.33
Benzene, tetraethyl-
AI3-09178
Tetraethylbenzene

33660-91-2
$C_{13}H_{20}O_3S.Na$
279.36
Benzenesulfonic acid, heptyl-, sodium salt (8CI, 9CI)
Heptylbenzenesulfonic acid sodium salt
Sodium heptylbenzenesulfonate

33665-90-6
$C_4H_5NO_4S$
163.15
Acesulfame
Acesulfamo (Spanish)
Acesulfamum (Latin)
Acetosulfam
3,4-Dihydro-6-methyl-1,2,3-oxathiazin-4-one-2,2-dioxide potassium salt
6-Methyl-3,4-dihydro-1,2,3-oxathiazin-4-one 2,2-dioxide
6-Methyl-1,2,3-oxathiazin-4(3H)-one 2,2-dioxide
1,2,3-Oxathiazin-4(3H)-one, 6-methyl-, 2,2-dioxide

33669-76-0
$C_{10}H_{20}O_2$
172.27
1,2-Cyclohexanediol, 1-methyl-4-(1-methylethyl)-(9CI)
AI3-24709
Isocarvomenthol, 1-hydroxy-(VAN)
p-Menthane-1,2-diol
NSC-96755

33671-46-4
$C_{16}H_{15}ClN_2OS$
318.81
Clotiazepam
5-(o-Chlorophenyl)-7-ethyl-1,3-dihydro-1-methyl-2H-thieno(2,3-E)-1,4-diazepin-2-one
5-(2-Chlorophenyl)-7-ethyl-1-methyl-1,3-dihydro-2H-thieno(2,3-E)(1,4)diazepin-2-one
Clotiazepamum (Latin)
DEA No. 2752
Rise
2H-Thieno(2,3e)(1,4)-diazepin-2-one, 5-(o-chlorophenyl)-7-ethyl-1,3-dihydro-1-methyl-(8CI)
2H-Thieno(2,3-e)(1,4)diazepin-2-one, 1,3-dihydro-5-(o-chlorophenyl)-7-ethyl-1-methyl-
Trecalmo
Y 6047

33684-08-1
$C_{10}H_{15}O_3P$
214.20
Phosphonic acid, methyl-, 1-methylethyl phenyl ester (9CI)
Phosphonic acid, methyl-, isopropyl phenyl ester (8CI)

33684-09-2
$C_{16}H_{36}N.C_4H_6NO_2$
342.57
Ammonium, tetrabutyl-, salt with 2,3-butanedione monooxime (1:1) (8CI)

33684-10-5
$C_{16}H_{36}N.C_7H_6NO_2$
378.60
Ammonium, tetrabutyl-, benzohydroxamate (8CI)

33684-11-6
$C_{16}H_{36}N.C_{10}H_{11}O_2$
405.67
Ammonium, tetrabutyl-, salt with 4'-hydroxy-3',5'-dimethylacetophenone (1:1) (8CI)

33692-99-8
$C_8H_{14}ClN_5$
215.66
1,3,5-Triazine-2,4-diamine, 6-chloro-N-methyl-N'-(1-methylpropyl)- (9CI)
GS 18182
s-Triazine, 2-(sec-butylamino)-4-chloro-6-(methylamino)-(8CI)

33693-04-8
$C_{10}H_{19}N_5O$
225.34
O(c(nc(nc1NC(C)(C)C)NCC)n1)C
s-Triazine, 2-tert-butylamino-4-ethylamino-6-methoxy
2-tert-Butylamino-4-ethylamino-6-methoxy-s-triazine
2-tert-Butylamino-4-ethylamino-6-methoxy-1,3,5-triazine
Caragard
GS 14259
2-Methoxy-4-tert-butylamino-6-aethylamino-s-triazin (German)
Terbumeton

33698-87-2
$C_5H_8O_2$
100.12
3-Pentenoic acid, (Z)-(9CI)
(Z)-3-Pentenoic acid

33700-25-3
$C_{38}H_{16}Cl_2N_2O_{14}S_4.4Na$
1015.67
Diphenaleno(1,9-ab:1',9'-lm)-triphenodioxazinetetrasulfonic acid, 8,19-dichloro-, tetrasodium salt (9CI)
8,19-Dichlorotetrasulfodiphenaleno(1,9-ab:1',9'-lm)triphenodioxazine, tetrasodium salt

33703-08-1
$C_{24}H_{46}O_4$
398.63
Diisononyl adipate
Diisononyl hexanedioate
Hexanedioic acid, diisononyl ester (9CI)

33704-59-5
$C_{14}H_{24}$
192.34
C(=C(C(C1C)(C)C)CCC2)(C1(C)C)C2
1H-Indene, 2,3,4,5,6,7-hexahydro-1,1,2,3,3-pentamethyl-(9CI)

33704-61-9
$C_{14}H_{22}O$
206.33
O=C(C(=C(C(C1C)(C)C)CC2)C1(C)C)C2
4H-Inden-4-one, 1,2,3,5,6,7-

hexahydro-1,1,2,3,3-pentamethyl- (9CI)

33719-74-3
$C_7H_6Cl_2O$
177.03
COc1cc(Cl)cc(Cl)c1
Benzene, 1,3-dichloro-5-methoxy- (9CI)
Anisole, 3,5-dichloro- (8CI)

33752-16-8
C_5HCl_4N
216.87
Pyridine, tetrachloro- (9CI)

33791-58-1
$C_{13}H_{16}O_2$
204.27
3a,4,5,6,7,7a-Hexahydro-4,7-methanoindenyl acrylate
Acrylic acid, 3a,4,5,6,7,7a-hexahydro-4,7-methanoindenyl ester
Dicyclopentadiene acrylate
2-Propenoic acid, 3a,4,5,6,7,7a-hexahydro-4,7-methano-1H-indenyl ester (9CI)

33796-87-1
$C_9H_{18}O_3$
174.24
Nonanoic acid, 3-hydroxy-, D- (8CI)

33806-58-5
$C_{18}H_{30}O_3$
294.43
2,5-Furandione, dihydro-3-(tetradecenyl)- (9CI)
Dihydro-3-(tetradecenyl)-2,5-furandione
Tetradecenylsuccinic anhydride

33813-20-6
$C_4H_4N_2S_3$
176.28
3H-Imidazo(2,1-c)-1,2,4-dithiazole-3-thione, 5,6-dihydro
5,6-Dihydro-3H-imidazo(2,1-c)-1,2,4-dithiazole-3-thione
Endodan
ETEM
Ethylene bisthiuram monosulfide
Ethylenethiocarbamyl sulfide
Ethylene thiuram monosulfide
Ethylene thiuram monosulphide
Ethylenethiuram sulfide
ETM
ETM (Heterocycle)
Hortocritt

33820-53-0
$C_{15}H_{23}N_3O_4$
309.41
CCCN(CCC)c1c(cc(cc1N(=O)=O)C(C)C)N(=O)=O
Aniline, 2,6-dinitro-N,N-dipropyl-p-isopropyl
Benzenamine, 4-(1-methylethyl)-2,6-dinitro-N,N-dipropyl- (9CI)
Cumidine, 2,6-dinitro-N,N-dipropyl-
2,6-Dinitro-N,N-dipropylcumidine
2,6-Dinitro-N,N-dipropyl-4-isopropylaniline
EL-179
Isopropalin
Isopropaline
4-Isopropyl-2,6-dinitro-N,N-dipropylaniline
4-(1-Methylethyl)-2,6-dinitro-N,N-dipropylbenzenamine
Paarlan

33857-23-7
$C_{12}H_{19}N_2O_4PS$
318.36
Phosphoramidothioic acid, isopropyl-, O-ethyl O-(2-nitro-p-tolyl) ester
Amiprophos
Bay-NTN 5006
NTN 5006
Phosphoramidothioic acid, (1-methylethyl)-, O-ethyl O-(4-methyl-2-nitrophenyl) ester (9CI)

33857-26-0
$C_{12}H_6Cl_2O_2$
253.08
Clc3ccc2Oc1cc(Cl)ccc1Oc2c3
Dibenzo-p-dioxin, 2,7-dichloro
DCDD
Dibenzo(b,e)(1,4)dioxin, 2,7-dichloro-
2,7-Dichlorodibenzo-p-dioxin
2,7-Dichlorodibenzodioxin
NCI-C03667

33857-28-2
$C_{12}H_5Cl_3O_2$
287.52
Dibenzo-p-dioxin, 2,3,7-trichloro
2,3,7-Trichlorodibenzo-p-dioxin

33877-87-1
$C_{10}H_{12}$
132.21
Azulene, 1,2,3,3a-tetrahydro- (8CI,9CI)

33878-50-1
$C_{18}H_{17}Cl_2NO_3$
366.26
CCOC(=O)C(C)N(C(=O)c1ccccc1)c2ccc(Cl)c(Cl)c2
Alanine, N-benzoyl-N-(3,4-dichlorophenyl)-, ethyl ester, L
L-N-Benzoyl-N-(3,4-dichlorophenyl)alanine ethyl ester
Karakhol
WL 17731

33880-83-0
$C_{15}H_{24}$
204.36
C(C(C(C=C)(CCC1C(=C)C)C)C1)(=C)C
Cyclohexane, 1-ethenyl-1-methyl-2,4-bis(1-methylethenyl)-, (1α,2β,4β)- (9CI)
β-Elemene

33931-68-9
$C_{10}H_{20}O_3$
188.27
Butyric acid, 1-ethoxybutyl ester (8CI)

33933-73-2
$C_8H_{12}S$
140.25
Thiophene, 2-methyl-5-propyl- (8CI,9CI)

33933-74-3
C_9H_{18}
126.24
3-Heptene, 4-ethyl- (8CI,9CI)

33956-01-3
$C_{36}H_{52}ClN_3O_3$
610.26
O=C(N=C(NN(C1=O)c(c(cc(c2)C)C)c2Cl)C1)C(Oc(cccc3CCCCCCCCCCCCCCC)c3)CC
Butanamide, N-(1-(2-chloro-4,6-dimethylphenyl)-4,5-dihydro-5-oxo-1H-pyrazol-3-yl)-2-(3-pentadecylphenoxy)-(9CI)

33956-61-5
$C_{44}H_{72}O_{12}$
793.16
Nonactin, 5,14,23,32-tetrademethyl-5,14,23,32-tetraethyl
S-3466-C
Tetranactin

33957-63-0
$C_9H_8N_4O$
188.16
1,3,5-Triazin-2(1H)-one, 4-amino-6-phenyl- (9CI)
Benzoguanide
s-Triazin-2-ol, 4-amino-6-phenyl- (6CI,7CI)
s-Triazin-2(1H)-one, 4-amino-6-phenyl- (8CI)

33962-13-9
$C_{14}H_{22}$
190.33
Benzene, 1,4-diethyl-2,3,5,6-tetramethyl- (7CI,8CI,9CI)

33967-19-0
$C_{10}H_{14}O$
150.22
Benzenemethanol, α,3,4-trimethyl- (9CI)
Benzyl alcohol, α,3,4-trimethyl- (8CI)

33979-03-2
$C_{12}H_4Cl_6$
360.88
Clc1cc(Cl)c(c(Cl)c1)c2c(Cl)cc(Cl)cc2Cl
2,2',4,4',6,6'-Hexachlorobiphenyl

34003-77-5
$C_8H_{14}O_2$
142.20
3(2H)-Furanone, 5-tert-butyldihydro- (8CI)

34006-76-3
$C_{13}H_{16}O_4$
236.27
1,2-Benzenedicarboxylic acid, butyl methyl ester
AI3-03342

34006-77-4
$C_{11}H_{12}O_4$
208.21
1,2-Benzenedicarboxylic acid, ethyl methyl ester (9CI)
Phthalic acid, ethyl methyl ester (8CI)

34010-15-6
$C_{14}H_{28}O$
212.38
OCCCCCCCCCC=CCC
11-Tetradecen-1-ol, (Z)-
AI3-35174
(Z)-11-Tetradecen-1-ol

34014-18-1
$C_9H_{16}N_4OS$
228.35
CNC(=O)N(C)c1nnc(s1)C(C)(C)C
Urea, 1-(5-(t-butyl)-1,3,4-thiadiazol-2-yl)-1,3-dimethyl
Brulan
1-(5-tert-Butyl-1,3,4-thiadiazol-2yl)-3-dimethylharnstoff (German)
1-(5-tert-Butyl-1,3,4-thiadiazol-2-yl)-1,3-dimethylurea
N-(5-(1,1-Dimethylaethyl)-1,3,4-thiadiazol-2-yl)-N,N'-dimethylharnstoff (German)
E-103
EI-103
EL-103
Graslan
Perflan
Perfmid
Preflan
Prefmid
Spike
Tebulan
Tebuthiuron
Tiurolan

34025-32-6
$C_8H_7ClO_2$
170.60
Benzeneacetaldehyde, 4-chloro-α-hydroxy- (9CI)
Mandelaldehyde, p-chloro- (8CI)

34040-64-7
$C_9H_9ClO_2$
184.62
Benzoic acid, 4-(chloromethyl)-, methyl ester (9CI)
p-Toluic acid, α-chloro-, methyl ester (8CI)

34061-80-8
$C_{10}H_{20}O$
156.27
2-Pentene, 5-(pentyloxy)- (9CI)
Ether, 3-pentenyl pentyl (8CI)

34067-75-9
C_8H_{14}
110.20
Cyclopentene, 3-propyl- (8CI, 9CI)

34090-00-1
$(C_4H_{10}O_2.C_2H_2O_4)x$
Polymer NA
Ethanedioic acid, Polymer with 1,4-butanediol (9CI)
1,4-Butanediol-oxalic acid copolymer
1,4-Butanediol-oxalic acid polymer
1,4-Butanediol, Polymer with ethanedioic acid (9CI)
Oxalic acid, Polyester with 1,4-butanediol (8CI)

34099-73-5
$C_2H_7BO_3$
89.89
Ethyl borate
Boric acid, ethyl ester
UN 1176 [Ethyl borate]

34101-86-5
$C_{18}H_{22}$
238.38
Benzene, 1,1'-(1,2-ethanediyl)bis(3,4-dimethyl- (9CI)
Bibenzyl, 3,3',4,4'-tetramethyl- (8CI)
3,3',4,4'-Tetramethylbibenzyl
3,3',4,4'-Tetramethyldiphenylethane
3,3',4,4'-Tetramethyldiphenyl-1,2-ethane

34114-36-8
$C_{21}H_{16}N_2O_5S$
408.42
O=C(c(c(C(=O)c1c(Nc(c(ccc2)C)c2)cc(S(=O)(=O)O)c3N)ccc4)c4)c13
2-Anthracenesulfonic acid, 1-amino-9,10-dihydro-9,10-dioxo-4-o-toluidino- (8CI)
2-Anthracenesulfonic acid, 1-amino-9,10-dihydro-4-((2-methylphenyl)amino)-9,10-dioxo-

34123-59-6
$C_{12}H_{18}N_2O$
206.32
CC(C)c1ccc(NC(=O)N(C)C)cc1
Urea, 1,1-dimethyl-3-(p-isopropylphenyl)
Alon
Arelon
Belgran
CGA-18731
Graminon
HOE 16410
IP 50
N-(4-Isopropylphenyl)-N',N'-dimethylharnstoff (German)
N-4-Isopropylphenyl-N,N-dimethylurea
N-(4-Isopropylphenyl)-N',N'-dimethylurea
Isoproturon
Tolkan
Tolken

34128-01-3
$C_6H_8O_7.3Na$
MW:261.10
Butanedioic acid, (carboxymethoxy)-, trisodium salt (9CI)
Carboxymethyloxysuccinate trisodium salt
Carboxymethyloxysuccinic acid, trisodium salt
Trisodium carboxymethyloxysuccinate

34128-99-9

$C_{15}H_{18}F_3N_3O_5$
377.30
2-Furanmethanamine, N-(2,6-dinitro-4-(trifluoromethyl)phenyl)tetrahydro-N-propyl- (9CI)
CGA 14397
Furfurylamine, tetrahydro-N-propyl-N-(α,α,α-trifluoro-2,6-dinitro-p-tolyl)- (8CI)
GS 39985

34129-07-2
$C_{14}H_{16}F_3N_3O_5$
363.28
2-Furanmethanamine, N-(2,6-dinitro-4-(trifluoromethyl)phenyl)-N-ethyltetrahydro- (9CI)
Furfurylamine, N-ethyltetrahydro-N-(α,α,α-trifluoro-2,6-dinitro-p-tolyl)- (8CI)
GS 38946

34137-09-2
$C_{60}H_{87}N_3O_{12}$
1042.35
Benzenepropanoic acid, 3,5-bis(1,1-dimethylethyl)-4-hydroxy-, (2,4,6-trioxo-1,3,5-triazine-1,3,5(2H,4H,6H)-triyl)tri-2,1-ethanediyl ester (9CI)
3,5-Di-tert-butyl-4-hydroxyhydrocinnamic acid, 1,3,5-tris(2-hydroxyethyl)-s-triazine-2,4,6(1H,3H,5H)trione triester

34139-62-3
$C_7H_5NO_4.C_6H_{13}N$
266.28
Benzoic acid, 3-nitro-, Compd. with cyclohexanamine (1:1) (9CI)
Benzoic acid, m-nitro-, Compd. with cyclohexylamine (1:1) (8CI)

34149-92-3
$(C_2H_4O.C_2H_3Cl)x$
Polymer
Vinylchloride-acetate copolymer
Acetic acid, vinyl ester, chloroethylene copolymer
Disposlips
Polyvinyl acetate chloride
Polyvinylchloride acetate
Vinyl chloride acetate copolymer
Vinyl chloride vinyl acetate copolymer
Vinyl chloride - vinyl acetate copolymers

34157-48-7
$C_8H_{10}Br_3N_3O$
403.92
3,4,5-Tribromo-N,N,α-trimethyl-1H-pyrazole-1-acetamide

34170-84-8
$C_6H_{13}O_5P.K$
234.24
Acetic acid, (diethoxyphosphinyl)-, potassium salt (9CI)
Acetic acid, phosphono-, P,P-diethyl ester, potassium salt (8CI)

34176-71-1
$C_{11}H_{18}N_2$
178.28
1H-Cyclodecapyrazole, 4,5,6,7,8,9,10,11-octahydro- (8CI)

34197-05-2
$C_{17}H_{17}Cl_3O$
343.69
Ethane, 2-(p-ethoxyphenyl)-2-p-tolyl-1,1,1-trichloro
Benzene, 1-ethoxy-4-(2,2,2-trichloro-1-(4-methylphenyl)-ethyl)- (9CI)
2-(p-Ethoxyphenyl)-2-p-tolyl-1,1,1-trichloroethane
Methylethoxychlor

34197-16-5
$C_{16}H_{15}Cl_3OS$
361.72
Ethane, 2-(p-methoxyphenyl)-2-(p-(methylthio)phenyl)-1,1,1-trichloro
Benzene, 1-methoxy-4-(2,2,2-trichloro-1-(4-(methylthio)-phenyl)ethyl)- (9CI)
Methoxy-methiochlor
2-(p-Methoxyphenyl)-2-(p-(methylthio)phenyl)-1,1,1-trichloroethane

34197-26-7
$C_{17}H_{19}NO_4$
301.34
COc1ccc(cc1)C(C(C)N(=O)=O)c2ccc(OC)cc2
Benzene, 1,1'-(2-nitropropylidene)bis(4-methoxy- (9CI)
Propane, 1,1-bis(p-methoxyphenyl)-2-nitro- (8CI)

34197-98-3
$C_9H_{10}ClNO_2$
199.64
Benzene, 1-(2-chloropropyl)-4-nitro- (8CI, 9CI)
2-Chloro-1-(p-nitrophenyl)-propane

34202-30-7
$C_3H_5AlO_5$
148.05
Aluminum, (acetato-O)(formato-O)hydroxy- (9CI)
(Acetato-O)(formato-O)-hydroxyaluminum

34202-69-2
$C_3F_6O.3H_2O$
220.09
O.O.O.FC(F)(F)C(=O)C(F)(F)F
2-Propanone, hexafluoro-, trihydrate
Acetone, hexafluoro-, trihydrate
GC 7787
Hexafluoroacetone trihydrate
HFA

34214-79-4
$C_{16}H_{33}Cl$
260.89
Hexadecane, chloro- (8CI,9CI)

34214-82-9
$C_{11}H_{23}Cl$
190.76
Undecane, chloro- (6CI, 8CI,9CI)

34214-84-1
$C_{13}H_{27}Cl$
218.81
Tridecane, chloro- (6CI, 8CI,9CI)
Chlorotridecane

34214-86-3
$C_{15}H_{31}Cl$
246.86
Pentadecane, chloro- (8CI,9CI)

34231-26-0
$C_{20}H_{21}NO_5$
355.38
9,10-Anthracenedione, 1-amino-4-hydroxy-2-((6-hydroxyhexyl)oxy)- (9CI)

34244-14-9
$C_{16}H_9Cl$
236.70
Pyrene, 1-chloro
1-Chloropyrene
3-Chloropyrene

34256-82-1
Unknown
Unknown
Acetochlor
Acetamide, 2-chloro-N-(ethoxymethyl)-N-(2-ethyl-6-methylphenyl)- (9CI)

Acetochlore (French)
o-Acetotoluidide, 2-chloro-N-(ethoxymethyl)-6'-ethyl-
Caswell No. 003B
2-Chloro-N-(ethoxymethyl)-6'-ethylacet-o-toluidide
2-Chloro-N-(ethoxymethyl)-6'-ethyl-o-acetotoluidide
2-Chloro-N-(ethoxymethyl)-N-(2-ethyl-6-methylphenyl)-acetamide
EPA Pesticide Chemical Code 121601

34262-88-9
$C_8H_6O_4.xCo$
Unknown
1,4-Benzenedicarboxylic acid, cobalt salt (9CI)
Cobalt 1,4-benzenedicarboxylate
Cobalt terephthalate

34262-89-0
$C_8H_8Cu_2O_4$
295.24
1,4-Benzenedicarboxylic acid, copper salt (9CI)
Terephthalic acid, copper salt (8CI)

34314-32-4
$C_{12}H_6O_3$
198.18
Naphthalenedicarboxylic acid, cyclic anhydride (9CI)
Naphthalenedicarboxylic anhydride (8CI)

34314-35-7
$C_9H_{19}Cl$
162.70
Nonane, chloro- (6CI,8CI, 9CI)

34314-83-5
C_5H_8O
84.12
Furan, 2,3-dihydro-4-methyl- (9CI)

34314-84-6
$C_7H_{12}O$
112.17
Furan, 2,3-dihydro-4-(1-methylethyl)- (9CI)

34333-27-2
$C_8H_{14}ClN_5$
215.66
1,3,5-Triazine-2,4-diamine, 6-chloro-N-(1,1-dimethylethyl)-N'-methyl- (9CI)
GS 18183

34364-42-6
$C_{27}H_{25}O_4P$
444.47
Phosphoric acid, (1-methyl-1-phenylethyl)phenyl diphenyl ester (9CI)

34367-95-8
$C_{18}H_{14}N_4S$
318.38
N(=C(N=NC(c(c(N1C)ccc2)c2)=C1c(cccc3)c3)SC=4)C4
1H-Indole, 1-methyl-2-phenyl-3-(2-thiazolylazo)- (9CI)
1-Methyl-2-phenyl-3-(2-thiazolylazo)-1H-indole

34372-18-4
$C_{19}H_{12}Br_8O_4$
943.53
Phenol, 4,4'-(1-methylethylidene)bis(2,3,5,6-tetrabromo-, diacetate (9CI)
Phenol, 4,4'-isopropylidenebis(2,3,5,6-tetrabromo-, diacetate (8CI)

34372-72-0
$C_{38}H_{34}N_2O_3$
566.69
O=C(OC(c(c(Oc1ccc(N(Cc(ccc c2)c2)Cc(cccc3)c3)c4)cc(N(CC)CC)c5)c5)(c14)c6cccc7)c67
Spiro(isobenzofuran-1(3H),9'-(9H)xanthen)-3-one, 2'-(bis-(phenylmethyl)amino)-6'-(diethylamino)- (9CI)

34375-28-5
$C_3H_9NO_2$
91.10
OCCN(O)C
2-((Hydroxymethyl)amino)-ethanol
Caswell No. 494C
EPA Pesticide Chemical Code 099001
Ethanol, 2-(hydroxymethyl-amino)- (9CI)
2-(Hydroxymethylamino)ethanol
Troysan 174

34397-99-4
$(C_2H_4O)xC_{19}H_{36}O_2$
Polymer
Poly(oxy-1,2-ethanediyl), α-(1-oxo-9-octadecenyl)-ω-methoxy-, (Z)- (9CI)
α-(1-Oxo-9-octadecenyl)-ω-methoxypoly(oxy-1,2-ethanediyl), (Z)

34408-25-8
$C_8H_5Cl_2NO_4$
250.03
O=C(OC)c(c(c(N(=O)=O)cc1Cl)Cl)c1
Methyl 2,5-dichloro-3-nitro-benzoate
Benzoic acid, 2,5-dichloro-3-nitro-, methyl ester (9CI)
Methyl 3-nitro-2,5-dichloro-benzoate

34413-35-9
$C_8H_{10}N_2$
134.17
n(c(c(nc1)CCC2)C2)c1
Quinoxaline, 5,6,7,8-tetra-hydro- (9CI)
5,6,7,8-Tetrahydroquinoxaline

34442-00-7
$C_2H_7AsO_4$
170.00
Arsenic acid, dimethyl ester (9CI)

34444-01-4
$C_{18}H_{18}N_6O_5S_2$
462.54
Cn1nnnc1SCC2=C(N4C(SC2)C(NC(=O)C(O)c3ccccc3)C4=O)C(O)=O
5-Thia-1-azabicyclo(4.2.0)oct-2-ene-2-carboxylic acid, 7-((hydroxyphenylacetyl)-amino)-3- (((1-methyl-1H-tetrazol-5-yl)thio)methyl)-8-oxo-, (6R-(6-α,7-β(R*)))
Cefadole
Cefamandol
Cefamandole
L-Cefamandole
Cephadole
Cephamandole
Mandokef

34449-89-3
$C_8H_{10}F_9NO_3S$
371.22
O=S(=O)(N(CCO)CC)C(F)(F)C(F)(F)C(F)(F)C(F)(F)F
1-Butanesulfonamide, N-ethyl-1,1,2,2,3,3,4,4,4-nonafluoro-N-(2-hydroxyethyl)- (9CI)
N-Ethylnonafluoro-N-(2-hydroxymethyl)-1-butanesulfonamide

34454-97-2
$C_7H_8F_9NO_3S$
357.19
O=S(=O)(N(CCO)C)C(F)(F)C(F)(F)C(F)(F)C(F)(F)F
1-Butanesulfonamide, 1,1,2,2,3,3,4,4,4-nonafluoro-N-(2-hydroxyethyl)-N-methyl- (9CI)
Nonafluoro-N-(2-hydroxyethyl)-N-methyl-1-butanesulfonamide

34455-03-3
$C_{10}H_{10}F_{13}NO_3S$
471.23
O=S(=O)(N(CCO)CC)C(F)(F)C(F)(F)C(F)(F)C(F)(F)C(F)(F)C(F)(F)F
1-Hexanesulfonamide, N-ethyl-1,1,2,2,3,3,4,4,5,5,6,6,6-tridecafluoro-N-(2-hydroxyethyl)- (9CI)
N-Ethyltridecafluoro-N-(2-hydroxyethyl)-1-hexanesulfonamide

34455-29-3
$C_{15}H_{19}F_{13}N_2O_4S$
570.36
O=C(O)CN(CCCNS(=O)(=O)CCC(F)(F)C(F)(F)C(F)(F)C(F)(F)C(F)(F)C(F)(F)F)(C)C
1-Propanaminium, N-(carboxymethyl)-N,N-dimethyl-3-(((3,3,4,4,5,5,6,6,7,7,8,8,8-tridecafluorooctyl)sulfonyl)amino)-, hydroxide, inner salt (9CI)
1-Propanaminium, N-(carboxymethyl)-N,N-dimethyl-3-((1,1,2,2-tetrahydroperfluorooctyl)sulfonylamino)-, hydroxide, inner salt

34464-38-5
$C_{10}H_{22}$
142.28
Isodecane (8CI,9CI)

34464-40-9
C_9H_{20}
128.29
Isononane [UN 1920]
UN 1920 [Nonanes]

34464-43-2
$C_{11}H_{24}$
156.31
Isoundecane (8CI,9CI)

34465-46-8
$C_{12}H_2Cl_6O_2$
390.87
Hexachlorodibenzo-4-dioxin
Hexachlorodibenzodioxin
Hexachlorodibenzo-p-dioxin

34484-77-0
$C_{13}H_{18}N_2O$
218.30
Cisanilide
Caswell No. 380A
cis-2,5-Dimethyl-N-phenyl-1-pyrrolidinecarboxamide
2,5-Dimethyl-1-pyrrolidinecarboxanilide
EPA Pesticide Chemical Code 375300
Rowtate
1-Pyrrolidinecarboxamide, 2,5-dimethyl-N-phenyl-, cis- (9CI)

34490-93-2
$C_{17}H_{13}O_5.Na$
320.28
Sodium fumarin
3-(α-Acetonylfurfuryl)-4-hydroxycoumarin sodium salt
3-(α-Acetonylfurfuryl)-4-hydroxycoumarin, sodium salt
2H-1-Benzopyran-2-one, 3-(1-(2-furanyl)-3-oxobutyl)-4-hydroxy-, sodium salt (9CI)
Caswell No. 005A
Coumarin, 3-(α-acetonylfurfuryl)-4-hydroxy-, sodium salt
EPA Pesticide Chemical Code 086004
Fumasol
3-(1-Furyl-2-acetylethyl)-4-hydroxycoumarin, sodium salt

34494-03-6
$C_3H_8NO_5P.Na$
192.12
Sodium glyphosate
Glycine, N-(phosphonomethyl)-, monosodium salt (9CI)
MON 0459

34508-68-4
C_3H_6ClN
91.53
2-Propanimine, N-chloro- (9CI)
Ethylidenimine, N-chloro-1-methyl-

34562-31-7
$C_{18}H_{25}N$
255.40
N(c(cccc1)c1)(C=C(C=C2CC)CC)C2CCC
Pyridine, 3,5-diethyl-1,2-dihydro-1-phenyl-2-propyl- (9CI)
1-Phenyl-3,5-diethyl-2-propyl-1,2-dihydropyridine

34581-41-4
C_4H_5Cl
88.54
1,2-Butadiene, 3-chloro- (9CI)
3-Chloro-1,2-butadiene

34586-49-7
$C_8H_9FO_2S$
188.22
O=S(=O)(F)c(c(ccc1)CC)c1
Benzenesulfonyl fluoride, 2-ethyl- (9CI)
2-Ethylbenzenesulfonyl fluoride

34590-94-8
$C_7H_{16}O_3$
148.20
COC(C)COCC(C)O
Dipropylene glycol monomethyl ether
Arcosolv
1,4-Dimethyl-3,6-dioxa-1-heptanol
Dipropylene glycol methyl ether (ACGIH,OSHA)
Dowanol DPM
Dowanol 50B
DPGME
1-(2-Methoxyisopropoxy)-2-propanol
1(or 2)-(2-Methoxymethylethoxy)propanol
PPG-2 Methyl Ether
Propanol, (2-methoxymethylethoxy)-
Propanol, 1(or 2)-(2-methoxymethylethoxy)- (9CI)
UCAR Solvent 2LM

34593-75-4
$C_{10}H_{12}Cl_2O$
219.11
Phenol, 2,6-dichloro-4-(1,1-dimethylethyl)- (9CI)
NSC-407752

34622-58-7
$C_{12}H_{16}ClNOS$
257.80
Carbamic acid, diethylthio-, S-(o-chlorobenzyl) ester
B 3356
Carbamothioic acid, diethyl-, S-((2-chlorophenyl)methyl) ester (9CI)
S-(2-Chlorobenzyl)-N,N-diethylthiolcarbamate
Diethylthiocarbamic acid S-(o-chlorobenzyl) ester
Lanray
Orbencarb
Orthobencarb

34643-46-4
$C_{11}H_{15}Cl_2O_2PS_2$
345.25
Phosphorodithioic acid, O-(2,4-dichlorophenyl) O-ethyl S-propyl ester
Bay NTN 8629
Bideron
O-(2,4-Dichlorophenyl) O-ethyl S-propylphosphorodithioate
Dichlorpropaphos
O-Ethyl-O-(2,4-dichlorophenyl)-S-n-propyl-dithiophosphate
NTN-8629
Prothiophos
Tokuthion
Toyodan
Toyothion

34651-95-1
Unknown
Unknown
Mono-(trichloro)tetra(mono-potassium dichloro)-penta-s-triazine-trione

34652-54-5
C_3H_6BrCl
157.44
Propane, bromochloro- (9CI)

34664-47-6
$C_{32}H_{26}CoN_{10}O_8S_2.Na$
824.62
Cobaltate(1-), bis(3-((4,5-dihydro-3-methyl-5-oxo-1-phenyl-1H-pyrazol-4-yl)azo)-4-hydroxybenzenesulfonamidato(2-))-, sodium, (OC-6-22')- (9CI)

34681-10-2
$C_7H_{14}N_2O_2S$
190.29
CNC(=O)ON=C(C)C(C)SC
2-Butanone, 3-(methylthio)-, O-(N-methylcarbamoyl) oxime
Afiline
Butocarboxim (German)
Butocarboxime
CO 755
Drawin 755
3-(Methylthio)-2-butanone O-((methylamino)carbonyl) oxime
3-(Methylthio)butanone O-methylcarbamoyloxime
3-(Methylthio)-2-butanone O-(methylcarbamoyl)oxime
3-(Methylthio)-O-((methylamino)carbonyl)oxime-2-butanone
2-Methylthio-O-(N-methylcarbamoyl)-butanonoxim-3 (German)

34681-23-7
$C_7H_{14}N_2O_4S$
222.25
Butoxycarboxim
AI3-29332-X
2-Butanone, 3-(methylsulfonyl)-, O-((methylamino)carbonyl)oxime (10%)
2-Butanone, 3-methylsulfonyl-, O-(N-methylcarbamoyl)oxime
Butoxicarboxim (German)
Butoxycarboxime (French)
Caswell No. 120A
CO 859
EPA Pesticide Chemical Code 113001
3-Mesylbutanone O-methylcarbamoyloxime
3-(Methylsulfonyl)-2-butanone O-((methylamino)carbonyl)-oxime (9CI)
3-Methylsulfonylbutanone O-methylcarbamoyloxime
3-(Methylsulfonyl)-2-butanone O-(methylcarbamoyl)oxime (8CI)
2-Methylsulfonyl-O-(N-methylcarbamoyl)-butanon-(3)-oxim (German)
3-Methylsulphonylbutanone O-methylcarbamoyloxime
Plant Pin
3-(Sulfonyl)-O-((methylamino)-carbonyl)oxime-2-butanone

34689-46-8
$C_7H_8O.Na$
131.13
Sodium cresylate
Caswell No. 757A
Cresylic acid, sodium salt
Cresylic acid, sodium salt solution
EPA Pesticide Chemical Code 357200
Methylphenol sodium salt
Phenol, methyl-, sodium salt (9CI)
Sodium cresolate

34690-00-1
$C_{17}H_{44}N_3O_{15}P_5$
685.40
O=P(O)(O)CN(CCCCCCN(CCCCCCN(CP(=O)(O)O)CP(=O)(O)O)CP(=O)(O)O)CP(=O)(O)O
Phosphonic acid, (((phosphonomethyl)imino)bis(6,1-hexanediylnitrilobis(methylene)))tetrakis- (9CI)
Bishexamethylenetriamine, pentamethylenepentaphosphonic acid

34732-09-7
$C_6H_2Cl_4O_2S$
279.96
Benzenesulfonyl chloride, 2,3,4-trichloro- (9CI)

34761-82-5
$C_8H_{10}N_2O_2$
166.18
Cc1cc(N)cc(C)c1N(=O)=O
3,5-Dimethyl-4-nitroaniline
Benzenamine, 3,5-dimethyl-4-nitro-

34777-33-8
$C_{16}H_{11}N$
217.26
1H-Benzo(c)carbazole (9CI)

34783-40-9
$C_9H_{12}ClO_4P$
250.62
COP(=O)(OC)OC1=C(Cl)C2CC=CC12
Heptenophos
5-(O,O-Dimethylphosphoryl)-6-chlorobicyclo(3.2.0)-hepta-1,5-diene
HOE 2982
Hostaquick
Ragadan

34787-01-4
$C_{15}H_{16}N_2O_6S_2$
384.43
CC3(C)SC2C(NC(=O)C(C(O)=O)c1ccsc1)C(=O)N2C3C(O)=O
Ticarcillin
Ticarcilina (Spanish)
Ticarcilline (French)
Ticarcillinum (Latin)

34807-41-5
$C_{38}H_{38}O_{10}$
654.76
CC5C(OC(=O)C=CC=Cc1ccccc1)C2(OC7(OC2C6C3OC3(CO)C(O)C4(O)C(C=C(C)C4=O)C56O7)c8ccccc8)C(C)=C
Daphnetoxin, 12-((1-oxo-5-phenyl-2,4-pentadienyl)-oxy)-, (12-β(E,E))
Meserein
Mezerein

34816-53-0
$C_{12}H_4Cl_4O_2$
321.96
Dibenzo-p-dioxin, 1,2,7,8-tetrachloro
Dibenzo(b,e)(1,4)dioxin, 1,2,7,8-tetrachloro-
1,2,7,8-Tetrachlorodibenzo-p-dioxin
2,3,6,7-Tetrachlorodibenzo-dioxin

34819-62-0
$C_4H_3Cl_3$
157.43
1,2-Butadiene, 4,4,4-trichloro- (9CI)
4,4,4-Trichloro-1,2-butadiene

34825-93-9
$C_7H_{11}Br$
175.07
Cyclohexene, 3-(bromomethyl)- (9CI)

34831-02-2
$C_6H_6CuNO_6.H$
252.67
Copper (2+) NTA

Copper nitrilotriacetic acid
Cuprate(1-), (N,N-bis(carboxy-methyl)glycinato(3-)-N,O,O', O'')-, hydrogen, (T-4)-
Nitrilotriacetic acid and its salts
NTA, copper(2+) hydrogen complex

34831-03-3
$C_6H_6NNiO_6.H$
247.81
Nickel (2+) NTA
Nickel nitrilotriacetic acid
Nickelate(1-), (N,N-bis(car-boxymethyl)glycinato(3-)-N,O,O',O'')-, hydrogen, (T-4)- (9CI)
Nitrilotriacetic acid and its salts
NTA, nickel(2+) hydrogen complex

34863-74-6
$C_{11}H_{15}ClN_2S.ClH$
279.22
Methanimidamide, N'-(4-chloro-2-methylphenyl)-N-methyl-N-((methylthio)-methyl)-, monohydrochloride (9CI)
Hokupanon

34870-88-7
$C_{27}H_{42}O_6$
462.63
1,2,4-Benzenetricarboxylic acid, monodecyl monooctyl ester (9CI)

34883-05-1
$C_{10}H_{14}O_2$
166.22
Phenol, 3-methoxy-2,4,6-tri-methyl- (9CI)

34883-39-1
$C_{12}H_8Cl_2$
223.10
Clc1ccc(Cl)c(c1)c2ccccc2
2,5-Dichlorobiphenyl
2,5-Dichloro-1,1'-biphenyl

34883-41-5
$C_{12}H_8Cl_2$
223.10
3,5-Dichlorobiphenyl
3,5-Dichloro-1,1'-biphenyl

34883-43-7
$C_{12}H_8Cl_2$
223.10
Clc1ccc(cc1)c2ccccc2Cl
Biphenyl, 2,4'-dichloro
1,1'-Biphenyl, 2,4'-dichloro- (9CI)
2,4'-Dichlorobiphenyl

34911-46-1
Unknown
Unknown
para-Hydroxy-2-oxo-phenyl-acethydroxymic acid chloride

34913-07-0
$C_{26}H_{32}N_2O$
388.56
Pyrimidine, 5-(4-(pentyl-oxy)phenyl)-2-(4-pentyl-phenyl)- (9CI)

34970-00-8
CHBrClI
255.28
Methane, bromochloroiodo- (9CI)

34973-41-6
$C_4H_3Cl_7$
299.24
Butane, 1,1,2,2,3,4,4-hepta-chloro- (9CI)

34992-00-2
Unknown
Unknown
Nonyl phenol sulfide solution

35029-96-0
$CH_4S.1/2Pb$
151.71
Lead (II) methylthiolate
Bis(methylthio)lead
Methanethiol, lead(2+) salt (9CI)
NSC-68072

35045-02-4
$C_8H_{13}N_3OS$
199.30
1,2,4-Triazin-5(2H)-one, 6-(1,1-dimethylethyl)-3-(methylthio)
Deaminated Sencor
Deaminometribuzin
6-(1,1-Dimethylethyl)-3-(methylthio)-1,2,4-triazin-5(2H)-one

35065-27-1
$C_{12}H_4Cl_6$
360.86
Clc1cc(Cl)c(cc1Cl)c2cc(Cl)c(Cl)cc2Cl
Biphenyl, 2,2',4,4',5,5'-hexa-chloro
2,2',4,4'5,5'-Hexachlorobi-phenyl
2,2',4,4',5,5'-Hexachloro-1,1'-biphenyl (9CI)
2,4,5,2',4',5'-Hexachlorobi-phenyl

35065-28-2
$C_{12}H_4Cl_6$
360.88
Clc1cc(Cl)c(cc1Cl)c2ccc(Cl)c(Cl)c2Cl
2,2',3',4,4',5-Hexachlorobi-phenyl
2,3,4,2',4',5'-Hexachlorobi-phenyl

35065-29-3
$C_{12}H_3Cl_7$
395.32
1,1'-Biphenyl, 2,2',3,4,4',5,5'-heptachloro- (9CI)
2,2',3,4,4',5,5'-Hepta-chlorobiphenyl
2,2',3,4,4',5,5'-PCB

35065-30-6
$C_{12}H_3Cl_7$
395.32
1,1'-Biphenyl, 2,2',3,3',4,4',5-heptachloro- (9CI)
2,2',3,3',4,4',5-Hepta-chlorobiphenyl
2,2',3,3',4,4',5-PCB

35075-24-2
$C_{10}H_{22}O_3$
190.29
2-Propanol, 1-(3-butoxy-propoxy)- (9CI)
1-(3-Butoxypropoxy)-2-pro-panol

35076-92-7
$C_6H_{10}Br_2$
241.95
BrC(CCC(Br)C1)C1
Cyclohexane, 1,4-dibromo- (9CI)
1,4-Dibromocyclohexane

35082-49-6
$C_{29}H_{26}O_{10}$
534.52
Cercosporin
NSC-153111

35087-77-5
$C_6H_{12}O_7.xK$
Unknown
D-Gluconic acid, potassium salt (9CI)
Potassium D-gluconate

35103-34-5
$C_{10}H_{13}NO_2$
179.21
Acetamide, N-((4-methoxy-phenyl)methyl)- (9CI)

Acetamide, N-(p-methoxy-
benzyl)-

35112-28-8
$C_8H_6Cl_2O_2$
205.04
Benzoic acid, 2,4-dichloro-, methyl ester (9CI)

35141-30-1
$C_{10}H_{27}N_3O_3Si$
265.41
1,2-Ethanediamine, N-(2-aminoethyl)-N'-(3-(trimethoxysilyl)propyl)- (9CI)
(3-Trimethoxysilylpropyl)diethylenetriamine

35148-19-7
$C_{14}H_{26}O_2$
226.36
O=C(OCCCCCCCCC=CCC)C
9-Dodecen-1-ol, acetate, (E)-
AI3-33973
(E)-9-Dodecen-1-ol acetate
(E)-9-Dodecenyl acetate

35153-15-2
$C_{14}H_{28}O$
212.38
OCCCCCCCCC=CCCCC
9-Tetradecen-1-ol, (Z)-
AI3-35291
(Z)-9-Tetradecen-1-ol

35187-24-7
$C_{21}H_{18}$
270.39
Benz(a)anthracene, 4,7,12-trimethyl
4,7,12-Trimethylbenz(a)anthracene

35187-27-0
$C_{21}H_{18}$
270.39
Benz(a)anthracene, 7,10,12-trimethyl
7,10,12-Trimethylbenz(a)anthracene

35187-28-1
$C_{20}H_{16}$
256.36
Benz(a)anthracene, 7,11-dimethyl
7,11-Dimethylbenz(a)anthracene
8,10-Dimethyl-1,2-benzanthracene

35194-22-0
$C_{14}H_{26}O$
210.36
11-Dodecen-2-one, 7,7-dimethyl- (9CI)

35194-39-9
$C_{11}H_{20}O_2$
184.28
8-Nonenoic acid, ethyl ester (7CI,9CI)
Ethyl 8-nonenoate

35200-02-3
$C_5H_{10}OS_2$
150.27
O(C(=S)SC)C(C)C
Carbonodithioic acid, S-methyl O-(1-methylethyl) ester (9CI)

35200-79-4
$C_7H_{12}O$
112.17
Cyclopropane, (2-methoxy-1-methylethenyl)-, (Z)- (9CI)

35200-80-7
$C_7H_{12}O$
112.17
Cyclopropane, (2-methoxy-1-methylethenyl)-, (E)- (9CI)

35203-06-6
$C_{10}H_{13}N$
147.21
Benzenamine, 2-ethyl-6-methyl-N-methylene- (9CI)
2-Ethyl-6-methyl-N-methylenebenzenamine

35203-08-8
$C_{11}H_{15}N$
161.24
N(c(c(ccc1)CC)c1CC)=C
Benzenamine, 2,6-diethyl-N-methylene- (9CI)
2,6-Diethyl-N-methylenebenzenamine

35243-89-1
$C_6H_{10}Br_2O_2$
273.95
O(C1COC(Br)C(Br)C)C1
Oxirane, ((1,2-dibromopropoxy)methyl)- (9CI)
((1,2-Dibromopropoxy)methyl)oxirane

35245-80-8
$C_{12}HCl_9O_2$
496.18
Phenol, 2,3,4,5-tetrachloro-6-(pentachlorophenoxy)
2-Hydroxy-nonachlorodiphenyl ether
Nonachloropredioxin
2,3,4,5-Tetrachloro-6-(pentachlorophenoxy)phenol

35254-70-7
$C_{10}H_5ClN_2$
188.60
Propanedinitrile, ((chlorophenyl)methylene)- (9CI)

35289-89-5
$C_{13}H_{10}N_2O_4$
242.24
Carbamic acid, phenyl-, 3-nitrophenyl ester (9CI)
Carbanilic acid, m-nitrophenyl ester
3-Nitrophenyl carbanilate
3-Nitrophenyl phenylcarbamate
m-Nitrophenyl phenylcarbamate
Phenol, m-nitro-, carbanilate (6CI)

35294-62-3
$C_{18}H_{13}N_5O_6S_2.Na$
482.42
7-Benzothiazolesulfonic acid, 2-(4-((hexahydro-2,4,6-trioxo-5-pyrimidinyl)azo)phenyl)-6-methyl-, monosodium salt (9CI)
7-Benzothiazolesulfonic acid, 6-methyl-2-(4-((2,4,6-trihydroxypyrimidin-5-yl)azo)phenyl)-, sodium salt

35296-72-1
$C_4H_{10}O$
74.12
Butanol (9CI) (VAN)

35301-43-0
$C_6H_{12}O$
100.16
Cyclobutanol, 2-ethyl- (9CI)

35306-33-3
$C_{20}H_{22}N_2O_2.ClH$
358.90
Gelsemine, monohydrochloride
Gesemine hydrochloride

35318-10-6
$C_{44}H_{88}N_2S_4$
773.45
N(C(=S)SSC(N(CCCCCCCCCCCCCCCCCC)C(C)C)=S)(CCCCCCCCCCCCCCCCCC)C(C)C
Thioperoxydicarbonic diamide (((H2N)C(S))2S2), N,N'-bis-

(1-methylethyl)-N,N'-dioctadecyl- (9CI)

35331-58-9
$C_{11}H_{28}O_4Si_4$
336.69
7,9,11,13-Tetraoxa-6,8,10,12-tetrasilaspiro(5.7)tridecane, 8,8,10,10,12,12-hexamethyl- (9CI)
Hexamethyl(silacyclohexyl)cyclotetrasiloxane

35331-89-6
$C_{14}H_{19}N$
201.31
Benzenamine, N-(2-ethyl-2-hexenylidene)- (9CI)
N-(2-Ethyl-2-hexenylidene)aniline
N-(2-Ethyl-2-hexenylidene)benzenamine

35340-00-2
$C_9H_{12}O_4$
184.19
3-Furancarboxylic acid, 5-(methoxymethyl)-2-methyl-, methyl ester (9CI)

35367-38-5
$C_{14}H_9ClF_2N_2O_2$
310.70
O=C(NC(=O)c(c(F)ccc1)c1F)Nc(ccc(c2)Cl)c2
Urea, 1-(p-chlorophenyl)-3-(2,6-difluorobenzoyl)
Benzamide, N-(((4-chlorophenyl)amino)carbonyl)-2,6-difluoro-
N-(((4-Chlorophenyl)amino)carbonyl)-2,6-difluorobenzamide
1-(4-Chlorophenyl)-3-(2,6-difluorobenzoyl)urea
Diflubenzuron
Difluron
Dimilin
DU 112307
ENT 29,054
OMS 1804
PDD 6040I
PH 60-40
Philips-Duphar PH 60-40
TH 6040
Thompson-Hayward TH6040

35400-43-2
$C_{12}H_{19}O_2PS_3$
322.46
CCCSP(=S)(OCC)Oc1ccc(SC)cc1
Phosphorodithioic acid, O-ethyl O-(4-(methylthio)-phenyl) S-propyl ester
Bay-NTN-9306
Bolstar
O-Ethyl O-(4-(methylmercapto)-phenyl)-S-n-propylphosphorothionothiolate
O-Ethyl O-(4-(methylthio)-phenyl)phosphorodithioic acid S-propyl ester
O-Ethyl O-(4-(methylthio)-phenyl) S-propyl phosphorodithioate
Helothion
NTN 9306
Sulprofos (ACGIH,OSHA)

35404-55-8
$C_5H_{11}N_3O_2$
145.14
β-Alanine, N-(aminoiminomethyl)-N-methyl- (9CI)

35438-85-8
$C_{15}H_{16}N_2O_2$
256.29
Benzene, 1-isocyanato-4-((4-isocyanatocyclohexyl)methyl)- (9CI)

35441-13-5
$C_{18}H_{18}Cl_2N_2O_5S_2.Na$
500.37
Ethanesulfonic acid, 2-((4-(3-(4,5-dichloro-2-methylphenyl)-4,5-dihydro-1H-pyrazol-1-yl)phenyl)sulfonyl)-, sodium salt (9CI)

35464-94-9
$(C_{12}H_{22}O_4.C_{10}H_{22}O_2)x$
Polymer NA
Dodecanedioic acid, Polymer with 1,10-decanediol (9CI)
Decamethylene glycol-decamethylene dicarboxylic acid copolymer
Decamethylene glycol-dodecanedioic acid copolymer
Decamethylene glycol-α,ω-dodecanedioic acid polymer
1,10-Decanediol-dodecanedioic acid copolymer
1,10-Decanediol-dodecanedioic acid polymer
1,10-Decanediol-1,12-dodecanedioic acid polymer
1,10-Decanediol, Polymer with dodecanedioic acid (9CI)

35465-71-5
$C_{16}H_{12}$
204.27
Naphthalene, phenyl- (9CI)

35471-38-6
Unknown
Unknown
Potassium 4-chloro-2-cyclopentylphenate

35471-49-9
$C_{13}H_{11}ClO.K$
257.78
Potassium 2-benzyl-4-chlorophenate
o-Benzyl-p-chlorophenol potassium salt
Caswell No. 083A
EPA Pesticide Chemical Code 062202
Phenol, 4-chloro-2-(phenylmethyl)-, potassium salt (9CI)
Potassium o-benzyl-p-chlorophenate
Potassium o-benzyl-p-chlorophenolate
Potassium 2-benzyl-4-chlorophenolate

35488-17-6
$C_7H_4Br_4O_2$
439.72
Phenol, 2,3,4,5-tetrabromo-6-methoxy- (7CI, 9CI)
Tetrabromoguaiacol

35493-90-4
$C_{15}H_{33}NO.C_8H_{10}O_5$
429.59
7-Oxabicyclo(2.2.1)heptane-2,3-dicarboxylic acid, Compd. with N,N-dimethyl-1-tridecanamine N-oxide (1:1) (9CI)
TD 1874

35503-54-9
$C_7H_{11}N_3O_4$
201.16
1,3,5-Triazin-2(1H)-one, (hydroxyethyl)bis(hydroxymethyl)- (9CI)
Dimethylol hydroxyethyltriazone

35554-08-6
$C_{10}H_{17}N_7O_4.2ClH$
372.26
1H,10H-Pyrrolo(1,2-c)purine-10,10-diol, 3a,4,8,9-tetrahydro-2,6-diamino-4-(((aminocarbonyl) oxy)-methyl)-, dihydrochloride, (3as-(3a-α,4-α,10ar*))
Biclorhidrato de saxotoxina (Spanish)
Clam Poison Dihydrochloride
Gonyaulax Catenella Poison Dihydrochloride
Gonyaulax Toxic Dihydrochloride
Mussel Poison Dihydrochloride
Paralytic Shellfish Poison Di-

hydrochloride
Saxitoxin Dihydrochloride
Saxitoxin Hydrochloride
STX Dihydrochloride

35554-44-0
$C_{14}H_{14}Cl_2N_2O$
297.20
Clc2ccc(C(Cn1ccnc1)OCC=C)c(Cl)c2
1H-Imidazole, 1-(2-(2,4-dichlorophenyl)-2-(2-propenyloxy)ethyl)
(+-)-1-(β-(Allyloxy)-2,4-dichlorophenethyl)imidazole
Chloramizol
Eniloconazol (SP)
1-(2-(2,4-Dichlorophenyl)-2-(2-propenyloxy)ethyl)-1H-imidazole
1-(2-(2,4-Dichlorphenyl)-2-(2-propenyloxy)aethyl)-1H-imidazol (German)
Fungaflor
Imazalil
R 23979

35572-78-2
$C_7H_7N_3O_4$
197.17
o-Toluidine, 3,5-dinitro
2-ADNT
2-Amino-4,6-dinitrotoluene
3,5-Dinitro-o-toluidine
2-Methyl-3,5-dinitrobenzenamine

35576-91-1
H_2N_2O
46.02
Nitrosamide (9CI)
Nitrosamine
NSC-223080

35597-43-4
$C_{11}H_{22}N_3O_6P$
323.33
CC(NC(=O)C(C)NC(=O)C(N)CCP(=O)CO)C(O)=O
L-Alanine, γ-(hydroxymethylphosphinyl)-L-α-aminobutyryl-L-alanyl
Antibiotic SF 1293
γ-(Hydroxymethylphosphinyl)-L-α-aminobutyryl-L-alanyl-L-alanine
Phosphinothricylalanylalanine
SF 1293

35597-44-5
$C_5H_{12}NO_4P$
181.12
Butanoic acid, 2-amino-4-(hydroxymethylphosphinyl)-, (S)- (9CI)
Phosphinothricin

35600-63-6
$C_{10}H_{12}ClNO_2$
213.67
Carbamic acid, (chloromethyl)phenyl-, ethyl ester (9CI)

35607-66-0
$C_{16}H_{17}N_3O_7S_2$
427.48
COC3(NC(=O)Cc1cccs1)C2SCC(=C(N2C3=O)C(O)=O)COC(N)=O
5-Thia-1-azabicyclo(4.2.0)oct-2-ene-2-carboxylic acid, 3-(((aminocarbonyl)oxy)-methyl)-7- methoxy-8-oxo-7-((2-thienylacetyl)amino)-, (6R-cis)
Cefoxitin
Cephoxitin
CFX
Rephoxitin

35632-99-6
$C_{34}H_{32}N_{12}O_6S_2$
768.76
Benzenesulfonic acid, 2,2'-(1,2-ethenediyl)bis(5-((4-(methylamino)-6-(phenylamino)-1,3,5-triazin-2-yl)-amino)- (9CI)

35633-50-2
$C_6H_{11}NO_2$
129.15
N#CCCOCCOC
Propanenitrile, 3-(2-methoxyethoxy)- (9CI)
3-(2-Methoxyethoxy)propanenitrile
Methoxyethoxypropionitrile

35656-51-0
$C_{12}H_7ClO_2$
218.64
Dibenzo(b,e)(1,4)dioxin, chloro-(9CI)
Chlorodibenzo(b,e)(1,4)dioxin
Chlorodibenzo-p-dioxin

35657-77-3
$C_{17}H_{44}N_3O_{15}P_5.xNa$
Unknown
Phosphonic acid, (((phosphonomethyl)imino)bis(6,1-hexanediylnitrilobis(methylene)))tetrakis-, sodium salt (9CI)
Dihexylenetriaminepentakismethylenephosphonic acid, sodium salt

35661-56-4
$C_9H_9Cl_2NO_2$
234.08
Ethanol, 2,2-dichloro-, phenylcarbamate (9CI)
2,2-Dichloroethyl N-phenylcarbamate

35674-56-7
$C_{16}H_{14}N.Cl$
255.75
Isoquinolinium, 2-(phenylmethyl)-, chloride
AI3-51472
2-(Phenylmethyl)isoquinolinium chloride

35687-41-3
$C_5H_7N_3O_3$
157.13
1H-Imidazole, 5-methoxy-1-methyl-3-nitro- (9CI)
5-Methoxy-1-methyl-4-nitroimidazole

35687-42-4
$C_5H_7N_3O_3$
157.11
1H-Imidazole, 4-methoxy-2-methyl-5-nitro- (9CI)

35687-44-6
$C_6H_9N_3O_3$
171.16
1H-Imidazole, 5-methoxy-1,2-dimethyl-4-nitro- (9CI)

35691-65-7
$C_6H_6Br_2N_2$
265.96
N#CC(Br)(CCC#N)CBr
Glutaronitrile, 2-bromo-2-(bromomethyl)
2-Bromo-2-(bromomethyl)-glutaronitrile
1,2-Dibromo-2,4-dicyanobutane
Methyldibromoglutaronitrile
Pentanedinitrile, 2-bromo-2-(bromomethyl)-

35692-98-9
$C_9H_{12}O_3$
168.19
2-Cyclohexene-1,4-dione, 2-hydroxy-3,5,5-trimethyl- (9CI)

35693-92-6
$C_{12}H_7Cl_3$
257.55
Clc1cc(Cl)c(c(Cl)c1)c2ccccc2
1,1'-Biphenyl, 2,4,6-trichloro-(9CI)
Biphenyl, 2,4,6-trichloro-
2,4,6-PCB
2,4,6-Trichlorobiphenyl

35693-99-3
$C_{12}H_6Cl_4$
291.98
Clc1ccc(Cl)c(c1)c2cc(Cl)ccc2Cl
1,1'-Biphenyl, 2,2',5,5'-tetrachloro
2,2',5,5'-TCB
2,2',5,5'-Tetrachlorobiphenyl
2,5,2',5'-Tetrachlorobiphenyl

35694-04-3
$C_{12}H_4Cl_6$
360.88
2,3,5,2',3',5'-Hexachlorobiphenyl
2,2',3,3',5,5'-Hexachlorobiphenyl

35694-06-5
$C_{12}H_4Cl_6$
360.88
Clc1ccc(c(Cl)c1)c2cc(Cl)c(Cl)c(Cl)c2Cl
1,1'-Biphenyl, 2,2',3,4,4',5-hexachloro- (9CI)
2,2',3,4,4',5-Hexachlorobiphenyl
2,2',3,4,4',5-PCB

35694-08-7
$C_{12}H_2Cl_8$
429.77
Clc1cc(c(Cl)c(Cl)c1Cl)c2cc(Cl)c(Cl)c(Cl)c2Cl
1,1'-Biphenyl, 2,2',3,3',4,4',5,5'-octachloro- (9CI)
2,2',3,3',4,4',5,5'-Octachlorobiphenyl
2,2',3,3',4,4',5,5'-PCB

35723-89-8
$C_{30}H_{63}N$
437.83
Isodecanamine, N,N-diisodecyl- (9CI)
N,N-Diisodecylisodecanamine
Tri(isodecyl)amine
N,N,N-Tri(isodecyl)amine

35818-31-6
$C_{16}H_{20}O_5$
292.36
1H-2-Benzopyran-1-one, 3,4-dihydro-6,8-dihydroxy-3-((tetrahydro-6-methyl-2H-pyran-2-yl) methyl)-, (2R-(2-α(R*),6-β))
Asperentin
Cladosporin
Isocoumarin, 3,4-dihydro-6,8-dihydroxy-3-(6-methyltetrahydro-2H-pyran-2-yl)-

35822-46-9
$C_{12}HCl_7O_2$
425.28
Dibenzo-p-dioxin, 1,2,3,4,6,7,8-heptachloro
Dibenzo(b,e)(1,4)dioxin, 1,2,3,4,6,7,8-heptachloro-
Heptachlorodibenzo-p-dioxin
1,2,3,4,6,7,8-Heptachlorodibenzodioxin
1,2,3,4,6,7,8-Heptachlorodibenzo-para-dioxin

35832-11-2
Unknown
Unknown
Triethylamine picloram

35835-94-0
$C_{20}H_{20}P.C_2H_3O_2$
350.40
Phosphonium, ethyltriphenyl-, acetate (9CI)
Ethyltriphenylphosphonium acetate

35860-86-7
Unknown
Unknown
Nonylphenoxypolyethoxyethanol-iodine complex

35869-50-2
$C_8H_8Cl_2O_3$
223.06
Phenol, 3,4-dichloro-2,6-dimethoxy- (9CI)

35869-60-4
$C_{22}H_{24}N_3O_2.CH_3O_4S$
473.53
1H-Benzimidazolium, 2-(7-(diethylamino)-2-oxo-2H-1-benzopyran-3-yl)-1,3-dimethyl-, methyl sulfate (9CI)

35869-64-8
$C_{40}H_{23}Cl_3N_8O_8$
849.98
2-Naphthalenecarboxamide, N,N'-(2-chloro-1,4-phenylene)bis(4-((4-chloro-2-nitrophenyl)azo)-3-hydroxy- (9CI)

35875-13-9
$C_{12}H_{24}O_3$
216.32
Dodecanoic acid, 6-hydroxy- (9CI)
6-Hydroxydodecanoic acid
6-Hydroxylauric acid

35884-66-3
$C_{84}H_{84}NiO_{12}P_4$
1468.17
Nickel, tetrakis(tris(methylphenyl) phosphite-P)- (9CI)
Nickel, tetrakis(tritolyl phosphite)-
Tetrakis(tris(methylphenyl) phosphite-P)nickel

35884-76-5
$C_8H_{15}O_6PS_2$
302.31
Butanedioic acid, ((dimethoxyphosphinothioyl)thio)-, monoethyl ester (9CI)
Succinic acid, mercapto-, monoethyl ester, S-ester with O,O-dimethyl phosphorodithioate (8CI)

35884-77-6
C_8H_9Br
185.07
Xylyl bromide [UN 1701]
Benzene, bromodimethyl-
α-Bromoxylene
Bromure de xylyle (French)
Bromuro de xililo (Spanish)
UN 1701 [Xylyl bromide]

35915-19-6
$C_7H_4Cl_2O_2$
191.01
Benzoic acid, 2,4(or 2,5)-dichloro- (9CI)
Dichlorobenzoic acid
2,4(or 2,5)-Dichlorobenzoic acid

35915-22-1
$C_5H_{10}O_2$
102.13
Butanoic acid, methyl- (9CI)

35946-91-9
$C_{12}H_{18}O$
178.27
Phenol, 2,5-bis(1-methylethyl)- (9CI)
Phenol, 2,5-diisopropyl-

35948-25-5
$C_{12}H_9O_2P$
216.18
O=P(Oc(c(c1cccc2)ccc3)c3)c12
6H-Dibenz(c,e)(1,2)oxaphosphorin, 6-oxide (9CI)
9,10-Dihydro-9-oxa-10-phosphaphenanthrene 10-oxide

36011-19-5
$C_{28}H_{32}NO_7$
494.61
Cytochalasin-E

36016-24-7
$C_{14}H_{12}ClNO_2$
261.71

Benzamide, N-(4-chloro-phenyl)-N-hydroxy-3-methyl- (9CI)
N-p-Chlorophenyl-m-methylbenzohydroxamic acid

36016-27-0
$C_{13}H_9Cl_2NO_2$
282.13
Benzamide, 3-chloro-N-(4-chlorophenyl)-N-hydroxy- (9CI)

36016-30-5
$C_{13}H_9ClN_2O_4$
292.68
Benzamide, N-(4-chloro-phenyl)-N-hydroxy-3-nitro- (9CI)
N-p-(Chlorophenyl)-m-nitro-benzohydroxamic acid

36038-53-6
$C_4H_2Cl_4$
191.87
1,3-Butadiene, 1,1,4,4-tetra-chloro- (9CI)

36059-21-9
$C_8H_6Br_4$
421.75
Benzene, dimethyl-, tetrabromo deriv. (9CI)

36060-61-4
$C_{12}H_{24}N_2O$
212.32
N(=C(N(C1)CCO)CCCCCCC)C1
Caprylyl hydroxyethyl imida-zoline
Caprylic acid, aminoethyl-ethanolamine amide-imida-zoline
2-Heptyl-4,5-dihydro-1H-imida-zole-1-ethanol
2-Heptyl-2-imidazoline-1-ethanol
2-(2-Heptylimidazolin-1-yl)-ethanol
1H-Imidazole-1-ethanol, 2-heptyl-4,5-dihydro- (9CI)

36069-45-1
$C_{29}H_{48}NO_3$
458.78
Muldamine

36088-22-9
$C_{12}H_3Cl_5O_2$
356.42
Dibenzo(b,e)(1,4)dioxin, penta-chloro- (9CI)
Pentachlorodibenzo(b,e)(1,4)diox-in
Pentachlorodibenzo-p-dioxin

36104-80-0
$C_{19}H_{18}ClN_3O_3$
371.80
Camazepam
Albego
B 5333
Camazepamum (Latin)
Carbamic acid, dimethyl-, 7-chloro-2,3-dihydro-1-methyl-2-oxo-5-phenyl-1H-1,4-benzodiazepin-3-yl ester
7-Chloro-1,3-dihydro-3-(N,N-di-methylcarbamoyl)-1-methyl-5-phenyl-2H-1,4-benzodia-zepin-2-one
7-Chloro-1,3-dihydro-3-hydroxy-1-methyl-5-phenyl-1,4-benzodiazepin-2-one dimethylcarbamate
DEA No. 2749
3-N,N-Dimethylcarbamoyloxy-7-chloro-5-phenyl-1-methyl-1,3-dihydro-2H-1,4-benzo-diazepin-2-one
SB 5833

36112-95-5
$C_{13}H_{14}O_3$
218.27
OCC(O)COc1cccc2ccccc12
1,2-Propanediol, 3-(1-naph-thyloxy)
1-(2,3-Dihydroxypropoxy)naph-thalene
Naphthalene, 1-(2,3-dihydroxy-propoxy)-
3-(α-Naphthoxy)-1,2-propane-diol
Propanolol glycol

36150-73-9
$C_{24}H_{39}NO_5$
421.64
1-Phenanthrenecarboxylic acid, tetradecahydro-9-hydroxy-7-(2-(2-(methyl-amino)ethoxy)-2- oxoethyl-idene)-1,4a,8-trimethyl-, methyl ester
Erythrophleine
Norcassamidine

36220-29-8
$C_9H_{11}ClO_2$
186.64
Ethanol, 2-(4-chloro-2-methyl-phenoxy)- (9CI)
MCPE

36236-41-6
$C_{18}H_{20}Cl_3NO_2$
388.71
Benzenemethanamine, 4-eth-oxy-N-(4-ethoxyphenyl)-α-(tri-chloromethyl)- (9CI)

36311-34-9
$C_{16}H_{34}O$
242.45
Isocetyl alcohol
Isohexadecanol (9CI)
Unimul-G-16

36322-90-4
$C_{15}H_{13}N_3O_4S$
331.37
CN2C(=C(O)c1ccccc1S2(=O)=O)C(=O)Nc3ccccn3
2H-1,2-Benzothiazine-3-car-boxamide, 4-hydroxy-2-methyl-N-2-pyridinyl-, 1,1-dioxide
CP 16171
Feldene
4-Hydroxy-2-methyl-N-(2-pyridyl)-2H-1,2-benzothiazin-3-caboxyamid-1,1-dioxid (German)
4-Hydroxy-2-methyl-N-(2-pyridyl)-2H-1,2-benzothiazine-3-carboxamide-1,1-dioxide
Piroxicam

36335-67-8
$C_{13}H_{21}N_2O_4PS$
332.39
Phosphoramidothioic acid, N -(sec-butyl)-, O-ethyl O-(6-nitro-m-tolyl) ester
Butamifos
Cremart
O-Ethyl O-(3-methyl-6-nitro-phenyl) N-sec-butylphos-phorothioamidate
O-Ethyl-O-(5-methyl-2-nitro-phenyl)(1-methylpropyl)phos-phoramidothioate
H 26905
Hercules 26905
Kremart
Metacrefos
Phosphoramidothioic acid, (1-methylpropyl)-, O-ethyl O-(5-methyl-2-nitrophenyl) ester (9CI)
S 28
S 28 (Pesticide)
S 2846

36351-18-5
$C_{27}H_{33}N_2.C_2H_2O_4$.1/2C2O4
519.61
Ethanaminium, N-(4-((4-(diethylamino)-phenyl)phenylmethyl-ene)-2,5-cyclohexadien-1-y lidene)-N-ethyl-, ethanedioate, ethane-dioate(2:1:2) (9CI)
Brilliant Green Oxalate
Ethanedioic acid, ion(2-), bis(N-(4-((4-(diethyl-amino)phenyl)phenyl-

methylene)-2,5-cyclohexadien-1-ylidene)-N-ethylethanaminium), ethanedioate (1:2) (9CI)

36355-01-8
$C_{12}H_4Br_6$
627.62
Brc1cc(Br)c(c(Br)c1)c2c(Br)cc(Br)cc2Br
Biphenyl, hexabromo
HBB
Hexabromobiphenyl
NCI-C53634
Polybrominated biphenyl

36388-36-0
$C_{48}H_{74}O_4$
715.11
1,2-Benzenedicarboxylic acid, bis((tetradecahydro-1,4a-dimethyl-7-(1-methylethyl)-1-phenanthrenyl)methyl) ester (9CI)
Phthalic acid, tetrahydroabietyl alcohol diester

36404-30-5
$C_7H_6Cl_2O$
177.03
Benzene, 1,2-dichloro-4-methoxy- (9CI)

36409-57-1
$C_{16}H_{30}O_7S.2Na$
412.46
Butanedioic acid, ((dodecyloxy)sulfonyl)-, disodium salt (9CI)
Butanedioic acid, sulfo-, C-dodecyl ester, disodium salt
Disodium lauryl sulfosuccinate
Succinic acid, sulfo-, monododecyl ester, disodium salt (8CI)

36413-60-2
$C_7H_{12}O_6$
192.17
Cyclohexanecarboxylic acid, 1,3,4,5-tetrahydroxy-, (1α,3α,4α,5β)- (9CI)
NSC-1115
Quinic acid (VAN)

36417-14-8
$C_3H_3Br_3$
278.77
1-Propene, 1,1,3-tribromo- (9CI)
Propene, 1,1,3-tribromo- (6CI)
1,1,3-Tribromo-1-propene

36429-48-8
$C_{14}H_{14}O_3$
230.26
Phenol, 3-(2-phenoxyethoxy)- (9CI)

36443-68-2
$C_{34}H_{50}O_8$
586.77
Benzenepropanoic acid, 3-(1,1-dimethylethyl)-4-hydroxy-5-methyl-, 1,2-ethanediylbis(oxy-2,1-ethanediyl) ester (9CI)

36445-71-3
$C_{22}H_{30}O_7S_2.2Na$
516.59
Benzenesulfonic acid, decyl-(sulfophenoxy)-, disodium salt (9CI)
Decyl(sulfophenoxy)benzenesulfonic acid, disodium salt

36452-21-8
$C_3H_3N_3O_3.2Na$
175.04
1,3,5-Triazine-2,4,6(1H,3H,5H)-trione, disodium salt (9CI)
Disodium cyanurate
Disodium 1,3,5-triazine-2,4,6-(1H,3H,5H)-trione

36478-76-9
N_2O_8U
394.02
Uranium, bis(nitrato-O,O')-dioxo-, (OC-6-11)
Uranyl nitrate

36483-57-5
$C_5H_9Br_3O$
324.84
1-Propanol, 2,2-dimethyl-, tribromo deriv. (9CI)
2,2-Dimethyl-1-propanol tribromo deriv.
Tribromoneopentyl alcohol

36483-60-0
$C_{12}H_4Br_6O$
643.59
Benzene, 1,1'-oxybis-, hexabromo deriv. (9CI)
Hexabromodiphenyl ether
Hexabromodiphenyl oxide
Hexabromophenoxybenzene
1,1'-Oxybisbenzene hexabromo deriv.

36486-76-7
$(C_4H_6O_2)x$
Polymer NA
2-Oxetanone, 4-methyl-, Homopolymer (9CI)
β-Butyrolactone homopolymer
β-Butyrolactone polymer
Poly-β-butyrolactone
Poly(β-methylpropiolactone)
Poly(β-methyl-β-propiolactone)

36501-84-5
$C_{22}H_{44}N_2PbS_4$
672.06
Lead, bis(dipentylcarbamodithioato-S,S')-, (T-4)- (9CI)
(T-4)-Bis(dipentylcarbamodithioato-S,S')lead
Lead diamyldithiocarbamate

36511-35-0
$C_{12}H_6Br_4O_2$
501.79
(1,1'-Biphenyl)-ar,ar'-diol, tetrabromo- (9CI)

36521-89-8
$C_{42}H_{80}O_7$
697.09
Sorbitan, dioctadecanoate
AI3-00978
Sorbitan distearate

36536-46-6
$C_4H_6O_2$
86.09
2-Oxetanone, 4-methyl-, (+-)- (9CI)

36541-18-1
$C_{12}H_{16}$
160.26
1H-Indene, 2,3-dihydrotrimethyl- (9CI)

36541-21-6
$C_{13}H_{12}$
168.24
Acenaphthylene, 1,2-dihydromethyl- (9CI)

36563-47-0
$C_{12}H_9BrO$
249.11
Benzene, bromophenoxy- (9CI)
Ether, bromophenyl phenyl

36596-36-8
$C_{41}H_{28}N_6O_{15}S_4.4C_6H_{15}NO_3$
1010.23
2-Naphthalenesulfonic acid, 7,7'-(carbonyldiimino)bis-(4-hydroxy-3-((6-sulfo-2-naphthalenyl)azo)-, Compd. with 2,2',2''-nitrilotris-(ethanol) (1:4) (9CI)

36614-38-7
$C_7H_{17}O_2PS_3$
260.39
COP(=S)(OC)SCCSC(C)C
Phosphorodithioic acid, S-2-(isopropylthio)ethyl O,O-dimethyl ester
O,O-Dimethyl-S-2-(isopropylthio)ethylphosphorodithioate
Hodson
Hosalon
Hosdon
Hosdon Granule
S-2-Isopropylthioethyl O,O-dimethyl phosphorodithioate
Isothioate
Isothionate
Phosphorodithioic acid, O,O-dimethyl S-(2-((1-methylethyl)-thio)ethyl) ester

36616-52-1
$C_{14}H_{16}Cl_2O_2$
287.20
Acetic acid, 2-chloro-2-(3-chloro-4-cyclohexylphenyl)
Chloro(3-chloro-4-cyclohexylphenyl)acetic acid
α,3-Dichloro-4-cyclohexylbenzeneacetic acid
Fenclorac
WHR 539

36617-02-4
C_9H_9Br
197.08
Benzene, (2-bromocyclopropyl)- (7CI,9CI)
(2-Bromocyclopropyl)-benzene
1-Bromo-2-phenylcyclopropane

36627-56-2
$C_{11}H_{16}N_2O$
192.26
CN(C)C(=O)Nc1cc(C)cc(C)c1
Urea, N'-(3,5-dimethylphenyl)-N,N-dimethyl-(9CI)

36631-30-8
$C_{39}H_{66}O_6$
630.95
1,2,4-Benzenetricarboxylic acid, triisodecyl ester (9CI)

36637-18-0
$C_{17}H_{28}N_2O$
276.47
CCCN(CC)C(CC)C(=O)Nc1c(C)cccc1C
Butanamide, N-(2,6-dimethylphenyl)-2-(ethylpropylamino)-, (+-)
N-(2,6-Dimethylphenyl)-2-(ethylpropylamino)butanamide
Duranest
2-(Ethylpropylamino)-2',6'-butyroxylidide
Etidocaine

36643-28-4
$C_{12}H_{27}Sn$
290.06
CCCC[Sn](CCCC)CCCC
Stannylium, tributyl- (9CI)
Tin(1+), tributyl-, ion (8CI)

36653-82-4
$C_{16}H_{34}O$
242.50
OCCCCCCCCCCCCCCCC
1-Hexadecanol
Adol
Adol 52
Adol 54
Adol 520
Alcohol C-16
Atalco C
Cachalot C-50
Cachalot C-51
Cachalot C-52
Cetaffine
Cetal
Cetalol CA
Cetyl alcohol
Cetylic alcohol
Cetylol
CO-1670
CO-1695
Crodacol-CAS
Crodacol-CAT
Cyclal cetyl alcohol
Dytol F-11
Epal 16NF
Ethal
Ethol
Hexadecanol
n-Hexadecanol
Hexadecan-1-ol
Hexadecyl alcohol
n-Hexadecyl alcohol
Lorol 24
Loxanol K
Loxanol K Extra
Palmityl alcohol
Product 308

36668-45-8
$C_3H_5BrCl_2$
191.88
Propane, 3-bromo-1,1-dichloro-(9CI)

36671-85-9
$(C_{21}H_{41}NO_2)x$
Polymer
Carbamic acid, octadecyl-, ethenyl ester, Homopolymer (9CI)
Polyvinyl octadecyl carbamate

36673-16-2
$C_{18}H_{42}N_2O_8Ti$
462.41
Titanium, bis((2,2',2''-nitrilotris(ethanolato))(1-)-N,O)bis-(2-propanolato)- (9CI)
Bis(bis(2-hydroxyethyl)aminoethyl) diisopropyl titanate
Diisopropoxybis(2-(bis(2-hydroxyethyl)amino)ethoxy)-titanium
Isopropyl triethanolamine titanate
NSC-5285

36675-34-0
$C_{18}H_{38}O_{13}$
462.49
Hexaglycerol (9CI)

36678-45-2
$C_4H_2Br_6$
529.48
BrC(Br)C(=C(Br)C(Br)Br)Br
1,1,2,3,4,4-Hexabromo-2-butene
2-Butene, hexabromo-
2-Butene, 1,1,2,3,4,4-hexabromo- (9CI)
Hexabromo-2-butene

36687-98-6
$C_6H_{12}O_2$
116.16
2-Butanone, 3-methoxy-3-methyl- (9CI)

36687-99-7
$C_7H_{14}O_2$
130.19
2-Butanone, 3-ethoxy-3-methyl- (6CI,7CI,9CI)

36711-58-7
Unknown
Unknown
Nitrilotriacetic acid, manganese salt
NTA, manganese salt

36712-20-6
$C_9H_{18}O$
142.24
Furan, 3-butyltetrahydro-2-methyl-, trans- (9CI)
trans-2-Methyl-3-butyltetrahydrofuran

36724-43-3
$C_6H_{14}O_7S_2$
262.30
O=S(=O)(CCO)COCS(=O)(=O)CCO
Ethanol, 2,2'-(oxybis(methylenesulfonyl))bis- (9CI)
2,2'-(Oxybis(methylenesulfonyl))bisethanol

36727-29-4
$C_9H_{17}ClO$
176.69
O=C(CC(CC(C)(C)C)C)Cl
Hexanoyl chloride, 3,5,5-trimethyl- (9CI)
3,5,5-Trimethylhexanoyl chloride

36727-72-7
$C_6H_{14}O_3S_2$
198.31
O(CSCCO)CSCCO
Ethanol, 2,2'-(oxybis(methylenethio))bis- (9CI)
2,2'-(Oxybis(methylenethio))-bisethanol

36728-72-0
$C_{29}H_{50}$
398.72
A'-Neo-30-norgammacerane (9CI)

36731-23-4
$C_{11}H_{16}O$
164.25
Benzene, (1,1-dimethylethyl)-methoxy- (9CI)

36734-19-7
$C_{13}H_{13}Cl_2N_3O_3$
330.19
CC(C)NC(=O)N1CC(=O)N(C1=O)c2cc(Cl)cc(Cl)c2
1-Imidazolidinecarboxamide, 3-(3,5-dichlorophenyl)-N-(1-methylethyl)-2,4-dioxo
Chipco 26019
3-(3,5-Dichlorophenyl)-N-(1-methylethyl)-2,4-dioxo-1-imidazolidinecarboxamide
FA 2071
Glycophen
Glycophene
Iprodione
1-Isopropyl carbamoyl-3-(3,5-dichlorophenyl)-hydantoin
LFA 2043
MRC 910
NRC 910
Promidione
ROP 500 F
Rovral
RP 26019

36735-22-5
$C_{17}H_{11}ClF_4N_2S$
386.79
Quazepam
2H-1,4-Benzodiazepine-2-thione, 7-chloro-5-(2-fluorophenyl)-1,3-dihydro-1-(2,2,2-trifluoroethyl)- (9CI)
7-Chloro-5-(o-fluorophenyl)-1,3-dihydro-1-(2,2,2-trifluoroethyl)-2H-1,4-benzodiazepine-2-thione
7-Chloro-5-(2-fluorophenyl)-1,3-dihydro-1-(2,2,2-trifluoroethyl)-2H-1,4-benzodiazepine-2-thione
DEA No. 2881
Dormalin
NSC-309702
Quazepamum (Latin)
Sch 16134

36749-13-0
$C_7H_{16}O$
116.20
Pentane, 3-ethoxy- (9CI)
Ether, ethyl 1-ethylpropyl

36756-79-3
$C_{16}H_{25}NOS$
279.48
CCC(C)N(C(C)CC)C(=O)SCc1ccccc1
Carbamic acid, di-sec-butylthio-, S-benzyl ester
S-Benzyl N,N-di-sec-butyl thiocarbamate
N,N-Di-sec-butyl-S-benzylthiocarbamate
Drepamon
M-3432
Tiocarbazil

36783-34-3
$C_{17}H_{11}F_3N_2O_2$
332.29
Floctafenic acid
Benzoic acid, 2-((8-(trifluoromethyl)-4-quinolinyl)amino)-(9CI)

36788-39-3
$C_{18}H_{39}O_9P$
430.54
Propanol, oxydi-, phosphite (3:1)
Oxydipropanol phosphite (3:1)
Phosphine, tris(dipropylene glycol)-
Phosphorous acid, tris(dipropylene glycol) ester
Tris(dipropylene glycol)phosphine
Tris(dipropylene glycol)phosphonate

36791-04-5
$C_8H_{12}N_4O_5$
244.18
Ribavirin
ICN-1229
NSC-163039
Ribamidyl
Ribavirina (Spanish)
Ribavirine (French)
Ribavirinum (Latin)
RTC
RTCA
1-β-D-Ribofuranosyl-1,2,4-triazole-3-carboxamide
1-β-D-Ribofuranosyl-1H-1,2,4-triazole-3-carboxamide
1,2,4-Triazole-3-carboxamide, 1-β-D-ribofuranosyl-
1H-1,2,4-Triazole-3-carboxamide, 1-β-D-ribofuranosyl-(9CI)
Tribavirin
Viramid
Virazole

36812-13-2
$C_{12}H_{18}O$
178.27
Phenol, (1,1-dimethylethyl)dimethyl- (9CI)
(1,1-Dimethylethyl)dimethylphenol
Xylenol,tert-butyl

36839-67-5
$C_6H_{14}O$
102.18
Pentane, 3-methoxy- (9CI)
Ether, 1-ethylpropyl methyl (6CI,7CI)
1-Ethylpropyl methyl ether
3-Methoxypentane

36876-13-8
$C_{18}H_{22}$
238.37
Diisopropylbiphenyl
1,1'-Biphenyl, ar,ar'-bis(1-methylethyl)- (9CI)
ar,ar'-Bis(1-methylethyl)-1,1'-biphenyl
Dicumyl
Diisopropyl biphenyl

36877-68-6
$C_3H_3N_3O_2$
113.06
1H-Imidazole, nitro- (9CI)

36878-20-3
$C_{30}H_{47}N$
421.70
Benzenamine, ar-nonyl-N-(nonylphenyl)- (9CI)
ar-Nonyl-N-(nonylphenyl)-benzenamine

36888-99-0
$C_{16}H_9N_5O_6$
367.25
O=C(NC(=O)C(=C(NC(c1cccc2)=C(C(=O)NC(=O)N3)C3=O)c12)C4=O)N4
2,4,6(1H,3H,5H)-Pyrimidinetrione, 5,5'-(1H-isoindole-1,3(2H)-diylidene)bis- (9CI)
1,3-Di(2,4,6-trioxohexahydro-

5-pyrimidinylidene)isoindole

36894-69-6
$C_{19}H_{24}N_2O_3$
328.45
CC(CCc1ccccc1)NCC(O)c2ccc(O)c(c2)C(N)=O
Benzamide, 2-hydroxy-5-(1-hydroxy-2-((1-methyl-3-phenylpropyl)amino)ethyl)
AH 5158
3-Carboxamido-4-hydroxy-α-((1-methyl-3-phenylpropyl-amino)methyl)benzyl alcohol
5-(1-Hydroxy-2-(1-methyl-3-phenylpropylamino)ethyl)-salicylamide
Labetalol
Normodyne
ScH 15719W

36897-88-8
$C_{24}H_{23}N_3O_8S_2$
545.58
O=C(c(c(C(=O)c1c(N)c(S(=O)(=O)O)cc2Nc(cc(S(=O)(=O)NCCO)c(c3C)C)c3)ccc4)c4)c12
2-Anthracenesulfonic acid, 1-amino-9,10-dihydro-4-((3-(((2-hydroxyethyl)amino)-sulfonyl)-4,5-dimethyl-phenyl)amino)-9,10-dioxo-(9CI)
1-(3,4-Dimethyl-5-sulfethanol-amide anilino)-4-amino-3-anthraquinonesulfonic acid

36911-94-1
$C_{22}H_{20}$
284.42
Benz(a)anthracene, 6,8-diethyl
6,8-Diethylbenz(a)anthracene

36911-95-2
$C_{22}H_{20}$
284.42
Benz(a)anthracene, 8,12-di-ethyl
8,12-Diethylbenz(a)anthracene

36917-36-9
$C_9H_{13}N$
135.21
Pyridine, 4-ethyl-2,6-di-methyl- (9CI)
2,6-Dimethyl-4-ethylpyri-dine
4-Ethyl-2,6-dimethylpyri-dine
2,6-Lutidine, 4-ethyl- (6CI, 7CI)

36936-60-4
$(C_2H_4O)x(C_2H_4O)x(C_2H_4O)x$ $C_6H_{15}NO_3$
Polymer
Poly(oxy-1,2-ethanediyl), α,α',α''-(nitrilotri-2,1-ethanediyl)tris(ω-hydroxy-(9CI)
Triethanolamine ethoxylated

36945-03-6
$C_{17}H_{18}ClN_3$
299.78
Lergotrile
2-Chloro-6-methylergoline-8β-acetonitrile
Ergoline-8-acetonitrile, 2-chloro-6-methyl-, (8β)-
Lergotrilo (Spanish)
Lergotrilum (Latin)
79907

36960-22-2
$C_5H_{10}O_2$
102.13
2-Butanone, 1-hydroxy-3-methyl- (6CI,9CI)
1-Hydroxy-3-methyl-2-but-anone

36968-27-1
$C_{25}H_{20}N_4O_4$
440.43
O=C(Nc(c(OC)ccc1)c1)c(c(O)c(N=Nc(ccc(C(=O)N)c2)c2)c(c3ccc4)c4)c3
2-Naphthalenecarboxamide, 4-((4-(aminocarbonyl)-phenyl)azo)-3-hydroxy-N-(2-methoxyphenyl)- (9CI)
4-((4-(Aminocarbonyl)phenyl)-azo)-3-hydroxy-N-(2-methoxy-phenyl)-2-naphthalenecarbox-amide
2-Hydrosy-1-((4-carboxamido-phenyl)azo)-N-(2-methoxy-phenyl)-3-naphthalenecarbox-amide
2-Naphthalenecarboxamide, 3-hydroxy-4-((4-phenylcar-boxyamide)azo)-N-(2-meth-oxyphenyl)-

36986-04-6
$C_{21}H_{21}ClN_5O_2.Cl$
446.31
Pyridinium, 1-(2-((4-((2-chloro-4-nitrophenyl)azo)-phenyl)ethylamino)ethyl)-, chloride (9CI)

37020-93-2
Unknown
Unknown
Mercury cyanide, solid
UN 1636

37069-54-8
$C_{30}H_{22}N_6O_6S_2$
626.64
O=S(=O)(O)c(c(ccc1N(N=CC=2c(cccc3)c3)N2)C=Cc(c(S(=O)(=O)O)cc(N(N=CC=4c(cccc5)c5)N4)c6)c6)c1
Blankophor BHC
Benzenesulfonic acid, 2,2'-(1,2-ethenediyl)bis(5-(4-phenyl-2H-1,2,3-triazol-2-yl)- (9CI)
2,2'-(1,2-Ethenediyl)bis(5-(4-phenyl-1,2,3-triazol-2-yl)-benzenesulfonic acid)

37070-83-0
$C_{16}H_{15}NO_4$
285.30
Benzoic acid, 3-(((phenylamino)carbon-yl)oxy)-ethyl ester (9CI)
3-Ethoxycarbonylphenyl phenylcarbamate
3-Carbethoxyphenyl N-phenylcarbamate

37070-85-2
$C_{14}H_{10}N_2O_2$
238.25
Benzonitrile, 4-(((phenyl-amino)carbonyl)oxy)-(9CI)
4-Cyanophenyl phenylcar-bamate
4-Cyanophenyl N-phenyl-carbamate

37070-86-3
$C_{15}H_{13}NO_3$
255.28
Ethanone, 1-(4-(((phenyl-amino)carbonyl)oxy)-phenyl)- (9CI)
4-Acetylphenyl phenylcar-bamate
4-Acetylphenyl N-phenyl-carbamate

37070-87-4
$C_{14}H_{11}NO_3$
241.25
Benzaldehyde, 3-(((phenyl-amino)carbonyl)oxy)-(9CI)
3-Formylphenyl phenyl-carbamate
3-Formylphenyl N-phenyl-carbamate

37076-88-3
$C_{14}H_{11}NO_3$
241.25
Benzaldehyde, 4-(((phenyl-amino)carbonyl)oxy)-(9CI)
4-Formylphenyl phenylcar-bamate
4-Formylphenyl N-phenyl-carbamate

37077-84-2
$C_3H_3Cl_3$
145.41
1-Propene, 2,3,3-trichloro- (9CI)
Propene, 2,3,3-trichloro-

37099-12-0
$(C_{14}H_{14}O_6)x$
Polymer NA
1,2-Benzenedicarboxylic acid, bis(oxiranylmethyl) ester, Homopolymer (9CI)
Denacol EX 721
DGF 25
ED 5661
Gly-Cel A 100
Shodine 500
Shodine 508

37102-63-9
Unknown
Unknown
Heptylamine 2,4-dichlorophenoxyacetate

37102-74-2
$C_9H_8O_4$
180.16
3-Methylphthalate
3-Methyl-1,2-benzenedicarboxylic acid

37138-23-1
$C_{32}H_{24}Cl_2N_{10}O_6S_2.2Na$
825.56
Benzenesulfonic acid, 2,2'-(1,2-ethenediyl)bis(5-((4-chloro-6-(phenylamino)-1,3,5-triazin-2-yl)amino)-, disodium salt (9CI)
4,4'-Bis(4-anilino-6-chloro-s-triazin-2-yl)amino)-2,2'-stilbenedisulfonic acid, disodium salt
4,4'-Bis(2-chloro-4-anilino-1,3,5-triazin-6-yl)amino-stilbene-2,2'-disulfonic acid, disodium salt

37139-99-4
$C_{27}H_{48}N.Cl$
422.13
Olealkonium chloride
Benzenemethanaminium, N,N-dimethyl-N-9-octadecenyl-, chloride
Benzenemethanaminium, N,N-dimethyl-N-9-octadecenyl-, chloride, (Z)- (9CI)
Benzyloleyldimethylammonium chloride
N,N-Dimethyl-N-9-octadecenyl-benzenemethanaminium chloride
Incroquat O-50
Jordaquat JO-50
Jordaquat ODBAC
Oleyl dimethyl benzyl ammonium chloride
N-Oleyl-N,N-dimethylbenzyl-ammonium chloride

37187-22-7
Unknown
Unknown
2,4-Pentanedione peroxide
Acetyl acetone peroxide, Maximum concentration 40% in solution (DOT)
Acetyl acetone peroxide, With more than 9% active oxygen (DOT)
Acetyl acetone peroxide, With not more than 9% by weight active oxygen (DOT)
Trigonox 40
UN 2080 (DOT)

37199-66-9
Unknown
Unknown
Potassium polysulfide
Caswell No. 701
EPA Pesticide Chemical Code 076703
Potassium sulfide (9CI)

37203-41-1
Unknown
Unknown
Dobane 055 (9CI)

37203-85-3
$C_6H_4Cl_2N_2O_2.C_6Cl_5NO_2$
502.33
Benzenamine, 2,6-dichloro-4-nitro-, Mixt. with pentachloronitrobenzene (9CI)

37206-01-2
$C_{18}H_{32}O_{14}$
472.45
Cellulose, carboxymethyl methyl ether (9CI)

37206-20-5
Unknown
Unknown
Methyl isobutyl ketone peroxide
Isobutyl methyl ketone peroxide
Isobutyl methyl ketone peroxide, No more than 62% in solution
Methyl isobutyl ketone peroxide, Solution with more than 9% by weight active oxygen
Methyl isobutyl ketone peroxide, Solution with not more than 9% by weight active oxygen
Methyl isobutyl ketone reaction product with hydrogen peroxide
4-Methyl-2-pentanone peroxide
2-Pentanone, 4-methyl-, peroxide (9CI)
2-Pentanone, 4-methyl-, peroxide, With more than 9% by weight active oxygen
2-Pentanone, 4-methyl-, peroxide, With not more than 9% by weight active oxygen
Peroxido de metilisobutilcetona (Spanish)
Peroxyde de methylisobutylcetone (French)
Trigonox HM 80
UN 2126

37220-82-9
$C_{21}H_{40}O_4$
356.55
9-Octadecenoic acid (Z)-, ester with 1,2,3-propanetriol (9CI)

37224-57-0
Unknown
Unknown
Chromium potassium zinc oxide (9CI)

37224-61-6
$C_{14}H_{20}N_2O_5$
296.31
Isooctyldinitrophenol, isomer unspecified
Dinitroisooctylphenol
Isooctyldinitrophenol
Phenol, isooctyldinitro- (9CI)

37227-61-5
Unknown
Unknown
Nickel alloy, Ni,Be
Beryllium-nickel alloy
Nickel-beryllium alloy

37237-76-6
$(C_3H_6O)x(C_3H_6O)x(C_3H_6O)x\ C_{12}H_{20}O_6$
Polymer
Poly(oxy(methyl-1,2-ethanediyl)), α,α',α''-1,2,3-propanetriyltris(ω-(oxiranylmethoxy)- (9CI)
α,α',α''-1,2,3-Propanetriyltris(ω-(2,3-epoxypropoxy)-poly(oxypropylene))

37243-36-0
$(C_3H_3N)x$
Polymer
Polyacrylonitrile
Acrylonitrile-cellulose copolymer
Acrylonitrile-cellulose co-

polymers
Acrylonitrile-cellulose graft copolymer
Acrylonitrile-cellulose polymer
Cellulose-polyacrylonitrile copolymer
Cellulose-polyacrylonitrile graft copolymer
Cellulose, Polymer with 2-propenenitrile
Fiber A
Orlon (Trademark)

37275-41-5
C_6H_{12}
84.16
Pentene, methyl- (9CI)

37275-48-2
$C_{10}H_8N_2$
156.17
Bipyridine (9CI)

37281-58-6
$(C_2H_4O)xC_{15}H_{24}O$
Polymer NA
Poly(oxy-1,2-ethanediyl), α-(tert-nonylphenyl)-ω-hydroxy- (9CI)
tert-Nonylphenyl undecaethylene glycol ether

37286-64-9
$(C_3H_6O)x.CH_4O$
Polymer
Poly(oxy(methyl-1,2-ethanediyl)), α-methyl-ω-hydroxy
Dowfroth 250
Jeffox Ol 2700
Polypropylene glycol methyl ether
Polypropylene glycol monomethylether
Propylene oxide-methanol Adduct
Ucon LB-1715

37294-49-8
$C_{14}H_{26}O_7S.2Na$
384.40
Disodium isodecyl sulfosuccinate
Butanedioic acid, sulfo-, C-isodecyl ester, disodium salt (9CI)
Butanedioic acid, sulfo-, 4-isodecyl ester, disodium salt
Sulfobutanedioic acid, 4-isodecyl ester, disodium salt

37299-86-8
$C_{29}H_{29}N_2O_5.Cl.2Na$
567.03
Ethanaminium, N-(9-(2,4-dicarboxyphenyl)-6-(diethylamino)-3H-xanthen-3-ylidene)-N-ethyl-, chloride, disodium salt
Acid Red 388
Rhodamine WT

37300-23-5
$CrO_4.Zn.H_4O_2Zn.CrO_3$
382.78
Chromic acid, zinc salt, Compd. with zinc hydroxide and chromium oxide (9:1)
Zinc Yellow

37304-37-3
Fe.Mo.Ni
210.48
Nickel alloy, base, Ni,Fe,Mo (9CI)
Iron alloy, nonbase, Ni,Fe,Mo (9CI)
Molybdenum alloy, nonbase, Ni,Fe,Mo (9CI)

37304-88-4
$C_{13}H_{25}N_3O_4.C_8H_{16}N_2O_3$
475.59
Bioban P 1487
Bioban P-1487
Morpholine, 4,4'-(2-ethyl-2-nitro-1,3-propanediyl)bis-, Mixt. with 4-(2-nitrobutyl)-morpholine (9CI)
P 1487

37317-41-2
Unknown
Unknown
Polychlorinated biphenyl (Kanechlor 500)
Kanechlor 500
KC-500

37337-13-6
Unknown
Unknown
Chromated copper arsenate

37339-32-5
$C_{22}H_{29}NO_2$
339.47
Oc(c(C(=NO)c(cccc1)c1)cc(c2)CCCCCCCCC)c2
Methanone, (2-hydroxy-5-nonylphenyl)phenyl-, oxime (9CI)
5-Nonyl-2-hydroxybenzophenoxime

37340-60-6
Unknown
Unknown
Poly(oxy-1,2-ethanediyl), α-(nonylphenyl)-ω-hydroxy-, phosphate, sodium salt (9CI)
Alcohol ethoxylate, phosphate ester, sodium salt
Ethoxylated nonyl phenol, polyphosphates, sodium salt
Nonyl phenol, ethoxylated, phosphated, sodium salt
Nonylphenol ethoxylated, phosphated, sodium salt
Nonylphenoxypoly(ethyleneoxy)ethyl ester of phosphoric acid, sodium salt
Poly(oxy-1,2-ethanediyl)-α-(nonylphenyl)-ω-hydroxy-phosphate, sodium salt
Poly(oxy-1,2-ethanediyl)-α-(nonylphenyl)-ω-hydroxy-, phosphate, sodium salt
Poly(oxy-1,2-ethanediyl), α-(nonylphenyl)-, ω-hydroxy, phosphate, sodium salt
Poly(oxy-1,2-ethanediyl), α-(nonylphenyl)-ω-hydroxy-, phosphate, sodium salt
Poly(oxy-1,2-ethanediyl), α-(nonylphenyl)-ω-hydroxy, phosphate, sodium salt

37341-11-0
$C_{10}H_{19}N_5S.C_{10}H_{13}ClN_2O$
454.00
Urea, N'-(3-chloro-4-methylphenyl)-N,N-dimethyl-, Mixt. with N-(1,1-dimethylethyl)-N'-ethyl-6-(methylthio)-1,3,5-triazine-2,4-diamine (9CI)
Igran Special

37350-58-6
$C_{15}H_{25}NO_3$
267.41
COCCc1ccc(OCC(O)CNC(C)C)cc1
2-Propanol, 1-(4-(2-methoxyethyl)phenoxy)-3-((1-methylethyl)amino)-, (+-)
CGP 2175
H 93/26
Metoprolol
(+-)-Metoprolol

37353-59-6
Unknown
Unknown
Cellulose, hydroxymethyl ether (9CI)
WP 40 (Polysaccharide)

37353-62-1
Unknown
Unknown
Kanechlor 200 (9CI)
KC 200

37353-63-2
Unknown

Unknown
Polychlorinated biphenyl (Kanechlor 300)
Kanechlor 300
KC-300

37380-42-0
Unknown
Unknown
XAD-4 Resin
Amberlite XAD 4

37437-20-0
$C_7H_5NS_2.C_6H_{13}N$
266.42
2(3H)-Benzothiazolethione, Compd. with cyclohexanamine (1:1) (9CI)
AI3-19517

37439-34-2
$C_5H_2Cl_3NO.Na$
221.42
2(1H)-Pyridinone, 3,5,6-trichloro-, sodium salt (9CI)
3,5,6-Trichloro-2-pyridinol, sodium salt
3,5,6-Trichloro-2(1H)-pyridinone sodium salt

37442-55-0
$C_{10}H_{10}O$
146.19
3-Buten-2-one, 1-phenyl- (9CI)
Benzyl vinyl ketone
1-Phenyl-3-buten-2-one

37475-88-0
$C_9H_{12}O_3S.H_3N$
217.28
Ammonium cumenesulfonate
Benzenesulfonic acid, (1-methylethyl)-, ammonium salt (9CI)
(1-Methylethyl)benzenesulfonic acid, ammonium salt

37482-11-4
$C_2H_6OS.Na$
101.12
Ethanol, 2-mercapto-, monosodium salt (9CI)
2-Hydroxyethyl sodium sulfide
α-Hydroxyethylsulfide, monosodium salt
2-Mercaptoethanol monosodium salt
Sodium β-hydroxyethyl sulfide

37491-68-2
$C_7H_9NO_2$
139.17
1,2-Benzenediol, 4-(aminomethyl)
4-(Aminomethyl)-1,2-benzenediol
3,4-Dihydroxybenzylamine

37517-28-5
$C_{22}H_{43}N_5O_{13}$
585.70
NCCC(O)C(=O)NC2CC(N)C(OC1OC(CN)C(O)C(O)C1O)C(O)C2OC3OC(CO)C(O)C(N)C3O
D-Streptamine, O-3-amino-3-deoxy-α-D-glucopyranosyl-(1-6)-O-(6-amino-6-deoxy-α-D- glucopyranosyl-(1,4))-N-(4-amino-2-hydroxy-1-oxobutyl)-2-deoxy-, (S)
Amicacin
Amikacin
1-N-(L(-)-γ-Amino-α-hydroxybutyryl)kanamycin A
1-N-(L-(-)-4-Amino-2-hydroxybutyryl)kanamycin A
Antibiotic BB-K 8
BB-K 8

37529-30-9
$C_{16}H_{27}N$
233.39
Benzenamine, 4-decyl- (9CI)

37574-47-3
$C_{20}H_{12}O$
268.32
O1C3C1c5cc2ccccc2c6ccc4cccc3c4c56
Benzo(a)pyrene 4,5-oxide
Benzo(a)pyrene 4,5-epoxide
Benzo(1,2)pyreno(4,5-b)oxirene, 3b,4a-dihydro-
Benz(a)pyrene 4,5-oxide
BP 4,5-epoxide
BP 4,5-oxide

37658-95-0
$C_7H_{14}N_2O_4$
190.20
Rhizobitoxine

37677-14-8
$C_{13}H_{20}O$
192.30
O=CC(CCC(=C1)CCC=C(C)C)C1
3-Cyclohexene-1-carboxaldehyde, 4-(4-methyl-3-pentenyl)- (9CI)
3-Cyclohexene-1-carboxaldehyde, 4-(5-methyl-3-penten-1-yl)-
1-Formyl-4-isohexenyl-4-cyclohexene
1-Formyl-4-ixohexenyl-4-cyclohexene
Isohexenyl cyclohexenyl carboxaldehyde
4-(4-Methylpent-3-enyl)cyclohex-3-ene-1-carboxaldehyde
1-(4-Methyl-3-pentenyl)-1-cyclohexene-4-carboxaldehyde
4-(4-Methyl-3-pentenyl)-3-cyclohexenecarboxaldehyde
4-(4-Methyl-3-pentenyl)-3-cyclohexene-1-carboxaldehyde
4-(4-Methyl-3-penten-1-yl)-3-cyclohexen-1-carboxaldehyde
4-(4-Methyl-3-penten-1-yl)-3-cyclohexene-1-carboxaldehyde
1-(4-Methyl-3-pentenyl)-4-formyl-1-cyclohexene
Myrac aldehyde

37680-65-2
$C_{12}H_7Cl_3$
257.55
Clc1ccc(Cl)c(c1)c2ccccc2Cl
1,1'-Biphenyl, 2,2',5-trichloro- (9CI)
2,2',5-PCB
2,2',5-Trichlorobiphenyl

37680-66-3
$C_{12}H_7Cl_3$
257.55
Clc1ccc(c(Cl)c1)c2ccccc2Cl
1,1'-Biphenyl, 2,2',4-trichloro- (9CI)
2,2',4-PCB
2,2',4-Trichlorobiphenyl

37680-68-5
$C_{12}H_7Cl_3$
257.55
1,1'-Biphenyl, 2,3',5'-trichloro- (9CI)
2,3',5'-PCB
2,3',5'-Trichlorobiphenyl

37680-69-6
$C_{12}H_7Cl_3$
257.55
1,1'-Biphenyl, 3,3',4-trichloro- (9CI)
3,3',4-PCB
3,3',4-Trichlorobiphenyl

37680-73-2
$C_{12}H_5Cl_5$
326.44
Clc1ccc(Cl)c(c1)c2cc(Cl)c(Cl)cc2Cl
2,4,5,2',5'-Pentachlorobiphenyl

37704-51-1
$C_9H_{10}N_2O_2$
178.19
Isoxazole, 5,5'-(1,3-propanediyl)bis- (9CI)

37751-39-6
$C_{17}H_{15}ClN_2O$

298.76
Ciclazindol
10-(m-Chlorophenyl)-2,3,4,10-tetrahydropyrimido(1,2-a)-indol-10-ol
Ciclazindolum (Latin)
Pyrimido(1,2-a)indol-10-ol, 10-(3-chlorophenyl)-2,3,4,10-tetrahydro-
WY-23,409

37764-25-3
$C_8H_{11}Cl_2NO$
208.10
O=C(N(CC=C)CC=C)C(Cl)Cl
Acetamide, N,N-diallyl-2,2-dichloro
Acetamide, 2,2-dichloro-N,N-di-2-propenyl- (9CI)
Compound R-25788
N,N-Diallyldichloroacetamide
Dichlormid
N,N-Diallyl-2,2-dichloroacetamide
R-25788
Stauffer R-25788

37788-55-9
$C_6H_{10}N_2O$
126.15
N(C=CN1CC(O)C)=C1
1H-Imidazole-1-ethanol, α-methyl- (9CI)
1H-Imidazole-1-(2-propanol)
α-Methyl-1H-imidazole-1-ethanol

37853-59-1
$C_{14}H_8Br_6O_2$
687.68
O(c(c(cc(c1)Br)Br)c1Br)CCOc(c(cc(c2)Br)Br)c2Br
Benzene, 1,1'-(1,2-ethanediylbis(oxy))bis(2,4,6-tribromo

37853-61-5
$C_{17}H_{16}Br_4O_2$
571.93
O(c(c(cc(c1)C(c(cc(c(OC)c2Br)Br)c2)(C)C)Br)c1Br)C
Benzene, 1,1'-(1-methylethylidene)bis(3,5-dibromo-4-methoxy- (9CI)
4,4'-Isopropylidenebis(2,6-dibromomethoxybenzene)

37871-00-4
$C_{12}HCl_7O_2$
425.31
Heptachlorobenzo-p-dioxin

37913-89-6
$C_9H_7Cl_3O_3.Na$
292.50
Propanoic acid, 2-(2,4,5-trichlorophenoxy)-, sodium salt (9CI)
Caswell No. 739O
EPA Pesticide Chemical Code 082504
Silvex, sodium salt
2-(2,4,5-Trichlorophenoxy)propionic acid, sodium salt

37921-74-7
$C_{45}H_{51}N_3O_6S$
761.96
1H-Indole-7-carboxylic acid, 3-(1-(7-((hexadecylsulfonyl)-amino)-1H-indol-3-yl)-3-oxo-1H,3H-naphtho(1,8-cd)-pyran-1-yl)- (9CI)

37924-13-3
$C_{14}H_{12}F_3NO_4S_2$
379.39
Cc1cc(ccc1NS(=O)(=O)C(F)(F)F)S(=O)(=O)c2ccccc2
Methanesulfonamide, N-(4-phenylsulfonyl-o-tolyl)-1,1,1-trifluoro
Destun
MBR 8251
Perfluidone
1,1,1-Trifluoro-N-(2-methyl-4-(phenylsulfonyl)phenyl)-methanesulfonamide

37956-57-3
$(C_3H_4O_3)x.xNa$
Polymer NA
2-Propenoic acid, 2-hydroxy-, Homopolymer, sodium salt (9CI)

37971-36-1
$C_7H_{11}O_9P$
270.13
O=C(O)C(P(=O)(O)O)(CC(=O)O)CCC(=O)O
2-Phosphonobutane-1,2,4-tricarboxylic acid
1,2,4-Butanetricarboxylic acid, 2-phosphono- (9CI)
2-Phosphonobutane-1,2,4-tricarbonic acid
2-Phosphono-1,2,4-butanetricarboxylic acid

38006-74-5
$C_{14}H_{16}F_{17}N_2O_2S.Cl$
634.77
1-Propanaminium, 3-(((heptadecafluorooctyl)sulfonyl)-amino)-N,N,N-trimethyl-, chloride (9CI)
3-(((Heptadecafluorooctyl)sulfonyl)amino)-N,N,N-trimethyl-1-propanaminium chloride

38011-25-5
Unknown
Unknown
Disodium dihydroxyethyl ethylenediaminediacetate

38049-26-2
$C_{10}H_{18}O$
154.25
Cyclohexanol, 2-methyl-5-(1-methylethenyl)-, (1α,2β,5α)- (9CI)
Carveol, dihydro-

38051-10-4
$C_{13}H_{24}Cl_6O_8P_2$
582.99
O=P(OCCCl)(OCCCl)OCC(CCl)(CCl)COP(=O)(OCCCl)OCCCl
Phosphoric acid, 2,2-bis-(chloromethyl)-1,3-propanediyl tetrakis(2-chloroethyl) ester (9CI)

38103-06-9
$C_{31}H_{20}O_8$
520.50
O=C(OC(=O)c1cc(Oc(ccc(c2)C(c(ccc(Oc(ccc(c3C(=O)O4)C4=O)c3)c5)c5)(C)C)c2)cc6)c16
1,3-Isobenzofurandione, 5,5'-((1-methylethylidene)bis(4,1-phenyleneoxy))bis- (9CI)
2,2-Bis(4-(3,4-dicarboxyphenoxy)phenyl)propane dianhydride

38116-59-5
$C_6H_4Cl_2N_2O_3$
223.00
2-Pyridinecarboxylic acid, 4-amino-3,5-dichloro-6-hydroxy- (9CI)

38122-80-4
$C_{53}H_{58}O_9$
839.04
2-Naphthalenecarboxylic acid, 1-hydroxy-4-(1-(4-hydroxy-3-(methoxycarbonyl)-1-naphthalenyl)-3-oxo-1H,3H-naphtho(1,8-cd)pyran-1-yl)-6-(octadecyloxy)- (9CI)
1-Hydroxy-4-(1-(4-hydroxy-3-(methoxycarbonyl)-1-naphthalenyl)-3-oxo-1H,3H-naphtho(1,8-cd)pyran-1-yl)-6-(octadecyloxy)-2-naphthalenecarboxylic acid

38134-94-0
$C_{29}H_{44}O_4$
456.67
O=C(O)c(c(O)c(c(c1)cc(OCCCCCCCCCCCCCCCCC)c2)c2)c1

2-Naphthalenecarboxylic acid, 1-hydroxy-6-(octadecyloxy)- (9CI)
1-Hydroxy-6-octadecyloxy-2-naphthoic acid

38171-97-0
$C_{16}H_{18}$
210.32
Naphthalene, 1-methyl-4-(1-methyl-2-butenyl)- (9CI)

38178-38-0
$C_{12}H_6Cl_2O_2$
253.08
Dibenzo-p-dioxin, 1,6-dichloro
Dibenzo(b,e)(1,4)dioxin, 1,6-dichloro-
1,6-Dichlorodibenzo-para-dioxin

38178-99-3
$C_{13}H_4Cl_6O$
388.89
1,2,4,5,7,8-Hexachloro(9H)-xanthene
1,2,4,5,7,8-HCX
9H-Xanthene, 1,2,4,5,7,8-hexachloro-

38185-06-7
$C_6H_3ClN_2O_7S.K$
321.70
Benzenesulfonic acid, 4-chloro-3,5-dinitro-, potassium salt (9CI)
4-Chloro-3,5-dinitrobenzenesulfonic acid potassium salt
NSC-123956

38194-50-2
$C_{20}H_{17}FO_3S$
356.43
CC2=C(CC(O)=O)c1cc(F)ccc1C2=Cc3ccc(cc3)S(C)=O
1H-Indene-3-acetic acid, 5-fluoro-2-methyl-1-((4-(methylsulfinyl)phenyl)-methylene)-, (Z)
Arthrocine
Clinoril
3-Oxo-2,3-dihydro-1H-indene-1-acetic acid
cis-5-Fluoro-2-methyl-1-((4-(methylsulfinyl)phenyl)-methylene)-1H-indene-3-acetic acid
(Z)-5-Fluoro-2-methyl-1-((p-(methylsulfinyl)phenyl)-methylene)-1H-indene-3-acetic acid
MK 231
Sulindac

38232-01-8
$C_{31}H_{62}O_2$
466.83
Hentriacontanoic acid (9CI)

38232-63-2
Hg_2N_6
485.24
Mercury-azide
Mercurous azide (DOT)

38239-27-9
C_3H_5N
55.07
Ethenamine, N-methylene- (9CI)

38260-54-7
$C_{10}H_{17}N_2O_4PS$
292.32
CCOc1cc(OP(=S)(OC)OC)nc(CC)n1
Phosphorothioic acid, O,O-dimethyl O-(6-ethoxy-2-ethyl-4-pyrimidinyl) ester
Ekamet
Ekamet G
Ekamet ULV
ENT 29,126
O-6-Ethoxy-2-ethylpyrimidin-4-yl O,O-dimethyl phosphorothioate
O-(6-Ethoxy-2-ethyl-4-pyrimidinyl) O,O-dimethyl phosphorothioate
Etrimfos
OMS 1806
San 197 I
Satisfar

38261-35-7
$C_9H_{10}N_4O$
190.18
1,3,5-Triazin-2(1H)-one, 6-amino-3,4-dihydro-4-phenyl- (9CI)
s-Triazin-2(1H)-one, tetrahydro-4-imino-6-phenyl- (8CI)

38274-67-8
$C_{13}H_{20}F_6O_2$
322.29
1,3-Dioxolane, 4-ethyl-5-hexyl-2,2-bis(trifluoromethyl)-, trans- (9CI)

38299-08-0
$C_6H_6N_4O.Na$
173.11
(1,2,4)Triazolo(1,5-a)pyrimidin-7-ol, 5-methyl-, sodium salt (9CI)

38304-52-8
$C_{22}H_{32}N_4O_8$
480.49
O=C(N(C(=O)C1(C)C)CC(OCC(O2)C2)CN(C(=O)C(N3CC(O4)C4)(C)C)C3=O)N1CC(O5)C5
2,4-Imidazolidinedione, 3,3'-(2-(oxiranylmethoxy)-1,3-propanediyl)bis(5,5-dimethyl-1-(oxiranylmethyl)- (9CI)
1,3-Bis(1-oxiranylmethyl-5,5-dimethylhydantoin-3-yl)-2-oxiranylmethoxypropane

38338-57-7
$C_{11}H_{13}Cl_2O_2PS$
311.17
Phosphonothioic acid, ethyl-, O-(1-(2,4-dichlorophenyl)-ethenyl) O-methyl ester (9CI)
WL 26738

38353-82-1
$C_{15}H_{18}N_2O_2$
258.31
O(c(cccc1)c1)CC(O)CNc(cccc2N)c2
2-Propanol, 1-((3-aminophenyl)amino)-3-phenoxy- (9CI)
1-((3-Aminophenyl)amino)-3-phenoxy-2-propanol

38355-75-8
$(C_{15}H_{16}O_2.C_4H_4O_4.C_3H_6O)x$
Polymer
2-Butenedioic acid (Z)-, Polymer with 4,4'-(1-methylethylidene)bis(phenol) and methyloxirane (9CI)
Bisphenol A, propylene oxide, fumaric acid polymer
2-Butenedioic acid, (Z)-, 4,4'-(1-methylethylidene)bisphenol, methyloxirane polymer

38379-99-6
$C_{12}H_5Cl_5$
326.44
Clc1ccc(Cl)c(c1)c2c(Cl)ccc(Cl)c2Cl
2,2',3,5',6-Pentachlorobiphenyl

38380-01-7
$C_{12}H_5Cl_5$
326.44
Clc1ccc(c(Cl)c1)c2cc(Cl)c(Cl)cc2Cl
1,1'-Biphenyl, 2,2',4,4',5-pentachloro- (9CI)
2,2',4,4',5-PCB
2,2',4,4',5-Pentachlorobiphenyl

38380-02-8

$C_{12}H_5Cl_5$
326.44
Clc1ccc(Cl)c(c1)c2ccc(Cl)c(Cl)c2Cl
1,1'-Biphenyl, 2,2',3,4,5'-pentachloro- (9CI)
2,2',3,4,5'-PCB
2,2',3,4,5'-Pentachlorobiphenyl

38380-03-9
$C_{12}H_5Cl_5$
326.44
2,3,3',4',6-Pentachlorobiphenyl

38380-04-0
$C_{12}H_4Cl_6$
360.88
Clc1cc(Cl)c(cc1Cl)c2c(Cl)ccc(Cl)c2Cl
1,1'-Biphenyl, 2,2',3,4',5',6-hexachloro- (9CI)
2,2',3,4',5',6-Hexachlorobiphenyl
2,2',3,4',5',6-PCB

38380-05-1
$C_{12}H_4Cl_6$
360.88
Clc1ccc(c(Cl)c1Cl)c2c(Cl)ccc(Cl)c2Cl
1,1'-Biphenyl, 2,2',3,3',4,6'-hexachloro- (9CI)
2,2',3,3',4,6'-Hexachlorobiphenyl
2,2',3,3',4,6'-PCB

38380-07-3
$C_{12}H_4Cl_6$
360.86
Clc1ccc(c(Cl)c1Cl)c2ccc(Cl)c(Cl)c2Cl
1,1'-Biphenyl, 2,2',3,3',4,4'-hexachloro
HCB
2,2',3,3',4,4'-Hexachlorobiphenyl
2,3,4,2',3',4'-Hexachlorobiphenyl

38380-08-4
$C_{12}H_4Cl_6$
360.86
1,1'-Biphenyl, 2,3,3',4,4',5-hexachloro
2,3,3',4,4',5-Hexachlorobiphenyl
2,3,3',4,4',5-Hexachloro-1,1'-biphenyl
2,3,4,5,3',4'-Hexachlorobiphenyl
3,4,2',3',4',5'-Hexachlorobiphenyl

38393-92-9
$C_{11}H_{14}O$
162.23
1H-Inden-1-ol, 2,3-dihydro-3,3-dimethyl- (9CI)
3,3-Dimethyl-1-indanol
1-Indanol, 3,3-dimethyl- (7CI)

38393-97-4
$C_{15}H_{22}$
202.34
1H-Indene, 5-(1,1-dimethylethyl)-2,3-dihydro-1,1-dimethyl- (9CI)

38411-22-2
$C_{12}H_4Cl_6$
360.86
Clc1ccc(Cl)c(c1Cl)c2c(Cl)ccc(Cl)c2Cl
1,1'-Biphenyl, 2,2',3,3',6,6'-hexachloro
2,3,6,2',3',6'-Hexachlorobiphenyl
2,2',3,3',6,6'-Hexachloro-1,1'-biphenyl

38411-25-5
$C_{12}H_3Cl_7$
395.32
1,1'-Biphenyl, 2,2',3,3',4,5,6'-heptachloro- (9CI)
2,2',3,3',4,5,6'-Heptachlorobiphenyl
2,2',3,3',4,5,6'-PCB

38420-60-9
$C_{10}H_{11}N_2O.HO_4S$
272.27
7-Benzofurandiazonium, 2,3-dihydro-2,2-dimethyl-, hydrogen sulfate (9CI)

38421-40-8
$C_{40}H_{42}O_3Si_4$
683.11
Tetrasiloxane, 1,3,5,7-tetramethyl-1,1,3,5,7,7-hexaphenyl- (9CI)

38421-90-8
$C_{12}H_{22}O_2$
198.31
O=C(OCCCCC=CCCCC)C
5-Decen-1-ol, acetate, (E)- (9CI)
AI3-35123
(E)-5-Decenyl acetate
(E)-5-Decen-1-yl acetate

38444-73-4
$C_{12}H_7Cl_3$
257.55
Clc1ccccc1c2c(Cl)cccc2Cl
1,1'-Biphenyl, 2,2',6-trichloro- (9CI)
2,2',6-PCB
2,2',6-Trichlorobiphenyl

38444-76-7
$C_{12}H_7Cl_3$
257.55
1,1'-Biphenyl, 2,3',6-trichloro- (9CI)
2,3',6-PCB
2,3',6-Trichlorobiphenyl

38444-77-8
$C_{12}H_7Cl_3$
257.55
Clc1ccc(cc1)c2c(Cl)cccc2Cl
1,1'-Biphenyl, 2,4',6-trichloro- (9CI)
2,4',6-PCB
2,4',6-Trichlorobiphenyl

38444-78-9
$C_{12}H_7Cl_3$
257.55
Clc1cccc(c1Cl)c2ccccc2Cl
1,1'-Biphenyl, 2,2',3-trichloro- (9CI)
2,2',3-PCB
2,2',3-Trichlorobiphenyl

38444-81-4
$C_{12}H_7Cl_3$
257.55
Clc1cccc(c1)c2cc(Cl)ccc2Cl
1,1'-Biphenyl, 2,3',5-trichloro- (9CI)
2,3',5-PCB
2,3',5-Trichlorobiphenyl

38444-84-7
$C_{12}H_7Cl_3$
257.55
Clc1cccc(c1)c2cccc(Cl)c2Cl
1,1'-Biphenyl, 2,3,3'-trichloro- (9CI)
2,3,3'-PCB
2,3,3'-Trichlorobiphenyl

38444-85-8
$C_{12}H_7Cl_3$
257.55
Clc1ccc(cc1)c2cccc(Cl)c2Cl
1,1'-Biphenyl, 2,3,4'-trichloro- (9CI)
2,3,4'-PCB
2,3,4'-Trichlorobiphenyl

38444-86-9
$C_{12}H_7Cl_3$
257.55
1,1'-Biphenyl, 2',3,4-trichloro- (9CI)
2',3,4-PCB
2',3,4-Trichlorobiphenyl

38444-87-0
$C_{12}H_7Cl_3$
257.55
Clc1cccc(c1)c2cc(Cl)cc(Cl)c2
1,1'-Biphenyl, 3,3',5-trichloro- (9CI)
3,3',5-PCB
3,3',5-Trichlorobiphenyl

38444-88-1
$C_{12}H_7Cl_3$
257.55
1,1'-Biphenyl, 3,4',5-trichloro- (9CI)
3,4',5-PCB
3,4',5-Trichlorobiphenyl

38444-90-5
$C_{12}H_7Cl_3$
257.55
Clc1ccc(cc1)c2ccc(Cl)c(Cl)c2
1,1'-Biphenyl, 3,4,4'-trichloro- (9CI)
3,4,4'-PCB
3,4,4'-Trichlorobiphenyl

38444-93-8
$C_{12}H_6Cl_4$
291.99
Clc1cccc(c1Cl)c2cccc(Cl)c2Cl
1,1'-Biphenyl, 2,2',3,3'-tetrachloro- (9CI)
2,2',3,3'-PCB
2,2',3,3'-Tetrachlorobiphenyl

38455-77-5
Unknown
Unknown
Tin(II) chromate

38471-49-7
$C_{15}H_{32}O_2$
244.42
Ethanol, 2-(tridecyloxy)- (9CI)

38483-28-2
$CH_2N_2O_6$
138.02
Methylene glycol dinitrate
Dinitrate de methylene glycol (French)
Dinitrato de metilenglicol (Spanish)
Methanediol, dinitrate
Methylene dinitrate

38512-20-8
Unknown
Unknown
Di-(1-naphthoyl)peroxide

38571-73-2
$C_6H_{11}Cl_3O_3$
237.52
ClCOCC(COCCl)OCCl
Propane, 1,2,3-tris(chloromethoxy)
Glycerol(tri(chloromethyl))ether
Tris-1,2,3-(chloromethoxy)propane

38613-77-3
$C_{68}H_{92}O_4P_2$
1035.42
Phosphonous acid, (1,1'-biphenyl)-4,4'-diylbis-, tetrakis(2,4-bis(1,1-dimethylethyl)phenyl) ester (9CI)

38620-92-7
$C_{17}H_{14}$
218.30
Naphthalene, (phenylmethyl)- (9CI)

38621-44-2
$C_{22}H_{42}O_6S$
434.64
9-Octadecenoic acid, (sulfooxy)-, 1-butyl ester (9CI)
1-Butyl (sulfooxy)-9-octadecenoate
(Sulfooxy)-9-octadecenoic acid, 1-butyl ester

38622-18-3
$C_{12}H_{12}N_2$
184.23
Hydrazine, diphenyl- (9CI)

38622-51-4
$C_{17}H_{22}$
226.36
Naphthalene, heptyl- (9CI)

38638-05-0
$C_{27}H_{33}O_4P$
452.53
Phosphoric acid, nonylphenyl diphenyl ester (9CI)
Nonylphenyl diphenyl phosphate

38640-62-9
$C_{16}H_{20}$
212.33
Diisopropylnaphthalene
Bis(isopropyl)naphthalene
Bis(1-methylethyl)naphthalene
Diisopropyl naphthalene
Naphthalene, bis(1-methylethyl)- (9CI)
Naphthalene, diisopropyl-

38641-16-6
$C_{22}H_{32}$
296.50
Naphthalene, dodecyl- (9CI)
Dodecylnaphthalene

38641-94-0
$C_3H_9N.C_3H_8NO_5P$
228.22
Glycine, N-(phosphonomethyl)-, Compd. with 2-propanamine (1:1)
Glycel
Glyphosate isopropylamine salt
Mono-isopropylammoniova sul (Czech)
Mon 39
Mon 139
Nitosorg
Roundup
Utal

38653-34-8
$C_9H_{14}O_2$
154.21
1,4-Benzodioxin, octahydro-2-methylene-, trans- (9CI)

38668-48-3
$C_{13}H_{21}NO_2$
223.31
OC(C)CN(c(ccc(c1)C)c1)CC(O)C
2-Propanol, 1,1'-((4-methylphenyl)imino)bis- (9CI)
N,N-Bis(2-hydroxypropyl)-p-toluidine
1,1'-((4-Methylphenyl)imino)-bis-2-propanol
1,1-(p-Tolylimino)dipropan-2-ol

38714-47-5
$CO_3.H_{12}N_4Zn$
193.50
Zinc(2+), tetraammine-, (T-4)-, carbonate (1:1) (9CI)
(T-4)-Tetraamminezinc(2+) carbonate (1:1)
Tetramminezinc(2+), carbonate

38721-71-0
$C_7H_5Cl_3$
195.47
Dichlorobenzyl chloride
Benzene, dichloro(chloromethyl)- (9CI)
Dichloro(chloromethyl)benzene
NSC-30678
Toluene, α, ar, ar-trichloro-

38727-55-8
$C_{16}H_{22}ClNO_3$
311.84
CCOC(=O)CN(C(=O)CCl)c1c(CC)cccc1CC
Glycine, N-(chloroacetyl)-N-(2,6-diethylphenyl)-, ethyl ester

Antor
Bay NNT 6867
N-Chloroacetyl-N-(2,6-diethylphenyl)glycine ethyl ester
Diethatyl ethyl
N-(Chloroacetyl)-N-(2,6-diethylphenyl)glycine ethyl ester
H 22234
Hercules 22234

38766-64-2
$C_{16}H_{16}Cl_3NO_2$
360.67
N-(α-Trichloromethyl-4-methoxybenzyl)-4-methoxyaniline
4-Methoxy-N-(4-methoxyphenyl)-α-(trichloromethyl)benzenemethanamine
N-(α-Trichloromethyl-p-methoxybenzyl)-p-methoxyaniline

38775-38-1
$C_{16}H_{33}ClO_2S$
324.96
O=S(=O)(CCCCCCCCCCCCCCCC)Cl
1-Hexadecanesulfonyl chloride (9CI)
Hexadecylsulfonyl chloride
NSC-93798

38821-53-3
$C_{16}H_{19}N_3O_4S$
349.44
CC1=C(N3C(SC1)C(NC(=O)C(N)C2=CCC=CC2)C3=O)C(O)=O
5-Thia-1-azabicyclo(4.2.0)oct-2-ene-2-carboxylic acid, 7-((amino-1,4-cyclohexadien-1-ylacetyl)amino)-3-methyl-8-oxo-, (6R-(6-α,7-β(R*)))
Cefradine
Cephradin
Cephradine
Sefril
Velosef

38827-35-9
$C_8H_8Cl_3NO_2S$
288.58
2,3,5-Trichloro-4-(propylsulfonyl)pyridine
Caswell No. 882FF
Dowicil A-40
EPA Pesticide Chemical Code 102801
Pyridine, 2,3,5-trichloro-4-(n-propylsulfonyl)-
2,3,5-Trichloro-4-(n-propylsulfonyl)-pyridine

38842-14-7
$C_{15}H_{10}F_5NO$
315.24
Benzamide, 2,3,4,5,6-pentafluoro-N-(2-phenylethyl)- (9CI)

38848-76-9
$C_{19}H_{40}N_2O_2$
328.53
CCCCCCCCCCCCCC(=O)[N-][N+](C)(C)CC(C)O
Hydrazinium, 1-(2-hydroxypropyl)-1,1-dimethyl-2-(1-oxotetradecyl)-, hydroxide, inner salt (9CI)

38888-98-1
$C_{14}H_{14}$
182.27
Benzene, (phenylethyl)- (9CI)
(Phenylethyl)benzene

38926-85-1
$C_7H_5Br_3O_2$
360.83
Phenol, 2,3,4-tribromo-6-methoxy- (9CI)
3,4,5-Tribromoguaiacol

38945-27-6
$C_6H_8O_7$
192.13
Butanedioic acid, (carboxymethoxy)- (9CI)

38954-75-5
$C_{17}H_{34}O_2$
270.46
O(C1COC1)CCCCCCCCCCCCCC
Oxirane, ((tetradecyloxy)methyl)- (9CI)
(Myristyloxymethyl)oxirane
((Tetradecyloxy)methyl)oxirane

38963-91-6
$C_{15}H_{26}O_3$
254.37
Cyclohexanecarboxylic acid, 4-(1,5-dimethyl-3-oxohexyl)-, (4(R)-cis)- (9CI)
cis-Dihydrotodomatuic acid

38964-22-6
$C_{12}H_6Cl_2O_2$
253.08
Dibenzo-p-dioxin, 2,8-dichloro
Dibenzo(b,e)(1,4)dioxin, 2,8-dichloro-
2,8-Dichlorodibenzodioxin
2,8-Dichlorodibenzo-para-dioxin

38998-75-3
$C_{12}HCl_7O$
409.31
Dibenzofuran, heptachloro- (9CI)

39001-02-0
$C_{12}Cl_8O$
443.76
Clc3c(Cl)c(Cl)c1c(oc2c(Cl)c(Cl)c(Cl)c(Cl)c12)c3Cl
Octachlorodibenzofuran

39004-94-9
$C_{13}H_{11}N_2O_7PS$
370.29
COP(=S)(Oc1ccc(cc1)N(=O)=O)Oc2ccc(cc2)N(=O)=O
Phosphorothioic acid, O,O-bis(p-nitrophenyl) O-methyl ester
Bis-methylparathion (Czech)
O,O-Bis-p-nitrofenyl-O-methylester kyseliny thiofosforecne (Czech)
O-Methyl-O,O-bis-(p-nitrofenyl)thiofosfat (Czech)

39028-58-5
$C_{10}H_{18}O_2$
170.25
2H-Pyran-3-ol, 6-ethenyltetrahydro-2,2,6-trimethyl-, trans- (9CI)
Linalool oxide C

39073-07-9
$C_{12}H_6Br_2O_2$
341.99
Dibenzo(b,e)(1,4)dioxin, 2,7-dibromo- (9CI)
Dibenzo-p-dioxin, 2,7-dibromo-

39076-02-3
$C_6H_{13}NO_2$
131.18
Carbamic acid, (1-methylpropyl)-, methyl ester (9CI)

39083-26-6
$C_4H_3Cl_3$
157.43
1,3-Butadiene, 1,2,3-trichloro-, (Z)- (9CI)

39098-01-6
$C_{12}H_{12}N_2O_3$
232.23
Benzamide, N-(1,1-dimethyl-2-propynyl)-4-nitro- (9CI)

39108-81-1
$C_{13}H_{15}NO_2$
217.26
Benzamide, N-(1,1-dimethyl-2-propynyl)-4-methoxy- (9CI)

39108-91-3
$C_{13}H_{15}NO$
201.26
Benzamide, N-(1,1-dimethyl-2-propynyl)-4-methyl- (9CI)

39148-24-8
$C_6H_{18}O_9P_3.Al$
354.13
Phosphonic acid, monoethyl ester, aluminum salt (3:1)
Aliette
Aluminum phosethyl
Aluminum tris(O-ethyl phosphonate)
Efosite Aluminum
Epal
Fosetyl Aluminum
LS 74783
Mikal
Phosethyl Aluminum
32545 RP

39156-41-7
$C_7H_{10}N_2O.H_2O_4S$
824.75
COc1ccc(N)cc1N.OS(O)(=O)=O
m-Phenylenediamine, 4-methoxy-, sulfate
Anisole, 2,4-diamino-, hydrogen sulfate
Anisole, 2,4-diamino-, sulfate
BASF Ursol SLA
1,3-Benzenediamine, 4-methoxy-, sulfate
1,3-Benzenediamine, 4-methoxy, sulfate (1:1) (9CI)
C.I. Oxidation Base 12A
C.I. 76051
2,4-DAA sulfate
2,4-Diaminoanisole sulphate
2,4-Diaminoanisole sulfate
2,4-Diamino-anisol sulphate
2,4-Diamino-1-methoxybenzene
1,3-Diamino-4-methoxybenzene sulphate
2,4-Diamino-1-methoxybenzene sulphate
2,4-Diaminosole sulphate
Durafur Brown MN
Fouramine BA
Fourrine 76
Fourrine SLA
Furro SLA
4-Methoxy-1,3-benzenediamine sulfate
4-Methoxy-1,3-benzenediamine sulfate (1:1)
4-Methoxy-1,3-benzenediamine sulphate
4-Methoxy-m-phenylenediamine sulfate
p-Methoxy-m-phenylenediamine sulphate
4-Methoxy-m-phenylenediamine sulphate
4-MMPD Sulphate
Nako TSA
NCI-C01989
Oxidation Base 12A
Pelagol BA
Pelagol Grey
Pelagol Grey SLA
Pelagol SLA
Renal SLA
Ursol SLA
Zoba SLE

39156-49-5
$C_{17}H_{28}O_3S$
312.47
Benzenesulfonic acid, 4-undecyl- (9CI)

39196-18-4
$C_9H_{18}N_2O_2S$
218.35
O=C(ON=C(C(C)(C)C)CSC)NC
2-Butanone, 3,3-dimethyl-1-(methylthio)-, o-((methylamino)carbonyl)oxime
Dacamox
Diamond Shamrock DS-15647
3,3-Dimethyl-1-(methylthio)-2-butanone-O-((methylamino)-carbonyl)oxime
DS-15647
ENT 27,851
RCRA waste number P045
Thiofanocarb
Thiofanox

39201-33-7
$C_5H_7NO_2$
113.12
OC(=O)CCCC#N
Butanoic acid, 4-cyano- (9CI)
Butyric acid, 4-cyano- (7CI)
4-Cyanobutanoic acid
4-Cyanobutyric acid
γ-Cyanobutyric acid

39224-65-2
$C_6H_3Cl_2NO_3$
207.99
Phenol, 4,5-dichloro-2-nitro- (9CI)

39227-28-6
$C_{12}H_2Cl_6O_2$
390.84
Clc3ccc2Oc1c(Cl)c(Cl)c(Cl)c(Cl)c1Oc2c3Cl
Dibenzo-p-dioxin, hexachloro
HCDD
Hexachlorodibenzo-p-dioxin

39227-53-7
$C_{12}H_7ClO_2$
218.64
Clc2cccc3Oc1ccccc1Oc23
Dibenzo-p-dioxin, 1-chloro
1-Chlorodibenzodioxin
1-Chlorodibenzo-p-dioxin
Dibenzo(b,e)(1,4)dioxin, 1-chloro-

39227-54-8
$C_{12}H_7ClO_2$
218.64
Clc3ccc2Oc1ccccc1Oc2c3
Dibenzo-p-dioxin, 2-chloro
2-Chlorodibenzo-para-dioxin
Dibenzo(b,e)(1,4)dioxin, 2-chloro-

39227-58-2
$C_{12}H_5Cl_3O_2$
287.52
Clc3cc(Cl)c2Oc1ccccc1Oc2c3Cl
Dibenzo-p-dioxin, 1,2,4-trichloro
Dibenzo(b,e)(1,4)dioxin, 1,2,4-trichloro-
1,2,4-Trichlorodibenzodioxin
1,2,4-Trichlorodibenzo-para-dioxin

39227-61-7
$C_{12}H_3Cl_5O_2$
356.40
Clc3ccc2Oc1c(Cl)c(Cl)c(Cl)c(Cl)c1Oc2c3
Dibenzo-p-dioxin, 1,2,3,4,7-pentachloro
1,2,3,4,7-Pentachlorodibenzo-dioxin
1,2,3,4,7-Pentachlorodibenzo-para-dioxin

39227-62-8
$C_{12}H_2Cl_6O_2$
390.84
Dibenzo-p-dioxin, 1,2,4,6,7,9-hexachloro
Dibenzo(b,e)(1,4)dioxin, 1,2,4,6,7,9-hexachloro-
1,2,4,6,7,9-Hexachlorodibenzo-dioxin
1,2,4,6,7,9-Hexachlorodibenzo-para-dioxin

39277-47-9
$C_{12}H_{14}Cl_2O_3.C_{12}H_{13}Cl_3O_3$
588.76
Acetic acid, 2,4-dichlorophenoxy-, butyl ester and 2,4,5-trichlorophenoxyacetic acid (45.5%: 48.2%)
Agent Orange
2,4-D n-butyl ester Mixed with 2,4,5-T n-butyl ester (1:1)
2,4,5-T n-butyl ester Mixed with 2,4-d n-butyl ester

39278-82-5
Unknown
Unknown
BP 1100 (9CI)

39283-72-2
$C_{10}H_{11}N_3OS.C_9H_8Cl_2O_3$
456.33
Propanoic acid, 2-(2,4-dichlorophenoxy)-, Mixt. with N-2-benzothiazolyl-N,N'-dimethylurea (9CI)
Tribunil-Combi

39292-53-0
$C_{11}H_{12}$
144.22
Naphthalene, dihydromethyl- (9CI)

39300-45-3
$C_{18}H_{24}N_2O_6$
364.44
CCCCCCCCc1cc(N(=O)=O)c(OC(=O)C=CC)c(c1)N(=O)=O
Crotonic acid, 2(or 4)-(1-methylheptyl)-4,6(or 2,6)-dinitrophenyl ester
Arathane
2-Butenoic acid, 2-(or 4)-iso-octyl-4,6(or 2,6)-dinitrophenyl ester (9CI)
Caprane
Capryldinitrophenyl crotonate
2-Capryl-4,6-dinitrophenyl crotonate
CR 1639
CR 1693
Crotonate de 2,4-dinitro 6-(1-methyl-heptyl)-phenyle (French)
Crotonic acid 2,4-dinitro-6-(1-methylheptyl)phenyl ester
Crotonic acid 2,4-dinitro-6-(2-octyl)phenyl ester
Crotothane
4,6-Dinitro-2-caprylphenyl crotonate
4,6-Dinitro-2-(2-capryl)phenyl crotonate
2,4-Dinitro-6-(1-methylheptyl)-phenylcrotonat (German)
Dinitro(1-methylheptyl)phenyl crotonate
4,6-Dinitro-2-(1-methylheptyl)-phenyl crotonate
2,4-Dinitro-6-(1-methylheptyl)-phenyl crotonate
2,4-Dinitro-6-(2-octyl)phenyl crotonate
Dinocap
Dinokap
DNOCP
DNOPC
DPC
ENT 24,727
Iscothan
Iscothane
Karathane
Karathane WD
Karathene
(6-(1-Methyl-heptyl)-2,4-dinitro-fenyl)-crotonaat (Dutch)
2-(1-Methylheptyl)-4,6-dinitro-fenylester kyseliny krotonove (Czech)
(6-(1-Methyl-heptyl)-2,3-dinitro-phenyl)-crotonat (German)
2-(1-Methylheptyl)-4,6-dinitro-phenyl crotonate
(6-(1-Metil-epitl)-2,4-dinitro-fenil)-crotonato (Italian)
Mildex
Phenol, 2-(1-methylheptyl)-4,6-dinitro-, crotonate (ester)

39300-88-4
Unknown
Unknown
Gum-Tara
NCI-C54364
Tara Gum

39316-51-3
Unknown
Unknown
Plurafac RA 30 (9CI)

39319-42-1
$C_{207}H_{309}N_{57}O_{57}S$
4539.80
α(1-39)-Corticotropin (Pig) , 31-L-serine-33-L-glutamine- (9CI)
α(1-39)-Corticotropin (Sheep)
α(1-39)-Corticotropin (Ox)

39327-16-7
$C_{13}H_9N$
179.22
Benzoquinoline (9CI)

39362-29-3
Unknown
Unknown
Corexit (9CI)

39362-66-8
C.Cu.Fe.Mn.Si
214.42
Steel, (API X52) (9CI)
API X 52
API 5L X52
API 5LX-52
API 5LX X52
API 5XL X52
KX 52
StE 360.7
X52

39390-00-6
Unknown
Unknown
Lead chloride silicate (9CI)

39390-54-0
Unknown
Unknown
CA 24 (9CI)

39390-62-0
Unknown
Unknown
Epoxide 8

39393-20-9
Unknown
Unknown
Superfloc 127 (9CI)

39393-37-8
Unknown
Unknown
Santicizer 711 (9CI)

39403-84-4
Unknown
Unknown
BP 1100X (9CI)
BP-1100X
Oil Dispersant BP 1100X

39407-03-9
$C_8H_{19}O_4P$
210.21
Phosphoric acid, octyl ester (9CI)
Octyl phosphate

39409-52-4
Unknown
Unknown
Santicizer 409 (9CI)

39409-64-8
Unknown
Unknown
Tris bis-bifluoroamino diethoxy propane
Tris, bis-bifluoroamino dietoxi propano (Spanish)
Tris, bis-fluoroamino diethoxy propane (French)

39413-47-3
Unknown
Unknown
Silicic acid, beryllium zinc salt
Beryllium zinc silicate
Zinc beryllium silicate

39425-24-6
Unknown
Unknown
Hexapon (9CI)

39430-27-8
$CH_4Ni_3O_7.4H_2O$
376.26
Nickel, (carbonato(2-))tetrahydroxytri-, tetrahydrate
Basic Nickel Carbonate

Nickel carbonate hydroxide

39450-05-0
Unknown
Unknown
Halowax 1099 (9CI)

39450-10-7
Unknown
Unknown
Palmotoxin-Bo

39450-11-8
Unknown
Unknown
Palmotoxin Go
Palmotoxin G0 (9CI)
Palmotoxin G(O)

39456-75-2
Unknown
Unknown
HOE 2874 (9CI)

39457-26-6
Unknown
Unknown
Bondolane M (9CI)

39464-64-7
Unknown
Unknown
Poly(oxy-1,2-ethanediyl), α-(dinonylphenyl)-ω-hydroxy-, phosphate (9CI)
Dinonylphenol, ethoxylate, phosphate
Dinonylphenol, ethoxylated, phosphated

39485-83-1
$C_{12}H_5Cl_5$
326.44
1,1'-Biphenyl, 2,2',4,4',6-pentachloro- (9CI)
2,2',4,4',6-PCB
2,2',4,4',6-Pentachlorobiphenyl

39515-41-8
$C_{22}H_{23}NO_3$
349.46
CC3(C)C(C(=O)OC(C#N)c2cccc(Oc1ccccc1)c2)C3(C)C
Cyclopropanecarboxylic acid, 2,2,3,3-tetramethyl-, cyano-(3-phenoxyphenyl)methyl ester
α-Cyano-3-phenoxybenzyl 2,2,3,3-tetramethyl-1-cyclopropanecarboxylate
Danitol
Danitrol
Fenpropanate
Fenpropathrin
Herald
Meothrin
Rody
S 3206
SD 41706
WL 41706
XE-938

39515-51-0
$C_{13}H_{10}O_2$
198.22
O=Cc(cc(Oc(cccc1)c1)cc2)c2
Benzaldehyde, 3-phenoxy- (9CI)
m-Phenoxybenzaldehyde
3-Phenoxybenzaldehyde

39542-65-9
$C_7H_6Cl_2O_2$
193.03
Phenol, 2,3-dichloro-4-methoxy- (9CI)
2,3-Dichlorohydroquinone monomethylether
2,3-Dichloro-4-methoxyphenol

39569-21-6
$C_7H_3Br_4Cl$
442.17
Benzene, 1,2,3,4-tetrabromo-5-chloro-6-methyl- (9CI)

39589-98-5
$C_{11}H_{14}O_4$
210.23
Bicyclo(2.2.1)hept-5-ene-2,3-dicarboxylic acid, dimethyl ester, (endo,endo)- (9CI)
Dimelone
Dimethyl carbate
5-Norbornene-2,3-dicarboxylic acid, dimethyl ester, cis-endo- (8CI)
NSC-46419
NSC-196235

39610-34-9
$(C_8H_{14}N_2O_2)x$
Polymer NA
Poly(imino(1-oxo-1,2-ethanediyl)imino(1-oxo-1,6-hexanediyl)) (9CI)
Poly(glycyl-ε-aminocaproic acid)

39624-86-7
$C_9H_{13}FNO_3PS$
265.26
CCOP(=S)(OCC)Oc1cccc(F)n1
Phosphorothioic acid, O,O-diethyl O-(6-fluoro-2-pyridyl) ester
O,O-Dimethyl S-9-thiabicyclo-(4.2.1)nonenyl phosphorodithioate (Isomeric mixture)
O,O-Diethyl O-(6-fluoro-2-pyridyl ester phosphorothioic) acid
DOWCO 275

39635-31-9
$C_{12}H_3Cl_7$
395.33
2,3,4,5,3',4',5'-Heptachlorobiphenyl

39638-32-9
$C_6H_{12}Cl_2O$
171.07
O(C(Cl)(C)C)C(Cl)(C)C
Bis(2-chloroisopropyl) ether
Bis(2-chloro-1-methylethyl)-ether
Dichloroisopropyl ether
2,2'-Oxybis(2-chloropropane)
Propane, 2,2'-oxybis(2-chloro- (9CI)

39660-14-5
$C_{10}H_6Cl_6O_2$
370.87
4,7-Methano-1H-indenediol, 4,5,6,7,8,8-hexachloro-3a,4,7,7a-tetrahydro- (9CI)

39735-13-2
$C_{23}H_{40}O_3S$
396.63
Benzenesulfonic acid, heptadecyl- (9CI)
Heptadecylbenzenesulfonic acid

39765-80-5
$C_{10}H_5Cl_9$
444.20
4,7-Methano-1H-indene, 1,2,3,4,5,6,7,8,8-nonachlor-2,3,3a,4,7,7a-hexahydro-, (1-α,2-β, 3-α,3a-α,4-β,7-β, 7a-α)
trans-Nonachlor

39801-14-4
$C_{10}HCl_{11}$
511.06
1,3,4-Metheno-1H-cyclobuta-(cd)pentalene, 1,1a,2,2,3,3a,4,5,5,5a,5-undecachloro-octahydro
Hydromirex
8-Monohydro mirex
Photomirex
1,2,3,4,5,5,6,7,9,10,10-Undecachloropentacyclo(5.3.0.$0^{2,6}.0^{3,9}.0^{4,8}$)decane

39817-09-9
$C_{19}H_{20}O_4$
312.37

Oxirane, 2,2'-(methylenebis-(phenyleneoxymethylene))-bis- (9CI)

40039-93-8
$(C_{15}H_{12}Br_4O_2.C_3H_5ClO)x$
Polymer
Phenol, 4,4'-(1-methylethylidene)bis(2,6-dibromo-, Polymer with (chloromethyl)oxirane (9CI)
Chloromethyloxirane, Polymer with 4,4'-(1-methylethylidene)bis(2,6-dibromophenol)
4,4'-(1-Methylethylidene)-bis(2,6-dibromophenol), (chloromethyl)oxirane polymer
4,4'-(1-Methylethylidene)bis-(2,6-dibromophenol), epichlorohydrin polymer
3,5,3',5'-Tetrabromobisphenol A, epichlorohydrin polymer

40086-66-6
C_7H_6BrNO
200.04
2-Bromoacetamidopyridine
Ethanone, 2-bromo-1-(2-pyridinyl)-

40088-45-7
$C_{12}H_6Br_4$
469.80
1,1'-Biphenyl, tetrabromo- (9CI)
Tetrabromobiphenyltetrabromo

40088-47-9
$C_{12}H_6Br_4O$
485.80
Benzene, 1,1'-oxybis-, tetrabromo deriv. (9CI)
1,1'-Oxybisbenzene tetrabromo deriv.
Tetrabromodiphenyl ether
Tetrabromodiphenyl oxide
Tetrabromophenoxybenzene

40091-57-4
$C_7H_{16}O_2$
132.20
2-Butanol, 4-(1-methylethoxy)- (9CI)

40120-74-9
$C_9H_{15}Cl_6O_4P$
430.91
1-Propanol, 1,3-dichloro-, phosphate (3:1) (9CI)

40137-60-8
$C_9H_{17}ClO_3$
208.68
O=C(OC(C)COCC(C)C)CCl
Acetic acid, chloro-, 1-methyl-2-(2-methylpropoxy)ethyl ester (9CI)

40164-67-8
Unknown
Unknown
N-((Acetylamino)methyl)-2-chloro-N-(2,6-diethylphenyl)acetamide
2-((Acetylamino)methyl)-2-chloro-N-(2,6-diethylphenyl)acetamide
Caswell No. 005B
EPA Pesticide Chemical Code 128001
Mon 4620

40164-69-0
$C_{13}H_{17}Cl_2NO$
274.18
O=C(N(c(c(ccc1)CC)c1CC)CCl)CCl
Acetamide, 2-chloro-N-(chloromethyl)-N-(2,6-diethylphenyl)- (9CI)

40169-27-5
$C_{18}H_{26}O_5Si_4$
434.75
Spiro(cyclotetrasiloxane-2,6'-(6H)dibenz(c,e)(1,2)-oxasilin), 4,4,6,6,8,8-hexamethyl- (9CI)

40186-70-7
$C_{12}H_3Cl_7$
395.32
1,1'-Biphenyl, 2,2',3,3',4,5',6-heptachloro- (9CI)
2,2',3,3',4,5',6-Heptachlorobiphenyl
2,2',3,3',4,5',6-PCB

40186-71-8
$C_{12}H_2Cl_8$
429.77
1,1'-Biphenyl, 2,2',3,3',4,5',6,6'-octachloro- (9CI)
2,2',3,3',4,5',6,6'-PCB

40186-72-9
$C_{12}HCl_9$
464.21
1,1'-Biphenyl, 2,2',3,3',4,4',5,5',6-nonachloro- (9CI)
2,2',3,3',4,4',5,5',6-Nonachlorobiphenyl
2,2',3,3',4,4',5,5',6-PCB

40188-83-8
$C_8H_5Cl_2NO_4$
250.03
O=C(OC)c(c(ccc1Cl)Cl)c1N(=O)=O
Benzoic acid, 3,6-dichloro-2-nitro-, methyl ester (9CI)
Methyl 2,5-dichloro-6-nitrobenzoate
Methyl 3,6-dichloro-2-nitrobenzoate

40216-08-8
$(C_{15}H_{16}O_2.C_6H_6O.C_3H_5ClO.CH_2O)x$
Polymer
Formaldehyde, Polymer with (chloromethyl)oxirane, 4,4'-(1-methylethylidene)bis-(phenol) and phenol (9CI)
Phenol, formaldehyde, biphenol A, epichlorohydrin polymer
Phenol, formaldehyde, bisphenol A, epichlorohydrin polymer
Phenol, Polymer with formaldehyde, bisphenol A and epichlorohydrin

40292-82-8
$C_{10}H_{19}ClO$
190.71
Neodecanoyl chloride (9CI)

40321-76-4
$C_{12}H_3Cl_5O_2$
356.40
Clc3cc2Oc1cc(Cl)c(Cl)c(Cl)c1Oc2cc3Cl
Dibenzo-p-dioxin, 1,2,3,7,8-pentachloro
1,2,3,7,8-Pentachlorodibenzo-p-dioxin

40482-18-6
$C_{12}H_{20}$
164.29
Cyclohexane, 1,1,4,4-tetramethyl-2,6-bis(methylene)- (9CI)
1,3-Dimethylene-2,2,5,5-tetramethylcyclohexane

40487-42-1
$C_{13}H_{19}N_3O_4$
281.35
CCC(CC)Nc1c(cc(C)c(C)c1N(=O)=O)N(=O)=O
Aniline, 3,4-dimethyl-2,6-dinitro-N-(1-ethylpropyl)
AC 92553
Benzenamine, 3,4-dimethyl-2,6-dinitro-N-(1-ethylpropyl)-
Benzenamine, N-(1-ethylpropyl)-3,4-dimethyl-2,6-dinitro- (9CI)
N-(1-Aethylpropyl)-3,4-dimethyl-2,6-dinitroanilin (German)
N-(1-Ethylpropyl)-3,4-dimethyl-2,6-dinitrobenzenamine
N-(1-Aethylpropyl)-2,6-dinitro-

3,4-xylidin (German)
Herbadox
Horbadox
Pay-Off
Pendimethalin
Pendimethaline
Penoxalin
Penoxaline
Penoxyn
N-(3-Pentyl)-3,4-dimethyl-2,6-dinitroaniline
Phenoxalin
Prowl
Stomp
Stomp 330D
Stomp 330E
Tendimethalin
3,4-Xylidine, 2,6-dinitro-N-(1-ethylpropyl)-

40529-66-6
$C_{14}H_{14}$
182.27
1,1'-Biphenyl, ethyl- (9CI)
Ethylbiphenyl
Ethyl-1,1'-biphenyl

40552-84-9
$C_{12}H_{14}O_3$
206.24
O=C(OC(c(cccc1)c1)C)CC(=O)C
Butanoic acid, 3-oxo-, 1-phenylethyl ester (9CI)
α-Methylbenzyl acetoacetate
1-Phenylethyl 3-oxobutanoate

40581-90-6
$C_{12}H_4Cl_4O_2$
321.97
Dibenzo(b,e)(1,4)dioxin, 1,2,6,7-tetrachloro- (9CI)
1,2,6,7-Tetrachlorodibenzo-(b,e)(1,4)dioxin
1,2,6,7-Tetrachlorodibenzo-p-dioxin

40581-91-7
$C_{12}H_4Cl_4O_2$
321.97
Dibenzo(b,e)(1,4)dioxin, 1,2,6,9-tetrachloro- (9CI)
1,2,6,9-Tetrachlorodibenzo-(b,e)(1,4)dioxin
1,2,6,9-Tetrachlorodibenzo-p-dioxin

40581-93-9
$C_{12}H_4Cl_4O_2$
321.97
Dibenzo(b,e)(1,4)dioxin, 1,4,6,9-tetrachloro- (9CI)
1,4,6,9-Tetrachlorodibenzo-(b,e)(1,4)dioxin
1,4,6,9-Tetrachlorodibenzo-p-dioxin

40581-94-0
$C_{12}H_4Cl_4O_2$
321.97
Dibenzo(b,e)(1,4)dioxin, 1,4,7,8-tetrachloro- (9CI)
1,4,7,8-Tetrachlorodibenzo-(b,e)(1,4)dioxin
1,4,7,8-Tetrachlorodibenzo-p-dioxin

40596-69-8
$C_{19}H_{34}O_3$
310.53
COC(C)(C)CCCC(C)CC=CC(C)=CC(=O)OC(C)C
2,4-Dodecadienoic acid, 11-methoxy-3,7,11-trimethyl-, 1-methylethyl ester, (E,E)
Altosid
Altosid IGR
Altosid SR 10
Apex
Diacon
Dianex
ENT 70,460
Isopropyl(2E,4E)-11-methoxy-3,7,11-trimethyl-2,4-dodecadienoate
Kabat
Manta
Methoprene
Minex
OMS 1697
Pharoid
Precor
ZR 515

40623-75-4
$(C_7H_{13}NO_4S.C_3H_4O_2)x$
Polymer NA
2-Propenoic acid, Polymer with 2-methyl-2-((1-oxo-2-propenyl)amino)-1-propanesulfonic acid (9CI)

40630-63-5
$C_8H_{17}FO_2S$
196.29
O=S(=O)(F)CCCCCCCC
1-Octanesulfonyl fluoride (9CI)

40648-26-8
$C_{11}H_{20}O$
168.28
Cyclohexene, 1-(1,1-dimethylethoxy)-2-methyl- (9CI)
1-tert-Butoxy-2-methylcyclohexene

40693-04-7
$C_{15}H_{18}O_2$
230.33
C=CCc1cccc(CC=C)c1OCC2CO2
Ether, 2,6-diallylphenyl 2,3-epoxypropyl
2,6-Diallylphenyl 2,3-epoxypropyl ether

40702-26-9
$C_{10}H_{16}O$
152.24
3-Cyclohexene-1-carboxaldehyde, 1,3,4-trimethyl- (9CI)

40703-79-5
$C_{15}H_{13}Br_2NO_2$
399.08
O=C(N(C(=O)C1C(CC2C3Br)C3Br)c(cccc4)c4)C12
4,7-Methano-1H-isoindole-1,3(2H)-dione, 5,6-dibromohexahydro-2-phenyl- (9CI)

40704-75-4
Unknown
Polymer
Duxon
N-(2-Hydroxypropyl)methacrylamide polymer
Poly(N-(2-hydroxypropyl)methacrylamide)

40709-82-8
C_4H_7ClOS
138.62
Sulfoxide, 2-chloroethyl vinyl
TL 906

40710-32-5
$C_{69}H_{138}O_2$
999.87
Nonahexacontanoic acid (9CI)

40710-42-7
$C_{41}H_{84}O$
593.13
1-Hentetracontanol (9CI)

40716-47-0
$C_{29}H_{12}Cl_8N_6O_2$
760.04
1H-Isoindol-1-one, 4,5,6,7-tetrachloro-3-((3-methyl-4-((4-((4,5,6,7-tetrachloro-1-oxo-1H-isoindol-3-yl)amino)phenyl)azo)phenyl)amino)- (9CI)

40723-63-5
$C_3H_4F_4$
116.06
Propane, 1,1,2,2-tetrafluoro- (9CI)

40780-64-1
Unknown

Unknown
Ethylbutyl acetate
UN 1177

40843-25-2
$C_{15}H_{12}Cl_2O_4$
327.17
Diclofop
Caswell No. 328B
2-(4-(2,4-Dichlorophenoxy)-phenoxy)propanoate
(RS)-2-(4-(2,4-Dichlorophen-oxy)phenoxy)propanoic acid
2-(4-(2,4-Dichlorophenoxy)-phenoxy)propanoic acid
Methyl (RS)-2-(4-(2,4-di-chlorophenoxy)phenoxy)-propionate
Methyl 2-(4-(2,4-dichloropheny-oxy)phenoxy)propanoate
Propanoic acid, 2-(4-(2,4-di-chlorophenoxy)phenoxy)-

40911-36-2
$C_{13}H_{11}Cl_2O_3P$
317.11
Dichloromethyl O,O-diphenyl phosphonate
(Dichloromethyl)phosphonic acid, diphenyl

40946-60-9
$C_{18}H_{12}Br_3O_4P$
562.99
Phenol, 4-bromo-, phos-phate (3:1) (9CI)
Tris(4-bromophenyl) phos-phate
Tri(p-bromophenyl)-phosphate
Tris(p-bromophenyl) phos-phate

40991-38-6
C_8H_7NOS
165.22
1,2-Benzisothiazole, 3-methoxy- (9CI)
3-Methoxy-1,2-benziso-thiazole

40991-93-3
$C_8H_{14}O_2$
142.20
Cyclopentanol, 2-methyl-, acetate, cis- (9CI)

41011-01-2
C_8H_6BrClO
233.50
2-Bromo-3'-chloroacetophen-one
α-Bromo-3-chloroacetophenone
2-Bromo-1-(3-chlorophenyl)-ethanone
m-Chlorophenacyl bromide
Ethanone, 2-bromo-1-(3-chloro-phenyl)-

41037-13-2
$C_{11}H_9NO_2$
187.19
Naphthalene, 1-methyl-3-nitro-(9CI)

41065-91-2
$C_8H_{16}O_2$
144.21
Hexanoic acid, 3-ethyl- (9CI)

41065-97-8
$C_7H_{14}O$
114.19
Hexanal, 4-methyl- (9CI)

41079-92-9
C_2H_6ClSn
184.23
Stannyl, chlorodimethyl- (9CI)

41096-46-2
$C_{17}H_{30}O_2$
266.42
Altozar
Caswell No. 456H
Dodeca-2,4-dienoic acid, 3,7,11-trimethyl-, ethyl ester, (2E,4E)-
2,4-Dodecadienoic acid, 3,7,11-trimethyl-, ethyl ester, (E,E)- (9CI)
ENT 70,459
EPA Pesticide Chemical Code 486300
Ethyl 3,7,11-trimethyldodeca-2,4-dienoate
Ethyl (E,E)-3,7,11-trimethyl-dodeca-2,4-dienoate
Ethyl (E,E)-3,7,11-trimethyl-2,4-dodecadienoate
Ethyl(2E,4E)-3,7,11-trimethyl-2,4-dodecadienoate
Ethyl (2E,4E)-3,7,11-trimethyl-dodeca-2-4-dienoate
(E,E)-Ethyl 3,7,11-trimethyl-2,4-dodecadienoate
Hydroprene
OMS 1696
(E,E)-3,7,11-Trimethyl-2,4-do-decadienoic acid, ethyl ester
ZR 512

41114-00-5
$C_{17}H_{34}O_2$
270.46
Pentadecanoic acid, ethyl ester (9CI)
NSC-137833

41122-70-7
$C_{19}H_{21}N$
263.38
N#Cc(ccc(c(ccc(c1)CCCCCC)c1)c2)c2
(1,1'-Biphenyl)-4-carbonitrile, 4'-hexyl- (9CI)
4-Cyano-4'-hexylbiphenyl
4'-Hexyl-(1,1'-biphenyl)-4-carbonitrile

41198-08-7
$C_{11}H_{15}BrClO_3PS$
373.65
CCCSP(=O)(OCC)Oc1ccc(Br)cc1Cl
Phosphorothioic acid, O-(4-bromo-2-chlorophenyl)-O-ethyl-S-propyl ester
O-(4-Bromo-2-chlorophenyl)-O-ethyl-S-propyl phosphoro-thioate
CGA 15324
Curacron
Polycron
Profenofos
Selecron

41261-95-4
$CrH_2O_4.xCr$
Unknown
Chromium chromate
Chromic acid, chromium salt

41295-28-7
$C_{16}H_{16}O_2$
240.30
(4-Methoxy-3-methylphenyl)-(3-methylphenyl)methanone (9CI)
4-Methoxy-3,3'-dimethylbenzo-phenone (8CI)
Methoxyphenone

41361-12-0
$C_9H_{20}N_4S$
216.33
Carbonothioic dihydrazide, (1-methylheptylidene)- (9CI)

41363-16-0
$C_{12}H_6Cl_4O_2$
323.99
Oc(cc(c(c1Cl)c(c(cc(O)c2)Cl)c2Cl)Cl)c1
(1,1'-Biphenyl)-4,4'-diol, 2,2',6,6'-tetrachloro-
AI3-31286
2,2',6,6'-Tetrachloro-(1,1'-bi-phenyl)-4,4'-diol

41372-08-1
$C_{10}H_{13}NO_4.3/2H_2O$
238.27
O.O.O.CC(N)(Cc1ccc(O)c(O)c1)C(O)=O.CC(N)(Cc1ccc(O)c(O)c1)C(O)=O
L-Tyrosine, 3-hydroxy-α-methyl-, sesquihydrate

Aldomet
Aldometil
Aldomin
AMD
Bayer 1440 L
Baypresol
Dopamet
Dopatec
Dopegyt
Hyperpax
Methyl Dopa Sesquihydrate
Medomet
Medopren
Methoplain
α-Methyl-l-3,4-dihydroxy-phenylalanine
MK.B51
MK-351
α-Medopa
Presinol
Presolisin
Sedometil
Sembrina

41394-05-2
$C_{10}H_{10}N_4O$
202.24
as-Triazin-5(4H)-one, 4-amino-3-methyl-6-phenyl
4-Amino-3-methyl-6-phenyl-1,2,4-triazin-5(4H)-one
Bay-DRW 1139
DRW 1139
Goltix
Herbrak
Metamiton
Metamitron (German)
Methiamitron (French)
3-Methyl-4-amino-6-phenyl-1,2,4-triazin(4H)-on (German)

41411-63-6
$C_{12}H_4Cl_6$
360.88
1,1'-Biphenyl, 2,3,4,4',5,6-hexachloro- (9CI)
2,3,4,4',5,6-Hexachlorobiphenyl
2,3,4,4',5,6-PCB

41411-64-7
$C_{12}H_3Cl_7$
395.32
1,1'-Biphenyl, 2,3,3',4,4',5,6-heptachloro- (9CI)
2,3,3',4,4',5,6-Heptachlorobiphenyl
2,3,3',4,4',5,6-PCB

41453-50-3
$C_7H_6O_4.1/2Pb$
257.72
Lead β-resorcylate
Benzoic acid, 2,4-dihydroxy-, lead(2+) salt (2:1) (9CI)
Bis(2,4-dihydroxyl)benzoatolead(II)

41464-39-5
$C_{12}H_6Cl_4$
291.99
Clc1ccc(Cl)c(c1)c2cccc(Cl)c2Cl
1,1'-Biphenyl, 2,2',3,5'-tetrachloro- (9CI)
2,2',3,5'-PCB
2,2',3,5'-Tetrachlorobiphenyl

41464-40-8
$C_{12}H_6Cl_4$
291.99
Clc1ccc(c(Cl)c1)c2cc(Cl)ccc2Cl
2,2',4,5'-Tetrachlorobiphenyl
2,2',4,5'-TCB

41464-41-9
$C_{12}H_6Cl_4$
291.99
1,1'-Biphenyl, 2,2',5,6'-tetrachloro- (9CI)
2,2',5,6'-PCB
2,2',5,6'-Tetrachlorobiphenyl

41464-42-0
$C_{12}H_6Cl_4$
291.99
1,1'-Biphenyl, 2,3',5,5'-tetrachloro- (9CI)
2,3',5,5'-PCB
2,3',5,5'-Tetrachlorobiphenyl

41464-43-1
$C_{12}H_6Cl_4$
291.99
1,1'-Biphenyl, 2,3,3',4'-tetrachloro- (9CI)
2,3,3',4'-PCB
2,3,3',4'-Tetrachlorobiphenyl

41464-46-4
$C_{12}H_6Cl_4$
291.99
1,1'-Biphenyl, 2,3',4',6-tetrachloro- (9CI)
2,3',4',6-PCB
2,3',4',6-Tetrachlorobiphenyl

41464-47-5
$C_{12}H_6Cl_4$
291.99
1,1'-Biphenyl, 2,2',3,6'-tetrachloro- (9CI)
2,2',3,6'-PCB
2,2',3,6'-Tetrachlorobiphenyl

41464-51-1
$C_{12}H_5Cl_5$
326.44
Clc1cc(Cl)c(cc1Cl)c2cccc(Cl)c2Cl
1,1'-Biphenyl, 2,2',3,4',5'-pentachloro- (9CI)
2,2',3,4',5'-PCB
2,2',3,4',5'-Pentachlorobiphenyl

41468-25-1
$C_8H_{10}NO_6P.H_2O$
265.18
Pyridoxal, 5-(dihydrogen phosphate), monohydrate
3-Hydroxy-5-(hydroxymethyl)-2-methylisonicotinaldehyde 5-phosphate monohydrate
3-Hydroxy-2-methyl-5-((phosphonooxy)methyl)-4-pyridinecarboxaldehyde monohydrate
Phosphopyridoxal monohydrate
4-Pyridinecarboxaldehyde, 3-hydroxy-2-methyl-5-((phosphonooxy)methyl)-, monohydrate
Pyridoxal-5-monophosphoric acid ester monohydrate
Pyridoxal-5-phosphate, monohydrate
Pyridoxal-5'-phosphate, monohydrate
Pyridoxyl phosphate monohydrate

41491-52-5
$C_{11}H_{12}Cl_3O_2PS$
345.61
Phosphonothioic acid, ethyl-, O-(2-chloro-1-(2,5-dichlorophenyl)ethenyl) O-methyl ester
AI3-27829

41541-11-1
$C_{18}H_{22}N_4O_5$
374.37
Ethanol, 2,2'-((4-((2-methoxy-4-nitrophenyl)azo)-3-methylphenyl)imino)bis- (9CI)

41541-13-3
$C_{17}H_{19}N_5O_6$
389.34
Ethanol, 2,2'-((4-((2,4-dinitrophenyl)azo)-3-methylphenyl)imino)bis- (9CI)
Disperse Violet 4K
Ethanol, 2,2'-(4-(2,4-dinitrophenylazo)-m-tolylimino)di-

41541-14-4
$C_{17}H_{20}N_4O_5$
360.35
Ethanol, 2,2'-((4-((2-methoxy-4-nitrophenyl)azo)phenyl)imino)bis- (9CI)
Ethanol, 2,2'-(p-(2-methoxy-4-nitrophenylazo)phenylimino)di-

41593-24-2
$C_{16}H_{12}$
204.27
Fluoranthene, dihydro- (9CI)

41593-31-1
$C_{18}H_{14}$
230.32
Chrysene, 1,2-dihydro
Chrysene, dihydro- (9CI)
Dihydrochrysene
1,2-Dihydrochrysene

41601-59-6
$C_4H_5Cl_3$
159.44
2-Butene, 1,1,4-trichloro- (9CI)
1,1,4-Trichloro-2-butene

41632-89-7
$C_6H_{14}O_2$
118.18
Propane, 1,1-dimethoxy-2 -methyl- (9CI)
1,1-Dimethoxy-2-methyl- propane
Isobutyraldehyde dimethyl acetal (6CI)
2-Methylpropanal dimethyl acetal

41637-90-5
$C_{19}H_{14}$
242.32
Chrysene, methyl- (9CI)

41637-92-7
$C_{20}H_{16}$
256.35
Chrysene, dimethyl- (9CI)

41638-13-5
$C_{12}H_{22}O_5$
246.30
Oxirane, 2,2'-(oxybis((methyl- 2,1-ethanediyl)oxymethyl- ene))bis- (9CI)
Dipropylene glycol diglycidyl ether

41638-55-5
$C_{16}H_{18}$
210.32
1,1'-Biphenyl, butyl- (9CI)
Butyl-1,1'-biphenyl
Butylphenyl

41638-56-6
$C_{18}H_{24}$
240.39
Naphthalene, octyl- (9CI)

41642-51-7
$C_{20}H_{19}N_7O_3$
405.37
O=C(Nc(c(N=Nc(c(C#N)cc(N(=O)=O)c1)c1C#N)ccc2N(CC)CC)c2)C
Acetamide, N-(2-((2,6-dicyano- 4-nitrophenyl)azo)-5-(di- ethylamino)phenyl)- (9CI)
N-(2-((2,6-Dicyano-4-nitro- phenyl)azo)-5-(diethylamino)- phenyl)acetamide
N-(2-((4-Nitro-2,6-dicyano- phenyl)azo)-5-(diethylamino)- phenyl)acetamide

41653-93-4
$C_6H_{10}O_2$
114.15
3-Pentenoic acid, 3- methyl-, (E)- (9CI)
(E)-3-Methyl-3-pentenoic acid
trans-δ3-3-Methylpentenoic acid

41654-04-0
$C_7H_{12}O_3$
144.17
Hexanoic acid, 5-methyl- 4-oxo- (9CI)
5-Methyl-4-oxohexanoic acid
4-Oxo-5-methylhexanoic acid

41654-09-5
$C_7H_{10}O_4$
158.15
2-Butenedioic acid, 2-ethyl- 3-methyl-, (Z)- (9CI)
Maleic acid, ethylmethyl-

41654-27-7
$C_{10}H_{14}O_3$
182.22
2-Cyclohexene-1,4-dione, 2-methoxy-3,5,5-tri- methyl- (9CI)

41663-84-7
$C_9H_6N_2O_4$
206.15
O=C(N(C(=O)c1ccc(N(=O)=O)c2)C)c12
1H-Isoindole-1,3(2H)-dione, 2-methyl-5-nitro-
AI3-28673
2-Methyl-5-nitro-1H-isoindole- 1,3(2H)-dione
N-Methyl-4-nitrophthalimide
4-Nitro-N-methylphthalimide

41669-40-3
$C_{14}H_{28}O_2.C_6H_{15}NO_3$
377.56
TEA-myristate
Caswell No. 887C
EPA Pesticide Chemical Code 079044
Myristic acid, triethanolamine salt
Tetradecanoic acid, Compd. with 2,2',2''-nitrilotris- (ethanol)
Tetradecanoic acid, Compd. with 2,2',2''-nitrilotris- (ethanol) (1:1) (9CI)
Triethanolamine myristate

41674-04-8
$C_{12}H_{11}N$
169.23
Aminobiphenyl

41699-09-6
$C_{23}H_{14}$
290.37
Benzo(ghi)perylene, methyl
Methyl-1,12-benzoperylene

41708-72-9
$C_{11}H_{16}N_2O$
192.29
CC(N)C(=O)Nc1c(C)cccc1C
2',6'-Propionoxylidide, 2-amino
Alanyl-2,6-xylidide
2-Amino-2',6'-propionoxylidide
Tocainide
Tonocard
W 36095

41708-76-3
$C_{15}H_{25}NO_6$
315.41
Butanoic acid, 2,3-dihydroxy- 2-(1-methylethyl)-, (2,3,5,7a- tetrahydro-1-hydroxy-1H- pyrrolizin-7-yl)methyl ester, N-oxide, (1R-(1-α,7(2R*,3S), 7-α,β))
Indi
Indicine N-oxide
NSC-132,319

41709-76-6
$C_{46}H_{30}Cl_2N_6O_4$
801.65
Chlorodiane blue
Biphenyl, 3,3'-dichloro-4,4'- bis((2-hydroxy-3-(N-phenyl- carbamyl)-1-naphthyl)azo)-
2-Naphthalenecarboxamide, 4,4'-((3,3'-dichloro(1,1'-bi- phenyl)-4,4'-diyl)bis(azo))- bis(3-hydroxy-N-phenyl- (9CI)

41772-23-0
$C_{10}H_9NO.ClH$
195.66
1-Naphthol, 2-amino-, hydro- chloride
2-Amino-1-naphthol hydro- chloride
1-Hydroxy-2-naphthylamine

41787-75-1
$C_{19}H_{23}ClO_7$
398.87
Propanoic acid, 3-chloro-2-hydroxy-2-methyl-, decahydro-8-hydroxy-3,6-bis-(methylene)-2-oxospiro-(azuleno(4,5-b)furan-9(2H),2'-oxiran)-4-yl ester, (3ar-(3a-α,4-α(S*),6a- α,8-β,9-α,9a-α,9b-β))
Acroptilin
Chlorohyssopifolin C

41814-78-2
$C_9H_7N_3S$
189.25
N(N=C(N1c(c2ccc3)c3C)S2)=C1
s-Triazolo(3,4-b)benzothiazole, 5-methyl
Beam
Bim
Blascide
EL-291
5-Methyl-1,2,4-triazole(3,4-b)-benzothiazole
Tricyclazole
Tricyclazone

41826-92-0
$C_{16}H_{22}O_6$
310.38
CCOc1cc(OCC)c(cc1OCC)C(=O)CCC(O)=O
Propionic acid, 3-(2,4,5-triethoxybenzoyl)
AA149
Colibil
Supacal
Trepibutone
3-(2,4,5-Triethoxybenzoyl)-propionic acid
2,4,5-Triethoxy-γ-oxobenzene-butanoic acid

41851-50-7
C_5H_5Cl
100.55
1,3-Cyclopentadiene, 5-chloro- (9CI)

41903-57-5
$C_{12}H_4Cl_4O_2$
321.98
2,3,7,8-Tetrachlorodibenzo-p-dioxin
Tetrachlorodibenzo-p-dioxin

41906-38-1
$C_8H_4Cl_2O_4$
235.02
Benzenedicarboxylic acid, dichloro- (9CI)

41977-34-8
$C_{11}H_{22}$
154.30
Cyclopropane, 1-butyl-1-methyl-2-propyl- (9CI)

41977-40-6
$C_{16}H_{32}$
224.43
Cyclopropane, 1-methyl-1-(1-methylethyl)-2-nonyl- (9CI)

41977-41-7
$C_{17}H_{34}$
238.46
Cyclopropane, 1-methyl-1-(2-methylpropyl)-2-nonyl- (9CI)

41977-45-1
$C_{12}H_{22}$
166.31
Bicyclo(4.1.0)heptane, 7-pentyl- (9CI)

41981-60-6
C_6H_9NS
127.20
Thiazole, 4-propyl- (9CI)

41981-63-9
$C_7H_{11}NS$
141.24
Thiazole, 2-methyl-4-propyl- (9CI)
2-Methyl-4-propyl-thiazole

41999-58-0
$C_5H_6O_7.3Na$
247.07
Propanedioic acid, (carboxy-methoxy)-, trisodium salt (9CI)

42045-86-3
$C_{23}H_{30}N_2O$
350.49
3-Methylfentanyl
DEA No. 9813
F 7209
Mefentanyl
N-(3-Methyl-1-(2-phenylethyl)-4-piperidinyl)-N-phenylpropanamide
N-(3-Methyl-1-(2-phenylethyl)-4-piperidyl)-N-phenylpropanamide
Propanamide, N-(3-methyl-1-(2-phenylethyl)-4-piperidinyl)-N-phenyl-

42058-59-3
$C_9H_{10}O_3$
166.18
Benzeneacetic acid, 3-hydroxy-, methyl ester (9CI)
Methyl m-hydroxyphenyl-acetate
Methyl 3-hydroxybenzene-acetate

42064-17-5
$C_8H_{16}O_5$
192.21
Acetic acid, (2-methoxyethoxy)-, 2-methoxyethyl ester (9CI)

42072-27-5
$C_4H_{10}NO_3PS$
183.16
O=C(N=P(OC)(OC)S)C
Phosphoramidothioic acid, acetyl-, O,O-dimethyl ester (9CI)

42087-80-9
$C_8H_6ClNO_4$
215.59
Benzoic acid, 4-chloro-2-nitro-, methyl ester (9CI)
NSC-17028

42131-42-0
$(C_2H_4O)xC_8H_{16}O_2$
Polymer
Poly(oxy-1,2-ethanediyl), α-(1-oxooctyl)-ω-hydroxy- (9CI)
Polyethylene glycol mono-caprylate

42135-22-8
$C_{13}H_7NO_3$
225.21
9H-Fluoren-9-one, 3-nitro
3-Nitrofluorenone
3-Nitro-9-fluorenone

42154-69-8
C_7H_{14}
98.19
3-Hexene, 2-methyl- (9CI)

42180-82-5
C_9H_7ClO
166.61
Benzofuran, 5-chloro-2-methyl- (9CI)

42200-33-9
$C_{17}H_{27}NO_4$
309.45
CC(C)(C)NCC(O)COc1cccc2CC(O)C(O)Cc12
2,3-Naphthalenediol, 5-

(3-((1,1-dimethylethyl)-amino)-2-hydroxypropoxy)-1,2,3,4-tetrahydro
1-(tert-Butylamino)-3-((5,6,7,8-tetrahydro-cis-6,7-dihydroxy-1-naphthyl)oxy)-2-propanol
Corgard
5-(3-((1,1-Dimethylethyl)-amino)-2-hydroxypropoxy)-1,2,3,4-tetrahydro-2,3-naphthalenediol
Nadolol
2,3-cis-1,2,3,4-Tetrahydro-5-((2-hydroxy-3-tert-butyl-amino)propoxy)-2,3-naphthalenediol
Solgol
SQ 11725

42205-08-3
$C_{12}H_{18}$
162.28
Benzene, 1,2,3-triethyl- (6CI,7CI,9CI)
1,2,3-Triethylbenzene

42217-02-7
$C_{20}H_{41}Cl$
317.00
Eicosane, 1-chloro- (9CI)

42255-14-1
$C_{12}HCl_9O_2$
496.18
Phenol, 2,3,4,6-tetrachloro-5-(pentachlorophenoxy)
Isoprediolxin
3-Hydroxy-nonachlorodiphenyl ether
2,3,4,6-Tetrachloro-5-(pentachlorophenoxy)phenol

42268-97-3
$C_{12}H_{10}F_{15}NO_3$
501.19
Octanamide, 2,2,3,3,4,4,5,5,6,6,7,7,8,8,8-pentadecafluoro-N,N-bis(2-hydroxyethyl)-(9CI)

42279-29-8
$C_{12}H_5Cl_5O$
342.42
Ether, pentachlorophenyl
Benzene, 1,1'-oxybis-, tetrachloro deriv. (9CI)
Pentachloro diphenyl ether
Pentachloro diphenyl oxide
Phenyl ether pentachloro
Phenyl ether, tetrachloro deriv. (8CI)

42286-46-4
$C_{20}H_{10}O_2$
282.30
O=C2C(=O)c4cc1ccccc1c5ccc3cccc2c3c45
Benzo(a)pyrene-4,5-dione
Benzo(a)pyrene-4,5-quinone

42296-74-2
C_6H_{10}
82.16
Hexadiene [UN 2458]
UN 2458 [Hexadienes]

42328-43-8
$C_7H_{14}O$
114.19
Oxirane, 2-methyl-2-(1-methylpropyl)- (9CI)
2-sec-Butyl-2-methyloxirane

42329-90-8
$C_8H_{16}O_2$
144.22
Hexanoic acid, 2,4-dimethyl-, (R-(R*,S*))-(9CI)

42343-35-1
$C_{26}H_{42}O_4$
418.62
1,2-Benzenedicarboxylic acid, isodecyl isooctyl ester (9CI)
NSC-6385
Phthalic acid, isodecyl isooctyl ester

42343-36-2
$C_{22}H_{34}O_4$
362.51
1,2-Benzenedicarboxylic acid, butyl isodecyl ester (9CI)
Phthalic acid, butyl isodecyl ester

42350-99-2
Unknown
Unknown
2-Chloro-4,6-di-tert-amylphenol

42373-04-6
$C_{19}H_{17}N_4S.Cl$
368.91
Thiazolium, 3-methyl-2-((1-methyl-2-phenyl-1H-indol-3-yl)azo)-, chloride
Basacryl Red GL
Basic Red 29
C.I. Basic Red 29
3-Methyl-2-((1-methyl-2-phenyl-1H-indol-3-yl)azo)-thiazolium chloride

42389-30-0
$C_6H_6ClN_3O_2$
187.57
O=N(=O)c(c(N)c(N)cc1Cl)c1
5-Chloro-3-nitro-1,2-benzenediamine
1,2-Benzenediamine, 5-chloro-3-nitro- (9CI)

42397-64-8
$C_{16}H_8N_2O_4$
292.26
Pyrene, 1,6-dinitro
1,6-Dinitropyrene

42397-65-9
$C_{16}H_8N_2O_4$
292.26
Pyrene, 1,8-dinitro
1,8-Dinitropyrene

42408-82-2
$C_{21}H_{29}NO_2$
327.51
Oc4ccc3CC1N(CCC2(CCCCC12O)c3c4)CC5CCC5
Morphinan-3,14-diol, 17-(cyclobutylmethyl)
l-BC 2627
Butorphanol
(-)-Butorphanol
17-(Cyclobutylmethyl)morphinan-3,14-diol

42474-44-2
$C_3H_8S_3$
140.29
Disulfide, methyl (methylthio)-methyl (9CI)
Methane, (methyldithio)(methylthio)-

42498-33-9
$C_{12}H_{17}NO$
163.26
Benzamide, N-(1,1-dimethylethyl)-3-methyl-(9CI)
N-tert-Butyl-3-methylbenzamide

42504-46-1
$C_{18}H_{30}O_3S.C_3H_9NO$
401.60
MIPA-Dodecylbenzenesulfonate
Benzenesulfonic acid, dodecyl-, Compd. with 1-amino-2-propanol (1:1) (9CI)
Dodecylbenzenesulfonic acid, Compd. with 1-amino-2-propanol (1:1)
Isopropanolamine dodecylbenzene sulfonate
Monoisopropanolamine dodecylbenzenesulfonate

42504-54-1
$C_{15}H_{16}$
196.29
Benzene, ethyl(phenyl-

methyl)- (9CI)
Ethyldiphenylmethane

42509-80-8
$C_9H_{17}ClN_3O_3PS$
313.77
CCOP(=S)(OCC)Oc1cn(C(C)C)c(Cl)n1
Phosphorothioic acid, O-(5-chloro-1-(1-methylethyl)-1H-1,2,4-triazol-3-yl) O,O-diethyl ester
A-12223
AI3-29128
CGA-12223
O-(5-Chloro-1-(1-methylethyl)-1H-1,2,4-triazol-3-yl) O,O-diethyl phosphorothioate
Ciba 12223
Isazofos
Isazophos
Miral
Miral 10 G

42534-61-2
Unknown
Unknown
d-cis-trans-Allethrin

42540-91-0
$C_8H_3N_2O_5$
207.12
1,2-Benzisoxazole-3-carboxylic acid, 6-nitro-, ion(1-) (9CI)
6-Nitrobenzisoxazole-3-carboxylate

42576-02-3
$C_{14}H_9Cl_2NO_5$
342.14
COC(=O)c2cc(Oc1ccc(Cl)cc1Cl)ccc2N(=O)=O
Benzoic acid, 5-(2,4-dichlorophenoxy)-2-nitro-, methyl ester
Bifenox
MC-4379
Methyl 5-(2,4-dichlorophenoxy)-2-nitrobenzoate
Modown

42576-07-8
$C_{20}H_{25}NO_2$
311.42
Benzeneethanol, α-(2-amino-1-methylethyl)-α-phenyl-, propanoate (ester), (S-(R*,S*))- (9CI)

42588-37-4
Unknown
Unknown
Kinoprene
Caswell No. 714AA
2,4-Dodecadienoic acid, 3,7,11-trimethyl-, 2-propynyl ester, (E,E)- (9CI)
Enstar
Enstar IGR
ENT 70,531
EPA Pesticide Chemical Code 107501
2-Kinoprene
(E,E)-2-Propynyl 3,7,11-trimethyl-2,4-dodecadienoate
2-Propynyl (E,E)-3,7,11-trimethyl-2,4-dodecadienoate
ZR77

42609-52-9
$C_{17}H_{20}N_2O$
268.39
Urea, 1-(α,α-dimethylbenyzl)-3-methyl-3-phenyl
1-(α,α-Dimethylbenzyl)-3-methyl-3-phenylurea

42615-29-2
Unknown
Unknown
Benzenesulfonic acid, alkyl deriv.
ABS
Alkylbenzenesulfonate
LAS
Linear alkylbenzene sulfonate
Linear alkylbenzene sulphonate

42721-99-3
$C_{23}H_{15}O_3.Na$
362.36
Sodium diphacinone
Caswell No. 394A
Diphacinone, monosodium salt
Diphacinone sodium salt
2-(Diphenylacetyl)-1,3-indandione sodium salt
EPA Pesticide Chemical Code 067705
1H-Indene-1,3(2H)-dione, 2-(diphenylacetyl)-, ion(1-), sodium (9CI)

42740-50-1
$C_{12}H_2Cl_8$
429.77
1,1'-Biphenyl, 2,2',3,3',4,4',5,6'-octachloro- (9CI)
2,2',3,3',4,4',5,6'-Octachlorobiphenyl
2,2',3,3',4,4',5,6'-PCB

42864-21-1
$C_9H_8Cl_3NO_2$
268.53
Ethanol, 2,2,2-trichloro-, phenylcarbamate (9CI)
Ethanol, 2,2,2-trichloro-, carbanilate (6CI)
2,2,2-Trichloroethyl N-phenylcarbamate

42874-03-3
$C_{15}H_{11}ClF_3NO_4$
361.72
CCOc2cc(Oc1ccc(cc1Cl)C(F)(F)F)ccc2N(=O)=O
Ether, 2-chloro-α,α,α-trifluoro-p-tolyl 3-ethoxy-4-nitrophenyl
2-Chloro-1-(3-ethoxy-4-nitrophenoxy)-4-trifluoromethylbenzene
2-Chloro-α,α,α-trifluoro-p-tolyl-3-ethoxy-4-nitrophenyl ether
Goal
Oxyfluorfen
Oxyfluorfene
RH-2915

42884-33-3
$C_{10}H_9NO$
159.20
1-Naphthol, 2-amino
2-Amino-1-naphthol
Aminonaphthalenol
Naphthalenol, amino- (9CI)

42966-64-3
$C_8H_{19}N$
129.25
2-Pentanamine, N-ethyl-4-methyl- (9CI)
Butylamine, N-ethyl-1,3-dimethyl- (6CI,7CI)
N-Ethyl-1,3-dimethylbutylamine

42975-18-8
$C_{12}H_{10}N_2O_3S.Na$
285.27
Benzenesulfonic acid, 4-(phenylazo)-, sodium salt (9CI)
Benzenesulfonic acid, p-(phenylazo)-, sodium salt
NSC-226920

42978-66-5
$C_{15}H_{24}O_6$
300.39
Acrylic acid, propylenebis(oxypropylene) ester
2-Propenoic acid, (1-methyl-1,2-ethanediyl)bis(oxy(methyl-2,1-ethanediyl)) ester
Tripropyleneglycol diacrylate

42981-76-0
$C_{18}H_{28}$
244.42
Naphthalene, 2,6-bis-(1,1-dimethylethyl)-1,2,3,4-tetrahydro- (9CI)
2,6-Di-tert-butyltetralin

43047-99-0
$C_{12}H_6Cl_2O$
237.08
Dibenzofuran, dichloro

43048-00-6
$C_{12}H_5Cl_3O$
271.53
Dibenzofuran, trichloro- (9CI)

43121-43-3
$C_{14}H_{16}ClN_3O_2$
293.78
CC(C)(C)C(=O)C(Oc1ccc(Cl)cc1)n2cncn2
2-Butanone, 1-(4-chlorophenoxy)-3,3-dimethyl-1-(1,2,4-triazol-1-yl)
Amiral
Azocene
Bay 6681 F
Bayleton
Bay-MEB-6447
1-(4-Chlorophenoxy)-3,3-dimethyl-1-(1H-1,2,4-triazol-1-yl)-2-butanone
1-(4-Chlorophenoxy)-3,3-dimethyl-1-(1,2,4-triazol-1-yl)-butan-2-one
MEB 6447
Triadimefon
Triadimefone
1H-1,2,4-Triazole, 1-((tert-butylcarbonyl-4-chlorophenoxy)methyl)-

43130-12-7
$C_{21}H_{29}N_3.HNO_3$
386.47
Benzenamine, 4,4'-carbonimidoylbis(N,N-diethyl-, mononitrate (9CI)
4,4'-Carbonimidoylbis(N,N-diethylbenzenamine) mononitrate
Ethyl auramine nitrate
Nitrate salt of ethyl auramine

43142-43-4
$C_9H_{12}O$
136.19
3,5-Nonadien-7-yn-2-ol, (E,E)- (9CI)

43143-11-9
$C_{10}H_8N_2O_2S_2.O_4S.Mg$
372.69
Sulfuric acid, magnesium salt (1:1), Compd. with 2,2'-dithiobis(pyridine) 1,1'-oxide
Magnesium sulfate adduct of 2,2-dithio-bis-pyridine 1-oxide
Omadine MDS

43145-54-6
C_9H_8O
132.16
Benzaldehyde, ethenyl- (9CI)
Benzaldehyde, vinyl-

43171-59-1
$C_7H_9ClN_2$
156.60
1,3-Benzenediamine, 4-chloro-2-methyl- (9CI)

43178-22-9
$C_{19}H_{14}$
242.32
Benz(a)anthracene, methyl- (9CI)

43180-81-0
$C_4H_2Cl_2O_4$
184.96
2-Butenedioic acid, 2,3-dichloro- (9CI)

43210-67-9
$C_{15}H_{13}N_3O_2S$
299.37
2-Benzimidazolecarbamic acid, 5-(phenylthio)-, methyl ester
Carbamic acid, (5-(phenylthio)-1H-benzimidazol-2-yl)-, methyl ester
Fenbendazol
Fenbendazole
HOE 881
Panacur

43212-86-8
$C_9H_{12}O$
136.20
2,4-Nonadien-6-yn-1-ol, (E,E)- (9CI)

43216-72-4
$C_7H_9ClN_2$
156.60
1,3-Benzenediamine, 4-chloro-6-methyl- (9CI)

43216-73-5
$C_7H_9ClN_2$
156.60
1,3-Benzenediamine, 2-chloro-4-methyl- (9CI)

43217-65-8
$C_{28}H_{50}O$
402.71
Cholestan-3-ol, 6-methyl-, (3β,5α,6β)- (9CI)
6β-Methylcholestanol
6β-Methyl-5α-cholestan-3β-ol

43222-48-6
$C_{16}H_{17}N_2.CH_3O_4S$
348.45
COS([O-])(=O)=O.Cn2c(cc(c1ccccc1)[n+]2C)c3ccccc3
1H-Pyrazolium, 1,2-dimethyl-3,5-diphenyl-, methyl sulfate
AC 84777
Avenge
Difenzoquat methyl sulfate
1,2-Dimethyl-3,5-diphenyl-1-h-pyrazolium methyl sulfate
Finaven
Mataven
Yeh-Yan-Ku

44648-02-4
$C_3H_8N_4O$
116.10
Urea, (imino(methylamino)-methyl)- (9CI)

44910-38-5
$C_9H_{21}Pb$
336.46
Plumbylium, tripropyl- (9CI)

44992-01-0
$C_8H_{16}NO_2.Cl$
193.67
Ethanaminium, N,N,N-trimethyl-2-((1-oxo-2-propenyl)oxy)-, chloride (9CI)

45115-34-2
Unknown
Unknown
Vinyl neodecanate

45294-18-6
$C_{24}H_{46}O_2$
366.63
O=C(OCCCCCCCCCCCCCCCCCCCC)C(=C)C
2-Propenoic acid, 2-methyl-, eicosyl ester (9CI)
Eicosyl methacrylate
Eicosyl 2-methyl-2-propenoate

45803-84-7
$C_8H_{12}Cl_2$
179.09
Cyclohexane, 1,2-dichloro-4-ethenyl- (9CI)

45955-66-6
$C_9H_{14}O_2$
154.21
Cyclopentaneacetic acid, ethenyl ester (9CI)

46005-09-8

$C_9H_{12}O_3$
168.19
1,2,4-Cyclopentanetrione, 3-butyl- (9CI)

46242-44-8
Unknown
Unknown
Nitrilotriacetic acid, antimony(3+) complex
NTA, antimony(3+) complex

46355-07-1
$C_9H_{13}O_4P$
216.17
Phosphoric acid, mono(1-methylethyl) monophenyl ester (9CI)
Isopropyl phenyl phosphate

46817-91-8
$C_{13}H_{19}NO_3$
237.33
CCOc1ccccc1OCC2CNCCO2
Morpholine, 2-((2-ethoxyphenoxy)methyl)
2-((2-Ethoxyphenoxy)methyl)-morpholine
2-(2-Ethoxyphenoxymethyl)-tetrahydro-1,4-oxazine
ICI-58834
Viloxazin
Viloxazine

47000-92-0
$C_{10}H_{11}F_3N_2O_3S$
296.29
CC(=O)Nc1ccc(C)c(NS(=O)(=O)C(F)(F)F)c1
Toluene-2,4-diamine, N^4-acetyl-N^2-trifluoromethylsulfonyl
Acetamide, N-(4-methyl-3-(((trifluoromethyl)sulfonyl)amino)-phenyl)-
Fluoridamid
MBR 6033
(N-4-Methyl-(((1,1,1-trifluoromethyl)sulfonyl)amino)-phenyl)acetamide
Sustar
Sustar 2S
3-Trifluoromethylsulfonamido-p-acetotoluidide

47221-31-8
$C_{18}H_{30}O_3S$
326.50
Benzenesulfonic acid, 2-dodecyl- (9CI)

47377-16-2
$C_{20}H_{34}O_3S$
354.55
O=S(=O)(O)c(ccc(c1)CCCCCCCCCCCCCC)c1
Benzenesulfonic acid, 4-tetradecyl- (9CI)
4-Tetradecylbenzenesulfonic acid

47747-56-8
$C_{24}H_{23}N_3O_6S$
481.43
Talampicillin
Talampicilina (Spanish)
Talampicilline (French)
Talampicillinum (Latin)
4-Thia-1-azabicyclo(3.2.0)-heptane-2-carboxylic acid, 6-((aminophenylacetyl)amino)-3,3-dimethyl-7-oxo-, 1,3-dihydro-3-oxo-1-isobenzofuranyl ester, (2S-(2α,5α,6β(S*)))-(9CI)

49584-26-1
$C_7H_4ClNO_2S$
201.63
Benzenesulfonyl chloride, 4-cyano- (9CI)
p-Cyanobenzenesulfonyl chloride
p-Cyanophenylsulfonyl chloride
4-Cyanobenzenesulfonyl chloride
Benzenesulfonyl chloride, p-cyano- (6CI)

49622-18-6
$C_{13}H_{28}$
184.37
Decane, 3,3,4-trimethyl- (9CI)
3,3,4-Trimethyldecane

49624-61-5
$C_5H_7ClN_4O$
174.57
1,3,5-Triazin-2(1H)-one, 4-chloro-6-(ethylamino)- (9CI)
GS 12515
s-Triazin-2-ol, 4-chloro-6-(ethylamino)-

49681-82-5
Unknown
Unknown
N,N-Dimethylamphetamine
DEA No. 1480

49690-94-0
$C_{12}H_7Br_3O$
406.90
Benzene, 1,1'-oxybis-, tribromo deriv. (9CI)
1,1'-Oxybisbenzene tribromo deriv.
Tribromodiphenyl ether
Tribromodiphenyl oxide
Tribromophenoxybenzene

49693-09-6
$C_{14}H_{10}Br_4O_4$
561.85
1,2-Benzenedicarboxylic acid, 3,4,5,6-tetrabromo-, di-2-propenyl ester (9CI)

49693-20-1
$C_{20}H_{10}Br_6N_2O_2$
789.72
1,4-Benzenedicarboxamide, N,N'-bis(2,4,6-tribromophenyl)- (9CI)

49784-44-3
$C_6H_6CdNO_6.H$
301.53
Cadmate(1-), (N,N-bis(carboxymethyl)glycinato(3-)-N,O,O',O'')-, hydrogen, (T-4)- (9CI)
Cadmium (2+) NTA

49794-90-3
$C_{18}H_{24}N_2O_6$
364.39
2,4-Dinitro-6-octylphenyl crotonate
2-Butenoic acid, 2,4-dinitro-6-octylphenyl ester (9CI)

49794-91-4
$C_{18}H_{24}N_2O_6$
364.39
2,6-Dinitro-4-octylphenyl crotonate
2-Butenoic acid, 2,6-dinitro-4-octylphenyl ester (9CI)

49813-61-8
C_6H_9N
95.14
1H-Pyrrole, dimethyl- (9CI)

49826-53-1
$C_{10}H_{14}$
134.22
Bicyclo(3.2.1)oct-2-ene, 3-methyl-4-methylene- (9CI)

49826-54-2
$C_{10}H_{14}$
134.22
Bicyclo(3.2.1)octane, 2,3-bis(methylene)- (9CI)

49833-96-7
$C_{10}H_{18}O$
154.25
4-Hepten-3-one, 5-ethyl-2-methyl- (9CI)

49839-35-2
$C_{14}H_{12}N_2O_4$
272.26
Carbamic acid, methylphenyl-, 4-nitrophenyl ester (9CI)
4-Nitrophenyl N-methyl-N-phenylcarbamate

49866-87-7
$C_{17}H_{17}N_2$
249.34
Difenzoquat
1,2-Dimethyl-3,5-diphenylpyrazolium ion
1,2-Dimethyl-3,5-diphenyl-1H-pyrazolium
1H-Pyrazolium, 1,2-dimethyl-3,5-diphenyl- (9CI)

50285-70-6
$C_7H_{15}N_3O_2$
173.25
CCN(CC)C(=O)N(CC)N=O
Urea, 3-nitroso-1,1,3-triethyl
Nitrosotriaethylharnstoff (German)
Nitrosotriethylurea
N-Nitrosotriethylurea
1,1,3-Triethyl-3-nitrosourea

50285-71-7
$C_5H_{11}N_3O_2$
145.19
CCN(N=O)C(=O)N(C)C
Urea, 1,1-dimethyl-3-ethyl-3-nitroso
1,1-Dimethyl-3-ethyl-3-nitrosourea
Nitrosoaethyldimethylharnstoff
Nitroso-1,1-dimethyl-3-ethylurea
Nitrosoethyldimethylurea
1-Nitroso-1-ethyl-3,3-dimethylurea

50285-72-8
$C_6H_{13}N_3O_2$
159.22
CCN(CC)C(=O)N(C)N=O
Urea, 1,1-diethyl-3-methyl-3-nitroso
1,1-Diethyl-3-methyl-3-nitrosourea
Nitroso-1,1-diethyl-3-methylurea
Nitrosomethyldiaethylharnstoff
Nitrosomethyldiethylurea
1-Nitroso-1-methyl-3,3-diethylurea

50317-11-8
$C_4H_8O_2$
88.11
2-Butene-1,2-diol (9CI)

50318-32-6
$C_3H_7O_2.F_6P$
220.05
Methylium, dimethoxy-, hexafluorophosphate(1-) (9CI)
Dimethoxycarbenium hexafluorophosphate
Dimethoxymethylium hexafluorophosphate(1-)

50370-12-2
$C_{16}H_{17}N_3O_5S$
363.42
CC1=C(N3C(SC1)C(NC(=O)C(N)c2ccc(O)cc2)C3=O)C(O)=O
5-Thia-1-azabicyclo(4.2.0)oct-2-ene-2-carboxylic acid, 7-((amino(4-hydroxyphenyl)acetyl) amino)-3-methyl-8-oxo-, (6R-(6-α,7-β(R*)))
BL-S 578
Bidocef
Duricef
S-578

50375-10-5
$C_7H_5Cl_3O$
211.47
Benzene, 1,2,4-trichloro-3-methoxy- (9CI)

50376-91-5
Unknown
Unknown
Copper etidronic acid complex

50397-64-3
$C_{10}H_{12}O_4S$
228.27
Acetic acid, ((4-methylphenyl)sulfonyl)-, methyl ester (9CI)
Methyl (p-tolylsulfonyl)acetate
Methyl (4-methylbenzenesulfonyl)acetate

50406-54-7
$C_3H_4N_2S$
100.14
1,2,3-Thiadiazole, 5-methyl- (7CI,9CI)
5-Methyl-1,2,3-thiadiazole

50464-96-5
$C_{13}H_{22}O_3$
226.32
2,5,10-Undecanetrione, 6,6-dimethyl- (9CI)

50471-44-8
$C_{12}H_9Cl_2NO_3$
286.12
2,4-Oxazolidinedione, 3-(3,5-dichlorophenyl)-5-methyl-5-vinyl
BAS 352 F
3-(3,5-Dichlorophenyl)-5-ethenyl-5-methyl-2,4-oxazolidinedione
3-(3,5-Dichlorphenyl)-5-methyl-5-vinyl-1,3-oxazolidin-2,4-dion (German)
Ornalin
Ronilan
Vinclozolin (German)
Vinclozoline

50473-86-4
$C_7H_{10}N_6$
178.20
Imidazo(5,1-f)(1,2,4)triazine-2,7-diamine, 4,5-dimethyl- (9CI)

50483-82-4
$C_8H_6N_2S$
162.21
1,2,4-Thiadiazole, 3-phenyl- (6CI,9CI)

50512-35-1
$C_{12}H_{18}O_4S_2$
290.42
CC(C)OC(=O)C(C(=O)OC(C)C)=C1SCCS1
Propanedioic acid, 1,3-dithiolan-2-ylidene-, bis(1-methylethyl) ester
Di-isopropyl 1,3-dithiolane-2-ylidenemalonate
Fudiolan
Fuji 1
Fujione
Fuji-one
IPT
IPT (Pesticide)
Isoprothiolane
NKK 100
NNF-109
SS 11946

50539-85-0
$C_{18}H_{21}NO_3S$
331.46
Carbamic acid, methyl((2-methylphenyl)thio)-, o-isopropoxyphenyl ester
o-Isopropoxyphenyl methyl((2-methylphenyl)thio)carbamate
N-(2-Toluenesulfenyl)propoxur

50563-36-5
$C_{13}H_{18}ClNO_2$
255.77
O=C(N(c(c(ccc1)C)c1C)CCOC)CCl
Acetamide, 2-chloro-N-(2,6-dimethylphenyl)-N-(2-meth-

oxyethyl)
A 4766
A 5089
2,6-Acetoxylidide, 2-chloro-N-(2-methoxyethyl)-
2-Chloro-N-(2,6-dimethylphenyl)-N-(2-methoxyethyl)-acetamide
2-Chloro-N-(2-methoxyethyl)-acet-2',6'-xylidide
Dimethachlor
Dimethachlore
2,6-Dimethyl-N-(2-methoxyethyl)chloroacetanilide
Teridox

50563-41-2
$C_{12}H_{16}ClNO_2$
241.74
o-Acetotoluidide, 2-chloro-N-(2-methoxyethyl)
Acetamide, 2-chloro-N-(2-methoxyethyl)-N-(2-methylphenyl)-
2-Chloro-N-(2-methoxyethyl)-o-acetotoluidide

50585-37-0
$C_{12}H_6Br_2O_2$
341.99
Dibenzo(b,e)(1,4)dioxin, 2,3-dibromo- (9CI)
2,3-Dibromodibenzo(b,e)(1,4)dioxin
2,3-Dibromodibenzo-p-dioxin

50585-38-1
$C_{12}H_6F_2O_2$
220.18
Dibenzo(b,e)(1,4)dioxin, 2,3-difluoro- (9CI)
2,3-Difluorodibenzo(b,e)(1,4)dioxin
2,3-Difluorodibenzo-p-dioxin

50585-39-2
$C_{12}H_6Cl_2O_2$
253.08
Dibenzo-p-dioxin, 1,3-dichloro
Dibenzo(b,e)(1,4)dioxin, 1,3-dichloro-
1,3-Dichlorodibenzo-para-dioxin

50585-40-5
$C_{12}H_4Br_2Cl_2O_2$
410.88
Dibenzo(b,e)(1,4)dioxin, 2,3-dibromo-7,8-dichloro- (9CI)
2,3-Dibromo-7,8-dichlorodibenzo(b,e)(1,4)dioxin
2,3-Dibromo-7,8-dichlorodibenzo-p-dioxin

50585-41-6
$C_{12}H_4Br_4O_2$
499.78
Dibenzo(b,e)(1,4)dioxin, 2,3,7,8-tetrabromo- (9CI)
Dibenzo-p-dioxin, 2,3,7,8-tetrabromo-
2,3,7,8-Tetrabromodibenzo-p-dioxin

50585-42-7
$C_{12}H_4Cl_2F_2O_2$
289.06
Dibenzo(b,e)(1,4)dioxin, 2,3-dichloro-7,8-difluoro- (9CI)
2,3-Dichloro-7,8-difluorodibenzo(b,e)(1,4)dioxin
2,3-Dichloro-7,8-difluorodibenzo-p-dioxin

50585-43-8
$C_{12}H_4Br_2F_2O_2$
377.97
Dibenzo(b,e)(1,4)dioxin, 2,3-dibromo-7,8-difluoro- (9CI)
2,3-Dibromo-7,8-difluorodibenzo(b,e)(1,4)dioxin
2,3-Dibromo-7,8-difluorodibenzo-p-dioxin

50585-46-1
$C_{12}H_4Cl_4O_2$
321.96
Dibenzo-p-dioxin, 1,3,7,8-tetrachloro
Dibenzo(b,e)(1,4)dioxin, 1,3,7,8-tetrachloro-
1,3,7,8-Tetrachlorodibenzodioxin
1,3,7,8-Tetrachlorodibenzo-para-dioxin

50598-50-0
$C_7H_{10}O_2$
126.16
2(5H)-Furanone, 3,5,5-trimethyl- (9CI)
2,4,4-Trimethyl-2-butenolide

50602-11-4
$C_{14}H_9Cl$
212.68
Anthracene, chloro- (9CI)

50623-57-9
$C_{13}H_{26}O_2$
214.35
Nonanoic acid, butyl ester (9CI)

50639-00-4
$C_8H_{16}O$
128.21
2-Hexen-1-ol, 2-ethyl- (9CI)

50639-02-6
$C_{12}H_{24}O$
184.32
5-Undecanone, 2-methyl- (9CI)

50642-02-9
Unknown
Unknown
Teepol CH 31 (9CI)

50642-03-0
$C_{18}H_{30}O_3S.(C_2H_4O)xC_{15}H_{24}O.Na$
Polymer NA
Benzenesulfonic acid, dodecyl-, sodium salt, Mixt. with α-(nonylphenyl)-ω-hydroxypoly(oxy-1,2-ethanediyl) (9CI)
Comprox
Teepol CH 53

50646-06-5
ClMn
90.39
Manganese chloride (MnCl) (6CI,7CI,9CI)
Manganese monochloride

50648-02-7
$C_{12}H_{12}Cd_3N_2O_{12}$
713.45
Nitrilotriacetic acid, tricadmium(2+) complex
Cadmium, bis(N,N-bis(carboxymethyl)glycinato(3-))tri- (9CI)
Cadmium nitrilotriacetic acid
NTA, tricadmium(2+) complex

50694-81-0
$C_7H_8Cl_2N_2$
191.05
1,3-Benzenediamine, 2,4-dichloro-6-methyl- (9CI)

50694-82-1
$C_7H_8Cl_2N_2$
191.05
1,3-Benzenediamine, 4,6-dichloro-2-methyl- (9CI)

50694-83-2
$C_7H_8Cl_2N_2$
191.0
1,3-Benzenediamine, 4,5-dichloro-6-methyl- (9CI)

50700-49-7
$C_8H_7NO_2$
149.16
Acetamide, N-(4-oxo-2,5-cyclohexadien-1-ylidene)
Acetimidoquinone
N-Acetyl-p-benzoquinone imine
N-Acetyl-p-quinonimine
2,5-Cyclohexadien-1-one, 4-acetylimino-

Napqi
N-(4-Oxo-2,5-cyclohexadien-ylidene)acetamide
N-(4-Oxo-2,5-cyclohexadien-1-ylidene)acetamide

50702-38-0
$C_6H_4ClIO_2S$
302.52
Benzenesulfonyl chloride, 3-iodo- (9CI)
3-Iodobenzenesulfonyl chloride

50723-80-3
Unknown
Unknown
Sodium bentazon
Caswell No. 509D
EPA Pesticide Chemical Code 103901
3-Isopropyl-1H-2,1,3-benzothiadiazin-4(3H)-one-2,2-dioxide, sodium salt

50746-55-9
$C_{13}H_{22}$
178.32
Bicyclo(3.1.1)heptane, 2,6,6-trimethyl-3-(2-propenyl)-, (1α,2β,3α,5α)- (9CI)

50782-69-9
$C_{11}H_{26}NO_2S$
236.44
Phosphonothioic acid, methyl-, S-(2-(diisopropylamino)ethyl) O-ethyl ester
S-(2-Diisopropylaminoethyl) O-ethyl methyl phosphonothiolate
EA 1701
Ethyl S-2-diisopropylaminoethyl methylphosphonothiolate
Ethyl-S-diisopropylaminoethyl methylthiophosphonate
O-Ethyl S-2-diisopropylaminoethyl methylphosphonothiote
O-Ethyl-S-2-diisopropylaminoethylester kyseliny methylthiofosfonove (Czech)
Ethyl S-dimethylaminoethyl methylphosphonothiolate
Methylphosphonothioic acid S-(2-(bis(methylethyl)amino)-ethyl) O-ethyl ester
Phosphonothioic acid, methyl-, S-(2-(bis(1-methylethyl)-amino)ethyl) O-ethyl ether
VX

50789-44-1
$C_{15}H_{14}O_3$
242.27
O=C(OCc(cc(Oc(cccc1)c1)cc2)c2)C
Benzenemethanol, 3-phenoxy-, acetate (9CI)
α-Hydroxy-3-phenoxytoluene, acetate
3-Phenoxybenzenemethanol acetate

50814-31-8
Unknown
Unknown
C.I. Direct Orange 46 (9CI)
Sirius Supra Orange 7GL

50815-77-5
$C4H_5CaCl_2NOS.C_4H_4CaCl_3NOS$
486.54
Calcium, dichloro(5-chloro-2-methyl-3(2H)-isothiazolone-O)-, Mixt. with dichloro(2-methyl-3(2H)-isothiazolone-O)calcium (9CI)
Euxyl K 100
RH 886

50816-31-4
$C_5H_{12}N_2O$
116.16
Diethylurea

50819-06-2
C_7H_{14}
98.19
Pentene, 2,2-dimethyl- (9CI)

50825-29-1
$C_7H_{12}O_2Pb$
335.36
Lead naphthenate
Cyclohexanecarboxylic acid, lead salt (9CI)
Lead cyclohexanecarboxylate

50851-28-0
$C_{27}H_{25}O_4P$
444.47
Phosphoric acid, ar'-(1-methylethyl)-(1,1'-biphenyl)yl diphenyl ester (9CI)

50854-94-9
$C_{17}H_{28}O_3S$
312.47
Benzenesulfonic acid, undecyl- (9CI)
Undecylbenzenesulfonic acid

50901-13-8
$C_5H_6Cl_2O_4$
201.01
Pentanedioic acid, 2,2-dichloro- (9CI)
α,α-Dichloroglutaric acid
2,2-Dichloroglutaric acid
Glutaric acid, 2,2-dichloro- (6CI)

50922-29-7
$CrO_4Zn.H_2O_2Zn$
280.76
Chromic acid, zinc salt, basic
Basic Zinc Chromate

50922-60-6
Unknown
Unknown
C.I. Disperse Blue 139 (9CI)
Resolin Navy Blue GLS
Samaron Navy Blue GR

50933-33-0
$C_{18}H_{32}O_2$
280.45
O=C(OCCCCCCC=CCCC=CCCCC)C
Gossyplure
7,11-Hexadecadien-1-ol, acetate (9CI)
(Z,E)-7,11-Hexadecadien-1-ol, acetate
Nomate PBW

50958-32-2
Unknown
Unknown
Teepol CH 610 (9CI)

50973-35-8
$(C_6H_6O_4S.CH_2O)x$
Polymer
Benzenesulfonic acid, hydroxy-, Polymer with formaldehyde (9CI)

50976-02-8
$C_{13}H_{14}O_2$
202.27
C=CC(=O)OC3=CC2C1CC(C=C1)C2C3
Acrylic acid, ((3a,4,7,7a-tetrahydro)4,7-methanoindenyl) ester
Acrylic acid, dicyclopentadienyl ester
DCPA
Dicyclopentadienyl acrylate
Dicyclopentenyl acrylate
2-Propenoic acid, 3a,4,7,7a-tetrahydro-4,7-methano-1H-indenyl ester

51000-52-3
$C_{12}H_{22}O_2$
198.31
Neodecanoic acid, ethenyl ester (9CI)
Ethenyl neodecanoate
Neodecanoic acid, vinyl ester

51004-63-8
Unknown
Unknown
ASA 3 (Antistatic agent)

51022-71-0
$C_{24}H_{36}O_3$
372.55
Nabilone
Cesamet
DEA No. 7379
9H-Dibenzo(b,d)pyran-9-one, 3-(1,1-dimethylheptyl)-6,6a,7,8,10,10a-hexahydro-1-hydroxy-6,6-dimethyl-, trans-(+-)-
(+)-trans-3-(1,1-Dimethylheptyl)-6,6a,7,8,10,10a-hexahydro 1-hydroxy-6,6-dimethyl-9H-dibenzo(b,d)pyran-9-one)
(+-)-3-(1,1-Dimethylheptyl-6,6aβ,7,8,10,10aα-hexahydro-1-hydroxy-6,6-dimethyl-9H-dibenzo(b,d)pyran-9-one
Lilly 109514

51023-22-4
$C_4H_5Cl_3$
159.44
Butene, trichloro
Trichlorbutylene
Trichlorobutene [UN 2322]
UN 2322 [Trichlorobutene]

51026-28-9
$C_3H_6NOS_2.K$
175.31
Potassium N-hydroxymethyl-N-methyldithiocarbamate
Bunema
Carbamic acid, N-hydroxy-methyl-n-methyldithio-, potassium salt

51037-30-0
$C_6H_6N_2O_3$
154.14
Cc1cnc(cn1=O)C(O)=O
2-Pyrazinecarboxylic acid, 5-methyl-, 4-oxide
Acipimox
5-Methyl-2-pyrazinecarboxylic acid 4-oxide

51083-28-4
$C_{17}H_{15}N_5O_3$
337.31
Butanamide, N-(2,3-dihydro-2-oxo-1H-benzimidazol-5-yl)-3-oxo-2-(phenylazo)- (9CI)

51109-97-8
Unknown
Unknown
Solvesso (9CI)

51116-73-5
$C_6H_{13}Br$
165.07
Pentane, 1-bromo-3-methyl- (9CI)

51149-70-3
$C_9H_{18}O_2$
158.24
2-Heptanone, 1-ethoxy- (9CI)
NSC-244920

51158-21-5
Unknown
Unknown
Corexit 8666 (9CI)
Esso Corexit 8666

51170-59-3
$C_9H_7Cl_3O_3.C_4H_{11}NO_2$
374.64
Propanoic acid, 2-(2,4,5-trichlorophenoxy)-, Compd. with 2,2'-iminobis(ethanol) (1:1) (9CI)
Caswell No. 739D
EPA Pesticide Chemical Code 082516
Silvex, diethanolamine salt
2-(2,4,5-Trichlorophenoxy)propionic acid, diethanolamine salt

51200-87-4
$C_5H_{11}NO$
101.17
O(CC(N1)(C)C)C1
Oxazolidine, 4,4-dimethyl
Dimethyl oxazolidine
4,4-Dimethyloxazolidine
Oxazolidine A

51207-31-9
$C_{12}H_4Cl_4O$
305.96
Clc3cc2oc1cc(Cl)c(Cl)cc1c2cc3Cl
Dibenzofuran, 2,3,7,8-tetrachloro
NCI-C56611
2,3,7,8-Tetrachlorodibenzofuran

51218-45-2
$C_{15}H_{22}ClNO_2$
283.83
CCc1cccc(C)c1N(C(C)COC)C(=O)CCl
o-Acetotoluidide, 2-chloro-6'-ethyl-N-(2-methoxy-1-methylethyl)
Acetamide, 2-chloro-N-(6-ethyl-o-tolyl)-N-(2-methoxy-1-methylethyl)-
2-Aethyl-6-methyl-N-(1-methyl-2-methoxyaethyl)-chloracetanilid (German)
Bicep
CGA-24705
α-Chlor-6'-aethyl-n-(2-methoxy-1-methylaethyl)-acet-o-toluidin (German)
2-Chloro-6'-ethyl-N-(2-methoxy-1-methylethyl)acet-o-toluidide
α-Chloro-2'-ethyl-6'-methyl-N-(1-methyl-2-methoxyethyl)-acetanilide
2-Chloro-N-(2-ethyl-6-methylphenyl)-N-(2-methoxy-1-methylethyl)acetamide
Codal
Cotoran Multi
Dual
2-Ethyl-6-methyl-1-N-(2-methoxy-1-methylethyl)chloroacetanilide
2-Etylo-6-metylo-N-(1'-metylo-2'-metoksyetylo)chloroacetanilid (Polish)
Metachlore
Metelilachlor
Metolachlor
Ontrack 8E
Primagram
Primextra

51219-00-2
$C_{13}H_{21}NO$
207.31
O(CC(Nc(c(ccc1)CC)c1C)C)C
2-Ethyl-6-methyl-N-(1'-methyl-2-methoxyethyl)-aniline
Benzenamine, 2-ethyl-N-(2-methoxy-1-methylethyl)-6-methyl- (9CI)

51230-49-0
$C_{12}H_7ClO$
202.64
2-Chlorodibenzofuran

51234-28-7
$C_{16}H_{12}ClNO_3$
301.74
CC(C(O)=O)c2ccc1oc(nc1c2)c3ccc(Cl)cc3
5-Benzoxazoleacetic acid, 2-(4-chlorophenyl)-α-methyl
Benoxaprofen
2-(4-Chlorophenyl)-α-methyl-5-benzoxazoleacetic acid
Compound 90459
Coxigon
Opren
Oraflex
Uniprofen

51235-04-2
$C_{12}H_{20}N_4O_2$
252.36
O=C(N=C(N(C1=O)C)N(C)C)N1

C(CCCC2)C2
s-Triazine-2,4(1H,3H)-dione, 3-cyclohexyl-6-(dimethylamino)-1-methyl
3-Cyclohexyl-6-(dimethylamino)-1-methyl-s-triazine-2,4(1H,3H)-dione
3-Cyclohexyl-6-(dimethylamino)-1-methyl-1,3,5-triazine-2,4(1H,3H)-dione
DPX 3674
Hexazinone
Velpar
Velpar Weed Killer

51247-87-1
$C_{17}H_{28}O_6$
328.41
2-Propenoic acid, 2-methyl-, (1-methyl-1,2-ethanediyl)bis(oxy(methyl-2,1-ethanediyl)) ester (9CI)
Tripropylene glycol dimethacrylate

51249-05-9
$C_{18}H_{37}NO_3P$
346.53
Phosphonic acid, 1-(butylamino)cyclohexyl-, dibutyl ester
Aminophon
Buminafos
1-(Butylamino)cyclohexylphosphonic acid dibutyl ester
CKB 1028A
O,O-Dibutyl 1-butylamino-cyclohexylphosphonate
Trakephon

51249-07-1
$C_{21}H_{25}N_5O_4$
411.43
3-Pyridinecarbonitrile, 1-(2-ethylhexyl)-1,2-dihydro-6-hydroxy-4-methyl-5-((2-nitrophenyl)azo)-2-oxo- (9CI)

51264-14-3
$C_{21}H_{19}N_3O_3S$
393.49
COc1cc(NS(C)(=O)=O)ccc1Nc3c2ccccc2nc4ccccc34
Methanesulfon-m-anisidide, 4'-(9-acridinylamino)
4'-(9-Acridinylamino)methanesulfon-m-anisidide
4'-(9-Acridinylamino)methanesulphon-m-anisidide
4'-(9-Acridinylamino)-3'-methoxymethanesulfonanilide
4'-(9-Acridinylamino)-methylsulfonyl-m-anisidine
AMSA
m-AMSA
Amsacrine
m-AMSA Methanesulfonate
Amsidine
Amsine
Methanesulfonamide, N-(4-(9-acridinylamino)-3-methoxyphenyl)-
Methanesulfonanilide, 4'-(9-acridinylamino)-3'-methoxy-
NSC-249992

51266-87-6
$C_6H_{14}O_5$
166.17
Propanediol, oxybis- (9CI) (VAN)

51273-71-3
FH.FNa
61.99
Hydrofluoric acid, Mixt. with sodium fluoride (9CI)

51276-47-2
$C_5H_{12}NO_4P$
181.13
Phosphinothricin
2-Amino-4-methylphosphinobutyric acid
Ammonium glufosinate

51282-49-6
$C_8H_6ClNO_4$
215.59
O=C(OC)c(c(N(=O)=O)ccc1Cl)c1
Benzoic acid, 5-chloro-2-nitro-, methyl ester (9CI)
Methyl 5-chloro-2-nitrobenzoate

51282-69-0
$C_{14}H_9ClFNO_5$
325.67
Benzoic acid, 5-(2-chloro-4-fluorophenoxy)-2-nitro-, methyl ester (9CI)
MC 6063

51287-84-4
$C_{26}H_{56}S_2Sn$
551.57
Stannane, bis(dodecylthio)dimethyl- (9CI)
Bis(dodecylthio)dimethylstannane

51308-54-4
$C_{21}H_{28}N_2S_2$
372.63
Carbonimidodithioic acid, 3-pyridinyl-, butyl (4-(1,1-dimethylethyl)phenyl)methyl ester
Buthiobate
Butyl 4-tert-butylbenzyl N-(3-pyridyl)dithiocarbonimidate
S-n-Butyl S'-p-tert-butylbenzyl N-3-pyridyldithiocarbonimidate
Butyl (4-(1,1-dimethylethyl)phenyl)methyl 3-pyridinylcarbonimidodithioate
Denmert
3-Pyridinylcarbonimidodithioic acid butyl (4-(1,1-dimethylethyl)phenyl)methyl ester
S-1358

51317-24-9
$C_6H_5NO_4.xPb$
Unknown
Lead mononitroresorcinate
Initiating Explosive Lead Mononitroresorcinate
Lead mononitroresorcinate (Dry)
Lead nitroresorcinate
Mononitroresorcinate de plomb (French)
NA 2811
Nitroresorcinato de plomo (Spanish)

51338-10-4
$C_{14}H_{10}Cl_2O_4$
313.14
Acetic acid, (4-(2,4-dichlorophenoxy)phenoxy)- (9CI)

51338-27-3
$C_{16}H_{14}Cl_2O_4$
341.20
COC(=O)C(C)Oc2ccc(Oc1ccc(Cl)cc1Cl)cc2
Propionic acid, 2-(4-(2,4-dichlorophenoxy)phenoxy)-, methyl ester
2-(4-(2,4-Dichlorophenoxy)-phenoxy)-methyl-propionate
Diclofop-Methyl
HOE 23408
Hoegrass
Hoelon
Hoelon 3EC
Illoxan
Iloxan
Methyl 2-(4-(2,4-dichlorophenoxy)phenoxy)propionate

51344-62-8
$(C_2H_4O)x(C_2H_4O)xC_{24}H_{39}NO_2$
Polymer
Dehydroabietylamine-ethylene oxide condensate
Caswell No. 277
Dehydroabietylamine, ethylene oxide adduct
Dehydroabietylamine-ethylene oxide condensate
EPA Pesticide Chemical Code 004203
Polyethylene oxide, dehydroabietylamine polymer
Poly(oxy-1,2-ethanediyl), α,α'-((((1,2,3,4,4a,9,10,

10a-octahydro-1,4a-dimethyl-7-(1-methylethyl)-1-phenanthrenyl)methyl)imino)di-2,1-ethanediyl)bis(ω-hydroxy-, (1R-(1α,4aβ,10aα))- (9CI)

51363-64-5
$C_{26}H_{47}O_4P$
454.63
Diisodecyl phenyl phosphate
Phosphoric acid, diisodecyl phenyl ester (9CI)

51365-70-9
$C_{23}H_{43}N_3O_3$
409.62
Hexanoic acid, 2-ethyl-, Compd. with 2,4,6-tris((dimethylamino)methyl)phenol (9CI)

51366-52-0
$C_{12}H_{13}N$
171.24
Quinoline, trimethyl- (9CI)

51388-00-2
$C_8H_{18}N_2$
142.23
1H-Azepine-1-ethanamine, hexahydro- (9CI)
Hexamethylenimine, 1-(2-aminoethyl)-

51410-44-7
$C_{10}H_{12}O_2$
164.22
COc1ccc(cc1)C(O)C=C
Benzyl alcohol, p-methoxy-α-vinyl
Estragole, 1'-hydroxy-
1'-Hydroxyestragole

51422-54-9
$C_8H_{18}O_2$
146.23
O(CCOC(C)(C)C)CC
Propane, 2-(2-ethoxyethoxy)-2-methyl- (9CI)
2-(2-Ethoxyethoxy)-2-methylpropane

51422-75-4
$C_6H_{10}BrCl$
197.50
Cyclohexane, 1-bromo-2-chloro-, cis- (9CI)

51461-71-3
$C_{12}H_{22}N_4O_3S$
302.37
Urea, N-(2,2-dimethoxyethyl)-N'-(5-(1,1-dimethylethyl)-1,3,4-thiadiazol-2-yl)-N-methyl- (9CI)
HCS 3510
VEL 3510

51479-36-8
$C_{16}H_{31}ClO_2$
290.87
Tetradecanoic acid, 2-chloroethyl ester (9CI)
NSC-406547

51481-10-8
$C_{15}H_{20}O_6$
296.35
CC4=CC3OC1C(O)CC(C)(C12CO2)C3(CO)C(O)C4=O
Trichothec-9-en-8-one, 12,13-epoxy-3,7,15-trihydroxy-, (3-α,7-α)
Dehydronivalenol
Deoxynivalenol
4-Deoxynivalenol
Desoxynivalenol
Don
Vomitoxin

51481-61-9
$C_{10}H_{16}N_6S$
252.38
CNC(NCCSCc1nc[nH]c1C)=NC#N
Guanidine, N-cyano-N'-methyl-N''-(2-(((5-methyl-1H-imidazol-4-yl)methyl)thio)ethyl)
Cimetidine
N-Cyano-N'-methyl-N''-(2-(((5-methyl-1H-imidazol-4-yl)-methyl)thio)ethyl)guanidine
1-Cyano-2-methyl-3-(2-(((5-methyl-4-imidazolyl)methyl)-thio)ethyl)guanidine
2-Cyano-1-methyl-3-(2-(((5-methylimidazol-4-yl)methyl)-thio)ethyl)guanidine
Eureceptor
FPF 1002
Gastromet
SKF 92334
Tagamet
Tametin
Tratul
Ulcedine
Ulcimet
Ulcomet

51481-65-3
$C_{21}H_{24}N_5O_8S_2.Na$
561.61
CC4(C)SC3C(NC(=O)C(NC(=O)N1CCN(C1=O)S(C)(=O)=O)c2ccccc2)C(=O)N3C4C(O)=O
4-Thia-1-azabicyclo(3.2.0)-heptane-2-carboxylic acid, 3,3-dimethyl-6-(((((3-(methylsulfonyl)- 2-oxo-1-imidazolidinyl)carbonyl)-amino)phenylacetyl)amino)-7-oxo-, sodium salt, (2s-(2-α,5-α,6-β)(S*))
Antibiotic Bay-f 1353
Mezlocillin

51487-69-5
$C_{11}H_{14}ClNO_4$
259.71
Phenol, 2-(2-chloro-1-methoxyethoxy)-, methylcarbamate
BAS 263
BAS 263I
2-(2-Chloro-1-methoxyethoxy)-phenol methylcarbamate
Cloethocarb
Cloetocarb
Lance

51496-03-8
$C_{12}H_{19}O_4P$
258.25
Phosphoric acid, bis(1-methylethyl) phenyl ester (9CI)

51555-31-8
$C_{15}H_{32}O_{11}$
388.41
Pentaglycerol (9CI)

51555-36-3
$(C_5H_8O_2.C_5H_8O)x$
Polymer NA
2-Propenoic acid, 2-methyl-, methyl ester, Polymer with 3-methyl-3-buten-2-one (9CI)
3-Buten-2-one, 3-methyl-, Polymer with methyl 2-methyl-2-propenoate (9CI)
3-Methyl-3-buten-2-one-methyl methacrylate copolymer
Methyl isopropenyl ketone-methyl methacrylate copolymer
Methyl isopropenyl ketone-methyl methacrylate polymer
Methyl methacrylate-methyl 1-methylvinyl ketone copolymer

51580-86-0
Unknown
Unknown
Sodium dichloroisocyanurate dihydrate
Sodium dichloro-s-triazinetrione dihydrate

51584-27-1
$C_{10}H_{15}OPSe$
261.16
Phosphinoselenoic acid,

(1,1-dimethylethyl)phenyl-, (R)- (9CI)

51584-28-2
$C_{10}H_{15}OPSe$
261.16
Phosphinoselenoic acid, (1,1-dimethylethyl)-phenyl-, (S)- (9CI)

51590-67-1
$C_4H_{10}OSn$
192.83
Stannane, butyloxo- (9CI)
Butyloxostannane
Monobutyltin oxide

51607-94-4
$C_{18}H_{32}O_2$
280.45
7,11-Hexadecadien-1-ol, acetate (Stereoisomer unspec.)
AI3-36135
7,11-Hexadecadien-1-ol, acetate, (Z,E)-

51630-58-1
$C_{25}H_{22}ClNO_3$
419.93
CC(C)C(C(=O)OC(C#N)c2cccc(Oc1ccccc1)c2)c3ccc(Cl)cc3
Benzeneacetic acid, 4-chloro-α-(1-methylethyl)-, cyano-(3-phenoxyphenyl)methyl ester
Belmark
α-Cyano-3-phenoxybenzyl 2-(4-chlorophenyl)isovalerate
α-Cyano-3-phenoxybenzyl-2-(4-chlorophenyl)-3-methylbutyrate
Cyano(3-phenoxyphenyl)methyl 4-chloro-α-(1-methylethyl)-benzeneacetate
Ectrin
Fenvalerate
Phenvalerate
Pydrin
S 5602
Sanmarton
SD 43775
Sumicidin
Sumifly
Sumipower
WL 43775

51632-16-7
$C_{13}H_{11}BrO$
263.13
O(c(cccc1CBr)c1)c(cccc2)c2
Benzene, 1-(bromomethyl)-3-phenoxy- (9CI)
1-(Bromomethyl)-3-phenoxybenzene
α-Bromo-3-phenoxytoluene
3-Phenoxybenzyl bromide

51707-55-2
$C_9H_8N_4OS$
220.27
Urea, 1-phenyl-3-(1,2,3-thiadiazol-5-yl)
Defolit
Dropp
N-Phenyl-N'-1,2,3-thiadiazol-5-yl-urea
SN 49537
(N-1,2,3-Thiadiazolyl-5)-N'-phenylurea
Thidiazuron

51762-05-1
$C_{16}H_{19}N_3O_5S$
365.44
COC1=C(N3C(SC1)C(NC(=O)C(N)C2C=CCC=C2)C3=O)C(O)=O
5-Thia-1-azabicyclo(4.2.0)oct-2-ene-2-carboxylic acid, 7-((amino-1,4-cyclohexadien-1- ylacetyl)amino)-3-methoxy-8-oxo-, (6R-(6-α,7-β-(R*)))
7-(D-2-Amino-2-(1,4-cyclohexadienyl)acetamide)-3-methoxy-3-cephem-4-carboxylic acid
Antibiotic CGP 9000
Cefroxadin
Cefroxadine
CGP 9000
CXD
Oraspor

51772-35-1
$C_{24}H_{29}N$
331.49
1-Naphthalenamine, N-((1,1,3,3-tetramethylbutyl)-phenyl)- (9CI)
tert-Octylphenyl-α-naphthylamine

51775-36-1
$C_{10}H_{11}Cl_7$
379.36
Bornane, 2,2,5-endo,6-exo,-8,9,10-heptachloro
Bicyclo(2.2.1)heptane, 2,2,5,6-tetrachloro-1,7,7-tris(chloromethyl)-, (5-endo,6-exo)-
2,2,5-endo,6-exo,8,9,10-Heptachlorobornane
5-endo,6-exo-2,2,5,6-Tetrachloro-1,7,7-tris(chloromethyl)-bicyclo(2.2.1)heptane
Toxaphene Toxicant B

51786-53-9
$C_8H_{11}N.ClH$
157.66
Cl.Cc1ccc(C)c(N)c1
2,5-Xylidine, hydrochloride
1-Amino-2,5-dimethylbenzene hydrochloride
3-Amino-1,4-dimethylbenzene hydrochloride
5-Amino-1,4-dimethylbenzene hydrochloride
2-Amino-4-methyltoluene hydrochloride
2-Amino-1,4-xylene hydrochloride
2,5-Dimethyl aniline hydrochloride
2,5-Dimethylaniline hydrochloride
2,5-Dimethylbenzenamine hydrochloride
5-Methyl-o-toluidine hydrochloride
6-Methyl-m-toluidine hydrochloride
para-Xylidine hydrochloride

51787-42-9
$C_{20}H_{17}N$
271.38
Benz(c)acridine, 7,9,11-trimethyl
1,3,10-Trimethyl-7,8-benzacridine (French)
7,9,11-Trimethylbenz(c)acridine

51787-43-0
$C_{20}H_{17}N$
271.38
Benz(a)acridine, 8,10,12-trimethyl
1,3,10-Trimethyl-5,6-benzacridine (French)

51787-44-1
$C_{21}H_{19}N$
285.41
Benz(c)acridine, 7,8,9,11-tetramethyl
1,3,4,10-Tetramethyl-7,8-benzacridine (French)
7,8,9,11-Tetramethylbenz(c)-acridine

51811-79-1
Unknown
Unknown
Poly(oxy-1,2-ethanediyl), α-(nonylphenyl)-ω-hydroxy-, phosphate (9CI)
Ethoxylated nonylphenol phosphate
Nonylphenol, ethoxylated and phosphated
Phosphated, ethoxylated nonylphenol
Phosphoric ester of poly(oxyethylene) nonylphenol ether

51845-86-4
Unknown
Unknown
Ethyl borate
UN 1176

51892-16-1
$C_6H_6N_2O$
122.13
3-Pyridinecarboxaldehyde, oxime, (E)- (9CI)

51908-16-8
$C_{12}H_4Cl_6$
360.88
1,1'-Biphenyl, 2,2',3,4',5,5'-hexachloro- (9CI)
2,2',3,4',5,5'-Hexachlorobiphenyl
2,3,5,2',4',5'-Hexachlorobiphenyl
PCB 146
2,2',3,4',5,5'-PCB

51908-64-6
C_4H_5Cl
88.54
1-Butyne, 4-chloro- (9CI)
4-Chloro-1-butyne

51913-96-3
$C_{18}H_{12}N_2$
256.31
Biquinoline (9CI)

51936-55-1
$C_{13}H_{12}Br_2Cl_6$
540.76
BrC(C(Br)CCC(C(C(=C(C12Cl)Cl)Cl)(C2(Cl)Cl)Cl)C1C3)C3
1,4-Methanobenzocyclooctene, 7,8-dibromo-1,2,3,4,11,11-hexachloro-1,4,4a,5,6,7,8,9,10,10a-decahydro- (9CI)

51937-92-9
$C_{14}H_8Cl_2FNO_5$
360.13
Benzoic acid, 5-(2,4-dichloro-6-fluorophenoxy)-2-nitro-, methyl ester (9CI)
MC 7181
Methyl 5-(2,4-dichloro-6-fluorophenoxy)-2-nitrobenzoate
Methyl 5-(2',4'-dichloro-6'-fluorophenoxy)-2-nitrobenzoate

51945-98-3
$C_7H_{12}O_2$
128.17
1,5-Heptadiene-3,4-diol (9CI)

51953-10-7
$C_{12}H_{22}O$
182.31
Naphthalene, 4a-ethoxydecahydro-, cis- (9CI)
cis-9-Ethoxydecalin

51962-63-1
$C_8H_{12}Cl_4$
250.00
Cyclohexane, 1,2-dichloro-4-(1,2-dichloroethyl)- (9CI)

51974-40-4
$C_{12}H_5Br_3O_2$
420.90
2,3,7-Tribromobenzo-4-dioxin
2,3,7-TBDD

51990-04-6
$C_9H_{19}NOS.C_8H_{11}Cl_2NO$
397.45
Eradicane
EPTC Plus Inert Herbicide Safener
EPTC Plus R-25788
R-25788 Plus EPTC

52006-63-0
C_6H_8S
112.20
CCc1ccsc1
Thiophene, ethyl- (9CI)

52078-56-5
$C_{23}H_{46}$
322.62
11-Tricosene (9CI)
NSC-66473

52106-86-2
$C_7H_7ClO.Na$
165.57
Phenol, 4-chloro-2-methyl-, sodium salt (9CI)
4-Chloro-2-methylphenol sodium salt

52112-04-6
$C_{21}H_{20}NP$
317.36
1,3-Azaphospholidine, 1,2,3-triphenyl- (9CI)

52166-72-0
$C_6H_4Cl_2O.Na$
185.99
Phenol, 2,5-dichloro-, sodium salt (9CI)
2,5-Dichlorophenol, sodium salt

52175-10-7
$C_5H_5N_4.xH_3O_4P$
807.14
Adenine, phosphate
1H-Purin-6-amine, phosphate

52181-51-8
$C_7H_4ClF_3$
180.56
Benzene, chloro(trifluoromethyl)
Chloro(trifluoromethyl)benzene
(Trifluoromethyl)chlorobenzene

52184-19-7
$C_{22}H_{29}N_3O_3$
383.47
O=N(=O)c(c(N=Nc(c(O)c(cc1C(CC)(C)C)C(CC)(C)C)c1)ccc2)c2
Phenol, 2,4-bis(1,1-dimethylpropyl)-6-((2-nitrophenyl)azo)- (9CI)

52196-74-4
$C_{10}H_{13}ClO_2$
200.66
O(c(c(cc(OCC)c1)Cl)c1)CC
Benzene, 2-chloro-1,4-diethoxy- (9CI)
Chloro-2,5-diethoxybenzene
1-Chloro-2,5-diethoxybenzene
2-Chloro-1,4-diethoxybenzene
2,5-Diethoxychlorobenzene
2,5-Diethoxy-1-chlorobenzene
NSC-89737

52198-64-8
$C_7H_{18}N_2$
130.22
1,3-Propanediamine, N-(1,1-dimethylethyl)- (9CI)

52207-99-5
$C_{18}H_{32}O_2$
280.45
(Z,Z)-7,11-Hexadecadien-1-ol, acetate
AI3-36282
7,11-Hexadecadien-1-ol, acetate, (Z,Z)-

52221-67-7
$C_4H_{12}O_4P.1/2C_2O_4$
199.12
Phosphonium, tetrakis-(hydroxymethyl)-, ethanedioate (2:1) (Salt) (9CI)

52232-24-3
Unknown
Unknown
Paraplex (9CI)

52235-18-4
$C_{27}H_{28}N_2O_4$
444.52
Pyrrolidine, 1-(2-(4-(1-(4-meth-

oxyphenyl)-2-nitro-2-phenylethenyl)phenoxy)ethyl)-, (Z)- (9CI)

52251-71-5
$C_{16}H_{14}$
206.29
Anthracene, 2-ethyl- (9CI)

52253-69-7
$C_9H_8N_2S.BrH.H_2O$
176.24
2-Thiazolamine, 4-phenyl-, monohydrobromide, monohydrate (9CI)
2-Amino-4-phenylthiazole hydrobromide monohydrate

52275-04-4
$C_{17}H_{22}O_2$
258.36
Benzaldehyde, 4-hydroxy-3,5-bis(3-methyl-2-butenyl)- (9CI)

52289-93-7
C_8H_9BrO
201.06
Benzene, 1-(bromomethyl)-2-methoxy- (9CI)
Anisole, o-(bromomethyl)-

52292-17-8
$(C_2H_4O)xC_{18}H_{38}O$
Polymer
Arosurf
Isosteareth-2
Isosteareth-3
Isosteareth-10
Isosteareth-12
Isosteareth-20
Isosteareth-22
Isosteareth-50
PEG-2 Isostearyl Ether
PEG-3 Isostearyl Ether
PEG-10 Isostearyl Ether
PEG-12 Isostearyl Ether
PEG-20 Isostearyl Ether
PEG-22 Isostearyl Ether
PEG-50 Isostearyl Ether
Polyethoxylated isooctadecanol
Polyethylene glycol (3) isostearyl ether
Polyethylene glycol (22) isostearyl ether
Polyethylene glycol (50) isostearyl ether
Polyethylene glycol 100 isostearyl ether
Polyethylene glycol 500 isostearyl ether
Polyethylene glycol 600 isostearyl ether
Polyethylene glycol 1000 isostearyl ether
Poly(oxy-1,2-ethanediyl), α-isooctadecyl-ω-hydroxy- (9CI)
Polyoxyethylene (2) isostearyl ether
Polyoxyethylene (3) isostearyl ether
Polyoxyethylene (10) isostearyl ether
Polyoxyethylene (12) isostearyl ether
Polyoxyethylene (20) isostearyl ether
Polyoxyethylene (22) isostearyl ether
Polyoxyethylene (50) isostearyl ether

52299-20-4
$C_5H_{13}NO_2$
119.16
OCNC(CO)(C)C
2-((Hydroxymethyl)amino)-2-methyl-1-propanol
2-((Hydroxymethyl)amino)-2-methyl-1-propanal
1-Propanol, 2-((hydroxymethyl)-amino)-2-methyl- (9CI)

52304-17-3
$C_7H_{15}NO_4$
177.19
O=C(OCC(C)C)N(CO)CO
Carbamic acid, bis(hydroxymethyl)-, 2-methylpropyl ester (9CI)
Dimethylol isobutyl carbamate
Isobutyl dimethylolcarbamate
Isobutyl N,N-dimethylolcarbamate

52314-67-7
$C_8H_9Cl_3O$
227.52
O=C(C(C1(C)C)C1C=C(Cl)Cl)Cl
Cyclopropanecarbonyl chloride, 3-(2,2-dichloroethenyl)-2,2-dimethyl- (9CI)
3-(2,2-Dichloroethenyl)-2,2-dimethylcyclopropanecarbonyl chloride
3-(2,2-Dichlorovinyl)-2,2-dimethylcyclopropanecarbonyl chloride

52315-07-8
$C_{22}H_{19}Cl_2NO_3$
416.32
CC1(C)C(C=C(Cl)Cl)C1C(=O)OC(C#N)c3cccc(Oc2ccccc2)c3
Cyclopropanecarboxylic acid, 3-(2,2-dichloroethenyl)-2,2-dimethyl-, cyano(3-phenoxyphenyl) methyl ester, (+-)
α-Cypermethrin
Alfamethrin
Alphamethrin
Ambush C
Ammo
Ammo Pesticide
Antiborer 3767
Ardap
Arrivo
Avicade
Barricade
CCN52
(+-)-α-Cyano-3-phenoxybenzyl 2,2-dimethyl-3-(2,2-dichlorovinyl)cyclopropane carboxylate
Cymbush
Cyomethrin
Cypercopal
Cyperkill
Cypermethrin
Cypermethrin-25EC
Cypermethrine
EXP 5598
Fendona
Flectron
FMC 30980
FMC 45497
FMC 45806
Folcord
Imperator
JF 5705F
Kafil super
NRDC 149
NRDC 160
NRDC 166
Polytrin
PP383
Ripcord
RU 27998
Sherpa
Siperin
Stockade
Toppel
WL 8517
WL 43467

52316-55-9
$C_9H_9N_3O_2.H_3O_4P$
289.17
Methyl (2-benzimidazole)-carbamate phosphate
Carbamic acid, (1H-benzimidazol-2-yl)-, methyl ester, phosphate (1:1)
Carbendazim phosphate
Caswell No. 525BB
Correx
EPA Pesticide Chemical Code 099102
Lignosan BLP
MBC-P
Methyl-2-benzimidazole-carbamate phosphate

52322-80-2
$C_{12}H_7Cl_3O$
273.55
Benzene, 1,2,4-trichloro-5-phenoxy- (9CI)

52326-66-6
$C_{38}H_{74}O_6$
627.00

Distearyl peroxydicarbonate
Distearyl peroxydicarbonate (Not more than 85% with stearyl alcohol)
Distearylperoxydicarbonate, With 15% stearyl alcohol
Peroxidicarbonato de diestearilo (Spanish)
Peroxydicarbonate d'octadecyle (French)
Peroxydicarbonic acid, dioctadecyl ester, Not more than 85% with stearyl alcohol
UN 2592

52338-87-1
$C_{11}H_{26}N_4O$
230.33
O=C(NCCCN(C)C)NCCCN(C)C
Urea, N,N'-bis(3-(dimethylamino)propyl)- (9CI)
N,N'-Bis(3-(dimethylamino)-propyl)urea

52341-32-9
$C_{21}H_{20}Cl_2O_3$
391.29
Cyclopropanecarboxylic acid, 3-(2,2-dichloroethenyl)-2,2-dimethyl-, (3-phenoxyphenyl)methyl ester, trans-(+-)- (9CI)
NRDC 146

52341-33-0
$C_{21}H_{20}Cl_2O_3$
391.31
Cyclopropanecarboxylic acid, 3-(2,2-dichloroethenyl)-2,2-dimethyl-, (3-phenoxyphenyl) methyl ester, cis-(+-)
(+-)-cis-FMC 33297
FMC 35171
NRDC 148
(+-)-cis-Permethrin
1RS,cis-Permethrin

52355-31-4
$C_{19}H_{36}O_2$
296.50
6-Octadecenoic acid, methyl ester (7CI,9CI)
Methyl 6-octadecenoate

52372-17-5
$C_9H_{17}NOS$
187.30
1-Piperidinecarbothioic acid, 2-methyl-, S-ethyl ester (9CI)
S 35 (pesticide)

52380-33-3
$C_{19}H_{36}O_2$
296.49
11-Octadecenoic acid, methyl ester (9CI)

52414-82-1
C_6H_9NS
127.20
N(C=C(S1)CCC)=C1
Thiazole, 5-propyl- (9CI)
5-Propylthiazole

52414-91-2
C_6H_9NS
127.20
Thiazole, 4-ethyl-5-methyl- (9CI)

52417-22-8
$C_{13}H_{10}N_2.ClH.H_2O$
248.73
9-Acridamine, monohydrochloride, monohydrate

52427-13-1
$C_{15}H_{24}O$
220.35
Phenol, 4-(1-ethyl-1-methylhexyl)- (9CI)

52463-83-9
$C_{18}H_{13}ClN_2O$
308.75
Pinazepam
2H-1,4-Benzodiazepin-2-one, 7-chloro-1,3-dihydro-5-phenyl-1-(2-propynyl)-
2H-1,4-Benzodiazepin-2-one, 1,3-dihydro-7-chloro-5-phenyl-1-(2-propynyl)-
7-Chloro-1,3-dihydro-5-phenyl-1-(2-propynyl)-2h-1,4-benzodiazepin-2-one
7-Chloro-1-propargyl-5-phenyl-2H-1,4-benzodiazepin-2-one
DEA No. 2883
Domar
Pinazepamum (Latin)
Z-905
Zami 905

52470-25-4
$CH_6N_4O_3$
122.08
Guanidine, nitrate (9CI) (VAN)

52485-79-7
$C_{29}H_{41}NO_4$
467.64
Buprenorphine
Buprenorfina (Spanish)
Buprenorphinum (Latin)
DEA No. 9064
6,14-Ethenomorphinan-7-methanol, 17-(cyclopropylmethyl)-α-(1,1-dimethylethyl)-4,5-epoxy-18,19-dihydro-3-hydroxy 6-methoxy-α-methyl-, (5-α,7-α-(S))-
RX 6029M
Temgesic

52495-71-3
$(C_2H_4O)x(C_2H_4O)x(C_2H_4O)x C_{15}H_{26}O_6$
Polymer NA
Poly(oxy-1,2-ethanediyl), α-hydro-ω-(oxiranylmethoxy)-, ether with 2-ethyl-2-(hydroxymethyl)-1,3-propanediol (3:1) (9CI)

52508-35-7
$C_{12}H_{17}O_7.Na$
296.25
Dikegulac
Atrinal
2,3':4,6-Bis-O-(1-methylethylidene)-α-L-xylo-2-hexulofuranosonic acid sodium salt
Caswell No. 098AA
Cutlass
Dikegulac (German)
Dikegulac sodium
EPA Pesticide Chemical Code 109601
Natrium-2,3:4,6-di-O-isopropyliden-2-keto-l-gulonat (German)
Ro 7-6145
Sodium 2,3:4,6-bis-O-(1-methylethylidine)-α-L-xylo-2-hexulofuranosonate
α-L-Xylo-2-hexulofuranosonic acid, 2,3:4,6-bis-O-(1-methylethylidene)-, sodium salt

52551-67-4
$C_{10}H_{14}N_2O_5$
242.26
O=N(=O)c(ccc(N(CCO)CCO)c1O)c1
Phenol, 2-(bis(2-hydroxyethyl)amino)-5-nitro
2-(Bis(2-hydroxyethyl)amino)-5-nitrophenol
HC Yellow No.4
NCI-C56019

52570-16-8
$C_{19}H_{17}NO_2$
291.37
Propanamide, 2-(2-naphthalenyloxy)-N-phenyl
MT 101
2-(2-Naphthalenyloxy)-N-phenylpropanamide
α-(β-Naphthoxy)propionanilide
α-(2-Naphthoxy)propionanilide
2-(2-Naphthyloxy)propionanilide
Naproanilide
Uribest

52588-78-0

$C_{13}H_{20}O_2$
208.30
3,4-Undecadiene-2,10-dione, 6,6-dimethyl- (9CI)

52591-22-7
$C_4HO_3.K.H_2O$
154.17
[K+].[O-]c1cc(=O)c1=O
3-Cyclobutene-1,2-dione, 3-hydroxy-, potassium salt, hydrate
1-Hydroxycyclobut-1-ene-3,4-dione potassium salt hydrate
Moniliformin

52613-22-6
$(C_{12}H_{12}.CH_2O)x$
Polymer
Formaldehyde, Polymer with dimethylnaphthalene (9CI)
Dimethylnaphthalene, formaldehyde polymer

52623-95-7
Unknown
Unknown
Poly(oxy-1,2-ethanediyl), α-((1,1,3,3-tetramethylbutyl)phenyl)-ω-hydroxy-, phosphate (9CI)

52628-25-8
Unknown
Unknown
Zinc ammonium chloride
Ammonium zinc chloride (9CI)

52628-37-2
$C_6H_4Br_3N$
329.81
Benzenamine, ar,ar,ar-tribromo- (9CI)

52642-16-7
$C_{11}H_9N$
155.19
Pyridine, phenyl- (9CI)

52645-53-1
$C_{21}H_{20}Cl_2O_3$
391.31
CC1(C)C(C=C(Cl)Cl)C1C(=O)OCc3cccc(Oc2ccccc2)c3
Cyclopropanecarboxylic acid, 3-(2,2-dichlorovinyl)-2,2-dimethyl-, 3-phenoxybenzyl ester, (+-)-, (cis,trans)
AI3-29158
Ambush
Ambushfog
Antiborer 3768
BW-21-Z
Chinetrin
Coopex
Corsair
Diffusil H
Dragnet
Ecsumin
Ectiban
Efmethrin
Exmin
Exsmin
FMC 33297
FMC 41655
ICI-PP 557
Indothrin
Ipitox
Kafil
Kestrel (Pesticide)
LE 79-519
MP79
NIA 33297
NRDC 143
Outflank
Outflank-Stockade
Peregin
Peregin W
Permasect
Permasect-25EC
Permethrin
Permetrin (Hungarian)
Permetrina (Portuguese)
Permitrene (Hungarian)
3-Phenoxybenzyl (+-)-3-(2,2-dichlorovinyl)-2,2-dimethylcyclopropanecarboxylate
3-Phenoxybenzyl dl-cis/trans-3-(2,2-dichlorovinyl)-2,2-dimethyl-1-cyclopropanecarboxylate
(3-Phenoxyphenyl)methyl 3-(2,2-dichlorethenyl)-2,2-dimethylcyclopropanecarboxylate
Picket
Picket G
Pounce
PP 557
Pramex
Qamlin
S-3151
SBP-1513
SBP-1513TEC
Stockade
Stomoxin
Stomoxin P
Talcord
Tornade
WL 43479

52663-57-7
$C_6H_{14}O_2.Na$
141.17
Ethanol, 2-butoxy-, sodium salt (9CI)
2-Butoxyethanol sodium salt
Sodium 2-butoxyethoxide

52663-58-8
$C_{12}H_6Cl_4$
291.99
2,3,4',6-Tetrachlorobiphenyl
2,3,4',6-TCBP

52663-59-9
$C_{12}H_6Cl_4$
291.99
Clc1ccc(c(Cl)c1Cl)c2ccccc2Cl
1,1'-Biphenyl, 2,2',3,4-tetrachloro- (9CI)
2,2',3,4-PCB
2,2',3,4-Tetrachlorobiphenyl

52663-60-2
$C_{12}H_5Cl_5$
326.44
Clc1cccc(c1Cl)c2c(Cl)ccc(Cl)c2Cl
2,2',3,3',6-Pentachlorobiphenyl

52663-61-3
$C_{12}H_5Cl_5$
326.44
Clc1ccc(Cl)c(c1)c2cc(Cl)cc(Cl)c2Cl
1,1'-Biphenyl, 2,2',3,5,5'-pentachloro- (9CI)
2,2',3,5,5'-PCB
2,2',3,5,5'-Pentachlorobiphenyl

52663-62-4
$C_{12}H_5Cl_5$
326.44
1,1'-Biphenyl, 2,2',3,3',4-pentachloro- (9CI)
2,2',3,3',4-PCB
2,2',3,3',4-Pentachlorobiphenyl

52663-63-5
$C_{12}H_4Cl_6$
360.88
1,1'-Biphenyl, 2,2',3,5,5',6-hexachloro- (9CI)
2,2',3,5,5',6-Hexachlorobiphenyl
2,2',3,5,5',6-PCB

52663-64-6
$C_{12}H_3Cl_7$
395.32
Clc1ccc(Cl)c(c1Cl)c2c(Cl)c(Cl)cc(Cl)c2Cl
1,1'-Biphenyl, 2,2',3,3',5,6,6'-heptachloro- (9CI)
2,2',3,3',5,6,6'-Heptachlorobiphenyl
2,2',3,3',5,6,6'-PCB

52663-66-8
$C_{12}H_4Cl_6$
360.88
Clc1cc(Cl)c(Cl)c(c1)c2ccc(Cl)c(Cl)c2Cl
1,1'-Biphenyl, 2,2',3,3',4,5'-hexachloro- (9CI)
2,2',3,3',4,5'-Hexachlorobiphenyl

2,2',3,3',4,5'-PCB

52663-67-9
$C_{12}H_3Cl_7$
395.32
1,1'-Biphenyl, 2,2',3,3',5,5',6-heptachloro- (9CI)
2,2',3,3',5,5',6-Heptachlorobiphenyl
2,2',3,3',5,5',6-PCB

52663-68-0
$C_{12}H_3Cl_7$
395.32
1,1'-Biphenyl, 2,2',3,4',5,5',6-heptachloro- (9CI)
2,2',3,4',5,5',6-Heptachlorobiphenyl
2,2',3,4',5,5',6-PCB

52663-69-1
$C_{12}H_3Cl_7$
395.32
1,1'-Biphenyl, 2,2',3,4,4',5',6-heptachloro- (9CI)
2,2',3,4,4',5',6-Heptachlorobiphenyl
2,2',3,4,4',5',6-PCB

52663-70-4
$C_{12}H_3Cl_7$
395.32
1,1'-Biphenyl, 2,2',3,3',4,5',6'-heptachloro- (9CI)
2,2',3,3',4,5',6'-Heptachlorobiphenyl
2,2',3,3',4,5',6'-PCB

52663-71-5
$C_{12}H_3Cl_7$
395.32
Clc1ccc(c(Cl)c1Cl)c2c(Cl)cc(Cl)c(Cl)c2Cl
1,1'-Biphenyl, 2,2',3,3',4,4',6-heptachloro- (9CI)
Heptachlorobiphenyl
2,2',3,3',4,4',6-PCB

52663-72-6
$C_{12}H_4Cl_6$
360.88
2,4,5,3',4',5'-Hexachlorobiphenyl
1,1'-Biphenyl, 2,3',4,4',5,5'-hexachloro-

52663-73-7
$C_{12}H_2Cl_8$
429.77
1,1'-Biphenyl, 2,2',3,3',4,5,6,6'-octachloro- (9CI)
2,2',3,3',4,5,6,6'-Octachlorobiphenyl
2,2',3,3',4,5,6,6'-PCB

52663-74-8
$C_{12}H_3Cl_7$
395.32
1,1'-Biphenyl, 2,2',3,3',4,5,5'-heptachloro- (9CI)
2,2',3,3',4,5,5'-Heptachlorobiphenyl
2,2',3,3',4,5,5'-PCB

52663-75-9
$C_{12}H_2Cl_8$
429.77
1,1'-Biphenyl, 2,2',3,3',4,5,5',6'-octachloro- (9CI)
2,2',3,3',4,5,5',6'-Octachlorobiphenyl
2,2',3,3',4,5,5',6'-PCB

52663-76-0
$C_{12}H_2Cl_8$
429.77
1,1'-Biphenyl, 2,2',3,4,4',5,5',6-octachloro- (9CI)
2,2',3,4,4',5,5',6-Octachlorobiphenyl
2,2',3,4,4',5,5',6-PCB

52663-77-1
$C_{12}HCl_9$
464.21
Clc1cc(Cl)c(Cl)c(c1Cl)c2c(Cl)c(Cl)c(Cl)c(Cl)c2Cl
1,1'-Biphenyl, 2,2',3,3',4,5,5',6,6'-nonachloro- (9CI)
2,2',3,3',4,5,5',6,6'-Nonachlorobiphenyl
2,2',3,3',4,5,5',6,6'-PCB

52663-78-2
$C_{12}H_2Cl_8$
429.77
Clc1ccc(c(Cl)c1Cl)c2c(Cl)c(Cl)c(Cl)c(Cl)c2Cl
1,1'-Biphenyl, 2,2',3,3',4,4',5,6-octachloro- (9CI)
2,2',3,3',4,4',5,6-Octachlorobiphenyl
2,2',3,3',4,4',5,6-PCB

52670-79-8
$C_{15}H_{26}NO_3P$
299.34
Phosphoramidic acid, methylphenyl-, dibutyl ester (9CI)
NSC-203106

52677-44-8
$C_{20}H_{12}CrN_4O_8S_2.Na$
575.43
Chromate(1-), (4-((4,5-dihydro-3-methyl-5-oxo-1-(3-sulfophenyl)-1H-pyrazol-4-yl)azo)-3-hydroxy-1-naphthalenesulfonato(4-))-, sodium (9CI)
3-Hydroxy-4-((3-methyl-1-(3-sulfophenyl)-5-hydroxy-4-pyrazolyl)azo)-1-naphthalenesulfonic acid, sodium salt, chromium complex

52698-46-1
$C_7H_{16}O_3$
148.20
Propane, 1,1,1-trimethoxy-2-methyl- (9CI)
Trimethyl orthoisobutyrate

52704-70-8
$C_{12}H_4Cl_6$
360.88
1,1'-Biphenyl, 2,2',3,3',5,6-hexachloro- (9CI)
2,2',3,3',5,6-Hexachlorobiphenyl
2,2',3,3',5,6-PCB

52704-98-0
Unknown
Unknown
Ammonium 2-phenylphenate

52712-04-6
$C_{12}H_4Cl_6$
360.88
1,1'-Biphenyl, 2,2',3,4,5,5'-hexachloro- (9CI)
2,2',3,4,5,5'-Hexachlorobiphenyl
2,2',3,4,5,5'-PCB

52712-05-7
$C_{12}H_3Cl_7$
395.32
1,1'-Biphenyl, 2,2',3,4,5,5',6-heptachloro- (9CI)
2,2',3,4,5,5',6-Heptachlorobiphenyl
2,2',3,4,5,5',6-PCB

52716-17-3
$C_{12}H_{15}ClO_3$
242.70
Acetic acid, (4-chlorophenoxy)-, butyl ester (9CI)
Acetic acid, (p-chlorophenoxy)-, butyl ester

52738-29-1
$C_{14}H_9NO_3$
239.23
9,10-Anthracenedione, aminohydroxy- (9CI)
Aminohydroxyanthraquinone

52740-16-6
$AsH_3O_3.Ca$
166.02

Calcium arsenite
Arsenic Compound
Arsenous acid, calcium salt
Arsonic acid, calcium salt (1:1)
Calcium arsenite (1:1)
Calcium arsenite, Solid [NA 1574]
Calcium meta-arsenite
Mono-calcium arsenite
NA 1574 [Calcium arsenite, Solid]

52744-13-5
$C_{12}H_4Cl_6$
360.88
Clc1cc(Cl)c(Cl)c(c1)c2c(Cl)ccc(Cl)c2Cl
1,1'-Biphenyl, 2,2',3,3',5,6'-hexachloro- (9CI)
2,2',3,3',5,6'-Hexachlorobiphenyl
2,2',3,3',5,6'-PCB

52748-69-3
$C_{21}H_{22}N_2O_2.HI$
462.33
Strychnidin-10-one, monohydriodide (9CI)
Strychnine hydriodide
4,6-Methano-6H, 14H-indolo(3,2,1-ij)oxepino(2,3,4-de)pyrrolo-(2,3-h)quinoline, strychnidin-10-one deriv. (9CI)

52756-22-6
$C_{19}H_{19}ClFNO_3$
363.82
Flamprop-Isopropyl
Isopropyl N-benzoyl-N-(3-chloro-4-fluorophenyl)-2-aminopropionate

52756-25-9
$C_{17}H_{15}ClFNO_3$
335.78
COC(=O)C(C)N(C(=O)c1ccccc1)c2ccc(F)c(Cl)c2
Propionic acid, 2-(N-(3-chloro-4-fluorophenyl)benzamido)-, methyl ester
Flamprop-Methyl
Mataven
Methyl N-benzoyl-N-(3-chloro-4-fluorophenyl)-2-aminopropionate
WL 29761

52783-43-4
$C_{19}H_{40}O$
284.53
Nonadecanol (9CI)

52784-49-3
$C_{20}H_{19}O_4P$
354.34
Phosphoric acid, 3-ethylphenyl diphenyl ester (9CI)

52793-97-2
$C_{20}H_{28}NO.CH_3O_4S$
409.54
Benzenaminium, N,N,N-trimethyl-4-((4,7,7-trimethyl-3-oxobicyclo(2.2.1)hept-2-ylidene)methyl)-, methyl sulfate (9CI)

52819-37-1
$C_{10}H_{10}Cl_3NO_2$
282.56
Carbamic acid, (2,4,5-trichlorophenyl)-, 1-methylethyl ester (9CI)

52819-39-3
$C_{10}H_9Cl_9$
448.26
Bicyclo(2.2.1)heptane, 2,3,3,5,6-pentachloro-7,7-bis(chloromethyl)-1-(dichloromethyl)-, (2-endo,5-exo,6-exo)- (9CI)
Toxaphene Toxicant C

52820-00-5
$C_{22}H_{19}Br_2NO_3$
505.24
Cyclopropanecarboxylic acid, 3-(2,2-dibromoethenyl)-2,2-dimethyl-, cyano(3-phenoxyphenyl) methyl ester
NRDC 156
NRDC 158
RU 22950

52900-12-6
$C_{30}H_{59}NO_4$
497.80
O=C(N(C(=O)CCCCCCCCCCCC C)(CCO)CCO)CCCCCCC CCCC
1-Tetradecanaminium, N,N-bis(2-hydroxyethyl)-1-oxo-N-(1-oxododecyl)-, hydroxide, inner salt (9CI)

52907-07-0
$C_{20}H_{20}Br_4N_2O_4$
671.99
4,7-Methano-1H-isoindole-1,3(2H)-dione, 2,2'-(1,2-ethanediyl)bis(5,6-dibromohexahydro- (9CI)
Ethylenebis(5,6-dibromonorbornane-2,3-dicarboximide)

52918-63-5
$C_{22}H_{19}Br_2NO_3$
505.24
CC1(C)C(C=C(Br)Br)C1C(=O)OC(C#N)c3cccc(Oc2ccccc2)c3
Cyclopropanecarboxylic acid, 3-(2,2-dibromoethenyl)-2,2-dimethyl-, cyano(3-phenoxyphenyl) methyl ester, (1R-(1-A(S*),3-α))
Butoflin
Butox
Decamethrin
Decamethrine
Decis
Dekametrin (Hungarian)
Deltamethrin
Deltamethrine
Esbecythrin
FMC 45498
K-Obiol
K-Othrin
NRDC 161
OMS 1988
Othrine
RU 22974

52933-01-4
Unknown
Unknown
Butyl acid phosphate
UN 1718

52942-64-0
$C_{11}H_{16}O_6$
244.27
Butanedioic acid, (ethoxymethylene)oxo-, diethyl ester
Diethylethoxymethyleneoxalacetate

53001-22-2
$C_4D_{10}O$
64.04
2-Propan-1,1,1,3,3,3-d6-ol-d, 2-(methyl-d3)- (9CI)
tert-Butyl-d9 alcohol-d

53014-37-2
$C_6H_3N_5O_8$
273.14
Aniline, tetranitro
Benzenamine, tetranitro- (9CI)
Tetranitroaniline [UN 0207]
UN 0207 [Tetranitroaniline]

53014-41-8
$C_7H_4Cl_4S$
261.99
Benzene, tetrachloro(methylthio)- (9CI)

53028-35-6
$C_{16}H_{16}N_2O_4.C_{13}H_{18}N_2O_2$
534.59
Carbamic acid, (3-methylphenyl)-, 3-((methoxycarbonyl)amino)phenyl ester, Mixt. with 3-cyclohexyl-6,7-dihydro-1H-cyclopentapyrimidine-2,4(3H,5H)-dione (9CI)

53042-79-8
Unknown
Unknown
(E,Z)-7,11-Hexadecadien-1-ol, acetate

53048-47-8
$C_{18}H_{34}O_2.CH_5N_3$
341.52
9-Octadecenoic acid (Z)-, Compd. with guanidine (1:1) (9CI)

53066-65-2
$C_{14}H_{13}Cl_2O_3PS$
363.20
Phosphorothioic acid, O,O-bis-(2-chlorophenyl) S-ethyl ester (9CI)

53066-66-3
$C_{14}H_{11}Cl_4O_3PS$
432.09
Phosphorothioic acid, O,O-bis-(2,4-dichlorophenyl) S-ethyl ester (9CI)

53066-68-5
$C_{14}H_{13}Cl_2O_3PS$
363.20
Phosphorothioic acid, O,O-bis-(4-chlorophenyl) S-ethyl ester (9CI)

53102-14-0
C_5H_5ClO
116.55
2-Cyclopenten-1-one, 3-chloro- (9CI)
3-Chloro-2-cyclopentenone
3-Chloro-2-cyclopenten-1-one

53108-47-7
Unknown
Unknown
Nitrilotriacetic acid, copper-(2+) complex sodium salt
NTA, copper(2+) complex sodium salt

53108-50-2
$C_6H_6CoNO_6.H$
248.05
Cobaltate(1-), (N,N-bis(carboxymethyl)glycinato(3-)-N,O,O',O'')-, hydrogen (9CI)
Cobalt nitrilotriacetic acid
Cobalt (2+) NTA
Cobaltate(1-), (N,N-bis(carboxymethyl)glycinato(3-)-N,O,O',O'')-, hydrogen, (T-4)-

53113-57-8
$C_6H_6NO_6Zn.H$
254.51
Zinc (2+) NTA
Zincate(1-), (N,N-bis(carboxymethyl)glycinato(3-)-N,O,O', O'')-, hydrogen, (T-4)-

53113-59-0
$C_6H_6NO_6Pb.H$
396.32
Plumbate(1-), (N,N-bis(carboxymethyl)glycinato(3-)-N,O,O',O''), hydrogen, (T-4)- (9CI)
Lead (2+) NTA

53113-61-4
$C_6H_6HgNO_6.H$
389.71
Mercurate(1-), (N,N-bis(carboxymethyl)glycinato(3-)-N,O,O',O'')-, hydrogen, (T-4)- (9CI)
Mercury (2+) NTA

53120-26-6
$C_{20}H_{36}O_2$
308.50
O=C(OCCC=CCCCCCCCCC=CCCCC)C
(E,Z)-3,13-Octadecadien-1-ol acetate
AI3-36727
3,13-Octadecadien-1-ol, acetate, (E,Z)-
(E,Z)-3,13-Octadecadienyl acetate

53120-27-7
$C_{20}H_{36}O_2$
308.50
O=C(OCCC=CCCCCCCCCC=CCCCC)C
(Z,Z)-3,13-Octadecadien-1-ol acetate
AI3-36728
Caswell No. 609AB
EPA Pesticide Chemical Code 117201
Exitlure
3,13-Octadecadien-1-ol, acetate
3,13-Octadecadien-1-ol, acetate, (Z,Z)-
(Z,Z) 3,13-Octadecadien-1-ol acetate
Z,Z-ODDA

53123-73-2
$C_{11}H_{11}N$
157.21
Quinoline, ethyl- (9CI)

53123-81-2
$C_{14}H_8O_8S_2$
368.34
Anthracenedisulfonic acid, 9,10-dihydro-9,10-dioxo- (9CI)

53135-94-7
$C_{12}H_{12}O_4S_2$
284.36
Naphthalene, 1,5-bis-(methylsulfonyl)- (9CI)
1,5-Bis(methylsulfonyl)-naphthalene

53135-95-8
$C_{11}H_9ClO_2S_2$
272.77
1-Naphthalenesulfonyl chloride, 5-(methylthio)- (9CI)
1-Methylthio-5-chlorosulfonylnaphthalene

53179-11-6
$C_{29}H_{33}ClN_2O_2$
477.09
CN(C)C(=O)C(CCN1CCC(O)(CC1)c2ccc(Cl)cc2)(c3ccccc3)c4ccccc4
1-Piperidinebutanamide, 4-(4-chlorophenyl)-4-hydroxy-N,N-dimethyl-α,α-diphenyl
4-(4-Chlorophenyl)-N,N-dimethyl-α,α-diphenyl-4-hydroxy-1-piperidinebutanamide
Loperamide

53188-07-1
$C_{14}H_{18}O_4$
250.29
2H-1-Benzopyran-2-carboxylic acid, 3,4-dihydro-6-hydroxy-2,5,7,8-tetramethyl-, (+-)- (9CI)

53202-98-5
$C_{38}H_{60}O_9$
660.89
Croton Factor F1
NSC-338250
12-o-Palmitoyl-16-hydroxyphorbol-13-acetate
Phorbol-13-acetate, 12-o-palmitoyl-16-hydroxy

53204-57-2
$C_{11}H_{12}$
144.22
1H-Indene, 1,2-dimethyl-, (S)- (9CI)
(+)-1,2-Dimethylindene

53220-22-7
$C_{30}H_{58}O_6$
514.79

O=C(OCCCCCCCCCCCCCCC)
OOC(=O)OCCCCCCCCCC
CCCCC
Dimyristyl peroxydicarbonate
Dimyristyl peroxydicarbonate, Not more than 22% in water
Dimyristyl peroxydicarbonate (Not more than 22% stable dispersion, in water)
Dimyristyl peroxydicarbonate, Technically pure
Ditetradecyl peroxydicarbonate
Peroxidicarbonato de dimiristilo (Spanish)
Peroxydicarbonate de dimyristyle (French)
Peroxydicarbonic acid, dimyristyl ester
Peroxydicarbonic acid, ditetradecyl ester (9CI)
Peroxydicarbonic acid, ditetradecyl ester, Not more than 22% in water
UN 2595
UN 2892

53223-75-9
$C_{21}H_{12}O$
280.33
12H-Dibenzo(b,h)fluoren-12-one (9CI)

53229-39-3
$C_7H_{14}O$
114.19
Oxirane, (1-methylbutyl)- (9CI)

53237-59-5
Unknown
Unknown
Urushiol (9CI)

53268-44-3
$C_{12}H_{22}O_4$
230.31
Propanedioic acid, (1,1-dimethylethyl)methyl-, diethyl ester (9CI)
Diethyl tert-butylmethylmalonate

53282-47-6
$C_{10}H_{16}$
136.24
Bicyclo(4.1.0)heptane, 7-(1-methylethylidene)- (9CI)

53299-53-9
C_9H_9ClO
168.62
Benzene, ((2-chloro-2-propenyl)oxy)- (9CI)
Ether, 2-chloroallyl phenyl (6CI)

53306-54-0
$C_{28}H_{46}O_4$
446.67
O=C(OCC(CCCCC)CCC)c(c(c
cc1)C(=O)OCC(CCCCC)
CCC)c1
1,2-Benzenedicarboxylic acid, bis(2-propylheptyl) ester (9CI)
Bis-(2-propylheptyl) phthalate
NSC-17071

53317-48-9
$C_4H_3Cl_3$
157.43
1,3-Butadiene, trichloro- (9CI)

53404-00-5
Unknown
Unknown
Silver thiuronium acrylate copolymer

53404-04-9
Unknown
Unknown
p-Octylphenoxypolyethoxyethanol-iodine complex

53404-09-4
$C_9H_7Cl_3O_3.C_6H_{15}NO_2$
402.71
Propanoic acid, 2-(2,4,5-trichlorophenoxy)-, Compd. with 1,1'-iminobis(2-propanol) (1:1) (9CI)
2-Propanol, 1,1'-iminobis-, 2-(2,4,5-trichlorophenoxy)propanoate (Salt) (9CI)

53404-10-7
$C_{17}H_{23}Cl_3O_3$
381.73
Propanoic acid, 2-(2,4,5-trichlorophenoxy)-, 2-ethyl-4-methylpentyl ester (9CI)

53404-13-0
$C_9H_7Cl_3O_3.C_3H_9NO$
344.62
Propanoic acid, 2-(2,4,5-trichlorophenoxy)-, Compd. with 1-amino-2-propanol (1:1) (9CI)
2-Propanol, 1-amino-, (2,4,5-trichlorophenoxy)propanoate (Salt) (9CI)

53404-14-1
$C_{17}H_{23}Cl_3O_3$
381.73
Propanoic acid, 2-(2,4,5-trichlorophenoxy)-, 1-methylheptyl ester (9CI)

53404-18-5
$C_{11}H_{16}O.K$
203.35
Potassium 4-tert-amylphenate
p-tert-Amylphenol, potassium salt
Caswell No. 050A
4-(1,1-Dimethylpropyl)phenol potassium salt
EPA Pesticide Chemical Code 064111
Phenol, 4-(1,1-dimethylpropyl)-, potassium salt (9CI)
Potassium p-tert-amylphenate
Potassium p-tert-amylphenolate

53404-19-6
$C_9H_{13}BrN_2O_2.Li$
268.05
Lithium bromacil
Bromacil lithium salt
Caswell No. 111A
EPA Pesticide Chemical Code 012302
2,4(1H,3H)-Pyrimidinedione, 5-bromo-6-methyl-3-(1-methylpropyl)-, lithium salt (9CI)

53404-20-9
Unknown
Unknown
Sodium 4-chloro-2-cyclopentylphenate

53404-21-0
Unknown
Unknown
Potassium 4-chloro-2-phenylphenate

53404-22-1
Unknown
Unknown
2-(m-Chlorophenoxy)propionic acid, sodium salt

53404-23-2
Unknown
Unknown
Diethanolamine 4-chlorophenoxyacetate

53404-24-3
Unknown
Unknown
Copper dehydroabietyl ammonium 2-ethylhexanoate

53404-28-7

Unknown
Unknown
Monoethanolamine dicamba

53404-30-1
Unknown
Unknown
Potassium 4,6-dichloro-2-phenylphenate

53404-31-2
$C_{15}H_{20}Cl_2O_4$
335.23
O=C(OCCOCCCC)C(Oc(c(cc(c1)Cl)Cl)c1)C
Butoxyethyl 2-(2,4-dichlorophenoxy)propionate
Caswell No. 320A
2-(2,4-Dichlorophenoxy)propionic acid 2-butoxyethanol ester
2-(2,4-Dichlorophenoxy)propionic acid, 2-butoxyethyl ester
EPA Pesticide Chemical Code 031453
Propanoic acid, 2-(2,4-dichlorophenoxy)-, 2-butoxyethyl ester (9CI)

53404-32-3
Unknown
Unknown
Dimethylamine 2-(2,4-dichlorophenoxy)propionate

53404-37-8
Unknown
Unknown
2-Ethyl-4-methylpentyl 2,4-dichlorophenoxyacetate

53404-47-0
Unknown
Unknown
Dodecylammonium methanearsonate

53404-51-6
Unknown
Unknown
Potassium ethylenediaminetetraacetate

53404-52-7
Unknown
Unknown
Tetra(ethanolamine) ethylenediaminetetraacetate

53404-54-9
Unknown
Unknown
Sodium m-(2-hydroxyethyl)-ethylenediaminetriacetate

53404-60-7
Unknown
Unknown
Sodium tetrahydro-3,5-dimethyl-2H-1,3,5-thiadiazine-2-thione

53404-62-9
Unknown
Unknown
Potassium N-(α-(nitroethyl)-benzyl)ethylenediamine

53404-67-4
Unknown
Unknown
Phenylmercuric ammonium acetate

53404-68-5
Unknown
Unknown
Phenylmercuric ammonium propionate

53404-73-2
$C_{18}H_{37}N.C_9H_7Cl_3O_3$
537.02
Propanoic acid, 2-(2,4,5-trichlorophenoxy)-, Compd. with (Z)-9-octadecen-1-amine (1:1) (9CI)
9-Octadecen-1-amine, (Z)-, 2-(2,4,5-trichlorophenoxy)propanoate (9CI)

53404-74-3
$C_9H_7Cl_3O_3.C_6H_{15}N$
370.71
Propanoic acid, 2-(2,4,5-trichlorophenoxy)-, Compd. with N,N-diethylethanamine (1:1) (9CI)
Ethanamine, N,N-diethyl-, 2-(2,4,5-trichlorophenoxy)propanoate

53404-75-4
$C_9H_{21}NO_3.C_9H_7Cl_3O_3$
460.79
Propanoic acid, 2-(2,4,5-trichlorophenoxy)-, Compd. with 1,1',1''-nitrilotris(2-propanol) (1:1) (9CI)
2-Propanol, 1,1',1''-nitrilotris-, 2-(2,4,5-trichlorophenoxy)propanoate (Salt) (9CI)

53404-76-5
$C_{17}H_{23}Cl_3O_3$
381.73
Propanoic acid, 2-(2,4,5-trichlorophenoxy)-, 2-ethylhexyl ester (9CI)
Propionic acid, 2-(2,4,5-trichlorophenoxy)-, 2-ethylhexyl ester

53404-81-2
Unknown
Unknown
Sodium N-cyclohexyl-N-palmitoyltaurate - iodine complex

53404-82-3
$C_{19}H_{38}O_4Sn$
449.22
Tributyltin isopropyl succinate
Caswell No. 867D
EPA Pesticide Chemical Code 083115
Stannane, (isopropylsuccinyloxy)tributyl-
Succinic acid, O-isopropyl-O'-tributylstannyl ester
Tributyltin isopropylsuccinate

53404-84-5
$C_{12}H_{27}N.C_8H_5Cl_3O_3$
440.83
Acetic acid, (2,4,5-trichlorophenoxy)-, Compd. with 1-dodecanamine (1:1) (9CI)
1-Dodecanamine, (2,4,5-trichlorophenoxy)acetate (9CI)

53404-85-6
$C_{13}H_{29}N.C_8H_5Cl_3O_3$
454.87
Acetic acid, (2,4,5-trichlorophenoxy)-, Compd. with 1-tridecanamine (1:1) (9CI)
1-Tridecanamine, (2,4,5-trichlorophenoxy)acetate (9CI)

53404-86-7
$C_8H_5Cl_3O_3.C_6H_{15}NO$
372.67
Acetic acid, (2,4,5-trichlorophenoxy)-, Compd. with 2-(diethylamino)ethanol (1:1) (9CI)
Caswell No. 881H
EPA Pesticide Chemical Code 082038
2,4,5-Trichlorophenoxyacetic acid, N,N-diethylethanolamine salt

53404-87-8
$C_{21}H_{44}N_2.C_8H_5Cl_3O_3$
580.08
Acetic acid, (2,4,5-trichlorophenoxy)-, Compd. with (Z)-N-9-octadecenyl-1,3-propanediamine (1:1) (9CI)
1,3-Propanediamine, N-9-octadecenyl-, (Z)-, mono((2,4,5-trichlorophenoxy)acetate) (9CI)

53404-88-9
$C_{20}H_{39}N.C_8H_5Cl_3O_3$
549.03
Acetic acid, (2,4,5-trichlorophenoxy)-, Compd. with (Z,Z)-N,N-dimethyl-9,12-octadecadien-1-amine (1:1) (9CI)
9,12-Octadecadien-1-amine, N,N-dimethyl-, (Z,Z)-, (2,4,5-trichlorophenoxy)-acetate (9CI)

53404-89-0
$C_{20}H_{41}N.C_8H_5Cl_3O_3$
551.04
Acetic acid, (2,4,5-trichlorophenoxy)-, Compd. with (Z)-N,N-dimethyl-9-octadecen-1-amine (1:1) (9CI)
9-Octadecen-1-amine, N,N-dimethyl-, (Z)-, (2,4,5-trichlorophenoxy)-acetate (9CI)

53404-90-3
Unknown
Unknown
Ammonium 2,3,6-trichlorophenylacetate

53417-29-1
$C_{11}H_{16}O_6$
244.24
O=C(OCC(CO)(CO)COC(=O)C=C)C=C
2-Propenoic acid, 2,2-bis-(hydroxymethyl)-1,3-propanediyl ester (9CI)
Pentaerythritol diacrylate

53422-16-5
$C_{19}H_{38}O_3.Li$
321.45
Octadecanoic acid, 12-hydroxy-, methyl ester, lithium salt (9CI)
Methyl 12-hydroxy stearate, lithium salt
Methyl 12-hydroxystearate, lithium salt

53422-49-4
$C_2H_4NO_3$
90.05
Azidoethyl nitrate
2-Azidoethanol nitrate
2-Azidoethyl nitrate
Ethanol, 2-azido-, nitrate (Ester)
Nitrate d'azidoethyle (French)
Nitrato de azidoetilo (Spanish)

53452-80-5
$C_7H_5Cl_3O$
211.47
Benzene, trichloromethoxy-(9CI)

53452-81-6
$C_7H_4Cl_4O$
245.92
Benzene, tetrachloromethoxy-(9CI)
Anisole, tetrachloro-

53466-68-5
Unknown
Unknown
2-Alkyl isoquinolinium bromide

53466-71-0
Unknown
Unknown
Polyoxyethylene sorbitol oleate - laurate

53466-84-5
$C_{16}H_{21}Cl_3O_4$
653.22
Propanoic acid, 2-(2,4,5-trichlorophenoxy)-, ester with 1(or 2)-(2-methylpropoxy)-propanol (9CI)

53466-85-6
Unknown
Unknown
Tributyltin monopropyleneglycol maleate

53466-86-7
$C_{15}H_{19}Cl_3O_4$
625.16
Acetic acid, (2,4,5-trichlorophenoxy)-, ester with 1(or 2)-(2-methylpropoxy)propanol (9CI)

53466-91-4
Unknown
Unknown
Heptadecyl hydroxyethyl imidazoline

53466-92-5
Unknown
Unknown
Heptadecyl hydroxyethyl-imidazolinium chloride

53467-00-8
Unknown
Unknown
Sodium dodecyl diphenyl oxide sulfonate

53467-01-9
Unknown
Unknown
Sodium dodecylbenzene-sulfonate-iodine complex

53469-21-9
Unknown
Unknown
Arochlor 1242
Aroclor 1242
Chlorierte biphenyle, Chlorgehalt 42% (German)
Chlorodiphenyl (42% Chlorine) (ACGIH,OSHA)
Chlorodiphenyl (42% cl)
Clorodifenili, cloro 42% (Italian)
Diphenyle chlore, 42% de chlore (French)
Gechloreerdedifenyl (Dutch)
Polychlorinated biphenyl (Aroclor 1242)

53494-70-5
$C_{12}H_8Cl_6O$
380.90
2,5,7-Metheno-3H-cyclopenta-(a)pentalen-3-one, 3b,4,5,6,6,6a-hexachlorodecahydro-, (2-α, 3a-β,3b-β,4-β,5-β,6a-β,7-α,7a-β,8R*)
Endrin Ketone
δ-Keto 153
δ-Ketoendrin
SD 2614

53496-16-5
$(C_2H_4O)xC_{15}H_{24}O$
190.33
Poly(oxy-1,2-ethanediyl), α-(4-tert-nonylphenyl)-ω-hydroxy- (9CI)
Polyethylene glycol mono-(4-tert-nonylphenyl)ether
Polyethylene glycol mono-(p-tert-nonylphenyl) ether
Polyoxyethylene p-tert-nonylphenyl ether

53498-30-9
$C_8H_{13}NS$
155.26

Thiazole, 4,5-dimethyl-2-(1-methylethyl)- (9CI)

53517-92-3
$C_7H_{14}N_2$
126.20
Pyrimidine, 1,4,5,6-tetrahydro-1,2,4-trimethyl-(9CI)
1,2,4-Trimethyl-1,4,5,6-tetrahydropyrimidine

53520-89-1
$C_3H_2F_6O$
168.04
1-Propanol, hexafluoro- (9CI)

53529-45-6
$C_{16}H_{35}O_9P_3$
464.42
Phosphonic acid, ethenyl-, bis(2-((butoxymethylphosphinyl)oxy)ethyl) ester
Fyrol 76

53535-26-5
$C_{19}H_{27}Cl_3O_5$
413.73
Propanoic acid, 2-(2,4,5-trichlorophenoxy)-, methyl-2-(methyl-2-(2-methylpropoxy)-ethoxy)ethyl ester (9CI)

53535-27-6
$C_6H_2Cl_4O.K$
271.00
Potassium tetrachlorophenate
Phenol, tetrachloro-, potassium salt (9CI)

53535-30-1
$C_{22}H_{33}Cl_3O_6$
499.86
Propanoic acid, 2-(2,4,5-trichlorophenoxy)-, methyl-2-(methyl-2-(methyl-2-(2-methylpropoxy)ethoxy)ethoxy)-ethyl ester (9CI)

53535-31-2
$C_{18}H_{25}Cl_3O_5$
427.75
Acetic acid, (2,4,5-trichlorophenoxy)-, methyl-2-(methyl-2-(2-methylpropoxy)ethoxy)ethyl ester (9CI)

53535-32-3
$C_{21}H_{31}Cl_3O_6$
485.83
Acetic acid, (2,4,5-trichlorophenoxy)-, methyl-2-(methyl-2-(methyl-2-(2-methylpropoxy)-ethoxy)ethoxy)ethyl ester (9CI)

53535-33-4
$C_7H_{16}O$
116.20
Heptanol (VAN)(9CI)

53535-36-7
Unknown
Unknown
N,N-Dimethyloleylamine 2,4-dichlorophenoxyacetate

53535-37-8
$C_{14}H_{31}N.C_8H_5Cl_3O_3$
468.88
Acetic acid, (2,4,5-trichlorophenoxy)-, Compd. with 1-tetradecanamine (1:1) (9CI)
1-Tetradecanamine, (2,4,5-trichlorophenoxy)-acetate (9CI)

53537-63-6
Unknown
Unknown
Diethanolamine 4-chloro-2-phenylphenate
Diethanolamine 4(or 6)-chloro-2-phenylphenate

53555-02-5
$C_{12}H_4Cl_4O_2$
321.96
Dibenzo-p-dioxin, 1,2,3,8-tetrachloro
Dibenzo(b,e)(1,4)dioxin, 1,2,3,8-tetrachloro-
1,2,3,8-Tetrachlorodibenzo-dioxin
1,2,3,8-Tetrachlorodibenzo-para-dioxin

53558-25-1
$C_{13}H_{12}N_4O_3$
272.29
O=C(NCc(cccn1)c1)Nc(ccc(N(=O)=O)c2)c2
Urea, 1-nitrophenyl-3-(3-pyridylmethyl)
DLP-87
DLP 787
N-(4-Nitrophenyl)-N'-(3-pyridinylmethyl)urea
N-3-Pyridylmethyl-N'-p-nitrophenylurea
1-(3-Pyridylmethyl)-3-(4-nitrophenyl)urea
Pyriminil
Pyriminyl
Pyrinuron
RH-787
Vacor

53563-67-0
$C_{11}H_{14}$
146.23
1H-Indene, 2,3-dihydrodi methyl- (9CI)
Indan, dimethyl-

53568-85-7
$C_{18}H_{18}O_3.C_{17}H_{25}ClO_3$
645.24
Aniten
Aniten M
9H-Fluorene-9-carboxylic acid, 9-hydroxy-, butyl ester, Mixt. with isooctyl (4-chloro-2-methylphenoxy)acetate (9CI)

53569-62-3
$CO_2.N_2O$
88.01
Carbon dioxide and nitrous oxide mixtures
Carbon dioxide, Mixture with nitrogen oxide (9CI)
Carbon dioxide-nitrous oxide, Mixture
Dioxyde de carbone et protoxyde d'azote en melange (French)
Mezclas de dioxido de carbonoy oxido nitroso (Spanish)
UN 1015 [Carbon dioxide and nitrous oxide mixtures]

53581-53-6
$C_{11}H_{16}BrNO_2.BrH$
355.06
4-Bromo-2,5-dimethoxy-amphetamine hydrobromide
Benzeneethanamine, 4-bromo-2,5-dimethoxy-α-methyl-, hydrobromide, (+-)- (9CI)
dl-4-Bromo-2,5-dimethoxy-amphetamine hydrobromide
4-Bromo-2,5-dimethoxy-α-methylphenethyl-amine hydrobromide
DL-4-Bromo-2,5-dimethoxy-α-methylphenethylamine hydrobromide
4-Bromo-2,5-DMA hydrobromide
DEA No. 7391
dl-2,5-Dimethoxy-4-bromo-amphetamine hydrobromide
DOB hydrobromide
Phenethylamine, 4-bromo-2,5-dimethoxy-α-methyl-, hydrobromide, DL-

53584-59-1
$C_{27}H_{46}$
370.66
A'-Neo-22,29,30-trinorgamma-

cerane, (17α)- (9CI)

53584-60-4
$C_{29}H_{50}$
398.72
A'-Neo-30-norgammacerane, (17α)- (9CI)

53584-62-6
$C_{31}H_{54}$
426.77
A'-Neo-30-norgammacerane, 22-ethyl- (9CI)
Homohopane

53606-41-0
$C_9H_9NO_5$
211.18
Phenol, 2-methoxy-5-nitro-, acetate (ester) (9CI)

53609-64-6
$C_6H_{14}N_2O_3$
162.22
CC(O)CN(CC(C)O)N=O
Dipropylamine, 2,2'-dihydroxy-N-nitroso
BHP
N-Bis(2-hydroxypropyl)nitrosamine
2,2'-Bishydroxypropylnitrosamine
DHPN
2,2'-Dihydroxy-di-n-propylnitrosoamine
Di(2-hydroxypropyl)nitrosamine
N,N-Di-(2-hydroxypropyl)-nitrosamine
Diisopropanolnitrosamine
DIPN
N-Nitrosobis(2-hydroxypropyl)-amine
N-Nitroso-N,N-di(2-hydroxypropyl)amine
2-Propanol, N-nitroso-1,1'-iminodi-
2-Propanol, 1,1'-nitrosoiminodi-

53633-54-8
$(C_8H_{15}NO_2.C_6H_9NO)x.xC_4H_{10}O_4S$
Polymer
Polyquaternium-11
2-Propenoic acid, 2-methyl-, 2-(dimethylamino)ethyl ester, Polymer with 1-ethenyl-2-pyrrolidinone, Compd. with diethyl sulfate (9CI)
Quaternium-23

53634-34-7
$C_{20}H_{21}ClOSi_2$
369.01
Disiloxane, 1-chloro-1,3-dimethyl-1,3,3-triphenyl- (9CI)

53664-71-4
Unknown
Unknown
Epi-Rez 505 (9CI)

53690-92-9
$C_{10}H_{16}O_2$
168.24
2-Cyclopenten-1-one, 4-butyl-3-methoxy- (9CI)

53742-07-7
$C_{12}HCl_9$
464.21
1,1'-Biphenyl, nonachloro- (9CI)
Nonachlorobiphenyl

53744-50-6
$C_{10}H_{12}O_3$
180.20
Benzenemethanol, 4-(acetyloxy)-α-methyl- (9CI)
Benzyl alcohol, p-hydroxy-α-methyl-, 4-acetate

53763-23-8
Unknown
Unknown
Disperse-Oil (9CI)

53772-82-0
$C_{23}H_{25}F_3N_2OS$
434.51
1-Piperazineethanol, 4-(3-(2-(trifluoromethyl)-9H-thioxanthen-9-ylidene)propyl)-, (Z)- (9CI)
α-Flupenthixol

53778-61-3
$C_6H_{12}O$
100.16
Oxetane, 2,3,4-trimethyl- (9CI)

53778-62-4
$C_6H_{12}O$
100.16
Oxetane, 2-ethyl-3-methyl- (9CI)

53778-73-7
$C_5H_{12}O_2$
104.15
O(CC(O)CC)C
2-Butanol, 1-methoxy- (9CI)
1-Methoxy-2-butanol

53780-34-0
$C_{11}H_{13}F_3N_2O_3S$
310.32
CC(=O)Nc1cc(NS(=O)(=O)C(F)(F)F)c(C)cc1C
Acetanilide, 2',4'-dimethyl-5-((trifluoromethyl)sulfonamido)
Acetamide, N-(2,4-dimethyl-5-(((trifluoromethyl)sulfonyl)-amino)phenyl)-
N-(2,4-Dimethyl-5-(((trifluoromethyl)sulfonyl)amino)-phenyl)acetamide
Embark
Embark Plant Growth Regulator
MBR 12325
Mefluidide
Vel 3973
Vistar
Vistar Herbicide

53780-36-2
Unknown
Unknown
Diethanolamine mefluidide

53783-86-1
$C_7H_{11}N$
109.17
Cyclobutanecarbonitrile, 3,3-dimethyl- (9CI)

53783-87-2
$C_7H_{10}O$
110.16
Bicyclo(2.2.1)hept-2-en-7-ol (9CI)
2-Norbornen-7-ol (6CI,7CI)

53783-88-3
$C_7H_{10}O$
110.16
1,3,4-Hexatriene, 3-methoxy- (9CI)

53809-87-3
$Fe_9Ni_9S_{16}$
1543.89
Iron nickel sulfide ($Fe_9Ni_9S_{16}$) (9CI)

53818-84-1
Unknown
Unknown
Nitrilotriacetic acid, tin(2+) salt
NTA, tin(2+) salt

53850-34-3
Unknown
Unknown
Thaumatin
Proteins, Thaumatins
Talin

53879-54-2
$(C_3H_6O)x(C_3H_6O)x(C_3H_6O)x$
$C_{15}H_{20}O_6$
Polymer
Poly(oxy(methyl-1,2-ethanediyl)), α-hydro-ω-((1-oxo-2-propenyl)oxy)-, ether with 2-ethyl-2-(hydroxymethyl)-1,3-propanediol (3:1) (9CI)
Trimethylolpropane propoxylate triacrylate

53894-31-8
$C_{11}H_{16}O_2$
180.25
Phenol, (1,1-dimethylethyl)-2-methoxy- (9CI)

53897-31-7
$C_7H_{14}O$
114.19
Oxirane, 2-methyl-2-(2-methylpropyl)- (9CI)

53897-32-8
$C_7H_{14}O$
114.19
Oxirane, 2-ethyl-3-propyl- (9CI)
Heptane, 3,4-epoxy-

53902-12-8
$C_{18}H_{17}NO_5$
327.36
COc2ccc(C=CC(=O)Nc1ccccc1C(O)=O)cc2OC
Benzoic acid, 2-((3-(3,4-dimethoxyphenyl)-1-oxo-2-propenyl)amino)
Anthranilic acid, N-(3,4-dimethoxycinnamoyl)-
N-(3,4-Dimethoxycinnamoyl)-anthranilic acid
N-(3',4'-Dimethoxycinnamoyl)-anthranilic acid
N-5'
Rizaben
Tranilast

53905-38-7
$C_{21}H_{36}O$
304.52
Benzene, 1-(8-methoxy-4,8-dimethylnonyl)-4-(1-methylethyl)
AI3-36206
Benzene, 1-(8-methoxy-4,8-dimethylnonyl)-4-(1-methylethyl)-
Caswell No. 584C
EPA Pesticide Chemical Code 119501
1-(8-Methoxy-4,8-dimethylnonyl)-4-(1-methylethyl)-benzene
Pro-Drone

53907-61-2
$C_8H_{14}O$
126.20
2-Pentenal, 2,4,4-trimethyl- (6CI,9CI)
2,4,4-Trimethyl-2-pentenal

53907-72-5
$C_8H_{16}O$
128.21
7-Octen-4-ol (9CI)

53907-91-8
$C_7H_{14}O_2$
130.19
1,4-Dioxane, 2-ethyl-5-methyl- (9CI)

53907-95-2
$C_7H_{16}O_2$
132.20
2-Propanol, 1-(1-methylpropoxy)- (9CI)

53939-27-8
$C_{14}H_{26}O$
210.36
O=CCCCCCCCC=CCCCC
(Z)-9-Tetradecenal
AI3-35584
Caswell No. 934A
9-Tetradecenal, (Z)-
(Z)-9-Tetradecen-1-al

53939-28-9
$C_{16}H_{30}O$
238.41
O=CCCCCCCCCCC=CCCCC
11-Hexadecenal
AI3-35937
Caswell No. 472C
EPA Pesticide Chemical Code 120001
Heliothis pheromone
Heliothis virescens
Hercon disrupt
11-Hexadecenal, (Z)-
(Z)-11-Hexadecenal

53951-50-1
$C_9H_{10}O$
134.18
Benzaldehyde, ethyl- (9CI)
Ethylbenzaldehyde

53973-98-1
$(C_{12}H_{16}M_2O_{15}S_2)x$
Polymer
Poligeenan
3,6-Anhydro-4-O-β-galactopyranosyl-α-D-galactopyranose 2,4'-bis(potassium/sodium sulfate) (13')-polysaccharide
3,6-Anhydro-4-O-β-D-galactopyranosyl-α-D-galactopyranose 2,4'-bis(potassium/sodium sulfate)-(1-3')-polysaccharide
Furose
Poligeenane (French)
Poligeenano (Spanish)
Poligeenanum (Latin)
Polygeenan
XLV

53978-04-4
$C_4H_3Cl_3$
157.43
1,3-Butadiene, 1,2,3-trichloro-, (E)- (9CI)

53980-88-4
$C_{21}H_{36}O_4$
352.51
Acrylinoleic acid
5(6)-Carboxy-4-hexyl-2-cyclohexene-1-octanoic acid
5(or 6)-Carboxy-4-hexyl-2-cyclohexene-1-octanoic acid
2-Cyclohexene-1-octanoic acid, 5(or 6)-carboxy-4-hexyl- (9CI)
C21-Dicarboxylic acid

53988-69-5
Unknown
Unknown
Gamlen Sea Clean (9CI)

53988-70-8
Unknown
Unknown
G.H. Woods Degreaser-Formula 11470 (9CI)

53988-71-9
Unknown
Unknown
Sugee 2 (9CI)

53994-73-3
$C_{15}H_{14}ClN_3O_4S$
367.83
NC(C(=O)NC2C1SCC(=C(N1C2=O)C(O)=O)Cl)c3ccccc3
5-Thia-1-azabicyclo(4.2.0)oct-2-ene-2-carboxylic acid, 7-((aminophenylacetyl)-amino)-3-chloro- 8-oxo-, (6R-(6-α,7-β(R*)))
CCL
Cefaclor
3-Chloro-7-D-(2-phenylglycinamido)-3-cephem-4-carboxylic acid
Lilly 99638
Panoral

54010-81-0
$C_{18}H_{21}NO_3$
299.36
Benzene, 1-ethoxy-4-(1-(4-methylphenyl)-2-nitropropyl)- (9CI)

54043-65-1
$C_{10}H_{20}N_2O$
184.27
OC(N=NC(C)(C)C)(CCCC1)C1
Cyclohexanol, 1-((1,1-dimethylethyl)azo)- (9CI)
1-((1,1-Dimethylethyl)azo)-cyclohexanol

54050-62-3
$C_{24}H_{46}O_4$
398.63
A(2)C
A2C
A2C Reagent
Membrane Mobility Agent A(2)C
2-(2-Methoxy)ethoxyethyl-8-(2-n-octylcyclopropyl)octanoate

54060-92-3
$C_{20}H_{24}N_3O.CH_3O_4S$
433.51
3H-Indolium, 2-(((4-methoxyphenyl)methylhydrazono)-methyl)-1,3,3-trimethyl-, methyl sulfate (9CI)

54063-14-8
$C_7H_{14}O_3$
146.19
1,3-Dioxan-5-ol, 4,4,5-trimethyl- (9CI)

54063-18-2
$C_7H_{14}O_3$
146.19
Ethene, (2-ethoxy-1-methoxyethoxy)- (9CI)

54075-76-2
$C_3H_9O.Cl_6Sb$
395.57
Trimethyloxonium hexachloroantimonate
Oxonium, trimethyl-, (OC-6-11)-hexachloroantimonate(1-) (9CI)

54091-06-4
$C_7H_6BrClO_2S$
269.55
4-Bromophenylchloromethyl sulfone
BPCMS

54105-66-7
$C_{17}H_{34}$
238.46
Cyclohexane, undecyl- (9CI)
Undecane, 1-cyclohexyl-

54116-90-4
$C_{14}H_{11}N$
193.24
Acridine, methyl- (9CI)

54120-62-6
$C_{11}H_{16}$
148.25
Benzene, ethyl-1,2,4-trimethyl- (9CI)

54120-64-8
$C_9H_8O_2$
148.16
1(3H)-Isobenzofuranone, 5-methyl- (9CI)

54135-80-7
$C_7H_5Cl_3O$
211.47
Benzene, 1,2,3-trichloro-4-methoxy- (9CI)

54135-81-8
$C_7H_5Cl_3O$
211.47
Benzene, 1,2,5-trichloro-3-methoxy- (9CI)

54135-82-9
$C_7H_5Cl_3O$
211.47
Benzene, 1,2,3-trichloro-5-methoxy- (9CI)

54150-69-5
$C_8H_{11}NO_2.ClH$
189.66
Cl.COc1ccc(N)c(OC)c1
Aniline, 2,4-dimethoxy-, hydrochloride
Benzenamine, 2,4-dimethoxy-, hydrochloride
2,4-Dimethoxyaniline hydrochloride
2,4-Dimethoxybenzenamine hydrochloride
2-Methoxy-p-anisidine hydrochloride
4-Methoxy-o-anisidine hydrochloride
NCI-C02255

54182-58-0
$C_{12}H_{54}Al_{16}O_{75}S_8$
2086.74
Sucralfate
Aluminum, hexadeca-mu-hydroxytetracosahydroxy-(mu8-(1,3,4,6-tetra-O-sulfo-β-D-fructofuranosyl α-D-glucopyranoside tetrakis-(hydrogen sulfato)(8-)))hexadeca-
Antepsin
Carafate
β-D-Fructofuranosyl-α-D-glucopyranoside octakis-(hydrogen sulfate) aluminum complex
Hexadeca-mu-hydroxytetracosahydroxy-(u8-(1,3,4,6-tetra-O-sulfo-β-D-fructofuranosyl-α-D-glucopyranoside tetrakis-(hydrogen sulfato)(8-))) hexadecaaluminum
α-D-Glucopyranoside, β-D-fructofuranosyl-, octakis-(hydrogen sulfate), aluminum complex
Sucralfato (Spanish)
Sucralfatum (Latin)
Sucrose octakis(hydrogen sulfate) aluminum complex
Ulcerban
Ulcerlmin
Ulcogant

54206-54-1
$C_6H_{14}O$
102.18
Butanol, 2,3-dimethyl- (9CI)

54208-63-8
$C_{19}H_{20}O_4$
312.37
O(C1COc(c(ccc2)Cc(c(OCC(O3)C3)ccc4)c4)c2)C1
Oxirane, 2,2'-(methylenebis-(2,1-phenyleneoxymethylene))bis- (9CI)
Methylenebis(o-phenol), 3-propylene oxide ether

54244-72-3
$C_9H_{14}O_2$
154.21
1,3-Cyclopentanedione, 4-butyl- (9CI)

54244-89-2
$C_{10}H_{18}O$
154.25
Furan, 4,5-diethyl-2,3-dihydro-2,3-dimethyl- (9CI)

54261-67-5
$C_{72}H_{116}O_4P_2S_4Zn$
1301.31
Phenol, dodecyl-, hydrogen phosphorodithioate, zinc salt (9CI)
Phosphorodithioic acid, O,O-bis(dodecylphenyl) ester, zinc

salt

54264-96-9
$C_{10}H_{14}O$
150.22
Benzenemethanol, 3-ethyl-α-methyl- (9CI)

54268-02-9
$C_3H_4Br_4$
359.70
Propane, 1,2,2,3-tetrabromo- (6CI,7CI,9CI)
1,2,2,3-Tetrabromopropane

54289-46-2
$C_{13}H_{15}N_5O_2S$
305.33
O=N(=O)C(SC(=N1)N=Nc(ccc(N(CC)CC)c2)c2)=C1
Benzenamine, N,N-diethyl-4-((5-nitro-2-thiazolyl)azo)- (9CI)

54322-31-5
$C_8H_9BrO_3S$
265.13
O=S(=O)(O)c(ccc(c1)CCBr)c1
Benzenesulfonic acid, 4-(2-bromoethyl)- (9CI)
Benzenesulfonic acid, p-(2-bromoethyl)-
4-(2-Bromoethyl)benzenesulfonic acid

54340-87-3
$C_{12}H_{16}$
160.26
1H-Indene, 2,3-dihydro-1,4,7-trimethyl- (9CI)
2,3-Dihydro-1,4,7-trimethyl-1H-indene
NSC-22050
1,4,7-Trimethyl-(2,3-dihydroindene)
1,4,7-Trimethylindan

54340-88-4
$C_{12}H_{16}$
160.26
1H-Indene, 2,3-dihydro-1,5,7-trimethyl- (9CI)

54340-89-5
$C_9H_{20}O_2$
160.26
2-Propanol, 1-(1,3-dimethylbutoxy)- (9CI)

54346-06-4
$C_{10}H_{14}O_3$
182.22
1(3H)-Isobenzofuranone, 3a,4,5,7a-tetrahydro-4-hydroxy-3a,7a-dimethyl-, (3aα,4β,7aα)-(.+-.)- (9CI)

54350-48-0
$C_{23}H_{30}O_3$
354.49
Etretinate
Ethyl etrinoate
Ethyl all-trans-9-(4-methoxy-2,3,6-trimethylphenyl)-3,7-dimethyl-2,4,6,8-nonatetraenoate
Ethyl (all-E)-9-(4-methoxy-2,3,6-trimethylphenyl)-3,7-dimethyl-2,4,6,8-nonatetraenoate
Etretinato (Spanish)
Etretinatum (Latin)
2,4,6,8-Nonanetetraenoic acid, 9-(4-methoxy-2,3,6-trimethylphenyl)-3,7-dimethyl-, ethyl ester, All-trans-
2,4,6,8-Nonatetraenoic acid, 9-(4-methoxy-2,3,6-trimethylphenyl)-3,7-dimethyl-, ethyl ester, (All-E)- (9CI)
2,4,6,8-Nonatetraenoic acid, 9-(4-methoxy-2,3,6-trimethylphenyl)-, ethyl ester, (All-E-)
NSC-297936
Ro 10-9359
Tegison
Tigason

54357-08-3
$C_{11}H_9NO_2$
187.19
Naphthalene, 2-methyl-6-nitro- (9CI)

54381-16-7
$C_{10}H_{16}N_2O_2.H_2O_4S$
294.32
Nc1ccc(cc1)N(CCO)CCO
N,N-Bis(2-hydroxyethyl)-p-phenylenediamiamine sulfate
2,2'-((4-Aminophenyl)imino)-bisethanol sulfate (Salt)
2,2'-((4-Aminophenyl)imino)-bis(ethanol) sulfate
N,N-Bis(2-hydroxyethyl)-4-aminoaniline sulfuric acid salt
N,N-Bis(β-hydroxyethyl)-p-phenylenediamine sulfate
Ethanol, 2,2'-((4-aminophenyl)-imino)bis-, sulfate (Salt)
Ethanol, 2,2'-((4-aminophenyl)-imino)bis-, sulfate (1:1) (Salt) (9CI)

54382-58-0
$C_{10}H_{18}O_2$
170.25
4-Isobenzofuranol, octahydro-3a,7a-dimethyl-, (3aα,4β,7aα)-(.+-.)- (9CI)

54385-63-6
$C_{10}H_{10}S$
162.26
Benzo(b)thiophene, ethyl- (9CI)

54392-02-8
Unknown
Unknown
KM 102 (9CI)

54392-15-3
Unknown
Unknown
Proxel CRL (9CI)

54395-52-7
$C_{33}H_{26}N_2O_6$
546.57
O=C(N(C(=O)c1cc(Oc(ccc(c2)C(c(ccc(Oc(ccc(c3C(=O)N4C)C4=O)c3)c5)c5)(C)C)c2)cc6)C)c16
1H-Isoindole-1,3(2H)-dione, 5,5'-((1-methylethylidene)-bis(4,1-phenyleneoxy))bis-(2-methyl- (9CI)
N,N'-Dimethyl-2,2-bis(4-(3,4-dicarboxyphenoxy)phenyl)propane diimide

54410-74-1
$C_{12}H_{18}$
162.28
Benzene, 1-methyl-2-(1-ethylpropyl)- (9CI)
1-(1-Ethylpropyl)-2-methylbenzene

54410-84-3
C_4H_6BrCl
169.45
2-Butene, 1-bromo-2-chloro- (9CI)

54410-98-9
$C_{12}H_{24}$
168.33
1-Nonene, 4,6,8-trimethyl- (6CI,9CI)
4,6,8-Trimethyl-1-nonene

54411-03-9
$C_{11}H_{20}O$
168.28
3-Decen-2-one, 3-methyl- (9CI)

54411-19-7
$C_{10}H_{13}Cl$
168.67
Benzene, chloro-1-methyl-4-(1-methylethyl)- (9CI)

54411-21-1

$C_{10}H_{13}Cl$
168.67
Benzene, (1-chloroethyl)-dimethyl- (9CI)
Xylene, (1-chloroethyl)- (7CI)

54423-73-3
$C_4H_{11}O_2P$
122.10
Phosphinic acid, (2-methylpropyl)- (9CI)
Isobutylphosphinic acid
Isobutylphosphonous acid

54446-78-5
$C_8H_{18}O_3$
162.23
Ethanol, 1-(2-butoxyethoxy)- (9CI)

54453-03-1
$C_{10}H_{12}CuN_2O_8.2H$
353.76
Copper ethylenediaminetetra-acetate
Acetic acid, (ethylenedinitrilo)tetra-, copper(II) complex
Caswell No. 240
Copper EDTA complex
Cuprate(2-), ((N,N'-1,2-ethanediylbis(N-(carboxymethyl)-glycinato))(4-)-N,N',O,O',ON,ON')-, dihydrogen, (OC-6-21)- (9CI)
EPA Pesticide Chemical Code 039105

54458-61-6
$C_9H_{14}O$
138.21
2-Cyclopenten-1-one, 2,3,4,5-tetramethyl- (9CI)

54460-46-7
$C_{20}H_{38}O_2$
310.58
CCCCCCCCCCCCCCCCOC(=O)C1CC1
Cyclopropanecarboxylic acid, hexadecyl ester
Cycloprate
Hexadecyl cyclopropanecarboxylate
Zardex
ZR-856

54518-04-6
$C_9H_{20}O_3$
176.26
Methanol, dibutoxy- (9CI)

54518-11-5
$C_{12}H_{18}O$
178.28
Benzeneethanol, α-methyl-3-(1-methylethyl)- (9CI)

54518-15-9
$C_7H_6Cl_2O$
177.03
Benzene, dichloromethoxy- (9CI)

54536-18-4
$C_{12}H_6Cl_2O_2$
253.08
Dibenzo(b,e)(1,4)dioxin, 1,2-dichloro- (9CI)
1,2-Dichlorodibenzo(b,e)(1,4)dioxin
1,2-Dichlorodibenzo-p-dioxin

54536-19-5
$C_{12}H_6Cl_2O_2$
253.08
Dibenzo(b,e)(1,4)dioxin, 1,4-dichloro- (9CI)
1,4-Dichlorodibenzo(b,e)(1,4)dioxin
1,4-Dichlorodibenzo-p-dioxin

54537-30-3
$C_8H_6O_4.Na$
189.12
Benzeneacetic acid, 4-hydroxy-α-oxo-, monosodium salt (9CI)

54548-50-4
C_7H_7ClO
142.59
m-Cresol, chloro
Chloro-m-cresol [UN 2669]
Phenol, 3-methyl-, monochloro deriv. (9CI)
UN 2669 [Chlorocresols, solid]

54549-72-3
$C_{11}H_{14}O_2$
178.23
Ethanone, 1-(4-(1-hydroxy-1-methylethyl)phenyl)- (9CI)
Acetophenone, 4'-(1-hydroxy-1-methylethyl)-

54549-80-3
C_9H_{18}
126.24
Cyclopentane, 2-ethyl-1,1-dimethyl- (9CI)

54549-81-4
$C_9H_{16}O$
140.23
Cyclopentanone, 2-methyl-3-(1-methylethyl)- (9CI)

54562-24-2
$C_8H_{12}O$
124.18
2-Cyclopenten-1-one, 2,3,5-trimethyl- (9CI)

54576-36-2
$C_{11}H_{16}S$
180.31
Benzene, 1-methyl-3-((2-methylpropyl)thio-)- (9CI)

54576-37-3
$C_{11}H_{16}S$
180.31
Benzene, 1-methyl-4-((2-methylpropyl)thio)- (9CI)
Sulfide, isobutyl p-tolyl (7CI)

54576-41-9
$C_{11}H_{16}S$
180.31
Benzene, 1-(butylthio)-3-methyl- (9CI)

54578-21-1
Unknown
Unknown
Separan MG 700 (9CI)

54578-28-8
Unknown
Unknown
Therminol 66 (9CI)

54579-28-1
Unknown
Unknown
CI Direct Orange 1
Sodium dichloro-s-triazinetrione (Dry, Containing more than 39% available chlorine)
UN 2465

54589-71-8
$C_{12}H_5Cl_3O$
271.53
Dibenzofuran, 2,4,8-trichloro- (9CI)
2,4,8-Trichlorodibenzofuran

54590-59-9
$(C_6H_{11}NO.C_2H_5NO_2)x$
Polymer NA
Glycine, Polymer with hexahydro-2H-azepin-2-one (9CI)
2H-Azepin-2-one, hexahydro-, Polymer with glycine (9CI)

Nylon 2-nylon 6 copolymer

54594-42-2
$C_{14}H_{24}O$
208.35
2(1H)-Naphthalenone, octahydro-4a-methyl-7-(1-methylethyl)-, (4aα, 7β,8aβ)- (9CI)

54616-10-3
$C_{12}H_{12}O_2S_2$
252.36
Naphthalene, 1-(methylsulfonyl)-5-(methylthio)- (9CI)

54630-50-1
$C_8H_{18}O$
130.23
2-Heptanol, 5-methyl- (7CI,9CI)
5-Methyl-2-heptanol

54644-40-5
$C_{14}H_{20}O_2$
220.31
Propanoic acid, 2,2-dimethyl-, 2,4,6-trimethylphenyl ester (9CI)

54661-98-2
$C_{15}H_{32}O_2$
244.42
Ethanol, 2-((1,1-dibutylpentyl)oxy)- (9CI)
Ethylene glycol mono-(1,1-dibutylpentyl) ether

54675-14-8
$C_{10}H_{15}N$
149.23
Benzenamine, 3,4-diethyl- (9CI)
3,4-Diethylaniline

54693-46-8
Unknown
Unknown
Diacetone alcohol peroxides (More than 57% in solution with more than 9% hydrogen peroxide, less than 26% diacetone alcohol and less than 9% water; total active oxygen content more than 9% by weight)
Diacetone alcohol peroxide
Diacetone alcohol peroxide, More than 57% in solution
Diacetone alcohol peroxide, Not more than 57% in solution
Diacetone alcohol peroxides (Not more than 57% in solution with not more than 9% hydrogen peroxide, not less than 26% diacetone alcohol and not less than 9% water, total active oxygen content not more than 9% by weight)
4-Hydroxy-4-methyl-2-pentanone peroxide
2-Pentanone, 4-hydroxy-4-methyl-, peroxide (9CI)
2-Pentanone, 4-hydroxy-4-methyl-, peroxide, More than 57% in solution
2-Pentanone, 4-hydroxy-4-methyl-, peroxide, Not more than 57% in solution
Peroxido de diacetonalcohol (Spanish)
Peroxyde de diacetone-alcool (French)
UN 2163

54698-11-2
C_9H_9NO
147.17
1H-Indole, 1-methoxy- (9CI)

54699-44-4
$C_{18}H_{24}O_4$
304.39
5-Isobenzofurancarboxylic acid, 1,3-dihydro-3-oxo,-nonyl ester (9CI)

54738-93-1
$C_{14}H_9NO_2$
223.22
Anthracene, nitro- (9CI)

54749-90-5
$C_9H_{16}ClN_3O_7$
313.73
OCC(O)C(O)C(O)C(NC(=O)N(CCCl)N=O)C=O
Glucopyranose, 2-deoxy-2-(3-(2-chloroethyl)-3-nitrosoureido)-, D
1-(2-Chloroethyl)-3-(D-glucopyranos-2-yl)-1-nitrosourea
2-((((2-Chloroethyl)nitrosoamino)carbonyl)amino)-2-deoxy-D-glucose
2-(3-(2-Chloroethyl)-3-nitrosoureido)-2-deoxy-D-glucosopyranose
2-(3-(2-Chloroethyl)-3-nitrosoureido)-D-gluco-pyranose
Chlorozotocin
CHLZ
CZT
DCNU
D-Glucopyranose, 2-((((2-chloroethyl)nitrosoamino)carbonyl)amino)-2-deoxy-
D-Glucose, 2-((((2-chloroethyl)nitrosoamino)carbonyl)amino)-2-deoxy- (9CI)
NSC-178248
NSC-D 254157
Urea, 1-(2-chloroethyl)-3-(D-glucopyranos-2-yl)-1-nitroso-

54755-20-3
$C_{11}H_9NO_2$
187.19
Naphthalene, 6-methyl-1-nitro- (9CI)

54755-21-4
$C_{11}H_9NO_2$
187.19
Naphthalene, 7-methyl-1-nitro- (9CI)

54771-30-1
$C_{54}H_{87}O_3P$
815.26
Phosphorous acid, dinonylphenyl bis(nonylphenyl) ester (9CI)

54773-19-2
$C_7H_3Cl_2F_3$
215.00
Benzene, 1,2-dichloro-3-(trifluoromethyl)- (9CI)

54774-27-5
$C_6H_{12}O$
100.16
5-Hexen-2-ol, (.+-.)- (9CI)

54774-28-6
$C_6H_{12}O_2$
116.16
2-Furanmethanol, tetrahydro-5-methyl-, trans- (9CI)

54774-89-9
$C_{14}H_{16}$
184.28
Naphthalene, 2-methyl-1-propyl- (9CI)

54774-94-6
$C_{10}H_{16}O_3$
184.24
Bicyclo(2.2.2)octane-1,4-diol, monoacetate (9CI)
4-Acetoxybicyclo(2.2.2)octan-1-ol

54789-11-6
$(C_5H_8O)_2$
84.12
3-Buten-2-one, 3-methyl-, dimer (6CI,9CI)

54789-15-0

$C_{14}H_{22}$
190.33
Benzene, 1-(1-ethylpropyl)-2-propyl- (9CI)

54789-29-6
$C_9H_{10}Cl_2$
189.09
Benzene, 1-(dichloromethyl)-4-ethyl- (9CI)
4-Ethylbenzal chloride

54789-45-6
$C_{13}H_{16}O$
188.27
3-Buten-2-one, 1-(2,3,6-trimethylphenyl)- (9CI)

54815-21-3
$C_{11}H_{12}O_3$
192.22
1,3-Benzodioxol-2-one, 5-(1,1-dimethylethyl)- (9CI)

54823-94-8
$C_{14}H_{26}$
194.36
Cyclohexane, 1-(cyclohexylmethyl)-2-methyl-, trans- (9CI)

54823-95-9
$C_{14}H_{26}$
194.36
Cyclohexane, 1-(cyclohexylmethyl)-3-methyl-, trans- (9CI)

54823-98-2
$C_{14}H_{26}$
194.36
Cyclohexane, 1-(cyclohexylmethyl)-4-methyl-, trans- (9CI)

54827-17-7
$C_{16}H_{20}N_2$
240.38
Nc(c(cc(c(cc(c(N)c1C)C)c1)c2)C)c2C
(1,1'-Biphenyl)-4,4-diamine, 3,3',5,5'-tetramethyl
3,3',5,5'-Tetramethylbenzidine
3,5,3',5'-Tetramethylbenzidine
TMB

54832-83-6
$C_{15}H_{28}$
208.39
1H-Indene, octahydro-2,2,4,4,7,7-hexamethyl-, trans- (9CI)

54833-23-7
$C_{21}H_{44}$
296.58
Eicosane, 10-methyl- (9CI)

54833-48-6
$C_{21}H_{44}$
296.58
Heptadecane, 2,6,10,15-tetramethyl- (9CI)

54849-38-6
$C_{31}H_{60}O_6S_3Sn$
743.72
Monomethyltin tris(isooctyl mercaptoacetate)
Acetic acid, 2,2',2''-((methylstannylidyne)tris(thio))tris-, triisooctyl ester (9CI)
Methyltin tris(isooctyl mercaptoacetate)
Methyltin S,S',S''-tris(isooctyl mercaptoacetate)
Methyltin tris(isooctyl thioglycolate)
Monomethyltin tris(isooctyl thioglycolate)
Stannane, tris(((isooctylthio)-acetyl)oxy)methyl-
Stannane methyltris((carboxymethyl)thio)tris isooctyl ester

54852-75-4
$C_{13}H_{26}O$
198.35
Cyclohexane, (2-(pentyloxy)ethyl)- (9CI)

54889-98-4
$C_{13}H_{18}O_3$
222.29
Methanol, (4-(1,1-dimethylethyl)phenoxy)-, acetate (9CI)

54932-78-4
$C_{14}H_{22}O$
206.33
Phenol, 4-(2,2,3,3-tetramethylbutyl)- (9CI)

54934-71-3
$C_{22}H_{38}$
302.54
Cyclopentane, 1,1'-(3-(2-cyclopentylethylidene)-1,5-pentanediyl)bis- (9CI)
NSC-175283

54934-92-8
$C_{15}H_{28}$
208.39
Cyclohexane, 1-(cyclohexylmethyl)-2-ethyl-, trans- (9CI)

54935-00-1
$C_8H_5Cl_3$
207.49
Benzene, 1,4-dichloro-2-(2-chloroethenyl)- (9CI)
Styrene, β,2,5-trichloro-

54946-52-0
Unknown
Unknown
3,4-Methylenedioxymethamphetamine
DEA No. 7405
MDMA

54964-75-9
$C_7H_{18}Pb$
309.42
Plumbane, butyltrimethyl- (9CI)
Lead, butyltrimethyl-

54964-83-9
$C_{20}H_{38}$
278.53
Naphthalene, decahydro-1,4-dimethyl-5-octyl- (7CI,9CI)

54965-61-6
$C_{16}H_{30}$
222.42
Cyclohexane, 1-(cyclohexylmethyl)-4-(1-methylethyl)- (9CI)
Methane, cyclohexyl(4-isopropylcyclohexyl)- (7CI)

54966-51-7
$C_{10}H_{12}O_3$
180.21
2-Butynoic acid, 4-cyclobutyl-4-oxo-, ethyl ester (9CI)

54972-97-3
$C_6H_{14}O$
102.18
Methyl amyl alcohol
Methyl-1-pentanol
1-Pentanol, methyl- (9CI)

55000-52-7
$C_{19}H_{40}$
268.53
Hexadecane, 2,6,10-trimethyl- (9CI)

55000-53-8
$C_{20}H_{28}$
268.45
Naphthalene, 1,4-dimethyl-

5-octyl- (7CI,9CI)

55012-69-6
$C_{12}H_{15}ClO$
210.71
1-Propanone, 2-chloro-1-(4-ethylphenyl)-2-methyl- (9CI)

55013-32-6
$C_9H_{16}O_2$
156.22
2(3H)-Furanone, 5-butyldihydro-4-methyl-, cis- (9CI)

55030-62-1
$C_{15}H_{32}$
212.42
Tridecane, 4,8-dimethyl- (9CI)

55030-65-4
$C_{15}H_{17}N$
211.31
Acridine, 1,2,3,4-tetrahydro-4,9-dimethyl- (7CI,9CI)

55044-46-7
$C_6H_2Cl_6$
286.80
Cyclobutane, 1,2-dichloro 3,4-bis(dichloromethylene)- (9CI)

55045-11-9
$C_{16}H_{34}$
226.45
Tridecane, 5-propyl- (9CI)

55045-14-2
$C_{16}H_{34}$
226.45
Tetradecane, 4-ethyl- (9CI)

55069-01-7
Unknown
Unknown
Eulan WA New (9CI)

55069-41-5
Unknown
Unknown
ADPA (9CI)

55072-57-6
Unknown
Unknown
Copper zinc hydroxide sulfate

55090-44-3
$C_{13}H_{28}N_2O$
228.43
CCCCCCCCCCCCN(C)N=O
Dodecylamine, N-methyl-N-nitroso
Laurylamine, N-methyl-N-nitroso-
N-Methyl-N-nitrosolaurylamine
N-Nitroso-N-methyl-N-dodecylamin (German)
Nitrosomethyl-n-dodecylamine
NMDDA

55103-65-6
$C_{12}H_{20}O$
180.29
2H-Cyclopentacycloocten 2-one, decahydro-3a-methyl-, trans- (9CI)

55103-68-9
$C_{13}H_{22}O$
194.32
2(1H)-Benzocycloocten-one, decahydro-10a-methyl-, trans- (9CI)

55114-29-9
$C_{12}H_{22}O_4$
230.31
Propanedioic acid, butylmethyl-, diethyl ester (9CI)
Diethyl butylmethylmalonate

55124-80-6
$C_{23}H_{48}$
324.63
Nonadecane, 2,6,10,14-tetramethyl- (9CI)

55158-44-6
Unknown
Unknown
Copper alloy, Cu,Be,Co
Beryllium-copper-cobalt alloy
Cobalt-beryllium copper

55162-35-1
$C_3H_5Br_2Cl$
236.34
Propane, 1,1-dibromo-2-chloro- (9CI)
1,1-Dibromo-2-chloropropane

55162-38-4
$C_{11}H_{23}Br$
235.21
Nonane, 2-bromo-5-ethyl- (7CI,9CI)
2-Bromo-5-ethylnonane

55162-41-9
$C_{11}H_{13}ClO_3$
228.68
Propanoic acid, 2-(4-chlorophenoxy)-2-methyl-, methyl ester (9CI)
Propionic acid, 2-(p-chlorophenoxy)-2-methyl-, methyl ester

55162-61-3
$C_{43}H_{88}$
605.18
Tetracontane, 3,5,24-trimethyl- (9CI)
3,5,24-Trimethyltetracontane

55179-31-2
$C_{20}H_{23}N_3O_2$
337.46
1H-1,2,4-Triazole-1-ethanol, β-((1,1'-biphenyl)-4-yloxy)-α-(1,1-dimethylethyl)
Baycor
Baymat-Spray
Biloxazol
β-((1,1'-Biphenyl)-4-yloxy)-α-(1,1-dimethylethyl)-1H-1,2,4-triazole-1-ethanol
Bitertanol
KWG 0599
Sibutol

55193-79-8
$C_{22}H_{44}O_2$
340.59
Octanoic acid, 1-methyltridecyl ester (9CI)

55199-72-9
$C_{27}H_{46}$
370.66
20,29,30-Trinorlupane, (17α)- (9CI)

55203-12-8
$C_5H_6O_7$
178.11
Propanedioic acid, (carboxymethoxy)
(Carboxymethoxy)malonic acid
(Carboxymethoxy)propanedioic acid
o-(Carboxymethyl)tartronic acid

55215-17-3
$C_{12}H_5Cl_5$
326.44
1,1'-Biphenyl, 2,2',3,4,6-pentachloro- (9CI)
2,2',3,4,6-PCB
2,2',3,4,6-Pentachlorobiphenyl

55215-18-4
$C_{12}H_4Cl_6$
360.88

Clc1cccc(c1Cl)c2cc(Cl)c(Cl)c(Cl)c2Cl
1,1'-Biphenyl, 2,2',3,3',4,5-hexachloro- (9CI)
2,2',3,3',4,5-Hexachlorobiphenyl
2,2',3,3',4,5-PCB

55219-65-3
$C_{14}H_{18}ClN_3O_2$
295.80
CC(C)C(O)C(Oc1ccc(Cl)cc1)n2cncn2
Ethanol, 2-(4-chlorophenoxy)-1-tert-butyl-2-(1H-1,2,4-triazole-1-yl)
Bayfidan
Bay KWG 0519
Baytan
2-(4-Chlorophenoxy)-1-tert-butyl-2-(1H-1,2,4-triazole-1-yl)ethanol
β-(4-Chlorophenoxy)-α-(1,1-dimethylethyl)-1H-1,2,4-triazole-1-ethanol
Summit
Triadimenol

55236-56-1
$C_{10}H_{14}FOP$
200.19
Phosphinic fluoride, (1,1-dimethylethyl)-phenyl- (9CI)
tert-Butylphenylphosphinoyl fluoride

55255-58-8
$C_{20}H_{32}$
272.48
Naphthalene, 1,2,3,4-tetrahydro-5,8-dimethyl-1-octyl- (7CI,9CI)

55256-17-2
C_6H_4ClF
130.55
Benzene, chlorofluoro- (6CI,7CI,9CI)

55256-32-1
Unknown
Unknown
N,N-Dimethyl oleyl-linoleyl amine 2,4-dichlorophenoxyacetate

55256-33-2
$C_{20}H_{41}N.C_{20}H_{39}N.C_8H_5Cl_3O_3.C_8H_5Cl_3O_3$
791.33
Acetic acid, (2,4,5-trichlorophenoxy)-, Compd. with (Z,Z)-N,N-dimethyl-9,12-octadecadien-1-amine (1:1), Mixt. with (Z)-N,N-dimethyl-9-octadecen-1-amine (2,4,5-trichlorophenoxy)-acetate (9CI)
9,12-Octadecadien-1-amine, N,N-dimethyl-, (Z,Z)-, (2,4,5-trichlorophenoxy)-acetate, Mixt. contg. (9CI)
9-Octadecen-1-amine, N,N-dimethyl-, (Z)-, (2,4,5-trichlorophenoxy)-acetate, Mixt. contg. (9CI)

55268-74-1
$C_{19}H_{24}N_2O_2$
312.45
O=C1CN(CC2N1CCc3ccccc23)C(=O)C4CCCCC4
4H-Pyrazino(2,1-a)isoquinolin-4-one, 2-(cyclohexylcarbonyl)-1,2,3,6,7,11b-hexahydro
Biltricide
Cesol
2-Cyclohexylcarbonyl-1,2,3,6,7,11b-hexahydro-4H-pyrazino(2,1-a)isoquinolin-4-one
Droncit
Embay 8440
Praziquantel
Pyquiton

55268-75-2
$C_{16}H_{16}N_4O_8S$
424.42
CON=C(C(=O)NC2C1SCC(=C(N1C2=O)C(O)=O)COC(N)=O)c3ccco3
5-Thia-1-azabicyclo(4.2.0)oct-2-ene-2-carboxylic acid, 3-(((aminocarbonyl)oxy)-methyl)-7-((2- furanyl-(methyoxyimino)acetyl)-amino)-8-oxo-, (6R-(6-α, 7-β(Z)))
Cefuroxim
Cefuroxime
Cephuroxime
CXM
Zinacef

55282-34-3
$C_{27}H_{54}$
378.73
Cyclohexane, 1,3,5-trimethyl-2-octadecyl- (9CI)
Octadecane, 1-(2,4,6-trimethylcyclohexyl)- (6CI)

55282-90-1
$C_9H_{14}O$
138.21
3,8-Nonadien-2-one, (E)-(9CI)

55283-68-6
$C_{13}H_{14}F_3N_3O_4$
333.30
O=N(=O)c(c(N(CC(=C)C)CC)c(N(=O)=O)cc1C(F)(F)F)c1
p-Toluidine, 2,6-dinitro-N-ethyl-N-(2-methyl-2-propenyl)-α,α,α-trifluoro
EL-161
Ethalflurlin
Ethalfluralin
N-Ethyl-N-(2-methyl-2-propenyl)-2,6-dinitro-4-(trifluoromethyl)benzenamine
Somilan
Sonalan
Sonalen

55285-05-7
$C_{16}H_{22}N_2O_4S$
338.46
Carbamic acid, N-methyl-N-(morpholinothio)-, 2,3-dihydro-2,2-dimethyl-7-benzofuranyl ester
N-Methyl-N-(morpholinothio)-carbamic acid 2,3-dihydro-2,2-dimethyl-7-benzofuranyl ester
N-(Morpholinosulfenyl)carbofuran

55285-14-8
$C_{20}H_{31}N_2O_3S$
379.59
CCCCN(CCCC)SN(C)C(=O)Oc2ccc1CC(C)(C)Oc1c2
Carbamic acid, ((dibutylamino)thio)methyl-, 2,2-dimethyl-2,3-dihydro-7-benzofuranyl ester
Advantage
Carbosulfan
((Dibutylamino)thio)methyl-carbamic acid, 2,2-dimethyl-2,3-dihydro-7-benzofuranyl ester
2,3-Dihydro-2,2-dimethyl-7-benzofuranyl (di-n-butyl-aminosulfenyl)methyl-carbamate
2,3-Dihydro-2,2-dimethyl-7-benzofuranyl((dibutyl-amino)thio) methyl carbamate
FMC 35001
Marshal
Posse

55290-64-7
$C_6H_{10}O_4S_2$
210.28
CC1=C(C)S(=O)(=O)CCS1(=O)=O
p-Dithiane, 2,3-dehydro-2,3-dimethyl-, tetroxide
2,3-Dihydro-5,6-dimethyl-1,4-dithiin 1,1,4,4-tetroxide
Dimethipin
1,4-Dithiin, 2,3-dihydro-5,6-dimethyl-, 1,1,4,4-tetraoxide
Harvade
N 252
Oxidimethiin

Tetrathiin
Tetrathiin (Desiccant)
UBI-N 252

55299-12-2
$C_{13}H_9ClO_2$
232.67
Methanone, diphenyl-, monochloro monohydroxy deriv. (9CI)

55312-69-1
$C_{12}H_5Cl_5$
326.44
1,1'-Biphenyl, 2,2',3,4,5-pentachloro- (9CI)
2,2',3,4,5-PCB
2,2',3,4,5-Pentachlorobiphenyl

55320-02-0
$C_{25}H_{40}O_6$
436.59
9,12,15-Octadecatrienoic acid, 2,3-bis(acetyloxy)-propyl ester, (Z,Z,Z)- (9CI)

55320-58-6
$C_8H_{16}O$
128.22
Hexanal, 5,5-dimethyl- (9CI)

55332-02-0
$C_{15}H_{26}O$
222.37
1(2H)-Naphthalenone, octahydro-4a,5-dimethyl-3-(1-methyl ethyl)-, (3α,4aα,5α,8aα)- (9CI)

55332-03-1
$C_{15}H_{26}O$
222.37
2(1H)-Naphthalenone, octahydro-4a,5-dimethyl-3-(1-methylethyl)-,(3α,4aα,5α,8aα)- (9CI)

55335-06-3
$C_7H_4Cl_3NO_3$
256.47
OC(=O)COc1nc(Cl)c(Cl)cc1Cl
Acetic acid, (3,5,6-trichloro-2-pyridyloxy)
DOWCO 233
Garlon
3,5,6-Trichloro-2-pyridyloxy-acetic acid
Triclopyr

55345-04-5
$C_{13}H_9NO_2$
211.23
Fluorene, nitro
9H-Fluorene, nitro- (9CI)
Nitrofluorene

55353-53-2
Unknown
Unknown
Advastab TM 181FS (9CI)

55373-89-2
$C_{26}H_{52}O_2$
396.70
Pentacosanoic acid, methyl ester (9CI)

55401-55-3
$C_{32}H_{66}$
450.88
Docosane, 11-decyl- (6CI, 7CI,9CI)

55402-04-5
$C_9H_{14}O_3$
170.21
3-Pentyn-2-one, 5,5-diethoxy- (9CI)

55402-13-6
$C_{11}H_{20}$
152.28
3-Octyne, 2,2,7-trimethyl- (9CI)

55406-53-6
$C_8H_{12}INO_2$
281.09
O=C(OCC#CI)NCCCC
3-Iodo-2-propynyl butylcarbamate
Carbamic acid, butyl-, 3-iodo-2-propynyl ester (9CI)
Caswell No. 501A
EPA Pesticide Chemical Code 107801
Troysan KK-108A
Troysan Polyphase Anti-Mildew
Woodlife

55449-70-2
$C_8H_{12}O$
124.18
Pentaleno(1,2-b)oxirene, octahydro-, (1aα,1bβ, 4aα,5aα)- (9CI)

55470-97-8
$C_{42}H_{86}$
591.15
Octadecane, 2,2,4,15,17, 17-hexamethyl-7,12-bis-(3,5,5-trimethylhexyl)- (6CI,9CI)

55488-87-4
$C_2H_7FeNO_4$
164.93
Ferric ammonium oxalate
Ammonium iron ethanedioate
Ethanedioic acid, ammonium iron salt (9CI)

55493-86-2
$C_{34}H_{51}NO$
489.79
1'H-Cholest-2-eno(3,2-b)-indole, 5'-methoxy-, (5α)- (9CI)
Cyclopenta(5,6)naphtho-(2,1-b)carbazole, 1'H-cholest-2-eno(3,2-b)indole deriv. (9CI)

55499-04-2
$C_{12}H_{26}$
170.34
Nonane, 2,2,3-trimethyl- (9CI)

55510-04-8
$C_4H_4N_6O_6$
232.11
Dinitroglycoluril

55511-98-3
$C_{10}H_{16}N_4O_2S$
256.36
2-Imidazolidinone, 3-(5-tert-butyl-1,3,4-thiadiazol-2-yl)-4-hydroxy-1-methyl
Buthidazole
1-(5-tert-Butyl-1,3,4-thiadiazol-2-yl)-4-hydroxy-1-methyl-2-imidazolidinone
2-Imidazolidinone, 3-(5-(1,1-dimethylethyl)-1,3,4-thiadiazol-2-yl)-4-hydroxy-1-methyl- (9CI)
Ravage
VEL-5026

55512-33-9
$C_{19}H_{23}ClN_2O_2S$
378.95
Carbonothioic acid, O-(6-chloro-3-phenyl-4-pyridazinyl) S-octyl ester
O-(6-Chloro-3-phenyl-4-pyridazinyl) S-octyl carbonothioate
CL 11344
Fenpyrate
Lentagran
P1 3419
Pyridate
Pyron
Tough

55554-09-1
$C_{16}H_{32}O_2$

256.43
Tetradecanoic acid, 2-methyl-, methyl ester (7CI,9CI)

55557-01-2
$C_6H_8N_2O_3$
156.13
N-Nitrosoguvacine
3-Pyridinecarboxylic acid, 1,2,5,6-tetrahydro-1-nitroso-
1,2,5,6-Tetrahydro-1-nitroso-3-pyridinecarboxylic acid

55557-02-3
$C_7H_{10}N_2O_3$
170.16
N-Nitrosoguvacoline
Nicotinic acid, 1,2,5,6-tetrahydro-1-nitroso-, methyl ester
Nitrosoguvacoline
1-Nitroso-1,2,5,6-tetrahydronicotinic acid methyl ester
3-Pyridinecarboxylic acid, 1,2,5,6-tetrahydro-1-nitroso-, methyl ester

55557-21-6
$C_{12}H_{22}O_4Si_2$
286.48
4H-Pyran-4-one, 5-((trimethylsilyl)oxy)-2-(((trimethylsilyl)oxy)methyl)- (9CI)

55566-30-8
$C_8H_{24}O_8P_2.O_4S$
406.32
OCP(CO)(CO)CO.OS(O)(=O)=O
Phosphonium, tetrakis-(hydroxymethyl)-, sulfate (2:1)
NCI-C55050
Pyroset TKO
Tetrakis(hydroxymethyl)phosphonium sulfate
THPS

55569-78-3
$C_6H_{12}S$
116.23
Thiophene, tetrahydrodimethyl- (7CI,9CI)

55573-38-1
$C_{11}H_{21}NOS$
215.35
1H-Azepine-1-carbothioic acid, hexahydro-3,6-dimethyl-, S-ethyl ester (9CI)

55591-12-3
$C_{14}H_{18}O_2$
218.30
1H-Indene-4-carboxylic acid, 2,3-dihydro-1,1-dimethyl-, ethyl ester (9CI)

55600-34-5
Unknown
Unknown
Clophen A-30

55617-85-1
$C_9H_7Cl_3O_3.C_2H_7N$
314.59
Propanoic acid, 2-(2,4,5-trichlorophenoxy)-, Compd. with N-methylmethanamine (1:1) (9CI)

55635-13-7
$C_{17}H_{24}NO_5.Na$
345.41
[Na+].CCCC([N]OCC=C)=C1C(=O)CC(C)(C)C(C(=O)OC)C1=O
Cyclohexanecarboxylic acid, 3-(1-(allyloxyamino)butylidene)-6,6-dimethyl-2,4-dioxo-, methyl ester, sodium salt
ADS
2-(1-Allyloxyaminobutylidene)-5,5-dimethylmethoxycarbonylcyclohexane-1,3-dione sodium salt
Alloxydim-Sodium
Bas 90210H
Clout
Fervin
Grasip
Grasipan
Graspaz
Kusagard
Nippon Soda
NP-48
NP-48 Na

55638-50-1
$C_{12}H_{22}$
166.31
3,5-Decadiene, 2,2-dimethyl-, (Z,Z)- (9CI)

55649-81-5
$C_{15}H_{14}N_4O$
268.32
1,2,3-Benzotriazin-4(3H)-one, 3-(4-(dimethylamino)phenyl)- (9CI)
1,2,3-Benzotriazin-4(3H)-one, 3-(p-dimethylaminophenyl)- (6CI)

55667-43-1
$C_6H_8Cl_2$
151.03
C(=CC=C(Cl)Cl)(C)C
1,3-Pentadiene, 1,1-dichloro-4-methyl- (9CI)
1,1-Dichloro-4-methyl-1,3-pentadiene

55669-88-0
$C_{12}H_{18}$
162.28
Benzene, 1,4-dimethyl-2-(2-methylpropyl)- (9CI)
1,4-Dimethyl-2-isobutylbenzene
1-Isobutyl-2,5-dimethylbenzene
p-Xylene, 2-isobutyl- (6CI)

55670-09-2
$C_7H_{12}O_2$
128.17
1-Butanol, 2-methylene-, acetate (6CI,9CI)
2-(Acetoxymethyl)-1-butene

55673-89-7
$C_{12}HCl_7O$
409.31
Dibenzofuran, 1,2,3,4,7,8,9-heptachloro- (9CI)

55682-73-0
$C_{12}H_{18}$
162.27
3,5-Decadiyne, 2,2-dimethyl- (9CI)

55682-80-9
$C_{13}H_{16}$
172.27
Naphthalene, 1,2-dihydro-1,4,6-trimethyl- (9CI)

55682-88-7
$C_{21}H_{36}O_2$
320.52
11,14,17-Eicosatrienoic acid, methyl ester (9CI)

55682-91-2
$C_{28}H_{56}O_2$
424.75
Heptacosanoic acid, methyl ester (9CI)

55682-92-3
$C_{29}H_{58}O_2$
438.78
Octacosanoic acid, methyl ester (9CI)

55683-10-8
$C_{13}H_{18}O_2$
206.29
Benzenepropanoic acid,

β,β,3,4-tetramethyl-
(9CI)

55683-21-1
$C_8H_{12}O$
124.18
2-Cyclopenten-1-one, 3,4,5-trimethyl- (9CI)

55684-92-9
$C_{12}H_3Cl_7O$
411.30
Ether, heptachlorodiphenyl
Heptachloro diphenyl ether

55684-94-1
$C_{12}H_2Cl_6O$
374.86
Dibenzofuran, hexachloro- (9CI)
Hexachlorodibenzofuran

55688-01-2
$C_{10}HBr_7$
680.45
Naphthalene, heptabromo- (9CI)

55702-45-9
$C_{12}H_7Cl_3$
257.55
Clc2ccc(Cl)c(c1ccccc1)c2Cl
1,1'-Biphenyl, 2,3,6-trichloro- (9CI)
2,3,6-PCB
2,3,6-Trichlorobiphenyl

55702-46-0
$C_{12}H_7Cl_3$
257.55
1,1'-Biphenyl, 2,3,4-trichloro- (9CI)
2,3,4-PCB
2,3,4-Trichlorobiphenyl

55702-54-0
$C_{11}H_{18}O$
166.27
4-Penten-2-one, 3-cyclohexyl- (9CI)

55702-60-8
$C_9H_{18}O$
142.24
2-Pentene, 1-ethoxy-4,4-dimethyl- (9CI)
Ether, 4,4-dimethyl-2-pentenyl ethyl (6CI)

55712-37-3
$C_{12}H_7Cl_3$
257.55
1,1'-Biphenyl, 2,3',4-trichloro- (9CI)
2,3',4-PCB
2,3',4-Trichlorobiphenyl

55720-44-0
$C_{12}H_7Cl_3$
257.55
1,1'-Biphenyl, 2,3,5-trichloro- (9CI)
2,3,5-PCB
2,3,5-Trichlorobiphenyl

55720-99-5
$C_{12}H_4Cl_6O$
376.88
Chlorinated diphenyl oxide
Hexachlorodiphenyl oxide

55722-26-4
$C_{12}H_2Cl_8$
429.77
1,1'-Biphenyl, octachloro- (9CI)
Biphenyl-, octachloro-
Octachlorobiphenyl

55723-93-8
$C_{18}H_{38}O$
270.50
7-Heptadecanol, 7-methyl- (9CI)

55723-99-4
$C_9H_{13}Cl$
156.65
Cyclohexane, (3-chloro-1-propynyl)- (9CI)

55724-04-4
$C_7H_{16}O_2$
132.20
2-Pentanol, 5-methoxy-2-methyl- (9CI)

55724-73-7
$C_8H_{16}O_3$
160.21
Butanoic acid, 4-butoxy- (9CI)

55738-54-0
$C_{11}H_{11}N_5O_4$
277.21
trans-2-((Dimethylamino)methylimino)-5-(2-(5-nitro-2-furyl)vinyl)-1,3,4-oxadiazole
1,3,4-Oxadiazole, 2-((dimethylamino)methylimino)-5-(2-(5-nitro-2-furyl)vinyl)-, (E)-

55739-99-6
Th
270
Thorium, Isotope of mass 270 (9CI)
Thorium-270

55755-19-6
$H_2O_3S_2.H_2O.2Na$
155.15
Thiosulfuric acid, disodium salt, monohydrate (9CI)

55759-85-8
$C_7H_{17}NO_2$
147.21
1-Propanamine, 3-(2-methoxy-1-methylethoxy)- (9CI)

55773-90-5
$C_{10}H_7Cl_5O_3$
352.43
Acetic acid, (pentachlorophenoxy)-, ethyl ester (9CI)

55777-84-9
C_6H_6BrN
172.02
Benzenamine, ar-bromo- (9CI)
Aniline, ar-bromo-

55785-58-5
$C_{15}H_{20}O_4$
264.32
Phomenone

55814-41-0
$C_{17}H_{19}NO_2$
269.37
Benzamide, 2-methyl-N-(3-(1-methylethoxy)phenyl)
Basitac
3'-Isopropoxy-2-methylbenzanilide
Mepronil
Mepronil (Pesticide)
2-Methyl-N-(3-(1-methylethoxy)phenyl)benzamide

55818-96-7
$C_4H_{12}O_4P.C_4H_{12}O_4P.C_2H_3O_2.1/3O_4P$
400.97
Phosphonium, tetrakis-(hydroxymethyl)-, acetate mixed with tetrakis-(hydroxymethyl) phosphonium dihydrogen phosphate (76:24)
Pyroset Flame Retardant TKP
Pyroset TKP

55836-33-4
$C_7H_8O_2.C_6H_4O_2$
232.24
2,5-Cyclohexadiene-1,4-dione, Compd. with 2-methyl-1,4-benzenediol

(1:1) (9CI)
1,4-Benzenediol, 2-methyl-, Compd. with 2,5-cyclohexadiene-1,4-dione (1:1) (9CI)

55838-67-0
Unknown
Unknown
Epoxide 7
EP 587

55840-82-9
Unknown
Unknown
C.I. Basic Blue 3 (VAN) (9CI)
Astrazon Blue BG
Astrazon Blue BGE/X
Atacryl Blue 3G
C.I. Basic Blue 4
Deorlene Blue 5G
Maxilon Blue 5G
Sevron Blue 5G
Sevron Blue 5GNF
Sumiacryl Blue 6G
Synacril Blue 5G

55852-95-4
$C_{12}H_{23}NOS$
229.38
1H-Azepine-1-carbothioic acid, hexahydro-, S-pentyl ester (9CI)

55861-78-4
$C_{10}H_{17}N_3O_2$
211.30
Urea, N,N-dimethyl-N'-(5-(1,1-dimethylethyl)-3-isoxazolyl)
3-(5-tert-Butylisoxazol-3-yl)-1,1-dimethylurea
N,N-Dimethyl-N'-(5-(1,1-dimethylethyl)-3-isoxazolyl)urea
N'-(5-(1,1-Dimethylethyl)-3-isoxazolyl)-N,N-dimethylurea
Isouron
Isoxyl
SSH 43

55864-04-5
$C_{21}H_{21}O_4P$
368.37
Phosphoric acid, 4-(1-methylethyl)phenyl diphenyl ester (9CI)
p-Cumenyl phenyl phosphate

55880-77-8
C_4HCl_5
226.32
1,3-Butadiene, pentachloro- (9CI)

55898-43-6
$C_{11}H_{20}O_4$
216.28
Propanedioic acid, methylpropyl-, diethyl ester (9CI)
Diethyl methylpropylmalonate
Malonic acid, methylpropyl-, diethyl ester (6CI)

55909-73-4
C_7H_7BrO
187.04
Phenol, bromomethyl- (9CI)

55936-40-8
$C_{12}H_9N_3O_3$
243.20
Phenol, 2-nitro-4-(phenylazo)- (9CI)

55955-78-7
$C_{15}H_{30}O_2$
242.41
Tridecanoic acid, 2-methyl-, methyl ester (9CI)

55956-21-3
$C_7H_{16}O_3$
148.20
1-Propanol, 2-(2-methoxy-1-methylethoxy)- (9CI)

55956-22-4
$C_7H_{16}O_3$
148.20
2-Propanol, 2-(2-methoxy-1-methylethoxy)- (7CI, 9CI)

55956-25-7
$C_9H_{18}O_3$
174.24
2-Propanol, 1-(1-methyl-2-(2-propenyloxy)-ethoxy)- (9CI)

55956-31-5
$C_{10}H_{21}N$
155.29
3-Octen-2-amine, N,N-dimethyl-, (E)- (9CI)

55956-37-1
$C_9H_{18}O$
142.24
2-Hepten-3-ol, 4,5-dimethyl- (9CI)

55956-43-9
C_9H_{12}
120.20
Cyclohexene, 3-(2-propynyl)- (9CI)

55957-10-3
Unknown
Unknown
Fyrquel 220 (9CI)

55962-27-1
Unknown
Unknown
Skydrol LD (9CI)

55965-84-9
$C_4H_5NOS.C_4H_4ClNOS$
264.76
3(2H)-Isothiazolone, 5-chloro-2-methyl-, Mixt. with 2-Methyl-3(2H)-isothiazolone
Bio-Perge
Isothiazolinone chloride
Kathon 886
Kathon Biocide
Kathon CG
Kathon LX
Kathon 886MW
Kathon RH 886
Kathon 886 W
Kathon WT
KKM 43

55976-10-8
$C_{12}H_{20}$
164.29
5-Decene, 4-ethynyl-, (E)- (9CI)

55976-13-1
$C_{11}H_{20}$
152.28
1,4-Undecadiene, (E)- (9CI)
NSC-244871

56009-20-2
$C_{20}H_{40}$
280.54
Cyclohexane, 1-(1,5-dimethylhexyl)-4-(4-methylpentyl)- (9CI)

56009-36-0
$C_8H_{14}O_3$
158.20
2-Pentenoic acid, 2-methoxy-4-methyl-, methyl ester (9CI)

56025-96-8
$C_{12}H_{20}O$
180.29
(1,1'-Bicyclohexyl)one
AI3-26971

56030-49-0
$C_{11}H_{18}$

150.27
Cyclohexene, 3-(3-methyl-1-butenyl)-, (E)- (9CI)

56030-54-7
$C_{22}H_{30}N_2O_2S$
386.55
Sufentanil
DEA No. 9740
N-(4-(Methoxymethyl)-1-(2-(2-thienyl)ethyl)-4-piperidinyl)-N-phenylpropanamide
N-(4-(Methoxymethyl)-1-(2-(2-thienyl)ethyl)-4-piperidyl)propionanilide
Propanamide, N-(4-(methoxymethyl)-1-(2-(2-thienyl)ethyl) 4-piperidinyl)-N-phenyl-
R 30,730
Sufentanilum (Latin)
Sulfentanil

56038-89-2
$C_{13}H_{21}N$
191.32
Benzenamine, N-(1-ethylpropyl)-4,5-dimethyl- (9CI)
N-(1-Ethylpropyl)-3,4-xylidine
N-(1-Ethylpropyl)-3,4-dimethylaniline

56046-62-9
$C_{12}H_{19}N_3O_3S$
285.35
O=S(=O)(NCCN(c(ccc(N=O)c1C)c1)CC)C
N-(2-(Ethyl(3-methyl-4-nitrosophenyl)amino)ethyl)-methanesulfonamide
Methanesulfonamide, N-(2-(ethyl(3-methyl-4-nitrosophenyl)amino)ethyl)- (9CI)
N-(2-(Methylsulfamido)ethyl)-N-ethyl-3-methyl-4-nitrosoaniline
4-Nitroso-N-ethyl-N-(β-methylsulfonamidoethyl)-m-toluidine

56051-60-6
$C_{14}H_{26}O_4$
258.36
Pentanedioic acid, 3-methyl-, dibutyl ester (9CI)
Glutaric acid, 3-methyl-, dibutyl ester (6CI)

56052-83-6
$C_7H_{14}O$
114.19
2-Hexene, 1-methoxy-, (E)- (9CI)

56052-85-8
$C_{10}H_{20}O$
156.27
2-Pentene, 5-(pentyloxy)-, (E)- (9CI)

56052-95-0
$C_7H_{14}O$
114.19
Oxirane, 2-ethyl-3-propyl-, trans- (9CI)
trans-3,4-Epoxyheptane

56070-16-7
$C_9H_{21}O_4PS_3$
320.43
CCOP(=S)(OCC)SCS(=O)(=O)C(C)(C)C
Terbufos sulfone
AC 94320
Phosphorodithioic acid, S-(((1,1-dimethylethyl)sulfonyl)-methyl)-O,O-diethyl ester (9CI)

56073-10-0
$C_{31}H_{23}BrO_3$
523.45
Oc5c(C1CC(Cc2ccccc12)c3ccc(cc3)c4ccc(Br)cc4)c(=O)oc6ccccc56
Coumarin, 3-(3-(4'-bromo-1,1'-biphenyl-4-yl)-1,2,3,4-tetrahydro-1-naphthyl)-4-hydroxy
Brodifacoum
Brodifakum (Czech)
3-(3-(4'-Bromobiphenyl-4-yl)-1,2,3,4-tetrahydronaphth-1-yl)-4-hydroxycoumarin
3-(3-(4'-Bromo-1,1'-biphenyl-4-yl)-1,2,3,4-tetrahydro-1-naphthyl)-4-hydroxycoumarin
Havoc
Klerat
PP 581
Ratak
Ratak Plus
Talon
Talon Rodenticide
Volid
WBA 8119

56090-54-1
$C_9H_{20}O_7$
240.25
Triglycerol (9CI)

56141-00-5
$C_{26}H_{23}Cl_4NO_6$
587.30
COc1c(Cl)ccc(Cl)c1C(=O)OCCN(CCOC(=O)c2c(Cl)ccc(Cl)c2OC)c3ccccc3
Benzoic acid, 3,6-dichloro-2-methoxy-, ester with N-phenyldiethanolamine (2:1)
Cambendichlor
Cambendichlore
N-Phenyldiethanolamine bis-(2-methoxy-3,6-dichlorobenzoate)
VEL-4207

56147-63-8
$C_{11}H_{14}$
146.23
1H-Indene, 2-ethyl-2,3-dihydro- (9CI)
Indan, 2-ethyl-

56157-92-7
$C_7H_6Cl_2O_2S$
225.09
Benzenesulfonyl chloride, 4-chloro-2-methyl- (9CI)
4-Chloro-2-methylphenylsulfonyl chloride

56157-93-8
$C_6H_5ClO_3S$
192.62
Benzenesulfonyl chloride, 3-hydroxy- (9CI)

56185-01-4
$C_{10}H_{16}ClN_2O_2P$
262.68
Phosphorodiamidic acid, tetramethyl-, 4-chlorophenyl ester (9CI)

56189-09-4
$C_{36}H_{70}O_6Pb_2$
1013.35
Lead stearate
Bis(octadecanoato)dioxodilead
Lead, bis(octadecanoato)dioxodi- (9CI)

56207-39-7
$C_7H_7N_3O_4$
197.17
Cc1c(N)c(ccc1N(=O)=O)N(=O)=O
o-Toluidine, 3,6-dinitro
2-Amino-3,6-dinitrotoluene
3,6-Dinitro-o-toluidine

56222-10-7
$C_9H_{10}N_2O_3$
194.19
Acetamide, N-((4-nitrophenyl)methyl)- (9CI)
Acetamide, N-p-nitrobenzyl- (6CI)
N-p-Nitrobenzylacetamide

56265-21-5
$C_{20}H_{26}O_2$

298.43
Benzene, 1,1'-(2-methyl-propylidene)bis(4-ethoxy-
AI3-29315

56265-22-6
$C_{19}H_{23}ClO_2$
318.87
Benzene, 1,1'-(2-chloropropylidene)bis(4-ethoxy
1,1'-(2-Chloropropylidene)bis-(4-ethoxybenzene)

56265-23-7
$C_{19}H_{22}Cl_2O_2$
353.31
Benzene, 1,1'-(2,2-dichloro-propylidene)bis(4-ethoxy
1,1'-(2,2-Dichloropropylidene)-bis(4-ethoxybenzene)

56265-24-8
$C_{20}H_{25}ClO_2$
332.88
Benzene, 1,1'-(2-chloro-2-methylpropylidene)bis(4-ethoxy- (9CI)

56265-26-0
$C_{19}H_{24}O$
268.40
Benzene, 1-ethoxy-4-(2-methyl-1-(4-methylphenyl)propyl)- (9CI)

56265-27-1
$C_{18}H_{21}ClO$
288.82
Benzene, 1-(2-chloro-1-(4-ethoxyphenyl)propyl)-4-methyl- (9CI)

56292-64-9
$C_{14}H_{26}$
226.45
Naphthalene, 1-(1,1-dimethylethyl)decahydro- (9CI)

56292-65-0
$C_{14}H_{30}$
198.40
Dodecane, 2,5-dimethyl- (7CI,9CI)

56292-66-1
$C_{15}H_{32}$
212.42
Tridecane, 2,5-dimethyl- (9CI)

56298-75-0
$C_{12}H_{16}$
160.26
1H-Indene, 1-ethyl-2,3-dihydro-1-methyl- (9CI)
1-Ethyl-1-methylindan

56312-55-1
$C_{11}H_{20}O$
168.28
5-Undecen-4-one (6CI,9CI)

56324-70-0
$C_{11}H_{18}$
150.27
1H-Indene, 1-ethylidene-octahydro-, trans- (9CI)

56348-72-2
$C_{12}H_6Cl_4O$
307.99
Benzene, 1,1'-oxybis(3,4-dichloro- (9CI)

56375-33-8
$C_4H_{10}N_2O$
102.16
Butylamine, N-nitroso
Butanamine, N-nitroso-
Butylnitrosamine
N-Nitrosobutylamine

56388-47-7
$C_{26}H_{40}$
352.60
Naphthalene, hexadecyl- (9CI)

56391-56-1
$C_{21}H_{41}N_5O_7$
475.67
CCNC2CC(N)C(OC1OC(=CCC1N)CN)C(O)C2OC3OCC(C)(O)C(NC)C3O
D-Streptamine, O-3-deoxy-4-c-methyl-3-(methylamino)-β-l-arabinopyranosyl(1-6)-O-(2,6- diamino-2,3,4,6-tetradeoxy-α-d-glycero-hex-4-enopyranosyl(1-4))-2-deoxy-N[1]-ethyl
1-N-Aethylsisomicin
1-N-Ethylsisomicin
Netilmicin
NTL
SCH 20569

56400-60-3
$C_2H_8N_2$.ClH
96.58
[Cl-].CNNC
Hydrazine, 1,2-dimethyl-, hydrochloride
sym-Dimethylhydrazine hydrochloride
1,2-Dimethylhydrazine hydrochloride
DMH

56428-00-3
$C_{14}H_{12}Cl_2O$
268.02
Bis(p-chlorobenzyl) ether
Benzene, 1,1'(oxybis(methylene))bis(4-chloro-

56438-07-4
$C_{16}H_{30}O_4$
286.42
1,1-Dodecanediol, diacetate (9CI)

56438-08-5
$C_{14}H_{26}O_2$
226.36
1-Dodecen-1-ol, acetate (6CI,9CI)

56444-79-2
$C_{27}H_{33}O_4P$
452.53
O=P(Oc(c(cc(c1)C)C)c1C)(Oc(c(cc(c2)C)C)c2C)Oc(c(cc(c3)C)C)c3C
Phenol, 2,4,6-trimethyl-, phosphate (3:1) (9CI)
2,4,6-Trimethylphenol phosphate (3:1)
Tri(2,4,6-trimethylphenyl) phosphate

56449-31-1
Unknown
Unknown
Direct Black (9CI)

56480-06-9
$C_{10}H_2Br_6$
601.58
Naphthalene, hexabromo
Hexabromonaphthalene

56507-37-0
$C_7H_{12}N_4O_2$
184.17
1,2,4-Triazine-3,5(2H,4H)-dione, 4-amino-6-(1,1-dimethylethyl)- (9CI)

56539-66-3
$C_6H_{14}O_2$
118.18
O(C(CCO)(C)C)C
3-Methyl-3-methoxybutanol
1-Butanol, 3-methoxy-3-methyl (9CI)
3-Methoxy-3-methyl-1-butanol

56549-12-3
$C_{10}H_{10}Cl_3O_2PS$
331.59
Phosphonothioic acid, methyl-, O-(2-chloro-1-(2,5-dichloro-

phenyl)ethenyl) O-methyl ester (9CI)
WL 25735

56554-35-9
$C_{18}H_{32}O$
264.46
9,17-Octadecadienal, (Z)- (9CI)

56554-64-4
$C_{40}H_{82}O_2$
595.10
Hexadecane, 1,1-bis(dodecyloxy)- (9CI)

56554-67-7
$C_{18}H_{32}O_3S$
272.41
Sulfuric acid, 5,8,11-heptadecatrienyl methyl ester (9CI)

56554-86-0
$C_{18}H_{34}O$
266.47
17-Octadecenal (9CI)

56554-96-2
$C_{18}H_{34}O$
266.47
2-Octadecenal (9CI)

56554-98-4
$C_{18}H_{34}O$
266.47
4-Octadecenal (9CI)

56554-99-5
$C_{18}H_{34}O$
266.47
3-Octadecenal (9CI)

56573-11-6
Unknown
Unknown
Indane (Alkane)

56573-85-4
Unknown
Unknown
Tributyltin chloride complex of ethylene oxide condensate of abietylamine
Caswell No. 867B
EPA Pesticide Chemical Code 083108
Tin-San
Tributyltin chloride complex of ethylene oxide condensate of abietylamine (Give equivalent tin)

56588-40-0
$C_{10}H_{12}O_2$
164.21
1H-Indene-1,2-diol, 2,3-dihydro-2-methyl-, cis- (9CI)

56590-81-9
Unknown
Unknown
Plurafac RA 40 (9CI)

56594-21-9
$C_{15}H_{10}ClNO_3$
287.70
9,10-Anthracenedione, 1-amino-4-chloro-2-(hydroxymethyl)- (9CI)

56594-22-0
$C_{15}H_{10}ClNO_4S$
335.77
Chlorosulfurous acid, (1-amino-9,10-dihydro-9,10-dioxo-2-anthracenyl)methyl ester (9CI)
1-Amino-2-((chlorosulfinyloxy)methyl)anthraquinone

56594-25-3
$C_{15}H_9Cl_2NO_2$
306.15
9,10-Anthracenedione, 1-amino-4-chloro-2-(chloromethyl)- (9CI)
1-Amino-4-chloro-2-(chloromethyl)anthraquinone

56594-27-5
$C_{16}H_{13}NO_3$
267.29
9,10-Anthracenedione, 1-amino-2-(methoxymethyl)- (9CI)

56594-28-6
$C_{17}H_{15}NO_3$
281.31
9,10-Anthracenedione, 1-amino-2-(ethoxymethyl)- (9CI)

56594-29-7
$C_{18}H_{17}NO_3$
295.34
9,10-Anthracenedione, 1-amino-2-(propoxymethyl)- (9CI)

56625-58-2
$C_{23}H_{27}NO_3$
365.48
9,10-Anthracenedione, 1-amino-2-((octyloxy)methyl)- (9CI)

56654-52-5
$C_9H_{19}N_3O_2$
201.31
CCCCNC(=O)N(CCCC)N=O
Urea, 1,3-dibutyl-3-nitroso
N,N'-Dibutyl-N-nitrosourea

56666-38-7
$C_{20}H_{36}O_2$
308.51
2H-Pyran, tetrahydro-2-(12-pentadecynyloxy)- (9CI)

56667-01-7
$C_{18}H_{22}$
238.38
1,1'-Biphenyl, 3,3',4,4',5,5'-hexamethyl- (9CI)
3,4,5,3',4',5'-Hexamethylbiphenyl

56667-10-8
$C_{11}H_{12}O$
160.22
2-Cyclopenten-1-ol, 1-phenyl- (7CI,9CI)

56680-68-3
$C_7H_6Cl_2O_2$
193.03
Phenol, 3,5-dichloro-4-methoxy- (9CI)

56680-89-8
$C_7H_6Cl_2O_2$
193.03
Phenol, 3,5-dichloro-2-methoxy- (9CI)

56682-87-2
$C_9H_6Cl_6$
326.87
Benzene, 1,3,5-tris(dichloromethyl)- (9CI)
1,3,5-Tris(dichloromethyl)benzene

56683-54-6
$C_{16}H_{32}O$
240.43
OCCCCCCCCCCC=CCCCC
11-Hexadecen-1-ol, (Z)-
AI3-35169
(Z)-11-Hexadecen-1-ol

56700-77-7
C_9H_{16}
124.23

1,3-Nonadiene, (E)- (9CI)
Cu-Be25
trans-1,3-Nonadiene
(E)-1,3-Nonadiene

56728-08-6
$C_{10}H_{13}NO_2$
179.21
Tricyclo(3.3.1.13,7)decane-2,6-dione, 4-amino- (9CI)

56728-10-0
C_9H_{18}
126.24
1-Hexene, 3,4,5-trimethyl- (9CI)

56771-50-7
$C_{15}H_{14}O$
210.28
Tricyclo(5.2.0.02,5)nona-3,8-dien-6-ol, 6-phenyl-, (1α,2β,5β,6β,7α)- (9CI)

56771-77-8
$C_7H_8O_4$
156.14
4-Hexenoic acid, 3-methyl-2,6-dioxo- (9CI)

56775-88-3
$C_{16}H_{17}BrN_2$
317.26
CN(C)CC=C(c1ccc(Br)cc1)c2cccnc2
2-Propen-1-amine, 3-(4-bromophenyl)-N,N-dimethyl-3-(3-pyridinyl)-, (Z)
Allylamine, 3-(p-bromophenyl)-N,N-dimethyl-3-(3-pyridyl)-
3-(4-Bromophenyl)-N,N-dimethyl-3-(3-pyridinyl)-2-propen-1-amine
(Z)-3-(4'-Bromophenyl)-3-(3''-pyridyl)dimethylallylamine
cis-H 102/09
Zimelidine
cis-Zimelidine
(Z)-Zimelidine

56776-27-3
$C_{24}H_{17}N_3O_3S.Na$
450.45
Benzenesulfonic acid, 5-(2H-naphtho(1,2-d)triazol-2-yl)-2-(2-phenylethenyl)-, sodium salt, (E)- (9CI)
NTS 1

56776-29-5
$C_{38}H_{40}N_{12}O_8S_2.2Na$
902.85
Benzenesulfonic acid, 2,2'-(1,2-ethenediyl)bis(5-((4-((2-hydroxyethyl)methylamino)-6-(phenylamino)-1,3,5-triazin-2-yl)amino)-, disodium salt, (E)- (9CI)
DASC 4

56776-30-8
$C_{40}H_{40}N_{12}O_8S_2.2Na$
926.87
Benzenesulfonic acid, 2,2'-(1,2-ethenediyl)bis(5-((4-(4-morpholinyl)-6-(phenylamino)-1,3,5-triazin-2-yl)amino)-, disodium salt, (E)- (9CI)
DASC 3

56795-65-4
$C_4H_{12}N_2.ClH$
124.64
Cl.CCCCNN
Hydrazine, butyl-, hydrochloride
n-Butylhydrazine hydrochloride

56802-99-4
Unknown
Unknown
Chlorinated trisodium phosphate

56803-37-3
$C_{22}H_{23}O_4P$
382.40
t-Butylphenyl diphenyl phosphate
tert-Butylphenyl diphenyl phosphate
Phosphoric acid, (1,1-dimethylethyl)phenyl diphenyl ester (9CI)

56805-23-3
$C_9H_{16}O$
140.23
3,6-Nonadien-1-ol, (E,Z)- (9CI)
trans,cis-3,6-Nonadien-1-ol

56827-79-3
C_4Cl_6
260.76
1,2-Butadiene, 1,1,3,4,4,4-hexachloro- (9CI)

56832-73-6
$C_{22}H_{14}$
278.36
Benzofluoranthene (9CI)

56842-14-9
$C_{10}H_8S$
160.24
Thiophene, phenyl- (9CI)

56842-43-4
$C_{17}H_{13}N$
231.29
Pyridine, diphenyl- (9CI)

56856-83-8
$C_4H_8N_2O_3$
132.14
CN(COC(C)=O)N=O
Acetic acid, methylnitrosaminomethyl ester
α-Acetoxy dimethylnitrosamine
Acetoxymethyl-methyl-nitrosamin (German)
Acetoxymethyl methylnitrosamine
N-α-Acetoxymethyl-N-methylnitrosamine
1-Acetoxy-N-nitrosodimethylamine
AMMN
ANN (German)
Dimethylamine, 1-acetoxy-N-nitroso-
DMN-OAC
MAMN
Methanol, (methylnitrosoamino)-, acetate (ester)
Methyl(acetoxymethyl)nitrosamine
Methylamine, N-acetoxymethyl-N-nitroso-
N-Nitroso-N-(acetoxy)methyl-N-methylamine
N-Nitroso-N-methyl-N-acetoxymethylamine

56860-81-2
$C_4H_3F_7O$
200.06
Propane, 3-(difluoromethoxy)-1,1,1,2,2-pentafluoro- (9CI)

56882-52-1
$C_9H_8Cl_2O_2$
219.07
Benzoic acid, 2,4-dichloro-, ethyl ester (9CI)

56894-91-8
$C_{10}H_{12}Cl_2O_2$
235.12
ClCOCc1ccc(COCCl)cc1
Benzene, 1,4-bis(chloromethoxymethyl)
Bis-1,4-(chloromethoxy)-p-xylene
1,4-Bis(chloromethoxymethyl)-benzene

56922-75-9
$C_8H_{14}O_2$
142.20
2-Hexen-1-ol, acetate, (Z)- (9CI)

56960-97-5
$C_{14}H_{12}Cl_2O$
267.15
1,2-Bis(p-chlorophenyl) ethanol
AI3-23251
Benzeneethanol, 4-chloro-α-(4-chlorophenyl)-
Ethanol, 1,2-bis(p-chlorophenyl)-

56961-05-8
$C_8H_9Cl_2N$
190.07
2,6-Dichloro-N,N-dimethylaniline
Benzenamine, 2,6-dichloro-N,N-dimethyl-

56961-07-0
$C_{13}H_{11}Br$
247.13
4'-Bromo-3-methylbiphenyl
1,1'-Biphenyl, 4'-bromo-3-methyl-

56961-11-6
$C_{12}H_{17}NO_2$
207.27
Carbanilic acid, N-ethyl-isopropyl ester
Carbamic acid, ethylphenyl-, 1-methylethyl ester

56961-20-7
$C_6H_3Cl_3O_2$
213.45
Oc1cc(Cl)c(Cl)c(Cl)c1O
1,2-Benzenediol, 3,4,5-trichloro- (9CI)

56961-21-8
$C_6H_3Cl_3O_3$
229.44
1,2,3-Benzenetriol, 4,5,6-trichloro
Pyrogallol, trichloro-
Trichloropyrogallol
Trichloro-1,2,3-trihydroxybenzene

56961-25-2
$C_7H_5Cl_2NO_2$
206.03
Benzoic acid, 4-amino-3,5-dichloro
4-Amino-3,5-dichlorobenzoic acid
M-6
NAB-930

56961-47-8
$C_{13}H_{10}Cl_2$
237.13
Chloro-(o-chlorophenyl)-phenylmethane
AI3-50309
Benzene, 1-chloro-2-(chlorophenylmethyl)-
α-(2-Chlorophenyl)benzyl chloride

56961-60-5
$C_{18}H_{13}N$
243.32
Benz(a)anthracen-8-amine
5-Amino-1:2-benzanthracene

56961-62-7
$C_{20}H_{16}$
256.36
Benz(a)anthracene, 8-ethyl
5-Ethyl-1,2-benzanthracene

56961-77-4
$C_6H_3BrCl_2$
225.90
Benzene, 1-bromo-2,3-dichloro- (9CI)

56974-57-3
$C_5H_{10}O_3.Na$
141.12
Propanoic acid, 3-hydroxy-2,2-dimethyl-, monosodium salt (9CI)

56975-84-9
$C_{27}H_{48}$
372.68
18,19-Dinorcholestane, 5,14-dimethyl-, (5β,8α,9β,10α,14β, 20S)- (9CI)

56977-47-0
Unknown
Unknown
Hydrofluorosilicic acid

56986-36-8
$C_7H_{14}N_2O_3$
174.23
CCCCN(COC(C)=O)N=O
Acetic acid, butylnitrosaminomethyl ester
Acetoxymethylbutylnitrosamine
N-(Acetoxy)methyl-N-n-butylnitrosamine
Bamn
Butyl acetoxymethylnitrosamine
N-Butyl-N-(acetoxymethyl)-nitrosamine
Butyl amine, N-(1-acetoxymethyl)-N-nitroso-
Butylnitrosaminomethyl acetate
Methanol, (butylnitrosoamino)-, acetate (ester)
Methylamine, 1-acetoxy-n-butyl-N-nitroso-
N-Nitroso-N-(1-acetoxymethyl)-butylamine

57030-15-6
$C_{27}H_{48}$
372.68
18,19-Dinorcholestane, 5,14-dimethyl-, (5β,8α,9β,10α,14β)- (9CI)

57036-00-7
Unknown
Unknown
Perko cleaner (9CI)

57055-38-6
$C_{20}H_{27}ClO_2$
334.89
1-Phenanthrenecarboxylic acid, chloro-1,2,3,4,4a,9,10,10a-octahydro-1,4a-dimethyl-7-(1-methylethyl)-, (1R-(1α,4aβ,10aα))- (9CI)

57055-39-7
$C_{20}H_{26}Cl_2O_2$
369.33
1-Phenanthrenecarboxylic acid, dichloro-1,2,3,4,4a,9,10,10a-octahydro-1,4a-dimethyl-7-(1-methylethyl)-, (1R-(1α,4aβ,10aα))- (9CI)

57057-83-7
$C_7H_5Cl_3O_2$
227.47
COc1c(O)cc(Cl)c(Cl)c1Cl
Phenol, 2-methoxy-trichloro
2-Methoxy-trichlorophenol
Trichloroguaiacol

57058-33-0
$C_9H_{10}ClNO$
183.64
Acetamide, N-((4-chlorophenyl)methyl)- (9CI)

57063-29-3
Unknown
Unknown
4,5-Dichloro-2-cyclohexyl-4-isothiazolin-3-one

57117-31-4
$C_{12}H_3Cl_5O$
340.40
Dibenzofuran, 2,3,4,7,8-pentachloro
2,3,4,7,8-Pentachlorodibenzofuran

57117-32-5
$C_{12}H_5Cl_3O$
271.53

Dibenzofuran, 2,3,8-trichloro- (9CI)
2,3,8-Trichlorodibenzofuran

57117-33-6
$C_{12}H_5Cl_3O$
271.53
Dibenzofuran, 2,3,6-tri- chloro- (9CI)

57117-34-7
$C_{12}H_5Cl_3O$
271.53
Dibenzofuran, 2,3,4-trichloro- (9CI)
2,3,4-Trichlorodibenzofuran

57117-35-8
$C_{12}H_4Cl_4O$
305.97
Dibenzofuran, 1,3,7,8-tetra- chloro- (9CI)
1,3,7,8-Tetrachlorodibenzo- furan

57117-36-9
$C_{12}H_4Cl_4O$
305.97
Dibenzofuran, 1,3,6,7-tetra- chloro- (9CI)
1,3,6,7-Tetrachlorodibenzo- furan

57117-37-0
$C_{12}H_4Cl_4O$
305.97
Dibenzofuran, 2,3,6,8-tetra- chloro- (9CI)
2,3,6,8-Tetrachlorodibenzo- furan

57117-38-1
$C_{12}H_4Cl_4O$
305.97
Dibenzofuran, 2,4,6,7-tetra- chloro- (9CI)
2,4,6,7-Tetrachlorodibenzo- furan

57117-39-2
$C_{12}H_4Cl_4O$
305.97
Dibenzofuran, 2,3,6,7-tetra- chloro- (9CI)
2,3,6,7-Tetrachlorodibenzo- furan

57117-40-5
$C_{12}H_4Cl_4O$
305.97
Dibenzofuran, 3,4,6,7-tetra- chloro- (9CI)
3,4,6,7-Tetrachlorodibenzo- furan

57117-41-6
$C_{12}H_3Cl_5O$
340.40
Dibenzofuran, 1,2,3,7,8-penta- chloro
1,2,3,7,8-Pentachlorodibenzo- furan

57117-42-7
$C_{12}H_3Cl_5O$
340.42
Dibenzofuran, 1,2,3,6,7-penta- chloro- (9CI)
1,2,3,6,7-Tentachlorodibenzo- furan

57117-43-8
$C_{12}H_3Cl_5O$
340.42
Dibenzofuran, 2,3,4,6,7-penta- chloro- (9CI)
2,3,4,6,7-Tentachlorodibenzo- furan

57117-44-9
$C_{12}H_2Cl_6O$
374.86
Dibenzofuran, 1,2,3,6,7,8-hexa- chloro- (9CI)
1,2,3,6,7,8-Texachlorodibenzo- furan

57138-85-9
$(C_6H_7N.CH_2O)x.xClH$
Polymer
Formaldehyde, Polymer with benzenamine, hydrochloride (VAN) (9CI)
Aniline, formaldehyde polymer, hydrochloride

57156-91-9
$C_{19}H_{30}O_2$
290.45
2,5-Octadecadiynoic acid, methyl ester (9CI)

57213-69-1
$C_7H_4Cl_3NO_3.C_6H_{15}N$
357.69
CCN(CC)CC.OC(=O)COc1nc(Cl) c(Cl)cc1Cl
Acetic acid, ((3,5,6-trichloro- 2-pyridinyl)oxy)-, Compd. with N,N-diethylethanamine (1:1)
N,N-Diethylethanamine Compd. with ((3,5,6-trichloro-2-pyrid- inyl)oxy)acetic acid (1:1)
Garlon 3A
M 3724
((3,5,6-Trichloro-2-pyridinyl)- oxy)acetic acid,- Compd. with N,N-diethylethanamine (1:1)
Triclopyr triethylamine
Triclopyr triethylamine salt

57219-31-5
Unknown
Unknown
Caradol 520 (9CI)

57230-48-5
$C_{10}H_{14}O_6$
230.24
Galactitol, 1,2:5,6-dianhydro-, diacetate
3,4-Diacetyl-1,2-5,6-dianhydro- dulcitol
3,4-Diacetyldianhydrogalactitol

57230-49-6
$C_{24}H_{26}O_6$
410.50
Galactitol, 1,2:5,6-dianhydro-, bis(benzenepropionate)
3,4-Di-β-phenylpropionyl-1,2- 5,6-dianhydro-dulcitol

57244-88-9
$C_9H_{10}O_4$
182.18
1,4-Benzenediol, 2-methoxy-, 4-acetate (9CI)

57274-46-1
$C_{20}H_{40}O_2$
312.54
Heptadecanoic acid, 15-methyl-, ethyl ester (9CI)

57289-07-3
$C_{17}H_{36}O$
256.48
Isoheptadecanol (9CI)

57289-16-4
$C_{13}H_{20}O_2$
208.30
2,6-Naphthalenedione, oc- tahydro-1,1,8a-tri- methyl-, cis- (9CI)

57289-26-6
$C_{13}H_{28}O$
200.37
1-Dodecanol, 2-methyl-, (S)- (9CI)
(-)-2-Methyl-1-dodecanol

57292-32-7
Unknown
Unknown
Aluminum sulfate solution
Aluminum sulfate solution (20% or less)

57308-11-9
Unknown
Unknown
Chloroalkylene-9 (9CI)

57321-63-8
$C_{12}H_7Cl_3O$
273.55
Benzene, 1,1'-oxybis-, trichloro deriv. (9CI)

57342-02-6
$C_{20}H_{32}O_2$
304.48
Ro 10-3108
6,7-Epoxy-1-(p-ethylphenoxy)-3-ethyl-7-methylnonane

57344-01-1
$(C_2H_4O)x(C_2H_4O)xC_8H_{19}O_4P$
Polymer
Poly(oxy-1,2-ethanediyl), α,α'-((octyloxy)phosphinylidene)-bis(ω-hydroxy- (9CI)

57344-02-2
$(C_2H_4O)xC_{16}H_{35}O_4P$
Polymer
Poly(oxy-1,2-ethanediyl), α-(bis(octyloxy)phosphinyl)-ω-hydroxy- (9CI)

57345-30-9
$C_{17}H_{22}O_2$
258.36
1-Phenanthrenecarboxylic acid, 1,2,3,4,4a,9,10,10a-octahydro-1,4a-dimethyl-, (1S-(1α,4aα,10aα))- (9CI)
5β-Podocarpa-8,11,13-trien-16-oic acid (7CI)

57352-34-8
$C_8H_{10}O_3S$
186.23
Benzenesulfonic acid, ethyl- (9CI)
Ethylbenzenesulfonic acid

57364-79-1
$C_{18}H_{22}$
210.32
Benzene, 1-ethyl-4-((4-ethylphenyl)ethyl)- (9CI)
4,4'-Diethyldiphenylethane

57373-19-0
$C_4H_4CaCl_3NOS$
260.59
5-Chloro-2-methyl-4-isothiazolin-3-one calcium(II) chloride

57373-20-3
$C_4H_5CaCl_2NOS$
226.14
2-Methyl-4-isothiazolin-3-one calcium chloride

57377-32-9
$C_{15}H_{22}O_5$
282.37
CC2CC1OC(=O)C(=C)C1CC3(C)C(O)OC(O)CC23
Furo(2',3':5,6)cyclohepta-(1,2-c)pyran-2(3H)-one, decahydro-5,7-dihydroxy-4a,9-dimethyl-3- methylene-, (3ar-(3a-α,4a-β,5-α,7-β,8a-α,9-α,10a-α))
Hymenovin
Hymenoxon
Hymenoxone

57383-80-9
$C_6H_4Br_2O$
251.92
Phenol, 2,3-dibromo- (9CI)
2,3-Dibromophenol

57425-31-7
Unknown
Unknown
ED 5662 (9CI)

57455-06-8
C_7H_7IO
234.04
Benzyl alcohol, m-iodo
Benzenemethanol, 3-iodo-
m-Iodobenzyl alcohol
3-Iodobenzyl alcohol

57465-28-8
$C_{12}H_5Cl_5$
326.44
3,4,5,3',4'-Pentachlorobiphenyl
3,3',4,4',5-Pentachloro-1,1'-biphenyl

57472-68-1
$C_{12}H_{18}O_5$
242.30
2-Propenoic acid, oxybis-(methyl-2,1-ethanediyl) ester
Dipropylene glycol diacrylate

57474-29-0
$C_{14}H_8N_2O_6$
300.22
Nifuroquine
Nifuroquina (Spanish)
Nifuroquinum (Latin)
Quinaldofur

57520-17-9
$C_{18}H_{41}N_7.3C_2H_4O_2$
535.84
Guanidine, 1,1'-(iminobis-(octamethylene))di-, triacetate
9-Aza-1,17-diguanidinoheptadecane triacetate
DF 125
Guanidine, N,N'''-(iminodi-8,1-octanediyl)bis-, triacetate
Guanoctine triacetate
Guazatine triacetate
Guazatin triacetate
Heptadecane, 9-aza-1,17-diguanidino-, triacetate
1,1'-(Iminobis(octamethylene))-diguanidine triacetate
Panoctine triacetate
SN 513

57531-37-0
$C_3H_2ClN_3O_2$
147.50
1H-Imidazole, 2-chloro-4-nitro- (9CI)

57531-38-1
$C_3H_2ClN_3O_2$
147.50
1H-Imidazole, 4-chloro-5-nitro- (9CI)
Imidazole, 4(or 5)-chloro-5(or 4)-nitro-

57583-34-3
$C_{31}H_{60}O_6S_3Sn$
743.78
Stannane, methyltris(2-ethylhexyloxycarbonylmethylthio)
Methyltris(2-ethylhexyloxycarbonylmethylthio)stannane
Tin, methyl-, tris(isooctyl thioglycollate)

57583-35-4
$C_{22}H_{44}O_4S_2Sn$
555.43
8-Oxa-3,5-dithia-4-stannatetradecanoic acid, 10-ethyl-4,4-dimethyl-7-oxo-, 2-ethylhexyl ester (9CI)

57608-19-2
Unknown
Unknown
Bio-Luvil (9CI)

57608-28-3
Unknown
Unknown
Manoxol DT (9CI)

57648-55-2
$C_8H_{16}O$

128.22
3-Octen-2-ol, (E)- (9CI)
trans-3-Octen-2-ol

57653-85-7
$C_{12}H_2Cl_6O_2$
390.84
Clc2cc1Oc3c(Oc1c(Cl)c2Cl)cc(Cl)c(Cl)c3Cl
Dibenzo-p-dioxin, 1,2,3,6,7,8-hexachloro
Dibenzo(b,e)(1,4)dioxin, 1,2,3,6,7,8-hexachloro-
1,2,3,6,7,8-Hexachloro-p-dioxin

57657-42-8
$C_9H_{13}N_7$
219.25
Rhoplex (9CI)
Primal (VAN)

57702-05-3
$C_{13}H_{26}O$
198.35
Tridecanone (9CI)

57716-72-0
$C_{15}H_{24}O$
220.35
Benzenemethanol, α-octyl- (9CI)
1-Nonanol, 1-phenyl-

57741-47-6
$C_{17}H_{11}ClF_3N_3O_4S.Na$
468.78
2-Naphthalenesulfonic acid, 6-amino-5-((4-chloro-2-(trifluoromethyl)phenyl)azo)-4-hydroxy-, monosodium salt (9CI)
6-Amino-5-((4-chloro-2-(trifluoromethyl)phenyl)azo)-4-hydroxy-2-naphthalenesulfonic acid, monosodium salt

57807-89-3
$C_{12}H_{25}ClO_2SSn$
387.56
Acetic acid, ((chlorodimethylstannyl)thio)-, isooctyl ester (9CI)

57808-36-3
$C_4H_5Cl_3$
159.44
2-Butene, 1,1,4-trichloro-, (E)- (9CI)
trans-1,1,4-Trichlorobut-2-ene

57827-84-6
$C_{22}H_{14}$
278.35
Benzochrysene (9CI)

57835-92-4
$C_{16}H_9NO_2$
247.25
4-Nitropyrene
Pyrene, 4-nitro-

57837-19-1
$C_{15}H_{21}NO_4$
279.37
COCC(=O)N(C(C)C(=O)OC)c1c(C)cccc1C
Alanine, N-(methoxyacetyl)-N-(2,6-xylyl)-, methyl ester, DL
DL-Alanine, N-(2,6-dimethylphenyl)-N-(methoxyacetyl)-, methyl ester (9CI)
Apron
Apron 2E
Apron FL
CG 117
CGA 48988
N-(2,6-Dimethylphenyl)-N-(methoxyacetyl)-alanine methyl ester
N-(2,6-Dimethylphenyl)-N-(methoxyacetyl)-dl-alanine methyl ester
Metalaxil
Metalaxyl
Ridomil
Ridomil 2E
Subdue
Subdue 2E
Subdue 5SP

57855-77-3
$C_{28}H_{44}O_3S.1/2Ca$
480.80
Naphthalenesulfonic acid, dinonyl-, calcium salt (9CI)
Calcium dinonylnaphthalenesulfonate
Dinonylnaphthalenesulfonic acid, calcium salt

57875-61-3
$C_{14}H_6N_2O_6$
298.20
9,10-Anthracenedione, 1,2-dinitro- (9CI)

57910-79-9
$C_8H_{18}N_2O$
158.23
OC(N=NC(C)(C)C)(CC)C
2-Butanol, 2-((1,1-dimethylethyl)azo)- (9CI)
2-((1,1-Dimethylethyl)azo)-2-butanol

57966-95-7
$C_7H_{10}N_4O_3$
198.21
O=C(NCC)NC(=O)C(=NOC)C#N
Acetamide, 2-cyano-N-((ethylamino)carbonyl)-2-(methoxyimino)
Curzate
2-Cyano-N-((ethylamino)carbonyl)-2-(methoxyimino)-acetamide
Cymoxanil
DPX 3217
DPX 3217M

58003-48-8
$C_{33}H_{48}O_2$
476.75
26,27-Dinorergost-5-en-3-ol, benzoate, (3β)- (9CI)

58011-68-0
$C_{19}H_{16}Cl_2N_2O_4S$
439.33
Methanone, (2,4-dichlorophenyl)(1,3-dimethyl-5-(((4-methylphenyl)sulfonyl)oxy)-1H-pyrazol-4- yl)
4-(2,4-Dichlorobenzoyl)-1,3-dimethyl-5-pyrazolyl p-toluenesulfonate
Pyrazolate
1H-Pyrazol-4-ol, 4-(2,4-dichlorobenzoyl)-1,3-dimethyl-, p-toluenesulfonate
SW-751

58033-85-5
Unknown
Unknown
C 180 (9CI)

58051-96-0
Unknown
Unknown
C.I. Disperse Red 135 (9CI)
Palanil Scarlet BRE
Sumikaron Red S-GG

58109-40-3
$C_{12}H_{10}I.F_6P$
426.08
Iodonium, diphenyl-, hexafluorophosphate(1-) (9CI)
Diphenyliodonium hexafluorophosphate
Diphenyliodonium hexafluorophosphate(1-)

58139-59-6
$C_6H_6N_4S_2$
198.28
Nc1nc(cs1)c2csc(N)n2
(4,4'-Bithiazole)-2,2'-diamine
4,4'-Bithiazole, 2,2'-diamino- (7CI)

58145-38-3
$(C_{19}H_{20}O_4)x$
Polymer NA
Oxirane, 2,2'-(methylenebis-(2,1-phenyleneoxymethyl-ene))bis-, Homopolymer (9CI)
XD 7818

58164-88-8
$C_3H_6O_3$.1/3Sb
129.64
Antimony lactate
Antimony lactate, Solid
Lactate d'antimoine (French)
Lactato de antimonio (Spanish)
Lactic acid, antimony salt
Propanoic acid, 2-hydroxy-, antimony(3+) salt (3:1)
Propanoic acid, 2-hydroxy-, trianhydride with antimonic acid
UN 1550 [Antimony lactate]

58170-30-2
C_8H_9BrO
201.06
Phenol, bromodimethyl- (9CI)

58170-32-4
$C_8H_7Br_3O$
358.85
Phenol, tribromodimethyl- (9CI)

58200-66-1
$C_{12}H_2Cl_6O_2$
390.86
Dibenzo(b,e)(1,4)dioxin, 1,2,3,4,6,7-hexachloro- (9CI)
1,2,3,4,6,7-Hexachloro-dibenzo(b,e)(1,4)dioxin
1,2,3,4,6,7-Hexachlorodibenzo-p-dioxin

58200-67-2
$C_{12}H_2Cl_6O_2$
390.86
Dibenzo(b,e)(1,4)dioxin, 1,2,3,4,6,8-hexachloro- (9CI)
1,2,3,4,6,8-Hexachloro-dibenzo(b,e)(1,4)dioxin
1,2,3,4,6,8-Hexachlorodibenzo-p-dioxin

58200-68-3
$C_{12}H_2Cl_6O_2$
390.86
Dibenzo(b,e)(1,4)dioxin, 1,2,3,4,6,9-hexachloro- (9CI)
1,2,3,4,6,9-Hexachlorodi-benzo(b,e)(1,4)dioxin
1,2,3,4,6,9-Hexachlorodibenzo-p-dioxin

58200-69-4
$C_{12}H_2Cl_6O_2$
390.86
Dibenzo(b,e)(1,4)dioxin, 1,2,3,6,8,9-hexachloro- (9CI)
1,2,3,6,8,9-Hexachlorodi-benzo(b,e)(1,4)dioxin
1,2,3,6,8,9-Hexachlorodibenzo-p-dioxin

58200-70-7
$C_{12}HCl_7O_2$
425.28
Clc3cc2Oc1c(Cl)c(Cl)c(Cl)c(Cl)c1Oc2c(Cl)c3Cl
Dibenzo-p-dioxin, 1,2,3,4,6,7,9-heptachloro
Dibenzo(b,e)(1,4)dioxin, 1,2,3,4,6,7,9-heptachloro-
1,2,3,4,6,7,9-Heptachlorodi-benzodioxin
1,2,3,4,6,7,9-Heptachlorodi-benzo-para-dioxin

58270-08-9
$C_9H_{15}Cl_2N_3O_2Zn$
333.51
Zinc, dichloro(4,4-dimethyl-5-((((methylamino)carbonyl)-oxy)imino)pentanenitrile)-, (T-4)- (9CI)
AC 85258
Ethienocarb

58327-09-6
$C_{17}H_{23}FO_3$
294.40
Acetic acid, (4-fluorophen-oxy)-, 3,3,5-trimethylcyclo-hexyl ester
(4-Fluorophenoxy)acetic acid 3,3,5-trimethylcyclohexyl ester

58334-79-5
$C_4H_2Cl_4$
191.87
1,3-Butadiene, tetrachloro-
Tetrachlorobutadiene

58349-01-2
$C_6H_3BrClNO_3$
252.45
Phenol, 4-bromo-2-chloro-6-nitro- (9CI)

58353-63-2
$C_{18}H_{29}NO_2$
291.43
Benzene, dodecylnitro- (9CI)
Dodecylnitrobenzene

58425-67-5
$C_{12}H_{18}O_3S$
242.34
Benzenesulfonic acid, hexyl- (7CI,9CI)
Hexylbenzenesulfonic acid

58429-99-5
$C_{20}H_{16}$
256.36
Benz(a)anthracene, 9,10-di-methyl
6,7-Dimethyl-1,2-benzanthra-cene
9,10-Dimethylbenz(a)anthracene

58443-82-6
$C_{12}H_{18}O$
178.28
Benzeneethanol, α-methyl-2-(1-methylethyl)- (9CI)

58447-69-1
$C_{11}H_{20}O_4$
216.28
Propanedioic acid, methyl(1-methylethyl)-, diethyl ester (9CI)

58447-70-4
$C_{12}H_{22}O_4$
230.31
Propanedioic acid, methyl(2-methylpropyl)-, diethyl ester (9CI)

58500-38-2
Unknown
Unknown
Silicic acid, beryllium salt (9CI)

58528-60-2
$C_{17}H_{18}Cl_2N_4O_4$
413.24
Ethanol, 2,2'-((4-((2,6-dichloro-4-nitrophenyl)azo)-3-methyl-phenyl)imino)bis- (9CI)
Ethanol, 2,2'-(4-(2,6-dichloro-4-nitrophenylazo)-m-tolyl-imino)di-

58543-15-0
Unknown
Unknown
MCS 1043 (9CI)

58548-38-2
$C_{13}H_{10}$
166.22
Acenaphthylene, methyl- (9CI)

58570-87-9
$C_{24}H_{27}O_4P$
410.45
Phosphoric acid, bis(1-methyl-

ethyl)phenyl diphenyl ester (9CI)

58609-76-0
$C_{10}H_{14}N_2O$
178.22
Urea, (1-methyl-1-phenylethyl)- (9CI)

58615-36-4
$C_{28}H_{18}$
354.46
Dibenzopyrene (9CI)

58629-01-9
$C_5H_8ClNO_2$
149.58
3-Chloro-4,4-dimethyl-2-oxazolidinone
Agent I
3-CDO

58654-67-4
$C_9H_{18}O$
142.27
O=C(CCC(CCC)C)C
2-Octanone, 5-methyl
Methyl 3-methylhexyl ketone
5-Methyl-2-octanone

58667-63-3
$C_{16}H_{13}ClFNO_3$
321.74
Flamprop
N-Benzoyl-N-(3-chloro-4-fluorophenyl)-DL-alanine (9CI)

58670-89-6
$C_{24}H_{50}O$
354.66
1-Tetradecanol, 2-decyl- (9CI)
Decyltetradecanol
2-Decyl-1-Tetradecanol

58679-08-6
$C_4H_3Cl_3$
157.43
1,2-Butadiene, 1,1,4-trichloro- (9CI)

58695-41-3
$C_8H_{15}NO_2$
157.24
OC(=O)CNC1CCCCC1
Acetic acid, cyclohexylamino
Cyclohexylamine acetate
Cyklohexylaminacetat (Czech)
Octan cyklohexylaminu (Czech)

58695-42-4
$C_6H_{13}NO_2$
131.18
N-Isobutyrylglycine
N-(1-Methylpropyl)glycine

58718-66-4
Unknown
Unknown
Halowax 1000 (9CI)

58718-67-5
Unknown
Unknown
Halowax 1001

58769-20-3
$C_{23}H_{24}O_4S$
396.53
Cyclopropanecarboxylic acid, 3-((dihydro-2-oxo-3(2H)-thienylidene)methyl)-2,2-dimethyl-, (5-(phenylmethyl)-3-furanyl)methyl ester, (1R-(1-α,3-α(E)))
ENT 29,117
Kadethrin
cis-Kadethrin
RU 15525
1R,cis-RU 15525
Spray-Tox

58782-15-3
$(C_{10}H_{10}O_4.(C_2H_4O)xH_2O)x$
Polymer
1,4-Benzenedicarboxylic acid, dimethyl ester, Polymer with α-hydro-ω-hydroxypoly(oxy-1,2-ethanediyl) (9CI)
Dimethyl terephthalate, polyethylene glycol polyester

58802-08-7
$C_{12}H_3Cl_5O_2$
356.40
Dibenzo-p-dioxin, 1,2,4,7,8-pentachloro
1,2,4,7,8-Pentachlorodibenzo-p-dioxin

58802-09-8
$C_{12}H_2Cl_6O_2$
390.86
Dibenzo(b,e)(1,4)dioxin, 1,2,4,6,8,9-hexachloro- (9CI)
1,2,4,6,8,9-Hexachlorodibenzo(b,e)(1,4)dioxin
1,2,4,6,8,9-Hexachlorodibenzo-p-dioxin

58802-14-5
$C_{12}H_5Cl_3O$
271.53
Dibenzofuran, 2,4,6-trichloro- (9CI)
2,4,6-Trichlorodibenzofuran

58802-15-6
$C_{12}H_3Cl_5O$
340.42
Dibenzofuran, 1,2,4,7,8-pentachloro- (9CI)
1,2,4,7,8-Pentachlorodibenzofuran

58802-16-7
$C_{12}H_3Cl_5O$
340.42
Dibenzofuran, 1,3,4,7,8-pentachloro- (9CI)
1,3,4,7,8-Pentachlorodibenzofuran

58802-17-8
$C_{12}H_5Cl_3O$
271.53
Dibenzofuran, 2,3,7-trichloro- (9CI)
2,3,7-Trichlorodibenzofuran

58802-18-9
$C_{12}H_5Cl_3O$
271.53
Dibenzofuran, 1,7,8-trichloro- (9CI)

58802-19-0
$C_{12}H_4Cl_4O$
305.98
2,4,6,8-Tetrachlorodibenzofuran
2,4,6,8-TCDF

58802-20-3
$C_{12}H_4Cl_4O$
305.97
Dibenzofuran, 1,2,7,8-tetrachloro- (9CI)
1,2,7,8-Tetrachlorodibenzofuran

58802-21-4
$C_{12}H_6Cl_2O$
237.08
Dibenzofuran, 3,7-dichloro- (9CI)
3,7-Dichlorodibenzofuran

58823-09-9
$C_{10}H_{20}Cl_4O_6P_2$
440.02
O=P(OCCCl)(OCCP(=O)(OCCCl)OCCCl)CCCl
Phosphonic acid, (2-(((2-chloroethoxy)(2-chloroethyl)phosphinyl)oxy)ethyl)-, bis(2-chloroethyl) ester (9CI)
(2-(((2-Chloroethoxy)(2-chloroethyl)phosphinyl)oxy)ethyl)phosphonic acid, bis(2-chloroethyl) ester

58841-70-6
$C_{24}H_{18}O$
322.41
Diphenyl, diphenyl ether
1,1'-Biphenyl, 4,4''-oxybis- (9CI)
4-Biphenylyl ether (6CI, 7CI)
4,4'-Diphenyldiphenyl ether
Dowathurm A

58842-20-9
$C_5H_8N_2O_2S$
160.18
SD 35651
AI3-29270
2H-1,3-Thiazine, tetrahydro-2-(nitromethylene)-

58849-75-5
$C_{17}H_{34}O_5S$
350.52
O=C(OC)C(S(=O)(=O)O)CCCCCCCCCCCCCC
Hexadecanoic acid, 2-sulfo-, 1-methyl ester (9CI)
Methyl palmitate α-sulfonic acid
1-Methyl 2-sulfohexadecanoate

58882-17-0
$C_{15}H_{23}NO_4$
281.35
Ethyl dihydroxypropyl PABA
Benzoic acid, 4-(bis(2-hydroxypropyl)amino)-, ethyl ester
4-(Bis(2-hydroxypropyl)amino)-benzoic acid, ethyl ester
Ethyl dihydroxypropyl p-aminobenzoate

58882-68-1
$C_7H_7NO_3$
153.13
Phenol, methyl-4-nitro- (9CI)

58933-55-4
HeO_2
36.00
Heliox
Helium-oxygen (Mixture)
NA 1980

58943-98-9
$C_{18}H_{18}F_{24}N_3O_6P_3$
921.22
1,3,5,2,4,6-Triazatriphosphorine, 2,2,4,4,6,6-hexahydro-2,2,4,4,6,6-hexakis(2,2,3,3-tetrafluoropropoxy)- (9CI)

58968-32-4
Unknown
Unknown
Lauropal (9CI)

58968-53-9
Unknown
Unknown
Polygard
Tri(mixed mono- and dinonylphenyl)phosphite
Tri(polynonylphenyl)phosphite

58969-15-6
Unknown
Unknown
Emkapyl 1839 (9CI)

58973-18-5
$C_6H_{10}O_3$
130.14
2-Butenoic acid, 2-methoxy-3-methyl- (9CI)

58984-19-3
Unknown
Unknown
Hexanedioic acid, ester with 2,2'-oxybis(ethanol) (9CI)

59010-86-5
$C_{17}H_{20}ClO_3PS_2$
402.90
RH 0994
AI3-29409
Phosphorothioic acid, O-(4-((4-chlorophenyl)thio)phenyl) O-ethyl S-propyl ester

59080-40-9
$C_{12}H_4Br_6$
627.62
Brc1cc(Br)c(cc1Br)c2cc(Br)c(Br)cc2Br
1,1'-Biphenyl, 2,2',4,4',5,5'-hexabromo
2,4,5,2',4',5'-Hexabromobiphenyl

59094-71-2
$C_{12}H_{16}O_2$
192.26
Pentanoic acid, (4-methylphenyl)- (9CI)

59104-79-9
$C_7H_{14}Br_2$
258.01
Heptane, 1,1-dibromo- (9CI)

59113-22-3
Unknown
Unknown
Varisoft 222 (9CI)

59113-36-9
$C_6H_{14}O_5$
166.17
Propanediol, oxybis- (9CI) (VAN)
Diglycerol

59116-88-0
$C_{14}H_8Cl_2$
247.12
Phenanthrene, dichloro- (9CI)

59118-78-4
$C_{20}H_{38}O_2S$
342.59
O=C(OCCS)CCCCCCCC=CCCCCCCC
9-Octadecenoic acid (Z)-, 2-mercaptoethyl ester (9CI)
2-Mercaptoethyl oleate

59118-79-5
$C_{61}H_{114}O_6S_3Sn$
1158.48
9-Octadecenoic acid (Z)-, (methylstannylidyne)tris-(thio-2,1-ethanediyl) ester (9CI)
Methyltin tris(2-mercaptoethyl oleate)

59128-97-1
$C_{17}H_{14}BrFN_2O_2$
377.20
CS 430
SY:10-Bromo-11b-(2-fluorophenyl)-2,3,7,11b-tetrahydrooxazolo(3,2-d)(1,4)benzodiazepin-6(5H)-one
CS-430
DEA No. 2771
Haloxazolam
Oxazolo(3,2-d)(1,4)benzodiazepin-6(5H)-one, 2,3,7,11b-tetrahydro-10-bromo-11b-(2-fluoro phenyl)-
Somelin

59145-63-0
Unknown
Unknown
4-Cyclododecyl-2,6-dimethylmorpholine benzoate

59177-47-8
$C_4H_4BrN_3O_2$
205.98
1H-Imidazole, 4-bromo-1-methyl-5-nitro- (9CI)
NSC-329117
RSU 3071

59204-74-9

$C_{10}H_{12}O_2$
164.22
Egomaketone

59291-64-4
$C_{12}H_4Cl_6$
360.88
1,1'-Biphenyl, 2,2',3,4,4',6'-hexachloro- (9CI)
2,2',3,4,4',6'-Hexachlorobiphenyl
2,2',3,4,4',6'-PCB

59355-75-8
Unknown
Unknown
MAPP (OSHA)
Methyl acetylene and propadiene mixtures
Methyl acetylene and propadiene mixture, Stabilized [UN 1060]
Methyl acetylene-propadiene mixture (ACGIH,OSHA)
Methyl acetylene propadiene, Stabilized
Methylacetylene et propadiene en melange (French)
Methylacetylene-propadiene, Stabilized
Mezclas estabilizadas de metilacetileno y propadieno (Spanish)
Propyne, Mixed with propadiene
UN 1060 [Methylacetylene and propadiene mixtures, Stabilized]

59401-04-6
$C_2H_7O_3PS$
142.12
Phosphorothioic acid, dimethyl ester (9CI)

59453-69-9
$C_{40}H_{44}N_{12}O_{10}S_2.2Na$
960.96
Benzenesulfonic acid, 2,2'-(1,2-ethenediyl)bis-(5-((4-(bis(2-hydroxyethyl)amino)-6-(phenylamino)-1,3,5-triazin-2-yl)amino)-, disodium salt, (E)- (9CI)
DASC 2

59467-70-8
$C_{18}H_{13}ClFN_3$
325.75
Midazolam
DEA No. 2884
4H-Imidazo(1,5-a)(1,4)benzodiazepine, 8-chloro-6-(2-fluorophenyl)-1-methyl-
Midazolamum (Latin)

59536-56-0
Unknown
Unknown
Dobanol 45-7, acetate (9CI)

59536-65-1
Unknown
Unknown
Firemaster BP-6
Biphenyl, hexabromo- (Technical grade)
Hexabromobiphenyl (Technical grade)
PBB
Polybrominated biphenyls

59607-71-5
$C_8H_6N_2O_3S$
210.22
N#CSc(c(N(=O)=O)cc(OC)c1)c1
Thiocyanic acid, 4-methoxy-2-nitrophenyl ester
4-Methoxy-2-nitrophenylthiocyanate

59652-20-9
$C_{22}H_{15}N$
293.38
Dibenz(a,j)acridine, 14-methyl
10-Methyl-3,4,5,6-dibenzacridine
14-Methyl dibenz(a,j)acridine

59652-21-0
$C_{22}H_{15}N$
293.38
Dibenz(c,h)acridine, 7-methyl
7-Methyldibenz(c,h)acridine
9-Methyl-3,4,5,6-dibenzacridine
10-Methyl-1,2:7,8-dibenzacridine

59653-29-1
$C_8H_{18}N_4S$
202.30
Carbonothioic dihydrazide, (1-methylhexylidene)- (9CI)

59669-26-0
$C_{10}H_{18}N_4O_4S_3$
354.45
Thiodicarb
AI3-29311
Bismethomyl thioether
Caswell No. 900AA
CGA 45156
Dicarbasulf
Dimethyl N,N'-(thiobis((methylimino)carbonyloxy))bis-(ethanimidothioate)
Dimethyl-N,N'-(thiobis-(((methylimino)carbonyl)-oxy))bis(ethanimidothioate)
EPA Pesticide Chemical Code 114501
Ethanimidothioic acid, N,N'-(thiobis((methylimino)-carbonyloxy))bis-, dimethyl ester
Larvin
Lepicron
Semevin
3,7,9,13-Tetramethyl-5,11-dioxa-2,8,14-trithia-4,7,9,12-tetra-azapentadeca-3,12-diene-6,10-dione
N,N'-(Thiobis((methylimino)-carbonyloxy))bisethanimidothioic acid dimethyl ester
UC 51762
UC 51769
UC 80502

59700-57-1
$C_3H_3Cl_5F_4N_3OP_3$
443.26
1,3,5,2,4,6-Triazatriphosphorine, 2,2,4,4,6-pentachloro-2,2,4,4,6,6-hexahydro-6-(2,2,3,3-tetrafluoropr opoxy)- (9CI)
Pentachloro(2,2,3,3-tetrafluoropropoxy)cyclotriphosphazene

59700-60-6
$C_{15}H_{15}ClF_{20}N_3O_5P_3$
825.62
1,3,5,2,4,6-Triazatriphosphorine, 2-chloro-2,2,4,4,6,6-hexahydro-2,4,4,6,6-pentakis(2,2,3,3-tetrafluor opropoxy)-(9CI)

59708-52-0
$C_{24}H_{30}N_2O_3$
394.50
Carfentanil
Carfentanila (Spanish)
Carfentanilum (Latin)
Carfentanyl
DEA No. 9743
Methyl 4-(N-(1-oxopropyl)-n-phenylamino)-1-(2-phenylethyl)-4-piperidinecarboxylate
Methyl 1-phenylethyl-4-(N-phenylpropionamido)isonipecotate
Methyl 4-(N-propionyl-n-phenylamino)-1-(2-phenylethyl)-4-piperidine-carboxylate
4-((1-Oxopropyl)phenylamino)-1-(2-phenylethyl)-4-piperidinecarboxylic acid methyl ester
4-Piperidinecarboxylic acid, 4((1-oxopropyl)phenylamino)-1-(2-phenylethyl)-, methyl ester
R-33799

59709-38-5
$C_{20}H_{20}BrClN_4O_6$
527.73
O=C(OC)CCN(c(ccc(N=Nc(c(cc

(N(=O)=O)c1)Cl)c1Br)c2)
c2)CCC(=O)OC
β-Alanine, N-(4-((2-bromo-6-chloro-4-nitrophenyl)azo)-phenyl)-N-(3-methoxy-3-oxo-propyl)-, methyl ester (9CI)
3,3'-((4-((2-Bromo-4-nitro-6-chlorophenyl)azo)phenyl)-imino)bis(propanoic acid), di-methyl ester

59720-42-2
Unknown
Unknown
5-Hydroxymethoxymethyl-1-aza-3,7-dioxabicyclo-(3.3.0)octane

59756-60-4
$C_{19}H_{14}F_3NO$
329.31
O=C(C(c(cccc1)c1)=CN(C=2)C)
C2c(cccc3C(F)(F)F)c3
Fluridone
Caswell No. 130C
EL 171
EPA Pesticide Chemical Code 112900
1-Methyl-3-phenyl-5-(3-(tri-fluoromethyl)phenyl)-4(1H)-pyridinone
1-Methyl-3-phenyl-5-(α,α,α-tri-fluoro-m-tolyl)-4-pyridone
4(1H)-Pyridinone, 1-methyl-3-phenyl-5-(3-(trifluoromethyl)-phenyl)-

59763-33-6
Unknown
Unknown
Dobane 83 (9CI)

59800-48-5
Unknown
Unknown
Varcum 5169 (9CI)

59808-78-5
$C_5H_6Cl_4$
207.91
Cyclopentane, tetrachloro-(9CI)
Tetrachlorocyclopentane

59865-13-3
$C_{62}H_{111}N_{11}O_{12}$
1202.84
Cyclosporin-A
Antibiotic S 7481F1
Ciclosporin
Cyclosporin
Cyclosporine
Cyclosporine A
OL 27-400
S 7481F1
Sandimmun
Sandimmune

59900-47-9
$C_{17}H_{19}ClO_2$
290.79
Benzene, 1,1'-(2-chloro-propylidene)bis(4-methoxy-(9CI)

59901-90-5
$C_{12}H_{12}O_5$
236.24
CC(=O)OC(C1CO1)c3ccc2OCOc
2c3
Benzyl alcohol, α-epoxyethyl-1,2-(methylenedioxy)-, acetate
1'-Acetoxysafrole-2',3'-oxide
1,3-Benzodioxole-5-methanol, α-oxiranyl-, acetate (9CI)
1,3-Benzodioxole-5-methanol, α-(oxiranyl)-, acetate
α-Epoxyethyl-1,2-(methylene-dioxy)benzyl alcohol acetate

59901-91-6
$C_{10}H_{10}O_4$
194.20
OC(C1CO1)c3ccc2OCOc2c3
1,3-Benzodioxole-5-methanol, α-epoxyethyl
1'-Hydroxysafrole-2',3'-oxide

59917-23-6
$H_3NO.HI$
160.94
Hydroxyl amine iodide
Hydroxylamine, hydriodide
Hydroxylamine iodide

59948-01-5
Unknown
Unknown
Terasil Black S-RL (9CI)

59985-42-1
$C_8H_8O_5.2K$
262.35
7-Oxabicyclo(2.2.1)hept-5-ene-2,3-dicarboxylic acid, di-potassium salt, (endo,endo)-(9CI)
Aquathol K

59993-86-1
$C_{11}H_{14}O_2.C_6H_{15}NO_3$
327.41
Benzoic acid, 4-(1,1-dimethyl-ethyl)-, Compd. with 2,2',2''-nitrilotris(ethanol) (1:1) (9CI)
p-t-Butylbenzoic acid, tri-ethanolamine salt
p-tert-Butylbenozic acid, tri-ethanolamine salt
2,2',2''-Nitrilo trisethanol 4-tert-butyl benzoate
Triethanolamine p-tert-butyl benzoate

59997-74-9
$C_{18}H_{37}NO_5S$
379.56
Hexadecanoic acid, 2-sulfo-, 1-(2-aminoethyl) ester (9CI)

59997-79-4
$C_{16}H_{33}NO_6S$
367.51
Dodecanoic acid, 2-sulfo-, 1-(2-(2-aminoethoxy)-ethyl) ester (9CI)

59997-80-7
$C_{20}H_{41}NO_6S$
423.62
Hexadecanoic acid, 2-sulfo-, 1-(2-(2-aminoe-thoxy)ethyl) ester (9CI)

59997-81-8
$C_{15}H_{31}NO_5S$
337.48
Dodecanoic acid, 2-sulfo-, 1-(2-(methylamino)-ethyl) ester (9CI)

59997-83-0
$C_{16}H_{33}NO_5S$
351.51
Dodecanoic acid, 2-sulfo-, 1-(2-(dimethylamino)-ethyl) ester (9CI)

59997-85-2
$C_{21}H_{43}NO_5S$
421.64
Ethanaminium, N,N,N-tri-methyl-2-((1-oxo-2-sulfo-hexadecyl)oxy)-, hydrox-ide, inner salt (9CI)

60029-23-4
$C_7H_{11}Cl_3N_2O_2$
261.55
ClC(Cl)(Cl)C(NC=O)N1CCOCC1
Formamide, N-(2,2,2-tri-chloro-1-(morpholinyl)ethyl)
N-(1-Formamido-2,2,2-trichlor-etyl)morfolin (Czech)
Trimorfamid
Trimorfamide
Trimorphamid
Trimorphamide
VUAgT 866
VUAgT 866/72

60034-45-9
$C_6H_9NO_6.Ca.Na$

254.20
Nitrilotriacetic acid and its salts
Calcium sodium nitrilotriacetic acid
Glycine, N,N-bis(carboxymethyl)-, calcium sodium salt (1:1:1)
NTA, calcium sodium salt (1:1:1)
Sodium (nitrilotriacetato)-calciate (7CI)

60044-33-9
$C_{15}H_{23}ClO$
254.80
Oc(c(cc(c1)CCCCCCCCC)Cl)c1
Phenol, 2-chloro-4-nonyl-
AI3-18979
2-Chloro-4-nonylphenol

60046-87-9
$C_{20}H_{40}O$
296.54
1-Hexadecen-3-ol, 3,7,11,15-tetramethyl- (6CI,7CI,9CI)
3,7,11,15-Tetramethyl-1-hexadecen-3-ol

60066-88-8
$C_{15}H_{22}O$
218.34
O=CC(=CCCC(=CCCC(C=C)=C)C)C
2,6,11-Dodecatrienal, 2,6-dimethyl-10-methylene- (9CI)
2,6-Dimethyl-10-methylene-2,6,11-dodecatrienal

60083-44-5
Unknown
Unknown
Ethyl 733 (9CI)

60102-37-6
$C_{19}H_{27}NO_7$
381.47
CC2CC1(OC1C)C(=O)OC3CCN(C)CC=C(COC(=O)C2(C)O)C3=O
Petasitenine
Fukinotoxin
Fukinotoxin (Neutral)
Petasitenine (Neutral)

60123-64-0
$C_{12}H_5Cl_5O$
342.43
Benzene, 1,2,4-trichloro-5-(2,4-dichlorophenoxy)- (9CI)

60123-65-1
$C_{12}H_5Cl_5O$
342.43
Benzene, 1,2,4-trichloro-5-(3,4-dichlorophenoxy)- (9CI)

60129-67-1
$C_{16}H_{17}N_5O_6$
375.31
Ethanol, 2,2'-((4-((2,4-dinitrophenyl)azo)phenyl)imino)bis- (9CI)
Ethanol, 2,2'-(p-(2,4-dinitrophenylazo)phenylimino)di-

60145-23-5
$C_{12}H_3Cl_7$
395.32
1,1'-Biphenyl, 2,2',3,4,4',5,6'-heptachloro- (9CI)
2,2',3,4,4',5,6'-Heptachlorobiphenyl
2,2',3,4,4',5,6'-PCB

60148-94-9
$C_9H_{18}O_2$
158.24
Heptanoic acid, 2,6-dimethyl-, (R)- (9CI)
(R)-(-)-2,6-Dimethylheptanoic acid

60153-49-3
$C_4H_7N_3O$
113.14
Propionitrile, 3-(methylnitrosamino)
3-Methylnitrosaminopropionitrile
MNPN
Propanenitrile, 3-(methylnitrosoamino)-

60168-88-9
$C_{17}H_{12}Cl_2N_2O$
331.21
n(cc(C(O)(c(ccc(c1)Cl)c1)c(c(ccc2)Cl)c2)cn3)c3
5-Pyrimidinemethanol, α-(2-chlorophenyl)-α-(4-chlorophenyl)
Bloc
(2-Chlorophenyl)-α-(4-chlorophenyl)-5-pyrimidinemethanol
α-(2-Chlorophenyl)-α-(4-chlorophenyl)-5-pyrimidinemethanol
EL 222
Fenarimol
Rimidin
Rubigan

60202-36-0
Unknown
Unknown
C.I. Direct Yellow 133 (9CI)

60207-31-0
$C_{12}H_{11}Cl_2N_3O_2$
300.13
Clc1ccc(c(Cl)c1)C3(Cn2cncn2)OCCO3
Azaconazole
Azaconazol (Spanish)
Azaconazolum (Latin)
1-((2-(2,4-Dichlorophenyl)-1,3-dioxolan-2-yl)methyl)-1H-1,2,4-triazole
R-28,644
1H-1,2,4-Triazole, 1-((2-(2,4-dichlorophenyl)-1,3-dioxolan-2-yl)methyl)

60207-90-1
$C_{15}H_{17}Cl_2N_3O_2$
342.25
1H-1,2,4-Triazole, 1-((2-(2,4-dichlorophenyl)-4-propyl-1,3-dioxolan-2-yl)methyl)
Banner
CGA-64250
CGD 92710F
Desmel
1-(2-(2,4-Dichlorophenyl)-4-propyl-1,3-dioxolan-2-ylmethyl)-1H-1,2,4-triazole
Proconazole
Propiconazole
Radar
Tilt

60211-57-6
$C_7H_6Cl_2O$
177.03
Benzenemethanol, 3,5-dichloro- (9CI)

60233-25-2
$C_{12}H_5Cl_5$
326.44
1,1'-Biphenyl, 2,2',3,4',6'-pentachloro- (9CI)
2,2',3,4',6'-PCB
2,2',3,4',6'-Pentachlorobiphenyl

60238-56-4
$C_{11}H_{15}Cl_2O_3PS_2$
361.25
Phosphorothioic acid, O-(dichloro(methylthio)phenyl) O,O-diethyl ester
Celathion
Chlorthiophos
CM-S 2957
O,O-Diethyl O-(dichloro-(methylthio)phenyl) phosphorothioate
O,O-Diethyl O-dichloro(methylthio)phenyl thiophosphate
ENT 27,635
OMS 1342
S 2957

60305-22-8

$C_{31}H_{54}$
426.77
A'-Neo-30-norgammacerane, 22-ethyl-, (17α,22R)- (9CI)

60305-23-9
$C_{31}H_{54}$
426.77
A'-Neo-30-norgammacerane, 22-ethyl-, (17α,22S)- (9CI)

60337-47-5
$C_{18}H_{12}N_3O_{10}P$
461.26
Phosphoric acid, tris(nitrophenyl) ester (9CI)

60382-88-9
$C_{28}H_{18}$
354.46
Dibenzofluoranthene (9CI)

60390-27-4
$C_{12}H_6Cl_2O$
237.08
Dibenzofuran, 2,6-dichloro- (9CI)
2,6-Dichlorodibenzofuran

60468-28-2
$C_{12}H_8Cl_5DO$
345.46
2,7:3,6-Dimethanonaphth-(2,3-b)oxirene-4-d, 3,5,6,9,9-pentachloro-1a,2,2a,3,6,6a,7,7a-octahydro-, (1aα,2β,2aα,3β,6β,6aα,7β,7aα)- (9CI)

60501-41-9
$C_{21}H_{40}O_2$
324.55
O(C1COCCCCCCCCC=CCCCC CCCC)C1
Oxirane, ((9-octadecenyloxy)-methyl)-, (Z)- (9CI)
(Z)-((9-Octadecenyloxy)-methyl)oxirane
(Oleyloxymethyl)oxirane

60529-17-1
Unknown
Unknown
Dobane IN (9CI)

60529-18-2
Unknown
Unknown
Dobane INQ (9CI)

60544-75-4
$C_8H_6N_2O_2$
162.14
1H-Indole, nitro- (9CI)

60568-05-0
$C_{14}H_{21}NO_3$
251.32
Furmecyclox
BAS 389
BAS 389F
Campogran
Caswell No. 907B
N-Cyclohexyl-N-methoxy-2,5-dimethyl-3-furan-carboxamide
EPA Pesticide Chemical Code 122601
Epic
3-Furancarboxamide, N-cyclohexyl-N-methoxy-2,5-dimethyl-
Furmetamide
Gus 215
Methyl N-cyclohexyl-2,5-dimethylfuran-3-carbohydroxamate
Xyligen B

60573-45-7
$C_{24}H_{46}O_5Si_6$
583.14
Hexasiloxane, 1,1,1,3,3,5,7,7,9,11,11,11-dodecamethyl-5,9-diphenyl- (9CI)

60573-46-8
$C_{42}H_{74}O_9Si_{10}$
1003.90
Decasiloxane, 1,1,1,3,3,5,7,7,9,11,11,13,15,15,17,19,19,19-octadecamethyl-5,9,13,1 7-tetraphenyl- (9CI)

60573-48-0
$C_{32}H_{44}O_6Si_6$
693.21
Cyclohexasiloxane, 2,2,4,6,8,8,10,12-octamethyl-4,6,10,12-tetraphenyl- (9CI)

60587-10-2
$C_{17}H_{38}O_4Si_5$
446.91
Pentasiloxane, 1,1,1,3,3,5,7,7,9,9,9-undecamethyl-5-phenyl- (9CI)

60597-20-8
$C_3H_{10}N_2$.ClH
110.61
Hydrazine, trimethyl-, hydrochloride
Trimethylhydrazine hydrochloride

60617-06-3
Unknown
Unknown
Corexit 9527

60617-40-5
$C_{33}H_{60}O_7Si_8$
793.53
Octasiloxane, 1,1,1,3,3,5,7,7,9,11,11,13,15,15,15-pentadecamethyl-5,9,13-triphenyl - (9CI)

60632-40-8
$C_8H_7BrO_3$
231.05
Benzaldehyde, 2-bromo-4-hydroxy-5-methoxy- (9CI)
Vanillin, 6-bromo- (6CI)

60676-86-0
O_2Si
60.09
Silica, Crystalline - Fused
Accusand
Fused Quartz
Fused Silica
Quartz Glass
Quartz Sand
SG-67
Quarzsand (German)
Silica, Fused (ACGIH,OSHA)
Silicon dioxide
Silicone dioxide
Siltex
Suprasil
Vitreous Quartz

60712-44-9
$C_7H_5Cl_3O_2$
227.47
Phenol, 3,4,6-trichloro-2-methoxy- (9CI)

60718-52-7
$C_7H_{11}N_3O$
153.19
1H-1,2,4-Triazole, 1-(2,2-dimethyl-1-oxopropyl)- (9CI)
1-Pivaloyl-1,2,4-triazole

60754-24-7
$C_5H_8N_2O_2$
128.12
Carbamic acid, cyanomethyl-, ethyl ester (9CI)
Ethyl cyanomethylcarbamate

60763-39-5
$C_{15}H_{17}O_4P$
292.27
Phosphoric acid, 1-methylethyl diphenyl ester (9CI)

60825-26-5
$C_8H_6Cl_3NO_3$
270.49
O=C(OC)COc(nc(c(c1)Cl)Cl)c1Cl
Acetic acid, ((3,5,6-trichloro-2-pyridinyl)oxy)-, methyl ester (9CI)
((3,5,6-Trichloro-2-pyridyl)-oxy)acetic acid, methyl ester

60826-62-2
$C_{13}H_{10}O$
182.22
Dibenzofuran, methyl- (9CI)

60845-51-4
$C_4H_5Cl_3$
159.44
1-Propene, 3,3-dichloro-2-(chloromethyl)- (9CI)

60846-21-1
$C_{16}H_{17}N_5O_7S_2$
455.44
5-Thia-1-azabicyclo(4.2.0)oct-2-ene-2-carboxylic acid, 3-((acetyloxy)methyl)-7-(((2-amino-4-thiazolyl)(methoxyimino)acetyl)amino)-8-oxo-, (6R-trans)- (9CI)

60848-01-3
$C_{21}H_{12}O$
280.33
7H-Dibenz(a,kl)anthracen-7-one (9CI)

60851-34-5
$C_{12}H_2Cl_6O$
374.84
Dibenzofuran, 2,3,4,6,7,8-hexachloro
2,3,4,6,7,8-Hexachlorodibenzo-furan

60913-86-2
$C_4H_8N_2O_4$
148.11
1,2-Hydrazinedicarboxylic acid, monoethyl ester (9CI)
NSC-106273

60916-92-9
$C_2H_3FO_2.H_3N$
95.07
Acetic acid, fluoro-, ammonium salt (9CI)
Ammonium fluoroacetate

60999-18-0
$C_7H_8N_2O_2$
152.14
Benzenamine, ar-methyl-ar-nitro- (9CI)

61017-62-7
Unknown
Unknown
Nitrilotriacetic acid, iron(2+) complex sodium salt (1:1:1)
NTA, iron(2+) complex sodium salt (1:1:1)

61031-72-9
$C_8H_6Cl_2O_2$
205.04
Benzeneacetic acid, α,α-dichloro- (9CI)
2,2-Dichloro-2-phenylacetic acid

61050-97-3
$C_{12}H_{15}N_3O_6$
297.25
1,3,5-Triazine-2,4,6(1H,3H,5H)-trione, 1,3,5-tris-(2-oxopropyl)- (9CI)

61073-10-7
$C_{13}H_{10}NO_2PS$
275.27
Phosphonothioic acid, phenyl-, O-(4-cyanophenyl) ester (9CI)

61089-87-0
$C_{19}H_{14}$
242.32
Benzofluorene (9CI)

61090-94-6
$C_9H_{10}NO_4P$
227.17
Phosphoric acid, dimethyl ester, ester with p-hydroxybenzonitrile
p-Cyanophenyl dimethyl phosphate
Phosphoric acid, 4-cyanophenyl dimethyl ester

61105-31-5
Fe.4H2O3Si.3/2Mg.Na
439.19
Crocidolite (9CI)

61128-00-5
$C_8H_2Cl_6$
310.82
Benzene, ethenyl-, hexachloro deriv. (9CI)

61128-87-8
$C_{15}H_{12}O$
208.26
Phenanthrene, methoxy- (9CI)

61141-66-0
$C_{16}H_{18}$
210.32
1,1'-Biphenyl, 3,4-diethyl- (9CI)

61141-83-1
C_8H_{16}
112.22
Cyclobutane, 1,2-diethyl- (6CI,9CI)

61142-36-7
C_9H_{16}
124.23
1,3-Hexadiene, 3-ethyl-2-methyl- (9CI)

61142-77-6
$C_6H_{12}O_2$
116.16
2-Furanol, tetrahydro-2,3-dimethyl-, trans- (9CI)

61167-23-5
$C_{11}H_8Cl_6O$
368.90
5,8-Epoxy-1,4-methanonaphthalene, 1,2,3,4,10,10-hexachloro-1,4,4a,5,6,7,8,8a-octahydro-, (1α,4α,4aβ,5α,8α,8aβ)- (9CI)
5,8-Epoxy-1,4-methanonaphthalene, 1,2,3,4,10,10-hexachloro-1,4,4a,5,6,7,8,8a-octahydro-
Oxadihydroaldrin

61197-73-7
$C_{23}H_{21}ClN_6O_3$
464.87
Loprazolam
6-(2-Chlorophenyl)-2,4-dihydro 2-((4-methyl-1-piperazinyl)-methylene)-8-nitro-1H-imidazo(1,2-a) (1,4)benzodiazepin-1-one
DEA No. 2773
Loprazolamum (Latin)
Triazulenone

61201-44-3
$C_2H_2NO_4$
104.04
Acetic acid, nitro-, ion(1-) (9CI)
Nitroacetate

61204-26-0
Unknown
Unknown
Bismuth chromate

61213-25-0
$C_{12}H_{10}Cl_2F_3NO$
312.13
2-Pyrrolidinone, 3-chloro-4-(chloromethyl)-1-(3-(trifluoromethyl)phenyl)
3-Chloro-4-(chloromethyl)-1-(3-(trifluoromethyl)phenyl)-2-pyrrolidinone
Fluorochloridone
Flurochloridone
R 40244
Racer
1-(m-Trifluoromethylphenyl)-3-chloro-4-chloromethyl-2-pyrrolidone

61215-89-2
$C_{21}H_{36}O_3S$
368.58
Benzenesulfonic acid, pentadecyl- (9CI)
Pentadecylbenzenesulfonic acid

61217-08-1
$C_{11}H_6Cl_6O_2$
382.88
2,7-Epoxy-3,6-methanonaphth(2,3-b)oxirene, 3,4,5,6,9,9-hexachloro-1a,2,2a,3,6,6a,7,7a-octahydro-, (1aα,2β,2aα,3β,6β,6aα,7β,7aα)- (9CI)
Oxadieldrin

61228-92-0
Unknown
Unknown
1,10(14)-Diepoxy-4(15),5-germecradiene-9-one
Caswell No. 333F
(1Z,5E)-1,10(14)-Diepoxy-4(15),5-germacradiene-9-one
EPA Pesticide Chemical Code 124801
Periplanone B

61235-00-5
$C_{28}H_{42}O_7$
490.40
Colletotrichin
4H-Pyran-3-carboxylic acid, 5-((decahydro-6-hydroxy-5-(3-hydroxy-3-methylbutyl)-5,8a-dimethyl-2-methylene-1-naphthalenyl)methyl)-6-methoxy-2-methyl-4-oxo-, methyl ester, (1α,4aα,5α,6α,8aβ)- (9CI)

61255-81-0
C_8HCl_7
345.26
Benzene, ethenyl-, heptachloro deriv. (9CI)

61288-13-9
Unknown
Unknown
Bromkal 80
Polybrominated biphenyls

61288-32-2
Unknown
Unknown
Pentabromprop (9CI)

61328-44-7
$C_{12}H_7Cl_3O$
273.55
Benzene, 1,2-dichloro-4-(2-chlorophenoxy)- (9CI)

61328-45-8
$C_{12}H_6Cl_4O$
307.98
Benzene, 5-(4-chlorophenoxy)-1,2,4-trichloro
2,4,4',5-Tetrachlorodiphenyl ether

61350-03-6
$C_8H_{16}F_2$
150.21
Octane, 1,1-difluoro- (9CI)
1,1-Difluorooctane

61443-77-4
$C_{18}H_{18}ClN_3O_2$
343.82
8-Hydroxyloxapine
Dibenz(b,f)(1,4)oxazepin-8-ol, 2-chloro-11-(4-methyl-1-piperazinyl)-

61443-78-5
$C_{17}H_{16}ClN_3O_2$
329.79
8-Hydroxyamoxapine
Dibenz(b,f)(1,4)oxazepin-8-ol, 2-chloro-11-(1-piperazinyl)-

61444-39-1
$C_{13}H_{24}O_2$
212.33
Heptanoic acid, 3-hexenyl ester, (Z)- (9CI)

61444-41-5
$C_{14}H_{26}O_2$
226.36
Octanoic acid, 3-hexenyl ester, (Z)- (9CI)

61465-79-0
$C_9H_9Cl_3$
223.53
Benzene, (1-methylethyl)-, trichloro deriv. (9CI)
Cumene, trichloro- (7CI)
Trichlorocumene

61465-81-4
$C_8H_7Cl_3O_2$
241.50
Benzene, trichlorodimethoxy- (9CI)

61467-64-9
$C_{19}H_{17}N_3.HNO_3$
350.35
Benzenamine, 4-((4-aminophenyl)(4-imino-2,5-cyclohexadien-1-ylidene)methyl)-, mononitrate (9CI)
Benzenamine, 4-((4-aminophenyl) (4-imino-2,5-cyclohexadien-1-ylidene)methyl)-, nitric acid salt
Fuchsin nitrate
p-Rosaniline nitrate
Parafuchsine, nitric acid salt

61523-34-0
$C_{16}H_{10}S$
234.32
Benzonaphthothiophene (9CI)

61556-82-9
$C_4H_{12}N_2.ClH$
124.64
Hydrazine, tetramethyl-, monohydrochloride
Tetramethylhydrazine hydrochloride

61578-04-9
$C_{18}H_{20}O_2$
268.36
O(C1COc(ccc(c2)C(c(cccc3)c3)(C)C)c2)C1
Oxirane, ((4-(1-methyl-1-phenylethyl)phenoxy)methyl)- (9CI)
p-Cumylphenyl glycidyl ether

61583-60-6
Unknown
Unknown
Molybdenum zinc oxide (9CI)
Basic Zinc Molybdate

61593-44-0
C_8HCl_7
345.26
Benzene, pentachloro(dichloroethenyl)- (9CI)

61593-45-1
C_8H_{16}
112.22
Cyclopentane, ethylmethyl-

(9CI)

61626-71-9
$C_5H_6Cl_2$
137.01
Pentadiene, dichloro- (9CI)

61639-90-5
$C_{12}H_5Cl_5O_2$
358.42
Phenol, 4,5-dichloro-2-(2,4,5-trichlorophenoxy)
4,5-Dichloro-2-(2,4,5-trichlorophenoxy)phenol

61653-33-6
$C_{16}H_{18}N_2$
238.32
Diazene, bis(4-ethylphenyl)- (9CI)
Azobenzene, 4,4'-diethyl- (6CI,7CI)

61665-19-8
C_8H_{16}
112.22
Pentene, trimethyl- (9CI)

61670-76-6
$C_8H_7ClO_3$
186.59
Benzaldehyde, chloro-4-hydroxy-3-methoxy- (9CI)

61702-44-1
$C_6H_7ClN_2.H_2O_4S$
240.68
Nc1ccc(N)c(Cl)c1.OS(O)(=O)=O
1,4-Benzenediamine, 2-chloro-, sulfate (1:1)

61703-05-7
Unknown
Unknown
C.I. Direct Black 114 (9CI)

61724-08-1
Unknown
Unknown
C.I. Acid Brown 100 (9CI)

61736-91-2
$C_{14}H_{10}O_6S_2$
338.36
Anthracene-1,5-disulfonic acid
A-1,5-DSA

61736-92-3
$C_{14}H_{10}O_6S_2$
338.36
1,8-Anthracenedisulfonic acid (9CI)
Anthracene-1,8-disulfonic acid
1,8-Disulfoanthracene

61788-32-7
Unknown
Unknown
Terphenyls, hydrogenated
Hydrogenated terphenyls (ACGIH,OSHA)

61788-33-8
Unknown
Unknown
Terphenyl, chlorinated
Kanechlor 500
Kanechlor C
PCT
Polychlorinated terphenyl

61788-72-5
Unknown
Unknown
Fatty acid, Tall Oil, Epoxidized, Octyl ester
Admex 741
Admex 746
Drapex 4.4
Octyl epoxytallate
Plastolein 9214
PX-806

61788-76-9
Unknown
Unknown
Alkanes, chloro
Chlorinated paraffins

61788-97-4
Unknown
Unknown
Epoxy resins
Condensation products, Epoxy
Epoxides, Polymers, Epoxy resins
Epoxy Compounds (VAN)
Ethers, cyclic, epoxides, Polymers
Plastics, Epoxy
Polyethers, Epoxy resins

61788-99-6
Unknown
Unknown
(Soya alkyl) dimethyl ethyl ammonium bromide
Quaternary ammonium compounds, ethyldimethylsoya alkyl, bromides

61789-22-8
Unknown
Unknown
Copper salts of the acids of tall oil
Caswell No. 254
Copper tallate
EPA Pesticide Chemical Code 023103
Fatty acids, tall-oil, Copper salts
Tall oil, copper salt

61789-36-4
Unknown
Unknown
Naphthenic acid, calcium salt

61789-39-7
Unknown
Unknown
1-Propanaminium, 3-amino-N-(carboxymethyl)-N,N-dimethyl-, N-coco acyl derivs., chlorides, sodium salts

61789-40-0
Unknown
Unknown
Cocamidopropyl betaine
CADG
N-(Carboxymethyl)-N,N-dimethyl-3-((1-oxococonut)-amino)-1-propanaminium hydroxide, inner salt
Cocamidopropyl dimethyl glycine
N-Cocamidopropyl-N,N-dimethylglycine, hydroxide, inner salt
Cocoamidopropylbetaine
N-(Cocoamidopropyl)-N,N-dimethyl-N-carboxymethyl ammonium, betaine
N-(3-Cocoamidopropyl)-N,N-dimethyl-N-carboxymethyl-ammonium hydroxide, inner salt
N-(3-Cocoamidopropyl)-N,N-dimethyl-N-carboxymethyl betaine
Coconut oil amidopropyl betaine
Cocoyl amide propylbetaine
Emcol NA-30
FMB CAP B
Incronam 30
Jortaine CAB-35
Jortaine CFA-35
1-Propanaminium, 3-amino-N-(carboxymethyl)-N,N-dimethyl-, N-coco acyl derivs., hydroxides, inner salts
1-Propanaminium, N-(carboxymethyl)-N,N-dimethyl-3-((1-oxococonut)amino)-, hydroxide, inner salt
Quaternary ammonium compounds, (carboxymethyl)-(3-cocoamidopropyl)dimethyl, hydroxides, inner salts
Unibetaine BA-35
Unibetaine BC-35
Unibetaine K

61789-51-3
Unknown
Unknown
Naphthenic acid, cobalt salt
Cobalt naphthenate
Cobalt naphthenate, Powder [UN 2001]
Naphtenate de cobalt (French)
UN 2001 [Cobalt naphthenates, powder]

61789-60-4
Unknown
Unknown
Pitch
Naval Stores, Pitch

61789-68-2
Unknown
Unknown
Alkyl bis(2-hydroxyethyl) benzyl ammonium chloride
Bis(hydroxyethyl)cocobenzyl ammonium chloride
Cocobis(hydroxyethyl)benzyl-ammonium, chloride
Quaternary ammonium compounds, benzylcoco alkyl-bis(hydroxyethyl), chlorides

61789-77-3
Unknown
Unknown
Dicocodimonium Chloride
Di(coco alkyl) dimethyl ammonium chloride
Dicoco dimethyl ammonium chloride
Dicocodimethylammonium chloride
Dimethyldicocoammonium chloride
Quaternary ammonium compounds, dicoco alkyl dimethyl, chlorides
Quaternary ammonium compounds, dicoco alkyldimethyl, chlorides
Quaternium-34

61789-85-3
Unknown
Unknown
Sulfonic acids, petroleum
Caswell No. 647B
EPA Pesticide Chemical Code 598500
Petrolatum acid sulfonate
Petroleum sulfonates

61789-87-5
Unknown
Unknown
Sulfonic acids, petroleum, magnesium salts

61789-97-7
Unknown
Unknown
Tallow
Beef Tallow
Mutton Tallow
Tallow, Beef
Tallow, Mutton

61790-12-3
Unknown
Unknown
Tall oil acid
Disproportionated tall oil fatty acid
Fatty acids, tall oil
Tall oil acids

61790-13-4
Unknown
Unknown
Naphthenic acid, sodium salt solution
Caswell No. 589E
EPA Pesticide Chemical Code 589600
Naphthathenic soap
Naphthenic acids, sodium salts

61790-14-5
$C_7H_{12}O_2.xPb$
1578.52
Naphthenic acid, lead salt
Cyclohexanecarboxylic acid, lead salt
Lead naphthenate

61790-19-0
Unknown
Unknown
Oils, Vegetable, Sulfated
Caswell No. 811
EPA Pesticide Chemical Code 079013
Mixed Vegetable Oil, Sulfated
Sulfonated Vegetable Oil

61790-23-6
Unknown
Unknown
Oils, sassafras, hydrogenated

61790-28-1
Unknown
Unknown
Tallow nitrile
Nitriles, tallow

61790-37-2
Unknown
Unknown
Tallow acid
Fatty acids, tallow
Tallow fatty acid

61790-41-8
Unknown
Unknown
Soytrimonium Chloride
Quaternary ammonium compounds, trimethylsoya alkyl, chlorides
Quaternium-9
(Soya alkyl) trimethyl ammonium chloride
N-(Soya alkyl)-N,N,N-trimethyl ammonium chloride
Soya trimethyl ammonium chloride

61790-53-2
Unknown
Unknown
Kieselguhr
Amorphous silica
Diatomaceous earth
Diatomaceous earth, Natural
Diatomaceous silica, Calcined
Diatomite
Silica, amorphous, diatomaceous earth (OSHA)
Silica, amorphous-diatomaceous earth (ACGIH)
Silica, amorphous-diatomaceous earth uncalcined; Containing < 1% quartz

61790-81-6
Unknown
Unknown
Ethoxylated lanolin
Caswell No. 427A
EPA Pesticide Chemical Code 031607
Ethylene oxide, lanolin adduct
Ivarlan 3406
Ivarlan 3407
Lanolin, ethoxylated
PEG-5 Lanolin
PEG-20 Lanolin
PEG-24 Lanolin
PEG-30 Lanolin
PEG-50 Lanolin
PEG-60 Lanolin
PEG-85 Lanolin
PEG-100 Lanolin
Polyethylene glycol (5) lanolin
Polyethylene glycol (24) lanolin
Polyethylene glycol (30) lanolin
Polyethylene glycol (50) lanolin
Polyethylene glycol (60) lanolin
Polyethylene glycol (85) lanolin
Polyethylene glycol (100) lanolin
Polyethylene glycol 1000 lanolin
Polyethylene glycol-27 lanolin
Polyethylene glycol-40 lanolin
Polyethylene glycol-75 lanolin
Polyoxyethylene (5) lanolin
Polyoxyethylene (20) lanolin
Polyoxyethylene (24) lanolin

Polyoxyethylene (30) lanolin
Polyoxyethylene (50) lanolin
Polyoxyethylene (60) lanolin
Polyoxyethylene (85) lanolin
Polyoxyethylene (100) lanolin

61791-00-2
Unknown
Unknown
Polyethylene glycol monoester of tall oil
Chemax E-400-MT
Chemax TO-10
Ethoxylated tall oil fatty acid
Ethoxylated tall oil fatty acids
Fatty acids, tall-oil, ethoxylated
Fatty acids, tall oil, monoesters with polyethylene glycol
PEG-4 Tallate
PEG-8 Tallate
PEG-10 Tallate
PEG-12 Tallate
PEG-16 Tallate
PEG-20 Tallate
Polyethylene glycol, monoester with tall oil acids
Polyethylene glycol monotallate
Polyethylene glycol (16) monotallate
Polyethylene glycol 200 monotallate
Polyethylene glycol 400 monotallate
Polyethylene glycol 500 monotallate
Polyethylene glycol 600 monotallate
Polyethylene glycol 1000 monotallate
Polyethylene glycol, tall oil ester
Polyethylene glycol, tall oil fatty acid polymer
Polyethylene glycol tallate
Polyoxyethylene (4) monotallate
Polyoxyethylene (8) monotallate
Polyoxyethylene (10) monotallate
Polyoxyethylene (12) monotallate
Polyoxyethylene (16) monotallate
Polyoxyethylene (20) monotallate
Tall oil acids, ethoxylated
Tall oil fatty acid, esters, ethoxylated
Tall oil fatty acid ethoxylate
Tall oil fatty acid, ethoxylated
Tall oil fatty acid monoester of polyethylene glycol
Tall oil fatty acid, polyethylene glycol ester
Tall oil fatty acids, polyethylene glycol ester
Tall oil monoester of polyethylene glycol
Unipeg-400 MOT

61791-14-8
Unknown
Unknown
Amines, coco alkyl, ethoxylated
Chemeen C-2
Chemeen C-5
Chemeen C-10
Chemeen C-15
Chemeen C 12G
Cocoamine, ethoxylated
(Coconut oil alkyl)amine, ethoxylated
Ethomeen C
Genamin C
2-Hydroxyethyl coco amine, ethoxylated
Kostat P 650/5
Nissan Nymeen F 215
Noramox C
Optamine PC 5
PEG-2 Cocamine
PEG-3 Cocamine
PEG-5 Cocamine
PEG-10 Cocamine
PEG-15 Cocamine
Polyethylene glycol (3) coconut amine
Polyethylene glycol (5) coconut amine
Polyethylene glycol (15) coconut amine
Polyethylene glycol 100 coconut amine
Polyethylene glycol 500 coconut amine
Polyoxyethylene (2) coconut amine
Polyoxyethylene (3) coconut amine
Polyoxyethylene (5) coconut amine
Polyoxyethylene (10) coconut amine
Polyoxyethylene (15) coconut amine
Primary coco amine ethylene oxide adduct
Unizeen C-2
Unizeen C-5
Unizeen C-10
Varonic K 205LC
Varonic K 215

61791-24-0
Unknown
Unknown
Ethomeen S/12
Ethomeen S/15

61791-34-2
Unknown
Unknown
Soyaethyl Morpholinium Ethosulfate
Morpholinium compounds, N-ethyl-N-soya alkyl, ethyl sulfates
Onium compounds, morpholinium, 4-ethyl-4-soya alkyl, Et sulfates
Quaternium-2
N-(Soya alkyl)-N-ethyl morpholinium ethyl sulfate
N-Soya-N-ethyl morpholinium ethosulfate

61791-36-4
Unknown
Unknown
1H-Imidazole, 4,5-dihydro-, 2-nortall-oil alkyl derivs.
Tall oil imidazoline
Tall oil, imidazoline deriv.

61791-39-7
Unknown
Unknown
Tall oil hydroxyethyl imidazoline
4,5-Dihydro-7-nortall oil-1H-imidazole-1-ethanol
1-(2-Hydroxyethyl)-2-(tall oil alkyl)-2-imidazoline
1H-Imidazole-1-ethanol, 4,5-dihydro-, 2-nortall oil
1H-Imidazole-1-ethanol, 4,5-dihydro-, 2-nortall-oil alkyl derivs.
2-Imidazoline, 1-(2-hydroxyethyl)-2-(tall oil alkyl)-
Miramine TOC
2-Nortall oil-1H-imidazole-1-ethanol, 4,5-dihydro-
2-(Tall oil alkyl)-1-(2-hydroxyethyl)-2-imidazoline

61791-44-4
Unknown
Unknown
Ethanol, 2,2'-iminobis-, N-tallow alkyl derivs.

61791-52-4
Unknown
Unknown
2-(Cocoalkyl)-1-benzyl-1-(2-hydroxyethyl)-2-imidazolinium chloride
Imidazolium compounds, 1-benzyl-4,5-dihydro-1-(hydroxyethyl)-2-norcoco alkyl, chlorides

61791-63-7
Unknown
Unknown
1-((Coco alkyl)amino)-3-aminopropane
Amines, N-coco alkyltrimethylenedi-
Cocopropylenediamine
N-Coco-1,3-propylenediamine

61791-64-8

Unknown
Unknown
1-((Coco alkyl)amino)-3-aminopropane acetates
Amines, N-coco alkyltrimethylenedi-, acetates
N-Coco-1,3-diaminopropane acetate

61791-67-1
Unknown
Unknown
1-((Soya alkyl)amino)-3-aminopropane
Amines, N-soya alkyltrimethylenedi-

61792-07-2
$C_{12}H_{13}Cl_3O_3$
312.80
Acetic acid, (2,4,5-trichlorophenoxy)-, 1-methylpropyl ester (9CI)

61814-27-5
Unknown
Unknown
C.I. Leuco Sulphur Black 2 (9CI)

61814-69-5
Unknown
Unknown
C.I. Direct Black 8 (9CI)

61840-22-0
Unknown
Unknown
Antiblaze 19 (9CI)

61840-33-3
Unknown
Unknown
Gly-Cel C 200 (9CI)

61847-60-7
Unknown
Unknown
C.I. Acid Red 348 (9CI)

61878-55-5
$C_6H_3I_3$
455.80
Benzene, triiodo- (9CI)

61878-56-6
$C_6H_4BrNO_2$
202.01
Benzene, bromonitro- (9CI)
Bromonitrobenzene

61878-58-8
C_7H_7I
218.04
Benzene, methyl-, monoiodo deriv. (9CI)

61898-58-6
$C_7H_{12}O_4$
160.17
Hexanedioic acid, 3-methyl-, (S)- (9CI)

61898-95-1
$C_9H_{12}Cl_2O_2$
223.10
O=C(OC)C(C1(C)C)C1C=C(Cl)Cl
Cyclopropanecarboxylic acid, 3-(2,2-dichloroethenyl)-2,2-dimethyl-, methyl ester (9CI)
Methyl 3-(2,2-dichlorovinyl)-2,2-dimethylcyclopropanecarboxylate

61901-21-1
Unknown
Unknown
C.I. Acid Brown 159 (9CI)

61902-16-7
Unknown
Unknown
C.I. Disperse Yellow 50 (9CI)
Foron Yellow SE-2GL

61931-22-4
Unknown
Unknown
C.I. Acid Red 361 (9CI)
Tectilon Red 2B

61931-82-6
$C_{18}H_{24}N_2$
268.39
N(c(ccc(N)c1)c1)(c(cccc2)c2)C(CC(C)C)C
1,4-Benzenediamine, N-(1,3-dimethylbutyl)-N-phenyl- (9CI)

61931-87-1
$C_{10}H_8O_7S_2.Na$
326.28
2,7-Naphthalenedisulfonic acid, 4-hydroxy-, monosodium salt (9CI)

61949-26-6
$C_7H_{16}O$
116.20
1-Hexanol, methyl- (9CI)

61949-76-6
$C_{21}H_{20}Cl_2O_3$
391.31
Cyclopropanecarboxylic acid, 3-(2,2-dichloroethenyl)-2,2-dimethyl-, (3-phenoxyphenyl) methyl ester, cis
cis-Permethrin

61949-77-7
$C_{21}H_{20}Cl_2O_3$
391.29
Cyclopropanecarboxylic acid, 3-(2,2-dichloroethenyl)-2,2-dimethyl-, (3-phenoxyphenyl)methyl ester, trans- (9CI)

61966-36-7
$C_7H_5Cl_3O_2$
227.48
Trichloroguaiacol
Trichloro-2-methoxyphenol

61967-93-9
Unknown
Unknown
C.I. Acid Blue 277 (9CI)
Erio Blue BGL

61968-26-1
Unknown
Unknown
C.I. Direct Yellow 132 (9CI)

61996-25-6
$C_5H_{10}O_3$
118.13
2-Butanone, 3-hydroxy-3-methoxy- (9CI)

62005-54-3
$C_9H_{20}O_4$
192.26
Ethane, 1,1-diethoxy-2-(2-methoxyethoxy)- (9CI)

62014-96-4
$C_{28}H_{48}O$
400.69
Cholest-8(14)-en-3-ol, 4-methyl-, (3β,4α,5α)- (9CI)

62015-39-8
Unknown
Unknown
Misoprostol

62016-34-6
$C_{11}H_{24}$
156.31
Octane, 2,3,7-trimethyl- (9CI)

62016-75-5
$C_{17}H_{35}Cl$
274.92
ClCCCCCCCCCCCCCCCCC
Heptadecane, 1-chloro- (9CI)
1-Chloroheptadecane

62046-37-1
$C_9H_9Cl_2NO_2$
234.09
Carbamic acid, N-methyl-, 3,4-dichlorobenzyl ester, Mixed with carbamic acid, N-methyl-, 2,3-dichlorobenzyl ester (4:1)
Benzenemethanol, 2,3(or 3,4)-dichloro-, methyl carbamate (9CI)
Chlorxylam
3,4-Dichlorobenzyl methylcarbamate mixed with 2,3-dichlorobenzyl methylcarbamate (80%:20%)
ENT 25,736
Rowmate
Sirmate
U-17004
UC 22,463

62064-85-1
$C_{15}H_{14}O_2$
226.27
Methanone, (4-hydroxy-3-methylphenyl)(3-methylphenyl)- (9CI)
Benzophenone, 4-hydroxy-3,3'-dimethyl-

62108-21-8
$C_{13}H_{28}$
184.37
Decane, 6-ethyl-2-methyl- (9CI)

62108-25-2
$C_{13}H_{28}$
184.37
Decane, 2,6,7-trimethyl- (9CI)

62108-31-0
$C_{13}H_{28}$
184.37
Heptane, 4-ethyl-2,2,6,6-tetramethyl- (9CI)

62108-37-6
$C_7H_{14}Si$
126.28
Silane, 1-butynyltrimethyl- (9CI)
1-Butynyltrimethylsilane
1-(Trimethylsilyl)-1-butyne

62127-42-8
$C_9H_{18}F_2$
164.24
Nonane, 1,1-difluoro- (9CI)

62168-26-7
$C_8H_{16}Br_2$
272.03
Octane, 1,1-dibromo- (9CI)
1,1-Dibromooctane

62168-27-8
$C_9H_{18}Br_2$
286.06
Nonane, 1,1-dibromo- (9CI)
1,1-Dibromononane

62180-90-9
$C_6Cl_6.C_6Cl_5NO_2$
580.11
Benzene, hexachloro-, Mixt. with pentachloronitrobenzene (9CI)

62185-54-0
$C_{13}H_{28}$
184.37
Nonane, 5-(1-methylpropyl)- (9CI)

62199-50-2
$C_{12}H_{24}$
168.32
Cyclopentane, 1-butyl-2-propyl- (9CI)

62207-76-5
$C_{16}H_{12}CoF_2N_2O_2$
361.20
Fluomine
Cobalt, bis(3-fluorosalicylaldehyde)ethylenediimine-
Cobalt, bis(3-fluorosalicylaldehyde)-ethylenediimine-
Cobalt, ((2,2'-(1,2-ethanediylbis(nitrilomethylidyne))bis-(6-fluorophenolato))(2-)-N,N',O,O')-
Cobalt, ((2,2'-(1,2-ethanediylbis(nitrilomethylidyne))bis-(6-fluorophenolato))(2-)-N,N',O,O')-, (SP-4-2)- (9CI)
Cobalt(II), N,N'-ethylenebis-(3-fluorosalicylideneiminato)-
N,N'-Ethylenebis(3-fluorosalicylideneiminato)cobalt (II)
Fluomine Dust

62237-99-4
$C_{13}H_{28}$
184.37
Decane, 2,2,7-trimethyl- (9CI)

62238-02-2
$C_9H_{20}O$
144.26
Pentane, 2-butoxy- (9CI)

62243-57-6
$C_7H_{12}O_2$
128.17
3-Hexenoic acid, 2-methyl-, (E)- (9CI)

62245-47-0
$C_{14}H_{10}N_2O_2$
238.23
2-Phenanthrenamine, 7-nitro- (9CI)

62251-96-1
$C_{18}H_{25}NO_4$
319.40
Coronatine
Cyclopropanecarboxylic acid, 2-ethyl-1-(((6-ethyl-2,3,3a,6,7,7a-hexahydro-1-oxo-1H-inden-4-yl)carbonyl)amino)-, (3aS-(3aα,4(1R*,2R*),6β,7aα))- (9CI)

62308-10-5
$C_{13}H_{12}N_4O_2$
256.27
Benzenamine, 3-methyl-4-((4-nitrophenyl)azo)- (9CI)

62338-00-5
C_8H_{12}
108.18
Cyclopentene, 3-ethylidene-1-methyl- (9CI)

62338-16-3
$C_{13}H_{28}$
184.37
Decane, 3,3,8-trimethyl- (9CI)

62338-24-3
$C_{12}H_{20}O$
180.29
Ethanone, 1-(4,5-diethyl-2-methyl-1-cyclopenten-1-yl)- (9CI)

62450-06-0
$C_{13}H_{13}N_3$
211.29
Cc2nc(N)c(C)c3c1ccccc1[nH]c23
5H-Pyrido(4,3-b)indole, 3-amino-1,4-dimethyl
3-Amino-1,4-dimethyl-γ-carboline

3-Amino-1,4-dimethyl-5H-pyrido(4,3-b)indole
5H-Pyrido(4,3-b)indol-3-amine, 1,4-dimethyl-
Trp-P-1
Trp-P-1
Tryptophan P1

62450-07-1
$C_{12}H_{11}N_3$
197.26
CC1=NC(=CC2C1Nc3ccccc23)N
5H-Pyrido(4,3-b)indole, 3-amino-1-methyl
3-Amino-1-methyl-γ-carboline
3-Amino-1-methyl-5H-pyrido-(4,3-b)indole
1-Methyl-3-amino-5H-pyrido-(4,3-b)indole
5H-Pyrido(4,3-b)indol-3-amine, 1-methyl-
TRP-P-2
TRP-P-2
Tryptophan P2

62470-53-5
$C_{12}H_4Cl_4O_2$
321.98
Clc3cc(Cl)c2Oc1c(Cl)cc(Cl)cc1Oc2c3
1,3,7,9-Tetrachlorodibenzo-p-dioxin

62470-54-6
$C_{12}H_4Cl_4O_2$
321.97
Dibenzo(b,e)(1,4)dioxin, 1,2,8,9-tetrachloro- (9CI)
1,2,8,9-Tetrachlorodibenzo-(b,e)(1,4)dioxin
1,2,8,9-Tetrachlorodibenzo-p-dioxin

62476-59-9
$C_{14}H_6ClF_3NO_5.Na$
383.65
[Na+].[O-]C(=O)c2cc(Oc1ccc(cc1Cl)C(F)(F)F)ccc2N(=O)=O
Benzoic acid, 5-(2-chloro-4-(trifluoromethyl)phenoxy)-2-nitro-, sodium salt
Acifluorfen
Acifluorfen sodium
Blazer
Blazer 2S
5-(2-Chloro-4-(trifluoromethyl)-phenoxy)-2-nitrobenzoic acid sodium salt
LS 80.1213
MC 10978
RH 6201
Sodium 5-(2-chloro-4-(trifluoro-methyl)phenoxy)-2-nitro-benzoate
Tackle
Tackle 2AS
Tackle 2S
Scifluorfen
Sodium salt of acifluorfen

62488-57-7
$C_8H_{14}N_4O_4$
230.26
s-Triazin-2(1H)-one, 5,6-dihydro-4-amino-1-β-D-ribo-furanosyl
4-Amino-5,6-dihydro-1-β-D-ribofuranosyl-s-triazin-2(1H)-one
5,6-Dihydro-4-amino-1-β-d-ribofuranosyl-s-triazin-2(1H)-one
5,6-Dihydro-5-azacytidine

62570-20-1
$C_{16}H_{17}N_5O_5$
359.31
Ethanol, 2-((4-((2,4-dinitro-phenyl)azo)phenyl)ethyl-amino)- (9CI)
C.I. Disperse Red 8
Ethanol, 2-(p-(2,4-dinitro-phenylazo)-N-ethylanilino)-

62571-86-2
$C_9H_{12}NO_3S$
214.28
CC(CS)C(=O)N1CCCC1C(O)=O
1-Pyrrolidinecarboxylic acid, 1-(D-3-mercapto-2-methyl-1-propionyl)-, l-(S,S)
Capoten
Captopril
Captopryl
Lopirin
1-(3-Mercapto-2-methyl-1-oxo-propyl)-l-proline
1-(D-3-Mercapto-2-methyl-1-oxopropyl)-l-proline (S,S)
D-3-Mercapto-2-methylpro-panoyl-l-proline
1-((2S)-3-Mercapto-2-methyl-propionyl)-l-proline
(2S)-1-(3-Mercapto-2-methyl-propionyl)-l-proline
D-2-Methyl-3-mercaptopro-panoyl-l-proline
l-Proline, 1-(3-mercapto-2-methyl-1-oxopropyl)-, (S)-
SQ 14,225

62587-63-7
$C_{18}H_{14}O.C_{12}H_{10}O$
416.52
1,1'-Biphenyl, phenoxy-, Mixt. with 1,1'-oxybis(benzene) (9CI)
Dowtherm G

62601-62-1
Unknown
Unknown 379.71
PCS (9CI)

62607-69-6
$C_{16}H_{26}O$
234.38
Benzenedecanol (9CI)

62610-39-3
Unknown
Unknown
Banvel (9CI)

62613-15-4
$C_{12}H_{10}I.AsF_6$
470.03
Iodonium, diphenyl-, hexa-fluoroarsenate(1-) (9CI)
Diphenyliodonium hexafluoro-arsenate
Diphenyliodonium hexafluoro-arsenate(1-)

62615-08-1
$C_{12}H_4Cl_4O$
305.97
Dibenzofuran, 1,2,3,8-tetra-chloro- (9CI)
1,2,3,8-Tetrachlorodibenzo-furan

62625-14-3
$C_6H_5ClN_2O_3.ClH$
225.04
Cl.Nc1cc(cc(Cl)c1O)N(=O)=O
Phenol, 2-amino-6-chloro-4-nitro-, hydrochloride
2-Amino-6-chloro-4-nitrophenol hydrochloride

62654-17-5
$C_6H_8N_2.C_2H_2O_4$
198.17
1,4-Benzenediamine ethane-dioate
1,4-Benzenediamine, ethane-dioate (1:1) (9CI)

62695-55-0
$C_{12}H_{24}O_4$
232.32
tert-Butyl peroxy-2-ethyl-hexanoate
tert-Butyl peroxyethylhexanoate
tert-Butyl peroxy-2-ethyl hex-anoate, Technically pure
tert-Butyl peroxy-2-ethylhexan-oate (Not more than 12% with 2,2-di-(tert-butylperoxy)-butane, not more than 14% with not less than 14% phlegmatizer and 60% inert inorganic solid)
tert-Butyl peroxy-2-ethylhexan-oate (Not more than 30% with 2,2-di-(tert-butylperoxy)-butane, not more than 35%, with not less than 35%

phlegmatizer)
tert-Butyl peroxy-2-ethylhexanoate (Not more than 50% with phlegmatizer)
Ethyl-2 peroxyhexanoate de tert-butyle (French)
Hexanoic acid, 2-((1,1-dimethylethyl)dioxy)ethyl ester
Peroxi-2-etilhexanoato de terc-butilo (Spanish)
UN 2143
UN 2886
UN 2887
UN 2888

62712-23-6
Unknown
Unknown
Zepel (9CI)

62716-20-5
$C_{21}H_{12}O$
280.33
5H-Naphth(3,2,1-de)anthracen-5-one (9CI)

62732-91-6
Unknown
Unknown
2-(2-Ethoxyethoxy)ethyl 2-benzimidazole carbamate

62763-89-7
$C_{10}H_9N.ClH$
179.64
Quinoline, 2-methyl-, hydrochloride (9CI)
2-Methylquinoline hydrochloride

62765-93-9
$C_8H_20N_2O.C_5H_10N_2$
258.38
NIAX catalyst ESN
NIAX ES-N
Niax ESN
Propanenitrile, 3-(dimethylamino)-, Mixt. with 2,2'-oxybis(N,N-dimethylethanamine) (9CI)

62796-65-0
$C_{12}H_6Cl_4$
291.99
1,1'-Biphenyl, 2,2',4,6-tetrachloro- (9CI)
2,2',4,6-PCB
2,2',4,6-Tetrachlorobiphenyl

62850-32-2
$C_{13}H_{19}NO_2S$
253.39
Carbamothioic acid, dimethyl-, S-(4-phenoxybutyl)-ester
BI-5452
Dimethylcarbamothioic acid S-(4-phenoxybutyl)ester
Fenothiocarb
KCO-3001
Panocon
Phenothiocarb
S-4-Phenoxybutyl dimethylthiocarbamate

62922-39-8
$C_{15}H_{19}Cl_3O_4$
369.68
2,4,5-T propylene glycol butyl ester
PGBE 2,4,5-T
2,4,5-T PGBE ester
2,4,5-Trichlorophenoxyacetic acid propylene glycol butyl ester

62936-23-6
$C_8H_7ClO_4$
202.59
Benzoic acid, 3-chloro-4-hydroxy-5-methoxy- (9CI)
NSC-45930
Vanillic acid, 5-chloro-

62936-24-7
$C_8H_7ClO_4$
202.59
Benzoic acid, 2-chloro-4-hydroxy-5-methoxy- (9CI)

62959-39-1
$C_{13}H_{12}N_2O_3S.Na$
299.29
Benzenesulfonic acid, 4-((4-methylphenyl)azo)-, sodium salt (9CI)

62959-40-4
$C_{14}H_{14}N_2O_3S.Na$
313.32
Benzenesulfonic acid, 4-((2,4-dimethylphenyl)azo)-, sodium salt (9CI)

62959-41-5
$C_{12}H_8Cl_2N_2O_3S.Na$
354.16
Benzenesulfonic acid, 4-((2,4-dichlorophenyl)azo)-, sodium salt (9CI)

62979-89-6
Unknown
Unknown
Nitrilotriacetic acid, calcium salt (2:3)
NTA, calcium salt (2:3)

62981-74-2
C_4H_5Cl
88.54
1-Butyne, 1-chloro- (7CI, 9CI)

62994-32-5
$C_{12}HBr_7O$
720.47
Dibenzofuran, heptabromo- (9CI)
Heptabromodibenzofuran

63017-87-8
$C_{11}H_9NO_2$
187.21
Naphthalene, 1-methyl-2-nitro
1-Methyl-2-nitronaphthalene

63018-79-1
$C_{20}H_{16}$
256.35
Benz(a)anthracene, 2,10-dimethyl- (9CI)

63018-94-0
$C_{17}H_{16}$
220.33
Anthracene, 2,9,10-trimethyl
2,9,10-Trimethylanthracene

63021-33-0
$C_{23}H_{17}N$
307.41
Dibenz(a,h)acridine, 1-ethyl
1'-Ethyl-1,2,5,6-dibenzacridine (French)

63021-35-2
$C_{23}H_{17}N$
307.41
Dibenz(a,j)acridine, 1-ethyl
1'-Ethyl-3,4,5,6-dibenzacridine (French)
1-Ethyl-dibenz(a,j)acridine

63021-85-2
$C_{18}H_{11}NO_2$
273.28
Chrysene, nitro- (9CI)

63021-86-3
$C_{16}H_9NO_2$
247.26
Pyrene, nitro
Nitropyrene

63028-27-3
$C_{15}H_{14}Cl_2$
265.18
1,1'-Biphenyl, 2,5-dichloro-4'-(1-methylethyl)- (9CI)

63035-28-9
$C_9H_8O_6$
212.16
1,2-Benzenedicarboxylic acid, 4-hydroxy-5-methoxy- (9CI)

63037-96-7
$C_{12}H_{10}O_4$
218.21
Acetic acid, ((7-hydroxy-1-naphthalenyl)oxy)- (9CI)

63041-47-4
$C_{21}H_{12}O$
280.33
13H-Dibenzo(a,g)fluoren-13-one (9CI)

63041-90-7
$C_{20}H_{11}NO_2$
297.32
O=N(=O)c2c1ccccc1c3ccc4cccc5ccc2c3c45
Benzo(a)pyrene, 6-nitro
6-Nitrobenz(a)pyrene
6-Nitrobenzo(a)pyrene

63075-06-9
$C_{12}H_{17}NO_2$
207.27
Carbamic acid, phenyl-, pentyl ester (9CI)
Carbanilic acid, pentyl ester (6CI,7CI)

63084-98-0
$C_6H_7NO.1/2H_2O_4S$
158.17
Phenol, 4-amino-, sulfate (2:1) (Salt) (9CI)
p-Aminophenol, hemisulfate
4-Aminophenol sulfate (2:1) (Salt)

63148-52-7
Unknown
Unknown
Siloxanes and silicones, di-Me, Me Ph

63148-53-8
Unknown
Unknown
Mentor 28

63148-56-1
Unknown
Unknown
Siloxanes and silicones, Me 3,3,3-trifluoropropyl

63148-57-2
Unknown
Unknown
Siloxanes and silicones, Me hydrogen
Polymethylhydrogensiloxane, trimethylsiloxy end blocked
Poly(methylhydrosiloxane)
Siloxanes and silicones, methyl hydrogen

63148-62-9
Unknown
Unknown
Siloxanes and silicones, di-Me
AF 75
AK 100 (siloxane)
Antaphron NM 42
Antifoam FD 62
Aquasil E
AV 1000
Baysilon
Baysilon M 100
Baysilon M 500
DC 360
Dimethylpolysiloxane hydrolyzate
Dow Corning 93-120
Dow Corning 200
Dow Corning 561
F 1-3563
F 111/5000
F 114 (silicone)
FG 10
Foamkill 8D
FZ 132
Gascon
HV 490
K 21 (silicone)
K 331
KE 77
KF 96
KF 96-100
KF 96-500
KM 9k
KO 08
KO 811
KR 220
KS 607A
KS 700
KS 705F
KS 770
KS 773
KS 774
L 43
L 45 (silicone)
L 546
Lukooil M 100
Lukooil M 200
Lukosan A 311
Lukosan M 02
Lukosan M 07
α-Methyl-ω-methoxypolydimethylsiloxane
NAE
ND 8 (silicone)
NMS 03
PMF 600
PMS (siloxane)
PMS 1.5
PMS 1000A
PMS 154A
PMS 200A
PNS 25
Polsil OM 1000
Polsil 350
Poly(dimethylsiloxane)
Poly(oxy(dimethylsilylene))
Polydimethyl silicone oil
Polydimethylsiloxane, methyl end-blocked
Polyoxy(dimethylsilylene), α-(trimethylsilyl)-ω-hydroxy
PS 197
Releasil 8
Rhodorsil CAF 3B
S DC 200
SAG 100
SAH 283
SAH 288
SF 96
SF 96-100
SF 97(50)
SGM 36
SH 200
SH 6188
SH 8708
Silak M 10
Silar 10:100
Silicone DC 200
Silicone DC 360
Silicone DC 360 Fluid
Silicone Oils
Silicone Release L 45
Silicone 360
Siligaz
Silikon Antifoam FD 62
Silol 350
Silol 5
Siloprene C 1
Siloprene C 18
Siloprene C 2
Siloxane and silicones, dimethyl
Siloxanes and silicones, dimethyl
Silpian E 2
Silpian 2
SM 2013
SM 2061
SM 2138
SM 5512
SM 8708
SP 2100
SWS 03314
Sylgard 1107-69
Sylgard 182-62B
Sylgard 184-31B
Sylgard 184-36B
Syltherm 800
Tegiloxan 50
Teginex FP 90
α-(Trimethylsilyl)poly(oxy(dimethylsilylene))-ω-methyl
α-(Trimethylsilyl)-ω-((trimethylsilyl)oxy)
TSF 431
TSF 451
TSF 451-1000
UC Liquid G
Ucarsil DJ
Union Carbide Liquid G
Viscasil
Viscasil 5000
Viscasil 10000
X 2-1163

X 20-201
XF 13-563
Y 9208
YF 3842

63148-65-2
$H_2.(C_8H_{14}O_2)x$
Polymer
Poly(2-propyl-m-dioxane-4,6-diylene)
Butvar
Polyvinylbutyral (Czech)
Polyvinyl Butyral Resins

63151-11-1
$C_3H_6Cl_2O$
128.99
1-Propanol, 2,2-dichloro- (7CI,9CI)

63231-51-6
Unknown
Unknown
Aerotex 3470

63231-67-4
Unknown
Unknown
Silochrome
Silica gel
Spherosil
Syloid
Syloid 63
Syloid 266

63283-80-7
$C_6H_{10}Cl_4O$
239.96
Dichloroisopropyl ether
Eter dicloroisopropilico (Spanish)
Ether dichloroisopropylique (French)

63302-49-8
$C_{30}H_{46}ClO_2P$
505.12
O(c(ccc(c1)CCCCCCCCC)c1)P(Oc(ccc(c2)CCCCCCCCC)c2)Cl
Phosphorochloridous acid, bis(4-nonylphenyl) ester (9CI)

63302-50-1
$C_7H_5NS_2.C_4H_{11}N$
240.38
2(3H)-Benzothiazolethione, Compd. with 2-methyl-2-propanamine (1:1) (9CI)

63302-94-3
$C_{36}H_{51}O_4P$
578.77
Phosphoric acid, bis(nonylphenyl) phenyl ester (9CI)
Bis(nonylphenyl) phenyl phosphate

63302-95-4
$C_{36}H_{35}O_4P$
562.65
Phosphoric acid, bis((1-methyl-1-phenylethyl)phenyl) phenyl ester (9CI)

63302-98-7
$C_{45}H_{45}O_4P$
680.82
Phenol, (1-methyl-1-phenylethyl)-, phosphate (9CI)
(1-Methyl-1-phenylethyl)phenol phosphate

63325-06-4
$C_{14}H_{13}F_{12}O_4P$
504.21
Phosphorane, phenyltetrakis(2,2,2-trifluoroethoxy)- (9CI)

63333-35-7
$C_{14}H_7Br_3F_3N_3O_4$
471.92
Bromethalin
AI3-29577-X
Benzenamine, 2,4-dinitro-N-methyl-n-(2,4,6-tribromophenyl)-6-(trifluoromethyl)-
Benzenamine, N-methyl-2,4-dinitro-N-(2,4,6-tribromophenyl)-6-(trifluoromethyl)-(9CI)
Bromethaline (French)
Caswell No. 561BB
4,6-Dinitro-N-methyl-N-(2,4,6-tribromophenyl)-α,α,α-trifluoro-o-toluidine
EL 614
EPA Pesticide Chemical Code 112802
N-Methyl-2,4-dinitro-N-(2,4,6-tribromophenyl)-6-(trifluoromethyl)benzenamine
N-Methyl-2,4-dinitro-N-2,4,6-tribromophenyl(-6)trifluoromethylbenzeneamin-E
o-Toluidine, 4,6-dinitro-N-methyl-N-(2,4,6-tribromophenyl)-α,α,α-trifluoro-
α,α,α-Trifluoro-N-methyl-4,6-dinitro-N-(2,4,6-tribromophenyl)-o-toluidine

63340-28-3
$C_{36}H_{43}O_4P$
570.71
Phosphoric acid, (1-methyl-1-phenylethyl)phenyl nonylphenyl phenyl ester (9CI)

63393-82-8
Unknown
Unknown
Alcohols, C12-15
C12-15 alcohol

63393-93-1
Unknown
Unknown
Isopropyl lanolate
Fatty acids, lanolin, iso-Pr esters
Fatty acids, lanolin, isopropyl esters
Lanolin fatty acids, isopropyl esters
Isopropyl lanolate
Isopropyl lanolin

63397-60-4
$C_{34}H_{64}O_8S_2Sn$
783.72
Propanoic acid, 3,3'-(bis((2-(isooctyloxy)-2-oxoethyl)-thio)stannylene)bis-, dibutyl ester (9CI)
Bis-β-carbobutoxyethyltin bis-isooctylthioglycolate

63428-83-1
$(C_6H_{11}NO)x$
Polymer
Nylon
Amilan
Ashlene
Caprolon
Enkalon
Grilon
Kapron
Mirlon
Perlon
Phrilon
Polyamid (German)
Silon
Trogamid T
Vydyne

63428-84-2
Unknown
Polymer NA
Polyamide fibers
Acids, polyamides (VAN)
Aeron (Fiber)
Agilon
Agilon D
Amino acids, polyamides (VAN)
Antron
Arenka 900
Arenka 930
Artslon
Banlon
Belima
Bri-Nylon
Cadon (Fiber)
Cantrece
Caprolan (polyamide)
Carboxylic acids, polyamides

Celon
Cerex
Chemlon VPK
Chemlon 67/16
Condensation products, polyamides (VAN)
Dederon
Dedotex
Durette
Elastik (Fiber)
Elastil
Elder (Fiber)
Ethilon
Flawil
Fluflon
Forlien
Fypro
Gofron
Helanca
HT 4 (VAN)
Kaprilon
Lactams, polyamides (VAN)
Lanastil
Lilion
M 3P
MED-T
MED 6
Meron (VAN)
N 40 (VAN)
N 20 (Polyamide)
Nevaflor
Nylon (VAN)
Nylon BCF 800
Nylsuisse
Ozhilon
PABH-T
Parel (Fiber)
Pelargon (VAN)
Pellon P 6
Polana
PRD 49
PRD 49-1
PRD 49III
Promilan
Pulon
Qiana
Radiatox
Rayona
Relon
Rilon
Shelon
Silon (Polyamide)
Speckelon
Stilofil
Stilon Standard A
Stilon Standard C
Sualen
Suturamid
Synthetic fibers, Polymeric, Polyamides
Terlon (Fiber)
Tine
Ultralon PA 6
Undekan (Fiber)
Vniivlon M
Vylor
Xylon
Yulon
23K (Fiber)

63439-57-6
Unknown
Unknown
Decapol A 33 (9CI)

63449-39-8
Unknown
Unknown
Paraffin waxes and hydrocarbon waxes, Chlorinated (C12, 60% chlorine)
Cerechlor 54
Cereclor
Cereclor 30
Cereclor 42
Cereclor 48
Cereclor 52
Cereclor 54
Cereclor 70
Cereclor 511
Cereclor 631
Cereclor 651
Cereclor 701
Cereclor 50LV
Cereclor S 42
Cereclor S52
Cereclor S70
Chlorcosane
Chlorez 700
Chlorez 700HMP
Chlorinated paraffins
Chlorinated paraffins (C12, 60% chlorine)
Chlorinated paraffins (C23, 43% chlorine)
Chlorowax
Chlorowax 40
Chlorowax 50
Chlorowax 70
Chlorowax 500C
Chlorowax 70S
Chlorowax S 70
Clorafin
Crechlor S 45
Flexchlor
NCI-C53587
Paraffin waxes and hydrocarbon waxes, Chlorinated (C23, 43% chlorine)
Paroil Chlorez
Unichlor
Unichlor 50

63449-41-2
Unknown
Unknown
Ammonium, alkyl(C14-16)dimethylbenzyl-, chlorides
Roccal
Tret-O-Lite XC 511

63450-73-7
$CH_5NO_7S_2.2Na$
253.16
Imidodisulfuric acid, methoxy-, disodium salt (9CI)
Disodium methoxyimidodisulfurate
Disodium O-methylhydroxylamine-N,N-disulfonate

63451-40-1
$C_{12}H_{10}N_2O.Na$
221.20
2,5-Cyclohexadien-1-one, 4-(phenylimino)-, oxime, sodium salt (9CI)

63452-61-9
$C_{12}H_{18}O$
178.27
Phenol, 4-(1,1-dimethylethyl)-2-ethyl- (9CI)

63455-63-0
$C_{12}H_{18}BrNO$
272.19
Ethanol, 2-((4-bromophenyl)butylamino)- (9CI)

63455-64-1
$C_{14}H_{23}NO_2$
237.34
Ethanol, 2-(2-(butylphenylamino)ethoxy)- (9CI)

63455-65-2
$C_{14}H_{15}NO_2$
229.28
1H-Indole-1-acetaldehyde, 2,3-dihydro-3,3-dimethyl-2-(2-oxoethylidene)- (9CI)

63460-04-8
$C_8H_{11}NO$
137.18
Benzenamine, methoxymethyl- (9CI)

63460-05-9
$C_{10}H_{14}BrN$
228.14
Benzenamine, bromo-N,N-diethyl- (9CI)

63460-06-0
$C_6H_3BrN_2O_4$
247.01
Benzene, bromodinitro- (9CI)

63460-07-1
$C_6H_5BrN_2O_2$
217.03
Benzenamine, ar-bromo-ar-nitro- (9CI)

63460-08-2
$C_{13}H_{12}N_2O$
212.25

Diazene, (methoxyphenyl)-phenyl- (9CI)

63460-09-3
$C_6H_4BrN_3O_4$
262.03
Benzenamine, ar-bromo-ar,ar-dinitro- (9CI)

63460-10-6
$C_{15}H_{21}N_3O_2$
275.35
Acetamide, N-(((2-cyano-ethyl)ethylamino)ethoxy-phenyl)- (9CI)

63460-11-7
$C_{14}H_8BrNO_3$
318.13
9,10-Anthracenedione, aminobromohydroxy- (9CI)

63493-28-7
$C_5H_{13}N$
87.16
NC(CCC)C
2-Pentanamine, (+-)- (9CI)
DL-2-Aminopentane
(+-)-2-Pentanamine

63494-59-7
$C_{12}H_{19}N_3O_3S$
285.35
O=S(=O)(NC)CCN(c(ccc(N=O)c1C)c1)CC
Ethanesulfonamide, 2-(ethyl-(3-methyl-4-nitrosophenyl)-amino)-N-methyl- (9CI)

63502-25-0
$(C_4H_{12}O_4P.CH_4N_2O.1/2O_4S)x$
Polymer
Phosphonium, tetrakis-(hydroxymethyl)-, sulfate (2:1) (Salt), Polymer with urea (9CI)
Tetrakis(hydroxymethyl)-phosphonium sulfate-urea condensation product
Tetrakis(hydroxymethyl)phos-phonium, sulfate (2:1), urea polymer

63505-64-6
$C_6H_5Br_2N$
250.93
Benzenamine, dibromo- (9CI)

63505-65-7
$C_{14}H_{19}N_3O_2$
261.33
Acetamide, N-(((2-cyano-ethyl)ethylamino)-methoxyphenyl)- (9CI)

63512-64-1
$C_{28}H_{44}$
380.66
Naphthalene, diisononyl- (9CI)
Diisononylnaphthalene

63610-08-2
$C_{18}H_{17}NO_3$
295.36
CCC(C(O)=O)c1ccc(cc1)N3Cc2ccccc2C3=O
Butyric acid, 2-(p-(1-oxo-2-isoindolinyl)phenyl)-, (+-)
4-(1,3-Dihydro-1-oxo-2H-iso-indol-2-yl)-α-ethyl-benzene-acetic acid
Indobufen
K-3920
1-Oxo-2-(p-((α-ethyl)carboxy-methyl)phenyl)isoindoline

63619-09-0
$C_4H_4Br_2N_2O$
255.88
Propanamide, dibromocyano- (9CI)

63634-21-9
$C_4H_4ClN_3O_2$
161.56
1H-Imidazole, 2-chloro-1-methyl-4-nitro
2-Chloro-1-methyl-4-nitro-1H-imidazole
1-Methyl-2-chloro-4-nitro-imidazole

63662-67-9
$C_4H_4ClN_3O_2$
161.53
1H-Imidazole, 4-chloro-2-methyl-5-nitro- (9CI)

63690-09-5
$C_{15}H_{16}N_4$
252.35
Benzenecarboximidamide, 4,4'-methylenebis
p,p'-Diamidinodiphenylmethane
Methylenediphenyl-4,4'-di-amidine

63709-64-8
$C_{13}H_7Cl_5O_2$
372.46
Benzene, 1,2,3-trichloro-5-(2,4-dichlorophenoxy)-4-meth-oxy- (9CI)

63721-05-1
$C_8H_{14}O_2$
142.20
O=C(OC)CC(C=C)(C)C
4-Pentenoic acid, 3,3-di-methyl-, methyl ester (9CI)
3,3-Dimethyl-4-pentenoic acid, methyl ester
Methyl 3,3-dimethyl-4-penteno-ate
Penten-4-oic acid, 3,3-dimethyl, methyl ester

63734-62-3
$C_{14}H_8ClF_3O_3$
316.66
O=C(O)c(cc(Oc(c(cc(c1)C(F)(F)F)Cl)c1)cc2)c2
Benzoic acid, 3-(2-chloro-4-(trifluoromethyl)phenoxy)- (9CI)
3-(2-Chloro-4-(trifluoromethyl)-phenoxy)benzoic acid

63843-89-0
$C_{42}H_{72}N_2O_5$
685.16
O=C(OC(CC(N(C1(C)C)C)(C)C)C1)C(C(=O)OC(CC(N(C2(C)C)C)(C)C)C2)(CCCC)Cc(cc(c(O)c3C(C)(C)C)C(C)(C)C)c3
Propanedioic acid, ((3,5-bis-(1,1-dimethylethyl)-4-hydroxyphenyl)methyl)-butyl-, bis(1,2,2,6,6- penta-methyl-4-piperidinyl) ester
Tinuvin 144

63848-94-2
Unknown
Unknown
Reofos 50 (9CI)

63868-82-6
$C_{24}H_{19}N_{12}O_{20}.Zr$
886.77
Picramic acid, zirconium salt (Wet)
Picramic acid, zirconium salt, Dry
UN 1517 [Zirconium picramate, Wetted with not less than 20 per cent water, by mass]
UN 0236 [Zirconium picramate, Dry or wetted with less than 20 per cent water, by mass]
Zirconium picramate, Dry or containing less than 20% water [UN 0236]
Zirconium picramate, Wet (with at least 20% water) [UN 1517]

63885-01-8
Unknown
Unknown
Zinc ammonium nitrite

UN 1512

63892-06-8
$C_8H_8N_2O_3$
180.16
Ethanone, 1-(2-amino-phenyl)-2-nitro- (9CI)
ω-Nitro-o-aminoaceto-phenone

63905-03-3
$C_{15}H_{22}Cl_2N_2O_3S$
381.31
Codeine
3-(Bis(2-chloroethyl)amino)-4-methylbenzenesulfonylmorpholine
DEA No. 9050
Morpholine, 4-(3-(bis(2-chloroethyl)amino)-4-methylbenzenesulfonyl)-

63906-56-9
$O_4S.Tl$
300.43
Thallium(II) sulfate (1:1)
Sulfuric acid, thallium(2+) salt

63907-12-0
$C_{26}H_{48}O_4$
424.74
Octadecanoic acid, 9,10-epoxy-, 2,3-epoxy-2-ethyl-hexyl ester

63907-41-5
Unknown
Unknown
Nitrogen monoxide, Mixed with nitrogen tetroxide
Azotu tlenki (Polish)
Nitric oxide and nitrogen tetroxide, Mixtures [UN 1975]
UN 1975 [Nitric oxide and dinitrogen tetroxide mixtures (Nitric oxide and nitrogen dioxide mixtures)]

63915-78-6
$C_{18}H_{26}O_6$
338.44
Sorbic acid, glycidyl ester dimer
Glycidyl sorbate dimer

63918-89-8
$C_8H_{16}Cl_2OS_2$
263.26
ClCCSCCOCCSCCCl
Ether, bis(2-chloroethyl-thioethyl)
Bis(β-chloroethylthioethyl) ether
Bis(2-chloroethylthioethyl) ether
2-2'-Di(3-chloroethylthio)-diethyl ether
Ethane, 1,1'-oxybis(2-(2-chloroethyl)thio-

63919-02-8
$C_{12}H_{14}O_3$
206.26
Ether, 2,3-epoxypropyl (2,3-epoxypropyl)phenyl
2,3-Epoxypropyl-phenyl glycidyl ether

63936-56-1
$C_{12}HBr_9O$
880.28
Benzene, pentabromo(tetra-bromophenoxy)- (9CI)
Nonabromodiphenyl ether
Nonabromodiphenyl oxide
Nonabromophenoxybenzene
Pentabromo(tetrabromophenoxy)benzene

63937-14-4
$C_6H_{11}O_7.Hg$
395.74
Mercurous gluconate, solid
Mercurous gluconate
Mercury gluconate
Mercury(I) gluconate
UN 1637 [Mercury gluconate]

63938-10-3
C_2HClF_4
136.48
Ethane, chlorotetrafluoro
Chlorotetrafluoroethane [UN 1021]
Monochlorotetrafluoroethane (DOT)
UN 1021 [Chlorotetrafluoroethane]

63958-66-7
$C_8H_6O_6$
198.13
4,5-Dihydroxyphthalic acid
1,2-Benzenedicarboxylic acid, 4,5-dihydroxy-
4,5-DHPA

63978-73-4
$C_{16}H_{30}O_3$
270.46
Tridecanoic acid, 2,3-epoxy-propyl ester
Dodecanoic acid, glycidyl ester
Glycidyl ester of dodecanoic acid

63989-69-5
$As_2Fe_2O_6.Fe_2O_3.5H_2O$
607.30
Ferric arsenite
Ferric arsenite, Basic
Ferric arsenite, Solid
Iron(III) o-arsenite pentahydrate
UN 1607 [Ferric arsenite]

63989-82-2
$C_7H_6N_2O_5$
198.15
p-Cresol, 3,5-dinitro
3,5-Dinitro-p-cresol

63990-84-1
$C_{22}H_{27}NO_2.ClH$
373.96
Acetophenone, 2-(6-(β-hydroxyphenethyl)-1-methyl 2-piperidyl)-, hydrochloride
2-(6-(β-Hydroxyphenethyl)-1-methyl-2-piperidyl)acetophenone hydrochloride
Lobeline hydrochloride
Lobron
Zoolobelin

64031-91-0
$C_{21}H_{14}$
266.34
Perylene, methyl- (9CI)

64036-91-5
$C_8H_{18}ClO_2S_2.Cl$
281.28
Sulfonium, bis(2-hydroxy-ethyl)-2-(2-chloroethylthio)-ethyl-, chloride
Bis(2-hydroxyethyl)-2-(2-chloroethylthio)ethylsulfonium, chloride
β-Chloroethyl β-(bis(β-hydroxyethyl)sulfonium)ethyl sulfide chloride
2-(2-Chloroethyl)thioethylbis-(2-hydroxyethyl)-, chloride
H-1TG

64037-54-3
$C_4H_6Cl_2$
125.00
3,4-Dichlorobutene-1
1-Butene, 3,4-dichloro-(Racemic mixture)
1,2-Dichloro-3-butene (Racemic mixture)
3,4-Dichlorobutene-1 (Racemic mixture)

64047-88-7
$C_6H_2Cl_2NO_3.Na$
229.98
Phenol, 2,4-dichloro-6-nitro-, sodium salt

64061-59-2
$C_{10}H_{11}NO_3$
193.20
O=N(=O)c(c(O)c(cc1)C=C(C)C)

cl
Phenol, 2-(2-methyl-1-propenyl)-6-nitro- (9CI)
2-(2-Methyl-1-propenyl)-6-nitrophenol

64070-12-8
$C_4H_4O_7Sb.K$
324.93
DL-Antimony potassium tartrate
Potassium antimonyl D,L-tartrate
DL-Tartaric acid, antimony potassium salt

64079-01-2
$C_9H_{14}O$
138.21
Furan, pentyl- (9CI)

64082-35-5
FeLiSi
90.88
Lithium-iron-silicon
Lithium ferrosilicon [UN 2830]
UN 2830 [Lithium ferrosilicon]

64091-90-3
$C_{10}H_{13}N_3O_2$
207.21
4-(Methylnitrosamino)-4-(3-pyridyl)butanal
Butanal, 4-(N-methyl-N-nitrosamino)-4-(3-pyridyl)-
γ-(Methylnitrosamino)-3-pyridinebutyraldehyde
4-(N-Methyl-N-nitrosamino)-4-(3-pyridyl)butanal
4-(N-Nitrosomethylamino)-4-(3-pyridyl)-1-butanal
4-(N-Nitroso-N-methylamino)-4-(3-pyridyl)butanal
NNA
3-Pyridinebutanal, γ-(methylnitrosoamino)-

64091-91-4
$C_{10}H_{13}N_3O_2$
207.26
Ketone, 3-pyridyl 3-(N-methyl-N-nitrosamino)propyl
1-Butanone, 4-(methylnitrosoamino)-1-(3-pyridinyl)-
4-(N-Methyl-N-nitrosamino)-1-(3-pyridyl)-1-butanone
4-(N-Methyl-N-nitrosoamino)-4-(3-pyridyl)-1-butanone
N-Methyl-N-nitroso-4-oxo-4-(3-pyridyl)butyl amine
4-(Nitrosoamino-N-methyl)-1-(3-pyridyl)-1-butanone
4-(N-Nitroso-N-methylamino)-1-(3-pyridyl)-1-butanone
NNK

64092-48-4
$C_{15}H_{13}ClNO_3.Na.2H_2O$
349.77
1H-Pyrrole-2-acetic acid, 5-(4-chlorobenzoyl)-1,4-dimethyl-, sodium salt, dihydrate
5-(4-Chlorobenzoyl)-1,4-dimethyl-1H-pyrrole-2-acetic acid sodium salt dihydrate
MCN 2783-21-98
Sodium 5-(4-chlorobenzoyl)-1,4-dimethyl-1H-pyrrole-2-acetate dihydrate
Sodium zomepirac
Zomax
Zomepirac
Zomepirac sodium
Zomepirac sodium salt

64126-85-8
$C_{12}H_6Cl_2O$
237.08
Dibenzofuran, 1,2-dichloro- (9CI)
1,2-Dichlorodibenzofuran

64126-86-9
$C_{12}H_6Cl_2O$
237.08
Dibenzofuran, 2,3-dichloro- (9CI)
2,3-Dichlorodibenzofuran

64126-87-0
$C_{12}H_4Cl_4O$
305.97
Dibenzofuran, 1,2,4,8-tetrachloro- (9CI)
1,2,4,8-Detrachlorodibenzofuran

64142-01-4
$C_{14}H_{15}NS$
229.35
2-Naphthalenecarbothioamide, N-(1-methylethyl)- (9CI)
N-Isopropyl-2-thionaphthoic amide

64176-42-7
Unknown
Unknown
Antiblaze 78 (9CI)

64176-84-7
Unknown
Unknown
Phosflex 300 (9CI)

64176-85-8
Unknown
Unknown
Phosflex 400 (9CI)

64216-20-2
$C_{17}H_{14}N_4OS_4$
418.56
Urea, N,N'-bis(2-benzothiazolylmercaptomethyl)- (9CI)

64275-73-6
$C_8H_{16}O$
128.21
OCCCCC=CCC
5-Octen-1-ol, (Z)- (9CI)
cis-5-Octen-1-ol
(Z)-5-Octen-1-ol

64330-03-6
$C_{17}H_{14}O_7$
330.31
Cyclopenta(c)furo(3',2':4,5)furo(2,3-h)(1)benzopyran-11(1H)-one, 2,3,6a,9a-tetrahydro-1,9a- dihydroxy-4-methoxy-, (6ar-cis)
Aflatoxicol M1

64341-49-7
C_4H_7BrO
151.01
3-Buten-2-ol, 1-bromo- (7CI,9CI)
1-Bromo-3-buten-2-ol

64346-32-3
$C_{10}H_{24}Pb$
351.50
Plumbane, butyltriethyl- (9CI)

64346-47-0
$C_{10}H_{18}N_2O_3$
214.25
O=C(NC(C)(C)C)NC(=CC(=O)OC)C
2-Butenoic acid, 3-((((1,1-dimethylethyl)amino)carbonyl)amino)-, methyl ester (9CI)
Methyl 3-((((1,1-dimethylethyl)amino)carbonyl)amino)-2-butenoate

64436-13-1
$C_5H_{11}AsO_2$
178.06
Arsenobetaine
Arsonium, (carboxymethyl)trimethyl-, hydroxide, inner salt (9CI)

64461-98-9
$C_{12}H_2Cl_6O_2$
390.84
Dibenzo-p-dioxin, 1,2,3,6,7,9-hexachloro
Dibenzo(b,e)(1,4)dioxin,

1,2,3,6,7,9-hexachloro-
1,2,3,6,7,9-Hexachlorodibenzo-dioxin
1,2,3,6,7,9-Hexachlorodibenzo-para-dioxin

64475-85-0
Unknown
Unknown
Mineral-spirits
Petroleum spirits

64501-00-4
$C_{12}H_6Cl_2O_2$
253.08
Dibenzo(b,e)(1,4)dioxin, dichloro- (9CI)
Dichloro dibenzo(b,e)(1,4)dioxin
Dichloro dibenzo-p-dioxin

64532-94-1
$C_{21}H_{21}O_4P$
368.37
Phosphoric acid, 2-(1-methylethyl)phenyl diphenyl ester (9CI)

64532-96-3
$C_{24}H_{27}O_4P$
410.45
Phosphoric acid, 4-hexylphenyl diphenyl ester (9CI)

64532-97-4
$C_{27}H_{33}O_4P$
452.53
Phosphoric acid, 4-nonylphenyl diphenyl ester (9CI)

64535-95-1
Be.Cu
72.56
Beryllium, Compd. with copper (9CI)
Copper, Compd. with beryllium (9CI)

64560-13-0
$C_{12}H_6Cl_2O$
237.08
Dibenzofuran, 4,6-dichloro- (9CI)

64560-14-1
$C_{12}H_5Cl_3O$
271.53
1,4,8-Trichlorodibenzofuran

64560-15-2
$C_{12}H_5Cl_3O$
271.53
Dibenzofuran, 1,2,6-trichloro- (9CI)
1,2,6-Trichlorodibenzofuran

64560-16-3
$C_{12}H_5Cl_3O$
271.53
Dibenzofuran, 1,3,7-trichloro- (9CI)
1,3,7-Trichlorodibenzofuran

64560-17-4
$C_{12}H_4Cl_4O$
305.97
Dibenzofuran, 1,3,7,9-tetrachloro- (9CI)
1,3,7,9-Tetrachlorodibenzofuran

64628-44-0
$C_{15}H_{10}ClF_3N_2O_3$
358.69
BAY-SIR 8514
AI3-29368
Benzamide, 2-chloro-N-(((4-(trifluoromethoxy)phenyl)amino)carbonyl)-
Caswell No. 217C
1-(o-Chlorobenzoyl)-3-(p-(trifluoromethoxy)phenyl)urea (8CI)
2-Chloro-N-(((4-trifluoromethoxy)phenyl)amino)carbonyl)benzamide
EPA Pesticide Chemical Code 118201
Trifumuron

64630-64-4
$C_{12}H_6Cl_5NO$
357.44
Benzenamine, 2,3,4-trichloro-6-(2,4-dichlorophenoxy)- (9CI)

64630-65-5
$C_{12}H_5Cl_6NO$
391.89
Benzenamine, 2,3,4,5-tetrachloro-6-(2,4-dichlorophenoxy)- (9CI)

64665-57-2
$C_7H_7N_3.Na$
156.12
1H-Benzotriazole, 4(or 5)-methyl-, sodium salt (9CI)
4(or 5)-Methyl-1H-benzotriazole sodium salt
Tolyltriazole, sodium salt

64667-33-0
$C_9H_{14}Cl_4O_2$
296.02
O=C(OC)CC(C(Cl)CC(Cl)(Cl)Cl)(C)C
Hexanoic acid, 4,6,6,6-tetrachloro-3,3-dimethyl-, methyl ester (9CI)
Hexanoic acid, 3,3-dimethyl-4,6,6,6-tetrachloro, methyl ester
Methyl 4,6,6,6-tetrachloro-3,3-dimethylhexanoate
4,6,6,6-Tetrachloro-3,3-dimethylhexanoic acid, methyl ester

64690-01-3
$C_{12}H_9N_9O_5S_2$
423.34
1,3-Benzenedisulfonyl diazide, 4-(((2-pyridinylamino)carbonyl)amino)- (9CI)

64700-56-7
$C_{13}H_{16}Cl_3NO_4$
356.65
CCCCOCCOC(=O)COc1nc(Cl)c(Cl)cc1Cl
Acetic acid, ((3,5,6-trichloro-2-pyridinyl)oxy)-, 2-butoxyethyl ester
Garlon 4
Garlon 4E
M 4021
((3,5,6-Trichloro-2-pyridinyl)-oxy)acetic acid 2-butoxyethyl ester

64726-91-6
$C_{14}H_{24}O_2$
224.34
O=C(OC(C=CCCCCCCCC)C1)C1
(R-(Z))-5-(1-Decenyl)dihydro-2-(3H)-furanone
AI3-38648
Caswell No. 275AA
(R,Z)-5-(1-Decenyl)dihydro-2(3H)-furanone
(R-(Z))-5-(1-Decenyl)dihydro-2(3H)-furanone
EPA Pesticide Chemical Code 116501
2(3H)-Furanone, 5-(1-decenyl)-dihydro-, (R-(Z))-
2(3H)-Furanone, 5-(1-decenyl)-dihydro-, (Z)-(R)-(-)-

64741-41-9
Unknown
Unknown
Heavy straight-run naphtha
Atmospheric gas oil (Petroleum)
Heavy straight run naphtha (Petroleum)
Naphtha, petroleum, heavy straight-run

64741-43-1
Unknown
Unknown
Straight-Run Gas Oil
Gas Oils, Petroleum, Straight-

run
Straight Run Gas Oil (Petroleum)

64741-44-2
Unknown
Unknown
Gas Oil, Blend

64741-45-3
Unknown
Unknown
Atmospheric Tower Residue (Reduced Crude Oil)
Residues, Petroleum, Atm. Tower

64741-46-4
Unknown
Unknown
Naphtha (Petroleum), Light straight-run
Light straight-run naphtha

64741-49-7
Unknown
Unknown
Mineral Oil, Petroleum condensates, Vacuum tower
Condensates (Petroleum), Vacuum tower (9CI)
Vacuum residuum

64741-50-0
Unknown
Unknown
Mineral oil, Petroleum distillates, Light paraffinic
Distillates (Petroleum), Light paraffinic (9CI)
Light paraffinic distillate

64741-51-1
Unknown
Unknown
Mineral oil, Petroleum distillates, Heavy paraffinic
Distillates (Petroleum), Heavy paraffinic (9CI)
Heavy paraffinic distillate

64741-52-2
Unknown
Unknown
Mineral oil, Petroleum distillates, Light naphthenic
Distillates (Petroleum), Light naphthenic (99CI)
Light naphthenic distillate
Light naphthenic distillates (Petroleum)

64741-53-3
Unknown
Unknown
Mineral oil, Petroleum distillates, Heavy naphthenic
Distillates (Petroleum), Heavy naphthenic (9CI)
Heavy naphthenic distillate
Heavy naphthenic distillates (Petroleum)

64741-54-4
Unknown
Unknown
Heavy catalytically cracked naphtha
Naphtha, petroleum, heavy catalytic cracked

64741-55-5
Unknown
Unknown
Naphtha (Petroleum), Light catalytic cracked
Light catalytically cracked naphtha

64741-56-6
Unknown
Unknown
Vacuum residue
Residues, Petroleum, Vacuum

64741-57-7
Unknown
Unknown
Heavy Vacuum Distillate (Heavy Vacuum Gas Oil)
Gas Oils, Petroleum, Heavy Vacuum
Vacuum Gas Oil (Petroleum)

64741-58-8
Unknown
Unknown
Light Vacuum Distillate (Light Vacuum Gas Oil)
Gas Oils, Petroleum, Light Vacuum
Light Gas Oil
Light Vacuum Gas Oil (Petroleum)

64741-59-9
Unknown
Unknown
Distillates (Petroleum), Light catalytic cracked
Light catalytically cracked distillate

64741-60-2
Unknown
Unknown
Intermediate Catalytically Cracked Distillate
Distillates, Petroleum, Intermediate catalytic cracked

64741-61-3
Unknown
Unknown
Distillates (Petroleum), Heavy catalytic cracked
Heavy catalytically cracked distillate

64741-62-4
Unknown
Unknown
Clarified slurry oil
Carbon black oil (Petroleum)
Catalytically cracked clarified oil
Clarified oils, Petroleum, Catalytic cracked
Recycle catalytic cracked slurry oil

64741-63-5
Unknown
Unknown
Light reformed naphtha
Naphtha, Petroleum, Light catalytic reformed
Naphthalene plant light gasoline
Platformate

64741-64-6
Unknown
Unknown
Full-range alkylate naphtha
Naphtha, Petroleum, Full-range alkylate

64741-66-8
Unknown
Unknown
Naphtha, Petroleum, Light alkylate

64741-68-0
Unknown
Unknown
Heavy reformed naphtha
Naphtha, Petroleum, Heavy catalytic reformed
Naphthalene plant heavy gasoline

64741-69-1
Unknown
Unknown
Light hydrocracked naphtha
Naphtha, Petroleum, Light Hydrocracked

64741-70-4
Unknown
Unknown
Isomerization naphtha

Naphtha, Petroleum, Isomerization

64741-72-6
Unknown
Unknown
Polymerization naphtha
Naphtha, Petroleum, Polymn.
(Petroleum) decene
(Petroleum) dodecene
(Petroleum) heptene
(Petroleum) nonene
(Petroleum) octene

64741-74-8
Unknown
Unknown
Light Thermally Cracked Naphtha
Light Coker Naphtha (Petroleum)
Light Thermal Cracked C4-C5 Naphtha and Gas Oil Distillate
Naphtha, Petroleum, Light Thermal Cracked

64741-79-3
Unknown
Unknown
Petroleum coke, Uncalcined (DOT)

64741-80-6
Unknown
Unknown
Thermally Cracked Residue
Residues, Petroleum, Thermal Cracked
Thermal Cracked Residuum (Petroleum)

64741-82-8
Unknown
Unknown
Light thermally cracked distillate
Distillates, Petroleum, Light thermal cracked

64741-83-9
Unknown
Unknown
Heavy thermally cracked naphtha
Naphtha, Petroleum, Heavy thermal cracked

64741-87-3
Unknown
Unknown
Naphtha, Petroleum, Sweetened
Sweetened hydrotreated light aromatic solvent naphtha

64741-88-4
Unknown
Unknown
Mineral oil, Petroleum distillates, Solvent-refined heavy paraffinic
Distillates (Petroleum), Solvent-refined heavy paraffinic (9CI)
Solvent-refined heavy paraffinic distillate

64741-89-5
Unknown
Unknown
Mineral oil, Petroleum distillates, Solvent-refined light paraffinic
Distillates (Petroleum), Solvent-refined light paraffinic (9CI)
Solvent-refined light paraffinic distillate

64741-91-9
Unknown
Unknown
Distillates, Petroleum, Solvent-refined middle

64741-95-3
Unknown
Unknown
Solvent deasphalted residual oil
Residual oils, Petroleum, Solvent deasphalted

64741-96-4
Unknown
Unknown
Mineral oil, Petroleum distillates, Solvent-refined heavy naphthenic
Distillates (Petroleum), Solvent-refined heavy naphthenic (9CI)
Solvent-refined heavy naphthenic distillate

64741-97-5
Unknown
Unknown
Mineral oil, Petroleum distillates, Solvent-refined light naphthenic
Distillates (Petroleum), Solvent-refined light naphthenic (9CI)
Solvent-refined light naphthenic distillate

64742-01-4
Unknown
Unknown
Residual oils, Petroleum, Solvent-refined

64742-03-6
Unknown
Unknown
Mineral oil, Petroleum extracts, Light naphthenic distillate solvent
Extracts (Petroleum), Light naphthenic distillate solvent (9CI)
Light naphthenic distillate, Solvent extract

64742-04-7
Unknown
Unknown
Mineral oil, Petroleum extracts, Heavy paraffinic distillate solvent
Extracts (Petroleum), Heavy paraffinic distillate solvent (9CI)
Heavy paraffinic distillate, Solvent extract

64742-05-8
Unknown
Unknown
Mineral oil, Petroleum extracts, Light paraffinic distillate solvent
Extracts (Petroleum), Light paraffinic distillate solvent (9CI)
Light paraffinic distillate, Solvent extract

64742-10-5
Unknown
Unknown
Mineral oil, Petroleum extracts, Residual oil solvent
Extracts (Petroleum), Residual oil solvent (9CI)
Residual oil solvent extract

64742-11-6
Unknown
Unknown
Mineral oil, Petroleum extracts, Heavy naphthenic distillate solvent
Extracts (Petroleum), Heavy naphthenic distillate solvent (9CI)
Heavy naphthenic distillate solvent extract

64742-16-1
Unknown
Unknown
Arien
Arkon M 100
Arkon P 70
Arkon 115
Arsolen

Caswell No. 647
Copar 100
EPA Pesticide Chemical Code 011401
Escorez 100
Escorez 1304
Escorez 1310
Escorez 1401
Escorez 2101
Escorez 2203
Escorez 5280
FR 40 (Petroleum resin)
FR 80 (Petroleum resin)
FR 100 (Petroleum resin)
FTR 6100
Geon D 100
H 130 (Resin)
Hercurez A 100
Hercurez AR 100
Hi-rez
Hiresin (petroleum resin)
Hydrocarbons, petroleum resins
LX 1065
Marukarez R 100A
Marukarez R 100B
Neopolymer
Neopolymer L
Neopolymer L 120
Neopolymer NP 150
Neopolymer 120
Neopolymer 140
Neopolymer 150
Neopolymer 160
Neopolymer 180
Neopolymer 170S
NK 1 (Resin)
Petcoal LX
Petcoal 140
Petrosin (VAN)
Petrosin K
Petrosin PR 120
Petrosin 80
Petrosin 150
Piccodiene 2025
Piccopale
Piccopale 100
Piccopale 100BHT
Piccopale 100SF
Piccopale 200HM
Pirolen 120A
Polyvel G
Polyvel GP 65
Polyvel M
Pyroplast (VAN)
Pyroplast 2
Quintol
Quintone A 100
Quintone C 200S
Quintone D 100
Quintone M 100
Quintone N 180
Quintone RX 05
Quintone U 185
SK 1000 (Petroleum resin)
SPP (Coating)
Statac B
SzF 2/3
SzF 8/9

64742-17-2
Unknown
Unknown
Mineral oil, Petroleum residual oils, Acid-treated
Acid-treated residual oil
Residual oils (Petroleum), Acid-treated (9CI)

64742-18-3
Unknown
Unknown
Mineral oil, Petroleum distillates, Acid-treated heavy naphthenic
Acid-treated heavy naphthenic distillate
Distillates (Petroleum), Acid-treated heavy naphthenic (9CI)

64742-19-4
Unknown
Unknown
Mineral oil, Petroleum distillates, Acid-treated light naphthenic
Acid-treated light naphthenic distillate
Distillates (Petroleum), Acid-treated light naphthenic (9CI)

64742-20-7
Unknown
Unknown
Mineral oil, Petroleum distillates, Acid-treated heavy paraffinic
Acid-treated heavy paraffinic distillate
Distillates (Petroleum), Acid-treated heavy paraffinic (9CI)

64742-21-8
Unknown
Unknown
Mineral oil, Petroleum distillates, Acid-treated light paraffinic
Acid-treated light paraffinic distillate
Distillates (Petroleum), Acid-treated light paraffinic (9CI)

64742-27-4
Unknown
Unknown
Distillates, Petroleum, Chemically neutralized heavy paraffinic

64742-28-5
Unknown
Unknown
Distillates, Petroleum, Chemically neutralized light paraffinic

64742-30-9
Unknown
Unknown
Distillates, Petroleum, Chemically neutralized middle

64742-31-0
Unknown
Unknown
Chemically neutralized kerosene
Chemically-neutralized light distillate
Distillate fuel oils, Light
Distillates, Petroleum, Chemically neutralized light

64742-34-3
Unknown
Unknown
Chemically neutralized heavy naphthenic distillate
Distillates, Petroleum, Chemically neutralized heavy naphthenic

64742-35-4
Unknown
Unknown
Distillates, Petroleum, Chemically neutralized light naphthenic

64742-36-5
Unknown
Unknown
Distillates, Petroleum, Clay treated heavy paraffinic

64742-37-6
Unknown
Unknown
Distillates, Petroleum, Clay-treated light paraffinic

64742-40-1
Unknown
Unknown
Neutralizing agents, Petroleum, Spent sodium hydroxide
Spent caustic
Spent caustic containing an average of 6.9 wt. % cresylate acid

64742-41-2
Unknown
Unknown
Residual oils, Petroleum, Clay-treated

64742-44-5
Unknown
Unknown
Petroleum distillates, Clay-treated heavy naphthenic

64742-45-6
Unknown
Unknown
Petroleum distillates, Clay-treated light naphthenic

64742-46-7
Unknown
Unknown
Distillates (Petroleum), Hydrotreated middle
Amoco NT-45 Process Oil
Kermac 600W (Mineral Seal Oil)

64742-47-8
Unknown
Unknown
Kerosene (Petroleum), Hydrotreated
Hydrotreated kerosene

64742-52-5
Unknown
Unknown
Mineral oil, Petroleum distillates, Hydrotreated heavy naphthenic
Distillates (Petroleum), Hydrotreated heavy naphthenic (9CI)
Hydrotreated heavy naphthenic distillate
Hydrotreated heavy naphthenic distillates (Petroleum)
Petroleum distillates, Hydrotreated heavy naphthenic

64742-53-6
Unknown
Unknown
Mineral oil, Petroleum distillates, Hydrotreated light naphthenic
Distillates (Petroleum), Hydrotreated light naphthenic (9CI)
Hydrotreated light naphthenic distillate
Hydrotreated light naphthenic distillates (Petroleum)

64742-54-7
Unknown
Unknown
Mineral oil, Petroleum distillates, Hydrotreated heavy paraffinic
Distillates (Petroleum), Hydrotreated heavy paraffinic (9CI)
Hydrotreated heavy paraffinic distillate

64742-55-8
Unknown
Unknown
Mineral oil, Petroleum distillates, Hydrotreated light paraffinic
Distillates (Petroleum), Hydrotreated light paraffinic (9CI)
Hydrotreated light paraffinic distillate

64742-56-9
Unknown
Unknown
Mineral oil, Petroleum distillates, Solvent-dewaxed light paraffinic
Distillates (Petroleum), Solvent-dewaxed light paraffinic (9CI)
Solvent-dewaxed light paraffinic distillate

64742-57-0
Unknown
Unknown
Residual oils, Petroleum, Hydrotreated

64742-61-6
Unknown
Unknown
Slack wax, Petroleum

64742-62-7
Unknown
Unknown
Residual oils, Petroleum, Solvent-dewaxed

64742-63-8
Unknown
Unknown
Mineral oil, Petroleum distillates, Solvent-dewaxed heavy naphthenic
Distillates (Petroleum), Solvent-dewaxed heavy naphthenic (9CI)
Solvent-dewaxed heavy naphthenic distillate

64742-64-9
Unknown
Unknown
Mineral oil, Petroleum distillates, Solvent-dewaxed light naphthenic
Distillates (Petroleum), Solvent-dewaxed light naphthenic (9CI)
Solvent-dewaxed light naphthenic distillate

64742-65-0
Unknown
Unknown
Mineral oil, Petroleum distillates, Solvent-dewaxed heavy paraffinic
Distillates (Petroleum), Solvent-dewaxed heavy paraffinic (9CI)
Petroleum distillates, Solvent-dewaxed heavy paraffinic
Solvent-dewaxed heavy paraffinic distillate

64742-68-3
Unknown
Unknown
Mineral oil, Petroleum naphthenic oils, Catalytic dewaxed heavy
Catalytic-Dewaxed heavy naphthenic distillate
Naphthenic oils (Petroleum), Catalytic dewaxed heavy (9CI)

64742-69-4
Unknown
Unknown
Mineral oil, Petroleum naphthenic oils, Catalytic dewaxed light
Catalytic-Dewaxed light naphthenic distillate
Naphthenic oils (Petroleum), Catalytic dewaxed light (9CI)

64742-70-7
Unknown
Unknown
Mineral oil, Petroleum paraffin oils, Catalytic dewaxed heavy
Catalytic-Dewaxed heavy paraffinic distillate
Paraffin oils (Petroleum), Catalytic dewaxed heavy (9CI)

64742-71-8
Unknown
Unknown
Mineral oil, Petroleum paraffin oils, Catalytic dewaxed light
Catalytic-Dewaxed light paraffinic distillate
Paraffin oils (Petroleum), Catalytic dewaxed light (9CI)

64742-75-2
Unknown
Unknown
Naphthenic oils, Petroleum, Complex dewaxed heavy

64742-76-3
Unknown
Unknown
Naphthenic oils, Petroleum, Complex dewaxed light

64742-80-9
Unknown
Unknown
Hydrodesulfurized Middle Distillate
Distillates, Petroleum, hydrodesulfurized middle

64742-81-0
Unknown
Unknown
Hydrodesulfurized Kerosene
Kerosine, Petroleum, hydrodesulfurized

64742-82-1
Unknown
Unknown
Hydrodesulfurized Heavy Naphtha
Naphtha, Petroleum, Hydrodesulfurized Heavy

64742-83-2
Unknown
Unknown
Light Steam-cracked Naphtha
Crude Butadiene (Petroleum)
Light Steam Cracked Aromatic Naphtha (C6) Concentrate (Petroleum)
Light Steam Cracked Naphtha Piperylene Concentrate (Petroleum)
Naphtha, Petroleum, Light Steam-cracked
Steam Cracked Narrow Cut Naphtha (Petroleum)

64742-88-7
Unknown
Unknown
Solvent naphtha, Petroleum, Medium aliph.

64742-90-1
Unknown
Unknown
Steam-cracked Residue
Residues, Petroleum, Steam-cracked
Steam Cracked Residuum Pitch (Petroleum)

64742-93-4
Unknown
Unknown
Asphalt, Oxidized
Asphalt oxide
Blown asphalt

64742-94-5
Unknown
Unknown
Heavy aromatic naphtha
Caswell No. 472A
EPA Pesticide Chemical Code 006602
Solvent naphtha, Petroleum, Heavy arom.

64742-95-6
Unknown
Unknown
Aromatic naphtha, Type I
Naphtha, Aromatic, High flash
Solvent naphtha, Petroleum, Light arom.

64743-02-8
Unknown
Unknown
Alkenes, C>10 α-
α Olefins (Petroleum), C10 cut
α Olefins (Petroleum), (C11-C12) cut
α Olefins (Petroleum), (C11-C14) cut
α Olefins (Petroleum), (C15-C20) cut
α Olefins (Petroleum), (C18-C20) cut

64771-71-7
$C_{12}H_{26}$
170.34
Paraffins, Petroleum, Normal C>10

64771-72-8
C_5H_{12}
72.15
Paraffins, Petroleum, Normal C5-20
Caswell No. 505A
EPA Pesticide Chemical Code 505200
Heavy normal paraffins concentrate (Petroleum)
Heavy normal paraffins (Petroleum)
Isoparaffinic hydrocarbons
Light normal paraffin concentrate (Petroleum)
Light normal paraffins (Petroleum)
Normal paraffins

64800-83-5
$C_{16}H_{18}$
210.32
Benzene, ethyl(phenylethyl)- (9CI)
Ethyl diphenylethane
Ethyl(phenylethyl)benzene

64819-51-8
$C_{11}H_{24}N_2O$
200.31
OC(N=NC(C)(C)C)(CCC(C)C)C
2-tert-Butylazo-2-hydroxy-5-methylhexane
2-((1,1-Dimethylethyl)azo)-5-methyl-2-hexanol
2-Hexanol, 2-((1,1-dimethylethyl)azo)-5-methyl- (9CI)
2-Hexanol-2-((1,1-dimethylethyl)azo)-5-methyl-

64828-44-0
$C_{15}H_{13}N$
207.27
Acridine, ethyl- (9CI)

64828-54-2
$C_{12}H_{11}N$
169.22
Pyridine, methylphenyl- (9CI)

64844-52-6
$C_{15}H_{13}N$
207.27
1H-Indole, methylphenyl- (9CI)

64855-18-1
$C_5H_8Cl_2O_2$
171.02
Propanoic acid, 3,3-dichloro-2,2-dimethyl- (9CI)

64859-47-8
$C_{11}H_{10}N_2$
170.20
Bipyridine, methyl- (9CI)

64859-69-4
Unknown
Unknown
Araldite EPN 1139 (9CI)

64890-90-0
$C_8H_{17}NO$
129.25
Pentanamide, N-propyl- (9CI)

64902-72-3
$C_{12}H_{12}ClN_5O_4S$
357.80
COc2nc(C)nc(NC(=O)NS(=O)(=O)c1ccccc1Cl)n2
Urea, 1-((o-chlorophenyl)-sulfonyl)-3-(4-methoxy-6-methyl-s-triazin-2-yl)
Benzenesulfonamide, 2-chloro-N-(((4-methoxy-6-methyl-

1,3,5-triazin-2-yl)amino)-
carbonyl)-
2-Chloro-N-((4-methoxy-6-methyl-1,3,5-triazin-2-yl)aminocarbonyl)-benzenesulfonamide
1-((o-Chlorophenyl)sulfonyl)-3-(4-methoxy-6-methyl-s-triazin-2-yl)urea
Chlorsulfon
Chlorsulfuron
DPX 4189
Glean
Glean 20DF
Telar
Vertavis

64919-15-9
$C_{14}H_{27}ClO_2$
262.82
Dodecanoic acid, 2-chloroethyl ester (9CI)
NSC-406536

64925-80-0
Unknown
Unknown
Phomopsin-A
NSC-381839
Phomopsin

65036-65-9
$C_8H_9BrO_3S.Na$
288.12
Benzenesulfonic acid, 4-(2-bromoethyl)-, sodium salt (9CI)

65072-04-0
Unknown
Unknown
Veriloid
Alkaloids, Veratrum
Alkavervir
American Hellebore
American Veratrum
Green Hellebore
Indian Poke
Veratrum Viride
Veratrum Viride Alkaloids Extract

65086-97-7
$C_9H_{14}N_2O_2.Na$
205.20
2,4(1H,3H)-Pyrimidinedione, 3-(1,1-dimethylethyl)-6-methyl-, sodium salt (9CI)
3-tert-Butyl-6-methyluracil, sodium salt

65105-00-2
$C_{22}H_{22}N_4O_6$
438.42
Carbamic acid, (3-isocyanatomethylphenyl)-, 1-methyl-1,3-propanediyl ester (VAN) (9CI)

65122-13-6
$C_9H_{22}Pb$
337.47
Plumbane, butyldiethylmethyl-(9CI)

65122-14-7
$C_8H_{20}Pb$
323.45
Plumbane, butylethyldimethyl-(9CI)

65122-21-6
$C_8H_{10}Cl_2$
177.07
Cyclohexene, 1,4-dichloro-4-ethenyl- (9CI)
Cyclohexene, 1,4-dichloro-4-vinyl- (6CI)

65125-87-3
$C_{17}H_{18}ClN_5O_6$
423.78
Ethanol, 2,2'-((4-((2-chloro-4,6-dinitrophenyl)azo)-3-methylphenyl)imino)bis-(9CI)

65150-94-9
$C_{18}H_4F_{33}I$
974.08
Octadecane, 1,1,1,2,2,3,3,4,4,5,5,6,6,7,7,8,8,9,9,10,10,11,11,12,12,13,13,14,14,15,15,16,16-tritriacontafluoro-18-iodo- (9CI)
1,1,2,2-Tetrahydroperfluorooctadecyl iodide

65151-10-2
$C_{14}H_{32}Pb$
407.61
Plumbane, tributylethyl- (9CI)

65195-55-3
$C_{48}H_{72}O_{14}$
873.20
Avermectin A1A, 5-O-demethyl
Antibiotic C 076B1a
Avermectin B1A
5-O-Demethylavermectin A1A

65206-90-8
$C_7H_{15}N_3O_2.ClH$
209.66
Carbamic acid, (aminoiminomethyl)methyl-, dimethyl deriv., ethyl ester, monohydrochloride (9CI)
Ethyl (dimethylamidino)methylcarbamate, hydrochloride

65208-42-6
$C_9H_{14}N_2O_2.Na$
205.20
2,4(1H,3H)-Pyrimidinedione, 6-methyl-3-(1-methylpropyl)-, sodium salt (9CI)
3-sec-Butyl-6-methyluracil, sodium salt

65256-17-9
Unknown
Unknown 430.55
Tribenzofluoranthene (9CI)

65256-18-0
Unknown
Unknown 480.61
Tribenzoperylene (9CI)

65256-40-8
$C_{32}H_{20}$
404.52
Dibenzoperylene (9CI)

65280-19-5
$C_9H_9ClO_3.C_9H_6ClNO_3S.Na$
467.28
3(2H)-Benzothiazoleacetic acid, 4-chloro-2-oxo-, Mixt. with sodium (4-chloro-2-methylphenoxy)acetate (9CI)
Leymin
Vuagt 210

65281-77-8
$C_{20}H_{26}Cl_2O_2$
369.33
1-Phenanthrenecarboxylic acid, 6,8-dichloro-1,2,3,4,4a,9,10,10a-octahydro-1,4a-dimethyl-7-(1-methylethyl)-, (1R-(1α,4aβ,10aα))- (9CI)

65332-44-7
$C_{16}H_{36}N_2S_2$
320.60
2-Propanamine, N,N'-(dithiodi-2,1-ethanediyl)bis(N-(1-methylethyl)- (9CI)
Triethylamine, 2,2'''-dithiobis(1',1''-dimethyl-

65431-33-6
Unknown
Unknown
Trypaflavine
Trypaflavin

65436-87-5
$C_7H_5Br_3O$
344.83

Phenol, methyl-, tribromo deriv. (9CI)

65455-72-3
$C_{35}H_{64}O_{11}$
660.89
3,6,9,12,15,18,21,24,27-Nonaoxanonacosan-1-ol, 29-(isononylphenoxy)- (9CI)

65489-80-7
$C_{12}H_6Br_2O$
325.99
Dibenzofuran, 2,7-dibromo- (9CI)
2,7-Dibromodibenzofuran

65510-44-3
$C_{12}H_5Cl_5$
326.44
1,1'-Biphenyl, 2,3',4,4',5'-pentachloro- (9CI)
2,3',4,4',5'-PCB
2,3',4,4',5'-Pentachlorobiphenyl

65510-45-4
$C_{12}H_5Cl_5$
326.44
Clc1ccc(c(Cl)c1)c2ccc(Cl)c(Cl)c2Cl
1,1'-Biphenyl, 2,2',3,4,4'-pentachloro- (9CI)
2,2',3,4,4'-PCB
2,2',3,4,4'-Pentachlorobiphenyl

65520-63-0
$C_{10}H_{20}O_5S.Na$
274.31
Nonanoic acid, 2-sulfo-, 1-methyl ester, sodium salt (9CI)

65520-64-1
$C_{17}H_{35}NO_5S$
365.54
Ethanaminium, N,N,N-trimethyl-2-((1-oxo-2-sulfododecyl)oxy)-, hydroxide, inner salt (9CI)

65520-65-2
$C_{16}H_{33}NO_6S$
367.51
Dodecanoic acid, 2-sulfo-, 1-(2-((2-hydroxyethyl)-amino)ethyl) ester (9CI)

65520-66-3
$C_{12}H_{24}O_5S.C_4H_{11}NO_2$
385.52
Dodecanoic acid, 2-sulfo-, Compd. with 2-(2-aminoethoxy)ethanol (1:1) (9CI)
Ethanol, 2-(2-aminoethoxy)-, 2-sulfododecanoate (Salt) (9CI)

65530-66-7
$(CF_2)xC_6H_9FO_2$
Polymer
Poly(difluoromethylene), α-fluoro-ω-(2-((2-methyl-1-oxo-2-propenyl)oxy)ethyl)- (9CI)

65601-40-3
$C_6H_6Cl_4F_8N_3O_2P_3$
538.85
1,3,5,2,4,6-Triazatriphosphorine, 2,2,4,4-tetrachloro-2,2,4,4,6,6-hexahydro-6,6-bis-(2,2,3,3-tetrafluor opropoxy)- (9CI)

65601-41-4
$C_9H_9Cl_3F_{12}N_3O_3P_3$
634.45
1,3,5,2,4,6-Triazatriphosphorine, 2,2,4-trichloro-2,2,4,4,6,6-hexahydro-4,6,6-tris(2,2,3,3-tetrafluoro propoxy)- (9CI)

65601-42-5
$C_{12}H_{12}Cl_2F_{16}N_3O_4P_3$
730.05
1,3,5,2,4,6-Triazatriphosphorine, 2,2-dichloro-2,2,4,4,6,6-hexahydro-4,4,6,6-tetrakis(2,2,3,3-tetrafluoropropoxy)- (9CI)

65636-26-2
$C_{28}H_{48}$
384.69
A'-Neo-28,30-dinorgammacerane, (17α)- (9CI)

65646-68-6
$C_{26}H_{33}NO_2$
391.60
Retinamide, N-(4-hydroxyphenyl)
Fenretinide
N-(4-Hydroxyphenyl)retinamide
Retinoic acid p-hydroxyphenylamide

65652-41-7
$C_{26}H_{31}O_4P$
438.50
Phosphoric acid, bis((1,1-dimethylethyl)phenyl) phenyl ester (9CI)
Di-tert-butylphenyl phenyl phosphate
Di(tert-butylphenyl) phenyl phosphate

65724-11-0
$C_{10}H_{13}Br$
213.14
p-Cymene, bromo
Bromo-p-cymeme

65724-12-1
$C_{10}H_{12}Cl_2$
203.12
p-Cymene, dichloro
Dichloro-p-cymene

65724-16-5
$C_7H_6Cl_2O_2$
193.03
Phenol, dichloro-2-methoxy- (9CI)

65752-48-9
$C_{18}H_{14}ClNO_2S$
343.83
4-Thiazolol, 2-(4-chlorophenyl)-5-(4-methylphenyl)-, acetate (ester) (9CI)

65755-17-1
$C_{21}H_{22}$
274.41
Chrysene, 1,2,3,4-tetrahydro-3,3,7-trimethyl- (9CI)

65777-08-4
$C_{18}H_{12}$
228.29
Benzophenanthrene (9CI)

65879-44-9
$C_7H_{11}ClN_2O$
174.63
2,4-Diamino-6-methylphenol hydrochloride
4,6-Diamino-2-methylphenol, hydrochloride
Phenol, 2,4-diamino-6-methyl-, hydrochloride (9CI)

65882-19-1
$C_8H_{15}NO_2$
157.21
Pentanamide, N-acetyl-N-methyl- (9CI)

65882-20-4
$C_{10}H_{19}NO_2$
185.27
Pentanamide, N-acetyl-N-propyl- (9CI)

65882-21-5
$C_8H_{15}NO_2$
157.21
Acetamide, N-acetyl-N-(1,1-dimethylethyl)-(9CI)

65882-22-6
$C_{18}H_{35}NO_2$
297.49
Hexadecanamide, N-acetyl- (9CI)

65882-23-7
$C_{20}H_{39}NO_2$
325.54
Octadecanamide, N-acetyl-(9CI)

65902-59-2
$C_3H_2BrN_3O_2$
191.98
1H-Imidazole, 2-bromo-4-nitro- (9CI)

65907-30-4
$C_{18}H_{26}N_2O_5S$
382.52
Benzofuran, 2,3-dihydro-2,2-dimethyl-7-(N-(N-methyl-N-butoxycarbonylamino-thio)-N- methylcarbamoyl-oxy)
CGA 73102
Deltanet
Furathiocarb
Promet
Promet 660SCO

65954-19-0
Unknown
Unknown
(Z)-4-Tridecen-1-yl acetate

65983-31-5
$C_{15}H_{20}O_3$
248.32
O=C(OCCOC(C(C(C(C12)C=C3)C3)C1)C2)C=C
2-Propenoic acid, 2-((3a,4,5,6,7,7a-hexahydro-4,7-methano-1H-inden-6-yl)-oxy)ethyl ester (9CI)
Dicyclopentyloxyethyl acrylate

65996-67-0
Unknown
Unknown
Iron, Furnace

65996-68-1
Unknown
Unknown
Gas, Blast furnace
Flue Gases, Ferrous metal, blast furnace

65996-69-2
Unknown
Unknown
Slags, Ferrous metal, Blast furnace
Blast furnace slag

65996-70-5
Unknown
Unknown
Dust, Ferrous metal, Blast furnace

65996-71-6
Unknown
Unknown
Slags, Steelmaking

65996-72-7
Unknown
Unknown
Dust, Steelmaking
Grinding dust

65996-74-9
Unknown
Unknown
Mill scale, Ferrous metal

65996-77-2
Unknown
Unknown
Coke
Catalytic Coke
Coke (Chaud) (French)
Coke, Coal
Coke, Hot
Coque Caliente (Spanish)

65996-78-3
Unknown
Unknown
Light oil, Coal, Coke-oven
Crude Light Oil (Coal)
Intermediate Light Oil (Coal)

65996-82-9
Unknown
Unknown
Coal Tar Neutral Oils
Chemical Oil (Coal)
High Flash Acid Fraction
Intermediate Naphthalene
Tar Oils, Coal

65996-84-1
Unknown
Unknown
Tar bases, Coal, Crude
Crude Tar Bases (Coal)

65996-85-2
Unknown
Unknown
Tar acids, Coal, Crude
Crude Tar Acids (Coal)

65996-89-6
Unknown
Unknown
Coal-Tar

65996-90-9
Unknown
Unknown
Coal-tars, Low temperature
Lignite coal tar

65996-91-0
Unknown
Unknown
Coal Tar Light Oil
Anthracene Oil
Caswell No. 052
Distillates, Coal Tar, Upper
EPA Pesticide Chemical Code 006101
Upper Coal Tar Distillate

65996-92-1
Unknown
Unknown
Coal Tar Distillate
Coal Tar Distillates
Destilados de Alquitran de Hulla (Spanish)
Distillates, Coal Tar
Goudron de Houille, Distillats de (French)
Middle Coal Tar Distillate
Middle Tar Distillate (Coal)
UN 1136 [Coal tar distillates, flammable]
UN 1137

65996-93-2
Unknown
Unknown
Coal-Tar Pitch
Coal Tar Pitch Volatiles (ACGIH,OSHA)
Pitch
Pitch, Coal Tar

65996-94-3
Unknown
Unknown
Phosphate rock and phosphorite, calcined

65997-15-1
Unknown
Unknown

Silicate, Portland Cement
Portland Cement (ACGIH, OSHA)

65997-17-3
Unknown
Unknown
Glass, Oxide, Chemicals
Glass
Glass enamel 19 E 110
Glassy sodium phosphate
Lead borosilicate glass enamel flux
Phosphorus furnace slag
Sodium calcium magnesium polyphosphate
Sodium calcium magnesium silica polyphosphate
Sodium calcium polyphosphate
Sodium calcium zinc silica polyphosphate
Sodium zinc polyphosphate
Sodium zinc potassium polyphosphate

65997-19-5
Unknown
Unknown
Steel manufacture, Chemicals

66017-91-2
$C_6H_{12}N_2O_3$
160.20
CCCN(COC(C)=O)N=O
Acetic acid, propylnitrosaminomethyl ester
Acetic acid, (nitrosopropylamino)methyl ester
Acetoxymethylpropylnitrosamine
N-(Acetoxy)methyl-N-n-propylnitrosamine
Methanol, (nitrosopropylamino)-, acetate (ester)
N-Nitroso-N-(1-acetoxymethyl)-propyl amine
(Nitrosopropylamino)methyl acetate
PAMN
Propyl acetoxymethylnitrosamine
N-Propyl-N-(acetoxymethyl)-nitrosamine
N-Nitroso-N-(acetoxy)methyl-N-n-propylamine
Propylamine, N-(1-acetoxymethyl)-N-nitroso-
Propylnitrosaminomethyl acetate

66028-01-1
$C_{20}H_{33}ClO_3$
356.93
O(CCOCCOc(ccc(c1)CCCCCCCC)c1)CCCl
Benzene, 1-(2-(2-(2-chloroethoxy)ethoxy)ethoxy)-4-octyl- (9CI)
1-Chloro-8-(p-octylphenoxy)-3,6-dioxaoctane

66046-78-4
$C_8H_2Br_4O_4.1/2C_2H_8N_2$
511.77
1,2-Benzenedicarboxylic acid, 3,4,5,6-tetrabromo-, Compd. with 1,2-ethanediamine (2:1) (9CI)
Ethylenediamine di(tetrabromophthalic acid)

66070-60-8
Unknown
Unknown
Soybean oil, Polymer with pentaerythritol and phthalic anhydride
1,2-Benzenedicarboxylic anhydride, pentaerythritol, soybean oil polymer
Pentaerythritol, phthalic anhydride, soya oil polymer
Pentaerythritol, phthalic anhydride, soybean oil polymer
Pentaerythritol, phthalic anhydride, soybean oil resin
Phthalic anhydride, pentaerythritol, soybean oil resin
Soya oil, pentaerythritol, phthalic anhydride polymer
Soya oil, pentaerythritol, phthalic anhydride resin
Soya oil, phthalic anhydride, pentaerythritol polymer
Soya oil, phthalic anhydride, pentaerythritol resin
Soybean oil modified, pentaerythritol, phthalic anhydride alkyd resin
Soybean oil, pentaerythritol, phthalic anhydride alkyd resin
Soybean oil, pentaerythritol, phthalic anhydride polymer
Soybean oil, pentaerythritol, phthalic anhydride resin
Soybean oil, phthalic anhydride, pentaerythritol alkyd resin
Soybean oil, phthalic anhydride, pentaerythritol polymer
Soybean oil, phthalic anhydride, pentaerythritol resin
Soybean oil, tetramethylolmethane, phthalic anhydride polymer

66070-61-9
Unknown
Unknown
Soybean oil, Polymer with glycerol and phthalic anhydride
1,2-Benzenedicarboxylic anhydride, glycerin, soybean oil polymer
Glycerine, soybean oil, phthalic anhydride, litharge polymer
Glycerol, phthalic anhydride, soybean oil polymer
Phthalic anhydride, glycerin, soybean oil polymer
Phthalic anhydride, glycerin, soybean oil resin
Phthalic anhydride, soya oil, glycerin polymer
Phthalic anhydride, soya oil, glycerin resin
Phthalic anhydride, soybean oil, glycerin polymer
1,2,3-Propanetriol, 1,3-isobenzofurandione, soybean oil polymer
Soya, glycerin, phthalic anhydride alkyd resin
Soya oil, glycerin, phthalic anhydride polymer
Soya oil, glycerol, phthalic anhydride polymer
Soya oil, glycerol, phthalic anhydride resin
Soya oil, glycerol, phthalic anhydride, Polymer
Soya oil, phthalic anhydride, glycerol polymer
Soya oil, phthalic anhydride, glycerol resin
Soybean oil, glycerin, phthalic anhydride polymer
Soybean oil, glycerin, phthalic anhydride resin
Soybean oil, glycerine, phthalic anhydride polymer
Soybean oil, glycerine, phthalic anhydrided polymer
Soybean oil, glycerol, phthalic anhydride alkyd resin
Soybean oil, glycerol, phthalic anhydride polymer
Soybean oil, glycerol, phthalic anhydride resin
Soybean oil modified, glycerin, phthalic anhydride alkyd resin
Soybean oil, phthalic anhydride glycerin polymer
Soybean oil, phthalic anhydride, glycerin polymer
Soybean oil, phthalic anhydride, glycerin resin
Soybean oil, phthalic anhydride, glycerine polymer
Soybean oil, phthalic anhydride, glycerol polymer
Soybean oil, phthalic anhydride, glycerol resin

66070-71-1
Unknown
Unknown
Fatty acids, tall-oil, Polymers with glycerol and phthalic anhydride
1,2-Benzenedicarboxylic anhydride, glycerin, tall oil fatty acids polymer
Glycerol, phthalic anhydride, tall oil acids polymer

Phthalic anhydride, glycerin, tall oil acids resin
Phthalic anhydride, tall oil fatty acids, glycerin polymer
Tall oil fatty acid, glycerin, phthalic anhydride polymer
Tall oil fatty acid, glycerine, phthalic anhydride resin
Tall oil fatty acid, glycerol, phthalic anhydride alkyd resin
Tall oil fatty acid, phthalic anhydride, glycerin polymer
Tall oil fatty acid, phthalic anhydride, glycerol polymer
Tall oil fatty acids, 1,2-benzenedicarboxylic anhydride, glycerin polymer
Tall oil fatty acids, glycerin, phthalic anhydride alkyd resin
Tall oil fatty acids, glycerin, phthalic anhydride polymer
Tall oil fatty acids, glycerine, phthalic anhydride polymer
Tall oil fatty acids, phthalic acid, glycerol polymer
Tall oil fatty acids, phthalic anhydride, glycerin polymer
Tall oil fatty acids, phthalic anhydride, glycerol polymer

66070-84-6
Unknown
Unknown
Fatty acids, tall-oil, Polymers with benzoic acid, pentaerythritol and phthalic anhydride
Benzoic acid, pentaerythritol, phthalic anhydride, tall fatty oil acids polymer
Benzoic acid, pentaerythritol, phthalic anhydride, tall oil fatty acid polymer
Benzoic acid, pentaerythritol, phthalic anhydride, tall oil fatty acids polymer
Benzoic acid, pentaerythritol, tall oil fatty acid, phthalic anhydride polymer
Benzoic acid, tall oil fatty acid, phthalic anhydride, pentaerythritol polymer
Benzoic acid, tall oil fatty acids, phthalic anhydride, pentaerythritol polymer
Tall oil acids, phthalic anhydride, benzoic acid, pentaerythritol resin
Tall oil fatty acid, benzoic acid, phthalic anhydride, pentaerythritol polymer
Tall oil fatty acid, pentaerythritol, benzoic acid, phthalic anhydride resin
Tall oil fatty acid pentaerythritol, phthalic anhydride alkyd resin benzoic acid modified
Tall oil fatty acid, pentaerythritol, phthalic anhydride, benzoic acid polymer
Tall oil fatty acids, pentaerythritol, benzoic acid, phthalic anhydride polymer
Tall oil fatty acids, pentaerythritol, phthalic anhydride, benzoic acid alkyd resin
Tall oil fatty acids, pentaerythritol, phthalic anhydride, benzoic acid polymer
Tall oil fatty acids, phthalic anhydride, benzoic acid, pentaerythritol polymer
Tall oil fatty acids, phthalic anhydride, pentaerythritol, benzoic acid polymer
Tall oil fatty acids, phthalic anhydride, pentaerythritol, benzoic acid resin
Tall oil fatty acids, phthalic anhydride, pentaerythritol, benzoic acid resin

66070-87-9
Unknown
Unknown
Coconut oil, Polymer with glycerol and phthalic anhydride
1,2-Benzenedicarboxylic anhydride, glycerin, coconut oil polymer
Coconut oil fatty acids, phthalic anhydride, glycerin polymer
Coconut oil, glycerin, phthalic anhydride polymer
Coconut oil, glycerin, phthalic anhydride resin
Coconut oil, glycerine, phthalic anhydride polymer
Coconut oil, glycerol, phthalic anhydride alkyd resin
Coconut oil, glycerol, phthalic anhydride polymer
Coconut oil, glycerol, phthalic anhydride resin
Coconut oil, phthalic anhydride, glycerin polymer
Coconut oil, phthalic anhydride, glycerol polyester
Coconut oil, phthalic anhydride, glycerol polymer
Coconut oil, phthalic anhydride, glycerol resin
Glycerine, phthalic anhydride, coconut oil fatty acid polymer
Phthalic anhydride, coconut oil, glycerin polymer
Phthalic anhydride, glycerin, coconut oil polymer
1,2,3-Propanetriol, 1,3-isobenzofurandione, coconut oil polymer

66070-93-7
Unknown
Unknown
Soybean oil, Polymer with glycerol, pentaerythritol and phthalic anhydride
1,2-Benzenedicarboxylic anhydride, glycerin, pentaerythritol, soybean oil polymer
Glycerin, pentaerythritol, phthalic anhydride, soybean oil polymer
Glycerol, pentaerythritol, phthalic anhydride, soybean oil polymer
Soya oil, pentaerythritol, glycerol, phthalic anhydride polymer
Soya oil, phthalic anhydride, glycerol, pentaerythritol polymer
Soya oil, phthalic anhydride, pentaerythritol, glycerin polymer
Soya oil, phthalic anhydride, pentaerythritol, glycerine resin
Soya oil, Polymer with pentaerythritol, glycerin and phthalic anhydride
Soybean oil acids, phthalic anhydride, pentaerythritol, glycerol resin
Soybean oil modified, glycerin, pentaerythritol, phthalic anhydride alkyd resin
Soybean oil, glycerin, pentaerythritol, phthalic anhydride polymer
Soybean oil, glycerin, pentaerythritol, phthalic anhydride resin
Soybean oil, glycerin, phthalic anhydride, pentaerythritol polymer
Soybean oil, glycerin, tetramethylolmethane, phthalic anhydride resin
Soybean oil, glycerine, pentaerythritol, phthalic anhydride polymer
Soybean oil, glycerol, pentaerythritol phthalic anhydride resin
Soybean oil, pentaerythritol, glycerin, phthalic anhydride alkyd resin
Soybean oil, pentaerythritol, glycerin, phthalic anhydride polymer
Soybean oil, pentaerythritol, glycerine, phthalic anhydride polymer
Soybean oil, pentaerythritol, glycerol, phthalic anhydride resin
Soybean oil, pentaerythritol, phthalic anhydride, benzoic acid, trimethylolethane resin
Soybean oil, pentaerythritol, phthalic anhydride, glycerin polymer
Soybean oil, pentaerythritol, phthalic anhydride, glycerin resin

Soybean oil, phthalic anhydride, glycerin, pentaerythritol polymer
Soybean oil, phthalic anhydride, glycerol, pentaerythritol polymer
Soybean oil, phthalic anhydride, glycerol, pentaerythritol resin
Soybean oil, phthalic anhydride, pentaerythritol, glycerin polymer
Soybean oil, phthalic anhydride, pentaerythritol, glycerin resin
Soybean oil, phthalic anhydride, pentaerythritol, glycerol polymer
Soybean oil, phthalic anhydride, pentaerythritol, glycerol resin

66071-18-9
Unknown
Unknown
Linseed oil, Polymer with glycerol, phthalic anhydride and tung oil
Chinawood oil, glycerol, linseed oil, phthalic anhydride polymer
Glycerol, phthalic anhydride, tung oil, linseed oil polymer
Linseed oil, tung oil, glycerin, phthalic anhydride polymer
Linseed oil, tung oil, glycerin, phthalic anhydride resin
Linseed oil, tung oil, phthalic anhydride, glycerol resin
Linseed, tung oil, phthalic anhydride, glycerol, alkyd resin
Phthalic anhydride, glycerol, linseed oil, tung oil polymer

66071-28-1
Unknown
Unknown
Fatty acids, dehydrated castor-oil, Polymers with glycerol, linseed oil, phthalic anhydride and rosin

66071-63-4
Unknown
Unknown
Linseed oil, Polymer with glycerol, pentaerythritol, phthalic anhydride and styrene
1,2-Benzenedicarboxylic anhydride, glycerin, pentaerythritol, linseed oil, styrene polymer

66104-24-3
$C_2H_2Be_3O_8$
181.07
Beryllium aluminum alloy
Beryllium, bis(carbonato-(2-))dihydroxytri-
Beryllium carbonate
Beryllium carbonate, Basic
Beryllium compound
Berylliumoxide carbonate
Bis(carbonato(2-))dihydroxytriberyllium

66108-37-0
$C_{11}H_{18}Br_2Cl_5O_4P$
582.31
O=P(OCC(CCl)(CBr)CBr)(OC(CCl)CCl)OC(CCl)CCl
2,2-Bis(bromomethyl)-3-chloropropyl bis(2-chloro-1-(chloromethyl)ethyl) phosphate
Bis(1,3-dichloro-2-propyl) 3-c hloro-2,2-dibromomethyl-1-propyl phosphate
Phosphoric acid, 2,2-bis(bromomethyl)-3-chloropropyl bis(2-chloro-1-(chloromethyl)ethyl) ester

66121-41-3
$C_{14}H_6Cl_2N_2O_4$
337.11
9,10-Anthracenedione, 1-amino-5,8-dichloro-4-nitro- (9CI)

66230-04-4
$C_{25}H_{22}ClNO_3$
419.93
Benzeneacetic acid, 4-chloro-α-(1-methylethyl)-, cyano (3-phenoxyphenyl)methyl ester, (S-(R*,R*))
Asana
Asana XL
(S)-α-Cyano-3-phenoxybenzyl-(S)-2-(4-chlorophenyl)-3-methylbutyrate
Esfenvalerate
Fenvalerate α
Fenvalerate A α
Halmark
OMS 3023
S 1844
S 5602 A Alpha
Sumi-Alfa
Sumi-Alpha
Sumicidin A Alpha

66241-11-0
Unknown
Unknown
C.I. Leuco Sulphur Black 1 (9CI)
C.I. Leuco Sulfur Black 1
C.I. 53185

66267-77-4
$C_{25}H_{22}ClNO_3$
419.90
Benzeneacetic acid, 4-chloro-α-(1-methylethyl)-, cyano-(3-phenoxyphenyl)methyl ester, (R-(R*,S*))- (9CI)
Fenvalerate β

66280-95-3
$C_6H_4Cl_2N_2O_2$
207.00
2-Pyridinecarboxylic acid, 4-aminodichloro- (9CI)

66289-74-5
$C_{14}H_{18}$
186.32
4,7-Methano-2,3,8-metheno-cyclopent(a)indene, dodecahydro-, stereoisomer
endo,endo-Dihydrodi(norbornadiene)
RJ 5
Shelloyne H

66290-87-7
$C_{13}H_{21}N_2O.I$
348.23
Pyridinium, 1-heptyl-3-((hydroxyimino)-methyl)-, iodide (9CI)

66291-32-5
$C_{16}H_{14}$
206.29
Phenanthrene, 3,9-dimethyl- (9CI)

66291-33-6
$C_{16}H_{14}$
206.29
Phenanthrene, 3,10-dimethyl- (9CI)

66330-88-9
Unknown
Unknown
Hydrothol-191 (9CI)

66332-96-5
$C_{17}H_{16}F_3NO_2$
323.34
Benzamide, N-(3-(1-methylethoxy)phenyl)-2-(trifluoromethyl)
Flutolanil
N-(3-(1-Methylethoxy)phenyl)-2-(trifluoromethyl)benzamide
Moncut
NNF-136

66357-35-5
$C_{13}H_{24}N_4O_3S.ClH$
352.93
CNC(NCCSCc1ccc(CN(C)C)o1)=CN(=O)=O
1,1-Ethenediamine, N-(2-(((5-((dimethylamino)methyl)-

2-furanyl)methyl)thio)ethyl)-N'-methyl- 2-nitro-, hydrochloride
AH 19065
Ranidil
Ranitidine hydrochloride
Zantac

66402-68-4
Unknown
Unknown
Ceramic Materials and Wares, Chemicals
Antimony oxide calcium titanate silicate ceramic opacifier
Barium, calcium, magnesium, strontium, aluminum silicate flux
Calcined bauxite
Calcined clay
Calcined clays
Calcined fireclay
Calcined kaolin
Calcined kaolin clay
Calcined Kentucky flint clay
Calcined lightweight aggregate
Calcined Missouri flint clay
Calcined semi-flint clay (Blum)
Calcined semi-flint clay (Harris)
Ceramic
Ceramic bonded alumina
Ceramic bonded silicon carbide
Clay bonded mordenite
Clay bonded natural zeolite
Cordierite
Cristobalite
Diatomaceous silica, Flux-calcined
Expanded clay, Lightweight aggregates
Fireclay, calcined
Fired clay
Kaolin, calcined
Mullite
Nickel oxide coated ceramic bonded zircon
Synthetic mordenite

66418-17-5
$C_{32}H_{16}Cu_2N_5O_{13}S_3.3Na$
970.73
Cuprate(3-), (mu-(4-((4'-((6-amino-1-hydroxy-3-sulfo-2-naphthalenyl)azo)-3,3'-dihydroxy(1,1'-biphenyl)-4-yl)-azo)-3-hydroxy-2,7-naphthalenedisulfonato(7-)))di-, trisodium (9CI)
Copper, (trihydrogen 7-amino-3-((3,3'-dihydroxy-4'-((2-hydroxy-3,6-disulfonato(4-)-1-naphthyl)azo)-4-biphenyl)azo)-4-hydroxy-2-naphthalenesulfonato(2-))di-, trisodium salt
(Trihydrogen-2-hydroxy-1-((((6-amino-1-hydroxy-3-sulfo-2-naphthyl)azo)-3,3'-dihydroxy-4,4'-biphenylene)azo)-3,6-naphthalenedisulfonato-)dicopper trisodium salt

66422-95-5
$C_8H_{12}N_2O_2.2ClH$
241.14
Ethanol, 2-(2,4-diamino-phenoxy)-, dihydrochloride
2-(2,4-Diaminophenoxy)ethanol dihydrochloride

66441-11-0
$C_{18}H_{16}ClNO_4S$
377.86
Propanoic acid, 2-(4-((6-chloro-2-benzothiazolyl)oxy)-phenoxy)-, ethyl ester
Ethyl 2-(4-((6-chloro-2-benzothiazolyl)oxy)phenoxy)propanoate
Fenthiaprop-ethyl
HOE 35 609
Joker
Taifun

66441-23-4
$C_{18}H_{16}ClNO_5$
361.78
Propanoic acid, 2-(4-((6-chloro-2-benzoxazolyl)oxy)phenoxy)-, ethyl ester (9CI)
Acclaim
Caswell No. 431C
EPA Pesticide Chemical Code 128701
Ethyl-2-((4-(6-chloro-2-benzoxazolyloxy))-phenoxy)propionate
(+-)-Ethyl 2-(4-((6-chloro-2-benzoxazolyl)oxy)phenoxy)propanoate
Fenoxaprop-ethyl
Fenoxaprop ethyl ester
Furore
HOE-A 25-01
HOE 33171
Option
Whip

66455-17-2
$C_{10}H_{22}O$
158.29
Alcohols, C9-11
C9-11 alcohol
Dobanol 911
Linevol 911
Neodol 91

66587-56-2
Unknown
Unknown
Alphanol 79
Linevol 79

66594-31-8
Unknown
Unknown
Pydraul 50E (9CI)
Cumylphenyl diphenyl phosphate

66594-32-9
Unknown
Unknown
Pydraul 115E (9CI)

66630-68-0
$C_{12}H_{12}Br_2N_2$
344.06
Dipyrido(1,2-a:2',1'-c)pyrazinediium, 6,7-dihydro-, dibromide, Mixt. with Cutrine Plus (9CI)

66651-97-6
$C_5H_3Cl_3NO_3PS$
294.48
2-Pyridinol, 3,5,6-trichloro-, dihydrogen phosphorothioate (ester) (9CI)

66733-21-9
$Al_2O_{18}Si_7.1/2Ca.7H_2O.1/2Na$
715.68
Erionite (Cakna $(Al_2Si_7O_{18})2.14H_{2O}$)

66793-76-8
$C_9H_{21}N$
143.27
1-Heptanamine, N-ethyl- (9CI)

66794-59-0
$C_{12}H_4Cl_4O$
305.97
Dibenzofuran, 1,4,6,7-tetrachloro- (9CI)
1,4,6,7-Tetrachlorodibenzofuran

66797-44-2
Unknown
Unknown
Kronitex 100 (9CI)

66822-98-8
$C_7H_{14}O_3$
146.19
1,3-Dioxolane, 2-methoxy-2-(1-methylethyl)- (9CI)

66841-25-6
$C_{22}H_{19}Br_4NO_3$
665.06
CC1(C)C(C(Br)C(Br)(Br)Br)C1C(=O)OC(C#N)c3cccc(Oc2ccccc2)c3
Cyclopropanecarboxylic acid, 2,2-dimethyl-3-(1,2,2,2-tetrabromoethyl)-, cyano-(3-phenoxy phenyl)methyl

ester
Cyano(3-phenoxyphenyl)methyl 2,2-dimethyl-3-(1,2,2,2-tetrabromoethyl)cyclopropanecarboxylate
HAG 107
RU 25472
RU 25474
Tralomethrin
Tralomethrine

66849-71-6
Unknown
Unknown
Trinitroethylnitrate

67028-17-5
$C_{12}H_5Cl_3O_2$
287.53
Dibenzo(b,e)(1,4)dioxin, 1,3,7-trichloro- (9CI)
1,3,7-Trichlorodibenzo-(b,e)(1,4)dioxin
1,3,7-Trichlorodibenzo-p-dioxin

67028-18-6
$C_{12}H_4Cl_4O_2$
321.97
Clc3ccc2Oc1cc(Cl)c(Cl)c(Cl)c1Oc2c3
Dibenzo(b,e)(1,4)dioxin, 1,2,3,7-tetrachloro- (9CI)
1,2,3,7-Tetrachlorodibenzo-(b,e)(1,4)dioxin
1,2,3,7-Tetrachlorodibenzo-p-dioxin

67028-19-7
$C_{12}H_3Cl_5O_2$
356.42
Dibenzo(b,e)(1,4)dioxin, 1,2,3,4,6-pentachloro- (9CI)
1,2,3,4,6-Pentachlorodibenzo-(b,e)(1,4)dioxin
1,2,3,4,6-Pentachlorodibenzo-p-dioxin

67069-15-2
$C_{32}H_{56}$
440.80
A'-Neo-30-norgammacerane, 22-propyl-, (17α,22S)- (9CI)

67069-16-3
$C_{33}H_{58}$
454.82
A'-Neo-30-norgammacerane, 22-butyl-, (17α,22S)- (9CI)

67069-25-4
$C_{32}H_{56}$
440.80
A'-Neo-30-norgammacerane, 22-propyl-, (17α,22R)- (9CI)

67069-26-5
$C_{33}H_{58}$
454.82
A'-Neo-30-norgammacerane, 22-butyl-, (17α,22R)- (9CI)

67124-09-8
$C_{15}H_{32}OS$
260.48
2-Propanol, 1-(tert-dodecylthio)- (9CI)
tert-Dodecyl 2-hydroxypropyl sulfide
1-(tert-Dodecylthio)-2-propanol

67129-08-2
$C_{14}H_{16}ClN_3O$
277.78
Cc1cccc(C)c1N(Cn2cccn2)C(=O)CCl
Acetamide, 2-chloro-N-(2,6-dimethylphenyl)-N-(1H-pyrazol-1-ylmethyl)
BAS 479H
Butisan S
2-Chloro-N-(2,6-dimethylphenyl)-N-(1H-pyrazol-1-ylmethyl)-acetamide
Metazachlor
Metazachlore

67254-74-4
Unknown
Unknown
Naphthenic oils

67254-79-9
Unknown
Unknown
Fatty acids
Acids, carboxylic (VAN)
Aliphatic compounds (VAN)
Carboxylic acids, fatty
KO (fatty acid)
KTIOL 15
KTIOL 77

67292-92-6
$C_8H_{15}N_3O_2$
185.21
CC(C)CNC(=O)N1CCNC1=O
Imidazolin-2-one, 1-isobutylcarbamoyl-
Imidazolidin-2-on-1-carbonsaeure-isobutylamid
1-Isobutylcarbamoyl-imidazolin-2-one
Isocarbamid
Merpelan AZ
MNF O 166

67323-56-2
$C_{12}H_4Cl_4O_2$
321.97
Dibenzo(b,e)(1,4)dioxin, 1,2,6,8-tetrachloro- (9CI)
1,2,6,8-Tetrachlorodibenzo(b,e-)(1,4)dioxin
1,2,6,8-Tetrachlorodibenzo-p-dioxin

67329-11-7
$C_5H_8Cl_2O_2$
171.02
Propanoic acid, 3-chloro-2-(chloromethyl)-2-methyl- (9CI)
Propionic acid, 2,2-bis(chloromethyl)-

67361-76-6
$C_{64}H_{120}O_6S_3Sn$
1200.56
9-Octadecenoic acid (Z)-, (butylstannylidyne)tris(thio-2,1-ethanediyl) ester (9CI)
Butyltintris(2-oleyoyloxyethylmercaptide)

67361-77-7
$C_{48}H_{92}O_4S_2Sn$
916.10
9-Octadecenoic acid (Z)-, (dibutylstannylene)bis(thio-2,1-ethanediyl) ester (9CI)
Dibutyltinbis(2-oleoyloxyethylmercaptide)

67375-30-8
$C_{22}H_{19}Cl_2NO_3$
416.32
Cyclopropanecarboxylic acid, 3-(2,2-dichloroethenyl)-2,2-dimethyl-, cyano(3-phenoxyphenyl) methyl ester, (1-α(S*),3-α)-(+-)
Alfoxylate
Alphacypermethrin
Concord
Fastac
Fendona
Renegade
WL 85871

67426-57-7
Unknown
Unknown
Kronitex 50 (9CI)

67426-58-8
Unknown
Unknown
Kronitex 300 (9CI)

67446-04-2
$C_7H_9O_2PS.K$
226.28
Phosphonothioic acid, phenyl-, O-methyl ester,

potassium salt (9CI)
O-Methyl potassium phenylphosphonothioate
Potassium O-methyl phenylphosphonothioate

67446-07-5
$C_{12}H_{22}O_2$
198.31
5-Decen-1-ol, acetate, (Z)- (9CI)

67452-27-1
$C_{12}H_{22}O_2$
198.31
4-Decen-1-ol, acetate, (Z)- (9CI)

67465-67-2
$C_{18}H_{23}N_3S$
313.50
10H-Pyrido(3,2-b)(1,4)benzothiazine, 10-(3-(diethylamino)propyl)
D 209

67481-22-5
$C_{12}H_3Cl_5O$
340.42
Dibenzofuran, 2,3,4,6,8-pentachloro- (9CI)
2,3,4,6,8-Pentachlorodibenzofuran

67485-29-4
$C_{25}H_{24}F_6N_4$
494.53
2(1H)-Pyrimidinone, tetrahydro-5,5-dimethyl-, (3-(4-(trifluoromethyl)phenyl)-1-(2-(4- (trifluoromethyl)phenyl)ethenyl)-2-propenylidene)hydrazone
AC 217300
Amdro
CL 217300
Combat
Hydramethylnon
Matox
Maxforce
Wipeout

67517-48-0
$C_{12}H_3Cl_5O$
340.42
Dibenzofuran, 1,2,3,4,8-pentachloro- (9CI)
1,2,3,4,8-Pentachlorodibenzofuran

67526-84-5
$C_{16}H_{11}N$
217.26
Benzocarbazole (9CI)

67546-51-4
$C_6H_8Cl_2$
151.04
1,5-Hexadiene, 1,6-dichloro- (7CI,9CI)
1,6-Dichloro-1,5-hexadiene

67562-39-4
$C_{12}HCl_7O$
409.31
Dibenzofuran, 1,2,3,4,6,7,8-heptachloro- (9CI)
1,2,3,4,6,7,8-Heptachlorodibenzofuran

67562-40-7
$C_{12}H_2Cl_6O$
374.86
Dibenzofuran, 1,2,4,6,7,8-hexachloro- (9CI)

67587-12-6
$C_9H_{17}NOS$
187.30
1-Piperidinecarbothioic acid, 4-methyl-, S-ethyl ester (9CI)

67597-34-6
$C_{29}H_{52}$
400.73
18,19-Dinorstigmastane, 5,14-dimethyl-, (5β,8α,9β, 10α,14β,20S,24XI)- (9CI)

67597-35-7
$C_{29}H_{52}$
400.73
18,19-Dinorstigmastane, 5,14-dimethyl-, (5β,8α,9β, 10α,14β,24XI)- (9CI)

67614-32-8
$C_{25}H_{22}ClNO_3$
419.90
Benzeneacetic acid, 4-chloro-α-(1-methylethyl)-, cyano-(3-phenoxyphenyl)methyl ester, (S-(R*,S*))- (9CI)

67614-33-9
$C_{25}H_{22}ClNO_3$
419.90
Benzeneacetic acid, 4-chloro-α-(1-methylethyl)-, cyano-(3-phenoxyphenyl)methyl ester, (R-(R*,R*))- (9CI)

67632-66-0
CCl_4O_4
217.82
Trichloromethyl perchlorate
Perchloric acid, trichloromethyl ester

67652-39-5
$C_{28}H_{35}NO_4.ClH$
486.04
4H-1-Benzopyran-4-one, 2-(4-(2-(dibutylamino)ethoxy)-3,5-dimethylbenzoyl)-, hydrochloride (9CI)

67696-25-7
$C_{22}H_{19}F_6O_2P$
460.36
Phosphorane, triphenylbis(2,2,2-trifluoroethoxy)-, (TB-5-11)- (9CI)

67700-45-2
Unknown
Unknown
Rosin, polymer with formaldehyde and phenol
Gum rosin modified phenol formaldehyde polymer
Paraformaldehyde, rosin, phenol polymer
Phenol, formaldehyde, gum rosin polymer
Phenol, formaldehyde, rosin polymer
Phenol, rosin, formaldehyde polymer
Phenol, rosin, formaldehyde resin
Rosin modified phenol, formaldehyde polymer
Rosin modified phenolic resin of phenol and formaldehyde
Rosin, formaldehyde, phenol polymer
Rosin, phenol, formaldehyde polymer
Rosin, phenol, formaldehyde reaction product

67700-99-6
$C_{33}H_{69}N$
479.92
Amines, di-C14-18-alkylmethyl
(C14-C18) Dialkylmethylamine
SDA 17-043-00

67701-06-8
$C_{15}H_{16}O_2$
228.29
Fatty acids, C14-18 and C16-18-unsatd.
C14-C18 and C16-C18 Alkylcarboxylic acid
(C14-C18) and (C16-C18) Alkylcarboxylic acid
(C14-C18) and (C16-C18)-Unsaturated alkyl carboxylic acid
(C14-C18) and (C16-C18)-Unsaturated alkylcarboxylic acid
(C14-C18) and C18 Unsaturated alkylcarboxylic acid

SDA 04-005-00

67704-68-1
$C_{11}H_{19}N_7$
249.37
Cyanamide, (4,6-bis((1-methylethyl)amino)-1,3,5-triazin-2-yl)methyl
Metazin
Metazine
Metazine (Pesticide)
Methazine

67708-83-2
$C_3H_5Br_2Cl$
236.33
Propane, dibromochloro- (9CI)

67711-86-8
O_7Pb_2SSi
586.55
Lead silicate sulfate (9CI)

67716-07-8
$C_{16}H_{26}O_3S$
298.45
Benzenepentanesulfonic acid, ε-pentyl- (9CI)

67725-14-8
Unknown
Unknown
Epi-Rez 5071 (9CI)

67730-10-3
$C_{10}H_8N_4$
184.22
Nc3ccc2nc1C=CC=Cn1c2n3
Dipyrido(1,2-a:3',2'-d)imidazole, 2-amino
2-Amino-dipyrido(1,2-a:3',2'-d)-imidazole
Dipyrido(1,2-a:3',2'-d)imidazol-2-amine
GLU-P-2

67730-11-4
$C_{11}H_{10}N_4$
198.25
CC1=CC=Cn2c1nc3ccc(N)nc23
Dipyrido(1,2-a:3',2'-d)imidazole, 2-amino-6-methyl
2-Amino-6-methyldipyrido(1,2-a:3',2'-d)imidazole
Dipyrido(1,2-a:3',2'-d)imidazol-2-amine, 6-methyl-
GLU-P-1
6-Me-GLU-P-2

67733-52-2
$C_{12}H_3Br_7$
706.48
1,1'-Biphenyl, 2,2',3,4,4',5,5'-heptabromo- (9CI)
2,2',3,4,4',5,5'-Heptabromobiphenyl
2,2',3,4,4',5,5'-PBB

67733-57-7
$C_{12}H_4Br_4O$
483.80
Dibenzofuran, 2,3,7,8-tetrabromo
2,3,7,8-Tetrabromodibenzofuran

67755-97-9
$C_5H_{10}O_2$
102.13
O=CC(O)C(C)C
Butanal, 2-hydroxy-3-methyl- (9CI)
2-Hydroxy-3-methylbutanal

67762-19-0
Unknown
Unknown
Poly(oxy-1,2-ethanediyl), α-sulfo-ω-hydroxy-, C10-16-alkyl ethers, ammonium salts
(C10-C16)Alcohol ethoxylate, sulfated, ammonium salt
(C10-C16)Alkyl alcohol, ethoxylate, sulfuric acid, ammonium salt
(C10-C16)-Alkyl alcohol ethoxylate sulfuric acid ammonium salt
(C13-C16)Alkyl ethoxylate sulfuric acid, ammonium salt
(C10-C16) Alkylethoxylate sulfuric acid, ammonium salt
SDA 15-067-01

67762-21-4
Unknown
Unknown
Poly(oxy-1,2-ethanediyl), α-sulfo-ω-hydroxy-, C10-16-alkyl ethers, magnesium salts
(C10-C16)-Alkyl alcohol ethoxylate sulfuric acid magnesium salt

67762-27-0
Unknown
Unknown
Cetearyl Alcohol
Alcohols, C16-18
(C16-C18) Alkyl alcohol
(C16-C18)-Alkyl alcohol
Almolan AE
Almolan Lis
Brookswax D
Brookswax R
Cetomacrogol Wax BP
Cetostearyl alcohol
Cetyl/stearyl alcohol
Cosmowax S
Crodacol CS-50
Incroquat CR CON
Incroquat CR Concentrate
Unette-O
Unette-W
Unicol CPS
Unicol 123
Unihydag WAX-O
Unihydag WAX-SX
Unimul-1002 Conc.
Unimulgade-F
Unimulgade-F Special
Unimulgade-1000NI
Upiwax 163
Upiwax 163 R

67762-39-4
Unknown
Unknown
Methyl esters of fatty acids (C8-C12)
(C6-C12) Alkylcarboxylic acid methyl ester
Caswell No. 568C
EPA Pesticide Chemical Code 079034
Fatty acids, C6-12, Me esters
Fatty acids, methyl esters

67762-41-8
Unknown
Unknown
Mixed fatty alcohols (C10-C16)
Alcohols, C10-16
C10-C16 Alkyl alcohol
(C10-C16) Alkyl alcohol
SDA 15-060-00

67762-72-5
Unknown
Unknown
2,5-Pyrrolidinedione, 1-(2-((2-((2-((2-aminoethyl)amino)ethyl)amino)ethyl)amino)ethyl)-, monopolyisobutenyl derivs.
Polyisobutenyl tetraethylenepentamine succinimide

67762-92-9
Unknown
Unknown
Silicone Y-6607
Siloxanes and silicones, dimethyl, (dimethylamino)-terminated

67763-87-5
Unknown
Unknown
Bleomycin, hydrochloride
Bleomycin chlorhydrate

67774-32-7
Unknown
Unknown

Firemaster FF-1
2,4,5,2',4',5'-Hexabromobiphenyl
PBB
Polybrominated biphenyl
Polybrominated biphenyl (FF-1)

67774-74-7
Unknown
Unknown
Benzene, C10-13-alkyl derivs.
Alkyl(C10-C13)benzene

67784-77-4
Unknown
Unknown
Quaternary ammonium compounds, bis(hydroxyethyl)-methyltallow alkyl, chlorides
N,N-Bis(2-hydroxyethyl)-N-methyl-N-tallow ammonium chloride
Methylbis(2-hydroxyethyl)-(tallowalkyl)ammonium chloride

67786-03-2
$C_{28}H_{28}O_6$
460.53
O(C1COc(ccc(c2)C(c(c(OCC(O3)C3)ccc4)c4)c(ccc(OCC(O5)C5)c6)c6)c2)C1
Oxirane, 2,2'-(((2-(oxiranylmethoxy)phenyl)methylene)-bis(4,1-phenyleneoxymethylene))bis- (9CI)

67801-06-3
$C_8H_{12}N_2O.2ClH$
225.14
m-Phenylenediamine, 4-ethoxy-, dihydrochloride
1,3-Benzenediamine, 4-ethoxy-, dihydrochloride (9CI)
2,4-Diaminoethoxybenzene dihydrochloride

67815-88-7
$C_{25}H_{45}NO_8S.2Na$
565.67
Butanedioic acid, sulfo-, 4-(1-methyl-2-((1-oxo-9-octadecenyl)amino)ethyl) ester, disodium salt (9CI)
Disodium (2-oleoylamido-1-methylethyl)sulfosuccinate

67845-79-8
$C_6H_7NO.1/2H_2O_4S$
158.16
Phenol, 2-amino-, sulfate (2:1) (Salt) (9CI)
o-Aminophenol, hemisulfate
2-Aminophenol sulfate (2:1) (Salt)

67851-29-0
$C_9H_9NO_5$
211.17
Phenol, 2-methoxy-4-nitro-, acetate (ester) (9CI)

67859-39-6
$C_{22}H_{44}O_6S.Na$
459.64
Octadecanoic acid, 10-(sulfooxy)-, 1-(2-methylpropyl) ester, sodium salt (9CI)
Isobutyl oleate, sulfate, sodium salt

67859-51-2
$C_{10}H_{12}N_2O_8Zn.2H_4N$
389.67
Zincate(2-), ((N,N'-1,2-ethanediylbis(N-(carboxymethyl)-glycinato))(4-)-N,N',O,O',ON,ON')-, diammonium, (OC-6-21-)- (9CI)
Ammonium zinc edetate
(Ethylenedinitrilo)tetraacetato zincate(2-), diammonium salt
Glycine, N,N'-1,2-ethanediyl-bis(N-carboxymethyl)-, diammonium zinc salt

67859-63-6
$C_{42}H_{80}O_4S_2Sn$
831.94
9-Octadecenoic acid (Z)-, (dimethylstannylene)bis(thio-2,1-ethanediyl) ester (9CI)
Dimethyltinbis(2-oleoyloxy-ethylmercaptide)

67859-64-7
$C_{42}H_{76}O_4S_2Sn$
827.91
9,12-Octadecadienoic acid (Z,Z)-, (dimethylstannylene)bis(thio-2,1-ethanediyl) ester (9CI)
Dimethyltinbis(2-linoleoyloxy-ethylmercaptide)

67860-04-2
$C_{19}H_{38}O$
282.51
O(C1CCCCCCCCCCCCCCCC)C1
Oxirane, heptadecyl- (9CI)
1,2-Epoxynonadecane
Heptadecyloxirane
1,2-Nonadecane oxide

67874-35-5
$C_{10}H_{24}N_2$
172.30
Isodecanediamine (9CI)
Isodecyldiamine

67874-38-8
$C_{26}H_{48}O_2$
392.67
9,12-Octadecadienoic acid (Z,Z)-, isooctyl ester (9CI)
Isooctyl linoleate

67874-55-9
$CH_5NO_7S_2.Na$
230.17
Imidodisulfuric acid, methoxy-, monosodium salt (9CI)
Monosodium methoxyimidodisulfurate
Monosodium O-methyl-hydroxylamine-N,N-disulfonate

67880-17-5
$C_2H_2N_8O_2$
170.04
Hydrazine dicarbonic acid diazide
1,2-Hydrazinedicarbonyl diazide

67881-24-7
$(C_2H_4O)xC_{18}H_{30}O$
Polymer NA
Poly(oxy-1,2-ethanediyl), α-(4-tert-dodecyl-phenyl)-ω-hydroxy- (9CI)

67883-07-2
$C_8H_9Cl_2NO$
206.06
Ethanamine, 2-(3,5-dichloro-phenoxy)- (9CI)

67883-08-3
$C_{10}H_{10}Cl_2O_3$
249.09
Butanoic acid, 3-(3,5-dichloro-phenoxy)- (9CI)

67888-96-4
$C_{12}H_5Br_5$
548.69
1,1'-Biphenyl, 2,2',4,5,5'-pentabromo- (9CI)
2,2',4,5,5'-PBB
2,2',4,5,5'-Pentabromo-biphenyl

67893-02-1
$C_{21}H_{34}O_2$
318.50
O=C(OC)C(C(C(C(C(=C1)CC(C2)C(C)C)C2)(CC3)C)C1)(C3)C
1-Phenanthrenecarboxylic acid, 1,2,3,4,4a,4b,5,6,7,8,10,10a-dodecahydro-1,4a-dimethyl-7-(1-methylethyl)-, methyl ester (9CI)

Methyl dihydroabietate

67905-27-5
$C_{28}H_{58}O.1/3Al$
419.76
1-Octacosanol, aluminum salt (9CI)
Aluminum n-octacosoxide

67905-28-6
$C_{26}H_{54}O.1/3Al$
391.71
1-Hexacosanol, aluminum salt (9CI)
Aluminum n-hexacosoxide

67905-29-7
$C_{24}H_{50}O.1/3Al$
363.65
1-Tetracosanol, aluminum salt (9CI)
Aluminum n-tetracosoxide

67905-30-0
$C_{22}H_{46}O.1/3Al$
335.60
1-Docosanol, aluminum salt (9CI)
Aluminum n-docosoxide

67905-31-1
$C_{20}H_{42}O.1/3Al$
307.54
1-Eicosanol, aluminum salt (9CI)
Aluminum n-eicosoxide

67905-32-2
$C_{14}H_{30}O.1/3Al$
223/38
1-Tetradecanol, aluminum salt (9CI)
Aluminum n-tetradecoxide

67905-86-6
$(C_{18}H_{37}N.C_4H_{13}N_3.C_2H_4Cl_2)x$
Polymer
N-Polyethylenepolyamine-N-oleylamine hydrochloride
Diethylenetriamine, ethylenedichloride, oleylamine polymer
1,2-Ethanediamine, N-(2-aminoethyl)-, Polymer with 1,2-dichloroethane and (Z)-9-octadecen-1-amine (9CI)

67923-88-0
$C_{16}H_{16}O_2$
240.30
9,10-Anthracenediol, ethyl-9,10-dihydro- (9CI)
Ethyl-9,10-dihydro-9,10-anthracenediol

67932-85-8
$C_{15}H_{18}ClNO_4$
311.76
Acetamide, N-(3-chloro-2,6-dimethylphenyl)-2-methoxy-N-(tetrahydro-2-oxo-3-furanyl)- (9CI)
CGA 80000

67952-57-2
$C_{10}H_{20}O_2$
172.27
O=C(OC(CCCC(C)C)C)C
2-Heptanol, 6-methyl-, acetate (9CI)
1,5-Dimethylhexyl acetate
6-Methyl-2-heptanol acetate

67953-04-2
$C_9H_{21}N$
143.27
NC(CCC(CCC)C)C
2-Octanamine, 5-methyl- (9CI)
2-Amino-5-methyloctane
1,4-Dimethylheptylamine
5-Methyl-2-octanamine

67953-32-6
$C_6H_{13}NO_3$
147.17
O=C(OCC(C)C)NCO
Isobutyl N-(hydroxymethyl) carbamate
Carbamic acid, (hydroxymethyl)-, 2-methylpropyl ester (9CI)
(Hydroxymethyl)carbamic acid, isobutyl ester

67953-54-2
$(C_4H_{13}N_3.C_3H_5ClO.C_2H_4N_4)x$
Polymer
Guanidine, cyano-, Polymer with N-(2-aminoethyl)-1,2-ethanediamine and (chloromethyl)oxirane (9CI)
Diethylenetriamine, cyanoguanidine, epichlorohydrin polymer

67953-76-8
$C_2H_8O_7P_2.xK$
UNKOWN
Phosphonic acid, (1-hydroxyethylidene)bis-, potassium salt (9CI)
(1-Hydroxyethylidene)bisphosphonic acid, potassium salt
Hydroxyethylidene diphosphonic acid, potassium salt
1-Hydroxylethanediphosphonic acid, potassium salt
Phosphonic acid, (1-hydroxyethylidene) bis-, potassium salt

67969-81-7
$C_{10}H_{19}ClO_3$
222.71
O=C(OC(C)COCC(C)C)C(Cl)C
Propanoic acid, 2-chloro-, 1-methyl-2-(2-methylpropoxy)ethyl ester (9CI)

67989-23-5
$C_{27}H_{42}O_6$
462.63
1,2,4-Benzenetricarboxylic acid, decyl octyl ester (9CI)

67990-32-3
$C_{13}H_4Br_6O_3$
687.60
O=C(Oc(c(cc(c1)Br)Br)c1Br)Oc(c(cc(c2)Br)Br)c2Br
Phenol, 2,4,6-tribromo-, carbonate (2:1) (9CI)
Bis(2,4,6-tribromophenyl) carbonate
2,4,6-Tribromophenol carbonate (2:1)

68002-90-4
$C_{13}H_{26}O_2$
214.35
Fatty acids, C10-16
(C10-C16)Alkylcarboxylic acid
(C10-C16) Carboxylic acid

68003-17-8
$C_{14}H_{28}O_6S_2.2Na$
402.48
1-Tetradecenedisulfonic acid, disodium salt (9CI)
Disodium 1-tetradecenedisulfonate

68006-83-7
$C_{12}H_{11}N_3$
197.26
Cc1cc2c(nc1N)[nH]c3ccccc23
9H-Pyrido(2,3-b)indole, 2-amino-3-methyl
2-Amino-3-methyl-α-carboline
2-Amino-3-methyl-1H-pyrido-(2,3-b)indole
2-Amino-3-methyl-9H-pyrido-(2,3-b)indole
Glob-P-1
3-Methyl-1H-pyrido(2,3-b)-indol-2-amine

68012-07-7
$(C_{12}H_{18}N_2O_4)x$
Polymer NA
2,4-Imidazolidinedione, 5-ethyl-5-methyl-1,3-bis(oxiranylmethyl)-, Homopolymer (9CI)
XU 238

68013-64-9
$CMnO_3.H$
115.95
Manganese, (carbonato(2-)-O)-, monohydrogen (9CI)

68015-98-5
$C_8H_{12}N_2O.H_2O_4S$
250.26
4-Ethoxy-m-phenylenediamine sulfate
1,3-Benzenediamine, 4-ethoxy-, sulfate (1:1) (9CI)
4-Ethoxy-1,3-benzenediamine sulfate
4-Ethoxy-1,3-benzenediamine sulfate (1:1)

68019-78-3
$C_4H_5N_3O_3$
143.10
1H-Imidazole, 4-methoxy-5-nitro- (9CI)
5-Methoxy-4-nitroimidazole

68037-39-8
Unknown
Unknown
Ethene, Homopolymer, chlorinated, chlorosulfonated
Polyethylene, chlorosulfonated

68037-49-0
$C_{14}H_{29}NaO_3S$
300.44
Sodium C14-17 sec alcohol sulfonate
(C10-C18)Alkylsulfonic acid, sodium salt
n-Alkyl(C10-C18)sulfonic acids, sodium salts
Sodium C14-17 alcohol sulfonate
Sulfonic acids, C10-18-alkane, sodium salts

68038-71-1
Unknown
Unknown
Bitoxibacillin
Agritol
Bacillus thuringensis
Bacillus thuringiensis
Bacillus thuringiensis Berliner
Bacillus thuringiensis (Berliner) var. aizawai
Bacillus thuringiensis (Berliner) var. israelensis
Bacillus thuringiensis (Berliner) var. kurstaki
Bacillus thuringiensis (Berliner) var. san diego
Bacillus thuringiensis (Berliner) var. tenebrionis
Bakthane
Biotrol
Bitoksybacillin
BTB
Caswell No. 066
Dipel
EPA Pesticide Chemical Code 006401
Larvatrol
Thuricide

68052-04-0
$C_{24}H_{46}O_4$
398.63
Hexanedioic acid, 2-ethylhexyl isodecyl ester (9CI)
2-Ethylhexyl isodecyl adipate
2-Ethylhexyl isodecyl hexanedioate

68052-23-3
$C_{22}H_{26}O_4$
354.45
O=C(OCC(C)(C)C(OC(=O)c(cccc1)c1)C(C)C)c(cccc2)c2
1,3-Pentanediol, 2,2,4-trimethyl-, dibenzoate (9CI)
2,2,4-Trimethyl-1,3-pentanediol, dibenzoate

68070-94-0
Unknown
Unknown
Dextrin, hydrogen 1-octenylbutanedioate (9CI)
Starch hydrogen 1-octenylsuccinate, dextrinized

68072-38-8
$(C_{15}H_{24}O.C_3H_6O_2)x$
Polymer
Oxiranemethanol, Polymer with nonylphenol (9CI)
Nonyl phenol, glycidyl polyether

68085-85-8
$C_{23}H_{19}ClF_3NO_3$
449.88
CC1(C)C(C=C(Cl)C(F)(F)F)C1C(=O)OC(C#N)c3cccc(Oc2ccccc2)c3
Cyclopropanecarboxylic acid, 3-(2-chloro-3,3,3-trifluoro-1-propenyl)-2,2-dimethyl-, cyano(3-phenoxyphenyl)-methyl ester
Cyhalothrin
Cyhalothrine
Grenade
ICI 146814
ICI-PP 563
PP 563

68122-86-1
Unknown
Unknown
Imidazolium compounds, 4,5-dihydro-1-methyl-2-nortallow alkyl-1-(2-tallow amidoethyl), Me sulfates
1-Methyl-1-(tallow alkyl amido)-2-nor(tallow alkyl)-2-imidazolinium, methylsulfate
1-Methyl-1-tallowalkylamidoethyl-2-tallowalkylimidazoline methosulfate
1-Methyl-1-(2-tallowamidoethyl)-2-tallowimidazolinium methylsulfate
1-(2-Tallow amidoethyl)-1-methyl-2-nor(tallow alkyl)-2-imidazolinium methyl sulfate
1-(2-Tallowamidoethyl)-1-methyl-2-tallowalkylimidazolinium methylsulfate

68127-33-3
$C_{21}H_{36}O_4.xK$
Unknown
2-Cyclohexene-1-octanoic acid, 5(or 6)-carboxy-4-hexyl-, potassium salt (9CI)
5(6)-Carboxy-4-hexyl-2-cyclohexene-1-octanoic acid, potassium salt

68131-39-5
Unknown
Unknown
C12-15 Pareth-12
Alcohols, C12-15, ethoxylated
C12-15 Pareth-2
C12-15 Pareth-3
C12-15 Pareth-4
C12-15 Pareth-5
C12-15 Pareth-7
C12-15 Pareth-9
Linear (C12-C15) alkyl alcohols, ethoxylated
Linear primary alcohol (C12-C15) ethoxylate
Linear primary(C12-C15)-alcohol, ethoxylate
Neodol-12
Pareth-25-2
Pareth-25-3
Pareth-25-4
Pareth-25-5
Pareth-25-7
Pareth-25-9
Pareth-25-12
Polyethoxylated (C12-C15) linear primary saturated alcohols
Polyethylene glycol, (C12-C15) alkyl ethers
Polyethylene glycol, linear (C12-C15)alkyl alcohols ether
Poly(ethylene oxide) ether with (C12-C15)linear primary alcohols

68131-40-8
Unknown
Unknown
Linear alchohol ethoxylated C12Eq
Alcohols, C11-15-secondary, ethoxylated
Alcohols, C11-15-secondary, ethoxylated
C11-15 Pareth-20
C11-15 Pareth-3
C11-15 Pareth-30
C11-15 Pareth-40
C11-15 Pareth-5
C11-15 Pareth-7
C11-15 Pareth-9
Linear random secondary alcohol (C11-C15) ethoxylate
Linear secondary alcohol (C11-C15) ethoxylate
Linear secondary(C11-C15)-alcohol, ethoxylate
Pareth 15
Pareth-15-20
Pareth-15-3
Pareth-15-30
Pareth-15-40
Pareth-15-5
Pareth-15-7
Pareth-15-9
Tergitol 15S
Tergitol 15-S-3
Tergitol 15-S-5
Tergitol 15-S-9
Tergitol 15-S-15
Tergitol 15-S-20
Tergitol 15-S-7 (Nonionic)
Tergitol 15-S-9 (Nonionic)
Tergitol 15-S-12 (Nonionic)

68131-74-8
Unknown
Unknown
Coal-Fly-Ash
Ashes (Residues)
Coal Ash

68132-21-8
Unknown
Unknown
Perilla Seed Oil
Oil, Misc.
Oils, Perilla
Perilla Oil
Shiso Oil

68134-06-5
$C_9H_{18}O_2$
158.24
O(C1COC(CC(C)C)C)C1
Oxirane, ((1,3-dimethylbutoxy)methyl)- (9CI)
((1,3-Dimethylbutoxy)methyl)-oxirane
Oxirane, ((1,3-dimethylbutyloxy)methyl)-

68134-07-6
$C_{11}H_{22}O_2$
186.29
O(C1COCCCCCC(C)C)C1
Oxirane, (((6-methylheptyl)oxy)methyl)- (9CI)
(((6-Methylheptyl)oxy)methyl)-oxirane
Oxirane, ((6-methylheptyloxy)-methyl)-

68137-05-3
C_7H_7ClO
142.59
Phenol, 2-chloromethyl- (9CI)

68137-08-6
$C_7H_7NO_3$
306.28
Phenol, 3(or 4)-methyl-2-nitro- (9CI)

68137-09-7
$C_7H_7NO_3$
306.28
Phenol, 2(or 4)-methyl-3-nitro- (9CI)

68153-30-0
Unknown
Unknown
Quaternary ammonium compounds, benzylbis(hydrogenated tallow alkyl)methyl, chlorides, Compds. with bentonite
Bis(hydrogenated tallow alkyl)-benzylmethylammonium bentonite
Methylbenzylbis(hydrogenated tallow)ammonium bentonite

68153-35-5
Unknown
Unknown
Ethanaminium, 2-amino-N-(2-aminoethyl)-N-(2-hydroxyethyl)-N-methyl-, N,N'-ditallow acyl derivs., Me sulfates (Salts)
N,N-Bis(2-tallowamidoethyl)-N-(2-hydroxyethyl)-N-methylammonium methylsulfate
N,N-Di(2-tallow amidoethyl)-N-(2-hydroxyethyl)-N-methyl ammonium methyl sulfate
N,N-Di(2-tallowamidoethyl)-N-(2-hydroxyethyl)-N-methylammonium methylsulfate
N-((Tallow alkyl)amidoethyl)-N-((tallow alkyl)ethyl)methyl ethylammonium, methosulfate

68153-81-1
Unknown
Unknown
Grease
Animal Grease, Inedible
Grease (Animal)

68155-37-3
Unknown
Unknown
1-(Alkyl amino)-3-aminopropane
N-(C12-C18)Alkyl propylenediamine
(C12-C18) N-Alkylpropylenediamine
Amines, N-C12-18-alkyltrimethylenedi-

68155-42-0
Unknown
Unknown
1-((Coco alkyl)amino)-3-aminopropane adipate
Adipic acid, N-cocoalkyl-1,3-propylenediamine salt
Adipic acid, N-(coconut oil alkyl)trimethylenediamine salt
Amines, N-coco alkyltrimethylenedi-, adipates

68155-43-1
Unknown
Unknown
1-((Coco alkyl)amino)-3-aminopropane hydroxyacetate
Amines, N-coco alkyltrimethylenedi-, glycolates
N-(Coconut oil alkyl)trimethylenediamine, hydroxyacetate
Hydroxyacetic acid, cocopropylenediamine salt

68171-29-9
$C_6H_{18}NO_{12}P_3.xNa$
Unknown
Ethanol, 2,2',2''-nitrilotris-, tris(dihydrogen phosphate) (ester), sodium salt (9CI)
Nitrilotris(ethyl phosphate), sodium salt

68186-36-7
Unknown
Unknown
Poly(oxy-1,2-ethanediyl), α-tridecyl-ω-hydroxy-, phosphate, potassium salt (9CI)
Poly(oxy-1,2-ethanediyl)-α-tridecyl-ω-hydroxy-, phosphated, potassium salt
Polyethylene glycol, tridecyl ether, phosphate, potassium salt
Tridecyl alcohol, ethoxylate, phosphate, potassium salt

Tridecyl alcohol, ethoxylated and phosphated, potassium salt
Tridecyl alcohol, ethoxylated, phosphate, potassium salt
Tridecyl alcohol, ethoxylated, phosphated, potassium salt
Tridecylalcohol, ethoxylated, phosphated, potassium salt

68187-41-7
$C_{14}H_{31}O_2PS_2$
326.50
Phosphorodithioic acid, O,O-di-C1-14-alkyl esters
Dithiophosphoric acid, O,O-di(C1-C14)alkyl ester

68187-58-6
Unknown
Unknown
Pitch, Petroleum, Arom.
Aromatic Petroleum Pitch
Petroleum Pitch

68187-63-3
Unknown
Unknown
3-Alkoxy-2-hydroxypropyl trimethyl ammonium chloride
3-(C12-C15) Alkoxy-2-hydroxypropyltrimethylammonium chloride
1-Propanaminium, 2-hydroxy-N,N,N-trimethyl-, 3-(C12-15-alkyloxy) derivs., chlorides

68187-76-8
Unknown
Unknown
Sulfonated Castor Oil, sodium salt
Castor Oil, Sulfate, sodium salt
Castor Oil, Sulfated, sodium salt
Turkey Red Oil, Sodium salt

68188-15-8
Unknown
Unknown
Slimes and sludges, Papermaking, Secondary
Activated sludge, Dried
Secondary sludge
Waste treatment plant sludge

68188-29-4
Unknown
Unknown
1-((Coco alkyl)amino)-3-aminopropane benzoate
Amines, N-coco alkyltrimethylenedi-, benzoates
Benzoic acid, N-cocoalkyltrimethylenediamine salt
N-Cocoalkyltrimethylendiamine benzoate
N-(Coconut oil alkyl)-trimethylenediamine, benzoic acid salt

68188-83-0
Unknown
Unknown
Rare earth oxides

68188-96-5
$C_6H_{20}N_2O_{12}P_4.4K$
592.51
Phosphonic acid, (1,2-ethanediylbis(nitrilobis(methylene)))tetrakis-, tetrapotassium salt (9CI)
Ethylene bis(nitrilodimethylene)tetraphosphonic acid, tetrapotassium salt

68189-23-1
$C_{23}H_{25}N_3$
343.45
N(N(c(cccc1)c1)c(cccc2)c2)=Cc(ccc(N(CC)CC)c3)c3
Benzaldehyde, 4-(diethylamino)-, diphenylhydrazone (9CI)
p-Diethylaminobenzaldehyde, 1,1-diphenylhydrazone

68191-07-1
$C_7H_6N_2O_5$
198.12
Phenol, 4-methyl-2,3-dinitro- (9CI)

68194-05-8
$C_{12}H_5Cl_5$
326.44
Clc1ccc(c(Cl)c1)c2c(Cl)ccc(Cl)c2Cl
1,1'-Biphenyl, 2,2',3,4',6-pentachloro- (9CI)
2,2',3,4',6-PCB
2,2',3,4',6-Pentachlorobiphenyl

68194-07-0
$C_{12}H_5Cl_5$
326.44
1,1'-Biphenyl, 2,2',3,4',5-pentachloro- (9CI)
2,2',3,4',5-PCB
2,2',3,4',5-Pentachlorobiphenyl

68194-08-1
$C_{12}H_4Cl_6$
360.88
1,1'-Biphenyl, 2,2',3,4',6,6'-hexachloro- (9CI)
2,2',3,4',6,6'-Hexachlorobiphenyl
2,2',3,4',6,6'-PCB

68194-12-7
$C_{12}H_5Cl_5$
326.44
Clc1cc(Cl)cc(c1)c2cc(Cl)c(Cl)cc2Cl
2,3',4,5,5'-Pentachlorobiphenyl

68194-13-8
$C_{12}H_4Cl_6$
360.88
1,1'-Biphenyl, 2,2',3,4',5,6-hexachloro- (9CI)
2,2',3,4',5,6-Hexachlorobiphenyl
2,2',3,4',5,6-PCB

68194-14-9
$C_{12}H_4Cl_6$
360.88
1,1'-Biphenyl, 2,2',3,4,5',6-hexachloro- (9CI)
2,2',3,4,5',6-Hexachlorobiphenyl
2,2',3,4,5',6-PCB

68194-16-1
$C_{12}H_3Cl_7$
395.32
1,1'-Biphenyl, 2,2',3,3',4,5,6-heptachloro- (9CI)
2,2',3,3',4,5,6-Heptachlorobiphenyl
2,2',3,3',4,5,6-PCB

68194-17-2
$C_{12}H_2Cl_8$
429.77
1,1'-Biphenyl, 2,2',3,3',4,5,5',6-octachloro- (9CI)
2,2',3,3',4,5,5',6-Octachlorobiphenyl
2,2',3,3',4,5,5',6-PCB

68201-79-6
$C_{44}H_{82}O_4$
675.13
O=C(OCC(C)(C)C(OC(=O)CCCCCCC=CCCCCCCCC)C(C)C)CCCCCCCC=CCCCCCCCC
9-Octadecenoic acid (Z)-, 2,2-dimethyl-1-(1-methylethyl)-1,3-propanediyl ester (9CI)
2,2,4-Trimethyl-1,3-pentanediol, dioleate

68201-84-3
$C_{18}H_{32}O_5S$
360.52
O=C(O)CCCCCCCC=CCC(S(=O)(=O)O)=CCCCCC
9,12-Octadecadienoic acid, 12-sulfo-, (?,Z)- (9CI)
12-Sulfolinoleic acid
(?,Z)-12-Sulfo-9,12-octadecadienoic acid

68213-23-0
Unknown
Unknown
Alcohols, C12-18, ethoxylated
C12-18 Alkyl alcohol ethoxylate
(C12-C18) Alkyl alcohol ethoxylate
Poly(oxy-1,2-ethanediyl), α-(C12-C18) alkyl-ω-hydroxy-

68215-98-5
$C_{14}H_{18}ClN_5O$
307.75
1,3,5-Triazine-2,4-diamine, 6-(5-chloro-2-methoxyphenyl)-N,N-diethyl- (9CI)

68228-10-4
$C_{13}H_{15}N$
185.26
Quinoline, diethyl- (9CI)
Diethylquinoline

68238-35-7
Unknown
Unknown
Keratin
Cheratina (Italian)
Detoxin

68239-80-5
$C_6H_7ClN_2.H_2O_4S$
240.66
4-Chloro-1,3-benzenediamine sulfate
1,3-Benzenediamine, 4-chloro-, sulfate (1:1) (9CI)
4-Chloro-1,3-benzenediamine sulfate (1:1)
m-Phenylenediamine,4-chloro-, sulfate

68239-82-7
$C_6H_7N_3O_2.H_2O_4S$
251.20
4-Nitro-1,2-benzenediamine sulfate
1,2-Benzenediamine, 4-nitro-, sulfate (1:1) (9CI)
4-Nitro-1,2-benzenediamine sulfate (1:1)
4-Nitro-o-phenylenediamine, sulfate

68239-83-8
$C_6H_7N_3O_2.H_2O_4S$
251.20
2-Nitro-1,4-benzenediamine sulfate
1,4-Benzenediamine, 2-nitro-, sulfate (1:1) (9CI)
2-Nitro-1,4-benzenediamine sulfate (1:1)
2-Nitro-p-phenylenediamine, sulfate

68258-91-3
$C_5H_4Cl_6$
276.80
Cyclopentane, hexachloro- (9CI)
Hexachlorocyclopentane

68278-98-8
$C_{14}H_{24}NS.Cl$
273.86
Pyridinium, 1-((octylthio)-methyl)-, chloride (9CI)

68279-00-5
$C_{13}H_{25}N_2S.Cl$
276.86
1H-Imidazolium, 1-methyl-3-((octylthio)methyl)-, chloride (9CI)

68279-02-7
$C_{17}H_{33}N_2S.Cl$
332.97
1H-Imidazolium, 1-((dodecylthio)methyl)-3-methyl-, chloride (9CI)

68279-54-9
$C_{16}H_{18}O_2$
242.32
Oc(c(c(c(O)c1CCCC2)ccc3CC)c3)c12
9,10-Anthracenediol, 6-ethyl-1,2,3,4-tetrahydro- (9CI)
6-Ethyl-1,2,3,4-tetrahydro-9,10-anthracenediol
5,6,7,8-Tetrahydro-2-ethylanthrahydroquinone

68298-46-4
$C_{10}H_{13}NO$
163.21
O(c(c(ccc1)C2)c1N)C2(C)C
7-Benzofuranamine, 2,3-dihydro-2,2-dimethyl- (9CI)
7-Amino-2,3-dihydro-2,2-dimethylbenzofuran
2,3-Dihydro-2,2-dimethyl-7-benzofuranamine

68298-47-5
$C_{10}H_{11}N_2O.1/2O_4S$
223.22
7-Benzofurandiazonium, 2,3-dihydro-2,2-dimethyl-, sulfate (2:1) (9CI)
2,3-Dihydro-2,2-dimethyl-7-benzofurandiazonium, sulfate (2:1)

68299-16-1
$C_{15}H_{30}O_3$
258.40
tert-Amyl peroxyneodecanoate
tert-Amyl peroxyneodecanoate (Not more than 75% with phlegmatiser)
1,1-Dimethylpropyl neodecaneperoxoate
Neodecaneperoxoic acid, 1,1-dimethylpropyl ester (9CI)
UN 2891

68308-34-9
Unknown
Unknown
Shale Oils, Crude
Blue Oil
Crude Shale Oils
Green Oil
Raw Shale Oil
Shale Oil [UN 1288]
UN 1288 [Shale oil]
Unfinished Lubricating Oil

68308-67-8
Unknown
Unknown
Quaternary ammonium compounds, ethyldimethylsoya alkyl, Et sulfates
Dimethylethylsoyaammonium ethosulfate
(Soya alkyl)ethyl dimethyl ammonium ethyl sulfates
Soyadimethylethylammonium, ethylsulfate

68309-34-2
Unknown
Unknown
2-(Tall oil alkyl)-1-benzyl-1-(2-hydroxyethyl)-2-imidazolinium chloride
1-Benzyl-1-(2-hydroxyethyl)-2-tall oil alkyl-2-imidazolinium chloride
1-Benzyl-1-(2-hydroxyethyl)-2-(tall oil alkyl)-2-imidazolinium chloride
Imidazolium compounds, 1-benzyl-4,5-dihydro-1-(hydroxyethyl)-2-nortall-oil alkyl, chlorides

68315-17-3
$C_{18}H_{32}NS.Cl$
329.97
Pyridinium, 1-((dodecylthio)methyl)-, chloride (9CI)

68318-35-4
$C_{36}H_{29}N_7O_{12}S_3.3Na$
916.79
2,7-Naphthalenedisulfonic acid, 4-amino-3-((4'-((2,4-dihydroxyphenyl)azo)-3,3'-dimethyl(1,1'-biphenyl)-

4-yl)azo)-5-hydroxy-6-((4-sulfophenyl)azo)-, trisodium salt (9CI)

68333-81-3
Unknown
Unknown
Alkanes, C4-C12

68334-13-4
Unknown
Unknown
Ethyl hexyl tallate
2-Ethylhexyl ester of tall oil acids
2-Ethylhexyl tallate
Fatty acids, tall-oil, 2-ethylhexyl esters

68334-30-5
Unknown
Unknown
Diesel Fuel [NA 1993]
Automotive Diesel Oil
Diesel Oil (Petroleum)
Diesel Test Fuel
Fuel Oil, Diesel (DOT)
Fuels, Diesel
NA 1993 [Diesel fuel]
Olej Napedowy III (Polish)

68334-37-2
Unknown
Unknown
Shale, Expanded, Aggregates
Expanded shale
Expanded shale, Lightweight aggregate
Haydite

68334-50-9
$C_{12}H_{10}O_2$
186.21
(1,1'-Biphenyl)diol (VAN) (9CI)

68334-67-8
$C_7H_7Cl_5O$
284.40
O(C(=C(C(=C(Cl)Cl)Cl)Cl)Cl)C(C)C
1,1,2,3,4-Pentachloro-4-(1-methylethoxy)-1,3-butadiene
1,3-Butadiene, 1,1,2,3,4-pentachloro-4-(1-methylethoxy)-(9CI)
Isopropoxy pentachlorobutadiene

68359-37-5
$C_{22}H_{18}Cl_2FNO_3$
434.31
CC1(C)C(C=C(Cl)Cl)C1C(=O)OC(C#N)c3ccc(F)c(Oc2ccccc2)c3
Cyclopropanecarboxylic acid, 2-(2,2-dichlorovinyl)-3,3-dimethyl-, ester with (4-fluoro-3- phenoxyphenyl)-hydroxyacetonitrile
Bay FCR 1272
Baythroid
Baythroid H
Cyfluthin
Cyfluthrin
Cyfluthrine
Cyfoxylate
Eulan SP
FCR 1272
Responsar
Solfac
Tempo

68389-89-9
Unknown
Unknown
Poly(oxy-1,2-ethanediyl), α-(2-(bis(2-aminoethyl)methylammonio)ethyl)-ω-hydroxy-, N,N'-bis(hydrogenated tallow acyl) derivs., Me sulfates (Salts)
Methyl, hydrogenated tallow diethylene triamine condensate, polyethoxylated, methyl sulfate

68391-01-5
Unknown
Unknown
Quaternary ammonium compounds, benzyl-C12-18-alkyldimethyl, chlorides
Hyamine 3500

68391-03-7
Unknown
Unknown
Alkyl trimethyl ammonium chloride
(C12-C18)Alkyl trimethylammonium chloride
(C12-C18)Alkyltrimethylammonium chloride
((C12-C18)Alkyl)trimethylammonium chloride
Quaternary ammonium compounds, C12-18-alkyltrimethyl, chlorides

68391-11-7
Unknown
Unknown
Alkyl pyridines
Paraldehyde and ammonia reaction product
Pyridine, alkyl derivs.
Pyridine bases
Pyridines, polyalkylated, higher boiling fraction
Pyridines, polyalkylated, lower boiling fraction
Pyridines, polyalkylated: polyalkylated pyridines

68391-67-3
$C_6H_9FeNO_6$
246.99
Iron (2+) NTA

68398-19-6
$C_{18}H_{22}$
238.37
Benzene, ethyl(phenylethyl)-, mono-ar-ethyl deriv. (9CI)
Diethyl diphenylethane

68400-67-9
$(C_9H_6N_2O_2.C_4H_{10}O_2.(C_3H_6O)x C_4H_{10}O)x$
Polymer
1,3-Butanediol, Polymer with α-butyl-ω-hydroxypoly(oxy-(methyl-1,2-ethanediyl)) and 1,3-diisocyanatomethylbenzene (9CI)
Polyurethane polymer

68400-78-2
$C_{10}H_{15}N$
149.23
Benzenamine, ar-(1-methylpropyl)- (9CI)
sec-Butylaniline
ar-(1-Methylpropyl)benzenamine

68400-79-3
$C_{40}H_{60}P_2S_5$
763.19
Thiodiphosphonic acid (((HS)HP(S))2S), bis(2,6,6-trimethylbicyclo(3.1.1)hept-2-enyl)-, bis(2,6,6-trimethylbicyclo(3.1.1)hept-2-enyl) ester (9CI)

68402-20-0
$C_{16}H_8Cl_2$
271.14
Pyrene, dichloro- (9CI)

68406-57-5
$C_{13}H_{17}N_3S$
247.35
1H-Benzimidazole-2-sulfenamide, N-cyclohexyl- (9CI)

68409-94-9
Unknown
Unknown
Residues (Coal), Solvent -refining (SRC), Vacuum-distn.

68410-00-4
Unknown
Unknown
Distillates, Petroleum, Crude oil

68410-05-9
Unknown
Unknown
Light Crude Oil Distillate
Distillates, Petroleum, Straight-run light

68410-07-1
Unknown
Unknown
Distillates (Coal), Solvent-efining (SRC), Heavy

68410-08-2
Unknown
Unknown
Distillates (Coal), Solvent-refining (SRC), Recycle

68410-09-3
Unknown
Unknown
Distillates, Coal, Solvent -refining (SRC), Wash
SRC wash solvent

68410-69-5
Unknown
Unknown
Poly(oxy-1,2-ethanediyl), α-(2-(bis(2-aminoethyl)methyl-ammonio)ethyl)-ω-hydroxy-, N,N'-ditallow acyl derivs., Me sulfates (Salts)
Methyl tallow diethylenetriamine condensate, polyethoxylated, methyl sulfate

68410-72-0
Unknown
Unknown
Residues, Coal, Solvent -refining (SRC) filtration
SRC mineral residue

68411-00-7
$C_{10}H_{20}$
140.27
Alkenes, C>8

68411-15-4
Unknown
Unknown
Calcium, Acetate octanoate stearate complexes
Acetic acid, stearic acid, caprylic acid, calcium salts

68411-30-3
Unknown
Unknown
Benzenesulfonic acid, linear alkyl-, sodium salt
Linear Alkylbenzenesulfonate, sodium salt
LAS-Na
LAS, sodium salt
Straight-chain alkyl benzene sulfonate

68411-81-4
Unknown
Unknown
2-Imidazolidinone, 4,5-dihydroxy-1,3-bis(hydroxymethyl)-, methylated
4,5-Dihydroxy-1,3-bis(hydroxymethyl)-2-imidazolidinone, methylated
Dihydroxydimethylolethyleneurea, methylated

68412-04-4
Unknown
Unknown
1,6-Octadiene, 7-methyl-3-methylene-, acetylated
Acetylated myrcene

68412-58-8
Unknown
Unknown
Phosphorodithioic acid, Mixed hexyl and iso-Pr esters, zinc salts
Phosphorodithioic acid, hexyl isopropyl esters, zinc salt

68412-60-2
Unknown
Unknown
Phosphoric acid, Mixed decyl and Et and octyl esters
Phosphoric anhydride esters with octyl alcohol, decyl alcohol, ethanol

68413-24-1
Unknown
Unknown
Cashew, Nutshell liq., Polymer with epichlorohydrin
Cashew nutshell oil, Polymer with (chloromethyl)oxirane

68413-48-9
$C_{28}H_{55}O_6PS_2$
582.85
O=C(OCCCC)CC(SP(OCC(CCCC)CC)(OCC(CCCC)CC)=S)C(=O)OCCCC
Butanedioic acid, ((bis((2-ethylhexyl)oxy)phosphinothioyl)thio)-, dibutyl ester (9CI)
Phosphorodithioic acid, O,O-di-(2-ethylhexyl)-S-(1,2-dicarbobutoxyethyl) ester

68413-71-8
$C_8H_6Br_4O$
437.75
O(c(c(cc(c1)Br)Br)c1Br)CCBr
Benzene, 1,3,5-tribromo-2-(2-bromoethoxy)- (9CI)
1,3,5-Tribromo-2-(2-bromoethoxy)benzene
1-(2,4,6-Tribromophenoxy)-2-bromoethane

68413-83-2
$C_8H_{17}NO_5$
207.22
O=C(OCCOC(C)C)N(CO)CO
Carbamic acid, bis(hydroxymethyl)-, 2-(1-methylethoxy)ethyl ester (9CI)
N,N-Dimethylol isopropoxyethyl carbamate

68424-92-0
Unknown
Unknown
Alkyl trimethyl ammonium bromide
(C14-C18) Alkyltrimethylammonium bromide
Quaternary ammonium compounds, C14-18-alkyltrimethyl, bromides

68424-94-2
Unknown
Unknown
Alkyl amino betaine
Betaines, coco alkyldimethyl

68425-29-6
Unknown
Unknown
Gasoline Blending Stock, Alkylates
Distillates, Petroleum, Naphtha Raffinate Pyrolyzate-derived, Gasoline-blending
Gasoline Blend Stock
Gasoline Blending Stock, Reformates

68425-61-6
$C_{16}H_{20}O_3S.C_6H_{13}N$
391.57
Naphthalenesulfonic acid, bis-(1-methylethyl)-, Compd. with cyclohexanamine (1:1) (9CI)
Bis(1-methylethyl)naphthalenesulfonic acid, cyclohexylamine salt

68439-45-2
Unknown
Unknown
Alcohols, C6-12, ethoxylated
C6-C12 Alkyl alcohol ethoxylate
(C6-C12) Alkylalcohol ethoxylate
C6-C12 (Alkyl) alcohol, ethoxylated

68439-49-6
Unknown
Unknown
Alcohols, C16-18, ethoxylated
Aliphatic (C16-C18)alcohol, ethoxylated
(C16-C18) Alkyl alcohol ethoxylate
Cetomacrogolum (Latin)
(C16-C18) Fatty alcohol, ethylene oxide reaction product
SDA 19-065-00

68439-50-9
Unknown
Unknown
Alcohols, C12-14, ethoxylated
Linear (C12 and C14) alkyl alcohols, ethoxylated

68439-51-0
Unknown
Unknown
Alcohols, C12-14, ethoxylated propoxylated
Linear (C12-C14) alkyl alcohols, ethoxylated, propoxylated

68441-14-5
Unknown
Unknown
1,3-Butadiene, 2-methyl-, Polymer with 2-methyl-1-propene, brominated
Brominated butyl rubber
Butyl rubber, brominated
Isobutylene, isoprene polymer, brominated

68442-12-6
Unknown
Unknown
9-Octadecenoic acid (Z)-, 2-mercaptoethyl ester, Reaction products with dichlorodimethylstannane, sodium sulfide(Na2S) and trichloromethylstannane

68442-22-8
Unknown
Unknown
Phosphorodithioic acid, Mixed O,O-bis(2-ethylhexyl and iso-Bu) esters, zinc salts

68442-97-7
Unknown
Unknown
1H-Imidazole-1-ethanamine, 4,5-dihydro-, 2-nortall-oil alkyl derivs.
1-Aminoethyl-2-tall oil alkyl-2-imidazoline
Tall oil, diethylenetriamine imidazoline

68443-05-0
Unknown
Unknown
Sodium sulfonate oleic acid
9-Octadecenoic acid (Z)-, sulfonated, sodium salts
Sulfonated oleic acid, sodium salt

68444-05-3
$C_{16}H_{26}N_2O_4$
310.38
O=C(N(C(C1=O)(CC(CC)C)CC)CC(O2)C2)N1CC(O3)C3
2,4-Imidazolidinedione, 5-ethyl-5-(2-methylbutyl)-1,3-bis(oxiranylmethyl)- (9CI)
1,3-Bis(oxiranylmethyl)-5-ethyl-5-(2-methylbutyl)hydantoin

68455-92-5
$C_{14}H_9NO_2$
223.22
Phenanthrene, nitro- (9CI)

68457-79-4
Unknown
Unknown
Phosphorodithioic acid, Mixed O,O-bis(iso-Bu and pentyl) esters, zinc salts
Dithiophosphoric acid, O,O'-isobutyl amyl ester, zinc salt
Zinc dialkyl dithiophosphorate

68458-86-6
Unknown
Unknown
Ketones, C14

68459-31-4
Unknown
Unknown
Fatty acids, C9-11-branched, glycidyl esters, Polymers with castor oil, formaldehyde, 6-phenyl-1,3,5-triazine-2,4-diamine and phthalic anhydride
Alkyd resin

68459-98-3
$C_6H_7ClN_2.H_2O_4S$
240.66
1,2-Benzenediamine, 4-chloro-, sulfate (1:1) (9CI)
4-Chloro-1,2-benzenediamine sulfate (1:1)
4-Chloro-o-phenylenediamine, monosulfate

68460-03-7
$C_9H_{15}Cl_6O_4P$
430.91
O=P(OCC(CCl)Cl)(OC(CCl)CCl)OC(CCl)CCl
Bis(2-chloro-1-(chloromethyl)-ethyl) 2,3-dichloropropyl phosphate
Bis(1,3-dichloro-2-propyl)-2,3-dichloro-1-propyl phosphate
Phosphoric acid, bis(2-chloro-1-(chloromethyl)ethyl) 2,3-dichloropropyl ester (9CI)

68475-57-0
Unknown
Unknown
Alkanes, C1-2

68475-58-1
C_5H_{14}
74.17
Ethane - propane mixture, Refrigerated liquid [NA 1961]
Alkanes, C2-3
Ethane propane mixture
Field ethane
NA 1961 [Ethane-Propane mixture, Refrigerated liquid (Cryogenic liquid)]

68475-59-2
C_4H_{10}
58.12
Alkanes, C3-4
Butane-propane mixture

68476-30-2
Unknown
Unknown
Fuel Oil, No. 2
API No.2 Fuel Oil
Gas Oil
Home Heating Oil No.2
#2 Home Heating Oils
Number 2 Burner fuel
Number 2 Fuel Oil

68476-31-3
Unknown
Unknown
Fuel Oil, No. 4
Caswell No. 333AB
Cat Cracker Feed Stock

Diesel Fuel Oil #2
EPA Pesticide Chemical Code 063514
No. 4 Fuel Oil
Residual Fuel Oils, Heavy

68476-32-4
Unknown
Unknown
Fuel Oil, Residues-straight-run gas oils, high-sulfur
Bunker "C"
High sulfur fuel oil

68476-33-5
Unknown
Unknown
Fuel Oil, Residual
Residual(Heavy) Fuel Oil

68476-34-6
Unknown
Unknown
Diesel Fuel No.2
Fuels, Diesel, No. 2
No. 2 Diesel Fuel

68476-54-0
Unknown
Unknown
Polymerization Feed
Hydrocarbons, C3-5, Polymn. Unit Feed

68476-78-8
Unknown
Unknown
Molasses
Beet Molasses
Beet Sugar Molasses
Molasses, Beet

68476-79-9
Unknown
Unknown
Naphtha, Coal, Solvent-refining
SRC naphtha

68476-85-7
Unknown
Unknown
L.P.G.
Gases de Petroleo (Spanish)
Gaz de Petrole (French)
L.P.G. (ACGIH,OSHA)
Liquefied Petroleum Gas (DOT,OSHA)
Petroleum Gas Liquefied
Petroleum Gases
Petroleum Gases, Liquefied
UN 1075 [Petroleum gases, Liquefied see also Liquefied petroleum gas]

68476-95-9
Unknown
Unknown
Shale, Expanded
Rotary kiln produced expanded shale lightweight aggregate

68476-96-0
Unknown
Unknown
Slags, Coal
Boiler slag
Coal slag
Electric utility boiler slag (Coal)

68477-31-6
Unknown
Unknown
Aromatic petroleum derivative solvent
Aromatic solvent (Petroleum)
Caswell No. 054
Distillates, Petroleum, Catalytic reformer fractionator residue, Low-boiling
EPA Pesticide Chemical Code 006501

68477-58-7
Unknown
Unknown
Distillates, Petroleum, Steam-cracked petroleum distillates, C5-18 fraction
(C5-C18) Partial fraction (Petroleum)

68477-83-8
Unknown
Unknown
Alkylation Feed
Gases, Petroleum, C3-5 Olefinic-Paraffinic Alkylation Feed
Light Olefin Feed

68477-99-6
Unknown
Unknown
Gases, Petroleum, Isomerized naphtha fractionater, C4-rich, Hydrogen sulfide-Free
Isomerization butane isomer hydrocarbon stream

68478-35-3
Unknown
Unknown
Acetic acid, C>19-alkyl esters
Aliphatic alcohols, acetate ester

68478-60-4
Unknown
Unknown
Dinaphtho(1,2,3-cd:3',2',1'-lm)perylene-5,10-dione, nitro derivs.

68479-98-1
$C_{11}H_{18}N_2$
178.27
Benzenediamine, ar,ar-diethyl-ar-methyl- (9CI)
ar,ar-Diethyl-ar-methyl-benzenediamine
Diethyltoluenediamine

68514-06-7
Unknown
Unknown
Vanillin Black Liquor
Sulfite Liquors and Cooking Liquors, Black, Vanillin

68514-95-4
Unknown
Unknown
Dialkyl dimethyl ammonium chloride
Quaternary ammonium compounds, di-C12-20-alkyldimethyl, chlorides

68515-39-9
Unknown
Unknown
1,2-Benzenedicarboxylic acid, tridecyl ester, Manuf. of, by-products from
Tridecyl recycle alcohol

68515-41-3
$C_{24}H_{38}O_4$
390.57
1,2-Benzenedicarboxylic acid, di-C7-9-branched and linear alkyl esters
Phthalic acid, dialkyl(C7-C9) ester

68515-42-4
$C_{24}H_{38}O_4$
390.57
1,2-Benzenedicarboxylic acid, di-C7-11-branched and linear alkyl esters
Phthalic acid, dialkyl(C7-C11) ester

68515-44-6
$C_{22}H_{34}O_4$
362.51
1,2-Benzenedicarboxylic acid, diheptyl ester, branched and linear
Phthalic acid, dialkyl(C7)

ester

68515-45-7
$C_{26}H_{42}O_4$
418.62
1,2-Benzenedicarboxylic acid, dinonyl ester, branched and linear
Phthalic acid, dialkyl(C9) ester

68515-47-9
$C_{34}H_{58}O_4$
530.84
1,2-Benzenedicarboxylic acid, di-C11-14-branched alkyl esters, C13-rich

68515-48-0
$C_{26}H_{42}O_4$
418.62
1,2-Benzenedicarboxylic acid, di-C8-10-branched alkyl esters, C9-rich

68515-49-1
$C_{30}H_{50}O_4$
474.73
1,2-Benzenedicarboxylic acid, di-C9-11-branched alkyl esters, C10-rich

68515-50-4
$C_{20}H_{30}O_4$
334.46
1,2-Benzenedicarboxylic acid, dihexyl ester, Branched and linear

68515-51-5
$C_{26}H_{42}O_4$
418.62
1,2-Benzenedicarboxylic acid, di-C6-10-alkyl esters
Di(C6-C10)alkyl phthalate

68516-01-8
Unknown
Unknown
Phosphorodithioic acid, mixed O,O-bis(iso-Bu and pentyl) esters
Dithiophosphoric acid, O,O'-isobutyl amyl esters

68516-06-3
Unknown
Unknown
Alkyl di(2-hydroxypropyl)-amine
1,1'-((Coconut oil alkyl)imino)-di-2-propanol
2-Propanol, 1,1'-iminobis-, N-coco alkyl derivs.

68516-18-7
$C_{22}H_{48}O_2$
344.63
Decene, hydroformylation products
Alkyl alcohol (C11)

68517-02-2
$C_{30}H_{32}O_6$
488.58
O(C1COc(ccc(c2)C(c(ccc(OCC(O3)C3)c4)c4)(c(ccc(OCC(O5)C5)c6)c6)CC)c2)C1
Oxirane, 2,2',2''-(propylidynetris(4,1-phenyleneoxymethylene))tris- (9CI)
Tris(4-hydroxyphenyl)propane-triglycidyl ether

68526-65-8
Unknown
Unknown
(Coco alkyl)amine mono-benzoate
Amines, coco alkyl, benzoates
Cocoamine, benzoic acid salt

68527-08-2
Unknown
Unknown
Alkenes, C>10 α-, Polymd.
Poly(α-olefins)

68527-62-8
$C_{29}H_{46}CaO_2$
466.77
Phenol, 2,2'-methylenebis-(4-(1,1,3,3-tetramethyl-butyl)-, calcium salt (9CI)
2,2'-Methylenebis(p-tert-octyl-phenol), calcium salt

68533-00-4
Unknown
Unknown
Residual Fuel Oils, Heavy

68540-40-9
$C_{23}H_{38}O_3.1/2Ca$
382.59
Benzoic acid, 5-hexadecyl-2-hydroxy-, calcium salt (2:1) (9CI)
5-Hexadecylsalicylic acid, calcium salt

68540-41-0
$C_{10}H_9NO_3S.H_3N$
240.27
1-Naphthalenesulfonic acid, 2-amino-, monoammonium salt (9CI)
2-Naphthylamine-1-sulfonic acid, ammonium salt

68551-07-5
$C_{13}H_{28}O$
200.37
Alcohols, C8-18
Fatty (C8-C18) alcohol

68551-12-2
Unknown
Unknown
Alcohols, C12-16, ethoxylated
Polyethylene glycol, dodecyl, tetradecyl, hexadecyl ether

68553-00-4
Unknown
Unknown
Oil, Fuel, No. 6
Fuel Oil, No. 6
No. 6 Fuel Oil

68553-81-1
Unknown
Unknown
Rice Bran Oil
Crude Rice Bran Oil
Crude Rice Oil
Oils, Rice Bran
Rice Bran

68554-00-7
Unknown
Unknown
Polyphosphoric acids, 2-ethoxyethyl esters
2-Ethoxyethanol, phosphated

68555-24-8
Unknown
Unknown
Tar Acids, Cresylic Residues
(C1-C3) Alkylphenol
Cresylic Acid Tar
Cresylic Acids Residue (Coal)

68555-86-2
$C_{21}H_{20}N_4O_5S.Na$
463.45
Benzenesulfonic acid, 4-((5-methoxy-4-((4-methoxy-phenyl)azo)-2-methylphenyl)-azo)-, sodium salt (9CI)
4-((5-Methoxy-4-((4-methoxy-phenyl)azo)-2-methylphenyl)-azo)benzenesulfonic acid, sodium salt

68558-73-6
$C_{18}H_{10}S$
258.34
Triphenyleno(1,12-bcd)thio-phene (9CI)
1,12-Epithiotriphenylene

NSC-334055

68568-63-8
$C_5H_{10}CaO_7$
222.21
Pentaric acid, calcium salt (9CI)
Calcium pentarate

68568-82-1
$C_{49}H_{79}N_3O_3.xCa$
Unknown
Phenol, 2,2'-((((2-hydroxy-5-octylphenyl)methyl)imino)-bis(2,1-ethanediylimino-methylene))bis(4-octyl-, calcium salt (9CI)
N,N-Bis-(2-(2-hydroxy-5-octyl-benzylamino)ethyl)-2-hydroxy-5-octylbenzylamine, calcium salt

68584-22-5
$C_{16}H_{26}O_3S$
298.45
Benzenesulfonic acid, C10-16-alkyl derivs.
(C10-C16)Alkyl benzene sulfonic acid
C10-C16 Alkylbenzenesulfonic acid
(C10-C16) Alkylbenzenesulfonic acid
(C10-C16) Saturated alkyl-benzenesulfonic acid
SDA 15-080-00

68585-34-2
Unknown
Unknown
Poly(oxy-1,2-ethanediyl), α-sulfo-ω-hydroxy-, C10-16-alkyl ethers, sodium salts
(C10-C16) Alcohol ethoxylate, sulfated, sodium salt
C10-C16 Alkyl (alcohol) ethoxylate sulfuric acid sodium salt
(C10-C16)Alkyl(alcohol)ethoxylate sulfuric acid, sodium salt
(C10-C16)Alkyl ethoxylate sulfuric acid, sodium salt
(C10-C16) Alkyl ethoxylate sulfuric acid, sodium salt
(C10-C16) Alkylethoxylate sulfuric acid, sodium salt
SDA 15-067-04

68586-19-6
$C_{16}H_{22}O_3$
262.35
2-Propenoic acid, 2-methyl-, 2-((2,3,3a,4,7,7a(or 3a,4,5,6,7,7a)-hexahydro-4,7-methano-1H-indenyl)oxy)-ethyl ester (9CI)
Dicyclopentenyloxyethyl methacrylate

68586-20-9
$C_{32}H_{43}N$
441.69
Benzenamine, ar-(1,7,7-trimethylbicyclo(2.2.1)hept-2-yl)-N-((1,7,7-trimethyl-bicyclo(2.2.1)hept-2-yl)-phenyl)- (9CI)
Di(isobornylphenyl)amine

68602-80-2
Unknown
Unknown
Chevron 100
(C12-C30) Aromatic Oil
Aromatic Oil Distillate
Aromatic Petroleum Distillate
Caswell No. 055
Distillates, Petroleum, C12-30-arom.
EPA Pesticide Chemical Code 006601

68603-15-6
$C_{10}H_{22}O$
158.29
Mixed fatty alcohols (C6-C12)
Alcohols, C6-12
(C6-C12) Alkyl alcohol
Caswell No. 456E
EPA Pesticide Chemical Code 079029
Fatty alcohols
SDA 13-060-00

68603-42-9
Unknown
Unknown
Coconut-Oil-Acid-Diethanolamine
Amides, Coco, N,N-bis-(hydroxyethyl)
Clindrol 200CGN
Clindrol 202CGN
Clindrol Superamide 100CG
Coconut Diethanolamide
Ethylan LD
NCI-C55312
Ninol 2012E

68603-84-9
$C_6H_{12}O_2$
116.16
Carboxylic acids, C5-9
(C5-C9) Monobasic acids

68606-11-1
Unknown
Unknown
Gasoline, Straight-run, Topping-plant
Raw, Straight run gasoline

68607-11-4
Unknown
Unknown
Petroleum products, Refinery gases
Refinery gas

68607-18-1
Unknown
Unknown
Polyphosphoric acids, zinc salts
Zinc polyphosphate

68607-28-3
Unknown
Unknown
(Oxydiethyleneglycol)bis(coco alkyl)dimethyl ammonium chloride
Caswell No. 627B
Dimethylcocoamine, bis(chloroethyl) ether, diquaternary ammonium salt
EPA Pesticide Chemical Code 069173
Oxydiethylenebis(alkyl*-dimethyl ammonium chloride) *(Derived from coconut oil fatty acids)
Quaternary ammonium compounds, (oxydi-2,1-ethane-diyl)bis(coco alkyldimethyl, dichlorides

68608-15-1
$C_4H_9NaO_3S$
160.17
Sulfonic acids, alkane, sodium salts
Paraffin sulfonate
Paraffin, Sulfonated, Sodium salt
Paraffin wax sulfonic acids, Sodium salts

68608-26-4
Unknown
Unknown
Sodium petroleum sulfonate
Mineral oil sulfonic acids, sodium salts
Oil soluble petroleum sulfonate, sodium salt
Oil soluble petroleum sulfonates, sodium salts
Petroleum sulfonic acid, monosodium salt
Petroleum sulfonic acid, sodium salt
Petroleumsulfonate, sodium salt
Sulfonated petroleum, sodium salt
Sulfonic acids, petroleum, sodium salts

68608-77-5
Unknown

Unknown
Benzenamine, 2-ethyl-N-(2-ethylphenyl)-, (tripropenyl) derivs.
N-(2-Ethylphenyl)-2-ethylbenzenamine, 1-propene trimer reaction product

68608-87-7
Unknown
Unknown
Benzenesulfonic acid, mono-C6-12-alkyl derivs., sodium salts
(C6-C12) Alkylbenzenesulfonic acid, sodium salt
(C6-C12)Alkylbenzenesulfonic acid, sodium salt

68609-96-1
Unknown
Unknown
Oxirane, mono((C8-10-alkyloxy)methyl) derivs.
Alkyl (C8,C10) glycidyl ether

68609-97-2
Unknown
Unknown
Oxirane, mono((C12-14-alkyloxy)methyl) derivs.
Alkyl (C12, C14) glycidyl ether

68610-00-4
Unknown
Unknown
Butoxypolypropoxypolyethoxyethanol iodine complex
Oxirane, methyl-, Polymer with oxirane, monobutyl ether, Compd. with iodine (9CI)

68610-06-0
Unknown
Unknown
Phenol, isobutylenated
Oil soluble polyolefin phenol

68610-22-0
Unknown
Unknown
Poly(oxy-1,2-ethanediyl), α-sulfo-ω-hydroxy-, C12-18-alkyl ethers, ammonium salts
(C12-C18) Alkyl ethoxylate, sulfate, ammonium salt
C12-C18 Alkyl ethoxylate sulfuric acid ammonium salt

68611-64-3
Unknown
Unknown
Urea, Reaction products with formaldehyde
Formaldehyde, Urea adduct

68628-60-4
$C_{18}H_{30}O_3S.Na$
349.49
Benzenesulfonic acid, 4-sec-dodecyl-, sodium salt (9CI)
Sodium 4-sec-dodecylbenzenesulfonate

68631-02-7
$C_{14}H_8Cl_6$
388.93
Benzene, 1,1'-(tetrachloroethylidene)bis(chloro- (9CI)
1,1'-Bis(chlorophenyl)-1,2,2,2-tetrachloroethylene
1,1'-(Tetrachloroethylidene)bis(chlorobenzene)

68647-44-9
$C_{18}H_{35}N_2O_3.HO.Na$
367.47
1H-Imidazolium, 1-(carboxymethyl)-4,5-dihydro-1(or 3)-(2-hydroxyethyl)-2-undecyl-, hydroxide, monosodium salt (9CI)
1H-Imidazole-1-ethanol, 4,5-dihydro-2-undecyl, carboxymethylated, sodium salt

68647-53-0
Unknown
Unknown
Imidazolium compounds, 1(or 3)-(carboxymethyl)-4,5-dihydro-1-(hydroxyethyl)-2-norcoco alkyl, hydroxides, monosodium salts
Coconut fatty acid, aminoethylethanolamine amide-imidazoline, carboxymethylated, sodium salt
Coconut fatty acid, aminoethylethanolamine amide-imidazoline, dicarboxymethylated, disodium salt
Coconut fatty acid, aminoethylethanolamine imidazoline, carboxymethylated, sodium salt
Coconut fatty acid, aminoethylethanolamine reaction product, dicarboxymethylated, disodium salt

68647-67-6
Unknown
Unknown
Sesquiterpenes and sesquiterpenoids, Guaiac wood-oil
Sesquiterpene mixture from guaiacwood oil

68648-86-2
$C_{16}H_{26}$
218.39
Benzene, C4-16-alkyl derivs.
Alkyl(C4-C16)benzene

68648-87-3
$C_{18}H_{30}$
246.44
Benzene, C10-16-alkyl derivs.
(C8-C16) Alkylbenzene

68648-91-9
$C_{26}H_{42}O_4$
418.62
1,2-Benzenedicarboxylic acid, di-C7-11-alkyl esters
Phthalic acid, (C7-C11) alkyl esters

68648-93-1
Unknown
Unknown
1,2-Benzenedicarboxylic acid, Mixed decyl and hexyl and octyl diesters
Mixed hexyl, octyl, decyl phthalates
Phthalic anhydride, hexyl, octyl, decyl esters

68649-43-4
Unknown
Unknown
Phosphorodithioic acid, O,O-dioctyl ester, branched

68679-99-2
$C_{14}H_8Cl_4$
318.03
Benzene, 1,1'-(dichloroethenylidene)bis(chloro-(VAN)(9CI)
1,1'-Bis(chlorophenyl)-2,2-dichloroethylene

68683-30-7
$C_{12}H_{13}ClO_3$
240.69
O=C(OC(c(cccc1)c1)C)C(C(=O)C)Cl
Butanoic acid, 2-chloro-3-oxo-, 1-phenylethyl ester (9CI)
α-Methylbenzyl 2-chloroacetoacetate

68760-70-3
$C_{12}H_{22}O_2$
198.31
6-Decen-1-ol, acetate, (Z)-(9CI)

68782-98-9
Unknown

Unknown
Extracts, Petroleum, Clarified oil solvent, Condensed-ring-arom.-contg.
Total clarified oil solvent extract (Petroleum)

68783-03-9
Unknown
Unknown
Extracts, Petroleum, Light clarified oil solvent, Condensed-ring-arom.-contg.
Light clarified oil solvent extract (Petroleum)

68783-78-8
Unknown
Unknown
Ditallowdimonium chloride
Dimethyl ditallow ammonium chloride
Dimethylditallowammonium chloride
Ditallow dimethyl ammonium chloride
Quaternary ammonium compounds, dimethyl ditallow alkyl, chlorides
Quaternary ammonium compounds, dimethylditallow alkyl, chlorides
Quaternium-48

68784-30-5
Unknown
Unknown
Phosphorodithioic acid, Mixed O,O-bis(sec-Bu and 1,3-dimethylbutyl) esters
Phosphorodithioic acid, O,O-(2-butyl, 4-methyl-2-pentyl) mixed esters

68784-31-6
Unknown
Unknown
Phosphorodithioic acid, Mixed O,O-bis(sec-Bu and 1,3-dimethylbutyl) esters, zinc salts
Phosphorodithioic acid, O,O-(2-butyl,4-methyl-2-phenyl) mixed esters, zinc salt

68784-32-7
Unknown
Unknown
Phosphorodithioic acid, Mixed O,O-bis(2-ethylhexyl and iso-Bu) esters
Phosphorodithioic acid, O,O-(isobutyl, 2-ethylhexyl) mixed esters

68784-33-8
Unknown
Unknown
Phosphorodithioic acid, Mixed O,O-bis(hexyl and iso-Bu) esters
Phosphorodithioic acid, O,O-(isobutyl, 1-hexyl) mixed esters

68784-34-9
Unknown
Unknown
Phosphorodithioic acid, Mixed O,O-bis(hexyl and iso-Bu) esters, zinc salts
Phosphorodithioic acid, O,O-(isobutyl, 1-hexyl) mixed esters, zinc salt

68795-14-2
$C_{12}H_3Br_5O$
562.68
Dibenzofuran, pentabromo- (9CI)
Pentabromodibenzofuran

68808-54-8
$C_{13}H_{13}N_3.C_2H_4O_2$
271.35
5H-Pyrido(4,3-b)indole, 3-amino-1,4-dimethyl-, acetate
3-Amino-1,4-dimethyl-5H-pyrido(4,3-b)indole acetate
5H-Pyrido(4,3-b)indol-3-amine, 1,4-dimethyl-, acetate
5H-Pyrido(4,3-b)indol-3-amine, 1,4-dimethyl-, monoacetate
TRP-P-1 (Acetate)

68813-94-5
Unknown
Unknown
Basic Zinc Sulfate
Caswell No. 928
EPA Pesticide Chemical Code 089101
Sulfuric acid, zinc salt, Basic (9CI)
Zinc sulfate, Basic

68815-21-4
Unknown
Unknown
Cresylate Spent Caustic
Cresylate Spent Caustic Solution
Tar Acids, Cresylic, Sodium Salts, Caustic Solns.

68815-49-6
Unknown
Unknown
Lithium, 12-hydroxyoctadecanoate sebacate complexes

68815-67-8
$C_{36}H_{58}O_2S$
554.92
Phenol, thiobis(tetrapropylene- (9CI)
Thiobis(tetrapropylenephenol)
Thiobis((tetrapropylene)-phenol)

68833-55-6
C_2HHg
225.62
Mercury acetylide
Acetiluro de mercurio (Spanish)
Acetylure de mercure (French)

68848-64-6
Unknown
Unknown
Lithium silicon
UN 1417

68849-24-1
$C_7H_3ClF_3NO_2$
225.55
Benzene, chloronitro(trifluoromethyl)- (9CI)

68855-54-9
Unknown
Unknown
Diatomite
Diatomaceous earth, Flux-calcined
Flux-calcined diatomaceous earth
Kieselguhr, Soda ash flux-calcined
Silica, Amorphous, Diatomaceous earth (Containing less than 1% crystalline silica)

68877-33-8
$C_{28}H_{23}N_9O_7S_2.2Na$
707.60
2,7-Naphthalenedisulfonic acid, 4-amino-3-((4-((2,4-diaminophenyl)azo)phenyl)-azo)-5-hydroxy-6-(phenyl-azo)-, disodium salt (9CI)

68877-63-4
$C_{21}H_{20}BrN_7O_6$
546.29
O=C(Nc(c(N=Nc(c(cc(N(=O)=O)c1)Br)c1N(=O)=O)cc(OC)c2N(CCC#N)CC=C)c2)C
Acetamide, N-(2-((2-bromo-4,6-dinitrophenyl)azo)-5-((2-cyanoethyl)-2-propenylamino)-4-methoxyphenyl)- (9CI)

68890-88-0
Unknown
Unknown
Poly(oxy-1,2-ethanediyl), α-sulfo-ω-hydroxy-, C10-12-alkyl ethers, ammonium salts
α-(C10-C12)Alkyl-ω-hydroxy-poly(oxy-1,2-ethanediyl), sulfate, ammonium salt

68890-99-3
$C_{21}H_{36}$
288.52
Benzene, mono-C10-16-alkyl derivs.
(C10-C16)Alkylbenzene

68891-38-3
Unknown
Unknown
Poly(oxy-1,2-ethanediyl), α-sulfo-ω-hydroxy-, C12-14-alkyl ethers, sodium salts
Linear(C12-C14)alkanol, ethoxylated, sulfated, sodium salt

68897-50-7
$C_{13}H_{18}N_4O_5$
310.29
Benzenamine, N-(1-ethyl propyl)-3,4-dimethyl-2,6-dinitro-N-nitroso- (9CI)

68908-87-2
Unknown
Unknown
Benzene, 1,3-dimethyl-, benzylated
m-Xylene, benzylated

68909-12-6
Unknown
Unknown
Bastnaesite, Acid-insol. fraction, Cerium-rich
Cerium concentrate

68909-13-7
Unknown
Unknown
Bastnaesite, calcined conc.
Calcined bastnasite

68909-93-3
Unknown
Unknown
Phosphorodithioic acid, Mixed O,O-bis(2-ethylhexyl and iso-Pr) esters, zinc salts

68911-49-9
Unknown
Unknown
Dried Blood
Animal Blood, Denatured
Blood, Glyoxal-denatured, Dried

68911-57-9
Unknown
Unknown
Distillates (Coal), Solvent-refining (SRC), Middle

68911-78-4
Unknown
Unknown
1-((Tallow alkyl)amino)-3-aminopropane diacetate
Amines, N-tallow alkyltri-methylenedi-, diacetates
1,3-Propanediamine, N-tallow-, diacetate

68915-31-1
Unknown
Unknown
Polyphosphoric acids, sodium salts
Sodium polyphosphate

68915-86-6
Unknown
Unknown
Oil, Edible
Oils, Raisin
Raisin Extract
Raisin Seed Oil

68915-97-9
Unknown
Unknown
Gas oils, Petroleum, Straight-run, high-boiling

68916-39-2
Unknown
Unknown
Hamamelis
NCI-C50544
Snapping Hazel
Spotted Alder
Striped Alder
Tobacco Wood
Winter Bloom
Witch Hazel

68916-91-6
Unknown
Unknown
Licorice, ext.
Glycyrrhiza
Glycyrrhiza extract
Glycyrrhizae (Latin)
Kanzo (Japanese)
Kanzou (Chinese)
Licorice
Licorice Extract
Licorice Root
Licorice Root Extract

68916-91-6
Unknown
Unknown
Glycyrrhiza-extract
Glycyrrhiza
Glycyrrhizae (Latin)
Kanzo (Japanese)
Kanzou (Chinese)
Licorice
Licorice Extract
Licorice Root
Licorice Root Extract

68918-07-0
Unknown
Unknown
Sulfonic acids, petrolatum, sodium salts
Petrolatum sulfonic acids, sodium salts

68919-37-9
Unknown
Unknown
Full-range reformed naphtha
Full range reformed naphtha (Petroleum)
Naphtha, Petroleum, Full-range reformed

68920-66-1
Unknown
Unknown
Alcohols, C16-18 and C18-unsatd., ethoxylated
(C16-C18) and (C18) Un-saturated alkylalcohol, ethoxylate

68920-70-7
$C_6H_{13}Cl$
120.62
Alkanes, C6-18, chloro
Chlorinated paraffins
Chlorinated n-paraffins (C6-C18)

68921-34-6
Unknown
Unknown
1,6-Octadiene, 7-methyl-2-methylene-, Reaction products with hydrochloric acid

68925-41-7
$C_5H_7ClN_4S$
190.63
2,4-Pyrimidinediamine, 6-chloro-5-(methylthio)-(9CI)

68928-76-7
$C_{22}H_{44}O_4Sn$
491.30
Stannane, dimethylbis((1-oxoneodecyl)oxy)- (9CI)
Dimethylbis((1-oxoneodecyl)-oxy)stannane
Dimethyltindineodecanoate

68928-80-3
$C_{12}H_3Br_7O$
722.48
Benzene, 1,1'-oxybis-, heptabromo deriv. (9CI)
Heptabromodiphenyl ether
Heptabromodiphenyl oxide
Heptabromophenoxybenzene
1,1'-Oxybisbenzene heptabromo deriv.

68937-40-6
Unknown
Unknown
Phenol, isobutylenated, phosphate (3:1)

68937-41-7
Unknown
Unknown
Phenol, isopropylated, phosphate (3:1)

68937-68-8
$C_3H_6O_2$
74.08
Carboxylic acids, C1-5
Top distillation cut by-product acids, monobasic (C1-C5)

68937-69-9
$C_{24}H_{46}O_6$
430.63
Carboxylic acids, C6-18 and C5-15-di-

68938-79-4
$C_7H_4Cl_2O_3.K.Na$
269.10
Benzoic acid, 3,6-dichloro-2-hydroxy-, potassium sodium salt (9CI)
3,6-Dichlorosalicylate, sodium, potassium salt

68938-81-8
$C_6H_4Cl_2O.K$
202.10
Phenol, 2,5-dichloro-, potassium salt (9CI)
2,5-Dichlorophenol potassium salt
Potassium 2,5-dichlorophenolate

68951-67-7
Unknown
Unknown
Alcohols, C14-15, ethoxylated

68952-33-0
Unknown
Unknown
Tar acids, Cresylic, C8-rich, phosphates
Phosphate esters of coal tar or petroleum derived cresylic acid

68952-35-2
Unknown
Unknown
Tar acids, Cresylic, Ph phosphates
Phosphate esters of coal tar or petroleum-derived cresylic acid

68952-95-4
Unknown
Unknown
Soap
Acidulated Soapstock
Caswell No. 741
EPA Pesticide Chemical Code 079009
Neutral Soap Stock
Neutral Soapstock
Soaps, Stocks, Vegetable-oil, Acidulated
Soapstock, Acidulated

68953-58-2
Unknown
Unknown
Quaternium-18 bentonite
Bis(hydrogenated tallow alkyl)-dimethylammonium bentonite
Clayamine ARO
Clayamine EP
Clayamine EPA
Clayamine #4
Dimethyl dihydrogenated tallow ammonium chloride, Reaction product with bentonite
Di(tallow alkyl) dimethyl ammonium bentonite
Quaternary ammonium compounds, bis(hydrogenated tallow alkyl)dimethyl, chlorides, Reaction products with bentonite
Quaternary ammonium compounds, bis(hydrogenated tallow alkyl)dimethyl, salts with bentonite
Tixogel VP

68953-95-7
Unknown
Unknown
Benzenesulfonic acid, mono-C9-17-alkyl derivs., sodium salts
(C9-C17) Branched alkylbenzenesulfonic acid, sodium salt

68955-06-6
$C_{12}H_{24}O$
184.36
Triisobutylene oxide (Obs.)
EP-1086

68955-35-1
Unknown
Unknown
Naphtha, petroleum, catalytic reformed
Catalytic reformed naphtha (Petroleum)

68956-68-3
Unknown
Unknown
Vegetable Oil
Edible Vegetable Oil
Oil, Edible
Oils, Vegetable
Salad Oil
Vegetable Oil Mist
Vegetable Oil Mist, Respirable fraction (OSHA)
Vegetable Oil Mist, Total dust (OSHA)
Viscoleo Oil

68956-68-3
Unknown
Unknown
Vegetable Oil
Edible Vegetable Oil
Oils, Vegetable
Salad Oil
Vegetable Oil Mist
Vegetable Oil Mist, Respirable fraction (OSHA)
Vegetable Oil Mist, Total dust (OSHA)
Viscoleo Oil

68956-68-3
Unknown
Unknown
Viscoleo-Oil

68956-70-7
Unknown
Unknown
Petroleum products, C5-12, Reclaimed, Wastewater treatment

68956-82-1
Unknown
Unknown

Cobalt Rosinate
Cobalt Resinate, Precipitated [UN 1318]
Resin acids and Rosin acids, Cobalt salts
Resinate de cobalt (French)
Resinato de cobalto (Spanish)
UN 1318 [Cobalt resinate, Precipitated]

68959-20-6
$C_{27}H_{60}NO_3Si.Cl$
510.31
Disiquonium Chloride
Caswell No. 331C
1-Decanaminium, N-decyl-N-methyl-N-(3-(trimethoxysilyl)propyl)-, chloride (9CI)
Didecylmethyl(3-(trimethoxysilyl)propyl)ammonium chloride.
Di-N-decylmethyl(3-trimethoxysilylpropyl)ammonium chloride
N,N-Didecyl-N-methyl-3-(trimethoxysilyl)propanediol
EPA Pesticide Chemical Code 169160
Trimethoxysilylpropyldidecylmethylammonium chloride

68959-23-9
$C_{15}H_{26}O_6$
302.37
O(C1COC(CCCCOCC(O2)C2)COCC(O3)C3)C1
Oxirane, 2,2',2''-(1,2,6-hexanetriyltris(oxymethylene))tris- (9CI)
1,2,6-Hexanetriol tris(glycidyl) ether

68966-50-7
$C_{34}H_{29}N_6NaO_{16}S_4$
928.89
2,7-Naphthalenedisulfonic acid, 3,3'-((3,3'-dimethoxy-(1,1'-biphenyl)-4,4'-diyl)bis(azo))bis(5-amino-4-hydroxy-, sodium salt (9CI)

68966-84-7
$C_9H_{14}N_2$
150.21
1,3-Benzenediamine, ar-ethyl-ar-methyl- (9CI)
ar-Ethyl-ar-methyl-1,3-benzenediamine

68967-09-9
$C_{17}H_{10}O$
230.27
Pyrenecarboxaldehyde (9CI)

68971-14-2
$C_{19}H_{24}N_6O_4S_3$
496.60
Benzenesulfonic acid, 4,4'-(carbonothioyldiimino)bis-, bis-((1-methylethylidene)hydrazide) (9CI)

68973-26-2
$CH_2O_3.2H2O.Mg$
86.34
Carbonic acid, magnesium salt (1:1), dihydrate (9CI)
Magnesium carbonate ($MgCO_3$) dihydrate

68974-78-7
$C_{36}H_{60}MgO_2S$
581.25
Phenol, thiobis((tetrapropenyl)-, magnesium salt (9CI)
Thiobis((tetrapropenyl)phenol) magnesium salt
Thiobis(tetrapropylenephenol), magnesium salt

68987-80-4
Unknown
Unknown
Oxirane, mono((C6-12-alkyloxy)methyl) derivs.
(C6-C12) Alkylglycidyl ether

68988-46-5
Unknown
Unknown
Phosphorodithioic acid, Mixed O,O-bis(iso-Bu and isooctyl and pentyl) esters, zinc salts
Phosphorodithioic acid, O,O-di(isobutyl, amyl, isooctyl) mixed esters, zinc salt

68991-48-0
Unknown
Unknown
Alcohols, C7-21, ethoxylated

69009-90-1
$C_{18}H_{22}$
238.37
1,1'-Biphenyl, bis(1-methylethyl)- (9CI)
Bis(1-methylethyl)-1,1'-biphenyl
Diisopropylbiphenyl

69011-63-8
Unknown
Unknown
Magnesium Dross
Magnesio, Escorias de, humedas o Calientes (Spanish)
Magnesium Dross, Hot
Magnesium Dross, Wet
Magnesium Dross (Wet or hot)
Scories de magnesium (Humides ou chaudes) (French)

69011-64-9
Unknown
Unknown
Babbitt, Dross
Caustic nickel skims (Secondary nonferrous plant)

69011-68-3
Unknown
Unknown
Brass, Dross

69011-84-3
Unknown
Unknown
Poly(oxy-1,2-ethanediyl), α-sulfo-ω-(octylphenoxy)-, branched, sodium salt
C8 Branched alkyl phenol ethoxylate sulfuric acid, sodium salt
C8 Branched alkylphenol ethoxylate sulfuric acid, sodium salt
SDA 22-101-04

69012-26-6
Unknown
Unknown
Slags, Brass-manufg.
Brass slag
Brass slag (Secondary nonferrous plant)
Cupola slag (Secondary nonferrous plant)
Slag, Skims and fines from brass furnace (Secondary nonferrous plant)

69012-56-2
Unknown
Unknown
Flue Dust, Brass-manufg.
Brass baghouse fume

69012-58-4
Unknown
Unknown
Flue Dust, Copper alloy-manufg., zinc oxide-contg.
Zinc flue dust (Secondary nonferrous plant)

69012-65-3
Unknown
Unknown
Fumes, Zinc
Zinc fume

69013-19-0
Unknown

Unknown
Alcohols, C8-22, ethoxylated
Linear (C8-C22) alkyl alcohol, Ethoxylated

69013-21-4
Unknown
Unknown
Fuel Oil, Pyrolysis
Pyrolysis Fuel Oil

69029-52-3
Unknown
Unknown
Lead Dross
Copper Dross (Lead refinery)
Decopperizing Dross (Secondary nonferrous plant)
Lead Dross (Containing 3% or more free acid)
Lead Kettle Dross (Secondary nonferrous plant)
Lead Refinery Copper Dross
Lead Refinery Pyrite Dross
Lead Refining Caustic Dross
Lead Scrap
Lead Skimmings
Lead Smelter Copper Dross
Lead Smelter Dross
Lead Smelter Lead Dross
NA 1794
Smelter Dross (Lead)
Test Lead Dross
UN 1794 [Lead sulfate with more than 3 per cent free acid]

69029-87-4
Unknown
Unknown
Slags, Type metal smelting

69036-12-0
$C_4H_7Br_2Cl$
250.36
Propane, 1,2-dibromo-1-chloro-2-methyl- (9CI)

69045-78-9
$C_6H_3Cl_4N$
230.90
Pyridine, 2-chloro-5-(trichloromethyl)- (9CI)

69045-83-6
$C_6H_2Cl_5N$
265.35
Pyridine, 2,3-dichloro-5-(trichloromethyl)- (9CI)

69078-83-7
$C_7H_{16}S_3$
196.40
Disulfide, methyl 2-methyl-1-(methylthio)-butyl (9CI)

69103-20-4
$C_{10}H_{16}O$
152.24
O(C1CC=C(C=C)C)C1(C)C
Oxirane, 2,2-dimethyl-3-(3-methyl-2,4-pentadienyl)- (9CI)
6,7-Epoxy-3,7-dimethyl-1,3-octadiene

69112-21-6
$C_6H_{10}O$
98.16
3-Hexenal, (E)
trans-3-Hexenal

69155-42-6
$C_{20}H_{46}O_7Si_4$
510.92
Tetrasiloxane, 1,1,1,3,5,7,7,7-octamethyl-3,5-bis(3-(oxiranylmethoxy)propyl)- (9CI)
1,1,1,3,5,7,7,7-Octamethyl-3,5-bis(6,7-epoxy-4-oxaheptyl)-tetrasiloxane

69158-26-5
$C_{10}H_6Cl_4O_4$
331.97
1,3-Benzenediol, 2,4,5,6-tetrachloro-, diacetate (9CI)

69227-21-0
Unknown
Unknown
Alcohols, C12-18, ethoxylated propoxylated
C12-18 Alkyl alcohol ethoxylate propoxylate
(C12-C18) Alkyl alcohol ethoxylate propoxylate
(C12-C18)Alkylalcohol ethoxylate propoxylate
SDA 16-070-00

69327-76-0
$C_{16}H_{23}N_3OS$
305.48
4H-1,3,5-Thiadiazin-4-one, 2-((1,1-dimethylethyl)imino)-tetrahydro-3-(1-methylethyl)-5-phenyl
Applaud
Buprofezin
Buprofezine
2-tert-Butylimino-3-isopropyl-5-phenylperhydro-1,3,5-thiadiazinan-4-one
2-tert-Butylimino-3-isopropyl-5-phenyl-3,4,5,6-tetrahydro-2H-1,3,5-thiadiazin-4-one
2-tert-Butylimino-3-isopropyl-5-phenyl-1,3,5-thiadiazinan-4-one
2-((1,1-Dimethylethyl)imino)-tetrahydro-3-(1-methylethyl)-5-phenyl-4H-1,3,5-thiadiazin-4-one
NNI 750
PP618

69329-95-9
$C_{16}H_{23}N_3O_2S$
321.43
4H-1,3,5-Thiadiazin-4-one, 2-((1,1-dimethylethyl)imino)-tetrahydro-5-(4-hydroxyphenyl)-3-(1-methylethyl)- (9CI)

69375-05-9
$C_{20}H_{41}N_5O_7$
463.66
Antibiotic XK 62-3
XK-62-3

69402-28-4
Unknown
Unknown
Montmorillonite, cuprian (9CI)

69409-94-5
$C_{26}H_{22}ClF_3N_2O_3$
502.95
CC(C)C(Nc1ccc(cc1Cl)C(F)(F)F)C(=O)OC(C#N)c3cccc(Oc2ccccc2)c3
DL-Valine, N-(2-chloro-4-(trifluoromethyl)phenyl)-, cyano(3-phenoxyphenyl)-methyl ester
N-(2-Chloro-4-(trifluoromethyl)-phenyl)-dl-valine cyano(3-phenoxyphenyl)methyl ester
Fluvalinate
Mavrik
Mavrik HR
Spur

69430-24-6
$(C_2H_6OSi)x$
Polymer
Cyclomethicone
Cyclopolydimethylsiloxane
Cyclosiloxanes, di-Me
Dimethylcyclopolysiloxane
Polydimethyl siloxy cyclics
Polydimethylcyclosiloxane
Rhodorsil Oils 70045
Sentry Cyclomethicone
Silbione
Silicone COM 10000
Silicone COM 16520
Silicone COM 20000
Silicone COM 27510
Silicone COM 29010
Unisil SF-V

69432-94-6
$C_5H_7D_3$

73.15
2-Butene-1,1,1-d3, 2-methyl-, (E)- (9CI)

69432-95-7
$C_6H_9D_3$
87.18
2-Pentene-1,1,1-d3, 2-methyl-, (E)- (9CI)

69432-96-8
$C_7H_{11}D_3$
101.21
2-Pentene-1,1,1-d3, 2,4-dimethyl-, (E)- (9CI)

69432-97-9
$C_6H_{12}O$
100.16
1-Butene, 1-methoxy-2-methyl-, (E)- (9CI)

69432-98-0
$C_6H_{12}O$
100.16
1-Butene, 1-methoxy-2-methyl-, (Z)- (9CI)

69433-00-7
$C_{12}H_3Cl_5O$
340.42
Dibenzofuran, 1,2,6,7,8-pentachloro- (9CI)

69462-12-0
$C_{16}H_{21}Cl_3O_3$
367.70
Acetic acid, (2,4,5-trichlorophenoxy)-, 2-ethyl-4-methylpentyl ester (9CI)

69462-13-1
Unknown
Unknown
Dimethylamine 2,3,6-trichlorophenylacetate

69472-19-1
$C_{19}H_{21}N_5O_2$
351.38
Propanenitrile, 3-(butyl(4-((4-nitrophenyl)azo)phenyl)amino)- (9CI)

69481-32-9
$C_9H_{12}N_2O$
164.20
Pyridine, diethylnitroso- (9CI)

69484-12-4
$C_9H_{13}BrN_2O_2.Na$
284.20
Bromacil, dimethylamine salt
2,4(1H,3H)-Pyrimidinedione, 5-bromo-6-methyl-3-(1-methylpropyl)-, sodium salt (9CI)
Sodium bromacil

69500-29-4
$C_{24}H_{27}O_4P$
410.45
Phosphoric acid, bis(2-(1-methylethyl)phenyl) phenyl ester (9CI)

69579-72-2
$C_{18}H_{12}N_2O_5$
336.31
2-Naphthalenecarboxylic acid, 5-((4-carboxyphenyl)azo)-6-hydroxy- (9CI)

69594-78-1
C_8H_7NO
133.14
1H-Indolol (9CI)

69622-82-8
$C_{18}H_{13}AlCl_6O_7$
581.00
Aluminum, hydroxybis-(2-(2,4,5-trichlorophenoxy)propanoato-O1,O2)- (9CI)
Propanoic acid, 2-(2,4,5-trichlorophenoxy)-, aluminum complex (9CI)

69644-64-0
$C_{17}H_{12}N_2O_3$
292.30
2-Naphthalenecarboxylic acid, 6-hydroxy-5-(phenylazo)- (9CI)
1-Phenylazo-2-naphthol-6-carboxylic acid

69644-65-1
$C_{17}H_{12}N_2O_6S$
372.36
2-Naphthalenecarboxylic acid, 6-hydroxy-5-((4-sulfophenyl)azo)- (9CI)

69644-66-2
$C_{17}H_{12}N_2O_6S$
372.36
Benzoic acid, 4-((2-hydroxy-6-sulfo-1-naphthalenyl)azo)- (9CI)

69645-07-4
$C_6H_6Cl_2$
149.02
1,3,5-Hexatriene, 1,6-dichloro- (9CI)

69668-83-3
$C_{10}H_{18}O_2$
170.25
3-Octen-1-ol, acetate, (Z)- (9CI)

69668-85-5
$C_9H_{16}O_2$
156.23
3-Octenoic acid, methyl ester, (Z)- (9CI)

69682-29-7
$C_{24}H_{27}O_4P$
410.45
Phosphoric acid, hexylphenyl diphenyl ester (9CI)

69698-57-3
$C_{12}H_3Cl_5O$
340.42
Dibenzofuran, 1,2,4,6,8-pentachloro- (9CI)
1,2,4,6,8-Pentachlorodibenzofuran

69698-58-4
$C_{12}HCl_7O$
409.31
Dibenzofuran, 1,2,3,4,6,8,9-heptachloro- (9CI)
1,2,3,4,6,8,9-Heptachlorodibenzofuran

69698-59-5
$C_{12}H_2Cl_6O$
374.86
Dibenzofuran, 1,2,4,6,8,9-hexachloro- (9CI)
1,2,4,6,8,9-Hexachlorodibenzofuran

69698-60-8
$C_{12}H_2Cl_6O$
374.86
Dibenzofuran, 1,2,3,4,6,8-hexachloro- (9CI)
1,2,3,4,6,8-Hexachlorodibenzofuran

69742-55-8
$C_{24}H_{12}Cl_4CuN_4$
565.78
Copper(1+), bis(2,9-dichloro-1,10-phenanthroline-N1,N10)-, (T-4)- (9CI)
1,10-Phenanthroline, 2,9-dichloro-, copper complex (9CI)

69742-90-1
$C_{15}H_{11}N_4O_9S.K$
463.45
2-Propanone, 1-(5-((2,4-dinitrophenyl)thio)-2,4-dinitro-2,4-cyclohexadien-1-yl)-, ion(1-), potassium (9CI)

69760-73-2
$C_{27}H_{44}$
368.65
Cholestadiene (9CI) (VAN)

69760-96-9
$C_{12}H_5Cl_3O_2$
287.53
Dibenzo(b,e)(1,4)dioxin, trichloro- (9CI)
Trichlorodibenzo(b,e)(1,4)dioxin
Trichlorodibenzo-p-dioxin

69782-90-7
$C_{12}H_4Cl_6$
360.88
1,1'-Biphenyl, 2,3,3',4,4',5'-hexachloro- (9CI)
2,3,3',4,4',5'-Hexachlorobiphenyl
2,3,3',4,4',5'-PCB

69782-91-8
$C_{12}H_3Cl_7$
395.32
1,1'-Biphenyl, 2,3,3',4',5,5',6-heptachloro- (9CI)
2,3,3',4',5,5',6-Heptachlorobiphenyl
2,3,3',4',5,5',6-PCB

69806-34-4
$C_{15}H_{11}ClF_3NO_4$
361.71
Haloxyfop
(RS)-2-(4-(3-Chloro-5-trifluoromethyl-2-pyridyloxy)phenoxy)propionic acid
Propanoic acid, 2-(4-((3-chloro-5-(trifluoromethyl)-2-pyridinyl)oxy)phenoxy)- (9CI)

69806-40-2
$C_{16}H_{13}ClF_3NO_4$
375.75
Propanoic acid, 2-(4-((3-chloro-5-(trifluoromethyl)-2-pyridinyl)oxy)phenoxy)-, methyl ester
2-(4-((3-Chloro-5-(trifluoromethyl)-2-pyridinyl)oxy)phenoxy)propanoic acid methyl ester
DOWCO 453
DOWCO 453ME
Gallant
Haloxyfop-Methyl
Verdict
Zellek

69806-50-4
$C_{19}H_{20}F_3NO_4$
383.40
CCCCOC(=O)C(C)Oc2ccc(Oc1ccc(cn1)C(F)(F)F)cc2
Propanoic acid, 2-(4-((5-(trifluoromethyl)-2-pyridinyl)oxy)phenoxy)-, butyl ester
Butyl 2-(4-(5-trifluoromethyl-2-pyridinyloxy)phenoxy)propanoate
Fluazifop-butyl
Fusilade
IH 773B
PP 009
Propionic acid, 2-(p-((5-(trifluoromethyl)-2-pyridyl)oxy)phenoxy)-, butyl ester
TF 1169

69834-19-1
$C_{36}H_{58}O$
506.86
Benzene, 1,1'-oxybis(dodecyl- (9CI)
Oxybis(dodecylbenzene)
1,1'-Oxybis(dodecylbenzene)

69845-51-8
$C_7H_4Cl_2O_4$
223.01
Benzoic acid, dichloro-3,4-dihydroxy- (9CI)

69845-52-9
$C_8H_7ClO_4$
202.59
Benzoic acid, chloro-4-hydroxy-3-methoxy- (9CI)

69882-11-7
$(C_6H_4Br_2O)x$
Polymer
Phenol, 2,4(or 2,6)-dibromo-, Homopolymer (9CI)
2,4(or 2,6)-Dibromophenol homopolymer

69911-61-1
$C_7H_3Cl_5$
264.36
Benzene, methyl-, pentachloro deriv. (9CI)

69913-55-9
Unknown
Unknown
Renex 714 (9CI)

69975-77-5
$C_{22}H_{18}O_8$
410.38
Orlandin
(8,8'-Bi-2H-1-benzopyran)-2,2'-dione, 7,7'-dihydroxy-4,4'-dimethoxy-5,5'-dimethyl-
Bis(8,8'-(7-hydroxy-4-methoxy-5-methylcoumarin))

69975-80-0
$C_3H_8NO_4P.H_3N$
170.11
Phosphonic acid, (aminocarbonyl-14C)-, monoethyl ester, monoammonium salt (9CI)

70017-56-0
Unknown
Unknown
N-(2-Methyl-1-naphthyl)maleimide

70021-42-0
$C_{20}H_{11}NO_2$
297.31
Benzo(a)pyrene, nitro- (9CI)

70021-47-5
$C_{14}H_{12}S$
212.32
Dibenzothiophene, dimethyl- (9CI)

70021-48-6
$C_{15}H_{14}S$
226.34
Dibenzothiophene, trimethyl- (9CI)

70021-98-6
$C_{20}H_{11}NO_2$
297.32
Benzo(a)pyrene, 3-nitro
3-Nitrobenz(a)pyrene

70021-99-7
$C_{20}H_{11}NO_2$
297.32
Benzo(a)pyrene, 1-nitro
1-Nitrobenzo(a)pyrene

70084-98-9
Unknown
Unknown
Terpenes and Terpenoids, C10-30, distn. residues

70096-14-9
$C_{20}H_{24}N_2O_3.C_4H_4O_4$
456.48
1H-Isoindol-1-one, 2,3-dihydro-3-(2-hydroxy-3-((1-methylethyl)amino)propoxy)-2-phenyl-, (R*,S*)-(+-)-, (E)-2-butenedioate (1:1) (Salt)

(9CI)

70116-00-6
$C_{20}H_{24}N_2O_3$
340.43
Quinidine-N-oxide

70124-77-5
$C_{26}H_{23}F_2NO_4$
451.50
CC(C)C(C(=O)OC(C#N)c2cccc(Oc1ccccc1)c2)c3ccc(OC(F)F)cc3
Benzeneacetic acid, 4-(difluoromethoxy)-α-(1-methylethyl)-, cyano(3-phenoxyphenyl) methyl ester
Aastar
AC 222705
(+-)-Cyano(3-phenoxyphenyl)methyl(+)-4-(difluoromethoxy)-α-(1-methylethyl)benzeneacetate
Cybolt
Cythrin
Flucythrinate
Pay-Off

70131-50-9
Unknown
Unknown
Bentonite, Acid-leached
Activated clay
Bleaching clay
Clay adsorbent

70152-47-5
Unknown
Unknown
Sodium cyanide, Solution
UN 1689

70192-84-6
$C_3H_3BrF_4$
194.95
Propane, 1-bromo-1,1,2,2-tetrafluoro- (9CI)

70210-28-5
$C_{38}H_{30}N_{10}O_9S.2Na$
848.70
Benzoic acid, 5-((4'-((6-amino-5-(1H-benzotriazol-5-ylazo)-1-hydroxy-3-sulfo-2-naphthalenyl)azo)-3,3'-dimethoxy-(1,1'-biphenyl)-4-yl)azo)-2-hydroxy-4-methyl-, disodium salt (9CI)
5-((4'-((6-Amino-5-(1H-benzotriazol-5-ylazo)-1-hydroxy-3-sulfo-2-naphthalenyl)azo)-3,3'-dimethoxy(1,1'-biphenyl)-4-yl)azo)-2-hydroxy-4-methylbenzoic acid, disodium salt

70210-46-7
$C_{27}H_{22}ClN_7O_{10}S_3.3Na$
805.09
2,7-Naphthalenedisulfonic acid, 5-((4-chloro-6-(methylphenylamino)-1,3,5-triazin-2-yl)amino)-4-hydroxy-3-((4-methyl-2-sulfophenyl)azo)-, trisodium salt (9CI)
5-((4-Chloro-6-(methylphenylamino)-1,3,5-triazin-2-yl)amino)-4-hydroxy-3-((4-methyl-2-sulfophenyl)azo)-2,7-naphthalenedisulfonic acid, trisodium salt

70225-91-1
Unknown
Unknown
Atlox 3409F (9CI)

70288-86-7
Unknown
Unknown
22,23-Dihydroavermectin B1
Hyvermectin
Ivermectin (9CI)
MK 933

70321-63-0
Unknown
Unknown
Lanolin oil
Oil, Misc.
Oils, Lanolin

70321-79-8
Unknown
Unknown
Creosote Oil (Derived from any source)
Creosote Oil, High-boiling Distillate

70321-86-7
$C_{30}H_{29}N_3O$
447.56
Oc(c(cc(c1)C(c(cccc2)c2)(C)C)C(c(cccc3)c3)(C)C)c1N(N=C(C=4C=CC=5)C5)N4
Phenol, 2-(2H-benzotriazol-2-yl)-4,6-bis(1-methyl-1-phenylethyl)- (9CI)

70323-51-2
Unknown
Unknown
Santicizer 143 (9CI)

70343-06-5
$C_7H_7N_3O_4$
197.17
Cc1ccc(N(=O)=O)c(N)c1N(=O)=O
m-Toluidine, 2,6-dinitro
3-Amino-2,4-dinitrotoluene
2,6-Dinitro-m-toluidine

70356-09-1
$C_{20}H_{22}O_3$
310.39
O=C(c(ccc(c1)C(C)(C)C)c1)CC(=O)c(ccc(OC)c2)c2
Butyl methoxydibenzoylmethane
1-(4-(1,1-Dimethylethyl)phenyl)-3-(4-methoxyphenyl)-1,3-propanedione
Parsol 1789
1,3-Propanedione, 1-(4-(1,1-dimethylethyl)phenyl)-3-(4-methoxyphenyl)- (9CI)

70362-46-8
$C_{12}H_6Cl_4$
291.99
1,1'-Biphenyl, 2,2',3,5-tetrachloro- (9CI)
2,2',3,5-PCB
2,2',3,5-Tetrachlorobiphenyl

70362-47-9
$C_{12}H_6Cl_4$
291.99
1,1'-Biphenyl, 2,2',4,5-tetrachloro- (9CI)
2,2',4,5-PCB
2,2',4,5-Tetrachlorobiphenyl

70362-48-0
$C_{12}H_6Cl_4$
291.99
1,1'-Biphenyl, 2,3',4',5'-tetrachloro- (9CI)
2,3',4',5'-PCB
2,3',4',5'-Tetrachlorobiphenyl

70362-49-1
$C_{12}H_6Cl_4$
291.99
1,1'-Biphenyl, 3,3',4,5-tetrachloro- (9CI)
3,3',4,5-PCB
3,3',4,5-Tetrachlorobiphenyl

70362-50-4
$C_{12}H_6Cl_4$
291.99
1,1'-Biphenyl, 3,4,4',5-tetrachloro- (9CI)
3,4,4',5-PCB
3,4,4',5-Tetrachlorobiphenyl

70424-68-9
$C_{12}H_5Cl_5$
326.44
1,1'-Biphenyl, 2,3,3',4',5-pentachloro- (9CI)
2,3,3',4',5-PCB

2,3,3',4',5-Pentachlorobiphenyl

70424-69-0
$C_{12}H_5Cl_5$
326.44
1,1'-Biphenyl, 2,3,3',4,5-pentachloro- (9CI)
2,3,3',4,5-PCB
2,3,3',4,5-Pentachlorobiphenyl

70431-21-9
Unknown
Unknown
Surfonic LF 17 (9CI)

70439-96-2
$C_7H_5Cl_4NO$
260.93
Benzenamine, 2,3,5,6-tetrachloro-4-methoxy- (9CI)

70528-90-4
$C_{15}H_{12}ClN_5O_4$
361.71
O=C(N(C(O)=C(N=Nc(c(N(=O)=O)cc(c1)Cl)c1)C=2C)CC)C2C#N
3-Pyridinecarbonitrile, 5-((4-chloro-2-nitrophenyl)azo)-1-ethyl-1,2-dihydro-6-hydroxy-4-methyl-2-oxo- (9CI)
5-((4-Chloro-2-nitrophenyl)azo)-1-ethyl-1,2-dihydro-6-hydroxy-4-methyl-2-oxo-3-pyridinecarbonitrile

70592-80-2
Unknown
Unknown
(C10-C16-Alkyl)dimethylamines, N-oxides
(C10-C16)Alkyldimethylamine oxide
Amines, C10-16-alkyldimethyl, N-oxides

70616-90-9
$C_{24}H_{16}Cl_2N_6O_{10}S_3.3Na$
784.46
1,5-Naphthalenedisulfonic acid, 2-((6-((4,6-dichloro-1,3,5-triazin-2-yl)methylamino)-1-hydroxy-3-sulfo-2-naphthalenyl)azo)-, trisodium salt (9CI)
2-((6-((4,6-Dichloro-1,3,5-triazin-2-yl)methylamino)-1-hydroxy-3-sulfo-2-naphthalenyl)azo)-1,5-naphthalenedisulfonic acid, trisodium salt

70648-13-4
$C_{12}H_5Cl_3O$
271.53
Dibenzofuran, 1,4,9-trichloro- (9CI)
1,4,9-Trichlorodibenzofuran

70648-14-5
$C_{12}H_6Cl_2O$
237.08
Dibenzofuran, 1,9-dichloro- (9CI)
1,9-Dichlorodibenzofuran

70648-15-6
$C_{12}H_3Cl_5O$
340.42
Dibenzofuran, 1,3,4,6,9-pentachloro- (9CI)
1,3,4,6,9-Pentachlorodibenzofuran

70648-16-7
$C_{12}H_4Cl_4O$
305.97
Dibenzofuran, 1,3,4,7-tetrachloro- (9CI)
1,3,4,7-Tetrachlorodibenzofuran

70648-18-9
$C_{12}H_4Cl_4O$
305.97
Dibenzofuran, 1,2,6,9-tetrachloro- (9CI)
1,2,6,9-Tetrachlorodibenzofuran

70648-19-0
$C_{12}H_4Cl_4O$
305.97
Dibenzofuran, 1,4,6,9-tetrachloro- (9CI)
1,4,6,9-Tetrachlorodibenzofuran

70648-20-3
$C_{12}H_3Cl_5O$
340.42
Dibenzofuran, 1,3,4,7,9-pentachloro- (9CI)
1,3,4,7,9-Pentachlorodibenzofuran

70648-21-4
$C_{12}H_3Cl_5O$
340.42
Dibenzofuran, 1,3,6,7,8-pentachloro- (9CI)
1,3,6,7,8-Pentachlorodibenzofuran

70648-22-5
$C_{12}H_4Cl_4O$
305.97
Dibenzofuran, 1,2,8,9-tetrachloro- (9CI)
1,2,8,9-Tetrachlorodibenzofuran

70648-23-6
$C_{12}H_3Cl_5O$
340.42
Dibenzofuran, 1,2,4,8,9-pentachloro- (9CI)
1,2,4,8,9-Pentachlorodibenzofuran

70648-24-7
$C_{12}H_3Cl_5O$
340.42
Dibenzofuran, 1,2,4,6,9-pentachloro- (9CI)
1,2,4,6,9-Pentachlorodibenzofuran

70648-25-8
$C_{12}HCl_7O$
409.31
Dibenzofuran, 1,2,3,4,6,7,9-heptachloro- (9CI)
1,2,3,4,6,7,9-Heptachlorodibenzofuran

70648-26-9
$C_{12}H_2Cl_6O$
374.84
Dibenzofuran, 1,2,3,4,7,8-hexachloro
1,2,3,4,7,8-Hexachlorodibenzofuran

70657-70-4
$C_6H_{12}O_3$
132.18
Acetic acid, 2-methoxypropyl ester
2-Methoxy-1-propyl acetate
2-Methoxypropylacetate-1

70693-06-0
$C_{10}H_{20}$
140.27
Aromatic hydrocarbons, C9-11

70693-50-4
$C_{30}H_{29}N_3O_3$
479.56
O=N(=O)c(cccc1)c1N=Nc(c(O)c(cc2C(c(cccc3)c3)(C)C)C(c(cccc4)c4)(C)C)c2
Phenol, 2,4-bis(1-methyl-1-phenylethyl)-6-((2-nitrophenyl)azo)- (9CI)
2-Nitro-2'-hydroxy-3',5'-bis-(α,α-dimethylbenzyl)azobenzene

70700-59-3
$C_{15}H_{26}NS.Cl$

287.89
Pyridinium, 3-methyl-1-((octylthio)methyl)-, chloride (9CI)

70700-60-6
$C_{19}H_{34}NS.Cl$
344.05
Pyridinium, 1-((dodecylthio)-methyl)-3-methyl-, chloride
1-((Dodecylthio)methyl)-3-methylpyridinium chloride
3-Methyl-n-dodecylthiomethyl-pyridinium chloride

70700-62-8
$C_{16}H_{28}NS.Cl$
301.92
Pyridinium, 3,5-dimethyl-1-((octylthio)methyl)-, chloride (9CI)

70700-63-9
$C_{20}H_{36}NS.Cl$
358.03
Pyridinium, 1-((dodecylthio)-methyl)-3,5-dimethyl-, chloride (9CI)

70709-94-3
$C_{17}H_{36}O_2$
272.47
Ethanol, 2-(pentadecyloxy)-(9CI)

70709-95-4
$C_{13}H_{28}O_2$
216.37
Ethanol, 2-((1-pentylhexyl)oxy)- (9CI)
Ethylene glycol mono-6-undecyl ether

70709-96-5
$C_{17}H_{36}O_2$
272.48
Ethanol, 2-((1-heptyloctyl)-oxy)- (9CI)
Ethylene glycol mono-8-pentadecyl ether

70709-97-6
$C_{12}H_{26}O_2$
202.34
Ethanol, 2-(1,1-dipropylbutoxy)- (9CI)
Ethylene glycol mono-(1,1-dipropylbutyl) ether

70709-98-7
$C_{18}H_{38}O_2$
286.50
Ethanol, 2-((1,1-dipentylhexyl)oxy)- (9CI)
Ethylene glycol mono-(1,1-dipentylhexyl) ether

70709-99-8
$C_{14}H_{28}O_2$
228.38
Ethanol, 2-((1-hexylcyclohexyl)oxy)- (9CI)
Ethylene glycol mono-(1-hexyl-1-cyclohexyl) ether

70710-00-8
$C_{18}H_{36}O_2$
284.49
Ethanol, 2-((1-decylcyclohexyl)oxy)- (9CI)
Ethylene glycol mono-(1-decyl-1-cyclohexyl) ether

70729-68-9
$C_{22}H_{42}O_7$
418.64
O=C(OCCOCCOCCOCCOC(=O)CCCCCC)CCCCCC
Heptanoic acid, oxybis(2,1-ethanediyloxy-2,1-ethanediyl) ester
TEGDH
Tetraethylene glycol-di-n-heptanoate

70750-47-9
Unknown
Unknown
(Coco alkyl) bis(2-hydroxyethyl) methyl ammonium chloride
N,N-Di(2-hydroxyethyl)-N-coco-N-methylammonium chloride
Quaternary ammonium compounds, coco alkylbis-(hydroxyethyl)methyl, chlorides

70776-26-0
$C_{16}H_{32}Cl_2S_2$
359.47
S(SCC(CCCCCC)Cl)CC(CCCCCC)Cl
Disulfide, bis(2-chlorooctyl) (9CI)
Bis(2-chlorooctyl) disulfide

70781-06-5
Unknown
Unknown
SN 38210 (9CI)

70781-07-6
Unknown
Unknown
SN 38212 (9CI)

70816-59-0
$C_2H_2N_4$
82.04
Tetrazine (French)
Tetracina (Spanish)
Tetrazine, Dry

70840-42-5
$C_9H_{11}Cl_3NO_3PS.C_4H_7Cl_2O_4P$
571.56
Phosphoric acid, 2,2-dichloroethenyl dimethyl ester, Mixt. with O,O-diethyl O-(3,5,6-trichloro-2-pyridinyl) phosphorothioate (9CI)

70848-82-7
$C_{12}H_8O_2$
184.19
Naphthalenedicarboxaldehyde (9CI)

70872-82-1
$C_{12}H_3Cl_5O$
340.42
Dibenzofuran, 1,2,6,7,9-pentachloro- (9CI)
1,2,6,7,9-Pentachlorodibenzofuran

70892-59-0
Unknown
Unknown
Montmorillonite ((Al1.33-1.67Mg0.33-0.67)(Ca0-1Na 0-1)0.33Si4(OH)2O10.x H2O), calcined
Montmorillonite clay, calcined

70910-35-9
Unknown
Unknown
Lead arsenite, Solid
UN 1618

70913-86-9
Unknown
Unknown
Alkanes, C18-70
(C18-C70) Paraffins

70942-15-3
$C_{20}H_{10}CrN_2O_8S_2.Na$
545.41
Chromate(1-), (3-hydroxy-4-((1-hydroxy-8-sulfo-2-naphthalenyl)azo)-1-naphthalenesulfonato(4-))-, sodium (9CI)

70958-50-8
$C_5H_7ClN_4S$
190.63
4,6-Pyrimidinediamine, 2-chloro-5-(methylthio)-

(9CI)

71000-82-3
CNO
42.01
Isocyanate (9CI)

71011-12-6
Unknown
Unknown
Alkanes, C12-13, chloro
Chlorinated paraffins

71011-26-2
Unknown
Unknown
Quaternary ammonium compounds, benzyl(hydrogenated tallow alkyl)dimethyl, chlorides, Compds. with hectorite
Dimethyl benzyl hydrogenated tallow ammonium chloride, Reaction product with hectorite

71011-27-3
Unknown
Unknown
Quaternary ammonium compounds, bis(hydrogenated tallow alkyl)dimethyl, chlorides, Compds. with hectorite
Dimethyl dihydrogenated tallow ammonium chloride, Reaction product with hectorite

71012-25-4
$C_{20}H_{13}N$
267.32
Dibenzocarbazole (9CI)

71033-08-4
$C_{35}H_{52}O_8$
600.79
O(C1COC(COCCCC)COc(ccc(c2)C(c(ccc(OCC(OCC(O3)C3)COCCCC)c4)c4)(C)C)c2)C1
Oxirane, 2,2'-((1-methylethylidene)bis(4,1-phenyleneoxy(1-(butoxymethyl)-2,1-ethanediyl)oxymethylene))bis- (9CI)
2,2-Bis(p-(2-glycidyloxy-3-butoxypropyloxy)phenyl)propane

71181-76-5
$C_{10}H_{10}F_{15}O_5P$
526.14
Phosphorane, pentakis-(2,2,2-trifluoroethoxy)- (9CI)

71195-58-9
Unknown
Unknown
Alfentanil
Alfentanilum (Latin)
DEA No. 9737

71206-09-2
$C_{13}H_{24}O_3$
228.33
Cardura E10
tert-Decanoic acid oxiranylmethyl ester

71242-00-7
$C_{17}H_{34}N_2O.ClH$
318.92
1H-Imidazole-1-ethanol, 2-dodecyl-4,5-dihydro-, monohydrochloride (9CI)

71245-27-7
Unknown
Unknown
Dechlorane 604 (9CI)

71261-64-8
$C_3H_7NO_2$
89.09
DL-Alanine-15N (9CI)
Alanine-15N (6CI,7CI)

71264-32-9
Unknown
Unknown
Nitrilotriacetic acid, diammonium salt
NTA, diammonium salt

71267-22-6
$C_{10}H_{11}N_3O$
189.24
2,3'-Bipyridine, 1,2,3,6-tetrahydro-1-nitroso
N'-Nitrosoanatabine
1,2,3,6-Tetrahydro-1-nitroso-2,3'-bipyridine

71277-90-2
$C_{22}H_{20}$
284.40
Chrysene, tetramethyl- (9CI)

71326-18-6
$C_{11}H_{24}N_2$
184.31
1,3-Propanediamine, N'-cyclohexyl-N,N-dimethyl- (9CI)

71328-89-7
Unknown
Unknown
Aroclor 1240 (9CI)

71342-62-6
C_4HClF_8
236.49
Butane, chlorooctafluoro- (9CI)

71484-80-5
$C_6H_6CuNO_6.H_4N$
269.69
Nitrilotriacetic acid, copper-(2+) complex ammonium salt
(N,N-Bis(carboxymethyl)glycinato(3-)-N,O,O',O'')cuprate-(-1), ammonium
Cuprate(1-), (N,N-bis(carboxymethyl)glycinato(3-)-N,O,O',O'')-, ammonium, (T-4)- (9CI)
NTA, copper(2+) complex ammonium salt

71489-58-2
$C_8H_4Cl_4$
241.93
Benzene, ethenyl-, tetrachloro deriv. (9CI)

71549-78-5
$C_{26}H_{42}O_4$
418.62
1,2-Benzenedicarboxylic acid, dinonyl ester, branched
Di-(C9-branched alkyl) phthalate

71566-41-1
$C_{41}H_{30}N_8O_{14}S_2.3Na$
991.79
Benzoic acid, 2-((2-amino-5-hydroxy-6-((4'-((2-hydroxy-6-sulfo-1-naphthalenyl)azo)-3,3'-dimethoxy(1,1'-biphenyl)-4-yl)azo)-7-sulfo-1-naphthalenyl)azo)-5-nitro-, trisodium salt (9CI)
2-((2-Amino-5-hydroxy-6-((4'-((2-hydroxy-6-sulfo-1-naphthalenyl)azo)-3,3'-dimethoxy-(1,1'-biphenyl)-4-yl)azo)-7-sulfo-1-naphthalenyl)azo)-5-nitrobenzoic acid, trisodium salt
3,3'-Dimethoxy-4-((8-((2-carboxy-4-nitrophenyl)azo)-7-amino-4-hydroxy-2-sulfonaphth-3-yl)azo)-4'-((2-hydroxy-6-sulfonaphth-1-yl)azo)-1,1'-biphenyl, trisodium salt

71607-70-0
$C_{18}H_{18}$

234.34
Phenanthrene, tetramethyl- (9CI)

71617-28-2
$C_{17}H_{17}Cl_2N_5O_4$
426.23
O=C(Nc(c(N=Nc(c(cc(N(=O)=O)c1)Cl)c1)cc(c2NCC(O)C)Cl)c2)C
Acetamide, N-(4-chloro-2-((2-chloro-4-nitrophenyl)azo)-5-((2-hydroxypropyl)amino)phenyl)- (9CI)
N-(4-Chloro-2-((2-chloro-4-nitrophenyl)azo)-5-((2-hydroxypropyl)amino)phenyl)-acetamide

71626-11-4
$C_{20}H_{23}NO_3$
325.44
Alanine, N-(2,6-dimethylphenyl)-N-(phenylacetyl)-, methyl ester, dl
Benalaxyl
Galben
M 9834

71665-99-1
$C_{12}H_4Cl_4O_2$
321.97
Dibenzo(b,e)(1,4)dioxin, 1,2,4,9-tetrachloro- (9CI)
1,2,4,9-Tetrachlorodibenzo-(b,e)(1,4)dioxin
1,2,4,9-Tetrachlorodibenzo-p-dioxin

71669-23-3
$C_{12}H_4Cl_4O_2$
321.97
Dibenzo(b,e)(1,4)dioxin, 1,2,7,9-tetrachloro- (9CI)
1,2,7,9-Tetrachlorodibenzo-(b,e)(1,4)dioxin
1,2,7,9-Tetrachlorodibenzo-p-dioxin

71669-24-4
$C_{12}H_4Cl_4O_2$
321.97
Dibenzo(b,e)(1,4)dioxin, 1,3,6,9-tetrachloro- (9CI)
1,3,6,9-Tetrachlorodibenzo-(b,e)(1,4)dioxin
1,3,6,9-Tetrachlorodibenzo-p-dioxin

71669-25-5
$C_{12}H_4Cl_4O_2$
321.97
Dibenzo(b,e)(1,4)dioxin, 1,2,3,6-tetrachloro- (9CI)
1,2,3,6-Tetrachlorodibenzo-(b,e)(1,4)dioxin
1,2,3,6-Tetrachlorodibenzo-p-dioxin

71669-26-6
$C_{12}H_4Cl_4O_2$
321.97
Dibenzo(b,e)(1,4)dioxin, 1,2,3,9-tetrachloro- (9CI)
1,2,3,9-Tetrachlorodibenzo-(b,e)(1,4)dioxin
1,2,3,9-Tetrachlorodibenzo-p-dioxin

71669-27-7
$C_{12}H_4Cl_4O_2$
321.97
Dibenzo(b,e)(1,4)dioxin, 1,2,4,6-tetrachloro- (9CI)
1,2,4,6-Tetrachlorodibenzo-(b,e)(1,4)dioxin
1,2,4,6-Tetrachlorodibenzo-p-dioxin

71669-28-8
$C_{12}H_4Cl_4O_2$
321.97
Dibenzo(b,e)(1,4)dioxin, 1,2,4,7-tetrachloro- (9CI)
1,2,4,7-Tetrachlorodibenzo-(b,e)(1,4)dioxin
1,2,4,7-Tetrachlorodibenzo-p-dioxin

71669-29-9
$C_{12}H_4Cl_4O_2$
321.97
Dibenzo(b,e)(1,4)dioxin, 1,2,4,8-tetrachloro- (9CI)
1,2,4,8-Tetrachlorodibenzo-(b,e)(1,4)dioxin
1,2,4,8-Tetrachlorodibenzo-p-dioxin

71697-59-1
$C_{22}H_{19}Cl_2NO_3$
416.32
Cyclopropanecarboxylic acid, 3-(2,2-dichloroethenyl)-2,2-dimethyl-, cyano(3-phenoxyphenyl) methyl ester, (1-α(S*),3-β)-(+-)
NRDC 159

71698-60-7
$C_{22}H_{20}ClF_3O_3$
424.85
Cyclopropanecarboxylic acid, 3-(2-chloro-3,3,3-trifluoro-1-propenyl)-2,2-dimethyl-, (3-phenoxyphenyl)methyl ester (9CI)

71700-95-3
Unknown
Unknown
Versatic 9-11 acid
Versatic 9-11
Versatic acid 911

71701-30-9
$C_{36}H_{28}N_4O_{10}S_3.2Na$
818.79
2,7-Naphthalenedisulfonic acid, 3-((3,3'-dimethyl-4'-((4-((phenylsulfonyl)oxy)phenyl)azo)(1,1'-biphenyl)-4-yl)azo)-4-hydroxy-, disodium salt (9CI)

71729-96-9
$C_{27}H_{55}N_2.Cl$
443.19
1H-Imidazolium, 1,1-didodecyl-dihydro-, chloride (9CI)

71732-95-1
$C_{17}H_{36}N_2O.ClH$
320.93
Dodecanamide, N-(3-(dimethylamino)propyl)-, monohydrochloride (9CI)

71732-96-2
$C_{21}H_{32}N.Cl$
333.94
Isoquinolinium, 2-dodecyl-, chloride (9CI)

71753-42-9
$C_{12}H_8Cl_4N_2$
322.01
Hydrazine, 1,2-bis(3,4-dichlorophenyl)- (9CI)
3,3',4,4'-Tetrachlorohydrazobenzene
3,4,3',4'-Tetrachlorohydrazobenzene

71764-17-5
$C_{32}H_{67}N.C_2H_4O_2$
525.94
1-Hexadecanamine, N-hexadecyl-, acetate (9CI)

71769-74-9
$C_9H_8N_4OS$
220.25
Photothidiazuron
N-Phenyl-N'-1,2,5-thiadiazol-3-ylurea
Urea, N-phenyl-N'-1,2,5-thiadiazol-3-yl-

71786-60-2
Unknown
Unknown
Alkyl-N,N-bis(2-hydroxyethyl)amine
N-(C12-C18)Alkyldiethanolamine

Ethanol, 2,2'-iminobis-, N-C12-18-alkyl derivs.

71808-64-5
$C_{11}H_{23}ClO_4Si$
282.84
Silane, (3-chloropropyl)-dimethoxy(3-(oxiranyl-methoxy)propyl)- (9CI)
(3-Glycidoxypropyl)(3-chloro-propyl)dimethoxysilane

71819-57-3
Unknown
Unknown
C.I. Acid Yellow 219 (9CI)

71859-30-8
$C_{12}H_4Cl_6O$
376.88
Benzene, 1,1'-oxybis(2,4,5-tri-chloro- (9CI)

71868-10-5
$C_{15}H_{21}NO_2S$
279.40
1-Propanone, 2-methyl-1-(4-(methylthio)phenyl)-2-(4-morpholinyl)- (9CI)

71872-22-5
Unknown
Unknown
C.I. Acid Green 108 (9CI)
Acidol Green M-FGL

71878-19-8
$(C_{35}H_{66}N_8)x$
Polymer
Poly((6-((1,1,3,3-tetramethyl-butyl)amino)-1,3,5-triazine-2,4-diyl)((2,2,6,6-tetramethyl-4- piperidinyl)imino)-1,6-hexanediyl((2,2,6,6-tetra-methyl-4-piperidinyl)imino))
Chimassorb 944
CR-144
Hals 3

71888-89-6
Unknown
Unknown
1,2-Benzenedicarboxylic acid, di-C6-8-branched alkyl esters, C7-rich

71902-14-2
Unknown
Unknown
C.I. Reactive Blue 170 (9CI)

71925-15-0
$C_{12}H_3Cl_5O_2$
356.42
Dibenzo(b,e)(1,4)dioxin, 1,2,3,6,7-pentachloro- (9CI)
1,2,3,6,7-Pentachlorodibenzo-(b,e)(1,4)dioxin
1,2,3,6,7-Pentachlorodibenzo-p-dioxin

71925-16-1
$C_{12}H_3Cl_5O_2$
356.42
Dibenzo(b,e)(1,4)dioxin, 1,2,3,6,8-pentachloro- (9CI)
1,2,3,6,8-Pentachlorodibenzo-(b,e)(1,4)dioxin
1,2,3,6,8-Pentachlorodibenzo-p-dioxin

71925-17-2
$C_{12}H_3Cl_5O_2$
356.42
Dibenzo(b,e)(1,4)dioxin, 1,2,3,7,9-pentachloro- (9CI)
1,2,3,7,9-Pentachlorodibenzo-(b,e)(1,4)dioxin
1,2,3,7,9-Pentachlorodibenzo-p-dioxin

71925-18-3
$C_{12}H_3Cl_5O_2$
356.42
Dibenzo(b,e)(1,4)dioxin, 1,2,3,8,9-pentachloro- (9CI)
1,2,3,8,9-Pentachlorodibenzo-(b,e)(1,4)dioxin
1,2,3,8,9-Pentachlorodibenzo-p-dioxin

71998-72-6
$C_{12}H_4Cl_4O$
305.98
1,3,6,8-Tetrachlorodibenzo-furan
1,3,6,8-TCDF

71998-73-7
$C_{12}H_4Cl_4O$
305.97
Dibenzofuran, 1,2,4,6-tetra-chloro- (9CI)
1,2,4,6-Tetrachlorodibenzo-furan

71998-74-8
$C_{12}H_3Cl_5O$
340.42
Dibenzofuran, 1,2,4,7,9-penta-chloro- (9CI)
1,2,4,7,9-Pentachlorodibenzo-furan

71998-75-9
$C_{12}H_2Cl_6O$
374.86
Dibenzofuran, 1,3,4,6,7,8-hexa-chloro- (9CI)
1,3,4,6,7,8-Hexachlorodibenzo-furan

71998-76-0
$C_{12}H_3Cl_5O_2$
356.42
Dibenzo(b,e)(1,4)dioxin, 1,2,4,6,8-pentachloro- (9CI)
1,2,4,6,8-Pentachlorodibenzo-(b,e)(1,4)dioxin
1,2,4,6,8-Pentachlorodibenzo-p-dioxin

72030-26-3
$C_5H_2Cl_6$
274.79
Cyclopentene, hexachloro- (9CI)
Hexachlorocyclopentene

72050-94-3
$C_{46}H_{90}N_4O_3$
747.22
L-Asparagine, N,N2-bis(3-(9-octadecenylamino)-propyl)-, (Z,Z)- (9CI)

72121-83-6
$C_{22}H_{23}O_4P$
382.40
O=P(Oc(cccc1)c1)(Oc(c(ccc2C)C)c2)Oc(c(ccc3C)C)c3
Phosphoric acid, bis(2,5-di-methylphenyl) phenyl ester (9CI)
Di(2,5-dimethylphenyl) phenyl phosphate
Di-p-xylyl phenyl phosphate

72175-27-0
$C_9H_{13}ClO$
172.65
2-Cyclohexen-1-one, 2(or 4)-chloro-3,5,5-trimethyl- (9CI)

72252-48-3
$C_{14}H_8ClF_3O_3.K$
355.76
Benzoic acid, 3-(2-chloro-4-(trifluoromethyl)phen-oxy)-, potassium salt (9CI)
Potassium 3-(2-chloro-4-(tri-fluoromethyl)phenoxy)benzo-ate

72254-06-9
Unknown
Unknown
Indenopyrene (9CI)

72254-58-1
$C_{12}H_{11}N_3.C_2H_4O_2$
257.32
CC(O)=O.Cc2nc(N)cc3[nH]c1cc

cccl c23
5H-Pyrido(4,3-b)indole, 3-amino-1-methyl-, acetate
3-Amino-1-methyl-5H-pyrido-(4,3-b)indole acetate
5H-Pyrido(4,3-b)indol-3-amine, 1-methyl-, monoacetate
TRP-P-2 (Acetate)

72269-41-1
Unknown
Unknown
2,5-Pyrrolidinedione, 1-(2-((2-((2-((2-aminoethyl)amino)-ethyl)amino)ethyl)amino)-ethyl)-, monopolyisobutenyl derivs., reaction products with molybdenum oxide (MoO_3), sulfurized
Tetraethylenepentamine poly-isobutylene succinimide, molybdenum complex, sulfur-ized

72269-48-8
$C_{15}H_{28}O_2$
240.39
O=C(OCCCC=CCCCCCCCC)C
(E)-4-Tridecen-l-yl acetate
AI3-34338
Caswell No. 456DD
EPA Pesticide Chemical Code 121902
4-Tridecen-1-ol, acetate, (E)-
E-4-Tridecenyl acetate
(E)-4-Tridecenyl acetate

72274-16-9
$C_{22}H_{17}ClFNO_3$
397.84
DL-Phenylalanine, N-ben-zoyl-N-(3-chloro-4-fluoro-phenyl)- (9CI)

72319-24-5
$C_{57}H_{64}O_8$
877.13
Oxirane, 2,2'-((1-methyl-ethylidene)bis(4,1-phenyl-eneoxy-3,1-propanediyloxy-4,1-phenylene(1-methyl-ethylidene)-4,1-phenylene-oxymethylene))bis- (9CI)

72378-89-3
Unknown
Unknown
Tin-sodium-tartrate
Sodium stannous tartrate

72382-90-2
$C_{36}H_{20}$
452.56
Dibenzo(a,o)naphtho-(1,2,3,4-rst)pentaphene (9CI)

72382-91-3
$C_{44}H_{24}$
552.68
Tetrabenzo(a,c,m,o)-naphtho(1,2,3,4-rst)-pentaphene (9CI)

72382-92-4
$C_{50}H_{26}$
626.76
Tetranaphtho(3,2,1-de: 1',2',3'-jk:3'',2'',1''-op: 1''',2''',3'''-uv)p entac-ene (9CI)

72391-46-9
$C_{13}H_{11}Cl_2NO_5$
332.15
Oxazolidine-5-carboxylic acid, 3-(3,5-dichlorophenyl)-2,4-dioxo-5-methyl-, ethyl ester
Chlozolinate
Dichlozolinate
Ethyl 3-(3,5-dichlorophenyl)-5-methyl-2,4-dioxo-5-oxazol-idine carboxylate
M 8164
Manderol
Serinal

72428-03-6
$C_8H_{10}O_2S$
170.23
Benzene, methyl(methylsulf-onyl)- (9CI)

72490-01-8
$C_{17}H_{19}NO_4$
301.34
Carbamic acid, (2-(4-phenoxy-phenoxy)ethyl)-, ethyl ester (9CI)
Caswell No. 652C
EPA Pesticide Chemical Code 125301
Ethyl 2-(4-phenoxyphenoxy)-ethylcarbamate
Ethyl (2-(4-phenoxyphenoxy)-ethyl)carbamate
Ethyl(2-(p-phenoxyphenoxy)-ethyl)carbamate
Fenoxycarb
Insegar
N-(2-(p-Phenoxyphenoxy)ethyl)-carbamic acid
(2-(4-Phenoxyphenoxy)ethyl)-carbamic acid ethyl ester
Ro 13-5223

72496-88-9
$C_{32}H_{28}CoN_8O_{10}S_2.Na$
830.63
Cobaltate(1-), bis(2-((5-(aminosulfonyl)-2-hydroxy-phenyl)azo)-3-oxo-N-phenyl-butanamidato- (2-))-, sodium (9CI)

72542-56-4
$C_{14}H_{23}N_3O_4PS_3$
424.51
Ethanimodithioic acid, N-((((((5,5-dimethyl-1,3,2-di-oxaphosphorinan-2-yl)(1,1-dimethylethyl)amino)thio)-methylamino)carbonyl)oxy)-, methyl ester, P-sulfide
AI3-29549

72623-82-6
Unknown
Unknown
Imidazolium compounds, 2-C13-17-alkyl-1-(2-C14-18-amidoethyl)-4,5-dihydro-3-methyl, Me sulfates
1-(2-(C14-C18)-Alkylamido-ethyl)-2-nor(C14-C18)alkyl-3-methylimidazolinium methyl sulfate

72629-49-3
Unknown
Unknown
Nitrilotriacetic acid, dilithium salt
NTA, dilithium salt

72668-27-0
$C_{27}H_{33}O_4P$
452.53
Phenol, 3-(1-methylethyl)-, phosphate (3:1) (9CI)
m-Cumenyl phosphate

72674-05-6
Unknown
Unknown
Sulfonic acid, α-alkene
AOS
α-Olefin sulfonate
α-Olefin sulphonate

72776-75-1
$C_{25}H_{18}O$
334.42
Methanone, bis((1,1'-biphenyl)-yl)- (9CI)

72776-77-3
$C_{15}H_{11}N$
205.25
Quinoline, phenyl- (9CI)

72828-64-9
$C_{23}H_{24}N_6O_4$
448.45
O=C(OCCN(c(ccc(N=Nc(c(C#N)

cc(N(=O)=O)c1)c1C#N)c2C)c2)CCCC)C
1,3-Benzenedicarbonitrile, 2-((4-((2-(acetyloxy)ethyl)-butylamino)-2-methyl-phenyl)azo)-5-nitro- (9CI)
N-Butyl-N-(2-acetoxyethyl)-4-((4-nitro-2,6-dicyanophenyl)-azo)-3-methylbenzeneamine

72918-21-9
$C_{12}H_2Cl_6O$
374.86
Dibenzofuran, 1,2,3,7,8,9-hexa-chloro- (9CI)
1,2,3,7,8,9-Hexachlorodibenzo-furan

72968-42-4
Unknown
Unknown
Gentian Extract
Bitter Root
Enzianwurzel
Pale Gentian
Yellow Gentian

72979-85-2
Unknown
Unknown
C.I. Reactive Red 133 (9CI)
Remazol Printing Rhodamine BB

73049-73-7
Unknown
Unknown
Tryptones
Hydrolyzed protein
Peptones

73090-68-3
Unknown
Unknown
tert-Butyl tetralin

73090-69-4
Unknown
Unknown
Chloro-4-tert-amylphenol

73131-17-6
$(C_2H_4O)xC_{20}H_{43}NO_4S$
Polymer NA
Poly(oxy-1,2-ethanediyl), α-sulfo-ω-(2-(hexa-decyldimethylammonio)-ethoxy)-, hydroxide, inner salt (9CI)

73138-29-1
Unknown
Unknown
Alkanes, C10-18
(C10-C18) Alkanes

73179-37-0
$C_{25}H_{29}O_4P$
424.48
O=P(Oc(c(ccc1)C)c1C)(Oc(c(cc2)C)c2C)Oc(c(cc(c3)C)C)c3C
Phosphoric acid, bis(2,6-di-methylphenyl) 2,4,6-tri-methylphenyl ester (9CI)
Di(2,6-dimethylphenyl) 2,4,6-trimethylphenyl phosphate

73179-38-1
$C_{26}H_{31}O_4P$
438.50
O=P(Oc(c(cc(c1)C)C)c1C)(Oc(cc(c2)C)C)c2C)Oc(c(ccc3C)C)c3
Phosphoric acid, 2,5-dimethyl-phenyl bis(2,4,6-trimethyl-phenyl) ester (9CI)

73179-40-5
$C_{20}H_{19}O_4P$
354.34
O=P(Oc(cccc1)c1)(Oc(cccc2)c2)Oc(c(ccc3C)C)c3
Phosphoric acid, 2,5-dimethyl-phenyl diphenyl ester (9CI)

73179-41-6
$C_{23}H_{25}O_4P$
396.42
O=P(Oc(cccc1)c1)(Oc(c(ccc2)C)c2C)Oc(c(cc(c3)C)C)c3C
Phosphoric acid, 2,6-di-methylphenyl phenyl 2,4,6-trimethylphenyl ester (9CI)
2,4,6-Trimethylphenyl 2,6-di-methylphenyl phenyl phos-phate

73179-42-7
$C_{23}H_{25}O_4P$
396.42
O=P(Oc(cccc1)c1)(Oc(c(cc(c2)C)C)c2C)Oc(c(ccc3C)C)c3
Phosphoric acid, 2,5-dimethyl-phenyl phenyl 2,4,6-trimethyl-phenyl ester (9CI)

73179-43-8
$C_{21}H_{21}O_4P$
368.37
O=P(Oc(cccc1)c1)(Oc(cccc2)c2)Oc(c(cc(c3)C)C)c3C
Phosphoric acid, diphenyl 2,4,6-trimethylphenyl ester (9CI)
Diphenyl 2,4,6-trimethylphenyl phosphate

73179-44-9
$C_{24}H_{27}O_4P$
410.45
O=P(Oc(cccc1)c1)(Oc(c(cc(c2)C)C)c2C)Oc(c(cc(c3)C)C)c3C
Phosphoric acid, phenyl bis-(2,4,6-trimethylphenyl) ester (9CI)
Di(2,4,6-trimethylphenyl) phenyl phosphate

73179-45-0
$C_{22}H_{23}O_4P$
382.40
O=P(Oc(cccc1)c1)(Oc(c(ccc2)C)c2C)Oc(c(ccc3C)C)c3
Phosphoric acid, 2,5-dimethyl-phenyl 2,6-dimethylphenyl phenyl ester (9CI)

73179-46-1
$C_{25}H_{29}O_4P$
424.48
O=P(Oc(c(ccc1)C)c1C)(Oc(c(cc(c2)C)C)c2C)Oc(c(ccc3C)C)c3
Phosphoric acid, 2,5-dimethyl-phenyl 2,6-dimethylphenyl 2,4,6-trimethylphenyl ester (9CI)

73179-47-2
$C_{25}H_{29}O_4P$
424.48
O=P(Oc(c(cc(c1)C)C)c1C)(Oc(c(ccc2C)C)c2)Oc(c(ccc3C)C)c3
Phosphoric acid, bis(2,5-di-methylphenyl) 2,4,6-trimethyl-phenyl ester (9CI)

73179-48-3
$C_{24}H_{27}O_4P$
410.45
O=P(Oc(c(ccc1)C)c1C)(Oc(c(ccc2)C)c2C)Oc(c(ccc3C)C)c3
Phosphoric acid, 2,5-dimethyl-phenyl bis(2,6-dimethyl-phenyl) ester (9CI)

73179-49-4
$C_{24}H_{27}O_4P$
410.45
O=P(Oc(c(ccc1)C)c1C)(Oc(c(ccc2C)C)c2)Oc(c(ccc3C)C)c3
Phosphoric acid, bis(2,5-di-methylphenyl) 2,6-dimethyl-phenyl ester (9CI)

73180-15-1
$C_6H_{12}S$
116.23
2H-Thiopyran, tetrahydro-methyl- (9CI)

73195-13-8
$C_{26}H_{31}O_4P$
438.50

O=P(Oc(c(ccc1)C)c1C)(Oc(c(cc(c2)C)C)c2C)Oc(c(cc(c3)C)C)c3C
Phosphoric acid, 2,6-dimethylphenyl bis(2,4,6-trimethylphenyl) ester (9CI)
Di(2,4,6-trimethylphenyl) 2,6-dimethylphenyl phosphate

73207-98-4
$C_9H_{22}NO_2PS$
239.31
Phosphonothioic acid, methyl-, S-(2-(bis(1-methylethyl)-amino)ethyl) ester (9CI)

73215-09-5
$C_{13}H_{18}N_4O_6$
326.31
Benzenamine, N-(1-ethylpropyl)-3,4-dimethyl-N,2,6-trinitro- (9CI)
N-Nitropendimethalin

73246-95-4
Unknown
Unknown
Benzaldehyde, 4-hydroxy-3-methoxy-, manuf. of, distn. residues
Vanillin Still Bottoms

73298-54-1
Unknown
Unknown
Calmodulin (ox brain)

73299-03-3
Unknown
Unknown
Benzothiadiazole (9CI)

73347-80-5
$C_{14}H_{12}O_2.2Na$
258.23
1,4-Dihydro-9,10-dihydroxy anthracene, disodium salt solution
9,10-Anthracenediol, 1,4-dihydro-, disodium salt (9CI)
1,4-Dihydro-9,10-anthracenediol disodium salt

73398-58-0
Unknown
Unknown
Amides, Vegetable-oil, N,N'-hexanediylbis-
1,6-Hexandiamine, Vegetable oil fatty acids diamide

73459-03-7
$C_{12}H_8O_3$
200.20
Cc2cc1occc1c3oc(=O)ccc23
2H-Furo(2,3-H)-1-benzopyran-2-one, 5-methyl
4-Hydroxy-6-methyl-5-benzofuranacrylic acid γ-lactone
5-Methylangelicin
5-Methyl-2H-furo(2,3-H)-1-benzopyran-2-one

73467-76-2
$C_{20}H_{12}$
252.32
Benzopyrene (9CI) (VAN)

73506-32-8
$H_4N_2.xH_2O_4Se$
Unknown
Hydrazine selenate
Hydrazine, selenate
Selenic acid, Compd. with hydrazine

73507-01-4
C_9H_{20}
128.26
Heptane, ethyl- (9CI)

73513-30-1
Unknown
Unknown
Methylpentaldehyde

73560-78-8
Unknown
Unknown
Dibenzofluorene (9CI)

73573-88-3
$C_{23}H_{34}O_5$
390.52
Compactin
7-(1,2,6,7,8,8a-Hexahydro-2-methyl-8- (2-methylbutyryloxy)naphthyl)-3-hydroxyheptan-5-olide
Mevastatin
Mevastatina (Spanish)
Mevastatine (French)
Mevastatinum (Latin)

73575-52-7
$C_{12}H_6Cl_4$
291.99
1,1'-Biphenyl, 2,3',4,5'-tetrachloro- (9CI)
2,3',4,5'-PCB
2,3',4,5'-Tetrachlorobiphenyl

73575-53-8
$C_{12}H_6Cl_4$
291.99
1,1'-Biphenyl, 2,3',4,5-tetrachloro- (9CI)
2,3',4,5-PCB

73575-57-2
$C_{12}H_5Cl_5$
326.44
1,1'-Biphenyl, 2,2',3,4,6'-pentachloro- (9CI)
2,2',3,4,6'-PCB
2,2',3,4,6'-Pentachlorobiphenyl

73602-65-0
$C_{16}H_{26}O_3S.Na$
321.44
Benzenesulfonic acid, 4-(1-methylnonyl)-, sodium salt (9CI)

73602-67-2
$C_{16}H_{26}O_3S.Na$
321.44
Benzenesulfonic acid, 4-(1-butylhexyl)-, sodium salt (9CI)

73622-98-7
$C_{10}H_{11}NO_4$
209.22
Carbanilic acid, 1-carboxyethyl ester
α-Carboxyethyl N-phenylcarbamate
Phenylcarbamic acid carboxyethyl ester
Propionic acid, 2-phenylcarbamoyloxy-

73727-39-6
C_3H_6BrClO
173.44
1-Propanol, 2-bromo-3-chloro- (9CI)
NSC-227854

73772-91-5
Unknown
Unknown
Nitrilotriacetic acid, magnesium salt
NTA, magnesium salt

73807-55-3
Unknown
Unknown
Allitin (9CI)

73908-22-2
C_9H_7ClO
166.61
1H-Inden-1-one, 2-chloro-2,3-dihydro-, (.+-.)- (9CI)

73908-23-3
C_8H_9BrO
201.07
Benzenemethanol, α-(bro-

momethyl)-, (R)- (9CI)

73908-26-6
$C_{15}H_{13}ClO$
244.72
Ethanone, 2-chloro-1-(4-methylphenyl)-2-phenyl-, (.+-.)- (9CI)

73908-28-8
$C_{10}H_{11}BrO$
227.11
1-Butanone, 2-bromo-1-phenyl-, (.+-.)- (9CI)

73908-29-9
$C_{10}H_{11}ClO$
182.65
1-Butanone, 2-chloro-1-phenyl-, (.+-.)- (9CI)

73926-79-1
Unknown
Unknown
Tin-potassium-tartrate
Stannous potassium tartrate

73928-09-3
$C_8H_{12}N_2O_4S_2$
264.34
Mannitol, 1,6-dideoxy-1,6-dithiocyanato-, (D)
1,6-Dithiocyanatomannitol
1,6-Dithiocyano-1,6-dideoxy-D-mannitol
Dithiocyanomannitol
DTM

73986-52-4
$C_{14}H_{20}Cl_2O$
275.24
Phenol, 2,6-dichloro-4-octyl
2,6-Dichloro-4-octylphenol-

74004-30-1
$C_2H_6S_2$
94.20
Ethanesulfenothioic acid (9CI)

74051-80-2
$C_{17}H_{29}NO_3S$
327.53
CCCC(=NOCC)C1=C(O)CC(CC(C)SCC)CC1=O
2-Cyclohexen-1-one, 2-(1-(ethoxyimino)butyl)-5-(2-(ethylthio)propyl)-3-hydroxy
Aljaden
Alloxol S
ARD 34/02
BAS 9052
BAS 9052H
BAS 90520H
Caswell No. 072A
Checkmate
2-Cyclohexen-1-one, 2-(1-(ethoxyimino)butyl)-5-(2-(ethylthio)propyl)-3-hydroxy- (9CI)
Cyethoxydim
EPA Pesticide Chemical Code 121001
2-((1-Ethoxyimino)butyl)-5-((ethylthio)propyl)-3-hydroxy-2-cyclohexen-1-one
2-(1-(Ethoxyimino)butyl)-5-(2-(ethylthio)propyl)-3-hydroxy-2-cyclohexen-1-one
(+-)-2-(1-(Ethoxyimino)butyl)-5-(2-(ethylthio)propyl)-3-hydroxy-2-cyclohexen-1-one
(+-)-(ZE)-2-(1-Ethoxyimino-butyl)-5-(2-(ethylthio)propyl)-3-hydroxycyclohex-2-enone
Expand
Fervinal
Nabu
NP 55
Poast
Sethoxydim
Sethoxydime (French)
SN 81742
Tritex-Extra

74082-93-2
$(C_6H_2Br_2O)x$
Polymer NA
Poly(oxy(dibromophenylene)) (9CI)
Fire Master TSA
Fire Master TSA-PO 64P
GLC 935P
PO 64P

74222-97-2
$C_{15}H_{16}N_4O_5S$
364.41
Benzoic acid, o-((3-(4,6-dimethyl-2-pyrimidinyl)-ureido)sulfonyl)-, methyl ester
Methyl 2-(((((4,6-dimethyl-2-pyrimidinyl)amino)carbonyl)amino)sulfonyl)benzoate
Sulfometuron methyl

74223-56-6
$C_{14}H_{14}N_4O_5S$
350.36
Sulfometuron
Benzoic acid, 2-(((((4,6-dimethyl-2-pyrimidinyl)amino)-carbonyl)amino)sulfonyl)-(9CI)
2-(((((4,6-Dimethyl-2-pyrimidinyl)amino)carbonyl)amino)-sulfonyl)benzoic acid
2-(3-(4,6-Dimethylpyrimidin-2-yl)ureidosulfonyl)benzoic acid
2-(3-(4,6-Dimethylpyrimidin-2-yl)ureidosulphonyl)benzoic acid

74223-64-6
$C_{14}H_{15}N_5O_6S$
381.37
COC(=O)c1ccccc1S(=O)(=O)NC(=O)Nc2nc(C)nc(OC)n2
Metsulfuron methyl
Ally
Caswell No. 419H
DPD 63760M
DPX 6376
DPX T6376
DPX-T 6376
EPA Pesticide Chemical Code 122010
Methyl-2-(((((4-methoxy-6-methyl-1,3,5-triazin-2-yl)-amino)carbonyl)amino)sulfonyl)benzoate
Metsulfuron methyl ester

74229-81-5
$C_{24}H_{22}$
310.44
Picene, 1,2,3,4-tetrahydro-1,2-dimethyl- (9CI)

74229-83-7
$C_{21}H_{22}$
274.41
Chrysene, 1,2,3,4-tetrahydro-3,4,7-trimethyl- (9CI)

74398-71-3
$(C_{66}H_{116}O_{12})x$
Polymer
9-Octadecenoic acid, 12-(oxiranylmethoxy)-, 1,2,3-propanetriyl ester, Homopolymer (9CI)

74472-33-6
$C_{12}H_6Cl_4$
291.99
1,1'-Biphenyl, 2,3,3',6-tetrachloro- (9CI)
2,3,3',6-PCB
2,3,3',6-Tetrachlorobiphenyl

74472-34-7
$C_{12}H_6Cl_4$
291.99
1,1'-Biphenyl, 2,3,4',5-tetrachloro- (9CI)
2,3,4',5-PCB
2,3,4',5-Tetrachlorobiphenyl

74472-36-9
$C_{12}H_5Cl_5$
326.44
1,1'-Biphenyl, 2,3,3',5,6-pentachloro- (9CI)
2,3,3',5,6-PCB
2,3,3',5,6-Pentachlorobiphenyl

74472-37-0
$C_{12}H_5Cl_5$
326.44
2,3,4,4'5-Pentachlorobiphenyl

74472-42-7
$C_{12}H_4Cl_6$
360.88
1,1'-Biphenyl, 2,3,3',4,4',6-hexachloro- (9CI)
2,3,3',4,4',6-Hexachlorobiphenyl
2,3,3',4,4',6-PCB

74472-46-1
$C_{12}H_4Cl_6$
360.88
Clc1cc(Cl)cc(c1)c2c(Cl)c(Cl)cc(Cl)c2Cl
1,1'-Biphenyl, 2,3,3',5,5',6-hexachloro- (9CI)
2,3,3',5,5',6-Hexachlorobiphenyl
2,3,3',5,5',6-PCB

74472-47-2
$C_{12}H_3Cl_7$
395.32
1,1'-Biphenyl, 2,2',3,4,4',5,6-heptachloro- (9CI)
2,2',3,4,4',5,6-Heptachlorobiphenyl
2,2',3,4,4',5,6-PCB

74472-48-3
$C_{12}H_3Cl_7$
395.32
1,1'-Biphenyl, 2,2',3,4,4',6,6'-heptachloro- (9CI)
2,2',3,4,4',6,6'-Heptachlorobiphenyl
2,2',3,4,4',6,6'-PCB

74472-50-7
$C_{12}H_3Cl_7$
395.32
1,1'-Biphenyl, 2,3,3',4,4',5',6-heptachloro- (9CI)
2,3,3',4,4',5',6-Heptachlorobiphenyl
2,3,3',4,4',5',6-PCB

74548-80-4
$C_{16}H_{14}Cl_3O_5P$
423.62
Phosphonic acid, (1-(acetyloxy)-2,2,2-trichloroethyl)-, diphenyl ester
(1-(Acetyloxy)-2,2,2-trichloroethyl)phosphonic acid diphenyl ester
Aphos

74562-99-5
$C_{26}H_{18}Cl_4O$
488.24
Benzene, 1,1',1'',1'''-(oxybis(methylidyne))tetrakis(4-chloro- (9CI)
Ether, bis(bis(p-chlorophenyl)methyl) (6CI)
1,1,1',1'-Tetra(p-chlorophenyl)dimethyl ether

74664-93-0
Unknown
Unknown
Alkanes, C14-30
Paraffinic hydrocarbons (C14-C30)

74665-17-1
Unknown
Unknown
Titanium, iso-Pr alc. triethanolamine complexes
Tetraisopropoxy titanate, Reaction products with triethanolamine

74712-19-9
$C_{15}H_{22}BrNO$
312.26
2-Bromo-3,3-dimethyl-N-N-(α-α-dimethylbenzyl) butyramide

74744-31-3
$C_{11}H_{18}$
150.26
5-Undecen-3-yne, (E)- (9CI)

74754-55-5
$C_6H_{11}NO_5$
177.15
Hexanoic acid, 6-(nitrooxy)- (9CI)

74798-20-2
$C_{15}H_{16}O_4$
260.29
7H-Furo(3,2-h)(2)benzopyran-3(2H)-one, 6,9-dihydro-7-hydroxy-7-methyl-2-(1-methylethylidene)-, (R)- (9CI)
Pergillin

74851-17-5
$C_{18}H_{24}O$
256.39
Bicyclo(3.1.1)hept-2-ene, 6,6-dimethyl-2-(2-(phenylmethoxy)ethyl)-, (1R)- (9CI)

74918-40-4
$C_{12}H_6Cl_2O$
237.08
Dibenzofuran, 3,6-dichloro- (9CI)
3,6-Dichlorodibenzofuran

74992-96-4
$C_{12}H_7ClO$
202.64
4-Chlorodibenzofuran

74992-97-5
$C_{12}H_6Cl_2O$
237.08
Dibenzofuran, 1,6-dichloro- (9CI)
1,6-Dichlorodibenzofuran

74992-98-6
$C_{12}H_6Cl_2O$
237.08
Dibenzofuran, 2,7-dichloro- (9CI)
2,7-Dichlorodibenzofuran

75013-55-7
$C_{12}H_{23}NOS$
229.38
1H-Azepine-1-carbothioic acid, hexahydro-, S-(1-ethylpropyl) ester (9CI)

75096-86-5
$C_{17}H_{25}ClN_2O_4$
356.84
Ethanaminium, 2-((((2,3-dihydro-2,2-dimethyl-7-benzofuranyl)oxy)carbonyl)methylamino)-N,N,N-trimethyl-2-oxo-, chloride
AI3-29783

75104-43-7
$C_{15}H_{17}N_3O_2$
271.32
5H-Pyrido(4,3-b)indol-3-amine, 1,4-dimethyl-, acetate (9CI)

75112-79-7
$C_9H_{10}O_2.Ag$
259.06
Benzenepropanoic acid, silver(1+) salt (9CI)
Silver 3-phenylpropanoate

75147-20-5
$C_8H_9Cl_5O_2$
314.42
3-Butenoic acid, 2,2,3,4,4-pentachloro-, butyl ester (9CI)
2,2,3,4,4-Pentachloro-3-butenoic acid, n-butyl ester

75150-13-9
$C_{10}H_{10}Br_2O_2$
322.00
Oxirane, ((2,4-dibromo-6-

methylphenoxy)methyl)- (9CI)
((2,4-Dibromo-6-methylphenoxy)methyl)oxirane

75181-94-1
$C_6H_2ClF_3$
166.53
Benzene, chlorotrifluoro- (9CI)

75198-38-8
$C_{12}H_2Cl_6O$
374.86
Dibenzofuran, 1,2,3,6,8,9-hexachloro- (9CI)
1,2,3,6,8,9-Hexachlorodibenzofuran

75217-43-5
$C_{17}H_{20}O_5$
304.35
1,3-Propanediol, 1-(3,4-dimethoxyphenyl)-2-phenoxy- (9CI)
1-(3,4-Dimethoxyphenyl)-2-phenoxy-1,3-propanediol
Veratrylglycerol β-phenyl ether

75217-44-6
$C_{17}H_{18}Cl_2O_5$
373.23
1,3-Propanediol, 2-(2,4-dichlorophenoxy)-1-(3,4-dimethoxyphenyl)- (9CI)

75248-87-2
$C_7H_4Cl_2O_2$
191.01
Benzoic acid, dichloro- (6CI,9CI)

75248-88-3
$C_8H_8Cl_2O_3$
223.06
Phenol, dichloro-2,6-dimethoxy- (9CI)

75315-44-5
$C_{11}H_{12}Cl_2O_4$
279.12
Ethanone, 1-(2,6-dichloro-3,4,5-trimethoxyphenyl)- (9CI)

75315-45-6
$C_{11}H_{12}Cl_2O_5$
295.12
Benzoic acid, 2,6-dichloro-3,4,5-trimethoxy-, methyl ester (9CI)

75315-46-7
$C_{11}H_{12}Cl_2O_4$
279.12
2-Propanone, 1-(2,6-dichloro-4-hydroxy-3,5-dimethoxyphenyl)- (9CI)

75315-50-3
$C_{10}H_8Cl_2O_6$
295.08
Benzeneacetic acid, 2,6-dichloro-4-hydroxy-3,5-dimethoxy-α-oxo- (9CI)

75315-51-4
$C_{10}H_{10}Cl_2O_6$
297.09
Benzeneacetic acid, 2,6-dichloro-α,4-dihydroxy-3,5-dimethoxy- (9CI)

75315-54-7
$C_{11}H_{14}Cl_2O_4$
281.14
Benzenemethanol, 2,6-dichloro-3,4,5-trimethoxy-α-methyl- (9CI)

75315-55-8
$C_{12}H_{14}Cl_2O_5$
309.15
Benzeneacetic acid, 2,6-dichloro-3,4,5-trimethoxy-, methyl ester (9CI)

75315-56-9
$C_{12}H_{14}Cl_2O_4$
293.15
2-Propanone, 1-(2,6-dichloro-3,4,5-trimethoxyphenyl)- (9CI)

75315-57-0
$C_{12}H_{14}Cl_2O_6$
325.15
Benzeneacetic acid, 2,6-dichloro-α-hydroxy-3,4,5-trimethoxy-, methyl ester (9CI)

75321-19-6
$C_{16}H_7N_3O_6$
337.26
Pyrene, 1,3,6-trinitro
1,3,6-Trinitropyrene

75321-20-9
$C_{16}H_8N_2O_4$
292.26
Pyrene, 1,3-dinitro
1,3-Dinitropyrene

75330-75-5
$C_{24}H_{36}O_5$
404.60
Butanoic acid, 2-methyl-, 1,2,3,7,8,8a-hexahydro-3,7-dimethyl-8-(2-(tetrahydro-4-hydroxy- 6-oxo-2H-pyran-2-yl)ethyl)-1-naphthalenyl ester, (1S-(1-α-(R*),3-α,7-β, 8-β-(2s*,4s*), 8a-β))
Mevinolin
MSD 803

75383-83-4
$C_{18}H_{22}O_5S$
350.44
1,3-Propanediol, 1-(3,4-dimethoxyphenyl)-2-(4-(methylthio)phenoxy)- (9CI)

75455-41-3
C_4H_7ClO
106.55
3-Buten-1-ol, 2-chloro- (6CI,7CI,9CI)

75536-53-7
$C_4H_6Cl_4O$
211.90
Butanol, tetrachloro- (9CI)
AA 81

75562-93-5
$C_7H_6Cl_2O_3$
209.03
Benzenediol, dichloromethoxy- (9CI) (VAN)

75602-99-2
Unknown
Unknown
LTX (9CI)

75625-24-0
$C_{10}H_2Br_6$
601.58
Naphthalene, 1,2,3,4,6,7-hexabromo

75627-02-0
$C_{12}H_2Cl_6O$
374.86
Dibenzofuran, 1,2,4,6,7,9-hexachloro- (9CI)
1,2,4,6,7,9-Hexachlorodibenzofuran

75673-43-7
$C_6H_{13}NO$
115.18
3,4,4-Trimethyloxazolidine

Caswell No. 892AA
EPA Pesticide Chemical Code 114802
3,4,4-Trimethyl-1-oxa-3-azacyclopentane

75675-48-8
$C_{22}H_{23}O_4P$
382.40
Phosphoric acid, butylphenyl diphenyl ester (9CI)

75701-74-5
$C_{15}H_{21}NO_4.C_4H_6MnN_2S_4.C_4H_6N_2S_4Zn$
820.40
DL-Alanine, N-(2,6-dimethylphenyl)-N-(methoxyacetyl)-, methyl ester, Mixt. with ((1,2- Ethanediylbis(carbamodithioato))(2-))manganese and ((1,2-ethanediylbis(carbamodithioato))(2-))zinc
Fubol
Metalaxyl-Mancozeb Mixt.
Ridomil Fitorex
Ridomil MZ

75736-33-3
$C_{15}H_{19}Cl_2N_3O$
328.27
1H-1,2,4-Triazole-1-ethanol, β-((2,4-dichlorophenyl)methyl)-α-(1,1-dimethylethyl)-, (R*,R*)-(+-)
1-(2,4-Dichlorophenyl)-4,4-dimethyl-2-(1,2,4-triazol-1-yl)-pentan-3-ol
Diclobutrazol
PP296
Vigil

75840-23-2
$C_{10}D_{12}$
144.28
Naphthalene-d8, 1,2,3,4-tetrahydro-1,2,3,4-d4- (9CI)

75881-81-1
Unknown
Unknown
Caradol 560 (9CI)

75882-11-0
Unknown
Unknown
Marlophen 830 (9CI)

76180-96-6
$C_{11}H_{10}N_4$
198.25
Cc3cc1c(ccc2[nH]cnc12)nc3N
3H-Imidazo(4,5-f)quinoline, 2-amino-3-methyl
2-Amino-3-methyl-3H-imidazo(4,5-f)quinoline
IQ
3-Methyl-3H-imidazo(4,5-f)-quinolin-2-amine

76229-76-0
$C_6H_{14}S_2$
150.31
Propane, 1-((2-(methylthio)-ethyl)thio)- (9CI)

76253-60-6
$C_{14}H_{10}Cl_4$
320.04
Benzene, dichloro((dichlorophenyl)methyl)methyl- (9CI)

76280-91-6
$C_{14}H_5Cl_6NO_3$
447.90
Benzoic acid, 6-(((2,3-dichlorophenyl)amino)carbonyl)-2,3,4,5-tetrachloro
6-(((2,3-Dichlorophenyl)amino)-carbonyl)-2,3,4,5-tetrachlorobenzoic acid
N-(2,3-Dichlorophenyl)-3,4,5,6-tetrachlorophthalamic acid
Shirahagen S
Techlofthalam

76330-06-8
$C_9H_8Cl_2O_4$
251.07
Benzaldehyde, 2,6-dichloro-4-hydroxy-3,5-dimethoxy- (9CI)

76341-69-0
$C_9H_9ClO_4$
216.62
Benzaldehyde, 2-chloro-4-hydroxy-3,5-dimethoxy- (9CI)

76379-66-3
$C_{12}H_{14}O$
174.24
5,7,11-Dodecatriyn-1-ol
Dodecto

76379-67-4
$C_{18}H_{22}O_2$
270.37
5,7,11,13-Octadecatrayne-1,18-diol
OCTD

76501-51-4
$C_{22}H_{26}$
290.45
Phenanthrene, octyl- (9CI)

76572-48-0
$C_9H_{15}NS$
169.28
Thiazole, 2-butyl-4,5-dimethyl- (9CI)

76578-14-8
$C_{19}H_{17}ClN_2O_4$
372.83
CCOC(=O)C(C)Oc3ccc(Oc2cnc1cc(Cl)ccc1n2)cc3
Propanoic acid, 2-(4-((6-chloro-2-quinoxalinyl)oxy)-phenoxy)-, ethyl ester
Assure
2-(4-((6-Chloro-2-quinoxalinyl)-oxy)phenoxy)propanoic acid ethyl ester
DPX-Y 6202
Ethyl 2-(4-(6-chloro-2-quinoxal-inyloxy)phenoxy)propanoate
Exp 3864
FBC 32197
NC 302
NCI-96683
Pilot
Quinofop-Ethyl
Quizalofop-Ethyl
Targa
Xylofop-Ethyl

76584-71-9
$C_{12}H_4Br_4O_2$
499.78
Dibenzo(b,e)(1,4)dioxin, 1,3,6,8-tetrabromo- (9CI)
1,3,6,8-Tetrabromodibenzo-(b,e)(1,4)dioxin
1,3,6,8-Tetrabromodibenzo-p-dioxin

76600-84-5
Unknown
Unknown
Antioxidant CD (9CI)

76602-24-9
$C_{12}H_{13}N$
171.24
Quinoline, ethylmethyl- (9CI)

76608-88-3
$C_{15}H_{25}N_3O$
263.36
1H-1,2,4-Triazole-1-ethanol, β-(cyclohexylmethylene)-α-(1,1-dimethylethyl)-, (E)- (9CI)
NTN 811

76621-12-0
$C_{12}H_5Cl_3O$
271.53
Dibenzofuran, 1,3,8-trichloro- (9CI)

1,3,8-Trichlorodibenzofuran

76723-60-9
Unknown
Unknown
Benzofluorenone (9CI)

76775-00-3
Unknown
Unknown
Kronitex TXP (9CI)

76943-21-0
$C_{18}H_{16}F_9O_3P$
482.29
Phosphorane, diphenyltris-(2,2,2-trifluoroethoxy)-(9CI)

77094-11-2
$C_{12}H_{12}N_4$
212.28
3H-Imidazo(4,5-f)quinoline, 2-amino-3,4-dimethyl
2-Amino-3,4-dimethylimidazo-(4,5-f)quinoline
3,4-Dimethyl-3H-imidazo-(4,5-f)quinolin-2-amine

77100-49-3
$C_8H_{15}NS_2.Na$
212.33
Carbamodithioic acid, cyclohexylmethyl-, sodium salt (9CI)

77102-93-3
$C_7H_6Cl_2O_2$
193.03
Phenol, 3,6-dichloro-2-methoxy-(9CI)

77102-94-4
$C_7H_6Cl_2O_2$
193.03
Phenol, 3,4-dichloro-2-methoxy-(9CI)

77182-82-2
$C_5H_{11}NO_4P.H_4N$
198.19
N.CP(O)(=O)CCC(N)C(O)=O
Butanoic acid, 2-amino-4-(hydroxymethylphosphinyl)-, monoammonium salt
2-Amino-4-(hydroxymethylphosphinyl)butanoic acid monoammonium salt
Ammonium (3-amino-3-carboxypropyl)methylphosphinate
Ammonium 2-amino-4-(hydroxymethylphosphinyl)-butanoate
Ammonium (dl-homoalanine-4-yl)methylphosphinate
Basta
Glufosinate-Ammonium
HOE 00661
HOE 39866
Rubout
Total

77311-02-5
C_5H_9NO
99.13
Oxazole, 2,5-dihydro-2,4-dimethyl- (9CI)

77327-07-2
$C_{28}H_{46}$
382.67
Ergosta-3,5-diene, (24XI)-(9CI)

77417-07-3
$C_{12}H_9N$
167.20
Naphthalenecarbonitrile, methyl- (9CI)

77439-76-0
$C_5H_3Cl_3O_3$
217.43
2(5H)-Furanone, 3-chloro-4-dichloromethyl-5-hydroxy
Chloro(dichloromethyl)-5-hydroxy-2(5H)-furanone
3-Chloro-4-dichloromethyl-5-hydroxy-2(5H)-furanone
MX

77468-36-1
$C_{16}H_9NO_2$
247.25
Fluoranthene, nitro- (9CI)

77468-37-2
$C_{12}H_{10}O$
170.21
Naphthalenecarboxaldehyde, methyl- (9CI)

77468-39-4
$C_{14}H_{10}O$
194.23
9H-Fluoren-9-one, methyl-(9CI)

77491-30-6
$C_{10}H_{20}NO_4PS.C_4H_7Cl_2O_4P$
502.32
2-Butenoic acid, 3-(((ethylamino)methoxyphosphinothioyl)oxy)-, 1-methylethyl ester, (E)-, Mixt. with 2,2-Dichloroethenyl dimethyl phosphate
Safrotin

77500-04-0
$C_{11}H_{11}N_5$
213.27
3H-Imidazo(4,5-f)quinoxaline, 2-amino-3,8-dimethyl
2-Amino-3,8-dimethylimidazo-(4,5-f)quinoxaline
2-Amino-3,8-dimethyl-3H-imidazo(4,5-f)quinoxaline
3,8-Dimethyl-3H-imidazo(4,5-f)quinoxalin-2-amine

77501-63-4
Unknown
Unknown
Lactofen
Benzoic acid, 5-(2-chloro-4-(trifluoromethyl)phenoxy)-2-nitro-, 2-ethoxy-1-methyl-2-oxoethyl ester,
(+-)-2-Ethoxy-1-methyl-2-oxoethyl 5-(2-chloro-4-(trifluoromethyl)phenoxy)-2 nitrobenzoate

77536-66-4
Unknown
Unknown
Asbestos, actinolite
Asbestos (ACGIH)
Actinolite asbestos

77536-67-5
Unknown
Unknown
Asbestos, anthophylite
Anthophylite
Anthophylite asbestos
Asbestos (ACGIH)
Azbolen asbestos
16 F
Ferroanthophyllite

77536-68-6
Unknown
Unknown
Asbestos, tremolite
Asbestos (ACGIH)
Fibrous tremolite
NCI-C08991
Tremolite asbestos
Tremolite (OSHA)

77630-51-4
$C_{32}H_{64}O_2$
480.86
Hentriacontanoic acid, methyl ester (9CI)

77732-09-3
$C_{14}H_{18}N_2O_4$
278.34
Acetamide, N-(2,6-dimethylphenyl)-2-methoxy-N-(2-oxo-

3-oxazolidinyl)
N-(2,6-Dimethylphenyl)-2-methoxy-N-(2-oxo-3-oxazolidinyl)-acetamide
M 10797
Oxadixyl
Pulsan
Recoil
Ripost
San 371
Sandofan
San 371F
Wakil

77753-24-3
$C_4H_5Cl_5$
230.35
Butane, 1,1,2,3,4-pentachloro- (9CI)

77907-22-3
Unknown
Unknown
C.I. Acid Yellow 237 (9CI)

77915-81-2
$C_{14}H_{12}ClNO_3$
277.71
Benzamide, N-(4-chlorophenyl)-N-hydroxy-3-methoxy- (9CI)
N-p-Chlorophenyl-m-methoxybenzohydroxamic acid

78051-43-1
$C_{11}H_{22}FN_2O_3PS_2$
344.40
Carbamic fluoride, (((5,5-dimethyl-1,3,2-dioxaphosphorinan-2-yl)(1,1-dimethylethyl)amino)thio)methyl-, P-sulfide (9CI)

78099-58-8
C_4HCl_3O
171.41
2-Cyclobuten-1-one, trichloro- (9CI)

78232-98-1
Unknown
Unknown
Advastab TM 692 (9CI)

78328-47-9
$C_{19}H_{14}$
242.32
Benzo(c)phenanthrene, methyl- (9CI)

78335-09-8
$C_{17}H_{11}N_3O_4S$
353.36
2-Naphthalenesulfonic acid, 5-((4-cyanophenyl)azo)-6-hydroxy- (9CI)

78335-10-1
$C_{16}H_{12}N_2O_{13}S_4$
568.54
1,3-Naphthalenedisulfonic acid, 8-((2,5-disulfophenyl)azo)-7-hydroxy- (9CI)

78335-11-2
$C_{18}H_{17}N_3O_7S_2$
451.48
1,3-Naphthalenedisulfonic acid, 8-((4-(2-aminoethyl)phenyl)azo)-7-hydroxy- (9CI)

78335-12-3
$C_{22}H_{19}N_3O_7S_2$
501.54
2,7-Naphthalenedisulfonic acid, 4-((4-(ethylamino)-1-naphthalenyl)azo)-5-hydroxy- (9CI)

78335-13-4
$C_{22}H_{15}N_5O_8S_2$
541.52
2,7-Naphthalenedisulfonic acid, 3-((4-nitrophenyl)azo)-6-(phenylazo)- (9CI)

78361-94-1
Unknown
Unknown
Naphthoquinoline (9CI)

78432-19-6
$C_{16}H_8N_2O_4$
292.26
Pyrene, dinitro
Dinitropyrene

78508-43-7
$C_{11}H_{15}BrN_2O$
271.16
Urea, N'-(4-bromo-3,5-dimethylphenyl)-N,N-dimethyl- (9CI)

78508-44-8
$C_{13}H_{12}N_2O_2$
228.24
Urea, (4-phenoxyphenyl)- (9CI)

78508-45-9
$C_8H_9FN_2O$
168.16
Urea, (4-fluoro-3-methylphenyl)- (9CI)

78508-46-0
$C_8H_9BrN_2O$
229.06
Urea, (4-bromo-3-methylphenyl)- (9CI)

78690-83-2
Unknown
Unknown
Santicizer 275 (9CI)

78744-33-9
$C_5H_7ClN_4O_2S$
222.63
2,4-Pyrimidinediamine, 6-chloro-5-(methylsulfonyl)- (9CI)
TE 194

78749-45-8
$C_{20}H_{26}O_5S$
378.49
1,3-Propanediol, 1-(3,4-dimethoxyphenyl)-2-(3,5-dimethyl-4-(methylthio)phenoxy)- (9CI)

78763-54-9
C_4H_9Sn
175.83
Tin(3+), butyl- (9CI)

78782-46-4
$C_8H_8Cl_2O_3$
223.06
3,5-Dichloro-2,6-dimethoxyphenol
3,5-Dichlorosyringol

78865-85-7
$C_{19}H_{42}NO_2S.Cl$
384.06
1-Hexanaminium, N,N-bis(2-hydroxyethyl)-N-((octylthio)methyl)-, chloride (9CI)

78865-87-9
$C_{23}H_{50}NO_2S.Cl$
440.17
1-Hexanaminium, N-((dodecylthio)methyl)-N,N-bis(2-hydroxyethyl)-, chloride (9CI)

78865-89-1
$C_{25}H_{54}NO_2S.Cl$
468.22
1-Dodecanaminium, N,N-bis-(2-hydroxyethyl)-N-((octylthio)methyl)-, chloride (9CI)

78865-90-4

$C_{27}H_{58}NO_2S.Cl$
496.28
1-Dodecanaminium, N-((decylthio)methyl)-N,N-bis(2-hydroxyethyl)-, chloride (9CI)

78982-40-8
$C_{12}H_8O_3$
200.20
4'-Methylangelicin
9-Methylangelicin

78995-10-5
Unknown
Unknown
N-(1-(2-Hydroxy-2-phenyl)-ethyl-4-piperidyl)-N-phenyl-propanamide, Its optical isomers, salts, and salts of isomers
DEA No. 9830
β-Hydroxyfentanyl

78995-14-9
$C_{23}H_{30}N_2O_2$
366.49
Propanamide, N-(1-(2-hydroxy-1-methyl-2-phenyl-ethyl)-3-methyl-4-piperidin-yl)-N-phenyl-
DEA No. 9831
F 7302
β-Hydroxy-3-methylfentanyl
N-(1-(2-Hydroxy-1-methyl-2-phenylethyl)-3-methyl-4-piperidinyl)-N-phenyl-propanamide
N-(3-Methyl-1-(2-hydroxy-2-phenyl)ethyl-4-piperidyl)-N-phenylpropanamide
Ohmefentanyl

79060-60-9
$C_{12}H_2Cl_6O$
374.86
Dibenzofuran, 1,2,3,4,6,7-hexachloro- (9CI)
1,2,3,4,6,7-Hexachlorodibenzo-furan

79064-73-6
$C_3H_9N_4OPS_2$
212.23
Thiourea, N,N''-(methyl-phosphinylidene)bis- (9CI)
Antipyrene T 1

79075-22-2
$C_{13}H_8O_3$
212.20
1H,3H-Naphtho(1,8-cd)pyran-1,3-dione, methyl- (9CI)

79075-27-7
$C_{16}H_{12}O$
220.27
Anthracenecarboxaldehyde, methyl- (9CI)

79147-47-0
$C_{14}H_{10}O$
194.23
Fluorenone, methyl- (9CI)

79188-95-7
$C_7H_5ClO_4$
188.57
Benzoic acid, chloro-3,4-di hydroxy- (9CI)

79241-46-6
$C_{19}H_{20}F_3NO_4$
383.40
Propanoic acid, 2-(4-((5-(tri-fluoromethyl)-2-pyridinyl)-oxy)phenoxy)-, butyl ester, (R)
Fluazifop-p-butyl
Fusilade 5
Fusilade 2000
Fusilade Super
PP 005

79419-43-5
$C_{15}H_{24}O_3S$
284.42
2-Cyclohexen-1-one, 5-(2-(ethyl-thio)propyl)-3-hydroxy-2-(1-oxobutyl)- (9CI)

79419-72-0
$C_{10}H_{22}N_2$
170.29
1,4-Butanediamine, N-cyclo-hexyl- (9CI)

79458-54-1
$C_4H_4Cl_6$
264.79
Butane, 1,1,1,4,4,4-hexa-chloro- (9CI)

79485-04-4
Unknown
Unknown
Retil (9CI)

79504-02-2
C_2BrCl_5
281.20
Ethane, bromopenta-chloro- (7CI,9CI)

79538-32-2
$C_{17}H_{14}ClF_7O_2$
418.76
Cyclopropanecarboxylic acid, 3-(2-chloro-3,3,3-trifluoro-1-propenyl)-2,2-dimethyl-, (2,3,5,6-tetrafluoro-4-methylphenyl)methyl ester, (1-α,3-α(Z))-(+-)
Force
Forza
PP993
Tefluthrin
Tefluthrine

79554-39-5
$C_{15}H_{24}$
204.36
Nonane, phenyl- (9CI)

79606-18-1
$C_{20}H_{26}$
378.49
Benzene, 1,1'-(1-pentyl-1,3-propanediyl)bis- (9CI)
1,3-Diphenyloctane

79704-88-4
Unknown
Unknown
α-Methylfentanyl
DEA No. 9814
α Methyl fentanyl
1-(1-Methyl-2-phenyl-ethyl)-4-(N-propanilido)piperidine
N-(1-(α-Methyl-β-phenyl)ethyl 4-piperidyl) propionanilide

79720-82-4
$C_{10}H_{18}ClNOS$
235.77
1H-Azepine-1-carbothioic acid, hexahydro-, S-(3-chloro-propyl) ester (9CI)

79746-00-2
$C_{15}H_{11}N_3O$
249.27
1H-1,2,4-Triazole, 1-benzo-yl-3-phenyl- (9CI)
1-Benzoyl-3-phenyl-1,2,4-triazole

79746-01-3
$C_{16}H_{13}N_3O$
263.30
Ethanone, 1-phenyl-2-(3-phenyl-1H-1,2,4-tri-azol-1-yl)- (9CI)

79787-65-8
$C_8H_{17}N_5$
183.23
1H-Azepine-1-carboximidamide, N-(aminoimino-methyl)hexahydro- (9CI)

79849-02-8

Unknown
Unknown
Nitrilotriacetic acid and its salts
NTA, lead(2+) salt (1:1)

79897-80-6
$C_{29}H_{48}$
396.70
Stigmasta-3,5-diene, (24XI)- (9CI)

79915-08-5
Unknown
Unknown
Nitrilotriacetic acid, lead(2+) potassium salt (1:1:1)
NTA, lead(2+) potassium salt (1:1:1)

79915-09-6
Unknown
Unknown
Nitrilotriacetic acid, lead(2+) salt (2:3)
NTA, lead(2+) salt (2:3)

79956-98-2
$C_6H_{14}O$
102.18
1-Butanol, dimethyl- (9CI)

80045-50-7
$C_{12}H_{12}BrNO$
266.13
Benzamide, 4-bromo-N-(1,1-dimethyl-2-propynyl)- (9CI)

80045-51-8
$C_{12}H_{12}FNO$
205.23
Benzamide, N-(1,1-dimethyl-2-propynyl)-4-fluoro- (9CI)

80045-52-9
$C_{15}H_{19}NO$
229.32
Benzamide, N-(1,1-dimethyl-2-propynyl)-4-(1-methylethyl)- (9CI)

80090-30-8
$C_{32}H_{36}N_2O_{10}$
608.65
Acetic acid, ((5-(3-((2-(3,4-dimethoxyphenyl)ethyl)((phenylmethoxy)carbonyl)amino)-2-hydroxypropoxy)-1,2,3,4-tetrahydro-2-oxo-8-quinolinyl)oxy)- (9CI)

80182-27-0
$C_{16}H_{13}NO_2$
251.28
Phenanthrene, dimethylnitro- (9CI)

80182-33-8
$C_{19}H_{13}NO_2$
287.31
Chrysene, methylnitro- (9CI)

80191-43-1
$C_{15}H_{11}NO_2$
237.25
Anthracene, methylnitro- (9CI)

80191-44-2
$C_{15}H_{11}NO_2$
237.25
Phenanthrene, methylnitro- (9CI)

80191-45-3
$C_{16}H_{13}NO_2$
251.28
Anthracene, dimethylnitro- (9CI)

80246-33-9
$C_{12}H_5Br_3O_2$
420.90
Dibenzo(b,e)(1,4)dioxin, 1,3,8-tribromo- (9CI)

80267-67-0
$C_{10}H_5NO_4$
203.15
Naphthalenedione, nitro- (9CI)

80387-97-9
$C_{25}H_{42}O_3S$
422.73
Acetic acid, (((3,5-bis(1,1-dimethylethyl)-4-hydroxyphenyl)methyl)thio)-, 2-ethylhexyl ester
(((3,5-Bis(1,1-dimethylethyl)-4-hydroxyphenyl)methyl)thio)acetic acid 2-ethylhexyl ester

80398-28-3
$C_{19}H_{10}O$
254.29
Benzo(cd)pyrenone (9CI)

80440-44-4
$C_{21}H_{12}O$
280.33
7H-Benzo(hi)chrysen-7-one (9CI)

80450-55-1
Unknown
Unknown
Glycoproteins, specific or class, emulsans
Emulsans
Glycoproteins, emulsans
Viscoemulsan

80455-52-3
Unknown
Unknown
Cyclopentaphenanthrene (9CI)

80508-23-2
Unknown
Unknown
N'-Nitrosonornicotine
Nitrosonornicotine
Pyridine, 3-(1-nitroso-2-pyrrolidinyl)-, (S)-

80547-56-4
$C_{22}H_{28}N_4O_2.2ClH$
453.38
1,4-Pentanediamine, N(1),N(1)-diethyl-N(4)-(1-nitro-9-acridinyl)-, dihydrochloride (9CI)

80789-74-8
$C_8H_5NO_5S.Na$
250.18
1H-Indole-5-sulfonic acid, 2,3-dihydro-2,3-dioxo-, monosodium salt (9CI)
5-Indolinesulfonic acid, 2,3-dioxo-, sodium salt (7CI)

80845-12-1
$C_{22}H_{19}Br_2NO_3$
505.20
Cyclopropanecarboxylic acid, 3-(2,2-dibromoethenyl)-2,2-dimethyl-, cyano(3-phenoxyphenyl)methyl ester, (1α(S*),3α)-(+-)- (9CI)
RU 43501

81325-79-3
$C_{17}H_{26}O_2$
262.40
Benzyl 2,2-dimethyloctanoate
BDOT
Octanoic acid, 2,2-dimethyl-, phenylmethyl ester

81325-80-6
$C_{16}H_{24}O_2$
248.37
Benzyl-2,2,4,4-tetramethylpentanoate
B-TMPN
Pentanoic acid, 2,2,4,4-tetramethyl-, phenylmethyl ester

81334-34-1
$C_{13}H_{15}N_3O_3$
261.28
Imazapyr
AC 243997
AC 243,997
Caswell No. 003F
2-(4,5-Dihydro-4-methyl-4-(1-methylethyl)-5-oxo-1H-imidazol-2-yl)-3-pyridinecarboxylic acid
EPA Pesticide Chemical Code 128821
2-(4-Isopropyl-4-methyl-5-oxo-2-imidazolin-2-yl)nicotinic acid

81335-37-7
$C_{17}H_{17}N_3O_3$
311.37
CC(C)C1(C)N=C(NC1=O)c3nc2ccccc2cc3C(O)=O
3-Quinolinecarboxylic acid, 2-(4,5-dihydro-4-methyl-4-(1-methylethyl)-5-oxo-1H-imidazol-2-yl)
AC 252214
2-(4,5-Dihydro-4-methyl-4-(1-methylethyl)-5-oxo-1H-imidazol-2-yl)-3-quinolinecarboxylic acid
Imazaquin
Scepter

81335-77-5
$C_{15}H_{19}N_3O_3$
289.34
Imazethapyr
AC 263499
2-(4,5-Dihydro-4-methyl-4-(1-methylethyl)-5-oxo-1H-imidazol-2-yl)-5-ethyl-3-pyridinecarboxylic acid
5-Ethyl-2-(4-isopropyl-4-methyl-5-oxo-2-imidazolin-2-yl)nicotinic acid
(RS)-5-Ethyl-2-(4-isopropyl-4-methyl-5-oxo-2-imidazolin-2-yl)nicotinic acid

81393-48-8
$C_{15}H_{21}ClN_2O_8$
392.78
Urea, N'-(4-chlorophenyl)-N-((β-D-glucopyranosyloxy)methyl)-N-methoxy- (9CI)

81412-56-8
Unknown
Unknown
Fatty acids, C9-11-branched, glycidyl esters

81510-83-0
Unknown
Unknown
2-(4,5-Dihydro-4-methyl-4-(1-methylethyl)-5-oxo-1H-imidazol-2-yl)-3-pyridnecarboxylic acid, monoisopropylamine
AC 252,925
Caswell No. 003D
EPA Pesticide Chemical Code 128829
Imazapyr-isopropylammonium

81546-39-6
$C_{27}H_{44}$
368.65
Cholestadiene (9CI) (VAN)

81598-29-0
$C_{12}H_{16}$
160.26
Naphthalene, ethyl-1,2,3,4-tetrahydro- (9CI)

81634-99-3
$C_{12}H_{22}O_2$
198.31
3-Decen-1-ol, acetate, (Z)-(9CI)

81638-37-1
$C_{12}H_6Cl_2O$
237.08
Dibenzofuran, 1,8-dichloro-(9CI)

81777-89-1
$C_{12}H_{14}ClNO_2$
239.72
3-Isoxazolidinone, 2-((2-chlorophenyl)methyl)-4,4-dimethyl
2-(2-Chlorobenzyl)-4,4-dimethyl-1,2-oxazolidin-3-one
2-((2-Chlorophenyl)methyl)-4,4-dimethyl-3-isoxazolidinone
Command
Dimethazone
FMC 57020
Gamit

81826-15-5
$C_{17}H_{19}NO_7$
349.34
1,3-Propanediol, 1-(3,4-dimethoxyphenyl)-2-(4-nitrophenoxy)- (9CI)

81846-81-3
Unknown
Unknown
1,1'-Biphenyl, butenylated

82010-82-0
Unknown
Unknown
Cuprous and cupric oxide, mixed

82027-59-6
Unknown
Unknown
Copper triethanolamine complex

82039-09-6
$C_{20}H_{11}NO_3$
313.31
Benzo(a)pyren-3-ol, 6-nitro-(9CI)

82039-10-9
$C_{20}H_{11}NO_3$
313.31
Benzo(a)pyren-1-ol, 6-nitro-(9CI)

82065-80-3
$C_{28}H_{54}O_6$
486.73
Diisotridecyl peroxydicarbonate (Technically pure)
Peroxydicarbonic acid, diisotridecyl ester
UN 2889

82078-98-6
Unknown
Unknown
Heptadecenyl imidazolinium chloride

82097-50-5
$C_{14}H_{16}ClN_5O_5S$
401.80
Benzenesulfonamide, 2-(2-chloroethoxy)-N-(((4-methoxy-6-methyl-1,3,5-triazin-2-yl)amino)carbonyl)- (9CI)

82291-34-7
$C_{12}H_3Cl_5O_2$
356.42
Dibenzo(b,e)(1,4)dioxin, 1,2,3,6,9-pentachloro- (9CI)
1,2,3,6,9-Pentachlorodibenzo-(b,e)(1,4)dioxin
1,2,3,6,9-Pentachlorodibenzo-p-dioxin

82291-35-8
$C_{12}H_3Cl_5O_2$
356.42
Dibenzo-p-dioxin, 1,2,4,6,7-pentachloro- (9CI)
1,2,4,6,7-Pentachlorodibenzo-(b,e)(1,4)dioxin
1,2,4,6,7-Pentachlorodibenzo-p-dioxin

82291-36-9
$C_{12}H_3Cl_5O_2$
356.42
Dibenzo(b,e)(1,4)dioxin, 1,2,4,6,9-pentachloro- (9CI)
1,2,4,6,9-Pentachlorodibenzo-(b,e)(1,4)dioxin
1,2,4,6,9-Pentachlorodibenzo-p-dioxin

82291-37-0
$C_{12}H_3Cl_5O_2$
356.42
Dibenzo(b,e)(1,4)dioxin, 1,2,4,7,9-pentachloro- (9CI)
1,2,4,7,9-Pentachlorodibenzo-(b,e)(1,4)dioxin
1,2,4,7,9-Pentachlorodibenzo-p-dioxin

82291-38-1
$C_{12}H_3Cl_5O_2$
356.42
Dibenzo(b,e)(1,4)dioxin, 1,2,4,8,9-pentachloro- (9CI)
1,2,4,8,9-Pentachlorodibenzo-(b,e)(1,4)dioxin
1,2,4,8,9-Pentachlorodibenzo-p-dioxin

82322-43-8
$C_{10}H_7NO_3$
189.16
Naphthalenol, nitro- (9CI)

82347-33-9
Unknown
Unknown
Monalube 29-78 (9CI)

82558-50-7
Unknown
Unknown
CCC(C)(CC)c2cc(NC(=O)c1c(OC)cccc1OC)on2
Isoxaben
Caswell No. 419F
Compound 121607
EPA Pesticide Chemical Code 125851
N-(3-(1-Ethyl-methylpropyl)-5-isoxazolyl)-2,6-dimethoxybenzamide
N-(3-(1-Ethyl-1-methylpropyl)-isoxazol-5-yl)-2,6-dimethoxybenzamide

82668-20-0
$C_8H_7ClO_3$
186.60
Benzaldehyde, 2-chloro-4-hydroxy-3-methoxy-(9CI)

82799-46-0
$C_{25}H_{32}OS$
380.59
9H-Thioxanthen-9-one, 2-dodecyl- (9CI)

82863-50-1
$C_{14}H_{18}.C_{10}H_{16}.C_7H_{14}$
420.72
4,7-Methano-2,3,8-methenocyclopent(a)indene, dodecahydro-, stereoisomer, Mixt. with methylcyclohexane and (3aα,4β,7β,7aα)-octahydro-4,7-methano-1H-indene (9CI)
JP 9

82911-58-8
$C_{12}H_4Cl_4O$
305.97
Dibenzofuran, 1,4,6,8-tetrachloro- (9CI)
1,4,6,8-Tetrachlorodibenzofuran

82911-59-9
$C_{12}H_5Cl_3O$
271.53
Dibenzofuran, 1,6,8-trichloro- (9CI)
2,4,9-Trichlorodibenzofuran

82911-60-2
$C_{12}H_5Cl_3O$
271.53
Dibenzofuran, 1,4,6-trichloro-(9CI)
1,4,6-Trichlorodibenzofuran

82911-61-3
$C_{12}H_5Cl_3O$
271.53
Dibenzofuran, 1,3,4-trichloro- (9CI)
1,3,4-Trichlorodibenzofuran

82987-03-9
$C_4H_{10}S$
90.19
Thiapentane (9CI)

83055-99-6
$C_{16}H_{18}N_4O_7S$
410.38
Benzoic acid, 2-((((((4,6-dimethoxy-2-pyrimidinyl)-amino)carbonyl)amino)sulfonyl)methyl)-, methyl ester (9CI)
DPX-F 5384

83242-23-3
$C_{22}H_{23}O_4P$
382.40
Phosphoric acid, 2-(1,1-dimethylethyl)phenyl diphenyl ester (9CI)

83293-82-7
$C_4H_5Cl_5$
230.35
Butane, 1,2,2,3,3-pentachloro- (9CI)
1,2,2,3,3-Pentachlorobutane

83381-96-8
Unknown
Unknown
Cyclopentapyrene (9CI)

83463-62-1
$C_2HBrClN$
154.40
Acetonitrile, bromochloro
Bromochloroacetonitrile
Bromochloromethyl cyanide

83484-75-7
$C_8H_3Cl_5$
276.38
Benzene, ethenyl-, pentachloro deriv. (9CI)

83484-79-1
$C_{23}H_{12}O$
304.35
6H-Cyclopenta(ghi)picen-6-one (9CI)

83536-56-5
$C_{13}H_9N$
179.22
Acenaphthylenecarbonitrile, 1,2-dihydro- (9CI)

83536-57-6
$C_{17}H_9N$
227.26
Pyrenecarbonitrile (9CI)

83542-86-3
Unknown
Unknown
N-cis-9-Octadecenyl-1,3-propanediamine monoglucomate

83589-41-7
Unknown
Unknown
Degreaser P (9CI)

83589-42-8
Unknown

Unknown
Dekarbon T (9CI)

83589-46-2
$C_{21}H_{12}O$
280.33
Dibenzofluorenone (9CI)

83589-63-3
Unknown
Unknown
Glacidet K (9CI)

83589-99-5
Unknown
Unknown
Skorexol (9CI)

83590-03-8
Unknown
Unknown
Syntol HD (9CI)

83601-83-6
Unknown
Unknown
Mefluidide, potassium salt

83622-91-7
$C_{21}H_{10}O$
278.31
1H-Cyclopenta(ghi)perylen-1-one (9CI)

83623-05-6
$C_5H_7ClN_4S$
190.63
Pyrimidinediamine, 2(or 6)-chloro-5-(methylthio)- (9CI)
UKJ 1506

83636-47-9
$C_{12}H_5Cl_3O$
271.53
Dibenzofuran, 1,2,3-trichloro- (9CI)
1,2,3-Trichlorodibenzofuran

83682-28-4
$C_6H_8Cl_6$
292.85
Hexane, 2,2,3,4,5,5-hexachloro- (9CI)

83682-29-5
$C_6H_8Cl_6$
292.85
Hexane, 1,2,2,5,5,6-hexachloro- (9CI)

83682-30-8
$C_6H_4Cl_8$
290.83
2-Butene, 1,1,4,4-tetrachloro-2,3-bis(dichloromethyl)- (9CI)

83682-31-9
$C_6H_6Cl_6$
290.83
3-Hexene, 2,2,3,4,5,5-hexachloro- (9CI)

83682-32-0
$C_3H_2Cl_2$
108.96
1,2-Propadiene, 1,3-dichloro- (9CI)

83682-33-1
$C_6H_8Cl_2$
151.04
1,5-Hexadiene, 3,4-dichloro- (9CI)

83682-34-2
C_6Cl_{14}
568.41
Hexane, tetradecachloro- (9CI)

83682-35-3
C_6H_7Cl
114.58
1,2,5-Hexatriene, 3-chloro- (9CI)

83682-36-4
C_7H_9Cl
128.60
5-Hepten-1-yne, 1-chloro- (9CI)

83682-37-5
$C_6H_6Cl_2$
149.02
1,3,5-Hexatriene, 3,4-dichloro- (6CI,9CI)

83682-38-6
$C_4H_2Cl_6$
262.78
1-Propene, 3,3,3-trichloro-2-(trichloromethyl)- (9CI)

83682-39-7
C_4HCl_7
297.22
1-Propene, 1,3,3,3-tetra chloro-2-(trichloromethyl)- (9CI)

83682-40-0
$C_5H_2Cl_8$
345.70
1-Butene, 1,1,4,4,4-pentachloro-2-(trichloromethyl)- (9CI)

83682-41-1
$C_4H_4Cl_2$
122.98
1,2-Butadiene, 4,4-dichloro- (9CI)

83682-42-2
$C_4H_4Cl_2$
122.98
1-Butyne, 4,4-dichloro- (9CI)

83682-43-3
$C_4H_3Cl_3$
157.43
1-Butyne, 4,4,4-trichloro- (9CI)

83682-44-4
$C_4H_4Cl_2$
122.98
1,2-Butadiene, 1,4-dichloro- (9CI)

83682-45-5
$C_4H_4Cl_2$
122.98
1-Butyne, 1,4-dichloro- (9CI)

83682-46-6
$C_4H_3Cl_3$
157.43
1-Butyne, 1,4,4-trichloro- (9CI)

83682-47-7
C_4Cl_6
260.76
1-Butyne, 1,3,3,4,4,4-hexachloro- (9CI)

83682-48-8
$C_8H_{12}Cl_2$
179.09
1,5-Hexadiene, 1,6-dichloro-2,5-dimethyl- (9CI)

83682-49-9
$C_8H_{12}Cl_2$
179.09
1,5-Hexadiene, 3,4-dichloro-2,5-dimethyl- (6CI,9CI)

83682-50-2
$C_8H_{10}Cl_4$
247.98
1,5-Hexadiene, 3,3,4,4-tetrachloro-2,5-dimethyl- (9CI)

83682-51-3
$C_8H_{12}Cl_2$
179.09
1,5-Hexadiene, 2,5-bis-(chloromethyl)- (9CI)

83682-52-4
$C_8H_{10}Cl_4$
247.98
1,5-Hexadiene, 3,4-dichloro-2,5-bis(chloromethyl)- (9CI)

83682-53-5
$C_8H_8Cl_6$
316.87
1,5-Hexadiene, 3,3,4,4-tetrachloro-2,5-bis(chloromethyl)- (9CI)

83682-54-6
$C_8H_{10}Cl_2$
177.07
1,3,5-Hexatriene, 1,6-dichloro-2,5-dimethyl- (9CI)

83682-55-7
$C_8H_8Cl_4$
245.96
1,3,5-Hexatriene, 3,4-dichloro-2,5-bis(chloromethyl)- (9CI)

83682-56-8
$C_8H_{10}Cl_2$
177.07
1,3,5-Hexatriene, 2,5-bis-(chloromethyl)- (9CI)

83682-57-9
$C_8H_6Cl_6$
314.86
1,3,5-Hexatriene, 3,4-dichloro-2,5-bis(dichloromethyl)- (9CI)

83682-58-0
$C_8H_4Cl_8$
383.75
1,3,5-Hexatriene, 3,4-dichloro-2,5-bis(trichloromethyl)- (9CI)

83682-59-1
$C_{12}H_{12}Cl_6$
368.95
Benzene, 1,3,5-tris(2,2-dichloroethyl)- (9CI)

83682-60-4
$C_{12}H_9Cl_9$
472.28
Benzene, 1,3,5-tris(2,2,2-trichloroethyl)- (9CI)

83682-61-5
$C_8H_{12}Cl_2$
193.12
1,5-Octadiene, 1,2-dichloro- (9CI)

83682-62-6
$C_8H_{12}Cl_4$
250.00
Cyclobutane, 1,2-bis-(1,2-dichloroethyl)- (9CI)

83682-63-7
$C_8H_9Cl_3$
211.52
Cyclohexene, 1,4-dichloro-4-(1-chloroethenyl)- (9CI)

83682-64-8
$C_8H_{10}Cl_6$
318.89
Cyclohexane, 1,2,4-trichloro-4-(1,1,2-trichloroethyl)- (9CI)

83682-65-9
$C_8H_{10}Cl_2$
177.07
1,4-Cyclooctadiene, 1,5-dichloro- (9CI)

83682-66-0
$C_8H_{12}Cl_2O$
195.09
1-Butene, 4,4'-oxybis-(3-chloro- (9CI)

83682-67-1
$C_8H_{14}O_3$
158.20
3-Buten-2-ol, 1,1'-oxybis- (9CI)

83682-68-2
$C_8H_{14}O_3$
158.20
3-Buten-1-ol, 2,2'-oxybis- (9CI)

83682-69-3
$C_4H_4Cl_6$
264.79
Butane, 1,1,2,2,3,3-hexachloro- (9CI)
1,1,2,2,3,3-Hexachlorobutane

83682-70-6
$C_4H_3Cl_7$
299.24
Butane, 1,1,1,2,2,3,3-heptachloro- (9CI)

83682-71-7
$C_4H_2Cl_6$
262.78
1-Butene, 2,3,3,4,4,4-hexachloro- (9CI)

83682-72-8
$C_3H_6Cl_2O$
128.99
1-Propanol, 3,3-dichloro- (7CI,9CI)

83690-98-6
$C_{12}H_4Cl_4O$
305.97
Dibenzofuran, 1,3,6,9-tetrachloro- (9CI)
1,3,6,9-Tetrachlorodibenzofuran

83704-21-6
$C_{12}H_4Cl_4O$
305.97
Dibenzofuran, 1,2,3,6-tetrachloro- (9CI)
1,2,3,6-Tetrachlorodibenzofuran

83704-22-7
$C_{12}H_4Cl_4O$
305.97
Dibenzofuran, 1,2,3,7-tetrachloro- (9CI)
1,2,3,7-Tetrachlorodibenzofuran

83704-23-8
$C_{12}H_4Cl_4O$
305.97
Dibenzofuran, 1,2,3,9-tetrachloro- (9CI)
1,2,3,9-Tetrachlorodibenzofuran

83704-24-9
$C_{12}H_4Cl_4O$
305.97
Dibenzofuran, 1,2,4,9-tetrachloro- (9CI)
1,2,4,9-Tetrachlorodibenzofuran

83704-25-0

$C_{12}H_4Cl_4O$
305.97
Dibenzofuran, 1,2,6,7-tetrachloro- (9CI)
1,2,6,7-Tetrachlorodibenzofuran

83704-26-1
$C_{12}H_4Cl_4O$
305.97
Dibenzofuran, 1,2,7,9-tetrachloro- (9CI)
1,2,7,9-Tetrachlorodibenzofuran

83704-27-2
$C_{12}H_4Cl_4O$
305.97
Dibenzofuran, 1,3,4,6-tetrachloro- (9CI)
1,3,4,6-Tetrachlorodibenzofuran

83704-28-3
$C_{12}H_4Cl_4O$
305.97
Dibenzofuran, 1,3,4,9-tetrachloro- (9CI)
1,3,4,9-Tetrachlorodibenzofuran

83704-29-4
$C_{12}H_4Cl_4O$
305.97
Dibenzofuran, 1,4,7,8-tetrachloro- (9CI)
1,4,7,8-Tetrachlorodibenzofuran

83704-30-7
$C_{12}H_4Cl_4O$
305.97
Dibenzofuran, 2,3,4,6-tetrachloro- (9CI)
2,3,4,6-Tetrachlorodibenzofuran

83704-31-8
$C_{12}H_4Cl_4O$
305.97
Dibenzofuran, 2,3,4,7-tetrachloro- (9CI)
2,3,4,7-Tetrachlorodibenzofuran

83704-32-9
$C_{12}H_4Cl_4O$
305.97
Dibenzofuran, 2,3,4,8-tetrachloro- (9CI)
2,3,4,8-Tetrachlorodibenzofuran

83704-33-0
$C_{12}H_4Cl_4O$
305.97
Dibenzofuran, 1,6,7,8-tetrachloro- (9CI)
1,6,7,8-Tetrachlorodibenzofuran

83704-34-1
$C_{12}H_5Cl_3O$
271.53
Dibenzofuran, 1,2,8-trichloro- (9CI)

83704-35-2
$C_{12}H_3Cl_5O$
340.42
Dibenzofuran, 1,4,6,7,8-pentachloro- (9CI)
1,4,6,7,8-Pentachlorodibenzofuran

83704-36-3
$C_{12}H_3Cl_5O$
340.42
Dibenzofuran, 1,3,4,6,7-pentachloro- (9CI)
1,3,4,6,7-Pentachlorodibenzofuran

83704-37-4
$C_{12}H_5Cl_3O$
271.53
Dibenzofuran, 1,2,7-trichloro- (9CI)

83704-38-5
$C_{12}H_5Cl_3O$
271.53
Dibenzofuran, 1,2,9-trichloro- (9CI)

83704-39-6
$C_{12}H_5Cl_3O$
271.53
Dibenzofuran, 1,3,6-trichloro- (9CI)
1,3,6-Trichlorodibenzofuran

83704-40-9
$C_{12}H_5Cl_3O$
271.53
Dibenzofuran, 1,3,9-trichloro- (9CI)

83704-41-0
$C_{12}H_5Cl_3O$
271.53
Dibenzofuran, 1,4,7-trichloro- (9CI)

83704-42-1
$C_{12}H_5Cl_3O$
271.53
Dibenzofuran, 2,4,7-trichloro- (9CI)
2,4,7-Trichlorodibenzofuran

83704-43-2
$C_{12}H_5Cl_3O$
271.53
Dibenzofuran, 3,4,6-trichloro- (9CI)

83704-44-3
$C_{12}H_5Cl_3O$
271.53
Dibenzofuran, 3,4,7-trichloro- (9CI)

83704-45-4
$C_{12}H_5Cl_3O$
271.53
Dibenzofuran, 2,6,7-trichloro- (9CI)
2,6,7-Trichlorodibenzofuran

83704-46-5
$C_{12}H_5Cl_3O$
271.53
Dibenzofuran, 1,6,7-trichloro- (9CI)

83704-47-6
$C_{12}H_3Cl_5O$
340.42
Dibenzofuran, 1,2,3,4,6-pentachloro- (9CI)
1,2,3,4,6-Pentachlorodibenzofuran

83704-48-7
$C_{12}H_3Cl_5O$
340.42
Dibenzofuran, 1,2,3,4,7-pentachloro- (9CI)
1,2,3,4,7-Pentachlorodibenzofuran

83704-49-8
$C_{12}H_3Cl_5O$
340.42
Dibenzofuran, 1,2,3,4,9-pentachloro- (9CI)
1,2,3,4,9-Pentachlorodibenzofuran

83704-50-1
$C_{12}H_3Cl_5O$
340.42
Dibenzofuran, 1,2,4,6,7-pentachloro- (9CI)
1,2,4,6,7-Pentachlorodibenzofuran

83704-51-2
$C_{12}H_3Cl_5O$
340.42

Dibenzofuran, 1,2,3,6,8-pentachloro- (9CI)
1,2,3,6,8-Pentachlorodibenzo furan

83704-52-3
$C_{12}H_3Cl_5O$
340.42
Dibenzofuran, 1,2,3,6,9-pentachloro- (9CI)
1,2,3,6,9-Pentachlorodibenzofuran

83704-53-4
$C_{12}H_3Cl_5O$
340.42
Dibenzofuran, 1,2,3,7,9-pentachloro- (9CI)
1,2,3,7,9-Pentachlorodibenzofuran

83704-54-5
$C_{12}H_3Cl_5O$
340.42
Dibenzofuran, 1,2,3,8,9-pentachloro- (9CI)
1,2,3,8,9-Pentachlorodibenzofuran

83704-55-6
$C_{12}H_3Cl_5O$
340.42
Dibenzofuran, 1,3,4,6,8-pentachloro- (9CI)
1,3,4,6,8-Pentachlorodibenzofuran

83710-07-0
$C_{12}H_4Cl_4O$
305.97
Dibenzofuran, 1,2,6,8-tetrachloro- (9CI)
1,2,6,8-Tetrachlorodibenzofuran

83719-40-8
$C_{12}H_4Cl_4O$
305.97
Dibenzofuran, 1,2,4,7-tetrachloro- (9CI)
1,2,4,7-Tetrachlorodibenzofuran

83733-82-8
$C_{13}H_{19}ClNO_3PS_2$
367.87
Phosphorodithioic acid, S-(((2-chlorophenyl)(1-oxobutyl)amino)methyl) O,O-dimethyl ester
S-(((2-Chlorophenyl)(1-oxobutyl)amino)methyl) O,O-dimethylphosphorodithioate
Fosmethilan
Fosmetilan
NE 79168
Nevifos
Nevifos 50
Neviki 79168
Phosmethylan

83929-87-7
Unknown
Unknown
C.I. Disperse Red 274 (9CI)

84002-64-2
$C_4H_7N_3O_9$
241.10
α-Methylglycerol trinitrate
1,2,3-Butanetriol, trinitrate

84082-38-2
Unknown
Unknown
Alkanes, C10-21, chloro
Chlorinated paraffins

84237-39-8
$C_{10}H_{13}N_3O$
191.21
N'-Nitrosoanabasine
(+-)-N-Nitrosoanabasine
(+-)-1-Nitrosoanabasine
(+-)-3-(1-Nitroso-2-piperidinyl)pyridine
Pyridine, 3-(1-nitroso-2-piperidinyl)-, (+-)-

84268-33-7
$C_{20}H_{23}N_3O_3$
353.40
Benzenepropanoic acid, 3-(2H-benzotriazol-2-yl)-5-(1,1-dimethylethyl)-4-hydroxy-, methyl ester (9CI)

84412-11-3
$C_{13}H_{10}O$
182.22
Acenaphthylenecarboxaldehyde, 1,2-dihydro- (9CI)

84455-05-0
C_6H_6ClNO
143.57
1H-Pyrrole, 1-acetyl-2-chloro- (9CI)

84455-06-1
C_6H_6BrNO
188.03
1H-Pyrrole, 1-acetyl-2-bromo- (9CI)

84540-57-8
$C_6H_{12}O_3$
132.16
Propanol, methoxy-, acetate (9CI)

84605-28-7
Unknown
Unknown
Phosphorodithioic acid, Mixed O,O-bis(1,3-dimethylbutyl and iso-Pr) esters
Phosphorodithioic acid, Mixed O,O'-bis(1,3-dimethylbutyl and 1-methylethyl)esters

84605-29-8
Unknown
Unknown
Phosphorodithioic acid, Mixed O,O-bis(1,3-dimethylbutyl and iso-Pr) esters, zinc salts
Phosphorodithioic acid, Mixed O,O'-bis(1,3-dimethylbutyl and 1-methyl) esters, zinc salts
Phosphorodithioic acid, Mixed O,O'-bis(1,3-dimethylbutyl and 1-methylethyl) esters, zinc salts

84665-39-4
$C_{22}H_{12}O$
292.34
Cyclopenta(cd)perylenone (9CI)

84665-41-8
$C_{24}H_{12}O$
316.36
Benzo(ghi)cyclopenta(cd) perylenone (9CI)

84704-01-8
$C_{12}H_{15}ClNO_3PS$
319.74
Phosphorodithioic acid, O-((4-chlorophenyl)-cyanomethyl) O,O-diethyl ester (9CI)

84761-80-8
$C_{12}H_4Br_4O$
483.78
Dibenzofuran, 1,2,7,8-tetrabromo- (9CI)
1,2,7,8-Tetrabromodibenzofuran

84761-81-9
$C_{12}H_5Br_3O$
404.88
Dibenzofuran, 1,2,8-tribromo- (9CI)
1,2,8-Tribromodibenzofuran

84761-82-0
$C_{12}H_5Br_3O$

404.88
Dibenzofuran, 2,3,8-tribromo- (9CI)
2,3,8-Tribromodibenzofuran

84761-86-4
$C_{12}H_7ClO$
202.64
1-Chlorodibenzofuran

84776-06-7
Unknown
Unknown
Alkanes, C10-32, chloro
Chlorinated paraffins

84776-07-8
Unknown
Unknown
Alkanes, C16-27, chloro
Chlorinated paraffins

84878-64-8
$C_8H_{13}N_2O_5P.2H2O$
284.19
3-Pyridinemethanol, 4-(aminomethyl)-5-hydroxy-6-methyl-, α-(dihydrogen phosphate), dihydrate (9CI)

84961-66-0
Unknown
Unknown
Tobacco Dust
Tobacco Leaf, Aqueous Extract

84987-77-9
$C_5H_7Cl_3O$
189.47
Ether, 2-chloroethyl 2,3-dichloroallyl
3-(2-Chloroethoxy)-1,2-dichloropropene
1-Propene, 3-(2-chloroethoxy)-1,3-dichloro-

85026-55-7
Unknown
Unknown
Rosin Oil
Rosin Soap (Disproportionated) Solution

85049-26-9
Unknown
Unknown
Alkanes, C16-35, chloro
Chlorinated paraffins

85255-90-9
Unknown
Unknown
Epi-Rez 5011 (9CI)

85269-46-1
$C_8H_{15}Cl_3$
217.57
Octane, 1,2,3-trichloro- (9CI)

85298-07-3
$C_8H_7Cl_3O_2$
241.50
Benzene, 1,2,5-trichloro-3,4-dimethoxy- (9CI)

85422-92-0
Unknown
Unknown
Paraffin oils and hydrocarbon oils, Chloro

85535-84-8
Unknown
Unknown
Alkanes, C10-13, chloro
Chlorinated paraffins

85535-85-9
Unknown
Unknown
Alkanes, C14-17, chloro
Chlorinated paraffins

85535-86-0
Unknown
Unknown
Alkanes, C18-28, chloro
Chlorinated paraffins

85536-22-7
Unknown
Unknown
Alkanes, C12-14, chloro
Chlorinated paraffins

85568-72-5
Unknown
Unknown
Procion Blue (9CI)

85681-73-8
Unknown
Unknown
Alkanes, C10-14, chloro
Chlorinated paraffins

85682-59-3
$C_{22}H_{19}Cl_2NO_3.C_{12}H_{14}Cl_3O_4P$
775.87
Cyclopropanecarboxylic acid, 3-(2,2-dichloroethenyl)-2,2-dimethyl-, cyano(3-phenoxyphenyl)methyl ester, Mixt. with 2-chloro-1-(2,4-dichlorophenyl)ethenyl diethyl phosphate (9CI)

85822-16-8
$C_7H_5ClO_3S$
204.63
Benzenesulfonyl chloride, 4-formyl- (9CI)

85847-73-0
Unknown
Unknown
2,3-Dichloro-4-(propylsulfonyl)pyridine

85878-62-2
$C_{14}H_7NO_3$
237.22
5H-Furo(3',2':6,7)(1)benzopyrano(3,4-c)pyridin-5-one
Pyrido(3,4-c)psoralen

85878-63-3
$C_{15}H_9NO_3$
251.24
Pyrido(3,4-c)-7-methylpsoralen
5H-Furo(3',2':6,7)(1)benzopyrano(3,4-c)pyridin-5-one, 7-methyl-
7-Methylpyrido(3,4-c)psoralen
NSC-376770

85897-29-6
$C_{13}H_7ClO$
214.65
9H-Fluoren-9-one, chloro- (9CI)

86006-42-0
$C_7H_6Br_2O$
265.93
Phenol, dibromomethyl- (9CI)

86006-43-1
$C_7H_5BrCl_2O$
255.93
Phenol, bromodichloromethyl- (9CI)

86006-44-2
$C_7H_5Br_2ClO$
300.38
Phenol, dibromochloromethyl- (9CI)

86072-07-3
$C_4H_4ClN_3O_2$
161.53
1H-Imidazole, 2-chloro-1-methyl-5-nitro- (9CI)

86220-42-0
$C_{66}H_{83}N_{17}O_{13}.xC_2H_4O_2.yH_2O$
Polymer
Nafarelin acetate
Luteinizing hormone-releasing factor (Pig), 6-(3-(2-naphthalenyl)-D-alanine)-, acetate (Salt), hydrate
5-Oxo-L-prolyl-L-histidyl-L-tryptophyl-L-seryl-L-tyrosyl-3-(2-naphthyl)-D-alanyl-L-leucyl-L-arginyl-L-prolylglycinamide acetate (Salt) hydrate
RS-94991-298
Synarel

86290-81-5
Unknown
Unknown
Gasoline
Automobiles, Fuel systems and fuels (VAN)
Benzine (Motor fuel)
Fuels, Gasoline
Herbicide ES
Hydrofining (VAN)
Metaforming
Motor Fuels (VAN)
Natural Gas Condensates, Gasoline
Nefras S 150/200
Petrol (VAN)
Synfuels (VAN)

86329-60-4
$C_{16}H_8Cl_2$
271.14
Fluoranthene, dichloro- (9CI)

86709-50-4
$C_{29}H_{46}$
394.68
Stigmasta-3,5,24(28)-triene (9CI)

86722-66-9
$C_{17}H_{16}N_2O_3$
296.31
Anthraquinone, 1-((2-hydroxyethyl)amino)-4-(methylamino)- (8CI)
Acetate Brilliant Blue 4B
Acetoquinone Light Pure Blue R
Altocyl Brilliant Blue B
Amacel Blue BNN
Amacel Brilliant Blue B
9,10-Anthracenedione, 1-((2-hydroxyethyl)amino)-4-(methylamino)- (9CI)
Artisil Blue BSG
Artisil Blue BSQ
C.I. Disperse Blue 3
C.I. Disperse Blue 41
C.I. 61505
Calcosyn Sapphire Blue R
Calcosyn Sapphire Blue 2GS
Celanthrene Brilliant Blue
Celanthrene Brilliant Blue FFS
Celliton Blue FFR
Celliton Fast Blue FBBN
Celliton Fast Blue FFR
Celliton Fast Blue FFRN
Celliton Fast Blue FFRS
Celutate Blue BLT
Celutate Blue RNH
Celutate Brilliant Blue B
Cibacet Blue BNG
Cibacet Blue F3R
Cibacet Brilliant Blue BG new
Cilla Fast Blue FFR
Diacelliton Fast Brilliant Blue B
Diacelliton Fast Brilliant Blue BF
Disperse Blue K
Disperse Blue 3
Dispersive blue K
Duranol Brilliant Blue B
Duranol Brilliant Blue BN
Duranol Printing Blue B
Eastman Blue BNN
Eastman Blue GBN
Fenacet Fast Blue FF
Fenacet Fast Blue FFN
1-(β-Hydroxyethylamino)-4-(methylamino)anthraquinone
4-((Hydroxyethyl)amino)-1-(methylamino)anthraquinone
Interchem Acetate Blue B
Interchem Acetate Blue NBN
Interchem Acetate Blue RBN
Interchem Acetate Blue WNBN
Kayalon Fast Blue FN
1-(Methylamino)-4-ethanolaminoanthraquinone
1-(Methylamino)-4-(β-hydroxyethylamino)anthraquinone
1-(Methylamino)-4-(2-hydroxyethylamino)anthraquinone
Microsetile Blue FF
Microsetile Blue FFR
Miketon Brilliant Blue B
Mireton Brilliant Blue B
Nacelan Blue KLT
NSC-64753
Nyloquinone Pure Blue
Nyloquinone Pure Blue R
Perliton Blue FFR
Serinyl Hosiery Blue
Serinyl Hosiery Blue BG
Serisol Brilliant Blue BG
Serisol Brilliant Blue BP
Serisol Brilliant Blue FF
Setacyl Blue BN
Setacyl Blue FG
Setacyl Blue RF
Setacyl Brilliant Blue
Setacyl Brilliant Blue BG
Supracet Brilliant Blue BG

86812-27-3
$C_{23}H_{40}O$
332.58
Phenol, heptadecyl- (9CI)

86825-83-4
$C_{15}H_{24}$
204.36
Naphthalene, hexahydrodimethylpropyl- (9CI)

86853-03-4
$C_{10}H_{16}$
136.24
Cyclohexene, 5-methyl-3-(1-methylethenyl)- (9CI)

86853-88-5
$C_{17}H_{10}O$
230.27
1H-Benz(de)anthracen-1-one (9CI)

86853-89-6
$C_{19}H_{10}O$
254.29
11H-Benz(bc)aceanthrylen-11-one (9CI)

86853-90-9
$C_{19}H_{10}O$
254.29
4H-Cyclopenta(def)triphenylen 4-one (9CI)

86853-91-0
$C_{18}H_{10}O$
242.28
4H-Cyclopenta(def)chrysen-4-one
4,5-Oxochrysene

86853-92-1
$C_{21}H_{10}O$
278.31
4H-Benzo(def)cyclopenta-(mno)chrysen-4-one (9CI)

86853-93-2
$C_{21}H_{12}O$
280.33
9H-Indeno(2,1-c)phenanthren-9-one (9CI)

86853-94-3
$C_{21}H_{12}O$
280.33
8H-Indeno(1,2-a)anthracen-8-one (9CI)

86853-95-4
$C_{21}H_{12}O$
280.33
7H-Indeno(1,2-a)phenanthren-7-one (9CI)

86853-96-5
$C_{21}H_{12}O$
280.33

13H-Indeno(1,2-l)phenanthren-13-one (9CI)

86853-97-6
$C_{21}H_{12}O$
280.33
7H-Dibenzo(c,g)fluoren-7-one (9CI)

86853-98-7
$C_{21}H_{12}O$
280.33
8H-Indeno(2,1-b)phenanthren-8-one (9CI)

86853-99-8
$C_{21}H_{12}O$
280.33
13H-Indeno(1,2-b)anthracen-13-one (9CI)

86854-00-4
$C_{21}H_{12}O$
280.33
12H-Indeno(1,2-b)phenanthren-12-one (9CI)

86854-01-5
$C_{21}H_{12}O$
280.33
13H-Dibenzo(a,i)fluoren-13-one (9CI)

86854-02-6
$C_{21}H_{12}O$
280.33
7H-Dibenzo(b,g)fluoren-7-one (9CI)

86854-03-7
$C_{21}H_{12}O$
280.33
13H-Indeno(1,2-c)phenanthren-13-one (9CI)

86854-04-8
$C_{21}H_{12}O$
280.33
13H-Indeno(2,1-a)anthracen-13-one (9CI)

86854-05-9
$C_{21}H_{12}O$
280.33
9H-Benzo(fg)naphthacen-9-one (9CI)

86854-06-0
$C_{21}H_{12}O$
280.33
13H-Dibenz(a,de)anthracen-13-one (9CI)

86854-07-1
$C_{23}H_{10}O$
302.33
1H-Benzo(ghi)cyclopenta(pqr)perylen-1-one (9CI)

86854-08-2
$C_{23}H_{12}O$
304.35
5H-Benzo(b)cyclopenta(def)chrysen-5-one (9CI)

86854-09-3
$C_{23}H_{12}O$
304.35
4H-Benzo(b)cyclopenta(mno)chrysen-4-one (9CI)

86854-10-6
$C_{23}H_{12}O$
304.35
4H-Benzo(c)cyclopenta(mno)chrysen-4-one (9CI)

86854-11-7
$C_{23}H_{12}O$
304.35
8H-Benzo(p)cyclopenta(def)chrysen-8-one (9CI)

86854-12-8
$C_{23}H_{12}O$
304.35
13H-Benzo(b)cyclopenta(def)triphenylen-13-one (9CI)

86854-13-9
$C_{23}H_{12}O$
304.35
4H-Benzo(b)cyclopenta(jkl)triphenylen-4-one (9CI)

86854-14-0
$C_{23}H_{12}O$
304.35
4H-Indeno(7,1,2-ghi)chrysen-4-one (9CI)

86854-15-1
$C_{23}H_{12}O$
304.35
13H-Cyclopenta(rst)pentaphen-13-one (9CI)

86854-16-2
$C_{23}H_{12}O$
304.35
4H-Cyclopenta(pqr)picen-4-one (9CI)

86854-17-3
$C_{23}H_{12}O$
304.35
13H-Dibenz(bc,j)aceanthrylen-13-one (9CI)

86854-18-4
$C_{23}H_{12}O$
304.35
13H-Dibenz(bc,l)aceanthrylen-13-one (9CI)

86854-19-5
$C_{23}H_{12}O$
304.35
13H-Indeno(2,1,7-qra)naphthacen-13-one (9CI)

86854-20-8
$C_{23}H_{12}O$
304.35
7H-Indeno(1,2-a)pyren-7-one (9CI)

86854-21-9
$C_{23}H_{12}O$
304.35
13H-Indeno(1,2-e)pyren-13-one (9CI)

86854-22-0
$C_{23}H_{12}O$
304.35
6H-Naphtho(2,1,8,7-defg)naphthacen-6-one (9CI)

86854-23-1
$C_{23}H_{12}O$
304.35
8H-Naphtho(3,2,1,8-defg)chrysen-8-one (9CI)

86854-24-2
$C_{23}H_{12}O$
304.35
13H-Dibenzo(def,qr)chrysen-13-one (9CI)

86854-25-3
$C_{25}H_{12}O$
328.37
13H-Benz(qr)indeno(6,7,1,2-defg)naphthacen-13-one (9CI)

86854-26-4
$C_{25}H_{14}O$
330.39
7H-Benz(5,6)indeno(1,2-a)phenanthren-7-one (9CI)

86862-68-2
$C_{21}H_{10}O$
278.31

3H-Indeno(2,1,7-cde)pyren-3-one (9CI)

86892-89-9
Unknown
Unknown
Nitrilotriacetic acid, disodium ammonium salt
NTA, disodium ammonium salt

86903-93-7
Unknown
Unknown
Epi-Rez 5077 (9CI)

87130-20-9
$C_{14}H_{21}NO_4$
267.32
Carbamic acid, (3,4-diethoxyphenyl)-, 1-methylethyl ester (9CI)

87175-02-8
Unknown
Unknown
Alkyl dimethyl dodecylbenzyl ammonium chloride

87237-48-7
$C_{19}H_{19}ClF_3NO_5$
433.84
Propanoic acid, 2-(4-((3-chloro-5-(trifluoromethyl)-2-pyridinyl)oxy)phenoxy)-, 2-ethoxyethyl ester
DOWCO 453
Dowco 453EE
Gallant
Haloxyfop-(2-ethoxyethyl)
Zellek

87257-05-4
$C_{17}H_{32}O_6$
332.44
Oxirane, 2,2'-(3,7,7,11-tetramethyl-2,5,9,12-tetraoxatridecane-1,13-diyl)bis- (9CI)

87259-53-8
$C_{10}H_{10}O$
146.19
Indenone, 2,3-dihydromethyl- (9CI)

87397-71-5
Unknown
Unknown
RDX 58456 (9CI)

87478-71-5
$C_3H_9N.HNO_3$
122.11
2-Propanamine, nitrate (9CI)

87495-30-5
Unknown
Unknown
Resorcine Blue (9CI) (VAN)

87625-62-5
$C_{20}H_{30}O_8$
398.50
Spiro(cyclopropane-1,5'-(5H)-inden)-3'(2'H)-one, 1',3'-α,4',7'-α-tetrahydro-7'-α-(β-d-glucopyranosyloxy)-4'-hydroxy-2',4',6'-trimethyl
Aquilide A
Ptaquiloside

87676-93-5
$C_{16}H_{13}F_2N_3O$
301.28
1H-1,2,4-Triazole-1-ethanol, α-(2-fluorophenyl)-α-(4-fluorophenyl)-, (+-)- (9CI)
Impact Sopra
PP 450

87818-31-3
$C_{18}H_{26}O_2$
274.44
7-Oxabicyclo(2.2.1)heptane, 1-methyl-4-(1-methylethyl)-2-((2-methylphenyl)methoxy)-, exo-(+-)
Cinmethylin
(+-)-exo-1-Methyl-4-(1-methylethyl)-2-((2-methylphenyl)methoxy)-7-oxabicyclo(2.2.1)heptane

87835-45-8
Unknown
Unknown
Sodium phosphide
UN 1432

87903-39-7
$C_7H_6O_4Pb$
361.32
Lead hydroxysalicylate
Hydroxy(2-hydroxybenzoato-O1,O2)lead
Lead (II) hydroxide salicylate
Lead, hydroxy(2-hydroxybenzoato-O1,O2)- (9CI)
Lead monobasic salicylate
Monobasic lead salicylate

87915-38-6
Unknown
Unknown
Blue Dextran
Blue Dextran 2000
Dextran Blue

87954-49-2
$C_2H_6O_4S$
126.13
Methanol, sulfonylbis- (9CI)

88083-39-0
$C_{15}H_{26}O_7S_3$
414.56
Butanoic acid, 2-(2-hydroxy-4-(methylthio)-1-oxobutoxy)-4-(methylthio)-, 1-carboxy-3-(methylthio)propyl ester (9CI)

88683-38-9
$C_{14}H_{17}NO_2S$
263.36
1H-Isoindole-1,3(2H)-dione, 2-((1-ethyl-2-methylpropyl)thio)- (9CI)

88813-63-2
$C_{14}H_{11}N$
193.24
Benzoquinoline, methyl- (9CI)

88899-62-1
$C_{20}H_{14}O_6$
350.33
Alteichin
3,10-Perylenedione, 1,2,12a,12b-tetrahydro-1,4,9,12a-tetrahydroxy-, (1α,12aβ,12bα)-(+)-

88927-42-8
$C_{10}H_8Cl_4O_3$
317.99
Acetic acid, (2,3,4,5-tetrachlorophenoxy)-, ethyl ester (9CI)

88997-61-9
Unknown
Unknown
Linco 4 (9CI)

88997-62-0
Unknown
Unknown
Magnus 101 (9CI)

89072-60-6
Unknown
Unknown
Cyclosol 63 (9CI)

89126-45-4
Unknown
Unknown
Azafluoranthene (9CI)

89126-46-5
Unknown
Unknown
Azapyrene (9CI)

89151-70-2
$C_{11}H_{15}NO$
177.25
Benzamide, 2,6-diethyl- (6CI,9CI)
2,6-Diethylbenzamide

89286-97-5
Unknown
Unknown
Permatox 100 (9CI)

89590-79-4
$C_{24}H_8Cl_{10}$
650.85
Quaterphenyl, decachloro- (9CI)

89590-80-7
$C_{24}H_{10}Cl_8$
581.97
Quaterphenyl, octachloro- (9CI)

89590-81-8
$C_{24}H_{12}Cl_6$
513.08
Quaterphenyl, hexachloro- (9CI)

89697-80-3
$(C_2H_4O)xC_{18}H_{31}NO_3S$
Polymer NA
Poly(oxy-1,2-ethanediyl), α-(2-(((4-(1,1-dimethylethyl)phenyl)sulfonyl)hexylamino)ethyl) -ω-hydroxy- (9CI)

89697-83-6
$(C_2H_4O)xC_{20}H_{35}NO_3S$
Polymer NA
Poly(oxy-1,2-ethanediyl), α-(2-(((4-(1,1-dimethylethyl)phenyl)sulfonyl)octylamino)ethyl) -ω-hydroxy- (9CI)

89697-89-2
$(C_2H_4O)xC_{22}H_{39}NO_3S$
Polymer NA
Poly(oxy-1,2-ethanediyl), α-(2-(decyl((4-(1,1-dimethylethyl)phenyl)sulfonyl)amino)ethyl) ω-hydroxy- (9CI)

89896-73-1
$C_7H_{12}O$
112.17
3-Heptenal (9CI)

89942-12-1
$C_7H_{13}Br$
177.09
1-Heptene, 1-bromo- (7CI)

90030-80-1
C_4H_9NOSe
166.08
Acetamide, 2-(ethylselenyl)- (7CI)

90077-73-9
$C_{14}H_9Cl$
212.68
Benzene, chloro(phenylethynyl)- (9CI)

90077-74-0
$C_{13}H_6Cl_2O$
249.10
9H-Fluoren-9-one, dichloro- (9CI)

90077-75-1
$C_{13}H_5Cl_3O$
283.54
9H-Fluoren-9-one, trichloro- (9CI)

90077-76-2
$C_{13}H_4Cl_4O$
317.98
9H-Fluoren-9-one, tetrachloro- (9CI)

90077-77-3
$C_{13}H_3Cl_5O$
352.43
9H-Fluoren-9-one, pentachloro- (9CI)

90077-78-4
C_6HBrCl_4
294.79
Benzene, bromotetrachloro- (9CI)

90077-79-5
$C_{12}H_4Cl_4$
289.97
Acenaphthylene, tetrachloro- (9CI)

90077-80-8
$C_8Cl_4N_2$
265.90
Benzenedicarbonitrile, tetrachloro- (9CI)

90282-99-8
$C_8H_9ClO_2$
172.61
Benzene, 1-chloro-2,3-dimethoxy- (9CI)

90283-00-4
$C_8H_8Cl_2O_2$
207.06
Benzene, 1,2-dichloro-3,4-dimethoxy- (9CI)

90283-01-5
$C_8H_8Cl_2O_2$
207.06
Benzene, 1,5-dichloro-2,3-dimethoxy- (9CI)

90283-02-6
$C_8H_8Cl_2O_2$
207.06
Benzene, 1,4-dichloro-2,3-dimethoxy- (9CI)

90370-29-9
$C_{14}H_{12}O_3$
228.26
2H-Furo(2,3-h)(1)benzopyran-2-one, 4,6,9-trimethyl
4,4',6-Trimethylangelicin
4,6,9-Trimethyl-2H-furo(2,3-h)-(1)benzopyran-2-one

90588-08-2
$(C_6H_{10}O_2.C_2H_4)x$
Polymer NA
1,3-Dioxepane, 2-methylene-, Polymer with ethene (9CI)
Ethene, Polymer with 2-methylene-1,3-dioxepane (9CI)
Ethylene-2-methylene-1,3-dioxepane copolymer

90736-23-5
Unknown
Unknown
para-Fluorofentanyl
DEA No. 9812
N-(4-Fluorophenyl)-N-(1-(2-phenethyl)-4-piperidinyl)-propanamide
N-(1-(2-Phenylethyl)-4-piperidyl)-N-(4-fluorophenyl)-propanamide its optical isomers, salts and salts of isomers

90853-11-5
$C_9H_{20}N_2$
156.26
1,4-Butanediamine, N-cyclopentyl- (9CI)

90853-12-6
$C_{10}H_{22}N_2$
170.29
Cyclohexanemethanamine, 3-amino-α,α,4-trimethyl- (9CI)

90853-13-7
$C_{10}H_{22}N_2$
170.29
1,3-Propanediamine, N-cyclohexyl-N'-methyl- (9CI)

90853-14-8
$C_8H_{19}N_3$
157.24
1-Piperazinepropanamine, γ-methyl- (9CI)

90853-15-9
$C_9H_{22}N_2O$
174.27
Ethanol, 2-((3-aminopropyl)-butylamino)- (9CI)

90853-16-0
$C_9H_{22}N_2O$
174.27
Ethanol, 2-((1,1-dimethyl-2-((1-methylethyl)amino)ethyl)-amino)- (9CI)

90853-17-1
$C_8H_{20}N_2O$
160.25
Ethanol, 2-((2-(dimethylamino)-1,1-dimethylethyl)amino)- (9CI)

90853-18-2
$C_8H_{18}N_2O_2$
174.23
Alanine, N-(3-(dimethylamino)-propyl)- (9CI)

90853-19-3
$C_7H_{16}N_2O_2$
160.20
Alanine, N-(2-(dimethylamino)-ethyl)- (9CI)

90853-20-6
$C_{12}H_{24}N_2O_2$
228.32
Alanine, N-(3-(cyclohexyl-amino)propyl)- (9CI)

90853-21-7
$C_{13}H_{26}N_2O_2$
242.35
Alanine, N-(3-(cyclohexyl-amino)propyl)-2-methyl- (9CI)

90853-22-8
$C_8H_{18}N_2O_2$
174.23
Alanine, N-(2-amino-2-methyl-propyl)-2-methyl- (9CI)

90853-23-9
$C_{12}H_{24}N_2O_2$
228.32
Alanine, N-(3-(2-methyl-1-piperidinyl)propyl)- (9CI)

90853-24-0
$C_{13}H_{26}N_2O_2$
242.35
Butanoic acid, 3-((3-(2-methyl-1-piperidinyl)propyl)amino)- (9CI)

90985-94-7
$C_{13}H_7Cl_5O_2$
372.46
Benzene, 1,3-dichloro-2-methoxy-5-(2,4,6-trichlorophenoxy)- (9CI)

90985-96-9
$C_{13}H_6Cl_6O_2$
406.91
Benzene, 1,2,3,5-tetrachloro-4-(3,5-dichloro-4-methoxyphenoxy)- (9CI)

90986-10-0
$C_{12}H_5Cl_5O_2$
358.43
Phenol, 2,6-dichloro-4-(2,4,6-trichlorophenoxy)- (9CI)

90986-11-1
$C_{12}H_4Cl_6O_2$
392.88
Phenol, 2,6-dichloro-4-(2,3,4,6-tetrachlorophenoxy)- (9CI)

91027-93-9
$C_5H_6ClN_3O_2$
175.56
1H-Imidazole, 5-chloro-1,2-dimethyl-4-nitro- (9CI)

91027-94-0
$C_5H_6ClN_3O_2$
175.56
1H-Imidazole, 4-chloro-1,2-dimethyl-5-nitro- (9CI)

91137-27-8
$C_{11}H_9NO_2$
187.19
Naphthalene, 1-methyl-5-nitro- (9CI)

91137-28-9
$C_{11}H_9NO_2$
187.19
Naphthalene, 2-methyl-7-nitro- (9CI)

91311-51-2
$C_{13}H_9NO$
195.22
Benzo(f)quinolinol (9CI)

91371-14-1
$C_{12}H_6Br_2O_2$
341.99
Dibenzo(b,e)(1,4)dioxin, 1,6-dibromo- (9CI)
Dibenzo-p-dioxin, 1,6-dibromo- (9)
1,6-Dibromodibenzo-p-dioxin

91455-17-3
$C_{18}H_{50}O_3Si_6$
483.11
Tetrasiloxane, 1,1,1,3,5,7,7,7-octamethyl-3,5-bis-(2-(trimethylsilyl)ethyl)- (9CI)

91538-83-9
$C_{12}H_2Cl_6O$
374.86
Dibenzofuran, 1,2,3,4,6,9-hexachloro- (9CI)

91538-84-0
$C_{12}H_2Cl_6O$
374.86
Dibenzofuran, 1,2,3,4,7,9-hexachloro- (9CI)
1,2,3,4,7,9-Hexachlorodibenzofuran

91724-16-2
As_2O_4Sr
301.46
Strontium arsenite
Arsenious acid, strontium salt
Arsenite de strontium (French)
Arsenito de estroncio (Spanish)
Strontium arsenite, Solid
UN 1691 [Strontium arsenite]

91741-91-2
$C_{18}H_{20}$
228.30
Chrysene, octahydro- (9CI)

91930-03-9
$C_7H_4Cl_2O_2$
191.01
Benzaldehyde, dichloro-2-hydroxy- (9CI)
Salicylaldehyde, dichloro-

92134-93-5
$C_4H_8N_2O_2S$
148.20
Thiazolidine-2-methanol, 3-nitroso
2-Hydroxymethyl-N-nitroso-thiazolidine
3-Nitrosothiazolidine-2-methan-ol

92170-50-8
$C_{37}H_{36}N_2O_9S_3.C_{16}H_{12}N_4O_9S_2.2H_3N.3Na$
1334.31
Aquashade
Benzenemethanaminium, N-ethyl-N-(4-((4-(ethyl((3-sulfo-phenyl)methyl)amino)phenyl)-(2-sulfophenyl(methylene)-2,5-cyclohexadien-1-ylidene)-3-sulfo-, hydroxide, inner salt, diammonium salt, Mixt. with 4,5-dihydro-5-oxo-1-(4-sulfo-phenyl)-4-((4-sulfophenyl)-azo)-1H-pyrazole-3-carboxylic acid trisodium salt (9CI)

92245-57-3
$C_{14}H_{14}N_2O_2$
242.28
Ethanol, 2-(4-(phenylazo)-phenoxy)- (9CI)
Ethanol, 2-(p-(phenylazo)-phenoxy)- (7CI)

92341-04-3
$C_{12}H_4Cl_4O$
305.97
Dibenzofuran, 1,3,4,8-tetra-chloro- (9CI)
1,3,4,8-Tetrachlorodibenzo-furan

92341-05-4
$C_{12}H_2Cl_6O$
374.86
Dibenzofuran, 1,3,4,6,7,9-hexa-chloro- (9CI)
1,3,4,6,7,9-Hexachlorodibenzo-furan

92341-06-5
$C_{12}H_2Cl_6O$
374.86
Dibenzofuran, 1,2,3,6,7,9-hexa-chloro- (9CI)
1,2,3,6,7,9-Hexachlorodibenzo-furan

92341-07-6
$C_{12}H_2Cl_6O$
374.86
Dibenzofuran, 1,2,3,4,8,9-hexa-chloro- (9CI)
1,2,3,4,8,9-Hexachlorodibenzo-furan

92366-34-2
$C_9H_{14}O$
138.21
2-Cyclopenten-1-one, tetra-methyl- (9CI)

92366-35-3
$C_{10}H_{12}Cl_2O$
219.11
Benzenemethanol, α,α,4-tri-methyl-, dichloro deriv. (9CI)

92387-48-9
$C_{16}H_{13}ClO_4$
304.73
1,2-Benzenedicarboxylic acid, mono(1-(4-chloro-phenyl)ethyl) ester (9CI)

92387-49-0
$C_{16}H_{13}BrO_4$
349.19
1,2-Benzenedicarboxylic acid, mono(1-(4-bromo-phenyl)ethyl) ester (9CI)

92474-39-0
Unknown
Unknown
Nitrilotriacetic acid, trisilver salt
NTA, trisilver salt

92988-11-9
Unknown
Unknown
Nitrilotriacetic acid, strontium sodium salt
NTA, strontium sodium salt

93037-15-1
$C_{22}H_{20}$
284.40
Benz(a)anthracene, 12-butyl- (9CI)

93746-34-0
$C_{10}H_{11}ClO_3.C_9H_6ClNO_3S.C_8H_6Cl_2O_3$
679.35
3(2H)-Benzothiazoleacetic acid, 4-chloro-2-oxo-, Mixt. with (+-)-2-(4-chloro-2-methyl-phenoxy)propanoic acid and (2,4-dichlorophenoxy)acetic acid (9CI)
DMA-sol

93763-70-3
Unknown
Unknown
Polytetrafluoroethylene Decomposition Products

93921-16-5
$C_{18}H_{16}ClNO_4S$
377.84
Propanoic acid, 2-(4-((6-chloro-2-benzothiazolyl)oxy)phen-oxy)-, ethyl ester, (+-)- (9CI)
Fenthiaprop-ethyl

94370-36-2
$C_{13}H_9ClFNO_2$
265.67
Benzamide, N-(4-chloro-phenyl)-3-fluoro-N-hydroxy- (9CI)

94483-57-5
$C_{11}H_{14}ClNO_3$
243.68
Carbamic acid, (3-chloro-4-methoxyphenyl)-, 1-methyl-ethyl ester (9CI)

94538-00-8
$C_{12}H_6Cl_2O$
237.09
Dibenzofuran, 1,3-di-chloro- (9CI)

94538-01-9
$C_{12}H_6Cl_2O$
237.08
Dibenzofuran, 1,4-dichloro- (9CI)
1,4-Dichlorodibenzofuran

94538-02-0
$C_{12}H_6Cl_2O$
237.08
Dibenzofuran, 1,7-dichloro- (9CI)
1,7-Dichlorodibenzofuran

94570-83-9
$C_{12}H_6Cl_2O$
237.09
Dibenzofuran, 3,4-di-chloro- (9CI)

94650-90-5
$C_6H_4Cl_2O_3$
195.00
Benzenetriol, dichloro- (9CI)

94650-91-6
$C_6H_3Cl_3O_3$

229.45
Benzenetriol, trichloro- (9CI)

94650-97-2
$C_5H_2Cl_2O_2$
164.98
3-Cyclopentene-1,2-dione, dichloro- (9CI)

94751-62-9
$C_4H_6N_2O_3S$
162.16
4-Isothiazolidinecarboxylic acid, 2-nitroso- (9CI)

94805-33-1
Unknown
Unknown
1-Octanethiol
Thiols

94818-85-6
$C_4H_7NO_3$
117.11
Oxetin
3-Amino-2-oxetane carboxylic acid
2-Oxetanecarboxylic acid, 3-amino-, (2R-cis)-

94888-09-2
$C_{12}H_5Cl_5O_2$
358.43
Phenol, 2,4-dichloro-6-(2,4,6-trichlorophenoxy)-(9CI)

94888-10-5
$C_{12}H_4Cl_6O_2$
392.88
Phenol, 2,4-dichloro-6-(2,3,4,6-tetrachlorophenoxy)-(9CI)

94888-11-6
$C_{12}H_4Cl_6O_2$
392.88
Phenol, 2,3,6-trichloro-4-(2,4,6-trichlorophenoxy)-(9CI)

94888-12-7
$C_{12}H_3Cl_7O_2$
427.32
Phenol, 2,3,4-trichloro-6-(2,3,4,6-tetrachlorophenoxy)-(9CI)

94888-13-8
$C_{12}H_3Cl_7O_2$
427.32
Phenol, 2,3,6-trichloro-4-(2,3,4,6-tetrachlorophenoxy)-(9CI)

94897-81-1
$C_{12}H_4Cl_6O_2$
392.88
Phenol, 2,3,4-trichloro-6-(2,4,6-trichlorophenoxy)-(9CI)

95032-59-0
Unknown
Unknown
TM (9CI)

95114-66-2
$C_{14}H_5Cl_{10}O_3PS$
638.76
Phosphorothioic acid, S-ethyl O,O-bis(pentachlorophenyl) ester (9CI)

95114-67-3
$C_{16}H_{19}O_3PS$
322.36
Phosphorothioic acid, S-ethyl O,O-bis(2-methylphenyl) ester (9CI)

95114-68-4
$C_{16}H_{19}O_3PS$
322.36
Phosphorothioic acid, S-ethyl O,O-bis(3-methylphenyl) ester (9CI)

95114-69-5
$C_{16}H_{19}O_3PS$
322.36
Phosphorothioic acid, S-ethyl O,O-bis(4-methylphenyl) ester (9CI)

95114-70-8
$C_{22}H_{31}O_3PS$
406.53
Phosphorothioic acid, O,O-bis-(2-(1,1-dimethylethyl)phenyl) S-ethyl ester (9CI)

95114-71-9
$C_{16}H_{17}Cl_2O_3PS$
391.25
Phosphorothioic acid, O,O-bis-(4-chloro-3-methylphenyl) S-ethyl ester (9CI)

95114-72-0
$C_{18}H_{23}O_3PS$
350.42
Phosphorothioic acid, O,O-bis-(2,3-dimethylphenyl) S-ethyl ester (9CI)

95114-73-1
$C_{18}H_{23}O_3PS$
350.42
Phosphorothioic acid, O,O-bis-(2,4-dimethylphenyl) S-ethyl ester (9CI)

95114-74-2
$C_{18}H_{23}O_3PS$
350.42
Phosphorothioic acid, O,O-bis-(2,5-dimethylphenyl) S-ethyl ester (9CI)

95114-75-3
$C_{18}H_{23}O_3PS$
350.42
Phosphorothioic acid, O,O-bis-(2,6-dimethylphenyl) S-ethyl ester (9CI)

95114-76-4
$C_{18}H_{23}O_3PS$
350.42
Phosphorothioic acid, O,O-bis-(3,4-dimethylphenyl) S-ethyl ester (9CI)

95114-77-5
$C_{18}H_{23}O_3PS$
350.42
Phosphorothioic acid, O,O-bis-(3,5-dimethylphenyl) S-ethyl ester (9CI)

95114-78-6
$C_{20}H_{27}O_3PS$
378.47
Phosphorothioic acid, S-ethyl O,O-bis(2,3,5-trimethyl-phenyl) ester (9CI)

95150-15-5
$C_{14}H_9Cl_6O_3PS$
500.98
Phosphorothioic acid, S-ethyl O,O-bis(2,4,5-trichloro-phenyl) ester (9CI)

95150-16-6
$C_{22}H_{31}O_3PS$
406.53
Phosphorothioic acid, O,O-bis-(4-(1,1-dimethylethyl)phenyl) S-ethyl ester (9CI)

95273-11-3
$C_{12}H_8N_2O_2$
212.20
9H-Carbazole, nitro- (9CI)
Carbazole, nitro-

95327-33-6
Unknown
Unknown
Belgard (9CI)

95461-54-4
C_9H_{18}
126.24
Hexene, trimethyl- (9CI)

95609-86-2
$C_{10}H_9NO$
159.18
2-Naphthalenol, amino- (9CI)

95686-15-0
Unknown
Unknown
Smithion (9CI)

95998-64-4
$C_{14}H_7ClF_6$
324.65
1,1'-Biphenyl, chloro-ar,ar'-bis-(trifluoromethyl)- (9CI)

95998-66-6
$C_{14}H_5Cl_3F_6$
393.54
1,1'-Biphenyl, trichloro-ar,ar'-bis(trifluoromethyl)-(9CI)

95998-67-7
$C_{14}H_4Cl_4F_6$
427.99
1,1'-Biphenyl, tetrachloro-ar,ar'-bis(trifluoromethyl)-(9CI)

95998-68-8
$C_{14}H_3Cl_5F_6$
462.43
1,1'-Biphenyl, pentachloro-ar,ar'-bis(trifluoromethyl)-(9CI)

95998-69-9
$C_{14}H_7Cl_2F_3O$
319.11
Methanone, (4-chlorophenyl)-(2-chloro-5-(trifluoromethyl)-phenyl)- (9CI)

95998-70-2
$C_{14}H_7Cl_2F_5$
341.11
Benzene, 1-chloro-2-((4-chloro-phenyl)difluoromethyl)-4-(tri-fluoromethyl)- (9CI)

96300-95-7
$C_{40}H_{46}O_8P_2$
716.76
Santicizer 148

96300-96-8
$C_{22}H_{23}O_4P.C_{18}H_{15}O_4P$
708.68
Phosphoric acid, 4-(1,1-di-methylethyl)phenyl diphenyl ester, Mixt. with triphenyl phosphate (9CI)
Santicizer 154

96300-97-9
$C_{21}H_{21}O_4P.C_{18}H_{15}O_4P$
694.66
Phosphoric acid, 2-(1-methyl-ethyl)phenyl diphenyl ester, Mixt. with triphenyl phos-phate (9CI)
Kronitex 200
Phosflex 31P

96320-70-6
Unknown
Unknown
Dalkon 11 (9CI)

96334-91-7
$C_{18}H_{19}NO$
265.36
1-Butanamine, 4-(9-anth-racenyloxy)- (9CI)

96420-93-8
Unknown
Unknown
QM 867 (9CI)

96881-25-3
Unknown
Unknown
2,4-Dichlorobenzoyl peroxide, More than 75% with water

97073-93-3
$C_{12}H_{11}N_3O_3S$
277.29
Carbamothioic acid, dimethyl-, O-(5-nitro-8-quinolinyl) ester (9CI)

97073-94-4
$C_{12}H_{11}ClN_2OS$
266.74
Carbamothioic acid, dimethyl-, O-(5-chloro-8-quinolinyl) ester (9CI)

97073-95-5
$C_{12}H_{12}N_2OS$
232.29
Carbamothioic acid, dimethyl-, O-8-quinolinyl ester (9CI)

97165-23-6
$C_{15}H_{20}O_2$
232.32
1H-Indene-1,7-dicarbox-aldehyde, 2,3,5,6-tetra-hydro-1,3,3,6-tetra-methyl-, (1S-cis)- (9CI)
Botrydienal

97190-65-3
$C_5H_8N_2O_2S$
160.18
2H-1,3-Thiazine, tetrahydro-2-(nitromethylene)-, (E)-(9CI)

97232-29-6
$C_{17}H_{14}$
218.30
Naphthalene, methylphenyl-(9CI)

97340-75-5
$C_{15}H_{13}N$
207.27
Acridine, 2,3-dimethyl- (9CI)

97419-16-4
$C_8H_{18}O_3$
162.23
Propane, 1,1,1-trimethoxy-2,2-dimethyl- (9CI)

97485-90-0
$C_{12}H_9N$
167.20
Azafluorene (9CI)

97502-49-3
Unknown
Unknown
Acid Red G (9CI) (VAN)

97534-02-6
$C_{13}H_7Cl_5O_2$
372.46
Benzene, 1,5-dichloro-2-meth-oxy-3-(2,4,6-trichlorophen-oxy)- (9CI)

97534-03-7
$C_{13}H_6Cl_6O_2$
406.91
Benzene, 1,2,3,5-tetrachloro-4-(3,5-dichloro-2-methoxy-phenoxy)- (9CI)

97534-04-8
$C_{13}H_6Cl_6O_2$
406.91
Benzene, 1,3,4-trichloro-2-methoxy-5-(2,4,6-tri-chlorophenoxy)- (9CI)

97534-05-9
$C_{13}H_6Cl_6O_2$
406.91
Benzene, 1,2,3-trichloro-4-methoxy-5-(2,4,6-trichlorophenoxy)- (9CI)

97534-06-0
$C_{13}H_5Cl_7O_2$
441.35
Benzene, 1,2,3,5-tetrachloro-4-(2,3,5-trichloro-4-methoxyphenoxy)- (9CI)

97534-07-1
$C_{13}H_5Cl_7O_2$
441.35
Benzene, 1,2,3,5-tetrachloro-4-(3,4,5-trichloro-2-methoxyphenoxy)- (9CI)

97553-43-0
Unknown
Unknown
Paraffins (Petroleum), Normal C > 10, chloro
Chlorinated paraffins

97659-46-6
Unknown
Unknown
Alkanes, C10-26, chloro
Chlorinated paraffins

98113-10-1
Unknown
Unknown
NP 9 (Surfactant)
Antarox CO 630
Nonoxynol-9

98318-97-9
$C_{10}H_5Cl_9$
444.22
4,7-Methano-1H-indene, 1,2,2,4,5,6,7,8,8-nonachloro-2,3,3a,4,7,7a-hexahydro-, (1α,3aα,4β,7β,7aα)- (9CI)

98404-93-4
$C_{13}H_{12}N_2O$
212.24
Benzamide, 4-cyano-N-(1,1-dimethyl-2-propynyl)- (9CI)

98526-74-0
$C_{77}H_{129}N_{27}O_{36}S_7$
2233.33
Metallothionein (Neurospora crassa copper-binding peptide moiety reduced), 12-L-threonine-14-L-alanine-16a-endo-L-glutamine-18-L-threonine-21-g lycine-24-de-L-serine- (9CI)
Metallothionein (Agaricus campestris bisporus copper-binding peptide moiety reduced)

98555-82-9
$C_7H_4Cl_4O_2S$
293.98
Benzene, 1,2,4,5-tetrachloro-3-(methylsulfonyl)- (9CI)
Sulfone, methyl 2,3,5,6-tetrachlorophenyl

98611-44-0
$C_{33}H_{38}ClNO_5$
564.13
Benzoic acid, 4-(((4-(nonyloxy)phenyl)methylene)amino)-, 2-(4-chlorophenyl)-1,3-dioxan-5-yl ester, (2α,5β(E))- (9CI)

98913-83-8
Unknown
Unknown
Pliabrac 521 (9CI)

99165-89-6
$C_3H_2Cl_2O_2$
140.95
2-Propenoic acid, dichloro- (9CI)

99165-90-9
$C_5H_3Cl_3O_3$
217.44
3-Pentenoic acid, 5,5,?-trichloro-2-oxo- (9CI)

99165-91-0
$C_5H_2Cl_4O_3$
251.88
3-Pentenoic acid, 5,5,5,?-tetrachloro-2-oxo- (9CI)

99165-92-1
$C_5H_3ClO_2S$
162.60
2-Thiophenecarboxylic acid, chloro- (9CI)

99165-93-2
$C_6H_3ClO_4S$
206.61
Thiophenedicarboxylic acid, chloro- (9CI)

99165-94-3
$C_5H_2Cl_4O_3$
251.88
3-Pentenoic acid, 3,4,5,5-tetrachloro-2-oxo- (9CI)

99165-95-4
$C_5HCl_5O_3$
286.32
3-Pentenoic acid, 3,4,5,5,5-pentachloro-2-oxo- (9CI)

99165-96-5
$C_7H_3Cl_3O_4$
257.46
Benzoic acid, 2,3,6-trichloro-4,5-dihydroxy- (9CI)

99165-97-6
$C_5H_3Cl_3O_3$
217.44
3-Pentenoic acid, 5,5,5-trichloro-2-oxo- (9CI)

99308-22-2
$C_6H_9Cl_3$
187.50
2-Hexene, 1,?,?-trichloro- (9CI)

99308-23-3
$C_9H_{14}Cl_4O$
280.02
Cyclohexane, dichloro(dichloropropoxy)- (9CI)

99308-24-4
$C_9H_{15}Cl_5O$
316.48
Hexane, trichloro-1-(dichloropropoxy)- (9CI)

99308-25-5
$C_9H_{14}Cl_6O$
350.93
Hexane, tetrachloro-1-(dichloropropoxy)- (9CI)

99308-26-6
$C_{12}H_{20}Cl_6O$
393.01
Nonane, tetrachloro-1-(dichloropropoxy)- (9CI)

99308-27-7
$C_{12}H_{20}Cl_6O_2$
409.01
Hexane, dichloro-1,6-bis(dichloropropoxy)- (9CI)

99308-28-8
$C_{12}H_{19}Cl_7O_2$
443.45
Hexane, trichloro-1,6-bis(dichloropropoxy)- (9CI)

99339-80-7
$_{15}H_{13}N$

207.27
Acridine, dimethyl- (9CI)

99342-08-2
$C_6H_{10}Cl_4O$
239.96
Propane, 1,1'-oxybis(dichloro- (9CI)

99422-01-2
$C_{14}H_{22}O_3S$
270.42
1,3-Cyclohexanedione, 5-(2-(ethylthio)propyl)-2-(1-oxopropyl)
5-((2-Ethylthio)propyl)-2-(1-oxopropyl)-1,3-cyclohexanedione
RE-45550

99554-33-3
$C_{29}H_{50}O_6$
494.71
Cyclohexanecarboxylic acid, 2-ethyl-2-(((1-oxononyl)oxy)-methyl)-1,3-propanediyl ester (9CI)

99562-17-1
$C_{25}H_{48}O_4$
412.65
Dodecanoic acid, 3-((2-ethyl-1-oxohexyl)oxy)-2,2-dimethyl-propyl ester (9CI)

99686-52-9
$C_{14}H_6Cl_2F_6$
359.10
1,1'-Biphenyl, 2,2'-dichloro-5,5'-bis(trifluoromethyl)- (9CI)

99744-82-8
$C_{12}H_{24}O_3S.Na$
271.38
Dodecenesulfonic acid, sodium salt (9CI)

99752-90-6
Unknown
Unknown
Preventol WB (9CI)

99849-00-0
$C_9H_{10}Cl_2O_3$
237.08
Benzene, 1,5-dichloro-2,3,4-trimethoxy- (6CI, 9CI)
1,3-Dichloro-4,5,6-trimethoxybenzene
4,6-Dichloro-1,2,3-trimethoxybenzene

100179-07-5
Unknown
Unknown
Phosflex Z (9CI)

100182-85-2
$C_{11}H_{13}NS$
191.29
Benzothiazole, butyl- (9CI)

100253-12-1
$C_{12}H_{16}NO_3PS$
285.30
Phosphorothioic acid, O-(cyanophenylmethyl) O,O-diethyl ester (9CI)
Phosphorothioic acid, α-cyanobenzyl O,O-diethyl ester (6CI)

100291-87-0
Unknown
Unknown
BNA 80 (9CI)

100428-67-9
$C_{10}H_{16}O_2$
168.24
2H-Pyran-2-one, tetrahydro-6-(2-pentenyl)-, (Z)- (9CI) (VAN)

100647-29-8
$C_{17}H_{10}O$
230.27
Benzo(b)fluorenone (9CI)

101038-68-0
$C_{12}H_{14}O_3$
206.24
Benzenehexanoic acid, β-oxo- (9CI)

101380-00-1
Unknown
Unknown
C.I. Direct Red 254 (9CI)

101657-77-6
$C_{19}H_{18}N_2O_2$
306.35
Cyanic acid, methylenebis-(2,6-dimethyl-4,1-phenylene) ester (9CI)

101714-96-9
$C_{12}H_7IO_2$
310.09
Dibenzo(b,e)(1,4)dioxin, 2-iodo- (9CI)
Dibenzo-p-dioxin, 2-iodo-

102040-44-8
$C_{23}H_{21}F_6O_3P$
490.38
Phosphorane, (2-methoxy-phenyl)diphenylbis-(2,2,2-trifluoroethoxy)- (9CI)

102040-45-9
$C_{14}H_{19}F_6O_4P$
396.27
Phosphorane, diethoxy-phenylbis(2,2,2-trifluoroethoxy)- (9CI)

102040-46-0
$C_{15}H_{21}F_6O_4P$
410.29
Phosphorane, diethoxy-(2-methylphenyl)bis-(2,2,2-trifluoroethoxy)- (9CI)

102040-47-1
$C_{13}H_{25}F_6O_5P$
406.30
Phosphorane, tris(1-methylethoxy)bis(2,2,2-trifluoroethoxy)- (9CI)

102040-48-2
$C_{24}H_{24}F_6NO_2P$
503.42
Benzenamine, 4-(diphenylbis(2,2,2-trifluoro-ethoxy)phosphoranyl)-N,N-dimethyl- (9CI)

102040-49-3
$C_{25}H_{25}F_6O_5P$
550.43
Phosphorane, tris(4-methoxyphenyl)bis(2,2,2-trifluoroethoxy)- (9CI)

102040-50-6
$C_{25}H_{25}F_6O_2P$
502.44
Phosphorane, tris(4-methylphenyl)bis(2,2,2-trifluoroethoxy)- (9CI)

102040-51-7
$C_{23}H_{21}F_6O_3P$
490.38
Phosphorane, (4-methoxy-phenyl)diphenylbis-(2,2,2-trifluoroethoxy)- (9CI)

102040-52-8
$C_{25}H_{25}F_6O_2P$
502.44

Phosphorane, tris(3-methylphenyl)bis(2,2,2-trifluoroethoxy)- (9CI)

102040-53-9
$C_{23}H_{21}F_6O_2P$
474.38
Phosphorane, (4-methylphenyl)diphenylbis-(2,2,2-trifluoroethoxy)-(9CI)

102040-54-0
$C_{22}H_{16}Cl_3F_6O_2P$
563.69
Phosphorane, tris-(4-chlorophenyl)bis-(2,2,2-trifluoroethoxy)-(9CI)

102040-55-1
$C_{24}H_{23}F_6O_4P$
520.41
Phosphorane, bis(4-methoxyphenyl)phenylbis-(2,2,2-trifluoroethoxy)-(9CI)

102040-56-2
$C_{22}H_{16}F_9O_2P$
514.33
Phosphorane, tris(4-fluorophenyl)bis(2,2,2-trifluoroethoxy)- (9CI)

102040-57-3
$C_{22}H_{16}Cl_3F_6O_2P$
563.69
Phosphorane, tris-(3-chlorophenyl)bis-(2,2,2-trifluoroethoxy)-(9CI)

102040-58-4
$C_{15}H_{15}F_{12}O_4P$
518.24
Phosphorane, (4-methylphenyl)tetrakis(2,2,2-trifluoroethoxy)- (9CI)

102040-59-5
$C_{15}H_{15}F_{12}O_5P$
534.24
Phosphorane, (4-methoxyphenyl)tetrakis(2,2,2-trifluoroethoxy)- (9CI)

102040-60-8
$C_{16}H_{18}F_{12}NO_4P$
547.27
Benzenamine, N,N-dimethyl-4-(tetrakis(2,2,2-trifluoroethoxy)phosphoranyl)- (9CI)

102040-61-9
$C_{15}H_{15}F_{12}O_4P$
518.24
Phosphorane, (3-methylphenyl)tetrakis(2,2,2-trifluoroethoxy)- (9CI)

102040-62-0
$C_{14}H_{12}ClF_{12}O_4P$
538.65
Phosphorane, (4-chlorophenyl)tetrakis(2,2,2-trifluoroethoxy)- (9CI)

102255-22-1
$C_9H_{21}O_2PS_2$
256.37
Phosphorodithioic acid, S-methyl O,O-bis(2-methylpropyl) ester (9CI)

102255-23-2
$CH_5O_2PS_2$
144.16
Phosphorodithioic acid, S-methyl ester (9CI)

102256-72-4
Unknown
Unknown
C 178 (9CI)

102640-14-2
Unknown
Unknown
C.I. Acid Red 413 (9CI)

102640-17-5
Unknown
Unknown
C.I. Reactive Red 185 (9CI)

102640-18-6
Unknown
Unknown
C.I. Reactive Yellow 155 (9CI)

102785-99-9
Unknown
Unknown
Siderite, magnesian (9CI)

102938-79-4
$C_{10}H_{15}Br$
215.13
Tricyclo(3.3.1.1(3,7))decane, bromo- (9CI)

103188-54-1
$C_{12}Cl_2F_{24}$
671.00
Decane, 2,9-bis(chlorodifluoromethyl)-1,1,1,2,3,3,4,4,5,5,6,6,7,7,8,8,9,10,10,10-eicosafluoro- (9CI)

103188-55-2
$C_{12}F_{26}$
638.09
Decane, 1,1,1,2,3,3,4,4,5,5,6,6,7,7,8,8,9,10,10,10-eicosafluoro-2,9-bis(trifluoromethyl)- (9CI)

103339-60-2
$C_4H_2Cl_2O$
136.96
2-Cyclobuten-1-one, dichloro-(9CI)

103339-61-3
$C_5H_6Cl_2O$
153.01
Cyclopentanone, dichloro-(9CI)

103339-62-4
$C_5H_3ClO_2$
130.53
3-Cyclopentene-1,2-dione, chloro- (9CI)

103339-63-5
$C_7H_7ClO_3$
174.58
Benzenediol, chloromethoxy-(9CI) (VAN)

103339-64-6
$C_7H_5ClO_3$
172.57
Benzaldehyde, chlorodihydroxy- (9CI)

103339-65-7
$C_7H_5ClO_4$
188.57
Benzoic acid, chlorodihydroxy-(9CI)

103354-08-1
$C_5HCl_3O_2$
199.42
3-Cyclopentene-1,2-dione, trichloro- (9CI)

103370-70-3
Unknown
Unknown
Benzophenanthrene, 9,10-dihydro- (9CI)

103433-72-3
$C_{11}H_{12}O$
160.22
Furan, 2,3-dihydro-2-(methylphenyl)- (9CI)

103453-97-0
$C_9H_{10}O$
134.18
Benzofuran, dihydro-2-methyl- (9CI)

103456-33-3
$C_{12}H_2Br_6O$
641.57
Dibenzofuran, hexabromo- (9CI)
Hexabromodibenzofuran

103456-36-6
$C_{12}H_3Br_5O_2$
578.67
Dibenzo(b,e)(1,4)dioxin, pentabromo- (9CI)
Pentabromodibenzo(b,e)(1,4)dioxin
Pentabromodibenzo-p-dioxin

103456-39-9
$C_{12}H_4Br_4O_2$
499.78
Dibenzo(b,e)(1,4)dioxin, tetrabromo- (9CI)
Tetrabromodibenzo(b,e)(1,4)dioxin
Tetrabromodibenzo-p-dioxin

103456-42-4
$C_{12}H_2Br_6O_2$
657.57
Dibenzo(b,e)(1,4)dioxin, hexabromo- (9CI)
Hexabromodibenzo(b,e)(1,4)-dioxin
Hexabromodibenzo-p-dioxin

103456-43-5
$C_{12}HBr_7O_2$
736.47
Dibenzo(b,e)(1,4)dioxin, heptabromo- (9CI)
Heptabromodibenzo(b,e)(1,4)dioxin
Heptabromodibenzo-p-dioxin

103528-31-0
$C_{10}H_{12}O$
148.20
Benzaldehyde, 3-propyl- (9CI)

103582-29-2
$C_{12}Br_8O$
799.36
Dibenzofuran, octabromo- (9CI)
Octabromodibenzofuran

103614-75-1
$C_{13}H_{19}NO_3S$
269.36
Formamide, N,N-dimethyl-1-((4-phenoxybutyl)sulfinyl)- (9CI)

103730-90-1
$C_4H_{11}Pb$
266.33
Plumbylium, ethyldimethyl- (9CI)

103737-38-8
Unknown
Unknown
UC 70480 (9CI)

103737-39-9
Unknown
Unknown
UC 70667 (9CI)

103837-23-6
$C_{16}H_{13}N$
219.28
1H-Pyrrole, diphenyl- (9CI)

103947-07-5
$C_{14}H_{19}N_7O_4S$
381.42
4-(4-(3,3-Dimethyl-1-triazene)-phenylsulfamide)-5,6-dimethoxypyrimidine
Aryltriazene methoxypyrimidine
4-ATMP

105554-30-1
$C_{12}H_{12}N_6O_{18}$
528.22
Hexamethylol benzene hexanitrate
Benzenehexamethanol, hexanitrate

105735-71-5
$C_{16}H_8N_2O_4$
292.24
3,7-Dinitrofluoranthene
Fluoranthene, 3,7-dinitro-

106232-85-3
Unknown
Unknown
Alkanes, C18-20, chloro
Chlorinated paraffins

106232-86-4
Unknown
Unknown
Alkanes, C22-40, chloro
Chlorinated paraffins

106807-78-7
$C_{19}H_{30}O_4$
322.44
Acetic acid, (2-(4-nonylphenoxy)ethoxy)- (9CI)

106939-89-3
$C_{14}H_{12}O_5S$
292.31
Benzenesulfonic acid, 2-formyl-, 4-methoxyphenyl ester (9CI)

106939-90-6
$C_{14}H_{12}O_4S$
276.31
Benzenesulfonic acid, 2-formyl-, 4-methylphenyl ester (9CI)

106939-91-7
$C_{13}H_{10}O_4S$
262.29
Benzenesulfonic acid, 2-formyl-, phenyl ester (9CI)

106939-92-8
$C_{13}H_9BrO_4S$
341.18
Benzenesulfonic acid, 2-formyl-, 4-bromophenyl ester (9CI)

106939-93-9
$C_{13}H_9NO_6S$
307.28
Benzenesulfonic acid, 2-formyl-, 4-nitrophenyl ester (9CI)

106939-94-0
$C_{14}H_{12}O_5S$
292.31
Benzenesulfonic acid, 4-formyl-, 4-methoxyphenyl ester (9CI)

106939-95-1
$C_{14}H_{12}O_4S$
276.31
Benzenesulfonic acid, 4-formyl-, 4-methylphenyl ester (9CI)

106939-96-2
$C_{13}H_{10}O_4S$

262.29
Benzenesulfonic acid, 4-formyl-, phenyl ester (9CI)

106939-97-3
$C_{13}H_9BrO_4S$
341.18
Benzenesulfonic acid, 4-formyl-, 4-bromophenyl ester (9CI)

106939-98-4
$C_{13}H_9NO_6S$
307.28
Benzenesulfonic acid, 4-formyl-, 4-nitrophenyl ester (9CI)

106946-84-3
Unknown
Unknown
Hercofloc 1021 (9CI)

106946-88-7
Unknown
Unknown
Percol 352 (9CI)

106946-89-8
Unknown
Unknown
Percol E 10 (9CI)

107207-47-6
$C_{12}HBrCl_6O$
453.76
Dibenzofuran, bromohexachloro- (9CI)
Bromohexachlorodibenzofuran

107227-56-5
$C_{12}H_4BrCl_3O$
350.43
Dibenzofuran, bromotrichloro- (9CI)
Bromotrichlorodibenzofuran

107227-75-8
$C_{12}H_4BrCl_3O_2$
366.42
Dibenzo(b,e)(1,4)dioxin, bromotrichloro- (9CI)
Bromotrichlorodibenzo-(b,e)(1,4)dioxin
Bromotrichlorodibenzo-p-dioxin

107348-42-5
$C_{12}H_{23}NOS$
229.38
1H-Azepine-1-carbothioic acid, hexahydro-, S-(3-methylbutyl) ester (9CI)

107348-43-6
$C_{10}H_{17}NOS$
199.31
1H-Azepine-1-carbothioic acid, hexahydro-, S-cyclopropyl ester (9CI)

107348-44-7
$C_9H_{16}ClNOS$
221.74
1H-Azepine-1-carbothioic acid, hexahydro-, S-(2-chloroethyl) ester (9CI)

107348-45-8
$C_9H_{17}NOS$
187.30
1-Piperidinecarbothioic acid, 3-methyl-, S-ethyl ester (9CI)

107348-46-9
$C_{10}H_{19}NOS$
201.33
1-Piperidinecarbothioic acid, 2,6-dimethyl-, S-ethyl ester (9CI)

107409-52-9
$C_7H_4Cl_4OS$
277.98
Benzene, 1,2,4,5-tetrachloro-3-(methylsulfinyl)- (9CI)

107534-96-3
$C_{16}H_{23}ClN_3O$
308.87
1H-1,2,4-Triazole-1-ethanol, α-(2-(4-chlorophenyl)ethyl)-α-(1,1-dimethylethyl)-, (+-)
Bay-HWG 1608
α-(2-(4-Chlorophenyl)ethyl)-α-(1,1-dimethylethyl)-1H-1,2,4-triazole-1-ethanol
Terbuconazole

108082-06-0
$C_5H_2Cl_4O_3$
251.87
2(5H)-Furanone, 3,4-dichloro-5-(dichloromethyl)-5-hydroxy
3,4-Dichloro-5-(dichloromethyl)-5-hydroxy-2-furanone

108544-90-7
$C_7H_5Cl_3O_2$
227.47
Hypochlorous acid, 2,3-dichloro-6-methoxyphenyl ester (9CI)

108544-91-8
$C_{10}H_{12}Cl_2O_4$
267.11
Benzeneethanol, 2,6-dichloro-4-hydroxy-3,5-dimethoxy- (9CI)

108544-93-0
$C_9H_{10}Cl_2O_3$
237.08
Benzene, 1,2-dichloro-3,4,5-trimethoxy- (9CI)

108544-94-1
$C_{10}H_{11}Cl_3O_3$
285.55
Benzene, 1,3-dichloro-2-(chloromethyl)-4,5,6-trimethoxy- (9CI)

108544-95-2
$C_{12}H_{12}Cl_2O_6$
323.13
Benzeneacetic acid, 2,6-dichloro-3,4,5-trimethoxy-α-oxo-, methyl ester (9CI)

108544-96-3
$C_{12}H_{16}Cl_2O_4$
295.16
Benzeneethanol, 2,6-dichloro-3,4,5-trimethoxy-α-methyl- (9CI)

108544-97-4
$C_8H_6Cl_2O_4$
237.04
Benzoic acid, 2,3-dichloro-4-hydroxy-5-methoxy- (9CI)

108544-98-5
$C_{10}H_{11}ClO_4$
230.65
Benzoic acid, 3-chloro-4,5-dimethoxy-, methyl ester (9CI)

108544-99-6
$C_{10}H_{10}Cl_2O_4$
265.09
Benzoic acid, 2,3-dichloro-4,5-dimethoxy-, methyl ester (9CI)

108545-00-2
$C_8H_9ClO_3$
188.61
Phenol, 4-chloro-2,6-dimethoxy- (9CI)

108545-01-3
$C_8H_6Cl_2O_3$

221.04
Benzaldehyde, 2,6-dichloro-4-hydroxy-3-methoxy- (9CI)

108545-02-4
$C_{10}H_{12}Cl_2O_3$
251.11
Phenol, 3,5-dichloro-4-ethyl-2,6-dimethoxy- (9CI)

108545-03-5
$C_{11}H_{14}Cl_2O_4$
281.14
Benzeneethanol, 2,6-dichloro-4-hydroxy-3,5-dimethoxy-α-methyl- (9CI)

108548-66-9
$C_{14}H_{11}ClO_2$
246.69
Benzoic acid, chloro-, phenylmethyl ester (9CI)
Benzoic acid, chloro-, benzyl ester (6CI)

108548-68-1
$C_8H_8O_4$
168.15
Benzoic acid, 3-hydroxymethoxy- (9CI)

108548-69-2
$C_8H_8O_4$
168.15
Benzoic acid, 4-hydroxymethoxy- (9CI)

108548-70-5
$C_{10}H_9Cl_3O_4$
299.54
2,4-Cyclohexadien-1-one, acetyl-2,6-dimethoxy-, trichloro deriv. (9CI)

108548-71-6
$C_8H_8Cl_2O_3$
223.06
Phenol, dichloro-2,5-dimethoxy- (9CI)

108548-72-7
$C_{14}H_9Cl_3O_2$
315.58
Benzoic acid, chloro-, (dichlorophenyl)methyl ester (9CI)

108548-73-8
$C_{10}H_{10}Cl_2O_4$
265.09
Ethanone, 1-(dichlorohydroxydimethoxyphenyl)- (9CI)

108572-08-3
$C_{10}H_{10}Cl_2O_5$
281.09
Benzeneacetic acid, dichloro-4-hydroxy-3,5-dimethoxy- (9CI)

108673-04-7
$C_{10}H_{10}Cl_2O_4$
265.09
2,4-Cyclohexadien-1-one, acetyldichloro-2,6-dimethoxy- (9CI)

112926-00-8
Unknown
Unknown
Preciptated silica
Silica, Amporphous-precipitated silica
Silica, Amporphous-silica gel

114790-09-9
$C_{16}H_9NO_2$
247.25
Acephenanthrylene, nitro- (9CI)

115044-73-0
$C_8H_3Cl_3N_2O_2$
265.47
Benzamide, 2,3,6-trichloro-5-cyano-4-hydroxy- (9CI)

115094-43-4
$C_{23}H_{33}N_5O_6SSi$
535.67
Adenosine, 5'-O-((1,1-dimethylethyl)dimethylsilyl)-, 2'-(4-methylbenzenesulfonate) (9CI)

115133-45-4
Unknown
Unknown
Acid Brilliant Yellow G (9CI)

115133-49-8
Unknown
Unknown
Basic Orange (9CI)

115340-67-5
$C_5H_3Cl_3O_3$
217.43
115340-67-5)
2-Butenoic acid, 3-formyl-2,4,4-trichloro-, (E)
(E)-2-Chloro-3-(dichloromethyl)-4-oxo-butenoic acid
(E)-3-Formyl-2,4,4-trichloro-2-butenoic acid

115384-94-6
$C_{10}H_7Cl_9$
446.24
4,7-Methano-1H-indene, 1,2,3,4,5,6,7,8,8-nonachlorooctahydro- (9CI)

116211-83-7
$C_4H_5NO_3$
115.08
2,5-Pyrrolidinedione, hydroxy- (9CI)

116211-84-8
$C_5H_7NO_3$
129.11
2,5-Pyrrolidinedione, hydroxymethyl- (9CI)

116211-85-9
C_9H_9NO
147.17
1H-Indolol, methyl- (9CI)

116211-86-0
$C_8H_4O_4$
164.12
1,3-Isobenzofurandione, hydroxy- (9CI)

116211-87-1
$C_8H_5N_3O_2$
175.13
Cinnoline, nitro- (9CI)

116211-88-2
$C_9H_6O_4$
178.14
1,3-Isobenzofurandione, hydroxymethyl- (9CI)

116211-89-3
$C_{14}H_{12}O_2$
212.25
9H-Xanthenol, methyl- (9CI)

116211-91-7
$C_{10}H_8N_2O_4$
220.17
1H-Indene, methyldinitro- (9CI)

116211-92-8
$C_6H_6N_2O_2$
138.11
Pyridine, methylnitro- (9CI)

116211-93-9
$C_8H_5NO_3S$

195.19
Benzo(b)thiopheneol, nitro- (9CI)

116211-95-1
$C_{11}H_6O_4$
202.17
Naphthalenetetrone, methyl- (9CI)

116211-97-3
$C_{12}H_9NO_4$
231.20
(1,1'-Biphenyl)diol, nitro- (9CI)

116211-98-4
$C_{14}H_9NO_3$
239.22
Phenanthrenol, nitro- (9CI)

116212-00-1
$C_{14}H_8N_2O_5$
284.22
Phenanthrenol, dinitro- (9CI)

116212-01-2
$C_{18}H_{11}NO_3$
289.28
Chrysenol, nitro- (9CI)

116212-02-3
$C_{18}H_9N_3O_2$
299.27
Benzo(h)naphtho(8,1,2-cde)cinnoline, nitro- (9CI)

116232-62-3
$C_{17}H_{10}O$
230.27
Benzo(a)fluorenone (9CI)

116232-63-4
$C_{13}H_{10}N_2O_2$
226.22
9H-Carbazole, methylnitro- (9CI)

116490-11-0
$C_{12}H_2Br_6O_2$
657.57
Dibenzo(b,e)(1,4)dioxin, 1,2,3,4,6,8-hexabromo- (9CI)
1,2,3,4,6,8-Hexabromodibenzo-(b,e)(1,4)dioxin
1,2,3,4,6,8-Hexabromodibenzo-p-dioxin

116530-07-5
$C_{11}H_9NO_2$
187.19
Naphthalene, 1-methyl-7-nitro- (9CI)

117148-85-3
Unknown
Unknown
Saytex 115 (9CI)

119620-42-7
$C_{13}H_8O_2$
196.21
9H-Fluoren-9-one, hydroxy- (9CI)

119973-28-3
$C_{25}H_{42}$
342.61
Chol-3-ene, 23-methyl-, (5α)- (9CI)

119973-29-4
$C_{26}H_{42}$
354.62
26,27-Dinorergosta-3,5-diene (9CI)

119973-30-7
$C_{26}H_{44}$
356.64
26,27-Dinorergost-3-ene, (5α)- (9CI)

119973-31-8
$C_{26}H_{40}$
352.60
26,27-Dinorergosta-3,5,22-triene, (22E)- (9CI)

120026-55-3
Unknown
Unknown
Surfonic JL 80X (9CI)

120056-15-7
$C_{30}H_{50}O$
426.73
Stigmasta-7,24(28)-dien-3-ol, 4-methyl-, (3β,5α)- (9CI) (VAN)

120710-23-8
$C_{22}H_{19}Br_2NO_3$
505.20
Cyclopropanecarboxylic acid, 3-(2,2-dibromoethenyl)-2,2-dimethyl-, cyano(3-phenoxyphenyl)methyl ester, (1α(R*),3α)-(+-)- (9CI)

120710-24-9
$C_{22}H_{19}Br_2NO_3$
505.20
Cyclopropanecarboxylic acid, 3-(2,2-dibromoethenyl)-2,2-dimethyl-, cyano(3-phenoxyphenyl)methyl ester, (1α(S*),3β)-(+-)- (9CI)

120710-25-0
$C_{22}H_{19}Br_2NO_3$
505.20
Cyclopropanecarboxylic acid, 3-(2,2-dibromoethenyl)-2,2-dimethyl-, cyano(3-phenoxyphenyl)methyl ester, (1α(R*),3β)-(+-)- (9CI)

977040-42-8
Unknown
Unknown
Whale Oil

1080 **(62-74-8)**
16842 **(4044-65-9)**
5107 **(2218-68-0)**
79907 **(36945-03-6)**
11A **(527-73-1)**
1212A **(9000-30-0)**
1371A **(26264-06-2)**
3A **(9003-53-6)**
5082A **(327-98-0)**
688A **(63-92-3)**
72-A34 **(33629-47-9)**
734571A **(4205-90-7)**
8000A **(9003-55-8)**
A 00 **(7429-90-5)**
A 033 **(8064-90-2)**
A 1 **(102-08-9)**
A 1 (Sorbent) **(1344-28-1)**
A 3-80 **(9003-53-6)**
A 15 (Polymer) **(9003-22-9)**
A 21LV **(9011-14-7)**
A 21 Lundbeck **(469-79-4)**
A 42 **(3244-90-4)**
A 50 **(8068-44-8)**
A 60-20R **(9002-88-4)**
A 60-70R **(9002-88-4)**
A 66 **(134-49-6)**
A 71 **(113-18-8)**
A 95 **(7429-90-5)**
A 99 **(7429-90-5)**
A 100 **(9004-66-4)**
A 100 (Pharmaceutical) **(9004-66-4)**
A 145 **(58-40-2)**
A 172 **(1067-53-4)**
A 348 **(115-67-3)**
A 361 **(1912-24-9)**
A 363 **(2032-59-9)**
A-399-Y4 **(606-58-6)**
A 432-130B **(9004-35-7)**
A 688 **(59-96-1)**
A 820 **(33629-47-9)**
A 884 **(300-42-5)**
A 995 **(7429-90-5)**
A 999 **(7429-90-5)**
A 1030 **(25038-54-4)**
A 1030N0 **(25038-54-4)**
A 1093 **(834-12-8)**
A 1100 **(919-30-2)**
A 1530 **(1309-64-4)**
A 1530 **(1314-60-9)**
A 1582 **(1309-64-4)**
A 1588LP **(1309-64-4)**
A 1798 **(8073-77-6)**
A 1803 **(54-91-1)**
A 2079 **(122-34-9)**
A 3322 **(91-80-5)**
A 3620 **(8066-11-3)**
A 3823A **(17090-79-8)**
A 4766 **(50563-36-5)**
A 4942 **(3778-73-2)**
A 5089 **(50563-36-5)**
A 6366 **(545-55-1)**
A 10846 **(67-68-5)**
A 11032 **(51-79-6)**
A 13397 **(2152-34-3)**
A 21960 **(609-15-4)**
A1-0109 P **(1344-28-1)**
A1-3916 P **(1344-28-1)**
A1-3945 E 1/16" **(1344-28-1)**
A1-3970 P **(1344-28-1)**
A1-4126 E 1/16" **(1344-28-1)**
A-1,5-DSA **(61736-91-2)**
A-20D **(9000-30-0)**
A-101 **(1088-11-5)**
A-139 **(800-24-8)**
A-500 **(144-80-9)**
A-502 **(57-68-1)**
A-980 **(101-27-9)**
A-1348 **(63-98-9)**
A-2371 **(18378-89-7)**
A-8103 **(54-91-1)**
A-12223 **(42509-80-8)**
A-91033 **(60-87-7)**
A7 Vapam **(137-42-8)**
A15 **(9003-22-9)**
A15-0 **(9003-22-9)**
A047 **(723-46-6)**
A65 **(302-41-0)**
A1030 **(105-60-2)**
A(S50154-9) **(4897-25-0)**
AH-42 **(91-80-5)**
AS-15 **(3810-74-0)**
AS-101 **(98-50-0)**
AS-17665 **(3570-75-0)**
9AA **(90-45-9)**
AA **(107-18-6)**
AA **(118-92-3)**
AA **(50-81-7)**
AA 81 **(75536-53-7)**
AA 1099 **(7429-90-5)**
AA-9 **(25155-30-0)**
AA-10 **(25155-30-0)**
AA149 **(41826-92-0)**
AA1199 **(7429-90-5)**
AAB **(60-09-3)**
2-AAF **(53-96-3)**
AAF **(53-96-3)**
AAN **(102-01-2)**
AAQ **(117-79-3)**
AAT **(56-38-2)**
AAT **(97-56-3)**
o-AAT **(97-56-3)**
AATP **(56-38-2)**
AAoC **(93-70-9)**
2-AB **(13952-84-6)**
AB-42 **(140-87-4)**
ABAR **(21609-90-5)**
ABG 3034 **(1214-39-7)**
AB-PC **(69-53-4)**
ABS **(11067-81-5)**
ABS **(11067-82-6)**
ABS **(42615-29-2)**
m-ABTF **(98-16-8)**
5-AC **(320-67-2)**
A(2)C **(54050-62-3)**
A2C **(54050-62-3)**
AC 8 **(9002-88-4)**
AC 394 **(9002-88-4)**
AC 528 **(78-34-2)**
AC 680 **(9002-88-4)**
AC 1075 **(69-74-9)**
AC 1220 **(9002-88-4)**
AC 3422 **(563-12-2)**
AC 3422 **(56-38-2)**
AC 3911 **(298-02-2)**

AC 4124	**(2463-84-5)**
AC 5223	**(2439-10-3)**
AC 5230	**(50-78-2)**
AC 5727	**(64-00-6)**
AC 18133	**(297-97-2)**
AC 18682	**(2275-18-5)**
AC 18706	**(116-01-8)**
AC 26,691	**(115-93-5)**
AC 38023	**(52-85-7)**
AC 38555	**(999-81-5)**
AC 47031	**(947-02-4)**
AC 47470	**(950-10-7)**
AC 52160	**(3383-96-8)**
AC 64475	**(21548-32-3)**
AC 84777	**(43222-48-6)**
AC 85258	**(58270-08-9)**
AC 92100	**(13071-79-9)**
AC 92553	**(40487-42-1)**
AC 94320	**(56070-16-7)**
AC 217300	**(67485-29-4)**
AC 222705	**(70124-77-5)**
AC 243,997	**(81334-34-1)**
AC 243997	**(81334-34-1)**
AC 252214	**(81335-37-7)**
AC 252,925	**(81510-83-0)**
AC 263499	**(81335-77-5)**
AC-1075	**(147-94-4)**
AC-12880	**(60-51-5)**
AC-18682	**(60-51-5)**
AC-18,737	**(2778-04-3)**
AC-47300	**(122-14-5)**
ACAR	**(510-15-6)**
ACC 3422	**(56-38-2)**
ACCEL R	**(103-34-4)**
AC-Di-Sol. NF	**(9004-32-4)**
AC GA	**(9002-88-4)**
ACHTL	**(1195-16-0)**
ACL-59	**(2244-21-5)**
ACL 60	**(2893-78-9)**
ACL 70	**(2782-57-2)**
ACL 85	**(87-90-1)**
ACN	**(2797-51-5)**
ACNQ	**(2797-51-5)**
ACP 322	**(132-66-1)**
ACP 322	**(132-67-2)**
ACP 6	**(9002-88-4)**
ACP 63303	**(1689-83-4)**
ACPC	**(52-52-8)**
ACP Grass Killer	**(650-51-1)**
ACP-M-728	**(133-90-4)**
ACPM-629	**(133-90-4)**
AC 8 (Polymer)	**(9002-88-4)**
AC-R-11	**(126-15-8)**
A2C Reagent	**(54050-62-3)**
ACS	**(60-32-2)**
ACS	**(9003-54-7)**
ACTH$^{1-24}$	**(16960-16-0)**
ACTH-α^{1-24}	**(16960-16-0)**
α^{1-24}-ACTH	**(16960-16-0)**
β^{1-24}-ACTH	**(16960-16-0)**
AD	**(50-76-0)**
AD 1	**(7429-90-5)**
AD 6	**(75-56-9)**
ADAB	**(539-17-3)**
ADC Auramine O	**(2465-27-2)**
ADC Brilliant Green Crystals	**(633-03-4)**
ADC Malachite Green Crystals	**(569-64-2)**
ADC Permanent Red Toner R	**(2814-77-9)**
ADC Rhodamine B	**(81-88-9)**
ADC Toluidine Red B	**(2425-85-6)**
ADE	**(73-24-5)**
ADH	**(11000-17-2)**
AD1M	**(7429-90-5)**
ADM	**(23214-92-8)**
ADMA 2	**(112-18-5)**
ADM hydrochloride	**(25316-40-9)**
2-ADNT	**(35572-78-2)**
4-ADNT	**(19406-51-0)**
ADO	**(7429-90-5)**
ADPA (9CI)	**(55069-41-5)**
ADR	**(25316-40-9)**
ADS	**(55635-13-7)**
AD 6 (Suspending agent)	**(75-56-9)**
AE	**(7429-90-5)**
AEP 1	**(9080-79-9)**
AEPD	**(115-70-8)**
AF	**(86-40-8)**
AF (Accelerator)	**(13733-91-0)**
AF 2 (Preservative)	**(3688-53-7)**
AF 5	**(9014-90-8)**
AF 72	**(9016-00-6)**
AF 75	**(63148-62-9)**
AF 75	**(9016-00-6)**
AF 10	**(25214-70-4)**
AF 101	**(330-54-1)**
AF 260	**(21645-51-2)**
AF-2	**(3688-53-7)**
A.F. Blue No. 1	**(2650-18-2)**
A.F. Blue No. 2	**(860-22-0)**
A.F. Green No. 1	**(4680-78-8)**
A.F. Green No. 2	**(5141-20-8)**
A.F. Orange No. 1	**(523-44-4)**
A.F. Orange No. 2	**(2646-17-5)**
A.F. Red No. 1	**(3564-09-8)**
A.F. Red No. 5	**(3118-97-6)**
A.F. Violet No. 1	**(1694-09-3)**
A.F. Yellow No. 2	**(85-84-7)**
A.F. Yellow No. 3	**(131-79-3)**
A.F. Yellow No. 4	**(1934-21-0)**
A.F. Yellow No. 5	**(2783-94-0)**
AFB1	**(1162-65-8)**
AFBI	**(1162-65-8)**
AFCF	**(25869-00-5)**
AFL 1081	**(640-19-7)**
A-Fax	**(9003-07-0)**
A-Fil Cream	**(13463-67-7)**
8 AG	**(134-58-7)**
AG 3	**(7440-44-0)**
AG 3 (Adsorbent)	**(7440-44-0)**
AG 5	**(7440-44-0)**
AG 5 (Adsorbent)	**(7440-44-0)**
AGC	**(15879-93-3)**
AGE	**(106-92-3)**
AGE (OSHA)	**(106-92-3)**
AGM-9	**(919-30-2)**
A-Gro	**(298-00-0)**
AH	**(59-33-6)**
AHA	**(546-88-3)**
AH-289 hydrochloride	**(14362-31-3)**
AH 501	**(1910-42-5)**
AH 5158	**(36894-69-6)**
AH 19065	**(66357-35-5)**
AHCTL	**(1195-16-0)**
AHR 85	**(532-03-6)**

AHR-438	(1665-48-1)	AI3-01046	(93-98-1)
AHR-619	(309-29-5)	AI3-01066	(13464-37-4)
AHR-712	(653-03-2)	AI3-01074	(10282-57-2)
AHR 2438B	(8061-51-6)	AI3-01091	(92-53-5)
AHR 2438B	(8062-15-5)	AI3-01165	(27152-57-4)
AI 3-22542	(134-62-3)	AI3-01169	(589-91-3)
AI 3-27985	(22431-62-5)	AI3-01174	(92-51-3)
AI 3-29024	(28434-00-6)	AI3-01234	(1454-85-9)
AI3-00033	(713-46-2)	AI3-01269	(2173-56-0)
AI3-00043	(132-65-0)	AI3-01270	(539-82-2)
AI3-00046	(78-59-1)	AI3-01358	(613-69-4)
AI3-00060	(98-28-2)	AI3-01391	(1129-50-6)
AI3-00101	(609-22-3)	AI3-01393	(550-44-7)
AI3-00115	(640-61-9)	AI3-01417	(94-69-9)
AI3-00179	(5328-01-8)	AI3-01432	(1521-38-6)
AI3-00183	(2050-76-2)	AI3-01463	(92-50-2)
AI3-00193	(6267-02-3)	AI3-01508	(496-03-7)
AI3-00203	(106-23-0)	AI3-01538	(538-51-2)
AI3-00277	(1314-13-2)	AI3-01565	(576-55-6)
AI3-00287	(2437-23-2)	AI3-01633	(555-43-1)
AI3-00310	(629-63-0)	AI3-01713	(2523-44-6)
AI3-00356	(151-21-3)	AI3-01744	(27136-73-8)
AI3-00358	(142-15-4)	AI3-01762	(2451-01-6)
AI3-00394	(77-94-1)	AI3-01772	(627-91-8)
AI3-00400	(140-04-5)	AI3-01812	(643-43-6)
AI3-00452	(538-68-1)	AI3-01876	(581-42-0)
AI3-00455	(136-81-2)	AI3-01932	(101-53-1)
AI3-00482	(4232-27-3)	AI3-01969	(102-13-6)
AI3-00489	(99-90-1)	AI3-01973	(2315-68-6)
AI3-00493	(615-43-0)	AI3-01979	(111-11-5)
AI3-00494	(89-59-8)	AI3-01980	(124-10-7)
AI3-00511	(607-85-2)	AI3-01982	(2917-73-9)
AI3-00516	(606-28-0)	AI3-01988	(106-06-9)
AI3-00584	(547-64-8)	AI3-01999	(77-90-7)
AI3-00586	(617-51-6)	AI3-02062	(495-40-9)
AI3-00638	(92-85-3)	AI3-02077	(634-36-6)
AI3-00641	(607-99-8)	AI3-02097	(93-96-9)
AI3-00645	(106-33-2)	AI3-02144	(2216-69-5)
AI3-00647	(111-81-9)	AI3-02166	(143-15-7)
AI3-00657	(111-62-6)	AI3-02178	(77-53-2)
AI3-00662	(106-79-6)	AI3-02183	(1119-49-9)
AI3-00665	(76-49-3)	AI3-02254	(2122-70-5)
AI3-00669	(111-82-0)	AI3-02257	(597-09-1)
AI3-00703	(623-30-3)	AI3-02268	(544-01-4)
AI3-00705	(134-85-0)	AI3-02270	(539-30-0)
AI3-00732	(1197-01-9)	AI3-02278	(620-23-5)
AI3-00733	(1632-73-1)	AI3-02282	(10277-04-0)
AI3-00736	(1195-79-5)	AI3-02332	(4376-18-5)
AI3-00737	(673-84-7)	AI3-02337	(519-73-3)
AI3-00789	(104-66-5)	AI3-02376	(13014-18-1)
AI3-00840	(483-65-8)	AI3-02418	(1515-72-6)
AI3-00842	(538-74-9)	AI3-02440	(81-14-1)
AI3-00862	(1137-42-4)	AI3-02453	(103-25-3)
AI3-00892	(501-52-0)	AI3-02463	(3391-10-4)
AI3-00897	(92-91-1)	AI3-02478	(28299-41-4)
AI3-00903	(660-60-6)	AI3-02479	(552-82-9)
AI3-00932	(7554-12-3)	AI3-02480	(106-65-0)
AI3-00957	(28804-88-8)	AI3-02489	(3739-67-1)
AI3-00959	(5341-95-7)	AI3-02545	(552-86-3)
AI3-00967	(1323-38-2)	AI3-02581	(95-75-0)
AI3-00969	(141-20-8)	AI3-02582	(13014-24-9)
AI3-00971	(106-12-7)	AI3-02693	(140-24-9)
AI3-00972	(1330-80-9)	AI3-02711	(103-60-6)
AI3-00978	(36521-89-8)	AI3-02743	(582-33-2)
AI3-00994	(112-89-0)	AI3-02820	(2136-89-2)
AI3-01023	(6531-86-8)	AI3-02940	(125-12-2)
AI3-01024	(124-06-1)	AI3-02952	(122-63-4)

AI3-02955	(94-18-8)
AI3-03103	(10248-74-5)
AI3-03113	(525-52-0)
AI3-03198	(2156-97-0)
AI3-03203	(2399-48-6)
AI3-03271	(85-97-2)
AI3-03273	(1817-74-9)
AI3-03277	(877-43-0)
AI3-03342	(34006-76-3)
AI3-03358	(115-70-8)
AI3-03389	(3878-55-5)
AI3-03442	(7756-96-9)
AI3-03443	(7779-77-3)
AI3-03502	(929-16-8)
AI3-03509	(112-39-0)
AI3-03520	(112-63-0)
AI3-03528	(141-19-5)
AI3-03564	(581-30-6)
AI3-03603	(6830-82-6)
AI3-03649	(618-80-4)
AI3-03666	(2051-95-8)
AI3-03698	(93-97-0)
AI3-03699	(88-65-3)
AI3-03709	(1016-05-3)
AI3-03775	(115-84-4)
AI3-03804	(479-27-6)
AI3-03827	(2499-59-4)
AI3-03945	(95-20-5)
AI3-03960	(110-41-8)
AI3-04026	(1011-12-7)
AI3-04095	(551-93-9)
AI3-04097	(104-21-2)
AI3-04110	(2555-49-9)
AI3-04168	(533-18-6)
AI3-04219	(102-25-0)
AI3-04220	(605-01-6)
AI3-04238	(593-08-8)
AI3-04250	(627-90-7)
AI3-04253	(637-27-4)
AI3-04318	(2065-23-8)
AI3-04341	(612-00-0)
AI3-04360	(501-65-5)
AI3-04363	(2756-56-1)
AI3-04487	(8047-99-2)
AI3-04490	(78-32-0)
AI3-04493	(609-31-4)
AI3-04494	(109-38-6)
AI3-04505	(13393-93-6)
AI3-04696	(99-88-7)
AI3-04702	(1541-81-7)
AI3-04979	(6975-71-9)
AI3-05001	(102-04-5)
AI3-05050	(504-57-4)
AI3-05084	(493-09-4)
AI3-05090	(90-97-1)
AI3-05526	(637-69-4)
AI3-05620	(1115-30-6)
AI3-05627	(10203-58-4)
AI3-05636	(617-35-6)
AI3-05639	(110-93-0)
AI3-05667	(2043-61-0)
AI3-05675	(2568-90-3)
AI3-05702	(495-76-1)
AI3-05710	(5466-77-3)
AI3-05775	(2201-24-3)
AI3-05777	(623-15-4)
AI3-05785	(87-41-2)
AI3-05827	(517-23-7)
AI3-05886	(120-21-8)
AI3-05904	(1806-54-8)
AI3-05972	(94-53-1)
AI3-05977	(4780-79-4)
AI3-05996	(6351-10-6)
AI3-06007	(818-38-2)
AI3-06008	(18707-60-3)
AI3-06011	(2050-60-4)
AI3-06014	(626-77-7)
AI3-06026	(1119-40-0)
AI3-06030	(540-07-8)
AI3-06080	(1732-10-1)
AI3-06164	(530-48-3)
AI3-06187	(1013-75-8)
AI3-06234	(504-01-8)
AI3-06279	(624-17-9)
AI3-06292	(617-48-1)
AI3-06325	(3937-56-2)
AI3-06331	(628-97-7)
AI3-06468	(554-95-0)
AI3-06520	(143-10-2)
AI3-06521	(112-88-9)
AI3-06523	(593-45-3)
AI3-06549	(533-60-8)
AI3-06556	(629-73-2)
AI3-06557	(111-88-6)
AI3-07023	(5500-21-0)
AI3-07025	(1823-91-2)
AI3-07159	(126-86-3)
AI3-07194	(18185-81-4)
AI3-07211	(623-12-1)
AI3-07328	(1423-46-7)
AI3-07380	(13361-34-7)
AI3-07400	(5330-17-6)
AI3-07501	(10332-32-8)
AI3-07552	(497-06-3)
AI3-07618	(5399-02-0)
AI3-07823	(583-04-0)
AI3-07842	(6259-76-3)
AI3-07848	(513-08-6)
AI3-07850	(126-71-6)
AI3-07854	(115-89-9)
AI3-07856	(2752-95-6)
AI3-07958	(110-36-1)
AI3-07959	(111-06-8)
AI3-07963	(110-33-8)
AI3-07964	(105-80-6)
AI3-07975	(7397-62-8)
AI3-08014	(8047-99-2)
AI3-08039	(621-62-5)
AI3-08042	(2568-30-1)
AI3-08092	(767-15-7)
AI3-08106	(3141-27-3)
AI3-08161	(1490-04-6)
AI3-08191	(538-75-0)
AI3-08196	(614-33-5)
AI3-08219	(2459-10-1)
AI3-08271	(496-46-8)
AI3-08446	(25735-29-9)
AI3-08497	(2455-24-5)
AI3-08507	(937-30-4)
AI3-08515	(110-30-5)
AI3-08532	(89-36-1)
AI3-08537	(592-20-1)
AI3-08540	(20893-30-5)
AI3-08686	(94-80-4)

AI3-08707	(765-43-5)	AI3-11170	(2052-07-5)
AI3-08751	(2439-35-2)	AI3-11181	(112-82-3)
AI3-08767	(2157-01-9)	AI3-11199	(629-30-1)
AI3-08826	(100-15-2)	AI3-11204	(611-70-1)
AI3-08832	(122-28-1)	AI3-11208	(4100-80-5)
AI3-08840	(96-97-9)	AI3-11230	(90-99-3)
AI3-08841	(603-62-3)	AI3-11234	(629-79-8)
AI3-08843	(552-32-9)	AI3-11240	(4265-25-2)
AI3-08854	(585-76-2)	AI3-11248	(1012-72-2)
AI3-08878	(636-98-6)	AI3-11264	(103-63-9)
AI3-08881	(1198-37-4)	AI3-11509	(815-17-8)
AI3-08882	(119-75-5)	AI3-11530	(621-77-2)
AI3-08884	(622-80-0)	AI3-11535	(931-20-4)
AI3-08885	(624-38-4)	AI3-11545	(1120-06-5)
AI3-08886	(100-23-2)	AI3-11583	(143-13-5)
AI3-08890	(501-60-0)	AI3-11586	(638-59-5)
AI3-08898	(96-73-1)	AI3-11591	(2021-28-5)
AI3-08905	(1155-00-6)	AI3-11735	(635-90-5)
AI3-08929	(76-84-6)	AI3-11743	(2067-33-6)
AI3-08954	(88-87-9)	AI3-11747	(590-90-9)
AI3-08977	(2473-03-2)	AI3-11798	(83-33-0)
AI3-08981	(95-12-5)	AI3-12032	(771-29-9)
AI3-09021	(824-78-2)	AI3-12065	(1821-12-1)
AI3-09026	(613-31-0)	AI3-12094	(13372-77-5)
AI3-09032	(589-87-7)	AI3-12116	(537-65-5)
AI3-09041	(74-39-5)	AI3-13058	(1117-86-8)
AI3-09044	(89-62-3)	AI3-13150	(613-90-1)
AI3-09046	(607-12-5)	AI3-13188	(4542-47-6)
AI3-09047	(119-42-6)	AI3-14148	(6833-13-2)
AI3-09061	(823-87-0)	AI3-14198	(2404-44-6)
AI3-09066	(2050-68-2)	AI3-14200	(3234-28-4)
AI3-09117	(1120-23-6)	AI3-14247	(140-77-2)
AI3-09125	(2664-42-8)	AI3-14319	(595-90-4)
AI3-09172	(6130-75-2)	AI3-14500	(142-30-3)
AI3-09173	(87-40-1)	AI3-14631	(593-85-1)
AI3-09178	(33637-20-6)	AI3-14650	(121-71-1)
AI3-09311	(886-77-1)	AI3-14655	(76-08-4)
AI3-09330	(1131-60-8)	AI3-14663	(6289-46-9)
AI3-09412	(575-89-3)	AI3-14664	(4705-34-4)
AI3-09491	(7388-44-5)	AI3-14675	(70-69-9)
AI3-09503	(105-62-4)	AI3-14677	(533-98-2)
AI3-09515	(13195-76-1)	AI3-14678	(75-81-0)
AI3-09529	(4542-57-8)	AI3-14682	(770-35-4)
AI3-09536	(1471-17-6)	AI3-14686	(86-28-2)
AI3-09664	(609-66-5)	AI3-14689	(14484-64-1)
AI3-10009	(583-53-9)	AI3-14764	(78-23-9)
AI3-10033	(1603-79-8)	AI3-14798	(14450-05-6)
AI3-10034	(635-46-1)	AI3-14852	(625-60-5)
AI3-10509	(1120-36-1)	AI3-14885	(611-19-8)
AI3-10513	(18835-33-1)	AI3-14886	(94-99-5)
AI3-10519	(7212-44-4)	AI3-14887	(102-47-6)
AI3-10523	(141-24-2)	AI3-14889	(122-01-0)
AI3-10532	(587-04-2)	AI3-14890	(89-75-8)
AI3-10570	(2243-35-8)	AI3-14898	(98-37-3)
AI3-10595	(459-60-9)	AI3-14905	(108-19-0)
AI3-10600	(628-39-7)	AI3-15013	(875-51-4)
AI3-10627	(1190-28-9)	AI3-15015	(12037-82-0)
AI3-11007	(4101-68-2)	AI3-15021	(1623-15-0)
AI3-11062	(504-02-9)	AI3-15029	(1120-48-5)
AI3-11086	(5798-75-4)	AI3-15040	(1779-48-2)
AI3-11096	(815-24-7)	AI3-15045	(3115-39-7)
AI3-11098	(112-17-4)	AI3-15046	(1623-14-9)
AI3-11101	(1975-78-6)	AI3-15051	(26658-09-3)
AI3-11112	(4593-90-2)	AI3-15053	(3991-73-9)
AI3-11124	(538-24-9)	AI3-15064	(824-72-6)
AI3-11163	(1131-62-0)	AI3-15067	(1609-21-8)
AI3-11164	(1667-01-2)	AI3-15074	(27900-75-0)

AI3-15076	(2016-42-4)	AI3-16611	(98-03-3)
AI3-15077	(2603-10-3)	AI3-16635	(2460-77-7)
AI3-15103	(574-42-5)	AI3-16644	(553-82-2)
AI3-15109	(6108-10-7)	AI3-16648	(492-86-4)
AI3-15119	(80-30-8)	AI3-16725	(1541-67-9)
AI3-15182	(84-54-8)	AI3-16727	(112-69-6)
AI3-15184	(504-53-0)	AI3-16771	(1330-78-5)
AI3-15228	(85-29-0)	AI3-16787	(77-08-7)
AI3-15229	(635-21-2)	AI3-16866	(534-26-9)
AI3-15237	(18266-55-2)	AI3-16897	(615-41-8)
AI3-15281	(97-59-6)	AI3-16899	(624-31-7)
AI3-15284	(2835-81-6)	AI3-16901	(97-65-4)
AI3-15292	(1002-16-0)	AI3-16904	(1087-21-4)
AI3-15299	(100-46-9)	AI3-16924	(103-62-8)
AI3-15320	(57-00-1)	AI3-16939	(121-53-9)
AI3-15321	(60-27-5)	AI3-16953	(926-39-6)
AI3-15325	(616-29-5)	AI3-16970	(1804-93-9)
AI3-15327	(103-49-1)	AI3-16972	(3138-42-9)
AI3-15332	(25167-81-1)	AI3-17002	(598-02-7)
AI3-15348	(1678-91-7)	AI3-17004	(3772-94-9)
AI3-15372	(1591-31-7)	AI3-17095	(622-38-8)
AI3-15376	(108-62-3)	AI3-17199	(4861-19-2)
AI3-15390	(645-00-1)	AI3-17201	(3921-30-0)
AI3-15398	(19525-59-8)	AI3-17229	(2492-26-4)
AI3-15403	(88-89-1)	AI3-17233	(3747-48-6)
AI3-15422	(25323-89-1)	AI3-17246	(830-81-9)
AI3-15483	(626-39-1)	AI3-17250	(2623-33-8)
AI3-15523	(95-15-8)	AI3-17266	(59-88-1)
AI3-15527	(645-13-6)	AI3-17279	(101-67-7)
AI3-15546	(141-94-6)	AI3-17283	(1333-13-7)
AI3-15587	(901-44-0)	AI3-17349	(1552-42-7)
AI3-15588	(116-37-0)	AI3-17378	(92-86-4)
AI3-15633	(598-92-5)	AI3-17381	(92-05-7)
AI3-15645	(1471-18-7)	AI3-17420	(505-10-2)
AI3-15687	(4813-57-4)	AI3-17422	(5335-05-7)
AI3-15694	(13402-02-3)	AI3-17591	(136-45-8)
AI3-15698	(999-55-3)	AI3-17608	(575-43-9)
AI3-15706	(689-12-3)	AI3-17609	(581-40-8)
AI3-15871	(3457-46-3)	AI3-17610	(582-16-1)
AI3-15885	(1689-78-7)	AI3-17611	(829-26-5)
AI3-15914	(2917-26-2)	AI3-17673	(2759-54-8)
AI3-15917	(590-67-0)	AI3-17738	(543-24-8)
AI3-15984	(95-14-7)	AI3-17741	(7463-22-1)
AI3-15989	(1604-34-8)	AI3-17757	(15096-41-0)
AI3-16044	(2189-60-8)	AI3-17824	(123-79-5)
AI3-16047	(107-40-4)	AI3-17837	(5414-19-7)
AI3-16063	(874-42-0)	AI3-17846	(78-33-1)
AI3-16095	(30667-99-3)	AI3-17853	(626-55-1)
AI3-16111	(87-89-8)	AI3-17876	(594-60-5)
AI3-16131	(612-25-9)	AI3-17947	(3467-59-2)
AI3-16135	(557-24-4)	AI3-17970	(536-66-3)
AI3-16183	(2580-77-0)	AI3-18009	(5339-85-5)
AI3-16253	(2051-24-3)	AI3-18010	(104-10-9)
AI3-16316	(5263-87-6)	AI3-18131	(615-54-3)
AI3-16362	(1142-19-4)	AI3-18146	(604-53-5)
AI3-16452	(127-52-6)	AI3-18152	(93-92-5)
AI3-16497	(110-87-2)	AI3-18153	(555-45-3)
AI3-16499	(142-68-7)	AI3-18160	(124-83-4)
AI3-16506	(556-08-1)	AI3-18168	(505-52-2)
AI3-16553	(5329-79-3)	AI3-18180	(112-86-7)
AI3-16560	(5332-24-1)	AI3-18185	(115-87-7)
AI3-16562	(112-20-9)	AI3-18209	(1121-55-7)
AI3-16570	(629-60-7)	AI3-18242	(1260-17-9)
AI3-16575	(102-86-3)	AI3-18245	(603-45-2)
AI3-16576	(3007-31-6)	AI3-18247	(3121-71-9)
AI3-16578	(2516-96-3)	AI3-18283	(3002-18-4)
AI3-16579	(98-33-9)	AI3-18299	(760-78-1)

AI3-18308 **(516-06-3)**
AI3-18377 **(3370-35-2)**
AI3-18429 **(542-90-5)**
AI3-18436 **(150-30-1)**
AI3-18442 **(3458-28-4)**
AI3-18470 **(107-95-9)**
AI3-18528 **(590-02-3)**
AI3-18544 **(122-70-3)**
AI3-18558 **(78-38-6)**
AI3-18786 **(624-89-5)**
AI3-18787 **(111-47-7)**
AI3-18857 **(922-80-5)**
AI3-18858 **(3006-15-3)**
AI3-18859 **(127-39-9)**
AI3-18864 **(544-02-5)**
AI3-18871 **(82-38-2)**
AI3-18876 **(57-87-4)**
AI3-18877 **(528-50-7)**
AI3-18904 **(5335-24-0)**
AI3-18979 **(60044-33-9)**
AI3-19022 **(120-46-7)**
AI3-19024 **(1570-65-6)**
AI3-19031 **(700-38-9)**
AI3-19045 **(3996-59-6)**
AI3-19099 **(138-52-3)**
AI3-19148 **(398-23-2)**
AI3-19232 **(100-48-1)**
AI3-19238 **(100-26-5)**
AI3-19250 **(632-51-9)**
AI3-19252 **(1070-83-3)**
AI3-19261 **(874-23-7)**
AI3-19279 **(98-31-7)**
AI3-19307 **(93-02-7)**
AI3-19423 **(608-66-2)**
AI3-19424 **(50-70-4)**
AI3-19425 **(87-79-6)**
AI3-19427 **(512-69-6)**
AI3-19441 **(16365-27-8)**
AI3-19476 **(100-84-5)**
AI3-19481 **(133-13-1)**
AI3-19501 **(130-13-2)**
AI3-19502 **(90-51-7)**
AI3-19517 **(37437-20-0)**
AI3-19536 **(2510-86-3)**
AI3-19577 **(7208-47-1)**
AI3-19626 **(3774-52-5)**
AI3-19668 **(14025-21-9)**
AI3-19737 **(151-13-3)**
AI3-19742 **(3278-35-1)**
AI3-19768 **(7620-77-1)**
AI3-19803 **(688-37-9)**
AI3-19804 **(142-17-6)**
AI3-19805 **(1555-53-9)**
AI3-19807 **(555-35-1)**
AI3-19928 **(580-13-2)**
AI3-19935 **(4221-03-8)**
AI3-19938 **(620-17-7)**
AI3-19939 **(126-81-8)**
AI3-19978 **(603-48-5)**
AI3-20047 **(551-45-1)**
AI3-20152 **(619-86-3)**
AI3-20196 **(114-38-5)**
AI3-20213 **(2345-34-8)**
AI3-20321 **(118-44-5)**
AI3-20480 **(593-50-0)**
AI3-20627 **(17849-38-6)**
AI3-20628 **(873-76-7)**
AI3-20685 **(110-03-2)**
AI3-20801 **(89-74-7)**
AI3-20871 **(51-14-9)**
AI3-20877 **(2444-36-2)**
AI3-20879 **(75-84-3)**
AI3-20881 **(119-56-2)**
AI3-20884 **(620-13-3)**
AI3-20950 **(230-17-1)**
AI3-20957 **(2150-38-1)**
AI3-21066 **(610-69-5)**
AI3-21153 **(118-41-2)**
AI3-21209 **(637-64-9)**
AI3-21213 **(93-04-9)**
AI3-21214 **(553-90-2)**
AI3-21247 **(1112-38-5)**
AI3-21349 **(934-00-9)**
AI3-21374 **(768-50-3)**
AI3-21419 **(877-65-6)**
AI3-21535 **(768-59-2)**
AI3-21536 **(89-95-2)**
AI3-21575 **(587-03-1)**
AI3-21616 **(613-33-2)**
AI3-21675 **(300-85-6)**
AI3-21892 **(5426-78-8)**
AI3-21918 **(529-20-4)**
AI3-21995 **(109-49-9)**
AI3-22019 **(7495-84-3)**
AI3-22030 **(625-99-0)**
AI3-22032 **(1119-44-4)**
AI3-22090 **(831-81-2)**
AI3-22124 **(2510-55-6)**
AI3-22131 **(141-10-6)**
AI3-22142 **(3452-97-9)**
AI3-22166 **(5397-01-3)**
AI3-22178 **(2425-01-6)**
AI3-22251 **(19343-78-3)**
AI3-22330 **(6936-40-9)**
AI3-22410 **(637-88-7)**
AI3-22613 **(5660-60-6)**
AI3-22668 **(3521-06-0)**
AI3-22671 **(4657-00-5)**
AI3-22766 **(1518-83-8)**
AI3-22781 **(587-56-4)**
AI3-23023 **(530-50-7)**
AI3-23031 **(610-96-8)**
AI3-23120 **(111-63-7)**
AI3-23126 **(107-54-0)**
AI3-23129 **(4501-58-0)**
AI3-23192 **(2107-76-8)**
AI3-23206 **(76-24-4)**
AI3-23227 **(119-70-0)**
AI3-23239 **(20925-85-3)**
AI3-23251 **(56960-97-5)**
AI3-23257 **(620-73-5)**
AI3-23286 **(870-23-5)**
AI3-23305 **(7057-92-3)**
AI3-23391 **(137-00-8)**
AI3-23399 **(830-09-1)**
AI3-23404 **(115-22-0)**
AI3-23412 **(151-19-9)**
AI3-23448 **(137-43-9)**
AI3-23452 **(617-62-9)**
AI3-23460 **(101-55-3)**
AI3-23491 **(2678-21-9)**
AI3-23514 **(57-48-7)**
AI3-23578 **(1126-46-1)**
AI3-23674 **(15448-99-4)**

AI3-23779 **(482-05-3)**
AI3-23843 **(78-39-7)**
AI3-23844 **(115-80-0)**
AI3-23867 **(2305-26-2)**
AI3-23868 **(579-07-7)**
AI3-23878 **(93-90-3)**
AI3-23975 **(4282-44-4)**
AI3-23986 **(463-40-1)**
AI3-23988 **(467-85-6)**
AI3-24008 **(885-82-5)**
AI3-24009 **(2432-12-4)**
AI3-24011 **(133-53-9)**
AI3-24040 **(107-85-7)**
AI3-24119 **(606-43-9)**
AI3-24120 **(604-35-3)**
AI3-24181 **(93-03-8)**
AI3-24199 **(541-35-5)**
AI3-24202 **(116-53-0)**
AI3-24210 **(698-71-5)**
AI3-24218 **(542-10-9)**
AI3-24251 **(106-30-9)**
AI3-24252 **(629-80-1)**
AI3-24259 **(127-43-5)**
AI3-24261 **(97-87-0)**
AI3-24280 **(583-61-9)**
AI3-24290 **(5780-07-4)**
AI3-24332 **(1445-45-0)**
AI3-24338 **(3744-02-3)**
AI3-24342 **(25154-55-6)**
AI3-24349 **(869-29-4)**
AI3-24356 **(624-54-4)**
AI3-24358 **(141-14-0)**
AI3-24379 **(497-03-0)**
AI3-24380 **(104-87-0)**
AI3-24382 **(99-36-5)**
AI3-24387 **(626-97-1)**
AI3-24427 **(2316-26-9)**
AI3-24476 **(621-32-9)**
AI3-24484 **(934-34-9)**
AI3-24486 **(99-82-1)**
AI3-24502 **(931-19-1)**
AI3-24563 **(3699-54-5)**
AI3-24564 **(6281-42-1)**
AI3-24571 **(614-78-8)**
AI3-24620 **(471-74-9)**
AI3-24709 **(33669-76-0)**
AI3-24787 **(1679-51-2)**
AI3-24875 **(1034-41-9)**
AI3-24890 **(94-04-2)**
AI3-24902 **(18479-57-7)**
AI3-24906 **(18479-51-1)**
AI3-24917 **(940-71-6)**
AI3-24920 **(2517-43-3)**
AI3-24973 **(4209-91-0)**
AI3-25024 **(542-28-9)**
AI3-25058 **(3917-15-5)**
AI3-25059 **(103-44-6)**
AI3-25090 **(505-32-8)**
AI3-25091 **(544-12-7)**
AI3-25132 **(78-80-8)**
AI3-25180 **(80-04-6)**
AI3-25204 **(1719-58-0)**
AI3-25222 **(589-29-7)**
AI3-25255 **(1122-61-8)**
AI3-25256 **(1124-33-0)**
AI3-25306 **(67-71-0)**
AI3-25310 **(126-39-6)**
AI3-25317 **(3766-27-6)**
AI3-25333 **(14861-06-4)**
AI3-25349 **(7783-28-0)**
AI3-25413 **(6294-34-4)**
AI3-25418 **(32360-05-7)**
AI3-25419 **(142-09-6)**
AI3-25443 **(6291-84-5)**
AI3-25450 **(2141-62-0)**
AI3-25516 **(87-10-5)**
AI3-25612 **(2703-13-1)**
AI3-25787 **(2665-30-7)**
AI3-26011 **(586-37-8)**
AI3-26062 **(7722-76-1)**
AI3-26087 **(123-81-9)**
AI3-26109 **(15843-02-4)**
AI3-26168 **(110-42-9)**
AI3-26171 **(111-59-1)**
AI3-26172 **(3658-80-8)**
AI3-26173 **(3012-65-5)**
AI3-26191 **(14832-14-5)**
AI3-26201 **(874-68-0)**
AI3-26247 **(9003-29-6)**
AI3-26248 **(9003-29-6)**
AI3-26249 **(9003-29-6)**
AI3-26250 **(9003-29-6)**
AI3-26251 **(9003-29-6)**
AI3-26252 **(9003-29-6)**
AI3-26253 **(9003-29-6)**
AI3-26254 **(9003-29-6)**
AI3-26255 **(9003-29-6)**
AI3-26256 **(9003-29-6)**
AI3-26275 **(77-76-9)**
AI3-26311 **(7390-81-0)**
AI3-26368 **(927-60-6)**
AI3-26413 **(88-97-1)**
AI3-26439 **(286-99-7)**
AI3-26465 **(508-32-7)**
AI3-26469 **(5502-88-5)**
AI3-26484 **(112-51-6)**
AI3-26487 **(118-69-4)**
AI3-26504 **(20020-02-4)**
AI3-26638 **(100-36-7)**
AI3-26640 **(51-80-9)**
AI3-26645 **(1838-19-3)**
AI3-26692 **(111-78-4)**
AI3-26693 **(931-88-4)**
AI3-26694 **(292-64-8)**
AI3-26696 **(1700-10-3)**
AI3-26709 **(328-39-2)**
AI3-26714 **(95-59-0)**
AI3-26793 **(103-67-3)**
AI3-26796 **(104-63-2)**
AI3-26821 **(541-48-0)**
AI3-26935 **(301-00-8)**
AI3-26938 **(328-50-7)**
AI3-26971 **(56025-96-8)**
AI3-26989 **(764-13-6)**
AI3-26997 **(121-06-2)**
AI3-27005 **(3734-49-4)**
AI3-27056 **(14979-34-1)**
AI3-27067 **(706-78-5)**
AI3-27096 **(2655-15-4)**
AI3-27133 **(95-32-9)**
AI3-27223 **(1031-47-6)**
AI3-27305 **(17702-57-7)**
AI3-27449 **(14816-20-7)**
AI3-27477 **(4301-50-2)**

AI3-27488	**(1563-38-8)**	AI3-29128	**(42509-80-8)**
AI3-27498	**(91-52-1)**	AI3-29158	**(52645-53-1)**
AI3-27522	**(2619-00-3)**	AI3-29270	**(58842-20-9)**
AI3-27531	**(1120-24-7)**	AI3-29311	**(59669-26-0)**
AI3-27537	**(499-74-1)**	AI3-29315	**(56265-21-5)**
AI3-27542	**(22941-83-9)**	AI3-29368	**(64628-44-0)**
AI3-27696	**(26419-73-8)**	AI3-29409	**(59010-86-5)**
AI3-27829	**(41491-52-5)**	AI3-29549	**(72542-56-4)**
AI3-27854	**(4104-14-7)**	AI3-29558	**(83-41-0)**
AI3-27887	**(31107-44-5)**	AI3-29783	**(75096-86-5)**
AI3-27990	**(26258-70-8)**	AI3-30202	**(689-89-4)**
AI3-28014	**(15438-71-8)**	AI3-30205	**(883-99-8)**
AI3-28035	**(4620-70-6)**	AI3-30436	**(4351-54-6)**
AI3-28052	**(7328-91-8)**	AI3-30512	**(5451-76-3)**
AI3-28072	**(3287-06-7)**	AI3-30528	**(627-58-7)**
AI3-28136	**(6175-49-1)**	AI3-30534	**(16529-56-9)**
AI3-28205	**(2163-79-3)**	AI3-30738	**(5454-28-4)**
AI3-28217	**(29590-42-9)**	AI3-30763	**(2512-29-0)**
AI3-28228	**(1741-41-9)**	AI3-30876	**(14167-18-1)**
AI3-28255	**(614-68-6)**	AI3-30956	**(106-02-5)**
AI3-28258	**(4044-65-9)**	AI3-30983	**(589-75-3)**
AI3-28269	**(5124-25-4)**	AI3-31017	**(2306-88-9)**
AI3-28301	**(629-08-3)**	AI3-31286	**(41363-16-0)**
AI3-28302	**(2244-07-7)**	AI3-31290	**(694-80-4)**
AI3-28402	**(592-43-8)**	AI3-31295	**(326-91-0)**
AI3-28403	**(111-66-0)**	AI3-31313	**(594-61-6)**
AI3-28404	**(112-95-8)**	AI3-31362	**(2150-93-8)**
AI3-28453	**(603-32-7)**	AI3-31503	**(2150-47-2)**
AI3-28462	**(91-67-8)**	AI3-31575	**(2239-78-3)**
AI3-28480	**(1604-11-1)**	AI3-31576	**(110-34-9)**
AI3-28514	**(626-23-3)**	AI3-31880	**(620-24-6)**
AI3-28517	**(21145-77-7)**	AI3-32117	**(625-38-7)**
AI3-28518	**(106-26-3)**	AI3-32389	**(588-30-7)**
AI3-28524	**(86-48-6)**	AI3-32462	**(112-11-8)**
AI3-28527	**(89-64-5)**	AI3-32576	**(110-06-5)**
AI3-28528	**(88-43-7)**	AI3-32578	**(629-45-8)**
AI3-28529	**(88-53-9)**	AI3-32895	**(7005-72-3)**
AI3-28531	**(98-48-6)**	AI3-33125	**(7304-99-6)**
AI3-28537	**(110-21-4)**	AI3-33228	**(32749-94-3)**
AI3-28570	**(1731-84-6)**	AI3-33229	**(630-19-3)**
AI3-28573	**(13171-00-1)**	AI3-33230	**(1121-60-4)**
AI3-28577	**(515-84-4)**	AI3-33242	**(630-18-2)**
AI3-28580	**(94-60-0)**	AI3-33266	**(14959-86-5)**
AI3-28585	**(108-85-0)**	AI3-33324	**(101-43-9)**
AI3-28589	**(544-10-5)**	AI3-33373	**(22775-37-7)**
AI3-28591	**(3386-33-2)**	AI3-33410	**(2136-79-0)**
AI3-28606	**(616-25-1)**	AI3-33474	**(16725-53-4)**
AI3-28607	**(1569-50-2)**	AI3-33573	**(30673-36-0)**
AI3-28609	**(625-31-0)**	AI3-33581	**(106-73-0)**
AI3-28612	**(4798-44-1)**	AI3-33584	**(109-19-3)**
AI3-28621	**(4938-52-7)**	AI3-33593	**(2445-76-3)**
AI3-28622	**(3521-91-3)**	AI3-33881	**(629-76-5)**
AI3-28673	**(41663-84-7)**	AI3-33971	**(16974-11-1)**
AI3-28714	**(94-34-8)**	AI3-33973	**(35148-19-7)**
AI3-28762	**(13601-19-9)**	AI3-33978	**(873-94-9)**
AI3-28792	**(2245-38-7)**	AI3-34338	**(72269-48-8)**
AI3-28793	**(590-66-9)**	AI3-34392	**(3681-71-8)**
AI3-28802	**(14481-29-9)**	AI3-34445	**(22047-25-2)**
AI3-28849	**(2207-01-4)**	AI3-34461	**(868-57-5)**
AI3-28850	**(6876-23-9)**	AI3-34601	**(21213-69-4)**
AI3-28854	**(16747-26-5)**	AI3-34794	**(928-97-2)**
AI3-28901	**(1113-68-4)**	AI3-34886	**(29804-22-6)**
AI3-28913	**(489-98-5)**	AI3-35092	**(544-00-3)**
AI3-28914	**(777-37-7)**	AI3-35104	**(1003-29-8)**
AI3-28936	**(535-15-9)**	AI3-35123	**(38421-90-8)**
AI3-28938	**(102-52-3)**	AI3-35155	**(7452-79-1)**
AI3-29087	**(1314-23-4)**	AI3-35169	**(56683-54-6)**

AI3-35174	**(34010-15-6)**	AI3-37199	**(939-97-9)**
AI3-35195	**(1560-88-9)**	AI3-37201	**(629-19-6)**
AI3-35251	**(1072-33-9)**	AI3-37210	**(628-99-9)**
AI3-35271	**(4706-81-4)**	AI3-37211	**(624-51-1)**
AI3-35291	**(35153-15-2)**	AI3-37212	**(5932-79-6)**
AI3-35349	**(27519-02-4)**	AI3-37214	**(589-62-8)**
AI3-35565	**(1560-89-0)**	AI3-37227	**(432-25-7)**
AI3-35584	**(53939-27-8)**	AI3-37252	**(764-01-2)**
AI3-35598	**(104-30-3)**	AI3-37268	**(625-25-2)**
AI3-35599	**(833-43-2)**	AI3-37707	**(563-79-1)**
AI3-35680	**(1653-30-1)**	AI3-37709	**(627-19-0)**
AI3-35817	**(141-12-8)**	AI3-37712	**(5026-76-6)**
AI3-35917	**(638-67-5)**	AI3-37714	**(2396-65-8)**
AI3-35937	**(53939-28-9)**	AI3-37786	**(6750-03-4)**
AI3-36005	**(7493-63-2)**	AI3-37787	**(1191-95-3)**
AI3-36027	**(3301-94-8)**	AI3-37790	**(6342-56-9)**
AI3-36028	**(705-86-2)**	AI3-38086	**(20680-10-8)**
AI3-36042	**(33467-76-4)**	AI3-38157	**(2179-60-4)**
AI3-36043	**(18409-17-1)**	AI3-38428	**(490-64-2)**
AI3-36062	**(14199-15-6)**	AI3-38563	**(589-35-5)**
AI3-36074	**(127-51-5)**	AI3-38565	**(6305-71-1)**
AI3-36117	**(925-78-0)**	AI3-38648	**(64726-91-6)**
AI3-36122	**(629-92-5)**	AI3-39164	**(496-15-1)**
AI3-36135	**(51607-94-4)**	AI3-39196	**(1120-72-5)**
AI3-36149	**(458-35-5)**	AI3-50012	**(98-66-8)**
AI3-36188	**(91-61-2)**	AI3-50132	**(514-73-8)**
AI3-36206	**(53905-38-7)**	AI3-50133	**(5490-27-7)**
AI3-36269	**(2363-89-5)**	AI3-50309	**(56961-47-8)**
AI3-36270	**(2463-63-0)**	AI3-50432	**(2465-59-0)**
AI3-36271	**(15121-84-3)**	AI3-50436	**(26628-22-8)**
AI3-36282	**(52207-99-5)**	AI3-50502	**(14697-50-8)**
AI3-36283	**(593-49-7)**	AI3-50606	**(50-35-1)**
AI3-36284	**(630-03-5)**	AI3-50705	**(10302-15-5)**
AI3-36320	**(100-27-6)**	AI3-50715	**(542-02-9)**
AI3-36432	**(1484-84-0)**	AI3-50866	**(622-46-8)**
AI3-36442	**(646-30-0)**	AI3-50950	**(2795-39-3)**
AI3-36443	**(373-49-9)**	AI3-50982	**(3397-62-4)**
AI3-36444	**(10030-73-6)**	AI3-50983	**(645-92-1)**
AI3-36448	**(506-21-8)**	AI3-51030	**(80-70-6)**
AI3-36449	**(1937-62-8)**	AI3-51074	**(3076-63-9)**
AI3-36450	**(1120-25-8)**	AI3-51088	**(717-74-8)**
AI3-36452	**(7132-64-1)**	AI3-51094	**(1973-05-3)**
AI3-36453	**(1731-92-6)**	AI3-51102	**(615-22-5)**
AI3-36454	**(1731-94-8)**	AI3-51156	**(993-13-5)**
AI3-36455	**(1120-28-1)**	AI3-51263	**(504-08-5)**
AI3-36456	**(929-77-1)**	AI3-51284	**(7450-69-3)**
AI3-36458	**(693-72-1)**	AI3-51294	**(934-75-8)**
AI3-36475	**(18435-45-5)**	AI3-51332	**(3389-71-7)**
AI3-36478	**(629-99-2)**	AI3-51333	**(15619-48-4)**
AI3-36479	**(629-94-7)**	AI3-51352	**(6004-38-2)**
AI3-36483	**(6765-39-5)**	AI3-51439	**(5962-23-2)**
AI3-36485	**(629-96-9)**	AI3-51456	**(1640-39-7)**
AI3-36489	**(661-19-8)**	AI3-51472	**(35674-56-7)**
AI3-36490	**(4181-95-7)**	AI3-51623	**(13464-10-3)**
AI3-36491	**(21964-49-8)**	AI3-51760	**(7738-94-5)**
AI3-36492	**(14167-59-0)**	AI3-51765	**(877-24-7)**
AI3-36493	**(7098-22-8)**	AI3-51821	**(500-44-7)**
AI3-36495	**(7194-85-6)**	AI3-51823	**(554-77-8)**
AI3-36496	**(3452-07-1)**	AI3-51991	**(2002-60-0)**
AI3-36497	**(1599-67-3)**	AI3-52142	**(56-05-3)**
AI3-36516	**(1193-81-3)**	AI3-52175	**(94-97-3)**
AI3-36531	**(12007-89-5)**	AI3-52207	**(5452-35-7)**
AI3-36578	**(544-60-5)**	AI3-52210	**(1122-62-9)**
AI3-36657	**(3978-81-2)**	AI3-52221	**(383-63-1)**
AI3-36727	**(53120-26-6)**	AI3-52225	**(598-56-1)**
AI3-36728	**(53120-27-7)**	AI3-52228	**(533-67-5)**
AI3-36742	**(301-02-0)**	AI3-52234	**(124-63-0)**

AI3-52239 (402-31-3)
AI3-52243 (471-47-6)
AI3-52244 (2002-59-7)
AI3-52248 (98-74-8)
AI3-52254 (98-59-9)
AI3-52262 (111-50-2)
AI3-52273 (2038-03-1)
AI3-52274 (140-31-8)
AI3-52275 (61-78-9)
AI3-52287 (873-55-2)
AI3-52308 (590-88-5)
AI3-52321 (541-88-8)
AI3-52341 (1132-21-4)
AI3-52355 (530-47-2)
AI3-52378 (1906-79-2)
AI3-52389 (630-06-8)
AI3-52402 (5536-61-8)
AI3-52409 (112-16-3)
AI3-52416 (13548-68-0)
AI3-52423 (75-39-8)
AI3-52448 (1072-98-6)
AI3-52469 (593-51-1)
AI3-52478 (591-49-1)
AI3-52539 (1562-94-3)
AI3-52555 (521-31-3)
AI3-52565 (28836-03-5)
AI3-52571 (56-03-1)
AI3-52581 (2219-31-0)
AI3-52586 (27241-31-2)
AI3-52592 (1249-84-9)
AI3-52603 (601-89-8)
AI3-52614 (112-67-4)
AI3-52615 (630-02-4)
AI3-52627 (405-50-5)
AI3-52643 (1202-34-2)
AI3-52657 (372-31-6)
AI3-52660 (141-52-6)
AI3-52667 (50-69-1)
AI3-52669 (499-81-0)
AI3-52671 (103-74-2)
AI3-52672 (505-48-6)
AI3-52685 (877-10-1)
AI3-52686 (120-29-6)
AI3-52691 (24313-88-0)
AI3-52698 (646-31-1)
AI3-52709 (112-85-6)
AI3-52759 (5421-66-9)
AI3-52863 (567-72-6)
AI3-52873 (1919-48-8)
AI3-52881 (21087-61-6)
AI3-60016 (1004-38-2)
AI3-60104 (30354-74-6)
AI3-60110 (1184-78-7)
AI3-60115 (1073-23-0)
AI3-60150 (2654-57-1)
AI3-60157 (3232-26-6)
AI3-60220 (87-82-1)
AI3-60245 (121-43-7)
AI3-60290 (1025-15-6)
AI3-60291 (839-90-7)
AI3-60313 (6291-87-8)
AI3-60335 (776-19-2)
AI3-60350 (5606-16-6)
AI3-60390 (143-66-8)
AI3-60391 (960-71-4)
AI3-61032 (493-77-6)
AI3-61038 (645-93-2)
AI3-61050 (608-25-3)
AI3-61053 (2428-04-8)
AI3-61061 (16867-03-1)
AI3-61075 (6088-51-3)
AI3-61104 (3232-84-6)
AI3-61301 (1967-25-5)
AI3-61314 (603-54-3)
AI3-61325 (2327-02-8)
AI3-61347 (701-82-6)
AI3-61351 (2989-98-2)
AI3-61362 (7160-01-2)
AI3-61395 (503-29-7)
AI3-61434 (4726-14-1)
AI3-61629 (22752-98-3)
AI3-61639 (52-51-7)
AI3-61640 (19757-13-2)
AI3-61817 (627-42-9)
AI3-61833 (16898-52-5)
AI3-61846 (87-88-7)
AI3-61848 (1561-86-0)
AI3-62005 (541-05-9)
AI3-62011 (6228-73-5)
AI3-62012 (3570-55-6)
AI3-62053 (6294-89-9)
AI3-62099 (480-96-6)
AI3-62131 (491-30-5)
AI3-62156 (1488-42-2)
AI3-62232 (546-88-3)
AI3-62444 (2832-19-1)
AI3-62516 (10191-18-1)
AI3-62519 (4839-46-7)
AI3-62521 (556-50-3)
AI3-62692 (26389-78-6)
AI3-62729 (5388-62-5)
AI3-62911 (333-27-7)
AI3-62912 (1493-13-6)
AI3-62933 (2657-00-3)
AI3-63017 (89-00-9)
AI3-63213 (90-66-4)
AI3-01094-X (10233-13-3)
AI3-04488-X (1333-07-9)
AI3-04543-X (25155-18-4)
AI3-07600-X (2917-26-2)
AI3-07871-X (2915-52-8)
AI3-14672-X (1118-58-7)
AI3-22033-X (1323-65-5)
AI3-24403-X (28804-88-8)
AI3-25173-X (112-02-7)
AI3-25192-X (25550-52-1)
AI3-25359-X (25549-16-0)
AI3-25550-X (16919-19-0)
AI3-25728-X (4784-77-4)
AI3-25962-X (15271-41-7)
AI3-26663-X (2664-42-8)
AI3-26730-X (27323-41-7)
AI3-27697-X (27554-26-3)
AI3-28310-X (28473-21-4)
AI3-29111-X (18428-88-1)
AI3-29162-X (9003-29-6)
AI3-29275-X (12057-74-8)
AI3-29332-X (34681-23-7)
AI3-29577-X (63333-35-7)
AL-100 (315-30-0)
ALCA (1317-25-5)
ALOIN (1415-73-2)
d-AM (51-64-9)
AMB (1397-89-3)

6-AMC (2642-98-0)
AMD (41372-08-1)
AMD (555-30-6)
AMMN (56856-83-8)
AMN (13256-07-0)
AMO 1618 (2438-53-1)
AMOCO 600 (9003-27-4)
3',5'-AMP (60-92-4)
5'-AMP (61-19-8)
5-AMP (61-19-8)
A5MP (61-19-8)
AMP (124-68-5)
AMP (61-19-8)
AMP-95 (124-68-5)
AMPC (26787-78-0)
AMPD (115-69-5)
2-AMPS (15214-89-8)
AMP (Nucleotide) (61-19-8)
AMS (7773-06-0)
AMSA (51264-14-3)
m-AMSA (51264-14-3)
m-AMSA Methanesulfonate (51264-14-3)
AMSR 3 (1344-00-9)
AN (103-84-4)
AN 23 (50-11-3)
AN 33501 (55-98-1)
ANA (86-87-3)
ANI (551-06-4)
ANIT (551-06-4)
ANN (German) (56856-83-8)
ANS (82-76-8)
ANU (86-87-3)
A-Ninopterin (54-62-6)
AO 4 (2082-79-3)
AO1 (119-47-1)
AO 1 (Antioxidant) (119-47-1)
AO 10 (9011-14-7)
AO 29 (128-37-0)
AO-40 (1709-70-2)
AO 754 (88-26-6)
AO 2246 (119-47-1)
AO 4K (128-37-0)
AO A1 (7429-90-5)
AOM (25843-45-2)
AOMB (583-39-1)
AOS (72674-05-6)
4-AP (504-24-5)
AP (94-78-0)
AP 50 (1309-64-4)
APAP (103-90-2)
APC (62-44-2)
APC (8003-03-0)
APCO 2330 (108-45-2)
APC (Pharmaceutical) (8003-03-0)
APFO (3825-26-1)
APGA (54-62-6)
APH (114-83-0)
API 5L X52 (39362-66-8)
API 5LX-52 (39362-66-8)
API 5LX X52 (39362-66-8)
API No.2 Fuel Oil (68476-30-2)
API X 52 (39362-66-8)
API 5XL X52 (39362-66-8)
APL (9002-61-3)
APL (Hormone) (9002-61-3)
A1-1401 P(MS) (1344-28-1)
APN (52-46-0)
APO (545-55-1)
APPA (732-11-6)
APV (111-90-0)
AR 1 (151-19-9)
AR 3 (7440-44-0)
AR-44 (57-64-7)
AR2 (7429-90-5)
ARD 13/02 (2302-17-2)
ARD 34/02 (74051-80-2)
ART 2 (7440-44-0)
A 15S (9003-22-9)
AS (9004-70-0)
AS 1 (7782-42-5)
AS17665 (531-82-8)
A.S.A. (50-78-2)
ASA (50-78-2)
ASA 3 (Antistatic agent) (9CI) (51004-63-8)
ASA-140 (712-68-5)
ASA 226 (500-42-5)
ASA Compound (62-44-2)
A.S.A. Empirin (50-78-2)
ASB 516 (9003-20-7)
ASC (121-60-8)
ASC-4 (87-10-5)
AS 61CL (9003-54-7)
ASM MB (583-39-1)
ASP 47 (3689-24-5)
ASP 51 (3244-90-4)
ASTA (50-18-0)
ASTA B518 (50-18-0)
A-Stoff (78-95-5)
3,A-T (61-82-5)
A.T. 10 (67-96-9)
A1-0104 T 3/16" (1344-28-1)
A1-1404 T 3/16" (1344-28-1)
A1-3438 T 1/8" (1344-28-1)
A1-3980 T 5/32" (1344-28-1)
A1-4028 T 3/16" (1344-28-1)
AT (61-82-5)
AT 7 (70-30-4)
AT-17 (70-30-4)
AT 20 (7782-42-5)
AT 36 (9003-07-0)
AT-90 (61-82-5)
AT 101 (652-67-5)
AT-290 (148-82-3)
AT 717 (9003-39-8)
o-AT (97-56-3)
ATA (61-82-5)
ATCP (1918-02-1)
ATEC (77-89-4)
ATIPI (56-65-5)
ATJ-S (7782-42-5)
ATJ-S Graphite (7782-42-5)
ATM 2 (Nylon) (25038-54-4)
ATM 2(Nylon) (105-60-2)
4-ATMP (103947-07-5)
5'-ATP (56-65-5)
ATP (56-65-5)
ATP (Nucleotide) (56-65-5)
AT Liquid (61-82-5)
AU 3 (7440-44-0)
AV 1000 (63148-62-9)
AV00 (7429-90-5)
AV000 (7429-90-5)
AVC/Dienestrol cream (97-59-6)
AWPA #1 (8001-58-9)

AW 15 (Polysaccharide)	**(9004-62-0)**
AY 6608	**(5534-95-2)**
AY-5312	**(3697-42-5)**
AY-6108	**(69-53-4)**
AY 21011	**(6673-35-4)**
AY 61123	**(637-07-0)**
AY 64043	**(525-66-6)**
AYAA	**(9003-20-7)**
AYAF	**(9003-20-7)**
AYJV	**(9003-20-7)**
AZ 2088	**(20091-61-6)**
5 AZC	**(320-67-2)**
5-AZCR	**(320-67-2)**
AZG	**(134-58-7)**
AZS	**(115-02-6)**
AZT	**(30516-87-1)**
Aacaptan	**(133-06-2)**
Aacifemine	**(50-27-1)**
Aafertis	**(14484-64-1)**
Acisal	**(50-78-2)**
Aalindan	**(58-89-9)**
Aamangan	**(12427-38-2)**
Aaprotect	**(137-30-4)**
Aarane	**(15826-37-6)**
Aararre	**(15826-37-6)**
Aastar	**(298-02-2)**
Aastar	**(70124-77-5)**
Aatack	**(137-26-8)**
Aaterra	**(2593-15-9)**
Aatrex	**(1912-24-9)**
Aatrex 4L	**(1912-24-9)**
Aatrex Nine-O	**(1912-24-9)**
Aatrex 80W	**(1912-24-9)**
Aavolex	**(137-30-4)**
Aazira	**(137-30-4)**
Abacil	**(18472-51-0)**
Abacin	**(8064-90-2)**
Abactrim	**(8064-90-2)**
Abadol	**(96-50-4)**
Abadole	**(96-50-4)**
Abat	**(3383-96-8)**
Abate	**(3383-96-8)**
Abathion	**(3383-96-8)**
Abavit	**(2279-64-3)**
Abbocillin	**(61-33-6)**
Abbocillin-DC	**(6130-64-9)**
Abbomeen E-2	**(4500-29-2)**
Abbomeen E-2 Aerosol	**(4500-29-2)**
Abbott 40566	**(1972-08-3)**
Abbott 44090	**(99-66-1)**
Abelmosco, Semillas (Spanish)	**(8015-62-1)**
Abensanil	**(103-90-2)**
Aberel	**(302-79-4)**
Abeson Nam	**(25155-30-0)**
Abesta	**(50-55-5)**
Abicel	**(9004-34-6)**
Abicol	**(50-55-5)**
Abies Alba Oil	**(8021-28-1)**
Abies Balsamea, Pinaceae	**(8021-28-1)**
Abies Excelsa Oil	**(8021-28-1)**
Abies Oil	**(8021-28-1)**
Abies Picea Oil	**(8021-28-1)**
δ 6,8(14)-Abietadienoic acid	**(79-54-9)**
Abietic acid	**(514-10-3)**
Abietic acid, dihydro-, triester with glycerol	**(125-93-9)**
Abietic acid, methyl ester	**(127-25-3)**
Abiguanil	**(57-67-0)**
Abilit	**(15676-16-1)**
Abiol	**(99-76-3)**
Abirol	**(72-63-9)**
Abminthic	**(514-73-8)**
Abol	**(23103-98-2)**
Abomasal Enzyme	**(9001-98-3)**
Abracol S.L.G.	**(31566-31-1)**
Abramant	**(1344-28-1)**
Abramax	**(1344-28-1)**
Abramycin	**(60-54-8)**
Abrarex	**(1344-28-1)**
Abrasin oil	**(8001-20-5)**
Abrasit	**(1344-28-1)**
Abricycline	**(60-54-8)**
Abril wax 10DS	**(110-30-5)**
Abroden	**(126-31-8)**
Abrodil	**(126-31-8)**
Abromeen E-25	**(2842-38-8)**
Abromine	**(107-43-7)**
Absolute ethanol	**(64-17-5)**
Absorbable Gelatin Sponge	**(9000-70-8)**
Abstensil	**(97-77-8)**
Abstinil	**(97-77-8)**
Abstinyl	**(97-77-8)**
Abufene	**(107-95-9)**
Acacia	**(9000-01-5)**
Acacia Dealbata Gum	**(9000-01-5)**
Acacia Gum	**(9000-01-5)**
Acacia Mollissima Tannin	**(1401-55-4)**
Acacia Senegal	**(9000-01-5)**
Acacia Syrup	**(9000-01-5)**
Acadyl	**(66-76-2)**
Acamol	**(103-90-2)**
Acaraben	**(510-15-6)**
Acaraben 4E	**(510-15-6)**
Acarac	**(33089-61-1)**
Acaracide	**(140-57-8)**
Acaralate	**(5836-10-2)**
Acardite	**(102-07-8)**
Acaricydol E 20	**(80-33-1)**
Acarin	**(115-32-2)**
Acarithion	**(786-19-6)**
Acarol	**(18181-80-1)**
Acaron	**(6164-98-3)**
Acavyl	**(66-76-2)**
Accel	**(2312-73-4)**
Accel BNS	**(95-31-8)**
Accelerate	**(129-67-9)**
Accelerator 552	**(98-77-1)**
Accelerator L	**(137-30-4)**
Accelerator OTOS	**(13752-51-7)**
Accelerator thiuram	**(137-26-8)**
Accelerine	**(138-89-6)**
Accent	**(142-47-2)**
Acclaim	**(66441-23-4)**
Acco Fast Red KB Base	**(95-79-4)**
Acco Naf-Sol AS-D	**(135-61-5)**
Acco Naf-Sol AS-KB	**(135-63-7)**
Acco Naf-Sol AS-phenyl	**(92-74-0)**
Acco Naphthol AS	**(92-77-3)**
Acco Naphthol AS-BO	**(132-68-3)**
Acco Naphthol AS-BR	**(91-92-9)**
Acco Naphthol AS-BS	**(135-65-9)**
Acco Naphthol AS-D	**(135-61-5)**
Acco Naphthol AS-KB	**(135-63-7)**
Acco Naphthol AS-phenyl	**(92-74-0)**
Acco Sulfur Blue B-CF	**(1327-57-7)**

Name	CAS
Acco Sulfur Blue GLP-CF	**(1327-57-7)**
Acco Sulfur Blue GLR-CF	**(1327-57-7)**
Acco Sulfur Blue 2R-CF	**(1327-57-7)**
Acco Sulfur Blue 4R-CF	**(1327-57-7)**
Acco Sulfur Blue 6R-CF	**(1327-57-7)**
Acco Sulfur Blue R-CF	**(1327-57-7)**
Accobond 3524	**(9003-08-1)**
Acconem	**(21548-32-3)**
Accosperse Cyan Blue GT	**(147-14-8)**
Accosperse Cyan Green G	**(1328-53-6)**
Accosperse Hansa Yellow G	**(2512-29-0)**
Accosperse Toluidine Red XL	**(2425-85-6)**
Accothion	**(122-14-5)**
Accothion o-analog	**(2255-17-6)**
Accucol	**(599-79-1)**
Accusand	**(60676-86-0)**
Accutane	**(4759-48-2)**
Accuzole	**(127-69-5)**
Acede cresylique (French)	**(1319-77-3)**
Acedicon	**(466-90-0)**
Acedikon (Czech)	**(466-90-0)**
Acedoben	**(556-08-1)**
Acedoben (Spanish)	**(556-08-1)**
Acedobene (French)	**(556-08-1)**
Acedobenum (Latin)	**(556-08-1)**
Acedoxin	**(71-63-6)**
Acedron	**(51-63-8)**
Aceite de alcanfor (Spanish)	**(8008-51-3)**
Aceite de colofonia (Spanish)	**(8002-16-2)**
Aceite de fusel (Spanish)	**(8013-75-0)**
Acemethadone	**(509-74-0)**
Acenaphthanthracene	**(5779-79-3)**
Acenaphth(1,2-b)anthracene	**(207-18-1)**
5-Acenaphthenamine	**(4657-93-6)**
Acenaphthene	**(83-32-9)**
5-Acenaphtheneamine	**(4657-93-6)**
Acenaphthenedione	**(82-86-0)**
Acenaphthene, 5-nitro	**(602-87-9)**
Acenaphthene-1-ol	**(6306-07-6)**
1-Acenaphthenol (8CI)	**(6306-07-6)**
Acenaphtho(1,2-b)phenanthrene	**(238-04-0)**
Acenaphtho(1,2-b)pyridine (8CI,9CI)	**(206-49-5)**
5-Acenaphthylenamine (8CI,9CI)	**(4523-49-3)**
5-Acenaphthylenamine, 1,2-dihydro-	**(4657-93-6)**
Acenaphthylene	**(208-96-8)**
Acenaphthylenecarbonitrile, 1,2-dihydro- (9CI)	**(83536-56-5)**
Acenaphthylenecarboxaldehyde, 1,2-dihydro- (9CI)	**(84412-11-3)**
Acenaphthylene, 1,2-dihydro-	**(83-32-9)**
Acenaphthylene, 1,2-dihydromethyl- (9CI)	**(36541-21-6)**
Acenaphthylene, 1,2-dihydro-5-nitro-	**(602-87-9)**
1,2-Acenaphthylenedione	**(82-86-0)**
Acenaphthylene, methyl- (9CI)	**(58548-38-2)**
Acenaphthylene, tetrachloro- (9CI)	**(90077-79-5)**
1-Acenaphthylenol, 1,2-dihydro- (9CI)	**(6306-07-6)**
Acenocoumarin	**(152-72-7)**
Acenocoumarol	**(152-72-7)**
Acenocumarol	**(152-72-7)**
Acenokumarin (Czech)	**(152-72-7)**
Acenterine	**(50-78-2)**
Aceothion	**(122-14-5)**
Acephat (German)	**(30560-19-1)**
Acephate	**(30560-19-1)**
Acephate-Met	**(10265-92-6)**
Acephenanthrylene	**(201-06-9)**
Acephenanthrylene, nitro- (9CI)	**(114790-09-9)**
Acepramin	**(60-32-2)**
Acepramine	**(60-32-2)**
Acepyrene	**(27208-37-3)**
Acepyrylene	**(27208-37-3)**
Acesal	**(50-78-2)**
Acesulfame	**(33665-90-6)**
Acesulfamo (Spanish)	**(33665-90-6)**
Acesulfamum (Latin)	**(33665-90-6)**
Acet-Theocin	**(58-55-9)**
Acetaal (Dutch)	**(105-57-7)**
Acetacid Red B	**(3567-69-9)**
Acetacid Red 2BR	**(915-67-3)**
Acetacid Red J	**(3761-53-3)**
Acetagesic	**(103-90-2)**
Acetal	**(50-78-2)**
Acetal [UN 1088]	**(105-57-7)**
Acetaldehidato amonico (Spanish)	**(75-39-8)**
Acetaldehyd (German)	**(75-07-0)**
Acetaldehyde (ACGIH,OSHA) [UN 1089]	**(75-07-0)**
Acetaldehyde, Homopolymer (9CI)	**(9002-91-9)**
Acetaldehyde, Polymers	**(9002-91-9)**
Acetaldehyde, amine salt	**(75-39-8)**
Acetaldehyde, amino-, dimethyl acetal (8CI)	**(22483-09-6)**
Acetaldehyde ammonia	**(75-39-8)**
Acetaldehyde, chloro	**(107-20-0)**
Acetaldehyde, chloro-, diethyl acetal (8CI)	**(621-62-5)**
Acetaldehyde, chloro-, dimethyl acetal (8CI)	**(97-97-2)**
Acetaldehyde, dibromo- (8CI,9CI)	**(3039-13-2)**
Acetaldehyde, 2,2-dichloro	**(79-02-7)**
Acetaldehyde, dichloro-	**(79-02-7)**
Acetaldehyde, dichloro-, diethyl acetal (8CI)	**(619-33-0)**
Acetaldehyde, diethyl acetal	**(105-57-7)**
Acetaldehyde, dihexyl acetal (8CI)	**(5405-58-3)**
Acetaldehyde, (1,3-dihydro-1,3,3-trimethyl-2H-indol-2-ylidene)- (9CI)	**(84-83-3)**
Acetaldehyde, dimethyl acetal	**(534-15-6)**
Acetaldehyde, (3,3-dimethylcyclohexylidene)-, (Z)- (9CI)	**(26532-24-1)**
Acetaldehyde, diphenyl	**(947-91-1)**
Acetaldehyde, 2-(2-ethoxyethoxy)ethyl 3,4-(methylene dioxy)phenyl acetal	**(51-14-9)**
Acetaldehyde, ethylmethyl-	**(96-17-3)**
Acetaldehyde, ethyl phenyl acetal (8CI)	**(5426-78-8)**
Acetaldehyde-N-formyl-N-methylhydrazone	**(16568-02-8)**
Acetaldehyde, hydroxy- (9CI)	**(141-46-8)**
Acetaldehyde, (p-hydroxyphenyl)- (8CI)	**(7339-87-9)**
Acetaldehyde, (methylamino)-, dimethyl acetal (8CI)	**(122-07-6)**
Acetaldehyde methyl ethyl acetyl	**(10471-14-4)**
Acetaldehyde-N-methyl-N-formylhydrazone	**(16568-02-8)**
Acetaldehyde, (methylthio)-, oxime (9CI)	**(10533-67-2)**
Acetaldehyde-oxime	**(107-29-9)**
Acetaldehyde oxime [UN 2332]	**(107-29-9)**
Acetaldehyde, phenyl-	**(122-78-1)**
Acetaldehyde, phenyl-, dimethyl acetal	**(101-48-4)**
Acetaldehyde propyl phenylethyl acetal	**(7493-57-4)**
Acetaldehyde, tetramer	**(108-62-3)**
Acetaldehyde, tribromo- (9CI)	**(115-17-3)**
Acetaldehyde, trichloro- (9CI)	**(75-87-6)**
Acetaldehyde, trimer	**(123-63-7)**
Acetal diethylique (French)	**(105-57-7)**
Acetaldol	**(107-89-1)**
Acetaldoxime	**(107-29-9)**
Acetale (Italian)	**(105-57-7)**
Acetalgin	**(103-90-2)**
3-Acetamido-5-(acetamidomethyl)-2,4,6-triiodobenzoic acid	**(440-58-4)**
Acetamide	**(60-35-5)**
Acetamide, N-acetyl- (9CI)	**(625-77-4)**
Acetamide, N-acetyl-N-(1,1-dimethylethyl)- (9CI)	**(65882-21-5)**
Acetamide, N-acetyl-N-hexyl- (9CI)	**(25457-47-0)**
Acetamide, N-acetyl-N-methyl-	**(1113-68-4)**

Acetamide, N-acetyl-N-(2-methyl-4-((2-methylphenyl)azo)phenyl) **(83-63-6)**
Acetamide, N-(4-(acetyloxy)phenyl)- (9CI) **(2623-33-8)**
Acetamide, N-acetyl-N-(2-phenylethyl)- (9CI) **(27179-64-2)**
Acetamide, N-acetyl-N-propyl- (9CI) **(1563-84-4)**
Acetamide, N-(4'-amino(1,1'-biphenyl)-4-yl)- **(3366-61-8)**
Acetamide, N-(4-(((aminoiminomethyl)amino)sulfonyl)phenyl)- (9CI) **(19077-97-5)**
Acetamide, N-(3-amino-4-methoxyphenyl)- (9CI) **(6375-47-9)**
Acetamide, N-(4-amino-6-methyl-1,3,5-triazin-2-yl)- (9CI) **(30355-53-4)**
Acetamide, N-(4-amino-6-methyl-s-triazin-2-yl)- (8CI) **(30355-53-4)**
Acetamide, 2-amino-N-phenyl- (9CI) **(555-48-6)**
Acetamide, N-(3-aminophenyl)- (9CI) **(102-28-3)**
Acetamide, N-(4-aminophenyl)- (9CI) **(122-80-5)**
Acetamide, 2-amino-N-phenyl-, monohydrochloride (9CI) **(4801-39-2)**
Acetamide, N-(3-aminophenyl)-, monohydrochloride (9CI) **(621-35-2)**
Acetamide, N-((4-aminophenyl)sulfonyl)- (9CI) **(144-80-9)**
Acetamide, N-(5-(aminosulfonyl)-3-methyl-1,3,4-thiadiazol-2(3H)-ylidene)- (9CI) **(554-57-4)**
Acetamide, N-(5-(aminosulfonyl)-1,3,4-thiadiazol-2-yl)- **(59-66-5)**
Acetamide, N-benzyl **(588-46-5)**
Acetamide, N-(1,1'-biphenyl)-4-yl- (9CI) **(4075-79-0)**
Acetamide, N-(5-(bis(2-(acetyloxy)ethyl)amino)-2-((2-bromo-4,6-dinitrophenyl)azo)-4-ethoxyphenyl)- (9CI) **(12239-34-8)**
Acetamide, N-(5-(bis(2-(acetyloxy)ethyl)amino)-2-((2-bromo-4,6-dinitrophenyl)azo)-4-methoxyphenyl)- (9CI) **(3618-72-2)**
Acetamide, N-(5-(bis(2-(acetyloxy)ethyl)amino)-2-((2-chloro-4,6-dinitrophenyl)azo)-4-methoxyphenyl)- (9CI) **(3618-73-3)**
Acetamide, N-(5-(bis(2-(acetyloxy)ethyl)amino)-2-((2-chloro-4-nitrophenyl)azo)phenyl)- (9CI) **(1533-78-4)**
Acetamide, N-(3-(bis(2-(acetyloxy)ethyl)amino)-4-methoxyphenyl)- (9CI) **(23128-51-0)**
Acetamide, N-(3-(bis(2-(acetyloxy)ethyl)amino)phenyl)- (9CI) **(27059-08-1)**
Acetamide, N-bromo- (9CI) **(79-15-2)**
Acetamide, 2-bromo-N-(2-(1,1-dimethylethyl)-6-methylphenyl)-N-(methoxymethyl)- (9CI) **(2163-81-7)**
Acetamide, N-(2-((2-bromo-4,6-dinitrophenyl)azo)-5-((2-cyanoethyl)ethylamino)-4-methoxyphenyl)- (9CI) **(22578-86-5)**
Acetamide, N-(2-((2-bromo-4,6-dinitrophenyl)azo)-5-((2-cyanoethyl)-2-propenylamino)-4-methoxyphenyl)- (9CI) **(68877-63-4)**
Acetamide, N-(4-bromophenyl)- **(103-88-8)**
Acetamide, N-(3-bromophenyl)- (9CI) **(621-38-5)**
Acetamide, N-(4-bromophenyl)-2-fluoro- (9CI) **(351-05-3)**
Acetamide, N-(butoxymethyl)-2-chloro-N-(2,6-diethylphenyl)- (9CI) **(23184-66-9)**
Acetamide, N-butyl- (8CI,9CI) **(1119-49-9)**
Acetamide, N-tert-butyl- (6CI,7CI,8CI) **(762-84-5)**
Acetamide, N-butyl-N-phenyl- (9CI) **(91-49-6)**
Acetamide, 2-chloro **(79-07-2)**
Acetamide, N-(4-(chloroacetyl)phenyl)- (9CI) **(140-49-8)**
Acetamide, 2-chloro-N-(chloromethyl)-N-(2,6-diethylphenyl)- (9CI) **(40164-69-0)**
Acetamide, N-(4-chloro-2-((2-chloro-4-nitrophenyl)azo)-5-((2-hydroxypropyl)amino)phenyl)- (9CI) **(71617-28-2)**
Acetamide, 2-chloro-N,N-diallyl **(93-71-0)**
Acetamide, 2-chloro-N,N-diethyl **(2315-36-8)**
Acetamide, 2-chloro-N-(2,6-diethylphenyl)-N-(methoxymethyl)- (9CI) **(15972-60-8)**
Acetamide, 2-chloro-N-(2-(1,1-dimethylethyl)-6-methylphenyl)- (9CI) **(3785-20-4)**
Acetamide, 2-chloro-N-(2,6-dimethylphenyl)-N-(2-methoxyethyl) **(50563-36-5)**
Acetamide, N-(3-chloro-2,6-dimethylphenyl)-2-methoxy-N-(tetrahydro-2-oxo-3-furanyl)- (9CI) **(67932-85-8)**
Acetamide, 2-chloro-N-(2,6-dimethylphenyl)-N-(1H-pyrazol-1-ylmethyl) **(67129-08-2)**
Acetamide, 2-chloro-N,N-di-2-propenyl- (9CI) **(93-71-0)**
Acetamide, 2-chloro-N-(ethoxymethyl)-N-(2-ethyl-6-methylphenyl)- (9CI) **(34256-82-1)**
Acetamide, 2-chloro-N-(6-ethyl-o-tolyl)-N-(2-methoxy-1-methylethyl)- **(51218-45-2)**
Acetamide, 2-chloro-N-(hydroxymethyl)- (8CI,9CI) **(2832-19-1)**
Acetamide, 2-chloro-N-(2-methoxyethyl)-N-(2-methylphenyl)- **(50563-41-2)**
Acetamide, N-(3-chloro-4-methoxyphenyl)- (9CI) **(7073-42-9)**
Acetamide, 2-chloro-N-(1-methylethyl)-N-phenyl- **(1918-16-7)**
Acetamide, 2-chloro-N-(1-methyl-2-propynyl)-N-phenyl- (9CI) **(21267-72-1)**
Acetamide, N-(2-((2-chloro-4-nitrophenyl)azo)-5-((2-(2,5-dioxo-1-pyrrolidinyl)ethyl)ethylamino)phenyl)- (9CI) **(29649-47-6)**
Acetamide, N-(4-chlorophenyl)- **(539-03-7)**
Acetamide, 2-chloro-N-phenyl- (9CI) **(587-65-5)**
Acetamide, N-(2-chlorophenyl)- (9CI) **(533-17-5)**
Acetamide, N-((4-chlorophenyl)methyl)- (9CI) **(57058-33-0)**
Acetamide, N,N'-(6-chloro-s-triazine-2,4-diyl)bis(N-ethyl- (8CI) **(4065-24-1)**
Acetamide, 2-cyano **(107-91-5)**
Acetamide, 2-cyano-2,2-dibromo **(10222-01-2)**
Acetamide, N-(2-((2-cyano-4,6-dinitrophenyl)azo)-5-(diethylamino)phenyl)- (9CI) **(24170-60-3)**
Acetamide, 2-cyano-N-((ethylamino)carbonyl)-2-(methoxyimino) **(57966-95-7)**
Acetamide, N-(3-((2-cyanoethyl)amino)-4-methoxyphenyl)- (9CI) **(26408-28-6)**
Acetamide, N-(((2-cyanoethyl)ethylamino)ethoxyphenyl)- (9CI) **(63460-10-6)**
Acetamide, N-(((2-cyanoethyl)ethylamino)methoxyphenyl)- (9CI) **(63505-65-7)**
Acetamide, N-(3-((2-cyanoethyl)ethylamino)-4-methoxyphenyl)- (9CI) **(19433-94-4)**
Acetamide, N-(3-((2-cyanoethyl)(2-hydroxyethyl)amino)-4-methoxyphenyl)- (9CI) **(22588-78-9)**
Acetamide, N-cyclohexyl **(1124-53-4)**
Acetamide, N,N-diallyl-2-chloro- (8CI) **(93-71-0)**
Acetamide, N,N-diallyl-2,2-dichloro **(37764-25-3)**
Acetamide, N-(4-((2,4-diaminophenyl)azo)phenyl)- (9CI) **(18371-12-5)**
Acetamide, 2,2-dibromo-2-cyano- (8CI,9CI) **(10222-01-2)**
Acetamide, 2,2-dichloro-N,N-di-2-propenyl- (9CI) **(37764-25-3)**
Acetamide, 2,2-dichloro-N-(β-hydroxy-α-(hydroxymethyl)-p-nitrophenethyl)- **(56-75-7)**
Acetamide, 2,2-dichloro-N-(β-hydroxy-α-(hydroxymethyl)-p-nitrophenethyl)-, D-(-)-threo **(56-75-7)**
Acetamide, 2,2-dichloro-N-(β-hydroxy-α-(hydroxymethyl)-p-nitrophenethyl)-, α-ester with sodium succinate **(982-57-0)**
Acetamide, 2,2-dichloro-N-(2-hydroxy-1-(hydroxymethyl)-2-(4-nitrophenyl)ethyl)-, (R-(R*,R*))- **(56-75-7)**
Acetamide, N-(3,4-dichlorophenyl)- **(2150-93-8)**
Acetamide, N-(2-((2,6-dicyano-4-nitrophenyl)azo)-5-(diethylamino)phenyl)- (9CI) **(41642-51-7)**
Acetamide, N,N-diethyl **(685-91-6)**
Acetamide, 2-(diethylamino)-N-(2,6-dimethylphenyl)- (9CI) **(137-58-6)**
Acetamide, N-(3-(diethylamino)-4-methoxyphenyl)- (9CI) **(19433-93-3)**
Acetamide, N-(3-(diethylamino)phenyl)- (9CI) **(6375-46-8)**
Acetamide, N-(2,5-dimethoxyphenyl)- (9CI) **(3467-59-2)**
Acetamide, N,N-dimethyl **(127-19-5)**
Acetamide, N,N-dimethyl-2,2-diphenyl **(957-51-7)**
Acetamide, N-(1,1-dimethylethyl)- (9CI) **(762-84-5)**
Acetamide, N-(2,4-dimethylphenyl)- (9CI) **(2050-43-3)**
Acetamide, N-(2,6-dimethylphenyl)-2-methoxy-N-(2-oxo-3-oxazolidinyl) **(77732-09-3)**
Acetamide, N-(2,4-dimethyl-5-(((trifluoromethyl)sulfonyl)amino)phenyl)- **(53780-34-0)**
Acetamide, N-(3-(((2,5-dioxo-1-pyrrolidinyl)ethyl)ethylamino)phenyl)- (9CI) **(27550-64-7)**
Acetamide, N-(3-ethoxyphenyl)- (9CI) **(591-33-3)**
Acetamide, N-(4-ethoxyphenyl)- (9CI) **(62-44-2)**
Acetamide, N-ethyl **(625-50-3)**
Acetamide, N-(3-(ethyl(phenylmethyl)amino)phenyl)- (9CI) **(29103-58-0)**
Acetamide, 2-(ethylselenyl)- (7CI) **(90030-80-1)**
Acetamide, N-fluoren-2-yl **(53-96-3)**
Acetamide, N-fluoren-4-yl **(28322-02-3)**
Acetamide, N-9H-fluoren-2-yl- (9CI) **(53-96-3)**
Acetamide, N-9H-fluoren-4-yl- (9CI) **(28322-02-3)**

Acetamide, N-9H-fluoren-2-yl-N-hydroxy- (9CI) **(53-95-2)**
Acetamide, 2-fluoro **(640-19-7)**
Acetamide, N-(3-fluorophenyl)- (9CI) **(351-28-0)**
Acetamide, N-(4-fluorophenyl)- (9CI) **(351-83-7)**
Acetamide, N-formyl-2-mercapto-N-methyl-, S-ester with O,O-dimethyl phosphorodithioate **(2540-82-1)**
Acetamide, N-hexyl- (6CI,7CI,8CI,9CI) **(7501-79-3)**
Acetamide, N-hydroxy- (9CI) **(546-88-3)**
Acetamide, N-(2-hydroxyethyl) **(142-26-7)**
Acetamide, 2,2'-((2-hydroxyethyl)imino)bis(N-(α,α-dimethyl-phenethyl)-N-methyl **(126-27-2)**
Acetamide, 2,2'-(2-hydroxyethyl)imino)bis(N-(1,1-dimethylphenyl-ethyl)-N-methyl- (9CI) **(126-27-2)**
Acetamide, N-hydroxy-N-(2-fluorenyl)- **(53-95-2)**
Acetamide, 2-(p-(2-hydroxy-3-(isopropylamino)propoxy)phenyl) **(29122-68-7)**
Acetamide, N-(4-(2-hydroxy-3-((1-methylethyl)amino)propoxy)-phenyl)- (9CI) **(6673-35-4)**
Acetamide, N-(2-hydroxy-5-methylphenyl)- (9CI) **(6375-17-3)**
Acetamide, N-(4-((2-hydroxy-5-methylphenyl)azo)phenyl) **(2832-40-8)**
Acetamide, 2-hydroxy-N-phenyl **(4746-61-6)**
Acetamide, N-(4-hydroxyphenyl)- **(103-90-2)**
Acetamide, N-(p-hydroxyphenyl)- **(103-90-2)**
Acetamide, N-(2-hydroxyphenyl)- (9CI) **(614-80-2)**
Acetamide, N-(3-hydroxyphenyl)- (9CI) **(621-42-1)**
Acetamide, N-(4-((4-hydroxyphenyl)azo)phenyl)- (9CI) **(5302-39-6)**
Acetamide, N-isopropyl-2-mercapto-, S-ester with O,O-diethyl phosphorodithioate **(2275-18-5)**
2-Acetamide-4-mercaptobutyric acid γ-thiolactone **(1195-16-0)**
Acetamide, 2-mercapto-N-(2-methoxyethyl)-, S-ester with O,O-dimethyl phosphorodithioate **(919-76-6)**
Acetamide, 2-mercapto-N-methyl-, S-ester with O,O-dimethyl phosphorothioate **(1113-02-6)**
Acetamide, N-(p-methoxybenzyl)- **(35103-34-5)**
Acetamide, N-(2-methoxyethyl)- (7CI,8CI,9CI) **(5417-42-5)**
Acetamide, N-(2-(5-methoxyindol-3-yl)ethyl) **(73-31-4)**
Acetamide, N-(2-(5-methoxy-1H-indol-3-yl)ethyl)- (9CI) **(73-31-4)**
Acetamide, N-(2-methoxy-5-methylphenyl)- (9CI) **(6962-44-3)**
Acetamide, N-(4-methoxyphenyl)- (9CI) **(51-66-1)**
Acetamide, N-((4-methoxyphenyl)methyl)- (9CI) **(35103-34-5)**
Acetamide, N-methyl **(79-16-3)**
Acetamide, N-(p-methylbenzyl)- (6CI,7CI,8CI) **(25079-96-3)**
Acetamide, N-methyl-N-nitroso **(7417-67-6)**
Acetamide, N-(2-methylphenyl)- (9CI) **(120-66-1)**
Acetamide, N-(3-methylphenyl)- (9CI) **(537-92-8)**
Acetamide, N-(4-methylphenyl)- (9CI) **(103-89-9)**
Acetamide, N-((4-methylphenyl)methyl)- (9CI) **(25079-96-3)**
Acetamide, N-(4-methyl-2-sulfamoyl-δ^2-1,3,4-thiadiazolin-5-yl-idene) **(554-57-4)**
Acetamide, N,N'-(6-methyl-s-triazine-2,4-diyl)bis- (8CI) **(30355-54-5)**
Acetamide, N-(4-methyl-3-(((trifluoromethyl)sulfonyl)-amino)phenyl)- **(47000-92-0)**
Acetamide, N-p-nitrobenzyl- (6CI) **(56222-10-7)**
Acetamide, N-(4-(5-nitro-2-furyl)-2-thiazolyl) **(531-82-8)**
Acetamide, N-(2-nitrophenyl)- (9CI) **(552-32-9)**
Acetamide, N-(3-nitrophenyl)- (9CI) **(122-28-1)**
Acetamide, N-(4-nitrophenyl)- (9CI) **(104-04-1)**
Acetamide, N-((4-nitrophenyl)methyl)- (9CI) **(56222-10-7)**
Acetamide, N-(4-oxo-2,5-cyclohexadien-1-ylidene) **(50700-49-7)**
Acetamide, 2-phenyl **(103-81-1)**
Acetamide, N-phenyl- **(103-84-4)**
Acetamide, N-(4-(phenylazo)phenyl)- (9CI) **(4128-71-6)**
Acetamide, N-(phenylmethyl)- (9CI) **(588-46-5)**
Acetamide, N-propyl- (8CI,9CI) **(5331-48-6)**
Acetamide, N-(5-sulfamoyl-1,3,4-thiadiazol-2-yl) **(59-66-5)**
Acetamide, N-sulfanilyl **(144-80-9)**
Acetamide, N-(tetrahydro-2-oxo-3-thienyl) **(1195-16-0)**
Acetamide, N-(5,6,7,9-tetrahydro-1,2,3,10-tetramethoxy-9-oxobenzo(α)heptalen-7-yl)- **(64-86-8)**
5-Acetamide-1,3,4-thiadiazole-2-sulfonamide **(59-66-5)**
Acetamide, thio **(62-55-5)**
1-Acetamido-4-ethoxybenzene **(62-44-2)**
Acetamide, N,N',N''-1,3,5-triazine-2,4,6-triyltris- (9CI) **(5637-82-1)**
Acetamide, N,N',N''-s-triazine-2,4,6-triyltris- (8CI) **(5637-82-1)**
Acetamide, N,N',N''-s-triazine-2,4,6-triyltris(2-chloro- (8CI) **(30360-27-1)**
Acetamide, N,N',N''-s-triazine-2,4,6-triyltris(2,2-dichloro- (8CI) **(30360-28-2)**
Acetamide, N,N',N''-s-triazine-2,4,6-triyltris(2,2,2-trichloro- (8CI) **(30360-29-3)**
Acetamide, 2,2,2-trichloro **(594-65-0)**
Acetamide, α-trichloro- **(594-65-0)**
Acetamide, N,N'-(6-(trichloromethyl)-s-triazine-2,4-diyl)bis- (8CI) **(30359-60-5)**
Acetamide, N,N'-(6-(trichloromethyl)-s-triazine-2,4-diyl)bis-(N-ethyl- (8CI) **(30359-61-6)**
Acetamide, 2,2,2-trichloro-N-(4-methyl-6-(trichloromethyl)-s-triazin-2-yl)- (8CI) **(30388-91-1)**
Acetamide, N-(3-(trifluoromethyl)phenyl)- (9CI) **(351-36-0)**
Acetamidine, N,N'-diphenyl **(621-09-0)**
Acetamidine, 2-chloro-, Compd. with 4,6-bis(chloro-methyl)-s-triazin-2-ol (1:1) (8CI) **(30886-00-1)**
Acetamidine, 2-chloro-, Compd. with 4,6-bis(chloro-methyl)-s-triazin-2-ol (6CI) **(30886-00-1)**
Acetamidine, 2,2-dichloro-, Compd. with 4,6-bis(dichloro-methyl)-s-triazin-2-ol (1:1) (8CI) **(30886-01-2)**
Acetamidine, 2,2-dichloro-, Compd. with 4,6-bis(dichloro-methyl)-s-triazin-2-ol (6CI) **(30886-01-2)**
Acetamidine, N-N'-o-phenylene- **(615-15-6)**
Acetamidine, 2,2,2-trichloro-, Compd. with 4,6-bis(trichloro-methyl)-s-triazin-2-ol (1:1) (8CI) **(30886-02-3)**
Acetamidine, 2,2,2-trichloro-, Compd. with 4,6-bis(trichloro-methyl)-s-triazin-2-ol (6CI) **(30886-02-3)**
Acetamidoacetic acid **(543-24-8)**
4-Acetamidoaniline **(122-80-5)**
p-Acetamidoaniline **(122-80-5)**
p-Acetamidoazobenzene **(4128-71-6)**
Acetamidobenzene **(103-84-4)**
4-Acetamidobenzenesulfonyl chloride **(121-60-8)**
p-Acetamidobenzenesulfonyl chloride **(121-60-8)**
4'-Acetamidobenzidine **(3366-61-8)**
2-Acetamidobenzoic acid **(89-52-1)**
4-Acetamidobenzoic acid **(556-08-1)**
o-Acetamidobenzoic acid **(89-52-1)**
4-Acetamidobiphenyl **(4075-79-0)**
Acetamidocyclohexane **(1124-53-4)**
Acetamidoethane **(625-50-3)**
2-Acetamidoethanol **(142-26-7)**
2-Acetamidofluorene **(53-96-3)**
Acetamidomalonic acid diethyl ester **(1068-90-2)**
2-Acetamido-4-mercaptobutyric acid thiolactone **(1195-16-0)**
L-α-Acetamido-β-mercaptopropionic acid **(616-91-1)**
2-Acetamido-2-methylpropane **(762-84-5)**
p-Acetamidonitrobenzene **(104-04-1)**
2-Acetamido-4-(5-nitro-2-furyl)thiazole **(531-82-8)**
p-Acetamidophenacyl chloride **(140-49-8)**
2-Acetamidophenol **(614-80-2)**
3-Acetamidophenol **(621-42-1)**
4-Acetamidophenol **(103-90-2)**
m-Acetamidophenol **(621-42-1)**
o-Acetamidophenol **(614-80-2)**
p-Acetamidophenol **(103-90-2)**
1-(4-Acetamidophenoxy)-3-isopropylamino-2-propanol **(6673-35-4)**
4-Acetamidophenylsulfonyl chloride **(121-60-8)**
p-Acetamidophenylsulfonyl chloride **(121-60-8)**
2-Acetamido-5-sulfonamido-1,3,4-thiadiazole **(59-66-5)**
Acetamidothiadiazolesulfonamide **(59-66-5)**

Name	CAS
α-Acetamido-γ-thiobutyrolactone	**(1195-16-0)**
3-Acetamidotoluene	**(537-92-8)**
p-Acetamidotoluene	**(103-89-9)**
Acetamine Diazo Black RD	**(119-90-4)**
Acetamine Diazo Navy RD	**(119-90-4)**
Acetamine Yellow CG	**(2832-40-8)**
Acetamine Yellow 2R	**(119-15-3)**
m-Acetaminoaniline	**(102-28-3)**
p-Acetaminobenzenesulfonyl chloride	**(121-60-8)**
p-Acetaminobenzoic acid	**(556-08-1)**
Acetaminofen	**(103-90-2)**
Acet-o-aminofenol (Czech)	**(614-80-2)**
p-Acetaminofenyl-2-hydroxyethylsulfon (Czech)	**(27375-52-6)**
p-Acetaminofenyl-β-hydroxyethylsulfon (Czech)	**(27375-52-6)**
2-Acetaminofluorene	**(53-96-3)**
2-Acetamino-4-(5-nitro-2-furyl)thiazole	**(531-82-8)**
4-Acetamino-2-nitrophenetole	**(1777-84-0)**
Acetaminophen	**(103-90-2)**
2-Acetaminophenol	**(614-80-2)**
p-Acetaminophenol	**(103-90-2)**
Aceto-m-aminotoluene	**(537-92-8)**
Acetamox	**(59-66-5)**
Acetanhydride	**(108-24-7)**
Acetanil	**(103-84-4)**
Acetanilid	**(103-84-4)**
Acetanilide	**(103-84-4)**
Acetanilide (8CI), 3',4'-dichloro-	**(2150-93-8)**
Acetanilide, 2-acetyl-	**(102-01-2)**
Acetanilide, 3'-amino	**(102-28-3)**
Acetanilide, 4'-amino	**(122-80-5)**
Acetanilide, 2-amino- (8CI)	**(555-48-6)**
Acetanilide, 3'-amino-4'-ethoxy	**(17026-81-2)**
Acetanilide, 3-amino-4-ethoxy-	**(17026-81-2)**
Acetanilide, 2-amino-, monohydrochloride	**(4801-39-2)**
Acetanilide, 4'-(p-aminophenyl)	**(3366-61-8)**
Acetanilide, 4',4'''-azobis	**(15446-39-6)**
Acetanilide, 5'-(bis(2-hydroxyethyl)amino)-2'-((2-methoxy-4-nitrophenyl)azo)-, diacetate (ester) (8CI)	**(1533-77-3)**
Acetanilide, 4'-bromo	**(103-88-8)**
Acetanilide, p-bromo-	**(103-88-8)**
Acetanilide, 3'-bromo- (6CI,7CI,8CI)	**(621-38-5)**
Acetanilide, 4'-bromo-2-fluoro	**(351-05-3)**
Acetanilide, N-(butoxymethyl)-2-chloro-2',6'-diethyl- (8CI)	**(23184-66-9)**
Acetanilide, N-butyl	**(91-49-6)**
Acetanilide, 2-chloro	**(587-65-5)**
Acetanilide, 3'-chloro	**(588-07-8)**
Acetanilide, 4'-chloro	**(539-03-7)**
Acetanilide, 2'-chloro- (8CI)	**(533-17-5)**
Acetanilide, 4'-(chloroacetyl)	**(140-49-8)**
Acetanilide, 2-chloro-N-(3-butynyl)	**(21267-72-1)**
Acetanilide, 2-chloro-2',6'-diethyl-N-(butoxymethyl)	**(23184-66-9)**
Acetanilide, 2-chloro-2',6'-diethyl-N-(methoxymethyl)	**(15972-60-8)**
Acetanilide, 2-chloro-N-isopropyl	**(1918-16-7)**
Acetanilide, 2-chloro-N-(1-methyl-2-propynyl)- (8CI)	**(21267-72-1)**
Acetanilide, 2'-((2-cyano-4,6-dinitrophenyl)azo)-5'-(diethyl-amino)-	**(24170-60-3)**
Acetanilide, 3-(N,N-di-β-acetoxyethyl)amino-	**(27059-08-1)**
Acetanilide, 4'-((2,4-diaminophenyl)azo)- (8CI)	**(18371-12-5)**
Acetanilide, 2',5'-dimethoxy- (8CI)	**(3467-59-2)**
Acetanilide, 2',4'-dimethyl-	**(2050-43-3)**
Acetanilide, 2',4'-dimethyl-5-((trifluoromethyl)sulfonamido)	**(53780-34-0)**
Acetanilide, 2-((2,4-dinitro-6-cyanophenyl)azo)-5-diethylamino	**(24170-60-3)**
Acetanilide, 3'-ethoxy-	**(591-33-3)**
Acetanilide, 4'-ethoxy-	**(62-44-2)**
Acetanilide, 4'-fluoro	**(351-83-7)**
Acetanilide, 3'-fluoro- (8CI)	**(351-28-0)**
Acetanilide, 2'-hydroxy	**(614-80-2)**
Acetanilide, 3'-hydroxy	**(621-42-1)**
Acetanilide, 4'-hydroxy	**(103-90-2)**
Acetanilide, 4'-hydroxy-, acetate (ester) (8CI)	**(2623-33-8)**
Acetanilide, 4'-(2-hydroxyethylsulfonyl)	**(27375-52-6)**
Acetanilide, 4'-(2-hydroxy-3-(isopropylamino)propoxy)	**(6673-35-4)**
Acetanilide, 4'-((p-hydroxyphenyl)azo)- (8CI)	**(5302-39-6)**
Acetanilide, 4'-methoxy-	**(51-66-1)**
Acetanilide, p-nitro	**(104-04-1)**
Acetanilide, 2'-nitro- (8CI)	**(552-32-9)**
Acetanilide, 3'-nitro- (8CI)	**(122-28-1)**
Acetanilide, 4'-phenyl	**(4075-79-0)**
Acetanilide, 4'-phenylazo	**(4128-71-6)**
Acetanilide-p-sulfonyl chloride	**(121-60-8)**
Acetanilide, 3-(trifluoromethyl)-	**(351-36-0)**
p-Acetanisidide	**(51-66-1)**
p-Acetanisidide, 2'-((2-bromo-4,6-dinitrophenyl)azo)-5'-((2-cyanoethyl)ethylamino)-	**(22578-86-5)**
p-Acetanisidide, 2'-((2-bromo-4,6-dinitrophenyl)azo)-5'-(N-(2-cyanoethyl)-N-ethylamino)-	**(22578-86-5)**
p-Acetanisidide, 3'-chloro- (8CI)	**(7073-42-9)**
p-Acetanisidide, 3'-((2-cyanoethyl)ethylamino)-	**(19433-94-4)**
p-Acetanisidide, 3'-(diethylamino)-	**(19433-93-3)**
p-Acetanisidine	**(51-66-1)**
Acetanisole	**(100-06-1)**
Acetaphos	**(2425-25-4)**
Acetate Blue B	**(2475-44-7)**
Acetate Blue G	**(2475-45-8)**
Acetate Brilliant Blue 4B	**(2475-46-9)**
Acetate Brilliant Blue 4B	**(86722-66-9)**
Acetate C 10	**(112-17-4)**
Acetate C-10	**(112-17-4)**
Acetate C-7	**(112-06-1)**
Acetate C-8	**(112-14-1)**
Acetate C-12	**(112-66-3)**
Acetate Fast Orange R	**(82-28-0)**
Acetate Fast Pink 3B	**(2872-48-2)**
Acetate Fast Red 2B	**(116-85-8)**
Acetate Fast Yellow G	**(2832-40-8)**
Acetate Fast Yellow 5RL	**(6300-37-4)**
Acetate Red Violet R	**(128-95-0)**
Acetate (VAN)	**(71-50-1)**
Acetate d'amyle (French)	**(628-63-7)**
Acetate cortisone	**(50-04-4)**
Acetate cotton	**(9004-35-7)**
Acetate de butyle (French)	**(123-86-4)**
Acetate de butyle secondaire (French)	**(105-46-4)**
Acetate de cellosolve (French)	**(111-15-9)**
Acetate de cuivre (French)	**(142-71-2)**
Acetate de dodemorphe (French)	**(31717-87-0)**
Acetate de l'ether monoethylique de l'ethylene-glycol (French)	**(111-15-9)**
Acetate de l'ether monomethylique de l'ethylene-glycol (French)	**(110-49-6)**
Acetate de methyle (French)	**(79-20-9)**
Acetate de methyle glycol (French)	**(110-49-6)**
Acetate de plomb (French)	**(301-04-2)**
Acetate de propyle normal (French)	**(109-60-4)**
Acetate de triphenyl-etain (French)	**(900-95-8)**
Acetate de vinyle (French)	**(108-05-4)**
Acetate ester of cellulose	**(9004-35-7)**
Acetate d'ethylglycol (French)	**(111-15-9)**
Acetate d'isobutyle (French)	**(110-19-0)**
Acetate d'isopropyle (French)	**(108-21-4)**
Acetate phenylmercurique (French)	**(62-38-4)**
Acetate-replacing factor	**(62-46-4)**
Acetato(chloromethoxypropyl)mercury	**(1319-86-4)**
Acetato di cellosolve (Italian)	**(111-15-9)**
Acetato di metil cellosolve (Italian)	**(110-49-6)**
Acetato di stagno trifenile (Italian)	**(900-95-8)**

(Acetato-o)ethylmercury **(109-62-6)**
(Acetato-O)(formato-O)hydroxyaluminum **(34202-30-7)**
Acetato(2-methoxyethyl)mercury **(151-38-2)**
(Acetato)phenylmercury **(62-38-4)**
(Acetato-o)(trimetaarsenito)dicopper **(12002-03-8)**
Acetatotriphenylstannane **(900-95-8)**
Acetazolamid **(59-66-5)**
Acetazolamide **(59-66-5)**
Acetazoleamide **(59-66-5)**
Aceto-caustin **(76-03-9)**
Acetdimethylamide **(127-19-5)**
Acetdron **(60-13-9)**
Acetein **(616-91-1)**
Acetene **(74-85-1)**
Aceteugenol **(93-28-7)**
Acethion **(919-54-0)**
Acethione **(919-54-0)**
Acethydroxamsaure (German) **(546-88-3)**
Acetic-acid **(64-19-7)**
Acetic acid (ACGIH,OSHA) **(64-19-7)**
Acetic acid (Aqueous solution) (DOT) **(64-19-7)**
Acetic acid, C>19-alkyl esters **(68478-35-3)**
Acetic acid, allyl ester **(591-87-7)**
Acetic acid, aluminum salt (9CI) **(139-12-8)**
Acetic acid amide **(60-35-5)**
Acetic acid, amide, N(2-chlorophenyl)- **(533-17-5)**
Acetic acid, amino-sec-butyl- **(73-32-5)**
Acetic acid, ((aminocarbonyl)amino)- **(462-60-2)**
Acetic acid, aminooxo- (9CI) **(471-47-6)**
Acetic acid, (p-aminophenyl) **(1197-55-3)**
Acetic acid, ammonium copper salt (9CI) **(23087-46-9)**
Acetic acid, ammonium salt **(631-61-8)**
Acetic acid, amyl ester **(628-63-7)**
Acetic acid, anhydride (9CI) **(108-24-7)**
Acetic acid anilide **(103-84-4)**
Acetic acid, barium salt **(543-80-6)**
Acetic acid, (benzamidooxy)- (8CI) **(5251-93-4)**
Acetic acid, ((benzoylamino)oxy)- (9CI) **(5251-93-4)**
Acetic acid, benzoyl-, ethyl ester **(94-02-0)**
Acetic acid, benzoyl-, methyl ester (8CI) **(614-27-7)**
Acetic acid, benzyl ester **(140-11-4)**
Acetic acid, beryllium salt **(543-81-7)**
Acetic acid, (4-(bis(2-chloroethyl)amino)phenyl)-, cholesteryl ester **(3546-10-9)**
Acetic acid, bis(p-chlorophenyl) **(83-05-6)**
Acetic acid, (((3,5-bis(1,1-dimethylethyl)-4-hydroxyphenyl)-methyl)thio)-, 2-ethylhexyl ester **(80387-97-9)**
Acetic acid, (2,4-bis(1,1-dimethylpropyl)phenoxy)- (9CI) **(13402-96-5)**
Acetic acid, bismuth(3+) salt **(22306-37-2)**
Acetic acid, bromo **(79-08-3)**
Acetic acid, bromo-, 2-butene-1,4-diyl ester (9CI) **(20679-58-7)**
Acetic acid, bromo-, 2-butenylene ester (8CI) **(20679-58-7)**
Acetic acid, bromo-, ethyl ester **(105-36-2)**
Acetic acid, bromo-, methyl ester **(96-32-2)**
Acetic acid, bromophenyl-, nitrile **(5798-79-8)**
Acetic acid, (2-butenylidene)- **(110-44-1)**
Acetic acid, 2-butoxy ester **(105-46-4)**
Acetic acid 2-(2-butoxyethoxy)ethyl ester **(124-17-4)**
Acetic acid, 2-butoxyethyl ester **(112-07-2)**
Acetic acid, (4-tert-butylcyclohexyl) ester **(32210-23-4)**
Acetic acid, 2-(sec-butyl)-4,6-dinitrophenyl ester **(2813-95-8)**
Acetic acid n-butyl ester **(123-86-4)**
Acetic acid, butyl ester **(123-86-4)**
Acetic acid, sec-butyl ester **(105-46-4)**
Acetic acid, tert-butyl ester **(540-88-5)**
Acetic acid, butylnitrosaminomethyl ester **(56986-36-8)**
Acetic acid, (p-tert-butylphenoxy)- (8CI) **(1798-04-5)**
Acetic acid, cadmium salt **(543-90-8)**
Acetic acid, calcium salt **(62-54-4)**
Acetic acid, ((carboxymethylimino)bis(ethylenenitrilo))tetra- **(67-43-6)**
Acetic acid, ((carboxymethylimino)bis(ethylenenitrilo))-tetra-, calcium trisodium salt **(12111-24-9)**
Acetic acid, ((carboxymethylimino)bis(ethylenenitrilo))-tetra-, pentasodium salt **(140-01-2)**
Acetic acid, (4-carboxyphenyl)- **(1679-64-7)**
Acetic acid, cedrol ester **(77-54-3)**
Acetic acid, cellulose ester **(9004-35-7)**
Acetic acid, chloride **(75-36-5)**
Acetic acid, chloro **(79-11-8)**
Acetic acid, chloro-, anhydride (9CI) **(541-88-8)**
Acetic acid, chloro-, benzyl ester **(140-18-1)**
Acetic acid, chloro-, bicyclo(2.2.1)hept-5-en-2-ylmethyl ester (9CI) **(28693-00-7)**
Acetic acid, chloro-, 2-butoxyethyl ester (9CI) **(5330-17-6)**
Acetic acid, chloro-, butyl ester (9CI) **(590-02-3)**
Acetic acid, 2-chloro-2-(3-chloro-4-cyclohexylphenyl) **(36616-52-1)**
Acetic acid, ((chlorodimethylstannyl)thio)-, isooctyl ester (9CI) **(57807-89-3)**
Acetic acid, chloro-, ethyl ester **(105-39-5)**
Acetic acid, chloro-, isopropyl ester **(105-48-6)**
Acetic acid, chloro-, methyl ester **(96-34-4)**
Acetic acid, chloro-, 1-methyl-2-(2-methylpropoxy)ethyl ester (9CI) **(40137-60-8)**
Acetic acid, (4-chloro-2-methylphenoxy)- **(94-74-6)**
Acetic acid, (4-chloro-2-methylphenoxy)-, ammonium salt **(19480-39-8)**
Acetic acid, (4-chloro-2-methylphenoxy)-, 2-butoxyethyl ester (9CI) **(19480-43-4)**
Acetic acid, (4-chloro-2-methylphenoxy)-, butyl ester **(1713-12-8)**
Acetic acid, (4-chloro-2-methylphenoxy)-, isooctyl ester (9CI) **(26544-20-7)**
Acetic acid, (4-chloro-2-methylphenoxy)-, sodium salt **(3653-48-3)**
Acetic acid, ((4-chloro-2-methyl)phenyl)thio **(94-76-8)**
Acetic acid, (6-chloro-2-methyl-5-pyrimidyl)-, ethyl ester **(14273-76-8)**
Acetic acid, chloro-, 5-norbornen-2-ylmethyl ester **(28693-00-7)**
Acetic acid, (4-chloro-2-oxobenzothiazol-3-yl) **(3813-05-6)**
Acetic acid, chlorooxo-, ethyl ester (9CI) **(4755-77-5)**
Acetic acid, (p-chlorophenoxy) **(122-88-3)**
Acetic acid, o-chlorophenoxy **(614-61-9)**
Acetic acid, (2-chlorophenoxy)- (9CI) **(614-61-9)**
Acetic acid, (3-chlorophenoxy)- (9CI) **(588-32-9)**
Acetic acid, (4-chlorophenoxy)- (9CI) **(122-88-3)**
Acetic acid, (m-chlorophenoxy)- (8CI) **(588-32-9)**
Acetic acid, (p-chlorophenoxy)-, butyl ester **(52716-17-3)**
Acetic acid, (4-chlorophenoxy)-, butyl ester (9CI) **(52716-17-3)**
Acetic acid, (p-chlorophenoxy)dimethyl- **(882-09-7)**
Acetic acid, (p-chlorophenoxy)dimethyl-, ethyl ester **(637-07-0)**
Acetic acid, (4-chlorophenoxy)-, ethyl ester (9CI) **(14426-42-7)**
Acetic acid, (p-chlorophenoxy)-, ethyl ester (8CI) **(14426-42-7)**
Acetic acid, (p-chlorophenyl) **(1878-66-6)**
Acetic acid, (m-chlorophenyl)- (8CI) **(1878-65-5)**
Acetic acid, (o-chlorophenyl)- (8CI) **(2444-36-2)**
Acetic acid, p-chlorophenyl ester **(876-27-7)**
Acetic acid, 4-chlorophenyl ester (9CI) **(876-27-7)**
Acetic acid, chloro-, phenyl ester (9CI) **(620-73-5)**
Acetic acid, chloro-, phenylmethyl ester (9CI) **(140-18-1)**
Acetic acid, ((2-chlorophenyl)thio)- (9CI) **(18619-18-6)**
Acetic acid, ((o-chlorophenyl)thio)- (8CI) **(18619-18-6)**
Acetic acid, chloro-, sodium salt **(3926-62-3)**
Acetic acid, ((4-chloro-o-tolyl)oxy) **(94-74-6)**
Acetic acid, ((4-chloro-o-tolyl)oxy)-, Compd. with 2,2'-Iminodiethanol (1:1) **(20405-19-0)**
Acetic acid, ((4-chloro-o-tolyl)oxy)-, amine salt **(19480-39-8)**
Acetic acid, ((4-chloro-o-tolyl)oxy)-, ammonium salt (8CI) **(19480-39-8)**
Acetic acid, ((4-chloro-o-tolyl)oxy)-, butyl ester **(1713-12-8)**
Acetic acid, ((4-chloro-o-tolyl)oxy)-, isooctyl ester **(26544-20-7)**
Acetic acid, ((4-chloro-o-tolyl)oxy)-, sodium salt **(3653-48-3)**
Acetic acid, ((4-chloro-o-tolyl)oxy)thio-, S-ethyl ester **(25319-90-8)**

Acetic acid, chloro-, vinyl ester **(2549-51-1)**
Acetic acid, chromium(3+) salt **(1066-30-4)**
Acetic acid, citronellyl ester **(150-84-5)**
Acetic acid, cobalt(2+) salt **(71-48-7)**
Acetic acid, cobalt(2+) salt, tetrahydrate **(6147-53-1)**
Acetic acid, copper(2+) salt **(142-71-2)**
Acetic acid, crotylidene- **(110-44-1)**
Acetic acid, cupric salt **(142-71-2)**
Acetic acid, cyano **(372-09-8)**
Acetic acid, cyano-, ethyl ester **(105-56-6)**
Acetic acid, cyano-, 2-ethylhexyl ester (9CI) **(13361-34-7)**
Acetic acid, cyano-, hydrazide **(140-87-4)**
Acetic acid cyanomethyl ester **(1001-55-4)**
Acetic acid, cyano-, methyl ester **(105-34-0)**
Acetic acid, cyclohexylamino **(58695-41-3)**
Acetic acid, (1,2-cyclohexylenedinitrilo)tetra **(482-54-2)**
Acetic acid, cyclohexyl ester **(622-45-7)**
Acetic acid, decyl ester (9CI) **(112-17-4)**
Acetic acid, dehydro- **(520-45-6)**
Acetic acid, diaminopropanoltetra- **(3148-72-9)**
Acetic acid, diazo-, ester with serine **(115-02-6)**
Acetic acid, dibromo- (9CI) **(631-64-1)**
Acetic acid, 1,2-dibromoethyl ester **(24442-57-7)**
Acetic acid, (3,5-dibromophenoxy)- (9CI) **(7507-35-9)**
Acetic acid, dichloro **(79-43-6)**
Acetic acid, o-(2,6-dichloroanilino)phenyl-, monosodium salt **(15307-79-6)**
Acetic acid 1,2-dichloroethyl ester **(10140-87-1)**
Acetic acid, dichloro-, ethyl ester (9CI) **(535-15-9)**
Acetic acid, (3,6-dichloro-2-methoxyphenyl)- (8CI) **(3004-74-8)**
Acetic acid, (2,3-dichloro-4-(2-methylenebutyryl)phenoxy) **(58-54-8)**
Acetic acid, (2,3-dichloro-4-(2-methylene-1-oxobutyl)phenoxy)- (9CI) **(58-54-8)**
Acetic acid, dichloro-, methyl ester **(116-54-1)**
Acetic acid, (2,4-dichlorophenoxy) **(94-75-7)**
Acetic acid, (3,4-dichlorophenoxy) **(588-22-7)**
Acetic acid, (2,3-dichlorophenoxy)- (9CI) **(2976-74-1)**
Acetic acid, (2,6-dichlorophenoxy)- (9CI) **(575-90-6)**
Acetic acid, (3,5-dichlorophenoxy)- (8CI,9CI) **(587-64-4)**
Acetic acid, (2,4-dichlorophenoxy)- Mixed with acetic acid, (2,4,5-trichlorophenoxy)-, (2:1) **(8015-35-8)**
Acetic acid, (2,4-dichlorophenoxy)- Mixed with (2,4,5-trichlorophenoxy)acetic acid (2:1) **(8015-35-8)**
Acetic acid, (2,4-dichlorophenoxy)-, ammonium salt **(2307-55-3)**
Acetic acid, (2,4-dichlorophenoxy)-, butoxyethyl ester **(1929-73-3)**
Acetic acid, (2,4-dichlorophenoxy)-, 2-butoxyethyl ester (8ci,9CI) **(1929-73-3)**
Acetic acid, (2,4-dichlorophenoxy)-, 2-butoxymethylethyl ester (9CI) **(1320-18-9)**
Acetic acid, (2,4-dichlorophenoxy)-, 2-butoxy-1-methylethyl ester (8CI,9CI) **(3966-11-8)**
Acetic acid, (2,4-dichlorophenoxy)-, butoxy propylene deriv (8CI) **(1320-18-9)**
Acetic acid, 2,4-dichlorophenoxy-, butoxypropyl ester **(1320-18-9)**
Acetic acid, 2,4-dichlorophenoxy-, butoxypropyl ester **(1928-45-6)**
Acetic acid, (2,4-dichlorophenoxy)-, n-butyl ester (8CI,9CI) **(94-80-4)**
Acetic acid, 2,4-dichlorophenoxy-, butyl ester and 2,4,5-trichlorophenoxyacetic acid (45.5%: 48.2%) **(39277-47-9)**
Acetic acid, 2,4-dichlorophenoxy-, 4-chloro-2-butenyl ester **(2971-38-2)**
Acetic acid, (2,4-dichlorophenoxy)-, Compd. with dimethylamine (1:1) **(2008-39-1)**
Acetic acid, (2,4'dichlorophenoxy)-, Compd. with N-methylmethanamine (1:1) (9CI) **(2008-39-1)**
Acetic acid, (2,4-dichlorophenoxy)-, Compd. with 1-tetradecanamine (1:1) (9CI) **(28685-18-9)**
Acetic acid, (2,4-dichlorophenoxy)-, Compd. with tetradecylamine (1:1) (8CI) **(28685-18-9)**
Acetic acid, (2,4-dichlorophenoxy)-, diethylamine salt **(20940-37-8)**
Acetic acid, (2,4-dichlorophenoxy)-, ethyl ester **(533-23-3)**
Acetic acid, (2,4-dichlorophenoxy)-, 2-ethylhexyl ester **(1928-43-4)**
Acetic acid, (2,4-dichlorophenoxy)-, hexyl ester (8CI,9CI) **(1917-95-9)**
Acetic acid, (2,4-dichlorophenoxy)-, isobutyl ester **(1713-15-1)**
Acetic acid, (2,4-dichlorophenoxy)-, isooctyl ester **(25168-26-7)**
Acetic acid, (2,4-dichlorophenoxy)-, isopropyl ester **(94-11-1)**
Acetic acid, (2,4-dichlorophenoxy)-, lithium salt **(3766-27-6)**
Acetic acid, 2,4-dichlorophenoxy-, lithium salt **(3766-27-6)**
Acetic acid, (2,4-dichlorophenoxy)-, methyl ester **(1928-38-7)**
Acetic acid, (2,4-dichlorophenoxy)-, 1-methylethyl ester (9CI) **(94-11-1)**
Acetic acid, (2,4-dichlorophenoxy)-, 2-methylpropyl ester (9CI) **(1713-15-1)**
Acetic acid, 2,4-dichlorophenoxy-, octyl ester **(1928-44-5)**
Acetic acid, (4-(2,4-dichlorophenoxy)phenoxy)- (9CI) **(51338-10-4)**
Acetic acid, (2,4-dichlorophenoxy)-, propyl ester (8CI,9CI) **(1928-61-6)**
Acetic acid, (2,4-dichlorophenoxy)-, sodium salt **(2702-72-9)**
Acetic acid, ((2,5-dichlorophenyl)thio)- (9CI) **(6274-27-7)**
Acetic acid, ((diethoxyphosphinothioyl)thio)-, ethyl ester (9CI) **(919-54-0)**
Acetic acid, (diethoxyphosphinyl)- (9CI) **(3095-95-2)**
Acetic acid, (diethoxyphosphinyl)-, ethyl ester **(867-13-0)**
Acetic acid, (diethoxyphosphinyl)-, potassium salt (9CI) **(34170-84-8)**
Acetic acid, ((diethoxyphosphinyl)thio)-, ethyl ester (9CI) **(2425-25-4)**
Acetic acid, diethyl- **(88-09-5)**
Acetic acid, diethylphosphono-, ethyl ester **(867-13-0)**
Acetic acid, difluoro **(381-73-7)**
Acetic acid, difluoro-, ethyl ester (9CI) **(454-31-9)**
Acetic acid, difluoro-, sodium salt (9CI) **(2218-52-2)**
Acetic acid, (3,4-dihydroxyphenyl) **(102-32-9)**
Acetic acid, (2,5-dihydroxyphenyl)- (8CI) **(451-13-8)**
Acetic acid dimer, sodium salt **(126-96-5)**
Acetic acid, (3,4-dimethoxyphenyl) **(93-40-3)**
Acetic acid, ((5-(3-((2-(3,4-dimethoxyphenyl)ethyl)((phenylmethoxy)carbonyl)amino)-2-hydroxypropoxy)-1,2,3,4-tetrahydro-2-oxo-8-quinolinyl)oxy)- (9CI) **(80090-30-8)**
Acetic acid, ((dimethoxyphosphinothioyl)thio)- **(1113-01-5)**
Acetic acid, ((dimethoxyphosphinyl)thio)-, ethyl ester (9CI) **(2088-72-4)**
Acetic acid, dimethyl- **(79-31-2)**
Acetic acid, dimethylamide **(127-19-5)**
Acetic acid, 1,3-dimethylbutyl ester **(108-84-9)**
Acetic acid, 2,6-dimethyl-m-dioxan-4-yl ester **(828-00-2)**
Acetic acid, O,O-dimethyldithiophosphoryl-, N-monomethylamide salt **(60-51-5)**
Acetic acid, (O,O-dimethyldithiophosphorylphenyl)-, ethyl ester **(2597-03-7)**
Acetic acid, 1,1-dimethylethyl ester (9CI) **(540-88-5)**
Acetic acid, (4-(1,1-dimethylethyl)phenoxy)- (9CI) **(1798-04-5)**
Acetic acid, 3,7-dimethyl-6-octen-1-yl ester **(150-84-5)**
Acetic acid, (2,4-dinitro-6-s-butylphenyl) ester **(2813-95-8)**
Acetic acid, (4,6-dinitro-2-s-butylphenyl) ester **(2813-95-8)**
Acetic acid, 2,2'-((dioctylstannylene)bis(thio))bis-, diisooctyl ester **(26401-97-8)**
Acetic acid, (2,4-di-tert-pentylphenoxy)- **(13402-96-5)**
Acetic acid, diphenyl **(117-34-0)**
Acetic acid, diphenyl-, 2-(diethylamino)ethyl ester **(64-95-9)**
Acetic acid, dipropyl- **(99-66-1)**
Acetic acid, (dodecahydro-7-hydroxy-1,4b,8,8-tetramethyl-10-oxo-2(1H)-phenanthrenylidene)-, 2-(dimethylamino)ethyl ester, (1R-(1α,2e,4aα,4bβ,7β,8aα,10aβ))- **(468-76-8)**
Acetic acid, dodecyl ester **(112-66-3)**
Acetic acid, ester with 2,6-dimethyl-m-dioxan-4-ol **(828-00-2)**
Acetic acid, ester with trichloroethanol **(515-84-4)**
Acetic acid, ethenyl ester **(108-05-4)**
Acetic acid ethenyl ester, Polymer with ethene (9CI) **(24937-78-8)**
Acetic acid ethenyl ester homopolymer (9CI) **(9003-20-7)**
Acetic acid ethenyl ester polymer with chlorethene (9CI) **(9003-22-9)**
Acetic acid, ethoxy **(627-03-2)**
Acetic acid 2-(2-ethoxyethoxy)ethyl ester **(112-15-2)**
Acetic acid, 2-ethoxyethyl ester **(111-15-9)**
Acetic acid, (ethylenedinitrilo)tetra **(60-00-4)**
Acetic acid, (ethylenedinitrilo)tetra-, calcium disodium salt **(62-33-9)**
Acetic acid, (ethylenedinitrilo)tetra-, copper(II) complex **(54453-03-1)**

Acetic acid, (ethylenedinitrilo)tetra-, disodium salt **(139-33-3)**
Acetic acid, (ethylenedinitrilo)tetra-, sodium salt, iron complex **(15708-41-5)**
Acetic acid, (ethylenedinitrilo)tetra-, tetrasodium salt **(64-02-8)**
Acetic acid, (ethylenedinitrilo)tetra-, trisodium salt **(150-38-9)**
Acetic acid, ethylene ether **(108-05-4)**
Acetic acid, ethyl ester **(141-78-6)**
Acetic acid α-ethylhexyl ester **(103-09-3)**
Acetic acid, 2-ethylhexyl ester **(103-09-3)**
Acetic acid, fluoro **(144-49-0)**
Acetic acid, fluoro-, ammonium salt (9CI) **(60916-92-9)**
Acetic acid, fluoro-, ethyl ester **(459-72-3)**
Acetic acid, fluoro-, (2-fluoroethyl) ester **(459-99-4)**
Acetic acid, (4-fluorophenoxy)-, 3,3,5-trimethylcyclohexyl ester **(58327-09-6)**
Acetic acid, (p-fluorophenyl)- (8CI) **(405-50-5)**
Acetic acid, fluoro-, sodium salt **(62-74-8)**
Acetic acid furfuryl ester **(623-17-6)**
Acetic acid, geraniol ester **(105-87-3)**
Acetic acid, glacial **(64-19-7)**
Acetic acid, glacial (DOT) **(64-19-7)**
Acetic acid, glacial, More than 80% acid, by weight [UN 2789] **(64-19-7)**
Acetic acid, heptyl ester **(112-06-1)**
Acetic acid, hexyl ester **(142-92-7)**
Acetic acid, hydroxy- **(79-14-1)**
Acetic acid, hydroxy-, Homopolymer (9CI) **(26124-68-5)**
Acetic acid, hydroxy-, butyl ester **(7397-62-8)**
Acetic acid 2-hydroxyethyl ester **(542-59-6)**
Acetic acid, 2-(hydroxyethylimino)di (9CI) **(93-62-9)**
Acetic acid, (4-hydroxy-3-methoxyphenyl)- (8CI) **(306-08-1)**
Acetic acid, hydroxy-, monopotassium salt (9CI) **(1932-50-9)**
Acetic acid, hydroxy-, monosodium salt **(2836-32-0)**
Acetic acid, ((6-hydroxy-2-naphthalenyl)oxy)- (9CI) **(10441-36-8)**
Acetic acid, ((7-hydroxy-1-naphthalenyl)oxy)- (9CI) **(63037-96-7)**
Acetic acid, ((6-hydroxy-2-naphthyl)oxy)- (8CI) **(10441-36-8)**
Acetic acid, (2-hydroxyphenoxy)- (9CI) **(6324-11-4)**
Acetic acid, (4-hydroxyphenoxy)- (9CI) **(1878-84-8)**
Acetic acid, (o-hydroxyphenoxy)- (8CI) **(6324-11-4)**
Acetic acid, (p-hydroxyphenyl) **(156-38-7)**
Acetic acid, (m-hydroxyphenyl)- (8CI) **(621-37-4)**
Acetic acid, (o-hydroxyphenyl)- (8CI) **(614-75-5)**
Acetic acid, hydroxy-, potassium salt (9CI) **(25904-89-6)**
Acetic acid, ((2-hydroxy-1,3-trimethylene)dinitrilo)tetra **(3148-72-9)**
Acetic acid, ((2-hydroxytrimethylene)dinitrilo)tetra- (8CI) **(3148-72-9)**
Acetic acid, iminodi **(142-73-4)**
Acetic acid, iminodi-, disodium salt **(928-72-3)**
Acetic acid, indolyl- **(87-51-4)**
Acetic acid, ion(1-) (8CI,9CI) (VAN) **(71-50-1)**
Acetic acid, iodo **(64-69-7)**
Acetic acid, iodo-, ethyl ester **(623-48-3)**
Acetic acid, isobornyl ester **(125-12-2)**
Acetic acid, isobutyl ester **(110-19-0)**
Acetic acid, isopentyl ester **(123-92-2)**
Acetic acid, isopropenyl ester **(108-22-5)**
Acetic acid, isopropyl- **(503-74-2)**
Acetic acid, isopropyl ester **(108-21-4)**
Acetic acid, lead salt **(15347-57-6)**
Acetic acid, lead(2+) salt **(301-04-2)**
Acetic acid, lead(4+) salt **(546-67-8)**
Acetic acid, lead(+2) salt trihydrate **(6080-56-4)**
Acetic acid linalool ester **(115-95-7)**
Acetic acid, lithium salt **(546-89-4)**
Acetic acid, magnesium salt **(142-72-3)**
Acetic acid, manganese(II) salt (2:1) **(638-38-0)**
Acetic acid, manganese(2+) salt, tetrahydrate **(6156-78-1)**
Acetic acid, p-menth-3-yl ester, dl **(16409-45-3)**
Acetic acid, mercapto **(68-11-1)**
Acetic acid, mercapto-, calcium salt (2:1) (9CI) **(814-71-1)**
Acetic acid, mercapto-, dodecyl ester **(3746-39-2)**
Acetic acid, mercapto-, dodecyl ester (9CI) **(3746-39-2)**
Acetic acid, mercapto-, 1,2-ethanediyl ester (9CI) **(123-81-9)**
Acetic acid, mercapto-, ethylene ester (8CI) **(123-81-9)**
Acetic acid, mercapto-, ethyl ester **(623-51-8)**
Acetic acid, mercapto-, ethyl ester, S-ester with O,O-diethyl phosphorodithioate **(919-54-0)**
Acetic acid, mercapto-, ethyl ester, S-ester with O,O-diethyl phosphorothioate **(2425-25-4)**
Acetic acid, mercapto-, ethyl ester, S-ester with O,O-dimethyl phosphorothioate **(2088-72-4)**
Acetic acid, mercapto-, 2-ethylhexyl ester **(7659-86-1)**
Acetic acid, mercapto-, isooctyl ester **(25103-09-7)**
Acetic acid, mercapto-, isooctyl ester, antimony(3+) salt (9CI) **(27288-44-4)**
Acetic acid, mercapto-, methyl ester **(2365-48-2)**
Acetic acid, mercapto-, monoammonium salt **(5421-46-5)**
Acetic acid, mercapto-, monosodium salt **(367-51-1)**
Acetic acid, mercaptophenyl-, ethyl ester, S-ester with O,O-dimethyl phosphorodithioate **(2597-03-7)**
Acetic acid, mercury(2+) salt **(1600-27-7)**
Acetic acid, mercury (1+) salt (8CI,9CI) **(631-60-7)**
Acetic acid, methoxy **(625-45-6)**
Acetic acid, 3-methoxybutyl ester **(4435-53-4)**
Acetic acid, (2-methoxyethoxy)- (8CI,9CI) **(16024-56-9)**
Acetic acid, 2-(2-methoxyethoxy)ethyl ester **(629-38-9)**
Acetic acid, (2-methoxyethoxy)-, 2-methoxyethyl ester (9CI) **(42064-17-5)**
Acetic acid, (2-methoxyethoxy)-, methyl ester (8CI,9CI) **(17640-28-7)**
Acetic acid 2-methoxyethyl ester **(110-49-6)**
Acetic acid, 2-methoxy-1-methylethyl ester **(108-65-6)**
Acetic acid, (o-methoxyphenoxy) **(1878-85-9)**
Acetic acid, methoxyphenoxy-, methyl ester (6CI,9CI) **(24607-12-3)**
Acetic acid, p-methoxyphenyl **(104-01-8)**
Acetic acid, (o-methoxyphenyl)- (8CI) **(93-25-4)**
Acetic acid, 2-methoxypropyl ester **(70657-70-4)**
Acetic acid, (methylamino)- **(107-97-1)**
Acetic acid, (methyl-ONN-azoxy)methyl ester **(592-62-1)**
Acetic acid, (3,4-methylenedioxy)benzyl ester **(326-61-4)**
Acetic acid, methyl ester **(79-20-9)**
Acetic acid, 1-methylethyl ester (9CI) **(108-21-4)**
Acetic acid, (methylimino)di- (8CI) **(4408-64-4)**
Acetic acid, 2-methyl-6-methylene-7-octen-2-yl ester **(1118-39-4)**
Acetic acid, (3-methyl-4-(methylsulfonyl)phenoxy)- (9CI) **(15267-77-3)**
Acetic acid, methylnitrosaminomethyl ester **(56856-83-8)**
Acetic acid, 2-methylphenyl ester (9CI) **(533-18-6)**
Acetic acid, 4-methylphenyl ester (9CI) **(140-39-6)**
Acetic acid, ((4-methylphenyl)sulfonyl)-, methyl ester (9CI) **(50397-64-3)**
Acetic acid, 2-methyl-2-propene-1,1-diol diester **(10476-95-6)**
Acetic acid, 2-methylpropyl ester **(110-19-0)**
Acetic acid, 1-methylpropyl ester (9CI) **(105-46-4)**
Acetic acid, 2,2',2''-((methylstannylidyne)tris(thio))tris-, triisooctyl ester (9CI) **(54849-38-6)**
Acetic acid, ((4-(methylsulfonyl)-m-tolyl)oxy)- (8CI) **(15267-77-3)**
Acetic acid, myrcenyl ester **(1118-39-4)**
Acetic acid, (2-naphthyloxy) **(120-23-0)**
Acetic acid, nickel(2+) salt **(373-02-4)**
Acetic acid, nitrilotri **(139-13-9)**
Acetic acid, nitrilotri-, disodium salt **(15467-20-6)**
Acetic acid, nitrilotri-, disodium salt, monohydrate (8CI) **(23255-03-0)**
Acetic acid, nitrilotri-, iron(III) chelate **(16448-54-7)**
Acetic acid, nitrilotri-, sodium salt **(10042-84-9)**
Acetic acid, nitrilotri-, tripotassium salt **(2399-85-1)**
Acetic acid, nitrilotri-, trisodium salt **(5064-31-3)**
Acetic acid, nitrilotri-, trisodium salt, monohydrate **(18662-53-8)**
Acetic acid, nitro-, ion(1-) (9CI) **(61201-44-3)**
Acetic acid, (2-nitrophenoxy)- (9CI) **(1878-87-1)**
Acetic acid, (o-nitrophenoxy)- (8CI) **(1878-87-1)**
Acetic acid, p-nitrophenyl ester **(830-03-5)**
Acetic acid, 2-nitrophenyl ester (9CI) **(610-69-5)**

Acetic acid, 3-nitrophenyl ester (9CI) **(1523-06-4)**
Acetic acid, 4-nitrophenyl ester (9CI) **(830-03-5)**
Acetic acid, m-nitrophenyl ester (8CI) **(1523-06-4)**
Acetic acid, (nitrosopropylamino)methyl ester **(66017-91-2)**
Acetic acid n-nonyl ester **(143-13-5)**
Acetic acid, nonyl ester (9CI) **(143-13-5)**
Acetic acid, (4-nonylphenoxy)- (9CI) **(3115-49-9)**
Acetic acid, (2-(4-nonylphenoxy)ethoxy)- (9CI) **(106807-78-7)**
Acetic acid, 1-octa-2,7-dienyl ester **(3491-27-8)**
Acetic acid, octyl ester **(112-14-1)**
Acetic acid, (p-octylphenoxy)- (8CI) **(15234-85-2)**
Acetic acid, 2,2',2''-((octylstannylidyne)tris(thio))tris-, triisooctyl ester (9CI) **(26401-86-5)**
Acetic acid, oxime **(546-88-3)**
Acetic acid, oxo-, methyl ester (9CI) **(922-68-9)**
Acetic acid, 2,2'-oxybis- (9CI) **(110-99-6)**
Acetic acid, oxydi **(110-99-6)**
Acetic acid, (pentachlorophenoxy)- (8CI,9CI) **(2877-14-7)**
Acetic acid, (pentachlorophenoxy)-, ethyl ester (9CI) **(55773-90-5)**
Acetic acid, 2-pentyl ester **(626-38-0)**
Acetic acid, pentyl ester **(628-63-7)**
Acetic acid, phenethyl ester **(103-45-7)**
Acetic acid, phenoxy **(122-59-8)**
Acetic acid, phenoxy-, ethyl ester **(2555-49-9)**
Acetic acid, phenoxy-, methyl ester (9CI) **(2065-23-8)**
Acetic acid, phenoxy-, sodium salt **(3598-16-1)**
Acetic acid, phenyl **(103-82-2)**
Acetic acid, phenyl ester **(122-79-2)**
Acetic acid, 2-phenylethyl ester **(103-45-7)**
Acetic acid, phenyl-, ethyl ester **(101-97-3)**
Acetic acid, 2-phenylhydrazide **(114-83-0)**
Acetic acid phenylhydrazone **(114-83-0)**
Acetic acid, phenyl-, isobutyl ester (8CI) **(102-13-6)**
Acetic acid, phenylmercury deriv. **(62-38-4)**
Acetic acid, phenyl-, methyl ester **(101-41-7)**
Acetic acid, phenylmethyl ester **(140-11-4)**
Acetic acid, phenyl-, phenethyl ester **(102-20-5)**
Acetic acid, phenyl-, sodium salt **(114-70-5)**
Acetic acid, phosphono **(4408-78-0)**
Acetic acid, phosphono-, P,P-diethyl ester (8CI) **(3095-95-2)**
Acetic acid, phosphono-, P,P-diethyl ester, potassium salt (8CI) **(34170-84-8)**
Acetic acid, ((phosphonomethyl)imino)di- (8CI) **(5994-61-6)**
Acetic acid, phosphono-, triethyl ester **(867-13-0)**
Acetic acid, potassium salt **(127-08-2)**
Acetic acid, 2-propenyl ester (9CI) **(591-87-7)**
Acetic acid n-propyl ester **(109-60-4)**
Acetic acid, propyl ester **(109-60-4)**
Acetic acid, propylnitrosaminomethyl ester **(66017-91-2)**
Acetic acid, sodium salt **(127-09-3)**
Acetic acid, sodium salt (2:1) (9CI) **(126-96-5)**
Acetic acid, sodium salt, Compd with acetic acid (1:1) **(126-96-5)**
Acetic acid solution, More than 80% acid, by weight [UN 2789] **(64-19-7)**
Acetic acid solution, More than 25% but not more than 80% acid, by weight [UN 2790] **(64-19-7)**
Acetic acid, stearic acid, caprylic acid, calcium salts **(68411-15-4)**
Acetic acid, strontium salt **(543-94-2)**
Acetic acid, sulfo **(123-43-3)**
Acetic acid, sulfo-, 1-dodecyl ester, sodium salt **(1847-58-1)**
Acetic acid, sulfo-, dodecyl ester, s-sodium salt (7CI) **(1847-58-1)**
Acetic acid, terephthaloyldi-, diethyl ester (8CI) **(93-94-7)**
Acetic acid, (2,3,4,5-tetrachlorophenoxy)-, ethyl ester (9CI) **(88927-42-8)**
Acetic acid, thallium(i) salt **(563-68-8)**
Acetic acid, thio **(507-09-5)**
Acetic acid, 2,2'-thiobis- (9CI) **(123-93-3)**
Acetic acid, thiocyanato-, isobornyl ester **(115-31-1)**
Acetic acid, thiocyanato-, 1,7,7-trimethylbicyclo(2,2,1)hept-2-yl ester, exo- (9CI) **(115-31-1)**
Acetic acid, thiodi **(123-93-3)**
Acetic acid, (p-tolyl) **(622-47-9)**
Acetic acid, m-tolyl- (8CI) **(621-36-3)**
Acetic acid, p-tolyl ester **(140-39-6)**
Acetic acid, o-tolyl ester (8CI) **(533-18-6)**
Acetic acid, tribromo- (9CI) **(75-96-7)**
Acetic acid, trichloro **(76-03-9)**
Acetic acid, trichloro-, Compd. with 3-(p-chlorophenyl)-1,1-dimethylurea (1:1) **(140-41-0)**
Acetic acid, trichloro-, Compd. with N'-(4-chlorophenyl)-N,N-dimethylurea (1:1) (9CI) **(140-41-0)**
Acetic acid, (2,4,5-trichlorophenoxy)-, Compd. with dimethylamine (1:1) (8CI) **(6369-97-7)**
Acetic acid, (2,4,5-trichlorophenoxy)-, Compd. with N-methylmethanamine (1:1) (9CI) **(6369-97-7)**
Acetic acid, trichloro-, Compd. with 1,1-dimethyl-3-phenylurea (1:1) **(4482-55-7)**
Acetic acid, trichloro-, Compd. with N,N-dimethyl-N'-phenylurea (1:1) (9CI) **(4482-55-7)**
Acetic acid, trichloro-, ammonium salt (8CI,9CI) **(7646-88-0)**
Acetic acid, trichloro-, anhydride (9CI) **(4124-31-6)**
Acetic acid, 2,2,2-trichloroethyl ester **(515-84-4)**
Acetic acid, trichloro-, ethyl ester (9CI) **(515-84-4)**
Acetic acid, α-(trichloromethyl)benzyl ester **(90-17-5)**
Acetic acid, trichloro-, methyl ester **(598-99-2)**
Acetic acid, trichloro-, mixed with urea, 1,1-dimethyl-3-phenyl- (1:1) **(4482-55-7)**
Acetic acid, (2,4,5-trichlorophenoxy)-, Compd. with Et3N **(2008-46-0)**
Acetic acid, (2,4,5-trichlorophenoxy)-, Compd. with 1-amino-2-propanol (1:1) (8CI) **(1319-72-8)**
Acetic acid, (2,4,5-trichlorophenoxy)-, Compd. with 2-(diethylamino)ethanol (1:1) (9CI) **(53404-86-7)**
Acetic acid, (2,4,5-trichlorophenoxy)-, Compd. with N,N-diethylethanamine (1:1) (9CI) **(2008-46-0)**
Acetic acid, (2,4,5-trichlorophenoxy)-, Compd. with N,N-dimethylmethanamine (1:1) (9CI) **(6369-96-6)**
Acetic acid, (2,4,5-trichlorophenoxy)-, Compd. with (Z,Z)-N,N-dimethyl-9,12-octadecadien-1-amine (1:1) (9CI) **(53404-88-9)**
Acetic acid, (2,4,5-trichlorophenoxy)-, Compd. with (Z,Z)-N,N-dimethyl-9,12-octadecadien-1-amine (1:1), Mixt. with (Z)-N,N-dimethyl-9-octadecen-1-amine (2,4,5-trichlorophenoxy)-acetate (9CI) **(55256-33-2)**
Acetic acid, (2,4,5-trichlorophenoxy)-, Compd. with (Z)-N,N-dimethyl-9-octadecen-1-amine (1:1) (9CI) **(53404-89-0)**
Acetic acid, (2,4,5-trichlorophenoxy)-, Compd. with 1-dodecanamine (1:1) (9CI) **(53404-84-5)**
Acetic acid, (2,4,5-trichlorophenoxy)-, Compd. with 2,2',2''-nitrilotriethanol (1:1) (8CI) **(3813-14-7)**
Acetic acid, (2,4,5-trichlorophenoxy)-, Compd. with 2,2',2''-nitrilotris(ethanol) (1:1) (9CI) **(3813-14-7)**
Acetic acid, (2,4,5-trichlorophenoxy)-, Compd. with (Z)-N-9-octadecenyl-1,3-propanediamine (1:1) (9CI) **(53404-87-8)**
Acetic acid, (2,4,5-trichlorophenoxy)-, Compd. with 1-tetradecanamine (1:1) (9CI) **(53535-37-8)**
Acetic acid, (2,4,5-trichlorophenoxy)-, Compd. with 1-tridecanamine (1:1) (9CI) **(53404-85-6)**
Acetic acid, (2,4,5-trichlorophenoxy)-, Compd. with triethylamine (1:1) (8CI) **(2008-46-0)**
Acetic acid, (2,4,5-trichlorophenoxy)-, Compd. with trimethylamine (1:1) (8CI) **(6369-96-6)**
Acetic acid, (2,4,5-trichlorophenoxy)-, ester with
Acetic acid, (2,4,5-trichlorophenoxy) **(93-76-5)**
Acetic acid, (2,4,6-trichlorophenoxy)- (9CI) **(575-89-3)**
Acetic acid, (2,3,5-trichlorophenoxy)- (8CI,9CI) **(33433-95-3)**
Acetic acid, (2,4,5-trichlorophenoxy)-, 3-(2-butoxyethoxy)-propyl ester (7CI,8CI) **(1928-58-1)**
Acetic acid, (2,4,5-trichlorophenoxy)-, 2-butoxyethyl ester **(2545-59-7)**

Acetic acid, (2,4,5-trichlorophenoxy)-, 3-butoxypropyl ester (8CI,9CI) **(1928-48-9)**
Acetic acid, (2,4,5-trichlorophenoxy)-, butyl ester **(93-79-8)**
1(or 2)-(2-methylpropoxy)propanol (9CI) **(53466-86-7)**
Acetic acid, (2,4,5-trichlorophenoxy)-, ethyl ester (8CI,9CI) **(1928-39-8)**
Acetic acid, (2,4,5-trichlorophenoxy)-, 2-ethylhexyl ester **(1928-47-8)**
Acetic acid, (2,4,5-trichlorophenoxy)-, 2-ethyl-4-methylpentyl ester (9CI) **(69462-12-0)**
Acetic acid, (2,4,5-trichlorophenoxy)-, isobutyl ester **(4938-72-1)**
Acetic acid, 2,4,5-trichlorophenoxy-, isooctyl ester **(25168-15-4)**
Acetic acid, (2,4,5-trichlorophenoxy)-, isopropyl ester (8CI) **(93-78-7)**
Acetic acid, (2,4,5-trichlorophenoxy)-, methyl ester (9CI) **(1928-37-6)**
Acetic acid, (2,4,5-trichlorophenoxy)-, 1-methylethyl ester (9CI) **(93-78-7)**
Acetic acid, (2,4,5-trichlorophenoxy)-, methyl-2-(methyl-2-(methyl-2-(2-methylpropoxy)ethoxy)ethoxy)ethyl ester (9CI) **(53535-32-3)**
Acetic acid, (2,4,5-trichlorophenoxy)-, methyl-2-(methyl-2-(2-methylpropoxy)ethoxy)ethyl ester (9CI) **(53535-31-2)**
Acetic acid, (2,4,5-trichlorophenoxy)-, 2-methylpropyl ester **(4938-72-1)**
Acetic acid, (2,4,5-trichlorophenoxy)-, 1-methylpropyl ester (9CI) **(61792-07-2)**
Acetic acid, (2,4,5-trichlorophenoxy)-, pentyl ester (8CI,9CI) **(120-39-8)**
Acetic acid, (2,4,5-trichlorophenoxy)-, sodium salt **(13560-99-1)**
Acetic acid, (2,3,6-trichlorophenyl) **(85-34-7)**
Acetic acid, ((3,5,6-trichloro-2-pyridinyl)oxy)-, Compd. with N,N-diethylethanamine (1:1) **(57213-69-1)**
Acetic acid, ((3,5,6-trichloro-2-pyridinyl)oxy)-, 2-butoxyethyl ester **(64700-56-7)**
Acetic acid, ((3,5,6-trichloro-2-pyridinyl)oxy)-, methyl ester (9CI) **(60825-26-5)**
Acetic acid, (3,5,6-trichloro-2-pyridyloxy) **(55335-06-3)**
Acetic acid, trichloro-, sodium salt **(650-51-1)**
Acetic acid, triethanolamine salt **(14806-72-5)**
Acetic acid, triethylene glycol diester **(111-21-7)**
Acetic acid, trifluoro **(76-05-1)**
Acetic acid, trifluoro-, anhydride **(407-25-0)**
Acetic acid, trifluoro-, ethyl ester (9CI) **(383-63-1)**
Acetic acid, trifluoro-, methyl ester (9CI) **(431-47-0)**
Acetic acid, trifluoro-, silver(1+) salt (9CI) **(2966-50-9)**
Acetic acid, trimethyl- **(75-98-9)**
Acetic acid, ureido- **(462-60-2)**
Acetic acid, vetiverol ester **(117-98-6)**
Acetic acid, vinyl ester **(108-05-4)**
Acetic acid, vinyl ester, chloroethylene copolymer **(34149-92-3)**
Acetic acid, vinyl ester, Polymer **(9003-20-7)**
Acetic acid, vinyl ester, Polymer with chloroethylene **(9003-22-9)**
Acetic acid, zinc(II) salt **(557-34-6)**
Acetic acid, zinc salt (8CI,9CI) **(557-34-6)**
Acetic acid, zinc salt, dihydrate **(5970-45-6)**
Acetic acid, zirconium salt (9CI) **(7585-20-8)**
Acetic aldehyde **(75-07-0)**
Acetic-anhydride **(108-24-7)**
Acetic anhydride (ACGIH,OSHA) [UN 1715] **(108-24-7)**
Acetic chloride **(75-36-5)**
Acetic-4-chloroanilide **(539-03-7)**
Acetic ether **(141-78-6)**
Acetic oxide **(108-24-7)**
Acetic peroxide **(79-21-0)**
Aceticyl **(50-78-2)**
Acetidin **(141-78-6)**
Acetile Diazo Black N **(539-17-3)**
Acetile Diazo Black R **(539-17-3)**
Acetilmetadol (Spanish) **(509-74-0)**
Acetilsalicilico **(50-78-2)**
Acetilum acidulatum **(50-78-2)**
Acetiluro de mercurio (Spanish) **(68833-55-6)**
Acetiluro de plata (Spanish) **(13092-75-6)**
Acetimidic acid **(60-35-5)**
Acetimidic acid, N-((methylcarbamoyl)oxy)thio-, methyl ester (8CI) **(16752-77-5)**
Acetimidic acid, thio-N-((methylcarbamoyl)oxy)-, methyl ester **(16752-77-5)**
Acetimidoquinone **(50700-49-7)**
Acetimidothioic acid, methyl-, N-(methylcarbamoyl) ester **(16752-77-5)**
Acetimidoyl chloride, N,N'-(6-chloro-s-triazine-2,4-diyl)-bis(2,2,2-trichloro- (7CI,8CI) **(10243-83-1)**
Acetimidoyl chloride, 2,2,2-trichloro-N-(4,6-dichloro-s-triazin-2-yl)- (7CI,8CI) **(10243-82-0)**
Acetin **(26446-35-5)**
Acetin, di **(25395-31-7)**
Acetin, mono **(26446-35-5)**
Acetin, tri **(102-76-1)**
Acetisal **(50-78-2)**
Aceto DIPP **(93-46-9)**
Aceto DNPT 40 **(101-25-7)**
Aceto DNPT 80 **(101-25-7)**
Aceto DNPT 100 **(101-25-7)**
Aceto HMT **(100-97-0)**
Aceto PBN **(135-88-6)**
Aceto Pan **(90-30-2)**
Aceto SDD 40 **(128-04-1)**
Aceto TETD **(137-26-8)**
Aceto TMTM **(97-74-5)**
Aceto ZDBD **(136-23-2)**
Aceto ZDED **(137-30-4)**
Aceto ZDMD **(137-30-4)**
Acetoacet-o-anisidin (Czech) **(92-15-9)**
Acetoacet-o-chloranilide **(93-70-9)**
Acetoacet-o-chloroanilide **(93-70-9)**
Acetoacet-4-chloro-2-methylanilide **(20139-55-3)**
Acetoacet-2,5-dimethoxyanilide **(6375-27-5)**
Acetoacet-2,4-dimethylphenyl **(97-36-9)**
Acetoacet-monomethylamide **(20306-75-6)**
Acetoacetamide, N,N-diethyl **(2235-46-3)**
Acetoacetamide, N,N-dimethyl **(2044-64-6)**
Acetoacetamidobenzene **(102-01-2)**
Acetoacetanilid **(102-01-2)**
Acetoacetanilide **(102-01-2)**
Acetoacetanilide, 2'-chloro **(93-70-9)**
Acetoacetanilide, 4'-chloro **(101-92-8)**
Acetoacetanilide, o-chloro- **(93-70-9)**
Acetoacetanilide, p-chloro- **(101-92-8)**
Acetoacetanilide, 4'-chloro-2',5'-dimethoxy- (8CI) **(4433-79-8)**
Acetoacetanilide, 2,2'-((3,3'-dichloro-4,4'-biphenylylene)diazo)bis **(6358-85-6)**
Acetoacetanilide, 2',5'-dimethoxy- (8CI) **(6375-27-5)**
Acetoacetanilide, 2',4'-dimethyl **(97-36-9)**
Acetoacetanilide, 4'-ethoxy- **(122-82-7)**
o-Acetoacetanisidide (8CI) **(92-15-9)**
o-Acetoacetanisidide, 4'-chloro **(20139-55-3)**
Acetoacetate **(541-50-4)**
Acetoacet-p-chloroanilide **(101-92-8)**
Acetoacetic acid **(541-50-4)**
Acetoacetic acid anilide **(102-01-2)**
Acetoacetic acid o-anisidide **(92-15-9)**
Acetoacetic acid, butyl ester **(591-60-6)**
Acetoacetic acid, tert-butyl ester (8CI) **(1694-31-1)**
Acetoacetic acid, 2-chloro-, ethyl ester (8CI) **(609-15-4)**
Acetoacetic acid, 2-ethyl- (8CI) **(4433-85-6)**
Acetoacetic acid, ethyl ester **(141-97-9)**
Acetoacetic acid, methyl ester **(105-45-3)**
Acetoacetic acid, p-phenetidide **(122-82-7)**
Acetoacetic acid, 4,4,4-trifluoro-, ethyl ester **(372-31-6)**
Acetoacetic anilide **(102-01-2)**
Acetoacetic ester **(141-97-9)**
Acetoacetic methyl ester **(105-45-3)**
o-Acetoacetochloranilide **(93-70-9)**
Acetoacetone **(123-54-6)**

p-Acetoacetophenetidide	**(122-82-7)**
p-Acetoacetophenetidide, 2,2''-((2,2'-dichloro-4,4'-biphenylylene)-bis(azo))bis- (8CI)	**(20139-66-6)**
Acetoacetophenone	**(93-91-4)**
o-Acetoacetotoluidide	**(93-68-5)**
p-Acetoacetotoluidide (8CI)	**(2415-85-2)**
2',4'-Acetoacetoxylidide	**(97-36-9)**
2,4-Acetoacetoxylidide	**(97-36-9)**
2',4'-Acetoacetoxylidide, 2,2''-((2,2',5,5'-tetrachloro-4,4'-biphenylene)bis(azo))bis-	**(22094-93-5)**
Acetoacet-p-phenetidide	**(122-82-7)**
Acetoacet-o-toluidide	**(93-68-5)**
Acetoacet-ortho-toluidide	**(93-68-5)**
Acetoaceto-m-xylidide	**(97-36-9)**
2-Acetoacetylaminoanisole	**(92-15-9)**
((Acetoacetyl)amino)benzene	**(102-01-2)**
2-Acetoacetylaminotoluene	**(93-68-5)**
Acetoacetylaniline	**(102-01-2)**
Acetoacetyl-o-aniside	**(92-15-9)**
Acetoacetyl-o-anisine	**(92-15-9)**
Acetoacetyl-2-chloroanilide	**(93-70-9)**
Acetoacetyl-4-chloroanilide	**(101-92-8)**
Acetoacetyl-2,5-dimethoxyanilide	**(6375-27-5)**
Acetoacetyl-2-methylanilide	**(93-68-5)**
Acetoacetyl-m-xylidide	**(97-36-9)**
Acetoacet-m-xylidide	**(97-36-9)**
p-Acetoaminoaniline	**(122-80-5)**
o-Acetoaminobenzoic acid	**(89-52-1)**
p-Acetoaminobenzoic acid	**(556-08-1)**
Acetoaminofluorene	**(53-96-3)**
Acetoanilide	**(103-84-4)**
Aceto-p-anisidide	**(51-66-1)**
Acetoarsenite de cuivre (French)	**(12002-03-8)**
Aceto azib	**(78-67-1)**
α-Acetobutyrolactone	**(517-23-7)**
Acetochlor	**(34256-82-1)**
Acetochlore (French)	**(34256-82-1)**
Acetocid	**(144-80-9)**
Acetoctan ethylnaty (Czech)	**(141-97-9)**
Acetofenon (Czech)	**(98-86-2)**
Acetoferrocene	**(1271-55-2)**
Acetofos	**(2425-25-4)**
Acetoguaiacon	**(498-02-2)**
Acetoguaiacone	**(498-02-2)**
Acetoguanamine	**(542-02-9)**
Acetoguanamine, phenyl-	**(1853-88-9)**
Acetoguanide	**(16352-06-0)**
Acetohexamide	**(968-81-0)**
Acetohydroxamic acid	**(546-88-3)**
Acetohydroxamic acid, N-fluoren-2-yl	**(53-95-2)**
Acetohydroximic acid	**(546-88-3)**
Acetoin	**(513-86-0)**
Acetol	**(105-57-7)**
Acetol	**(116-09-6)**
Acetol	**(50-78-2)**
Acetomethoxan	**(828-00-2)**
Acetomethoxane	**(828-00-2)**
Acetomorfine	**(561-27-3)**
Acetomorphine	**(561-27-3)**
Aceton (German, Dutch, Polish)	**(67-64-1)**
Acetonanil	**(147-47-7)**
Acetonanil	**(26780-96-1)**
Acetonanyl	**(147-47-7)**
1-Acetonaphthalene	**(941-98-0)**
β-Acetonaphthalene	**(93-08-3)**
1'-Acetonaphthone	**(941-98-0)**
1-Acetonaphthone	**(941-98-0)**
2'-Acetonaphthone	**(93-08-3)**
2-Acetonaphthone	**(93-08-3)**
Acetonaphthone	**(93-08-3)**
α-Acetonaphthone	**(941-98-0)**
β-Acetonaphthone	**(93-08-3)**
2'-Acetonaphthone, 5',6',7',8'-tetrahydro-3',5',5',6',8',8'-hexamethyl- (VAN) (8CI)	**(1506-02-1)**
Acetonbromoform	**(76-08-4)**
Acetoncianhidrinei (Romanian)	**(75-86-5)**
Acetoncianidrina (Italian)	**(75-86-5)**
Acetoncyaanhydrine (Dutch)	**(75-86-5)**
Acetoncyanhydrin (German)	**(75-86-5)**
Acetone (ACGIH,OSHA) [UN 1090]	**(67-64-1)**
Acetone, acetonyl-	**(110-13-4)**
Acetone, acetyl-	**(123-54-6)**
Acetone anil	**(147-47-7)**
Acetone, bis(ethyl sulfone)	**(115-24-2)**
Acetone-bromoform	**(76-08-4)**
Acetone O-carbaniloyloxime	**(2828-42-4)**
Acetone, chloro-	**(78-95-5)**
Acetone chloroform	**(57-15-8)**
Acetone, 1-chloro-1,1,3,3,3-pentafluoro-	**(79-53-8)**
Acetonecyanhydrine (French)	**(75-86-5)**
Acetone cyanohydrin (DOT)	**(75-86-5)**
Acetone cyanohydrin, Stabilized [UN 1541]	**(75-86-5)**
Acetone, cyclic (hydroxymethyl)ethylene acetal	**(100-79-8)**
Acetone-d6 (8CI)	**(666-52-4)**
Acetone, diethyl acetal	**(126-84-1)**
Acetone diethyl ketal	**(126-84-1)**
Acetone diethylsulfone	**(115-24-2)**
Acetone, dimethyl acetal (8CI)	**(77-76-9)**
Acetone dimethyl ketal	**(77-76-9)**
Acetone, dimethyl mercaptole (8CI)	**(6156-18-9)**
Acetone, hexachloro-	**(116-16-5)**
Acetone, hexafluoro-	**(684-16-2)**
Acetone, hexafluoro-, sesquihydrate	**(13098-39-0)**
Acetone, hexafluoro-, trihydrate	**(34202-69-2)**
Acetone, 3-hydroxyimino-	**(306-44-5)**
Acetone, isopropylidene-	**(141-79-7)**
Acetone, methyl-	**(78-93-3)**
Acetone methylallyl	**(25659-22-7)**
Acetone, methylene-	**(78-94-4)**
Acetone, monochloropentafluoro-	**(79-53-8)**
Acetone, oxime	**(127-06-0)**
Acetone oxime N-phenylcarbamate	**(2828-42-4)**
Acetone, oxime, O-(phenylcarbamate)	**(2828-42-4)**
Acetone oxime phenylurethane	**(2828-42-4)**
Acetone, O-(phenylcarbamoyl)oxime (8CI)	**(2828-42-4)**
Acetone, semicarbazone	**(110-20-3)**
Acetone, thiosemicarbazone	**(1752-30-3)**
Acetonic acid	**(50-21-5)**
Acetonitril (German, Dutch)	**(75-05-8)**
Acetonitrile (ACGIH,DOT,OSHA)	**(75-05-8)**
Acetonitrile, (4-biphenylyl)-	**(31603-77-7)**
Acetonitrile, ((4,6-bis(ethylamino)-s-triazin-2-yl)thio)- (7CI,8CI)	**(30360-89-5)**
Acetonitrile, bromo- (8CI,9CI)	**(590-17-0)**
Acetonitrile, bromochloro	**(83463-62-1)**
Acetonitrile, (p-bromophenyl)	**(16532-79-9)**
Acetonitrile, bromophenyl	**(5798-79-8)**
Acetonitrile, chloro	**(107-14-2)**
Acetonitrile, (p-chlorophenyl)	**(140-53-4)**
Acetonitrile, dibromo	**(3252-43-5)**
Acetonitrile, dichloro	**(3018-12-0)**
Acetonitrile, (dimethylamino)	**(926-64-7)**
Acetonitrile, diphenyl	**(86-29-3)**
Acetonitrile, ((4-(ethylamino)-6-(isopropylamino)-s-triazin-2-yl)-thio)- (7CI,8CI)	**(30360-90-8)**

Acetonitrile, hydroxy **(107-16-4)**
Acetonitrile, (p-hydroxyphenyl) **(14191-95-8)**
Acetonitrile, hydroxyphenyl- **(532-28-5)**
Acetonitrile, (p-methoxypenyl) **(104-47-2)**
Acetonitrile, ((4-methyl-6-(trichloromethyl)-s-triazin-2-yl)-imino)di- (8CI) **(30388-90-0)**
Acetonitrile, morpholino- **(5807-02-3)**
Acetonitrile, (1-naphthyl) **(132-75-2)**
Acetonitrile, 2,2',2''-nitrilotris- (9CI) **(7327-60-8)**
Acetonitrile, nitrilotri- (8CI) **(7327-60-8)**
Acetonitrile, (o-nitrophenyl) **(610-66-2)**
Acetonitrile, phenyl **(140-29-4)**
Acetonitrile, trichloro **(545-06-2)**
Acetonitrile, trichloro-, trimer **(6542-67-2)**
Acetonitrile, trinitro- **(630-72-8)**
Acetonitrile, vinyl- **(109-75-1)**
Acetonkyanhydrin (Czech) **(75-86-5)**
Acetonoxime **(127-06-0)**
Acetonthiosemikarbazon (Czech) **(1752-30-3)**
Acetonyl **(50-78-2)**
Acetonyl acetone **(110-13-4)**
3-(Acetonylbenzyl)-4-hydroxycoumarin **(81-81-2)**
3-(α-Acetonylbenzyl)-4-hydroxycoumarin **(81-81-2)**
3-(α-Acetonylbenzyl)-4-hydroxy-coumarin sodium salt **(129-06-6)**
Acetonyl bromide **(598-31-2)**
Acetonyl chloride **(78-95-5)**
3-(α-Acetonyl-4-chlorobenzyl)-4-hydroxycoumarin **(81-82-3)**
3-(α-Acetonyl-p-chlorobenzyl)-4-hydroxycoumarin **(81-82-3)**
3-(α-Acetonylfurfuryl)-4-hydroxycoumarin **(117-52-2)**
3-(α-Acetonylfurfuryl)-4-hydroxycoumarin sodium salt **(34490-93-2)**
3-(α-Acetonylfurfuryl)-4-hydroxycoumarin, sodium salt **(34490-93-2)**
3-(α-Acetonyl-4-nitrobenzyl)-4-hydroxycoumarin **(152-72-7)**
3-(α-Acetonyl-p-nitrobenzyl)-4-hydroxy-coumarin **(152-72-7)**
Acetophen **(50-78-2)**
p-Acetophenetide **(62-44-2)**
p-Acetophenetide, 3'-nitro- **(1777-84-0)**
m-Acetophenetidide **(591-33-3)**
p-Acetophenetidide **(62-44-2)**
para-Acetophenetidide **(62-44-2)**
p-Acetophenetidide, 3-nitro **(1777-84-0)**
Acetophenetidin **(62-44-2)**
Acetophenetidine **(62-44-2)**
p-Acetophenetidine **(62-44-2)**
Acetophenetin **(62-44-2)**
Acetophenon **(98-86-2)**
Acetophenone **(98-86-2)**
Acetophenone, 3'-amino **(99-03-6)**
Acetophenone, m-amino- **(99-03-6)**
Acetophenone, p-amino **(99-92-3)**
Acetophenone, p-amino- **(99-92-3)**
Acetophenone, 2'-amino- (8CI) **(551-93-9)**
Acetophenone, 4'-amino- (8CI) **(99-92-3)**
Acetophenone, 2-bromo **(70-11-1)**
Acetophenone, 3'-bromo- (8CI) **(2142-63-4)**
Acetophenone, 4'-bromo- (8CI) **(99-90-1)**
Acetophenone, 4'-tert-butyl- (8CI) **(943-27-1)**
Acetophenone, 4'-tert-butyl-2',6'-dimethyl-3',5'-dinitro- (8CI) **(81-14-1)**
Acetophenone, 2-chloro **(532-27-4)**
Acetophenone, 4'-chloro- **(99-91-2)**
Acetophenone, 2-chloro-m-nitro **(99-47-8)**
Acetophenone, 2,4'-dibromo **(99-73-0)**
Acetophenone, 2',4'-dichloro- (8CI) **(2234-16-4)**
Acetophenone, 2',4'-dihydroxy **(89-84-9)**
Acetophenone, 2',5'-dihydroxy **(490-78-8)**
Acetophenone, 2',6'-dihydroxy-4'-methoxy- (8CI) **(7507-89-3)**
Acetophenone, 2',3'-dihydroxy-4'-methoxy- (6CI,7CI,8CI) **(708-53-2)**
Acetophenone, 2',6'-dimethoxy **(2040-04-2)**
Acetophenone, 3',4'-dimethoxy- (8CI) **(1131-62-0)**
Acetophenone, 3,5-dimethoxy-4-hydroxy **(2478-38-8)**
Acetophenone, 2',4'-dimethyl- (8CI) **(89-74-7)**
Acetophenone, 2',6'-dimethyl- (8CI) **(2142-76-9)**
Acetophenone, 2',3'-dimethyl- (6CI,7CI,8CI) **(2142-71-4)**
Acetophenone, 2'-ethyl- (8CI) **(2142-64-5)**
Acetophenone, 3'-ethyl- (8CI) **(22699-70-3)**
Acetophenone, 4'-ethyl- (8CI) **(937-30-4)**
Acetophenone, 3'-fluoro- (8CI) **(455-36-7)**
Acetophenone, 4'-fluoro- (8CI) **(403-42-9)**
Acetophenone, 4'-hydroxy- **(99-93-4)**
Acetophenone, o-hydroxy **(118-93-4)**
Acetophenone, p-hydroxy **(99-93-4)**
Acetophenone, 2'-hydroxy- (8CI) **(118-93-4)**
Acetophenone, 3'-hydroxy- (8CI) **(121-71-1)**
Acetophenone, 2-hydroxy-, acetate (8CI) **(2243-35-8)**
Acetophenone, 4'-hydroxy-3',5'-dimethoxy- (8CI) **(2478-38-8)**
Acetophenone, 4'-hydroxy-3'-methoxy **(498-02-2)**
Acetophenone, 2'-hydroxy-5'-methoxy-4'-methyl- (6CI,7CI,8CI) **(4223-84-1)**
Acetophenone, 4'-(1-hydroxy-1-methylethyl)- **(54549-72-3)**
Acetophenone, 2-(6-(β-hydroxyphenethyl)-1-methyl-2-piperidyl)-, hydrochloride **(63990-84-1)**
Acetophenone, 2-hydroxy-2-phenyl- **(119-53-9)**
Acetophenone, 4'-isopropyl- (8CI) **(645-13-6)**
Acetophenone, 2'-isopropyl- (6CI,7CI,8CI) **(2142-65-6)**
Acetophenone, 4'-methoxy **(100-06-1)**
Acetophenone, 3'-methoxy- (8CI) **(586-37-8)**
Acetophenone, 4'-methyl **(122-00-9)**
Acetophenone, 2'-methyl- (8CI) **(577-16-2)**
Acetophenone, 3'-methyl- (8CI) **(585-74-0)**
Acetophenone, 2'-nitro **(577-59-3)**
Acetophenone, 3'-nitro **(121-89-1)**
Acetophenone, 4'-nitro **(100-19-6)**
Acetophenone, 4'-(p-nitrophenyl)- (8CI) **(135-69-3)**
Acetophenone, 2,2,2',4',5'-pentachloro- **(1203-86-7)**
Acetophenone, 2-phenyl **(451-40-1)**
Acetophenone, 4'-phenyl- (8CI) **(92-91-1)**
Acetophenone, 2,2,2-trifluoro- (8CI) **(434-45-7)**
Acetophenone, 2',3',4'-trihydroxy **(528-21-2)**
Acetophenone, 2',4',6'-trimethyl- (8CI) **(1667-01-2)**
Acetophos **(2425-25-4)**
Acetopropionic acid **(123-76-2)**
Acetopropyl acetate **(5185-97-7)**
Acetopurpurine 8B **(6548-29-4)**
3-Acetopyridine **(350-03-8)**
Acetopyruvate **(5699-58-1)**
Acetoquat CPB **(140-72-7)**
Acetoquat CPC **(123-03-5)**
Acetoquat CTAB **(57-09-0)**
Acetoquinone Blue L **(2475-45-8)**
Acetoquinone Blue R **(2475-45-8)**
Acetoquinone Light Gooseberry RL **(116-85-8)**
Acetoquinone Light Heliotrope NL **(128-95-0)**
Acetoquinone Light Orange JL **(82-28-0)**
Acetoquinone Light Pure Blue R **(2475-46-9)**
Acetoquinone Light Pure Blue R **(86722-66-9)**
Acetoquinone Light Violet N **(1220-94-6)**
Acetoquinone Light Yellow 2RZ **(119-15-3)**
Acetoquinone Light Yellow **(2832-40-8)**
Acetoquinone Light Yellow 4JLZ **(2832-40-8)**
Acetorfin (Czech) **(62-67-9)**
Acetorfina **(25333-77-1)**
Acetorphine **(25333-77-1)**
Acetorphine **(62-67-9)**
Acetorphinum **(25333-77-1)**
Acetosal **(50-78-2)**
Acetosalic acid **(50-78-2)**

Acetosalin **(50-78-2)**
Acetose **(9004-35-7)**
Acetospan **(76-25-5)**
Acetosulfam **(33665-90-6)**
Acetosulfamin **(144-80-9)**
Acetosulfamine **(144-80-9)**
Acetosyringone **(2478-38-8)**
2-Acetothienone **(88-15-3)**
Acetothioamide **(62-55-5)**
2-Acetothiophene **(88-15-3)**
Acetotoluide **(537-92-8)**
4-Acetotoluidide **(103-89-9)**
m-Acetotoluidide **(537-92-8)**
o-Acetotoluidide **(120-66-1)**
p-Acetotoluidide **(103-89-9)**
o-Acetotoluidide, 2-bromo-6'-tert-butyl-N-(methoxymethyl)- (8CI) **(2163-81-7)**
o-Acetotoluidide, 6'-tert-butyl-2-chloro **(3785-20-4)**
o-Acetotoluidide, 2-chloro-N-(ethoxymethyl)-6'-ethyl- **(34256-82-1)**
o-Acetotoluidide, 2-chloro-6'-ethyl-N-(2-methoxy-1-methylethyl) **(51218-45-2)**
o-Acetotoluidide, 2-chloro-N-(2-methoxyethyl) **(50563-41-2)**
m-Acetotoluidide, 6'-hydroxy- (8CI) **(6375-17-3)**
m-Acetotoluidide, α,α,α-trifluoro **(351-36-0)**
Acetovanillone **(498-02-2)**
Acetovanilone **(498-02-2)**
Acetovanyllon **(498-02-2)**
Acetoxime **(127-06-0)**
Acetoxon **(2425-25-4)**
2-Acetoxy-3-(N,N-diethylcarboxamido)-9,10-dimethoxy-1,2 ,3,4,6,7-hexahydro-11bh-benzopyridocoline **(63-12-7)**
2-Acetoxybenzoic acid **(50-78-2)**
o-Acetoxybenzoic acid **(50-78-2)**
4-Acetoxybicyclo(2.2.2)octan-1-ol **(54774-94-6)**
p-Acetoxychlorobenzene **(876-27-7)**
17-Acetoxy-6-chloro-6-dehydroprogesterone **(302-22-7)**
17-α-Acetoxy-6-chloro-6,7-dehydroprogesterone **(302-22-7)**
17-α-Acetoxy-6-chloro-6-dehydroprogesterone **(302-22-7)**
17-α-Acetoxy-6-chloro-4,6-pregnadiene-3,20-dione **(302-22-7)**
17-α-Acetoxy-6-chloropregna-4,6-diene-3,20-dione **(302-22-7)**
17-α-Acetoxy-6-dehydro-6-methylprogesterone **(595-33-5)**
2-Acetoxy-3-diethylcarbamyl-9,10-dimethoxy-1,2,3,4,6,7-hexahydro-11b-benzo(a)quinolizine **(63-12-7)**
3-Acetoxy-6-dimethylamino-4,4-diphenylheptane **(509-74-0)**
5-Acetoxy-2-dimethylamino-4,4-diphenylheptane **(509-74-0)**
6-Acetoxy-2,4-dimethyl-m-dioxane **(828-00-2)**
α-Acetoxy dimethylnitrosamine **(56856-83-8)**
Acetoxyethane **(141-78-6)**
Acetoxyethyl chloride **(542-58-5)**
1-Acetoxyethylene **(108-05-4)**
1-Acetoxyhexadecane **(629-70-9)**
6-Acetoxy-2-hexanone **(4305-26-4)**
21-Acetoxy-17,α-hydroxypregn-4-ene-3,11,20-trione **(50-04-4)**
21-Acetoxy-17,α-hydroxy-3,11,20-triketopregnene-4 **(50-04-4)**
Acetoxyl **(94-36-0)**
2',4'-Acetoxylidide **(2050-43-3)**
2,6-Acetoxylidide, 2-chloro-N-(2-methoxyethyl)- **(50563-36-5)**
2',6'-Acetoxylidide, 2-(diethylamino) **(137-58-6)**
2',4'-Acetoxylidine **(2050-43-3)**
(Acetoxymercuri)benzene **(62-38-4)**
1-Acetoxy-2-methoxy-4-allylbenzene **(93-28-7)**
2-(Acetoxymethyl)-1-butene **(55670-09-2)**
Acetoxymethylbutylnitrosamine **(56986-36-8)**
N-(Acetoxy)methyl-N-n-butylnitrosamine **(56986-36-8)**
2-Acetoxymethylfuran **(623-17-6)**
Acetoxymethyl-methyl-nitrosamin (German) **(56856-83-8)**
Acetoxymethyl methylnitrosamine **(56856-83-8)**
N-α-Acetoxymethyl-N-methylnitrosamine **(56856-83-8)**
17-Acetoxy-6-methylpregna-4,6-diene-3,20-dione **(595-33-5)**
17-α-Acetoxy-6-methyl-4,6-pregnadiene-3,20-dione **(595-33-5)**
17-α-Acetoxy-6-methylpregna-4,6-diene-3,20-dione **(595-33-5)**
17-α-Acetoxy-6-α-methylpregn-4-ene-3,20-dione **(71-58-9)**
17-Acetoxy-6-α-methylprogesterone **(71-58-9)**
17-α-Acetoxy-6-α-methylprogesterone **(71-58-9)**
Acetoxymethylpropylnitrosamine **(66017-91-2)**
N-(Acetoxy)methyl-N-n-propylnitrosamine **(66017-91-2)**
p-Acetoxynitrobenzene **(830-03-5)**
1-Acetoxy-N-nitrosodimethylamine **(56856-83-8)**
17-Acetoxy-19-nor-17-α-pregn-4-en-20-yn-3-one **(51-98-9)**
17-β-Acetoxy-19-nor-17-α-pregn-4-en-20-yn-3-one **(51-98-9)**
1-Acetoxyoctadiene **(3491-27-8)**
2-Acetoxypentane **(626-38-0)**
4-Acetoxy-3-pentyltetrahydropyran **(18871-14-2)**
3-Acetoxyphenol **(102-29-4)**
Acetoxyphenylmercury **(62-38-4)**
1-Acetoxypropane **(109-60-4)**
2-Acetoxypropane **(108-21-4)**
2-Acetoxy-1,2,3-propanetricarboxylic acid tributyl ester **(77-90-7)**
1-Acetoxy-2-propanone **(592-20-1)**
3-Acetoxypropene **(591-87-7)**
1'-Acetoxysafrole-2',3'-oxide **(59901-90-5)**
4-Acetoxytoluene **(140-39-6)**
α-Acetoxytoluene **(140-11-4)**
o-Acetoxytoluene **(533-18-6)**
p-Acetoxytoluene **(140-39-6)**
Acetoxytriphenyllead **(1162-06-7)**
Acetoxy-triphenyl-stannan (German) **(900-95-8)**
Acetoxy-triphenylstannane **(900-95-8)**
Acetozalamide **(59-66-5)**
Acetparamin **(122-80-5)**
Acet-p-phenalide **(62-44-2)**
Aceto-para-phenalide **(62-44-2)**
Aceto-para-phenetidide **(62-44-2)**
Acet-p-phenetidin **(62-44-2)**
Acetphenetidin **(62-44-2)**
p-Acetphenetidin **(62-44-2)**
Aceto-4-phenetidine **(62-44-2)**
Acetpyrogall **(525-52-0)**
Aceturic acid **(543-24-8)**
Acetyl 35 **(9004-35-7)**
Acetyl Red B **(3567-66-6)**
Acetyl Red G **(3734-67-6)**
Acetyl Red J **(3734-67-6)**
Acetyl Rose 2GL **(3734-67-6)**
Acetylacetanilide **(102-01-2)**
α-Acetylacetanilide **(102-01-2)**
Acetylacetone **(123-54-6)**
Acetylacetone cobalt(II) **(14024-48-7)**
Acetyl acetone peroxide, Maximum concentration 40% in solution (DOT) **(37187-22-7)**
Acetyl acetone peroxide, With more than 9% active oxygen (DOT) **(37187-22-7)**
Acetyl acetone peroxide, With not more than 9% by weight active oxygen (DOT) **(37187-22-7)**
2-Acetylacetophenone **(93-91-4)**
α-Acetylacetophenone **(93-91-4)**
N-(Acetylacetyl)aniline **(102-01-2)**
Acetyladriamycin **(20830-81-3)**
Acetylalanine **(97-69-8)**
N-Acetyl-S-alanine **(97-69-8)**
N-Acetylalanine **(97-69-8)**
Acetylaldehyde **(75-07-0)**
Acetylaminoacetic acid **(543-24-8)**
3-(Acetylamino)-5-((acetylamino)methyl)-2,4,6-triiodobenzoic acid **(440-58-4)**
3-Acetylaminoaniline **(102-28-3)**
4-(Acetylamino)aniline **(122-80-5)**

m-(Acetylamino)aniline **(102-28-3)**
p-(Acetylamino)aniline **(122-80-5)**
4-Acetylaminoazobenzene **(4128-71-6)**
1-Acetyl-2-aminobenzene **(551-93-9)**
Acetylaminobenzene **(103-84-4)**
p-Acetylaminobenzenesulfochloride **(121-60-8)**
N-Acetyl-4-aminobenzenesulfonamide **(144-80-9)**
4-(Acetylamino)benzenesulfonic acid **(121-62-0)**
4-(Acetylamino)benzenesulfonyl chloride **(121-60-8)**
p-Acetylaminobenzenesulfonyl chloride **(121-60-8)**
2-(Acetylamino)benzoic acid **(89-52-1)**
4-(Acetylamino)benzoic acid **(556-08-1)**
4-Acetylaminobenzoic acid **(556-08-1)**
N-Acetyl-p-aminobenzoic acid **(556-08-1)**
N-Acetylaminobenzoic acid **(89-52-1)**
p-Acetylaminobenzoic acid **(556-08-1)**
4-Acetylaminobiphenyl **(4075-79-0)**
2-Acetylaminoethanol **(142-26-7)**
S-(2-(Acetylamino)ethyl) O,O-dimethyl phosphorodithioate **(13265-60-6)**
o-Acetylaminofenol (Czech) **(614-80-2)**
2-Acetylamino-fluoren (German) **(53-96-3)**
4-Acetylaminofluoren (German) **(28322-02-3)**
2-(Acetylamino)fluorene **(53-96-3)**
4-Acetylaminofluorene **(28322-02-3)**
N-Acetyl-2-aminofluorene **(53-96-3)**
2-Acetylaminofluorine (OSHA) **(53-96-3)**
2-((Acetylamino)methyl)-2-chloro-N-(2,6-diethylphenyl)-acetamide **(40164-67-8)**
N-((Acetylamino)methyl)-2-chloro-N-(2,6-diethylphenyl)-acetamide **(40164-67-8)**
2-Acetylamino-4-(5-nitro-2-furyl)thiazole **(531-82-8)**
p-(Acetylamino)phenacyl chloride **(140-49-8)**
2-(Acetylamino)phenol **(614-80-2)**
3-(Acetylamino)phenol **(621-42-1)**
N-Acetyl-2-aminophenol **(614-80-2)**
N-Acetyl-p-aminophenol **(103-90-2)**
m-(Acetylamino)phenol **(621-42-1)**
o-(Acetylamino)phenol **(614-80-2)**
p-Acetylaminophenol **(103-90-2)**
N^1-Acetyl-4-aminophenylsulfonamide **(144-80-9)**
2-Acetylamino-1,3,4-thiadiazole-5-sulfonamide **(59-66-5)**
4-(Acetylamino)toluene **(103-89-9)**
N-Acetyl-N-(3-amino-2,4,6-triiodophenyl)-β-aminoisobutyric acid **(16034-77-8)**
3-(Acetyl-(3-amino-2,4,6-triiodophenyl)amino)-2-methyl-propanoic acid **(16034-77-8)**
N-Acetyl-N-(3-amino-2,4,6-triiodophenyl)-2-methyl-β-alanine **(16034-77-8)**
Acetyl anhydride **(108-24-7)**
2-Acetylaniline **(551-93-9)**
3-Acetylaniline **(99-03-6)**
4-Acetylaniline **(99-92-3)**
Acetylaniline **(103-84-4)**
N-Acetylaniline **(103-84-4)**
m-Acetylaniline **(99-03-6)**
o-Acetylaniline **(551-93-9)**
p-Acetylaniline **(99-92-3)**
Acetyl-p-anisidine **(51-66-1)**
4-Acetylanisole **(100-06-1)**
p-Acetylanisole **(100-06-1)**
Acetylanthranilic acid **(89-52-1)**
N-Acetylanthranilic acid **(89-52-1)**
Acetylasulam **(1773-37-1)**
Acetylated myrcene **(68412-04-4)**
Acetylbenzene **(98-86-2)**
1-(p-Acetylbenzenesulfonyl)-3-cyclohexylurea **(968-81-0)**
N-Acetylbenzidine **(3366-61-8)**
N-Acetyl-p-benzoquinone imine **(50700-49-7)**
Acetylbenzoyl **(579-07-7)**
Acetylbenzoylmethane **(93-91-4)**
Acetyl benzoyl peroxide, Maximum concentration 45% in solution (DOT) **(644-31-5)**
Acetyl benzoyl peroxide, More than 40% in solution (DOT) **(644-31-5)**
Acetyl benzoyl peroxide, Solid (DOT) **(644-31-5)**
Acetyl benzoyl peroxide, Solution, Not over 40% peroxide (DOT) **(644-31-5)**
N-Acetylbenzylamine **(588-46-5)**
4-Acetylbiphenyl **(92-91-1)**
p-Acetylbiphenyl **(92-91-1)**
1-Acetyl-3,3-bis(4-(acetyloxy)phenyl)-1,3-dihydro-2H-indol-2-one **(18869-73-3)**
1-Acetyl-3,3-bis(p-hydroxyphenyl)oxindole diacetate **(18869-73-3)**
Acetyl bromide [UN 1716] **(506-96-7)**
N-Acetyl-N-3-butinyl-2-chloranilin (Czech) **(21267-72-1)**
o-Acetyl-2-sec-butyl-4,6-dinitrophenol **(2813-95-8)**
N-Acetylbutyramide **(22534-71-0)**
2-Acetyl-γ-butyrolactone **(517-23-7)**
2-Acetylbutyrolactone **(517-23-7)**
α-Acetyl-γ-butyrolactone **(517-23-7)**
α-Acetylbutyrolactone **(517-23-7)**
9-Acetylcedr-8-ene **(32388-55-9)**
9-Acetyl-8-cedrene **(32388-55-9)**
Acetylcellulose **(9004-35-7)**
Acetyl chloride [UN 1717] **(75-36-5)**
Acetyl chloride, chloro **(79-04-9)**
Acetyl chloride, dichloro **(79-36-7)**
Acetyl chloride, fluoro **(359-06-8)**
Acetyl chloride, phenyl **(103-80-0)**
Acetyl chloride, trichloro **(76-02-8)**
Acetyl chloride, trifluoro- (9CI) **(354-32-5)**
Acetyl chloride, trimethyl **(3282-30-2)**
3-(1-Acetyl-2-(p-chlorophenyl)ethyl)-4-hydroxycoumarin **(81-82-3)**
Acetylcitric acid tributyl ester **(77-90-7)**
Acetylcolamine **(142-26-7)**
Acetyl-o-cresol **(533-18-6)**
1-Acetyl-1-cycloheptene **(14377-11-8)**
Acetyl cyclohexanepersulfonate **(3179-56-4)**
Acetyl cyclohexanesulfonyl peroxide **(3179-56-4)**
Acetyl cyclohexanesulfonyl peroxide (More than 82%, Wetted with less than 12% water) **(3179-56-4)**
Acetyl cyclohexanesulfonyl peroxide, More than 82% Wetted with less than 12% water **(3179-56-4)**
Acetyl cyclohexanesulfonyl peroxide (Not more than 32% in solution) **(3179-56-4)**
Acetyl cyclohexanesulfonyl peroxide, Not more than 32% in solution **(3179-56-4)**
Acetyl cyclohexanesulfonyl peroxide (Not more than 82%, Wetted with less than 12% water) **(3179-56-4)**
Acetyl cyclohexanesulfonyl peroxide, Not more than 82% Wet with not less than 12% water **(3179-56-4)**
2-Acetylcyclohexanone **(874-23-7)**
N-Acetylcyclohexylamine **(1124-53-4)**
Acetyl cyclohexylsulfonyl peroxide **(3179-56-4)**
Acetylcysteine **(616-91-1)**
N-Acetyl-L-cysteine **(616-91-1)**
N-Acetylcysteine **(616-91-1)**
Acetyldihydrocodeine **(3861-72-1)**
Acetyldihydrocodeinone **(466-90-0)**
3-Acetyl-2(3H)-4,5-dihydrofuranone **(517-23-7)**
3-Acetyldihydro-2(3H)-furanone **(517-23-7)**
Acetyldihydrokodein (Czech) **(3861-72-1)**
Acetyldihydrokodeinon (Czech) **(466-90-0)**
O-Acetyl-6-dimethylamino-4,4-diphenyl-3-heptanol **(509-74-0)**
Acetylen **(74-86-2)**
Acetylene Black **(1333-86-4)**
Acetylene, Liquid [UN 1001] **(74-86-2)**
Acetylene [UN 1001] **(74-86-2)**
Acetylene carbamide **(496-46-8)**

Name	CAS No.
Acetylenecarboxylic acid	**(471-25-0)**
Acetylenecarboxylic acid methyl ester	**(922-67-8)**
Acetylene, (o-chlorophenyl)phenyl- (6CI,7CI,8CI)	**(10271-57-5)**
Acetylene, dibromo-	**(624-61-3)**
Acetylenedicarboxylic acid (8CI)	**(142-45-0)**
Acetylenedicarboxylic acid, dimethyl ester	**(762-42-5)**
trans-Acetylene dichloride	**(156-60-5)**
Acetylene dichloride (OSHA)	**(540-59-0)**
Acetylene, dichloro	**(7572-29-4)**
Acetylene, diiodo-	**(624-74-8)**
Acetylene, diphenyl- (8CI)	**(501-65-5)**
Acetylene, dissolved [UN 1001]	**(74-86-2)**
Acetylenediurea	**(496-46-8)**
Acetylenediureine	**(496-46-8)**
Acetylene, methyl-	**(74-99-7)**
Acetylene, phenyl-	**(536-74-3)**
Acetylenepinacol	**(142-30-3)**
Acetylene tetrabromide (ACGIH,OSHA) [UN 2504]	**(79-27-6)**
Acetylene tetrachloride	**(79-34-5)**
Acetylene trichloride	**(79-01-6)**
Acetyleneurea	**(496-46-8)**
N-Acetyl ethanolamine	**(142-26-7)**
Acetyl ether	**(108-24-7)**
Acetyl ethylene	**(78-94-4)**
Acetyleugenol	**(93-28-7)**
N-Acetyl-m-fenylendiamin (Czech)	**(102-28-3)**
N-Acetyl-p-fenylendiamin (Czech)	**(122-80-5)**
1-Acetylferrocene	**(1271-55-2)**
Acetylferrocene	**(1271-55-2)**
N-Acetyl-N-2-fluorenylhydroxylamine	**(53-95-2)**
Acetyl fluoride, difluoro(fluorosulfonyl)- (9CI)	**(677-67-8)**
Acetylformaldehyde	**(78-98-8)**
Acetylformic acid	**(127-17-3)**
Acetylformyl	**(78-98-8)**
2-Acetylfuran	**(1192-62-7)**
Acetylfuran	**(1192-62-7)**
Acetylglycine	**(543-24-8)**
N-Acetylglycine	**(543-24-8)**
Acetylglycocoll	**(543-24-8)**
6-Acetyl-1,1,2,4,4,7-hexamethyl-1,2,3,4-tetrahydronaphthalene	**(1506-02-1)**
7-Acetyl-1,1,3,4,4,6-hexamethyltetrahydronaphthalene	**(1506-02-1)**
Acetyl hexamethyl tetralin	**(1506-02-1)**
N-Acetylhomocysteine thiolactone	**(1195-16-0)**
N-Acetylhomocysteinthiolakton (German)	**(1195-16-0)**
Acetyl hydroperoxide	**(79-21-0)**
2-Acetylhydroquinone	**(490-78-8)**
Acetylhydroquinone	**(490-78-8)**
Acetylhydroxamic acid	**(546-88-3)**
2-Acetyl-4-hydroxybutyric acid γ-lactone	**(517-23-7)**
α-Acetyl-γ-hydroxybutyric acid γ-lactone	**(517-23-7)**
3-Acetyl-4-hydroxy-6-methyl-2H-pyran-2-one	**(520-45-6)**
2-Acetylimino-3-methyl-δ^4-1,3,4-thiadiazoline-5-sulfonamide	**(554-57-4)**
5-Acetylimino-4-methyl-δ^2-1,3,4-thiadiazoline-2-sulfonamide	**(554-57-4)**
Acetylin	**(50-78-2)**
Acetyl iodide [UN 1898]	**(507-02-8)**
N-Acetyl-N'-isonicotinyl hydrazide	**(17433-31-7)**
Acetyllandromedol	**(4720-09-6)**
Acetyl mercaptan	**(507-09-5)**
N-Acetyl-3-mercaptoalanine	**(616-91-1)**
Acetylmethadol	**(509-74-0)**
Acetylmethadolum (Latin)	**(509-74-0)**
N-Acetyl-5-methoxytryptamine	**(73-31-4)**
N-Acetyl-N-methylacetamide	**(1113-68-4)**
Acetyl methyl bromide	**(598-31-2)**
Acetyl methyl carbinol [UN 2621]	**(513-86-0)**
o-Acetyl-β-methylcholine chloride	**(62-51-1)**
2-Acetyl-5-methyl-furan	**(1193-79-9)**
N-Acetyl-N-(2-methyl-4-((2-methylphenyl)azo)phenyl)acetamide	**(83-63-6)**
3-Acetyl-6-methyl-2H-pyran-2,4(3H)-dione	**(520-45-6)**
3-Acetyl-6-methyl-2H-pyran-2,4(3H)-dione. enol form	**(520-45-6)**
3-Acetyl-6-methyl-2,4-pyrandione	**(520-45-6)**
3-Acetyl-6-methylpyrandione-2,4	**(520-45-6)**
2-Acetyl-3-methylpyrazine	**(23787-80-6)**
N-Acetylmorfolin (Czech)	**(1696-20-4)**
4-Acetylmorpholine	**(1696-20-4)**
N-Acetylmorpholine	**(1696-20-4)**
1-Acetylnaphthalene	**(941-98-0)**
2-Acetylnaphthalene	**(93-08-3)**
β-Acetylnaphthalene	**(93-08-3)**
N-Acetyl-m-nitroaniline	**(122-28-1)**
m-Acetylnitrobenzene	**(121-89-1)**
p-Acetylnitrobenzene	**(100-19-6)**
4-Acetyl-4'-nitrobiphenyl	**(135-69-3)**
Acetylon Fast Blue G	**(2475-45-8)**
Acetylon Fast Pink B	**(116-85-8)**
Acetylon Fast Red Violet R	**(128-95-0)**
Acetyl oxide	**(108-24-7)**
Acetyloxirane	**(4401-11-0)**
(3β,6β)-6-(Acetyloxy)-3-(β-D-glucopyranosyloxy)-8,14-dihydroxybufa-4,20,22-trienolide	**(507-60-8)**
3-β,6-β-6-Acetyloxy-3-(β-D-glucopyranosyloxy)-8,14-dihydroxybufa-4,20,22-trienolide	**(507-60-8)**
6 β-(Acetyloxy)-3-β-(β-D-glucopyranosyloxy)-8,14-dihydroxybufa-4,20,22-trienolide	**(507-60-8)**
2-(Acetyloxy)-N,N-diethyl-1,3,4,6,7,11b-hexahydro-9,10-dimethoxy-2H-benzo(a)quinolizine-3-carboxamide	**(63-12-7)**
2-(Acetyloxy)benzoic acid	**(50-78-2)**
4-(Acetyloxy)benzoic acid	**(2345-34-8)**
17-(Acetyloxy)-6-chloropregna-4,6-diene-3,20-dione	**(302-22-7)**
4-Acetyloxy-12,13-epoxy-3,7,15-trihydroxy-(3-α,4-β,7-β)-trichothec-9-en-8-one	**(23255-69-8)**
(6-α)-17-(Acetyloxy)-6-methylpreg-4-ene-3,20-dione	**(71-58-9)**
(17-α)-17-(Acetyloxy)-19-norpregn-4-en-20-yn-3-one	**(51-98-9)**
17-Acetyloxy(17-α)-19-norpregn-4-estren-17-β-ol-acetate-3-one	**(51-98-9)**
2-(Acetyloxy)-1-phenylethanone	**(2243-35-8)**
2-(Acetyloxy)-1,2,3-propanetricarboxylic acid, tributyl ester	**(77-90-7)**
1-(Acetyloxy)-2-propanone	**(592-20-1)**
(1-(Acetyloxy)-2,2,2-trichloroethyl)phosphonic acid diphenyl ester	**(74548-80-4)**
Acetyl-peroxide	**(110-22-5)**
Acetyl peroxide, More than 25% in solution (DOT)	**(110-22-5)**
Acetyl peroxide, Solid (DOT)	**(110-22-5)**
Acetyl peroxide (Solution)	**(110-22-5)**
Acetyl peroxide solution, Not over 25% peroxide (DOT)	**(110-22-5)**
Acetylphenetidin	**(62-44-2)**
N-Acetyl-p-phenetidine	**(62-44-2)**
2-Acetylphenol	**(118-93-4)**
3-Acetylphenol	**(121-71-1)**
4-Acetylphenol	**(99-93-4)**
Acetyl phenol	**(122-79-2)**
m-Acetylphenol	**(121-71-1)**
o-Acetylphenol	**(118-93-4)**
p-Acetylphenol	**(99-93-4)**
Acetyl-p-phenylenediamine	**(122-80-5)**
N-Acetyl-m-phenylenediamine	**(102-28-3)**
1-Acetyl-2-phenylhydrazine	**(114-83-0)**
Acetylphenylhydrazine	**(114-83-0)**
N-Acetyl-N'-phenylhydrazine	**(114-83-0)**
β-Acetylphenylhydrazine	**(114-83-0)**
4-Acetylphenyl N-phenylcarbamate	**(37070-86-3)**
4-Acetylphenyl phenylcarbamate	**(37070-86-3)**
1-((p-Acetylphenyl)sulfonyl)-3-cyclohexylurea	**(968-81-0)**
13-O-Acetylphorbol 12-myristate	**(16561-29-8)**
Acetylphosphoramidothioic acid O,S-dimethyl ester	**(30560-19-1)**

Name	CAS
1-Acetyl-2-picolinolhydrazine	(17433-31-7)
1-Acetyl-2-picolinoylhydrazine	(17433-31-7)
n-Acetylpiperidin (German)	(618-42-8)
1-Acetylpiperidine	(618-42-8)
2-Acetyl propane	(563-80-4)
Acetyl 2-propanone	(123-54-6)
β-Acetylpropionic acid	(123-76-2)
Acetylpropionyl	(600-14-6)
3-Acetylpropyl acetate	(5185-97-7)
γ-Acetylpropyl acetate	(5185-97-7)
Acetylpropylorvinol	(25333-77-1)
2-Acetylpyrazine	(22047-25-2)
Acetylpyrazine	(22047-25-2)
3-Acetylpyridine	(350-03-8)
4-Acetylpyridine	(1122-54-9)
β-Acetylpyridine	(350-03-8)
2-Acetylpyrrole	(1072-83-9)
Acetylpyruvic acid	(5699-58-1)
N-Acetyl-p-quinonimine	(50700-49-7)
4-Acetylresorcinol	(89-84-9)
Acetylresorcinol	(102-29-4)
Acetylsal	(50-78-2)
Acetylsalicylic acid	(530-75-6)
Acetylsalicylic acid (ACGIH,OSHA)	(50-78-2)
Acetylsalicylsalicylic acid	(530-75-6)
Acetylsalicylsaure (German)	(50-78-2)
N'-Acetylsulfanilamide	(144-80-9)
N^1-Acetylsulfanilamide	(144-80-9)
N-Acetylsulfanilamide	(144-80-9)
N-Acetylsulfanilamine	(144-80-9)
Acetylsulfanilyl chloride	(121-60-8)
N(4)-Acetylsulfanilyl chloride	(121-60-8)
N-Acetylsulfanilyl chloride	(121-60-8)
3-Acetyltetrahydro-2-furanone	(517-23-7)
1-Acetyl-1,2,3,4-tetrahydropyridine	(19615-27-1)
7-α-Acetylthio-3-oxo-17-α-pregn-4-ene-21,17-β-carbolactone	(52-01-7)
7-α-Acetylthio-3-oxo-17-β-pregn-4-ene-21,17-β-carbolactone	(52-01-7)
2-Acetylthiophene	(88-15-3)
1-Acetyl-2-thiourea	(591-08-2)
Acetyl thiourea	(591-08-2)
2-Acetyltoluene	(577-16-2)
o-Acetyltoluene	(577-16-2)
p-Acetyltoluene	(122-00-9)
N-Acetyl-p-toluidide	(103-89-9)
Acetyl-o-toluidine	(120-66-1)
Acetyl-p-toluidine	(103-89-9)
N-Acetyl-m-toluidine	(537-92-8)
1-Acetyl-1,2,4-triazole	(15625-88-4)
1-Acetyl-1H-1,2,4-triazole	(15625-88-4)
1-Acetyltriazole	(15625-88-4)
N-Acetyl-1,2,4-triazole	(15625-88-4)
2-Acetyltributylcitrate	(77-90-7)
Acetyl tributyl citrate	(77-90-7)
Acetyl triethyl citrate	(77-89-4)
N-Acetyl-N-(2,4,6-triiodo-3-aminophenyl)-β-aminoisobutyric acid	(16034-77-8)
N-Acetyl trimethylcolchicinic acid methylether	(64-86-8)
Acetylure d'argent (French)	(13092-75-6)
Acetylure de mercure (French)	(68833-55-6)
Acetylureum	(63-98-9)
N-Acetylxenylamin (Czech)	(4075-79-0)
Achilleic acid	(499-12-7)
Achletin	(133-67-5)
Achro	(64-75-5)
Achrocidin	(62-44-2)
Achromycin	(60-54-8)
Achromycin	(64-75-5)
Achromycin V	(64-75-5)
Achromycin hydrochloride	(64-75-5)
Achylin	(138-15-8)
Acid-Spar	(7789-75-5)
Acicontral	(813-94-5)
Acid	(50-37-3)
Acid Black 52	(5610-64-0)
Acid Alizarine Pure Blue R	(4368-56-3)
Acid Alizarine Violet	(2092-55-9)
Acid Alizarine Violet B	(2092-55-9)
Acid Alizarine Violet N	(2092-55-9)
Acid Amaranth	(915-67-3)
Acid Amaranth I	(915-67-3)
Acid Ámaranth N	(915-67-3)
Acid Anthracene Red G	(10169-02-5)
Acid Anthracene Red GA-CF	(10169-02-5)
Acid Anthracene Yellow GR	(6375-55-9)
Acid Black 1	(1064-48-8)
Acid Black 10A	(1064-48-8)
Acid Black 10B	(1064-48-8)
Acid Black 12B	(1064-48-8)
Acid Black 10BA	(1064-48-8)
Acid Black 10BN	(1064-48-8)
Acid Black 4BN	(1064-48-8)
Acid Black 4BNU	(1064-48-8)
Acid Black BRX	(1064-48-8)
Acid Black BX	(1064-48-8)
Acid Black Base M	(1064-48-8)
Acid Black H	(1064-48-8)
Acid Black JVS	(1064-48-8)
Acid Blue 1	(129-17-9)
Acid Blue 41	(2666-17-3)
Acid Blue 62	(4368-56-3)
Acid Blue 9	(2650-18-2)
Acid Blue 90	(6104-58-1)
Acid Blue 92	(3861-73-2)
Acid Blue A	(3861-73-2)
Acid Blue Black 10B	(1064-48-8)
Acid Blue Black B	(1064-48-8)
Acid Blue Black BG	(1064-48-8)
Acid Blue Black Double 600	(1064-48-8)
Acid Blue V	(129-17-9)
Acid Blue W	(860-22-0)
Acid Bright Azure Z	(129-17-9)
Acid Bright Orange Zh	(1934-20-9)
Acid Bright Red	(3734-67-6)
Acid Brilliant Blue VF	(129-17-9)
Acid Brilliant Blue Z	(129-17-9)
Acid Brilliant Green SF	(5141-20-8)
Acid Brilliant Orange Zh	(1934-20-9)
Acid Brilliant Pink B	(81-88-9)
Acid Brilliant Red	(3734-67-6)
Acid Brilliant Rubine A2G Conc.	(3567-69-9)
Acid Brilliant Rubine 2G	(3567-69-9)
Acid Brilliant Rubine 2GT	(3567-69-9)
Acid Brilliant Scarlet 3R	(2611-82-7)
Acid Brilliant Sky Blue Z	(129-17-9)
Acid Brilliant Yellow G (9CI)	(115133-45-4)
Acid Cardinal G	(1658-56-6)
Acid Chrome Blue BA	(3567-69-9)
Acid Chrome Blue BA-CF	(3567-69-9)
Acid Chrome Blue FBS	(3567-69-9)
Acid Chrome Blue 2R	(3567-69-9)
Acid Chrome Violet K	(2092-55-9)
Acid Chrome Violet N	(2092-55-9)
Acid Fast Orange Egg	(1936-15-8)
Acid Fast Orange G	(1936-15-8)
Acid Fast Red Egg	(3734-67-6)

Acid Fast Red FB	**(3567-69-9)**
Acid Fast Red 3G	**(3734-67-6)**
Acid Fast Violet 5BN	**(1694-09-3)**
Acid Fast Yellow AG	**(6373-74-6)**
Acid Fast Yellow E5R	**(6373-74-6)**
Acid Fast Yellow MR	**(6375-55-9)**
Acid Fuchsin Fast B	**(3567-66-6)**
Acid Fuchsine	**(3244-88-0)**
Acid Fuchsine D	**(3567-66-6)**
Acid Fuchsine FB	**(3244-88-0)**
Acid Fuchsine N	**(3244-88-0)**
Acid Fuchsine O	**(3244-88-0)**
Acid Fuchsine S	**(3244-88-0)**
Acid Green	**(4680-78-8)**
Acid Green 25	**(4403-90-1)**
Acid Green 3	**(4680-78-8)**
Acid Green 5	**(5141-20-8)**
Acid Green A	**(5141-20-8)**
Acid Green B	**(4680-78-8)**
Acid Green 2G	**(4680-78-8)**
Acid Green G	**(4680-78-8)**
Acid Green L	**(4680-78-8)**
Acid Green S	**(4680-78-8)**
Acid IV	**(131-27-1)**
Acid Leather Blue IC	**(860-22-0)**
Acid Leather Blue IGW	**(1064-48-8)**
Acid Leather Blue R	**(3861-73-2)**
Acid Leather Blue V	**(129-17-9)**
Acid Leather Brown 2G	**(1300-73-8)**
Acid Leather Dark Blue G	**(1064-48-8)**
Acid Leather Fast Blue Black G	**(1064-48-8)**
Acid Leather Green F	**(4680-78-8)**
Acid Leather Green 3G	**(4680-78-8)**
Acid Leather Light Brown G	**(6373-74-6)**
Acid Leather Magenta A	**(3244-88-0)**
Acid Leather Orange Extra	**(633-96-5)**
Acid Leather Orange Extra G	**(633-96-5)**
Acid Leather Orange Extra PRW	**(633-96-5)**
Acid Leather Orange I	**(523-44-4)**
Acid Leather Orange KG	**(1936-15-8)**
Acid Leather Orange PGW	**(1936-15-8)**
Acid Leather Red BG	**(6459-94-5)**
Acid Leather Red 12BW	**(915-67-3)**
Acid Leather Red GR	**(3567-65-5)**
Acid Leather Red IBW	**(2302-96-7)**
Acid Leather Red KG	**(3734-67-6)**
Acid Leather Red KPR	**(3761-53-3)**
Acid Leather Red P2R	**(3761-53-3)**
Acid Leather Red ROC	**(1658-56-6)**
Acid Leather Rubine S	**(915-67-3)**
Acid Leather Scarlet G	**(3567-65-5)**
Acid Leather Scarlet IRW	**(3761-53-3)**
Acid Leather Yellow CRS	**(6375-55-9)**
Acid Leather Yellow PGW	**(547-57-9)**
Acid Leather Yellow PRW	**(587-98-4)**
Acid Leather Yellow R	**(587-98-4)**
Acid Leather Yellow T	**(1934-21-0)**
Acid Light Orange G	**(1936-15-8)**
Acid Light Orange J	**(1936-15-8)**
Acid Light Orange JA Export	**(1936-15-8)**
Acid Light Orange SX	**(1936-15-8)**
Acid Magenta	**(3244-88-0)**
Acid Magenta O	**(3244-88-0)**
Acid Naftol Red G	**(3734-67-6)**
Acid Orange	**(633-96-5)**
Acid Orange 10	**(1936-15-8)**
Acid Orange 12	**(1934-20-9)**
Acid Orange 24	**(1300-73-8)**
Acid Orange 24	**(1320-07-6)**
Acid Orange 7	**(633-96-5)**
Acid Orange A	**(633-96-5)**
Acid Orange 2G	**(1936-15-8)**
Acid Orange G	**(1936-15-8)**
Acid Orange GG	**(1936-15-8)**
Acid Orange I	**(523-44-4)**
Acid Orange II	**(633-96-5)**
Acid Orange No. 3	**(6373-74-6)**
Acid Phloxine GA	**(3734-67-6)**
Acid Phosphine CL	**(523-44-4)**
Acid Phosphine G New	**(547-57-9)**
Acid Ponceau 4R	**(2611-82-7)**
Acid Ponceau R	**(3761-53-3)**
Acid Ponceau 2RL	**(3761-53-3)**
Acid Ponceau Special	**(3761-53-3)**
Acid Red 1	**(3734-67-6)**
Acid Red 13	**(2302-96-7)**
Acid Red 14	**(3567-69-9)**
Acid Red 18	**(2611-82-7)**
Acid Red 25	**(5858-93-5)**
Acid Red 26	**(3761-53-3)**
Acid Red 33	**(3567-66-6)**
Acid Red 37	**(915-67-3)**
Acid Red 41	**(5850-44-2)**
Acid Red 85	**(3567-65-5)**
Acid Red 88	**(1658-56-6)**
Acid Red 97	**(10169-02-5)**
Acid Red 114	**(6459-94-5)**
Acid Red 388	**(37299-86-8)**
Acid Red AV	**(1658-56-6)**
Acid Red Alizarine	**(130-22-3)**
Acid Red 2A	**(3567-66-6)**
Acid Red B	**(3567-66-6)**
Acid Red 2C	**(3567-69-9)**
Acid Red 2G	**(3734-67-6)**
Acid Red G	**(1658-56-6)**
Acid Red GA	**(3734-67-6)**
Acid Red G (9CI) (VAN)	**(97502-49-3)**
Acid Red PG	**(3567-65-5)**
Acid Red 4ZhM	**(6408-31-7)**
Acid Rose AV	**(1658-56-6)**
Acid Rose 2GL	**(3734-67-6)**
Acid Rosein	**(3244-88-0)**
Acid Rubin	**(3244-88-0)**
Acid Rubine	**(3567-69-9)**
Acid Rubine Extra	**(3567-69-9)**
Acid Scarlet	**(3761-53-3)**
Acid Scarlet 2B	**(3761-53-3)**
Acid Scarlet 2BN	**(3761-53-3)**
Acid Scarlet GNA	**(3257-28-1)**
Acid Scarlet JN Extra Pure A	**(3257-28-1)**
Acid Scarlet 2R	**(3761-53-3)**
Acid Scarlet 3R	**(2611-82-7)**
Acid Scarlet 4R	**(2611-82-7)**
Acid Scarlet 2RL	**(3761-53-3)**
Acid Scarlet 2RN	**(3761-53-3)**
Acid Scarlet 3RZ	**(2611-82-7)**
Acid Scarlet 2R for Lakes	**(3761-53-3)**
Acid Scarlet 2R for Lakes Bluish	**(3761-53-3)**
Acid Sky Blue A	**(3844-45-9)**
Acid, Tannic	**(1401-55-4)**
Acid Violet	**(1694-09-3)**
Acid Violet 17	**(4129-84-4)**
Acid Violet 49	**(1694-09-3)**
Acid Violet 5B	**(1694-09-3)**

Acid Violet 6B **(1694-09-3)**
Acid Violet 5BN **(1694-09-3)**
Acid Violet 4BNS **(1694-09-3)**
Acid Violet S **(1694-09-3)**
Acid Wool Blue RL **(3861-73-2)**
Acid Yellow **(547-57-9)**
Acid Yellow 3 **(8004-92-0)**
Acid Yellow 23 **(1934-21-0)**
Acid Yellow 36 **(587-98-4)**
Acid Yellow 42 **(6375-55-9)**
Acid Yellow 73 **(2321-07-5)**
Acid Yellow D **(554-73-4)**
Acid Yellow E **(6373-74-6)**
Acid Yellow K **(6375-55-9)**
Acid Yellow T **(1934-21-0)**
Acid Yellow TRA **(2783-94-0)**
Acid acetate **(126-96-5)**
Acidal Black 10B **(1064-48-8)**
Acidal Bright Ponceau 3R **(2611-82-7)**
Acidal Brilliant Red 2G **(3734-67-6)**
Acidal Fast Orange **(1936-15-8)**
Acidal Fuchsine **(3244-88-0)**
Acidal Green G **(4680-78-8)**
Acidal Light Green SF **(5141-20-8)**
Acidal Magenta **(3244-88-0)**
Acidal Navy Blue 3BR **(1064-48-8)**
Acidal Ponceau G **(3761-53-3)**
Acidal Red E **(2302-96-7)**
Acidalin **(138-15-8)**
Acid amide **(98-92-0)**
Acid ammonium carbonate **(1066-33-7)**
Acid ammonium fluoride **(1341-49-7)**
Acid ammonium sulfate **(7803-63-6)**
Acid butyl phosphate **(12788-93-1)**
Acid copper arsenite **(10290-12-7)**
Acide Arsenique Liquide (French) **(7778-39-4)**
Acide L-azetidine-2-carboxylic (French) **(2133-34-8)**
Acide acetique (French) **(64-19-7)**
Acide acetylsalicylique (French) **(50-78-2)**
Acide anisique (French) **(119-36-8)**
Acide p-arsanilique (French) **(98-50-0)**
Acide arsenieux (French) **(1327-53-3)**
Acide arsenique liquide (French) **(7778-39-4)**
Acide ascorbique (French) **(50-81-7)**
Acide azidodithiocarbonique (French) **(4472-06-4)**
Acide benzoique (French) **(65-85-0)**
L'Acide (benzoyl-3-phenyl)-2-propionique (French) **(22071-15-4)**
Acide bromacetique (French) **(79-08-3)**
Acide bromhydrique (French) **(10035-10-6)**
Acide cacodylique (French) **(75-60-5)**
Acide carbolique (French) **(108-95-2)**
Acide chloracetique (French) **(79-11-8)**
Acide chlorhydrique (French) **(7647-01-0)**
Acide chlorhydrique et acide nitrique, melange d' (French) **(8007-56-5)**
Acide 2-(4-chloro-2-methyl-phenoxy)propionique (French) **(93-65-2)**
Acide chloro-3 peroxybenzoique (French) **(937-14-4)**
Acide o-chlorophenoxyacetique (French) **(614-61-9)**
Acide (m-chlorophenoxy)-2 methyl-2 propionique (French) **(17413-73-9)**
Acide (p-chlorophenoxy)-2 methyl-2 propionique (French) **(882-09-7)**
Acide chloroplatinique (French) **(16941-12-1)**
Acide chromique (French) **(7738-94-5)**
Acide chrysaminique (French) **(517-92-0)**
Acide cromoglicique (French) **(16110-51-3)**
Acide cyanacetique (French) **(372-09-8)**
Acide cyanhydrique (French) **(74-90-8)**
Acide 2,4-dichloro phenoxyacetique (French) **(94-75-7)**
Acide 2-(2,4-dichloro-phenoxy) propionique (French) **(120-36-5)**
Acide difluorophosphorique (French) **(13779-41-4)**
Acide dimethylarsinique (French) **(75-60-5)**
Acide diphenylhydroxyacetique (French) **(76-93-7)**
Acide ethylenediaminetetracetique (French) **(60-00-4)**
Acide 1-etil-7-metil-1,8-naftiridin-4-one-3-carbossilico (Italian) **(389-08-2)**
Acide fluorhydrique (French) **(7664-39-3)**
Acide fluorosilicique (French) **(16961-83-4)**
Acide fluosilicique (French) **(16961-83-4)**
Acide formique (French) **(64-18-6)**
Acide fulminique (French) **(506-85-4)**
Acide gibberellique (French) **(77-06-5)**
Acide hyponitreux (French) **(14448-38-5)**
Acide (isobutyl-4 phenyl)-2 propionique (French) **(15687-27-1)**
Acide isophtalique (French) **(121-91-5)**
Acide isothiocyanique (French) **(3129-90-6)**
Acide isovanillique (French) **(645-08-9)**
Acide mefenamique (French) **(61-68-7)**
Acide methazoique (French) **(5653-21-4)**
Acide o-methoxyphenoxyacetique (French) **(1878-85-9)**
Acide methyl-o-benzoique (French) **(119-36-8)**
Acide monochloracetique (French) **(79-11-8)**
Acide-monofluoracetique (French) **(144-49-0)**
Acide nalidixico (Italian) **(389-08-2)**
Acide nalidixique (French) **(389-08-2)**
Acide β-naphthylacetique (French) **(581-96-4)**
Acide naphthyloxyacetique (French) **(120-23-0)**
Acide iso-nicotinique (French) **(55-22-1)**
Acide nicotinique (French) **(59-67-6)**
Acide nitrique (French) **(7697-37-2)**
Acide nitrobenzenesulfonique (French) **(31212-28-9)**
Acide oxalique (French) **(144-62-7)**
Acide oxiniacique (French) **(2398-81-4)**
Acide peracetique (French) **(79-21-0)**
Acide peroxyacetique (French) **(79-21-0)**
Acide phenoxyacetique (French) **(122-59-8)**
Acide phosphorique (French) **(7664-38-2)**
Acide phtalique (French) **(88-99-3)**
Acide picolique (French) **(98-98-6)**
Acide picramique (French) **(96-91-3)**
Acide picrique (French) **(88-89-1)**
Acide pidolique (French) **(98-79-3)**
Acide pipecolique (French) **(535-75-1)**
Acide piperidine-carboxylique-2 (French) **(535-75-1)**
Acide propionique (French) **(79-09-4)**
L'Acide ricinoleique (French) **(141-22-0)**
Acide sulfhydrique (French) **(7783-06-4)**
Acide sulfurique (French) **(7664-93-9)**
Acide terephtalique (French) **(100-21-0)**
Acide thioglycolique (French) **(68-11-1)**
Acide trichloracetique (French) **(76-03-9)**
Acide trichlorobenzoique (French) **(50-31-7)**
Acide 2,4,5-trichloro phenoxyacetique (French) **(93-76-5)**
Acide 2-(2,4,5-trichloro-phenoxy) propionique (French) **(93-72-1)**
Acide orthovanillique (French) **(877-22-5)**
Acide vanillique (French) **(121-34-6)**
Acid, hyaluronic **(9004-61-9)**
Acidic Metanil Yellow **(587-98-4)**
Acidine Orange GN **(1934-20-9)**
Acidine Red G **(3734-67-6)**
Acidine Scarlet GD **(3567-65-5)**
Acid lead arsenate **(7784-40-9)**
Acid lead orthoarsenate **(7784-40-9)**
Acid, lysergic **(82-58-6)**
Acid metanil Yellow **(587-98-4)**
Acido acetico (Italian) **(64-19-7)**
Acido o-acetil-benzoico (Italian) **(50-78-2)**
Acido acetilsalicilico (Italian) **(50-78-2)**

Acido ascorbico (Spanish) (50-81-7)
Acido azidoditiocarbonico (Spanish) (4472-06-4)
Acido bromidrico (Italian) (10035-10-6)
Acido cianidrico (Italian) (74-90-8)
Acido cloridrico (Italian) (7647-01-0)
Acido 2-(4-cloro-2-metil-fenossi)-propionico (Italian) (93-65-2)
Acido m-clorobenzoico (Italian) (535-80-8)
Acido p-clorobenzoico (Italian) (74-11-3)
Acido 3-cloroperoxibenzoico (Spanish) (937-14-4)
Acido cloroplatinico (Spanish) (16941-12-1)
Acido crisaminico (Spanish) (517-92-0)
Acido cromico (Spanish) (7738-94-5)
Acido cromoglicico (Spanish) (16110-51-3)
Acido(2,4-dicloro-fenossi)-acetico (Italian) (94-75-7)
Acido 2-(2,4-dicloro-fenossi)-propionico (Italian) (120-36-5)
Acido (3,6-dicloro-2-metossi)-benzoico (Italian) (1918-00-9)
Acido difluofosforico (Spanish) (13779-41-4)
Acido 5-fenil-5-etilbarbiturico (Italian) (50-06-6)
Acido fluoridrico (Italian) (7664-39-3)
Acido fluosilicico (Italian) (16961-83-4)
Acido formico (Italian) (64-18-6)
Acido fosforico (Italian) (7664-38-2)
Acido fulminico (Spanish) (506-85-4)
Acidogen (138-15-8)
Acidogen nitrate (124-47-0)
Acido hiponitroso (Spanish) (14448-38-5)
Acido m-idrossibenzoico (Italian) (99-06-9)
Acido ippurico (Italian) (495-69-2)
Acido isotiocianico (Spanish) (3129-90-6)
Acidol Green M-FGL (71872-22-5)
Acidol Red E (2302-96-7)
Acido metazoico (Spanish) (5653-21-4)
Acido 1-metil-5-(p-tolnil)-pirrol-2-acetico (Spanish) (26171-23-3)
Acidomonocloroacetico (Italian) (79-11-8)
Acido monofluoroacetio (Italian) (144-49-0)
Acido nitrico (Italian) (7697-37-2)
Acido nitrobencenosulfonico (Spanish) (31212-28-9)
Acido ortocresotinico (Italian) (83-40-9)
Acido 3-ossi-5-metil-benzoico (Italian) (83-40-9)
Acido ossalico (Italian) (144-62-7)
Acido oxiniacico (Spanish) (2398-81-4)
Acido peroxiacetico (Spanish) (79-21-0)
Acido picrico (Italian) (88-89-1)
Acido pidolico (Spanish) (98-79-3)
Acidoride (138-15-8)
Acido salicilico (Italian) (69-72-7)
Acido solforico (Italian) (7664-93-9)
Acidothyn (138-15-8)
Acido (2,4,5-tricloro-fenossi)-acetico (Italian) (93-76-5)
Acido 2-(2,4,5-tricloro-fenossi)-propionico (Italian) (93-72-1)
Acido tricloroacetico (Italian) (76-03-9)
Acid potassium sulfate (7646-93-7)
Acid potassium tartrate (868-14-4)
Acids, carboxylic (VAN) (67254-79-9)
Acids, polyamides (VAN) (63428-84-2)
Acidulated Soapstock (68952-95-4)
Acidulen (138-15-8)
Acidulin (138-15-8)
Acidum acetylsalicylicum (50-78-2)
Acidum arcorbicum (Latin) (50-81-7)
Acidum ascorbinicum (50-81-7)
Acidum cromoglicicum (Latin) (16110-51-3)
Acidum nicotinicum (59-67-6)
Acidum oxiniacicum (Latin) (2398-81-4)
Acidum pidolicum (Latin) (98-79-3)
Acifloctin (124-04-9)
Acifluorfen (62476-59-9)
Acifluorfen sodium (62476-59-9)
Acigena (70-30-4)
Aciglumin (138-15-8)
Acigluminum (138-15-8)
Acilan Chromotrope RR (4197-07-3)
Acilan Fast Navy Blue R (3861-73-2)
Acilan Green B (4680-78-8)
Acilan Green SFG (5141-20-8)
Acilan Naphthol Red G (3734-67-6)
Acilan Naphtol Red G (3734-67-6)
Acilan Orange G (1934-20-9)
Acilan Orange GX (1936-15-8)
Acilan Orange II (633-96-5)
Acilan Ponceau 4GBL (1934-20-9)
Acilan Ponceau 6R (5850-44-2)
Acilan Ponceau RRL (3761-53-3)
Acilan Red E (2302-96-7)
Acilan Red SE (915-67-3)
Acilan Scarlet V3R (2611-82-7)
Acilan Turquoise Blue AE (2650-18-2)
Acilan Violet S4BN (1694-09-3)
Acilan Yellow GG (1934-21-0)
Aciletten (77-92-9)
Acillin (69-53-4)
Acillin (7177-48-2)
Acimetion (59-51-8)
Acimetten (50-78-2)
Acinetten (124-04-9)
Acintene A (80-56-8)
Acintene DP (138-86-3)
Acintene DP Dipentene (138-86-3)
Acintene O (104-46-1)
Acintol C (8002-26-4)
Acipen V (87-08-1)
Aciphenochinoline (132-60-5)
Aciphenochinolinium (132-60-5)
Acipimox (51037-30-0)
Acid-treated heavy naphthenic distillate (64742-18-3)
Acid-treated heavy paraffinic distillate (64742-20-7)
Acid-treated light naphthenic distillate (64742-19-4)
Acid-treated light paraffinic distillate (64742-21-8)
Acid-treated residual oil (64742-17-2)
Acket (65-45-2)
Aclor (138-15-8)
Acme Amine 4 (94-75-7)
Acme LV 4 (94-75-7)
Acme MCPA Amine 4 (94-74-6)
Acme Yellow Acid Yellow RS (547-57-9)
Acme butyl ester 4 (94-75-7)
Acna Black DF Base (101-54-2)
Acna Naphthol C (92-77-3)
Acna Naphthol CA (137-52-0)
Acna Naphthol E (135-61-5)
Acna Naphthol F (132-68-3)
Acna Naphthol G (91-96-3)
Acna Naphthol M (135-65-9)
Acna Naphthol O (135-62-6)
Acna Naphthol OF (92-74-0)
Acnegel (94-36-0)
Acnestrol (56-53-1)
Acocantherin (630-60-4)
Acon (68-26-8)
Aconitane (302-27-2)
Aconitic acid (499-12-7)
Aconitin Cristallisat (German) (302-27-2)
Aconitine (302-27-2)
Aconitine (Crystalline) (302-27-2)

Name	CAS No.
Acovenoside B	**(2624-17-1)**
Acquinite	**(107-02-8)**
Acquinite	**(76-06-2)**
Acraldehyde	**(107-02-8)**
Acraldehydeacroleina (Italian)	**(107-02-8)**
Acramine Yellow	**(134-50-9)**
Acrawax CT	**(110-30-5)**
Acrec	**(973-21-7)**
Acrex	**(973-21-7)**
Acrichine	**(69-05-6)**
Acrichine	**(83-89-6)**
Acricid	**(485-31-4)**
9-Acridamine, monohydrochloride, monohydrate	**(52417-22-8)**
Acridan	**(92-81-9)**
Acridan, 9,9-dimethyl- (8CI)	**(6267-02-3)**
Acridane	**(92-81-9)**
9-Acridanone	**(578-95-0)**
Acridanone	**(578-95-0)**
2-Acridinamine (9CI)	**(581-28-2)**
3-Acridinamine (9CI)	**(581-29-3)**
9-Acridinamine (9CI)	**(90-45-9)**
9-Acridinamine, monohydrochloride (9CI)	**(134-50-9)**
9-Acridinamine, 1,2,3,4-tetrahydro- (9CI)	**(321-64-2)**
Acridine Orange	**(494-38-2)**
Acridine Orange	**(65-61-2)**
Acridine Orange Base	**(494-38-2)**
Acridine Orange Free Base	**(494-38-2)**
Acridine Orange NO	**(494-38-2)**
Acridine Orange No	**(65-61-2)**
Acridine Orange R	**(65-61-2)**
Acridine Red	**(2465-29-4)**
Acridine Red 3B	**(2465-29-4)**
Acridine Red, Hydrochloride	**(2465-29-4)**
Acridine [UN 2713]	**(260-94-6)**
Acridine, 2-amino	**(581-28-2)**
Acridine, 3-amino	**(581-29-3)**
Acridine, 9-amino	**(90-45-9)**
Acridine, 1-amino-9-((3-(dimethylamino)propyl)amino)-, dihydrochloride (8CI)	**(19395-62-1)**
Acridine, 9-amino-, hydrochloride	**(134-50-9)**
Acridine, 9-(4-bis(2-chloroethyl)amino-1-methylbutylamino)-6-chloro-2-methoxy-, dihydrochloride	**(4213-45-0)**
Acridine, 3,6-bis(dimethylamino)	**(494-38-2)**
Acridine, 3,6-bis(dimethylamino)-, monohydrochloride	**(65-61-2)**
Acridine, 6-chloro-9-((4-(diethylamino)-1-methylbutyl)-amino)-2-methoxy	**(83-89-6)**
Acridine, 6-chloro-9-((4-(diethylamino)-1-methylbutyl)-amino)-2-methoxy-, dihydrochloride	**(69-05-6)**
3,6-Acridinediamine (9CI)	**(92-62-6)**
Acridinediamine, N,N,N^{1},N^{1}-tetramethyl- (9CI)	**(494-38-2)**
3,6-Acridinediamine, N,N,N',N'-tetramethyl-, monohydrochloride	**(65-61-2)**
3,6-Acridinediamine, dihydrochloride (9CI)	**(531-73-7)**
1,9-Acridinediamine, N(9)-(3-(dimethylamino)propyl)-, dihydrochloride (9CI)	**(19395-62-1)**
3,6-Acridinediamine, monohydrochloride (9CI)	**(952-23-8)**
3,6-Acridinediamine sulfate	**(553-30-0)**
3,6-Acridinediamine, sulfate (1:1) (9CI)	**(553-30-0)**
3,6-Acridinediamine sulphate	**(553-30-0)**
Acridine, 3,6-diamino	**(92-62-6)**
Acridine, 3,6-diamino-, dihydrochloride	**(531-73-7)**
Acridine, 3,6-diamino-, monohydrochloride	**(952-23-8)**
Acridine, 3,6-diamino-, sulfate (1:1)	**(553-30-0)**
Acridine, 3,6-diamino-, sulfate (2:1)	**(1811-28-5)**
Acridine, 9,10-dihydro- (9CI)	**(92-81-9)**
Acridine, 9,10-dihydro-9,9-dimethyl- (9CI)	**(6267-02-3)**
Acridine, 2,3-dimethyl- (9CI)	**(97340-75-5)**
Acridine, dimethyl- (9CI)	**(99339-80-7)**
Acridine, 9-((3-(dimethylamino)propyl)amino)-3-nitro	**(6237-24-7)**
Acridine, ethyl- (9CI)	**(64828-44-0)**
Acridine, methyl- (9CI)	**(54116-90-4)**
Acridine mustard	**(146-59-8)**
Acridine, 1,2,3,4-tetrahydro-9-amino	**(321-64-2)**
Acridine, 1,2,3,4-tetrahydro-4,9-dimethyl- (7CI,9CI)	**(55030-65-4)**
Acridinium, 3,6-diamino-10-methyl-, chloride	**(86-40-8)**
Acridinium, 3,6-diamino-10-methyl-, chloride mixed with 3,6-acridinediamine	**(8048-52-0)**
Acridinium, 3,6-diamino-10-methyl-, chloride, monohydrochloride, mixt. with 3,6-acridinediamine monohydrochloride (9CI)	**(8018-07-3)**
9(10H)-Acridinone (9CI)	**(578-95-0)**
Acridinorange	**(494-38-2)**
4'-(9-Acridinylamino)methanesulfon-m-anisidide	**(51264-14-3)**
4'-(9-Acridinylamino)methanesulphon-m-anisidide	**(51264-14-3)**
4'-(9-Acridinylamino)-3'-methoxymethanesulfonanilide	**(51264-14-3)**
4'-(9-Acridinylamino)-methylsulfonyl-m-anisidine	**(51264-14-3)**
Acridogen	**(138-15-8)**
9(10H)-Acridone	**(578-95-0)**
9-Acridone	**(578-95-0)**
Acridone	**(578-95-0)**
Acridoride	**(138-15-8)**
Acriflavin	**(8048-52-0)**
Acriflavine	**(8048-52-0)**
Acriflavine	**(86-40-8)**
Acriflavine mixture with proflavine	**(8048-52-0)**
Acriflavine neutral	**(86-40-8)**
Acriflavinii chloridum	**(8048-52-0)**
Acriflavinii chloridum (Latin)	**(8018-07-3)**
Acriflavinio cloruro	**(8018-07-3)**
Acriflavinium chloride	**(8048-52-0)**
Acriflavinium chloride (8CI)	**(8018-07-3)**
Acriflavon	**(8048-52-0)**
Acriflavon	**(86-40-8)**
Acrilafil	**(9003-54-7)**
Acrinamine	**(83-89-6)**
Acriquine	**(69-05-6)**
Acriquine	**(83-89-6)**
Acrisin FS 017	**(9011-05-6)**
Acritet	**(107-13-1)**
Acroart	**(9002-88-4)**
Acroleic acid	**(79-10-7)**
trans-Acrolein	**(107-02-8)**
Acrolein (ACGIH,OSHA)	**(107-02-8)**
Acrolein, Inhibited [UN 1092]	**(107-02-8)**
Acroleina (Italian)	**(107-02-8)**
Acrolein, 2-chloro	**(683-51-2)**
Acrolein diacetate	**(869-29-4)**
Acrolein dimer	**(100-73-2)**
Acrolein dimer, Stabilized	**(100-73-2)**
Acroleine (Dutch, French)	**(107-02-8)**
Acroleine dimere stabilisee (French)	**(100-73-2)**
Acrolein, 2-ethyl-3-propyl-	**(645-62-5)**
Acrolein, 2-methyl-	**(78-85-3)**
Acrolein, 3-phenyl-	**(104-55-2)**
Acromona	**(443-48-1)**
Acromycine	**(7008-42-6)**
Acronal S 320 D	**(9011-14-7)**
Acronine	**(7008-42-6)**
Acronize	**(57-62-5)**
Acronycine	**(7008-42-6)**
Acroptilin	**(41787-75-1)**
Acrowax C	**(110-30-5)**
Acryl Brilliant Green B	**(569-64-2)**
Acrylaldehyd (German)	**(107-02-8)**
Acrylaldehyde	**(107-02-8)**
Acrylamide (ACGIH,OSHA) [UN 2074]	**(79-06-1)**

Acrylamide Yellow G **(2512-29-0)**
Acrylamide, N-butoxymethyl **(1852-16-0)**
Acrylamide, N-tert-butyl **(107-58-4)**
Acrylamide, N,N-dimethyl **(2680-03-7)**
Acrylamide, dimethylaminoethyl methacrylate copolymer, dimethyl sulfate quaternized **(26006-22-4)**
Acrylamide, N-(5,5-dimethylhexyl) **(4223-03-4)**
Acrylamide, N-(1,1-dimethyl-3-oxobutyl) **(2873-97-4)**
Acrylamide, ethyl acrylate, methylolacrylamide, methyl methacrylate polymer **(30394-81-1)**
Acrylamide, 2-(2-furyl)-3-(5-nitro-2-furyl) **(3688-53-7)**
Acrylamide, N-(hydroxymethyl) **(924-42-5)**
Acrylamide, N-(isobutoxymethyl) **(16669-59-3)**
Acrylamide, N-isopropyl **(2210-25-5)**
Acrylamide, methacryloyloxyethyltrimethylammonium methylsulfate copolymer **(26006-22-4)**
Acrylamide, N-methyl **(1187-59-3)**
Acrylamide, N,N'-methylenebis **(110-26-9)**
Acrylamide, Polymers **(9003-05-8)**
Acrylamide, N,N,N-trimethyl-2-((2-methyl-1-oxo-2-propenyl)-oxy)ethanaminium methyl sulfate polymer **(26006-22-4)**
2-Acrylamido-2-methylpropanesulfonate **(15214-89-8)**
2-Acrylamido-2-methylpropanesulfonic acid sodium salt **(5165-97-9)**
Acrylanilide, 3',4'-dichloro-2-methyl **(2164-09-2)**
Acrylate de methyle (French) **(96-33-3)**
Acrylate d'ethyle (French) **(140-88-5)**
Acrylic Resin **(9003-01-4)**
Acrylic-acid **(79-10-7)**
Acrylic acid (ACGIH,DOT,OSHA) **(79-10-7)**
Acrylic acid, Inhibited (DOT) **(79-10-7)**
Acrylic acid, allyl ester (8CI) **(999-55-3)**
Acrylic acid, 3-p-anisoyl-3-bromo-, sodium salt, (E) **(21739-91-3)**
Acrylic acid, butyl ester **(141-32-2)**
Acrylic acid, tert-butyl ester (8CI) **(1663-39-4)**
Acrylic acid, 3-(N-butyl-1-hydroxyformimidoyl)-, cyclic anhydride (7CI,8CI) **(27396-39-0)**
Acrylic acid, 3-(N-butyl-1-hydroxyformimidoyl)-, γ-lactone (6CI) **(27396-39-0)**
Acrylic acid, 3-carbamoyl-, (Z)- **(557-24-4)**
Acrylic acid chloride **(814-68-6)**
Acrylic acid, 2-chloro **(598-79-8)**
Acrylic acid, 3-chloro-, (E) **(2345-61-1)**
Acrylic acid, chloro- (8CI) **(26952-44-3)**
Acrylic acid, 3-chloro-, (Z)- (8CI) **(1609-93-4)**
Acrylic acid β-chloroethyl ester **(2206-89-5)**
Acrylic acid, 2-chloroethyl ester **(2206-89-5)**
Acrylic acid, 2-chloro-, methyl ester **(80-63-7)**
Acrylic acid, 2-cyanoethyl ester **(106-71-8)**
Acrylic acid, 2-cyano-, methyl ester **(137-05-3)**
Acrylic acid, decyl ester **(2156-96-9)**
Acrylic acid, 2,3-dibromopropyl ester **(19660-16-3)**
Acrylic acid, dicyclopentadienyl ester **(50976-02-8)**
Acrylic acid, diester with tetraethylene glycol **(17831-71-9)**
Acrylic acid, diester with triethylene glycol **(1680-21-3)**
Acrylic acid, 2-(diethylamino)ethyl ester **(2426-54-2)**
Acrylic acid, N,N-diethylaminoethyl ester **(2426-54-2)**
Acrylic acid, 2,2-dimethyl-1,3-propanediol diester **(2223-82-7)**
Acrylic acid, 2,2-dimethyltrimethylene ester **(2223-82-7)**
Acrylic acid, 2,2-dinitropropyl ester (8CI) **(17977-09-2)**
Acrylic acid, 2,3-epoxy-2-methylpropyl ester (6CI,7CI,8CI) **(19900-46-0)**
Acrylic acid, 2,3-epoxypropyl ester **(106-90-1)**
Acrylic acid, ester with hydracrylonitrile **(106-71-8)**
Acrylic acid, 4-ester with 2-hydroxy-4-(2-hydroxyethoxy)-benzophenone **(16432-81-8)**
Acrylic acid, 2-ethoxyethanol diester **(4074-88-8)**
Acrylic acid, 2-ethoxyethanol ester **(106-74-1)**
Acrylic acid, 2-ethoxyethyl ester **(106-74-1)**
Acrylic acid, 2-ethylbutyl ester **(3953-10-4)**
Acrylic acid, ethyl ester **(140-88-5)**
Acrylic acid, 2-ethylhexyl ester **(103-11-7)**
Acrylic acid, glacial **(79-10-7)**
Acrylic acid glycidyl ester **(106-90-1)**
Acrylic acid, hexadecyl ester **(13402-02-3)**
Acrylic acid, 3a,4,5,6,7,7a-hexahydro-4,7-methanoindenyl ester **(33791-58-1)**
Acrylic acid, hexamethylene ester **(13048-33-4)**
Acrylic acid, hexyl ester **(2499-95-8)**
Acrylic acid homopolymer **(9003-01-4)**
Acrylic acid, 2-hydroxyethyl ester **(818-61-1)**
Acrylic acid, 2-hydroxypropyl ester **(999-61-1)**
Acrylic acid, hydroxypropyl ester **(25584-83-2)**
Acrylic acid, ion(1-) (8CI) **(10344-93-1)**
Acrylic acid, isobutyl ester **(106-63-8)**
Acrylic acid, isodecyl ester **(1330-61-6)**
Acrylic acid, isopropyl ester (8CI) **(689-12-3)**
Acrylic acid, 2-methoxyethyl ester **(3121-61-7)**
Acrylic acid, 2-methyl- **(79-41-4)**
Acrylic acid, 3-methyl- **(3724-65-0)**
Acrylic acid, 2-methyl-, dodecyl ester **(142-90-5)**
Acrylic acid, methyl ester **(96-33-3)**
Acrylic acid, 2-methyl-, methyl ester **(80-62-6)**
Acrylic acid, 1-methyltrimethylene ester (8CI) **(19485-03-1)**
Acrylic acid, monoester with 1,2-propanediol **(25584-83-2)**
Acrylic acid, 5-norbornen-2-methyl ester **(95-39-6)**
Acrylic acid, 5-norbornen-2-ylmethyl ester **(95-39-6)**
Acrylic acid, octadecyl ester **(4813-57-4)**
Acrylic acid, octyl ester (8CI) **(2499-59-4)**
Acrylic acid, oxybis(ethyleneoxyethylene) ester **(17831-71-9)**
Acrylic acid, oxydiethylene ester (8CI) **(4074-88-8)**
Acrylic acid, pentaerithritol triester **(3524-68-3)**
Acrylic acid polymer **(9003-01-4)**
Acrylic acid, Polymers **(9003-01-4)**
Acrylic acid, propylenebis(oxypropylene) ester **(42978-66-5)**
Acrylic acid, propyl ester (8CI) **(925-60-0)**
Acrylic acid resin **(9003-01-4)**
Acrylic acid, sodium salt **(7446-81-3)**
Acrylic acid, ((3a,4,7,7a-tetrahydro)4,7-methanoindenyl) ester **(50976-02-8)**
Acrylic acid, trichloro- (8CI) **(2257-35-4)**
Acrylic acid, tridecyl ester **(3076-04-8)**
Acrylic acid, triester with 2-ethyl-2-(hydroxymethyl)-1,3-propanediol **(15625-89-5)**
Acrylic acid, 1,1,1-(trihydroxymethyl)propane triester **(15625-89-5)**
Acrylic acid, zinc salt **(14643-87-9)**
Acrylic aldehyde **(107-02-8)**
Acrylic amide **(79-06-1)**
Acrylic polymer **(9003-01-4)**
Acrylinoleic acid **(53980-88-4)**
Acrylite **(9011-14-7)**
Acrylnitril (German, Dutch) **(107-13-1)**
Acryloid A-15 **(9011-14-7)**
Acryloid B 66 **(25608-33-7)**
Acryloid K 125 **(25608-33-7)**
Acrylon **(107-13-1)**
Acrylonitrile (ACGIH,DOT,OSHA) **(107-13-1)**
Acrylonitrile, Inhibited [UN 1093] **(107-13-1)**
Acrylonitrile, Polymers (8CI) **(25014-41-9)**
Acrylonitrile, Polymer with 1,3-butadiene and styrene **(9003-56-9)**
Acrylonitrile, Polymer with styrene **(9003-54-7)**
Acrylonitrile-butadiene-styrene copolymer **(9003-56-9)**
Acrylonitrile-cellulose copolymer **(37243-36-0)**
Acrylonitrile-cellulose copolymers **(37243-36-0)**
Acrylonitrile-cellulose graft copolymer **(37243-36-0)**
Acrylonitrile-cellulose polymer **(37243-36-0)**
Acrylonitrile homopolymer **(25014-41-9)**
Acrylonitrile monomer **(107-13-1)**

Acrylonitrile-styrene resin **(9003-54-7)**
Acrylonitrile-styrene copolymer **(9003-54-7)**
Acrylonitrile-styrene polymer **(9003-54-7)**
Acrylonitrile, 2,3,3-trichloro **(16212-28-5)**
Acrylonitrile, trichloro- (8CI) **(16212-28-5)**
Acrylophenone, 3-phenyl- **(94-41-7)**
Acryloyl-chloride **(814-68-6)**
2-(Acryloyloxy)ethanol **(818-61-1)**
β-(Acryloyloxy)propanoic acid **(24615-84-7)**
3-Acryloyloxypropionic acid **(24615-84-7)**
Acrylsaeureaethylester (German) **(140-88-5)**
Acrylsaeuremethylester (German) **(96-33-3)**
Acrylyl chloride **(814-68-6)**
Acrypet **(9011-14-7)**
Acrypet M 001 **(9011-14-7)**
Acrypet V **(9011-14-7)**
Acrypet VH **(9011-14-7)**
Acrysol A 1 **(9003-01-4)**
Acrysol A 3 **(9003-01-4)**
Acrysol A 5 **(9003-01-4)**
Acrysol AC 5 **(9003-01-4)**
Acrysol ASE **(9011-14-7)**
Acrysol ASE-75 **(9003-01-4)**
Acrysol WS-24 **(9003-01-4)**
Act **(50-76-0)**
Acti-Aid **(66-81-9)**
Acti-Chlore **(127-65-1)**
Act D **(50-76-0)**
Acto-D **(50-76-0)**
Acti-Flow 68-SB **(8002-43-5)**
Actamer **(97-18-7)**
Actase **(9001-90-5)**
Actedron **(300-62-9)**
Actelic **(29232-93-7)**
Actellic **(29232-93-7)**
Actellifog **(29232-93-7)**
Actholain **(16960-16-0)**
Acticarbone **(7440-44-0)**
Acticel **(7631-86-9)**
Actidil **(6138-79-0)**
Actidion **(66-81-9)**
Actidione **(66-81-9)**
Actidione PM **(66-81-9)**
Actidione TGF **(66-81-9)**
Actidone **(66-81-9)**
Actinidiolide, dihydro- **(17092-92-1)**
Actinium **(7440-34-8)**
Actinol P 3035 **(9003-11-6)**
Actinolite (8CI,9CI) (VAN) **(12172-67-7)**
Actinolite asbestos **(77536-66-4)**
Actinomycin 1048A **(12623-78-8)**
Actinomycin 7 **(50-76-0)**
Actinomycin A IV **(50-76-0)**
Actinomycin C1 **(50-76-0)**
Actinomycin-D **(50-76-0)**
Actinomycin I **(50-76-0)**
Actinomycin IV **(50-76-0)**
Actinomycin I_1 **(50-76-0)**
Actinomycin-S **(12623-78-8)**
Actinomycin X 1 **(50-76-0)**
Actinomycin 11 cosmegen **(50-76-0)**
Actinomycindioic D acid, dilactone **(50-76-0)**
Actinon **(10043-92-2)**
Actispray **(66-81-9)**
Activated Attapulgite **(12174-11-7)**
Activated Carbon (DOT) **(7440-44-0)**
Activated aluminum oxide **(1344-28-1)**
Activated clay **(70131-50-9)**
Activated sludge, Dried **(68188-15-8)**
Active acetyl acetate **(141-97-9)**
Active dicumyl peroxide **(80-43-3)**
Active valeric acid **(116-53-0)**
Activol **(123-30-8)**
Activol **(77-06-5)**
Activol GA **(77-06-5)**
Actor Q **(105-11-3)**
Actox 14 **(1314-13-2)**
Actox 16 **(1314-13-2)**
Actox 216 **(1314-13-2)**
Actrapid **(9004-10-8)**
Actril **(1689-83-4)**
Actybaryte **(7727-43-7)**
Actylol **(97-64-3)**
Acutox **(87-86-5)**
Acylpyrin **(50-78-2)**
Acytol **(97-64-3)**
Adacene 12 **(112-41-4)**
Adakane 12 **(112-40-3)**
Adalat **(21829-25-4)**
Adalin **(77-65-6)**
1-Adamantamine **(768-94-5)**
1-Adamantanamine **(768-94-5)**
Adamantane (8CI) **(281-23-2)**
Adamantane, 1,3-dimethyl- (8CI) **(702-79-4)**
Adamsite **(578-94-9)**
Adanon **(76-99-3)**
Adbond 1000 Clear **(9006-03-5)**
Adc Pigment Yellow G **(2512-29-0)**
Adchem GMO **(25496-72-4)**
Addex-THAM **(77-86-1)**
Addisomnol **(77-65-6)**
Additin 30 **(90-30-2)**
Addukt hexachlorcyklopentadienu s cyklopentadienem (Czech) **(3734-48-3)**
Adeka CM 294 **(9003-11-6)**
Adeka CR 10 **(9006-03-5)**
Adeka CR 150 **(9006-03-5)**
Adeka CR 20 **(9006-03-5)**
Adeka CR 40 **(9006-03-5)**
Adeka CR 5 **(9006-03-5)**
Adeka Carpol GH 10 **(9003-11-6)**
Adeka Carpol MH 500 **(9003-11-6)**
Adeka Carpol PH 2000 **(9003-11-6)**
Adeka GH-200 **(9082-00-2)**
Adekacizer P 200 **(27941-08-8)**
Adelfan **(50-55-5)**
Adelphane **(50-55-5)**
Adelphin **(50-55-5)**
Adelphin-esidrex-K **(50-55-5)**
Ademine **(396-01-0)**
Ademol **(148-56-1)**
Adenex **(50-81-7)**
Adenine **(73-24-5)**
Adenine, N-benzyl- (8CI) **(1214-39-7)**
Adenine, N-benzyl-9-(tetrahydro-2H-pyran-2-yl)- (8CI) **(2312-73-4)**
Adenine, N-furfuryl **(525-79-1)**
Adenine, phosphate **(52175-10-7)**
Adenine riboside **(58-61-7)**
Adeninimine **(73-24-5)**
Adenock **(315-30-0)**
Adenohypophyseal Growth Hormone **(9002-72-6)**
Adenohypophyseal luteotropin **(9002-62-4)**
Adenosin (German) **(58-61-7)**
Adenosine **(58-61-7)**
β-Adenosine **(58-61-7)**

β-D-Adenosine	(58-61-7)
Adenosine, cyclic 3',5'-(hydrogenphosphate)	(60-92-4)
Adenosine 3',5'-cyclic monophosphate	(60-92-4)
Adenosine cyclic monophosphate	(60-92-4)
Adenosine cyclic 3',5'-phosphate	(60-92-4)
Adenosine 3',5'-cyclophosphate	(60-92-4)
Adenosine, 2'-deoxy-, 5'-(dihydrogen phosphate) (8CI)	(653-63-4)
Adenosine, N,6-didehydro-1,6-dihydro-1-methyl- (9CI)	(15763-06-1)
Adenosine, 5'-O-((1,1-dimethylethyl)dimethylsilyl)-, 2'-(4-methylbenzenesulfonate) (9CI)	(115094-43-4)
Adenosine, 1-methyl- (VAN) (8CI)	(15763-06-1)
Adenosine 3',5'-monophosphate	(60-92-4)
Adenosine 5'-monophosphate	(61-19-8)
Adenosine-5'-monophosphoric acid	(61-19-8)
Adenosine-5-monophosphoric acid	(61-19-8)
Adenosine 3',5'-phosphate	(60-92-4)
Adenosine 5'-phosphate	(61-19-8)
Adenosine phosphate	(61-19-8)
Adenosine 5'-phosphoric acid	(61-19-8)
Adenosine, 5'-O-phosphorylcytidylyl-(3'-5')- (8CI)	(15648-73-4)
Adenosine 5'-(tetrahydrogen triphosphate)	(56-65-5)
Adenosine 5'-(trihydrogen diphosphate), 2'-(dihydrogen phosphate), 5'-5'-ester with 1,4-dihydro-1-β-D-ribofuranosyl-3-pyridinecarboxamide (9CI)	(53-57-6)
Adenosine 5'-triphosphate	(56-65-5)
Adenosine triphosphate	(56-65-5)
Adenosine 5'-triphosphoric acid	(56-65-5)
Adenovite	(61-19-8)
Adenyl	(61-19-8)
5'-Adenylic acid	(61-19-8)
Adenylic acid	(61-19-8)
tert-Adenylic acid	(61-19-8)
5'-Adenylic acid, 2'-deoxy- (9CI)	(653-63-4)
Adenylpyrophosphoric acid	(56-65-5)
Adephos	(56-65-5)
Adeps Lane	(8006-54-0)
Adepsine Oil	(8012-95-1)
Adergon	(548-62-9)
Adermine	(65-23-6)
Adermine hydrochloride	(58-56-0)
Adetol	(56-65-5)
Adhere	(137-05-3)
Adiaben	(94-20-2)
Adiazine	(68-35-9)
Adilactetten	(124-04-9)
Adimoll DN	(151-32-6)
Adimoll DO	(123-79-5)
Adine 0102	(13654-09-6)
Adinol T	(137-20-2)
Adipamide	(628-94-4)
Adipan	(300-62-9)
Adipan	(60-13-9)
Adipan hexamethylendiaminu (Czech)	(3323-53-3)
Adiparthrol	(60-13-9)
Iso-Adipate 2/043700	(6938-94-9)
Adiphenin	(64-95-9)
Adiphenine	(64-95-9)
Adipic-acid	(124-04-9)
Adipic acid, Compd. with 1,6-hexanediamine	(3323-53-3)
Adipic acid, Compd. with 1,6-hexanediamine (1:1)	(3323-53-3)
Adipic acid amide	(628-94-4)
Adipic acid, bis(2-(2-butoxyethoxy)ethyl) ester	(141-17-3)
Adipic acid, bis(2-butyoxyethyl) ester	(141-18-4)
Adipic acid, bis(2,3-epoxypropyl) ester (8CI)	(2754-17-8)
Adipic acid, bis(2-(2-ethylbutoxy)ethyl) ester	(7790-07-0)
Adipic acid, bis(2-ethylbutyl) ester	(10022-60-3)
Adipic acid, 2-ethylhexyl ester	(4337-65-9)
Adipic acid, bis(2-ethylhexyl) ester	(103-23-1)
Adipic acid, bis(2-(hexyloxy)ethyl) ester	(110-32-7)
Adipic acid, bis(1-methylheptyl) ester (8CI)	(108-63-4)
Adipic acid, N-cocoalkyl-1,3-propylenediamine salt	(68155-42-0)
Adipic acid, N-(coconut oil alkyl)trimethylenediamine salt	(68155-42-0)
Adipic acid, cyclic tetramethylene ester	(777-95-7)
Adipic acid, decyl octyl ester	(110-29-2)
Adipic acid, diallyl ester	(2998-04-1)
Adipic acid diamide	(628-94-4)
Adipic acid, dibutoxyethyl ester	(141-18-4)
Adipic acid, dibutyl ester	(105-99-7)
Adipic acid, dicyclohexyl ester	(849-99-0)
Adipic acid, didecyl ester (8CI)	(105-97-5)
Adipic acid, di(2-(2-ethylbutoxy)ethyl) ester	(7790-07-0)
Adipic acid, di(2-ethylbutyl) ester	(10022-60-3)
Adipic acid, diethyl ester	(141-28-6)
Adipic acid, di(2-hexyloxyethyl) ester	(110-32-7)
Adipic acid, diisobutyl ester	(141-04-8)
Adipic acid, diisodecyl ester	(27178-16-1)
Adipic acid, diisooctyl ester	(1330-86-5)
Adipic acid, diisopropyl ester	(6938-94-9)
Adipic acid, dimethyl ester	(627-93-0)
Adipic acid dinitrile	(111-69-3)
Adipic acid, dinonyl ester (8CI)	(151-32-6)
Adipic acid, dioctyl ester (8CI)	(123-79-5)
Adipic acid, dipentyl ester (8CI)	(14027-78-2)
Adipic acid, dipropyl ester	(106-19-4)
Adipic acid, disodium salt	(7486-38-6)
Adipic acid, mono(2-ethylhexyl) ester (8CI)	(4337-65-9)
Adipic acid, monomethyl ester (8CI)	(627-91-8)
Adipic acid nitrile	(111-69-3)
Adipic diamide	(628-94-4)
Adipic ketone	(120-92-3)
Adipinic acid	(124-04-9)
Adiplon	(144-83-2)
Adipodinitrile	(111-69-3)
Adipoin	(533-60-8)
Adipol 10A	(1330-86-5)
Adipol BCA	(141-18-4)
Adipol 2EH	(103-23-1)
Adipol ODY	(110-29-2)
Adiponitrile	(111-69-3)
Adiponitrile (DOT)	(111-69-3)
Adipoyl chloride	(111-50-2)
Adjudets	(51-63-8)
Admer PB 02	(9003-07-0)
Admex 741	(61788-72-5)
Admex 746	(61788-72-5)
Admul	(31566-31-1)
Adnephrine	(51-43-4)
Adobacillin	(69-53-4)
Adogen 73	(301-02-0)
Adogen 360	(102-87-4)
Adogen 444	(112-02-7)
Adogen 471	(8030-78-2)
Adogenen 142	(124-30-1)
Adol	(112-92-5)
Adol	(143-28-2)
Adol	(36653-82-4)
Adol 34	(143-28-2)
Adol 52	(36653-82-4)
Adol 54	(36653-82-4)
Adol 68	(112-92-5)
Adol 80	(143-28-2)
Adol 85	(143-28-2)
Adol 90	(143-28-2)
Adol 320	(143-28-2)

Name	CAS No.
Adol 330	**(143-28-2)**
Adol 340	**(143-28-2)**
Adol 520	**(36653-82-4)**
Adonal	**(50-06-6)**
Adonite	**(488-81-3)**
Adonitol	**(488-81-3)**
Adopol	**(144-14-9)**
Adorm	**(52-31-3)**
Adran	**(15687-27-1)**
Adrenal	**(51-43-4)**
Adrenalex	**(53-06-5)**
Adrenalin	**(51-43-4)**
l-Adrenalin	**(51-43-4)**
Adrenalin bitartrate	**(51-42-3)**
Adrenalin chloride	**(55-31-2)**
(-)-Adrenaline	**(51-43-4)**
Adrenaline	**(51-43-4)**
DL-Adrenaline	**(329-65-7)**
l-Adrenaline	**(51-43-4)**
(-)-Adrenaline acid tartrate	**(51-42-3)**
Adrenaline acid tartrate	**(51-42-3)**
(-)-Adrenaline bitartrate	**(51-42-3)**
Adrenaline bitartrate	**(51-42-3)**
l-Adrenaline bitartrate	**(51-42-3)**
l-Adrenaline d-bitartrate	**(51-42-3)**
Adrenaline chloride	**(55-31-2)**
l-Adrenaline chloride	**(55-31-2)**
(+-)-Adrenaline hydrochloride	**(329-63-5)**
(-)-Adrenaline hydrochloride	**(55-31-2)**
Adrenaline hydrochloride	**(55-31-2)**
Adrenaline hydrochloride, (-)-	**(55-31-2)**
dl-Adrenaline hydrochloride	**(329-63-5)**
l-Adrenaline hydrochloride	**(55-31-2)**
(-)-Adrenaline hydrogen tartrate	**(51-42-3)**
Adrenaline hydrogen tartrate	**(51-42-3)**
l-Adrenaline hydrogen tartrate	**(51-42-3)**
(-)-Adrenaline tartrate	**(51-42-3)**
Adrenaline tartrate	**(51-42-3)**
l-Adrenaline tartrate	**(51-42-3)**
Adrenalin hydrochloride	**(55-31-2)**
Adrenalin in oil	**(51-43-4)**
Adrenalin-medihaler	**(51-43-4)**
Adrenamine	**(51-43-4)**
Adrenan	**(51-43-4)**
Adrenapax	**(51-43-4)**
Adrenasol	**(51-43-4)**
Adrenatrate	**(51-42-3)**
Adrenatrate	**(51-43-4)**
Adrenine	**(51-43-4)**
Adrenodis	**(51-43-4)**
Adrenohorma	**(51-43-4)**
Adrenor	**(51-41-2)**
Adrenosan	**(51-43-4)**
Adrenutol	**(51-43-4)**
Adreson	**(50-04-4)**
Adriacin	**(25316-40-9)**
Adriamycin	**(23214-92-8)**
Adriamycin	**(25316-40-9)**
Adriamycin, hydrochloride	**(25316-40-9)**
Adriamycin semiquinone	**(23214-92-8)**
Adrianol	**(61-76-7)**
Adriblastin	**(23214-92-8)**
Adriblastin	**(25316-40-9)**
Adriblastina	**(23214-92-8)**
Adriblastina	**(25316-40-9)**
Adrin	**(51-43-4)**
Adrine	**(51-43-4)**
Adrizine	**(51-63-8)**
Adroidin	**(434-07-1)**
Adronal	**(108-93-0)**
Adroyd	**(434-07-1)**
Adrucil	**(51-21-8)**
Adsorbonac	**(62-33-9)**
Adulsin	**(9004-67-5)**
Adumbran	**(604-75-1)**
Advantage	**(55285-14-8)**
Advastab 47	**(131-53-3)**
Advastab 52	**(77-58-7)**
Advastab 401	**(128-37-0)**
Advastab 405	**(119-47-1)**
Advastab 406	**(90-66-4)**
Advastab 800	**(123-28-4)**
Advastab 802	**(693-36-7)**
Advastab DBTM	**(78-04-6)**
Advastab 17 MO	**(26401-97-8)**
Advastab PS 802	**(693-36-7)**
Advastab T290	**(78-04-6)**
Advastab T340	**(78-04-6)**
Advastab TM 180 (9CI)	**(12750-71-9)**
Advastab TM 692 (9CI)	**(78232-98-1)**
Advastab TM 181FS (9CI)	**(55353-53-2)**
Advawachs 280	**(110-30-5)**
Advawax	**(110-30-5)**
Advawax 140	**(31566-31-1)**
Advawax 275	**(110-30-5)**
Advawax 280	**(110-30-5)**
Adynol	**(56-65-5)**
Aedelforsite	**(13983-17-0)**
Aenh (German)	**(759-73-9)**
Aephenal	**(50-06-6)**
Aero-Cyanamid	**(156-62-7)**
Aero	**(108-78-1)**
Aero Cyanamid Granular	**(156-62-7)**
Aero Cyanamid Special Grade	**(156-62-7)**
Aero Cyanate	**(590-28-3)**
Aero Cyanate Weedkiller	**(590-28-3)**
Aero Liquid HCN	**(74-90-8)**
Aerodag G	**(7782-42-5)**
Aeroflat 238	**(33619-92-0)**
Aerol 1	**(52-68-6)**
Aerolite 300	**(9011-05-6)**
Aerolite A 300	**(9011-05-6)**
Aerolite FFD	**(9011-05-6)**
Aerol 1 (Pesticide)	**(52-68-6)**
Aeron (Fiber)	**(63428-84-2)**
Aeroseb-HC	**(50-23-7)**
Aeroseb-dex	**(50-02-2)**
Aerosil	**(7631-86-9)**
Aerosol GPG	**((577-11-7)**
Aerosol OT	**(577-11-7)**
Aerosol OT 75	**(577-11-7)**
Aerosol OT-B	**(577-11-7)**
Aerotex 3470	**(63231-51-6)**
Aerotex Reactant No. 100	**(136-84-5)**
Aerotex glyoxal 40	**(107-22-2)**
Aerothene MM	**(75-09-2)**
Aerothene tt	**(71-55-6)**
Aeroxanthate	**(2720-73-2)**
Aeroxanthate 343	**(140-93-2)**
Aesculetin	**(305-01-1)**
Aesculetin dimethyl ether	**(120-08-1)**
Aethaldiamin (German)	**(107-15-3)**
Aethanethiol (German)	**(75-08-1)**
Aethanol (German)	**(64-17-5)**

Aethanolamin (German) (141-43-5)
Aethazol (94-19-9)
Aether (60-29-7)
Aether oenanthicus (106-30-9)
2-Aethinylbutanol (77-75-8)
Aethinyl-cyclohexyl-carbamat (German) (126-52-3)
Aethinyoestradiol (German) (57-63-6)
Aethionin (67-21-0)
Aethoheptazin (77-15-6)
Aethon (122-51-0)
Aethosuximide (German) (77-67-8)
2-(2-Aethoxy-aethoxy)-aethy-3,6,9-trioxa-undecan (German) (51-14-9)
2-Aethoxy-aethylacetat (German) (111-15-9)
3-(Aethoxycarbonylaminophenyl)-N-phenyl-carbamat (German) (13684-56-5)
p-Aethoxyphenylharnstoff (German) (150-69-6)
5-Aethoxy-3-trichlormethyl-1,2,4-thiadiazol (German) (2593-15-9)
S-Aethyl-N,N-dipropylthiolcarbamat (German) (759-94-4)
N-Aethylacetamid (German) (625-50-3)
Aethylacetat (German) (141-78-6)
Aethylacrylat (German) (140-88-5)
Aethyl-aethanol-nitrosoamin (German) (13147-25-6)
Aethylalkohol (German) (64-17-5)
Aethylamine (German) (75-04-7)
2-Aethylamino-4-sek.butylamino-6-chlor-1,3,5-triazin (German) (7286-69-3)
4-Aethylamino-2-tert-butylamino-6-methylthio-s-triazin (German) (886-50-0)
2-Aethylamino-4-chlor-6-isopropylamino-1,3,5-triazin (German) (1912-24-9)
2-Aethylamino-4-isopropylamino-6-chlor-1,3,5-triazin (German) (1912-24-9)
Aethylanilin (German) (103-69-5)
Aethylbenzol (German) (100-41-4)
Aethylbutylketon (German) (106-35-4)
Aethyl-n-butyl-nitrosoamin (German) (4549-44-4)
Aethyl-t-butyl-nitrosoamin (German) (3398-69-4)
Aethylcarbamat (German) (51-79-6)
Aethylchlorid (German) (75-00-3)
Aethyl-chlorvynol (113-18-8)
1-Aethyl-cyclohexanol-(1) (German) (1940-18-7)
Aethyl-2-(3',5'-dijoD-4'-oxybenzoyl)-3 cumaron (German) (68-90-6)
O-Aethyl-S,S-diphenyl-dithiophosphat (German) (17109-49-8)
1,1'-Aethylen-2,2'-bipyridinium-dibromid (German) (85-00-7)
Aethylenbromid (German) (106-93-4)
Aethylenchlorid (German) (107-06-2)
Aethylenechlorhydrin (German) (107-07-3)
Aethylenediamin (German) (107-15-3)
Aethylenglykolaetheracetat (German) (111-15-9)
Aethylenglykolmethylaetheracetat (German) (110-49-6)
Aethylenglykol-monomethylaether (German) (109-86-4)
Aethylenimin (German) (151-56-4)
Aethylenoxid (German) (75-21-8)
Aethylensulfid (German) (420-12-2)
N-Aethylformamid (German) (627-45-2)
Aethylformiat (German) (109-94-4)
S-Aethyl-N-hexahydro-1H-azepinthiolcarbamat (German) (2212-67-1)
2-Aethylhexanol (German) (104-76-7)
Aethylidenchlorid (German) (75-34-3)
Aethylis (75-00-3)
Aethylis chloridum (75-00-3)
2-(O-Aethyl-N-isopropylamidothiophosphoryloxy)-benzo-saeure-isopropylester (German) (25311-71-1)
Aethyl-isopropyl-nitrosoamin (German) (16339-04-1)
Aethylmercaptan (German) (75-08-1)
Aethylmethylketon (German) (78-93-3)
2-Aethyl-6-methyl-N-(1-methyl-2-methoxyaethyl)-chloracetanilid (German) (51218-45-2)
O-Aethyl-O-(3-methyl-4-methylthiophenyl)-isopropylamido-phosphorsaeureester (German) (22224-92-6)
1-Aethyl-7-methyl-1,8-naphthyridin-4-on-3-karbonsaeure (German) (389-08-2)
O-Aethyl-O-(4-nitro-phenyl)-phenyl-monothiophosphonat (German) (2104-64-5)
Aethylnitroso-harnstoff (German) (759-73-9)
Aethylnitrosourethan (German) (614-95-9)
3-Aethyl-pentanol-(3) (German) (597-49-9)
5-Aethyl-5-pentyl-(2')-barbitursaeure (German) (115-58-2)
O-Aethyl-S-phenyl-aethyl-dithiophosphonat (German) (944-22-9)
5-Aethyl-5-phenyl-hexahydropyrimidin-4,6-dion (German) (125-33-7)
4-Aethyl-1-phospha-2,6,7-trioxabicyclo(2.2.2)octan (German) (824-11-3)
4-Aethyl-1-phospha-2,6,7-trioxabicyclo(2.2.2)octan-1-oxid (German) (1005-93-2)
N-Aethylpiperidin (German) (766-09-6)
N-(1-Aethylpropyl)-3,4-dimethyl-2,6-dinitroanilin (German) (40487-42-1)
N-(1-Aethylpropyl)-2,6-dinitro-3,4-xylidin (German) (40487-42-1)
Aethylpropylketon (German) (589-38-8)
Aethylrhodanid (German) (542-90-5)
1-N-Aethylsisomicin (56391-56-1)
O-Aethyl-O-(2,4,5-trichlorphenyl)-aethylthionophosphonat (German) (327-98-0)
Aethylurethan (German) (51-79-6)
Aethyl-vinyl-nitrosoamin (German) (13256-13-8)
Aetina (536-33-4)
Aetiva (536-33-4)
Afi-Phyllin (479-18-5)
Afi-Tiazin (92-84-2)
Afalon (330-55-2)
Afalon Inuron (330-55-2)
Afalonu (Polish) (330-55-2)
Afastogen Blue 5040 (3468-11-9)
Afatin (51-63-8)
Afaxin (68-26-8)
Afcolac B 101 (9003-55-8)
Afcolene (9003-53-6)
Afcolene (9003-54-7)
Afcolene 666 (9003-53-6)
Afcolene S 100 (9003-53-6)
Afesin (1746-81-2)
Afibrin (60-32-2)
Aficida (23103-98-2)
Aficide (58-89-9)
Afiline (34681-10-2)
Afko-Hist (91-84-9)
Afko-sal (65-45-2)
Aflatoxicol M1 (64330-03-6)
Aflatoxin (1402-68-2)
Aflatoxin B (1162-65-8)
Aflatoxin B1 (1162-65-8)
Aflatoxin B2 (7220-81-7)
Aflatoxin G1 (1165-39-5)
Aflatoxin G2 (7241-98-7)
Aflatoxin M1 (6795-23-9)
Aflatoxin M2 (6885-57-0)
Aflix (2540-82-1)
Afolat (Czech) (52-46-0)
Afos (2595-54-2)
Afsillin (6130-64-9)
Afugan (13457-18-6)
Ag 1500 (7782-42-5)
Agar-Agar (9002-18-0)
Agar-Agar Gum (9002-18-0)
Agaldog (84-17-3)
Agalite (14807-96-6)
Agallol (123-88-6)
Agallolat (123-88-6)
Agalol (123-88-6)
Agar (9002-18-0)
Agar Agar Flake (9002-18-0)

Agarin	(2763-96-4)	Agrisil	(327-98-0)
Agaritine	(2757-90-6)	Agrisol G-20	(58-89-9)
Agate	(14808-60-7)	Agristrep	(3810-74-0)
Agedoite	(70-47-3)	Agritan	(50-29-3)
Ageflex AMA	(96-05-9)	Agritol	(68038-71-1)
Ageflex BGE	(2426-08-6)	Agritox	(327-98-0)
Ageflex EGDM	(97-90-5)	Agritox	(94-74-6)
Ageflex FA-10	(1330-61-6)	Agriya 1050	(122-14-5)
Ageflex FA-2	(2426-54-2)	Agrizan	(1332-40-7)
Ageflex FM 246	(142-90-5)	Agroceres	(76-44-8)
Ageflex FM-1	(2867-47-2)	Agrocide	(58-89-9)
Ageflex FM-10	(29964-84-9)	Agrocide 2	(58-89-9)
Ageflex FM-4	(3775-90-4)	Agrocide 7	(58-89-9)
Ageflex n-HA	(2499-95-8)	Agrocide 6G	(58-89-9)
Ageflex PGE	(122-60-1)	Agrocide III	(58-89-9)
Agel TG 37	(8004-09-9)	Agrocide WP	(58-89-9)
Agel TG 67	(8004-09-9)	Agroform	(9011-05-6)
Agelon	(8073-77-6)	Agroforotox	(52-68-6)
Agelon 1798	(8073-77-6)	Agromicina	(60-54-8)
Agenap	(1338-24-5)	Agronaa	(86-87-3)
Agene	(10025-85-1)	Agronexit	(58-89-9)
Agent 504	(112-30-1)	Agrosan	(62-38-4)
Agent AT 717	(9003-39-8)	Agrosan GN 5	(62-38-4)
Agent Blue	(75-60-5)	Agrosand	(62-38-4)
Agent I	(58629-01-9)	Agrosol	(502-39-6)
Agent Orange	(39277-47-9)	Agrosol S	(133-06-2)
Agerite	(103-16-2)	Agrotect	(94-75-7)
Agerite	(135-88-6)	Agrothion	(122-14-5)
Agerite	(74-31-7)	Agrox 2-Way and 3-Way	(133-06-2)
Agerite 150	(101-73-5)	Agroxon	(94-74-6)
Agerite Alba	(103-16-2)	Agroxone	(94-74-6)
Agerite DPPD	(74-31-7)	Agroxone 3	(3653-48-3)
Agerite Iso	(101-73-5)	Agrypnal	(50-06-6)
Agerite MA	(26780-96-1)	Agstone	(1317-65-3)
Agerite Powder	(135-88-6)	Aguathol	(129-67-9)
Agerite Resin D	(147-47-7)	Ahco Direct Black GX	(1937-37-7)
Agerite White	(93-46-9)	Ahco Direct Black RW	(2429-83-6)
Agidol	(128-37-0)	Ahcocid Carmine 2G	(3734-67-6)
Agidol 1	(128-37-0)	Ahcocid Fast Scarlet R	(3761-53-3)
Agilene	(9002-88-4)	Ahcoquinone Blue ASTB Base	(128-85-8)
Agilon	(63428-84-2)	Ahcoquinone Blue IR Base	(81-48-1)
Agilon D	(63428-84-2)	Ahcoquinone Red S	(130-22-3)
Agiolan	(68-26-8)	Ahcovat Blue BCF	(130-20-1)
Aglaiene	(3856-25-5)	Ahcovat Brilliant Violet 2R	(1324-55-6)
Aglicid	(64-77-7)	Ahcovat Brilliant Violet 4R	(1324-55-6)
Aglypt	(21087-63-8)	Ahcovat Brown BR	(2475-33-4)
Agnin	(8006-54-0)	Ahcovat Golden Orange G	(128-70-1)
Agnolin	(8006-54-0)	Ahcovat Jade Green B	(128-58-5)
Agnolin NO 1	(8006-54-0)	Ahcovat Jade Green BDA	(128-58-5)
Agofollin	(113-38-2)	Ahcovat Navy Blue BR	(1324-54-5)
Agontan	(66-02-4)	Ahcovat Olive ARN	(2379-81-9)
Agoral	(77-09-8)	Ahcovat Olive R	(2379-81-9)
Agostilben	(56-53-1)	Ahcovat Pink FFD	(2379-74-0)
Agotan	(132-60-5)	Ahcovat Printing Golden Yellow	(128-66-5)
Agovirin	(57-85-2)	Ahcovat Printing Jade Green B	(128-58-5)
Agral 90	(9016-45-9)	Ahcovat Printing Jade Green BDA	(128-58-5)
Agramed	(87-10-5)	Ahcovat Printing Navy Blue XSA	(1324-54-5)
Agrazine	(92-84-2)	Ahcovat Printing Orange R	(3263-31-8)
Agreflan	(1582-09-8)	Ahcovat Printing Pink FF	(2379-74-0)
Agri-Mycin	(3810-74-0)	Ahcovat Rubine R	(4203-77-4)
Agria 1050	(122-14-5)	Ahydol (Russian)	(1709-70-2)
Agricide Maggot Killer (F)	(8001-35-2)	Ahygroscopin-B	(606-58-6)
Agricultural limestone	(1317-65-3)	Aibn	(78-67-1)
Agridip	(56-72-4)	Aiglonyl	(15676-16-1)
Agriflan 24	(1582-09-8)	Aimco Systox	(301-12-2)
Agrimycin 17	(57-92-1)	Aimsan	(2597-03-7)
Agrion	(2702-72-9)	Air-Flo Green	(10290-12-7)

Name	CAS No.
Airbron	**(616-91-1)**
Airedale Black 2BG	**(1064-48-8)**
Airedale Black BHD	**(2429-73-4)**
Airedale Black ED	**(1937-37-7)**
Airedale Black RWD	**(2429-83-6)**
Airedale Blue 2BD	**(2602-46-2)**
Airedale Blue D	**(2429-74-5)**
Airedale Blue FFD	**(2610-05-1)**
Airedale Blue IN	**(860-22-0)**
Airedale Blue RL	**(3861-73-2)**
Airedale Brown BSD	**(2429-81-4)**
Airedale Brown MD	**(2429-82-5)**
Airedale Carmoisine	**(3567-69-9)**
Airedale Green BWD	**(3626-28-6)**
Airedale Orange II	**(633-96-5)**
Airedale Red A	**(1658-56-6)**
Airedale Red KD	**(2610-11-9)**
Airedale Red PGM	**(3567-65-5)**
Airedale Scarlet 3BD	**(6358-29-8)**
Airedale Scarlet GM	**(10169-02-5)**
Airedale Violet ND	**(2586-60-9)**
Airedale Yellow E	**(6373-74-6)**
Airedale Yellow 3GM	**(6375-55-9)**
Airedale Yellow RD	**(1325-37-7)**
Airedale Yellow T	**(1934-21-0)**
Airone	**(12071-83-9)**
Aiselazine	**(304-20-1)**
Aisemide	**(54-31-9)**
Aitk	**(57-06-7)**
Aizen Acid Violet 5BH	**(1694-09-3)**
Aizen Amaranth	**(915-67-3)**
Aizen Auramine	**(2465-27-2)**
Aizen Auramine Conc. SFA	**(2465-27-2)**
Aizen Auramine OH	**(2465-27-2)**
Aizen Brilliant Acid Pure Blue VH	**(129-17-9)**
Aizen Brilliant Blue FCF	**(2650-18-2)**
Aizen Brilliant Scarlet 3RH	**(2611-82-7)**
Aizen Cathilon Orange GL	**(3056-93-7)**
Aizen Cathilon Orange GLH	**(3056-93-7)**
Aizen Cathilon Red GTLH	**(14097-03-1)**
Aizen Chrome Violet BH	**(2092-55-9)**
Aizen Crystal Violet	**(548-62-9)**
Aizen Crystal Violet Extra Pure	**(548-62-9)**
Aizen Diamond Green GH	**(633-03-4)**
Aizen Direct Black BH	**(2429-73-4)**
Aizen Direct Blue 2BH	**(2602-46-2)**
Aizen Direct Brown MH	**(2429-82-5)**
Aizen Direct Dark Green BH	**(3626-28-6)**
Aizen Direct Deep Black EH	**(1937-37-7)**
Aizen Direct Deep Black GH	**(1937-37-7)**
Aizen Direct Deep Black RH	**(1937-37-7)**
Aizen Direct Sky Blue 5BH	**(2429-74-5)**
Aizen Direct Violet LNH	**(6426-67-1)**
Aizen Eosine GH	**(17372-87-1)**
Aizen Eosine GH	**(548-26-5)**
Aizen Erythrosine	**(16423-68-0)**
Aizen Food Blue No. 2	**(3844-45-9)**
Aizen Food Green No. 3	**(2353-45-9)**
Aizen Food Orange No. 1	**(523-44-4)**
Aizen Food Orange No. 2	**(2646-17-5)**
Aizen Food Red No. 5	**(3118-97-6)**
Aizen Food Violet No. 1	**(1694-09-3)**
Aizen Food Yellow No. 5	**(2783-94-0)**
Aizen Magenta	**(632-99-5)**
Aizen Malachite Green	**(569-64-2)**
Aizen Malachite Green Crystals	**(569-64-2)**
Aizen Metanil Yellow	**(587-98-4)**

Name	CAS No.
Aizen Methylene Blue BH	**(61-73-4)**
Aizen Methylene Blue FZ	**(61-73-4)**
Aizen Naphthol Orange I	**(523-44-4)**
Aizen Orange I	**(523-44-4)**
Aizen Ponceau RH	**(3761-53-3)**
Aizen Primula Brown BRLH	**(16071-86-6)**
Aizen Primula Brown PLH	**(16071-86-6)**
Aizen Primula Red 4BH	**(2610-11-9)**
Aizen Rhodamine B Base	**(509-34-2)**
Aizen Rhodamine BH	**(81-88-9)**
Aizen Rhodamine BHC	**(81-88-9)**
Aizen Tartrazine	**(1934-21-0)**
Aizen Uranine	**(518-47-8)**
Aizen Victoria Blue BOH	**(2185-86-6)**
Ajax GMO	**(25496-72-4)**
Ajinomoto	**(142-47-2)**
Akar	**(510-15-6)**
Akar 50	**(510-15-6)**
Akar 338	**(510-15-6)**
Akarithion	**(786-19-6)**
Akaritox	**(116-29-0)**
Aketdrin	**(60-13-9)**
Akhnot	**(481-39-0)**
Akiriku Rhodamine B	**(81-88-9)**
Aklomix-3	**(121-19-7)**
Akotin	**(59-67-6)**
Akro-Mag	**(1309-48-4)**
Akro-Zinc Bar 85	**(1314-13-2)**
Akro-Zinc Bar 90	**(1314-13-2)**
Akrichin	**(83-89-6)**
Akrichin (Czech)	**(69-05-6)**
Akridin (Czech)	**(260-94-6)**
Akrolein (Czech)	**(107-02-8)**
Akroleina (Polish)	**(107-02-8)**
Akrylamid (Czech)	**(79-06-1)**
Akrylanem etylu (Polish)	**(140-88-5)**
Akrylonitril (Czech)	**(107-13-1)**
Akrylonitryl (Polish)	**(107-13-1)**
Aktamin	**(51-41-2)**
Aktedrin	**(60-13-9)**
Aktidion (Czech)	**(66-81-9)**
Aktikon	**(1912-24-9)**
Aktikon PK	**(1912-24-9)**
Aktinit A	**(1912-24-9)**
Aktinit PK	**(1912-24-9)**
Aktinit S	**(122-34-9)**
Aktivex	**(123-03-5)**
Aktivin	**(127-65-1)**
Akton	**(1757-18-2)**
Akulon	**(105-60-2)**
Akulon	**(25038-54-4)**
Akulon M 2W	**(105-60-2)**
Akulon M 2W	**(25038-54-4)**
Akuripetto VH	**(9011-14-7)**
Akyporox O 50	**(9004-96-0)**
Akyporox S 100	**(9004-99-3)**
Akyposal ALS 33	**(2235-54-3)**
Akyposal SDS	**(151-21-3)**
Akyposal TLS	**(139-96-8)**
Akzo Chemie Maneb	**(12427-38-2)**
Al-Alchili (Italian)	**(1191-15-7)**
Al-Diisobutyl	**(1191-15-7)**
Al-Phos	**(20859-73-8)**
Ala Tet	**(64-75-5)**
Alabaster No. 3	**(2783-94-0)**
Alachlor	**(15972-60-8)**
Alachlore	**(15972-60-8)**

Alacid **(9000-71-9)**
Alacine **(156-51-4)**
Alamine 6 **(143-27-1)**
Alamine 7 **(124-30-1)**
Alamine 11 **(112-90-3)**
Alamine 304 **(102-87-4)**
Alamine 308 **(1116-76-3)**
Alamine 336 **(1116-76-3)**
β-Alaminenitrile **(151-18-8)**
Alanap-3 **(132-67-2)**
Alanap **(132-66-1)**
Alanap 1 **(132-66-1)**
Alanap 10G AT **(132-66-1)**
Alanape **(132-66-1)**
Alanate **(9000-71-9)**
Alane **(7784-21-6)**
Alanex **(15972-60-8)**
Alaninamide, N-carboxy-β-alanyl-l-tryptophyl-l-methionyl-l-aspartylphenyl-, N-tert-butyl ester, L **(5534-95-2)**
(S)-Alanine **(56-41-7)**
Alanine **(56-41-7)**
Alanine, β **(107-95-9)**
L(+)-Alanine **(56-41-7)**
L-(+)-Alanine **(56-41-7)**
L-Alanine **(56-41-7)**
L-α-Alanine **(56-41-7)**
α-Alanine **(56-41-7)**
(L)-Alanine (9CI) **(56-41-7)**
β-Alanine (9CI) **(107-95-9)**
Alanine, L- (8CI) **(56-41-7)**
DL-Alanine-15N (9CI) **(71261-64-8)**
Alanine-15N (6CI,7CI) **(71261-64-8)**
L-Alanine, N-acetyl- (9CI) **(97-69-8)**
Alanine, N-acetyl-, L- (8CI) **(97-69-8)**
β-Alanine, N-(aminoiminomethyl)-N-methyl- (9CI) **(35404-55-8)**
Alanine, N-(2-amino-2-methylpropyl)-2-methyl- (9CI) **(90853-22-8)**
Alanine, N-benzoyl-N-(3,4-dichlorophenyl)-, ethyl ester, DL- **(22212-55-1)**
Alanine, N-benzoyl-N-(3,4-dichlorophenyl)-, ethyl ester, L **(33878-50-1)**
Alanine, N-benzoyl-, methyl ester **(7244-67-9)**
L-Alanine, N-benzoyl-, methyl ester (9CI) **(7244-67-9)**
Alanine, N-benzoyl-, methyl ester, L- (8CI) **(7244-67-9)**
Alanine, 3-(p-(bis(2-chloroethyl)amino)phenyl)-, D **(13045-94-8)**
Alanine, 3-(p-(bis(2-chloroethyl)amino)phenyl)-, DL **(531-76-0)**
Alanine, 3-(p-(bis(2-chloroethyl)amino)phenyl)-, L **(148-82-3)**
β-Alanine, N-(4-((2-bromo-6-chloro-4-nitrophenyl)azo)phenyl)-N-(3-methoxy-3-oxopropyl)-, methyl ester (9CI) **(59709-38-5)**
β-Alanine, N-(2-carboxyethyl)- (9CI) **(505-47-5)**
β-Alanine, N-(2-carboxyethyl)-N-dodecyl-, disodium salt (9CI) **(3655-00-3)**
β-Alanine, N-(2-carboxyethyl)-N-dodecyl-, monosodium salt (9CI) **(14960-06-6)**
Alanine, 3-((carboxymethyl)thio)-, L **(638-23-3)**
Alanine, N-((5-chloro-8-hydroxy-3-methyl-1-oxo-7-isochromanyl)carbonyl)-3-phenyl-, (-) **(303-47-9)**
Alanine, N-((5-chloro-8-hydroxy-3-methyl-1-oxo-7-isochromanyl)carbonyl)-3-phenyl-, ethyl ester, L- **(4865-85-4)**
β-Alanine, N-cyclohexyl- (8CI,9CI) **(26872-84-4)**
Alanine, N-(3-(cyclohexylamino)propyl)- (9CI) **(90853-20-6)**
Alanine, N-(3-(cyclohexylamino)propyl)-2-methyl- (9CI) **(90853-21-7)**
β-Alanine, N-(2,4-dihydroxy-3,3-dimethyl-1-oxobutyl)-, monosodium salt, (R)- (9CI) **(867-81-2)**
β-Alanine, N-(2,4-dihydroxy-3,3-dimethyl-1-oxobutyl)-, (R)- (9CI) **(79-83-4)**
Alanine, 3-(3,4-dihydroxyphenyl)-, (-)- **(59-92-7)**
Alanine, 3-(3,4-dihydroxyphenyl)-, L **(59-92-7)**
Alanine, 3-(3,4-dihydroxyphenyl)-2-methyl-, l-(-) **(555-30-6)**
Alanine, N-(2-(dimethylamino)ethyl)- (9CI) **(90853-19-3)**
Alanine, N-(3-(dimethylamino)propyl)- (9CI) **(90853-18-2)**
DL-Alanine, N-(2,6-dimethylphenyl)-N-(methoxyacetyl)-, methyl ester (9CI) **(57837-19-1)**
DL-Alanine, N-(2,6-dimethylphenyl)-N-(methoxyacetyl)-, methyl ester, Mixt. with ((1,2-ethanediylbis(carbamodithioato))(2-))manganese and ((1,2-ethanediylbis(carbamodithioato))(2-))zinc **(75701-74-5)**
Alanine, N-(2,6-dimethylphenyl)-N-(phenylacetyl)-, methyl ester, dl **(71626-11-4)**
β-Alanine, N-dodecyl- (9CI) **(1462-54-0)**
Alanine, 3-(p-fluorophenyl) **(60-17-3)**
Alanine, 3-(p-fluorophenyl)-, DL **(51-65-0)**
Alanine, 3-(4-(4-hydroxy-3-iodophenoxy)-3,5-diiodophenyl)-, L **(6893-02-3)**
Alanine, N-((8-hydroxy-3-methyl-1-oxo-7-isochromanyl)carbonyl)-3-phenyl-, (-) **(4825-86-9)**
L-Alanine, γ-(hydroxymethylphosphinyl)-L-α-aminobutyryl-L-alanyl **(35597-43-4)**
Alanine, 3-indol-3-yl- **(73-22-3)**
Alanine, N-(methoxyacetyl)-N-(2,6-xylyl)-, methyl ester, DL **(57837-19-1)**
Alanine, N-(3-(2-methyl-1-piperidinyl)propyl)- (9CI) **(90853-23-9)**
Alanine, 3-(methylthio)-, L **(7728-98-5)**
Alanine nitrogen mustard **(148-82-3)**
Alanine, 3-phenyl- **(63-91-2)**
L-Alanine, phenyl- **(63-91-2)**
Alanine, phenyl- (8CI) **(63-91-2)**
Alanine, phenyl-, DL- (8CI) **(150-30-1)**
Alanine, phenyl-, D **(673-06-3)**
Alanine, phenyl-, L **(63-91-2)**
Alanine, 3-selenyl-, (+-) **(18312-66-8)**
DL-Alanine, 3-selenyl- (9CI) **(18312-66-8)**
Alanine, 3-selenyl-, DL (8CI) **(18312-66-8)**
β-Alaninol **(156-87-6)**
Alanox **(15972-60-8)**
Alantan **(97-59-6)**
Alanyl-2,6-xylidide **(41708-72-9)**
Alapurin **(8006-54-0)**
Alar **(1596-84-5)**
Alar-85 **(1596-84-5)**
Alaren **(9000-71-9)**
Alatate **(9000-71-9)**
Alathon **(25038-59-9)**
Alathon **(9002-88-4)**
Alathon 5B **(9002-88-4)**
Alathon 14 **(9002-88-4)**
Alathon 15 **(9002-88-4)**
Alathon 71XHN **(9002-88-4)**
Alathon 1560 **(9002-88-4)**
Alathon 6600 **(9002-88-4)**
Alathon 7026 **(9002-88-4)**
Alathon 7040 **(9002-88-4)**
Alathon 7050 **(9002-88-4)**
Alathon 7140 **(9002-88-4)**
Alathon 7511 **(9002-88-4)**
Alatox 480 **(15972-60-8)**
Alaun (German) **(7429-90-5)**
Alazin **(156-51-4)**
Alazine **(156-51-4)**
Alba-Dome **(103-16-2)**
Albagel Premium USP 4444 **(1302-78-9)**
Albalith **(1314-98-3)**
Albalon Liquifilm **(550-99-2)**
Albamine **(144-80-9)**
Albego **(36104-80-0)**
Albemap **(51-63-8)**
Albexan **(63-74-1)**
Albigen A **(9003-39-8)**
Albiotic **(154-21-2)**
Alboline **(8012-95-1)**
Albone **(7722-84-1)**

Albone 35	**(7722-84-1)**
Albone 35CG	**(7722-84-1)**
Albone 50	**(7722-84-1)**
Albone 50CG	**(7722-84-1)**
Albone 70	**(7722-84-1)**
Albone 70CG	**(7722-84-1)**
Alboral	**(439-14-5)**
Albosal	**(63-74-1)**
Albucid	**(144-80-9)**
Albumin, Blood serum, Labeled with iodine-125	**(9048-46-8)**
Albumin, Blood serum, Labeled with iodine-131	**(9048-46-8)**
Albumin, Iodinated I 125 serum	**(9048-46-8)**
Albumin, Iodinated I 131 serum	**(9048-46-8)**
Albumins, Blood serum	**(9048-46-8)**
Albumotope I-125	**(9048-46-8)**
Albumotope I-131	**(9048-46-8)**
Albutest	**(115-39-9)**
Alcalase	**(9014-01-1)**
Alcanfor	**(464-49-3)**
Alcapton	**(451-13-8)**
Alchloquin	**(130-26-7)**
Alcide	**(10049-04-4)**
Alcloxa	**(1317-25-5)**
Alcloxum (Latin)	**(1317-25-5)**
Alcon-Efrin	**(61-76-7)**
Alcoa 331	**(21645-51-2)**
Alcoa C 30BF	**(21645-51-2)**
Alcoa F 1	**(1344-28-1)**
Alcoa Sodium Fluoride	**(7681-49-4)**
Alcobam NM	**(128-04-1)**
Alcobam ZM	**(137-30-4)**
Alcobon	**(2022-85-7)**
Alcogum	**(9003-01-4)**
Alcohol	**(64-17-5)**
Alcohol, Anhydrous	**(64-17-5)**
Alcohol C-8	**(111-87-5)**
Alcohol C-9	**(143-08-8)**
Alcohol C-10	**(112-30-1)**
Alcohol C-11	**(112-42-5)**
Alcohol C-12	**(112-53-8)**
Alcohol C-12	**(27342-88-7)**
Alcohol C-16	**(36653-82-4)**
Alcohol, dehydrated	**(64-17-5)**
Alcohol ethoxylate, phosphate ester, sodium salt	**(37340-60-6)**
Alcohols, C6-12	**(68603-15-6)**
Alcohols, C8-18	**(68551-07-5)**
Alcohols, C9-11	**(66455-17-2)**
Alcohols, C10-16	**(67762-41-8)**
Alcohols, C12-15	**(63393-82-8)**
Alcohols, C16-18	**(67762-27-0)**
Alcohols, C6-12, ethoxylated	**(68439-45-2)**
Alcohols, C7-21, ethoxylated	**(68991-48-0)**
Alcohols, C8-22, ethoxylated	**(69013-19-0)**
Alcohols, C11-15-secondary, ethoxylated	**(68131-40-8)**
Alcohols, C12-14, ethoxylated	**(68439-50-9)**
Alcohols, C12-14, ethoxylated propoxylated	**(68439-51-0)**
Alcohols, C12-15, ethoxylated	**(68131-39-5)**
Alcohols, C12-16, ethoxylated	**(68551-12-2)**
Alcohols, C12-18, ethoxylated	**(68213-23-0)**
Alcohols, C12-18, ethoxylated propoxylated	**(69227-21-0)**
Alcohols, C14-15, ethoxylated	**(68951-67-7)**
Alcohols, C16-18, ethoxylated	**(68439-49-6)**
Alcohols, C16-18 and C18-unsatd., ethoxylated	**(68920-66-1)**
Alcoid	**(357-56-2)**
Alcojel	**(67-63-0)**
Alcolec S	**(8002-43-5)**
Alcool allilco (Italian)	**(107-18-6)**
Alcool allylique (French)	**(107-18-6)**
Alcool amilico (Italian)	**(123-51-3)**
Alcool amylique (French)	**(71-41-0)**
Alcool butylique (French)	**(71-36-3)**
Alcool butylique secondaire (French)	**(78-92-2)**
Alcool butylique tertiaire (French)	**(75-65-0)**
Alcool ethylique (French)	**(64-17-5)**
Alcool etilico (Italian)	**(64-17-5)**
l'Alcool N-heptylique primaire (French)	**(111-70-6)**
Alcool isoamylique (French)	**(123-51-3)**
Alcool isobutylique (French)	**(78-83-1)**
Alcool isopropilico (Italian)	**(67-63-0)**
Alcool isopropylique (French)	**(67-63-0)**
Alcool methyl amylique (French)	**(108-11-2)**
Alcool methylique (French)	**(67-56-1)**
Alcool metilico (Italian)	**(67-56-1)**
Alcool propilico (Italian)	**(71-23-8)**
Alcool propylique (French)	**(71-23-8)**
Alcophobin	**(97-77-8)**
Alcopol O	**(577-11-7)**
Alcosolve	**(67-63-0)**
Alcotex 88/05	**(9002-89-5)**
Alcotex 88/10	**(9002-89-5)**
Alcowax 6	**(9002-88-4)**
Aldo-28	**(31566-31-1)**
Aldo-72	**(31566-31-1)**
Aldacol Q	**(9003-39-8)**
Aldactazide	**(52-01-7)**
Aldactide	**(52-01-7)**
Aldactone	**(52-01-7)**
Aldactone A	**(52-01-7)**
Aldanil	**(149-44-0)**
Aldecarb	**(116-06-3)**
Aldehydate d'ammoniaque (French)	**(75-39-8)**
Aldehyde B	**(103-95-7)**
Aldehyde C-6	**(66-25-1)**
Aldehyde C-8	**(124-13-0)**
Aldehyde C-9	**(124-19-6)**
Aldehyde C10	**(112-31-2)**
Aldehyde C-12	**(110-41-8)**
Aldehyde C-14	**(104-67-6)**
Aldehyde C-16	**(77-83-8)**
Aldehyde C-18	**(104-61-0)**
Aldehyde C-12, lauric	**(112-54-9)**
Aldehyde C-14, Myristic	**(124-25-4)**
Aldehyde C-14 Peach	**(104-67-6)**
Aldehyde C-11, undecylenic	**(112-45-8)**
Aldehyde C-11, undecylic	**(112-44-7)**
Aldehyde M.N.A.	**(110-41-8)**
Aldehyde acetique (French)	**(75-07-0)**
Aldehyde acrylique (French)	**(107-02-8)**
Aldehyde ammonia	**(75-39-8)**
Aldehyde butyrique (French)	**(123-72-8)**
Aldehydecollidine	**(104-90-5)**
Aldehyde crotonique (French)	**(123-73-9)**
Aldehyde 2-ethylbutyrique (French)	**(97-96-1)**
Aldehyde formique (French)	**(50-00-0)**
Aldehyde propionique (French)	**(123-38-6)**
Aldehyd hydratropovy (Czech)	**(93-53-8)**
2,5-Aldehydine	**(104-90-5)**
Aldehydine	**(104-90-5)**
Aldehyd mravenci (Czech)	**(50-00-0)**
Aldehyd skoricovy (Czech)	**(104-55-2)**
Aldeide acetica (Italian)	**(75-07-0)**
Aldeide acrilica (Italian)	**(107-02-8)**
Aldeide butirrica (Italian)	**(123-72-8)**
Aldeide formica (Italian)	**(50-00-0)**

Name	CAS No.
Alderlin hydrochloride	**(51-02-5)**
Aldicarb	**(116-06-3)**
Aldicarbe (French)	**(116-06-3)**
Aldicarb oxime	**(1646-75-9)**
Aldicarb sulfone	**(1646-88-4)**
Aldicarb sulfoxide	**(1646-87-3)**
Aldifen	**(51-28-5)**
Aldinamid	**(98-96-4)**
Aldinamide	**(98-96-4)**
Aldo	**(52-86-8)**
Aldo 33	**(123-94-4)**
Aldo 40	**(25496-72-4)**
Aldo 75	**(123-94-4)**
Aldo HMO	**(111-03-5)**
Aldo HMS	**(31566-31-1)**
Aldo MO	**(111-03-5)**
Aldo MO-FG	**(25496-72-4)**
Aldo MS	**(31566-31-1)**
Aldo MSA	**(31566-31-1)**
Aldo MSD	**(123-94-4)**
Aldo MSLG	**(123-94-4)**
Aldo MSLG	**(31566-31-1)**
Aldo To	**(122-32-7)**
Aldol [UN 2839]	**(107-89-1)**
Aldomet	**(41372-08-1)**
Aldomet	**(555-30-6)**
Aldometil	**(41372-08-1)**
Aldometil	**(555-30-6)**
Aldomin	**(41372-08-1)**
Aldomin	**(555-30-6)**
Aldomycin	**(59-87-0)**
Aldosperse L 9	**(9002-92-0)**
Aldoxime	**(107-29-9)**
Aldoxycarb	**(1646-88-4)**
Aldoxycarbe (French)	**(1646-88-4)**
Aldrex	**(309-00-2)**
Aldrex 30	**(309-00-2)**
Aldrex 30 E.C.	**(309-00-2)**
Aldrich	**(91-10-1)**
Aldrin (ACGIH,DOT,OSHA)	**(309-00-2)**
Aldrin, Cast solid (DOT)	**(309-00-2)**
Aldrin, Liquid [NA 2762]	**(309-00-2)**
Aldrin, Solid [NA 2761]	**(309-00-2)**
Aldrine (French)	**(309-00-2)**
Aldrite	**(309-00-2)**
Aldrosol	**(309-00-2)**
Aldyl A	**(9002-88-4)**
Alentin	**(339-43-5)**
Alentol	**(60-13-9)**
Alepsin	**(630-93-3)**
Aleryl	**(58-73-1)**
Alesten	**(144-80-9)**
Aleviatin	**(57-41-0)**
Alexan	**(147-94-4)**
Alfa-Tox	**(333-41-5)**
Alfacillin	**(132-93-4)**
Alfacron	**(18181-70-9)**
Alfamat	**(15879-93-3)**
Alfamethrin	**(52315-07-8)**
Alfanaftilamina (Italian)	**(134-32-7)**
Alfenol 3	**(9002-93-1)**
Alfenol 9	**(9002-93-1)**
Alfentanil	**(71195-58-9)**
Alfentanilum (Latin)	**(71195-58-9)**
Alfeprol (Russian)	**(13655-52-2)**
Alficetyn	**(56-75-7)**
Alfimid	**(77-21-4)**
Alfine	**(9003-17-2)**
Alflorone	**(127-31-1)**
Alfocillin	**(132-93-4)**
Alfol 8	**(111-87-5)**
Alfol 12	**(112-53-8)**
Alfoxylate	**(67375-30-8)**
Alfucin	**(59-87-0)**
Algo-Dex	**(51-63-8)**
Algamon	**(65-45-2)**
Algaroba	**(9000-40-2)**
Algeon 22	**(75-45-6)**
Algiamida	**(65-45-2)**
Algil	**(50-13-5)**
L'-Algiline	**(9005-38-3)**
Algimycin	**(62-38-4)**
Algin	**(9005-38-3)**
Algin (Polysaccharide)	**(9005-38-3)**
Alginate KMF	**(9005-38-3)**
Alginic-acid	**(9005-32-7)**
Alginic acid, ammonium salt (9CI)	**(9005-34-9)**
Alginic acid, calcium salt	**(9005-35-0)**
Alginic acid, potassium salt (9CI)	**(9005-36-1)**
Alginic acid, sodium salt	**(9005-38-3)**
Algipon L-1168	**(9005-38-3)**
Algistat	**(117-80-6)**
Algocor	**(68-90-6)**
Algofrene 22	**(75-45-6)**
Algofrene Type 1	**(75-69-4)**
Algofrene Type 2	**(75-71-8)**
Algofrene Type 5	**(75-43-4)**
Algofrene Type 6	**(75-45-6)**
Algofrene Type 67	**(75-37-6)**
Algol Orange RF	**(3263-31-8)**
Algosediv	**(50-35-1)**
Algotropyl	**(103-90-2)**
Algrain	**(64-17-5)**
Algylen	**(79-01-6)**
Alicop	**(112-00-5)**
Alicyanate	**(590-28-3)**
Alidine	**(144-14-9)**
Alidochlor	**(93-71-0)**
Alidochlore	**(93-71-0)**
Aliette	**(39148-24-8)**
Alimemazine	**(84-96-8)**
Alimet	**(583-91-5)**
Alimezine	**(84-96-8)**
Alindor	**(50-33-9)**
Alipal CO 430	**(9014-90-8)**
Alipal CO 433	**(9014-90-8)**
Alipal CO 436	**(9051-57-4)**
Alipal EP	**(9051-57-4)**
Alipal EP 110	**(9051-57-4)**
Alipal EP 120	**(9051-57-4)**
Aliphatic (C16-C18)alcohol, ethoxylated	**(68439-49-6)**
Aliphatic alcohols, acetate ester	**(68478-35-3)**
Aliphatic compounds (VAN)	**(67254-79-9)**
Aliporina	**(50-59-9)**
Alipur-O	**(2163-69-1)**
Alipur	**(2163-69-1)**
Alipur	**(8015-55-2)**
Aliquat 203	**(7173-51-5)**
Aliquat 206	**(1812-53-9)**
Aliquat 207	**(107-64-2)**
Aliquat 336	**(5137-55-3)**
Aliquat 4	**(112-00-5)**
Aliquat 6	**(112-02-7)**
Aliquat 7	**(112-03-8)**

Aliquat 336N (5137-55-3)
Aliquat 336-PTC (5137-55-3)
Aliseum (439-14-5)
Alisobumal (77-26-9)
Alisobumalum (77-26-9)
Alithon 7050 (9002-88-4)
Alizanthrene Blue RC (130-20-1)
Alizanthrene Navy Blue R (1324-54-5)
Alizanthrene Navy Blue RT (1324-54-5)
Alizarin (72-48-0)
Alizarin B (72-48-0)
Alizarin Brilliant Blue BS (128-86-9)
Alizarin Carmine (Biological stain) (130-22-3)
Alizarin Red (72-48-0)
Alizarin Red S (130-22-3)
Alizarin S (130-22-3)
Alizarin Yellow C (528-21-2)
Alizarina (72-48-0)
Alizarinbordeaux (81-61-8)
Alizarine (72-48-0)
Alizarine 3B (72-48-0)
Alizarine B (72-48-0)
Alizarine Blue A (2666-17-3)
Alizarine Blue AR (2666-17-3)
Alizarine Blue BL (4474-24-2)
Alizarine Bordeaux (81-61-8)
Alizarine Bordeaux B (81-61-8)
Alizarine Brilliant Sapphire R (4368-56-3)
Alizarine Brilliant Sky Blue R (4368-56-3)
Alizarine Carmine Indicator (130-22-3)
Alizarine Chrome Red G (10169-02-5)
Alizarine Cyanine Green Base (128-80-3)
Alizarine Cyanol Grey G (VAN) (116-81-4)
Alizarine Direct Blue AR (2666-17-3)
Alizarine Direct Blue ARA (2666-17-3)
Alizarine Direct Pure Blue R (4368-56-3)
Alizarine Fast Blue RFE (4368-56-3)
Alizarine Green G Base (128-80-3)
Alizarine Indicator (72-48-0)
Alizarine Irisol R Base (81-48-1)
Alizarine L Paste (72-48-0)
Alizarine Lake Red IPX (72-48-0)
Alizarine Lake Red 2P (72-48-0)
Alizarine Lake Red 3P (72-48-0)
Alizarine NAC (72-48-0)
Alizarine Paste 20% bluish (72-48-0)
Alizarine Pure Blue B Base (128-85-8)
Alizarine Red (72-48-0)
Alizarine Red A (130-22-3)
Alizarine Red AS (130-22-3)
Alizarine Red B (72-48-0)
Alizarine Red B2 (72-48-0)
Alizarine Red IP (72-48-0)
Alizarine Red IPP (72-48-0)
Alizarine Red Indicator (130-22-3)
Alizarine Red L (72-48-0)
Alizarine Red S (Biological stain) (130-22-3)
Alizarine Red S Sodium salt (130-22-3)
Alizarine Red SW (130-22-3)
Alizarine Red SZ (130-22-3)
Alizarine Red W (130-22-3)
Alizarine Red WA (130-22-3)
Alizarine Red WS (130-22-3)
Alizarine Red for Wool (130-22-3)
Alizarine S (130-22-3)
Alizarine S Extra Conc. A Export (130-22-3)
Alizarine S Extra Pure A (130-22-3)
Alizarine Sapphire AR (2666-17-3)
Alizarine Sky Blue R (4368-56-3)
Alizarine Supra Blue R (4368-56-3)
Alizarine Supra Sky RA (4368-56-3)
Alizarine Violet 3B Base (81-48-1)
Alizarine Violet N (2092-55-9)
Alizarine Yellow (476-66-4)
Alizarine Yellow C (528-21-2)
Alizarinrot-S (German) (130-22-3)
Alizarinsulfonate (130-22-3)
Aljaden (74051-80-2)
Alk-Enzyme (9014-01-1)
Alkabutazona (50-33-9)
Alkacitron (144-33-2)
Alkali Resistant Red Dark (6471-49-4)
Alkaloid H 3, From colchicum antumnale (477-30-5)
Alkaloid V (4449-51-8)
Alkaloids, Ergot (12126-57-7)
Alkaloids, Veratrum (65072-04-0)
Alkamid (105-60-2)
Alkamid (25038-54-4)
Alkanate 3SL3 (9004-82-4)
Alkanes, C1-2 (68475-57-0)
Alkanes, C10-18 (73138-29-1)
Alkanes, C14-30 (74664-93-0)
Alkanes, C18-70 (70913-86-9)
Alkanes, C2-3 (68475-58-1)
Alkanes, C3-4 (68475-59-2)
Alkanes, C4-C12 (68333-81-3)
Alkanes, C6-18, chloro (68920-70-7)
Alkanes, C10-13, chloro (85535-84-8)
Alkanes, C10-14, chloro (85681-73-8)
Alkanes, C10-21, chloro (84082-38-2)
Alkanes, C10-26, chloro (97659-46-6)
Alkanes, C10-32, chloro (84776-06-7)
Alkanes, C12-13, chloro (71011-12-6)
Alkanes, C12-14, chloro (85536-22-7)
Alkanes, C14-17, chloro (85535-85-9)
Alkanes, C16-27, chloro (84776-07-8)
Alkanes, C16-35, chloro (85049-26-9)
Alkanes, C18-20, chloro (106232-85-3)
Alkanes, C18-28, chloro (85535-86-0)
Alkanes, C22-40, chloro (106232-86-4)
Alkanes, chloro (61788-76-9)
Alkapol PEG-200 (25322-68-3)
Alkapol PEG-300 (25322-68-3)
Alkapol PEG-600 (25322-68-3)
Alkapol PPG-1200 (25322-69-4)
Alkapol PPG-2000 (25322-69-4)
Alkapol PPG-4000 (25322-69-4)
Alkapol PEG-6000 (25322-68-3)
Alkapol PEG-8000 (25322-68-3)
Alkarau (50-55-5)
Alkarsodyl (124-65-2)
Alkaserp (50-55-5)
Alkathene (9002-88-4)
Alkathene 17/04/00 (9002-88-4)
Alkathene 22 300 (9002-88-4)
Alkathene 200 (9002-88-4)
Alkathene ARN 60 (9002-88-4)
Alkathene WJG 11 (9002-88-4)
Alkathene WNG 14 (9002-88-4)
Alkathene XDG 33 (9002-88-4)
Alkathene XJK 25 (9002-88-4)
Alk-aubs (97-77-8)
Alkavervir (65072-04-0)
Alkazene 32 (606-07-5)

Alkazene 42	**(30812-87-4)**
Alkenes, C>10 α-	**(64743-02-8)**
Alkenes, C>8	**(68411-00-7)**
Alkenes, C>10 α-, Polymd.	**(68527-08-2)**
Alkeran	**(148-82-3)**
Alkiron	**(56-04-2)**
Alkofen BP	**(128-37-0)**
Alkofen MBP	**(1817-68-1)**
Alkohol (German)	**(64-17-5)**
Alkohol skoricovy (Czech)	**(104-54-1)**
Alkoholu etylowego (Polish)	**(64-17-5)**
Alkotex	**(9002-89-5)**
Alkovert	**(83-86-3)**
3-Alkoxy-2-hydroxypropyl trimethyl ammonium chloride	**(68187-63-3)**
Alkron	**(56-38-2)**
Alkyd resin	**(68459-31-4)**
Alkyl(C10-C13)benzene	**(67774-74-7)**
Alkyl(C4-C16)benzene	**(68648-86-2)**
Alkyl (C12, C14) glycidyl ether	**(68609-97-2)**
Alkyl (C8,C10) glycidyl ether	**(68609-96-1)**
n-Alkyl(C10-C18)sulfonic acids, sodium salts	**(68037-49-0)**
Alkyl(C8H17 to C18H37) dimethyl 3,4-dichlorobenzyl ammonium chloride	**(8023-53-8)**
Alkyl(C-13) polyethoxylates(ethoxy-6)	**(24938-91-8)**
Alkyl(C9-15)tolyl methyltrimethyl ammonium chloride	**(1399-80-0)**
Alkyl alcohol (C11)	**(68516-18-7)**
Alkyl* amine 2,4-dichlorophenoxyacetate *(100% C12)	**(2212-54-6)**
1-(Alkyl amino)-3-aminopropane	**(68155-37-3)**
Alkyl amino betaine	**(68424-94-2)**
Alkylation Feed	**(68477-83-8)**
Alkylbenzenesulfonate	**(42615-29-2)**
Alkyl-N,N-bis(2-hydroxyethyl)amine	**(71786-60-2)**
Alkyl bis(2-hydroxyethyl) benzyl ammonium chloride	**(61789-68-2)**
Alkyl di(2-hydroxypropyl)amine	**(68516-06-3)**
Alkyl dimethylbenzyl ammonium chloride	**(8001-54-5)**
Alkyl dimethyl 3,4-dichlorobenzyl ammonium chloride	**(8023-53-8)**
Alkyl dimethyl dodecylbenzyl ammonium chloride	**(87175-02-8)**
Alkyl dimethyl 1-napthylmethyl ammonium chloride	**(1733-96-6)**
Alkyldimethyl(phenylmethyl)quaternary ammonium chlorides	**(8001-54-5)**
2-Alkyl isoquinolinium bromide	**(53466-68-5)**
Alkyl pyridines	**(68391-11-7)**
Alkyl trimethyl ammonium bromide	**(68424-92-0)**
Alkyl trimethyl ammonium chloride	**(68391-03-7)**
Allantoin (8CI)	**(97-59-6)**
Allantol	**(97-59-6)**
Allantoxaidin	**(71-33-0)**
Allantoxaidine	**(71-33-0)**
Allbarbital	**(52-43-7)**
Allbri Aluminum Paste and Powder	**(7429-90-5)**
Allbri Natural Copper	**(7440-50-8)**
Alledryl	**(58-73-1)**
(+)-Allelrethonyl (+)-cis,trans-chrysanthemate	**(584-79-2)**
Allene	**(463-49-0)**
Allene, methyl-	**(590-19-2)**
Alleract	**(6138-79-0)**
Allerclor	**(113-92-8)**
Allercorb	**(50-81-7)**
Allergan	**(147-24-0)**
Allergan B	**(58-73-1)**
Allergeval	**(58-73-1)**
Allergical	**(58-73-1)**
Allergican	**(132-22-9)**
Allergin	**(113-92-8)**
Allergin	**(58-73-1)**
Allergina	**(58-73-1)**
Allergisan	**(113-92-8)**
Allergisan	**(132-22-9)**

Allergival	**(147-24-0)**
Allergival	**(58-73-1)**
Alleron	**(56-38-2)**
Allethrin	**(584-79-2)**
d-Allethrin	**(584-79-2)**
d-cis-trans-Allethrin	**(42534-61-2)**
d-t-Allethrin	**(28434-00-6)**
d-trans Allethrin	**(584-79-2)**
d-trans-Allethrin	**(28057-48-9)**
trans-(+)-Allethrin	**(28434-00-6)**
Allethrin (DOT)	**(584-79-2)**
Allethrin I	**(584-79-2)**
Allethrolon	**(551-45-1)**
Allethrolone (VAN)	**(551-45-1)**
d-Allethrolone chrysanthemumate	**(28434-00-6)**
(+)-Allethronyl (+)-trans-chrysanthemumate	**(28434-00-6)**
Alletone	**(3658-77-3)**
Allidochlor	**(93-71-0)**
Allied GC 9160	**(4234-79-1)**
Allied GC-6506	**(3254-63-5)**
Allied PE 617	**(9002-88-4)**
Allile (cloruro di) (Italian)	**(107-05-1)**
Allil-glicidil-etere (Italian)	**(106-92-3)**
1-Allilossi-2,3 epossipropano (Italian)	**(106-92-3)**
Allilowy alkohol (Polish)	**(107-18-6)**
Allional	**(77-02-1)**
Allisan	**(99-30-9)**
Allitin (9CI)	**(73807-55-3)**
Allium sativum	**(8000-78-0)**
Allobarbital	**(52-43-7)**
Allobarbitone	**(52-43-7)**
Allocaine	**(51-05-8)**
Allocaine	**(59-46-1)**
Allochrysoketone	**(6051-98-5)**
Allodene	**(300-62-9)**
α-D-Allofuranose, 1,2:5,6-bis-O-(1-methylethylidene)-, 4-methylbenzenesulfonate (9CI)	**(13964-21-1)**
Allofuranose, 1,2:5,6-di-O-isopropylidene-, p-toluene-sulfonate, α-D- (8CI)	**(13964-21-1)**
Allogel	**(9004-35-7)**
Allomaleic acid	**(110-17-8)**
Allonal	**(77-02-1)**
Alloocimene	**(673-84-7)**
Allo-ocimenol	**(78-70-6)**
Allophanamide	**(108-19-0)**
Allophanic acid amide	**(108-19-0)**
Allophanic acid, 4,4'-o-phenylenebis(3-thio-, diethyl ester	**(23564-06-9)**
Allophanic acid, 4,4'-o-phenylenebis(3-thio-, dimethyl ester	**(23564-05-8)**
Allophanimidic acid (VAN)	**(108-19-0)**
Allopregnane	**(641-85-0)**
Alloprene	**(9006-03-5)**
Allopurinol	**(315-30-0)**
Allorphine	**(62-67-9)**
Allotropal	**(77-75-8)**
Alloxan	**(50-71-5)**
Alloxane	**(50-71-5)**
Alloxan hydrate	**(3237-50-1)**
Alloxan monohydrate	**(2244-11-3)**
Alloxan monohydrate	**(3237-50-1)**
Alloxantin	**(76-24-4)**
Alloxantin, dihydrate	**(76-24-4)**
Alloxol S	**(74051-80-2)**
Alloxydim-Sodium	**(55635-13-7)**
Allozym	**(315-30-0)**
Alltex	**(8001-35-2)**
Alltox	**(8001-35-2)**
Alluminio(cloruro di) (Italian)	**(7446-70-0)**

Alluminio diisobutil-monocloruro (Italian) **(1779-25-5)**
Allura Red **(25956-17-6)**
Allura Red AC **(25956-17-6)**
Allural **(315-30-0)**
Ally **(74223-64-6)**
Allybarbitural **(52-43-7)**
Allyl Al **(107-18-6)**
Allyl acetate [UN 2333] **(591-87-7)**
Allylacetic acid **(591-80-0)**
Allylacetone **(109-49-9)**
Allylacetonitrile **(592-51-8)**
Allyl acrylate **(999-55-3)**
Allyl adipate **(2998-04-1)**
Allyl-alcohol **(107-18-6)**
Allyl alcohol (ACGIH,OSHA) [UN 1098] **(107-18-6)**
Allyl aldehyde **(107-02-8)**
Allylalkohol (German) **(107-18-6)**
Allylamine, N,N,1-trimethyl-3,3-di-2-thienyl- **(524-84-5)**
Allylamine, N,N-diethyl-3,3-di-2-thienyl-1-methyl- **(86-14-6)**
Allylamine [UN 2334] **(107-11-9)**
Allylamine, 3-(p-bromophenyl)-N,N-dimethyl-3-(3-pyridyl)- **(56775-88-3)**
Allylamine, 3,3-di-2-thienyl-N,N,1-trimethyl- **(524-84-5)**
Allylamine, N-methyl-N-nitroso **(4549-43-3)**
p-Allylanisole **(140-67-0)**
Allylbarbital **(77-26-9)**
Allylbarbitone **(77-26-9)**
Allylbarbituric acid **(77-26-9)**
Allylbenzene **(300-57-2)**
5-Allyl-1,3-benzodioxole **(94-59-7)**
Allyl bromide [UN 1099] **(106-95-6)**
5-Allyl-5-sec-butylbarbituric acid **(115-44-6)**
Allyl caproate **(123-68-2)**
Allyl carbamate **(2114-11-6)**
Allylcarbamide **(557-11-9)**
Allyl carbonate **(15022-08-9)**
Allylcatechol methylene ether **(94-59-7)**
4-Allylcatechol-2-methyl ether **(97-53-0)**
Allylchlorid (German) **(107-05-1)**
Allyl chloride (ACGIH,OSHA) [UN 1100] **(107-05-1)**
Allyl chlorocarbonate **(2937-50-0)**
Allyl chlorocarbonate (DOT) **(2937-50-0)**
Allyl chloroformate [UN 1722] **(2937-50-0)**
Allyl cinerin **(584-79-2)**
Allyl cyanide **(109-75-1)**
Allyl cyclohexanepropionate **(2705-87-5)**
3-Allylcyclohexyl propionate **(2705-87-5)**
N-Allyl-7,8-dehydro-4,5-epoxy-3,6-dihydroxymorphinan **(62-67-9)**
N-Allyl-N-desmethylmorphine **(62-67-9)**
Allyl diglycol carbonate **(142-22-3)**
l-N-Allyl-7,8-dihydro-14-hydroxynormorphinone **(465-65-6)**
1-Allyl-3,4-dimethoxybenzene **(93-15-2)**
4-Allyl-1,2-dimethoxybenzene **(93-15-2)**
Allyldioxybenzene methylene ether **(94-59-7)**
Allyl-disulfide **(2179-57-9)**
Allyl disulphide **(2179-57-9)**
Allyle (chlorure d') (French) **(107-05-1)**
Allylene **(74-99-7)**
sym-Allylene **(463-49-0)**
17-Allyl-4,5-α-epoxy-3,14-dihydroxymorphinan-6-one **(465-65-6)**
Allyl 2,3-epoxypropyl ether **(106-92-3)**
Allyl-9,10-epoxystearate **(123-36-4)**
Allylester kyseliny chlormravenci (Czech) **(2937-50-0)**
Allylester kyseliny kapronove (Czech) **(123-68-2)**
Allylester kyseliny methakrylove (Czech) **(96-05-9)**
Allylether **(557-40-4)**
Allyl ether of propylene glycol **(1331-17-5)**
Allylether propylenglykolu (Czech) **(1331-17-5)**
Allyl ethyl ether [UN 2335] **(557-31-3)**
α-Allyl glycerol ether **(123-34-2)**
Allylglycidaether (German) **(106-92-3)**
Allyl glycidyl ether (ACGIH,DOT,OSHA) **(106-92-3)**
4-Allylguaiacol **(97-53-0)**
p-Allylguaiacol **(97-53-0)**
Allyl hexahydrophenylpropionate **(2705-87-5)**
Allyl hexanoate **(123-68-2)**
Allyl homolog of cinerin I **(584-79-2)**
4-Allyl-1-hydroxy-2-methoxybenzene **(97-53-0)**
Allyl (hydroxymethyl)carbamate **(24935-97-5)**
d,l-2-Allyl-4-hydroxy-3-methyl-2-cyclopenten-1-one-d,l-chrysanthemum monocarboxylate **(584-79-2)**
N-Allyl-3-hydroxymorphinan **(152-02-3)**
l-N-Allyl-14-hydroxynordihydromorphinone **(465-65-6)**
Allylic alcohol **(107-18-6)**
Allylidene acetate **(869-29-4)**
Allylidene diacetate **(869-29-4)**
Allyl iodide [UN 1723] **(556-56-9)**
Allyl α-ionone **(79-78-7)**
Allylisobutylbarbital **(77-26-9)**
Allylisobutylbarbiturate **(77-26-9)**
5-Allyl-5-isobutylbarbituric acid **(77-26-9)**
5-Allyl-5-isopropylbarbiturate **(77-02-1)**
5-Allyl-5-isopropylbarbituric acid **(77-02-1)**
Allylisopropylbarbituric acid **(77-02-1)**
Allylisopropylmalonylurea **(77-02-1)**
Allyl isorhodanide **(57-06-7)**
Allyl isosulfocyanate **(57-06-7)**
Allyl isosulphocyanate **(57-06-7)**
Allyl isothiocyanate **(57-06-7)**
Allyl isothiocyanate, Stabilized [UN 1545] **(57-06-7)**
Allylisothiokyanat (Czech) **(57-06-7)**
Allyl isovalerate **(2835-39-4)**
Allyl isovalerianate **(2835-39-4)**
3-Allyl-4-keto-2-methylcyclopentenyl chrysanthemummono-carboxylate **(584-79-2)**
Allylkyanid (Czech) **(109-75-1)**
Allyl methacrylate **(96-05-9)**
4-Allyl-1-methoxybenzene **(140-67-0)**
p-Allylmethoxybenzene **(140-67-0)**
5-Allyl-1-methoxy-2,3-(methylenedioxy)benzene **(607-91-0)**
4-Allyl-2-methoxyphenol **(97-53-0)**
4-Allyl-2-methoxyphenol acetate **(93-28-7)**
4-Allyl-2-methoxyphenol formate **(10031-96-6)**
4-Allyl-2-methoxyphenyl acetate **(93-28-7)**
5-Allyl-5-(1-methylbutyl)barbituric acid **(76-73-3)**
5-Allyl-5-(1-methylbutyl)malonylurea **(76-73-3)**
Allyl 3-methylbutyrate **(2835-39-4)**
Allylmethyl cyanide **(592-51-8)**
1-Allyl-3,4-methylenedioxybenzene **(94-59-7)**
4-Allyl-1,2-methylenedioxybenzene **(94-59-7)**
(+-)-5-Allyl-1-methyl-5-(1-methyl-2-pentynyl)barbituric acid **(151-83-7)**
5-Allyl-1-methyl-5-(1-methyl-2-pentynyl)barbituric acid sodium salt **(309-36-4)**
Allylmethylnitrosamine **(4549-43-3)**
Allyl N-methylolcarbamate **(24935-97-5)**
DL-3-Allyl-2-methyl-4-oxocyclopent-2-enyl DL-cis trans chrysanthemate **(584-79-2)**
3-Allyl-2-methyl-4-oxo-2-cyclopenten-1-yl chrysanthemate **(584-79-2)**
5-Allyl-5-(1-methylpropyl) barbituric acid **(115-44-6)**
5-Allyl-5-(2'-methyl-n-propyl) barbituric acid **(77-26-9)**
Allyl methyl sulfide **(10152-76-8)**
Allyl monosulfide **(592-88-1)**
Allyl mustard oil **(57-06-7)**
Allylnitrile **(109-75-1)**
N-Allylnormorphine **(62-67-9)**
2-(1-Allyloxyaminobutylidene)-5,5-dimethylmethoxycarbonyl-

Name	CAS
cyclohexane-1,3-dione sodium salt	**(55635-13-7)**
(+-)-1-(β-(Allyloxy)-2,4-dichlorophenethyl)imidazole	**(35554-44-0)**
1-Allyloxy-2,3-epoxy-propaan (Dutch)	**(106-92-3)**
1-Allyloxy-2,3-epoxypropan (German)	**(106-92-3)**
1-(Allyloxy)-2,3-epoxypropane	**(106-92-3)**
Allyloxy(hydroxymethyl)benzene	**(28655-62-1)**
1-(o-(Allyloxy)phenoxy)-3-(isopropylamino)-2-propanol	**(6452-71-7)**
1-Allyloxy-2,3-propanediol	**(123-34-2)**
3-Allyloxy-1,2-propanediol	**(123-34-2)**
Allyl pentareythritol	**(1471-18-7)**
2-Allylphenol	**(1745-81-9)**
1-(o-Allylphenoxy)-3-(isopropylamino)-2-propanol	**(13655-52-2)**
Allyl phosphate	**(1623-19-4)**
Allyl phosphorodithioate, (C3H5O)2(HS)PS (8CI)	**(5851-14-9)**
Allyl phosphorodithioate (6CI,7CI)	**(5851-14-9)**
Allyl phthalate, Homopolymer	**(25053-15-0)**
Allylprodine	**(25384-17-2)**
1-Allylpropene	**(592-45-0)**
Allylpropenyl	**(592-45-0)**
Allyl propyl disulfide (ACGIH,OSHA)	**(2179-59-1)**
Allylpropymal	**(77-02-1)**
m-Allylpyrocatechin methylene ether	**(94-59-7)**
4-Allylpyrocatechol formaldehyde acetal	**(94-59-7)**
Allylpyrocatechol methylene ether	**(94-59-7)**
Allylrethronyl dl-cis-trans-chrysanthemate	**(584-79-2)**
Allylsenevol	**(57-06-7)**
Allylsenfoel (German)	**(57-06-7)**
Allyl sevenolum	**(57-06-7)**
Allyl sucrose	**(12002-22-1)**
Allyl-sulfide	**(592-88-1)**
12-Allyl-7,7a,8,9-tetrahydro-3,7a-dihydroxy-4ah-8,9c-imino-ethanophenanthro(4,5-bcd)furanone	**(465-65-6)**
Allylthiocarbamide	**(109-57-9)**
Allyl thiocarbonimide	**(57-06-7)**
Allylthiomocovina (Czech)	**(109-57-9)**
1-Allyl-2-thiourea	**(109-57-9)**
1-Allylthiourea	**(109-57-9)**
Allylthiourea	**(109-57-9)**
N-Allylthiourea	**(109-57-9)**
Allyl trichloride	**(96-18-4)**
Allyl trichlorosilane (DOT)	**(107-37-9)**
Allyltrichlorosilane, Stabilized [UN 1724]	**(107-37-9)**
Allyltrimethylammonium chloride	**(1516-27-4)**
1-Allylurea	**(557-11-9)**
Allylurea	**(557-11-9)**
N-Allylurea	**(557-11-9)**
4-Allylveratrole	**(93-15-2)**
Allyl vinyl ether	**(3917-15-5)**
Allypropymal	**(77-02-1)**
Almazine	**(846-49-1)**
Almederm	**(70-30-4)**
Almefrin	**(61-76-7)**
Almefrol	**(59-02-9)**
Almelose	**(9000-11-7)**
Almite	**(1344-28-1)**
Almocarpine	**(54-71-7)**
Almocarpine	**(92-13-7)**
Almolan AE	**(67762-27-0)**
Almolan Lis	**(67762-27-0)**
Almond Artificial Essential Oil	**(100-52-7)**
Alnasid	**(51-65-0)**
Alnox	**(52-43-7)**
Alobarbital	**(52-43-7)**
Alocaine	**(51-05-8)**
Alochlor	**(15972-60-8)**
Alodan (Gerot)	**(50-13-5)**
Aloe Extract #103	**(8006-54-0)**
Alon	**(1344-28-1)**
Alon	**(34123-59-6)**
Alon C	**(1344-28-1)**
Alondra	**(53-33-8)**
Alondra-F	**(1524-88-5)**
Aloperidin	**(52-86-8)**
Aloperidolo	**(52-86-8)**
Aloperidon	**(52-86-8)**
Alositol	**(315-30-0)**
Aloxite	**(1344-28-1)**
Alpen	**(132-93-4)**
Alpen	**(69-53-4)**
Alperidine	**(25384-17-2)**
Alperox C	**(105-74-8)**
Alphacroic Violet B	**(2092-55-9)**
Alphacypermethrin	**(67375-30-8)**
Alphadrol	**(53-34-9)**
Alphaketoglutaric acid	**(328-50-7)**
Alphalin	**(68-26-8)**
Alpha medopa	**(555-30-6)**
Alphameprodine	**(468-51-9)**
Alphamethrin	**(52315-07-8)**
Alphanaphthyl thiourea	**(86-88-4)**
Alphanaphtyl thiouree (French)	**(86-88-4)**
Alphanol 79	**(66587-56-2)**
Alphaprodine	**(77-20-3)**
Alphasol OT	**(577-11-7)**
Alphaspra	**(86-87-3)**
Alphasterol	**(68-26-8)**
Alphatron	**(10043-92-2)**
Alphazole	**(127-69-5)**
Alphazurine	**(2650-18-2)**
Alphazurine FG	**(2650-18-2)**
Alphazurine FGND	**(2650-18-2)**
Alphazurine 2G	**(129-17-9)**
Alphazurine (Indicator)	**(2650-18-2)**
Alphex Fit 221	**(9002-88-4)**
Alphozone	**(123-23-9)**
Alpinyl	**(103-90-2)**
Alprazolam	**(28981-97-7)**
Alprazolamum (Latin)	**(28981-97-7)**
Alprenolol	**(13655-52-2)**
Alqoverin	**(50-33-9)**
Alrato	**(86-88-4)**
Alrheumat	**(22071-15-4)**
Alrheumun	**(22071-15-4)**
Alserin	**(50-55-5)**
Altabactina	**(139-91-3)**
Altafur	**(139-91-3)**
Altan	**(117-10-2)**
Altax	**(120-78-5)**
Altazine Black BH	**(2429-73-4)**
Altco Sperse Fast Yellow GFN New	**(2832-40-8)**
Alteichin	**(88899-62-1)**
Altersolanol A	**(22268-16-2)**
Alterungsschutz HS	**(2517-16-0)**
Altezol	**(80-35-3)**
Altheine	**(70-47-3)**
Altochrome Milling Scarlet G	**(3567-65-5)**
Altochrome Scarlet G	**(10169-02-5)**
Altocyl Brilliant Blue B	**(2475-46-9)**
Altocyl Brilliant Blue B	**(86722-66-9)**
Altosid	**(40596-69-8)**
Altosid IGR	**(40596-69-8)**
Altosid SR 10	**(40596-69-8)**
Altowhites	**(1332-58-7)**
Altox	**(309-00-2)**

Altozar **(41096-46-2)**
Altrad **(50-28-2)**
Altretamine **(645-05-6)**
Altulor M 70 **(9011-14-7)**
Alugel 34TN **(637-12-7)**
Aluline **(315-30-0)**
Alum **(10043-01-3)**
Alum **(10043-67-1)**
Alum **(7784-25-0)**
Alum **(7784-28-3)**
Alum, N.F. **(10043-67-1)**
Alum ammonium **(7784-25-0)**
Alum, ammonium **(7784-26-1)**
Alumigel **(21645-51-2)**
Alumina **(1344-28-1)**
β-Alumina **(12005-48-0)**
β-Alumina **(1344-28-1)**
γ-Alumina **(1344-28-1)**
α-Alumina (ACGIH,OSHA) **(1344-28-1)**
Alumina Fibre **(7429-90-5)**
Alumina hydrate **(21645-51-2)**
Alumina hydrated **(21645-51-2)**
Aluminate(1-), dihydrobis(2-methoxyethanolato-O,O')-, sodium (9CI) **(22722-98-1)**
Aluminate(12-), hexaoxotris(sulfato(2-))di-, calcium (1:6) (9CI) **(12004-14-7)**
Aluminate(12-), hexaoxotris(sulfato(2-))di-, calcium (1:6), hydrate (9CI) **(11070-82-9)**
Aluminatesilicate (9CI) **(1327-36-2)**
Aluminate, sodium (9CI) **(12005-48-0)**
Aluminate, sodium (9CI) **(1302-42-7)**
Aluminate(1-), tetrahydro-, calcium (8CI) **(16941-10-9)**
Aluminate(1-), tetrahydro-, calcium (2:1), (T-4)- (9CI) **(16941-10-9)**
Aluminate (1-), tetrahydro-, lithium **(16853-85-3)**
Aluminate(1-), tetrahydro-, lithium, (T-4)- (9CI) **(16853-85-3)**
Aluminate (1-), tetrahydro-, sodium **(13770-96-2)**
Aluminate(1-), tetrahydro-, sodium, (T-4)- (9CI) **(13770-96-2)**
Alumina trihydrate **(21645-51-2)**
α-Alumina trihydrate **(21645-51-2)**
Aluminic acid **(21645-51-2)**
Aluminite 37 **(1344-28-1)**
Aluminophosphoric acid **(7784-30-7)**
Aluminosilcate **(1302-76-7)**
Aluminosilicate **(1327-36-2)**
Aluminosilicic acid **(1335-30-4)**
Aluminosilicic acid (Unspecified), calcium sodium salt, hydrate **(1344-01-0)**
Aluminosilicic acid, calcium sodium salt **(1344-01-0)**
Aluminosilicic acid, calcium sodium salt, hydrate **(1344-01-0)**
Aluminosilicic acid, sodium salt **(1344-00-9)**
Aluminium **(7429-90-5)**
Aluminium Alloy, Al,Be **(12770-50-2)**
Aluminium Bronze **(7429-90-5)**
Aluminium Flake **(7429-90-5)**
Aluminium Fosfide (Dutch) **(20859-73-8)**
Aluminium acetylsalicylate **(23413-80-1)**
Aluminiumchlorid (German) **(7446-70-0)**
Aluminium chlorohydroxyallantoinate **(1317-25-5)**
Aluminium fluorure (French) **(7784-18-1)**
Aluminium hydroxide **(21645-51-2)**
Aluminium phosphide **(20859-73-8)**
Aluminium stearate **(637-12-7)**
Aluminum 27 **(7429-90-5)**
Aluminum A00 **(7429-90-5)**
Aluminum Alloy, Al,Be (9CI) **(12770-50-2)**
Aluminum Alloy, base, Al 60-66,Si 25-30,Ni 5-7,Al2O3 3-4 (SAS 1) (9CI) **(12743-20-3)**
Aluminum (ACGIH) **(7429-90-5)**
Aluminum Aspirin **(23413-80-1)**
Aluminum, Compd. with zirconium (3:1) (9CI) **(12004-83-0)**
Aluminum Hydroxide Gel **(21645-51-2)**
Aluminum, Metallic, Powder (DOT) **(7429-90-5)**
Aluminum Powder **(7429-90-5)**
Aluminum Powder, Coated [UN 1309] **(7429-90-5)**
Aluminum, Powder, Pyrophoric [UN 1383] **(7429-90-5)**
Aluminum, Powder, Uncoated, Non-pyrophoric [UN 1396] **(7429-90-5)**
Aluminum Pyro Powders (OSHA) **(7429-90-5)**
Aluminum Welding Fumes (OSHA) **(7429-90-5)**
Aluminum acetate **(139-12-8)**
Aluminum, (acetato-O)(formato-O)hydroxy- (9CI) **(34202-30-7)**
Aluminum acetylacetonate **(13963-57-0)**
Aluminum(III) acetylacetonate **(13963-57-0)**
Aluminum acid phosphate **(7784-30-7)**
Aluminum alum **(10043-01-3)**
Aluminum ammonium alum **(7784-25-0)**
Aluminum ammonium disulfate **(7784-25-0)**
Aluminum ammonium sulfate **(7784-25-0)**
Aluminum ammonium sulfate (1:1:2) dodecahydrate **(7784-26-1)**
Aluminum benzoate **(555-32-8)**
Aluminum beryllium alloy **(12770-50-2)**
Aluminum, bis(acetato-O)hydroxy- (9CI) **(142-03-0)**
Aluminum, bis(2-(acetyloxy)benzoato-O[1])hydroxy- (9CI) **(23413-80-1)**
Aluminum bis(acetylsalicylate) **(23413-80-1)**
Aluminum, bis(2-butanolato)(ethyl 3-oxobutanoato-O1',O3)-, (T-4)- (9CI) **(24772-51-8)**
Aluminum borate **(11121-16-7)**
Aluminum bromide **(7727-15-3)**
Aluminum bromide, Anhydrous [UN 1725] **(7727-15-3)**
Aluminum bromide, Solution [UN 2580] **(7727-15-3)**
Aluminum n-butoxide **(3085-30-1)**
Aluminum sec-butoxide **(2269-22-9)**
Aluminum calcium hydride (6CI) **(16941-10-9)**
Aluminum calcium oxide sulfate, hydrate (8CI) **(11070-82-9)**
Aluminum calcium sodium silicate **(1344-01-0)**
Aluminum caprylate **(6028-57-5)**
Aluminum carbide [UN 1394] **(12656-43-8)**
Aluminum chlorate **(15477-33-5)**
Aluminum chlorhydrate **(12042-91-0)**
Aluminum chlorhydrol **(12042-91-0)**
Aluminum chlorhydroxide **(12042-91-0)**
Aluminum chlorhydroxy allantoinate **(1317-25-5)**
Aluminum chloride (1:3) **(7446-70-0)**
Aluminum-chloride **(7446-70-0)**
Aluminum chloride, Anhydrous [UN 1726] **(7446-70-0)**
Aluminum chloride, Basic (9CI) **(1327-41-9)**
Aluminum chloride, Solution [UN 2581] **(7446-70-0)**
Aluminum chloride, hexahydrate **(7784-13-6)**
Aluminum(III) chloride, hexahydrate **(7784-13-6)**
Aluminum, chlorobis(2-methylpropyl)- (9CI) **(1779-25-5)**
Aluminum, chlorodiethyl **(96-10-6)**
Aluminum, chlorodiisobutyl **(1779-25-5)**
Aluminum, chloro((2,5-dioxo-4-imidazolidinyl)ureato)tetrahydroxydi- (9CI) **(1317-25-5)**
Aluminum chlorohydroxide **(12042-91-0)**
Aluminum-chloride-hydroxide **(12042-91-0)**
Aluminum-chloride-hydroxide **(1327-41-9)**
Aluminum chlorohydroxy allantoinate **(1317-25-5)**
Aluminum, chlorotetrahydroxy((4,5-dihydro-2-hydroxy-5-oxo-1H-imidazol-4-yl)ureato)di- **(1317-25-5)**
Aluminum, chlorotetrahydroxy((2-hydroxy-5-oxo-2-imidazolin-4-yl)ureato)di- **(1317-25-5)**
Aluminum n-decoxide **(26303-54-8)**
Aluminum dehydrated **(7429-90-5)**
Aluminum diacetate **(142-03-0)**
Aluminum di-sec-butoxide acetoacetic ester chelate **(24772-51-8)**
Aluminum, dichloroethyl **(563-43-9)**

Aluminum, di-mu-chlorotetrachlorodi- (9CI) **(13845-12-0)**
Aluminum, diethylhydro- (8CI,9CI) **(871-27-2)**
Aluminum, diethyliodo- (9CI) **(2040-00-8)**
Aluminum, dihydro(2-methyl-1,3-butadienyl)- (9CI) **(24683-32-7)**
Aluminum, dihydroxy(stearato) **(7047-84-9)**
Aluminum, diisobutylhydro **(1191-15-7)**
Aluminum distearate (ACGIH) **(300-92-5)**
Aluminum n-docosoxide **(67905-30-0)**
Aluminum n-dodecoxide **(14624-15-8)**
Aluminum n-eicosoxide **(67905-31-1)**
Aluminum ethoxide **(555-75-9)**
Aluminum, ethoxydiethyl- (9CI) **(1586-92-1)**
Aluminum ethylate **(555-75-9)**
Aluminum, (ethyl 3-oxobutanoato-O1',O3)bis (2-propanolato)-, (T-4)- (9CI) **(14782-75-3)**
Aluminum-fluoride **(7784-18-1)**
Aluminum formate **(7360-53-4)**
Aluminum glycinate **(13682-92-3)**
Aluminum, (glycinato-N,O)dihydroxy- **(13682-92-3)**
Aluminum, (glycinato-N,O)dihydroxy-, (T-4)- (9CI) **(13682-92-3)**
Aluminum hexachloride **(13845-12-0)**
Aluminum n-hexacosoxide **(67905-28-6)**
Aluminum, hexadeca-mu-hydroxytetracosahydroxy(mu8-(1,3,4,6-tetra-O-sulfo-β-D-fructofuranosyl α-D-glucopyranoside tetrakis(hydrogen sulfato)(8-)))hexadeca- **(54182-58-0)**
Aluminum hexadecanoate **(555-35-1)**
Aluminum n-hexadecoxide **(19141-82-3)**
Aluminum n-hexoxide **(23275-26-5)**
Aluminum hydrate **(21645-51-2)**
Aluminum hydride [UN 2463] **(7784-21-6)**
Aluminum, hydrobis(2-methylpropyl)- (9CI) **(1191-15-7)**
Aluminum, hydrodiisobutyl- (8CI) **(1191-15-7)**
Aluminum(III) hydroxide **(21645-51-2)**
Aluminum-hydroxide **(21645-51-2)**
Aluminum hydroxide ($Al(OH)_3$), Mixt. contg. (9CI) **(8047-42-5)**
Aluminum hydroxide chloride **(12042-91-0)**
Aluminum hydroxide distearate **(300-92-5)**
Aluminum hydroxyacetate **(142-03-0)**
Aluminum hydroxychloride **(12042-91-0)**
Aluminum hydroxydistearate **(300-92-5)**
Aluminum, hydroxybis(octadecanoato-O)- (9CI) **(300-92-5)**
Aluminum, hydroxybis(salicylic acid acetato) **(23413-80-1)**
Aluminum, hydroxybis(stearato) **(300-92-5)**
Aluminum, hydroxybis(2-(2,4,5-trichlorophenoxy) propanoato-O1,O2)- (9CI) **(69622-82-8)**
Aluminum isopropoxide **(555-31-7)**
Aluminum(II) isopropylate **(555-31-7)**
Aluminum lactate **(18917-91-4)**
Aluminum lithium hydride **(16853-85-3)**
Aluminum metal (OSHA) **(7429-90-5)**
Aluminum monophosphate **(7784-30-7)**
Aluminum monophosphide **(20859-73-8)**
Aluminum(111) nitrate (1:3) **(13473-90-0)**
Aluminum nitrate [UN 1438] **(13473-90-0)**
Aluminum n-octacosoxide **(67905-27-5)**
Aluminum, (octadecanoato-O)oxo- (9CI) **(13419-15-3)**
Aluminum n-octadecoxide **(3985-81-7)**
Aluminum octanoate **(6028-57-5)**
Aluminum n-octoxide **(14624-13-6)**
Aluminum oleate **(688-37-9)**
Aluminum oxide **(1302-74-5)**
Aluminum oxide **(1344-28-1)**
Aluminum oxide (2:3) **(1344-28-1)**
α-Aluminum oxide **(1344-28-1)**
β-Aluminum oxide **(1344-28-1)**
γ-Aluminum oxide **(1344-28-1)**
Aluminum oxide (Al_2O_3), alloy, Al 60-66,Si 25-30,Ni 5-7,Al_2O_3 3-4 (SAS 1) (9CI) **(12743-20-3)**
Aluminum oxide-$3H_2O$ **(21645-51-2)**
Aluminum oxide, alloy, Al 60-66,Si 25-30,Ni 5-7,Al_2O_3 3-4 (SAS 1) **(12743-20-3)**
Aluminum oxide hydrate **(21645-51-2)**
Aluminum oxide silicate **(1302-76-7)**
Aluminum oxide silicate (9CI) **(12068-56-3)**
Aluminum oxide trihydrate **(21645-51-2)**
Aluminum palmitate **(555-35-1)**
Aluminum phenoxide **(15086-27-8)**
Aluminum phosethyl **(39148-24-8)**
Aluminum phosphate **(7784-30-7)**
Aluminum phosphate (1:10) **(7784-30-7)**
Aluminum phosphide [UN 1397] **(20859-73-8)**
Aluminum-phosphide **(20859-73-8)**
Aluminum potassium alum **(10043-67-1)**
Aluminum potassium disulfate **(10043-67-1)**
Aluminum potassium sulfate **(10043-67-1)**
Aluminum potassium sulfate alum **(10043-67-1)**
Aluminum sesquioxide **(1344-28-1)**
Aluminum silicate **(12068-56-3)**
Aluminum silicate **(1327-36-2)**
Aluminum silicate **(1335-30-4)**
Aluminum(III) silicate (2:1) **(1302-76-7)**
Aluminum sodium disulfate dodecahydrate **(7784-28-3)**
Aluminum sodium fluoride **(15096-52-3)**
Aluminum sodium hydride **(13770-96-2)**
Aluminum sodium oxide **(1302-42-7)**
Aluminum sodium oxide phosphate **(1344-06-5)**
Aluminum sodium phosphate **(7785-88-8)**
Aluminum sodium silicate **(1344-00-9)**
Aluminum sodium sulfate **(10102-71-3)**
Aluminum stearate **(637-12-7)**
Aluminum stearate (ACGIH) **(7047-84-9)**
Aluminum stearate palmitate **(555-35-1)**
Aluminum subacetate **(142-03-0)**
Aluminum sulfate **(10043-01-3)**
Aluminum sulfate (2:3) **(10043-01-3)**
Aluminum sulfate, Solution (DOT) **(10043-01-3)**
Aluminum sulfate solution **(57292-32-7)**
Aluminum sulfate solution (20% or less) **(57292-32-7)**
Aluminum sulphate **(10043-01-3)**
Aluminum n-tetracosoxide **(67905-29-7)**
Aluminum n-tetradecoxide **(67905-32-2)**
Aluminum triacetylacetonate **(13963-57-0)**
Aluminum tribromide **(7727-15-3)**
Aluminum, tribromotrimethyldi **(12263-85-3)**
Aluminum, tributyl- (9CI) **(1116-70-7)**
Aluminum trichloride **(7446-70-0)**
Aluminum trichloride hexahydrate **(7784-13-6)**
Aluminum, trichlorotriethyldi **(12075-68-2)**
Aluminum, trichlorotrimethyldi- (8CI,9CI) **(12542-85-7)**
Aluminum, tridocosyl- (9CI) **(6651-25-8)**
Aluminum, tridodecyl- (9CI) **(1529-59-5)**
Aluminum, trieicosyl- (9CI) **(1529-57-3)**
Aluminum triethoxide **(555-75-9)**
Aluminum, triethyl **(97-93-8)**
Aluminum trifluoride **(7784-18-1)**
Aluminum, trihexacosyl- (8CI) **(10449-71-5)**
Aluminum, trihexadecyl- (8CI,9CI) **(1726-65-4)**
Aluminum, trihexyl- (9CI) **(1116-73-0)**
Aluminum trihydrat **(21645-51-2)**
Aluminum trihydrate **(21645-51-2)**
Aluminum trihydride **(7784-21-6)**
α-Aluminum trihydride **(7784-21-6)**
Aluminum trihydroxide **(21645-51-2)**
Aluminum, triisobutyl **(100-99-2)**

Name	CAS
Aluminum, trimethyl	(75-24-1)
Aluminum trinitrate	(13473-90-0)
Aluminum, trioctacosyl- (8CI)	(6651-27-0)
Aluminum, trioctadecyl- (9CI)	(3041-23-4)
Aluminum, trioctyl- (9CI)	(1070-00-4)
Aluminum trioleate	(688-37-9)
Aluminum, tripropyl- (8CI,9CI)	(102-67-0)
Aluminum tris(O-ethyl phosphonate)	(39148-24-8)
Aluminum tris(acetylacetonate)	(13963-57-0)
Aluminum, tris(decyl)- (9CI)	(1726-66-5)
Aluminum, tris(2-hydroxypropanoato-O^1,O^2)- (9CI)	(18917-91-4)
Aluminum, tris(lactato)	(18917-91-4)
Aluminum, tris(2-methylpropyl)- (9CI)	(100-99-2)
Aluminum, tris(2,4-pentanedionato)	(13963-57-0)
Aluminum, tris(2,4-pentanedionato-O,O')-, (OC-6-11)- (9CI)	(13963-57-0)
Aluminum tristearate (ACGIH)	(637-12-7)
Aluminum, tris(tetradecyl)-	(1529-58-4)
Aluminum, tris(tetrahydroborato(1-)-H,H')-, (OC-6-11)- (9CI)	(13771-22-7)
Aluminum trisulfate	(10043-01-3)
Aluminum, tritetracosyl- (9CI)	(6651-26-9)
Aluminum, tritetradecyl- (9CI)	(1529-58-4)
Alumite	(1344-28-1)
Alumite (Oxide)	(1344-28-1)
Alum, potassium	(10043-67-1)
Alundum	(1344-28-1)
Alundum 600	(1344-28-1)
Alunex	(113-92-8)
Aluphos	(7784-30-7)
Alurate	(77-02-1)
Alurate elixir verdum	(77-02-1)
Alurene	(58-94-6)
Alusal	(21645-51-2)
Alusil ET	(1344-00-9)
Alutyl	(132-60-5)
Aluzine	(54-31-9)
Alvedon	(103-90-2)
Alveograf	(1306-06-5)
Alverina (Spanish)	(150-59-4)
Alverine	(150-59-4)
Alverinum (Latin)	(150-59-4)
Alvinol	(113-18-8)
Alvison 8	(28345-91-7)
Alvit	(60-57-1)
Alvyl	(9002-89-5)
Alysine	(54-21-7)
Alzodef	(156-62-7)
Am-Fol	(7664-41-7)
Ami-Pilo	(54-71-7)
Amabevan	(121-59-5)
Amacel Blue BNN	(2475-46-9)
Amacel Blue BNN	(86722-66-9)
Amacel Blue GG	(2475-45-8)
Amacel Brilliant Blue B	(2475-46-9)
Amacel Brilliant Blue B	(86722-66-9)
Amacel Cerise B	(2872-48-2)
Amacel Developed Navy SD	(119-90-4)
Amacel Heliotrope R	(128-95-0)
Amacel Pink B	(116-85-8)
Amacel Pure Blue B	(2475-45-8)
Amacel Violet 6B	(1220-94-6)
Amacel Yellow G	(2832-40-8)
Amacel Yellow RR	(119-15-3)
Amacid Amaranth	(915-67-3)
Amacid Black 10BR	(1064-48-8)
Amacid Blue FG Conc	(2650-18-2)
Amacid Blue V	(129-17-9)
Amacid Brilliant Blue	(860-22-0)
Amacid Brilliant Orange	(1934-20-9)
Amacid Carmoisine B	(3567-69-9)
Amacid Chrome Blue R	(3567-69-9)
Amacid Fast Blue R	(3861-73-2)
Amacid Fast Red A	(1658-56-6)
Amacid Fast Yellow RS	(6375-55-9)
Amacid Fuchsine 4B	(3567-66-6)
Amacid Green B	(4680-78-8)
Amacid Green G	(5141-20-8)
Amacid Lake Scarlet 2R	(3761-53-3)
Amacid Milling Red PGS	(3567-65-5)
Amacid Milling Scarlet G	(10169-02-5)
Amacid Orange Y	(633-96-5)
Amacid Phloxine	(3734-67-6)
Amacid Scarlet 2R	(3761-53-3)
Amacid Yellow M	(587-98-4)
Amacid Yellow T	(1934-21-0)
Amadil	(103-90-2)
Amaizo W 13	(9005-25-8)
Amal	(57-43-2)
Amalox	(1314-13-2)
Amanil Black GL	(1937-37-7)
Amanil Black WD	(1937-37-7)
Amanil Blue 2BX	(2602-46-2)
Amanil Brown MR	(2429-82-5)
Amanil Chloramine Red 8BS	(6548-29-4)
Amanil Chrome Navy Blue B	(7082-31-7)
Amanil Developed Black BHSW	(2429-73-4)
Amanil Fast Brown BRL	(16071-86-6)
Amanil Fast Brown HP	(2429-81-4)
Amanil Fast Red 8BL	(2610-11-9)
Amanil Fast Red 8BLW	(2610-11-9)
Amanil Fast Scarlet 3B	(6358-29-8)
Amanil Fast Violet N	(2586-60-9)
Amanil Fast Yellow AN	(1325-37-7)
Amanil Green LT	(3626-28-6)
Amanil Naphthol AS-BO	(132-68-3)
Amanil Naphthol AS-BR	(91-92-9)
Amanil Naphthol AS-BS	(135-65-9)
Amanil Naphthol AS-D	(135-61-5)
Amanil Naphthol AS-EL	(137-52-0)
Amanil Naphthol AS-G	(91-96-3)
Amanil Naphthol AS-KB	(135-63-7)
Amanil Naphthol AS-OL	(135-62-6)
Amanil Naphthol AS	(92-77-3)
Amanil Naphthol AS-phenyl	(92-74-0)
Amanil Purpurine 4B	(992-59-6)
Amanil Rayon Brown B	(2429-81-4)
Amanil Sky Blue	(2429-74-5)
Amanil Sky Blue	(72-57-1)
Amanil Sky Blue 6B	(2610-05-1)
Amanil Sky Blue FF	(2610-05-1)
Amanil Sky Blue R	(72-57-1)
Amanil Supra Blue 9GL	(28407-37-6)
Amanil Supra Blue 9GL	(28407-37-6)
Amanil Supra Brown LBL	(16071-86-6)
Amanin	(21150-21-0)
Amanine	(21150-21-0)
β-Amanitin	(21150-22-1)
α-Amanitin (8CI,9CI)	(23109-05-9)
α-Amanitine	(23109-05-9)
γ-Amanitine	(13567-11-8)
α-Amanitin, 1-l-aspartic acid- (9CI)	(21150-22-1)
α-Amanitin, 1-l-aspartic acid-4-(2-mercapto-l-tryptophan)-	(21150-21-0)
Amanozina (Spanish)	(537-17-7)
Amanozine	(537-17-7)
Amanozinum (Latin)	(537-17-7)

Amantadine	(768-94-5)
Amanthrene Blue BCL	(130-20-1)
Amanthrene Brilliant Green J	(128-58-5)
Amanthrene Brilliant Green JP	(128-58-5)
Amanthrene Brilliant Violet RR	(1324-55-6)
Amanthrene Brown BR	(2475-33-4)
Amanthrene Golden Orange G	(128-70-1)
Amanthrene Golden Yellow	(128-66-5)
Amanthrene Green JF	(128-58-5)
Amanthrene Navy Blue BN	(1324-54-5)
Amanthrene Olive R	(2379-81-9)
Amanthrene Orange R	(3263-31-8)
Amanthrene Pink FF	(2379-74-0)
Amanthrene Pink FFD	(2379-74-0)
Amanthrene Pink FFWP	(2379-74-0)
Amanthrene Supra Green JF	(128-58-5)
Amanthrene Supra Navy Blue BN	(1324-54-5)
Amanthrene Supra Navy Blue BNR	(1324-54-5)
Amaplast Green OZ	(128-80-3)
Amaplast Red Violet P 2R	(128-95-0)
Amarant (Czech)	(915-67-3)
Amaranth	(915-67-3)
Amaranth A	(915-67-3)
Amaranth B	(915-67-3)
Amaranth BPC	(915-67-3)
Amaranth Extra	(915-67-3)
Amaranth Lake	(915-67-3)
Amaranth S	(915-67-3)
Amaranth S Specially Pure	(915-67-3)
Amaranth USP	(915-67-3)
Amaranth WD	(915-67-3)
Amaranthe	(915-67-3)
Amaranthe USP (Biological stain)	(915-67-3)
Amarex	(9006-42-2)
Amarine	(6199-67-3)
Amarthol AS-BO	(132-68-3)
Amarthol AS-BR	(91-92-9)
Amarthol AS-BS	(135-65-9)
Amarthol AS-D	(135-61-5)
Amarthol AS-G	(91-96-3)
Amarthol AS-OL	(135-62-6)
Amarthol AS	(92-77-3)
Amarthol AS-phenyl	(92-74-0)
Amarthol Fast Blue BB Base	(120-00-3)
Amarthol Fast Orange GC Base	(141-85-5)
Amarthol Fast Orange GC Salt	(141-85-5)
Amarthol Fast Orange R Base	(99-09-2)
Amarthol Fast Red B Base	(97-52-9)
Amarthol Fast Red GL Base	(89-62-3)
Amarthol Fast Red GL Salt	(89-62-3)
Amarthol Fast Red TR Base	(3165-93-3)
Amarthol Fast Red TR Base	(95-69-2)
Amarthol Fast Red TR Salt	(3165-93-3)
Amarthol Fast Scarlet G Base	(99-55-8)
Amarthol Fast Scarlet GG Base	(95-82-9)
Amarthol Fast Scarlet GGS Base	(95-82-9)
Amarthol Fast Scarlet G Salt	(99-55-8)
Amasust	(57-43-2)
Amatin	(118-74-1)
Amax	(102-77-2)
Amaze	(25311-71-1)
Amazon Yellow X2485	(6358-85-6)
Ambam	(3566-10-7)
Ambazon	(539-21-9)
Ambazone	(539-21-9)
Amben	(133-90-4)
Amben	(150-13-0)
Ambenyl	(147-24-0)
Amber Lanolin	(8006-54-0)
Amber acid	(110-15-6)
Amberlite 200 (9CI)	(12626-25-4)
Amberlite XAD 4	(37380-42-0)
Amberol ST 140F	(21645-51-2)
Ambeside	(63-74-1)
Ambiben	(133-90-4)
Ambilhar	(61-57-4)
Ambinon	(9002-61-3)
Amblosin	(69-53-4)
Ambochlorin	(305-03-3)
Amboclorin	(305-03-3)
Ambofen	(56-75-7)
Ambox	(485-31-4)
Ambracyn	(64-75-5)
Ambramicina	(60-54-8)
Ambramycin	(60-54-8)
Ambretta, Semi (Italian)	(8015-62-1)
Ambrette	(8015-62-1)
Ambrette, Graines (French)	(8015-62-1)
Ambrette Oil	(8015-62-1)
Ambrette Seed Absolute	(8015-62-1)
Ambrette Seed Liquid	(8015-62-1)
Ambrette Seed Oil	(8015-62-1)
Ambrosa-2,11(13)-dien-12-oic acid, 6-α,8-β-dihydroxy-4-oxo-, 12,8-lactone	(6754-13-8)
Ambush	(116-06-3)
Ambush	(52645-53-1)
Ambush C	(52315-07-8)
Ambushfog	(52645-53-1)
Ambythene	(9002-88-4)
Amcap	(7177-48-2)
Amchem 68-250	(16672-87-0)
Amchem 70-25	(33629-47-9)
Amchem A-280	(33629-47-9)
Amchem 3-CP	(101-10-0)
Amchem Grass Killer	(76-03-9)
Amchem 2,4,5-tp	(93-72-1)
Amchlor	(12125-02-9)
Amcide	(7773-06-0)
Amcill	(69-53-4)
Amcill	(7177-48-2)
Amco	(9003-07-0)
Amdex	(51-63-8)
Amdon	(1918-02-1)
Amdon Grazon	(1918-02-1)
Amdram	(300-42-5)
Amdro	(67485-29-4)
Ameban	(121-59-5)
Amebarsone	(121-59-5)
Amebicide	(316-42-7)
Amebil	(130-26-7)
Amechol	(62-51-1)
Amedel	(54-91-1)
Amedrine	(300-42-5)
Ameisenatod	(58-89-9)
Ameisenmittel Merck	(58-89-9)
Ameisensaeure (German)	(64-18-6)
Amenoron	(57-63-6)
Amenorone	(57-63-6)
Amepromat	(57-53-4)
Amercide	(133-06-2)
Amerfil	(9003-07-0)
Americaine	(94-09-7)
American CL-26691	(115-93-5)
American Cyanamid 3422	(56-38-2)

American Cyanamid 3,911	(298-02-2)
American Cyanamid 4,049	(121-75-5)
American Cyanamid 4,124	(2463-84-5)
American Cyanamid 5223	(2439-10-3)
American Cyanamid 12,008	(78-52-4)
American Cyanamid 12,503	(119-12-0)
American Cyanamid 12880	(60-51-5)
American Cyanamid 18133	(297-97-2)
American Cyanamid 18682	(2275-18-5)
American Cyanamid 18706	(116-01-8)
American Cyanamid 38023	(52-85-7)
American Cyanamid 43073	(2275-23-2)
American Cyanamid 47031	(947-02-4)
American Cyanamid AC 47,031	(947-02-4)
American Cyanamid AC 52,160	(3383-96-8)
American Cyanamid CL-26,691	(115-93-5)
American Cyanamid CL-38,023	(52-85-7)
American Cyanamid CL-47,300	(122-14-5)
American Cyanamid CL-47031	(947-02-4)
American Cyanamid CL-47470	(950-10-7)
American Cyanamid CL-52160	(3383-96-8)
American Cyanamid E.I. 52,160	(3383-96-8)
American Hellebore	(65072-04-0)
American Penicillin	(69-57-8)
American Veratrum	(65072-04-0)
American Vermilion	(2814-77-9)
Americium (9CI)	(7440-35-9)
Americium, Isotope of mass 241 (8CI,9CI)	(14596-10-2)
Amerol	(61-82-5)
Ameroxol OE 2	(9004-98-2)
Ameroxol OE 10	(9004-98-2)
Ameroxol OE 20	(9004-98-2)
Amet (German)	(440-58-4)
Amethocaine hydrochloride	(136-47-0)
Amethopterin	(59-05-2)
Amethopterin, 3',5'-dichloro-	(528-74-5)
Amethopterine	(59-05-2)
Amethyst	(14808-60-7)
Ametox	(10102-17-7)
Ametrex	(834-12-8)
Ametriodinic acid	(440-58-4)
Ametryn	(834-12-8)
Ametryne	(834-12-8)
Ametycin	(50-07-7)
Ametycine	(50-07-7)
Amex	(33629-47-9)
Amex 820	(33629-47-9)
Amfepramone	(90-84-6)
Amfetamina	(60-13-9)
Amfetamina (Italian)	(60-15-1)
Amfetamine	(60-13-9)
d-Amfetasul	(51-63-8)
Amfipen	(69-53-4)
Amianthus	(1332-21-4)
Amiben	(133-90-4)
Amiben DS	(133-90-4)
Amibiarson	(121-59-5)
Amibin	(133-90-4)
Amicacin	(37517-28-5)
Amical 48	(20018-09-1)
Amicar	(60-32-2)
Amicide	(7773-06-0)
Amicin	(527-73-1)
Amid-Thin	(86-86-2)
Amid-Thin W	(86-86-2)
Amidazin	(536-33-4)
Amidazine	(536-33-4)
Amidazophen	(58-15-1)
Amidazophene	(58-15-1)
Amide PP	(98-92-0)
Amides, Coco, N,N-bis(hydroxyethyl)	(68603-42-9)
Amides, Vegetable-oil, N,N'-hexanediylbis-	(73398-58-0)
Amidine Blue 4B	(72-57-1)
4-Amidino-1-(nitrosaminoamidino)-1-tetrazene	(109-27-3)
1-Amidinohydrazono-4-thiosemicarbazono-2,5-cyclohexadiene	(539-21-9)
N^1-Amidinosulfanilamide	(57-67-0)
Amidiphos	(919-76-6)
Amidithion	(919-76-6)
Amid kyseliny akrylove (Czech)	(79-06-1)
Amid kyseliny benzoove (Czech)	(55-21-0)
Amid kyseliny fluoroctove (Czech)	(640-19-7)
Amid kyseliny kyanoctove (Czech)	(107-91-5)
Amid kyseliny methakrylove (Czech)	(79-39-0)
Amid kyseliny mravenci (Czech)	(75-12-7)
Amid kyseliny 1-naftyloctove (Czech)	(86-86-2)
Amid kyseliny nikotinove (Czech)	(98-92-0)
Amid kyseliny octove (Czech)	(60-35-5)
Amid kyseliny propionove (Czech)	(79-05-0)
Amid kyseliny salicylove (Czech)	(65-45-2)
Amid kyseliny stavelove (Czech)	(471-46-5)
Amid kyseliny trichloroctove (Czech)	(594-65-0)
Amido-G-acid	(86-65-7)
Amido Naphthol Red 2G	(3734-67-6)
Amido Naphthol Red G	(3734-67-6)
Amido Naphthol Red GA	(3734-67-6)
Amido Red 2G	(3734-67-6)
Amido Yellow E	(6373-74-6)
Amido Yellow EA	(6373-74-6)
Amido Yellow EA-CF	(6373-74-6)
o-Amidoazotoluol (German)	(97-56-3)
ortho-Amidobenzoic acid	(118-92-3)
Amidocyanogen	(420-04-2)
Amidofebrin	(58-15-1)
Amidofen	(58-15-1)
Amidofos	(299-86-5)
Amidol	(137-09-7)
Amidol	(545-90-4)
Amidolacetate	(509-74-0)
Amidon	(54-85-3)
Amidon	(76-99-3)
Amidone	(76-99-3)
Amidophen	(58-15-1)
Amidophenazone	(58-15-1)
Amidophos	(299-86-5)
Amidopyrazoline	(58-15-1)
Amidopyrin	(58-15-1)
Amidopyrine	(58-15-1)
Amidosal	(65-45-2)
Amidosulfonic acid	(5329-14-6)
Amidosulfuric acid	(5329-14-6)
Amidotrizoic acid	(117-96-4)
Amidourea hydrochloride	(563-41-7)
Amidox	(94-75-7)
Amidoxal	(127-69-5)
Amidryl	(58-73-1)
Amifenazol	(490-55-1)
Amifur	(59-87-0)
α-Amino-β-hydroxypropionic acid	(56-45-1)
Amikacin	(37517-28-5)
Amikar	(60-32-2)
Amikol 65	(9011-05-6)
Amilan	(63428-83-1)
Amilan CM 1001	(105-60-2)
Amilan CM 1001	(25038-54-4)

Amilan CM 1001C (25038-54-4)
Amilan CM 1001G (25038-54-4)
Amilan CM 1001C (105-60-2)
Amilan CM 1001G (105-60-2)
Amilan CM 1011 (105-60-2)
Amilan CM 1011 (25038-54-4)
Amilan CM 1031 (25038-54-4)
Amilar (25038-59-9)
Amilnitrit (110-46-3)
Amilorida (Spanish) (2609-46-3)
Amiloride (2609-46-3)
Amiloride hydrochloride (2016-88-8)
Amiloridum (Latin) (2609-46-3)
Amilphenol (80-46-6)
b-Amin (59-43-8)
Aminacrin (90-45-9)
Aminacrine (90-45-9)
Aminacrine hydrochloride (134-50-9)
Aminarson (121-59-5)
Aminarsone (121-59-5)
Aminasine (50-53-3)
Aminate base (79-17-4)
Aminazin (50-53-3)
Aminazine (50-53-3)
Aminazin monohydrochloride (69-09-0)
Amine 220 (95-38-5)
Amine D (1446-61-3)
Amines, C10-16-alkyldimethyl, N-oxides (70592-80-2)
Amines, N-C12-18-alkyltrimethylenedi- (68155-37-3)
Amines, coco alkyl, benzoates (68526-65-8)
Amines, coco alkyl, ethoxylated (61791-14-8)
Amines, N-coco alkyltrimethylenedi- (61791-63-7)
Amines, N-coco alkyltrimethylenedi-, acetates (61791-64-8)
Amines, N-coco alkyltrimethylenedi-, adipates (68155-42-0)
Amines, N-coco alkyltrimethylenedi-, benzoates (68188-29-4)
Amines, N-coco alkyltrimethylenedi-, glycolates (68155-43-1)
Amines, di-C14-18-alkylmethyl (67700-99-6)
Amines, N-soya alkyltrimethylenedi- (61791-67-1)
Amines, N-tallow alkyltrimethylenedi-, diacetates (68911-78-4)
Amine 2,4,5-t for rice (93-76-5)
Aminic acid (64-18-6)
Aminicotin (98-92-0)
4'-Amino-Dab (539-17-3)
2-Amino-6-MP (154-42-7)
4-Amino-PGA (54-62-6)
Amino Triazole Weedkiller 90 (61-82-5)
5-Aminoacenaphthene (4657-93-6)
Aminoacetaldehyde dimethyl acetal (22483-09-6)
2-Amino-4-acetaminifenetol (Czech) (17026-81-2)
3-Aminoacetanilid (Czech) (102-28-3)
4'-Aminoacetanilid (Czech) (122-80-5)
2-Aminoacetanilide (555-48-6)
3'-Aminoacetanilide (102-28-3)
4'-Aminoacetanilide (122-80-5)
4-Aminoacetanilide (122-80-5)
m-Aminoacetanilide (102-28-3)
p-Aminoacetanilide (122-80-5)
2-Aminoacetanilide hydrochloride (4801-39-2)
3-Aminoacetanilide, hydrochloride (621-35-2)
Aminoacetic acid (56-40-6)
Aminoacetic anilide (555-48-6)
3-Aminoacetofenon (Czech) (99-03-6)
p-Aminoacetofenonu (Polish) (99-92-3)
2'-Aminoacetophenone (551-93-9)
2-Aminoacetophenone (551-93-9)
3'-Aminoacetophenone (99-03-6)
4'-Aminoacetophenone (99-92-3)
β-Aminoacetophenone (99-03-6)
m-Aminoacetophenone (99-03-6)
o-Aminoacetophenone (551-93-9)
p-Aminoacetophenone (99-92-3)
m-Aminoacetylbenzene (99-03-6)
o-Aminoacetylbenzene (551-93-9)
p-Aminoacetylbenzene (99-92-3)
Amino acids, polyamides (VAN) (63428-84-2)
2-Aminoacridine (581-28-2)
3-Aminoacridine (581-29-3)
5-Aminoacridine (90-45-9)
9-Aminoacridine (90-45-9)
2-Aminoacridine (European) (581-29-3)
3-Aminoacridine (European) (581-28-2)
5-Aminoacridine hydrochloride (134-50-9)
9-Aminoacridine hydrochloride (134-50-9)
Aminoacridine hydrochloride (134-50-9)
9-Aminoacridine monohydrochloride (134-50-9)
1-Aminoadamantane (768-94-5)
1-Aminoadamatane (768-94-5)
2-Aminoaethanol (German) (141-43-5)
β-Aminoaethyl-morpholin (German) (2038-03-1)
2-Aminoakridin (Czech) (581-28-2)
3-Aminoakridin (Czech) (581-29-3)
9-Aminoakridin (Czech) (90-45-9)
m-Aminoaline (108-45-2)
5-Amino-2-(p-aminoanilino)benzenesulfonic acid (119-70-0)
4-Amino-N-(aminoiminomethyl)benzenesulfonamide (57-67-0)
Amino-1 aminomethyl-1 cyclohexane (5062-67-9)
1-Amino-2-(aminomethyl)cyclopentane (21544-02-5)
4-Amino-N-(4-aminophenyl)benzamide (785-30-8)
1-Amino-4-anilino-2-anthraquinonesulfonic acid (2786-71-2)
4-Amino-5-anilino-1,8-dihydroxyanthraquinone (20241-77-4)
2-Amino-4-anilino-6-methyl-s-triazine (7426-35-9)
2-Aminoaniline (95-54-5)
3-Aminoaniline (108-45-2)
4-Aminoaniline (106-50-3)
p-Aminoaniline (106-50-3)
3-Aminoaniline dihydrochloride (541-69-5)
4-Aminoaniline dihydrochloride (624-18-0)
m-Aminoaniline dihydrochloride (541-69-5)
p-Aminoaniline dihydrochloride (624-18-0)
6-(p-Aminoanilino)metanilic acid (119-70-0)
2-(p-Aminoanilino)-5-nitrobenzenesulfonic acid (91-29-2)
4-(p-Aminoanilino)-3-sulfoaniline (119-70-0)
2-Aminoanisole (90-04-0)
3-Aminoanisole (536-90-3)
4-Aminoanisole (104-94-9)
Aminoanisole (29191-52-4)
m-Aminoanisole (536-90-3)
o-Aminoanisole (90-04-0)
p-Aminoanisole (104-94-9)
2-Aminoanisole hydrochloride (134-29-2)
4-Aminoanisole hydrochloride (20265-97-8)
o-Aminoanisole hydrochloride (134-29-2)
p-Aminoanisole hydrochloride (20265-97-8)
4-Aminoanisole-3-sulfonic acid (13244-33-2)
2-Aminoanthracene (613-13-8)
β-Aminoanthracene (613-13-8)
1-Amino-9,10-anthracenedione (82-45-1)
2-Amino-9,10-anthracenedione (117-79-3)
Amino-9,10-anthracenedione (25620-59-1)
1-Aminoanthrachinon (Czech) (82-45-1)
1-Amino-9,10-anthraquinone (82-45-1)
1-Aminoanthraquinone (82-45-1)
2-Amino-9,10-anthraquinone (117-79-3)
2-Aminoanthraquinone (117-79-3)

Aminoanthraquinone **(117-79-3)**
α-Aminoanthraquinone **(82-45-1)**
β-Aminoanthraquinone **(117-79-3)**
1-Aminoanthraquinone-2-carboxaldehyde **(6363-87-7)**
4-Amino-1-arabinofuranosyl-2-oxo-1,2-dihydropyrimidin (Czech) **(147-94-4)**
4-Amino-1-arabinofuranosyl-2-oxo-1,2-dihydropyrimidine **(147-94-4)**
4-Amino-1-β-D-arabinofuranosyl-2(1H)-pyrimidinon (Czech) **(147-94-4)**
Aminoarson **(121-59-5)**
4-Amino-1,1'-azobenzene **(60-09-3)**
4-Aminoazobenzene **(60-09-3)**
Aminoazobenzene **(60-09-3)**
p-Aminoazobenzene **(60-09-3)**
4-Aminoazobenzene-3,4'-disulfonic acid **(101-50-8)**
Aminoazobenzene (indicator) **(60-09-3)**
4'-Aminoazobenzene-4-sulfonic acid **(104-23-4)**
4-Aminoazobenzene-4'-sulphonic acid **(104-23-4)**
4-Aminoazobenzol **(60-09-3)**
p-Aminoazobenzol **(60-09-3)**
1,2-Aminoazophenylene **(95-14-7)**
2-Amino-5-azotoluene **(97-56-3)**
4'-Amino-2,3'-azotoluene **(97-56-3)**
4'-Amino-2:3'-azotoluene **(97-56-3)**
Aminoazotoluene **(97-56-3)**
o-Aminoazotoluene **(97-56-3)**
Aminoazotoluene (indicator) **(97-56-3)**
o-Aminoazotolueno (Spanish) **(97-56-3)**
o-Aminoazotoluol **(97-56-3)**
10-Aminobenz(a)acridine **(18936-75-9)**
4-Aminobenzaldehyde **(556-18-3)**
p-Aminobenzaldehyde **(556-18-3)**
m-Aminobenzal fluoride **(98-16-8)**
2-Aminobenzamide **(88-68-6)**
4-Aminobenzamide **(2835-68-9)**
o-Aminobenzamide **(88-68-6)**
10-Amino-1,2-benzanthracene **(2381-18-2)**
5-Amino-1:2-benzanthracene **(56961-60-5)**
Aminobenzene **(62-53-3)**
(+-)-α-Aminobenzeneacetic acid **(2835-06-5)**
(R)-α-Aminobenzeneacetic acid **(875-74-1)**
4-Aminobenzeneacetic acid **(1197-55-3)**
4-Aminobenzenearsonic acid **(98-50-0)**
p-Aminobenzenearsonic acid **(98-50-0)**
p-Aminobenzeneazodimethylaniline **(539-17-3)**
2-Aminobenzeneethanol **(5339-85-5)**
4-Aminobenzeneethanol **(104-10-9)**
(S)-α-Aminobenzenepropanoic acid **(63-91-2)**
p-Aminobenzenesulfamide **(63-74-1)**
3-(p-Aminobenzenesulfamido)-6-methoxypyridazine **(80-35-3)**
p-Aminobenzenesulfonacetamide **(144-80-9)**
3-Aminobenzenesulfonamide **(98-18-0)**
4-Aminobenzenesulfonamide **(63-74-1)**
m-Aminobenzenesulfonamide **(98-18-0)**
p-Aminobenzenesulfonamide **(63-74-1)**
5-(p-Aminobenzenesulfonamido)-3,4-dimethylisoxazole **(127-69-5)**
2-(p-Aminobenzenesulfonamido)-4,6-dimethylpyrimidine **(57-68-1)**
2-(p-Aminobenzenesulfonamido)-5-methylthiadiazole **(144-82-1)**
3-(p-Aminobenzenesulfonamido)-2-phenylpyrazole **(526-08-9)**
2-p-Aminobenzenesulfonamidoquinoxaline **(59-40-5)**
2-(p-Aminobenzenesulfonamido)thiazole **(72-14-0)**
1-Aminobenzene-3-sulfonic acid **(121-47-1)**
2-Aminobenzenesulfonic acid **(88-21-1)**
3-Amino-benzenesulfonic acid **(121-47-1)**
4-Aminobenzenesulfonic acid **(121-57-3)**
m-Aminobenzenesulfonic acid **(121-47-1)**
o-Aminobenzenesulfonic acid **(88-21-1)**
p-Aminobenzenesulfonic acid **(121-57-3)**
p-Aminobenzenesulfonoacetamide **(144-80-9)**
N-(4-Aminobenzenesulfonyl)-N'-butylurea **(339-43-5)**
p-Aminobenzenesulfonylguanidine **(57-67-0)**
m-Aminobenzenesulphonamide **(98-18-0)**
5-(p-Aminobenzenesulphonamide)-3,4-dimethylisoxazole **(127-69-5)**
3-p-Aminobenzenesulphonamido-6-methoxypyridazine **(80-35-3)**
5-(p-Aminobenzenesulphonamido)-3,4-dimethylisoxazole **(127-69-5)**
2-(p-Aminobenzenesulphonamido)pyridine **(144-83-2)**
2-p-Aminobenzenesulphonamidoquinoxaline **(59-40-5)**
2-(p-Aminobenzenesulphonamido)thiazole **(72-14-0)**
N-p-Aminobenzenesulphonylguanidine monohydrate **(57-67-0)**
2-Aminobenzenethiol **(137-07-5)**
4-Aminobenzenethiol **(1193-02-8)**
p-Aminobenzenethiol **(1193-02-8)**
2-Aminobenzimidazole **(934-32-7)**
Aminobenzoate potassium **(138-84-1)**
2-Aminobenzoic acid **(118-92-3)**
3-Aminobenzoic acid **(99-05-8)**
4-Aminobenzoic acid **(150-13-0)**
Aminobenzoic acid **(150-13-0)**
γ-Aminobenzoic acid **(150-13-0)**
m-Aminobenzoic acid **(99-05-8)**
o-Aminobenzoic acid **(118-92-3)**
ortho-Aminobenzoic acid **(118-92-3)**
p-Aminobenzoic acid **(150-13-0)**
p-Aminobenzoic acid amide **(2835-68-9)**
p-Aminobenzoic acid butyl ester **(94-25-7)**
4-Aminobenzoic acid diethylaminoethyl ester **(59-46-1)**
p-Aminobenzoic acid 2-diethylaminoethyl ester **(59-46-1)**
4-Aminobenzoic acid 2-(diethylamino)ethyl ester hydrochloride **(51-05-8)**
p-Aminobenzoic acid 2-diethylaminoethyl ester hydrochloride **(51-05-8)**
4-Aminobenzoic acid, ethyl ester **(94-09-7)**
m-Aminobenzoic acid, ethyl ester **(582-33-2)**
p-Aminobenzoic acid ethyl ester **(94-09-7)**
3-Aminobenzoic acid, ethyl ester, methanesulfonate **(582-33-2)**
2-Aminobenzoic acid methyl ester **(134-20-3)**
o-Aminobenzoic acid methyl ester **(134-20-3)**
p-Aminobenzoic acid methyl ester **(619-45-4)**
p-Aminobenzoic acid monoglyceryl ester **(136-44-7)**
2-Aminobenzoic acid 3-phenyl-2-propenyl ester **(87-29-6)**
p-Aminobenzoic acid propyl ester **(94-12-2)**
p-Aminobenzoic acid sodium salt **(555-06-6)**
p-Aminobenzoic diethylaminoethylamide **(51-06-9)**
6-(4'-Aminobenzol-sulfonamido)-2,4-dimethylpyrimidin (German) **(57-68-1)**
(p-Aminobenzolsulfonyl)-2-amino-4,6-dimethylpyrimidin (German) **(57-68-1)**
4-Amino-benzolsulfonyl-methylcarbamat (German) **(3337-71-1)**
2-Aminobenzonitrile **(1885-29-6)**
3-Aminobenzonitrile **(2237-30-1)**
4-Aminobenzonitrile **(873-74-5)**
m-Aminobenzonitrile **(2237-30-1)**
o-Aminobenzonitrile **(1885-29-6)**
p-Aminobenzonitrile **(873-74-5)**
p-Aminobenzophenone **(1137-41-3)**
2-Aminobenzothiazole **(136-95-8)**
2-Aminobenzothiazole-6-methyl sulfone **(17557-67-4)**
2-Amino-6-benzothiazolesulfonic acid **(21951-32-6)**
2-Aminobenzothiazolyl-6 methyl sulfone **(17557-67-4)**
5-((4'-((6-Amino-5-(1H-benzotriazol-5-ylazo)-1-hydroxy-3-sulfo-2-naphthalenyl)azo)-3,3'-dimethoxy(1,1'-biphenyl)-4-yl)azo)-2-hydroxy-4-methylbenzoic acid, disodium salt **(70210-28-5)**
3-Aminobenzotrifluoride **(98-16-8)**
m-Aminobenzotrifluoride **(98-16-8)**
o-Aminobenzotrifluoride **(88-17-5)**
N-(p-Aminobenzoyl)aminoacetic acid **(61-78-9)**
2-Amino-5-benzoylaminohydroquinone diethyl ether **(120-00-3)**
p-Aminobenzoyldiethylaminoethanol **(59-46-1)**
p-Aminobenzoyldiethylaminoethanol hydrochloride **(51-05-8)**
o-Aminobenzoylformic anhydride **(91-56-5)**

N-(4-Aminobenzoyl)glycine **(61-78-9)**
N-(p-Aminobenzoyl)glycine **(61-78-9)**
N-(para-Aminobenzoyl)glycine **(61-78-9)**
2-(p-Aminobenzoyloxy)benzoic acid **(23787-97-5)**
2-Aminobenzthiazole **(136-95-8)**
4-(4-Aminobenzyl)aniline **(101-77-9)**
Aminobenzylpenicillin **(69-53-4)**
D-(-)-α-Aminobenzylpenicillin **(69-53-4)**
Aminobenzylpenicillin trihydrate **(7177-48-2)**
α-Aminobenzylpenicillin trihydrate **(7177-48-2)**
2-Aminobifenyl (Czech) **(90-41-5)**
4-Aminobifenyl (Czech) **(92-67-1)**
2-Aminobiphenyl **(90-41-5)**
4-Aminobiphenyl **(92-67-1)**
Aminobiphenyl **(41674-04-8)**
o-Aminobiphenyl **(90-41-5)**
p-Aminobiphenyl **(92-67-1)**
4-Aminobiphenyl ether **(139-59-3)**
3-Amino-(1,1'-biphenyl)-4-ol **(1134-36-7)**
5-Amino-1-bis(dimethylamide)phosphoryl-3-phenyl-1,2,4-triazole **(1031-47-6)**
5-Amino-1-bis(dimethylamido)phosphoryl-3-phenyl-1,2,4-triazole **(1031-47-6)**
5-Amino-1-(bis(dimethylamino)phosphinyl)-3-phenyl-1,2,4-triazole **(1031-47-6)**
5-Amino-1,3-bis(2-ethylhexyl)-5-methylhexahydropyrimidine **(141-94-6)**
Aminobis(propylamine) **(56-18-8)**
1-Aminobiurea **(4381-07-1)**
1-Amino-2-brom-4-hydroxyanthrachinon (Czech) **(116-82-5)**
1-Amino-4-bromo-2-anthraquinonesulfonic acid, sodium salt acid **(116-81-4)**
1-Amino-2-bromo-4-hydroxyanthraquinone **(116-82-5)**
4-Amino-3-bromopyridine **(13534-98-0)**
1-Amino-butaan (Dutch) **(109-73-9)**
1-Aminobutan (German) **(109-73-9)**
(RS)-2-Aminobutane **(13952-84-6)**
1-Aminobutane **(109-73-9)**
2-Aminobutane **(13952-84-6)**
2-Aminobutane Base **(13952-84-6)**
(+-)-2-Aminobutanoic acid **(2835-81-6)**
3-Aminobutanoic acid **(541-48-0)**
4-Aminobutanoic acid **(56-12-2)**
2-Amino-1-butanol **(96-20-8)**
2-Aminobutan-1-ol **(96-20-8)**
γ-Aminobuttersaeure (German) **(56-12-2)**
2-Amino-n-butyl alcohol **(96-20-8)**
4-Amino-N-((butylamino)carbonyl)benzenesulfonamide **(339-43-5)**
1-Amino-4-butylbenzene **(104-13-2)**
p-Aminobutylbenzene **(104-13-2)**
(4-Aminobutyl)diethoxymethylsilane **(3037-72-7)**
4-Amino-6-tert-butyl-3-mercapto-1,2,4-triazine-5(4H)-one **(33509-43-2)**
δ-Aminobutylmethyldiethoxysilane **(3037-72-7)**
4-Amino-6-tert-butyl-3-methylthio-as-triazin-5-one **(21087-64-9)**
4-Amino-6-tert-butyl-3-(methylthio)-1,2,4-triazin-5-one **(21087-64-9)**
3-Aminobutyric acid **(541-48-0)**
4-Aminobutyric acid **(56-12-2)**
γ-Amino-n-butyric acid **(56-12-2)**
γ-Aminobutyric acid **(56-12-2)**
4-Aminobutyric acid lactam **(616-45-5)**
γ-Aminobutyric acid lactam **(616-45-5)**
γ-Aminobutyric lactam **(616-45-5)**
γ-Aminobutyrolactam **(616-45-5)**
Aminocaine **(51-05-8)**
2-Aminocaproic acid **(327-57-1)**
6-Aminocaproic acid **(60-32-2)**
Aminocaproic acid **(60-32-2)**
α-Aminocaproic acid **(327-57-1)**
ε-Aminocaproic acid **(60-32-2)**
ω-Aminocaproic acid **(60-32-2)**
6-Aminocaproic acid lactam **(105-60-2)**
Aminocaproic lactam **(105-60-2)**
6-Aminocapronitrile **(2432-74-8)**
ω-Aminocapronitrile **(2432-74-8)**
Aminocarb **(2032-59-9)**
Aminocarbe (French) **(2032-59-9)**
2-Amino-α-carboline **(26148-68-5)**
Amino-α-carboline **(26148-68-5)**
(3-((Aminocarbonyl)amino)-2-methoxypropyl)chloromercury **(62-37-3)**
(4-((Aminocarbonyl)amino)phenyl)arsonic acid **(121-59-5)**
N-(Aminocarbonyl)benzeneacetamide **(63-98-9)**
2-(Aminocarbonyl)benzoic acid **(88-97-1)**
N-(Aminocarbonyl)-2-bromo-2-ethylbutanamide **(77-65-6)**
N-(Aminocarbonyl)glycine **(462-60-2)**
4-((5-(Aminocarbonyl)-2-methylphenyl)azo)-3-hydroxy-N-phenyl-2-naphthalenecarboxamide **(16403-84-2)**
2-((Aminocarbonyl)oxy)-N,N,N-trimethylethanaminium chloride **(51-83-2)**
4-((4-(Aminocarbonyl)phenyl)azo)-N-(2-ethoxyphenyl)-3-hydroxy-2-naphthalenecarboxamide **(2786-76-7)**
4-((4-(Aminocarbonyl)phenyl)azo)-3-hydroxy-N-(2-methoxyphenyl)-2-naphthalenecarboxamide **(36968-27-1)**
1-Amino-2-carboxybenzene **(118-92-3)**
1-Amino-4-carboxybenzene **(150-13-0)**
1-Amino-2-carboxylate-4-nitro-anthraquinone **(82-24-6)**
3-Amino-N-(α-carboxyphenethyl)succinamic acid N-methyl ester, stereoisomer **(22839-47-0)**
2-Amino-4-carboxythiazoline **(2150-55-2)**
Aminocardol **(317-34-0)**
4-Amino-6-p-chloroanilino-1,2-dwuhydro-2,2-dwumethylo-1,3,5-trojazyna (Polish) **(516-21-2)**
2-Amino-4-chloroaniline **(95-83-0)**
2-Amino-4-(p-chloroanilino)-s-triazine **(500-42-5)**
1-Amino-5-chloroanthraquinone **(117-11-3)**
1-Amino-2-chlorobenzene **(95-51-2)**
1-Amino-3-chlorobenzene **(108-42-9)**
1-Amino-4-chlorobenzene **(106-47-8)**
m-Aminochlorobenzene **(108-42-9)**
3-Amino-4-chlorobenzenesulfonic acid **(98-36-2)**
5-Amino-2-chlorobenzenesulfonic acid **(88-43-7)**
2-Amino-5-chlorobenzoic acid **(635-21-2)**
2-Amino-5-chlorobenzophenone **(719-59-5)**
2-Amino-4-chlorobenzothiazole **(19952-47-7)**
1-Amino-4-chloro-2-(chloromethyl)anthraquinone **(56594-25-3)**
4-Amino-5-chloro-N-(2-(diethylamino)ethyl)-2-methoxybenzamide **(364-62-5)**
5-Amino-4-chloro-2,3-dihydro-3-oxo-2-phenylpyridazine **(1698-60-8)**
2-Amino-4-chloro-6-ethylamino-s-triazine **(1007-28-9)**
2-Amino-5-chloro-4-ethylbenzenesulfonic acid **(88-56-2)**
1-Amino-2-chloro-4-hydroxy-9,10-anthracenedione **(2478-67-3)**
1-Amino-2-chloro-4-hydroxyanthraquinone **(2478-67-3)**
3-Amino-5-chloro-2-hydroxybenzenesulfonic acid **(88-23-3)**
1-Amino-2-chloro-6-methylbenzene **(87-60-5)**
1-Amino-3-chloro-2-methylbenzene **(87-60-5)**
1-Amino-3-chloro-4-methylbenzene **(95-74-9)**
1-Amino-3-chloro-6-methylbenzene **(95-79-4)**
2-Amino-5-chloro-4-methylbenzenesulfonic acid **(88-53-9)**
2-Amino-4-chloro-6-methylpyrimidine **(5600-21-5)**
2-Amino-3-chloro-1,4-naphthoquinone **(2797-51-5)**
1-Amino-2-chloro-4-nitrobenzene **(121-87-9)**
2-Amino-4-chloro-5-nitrophenol **(6358-07-2)**
2-Amino-6-chloro-4-nitrophenol **(6358-09-4)**
2-Amino-6-chloro-4-nitrophenol hydrochloride **(62625-14-3)**
(2-Amino-5-chlorophenyl)phenylmethanone **(719-59-5)**
5-Amino-4-chloro-2-phenyl-3(2H)-pyridazinone **(1698-60-8)**
5-Amino-4-chloro-2-phenylpyridazin-3(2H)-one **(1698-60-8)**
1-Amino-2-((chlorosulfinyloxy)methyl)anthraquinone **(56594-22-0)**
2-Amino-3-chlorotoluene **(87-63-8)**
2-Amino-4-chlorotoluene **(95-79-4)**
2-Amino-5-chlorotoluene **(95-69-2)**
2-Amino-6-chlorotoluene **(87-60-5)**

4-Amino-2-chlorotoluene (95-74-9)
2-Amino-5-chlorotoluene hydrochloride (3165-93-3)
6-Amino-4-chloro-m-toluenesulfonic acid (88-51-7)
6-Amino-5-((4-chloro-2-(trifluoromethyl)phenyl)azo)-4-hydroxy-2-naphthalenesulfonic acid, monosodium salt (57741-47-6)
6-Aminochrysene (2642-98-0)
o-Aminocinnamic acid lactam (59-31-4)
5-Amino-o-cresol (2835-95-2)
m-Amino-p-cresol, methyl ester (120-71-8)
3-Amino-p-cresol methyl ether (120-71-8)
4-Aminocumene (99-88-7)
4-Amino-5-cyano-7-(D-ribofuranosyl)-7H-pyrrolo(2,3-d)pyrimidine (606-58-6)
1-Aminocycloheptanecarboxylic acid (6949-77-5)
7-(D-2-Amino-2-(1,4-cyclohexadienyl)acetamide)-3-methoxy-3-cephem-4-carboxylic acid (51762-05-1)
Aminocyclohexane (108-91-8)
2-((4-Aminocyclohexyl)methyl)cyclohexanamine (24650-10-0)
1-Amino-1-cyclopentanecarboxylic acid (52-52-8)
1-Aminocyclopentane-1-carboxylic acid (52-52-8)
2-Aminocyclopentanemethanamine (21544-02-5)
2-Aminocyclopentanemethylamine (21544-02-5)
1-Aminodecane (2016-57-1)
4-Amino-1-(2-deoxy-β-d-erythro-pentofuranosyl)-s-triazin-2(1H)-one (2353-33-5)
D-2-Amino-2-deoxygalactose (7535-00-4)
2-Amino-2-deoxy-d-glucose (3416-24-8)
4-Amino-4-deoxy-N^{10}-methylpteroylglutamate (59-05-2)
4-Amino-4-deoxy-N^{10}-methylpteroylglutamic acid (59-05-2)
4-Amino-4-deoxypteroylglutamate (54-62-6)
Aminodiacetic acid (142-73-4)
1-Aminodiamantane (768-94-5)
4-Amino-N-(diaminomethylene)benzenesulfonamide (57-67-0)
Aminodibenzanthrone (26763-69-9)
1-Amino-2,4-dibromanthrachinon (Czech) (81-49-2)
1-Amino-2,4-dibromoanthraquinone (81-49-2)
1-Amino-3,4-dichlorobenzene (95-76-1)
4-Amino-2,5-dichlorobenzenesulfonic acid (88-50-6)
3-Amino-2,5-dichlorobenzoic acid (133-90-4)
4-Amino-3,5-dichlorobenzoic acid (56961-25-2)
3-Amino-2,5-dichlorobenzoic acid, ammonium salt (1076-46-6)
3-Amino-2,5-dichlorobenzoic acid methyl ester (7286-84-2)
2-Amino-4,6-dichlorophenol (527-62-8)
2-Amino-4,6-dichloropyrimidine (56-05-3)
2-Amino-4,6-dichloro-s-triazine (933-20-0)
N-(4-Amino-2,5-diethoxyphenyl)benzamide (120-00-3)
1-Amino-2-(N,N-diethylamino)ethane (100-36-7)
1-Amino-2-(diethylamino)ethane (100-36-7)
p-Amino-N-(2-diethylaminoethyl)benzamide (51-06-9)
1-Amino-3-(diethylamino)propane (104-78-9)
p-Aminodiethylaniline (93-05-0)
p-Amino diethylaniline hydrochloride (2198-58-5)
4-Aminodifenil (Spanish) (92-67-1)
p-Aminodifenylamin (Czech) (101-54-2)
4-Aminodifenylether (Czech) (139-59-3)
2-Amino-4,6-difluoro-1,3,5-triazine (1652-36-4)
2-Amino-4,6-difluorotriazine (1652-36-4)
4-Amino-5,6-dihydro-1-β-D-ribofuranosyl-s-triazin-2(1H)-one (62488-57-7)
7-Amino-2,3-dihydro-2,2-dimethylbenzofuran (68298-46-4)
1-Amino-9,10-dihydro-9,10-dioxo-4-(phenylamino)-2-anthracenesulfonic acid (2786-71-2)
2-Amino-1,5-dihydro-1-methyl-4H-imidazol-4-one (60-27-5)
5-Amino-2,3-dihydro-1,4-phthalazinedione (521-31-3)
2-Amino-1,7-dihydro-6H-purin-6-thion (Czech) (154-42-7)
5-Amino-1,4-dihydro-7H-1,2,3-triazolo(4,5-d)pyrimidin-7-one (134-58-7)
5-Amino-1,6-dihydro-7H-v-triazolo(4,5-d)pyrimidin-7-one (134-58-7)
l-2-Amino-1-(3,4-dihydroxyphenyl)ethanol (51-41-2)
2-Amino-3-(3,4-dihydroxyphenyl)propanoic acid (59-92-7)
2-Amino-4,6-dihydroxypyrimidine (56-09-7)
2-Amino-1-(2,5-dimethoxyphenyl)-1-propanol hydrochloride (61-16-5)
6-Amino-2,4-dimethoxypyrimidine (3289-50-7)
1-(4-Amino-6,7-dimethoxy-2-quinazolinyl)-4-(2-furanyl-carbonyl)piperazine (19216-56-9)
4'-Amino-N,N-dimethyl-4-aminoazobenzene (539-17-3)
1-Amino-3-dimethylaminopropane (109-55-7)
p-Aminodimethylaniline (105-10-2)
4'-Amino-2,3'-dimethylazobenzene (97-56-3)
4-Amino-2',3-dimethylazobenzene (97-56-3)
1-Amino-2,4-dimethylbenzene (95-68-1)
1-Amino-2,5-dimethylbenzene (95-78-3)
3-Amino-1,4-dimethylbenzene (95-78-3)
4-Amino-1,3-dimethylbenzene (95-68-1)
Aminodimethylbenzene (1300-73-8)
1-Amino-2,4-dimethylbenzene hydrochloride (21436-96-4)
1-Amino-2,5-dimethylbenzene hydrochloride (51786-53-9)
3-Amino-1,4-dimethylbenzene hydrochloride (51786-53-9)
4-Amino-1,3-dimethylbenzene hydrochloride (21436-96-4)
5-Amino-1,4-dimethylbenzene hydrochloride (51786-53-9)
2-Amino-3,3-dimethylbutane (3850-30-4)
3-Amino-1,4-dimethyl-γ-carboline (62450-06-0)
2-Aminodimethylethanol (124-68-5)
4-Amino-6-(1,1-dimethylethyl)-3-(methylthio)-1,2,4-triazin-5(4H)-one (21087-64-9)
2-Amino-3,4-dimethylimidazo(4,5-f)quinoline (77094-11-2)
2-Amino-3,8-dimethyl-3H-imidazo(4,5-f)quinoxaline (77500-04-0)
2-Amino-3,8-dimethylimidazo(4,5-f)quinoxaline (77500-04-0)
4-Amino-N-(3,4-dimethyl-5-isoxazolyl)benzenesulphonamide (127-69-5)
5-Amino-1,2-dimethyl-4-nitroimidazole (21677-57-6)
6-Amino-2-(2,4-dimethylphenyl)-1h-benz(de)isoquinoline-1,3(2h)-dione (2478-20-8)
4-Amino-N-(2,4-dimethylphenyl)naphthalene-1,8-dicarboximide (2478-20-8)
2-Amino-4,6-dimethylpyridine (5407-87-4)
3-Amino-1,4-dimethyl-5H-pyrido(4,3-b)indole (62450-06-0)
3-Amino-1,4-dimethyl-5H-pyrido(4,3-b)indole acetate (68808-54-8)
2-Amino-4,6-dimethylpyrimidine (767-15-7)
2-Amino-3,6-dinitrotoluene (56207-39-7)
2-Amino-4,6-dinitrotoluene (35572-78-2)
3-Amino-2,4-dinitrotoluene (70343-06-5)
3-Amino-2,6-dinitrotoluene (19406-51-0)
4-Amino-3,5-dinitrotoluene (6393-42-6)
5-Amino-2,4-dinitrotoluene (5267-27-6)
1-Amino-9,10-dioxo-9,10-dihydro-2-anthracenecarboxylic acid (82-24-6)
2-Amino-4,6-dioxypyrimidine (56-09-7)
2-Aminodiphenyl (90-41-5)
o-Aminodiphenyl (90-41-5)
p-Aminodiphenyl (92-67-1)
4-Aminodiphenyl (ACGIH,OSHA) (92-67-1)
p-Aminodiphenylamine (101-54-2)
4-Aminodiphenyl ether (139-59-3)
p-Aminodiphenylimide (60-09-3)
2-Amino-dipyrido(1,2-a:3',2'-d)-imidazole (67730-10-3)
4-Amino-3,4'-disulfoazobenzene (101-50-8)
Aminodur (317-34-0)
2-Aminoetanolo (Italian) (141-43-5)
1-Aminoethane (75-04-7)
Aminoethane (75-04-7)
2-Aminoethanesulfonic acid (107-35-7)
2-Aminoethanethiol (60-23-1)
1-Amino-ethanol (75-39-8)
1-Aminoethanol (75-39-8)
2-Aminoethanol (OSHA) (141-43-5)
2-Aminoethanol hydrochloride (2002-24-6)
β-Aminoethanol hydrochloride (2002-24-6)
2-Aminoethanol, hydrogen sulfate (ester) (926-39-6)
2-Aminoethanol mononitrate (20748-72-5)

2-Aminoethanol, nitrate (20748-72-5)
2-Aminoethanol nitrate (Salt) (20748-72-5)
2-Aminoethanol phosphate (Salt) (29868-05-1)
2-Aminoethanol sulfite (2:1) (Salt) (15535-29-2)
3-Amino-4-ethoxyacetanilide (17026-81-2)
3-Amino-4-ethoxyanilid kyseliny octove (Czech) (17026-81-2)
4-Aminoethoxybenzene (156-43-4)
2-Amino-6-ethoxybenzothiazole (94-45-1)
2-Aminoethoxyethanol (929-06-6)
2-(2-Aminoethoxy)ethanol (DOT) (929-06-6)
α-Aminoethyl alcohol (75-39-8)
β-Aminoethyl alcohol (141-43-5)
N-(2-((2-((2-((2-Aminoethyl)amino)ethyl)amino)ethyl)amino)-ethyl)nonanamide (26392-63-2)
1-Amino-4-ethylbenzene (589-16-2)
α-Aminoethylbenzene (98-84-0)
β-Aminoethylbenzene (64-04-0)
o-Aminoethylbenzene (578-54-1)
α-(1-Aminoethyl)benzenemethanol hydrochloride (154-41-6)
2-Amino-N-ethylbenzenesulfonanilide (81-10-7)
dl-α-(1-Aminoethyl)benzyl alcohol (14838-15-4)
α-(1-Aminoethyl)benzyl alcohol hydrochloride (154-41-6)
1-Amino-2-ethylbutane (617-79-8)
(2-Aminoethyl)carbamic acid (109-58-0)
3-Amino-9-ethylcarbazole (132-32-1)
3-Amino-N-ethylcarbazole (132-32-1)
3-Amino-9-ethylcarbazole hydrochloride (6109-97-3)
Aminoethyldiethanolamine, aminoethylethanolamine solution (3197-06-6)
N-(2-Aminoethyl)-N,N-diethylamine (100-36-7)
4-(2-Aminoethyl)diethylenetriamine (4097-89-6)
α-(1-Aminoethyl)-2,5-dimethoxybenzyl alcohol hydrochloride (61-16-5)
Aminoethylene (151-56-4)
Aminoethylethandiamine (111-40-0)
Aminoethyl ethanolamine (111-41-1)
N-Aminoethylethanolamine (111-41-1)
N-(2-Aminoethyl)-N-ethyl-m-toluidine (19248-13-6)
β-Aminoethylglyoxaline (51-45-6)
1-(β-Aminoethyl)-2-heptadecenylimidazoline (27476-93-3)
1-Amino-2-ethylhexan (Czech) (104-75-6)
1-α-(1-Aminoethyl)-m-hydroxybenzyl alcohol (54-49-9)
3-(β-Aminoethyl)-5-hydroxyindole (50-67-9)
β-Aminoethylimidazole (51-45-6)
1-(2-Aminoethyl)-2-imidazolidinone (6281-42-1)
1-(2-Aminoethyl)-2-imidazolidone (6281-42-1)
3-(2-Aminoethyl)indole (61-54-1)
(Amino-2 ethyl)-3 indole (French) (61-54-1)
3-(2-Aminoethyl)indol-5-ol (50-67-9)
2-Aminoethyl mercaptan (60-23-1)
N-2-Aminoethylmorfolin (Czech) (2038-03-1)
4-(2-Aminoethyl)-morpholine (2038-03-1)
N-2-Aminoethylmorpholine (2038-03-1)
N-Aminoethylmorpholine (2038-03-1)
p-β-Aminoethylphenol (51-67-2)
2-Amino-N-ethyl-N-phenylbenzenesulfonamide (81-10-7)
1-(2-Aminoethyl)piperazine (140-31-8)
1-Aminoethylpiperazine (140-31-8)
Aminoethylpiperazine (140-31-8)
N-(2-Aminoethyl)piperazine (140-31-8)
N-(Aminoethyl)piperazine (140-31-8)
N-(β-Aminoethyl)piperazine (140-31-8)
N-Aminoethylpiperazine [UN 2815] (140-31-8)
N-(2-Aminoethyl)-1,4-piperazinediethanamine (31295-54-2)
N-(2-Aminoethyl)-N-(2-(1-piperazinyl)ethyl)-1,2-ethanediamine (31295-49-5)
N-(2-Aminoethyl)-1,3-propanediamine (13531-52-7)
2-Amino-2-ethyl-1,3-propanediol (115-70-8)
4-(2-Aminoethyl)pyrocatechol (51-61-6)
1-Aminoethyl-2-tall oil alkyl-2-imidazoline (68442-97-7)
2-Amino-4-(ethylthio)butyric acid (67-21-0)
D-2-Amino-4-(ethylthio)butyric acid (535-32-0)
DL-2-Amino-4-(ethylthio)butyric acid (67-21-0)
L-2-Amino-4-(ethylthio)butyric acid (13073-35-3)
N-(2-(4-Amino-N-ethyl-m-toluidino)ethyl)methane-sulfonamide sulfate (2:3) (25646-71-3)
3-Amino-α-ethyl-2,4,6-triiodohydrocinnamic acid (96-83-3)
Aminofenazone (Italian) (58-15-1)
p-Aminofenetol (Czech) (156-43-4)
5-Amino-3-fenil-1-bis(-dimetilamino)-fosforil-1,2,4-triazolo (Italian) (1031-47-6)
Aminofenitrothion (13306-69-9)
m-Aminofenol (Czech) (591-27-5)
p-Aminofenol (Czech) (123-30-8)
5-Amino-3-fenyl-1-bis(dimethyl-amino)-fosforyl-1,2,4-triazool (Dutch) (1031-47-6)
2-Amino-fenylethan (Czech) (64-04-0)
Aminofilina (Spanish) (317-34-0)
Aminofluoren (German) (153-78-6)
2-Aminofluorene (153-78-6)
4-Amino-5-fluoro-2(1H)-pyrimidinone (2022-85-7)
4-Aminofolic acid (54-62-6)
Aminoform (100-97-0)
Aminoformamidine (113-00-8)
Aminoformic acid (463-77-4)
2-Aminoglutaramic acid (56-85-9)
l-2-Aminoglutaramidic acid (56-85-9)
α-Aminoglutaric acid (56-86-0)
l-2-Aminoglutaric acid (56-86-0)
1-Aminoglycerol (616-30-8)
Aminoglycol (115-69-5)
Aminoguanidine (79-17-4)
Aminoguanidine hydrochloride (1937-19-5)
1-Aminoheptane (111-68-2)
dl-2-Aminoheptane (123-82-0)
Aminohexahydrobenzene (108-91-8)
1-Aminohexane (111-26-2)
6-Aminohexanenitrile (2432-74-8)
(S)-2-Aminohexanoic acid (327-57-1)
2-Aminohexanoic acid (327-57-1)
6-Aminohexanoic acid (60-32-2)
ε-Aminohexanoic acid (60-32-2)
ω-Aminohexanoic acid (60-32-2)
6-Aminohexanoic acid cyclic lactam (105-60-2)
6-Aminohexanoic acid homopolymer (25038-54-4)
(6-Aminohexyl)carbamic acid (143-06-6)
N-(6-Aminohexyl)-1,6-hexanediamine (143-23-7)
Aminohippuric Acid (61-78-9)
4-Aminohippuric acid (61-78-9)
Aminohippuric acid (61-78-9)
p-Aminohippuric acid (61-78-9)
para-Aminohippuric acid (61-78-9)
α-Aminohydrocinnamic acid (63-91-2)
Aminohydroquinone dimethyl ether (102-56-7)
2-Amino-6-hydroxy-4(1H)-pyrimidinone (56-09-7)
1-Amino-4-hydroxyanthraquinone (116-85-8)
Aminohydroxyanthraquinone (52738-29-1)
1-((4-((1-Amino-4-hydroxy-2-anthraquinonyl)oxy)phenyl)methyl)-hexahydro-2H-azepin-2-one (19014-53-0)
4-Amino-4'-hydroxyazobenzene (103-18-4)
2-Amino-1-hydroxybenzene (95-55-6)
3-Amino-1-hydroxybenzene (591-27-5)
4-Amino-1-hydroxybenzene (123-30-8)
(R)-α-Amino-4-hydroxybenzeneacetic acid (22818-40-2)
3-Amino-4-hydroxybenzenesulfonamide (98-32-8)
3-Amino-4-hydroxybenzenesulfonic acid (98-37-3)
2-Amino-3-hydroxybenzoic acid (548-93-6)

4-Amino-2-hydroxybenzoic acid **(65-49-6)**
α-Amino-p-hydroxybenzylpenicillin **(26787-78-0)**
1-N-(L(-)-γ-Amino-α-hydroxybutyryl)kanamycin A **(37517-28-5)**
1-N-(L-(-)-4-Amino-2-hydroxybutyryl)kanamycin A **(37517-28-5)**
2-Amino-3-hydroxy-1,7-dihydro-7-methyl-6H-purin-6-one **(30345-27-8)**
2-Amino-3-hydroxy-1,7-dihydro-8-methyl-6H-purin-6-one **(30345-28-9)**
2-Amino-4-((2-hydroxyethyl)sulfonyl)phenol **(17601-96-6)**
4-Amino-N-(3-((2-hydroxyethyl)sulfonyl)phenyl)benzamide **(20241-68-3)**
4-Amino-N-(3-(2-hydroxyethyl)sulfonylphenyl)benzamide **(20241-68-3)**
5-Amino-4-hydroxy-3-((2-hydroxy-5-nitrophenyl)azo)-2,7-naphthalenedisulfonic acid **(13301-33-2)**
2-((2-Amino-5-hydroxy-6-((4'-((2-hydroxy-6-sulfo-1-naphthalenyl) azo)-3,3'-dimethoxy-(1,1'-biphenyl)-4-yl)azo)-7-sulfo-1-naphthalenyl)azo)-5-nitrobenzoic acid, trisodium salt **(71566-41-1)**
1-Amino-2-(hydroxymethyl)anthraquinone **(24094-44-8)**
3-Amino-4-hydroxy-N-methylbenzenesulfonamide **(80-23-9)**
2-Amino-4-(hydroxymethylphosphinyl)butanoic acid monoammonium salt **(77182-82-2)**
2-Amino-2-(hydroxymethyl)-1,3-propanediol **(77-86-1)**
2-Amino-2-(hydroxymethyl)propane-1,3-diol **(77-86-1)**
4-Amino-5-hydroxy-1,3-naphthalenedisulfonic acid **(82-47-3)**
4-Amino-5-hydroxy-2,7-naphthalenedisulfonic acid **(90-20-0)**
4-Amino-3-hydroxy-1-naphthalenesulfonic acid **(116-63-2)**
6-Amino-4-hydroxy-2-naphthalenesulfonic acid **(90-51-7)**
7-Amino-4-hydroxy-2-naphthalenesulfonic acid **(87-02-5)**
6-Amino-4-hydroxy-2-naphthalenesulfonic acid (γ acid) **(90-51-7)**
3-Amino-4-hydroxynitrobenzene **(99-57-0)**
3-Amino-2-hydroxy-5-nitro-benzenesulfonic acid **(96-67-3)**
3-Amino-4-hydroxy-5-nitrobenzenesulfonic acid **(96-93-5)**
1-Amino-4-hydroxy-2-phenoxyanthraquinone **(17418-58-5)**
6-(D-(-)-α-Amino-p-hydroxyphenylacetamido)penicillanic acid **(26787-78-0)**
dl-2-Amino-1-hydroxy-1-phenylpropane **(14838-15-4)**
threo-2-Amino-1-hydroxy-1-phenylpropane **(492-39-7)**
(S)-2-Amino-3-hydroxypropanoic acid **(56-45-1)**
2-Amino-3-hydroxypropanoic acid **(56-45-1)**
2-Amino-3-hydroxypyridine **(16867-03-1)**
4-Amino-2-hydroxypyrimidine **(71-30-7)**
5-((4'-((7-Amino-1-hydroxy-3-sulfo-2-naphthyl)azo)-4-biphenyl)-azo)-salicylic acid **(2429-82-5)**
7-Amino-4-hydroxy-3-(4-(4-(sulfophenyl)azo)phenylazo)-2-naphthalenesulfonic acid, disodium salt **(6300-50-1)**
4-Amino-2-hydroxytoluene **(2835-95-2)**
5-Amino-7-hydroxy-1H-v-triazolo(d)pyrimidine **(134-58-7)**
2-Aminohypoxanthine **(73-40-5)**
Aminoiminomethanesulfinic acid **(1758-73-2)**
N-(4-(((Aminoiminomethyl)amino)sulfonyl)phenyl)acetamide **(19077-97-5)**
N-(Aminoiminomethyl)-N-methylglycine **(57-00-1)**
(Aminoiminomethyl)urea **(141-83-3)**
(Aminoiminomethyl)urea phosphate **(17675-60-4)**
(Aminoiminomethyl)urea sulfate (2:1) **(591-01-5)**
α'-Amino-3-indolepropionic acid **(73-22-3)**
m-Aminoiodobenzene **(626-01-7)**
2-Aminoisobutane **(75-64-9)**
β-Aminoisobutanol **(124-68-5)**
α-Aminoisobutyronitrile **(19355-69-2)**
α-Aminoisocaproic acid **(61-90-5)**
3-Aminoisonaphthoic acid **(5959-52-4)**
α-Aminoisopropyl alcohol **(78-96-6)**
2-Aminoisopropylbenzene **(643-28-7)**
4-Amino-1-isopropylbenzene **(99-88-7)**
o-Aminoisopropylbenzene **(643-28-7)**
D-4-Amino-3-isossazolidone (Italian) **(68-41-7)**
DL-α-Aminoisovaleric acid **(516-06-3)**
L(+)-α-Aminoisovaleric acid **(72-18-4)**
D-4-Amino-3-isoxazolidinone **(68-41-7)**
D-4-Amino-3-isoxazolidone **(68-41-7)**
Aminokapron **(60-32-2)**
Aminolevulinic Acid **(106-60-5)**
2-Amino-6-mercaptopurine **(154-42-7)**
2-Amino-5-mercapto-1,3,4-thiadiazole **(2349-67-9)**
5-Amino-2-mercapto-1,3,4-thiadiazole **(2349-67-9)**
Aminomercuric chloride **(10124-48-8)**
2-Amino-6-merkaptopurin (Czech) **(154-42-7)**
2-Aminomesitylene **(88-05-1)**
Aminomesitylene **(88-05-1)**
2-Aminomesitylene hydrochloride **(6334-11-8)**
Aminomesitylene hydrochloride **(6334-11-8)**
Aminomethanamidine **(113-00-8)**
Aminomethane **(74-89-5)**
3-Amino-4-methoxyacetanilide **(6375-47-9)**
3-Amino-4-methoxy benzanilide **(120-35-4)**
1-Amino-2-methoxybenzene **(90-04-0)**
1-Amino-4-methoxybenzene **(104-94-9)**
1-Amino-4-methoxybenzene hydrochloride **(20265-97-8)**
2-Amino-5-methoxy benzenesulfonic acid **(13244-33-2)**
2-Amino-5-methoxybenzenesulfonic acid, monosodium salt **(19433-86-4)**
2-Amino-4-methoxybenzothiazole **(5464-79-9)**
1-Amino-2-methoxy-5-methylbenzene **(120-71-8)**
4-Amino-5-methoxy-2-methylbenzenesulfonic acid **(6471-78-9)**
7-Amino-9-α-methoxymitosane **(50-07-7)**
2-Amino-1-methoxy-4-nitrobenzene **(99-59-2)**
3-Amino-4-methoxynitrobenzene **(99-59-2)**
N-(3-Amino-4-methoxyphenyl)acetamide **(6375-47-9)**
3-((4-Amino-3-methoxyphenyl)azo)benzenesulfonic acid, sodium salt **(6300-07-8)**
3-(4-Amino-3-methoxyphenylazo)benzenesulfonic acid, sodium salt **(6300-07-8)**
4-Amino-N-(6-methoxy-3-pyridazinyl)-benzenesulfonamide (9CI) **(80-35-3)**
3-Amino-4-methoxytoluene **(120-71-8)**
4-Amino-1-methylaminoanthraquinone **(1220-94-6)**
1-(Aminomethyl)-2-aminocyclopentane **(21544-02-5)**
DL-α-Amino-β-methylaminopropionic acid **(17463-44-4)**
4-Amino-2-methylaniline **(95-70-5)**
2-Amino-4-methylanisole **(120-71-8)**
1-Amino-2-methylanthraquinone **(82-28-0)**
3-Amino-4-methylbenzamide **(19406-86-1)**
1-Amino-2-methylbenzene **(95-53-4)**
2-Amino-1-methylbenzene **(95-53-4)**
3-Amino-1-methylbenzene **(108-44-1)**
4-Amino-1-methylbenzene **(106-49-0)**
4-(Aminomethyl)-1,2-benzenediol **(37491-68-2)**
1-Amino-2-methylbenzene hydrochloride **(636-21-5)**
2-Amino-1-methylbenzene hydrochloride **(636-21-5)**
2-Amino-5-methylbenzenesulfonic acid **(88-44-8)**
4-Amino-3-methylbenzenesulfonic acid **(98-33-9)**
3-Amino-4-methylbenzenesulfonylcyclohexylurea **(565-33-3)**
2-Amino-4-methylbenzothiazole **(1477-42-5)**
3-Amino-α-methylbenzyl alcohol **(2454-37-7)**
m-Amino-α-methylbenzyl alcohol **(2454-37-7)**
1-Amino-3-methylbutane **(107-85-7)**
3-Amino-3-methyl-1-butyne **(2978-58-7)**
2-Amino-3-methyl-α-carboline **(68006-83-7)**
3-Amino-1-methyl-γ-carboline **(62450-07-1)**
1-Aminomethylcyclohexylamine **(5062-67-9)**
2-(Aminomethyl)cyclopentylamine **(21544-02-5)**
4-Amino-3-methyl-N,N-diethylaniline hydrochloride **(2051-79-8)**
l-α-(Aminomethyl)-3,4-dihydroxybenzyl alcohol **(51-41-2)**
2-Amino-6-methyldipyrido(1,2-a:3',2'-d)imidazole **(67730-11-4)**
l-α-Amino-β-methylenecyclopropanepropionic acid **(156-56-9)**
α-Amino-2-methylenecyclopropanepropionic acid **(156-56-9)**
α-Aminomethylenecyclopropanepropionic acid **(156-56-9)**
α-Amino-β-(2-methylenecyclopropyl)propionic acid **(156-56-9)**
2-Amino-4,5-methylenehex-5-enoic acid **(156-56-9)**
4-Amino-10-methylfolic acid **(59-05-2)**

Name	CAS No.
α-(Aminomethyl)-p-hydroxybenzyl alcohol	**(104-14-3)**
5-Aminomethyl-3-hydroxyisoxazole	**(2763-96-4)**
4-(Aminomethyl)-5-hydroxy-6-methyl-3-pyridinemethanol dihydrochloride	**(524-36-7)**
2-Amino-3-methyl-3H-imidazo(4,5-f)quinoline	**(76180-96-6)**
5-(Aminomethyl)-3-isoxazolol	**(2763-96-4)**
5-(Aminomethyl)-3(2H)-isoxazolone	**(2763-96-4)**
4-Amino-N-(5-methyl-3-isoxazolyl)benzenesulfonamide	**(723-46-6)**
5-Aminomethyl-3-isoxyzole	**(2763-96-4)**
l-α-Amino-γ-methylmercaptobutyric acid	**(63-68-3)**
2-Amino-5-methyloctane	**(67953-04-2)**
2-Amino-2-methylol-1,3-propanediol	**(77-86-1)**
2-Amino-3-methylpentanoic acid	**(73-32-5)**
2-Amino-4-methylpentanoic acid	**(61-90-5)**
5-Amino-2-methylphenol	**(2835-95-2)**
4-((4-Amino-3-methylphenyl)amino)phenol	**(6219-89-2)**
2-Amino-4-methyl-6-phenylamino-1,3,5-triazine	**(7426-35-9)**
2-((4-Amino-3-methylphenyl)ethylamino)ethanol sulfate	**(25646-77-9)**
5-Amino-3-methyl-1-phenylpyrazole	**(1131-18-6)**
8-Amino-2-methyl-4-phenyl-1,2,3,4-tetrahydroisoquinoline	**(24526-64-5)**
4-Amino-3-methyl-6-phenyl-1,2,4-triazin-5(4H)-one	**(41394-05-2)**
2-Amino-4-methylphosphinobutyric acid	**(51276-47-2)**
1-Aminomethylphosphonic acid	**(1066-51-9)**
1-Amino-2-methylpropane	**(78-81-9)**
2-Amino-2-methylpropane	**(75-64-9)**
2-Amino-2-methyl-1,3-propanediol	**(115-69-5)**
2-Amino-2-methylpropanenitrile	**(19355-69-2)**
2-Amino-2-methyl-1-propanol	**(124-68-5)**
2-Amino-2-methylpropan-1-ol	**(124-68-5)**
2-Amino-2-methylpropanol	**(124-68-5)**
2-Amino-2-methyl-1-propanol hydrochloride	**(3207-12-3)**
(-)-α-(Aminomethyl)protocatechuyl alcohol	**(51-41-2)**
4-Amino-N^{10}-methylpteroylglutamic acid	**(59-05-2)**
2-Amino-5-methylpyridine	**(1603-41-4)**
2-Amino-6-methylpyridine	**(1824-81-3)**
2-Amino-3-methyl-1H-pyrido(2,3-b)indole	**(68006-83-7)**
2-Amino-3-methyl-9H-pyrido(2,3-b)indole	**(68006-83-7)**
3-Amino-1-methyl-5H-pyrido(4,3-b)indole	**(62450-07-1)**
3-Amino-1-methyl-5H-pyrido(4,3-b)indole acetate	**(72254-58-1)**
2-Amino-4-methylpyrimidine	**(108-52-1)**
3-((4-Amino-2-methyl-5-pyrimidinyl)methyl)-5-(2-hydroxyethyl)-4-methylthiazolium chloride	**(59-43-8)**
2-Amino-4-(methylselenyl)butyric acid	**(1464-42-2)**
2-Amino-4-(methylsulfonyl)phenol	**(98-30-6)**
l(-)-Amino-γ-methylthiobutyric acid	**(63-68-3)**
4-Amino-3-methylthio-6-phenyl-1,2,4-triazine-5-one	**(21087-63-8)**
4-Amino-3-methyltoluene	**(95-68-1)**
2-Amino-4-methyltoluene hydrochloride	**(51786-53-9)**
4-Amino-3-methyltoluene hydrochloride	**(21436-96-4)**
α-Amino-β-methylvaleric acid	**(73-32-5)**
α-Amino-γ-methylvaleric acid	**(61-90-5)**
Aminomocovina (Czech)	**(57-56-7)**
1-Aminonaftalen (Czech)	**(134-32-7)**
2-Aminonaftalen (Czech)	**(91-59-8)**
1-Aminonaphthalene	**(134-32-7)**
2-Aminonaphthalene	**(91-59-8)**
3-Amino-2-naphthalenecarboxylic acid	**(5959-52-4)**
2-Amino-1,5-naphthalenedisulfonic acid	**(117-62-4)**
2-Amino-4,8-naphthalenedisulfonic acid	**(131-27-1)**
3-Amino-1,5-naphthalenedisulfonic acid	**(131-27-1)**
7-Amino-1,3-naphthalenedisulfonic acid	**(86-65-7)**
7-Amino-1,5-naphthalenedisulfonic acid	**(131-27-1)**
1-Amino-naphthalene hydrochloride	**(552-46-5)**
2-Aminonaphthalene hydrochloride	**(612-52-2)**
1-Amino-8-naphthalene sulfonate	**(82-75-7)**
2-Aminonaphthalene-6-sulfonate	**(93-00-5)**
1-Amino-2-naphthalenesulfonic acid	**(81-06-1)**
1-Amino-6-naphthalenesulfonic acid	**(119-79-9)**
1-Amino-7-naphthalenesulfonic acid	**(119-28-8)**
1-Aminonaphthalene-4-sulfonic acid	**(84-86-6)**
1-Aminonaphthalene-7-sulfonic acid	**(119-28-8)**
1-Aminonaphthalene-8-sulfonic acid	**(82-75-7)**
2-Amino-1-naphthalenesulfonic acid	**(81-16-3)**
2-Amino-6-naphthalenesulfonic acid	**(93-00-5)**
2-Aminonaphthalene-6-sulfonic acid	**(93-00-5)**
4-Amino-1-naphthalenesulfonic acid	**(84-86-6)**
5-Amino-1-naphthalenesulfonic acid	**(84-89-9)**
5-Amino-2-naphthalenesulfonic acid	**(119-79-9)**
6-Amino-2-naphthalenesulfonic acid	**(93-00-5)**
6-Aminonaphthalene-2-sulfonic acid	**(93-00-5)**
8-Amino-1-naphthalenesulfonic acid	**(82-75-7)**
8-Amino-2-naphthalenesulfonic acid	**(119-28-8)**
8-Aminonaphthalene-2-sulfonic acid	**(119-28-8)**
4-Amino-1-naphthalene sulfonic acid, sodium salt	**(130-13-2)**
1-Amino-2-naphthalenol	**(2834-92-6)**
Aminonaphthalenol	**(42884-33-3)**
3-Amino-2-naphthoic acid	**(5959-52-4)**
1-Amino-2-naphthol	**(2834-92-6)**
2-Amino-1-naphthol	**(42884-33-3)**
4-Amino-1-naphthol	**(2834-90-4)**
8-Amino-1-naphthol-5,7-disulfonic acid	**(82-47-3)**
1-Amino-2-naphthol hydrochloride	**(1198-27-2)**
2-Amino-1-naphthol hydrochloride	**(41772-23-0)**
1-Amino-2-naphthol-4-sulfonic acid	**(116-63-2)**
Aminonaphthol sulfonic acid, γ-	**(90-51-7)**
Aminonaphthol sulfonic acid J	**(87-02-5)**
2-Amino-6-naphthylsulfonic acid	**(93-00-5)**
4-Amino-1-napthalene sulfonic acid, sodium salt	**(130-13-2)**
6-Aminonicotinic acid	**(3167-49-5)**
2-Amino-4-nitroaniline	**(99-56-9)**
4-Amino-2-nitroaniline	**(5307-14-2)**
2-Amino-5-nitroanisol (Czech)	**(97-52-9)**
2-Amino-4-nitroanisole	**(99-59-2)**
2-Amino-5-nitroanisole	**(97-52-9)**
1-Amino-2-nitrobenzene	**(88-74-4)**
1-Amino-3-nitrobenzene	**(99-09-2)**
1-Amino-4-nitrobenzene	**(100-01-6)**
3-Aminonitrobenzene	**(626-01-7)**
m-Aminonitrobenzene	**(99-09-2)**
p-Aminonitrobenzene	**(100-01-6)**
2-Amino-5-nitrobenzenesulfonic acid ammonium salt	**(4346-51-4)**
2-Amino-4-nitro-benzoic acid	**(619-17-0)**
2-Amino-5-nitrobenzoic acid	**(616-79-5)**
Aminonitrodurene	**(13171-61-4)**
2-Amino-4-nitrofenol (Czech)	**(99-57-0)**
4-Amino-2-nitrofenol (Czech)	**(119-34-6)**
2-Amino-5-(5-nitro-2-furyl)-1,3,4-thiadiazole	**(712-68-5)**
5-Amino-2-(5-nitro-2-furyl)-1,3,4-thiadiazole	**(712-68-5)**
1-Amino-2-nitro-4-methylbenzene	**(89-62-3)**
2-Amino-4-nitrophenol	**(99-57-0)**
2-Amino-5-nitrophenol	**(121-88-0)**
4-Amino-2-nitrophenol	**(119-34-6)**
2-((4-Amino-2-nitrophenyl)amino)ethanol	**(2871-01-4)**
4-Amino-4'-nitro-2,2'-stilbenedisulfonic acid	**(119-72-2)**
4-Amino-4'-nitro-2,2'-stilbenedisulfonic acid, disodium salt	**(6634-82-8)**
5-Amino-2-(2-(4-nitro-2-sulfophenyl)ethenyl)benzenesulfonic acid, disodium salt	**(6634-82-8)**
2-Amino-5-nitrothiazole	**(121-66-4)**
Aminonitrothiazole	**(121-66-4)**
Aminonitrothiazolum	**(121-66-4)**
2-Amino-4-nitrotoluene	**(99-55-8)**
2-Amino-6-nitrotoluene	**(603-83-8)**
3-Amino-4-nitrotoluene	**(578-46-1)**
4-Amino-2-nitrotoluene	**(119-32-4)**

4-Amino-3-nitrotoluene **(89-62-3)**
1-Aminooctane **(111-86-4)**
2-Aminooctane **(693-16-3)**
3-Amino-2-oxetane carboxylic acid **(94818-85-6)**
5-Amino-4-oxopentanoic acid **(106-60-5)**
1-Amino-4-oxyanthraquinone (Russian) **(116-85-8)**
Aminopar **(65-49-6)**
Aminoparathion **(3735-01-1)**
p-Aminoparathion **(3735-01-1)**
D-(-)-α-Aminopenicillin **(69-53-4)**
Aminopentafluorobenzene **(771-60-8)**
1-Aminopentane **(110-58-7)**
3-Aminopentane **(616-24-0)**
DL-2-Aminopentane **(63493-28-7)**
2-Aminopentanedioic acid **(56-86-0)**
2-Aminopentanedioic acid hydrochloride **(138-15-8)**
Aminophen **(62-53-3)**
Aminophenazone **(58-15-1)**
p-Aminophenethyl alcohol **(104-10-9)**
1-(p-Aminophenethyl)-4-phenylisonipecotic acid ethyl ester **(144-14-9)**
1-(p-Aminophenethyl)-4-phenylpiperidine-4-carboxylic acid ethyl ester **(144-14-9)**
4-Aminophenetole **(156-43-4)**
p-Aminophenetole **(156-43-4)**
2-Aminophenol **(95-55-6)**
3-Aminophenol **(591-27-5)**
4-Aminophenol **(123-30-8)**
m-Aminophenol [UN 2512] **(591-27-5)**
o-Aminophenol [UN 2512] **(95-55-6)**
p-Aminophenol [UN 2512] **(123-30-8)**
o-Aminophenol, hemisulfate **(67845-79-8)**
p-Aminophenol, hemisulfate **(63084-98-0)**
2-Aminophenol hydrochloride **(51-19-4)**
4-Aminophenol hydrochloride **(51-78-5)**
o-Aminophenol hydrochloride **(51-19-4)**
p-Aminophenol hydrochloride **(51-78-5)**
Aminophenol sulfamide **(98-32-8)**
4-Aminophenol sulfate (2:1) (Salt) **(63084-98-0)**
2-Aminophenol sulfate (2:1) (Salt) **(67845-79-8)**
2-Aminophenol-4-sulfonamide **(98-32-8)**
o-Aminophenol-p-sulfonamide **(98-32-8)**
2-Aminophenol-4-sulfonic acid **(98-37-3)**
o-Aminophenol-p-sulfonic acid **(98-37-3)**
Aminophenurobutane **(339-43-5)**
2-Amino-5-phenyl-4(5H)-oxazolone **(2152-34-3)**
2-Amino-N-phenylacetamide **(555-48-6)**
N-(4-Aminophenyl)acetamide **(122-80-5)**
N-(p-Aminophenyl)acetamide **(122-80-5)**
N-(3-Aminophenyl)acetamide monohydrochloride **(621-35-2)**
7-(D-α-Aminophenyl-acetamido)cephalosporanic acid **(3577-01-3)**
7-(D-α-Aminophenylacetamido)desacetoxycephalosporanic acid **(15686-71-2)**
7-(D-2-Amino-2-phenylacetamido)-3-methyl-δ^{3}-cephem-4-carboxylic acid **(15686-71-2)**
6-(D(-)-α-Aminophenylacetamido)penicillanic acid **(69-53-4)**
4'-(p-Aminophenyl)acetanilide **(3366-61-8)**
(p-Aminophenyl)acetic acid **(1197-55-3)**
4-Aminophenylacetic acid **(1197-55-3)**
1-Amino-4-(phenylamino)-9,10-anthracenedione **(4395-65-7)**
2-(4-Aminophenyl)-5-aminobenzimidazole **(7621-86-5)**
1-((3-Aminophenyl)amino)-3-phenoxy-2-propanol **(38353-82-1)**
Aminophenylarsine acid **(98-50-0)**
p-Aminophenylarsine acid **(98-50-0)**
p-Aminophenylarsinic acid **(98-50-0)**
4-Aminophenylarsonic acid **(98-50-0)**
p-Aminophenylarsonic acid **(98-50-0)**
(4-Aminophenyl)arsonic acid sodium salt **(127-85-5)**
p-((p-Aminophenyl)azo)benzenesulfonic acid **(104-23-4)**
4-((4-Aminophenyl)azo)-1-naphthalenamine **(6054-48-4)**
2,2'-((p-((p-Aminophenyl)azo)phenyl)imino)diethanol **(20721-50-0)**
2,2'-((4-((4-Aminophenyl)azo)phenyl)imino)bis(ethanol) **(20721-50-0)**
N-(4-Aminophenyl)-1,4-benzenediamine **(537-65-5)**
2-(4-Aminophenyl)-1H-benzimidazol-5-amine **(7621-86-5)**
5-Amino-3-phenyl-1-bis (dimethyl-amino)-phosphoryle-1,2,4-triazole (French) **(1031-47-6)**
5-Amino-3-phenyl-1-bis(dimethylamino)-phosphoryl-1H-1,2,4-triazol (German) **(1031-47-6)**
5-Amino-1-phenyl-4-((2,5-dichloro-4-sulfophenyl)azo)-3-methyl-pyrazole **(12239-15-5)**
O-(4-Aminophenyl) O,O-diethyl phosphorothioate **(3735-01-1)**
1-Amino-1-phenylethane **(98-84-0)**
1-Amino-2-phenylethane **(64-04-0)**
2-Amino-1-phenylethane **(64-04-0)**
1-(3-Aminophenyl)ethanol **(2454-37-7)**
2-(4-Aminophenyl)ethanol **(104-10-9)**
2-(p-Aminophenyl)ethanol **(104-10-9)**
2-Amino-1-phenyl-1-ethanol **(7568-93-6)**
1-(2-Aminophenyl)ethanone **(551-93-9)**
4-Aminophenyl ether **(101-80-4)**
p-Aminophenyl ether **(101-80-4)**
N-β-(p-Aminophenyl)ethylnormeperidine **(144-14-9)**
N-(β-(p-Aminophenyl)ethyl)-4-phenyl-4-carbethoxypiperidine **(144-14-9)**
1-(2-(4-Aminophenyl)ethyl)-4-phenyl-4-piperidinecarboxylic acid ethyl ester **(144-14-9)**
4-((4-Aminophenyl)(4-imino-2,5-cyclohexadien-1-ylidene)-methyl)-2-methylbenzenamine **(3248-93-9)**
α-(4-Aminophenyl)-α-(4-imino-2,5-cyclohexadien-1-ylidene)-4-toluidine monohydrochloride **(479-73-2)**
2,2'-((4-Aminophenyl)imino)bis(ethanol) sulfate **(54381-16-7)**
2,2'-((4-Aminophenyl)imino)bisethanol sulfate (Salt) **(54381-16-7)**
p-Aminophenyl iodide **(540-37-4)**
p-Aminophenylmercaptan **(1193-02-8)**
3-Aminophenylmethane **(108-44-1)**
2-(p-Aminophenyl)-6-methylbenzothiazole **(92-36-4)**
2-(p-Aminophenyl)-6-methylbenzothiazolyl-7-sulfonic acid **(130-17-6)**
(m-Aminophenyl)methyl carbinol **(2454-37-7)**
4-Aminophenyl phenyl ether **(139-59-3)**
p-Aminophenyl phenyl ether **(139-59-3)**
2-Amino-1-phenylpropane **(60-15-1)**
β-(4-Aminophenyl)propane **(99-88-7)**
d-2-Amino-1-phenylpropane **(51-64-9)**
(+-)-2-Amino-1-phenylpropane sulfate **(60-13-9)**
DL-2-Amino-3-phenylpropanoic acid **(150-30-1)**
(+-)-2-Amino-1-phenyl-1-propanol hydrochloride **(154-41-6)**
1-(4-Aminophenyl)-1-propanone **(70-69-9)**
DL-α-Amino-β-phenylpropionic acid **(150-30-1)**
α-Amino-β-phenylpropionic acid **(63-91-2)**
4-Amino-N-(1-phenyl-1H-pyrazol-5-yl)benzenesulfonamide **(526-08-9)**
4-Aminophenylsulfonamide **(63-74-1)**
p-Aminophenylsulfonamide **(63-74-1)**
5-(4-Aminophenylsulfonamido)-3,4-dimethylisoxazole **(127-69-5)**
o-Aminophenylsulfonic acid **(88-21-1)**
p-Aminophenylsulfonic acid **(121-57-3)**
N-((4-Aminophenyl)sulfonyl)acetamide **(144-80-9)**
2-((3-Aminophenyl)sulfonyl)ethanol **(5246-57-1)**
5-(4-Aminophenylsulphonamido)-3,4-dimethylisoxazole **(127-69-5)**
3-(para-Aminophenylsulphonamido)-5-methylisoxazole **(723-46-6)**
5-Amino-3-phenyl-1,2,4-thiadiazole **(17467-15-1)**
2-Amino-4-phenylthiazole hydrobromide monohydrate **(52253-69-7)**
5-Amino-3-phenyl-1,2,4-triazole-1-yl-N,N,N',N'-tetramethyl-phosphodiamide **(1031-47-6)**
5-Amino-3-phenyl-1,2,4-triazolyl-1-bis(dimethylamido)pheosphate **(1031-47-6)**
5-Amino-3-phenyl-1,2,4-triazolylbis(dimethylamino)-phosphinoxid (German) **(1031-47-6)**
5-Amino-3-phenyl-1,2,4-triazolyl-N,N,N'N'-tetramethyl-

phosphonamide **(1031-47-6)**
p-(5-Amino-3-phenyl-1H-1,2,4-triazol-1-yl)-N,N,N',N'-tetra-methylphosphonic diamide **(1031-47-6)**
Aminophon **(51249-05-9)**
3-Aminophthalhydrazide **(521-31-3)**
3-Aminophthalic acid hydrazide **(521-31-3)**
3-Aminophthalic hydrazide **(521-31-3)**
Aminophylline **(317-34-0)**
6-Amino-3-picoline **(1603-41-4)**
Aminopielik D **(8068-77-7)**
2-Amino-propaan (Dutch) **(75-31-0)**
2-Aminopropan (German) **(75-31-0)**
1-Aminopropane **(107-10-8)**
2-Aminopropane **(75-31-0)**
1-Aminopropane-1,3-dicarboxylic acid **(56-86-0)**
3-Amino-1,2-propanediol **(616-30-8)**
3-Aminopropanenitrile **(151-18-8)**
2-Amino-propano (Italian) **(75-31-0)**
3-Aminopropanoic acid **(107-95-9)**
1-Amino-2-propanol **(78-96-6)**
1-Aminopropan-2-ol **(78-96-6)**
3-Amino-1-propanol **(156-87-6)**
3-Aminopropanol **(156-87-6)**
γ-Aminopropanol **(156-87-6)**
3-Aminopropene **(107-11-9)**
p-Aminopropiofenon (Czech) **(107-05-1)**
2-Aminopropionic acid **(56-41-7)**
3-Aminopropionic acid **(107-95-9)**
L-2-Aminopropionic acid **(56-41-7)**
L-S-Aminopropionic acid **(56-41-7)**
L-α-Aminopropionic acid **(56-41-7)**
α-Aminopropionic acid **(56-41-7)**
β-Aminopropionic acid **(107-95-9)**
3-Aminopropionitrile **(151-18-8)**
β-Aminopropionitrile **(151-18-8)**
2-Amino-2',6'-propionoxylidide **(41708-72-9)**
4'-Aminopropiophenone **(70-69-9)**
4-Aminopropiophenone **(70-69-9)**
p-Aminopropiophenone **(70-69-9)**
3-Aminopropyl alcohol **(156-87-6)**
1-Amino-4-propylbenzene **(2696-84-6)**
β-Aminopropylbenzene **(60-15-1)**
N-(3-Aminopropyl) cyclohexylamine **(3312-60-5)**
Aminopropyldiethanolamine (DOT) **(4985-85-7)**
(3-Aminopropyl)diethoxymethylsilane **(3179-76-8)**
3-Aminopropylene **(107-11-9)**
N-(3-Aminopropyl)ethylenediamine **(13531-52-7)**
(3-Aminopropyl)methylamine **(6291-84-5)**
N-(3-Aminopropyl)morfolin (Czech) **(123-00-2)**
4-Aminopropylmorpholine **(123-00-2)**
N-(3-Aminopropyl)morpholine **(123-00-2)**
N-Aminopropylmorpholine (DOT) **(123-00-2)**
1-(3-Aminopropyl)pyrrolidine **(23159-07-1)**
N-(3-Aminopropyl)pyrrolidine **(23159-07-1)**
(3-Aminopropyl)triethoxysilane **(919-30-2)**
(γ-Aminopropyl)triethoxysilane **(919-30-2)**
(3-Aminopropyl)trimethoxysilane **(13822-56-5)**
(γ-Aminopropyl)trimethoxysilane **(13822-56-5)**
Aminopteridine **(54-62-6)**
2-Amino-4,7(3H,8H)-pteridinedione **(529-69-1)**
Aminopterin **(54-62-6)**
4-Aminopteroylglutamic acid **(54-62-6)**
2-Aminopurine **(452-06-2)**
6-Amino-1H-purine **(73-24-5)**
6-Amino-3H-purine **(73-24-5)**
6-Amino-9H-purine **(73-24-5)**
6-Aminopurine **(73-24-5)**
2-Amino-6-purinethiol **(154-42-7)**
2-Aminopurine-6-thiol **(154-42-7)**
2-Aminopurine-6(1H)-thione **(154-42-7)**
2-Aminopurin-6-thiol (Czech) **(154-42-7)**
1-Aminopyrene **(1606-67-3)**
3-Aminopyrene **(1606-67-3)**
3-Aminopyridine **(462-08-8)**
4-Aminopyridine **(504-24-5)**
Amino-2 pyridine **(504-29-0)**
Amino-3 pyridine **(462-08-8)**
Amino-4 pyridine **(504-24-5)**
α-Aminopyridine **(504-29-0)**
β-Aminopyridine **(462-08-8)**
γ-Aminopyridine **(504-24-5)**
2-Aminopyridine (ACGIH,OSHA) **(504-29-0)**
m-Aminopyridine [UN 2671] **(462-08-8)**
o-Aminopyridine [UN 2671] **(504-29-0)**
p-Aminopyridine [UN 2671] **(504-24-5)**
6-Amino-3-pyridinecarboxylic acid **(3167-49-5)**
2-Amino-3-pyridinol **(16867-03-1)**
2-Amino-9H-pyrido(2,3-b)indole **(26148-68-5)**
2-Aminopyrimidine **(109-12-6)**
2-Amino-4,6-pyrimidinedione **(56-09-7)**
4-Amino-2(1H)-pyrimidinone **(71-30-7)**
Aminopyrine **(58-15-1)**
2-Aminoquinoline **(580-22-3)**
3-Aminoquinoline **(580-17-6)**
5-Aminoquinoline **(611-34-7)**
6-Aminoquinoline **(580-15-4)**
8-Aminoquinoline **(578-66-5)**
5-Amino-2-β-D-ribofuranosyl-as-triazin-3(2H)-one **(3131-60-0)**
6-Amino-9-β-D-ribofuranosyl-9H-purine **(58-61-7)**
4-Amino-1-β-D-ribofuranosyl-2(1H)-pyrimidinone **(65-46-3)**
4-Amino-7-β-D-ribofuranosyl-7H-pyrrolo(2,3-d)pyrimidine-5-carbonitrile **(606-58-6)**
4-Amino-1-β-D-ribofuranosyl-D-triazin-2(1H)-one **(320-67-2)**
4-Aminosalicylic acid **(65-49-6)**
Aminosalicylic acid **(65-49-6)**
p-Aminosalicylic acid **(65-49-6)**
para-Amino salicylic acid **(65-49-6)**
Aminosidin **(7542-37-2)**
Aminosidine **(7542-37-2)**
Aminosidine I **(7542-37-2)**
Aminosin **(109-57-9)**
o-Aminosulfanilic acid **(88-63-1)**
1-Amino-4-sulfo-2-naphthol **(116-63-2)**
2-Amino-5-sulfobenzoic acid **(3577-63-7)**
1-Amino-4-sulfonaphthalene **(84-86-6)**
1-Amino-6-sulfonaphthalene **(119-79-9)**
1-Amino-7-sulfonaphthalene **(119-28-8)**
2-Amino-6-sulfonaphthalene **(93-00-5)**
Aminosulfonic acid **(5329-14-6)**
5-(Aminosulfonyl)-4-chloro-2-((2-furanylmethyl)amino)benzoic acid **(54-31-9)**
5-(Aminosulfonyl)-N-((1-ethyl-2-pyrrolidinyl)methyl)-2-methoxy-benzamide **(15676-16-1)**
O-(4-(Aminosulfonyl)phenyl) O,O-dimethyl phosphorothioate **(115-93-5)**
N-(5-(Aminosulfonyl)-1,3,4-thiadiazol-2-yl)acetamide **(59-66-5)**
2-Amino-4-((2-(sulfooxy)ethyl)sulfonyl)phenol **(4726-22-1)**
2-Amino-4-sulfophenol **(98-37-3)**
4-(4-Amino-3-sulfophenylazo)benzenesulfonic acid **(101-50-8)**
Aminosumithion **(13306-69-9)**
9-Amino-1,2,3,4-tetrahydroacridine **(321-64-2)**
5-Amino-6,7,8,9-tetrahydroacridine (European) **(321-64-2)**
5-Amino-1H-tetrazole **(4418-61-5)**
5-Aminotetrazole **(4418-61-5)**
Aminotetrazole **(4418-61-5)**
2-Amino-1,3,4-thiadiazole-5-thiol **(2349-67-9)**

5-Amino-1,3,4-thiadiazole-2-thiol **(2349-67-9)**
2-Amino-δ(2)-1,3,4-thiadiazoline-5-thione **(2349-67-9)**
5-Amino-1,3,4-thiadiazoline-2-thione **(2349-67-9)**
2-Aminothiazole **(96-50-4)**
Aminothiazole **(96-50-4)**
2-Amino-2-thiazoline **(1779-81-3)**
2-Amino-2-thiazoline-4-carboxylic acid **(2150-55-2)**
4-Amino-N-2-thiazolylbenzenesulfonamide **(72-14-0)**
2-Aminothiophenol **(137-07-5)**
4-Aminothiophenol **(1193-02-8)**
o-Aminothiophenol **(137-07-5)**
p-Aminothiophenol **(1193-02-8)**
N-Aminothiourea **(79-19-6)**
3-Aminotoluen (Czech) **(108-44-1)**
4-Aminotoluen (Czech) **(106-49-0)**
2-Aminotoluene **(95-53-4)**
3-Aminotoluene **(108-44-1)**
4-Aminotoluene **(106-49-0)**
Aminotoluene **(100-46-9)**
α-Amino toluene **(100-46-9)**
α-Aminotoluene **(100-46-9)**
m-Aminotoluene **(108-44-1)**
o-Aminotoluene **(95-53-4)**
ω-Aminotoluene **(100-46-9)**
p-Aminotoluene **(106-49-0)**
2-Aminotoluene hydrochloride **(636-21-5)**
4-Aminotoluene hydrochloride **(540-23-8)**
o-Aminotoluene hydrochloride **(636-21-5)**
2-Amino-5-toluenesulfonic acid **(98-33-9)**
4-Amino-m-toluenesulfonic acid **(98-33-9)**
4-Aminotoluene-3-sulfonic acid **(88-44-8)**
6-Amino-m-toluenesulfonic acid **(88-44-8)**
4-Amino-m-toluenesulfonic acid (SO_3H=1) **(98-33-9)**
4-Amino-meta-toluenesulfonic acid (SO_3H=1) **(98-33-9)**
p-Amino-α-toluic acid **(1197-55-3)**
3-Amino-p-toluidine **(95-80-7)**
5-Amino-o-toluidine **(95-80-7)**
1-Amino-4-p-toluidinoanthraquinone **(4395-65-7)**
1-(3-Amino-p-tolylsulfonyl)-3-cyclohexylurea **(565-33-3)**
Aminotriacetic acid **(139-13-9)**
6-Amino-1,3,5-triazine-2,4(1H,3H)-dithione **(2770-75-4)**
2-Amino-1,3,4-triazole **(61-82-5)**
2-Aminotriazole **(61-82-5)**
3-Amino-1H-1,2,4-triazole **(61-82-5)**
3-Amino-s-triazole **(61-82-5)**
3-Aminotriazole **(61-82-5)**
Aminotriazole **(61-82-5)**
3-Amino-1,2,4-triazole (ACGIH) **(61-82-5)**
Aminotriazole (plant regulator) **(61-82-5)**
5-Amino-1H-v-triazolo(d)pyrimidin-7-ol **(134-58-7)**
Aminotriazol-spritzpulver **(61-82-5)**
3-Amino-1-(2,4,6-trichlorophenyl)-2-pyrazolin-5-one **(27241-31-2)**
3-Amino-1-(2,4,6-trichlorophenyl)-5-pyrazolone **(27241-31-2)**
4-Amino-3,5,6-trichloro-2-picolinic acid **(1918-02-1)**
4-Amino-3,5,6-trichloropicolinic acid **(1918-02-1)**
4-Amino-3,5,6-trichloropicolinic acid potassium salt **(2545-60-0)**
4-Amino-3,5,6-trichloro-2-pyridinecarbonitrile **(14143-60-3)**
4-Amino-3,5,6-trichlorpicolinsaeure (German) **(1918-02-1)**
1-Aminotricyclo(3.3.1.1^{3,7})decane **(768-94-5)**
2-Amino-4-(trifluoromethyl)-5-thiazolecarboxylic acid ethyl ester **(344-72-9)**
30-Amino-3,14,25-trihydroxy-3,9,14,20,25-pentaazatriacontane-2,10,13,21,24-pentaone **(70-51-9)**
2-(3-Amino-2,4,6-triiodobenzyl)butyric acid **(96-83-3)**
3-(3-Amino-2,4,6-triiodophenyl)-2-ethylpropanoic acid **(96-83-3)**
β-(3-Amino-2,4,6-triiodophenyl)-α-ethylpropionic acid **(96-83-3)**
1-Amino-2,4,6-trimethylbenzen (Czech) **(88-05-1)**
1-Amino-2,4,5-trimethylbenzene **(137-17-7)**
2-Amino-1,3,5-trimethylbenzene **(88-05-1)**
1-Amino-2,4,5-trimethylbenzene hydrochloride **(21436-97-5)**
2-Amino-1,3,5-trimethylbenzene hydrochloride **(6334-11-8)**
4-Amino-a,a,4-trimethylcyclohexanemethamine **(80-52-4)**
Aminotri(methylenephosphonic acid) **(6419-19-8)**
Aminotri(methylenephosphonic acid), pentasodium salt **(2235-43-0)**
Aminotri(methylene phosphonic acid), sodium salt **(20592-85-2)**
Aminotrimethylolmethane **(77-86-1)**
Aminotri(methylphosphonic acid) **(6419-19-8)**
Aminotris(hydroxymethyl)methane **(77-86-1)**
Aminotris(methanephosphonic acid) **(6419-19-8)**
Amino tris(methylene phosphonic acid), potassium salt **(27794-93-0)**
Aminotris(methylphosphonic acid) **(6419-19-8)**
Aminotris(methylphosphonic acid), pentasodium salt **(2235-43-0)**
Aminotris(methylphosphonic acid), sodium salts **(20592-85-2)**
11-Aminoundecanoic acid **(2432-99-7)**
Aminoundecanoic acid **(2432-99-7)**
11-Aminoundecylic acid **(2432-99-7)**
6-Aminouracil **(873-83-6)**
Aminouracil mustard **(66-75-1)**
Aminourea **(57-56-7)**
Aminourea hydrochloride **(563-41-7)**
2-Aminovaleric acid **(760-78-1)**
Aminox **(65-49-6)**
2-Amino-1,4-xylene **(95-78-3)**
4-Amino-1,3-xylene **(95-68-1)**
2-Amino-1,4-xylene hydrochloride **(51786-53-9)**
4-Amino-1,3-xylene hydrochloride **(21436-96-4)**
4-Amino-N-(2',4'-xylyl)-1,8-naphthalimide **(2478-20-8)**
Aminoxyscopolamine hydrobromide **(6106-81-6)**
Aminozide **(1596-84-5)**
1,2,-Aminozophenylene **(95-14-7)**
1,2-Aminozophenylene **(95-14-7)**
Aminutrin **(56-87-1)**
Aminzol soluble **(121-66-4)**
Amipenix S **(69-53-4)**
Amiphenazol **(490-55-1)**
Amiphenazole **(490-55-1)**
Amiphos **(13265-60-6)**
Amipramizide **(2016-88-8)**
Amiprol **(439-14-5)**
Amiprophos **(33857-23-7)**
Amiral **(43121-43-3)**
Amid-sal **(65-45-2)**
Amisol LDE **(142-78-9)**
Amisol LME **(142-78-9)**
Amistura P **(54-71-7)**
Amital **(57-43-2)**
Amitol **(61-82-5)**
Amiton **(78-53-5)**
Amiton oxalate **(3734-97-2)**
Amitraz **(33089-61-1)**
Amitraze (French) **(33089-61-1)**
Amitraz estrella **(33089-61-1)**
Amitrene **(51-63-8)**
Amitril **(61-82-5)**
Amitril T.L. **(61-82-5)**
Amitriptylin (German) **(50-48-6)**
Amitriptyline **(50-48-6)**
Amitrol **(61-82-5)**
Amitrol 90 **(61-82-5)**
Amitrol-T **(61-82-5)**
Amitrole (ACGIH,OSHA) **(61-82-5)**
Amitryptyline **(50-48-6)**
Amitryptyline, demethyl- **(72-69-5)**
Amixicotyn **(98-92-0)**
Amizol **(61-82-5)**

Amizol D (61-82-5)
Amizol F (61-82-5)
Amizol dp nau (61-82-5)
Ammat (7773-06-0)
Ammate (7773-06-0)
Ammate X (7773-06-0)
Ammelide (108-78-1)
Ammeline (645-92-1)
Ammidin (482-44-0)
Amminosidin (7542-37-2)
Ammo (52315-07-8)
Ammo Pesticide (52315-07-8)
Ammoform (100-97-0)
Ammoidin (298-81-7)
Ammoneric (12125-02-9)
Ammonia (ACGIH,OSHA) (8CI,9CI) (7664-41-7)
Ammonia, Anhydrous [UN 1005] (7664-41-7)
Ammonia Gas (7664-41-7)
Ammonia Solution, Strong (7664-41-7)
Ammonia Water 29% (1336-21-6)
Ammonia aqueous (1336-21-6)
Ammoniac (French) (7664-41-7)
Ammoniaca (Italian) (7664-41-7)
Ammoniak (German) (7664-41-7)
Ammonia solution, Containing more than 50% ammonia (DOT) (7664-41-7)
Ammonia solution, Containing more than 44% ammonia [UN 1005] (7664-41-7)
Ammonia solution, Containing 44% or less ammonia (1336-21-6)
Ammonia solution, More than 10% and not more than 35% ammonia [UN 2672] (7664-41-7)
Ammonia solution, More than 35% and not more than 50% ammonia [UN 2073] (7664-41-7)
Ammoniated mercury (10124-48-8)
Ammonio (bicromato di) (Italian) (7789-09-5)
Ammonio (dicromato di) (Italian) (7789-09-5)
Ammonioformaldehyde (100-97-0)
Ammonium-aethyl-carbamoyl-phosphonat (German) (25954-13-6)
Ammonium-carbonate (506-87-6)
Ammonium-chloroplatinate (16919-58-7)
Ammonium-ferric-cyano-ferrate(II) (25869-00-5)
Ammonium-formate (540-69-2)
Ammonium-hydrogen-fluoride (1341-49-7)
Ammonium 2,4-D (2307-55-3)
Ammonium(I) nitrate (1:1) (6484-52-2)
Ammonium acetate (631-61-8)
Ammonium acid arsenate (7784-44-3)
Ammonium acid phosphate (7722-76-1)
Ammonium acid sulfate (7803-63-6)
Ammonium alginate (9005-34-9)
Ammonium, alkyl(C8-C18)dimethyl 3,4-dichlorobenzyl-, chloride (8023-53-8)
Ammonium, alkyl(C14-16)dimethylbenzyl-, chlorides (63449-41-2)
Ammonium, N-alkyl(C9-15)tolyl methyltrimethyl-, chloride (1399-80-0)
Ammonium, alkyldimethylbenzyl-, chloride (8001-54-5)
Ammonium, allyltrimethyl-, chloride (1516-27-4)
Ammonium alum (7784-25-0)
Ammonium aluminum alum (7784-25-0)
Ammonium aluminum sulfate (7784-25-0)
Ammonium amidosulfonate (7773-06-0)
Ammonium amidosulphate (7773-06-0)
Ammonium (3-amino-3-carboxypropyl)methylphosphinate (77182-82-2)
Ammonium aminoformate (1111-78-0)
Ammonium 2-amino-4-(hydroxymethylphosphinyl)butanoate (77182-82-2)
Ammonium arsenate, Solid (DOT) (7784-44-3)
Ammonium arsenate [UN 1546] (7784-44-3)
Ammonium azide (12164-94-2)
Ammonium benzoate (1863-63-4)
Ammonium, benzyldiethyl((2,6-xylylcarbamoyl)methyl)-, benzoate (8CI) (3734-33-6)
Ammonium, benzyldimethyldodecyl-, chloride (139-07-1)
Ammonium, benzyldimethylhexadecyl-, chloride (122-18-9)
Ammonium, benzyldimethyloctadecyl-, chloride (122-19-0)
Ammonium, benzyldimethyloctyl-, chloride (959-55-7)
Ammonium, benzyldimethyl(2-phenoxyethyl) (7181-73-9)
Ammonium, benzyldimethyltetradecyl-, chloride (139-08-2)
Ammonium, benzyldimethyl(2-(2-(p-(1,1,3,3-tetramethylbutyl)-phenoxy)ethoxy)ethyl)-, chloride (121-54-0)
Ammonium, benzyldodecyldimethyl-, bromide (7281-04-1)
Ammonium, benzyldodecyldimethyl-, chloride (139-07-1)
Ammonium, benzyltrimethyl-, bromide (5350-41-4)
Ammonium, benzyltrimethyl-, chloride (56-93-9)
Ammonium, benzyltrimethyl-, hydroxide (100-85-6)
Ammonium, benzyltrimethyl-, iodide (4525-46-6)
Ammonium bicarbonate (1:1) (1066-33-7)
Ammoniumbichromaat (Dutch) (7789-09-5)
Ammonium bichromate (7789-09-5)
Ammonium bifluoride (1341-49-7)
Ammonium bifluoride, Solid (DOT) (1341-49-7)
Ammonium bifluoride, Solution (DOT) (1341-49-7)
Ammonium biphosphate (7722-76-1)
Ammonium, (4-(bis(p-(dimethylamino)phenyl)methylene)-2,5-cyclohexadien-1-ylidene) dimethyl-, chloride (548-62-9)
Ammonium, bis(2-hydroxyethyl)(2-(N-(2-hydroxyethyl)octa-decanamido)ethyl)methyl-, methyl sulfate, stearate (Ester) (8CI) (13441-22-0)
Ammonium bisulfate (7803-63-6)
Ammonium bisulfide (12124-99-1)
Ammonium bisulfite (10192-30-0)
Ammonium bisulfite, Solid (DOT) (10192-30-0)
Ammonium bisulfite, Solution (DOT) (10192-30-0)
Ammonium borate (12007-89-5)
Ammonium boron oxide (9CI) (12007-89-5)
Ammonium bromate (13843-59-9)
Ammonium bromide (9CI) (12124-97-9)
Ammonium butanoate (14287-04-8)
Ammonium butyrate (14287-04-8)
Ammonium calcium nitrate (15245-12-2)
Ammonium carbamate (DOT) (1111-78-0)
Ammonium carbazoate (131-74-8)
Ammoniumcarbonat (German) (506-87-6)
Ammonium carbonate (1066-33-7)
Ammonium carbonate (DOT) (506-87-6)
Ammonium, (3-carboxy-2-hydroxypropyl)trimethyl-, hydroxide, inner salt, l (541-15-1)
Ammonium, (carboxymethyl)dimethyl-cis-9-octadecenyl-, hydroxide, inner salt (871-37-4)
Ammonium, (carboxymethyl)hexadecyldimethyl-, hydroxide, inner salt (693-33-4)
Ammonium, (carboxymethyl)trimethyl-, chloride (590-46-5)
Ammonium, (9-(o-carboxyphenyl)-6-(diethylamino)-3H-xanthen-3-ylidene)diethyl-, chloride (81-88-9)
Ammonium, cetyldiethylethyl-, bromide (13316-70-6)
Ammonium chloramben (1076-46-6)
Ammonium chlorate (10192-29-7)
Ammoniumchlorid (German) (12125-02-9)
Ammonium chloride (OSHA) (12125-02-9)
Ammonium, (2-chloroethyl)trimethyl-, chloride (999-81-5)
Ammonium, (3-chloro-2-hydroxypropyl)trimethyl-, chloride (3327-22-8)
Ammonium, (2-(p-((2-chloro-4-nitrophenyl)azo)phenethylamino)-ethyl)trimethyl- (14097-03-1)
Ammonium chromate (7788-98-9)
Ammonium chromate(VI) (7788-98-9)
Ammonium citrate (7632-50-0)
Ammonium citrate, dibasic (3012-65-5)
Ammonium copper acetate (23087-46-9)

Ammonium copper carbonate **(33113-08-5)**
Ammonium cumenesulfonate **(37475-88-0)**
Ammonium decaborate **(12007-89-5)**
Ammonium, decamethylenebis(trimethyl- **(156-74-1)**
Ammonium, decyldimethyloctyl, chloride **(32426-11-2)**
Ammonium, decyltrimethyl-, bromide **(2082-84-0)**
Ammonium, decyltrimethyl-, chloride (8CI) **(10108-87-9)**
Ammonium diacid phosphate **(7722-76-1)**
Ammonium, (3,4-dichlorobenzyl)dodecyldimethyl-, chloride **(102-30-7)**
Ammonium 2,4-dichlorophenoxyacetate **(2307-55-3)**
Ammoniumdichromaat (Dutch) **(7789-09-5)**
Ammoniumdichromat (German) **(7789-09-5)**
Ammonium (dichromate d') (French) **(7789-09-5)**
Ammonium dichromate [UN 1439] **(7789-09-5)**
Ammonium dichromate (VI) **(7789-09-5)**
Ammonium, didecyldimethyl-, bromide (8CI) **(2390-68-3)**
Ammonium, didecyldimethyl-, chloride **(7173-51-5)**
Ammonium, didodecyldimethyl-, bromide (8CI) **(3282-73-3)**
Ammonium, (4-(p-(diethylamino)-α-phenylbenzylidene)-2,5-cyclohexadien-1-ylidene) diethyl-, sulfate (1:1) **(633-03-4)**
Ammonium, (4-(α-(p-(diethylamino)phenyl)-2,4-disulfobenzylidene)-2,5-cyclohexadien-1- ylidene)diethyl-, hydroxide, monosodium salt **(129-17-9)**
Ammonium, (6-(diethylamino)-3H-xanthen-3-ylidene)diethyl-, chloride **(2150-48-3)**
Ammonium, diethyl(6-(diethylamino)-3H-xanthen-3-ylidene)-, chloride **(2150-48-3)**
Ammonium, diethyl(2-hydroxyethyl)methyl-, bromide, xanthene-9-carboxylate **(53-46-3)**
Ammonium, (2-(O,O-diethylphosphorothio)ethyl)trimethyl-, iodide **(513-10-0)**
Ammonium difluoride **(1341-49-7)**
Ammonium, diheptadecyldimethyl-, chloride (8CI) **(1118-41-8)**
Ammonium dihexadecyldimethyl-, chloride **(1812-53-9)**
Ammonium dihydrogen orthophosphate **(7722-76-1)**
Ammonium dihydrogen phosphate **(7722-76-1)**
Ammonium dihydrophosphate **(7722-76-1)**
Ammonium, diisopropyl(2-hydroxyethyl)methyl-, bromide, xanthene-9-carboxylate **(50-34-0)**
Ammonium, (4-(p-(dimethylamino)-α-(p-(ethyl(m-sulfobenzyl)-amino)phenyl)benzylidene)-2,5-cyclohexadien-1-ylidene)-ethyl(m-sulfobenzyl)-, hydroxide, inner salt, sodium salt **(1694-09-3)**
Ammonium, (4-(p-(dimethylamino)-α-phenylbenzylidene)-2,5-cyclohexadien-1-ylidene)- dimethyl-, chloride **(569-64-2)**
Ammonium, (4-(p-(dimethylamino)-α-phenylbenzylidene)-2,5-cyclohexadien-1-ylidene)- dimethyl-, oxalate (2:1), oxalate (1:1) **(2437-29-8)**
Ammonium dimethylbenzenesulfonate **(26447-10-9)**
Ammonium, dimethyldioctadecyl-, chloride **(107-64-2)**
Ammonium, dimethylditetradecyl-, chloride (8CI) **(10108-91-5)**
Ammonium disulfatonickelate(II) **(15699-18-0)**
Ammonium dithiocarbamate **(513-74-6)**
Ammonium dodecanoate **(2437-23-2)**
Ammonium dodecylbenzenesulfonate **(1331-61-9)**
Ammonium dodecyl sulfate **(2235-54-3)**
Ammonium n-dodecyl sulfate **(2235-54-3)**
Ammonium, dodecyltrimethyl-, chloride (8CI) **(112-00-5)**
Ammonium, dodecyltrimethyl-, methyl sulfate **(13623-06-8)**
Ammonium, (2,3-epoxypropyl)trimethyl-, chloride **(3033-77-0)**
Ammonium ethanedioate **(14258-49-2)**
Ammonium, (ethylbenzyl)hexadecyldimethyl-, chloride (VAN)(8CI) **(29656-52-8)**
Ammonium, ethyl carbamoylphosphonate **(25954-13-6)**
Ammonium ethyl carbamoylphosphonate solution **(25954-13-6)**
Ammonium ethylenediaminetetraacetate **(7379-26-2)**
Ammonium, ethyl(4-(p-(ethyl(m-sulfobenzyl)amino)-α-phenylbenzylidene)- 2,5-cyclohexadien-1-ylidene)(m-sulfobenzyl)-, hydroxide, inner salt, sodium salt **(4680-78-8)**
Ammonium, ethyl(4-(α-(p-(ethyl(m-sulfobenzyl)amino)phenyl)-4-hydroxy-3-sulfobenzylidene)-2,5-cyclohexadien-1-ylidene)(m-sulfobenzyl)-, hydroxide, inner salt, disodium salt **(2353-45-9)**
Ammonium, ethyl(4-(p-(ethyl(m-sulfobenzyl)amino)-α-(o-sulfophenyl)benzylidene)- 2,5-cyclohexadien-1-ylidene)(m-sulfobenzyl)-, hydroxide, inner salt, diammonium salt **(2650-18-2)**
Ammonium, ethyl(4-(p-(ethyl(m-sulfobenzyl)amino)-α-(o-sulfophenyl)benzylidene)- 2,5-cyclohexadien-1-ylidene)(m-sulfobenzyl)-, hydroxide, inner salt, disodium salt **(3844-45-9)**
Ammonium, ethyl(4-(p-(ethyl(m-sulfobenzyl)amino)-α-(p-sulfophenyl)benzylidene)- 2,5-cyclohexadien-1-ylidene)(m-sulfobenzyl)-, hydroxide, inner salt, disodium salt **(5141-20-8)**
Ammonium, ethylhexadecyldimethyl-, bromide **(124-03-8)**
Ammonium ferric hexacyanoferrate **(25869-00-5)**
Ammonium ferric oxalate **(14221-47-7)**
Ammonium ferrioxalate **(14221-47-7)**
Ammonium ferrocyanide **(14481-29-9)**
Ammonium ferrous sulfate **(10045-89-3)**
Ammonium fluoborate (DOT) **(13826-83-0)**
Ammonium fluoride [UN 2505] **(12125-01-8)**
Ammonium fluoride comp. with hydrogen fluoride (1:1) **(1341-49-7)**
Ammonium fluoroacetate **(60916-92-9)**
Ammonium fluoroborate **(13826-83-0)**
Ammonium fluorosilicate [UN 2854] **(16919-19-0)**
Ammonium fluorure (French) **(12125-01-8)**
Ammonium fluosilicate **(16919-19-0)**
Ammonium fluozirconate **(16919-31-6)**
Ammonium D-gluconate **(10361-31-6)**
Ammonium gluconate **(10361-31-6)**
Ammonium glufosinate **(51276-47-2)**
Ammoniumglutaminat (German) **(7558-63-6)**
Ammonium heptamolybdate **(12027-67-7)**
Ammonium hexachloroplatinate(IV) **(16919-58-7)**
Ammonium hexacyanoferrate (II) **(14481-29-9)**
Ammonium hexadecanoate **(593-26-0)**
Ammonium hexadecyl sulfate **(4696-47-3)**
Ammonium, hexadecyltrimethyl **(6899-10-1)**
Ammonium, hexadecyltrimethyl-, borohydride **(19710-01-1)**
Ammonium, hexadecyltrimethyl-, bromide **(57-09-0)**
Ammonium, hexadecyltrimethyl-, chloride **(112-02-7)**
Ammonium hexafluorosilicate **(16919-19-0)**
Ammonium hexafluoro zirconate **(16919-31-6)**
Ammonium, hexamethylenebis(fluoren-9-yldimethyl- **(4844-10-4)**
Ammonium (dl-homoalanine-4-yl)methylphosphinate **(77182-82-2)**
Ammonium hydrofluoride **(1341-49-7)**
Ammonium hydrogen bifluoride **(1341-49-7)**
Ammonium hydrogen carbonate **(1066-33-7)**
Ammonium hydrogen difluoride **(1341-49-7)**
Ammonium hydrogen fluoride, Solid [UN 1727] **(1341-49-7)**
Ammonium hydrogen fluoride, Solution [UN 2817] **(1341-49-7)**
Ammonium hydrogen phosphate **(7783-28-0)**
Ammonium hydrogen phosphate solution **(7783-28-0)**
Ammonium hydrogen sulfate [UN 2506] **(7803-63-6)**
Ammonium hydrogen sulfide **(12124-99-1)**
Ammonium hydrogen sulfite **(10192-30-0)**
Ammonium hydrosulfide **(12124-99-1)**
Ammonium hydrosulfide, Solution (DOT) **(12124-99-1)**
Ammonium hydroxide **(1336-21-6)**
Ammonium hydroxide, Containing less than 12% ammonia **(1336-21-6)**
Ammonium hydroxide, Containing not less than 12% but not more than 44% ammonia **(1336-21-6)**
Ammonium, (5-hydroxycarvacryl)trimethyl-, chloride, 1-piperidinecarboxylate **(2438-53-1)**
Ammonium, (2-hydroxyethyl)diisopropylmethyl-, xanthene-9-carboxylate (ester) **(298-50-0)**
Ammonium, (2-hydroxyethyl)trimethyl-, chloride **(67-48-1)**
Ammonium, (m-hydroxyphenyl)trimethyl-, bromide,

decamethylenebis(methylcarbamate) (56-94-0)
Ammonium, (m-hydroxyphenyl)trimethyl-, bromide, dimethylcarbamate (114-80-7)
Ammonium, (m-hydroxyphenyl)trimethyl-, dimethylcarbamate (ester) (59-99-4)
Ammonium, (m-hydroxyphenyl)trimethyl-, methylcarbamate (ester) (8CI) (17752-10-2)
Ammonium, (2-hydroxypropyl)trimethyl-, chloride, acetate (62-51-1)
Ammonium hypophosphite (7803-65-8)
Ammonium hyposulfite (7783-18-8)
Ammonium iodide (9CI) (12027-06-4)
Ammonium ion (8CI,9CI) (14798-03-9)
Ammonium ion, ion (NH4(+1)) (14798-03-9)
Ammonium iron(III) citrate (1185-57-5)
Ammonium iron ethanedioate (55488-87-4)
Ammonium iron (III) hexacyanoferrate (25869-00-5)
Ammonium iron sulfate (10045-89-3)
Ammonium iron sulfate (2:2:1) (10045-89-3)
Ammonium isobutyrate (22228-82-6)
Ammonium laurate (2437-23-2)
Ammonium laureth sulfate (32612-48-9)
Ammonium lauryl benzene sulfonate (1331-61-9)
Ammonium lauryl ether sulfate (32612-48-9)
Ammonium lauryl sulfate (2235-54-3)
Ammonium mercaptan (12124-99-1)
Ammonium mercaptoacetate (5421-46-5)
Ammonium, (2-mercaptoethyl)trimethyl-, iodide, S-ester with O,O-diethylphosphorothioate (513-10-0)
Ammonium metavanadate [UN 2859] (7803-55-6)
Ammonium 2-methylpropanoate (22228-82-6)
Ammonium, methyltrioctyl-, chloride (5137-55-3)
Ammonium molybdate (12027-67-7)
Ammonium molybdate (13106-76-8)
Ammonium monobasic phosphate (7722-76-1)
Ammonium monohydrogen citrate (3012-65-5)
Ammonium monohydrogen orthophosphate (7783-28-0)
Ammonium monohydrogen sulfate (7803-63-6)
Ammonium monosulfide (12135-76-1)
Ammonium monosulfite (10192-30-0)
Ammonium muriate (12125-02-9)
Ammonium myreth sulfate (27731-61-9)
Ammonium myristyl ether sulfate (27731-61-9)
Ammonium 1-naphthaleneacetate (25545-89-5)
Ammonium nickel sulfate (15699-18-0)
Ammonium nitrate (6484-52-2)
Ammonium nitrate, No organic coating (DOT) (6484-52-2)
Ammonium nitrate, Organic coating (DOT) (6484-52-2)
Ammonium nitrate, Solution (containing not less than 15% water) (DOT) (6484-52-2)
Ammonium nitrate, With more than 0.2% combustible substances [UN 0222] (6484-52-2)
Ammonium nitrate, With not more than 0.2% combustible substances [UN 1942] (6484-52-2)
Ammonium nitrate sulfate (8CI,9CI) (12436-94-1)
UN 0222 [Ammonium nitrate, with more than 0.2 per cent combustible substances, including any organic substance calculated as carbon, to the exclusion of any other added substance] (6484-52-2)
UN 1942 [Ammonium nitrate, with not more than 0.2 per cent of combustible substances, including any organic substance calculated as carbon, to the exclusion of any other added substance] (6484-52-2)
Ammonium nitrite (DOT) (13446-48-5)
Ammonium N-nitrosophenylhydroxylamine (135-20-6)
Ammonium nonoxynol-4 sulfate (31691-97-1)
Ammonium nonoxynol-4-sulfate (9051-57-4)
Ammonium 9-octadecenoate (544-60-5)
Ammonium oleate (544-60-5)
Ammonium oxalate (14258-49-2)
Ammonium oxalate (5972-73-6)
Ammonium oxalate [NA 2449] (1113-38-8)
Ammonium oxalate ((NH4)2C2O4) monohydrate (6009-70-7)
Ammonium oxalate monohydrate (6009-70-7)
Ammonium paramolybdate (12027-67-7)
Ammonium paramolybdate (13106-76-8)
Ammonium pentaborate (12007-89-5)
Ammonium pentadecafluorooctanate (3825-26-1)
Ammonium perchlorate (7790-98-9)
Ammonium perchlorate, Average particle size less than 45 microns (DOT) (7790-98-9)
Ammonium perfluorocaprilate (3825-26-1)
Ammonium perfluorocaprylate (3825-26-1)
Ammonium perfluorooctanoate (3825-26-1)
Ammonium permanganate (13446-10-1)
Ammonium peroxydisulfate (7727-54-0)
Ammonium persulfate (ACGIH,dot) (7727-54-0)
Ammonium persulphate [UN 1444] (7727-54-0)
Ammonium, (v-phenenyltris(oxyethylene)tris(triethyl-, triiodide (65-29-2)
Ammonium 2-phenylphenate (52704-98-0)
Ammonium phosphate (7722-76-1)
Ammonium phosphate (7783-28-0)
Ammonium phosphate, dibasic (7783-28-0)
Ammonium orthophosphate dihydrogen (7722-76-1)
Ammonium phosphate, monobasic (7722-76-1)
Ammonium phosphate sulfate (9CI) (12593-60-1)
Ammonium phosphinate (7803-65-8)
Ammonium picrate, Dry or containing, by weight, less than 10% water [UN 0004] (131-74-8)
Ammonium picrate (Wet) (131-74-8)
Ammonium picrate, Wet with 10% or more water [UN 1310] (131-74-8)
Ammonium picronitrate (131-74-8)
Ammonium, (5-(piperidinocarbonyloxy)carvacryl)trimethyl-, chloride (2438-53-1)
Ammonium platinic chloride (16919-58-7)
Ammonium polymannurate (9005-34-9)
Ammonium polyphosphate solution (14728-39-3)
Ammonium polysulfide (9080-17-5)
Ammonium polysulfide solution [UN 2818] (9080-17-5)
Ammonium primary phosphate (7722-76-1)
Ammonium 2-propenoate (10604-69-0)
Ammonium rhodanate (1762-95-4)
Ammonium rhodanide (1762-95-4)
Ammonium saccharin (6381-61-9)
Ammonium saltpeter (6484-52-2)
Ammoniumsalz der amidosulfonsaure (German) (7773-06-0)
Ammonium silicofluoride (16919-19-0)
Ammonium silicon fluoride (16919-19-0)
Ammonium stearate (ACGIH) (1002-89-7)
Ammonium sulfamate (7773-06-0)
Ammonium sulfate (2:1) (7783-20-2)
Ammonium sulfate nitrate (12436-94-1)
Ammonium sulfhydrate (12124-99-1)
Ammonium sulfide (9080-17-5)
Ammonium sulfide (Solution) (12135-76-1)
Ammonium sulfide, Solution [UN 2683] (12135-76-1)
Ammonium sulfite (10196-04-0)
Ammonium sulfocarbamate (513-74-6)
Ammonium sulfocyanate (1762-95-4)
Ammonium sulfocyanide (1762-95-4)
Ammonium sulphamate (7773-06-0)
Ammonium sulphamidate (7773-06-0)
Ammonium sulphate (7783-20-2)
Ammonium sulphide, Solution (DOT) (12135-76-1)
Ammonium d-tartrate (3164-29-2)

Ammonium tartrate (14307-43-8)
Ammonium tartrate (3164-29-2)
Ammonium, tetrabutyl-, Salt with p-nitrophenol (1:1) (8CI) (3002-48-0)
Ammonium, tetrabutyl-, benzohydroxamate (8CI) (33684-10-5)
Ammonium, tetrabutyl-, salt with 2,3-butanedione monooxime (1:1) (8CI) (33684-09-2)
Ammonium, tetrabutyl-, salt with 4'-hydroxy-3',5'-dimethylacetophenone (1:1) (8CI) (33684-11-6)
Ammonium, tetrabutyl-, salt with 4-nitropyrocatechol (1:1) (8CI) (11072-43-8)
Ammonium, tetradecyltrimethyl-, bromide (1119-97-7)
Ammonium, tetraethyl-, bromide (71-91-0)
Ammonium, tetraethyl-, chloride (56-34-8)
Ammonium, tetraethyl-, iodide (68-05-3)
Ammonium, tetraethyl-, perchlorate, Dry (2567-83-1)
Ammonium tetrafluoroborate (13826-83-0)
Ammonium tetrafluoroborate(1-) (13826-83-0)
Ammonium, tetramethyl-, bromide (64-20-0)
Ammonium, tetramethyl-, chloride (75-57-0)
Ammonium, tetramethyl-, hydroxide (75-59-2)
Ammonium, tetrapropyl-, iodide (631-40-3)
Ammonium thiocyanate (463-56-9)
Ammoniumthiocyanate (1762-95-4)
Ammonium thioglycolate (5421-46-5)
Ammonium thioglycollate (5421-46-5)
Ammonium thiosulfate (7783-18-8)
Ammonium thiosulfate solution (10103-43-2)
Ammonium trichloroacetate (7646-88-0)
Ammonium 2,3,6-trichlorophenylacetate (53404-90-3)
Ammonium, triethylhexadecyl-, bromide (8CI) (13316-70-6)
Ammonium, trimethylbenzyl-, iodide (4525-46-6)
Ammonium, trimethyl(1-methyl-2-phenothiazin-10-ylethyl)-, methyl sulfate (58-34-4)
Ammonium, trimethyloctadecyl-, chloride (112-03-8)
Ammonium, trimethyloctyl-, chloride (8CI) (10108-86-8)
Ammonium, trimethyltallow alkyl-, chlorides (8030-78-2)
Ammonium, trimethyltetradecyl-, bromide (1119-97-7)
Ammonium, trimethyltetradecyl-, chloride (8CI) (4574-04-3)
Ammonium, trimethylvinyl-, hydroxide (463-88-7)
Ammonium trioxalatoferrate(III) (14221-47-7)
Ammonium, tris(2-hydroxyethyl)(phenylmercurio)-, lactate (23319-66-6)
Ammonium vanadate (11115-67-6)
Ammonium vanadate (7803-55-6)
Ammonium xylenesulfonate (26447-10-9)
Ammonium zinc chloride (9CI) (52628-25-8)
Ammonium zinc edetate (67859-51-2)
Ammonium zinc sulfate (7783-24-6)
Ammonium zinc sulfate hexahydrate (7783-24-6)
Ammonium zirconium fluoride (16919-31-6)
Ammonium-picrate (131-74-8)
Ammonium-sulfide (12124-99-1)
Ammonium-vanadium-oxide (11115-67-6)
Ammonyl BR 1244 (7281-04-1)
Ammonyx (8001-54-5)
Ammonyx 4 (122-19-0)
Ammonyx 485 (122-19-0)
Ammonyx 490 (122-19-0)
Ammonyx 4002 (122-19-0)
Ammonyx CA Special (122-19-0)
Ammonyx CPC (123-03-5)
Ammonyx DME (124-03-8)
Ammonyx Lo (1643-20-5)
Ammophyllin (317-34-0)
Amnicotin (98-92-0)
Amnosed (52-31-3)
Amnucol (9005-38-3)
Amobam (3566-10-7)
Amobarbital (57-43-2)
Amobarbitone (57-43-2)
Amoben (133-90-4)
Amoco 1010 (9003-07-0)
Amoco 610A4 (9002-88-4)
Amoco 15H (9003-29-6)
Amoco H 300 (9003-29-6)
Amoco NT-45 Process Oil (64742-46-7)
Amoenol (130-26-7)
Amoglandin (551-11-1)
Amoil (131-18-0)
Amokin (54-05-7)
Amolin (26787-78-0)
Amoniaco (Spanish) (7664-41-7)
Amoniak (Polish) (7664-41-7)
Amonyx AO (1643-20-5)
Amopenixin (26787-78-0)
Amorphous crocidolite asbestos (12001-28-4)
Amorphous silica (61790-53-2)
Amorphous silica dust (7631-86-9)
Amosene (57-53-4)
Amosite (Obs.) (1332-21-4)
Amosite asbestos (12172-73-5)
Amospan (57-43-2)
Amosyt (523-87-5)
Amotin (Russian) (23360-92-1)
Amotril (637-07-0)
Amotril S (637-07-0)
Amoxi (26787-78-0)
Amoxicillin (26787-78-0)
Amoxil (26787-78-0)
Amoxipen (26787-78-0)
Amoxone (94-75-7)
Amoxycillin (26787-78-0)
Ampi-Bol (69-53-4)
Ampacet C/A (9004-35-7)
Ampacet E/C (9004-57-3)
Ampazine (58-40-2)
Amperil (69-53-4)
Amperil (7177-48-2)
Amphaetamin (60-13-9)
Amphaetex (51-63-8)
Amphamed (60-13-9)
Amphamine sulfate (60-13-9)
Amphatamin (60-13-9)
Amphate (60-13-9)
Amphedrine (51-63-8)
Amphedrine (60-13-9)
Amphedroxy (300-42-5)
Amphedroxyn (300-42-5)
Amphenicol (56-75-7)
Amphepramone (90-84-6)
Ampherex (51-63-8)
(+)-Amphetamine (51-64-9)
Amphetamine (60-15-1)
d-Amphetamine (51-64-9)
dl-Amphetamine (300-62-9)
Amphetamine, 2,5-dimethoxy-, hydrochloride (24973-25-9)
(+)-Amphetamine sulfate (51-63-8)
(+-)-Amphetamine sulfate (60-13-9)
Amphetamine sulfate (60-13-9)
d-Amphetamine sulfate (51-63-8)
dl-Amphetamine sulfate (60-13-9)
Amphetamine sulphate (60-13-9)
Amphetaminum (60-13-9)
Amphetasul (51-63-8)
Amphex (51-63-8)

Name	CAS No.
Amphezamin	(60-13-9)
Amphibole	(1332-21-4)
Amphiboles	(1318-09-8)
Amphicol	(56-75-7)
Amphoids-S	(60-13-9)
Amphojel	(21645-51-2)
Amphomoronal	(1397-89-3)
Amphoteracin B	(1397-89-3)
Amphotericin β	(1397-89-3)
Amphotericin-B	(1397-89-3)
Amphotericine B	(1397-89-3)
Amphozone	(1397-89-3)
Ampichel	(7177-48-2)
Ampicillin	(69-53-4)
D-(-)-Ampicillin	(69-53-4)
D-Ampicillin	(69-53-4)
Ampicillin A	(69-53-4)
Ampicillin acid	(69-53-4)
Ampicillin anhydrate	(69-53-4)
Ampicillin trihydrate	(7177-48-2)
Ampicin	(69-53-4)
Ampikel	(69-53-4)
Ampikel	(7177-48-2)
Ampimed	(69-53-4)
Ampin-Penicillin	(6130-64-9)
Ampinova	(7177-48-2)
Ampipenin	(69-53-4)
Ampliactil	(50-53-3)
Ampliactil monohydrochloride	(69-09-0)
Amplicitil	(50-53-3)
Ampligram	(50-59-9)
Amplin	(7177-48-2)
Amplisom	(69-53-4)
Amplital	(69-53-4)
Amplivix	(68-90-6)
Ampol C 60	(9003-07-0)
Amprolene	(75-21-8)
Amptrerex	(51-63-8)
Ampy-penyl	(69-53-4)
Ampyrox	(155-41-9)
Amsacrine	(51264-14-3)
Amsco H-J	(8030-30-6)
Amsco H-SB	(8030-30-6)
Amsco tetramer	(6842-15-5)
Amseclor	(56-75-7)
Amsidine	(51264-14-3)
Amsine	(51264-14-3)
Amsonic acid	(81-11-8)
Amsustain	(51-63-8)
Amsustain	(51-64-9)
Amthio	(1762-95-4)
Amudane	(126-07-8)
Amuno	(53-86-1)
Amvisc	(9004-61-9)
Amybal	(57-43-2)
Amycin	(60-54-8)
Amycin, hydrochloride	(64-75-5)
Amygdalic acid	(90-64-2)
Amygdalin	(29883-15-6)
D-Amygdalin	(29883-15-6)
R-Amygdalin	(29883-15-6)
Amygdalinic acid	(90-64-2)
Amygdalonitrile	(532-28-5)
Amygdaloside	(29883-15-6)
Amyl Zimate	(137-30-4)
tert-Amyl acetate	(625-16-1)
n-Amyl acetate (ACGIH,OSHA)	(628-63-7)
sec-Amyl acetate (ACGIH,OSHA) [UN 1104]	(626-38-0)
Amyl acetate [UN 1104]	(628-63-7)
Amyl acetic ester	(628-63-7)
Amyl acetic ether	(628-63-7)
3-Amyl-4-acetoxytetrahydropyran	(18871-14-2)
Amyl acid phosphate	(12789-46-7)
Amyl alcohol	(71-41-0)
n-Amyl alcohol [UN 1105]	(71-41-0)
sec-Amyl alcohol [UN 1105]	(6032-29-7)
tert-Amyl alcohol [UN 1105]	(75-85-4)
Amyl alcohol, normal	(71-41-0)
Amyl aldehyde	(110-62-3)
n-Amylalkohol (Czech)	(71-41-0)
Amylamine	(110-58-7)
n-Amylamine	(110-58-7)
sec-Amylamine	(625-30-9)
p-tert-Amylaniline	(2049-92-5)
2-Amylanthraquinone	(13936-21-5)
Amylazetat (German)	(628-63-7)
Amylbarbitone	(57-43-2)
Amylbenzene	(538-68-1)
n-Amylbenzene	(538-68-1)
sec-Amylbenzene	(29316-05-0)
tert-Amylbenzene	(2049-95-8)
Amyl benzoate	(2049-96-9)
Amyl bromide	(110-53-2)
n-Amyl bromide	(110-53-2)
Amyl butyrate	(540-18-1)
n-Amyl butyrate [UN 2620]	(540-18-1)
γ-n-Amylbutyrolactone	(104-61-0)
Amyl caproate	(540-07-8)
n-Amyl caproate	(540-07-8)
Amyl capronate	(540-07-8)
Amylcarbinol	(111-27-3)
Amyl chloride	(594-36-5)
n-Amyl chloride	(543-59-9)
tert-Amyl chloride	(594-36-5)
Amyl chloride [UN 1107]	(543-59-9)
α-Amylcinnamaldehyde	(122-40-7)
Amylcinnamaldehyde	(122-40-7)
α-Amyl cinnamaldehyde	(122-40-7)
Amylcinnamic acid aldehyde	(122-40-7)
Amyl cinnamic aldehyde	(122-40-7)
Amylcinnamic aldehyde	(122-40-7)
α-Amyl cinnamic aldehyde	(122-40-7)
4-tert-Amylcyclohexanone	(16587-71-6)
Amyl dimethyl PABA	(14779-78-3)
Amyl-p-dimethylaminobenzoate	(14779-78-3)
2-sec-Amyl-4,6-dinitrophenol	(4097-36-3)
Amyleine	(644-26-8)
Amylene	(25377-72-4)
Amylene	(513-35-9)
α-Amylene	(109-67-1)
α-n-Amylene	(109-67-1)
β-Amylene-cis	(627-20-3)
β-Amylene-trans	(646-04-8)
β-n-Amylene	(109-68-2)
cis-β-Amylene	(627-20-3)
cis-β-n-Amylene	(627-20-3)
tert-Amylene	(26760-64-5)
trans-β-Amylene	(646-04-8)
trans-β-n-Amylene	(646-04-8)
n-Amylene [UN 1108]	(25377-72-4)
Amylene dichloride	(507-45-9)
2,4-Amyleneglycol	(625-69-4)
Amylene hydrate	(75-85-4)
Amylene, normal (DOT)	(25377-72-4)

Amylester kyseliny dusicne (Czech) **(1002-16-0)**
2-Amylester kyseliny octove (Czech) **(626-38-0)**
Amylester kyseliny octove (Czech) **(628-63-7)**
sek.Amylester kyseliny octove (Czech) **(626-38-0)**
Amylester kyseliny salicylove (Czech) **(2050-08-0)**
Amyl ether **(693-65-2)**
n-Amyl ether **(693-65-2)**
Amylethylcarbinol **(589-98-0)**
Amyl ethyl ketone **(106-68-3)**
Amylethylmethylcarbinol **(5340-36-3)**
n-Amyl fluoride **(592-50-7)**
n-Amyl formate **(638-49-3)**
Amyl formate [UN 1109] **(638-49-3)**
2-Amylfuran **(3777-69-3)**
Amyl hexanoate **(540-07-8)**
Amyl hexoate **(540-07-8)**
t-Amyl hydroperoxide **(3425-61-4)**
tert-Amyl hydroperoxide **(3425-61-4)**
Amyl hydrosulfide **(110-66-7)**
3-Amyl-1-hydroxy-6,6,9-trimethyl-6H-dibenzo(b,d)pyran **(521-35-7)**
Amyl iodide **(628-17-1)**
n-Amyl iodide **(628-17-1)**
sec-Amyl iodide **(637-97-8)**
Amyl ketone **(927-49-1)**
n-Amyl mercaptan **(110-66-7)**
Amyl mercaptan [UN 1111] **(110-66-7)**
Amyl methyl alcohol **(105-30-6)**
Amyl methyl carbinol **(543-49-7)**
Amyl-methyl-cetone (French) **(110-43-0)**
n-Amyl methyl ketone **(110-43-0)**
Amyl methyl ketone [UN 1110] **(110-43-0)**
N-Amyl-N-methylnitrosamine **(13256-07-0)**
Amyl methyl sulfide **(1741-83-9)**
Amyl nitrate **(1002-16-0)**
Amyl nitrit **(110-46-3)**
Amyl nitrite **(110-46-3)**
n-Amyl nitrite **(463-04-7)**
Amyl nitrite [UN 1113] **(463-04-7)**
Amyl p-nitrobenzoate **(14309-42-3)**
Amylobarbital **(57-43-2)**
Amylobarbitone **(57-43-2)**
Amylocaine **(644-26-8)**
Amylodextrin (9CI) **(9005-84-9)**
Amylodextrins **(9005-84-9)**
Amylofene **(50-06-6)**
Amylol **(71-41-0)**
Amylomaize VII **(9005-25-8)**
Amylopectin **(9037-22-3)**
Amylopectin (9CI) **(9037-22-3)**
β-Amylose **(9004-34-6)**
Amylowy alkohol (Polish) **(123-51-3)**
tert-Amyl peroxybenzoate **(4511-39-1)**
tert-Amyl peroxy-2-ethylhexanoate, Technically pure (DOT) **(686-31-7)**
tert-Amyl peroxyneodecanoate **(68299-16-1)**
tert-Amyl peroxyneodecanoate (Not more than 75% with phlegmatiser) **(68299-16-1)**
4-n-Amylphenol **(14938-35-3)**
4-tert-Amylphenol **(80-46-6)**
o-Amyl phenol **(136-81-2)**
o-Amylphenol **(136-81-2)**
o-tert-Amylphenol **(3279-27-4)**
p-tert-Amylphenol **(80-46-6)**
Amyl phenol 4T **(80-46-6)**
p-tert-Amylphenol, potassium salt **(53404-18-5)**
p-tert-Amylphenol sodium salt **(31366-95-7)**
α-Amyl-β-phenylacrolein **(122-40-7)**
3-sec-Amylphenyl N-methylcarbamate **(2282-34-0)**
Amyl phthalate **(131-18-0)**
Amyl potassium xanthate **(2720-73-2)**
6-Amyl-α-pyrone **(27593-23-3)**
5-n-Amylresorcinol **(500-66-3)**
Amyl salicylate **(2050-08-0)**
Amyl sulfhydrate **(110-66-7)**
Amyl thioalcohol **(110-66-7)**
Amyl toluene **(1320-01-0)**
Amyltoluene **(1320-01-0)**
Amyl 2,4,5-trichlorophenoxyacetate **(120-39-8)**
Amyl trichlorosilane **(107-72-2)**
Amyltrichlorosilane [UN 1728] **(107-72-2)**
Amylum **(9005-25-8)**
Amyl-δ-valerolactone **(705-86-2)**
Amylvinylcarbinol **(3391-86-4)**
Amyl xylyl ether **(1320-21-4)**
Amyris-Oil **(8015-65-4)**
Amytal **(57-43-2)**
Amytriptiline **(50-48-6)**
A1030n0 **(105-60-2)**
(-)-Anabasin **(494-52-0)**
Anabasin **(494-52-0)**
Anabasine **(494-52-0)**
Anabasine, 1-nitroso **(1133-64-8)**
Anabazin **(494-52-0)**
Anabol **(10418-03-8)**
Anabolin **(72-63-9)**
Anac 110 **(7440-50-8)**
Anacardiol **(13898-68-5)**
Anacardol, tetrahydro- **(501-24-6)**
Anacardone **(59-26-7)**
Anacel **(136-47-0)**
Anacetin **(56-75-7)**
Anacobin **(68-19-9)**
Anacordone **(59-26-7)**
Anadolor **(51-05-8)**
Anadomis Green **(1308-38-9)**
Anadonis Green **(1308-38-9)**
Anadrol **(434-07-1)**
Anadroyd **(434-07-1)**
Anaesthesin **(94-09-7)**
Anaesthetic ether **(60-29-7)**
Anafebrina **(58-15-1)**
Anaflex **(9011-05-6)**
Anaflon **(103-90-2)**
Anafranil **(303-49-1)**
Anagiardil **(443-48-1)**
Anahist **(63-56-9)**
Analgesine **(60-80-0)**
Analgine **(9005-34-9)**
Analgizer **(76-38-0)**
O-Analog of Dimethoate **(1113-02-6)**
Anamenth **(79-01-6)**
Anamid **(65-45-2)**
Anapac **(62-44-2)**
Anapolon **(434-07-1)**
Anapral **(24815-24-5)**
Anaprel **(24815-24-5)**
Anara **(60-13-9)**
Anarcon **(62-67-9)**
Anasteron **(434-07-1)**
Anasteronal **(434-07-1)**
Anasterone **(434-07-1)**
Anastress **(57-53-4)**
(-)-Anatabine **(581-49-7)**
Anatabine **(581-49-7)**
Anatase (TiO_2) (9CI) **(1317-70-0)**

Anatase titanium dioxide **(1317-70-0)**
Anatensol **(69-23-8)**
Anathylmon **(57-53-4)**
Anatola **(68-26-8)**
Anatola A **(68-26-8)**
Anatran **(133-67-5)**
Anautine **(523-87-5)**
Anavar **(53-39-4)**
Anayodin **(7681-82-5)**
Anazolene, Sodium **(3861-73-2)**
Ancamine TL **(101-77-9)**
Anchoic acid **(123-99-9)**
Anchol **(519-95-9)**
Anchred Standard **(1309-37-1)**
Ancobon **(2022-85-7)**
Ancolan **(569-65-3)**
Ancolon **(569-65-3)**
Ancor EN 80/150 **(7439-89-6)**
Ancortone **(53-03-2)**
Ancotil **(2022-85-7)**
Ancrack **(132-66-1)**
Ancylol **(305-85-1)**
Ancymidol **(12771-68-5)**
Ancymidole **(12771-68-5)**
Andaksin **(57-53-4)**
Andaxin **(57-53-4)**
Andhist **(63-56-9)**
Andramine **(523-87-5)**
Andrazide **(54-85-3)**
Andrez **(9003-55-8)**
Androfluorene **(76-43-7)**
Androfluorone **(76-43-7)**
Androgen **(57-85-2)**
Androlin **(58-22-0)**
Andromedotoxin **(4720-09-6)**
Andrometh **(58-18-4)**
Andronaq **(58-22-0)**
Androsan **(57-85-2)**
Androsan **(58-18-4)**
Androsan (tablets) **(58-18-4)**
Androsta-1,4-dien-3-one, 17-β-hydroxy-17-α-methyl **(72-63-9)**
Androsta-1,4-dien-3-one, 17-β-hydroxy-17-methyl- (8CI) **(72-63-9)**
Androsta-1,4-dien-3-one, 17-hydroxy-17-methyl-, (17-β)- (9CI) **(72-63-9)**
Androstanazol (VAN) **(10418-03-8)**
Androstanazole (VAN) **(10418-03-8)**
5-α-Androstane-2-α-carbonitrile, 4-α,5-epoxy-17-β-hydroxy-3-oxo- **(13647-35-3)**
Androstane-2-carbonitrile, 4,5-epoxy-17-hydroxy-3-oxo-, (2-α,4-α,5-α,17-β)- **(13647-35-3)**
5α-Androstan-3α,17β-diol, 2β,16β-dipipecolinio-, dibromide, diacetate **(15500-66-0)**
Androstano(2,3-c)(1,2,5)oxadiazol-17-ol, 17-methyl-, (5-α,17-β)- **(434-07-1)**
5-α,17-β-Androstan-3-one, 17-hydroxy-2-(hydroxymethylene)-17-methyl **(434-07-1)**
Androstan-3-one, 17-hydroxy-2-(hydroxymethylene)-17-methyl-, (5-α,17-β)- **(434-07-1)**
5-α-Androstan-3-one, 17-β-hydroxy-2-(hydroxymethylene)-17-methyl- (8CI) **(434-07-1)**
5-α-Androstan-3-one, 17-β-hydroxy-2-α-methyl **(58-19-5)**
Androstan-3-one, 17-hydroxy-2-methyl-, (2-α,5-α,17-β)- **(58-19-5)**
Androsten **(58-18-4)**
Androst-2-ene-2-carbonitrile, 4,5-epoxy-3,17-dihydroxy-, (4α,5α,17β)- **(13647-35-3)**
4-Androstene-17-α-methyl-17-β-ol-3-one **(58-18-4)**
δ^4-Androstene-17-β-propionate-3-one **(57-85-2)**
2'H-5α-Androst-2-eno(3,2-c)pyrazol-17β-ol, 17-methyl- **(10418-03-8)**
2'H-Androst-2-eno(3,2-c)pyrazol-17-ol, 17-methyl-, (5α,17β)- (9CI) **(10418-03-8)**
Androst-5-en-3-β-ol, 17-β-((3-(dimethylamino)propyl)methyl-amino)-, dihydrochloride **(1249-84-9)**
Androst-5-en-3-ol, 17-((3-(dimethylamino)propyl)methyl-amino)-, dihydrochloride, (3β,17β)- **(1249-84-9)**
Androst-4-en-17β-ol-3-one **(58-22-0)**
δ^4-Androsten-17(β)-ol-3-one **(58-22-0)**
Androst-4-en-3-one, 7-β,17-α-dimethyl-17-β-hydroxy **(17021-26-0)**
Androst-4-en-3-one, 9-fluoro-11-β,17-β-dihydroxy-17-methyl **(76-43-7)**
Androst-4-en-3-one, 9-fluoro-11,17-dihydroxy-17-methyl-, (11-β,17-β)- (9CI) **(76-43-7)**
Androst-4-en-3-one, 17-β-hydroxy- **(58-22-0)**
Androst-4-en-3-one, 17-hydroxy-, (17-β)- **(58-22-0)**
Androst-4-en-3-one, 17-β-hydroxy-7-β,17-dimethyl- **(17021-26-0)**
Androst-4-en-3-one, 17-hydroxy-7,17-dimethyl-, (7-β,17-β)- (9CI) **(17021-26-0)**
Androst-4-en-3-one, 17-β-hydroxy-17-methyl **(58-18-4)**
Androst-4-en-3-one, 17-hydroxy-17-methyl-, (17-β)- (9CI) **(58-18-4)**
Androst-4-en-3-one, 17-β-hydroxy-6-α-methyl-17-(1-propynyl) **(79-64-1)**
Androst-4-en-3-one, 17-hydroxy-6-methyl-17-(1-propynyl)-, (6-α,17-β)- **(79-64-1)**
Androst-4-en-3-one, 17-((1-oxoheptyl)oxy)-, (17-β)- **(315-37-7)**
Androst-4-en-3-one, 17-(1-oxopropoxy)-(17-β)- **(57-85-2)**
Androsterolo **(76-43-7)**
Androtardyl **(315-37-7)**
Androtest P **(57-85-2)**
Androteston **(57-85-2)**
Andrusol **(58-22-0)**
Andrusol-P **(57-85-2)**
Andur **(9009-54-5)**
Anecotan **(76-38-0)**
Anectine **(306-40-1)**
Anelix **(103-90-2)**
Anelmid **(514-73-8)**
Anemolin **(26787-78-0)**
Anertan **(57-85-2)**
Anertan **(58-18-4)**
Anertan (tablets) **(58-18-4)**
Anerval **(50-33-9)**
Anestacon **(137-58-6)**
Anesthenyl **(109-87-5)**
Anesthesia ether **(60-29-7)**
Anesthesin **(94-09-7)**
Anesthesol **(51-05-8)**
Anesthetic Compound No. 347 **(13838-16-9)**
Anesthetic ether **(60-29-7)**
Anesthone **(94-09-7)**
Anestil **(51-05-8)**
Anethaine **(136-47-0)**
Anethol **(104-46-1)**
trans-Anethol **(4180-23-8)**
Anethole **(104-46-1)**
trans-Anethole **(4180-23-8)**
Aneural **(57-53-4)**
Aneurine **(59-43-8)**
Aneurine hydrochloride **(67-03-8)**
Aneurol **(57-53-4)**
Aneusral **(57-53-4)**
Aneuxal **(57-53-4)**
Aneuxral **(57-53-4)**
Anex **(2272-40-4)**
Anexol **(127-65-1)**
Anfetamina **(60-13-9)**
Anflagen **(15687-27-1)**
Anfram 3PB **(126-72-7)**
Angecin **(523-50-2)**
Angel Dust **(956-90-1)**

Angelecin (523-50-2)
Angelica-Root-Oil (8015-64-3)
Angelica-Seed-Oil (8015-64-3)
Angelic acid (565-63-9)
Angelic acid nitrile (20068-02-4)
Angelica lactone (1333-38-6)
Angelicin (523-50-2)
Angelicin (83-46-5)
Angelicin (coumarin deriv) (523-50-2)
Angelicin (steroid) (83-46-5)
Angeliconitrile (20068-02-4)
trans-Angeliconitrile (20068-02-4)
Angelika Oel (German) (8015-64-3)
22-β-Angeloyloxyoleanolic acid (467-81-2)
Angiazol (54-95-5)
Angibid (55-63-0)
Angicap (78-11-5)
Angiflan (8048-52-0)
Anginal (58-32-2)
Anginin (1882-26-4)
Anginine (1882-26-4)
Anginine (55-63-0)
Anginon (539-21-9)
Angiokapsul (637-07-0)
Angiolingual (55-63-0)
Angioton (54-95-5)
Angiotonin (54-95-5)
Angitet (78-11-5)
Anglislite (7446-14-2)
Angorin (55-63-0)
Anguifugan (514-73-8)
Anhiba (103-90-2)
Anhistabs (91-84-9)
Anhistol (91-84-9)
Anhydride acetique (French) (108-24-7)
Anhydride arsenieux (French) (1327-53-3)
Anhydride arsenique (French) (1303-28-2)
Anhydride carbonique (French) (124-38-9)
Anhydride chromique (French) (1333-82-0)
Anhydride phtalique (French) (85-44-9)
Anhydride vanadique (French) (1314-62-1)
Anhydrid kyseliny citrakonove (Czech) (616-02-4)
Anhydrid kyseliny ftalove (Czech) (85-44-9)
Anhydrid kyseliny glutarove (Czech) (108-55-4)
Anhydrid kyseliny krotonove (Czech) (623-68-7)
Anhydrid kyseliny maleinove (Czech) (108-31-6)
Anhydrid kyseliny maselne (Czech) (106-31-0)
Anhydrid kyseliny octove (Czech) (108-24-7)
Anhydrid kyseliny propionove (Czech) (123-62-6)
Anhydrid kyseliny tetrahydroftalove (Czech) (85-43-8)
Anhydrid kyseliny trifluoroctove (Czech) (407-25-0)
Anhydro-4,4'-bis(diethylamino)triphenylmethanol-2',4''-disulphonic acid, monosodium salt (129-17-9)
Anhydroformaldehyde aniline (VAN) (91-78-1)
3,6-Anhydro-D-galactan (9000-07-1)
3,6-Anhydro-4-O-β-D-galactopyranosyl-α-D-galactopyranose 2,4'-bis(potassium/sodium sulfate)-(1-3')-polysaccharide (53973-98-1)
3,6-Anhydro-4-O-β-galactopyranosyl-α-D-galactopyranose 2,4'-bis(potassium/sodium sulfate) (13')-polysaccharide (53973-98-1)
1,4-Anhydro-D-glucitol 6-dodecanoate (5959-89-7)
Anhydroglucochloral (15879-93-3)
10-(1',5'-Anhydroglucosyl)aloe-emodin-9-anthrone (1415-73-2)
Anhydrohexitol Sesquioleate (8007-43-0)
Anhydrohydroxynorprogesterone (68-22-4)
Anhydrol (64-17-5)
Anhydron (2259-96-3)
Anhydrone (10034-81-8)
Anhydrosorbitol stearate (1338-41-6)
Anhydrosorbitol tristearate (26658-19-5)
Anhydro-o-sulfaminebenzoic acid (81-07-2)
Anhydro trimellic acid (552-30-7)
Anhydrous ammonia (DOT) (7664-41-7)
Anhydrous borax (1330-43-4)
Anhydrous chloral (75-87-6)
Anhydrous chlorobutanol (57-15-8)
Anhydrous citric acid (77-92-9)
Anhydrous dextrose (50-99-7)
Anhydrous hydrazine [UN 2029] (302-01-2)
Anhydrous hydriodic acid (10034-85-2)
Anhydrous hydrobromic acid (10035-10-6)
Anhydrous hydrofluoric acid (DOT) (7664-39-3)
Anhydrous iron oxide (1309-37-1)
Anhydrous lanolin (8006-54-0)
Anhydrous lanum (8006-54-0)
Anhydrous oxide of iron (1309-37-1)
Anhydrous sodium acetate (127-09-3)
Anhydrous sodium arsanilate (127-85-5)
Anhydrous sodium sulfite (7757-83-7)
Anhydrous tetrasodium pyrophosphate (7722-88-5)
Anicon Kombi (94-74-6)
Anicon M (94-74-6)
Anicon P (7085-19-0)
Anidride acetica (Italian) (108-24-7)
Anidride cromica (Italian) (1333-82-0)
Anidride ftalica (Italian) (85-44-9)
Anilazin (101-05-3)
Anilazine (101-05-3)
Anileridina (Spanish) (144-14-9)
Anileridine (144-14-9)
Anileridinum (Latin) (144-14-9)
7-Anilino-3-diethylamino-6-methylfluoran (29512-49-0)
2-Anilino-4,6-dimethoxy-s-triazine (30358-01-1)
6-Anilino-2,4-diphenoxy-s-triazine (1973-08-6)
Anilid kyseliny acetoctove (Czech) (102-01-2)
Anilid kyseliny salicylove (Czech) (87-17-2)
7-Anilino-4-hydroxy-2-naphthalenesulfonic acid (119-40-4)
Anilin (Czech) (62-53-3)
Anilina (Italian, Polish) (62-53-3)
1-Anilino-8-naphthalenesulfonate (82-76-8)
1-Anilino-8-naphthalenesulfonic acid (82-76-8)
8-Anilino-1-naphthalenesulfonic acid (82-76-8)
4-(α-(4-Anilino-1-naphthyl)-p-dimethylamino)benzylidene-2,5-cyclohexadien-1-ylidenedimethyl ammonium chloride (2580-56-5)
1-Anilino-8-napthalenesulfonate (82-76-8)
Aniline (ACGIH,OSHA) [UN 1547] (62-53-3)
Aniline Carmine Powder (860-22-0)
Aniline Green (569-64-2)
Aniline Green (633-03-4)
Aniline, Polymer with formaldehyde (8CI) (25214-70-4)
Aniline Red (632-99-5)
Aniline Violet (548-62-9)
Aniline Violet Pyoktanine (548-62-9)
Aniline Yellow (60-09-3)
Aniline, N-acetoxyethyl-N-cyanoethyl- (22031-33-0)
Aniline, N-acetyl- (103-84-4)
Aniline-ω-acid (103-06-0)
p-Anilinearsonic acid (98-50-0)
Aniline, 4,4'-azodi (538-41-0)
Aniline, N-benzyl-N-dimethylaminoethyl-, hydrochloride (2045-52-5)
Aniline, N-benzylidene- (8CI) (538-51-2)
Aniline, N,N-bis(2-chloroethyl) (553-27-5)
Aniline, 3,5-bis(trifluoromethyl)- (328-74-5)
Aniline, ar-bromo- (55777-84-9)
Aniline, p-bromo (106-40-1)

Name	CAS No.
Aniline, m-bromo- (8CI)	**(591-19-5)**
Aniline, o-bromo- (8CI)	**(615-36-1)**
Aniline, p-bromo-N,N-diethyl- (8CI)	**(2052-06-4)**
Aniline, 2-bromo-6-chloro-4-nitro- (8CI)	**(99-29-6)**
Aniline, 2-bromo-4,6-dichloro- (8CI)	**(697-86-9)**
Aniline, 4-bromo-3,5-dichloro- (8CI)	**(1940-29-0)**
Aniline, 2-bromo-4,6-dinitro	**(1817-73-8)**
Aniline, 4-butyl	**(104-13-2)**
Aniline, n-butyl	**(1126-78-9)**
Aniline, N-sec-butyl-4-tert-butyl-2,6-dinitro	**(33629-47-9)**
Aniline, p-butyl-N-(p-methoxybenzylidene)	**(26227-73-6)**
Aniline, N-tert-butyl-p-nitro- (7CI,8CI)	**(4138-38-9)**
Aniline-3-carboxylic acid	**(99-05-8)**
Aniline chloride	**(142-04-1)**
Aniline, 4-chloro-	**(106-47-8)**
Aniline, m-chloro	**(108-42-9)**
Aniline, o-chloro	**(95-51-2)**
Aniline, p-chloro	**(106-47-8)**
Aniline, 4-chloro-2,5-dimethoxy	**(6358-64-1)**
Aniline, o-chloro-N,N-dimethyl-	**(698-01-1)**
Aniline, 6-chloro-2,4-dinitro	**(3531-19-9)**
Aniline, 4-chloro-2,6-dinitro- (8CI)	**(5388-62-5)**
Aniline, 3-chloro-2,6-dinitro- (7CI,8CI)	**(10250-71-2)**
Aniline, N-(2-chloroethyl)-N-ethyl	**(92-49-9)**
Aniline, 3-chloro-4-fluoro-	**(367-21-5)**
Aniline, p-chloro-, hydrochloride	**(20265-96-7)**
Aniline, m-chloro-, hydrochloride (8CI)	**(141-85-5)**
Aniline, 2-chloro-4-nitro	**(121-87-9)**
Aniline, 2-chloro-5-nitro	**(6283-25-6)**
Aniline, 4-chloro-2-nitro	**(89-63-4)**
Aniline, 4-chloro-3-nitro	**(635-22-3)**
Aniline, 2-chloro-6-nitro- (8CI)	**(769-11-9)**
Aniline, 3-chloro-4-nitro- (8CI)	**(825-41-2)**
Aniline, 3-chloro-5-nitro- (8CI)	**(5344-44-5)**
Aniline, 5-chloro-2-nitro- (8CI)	**(1635-61-6)**
Aniline, p-cyano	**(873-74-5)**
Aniline, N-(2-cyanoethyl)-N-ethyl-	**(148-87-8)**
Aniline, N-(β-cyanoethyl)-N-(β-hydroxyethyl)-	**(92-64-8)**
Aniline, N-(2-cyanoethyl)-N-methyl-4-((p-nitrophenyl)azo)-	**(31464-38-7)**
Aniline, 2-cyano-4-nitro	**(17420-30-3)**
Aniline, 2,6-dibromo-4-nitro- (8CI)	**(827-94-1)**
Aniline, 2,6-di-tert-butyl-4-nitro- (6CI,7CI,8CI)	**(5180-59-6)**
Aniline, 2,4-dichloro	**(554-00-7)**
Aniline, 2,5-dichloro	**(95-82-9)**
Aniline, 3,4-dichloro	**(95-76-1)**
Aniline, 3,5-dichloro-	**(626-43-7)**
Aniline, 2,3-dichloro- (8CI)	**(608-27-5)**
Aniline, N-((dichlorofluoromethyl)thio)-N-((dimethylamino)-sulfonyl)-	**(1085-98-9)**
Aniline, dichloro- (mixed isomers)	**(27134-27-6)**
Aniline, 2,5-dichloro-4-nitro	**(6627-34-5)**
Aniline, 2,6-dichloro-4-nitro	**(99-30-9)**
Aniline, 2,4-dichloro-6-nitro- (8CI)	**(2683-43-4)**
Aniline, 4,5-dichloro-2-nitro- (8CI)	**(6641-64-1)**
Aniline, N,N-dicyanoethyl-	**(1555-66-4)**
Aniline, 2,6-diethyl	**(579-66-8)**
Aniline, N,N-diethyl	**(91-66-7)**
Aniline, 2,4-diethyl- (8CI)	**(14719-47-2)**
Aniline, 3,5-diethyl- (8CI)	**(1701-68-4)**
Aniline, N,N-diethyl-p-((p-nitrophenyl)azo)- (8CI)	**(3025-52-3)**
Aniline, N,N-diethyl-p-(phenylazo)-	**(2481-94-9)**
Aniline, 2,4-difluoro	**(367-25-9)**
Aniline, 2,6-diisopropyl	**(24544-04-5)**
Aniline, 2,4-dimethoxy	**(2735-04-8)**
Aniline, 2,5-dimethoxy	**(102-56-7)**
Aniline, 2,4-dimethoxy-, hydrochloride	**(54150-69-5)**
Aniline, 2,4-dimethyl-	**(95-68-1)**
Aniline, 2,5-dimethyl-	**(95-78-3)**
Aniline, 2,6-dimethyl-	**(87-62-7)**
Aniline, 3,4-dimethyl	**(95-64-7)**
Aniline, N,N-dimethyl	**(121-69-7)**
Aniline, 4-(p-dimethylaminophenylazo)	**(539-17-3)**
Aniline, N,N-dimethyl-4,4'-azodi- (7CI)	**(539-17-3)**
Aniline, 3,4-dimethyl-2,6-dinitro-N-(1-ethylpropyl)	**(40487-42-1)**
Aniline, N,N'-dimethyl-4,4'-methylenedi-	**(1807-55-2)**
Aniline, N,N-dimethyl-p-(2'-methylphenylazo)-	**(3731-39-3)**
Aniline, N,N-dimethyl-p-(3'-methylphenylazo)-	**(55-80-1)**
Aniline, N,N-dimethyl-p-nitro- (8CI)	**(100-23-2)**
Aniline, N,N-dimethyl-p-((o-nitrophenyl)azo)	**(3010-38-6)**
Aniline, N,N-dimethyl-p-nitroso	**(138-89-6)**
Aniline, N,N-dimethyl-p-phenylazo	**(60-11-7)**
Aniline, N,N-dimethyl-4-(o-tolylazo)	**(3731-39-3)**
Aniline, N,N-dimethyl-p-(m-tolylazo)	**(55-80-1)**
Aniline, 2,4-dinitro	**(97-02-9)**
Aniline, 2,6-dinitro	**(606-22-4)**
Aniline, 3,5-dinitro	**(618-87-1)**
Aniline, 2,6-dinitro-N,N-dipropyl-p-isopropyl	**(33820-53-0)**
Aniline, 2,6-dinitro-N,N-dipropyl-4-(methylsulfonyl)-	**(4726-14-1)**
Aniline, 2,6-dinitro-4-methyl-	**(6393-42-6)**
Aniline, dinitro- (mixed isomers)	**(26471-56-7)**
Aniline, N,p-dinitroso-N-methyl	**(99-80-9)**
Aniline, p-(2,3-epoxypropoxy)-N,N-bis(2,3-epoxypropyl)	**(5026-74-4)**
Aniline, N,N-bis(2,3-epoxypropyl)	**(2095-06-9)**
Aniline, p-ethoxy-	**(156-43-4)**
Aniline, 2-ethyl	**(578-54-1)**
Aniline, 4-ethyl	**(589-16-2)**
Aniline, N-ethyl	**(103-69-5)**
Aniline, m-ethyl	**(587-02-0)**
Aniline, o-ethyl- (8CI)	**(578-54-1)**
Aniline, N-ethyl-N-(2-cyanoethyl)	**(148-87-8)**
Aniline, N-(2-ethylhexyl)-	**(10137-80-1)**
Aniline, N-ethyl-N-methyl- (8CI)	**(613-97-8)**
Aniline, N-ethyl-p-nitro- (8CI)	**(3665-80-3)**
Aniline, N-ethyl-N-nitroso	**(612-64-6)**
Aniline, p-fluoren-9-yl-N,N-dimethyl- (8CI)	**(32377-15-4)**
Aniline, 2-fluoro	**(348-54-9)**
Aniline, 3-fluoro	**(372-19-0)**
Aniline, 4-fluoro	**(371-40-4)**
Aniline, 4-fluoro-3-nitro	**(364-76-1)**
Aniline-formaldehyde Condensate	**(25214-70-4)**
Aniline-formaldehyde Polymer	**(25214-70-4)**
Aniline, formaldehyde polymer, hydrochloride	**(57138-85-9)**
Aniline, N-formyl	**(103-70-8)**
Aniline, hexahydro-	**(108-91-8)**
Aniline, (p-hexyl)	**(33228-45-4)**
Aniline hydrochloride [UN 1548]	**(142-04-1)**
Aniline, N-hydroxy-	**(100-65-2)**
Aniline, N-(2-hydroxyethyl)-	**(122-98-5)**
Aniline, N-(β-hydroxyethyl)-	**(122-98-5)**
Aniline, 4,4'-imidocarbonylbis(N,N-diethyl-, monohydrochloride (8CI)	**(6358-36-7)**
Aniline, 4,4'-(imidocarbonyl)bis(N,N-dimethyl	**(492-80-8)**
Aniline, 4,4'-(imidocarbonyl)bis(N,N-dimethyl-, hydrochloride	**(2465-27-2)**
Aniline, 4,4'-iminodi-	**(537-65-5)**
Aniline, 4-iodo-	**(540-37-4)**
Aniline, m-iodo	**(626-01-7)**
Aniline, p-iodo	**(540-37-4)**
Aniline, n-isopropyl	**(768-52-5)**
Aniline, o-isopropyl	**(643-28-7)**
Aniline, p-isopropyl-	**(99-88-7)**
Aniline, N-isopropyl-4,4'-methylenedi- (8CI)	**(10029-31-9)**
Aniline, p-methoxy-	**(104-94-9)**
Aniline, 2-methoxy-4-nitro-	**(97-52-9)**
Aniline, 2-methoxy-5-nitro-	**(99-59-2)**

Aniline, 4-methoxy-2-nitro	**(96-96-8)**
Aniline, 2-methyl-	**(95-53-4)**
Aniline, 3-methyl-	**(108-44-1)**
Aniline, N-methyl	**(100-61-8)**
Aniline, p-methyl-	**(106-49-0)**
Aniline, N-methyl-N,p-dinitroso	**(99-80-9)**
Aniline, 4,4'-methylenebis(2-chloro-	**(101-14-4)**
Aniline, 4,4'-methylenebis(N,N-dimethyl	**(101-61-1)**
Aniline, 4,4'-methylenebis(N-methyl	**(1807-55-2)**
Aniline, 2',4-methylenedi	**(1208-52-2)**
Aniline, 2,2'-methylenedi-	**(6582-52-1)**
Aniline, 4,4'-methylenedi	**(101-77-9)**
Aniline, 4,4'-methylenedi-, dihydrochloride	**(13552-44-8)**
Aniline, 4,4'-methylenebis(2-methyl	**(838-88-0)**
Aniline, 2-methyl-6-ethyl	**(24549-06-2)**
Aniline, 4,4',4''-methylidynetris(N,N-dimethyl- (8CI)	**(603-48-5)**
Aniline, 2-methyl-4-nitro-	**(99-52-5)**
Aniline, N-methyl-p-nitro- (8CI)	**(100-15-2)**
Aniline, N-(2-methyl-2-nitropropyl)-p-nitroso	**(24458-48-8)**
Aniline, N-methyl-N-nitroso	**(614-00-6)**
Aniline, 4-(methylsulfonyl)-2,6-dinitro-N,N-dipropyl-	**(4726-14-1)**
Aniline, N-methyl-N,2,4,6-tetranitro	**(479-45-8)**
Aniline mustard	**(553-27-5)**
1-Aniline-8-naphthalene sulfonate	**(82-76-8)**
Aniline, 4-nitro-	**(100-01-6)**
Aniline, N-nitro-	**(645-55-6)**
Aniline, m-nitro	**(99-09-2)**
Aniline, o-nitro	**(88-74-4)**
Aniline, p-nitro	**(100-01-6)**
Aniline, 4-((4-nitro-2-chlorophenyl)azo)-N-hydroxyethyl-N-cyanoethyl	**(6657-33-6)**
Aniline nitro nerol acid	**(91-29-2)**
Aniline, 4-nitro-N-trityl-	**(20222-29-1)**
Aniline, p-nitro-N-trityl- (7CI,8CI)	**(20222-29-1)**
Aniline, p-octyl- (8CI)	**(16245-79-7)**
Aniline oil	**(62-53-3)**
Aniline oil, Liquid [UN 1547]	**(62-53-3)**
Aniline, 4,4'-oxybis(2-chloro-	**(28434-86-8)**
Aniline, 4,4'-oxydi	**(101-80-4)**
Aniline, 2,3,4,5,6-pentachloro	**(527-20-8)**
Aniline, 2,3,4,5,6-pentafluoro	**(771-60-8)**
Aniline, 2,3,4,5,6-pentanitro-, Dry	**(21985-87-5)**
Aniline, p-phenoxy	**(139-59-3)**
Aniline, N-phenyl-	**(122-39-4)**
Aniline, N-(phenylazo)-	**(136-35-6)**
Aniline, p-(phenylazo)	**(60-09-3)**
Aniline, 4-propyl	**(2696-84-6)**
Aniline, p-propyl- (8CI)	**(2696-84-6)**
"Aniline salt"	**(142-04-1)**
Aniline sulfate	**(542-16-5)**
p-Anilinesulfonamide	**(63-74-1)**
Aniline-4-sulfonic acid	**(121-57-3)**
Aniline-ω-sulfonic acid	**(103-06-0)**
Aniline-p-sulfonic acid	**(121-57-3)**
m-Anilinesulfonic acid	**(121-47-1)**
Aniline-p-sulfonic amide	**(63-74-1)**
Aniline, 4,4'-sulfonyldi	**(80-08-0)**
Aniline-p-sulphonic acid	**(121-57-3)**
Aniline, 2,3,4,5-tetrachloro- (8CI)	**(634-83-3)**
Aniline, 2,3,5,6-tetrachloro- (8CI)	**(3481-20-7)**
Aniline, 2,3,4,5-tetrafluoro- (8CI)	**(5580-80-3)**
Aniline, 2,3,5,6-tetrafluoro- (8CI)	**(700-17-4)**
Aniline, 2,3,5,6-tetramethyl-4-nitro- (6CI,7CI,8CI)	**(13171-61-4)**
Aniline, 2,3,4,6-tetranitro	**(3698-54-2)**
Aniline, N,2,4,6-tetranitro-	**(4591-46-2)**
Aniline, tetranitro	**(53014-37-2)**
Aniline, 4,4'-thiodi	**(139-65-1)**
Aniline, 2,4,6-tribromo	**(147-82-0)**
Aniline, 2,4,6-trichloro	**(634-93-5)**
Aniline, 2,3,4-trichloro- (8CI)	**(634-67-3)**
Aniline, 2,3,5-trichloro- (8CI)	**(18487-39-3)**
Aniline, 2,4,5-trichloro- (8CI)	**(636-30-6)**
Aniline, 3,4,5-trichloro- (8CI)	**(634-91-3)**
Aniline, 2,4,5-trimethyl	**(137-17-7)**
Aniline, 2,4,6-trimethyl	**(88-05-1)**
Aniline, 2,4,5-trimethyl, hydrochloride	**(21436-97-5)**
Aniline, 2,4,6-trimethyl, hydrochloride	**(6334-11-8)**
Aniline, trinitro	**(26952-42-1)**
Aniline, 2,4,6-trinitro- (8CI)	**(489-98-5)**
4-Anilino-5-nitrochrysazin	**(20241-76-3)**
Anilinium chloride	**(142-04-1)**
Anilinlost (German)	**(553-27-5)**
p-Anilinoaniline	**(101-54-2)**
Anilinobenzene	**(122-39-4)**
2-Anilinobenzoic acid	**(91-40-7)**
o-Anilinobenzoic acid	**(91-40-7)**
Anilinoethane	**(103-69-5)**
2-Anilinoethanol	**(122-98-5)**
Anilinomethane	**(100-61-8)**
Anilinomethanesulfonate	**(103-06-0)**
Anilinomethanesulfonate sodium salt	**(26021-90-9)**
Anilinomethanesulfonic acid	**(103-06-0)**
Anilinomethanesulfonic acid, monosodium salt	**(26021-90-9)**
Anilinomethanesulfonic acid, sodium salt	**(26021-90-9)**
1-Anilinonaphthalene	**(90-30-2)**
2-Anilinonaphthalene	**(135-88-6)**
Anilinonaphthalene	**(135-88-6)**
8-Anilinonaphthalene-1-sulfonate	**(82-76-8)**
Anilinonaphthalenesulfonic acid	**(82-76-8)**
4-Anilinophenol	**(122-37-2)**
m-Anilinophenol	**(101-18-8)**
p-Anilinophenol	**(122-37-2)**
4-((4-Anilino-5-sulfo-1-naphthyl)azo)-5-hydroxy-2,7-naphthalene-disulfonic acid trisodium	**(3861-73-2)**
Anilino-2-sulfonic acid	**(88-21-1)**
Anilino-o-sulfonic acid	**(88-21-1)**
Anilino-o-sulphonic acid	**(88-21-1)**
Anilotic acid	**(96-97-9)**
Animag	**(1309-48-4)**
Animal Blood, Denatured	**(68911-49-9)**
Animal Coniine	**(462-94-2)**
Animal Grease, Inedible	**(68153-81-1)**
Animal Oil	**(8001-85-2)**
Animal galactose factor	**(65-86-1)**
Animert	**(2227-13-6)**
Animert V-10	**(2227-13-6)**
Animert V-101	**(2227-13-6)**
Animert V-10K	**(2227-13-6)**
Anion, Superoxide	**(11062-77-4)**
Aniphor	**(77-75-8)**
Anis Oel (German)	**(8007-70-3)**
2-Anisaldehyde	**(135-02-4)**
m-Anisaldehyde	**(591-31-1)**
o-Anisaldehyde	**(135-02-4)**
p-Anisaldehyde	**(123-11-5)**
m-Anisaldehyde, 4-hydroxy-	**(121-33-5)**
m-Anisaldehyde, 2-hydroxy- (8CI)	**(148-53-8)**
p-Anisaldehyde, 3-hydroxy- (8CI)	**(621-59-0)**
n-Anisamide, 4-amino-5-chloro-N-(2-(diethylamino)ethyl)	**(364-62-5)**
o-Anisamide, 4-amino-5-chloro-N-(2-(diethylamino)ethyl)- (8CI)	**(364-62-5)**
o-Anisamide, N-((1-ethyl-2-pyrrolidinyl)methyl)-5-sulfamoyl	**(15676-16-1)**
p-Anisanilide, 3-nitro- (8CI)	**(97-32-5)**
Anise LS	**(8001-21-6)**
Anise-Oil	**(8007-70-3)**

Anise alcohol	(105-13-5)
Anise camphor	(104-46-1)
Aniseed Oil	(8007-70-3)
Anisene	(569-57-3)
2-Anisic acid	(579-75-9)
4-Anisic acid	(100-09-4)
m-Anisic acid	(586-38-9)
o-Anisic acid	(119-36-8)
o-Anisic acid	(579-75-9)
p-Anisic acid	(100-09-4)
Anisic acid (8CI)	(1335-08-6)
o-Anisic acid, 3,6-dichloro-	(1918-00-9)
o-Anisic acid, 3,6-dichloro-, Compd. with dimethylamine (1:1)	(2300-66-5)
o-Anisic acid, 3,6-dichloro-, Compd. with 2,2'-iminodiethanol	(25059-78-3)
o-Anisic acid, 3,6-dichloro-, Compd. with 2,2'-iminodiethanol (1:1)	(25059-78-3)
p-Anisic acid, 3,5-dichloro-, methyl ester (6CI,8CI)	(24295-27-0)
o-Anisic acid, 3,6-dichloro-, sodium salt	(1982-69-0)
m-Anisic acid, 2-hydroxy	(877-22-5)
m-Anisic acid, 4-hydroxy-	(121-34-6)
p-Anisic acid, 3-hydroxy	(645-08-9)
p-Anisic acid, 2-hydroxy- (8CI)	(2237-36-7)
m-Anisic acid, 4'-hydroxy-4,5'-oxydi- (8CI)	(2555-99-9)
p-Anisic acid, methyl ester	(121-98-2)
o-Anisic acid, 3,5,6-trichloro	(2307-49-5)
Anisic alcohol	(105-13-5)
Anisic aldehyde	(123-11-5)
Anisicaldehyde dimethylacetal	(2186-92-7)
1-p-Anisidino-4-hydroxyanthraquinone	(23552-76-3)
2-Anisidine	(90-04-0)
4-Anisidine	(104-94-9)
Anisidine	(29191-52-4)
m-Anisidine	(536-90-3)
p-Anisidine	(104-94-9)
p-Anisidine (ACGIH,OSHA)	(104-94-9)
o-Anisidine (ACGIH,OSHA) [UN 2431]	(90-04-0)
Anisidine (o-,p-Isomers)	(29191-52-4)
o-Anisidine, acetoacetyl	(92-15-9)
p-Anisidine, 3-chloro	(5345-54-0)
2-Anisidine hydrochloride	(134-29-2)
4-Anisidine hydrochloride	(20265-97-8)
o-Anisidine, hydrochloride	(134-29-2)
p-Anisidine, hydrochloride	(20265-97-8)
Anisidine, isomers (8CI)	(29191-52-4)
o-Anisidine, 5-methyl	(120-71-8)
p-Anisidine, 2-methyl	(102-50-1)
p-Anisidine, N-methyl- (8CI)	(5961-59-1)
p-Anisidine monohydrochloride	(20265-97-8)
o-Anisidine nitrate	(99-59-2)
o-Anisidine, 4-nitro	(97-52-9)
o-Anisidine, 5-nitro	(99-59-2)
m-Anisidine, N-phenyl- (8CI)	(101-16-6)
p-Anisidine, N-phenyl- (8CI)	(1208-86-2)
o-Anisidinomethanesulfonic acid	(93-13-0)
o-Anisidyl-N-methanesulfonic acid	(93-13-0)
Anisindione	(117-37-3)
Anisin indandione	(117-37-3)
p-Anisol alcohol	(105-13-5)
Anisole [UN 2222]	(100-66-3)
Anisole, p-allyl	(140-67-0)
Anisole, p-amino-	(104-94-9)
Anisole, 2-amino-5-nitro-	(97-52-9)
Anisole, 3,3'-azodi- (7CI)	(6319-23-9)
Anisole, 2,2'-azodi- (6CI,7CI)	(613-55-8)
Anisole, p-benzyl- (8CI)	(834-14-0)
Anisole, o-bromo	(578-57-4)
Anisole, p-bromo	(104-92-7)
Anisole, m-bromo- (8CI)	(2398-37-0)
Anisole, o-(bromomethyl)-	(52289-93-7)
Anisole, 6-t-butyl-3-methyl-2,4-dinitro	(83-66-9)
Anisole, o-chloro- (8CI)	(766-51-8)
Anisole, p-chloro- (8CI)	(623-12-1)
Anisole, 2,4-diamino	(615-05-4)
Anisole, 2,4-diamino-, hydrogen sulfate	(39156-41-7)
Anisole, 2,4-diamino-, sulfate	(39156-41-7)
Anisole, 2,4-dichloro- (8CI)	(553-82-2)
Anisole, 3,5-dichloro- (8CI)	(33719-74-3)
Anisole, 2,4-dinitro-	(119-27-7)
Anisole, p-(2,3-epoxypropoxy)-	(2211-94-1)
Anisole, p-fluoren-9-yl- (6CI)	(21846-08-2)
Anisole, m-fluoro- (8CI)	(456-49-5)
Anisole, p-fluoro- (8CI)	(459-60-9)
Anisole, p-methoxy-	(150-78-7)
Anisole, p-methyl	(104-93-8)
Anisole, m-methyl- (8CI)	(100-84-5)
Anisole, methyl- (8CI)	(26897-24-5)
Anisole, p-(methylthio)- (8CI)	(1879-16-9)
Anisole, m-nitro	(555-03-3)
Anisole, o-nitro	(91-23-6)
Anisole, p-nitro	(100-17-4)
Anisole, 2,3,4,5,6-pentachloro	(1825-21-4)
Anisole, o-phenyl	(86-26-0)
Anisole, p-phenyl	(613-37-6)
Anisole, p-(phenylazo)-	(2396-60-3)
Anisole, o-(phenylazo)- (6CI,7CI)	(6319-21-7)
Anisole, p-propenyl	(104-46-1)
Anisole, p-propenyl-, trans	(4180-23-8)
Anisole, p-propenyl-, (E)- (8CI)	(4180-23-8)
Anisole, p-propyl	(104-45-0)
Anisole, p-styryl- (8CI)	(1142-15-0)
Anisole, tetrachloro-	(53452-81-6)
Anisole, 2,3,4,5-tetrachloro- (8CI)	(938-86-3)
Anisole, 2,3,4,6-tetrachloro- (8CI)	(938-22-7)
Anisole, 2,3,5,6-tetrachloro- (8CI)	(6936-40-9)
Anisole, 2,3,5,6-tetrachloro-4-nitro	(2438-88-2)
Anisole, o-(o-tolylazo)- (7CI)	(29268-78-8)
Anisole, 2,4,5-trichloro- (8CI)	(6130-75-2)
Anisole, 2,4,6-trichloro- (8CI)	(87-40-1)
Anisole, 2,4,6-trinitro	(606-35-9)
Anisole, p-vinyl- (8CI)	(637-69-4)
Anisomycin	(22862-76-6)
Anisopyradamine	(59-33-6)
Anisotropine methobromide	(80-50-2)
Anisotropine methylbromide	(80-50-2)
(E)-3-p-Anisoyl-3-bromoacrylic acid sodium salt	(21739-91-3)
Anisoyl-chloride	(100-07-2)
Anisoyl chloride [UN 1729]	(100-07-2)
Anistadin	(133-67-5)
2-(p-Anisyl)acetic acid	(104-01-8)
Anisylacetonitrile	(104-47-2)
Anisyl alcohol	(105-13-5)
m-Anisylamine	(536-90-3)
o-Anisylamine	(90-04-0)
p-Anisylamine	(104-94-9)
o-Anisylamine hydrochloride	(134-29-2)
p-Anisylamine hydrochloride	(20265-97-8)
Anisyl bromide	(104-92-7)
Anisyl bromide	(578-57-4)
Anisyl formate	(104-01-8)
2-p-Anisyl-1,3-indandione	(117-37-3)
p-Anisyl phenylcarbamate	(19219-48-8)
Aniten	(53568-85-7)
Aniten M	(53568-85-7)
Ankilostin	(127-18-4)

Annidalin (552-22-7)
Annogen (127-52-6)
(6)Annulene (71-43-2)
(8)Annulene (629-20-9)
Anodynin (60-80-0)
Anodynine (60-80-0)
Anodynon (75-00-3)
Anofex (50-29-3)
Anol (108-93-0)
Anone (108-94-1)
Anoprolin (315-30-0)
Anorexide (300-62-9)
Anorexine (60-13-9)
Anovigam (938-73-8)
Anovlar (57-63-6)
Anozol (84-66-2)
Anparton (637-07-0)
Anprolene (75-21-8)
Anproline (75-21-8)
Anpuzone (50-33-9)
Anquil (50-55-5)
Ansadol (87-17-2)
Ansar (75-60-5)
Ansar 138 (75-60-5)
Ansar 160 (124-65-2)
Ansar 170 (2163-80-6)
Ansar 184 (144-21-8)
Ansar 529 (2163-80-6)
Ansar 560 (124-65-2)
Ansar 8100 (144-21-8)
Ansar DSMA Liquid (144-21-8)
Ansar 170 H.C. (2163-80-6)
Ansar 529 H.C. (2163-80-6)
Ansar 170l (2163-80-6)
Anshist (63-56-9)
Ansiatan (57-53-4)
Ansibase Red KB (95-79-4)
Ansibases Orange GC (141-85-5)
Ansibases Red RL (99-52-5)
Ansietan (57-53-4)
Ansieten (27223-35-4)
Ansil (57-53-4)
Ansilan (2898-12-6)
Ansiolin (439-14-5)
Ansiolisina (439-14-5)
Ansiolisina (604-75-1)
Ansiowas (57-53-4)
Ansioxacepam (604-75-1)
Ansul ether 161 (112-49-2)
Ansul ether 181AT (143-24-8)
Antabus (97-77-8)
Antabuse (97-77-8)
Antadix (97-77-8)
Antadol (50-33-9)
Antaenyl (97-77-8)
Antaethan (97-77-8)
Antaethyl (97-77-8)
Antaetil (97-77-8)
Antage W 400 (119-47-1)
Antage w 500 (88-24-4)
Antagonate (113-92-8)
Antagothyroid (141-90-2)
Antagothyroil (141-90-2)
Antak (112-30-1)
Antalcol (97-77-8)
Antalergan (91-84-9)
Antalin (62-33-9)
Antalka (138-15-8)
Antallergan (91-84-9)
Antamine (91-84-9)
Antan (835-31-4)
Antapentan (634-03-7)
Antaphron NM 42 (63148-62-9)
Antar (629-82-3)
Antarox A-200 (9002-93-1)
Antarox CA 620 (9036-19-5)
Antarox CO 630 (98113-10-1)
Anteisopentadecanoic acid (5502-94-3)
Antemoqua (50-67-9)
Antemovis (50-67-9)
Antene (137-30-4)
Antepsin (54182-58-0)
Antergan (961-71-7)
Antergan hydrochloride (2045-52-5)
Antergyl (555-06-6)
Anterior Pituitary Growth Hormone (9002-72-6)
Anterior Pituitary Luteotropin (9002-62-4)
Antetan (97-77-8)
Antethyl (97-77-8)
Antetil (97-77-8)
Anteyl (97-77-8)
Anth(2,1-a)anthrene (217-54-9)
Anthanthren (German) (191-26-4)
Anthanthrene (191-26-4)
Anthcoquinone Cyanine Green Base (128-80-3)
Anthio (2540-82-1)
Anthion (7727-21-1)
Anthiphen (97-23-4)
Anthisan (91-84-9)
Anthisan maleate (59-33-6)
Anthium dioxcide (10049-04-4)
Anthon (52-68-6)
Anthonaphthol AS-BS (135-65-9)
Anthonaphthol AS-D (135-61-5)
Anthonaphthol AS (92-77-3)
Anthonaphthol M3B (132-68-3)
Anthonaphthol MF (135-62-6)
Anthophylite (77536-67-5)
Anthophylite asbestos (77536-67-5)
Anthophyllite (17068-78-9)
Anthra-Derm (1143-38-0)
Anthra Red G (10169-02-5)
(Anthra-2',1')-1,2-anthracene (217-54-9)
2',1'-Anthra-1,2-anthracene (217-54-9)
Anthra(9,1,2-cde)benzo(rst)pentaphene-5,10-dione, amino- (9CI) (26763-69-9)
Anthra(9,1,2-cde)benzo(rst)pentaphene-5,10-dione, 16-nitro- (9CI) (128-60-9)
Anthra(9,1,2-cde)benzo(rst)pentaphene-5,10-dione, 3,12,16,17-tetrachloro- (9CI) (6373-20-2)
Anthracen (German) (120-12-7)
2-Anthracenamine (613-13-8)
Anthracene (120-12-7)
Anthracene Oil (65996-91-0)
9-Anthracenecarbonitrile (9CI) (1210-12-4)
Anthracenecarbonitrile (9CI) (28805-75-6)
Anthracenecarboxaldehyde (9CI) (31671-77-9)
2-Anthracenecarboxaldehyde, 1-amino-9,10-dihydro-9,10-dioxo- (9CI) (6363-87-7)
Anthracenecarboxaldehyde, methyl- (9CI) (79075-27-7)
Anthracene-9-carboxylic acid (723-62-6)
2-Anthracenecarboxylic acid, 7-α-D-glucopyranosyl-9,10-dihydro-3,5,6,8-tetrahydroxy-1-methyl-9,10-dioxo- (1260-17-9)
2-Anthracenecarboxylic acid, 7-β-D-glucopyranosyl-9,10-dihydro-3,5,6,8-tetrahydroxy-1-methyl-9,10-dioxo- (9CI) (1260-17-9)

2-Anthracenecarboxylic acid, 9,10-dihydro-1-amino-9,10-dioxo **(82-24-6)**
2-Anthracenecarboxylic acid, 9,10-dihydro-4,5-dihydroxy-9,10-dioxo- (9CI) **(478-43-3)**
Anthracene, chloro- (9CI) **(50602-11-4)**
2,3-Anthracenedicarboximide, 1,4-diamino-N-butyl-9,10-dihydro-9,10-dioxo- (8CI) **(3176-88-3)**
2,3-Anthracenedicarboximide, 1,4-diamino-9,10-dihydro-9,10-dioxo- (8CI) **(128-81-4)**
1,2-Anthracenedicarboxylic acid, 7-acetyl-6-ethyl-9,10-dihydro-3,5,8-trihydroxy-9,10-dioxo- (9CI) **(6219-66-5)**
Anthracene, 9,10-dichloro- (9CI) **(605-48-1)**
Anthracene, 9,10-dihydro- (9CI) **(613-31-0)**
Anthracene, 9,10-dihydro-9,10-dioxo- **(84-65-1)**
Anthracene, 9,10-dimethyl **(781-43-1)**
Anthracene, dimethyl **(29063-00-1)**
Anthracene, 2,6-dimethyl- (8CI,9CI) **(613-26-3)**
Anthracene, dimethylnitro- (9CI) **(80191-45-3)**
9,10-Anthracenediol (9CI) **(4981-66-2)**
9,10-Anthracenediol, 1,4-dihydro-, disodium salt (9CI) **(73347-80-5)**
9,10-Anthracenediol, ethyl-9,10-dihydro- (9CI) **(67923-88-0)**
9,10-Anthracenediol, 6-ethyl-1,2,3,4-tetrahydro- (9CI) **(68279-54-9)**
9,10-Anthracenedione **(84-65-1)**
9,10-Anthracenedione, 1-amino- **(82-45-1)**
9,10-Anthracenedione, 2-amino- (9CI) **(117-79-3)**
9,10-Anthracenedione, amino- (9CI) **(25620-59-1)**
9,10-Anthracenedione, 1-amino-2-bromo-4-hydroxy- **(116-82-5)**
9,10-Anthracenedione, aminobromohydroxy- (9CI) **(63460-11-7)**
9,10-Anthracenedione, 1-amino-5-chloro- **(117-11-3)**
9,10-Anthracenedione, 1-amino-4-chloro-2-(chloromethyl)- (9CI) **(56594-25-3)**
9,10-Anthracenedione, 1-amino-2-chloro-4-hydroxy- (9CI) **(2478-67-3)**
9,10-Anthracenedione, 1-amino-4-chloro-2-(hydroxymethyl)-(9CI) **(56594-21-9)**
9,10-Anthracenedione, 1-amino-4-chloro-2-methyl- (9CI) **(3225-97-6)**
9,10-Anthracenedione, 1-amino-2,4-dibromo- **(81-49-2)**
9,10-Anthracenedione, 1-amino-5,8-dichloro-4-nitro- (9CI) **(66121-41-3)**
9,10-Anthracenedione, 1-amino-4,5-dihydroxy-8-(phenylamino)- (9CI) **(20241-77-4)**
9,10-Anthracenedione, 1-amino-2-(ethoxymethyl)- (9CI) **(56594-28-6)**
9,10-Anthracenedione, 1-amino-4-hydroxy- (9CI) **(116-85-8)**
9,10-Anthracenedione, aminohydroxy- (9CI) **(52738-29-1)**
9,10-Anthracenedione, 1-amino-4-hydroxy-2-(3-hydroxybutoxy)- (9CI) **(3224-15-5)**
9,10-Anthracenedione, 1-amino-4-hydroxy-2-((6-hydroxyhexyl)oxy)- (9CI) **(34231-26-0)**
9,10-Anthracenedione, 1-amino-2-(hydroxymethyl)- (9CI) **(24094-44-8)**
9,10-Anthracenedione, 1-amino-4-hydroxy-2-((α-(2-oxo-1-pyrrolidinyl)-p-tolyl)oxy)- **(25177-16-6)**
9,10-Anthracenedione, 1-amino-2-(methoxymethyl)- (9CI) **(56594-27-5)**
9,10-Anthracenedione, 1-amino-4-((methoxyphenyl)amino)-(9CI) **(27341-33-9)**
9,10-Anthracenedione, 1-amino-2-methyl- (9CI) **(82-28-0)**
9,10-Anthracenedione, 1-amino-4-(methylamino)- (9CI) **(1220-94-6)**
9,10-Anthracenedione, 1-amino-2-((octyloxy)methyl)- (9CI) **(56625-58-2)**
9,10-Anthracenedione, 1-amino-4-(phenylamino)- (9CI) **(4395-65-7)**
9,10-Anthracenedione, 1-amino-2-(propoxymethyl)- (9CI) **(56594-29-7)**
9,10-Anthracenedione, 1,5-bis(cyclohexylamino)- (9CI) **(15958-68-6)**
9,10-Anthracenedione, 1,4-bis((2,6-diethylphenyl)amino)- (9CI) **(20241-74-1)**
9,10-Anthracenedione, 1,5-bis(2,4-dinitrophenoxy)-4,8-dinitro-(9CI) **(116-78-9)**
9,10-Anthracenedione, 1,4-bis(ethylamino)- (9CI) **(6994-46-3)**
9,10-Anthracenedione, 1,4-bis((2-hydroxyethyl)amino)- (9CI) **(4471-41-4)**
9,10-Anthracenedione, 1,4-bis(methylamino)- (9CI) **(2475-44-7)**
9,10-Anthracenedione, 1,4-bis((1-methylethyl)amino)- (9CI) **(14233-37-5)**
9,10-Anthracenedione, 1,4-bis((4-methylphenyl)amino)- (9CI) **(128-80-3)**
9,10-Anthracenedione, 1,5(or 1,8)-bis((4-methylphenyl)amino)-(9CI) **(8005-40-1)**
9,10-Anthracenedione, 1,4-bis((2,4,6-trimethylphenyl)amino)- (9CI) **(116-75-6)**
9,10-Anthracenedione, 1-bromo- (9CI) **(632-83-7)**
9,10-Anthracenedione, 1-chloro- **(82-44-0)**
9,10-Anthracenedione, 2-chloro- **(131-09-9)**
9,10-Anthracenedione, 1-chloro-2-methyl- (9CI) **(129-35-1)**
9,10-Anthracenedione, 1-chloro-5-nitro- **(129-40-8)**
9,10-Anthracenedione, 1-(cyclohexylamino)- (9CI) **(1096-48-6)**
9,10-Anthracenedione, 1,5-diamino- **(129-44-2)**
9,10-Anthracenedione, 1,8-diamino- **(129-42-0)**
9,10-Anthracenedione, 1,4-diamino- (9CI) **(128-95-0)**
9,10-Anthracenedione, 1,5-diamino- (9CI) **(129-42-0)**
9,10-Anthracenedione, 1,8-diaminobromo-4,5-dihydroxy- (9CI) **(27733-08-0)**
9,10-Anthracenedione, 1,5-diaminochloro-4,8-dihydroxy **(12217-79-7)**
9,10-Anthracenedione, 1,4-diamino-2,3-dichloro- (9CI) **(81-42-5)**
9,10-Anthracenedione, 1,4-diamino-2,3-dihydro- **(81-63-0)**
9,10-Anthracenedione, 1,5-diamino-4,8-dihydroxy- **(145-49-3)**
9,10-Anthracenedione, 1,8-diamino-4,5-dihydroxy- **(128-94-9)**
9,10-Anthracenedione, 4,8-diamino-1,5-dihydroxy-2-(4-hydroxy-3-methylphenyl)- (9CI) **(4702-65-2)**
9,10-Anthracenedione, 1,5-diamino-4,8-dihydroxy-(4-hydroxyphenyl)- (9CI) **(31529-83-6)**
9,10-Anthracenedione, 4,8-diamino-1,5-dihydroxy-2-(4-hydroxyphenyl)- (9CI) **(7098-08-0)**
9,10-Anthracenedione, 1,5-diamino-4,8-dihydroxy(4-methoxyphenyl)- (9CI) **(31288-44-5)**
9,10-Anthracenedione, 1,4-diamino-2-methoxy- (9CI) **(2872-48-2)**
9,10-Anthracenedione, 1,4-diamino-5-nitro- (9CI) **(82-33-7)**
9,10-Anthracenedione, 1,5-dichloro- **(82-46-2)**
9,10-Anthracenedione, 1,8-dichloro- **(82-43-9)**
1,4-Anthracenedione, 2,3-dihydro-9,10-dihydroxy- (9CI) **(17648-03-2)**
9,10-Anthracenedione, 1,2-dihydroxy- **(72-48-0)**
9,10-Anthracenedione, 1,4-dihydroxy- **(81-64-1)**
9,10-Anthracenedione, 1,5-dihydroxy- **(117-12-4)**
9,10-Anthracenedione, 1,8-dihydroxy- **(117-10-2)**
9,10-Anthracenedione, 2,6-dihydroxy- **(84-60-6)**
9,10-Anthracenedione, dihydroxy- (9CI) **(1322-60-7)**
9,10-Anthracenedione, 1,5-dihydroxy-4,8-bis(methylamino)- (9CI) **(3860-63-7)**
9,10-Anthracenedione, 1,5-dihydroxy-4,8-dinitro- (9CI) **(128-91-6)**
9,10-Anthracenedione, 1,8-dihydroxy-4,5-dinitro- (9CI) **(81-55-0)**
9,10-Anthracenedione, 1,8-dihydroxy-4-((4-(2-hydroxyethyl)-phenyl)amino)-5-nitro- **(15791-78-3)**
9,10-Anthracenedione, 1,8-dihydroxy-3-methyl- (9CI) **(481-74-3)**
9,10-Anthracenedione, 1,8-dihydroxy-4-nitro-5-(phenylamino)-(9CI) **(20241-76-3)**
9,10-Anthracenedione, 1,8-dihydroxy-2,4,5,7-tetranitro- (9CI) **(517-92-0)**
9,10-Anthracenedione, 1,5-dimethoxy- **(6448-90-4)**
9,10-Anthracenedione, 1,2-dinitro- (9CI) **(57875-61-3)**
9,10-Anthracenedione, 1,5-diphenoxy- **(82-21-3)**
9,10-Anthracenedione, 1,8-diphenoxy- (9CI) **(82-17-7)**
9,10-Anthracenedione, 2-ethyl **(84-51-5)**
9,10-Anthracenedione, 6-ethyl-1,2,3,4-tetrahydro- **(15547-17-8)**
9,10-Anthracenedione, 1-hydroxy- **(129-43-1)**
9,10-Anthracenedione, 2-hydroxy- (9CI) **(605-32-3)**
9,10-Anthracenedione, hydroxy- (9CI) **(27938-76-7)**
9,10-Anthracenedione, 1-((2-hydroxyethyl)amino)- (9CI) **(4465-58-1)**
9,10-Anthracenedione, 1-((2-hydroxyethyl)amino)-4-(methylamino)- (9CI) **(2475-46-9)**
9,10-Anthracenedione, 1-((2-hydroxyethyl)amino)-4-(methylamino)- (9CI) **(86722-66-9)**
9,10-Anthracenedione, 1-hydroxy-4-((4-methoxyphenyl)-amino)- (9CI) **(23552-76-3)**
9,10-Anthracenedione, 1-hydroxy-4-((4-((methylsulfonyl)oxy)-phenyl)amino)- (9CI) **(1594-08-7)**
9,10-Anthracenedione, 1-methoxy- (9CI) **(82-39-3)**
9,10-Anthracenedione, 2-methoxy- (9CI) **(3274-20-2)**
9,10-Anthracenedione, 2-methyl- (9CI) **(84-54-8)**
9,10-Anthracenedione, 1-(methylamino)- (9CI) **(82-38-2)**
9,10-Anthracenedione, 1-(methylamino)-4-((4-methylphenyl)-

amino)- (9CI) **(128-85-8)**
9,10-Anthracenedione, 2-methyl-1-nitro- (9CI) **(129-15-7)**
9,10-Anthracenedione, 1-nitro- **(82-34-8)**
9,10-Anthracenedione, 2-pentyl- (9CI) **(13936-21-5)**
9,10-Anthracenedione, 1,4,5,8-tetraamino **(2475-45-8)**
9,10-Anthracenedione, 1,2,5,8-tetrahydroxy- (9CI) **(81-61-8)**
9,10-Anthracenedione, 1,2,4-trihydroxy- **(81-54-9)**
9,10-Anthracenedione, 1,3,8-trihydroxy-6-methyl- (9CI) **(518-82-1)**
Anthracene, 9,10-diphenyl- (9CI) **(1499-10-1)**
Anthracene-1,5-disulfonic acid **(61736-91-2)**
Anthracene-1,8-disulfonic acid **(61736-92-3)**
1,8-Anthracenedisulfonic acid (9CI) **(61736-92-3)**
2,6-Anthracenedisulfonic acid, 4,8-diamino-9,10-dihydro-1,5-dihydroxy-9,10-dioxo- (9CI) **(128-86-9)**
2,6-Anthracenedisulfonic acid, 9,10-dihydro-1,5-dihydroxy-9,10-dioxo-4,8-bis((sulfomethyl)amino)-, tetrasodium salt (9CI) **(6370-93-0)**
1,5-Anthracenedisulfonic acid, 9,10-dihydro-9,10-dioxo **(117-14-6)**
1,8-Anthracenedisulfonic acid, 9,10-dihydro-9,10-dioxo **(82-48-4)**
Anthracenedisulfonic acid, 9,10-dihydro-9,10-dioxo- (9CI) **(53123-81-2)**
1,5-Anthracenedisulfonic acid, 9,10-dihydro-9,10-dioxo-, disodium salt **(853-35-0)**
2,7-Anthracenedisulfonic acid, 9,10-dihydro-9,10-dioxo-, disodium salt (9CI) **(853-67-8)**
Anthracene, 9-ethenyl- (9CI) **(2444-68-0)**
Anthracene, 2-ethyl- (9CI) **(52251-71-5)**
Anthracene, 9-methoxy- (8CI,9CI) **(2395-96-2)**
Anthracene, 2-methyl **(613-12-7)**
Anthracene, 9-methyl **(779-02-2)**
Anthracene, 1-methyl- (9CI) **(610-48-0)**
Anthracene, methyl- (8CI,9CI) **(26914-18-1)**
Anthracene, methylnitro- (9CI) **(80191-43-1)**
Anthracene, 9-nitro **(602-60-8)**
Anthracene, nitro- (9CI) **(54738-93-1)**
Anthracene, 9-phenyl- (9CI) **(602-55-1)**
2-Anthracenesulfonic acid, 4-((4-(acetylamino)phenyl)amino)-1-amino-9,10-dihydro-9,10-dioxo- (9CI) **(6247-34-3)**
2-Anthracenesulfonic acid, 4-((4-(acetylamino)phenyl)amino)-1-amino-9,10-dihydro-9,10-dioxo-, monosodium salt (9CI) **(6424-85-7)**
2-Anthracenesulfonic acid, 1-amino-4-((3-amino-4-sulfophenyl)-amino)-9,10-dihydro-9,10-dioxo- (9CI) **(128-99-4)**
2-Anthracenesulfonic acid, 1-amino-4-bromo-9,10-dihydro-9,10-dioxo- (9CI) **(116-81-4)**
2-Anthracenesulfonic acid, 1-amino-4-(cyclohexylamino)-9,10-dihydro-9,10-dioxo- (9CI) **(5617-28-7)**
2-Anthracenesulfonic acid, 1-amino-9,10-dihydro-9,10-dioxo-4-(phenylamino)- (9CI) **(2786-71-2)**
2-Anthracenesulfonic acid, 1-amino-9,10-dihydro-9,10-dioxo-4-(phenylamino)-, monosodium salt (9CI) **(6408-78-2)**
2-Anthracenesulfonic acid, 1-amino-9,10-dihydro-9,10-dioxo-4-o-toluidino- (8CI) **(34114-36-8)**
2-Anthracenesulfonic acid, 1-amino-9,10-dihydro-4-((3-(((2-hydroxyethyl)amino)sulfonyl)-4,5-dimethylphenyl)amino)-9,10-dioxo- (9CI) **(36897-88-8)**
2-Anthracenesulfonic acid, 1-amino-9,10-dihydro-4-((5-(((2-hydroxyethyl)amino)sulfonyl)-3,4-dimethylphenyl) amino)-9,10-dioxo-, monosodium salt (9CI) **(25797-81-3)**
2-Anthracenesulfonic acid, 1-amino-9,10-dihydro-4-(m-((2-hydroxy-ethyl)sulfonyl)anilino- 9,10-dioxo-, hydrogen sulfate (ester), disodium salt **(2580-78-1)**
2-Anthracenesulfonic acid, 1-amino-9,10-dihydro-4-(p-(N-methyl-acetamido)anilino)-9,10-dioxo-, monosodium salt **(2666-17-3)**
2-Anthracenesulfonic acid, 1-amino-9,10-dihydro-4-((2-methyl-phenyl)amino)-9,10-dioxo- **(34114-36-8)**
2-Anthracenesulfonic acid, 9,10-dihydro-1-amino-4-(cyclohexyl-amino)-9,10-dioxo-, monosodium salt **(4368-56-3)**
2-Anthracenesulfonic acid, 9,10-dihydro-3,4-dihydroxy-9,10-dioxo-, monosodium salt **(130-22-3)**
1-Anthracenesulfonic acid, 9,10-dihydro-9,10-dioxo- (9CI) **(82-49-5)**
2-Anthracenesulfonic acid, 9,10-dihydro-9,10-dioxo-, sodium salt **(131-08-8)**
1-Anthracenesulfonic acid, 9,10-dihydro-9,10-dioxo-, sodium salt (8CI,9CI) **(128-56-3)**
Anthracene, tetradecahydro- (9CI) **(6596-35-6)**
Anthracene, tetrahydro- (8CI,9CI) **(30551-09-8)**
Anthracene, 2,9,10-trimethyl **(63018-94-0)**
Anthracene, trimethyl- (8CI,9CI) **(27358-28-7)**
1,8,9-Anthracenetriol **(480-22-8)**
1,8,9-Anthracenetriol, 3-methyl **(491-59-8)**
Anthraceno(2,1-a)anthracene **(217-54-9)**
Anthraceno(2',1':1,2)anthracene **(217-54-9)**
9(10H)-Anthracenone **(90-44-8)**
9(10H)-Anthracenone, 1,8-dihydroxy- **(1143-38-0)**
Anthrachinon-1,5-disulfonan disodny (Czech) **(853-35-0)**
Anthrachinon-1,5-disulfonan sodny (Czech) **(853-35-0)**
9,10-Anthrachinon (Czech) **(84-65-1)**
Anthrachinon-1-sulfonan sodny (Czech) **(128-56-3)**
Anthracin **(120-12-7)**
2-Anthracylamine **(613-13-8)**
Anthra(2,1-d:6,5-d')bisthiazole-6,12-dione, 2,8-diphenyl **(129-09-9)**
Anthra(2,1,9-def:6,5,10-d'e'f')diisoquinoline-1,3,8,10(2H,9H)-tetrone (9CI) **(81-33-4)**
Anthra(2,1,9-def:6,5,10-d'e'f')diisoquinoline-1,3,8,10(2H,9H)-tetrone, 2,9-dimethyl- (9CI) **(5521-31-3)**
Anthra(2,1,9-def:6,5,10-d'e'f')diisoquinoline-1,3,8,10(2H,9H)-tetrone, 2,9-bis(3,5-dimethylphenyl)- (9CI) **(4948-15-6)**
Anthra(2,1,9-def:6,5,10-d'e'f')diisoquinoline-1,3,8,10(2H,9H)-tetrone, 2,9-bis(4-ethoxyphenyl)- (9CI) **(24108-89-2)**
Anthra(2,1,9-def:6,5,10-d'e'f')diisoquinoline-1,3,8,10(2H,9H)-tetrone, 2,9-bis(4-methoxyphenyl)- (9CI) **(6424-77-7)**
Anthra(2,1,9-def:6,5,10-d'e'f')diisoquinoline-1,3,8,10(2H,9H)-tetrone, 2,9-bis(4-(phenylazo)phenyl)- (9CI) **(3049-71-6)**
Anthra(2,1,9-def)isoquinoline (7CI,8CI,9CI) **(189-90-2)**
Anthradione **(84-65-1)**
Anthra(2,3-d)oxazole-5,10-dione, 2-(1-amino-2-anthraquinonyl)-(8CI) **(2379-79-5)**
Anthra(2,3-d)oxazole-5,10-dione, 2-(1-amino-9,10-dihydro-9,10-dioxo-2-anthracenyl)- (9CI) **(2379-79-5)**
Anthraflavic acid **(84-60-6)**
Anthraflavin **(84-60-6)**
Anthralan Blue B **(2666-17-3)**
Anthralan Yellow RRT **(6373-74-6)**
Anthraldehyde **(31671-77-9)**
2-Anthraldehyde, 1-amino-9,10-dihydro-9,10-dioxo- (8CI) **(6363-87-7)**
Anthralin **(1143-38-0)**
2-Anthramine **(613-13-8)**
Anthra(2,1,9-mna)naphth(2,3-h)acridine-5,10,15(16H)-trione, 3-((9,10-dihydro-9,10-dioxo-1-anthracenyl)amino)- (9CI) **(4395-53-3)**
Anthra(2,1,9-mna)naphth(2,3-h)acridine-5,10,15-trione **(3271-76-9)**
Anthranilamide **(88-68-6)**
Anthranilic-acid **(118-92-3)**
o-Anthranilic acid **(118-92-3)**
Anthranilic acid, N-acetyl **(89-52-1)**
Anthranilic acid, N-acetyl-, methyl ester (8CI) **(2719-08-6)**
Anthranilic acid, allyl ester (8CI) **(7493-63-2)**
Anthranilic acid, N,N'-1,5-anthraquinonylenedi- (8CI) **(81-78-7)**
Anthranilic acid, butyl ester (8CI) **(7756-96-9)**
Anthranilic acid, 5-chloro- (8CI) **(635-21-2)**
Anthranilic acid, 4-chloro-N-furfuryl-5-sulfamoyl **(54-31-9)**
Anthranilic acid, cinnamyl ester **(87-29-6)**
Anthranilic acid, cyclohexyl ester (8CI) **(7779-16-0)**
Anthranilic acid, N-(3,4-dimethoxycinnamoyl)- **(53902-12-8)**
Anthranilic acid, 1,5-dimethyl-1-vinyl-4-hexenyl ester **(7149-26-0)**
Anthranilic acid, 3-hydroxy- **(548-93-6)**
Anthranilic acid, linalyl ester **(7149-26-0)**

Name	CAS No.
Anthranilic acid, methyl ester	**(134-20-3)**
Anthranilic acid, N-methyl-, methyl ester	**(85-91-6)**
Anthranilic acid, 4-nitro	**(619-17-0)**
Anthranilic acid, 5-nitro- (8CI)	**(616-79-5)**
Anthranilic acid, phenethyl ester	**(133-18-6)**
Anthranilic acid, N-phenyl	**(91-40-7)**
Anthranilic acid, N-(2,3-xylyl)	**(61-68-7)**
Anthranilimidic acid	**(88-68-6)**
Anthranilonitrile	**(1885-29-6)**
m-Anthranilonitrile	**(2237-30-1)**
Anthranthrene	**(191-26-4)**
Anthrapole AZ	**(136-60-7)**
9,10-Anthraquinone	**(84-65-1)**
Anthraquinone	**(84-65-1)**
Anthraquinone Blue	**(81-77-6)**
Anthraquinone Deep Blue	**(81-77-6)**
Anthraquinone Green G Base	**(128-80-3)**
Anthraquinone, 1-amino	**(82-45-1)**
Anthraquinone, 2-amino	**(117-79-3)**
Anthraquinone, 1-amino-2-bromo-4-hydroxy	**(116-82-5)**
Anthraquinone, 1-amino-5-chloro	**(117-11-3)**
Anthraquinone, 1-amino-4-chloro-2-methyl- (8CI)	**(3225-97-6)**
Anthraquinone, 1-amino-2,4-dibromo	**(81-49-2)**
Anthraquinone, 1-amino-4,5-dihydroxy-8-(phenylamino)-	**(20241-77-4)**
Anthraquinone, 1-amino-2-(p-((hexahydro-2-oxo-1H-azepin-1-yl)-methyl)phenoxy)-4-hydroxy- (8CI)	**(19014-53-0)**
Anthraquinone, 1-amino-4-hydroxy	**(116-85-8)**
Anthraquinone, 1-amino-4-hydroxy-	**(116-85-8)**
Anthraquinone, 1-amino-4-hydroxy-2-(3-hydroxybutoxy)-	**(3224-15-5)**
Anthraquinone, 1-amino-2-(hydroxymethyl)- (8CI)	**(24094-44-8)**
Anthraquinone, 1-amino-4-hydroxy-2-((α-(2-oxo-1-pyrrolidinyl)-p-tolyl)oxy)- (8CI)	**(25177-16-6)**
Anthraquinone, 1-amino-4-hydroxy-2-phenoxy	**(17418-58-5)**
Anthraquinone, 1-amino-2-methyl	**(82-28-0)**
Anthraquinone, 1-amino-4-methylamino	**(1220-94-6)**
Anthraquinone, 1-anilino-4,5-dihydroxy-8-nitro-	**(20241-76-3)**
Anthraquinone, 1,4-bis((2-hydroxyethyl)amino)- (8CI)	**(4471-41-4)**
Anthraquinone, 1,4-bis((2-hydroxyethyl)amino)-5,8-dihydroxy	**(3179-90-6)**
Anthraquinone, 1,4-bis(isopropylamino)- (8CI)	**(14233-37-5)**
Anthraquinone, 1,4-bis(methylamino)	**(2475-44-7)**
Anthraquinone, 1,4-bis(p-tolylamino)	**(128-80-3)**
Anthraquinone, 1,4-bis(2,4,6-trimethylanilino)- (8CI)	**(116-75-6)**
Anthraquinone, 1-bromo- (8CI)	**(632-83-7)**
Anthraquinone, 2-bromo-1,5-diamino-4,8-dihydroxy	**(27312-17-0)**
2-Anthraquinonecarboxaldehyde, 1-amino-	**(6363-87-7)**
Anthraquinone, 1-chloro	**(82-44-0)**
Anthraquinone, 2-chloro	**(131-09-9)**
Anthraquinone, 1-chloro-2-methyl- (8CI)	**(129-35-1)**
Anthraquinone, 1-chloro-5-nitro	**(129-40-8)**
Anthraquinone, 1,4-diamino	**(128-95-0)**
Anthraquinone, 1,5-diamino	**(129-44-2)**
Anthraquinone, 1,8-diamino	**(129-42-0)**
Anthraquinone, 1,5-diamino-4,8-dihydroxy	**(145-49-3)**
Anthraquinone, 1,8-diamino-4,5-dihydroxy	**(128-94-9)**
Anthraquinone, 1,5-diamino-4,8-dihydroxy-3-(p-methoxyphenyl)	**(4702-64-1)**
Anthraquinone, 1,4-diamino-2-methoxy	**(2872-48-2)**
Anthraquinone, 1,4-diamino-5-nitro	**(82-33-7)**
Anthraquinone, 1,5-dichloro	**(82-46-2)**
Anthraquinone, 1,8-dichloro	**(82-43-9)**
Anthraquinone, 2,3-dihydro-1,4-diamino	**(81-63-0)**
Anthraquinone, 1,2-dihydroxy	**(72-48-0)**
Anthraquinone, 1,4-dihydroxy	**(81-64-1)**
Anthraquinone, 1,5-dihydroxy	**(117-12-4)**
Anthraquinone, 1,8-dihydroxy	**(117-10-2)**
Anthraquinone, 2,6-dihydroxy	**(84-60-6)**
Anthraquinone, dihydroxy- (8CI)	**(1322-60-7)**
Anthraquinone, 1,8-dihydroxy-4,5-dinitro	**(81-55-0)**
Anthraquinone, 1,5-dihydroxy-4,8-dinitro- (8CI)	**(128-91-6)**
Anthraquinone, 1,8-dihydroxy-4-(p-(2-hydroxyethyl)anilino)-5-nitro	**(15791-78-3)**
Anthraquinone, 1,8-dihydroxy-3-methyl	**(481-74-3)**
Anthraquinone, 1,8-dihydroxy-2,4,5,7-tetranitro- (8CI)	**(517-92-0)**
Anthraquinone, 1,5-dimethoxy	**(6448-90-4)**
1,2-Anthraquinonediol	**(72-48-0)**
Anthraquinone, 1,5-diphenoxy	**(82-21-3)**
1,5-Anthraquinonedisulfonic acid	**(117-14-6)**
1,8-Anthraquinonedisulfonic acid	**(82-48-4)**
2,7-Anthraquinonedisulfonic acid, disodium salt	**(853-67-8)**
Anthraquinone-2,7-disulfonic acid, disodium salt	**(853-67-8)**
Anthraquinone, 1,4-di-p-toluidino- (8CI)	**(128-80-3)**
Anthraquinone, 2-ethyl-	**(84-51-5)**
Anthraquinone, 2-ethyl-5,6,7,8-tetrahydro	**(15547-17-8)**
Anthraquinone, 1-hydroxy	**(129-43-1)**
Anthraquinone, 2-hydroxy	**(605-32-3)**
Anthraquinone, hydroxy- (8CI)	**(27938-76-7)**
Anthraquinone, 1-((2-hydroxyethyl)amino)-4-(methylamino)	**(2475-46-9)**
Anthraquinone, 1-((2-hydroxyethyl)amino)-4-(methylamino)-(8CI)	**(86722-66-9)**
Anthraquinone, 1-hydroxy-4-(p-hydroxyanilino)-, 4-methane-sulfonate (Ester)	**(1594-08-7)**
Anthraquinone, 1-hydroxy-4-(p-toluidino)	**(81-48-1)**
Anthraquinone, 4,5'-iminobis(4-benzamido	**(128-89-2)**
Anthraquinone, 1-methoxy	**(82-39-3)**
Anthraquinone, 2-methoxy- (8CI)	**(3274-20-2)**
Anthraquinone, 2-methyl- (8CI)	**(84-54-8)**
Anthraquinone, 1-(methylamino)- (8CI)	**(82-38-2)**
Anthraquinone, 1-(methylamino)-4-p-toluidino- (8CI)	**(128-85-8)**
Anthraquinone, 2-methyl-1-nitro	**(129-15-7)**
Anthraquinone, 6-methyl-1,3,8-trihydroxy	**(518-82-1)**
Anthraquinone, 1-nitro	**(82-34-8)**
9,10-Anthraquinone-2-sodium sulfonate	**(131-08-8)**
2-Anthraquinonesulfonate sodium	**(131-08-8)**
9,10-Anthraquinone-1-sulfonate sodium salt	**(128-56-3)**
Anthraquinone-2-sulfonate sodium salt	**(131-08-8)**
2-Anthraquinonesulfonic acid, 1-amino-4-(cyclohexylamino)-	**(5617-28-7)**
1-Anthraquinonesulfonic acid sodium salt	**(128-56-3)**
2-Anthraquinonesulfonic acid sodium salt	**(131-08-8)**
Anthraquinone, 1,2,5,8-tetrahydroxy	**(81-61-8)**
Anthraquinone, 1,4,5,8-tetramino-	**(2475-45-8)**
Anthraquinone, 1,2,4-trihydroxy	**(81-54-9)**
Anthraquinone, 1,3,8-trihydroxy-6-methyl-	**(518-82-1)**
α-Anthraquinonylamine	**(82-45-1)**
β-Anthraquinonylamine	**(117-79-3)**
1,4-Anthraquinonyldiamine	**(128-95-0)**
1,5-Anthraquinonyldiamine	**(129-42-0)**
1,5-Anthraquinonyldiamine	**(129-44-2)**
1,8-Anthraquinonyldiamine	**(129-42-0)**
N,N'-(1,5-Anthraquinonylene)dianthranilic acid	**(81-78-7)**
2,2'-(1,4-Anthraquinonylenediimino)bis(5-methylbenzene-sulfonic acid) disodium salt	**(4403-90-1)**
Anthra(1,2,3,4-rst)pentaphene (8CI,9CI)	**(31541-07-8)**
Anthrarufin	**(117-12-4)**
Anthrarufin, 4,8-diamino-	**(145-49-3)**
Anthrarufin, 4,8-dinitro-	**(128-91-6)**
Anthrasorb	**(7440-44-0)**
1,8,9-Anthratriol	**(480-22-8)**
Anthravat Golden Yellow	**(128-66-5)**
Anthravat Navy Blue BR	**(1324-54-5)**
5,9,14,18-Anthrazinetetrone, chloro-6,15-dihydro- (9CI)	**(1324-27-2)**
5,9,14,18-Anthrazinetetrone, 7,16-dichloro-6,15-dihydro- (8CI,9CI)	**(130-20-1)**
5,9,14,18-Anthrazinetetrone, 6,15-dihydro	**(81-77-6)**
Anthrogon	**(9002-68-0)**
9-Anthroic acid	**(723-62-6)**
2-Anthroic acid, 9,10-dihydro-4,5-dihydroxy-9,10-dioxo- (8CI)	**(478-43-3)**

Name	CAS No.
1-Anthroic acid, 9,10-dihydro-2,5,7,8-tetrahydroxy-4-methyl-9,10-dioxo-6-(2,3,4,5-tetrahydroxyhexanoyl)- (8CI)	**(1260-17-9)**
Anthrone	**(90-44-8)**
Anthrone, 1,8-dihydroxy	**(1143-38-0)**
9-Anthronitrile (8CI)	**(1210-12-4)**
Anthronitrile (8CI)	**(28805-75-6)**
Anthropodeoxycholic acid	**(474-25-9)**
Anthropodesoxycholic acid	**(474-25-9)**
Anthropododesoxycholic acid	**(474-25-9)**
Anthrosin BRX	**(1658-56-6)**
2-Anthrylamine	**(613-13-8)**
Anti-Chromotrichia Factor	**(150-13-0)**
Anti-Ethyl	**(97-77-8)**
Anti-Germ 77	**(121-54-0)**
Anti-Infective Vitamin	**(68-26-8)**
Anti-Inflammatory Hormone	**(50-23-7)**
Anti OX	**(119-47-1)**
Anti-Pellagra Vitamin	**(59-67-6)**
Anti-Rust	**(7632-00-0)**
Anti-Stress	**(77-75-8)**
Antiaethan	**(97-77-8)**
Antiarigenin + L-rhamnose (German)	**(639-13-4)**
β-Antiarin	**(639-13-4)**
Antiasthmatique	**(51-43-4)**
Antibason	**(56-04-2)**
Antibiocin	**(132-98-9)**
Antibiotic 1037	**(606-58-6)**
Antibiotic 503-3	**(7542-37-2)**
Antibiotic A-399-Y4	**(606-58-6)**
Antibiotic N-329 B	**(2001-95-8)**
Antibiotic BB-K 8	**(37517-28-5)**
Antibiotic Bay-f 1353	**(51481-65-3)**
Antibiotic C 076B1A	**(65195-55-3)**
Antibiotic CGP 9000	**(51762-05-1)**
Antibiotic 2230D	**(7542-37-2)**
Antibiotic E212	**(606-58-6)**
Antibiotic FN 1636	**(63-91-2)**
Antibiotic LA 7017	**(18378-89-7)**
Antibiotic PA-106	**(22862-76-6)**
Antibiotic S 7481F1	**(59865-13-3)**
Antibiotic SF 1293	**(35597-43-4)**
Antibiotic SF 767B	**(7542-37-2)**
Antibiotic U 18496	**(320-67-2)**
Antibiotic XK 62-3	**(69375-05-9)**
Antiblaze 19 (9CI)	**(61840-22-0)**
Antiblaze 78 (9CI)	**(64176-42-7)**
Antiborer 3767	**(52315-07-8)**
Antiborer 3768	**(52645-53-1)**
Antibulit	**(7681-49-4)**
Anticanitic Vitamin	**(150-13-0)**
Anticarie	**(118-74-1)**
Antichlor	**(10102-17-7)**
Antideprin	**(50-49-7)**
Antidiuretic hormone	**(11000-17-2)**
Antidurol	**(50-13-5)**
Antiegene MB	**(583-39-1)**
Antierythrite	**(149-32-6)**
Antietanol	**(97-77-8)**
Antietil	**(97-77-8)**
Antifebrin	**(103-84-4)**
Antifoam A	**(8050-81-5)**
Antifoam FD 62	**(63148-62-9)**
Antifolan	**(59-05-2)**
Antiformin	**(7681-52-9)**
Antigene RDF	**(26780-96-1)**
Antigestil	**(56-53-1)**
Antihemorrhagic vitamin	**(84-80-0)**

Name	CAS No.
Antihist	**(59-33-6)**
Antiknock-33	**(12108-13-3)**
Antikol	**(97-77-8)**
Antilepsin	**(630-93-3)**
Antilipid	**(637-07-0)**
Antilipide	**(637-07-0)**
Antilirium	**(57-64-7)**
Antimalarina	**(83-89-6)**
Antimicina	**(54-85-3)**
Antimigrant C 45	**(9005-38-3)**
Antimilace	**(108-62-3)**
Antimilace	**(9002-91-9)**
Antiminth	**(15686-83-6)**
Antimit	**(55-86-7)**
Antimoine fluorure (French)	**(7783-56-4)**
Antimoine (pentachlorure d') (French)	**(7647-18-9)**
Antimoine (trichlorure d') (French)	**(10025-91-9)**
Antimol	**(532-32-1)**
Antimonate(2)-, bis(mu-tartrato(4-))di-, dipotassium, trihydrate	**(28300-74-5)**
Antimonate, potassium, (OC-6-11)- (9CI)	**(12208-13-8)**
Antimonial saffron	**(1315-04-4)**
Antimonic "acid"	**(1314-60-9)**
Antimonic acid, monopotassium salt (8CI)	**(12208-13-8)**
Antimonic oxide	**(1314-60-9)**
Antimonic sulfide	**(1315-04-4)**
Antimonio (pentacloruro di) (Italian)	**(7647-18-9)**
Antimonio (tricloruro di) (Italian)	**(10025-91-9)**
Antimonious oxide	**(1309-64-4)**
Antimonous chloride (DOT)	**(10025-91-9)**
Antimonous fluoride	**(7783-56-4)**
Antimonous sulfide	**(1345-04-6)**
Antimonpentachlorid (German)	**(7647-18-9)**
Antimontrichlorid (German)	**(10025-91-9)**
Antimonwasserstoffes (German)	**(7803-52-3)**
Antimony (ACGIH,OSHA)	**(7440-36-0)**
Antimony Black	**(7440-36-0)**
Antimony Butter	**(10025-91-9)**
Antimony, Compd. with lead (1:1) (9CI)	**(12266-38-5)**
Antimony Glance	**(1345-04-6)**
Antimony, Isotope of mass 125 (8CI,9CI)	**(14234-35-6)**
Antimony, Isotope of mass 127 (8CI,9CI)	**(13968-50-8)**
Antimony Orange	**(1345-04-6)**
Antimony, Powder [UN 2871]	**(7440-36-0)**
Antimony Red	**(1315-04-4)**
Antimony, Regulus	**(7440-36-0)**
Antimony Sulfide Golden	**(1315-04-4)**
Antimony (V) chloride	**(7647-18-9)**
Antimony (V) fluoride	**(7783-70-2)**
Antimony(V) pentafluoride	**(7783-70-2)**
Antimony bromide	**(7789-61-9)**
Antimony bromide oxide (8CI,9CI)	**(12323-32-9)**
Antimony (III) chloride	**(10025-91-9)**
Antimony chloride (DOT)	**(10025-91-9)**
Antimony diamyldithiocarbamate	**(15890-25-2)**
Antimony O,O'-di-2-ethylhexylphosphorodithioate	**(15874-52-9)**
Antimony O,O'-di-n-propyl phosphorodithioate	**(15874-48-3)**
Antimony fluoride	**(7783-70-2)**
Antimony(III) fluoride (1:3)	**(7783-56-4)**
Antimony hydride	**(7803-52-3)**
Antimony iodide	**(7790-44-5)**
Antimony iodide oxide (8CI,9CI)	**(12196-43-9)**
Antimony lactate	**(58164-88-8)**
Antimony lactate, Solid	**(58164-88-8)**
Antimony lead oxide (9CI)	**(13510-89-9)**
Antimonyl potassium tartrate	**(28300-74-5)**
Antimony(3+) oxide	**(1309-64-4)**
Antimony-oxide	**(1309-64-4)**

Antimony oxide (9CI) **(1332-81-6)**
Antimony oxide (VAN)(9CI) **(1327-33-9)**
Antimony oxide calcium titanate silicate ceramic opacifier **(66402-68-4)**
Antimony pentachloride (DOT) **(7647-18-9)**
Antimony pentachloride, Liquid [UN 1730] **(7647-18-9)**
Antimony pentachloride, Solution [UN 1731] **(7647-18-9)**
Antimony pentafluoride [UN 1732] **(7783-70-2)**
Antimony pentaoxide **(1314-60-9)**
Antimony pentasulfide **(1315-04-4)**
Antimony-pentoxide **(1314-60-9)**
Antimony perchloride **(7647-18-9)**
Antimony peroxide **(1309-64-4)**
DL-Antimony potassium tartrate **(64070-12-8)**
L-Antimony potassium tartrate **(11071-15-1)**
Antimony potassium tartrate, Solid (DOT) **(28300-74-5)**
Antimony potassium tartrate [UN 1551] **(28300-74-5)**
Antimony sesquioxide **(1309-64-4)**
Antimony sulfide **(1345-04-6)**
Antimony sulfide (Sb_2S_5) (8CI,9CI) **(1315-04-4)**
Antimony sulfide, Solid **(12627-52-0)**
Antimony sulfide, Solid (DOT) **(1345-04-6)**
Antimony tribromide **(7789-61-9)**
Antimony tribromide, Solid (DOT) **(7789-61-9)**
Antimony tribromide, Solution [NA 1549] **(7789-61-9)**
Antimony trichloride **(10025-91-9)**
Antimony trichloride, Liquid [UN 1733] **(10025-91-9)**
Antimony trichloride, Solid (DOT) **(10025-91-9)**
Antimony trichloride, Solution (DOT) **(10025-91-9)**
Antimony trifluoride **(7783-56-4)**
Antimony trifluoride, Solid [NA 1549] **(7783-56-4)**
Antimony trifluoride, Solution (DOT) **(7783-56-4)**
Antimony trihydride **(7803-52-3)**
Antimony triiodide **(7790-44-5)**
Antimony trioxide **(1309-64-4)**
Antimony, tris(dipentylcarbamodithioato-S,S')-, (OC-6-11)- (9CI) **(15890-25-2)**
Antimony tris(ethylene glycoxide) **(29736-75-2)**
Antimony tris(isooctyloxycarbonylmethylmercaptide) **(27288-44-4)**
Antimony-trisulfide **(1345-04-6)**
Antimony trisulfide colloid **(1345-04-6)**
Antimony white **(1309-64-4)**
Antimoonpentachloride (Dutch) **(7647-18-9)**
Antimoontrichlride (Dutch) **(10025-91-9)**
Antimucin WDR **(62-38-4)**
Antimycin **(518-75-2)**
Antimycin A **(1397-94-0)**
Antinonin **(534-52-1)**
Antinonnin **(534-52-1)**
Antio **(2540-82-1)**
Antiok S **(693-36-7)**
Antioxidant 1 **(119-47-1)**
Antioxidant 4 **(128-37-0)**
Antioxidant 29 **(128-37-0)**
Antioxidant 30 **(128-37-0)**
Antioxidant 116 **(135-88-6)**
Antioxidant 330 **(1709-70-2)**
Antioxidant 425 **(88-24-4)**
Antioxidant 754 **(88-26-6)**
Antioxidant 1076 **(2082-79-3)**
Antioxidant 2246 **(119-47-1)**
Antioxidant 4020 **(793-24-8)**
Antioxidant AS **(123-28-4)**
Antioxidant BKF **(119-47-1)**
Antioxidant CD (9CI) **(76600-84-5)**
Antioxidant D **(100-64-1)**
Antioxidant DBPC **(128-37-0)**
Antioxidant E 702 **(118-82-1)**
Antioxidant HS **(26780-96-1)**
Antioxidant HSL **(26780-96-1)**
Antioxidant 4K **(128-37-0)**
Antioxidant KB **(128-37-0)**
Antioxidant LTDP **(123-28-4)**
Antioxidant MB (Czech) **(583-39-1)**
Antioxidant NG-2246 **(119-47-1)**
Antioxidant No. 33 **(96-76-4)**
Antioxidant PBN **(135-88-6)**
Antioxine **(499-75-2)**
Antioxyne B **(25013-16-5)**
Antiperz **(650-51-1)**
Antiphen **(97-23-4)**
Antipiricullin **(1397-94-0)**
Antipirin **(60-80-0)**
Antiprex 461 **(9003-01-4)**
Antiprex A **(9003-01-4)**
Antipyonin **(1303-96-4)**
Antipyrene T 1 **(79064-73-6)**
Antipyrin **(60-80-0)**
Antipyrine **(60-80-0)**
Antipyrine, 4-(dimethylamino) **(58-15-1)**
Antiren **(110-85-0)**
Antisacer **(57-41-0)**
Antisacer **(630-93-3)**
Antisal 1a **(108-88-3)**
Antiscorbic vitamin **(50-81-7)**
Antiscorbutic vitamin **(50-81-7)**
Antisepsin **(103-88-8)**
Antiseptol **(121-54-0)**
Antisol 1 **(127-18-4)**
Antisterility vitamin **(59-02-9)**
Antistominum **(58-73-1)**
Antistrept **(63-74-1)**
Antitanil **(67-96-9)**
Antitrombosin **(66-76-2)**
Antituberkulosum **(54-85-3)**
Antiverm **(92-84-2)**
Antivitium **(97-77-8)**
Antixerophthalmic Vitamin **(68-26-8)**
Antodyn **(538-43-2)**
Antodyne **(538-43-2)**
Antofin **(62-67-9)**
Antol **(105-36-2)**
Antomin **(58-73-1)**
Antor **(38727-55-8)**
Antora **(78-11-5)**
Antorfin **(62-67-9)**
Antorphine **(62-67-9)**
Antox **(1309-64-4)**
Antoxylic acid **(98-50-0)**
Antozite 1 **(103-96-8)**
Antracol **(12071-83-9)**
Antraderm **(480-22-8)**
Antrancine 12 **(25013-16-5)**
Antrancine 8 **(128-37-0)**
Antrapurol **(117-10-2)**
Antron **(63428-84-2)**
Anti-tetany substance 10 **(67-96-9)**
Antu (ACGIH,OSHA) **(86-88-4)**
Antuitrin S **(9002-61-3)**
Anturan **(57-96-5)**
Anturane **(57-96-5)**
Anturat **(86-88-4)**
Antymon (Polish) **(7440-36-0)**
Antymonowodor (Polish) **(7803-52-3)**
Antyperz **(650-51-1)**
Antywylegacz **(999-81-5)**

Anullex PBA 15 (85-60-9)
Anural (57-53-4)
Anuspiramin (50-33-9)
Anxietil (57-53-4)
Anxiolit (604-75-1)
Anxon (27223-35-4)
Anyvim (62-53-3)
Anzief (315-30-0)
Aoral (68-26-8)
Apachlor (470-90-6)
Apacil (65-49-6)
Apadodine (2439-10-3)
Apadodine (31717-87-0)
Apadon (103-90-2)
Apadrin (6923-22-4)
Apamid (103-90-2)
Apamide (103-90-2)
Apamidon (13171-21-6)
Apamine (300-42-5)
Aparasin (58-89-9)
Apas (65-49-6)
Apascil (57-53-4)
Apasil (57-53-4)
Apatate drape (59-43-8)
Apate Drops (67-03-8)
Apatite, hydroxy (1306-06-5)
Apaurin (439-14-5)
Apavap (62-73-7)
Apavinphos (298-01-1)
Apavinphos (7786-34-7)
Apelagrin (59-67-6)
Apergel (9000-11-7)
Apesan (78-44-4)
Apetain (51-63-8)
Apex (40596-69-8)
Apex 4 (123-95-5)
Apex 462-5 (126-72-7)
Apexol (68-26-8)
Apeyel (9000-11-7)
Aphalon (330-55-2)
Aphamite (56-38-2)
Aphenylbarbit (50-06-6)
Aphenyletten (50-06-6)
Aphidan (5827-05-4)
Apholate (52-46-0)
Aphos (74548-80-4)
Aphosal (15879-93-3)
Aphox (23103-98-2)
Aphoxide (545-55-1)
Aphrodine (146-48-5)
Aphrodine hydrochloride (65-19-0)
Aphrosol (146-48-5)
Aphtasolon (50-02-2)
Aphtiria (58-89-9)
Apirelina (60-80-0)
Apirolio 1431 C (25323-68-6)
Apirolio 1476 C (25429-29-2)
Apl-Luster (148-79-8)
Aplakil (604-75-1)
Aplidal (58-89-9)
Apocid Milling Red G (3567-65-5)
Apocid Orange 2G (1936-15-8)
Apocynin (498-02-2)
Apocynine (498-02-2)
Apocynol (2480-86-6)
Apoidina (9002-61-3)
Apolan (637-07-0)
Apolon B_6 (54-47-7)
Apomine Black GX (1937-37-7)
Apomorfin (58-00-4)
Apomorphine (58-00-4)
Apomorphine chloride (314-19-2)
(-)-Apomorphine hydrochloride (314-19-2)
Apomorphine hydrochloride (314-19-2)
Apomorphinium chloride (314-19-2)
(-)-Apomorphinium hydrochloride (314-19-2)
Aponorin (133-67-5)
Apopen (87-08-1)
Apoplon (50-55-5)
Apormorphine (58-00-4)
6a-β-Aporphine-10,11-diol (58-00-4)
6a-β-Aporphine-10,11-diol, hydrochloride (314-19-2)
Aposulfatrim (8064-90-2)
Apothyrin (66-02-4)
Apozepam (439-14-5)
Appetrol-sr (57-53-4)
Appex (22248-79-9)
Appl-Set (86-87-3)
Applaud (69327-76-0)
Appresinum (304-20-1)
Aprelazine (304-20-1)
Apresazide (304-20-1)
Apresine (304-20-1)
Apresolin (304-20-1)
Apresolin (86-54-4)
Apresoline (304-20-1)
Apresoline (86-54-4)
Apresoline-Esidrix (304-20-1)
Apresoline HCl (304-20-1)
Apresoline hydrochloride (304-20-1)
Apressin (304-20-1)
Apressin (86-54-4)
Apressoline (304-20-1)
Aprezolin (304-20-1)
Aprezolin (86-54-4)
Apridol (77-75-8)
Aprinox (73-48-3)
Aprobarbital (77-02-1)
Aprobarbitone (77-02-1)
Aprobit (60-87-7)
Aprocarb (114-26-1)
Aprol 161 (5340-36-3)
Apron (57837-19-1)
Apron 2E (57837-19-1)
Apron FL (57837-19-1)
Apropin (8073-77-6)
Aprozal (77-02-1)
Apsical (50-55-5)
Apsin VK (132-98-9)
Aptal (59-50-7)
Apurin (315-30-0)
Apurina (57-66-9)
Apurol (315-30-0)
Apyonine Auramine Base (492-80-8)
Aqua Cera (106-11-6)
Aqua Fortis (7697-37-2)
Aqua-Kleen (94-75-7)
Aqua Mephyton (84-80-0)
Aqua Regia (8007-56-5)
Aqua-Vex (93-72-1)
Aqua ammonia (1336-21-6)
Aqua ammonia, Solution (1336-21-6)
Aquacal (1592-23-0)
Aquacat (7440-48-4)

Aquachloral (302-17-0)
Aquacide (85-00-7)
Aquacillin (6130-64-9)
Aquacrine (53-16-7)
Aquacycline (2058-46-0)
Aquadag (7782-42-5)
Aquafil (7631-86-9)
Aquakay (58-27-5)
Aqualin (107-02-8)
Aqualine (107-02-8)
Aqualine Blue (147-14-8)
Aquamoline BC (64-02-8)
Aquamollin (64-02-8)
Aquamox (73-49-4)
Aquamycetin (56-75-7)
Aquapel (Polysaccharide) (9005-25-8)
Aquaplast (9004-32-4)
Aquaplast, Caprolactone (24980-41-4)
Aquarex ME (151-21-3)
Aquarex methyl (151-21-3)
Aquarills (58-93-5)
Aquarius (58-93-5)
Aquashade (92170-50-8)
Aquasil E (63148-62-9)
Aquasuspen (6130-64-9)
Aquasynth (68-26-8)
Aquatag (91-33-8)
Aquatensen (135-07-9)
Aquathol (145-73-3)
Aquathol K (59985-42-1)
Aquatin (639-58-7)
Aquatin 20 EC (639-58-7)
Aquaviron (57-85-2)
Aquazine (122-34-9)
Aquilide A (87625-62-5)
Aquinite (76-06-2)
Aquinone (58-27-5)
Aquirel (2259-96-3)
Ara-ATP (56-65-5)
Ara-C (147-94-4)
Ara-C (69-74-9)
Ara-Cytidine (147-94-4)
Arab Rat Deth (81-81-2)
Arabic-Gum (9000-01-5)
D-Arabino-hexonic acid, 3-deoxy- (8CI,9CI) (1518-59-8)
D-Arabino-2-hexulosonic acid (9CI) (669-90-9)
Arabinocytidine (147-94-4)
9-β-D-Arabinofuranosyladenine 5'-triphosphate (56-65-5)
1-β-D-Arabinofuranosyl-4-amino-2(1H)pyrimidinone (147-94-4)
1-(β-D-Arabinofuranosyl)cytosine (147-94-4)
1-Arabinofuranosylcytosine (147-94-4)
1-β-Arabinofuranosylcytosine (147-94-4)
1-β-D-Arabinofuranosylcytosine hydrochloride (69-74-9)
Arabinofuranosylcytosine hydrochloride (69-74-9)
1-β-D-Arabinofuranosylcytosine monohydrochloride (69-74-9)
D-Arabino-hexose, 2-deoxy (154-17-6)
L-Arabinopyranose (9CI) (87-72-9)
L-Arabinose (147-81-9)
Arabinose (9CI) (147-81-9)
Arabinosylcytosine (147-94-4)
β-D-Arabinosylcytosine (147-94-4)
Arabinosylcytosine hydrochloride (69-74-9)
Arabitin (147-94-4)
Arabitin hydrochloride (69-74-9)
Araboascorbic acid (89-65-6)
D-Araboascorbic acid (89-65-6)
Aracet APV (9002-89-5)
Arachic acid (506-30-9)
Arachic alcohol (629-96-9)
Arachidic acid (506-30-9)
Arachidic alcohol (629-96-9)
Arachidonate (506-32-1)
Arachidonic-acid (506-32-1)
Arachidyl alcohol (629-96-9)
Arachis Oil (8002-03-7)
Aracid (80-38-6)
Aracide (140-57-8)
Aractidine (147-94-4)
Aracytidine hydrochloride (69-74-9)
Aracytin (147-94-4)
Aracytin hydrochloride (69-74-9)
Araldit DY 026 (2425-79-8)
Araldite Accelerator 062 (103-83-3)
Araldite B (25085-99-8)
Araldite EPN 1139 (9CI) (64859-69-4)
Araldite Ere 1359 (101-90-6)
Araldite F (25085-99-8)
Araldite HT 907 (85-42-7)
Araldite HY 951 (112-24-3)
Araldite Hardener 972 (101-77-9)
Araldite Hardener HY 951 (112-24-3)
Araldite PT 810 (28825-96-9)
Araldite PT 816 (28825-96-9)
Araldite TGIC (28825-96-9)
Aralen (54-05-7)
Aralkonium chloride (102-30-7)
Aralo (56-38-2)
Aramine (54-49-9)
Aramite (140-57-8)
Aramite-15W (140-57-8)
Aramitearamite-15W (140-57-8)
Arancio cromo (Italian) (18454-12-1)
Aranox (100-93-6)
Arasan (137-26-8)
Arasan 70 (137-26-8)
Arasan 75 (137-26-8)
Arasan-M (137-26-8)
Arasan 42-S (137-26-8)
Arasan-SF (137-26-8)
Arasan-SF-X (137-26-8)
Aratan (123-88-6)
Arathane (39300-45-3)
Aratone (3035-45-8)
Aratron (140-57-8)
Arbitex (58-89-9)
Arbocel (9004-34-6)
Arbocel BC 200 (9004-34-6)
Arbocell B 600/30 (9004-34-6)
Arbogal (122-14-5)
Arboricid (93-79-8)
Arborol (534-52-1)
Arbotect (148-79-8)
Arbotect (28558-32-9)
Arbotect 20-S (28558-32-9)
Arbotect S (28558-32-9)
Arc 1G2 (911-65-9)
Arcacil (132-98-9)
Arcasin (132-98-9)
Arco-cee (50-81-7)
Archidyn (13292-46-1)
Arcoban (57-53-4)
Arcosolv (34590-94-8)
Arcotrate (78-11-5)
Arcton (75-46-7)

Arcton 0 (75-73-0)
Arcton 3 (75-72-9)
Arcton 4 (75-45-6)
Arcton 6 (75-71-8)
Arcton 7 (75-43-4)
Arcton 9 (75-69-4)
Arcton 12 (75-71-8)
Arcton 22 (75-45-6)
Arcton 33 (76-14-2)
Arcton 63 (76-13-1)
Arcton 114 (76-14-2)
Arctuvin (123-31-9)
Arcum R-S (50-55-5)
Ardall (54-21-7)
Ardap (52315-07-8)
Ardex (51-63-8)
Aredion (116-29-0)
Areginal (109-94-4)
Arelon (34123-59-6)
Arenka 900 (63428-84-2)
Arenka 930 (63428-84-2)
Aresin (1746-81-2)
Aresol (93-14-1)
Arest (12771-68-5)
Aretan (123-88-6)
Aretan 6 (123-88-6)
Aretit (2813-95-8)
Aretit (88-85-7)
Aretit (The phenol) (2813-95-8)
Arezin (1746-81-2)
Arezine (1746-81-2)
Arficin (13292-46-1)
Argamine (1119-34-2)
Argent fluorure (French) (7783-95-1)
Argentic fluoride (7783-95-1)
Argentum (7440-22-4)
Argezin (1912-24-9)
(L)-Arginine (74-79-3)
Arginine (74-79-3)
Arginine Glutamate (4320-30-3)
Arginine, L (74-79-3)
L-Arginine L-glutamate (1:1) (4320-30-3)
Arginine hydrochloride (1119-34-2)
L-Arginine hydrochloride (1119-34-2)
Arginine, hydrochloride, L- (1119-34-2)
Arginine monohydrochloride (1119-34-2)
L-Arginine, monohydrochloride (1119-34-2)
Arginine, monohydrochloride, L (1119-34-2)
Arginylglutamate (4320-30-3)
Argivene (1119-34-2)
Argon-40 (7440-37-1)
Argo Brand Corn Starch (9005-25-8)
Argon, Compressed [UN 1006] (7440-37-1)
Argon (DOT) (7440-37-1)
Argon, Refrigerated liquid [UN 1951] (7440-37-1)
Arheol (8006-87-9)
Ariavit Red 2G (3734-67-6)
Aribine (486-84-0)
Arichin (69-05-6)
Arien (64742-16-1)
Arilat (63-25-2)
Arilate (17804-35-2)
Arilate (63-25-2)
Ariotox (108-62-3)
Arisan (3766-60-7)
Aristocort (124-94-7)
Aristocort acetonide (76-25-5)
Aristoderm (76-25-5)
Aristogel (76-25-5)
Aristol (552-22-7)
Aristophyllin (479-18-5)
Arizole (104-46-1)
Arizole (8002-09-3)
Arklone P (76-13-1)
Arkofix NG (1854-26-8)
Arkon 115 (64742-16-1)
Arkon M 100 (64742-16-1)
Arkon P 70 (64742-16-1)
Arkopal N-090 (9016-45-9)
Arkopon (97-80-3)
Arkotine (50-29-3)
Arkozal (64-77-7)
Arlacel 40 (26266-57-9)
Arlacel 60 (1338-41-6)
Arlacel 80 (1338-43-8)
Arlacel 83 (8007-43-0)
Arlacel 85 (26266-58-0)
Arlacel 161 (31566-31-1)
Arlacel 165 (123-94-4)
Arlacel 169 (31566-31-1)
Arlacel C (8007-43-0)
Arlacide A (56-95-1)
Arlacide G (18472-51-0)
Arlanthol ASG (91-96-3)
Arlanthrene Golden Yellow (128-66-5)
Arlanthrene Violet 4R (1324-55-6)
Arlatone UVB (21245-02-3)
Arlocyanine Blue PS (147-14-8)
Arlosol Green B (128-80-3)
Arlosol Green BS (128-80-3)
Arlosol Green BSS (128-80-3)
Arlosol Yellow S (8003-22-3)
Armac 16D (2016-52-6)
Armac 18D (2190-04-7)
Armac OD (2190-04-7)
Armacide (54-85-3)
Armazal (140-87-4)
Armazide (54-85-3)
Armco iron (7439-89-6)
Armeen 1-7 (123-82-0)
Armeen 8 (111-86-4)
Armeen 8D (111-86-4)
Armeen 16D (143-27-1)
Armeen 118D (124-30-1)
Armeen DM-12D (112-18-5)
Armeen DM 14D (112-75-4)
Armeen DM 16D (112-69-6)
Armeen O (112-90-3)
Armenian Bole (1309-37-1)
Armine DM14D (112-75-4)
Armofos (7758-29-4)
Armol (55-55-0)
Armoslip CP (301-02-0)
Armostat 801 (31566-31-1)
Armotan MO (1338-43-8)
Armotan MS (1338-41-6)
Armotan PML-20 (9005-64-5)
Armotan PMO-20 (9005-65-6)
Armowax EBS-P (110-30-5)
Armstrong's Acid (81-04-9)
Armstrong's S Acid (81-04-9)
Arnaudon's Green (7789-04-0)
Arnaudon's Green (Hemiheptahydrate) (7789-04-0)
Arneel 8 (124-12-9)

Name	CAS No.
Arnite A	**(25038-59-9)**
Arnite A 200	**(25038-59-9)**
Arnite A-049000	**(25038-59-9)**
Arnite FP 800	**(25038-59-9)**
Arnite G	**(25038-59-9)**
Arnite G 600	**(25038-59-9)**
Aro	**(1333-86-4)**
Aroall	**(54-21-7)**
Arochlor 1242	**(53469-21-9)**
Arochlor 1254	**(11097-69-1)**
Arochlor 1260	**(11096-82-5)**
Arochlor 5460	**(11126-42-4)**
Aroclor	**(12767-79-2)**
Aroclor	**(1336-36-3)**
Aroclor 1016	**(12674-11-2)**
Aroclor 1221	**(11104-28-2)**
Aroclor 1221	**(1336-36-3)**
Aroclor 1232	**(11141-16-5)**
Aroclor 1232	**(1336-36-3)**
Aroclor 1240 (9CI)	**(71328-89-7)**
Aroclor 1242	**(1336-36-3)**
Aroclor 1242	**(53469-21-9)**
Aroclor 1248	**(12672-29-6)**
Aroclor 1248	**(1336-36-3)**
Aroclor 1254	**(11097-69-1)**
Aroclor 1254	**(1336-36-3)**
Aroclor 1254	**(27323-18-8)**
Aroclor 1260	**(11096-82-5)**
Aroclor 1260	**(1336-36-3)**
Aroclor 1262	**(1336-36-3)**
Aroclor 1268	**(11100-14-4)**
Aroclor 1268	**(1336-36-3)**
Aroclor 2565	**(1336-36-3)**
Aroclor 4465	**(11120-29-9)**
Aroclor 4465	**(1336-36-3)**
Aroclor 5442	**(12642-23-8)**
Aroclor 5442	**(1336-36-3)**
Aroclor 5460	**(11126-42-4)**
Aroflow	**(1333-86-4)**
Arogen	**(1333-86-4)**
Arolon	**(9003-01-4)**
(+)-Aromadendrene	**(489-39-4)**
Aromadendrene, (+)-	**(489-39-4)**
Aromadendrene (VAN)	**(489-39-4)**
Aromantrene Olive FR	**(2379-81-9)**
Aromatic Ammonia, Vaporole	**(7664-41-7)**
Aromatic Castor Oil	**(8001-79-4)**
Aromatic Oil Distillate	**(68602-80-2)**
Aromatic Petroleum Distillate	**(68602-80-2)**
Aromatic Petroleum Pitch	**(68187-58-6)**
Aromatic hydrocarbons, C9-11	**(70693-06-0)**
Aromatic naphtha, Type I	**(64742-95-6)**
Aromatic petroleum derivative solvent	**(68477-31-6)**
Aromatic solvent (Petroleum)	**(68477-31-6)**
Aromex	**(1333-86-4)**
Aromox DMMC-W	**(1643-20-5)**
Aron	**(9003-01-4)**
Aron A 10H	**(9003-01-4)**
Arosol	**(122-99-6)**
Arosurf	**(52292-17-8)**
Arosurf 1855E40	**(9004-99-3)**
Arosurf TA 100	**(107-64-2)**
Arotone	**(1333-86-4)**
Arovel	**(1333-86-4)**
Arpon	**(57-53-4)**
Arprocarb	**(114-26-1)**
Arquad 12	**(112-00-5)**
Arquad 12D	**(112-00-5)**
Arquad 12/50	**(112-00-5)**
Arquad 16-29	**(112-02-7)**
Arquad 16-50	**(112-02-7)**
Arquad 16/28	**(112-02-7)**
Arquad 18	**(112-03-8)**
Arquad 18-50	**(112-03-8)**
Arquad DM14B-90	**(139-08-2)**
Arquad DM18B-90	**(122-19-0)**
Arquad DMMCB-75	**(8001-54-5)**
Arquad R 40	**(107-64-2)**
Arquad T	**(8030-78-2)**
Arquad T-50	**(8030-78-2)**
Arresin	**(1746-81-2)**
Arrhenal	**(144-21-8)**
Arrivo	**(52315-07-8)**
Arrow	**(1333-86-4)**
Arrowroot Starch	**(9005-25-8)**
Arsambide	**(121-59-5)**
Arsamin	**(127-85-5)**
Arsan	**(75-60-5)**
Arsanilic acid-100	**(98-50-0)**
4-Arsanilic acid	**(98-50-0)**
Arsanilic-acid	**(98-50-0)**
p-Arsanilic acid	**(98-50-0)**
Arsanilic acid, N-carbamoyl	**(121-59-5)**
Arsanilic acid, monosodium salt	**(127-85-5)**
Arsanilic acid sodium salt	**(127-85-5)**
Arsecodile	**(124-65-2)**
Arsen (German,Polish)	**(7440-38-2)**
Arsenate	**(7778-39-4)**
Arsenate (8CI,9CI)	**(15584-04-0)**
Arsenate of iron, ferric	**(10102-49-5)**
Arsenate of iron, ferrous	**(10102-50-8)**
Arsenate of lead	**(7784-40-9)**
Arsenenic acid, calcium salt (9CI)	**(15194-99-7)**
Arsenenous acid, potassium salt	**(10124-50-2)**
Arsenenous acid, potassium salt (9CI)	**(13464-35-2)**
Arsenenous acid, sodium salt (9CI)	**(7784-46-5)**
Arsenenous acid, strontium salt, tetrahydrate	**(10378-48-0)**
Arseniate de calcium (French)	**(7778-44-1)**
Arseniate de magnesium (French)	**(10103-50-1)**
Arseniate de plomb (French)	**(7645-25-2)**
Arsenic-75	**(7440-38-2)**
Arsenic (ACGIH,OSHA) [UN 1558]	**(7440-38-2)**
Arsenic Black	**(7440-38-2)**
Arsenic Butter	**(7784-34-1)**
Arsenic Compound	**(52740-16-6)**
Arsenic, Metallic (DOT)	**(7440-38-2)**
Arsenic, Solid (DOT)	**(7440-38-2)**
Arsenic (V) oxide	**(1303-28-2)**
Arsenic, White, Solid (DOT)	**(1327-53-3)**
Arsenic Yellow	**(1303-33-9)**
Arsenic acid	**(1303-28-2)**
Arsenic-acid	**(7778-39-4)**
Arsenic acid (9CI)	**(7778-39-4)**
Arsenic acid (H_3AsO_4), cobalt(2+) salt (2:3) (9CI)	**(24719-19-5)**
Arsenic acid, Liquid (DOT)	**(7778-39-4)**
Arsenic acid, Liquid [UN 1553]	**(7778-39-4)**
Arsenic acid, Solid (DOT)	**(7778-39-4)**
Arsenic acid, Solid [UN 1554]	**(7778-39-4)**
Arsenic acid, Solution (DOT)	**(7778-39-4)**
Arsenic acid anhydride	**(1303-28-2)**
Arsenic acid, calcium salt(2:3)	**(7778-44-1)**
Arsenic acid, calcium salt (8CI)	**(15194-99-7)**
Arsenic acid, diammonium salt	**(7784-44-3)**
Arsenic acid, dimethyl ester (9CI)	**(34442-00-7)**

Arsenic acid, disodium salt (7778-43-0)
Arsenic acid, disodium salt, heptahydrate (10048-95-0)
Arsenic acid, ion(3-) (15584-04-0)
Arsenic acid, lead salt (7645-25-2)
Arsenic acid, lead(2+) salt (2:3) (3687-31-8)
Arsenic acid, lead(2+) salt(1:1) (7784-40-9)
Arsenic acid, magnesium salt (10103-50-1)
Arsenic acid, monopotassium salt (7784-41-0)
Arsenic acid, sodium salt (7631-89-2)
Arsenic acid, trisodium salt (8CI,9CI) (13464-38-5)
Arsenic acid, zinc salt (1303-39-5)
Arsenical Dust (8028-73-7)
Arsenical Flue Dust (8028-73-7)
Arsenicals (7440-38-2)
Arsenic anhydride (1303-28-2)
Arsenic blanc (French) (1327-53-3)
Arsenic(II) bromide (7784-33-0)
Arsenic bromide, Solid (DOT) (7784-33-0)
Arsenic bromide [UN 1555] (7784-33-0)
Arsenic(III) chloride (7784-34-1)
Arsenic chloride (DOT) (7784-34-1)
Arsenic chloride, Liquid (DOT) (7784-34-1)
Arsenic compound (10103-62-5)
Arsenic compound (13464-35-2)
Arsenic compound (15194-98-6)
Arsenic compound (27152-57-4)
Arsenic dichloroethane (598-14-1)
Arsenic disulfide (1303-32-8)
Arsenic fluoride (7784-35-2)
Arsenic hydrid (7784-42-1)
Arsenic hydride (7784-42-1)
Arsenic iodide, Solid [NA 1557] (7784-45-4)
Arsenic (III) oxide (1327-53-3)
Arsenic oxide (1303-28-2)
Arsenic oxide (1327-53-3)
Arsenic pentaoxide (1303-28-2)
Arsenic-pentoxide (1303-28-2)
Arsenic pentoxide, Solid (DOT) (1303-28-2)
Arsenic pentoxide [UN 1559] (1303-28-2)
Arsenic selenide (9CI) (1303-36-2)
Arsenic sesquioxide (1327-53-3)
Arsenic sesquisulfide (1303-33-9)
Arsenic sesquisulphide (1303-33-9)
Arsenic sulfide, Solid [UN 1557] (1303-33-9)
Arsenic sulfide Yellow (1303-33-9)
Arsenic sulphide [NA 1557] (1303-33-9)
Arsenic tersulphide (1303-33-9)
Arsenic tribromide (7784-33-0)
Arsenic trichloride, Liquid (DOT) (7784-34-1)
Arsenic trichloride [UN 1560] (7784-34-1)
Arsenic trifluoride (7784-35-2)
Arsenic trihydride (7784-42-1)
Arsenic triiodide (7784-45-4)
Arsenic trioxide (ACGIH) [UN 1561] (1327-53-3)
Arsenic trioxide, Solid (DOT) (1327-53-3)
Arsenic trisulfide (DOT) (1303-33-9)
Arsenicum album (1327-53-3)
Arsenigen saure (German) (1327-53-3)
Arsenious acid (1327-53-3)
Arsenious acid (H_3AsO_3), trisodium salt (8CI) (13464-37-4)
Arsenious acid, Solid (DOT) (1327-53-3)
Arsenious acid, calcium salt (27152-57-4)
Arsenious acid, copper(II) salt (1:1) (10290-12-7)
Arsenious acid, monosodium salt (7784-46-5)
Arsenious acid, potassium salt (10124-50-2)
Arsenious acid, sodium salt (7784-46-5)
Arsenious acid sodium salt (Na_3AsO_3) (13464-37-4)
Arsenious acid, strontium salt (10378-48-0)
Arsenious acid, strontium salt (91724-16-2)
Arsenious acid, strontium salt, tetrahydrate (10378-48-0)
Arsenious chloride (7784-34-1)
Arsenious oxide (1327-53-3)
Arsenious sulphide (1303-33-9)
Arsenious trioxide (1327-53-3)
Arsenite (1327-53-3)
Arsenite (15502-74-6)
Arsenite de potassium (French) (10124-50-2)
Arsenite de potassium (French) (13464-35-2)
Arsenite de sodium (French) (7784-46-5)
Arsenite de strontium (French) (91724-16-2)
Arsenito de estroncio (Spanish) (91724-16-2)
Arsenito potasico (Spanish) (13464-35-2)
Arseniuretted hydrogen (7784-42-1)
Arsenobetaine (64436-13-1)
Arsenolite (1327-53-3)
Arsenotrithious acid, tripropyl ester (9CI) (5582-57-0)
Arsenous acid (1327-53-3)
Arsenous acid anhydride (1327-53-3)
Arsenous acid, calcium salt (52740-16-6)
Arsenous acid, calcium salt (2:3) (27152-57-4)
Arsenous acid, trisodium salt (9CI) (13464-37-4)
Arsenous anhydride (1327-53-3)
Arsenous bromide (7784-33-0)
Arsenous chloride (7784-34-1)
Arsenous fluoride (7784-35-2)
Arsenous hydride (7784-42-1)
Arsenous iodide (7784-45-4)
Arsenous oxide (1327-53-3)
Arsenous oxide anhydride (1327-53-3)
Arsenous sulfide (1303-33-9)
Arsenous tribromide (7784-33-0)
Arsenous trichloride (9CI) (7784-34-1)
Arsenous triiodide (9CI) (7784-45-4)
Arsenous-trifluoride (7784-35-2)
Arsenowodor (Polish) (7784-42-1)
Arsentrioxide (1327-53-3)
Arsenwasserstoff (German) (7784-42-1)
Arsicodile (124-65-2)
Arsine (ACGIH,OSHA) [UN 2188] (7784-42-1)
Arsine, butylphenyl(phenylthio)- (8CI) (24582-61-4)
Arsine, chlorodiphenyl (712-48-1)
Arsine, (2-chlorovinyl)dichloro- (541-25-3)
Arsine, dichloro(2-chlorovinyl) (541-25-3)
Arsine, dichloroethyl (598-14-1)
Arsine, dichloromethyl- (593-89-5)
Arsine, dichlorophenyl (696-28-6)
Arsine, diethyl- (692-42-2)
Arsine, ethoxydiphenyl- (8CI) (24582-55-6)
Arsine, ethyl(isopropylthio)phenyl- (8CI) (23486-02-4)
Arsine, ethylmethoxyphenyl- (8CI) (24582-56-7)
Arsine, (ethylthio)diphenyl- (8CI) (24582-60-3)
Arsine, methoxydiphenyl- (8CI) (24582-54-5)
Arsine, methylthioxo (2533-82-6)
Arsine oxide, dimethylhydroxy (75-60-5)
Arsine oxide, dimethylhydroxy-, sodium salt (124-65-2)
Arsine oxide, triphenyl- (9CI) (1153-05-5)
Arsine, triphenyl- (9CI) (603-32-7)
Arsinette (3687-31-8)
Arsinette (7784-40-9)
Arsinic acid, dimethyl- (9CI) (75-60-5)
Arsinic acid, dimethyl-, iron(3+) salt (5968-84-3)
Arsinic acid, dimethyl-, sodium salt (9CI) (124-65-2)
Arsinosolvin (127-85-5)
Arsinothious acid, ethylphenyl-, 1-methylethyl ester (9CI) (23486-02-4)

Arsinous acid, diphenyl-, ethyl ester (9CI) **(24582-55-6)**
Arsinous acid, diphenyl-, methyl ester (9CI) **(24582-54-5)**
Arsinous acid, ethylphenyl-, methyl ester (9CI) **(24582-56-7)**
Arsinous chloride, diphenyl- (9CI) **(712-48-1)**
Arsinyl **(144-21-8)**
Arsodent **(1327-53-3)**
Arsolen **(64742-16-1)**
Arsonate Liquid **(2163-80-6)**
Arsonic acid, (4-aminophenyl)- (9CI) **(98-50-0)**
Arsonic acid, (4-aminophenyl)-, monosodium salt (9CI) **(127-85-5)**
Arsonic acid, calcium salt (1:1) **(52740-16-6)**
Arsonic acid, copper(2+) salt (1:1) (9CI) **(10290-12-7)**
Arsonic acid, (4-hydroxy-3-nitrophenyl)- **(121-19-7)**
Arsonic acid, methyl-, disodium salt **(144-21-8)**
Arsonic acid, methyl-, monosodium salt **(2163-80-6)**
Arsonic acid, potassium salt **(10124-50-2)**
Arsonium, (carboxymethyl)trimethyl-, hydroxide, inner salt (9CI) **(64436-13-1)**
Arsonodithious acid, phenyl-, diethyl ester (9CI) **(5582-58-1)**
Arsonodithious acid, phenyl-, diphenyl ester (9CI) **(24582-59-0)**
p-Arsonophenylurea **(121-59-5)**
Arsonous acid, phenyl-, diethyl ester (9CI) **(3141-11-5)**
Arsonous acid, phenyl-, dimethyl ester (9CI) **(24582-52-3)**
Arsonous dichloride, (2-chloroethenyl)- (9CI) **(541-25-3)**
Arsonous dichloride, ethyl- (9CI) **(598-14-1)**
Arsonous dichloride, methyl- (9CI) **(593-89-5)**
Arsonous dichloride, phenyl- (9CI) **(696-28-6)**
Arsycodile **(124-65-2)**
Arsynal **(144-21-8)**
Artam **(132-60-5)**
Arterenol **(51-41-2)**
l-Arterenol **(51-41-2)**
dl-Arterenol hydrochloride **(55-27-6)**
Arterioflexin **(637-07-0)**
Arterocyn **(333-20-0)**
Arterosol **(637-07-0)**
Artes **(637-07-0)**
Artevil **(637-07-0)**
Artexin **(132-60-5)**
D-Arthin **(50-14-6)**
Artho LM **(86-85-1)**
Arthodibrom **(300-76-5)**
Arthrochin **(54-05-7)**
Arthrocine **(38194-50-2)**
Artic **(74-87-3)**
Artificial Almond Oil **(100-52-7)**
Artificial Ant Oil **(98-01-1)**
Artificial Barite **(7727-43-7)**
Artificial Cinnamon Oil **(8007-80-5)**
Artificial Essential Oil of Almond **(100-52-7)**
Artificial Heavy Spar **(7727-43-7)**
Artificial Mustard Oil **(57-06-7)**
Artificial Oil of Ants **(98-01-1)**
Artificial Oil of Mustard **(57-06-7)**
Artificial Silk Black G **(6428-31-5)**
Artificial Silk Black GN **(6428-31-5)**
Artificial Silk Black GR **(6428-31-5)**
Artificial Sweetening Substanz Gendorf 450 **(128-44-9)**
Artisil Blue BGL **(12222-78-5)**
Artisil Blue BRP **(2475-44-7)**
Artisil Blue BSG **(2475-46-9)**
Artisil Blue BSG **(86722-66-9)**
Artisil Blue BSQ **(2475-46-9)**
Artisil Blue BSQ **(86722-66-9)**
Artisil Blue SAP **(2475-45-8)**
Artisil Blue SAP Conc **(2475-45-8)**
Artisil Brilliant Rose 5BP **(2872-48-2)**
Artisil Direct Red 3BP **(116-85-8)**
Artisil Direct Violet 2RP **(128-95-0)**
Artisil Direct Yellow G **(2832-40-8)**
Artisil Orange 3RP **(82-28-0)**
Artisil Red 3BP **(116-85-8)**
Artisil Violet 2RP **(128-95-0)**
Artisil Yellow G **(2832-40-8)**
Artisil Yellow 2GN **(2832-40-8)**
Arto-Espasmol **(456-59-7)**
Artolon **(57-53-4)**
Artomycin **(64-75-5)**
Artosin **(64-77-7)**
Artozin **(64-77-7)**
Artracin **(53-86-1)**
Artrichin **(54-05-7)**
Artril 300 **(15687-27-1)**
Artrinovo **(53-86-1)**
Artriona **(50-04-4)**
Artrivia **(53-86-1)**
Artrizin **(50-33-9)**
Artrizone **(50-33-9)**
Artroflog **(129-20-4)**
Artropan **(50-33-9)**
Artslon **(63428-84-2)**
Arumel **(51-21-8)**
Arusal **(78-44-4)**
Arvynol **(113-18-8)**
Arwood Copper **(7440-50-8)**
Arylam **(63-25-2)**
Arylamide Yellow G **(2512-29-0)**
Aryltriazene methoxypyrimidine **(103947-07-5)**
Asabaine **(53-46-3)**
Asagraea Officinalis **(8051-02-3)**
Asagran **(50-78-2)**
Asahifron 113 **(76-13-1)**
Asahisol 1527 **(9003-20-7)**
Asamid **(77-67-8)**
Asana **(66230-04-4)**
Asana XL **(66230-04-4)**
Asaraldehyde **(4460-86-0)**
Asaron **(2883-98-9)**
Asaronaldehyde **(4460-86-0)**
Asarone **(2883-98-9)**
Asarone, trans- **(2883-98-9)**
α-Asarone **(2883-98-9)**
trans-Asarone **(2883-98-9)**
Asarum Camphor **(2883-98-9)**
Asatard **(50-78-2)**
Asazol **(2163-80-6)**
Asbest (German) **(1332-21-4)**
Asbestine **(14807-96-6)**
7-45 Asbestos **(12001-29-5)**
Asbestos (ACGIH,DOT,OSHA) **(1332-21-4)**
Asbestos (ACGIH) **(12001-28-4)**
Asbestos (ACGIH) **(12001-29-5)**
Asbestos (ACGIH) **(12172-73-5)**
Asbestos (ACGIH) **(77536-66-4)**
Asbestos (ACGIH) **(77536-67-5)**
Asbestos (ACGIH) **(77536-68-6)**
Asbestos Fiber **(1332-21-4)**
Asbestos Fibre **(1332-21-4)**
Asbestos, White (DOT) **(12001-29-5)**
Asbestos, actinolite **(77536-66-4)**
Asbestos, amosite **(12172-73-5)**
Asbestos, anthophylite **(77536-67-5)**
Asbestos, chrysotile **(12001-29-5)**
Asbestos, crocidolite **(12001-28-4)**
Asbestos, tremolite **(77536-68-6)**

Ascabin (120-51-4)
Ascabiol (120-51-4)
Ascaridol (512-85-6)
Ascaridole (DOT) (512-85-6)
Ascarisin (512-85-6)
Ascarol 2E (18181-80-1)
Ascaryl (136-77-6)
Ascor-B.I.D. (50-81-7)
Ascoltin (50-81-7)
Ascophen (8003-03-0)
Ascorb (50-81-7)
Ascorbajen (50-81-7)
Ascorbate (50-81-7)
Ascorbicab (50-81-7)
Ascorbic acid (50-81-7)
L(+)-Ascorbic acid (50-81-7)
L-Ascorbic acid (8CI,9CI) (50-81-7)
Ascorbic acid calcium salt (5743-27-1)
L-Ascorbic acid, calcium salt (2:1) (9CI) (5743-27-1)
L-Ascorbic acid, monosodium salt (134-03-2)
L-Ascorbic acid, 6-palmitate (137-66-6)
Ascorbic acid sodium salt (134-03-2)
l-Ascorbic acid sodium salt (134-03-2)
Ascorbicap (50-81-7)
Ascorbicin (134-03-2)
Ascorbin (134-03-2)
Ascorbin (50-81-7)
Ascorbutina (50-81-7)
Ascorbyl palmitate (137-66-6)
Ascorin (50-81-7)
Ascorteal (50-81-7)
Ascorvit (50-81-7)
Ascoserp (50-55-5)
Ascoserpina (50-55-5)
Ascotoxin (20350-15-6)
Asculetine (305-01-1)
Ascumar (152-72-7)
Asebotoxin (4720-09-6)
Asepsin (103-88-8)
Asepta Herban (18530-56-8)
Aseptoform (99-76-3)
Aseptoform E (120-47-8)
Aseptoform P (94-13-3)
Aseptoform butyl (94-26-8)
Asex (7775-09-9)
Ashes (Residues) (68131-74-8)
Ashland 10303 (17341-40-1)
Ashlene (63428-83-1)
Asidon 3 (50-35-1)
Asilan (3337-71-1)
Asmadion (50-35-1)
Asmatane Mist (51-42-3)
Asmatane mist (51-43-4)
Asmaval (50-35-1)
Asozin (2533-82-6)
Aspalon (50-78-2)
(-)-Asparagine (70-47-3)
L-(+)-Asparagine (70-47-3)
Asparagine, L- (8CI) (70-47-3)
L-Asparagine (9CI) (70-47-3)
Asparagine (VAN) (70-47-3)
L-β-Asparagine (VAN) (70-47-3)
Asparagine, N(2)-L-α-aspartyl-, L- (8CI) (13433-11-9)
L-Asparagine, N(2)-L-α-aspartyl- (9CI) (13433-11-9)
L-Asparagine, N,N2-bis(3-(9-octadecenylamino)propyl)-, (Z,Z)- (9CI) (72050-94-3)
Aspartame (22839-47-0)
(l)-Aspartic acid (56-84-8)
Aspartic acid, L (56-84-8)
Aspartic acid, N-carboxy-, N-benzyl ester, L- (8CI) (1152-61-0)
L-Aspartic acid, N-(3-carboxy-1-oxo-2-sulfopropyl)-N-octadecyl-, tetrasodium salt (9CI) (3401-73-8)
L-Aspartic acid, N-(3-carboxy-1-oxosulfopropyl)-N-octadecyl-, tetrasodium salt (3401-73-8)
L-Aspartic acid, N-((phenylmethoxy)carbonyl)- (9CI) (1152-61-0)
Aspartylphenylalanine methyl ester (22839-47-0)
Asperentin (35818-31-6)
Aspergillin (67-99-2)
Aspergum (50-78-2)
Asphalen Sulphur Black C (1326-82-5)
Asphalen Sulphur Black S (1326-82-5)
Asphalt (8052-42-4)
Asphalt, At or above its flashpoint [NA 1999] (8052-42-4)
Asphalt, Cut back (DOT) (8052-42-4)
Asphalt, Fumes (ACGIH) (8052-42-4)
Asphalt, Oxidized (64742-93-4)
Asphalt, Petroleum (8052-42-4)
Asphalt oxide (64742-93-4)
Asphaltum (8052-42-4)
Aspiral (110-46-3)
Aspirdrops (50-78-2)
Aspirin (50-78-2)
Aspirin (OSHA) (50-78-2)
Aspirin, Phenacetin and caffeine (8003-03-0)
Aspirin aluminium (23413-80-1)
Aspirine (50-78-2)
Aspon-chlordane (57-74-9)
Aspon (3244-90-4)
Aspor (12122-67-7)
Asporum (12122-67-7)
Aspro (50-78-2)
Aspron (561-27-3)
Assiflavine (8048-52-0)
Assival (439-14-5)
Assugrin (139-05-9)
Assure (76578-14-8)
Assurgrin Feinsuss (139-05-9)
Assurgrin Vollsuss (139-05-9)
Asta Z 4942 (3778-73-2)
Astedin (60-13-9)
Asteric (50-78-2)
Asthenthilo (71-63-6)
Asthma Meter Mist (51-43-4)
Asthma-Nefrin (51-43-4)
Astmahalin (51-43-4)
Astmamasit (479-18-5)
Astminhal (51-43-4)
Astra 1512 (721-50-6)
Astra 1515 (721-50-6)
Astra Chrysoidine R (532-82-1)
Astra Diamond Green GX (633-03-4)
Astra Fuchsine B (632-99-5)
Astra Malachite Green (569-64-2)
Astra Malachite Green B (569-64-2)
Astra Malachite Green BXX (569-64-2)
Astracillin (132-93-4)
Astrafer (9004-51-7)
Astrazolo (127-69-5)
Astrazon Blue BG (55840-82-9)
Astrazon Blue BGE/X (55840-82-9)
Astrazon Blue FGGL (12270-13-2)
Astrazon Blue FGL (12217-41-3)
Astrazon Blue FRR (12235-47-1)
Astrazon Orange (3056-93-7)

Astrazon Orange G	(3056-93-7)	Athebrate	(637-07-0)
Astrazon Red GTL	(14097-03-1)	Atheromide	(637-07-0)
Astreptine	(63-74-1)	Atheropront	(637-07-0)
Astress	(604-75-1)	Athranid-wirkstoff	(637-07-0)
Astringen	(12042-91-0)	Athrombin	(129-06-6)
Astrobain	(630-60-4)	Athrombin-K	(81-81-2)
Astrobot	(62-73-7)	Athrombine-K	(81-81-2)
Astrocar	(59-26-7)	Athrombon	(83-12-5)
Astrocid	(63-74-1)	Athyl-Gusathion	(2642-71-9)
Astrophyllin	(479-18-5)	Athylen (German)	(74-85-1)
Astrumal	(7778-74-7)	Athylenglykol (German)	(107-21-1)
Asturidon	(143-81-7)	Athylenglykol-monoathylather (German)	(110-80-5)
Asuccin	(110-15-6)	Athyromazole	(22232-54-8)
Asucrol	(94-20-2)	Atic Vat Blue XRN	(81-77-6)
Asugryn	(139-05-9)	Atic Vat Brilliant Purple 4R	(1324-55-6)
Asulam	(3337-71-1)	Atic Vat Jade Green XBN	(128-58-5)
Asulam sodium	(2302-17-2)	Atic Vat Olive R	(2379-81-9)
Asulam, sodium salt	(2302-17-2)	Atic Vat Printing Jade Green XBN	(128-58-5)
Asulam-sodium	(2302-17-2)	Atigoa	(132-60-5)
Asulfidine	(599-79-1)	Atilen	(439-14-5)
Asulfox F	(3337-71-1)	Atiran	(123-88-6)
Asulox	(2302-17-2)	Ativan	(846-49-1)
Asulox 40	(3337-71-1)	Atlacide	(7775-09-9)
Asulox, sodium salt	(2302-17-2)	Atladiol	(979-32-8)
Asunthol	(56-72-4)	Atlantic	(1333-86-4)
Asuntol	(56-72-4)	Atlantic Artificial Silk Black G	(6428-31-5)
Asymmetrical trimethylbenzene	(95-63-6)	Atlantic Black BD	(1937-37-7)
Atabrine	(69-05-6)	Atlantic Black C	(1937-37-7)
Atabrine	(83-89-6)	Atlantic Black E	(1937-37-7)
Atabrine dihydrochloride	(69-05-6)	Atlantic Black EA	(1937-37-7)
Atabrine hydrochloride	(69-05-6)	Atlantic Black GAC	(1937-37-7)
Atacryl Blue 3G	(55840-82-9)	Atlantic Black GG	(1937-37-7)
Atactic Poly(acrylic acid)	(9003-01-4)	Atlantic Black GXCW	(1937-37-7)
Atactic Polypropylene	(9003-07-0)	Atlantic Black GXOO	(1937-37-7)
Atactic Polystyrene	(9003-53-6)	Atlantic Black RW	(2429-83-6)
Atactic butadiene polymer	(9003-17-2)	Atlantic Black SD	(1937-37-7)
Atalco C	(36653-82-4)	Atlantic Blue 2B	(2602-46-2)
Atalco O	(143-28-2)	Atlantic Brown BCW	(2429-81-4)
Atalco S	(112-92-5)	Atlantic Brown BP	(2429-81-4)
Atara	(68-88-2)	Atlantic Brown D 3Y	(2586-58-5)
Atarax	(68-88-2)	Atlantic Brown M	(2429-82-5)
Ataraxine	(57-53-4)	Atlantic Congo Red	(573-58-0)
Ataraxoid	(68-88-2)	Atlantic Dark Green	(3626-28-6)
Atarazoid	(68-88-2)	Atlantic Fast Brown BRL	(16071-86-6)
Atars	(68-26-8)	Atlantic Green WT	(3626-28-6)
Atav	(68-26-8)	Atlantic Resin Fast Blue	(2610-05-1)
Atazina	(68-88-2)	Atlantic Resin Fast Brown BRL	(16071-86-6)
Atazinax	(1912-24-9)	Atlantic Scarlet 3B	(6358-29-8)
Atcotibine	(54-85-3)	Atlantic Sky Blue A	(2429-74-5)
Atebrin	(69-05-6)	Atlantic Sky Blue 6B	(2610-05-1)
Atebrin	(83-89-6)	Atlantic Sky Blue FF	(2610-05-1)
Atebrine	(69-05-6)	Atlantic Stilbene Yellow GA	(1325-37-7)
Atebrine	(83-89-6)	Atlantic Violet N	(2586-60-9)
Atebrin hydrochloride	(69-05-6)	Atlantichrome Violet B	(2092-55-9)
Ateculon	(637-07-0)	Atlas "A"	(7784-46-5)
Atemorin	(77-75-8)	Atlas G-263	(78-21-7)
Atempol	(77-75-8)	Atlas G 924	(1323-39-3)
Atenolol	(29122-68-7)	Atlas G-2124	(141-20-8)
Atensine	(439-14-5)	Atlas G-2133	(9002-92-0)
d-Ate phenyl 747	(51-63-8)	Atlas G-2142	(9004-96-0)
Aterax	(68-88-2)	Atlas G-2144	(9004-96-0)
Aterian	(57-67-0)	Atlas G 2146	(106-11-6)
Ateriosan	(637-07-0)	Atlas G-3705	(9002-92-0)
Aterocyn	(333-20-0)	Atlas White Titanium Dioxide	(13463-67-7)
Atgard	(62-73-7)	Atlasetox	(8065-62-1)
Atgard C	(62-73-7)	Atlatest	(315-37-7)
Atgard V	(62-73-7)	Atlox 1087	(9005-65-6)

Name	CAS No.
Atlox 3409F (9CI)	**(70225-91-1)**
Atlox 8916TF	**(9005-65-6)**
Atmonil	**(548-62-9)**
Atmos 150	**(31566-31-1)**
Atmospheric Tower Residue (Reduced Crude Oil)	**(64741-45-3)**
Atmospheric gas oil (petroleum)	**(64741-41-9)**
Atmul 124	**(31566-31-1)**
Atmul 67	**(31566-31-1)**
Atmul 84	**(31566-31-1)**
Atocin	**(132-60-5)**
Atofan	**(132-60-5)**
Atomit	**(471-34-1)**
Atophan	**(132-60-5)**
Atorel	**(58-63-9)**
Atosil	**(58-33-3)**
Atosil	**(60-87-7)**
Atoxan	**(63-25-2)**
Atoxicocaine	**(51-05-8)**
Atoxyl	**(127-85-5)**
Atoxyl	**(98-50-0)**
Atoxylic acid	**(98-50-0)**
Atranex	**(1912-24-9)**
Atrasine	**(1912-24-9)**
Atratol	**(7775-09-9)**
Atratol A	**(1912-24-9)**
Atraton	**(1610-17-9)**
Atratone	**(1610-17-9)**
Atraxin	**(57-53-4)**
Atrazin	**(1912-24-9)**
Atrazine (ACGIH,OSHA)	**(1912-24-9)**
Atrazine-Gesagard Mixt.	**(8073-77-6)**
Atrazine-Prometryn Mixt.	**(8073-77-6)**
Atrazine-Prometryne Mixt.	**(8073-77-6)**
Atred	**(1912-24-9)**
Atrex	**(1912-24-9)**
Atrinal	**(52508-35-7)**
Atriphos	**(56-65-5)**
Atrivyl	**(443-48-1)**
Atrochin	**(51-34-3)**
Atrolactic acid	**(515-30-0)**
Atrolen	**(637-07-0)**
Atromid	**(637-07-0)**
Atromid S	**(637-07-0)**
Atromida	**(637-07-0)**
Atromidin	**(637-07-0)**
Atropette	**(55-48-1)**
Atropin (German)	**(51-55-8)**
Atropina (Italian)	**(51-55-8)**
(-)-Atropine	**(101-31-5)**
Atropine	**(51-55-8)**
Atropine methobromide	**(2870-71-5)**
Atropine methylbromide	**(2870-71-5)**
Atropine sulfate	**(55-48-1)**
Atropine, sulfate (2:1)	**(55-48-1)**
Atropine sulphate	**(55-48-1)**
Atropin-flexiolen	**(51-55-8)**
Atropinium sulfate	**(55-48-1)**
Atropinol	**(51-55-8)**
Atropinsal	**(55-48-1)**
Atropin siran (Czech)	**(55-48-1)**
Atropinsulfat (German)	**(55-48-1)**
Atropiny siarczan (Polish)	**(55-48-1)**
Atropisal	**(55-48-1)**
Atropisol	**(51-55-8)**
Atropisol	**(55-48-1)**
Atroponitrile (6CI,7CI,8CI)	**(495-10-3)**
Atroquin	**(51-34-3)**
Atroton	**(1610-17-9)**
Atrovis	**(637-07-0)**
Attac 4-2	**(8001-35-2)**
Attac 4-4	**(8001-35-2)**
Attac 6	**(8001-35-2)**
Attac 6-3	**(8001-35-2)**
Attac 8	**(8001-35-2)**
Attaclay	**(12174-11-7)**
Attaclay X 250	**(12174-11-7)**
Attacote	**(12174-11-7)**
Attagel	**(12174-11-7)**
Attagel 40	**(12174-11-7)**
Attagel 50	**(12174-11-7)**
Attagel 150	**(12174-11-7)**
Attapulgite	**(12174-11-7)**
Attapulgite (VAN)	**(12174-11-7)**
Attasorb	**(12174-11-7)**
Atul Acid Black 10BX	**(1064-48-8)**
Atul Acid Black BX	**(1064-48-8)**
Atul Acid Crystal Orange G	**(1936-15-8)**
Atul Acid Crystal Red	**(3567-69-9)**
Atul Acid Fast Red A	**(1658-56-6)**
Atul Acid Geranine G	**(3734-67-6)**
Atul Acid Orange II	**(633-96-5)**
Atul Acid Scarlet 3R	**(2611-82-7)**
Atul Brilliant Oil Yellow G	**(1689-82-3)**
Atul Congo Red	**(573-58-0)**
Atul Crystal Red F	**(3567-69-9)**
Atul Developed Black BT	**(2429-73-4)**
Atul Direct Black E	**(1937-37-7)**
Atul Direct Blue 2B	**(2602-46-2)**
Atul Direct Brown CN	**(2586-58-5)**
Atul Direct Brown MR	**(2429-82-5)**
Atul Direct Dark Green P	**(3626-28-6)**
Atul Direct Red 4B	**(992-59-6)**
Atul Direct Sky Blue	**(2429-74-5)**
Atul Direct Sky Blue FB	**(2610-05-1)**
Atul Direct Violet N	**(2586-60-9)**
Atul Fast Yellow R	**(60-11-7)**
Atul Indigo Carmine	**(860-22-0)**
Atul Oil Orange T	**(2646-17-5)**
Atul Oil Red G	**(85-86-9)**
Atul Orange R	**(842-07-9)**
Atul Pigment Red RS Sodium salt	**(1248-18-6)**
Atul Scarlet F	**(2611-82-7)**
Atul Sulfur Navy Blue	**(1327-57-7)**
Atul Sulphur Black GXE	**(1326-82-5)**
Atul Sulphur Black GXR	**(1326-82-5)**
Atul Sulphur Black RP	**(1326-82-5)**
Atul Sunset Yellow FCF	**(2783-94-0)**
Atul Tartrazine	**(1934-21-0)**
Atul Vulcan Fast Pigment Orange G	**(3520-72-7)**
Atussil	**(77-23-6)**
Atysmal	**(77-67-8)**
Aubepine	**(123-11-5)**
Aubygel GS	**(9000-07-1)**
Aubygel X 52	**(9062-07-1)**
Aubygum DM	**(9000-07-1)**
Aules	**(137-26-8)**
Auligen	**(502-55-6)**
Aulinogen	**(502-55-6)**
Auramine	**(492-80-8)**
Auramine A1	**(2465-27-2)**
Auramine Base	**(492-80-8)**
Auramine Conc. Specially soluble in spirit	**(2465-27-2)**
Auramine Extra	**(2465-27-2)**
Auramine Extra Conc. A	**(2465-27-2)**

Auramine FA	(2465-27-2)	Avantyl	(72-69-5)
Auramine (Free base)	(492-80-8)	Avazyme	(9004-07-3)
Auramine Hydrochloride	(2465-27-2)	δ(5)-Avenasterol	(18472-36-1)
Auramine Lake Yellow O	(2465-27-2)	Avenge	(43222-48-6)
Auramine N	(2465-27-2)	Aventox	(1912-26-1)
Auramine N Base	(492-80-8)	Aventyl	(72-69-5)
Auramine O	(2465-27-2)	Avermectin B1A	(65195-55-3)
Auramine OAF	(492-80-8)	Avermectin A1A, 5-O-demethyl	(65195-55-3)
Auramine O Base	(492-80-8)	Avermin	(548-62-9)
Auramine O (Biological stain)	(2465-27-2)	Aversan	(97-77-8)
Auramine O Extra Conc. A export	(2465-27-2)	Averzan	(97-77-8)
Auramine ON	(2465-27-2)	Aviamide-6	(25038-54-4)
Auramine OO	(2465-27-2)	Avibest	(1343-88-0)
Auramine OO	(492-80-8)	Avibest C	(12001-29-5)
Auramine OOO	(2465-27-2)	Avibon	(68-26-8)
Auramine OS	(2465-27-2)	Avicade	(52315-07-8)
Auramine Pure	(2465-27-2)	Avicel	(9004-34-6)
Auramine SP	(2465-27-2)	Avicel 101	(9004-34-6)
Auramine SS	(492-80-8)	Avicel 102	(9004-34-6)
Auramine Yellow	(2465-27-2)	Avicel PH 101	(9004-34-6)
Auranile	(57-41-0)	Avicel PH 105	(9004-34-6)
Auranile	(630-93-3)	Avicol	(82-68-8)
Aurantia	(131-73-7)	Avil	(86-21-5)
Aureine	(130-01-8)	Avil-Retard	(132-20-7)
Auremine	(492-80-8)	Aviocaffaro	(1332-65-6)
Aureocina	(57-62-5)	Aviocaffaro PF	(1332-65-6)
Aureofungin	(8065-41-6)	Aviomarin	(523-87-5)
Aureolic acid	(18378-89-7)	Avirol 101	(151-21-3)
Aureomycin	(57-62-5)	Avirol 118 Conc	(151-21-3)
Aureomycin A-377	(57-62-5)	Avirol 100E	(9004-82-4)
Aureomykoin	(57-62-5)	Avirosan	(22936-75-0)
Aureotan	(12192-57-3)	Avirosan	(24151-93-7)
Auric chloride	(13453-07-1)	Avisun	(9003-07-0)
Aurin	(603-45-2)	Avisun 101	(9003-07-0)
Auripigment	(1303-33-9)	Avisun 12-270A	(9003-07-0)
Aurlelic acid	(18378-89-7)	Avisun 12-407A	(9003-07-0)
Auromyose	(12192-57-3)	Avita	(68-26-8)
Auropan	(57-33-0)	Avitex C	(1120-01-0)
Aurora Yellow	(1306-23-6)	Avitex SF	(1120-01-0)
Aurotan	(12192-57-3)	Avitol	(68-26-8)
1-Aurothio-D-glucopyranose	(12192-57-3)	Avitrol	(504-24-5)
Aurothioglucose	(12192-57-3)	Avitrol 100	(1124-33-0)
Aurumine	(12192-57-3)	Avitrol 200	(504-24-5)
Austiox	(13463-67-7)	Avivan SO 6	(9005-00-9)
Austracil	(56-75-7)	Avlochlor	(54-05-7)
Austracol	(56-75-7)	Avloclor	(54-05-7)
Australian Gum	(9000-01-5)	Avlon	(8048-52-0)
Australol	(99-89-8)	Avlon	(86-40-8)
Austranal	(504-17-6)	Avloprocil	(6130-64-9)
Austrapen	(69-53-4)	Avlosulfon	(80-08-0)
Austrapine	(50-55-5)	Avlosulphone	(80-08-0)
Austrian cinnabar	(18454-12-1)	Avlothane	(67-72-1)
Austrominal	(50-06-6)	Avolin	(131-11-3)
Austrovit PP	(98-92-0)	Avomine	(60-87-7)
Autan	(134-62-3)	Avon Green A-4379	(633-03-4)
Autarite	(7789-80-2)	Axerophthol	(68-26-8)
Authron	(12192-57-3)	Axiom	(1757-18-2)
Automin	(58-73-1)	Axion	(13422-51-0)
Automobiles, Fuel systems and fuels (VAN)	(86290-81-5)	Axiten	(64-55-1)
Automotive Diesel Oil	(68334-30-5)	Axlon	(13422-51-0)
Auxietil	(57-53-4)	Axuris	(548-62-9)
Auxinutril	(115-77-5)	Ayermate	(57-53-4)
Avadex	(2303-16-4)	Ayerst 62013	(1954-28-5)
Avadex BW	(2303-17-5)	Ayfivin	(1405-87-4)
Avagal	(53-46-3)	10-Azaanthracene	(260-94-6)
Avantin	(67-63-0)	9-Azaanthracene	(260-94-6)
Avantine	(67-63-0)	5-Aza-10-arsenaanthracene chloride	(578-94-9)

1-Azabenz(a)anthracene **(84-56-0)**
12-Azabenz(a)anthracene **(225-51-4)**
7-Azabenz(a)anthracene **(225-11-6)**
Azabenzene **(110-86-1)**
2-Azabenzo(b)pyrene **(189-90-2)**
3-Azabenzonitrile **(100-54-9)**
3-Azabicyclo(3.1.0)hexane-2,4-dione, 3-(3,5-dichlorophenyl)-1,5-dimethyl- **(32809-16-8)**
1-Azabicyclo(4.3.0)nonane **(13618-93-4)**
1-Azabicyclo(3.3.0)octane **(643-20-9)**
1-Azabicyclo(2.2.2)octane (9CI) **(100-76-5)**
8-Azabicyclo(3.2.1)octane-2-carboxylic acid, 3-(benzoyloxy)-8-methyl-, methyl ester, (1R-(exo,exo))- (9CI) **(50-36-2)**
8-Azabicyclo(3.2.1)octanol, 8-methyl- **(120-29-6)**
8-Azabicyclo(3.2.1)octan-3-ol, 8-methyl-, endo- (9CI) **(120-29-6)**
1-Azabicyclo(2.2.2)octan-3-one (9CI) **(3731-38-2)**
1-Azachrysene **(218-08-6)**
2-Azachrysene **(218-02-0)**
Azacitidine **(320-67-2)**
Azaconazol (Spanish) **(60207-31-0)**
Azaconazole **(60207-31-0)**
Azaconazolum (Latin) **(60207-31-0)**
Azacosterol dihydrochloride **(1249-84-9)**
Azacosterol hydrochloride **(1249-84-9)**
Azacyclodecane **(4396-27-4)**
1-Azacycloheptane **(111-49-9)**
Azacycloheptane **(111-49-9)**
1-Aza-2-cycloheptanone **(105-60-2)**
2-Azacycloheptanone **(105-60-2)**
Azacyclohexane **(110-89-4)**
1-Aza-2,4-cyclopentadiene **(109-97-7)**
Azacyclopentane **(123-75-1)**
2-Azacyclopropa(cd)pentalene, octahydro-2-methyl- (8CI) **(16967-61-6)**
Azacyclopropane **(151-56-4)**
2-Azacyclotridecanone **(947-04-6)**
Azacyclotridecan-2-one **(947-04-6)**
5'-Azacytidine **(320-67-2)**
5-Azacytidine **(320-67-2)**
6-Azacytidine **(3131-60-0)**
Azacytidine **(320-67-2)**
5-Azacytosine **(931-86-2)**
5-Aza-2'-deoxycytidine **(2353-33-5)**
5-Azadeoxycytidine **(2353-33-5)**
7-Azadibenz(a,h)anthracene **(226-36-8)**
14-Azadibenz(a,j)anthracene **(224-53-3)**
7-Azadibenz(a,j)anthracene **(224-42-0)**
5-Azadibenzo(a,e)cycloheptatriene **(256-96-2)**
7-Aza-7H-dibenzo(c,g)fluorene **(194-59-2)**
Azadieno **(33089-61-1)**
9-Aza-1,17-diguanidinoheptadecane triacetate **(57520-17-9)**
1-Aza-5-ethyl-3,7-dioxabicyclo(3.3.0)octane **(7747-35-5)**
1-Azafluoranthene **(206-56-4)**
3-Azafluoranthene **(206-55-3)**
7-Azafluoranthene **(206-49-5)**
Azafluoranthene (9CI) **(89126-45-4)**
4-Azafluorene **(244-99-5)**
9-Azafluorene **(86-74-8)**
Azafluorene (9CI) **(97485-90-0)**
8-Azaguanine **(134-58-7)**
Azaguanine **(134-58-7)**
Azaguanine-8 **(134-58-7)**
1-Azaindene **(120-72-9)**
2-Azaindole **(271-44-3)**
3-Azaindole **(51-17-2)**
4-Azaindole **(272-49-1)**
5-Azaindole **(271-34-1)**
6-Azaindole **(271-29-4)**
7-Azaindoline **(10592-27-5)**
Azak **(1918-11-2)**
Azamin 4B **(992-59-6)**
Azamun (Czech) **(446-86-6)**
Azan **(134-58-7)**
1-Azanaphthalene **(91-22-5)**
2-Azanaphthalene **(119-65-3)**
Azanil Red Salt TRD **(3165-93-3)**
Azanin **(446-86-6)**
Azanol Fast Acid Black 10B **(1064-48-8)**
3-Azapentane-1,5-diamine **(111-40-0)**
1-Azaphenanthrene **(85-02-9)**
4-Azaphenanthrene **(230-27-3)**
5-Azaphenanthrene **(229-87-8)**
9-Azaphenanthrene **(229-87-8)**
1,3-Azaphospholidine, 1,2,3-triphenyl- (9CI) **(52112-04-6)**
Azapicyl **(17433-31-7)**
Azaplant **(61-82-5)**
Azaplant Kombi **(61-82-5)**
1-Azapyrene **(313-80-4)**
4-Azapyrene **(194-03-6)**
Azapyrene (9CI) **(89126-46-5)**
Azarylaldehyde **(4460-86-0)**
Azaserin **(115-02-6)**
Azaserine **(115-02-6)**
l-Azaserine **(115-02-6)**
1-Aza-2-silacyclopentane, 2,2-diethoxy-1-(trimethylsilyl)- (9CI) **(21297-72-3)**
Azasterol **(1249-84-9)**
Azathioprine **(446-86-6)**
Azatioprin **(446-86-6)**
5-Azauracil **(71-33-0)**
6-Azauracil-β-D-riboside **(54-25-1)**
6-Azauracilribosid (Czech) **(54-25-1)**
6-Azauracilriboside **(54-25-1)**
6-Azauridine **(54-25-1)**
Azauridine **(54-25-1)**
8-Azaxanthine **(1468-26-4)**
Azbllen asbestos **(17068-78-9)**
Azbolen asbestos **(77536-67-5)**
Azdel **(9003-07-0)**
Azdid **(50-33-9)**
Azecine, decahydro- (7CI,8CI,9CI) **(4396-27-4)**
Azelaaldehydic acid, ethyl ester (7CI,8CI) **(3433-16-7)**
Azelaic-acid **(123-99-9)**
Azelaic acid, Technical grade **(123-99-9)**
Azelaic acid, bis(2-ethylhexyl) ester **(103-24-2)**
Azelaic acid, dibutyl ester (8CI) **(2917-73-9)**
Azelaic acid, diethyl ester **(624-17-9)**
Azelaic acid, di(2-ethylhexyl)ester **(103-24-2)**
Azelaic acid, dihexyl ester **(109-31-9)**
Azelaic acid, diisodecyl ester **(28472-97-1)**
Azelaic acid, dimethyl ester (8CI) **(1732-10-1)**
Azelaic semialdehyde monoethyl ester **(3433-16-7)**
1H-Azepine, 1-acetylhexahydro- (8CI,9CI) **(5809-41-6)**
1H-Azepine, 1-(3-aminopropyl)hexahydro- (8CI) **(3437-33-0)**
1H-Azepine-1-carbothioic acid, hexahydro-, S-allyl ester (8CI) **(16516-78-2)**
1H-Azepine-1-carbothioic acid, hexahydro-, S-benzyl ester (8CI) **(6996-88-9)**
1H-Azepine-1-carbothioic acid, hexahydro-, S-(2-chloroethyl) ester (9CI) **(107348-44-7)**
1H-Azepine-1-carbothioic acid, hexahydro-, S-(3-chloropropyl) ester (9CI) **(79720-82-4)**
1H-Azepine-1-carbothioic acid, hexahydro-, S-cyclopropyl ester (9CI) **(107348-43-6)**
1H-Azepine-1-carbothioic acid, hexahydro-3,6-dimethyl-, S-ethyl ester (9CI) **(55573-38-1)**
1H-Azepine-1-carbothioic acid, hexahydro-, S-ethyl ester **(2212-67-1)**
1H-Azepine-1-carbothioic acid, hexahydro-, S-(1-ethylpropyl)

Azo-55TT **(1314-13-2)**
Azo-66TT **(1314-13-2)**
Azo-77TT **(1314-13-2)**
Azo Acid Red GS **(1658-56-6)**
Azo Dark Blue C 2B **(1064-48-8)**
Azo Dark Blue HR **(1064-48-8)**
Azo Dark Blue S **(1064-48-8)**
Azo Dark Blue SH **(1064-48-8)**
Azo Fuchsine **(3567-66-6)**
Azo-Gantanol **(723-46-6)**
Azo Gantrisin **(127-69-5)**
Azo Gantrisin **(136-40-3)**
Azo Gastanol **(136-40-3)**
Azo Geranine 2G **(3734-67-6)**
Azo Geranine 2GA **(3734-67-6)**
Azo Grenadine **(3567-66-6)**
Azo Magenta G **(3567-66-6)**
Azo-Mandelamine **(136-40-3)**
Azo Milling Red G **(10169-02-5)**
Azo Phloxine GA **(3734-67-6)**
Azo Phloxine GA-CF **(3734-67-6)**
Azo Red R **(915-67-3)**
Azo Rhodine 2G **(3734-67-6)**
Azo Rubin Extra **(3567-69-9)**
Azo Rubin S **(3567-69-9)**
Azo Rubin XX **(3567-69-9)**
Azo Rubine **(3567-69-9)**
S-Azo Rubine **(915-67-3)**
Azo Rubine AF **(3567-69-9)**
Azo Rubine (Biological stain) **(3567-69-9)**
Azo Rubine Extra LC **(3567-69-9)**
Azo Rubine LZ **(3567-69-9)**
Azo Rubine S **(3567-69-9)**
Azo Rubine S **(915-67-3)**
Azo Rubine SF **(915-67-3)**
Azo Rubine S.Fq **(915-67-3)**
Azo Rubine S Specially Pure **(3567-69-9)**
Azo Rubine XX **(3567-69-9)**
Azo Rubine for Food **(3567-69-9)**
Azo Ruby S **(915-67-3)**
Azo-Standard **(136-40-3)**
Azo-Stat **(136-40-3)**
Azoamine Pink O **(97-52-9)**
Azoamine Red A **(89-62-3)**
Azoamine Red zh **(100-01-6)**
Azoamine Scarlet **(99-59-2)**
Azoamine Scarlet K **(99-59-2)**
Azobase DCA **(95-82-9)**
Azobase MNA **(99-09-2)**
Azobase NAT **(89-62-3)**
Azobenzeen (Dutch) **(103-33-3)**
Azobenzene **(103-33-3)**
Azobenzene, 4-amino- **(60-09-3)**
Azobenzene, 4-amino-4'-dimethylamino- **(539-17-3)**
Azobenzene, 4-amino-4'-hydroxy- **(103-18-4)**
Azobenzene, 4-chloro- (8CI) **(4340-77-6)**
Azobenzene-2,4-diamine **(495-54-5)**
Azobenzene, 4,4'-dichloro **(1602-00-2)**
Azobenzene, 3,3'-dichloro- (8CI) **(15426-14-9)**
Azobenzene, 4,4'-diethyl- (6CI,7CI) **(61653-33-6)**
Azobenzene, 2,2'-dimethoxy- (8CI) **(613-55-8)**
Azobenzene, 3,3'-dimethoxy- (8CI) **(6319-23-9)**
Azobenzene, 4,4'-dimethoxy- (8CI) **(501-58-6)**
Azobenzene, 2,4-dimethyl- (8CI) **(29418-21-1)**
Azobenzene, 2,6-dimethyl- (6CI,7CI,8CI) **(17590-87-3)**
Azobenzene, p-dimethylamino- **(60-11-7)**
Azobenzene-3,4'-disulfonic acid, 4-amino- **(101-50-8)**
Azobenzene, 2,2',4,4',6,6'-hexamethyl- (7CI,8CI) **(5692-66-0)**
Azobenzene, 2-methoxy- (8CI) **(6319-21-7)**
Azobenzene, 4-methoxy- (8CI) **(2396-60-3)**
Azobenzene, 2-methoxy-2'-methyl- (8CI) **(29268-78-8)**
Azobenzene, 2-methoxy-6-methyl- (8CI) **(29268-77-7)**
Azobenzene, 4'-methoxy-4-nitro **(29418-59-5)**
Azobenzene, 4-methyl- (8CI) **(949-87-1)**
Azobenzene, 2-methyl- (6CI,7CI,8CI) **(6676-90-0)**
Azobenzene, 2-methyl-4'-nitro- (7CI,8CI) **(7030-18-4)**
Azobenzene, 4-nitro **(2491-52-3)**
Azobenzene, oxide **(495-48-7)**
Azobenzene, 4-phenyl- (6CI,8CI) **(7466-42-4)**
Azobenzene, 3,3',4,4'-tetrachloro **(14047-09-7)**
Azobenzene, 2,2',4,4'-tetramethyl- (8CI) **(29418-25-5)**
Azobenzene, 2,2',5,5'-tetramethyl- (8CI) **(6311-44-0)**
Azobenzene, 2,2',4-trimethyl- (8CI) **(29418-23-3)**
Azobenzene, 2,4,4'-trimethyl- (8CI) **(29418-24-4)**
Azobenzene, 2,2',6-trimethyl- (7CI,8CI) **(6319-26-2)**
Azobenzene, 2,4,6-trimethyl- (6CI,7CI,8CI) **(29418-26-6)**
Azobenzide **(103-33-3)**
4-Azobenzoate **(586-91-4)**
p-Azobenzoate **(586-91-4)**
para-Azobenzoate **(586-91-4)**
Azobenzol **(103-33-3)**
2,2'-Azobiphenyl **(230-17-1)**
Azobisbenzene **(103-33-3)**
1,1'-Azobiscarbamide **(123-77-3)**
Azobiscarbonamide **(123-77-3)**
Azobiscarboxamide **(123-77-3)**
1,1'-Azobis(N-chloroformamidine) **(502-98-7)**
4,4'-Azobis(4-cyanovaleric acid) **(2638-94-0)**
Azobis(cyanovaleric acid) **(2638-94-0)**
1,1'-Azobis(formamide) **(123-77-3)**
Azobisisobutylonitrile **(78-67-1)**
α,α'-Azobisisobutylonitrile **(78-67-1)**
2,2'-Azobis(isobutyronitrile) **(78-67-1)**
Azobisisobutyronitrile **(78-67-1)**
2,2'-Azobis(2-methylbutanenitrile) **(13472-08-7)**
2,2'-Azobis(2-methylpropionamidine) dihydrochloride **(2997-92-4)**
2,2'-Azobis(2-methylpropionitrile) **(78-67-1)**
Azobutyl **(50-33-9)**
Azocard Black EW **(1937-37-7)**
Azocard Black RW **(2429-83-6)**
Azocard Blue 2B **(2602-46-2)**
Azocard Blue 6B **(2610-05-1)**
Azocard Blue BH **(2429-73-4)**
Azocard Brown M **(2429-82-5)**
Azocard Dark Green B **(3626-28-6)**
Azocard Red 4B **(992-59-6)**
Azocard Red Congo **(573-58-0)**
Azocard Violet N **(2586-60-9)**
Azocene **(43121-43-3)**
Azocine, 1-(2-guanidinoethyl)octahydro- **(55-65-2)**
Azocine, octahydro **(1121-92-2)**
p,p'-Azodiacetanilide **(15446-39-6)**
4,4'-Azodianiline **(538-41-0)**
4,4'-Azodianisole **(501-58-6)**
Azodibenzene **(103-33-3)**
Azodibenzeneazofume **(103-33-3)**
Azodicarbamide **(123-77-3)**
Azodicarboamide **(123-77-3)**
Azodicarbonamide **(123-77-3)**
Azodicarboxamide **(123-77-3)**
Azodicarboxylic acid diamide **(123-77-3)**
2,2'-Azodi-(2,4-dimethylvaleronitrile) [UN 2953] **(4419-11-8)**
Azodi-(1,1'-hexahydrobenzonitrile) [UN 2954] **(2094-98-6)**
2,2'-Azodiisobutyronitrile **(78-67-1)**

Name	CAS
α,α'-Azodiisobutyronitrile	(78-67-1)
Azodiisobutyronitrile [UN 2952]	(78-67-1)
Azodine	(136-40-3)
Azodium	(136-40-3)
Azodox-55	(1314-13-2)
Azodox-55TT	(1314-13-2)
Azodrin-71	(6923-22-4)
Azodrin Insecticide	(6923-22-4)
Azodrin (OSHA)	(6923-22-4)
Azodyne	(136-40-3)
Azoene Fast Blue BB Base	(120-00-3)
Azoene Fast Blue Base	(119-90-4)
Azoene Fast Blue Salt	(119-90-4)
Azoene Fast Orange GR Base	(88-74-4)
Azoene Fast Orange GR Salt	(88-74-4)
Azoene Fast Red B Base	(97-52-9)
Azoene Fast Red 3GL Base	(89-63-4)
Azoene Fast Red GL Base	(99-52-5)
Azoene Fast Red 3GL Salt	(89-63-4)
Azoene Fast Red KB Base	(95-79-4)
Azoene Fast Red Red GL Salt	(89-62-3)
Azoene Fast Red TR Base	(95-69-2)
Azoene Fast Red TR Salt	(3165-93-3)
Azoene Fast Scarlet 2G Base	(95-82-9)
Azoene Fast Scarlet GC Base	(99-55-8)
Azoene Fast Scarlet GC Salt	(99-55-8)
Azofene	(2310-17-0)
Azofenol 4K	(6300-37-4)
Azofix Blue B Salt	(119-90-4)
Azofix Orange GC Salt	(141-85-5)
Azofix Orange GR Salt	(88-74-4)
Azofix Red GG Salt	(100-01-6)
Azofix Red 3GL Salt	(89-63-4)
Azofix Red GL Salt	(89-62-3)
Azofix Scarlet G Salt	(99-55-8)
Azofos	(298-00-0)
Azofume	(103-33-3)
Azogen Developer A	(135-19-3)
Azogen Developer H	(95-80-7)
Azogene Ecarlate R	(99-59-2)
Azogene Fast Blue B	(119-90-4)
Azogene Fast Blue B Salt	(119-90-4)
Azogene Fast Orange GC Base	(141-85-5)
Azogene Fast Orange GCN Base	(141-85-5)
Azogene Fast Orange GC Salt	(141-85-5)
Azogene Fast Orange GEN Salt	(141-85-5)
Azogene Fast Orange GR	(88-74-4)
Azogene Fast Red NRL Salt	(99-52-5)
Azogene Fast Red RL	(99-52-5)
Azogene Fast Red TR	(3165-93-3)
Azogene Fast Red TR	(95-69-2)
Azogene Fast Scarlet G	(99-55-8)
Azogene Fast Scarlet GGC	(95-82-9)
Azogene Fast Scarlet GG (Free Base)	(95-82-9)
Azoground AS	(92-77-3)
Azoground BS	(135-65-9)
Azoground D	(135-61-5)
Azoground OL	(135-62-6)
Azoic Coupling Component 17	(135-65-9)
Azoic Coupling Component 18	(135-61-5)
Azoic Coupling Component 3	(91-92-9)
Azoic Diazo Component 3	(95-82-9)
Azoic Diazo Component 6	(88-74-4)
Azoic Diazo Component 8	(89-62-3)
Azoic Diazo Component 9	(89-63-4)
Azoic Diazo Component 32	(95-79-4)
Azoic Diazo Component 37	(100-01-6)
Azoic Diazo Component 46	(87-60-5)
Azoic Diazo Component 11 Base	(3165-93-3)
Azoic Diazo Component 11, Base	(95-69-2)
Azoic Diazo Component 13, Base	(99-59-2)
Azoic Red 36	(120-71-8)
Azoimide	(7782-79-8)
Azoksodon	(2152-34-3)
Azol	(123-30-8)
Azolamide	(30979-48-7)
Azolan	(61-82-5)
Azole	(109-97-7)
Azole	(61-82-5)
Azolid	(50-33-9)
Azolidine	(123-75-1)
Azolitmin (8CI,9CI)	(1395-18-2)
Azolmetazin	(57-68-1)
Azomesitylene	(5692-66-0)
Azomethane	(503-28-6)
Azomine	(136-40-3)
Azomine Black BH	(2429-73-4)
Azomine Black EWO	(1937-37-7)
Azomine Blue 2B	(2602-46-2)
Azomine Brown M	(2429-82-5)
Azomine Yellow R	(1325-37-7)
Azomycin	(527-73-1)
Azonaphthol Red J	(3734-67-6)
Azonaphtol A	(92-77-3)
Azonaphtol AN	(132-68-3)
Azonaphtol MNA	(135-65-9)
Azonaphtol OA	(135-62-6)
Azonaphtol OT	(135-61-5)
8-Azoniabicyclo(3.2.1)octane, 8,8-dimethyl-3-((1-oxo-2-propylpentyl)oxy)-, bromide, endo-	(80-50-2)
Azophen	(60-80-0)
Azophene	(60-80-0)
Azophenylene	(92-82-0)
Azophloxin	(3734-67-6)
Azophloxine	(3734-67-6)
Azophos	(298-00-0)
Azopyrin	(599-79-1)
Azoresorcin	(550-82-3)
Azorubin	(3567-69-9)
Azorubin S	(915-67-3)
Azosalt R	(101-54-2)
Azoseptale	(72-14-0)
Azossibenzene (Italian)	(495-48-7)
Azosulfizin	(127-69-5)
Azote (French)	(10102-44-0)
Azothioprin (Czech)	(446-86-6)
Azothioprine	(446-86-6)
Azotic acid	(7697-37-2)
Azoto (Italian)	(10102-44-0)
Azotol A	(92-77-3)
Azotol ANF	(132-68-3)
Azotol DA	(91-92-9)
Azotol KHA	(137-52-0)
Azotol MNA	(135-65-9)
Azotol NMA	(135-65-9)
Azotol OA	(135-62-6)
Azotol OF	(92-74-0)
Azotol OT	(135-61-5)
Azotol XA	(137-52-0)
4,4'-Azotoluene	(501-60-0)
ar,ar'-Azotoluene (8CI)	(26444-20-2)
o,o'-Azotoluene (8CI)	(584-90-7)
p,p'-Azotoluene (8CI)	(501-60-0)
o,p'-Azotoluene (7CI,8CI)	(29418-22-2)

Azotowy kwas (Polish) **(7697-37-2)**
Azotox **(50-29-3)**
Azotoyperite **(55-86-7)**
Azotrex **(136-40-3)**
Azoture d'ammonium (French) **(12164-94-2)**
Azoture de benzoyle (French) **(582-61-6)**
Azoture de tert-butoxycarbonyle (French) **(1070-19-5)**
Azoture de chlore (French) **(13973-88-1)**
Azoture de plomb (French) **(13424-46-9)**
Azoture de sodium (French) **(26628-22-8)**
Azoture d'iode (French) **(14696-82-3)**
Azotu tlenki (Polish) **(63907-41-5)**
Azovan Blue **(314-13-6)**
Azoxodone **(2152-34-3)**
Azoxyaethan (German) **(16301-26-1)**
Azoxybenzeen (Dutch) **(495-48-7)**
Azoxybenzene **(495-48-7)**
Azoxybenzene, 4,4'-dichloro **(614-26-6)**
Azoxybenzene, 4,4'-dimethoxy- (8CI) **(1562-94-3)**
Azoxybenzene, 3,3',4,4'-tetrachloro **(21232-47-3)**
Azoxybenzide **(495-48-7)**
Azoxybenzol (German) **(495-48-7)**
Azoxydibenzene **(495-48-7)**
Azoxyethane **(16301-26-1)**
Azoxymethane **(25843-45-2)**
Aztec BPO **(94-36-0)**
Azulene **(275-51-4)**
Azulene, 1,2,3,4,5,6,7,8-octahydro-1,4-dimethyl-7-(1-methylethylidene)-, (1S,cis)- (9CI) **(88-84-6)**
Azulene, 1,2,3,3a-tetrahydro- (8CI,9CI) **(33877-87-1)**
6-Azulenol, 1,2,3,3a,4,5,6,8a-octahydro-4,8-dimethyl-2-(1-methylethylidene)-, acetate **(117-98-6)**
6-Azulenol, 1,2,3,3a,4,5,6,8a-octahydro-2-isopropylidene-4,8-dimethyl-, acetate **(117-98-6)**
Azulfidine **(599-79-1)**
Azunthol **(56-72-4)**
6-Azur **(54-25-1)**
Azur **(54-25-1)**
Azuren **(54-85-3)**
6-Azuridine **(54-25-1)**
Azurro Diretto 3B **(72-57-1)**
2B **(122-19-0)**
70-314B **(33629-47-9)**
B 7 **(9003-17-2)**
B 9 **(14807-96-6)**
B-9 **(141-03-7)**
B-9 **(1596-84-5)**
B 10 **(9004-32-4)**
B 10 (Polysaccharide) **(9004-32-4)**
B 11 **(9003-17-2)**
B-12 **(68-19-9)**
B 13 **(14807-96-6)**
B 13 (Mineral) **(14807-96-6)**
B-28 **(134-58-7)**
B32 **(70-30-4)**
B-35 **(25038-54-4)**
B 75 **(9004-66-4)**
B/77 **(116-01-8)**
B-203 **(25038-54-4)**
B 208 **(340-57-8)**
B 208-Tropon **(340-57-8)**
B-216 **(25038-54-4)**
B-300 **(25038-54-4)**
B-350 **(25038-54-4)**
B 360 **(70-70-2)**
B 404 **(56-38-2)**
B-436 **(390-64-7)**
B-500 **(91-22-5)**
B 518 **(6055-19-2)**
B 518 **(50-18-0)**
B-622 **(101-05-3)**
B 673 **(53-34-9)**
B 995 **(1596-84-5)**
B-1,776 **(78-48-8)**
B-1843 **(1113-14-0)**
B 3000 **(9003-17-2)**
B-3015 **(28249-77-6)**
B 3356 **(34622-58-7)**
B-4130 **(440-58-4)**
B 5333 **(36104-80-0)**
B 29493 **(55-38-9)**
B 37344 **(2032-65-7)**
B 77488 **(14816-18-3)**
B-10094 **(13067-93-1)**
B-33172 **(3878-19-1)**
B Rose Liquid **(1390-65-4)**
6-BA **(1214-39-7)**
BA **(56-55-3)**
BA 51-090462 **(191-30-0)**
BA 59 **(79-94-7)**
BA 2666 **(125-71-3)**
BA 2726 **(1067-33-0)**
BA 2773 **(102-32-9)**
BA 5968 **(304-20-1)**
BA 5968 **(86-54-4)**
BA 32644 **(61-57-4)**
BA 32644 Ciba **(61-57-4)**
BA (Growth stimulant) **(1214-39-7)**
BAA **(91-49-6)**
BAB **(2079-00-7)**
BABS **(2079-00-7)**
6-BAP **(1214-39-7)**
BAP (Growth stimulant) **(1214-39-7)**
BAPN **(151-18-8)**
BAS-083 **(24307-26-4)**
BAS 083W **(24307-26-4)**
BAS 238F **(31717-87-0)**
BAS 263 **(51487-69-5)**
BAS 263I **(51487-69-5)**
BAS-290-H **(21267-72-1)**
BAS 305 **(7055-03-0)**
BAS 351-H **(25057-89-0)**
BAS 351-07H **(25057-89-0)**
BAS 352 F **(50471-44-8)**
BAS 389F **(60568-05-0)**
BAS 389 **(60568-05-0)**
BAS 392-H **(33245-39-5)**
BAS 479H **(67129-08-2)**
BAS 2205-F **(24602-86-6)**
BAS 2382 F **(31717-87-0)**
BAS 2900H **(21267-72-1)**
BAS 2903H **(21267-72-1)**
BAS-3050 **(7055-03-0)**
BAS 3050F **(7055-03-0)**
BAS-3170 **(15310-01-7)**
BAS 3170F **(15310-01-7)**
BAS 3220 **(23564-06-9)**
BAS-3460 **(10605-21-7)**
BAS 3510H **(25057-89-0)**
BAS 3510 **(25057-89-0)**
BAS 3512H **(25057-89-0)**
BAS 3517H **(25057-89-0)**
BAS 3920 **(33245-39-5)**
BAS 3922 **(33245-39-5)**

Name	CAS
BAS-08300W	(24307-26-4)
BAS 08301W	(24307-26-4)
BAS 08305 W	(24307-26-4)
BAS 08306 W	(24307-26-4)
BAS 08307 W	(24307-26-4)
BAS 9052	(74051-80-2)
BAS 9052H	(74051-80-2)
BAS 32500F	(23564-05-8)
BAS 67054	(10605-21-7)
BAS 90520H	(74051-80-2)
BAS85559X	(24307-26-4)
BASF	(9011-05-6)
BASF 238	(110-91-8)
BASF-Grunkupfer	(1332-40-7)
BASF III	(9003-53-6)
BASF-Maneb Spritzpulver	(12427-38-2)
BASF Ursol D	(106-50-3)
BASF Ursol EG	(591-27-5)
BASF Ursol Ern	(90-15-3)
BASF Ursol 3GA	(95-55-6)
BASF Ursol P Base	(123-30-8)
BASF Ursol SLA	(39156-41-7)
BASFapon B	(127-20-8)
BASF mehltaumittel	(31717-87-0)
BAU	(7440-44-0)
BAY 86791	(21085-19-8)
BAY-SIR 8514	(64628-44-0)
BB-8	(27858-07-7)
BBC	(17804-35-2)
BBC	(5798-79-8)
BBC 12	(96-12-8)
BBCE	(111-94-4)
BBH	(58-89-9)
BB-K 8	(37517-28-5)
BBM	(85-60-9)
BBN	(3817-11-6)
BBN	(5798-79-8)
BBNOH	(3817-11-6)
BBP	(85-68-7)
BC 20	(9011-05-6)
BC 40	(9011-05-6)
BC 77	(9011-05-6)
BCS-3	(2079-00-7)
l-BC 2627	(42408-82-2)
BCF	(353-59-3)
BCF-Bushkiller	(93-76-5)
BCM	(10605-21-7)
BCM	(551-74-6)
BCM	(576-68-1)
BCME	(542-88-1)
BCNU	(154-93-8)
BCPC	(2164-13-8)
BCPE	(80-06-8)
BC 20 (Polymer)	(9011-05-6)
BCS Copper Fungicide	(7758-98-7)
4311/B Ciba	(113-45-1)
BDCM	(75-27-4)
(BDH)	(77-75-8)
BDH 1298	(595-33-5)
BDH 29-790	(9003-53-6)
BDMA	(103-83-3)
BDOT	(81325-79-3)
BE	(540-51-2)
BEHA	(103-23-1)
BEHP	(117-81-7)
BEK	(502-55-6)
BEN-P	(1538-09-6)

Name	CAS
BEP	(115-84-4)
BES	(10191-18-1)
BETZ 402	(8061-51-6)
BEXIDE	(502-55-6)
BEXT	(502-55-6)
BFP	(115-26-4)
BFPO	(115-26-4)
BFV	(50-00-0)
BG 6080	(7440-44-0)
t-BGE	(7665-72-7)
BGE (OSHA)	(2426-08-6)
2-tert-BHA	(121-00-6)
3-BHA	(121-00-6)
3-tert-BHA	(88-32-4)
BHA	(25013-16-5)
BHBN	(3817-11-6)
BHC	(58-89-9)
BHC	(608-73-1)
BHC	(66-76-2)
α-BHC	(319-84-6)
β-BHC	(319-85-7)
δ-BHC	(319-86-8)
γ-BHC	(58-89-9)
BH 2,4-D	(94-75-7)
BH 2,4-DP	(120-36-5)
BH Dalapon	(75-99-0)
BHFT	(73-48-3)
BH MCPA	(94-74-6)
BH Mecoprop	(7085-19-0)
BH Mecoprop	(93-65-2)
BHP	(53609-64-6)
BHT	(128-37-0)
BHT (Food grade)	(128-37-0)
B-Herbatox	(7775-09-9)
BI-5452	(62850-32-2)
BI-58	(60-51-5)
BIC	(111-36-4)
BICP	(1967-16-4)
BI 58 EC	(60-51-5)
B-I-K	(57-13-6)
BIPC	(1967-16-4)
BIPC (The herbicide)	(1967-16-4)
8014 BIS HC	(60-51-5)
1,2-BITDO	(936-16-3)
BKF	(119-47-1)
B-K Liquid	(7681-52-9)
B-K Powder	(7778-54-3)
BL 15	(9004-62-0)
BL 2487	(14168-01-5)
BL 9	(9002-92-0)
BLA	(1335-32-6)
BLA-S	(2079-00-7)
BL 9EX	(9002-92-0)
BL H368	(73-48-3)
BLM	(11056-06-7)
BLP 1410	(26787-78-0)
BL-S 578	(50370-12-2)
80BM	(25608-33-7)
BM 1	(129-20-4)
BMC	(10605-21-7)
BMOO	(532-34-3)
BNA 80 (9CI)	(100291-87-0)
BNB	(117-26-0)
BNM	(17804-35-2)
BNP	(117-27-1)
BNP	(504-88-1)
BNP 30	(88-85-7)

BNS (102-96-5)
1,2-BNT (205-43-6)
2,3-BNT (243-46-9)
BNU (869-01-2)
B-Nine (1596-84-5)
BO 714 (1195-16-0)
BOH (109-84-2)
3,4-BP (50-32-8)
B(e)P (192-97-2)
BP (50-32-8)
BP 5015 (54-85-3)
BPA (2312-73-4)
BP 1100 (9CI) (39278-82-5)
BPCMS (54091-06-4)
BPE-I (9002-88-4)
BP-KLP (9003-53-6)
BPL (57-57-8)
BPMC (3766-81-2)
BPPS (2312-35-8)
BP-1100X (39403-84-4)
BP 1100X (9CI) (39403-84-4)
BP 4,5-epoxide (37574-47-3)
B-Picoline (108-99-6)
BP 4,5-oxide (37574-47-3)
BP-1,6-quinone (3067-13-8)
BP-3,6-quinone (3067-14-9)
BP-6,12-quinone (3067-12-7)
BQC Reagent (537-45-1)
BRL (69-53-4)
BRL 1341 (69-53-4)
BRL 152 (132-93-4)
BRL 1621 (61-72-3)
BRL 1702 (3116-76-5)
BRL 2039 (5250-39-5)
BRL 2333 (26787-78-0)
BR 55N (1163-19-5)
BS (123-95-5)
BS 479 (621-42-1)
BS 572 (456-59-7)
BSA (98-10-2)
BSB-S-E (9003-53-6)
BSB-S 40 (9003-53-6)
BSC-Refine D (98-09-9)
BSF (71-67-0)
BSF simes (71-67-0)
BSP (71-67-0)
BSP 5000 (9003-11-6)
BSP sodium (71-67-0)
B-Selektonon (94-75-7)
B-Selektonon M (94-74-6)
B-Stoff (598-31-2)
BT (16227-10-4)
BT 31 (78-04-6)
BT 324 (136-84-5)
BT 4651 (147-14-8)
BTB (68038-71-1)
BTC (8001-54-5)
BTC 50 (8001-54-5)
BTC 50 USP (8001-54-5)
BTC 65 (8001-54-5)
BTC 65 USP (8001-54-5)
BTC 100 (8001-54-5)
BTC 812 (10361-16-7)
BTC 824 (8001-54-5)
BTC 835 (8001-54-5)
BTC 1010 (7173-51-5)
BTC 2565 (8001-54-5)
BTC 8248 (8001-54-5)
BTC 8249 (8001-54-5)
BTC E-8358 (8001-54-5)
BTCO 1010 (7173-51-5)
1,3-BTDZD (1615-06-1)
BTFMEA (75-23-0)
B-TGDR (789-61-7)
BTKH (538-28-3)
BTM (56-93-9)
B-TMPN (81325-80-6)
BTO (56-35-9)
BTS (127-17-3)
BTS 14639 (2655-19-8)
BTS 27,419 (33089-61-1)
BTX (23509-16-2)
B.T.Z. (50-33-9)
B-Twelve (68-19-9)
B-Twelve ORA (68-19-9)
BU 700 (9011-05-6)
BUCB (112-34-5)
BUCS (111-76-2)
BU2AE (102-81-8)
BUKS (128-37-0)
BUX (2282-34-0)
BUX (672-04-8)
BUX (8065-36-9)
BUX Ten (2282-34-0)
BUX-Ten (8065-36-9)
BV 201 (2860-64-2)
BV 207 (2884-69-7)
B-W (6834-92-0)
BW 47-83 (82-92-8)
BW 50-63 (58-14-0)
BW 56-158 (315-30-0)
BW 56-72 (738-70-5)
BW 57-322 (446-86-6)
BW 5071 (154-42-7)
BW-A 509U (30516-87-1)
BW-21-Z (52645-53-1)
BZ 55 (339-43-5)
BZ 55 (64-77-7)
BZCF (501-53-1)
BZF-60 (94-36-0)
BZI (51-17-2)
BZQ (63-12-7)
BZT (121-54-0)
276-Ba (10262-69-8)
36278-Ba (23239-41-0)
Ba 2797 (12111-24-9)
Ba 2821 (405-50-5)
Ba-20684 (911-65-9)
Ba 30920 (16960-16-0)
Ba 36278 (23239-41-0)
Ba 36716 (16960-16-0)
B(a)P (50-32-8)
Baam (33089-61-1)
Babbitt, Dross (69011-64-9)
Babrocid (59-87-0)
Baci-Jel (1405-87-4)
Bacfor BL (7281-04-1)
Baciguent (1405-87-4)
Baciliquin (1405-87-4)
Bacillin (54-85-3)
Bacillol (1319-77-3)
Bacillopeptidase A (9014-01-1)
Bacillopeptidase B (9014-01-1)
Bacillus-Subtilis-Carlsberg (9014-01-1)

Bacillus thuringensis	**(68038-71-1)**
Bacillus thuringiensis	**(68038-71-1)**
Bacillus thuringiensis (Berliner) var. aizawai	**(68038-71-1)**
Bacillus thuringiensis (Berliner) var. israelensis	**(68038-71-1)**
Bacillus thuringiensis (Berliner) var. kurstaki	**(68038-71-1)**
Bacillus thuringiensis (Berliner) var. san diego	**(68038-71-1)**
Bacillus thuringiensis (Berliner) var. tenebrionis	**(68038-71-1)**
Bacillus thuringiensis berliner	**(68038-71-1)**
Bacitek Ointment	**(1405-87-4)**
Bacitracin	**(1405-87-4)**
Baco AF 260	**(21645-51-2)**
Bacteramid	**(63-74-1)**
Bacterial Vitamin H1	**(150-13-0)**
Bactesid	**(63-74-1)**
Bactesulf	**(127-69-5)**
Bactigras	**(56-95-1)**
Bactine	**(25155-18-4)**
Bactol	**(130-26-7)**
Bactolatex	**(9003-53-6)**
Bactramin	**(8064-90-2)**
Bactrim	**(723-46-6)**
Bactrim	**(8064-90-2)**
Bactrim DS	**(8064-90-2)**
Bactrin	**(8064-90-2)**
Bactrol	**(132-27-4)**
Bactromin	**(8064-90-2)**
Badil	**(548-62-9)**
Bafhameritin-M	**(61-68-7)**
Bagasse	**(9006-97-7)**
Bagodryl	**(58-73-1)**
Bagolax	**(9004-67-5)**
Bahama Blue BC	**(147-14-8)**
Bahama Blue BNC	**(147-14-8)**
Bahama Blue Lake NCNF	**(147-14-8)**
Bahama Blue WD	**(147-14-8)**
Bakelite AYAA	**(9003-20-7)**
Bakelite AYAF	**(9003-20-7)**
Bakelite AYAT	**(9003-20-7)**
Bakelite DFD 330	**(9002-88-4)**
Bakelite DHDA 4080	**(9002-88-4)**
Bakelite DYNH	**(9002-88-4)**
Bakelite LP 70	**(9003-22-9)**
Bakelite LP 90	**(9003-20-7)**
Bakelite RMD 4511	**(9003-54-7)**
Bakelite SMD 3500	**(9003-53-6)**
Bakelite VLFV	**(9003-22-9)**
Bakelite VMCC	**(9003-22-9)**
Bakelite VSJD 10	**(9003-22-9)**
Bakelite VYHD	**(9003-22-9)**
Bakelite VYHH	**(9003-22-9)**
Bakelite VYHN	**(9003-22-9)**
Bakelite VYNS	**(9003-22-9)**
Bakelite VYNW	**(9003-22-9)**
Baker's Antifol	**(134-62-3)**
Bakers P-6	**(140-04-5)**
Baker's P and S Liquid and Ointment	**(108-95-2)**
Baking soda	**(144-55-8)**
Bakontal	**(7727-43-7)**
Baktar	**(8064-90-2)**
Bakthane	**(68038-71-1)**
Baktol	**(59-50-7)**
Baktolan	**(59-50-7)**
Bal	**(59-52-9)**
Balab 615	**(9003-11-6)**
Balan	**(1861-40-1)**
Balfin	**(1861-40-1)**
Bali Viscose Black G	**(6428-31-5)**
Bali Viscose Black N	**(6428-31-5)**
Balmadren	**(51-43-4)**
Balsam Canada	**(8007-47-4)**
Balsam Fir Oil	**(8021-28-1)**
Balsam Gurjun	**(8030-55-5)**
Balsam of Fir	**(8007-47-4)**
Balsams, Canada	**(8007-47-4)**
Bamd 400	**(57-53-4)**
Bamn	**(56986-36-8)**
Bamo 400	**(57-53-4)**
Ban-HOE	**(122-42-9)**
Ban-HOE	**(2164-08-1)**
Banabin	**(80-77-3)**
Banabin-Sintyal	**(80-77-3)**
Banafine	**(1861-40-1)**
Banamite	**(25939-05-3)**
Banana Oil	**(123-92-2)**
Banasil	**(50-55-5)**
Banex	**(1918-00-9)**
Banex	**(2300-66-5)**
Bangton	**(133-06-2)**
Banirex N	**(8061-51-6)**
Banisil	**(50-55-5)**
Banlen	**(1918-00-9)**
Banlon	**(63428-84-2)**
Banminth	**(15686-83-6)**
Banner	**(60207-90-1)**
Banocide	**(1642-54-2)**
Banol	**(671-04-5)**
Banol Tuco Sok	**(671-04-5)**
Bant	**(25265-76-3)**
Banthin	**(53-46-3)**
Banthine	**(53-46-3)**
Banthine bromide	**(53-46-3)**
Banthionine	**(59-51-8)**
Bantrol	**(1689-83-4)**
Bantu	**(86-88-4)**
Banvel	**(1918-00-9)**
Banvel 720	**(8068-77-7)**
Banvel (9CI)	**(62610-39-3)**
Banvel CST	**(1918-00-9)**
Banvel 72D	**(8068-77-7)**
Banvel D	**(1918-00-9)**
Banvel D	**(2300-66-5)**
Banvel-2,4-D mixture	**(8068-77-7)**
Banvel Herbicide	**(1918-00-9)**
Banvel II Herbicide	**(1918-00-9)**
Banvel 4S	**(1918-00-9)**
Banvel 4S	**(2300-66-5)**
Banvel T	**(2307-49-5)**
Banvel 4WS	**(1918-00-9)**
Baptitoxin	**(485-35-8)**
Baptitoxine	**(485-35-8)**
Baracoumin	**(66-76-2)**
Baramine	**(58-73-1)**
Barazae	**(127-69-5)**
Barballyl	**(52-43-7)**
Barbaloin	**(1415-73-2)**
Barbamate	**(101-27-9)**
Barbamil	**(57-43-2)**
Barbamyl	**(57-43-2)**
Barbamyl acid	**(57-43-2)**
Barban	**(101-27-9)**
Barbanate	**(101-27-9)**
Barbane	**(101-27-9)**
Barbapil	**(50-06-6)**
Barbasco	**(83-79-4)**

Barbellen **(50-06-6)**
Barbellon **(50-06-6)**
Barbenyl **(50-06-6)**
Barbidal **(52-43-7)**
Barbidorm **(56-29-1)**
Barbilehae (Barbilettae) **(50-06-6)**
Barbinal **(50-06-6)**
Barbiphen **(50-06-6)**
Barbiphenyl **(50-06-6)**
Barbipil **(50-06-6)**
Barbita **(50-06-6)**
Barbital **(57-44-3)**
Barbitone **(57-44-3)**
Barbituric-acid **(67-52-7)**
Barbituric acid, 5-allyl-5-sec-butyl **(115-44-6)**
Barbituric acid, 5-allyl-5-isobutyl- **(77-26-9)**
Barbituric acid, 5-allyl-5-isopropyl **(77-02-1)**
Barbituric acid, 5-allyl-5-(1-methylbutyl) **(76-73-3)**
Barbituric acid, 5-allyl-1-methyl-5-(1-methyl-2-pentynyl)-, sodium deriv. **(309-36-4)**
Barbituric acid, 5-allyl-1-methyl-5-(1-methyl-2-pentynyl)-, sodium salt **(309-36-4)**
Barbituric acid, 5-butyl-5-ethyl **(77-28-1)**
Barbituric acid, 5-sec-butyl-5-ethyl **(125-40-6)**
Barbituric acid, 5-sec-butyl-5-ethyl-, sodium salt **(143-81-7)**
Barbituric acid, 5-(1-cyclohepten-1-yl)-5-ethyl **(509-86-4)**
Barbituric acid, 5-(1-cyclohexen-1-yl)-1,5-dimethyl **(56-29-1)**
Barbituric acid, 5-(1-cyclohexen-1-yl)-5-ethyl **(52-31-3)**
Barbituric acid, 5,5-diallyl **(52-43-7)**
Barbituric acid, 5,5-diethyl **(57-44-3)**
Barbituric acid, 5,5-diethyl-1-methyl **(50-11-3)**
Barbituric acid, 5,5-dihydroxy **(3237-50-1)**
Barbituric acid, 1,3-dimethyl-5-ethyl **(7391-61-9)**
Barbituric acid, 5-ethyl- (8CI) **(2518-72-1)**
Barbituric acid, 5-ethyl-5-isopentyl **(57-43-2)**
Barbituric acid, 5-ethyl-5-(1-methyl-1-butenyl)- (8CI) **(125-42-8)**
Barbituric acid, 5-ethyl-5-(1-methylbutyl) **(76-74-4)**
Barbituric acid, 5-ethyl-5-(1-methylbutyl)-, sodium salt **(57-33-0)**
Barbituric acid, 5-ethyl-5-(1-methylbutyl)-2-thio- **(76-75-5)**
Barbituric acid, 5-ethyl-1-methyl-5-phenyl **(115-38-8)**
Barbituric acid, 5-ethyl-5-(3 or 6-oxo-1-cyclohexen-1-yl) **(25104-37-4)**
Barbituric acid, 5-ethyl-5-pentyl- (8CI) **(115-58-2)**
Barbituric acid, 5-ethyl-5-phenyl **(50-06-6)**
Barbituric acid, 5-ethyl-5-phenyl-, sodium salt **(57-30-7)**
Barbituric acid, 5-oxo- **(50-71-5)**
Barbituric acid, 2-thio **(504-17-6)**
Barbivis **(50-06-6)**
Barbonal **(50-06-6)**
Barbophen **(50-06-6)**
Barbosec **(76-73-3)**
Bardac 22 **(7173-51-5)**
Bardiol **(50-28-2)**
Bardorm **(50-06-6)**
Bareco Polywax 2000 **(9002-88-4)**
Bareco Wax C 7500 **(9002-88-4)**
Barex 210 Resin **(25014-41-9)**
Barhist **(135-23-9)**
Baridium **(136-40-3)**
Baridol **(7727-43-7)**
Bario (perossido di) (Italian) **(1304-29-6)**
Barite **(7727-43-7)**
Baritop **(7727-43-7)**
Baritrate **(78-11-5)**
Barium (ACGIH,OSHA) [UN 1400] **(7440-39-3)**
Barium, Alloys, Non-pyrophoric (DOT) **(7440-39-3)**
Barium, Alloys, Pyrophoric [UN 1854] **(7440-39-3)**
Barium Lithol **(1103-38-4)**
Barium Lithol Red **(1103-38-4)**
Barium, Metal, Non-pyrophoric (DOT) **(7440-39-3)**
Barium acetate **(543-80-6)**
Barium-azide **(18810-58-7)**
Barium azide, Dry or containing less than 50% water [UN 0224] **(18810-58-7)**
Barium azide, Wet, 50% or more water [UN 1571] **(18810-58-7)**
Barium benzoate **(533-00-6)**
Barium binoxide **(1304-29-6)**
Barium cadmium laurate **(15337-60-7)**
Barium cadmium stearate **(1191-79-3)**
Barium, calcium, magnesium, strontium, aluminum silicate flux **(66402-68-4)**
Barium carbonate **(513-77-9)**
Barium carbonate (1:1) **(513-77-9)**
Barium chlorate [UN 1445] **(13477-00-4)**
Barium chlorate, Wet (DOT) **(13477-00-4)**
Barium-chloride **(10361-37-2)**
Barium chloride, dihydrate **(10326-27-9)**
Barium chromate **(10294-40-3)**
Barium chromate (1:1) **(10294-40-3)**
Barium chromate(VI) **(10294-40-3)**
Barium chromate oxide **(10294-40-3)**
Barium-cyanide **(542-62-1)**
Barium cyanide, Solid (DOT) **(542-62-1)**
Barium cyanide [UN 1565] **(542-62-1)**
Barium decanoate **(13098-41-4)**
Barium diacetate **(543-80-6)**
Barium dichloride **(10361-37-2)**
Barium dichloride dihydrate **(10326-27-9)**
Barium dichromate **(10031-16-0)**
Barium dicyanide **(542-62-1)**
Barium dihydroxide **(17194-00-2)**
Barium dinitrate **(10022-31-8)**
Barium dinonylnaphthalenesulfonate **(25619-56-1)**
Barium dioxide **(1304-29-6)**
Barium distearate **(6865-35-6)**
Barium dodecanoate **(4696-57-5)**
Barium 2-ethylhexanoate **(2457-01-4)**
Barium-fluoride **(7787-32-8)**
Barium fluorosilicate **(17125-80-3)**
Barium fluosilicate **(17125-80-3)**
Barium hexafluorosilicate **(17125-80-3)**
Barium hexafluorosilicate(2-) **(17125-80-3)**
Barium hydroxide (9CI) **(17194-00-2)**
Barium hydroxide lime **(17194-00-2)**
Barium metaborate **(13701-59-2)**
Barium monoxide **(1304-28-5)**
Barium myristate **(10196-66-4)**
Barium(II) nitrate (1:2) **(10022-31-8)**
Barium nitrate [UN 1446] **(10022-31-8)**
Barium oxide [UN 1884] **(1304-28-5)**
Barium perchlorate **(13465-95-7)**
Barium permanganate [UN 1448] **(7787-36-2)**
Bariumperoxid (German) **(1304-29-6)**
Barium peroxide [UN 1449] **(1304-29-6)**
Bariumperoxyde (Dutch) **(1304-29-6)**
Barium protoxide **(1304-28-5)**
Bariumsilicofluorid (German) **(17125-80-3)**
Barium-silicofluoride **(17125-80-3)**
Barium silicon fluoride **(17125-80-3)**
Barium stearate **(6865-35-6)**
Barium sulfate (1:1) **(7727-43-7)**
Barium-sulfate **(7727-43-7)**
Barium sulfate (ACGIH,OSHA) **(7727-43-7)**
Barium sulphate **(7727-43-7)**
Barium superoxide **(1304-29-6)**
Barium tetradecanoate **(10196-66-4)**
Barizon **(3766-81-2)**

Name	CAS
Barlene 125	**(112-18-5)**
Barolub FTO	**(106-14-9)**
Baron	**(136-25-4)**
Baros Camphor	**(507-70-0)**
Barosperse	**(7727-43-7)**
Barotrast	**(7727-43-7)**
Barpental	**(57-33-0)**
Barquat CME-A	**(78-21-7)**
Barquat CT 29	**(112-02-7)**
Barquat MB-50	**(8001-54-5)**
Barquat MB-80	**(8001-54-5)**
Barquat SB-25	**(122-19-0)**
Barquinol	**(130-26-7)**
Barricade	**(52315-07-8)**
Barseb HC	**(50-23-7)**
Barthrin	**((70-43-9)**
Bartol	**(50-06-6)**
Baryl	**(140-04-5)**
Baryta	**(1304-28-5)**
Baryta White	**(7727-43-7)**
Baryta Yellow	**(10294-40-3)**
Barytes	**(7727-43-7)**
Barytes 22	**(7727-43-7)**
Baryum fluorure (French)	**(7787-32-8)**
Bas 90210H	**(55635-13-7)**
Basacryl Blue X 3GL	**(12270-13-2)**
Basacryl Red GL	**(42373-04-6)**
Basacryl Red X-BL	**(12270-25-6)**
Basagran	**(25057-89-0)**
Basalin	**(33245-39-5)**
Basamaize	**(21267-72-1)**
Basamid	**(533-74-4)**
Basamid-Fluid	**(137-42-8)**
Basamid G	**(533-74-4)**
Basamid-Granular	**(533-74-4)**
Basamid P	**(533-74-4)**
Basamid-Puder	**(533-74-4)**
Basanite	**(88-85-7)**
Base 661	**(9003-55-8)**
R-Base (Czech)	**(86-72-6)**
Base LP 12	**(9002-92-0)**
Base Oil	**(8002-05-9)**
Basecil	**(56-04-2)**
Basedol	**(96-50-4)**
Basergin	**(60-79-7)**
Basethyrin	**(56-04-2)**
Basfapon	**(75-99-0)**
Basfapon B	**(75-99-0)**
Basfapon/basfapon N	**(75-99-0)**
Basic Aluminum Acetate	**(142-03-0)**
Basic Aluminum Chlorate	**(12042-91-0)**
Basic Bismuth Nitrate	**(1304-85-4)**
Basic Blue 41	**(12270-13-2)**
Basic Blue 9	**(61-73-4)**
Basic Bluek	**(2185-86-6)**
Basic Bright-Green Sulfate	**(633-03-4)**
Basic Bright Green	**(633-03-4)**
Basic Brilliant Green	**(633-03-4)**
Basic Copper Carbonate	**(12069-69-1)**
Basic Copper(II) Carbonate	**(12069-69-1)**
Basic Copper Chloride	**(1332-40-7)**
Basic Copper Chloride	**(1332-65-6)**
Basic Copper Sulfate	**(1332-03-2)**
Basic Copper Sulfate	**(1344-73-6)**
Basic Copper Sulfate	**(7758-98-7)**
Basic Copper TS-53	**(1332-03-2)**
Basic Cupric Carbonate	**(12069-69-1)**
Basic Fuchsine	**(632-99-5)**
Basic Green 1	**(633-03-4)**
Basic Green 4	**(569-64-2)**
Basic Green V	**(633-03-4)**
Basic Lead Acetate	**(1335-32-6)**
Basic Lead Carbonate	**(1319-46-6)**
Basic Lead Chromate	**(18454-12-1)**
Basic Magenta	**(632-99-5)**
Basic Magenta E-200	**(632-99-5)**
Basic Magnesium Carbonate	**(7760-50-1)**
Basic Mercuric Sulfate	**(1312-03-4)**
Basic Nickel Carbonate	**(39430-27-8)**
Basic Nickel(II) Carbonate	**(12607-70-4)**
Basic Orange 21	**(3056-93-7)**
Basic Orange (9CI)	**(115133-49-8)**
Basic Orange 3RN	**(65-61-2)**
Basic Orange 3rn	**(494-38-2)**
Basic Parafuchsine	**(569-61-9)**
Basic Red 1	**(989-38-8)**
Basic Red 18	**(14097-03-1)**
Basic Red 2	**(477-73-6)**
Basic Red 22	**(12221-52-2)**
Basic Red 29	**(42373-04-6)**
Basic Red 46	**(12221-69-1)**
Basic Rhodamine Yellow	**(989-38-8)**
Basic Rhodaminic Yellow	**(989-38-8)**
Basic Violet 10	**(81-88-9)**
Basic Violet 3	**(548-62-9)**
Basic Violet BN	**(548-62-9)**
Basic Violet K	**(8004-87-3)**
Basic Zinc Chromate	**(13530-65-9)**
Basic Zinc Chromate	**(50922-29-7)**
Basic Zinc Chromate X-2259	**(13530-65-9)**
Basic Zinc Molybdate	**(61583-60-6)**
Basic Zinc Sulfate	**(68813-94-5)**
Basic Zirconium Chloride	**(7699-43-6)**
0% Basicity Chrome Alum	**(10141-00-1)**
Basil LS	**(8001-21-6)**
Basinex	**(2517-16-0)**
Basinex	**(75-99-0)**
Basitac	**(55814-41-0)**
Basle Green	**(12002-03-8)**
Basofortina	**(113-42-8)**
Basolan	**(60-56-0)**
Basolest	**(22232-54-8)**
Bassa	**(3766-81-2)**
Basta	**(77182-82-2)**
Bastnaesite, Acid-insol. fraction, Cerium rich	**(68909-12-6)**
Bastnaesite, calcined conc.	**(68909-13-7)**
Basudin	**(333-41-5)**
Basudin 10 G	**(333-41-5)**
Batasan	**(900-95-8)**
Batazina	**(122-34-9)**
Bathyran	**(504-17-6)**
Batidrol	**(480-22-8)**
Batrachotoxin	**(23509-16-2)**
Batrachotoxinin A, 20-(2,4-dimethyl-1H-pyrrole-3-carboxylate)	**(23509-16-2)**
Batridol	**(1143-38-0)**
Batrilex	**(82-68-8)**
Bauxite	**(1318-16-7)**
Bauxite Residue	**(1309-37-1)**
Bavistin	**(10605-21-7)**
Bax	**(147-24-0)**
Bay 21/199	**(56-72-4)**
Bay 1040	**(21829-25-4)**
Bay 2353	**(50-65-7)**
Bay 3231	**(68-76-8)**

Bay 4934 **(2533-82-6)**
Bay 5072 **(140-56-7)**
Bay 5097 **(23593-75-1)**
Bay 5212 **(731-27-1)**
Bay 5621 **(14816-18-3)**
Bay 5712a **(731-27-1)**
Bay 5821 **(13593-03-8)**
Bay 9010 **(114-26-1)**
Bay 9026 **(2032-65-7)**
Bay 9027 **(86-50-0)**
Bay 10756 **(8065-48-3)**
Bay 11405 **(298-00-0)**
Bay 11678 **(950-35-6)**
Bay 15080 **(495-73-8)**
Bay 15203 **(8022-00-2)**
Bay 15203 **(867-27-6)**
Bay 15922 **(52-68-6)**
Bay 16255 **(2642-71-9)**
Bay 17147 **(86-50-0)**
Bay 18436 **(919-86-8)**
Bay 19639 **(298-04-4)**
Bay 21097 **(301-12-2)**
Bay 22555 **(140-56-7)**
Bay 23129 **(640-15-3)**
Bay 23323 **(2497-07-6)**
Bay 25141 **(115-90-2)**
Bay 25634 **(5836-29-3)**
Bay 29493 **(55-38-9)**
Bay 30130 **(709-98-8)**
Bay 30686 **(93-75-4)**
Bay 32651 **(3566-00-5)**
Bay 33051 **(2597-03-7)**
Bay 33172 **(3878-19-1)**
Bay 34727 **(2636-26-2)**
Bay 36205 **(2439-01-2)**
Bay 36743 **(2984-64-7)**
Bay 37289 **(327-98-0)**
Bay 37344 **(2032-65-7)**
Bay 38156 **(333-43-7)**
Bay 38819 **(4104-14-7)**
Bay 39007 **(114-26-1)**
Bay 39731 **(2631-40-5)**
Bay 41831 **(122-14-5)**
Bay 42247 **(2255-17-6)**
Bay 44646 **(2032-59-9)**
Bay 45432 **(1113-02-6)**
Bay 46131 **(12071-83-9)**
Bay 47531 **(1085-98-9)**
Bay 49854 **(731-27-1)**
Bay 60618 **(1929-88-0)**
Bay 61597 **(21087-64-9)**
Bay 68138 **(22224-92-6)**
Bay 70143 **(1563-66-2)**
Bay 70533 **(14437-17-3)**
Bay 71628 **(10265-92-6)**
Bay 74283 **(18691-97-9)**
Bay 77049 **(13593-03-8)**
Bay 77488 **(14816-18-3)**
Bay 78418 **(17109-49-8)**
Bay 79758 **(21087-63-8)**
Bay 79770 **(20856-57-9)**
Bay 94871 **(30979-48-7)**
Bay 105807 **(2631-40-5)**
Bay-19149 **(62-73-7)**
Bay-92114 **(25311-71-1)**
Bay A 139 **(800-24-8)**
Bay A 1040 **(21829-25-4)**
Bay B 5097 **(23593-75-1)**
Bay-B 5097 **(23593-75-1)**
Bay DIC 1468 **(21087-64-9)**
Bay-DRW 1139 **(41394-05-2)**
Bay-E-393 **(3689-24-5)**
Bay E-601 **(298-00-0)**
Bay E-605 **(56-38-2)**
Bay ENE 11183 B **(5836-29-3)**
Bay 6681 F **(43121-43-3)**
Bay FCR 1272 **(68359-37-5)**
Bay-HWG 1608 **(107534-96-3)**
Bay-Hox-1901 **(29973-13-5)**
Bay KWG 0519 **(55219-65-3)**
Bay Kue 2079A **(33439-45-1)**
Bay-MEB-6447 **(43121-43-3)**
Bay NNT 6867 **(38727-55-8)**
Bay-NTN-9306 **(35400-43-2)**
Bay NTN 8629 **(34643-46-4)**
Bay-NTN 5006 **(33857-23-7)**
Bay S 2758 **(3566-00-5)**
Bay S 276 **(298-04-4)**
Bay S 5660 **(122-14-5)**
Bay SRA 3886 **(22224-92-6)**
Bay SRA 7502 **(14816-18-3)**
Bay-SRA-12869 **(25311-71-1)**
Baycarb **(3766-81-2)**
Baychrom A **(10101-53-8)**
Baychrom F **(10101-53-8)**
Baycid **(55-38-9)**
Bayclean **(8001-54-5)**
Baycor **(55179-31-2)**
Bayer 21/116 **(8022-00-2)**
Bayer 21/199 **(56-72-4)**
Bayer 22/190 **(500-28-7)**
Bayer 22/190 **(2463-84-5)**
Bayer 25 634 **(5836-29-3)**
Bayer 25/154 **(919-86-8)**
Bayer 73 **(1420-04-8)**
Bayer 73 **(50-65-7)**
Bayer 1219 **(84-96-8)**
Bayer 1362 **(653-03-2)**
Bayer 1440 L **(41372-08-1)**
Bayer 1440 L **(555-30-6)**
Bayer 2353 **(50-65-7)**
Bayer 3231 **(68-76-8)**
Bayer 4935 **(93-75-4)**
Bayer 4964 **(2439-01-2)**
Bayer 5072 **(140-56-7)**
Bayer 5080 **(2032-59-9)**
Bayer 5081 **(327-98-0)**
Bayer 5312 **(536-33-4)**
Bayer 5360 **(443-48-1)**
Bayer 5712a **(731-27-1)**
Bayer 6159H **(21087-64-9)**
Bayer 6443H **(21087-64-9)**
Bayer 8169 **(8065-48-3)**
Bayer 8169 **(298-03-3)**
Bayer 9007 **(55-38-9)**
Bayer 9027 **(86-50-0)**
Bayer 15080 **(495-73-8)**
Bayer 15922 **(52-68-6)**
Bayer 16259 **(2642-71-9)**
Bayer 17147 **(86-50-0)**
Bayer 18510 **(2597-03-7)**
Bayer 19149 **(62-73-7)**
Bayer 19639 **(298-04-4)**
Bayer 20315 **(17040-19-6)**

Bayer 21097	**(301-12-2)**
Bayer 22555	**(140-56-7)**
Bayer 23129	**(640-15-3)**
Bayer 25141	**(115-90-2)**
Bayer 25648	**(1420-04-8)**
Bayer 29493	**(55-38-9)**
Bayer 29952	**(2703-13-1)**
Bayer 30686	**(93-75-4)**
Bayer 31686	**(93-75-4)**
Bayer 32651	**(3566-00-5)**
Bayer 33172	**(3878-19-1)**
Bayer 34727	**(2636-26-2)**
Bayer 36205	**(2439-01-2)**
Bayer 36743	**(2984-64-7)**
Bayer 37289	**(327-98-0)**
Bayer 37344	**(2032-65-7)**
Bayer 38156	**(333-43-7)**
Bayer 38819	**(4104-14-7)**
Bayer 39007	**(114-26-1)**
Bayer 39731	**(2631-40-5)**
Bayer 41367C	**(3766-81-2)**
Bayer 41637	**(3766-81-2)**
Bayer 41831	**(122-14-5)**
Bayer 44646	**(2032-59-9)**
Bayer 45,432	**(1113-02-6)**
Bayer 46131	**(12071-83-9)**
Bayer 47531	**(1085-98-9)**
Bayer 49854	**(731-27-1)**
Bayer 70533	**(14437-17-3)**
Bayer 71628	**(10265-92-6)**
Bayer 77488	**(14816-18-3)**
Bayer 78418	**(17109-49-8)**
Bayer 94337	**(21087-64-9)**
Bayer 96610	**(27355-22-2)**
Bayer A 139	**(800-24-8)**
Bayer E-605	**(56-38-2)**
Bayer E-838	**(299-45-6)**
Bayer-E 393	**(3689-24-5)**
Bayer E39 Soluble	**(800-24-8)**
Bayer L 13/59	**(52-68-6)**
Bayer S 4400	**(327-98-0)**
Bayer S 5660	**(122-14-5)**
Bayer S-1752	**(55-38-9)**
Bayer S767	**(115-90-2)**
Bayer SS2074	**(2439-01-2)**
Bayeritian	**(13463-67-7)**
Bayertitan	**(13463-67-7)**
Bayfidan	**(55219-65-3)**
Baygon (OSHA)	**(114-26-1)**
Bayleton	**(43121-43-3)**
Bayluscid	**(1420-04-8)**
Bayluscid	**(50-65-7)**
Bayluscide	**(1420-04-8)**
Baymat-Spray	**(55179-31-2)**
Baymix	**(56-72-4)**
Baymix 50	**(56-72-4)**
Bayol 55	**(8012-95-1)**
Bayol F	**(8012-95-1)**
Baypresol	**(41372-08-1)**
Baypresol	**(555-30-6)**
Bayrites	**(7727-43-7)**
Bayrusil	**(13593-03-8)**
Baysilon	**(63148-62-9)**
Baysilon M 100	**(63148-62-9)**
Baysilon M 500	**(63148-62-9)**
Baytan	**(123-88-6)**
Baytan	**(55219-65-3)**
Baytex	**(55-38-9)**
Baythion	**(14816-18-3)**
Baythroid	**(68359-37-5)**
Baythroid H	**(68359-37-5)**
Baytitan	**(13463-67-7)**
Baze michlerova (Czech)	**(101-61-1)**
Bazinon	**(333-41-5)**
Bazuden	**(333-41-5)**
B(b)F	**(205-99-2)**
B(c)AC	**(225-51-4)**
Be 724-A	**(73-48-3)**
Beacillin	**(1538-09-6)**
Beam	**(41814-78-2)**
Beamette	**(9003-07-0)**
Bean Seed Protectant	**(133-06-2)**
Bearberyy LS	**(8001-21-6)**
Beatine	**(67-03-8)**
Beaver White 200	**(14807-96-6)**
Bebenil	**(7055-03-0)**
Becafurazone	**(59-87-0)**
Becaptan	**(60-23-1)**
Becilan	**(58-56-0)**
Beckamine 21-511	**(9011-05-6)**
Beckamine NF 5	**(9011-05-6)**
Beckamine P 136	**(9011-05-6)**
Beckamine P 138	**(9011-05-6)**
Beckamine P 138-60	**(9011-05-6)**
Beckamine P 196M	**(9011-05-6)**
Becorel	**(434-07-1)**
Bedome	**(67-03-8)**
Beechwood Cresoate	**(8021-39-4)**
Beef Tallow	**(61789-97-7)**
Beesix	**(65-23-6)**
Beeswax	**(8012-89-3)**
Beeswax Absolute	**(8012-89-3)**
Beeswax Oil, Absolute	**(8012-89-3)**
Beeswax (White)	**(8012-89-3)**
Beeswax White	**(8012-89-3)**
Beeswax Yellow	**(8012-89-3)**
Beet-Kleen	**(101-21-3)**
Beet-Kleen	**(101-42-8)**
Beet-Kleen	**(122-42-9)**
Beet Molasses	**(68476-78-8)**
Beet Sugar Molasses	**(68476-78-8)**
Beetle 55	**(9011-05-6)**
Beetle 60	**(9011-05-6)**
Beetle 65	**(9011-05-6)**
Beetle 80	**(9011-05-6)**
Beetle 212-9	**(9011-05-6)**
Beetle BE 685	**(9011-05-6)**
Beetle BU 700	**(9011-05-6)**
Beetle XB 1050	**(9011-05-6)**
Beet sugar	**(57-50-1)**
Beflavine	**(83-88-5)**
Begiolan	**(67-03-8)**
Behenamide	**(3061-75-4)**
Behenic acid	**(112-85-6)**
Behenic acid amide	**(3061-75-4)**
Behenic acid, lithium salt	**(4499-91-6)**
Behenic acid, methyl ester	**(929-77-1)**
Behenic alcohol	**(661-19-8)**
Behenyl alcohol	**(661-19-8)**
Beih	**(51-12-7)**
Beivon	**(59-43-8)**
Bekadid	**(125-29-1)**
Belacid Milling Red G	**(10169-02-5)**
Belacid Milling Yellow R	**(6375-55-9)**

Name	CAS No.
Belacid Phloxine G	(3734-67-6)
Belamine Black GX	(1937-37-7)
Belamine Blue 2B	(2602-46-2)
Belamine Diazo Black BH	(2429-73-4)
Belamine Fast Brown BP	(2429-81-4)
Belamine Fast Brown BRLL	(16071-86-6)
Belamine Fast Brown M	(2429-82-5)
Belamine Fast Red 8 BL	(2610-11-9)
Belamine Sky Blue A	(2429-74-5)
Belamine Sky Blue FF	(2610-05-1)
Belanthrene Jade Green	(128-58-5)
Beldavrin	(114-49-8)
Belgard (9CI)	(95327-33-6)
Belgran	(34123-59-6)
Belima	(63428-84-2)
Bell Mine	(1305-62-0)
Bell Mine Pulverized Limestone	(1317-65-3)
Belloid FR	(2717-15-9)
Belmark	(51630-58-1)
Belt	(57-74-9)
Belustine	(13010-47-4)
Bemaco	(54-05-7)
Bemaphate	(54-05-7)
Bemasulph	(54-05-7)
Ben-30	(3813-05-6)
Ben-Allergin	(58-73-1)
Ben-Cornox	(3813-05-6)
Ben-Hex	(58-89-9)
Ben-U-Ron	(103-90-2)
Bena	(147-24-0)
Bena	(58-73-1)
Benachlor	(58-73-1)
Benactizina (Italian)	(302-40-9)
Benactyzin	(302-40-9)
Benactyzine	(302-40-9)
Benadon	(58-56-0)
Benadon	(58-73-1)
Benadrin	(58-73-1)
Benadryl	(147-24-0)
Benadryl	(58-73-1)
Benadryl hydrochloride	(147-24-0)
Benalan	(1861-40-1)
Benalaxyl	(71626-11-4)
Benanserin	(441-91-8)
Benapon	(58-73-1)
Benaquin	(54-05-7)
Benaspir	(50-78-2)
Benazalox	(3813-05-6)
Benazol P	(2440-22-4)
Benazolin	(3813-05-6)
Benazoline	(3813-05-6)
Benazyl	(50-55-5)
Bencarbate	(22781-23-3)
Benchinox	(495-73-8)
Bencidal Black E	(1937-37-7)
Bencidal Black RW	(2429-83-6)
Bencidal Blue 2B	(2602-46-2)
Bencidal Blue 3B	(72-57-1)
Bencidal Dark Green B	(3626-28-6)
Bencidal Fast Black G	(6428-31-5)
Bencidal Fast Brown M	(2429-82-5)
Bencidal Fast Violet N	(2586-60-9)
Bencidal Fast Yellow X	(1325-37-7)
Bencidal Navy Blue BH	(2429-73-4)
Bencidal Purple 4B	(992-59-6)
Bendectin	(8064-77-5)
Bendex	(13356-08-6)
Bendigon	(50-55-5)
Bendiocarb	(22781-23-3)
Bendiocarbe	(22781-23-3)
Bendioxide	(25057-89-0)
Bendopa	(59-92-7)
Bendralan	(132-93-4)
Bendrofluazide	(73-48-3)
Bendroflumethiazide	(73-48-3)
Bendylate	(147-24-0)
Benecid	(57-66-9)
Benefex	(1861-40-1)
Benefin	(1861-40-1)
Benemid	(57-66-9)
Benerva	(67-03-8)
Benesal	(65-45-2)
Benfluralin	(1861-40-1)
Benfos	(62-73-7)
Bengal	(9002-18-0)
Bengal Gelatin	(9002-18-0)
Bengal Isinglass	(9002-18-0)
Benguinox	(495-73-8)
Benhexol	(58-89-9)
Benicot	(98-92-0)
Benirol	(8001-54-5)
Benjamin Gum	(9000-05-9)
Benkfuran	(67-20-9)
Benlat	(17804-35-2)
Benlate	(17804-35-2)
Benlate 50	(17804-35-2)
Benlate 50 W	(17804-35-2)
Bennie	(60-13-9)
Benocten	(147-24-0)
Benodanil	(15310-01-7)
Benodin	(58-73-1)
Benodine	(58-73-1)
Benomyl (ACGIH,OSHA)	(17804-35-2)
Benomyl 50W	(17804-35-2)
Benopan	(3813-05-6)
Benoquin	(103-16-2)
Benovocylin	(50-50-0)
Benoxaprofen	(51234-28-7)
Benoxyl	(94-36-0)
Benozil	(1172-18-5)
Benzo-α-pyrone	(91-64-5)
Benquinox	(495-73-8)
Bensecal	(3813-05-6)
Bensulfoid	(7704-34-9)
Bensulide	(741-58-2)
Bensylyt NEN	(63-92-3)
Bensylyte	(59-96-1)
Bentanex	(13684-56-5)
Bentazon	(25057-89-0)
Bentazone	(25057-89-0)
Benthiocarb	(28249-77-6)
Bentone	(1332-58-7)
Bentone 38 (8CI,9CI)	(12001-31-9)
Bentonite	(1302-78-9)
Bentonite 2073	(1302-78-9)
Bentonite, Acid-leached	(70131-50-9)
Bentonite Magma	(1302-78-9)
Bentox 10	(58-89-9)
Bentride	(73-48-3)
Bentrol	(1689-83-4)
Bentylol	(77-19-0)
Benuron	(73-48-3)
Benuryl	(57-66-9)
Benvil	(4268-36-4)

Benylan (58-73-1)
Benylate (120-51-4)
Benz-O-Chlor (510-15-6)
Benz(a)aceanthrylene (203-33-8)
Benz(a)acridin-10-amine (9CI) (18936-75-9)
Benz(a)acridine (225-11-6)
Benz(a)acridine, 10-amino (18936-75-9)
Benz(a)acridine, 8,10-dimethyl (53-69-0)
Benz(a)acridine, 8,12-dimethyl (3518-05-6)
Benz(a)acridine, 9,12-dimethyl (17401-48-8)
Benz(a)acridine, 8,10,12-trimethyl (51787-43-0)
Benz(a)anthracen-7-amine (2381-18-2)
Benz(a)anthracen-8-amine (56961-60-5)
1,2-Benz(a)anthracene (56-55-3)
Benz(a)anthracene (56-55-3)
Benz(a)anthracene, 7-bromomethyl-12-methyl (16238-56-5)
Benz(a)anthracene, 12-butyl- (9CI) (93037-15-1)
Benz(a)anthracene, 6,8-diethyl (36911-94-1)
Benz(a)anthracene, 7,12-diethyl (16354-52-2)
Benz(a)anthracene, 8,12-diethyl (36911-95-2)
Benz(a)anthracene, 4,5-dimethyl (18429-70-4)
Benz(a)anthracene, 6,12-dimethyl (568-81-0)
Benz(a)anthracene, 7,11-dimethyl (35187-28-1)
Benz(a)anthracene, 7,12-dimethyl (57-97-6)
Benz(a)anthracene, 8,12-dimethyl (20627-31-0)
Benz(a)anthracene, 9,10-dimethyl (58429-99-5)
Benz(a)anthracene-d16, 7,12-dimethyl (32976-87-7)
Benz(a)anthracene, 2,10-dimethyl- (9CI) (63018-79-1)
Benz(a)anthracene, 2,9-dimethyl- (9CI) (572-89-4)
Benz(a)anthracene-5,6-dione (18508-00-4)
Benz(a)anthracene-7,12-dione (9CI) (2498-66-0)
Benz(a)anthracene-5,6-epoxide (790-60-3)
Benz(a)anthracene, 5,6-epoxy-5,6-dihydro (962-32-3)
Benz(a)anthracene, 12-ethyl (18868-66-1)
Benz(a)anthracene, 7-ethyl (3697-30-1)
Benz(a)anthracene, 8-ethyl (56961-62-7)
Benz(a)anthracene, 6-fluoro-7-methyl (2541-68-6)
Benz(a)anthracene, 5-fluoro-7-methyl- (9CI) (2498-63-7)
Benz(a)anthracene, 1-methyl (2498-77-3)
Benz(a)anthracene, 10-methyl (2381-15-9)
Benz(a)anthracene, 11-methyl (6111-78-0)
Benz(a)anthracene, 12-methyl (2422-79-9)
Benz(a)anthracene, 2-methyl (2498-76-2)
Benz(a)anthracene, 3-methyl (2498-75-1)
Benz(a)anthracene, 4-methyl (316-49-4)
Benz(a)anthracene, 7-methyl (2541-69-7)
Benz(a)anthracene, 8-methyl (2381-31-9)
Benz(a)anthracene, 9-methyl (2381-16-0)
Benz(a)anthracene, methyl- (9CI) (43178-22-9)
Benz(a)anthracene, 7-nitro- (20268-51-3)
Benz(a)anthracene 5,6-oxide (790-60-3)
Benz(a)anthracene, 5,6-oxide (962-32-3)
Benz(a)anthracene, 4,7,12-trimethyl (35187-24-7)
Benz(a)anthracene, 6,7,12-trimethyl (20627-33-2)
Benz(a)anthracene, 6,7,8-trimethyl (20627-32-1)
Benz(a)anthracene, 6,8,12-trimethyl (20627-34-3)
Benz(a)anthracene, 7,10,12-trimethyl (35187-27-0)
Benz(a)anthra-5,6-oxide (962-32-3)
Benzabar (50-31-7)
Benzac (1338-32-5)
Benzac (50-31-7)
Benzac (94-36-0)
Benzac 1281 (3426-62-8)
Benzac 354 (1338-32-5)
Benzac-1281 (50-31-7)
1,2-Benzacenaphthene (206-44-0)
Benzacillin (1538-09-6)
1,2-Benzacridine (225-11-6)
3,4-Benzacridine (225-51-4)
7,8-Benzacridine (French) (225-51-4)
Benzadone Blue RS (81-77-6)
Benzadone Brilliant Purple 2R (1324-55-6)
Benzadone Brilliant Purple 4R (1324-55-6)
Benzadone Brown BR (2475-33-4)
Benzadone Gold Orange G (128-70-1)
Benzadone Golden Yellow (128-66-5)
Benzadone Jade Green B (128-58-5)
Benzadone Jade Green X (128-58-5)
Benzadone Jade Green XBN (128-58-5)
Benzadone Jade Green XN (128-58-5)
Benzadone Olive R (2379-81-9)
Benzadox (5251-93-4)
Benzafinyl (60-13-9)
Benzak (Czech) (50-31-7)
Benzaknew (94-36-0)
Benzal Green (569-64-2)
Benzalaceton (German) (122-57-6)
Benzalacetone (122-57-6)
2-Benzalacetophenone (94-41-7)
Benzalacetophenone (94-41-7)
Benzal alcohol (100-51-6)
Benzalaniline (538-51-2)
N-Benzalaniline (538-51-2)
Benzal chloride (98-87-3)
Benzaldehyde (100-52-7)
Benzaldehyde (DOT) (100-52-7)
Benzaldehyde FFC (100-52-7)
Benzaldehyde Green (569-64-2)
Benzaldehyde Green (633-03-4)
Benzaldehyde, 4-amino (556-18-3)
Benzaldehyde anil (538-51-2)
Benzaldehyde, 3-bromo- (9CI) (3132-99-8)
Benzaldehyde, 2-bromo-4-hydroxy-5-methoxy- (9CI) (60632-40-8)
Benzaldehyde, 3-bromo-4-hydroxy-5-methoxy- (9CI) (2973-76-4)
Benzaldehyde, 4-butyl- (9CI) (1200-14-2)
Benzaldehyde, α-chloro- (98-88-4)
Benzaldehyde, o-chloro (89-98-5)
Benzaldehyde, p-chloro (104-88-1)
Benzaldehyde, 2-chloro- (9CI) (89-98-5)
Benzaldehyde, 3-chloro- (9CI) (587-04-2)
Benzaldehyde, m-chloro- (8CI) (587-04-2)
Benzaldehyde, chlorodihydroxy- (9CI) (103339-64-6)
Benzaldehyde, 3-chloro-4-hydroxy- (9CI) (2420-16-8)
Benzaldehyde, 2-chloro-4-hydroxy-3,5-dimethoxy- (9CI) (76341-69-0)
Benzaldehyde, 2-chloro-4-hydroxy-3-methoxy- (9CI) (82668-20-0)
Benzaldehyde, 2-chloro-4-hydroxy-5-methoxy- (9CI) (18268-76-3)
Benzaldehyde, 3-chloro-4-hydroxy-5-methoxy- (9CI) (19463-48-0)
Benzaldehyde, chloro-4-hydroxy-3-methoxy- (9CI) (61670-76-6)
Benzaldehyde cyanohydrin (532-28-5)
Benzaldehyde, 3,5-dibromo-4-hydroxy- (8CI,9CI) (2973-77-5)
Benzaldehyde, 3,5-dibromo-4-hydroxy-, (2,4-dinitrophenyl)oxime (13181-17-4)
Benzaldehyde, 2,4-dichloro- (9CI) (874-42-0)
Benzaldehyde, 2,6-dichloro- (9CI) (83-38-5)
Benzaldehyde, 3,4-dichloro- (9CI) (6287-38-3)
Benzaldehyde, dichloro- (9CI) (31155-09-6)
Benzaldehyde, dichloro-2-hydroxy- (9CI) (91930-03-9)
Benzaldehyde, 3,5-dichloro-4-hydroxy- (8CI,9CI) (2314-36-5)
Benzaldehyde, 2,6-dichloro-4-hydroxy-3,5-dimethoxy- (9CI) (76330-06-8)
Benzaldehyde, 2,3-dichloro-4-hydroxy-5-methoxy- (9CI) (18268-69-4)
Benzaldehyde, 2,6-dichloro-4-hydroxy-3-methoxy- (9CI) (108545-01-3)
Benzaldehyde, 4-(diethylamino)- (9CI) (120-21-8)
Benzaldehyde, p-(diethylamino)- (8CI) (120-21-8)
Benzaldehyde, 4-(diethylamino)-, diphenylhydrazone (9CI) (68189-23-1)
Benzaldehyde, 4-(diethylamino)-2-hydroxy- (9CI) (17754-90-4)

Benzaldehyde, 2,4-dihydroxy- **(95-01-2)**
Benzaldehyde, 3,4-dihydroxy- **(139-85-5)**
Benzaldehyde, 3,4-dimethoxy- **(120-14-9)**
Benzaldehyde, 2,5-dimethoxy- (9CI) **(93-02-7)**
Benzaldehyde, 3,5-dimethoxy-4-hydroxy **(134-96-3)**
Benzaldehyde, 2,5-dimethyl- (8CI,9CI) **(5779-94-2)**
Benzaldehyde, 3,4-dimethyl- (8CI,9CI) **(5973-71-7)**
Benzaldehyde, dimethyl- (8CI,9CI) **(28351-09-9)**
Benzaldehyde, p-(dimethylamino) **(100-10-7)**
Benzaldehyde, (dimethylamino)- (9CI) **(28602-27-9)**
Benzaldehyde, (4-(dimethylamino)-6-(trichloromethyl)-s-triazin-2-yl)hydrazone (8CI) **(30359-62-7)**
Benzaldehyde, 4-(1,1-dimethylethyl)- **(939-97-9)**
Benzaldehyde, ethenyl- (9CI) **(43145-54-6)**
Benzaldehyde, p-ethoxy **(10031-82-0)**
Benzaldehyde, 2-ethoxy- (9CI) **(613-69-4)**
Benzaldehyde, o-ethoxy- (8CI) **(613-69-4)**
Benzaldehyde, 3-ethoxy-4-hydroxy **(121-32-4)**
Benzaldehyde, 4-ethyl- (9CI) **(4748-78-1)**
Benzaldehyde, ethyl- (9CI) **(53951-50-1)**
Benzaldehyde, p-ethyl- (8CI) **(4748-78-1)**
Benzaldehyde, m-hydroxy **(100-83-4)**
Benzaldehyde, o-hydroxy- **(90-02-8)**
Benzaldehyde, p-hydroxy **(123-08-0)**
Benzaldehyde, 3-hydroxy- (9CI) **(100-83-4)**
Benzaldehyde, 4-hydroxy- (9CI) **(123-08-0)**
Benzaldehyde, 4-hydroxy-3,5-bis(3-methyl-2-butenyl)- (9CI) **(52275-04-4)**
Benzaldehyde, 2-hydroxy-3-methoxy **(148-53-8)**
Benzaldehyde, 3-hydroxy-4-methoxy **(621-59-0)**
Benzaldehyde, 4-hydroxy-3-methoxy- **(121-33-5)**
Benzaldehyde, 4-hydroxy-3-methoxy-, manuf. of, distn. residues **(73246-95-4)**
Benzaldehyde, p-isopropyl **(122-03-2)**
Benzaldehyde, 2-methoxy- (9CI) **(135-02-4)**
Benzaldehyde, 3-methoxy- (9CI) **(591-31-1)**
Benzaldehyde, methyl- **(1334-78-7)**
Benzaldehyde, 2-methyl- (9CI) **(529-20-4)**
Benzaldehyde, 3-methyl- (9CI) **(620-23-5)**
Benzaldehyde, 4-methyl- (9CI) **(104-87-0)**
Benzaldehyde, 3,4-(methylenedioxy)- **(120-57-0)**
Benzaldehyde, 4-(1-methylethyl)- (9CI) **(122-03-2)**
Benzaldehyde, 4-methyl-, oxime, (E)- (9CI) **(3717-15-5)**
Benzaldehyde, m-nitro **(99-61-6)**
Benzaldehyde, o-nitro **(552-89-6)**
Benzaldehyde, p-nitro **(555-16-8)**
Benzaldehyde, 3-nitro- (9CI) **(99-61-6)**
Benzaldehyde, 4-nitro- (9CI) **(555-16-8)**
Benzaldehyde, oxime, (Z)- (8CI,9CI) **(622-32-2)**
Benzaldehyde, pentachloro-, (4-(dimethylamino)-6-(trichloromethyl)-s-triazin-2-yl)hydrazone (8CI) **(30359-63-8)**
Benzaldehyde, 4-pentyl- (9CI) **(6853-57-2)**
Benzaldehyde, 3-phenoxy- (9CI) **(39515-51-0)**
Benzaldehyde, 3-(((phenylamino)carbonyl)oxy)- (9CI) **(37070-87-4)**
Benzaldehyde, 4-(((phenylamino)carbonyl)oxy)- (9CI) **(37076-88-3)**
Benzaldehyde, 3-propyl- (9CI) **(103528-31-0)**
Benzaldehyde, tetrahydro- **(1321-16-0)**
Benzaldehyde, 2,3,6-trichloro **(4659-47-6)**
Benzaldehyde, 4-(trifluoromethyl)- (9CI) **(455-19-6)**
Benzaldehyde, 2,4,5-trimethoxy **(4460-86-0)**
Benzaldehyde, 3,4,5-trimethoxy **(86-81-7)**
Benzaldehyde, 2,4,6-trimethyl **(487-68-3)**
Benzaldehyde, 2,4,5-trimethyl- (8CI,9CI) **(5779-72-6)**
Benzaldehyde, vinyl- **(43145-54-6)**
Benzaldehydkyanhydrin (Czech) **(532-28-5)**
Benzalin **(146-22-5)**
Benzalkonium bromide **(7281-04-1)**
Benzalkonium chloride **(8001-54-5)**
2-Benzamido-4-chloro-5-nitrophenol **(5099-06-9)**
Benzamide **(55-21-0)**
Benzamide, o-amino **(88-68-6)**
Benzamide, 2-amino- (9CI) **(88-68-6)**
Benzamide, 4-amino- (9CI) **(2835-68-9)**
Benzamide, p-amino- (8CI) **(2835-68-9)**
Benzamide, 4-amino-N-(4-aminophenyl)- (9CI) **(785-30-8)**
Benzamide, 4-amino-5-chloro-N-(2-(diethylamino)ethyl)-2-methoxy- (9CI) **(364-62-5)**
Benzamide, 3-amino-2,5-dichloro- (8CI,9CI) **(2280-93-5)**
Benzamide, N-(4-amino-2,5-diethoxyphenyl)- (9CI) **(120-00-3)**
Benzamide, p-amino-N-(2-(diethylamino)ethyl) **(51-06-9)**
Benzamide, 4-amino-N-(2-(diethylamino)ethyl)- (9CI) **(51-06-9)**
Benzamide, 4-amino-N-(3-((2-hydroxyethyl)sulfonyl)phenyl)-(9CI) **(20241-68-3)**
Benzamide, 3-amino-4-methyl- (9CI) **(19406-86-1)**
Benzamide, 5-(aminosulfonyl)-N-((1-ethyl-2-pyrrolidinyl)methyl)-2-methoxy- (9CI) **(15676-16-1)**
Benzamide, N-(5-(bis(2-(acetyloxy)ethyl)amino)-2-((4-nitrophenyl)azo)phenyl)- (9CI) **(29765-00-2)**
Benzamide, 3-(((2,4-bis(1,1-dimethylpropyl)phenoxy)acetyl)-amino)-N-(4,5-dihydro-5-oxo-1-(2,4,6-trichlorophenyl)-1H-pyrazol-3-yl)- (9CI) **(31188-91-7)**
Benzamide, 4-(((2,4-bis(1,1-dimethylpropyl)phenoxy)acetyl)-amino) N-(4,5-dihydro-5-((4-methoxyphenyl)azo)-5-oxo-1-(2,4,6-trichlorophenyl)-1H-pyrazol-3-yl)- (9CI) **(28279-36-9)**
Benzamide, 5-bromo-N,2-dihydroxy- (9CI) **(5798-94-7)**
Benzamide, 5-bromo-N-(4-bromophenyl)-2-hydroxy- (9CI) **(87-12-7)**
Benzamide, 4-bromo-N-(1,1-dimethyl-2-propynyl)- (9CI) **(80045-50-7)**
Benzamide, 2-bromo-N-phenyl- **(10282-57-2)**
Benzamide, 2-chloro- (9CI) **(609-66-5)**
Benzamide, 4-chloro- (9CI) **(619-56-7)**
Benzamide, o-chloro- (8CI) **(609-66-5)**
Benzamide, p-chloro- (8CI) **(619-56-7)**
Benzamide, 5-chloro-N-(2-chloro-4-nitrophenyl)-2-hydroxy- **(50-65-7)**
Benzamide, 3-chloro-N-(4-chlorophenyl)-N-hydroxy- (9CI) **(36016-27-0)**
Benzamide, 4-chloro-N-(1,1-dimethyl-2-propynyl)- (9CI) **(24911-15-7)**
Benzamide, p-chloro-N-(1,1-dimethyl-2-propynyl)- (8CI) **(24911-15-7)**
Benzamide, N-(5-chloro-2-hydroxy-4-nitrophenyl)- (9CI) **(5099-06-9)**
Benzamide, 3,3'-((2-chloro-5-methyl-1,4-phenylene)bis(imino-(1-acetyl-2-oxo-2,1-ethanediyl)azo))bis(4-chloro-N-(3-chloro-2-methylphenyl)- (9CI) **(5580-57-4)**
Benzamide, N-(2-chlorophenyl) **(1020-39-9)**
Benzamide, 2-chloro-N-phenyl- (9CI) **(6833-13-2)**
Benzamide, N-(((4-chlorophenyl)amino)carbonyl)-2,6-difluoro- **(35367-38-5)**
Benzamide, N-(4-chlorophenyl)-3-fluoro-N-hydroxy- (9CI) **(94370-36-2)**
Benzamide, N-(4-chlorophenyl)-N-hydroxy- (9CI) **(1528-82-1)**
Benzamide, N-(4-chlorophenyl)-N-hydroxy-3-methoxy- (9CI) **(77915-81-2)**
Benzamide, N-(4-chlorophenyl)-N-hydroxy-3-methyl- (9CI) **(36016-24-7)**
Benzamide, N-(4-chlorophenyl)-N-hydroxy-3-nitro- (9CI) **(36016-30-5)**
Benzamide, 2-chloro-N-(((4-(trifluoromethoxy)phenyl)amino)-carbonyl)- **(64628-44-0)**
Benzamide, 4-cyano-N-(1,1-dimethyl-2-propynyl)- (9CI) **(98404-93-4)**
Benzamide, 3,5-dibromo-N-(4-bromophenyl)-2-hydroxy- **(87-10-5)**
Benzamide, 3,5-dibromo-N-(4-bromophenyl)-2-hydroxy- (9CI) **(87-10-5)**
Benzamide, 3,5-dibromo-2-hydroxy-N-phenyl- (9CI) **(2577-72-2)**
Benzamide, 3,5-dibromo-2-hydroxy-N-(3-(trifluoromethyl)-phenyl)- (9CI) **(4776-06-1)**
Benzamide, N,N-dibutyl- (6CI,8CI,9CI) **(25033-65-2)**
Benzamide, 2,6-dichloro- (9CI) **(2008-58-4)**
Benzamide, 2,5-dichloro- (8CI,9CI) **(5980-26-7)**
Benzamide, 3,5-dichloro- (8CI,9CI) **(5980-23-4)**
Benzamide, 3,5-dichloro-N-(1,1-dimethyl-2-propynyl) **(23950-58-5)**
Benzamide, 2,6-dichloro-N-(hydroxymethyl)thio **(1953-89-5)**
Benzamide, N-(3,4-dichlorophenyl)- (9CI) **(10286-75-6)**
Benzamide, N-(3,4-dichlorophenyl)-N-((dimethylamino)carbonyl) **(3134-12-1)**
Benzamide, 2,6-dichlorothio **(1918-13-4)**
Benzamide, N,N-diethyl **(1696-17-9)**

Benzamide, 2,6-diethyl- (6CI,9CI) **(89151-70-2)**
Benzamide, N,N-diethyl-3-ethoxy-4-hydroxy **(13898-68-5)**
Benzamide, N,N-diethyl-3-methyl- **(134-62-3)**
Benzamide, N,N-dimethyl **(611-74-5)**
Benzamide, N-(p-(2-(dimethylamino)ethoxy)benzyl)-3,4,5-trimethoxy- **(138-56-7)**
Benzamide, N-((4-(2-(dimethylamino)ethoxy)phenyl)methyl)-3,4,5-trimethoxy- **(138-56-7)**
Benzamide, N-(1,1-dimethylethyl)-3-methyl- (9CI) **(42498-33-9)**
Benzamide, 3,3'-((2,5-dimethyl-1,4-phenylene)bis(imino-(1-acetyl-2-oxo-2,1-ethanediyl)azo))bis(4-chloro-N-(5-chloro-2-methylphenyl)- (9CI) **(5280-80-8)**
Benzamide, N-(1,1-dimethyl-2-propynyl)- (8CI,9CI) **(33244-86-9)**
Benzamide, N-(1,1-dimethyl-2-propynyl)-4-fluoro- (9CI) **(80045-51-8)**
Benzamide, N-(1,1-dimethyl-2-propynyl)-4-methoxy- (9CI) **(39108-81-1)**
Benzamide, N-(1,1-dimethyl-2-propynyl)-4-methyl- (9CI) **(39108-91-3)**
Benzamide, N-(1,1-dimethyl-2-propynyl)-4-(1-methylethyl)- (9CI) **(80045-52-9)**
Benzamide, N-(1,1-dimethyl-2-propynyl)-4-nitro- (9CI) **(39098-01-6)**
Benzamide, N,N'-(dithiodi-2,1-phenylene)bis- **(135-57-9)**
Benzamide, o-ethoxy **(938-73-8)**
Benzamide, 2-ethoxy- (9CI) **(938-73-8)**
Benzamide, N,N'-(ethoxymethylsilylene)bis(N-methyl- (9CI) **(16230-35-6)**
Benzamide, 2-fluoro-N-phenyl- (9CI) **(1747-80-4)**
Benzamide, 2-hydroxy- **(65-45-2)**
Benzamide, N-hydroxy- **(495-18-1)**
Benzamide, o-hydroxy- **(65-45-2)**
Benzamide, 2-hydroxy-5-(1-hydroxy-2-((1-methyl-3-phenylpropyl)-amino)ethyl) **(36894-69-6)**
Benzamide, 4-methoxy-3-nitro-N-phenyl- (9CI) **(97-32-5)**
Benzamide, N-methyl **(613-93-4)**
Benzamide, 2-methyl- (9CI) **(527-85-5)**
Benzamide, 4-methyl- (9CI) **(619-55-6)**
Benzamide, 2-methyl-3,5-dinitro- **(148-01-6)**
Benzamide, N-(3-(1-methylethoxy)phenyl)-2-(trifluoromethyl) **(66332-96-5)**
Benzamide, N-(1-methylethyl)-4-((2-methylhydrazino)methyl)-(9CI) **(671-16-9)**
Benzamide, N-(1-methylethyl)-4-((2-methylhydrazino)-methyl)-, monohydrochloride **(366-70-1)**
Benzamide, (p-(N'-methylhydrazinomethyl)-N-isopropyl)- **(366-70-1)**
Benzamide, 2-methyl-N-(3-(1-methylethoxy)phenyl) **(55814-41-0)**
Benzamide, 4-methyl-3-nitro- (9CI) **(19013-11-7)**
Benzamide, 2-methyl-N-phenyl **(7055-03-0)**
Benzamide, o-nitro **(610-15-1)**
Benzamide, p-nitro **(619-80-7)**
Benzamide, 3-nitro- (9CI) **(645-09-0)**
Benzamide, m-nitro- (8CI) **(645-09-0)**
Benzamide, 4-nitro-N-(4-nitrophenyl)- (9CI) **(6333-15-9)**
Benzamide, 2,3,4,5,6-pentafluoro-N-(2-phenylethyl)- (9CI) **(38842-14-7)**
Benzamide, N-phenyl- (9CI) **(93-98-1)**
Benzamide, N,N'-(10,15,16,17-tetrahydro-5,10,15,17-tetraoxo-5H-dinaphtho(2,3-a:2',3'-i)carbazole-4,9-diyl)bis- (9CI) **(131-92-0)**
Benzamide, thio **(2227-79-4)**
Benzamide, N,N',N''-1,3,5-triazine-2,4,6-triyltris- (9CI) **(5637-84-3)**
Benzamide, N,N',N''-s-triazine-2,4,6-triyltris- (8CI) **(5637-84-3)**
Benzamide, 2,3,6-trichloro-5-cyano-4-hydroxy- (9CI) **(115044-73-0)**
Benzamidine, 4,4'-(pentamethylenedioxy)di-, bis(β-hydroxy-ethanesulfonate) **(140-64-70)**
Benzamidoacetic acid **(495-69-2)**
(Benzamidooxy)acetic acid **(5251-93-4)**
Benzamidooxyacetic acid **(5251-93-4)**
Benzamidoxime, 2,6-dichloro- (8CI) **(23505-21-7)**
Benzamil Black E **(1937-37-7)**
Benzamil Supra Brown BRLL **(16071-86-6)**
Benzaminblau 3B **(72-57-1)**
Benzamine Blue **(72-57-1)**
Benzamine Blue 3B **(72-57-1)**
Benzamine, N,N,2-trimethyl- (9CI) **(609-72-3)**
Benzamphetamine **(60-13-9)**
Benzanil Black BH **(2429-73-4)**
Benzanil Black RW **(2429-83-6)**
Benzanil Blue 2B **(2602-46-2)**
Benzanil Blue 3BN **(72-57-1)**
Benzanil Blue R **(72-57-1)**
Benzanil Brown BS **(2429-81-4)**
Benzanil Brown M **(2429-82-5)**
Benzanil Dark Green BW **(3626-28-6)**
Benzanil Fast Black G **(6428-31-5)**
Benzanil Fast Red K **(2610-11-9)**
Benzanil Purpurine 4B **(992-59-6)**
Benzanil Scarlet 3B **(6358-29-8)**
Benzanil Sky Blue **(2429-74-5)**
Benzanil Sky Blue FF **(2610-05-1)**
Benzanil Supra Blue 2GN **(2610-05-1)**
Benzanil Supra Brown BRLN **(16071-86-6)**
Benzanil Violet BXN **(6426-67-1)**
Benzanil Violet N **(2586-60-9)**
Benzanil Yellow R **(1325-37-7)**
Benzanilide **(93-98-1)**
Benzanilide, 4'-amino-2',5'-diethoxy- (8CI) **(120-00-3)**
Benzanilide, 3-amino-4-methoxy **(120-35-4)**
Benzanilide, 2'-chloro- (8CI) **(1020-39-9)**
Benzanilide, 2-chloro- (8CI) **(6833-13-2)**
Benzanilide, 5'-chloro-2'-hydroxy-4'-nitro- **(5099-06-9)**
Benzanilide, 4,4'-diamino- (8CI) **(785-30-8)**
Benzanilide, 3',4'-dichloro- (8CI) **(10286-75-6)**
Benzanilide, 2',2'''-dithiobis **(135-57-9)**
Benzanilide, 2-fluoro- (8CI) **(1747-80-4)**
Benzanilide, 2-iodo **(15310-01-7)**
Benzanol Brilliant Scarlet 3B **(6358-29-8)**
Benzanol Brown M **(2429-82-5)**
Benz(3,4)anthra(1,2-b)oxirene (9CI) **(790-60-3)**
Benz(3,4)anthra(1,2-b)oxirene, 1a,11b-dihydro- (9CI) **(962-32-3)**
1,2-Benzanthracene **(56-55-3)**
1,2:5,6-Benzanthracene **(53-70-3)**
2,3-Benzanthracene **(92-24-0)**
Benzanthracene **(56-55-3)**
1,2-Benzanthrazen (German) **(56-55-3)**
1,2-Benzanthrene **(56-55-3)**
2,3-Benzanthrene **(92-24-0)**
Benzanthrene **(56-55-3)**
Benzanthrenone **(82-05-3)**
Benzanthrone **(82-05-3)**
Benzantine **(58-73-1)**
Benz(a)oxireno(c)anthracene **(790-60-3)**
Benz(a)phenanthrene **(218-01-9)**
3,4-Benz(a)pyrene **(50-32-8)**
Benz(a)pyrene **(50-32-8)**
Benz(a)pyrene 4,5-oxide **(37574-47-3)**
Benzar **(3813-05-6)**
Benzathine benzylpenicillin **(1538-09-6)**
Benzathine penicillin **(1538-09-6)**
Benzathine penicillin G **(1538-09-6)**
Benzathonium chloride **(121-54-0)**
Benzazide **(582-61-6)**
Benzazimide **(90-16-4)**
Benzazimidone **(90-16-4)**
1-Benzazine **(91-22-5)**
2-Benzazine **(119-65-3)**
1-Benzazole **(120-72-9)**
Benz(b)anthracene **(92-24-0)**
11H-Benz(bc)aceanthrylen-11-one (9CI) **(86853-89-6)**
Benz(c)acridine **(225-51-4)**
Benz(c)acridine, 7,10-dimethyl **(2381-40-0)**
Benz(c)acridine, 7,11-dimethyl **(32740-01-5)**

Benz(c)acridine, 7,9-dimethyl **(963-89-3)**
Benz(c)acridine, 5-methyl- (8CI,9CI) **(3519-87-7)**
Benz(c)acridine, 7,8,9,11-tetramethyl **(51787-44-1)**
Benz(c)acridine, 7,9,11-trimethyl **(51787-42-9)**
1,2-Benzcarbazole **(239-01-0)**
2,3-Benzcarbazole **(243-28-7)**
Benzchinamide **(63-12-7)**
Benzchinamidum **(63-12-7)**
Benzo-chinon (German) **(106-51-4)**
3,4-Benzchrysene **(213-46-7)**
Benzcurine iodide **(65-29-2)**
7H-Benz(de)anthracene-7-one **(82-05-3)**
7H-Benz(de)anthracen-7-one **(82-05-3)**
1H-Benz(de)anthracen-1-one (9CI) **(86853-88-5)**
7H-Benz(de)anthracen-7-one, 2-methyl- (7CI,8CI,9CI) **(82-03-1)**
15,16-Benzdehydrocholanthrene **(203-20-3)**
1H-Benz(de)isoquinoline-1,3(2H)-dione (9CI) **(81-83-4)**
1H-Benz(de)isoquinoline-1,3(2H)-dione, 6-amino-2-(2,4-dimethylphenyl)- (9CI) **(2478-20-8)**
1H-Benz(de)isoquinoline-1,3(2H)-dione, 2-butyl-6-(butylamino)- (9CI) **(19125-99-6)**
Benz(e)acenaphthylene **(201-06-9)**
3,4-Benz(e)acephenanthrylene **(205-99-2)**
Benz(e)acephenanthrylene **(205-99-2)**
Benz(e)-as-indacene-7,10-dione, 1,2,3,4,5,6,6a,6b,8,9,10a,10b-dodecahydro- (8CI) **(31991-61-4)**
Benzebar **(60-13-9)**
Benzedrex **(101-40-6)**
Benzedrina **(60-13-9)**
(+-)-Benzedrine **(300-62-9)**
Benzedrine **(300-62-9)**
Benzedrine **(60-13-9)**
Benzedrine **(60-15-1)**
dl-Benzedrine **(300-62-9)**
d-Benzedrine sulfate **(51-63-8)**
Benzedryna **(60-13-9)**
Benzeen (Dutch) **(71-43-2)**
Benzeethanamine, α,α-dimethyl (9CI) **(122-09-8)**
Benzehist **(147-24-0)**
Benzen (Polish) **(71-43-2)**
Benzenacetic acid **(103-82-2)**
Benzenamine **(62-53-3)**
Benzenamine, N-(2-acetoxy)ethyl-N-(2-cyano)ethyl-4-(((2,6-dichloro-4-nitro)phenyl)azo)- **(5261-31-4)**
Benzenamine, ar-amino- **(25265-76-3)**
Benzenamine, 4-((4-amino-3-methylphenyl)(4-imino-3-methyl-2,5-cyclohexadien-1-ylidene)methyl)-2-methyl-, monohydrochloride (9CI) **(3248-91-7)**
Benzenamine, 4-((4-aminophenyl)azo)-N,N-dimethyl- (9CI) **(539-17-3)**
Benzenamine, 4-((4-aminophenyl) (4-imino-2,5-cyclohexadien-1-ylidene)methyl)-, nitric acid salt **(61467-64-9)**
Benzenamine, 4-((4-aminophenyl)(4-imino-2,5-cyclohexadien-1-ylidene)methyl), monohydrochloride **(569-61-9)**
Benzenamine, 4-((4-aminophenyl)(4-imino-2,5-cyclohexadien-1-ylidene)methyl)-, mononitrate (9CI) **(61467-64-9)**
Benzenamine, 4-((4-aminophenyl)(4-imino-2,5-cyclohexadien-1-ylidene)methyl)-2-methyl- **(3248-93-9)**
Benzenamine, 4-((4-aminophenyl)(4-imino-2,5-cyclohexadien-1-ylidene)methyl)-2-methyl-, monohydrochloride **(632-99-5)**
Benzenamine, 2-((4-aminophenyl)methyl)- (9CI) **(1208-52-2)**
Benzenamine, 2,4-bis((4-aminophenyl)methyl)- (9CI) **(25834-80-4)**
Benzenamine, N,N-bis(2-chloroethyl)-4-methyl-2,6-dinitro- **(26389-78-6)**
Benzenamine, 2,6-bis(1,1-dimethylethyl)-4-nitro- (9CI) **(5180-59-6)**
Benzenamine, 2,6-bis(1-methylethyl)- **(24544-04-5)**
Benzenamine, 3,5-bis(trifluoromethyl)- (9CI) **(328-74-5)**
Benzenamine, 4-bromo- **(106-40-1)**
Benzenamine, 2-bromo- (9CI) **(615-36-1)**
Benzenamine, 3-bromo- (9CI) **(591-19-5)**
Benzenamine, ar-bromo- (9CI) **(55777-84-9)**
Benzenamine, 2-bromo-6-chloro-4-nitro- (9CI) **(99-29-6)**
Benzenamine, 2-bromo-4,6-dichloro- (9CI) **(697-86-9)**
Benzenamine, 4-bromo-3,5-dichloro- (9CI) **(1940-29-0)**
Benzenamine, 4-bromo-N,N-diethyl- (9CI) **(2052-06-4)**
Benzenamine, bromo-N,N-diethyl- (9CI) **(63460-05-9)**
Benzenamine, ar-bromo-ar,ar-dinitro- (9CI) **(63460-09-3)**
Benzenamine, 4-bromo-, hydrochloride (9CI) **(624-19-1)**
Benzenamine, 4-bromo-3-methyl- (9CI) **(6933-10-4)**
Benzenamine, 4-bromo-2-nitro- (9CI) **(875-51-4)**
Benzenamine, ar-bromo-ar-nitro- (9CI) **(63460-07-1)**
Benzenamine, 4-butyl- (9CI) **(104-13-2)**
Benzenamine, n-butyl- (9CI) **(1126-78-9)**
Benzenamine, 4,4'-carbonimidoylbis(N,N-diethyl-, monohydrochloride (9CI) **(6358-36-7)**
Benzenamine, 4,4'-carbonimidoylbis(N,N-diethyl-, mononitrate (9CI) **(43130-12-7)**
Benzenamine, 4,4'-carbonimidoylbis(N,N-dimethyl- (9CI) **(492-80-8)**
Benzenamine, 4,4'-carbonimidoylbis(N,N-dimethyl-, monohydrochloride (9CI) **(2465-27-2)**
Benzenamine, 2-chloro- (9CI) **(95-51-2)**
Benzenamine, 3-chloro- (9CI) **(108-42-9)**
Benzenamine, chloro- (9CI) **(27134-26-5)**
Benzenamine, 2-chloro-N,N-dimethyl- **(698-01-1)**
Benzenamine, 3-chloro-N,N-dimethyl- (9CI) **(6848-13-1)**
Benzenamine, 3-chloro-N-(4-chlorophenyl)- (9CI) **(15979-79-0)**
Benzenamine, 2-chloro-4,6-dinitro- **(3531-19-9)**
Benzenamine, 3-chloro-2,6-dinitro- (9CI) **(10250-71-2)**
Benzenamine, 4-chloro-2,6-dinitro- (9CI) **(5388-62-5)**
Benzenamine, N-(2-chloroethyl)-N-ethyl- (9CI) **(92-49-9)**
Benzenamine, 3-chloro-4-fluoro- (9CI) **(367-21-5)**
Benzenamine, 4-chloro-, hydrochloride **(20265-96-7)**
Benzenamine, 2-chloro-, hydrochloride (9CI) **(137-04-2)**
Benzenamine, 3-chloro-, hydrochloride (9CI) **(141-85-5)**
Benzenamine, 3-chloro-4-methoxy- (9CI) **(5345-54-0)**
Benzenamine, 2-chloro-4-methyl- **(615-65-6)**
Benzenamine, 4-chloro-3-methyl- (9CI) **(7149-75-9)**
Benzenamine, 4-chloro-2-methyl-, hydrochloride **(3165-93-3)**
Benzenamine, 5-chloro-2-methyl-, hydrochloride (9CI) **(6259-42-3)**
Benzenamine, 2-chloro-6-nitro- (9CI) **(769-11-9)**
Benzenamine, 3-chloro-4-nitro- (9CI) **(825-41-2)**
Benzenamine, 3-chloro-5-nitro- (9CI) **(5344-44-5)**
Benzenamine, 4-chloro-2-nitro- (9CI) **(89-63-4)**
Benzenamine, 5-chloro-2-nitro- (9CI) **(1635-61-6)**
Benzenamine, 3-chloro-N-(4-nitrophenyl)- (9CI) **(15979-85-8)**
Benzenamine, 4-(((2-chloro-4-nitro)phenyl)azo)-N-(2-cyanoethyl)-N-(2-hydroxyethyl)- **(6657-33-6)**
Benzenamine, 4-(((2-chloro-4-nitro)phenyl)azo)-3-methyl-N-(2-cyanoethyl)-N-ethyl- **(16586-43-9)**
Benzenamine, 3-chloro-N-phenyl- (9CI) **(101-17-7)**
Benzenamine, 4-chloro-N-phenyl- (9CI) **(1205-71-6)**
Benzenamine, N-(3-chlorophenyl)-2,4-dinitro- (9CI) **(16220-58-9)**
Benzenamine, 4-chloro-2-(trifluoromethyl)- (9CI) **(445-03-4)**
Benzenamine, 4-decyl- (9CI) **(37529-30-9)**
Benzenamine, 2,4-dibromo- (9CI) **(615-57-6)**
Benzenamine, dibromo- (9CI) **(63505-64-6)**
Benzenamine, 2,6-dibromo-4-nitro- (9CI) **(827-94-1)**
Benzenamine, N,N-dibutyl- (9CI) **(613-29-6)**
Benzenamine, 2,3-dichloro- (9CI) **(608-27-5)**
Benzenamine, 2,4-dichloro- (9CI) **(554-00-7)**
Benzenamine, 2,5-dichloro- (9CI) **(95-82-9)**
Benzenamine, 2,6-dichloro- (9CI) **(608-31-1)**
Benzenamine, 3,4-dichloro- (9CI) **(95-76-1)**
Benzenamine, 3,5-dichloro- (9CI) **(626-43-7)**
Benzenamine, ar,ar-dichloro- (9CI) **(27134-27-6)**
Benzenamine, 2,6-dichloro-N,N-dimethyl- **(56961-05-8)**

Benzenamine, 2,6-dichloro-N-2-imidazolidinylidene- (9CI) **(4205-90-7)**
Benzenamine, 2,4-dichloro-6-nitro- (9CI) **(2683-43-4)**
Benzenamine, 2,6-dichloro-4-nitro- (9CI) **(99-30-9)**
Benzenamine, 4,5-dichloro-2-nitro- (9CI) **(6641-64-1)**
Benzenamine, 2,6-dichloro-4-nitro-, Mixt. with pentachloronitrobenzene (9CI) **(37203-85-3)**
Benzenamine, 2,4-diethyl- (9CI) **(14719-47-2)**
Benzenamine, 2,6-diethyl- (9CI) **(579-66-8)**
Benzenamine, 3,4-diethyl- (9CI) **(54675-14-8)**
Benzenamine, 3,5-diethyl- (9CI) **(1701-68-4)**
Benzenamine, N,N-diethyl- (9CI) **(91-66-7)**
Benzenamine, N,N-diethyl-3-methyl- (9CI) **(91-67-8)**
Benzenamine, 2,6-diethyl-N-methylene- (9CI) **(35203-08-8)**
Benzenamine, N,N-diethyl-4-((4-nitrophenyl)azo)- (9CI) **(3025-52-3)**
Benzenamine, N,N-diethyl-4-((5-nitro-2-thiazolyl)azo)- (9CI) **(54289-46-2)**
Benzenamine, N,N-diethyl-4-(phenylazo) **(2481-94-9)**
Benzenamine, 2,4-difluoro- (9CI) **(367-25-9)**
Benzenamine, 2,5-dimethoxy- (9CI) **(102-56-7)**
Benzenamine, 2,4-dimethoxy-, hydrochloride **(54150-69-5)**
Benzenamine, 2,3-dimethyl- (9CI) **(87-59-2)**
Benzenamine, 3,4-dimethyl- (9CI) **(95-64-7)**
Benzenamine, N,4-dimethyl- (9CI) **(623-08-5)**
Benzenamine, N,N,-dimethyl- (9CI) **(121-69-7)**
Benzenamine, 3,4-dimethyl-2,6-dinitro-N-(1-ethylpropyl)- **(40487-42-1)**
Benzenamine, 4-(1,1-dimethylethyl)-N-(1-methylpropyl)-2,6-dinitro- (9CI) **(33629-47-9)**
Benzenamine, N-(1,1-dimethylethyl)-4-nitro- (9CI) **(4138-38-9)**
Benzenamine, N,N-dimethyl-4-((2-methylphenyl)azo)- **(3731-39-3)**
Benzenamine, N,N-dimethyl-4-((3-methylphenyl)azo)- (9CI) **(55-80-1)**
Benzenamine, 3,5-dimethyl-4-nitro- **(34761-82-5)**
Benzenamine, 2,6-dimethyl-4-nitro- (9CI) **(16947-63-0)**
Benzenamine, N,2-dimethyl-4-nitro- (9CI) **(10439-77-7)**
Benzenamine, N,N-dimethyl-4-nitro- (9CI) **(100-23-2)**
Benzenamine, N,N-dimethyl-4-((4-nitrophenyl)azo)- (9CI) **(2491-74-9)**
Benzenamine, N,N-dimethyl-4-nitroso- (9CI) **(138-89-6)**
Benzenamine, N,N-dimethyl-4-(phenylazo)- (9CI) **(60-11-7)**
Benzenamine, 4-((2,4-dimethylphenyl)azo)-2-methyl- (9CI) **(102-63-6)**
Benzenamine, 4-(1,1-dimethylpropyl)- (9CI) **(2049-92-5)**
Benzenamine, N,N-dimethyl-4-(tetrakis(2,2,2-trifluoroethoxy)-phosphoranyl)- (9CI) **(102040-60-8)**
Benzenamine, 2,4-dinitro- (9CI) **(97-02-9)**
Benzenamine, 2,6-dinitro- (9CI) **(606-22-4)**
Benzenamine, ar,ar-dinitro- (9CI) **(26471-56-7)**
Benzenamine, 2,6-dinitro-N,N-dipropyl-4-(trifluoromethyl)- (9CI) **(1582-09-8)**
Benzenamine, 2,4-dinitro-N-methyl-N-(2,4,6-tribromophenyl)-6-(trifluoromethyl)- **(63333-35-7)**
Benzenamine, 2,4-dinitro-N-(3-nitrophenyl)- (9CI) **(970-91-2)**
Benzenamine, 2,4-dinitro-N-(4-nitrophenyl)- (9CI) **(970-76-3)**
Benzenamine, 2,6-dinitro-N-propyl-4-(trifluoromethyl)- (9CI) **(2077-99-8)**
Benzenamine, 2,6-dinitro-4-(trifluoromethyl)- (9CI) **(445-66-9)**
Benzenamine, 2,4-dinitro-N-(3-(trifluoromethyl)phenyl)- (9CI) **(1869-67-6)**
Benzenamine, 2,4-dinitro-N-(4-(trifluoromethyl)phenyl)- (9CI) **(13744-79-1)**
Benzenamine, N,N-diphenyl- (9CI) **(603-34-9)**
Benzenamine, 4-(diphenylbis(2,2,2-trifluoroethoxy)-phosphoranyl)-N,N-dimethyl- (9CI) **(102040-48-2)**
Benzenamine, 4-dodecyl- (9CI) **(104-42-7)**
Benzenamine, 3-ethoxy- (9CI) **(621-33-0)**
Benzenamine, 4-ethoxy- (9CI) **(156-43-4)**
Benzenamine, 3-ethoxy-N,N-diethyl- (9CI) **(1864-92-2)**
Benzenamine, 2-ethyl- (9CI) **(578-54-1)**
Benzenamine, 3-ethyl- (9CI) **(587-02-0)**
Benzenamine, N-ethyl- (9CI) **(103-69-5)**
Benzenamine, 2-ethyl-N-(2-ethylphenyl)-, (tripropenyl) derivs. **(68608-77-5)**
Benzenamine, N-(2-ethyl-2-hexenylidene)- (9CI) **(35331-89-6)**
Benzenamine, N-(2-ethylhexyl)- (9CI) **(10137-80-1)**
Benzenamine, 2-ethyl-N-(2-methoxy-1-methylethyl)-6-methyl-(9CI) **(51219-00-2)**
Benzenamine, 2-ethyl-6-methyl- **(24549-06-2)**
Benzenamine, N-ethyl-2-methyl- (9CI) **(94-68-8)**
Benzenamine, N-ethyl-3-methyl- (9CI) **(102-27-2)**
Benzenamine, N-ethyl-N-methyl- (9CI) **(613-97-8)**
Benzenamine, 2-ethyl-6-methyl-N-methylene- (9CI) **(35203-06-6)**
Benzenamine, N-ethyl-4-nitro- (9CI) **(3665-80-3)**
Benzenamine, N-ethyl-N-nitroso- (9CI) **(612-64-6)**
Benzenamine, N-(1-ethylpropyl)-4,5-dimethyl- (9CI) **(56038-89-2)**
Benzenamine, N-(1-ethylpropyl)-3,4-dimethyl-2,6-dinitro- (9CI) **(40487-42-1)**
Benzenamine, N-(1-ethylpropyl)-3,4-dimethyl-2,6-dinitro-N-nitroso- (9CI) **(68897-50-7)**
Benzenamine, N-(1-ethylpropyl)-3,4-dimethyl-N,2,6-trinitro-(9CI) **(73215-09-5)**
Benzenamine, 4-(9H-fluoren-9-yl)-N,N-dimethyl- (9CI) **(32377-15-4)**
Benzenamine, 4-fluoro- (9CI) **(371-40-4)**
Benzenamine, 4-fluoro-3-nitro- **(364-76-1)**
Benzenamine, 4-hexyl- (9CI) **(33228-45-4)**
Benzenamine, hydrobromide (9CI) **(542-11-0)**
Benzenamine, hydrochloride **(142-04-1)**
Benzenamine, 4,4'-iminobis- **(537-65-5)**
Benzenamine, 2-iodo- (9CI) **(615-43-0)**
Benzenamine, 3-iodo- (9CI) **(626-01-7)**
Benzenamine, 4-iodo- (9CI) **(540-37-4)**
Benzenamine, N,N'-methanetetraylbis(2,6-bis(1-methylethyl)-(9CI) **(2162-74-5)**
Benzenamine, N,N'-methanetetraylbis(2-methyl- (9CI) **(1215-57-2)**
Benzenamine, 2-methoxy- (9CI) **(90-04-0)**
Benzenamine, 3-methoxy- (9CI) **(536-90-3)**
Benzenamine, 4-methoxy- (9CI) **(104-94-9)**
Benzenamine, ar-methoxy- (9CI) **(29191-52-4)**
Benzenamine, 2-methoxy-, hydrochloride (9CI) **(134-29-2)**
Benzenamine, 4-methoxy-, hydrochloride (9CI) **(20265-97-8)**
Benzenamine, 4-methoxy-N-(4-methoxyphenyl)- **(101-70-2)**
Benzenamine, 3-methoxy-4-methyl **(16452-01-0)**
Benzenamine, 4-methoxy-2-methyl- **(102-50-1)**
Benzenamine, 2-methoxy-5-methyl- (9CI) **(120-71-8)**
Benzenamine, 4-methoxy-N-methyl- (9CI) **(5961-59-1)**
Benzenamine, methoxymethyl- (9CI) **(63460-04-8)**
Benzenamine, 2-methoxy-4-nitro- (9CI) **(97-52-9)**
Benzenamine, 2-methoxy-5-nitro- (9CI) **(99-59-2)**
Benzenamine, 2-methoxy-4-((4-nitrophenyl)azo)- (9CI) **(101-52-0)**
Benzenamine, 3-methoxy-N-phenyl- (9CI) **(101-16-6)**
Benzenamine, 4-methoxy-N-phenyl- (9CI) **(1208-86-2)**
Benzenamine, 2-methyl- (9CI) **(95-53-4)**
Benzenamine, ar-methyl- (9CI) **(26915-12-8)**
Benzenamine, 4-(6-methyl-2-benzothiazolyl)- (9CI) **(92-36-4)**
Benzenamine, 4-methyl-2,6-dinitro- **(6393-42-6)**
Benzenamine, N-methyl-N,4-dinitroso- **(99-80-9)**
Benzenamine, N-methyl-2,4-dinitro-N-(2,4,6-tribromophenyl)-6-(trifluoromethyl)- (9CI) **(63333-35-7)**
Benzenamine, 4,4'-methylenebis- **(101-77-9)**
Benzenamine, 2,2'-methylenebis- (9CI) **(6582-52-1)**
Benzenamine, 4,4'-methylenebis-, Polymer with (chloromethyl)oxirane (9CI) **(28390-91-2)**
Benzenamine, 4,4'-methylenebis(2-chloro **(101-14-4)**
Benzenamine, 4,4'-methylenebis(N,N-diethyl- **(135-91-1)**
Benzenamine, 4,4'-methylenebis(N,N-diethyl- (9CI) **(135-91-1)**
Benzenamine, 4,4'-methylenebis-, dihydrochloride **(13552-44-8)**
Benzenamine, 4-4'-methylenebis(N,N-dimethyl)- (9CI) **(101-61-1)**
Benzenamine, 4,4'-methylenebis(2-methyl- **(838-88-0)**
Benzenamine, 4,4'-methylenebis(N-methyl- (9CI) **(1807-55-2)**
Benzenamine, 4,4'-methylenebis(N-(1-methylpropyl)- (9CI) **(5285-60-9)**
Benzenamine, 2-(1-methylethyl)- (9CI) **(643-28-7)**
Benzenamine, 4-(1-methylethyl)- (9CI) **(99-88-7)**
Benzenamine, 4-(1-methylethyl)-2,6-dinitro-N,N-dipropyl- (9CI) **(33820-53-0)**
Benzenamine, N-(1-methylethyl)-4-nitro- (9CI) **(25186-43-0)**
Benzenamine, 4-(1-methylethyl)-N-phenyl- (9CI) **(5650-10-2)**

Benzenamine, 2-methyl-, hydrochloride **(636-21-5)**
Benzenamine, 4-methyl-, hydrochloride (9CI) **(540-23-8)**
Benzenamine, 4,4',4''-methylidynetris(N,N-diethyl-3-methyl- (9CI) **(4482-70-6)**
Benzenamine, 4,4',4''-methylidynetris(N,N-dimethyl- (9CI) **(603-48-5)**
Benzenamine, 2-methyl-3-nitro- **(603-83-8)**
Benzenamine, 2-methyl-4-nitro- **(99-52-5)**
Benzenamine, 2-methyl-5-nitro- **(99-55-8)**
Benzenamine, 4-methyl-3-nitro- **(119-32-4)**
Benzenamine, 2-methyl-6-nitro- (9CI) **(570-24-1)**
Benzenamine, 3-methyl-4-nitro- (9CI) **(611-05-2)**
Benzenamine, 4-methyl-2-nitro- (9CI) **(89-62-3)**
Benzenamine, N-methyl-4-nitro- (9CI) **(100-15-2)**
Benzenamine, ar-methyl-ar-nitro- (9CI) **(60999-18-0)**
Benzenamine, 3-methyl-N-(4-nitrophenyl)- (9CI) **(15979-82-5)**
Benzenamine, 3-methyl-4-((4-nitrophenyl)azo)- (9CI) **(62308-10-5)**
Benzenamine, N-(2-methyl-2-nitropropyl)-p-nitroso- (9CI) **(24458-48-8)**
Benzenamine, N-methyl-N-nitroso- **(614-00-6)**
Benzenamine, 3-methyl-N-phenyl- (9CI) **(1205-64-7)**
Benzenamine, 4-methyl-N-phenyl- (9CI) **(620-84-8)**
Benzenamine, N-methyl-N-phenyl- (9CI) **(552-82-9)**
Benzenamine, N-(3-methylphenyl)-2,4-dinitro- (9CI) **(964-79-4)**
Benzenamine, 4-(1-methyl-1-phenylethyl)-N-(4-(1-methyl-1-phenylethyl)phenyl)- (9CI) **(10081-67-1)**
Benzenamine, ar-(1-methylpropyl)- (9CI) **(68400-78-2)**
Benzenamine, 4-(methylsulfonyl)-2,6-dinitro-N,N-dipropyl- **(4726-14-1)**
Benzenamine, 4-(methylsulfonyl)-N-phenyl- (9CI) **(15979-81-4)**
Benzenamine, N-methyl-N,2,4,6-tetranitro- (9CI) **(479-45-8)**
Benzenamine, N-nitro- **(645-55-6)**
Benzenamine, 3-nitro- (9CI) **(99-09-2)**
Benzenamine, 4-nitro- (9CI) **(100-01-6)**
Benzenamine, 4-nitro-, monohydrochloride (9CI) **(15873-51-5)**
Benzenamine, 3-nitro-N-(4-nitrophenyl)- (9CI) **(15979-87-0)**
Benzenamine, 4-nitro-N-phenyl- **(836-30-6)**
Benzenamine, 2-nitro-N-phenyl- (9CI) **(119-75-5)**
Benzenamine, 3-nitro-N-phenyl- (9CI) **(4531-79-7)**
Benzenamine, 4-((4-nitrophenyl)azo)- (9CI) **(730-40-5)**
Benzenamine, 4-(2-(4-nitrophenyl)ethenyl)- (9CI) **(4629-58-7)**
Benzenamine, N-(4-nitrophenyl)-3-(trifluoromethyl)- (9CI) **(369-90-4)**
Benzenamine, 4-nitroso-N-phenyl- (9CI) **(156-10-5)**
Benzenamine, N-nitroso-N-phenyl- (9CI) **(86-30-6)**
Benzenamine, 4-nitro-N-(triphenylmethyl)- (9CI) **(20222-29-1)**
Benzenamine, 4-nonyl-N-(4-nonylphenyl)- (9CI) **(24925-59-5)**
Benzenamine, ar-nonyl-N-(nonylphenyl)- (9CI) **(36878-20-3)**
Benzenamine, 4-octyl- (9CI) **(16245-79-7)**
Benzenamine, 4-octyl-N-(4-octylphenyl)- (9CI) **(101-67-7)**
Benzenamine, 4-octyl-N-phenyl- (9CI) **(4175-37-5)**
Benzenamine, 4,4'-oxybis- **(101-80-4)**
Benzenamine, 2,3,4,5,6-pentachloro- (9CI) **(527-20-8)**
Benzenamine, 2,3,4,5,6-pentafluoro- (9CI) **(771-60-8)**
Benzenamine, 2,3,4,5,6-pentanitro- **(21985-87-5)**
Benzenamine, 4-phenoxy- (9CI) **(139-59-3)**
Benzenamine, N-phenyl- (9CI) **(122-39-4)**
Benzenamine, N-(5-(phenylamino)-2,4-pentadienylidene)-, monohydrochloride (9CI) **(1497-49-0)**
Benzenamine, 4-(phenylazo)- (9CI) **(60-09-3)**
Benzenamine, N-(phenylmethylene)- (9CI) **(538-51-2)**
Benzenamine, N-phenyl-4-((4-(phenylamino)phenyl)(4-(phenylimino)-2,5-cyclohexadien-1-y-lidene)methyl)-, monohydrochloride (9CI) **(2152-64-9)**
Benzenamine, N-propyl- **(622-80-0)**
Benzenamine, 4-propyl- (9CI) **(2696-84-6)**
Benzenamine, sulfate (2:1) (9CI) **(542-16-5)**
Benzenamine, 4,4'-sulfonylbis- (9CI) **(80-08-0)**
Benzenamine, 2,3,4,5-tetrachloro- (9CI) **(634-83-3)**
Benzenamine, 2,3,5,6-tetrachloro- (9CI) **(3481-20-7)**
Benzenamine, 2,3,4,5-tetrachloro-6-(2,4-dichlorophenoxy)- (9CI) **(64630-65-5)**
Benzenamine, 2,3,5,6-tetrachloro-4-methoxy- (9CI) **(70439-96-2)**
Benzenamine, 2,3,4,5-tetrafluoro- (9CI) **(5580-80-3)**
Benzenamine, 2,3,5,6-tetrafluoro- (9CI) **(700-17-4)**
Benzenamine, 2,3,5,6-tetramethyl-4-nitro- (9CI) **(13171-61-4)**
Benzenamine, N,2,4,6-tetranitro- **(4591-46-2)**
Benzenamine, tetranitro- (9CI) **(53014-37-2)**
Benzenamine, 4,4'-thiobis- (9CI) **(139-65-1)**
Benzenamine, 2,4,6-tribromo- (9CI) **(147-82-0)**
Benzenamine, ar,ar,ar-tribromo- (9CI) **(52628-37-2)**
Benzenamine, 2,3,4-trichloro- (9CI) **(634-67-3)**
Benzenamine, 2,3,5-trichloro- (9CI) **(18487-39-3)**
Benzenamine, 2,4,5-trichloro- (9CI) **(636-30-6)**
Benzenamine, 2,4,6-trichloro- (9CI) **(634-93-5)**
Benzenamine, 3,4,5-trichloro- (9CI) **(634-91-3)**
Benzenamine, 2,3,4-trichloro-6-(2,4-dichlorophenoxy)- (9CI) **(64630-64-4)**
Benzenamine, 2-(trifluoromethyl)- (9CI) **(88-17-5)**
Benzenamine, 3,4,5-trimethoxy- (9CI) **(24313-88-0)**
Benzenamine, 2,4,5-trimethyl- **(137-17-7)**
Benzenamine, N,N,4-trimethyl- **(99-97-8)**
Benzenamine, 2,4,6-trimethyl- (9CI) **(88-05-1)**
Benzenamine, ar-(1,7,7-trimethylbicyclo(2.2.1)hept-2-yl)-N-((1,7,7-trimethylbicyclo(2.2.1)hept-2-yl)phenyl)- (9CI) **(68586-20-9)**
Benzenamine, 2,4,5-trimethyl-, hydrochloride (9CI) **(21436-97-5)**
Benzenamine, 2,4,6-trimethyl-, hydrochloride (9CI) **(6334-11-8)**
Benzenamine, N,N,2-trimethyl-4-((3-nitrophenyl)azo)- (9CI) **(4313-14-8)**
Benzenamine, N,N,2-trimethyl-4-((4-nitrophenyl)azo)- (9CI) **(4313-13-7)**
Benzenamine, 2,4,6-trinitro- (9CI) **(489-98-5)**
Benzenamine, trinitro- (9CI) **(26952-42-1)**
Benzenaminium, 3-(((dimethylamino)carbonyl)oxy)-N,N,N-trimethyl- (9CI) **(59-99-4)**
Benzenaminium, 3-(((dimethylamino)carbonyl)oxy)-N,N,N-trimethyl-, bromide (9CI) **(114-80-7)**
Benzenaminium, 4-((4-(dimethylamino)phenyl)(4-(dimethyliminio)-2,5-cyclohexadien-1-ylidene)methyl)-N,N,N-trimethyl-, dichloride (9CI) **(82-94-0)**
Benzenaminium, N,N,N-trimethyl-3-(((methylamino)carbonyl)-oxy)- (9CI) **(17752-10-2)**
Benzenaminium, N,N,N-trimethyl-4-((4,7,7-trimethyl-3-oxo-bicyclo(2.2.1)hept-2-ylidene)methyl)-, methyl sulfate (9CI) **(52793-97-2)**
1,2-Benzendikarbonitril (Czech) **(91-15-6)**
1,3-Benzendikarbonitril (Czech) **(626-17-5)**
Benzene (ACGIH,OSHA) [UN 1114] **(71-43-2)**
Benzene, C10-13-alkyl derivs. **(67774-74-7)**
Benzene, C10-16-alkyl derivs. **(68648-87-3)**
Benzene, C4-16-alkyl derivs. **(68648-86-2)**
Benzeneacetaldehyde **(122-78-1)**
Benzeneacetaldehyde, 4-chloro-α-hydroxy- (9CI) **(34025-32-6)**
Benzeneacetaldehyde, α-ethylidene- (9CI) **(4411-89-6)**
Benzeneacetaldehyde, 4-hydroxy- (9CI) **(7339-87-9)**
Benzeneacetaldehyde, α-methyl **(93-53-8)**
Benzeneacetaldehyde, 4-methyl- (9CI) **(104-09-6)**
Benzeneacetaldehyde, α-phenyl- (9CI) **(947-91-1)**
Benzeneacetamide (9CI) **(103-81-1)**
Benzeneacetamide, N-(aminocarbonyl)- (9CI) **(63-98-9)**
Benzeneacetamide, N,N-dimethyl-α-phenyl- (9CI) **(957-51-7)**
Benzeneacetamide, 4-(2-hydroxy-3-((1-methylethyl)amino)-propoxy)- (9CI) **(29122-68-7)**
Benzeneacetamide, N-methyl- (9CI) **(6830-82-6)**
Benzeneacetic acid **(103-82-2)**
Benzeneacetic acid, .α.-(acetyloxy)-.α.-methyl-, (S)- (9CI) **(10487-92-0)**
Benzeneacetic acid, 4-amino- (9CI) **(1197-55-3)**
Benzeneacetic acid, α-amino-, (+-)- (9CI) **(2835-06-5)**
Benzeneacetic acid, α-amino-, (R)- (9CI) **(875-74-1)**
Benzeneacetic acid, α-amino-4-hydroxy-, (R)- (9CI) **(22818-40-2)**
Benzeneacetic acid, 4-bromo-α-(4-bromophenyl)-α-hydroxy-, 1-methylethyl ester **(18181-80-1)**
Benzeneacetic acid, 2-chloro- (9CI) **(2444-36-2)**

Benzeneacetic acid, 3-chloro- (9CI) **(1878-65-5)**
Benzeneacetic acid, 4-chloro- (9CI) **(1878-66-6)**
Benzeneacetic acid, 4-chloro-α-(4-chlorophenyl)- (9CI) **(83-05-6)**
Benzeneacetic acid, 4-chloro-α-(4-chlorophenyl)-α-hydroxy-, ethyl ester **(510-15-6)**
Benzeneacetic acid, 4-chloro-α-hydroxy- (9CI) **(492-86-4)**
Benzeneacetic acid, 4-chloro-α-(1-methylethyl)-, cyano (3-phenoxyphenyl)methyl ester, (S-(R*,R*)) **(66230-04-4)**
Benzeneacetic acid, 4-chloro-α-(1-methylethyl)-, cyano-(3-phenoxyphenyl)methyl ester **(51630-58-1)**
Benzeneacetic acid, 4-chloro-α-(1-methylethyl)-, cyano-(3-phenoxyphenyl)methyl ester, (R-(R*,R*))- (9CI) **(67614-33-9)**
Benzeneacetic acid, 4-chloro-α-(1-methylethyl)-, cyano-(3-phenoxyphenyl)methyl ester, (R-(R*,S*))- (9CI) **(66267-77-4)**
Benzeneacetic acid, 4-chloro-α-(1-methylethyl)-, cyano-(3-phenoxyphenyl)methyl ester, (S-(R*,S*))- (9CI) **(67614-32-8)**
Benzeneacetic acid, α-cyclohexyl-α-hydroxy-, 4-(diethyl-amino)-2-butynyl ester **(5633-20-5)**
Benzeneacetic acid, .α.,.α.-dichloro- (9CI) **(61031-72-9)**
Benzeneacetic acid, 2,6-dichloro-.α.,4-dihydroxy-3,5-dimethoxy- (9CI) **(75315-51-4)**
Benzeneacetic acid, 2,4-dichloro-, ethyl ester (9CI) **(533-23-3)**
Benzeneacetic acid, dichloro-4-hydroxy-3,5-dimethoxy- (9CI) **(108572-08-3)**
Benzeneacetic acid, 2,6-dichloro-4-hydroxy-3,5-dimethoxy-.α.-oxo- (9CI) **(75315-50-3)**
Benzeneacetic acid, 2,6-dichloro-.α.-hydroxy-3,4,5-tri-methoxy-, methyl ester (9CI) **(75315-57-0)**
Benzeneacetic acid, 3,6-dichloro-2-methoxy- (9CI) **(3004-74-8)**
Benzeneacetic acid, 2-((2,6-dichlorophenyl)amino)-, monosodium salt **(15307-79-6)**
Benzeneacetic acid, 2,6-dichloro-3,4,5-trimethoxy-, methyl ester (9CI) **(75315-55-8)**
Benzeneacetic acid, 2,6-dichloro-3,4,5-trimethoxy-.α.-oxo-, methyl ester (9CI) **(108544-95-2)**
Benzeneacetic acid, 4-(difluoromethoxy)-α-(1-methylethyl)-, cyano(3-phenoxyphenyl) methyl ester **(70124-77-5)**
Benzeneacetic acid, 2,5-dihydroxy- (9CI) **(451-13-8)**
Benzeneacetic acid, 3,4-dihydroxy- (9CI) **(102-32-9)**
Benzeneacetic acid, α,4-dihydroxy-3-methoxy- (9CI) **(55-10-7)**
Benzeneacetic acid, 3,4-dimethoxy- (9CI) **(93-40-3)**
Benzeneacetic acid, α-((dimethoxyphosphinothioyl)thio)-, ethyl ester **(2597-03-7)**
Benzeneacetic acid, α,α-dimethyl- (9CI) **(826-55-1)**
Benzeneacetic acid, 2,4-dinitro- (9CI) **(643-43-6)**
Benzeneacetic acid, α-((3-ethoxy-1-methyl-3-oxo-1-propenyl)-amino)-, monopotassium salt, (R)- (9CI) **(961-69-3)**
Benzeneacetic acid, ethyl ester (9CI) **(101-97-3)**
Benzeneacetic acid, 4-fluoro- (9CI) **(405-50-5)**
Benzeneacetic acid, 2-formyl-, methyl ester **(5894-79-1)**
Benzeneacetic acid, α-formyl-, methyl ester (9CI) **(5894-79-1)**
Benzeneacetic acid, 2-hydroxy- (9CI) **(614-75-5)**
Benzeneacetic acid, 3-hydroxy- (9CI) **(621-37-4)**
Benzeneacetic acid, 4-hydroxy- (9CI) **(156-38-7)**
Benzeneacetic acid, α-hydroxy-, (+-)- (9CI) **(611-72-3)**
Benzeneacetic acid, 4-hydroxy-, ethyl ester (9CI) **(17138-28-2)**
Benzeneacetic acid, 4-hydroxy-3-methoxy- (9CI) **(306-08-1)**
Benzeneacetic acid, α-(hydroxymethyl)- (9CI) **(529-64-6)**
Benzeneacetic acid, α-hydroxy-α-methyl- (9CI) **(515-30-0)**
Benzeneacetic acid, 4-hydroxy-, methyl ester **(14199-15-6)**
Benzeneacetic acid, 3-hydroxy-, methyl ester (9CI) **(42058-59-3)**
Benzeneacetic acid, α-(hydroxymethyl)-, 9-methyl-3-oxa-9-aza-tricyclo(3.3.1.0(2,4))non-7-yl ester, N-oxide, hydrobromide, (7(S)-(1α,2β,4β,5α,7β))- **(6106-81-6)**
Benzeneacetic acid, α-(hydroxymethyl)-, 9-methyl-3-oxa-9-aza-tricyclo(3.3.1.02,4)non-7-yl ester, N-oxide, hydrobromide, (7(S)-(1α,2β,4β,5α,7β))- (9CI) **(6106-81-6)**
Benzeneacetic acid, 4-hydroxy-α-oxo-, monosodium salt (9CI) **(54537-30-3)**
Benzeneacetic acid, α-hydroxy-α-phenyl- (9CI) **(76-93-7)**
Benzeneacetic acid, α-hydroxy-α-phenyl-, 2-(diethylamino)ethyl ester (9CI) **(302-40-9)**
Benzeneacetic acid, α-hydroxy-α-phenyl-, 1-ethyl-3-piperidinyl ester (9CI) **(3567-12-2)**
Benzeneacetic acid, α-hydroxy-, 3,3,5-trimethylcyclohexyl ester (9CI) **(456-59-7)**
Benzeneacetic acid, 2-methoxy- (9CI) **(93-25-4)**
Benzeneacetic acid, 4-methoxy- (9CI) **(104-01-8)**
Benzeneacetic acid, 3-methyl- (9CI) **(621-36-3)**
Benzeneacetic acid, 4-methyl- (9CI) **(622-47-9)**
Benzeneacetic acid, α-methyl- (9CI) **(492-37-5)**
Benzeneacetic acid, methyl ester (9CI) **(101-41-7)**
Benzeneacetic acid, α-methyl-4-(2-methylpropyl)- **(15687-27-1)**
Benzeneacetic acid, 2-methylpropyl ester (9CI) **(102-13-6)**
Benzeneacetic acid, α-oxo-, ethyl ester (9CI) **(1603-79-8)**
Benzeneacetic acid, α-phenyl- **(117-34-0)**
Benzeneacetic acid, α-phenyl-, 2-(diethylamino)ethyl ester, (9CI) **(64-95-9)**
Benzeneacetic acid, 2-phenylethyl ester (9CI) **(102-20-5)**
Benzeneacetic acid, α-(phenylmethyl)- **(3333-15-1)**
Benzeneacetic acid, potassium salt (9CI) **(13005-36-2)**
Benzeneacetic acid, sodium salt (9CI) **(114-70-5)**
Benzeneacetic acid, 2,3,6-trichloro- (9CI) **(85-34-7)**
Benzeneacetic acid, 2,3,6-trichloro-, sodium salt (9CI) **(2439-00-1)**
Benzeneacetonitrile (9CI) **(140-29-4)**
Benzeneacetonitrile, 4-bromo- (9CI) **(16532-79-9)**
Benzeneacetonitrile, α-bromo- (9CI) **(5798-79-8)**
Benzeneacetonitrile, 4-chloro- **(140-53-4)**
Benzeneacetonitrile, 4-chloro-α-(1-methylethyl)- (9CI) **(2012-81-9)**
Benzeneacetonitrile, α-(((diethoxyphosphinothioyl)oxy)imino)- **(14816-18-3)**
Benzeneacetonitrile, α-((6-O-β-D-glucopyranosyl-β-D-gluco-pyranosyl)oxy)-, (R)- **(29883-15-6)**
Benzeneacetonitrile, 4-hydroxy- (9CI) **(14191-95-8)**
Benzeneacetonitrile, α-(hydroxyimino)- (9CI) **(825-52-5)**
Benzeneacetonitrile, 4-methoxy- **(104-47-2)**
Benzeneacetonitrile, α-methyl- (9CI) **(1823-91-2)**
Benzeneacetonitrile, .α.-methylene- (9CI) **(495-10-3)**
Benzeneacetonitrile, 2-nitro- (9CI) **(610-66-2)**
Benzeneacetonitrile, α-oxo- (9CI) **(613-90-1)**
Benzeneacetonitrile, α-phenyl- (9CI) **(86-29-3)**
Benzene, (acetoxymercuri)- **(62-38-4)**
Benzene, (acetoxymercurio)- **(62-38-4)**
Benzene, acetyl- **(98-86-2)**
Benzeneacetyl chloride (9CI) **(103-80-0)**
Benzene, allyl (8CI) **(300-57-2)**
Benzene, 4-allyl-1,2-dimethoxy **(93-15-2)**
Benzene, 5-allyl-1-methoxy-2,3-(methylenedioxy) **(607-91-0)**
Benzene, 4-allyl-1,2-(methylenedioxy) **(94-59-7)**
Benzeneamine, 5-acetamino-4-((2,4-dinitro-6-bromophenyl)-azo)-2-methoxy-N,N-bis((2-acetoxy)ethyl)- **(3618-72-2)**
Benzeneamine, 4-chloro **(106-47-8)**
Benzeneamine, 3,5-dinitro- **(618-87-1)**
Benzene, amino **(62-53-3)**
Benzeneazo-β-naphthol **(842-07-9)**
1-Benzene-azo-β-naphthylamine **(85-84-7)**
Benzene, anilino- **(122-39-4)**
Benzenearsonic-acid **(98-05-5)**
Benzenearsonic acid, p-amino- **(98-50-0)**
Benzenearsonic acid, 4-hydroxy-3-nitro **(121-19-7)**
Benzenearsonic acid, p-ureido- **(121-59-5)**
Benzenearsonous acid, diethyl ester (6CI,7CI,8CI) **(3141-11-5)**
Benzenearsonous acid, dimethyl ester (8CI) **(24582-52-3)**
Benzenearsonous acid, dithio-, diethyl ester (7CI,8CI) **(5582-58-1)**
Benzenearsonous acid, dithio-, diphenyl ester (7CI,8CI) **(24582-59-0)**
Benzene azimide **(95-14-7)**
4-Benzeneazoaniline **(60-09-3)**
Benzeneazobenzene **(103-33-3)**

Benzene-1-azobenzene-4-azo-o-cresol **(6300-37-4)**
Benzeneazobenzeneazo-β-naphthol **(85-86-9)**
Benzene, azodi **(103-33-3)**
Benzeneazodimethylaniline **(60-11-7)**
1-Benzeneazo-2-naphthol **(842-07-9)**
Benzene-1-azo-2-naphthol **(842-07-9)**
1-Benzeneazo-2-naphthylamine **(85-84-7)**
p-Benzeneazophenol **(1689-82-3)**
Benzeneazoresorcinol **(2051-85-6)**
Benzene, azoxydi- **(495-48-7)**
Benzene, benzoyl- **(119-61-9)**
Benzene, benzyl- **(101-81-5)**
Benzene, (1,1'-bicyclohexyl)-4-yl- (9CI) **(20273-27-2)**
Benzene, 1,4-bis(chloromethoxymethyl) **(56894-91-8)**
Benzene, 1,2-bis(chloromethyl)- **(612-12-4)**
Benzene, 1,3-bis(chloromethyl)- **(626-16-4)**
Benzene, bis(chloromethyl)- (9CI) **(28347-13-9)**
Benzene, 1,2-bis(dichloromethyl)- (9CI) **(25641-99-0)**
Benzene, 1,4-bis(dimethylamino)- **(100-22-1)**
Benzene, 1,4-bis(1,1-dimethylethyl)- (9CI) **(1012-72-2)**
Benzene, bis(isocyanatomethyl)- (9CI) **(25854-16-4)**
Benzene, 1,3-bis(1-isocyanato-1-methylethyl)- (9CI) **(2778-42-9)**
Benzene, 1,4-bis(1-isocyanato-1-methylethyl)- (9CI) **(2778-41-8)**
Benzene, 1,4-bis(1-methylethyl)- (9CI) **(100-18-5)**
Benzene, 1,3-bis(1-methylpropyl)- (9CI) **(1079-96-5)**
Benzene, 1,4-bis(1-methylpropyl)- (9CI) **(1014-41-1)**
Benzene, 1,3-bis(trifluoromethyl)- (9CI) **(402-31-3)**
Benzene, bromo **(108-86-1)**
Benzene, 1-bromo-3-chloro **(108-37-2)**
Benzene, 1-bromo-4-chloro **(106-39-8)**
Benzene, 1-bromo-2-chloro- (9CI) **(694-80-4)**
Benzene, bromochloro- (8CI,9CI) **(28906-38-9)**
Benzene, (2-bromocyclopropyl)- (7CI,9CI) **(36617-02-4)**
Benzene, 1-bromo-4-cyclopropyl- (6CI,7CI,8CI,9CI) **(1124-14-7)**
Benzene, 1-bromo-2,3-dichloro- (9CI) **(56961-77-4)**
Benzene, 1-bromo-3,5-dichloro- (9CI) **(19752-55-7)**
Benzene, 2-bromo-1,3-dichloro- (9CI) **(19393-92-1)**
Benzene, 4-bromo-1,2-dichloro- (9CI) **(18282-59-2)**
Benzene, 2-bromo-1,4-dichloro- (8CI,9CI) **(1435-50-3)**
Benzene, bromodimethyl- **(35884-77-6)**
Benzene, 2-bromo-1,4-dimethyl- (9CI) **(553-94-6)**
Benzene, 4-bromo-1,2-dinitro- **(610-38-8)**
Benzene, bromodinitro- (9CI) **(63460-06-0)**
Benzene, 1-bromo-2-ethenyl- (9CI) **(2039-88-5)**
Benzene, 1-bromo-4-ethenyl- (9CI) **(2039-82-9)**
Benzene, (1-bromoethyl)- (9CI) **(585-71-7)**
Benzene, (2-bromoethyl)- (9CI) **(103-63-9)**
Benzene, 1-bromo-4-ethyl- (9CI) **(1585-07-5)**
Benzene, 1-bromo-2-ethyl- (8CI,9CI) **(1973-22-4)**
Benzene, 1-(2-bromoethyl)-4-chloro- (6CI,7CI,8CI,9CI) **(6529-53-9)**
Benzene, 1-bromo-4-fluoro- (9CI) **(460-00-4)**
Benzene, 1-bromo-2-iodo- **(583-55-1)**
Benzene, 1-bromo-2-iodo- (9CI) **(583-55-1)**
Benzene, 1-bromo-3-iodo- (9CI) **(591-18-4)**
Benzene, 1-bromo-4-iodo- (9CI) **(589-87-7)**
Benzene, 1-bromo-2-methoxy- (9CI) **(578-57-4)**
Benzene, 1-bromo-3-methoxy- (9CI) **(2398-37-0)**
Benzene, 1-bromo-4-methoxy- (9CI) **(104-92-7)**
Benzene, (bromomethyl)- **(100-39-0)**
Benzene, 1-bromo-3-methyl- **(591-17-3)**
Benzene, 1-bromo-2-methyl-(9CI) **(95-46-5)**
Benzene, bromomethyl- (9CI) **(28807-97-8)**
Benzene, 1-(bromomethyl)-2-chloro- **(611-17-6)**
Benzene, 1-bromo-2-(1-methylethyl)- (9CI) **(7073-94-1)**
Benzene, 1-bromo-4-(1-methylethyl)- (9CI) **(586-61-8)**
Benzene, 1-(bromomethyl)-2-methoxy- (9CI) **(52289-93-7)**
Benzene, (bromomethyl)methyl- (9CI) **(28777-60-8)**
Benzene, 1-(bromomethyl)-2-methyl- (9CI) **(89-92-9)**
Benzene, 1-(bromomethyl)-3-methyl- (9CI) **(620-13-3)**
Benzene, 1-(bromomethyl)-4-methyl- (9CI) **(104-81-4)**
Benzene, 2-bromo-4-methyl-1-(1-methylethyl)- (9CI) **(4478-10-8)**
Benzene, 1-(bromomethyl)-3-phenoxy- (9CI) **(51632-16-7)**
Benzene, 1-bromo-2-nitro **(577-19-5)**
Benzene, 1-bromo-3-nitro **(585-79-5)**
Benzene, 1-bromo-4-nitro **(586-78-7)**
Benzene, bromonitro- (9CI) **(61878-56-6)**
Benzene, (2-bromo-2-nitroethenyl)- (9CI) **(7166-19-0)**
Benzene, bromopentafluoro- (9CI) **(344-04-7)**
Benzene, 1-bromo-4-phenoxy- (9CI) **(101-55-3)**
Benzene, bromophenoxy- (9CI) **(36563-47-0)**
Benzene, (3-bromopropyl)- (9CI) **(637-59-2)**
Benzene, bromotetrachloro- (9CI) **(90077-78-4)**
Benzene, 1-bromo-3-(trifluoromethyl)- **(401-78-5)**
Benzene, 1,1'-(1,3-butadiene-1,4-diyl)bis- (9CI) **(886-65-7)**
Benzenebutanal (9CI) **(18328-11-5)**
Benzenebutanoic acid (9CI) **(1821-12-1)**
Benzenebutanoic acid, 4-amino- (9CI) **(15118-60-2)**
Benzenebutanoic acid, 4-(bis(2-chloroethyl)amino)- **(305-03-3)**
Benzenebutanoic acid, 2,5-dimethyl- (9CI) **(1453-06-1)**
Benzenebutanoic acid, .β.-hydroxy- (9CI) **(6828-41-7)**
Benzenebutanoic acid, 4-hydroxy-γ-(4-hydroxyphenyl)-γ-methyl- (9CI) **(126-00-1)**
Benzenebutanoic acid, γ-methyl- (9CI) **(16433-43-5)**
Benzenebutanoic acid, 4-methyl-, methyl ester (9CI) **(24306-23-8)**
Benzenebutanoic acid, .β.-oxo- (9CI) **(25832-09-1)**
Benzenebutanoic acid, γ-oxo- (9CI) **(2051-95-8)**
Benzene, 2-butenyl- **(1560-06-1)**
Benzene, 3-butenyl- **(768-56-9)**
Benzene, 1-butenyl-, (E)- (9CI) **(1005-64-7)**
Benzene, 1-tert-butoxy-3-methoxy- (8CI) **(15359-99-6)**
Benzene, butoxymethyl- (9CI) **(29225-54-5)**
Benzene, butyl **(104-51-8)**
Benzene, sec-butyl **(135-98-8)**
Benzene, tert-butyl **(98-06-6)**
Benzene, (1-butyldecyl)- (9CI) **(4534-56-9)**
Benzene, 5-tert-butyl-1,3-dimethyl- **(98-19-1)**
Benzene, 1-tert-butyl-3,5-dimethyl-2,4,6-trinitro- **(81-15-2)**
Benzene, 1-tert-butyl-2,6-dinitro-3,4,5-trimethyl- **(145-39-1)**
Benzene, 2-tert-butyl-4-ethyl-1-hydroxy- **(96-70-8)**
Benzene, (1-butylheptyl)- (9CI) **(4537-15-9)**
Benzene, (1-butylhexadecyl)- (9CI) **(2400-04-6)**
Benzene, (1-butylhexyl)- (9CI) **(4537-11-5)**
Benzene, 1-tert-butyl-2-hydroxy-4-methyl- **(88-60-8)**
Benzene, (1-butylnonyl)- (9CI) **(4534-50-3)**
Benzene, (1-butyloctyl)- (9CI) **(2719-63-3)**
Benzene, 1-(butylthio)-3-methyl- (9CI) **(54576-41-9)**
Benzene carbaldehyde **(100-52-7)**
Benzenecarbinol **(100-51-6)**
Benzenecarbonal **(100-52-7)**
Benzenecarbonyl chloride **(98-88-4)**
Benzenecarboperoxoic acid, 3-chloro- (9CI) **(937-14-4)**
Benzenecarboperoxoic acid, 1,1-dimethylpropyl ester (9CI) **(4511-39-1)**
Benzenecarboperoxoic acid, 2-methyl-, 1,1-dimethylethyl ester (9CI) **(22313-62-8)**
Benzenecarboperoxoic acid, 1,1,4,4-tetramethyl-1,4-butanediyl ester (9CI) **(2618-77-1)**
Benzenecarbothioamide **(2227-79-4)**
Benzenecarbothioamide, 2,6-dichloro- (9CI) **(1918-13-4)**
Benzenecarboximidamide, 2,6-dichloro-N-hydroxy- (9CI) **(23505-21-7)**
Benzenecarboximidamide, 4,4'-methylenebis **(63690-09-5)**
Benzenecarboxylic acid **(65-85-0)**
Benzene chloramine **(127-52-6)**
Benzene chloride **(108-90-7)**
Benzene, chloro **(108-90-7)**

Benzene, (3-chloro-2-butenyl)-, (Z)- (9CI)	**(16608-68-7)**
Benzene, 1-chloro-2-(chloromethyl)- (9CI)	**(611-19-8)**
Benzene, 1-chloro-3-(chloromethyl)- (9CI)	**(620-20-2)**
Benzene, 1-chloro-2-(4-chlorophenoxy)- (9CI)	**(6903-65-7)**
Benzene, 4-chloro-1-(4-chlorophenoxy)-2-nitro-	**(135-12-6)**
Benzene, 1-chloro-2-((4-chlorophenyl)difluoromethyl)-4-(trifluoromethyl)- (9CI)	**(95998-70-2)**
Benzene, 1-chloro-2-(chlorophenylmethyl)-	**(56961-47-8)**
Benzene, 1-chloro-4-(chlorophenylmethyl)- (9CI)	**(134-83-8)**
Benzene, 1-chloro-2-(2,2-dichloro-1-(4-chlorophenyl)ethenyl)- (9CI)	**(3424-82-6)**
Benzene, 1-chloro-2-(2,2-dichloro-1-(4-chlorophenyl)ethyl)-	**(53-19-0)**
Benzene, 2-chloro-1,4-diethoxy- (9CI)	**(52196-74-4)**
Benzene, 1-chloro-2,5-diethoxy-4-nitro- (9CI)	**(91-43-0)**
Benzene, 1-chloro-2,3-dimethoxy- (9CI)	**(90282-99-8)**
Benzene, 2-chloro-1,4-dimethoxy- (8CI,9CI)	**(2100-42-7)**
Benzene, 4-chloro-1,2-dimethoxy- (8CI,9CI)	**(16766-27-1)**
Benzene, chlorodimethyl- (9CI)	**(25323-41-5)**
Benzene, (2-chloro-1,1-dimethylethyl)- (9CI)	**(515-40-2)**
Benzene, 1-chloro-2,4-dinitro	**(97-00-7)**
Benzene, 2-chloro-1,3-dinitro	**(606-21-3)**
Benzene, chlorodinitro- (Mixed isomers)	**(25567-67-3)**
Benzene, 2-chloro-1,3-dinitro-5-(trifluoromethyl)-	**(393-75-9)**
Benzene, 1,1'-(1-chloro-1,2-ethanediyl)bis- (9CI)	**(4714-14-1)**
Benzene, 1,1'-(1-chloro-1,2-ethenediyl)bis- (9CI)	**(1460-06-6)**
Benzene, (2-chloroethenyl)- (9CI)	**(622-25-3)**
Benzene, 1-chloro-3-ethenyl- (9CI)	**(2039-85-2)**
Benzene, 1-chloro-4-ethenyl- (9CI)	**(1073-67-2)**
Benzene, 1,1',1''-(1-chloro-1-ethenyl-2-ylidene)tris(4-methoxy)-	**(569-57-3)**
Benzene, (2-chloroethoxy)- (9CI)	**(622-86-6)**
Benzene, 1-(2-(2-(2-chloroethoxy)ethoxy)ethoxy)-4-octyl- (9CI)	**(66028-01-1)**
Benzene, 1-(2-chloro-1-(4-ethoxyphenyl)propyl)-4-methyl- (9CI)	**(56265-27-1)**
Benzene, (1-chloroethyl)- (9CI)	**(672-65-1)**
Benzene, (2-chloroethyl)- (9CI)	**(622-24-2)**
Benzene, 1-chloro-2-ethyl- (8CI,9CI)	**(89-96-3)**
Benzene, (1-chloroethyl)dimethyl- (9CI)	**(54411-21-1)**
Benzene, ((2-chloroethyl)sulfonyl)- (9CI)	**(938-09-0)**
Benzene, ((2-chloroethyl)thio)- (9CI)	**(5535-49-9)**
Benzene, 1-chloro-2-fluoro- (9CI)	**(348-51-6)**
Benzene, 1-chloro-3-fluoro- (9CI)	**(625-98-9)**
Benzene, 1-chloro-4-fluoro- (9CI)	**(352-33-0)**
Benzene, chlorofluoro- (6CI,7CI,9CI)	**(55256-17-2)**
Benzene, 1-chloro-3-fluoro-2-methyl- (9CI)	**(443-83-4)**
Benzene, 2-chloro-1-fluoro-4-nitro	**(350-30-1)**
Benzene, 1-chloro-2-iodo- (9CI)	**(615-41-8)**
Benzene, 1-chloro-3-iodo- (9CI)	**(625-99-0)**
Benzene, 1-chloro-4-iodo- (9CI)	**(637-87-6)**
Benzene, (chloromercuri)-	**(100-56-1)**
Benzene, 1-chloro-2-methoxy- (9CI)	**(766-51-8)**
Benzene, 1-chloro-3-methoxy- (9CI)	**(2845-89-8)**
Benzene, 1-chloro-4-methoxy- (9CI)	**(623-12-1)**
Benzene, (chloromethyl)-	**(100-44-7)**
Benzene, 1-chloro-3-methyl-	**(108-41-8)**
Benzene, 1-chloro-4-methyl-	**(106-43-4)**
Benzene, 1-chloro-2-methyl- (9CI)	**(95-49-8)**
Benzene, chloromethyl- (9CI)	**(25168-05-2)**
Benzene, 1,1'-(chloromethylene)bis- (9CI)	**(90-99-3)**
Benzene, (chloromethyl)ethenyl- (9CI)	**(30030-25-2)**
Benzene, (chloromethyl)ethyl- (9CI)	**(26968-58-1)**
Benzene, 1-chloro-2-(1-methylethyl)- (9CI)	**(2077-13-6)**
Benzene, 1-chloro-4-(1-methylethyl)- (9CI)	**(2621-46-7)**
Benzene, 1-(chloromethyl)-2-fluoro- (9CI)	**(345-35-7)**
Benzene, 1-(chloromethyl)-3-fluoro- (9CI)	**(456-42-8)**
Benzene, 1-(chloromethyl)-4-fluoro- (9CI)	**(352-11-4)**
Benzene, 1-(chloromethyl)-2-methyl- (9CI)	**(552-45-4)**
Benzene, 1-(chloromethyl)-3-methyl- (9CI)	**(620-19-9)**
Benzene, 1-(chloromethyl)-4-methyl- (9CI)	**(104-82-5)**
Benzene, 2-chloro-1-methyl-4-(1-methylethyl)- (9CI)	**(4395-79-3)**
Benzene, chloro-1-methyl-4-(1-methylethyl)- (9CI)	**(54411-19-7)**
Benzene, 1-chloro-2-methyl-3-nitro-	**(83-42-1)**
Benzene, 4-chloro-1-methyl-3-nitro-	**(89-59-8)**
Benzene, 1-(chloromethyl)-2-nitro- (9CI)	**(612-23-7)**
Benzene, 1-(chloromethyl)-3-nitro- (9CI)	**(619-23-8)**
Benzene, 1-(chloromethyl)-4-nitro- (9CI)	**(100-14-1)**
Benzene, 1-chloro-4-methyl-2-nitro- (9CI)	**(89-60-1)**
Benzene, 4-chloro-1-methyl-2-nitro- (9CI)	**(89-59-8)**
Benzene, 1,1'-(2-chloro-2-methylpropylidene)bis(4-ethoxy- (9CI)	**(56265-24-8)**
Benzene, 1-chloro-4-(methylsulfinyl)- (9CI)	**(934-73-6)**
Benzene, 1-chloro-4-(methylsulfonyl)- (9CI)	**(98-57-7)**
Benzene, 1-chloro-4-(methylthio)- (9CI)	**(123-09-1)**
Benzene, 1-chloro-2-nitro	**(88-73-3)**
Benzene, 1-chloro-3-nitro	**(121-73-3)**
Benzene, 1-chloro-4-nitro	**(100-00-5)**
Benzene, chloronitro- (Mixed isomers)	**(25167-93-5)**
Benzene, 4-chloro-2-nitro-1-(4-nitrophenoxy)- (9CI)	**(20115-34-8)**
Benzene, 1-chloro-4-nitroso- (9CI)	**(932-98-9)**
Benzene, 1-chloro-4-nitro-2-(trifluoromethyl)- (9CI)	**(777-37-7)**
Benzene, chloronitro(trifluoromethyl)- (9CI)	**(68849-24-1)**
Benzene, chloropentafluoro	**(344-07-0)**
Benzene, 1-chloro-4-phenoxy- (9CI)	**(7005-72-3)**
Benzene, 5-(4-chlorophenoxy)-1,2,4-trichloro	**(61328-45-8)**
Benzene, 1-chloro-2-(phenylethynyl)- (9CI)	**(10271-57-5)**
Benzene, chloro(phenylethynyl)- (9CI)	**(90077-73-9)**
Benzene, 1-chloro-2-(phenylmethoxy)-	**(949-38-2)**
Benzene, 1-chloro-4-(phenylmethyl)- (9CI)	**(831-81-2)**
Benzene, ((2-chloro-2-propenyl)oxy)- (9CI)	**(53299-53-9)**
Benzene, (3-chloropropyl)- (8CI,9CI)	**(104-52-9)**
Benzene, 1,1'-(2-chloropropylidene)bis(4-ethoxy	**(56265-22-6)**
Benzene, 1,1'-(2-chloropropylidene)bis(4-methoxy- (9CI)	**(59900-47-9)**
Benzene, 1-(2-chloropropyl)-4-nitro- (8CI,9CI)	**(34197-98-3)**
Benzene, 1-chloro-2-(trichloromethyl)- (9CI)	**(2136-89-2)**
Benzene, 1-chloro-4-(trichloromethyl)- (9CI)	**(5216-25-1)**
Benzene, 1-chloro-4-(2,2,2-trichloro-1-(4-methylphenyl)ethyl)- (9CI)	**(17925-97-2)**
Benzene, chlorotrifluoro- (9CI)	**(75181-94-1)**
Benzene, chloro(trifluoromethyl)	**(52181-51-8)**
Benzene, 1-chloro-2-(trifluoromethyl)- (9CI)	**(88-16-4)**
Benzene, 1-chloro-3-(trifluoromethyl)- (9CI)	**(98-15-7)**
Benzene, 5-chloro-1,2,3-trimethoxy- (8CI,9CI)	**(2675-80-1)**
Benzene, 1-chloro-4-(trimethyl)- (9CI)	**(98-56-6)**
Benzene, 2-chloro-1,3,5-trimethyl- (9CI)	**(1667-04-5)**
Benzene, 2-chloro-1,3,5-trinitro	**(88-88-0)**
Benzene, chlorotrinitro	**(28260-61-9)**
Benzene, cyano-	**(100-47-0)**
Benzene, (1,4-cyclohexadien-1-yl)- (8CI)	**(13703-52-1)**
Benzene, 1,4-cyclohexadien-1-yl- (9CI)	**(13703-52-1)**
Benzene, 2,5-cyclohexadien-1-yl- (6CI,7CI,8CI,9CI)	**(4794-05-2)**
Benzene, (3-cyclohexen-1-yl)-	**(4994-16-5)**
Benzene, cyclohexenyl	**(31017-40-0)**
Benzene, 3-cyclohexen-1-yl- (8CI,9CI)	**(4994-16-5)**
Benzene, cyclohexyl	**(827-52-1)**
Benzene, 1,1'-cyclohexylidenebis- (9CI)	**(21113-55-3)**
Benzenedecanoic acid (9CI)	**(18017-73-7)**
Benzenedecanol (9CI)	**(62607-69-6)**
Benzene, decyl- (9CI)	**(104-72-3)**
1,2-Benzenediacetic acid (9CI)	**(7500-53-0)**
o-Benzenediacetic acid (8CI)	**(7500-53-0)**
Benzene, diacetyl- (8CI)	**(30773-71-8)**
Benzene, m-diacetyl- (8CI)	**(6781-42-6)**
Benzene, p-diacetyl- (8CI)	**(1009-61-6)**
1,2-Benzenediamine	**(95-54-5)**
1,3-Benzenediamine	**(108-45-2)**
1,4-Benzenediamine	**(106-50-3)**
m-Benzenediamine	**(108-45-2)**

o-Benzenediamine **(95-54-5)**
p-Benzenediamine **(106-50-3)**
Benzenediamine (9CI) **(25265-76-3)**
1,4-Benzenediamine, N-(4-aminophenyl)- (9CI) **(537-65-5)**
1,3-Benzenediamine, 4-((4-aminophenyl)azo)- (9CI) **(6364-34-7)**
1,4-Benzenediamine, N,N'-bis(1-methylethyl)- (9CI) **(4251-01-8)**
1,4-Benzenediamine, N,N'-bis(1-methylheptyl)- (9CI) **(103-96-8)**
1,2-Benzenediamine, 4-butyl- (9CI) **(3663-23-8)**
1,2-Benzenediamine, 4-chloro- (9CI) **(95-83-0)**
1,3-Benzenediamine, 4-chloro- (9CI) **(5131-60-2)**
1,4-Benzenediamine, 2-chloro-, dihydrochloride (9CI) **(615-46-3)**
1,3-Benzenediamine, 2-chloro-4-methyl- (9CI) **(43216-73-5)**
1,3-Benzenediamine, 4-chloro-2-methyl- (9CI) **(43171-59-1)**
1,3-Benzenediamine, 4-chloro-6-methyl- (9CI) **(43216-72-4)**
1,2-Benzenediamine, 5-chloro-3-nitro- (9CI) **(42389-30-0)**
1,4-Benzenediamine, 2-chloro-, sulfate (1:1) **(61702-44-1)**
1,2-Benzenediamine, 4-chloro-, sulfate (1:1) (9CI) **(68459-98-3)**
1,3-Benzenediamine, 4-chloro-, sulfate (1:1) (9CI) **(68239-80-5)**
1,4-Benzenediamine, 2,6-dichloro- **(609-20-1)**
1,3-Benzenediamine, 2,4-dichloro-6-methyl- (9CI) **(50694-81-0)**
1,3-Benzenediamine, 4,5-dichloro-6-methyl- (9CI) **(50694-83-2)**
1,3-Benzenediamine, 4,6-dichloro-2-methyl- (9CI) **(50694-82-1)**
1,4-Benzenediamine, N,N'-dicyclohexyl- **(4175-38-6)**
1,3-Benzenediamine, N^3,N^3-diethyl-2,4-dinitro-6-(trifluoromethyl)- **(29091-05-2)**
1,4-Benzenediamine, N4,N4-diethyl-2-methyl- (9CI) **(148-71-0)**
Benzenediamine, ar,ar-diethyl-ar-methyl- (9CI) **(68479-98-1)**
1,4-Benzenediamine dihydrochloride **(624-18-0)**
m-Benzenediamine dihydrochloride **(541-69-5)**
p-Benzenediamine dihydrochloride **(624-18-0)**
1,4-Benzenediamine, N,N-dimethyl- (9CI) **(105-10-2)**
1,4-Benzenediamine, N'-(4-(dimethylamino)phenyl)-N,N-dimethyl- (9CI) **(637-31-0)**
1,4-Benzenediamine, N-(1,3-dimethylbutyl)-N'-phenyl- (9CI) **(793-24-8)**
1,4-Benzenediamine, N-(1,3-dimethylbutyl)-N-phenyl- (9CI) **(61931-82-6)**
1,4-Benzenediamine, N-(1,4-dimethylpentyl)-N'-phenyl- (9CI) **(3081-01-4)**
1,4-Benzenediamine, N-(2,4-dinitrophenyl)- (9CI) **(6373-73-5)**
1,4-Benzenediamine, N,N'-di-sec-octyl- (9CI) **(28633-36-5)**
1,4-Benzenediamine ethanedioate **(62654-17-5)**
1,4-Benzenediamine, ethanedioate (1:1) (9CI) **(62654-17-5)**
1,2-Benzenediamine, 4-ethoxy- (9CI) **(1197-37-1)**
1,3-Benzenediamine, 4-ethoxy-, dihydrochloride (9CI) **(67801-06-3)**
1,3-Benzenediamine, 4-ethoxy-, sulfate (1:1) (9CI) **(68015-98-5)**
1,3-Benzenediamine, ar-ethyl-ar-methyl- (9CI) **(68966-84-7)**
1,3-Benzenediamine hydrochloride **(541-69-5)**
1,3-Benzenediamine, 4-methoxy- (9CI) **(615-05-4)**
1,4-Benzenediamine, 2-methoxy- (9CI) **(5307-02-8)**
1,3-Benzenediamine, 4-methoxy-, dihydrochloride **(614-94-8)**
1,3-Benzenediamine, 4,4'-((4-methoxy-1,3-phenylene)bis-(azo))bis- (9CI) **(6358-83-4)**
1,3-Benzenediamine, 4-methoxy-, sulfate **(39156-41-7)**
1,3-Benzenediamine, 4-methoxy, sulfate (1:1) (9CI) **(39156-41-7)**
1,2-Benzenediamine, 4-methyl- **(496-72-0)**
1,3-Benzenediamine, 2-methyl- **(823-40-5)**
1,4-Benzenediamine, 2-methyl- **(95-70-5)**
Benzenediamine, ar-methyl- **(25376-45-8)**
1,2-Benzenediamine, 3-methyl- (9CI) **(2687-25-4)**
1,3-Benzenediamine, 5-methyl- (9CI) **(108-71-4)**
1,4-Benzenediamine, 2-methyl-, dihydrochloride (9CI) **(615-45-2)**
Benzenediamine, ar-(1-methylethyl)- (9CI) **(31626-02-5)**
1,4-Benzenediamine, N-(1-methylethyl)-N-phenyl- (9CI) **(3085-82-3)**
1,4-Benzenediamine, N-(1-methylheptyl)-N'-phenyl- (9CI) **(15233-47-3)**
1,3-Benzenediamine, 4-methyl-, monohydrochloride (9CI) **(5459-85-8)**
1,3-Benzenediamine, 4-methyl-6-(phenylazo)- (9CI) **(5042-54-6)**
1,3-Benzenediamine, 4,4'-((4-methyl-1,3-phenylene)bis(azo))-bis(6-methyl-, dihydrochloride (9CI) **(5421-66-9)**
1,4-Benzenediamine, N-(1-methylpropyl)-N'-phenyl- (9CI) **(788-17-0)**
1,4-Benzenediamine, 2-methyl-, sulfate (1:1) (9CI) **(615-50-9)**
1,4-Benzenediamine, 2-methyl-, sulfate (9CI) **(6369-59-1)**
1,3-Benzenediamine, 4-nitro- **(5131-58-8)**
1,3-Benzenediamine, 5-nitro- (9CI) **(5042-55-7)**
1,2-Benzenediamine, 4-nitro-, dihydrochloride (9CI) **(6219-77-8)**
1,4-Benzenediamine, 2-nitro-, dihydrochloride (9CI) **(18266-52-9)**
1,4-Benzenediamine, N-(4-nitrophenyl)- (9CI) **(6149-34-4)**
1,2-Benzenediamine, 4-nitro-, sulfate (1:1) (9CI) **(68239-82-7)**
1,4-Benzenediamine, 2-nitro-, sulfate (1:1) (9CI) **(68239-83-8)**
1,4-Benzenediamine, N-phenyl- (9CI) **(101-54-2)**
1,3-Benzenediamine, 4,4'-(1,3-phenylenebis(azo))bis **(1052-38-6)**
1,3-Benzenediamine, sulfate (1:1) (9CI) **(541-70-8)**
1,4-Benzenediamine, sulfate (1:1) (9CI) **(16245-77-5)**
1,4-Benzenediamine, N,N,N',N'-tetrakis(4-aminophenyl)- (9CI) **(3283-07-6)**
1,4-Benzenediamine, N,N,N',N'-tetrakis(4-(diethylamino)phenyl)-(9CI) **(3956-73-8)**
1,4-Benzenediamine, N,N,N',N'-tetramethyl- **(100-22-1)**
1,3-Benzenediamine, 2,4,6-trinitro- (9CI) **(1630-08-6)**
Benzene, diamino- **(25265-76-3)**
Benzene, 1,4-diazido- **(2294-47-5)**
Benzene, p-diazido- **(2294-47-5)**
Benzene, 1,1'-(diazomethylene)bis- **(883-40-9)**
Benzenediazonium, chloride **(100-34-5)**
Benzenediazonium chloride, Dry (DOT) **(100-34-5)**
Benzenediazonium, 4-chloro-2-nitro- (9CI) **(27165-22-6)**
Benzenediazonium, 4-chloro-2-nitro-, tetrachlorozincate(2-) (2:1) (9CI) **(14263-89-9)**
Benzenediazonium, 2,5-diethoxy-4-(4-morpholinyl)-, (T-4)-tetrachlorozincate(2-) (2:1) (9CI) **(6023-29-6)**
Benzenediazonium, 4-(diethylamino)-, (T-4)-tetrachlorozincate(2-) (2:1) (9CI) **(5149-85-9)**
Benzenediazonium, 4,4'-(1,2-ethenediyl)bis(3-sulfo-, dichloride (9CI) **(13954-62-6)**
Benzenediazonium, hexafluorophosphate(1-) (9CI) **(369-58-4)**
Benzenediazonium, 2-methoxy-5-nitro- (9CI) **(27165-17-9)**
Benzenediazonium, 2-methoxy-4-(phenylamino)-, Salt with 2,4,6-trimethylbenzenesulfonic acid (1:1) polymer with 1,1'-oxybis(4-(methoxymethyl)benzene) (9CI) **(32761-96-9)**
Benzenediazonium, 4-(phenylamino)-, sulfate (1:1) (9CI) **(4477-28-5)**
Benzenediazosulfonic acid, p-(dimethylamino)-, sodium salt **(140-56-7)**
Benzene, dibromo **(26249-12-7)**
Benzene, m-dibromo **(108-36-1)**
Benzene, p-dibromo **(106-37-6)**
Benzene, 1,2-dibromo- (9CI) **(583-53-9)**
Benzene, 1,3-dibromo- (9CI) **(108-36-1)**
Benzene, 1,4-dibromo- (9CI) **(106-37-6)**
Benzene, o-dibromo- (8CI) **(583-53-9)**
Benzene, (1,2-dibromoethyl) **(93-52-7)**
Benzene, dibromoethyl **(30812-87-4)**
Benzene, 2,4-dibromo-1-fluoro- (8CI,9CI) **(1435-53-6)**
Benzene, 1,3-dibromo-5-methyl- (9CI) **(1611-92-3)**
Benzene, 1,4-dibromo-2-methyl- (9CI) **(615-59-8)**
Benzene, 2,4-dibromo-1-methyl- (9CI) **(31543-75-6)**
Benzene, p-di-sec-butyl- (8CI) **(1014-41-1)**
Benzene, p-di-tert-butyl- (8CI) **(1012-72-2)**
Benzene, m-di-sec-butyl- (6CI,7CI,8CI) **(1079-96-5)**
1,3-Benzenedicarbonitrile **(626-17-5)**
1,3-Benzenedicarbonitrile, 2-((4-((2-(acetyloxy)ethyl)butylamino)-2-methylphenyl)azo)-5-nitro- (9CI) **(72828-64-9)**
1,3-Benzenedicarbonitrile, 4-hydroxy-2,5,6-trichloro **(28343-61-5)**
1,3-Benzenedicarbonitrile, 2,4,5,6-tetrachloro- **(1897-45-6)**
Benzenedicarbonitrile, tetrachloro- (9CI) **(90077-80-8)**
1,3-Benzenedicarbonyl chloride **(99-63-8)**
1,4-Benzenedicarbonyl chloride **(100-20-9)**
1,4-Benzenedicarbonyl dichloride **(100-20-9)**
1,2-Benzenedicarbonyl dichloride (9CI) **(88-95-9)**
1,4-Benzenedicarbonyl dichloride, 2,3,5,6-tetrachloro- (9CI) **(719-32-4)**

1,2-Benzenedicarboperoxoic acid, bis(1,1-dimethylethyl) ester (9CI) **(2155-71-7)**
p-Benzenedicarboxaldehyde **(623-27-8)**
1,4-Benzenedicarboxaldehyde (9CI) **(623-27-8)**
Benzenedicarboxaldehyde (8CI,9CI) **(30025-33-3)**
1,2-Benzenedicarboxamide **(88-96-0)**
1,4-Benzenedicarboxamide, N,N'-bis(2,4,6-tribromophenyl)- (9CI) **(49693-20-1)**
1,3-Benzenedicarboxamide, 2,4,5,6-tetrachloro- (9CI) **(1786-86-3)**
1,2-Benzenedicarboxylic acid **(88-99-3)**
1,4-Benzenedicarboxylic acid **(100-21-0)**
Benzene-1,2-dicarboxylic acid **(88-99-3)**
Benzene-1,3-dicarboxylic acid **(121-91-5)**
m-Benzenedicarboxylic acid **(121-91-5)**
o-Benzenedicarboxylic acid **(88-99-3)**
p-Benzenedicarboxylic acid **(100-21-0)**
Benzenedicarboxylic acid (8CI,9CI) **(29010-86-4)**
1,2-Benzenedicarboxylic acid, Mixed decyl and hexyl and octyl diesters **(68648-93-1)**
1,4-Benzenedicarboxylic acid, Polymer with 1,4-butanediol (9CI) **(26062-94-2)**
1,2-Benzenedicarboxylic acid, 3-amino-, cyclic hydrazide **(521-31-3)**
1,2-Benzenedicarboxylic acid anhydride **(85-44-9)**
1,2-Benzenedicarboxylic acid, bi(2-methoxyethyl) ester (9CI) **(117-82-8)**
1,2-Benzenedicarboxylic acid, bis(2-ethylhexyl) ester **(117-81-7)**
1,4-Benzenedicarboxylic acid, bis(2-ethylhexyl)ester (9CI) **(6422-86-2)**
1,4-Benzenedicarboxylic acid, bis(2-hydroxyethyl) ester (9CI) **(959-26-2)**
1,2-Benzenedicarboxylic acid, bis(1-methylethyl) ester **(605-45-8)**
1,2-Benzenedicarboxylic acid, bis(4-methylpentyl) ester (9CI) **(146-50-9)**
1,4-Benzenedicarboxylic acid, 2,5-bis((4-methylphenyl)amino)- (9CI) **(10291-28-8)**
1,3-Benzenedicarboxylic acid, bis(oxiranylmethyl) ester (9CI) **(7195-43-9)**
1,2-Benzenedicarboxylic acid, bis(oxiranylmethyl) ester, Homopolymer (9CI) **(37099-12-0)**
1,4-Benzenedicarboxylic acid, 2,5-bis(phenylamino)- (9CI) **(10109-95-2)**
1,2-Benzenedicarboxylic acid, bis(phenylmethyl) ester (9CI) **(523-31-9)**
1,2-Benzenedicarboxylic acid, bis(2-propylheptyl) ester (9CI) **(53306-54-0)**
1,2-Benzenedicarboxylic acid, bis((tetradecahydro-1,4a-dimethyl-7-(1-methylethyl)-1-phenanthrenyl)methyl) ester (9CI) **(36388-36-0)**
1,2-Benzenedicarboxylic acid, 2-butoxyethyl butyl ester (9CI) **(33374-28-6)**
1,2-Benzenedicarboxylic acid, butyl cyclohexyl ester (9CI) **(84-64-0)**
1,2-Benzenedicarboxylic acid, butyl 2-ethylhexyl ester (9CI) **(85-69-8)**
1,2-Benzenedicarboxylic acid, butyl isodecyl ester (9CI) **(42343-36-2)**
1,2-Benzenedicarboxylic acid, butyl methyl ester **(34006-76-3)**
1,2-Benzenedicarboxylic acid, butyl 2-methylpropyl ester (9CI) **(17851-53-5)**
1,2-Benzenedicarboxylic acid, butyl octyl ester (9CI) **(84-78-6)**
1,2-Benzenedicarboxylic acid, butyl phenylmethyl ester **(85-68-7)**
1,2-Benzenedicarboxylic acid, 4-chloro- (9CI) **(89-20-3)**
1,4-Benzenedicarboxylic acid, cobalt salt (9CI) **(34262-88-9)**
1,4-Benzenedicarboxylic acid, copper salt (9CI) **(34262-89-0)**
1,2-Benzenedicarboxylic acid, decyl octyl ester **(119-07-3)**
1,2-Benzenedicarboxylic acid, di-C6-10-alkyl esters **(68515-51-5)**
1,2-Benzenedicarboxylic acid, di-C7-11-alkyl esters **(68648-91-9)**
1,2-Benzenedicarboxylic acid, di-C11-14-branched alkyl esters, C13-rich **(68515-47-9)**
1,2-Benzenedicarboxylic acid, di-C6-8-branched alkyl esters, C7-rich **(71888-89-6)**
1,2-Benzenedicarboxylic acid, di-C8-10-branched alkyl esters, C9-rich **(68515-48-0)**
1,2-Benzenedicarboxylic acid, di-C9-11-branched alkyl esters, C10-rich **(68515-49-1)**
1,2-Benzenedicarboxylic acid, di-C7-11-branched and linear alkyl esters **(68515-42-4)**
1,2-Benzenedicarboxylic acid, di-C7-9-branched and linear alkyl esters **(68515-41-3)**
1,4-Benzenedicarboxylic acid, dibutyl ester **(1962-75-0)**
Benzene-o-dicarboxylic acid di-n-butyl ester **(84-74-2)**
o-Benzenedicarboxylic acid, dibutyl ester **(84-74-2)**
Benzenedicarboxylic acid, dichloro- (9CI) **(41906-38-1)**
1,2-Benzenedicarboxylic acid, dicyclohexyl ester **(84-61-7)**
1,2-Benzenedicarboxylic acid, didodecyl ester (9CI) **(2432-90-8)**
1,2-Benzenedicarboxylic acid, diethyl ester **(84-66-2)**
1,4-Benzenedicarboxylic acid, diethyl ester **(636-09-9)**
1,2-Benzenedicarboxylic acid, diheptyl ester (9CI) **(3648-21-3)**
1,2-Benzenedicarboxylic acid, diheptyl ester, branched and linear **(68515-44-6)**
1,2-Benzenedicarboxylic acid, dihexyl ester **(84-75-3)**
1,2-Benzenedicarboxylic acid, dihexyl ester, Branched and linear **(68515-50-4)**
1,3-Benzenedicarboxylic acid, dihydrazide (9CI) **(2760-98-7)**
1,2-Benzenedicarboxylic acid, 4,5-dihydroxy- **(63958-66-7)**
1,2-Benzenedicarboxylic acid, diisodecyl ester (9CI) **(26761-40-0)**
1,2-Benzenedicarboxylic acid, diisononyl ester (9CI) **(28553-12-0)**
1,2-Benzenedicarboxylic acid, diisooctyl ester **(27554-26-3)**
1,2-Benzenedicarboxylic acid, dimethyl ester **(131-11-3)**
1,3-Benzenedicarboxylic acid, dimethyl ester **(1459-93-4)**
1,4-Benzenedicarboxylic acid, dimethyl ester (9CI) **(120-61-6)**
Benzenedicarboxylic acid, dimethyl ester (8CI,9CI) **(26638-01-7)**
1,4-Benzenedicarboxylic acid, dimethyl ester, Polymer with α-hydro-ω-hydroxypoly(oxy-1,2-ethanediyl) (9CI) **(58782-15-3)**
1,2-Benzenedicarboxylic acid, 2,2-dimethyl-1-(1-methylethyl)-3-(2-methyl-1-oxopropoxy)propyl phenylmethyl ester (9CI) **(16883-83-3)**
1,2-Benzenedicarboxylic acid, dinonyl ester, branched **(71549-78-5)**
1,2-Benzenedicarboxylic acid, dinonyl ester, branched and linear **(68515-45-7)**
1,2-Benzenedicarboxylic acid, dioctadecyl ester (9CI) **(14117-96-5)**
1,2-Benzenedicarboxylic acid, dioctyl ester **(117-84-0)**
o-Benzenedicarboxylic acid, dioctyl ester **(117-84-0)**
1,2-Benzenedicarboxylic acid, dipentyl ester **(131-18-0)**
1,2-Benzenedicarboxylic acid, diphenyl ester (9CI) **(84-62-8)**
1,3-Benzenedicarboxylic acid, diphenyl ester (9CI) **(744-45-6)**
1,4-Benzenedicarboxylic acid, diphenyl ester (9CI) **(1539-04-4)**
1,2-Benzenedicarboxylic acid, di-2-propenyl ester **(131-17-9)**
1,3-Benzenedicarboxylic acid, di-2-propenyl ester (9CI) **(1087-21-4)**
1,3-Benzenedicarboxylic acid, di-2-propenyl ester, Homopolymer (9CI) **(25035-78-3)**
1,2-Benzenedicarboxylic acid, di-2-propenyl ester, Homopolymer (9CI) **(25053-15-0)**
1,2-Benzenedicarboxylic acid, dipropyl ester **(131-16-8)**
1,2-Benzenedicarboxylic acid, ditridecyl ester **(119-06-2)**
1,2-Benzenedicarboxylic acid, 2-ethoxy-2-oxoethyl-, ethyl ester (9CI) **(84-72-0)**
1,2-Benzenedicarboxylic acid, ethyl methyl ester (9CI) **(34006-77-4)**
1,2-Benzenedicarboxylic acid, 3-fluoro- (9CI) **(1583-67-1)**
1,2-Benzenedicarboxylic acid, 4-hydroxy- (9CI) **(610-35-5)**
1,3-Benzenedicarboxylic acid, 4-hydroxy- (9CI) **(636-46-4)**
1,2-Benzenedicarboxylic acid, 4-hydroxy-5-methoxy- (9CI) **(63035-28-9)**
1,3-Benzenedicarboxylic acid, 4-hydroxy-5-methoxy- (9CI) **(2134-91-0)**
1,2-Benzenedicarboxylic acid, isodecyl isooctyl ester (9CI) **(42343-35-1)**
1,2-Benzenedicarboxylic acid, isodecyl octyl ester (9CI) **(1330-96-7)**
1,2-Benzenedicarboxylic acid, lead(2+) salt (1:1) (9CI) **(6838-85-3)**
1,2-Benzenedicarboxylic acid, 4-methyl- (9CI) **(4316-23-8)**
1,2-Benzenedicarboxylic acid, methyl- (9CI) **(30497-87-1)**
1,2-Benzenedicarboxylic acid, 4-methyl-, dimethyl ester (9CI) **(20116-65-8)**
1,4-Benzenedicarboxylic acid, 2-methyl-, dimethyl ester (9CI) **(14186-60-8)**
1,2-Benzenedicarboxylic acid, mono(1-(4-bromophenyl)-ethyl) ester (9CI) **(92387-49-0)**
1,2-Benzenedicarboxylic acid, monobutyl ester (9CI) **(131-70-4)**
1,2-Benzenedicarboxylic acid, mono(1-(4-chlorophenyl)-ethyl) ester (9CI) **(92387-48-9)**
1,2-Benzenedicarboxylic acid, mono(1-(4-(1,1-dimethylethyl)-phenyl)ethyl) ester (9CI) **(33533-56-1)**
1,2-Benzenedicarboxylic acid, monododecyl ester (9CI) **(21577-80-0)**
1,2-Benzenedicarboxylic acid, monoisodecyl ester (9CI) **(31047-64-0)**
1,2-Benzenedicarboxylic acid, monoisooctyl ester (9CI) **(30849-48-0)**
1,2-Benzenedicarboxylic acid, mono(1-(4-methoxyphenyl)-ethyl) ester (9CI) **(33533-57-2)**
1,2-Benzenedicarboxylic acid, monomethyl ester (9CI) **(4376-18-5)**

1,4-Benzenedicarboxylic acid, monomethyl ester (9CI) **(1679-64-7)**
1,2-Benzenedicarboxylic acid, mono(1-(4-methylphenyl)-ethyl) ester (9CI) **(23005-56-3)**
1,2-Benzenedicarboxylic acid, mono(1-phenylethyl) ester (9CI) **(33533-53-8)**
1,2-Benzenedicarboxylic acid, mono(phenylmethyl) ester (9CI) **(2528-16-7)**
1,2-Benzenedicarboxylic acid, monopotassium salt **(877-24-7)**
1,3-Benzenedicarboxylic acid, 5-nitro- (9CI) **(618-88-2)**
1,3-Benzenedicarboxylic acid, 5-nitro-, dimethyl ester (9CI) **(13290-96-5)**
1,4-Benzenedicarboxylic acid, 2-nitro-, dimethyl ester (9CI) **(5292-45-5)**
1,4-Benzenedicarboxylic acid, potassium salt (9CI) **(13427-80-0)**
1,2-Benzenedicarboxylic acid, 4-sulfo- (9CI) **(89-08-7)**
1,3-Benzenedicarboxylic acid, 5-sulfo-, 1,3-dimethyl ester (9CI) **(138-25-0)**
1,3-Benzenedicarboxylic acid, 5-sulfo-, 1,3-bis(2-hydroxy-ethyl) ester, monosodium salt (9CI) **(24019-46-3)**
Benzene-1,3-dicarboxylic acid, 5-sulfo-, monosodium salt **(6362-79-4)**
1,2-Benzenedicarboxylic acid, 4-sulfo-, monosodium salt (9CI) **(33562-89-9)**
1,2-Benzenedicarboxylic acid, 3,4,5,6-tetrabromo-, Compd. with 1,2-ethanediamine (2:1) (9CI) **(66046-78-4)**
1,2-Benzenedicarboxylic acid, 3,4,5,6-tetrabromo-, di-2-propenyl ester (9CI) **(49693-09-6)**
1,2-Benzenedicarboxylic acid, 3,4,5,6-tetrachloro- (9CI) **(632-58-6)**
1,4-Benzenedicarboxylic acid, 2,3,5,6-tetrachloro- (9CI) **(2136-79-0)**
1,4-Benzenedicarboxylic acid, 2,3,5,6-tetrachloro-, dimethyl ester **(1861-32-1)**
1,4-Benzenedicarboxylic acid, 2,3,5,6-tetrachloro-, monomethyl ester (9CI) **(887-54-7)**
1,2-Benzenedicarboxylic acid, tridecyl ester, Manuf. of, by-products from **(68515-39-9)**
1,2-Benzenedicarboxylic anhydride, glycerin, coconut oil polymer **(66070-87-9)**
1,2-Benzenedicarboxylic anhydride, glycerin, penta-erythritol, linseed oil, styrene polymer **(66071-63-4)**
1,2-Benzenedicarboxylic anhydride, glycerin, penta-erythritol, soybean oil polymer **(66070-93-7)**
1,2-Benzenedicarboxylic anhydride, glycerin, soybean oil polymer **(66070-61-9)**
1,2-Benzenedicarboxylic anhydride, glycerin, tall oil fatty acids polymer **(66070-71-1)**
1,2-Benzenedicarboxylic anhydride, pentaerythritol, soybean oil polymer **(66070-60-8)**
Benzene, 1,2-dichloro- **(95-50-1)**
Benzene, m-dichloro **(541-73-1)**
Benzene, o-dichloro **(95-50-1)**
Benzene, 1,3-dichloro- (9CI) **(541-73-1)**
Benzene, 1,4-dichloro- (9CI) **(106-46-7)**
Benzene, p-dichloro (8CI) **(106-46-7)**
Benzene, dichloro- (8CI,9CI) **(25321-22-6)**
Benzene, 1,4-dichloro-2-(2-chloroethenyl)- (9CI) **(54935-00-1)**
Benzene, 1,2-dichloro-4-(chloromethyl)- (9CI) **(102-47-6)**
Benzene, 2,4-dichloro-1-(chloromethyl)- (9CI) **(94-99-5)**
Benzene, dichloro(chloromethyl)- (9CI) **(38721-71-0)**
Benzene, 1,3-dichloro-2-(chloromethyl)-4,5,6-trimethoxy- (9CI) **(108544-94-1)**
Benzene, 1,2-dichloro-4-(2-chlorophenoxy)- (9CI) **(61328-44-7)**
Benzene, 1,3-dichloro-2-(dichloromethyl)- (9CI) **(81-19-6)**
Benzene, 2,4-dichloro-1-(dichloromethyl)- (9CI) **(134-25-8)**
Benzene, dichloro((dichlorophenyl)methyl)methyl- (9CI) **(76253-60-6)**
Benzene, 1,4-dichloro-2,5-dimethoxy **(2675-77-6)**
Benzene, 1,2-dichloro-3,4-dimethoxy- (9CI) **(90283-00-4)**
Benzene, 1,4-dichloro-2,3-dimethoxy- (9CI) **(90283-02-6)**
Benzene, 1,5-dichloro-2,3-dimethoxy- (9CI) **(90283-01-5)**
Benzene, 1,2-dichloro-4,5-dimethoxy- (8CI,9CI) **(2772-46-5)**
Benzene, 1,4-dichloro-2,5-dimethyl- (9CI) **(1124-05-6)**
Benzene, 2,4-dichloro-1,3-dimethyl-5-hydroxy- **(133-53-9)**
Benzene, 2,4-dichloro-1,3-dinitro-5-(trifluoromethyl)- (9CI) **(29091-09-6)**
Benzene, 1,1'-(1,2-dichloro-1,2-ethanediyl)bis- (9CI) **(5963-49-5)**
Benzene, 1,1'-(1,2-dichloro-1,2-ethenediyl)bis-, (E)- (9CI) **(951-86-0)**
Benzene, 1,2-dichloro-3-ethenyl- (9CI) **(2123-28-6)**
Benzene, 1,3-dichloro-2-ethenyl- (9CI) **(28469-92-3)**
Benzene, 1,4-dichloro-2-ethenyl- (9CI) **(1123-84-8)**
Benzene, 2,4-dichloro-1-ethenyl- (9CI) **(2123-27-5)**
Benzene, 1,1'-(dichloroethenylidene)bis(chloro- (VAN)(9CI) **(68679-99-2)**
Benzene, (1,2-dichloroethyl)- (8CI,9CI) **(1074-11-9)**
Benzene, 1,1'-(2,2-dichloroethylidene)bis(4-ethoxy **(7388-32-1)**
Benzene, 1,2-dichloro-3-methoxy- (9CI) **(1984-59-4)**
Benzene, 1,2-dichloro-4-methoxy- (9CI) **(36404-30-5)**
Benzene, 1,3-dichloro-2-methoxy- (9CI) **(1984-65-2)**
Benzene, 1,3-dichloro-5-methoxy- (9CI) **(33719-74-3)**
Benzene, 1,4-dichloro-2-methoxy- (9CI) **(1984-58-3)**
Benzene, 2,4-dichloro-1-methoxy- (9CI) **(553-82-2)**
Benzene, dichloromethoxy- (9CI) **(54518-15-9)**
Benzene, 1,3-dichloro-2-methoxy-5-(2,4,6-trichlorophenoxy)-(9CI) **(90985-94-7)**
Benzene, 1,5-dichloro-2-methoxy-3-(2,4,6-trichlorophenoxy)-(9CI) **(97534-02-6)**
Benzene, (dichloromethyl) **(98-87-3)**
Benzene, 1,2-dichloro-3-methyl- (9CI) **(32768-54-0)**
Benzene, 1,2-dichloro-4-methyl- (9CI) **(95-75-0)**
Benzene, 1,3-dichloro-2-methyl- (9CI) **(118-69-4)**
Benzene, 1,3-dichloro-5-methyl- (9CI) **(25186-47-4)**
Benzene, 1,4-dichloro-2-methyl- (9CI) **(19398-61-9)**
Benzene, 2,4-dichloro-1-methyl- (9CI) **(95-73-8)**
Benzene, dichloromethyl- (9CI) **(29797-40-8)**
Benzene, 1,1'-(dichloromethylene)bis- (9CI) **(2051-90-3)**
Benzene, 1-(dichloromethyl)-4-ethyl- (9CI) **(54789-29-6)**
Benzene, 1,2-dichloro-3-nitro **(3209-22-1)**
Benzene, 1,4-dichloro-2-nitro **(89-61-2)**
Benzene, 2,4-dichloro-1-nitro **(611-06-3)**
Benzene, dichloronitro- **(27900-75-0)**
Benzene, 1,3-dichloro-5-nitro- (9CI) **(618-62-2)**
Benzene, 1,2-dichloro-4-nitro (8CI,9CI) **(99-54-7)**
Benzene, 1,3-dichloro-2-nitro- (8CI,9CI) **(601-88-7)**
Benzene, 2,4-dichloro-1-(4-nitrophenoxy)- **(1836-75-5)**
Benzene, 1,1'-(2,2-dichloropropylidene)bis(4-ethoxy **(56265-23-7)**
Benzene, 1,2-dichloro-4-(trichloromethyl)- (9CI) **(13014-24-9)**
Benzene, 2,4-dichloro-1-(trichloromethyl)- (9CI) **(13014-18-1)**
Benzene, 1,2-dichloro-3-(trifluoromethyl)- (9CI) **(54773-19-2)**
Benzene, 1,2-dichloro-4-(trifluoromethyl)- (9CI) **(328-84-7)**
Benzene, 2,4-dichloro-1-(trifluoromethyl)- (9CI) **(320-60-5)**
Benzene, dichloro(trifluoromethyl)- (9CI) **(30498-35-2)**
Benzene, 1,2-dichloro-3,4,5-trimethoxy- (9CI) **(108544-93-0)**
Benzene, 1,5-dichloro-2,3,4-trimethoxy- (6CI,9CI) **(99849-00-0)**
Benzene, 1,4-diethenyl- (9CI) **(105-06-6)**
Benzene, diethyl **(25340-17-4)**
Benzene, m-diethyl **(141-93-5)**
Benzene, o-diethyl **(135-01-3)**
Benzene, p-diethyl- **(105-05-5)**
Benzene, 1,4-diethyl- (9CI) **(105-05-5)**
Benzene, 1,3-diethyl-5-methyl- **(2050-24-0)**
Benzene, diethylmethyl- (9CI) **(25550-13-4)**
Benzene, 1,4-diethyl-2,3,5,6-tetramethyl- (7CI,8CI,9CI) **(33962-13-9)**
Benzene, 1,4-difluoro- **(540-36-3)**
Benzene, m-difluoro **(372-18-9)**
Benzene, o-difluoro **(367-11-3)**
Benzene, p-difluoro **(540-36-3)**
Benzene, m-dihydroxy- **(108-46-3)**
Benzene, o-dihydroxy- **(120-80-9)**
Benzene, p-dihydroxy- **(123-31-9)**
Benzene, 1,2-diiodo- (9CI) **(615-42-9)**
Benzene, 1,3-diiodo- (9CI) **(626-00-6)**
Benzene, 1,4-diiodo- (9CI) **(624-38-4)**
Benzene, o-diiodo- (8CI) **(615-42-9)**
Benzene, p-diiodo- (8CI) **(624-38-4)**
Benzene, 1-((diiodomethyl)sulfonyl)-4-methyl- (9CI) **(20018-09-1)**
Benzene, 2,4-diisocyanato-1-methyl **(584-84-9)**
Benzene, 2,6-diisocyanato-1-methyl **(91-08-7)**

Benzene-, 1,3-diisocyanatomethyl- **(26471-62-5)**
Benzene, diisocyanatomethyl- (9CI) **(1321-38-6)**
Benzene, diisopropyl **(25321-09-9)**
Benzene, m-diisopropyl **(99-62-7)**
Benzene, o-diisopropyl **(577-55-9)**
Benzene, p-diisopropyl **(100-18-5)**
Benzene, 1,4-diisothiocyanato- **(4044-65-9)**
1,4-Benzenedimethanol (9CI) **(589-29-7)**
1,4-Benzenedimethanol, α,α,α',α'-tetramethyl- (9CI) **(2948-46-1)**
Benzene, 1,2-dimethoxy- **(91-16-7)**
Benzene, 1,3-dimethoxy- **(151-10-0)**
Benzene, 1,4-dimethoxy- **(150-78-7)**
Benzene, m-dimethoxy **(151-10-0)**
Benzene, o-dimethoxy **(91-16-7)**
Benzene, p-dimethoxy **(150-78-7)**
Benzene, dimethoxy- (9CI) **(27598-81-8)**
Benzene, (2,2-dimethoxyethyl)- (9CI) **(101-48-4)**
Benzene, 1,2-dimethoxy-4-methyl- (9CI) **(494-99-5)**
Benzene, 1,3-dimethoxy-2-methyl- (9CI) **(5673-07-4)**
Benzene, 1-(dimethoxymethyl)-4-methoxy- (9CI) **(2186-92-7)**
Benzene, 1,2-dimethoxy-4-propenyl **(93-16-3)**
Benzene, dimethyl- **(1330-20-7)**
Benzene, 1,3-dimethyl-, benzylated **(68908-87-2)**
Benzene, 1,4-dimethyl-2,5-bis(1-methylethyl)- (9CI) **(10375-96-9)**
Benzene, 1,5-dimethyl-2,4-bis(1-methylethyl)- (9CI) **(5186-68-5)**
Benzene, 1,1'-(3,3-dimethyl-1-butenylidene)bis- (9CI) **(23586-64-3)**
Benzene, (1,1-dimethylbutyl)- (6CI,7CI,8CI,9CI) **(1985-57-5)**
Benzene, (2,2-dimethylbutyl)- (6CI,7CI,8CI,9CI) **(28080-86-6)**
Benzene, dimethyl-, dibromo deriv. (9CI) **(28805-90-5)**
Benzene, 1,1'-(1,2-dimethyl-1,2-ethanediyl)bis-, (R*,S*)- (9CI) **(4613-11-0)**
Benzene, 1-(1,1-dimethylethoxy)-3-methoxy- (9CI) **(15359-99-6)**
Benzene, 1-(1,1-dimethylethyl)-3,5-dimethyl-2,4,6-trinitro **(81-15-2)**
Benzene, 1-(1,1-dimethylethyl)-2,6-dinitro-3,4,5-trimethyl **(145-39-1)**
Benzene, 1-(1,1-dimethylethyl)-4-ethenyl- (9CI) **(1746-23-2)**
Benzene, 1-(1,1-dimethylethyl)-3-ethyl-5-methyl- (9CI) **(6630-01-9)**
Benzene, (1,1-dimethylethyl)methoxy- (9CI) **(36731-23-4)**
Benzene, (1,1-dimethylethyl)methyl- (9CI) **(27138-21-2)**
Benzene, 5-(1,1-dimethylethyl)-1,2,3-trimethyl- (9CI) **(98-23-7)**
Benzene, 1,3-dimethyl-5-(1-methylethyl)- (9CI) **(4706-90-5)**
Benzene, 2,4-dimethyl-1-(1-methylethyl)- (9CI) **(4706-89-2)**
Benzene, 1,4-dimethyl-2-(2-methylpropyl)- (9CI) **(55669-88-0)**
Benzene, 2,4-dimethyl-1-(1-methylpropyl)- (9CI) **(1483-60-9)**
Benzene, 1,4-dimethyl-2-nitro- **(89-58-7)**
Benzene, 1,2-dimethyl-3-nitro- (9CI) **(83-41-0)**
Benzene, 1,2-dimethyl-4-nitro- (9CI) **(99-51-4)**
Benzene, 1,3-dimethyl-5-nitro- (9CI) **(99-12-7)**
Benzene, 2,4-dimethyl-1-nitro- (9CI) **(89-87-2)**
Benzene, 1,3-dimethyl-5-(4-nitrophenoxy)- (9CI) **(1630-17-7)**
Benzene, dimethyl(pentyloxy)- (9CI) **(1320-21-4)**
Benzene, 1,2-dimethyl-4-(1-phenylethyl)- (9CI) **(6196-95-8)**
Benzene, 1,2-dimethyl-4-(phenylmethyl)- (9CI) **(13540-56-2)**
Benzene, (2,2-dimethylpropyl)- (9CI) **(1007-26-7)**
Benzene, 1,1'-(2,2-dimethylpropylidene)bis(4-ethoxy- (9CI) **(27955-87-9)**
Benzene, 1,1'-(2,2-dimethylpropylidene)bis(4-methoxy- (9CI) **(4741-74-6)**
Benzene, dimethyl-, tetrabromo deriv. (9CI) **(36059-21-9)**
Benzene, 1,3-dinitro- **(99-65-0)**
Benzene, dinitro **(25154-54-5)**
Benzene, m-dinitro **(99-65-0)**
Benzene, o-dinitro **(528-29-0)**
Benzene, p-dinitro **(100-25-4)**
Benzene, 1,2-dinitro- (9CI) **(528-29-0)**
Benzene, 2,4-dinitro-1-fluoro **(70-34-8)**
Benzene, o-dinitroso **(7617-57-4)**
Benzene, p-dinitroso **(105-12-4)**
Benzene, 1,4-dinitroso- (9CI) **(105-12-4)**
Benzene, 4,6-dinitro-1,2,3-trichloro **(6379-46-0)**
Benzene, dinitrotrichloro **(8003-46-1)**
1,2-Benzenediol **(120-80-9)**
1,3-Benzenediol **(108-46-3)**
1,4-Benzenediol **(123-31-9)**
m-Benzenediol **(108-46-3)**
o-Benzenediol **(120-80-9)**
p-Benzenediol **(123-31-9)**
1,2-Benzenediol, 4-(2-amino-1-hydroxyethyl)-, hydrochloride, (+-)- (9CI) **(55-27-6)**
1,2-Benzenediol, 4-(2-amino-1-hydroxyethyl)-, (R)- (9CI) **(51-41-2)**
1,2-Benzenediol, 4-(aminomethyl) **(37491-68-2)**
1,4-Benzenediol, 2,6-bis(1,1-dimethylpropyl)- (9CI) **(2349-85-1)**
1,2-Benzenediol, 3-bromo- (9CI) **(14381-51-2)**
1,2-Benzenediol, 4-chloro- **(2138-22-9)**
1,3-Benzenediol, 4,4'-((5-chloro-2-hydroxy-1,3-phenylene)bis-(methylene))bis- (9CI) **(31265-39-1)**
Benzenediol, chloromethoxy- (9CI) (VAN) **(103339-63-5)**
1,2-Benzenediol, 4,5-dichloro- **(3428-24-8)**
1,2-Benzenediol, 3,5-dichloro- (9CI) **(13673-92-2)**
1,2-Benzenediol, 3,6-dichloro- (9CI) **(3938-16-7)**
1,2-Benzenediol, dichloro- (9CI) **(25167-85-5)**
1,4-Benzenediol, 2,3-dichloro- (9CI) **(608-44-6)**
1,4-Benzenediol, 2,5-dichloro- (9CI) **(824-69-1)**
Benzenediol, dichloromethoxy- (9CI) (VAN) **(75562-93-5)**
1,3-Benzenediol, dihydro- **(504-02-9)**
1,2-Benzenediol, 3,4-dimethyl- **(2785-76-4)**
1,2-Benzenediol, 3,5-dimethyl- (9CI) **(2785-75-3)**
1,3-Benzenediol, 4,5-dimethyl- (9CI) **(527-55-9)**
1,2-Benzenediol, 4-(1,1-dimethylethyl)- **(98-29-3)**
1,2-Benzenediol, (1,1-dimethylethyl)- (9CI) **(27213-78-1)**
1,4-Benzenediol, 2-(1,1-dimethylethyl)- (9CI) **(1948-33-0)**
1,3-Benzenediol, 5-(2-((1,1-dimethylethyl)amino)-1-hydroxy-ethyl)- (9CI) **(23031-25-6)**
1,3-Benzenediol, 2,4-dinitro- (9CI) **(519-44-8)**
1,2-Benzenediol, 3-ethyl- **(933-99-3)**
1,3-Benzenediol, 4-ethyl- (9CI) **(2896-60-8)**
1,3-Benzenediol, hexahydro- **(504-01-8)**
1,2-Benzenediol, 4-(1-hydroxy-2-(methylamino)ethyl)-, hydrochloride, (+-)- (9CI) **(329-63-5)**
1,2-Benzenediol, 4-(1-hydroxy-2-(methylamino)ethyl)-, hydrochloride, (R)- (9CI) **(55-31-2)**
1,2-Benzenediol, 4-(1-hydroxy-2-(methylamino)ethyl)-, (R)- (9CI) **(51-43-4)**
1,2-Benzenediol, 4-(1-hydroxy-2-((1-methylethyl)amino)ethyl)-, hydrochloride **(51-30-9)**
1,2-Benzenediol, 3-methoxy- (9CI) **(934-00-9)**
1,4-Benzenediol, 2-methoxy-, 4-acetate (9CI) **(57244-88-9)**
1,2-Benzenediol, 3-methyl- **(488-17-5)**
1,4-Benzenediol, 2-methyl- **(95-71-6)**
1,2-Benzenediol, 4-methyl- (9CI) **(452-86-8)**
1,3-Benzenediol, 2-methyl- (9CI) **(608-25-3)**
1,3-Benzenediol, 4-methyl- (9CI) **(496-73-1)**
1,4-Benzenediol, 2-methyl-, Compd. with 2,5-cyclo-hexadiene-1,4-dione (1:1) (9CI) **(55836-33-4)**
1,2-Benzenediol, 4-(1-methylethyl)- **(2138-43-4)**
1,2-Benzenediol, 3-(1-methylethyl)- (9CI) **(2138-48-9)**
1,3-Benzenediol, 2-(3-methyl-6-(1-methylethenyl)-2-cyclo-hexen-1-yl)-5-pentyl-, (1R-trans)- **(13956-29-1)**
1,3-Benzenediol, monoacetate **(102-29-4)**
1,3-Benzenediol, 2-nitro- (9CI) **(601-89-8)**
1,3-Benzenediol, 4-((4-nitrophenyl)azo)- (9CI) **(74-39-5)**
1,3-Benzenediol, 5-pentyl- (9CI) **(500-66-3)**
1,2-Benzenediol, 3-phenyl- **(1133-63-7)**
1,2-Benzenediol, 4-phenyl- **(92-05-7)**
1,3-Benzenediol, 4-(phenylazo) **(2051-85-6)**
1,2-Benzenediol, 3,4,5,6-tetrabromo- **(488-47-1)**
1,3-Benzenediol, 2,4,5,6-tetrachloro-, diacetate (9CI) **(69158-26-5)**
1,2-Benzenediol, 3,4,5-tribromo- (9CI) **(2747-17-3)**
1,2-Benzenediol, 3,4,5-trichloro- (9CI) **(56961-20-7)**

Name	CAS No.
1,2-Benzenediol, 3,4,6-trichloro- (9CI)	**(32139-72-3)**
1,4-Benzenediol, 2,3,5-trichloro- (9CI)	**(608-94-6)**
1,4-Benzenediol, 2,3,5-trimethyl- (9CI)	**(700-13-0)**
1,3-Benzenediol, 2,4,6-trinitro- (9CI)	**(82-71-3)**
1,3-Benzenediol, 2,4,6-trinitro-, lead(2+) salt (1:1) (9CI)	**(15245-44-0)**
1,3-Benzenediol, 2,4,6-trinitro-, magnesium salt (1:1) (9CI)	**(13255-27-1)**
1,4-Benzenedipropanoic acid, β,β'-dioxo-, diethyl ester (9CI)	**(93-94-7)**
Benzene, 1,4-dipropyl- (9CI)	**(4815-57-0)**
Benzene, p-dipropyl- (8CI)	**(4815-57-0)**
1,3-Benzenedisulfonamide, 4,5-dichloro-	**(120-97-8)**
m-Benzenedisulfonamide, 4,5-dichloro	**(120-97-8)**
1,3-Benzenedisulfonic acid	**(98-48-6)**
m-Benzenedisulfonic acid	**(98-48-6)**
1,3-Benzenedisulfonic acid, 4,6-diamino- (9CI)	**(137-50-8)**
1,4-Benzenedisulfonic acid, 2,5-dihydroxy-, dipotassium salt (9CI)	**(15763-57-2)**
p-Benzenedisulfonic acid, 2,5-dihydroxy-, dipotassium salt (8CI)	**(15763-57-2)**
1,3-Benzenedisulfonic acid, 4,5-dihydroxy-, disodium salt	**(149-45-1)**
m-Benzenedisulfonic acid, 4,5-dihydroxy-, disodium salt	**(149-45-1)**
1,4-Benzenedisulfonic acid, 2-((4-((4-((1-hydroxy-6-(phenylamino)-3-sulfo-2-naphthalenyl)azo)-1-naphthalenyl)azo)-6-sulfo-1-naphthalenyl)azo)-, tetrasodium salt (9CI)	**(2503-73-3)**
1,3-Benzenedisulfonyl diazide, 4-(((2-pyridinylamino)-carbonyl)amino)- (9CI)	**(64690-01-3)**
Benzene, divinyl	**(1321-74-0)**
Benzene, m-divinyl	**(108-57-6)**
Benzene, dodecyl	**(123-01-3)**
Benzene, 1-dodecyl-4-methyl- (9CI)	**(104-41-6)**
Benzene, dodecylnitro- (9CI)	**(58353-63-2)**
Benzene, dodecylphenoxy- (9CI)	**(25619-63-0)**
Benzene, eicosyl- (9CI)	**(2398-68-7)**
Benzene, (epoxyethyl)	**(96-09-3)**
Benzene, (2,3-epoxypropoxy)-	**(122-60-1)**
Benzene, 1-(2,3-epoxypropoxy)-4-methoxy	**(2211-94-1)**
Benzeneethanamine	**(64-04-0)**
Benzeneethanamine, 4-bromo-2,5-dimethoxy-α-methyl-, hydrobromide, (+-)-	**(9CI)**
Benzeneethanamine, 2-chloro-α,α-dimethyl-	**(10389-73-8)**
Benzeneethanamine, N-(3-chloropropyl)-α-methyl- (9CI)	**(17243-57-1)**
Benzeneethanamine, 2,5-dimethoxy-α-methyl-, hydrochloride (9CI)	**(24973-25-9)**
Benzeneethanamine, N,N-dimethyl-α-phenyl-, hydrochloride, (R)-, (9CI)	**(14148-99-3)**
Benzeneethanamine, N,α-dimethyl-N-(phenylmethyl)-, (+)- (9CI)	**(156-08-1)**
Benzeneethanamine, N,α-dimethyl-N-(phenylmethyl)-, hydrochloride, (+)- (9CI)	**(5411-22-3)**
Benzeneethanamine, 4-methoxy-α-methyl-, (+-)- (9CI)	**(23239-32-9)**
Benzeneethanamine, α-methyl-, (S)-, sulfate (2:1)	**(51-63-8)**
Benzeneethanamine, sulfate (2:1) (9CI)	**(5471-08-9)**
Benzeneethanamine, 3,4,5-trimethoxy-	**(54-04-6)**
Benzene, 1,1'-(1,2-ethanediyl)bis(3,4-dimethyl- (9CI)	**(34101-86-5)**
Benzene, 1,2-(1,2-ethanediylbis(oxy))-	**(493-09-4)**
Benzene, 1,1'-(1,2-ethanediylbis(oxy))bis- (9CI)	**(104-66-5)**
Benzene, 1,1'-(1,2-ethanediylbis(oxy))bis(2,4,6-tribromo	**(37853-59-1)**
Benzeneethanol, 2-amino- (9CI)	**(5339-85-5)**
Benzeneethanol, 4-amino- (9CI)	**(104-10-9)**
Benzeneethanol, α-(2-amino-1-methylethyl)-α-phenyl-, propanoate (ester), (S-(R*,S*))- (9CI)	**(42576-07-8)**
Benzeneethanol, 4-chloro-α-(4-chlorophenyl)-	**(56960-97-5)**
Benzeneethanol, 4-chloro-β-(4-chlorophenyl)	**(2642-82-2)**
Benzeneethanol, 2,6-dichloro-4-hydroxy-3,5-dimethoxy- (9CI)	**(108544-91-8)**
Benzeneethanol, 2,6-dichloro-4-hydroxy-3,5-dimethoxy-.α.-methyl- (9CI)	**(108545-03-5)**
Benzeneethanol, 2,6-dichloro-3,4,5-trimethoxy-.α.-methyl- (9CI)	**(108544-96-3)**
Benzeneethanol, α-(2-(dimethylamino)-1-methylethyl)-α-phenyl-, propanoate (Ester), (R-(R*,S*))-	**(2338-37-6)**
Benzeneethanol, β-(2-(dimethylamino)propyl)-α-ethyl-β-phenyl- (9CI)	**(545-90-4)**
Benzeneethanol, β-(2-(dimethylamino)propyl)-α-ethyl-β-phenyl-, acetate (ester)	**(509-74-0)**
Benzeneethanol, 4-hydroxy- (9CI)	**(501-94-0)**
Benzeneethanol, β-methyl- (9CI)	**(1123-85-9)**
Benzeneethanol, .α.-methyl-2-(1-methylethyl)- (9CI)	**(58443-82-6)**
Benzeneethanol, .α.-methyl-3-(1-methylethyl)- (9CI)	**(54518-11-5)**
Benzeneethanol, 2-nitro-	**(15121-84-3)**
Benzeneethanol, 4-nitro- (9CI)	**(100-27-6)**
Benzeneethanol, 2-nitro-, acetate (ester)	**(833-43-2)**
Benzeneethanol, 4-nitro-, acetate (ester) (9CI)	**(104-30-3)**
Benzeneethanol, .α.,.α.,.β.-trimethyl- (9CI)	**(3280-08-8)**
Benzene, 1,1',1'',1'''-(1,2-ethendiylidene)tetrakis-	**(632-51-9)**
Benzene, 1,1'-(1,2-ethenediyl)bis- (9CI)	**(588-59-0)**
Benzene, 1,1'-(1,2-ethenediyl)bis-, (E)- (9CI)	**(103-30-0)**
Benzene, 1,1'-(1,2-ethenediyl)bis-, (Z)- (9CI)	**(645-49-8)**
Benzene, 1,1'-(1,2-ethenediyl)bis(4-methoxy- (9CI)	**(4705-34-4)**
Benzene, 1,1'-(1,2-ethenediyl)bis(2-nitro-	**(6275-02-1)**
Benzene, 1,1',1'',1'''-(1,2-ethenediylidene)tetrakis- (9CI)	**(632-51-9)**
Benzene, ethenyl-, Polymer with 1,3-butadiene	**(9003-55-8)**
Benzene, ethenyl-, Polymer with butyl 2-methyl-2-propenoate	**(25213-39-2)**
Benzene, ethenyl-, Polymer with 3-methyl-3-buten-2-one (9CI)	**(25191-48-4)**
Benzene, 1-ethenyl-3,5-dimethyl- (9CI)	**(5379-20-4)**
Benzene, 2-ethenyl-1,3-dimethyl- (9CI)	**(2039-90-9)**
Benzene, 4-ethenyl-1,2-dimethyl- (9CI)	**(27831-13-6)**
Benzene, ethenyl-, dimethyl deriv. (9CI)	**(27576-03-0)**
Benzene, 1-ethenyl-2-ethyl- (9CI)	**(7564-63-8)**
Benzene, 1-ethenyl-4-ethyl- (9CI)	**(3454-07-7)**
Benzene, ethenylethyl- (9CI)	**(28106-30-1)**
Benzene, ethenyl-, heptachloro deriv. (9CI)	**(61255-81-0)**
Benzene, ethenyl-, hexachloro deriv. (9CI)	**(61128-00-5)**
Benzene, ethenyl-, Homopolymer (9CI)	**(9003-53-6)**
Benzene, 1,1'-ethenylidenebis- (9CI)	**(530-48-3)**
Benzene, 1-ethenyl-4-methoxy- (9CI)	**(637-69-4)**
Benzene, 1-ethenyl-2-methyl- (9CI)	**(611-15-4)**
Benzene, 1-ethenyl-3-methyl- (9CI)	**(100-80-1)**
Benzene, 1-ethenyl-4-methyl- (9CI)	**(622-97-9)**
Benzene, ethenyl-, monomethyl deriv. (9CI)	**(1319-73-9)**
Benzene, ethenyl-, pentachloro deriv. (9CI)	**(83484-75-7)**
Benzene, ethenyl-, tetrachloro deriv. (9CI)	**(71489-58-2)**
Benzene, ethenyl-, trimer (9CI)	**(28213-80-1)**
Benzene, 2-ethenyl-1,3,5-trimethyl- (9CI)	**(769-25-5)**
Benzene, ethoxy-	**(103-73-1)**
Benzene, (1-ethoxyethoxy)- (9CI)	**(5426-78-8)**
Benzene, (1-ethoxyethyl)- (9CI)	**(3299-05-6)**
Benzene, (ethoxymethyl)- (9CI)	**(539-30-0)**
Benzene, 1-ethoxy-3-methyl- (9CI)	**(621-32-9)**
Benzene, 1-ethoxy-4-(2-methyl-1-(4-methylphenyl)propyl)- (9CI)	**(56265-26-0)**
Benzene, 1-ethoxy-4-(1-(4-methylphenyl)-2-nitropropyl)- (9CI)	**(54010-81-0)**
Benzene, 1-ethoxy-4-nitro	**(100-29-8)**
Benzene, ethoxynonyl- (9CI)	**(28679-13-2)**
Benzene, 1-ethoxy-4-(2,2,2-trichloro-1-(4-methylphenyl)-ethyl)- (9CI)	**(34197-05-2)**
Benzene, ethyl	**(100-41-4)**
Benzene, (1-ethylallyl)-	**(19947-22-9)**
Benzene, (1-ethylbutyl)- (9CI)	**(4468-42-2)**
Benzene, (2-ethylbutyl)- (6CI,7CI,8CI)	**(19219-85-3)**
Benzene, (1-ethyldecyl)- (9CI)	**(2400-00-2)**
Benzene, 1-ethyl-2,3-dimethyl- (9CI)	**(933-98-2)**
Benzene, 1-ethyl-2,4-dimethyl- (9CI)	**(874-41-9)**
Benzene, 1-ethyl-3,5-dimethyl- (9CI)	**(934-74-7)**
Benzene, 2-ethyl-1,3-dimethyl- (9CI)	**(2870-04-4)**
Benzene, 2-ethyl-1,4-dimethyl- (9CI)	**(1758-88-9)**
Benzene, 4-ethyl-1,2-dimethyl- (9CI)	**(934-80-5)**
Benzene, ethyldimethyl- (9CI)	**(29224-55-3)**
Benzene, (1-ethyldodecyl)- (9CI)	**(4534-58-1)**

Benzene, 1-ethyl-4-((4-ethylphenyl)ethyl)- (9CI) **(57364-79-1)**
Benzene, 1-ethyl-3-hydroxy- **(620-17-7)**
Benzene, 1,1'-ethylidenebis- (9CI) **(612-00-0)**
Benzene, 1,1'-ethylidenebis(4-ethyl- (9CI) **(10224-91-6)**
Benzene, 1-ethyl-2-methyl- (9CI) **(611-14-3)**
Benzene, 1-ethyl-3-methyl- (9CI) **(620-14-4)**
Benzene, 1-ethyl-4-methyl- (9CI) **(622-96-8)**
Benzene, ethylmethyl- (9CI) **(25550-14-5)**
Benzene, 1-ethyl-3-(1-methylethyl)- (9CI) **(4920-99-4)**
Benzene, 1-ethyl-4-(1-methylethyl)- (9CI) **(4218-48-8)**
Benzene, 1-ethyl-4-nitro- (9CI) **(100-12-9)**
Benzene, (1-ethylnonyl)- (9CI) **(4536-87-2)**
Benzene, (1-ethyloctadecyl)- (9CI) **(2400-02-4)**
Benzene, (1-ethyloctyl)- (9CI) **(4621-36-7)**
Benzene, ethyl(phenylethyl)- (9CI) **(64800-83-5)**
Benzene, ethyl(phenylethyl)-, mono-ar-ethyl deriv. (9CI) **(68398-19-6)**
Benzene, ethyl(phenylmethyl)- (9CI) **(42504-54-1)**
Benzene, (1-ethyl-1-propenyl)-, (E)- (9CI) **(4165-86-0)**
Benzene, (1-ethyl-2-propenyl)- (9CI) **(19947-22-9)**
Benzene, (1-ethyl-1-propenyl)-, (Z)- (9CI) **(4165-78-0)**
Benzene, (1-ethylpropyl)- (9CI) **(1196-58-3)**
Benzene, 1-(1-ethylpropyl)-4-methyl- (9CI) **(22975-58-2)**
Benzene, 1-(1-ethylpropyl)-2-propyl- (9CI) **(54789-15-0)**
Benzene, (ethylthio)- (9CI) **(622-38-8)**
Benzene, ethyl-1,2,4-trimethyl- (9CI) **(54120-62-6)**
Benzene, 1-ethyl-2,4,5-trimethyl- (8CI,9CI) **(17851-27-3)**
Benzene, (1-ethylundecyl)- (9CI) **(4534-52-5)**
Benzene, 1,1'-(1,2-ethynediyl)bis- (9CI) **(501-65-5)**
Benzene, ethynyl **(536-74-3)**
Benzene, fluoro **(462-06-6)**
Benzene, 1-fluoro-3-bromo **(1073-06-9)**
Benzene, 1-fluoro-4-iodo- (9CI) **(352-34-1)**
Benzene, 1-fluoro-3-methoxy- (9CI) **(456-49-5)**
Benzene, 1-fluoro-4-methoxy- (9CI) **(459-60-9)**
Benzene, (fluoromethyl)- (9CI) **(350-50-5)**
Benzene, 1-fluoro-2-methyl- (9CI) **(95-52-3)**
Benzene, 1-fluoro-3-methyl- (9CI) **(352-70-5)**
Benzene, 1-fluoro-4-methyl- (9CI) **(352-32-9)**
Benzene, fluoromethyl- (9CI) **(25496-08-6)**
Benzene, 1-fluoro-4-nitro **(350-46-9)**
Benzene, 1-fluoro-3-nitro- (9CI) **(402-67-5)**
Benzene, 1-fluoro-2-(trifluoromethyl)- (9CI) **(392-85-8)**
Benzeneformic acid **(65-85-0)**
Benzeneglyoxylic-acid **(611-73-4)**
Benzene, heptadecyl- (9CI) **(14752-75-1)**
Benzene, 1-heptenyl- (9CI) **(829-99-2)**
Benzene, 2-heptenyl- (9CI) **(26447-63-2)**
Benzene, heptyl- (9CI) **(1078-71-3)**
Benzene, hexabromo- (9CI) **(87-82-1)**
1,2,3,4,5,6-Benzenehexacarboxylic acid **(517-60-2)**
Benzenehexacarboxylic acid (9CI) **(517-60-2)**
Benzene hexachloride **(58-89-9)**
Benzene hexachloride **(608-73-1)**
α-Benzenehexachloride **(319-84-6)**
γ-Benzene hexachloride **(58-89-9)**
trans-α-Benzenehexachloride **(319-85-7)**
Benzene hexachloride-α-isomer **(319-84-6)**
Benzene, hexachloro **(118-74-1)**
Benzene, hexachloro-, Mixt. with pentachloronitrobenzene (9CI) **(62180-90-9)**
Benzene, hexadecyl- (9CI) **(1459-09-2)**
Benzene, hexaethyl **(604-88-6)**
Benzene, hexafluoro **(392-56-3)**
Benzene, hexahydro- **(110-82-7)**
Benzene, hexaiodo- (9CI) **(608-74-2)**
Benzenehexamethanol, hexanitrate **(105554-30-1)**
Benzene, hexamethyl **(87-85-4)**
Benzenehexanoic acid, .β.-oxo- (9CI) **(101038-68-0)**

Benzene, hexyl- (9CI) **(1077-16-3)**
Benzene, (1-hexylheptyl)- (9CI) **(2400-01-3)**
Benzene, (1-hexyloctyl)- (9CI) **(4534-54-7)**
Benzene, (1-hexyltetradecyl)- (9CI) **(2398-64-3)**
Benzene, hydrazodi- **(122-66-7)**
Benzeneiodide **(591-50-4)**
Benzene, iodo **(591-50-4)**
Benzene, (iodomethyl)- (9CI) **(620-05-3)**
Benzene, 1-iodo-2-methyl- (9CI) **(615-37-2)**
Benzene, 1-iodo-3-methyl- (9CI) **(625-95-6)**
Benzene, 1-iodo-4-methyl- (9CI) **(624-31-7)**
Benzene, 1-iodo-3-nitro- (9CI) **(645-00-1)**
Benzene, 1-iodo-4-nitro- (9CI) **(636-98-6)**
Benzene, isobutyl **(538-93-2)**
Benzene, isocyanato **(103-71-9)**
Benzene, 2-isocyanato-1,3-bis(1-methylethyl)- (9CI) **(28178-42-9)**
Benzene, 1-isocyanato-4-((4-isocyanatocyclohexyl)methyl)- (9CI) **(35438-85-8)**
Benzene, 1-isocyanato-2-((4-isocyanatophenyl)methyl)- (9CI) **(5873-54-1)**
Benzene, 1-isocyanato-2-methyl- **(614-68-6)**
Benzene, 1-(1-isocyanato-1-methylethyl)-3-(1-methylethenyl)- (9CI) **(2094-99-7)**
Benzene, 1-(1-isocyanato-1-methylethyl)-4-(1-methylethenyl)- (9CI) **(2889-58-9)**
Benzene, 1-isocyano-4-methyl- (9CI) **(7175-47-5)**
Benzene, isopentyl- (8CI) **(2049-94-7)**
Benzene, isopropyl **(98-82-8)**
Benzene, 1-isopropyl-4-methyl- **(99-87-6)**
Benzene, (1-isopropyl-2-methylpropyl)- (8CI) **(21777-84-4)**
Benzene-1-isothiocyanate **(103-72-0)**
Benzene, isothiocyanato- **(103-72-0)**
Benzenemethanamine (9CI) **(100-46-9)**
Benzenemethanamine, N,N-bis(phenylmethyl)-, hydrochloride (9CI) **(7673-07-6)**
Benzenemethanamine, N-(2-chloroethyl)-N-(1-methyl-2-phenoxyethyl)- **(59-96-1)**
Benzenemethanamine, N-(2-chloroethyl)-N-(1-methyl-2-phenoxyethyl)-, hydrochloride **(63-92-3)**
Benzenemethanamine, N,N-dimethyl- (9CI) **(103-83-3)**
Benzenemethanamine, 4-ethoxy-N-(4-ethoxyphenyl)-α-(trichloromethyl)- (9CI) **(36236-41-6)**
Benzenemethanamine, N-ethyl-N-(4-(1H-1,2,4-triazol-3-ylazo)-phenyl)- (9CI) **(13486-13-0)**
Benzenemethanamine, hydrochloride **(3287-99-8)**
Benzenemethanamine, N-methyl- (9CI) **(103-67-3)**
Benzenemethanamine, α-methyl-, (+-)- (9CI) **(618-36-0)**
Benzenemethanamine, α-methyl-, (R)- (9CI) **(3886-69-9)**
Benzenemethanamine, α-methyl-, (S)- (9CI) **(2627-86-3)**
Benzenemethanamine, N-(phenylmethyl)- (9CI) **(103-49-1)**
Benzenemethanamine, N-(phenylmethyl)-, hydrochloride (9CI) **(20455-68-9)**
Benzenemethanamine, N,N,α-trimethyl- (9CI) **(2449-49-2)**
Benzenemethanaminium, N-decyl-N,N-dimethyl-, chloride (9CI) **(965-32-2)**
Benzenemethanaminium, 3,4-dichloro-n-dodecyl-N,N-dimethyl-, chloride (9CI) **(102-30-7)**
Benzenemethanaminium, N-(4-((4-(diethylamino)-2-methylphenyl)-(4-(ethyl((3-sulfophenyl)methyl)amino)phenyl)methylene)-2,5-cyclohexadien-1-ylidene)-N-ethyl-3-sulfo-, hydroxide, inner salt, sodium salt (9CI) **(5863-46-7)**
Benzenemethanaminium, N-(4-((4-(diethylamino)phenyl)(4-(ethyl-((3-sulfophenyl)methyl)amino)-2-methylphenyl)methylene)-3-methyl-2,5-cyclohexadien-1-ylidene)-N-ethyl-3-sulfo-, hydroxide, inner salt, sodium salt (9CI) **(6505-30-2)**
Benzenemethanaminium, N-(4-((4-(diethylamino)phenyl)(4-(ethyl-((3-sulfophenyl)methyl)amino)phenyl)methylene)-2,5-cyclohexadien-1-ylidene)-N-ethyl-3-sulfo-, hydroxide, inner salt, sodium salt (9CI) **(4129-84-4)**
Benzenemethanaminium, N,N-dimethyl-N-(2-(2-(methyl-4-(1,1,3,3-tetramethylbutyl)phenoxy)ethoxy)ethyl)-, chloride **(25155-18-4)**

Benzenemethanaminium, N,N-dimethyl-N-(2-(2-(methyl-4-(1,1,3,3-tetramethylbutyl)phenoxy)ethoxy)ethyl)-, chloride (25% Aqueous) **(25155-18-4)**
Benzenemethanaminium, N,N-dimethyl-N-9-octadecenyl-, chloride **(37139-99-4)**
Benzenemethanaminium, N,N-dimethyl-N-9-octadecenyl-, chloride, (Z)- (9CI) **(37139-99-4)**
Benzenemethanaminium, N,N-dimethyl-N-octadecyl-, chloride (9CI) **(122-19-0)**
Benzenemethanaminium, N,N-dimethyl-N-octyl-, chloride (9CI) **(959-55-7)**
Benzenemethanaminium, N-(2-((2,6-dimethylphenyl)amino)-2-oxoethyl)-N,N-diethyl-, benzoate (9CI) **(3734-33-6)**
Benzenemethanaminium, N,N-dimethyl-N-tetradecyl-, chloride (9CI) **(139-08-2)**
Benzenemethanaminium, N-dodecyl-N,N-dimethyl-, bromide (9CI) **(7281-04-1)**
Benzenemethanaminium, N-dodecyl-N,N-dimethyl-, chloride (9CI) **(139-07-1)**
Benzenemethanaminium, N-dodecyl-ar-ethyl-N,N-dimethyl-, chloride (9CI) **(27479-28-3)**
Benzenemethanaminium, 4-dodecyl-N,N,N-trimethyl-, chloride **(1330-85-4)**
Benzenemethanaminium, ar-dodecyl-N,N,N-trimethyl-, chloride (9CI) **(1330-85-4)**
Benzenemethanaminium, ar-ethenyl-N,N,N-trimethyl-, chloride (9CI) **(26616-35-3)**
Benzenemethanaminium, N-(4-((4-((4-ethoxyphenyl)amino)-phenyl)(4-(ethyl((3-sulfophenyl)methyl)amino)-2-methyl-phenyl)methylene)-3-methyl-2,5-cyclohexadien-1-ylidene)-N-ethyl-3-sulfo-, hydroxide, inner salt, monosodium salt (9CI) **(6104-58-1)**
Benzenemethanaminium, N-ethyl-N,N-dimethyl-, chloride (9CI) **(5197-80-8)**
Benzenemethanaminium, ar-ethyl-N,N-dimethyl-N-tetra-decyl-, chloride (9CI) **(27479-29-4)**
Benzenemethanaminium, N-ethyl-N-(4-((4-(ethyl((3-sulfophenyl)-methyl)amino)phenyl)(2-sulfophenyl(methylene)-2,5-cyclo-hexadien-1-ylidene)-3-sulfo-, hydroxide, inner salt, diammonium salt, Mixt. with 4,5-dihydro-5-oxo-1-(4-sulfophenyl)-4-((4-sulfophenyl)azo) **(92170-50-8)**
Benzenemethanaminium, N-ethyl-N-(4-((4-(ethyl((3-sulfophenyl)-methylamino)phenyl)(2-sulfophenyl)methylene)-2,5-cyclo-hexadien-1-ylidene)-3-sulfo-, hydroxide, inner salt, aluminum salt (3:2) (9CI) **(15792-67-3)**
Benzenemethanaminium, N-hexadecyl-N,N-dimethyl-, chloride **(122-18-9)**
Benzenemethanaminium, N,N,N-trimethyl-, bromide **(5350-41-4)**
Benzenemethanaminium, N,N,N-trimethyl-, chloride (9CI) **(56-93-9)**
Benzenemethanaminium, N,N,N-trimethyl-, hydroxide (9CI) **(100-85-6)**
Benzenemethanaminium, N,N,N-trimethyl-, iodide (9CI) **(4525-46-6)**
Benzenemethaneamine, α-methyl- **(98-84-0)**
Benzenemethanesulfonic acid (9CI) **(100-87-8)**
Benzenemethanoic acid **(65-85-0)**
Benzenemethanol **(100-51-6)**
Benzenemethanol, 4-(acetyloxy)-α-methyl- (9CI) **(53744-50-6)**
Benzenemethanol, 4-amino-α,α-bis(4-aminophenyl)- (9CI) **(467-62-9)**
Benzenemethanol, α-(1-aminoethyl)-, (S-(R*,R*))- (9CI) **(492-39-7)**
Benzenemethanol, α-(1-aminoethyl)-, hydrochloride, (R*,S*)-, (+-) **(154-41-6)**
Benzenemethanol, α-(1-aminoethyl)-3-hydroxy-, (R-(R*,S*))- (9CI) **(54-49-9)**
Benzenemethanol, 3-amino-α-methyl- (9CI) **(2454-37-7)**
Benzenemethanol, 3,5-bis(1,1-dimethylethyl)-4-hydroxy- **(88-26-6)**
Benzenemethanol, .α.-(bromomethyl)-, (R)- (9CI) **(73908-23-3)**
Benzenemethanol, 2-chloro- **(17849-38-6)**
Benzenemethanol, 4-chloro- (9CI) **(873-76-7)**
Benzenemethanol, 4-chloro-α-(4-chlorophenyl)- **(90-97-1)**
Benzenemethanol, 4-chloro-α-(4-chlorophenyl)-α-(trichloromethyl)- **(115-32-2)**
Benzenemethanol, 2-chloro-α-(4-chlorophenyl)-α-(trichloro-methyl)- (9CI) **(10606-46-9)**
Benzenemethanol, 2-chloro-α-(dichloromethyl)- **(27683-60-9)**
Benzenemethanol, 2-chloro-α-methyl- (9CI) **(13524-04-4)**
Benzenemethanol, 4-chloro-α-methyl- (9CI) **(3391-10-4)**
Benzenemethanol, 4-chloro-α-phenyl- (9CI) **(119-56-2)**
Benzenemethanol, 2,6-dichloro- (9CI) **(15258-73-8)**
Benzenemethanol, 3,4-dichloro- (9CI) **(1805-32-9)**
Benzenemethanol, 3,5-dichloro- (9CI) **(60211-57-6)**
Benzenemethanol, 3,4-dichloro-, methylcarbamate **(1966-58-1)**
Benzenemethanol, 2,6-dichloro-3,4,5-trimethoxy-.α.-methyl- (9CI) **(75315-54-7)**
Benzenemethanol, 3,4-dimethoxy- (9CI) **(93-03-8)**
Benzenemethanol, 2,4-dimethyl- (9CI) **(16308-92-2)**
Benzenemethanol, 3,4-dimethyl- (9CI) **(6966-10-5)**
Benzenemethanol, 3,5-dimethyl- (9CI) **(27129-87-9)**
Benzenemethanol, α,α-dimethyl- (9CI) **(617-94-7)**
Benzenemethanol, 4-(1,1-dimethylethyl)- **(877-65-6)**
Benzenemethanol, α,α-diphenyl- (9CI) **(76-84-6)**
Benzenemethanol, 4-ethyl- **(768-59-2)**
Benzenemethanol, α-ethyl- (9CI) **(93-54-9)**
Benzenemethanol, α-ethyl-2-methoxy- (9CI) **(7452-01-9)**
Benzenemethanol, 3-ethyl-α-methyl- (9CI) **(54264-96-9)**
Benzenemethanol, 2-hydroxy- (9CI) **(90-01-7)**
Benzenemethanol, 3-hydroxy- (9CI) **(620-24-6)**
Benzenemethanol, 4-hydroxy- (9CI) **(623-05-2)**
Benzenemethanol, 4-hydroxy-3,5-dimethoxy- (9CI) **(530-56-3)**
Benzenemethanol, 4-hydroxy-3-methoxy-α-methyl- (9CI) **(2480-86-6)**
Benzenemethanol, 3-hydroxy-α-((methylamino)methyl)-, (R)- (9CI) **(59-42-7)**
Benzenemethanol, 3-hydroxy-α-((methylamino)methyl)-, hydro-chloride, (-)- **(61-76-7)**
Benzenemethanol, 4-hydroxy-α-(1-((1-methyl-3-phenylpropyl)-amino)ethyl)- **(447-41-6)**
Benzenemethanol, 3-iodo- **(57455-06-8)**
Benzenemethanol, ar-methoxy- (9CI) **(1331-81-3)**
Benzenemethanol, 4-methoxy-, acetate (9CI) **(104-21-2)**
Benzenemethanol, 3-methyl- **(587-03-1)**
Benzenemethanol, α-methyl- **(98-85-1)**
Benzenemethanol, 2-methyl- (9CI) **(89-95-2)**
Benzenemethanol, α-methyl-, acetate (9CI) **(93-92-5)**
Benzenemethanol, α-(1-(methylamino)ethyl)-, (R-(R*,S*))- **(299-42-3)**
Benzenemethanol, 5-(1-methyl-1-(4-(oxiranylmethoxy)phenyl)-ethyl)-2-(oxiranylmethoxy)- (9CI) **(3188-83-8)**
Benzenemethanol, 2-nitro- (9CI) **(612-25-9)**
Benzenemethanol, .α.-nonyl- (9CI) **(21078-95-5)**
Benzenemethanol, α-octyl- (9CI) **(57716-72-0)**
Benzenemethanol, 2,3(or 3,4)-dichloro-, methyl carbamate (9CI) **(62046-37-1)**
Benzenemethanol, 2,3,4,5,6-pentachloro- (9CI) **(16022-69-8)**
Benzenemethanol, 3-phenoxy- (9CI) **(13826-35-2)**
Benzenemethanol, 3-phenoxy-, acetate (9CI) **(50789-44-1)**
Benzenemethanol, 4-(phenylamino)-α,α-bis(4-(phenylamino)-phenyl)- (9CI) **(23681-60-9)**
Benzenemethanol, ar-(2-propenyloxy)- **(28655-62-1)**
Benzenemethanol, (2-propenyloxy)- (9CI) **(28655-62-1)**
Benzenemethanol, α-(trichloromethyl)- (9CI) **(2000-43-3)**
Benzenemethanol, α-(trichloromethyl)-, acetate **(90-17-5)**
Benzenemethanol, α,α,4-trimethyl- **(1197-01-9)**
Benzenemethanol, α,2,4-trimethyl- (9CI) **(5379-19-1)**
Benzenemethanol, α,3,4-trimethyl- (9CI) **(33967-19-0)**
Benzenemethanol, α,α,4-trimethyl-, dichloro deriv. (9CI) **(92366-35-3)**
Benzene, methoxy **(100-66-3)**
Benzene, 1-(8-methoxy-4,8-dimethylnonyl)-4-(1-methylethyl) **(53905-38-7)**
Benzene, 1-(8-methoxy-4,8-dimethylnonyl)-4-(1-methylethyl)- **(53905-38-7)**
Benzene, 1-methoxy-2,4-dinitro **(119-27-7)**
Benzene, (1-methoxyethyl)- (9CI) **(4013-34-7)**
Benzene, (methoxymethyl)- (9CI) **(538-86-3)**
Benzene, 1-methoxy-3-methyl- (9CI) **(100-84-5)**
Benzene, 1-methoxy-4-methyl- (9CI) **(104-93-8)**
Benzene, methoxymethyl- (9CI) **(26897-24-5)**
Benzene, 1-methoxy-4-methyl-2-nitro- (9CI) **(119-10-8)**
Benzene, 1-methoxy-4-(methylthio)- (9CI) **(1879-16-9)**

Name	CAS
Benzene, 1-methoxy-2-nitro- (9CI)	**(91-23-6)**
Benzene, 1-methoxy-4-nitro- (9CI)	**(100-17-4)**
Benzene, 1-methoxy-4-(2-phenylethenyl)- (9CI)	**(1142-15-0)**
Benzene, 1-methoxy-4-(phenylmethyl)- (9CI)	**(834-14-0)**
Benzene, (3-methoxy-1-propenyl)- (9CI)	**(16277-67-1)**
Benzene, 1-methoxy-4-(1-propenyl)-, (E)- (9CI)	**(4180-23-8)**
Benzene, 1-methoxy-2-(2,2,2-trichloro-1-(4-methoxyphenyl)ethyl)- (9CI)	**(30667-99-3)**
Benzene, 1-methoxy-4-(2,2,2-trichloro-1-(4-(methylthio)-phenyl)ethyl)- (9CI)	**(34197-16-5)**
Benzene, 1-methoxy-4-(trimethoxymethyl)- (9CI)	**(4316-33-0)**
Benzene, 2-methoxy-1,3,5-trinitro- (9CI)	**(606-35-9)**
Benzene, methyl-	**(108-88-3)**
Benzene, (2-methylallyl)- (8CI)	**(3290-53-7)**
Benzene, methylbis(phenylmethyl)- (9CI)	**(26898-17-9)**
Benzene, (3-methyl-2-butenyl)- (9CI)	**(4489-84-3)**
Benzene, (3-methylbutyl)- (9CI)	**(2049-94-7)**
Benzene, (1-methylbutyl)- (8CI,9CI)	**(2719-52-0)**
Benzene, methyl-d3- (9CI)	**(1124-18-1)**
Benzene-d5, methyl-d3- (9CI)	**(2037-26-5)**
Benzene, (1-methyldecyl)- (9CI)	**(4536-88-3)**
Benzene, 1-methyl-2,4-dinitro-	**(121-14-2)**
Benzene, methyldinitro-	**(25321-14-6)**
Benzene, 1-methyl-2,3-dinitro- (9CI)	**(602-01-7)**
Benzene, 1-methyl-3,5-dinitro- (9CI)	**(618-85-9)**
Benzene, 2-methyl-1,3-dinitro- (9CI)	**(606-20-2)**
Benzene, 2-methyl-1,4-dinitro- (9CI)	**(619-15-8)**
Benzene, 4-methyl-1,2-dinitro- (9CI)	**(610-39-9)**
Benzene, (1-methyldodecyl)- (9CI)	**(4534-53-6)**
Benzene, 1,1'-methylenebis(chloro- (9CI)	**(25249-39-2)**
Benzene, 1,1'-methylenebis(4-isocyanato- (9CI)	**(101-68-8)**
Benzene, 1,1'-methylenebis(isocyanato- (9CI)	**(26447-40-5)**
Benzene, 1,1'-methylenebis(4-nitro- (9CI)	**(1817-74-9)**
Benzene, 1,2-methylenedioxy	**(274-09-9)**
Benzene, 1,2-methylenedioxy-4-allyl-	**(94-59-7)**
Benzene, 1,2-(methylenedioxy)-4-(2-(octylsulfinyl)propyl)	**(120-62-7)**
Benzene, 1,2-(methylenedioxy)-4-propenyl	**(120-58-1)**
Benzene, 1,2-methylenedioxy-4-propyl	**(94-58-6)**
Benzene, 1,1'-(1-methyl-1,2-ethanediyl)bis- (9CI)	**(5814-85-7)**
Benzene, 1,1'-(1-methyl-1,2-ethenediyl)bis- (9CI)	**(779-51-1)**
Benzene, 1,1'-(1-methyl-1,2-ethenediyl)bis-, (E)- (9CI)	**(833-81-8)**
Benzene, (1-methylethenyl)-	**(98-83-9)**
Benzene, (1-methylethenyl)-, dimer (9CI)	**(6144-04-3)**
Benzene, 1-(1-methylethenyl)-4-(1-methylethyl)- (9CI)	**(2388-14-9)**
Benzene, (1-methylethyl)- (9CI)	**(98-82-8)**
Benzene, 1,1'-(1-methylethylidene)bis- (9CI)	**(778-22-3)**
Benzene, 1,1'-(1-methylethylidene)bis(3,5-dibromo-4-(2,3-dibromopropoxy)- (9CI)	**(21850-44-2)**
Benzene, 1,1'-(1-methylethylidene)bis(3,5-dibromo-4-methoxy- (9CI)	**(37853-61-5)**
Benzene, 1,1'-(1-methylethylidene)bis(3,5-dibromo-4-(2-propenyloxy)- (9CI)	**(25327-89-3)**
Benzene, 1,1'-(1-methylethylidene)bis(4-(2-propenyloxy)- (9CI)	**(3739-67-1)**
Benzene, 1-(1-methylethyl)-4-nitro- (9CI)	**(1817-47-6)**
Benzene, 1-methyl-2-(1-ethylpropyl)- (9CI)	**(54410-74-1)**
Benzene, (1-methylethyl)-, trichloro deriv. (9CI)	**(61465-79-0)**
Benzene, 1,1',1''-methylidynetris- (9CI)	**(519-73-3)**
Benzene, 1,1',1''-methylidynetris(4-isocyanato- (9CI)	**(2422-91-5)**
Benzene, methyl(1-methylethyl)- (9CI)	**(25155-15-1)**
Benzene, (2-methyl-1-(1-methylethyl)propyl)- (9CI)	**(21777-84-4)**
Benzene, 1-methyl-4-(2-methylpropyl)- (9CI)	**(5161-04-6)**
Benzene, 1-methyl-3-((2-methylpropyl)thio)- (9CI)	**(54576-36-2)**
Benzene, 1-methyl-4-((2-methylpropyl)thio)- (9CI)	**(54576-37-3)**
Benzene, 1-methyl-4-(methylsulfonyl)- (9CI)	**(3185-99-7)**
Benzene, methyl(methylsulfonyl)- (9CI)	**(72428-03-6)**
Benzene, methyl-, monoiodo deriv. (9CI)	**(61878-58-8)**
Benzene, methylnitro-	**(1321-12-6)**
Benzene, 1-methyl-2-(4-nitrophenoxy)-	**(2444-29-3)**
Benzene, 1-methyl-3-(4-nitrophenoxy)- (9CI)	**(2303-25-5)**
Benzene, (1-methylnonadecyl)- (9CI)	**(2398-66-5)**
Benzene, (1-methylnonyl)- (9CI)	**(4537-13-7)**
Benzene, methyl-, pentachloro deriv. (9CI)	**(69911-61-1)**
Benzene, (1-methylpentyl)- (9CI)	**(6031-02-3)**
Benzene, methylpentyl- (9CI)	**(1320-01-0)**
Benzene, 1-methyl-3-phenoxy	**(3586-14-9)**
Benzene, 1-methyl-3-(2-phenylethenyl)-, (E)- (9CI)	**(14064-48-3)**
Benzene, 1,1'-(1-methyl-1,3-propanediyl)bis- (9CI)	**(1520-44-1)**
Benzene, 1,1'-(3-methyl-1-propene-1,3-diyl)bis- (9CI)	**(7614-93-9)**
Benzene, (2-methyl-1-propenyl)- (9CI)	**(768-49-0)**
Benzene, (2-methyl-2-propenyl)- (9CI)	**(3290-53-7)**
Benzene, (2-methylpropenyl)- (8CI)	**(768-49-0)**
Benzene, 1-methyl-2-(2-propenyl)- (9CI)	**(1587-04-8)**
Benzene, 1-((2-methyl-2-propenyl)oxy)-2-nitro- (9CI)	**(13414-54-5)**
Benzene, (1-methyl-1-propoxyethyl)- (9CI)	**(24142-77-6)**
Benzene, 1-methyl-2-propyl- (9CI)	**(1074-17-5)**
Benzene, 1-methyl-3-propyl- (9CI)	**(1074-43-7)**
Benzene, 1-methyl-4-propyl- (9CI)	**(1074-55-1)**
Benzene, 1,1'-(2-methylpropylidene)bis(4-ethoxy-	**(56265-21-5)**
Benzene, (methylsulfinyl)- (9CI)	**(1193-82-4)**
Benzene, (methylsulfonyl)- (9CI)	**(3112-85-4)**
Benzene, (methylthio)	**(100-68-5)**
Benzene, ((methylthio)methyl)- (9CI)	**(766-92-7)**
Benzene, methyl-, tribromo deriv. (9CI)	**(27476-22-8)**
Benzene, (1-methyltridecyl)- (9CI)	**(4534-59-2)**
Benzene, methyl-, trifluoro deriv. (9CI)	**(27359-10-0)**
Benzene, 1-methyl-2,3,4-trinitro-	**(602-29-9)**
Benzene, 1-methyl-2,3,6-trinitro-	**(18292-97-2)**
Benzene, 1-methyl-2,4,5-trinitro-	**(610-25-3)**
Benzene, 2-methyl-1,3,5-trinitro-	**(118-96-7)**
Benzene, (1-methylundecyl)- (9CI)	**(2719-61-1)**
Benzene, mono-C10-16-alkyl derivs.	**(68890-99-3)**
Benzenenamine, N-methyl- (9CI)	**(100-61-8)**
Benzene, 1,2-(1,8-naphthylene)-	**(206-44-0)**
Benzene, neopentyl- (8CI)	**(1007-26-7)**
Benzenenitrile	**(100-47-0)**
Benzene, nitro	**(98-95-3)**
Benzene, (1-nitroethyl)- (6CI,7CI,8CI,9CI)	**(7214-61-1)**
Benzene, 2-nitro-5-(p-nitrophenyl)-1,3-diphenyl- (7CI,8CI)	**(4170-07-4)**
Benzene, nitropentachloro-	**(82-68-8)**
Benzene, 1-nitro-4-phenoxy- (9CI)	**(620-88-2)**
Benzene, (1-nitropropyl)- (7CI,8CI,9CI)	**(5279-14-1)**
Benzene, 1,1'-(2-nitropropylidene)bis(4-ethoxy- (9CI)	**(26258-70-8)**
Benzene, 1,1'-(2-nitropropylidene)bis(4-methoxy- (9CI)	**(34197-26-7)**
Benzene, nitroso- (9CI)	**(586-96-9)**
Benzene, 3-nitro-1,2,4,5-tetrachloro-	**(117-18-0)**
Benzene, 5-nitro-1,2,3,4-tetrachloro-	**(879-39-0)**
Benzene, 4-nitro-1,2,3-trichloro	**(17700-09-3)**
Benzene, 1-nitro-2-(trifluoromethyl)- (9CI)	**(384-22-5)**
Benzene, 1-nitro-4-(trifluoromethyl)- (9CI)	**(402-54-0)**
Benzene, nonadecyl- (9CI)	**(29136-19-4)**
Benzene, nonyl- (9CI)	**(1081-77-2)**
Benzene, octadecyl- (9CI)	**(4445-07-2)**
Benzene, octyl- (9CI)	**(2189-60-8)**
Benzene, (1-octyldodecyl)- (9CI)	**(2398-65-4)**
Benzene, 1,1'-oxybis-, Polymer with formaldehyde (9CI)	**(26007-63-6)**
Benzene, 1,1'-oxybis(4-chloro- (9CI)	**(2444-89-5)**
Benzene, 1,1'-oxybis(2-chloro-4-nitro- (9CI)	**(13867-27-1)**
Benzene, 1,1'-oxybis(2,4-dichloro-	**(28076-73-5)**
Benzene, 1,1'-oxybis(3,4-dichloro- (9CI)	**(56348-72-2)**
Benzene, 1,1'-oxybis(dodecyl- (9CI)	**(69834-19-1)**
Benzene, 1,1'-oxybis-, heptabromo deriv. (9CI)	**(68928-80-3)**
Benzene, 1,1'-oxybis-, hexabromo deriv. (9CI)	**(36483-60-0)**
Benzene, 1,1'-oxybis-, hexachloro deriv. (9CI)	**(31242-93-0)**
Benzene, 1,1'-oxybis(methyl-	**(28299-41-4)**

Benzene, 1,1'-(oxybis(methylene))bis(4-chloro- **(56428-00-3)**
Benzene, 1,1',1'',1'''-(oxybis(methylidyne))tetrakis(4-chloro- (9CI) **(74562-99-5)**
Benzene, 1,1'-oxybis(4-nitro- (9CI) **(101-63-3)**
Benzene, 1,1'-oxybis-, octabromo deriv. (9CI) **(32536-52-0)**
Benzene, 1,1'-oxybis(2,3,4,5,6-pentabromo- (9CI) **(1163-19-5)**
Benzene, 1,1'-oxybis-, pentabromo deriv. (9CI) **(32534-81-9)**
Benzene, 1,1'-oxybis(2,3,4,5,6-pentachloro- (9CI) **(31710-30-2)**
Benzene, 1,1'-oxybis-, tetrabromo deriv. (9CI) **(40088-47-9)**
Benzene, 1,1'-oxybis-, tetrachloro deriv. (9CI) **(42279-29-8)**
Benzene, 1,1'-oxybis-, tribromo deriv. (9CI) **(49690-94-0)**
Benzene, 1,1'-oxybis(2,4,5-trichloro- (9CI) **(71859-30-8)**
Benzene, 1,1'-oxybis-, trichloro deriv. (9CI) **(57321-63-8)**
Benzene, 1,1'-(oxydiethylidene)bis- (9CI) **(93-96-9)**
Benzene, 1,1',1'',1'''-(oxydimethylidyne)tetrakis- (9CI) **(574-42-5)**
Benzene, pentabromo- (8CI,9CI) **(608-90-2)**
Benzene, pentabromoethyl- (9CI) **(85-22-3)**
Benzene, pentabromomethyl- (9CI) **(87-83-2)**
Benzene, pentabromo(tetrabromophenoxy)- (9CI) **(63936-56-1)**
Benzenepentacarboxylate **(1585-40-6)**
Benzenepentacarboxylic acid (9CI) **(1585-40-6)**
Benzene, pentachloro **(608-93-5)**
Benzene, pentachloro(1,2-dichloroethenyl)-, (E)- (9CI) **(29086-38-2)**
Benzene, pentachloro(dichloroethenyl)- (9CI) **(61593-44-0)**
Benzene, pentachloro(1,2-dichloroethenyl)-, (Z)- (9CI) **(29086-39-3)**
Benzene, pentachloroethyl- (6CI,7CI,9CI) **(606-07-5)**
Benzene, pentachlorofluoro- (8CI,9CI) **(319-87-9)**
Benzene, pentachloromethoxy- (9CI) **(1825-21-4)**
Benzene, pentachloromethyl- (9CI) **(877-11-2)**
Benzene, pentachloro(methylthio)- (9CI) **(1825-19-0)**
Benzene, pentachloronitro **(82-68-8)**
Benzene, pentadecyl- (9CI) **(2131-18-2)**
Benzene, pentaethyl- (9CI) **(605-01-6)**
Benzene, pentafluoro- (9CI) **(363-72-4)**
Benzene, pentaiodo- (8CI,9CI) **(608-96-8)**
Benzene, pentamethyl- (9CI) **(700-12-9)**
Benzenepentanesulfonic acid, ε-pentyl- (9CI) **(67716-07-8)**
Benzenepentanoic acid (9CI) **(2270-20-4)**
Benzene, sec-pentyl **(29316-05-0)**
Benzene, tert-pentyl **(2049-95-8)**
Benzene, pentyl- (9CI) **(538-68-1)**
Benzene, (1-pentylheptyl)- (9CI) **(2719-62-2)**
Benzene, (1-pentylhexyl)- (9CI) **(4537-14-8)**
Benzene, (1-pentylnonyl)- (9CI) **(4534-55-8)**
Benzene, (1-pentyloctyl)- (9CI) **(4534-49-0)**
Benzene, 1,1'-(1-pentyl-1,3-propanediyl)bis- (9CI) **(79606-18-1)**
Benzene, phenoxy- **(101-84-8)**
Benzene, (phenylamino)- **(122-39-4)**
Benzene, (phenylethyl)- (9CI) **(38888-98-1)**
Benzene, ((phenylmethyl)thio)- (9CI) **(831-91-4)**
Benzenephosphonic acid **(1571-33-1)**
Benzenephosphonic acid, dioctyl ester **(1754-47-8)**
Benzenephosphonic acid, thiono-, ethyl-p-nitrophenyl ester **(2104-64-5)**
Benzene phosphorus dichloride **(824-72-6)**
Benzene phosphorus dichloride (DOT) **(644-97-3)**
Benzene phosphorus thiodichloride (DOT) **(3497-00-5)**
Benzenepropanal, 4-(1,1-dimethylethyl)-α-methyl- **(80-54-6)**
Benzenepropanal, α-methyl-2-(1-methylethyl)- (9CI) **(6502-20-1)**
Benzenepropanamide, N,N'-1,6-hexanediylbis(3,5-bis(1,1-dimethylethyl)-4-hydroxy- (9CI) **(23128-74-7)**
Benzenepropanamide, 2-hydroxy- (9CI) **(22367-76-6)**
Benzenepropanamine, N-ethyl-N-(3-phenylpropyl)- **(150-59-4)**
Benzenepropanamine, N-(1-methyl-2-phenylethyl)-γ-phenyl- **(390-64-7)**
Benzene, 1,1'-(1,3-propanediyl)bis- (9CI) **(1081-75-0)**
Benzenepropanenitrile, α,4-dichloro- (9CI) **(17849-64-8)**
Benzenepropanenitrile, α,3-dichloro-2-methyl- **(21342-85-8)**
Benzenepropanoic acid (9CI) **(501-52-0)**
Benzenepropanoic acid, α-amino-, (S)- **(63-91-2)**
Benzenepropanoic acid, 3-(2H-benzotriazol-2-yl)-5-(1,1-dimethylethyl)-4-hydroxy-, methyl ester (9CI) **(84268-33-7)**
Benzenepropanoic acid, 3,5-bis(1,1-dimethylethyl)-4-hydroxy-, 2,2-bis((3-(3,5-bis(1,1-dimethylethyl)-4-hydroxyphenyl)-1-oxopropoxy)methyl)-1,3-propanediyl ester (9CI) **(6683-19-8)**
Benzenepropanoic acid, 3,5-bis(1,1-dimethylethyl)-4-hydroxy-, 2-(3-(3,5-bis(1,1-dimethylethyl)-4-hydroxyphenyl)-1-oxopropyl)hydrazide (9CI) **(32687-78-8)**
Benzenepropanoic acid, 3,5-bis(1,1-dimethylethyl)-4-hydroxy-, hydrazide (9CI) **(32687-77-7)**
Benzenepropanoic acid, 3,5-bis(1,1-dimethylethyl)-4-hydroxy-, methyl ester (9CI) **(6386-38-5)**
Benzenepropanoic acid, 3,5-bis(1,1-dimethylethyl)-4-hydroxy-, octadecyl ester (9CI) **(2082-79-3)**
Benzenepropanoic acid, 3,5-bis(1,1-dimethylethyl)-4-hydroxy-, (2,4,6-trioxo-1,3,5-triazine-1,3,5(2H,4H,6H)-triyl)tri-2,1-ethanediyl ester (9CI) **(34137-09-2)**
Benzenepropanoic acid, α,4-dichloro- (9CI) **(14437-20-8)**
Benzenepropanoic acid, 3,4-dihydroxy- (9CI) **(1078-61-1)**
Benzenepropanoic acid, β,β-dimethyl- (9CI) **(1010-48-6)**
Benzenepropanoic acid, 3-(((dimethylamino)methylene)amino)-2,4,6-triiodo- (9CI) **(5587-89-3)**
Benzenepropanoic acid, 3-(1,1-dimethylethyl)-4-hydroxy-5-methyl-, 1,2-ethanediylbis(oxy-2,1-ethanediyl) ester (9CI) **(36443-68-2)**
Benzenepropanoic acid, β-ethyl- (9CI) **(5669-17-0)**
Benzenepropanoic acid, ethyl ester (9CI) **(2021-28-5)**
Benzenepropanoic acid, 4-hydroxy- (9CI) **(501-97-3)**
Benzenepropanoic acid, 4-hydroxy-3-methoxy- (9CI) **(1135-23-5)**
Benzenepropanoic acid, 4-hydroxy-α-oxo- (9CI) **(156-39-8)**
Benzenepropanoic acid, β-methyl- **(4593-90-2)**
Benzenepropanoic acid, 2-methyl- (9CI) **(22084-89-5)**
Benzenepropanoic acid, methyl ester (9CI) **(103-25-3)**
Benzenepropanoic acid, α-oxo- (9CI) **(156-06-9)**
Benzenepropanoic acid, β-oxo-, ethyl ester (9CI) **(94-02-0)**
Benzenepropanoic acid, β-oxo-, methyl ester (9CI) **(614-27-7)**
Benzenepropanoic acid, α-phenyl- (9CI) **(3333-15-1)**
Benzenepropanoic acid, silver(1+) salt (9CI) **(75112-79-7)**
Benzenepropanoic acid, .β.,.β.,3,4-tetramethyl- (9CI) **(55683-10-8)**
3-Benzenepropanol **(122-97-4)**
Benzenepropanol, 4-hydroxy-3-methoxy- (9CI) **(2305-13-7)**
Benzenepropanol, γ-phenyl- (9CI) **(20017-67-8)**
Benzene, propenyl **(637-50-3)**
Benzene, 2-propenyl- (9CI) **(300-57-2)**
Benzenepropionic acid **(501-52-0)**
Benzene, (2-(1-propoxyethoxy)ethyl)- (9CI) **(7493-57-4)**
Benzene, propyl **(103-65-1)**
Benzene, (1-propylbutyl)- (9CI) **(2132-86-7)**
Benzene, (1-propyldecyl)- (9CI) **(4534-51-4)**
Benzene, (1-propylheptadecyl)- (9CI) **(2400-03-5)**
Benzene, (1-propylheptyl)- (9CI) **(4537-12-6)**
Benzene, (1-propylnonyl)- (9CI) **(2719-64-4)**
Benzene, (1-propyloctyl)- (9CI) **(4536-86-1)**
Benzene, (1-propylundecyl)- (9CI) **(4534-57-0)**
Benzeneselenenyl chloride (8CI,9CI) **(5707-04-0)**
Benzeneseleninic acid (8CI,9CI) **(6996-92-5)**
Benzenesulfenyl chloride, 2-nitro- (9CI) **(7669-54-7)**
Benzenesulfenyl chloride, o-nitro- (8CI) **(7669-54-7)**
Benzenesulfinic acid (8CI,9CI) **(618-41-7)**
Benzenesulfinic acid, 4-methyl- (9CI) **(536-57-2)**
Benzenesulfinic acid, 4-methyl-, sodium salt (9CI) **(824-79-3)**
Benzenesulfinic acid, 4-methyl-, zinc salt (9CI) **(24345-02-6)**
Benzenesulfinic acid, sodium salt (9CI) **(873-55-2)**
Benzenesulfochloramide sodium **(127-52-6)**
Benzenesulfohydrazide **(80-17-1)**
Benzenesulfonamide **(98-10-2)**
Benzenesulfonamide, 4-acetyl-N-((cyclohexylamino)carbonyl)- **(968-81-0)**

Benzenesulfonamide, 3-amino- **(98-18-0)**
Benzenesulfonamide, p-amino- **(63-74-1)**
Benzenesulfonamide, 3-amino-N-((cyclohexylamino)carbonyl)-4-methyl- **(565-33-3)**
Benzenesulfonamide, 4-amino-N-(diaminomethylene)- **(57-67-0)**
Benzenesulfonamide, 4-amino-N-(3,4-dimethyl-5-isoxazolyl)- **(127-69-5)**
Benzenesulfonamide, 4-amino-N-(4,6-dimethyl-2-pyrimidinyl)- **(57-68-1)**
Benzenesulfonamide, 4-amino-N-(1-ethyl-1,2-dihydro-2-oxo-4-pyrimidinyl)- (9CI) **(17784-12-2)**
Benzenesulfonamide, 2-amino-N-ethyl-N-phenyl- (9CI) **(81-10-7)**
Benzenesulfonamide, 4-amino-N-(5-ethyl-1,3,4-thiadiazol-2-yl)- (9CI) **(94-19-9)**
Benzenesulfonamide, 3-amino-4-hydroxy- (9CI) **(98-32-8)**
Benzenesulfonamide, 3-amino-4-hydroxy-N-methyl- (9CI) **(80-23-9)**
Benzenesulfonamide, 4-amino-N-(5-methyl-3-isoxazolyl)- **(723-46-6)**
Benzenesulfonamide, 4-amino-N-(4-methyl-2-pyrimidinyl)-, monosodium salt (9CI) **(127-58-2)**
Benzenesulfonamide, 4-amino-N-(5-methyl-1,3,4-thiadiazol-2-yl)- **(144-82-1)**
Benzenesulfonamide, 4-amino-N-(1-phenyl-1H-pyrazol-5-yl) **(526-08-9)**
Benzenesulfonamide, 4-amino-N-pyrazinyl- (9CI) **(116-44-9)**
Benzenesulfonamide, 4-amino-N-(2-pyridinyl) **(144-83-2)**
Benzenesulfonamide, 4-amino-N-2-pyrimidinyl- **(68-35-9)**
Benzenesulfonamide, n-butyl **(3622-84-2)**
Benzenesulfonamide, N-((butylamino)carbonyl)-4-methyl- **(64-77-7)**
Benzenesulfonamide, p-chloro **(98-64-6)**
Benzenesulfonamide, 4-chloro-N-(4,6-dichloro-1,3,5-triazin-2-yl)- (9CI) **(17752-71-5)**
Benzenesulfonamide, p-chloro-N-(4,6-dichloro-s-triazin-2-yl)- (8CI) **(17752-71-5)**
Benzenesulfonamide, 2-chloro-5-(2,3-dihydro-1-hydroxy-3-oxo-1H-isoindol-1-yl)- (9CI) **(77-36-1)**
Benzenesulfonamide, 4-chloro-N,N-dimethyl- (9CI) **(7463-22-1)**
Benzenesulfonamide, 4-chloro-N,N-dimethyl-3-nitro- (9CI) **(137-47-3)**
Benzenesulfonamide, 2-(2-chloroethoxy)-N-(((4-methoxy-6-methyl-1,3,5-triazin-2-yl)amino)carbonyl)- (9CI) **(82097-50-5)**
Benzenesulfonamide, 2-chloro-5-(1-hydroxy-3-oxo-1-isoindolinyl) **(77-36-1)**
Benzenesulfonamide, 2-chloro-N-(((4-methoxy-6-methyl-1,3,5-triazin-2-yl)amino)carbonyl)- **(64902-72-3)**
Benzenesulfonamide, N-chloro-4-methyl-, sodium salt (9CI) **(127-65-1)**
Benzenesulfonamide, 4-chloro-3-nitro- (9CI) **(97-09-6)**
Benzenesulfonamide, N-(4-chlorophenyl)- (9CI) **(4750-28-1)**
Benzenesulfonamide, 4-(3-(4-chlorophenyl)-4,5-dihydro-1H-pyrazol-1-yl)- (9CI) **(2744-49-2)**
Benzenesulfonamide, 4-chloro-N-((propylamino)carbonyl)- **(94-20-2)**
Benzenesulfonamide, N-chloro-, sodium salt (9CI) **(127-52-6)**
Benzenesulfonamide, N-cyclohexyl-4-methyl- (9CI) **(80-30-8)**
Benzenesulfonamide, N-(4,6-dichloro-1,3,5-triazin-2-yl)- (9CI) **(30369-89-2)**
Benzenesulfonamide, N-(4,6-dichloro-s-triazin-2-yl)- (8CI) **(30369-89-2)**
Benzenesulfonamide, 3-((4,5-dihydro-3-methyl-5-oxo-1-phenyl-1H-pyrazol-4-yl)azo)-4-hydroxy- (9CI) **(5264-47-1)**
Benzenesulfonamide, 3-(4,5-dihydro-3-methyl-5-oxo-1H-pyrazol-1-yl)- (9CI) **(89-29-2)**
Benzenesulfonamide, N,4-dimethyl- (9CI) **(640-61-9)**
Benzenesulfonamide, N-ethyl-2-methyl- (9CI) **(1077-56-1)**
Benzenesulfonamide, N-ethyl-4-methyl- (9CI) **(80-39-7)**
Benzenesulfonamide, N-ethyl-2(or 4)-methyl- **(8047-99-2)**
Benzenesulfonamide, N-(((hexahydro-1H-azepin-1-yl)-amino)-carbonyl)-4-methyl- **(1156-19-0)**
Benzenesulfonamide, p-hydroxy-, O-ester with O,O-dimethyl phosphorothioate **(115-93-5)**
Benzenesulfonamide, 4-hydroxy-3-((2-hydroxy-1-naphthalenyl)-azo)- (9CI) **(16432-45-4)**
Benzenesulfonamide, p-iodo **(825-86-5)**
Benzenesulfonamide, N-(2-mercaptoethyl)-, S-ester with O,O-diisopropylphosphorodithioate **(741-58-2)**
Benzenesulfonamide, 2(and 4)-methyl- **(1333-07-9)**
Benzenesulfonamide, 2-methyl- **(88-19-7)**
Benzenesulfonamide, 3-methyl- (9CI) **(1899-94-1)**
Benzenesulfonamide, ar-methyl- (9CI) **(1333-07-9)**
Benzenesulfonamide, m-(3-methyl-5-oxo-2-pyrazolin-1-yl)- (8CI) **(89-29-2)**
Benzenesulfonamide, 4-methyl-N-(4-(phenylamino)phenyl)- (9CI) **(100-93-6)**
Benzenesulfonamide, 4-methyl-N-(phenylsulfonyl)- (9CI) **(14706-41-3)**
Benzenesulfonamide, p-nitro **(6325-93-5)**
Benzenesulfonamide, 3-nitro-N-phenyl-4-(phenylamino)- (9CI) **(5124-25-4)**
Benzenesulfonamide, N-phenyl-N-((trichloromethyl)thio)- (9CI) **(2280-49-1)**
Benzenesulfonanilide, 2-amino-N-ethyl- (8CI) **(81-10-7)**
Benzenesulfonanilide, 4'-chloro- (8CI) **(4750-28-1)**
Benzenesulfonate de 4-chlorophenyle (French) **(80-38-6)**
Benzene sulfonechloride **(98-09-9)**
Benzenesulfonic-acid **(98-11-3)**
Benzenesulfonic acid, C10-16-alkyl derivs. **(68584-22-5)**
Benzenesulfonic acid, 4-acetamido-, chloride **(121-60-8)**
Benzenesulfonic acid, 4-(acetylamino)- (9CI) **(121-62-0)**
Benzenesulfonic acid, alkyl deriv. **(42615-29-2)**
Benzenesulfonic acid, 3-amino- **(121-47-1)**
Benzenesulfonic acid, o-amino **(88-21-1)**
Benzenesulfonic acid, 4-amino- (9CI) **(121-57-3)**
Benzenesulfonic acid, 5-amino-2-(p-aminoanilino)- (8CI) **(119-70-0)**
Benzenesulfonic acid, 5-amino-2-((4-aminophenyl)amino)- (9CI) **(119-70-0)**
Benzenesulfonic acid, 2-amino-5-((4-amino-3-sulfophenyl)(4-imino-3-sulfo-2,5-cyclohexadien- 1-ylidene)methyl)-3-methyl-, disodium salt **(3244-88-0)**
Benzenesulfonic acid, 2-(p-aminoanilino)-5-nitro- (8CI) **(91-29-2)**
Benzenesulfonic acid, 6-amino-3,4'-azodi- **(101-50-8)**
Benzenesulfonic acid, 3-amino-4-chloro- (9CI) **(98-36-2)**
Benzenesulfonic acid, 5-amino-2-chloro- (9CI) **(88-43-7)**
Benzenesulfonic acid, 2-amino-5-chloro- (8CI,9CI) **(133-74-4)**
Benzenesulfonic acid, 2-amino-5-chloro-4-ethyl- (9CI) **(88-56-2)**
Benzenesulfonic acid, 3-amino-5-chloro-2-hydroxy- (9CI) **(88-23-3)**
Benzenesulfonic acid, 2-amino-5-chloro-4-methyl- **(88-53-9)**
Benzenesulfonic acid, 4-amino-2,5-dichloro- (9CI) **(88-50-6)**
Benzenesulfonic acid, ((1-amino-9,10-dihydro-4-hydroxy-9,10-dioxo-2-anthracenyl)oxy)(1,1-dimethylpropyl)- (9CI) **(28517-81-9)**
Benzenesulfonic acid, 3-amino-4-hydroxy- (9CI) **(98-37-3)**
Benzenesulfonic acid, 3-amino-2-hydroxy-5-nitro- **(96-67-3)**
Benzenesulfonic acid, 3-amino-4-hydroxy-5-nitro **(96-93-5)**
Benzenesulfonic acid, 2-amino-5-methoxy **(13244-33-2)**
Benzenesulfonic acid, 4-amino-5-methoxy-2-methyl- (9CI) **(6471-78-9)**
Benzenesulfonic acid, 2-amino-5-methoxy-, monosodium salt (9CI) **(19433-86-4)**
Benzenesulfonic acid, 3-((4-amino-3-methoxyphenyl)azo)-, monosodium salt (9CI) **(6300-07-8)**
Benzenesulfonic acid, 2-amino-5-methyl- **(88-44-8)**
Benzenesulfonic acid, 4-amino-3-methyl- (9CI) **(98-33-9)**
Benzenesulfonic acid, 4-((5-amino-3-methyl-1-phenyl-1H-pyrazol-4-yl)azo)-2,5-dichloro- (9CI) **(12239-15-5)**
Benzenesulfonic acid, 2-amino-5-nitro **(96-75-3)**
Benzenesulfonic acid, 2-amino-5-nitro-, ammonium salt **(4346-51-4)**
Benzenesulfonic acid, 5-amino-2-(2-(4-nitro-2-sulfophenyl)-ethenyl)-, disodium salt (9CI) **(6634-82-8)**
Benzenesulfonic acid, 2-((4-aminophenyl)amino)-5-nitro- (9CI) **(91-29-2)**
Benzenesulfonic acid, p-((p-aminophenyl)azo) **(104-23-4)**
Benzenesulfonic acid, 4-((4-aminophenyl)azo)-, monosodium salt (9CI) **(2491-71-6)**
Benzenesulfonic acid, 4-((4-aminophenyl)azo)-, sodium salt **(2491-71-6)**
Benzenesulfonic acid, 5-amino-2-(2-phenylethenyl)- (9CI) **(6265-01-6)**
Benzenesulfonic acid, 5-((4-(aminosulfonyl)-2-nitrophenyl)-amino)-2-((4-((4-(aminosulfonyl)-2-nitrophenyl)amino)-phenyl)amino)- (9CI) **(12239-00-8)**
Benzenesulfonic acid, 2-amino-5-((4-sulfophenyl)azo)- **(101-50-8)**
Benzenesulfonic acid, 4-(4-amino-3-sulfophenylazo) **(101-50-8)**
Benzenesulfonic acid, 2-anilino-5-(2,4-dinitroanilino)-, monosodium salt **(6373-74-6)**
Benzenesulfonic acid, p-((p-anilinophenyl)azo)-, monosodium salt **(554-73-4)**

Benzenesulfonic acid, p-((p-anilinophenyl)azo)-, sodium salt **(554-73-4)**
Benzenesulfonic acid, 2,2'-(1,4-anthraquinonylenediimino)bis-(5-methyl-, disodium salt **(4403-90-1)**
Benzenesulfonic acid, 5-benzoyl-4-hydroxy-2-methoxy **(4065-45-6)**
Benzenesulfonic acid, 2,2'-((1,1'-biphenyl)-4,4'-diyldi-2,1-ethenediyl)bis-, disodium salt **(27344-41-8)**
Benzenesulfonic acid, 2,2'-(4,4'-biphenylylenedivinylene)di-, disodium salt **(27344-41-8)**
Benzenesulfonic acid, 3,5-bis(methoxycarbonyl)-, sodium salt **(3965-55-7)**
Benzenesulfonic acid, p-(2-bromoethyl)- **(54322-31-5)**
Benzenesulfonic acid, 4-(2-bromoethyl)- (9CI) **(54322-31-5)**
Benzenesulfonic acid, 4-(2-bromoethyl)-, sodium salt (9CI) **(65036-65-9)**
Benzenesulfonic acid, 4-butyl- (9CI) **(18521-59-0)**
Benzenesulfonic acid, p-butyl- (8CI) **(18521-59-0)**
Benzenesulfonic acid butyl amide **(3622-84-2)**
Benzenesulfonic acid, butyl ester (8CI,9CI) **(80-44-4)**
Benzenesulfonic acid, 4-(1-butylhexyl)-, sodium salt (9CI) **(73602-67-2)**
Benzenesulfonic acid, butyl-, sodium salt (9CI) **(26746-29-2)**
Benzenesulfonic acid, 4,4'-(carbonothioyldiimino)bis-, bis((1-methylethylidene)hydrazide) (9CI) **(68971-14-2)**
Benzenesulfonic acid, 3,3'-(carbonylbis(imino(3-methoxy-4,1-phenylene)azo))bis- (9CI) **(25712-08-7)**
Benzenesulfonic acid, 3,3'-(carbonylbis(imino(3-methoxy-4,1-phenylene)azo))bis-, disodium salt (9CI) **(10114-86-0)**
Benzenesulfonic acid, 2,2'-(carbonylbis(imino(2-methyl-4,1-phenylene)azo(2-methyl-4,1-phenylene)azo))bis-(5-((4-sulfophenyl)azo)-, tetrasodium salt (9CI) **(6854-81-5)**
Benzenesulfonic (acid) chloride **(98-09-9)**
Benzenesulfonic acid, 4-chloro- **(98-66-8)**
Benzenesulfonic acid, 5-chloro-2-(4-chloro-2-(3-(3,4-dichlorophenyl)ureido)phenoxy)-, sodium salt **(3567-25-7)**
Benzenesulfonic acid, 4-chloro-, 4-chlorophenyl ester **(80-33-1)**
Benzenesulfonic acid, p-chloro-, 2-(4,6-dichloro-s-triazin-2-yl)-hydrazide (8CI) **(30357-78-9)**
Benzenesulfonic acid, 4-chloro-3-(4,5-dihydro-3-methyl-5-oxo-4-(phenylazo)-1H-pyrazol-1-yl)-, monosodium salt (9CI) **(6359-90-6)**
Benzenesulfonic acid, 5-chloro-3-[[4,5-dihydro-3-methyl-5-oxo-1-(3-sulfophenyl)-1H-pyrazol-4-yl]azo]-2-hydroxy-, chromium complex **(6408-31-7)**
Benzenesulfonic acid, 4-chloro-3,5-dinitro-, potassium salt (9CI) **(38185-06-7)**
Benzenesulfonic acid, 4-chloro-2-((2-hydroxy-1-naphthalenyl)azo)-5-methyl-, monosodium salt (9CI) **(5850-90-8)**
Benzenesulfonic acid, 5-chloro-2-((2-hydroxy-1-naphthalenyl)azo)-4-methyl-, barium salt **(5160-02-1)**
Benzenesulfonic acid, 5-chloro-4-methyl-2-nitro- (9CI) **(6973-13-3)**
Benzenesulfonic acid, 2-chloro-5-nitro- (9CI) **(96-73-1)**
Benzenesulfonic acid, p-((p-chlorophenyl)azo)-, sodium salt **(2777-05-1)**
Benzenesulfonic acid, 4-((4-chlorophenyl)azo)-, sodium salt (9CI) **(2777-05-1)**
Benzenesulfonic acid, 4-chlorophenyl ester **(80-38-6)**
Benzenesulfonic acid, p-chlorophenyl ester **(80-38-6)**
Benzenesulfonic acid, p-chloro-, sodium salt **(5138-90-9)**
Benzenesulfonic acid, p-decyl- **(140-60-3)**
Benzenesulfonic acid, 4-decyl- (9CI) **(140-60-3)**
Benzenesulfonic acid, decyl-, sodium salt **(1322-98-1)**
Benzenesulfonic acid, p-decyl-, sodium salt **(2627-06-7)**
Benzenesulfonic acid, 4-decyl-, sodium salt (9CI) **(2627-06-7)**
Benzenesulfonic acid, decyl(sulfophenoxy)-, disodium salt (9CI) **(36445-71-3)**
Benzenesulfonic acid, 2,4-diamino **(88-63-1)**
Benzenesulfonic acid, 4-((2,4-diaminophenyl)azo)-, monosodium salt **(10190-66-6)**
Benzenesulfonic acid, p-((2,4-diaminophenyl)azo)-, monosodium salt (8CI) **(10190-66-6)**
Benzenesulfonic acid, 3,5-dicarboxy-, sodium salt **(6362-79-4)**
Benzenesulfonic acid, 2,5-dichloro-4-(4,5-dihydro-3-methyl-5-oxo-4-(phenylazo)-1H-pyrazol-1-yl)-, sodium salt (9CI) **(6359-97-3)**
Benzenesulfonic acid, 2,5-dichloro-4-(4,5-dihydro-3-methyl-5-oxo-1H-pyrazol-1-yl)- **(84-57-1)**
Benzenesulfonic acid, 2,5-dichloro-4-(4,5-dihydro-3-methyl-5-oxo-4-((4-sulfophenyl)azo)-1H-pyrazol-1-yl)-, disodium salt (9CI) **(6359-98-4)**
Benzenesulfonic acid, 2,5-dichloro-4-(4-((2-((ethylphenylamino)-sulfonyl)phenyl)azo)-4,5-dihydro-3-methyl-5-oxo-1H-pyrazol-1-yl)-, sodium salt (9CI) **(12217-38-8)**
Benzenesulfonic acid, 4,5-dichloro-2-((2-hydroxy-1-naphthalenyl)-azo)-, monosodium salt (9CI) **(5850-81-7)**
Benzenesulfonic acid, 2,5-dichloro-4-(3-methyl-5-oxo-2-pyrazolin-1-yl) **(84-57-1)**
Benzenesulfonic acid, 4-((2,4-dichlorophenyl)azo)-, sodium salt (9CI) **(62959-41-5)**
Benzenesulfonic acid, 2,4-dichlorophenyl ester **(97-16-5)**
Benzenesulfonic acid, 4-((4,6-dichloro-1,3,5-triazin-2-yl)amino)-(9CI) **(16110-89-7)**
Benzenesulfonic acid, 2-(4,6-dichloro-s-triazin-2-ylamino)-4-(4-amino-3-sulfo-1- anthraquinonylamino)-, disodium salt **(4499-01-8)**
Benzenesulfonic acid, 2-(4,6-dichloro-s-triazin-2-yl)hydrazide (6CI,8CI) **(18237-29-1)**
Benzenesulfonic acid, 3-(diethylamino)-, sodium salt **(5123-63-7)**
Benzenesulfonic acid, 3,3'-((9,10-dihydro-9,10-dioxo-1,4-anthracenediyl)diimino)bis(2,4,6-trimethyl-, disodium salt (9CI) **(4474-24-2)**
Benzenesulfonic acid, 4-(4,5-dihydro-3-methyl-4-((4-methyl-3-((phenylamino)sulfonyl)phenyl)azo)-5-oxo-1H-pyrazol-1-yl)-, monosodium salt (9CI) **(6359-85-9)**
Benzenesulfonic acid, 4-(4,5-dihydro-3-methyl-5-oxo-1H-pyrazol-1-yl)- (9CI) **(89-36-1)**
Benzenesulfonic acid, 2,5-dihydroxy-, monopotassium salt (9CI) **(21799-87-1)**
Benzenesulfonic acid, 2,5-dihydroxy-, monosodium salt (9CI) **(10021-55-3)**
Benzenesulfonic acid, 3-((1,5-dihydroxy-2-naphthalenyl)-azo)-4-hydroxy-, monosodium salt (9CI) **(2052-25-7)**
Benzenesulfonic acid, 4-((2,4-dihydroxyphenyl)azo)-, monosodium salt (9CI) **(547-57-9)**
Benzenesulfonic acid, p-((2,4-dihydroxyphenyl)azo)-, sodium salt **(547-57-9)**
Benzenesulfonic acid, 4-(4-((2,5-dimethoxy-4-((2-(sulfooxy)ethyl)-sulfonyl)phenyl)azo)-4,5-dihydro-3-methyl-5-oxo-1H-pyrazol-1-yl)-, dipotassium salt (9CI) **(20317-19-5)**
Benzenesulfonic acid, 2,4-dimethyl- **(88-61-9)**
Benzenesulfonic acid, 2,5-dimethyl- (9CI) **(609-54-1)**
Benzenesulfonic acid, p-((p-(dimethylamino)phenyl)azo)-, sodium salt **(547-58-0)**
Benzenesulfonic acid, dimethyl-, ammonium salt (9CI) **(26447-10-9)**
Benzenesulfonic acid, p-(1,1-dimethyldecyl)- (8CI) **(24271-16-7)**
Benzenesulfonic acid, p-(2,2-dimethyldecyl)- (8CI) **(24271-17-8)**
Benzenesulfonic acid, p-(3,3-dimethyldecyl)- (8CI) **(24271-18-9)**
Benzenesulfonic acid, p-(9,9-dimethyldecyl)- (8CI) **(24271-19-0)**
Benzenesulfonic acid, 4-((3-((dimethylphenyl)azo)-2,4-dihydroxyphenyl)azo)-, monosodium salt (9CI) **(1320-07-6)**
Benzenesulfonic acid, 4-((2,4-dimethylphenyl)azo)-, sodium salt (9CI) **(62959-40-4)**
Benzenesulfonic acid, dimethyl-, potassium salt (9CI) **(30346-73-7)**
Benzenesulfonic acid, dimethyl-, sodium salt **(1300-72-7)**
Benzenesulfonic acid, 2,5-dimethyl-, sodium salt (9CI) **(827-19-0)**
Benzenesulfonic acid, 5-((2,4-dinitrophenyl)amino)-2-(phenylcamino)-, monosodium salt **(6373-74-6)**
Benzenesulfonic acid, 2-dodecyl- (9CI) **(47221-31-8)**
Benzenesulfonic acid, 3-dodecyl- (9CI) **(16577-13-2)**
Benzenesulfonic acid, 4-dodecyl- (9CI) **(121-65-3)**
Benzenesulfonic acid, m-dodecyl- (8CI) **(16577-13-2)**
Benzenesulfonic acid, dodecyl (8CI,9CI) **(27176-87-0)**
Benzenesulfonic acid, dodecyl-, Compd. with 2-aminoethanol (1:1) (8CI,9CI) **(26836-07-7)**
Benzenesulfonic acid, dodecyl-, Compd. with 1-amino-2-propanol (1:1) (9CI) **(42504-46-1)**
Benzenesulfonic acid, dodecyl-, Compd. with 2,2'-iminobis-(ethanol) (1:1) (9CI) **(26545-53-9)**

Benzenesulfonic acid, dodecyl-, Compd. with 2,2',2''-nitrilo-tris(ethanol) (1:1) (9CI) **(27323-41-7)**
Benzenesulfonic acid, dodecyl-, Compd. with 2-propanamine (1:1) (9CI) **(26264-05-1)**
Benzenesulfonic acid, dodecyl-, 2-aminoethanol salt (1:1) **(26836-07-7)**
Benzenesulfonic acid, dodecyl-, ammonium salt (9CI) **(1331-61-9)**
Benzenesulfonic acid, dodecyl-, calcium salt (9CI) **(26264-06-2)**
Benzenesulfonic acid, dodecyl-, potassium salt (9CI) **(27177-77-1)**
Benzenesulfonic acid, dodecyl-, sodium salt **(25155-30-0)**
Benzenesulfonic acid, 4-dodecyl-, sodium salt (9CI) **(2211-98-5)**
Benzenesulfonic acid, 4-sec-dodecyl-, sodium salt (9CI) **(68628-60-4)**
Benzenesulfonic acid, dodecyl-, sodium salt, Mixt. with α-(nonylphenyl)-ω-hydroxypoly(oxy-1,2-ethanediyl) (9CI) **(50642-03-0)**
Benzenesulfonic acid, dodecyl(sulfophenoxy)- (9CI) **(30260-72-1)**
Benzenesulfonic acid, dodecyl (sulfophenoxy)-, disodium salt **(28519-02-0)**
Benzenesulfonic acid, dodecyl(sulfophenoxy)-, disodium salt (9CI) **(28519-02-0)**
Benzenesulfonic acid, 2,2'-(1,2-ethenediyl)bis(5-amino-, disodium salt (9CI) **(7336-20-1)**
Benzenesulfonic acid, 2,2'-(1,2-ethenediyl)bis(5-((4-((3-amino-3-oxopropyl)(2-hydroxyethyl)amino)-6-(phenylamino)-1,3,5-triazin-2-yl)amino)-, disodium salt (9CI) **(27344-06-5)**
Benzenesulfonic acid, 2,2'-(1,2-ethenediyl)bis(5-amino-, sodium salt (9CI) **(25394-13-2)**
Benzenesulfonic acid, 2,2'-(1,2-ethenediyl)bis(5-((4-chloro-6-(phenylamino)-1,3,5-triazin-2-yl)amino)-, disodium salt (9CI) **(37138-23-1)**
Benzenesulfonic acid, 2,2'-(1,2-ethenediyl)bis(5-((4-ethoxyphenyl)-azo)-, disodium salt (9CI) **(2870-32-8)**
Benzenesulfonic acid, 2,2'-(1,2-ethenediyl)bis(5-((4-(bis-(2-hydroxyethyl)amino)-6-(phenyla mino)-1,3,5-triazin-2-yl)-amino)-, disodium salt, (E)- (9CI) **(59453-69-9)**
Benzenesulfonic acid, 2,2'-(1,2-ethenediyl)bis(5-((4-(bis(2-hydroxy-ethyl)amino)-6-(phenylamino)-1,3,5-triazin-2-yl)amino)- (9CI) **(4404-43-7)**
Benzenesulfonic acid, 2,2'-(1,2-ethenediyl)bis(5-((4-(bis(2-hydroxy-ethyl)amino)-6-((4-sulfophenyl)amino)-1,3,5-triazin-2-yl)-amino)-, tetrasodium salt (9CI) **(16470-24-9)**
Benzenesulfonic acid, 2,2'-(1,2-ethenediyl)bis(5-((4-((2-hydroxy-ethyl)methylamino)-6-(phenylamino)-1,3,5-triazin-2-yl)amino)-, disodium salt, (E)- (9CI) **(56776-29-5)**
Benzenesulfonic acid, 2,2'-(1,2-ethenediyl)bis(5-((4-hydroxyphenyl)-azo)- (9CI) **(91-34-9)**
Benzenesulfonic acid, 2,2'-(1,2-ethenediyl)bis(5-((4-hydroxyphenyl)-azo)-, disodium salt (9CI) **(3051-11-4)**
Benzenesulfonic acid, 2,2'-(1,2-ethenediyl)bis(5-((4-((2-hydroxy-propyl)amino)-6-(phenylamino)-1,3,5-triazin-2-yl)amino)-, disodium salt (9CI) **(32694-95-4)**
Benzenesulfonic acid, 2,2'-(1,2-ethenediyl)bis(5-((4-(methylamino)-6-(phenylamino)-1,3,5-triazin-2-yl)amino)- (9CI) **(35632-99-6)**
Benzenesulfonic acid, 2,2'-(1,2-ethenediyl)bis(5-((4-((2-methyl-phenyl)amino)-6-(4-morpholinyl)-1,3,5-triazin-2-yl)amino)-, disodium salt **(24019-80-5)**
Benzenesulfonic acid, 2,2'-(1,2-ethenediyl)bis(5-((4-(4-morpholinyl)-6-(phenylamino)-1,3,5-triazin-2-yl)amino)-, disodium salt, (E)- (9CI) **(56776-30-8)**
Benzenesulfonic acid, 2,2'-(1,2-ethenediyl)bis(5-nitro- **(128-42-7)**
Benzenesulfonic acid, 2,2'-(1,2-ethenediyl)bis(5-nitro-, disodium salt (9CI) **(3709-43-1)**
Benzenesulfonic acid, 2,2'-(1,2-ethenediyl)bis(5-nitro-, sodium salt (9CI) **(15883-59-7)**
Benzenesulfonic acid, 2,2'-(1,2-ethenediyl)bis(5-(4-phenyl-2H-1,2,3-triazol-2-yl)- (9CI) **(37069-54-8)**
Benzenesulfonic acid, 2,2'-(1,2-ethenediyl)bis(5-((4-((4-sulfo-phenyl)azo)phenyl)-NNO-azoxy)-, tetrasodium salt (9CI) **(32651-66-4)**
Benzenesulfonic acid, ethenyl-, Homopolymer, sodium salt **(9080-79-9)**
Benzenesulfonic acid, 4-ethenyl-, sodium salt (9CI) **(2695-37-6)**
Benzenesulfonic acid, ethenyl-, sodium salt (9CI) **(27457-28-9)**
Benzenesulfonic acid, p-ethyl- **(98-69-1)**
Benzenesulfonic acid, 4-ethyl- (9CI) **(98-69-1)**
Benzenesulfonic acid, ethyl- (9CI) **(57352-34-8)**
Benzenesulfonic acid, 4-(1-ethyldecyl)- (9CI) **(18777-54-3)**
Benzenesulfonic acid, 2,2'-(1,2-ethylenediyl)bis(5-amino- (9CI) **(81-11-8)**
Benzenesulfonic acid, ethyl ester (8CI,9CI) **(515-46-8)**
Benzenesulfonic acid, 3-((ethylphenylamino)methyl)- (9CI) **(101-11-1)**
Benzenesulfonic acid, p-ethyl-, sodium salt **(14995-38-1)**
Benzenesulfonic acid, 4-ethyl-, sodium salt (9CI) **(14995-38-1)**
Benzenesulfonic acid, ethyl-, sodium salt (9CI) **(30995-65-4)**
Benzenesulfonic acid, 2-formyl- (9CI) **(91-25-8)**
Benzenesulfonic acid, 2-formyl-, 4-bromophenyl ester (9CI) **(106939-92-8)**
Benzenesulfonic acid, 4-formyl-, 4-bromophenyl ester (9CI) **(106939-97-3)**
Benzenesulfonic acid, 2-formyl-, 4-methoxyphenyl ester (9CI) **(106939-89-3)**
Benzenesulfonic acid, 4-formyl-, 4-methoxyphenyl ester (9CI) **(106939-94-0)**
Benzenesulfonic acid, 2-formyl-, 4-methylphenyl ester (9CI) **(106939-90-6)**
Benzenesulfonic acid, 4-formyl-, 4-methylphenyl ester (9CI) **(106939-95-1)**
Benzenesulfonic acid, 2-formyl-, 4-nitrophenyl ester (9CI) **(106939-93-9)**
Benzenesulfonic acid, 4-formyl-, 4-nitrophenyl ester (9CI) **(106939-98-4)**
Benzenesulfonic acid, 2-formyl-, phenyl ester (9CI) **(106939-91-7)**
Benzenesulfonic acid, 4-formyl-, phenyl ester (9CI) **(106939-96-2)**
Benzenesulfonic acid, 2-formyl-, sodium salt (9CI) **(1008-72-6)**
Benzenesulfonic acid, o-formyl-, sodium salt (8CI) **(1008-72-6)**
Benzenesulfonic acid, heptadecyl- (9CI) **(39735-13-2)**
Benzenesulfonic acid, heptyl-, sodium salt (8CI,9CI) **(33660-91-2)**
Benzenesulfonic acid, p-hexadecyl- (8CI) **(16722-32-0)**
Benzenesulfonic acid, hexyl- (7CI,9CI) **(58425-67-5)**
Benzenesulfonic acid, hydrazide **(80-17-1)**
Benzenesulfonic acid, hydroxy **(1333-39-7)**
Benzenesulfonic acid, p-hydroxy **(98-67-9)**
Benzenesulfonic acid, 2-hydroxy- (9CI) **(609-46-1)**
Benzenesulfonic acid, o-hydroxy- (8CI) **(609-46-1)**
Benzenesulfonic acid, hydroxy-, Polymer with formaldehyde (9CI) **(50973-35-8)**
Benzenesulfonic acid, p-(4-((4-((2-hydroxyethyl)sulfonyl)-2,5-dimethoxyphenyl)azo)-3-methyl-5-oxo-2-pyrazolin-1-yl)-, hydrogen sulfate, dipotassium salt **(20317-19-5)**
Benzenesulfonic acid, 4-hydroxy-3-((2-hydroxy-1-naphthalenyl) azo)-, monosodium salt **(2092-55-9)**
Benzenesulfonic acid, hydroxy-, monosodium salt (9CI) **(1300-51-2)**
Benzenesulfonic acid, 2-((2-hydroxy-1-naphthalenyl)azo)- (9CI) **(6616-62-2)**
Benzenesulfonic acid, 3-((2-hydroxy-1-naphthalenyl)azo)- (9CI) **(22080-08-6)**
Benzenesulfonic acid, 4-((2-hydroxy-1-naphthalenyl)azo)- (9CI) **(573-89-7)**
Benzenesulfonic acid, 4-((4-hydroxy-1-naphthalenyl)azo)- (9CI) **(574-69-6)**
Benzenesulfonic acid, 4-((2-hydroxy-1-naphthalenyl)azo)-3-methyl-, monosodium salt (9CI) **(5850-86-2)**
Benzenesulfonic acid, 4-((2-hydroxy-1-naphthalenyl)azo)-, monosodium salt **(633-96-5)**
Benzenesulfonic acid, 4-((4-((2-hydroxy-1-naphthalenyl)azo)-phenyl)azo)-, monosodium salt (9CI) **(6406-56-0)**
Benzenesulfonic acid, 2-((2-hydroxy-1-naphthalenyl)azo)-5-((4-sulfophenyl)azo)-, disodium salt (9CI) **(4196-99-0)**
Benzenesulfonic acid, m-((2-hydroxy-1-naphthyl)azo)- (8CI) **(22080-08-6)**
Benzenesulfonic acid, o-((2-hydroxy-1-naphthyl)azo)- (8CI) **(6616-62-2)**
Benzenesulfonic acid, p-((4-hydroxy-1-naphthyl)azo)-, sodium salt **(523-44-4)**
Benzenesulfonic acid, 4-((4-hydroxyphenyl)azo)-, monosodium salt (9CI) **(2623-36-1)**
Benzenesulfonic acid, p-((p-hydroxyphenyl)azo)-, monosodium salt (8CI) **(2623-36-1)**
Benzenesulfonic acid, p-((p-hydroxyphenyl)azo)-, sodium salt **(2623-36-1)**
Benzenesulfonic acid, 4-hydroxy-, zinc salt (2:1) **(127-82-2)**
Benzenesulfonic acid, p-hydroxy-, zinc salt (2:1) **(127-82-2)**
Benzenesulfonic acid, linear alkyl-, sodium salt **(68411-30-3)**
Benzenesulfonic acid, 4-((5-methoxy-4-((4-methoxyphenyl)-azo)-2-methylphenyl)azo)-, sodium salt (9CI) **(68555-86-2)**
Benzenesulfonic acid, 2-methyl- (9CI) **(88-20-0)**
Benzenesulfonic acid, 4-methyl-, 2-(aminocarbonyl)hydrazide (9CI) **(10396-10-8)**

Benzenesulfonic acid, 4-methyl-, copper(2+) salt (9CI) **(7144-37-8)**
Benzenesulfonic acid, 4-methyl-, cyclohexyl ester (9CI) **(953-91-3)**
Benzenesulfonic acid, methyl ester **(80-18-2)**
Benzenesulfonic acid, 4-(1-methylethyl)- (9CI) **(16066-35-6)**
Benzenesulfonic acid, (1-methylethyl)-, ammonium salt (9CI) **(37475-88-0)**
Benzenesulfonic acid, (1-methylethyl)-, sodium salt (9CI) **(28348-53-0)**
Benzenesulfonic acid, 2-methyl-4-((4-((4-((3-methylphenyl)amino)phenyl)(4-((3-methylphenyl)imino)-2,5-cyclohexadien-1-ylidene)methyl)phenyl)amino)- (9CI) **(6417-46-5)**
Benzenesulfonic acid, 4-((4-((2-methyl-4-(((4-methylphenyl)sulfonyl)oxy)phenyl)azo)phenyl)amino)-3-nitro-, monosodium salt (9CI) **(12220-06-3)**
Benzenesulfonic acid, 2-methyl-5-nitro **(121-03-9)**
Benzenesulfonic acid, 2-methyl-5-nitro-, sodium salt (9CI) **(5258-64-0)**
Benzenesulfonic acid, 4-(1-methylnonyl)-, sodium salt (9CI) **(73602-65-0)**
Benzenesulfonic acid, p-(3-methyl-5-oxo-2-pyrazolin-1-yl)- (8CI) **(89-36-1)**
Benzenesulfonic acid, 4-((1-(((2-methylphenyl)amino)carbonyl)-2-oxopropyl)azo)-3-nitro-, calcium salt (2:1) (9CI) **(12286-66-7)**
Benzenesulfonic acid, 4-((4-methylphenyl)azo)-, sodium salt (9CI) **(62959-39-1)**
Benzenesulfonic acid, 4-methyl-, 2-phenylethyl ester (9CI) **(4455-09-8)**
Benzenesulfonic acid, methyl-, potassium salt (9CI) **(30526-22-8)**
Benzenesulfonic acid, 4-((2-methyl-2-propenyl)oxy)-, sodium salt (9CI) **(1208-67-9)**
Benzenesulfonic acid, 4-methyl-, sodium salt (9CI) **(657-84-1)**
Benzenesulfonic acid, methyl-, sodium salt (9CI) **(12068-03-0)**
Benzenesulfonic acid, 4-(1-methyltridecyl)-, sodium salt (9CI) **(13419-31-3)**
Benzenesulfonic acid, p-(1-methyltridecyl)-, sodium salt (8CI) **(13419-31-3)**
Benzenesulfonic acid, 4-(1-methylundecyl)-, sodium salt (9CI) **(2211-99-6)**
Benzenesulfonic acid, p-(1-methylundecyl)-, sodium salt (8CI) **(2211-99-6)**
Benzenesulfonic acid, mono-C6-12-alkyl derivs., sodium salts **(68608-87-7)**
Benzenesulfonic acid, mono-C9-17-alkyl derivs., sodium salts **(68953-95-7)**
Benzenesulfonic acid, 5-(2H-naphtho(1,2-d)triazol-2-yl)-2-(2-phenylethenyl)-, sodium salt **(6416-68-8)**
Benzenesulfonic acid, 5-(2H-naphtho(1,2-d)triazol-2-yl)-2-(2-phenylethenyl)-, sodium salt, (E)- (9CI) **(56776-27-3)**
Benzenesulfonic acid, 3-nitro- **(98-47-5)**
Benzenesulfonic acid, m-nitro **(98-47-5)**
Benzenesulfonic acid, nitro- **(31212-28-9)**
Benzenesulfonic acid, 2-nitro- (9CI) **(80-82-0)**
Benzenesulfonic acid, 4-nitro- (9CI) **(138-42-1)**
Benzenesulfonic acid, o-nitro- (8CI) **(80-82-0)**
Benzenesulfonic acid, p-nitro- (8CI) **(138-42-1)**
Benzenesulfonic acid, 4-nitro-, methyl ester (9CI) **(6214-20-6)**
Benzenesulfonic acid, 4-((4-nitrophenyl)azo)-, sodium salt (9CI) **(2491-72-7)**
Benzenesulfonic acid, 5-nitro-2-(2-phenylethenyl)-, sodium salt (9CI) **(10359-69-0)**
Benzenesulfonic acid, m-nitro-, sodium salt **(127-68-4)**
Benzenesulfonic acid, 4-octyl- (9CI) **(17012-98-5)**
Benzenesulfonic acid, octyl- (9CI) **(25321-43-1)**
Benzenesulfonic acid, p-octyl- (8CI) **(17012-98-5)**
Benzenesulfonic acid, 4-octyl-, sodium salt (9CI) **(6149-03-7)**
Benzenesulfonic acid, p-octyl-, sodium salt (8CI) **(6149-03-7)**
Benzenesulfonic acid, 4,4'-oxybis-, dihydrazide **(80-51-3)**
Benzenesulfonic acid, oxybis-, dihydrazide (9CI) **(80-51-3)**
Benzenesulfonic acid, oxybis(dodecyl- (9CI) **(30260-73-2)**
Benzenesulfonic acid, 2,2'(or 3,3')-oxybis(5(or 2)-dodecyl-, disodium salt (9CI) **(25167-32-2)**
Benzenesulfonic acid, pentadecyl- (9CI) **(61215-89-2)**
Benzenesulfonic acid, 4-(1-pentylheptyl)-, sodium salt (9CI) **(2212-52-4)**
Benzenesulfonic acid, p-(1-pentylheptyl)-, sodium salt (8CI) **(2212-52-4)**
Benzenesulfonic acid, 3-((4-(phenylamino)phenyl)azo)- (9CI) **(4005-68-9)**
Benzenesulfonic acid, 3-((4-(phenylamino)phenyl)azo)-, monosodium salt **(587-98-4)**
Benzenesulfonic acid, 4-((4-(phenylamino)phenyl)azo)-, monosodium salt (9CI) **(554-73-4)**
Benzenesulfonic acid, ((4-((4-(phenylamino)phenyl)(4-(phenylimino)-2,5-cyclohexadien-1-ylidene)methyl)phenyl)amino)- (9CI) **(1324-76-1)**
Benzenesulfonic acid, 4-(phenylazo)- (9CI) **(2484-88-0)**
Benzenesulfonic acid, p-(phenylazo)-, sodium salt **(42975-18-8)**
Benzenesulfonic acid, 4-(phenylazo)-, sodium salt (9CI) **(42975-18-8)**
Benzenesulfonic acid, 4-propyl- (9CI) **(15592-74-2)**
Benzenesulfonic acid, p-propyl- (7CI,8CI) **(15592-74-2)**
Benzenesulfonic acid, propyl ester (9CI) **(80-42-2)**
Benzenesulfonic acid, sodium salt **(515-42-4)**
Benzenesulfonic acid, 4-tetradecyl- (9CI) **(47377-16-2)**
Benzenesulfonic acid, tetradecyl- (9CI) **(30776-59-1)**
Benzenesulfonic acid, 4-tetradecyl-, sodium salt (9CI) **(1797-33-7)**
Benzenesulfonic acid, tetradecyl-, sodium salt (9CI) **(28348-61-0)**
Benzenesulfonic acid, tetrapropylene **(11067-81-5)**
Benzenesulfonic acid, tetrapropylene-, sodium salt **(11067-82-6)**
Benzenesulfonic acid, tridecyl- (9CI) **(25496-01-9)**
Benzenesulfonic acid, tridecyl-, sodium salt (9CI) **(26248-24-8)**
Benzenesulfonic acid, 4-undecyl- (9CI) **(39156-49-5)**
Benzenesulfonic acid, undecyl- (9CI) **(50854-94-9)**
Benzenesulfonic acid, undecyl-, sodium salt (9CI) **(27636-75-5)**
Benzenesulfonic acid, 3,3'-(ureylenebis((3-methoxy-p-phenylene)-azo))di **(25712-08-7)**
Benzenesulfonic chloride **(98-09-9)**
Benzenesulfonic hydrazide **(80-17-1)**
Benzenesulfonohydrazide **(80-17-1)**
n-Benzenesulfonylaziridine **(10302-15-5)**
Benzenesulfonyl-chloride **(98-09-9)**
Benzenesulfonyl chloride, 4-(acetylamino)- (9CI) **(121-60-8)**
Benzenesulfonyl chloride, 4-(acetylamino)-2,5-dimethoxy- (9CI) **(13279-58-8)**
Benzenesulfonyl chloride, 3-bromo- (9CI) **(2905-24-0)**
Benzenesulfonyl chloride, 4-bromo- (9CI) **(98-58-8)**
Benzenesulfonyl chloride, m-bromo- (8CI) **(2905-24-0)**
Benzenesulfonyl chloride, p-bromo- (8CI) **(98-58-8)**
Benzenesulfonyl chloride, p-chloro **(98-60-2)**
Benzenesulfonyl chloride, 3-chloro- (9CI) **(2888-06-4)**
Benzenesulfonyl chloride, m-chloro- (8CI) **(2888-06-4)**
Benzenesulfonyl chloride, 4-chloro-2-methyl- (9CI) **(56157-92-7)**
Benzenesulfonyl chloride, 2-chloro-5-nitro- (9CI) **(4533-95-3)**
Benzenesulfonyl chloride, 4-chloro-3-nitro- (9CI) **(97-08-5)**
Benzenesulfonyl chloride, 4-chloro-2-nitro- (8CI,9CI) **(4533-96-4)**
Benzenesulfonyl chloride, 4-cyano- (9CI) **(49584-26-1)**
Benzenesulfonyl chloride, p-cyano- (6CI) **(49584-26-1)**
Benzenesulfonyl chloride, 3,4-dichloro- (9CI) **(98-31-7)**
Benzenesulfonyl chloride, 2,4-dimethyl- (9CI) **(609-60-9)**
Benzenesulfonyl chloride, p-fluoro- **(349-88-2)**
Benzenesulfonyl chloride, 4-fluoro- (9CI) **(349-88-2)**
Benzenesulfonyl chloride, 2-formyl- (9CI) **(21639-41-8)**
Benzenesulfonyl chloride, 4-formyl- (9CI) **(85822-16-8)**
Benzenesulfonyl chloride, o-formyl- (8CI) **(21639-41-8)**
Benzenesulfonyl chloride, 3-hydroxy- (9CI) **(56157-93-8)**
Benzenesulfonyl chloride, 3-iodo- (9CI) **(50702-38-0)**
Benzenesulfonyl chloride, 4-iodo- (9CI) **(98-61-3)**
Benzenesulfonyl chloride, p-iodo- (8CI) **(98-61-3)**
Benzenesulfonyl chloride, 4-methoxy- (9CI) **(98-68-0)**
Benzenesulfonyl chloride, p-methoxy- (8CI) **(98-68-0)**
Benzenesulfonyl chloride, 2-methyl- (9CI) **(133-59-5)**
Benzenesulfonyl chloride, 4-methyl- (9CI) **(98-59-9)**
Benzenesulfonyl chloride, 2-methyl-5-nitro- **(121-02-8)**
Benzenesulfonyl chloride, 2-nitro- (9CI) **(1694-92-4)**
Benzenesulfonyl chloride, 3-nitro- (9CI) **(121-51-7)**
Benzenesulfonyl chloride, 4-nitro- (9CI) **(98-74-8)**
Benzenesulfonyl chloride, m-nitro- (8CI) **(121-51-7)**
Benzenesulfonyl chloride, o-nitro- (8CI) **(1694-92-4)**
Benzenesulfonyl chloride, p-nitro- (8CI) **(98-74-8)**
Benzenesulfonyl chloride, 4,4'-oxybis- (9CI) **(121-63-1)**
Benzenesulfonyl chloride, 4,4'-oxydi- (8CI) **(121-63-1)**
Benzenesulfonyl chloride, 2,3,4-trichloro- (9CI) **(34732-09-7)**
Benzenesulfonyl chloride, 2,4,5-trichloro- (9CI) **(15945-07-0)**

Name	CAS No.
Benzenesulfonyl chloride, 2,4,6-trimethyl	**(773-64-8)**
Benzenesulfonyl fluoride, 2-ethyl- (9CI)	**(34586-49-7)**
Benzenesulfonyl fluoride, 4-ethyl- (9CI)	**(455-20-9)**
Benzenesulfonyl hydrazide	**(80-17-1)**
Benzenesulfonyl hydrazine	**(80-17-1)**
Benzenesulfonyl isocyanate, 4-methyl- (9CI)	**(4083-64-1)**
Benzenesulfo-sodium chloramide	**(127-52-6)**
Benzenesulphonamide	**(98-10-2)**
Benzenesulphonic acid, 2,4-dichlorophenyl ester	**(97-16-5)**
Benzene sulphonohydrazide	**(80-17-1)**
Benzene sulphonyl chloride [UN 2225]	**(98-09-9)**
Benzene, 1,2,4,5-tetrabromo- (9CI)	**(636-28-2)**
Benzene, 1,2,3,4-tetrabromo- (8CI,9CI)	**(22311-25-7)**
Benzene, 1,2,3,5-tetrabromo- (8CI,9CI)	**(634-89-9)**
Benzene, 1,2,3,4-tetrabromo-5-chloro-6-methyl- (9CI)	**(39569-21-6)**
Benzene, 1,2,3,4-tetrabromo-5,6-dimethyl- (9CI)	**(2810-69-7)**
1,2,4,5-Benzenetetracarboxylic acid	**(89-05-4)**
1,2,3,4-Benzenetetracarboxylic acid (8CI,9CI)	**(476-73-3)**
1,2,3,5-Benzenetetracarboxylic acid (8CI,9CI)	**(479-47-0)**
1,2,4,5 Benzenetetracarboxylic 1,2:4,5 dianhydride	**(89-32-7)**
Benzene tetrachloride	**(1782-00-9)**
Benzene, 1,2,3,4-tetrachloro	**(634-66-2)**
Benzene, 1,2,3,5-tetrachloro	**(634-90-2)**
Benzene, 1,2,4,5-tetrachloro	**(95-94-3)**
Benzene, tetrachloro- (9CI)	**(12408-10-5)**
Benzene, 1,2,3,5-tetrachloro-4-(3,5-dichloro-2-methoxyphenoxy)- (9CI)	**(97534-03-7)**
Benzene, 1,2,3,5-tetrachloro-4-(3,5-dichloro-4-methoxyphenoxy)- (9CI)	**(90985-96-9)**
Benzene, 1,2,3,4-tetrachloro-5,6-dimethoxy- (8CI,9CI)	**(944-61-6)**
Benzene, 1,2,3,5-tetrachloro-4,6-dimethoxy- (8CI,9CI)	**(944-77-4)**
Benzene, 1,2,4,5-tetrachloro-3,6-dimethoxy- (8CI,9CI)	**(944-78-5)**
Benzene, 1,2,4,5-tetrachloro-3,6-dimethyl- (9CI)	**(877-10-1)**
Benzene, 1,1'-(1,1,2,2-tetrachloro-1,2-ethanediyl)bis- (9CI)	**(13700-81-7)**
Benzene, 1,1'-(tetrachloroethylidene)bis(chloro- (9CI)	**(68631-02-7)**
Benzene, 1,2,3,4-tetrachloro-5-methoxy- (9CI)	**(938-86-3)**
Benzene, 1,2,3,5-tetrachloro-4-methoxy- (9CI)	**(938-22-7)**
Benzene, 1,2,4,5-tetrachloro-3-methoxy- (9CI)	**(6936-40-9)**
Benzene, tetrachloromethoxy- (9CI)	**(53452-81-6)**
Benzene, 1,2,4,5-tetrachloro-3-methoxy-6-nitro- (9CI)	**(2438-88-2)**
Benzene, 1,2,3,4-tetrachloro-5-methyl- (9CI)	**(1006-32-2)**
Benzene, 1,2,4,5-tetrachloro-3-methyl- (9CI)	**(1006-31-1)**
Benzene, 1,2,4,5-tetrachloro-3-(methylsulfinyl)- (9CI)	**(107409-52-9)**
Benzene, 1,2,4,5-tetrachloro-3-(methylsulfonyl)- (9CI)	**(98555-82-9)**
Benzene, tetrachloro(methylthio)- (9CI)	**(53014-41-8)**
Benzene, 1,2,3,4-tetrachloro-5-nitro	**(879-39-0)**
Benzene, 1,2,4,5-tetrachloro-3-nitro	**(117-18-0)**
Benzene, tetrachloronitro- (9CI)	**(28804-67-3)**
Benzene, 1,2,3,5-tetrachloro-4-nitro- (8CI,9CI)	**(3714-62-3)**
Benzene, 1,2,3,5-tetrachloro-4-(2,3,5-trichloro-4-methoxy-phenoxy)- (9CI)	**(97534-06-0)**
Benzene, 1,2,3,5-tetrachloro-4-(3,4,5-trichloro-2-methoxy-phenoxy)- (9CI)	**(97534-07-1)**
Benzene, tetradecyl- (9CI)	**(1459-10-5)**
Benzene, tetraethyl-	**(33637-20-6)**
Benzene, 1,2,3,4-tetrafluoro- (8CI,9CI)	**(551-62-2)**
Benzene, 1,2,3,5-tetrafluoro- (8CI,9CI)	**(2367-82-0)**
Benzene, 1,2,4,5-tetrafluoro- (8CI,9CI)	**(327-54-8)**
Benzene, tetrafluoro- (8CI,9CI)	**(28016-01-5)**
Benzenetetrahydride	**(110-83-8)**
Benzene, tetrahydro-	**(110-83-8)**
Benzene, 1,2,3,4-tetraiodo- (8CI,9CI)	**(634-68-4)**
Benzene, 1,2,3,5-tetraiodo- (7CI,8CI,9CI)	**(634-92-4)**
Benzene, 1,2,4,5-tetraiodo- (7CI,8CI,9CI)	**(636-31-7)**
Benzene, 1,2,3,4-tetramethyl	**(488-23-3)**
Benzene, 1,2,3,5-tetramethyl	**(527-53-7)**
Benzene, 1,2,4,5-tetramethyl	**(95-93-2)**
Benzene, tetramethyl- (8CI,9CI)	**(25619-60-7)**
1,2,4,5-Benzenetetrol, 3-methyl- (9CI)	**(700-19-6)**
Benzene, 1,1'-(thiobis(methylene))bis- (9CI)	**(538-74-9)**
Benzenethiol (DOT)	**(108-98-5)**
Benzenethiol, o-amino	**(137-07-5)**
Benzenethiol, p-amino	**(1193-02-8)**
Benzenethiol, 3-amino- (9CI)	**(22948-02-3)**
Benzenethiol, 4-amino- (9CI)	**(1193-02-8)**
Benzenethiol, m-amino- (8CI)	**(22948-02-3)**
Benzenethiol, p-chloro	**(106-54-7)**
Benzenethiol, 3-chloro- (9CI)	**(2037-31-2)**
Benzenethiol, m-chloro- (8CI)	**(2037-31-2)**
Benzenethiol, 2,6-dichloro- (8CI,9CI)	**(24966-39-0)**
Benzenethiol, dimethyl-	**(25550-52-1)**
Benzenethiol, p-(dimethylamino)	**(4946-22-9)**
Benzenethiol, 3-methyl- (9CI)	**(108-40-7)**
Benzenethiol, 4-nitro-	**(1849-36-1)**
Benzenethiol, p-nitro	**(1849-36-1)**
Benzenethiol, 2-nitro- (9CI)	**(4875-10-9)**
Benzenethiol, o-nitro- (8CI)	**(4875-10-9)**
Benzenethiol, pentachloro	**(133-49-3)**
Benzenethiol, pentafluoro	**(771-62-0)**
Benzenethiol, sodium salt (9CI)	**(930-69-8)**
Benzenethiol, 2,4,5-trichloro	**(3773-14-6)**
1,3,5-Benzenetriamine, 2,4,6-trinitro	**(3058-38-6)**
Benzene, 1,2,4-tribromo- (9CI)	**(615-54-3)**
Benzene, 1,3,5-tribromo- (9CI)	**(626-39-1)**
Benzene, 1,2,3-tribromo- (8CI,9CI)	**(608-21-9)**
Benzene, tribromo- (8CI,9CI)	**(28779-08-0)**
Benzene, 1,3,5-tribromo-2-(2-bromoethoxy)- (9CI)	**(68413-71-8)**
Benzene, 1,3,5-tribromo-2-methoxy- (9CI)	**(607-99-8)**
Benzene, 1,2,4-tribromo-3,5,6-trichloro-	**(13075-01-9)**
Benzene, 1,3,5-tri-tert-butyl- (8CI)	**(1460-02-2)**
Benzene, 1,2,4-tri-tert-butyl- (7CI,8CI)	**(1459-11-6)**
Benzene, 1,2,4-tributyl- (8CI,9CI)	**(14800-16-9)**
1,3,5-Benzenetricarbonyl chloride	**(4422-95-1)**
1,3,5-Benzenetricarbonyl trichloride (9CI)	**(4422-95-1)**
1,2,4-Benzenetricarboxylic acid	**(528-44-9)**
Benzene 1,2,3-tricarboxylic acid	**(569-51-7)**
1,2,3-Benzenetricarboxylic acid (9CI)	**(569-51-7)**
1,3,5-Benzenetricarboxylic acid (9CI)	**(554-95-0)**
1,2,4-Benzenetricarboxylic acid 1,2-anhydride	**(552-30-7)**
1,2,4-Benzenetricarboxylic acid anhydride	**(552-30-7)**
1,2,4-Benzenetricarboxylic acid, cyclic 1,2-anhydride	**(552-30-7)**
1,2,4-Benzenetricarboxylic acid, decyl octyl ester (9CI)	**(67989-23-5)**
1,2,4-Benzenetricarboxylic acid, monodecyl monooctyl ester (9CI)	**(34870-88-7)**
1,2,4-Benzenetricarboxylic acid, trihexyl ester (9CI)	**(1528-49-0)**
1,2,4-Benzenetricarboxylic acid, triisodecyl ester (9CI)	**(36631-30-8)**
1,2,4-Benzenetricarboxylic acid, triisooctyl ester (9CI)	**(27251-75-8)**
1,2,4-Benzenetricarboxylic acid, trimethyl ester	**(2459-10-1)**
Benzene tricarboxylic acid, trioctyl ester	**(89-04-3)**
1,2,4-Benzenetricarboxylic acid, trioctyl ester (9CI)	**(89-04-3)**
1,2,4-Benzenetricarboxylic acid, tri-2-propenyl ester (9CI)	**(2694-54-4)**
1,2,4-Benzenetricarboxylic acid, tris(2-ethylhexyl) ester (9CI)	**(3319-31-1)**
1,2,4-Benzenetricarboxylic anhydride	**(552-30-7)**
Benzene, 1,2,3-trichloro	**(87-61-6)**
Benzene, 1,2,4-trichloro	**(120-82-1)**
Benzene, 1,3,5-trichloro	**(108-70-3)**
Benzene, trichloro	**(12002-48-1)**
Benzene, 1,2,4-trichloro-5-((4-chlorophenyl)sulfonyl)-	**(116-29-0)**
Benzene, 1,2,4-trichloro-5-((4-chlorophenyl)thio)-	**(2227-13-6)**
Benzene, 1,2,4-trichloro-5-(2,4-dichlorophenoxy)- (9CI)	**(60123-64-0)**
Benzene, 1,2,4-trichloro-5-(3,4-dichlorophenoxy)- (9CI)	**(60123-65-1)**
Benzene, 1,2,3-trichloro-5-(2,4-dichlorophenoxy)-4-methoxy- (9CI)	**(63709-64-8)**
Benzene, 1,2,5-trichloro-3,4-dimethoxy- (9CI)	**(85298-07-3)**

Benzene, trichlorodimethoxy- (9CI) **(61465-81-4)**
Benzene, 1,2,3-trichloro-4,5-dimethoxy- (8CI,9CI) **(16766-29-3)**
Benzene, 1,2,4-trichloro-3,5-dinitro- (9CI) **(2678-21-9)**
Benzene, 1,1'-(2,2,2-trichloroethylidene)bis(4-chloro- **(50-29-3)**
Benzene, 1,1'-(2,2,2-trichloroethylidene)bis(chloro- (9CI) **(33086-18-9)**
Benzene, 1,1'-(2,2,2-trichloroethylidene)bis(4-fluoro- **(475-26-3)**
Benzene, 1,1'-(2,2,2-trichloroethylidene)bis(4-methoxy- **(72-43-5)**
Benzene, 1,1'-(2,2,2-trichloroethylidene)bis(4-(methylthio)- (9CI) **(19679-38-0)**
Benzene, 1,2,3-trichloro-4-methoxy- (9CI) **(54135-80-7)**
Benzene, 1,2,3-trichloro-5-methoxy- (9CI) **(54135-82-9)**
Benzene, 1,2,4-trichloro-3-methoxy- (9CI) **(50375-10-5)**
Benzene, 1,2,4-trichloro-5-methoxy- (9CI) **(6130-75-2)**
Benzene, 1,2,5-trichloro-3-methoxy- (9CI) **(54135-81-8)**
Benzene, 1,3,5-trichloro-2-methoxy- (9CI) **(87-40-1)**
Benzene, trichloromethoxy- (9CI) **(53452-80-5)**
Benzene, 1,2,3-trichloro-4-methoxy-5-(2,4,6-trichloro-phenoxy)- (9CI) **(97534-05-9)**
Benzene, 1,3,4-trichloro-2-methoxy-5-(2,4,6-trichloro-phenoxy)- (9CI) **(97534-04-8)**
Benzene, (trichloromethyl)- **(98-07-7)**
Benzene, 1,2,4-trichloro-3-methyl- **(2077-46-5)**
Benzene, 1,2,3-trichloro-4-methyl- (9CI) **(7359-72-0)**
Benzene, 1,2,4-trichloro-5-methyl- (9CI) **(6639-30-1)**
Benzene, 1,3,5-trichloro-2-methyl- (9CI) **(23749-65-7)**
Benzene, trichloromethyl- (9CI) **(30583-33-6)**
Benzene, 1,2,4-trichloro-5-(methylsulfonyl)- (9CI) **(4163-80-8)**
Benzene, 1,2,4-trichloro-5-nitro **(89-69-0)**
Benzene, 1,3,5-trichloro-2-nitro- (9CI) **(18708-70-8)**
Benzene, 1,3,5-trichloro-2-(4-nitrophenoxy)- **(1836-77-7)**
Benzene, 1,2,4-trichloro-5-phenoxy- (9CI) **(52322-80-2)**
Benzene, 1,2,3-trichloro-4,5,6-trimethoxy- (8CI,9CI) **(2539-13-1)**
Benzene, 1,3,5-trichloro-2,4,6-trinitro **(2631-68-7)**
Benzene, tridecyl- (9CI) **(123-02-4)**
Benzene, 1,3,5-triethyl- (9CI) **(102-25-0)**
Benzene, 1,2,4-triethyl- (8CI,9CI) **(877-44-1)**
Benzene, 1,2,3-triethyl- (6CI,7CI,9CI) **(42205-08-3)**
Benzene, triethyl- (Mixed isomers) **(25340-18-5)**
Benzene, 1,3,5-trifluoro- (8CI,9CI) **(372-38-3)**
Benzene, 1,2,3-trifluoro- (6CI,7CI,8CI,9CI) **(1489-53-8)**
Benzene, 1,2,3-trihydroxy- **(87-66-1)**
Benzene, 1,3,5-trihydroxy- **(108-73-6)**
Benzene, trihydroxy **(108-73-6)**
Benzene, triiodo- (9CI) **(61878-55-5)**
Benzene, 1,2,3-triiodo- (7CI,8CI,9CI) **(608-29-7)**
Benzene, 1,3,5-triiodo- (6CI,7CI,8CI,9CI) **(626-44-8)**
Benzene, 1,3,5-triisopropyl- (8CI) **(717-74-8)**
Benzene, 1,3,5-trimethoxy **(621-23-8)**
Benzene, 1,2,3-trimethoxy- (9CI) **(634-36-6)**
Benzene, 1,2,4-trimethoxy-5-propenyl-, (E) **(2883-98-9)**
Benzene, 1,2,4-trimethoxy-5-propenyl-, trans- **(2883-98-9)**
Benzene, 1,2,3-trimethyl **(526-73-8)**
Benzene, 1,2,4-trimethyl **(95-63-6)**
Benzene, 1,2,5-trimethyl- **(95-63-6)**
Benzene, 1,3,5-trimethyl- **(108-67-8)**
Benzene, trimethyl- (Mixed isomers) **(25551-13-7)**
Benzene, 1,3,5-trimethyl-2,4-dinitro- (9CI) **(608-50-4)**
Benzene, (1,2,2-trimethylpropyl)- (6CI,8CI,9CI) **(19262-20-5)**
Benzene, 1,3,5-trimethyl-2,4,6-tris(3,5-di-t-butyl-4-hydroxy-benzyl) **(1709-70-2)**
Benzene, 1,3,5-trinitro **(99-35-4)**
1,2,3-Benzenetriol **(87-66-1)**
1,2,4-Benzenetriol **(533-73-3)**
1,3,5-Benzenetriol **(108-73-6)**
Benzene-1,3,5-triol **(108-73-6)**
Benzene-s-triol **(108-73-6)**
Benzenetriol, dichloro- (9CI) **(94650-90-5)**
1,2,4-Benzenetriol, phenyl **(29222-39-7)**
1,2,3-Benzenetriol, triacetate (9CI) **(525-52-0)**
1,2,3-Benzenetriol, 4,5,6-trichloro **(56961-21-8)**
Benzenetriol, trichloro- (9CI) **(94650-91-6)**
Benzene, 1,3,5-tris(2-chloroethyl)- (8CI) **(18226-46-5)**
Benzene, 1,3,5-tris(2,2-dichloroethyl)- (9CI) **(83682-59-1)**
Benzene, 1,3,5-tris(dichloromethyl)- (9CI) **(56682-87-2)**
Benzene, 1,2,4-tris(1,1-dimethylethyl)- (9CI) **(1459-11-6)**
Benzene, 1,3,5-tris(1,1-dimethylethyl)- (9CI) **(1460-02-2)**
Benzene, 1,3,5-tris((2-ethyl-aziridinyl)-carbonyl) **(7722-73-8)**
Benzene, 1,3,5-tris(1-methylethyl)- (9CI) **(717-74-8)**
Benzene, 1,3,5-tris(2,2,2-trichloroethyl)- (9CI) **(83682-60-4)**
Benzene, undecyl- (9CI) **(6742-54-7)**
Benzene, vinyl- **(100-42-5)**
Benzenol **(108-95-2)**
Benzenosulfochlorek (Polish) **(98-09-9)**
Benzenosulphochloride **(98-09-9)**
Benzenyl chloride **(98-07-7)**
Benzenyl fluoride **(98-08-8)**
Benzenyl trichloride **(98-07-7)**
Benzerythrene **(135-70-6)**
Benzethacil **(1538-09-6)**
Benzethidin **(3691-78-9)**
Benzethidine **(3691-78-9)**
Benzethidinum (Latin) **(3691-78-9)**
Benzethonium **(121-54-0)**
Benzethonium chloride **(121-54-0)**
Benzethonium chloride 1622 **(121-54-0)**
Benzetidina (Spanish) **(3691-78-9)**
Benzetonium chloride **(121-54-0)**
Benzfetamine **(5411-22-3)**
1,2-Benzfluoranthene **(203-33-8)**
10,11-Benzfluoranthene **(205-82-3)**
2,3-Benzfluoranthene **(205-99-2)**
3,4-Benzfluoranthene **(205-99-2)**
Benzo-12,13-fluoranthene **(205-82-3)**
1,2-Benzfluoranthrene **(203-33-8)**
Benzo-gynoestryl **(50-50-0)**
Benzhexol **(144-11-6)**
Benzhormovarine **(50-50-0)**
Benzhydramine **(58-73-1)**
Benzhydramine hydrochloride **(147-24-0)**
Benzhydraminum **(58-73-1)**
Benzhydrazide **(613-94-5)**
Benzhydril **(58-73-1)**
Benzhydroamina **(58-73-1)**
Benzhydrol **(91-01-0)**
Benzhydrol, 4,4'-bis-(dimethylamino) **(119-58-4)**
Benzhydrol, p-chloro- **(119-56-2)**
Benzhydrol, 4-chloro- (8CI) **(119-56-2)**
Benzhydrol, 4,4'-dichloro-α-(ethoxymethyl) **(6012-83-5)**
Benzhydrol, 4,4'-dichloro-α-methyl **(80-06-8)**
Benzhydrol, 2,4'-dichloro-α-(trichloromethyl)- **(10606-46-9)**
Benzhydrol, 4,4'-dichloro-α-(trichloromethyl) **(115-32-2)**
Benzhydrol ether **(574-42-5)**
Benzhydrol, α-(3-piperidinopropyl)- **(972-02-1)**
Benzhydryl **(58-73-1)**
Benzhydryl alcohol **(91-01-0)**
Benzhydryl chloride **(90-99-3)**
o-Benzhydryldimethylaminoethanol **(58-73-1)**
o-Benzhydryldimethylaminoethanol 8-chlorotheophyllinate **(523-87-5)**
Benzhydryl-β-dimethylaminoethylether 8-chlorotheophylline **(523-87-5)**
Benzhydryl ether **(574-42-5)**
(N-Benzhydryl)(N'-methyl)diethylenediamine **(82-92-8)**
N-Benzhydryl-N-methyl piperazine **(82-92-8)**
N-(Benzhydryloksy-etylo)dwumetyloamina (Polish) **(58-73-1)**
2-(Benzhydryloxy)-N,N-dimethylethylamine **(58-73-1)**
2-(Benzhydryloxy)-N,N-dimethylethylamine, Compd. with

8-Chlorotheophylline **(523-87-5)**
2-(Benzhydryloxy)-N,N-dimethylethylamine hydrochloride **(147-24-0)**
Benzidam **(62-53-3)**
Benzidene Yellow ABZ-245 **(6358-85-6)**
Benzidene Yellow WD-266 (Water dispersible) **(6358-85-6)**
Benzidene Yellow YB-1 **(6358-85-6)**
Benzidin (Czech) **(92-87-5)**
Benzidina (Italian) **(92-87-5)**
Benzidine (ACGIH,OSHA) [UN 1885] **(92-87-5)**
Benzidine Lacquer Yellow G **(6358-85-6)**
Benzidine, N,N,N',N'-tetramethyl **(366-29-0)**
Benzidine Orange **(3520-72-7)**
Benzidine Orange 45-2850 **(3520-72-7)**
Benzidine Orange 45-2880 **(3520-72-7)**
Benzidine Orange Toner **(3520-72-7)**
Benzidine Orange wd 265 **(3520-72-7)**
Benzidine Yellow **(6358-85-6)**
Benzidine Yellow 45-2650 **(6358-85-6)**
Benzidine Yellow 45-2680 **(6358-85-6)**
Benzidine Yellow 45-2685 **(6358-85-6)**
Benzidine Yellow E **(6358-85-6)**
Benzidine Yellow G **(6358-85-6)**
Benzidine Yellow GF **(6358-85-6)**
Benzidine Yellow GR **(6358-85-6)**
Benzidine Yellow GT **(6358-85-6)**
Benzidine Yellow GTR **(6358-85-6)**
Benzidine Yellow HG **(6358-85-6)**
Benzidine Yellow HG PLV **(6358-85-6)**
Benzidine Yellow Toner **(6358-85-6)**
Benzidine Yellow Toner YA-8081 **(6358-85-6)**
Benzidine Yellow Toner YT-378 **(6358-85-6)**
3,3'-Benzidinedicarboxylic acid **(2130-56-5)**
Benzidine, 2,2'-dichloro **(84-68-4)**
Benzidine, 3,3'-dichloro **(91-94-1)**
Benzidine, 3,3'-dichloro-, dihydrochloride **(612-83-9)**
Benzidine, dihydrochloride **(531-85-1)**
Benzidine, 3,3'-dihydroxy- **(2373-98-0)**
Benzidine, 3,3'-dimethoxy **(119-90-4)**
Benzidine, 3,3'-dimethoxy-, dihydrochloride **(20325-40-0)**
Benzidine, 3,3'-dimethyl **(119-93-7)**
Benzidine, 3,3'-dimethyl-, dihydrochloride **(612-82-8)**
Benzidine, 2,2'-disulfo- **(117-61-3)**
2,2'-Benzidinedisulfonic acid **(117-61-3)**
Benzidine hydrochloride **(531-85-1)**
Benzidine-sulfate **(531-86-2)**
Benzidine, 2,2',5,5'-tetrachloro **(15721-02-5)**
Benzies **(60-13-9)**
Benzil **(134-81-6)**
Benzilan **(510-15-6)**
Benzilate of pseudotropanol **(537-26-8)**
Benzile (cloruro di) (Italian) **(100-44-7)**
Benzilic-acid **(76-93-7)**
Benzilic acid, 4,4'-dibromo-, isopropyl ester **(18181-80-1)**
Benzilic acid, 4,4'-dichloro-, ethyl ester **(510-15-6)**
Benzilic acid, 4,4'-dichloro-, isopropyl ester **(5836-10-2)**
Benzilic acid β-diethylaminoethyl ester **(302-40-9)**
Benzilic acid, 2-(diethylamino)ethyl ester **(302-40-9)**
Benzilic acid, 1-ethyl-3-piperidyl ester (8CI) **(3567-12-2)**
Benzilic acid, 1-methyl-3-piperidyl ester **(3321-80-0)**
1H-Benzimidazol-5-amine, 2-(4-aminophenyl)- (9CI) **(7621-86-5)**
Benzimidazole **(51-17-2)**
o-Benzimidazole **(51-17-2)**
1H-Benzimidazole (9CI) **(51-17-2)**
Benzimidazole, 2-amino **(934-32-7)**
2-Benzimidazolecarbamic acid, 5-benzoyl-, methyl ester **(31431-39-7)**
2-Benzimidazolecarbamic acid, 1-(butylcarbamoyl)-, methyl ester **(17804-35-2)**
2-Benzimidazolecarbamic acid, 5-butyl-, methyl ester **(14255-87-9)**
2-Benzimidazolecarbamic acid, methyl ester **(10605-21-7)**
Benzimidazole-2-carbamic acid, methyl ester **(10605-21-7)**
2-Benzimidazolecarbamic acid, 5-(phenylthio)-, methyl ester **(43210-67-9)**
2-Benzimidazolecarbamic acid, 5-(2-thenoyl)-, methyl ester **(31430-18-9)**
2-Benzimidazolecarbamic acid, 5-(2-thienoyl)-, methyl ester **(31430-18-9)**
2-Benzimidazolecarbamic acid, 5-(2-thienylcarbonyl)-, methyl ester **(31430-18-9)**
1-Benzimidazolecarboxylic acid, 5,6-dichloro-2-(trifluoromethyl)-, phenyl ester **(14255-88-0)**
1H-Benzimidazole-1-carboxylic acid, 5,6-dichloro-2-(trifluoromethyl)-, phenyl ester **(14255-88-0)**
1H-Benzimidazole, 5,6-dichloro-2-(trifluoromethyl) **(2338-25-2)**
Benzimidazole, 4,5-dichloro-2-(trifluoromethyl) **(3615-21-2)**
Benzimidazole, 1-(2-(diethylamino)ethyl)-2-(p-ethoxybenzyl)-5-nitro- **(911-65-9)**
Benzimidazole, 5,6-dimethyl **(582-60-5)**
1H-Benzimidazole-1-ethanamine, 2-((4-ethoxyphenyl)methyl)-N,N-diethyl-5-nitro- (9CI) **(911-65-9)**
1H-Benzimidazole, 2,2'-(2,5-furandiyl)bis(1-methyl- (9CI) **(4751-43-3)**
1H-Benzimidazole, 2-(2-furanyl)- **(3878-19-1)**
Benzimidazole, 2-(2-furyl) **(3878-19-1)**
Benzimidazole, 2-isopropyl **(5851-43-4)**
Benzimidazole, 2-methyl **(615-15-6)**
Benzimidazole, 6-nitro **(94-52-0)**
1H-Benzimidazole-2-sulfenamide, N-cyclohexyl- (9CI) **(68406-57-5)**
1H-Benzimidazole-5-sulfonic acid, 2-phenyl- (9CI) **(27503-81-7)**
1H-Benzimidazole, 2-(4-thiazolyl)- **(148-79-8)**
Benzimidazole, 2-(4-thiazolyl) **(148-79-8)**
Benzimidazole, 2-(4-thiazolyl)-, monophosphinate **(28558-32-9)**
2-Benzimidazolethiol **(583-39-1)**
2-Benzimidazoleurea **(24370-25-0)**
2-Benzimidazolinone, 1-(1-(3-cyano-3,3-diphenylpropyl)-4-piperidyl)-3-propionyl- **(15301-48-1)**
2-Benzimidazolinone, 1-(1-(3-(p-fluorobenzoyl)propyl)-1,2,3,6-tetrahydro-4-pyridyl) **(548-73-2)**
2-Benzimidazolinthion (Czech) **(583-39-1)**
1H-Benzimidazolium, 2-(7-(diethylamino)-2-oxo-2H-1-benzopyran-3-yl)-1,3-dimethyl-, chloride (9CI) **(29556-33-0)**
1H-Benzimidazolium, 2-(7-(diethylamino)-2-oxo-2H-1-benzopyran-3-yl)-1,3-dimethyl-, methyl sulfate (9CI) **(35869-60-4)**
N-2-(Benzimidazolyl) carbamate **(10605-21-7)**
1H-Benzimidazol-2-ylcarbamic acid methyl ester **(10605-21-7)**
2-(1H-Benzimidazol-2-yl)phenol **(2963-66-8)**
4-(2-Benzimidazolyl)thiazole **(148-79-8)**
Benzimidazol-2-ylurea **(24370-25-0)**
Benziminazole **(51-17-2)**
Benzin **(8030-30-6)**
Benzin B70 **(8030-30-6)**
Benzin (Obs.) **(71-43-2)**
7H-Benz(5,6)indeno(1,2-a)phenanthren-7-one (9CI) **(86854-26-4)**
1H-Benz(e)indolium, 2-(7(1,1-dimethyl-3-(4-sulfobutyl)benz(E)-indolin-2-ylidene)- 1,3,5-heptatrienyl)-1,1-dimethyl-3-(4-sulfobutyl)-, hydroxide, inner salt, sodium salt **(3599-32-4)**
1-Benzine **(91-22-5)**
Benzine (Light petroleum distillate) **(8032-32-4)**
Benzine (Motor fuel) **(86290-81-5)**
Benzine (Obs.) **(71-43-2)**
Benzinoform **(56-23-5)**
Benzinol **(79-01-6)**
Benziodaron **(68-90-6)**
Benziodarone **(68-90-6)**
1,2-Benzisothiazol-3(2H)-one (9CI) **(2634-33-5)**
1,2-Benzisothiazol-3(2H)-one 1,1-dioxide **(81-07-2)**
1,2-Benzisothiazol-3(2H)-one, 1,1-dioxide, ammonium salt (9CI) **(6381-61-9)**
1,2-Benzisothiazol-3(2H)-one, 1,1-dioxide, calcium salt (9CI) **(6485-34-3)**
1,2-Benzisothiazol-3(2H)-one, 2-methyl-, 1,1-dioxide (9CI) **(15448-99-4)**
1,2-Benzisothiazole (8CI,9CI) **(272-16-2)**

1,2-Benzisothiazole, 3-methoxy- (9CI) **(40991-38-6)**
1,2-Benzisothiazolin-3-one (8CI) **(2634-33-5)**
1,2-Benzisothiazolin-3-one, 1,1-dioxide **(81-07-2)**
3-Benzisothiazolinone 1,1-dioxide **(81-07-2)**
1,2-Benzisothiazolin-3-one, 1,1-dioxide, ammonium salt **(6381-61-9)**
1,2-Benzisothiazolin-3-one, 1,1-dioxide, calcium salt **(6485-34-3)**
1,2-Benzisothiazolin-3-one, 1,1-dioxide, sodium salt **(128-44-9)**
1,2-Benzisothiazolin-3-one, 2-methyl-, 1,1-dioxide (8CI) **(15448-99-4)**
Benzisotriazole **(95-14-7)**
1,2-Benzisoxazole-3-carboxylic acid, 6-nitro-, ion(1-) (9CI) **(42540-91-0)**
Benzitramide **(15301-48-1)**
Benz(j)aceanthrylene, 1,2-dihydro- **(479-23-2)**
Benz(j)aceanthrylene, 1,2-dihydro-3-methyl- **(56-49-5)**
Benz(j)fluoranthene **(205-82-3)**
Benz(k)acephenanthrene **(5779-79-3)**
Benz(k)acephenanthrylene, 4,5-dihydro **(5779-79-3)**
9',10'-Benzo-meso-benzanthrone **(5623-32-5)**
Benzmide, 2-nitro- (9CI) **(610-15-1)**
Benzo Azurine G **(2429-71-2)**
Benzo Black Blue BH **(2429-73-4)**
Benzo Black Blue FBH **(2429-73-4)**
Benzo Blue **(72-57-1)**
Benzo Blue 3B **(72-57-1)**
Benzo Blue BBA-CF **(2602-46-2)**
Benzo Blue BBN-CF **(2602-46-2)**
Benzo Blue 3Bs **(72-57-1)**
Benzo Blue GS **(2602-46-2)**
Benzo Brilliant Blue 6BS **(2610-05-1)**
Benzo Brilliant Red 8BS **(6548-29-4)**
Benzo Brown D 3GA-CF **(2586-58-5)**
Benzo Brown M **(2429-82-5)**
Benzo Chrome Black Blue BS **(7082-31-7)**
Benzo Congo Red **(573-58-0)**
Benzo Dark Green B **(3626-28-6)**
Benzo Dark Green BA-CF **(3626-28-6)**
Benzo Deep Black E **(1937-37-7)**
Benzo Deep Black RW **(2429-83-6)**
Benzo Deep Brown NZ **(2429-81-4)**
Benzo Fast Black G **(6428-31-5)**
Benzo Fast Red 8BL **(2610-11-9)**
Benzo Fast Yellow A **(1325-37-7)**
Benzo Flex 2-45 **(120-55-8)**
Benzo Leather Black E **(1937-37-7)**
Benzo Red 3B **(6358-29-8)**
Benzo Sky Blue A-CF **(2429-74-5)**
Benzo Sky Blue S **(2429-74-5)**
Benzo Violet N **(2586-60-9)**
Benzo Viscose Yellow 5GL **(10190-68-8)**
Benzo(a)anthracene **(56-55-3)**
Benzo(a)anthracene 5,6-oxide **(962-32-3)**
11H-Benzo(a)carbazole **(239-01-0)**
4,5-Benzoacenaphthylene **(201-06-9)**
Benzoacenaphthylene **(201-06-9)**
Benzo(a)chrysene **(213-46-7)**
Benzo(a)coronene (8CI,9CI) **(190-70-5)**
3,4-Benzoacridine **(225-51-4)**
Benzo(a)dibenzothiophene **(239-35-0)**
Benzo(a)fluoranthene **(203-33-8)**
11H-Benzo(a)fluorene **(238-84-6)**
Benzo(a)fluorene **(238-84-6)**
Benzo(a)fluorene (9CI) (VAN) **(30777-18-5)**
Benzo(a)fluorenone (9CI) **(116232-62-3)**
11H-Benzo(a)fluoren-11-one (8CI,9CI) **(479-79-8)**
Benzo(a)heptalen-9(5H)-one, 7-acetamido-6,7-dihydro-1,2,3,10-tetra-methoxy- **(64-86-8)**
Benzo(a)heptalen-9(5H)-one, 6,7-dihydro-1,2,3,10-tetramethoxy-7-(methylamino)-, (S)- **(477-30-5)**
Benzoan sodny (Czech) **(532-32-1)**
1,2-Benzoanthracene **(56-55-3)**
Benzoanthracene **(56-55-3)**
Benzoanthrone **(82-05-3)**
Benzo(a)phenanthrene **(218-01-9)**
Benzo(a)phenanthrene **(56-55-3)**
Benzo(a)pyrazine **(91-19-0)**
Benzo(a)pyrene **(50-32-8)**
Benzo(a)pyrene (OSHA) **(50-32-8)**
Benzo(a)pyrene, 6-bromo **(21248-00-0)**
Benzo(a)pyrene, 1,2-dimethyl **(16757-85-0)**
Benzo(a)pyrene, 1,3-dimethyl **(16757-86-1)**
Benzo(a)pyrene, 1,4-dimethyl **(16757-88-3)**
Benzo(a)pyrene, 1,6-dimethyl **(16757-90-7)**
Benzo(a)pyrene, 2,3-dimethyl **(16757-87-2)**
Benzo(a)pyrene, 3,12-dimethyl **(16757-84-9)**
Benzo(a)pyrene, 3,6-dimethyl **(16757-91-8)**
Benzo(a)pyrene, 4,5-dimethyl **(16757-89-4)**
1,6-Benzo(a)pyrenedione **(3067-13-8)**
3,6-Benzo(a)pyrenedione **(3067-14-9)**
6,12-Benzo(a)pyrenedione **(3067-12-7)**
Benzo(a)pyrene-1,6-dione **(3067-13-8)**
Benzo(a)pyrene-3,6-dione **(3067-14-9)**
Benzo(a)pyrene-4,5-dione **(42286-46-4)**
Benzo(a)pyrene-6,12-dione **(3067-12-7)**
Benzo(a)pyrene 4,5-epoxide **(37574-47-3)**
Benzo(a)pyrene, 11-methyl **(16757-80-5)**
Benzo(a)pyrene, 12-methyl **(4514-19-6)**
Benzo(a)pyrene, 2-methyl **(16757-82-7)**
Benzo(a)pyrene, 3-methyl **(16757-81-6)**
Benzo(a)pyrene, 4-methyl **(16757-83-8)**
Benzo(a)pyrene, 6-methyl **(2381-39-7)**
Benzo(a)pyrene, 1-nitro **(70021-99-7)**
Benzo(a)pyrene, 3-nitro **(70021-98-6)**
Benzo(a)pyrene, 6-nitro **(63041-90-7)**
Benzo(a)pyrene, nitro- (9CI) **(70021-42-0)**
Benzo(a)pyrene 4,5-oxide **(37574-47-3)**
Benzo(a)pyrene 6,12-quinone **(3067-12-7)**
Benzo(a)pyrene-1,6-quinone **(3067-13-8)**
Benzo(a)pyrene-3,6-quinone **(3067-14-9)**
Benzo(a)pyrene-4,5-quinone **(42286-46-4)**
Benzo(a)pyrene, 1,3,6-trimethyl **(16757-92-9)**
Benzo(a)pyren-1-ol, 6-nitro- (9CI) **(82039-10-9)**
Benzo(a)pyren-3-ol, 6-nitro- (9CI) **(82039-09-6)**
Benzo(a)pyrimidine **(253-82-7)**
2H-Benzo(a)quinolizine-3-carboxamide, N,N-diethyl-1,3,4,6,7,11b-hexahydro-2-hydroxy-9,10-dimethoxy-, acetate (ester) **(63-12-7)**
2H-Benzo(a)quinolizine-3-carboxamide, 2-(acetyloxy)-N,N-diethyl-1,3,4,6,7,11b-hexahydro-9,10-dimethoxy- **(63-12-7)**
2H-Benzo(a)quinolizine, 3-ethyl-1,3,4,6,7,11b-hexahydro-9,10-dimethoxy-2-((1,2,3,4- tetrahydro-6,7-dimethoxy-1-isoquinolyl)methyl) **(483-18-1)**
Benzoaric acid **(476-66-4)**
Benzoate **(65-85-0)**
Benzoate de denatonium (French) **(3734-33-6)**
Benzoate d'oestradiol (French) **(50-50-0)**
Benzoate d'oestrone (French) **(2393-53-5)**
Benzoate of soda **(532-32-1)**
Benzoate sodium **(532-32-1)**
Benzoato de denatonio (Spanish) **(3734-33-6)**
1-Benzoazo-2-naphthol **(842-07-9)**
Benzoazurine G **(2429-71-2)**
5H-Benzo(b)carbazole (9CI) **(243-28-7)**
Benzo(b)chrysene **(214-17-5)**
5H-Benzo(b)cyclopenta(def)chrysen-5-one (9CI) **(86854-08-2)**
13H-Benzo(b)cyclopenta(def)triphenylen-13-one (9CI) **(86854-12-8)**

4H-Benzo(b)cyclopenta(jkl)triphenylen-4-one (9CI) **(86854-13-9)**
4H-Benzo(b)cyclopenta(mno)chrysen-4-one (9CI) **(86854-09-3)**
Benzo(b)fluoranthene **(205-99-2)**
11H-Benzo(b)fluorene **(243-17-4)**
Benzo(b)fluorene **(30777-19-6)**
11H-Benzo(b)fluoren-11-one **(3074-03-1)**
Benzo(b)fluoren-11-one **(3074-03-1)**
Benzo(b)fluorenone (9CI) **(100647-29-8)**
Benzo(b)furan **(271-89-6)**
Benzoblau 3B **(72-57-1)**
Benzo(b)naphthacene **(135-48-8)**
Benzo(b)naphtho(1,2-d)furan **(205-39-0)**
Benzo(b)naphtho(1,2-d)thiophene **(205-43-6)**
Benzo(b)naphtho(2,3-d)thiophene **(243-46-9)**
Benzo(b)naphtho(2,1-d)thiophene (9CI) **(239-35-0)**
Benzo(b)perylene (8CI,9CI) **(197-70-6)**
Benzo(b)phenanthrene **(56-55-3)**
Benzo(b)pyridine **(91-22-5)**
Benzo(b)quinoline **(260-94-6)**
Benzo(b)thien-4-yl methylcarbamate **(1079-33-0)**
Benzo(b)thiophen-3(2H)-one, 5-chloro-2-(5-chloro-7-methyl-3-oxo-benzo(b)thien-2(3H)-ylidene)-7-methyl- (9CI) **(5462-29-3)**
Benzo(b)thiophen-3(2H)-one, 6-chloro-2-(6-chloro-4-methyl-3-oxo-benzo(b)thien-2(3H)-ylidene)-4- ethyl **(2379-74-0)**
Benzo(b)thiophen-3(2H)-one, 4,7-dichloro-2-(4,7-dichloro-3-oxo-benzo(b)thien-2(3H)-ylidene)- (9CI) **(14295-43-3)**
Benzo(b)thiophene (9CI) **(95-15-8)**
Benzo(b)thiophene, dimethyl- (8CI,9CI) **(30027-44-2)**
Benzo(b)thiophene, ethyl- (9CI) **(54385-63-6)**
Benzo(b)thiophene, methyl- (9CI) **(31393-23-4)**
Benzo(b)thiophene-4-ol **(3610-02-4)**
Benzo(b)thiophene-4-ol, methylcarbamate **(1079-33-0)**
Benzo(b)thiopheneol, nitro- (9CI) **(116211-93-9)**
Benzo(b)triphenylene **(215-58-7)**
Benzocaine **(94-09-7)**
2,3-Benzocarbazole **(243-28-7)**
Benzocarbazole **(243-28-7)**
Benzocarbazole (9CI) **(67526-84-5)**
1H-Benzo(c)carbazole (9CI) **(34777-33-8)**
Benzo(1,2-c:4,5-c')diacridine-6,9,15,18(5H,14H)-tetrone (9CI) **(4424-87-7)**
1H,3H-Benzo(1,2-c:4,5-c')difuran-1,3,5,7-tetrone **(89-32-7)**
Benzo(c)chrysene **(194-69-4)**
Benzo(c)cinnoline (9CI) **(230-17-1)**
4H-Benzo(c)cyclopenta(mno)chrysen-4-one (9CI) **(86854-10-6)**
6H-Benzo[cd]pyren-6-one (6CI, 7CI, 8CI, 9CI) **(3074-00-8)**
Benzo(cd)pyrenone (9CI) **(80398-28-3)**
7H-Benzo(c)fluorene **(205-12-9)**
7H-Benzo(c)fluorene, 7-phenyl- (8CI,9CI) **(32377-10-9)**
7H-Benzo(c)fluoren-7-one (8CI,9CI) **(6051-98-5)**
Benzochinamide **(63-12-7)**
1,4-Benzochinondioxim (Czech) **(105-11-3)**
2,3-Benzochrysene **(214-17-5)**
Benzochrysene (9CI) **(57827-84-6)**
3,4-Benzocinnoline **(230-17-1)**
Benzo(c)phenanthrene **(195-19-7)**
Benzo(c)phenanthrene, methyl- (9CI) **(78328-47-9)**
Benzo(c)picene (8CI,9CI) **(217-37-8)**
Benzo(c)pyridine **(119-65-3)**
Benzo(c)quinoline **(229-87-8)**
Benzo(c)tetraphene **(214-17-5)**
Benzocyclobutene, 1,2-dihydro- **(694-87-1)**
1H-Benzocycloheptene, 2,4aβ,5,6,7,8-hexahydro-3,5,5,9-tetra-methyl-, (+)- (8CI) **(1461-03-6)**
1H-Benzocycloheptene, 2,4a,5,6,7,8-hexahydro-3,5,5,9-tetra-methyl-, (R)- (9CI) **(1461-03-6)**
2(1H)-Benzocyclooctenone, decahydro-10a-methyl-, trans- (9CI) **(55103-68-9)**
7H-Benzo(de)anthracen-7-one **(82-05-3)**
4H-Benzo(def)carbazole (8CI,9CI) **(203-65-6)**
Benzo(d,e,f)chrysene **(50-32-8)**
4H-Benzo(def)cyclopenta(mno)chrysen-4-one (9CI) **(86853-92-1)**
Benzo(def)dibenzothiophene **(30796-92-0)**
Benzo(def)phenanthrene **(129-00-0)**
7H-Benzo(de)naphthacen-7-one (8CI,9CI) **(5623-32-5)**
Benzo(6,7)-1,4-diazepino-(5,4-b)-oxazol-6-one, 10-chloro-2,3,5,6,7,11b-hexahydro-2-methyl-11b-phenyl- **(24143-17-7)**
3H-1,4-Benzodiazepin-2-amine, 7-chloro-N-methyl-5-phenyl-, 4-oxide (9CI) **(58-25-3)**
Benzodiazepine **(12794-10-4)**
1H-1,5-Benzodiazepine-2,4(3H,5H)-dione, 7-chloro-1-methyl-5-phenyl **(22316-47-8)**
1H-1,4-Benzodiazepine-3-carboxylic acid, 7-chloro-2,3-dihydro-2-oxo-5-phenyl- **(23887-31-2)**
1H-1,4-Benzodiazepine-3-carboxylic acid, 7-chloro-5-(2-fluorophenyl)-2,3-dihydro-2-oxo-, ethyl ester **(29177-84-2)**
1H-1,4-Benzodiazepine, 7-chloro-2,3-dihydro-1-methyl-5-phenyl- **(2898-12-6)**
3H-1,4-Benzodiazepine, 7-chloro-2-(methylamino)-5-phenyl-, 4-oxide **(58-25-3)**
1H-1,4-Benzodiazepine, 2,3-dihydro-7-chloro-1-methyl-5-phenyl- **(2898-12-6)**
2H-1,4-Benzodiazepine-2-thione, 7-chloro-5-(2-fluorophenyl)-1,3-dihydro-1-(2,2,2-trifluoroethyl)- (9CI) **(36735-22-5)**
2H-1,4-Benzodiazepin-2-one, 7-bromo-1,3-dihydro-5-(2-pyridinyl)- (9CI) **(1812-30-2)**
2H-1,4-Benzodiazepin-2-one, 7-bromo-1,3-dihydro-5-(2-pyridyl)- (8CI) **(1812-30-2)**
2H-1,4-Benzodiazepin-2-one, 7-chloro-5-(2-chloro-phenyl)-1,3-dihydro- (9CI) **(2894-67-9)**
2H-1,4-Benzodiazepin-2-one, 7-chloro-5-(o-chloro-phenyl)-1,3-dihydro- (8CI) **(2894-67-9)**
2H,1,4-Benzodiazepin-2-one, 7-chloro-5-(o-chloro-phenyl)-1,3-dihydro-3-hydroxy **(846-49-1)**
2H-1,4-Benzodiazepin-2-one, 7-chloro-5-(2-chloro-phenyl)-1,3-dihydro-3-hydroxy-1-methyl- **(848-75-9)**
2H-1,4-Benzodiazepin-2-one, 7-chloro-5-(1-cyclo-hexen-1-yl)-1,3-dihydro-1-methyl- **(10379-14-3)**
2H-1,4-Benzodiazepin-2-one, 7-chloro-1-(cyclopropyl-methyl)-1,3-dihydro-5-phenyl- (8CI,9CI) **(2955-38-6)**
2H-1,4-Benzodiazepin-2-one, 7-chloro-1-(2-(diethylamino)-ethyl)-5-(o-fluorophenyl)-, 1,3-dihydro-, dihydrochloride **(1172-18-5)**
2H-1,4-Benzodiazepin-2-one, 7-chloro-1,3-dihydro-3-hydroxy-1-methyl-5-phenyl **(846-50-4)**
2H-1,4-Benzodiazepin-2-one, 7-chloro-1,3-dihydro-3-hydroxy-5-phenyl **(604-75-1)**
2H-1,4-Benzodiazepin-2-one, 7-chloro-1,3-dihydro-1-methyl-5-phenyl **(439-14-5)**
2H-1,4-Benzodiazepin-2-one, 7-chloro-1,3-dihydro-5-phenyl **(1088-11-5)**
2H-1,4-Benzodiazepin-2-one, 7-chloro-1,3-dihydro-5-phenyl-1-(2-propynyl)- **(52463-83-9)**
2H-1,4-Benzodiazepin-2-one, 7-chloro-5-(o-fluorophenyl)-1,3-dihydro-1-methyl- **(3900-31-0)**
2H-1,4-Benzodiazepin-2-one, 7-chloro-5-(2-fluorophenyl)-1,3-dihydro-1-methyl- (9CI) **(3900-31-0)**
2-H-1,4-Benzodiazepin-2-one, 5-(o-chlorophenyl)-1,3-dihydro-7-nitro **(1622-61-3)**
2H-1,4-Benzodiazepin-2-one, 1,3-dihydro-7-bromo-5-(2-pyridyl)- **(1812-30-2)**
2H-1,4-Benzodiazepin-2-one, 1,3-dihydro-7-chloro-5-(2-chlorophenyl)- **(2894-67-9)**
2H-1,4-Benzodiazepin-2-one, 1,3-dihydro-7-chloro-5-(o-chloro-phenyl)-3-hydroxy-1-methyl- **(848-75-9)**
2H-1,4-Benzodiazepin-2-one, 1,3-dihydro-7-chloro-1-(cyclopropyl-methyl)-5-phenyl- **(2955-38-6)**
2H-1,4-Benzodiazepin-2-one, 1,3-dihydro-7-chloro-1-(2-(diethyl-amino)ethyl)-5-(o-fluorophenyl) **(17617-23-1)**
2H-1,4-Benzodiazepin-2-one, 1,3-dihydro-7-chloro-5-(o-fluoro-phenyl)-1-methyl **(3900-31-0)**

2H-1,4-Benzodiazepin-2-one, 1,3-dihydro-7-chloro-3-hydroxy-1-methyl-5-phenyl- **(846-50-4)**
2H-1,4-Benzodiazepin-2-one, 1,3-dihydro-7-chloro-5-phenyl- **(1088-11-5)**
2H-1,4-Benzodiazepin-2-one, 1,3-dihydro-7-chloro-5-phenyl-1-(2-propynyl)- **(52463-83-9)**
2H-1,4-Benzodiazepin-2-one, 1,3-dihydro-7-chloro-5-phenyl-1-(2,2,2-trifluoroethyl)- **(23092-17-3)**
2H-1,4-Benzodiazepin-2-one, 1,3-dihydro-5-(2-fluorophenyl)-1-methyl-7-nitro- **(1622-62-4)**
2H-1,4-Benzodiazepin-2-one, 1,3-dihydro-1-methyl-7-nitro-5-phenyl- **(2011-67-8)**
2H-1,4-Benzodiazepin-2-one, 1,3-dihydro-7-nitro-5-phenyl **(146-22-5)**
2H-1,4-Benzodiazepin-2-one, 5-(2-fluorophenyl)-1,3-dihydro-1-methyl-7-nitro- **(1622-62-4)**
1,3-Benzodiazine **(253-82-7)**
1,4-Benzodiazine **(91-19-0)**
2,3-Benzodiazine **(253-52-1)**
1,3-Benzodiazole **(51-17-2)**
1,2-Benzodihydropyrone **(119-84-6)**
1,4-Benzodioxan (8CI) **(493-09-4)**
1,4-Benzodioxane **(493-09-4)**
4H-(1.3.2)Benzodioxaphosphorine, 2-methoxy-, 2-sulfide **(3811-49-2)**
1,4-Benzodioxin, 2,3-dihydro- (9CI) **(493-09-4)**
1,4-Benzodioxin, octahydro-2-methylene-, trans- (9CI) **(38653-34-8)**
1,3-Benzodioxole, 5-allyl- **(94-59-7)**
1,3-Benzodioxole, 6-(bis(2-(2-butoxyethoxy)ethoxy))methyl **(5281-13-0)**
1,3-Benzodioxole, 5-((2-(2-butoxyethoxy)ethoxy)methyl)-6-propyl- **(51-03-6)**
1,3-Benzodioxole-5-carboxaldehyde **(120-57-0)**
1,3-Benzodioxole-5-carboxaldehyde, 7-methoxy- **(5780-07-4)**
1,3-Benzodioxole-5-carboxylic acid (9CI) **(94-53-1)**
1,3-Benzodioxole, 2,2-dimethyl-4-(N-methylaminocarboxylato)- **(22781-23-3)**
1,3-Benzodioxole, 2,2-dimethyl-4-(N-methylcarbamato)- **(22781-23-3)**
1,3-Benzodioxole-5-ethanamine, α-methyl- (9CI) **(4764-17-4)**
1,3-Benzodioxole, 5-(1-(2-(2-ethoxyethoxy)ethoxy)ethoxy)- **(51-14-9)**
1,3-Benzodioxole, 5-((5-(3-ethyl-3-methyloxiranyl)-3-methyl-2-pentenyl)oxy)- **(21213-69-4)**
1,3-Benzodioxole-5-methanol (9CI) **(495-76-1)**
1,3-Benzodioxole-5-methanol, α-epoxyethyl **(59901-91-6)**
1,3-Benzodioxole-5-methanol, α-ethenyl **(5208-87-7)**
1,3-Benzodioxole-5-methanol, α-(oxiranyl)-, acetate **(59901-90-5)**
1,3-Benzodioxole-5-methanol, α-oxiranyl-, acetate (9CI) **(59901-90-5)**
1,3-Benzodioxole-5-(2-propen-1-ol) **(5208-87-7)**
1,3-Benzodioxole, 5-(1-propenyl)- **(120-58-1)**
(1,3)-Benzodioxolo(5,6-c)-1,3-dioxolo(4,5-i)phenanthridinium, 13-methyl- (9CI) **(2447-54-3)**
1,3-Benzodioxol-4-ol, 2,2-dimethyl-, methylcarbamate **(22781-23-3)**
1,3-Benzodioxol-2-one, 5-(1,1-dimethylethyl)- (9CI) **(54815-21-3)**
2-(1,3-Benzodioxol-5-yl)ethyl octyl sulfoxide **(120-62-7)**
1,3-Benzodioxol-5-yl-1-oxo-2,4-pentadienyl-piperine **(94-62-2)**
Benzododecinium bromide **(7281-04-1)**
Benzo(d)pyridazine **(253-52-1)**
Benzoepin **(115-29-7)**
α-Benzoepin **(959-98-8)**
β-Benzoepin **(33213-65-9)**
Benzoesaeure (German) **(65-85-0)**
Benzoesaeure (na-salz) (German) **(532-32-1)**
Benzoestrofol **(50-50-0)**
Benzo(fgh)trinaphthylene (8CI,9CI) **(31541-03-4)**
9H-Benzo(fg)naphthacen-9-one (9CI) **(86854-05-9)**
Benzoflex S-552 **(4196-86-5)**
Benzoflex P 200 **(9004-86-8)**
Benzoflex P-600 **(9004-86-8)**
Benzoflex T 150 **(120-56-9)**
1,2-Benzofluoranthene **(203-33-8)**
10,11-Benzofluoranthene **(205-82-3)**
11,12-Benzofluoranthene **(207-08-9)**
2,13-Benzofluoranthene **(203-12-3)**
2,3-Benzofluoranthene **(205-99-2)**
3,4-Benzofluoranthene **(205-99-2)**
7,10-Benzofluoranthene **(203-12-3)**
7,8-Benzofluoranthene **(205-82-3)**
8,9-Benzofluoranthene **(207-08-9)**
Benzo(e)fluoranthene **(205-99-2)**
Benzofluoranthene (9CI) **(56832-73-6)**
2,3-Benzofluoranthrene **(205-99-2)**
1,2-Benzofluorene **(238-84-6)**
2,3-Benzofluorene **(243-17-4)**
3,4-Benzofluorene **(205-12-9)**
Benzofluorene (9CI) **(61089-87-0)**
Benzofluorenone (9CI) **(76723-60-9)**
Benzofoline **(50-50-0)**
Benzoform Black BCN-CF **(1937-37-7)**
Benzoform Black RRA-CF **(6428-31-5)**
Benzo(f)quinoline **(85-02-9)**
Benzo(f)quinoline, 3-methyl- (9CI) **(85-06-3)**
Benzo(f)quinolinol (9CI) **(91311-51-2)**
Benzofume **(103-33-3)**
Benzofur D **(106-50-3)**
Benzofur GG **(95-55-6)**
Benzofur MT **(95-80-7)**
Benzofur P **(123-30-8)**
2,3-Benzofuran **(271-89-6)**
Benzofuran **(271-89-6)**
5-Benzofuranacrylic acid, 4-hydroxy-β,2-dimethyl-, δ-lactone **(4063-41-6)**
5-Benzofuranacrylic acid, 6-hydroxy-7-methoxy-, δ-lactone **(298-81-7)**
5-Benzofuranacrylic acid, 6-hydroxy-7-((3-methyl-2-butenyl)oxy)-, δ-lactone **(482-44-0)**
7-Benzofuranamine, 2,3-dihydro-2,2-dimethyl- (9CI) **(68298-46-4)**
Benzofuran, 5-chloro-2-methyl- (9CI) **(42180-82-5)**
7-Benzofurandiazonium, 2,3-dihydro-2,2-dimethyl-, hydrogen sulfate (9CI) **(38420-60-9)**
7-Benzofurandiazonium, 2,3-dihydro-2,2-dimethyl-, sulfate (2:1) (9CI) **(68298-47-5)**
Benzofuran, 2,3-dihydro- (9CI) **(496-16-2)**
Benzofuran, 2,3-dihydro-2,2-dimethyl-7-(N-(N-methyl-N-butoxycarbonylaminothio)-N- methylcarbamoyloxy) **(65907-30-4)**
Benzofuran, 2,3-dihydro-2,2-dimethyl-7-nitro- (9CI) **(13414-55-6)**
Benzofuran, dihydro-2-methyl- (9CI) **(103453-97-0)**
Benzofuran, 3-(3,5-diiodo-4-hydroxybenzoyl)-2-ethyl- **(68-90-6)**
Benzofuran, 2,3-dimethyl- (9CI) **(3782-00-1)**
Benzofuran, 4,7-dimethyl- (8CI,9CI) **(28715-26-6)**
Benzofuran, dimethyl- (8CI,9CI) **(25586-39-4)**
3,7-Benzofurandiol, 2,3-dihydro-2,2-dimethyl-, 7-(methylcarbamate) (9CI) **(16655-82-6)**
Benzofuran, 2-methyl- **(4265-25-2)**
Benzofuran, methyl- (8CI,9CI) **(25586-38-3)**
7-Benzofuranol, 2,3-dihydro-2,2-dimethyl- **(1563-38-8)**
5-Benzofuranol, 2-ethoxy-2,3-dihydro-3,3-dimethyl-, methanesulfonate, (+-) **(26225-79-6)**
3(2H)-Benzofuranone (8CI,9CI) **(7169-34-8)**
2(3H)-Benzofuranone, 3,3-dimethyl- (8CI,9CI) **(13524-76-0)**
3(2H)-Benzofuranone, 4,7-dimethyl- (8CI,9CI) **(20895-45-8)**
3(2H)-Benzofuranone, 2,2-dimethyl-7-(((methylamino)-carbonyl)oxy)- (9CI) **(16709-30-1)**
3(2H)-Benzofuranone, 6-methyl- (7CI,8CI,9CI) **(20895-41-4)**
3(2H)-Benzofuranone, 7-methyl- (7CI,8CI,9CI) **(669-04-5)**
2(4H)-Benzofuranone, 5,6,7,7a-tetrahydro-6-hydroxy-4,4,7a-trimethyl-, cis- (8CI,9CI) **(1133-03-5)**
2(4H)-Benzofuranone, 5,6,7,7a-tetrahydro-4,4,7a-trimethyl-, (R)- (9CI) **(17092-92-1)**
2(4H)-Benzofuranone, 5,6,7,7a-tetrahydro-4,4,7a-trimethyl-, (S)- **(17092-92-1)**
2(4H)-Benzofuranone, 5,6,7,7a-tetrahydro-4,4,7a-trimethyl-(VAN) (8CI) **(17092-92-1)**
2(4H)-Benzofuranone, 5,6,7,7a-tetrahydro-4,4,7a-trimethyl-

(VAN)(9CI) **(15356-74-8)**
Benzofurazan, dinitro-, ion(1-), 1-oxide, potassium **(29267-75-2)**
Benzofurazan, 4,6-dinitro-, 1-oxide, potassium salt (8CI) **(29267-75-2)**
4-Benzofurazanol, 1,4-dihydro-5,7-dinitro-, 3-oxide, ion(1-), potassium (9CI) **(29267-75-2)**
Benzofurazan N-oxide **(480-96-6)**
Benzofurazan oxide **(480-96-6)**
Benzofurazan, 1-oxide (9CI) **(480-96-6)**
Benzofurfuran **(271-89-6)**
Benzofuroline **(10453-86-8)**
Benzofuroxan **(480-96-6)**
Benzofuroxane **(480-96-6)**
Benzo(g)-1,3-benzodioxolo(5,6-a)quinolizinium, 5,6-dihydro-9,10-dimethoxy-, sulfate (2:1) **(316-41-6)**
Benzo(g)chrysene **(196-78-1)**
Benzo(ghi)cyclopenta(cd)perylenone (9CI) **(84665-41-8)**
1H-Benzo(ghi)cyclopenta(pqr)perylen-1-one (9CI) **(86854-07-1)**
Benzo(ghi)fluoranthene **(203-12-3)**
Benzo(ghi)perylene **(191-24-2)**
Benzo(ghi)perylene, methyl **(41699-09-6)**
Benzoglyoxaline **(51-17-2)**
Benzoguanamine **(91-76-9)**
Benzoguanide **(33957-63-0)**
Benzoguanimine **(91-76-9)**
γ-Benzohexachloride **(58-89-9)**
7H-Benzo(hi)chrysen-7-one (9CI) **(80440-44-4)**
Benzo(h)naphtho(8,1,2-cde)cinnoline, nitro- (9CI) **(116212-02-3)**
Benzo(h)naphtho(1,2,3,4-rst)pentaphene (8CI,9CI) **(31541-02-3)**
Benzo(h)quinoline **(230-27-3)**
Benzohydrazide **(613-94-5)**
Benzohydrazine **(613-94-5)**
Benzohydrol **(91-01-0)**
Benzohydrol ether **(574-42-5)**
Benzohydroquinone **(123-31-9)**
Benzohydroxamate **(495-18-1)**
Benzohydroxamic-acid **(495-18-1)**
Benzohydroxamic acid, N-(p-chlorophenyl)- (8CI) **(1528-82-1)**
2-(Benzohydryloxy)-N,N-dimethylethylamine **(58-73-1)**
Benzoic-1'-13C acid **(3880-99-7)**
Benzoic Methyl Violet Lake **(1325-82-2)**
Benzoic-acid **(65-85-0)**
Benzoic acid, Compd. with cyclohexanamine (1:1) (9CI) **(3129-92-8)**
Benzoic acid N,N-diethylamide **(1696-17-9)**
Benzoic acid, 2-((6-O-β-D-xylopyranosyl-β-D-glucopyranosyl)oxy)-, methyl ester (9CI) **(490-67-5)**
Benzoic acid, p-acetamido- (8CI) **(556-08-1)**
Benzoic acid, p-acetonyl- (8CI) **(15482-54-9)**
Benzoic acid, 4-acetyl- (9CI) **(586-89-0)**
Benzoic acid, p-acetyl- (8CI) **(586-89-0)**
Benzoic acid, 2-(acetylamino)- (9CI) **(89-52-1)**
Benzoic acid, 4-(acetylamino)- (9CI) **(556-08-1)**
Benzoic acid, 3-(acetylamino)-5-((acetylamino)methyl)-2,4,6-triiodo **(440-58-4)**
Benzoic acid, 2-(acetylamino)-, methyl ester (9CI) **(2719-08-6)**
Benzoic acid, 4-(acetyloxy)- (9CI) **(2345-34-8)**
Benzoic acid, 2-(acetyloxy)-, Mixed with 3,7-dihydro-1,3,7-trimethyl-1H-purine-2,6-dione, and N-(4-ethoxyphenyl)-acetamide **(8003-03-0)**
Benzoic acid, allyl ester (8CI) **(583-04-0)**
Benzoic acid, aluminum salt (9CI) **(555-32-8)**
Benzoic acid amide **(55-21-0)**
Benzoic acid, 4-amino- **(150-13-0)**
Benzoic acid, m-amino **(99-05-8)**
Benzoic acid, o-amino- **(118-92-3)**
Benzoic acid, p-amino **(150-13-0)**
Benzoic acid, amino- (9CI) **(1321-11-5)**
Benzoic acid, 5-((4'-((6-amino-5-(1H-benzotriazol-5-ylazo)-1-hydroxy-3-sulfo-2-naphthalenyl)azo)-3,3'-dimethoxy(1,1'-biphenyl)-4-yl)azo)-2-hydroxy-4-methyl-, disodium salt (9CI) **(70210-28-5)**
Benzoic acid, 5-((4-aminobenzoyl)amino)-2-hydroxy-3-methyl-(9CI) **(6265-15-2)**
Benzoic acid, p-amino-, butyl ester **(94-25-7)**
Benzoic acid, 2-amino-, butyl ester (9CI) **(7756-96-9)**
Benzoic acid, 2-(aminocarbonyl)- (9CI) **(88-97-1)**
Benzoic acid, 2-((2-amino-6-((4'-((3-carboxy-4-hydroxyphenyl)azo)-3,3'-dimethoxy(1,1'-biphenyl)-4-yl)azo)-5-hydroxy-7-sulfo-1-naphthalenyl)azo)-5-nitro-, trisodium salt (9CI) **(6739-62-4)**
Benzoic acid, 4-amino-2-chloro **(2457-76-3)**
Benzoic acid, 2-amino-5-chloro- (9CI) **(635-21-2)**
Benzoic acid, 4-amino-3-chloro- (8CI,9CI) **(2486-71-7)**
Benzoic acid, 4-amino-2-chloro-, 2-(diethylamino)ethyl ester **(133-16-4)**
Benzoic acid, 2-amino-, cyclohexyl ester (9CI) **(7779-16-0)**
Benzoic acid, 3-amino-2,5-dichloro **(133-90-4)**
Benzoic acid, 4-amino-3,5-dichloro **(56961-25-2)**
Benzoic acid, 3-amino-2,5-dichloro-, methyl ester (9CI) **(7286-84-2)**
Benzoic acid, p-amino-, 2-(diethylamino)ethyl ester **(59-46-1)**
Benzoic acid, 4-amino-, 2-(diethylamino)ethyl ester, monohydrochloride **(51-05-8)**
Benzoic acid, p-amino-, 2-(diethylamino)ethyl ester, monohydrochloride **(51-05-8)**
Benzoic acid, 2-amino-, ethyl ester **(582-33-2)**
Benzoic acid, o-amino-, ethyl ester **(87-25-2)**
Benzoic acid, p-amino-, ethyl ester **(94-09-7)**
Benzoic acid, 3-amino-, ethyl ester (9CI) **(582-33-2)**
Benzoic acid, m-amino-, ethyl ester (8CI) **(582-33-2)**
Benzoic acid, 4-amino-, heptyl ester (9CI) **(14309-40-1)**
Benzoic acid, p-amino-, heptyl ester (8CI) **(14309-40-1)**
Benzoic acid, 2-amino-3-hydroxy **(548-93-6)**
Benzoic acid, 4-amino-2-hydroxy- **(65-49-6)**
Benzoic acid, 2-((2-amino-5-hydroxy-6-((4'-((2-hydroxy-6-sulfo-1-naphthalenyl)azo)-3,3'-dimethoxy(1,1'-biphenyl)-4-yl)azo)-7-sulfo-1-naphthalenyl)azo)-5-nitro-, trisodium salt (9CI) **(71566-41-1)**
Benzoic acid, 5-((4'-((8-amino-1-hydroxy-7-((4-nitrophenyl)-azo)-3,6-disulfo-2-naphthalenyl)azo)(1,1'-biphenyl)-4-yl)-azo)-2-hydroxy-, trisodium salt (9CI) **(5422-17-3)**
Benzoic acid, 3-amino-2-hydroxy-5-sulfo- (9CI) **(6201-86-1)**
Benzoic acid, 5-amino-2-hydroxy-3-sulfo- (9CI) **(6201-87-2)**
Benzoic acid, 5-((4'-((2-amino-8-hydroxy-6-sulfo-1-naphthalenyl)azo)(1,1'-biphenyl)-4-yl)azo)-2-hydroxy-, disodium salt (9CI) **(2429-84-7)**
Benzoic acid, 5-((4'-((7-amino-1-hydroxy-3-sulfo-2-naphthalenyl)azo)(1,1'-biphenyl)-4-yl)azo)-2-hydroxy-, disodium salt (9CI) **(2429-82-5)**
Benzoic acid, 5-((4-((4-((6-amino-1-hydroxy-3-sulfo-2-naphthalenyl)azo)-1-naphthalenyl)azo)-6-sulfo-1-naphthalenyl)azo)-2-hydroxy-, trisodium salt (9CI) **(3841-15-4)**
Benzoic acid, 4-amino-3-methyl- (9CI) **(2486-70-6)**
Benzoic acid, p-amino-, methyl ester **(619-45-4)**
Benzoic acid, 2-amino-, methyl ester (9CI) **(134-20-3)**
Benzoic acid, 2-amino-, 2-methylpropyl ester **(7779-77-3)**
Benzoic acid, 4-amino-, monopotassium salt (9CI) **(138-84-1)**
Benzoic acid, p-amino-, monosodium salt **(555-06-6)**
Benzoic acid, 4-amino-, monosodium salt (9CI) **(555-06-6)**
Benzoic acid, 2-amino-4-nitro- (9CI) **(619-17-0)**
Benzoic acid, 2-amino-5-nitro- (9CI) **(616-79-5)**
Benzoic acid, 4-amino-, octyl ester (9CI) **(14309-41-2)**
Benzoic acid, 4-amino-, pentyl ester (9CI) **(13110-37-7)**
Benzoic acid, p-amino-, pentyl ester (8CI) **(13110-37-7)**
Benzoic acid, 5-((4-aminophenyl)azo)-2-hydroxy-, monosodium salt (9CI) **(6470-98-0)**
Benzoic acid, 2-amino-, 2-phenylethyl ester **(133-18-6)**
Benzoic acid, 2-amino-, 3-phenyl-2-propenyl ester **(87-29-6)**
Benzoic acid, 2-amino-, 2-propenyl ester (9CI) **(7493-63-2)**

Benzoic acid, p-amino-, propyl ester (9CI) **(94-12-2)**
Benzoic acid, 4-amino-, propyl ester (7CI,8CI) **(94-12-2)**
Benzoic acid, 2-amino-5-sulfo- (9CI) **(3577-63-7)**
Benzoic acid, 5-((4'-((1-amino-4-sulfo-2-naphthalenyl)azo)-(1,1'-biphenyl)-4-yl)azo)-2-hydroxy-, disodium salt (9CI) **(2429-79-0)**
Benzoic acid, 5-(aminosulfonyl)-4-chloro-2-((2-furanylmethyl)-amino)- (9CI) **(54-31-9)**
Benzoic acid, ammonium salt **(1863-63-4)**
Benzoic acid, anhydride (9CI) **(93-97-0)**
Benzoic acid anilide **(93-98-1)**
Benzoic acid azide **(582-61-6)**
Benzoic acid, barium salt (9CI) **(533-00-6)**
Benzoic acid, 2-benzoyl- **(85-52-9)**
Benzoic acid, o-benzoyl **(85-52-9)**
Benzoic acid, benzoyl- (9CI) **(27458-06-6)**
Benzoic acid, 2-benzoyl-, methyl ester (9CI) **(606-28-0)**
Benzoic acid, o-benzoyl-, methyl ester (8CI) **(606-28-0)**
Benzoic acid, benzyl ester **(120-51-4)**
Benzoic acid, 4-(((1,1'-biphenyl)-4-ylmethylene)amino)-, ethyl ester (9CI) **(3782-80-7)**
Benzoic acid, 3,5-bis(acetylamino)-2,4,6-triiodo- (9CI) **(117-96-4)**
Benzoic acid, 3,5-bis(1,1-dimethylethyl)-4-hydroxy-, 2,4-bis-(1,1-dimethylethyl)phenyl ester (9CI) **(4221-80-1)**
Benzoic acid, 4-(bis(2-hydroxypropyl)amino)-, ethyl ester **(58882-17-0)**
Benzoic acid, p-bromo **(586-76-5)**
Benzoic acid, 2-bromo- (9CI) **(88-65-3)**
Benzoic acid, 3-bromo- (9CI) **(585-76-2)**
Benzoic acid, 4-bromo- (9CI) **(586-76-5)**
Benzoic acid, m-bromo- (8CI) **(585-76-2)**
Benzoic acid, o-bromo- (8CI) **(88-65-3)**
Benzoic acid, bromochloro- (8CI,9CI) **(25638-14-6)**
Benzoic acid, 4-bromo-, ethyl ester (9CI) **(5798-75-4)**
Benzoic acid, p-bromo-, ethyl ester (8CI) **(5798-75-4)**
Benzoic acid, 4-(bromomethyl)- (9CI) **(6232-88-8)**
Benzoic acid, 2-bromo-, methyl ester (9CI) **(610-94-6)**
Benzoic acid, 3-bromo-, methyl ester (9CI) **(618-89-3)**
Benzoic acid, 4-bromo-, methyl ester (9CI) **(619-42-1)**
Benzoic acid, o-bromo-, methyl ester (8CI) **(610-94-6)**
Benzoic acid, p-bromo-, methyl ester (8CI) **(619-42-1)**
Benzoic acid, p-tert-butyl **(98-73-7)**
Benzoic acid, 4-butyl- (9CI) **(20651-71-2)**
Benzoic acid, 4-(butylamino)-, 2-(dimethylamino)ethyl ester, monohydrochloride **(136-47-0)**
Benzoic acid, p-(butylamino)-, 2-(dimethylamino)ethyl ester, monohydrochloride **(136-47-0)**
Benzoic acid n-butyl ester **(136-60-7)**
Benzoic acid, butyl ester **(136-60-7)**
Benzoic acid, cadmium salt (9CI) **(3026-22-0)**
Benzoic acid, calcium salt (9CI) **(2090-05-3)**
Benzoic acid, o-carbamoyl- **(88-97-1)**
Benzoic acid, 3-(4-carboxy-2-methoxyphenoxy)-4-hydroxy-5-methoxy- (9CI) **(2555-99-9)**
Benzoic acid, 2-(((2-carboxyphenyl)amino)sulfonyl)-5-((4,5-dihydro-3-methyl-5-oxo-1-phenyl-1H-pyrazol-4-yl)azo)-, disodium salt (9CI) **(10482-43-6)**
Benzoic acid, chloride **(98-88-4)**
Benzoic acid, chlorinated **(12002-27-6)**
Benzoic acid, 2-chloro- **(118-91-2)**
Benzoic acid, 3-chloro- **(535-80-8)**
Benzoic acid, m-chloro **(535-80-8)**
Benzoic acid, o-chloro **(118-91-2)**
Benzoic acid, p-chloro **(74-11-3)**
Benzoic acid, 4-chloro- (9CI) **(74-11-3)**
Benzoic acid, chloro- (8CI,9CI) **(26264-09-5)**
Benzoic acid, chloro-, benzyl ester (6CI) **(108548-66-9)**
Benzoic acid, chloro-, (dichlorophenyl)methyl ester (9CI) **(108548-72-7)**
Benzoic acid, chloro-3,4-dihydroxy- (9CI) **(79188-95-7)**
Benzoic acid, chlorodihydroxy- (9CI) **(103339-65-7)**
Benzoic acid, 2-chloro-4,5-dimethoxy-, methyl ester (9CI) **(30714-88-6)**
Benzoic acid, 3-chloro-4,5-dimethoxy-, methyl ester (9CI) **(108544-98-5)**
Benzoic acid, 4-chloro-3,5-dinitro- (9CI) **(118-97-8)**
Benzoic acid, 5-(2-chloro-4-fluorophenoxy)-2-nitro-, methyl ester (9CI) **(51282-69-0)**
Benzoic acid, 4-chloro-, hydrazide (9CI) **(536-40-3)**
Benzoic acid, p-chloro-, hydrazide (8CI) **(536-40-3)**
Benzoic acid, 5-chloro-2-hydroxy- **(321-14-2)**
Benzoic acid, 3-chloro-4-hydroxy- (9CI) **(3964-58-7)**
Benzoic acid, 2-chloro-4-hydroxy-5-methoxy- (9CI) **(62936-24-7)**
Benzoic acid, 3-chloro-4-hydroxy-5-methoxy- (9CI) **(62936-23-6)**
Benzoic acid, chloro-4-hydroxy-3-methoxy- (9CI) **(69845-52-9)**
Benzoic acid, 2-chloro-5-((2-hydroxy-1-naphthalenyl)azo)-4-sulfo-, calcium sodium salt (2:1:2) (9CI) **(5850-80-6)**
Benzoic acid, p-(chloromercuri)- **(59-85-8)**
Benzoic acid, 4-chloro-, mercury(2+) salt (9CI) **(15516-76-4)**
Benzoic acid, p-chloro-, mercury(2+) salt (8CI) **(15516-76-4)**
Benzoic acid, m-chloro-, methyl ester **(2905-65-9)**
Benzoic acid, 2-chloro-, methyl ester (9CI) **(610-96-8)**
Benzoic acid, 3-chloro-, methyl ester (9CI) **(2905-65-9)**
Benzoic acid, 4-chloro-, methyl ester (9CI) **(1126-46-1)**
Benzoic acid, o-chloro-, methyl ester (8CI) **(610-96-8)**
Benzoic acid, p-chloro-, methyl ester (8CI) **(1126-46-1)**
Benzoic acid, 4-(chloromethyl)-, methyl ester (9CI) **(34040-64-7)**
Benzoic acid, 2-chloro-4-nitro **(99-60-5)**
Benzoic acid, 2-chloro-5-nitro- (9CI) **(2516-96-3)**
Benzoic acid, 2-chloro-3-nitro- (8CI,9CI) **(3970-35-2)**
Benzoic acid, 4-chloro-2-nitro-, methyl ester (9CI) **(42087-80-9)**
Benzoic acid, 5-chloro-2-nitro-, methyl ester (9CI) **(51282-49-6)**
Benzoic acid, chloro-, phenylmethyl ester (9CI) **(108548-66-9)**
Benzoic acid, 2-(chlorosulfonyl)-, methyl ester (9CI) **(26638-43-7)**
Benzoic acid, 3-(2-chloro-4-(trifluoromethyl)phenoxy)- (9CI) **(63734-62-3)**
Benzoic acid, 5-(2-chloro-4-(trifluoromethyl)phenoxy)-2-nitro-, 2-ethoxy-1-methyl-2-oxoethyl ester, **(77501-63-4)**
Benzoic acid, 5-(2-chloro-4-(trifluoromethyl)phenoxy)-2-nitro-, sodium salt **(62476-59-9)**
Benzoic acid, 3-(2-chloro-4-(trifluoromethyl)phenoxy)-, potassium salt (9CI) **(72252-48-3)**
Benzoic acid, N-cocoalkyltrimethylenediamine salt **(68188-29-4)**
Benzoic acid, 3-cyano- (9CI) **(1877-72-1)**
Benzoic acid, 4-cyano- (9CI) **(619-65-8)**
Benzoic acid, m-cyano- (8CI) **(1877-72-1)**
Benzoic acid, p-cyano- (8CI) **(619-65-8)**
Benzoic acid, 3,5-diacetamido-2,4,6-triiodo **(117-96-4)**
Benzoic acid, 5-((4'-((2,6-diamino-3-((8-hydroxy-3,6-disulfo-7-((4-sulfo-1-naphthalenyl)azo)2-naphthalenyl)azo)-5-methyl phenyl)azo)(1,1'-biphenyl)-4-yl)azo)-2-hydroxy-, tetrasodium salt **(2429-81-4)**
Benzoic acid, 5-((4'-((2,6-diamino-3-methyl-5-sulfophenyl)azo)-3,3'-dimethyl(1,1'-biphenyl)-4-yl)azo)-2-hydroxy-, disodium salt (9CI) **(6637-88-3)**
Benzoic acid, 5-((4'-((2,6-diamino-3-methyl-5-((4-sulfophenyl)-azo)phenyl)azo)(1,1'-biphenyl)- 4-yl)azo)-2-hydroxy-, disodium salt **(2586-58-5)**
Benzoic acid, 5-((4'-((2,6-diamino-3-methyl-5-((4-sulfophenyl)-azo)phenyl)azo)(1,1'-biphenyl)-4-yl)azo)-2-hydroxy-3-methyl-, disodium salt (9CI) **(6360-54-9)**
Benzoic acid, 2,4-dibromo- (8CI,9CI) **(611-00-7)**
Benzoic acid, 3,5-di-tert-butyl-4-hydroxy **(1421-49-4)**
Benzoic acid, 2,3-dichloro **(50-45-3)**
Benzoic acid, 2,4-dichloro **(50-84-0)**
Benzoic acid, 2,5-dichloro **(50-79-3)**
Benzoic acid, 2,6-dichloro **(50-30-6)**
Benzoic acid, 3,4-dichloro **(51-44-5)**
Benzoic acid, 3,5-dichloro **(51-36-5)**
Benzoic acid, 2,4(or 2,5)-dichloro- (9CI) **(35915-19-6)**

Benzoic acid, dichloro- (6CI,9CI)

Benzoic acid, dichloro- (6CI,9CI) **(75248-87-2)**
Benzoic acid, dichloro-3,4-dihydroxy- (9CI) **(69845-51-8)**
Benzoic acid, 2,3-dichloro-4,5-dimethoxy-, methyl ester (9CI) **(108544-99-6)**
Benzoic acid, 2,4-dichloro-, ethyl ester (9CI) **(56882-52-1)**
Benzoic acid, 5-(2,4-dichloro-6-fluorophenoxy)-2-nitro-, methyl ester (9CI) **(51937-92-9)**
Benzoic acid, 3,5-dichloro-4-hydroxy **(3336-41-2)**
Benzoic acid, 3,6-dichloro-2-hydroxy- (9CI) **(3401-80-7)**
Benzoic acid, 2,6-dichloro-4-hydroxy-3,5-dimethoxy- (8CI,9CI) **(20624-96-8)**
Benzoic acid, 2,3-dichloro-4-hydroxy-5-methoxy- (9CI) **(108544-97-4)**
Benzoic acid, 3,6-dichloro-2-hydroxy-, potassium sodium salt (9CI) **(68938-79-4)**
Benzoic acid, 3,5-dichloro-2-methoxy- **(22775-37-7)**
Benzoic acid, 3,6-dichloro-2-methoxy **(1918-00-9)**
Benzoic acid, 3,6-dichloro-2-methoxy-, Mixt. with (2,4-dichlorophenoxy)acetic acid **(8068-77-7)**
Benzoic acid, 3,6-dichloro-2-methoxy-, Compd. with 2,2'-iminobis(ethanol) (1:1) **(25059-78-3)**
Benzoic acid, 3,6-dichloro-2-methoxy-, Compd. with N-methylmethanamine (1:1) **(2300-66-5)**
Benzoic acid, 3,6-dichloro-2-methoxy-, ester with N-phenyldiethanolamine (2:1) **(56141-00-5)**
Benzoic acid, 3,5-dichloro-4-methoxy-, methyl ester (9CI) **(24295-27-0)**
Benzoic acid, 3,6-dichloro-2-methoxy-, sodium salt (9CI) **(1982-69-0)**
Benzoic acid, 2,5-dichloro-4-methyl- (9CI) **(21460-88-8)**
Benzoic acid, 2,4-dichloro-, methyl ester (9CI) **(35112-28-8)**
Benzoic acid, 2,5-dichloro-, methyl ester (9CI) **(2905-69-3)**
Benzoic acid, 3,4-dichloro-, 3-(2-methylpiperidino)propyl ester **(3478-94-2)**
Benzoic acid, 2,5-dichloro-3-nitro **(88-86-8)**
Benzoic acid, 2,5-dichloro-3-nitro-, methyl ester (9CI) **(34408-25-8)**
Benzoic acid, 3,6-dichloro-2-nitro-, methyl ester (9CI) **(40188-83-8)**
Benzoic acid, 5-(2,4-dichlorophenoxy)-2-nitro-, methyl ester **(42576-02-3)**
Benzoic acid, 6-(((2,3-dichlorophenyl)amino)carbonyl)-2,3,4,5-tetrachloro **(76280-91-6)**
Benzoic acid, p-(dichlorosulfamoyl) **(80-13-7)**
Benzoic acid, 2,6-dichloro-3,4,5-trimethoxy-, methyl ester (9CI) **(75315-45-6)**
Benzoic acid, diester with diethylene glycol **(120-55-8)**
Benzoic acid diester with polyethylene glycol 600 **(9004-86-8)**
Benzoic acid diethylamide **(1696-17-9)**
Benzoic acid, 2-(4-(diethylamino)-2-hydroxybenzoyl)- (9CI) **(5809-23-4)**
Benzoic acid, 2,6-difluoro **(385-00-2)**
Benzoic acid, 2,2'-((9,10-dihydro-9,10-dioxo-1,5-anthracenediyl)-diimino)bis- (9CI) **(81-78-7)**
Benzoic acid, 5-((4'-((4,5-dihydro-3-methyl-5-oxo-1-(4-sulfophenyl)-1H-pyrazol-4-yl)azo)(1,1'-biphenyl)-4-yl)azo)-2-hydroxy-, disodium salt (9CI) **(13164-93-7)**
Benzoic acid, 2-((3-(((2,3-dihydro-2-oxo-1H-benzimidazol-5-yl)-amino)carbonyl)-2-hydroxy-1-naphthalenyl)azo)-, butyl ester (9CI) **(31778-10-6)**
Benzoic acid, 2-((3-(((2,3-dihydro-2-oxo-1H-benzimidazol-5-yl)-amino)carbonyl)-2-hydroxy-1-naphthalenyl)azo)-, methyl ester (9CI) **(6985-92-8)**
Benzoic acid, 2-((1-(((2,3-dihydro-2-oxo-1H-benzimidazol-5-yl)-amino)carbonyl)-2-ox opropyl)azo)- (9CI) **(31837-42-0)**
Benzoic acid, 2,5-dihydroxy- **(490-79-9)**
Benzoic acid, 3,4-dihydroxy- **(99-50-3)**
Benzoic acid, 2,3-dihydroxy- (9CI) **(303-38-8)**
Benzoic acid, 2,4-dihydroxy- (9CI) **(89-86-1)**
Benzoic acid, 2,6-dihydroxy- (9CI) **(303-07-1)**
Benzoic acid, 3,5-dihydroxy- (9CI) **(99-10-5)**
Benzoic acid, 2,4-dihydroxy-3,6-dimethyl-, methyl ester (9CI) **(4707-47-5)**
Benzoic acid, 2,4-dihydroxy-, lead(2+) salt (2:1) (9CI) **(41453-50-3)**
Benzoic acid, 3,4-dihydroxy-5-methoxy- **(3934-84-7)**
Benzoic acid, 2,3-dihydroxy-4-methyl- (9CI) **(3929-89-3)**
Benzoic acid, 2,5-dihydroxy-3-methyl- (9CI) **(5981-39-5)**
Benzoic acid, 2,4-dihydroxy-, methyl ester **(2150-47-2)**
Benzoic acid, 2,3-dihydroxy-4-(1-methylethyl)- (9CI) **(19420-61-2)**
Benzoic acid, 2,4-dihydroxy-6-propyl- (9CI) **(4707-50-0)**
Benzoic acid, 5-((4'-((2,4-dihydroxy-3-((4-sulfophenyl)azo)-phenyl)azo)(1,1'-biphenyl)-4-yl)azo)-2-hydroxy-, disodium salt (9CI) **(2893-80-3)**
Benzoic acid, 3,5-diiodo-2-hydroxy- **(133-91-5)**
Benzoic acid, 2,3-dimethoxy- (9CI) **(1521-38-6)**
Benzoic acid, 2,4-dimethoxy- (9CI) **(91-52-1)**
Benzoic acid, 2,6-dimethoxy- (9CI) **(1466-76-8)**
Benzoic acid, 3,5-dimethoxy- (9CI) **(1132-21-4)**
Benzoic acid, 4-((2,4-dimethoxy-6-methylbenzoyl)oxy)-2-methoxy-6-methyl-, 3-methoxy-4-(methoxycarbonyl)-5-methylphenyl ester (9CI) **(19314-74-0)**
Benzoic acid, 3,4-dimethoxy-, methyl ester (9CI) **(2150-38-1)**
Benzoic acid, 2-((3-(3,4-dimethoxyphenyl)-1-oxo-2-propenyl)-amino) **(53902-12-8)**
Benzoic acid, 2-((((((4,6-dimethoxy-2-pyrimidinyl)amino)-carbonyl)amino)sulfonyl)methyl)-, methyl ester (9CI) **(83055-99-6)**
Benzoic acid, 2,3-dimethyl **(603-79-2)**
Benzoic acid, 2,6-dimethyl **(632-46-2)**
Benzoic acid, 3,4-dimethyl **(619-04-5)**
Benzoic acid, 3,5-dimethyl **(499-06-9)**
Benzoic acid, 2,4-dimethyl- (9CI) **(611-01-8)**
Benzoic acid, 2,5-dimethyl- (9CI) **(610-72-0)**
Benzoic acid, dimethyl- (8CI,9CI) **(30587-19-0)**
Benzoic acid, 4-(dimethylamino)- **(9CI)**
Benzoic acid, 3-(dimethylamino)- (9CI) **(99-64-9)**
Benzoic acid, m-(dimethylamino)- (8CI) **(99-64-9)**
Benzoic acid, p-(dimethylamino) (8CI) **(619-84-1)**
Benzoic acid, 2-(4-(dimethylamino)benzoyl)- (9CI) **(21528-31-4)**
Benzoic acid, 4-(dimethylamino)-, ethyl ester (9CI) **(10287-53-3)**
Benzoic acid, 4-(dimethylamino)-, 2-ethylhexyl ester (9CI) **(21245-02-3)**
Benzoic acid, p-dimethylamino-, pentyl ester **(14779-78-3)**
Benzoic acid, 2-((4-dimethylamino)phenylazo) **(493-52-7)**
Benzoic acid, m-((p-(dimethylamino)phenyl)azo) **(20691-84-3)**
Benzoic acid, 4-((4-(dimethylamino)phenyl)azo)- (9CI) **(6268-49-1)**
Benzoic acid, p-((p-(dimethylamino)phenyl)azo)- (8CI) **(6268-49-1)**
Benzoic acid, (1,1-dimethylethyl)- (9CI) **(1320-16-7)**
Benzoic acid, 4-(1,1-dimethylethyl)-, Compd. with 2,2',2''-nitrilotris(ethanol) (1:1) (9CI) **(59993-86-1)**
Benzoic acid, 2,3-dimethyl-, methyl ester (8CI,9CI) **(15012-36-9)**
Benzoic acid, 2,4-dimethyl-, methyl ester (8CI,9CI) **(23617-71-2)**
Benzoic acid, 3,5-dimethyl-, methyl ester (8CI,9CI) **(25081-39-4)**
Benzoic acid, 2-((2,3-dimethylphenyl)amino)- (9CI) **(61-68-7)**
Benzoic acid, 2-(((((4,6-dimethyl-2-pyrimidinyl)amino)-carbonyl)amino)sulfonyl)- (9CI) **(74223-56-6)**
Benzoic acid, o-((3-(4,6-dimethyl-2-pyrimidinyl)ureido)-sulfonyl)-, methyl ester **(74222-97-2)**
Benzoic acid, 2,4-dinitro **(610-30-0)**
Benzoic acid, 2,5-dinitro **(610-28-6)**
Benzoic acid, 3,4-dinitro **(528-45-0)**
Benzoic acid, 3,5-dinitro **(99-34-3)**
Benzoic acid, 3,5-dinitro-, Compd. with cyclohexanamine (1:1) (9CI) **(5473-16-5)**
Benzoic acid, 3,5-dinitro-, Compd. with cyclohexylamine (1:1) (8CI) **(5473-16-5)**
Benzoic acid, 4-((dipropylamino)sulfonyl)- (9CI) **(57-66-9)**
Benzoic acid, p-(dipropylsulfamoyl) **(57-66-9)**

Benzoic acid, 3,3'-((3,7-disulfo-1,5-naphthalenediyl)bis(azo-(6-hydroxy-3,1-phenylene)azo(6(or 7)-sulfo-4,1-naphthalenediyl)-azo-(1,1'-biphenyl)-4,4'-diylazo))bis(6-hydroxy-, hexasodium salt **(8014-91-3)**
Benzoic acid, 3,3'-((3,7-disulfo-1,5-naphthalenediyl)bis(azo-(6-hydroxy-3,1-phenylene)azo(6(or-biphenyl)-4,4'-diylazo)) bis(6-hydroxy-, hexasodium salt **(8014-91-3)**
Benzoic acid estradiol **(50-50-0)**
Benzoic acid, 4-ethenyl- (9CI) **(1075-49-6)**
Benzoic acid, 3-ethoxy- (9CI) **(621-51-2)**

Benzoic acid, 4-ethoxy- (9CI) **(619-86-3)**
Benzoic acid, m-ethoxy- (8CI) **(621-51-2)**
Benzoic acid, p-ethoxy- (8CI) **(619-86-3)**
Benzoic acid, 2-((ethoxy((1-methylethyl)amino)phosphinothioyl)-oxy)-, 1-methyl ester (9CI) **(25311-71-1)**
Benzoic acid, 2-((ethoxy((1-methylethyl)amino)phosphinyl)oxy)-, 1-methylethyl ester (9CI) **(31120-85-1)**
Benzoic acid, ethyl- (6CI,7CI,8CI,9CI) **(28134-31-8)**
Benzoic acid, o-(6-(ethylamino)-3-(ethylimino)-2,7-dimethyl-3H-xanthen-9-yl)-, ethyl ester, monohydrochloride **(989-38-8)**
Benzoic acid, ethyl ester **(93-89-0)**
Benzoic acid, 2-ethylhexyl ester (9CI) **(5444-75-7)**
Benzoic acid, 4-ethyl-, methyl ester (9CI) **(7364-20-7)**
Benzoic acid, p-ethyl-, methyl ester (8CI) **(7364-20-7)**
Benzoic acid, o-fluoro **(445-29-4)**
Benzoic acid, 3-fluoro- (9CI) **(455-38-9)**
Benzoic acid, 4-fluoro- (9CI) **(456-22-4)**
Benzoic acid, m-fluoro- (8CI) **(455-38-9)**
Benzoic acid, p-fluoro- (8CI) **(456-22-4)**
Benzoic acid, 2-fluoro-4-nitro- (8CI,9CI) **(403-24-7)**
Benzoic acid, 3-formyl- (9CI) **(619-21-6)**
Benzoic acid, 4-formyl-, methyl ester (9CI) **(1571-08-0)**
Benzoic acid, 5-hexadecyl-2-hydroxy-, calcium salt (2:1) (9CI) **(68540-40-9)**
Benzoic acid, hexyl ester **(6789-88-4)**
Benzoic acid, hydrazide **(613-94-5)**
Benzoic acid, 2-hydroxy- **(69-72-7)**
Benzoic acid, m-hydroxy **(99-06-9)**
Benzoic acid, p-hydroxy **(99-96-7)**
Benzoic acid, 3-hydroxy- (9CI) **(99-06-9)**
Benzoic acid, 4-hydroxy- (9CI) **(99-96-7)**
Benzoic acid, p-hydroxy-, benzyl ester (8CI) **(94-18-8)**
Benzoic acid, 4-hydroxy-, butyl ester (9CI) **(94-26-8)**
Benzoic acid, p-hydroxy-, butyl ester (8CI) **(94-26-8)**
Benzoic acid, 2-hydroxy-, 2-carboxyphenyl ester **(552-94-3)**
Benzoic acid, 2-hydroxydiiodo- (9CI) **(1321-04-6)**
Benzoic acid, 4-hydroxy-3,5-dimethoxy **(530-57-4)**
Benzoic acid, 2-hydroxy-, 4-(1,1-dimethylethyl)phenyl ester **(87-18-3)**
Benzoic acid, 2-hydroxy-3,5-dinitro- (9CI) **(609-99-4)**
Benzoic acid, 2-hydroxy-3,5-dinitro-, methyl ester (9CI) **(22633-33-6)**
Benzoic acid, 2-((2-hydroxy-3,6-disulfo-1-naphthalenyl)azo)-, barium salt (1:2) (9CI) **(1325-16-2)**
Benzoic acid, 2-((2-hydroxy-3,6-disulfo-1-naphthalenyl)azo)-, barium salt (2:3) (9CI) **(15782-06-6)**
Benzoic acid, p-hydroxy-, ethyl ester **(120-47-8)**
Benzoic acid, 3-hydroxy-, ethyl ester (9CI) **(7781-98-8)**
Benzoic acid, m-hydroxy-, ethyl ester (8CI) **(7781-98-8)**
Benzoic acid, 2-(1-hydroxyethyl)-, γ-lactone **(3453-64-3)**
Benzoic acid, 2-hydroxy-, hexyl ester **(6259-76-3)**
Benzoic acid, 2-hydroxy-5-((4-((4-((8-hydroxy-7-((4-((8-hydroxy-3,6-disulfo-1-naphthalenyl)azo)-2-methoxy-5-methylphenyl)azo)-3,6-disulfo-1-naphthalenyl)amino)-6-(phenylamino)-1,3,5-triazin-2-yl)amino)phenyl)azo)-, pentasodium salt (9CI) **(6388-26-7)**
Benzoic acid, 2-hydroxy-5-((4'-((1-hydroxy-7-(phenylamino)-3-sulfo-2-naphthalenyl)azo)(1,1'-biphenyl)-4-yl)azo)-, disodium salt (9CI) **(3476-90-2)**
Benzoic acid, 3-((1-hydroxy-6-((((5-hydroxy-6-(phenylazo)-7-sulfo-2-naphthalenyl)amino)carbonyl)amino)-3-sulfo-2-naphthalenyl)-azo)-, trisodium salt (9CI) **(6420-40-2)**
Benzoic acid, (4-(hydroxyimino)-2,5-cyclohexadien-1-ylidene) hydrazide **(495-73-8)**
Benzoic acid, (4-(hydroxyimino)-2,5-cyclohexadien-1-ylidene)-hydrazide, Mixt. with methylthioxoarsine (9CI) **(8066-69-1)**
Benzoic acid, 2-hydroxy-, ion(1-) (9CI) **(63-36-5)**
Benzoic acid, 4-hydroxy-, ion(1-) (9CI) **(456-23-5)**
Benzoic acid, p-hydroxy-, ion(1-) (8CI) **(456-23-5)**
Benzoic acid, 2-hydroxy-3-methoxy- **(877-22-5)**
Benzoic acid, 3-hydroxy-4-methoxy- **(645-08-9)**
Benzoic acid, 4-hydroxy-3-methoxy- **(121-34-6)**
Benzoic acid, 2-hydroxy-4-methoxy- (9CI) **(2237-36-7)**
Benzoic acid, 3-hydroxymethoxy- (9CI) **(108548-68-1)**
Benzoic acid, 4-hydroxymethoxy- (9CI) **(108548-69-2)**
Benzoic acid, 4-hydroxy-3-methoxy-, ion(1-) (9CI) **(6746-48-1)**
Benzoic acid, 2-hydroxy-5-((4-((((2-methoxy-4-((3-sulfo-phenyl)azo)phenyl)amino)carbonyl)amino)phenyl)azo)-, disodium salt (9CI) **(7248-45-5)**
Benzoic acid, 2-(hydroxymethyl)- (9CI) **(612-20-4)**
Benzoic acid, 2-hydroxy-3-methyl- (9CI) **(83-40-9)**
Benzoic acid, 2-hydroxy-4-methyl- (9CI) **(50-85-1)**
Benzoic acid, 2-hydroxy-5-methyl- (9CI) **(89-56-5)**
Benzoic acid, 2-hydroxy-6-methyl- (9CI) **(567-61-3)**
Benzoic acid, p-hydroxy-, methyl ester **(99-76-3)**
Benzoic acid, 2-hydroxy-, 1-methylethyl ester **(607-85-2)**
Benzoic acid, 2-hydroxy-3-methyl-, methyl ester **(23287-26-5)**
Benzoic acid, 2-hydroxy-, monosodium salt (9CI) **(54-21-7)**
Benzoic acid, 2-((2-hydroxy-1-naphthalenyl)azo)- (9CI) **(29128-56-1)**
Benzoic acid, 3-((2-hydroxy-1-naphthalenyl)azo)- (9CI) **(32624-41-2)**
Benzoic acid, 4-((2-hydroxy-1-naphthalenyl)azo)- (9CI) **(32624-40-1)**
Benzoic acid, 4-((4-hydroxy-1-naphthalenyl)azo)- (9CI) **(21184-58-7)**
Benzoic acid, p-((2-hydroxy-1-naphthyl)azo)- (8CI) **(32624-40-1)**
Benzoic acid, p-((4-hydroxy-1-naphthyl)azo)- (8CI) **(21184-58-7)**
Benzoic acid, m-((2-hydroxy-1-naphthyl)azo)- (7CI,8CI) **(32624-41-2)**
Benzoic acid, 2-hydroxy-5-nitro- (9CI) **(96-97-9)**
Benzoic acid, 4-hydroxy-3-nitro- (9CI) **(616-82-0)**
Benzoic acid, 3-hydroxy-4-nitro- (8CI,9CI) **(619-14-7)**
Benzoic acid, 4-hydroxy-3-nitro-, methyl ester (9CI) **(99-42-3)**
Benzoic acid, 2-(6-hydroxy-3-oxo-3H-xanthen-9-yl)- (9CI) (VAN) **(518-45-6)**
Benzoic acid, o-(6-hydroxy-3-oxo-3H-xanthen-9-yl)- (8CI) (VAN) **(518-45-6)**
Benzoic acid, 2-(6-hydroxy-3-oxo-3H-xanthen-9-yl)- (VAN) **(2321-07-5)**
Benzoic acid, m-hydroxyphenyl ester **(136-36-7)**
Benzoic acid, 4-hydroxy-, phenylmethyl ester (9CI) **(94-18-8)**
Benzoic acid, 4-hydroxy-, propyl ester (9CI) **(94-13-3)**
Benzoic acid, p-hydroxy-, propyl ester (6CI,8CI) **(94-13-3)**
Benzoic acid, 2-hydroxy-, strontium salt (2:1) **(526-26-1)**
Benzoic acid, 4-((2-hydroxy-6-sulfo-1-naphthalenyl)azo)- (9CI) **(69644-66-2)**
Benzoic acid, 2-hydroxy-5-((5-sulfo-2-naphthalenyl)azo)-, disodium salt (9CI) **(10114-96-2)**
Benzoic acid, 2-hydroxy-5-((8-sulfo-2-naphthalenyl)azo)-, disodium salt (9CI) **(10114-97-3)**
Benzoic acid, 2-hydroxy-, 4-(1,1,3,3-tetramethylbutyl)phenyl ester (9CI) **(2553-08-4)**
Benzoic acid, 2-hydroxy-, 3,3,5-trimethylcyclohexyl ester (9CI) **(118-56-9)**
Benzoic acid, ion(1-) (8CI,9CI) **(766-76-7)**
Benzoic acid, o-iodo **(88-67-5)**
Benzoic acid, p-iodo **(619-58-9)**
Benzoic acid, 3-iodo- (9CI) **(618-51-9)**
Benzoic acid, m-iodo- (8CI) **(618-51-9)**
Benzoic acid, o-isopropyl- (8CI) **(2438-04-2)**
Benzoic acid, p-isopropyl- (8CI) **(536-66-3)**
Benzoic acid, isopropyl ester **(939-48-0)**
Benzoic acid, magnesium salt (9CI) **(553-70-8)**
Benzoic acid, o-mercapto **(147-93-3)**
Benzoic acid, 2-methoxy- **(119-36-8)**
Benzoic acid, 2-methoxy- (9CI) **(579-75-9)**
Benzoic acid, 3-methoxy- (9CI) **(586-38-9)**
Benzoic acid, methoxy- (9CI) **(1335-08-6)**
Benzoic acid, p-methoxy-, methyl ester **(121-98-2)**
Benzoic acid, 2-(((3-methoxyphenyl)amino)carbonyl)- (9CI) **(19336-97-1)**
Benzoic acid, 2-methoxy-3,5,6-trichloro- **(2307-49-5)**
Benzoic acid, methyl- (9CI) **(25567-10-6)**
Benzoic acid, 2-methyl-3,5-dinitro- (9CI) **(28169-46-2)**
Benzoic acid, methylenebis(2-hydroxy- (9CI) **(27496-82-8)**
Benzoic acid, 3,4-(methylenedioxy)- **(94-53-1)**
Benzoic acid, methyl ester **(93-58-3)**
Benzoic acid, 2-(1-methylethyl)- (9CI) **(2438-04-2)**

Benzoic acid, 4-(1-methylethyl)- (9CI) **(536-66-3)**
Benzoic acid, 3-methyl-, methyl ester (9CI) **(99-36-5)**
Benzoic acid, methyl-, methyl ester (9CI) **(25567-11-7)**
Benzoic acid, 2-methyl-3-nitro- (9CI) **(1975-50-4)**
Benzoic acid, 2-methyl-5-nitro- (9CI) **(1975-52-6)**
Benzoic acid, 2-methyl-6-nitro- (9CI) **(13506-76-8)**
Benzoic acid, 3-methyl-2-nitro- (9CI) **(5437-38-7)**
Benzoic acid, 3-methyl-4-nitro- (9CI) **(3113-71-1)**
Benzoic acid, 4-methyl-3-nitro- (9CI) **(96-98-0)**
Benzoic acid, 5-methyl-2-nitro- (9CI) **(3113-72-2)**
Benzoic acid nitrile **(100-47-0)**
Benzoic acid, m-nitro **(121-92-6)**
Benzoic acid, o-nitro **(552-16-9)**
Benzoic acid, p-nitro **(62-23-7)**
Benzoic acid, 3-nitro- (9CI) **(121-92-6)**
Benzoic acid, 3-nitro-, Compd. with cyclohexanamine (1:1) (9CI) **(34139-62-3)**
Benzoic acid, m-nitro-, Compd. with cyclohexylamine (1:1) (8CI) **(34139-62-3)**
Benzoic acid, p-nitro-, ethyl ester **(99-77-4)**
Benzoic acid, p-nitro-, methyl ester **(619-50-1)**
Benzoic acid, 4-nitro-, pentyl ester (9CI) **(14309-42-3)**
Benzoic acid, 4-nitro-, propyl ester (9CI) **(94-22-4)**
Benzoic acid, p-nitro-, propyl ester (8CI) **(94-22-4)**
Benzoic acid, 3-nitro-, sodium salt (9CI) **(827-95-2)**
Benzoic acid, 4-(((4-(nonyloxy)phenyl)methylene)amino)-, 2-(4-chlorophenyl)-1,3-dioxan-5-yl ester, (2.α.,5.β.(E))- (9CI) **(98611-44-0)**
Benzoic acid, 4-((octadecylamino)carbonyl)-, methyl ester (9CI) **(7333-86-0)**
Benzoic acid, 4-((octadecylamino)carbonyl)-, monosodium salt (9CI) **(5994-45-6)**
Benzoic acid, 4-(octyloxy)- (9CI) **(2493-84-7)**
Benzoic acid, p-(octyloxy)- (8CI) **(2493-84-7)**
Benzoic acid, (4-oxo-2,5-cyclohexadien-1-ylidene)hydrazide, oxime **(495-73-8)**
Benzoic acid, 4-(1-oxopropyl)- (9CI) **(4219-55-0)**
Benzoic acid, 4-(2-oxopropyl)- (9CI) **(15482-54-9)**
Benzoic acid, pentaerythritol, phthalic anhydride, tall fatty oil acids polymer **(66070-84-6)**
Benzoic acid, pentaerythritol, phthalic anhydride, tall oil fatty acid polymer **(66070-84-6)**
Benzoic acid, pentaerythritol, phthalic anhydride, tall oil fatty acids polymer **(66070-84-6)**
Benzoic acid, pentaerythritol, tall oil fatty acid, phthalic anhydride polymer **(66070-84-6)**
Benzoic acid, pentyl ester **(2049-96-9)**
Benzoic acid, peroxide **(94-36-0)**
Benzoic acid, 3-phenoxy- **(3739-38-6)**
Benzoic acid, m-phenoxy **(3739-38-6)**
Benzoic acid, 2-(phenylamino)- (9CI) **(91-40-7)**
Benzoic acid, 3-(((phenylamino)carbonyl)oxy)-, ethyl ester (9CI) **(37070-83-0)**
Benzoic acid, 2-(phenylazo)- (9CI) **(3682-56-2)**
Benzoic acid, 4-(phenylazo)- (9CI) **(1562-93-2)**
Benzoic acid, o-(phenylazo)- (8CI) **(3682-56-2)**
Benzoic acid, p-(phenylazo)- (8CI) **(1562-93-2)**
Benzoic acid, 2-(phenylazo)-, ethyl ester (9CI) **(18277-91-3)**
Benzoic acid, o-(phenylazo)-, ethyl ester (7CI,8CI) **(18277-91-3)**
Benzoic acid, p-((p-phenylbenzylidene)amino)-, ethyl ester (7CI,8CI) **(3782-80-7)**
Benzoic acid, phenyl ester **(93-99-2)**
Benzoic acid, 2-(phenylmethyl)- (9CI) **(612-35-1)**
Benzoic acid, phenylmethyl ester **(120-51-4)**
Benzoic acid, potassium salt (9CI) **(582-25-2)**
Benzoic acid, 2-propenyl ester (9CI) **(583-04-0)**
Benzoic acid, p-propionyl- (7CI,8CI) **(4219-55-0)**
Benzoic acid, propyl ester (9CI) **(2315-68-6)**
Benzoic acid, sodium salt **(532-32-1)**
Benzoic acid, 2-sulfo- (9CI) **(632-25-7)**
Benzoic acid, 3-sulfo- (9CI) **(121-53-9)**
Benzoic acid, m-sulfo- (8CI) **(121-53-9)**
Benzoic acid, 2-sulfo-, cyclic anhydride **(81-08-3)**
Benzoic acid, o-sulfo-, cyclic anhydride (8CI) **(81-08-3)**
Benzoic acid, tall oil fatty acid, phthalic anhydride, pentaerythritol polymer **(66070-84-6)**
Benzoic acid, tall oil fatty acids, phthalic anhydride, pentaerythritol polymer **(66070-84-6)**
Benzoic acid, 2-(2,4,5,7-tetrabromo-6-hydroxy-3-oxo-3H-xanthen-9-yl)-, disodium salt **(548-26-5)**
Benzoic acid, 2,3,4,5-tetrachloro **(50-74-8)**
Benzoic acid, tetraester with pentaerythritol **(4196-86-5)**
Benzoic acid, 2,3,5-trichloro **(50-73-7)**
Benzoic acid, 2,3,6-trichloro **(50-31-7)**
Benzoic acid, 2,4,5-trichloro **(50-82-8)**
Benzoic acid, 2,4,6-trichloro **(50-43-1)**
Benzoic acid, trichloro- (8CI,9CI) **(1319-85-3)**
Benzoic acid, 2,3,6-trichloro-, Compd. with Dimethylamine (1:1) **(3426-62-8)**
Benzoic acid, 2,3,6-trichloro-4,5-dihydroxy- (9CI) **(99165-96-5)**
Benzoic acid, 2,3,5-trichloro-6-hydroxy-, pentyl ester (9CI) **(30431-53-9)**
Benzoic acid, 2,3,5-trichloro-6-methoxy- (9CI) **(2307-49-5)**
Benzoic acid, 3-(trifluoromethyl)- (9CI) **(454-92-2)**
Benzoic acid, 4-(trifluoromethyl)- (9CI) **(455-24-3)**
Benzoic acid, 2-((8-(trifluoromethyl)-4-quinolinyl)amino)- (9CI) **(36783-34-3)**
Benzoic acid, 2-(8'-trifluoromethyl-4'-quinolylamino)-, 2,3-dihydroxypropyl ester **(23779-99-9)**
Benzoic acid, 3,4,5-trihydroxy- **(149-91-7)**
Benzoic acid, 2,3,4-trihydroxy- (9CI) **(610-02-6)**
Benzoic acid, 3,4,5-trihydroxy-, ethyl ester (9CI) **(831-61-8)**
Benzoic acid, 3,4,5-trihydroxy-, methyl ester **(99-24-1)**
Benzoic acid, 3,4,5-trihydroxy-, propyl ester **(121-79-9)**
Benzoic acid, 2,3,5-triiodo **(88-82-4)**
Benzoic acid, 2,4,5-trimethoxy- **(490-64-2)**
Benzoic acid, 3,4,5-trimethoxy- (9CI) **(118-41-2)**
Benzoic acid, 2,3,4-trimethoxy- (8CI,9CI) **(573-11-5)**
Benzoic acid, 2,4,5-trimethyl **(528-90-5)**
Benzoic acid, 2,4,6-trimethyl **(480-63-7)**
Benzoic acid, 2,3,4-trimethyl- (8CI,9CI) **(1076-47-7)**
Benzoic acid, 2,3,6-trimethyl- (8CI,9CI) **(2529-36-4)**
Benzoic acid, trinitro- (Dry) **(129-66-8)**
Benzoic acid, trinitro- (10% to 30% water) **(129-66-8)**
Benzoic acid, p-vinyl- (8CI) **(1075-49-6)**
Benzoic acid, zinc salt (9CI) **(553-72-0)**
Benzoic aldehyde **(100-52-7)**
Benzoic anhydride (8CI) **(93-97-0)**
Benzoic-carboxy-13C acid (6CI,7CI,8CI,9CI) **(3880-99-7)**
Benzoic ether **(93-89-0)**
Benzoic hydrazide **(613-94-5)**
o-Benzoic sulfimide **(81-07-2)**
Benzoic sulphimide **(81-07-2)**
o-Benzoic sulphimide **(81-07-2)**
Benzoic trichloride **(98-07-7)**
Benzoimidazole **(51-17-2)**
Benzoin **(119-53-9)**
Benzoin **(9000-05-9)**
Benzoin Gum **(9000-05-9)**
Benzoin Gum, Sumatra **(9000-05-9)**
Benzoin Malasia **(9000-05-9)**
Benzoin Resin **(9000-05-9)**
Benzoin Resinoid **(9000-05-9)**
Benzoin Resinoid, Siam **(9000-05-9)**
Benzoin Siam **(9000-05-9)**
Benzoin Sumatra **(9000-05-9)**
Benzoinoxim (Czech) **(441-38-3)**
Benzoin, α-oxime **(441-38-3)**
Benzoin, oxime **(441-38-3)**
α-Benzoin oxime **(441-38-3)**
3,4-Benzoisoquinoline **(229-87-8)**
1,2-Benzoisothiazoline 1,1-dioxide **(936-16-3)**

Benzo(j)fluoranthene **(205-82-3)**
Benzo(jk)fluorene **(206-44-0)**
Benzoketotriazine **(90-16-4)**
11,12-Benzo(k)fluoranthene **(207-08-9)**
Benzo(k)fluoranthene **(207-08-9)**
Benzokoll **(555-48-6)**
Benzol [UN 1114] **(71-43-2)**
Benzole **(71-43-2)**
Benzolene **(71-43-2)**
Benzo(l)fluoranthene **(205-82-3)**
Benzoline **(8032-32-4)**
Benzo(lmn)phenanthridine **(194-03-6)**
Benzolo (Italian) **(71-43-2)**
Benzolone **(60-13-9)**
Benzo(l)phenanthrene **(217-59-4)**
Benzomarc **(3134-12-1)**
Benzo(mno)fluoranthene **(203-12-3)**
1H-Benzonaphthene **(203-80-5)**
Benzonaphthothiophene (9CI) **(61523-34-0)**
Benzone **(50-33-9)**
Benzonitrile [UN 2224] **(100-47-0)**
Benzonitrile, m-amino **(2237-30-1)**
Benzonitrile, 2-amino- (9CI) **(1885-29-6)**
Benzonitrile, 3-amino- (9CI) **(2237-30-1)**
Benzonitrile, 4-amino- (9CI) **(873-74-5)**
Benzonitrile, p-amino- (8CI) **(873-74-5)**
Benzonitrile, 2-((4-(bis(2-(acetyloxy)ethyl)amino)phenyl)azo)-5-nitro- (9CI) **(30124-94-8)**
Benzonitrile, 4-bromo **(623-00-7)**
Benzonitrile, m-bromo **(6952-59-6)**
Benzonitrile, 2-bromo- (9CI) **(2042-37-7)**
Benzonitrile, o-bromo- (8CI) **(2042-37-7)**
Benzonitrile, 3-bromo-5-chloro-4-hydroxy- (7CI,8CI,9CI) **(1689-86-7)**
Benzonitrile, o-chloro **(873-32-5)**
Benzonitrile, p-chloro **(623-03-0)**
Benzonitrile, 2-chloro-6-methyl- (9CI) **(6575-09-3)**
Benzonitrile, 2-((4-((2-cyanoethyl)ethylamino)phenyl)azo)-5-nitro **(16889-10-4)**
Benzonitrile, 3,5-dibromo-4-hydroxy **(1689-84-5)**
Benzonitrile, 3,5-dibromo-4-octanoyloxy **(1689-99-2)**
Benzonitrile, 2,6-dichloro **(1194-65-6)**
Benzonitrile, 3,5-diiodo-4-hydroxy **(1689-83-4)**
Benzonitrile, 3,5-diiodo-4-hydroxy-, octanoate **(3861-47-0)**
Benzonitrile, 3,5-diiodo-4-octanoyloxy- **(3861-47-0)**
Benzonitrile, p-hydroxy **(767-00-0)**
Benzonitrile, 2-hydroxy- (9CI) **(611-20-1)**
Benzonitrile, 3-hydroxy- (9CI) **(873-62-1)**
Benzonitrile, m-hydroxy- (8CI) **(873-62-1)**
Benzonitrile, p-hydroxy-, O-ester with O-ethyl phenylphosphonothioate **(13067-93-1)**
Benzonitrile, 4-hydroxy-3,5-diiodo- **(1689-83-4)**
Benzonitrile, 3-nitro- **(619-24-9)**
Benzonitrile, m-nitro **(619-24-9)**
Benzonitrile, p-nitro **(619-72-7)**
Benzonitrile, 4-nitro- (9CI) **(619-72-7)**
Benzonitrile, pentachloro- **(20925-85-3)**
Benzonitrile, 4-(((phenylamino)carbonyl)oxy)- (9CI) **(37070-85-2)**
Benzoparadiazine **(91-19-0)**
8H-Benzo(p)cyclopenta(def)chrysen-8-one (9CI) **(86854-11-7)**
Benzopenicillin **(61-33-6)**
Benzoperoxide **(94-36-0)**
1,12-Benzoperylene **(191-24-2)**
2,3-Benzoperylene **(197-70-6)**
Benzoperylene **(11057-45-7)**
1,2-Benzophenanthrene **(218-01-9)**
2,3-Benzophenanthrene **(56-55-3)**
3,4-Benzophenanthrene **(195-19-7)**
9,10-Benzophenanthrene **(217-59-4)**
Benzophenanthrene (9CI) **(65777-08-4)**
Benzophenanthrene, 9,10-dihydro- (9CI) **(103370-70-3)**
Benzophenone **(119-61-9)**
Benzophenone 4 **(4065-45-6)**
Benzophenone-2 **(131-55-5)**
Benzophenone-3 **(131-57-7)**
Benzophenone-6 **(131-54-4)**
Benzophenone-8 **(131-53-3)**
Benzophenone, 4-amino **(1137-41-3)**
Benzophenone, 4,4'-bis(diethylamino)- (8CI) **(90-93-7)**
Benzophenone, 4,4'-bis(dimethylamino) **(90-94-8)**
Benzophenone-2-carboxylic acid **(85-52-9)**
Benzophenone, 4-chloro- (8CI) **(134-85-0)**
Benzophenone, 5-chloro-2-hydroxy- (8CI) **(85-19-8)**
Benzophenone, 5-chloro-2-methyl- (8CI) **(33184-55-3)**
Benzophenone, 4,4'-dichloro **(90-98-2)**
Benzophenone, 2,4'-dichloro- (8CI) **(85-29-0)**
Benzophenone, 2,4-dihydroxy **(131-56-6)**
Benzophenone, 2,2'-dihydroxy-4,4'-dimethoxy **(131-54-4)**
Benzophenone, 2,2'-dihydroxy-4-methoxy **(131-53-3)**
Benzophenone, 4,4'-dimethoxy- (8CI) **(90-96-0)**
Benzophenone, 3,3'-dimethyl- (7CI,8CI) **(2852-68-8)**
Benzophenone, 4-(dimethylamino)- (8CI) **(530-44-9)**
Benzophenone, 4-ethyl- (8CI) **(18220-90-1)**
Benzophenone, 2-hydroxy- (8CI) **(117-99-7)**
Benzophenone, 4-hydroxy- (8CI) **(1137-42-4)**
Benzophenone, 4-hydroxy-3,3'-dimethyl- **(62064-85-1)**
Benzophenone, 2-hydroxy-4-methoxy **(131-57-7)**
Benzophenone, 2-hydroxy-4-(octyloxy) **(1843-05-6)**
Benzophenone, 4-methyl **(134-84-9)**
p-Benzophenone, methyl- **(134-84-9)**
Benzophenone, 4-nitro- **(1144-74-7)**
3,3',4,4'-Benzophenonetetracarboxylic acid dianhydride **(2421-28-5)**
Benzophenone, 2,2',4,4'-tetrahydroxy **(131-55-5)**
Benzophenone, 2,3,4-trihydroxy- (8CI) **(1143-72-2)**
Benzophosphate **(2310-17-0)**
3,4-Benzopirene (Italian) **(50-32-8)**
Benzo(p)naphtho(1,8,7-ghi)chrysene (7CI,8CI,9CI) **(385-14-8)**
Benzo(pqr)picene (8CI,9CI) **(189-96-8)**
Benzopurpurin 4B **(992-59-6)**
Benzopurpurine 4B **(992-59-6)**
Benzopurpurine 4BKX **(992-59-6)**
Benzopurpurine 4BX **(992-59-6)**
1H-2-Benzopyran-1,3(4H)-dione (9CI) **(703-59-3)**
2H-1-Benzopyran-2-carboxylic acid, 3,4-dihydro-6-hydroxy-2,5,7,8-tetramethyl-, (+-)- (9CI) **(53188-07-1)**
3H-2-Benzopyran-7-carboxylic acid, 4,6-dihydro-8-hydroxy-3,4,5-trimethyl-6-oxo-, (3R-trans) **(518-75-2)**
4H-1-Benzopyran-2-carboxylic acid, 5,5'-((2-hydroxy-1,3-propanediyl)bis(oxy))bis(4-oxo- **(16110-51-3)**
4H-1-Benzopyran-2-carboxylic acid, 5,5'-((2-hydroxytrimethylene)dioxy)bis(4-oxo- **(16110-51-3)**
4H-1-Benzopyran-2-carboxylic acid, 5,5'-((2-hydroxytrimethylene)-dioxy)bis(4-oxo-, disodium salt **(15826-37-6)**
2H-1-Benzopyran, 7-hydroxy-4-methyl-2-oxo- **(90-33-5)**
(1)Benzopyrano(3,4-b)furo(2,3-h)(1)benzopyran-6(6aH)-one, 1,2,12,12a-tetrahydro-8,9-dimethoxy-2-(1-methylethyl)-, (2R-(2α,6aα,12aα))- (9CI) **(6659-45-6)**
(1)Benzopyrano(3,4-b)furo(2,3-h)(1)benzopyran-6(6aαH)-one, 1,2,12,12aα-tetrahydro-2α-isopropyl-8,9-dimethoxy- (8CI) **(6659-45-6)**
(1)Benzopyrano(3,4-b)furo(2,3-h)(1)benzopyran-6(6ah)-one, 1,2,12,12a-tetrahydro-2-α- isopropenyl-8,9-dimethoxy **(83-79-4)**
(1)Benzopyrano(5,4,3-cde)(1)benzopyran-5,10-dione, 2,3,7,8-tetrahydroxy **(476-66-4)**
(2)Benzopyrano(6,5,4-def)(2)benzopyran-1,3,6,8-tetrone (9CI) **(81-30-1)**
2H-1-Benzopyran-6-ol, 3,4-dihydro-2,5,7,8-tetramethyl-2-(4,8,12-trimethyltridecyl)-, (2R-(2R*(4R*,8R*))) **(59-02-9)**

2H-1-Benzopyran-6-ol, 3,4-dihydro-2,5,7,8-tetramethyl-2-(4,8,12-trimethyltridecyl)-, acetate, (2R-(2R*(4R*,8R*)))- (9CI) **(58-95-7)**
2H-1-Benzopyran-2-one **(91-64-5)**
4H-1-Benzopyran-4-one, 3-((6-O-(6-deoxy-α-l-mannopyranosyl)-β-d-galactopyranosyl) oxy)-7-((6-deoxy-α-l-mannopyranosyl)oxy)-5-hydroxy-2-(4-hydroxyphenyl) **(301-19-9)**
2H-1-Benzopyran-2-one, 3-(1-(4-chlorophenyl)-3-oxobutyl)-4-hydroxy- (9CI) **(81-82-3)**
4H-1-Benzopyran-4-one, 2-(4-(2-(dibutylamino)ethoxy)-3,5-dimethylbenzoyl)-, hydrochloride (9CI) **(67652-39-5)**
2H-1-Benzopyran-2-one, 7-(diethylamino)-4-methyl- (9CI) **(91-44-1)**
1H-2-Benzopyran-1-one, 3,4-dihydro-6,8-dihydroxy-3-((tetrahydro-6-methyl-2H-pyran-2-yl) methyl)-, (2r-(2-α(R*),6-β)) **(35818-31-6)**
2H-1-Benzopyran-2-one, 6,7-dihydroxy- (9CI) **(305-01-1)**
2H-1-Benzopyran-2-one, 5,7-dihydroxy-4-methyl- (9CI) **(2107-76-8)**
2H-1-Benzopyran-2-one, 6,7-dihydroxy-4-methyl- (9CI) **(529-84-0)**
2H-1-Benzopyran-2-one, 7,8-dihydroxy-4-methyl- (9CI) **(2107-77-9)**
4H-1-Benzopyran-4-one, 2-(3,4-dihydroxyphenyl)-2,3-dihydro-3,5,7-trihydroxy- **(480-18-2)**
4H-1-Benzopyran-4-one, 2-(3,4-dihydroxyphenyl)-2,3-dihydro-3,5,7-trihydroxy-, (2R-trans) **(480-18-2)**
4H-1-Benzopyran-4-one, 2-(3,4-dihydroxyphenyl)-3,5,7-trihydroxy- (9CI) **(117-39-5)**
Benzopyran-2-one, 6,7-dimethoxy- (9CI) **(120-08-1)**
2H-1-Benzopyran-2-one, 7-(dimethylamino)-4-methyl- (9CI) **(87-01-4)**
2H-1-Benzopyran-2-one, 3-(1-(2-furanyl)-3-oxobutyl)-4-hydroxy-(9CI) **(117-52-2)**
2H-1-Benzopyran-2-one, 3-(1-(2-furanyl)-3-oxobutyl)-4-hydroxy-, sodium salt (9CI) **(34490-93-2)**
2H-1-Benzopyran-2-one, 7-hydroxy- **(93-35-6)**
4H-1-Benzopyran-4-one, 5-hydroxy-7-methoxy-2-methyl- (9CI) **(480-34-2)**
2H-1-Benzopyran-2-one, 4-hydroxy-3-(1-(4-nitrophenyl)-3-oxobutyl)- **(152-72-7)**
2H-1-Benzopyran-2-one, 4-hydroxy-3-(3-oxo-1-phenylbutyl)- (9CI) **(81-81-2)**
2H-1-Benzopyran-2-one, 4-hydroxy-3-(3-oxo-1-phenylbutyl)-, sodium salt (9CI) **(129-06-6)**
2H-1-Benzopyran-2-one, 4-hydroxy-3-(1-phenylpropyl)- (9CI) **(435-97-2)**
2H-1-Benzopyran-2-one, 4-hydroxy-3-(1,2,3,4-tetrahydro-1-naphthalenyl)- (9CI) **(5836-29-3)**
2H-1-Benzopyran-2-one, 6-methyl- **(92-48-8)**
2H-1-Benzopyran-2-one, 3,3'-methylenebis(4-hydroxy- (9CI) **(66-76-2)**
2H-1-Benzopyran-2-one, 7-(2H-naphtho(1,2-d)triazol-2-yl)-3-phenyl- (9CI) **(3333-62-8)**
4H-1-Benzopyran-4-one, 3,5,7-trihydroxy-2-(4-hydroxyphenyl)- **(520-18-3)**
2H-1-Benzopyran, 2-oxo- **(91-64-5)**
2H-1-Benzopyran-3,5,7-triol, 2-(3,4-dihydroxyphenyl)-3,4-dihydro-, (2R-trans) **(154-23-4)**
1,2-Benzopyrene **(192-97-2)**
3,4-Benzopyrene **(50-32-8)**
4,5-Benzopyrene **(192-97-2)**
6,7-Benzopyrene **(50-32-8)**
Benzo(e)pyrene **(192-97-2)**
Benzopyrene (9CI) (VAN) **(73467-76-2)**
6,12-Benzopyrene quinone **(3067-12-7)**
Benzo(1,2)pyreno(4,5-b)oxirene, 3b,4a-dihydro- **(37574-47-3)**
2H-1-Benzopyren-2-one, 7-hydroxy-4-methyl- **(90-33-5)**
5,6-Benzopyrimidine **(253-82-7)**
1,2-Benzopyrone **(91-64-5)**
2,3-Benzopyrrole **(120-72-9)**
Benzopyrrole **(120-72-9)**
Benzoquin **(103-16-2)**
Benzoquinamida (Spanish) **(63-12-7)**
Benzoquinamide **(63-12-7)**
1,4-Benzoquine **(106-51-4)**
Benzoquinol **(123-31-9)**
Benzoquinol (VAN) **(4323-21-1)**
2,3-Benzoquinoline **(260-94-6)**
3,4-Benzoquinoline **(229-87-8)**
5,6-Benzoquinoline **(85-02-9)**
7,8-Benzoquinoline **(230-27-3)**
α-Benzoquinoline **(230-27-3)**
Benzoquinoline (9CI) **(39327-16-7)**
Benzoquinoline, methyl- (9CI) **(88813-63-2)**
1,2-Benzoquinone **(583-63-1)**
1,4-Benzoquinone **(106-51-4)**
o-Benzoquinone **(583-63-1)**
p-Benzoquinone **(106-51-4)**
p-Benzoquinone, Compd. with Hydroquinone **(106-34-3)**
p-Benzoquinone (OSHA) **(106-51-4)**
Benzoquinone [UN 2587] **(106-51-4)**
Benzoquinone [UN 2587] **(583-63-1)**
p-Benzoquinone amidinohydrazone thiosemicarbazone **(539-21-9)**
Benzoquinone aziridine **(800-24-8)**
1,4-Benzoquinone N'-benzoylhydrazone oxime **(495-73-8)**
p-Benzoquinone, 2,5-bis(1-aziridinyl)-3,6-bis(2-methoxyethoxy) **(800-24-8)**
p-Benzoquinone, 2,6-di-tert-butyl **(719-22-2)**
p-Benzoquinone, 2,5-di-tert-butyl- (8CI) **(2460-77-7)**
p-Benzoquinone, 2,5-dichloro-3,6-dihydroxy- (8CI) **(87-88-7)**
1,4-Benzoquinone dioxime **(105-11-3)**
p-Benzoquinone, dioxime **(105-11-3)**
Benzoquinone guanylhydrazone thiosemicarbazone **(539-21-9)**
p-Benzoquinone, 2-methyl **(553-97-9)**
p-Benzoquinone, monooxime (8CI) **(637-62-7)**
p-Benzoquinone oxime benzoylhydrazone **(495-73-8)**
p-Benzoquinone, 2,3,5,6-tetraazido- **(22826-61-5)**
1,4-Benzoquinone, 2,3,5,6-tetrachloro- **(118-75-2)**
p-Benzoquinone, 2,3,5,6-tetrachloro **(118-75-2)**
o-Benzoquinone, 3,4,5,6-tetrachloro- (8CI) **(2435-53-2)**
p-Benzoquinone, 2,3,5-trimethyl- (8CI) **(935-92-2)**
p-Benzoquinone, 2,3,5-tris(1-aziridinyl) **(68-76-8)**
p-Benzoquinone, tris(1-aziridinyl)- **(68-76-8)**
Benzo(rst)pentaphene **(189-55-9)**
Benzo(rst)phenanthro(10,1,2-cde)pentaphene-9,18-dione, bromo- (9CI) **(1324-17-0)**
Benzo(rst)phenanthro(10,1,2-cde)pentaphene-9,18-dione, dichloro **(1324-55-6)**
o-Benzosulfimide **(81-07-2)**
Benzosulfonamide **(98-10-2)**
Benzosulfonazole **(95-16-9)**
Benzosulphimide **(81-07-2)**
o-Benzosulphimide **(81-07-2)**
3,4-Benzotetracene **(214-17-5)**
3,4-Benzotetraphene **(214-17-5)**
1H-2,1,3-Benzothiadiazin-4(3H)-one, 3-isopropyl-, 2,2-dioxide **(25057-89-0)**
1H-2,1,3-Benzothiadiazin-4(3H)-one, 3-methyl-, 2,2-dioxide (8CI,9CI) **(2225-40-3)**
2H-1,2,4-Benzothiadiazine-7-sulfonamide, 3-benzyl-3,4-dihydro-6-(trifluoromethyl)-, 1,1-dioxide **(73-48-3)**
2H-1,2,4-Benzothiadiazine-7-sulfonamide, 3-((benzylthio)-menthyl)-6-chloro-, 1,1-dioxide **(91-33-8)**
2H-1,2,4-Benzothiadiazine-7-sulfonamide, 6-chloro-3-(chloromethyl)-3,4-dihydro-, 1,1-dioxide **(135-07-9)**
2H-1,2,4-Benzothiadiazine-7-sulfonamide, 6-chloro-3-(dichloromethyl)-3,4-dihydro-, 1,1-dioxide **(133-67-5)**
2H-1,2,4-Benzothiadiazine-7-sulfonamide, 6-chloro-3,4-dihydro-, 1,1-dioxide **(58-93-5)**
2H-1,2,4-Benzothiadiazine-7-sulfonamide, 6-chloro-3,4-dihydro-2-methyl-3-(((2,2,2-trifluoroethyl)thio)methyl)-, 1,1-dioxide **(346-18-9)**
2H-1,2,4-Benzothiadiazine-7-sulfonamide, 6-chloro-, 1,1-dioxide **(58-94-6)**
2H-1,2,4-Benzothiadiazine-7-sulfonamide, 6-chloro-3-(((phenylmethyl)thio)methyl)-, dioxide, **(91-33-8)**
2H-1,2,4-Benzothiadiazine-7-sulfonamide, 3,4-dihydro-6-chloro-3-(5-norbornen-2-yl)-, 1,1-dioxide **(2259-96-3)**
2H-1,2,4-Benzothiadiazine-7-sulfonamide, 3,4-dihydro-6-(trifluoromethyl)-, 1,1-dioxide **(135-09-1)**

Name	CAS No.
4H-1,2,4-Benzothiadiazine-7-sulfonamide, 6-(trifluoromethyl)-, 1,1-dioxide	**(148-56-1)**
Benzothiadiazole (9CI)	**(73299-03-3)**
2,1,3-Benzothiadiazole, 1,3-dihydro-, 2,2-dioxide	**(1615-06-1)**
1H,3H-2,1,3-Benzothiadiazole 2,2-dioxide	**(1615-06-1)**
2,1,3-Benzothiadiazole, 4,5,7-trichloro	**(1982-55-4)**
Benzothiamide	**(2227-79-4)**
Benzothiazide	**(91-33-8)**
2H-1,4-Benzothiazin-3(4H)-one (8CI,9CI)	**(5325-20-2)**
2H-1,2-Benzothiazine-3-carboxamide, 4-hydroxy-2-methyl-N-2-pyridinyl-, 1,1-dioxide	**(36322-90-4)**
2-Benzothiazolamine, 5,6-dimethyl- (9CI)	**(29927-08-0)**
2-Benzothiazolamine, 6-ethoxy- (9CI)	**(94-45-1)**
2-Benzothiazolamine, N-ethyl- (9CI)	**(28291-69-2)**
2-Benzothiazolamine, 6-(methylsulfonyl)- (9CI)	**(17557-67-4)**
2-Benzothiazolamine, 6-nitro	**(6285-57-0)**
Benzothiazole	**(95-16-9)**
3(2H)-Benzothiazoleacetic acid, 4-chloro-2-oxo- (9CI)	**(3813-05-6)**
3(2H)-Benzothiazoleacetic acid, 4-chloro-2-oxo-, Mixt. with sodium (4-chloro-2-methylphenoxy)acetate (9CI)	**(65280-19-5)**
3(2H)-Benzothiazoleacetic acid, 4-chloro-2-oxo-, Mixt. with (+-)-2-(4-chloro-2-methylphenoxy)propanoic acid and (2,4-dichlorophenoxy)acetic acid (9CI)	**(93746-34-0)**
Benzothiazole, 2-amino	**(136-95-8)**
Benzothiazole, 2-amino-4-chloro	**(19952-47-7)**
Benzothiazole, 2-amino-5,6-dichloro	**(24072-75-1)**
Benzothiazole, 2-amino-5,6-dimethyl- (8CI)	**(29927-08-0)**
Benzothiazole, 2-amino-6-ethoxy- (8CI)	**(94-45-1)**
Benzothiazole, 2-amino-4-methoxy	**(5464-79-9)**
Benzothiazole, 2-amino-6-methoxy	**(1747-60-0)**
Benzothiazole, 2-amino-4-methyl	**(1477-42-5)**
Benzothiazole, 2-(p-aminophenyl)-6-methyl	**(92-36-4)**
Benzothiazole, butyl- (9CI)	**(100182-85-2)**
Benzothiazole, 2-((chloromethyl)thio)- (9CI)	**(28908-00-1)**
Benzothiazole, 2-((p-(N-(2-cyanoethyl)-N-ethylamino)phenylazo)-6-nitro-	**(25510-81-0)**
Benzothiazole, 2-(p-(dimethylamino)styryl)	**(1628-58-6)**
Benzothiazole disulfide	**(120-78-5)**
Benzothiazole, 2,2'-dithiobis	**(120-78-5)**
Benzothiazole, 2-(ethylamino)- (8CI)	**(28291-69-2)**
Benzothiazole, 2-hydrazino	**(615-21-4)**
Benzothiazole, 2-methyl	**(120-75-2)**
Benzothiazole, 4-methyl- (8CI,9CI)	**(3048-48-4)**
Benzothiazole, 2-(methylamino)	**(16954-69-1)**
Benzothiazole, 2-(methylthio)- (9CI)	**(615-22-5)**
Benzothiazole, 2-(1-methylureido)-	**(1929-88-0)**
Benzothiazole, 2-(morpholinodithio)- (8CI)	**(95-32-9)**
Benzothiazole, 2-(morpholinothio)	**(102-77-2)**
Benzothiazole, 2-(4-morpholinyldithio)- (9CI)	**(95-32-9)**
2-Benzothiazolesulfenamide, N-tert-butyl	**(95-31-8)**
2-Benzothiazolesulfenamide, N-cyclohexyl	**(95-33-0)**
2-Benzothiazolesulfenamide, N,N-dicyclohexyl	**(4979-32-2)**
2-Benzothiazolesulfenamide, N,N-diisopropyl	**(95-29-4)**
Benzothiazolesulfenamide, N-(1,1-dimethylethyl)- (9CI)	**(95-31-8)**
2-Benzothiazolesulfonamide, 6-ethoxy	**(452-35-7)**
6-Benzothiazolesulfonic acid, 2-amino- (9CI)	**(21951-32-6)**
7-Benzothiazolesulfonic acid, 2-(p-aminophenyl)-6-methyl	**(130-17-6)**
7-Benzothiazolesulfonic acid, 2-(4-((2-cyano-3-(4-(methyl-(2-sulfoethyl)amino)phenyl)-1-oxo-2-propenyl)amino)-phenyl)-6-methyl-, disodium salt (9CI)	**(2498-95-5)**
7-Benzothiazolesulfonic acid, 2-(p-(α-cyano-p-(methyl(2-sulfo-ethyl)amino)cinnamamido)phenyl)-6-methyl-, disodium salt	**(2498-95-5)**
7-Benzothiazolesulfonic acid, 2-(2-(2,3-dihydro-1,3-dioxo-1H-inden-2-yl)-6-quinolinyl)-6-methyl-, sodium salt (9CI)	**(4121-67-9)**
7-Benzothiazolesulfonic acid, 2-(4-((4,5-dihydro-3-methyl-5-oxo-1H-pyrazol-4-yl)azo)phenyl)-6-methyl-, monosodium salt (9CI)	**(21493-04-9)**
7-Benzothiazolesulfonic acid, 2-(4-((hexahydro-2,4,6-trioxo-5-pyrimidinyl)azo)phenyl)-6-methyl-, monosodium salt (9CI)	**(35294-62-3)**
7-Benzothiazolesulfonic acid, 2-(4-((1-(((2-methoxyphenyl)-amino)carbonyl)-2-oxopropyl)azo)-3-sulfophenyl)-6-methyl-, disodium salt (9CI)	**(10190-68-8)**
7-Benzothiazolesulfonic acid, 6-methyl-2-(2-methyl-6-quinolinyl)- (9CI)	**(3181-86-0)**
7-Benzothiazolesulfonic acid, 6-methyl-2-(2-methyl-6-quinolyl)-	**(3181-86-0)**
7-Benzothiazolesulfonic acid, 6-methyl-2-(4-((2,4,6-trihydroxy-pyrimidin-5-yl)azo)phenyl)-, sodium salt	**(35294-62-3)**
2-Benzothiazolesulfonic acid, sodium salt (9CI)	**(21465-51-0)**
2-Benzothiazolethiol	**(149-30-4)**
2-Benzothiazolethiol, potassium salt (8CI)	**(7778-70-3)**
2-Benzothiazolethiol, sodium deriv	**(2492-26-4)**
2-Benzothiazolethiol, sodium salt	**(2492-26-4)**
Benzothiazolethiol, sodium salt	**(2492-26-4)**
2-Benzothiazolethiol, zinc salt (2:1)	**(155-04-4)**
2(3H)-Benzothiazolethione, Compd. with cyclohexanamine (1:1) (9CI)	**(37437-20-0)**
2(3H)-Benzothiazolethione, Compd. with 2-methyl-2-propan-amine (1:1) (9CI)	**(63302-50-1)**
2(3H)-Benzothiazolethione, copper salt (9CI)	**(32510-27-3)**
2(3H)-Benzothiazolethione, potassium salt (9CI)	**(7778-70-3)**
2(3H)-Benzothiazolethione, sodium salt (9CI)	**(2492-26-4)**
3-Benzothiazolineacetic acid, 4-chloro-2-oxo- (8CI)	**(3813-05-6)**
2-Benzothiazolinone (8CI)	**(934-34-9)**
Benzothiazolium, 2-((4-((2,3-dimethoxypropyl)methylamino)-phenyl)azo)-6-methoxy-3-methyl-, chloride	**(12270-13-2)**
Benzothiazolium, 2-(4-(dimethylamino)phenyl)-3,6-dimethyl-, chloride (9CI)	**(2390-54-7)**
Benzothiazolium, 2-(p-(dimethylamino)phenyl)-3,6-dimethyl-, chloride (8CI)	**(2390-54-7)**
Benzothiazolium, 3-ethyl-2-(5-(3-ethyl-2(3H)-benzo-thiazolylidene)-1,3-pentadienyl)-, iodide (9CI)	**(514-73-8)**
Benzothiazolium, 3-ethyl-2-(5-(3-ethyl-2-benzo-thiazolinylidene)-1,3-pentadienyl)-, iodide (8CI)	**(514-73-8)**
Benzothiazolium, 2-((4-(ethyl(2-hydroxyethyl)amino)phenyl)-azo)-6-methoxy-3-methyl-, methyl sulfate (Salt) (9CI)	**(12270-13-2)**
2(3H)-Benzothiazolone (9CI)	**(934-34-9)**
2-Benzothiazolyl disulfide	**(120-78-5)**
Benzothiazolyl disulfide	**(120-78-5)**
N-(2-Benzothiazolyl)-N'-methylurea	**(1929-88-0)**
2-Benzothiazolyl morpholino disulfide	**(95-32-9)**
2-Benzothiazolyl N-morpholino sulfide	**(102-77-2)**
2-(2-Benzothiazolyl)phenol	**(3411-95-8)**
o-(2-Benzothiazolyl)phenol	**(3411-95-8)**
2-Benzothiazolylsulfenyl morpholine	**(102-77-2)**
4-(2-Benzothiazolylthio)morpholine	**(102-77-2)**
Benzothiazyl disulfide	**(120-78-5)**
4-Benzothienylester kyseliny methylkarbaminove (Czech)	**(1079-33-0)**
4-Benzothienyl methylcarbamate	**(1079-33-0)**
Benzothioamide	**(2227-79-4)**
Benzothiofuran	**(95-15-8)**
Benzothiophen	**(95-15-8)**
1-Benzothiophene	**(95-15-8)**
2,3-Benzothiophene	**(95-15-8)**
Benzothiophene (9CI) (VAN)	**(11095-43-5)**
1-Benzothiophene-4-ol	**(3610-02-4)**
2H-1-Benzothiopyran, 3,4-dihydro-2-phenyl- (9CI)	**(5961-99-9)**
4H-1-Benzothiopyran-4-one, 2,3-dihydro- (9CI)	**(3528-17-4)**
1,2,3-Benzotriazin-4(1H)-one	**(90-16-4)**
1,2,3-Benzotriazin-4(3H)-one, 3-(chloromethyl)- (9CI)	**(24310-41-6)**
1,2,3-Benzotriazin-4(3H)-one, 3-(4-(dimethylamino)phenyl)-(9CI)	**(55649-81-5)**
1,2,3-Benzotriazin-4(3H)-one, 3-(p-dimethylaminophenyl)- (6CI)	**(55649-81-5)**
1,2,3-Benzotriazin-4(3H)-one, 3-(mercaptomethyl)-, O,O-dimethyl phosphorodithioate	**(86-50-0)**
Benzotriazine derivative of a methyl dithiophosphate	**(86-50-0)**

Benzotriazine derivative of an ethyl dithiophosphate **(2642-71-9)**
Benzotriazinedithiophosphoric acid dimethoxy ester **(86-50-0)**
3H-1,2,3-Benzotriazin-4-one **(90-16-4)**
1,2,3-Benzotriazole **(95-14-7)**
1H-1,2,3-Benzotriazole **(95-14-7)**
1H-Benzotriazole **(95-14-7)**
Benzotriazole **(95-14-7)**
Benzotriazole (VAN) **(95-14-7)**
1H-Benzotriazole (VAN) (9CI) **(95-14-7)**
1H-Benzotriazole, 5-chloro- (9CI) **(94-97-3)**
1H-Benzotriazole, 1-methyl **(13351-73-0)**
1H-Benzotriazole, 5-methyl **(136-85-6)**
1H-Benzotriazole, methyl **(29385-43-1)**
1H-Benzotriazole, 4(or 5)-methyl-, sodium salt (9CI) **(64665-57-2)**
1H-Benzotriazole, silver(1+) salt (9CI) **(22257-44-9)**
2-(2H-Benzotriazol-2-yl)-4-(1,1,3,3-tetramethylbutyl)phenol **(3147-75-9)**
Benzotrichloride [UN 2226] **(98-07-7)**
Benzotrifluoride [UN 2338] **(98-08-8)**
Benzotrifluoride, 4-chloro-3,5-dinitro- **(393-75-9)**
Benzotrifluoride, 4-chloro-3-nitro- **(121-17-5)**
2,5,8-Benzotrioxacycloundecin-1,9-dione, 3,4,6,7-tetrahydro- (9CI) **(13988-26-6)**
Benzotropine **(86-13-5)**
Benzoxacyclotetradec-11-en-1-one, 14,16-dihydroxy-3-methyl-7-oxo-, trans- **(17924-92-4)**
1H-2-Benzoxacyclotetradecin-1-one, 3,4,5,6,7,8,9,10,11,12-decahydro-7,14,16-trihydroxy-3- methyl-, (3S,7X) **(26538-44-3)**
1H-2-benzoxacyclotetradecin-1-one, 3,4,5,6,7,8,9,10-octahydro-14,16-dihydroxy-3-methyl- 7-oxo-, (E) **(17924-92-4)**
Benzoxale **(91-81-6)**
3H-2,1-Benzoxathiol-3-one, 1,1-dioxide (9CI) **(81-08-3)**
2-Benzoxaxolol **(59-49-4)**
3,1-Benzoxazepine, 2-phenyl- (8CI,9CI) **(14300-21-1)**
2H-3,1-benzoxazine-2,4(1H)-dione **(118-48-9)**
1H-2,5-Benzoxazocine, 3,4,5,6,7-tetrahydro-5-methyl-1-phenyl **(13669-70-0)**
Benzoxazole **(273-53-0)**
5-Benzoxazoleacetic acid, 2-(4-chlorophenyl)-α-methyl **(51234-28-7)**
Benzoxazole, 5-chloro-2-methyl **(19219-99-9)**
Benzoxazole, 2,2'-(1,2-ethenediyl)bis(5-methyl- (9CI) **(1041-00-5)**
Benzoxazole, 2,2'-(1,2-ethenediyl)bis(5-methyl-, (E)- (9CI) **(17233-65-7)**
Benzoxazole, 2,2'-(1,2-ethenediyldi-4,1-phenylene)bis- (9CI) **(1533-45-5)**
Benzoxazole, 2-hydroxy- **(59-49-4)**
Benzoxazole, 2-(isobutylamino)- (8CI) **(28291-83-0)**
Benzoxazole, 2,2'-(2,5-thiophenediyl)bis- (9CI) **(2866-43-5)**
Benzoxazole, 2,2'-(2,5-thiophenediyl)bis(5-(1,1-dimethylethyl)- (9CI) **(7128-64-5)**
2-Benzoxazolinone **(59-49-4)**
Benzoxazolinone **(59-49-4)**
S-((3-Benzoxazolinyl-6-chloro-2-oxo)methyl) O,O-diethyl phosphorodithioate **(2310-17-0)**
2(3H)-Benzoxazolone **(59-49-4)**
Benzoxazolone **(59-49-4)**
2(3H)-Benzoxazolone, 6-((2-hydroxyethyl)sulfonyl)- (9CI) **(5031-74-3)**
1-Benzoxepin, 2,3,4,5-tetrahydro- (8CI,9CI) **(6169-78-4)**
Benzo(xyz)heptaphene (8CI,9CI) **(189-43-5)**
Benzoyl **(94-36-0)**
p-Benzoyl-N,N-dimethylaniline **(530-44-9)**
Benzoylacetic acid ethyl ester **(94-02-0)**
Benzoyl-aceton (German) **(93-91-4)**
Benzoylacetone **(93-91-4)**
2-Benzoylacetophenone **(120-46-7)**
ω-Benzoylacetophenone **(120-46-7)**
Benzoylacetyl **(579-07-7)**
Benzoyl alcohol **(100-51-6)**
Benzoylamide **(55-21-0)**
o-(Benzoylamino)phenyl disulfide **(135-57-9)**
Benzoyl anhydride **(93-97-0)**
N-Benzoylaniline **(93-98-1)**
Benzoyl azide **(582-61-6)**
Benzoylbenzene **(119-61-9)**
N-2 (5-Benzoyl-benzimidazole) carbamate de methyle (French) **(31431-39-7)**
5-Benzoyl-2-benzimidazolecarbamic acid methyl ester **(31431-39-7)**
N-(Benzoyl-5, benzimidazolyl)-2, carbamate de methyle (French) **(31431-39-7)**
(5-Benzoyl-1H-benzimidazol-2-yl)-carbamic acid methyl ester **(31431-39-7)**
2-Benzoylbenzoic acid **(85-52-9)**
2-Benzoylbenzoic acid, methyl ester **(606-28-0)**
Benzoyl bromide (9CI) **(618-32-6)**
Benzoyl-chloride **(98-88-4)**
Benzoyl chloride [UN 1736] **(98-88-4)**
Benzoyl chloride, 4-chloro- **(122-01-0)**
Benzoyl chloride, 3-chloro- (9CI) **(618-46-2)**
Benzoyl chloride, m-chloro- (8CI) **(618-46-2)**
Benzoyl chloride, 2,4-dichloro- **(89-75-8)**
Benzoyl chloride, 2,3-dichloro- (9CI) **(2905-60-4)**
Benzoyl chloride, 2,5-dichloro- (9CI) **(2905-61-5)**
Benzoyl chloride, 3,5-dichloro- (9CI) **(2905-62-6)**
Benzoyl chloride, dichloro- (9CI) **(25134-08-1)**
Benzoyl chloride, o-fluoro **(393-52-2)**
Benzoyl chloride, methoxy- (9CI) **(100-07-2)**
Benzoyl chloride, 4-methyl- (9CI) **(874-60-2)**
Benzoyl chloride, m-nitro **(121-90-4)**
Benzoyl chloride, o-nitro **(610-14-0)**
Benzoyl chloride, p-nitro **(122-04-3)**
Benzoyl chloride, 2-nitro- (9CI) **(610-14-0)**
Benzoyl chloride, 3-nitro- (9CI) **(121-90-4)**
Benzoyl chloride, (2,4,6-trichlorophenyl)hydrazone **(25939-05-3)**
Benzoyl chloride, α-(2,4,6-trichlorophenyl)hydrazono **(25939-05-3)**
N-Benzoyl-N-(3-chloro-4-fluorophenyl)-DL-alanine (9CI) **(58667-63-3)**
2-Benzoyl-4-chlorophenol **(85-19-8)**
Benzoyl cyanide O-(diethoxyphosphinothioyl)oxime **(14816-18-3)**
L-N-Benzoyl-N-(3,4-dichlorophenyl)alanine ethyl ester **(33878-50-1)**
N-Benzoyl-N-(3,4-dichlorophenyl)-DL-alanine ethyl ester **(22212-55-1)**
Benzoyldiethylamine **(1696-17-9)**
Benzoyl ecgonine **(519-09-5)**
Benzoylecgonine **(519-09-5)**
N-Benzoylferrioxamine B **(70-51-9)**
Benzoylglycine **(495-69-2)**
3-Benzoylhydratropic acid **(22071-15-4)**
m-Benzoylhydratropic acid **(22071-15-4)**
Benzoyl hydrazide **(613-94-5)**
Benzoyl hydrazine **(613-94-5)**
Benzoylhydroxamic acid **(495-18-1)**
Benzoyl methide **(98-86-2)**
3-Benzoyl-α-methylbenzeneacetic acid **(22071-15-4)**
Benzoylmethylecgonine **(50-36-2)**
Benzoyl methyl ketone **(579-07-7)**
3-(Benzoyloxy)estra-1,3,5(10)-trien-17-one **(2393-53-5)**
Benzoyloxytributylstannane **(4342-36-3)**
Benzoylperoxid (German) **(94-36-0)**
Benzoyl-peroxide **(94-36-0)**
Benzoyl peroxide (ACGIH,DOT,OSHA) **(94-36-0)**
Benzoyl peroxide, More than 72% but less than 95% as a paste (DOT) **(94-36-0)**
Benzoyl peroxide, More than 77% but less than 95% with water (DOT) **(94-36-0)**
Benzoyl-peroxide, More than 52% with inert solid (DOT) **(94-36-0)**
Benzoyl peroxide, Not less than 30% but not more than 52% with inert solid [UN 2089] **(94-36-0)**
Benzoyl peroxide, Not more than 72% as a paste [UN 2087] **(94-36-0)**
Benzoyl peroxide, Not more than 77% with water [UN 2090] **(94-36-0)**
Benzoyl peroxide, Technically pure (DOT) **(94-36-0)**
Benzoylperoxyde (Dutch) **(94-36-0)**
o-Benzoylphenol **(117-99-7)**
Benzoylphenylcarbinol **(119-53-9)**

1-Benzoyl-1-phenylethene (94-41-7)
2-(3-Benzoylphenyl)propionic acid (22071-15-4)
2-(m-Benzoylphenyl)propionic acid (22071-15-4)
1-Benzoyl-3-phenyl-1,2,4-triazole (79746-00-2)
Benzoylprop ethyl (22212-55-1)
Benzoylpseudotropeine (537-26-8)
Benzoylpseudotropine (537-26-8)
Benzoyl-psi-tropeine (537-26-8)
4-Benzoylpyridine (14548-46-0)
β-Benzoylstyrene (94-41-7)
o-Benzoyl sulfimide (81-07-2)
o-Benzoyl sulphimide (81-07-2)
Benzoyl superoxide (94-36-0)
o-Benzoyltropine (537-26-8)
1,12-Benzperylene (191-24-2)
1,2-Benzphenanthrene (218-01-9)
2,3-Benzphenanthrene (56-55-3)
3,4-Benzphenanthrene (195-19-7)
9,10-Benzphenanthrene (217-59-4)
Benzphetamine (156-08-1)
Benzphetamine chloride (5411-22-3)
Benzphetamine hydrochloride (5411-22-3)
Benzphos (2310-17-0)
Benzpropamine (60-13-9)
3,4-Benzpyren (German) (50-32-8)
1,2-Benzpyrene (192-97-2)
3,4-Benzpyrene (50-32-8)
13H-Benz(qr)indeno(6,7,1,2-defg)naphthacen-13-one (9CI) (86854-25-3)
Benzquinamide (63-12-7)
Benzquinamidum (Latin) (63-12-7)
Benzo-2-sulphimide (81-07-2)
Benzo-sulphinide (81-07-2)
1,2-Benzo-9-thiafluorene (239-35-0)
Benzthiazide (91-33-8)
3-(2-Benzthiazolyl)-1-methylharnstoff (German) (1929-88-0)
N-(2-Benzthiazolyl)-N'-methylharnstoff (German) (1929-88-0)
Benzthiazuron (1929-88-0)
Benztriazole (95-14-7)
Benztropine (86-13-5)
Benzulfide (741-58-2)
Benzuride (3134-12-1)
Benzuron (3134-12-1)
Benzydroflumethiazide (73-48-3)
Benzydyna (Polish) (92-87-5)
Benzyfuroline (10453-86-8)
Benzyhydrylcyanide (86-29-3)
Benzyl Blue R (3861-73-2)
Benzyl Fast Blue R (3861-73-2)
Benzyl Fast Red BG (6459-94-5)
Benzyl Fast Red GRG (3567-65-5)
Benzyl Fast Yellow RS (6375-55-9)
Benzyl Mustard Oil (622-78-6)
Benzyl Red BR (6459-94-5)
Benzyl Red GR (3567-65-5)
Benzyl Red MG (10169-02-5)
Benzyl Red ROC (1658-56-6)
Benzyl Red S (1658-56-6)
Benzyl Reg GS (10169-02-5)
Benzyl Scarlet 3BS (6358-29-8)
Benzyl Violet (1694-09-3)
Benzyl Violet 3B (1694-09-3)
Benzyl Violet 4B (1694-09-3)
Benzylacetamide (588-46-5)
N-Benzylacetamide (588-46-5)
Benzyl acetate (140-11-4)
Benzylacetic acid (501-52-0)
Benzylacetone (2550-26-7)
6-Benzyladenine (1214-39-7)
Benzyladenine (1214-39-7)
N^6-Benzyladenine (1214-39-7)
N-Benzyladenine (1214-39-7)
N6-Benzyladenine (1214-39-7)
N-benzyladine (1214-39-7)
Benzyl-alcohol (100-51-6)
Benzyl alcohol, α-(1-aminoethyl)-2,5-dimethoxy-, hydrochloride (61-16-5)
Benzyl alcohol, α-(1-aminoethyl)-, hydrochloride, (+-) (154-41-6)
Benzyl alcohol, α-(1-aminoethyl)-m-hydroxy-, (-) (54-49-9)
Benzyl alcohol, α-(aminomethyl) (7568-93-6)
Benzyl alcohol, m-amino-α-methyl- (8CI) (2454-37-7)
Benzyl alcohol, α-(aminomethyl)-3,4-dihydroxy-, (-) (51-41-2)
Benzyl alcohol, α-(aminomethyl)-3,4-dihydroxy-, hydrochloride-, (+-) (55-27-6)
Benzyl alcohol, α-(aminomethyl)-p-hydroxy (104-14-3)
Benzyl alcohol benzoic ester (120-51-4)
Benzyl alcohol, α-((t-butylamino)methyl)-3,5-dihydroxy (23031-25-6)
Benzyl alcohol, p-chloro- (8CI) (873-76-7)
Benzyl alcohol, o-chloro-α-(dichloromethyl)- (27683-60-9)
Benzyl alcohol, p-chloro-α-methyl- (8CI) (3391-10-4)
Benzyl alcohol, cinnamate (103-41-3)
Benzyl alcohol, cinnamic ester (103-41-3)
Benzyl alcohol, 3,5-di-tert-butyl-4-hydroxy (88-26-6)
Benzyl alcohol, 2,4-dichloro (1777-82-8)
Benzyl alcohol, 2,6-dichloro (15258-73-8)
Benzyl alcohol, 3,4-dichloro- (8CI) (1805-32-9)
Benzyl alcohol, 2,4-dichloro-α-(chloromethylene)-, diethyl phosphate (470-90-6)
Benzyl alcohol, 3,4-dihydroxy-α-((isopropylamino)methyl)-, hydrochloride (51-30-9)
Benzyl alcohol, 3,4-dihydroxy-α-((methylamino)methyl)-, (+-) (329-65-7)
Benzyl alcohol, 3,4-dihydroxy-α-((methylamino)methyl)-, (-) (51-43-4)
Benzyl alcohol, 3,4-dihydroxy-α-((methylamino)methyl)-, hydrochloride, (+-) (329-63-5)
Benzyl alcohol, 3,4-dihydroxy-α-((methylamino)methyl)-, hydrochloride, (-) (55-31-2)
Benzyl alcohol, 3,4-dihydroxy-α-((methylamino)methyl)-, (-)-, tartrate (1:1), (+) (51-42-3)
Benzyl alcohol, α,α-dimethyl (617-94-7)
Benzyl alcohol, 2,4-dimethyl- (8CI) (16308-92-2)
Benzyl alcohol, 3,4-dimethyl- (8CI) (6966-10-5)
Benzyl alcohol, 3,5-dimethyl- (8CI) (27129-87-9)
Benzyl alcohol, α-epoxyethyl-1,2-(methylenedioxy)-, acetate (59901-90-5)
Benzyl alcohol, α-ethyl (93-54-9)
Benzyl alcohol, α-ethyl-o-methoxy- (8CI) (7452-01-9)
Benzyl alcohol, formate (104-57-4)
Benzyl alcohol, hexahydro- (100-49-2)
Benzyl alcohol, o-hydroxy (90-01-7)
Benzyl alcohol, m-hydroxy- (8CI) (620-24-6)
Benzyl alcohol, p-hydroxy- (8CI) (623-05-2)
Benzyl alcohol, 4-hydroxy-3,5-dimethoxy- (8CI) (530-56-3)
Benzyl alcohol, o-hydroxy-, o-glucoside (138-52-3)
Benzyl alcohol, p-hydroxy-α-methyl-, 4-acetate (53744-50-6)
Benzyl alcohol, m-hydroxy-α-((methylamino)methyl)-, (-) (59-42-7)
Benzyl alcohol, m-hydroxy-α-((methylamino)methyl)-, hydrochloride, (-) (61-76-7)
Benzyl alcohol, p-hydroxy-α-(1-((1-methyl-2-phenoxyethyl)amino)-ethyl) (395-28-8)
Benzyl alcohol, p-hydroxy-α-(1-((1-methyl-3-phenylpropyl)amino)-ethyl)- (447-41-6)
Benzyl alcohol, m-iodo (57455-06-8)
Benzyl alcohol, p-isopropyl- (536-60-7)
Benzyl alcohol, p-methoxy (105-13-5)
Benzyl alcohol, p-methoxy-, acetate (8CI) (104-21-2)
Benzyl alcohol, p-methoxy-α-vinyl (51410-44-7)
Benzyl alcohol, α-methyl (98-85-1)

Benzyl alcohol, p-methyl **(589-18-4)**
Benzyl alcohol, o-methyl- (8CI) **(89-95-2)**
Benzyl alcohol, α-methyl-, acetate (8CI) **(93-92-5)**
Benzyl alcohol, 3,4-(methylenedioxy)- **(495-76-1)**
Benzyl alcohol, o-nitro- (8CI) **(612-25-9)**
Benzyl alcohol, p,α-dimethyl **(536-50-5)**
Benzyl alcohol, 2,3,4,5,6-pentachloro- (8CI) **(16022-69-8)**
Benzyl alcohol, m-phenoxy **(13826-35-2)**
Benzyl alcohol, 2,4,5-trichloro-α-(chloromethylene)-, dimethyl phosphate **(961-11-5)**
Benzyl alcohol, α-(trichloromethyl) **(2000-43-3)**
Benzyl alcohol, α-(trichloromethyl)-, acetate **(90-17-5)**
Benzyl alcohol, 2,4,6-trimethyl **(4170-90-5)**
Benzyl alcohol, α,2,4-trimethyl- (8CI) **(5379-19-1)**
Benzyl alcohol, α,3,4-trimethyl- (8CI) **(33967-19-0)**
Benzylamine (8CI) **(100-46-9)**
Benzylamine, N-(2-chloroethyl)-N-(1-methyl-2-phenoxyethyl) **(59-96-1)**
Benzylamine, N-(2-chloroethyl)-N-(1-methyl-2-phenoxyethyl)-, hydrochloride **(63-92-3)**
Benzylamine, N,N-dimethyl **(103-83-3)**
Benzylamine, dimethyl-, hydrochloride **(1875-92-9)**
Benzylamine, N-ethyl-N-phenyl **(92-59-1)**
Benzylamine, hydrochloride **(3287-99-8)**
Benzylamine, α-methyl **(98-84-0)**
Benzylamine, N-methyl- (8CI) **(103-67-3)**
Benzylamine, N-methyl-N-nitroso **(937-40-6)**
Benzylamine, N,N,α-trimethyl- (8CI) **(2449-49-2)**
2-(Benzylamino)ethanol **(104-63-2)**
2-Benzylaminoethanol **(104-63-2)**
Benzylaminoethanol **(104-63-2)**
Benzyl-6-aminopenicillinic acid **(61-33-6)**
6-(Benzylamino)purine **(1214-39-7)**
6-(N-Benzylamino)purine **(1214-39-7)**
Benzylaminopurine **(1214-39-7)**
N^6-(Benzylamino)purine **(1214-39-7)**
6-Benzylamino-9-tetrahydropyran-2-yl-9H-purine **(2312-73-4)**
Benzylammonium chloride **(3287-99-8)**
Benzylbenzene **(101-81-5)**
Benzyl benzenecarboxylate **(120-51-4)**
Benzyl benzoate **(120-51-4)**
N-Benzylbenzylamine **(103-49-1)**
Benzyl bromide [UN 1737] **(100-39-0)**
Benzyl bromide, 2-chloro **(611-17-6)**
Benzyl bromoacetate **(5437-45-6)**
Benzyl n-butanoate **(103-37-7)**
Benzyl-t-butanol **(103-05-9)**
Benzyl-butylester kyseliny ftalove (Czech) **(85-68-7)**
Benzyl butyl phthalate **(85-68-7)**
Benzyl n-butyl phthalate **(85-68-7)**
Benzyl butyrate **(103-37-7)**
Benzyl n-butyrate **(103-37-7)**
1-(2-(Benzylcarbamoyl)ethyl)-2-isonicotinoylhydrazine **(51-12-7)**
N^1-β-Benzylcarbamoylethyl-N^2-isonicotinoylhydrazine **(51-12-7)**
N-(2-(Benzylcarbamyl)ethylamino)isonicotinamide **(51-12-7)**
2-(2-(Benzylcarbamyl)ethyl)hydrazide isonicotinic acid **(51-12-7)**
Benzyl carbinol **(60-12-8)**
Benzylcarbinyl acetate **(103-45-7)**
Benzylcarbinyl anthranilate **(133-18-6)**
Benzylcarbinyl propionate **(122-70-3)**
Benzylcarbinyl α-toluate **(102-20-5)**
Benzylcarbonyl chloride **(501-53-1)**
N(1)-Benzyl-3-carboxylpyridinium chloride **(16214-98-5)**
1-Benzyl-3-carboxypyridinium chloride **(16214-98-5)**
N-Benzyl-3-carboxypyridinium chloride **(16214-98-5)**
N-Benzyl-3-carboxypyridinum chloride **(16214-98-5)**
Benzyl cellosolve **(622-08-2)**
Benzylcelosolv (Czech) **(622-08-2)**
Benzylchlorid (German) **(100-44-7)**
Benzyl chloride (ACGIH,DOT,OSHA) **(100-44-7)**
Benzyl chloride unstabilized [UN 1738] **(100-44-7)**
Benzyl chloroacetate **(140-18-1)**
Benzyl-α-chloroacetate **(140-18-1)**
Benzyl chlorocarbonate (DOT) **(501-53-1)**
2-(N-Benzyl-2-chloroethylamino)-1-phenoxypropane **(59-96-1)**
2-(N-Benzyl-2-chloroethylamino)-1-phenoxypropane hydrochloride **(63-92-3)**
Benzyl(2-chloroethyl)-(1-methyl-2-phenoxyethyl)amine **(59-96-1)**
Benzyl(2-chloroethyl)(1-methyl-2-phenoxyethyl)amine hydrochloride **(63-92-3)**
Benzyl chloroformate [UN 1739] **(501-53-1)**
2-Benzyl-4-chlorophenol **(120-32-1)**
Benzylchlorophenol **(120-32-1)**
o-Benzyl-p-chlorophenol **(120-32-1)**
o-Benzyl-p-chlorophenol potassium salt **(35471-49-9)**
2-Benzyl-4-chlorophenol, sodium salt **(3184-65-4)**
o-Benzyl-p-chlorophenol sodium salt **(3184-65-4)**
Benzyl o-chlorophenyl ether **(949-38-2)**
Benzyl cinnamate **(103-41-3)**
Benzyl cyanide **(140-29-4)**
S-Benzyl N,N-di-sec-butyl thiocarbamate **(36756-79-3)**
Benzyl dichloride **(98-87-3)**
N-Benzyl-N-(dichloro-3,4-phenyl)-N',N'-dimethylurea **(3134-12-1)**
N-Benzyldiethylamine **(772-54-3)**
S-Benzyl-O,O-diethylester kyseliny thiofosforecne (Czech) **(13286-32-3)**
Benzyldiethyl((2,6-xylylcarbamoyl)methyl)ammonium benzoate **(3734-33-6)**
3-Benzyl-3,4-dihydro-6-(trifluoromethyl)-2H-1,2,4-benzothiadiazine-7-sulfonamide 1,1-dioxide **(73-48-3)**
Benzyl-N,N-dimethylamine **(103-83-3)**
N-Benzyldimethylamine **(103-83-3)**
Benzyl dimethylamine (DOT) **(103-83-3)**
Benzyldimethylamine [UN 2619] **(103-83-3)**
Benzyldimethylamine methiodide **(4525-46-6)**
2-(Benzyl(2-dimethylaminoethyl)amino)pyridine **(91-81-6)**
2-(N-Benzyl-N-(2-dimethylaminoethyl)amino)pyridine **(91-81-6)**
2-(Benzyl(2-(dimethylamino)ethyl)amino)pyridine hydrochloride **(154-69-8)**
N-Benzyl-N-dimethylaminoethyl α-aminopyridine hydrochloride **(154-69-8)**
2-(Benzyl(2-(dimethylamino)ethyl)amino)pyridine monohydrochloride **(154-69-8)**
Benzyldimethyl carbinyl acetate **(151-05-3)**
Benzyldimethylcetylammonium chloride **(122-18-9)**
Benzyldimethyldodecylammonium bromide **(7281-04-1)**
2-Benzyl-1,3-dimethylguanidine **(55-73-2)**
Benzyldimethylhexadecylammonium chloride **(122-18-9)**
Benzyl 2,2-dimethyloctanoate **(81325-79-3)**
N-Benzyl-N',N'-dimethyl-N-phenylethylenediamine **(961-71-7)**
N-Benzyl-N',N'-dimethyl-N-2-pyridylethylenediamine **(91-81-6)**
N-Benzyl-N',N'-dimethyl-N-2-pyridyl-ethylenediamine hydrochloride **(154-69-8)**
Benzyldimethylstearylammonium chloride **(122-19-0)**
Benzyldimethyl(2-(2-(p-(1,1,3,3-tetramethylbutyl)phenoxy)ethoxy)-ethyl)ammonium chloride **(121-54-0)**
Benzyldimethyl-p-(1,1,3,3-tetramethylbutyl)phenoxyethoxy-ethylammonium chloride **(121-54-0)**
Benzyldodecyldimethylammonium bromide **(7281-04-1)**
Benzyle (chlorure de) (French) **(100-44-7)**
Benzylene chloride **(98-87-3)**
Benzylester kyseliny benzoove (Czech) **(120-51-4)**
Benzylester kyseliny maselne (Czech) **(103-37-7)**
Benzylester kyseliny mravenci (Czech) **(104-57-4)**
Benzylester kyseliny octove (Czech) **(140-11-4)**
Benzylester kyseliny skoricove (Czech) **(103-41-3)**
Benzyl ethanoate **(140-11-4)**
Benzyl ethanolamine **(104-63-2)**
Benzylethanolamine **(104-63-2)**
N-Benzylethanolamine **(104-63-2)**

Benzyl-ether (103-50-4)
3-((p-(Benzylethylamino)phenyl)azo)-1H-1,2,4-triazole (13486-13-0)
1-Benzyl-1-ethylaziridinium 2,4,6-trinitrobenzenesulfonate (3806-34-6)
Benzyl ethyl ether (539-30-0)
Benzyl ethyl oxide (539-30-0)
Benzylethylphenylamine (92-59-1)
Benzylets (120-51-4)
9-Benzylfluorene (1572-46-9)
Benzyl fluoride (350-50-5)
Benzyl formate (104-57-4)
5-Benzyl-3-furylmethyl(+)-trans-chrysanthemate (28434-01-7)
5-Benzyl-3-furylmethyl(+-)-cis,trans-chrysanthemate (10453-86-8)
(5-Benzyl-3-furyl) methyl-2,2-dimethyl-3-(2-methylpropenyl)-cyclopropanecarboxylate (10453-86-8)
5-Benzyl-3-furylmethyl (+)-trans-ethanochrysanthemate (22431-62-5)
Benzyl hemiformal (14548-60-8)
Benzylhydroflumethiazide (73-48-3)
Benzyl hydroquinone (103-16-2)
Benzylhydrosulfide (100-53-8)
Benzyl 4-hydroxybenzoate (94-18-8)
Benzyl o-hydroxybenzoate (118-58-1)
Benzyl p-hydroxybenzoate (94-18-8)
1-Benzyl-1-(2-hydroxyethyl)-2-(tall oil alkyl)-2-imidazolinium chloride (68309-34-2)
1-Benzyl-1-(2-hydroxyethyl)-2-tall oil alkyl-2-imidazolinium chloride (68309-34-2)
N-Benzylidenaniline (538-51-2)
Benzylideneacetaldehyde (104-55-2)
Benzylidene acetone (122-57-6)
2-Benzylideneacetophenone (94-41-7)
Benzylideneacetophenone (94-41-7)
α-Benzylideneacetophenone (94-41-7)
Benzylideneaniline (538-51-2)
N-Benzylideneaniline (538-51-2)
1,1'-(Benzylidenebis((2-methoxy-p-phenylene)(azo))di-2-naphthol) (6483-64-3)
3-Benzylidene-2-bornanone (15087-24-8)
3-Benzylidene camphor (15087-24-8)
Benzylidene chloride [UN 1886] (98-87-3)
2-Benzylideneheptanal (122-40-7)
Benzylidyne chloride (98-07-7)
Benzylidyne fluoride (98-08-8)
Benzyl iodide [UN 2653] (620-05-3)
Benzyl isoamyl ether (122-73-6)
N-Benzyl-β-(isonicotinoylhydrazine)propionamide (51-12-7)
N-Benzyl-β-(isonicotinylhydrazino)propionamide (51-12-7)
Benzyl isopentyl ether (122-73-6)
Benzyl-isothiocyanate (622-78-6)
Benzylisothiourea hydrochloride (538-28-3)
2-Benzylisothiouronium chloride (538-28-3)
Benzylisothiouronium chloride (538-28-3)
Benzyl ketone (102-04-5)
Benzylkyanid (Czech) (140-29-4)
Benzyl-lauryldimethylammonium chloride (139-07-1)
Benzyl mercaptan (100-53-8)
Benzyl methacrylate (2495-37-6)
Benzyl methanoate (104-57-4)
Benzylmethylamine (103-67-3)
N-Benzylmethylamine (103-67-3)
Benzyl methyl ether (538-86-3)
1-Benzyl-2-methylhydrazine (10309-79-2)
Benzyl methyl ketone (103-79-7)
Benzyl monochloracetate (140-18-1)
Benzyl monosulfide (538-74-9)
Benzylmorphine (14297-87-1)
Benzyl nitrile (140-29-4)
Benzyloleyldimethylammonium chloride (37139-99-4)
Benzyl oxide (Czech) (103-50-4)
Benzyloxycarbonyl chloride (501-53-1)
2-Benzyloxyethanol (622-08-2)
4-Benzyloxyphenol (103-16-2)
p-Benzyloxyphenol (103-16-2)
3-(Benzyloxy)propionitrile (6328-48-9)
Benzyl paraben (94-18-8)
Benzylparaben (94-18-8)
Benzyl parahydroxybenzoate (94-18-8)
Benzyl parasept (94-18-8)
Benzyl-parasept (94-18-8)
(5R,6R)-Benzylpenicillin (61-33-6)
Benzylpenicillin (61-33-6)
Benzylpenicillin G (61-33-6)
Benzylpenicillin benzathine (1538-09-6)
Benzylpenicillin dibenzylethylenediamine salt (1538-09-6)
Benzylpenicillinic acid (61-33-6)
Benzylpenicillinic acid potassium salt (113-98-4)
Benzylpenicillinic acid procaine salt (6130-64-9)
Benzylpenicillinic acid sodium salt (69-57-8)
Benzylpenicillin novocaine salt (54-35-3)
Benzylpenicillin potassium (113-98-4)
Benzylpenicillin potassium salt (113-98-4)
Benzylpenicillin procaine (54-35-3)
Benzylpenicillin procaine monohydrate (6130-64-9)
Benzylpenicillin procaine salt (6130-64-9)
Benzylpenicillin sodium (69-57-8)
Benzylpenicillin sodium salt (69-57-8)
2-Benzylphenol (28994-41-4)
4-Benzylphenol (101-53-1)
N-Benzyl-N-phenoxyisopropyl-β-chlorethylamine hydrochloride (63-92-3)
Benzyl γ-phenylacrylate (103-41-3)
Benzyl phenylformate (120-51-4)
Benzyl phenyl ketone (451-40-1)
N-Benzylpicolinonium chloride (26747-91-1)
Benzyl propionate (122-63-4)
2-Benzylpyridine (101-82-6)
4-Benzylpyridine (2116-65-6)
Benzyl-(α-pyridyl)-dimethylaethylendiamin (German) (91-81-6)
N-Benzyl-N-α-pyridyl-N',N'-dimethyl-aethylendiamin-hydrochlorid (German) (154-69-8)
N-Benzyl-N-(2-pyridyl)-N',N'-dimethyl ethylenediamine (91-81-6)
1-Benzylquinolinium chloride (15619-48-4)
Benzylquinolinium chloride (15619-48-4)
N-Benzylquinolinium chloride (15619-48-4)
Benzylrodiuran (73-48-3)
Benzyl salicylate (118-58-1)
Benzylsenfoel (German) (622-78-6)
Benzylstearyldimethylammonium chloride (122-19-0)
Benzyl sulfide (538-74-9)
Benzyl-α-sulfonic acid (100-87-8)
Benzylsulfonic acid (100-87-8)
Benzyl tegosept (94-18-8)
Benzyl-2,2,4,4-tetramethylpentanoate (81325-80-6)
Benzyl-thiocyanate (3012-37-1)
Benzyl thioether (538-74-9)
Benzylthiol (100-53-8)
3-((Benzylthio)methyl)-6-chloro-1,2,4-benzothiadiazine-7-sulfonamide 1,1-dioxide (91-33-8)
3-Benzylthiomethyl-6-chloro-2H-1,2,4-benzothiadiazine-7-sulfonamide 1,1-dioxide (91-33-8)
3-Benzylthiomethyl-6-chloro-7-sulfamoyl-1,2,4-benzothiadiazine 1,1-dioxide (91-33-8)
Benzyl thiopseudourea hydrochloride (538-28-3)
Benzylthiuronium chloride (538-28-3)
S-Benzylthiuronium chloride (538-28-3)
Benzyl trichloride (98-07-7)
3-Benzyl-6-trifluoromethyl-7-sulfamoyl-3,4-dihydro-1,2,4-benzo-

thiadiazine, 1,1-dioxide (73-48-3)
N-Benzyl-N,N,N-trimethylammonium bromide (5350-41-4)
Benzyltrimethylammonium bromide(btm) (5350-41-4)
Benzyltrimethylammonium chloride (56-93-9)
Benzyltrimethylammonium hydroxide (100-85-6)
Benzyl-trimethylammonium iodide (4525-46-6)
Benzyl vinyl ketone (37442-55-0)
Benzylyt (59-96-1)
Benzylyt (63-92-3)
3,4-Benzypyrene (50-32-8)
Benzytol (88-04-0)
Beosit (115-29-7)
Bepanthen (81-13-0)
Bepanthene (81-13-0)
Bepantol (81-13-0)
Beparon (71-91-0)
Bephenium (7181-73-9)
Bequin (67-03-8)
Berberine sulfate (316-41-6)
Berberine sulfate (2:1) (316-41-6)
Berberin sulfate (316-41-6)
Berbinium, 7,8,13,13a-tetradehydro-9,10-dimethoxy-2,3-(methylene-dioxy)-, sulfate (2:1) (316-41-6)
Bercema (12122-67-7)
Bercema (74-83-9)
Bercema Fertam 50 (14484-64-1)
Bercema NMC50 (63-25-2)
Berelex (77-06-5)
Bergamiol (115-95-7)
Bergamot-Oil-Rectified (8007-75-8)
Bergamotte Oel (German) (8007-75-8)
Bergaptan (484-20-8)
Bergapten (484-20-8)
Bergaptene (484-20-8)
Berin (67-03-8)
Berkendyl (127-65-1)
Berkflam B 10 (13654-09-6)
Berkflam B 10E (1163-19-5)
Berkfurin (67-20-9)
Berkomine (50-49-7)
Berkozide (73-48-3)
Berlin White (1319-46-6)
Berlophen (94-19-9)
Bermat (6164-98-3)
Bermuda Blue (147-14-8)
Bernarenin (51-43-4)
Bernice (50-36-2)
Bernies (50-36-2)
Bernocaine (51-05-8)
Bernsteinsaeure-2,2-dimethylhydrazid (German) (1596-84-5)
Bernsteinsaure (German) (110-15-6)
Bernsteinsaure-anhydrid (German) (108-30-5)
Berol 043 (9005-00-9)
Berol 08 (9005-00-9)
Berol 370 (9003-11-6)
Berol 452 (151-21-3)
Berol 478 (577-11-7)
Berol TVM 370 (9003-11-6)
Beromycin (132-98-9)
Beromycin (87-08-1)
Beromycin 400 (132-98-9)
Beromycin (Penicillin) (132-98-9)
Beronald (54-31-9)
Bertholite (7782-50-5)
Berthollet Salt (3811-04-9)
Berthollet's Salt (3811-04-9)
Bertrandite (12161-82-9)
Berubigen (68-19-9)
Beryl (1302-52-9)
Beryl Ore (1302-52-9)
Beryllia (1304-56-9)
Beryllium-9 (7440-41-7)
Beryllium (ACGIH,OSHA) (7440-41-7)
Beryllium, Metal powder [UN 1567] (7440-41-7)
Beryllium acetate (543-81-7)
Beryllium acetate, basic (19049-40-2)
Beryllium acetate normal (543-81-7)
Beryllium-aluminium alloy (12770-50-2)
Beryllium aluminium silicate (1302-52-9)
Beryllium aluminosilicate (1302-52-9)
Beryllium aluminum alloy (66104-24-3)
Beryllium aluminum silicate (1302-52-9)
Beryllium, bis(carbonato(2-))dihydroxytri- (66104-24-3)
Beryllium carbonate (66104-24-3)
Beryllium carbonate (1:1) (13106-47-3)
Beryllium carbonate, Basic (66104-24-3)
Beryllium chloride (DOT) (7787-47-5)
Beryllium, Compd. with copper (9CI) (64535-95-1)
Beryllium compound (66104-24-3)
Beryllium-copper-cobalt alloy (55158-44-6)
Beryllium-copper alloy (11133-98-5)
Beryllium dichloride (7787-47-5)
Beryllium difluoride (7787-49-7)
Beryllium dihydroxide (13327-32-7)
Beryllium dinitrate (13597-99-4)
Beryllium fluoride (DOT) (7787-49-7)
Beryllium, hexakis(mu-(acetato-O:O'))-mu[4]-oxotetra- (9CI) (19049-40-2)
Beryllium, hexakis(mu-acetato)-mu[4]-oxotetra (19049-40-2)
Beryllium hydrate (13327-32-7)
Beryllium hydrogen phosphate (1:1) (13598-15-7)
Beryllium-hydroxide (13327-32-7)
Beryllium monoxide (1304-56-9)
Beryllium-nickel alloy (37227-61-5)
Beryllium nitrate (7787-55-5)
Beryllium nitrate [UN 2464] (13597-99-4)
Beryllium-oxide (1304-56-9)
Beryllium oxide acetate (19049-40-2)
Berylliumoxide carbonate (66104-24-3)
Beryllium oxyacetate (19049-40-2)
Beryllium phosphate (BeHPO4) (13598-15-7)
Beryllium orthosilicate (15191-85-2)
Beryllium silicate (15191-85-2)
Beryllium silicate hydrate (12161-82-9)
Beryllium silicic acid (15191-85-2)
Beryllium sulfate (1:1) (13510-49-1)
Beryllium sulfate, tetrahydrate (1:1:4) (7787-56-6)
Beryllium sulphate (13510-49-1)
Beryllium sulphate tetrahydrate (7787-56-6)
Beryllium zinc silicate (39413-47-3)
Besan (90-82-4)
Betabion (59-43-8)
Betabion hydrochloride (67-03-8)
Betacarotene (7235-40-7)
Betacaroteno (Spanish) (7235-40-7)
Betacarotenum (INN-Latin) (7235-40-7)
Betacide P (94-13-3)
Betadine (25655-41-8)
Betafedrina (51-63-8)
Betafedrine (51-63-8)
Betafen (60-13-9)
Betainchloralum (2218-68-0)
Betaine (107-43-7)
Betaine (19223-55-3)
Betaine cephaloridine (50-59-9)

Betaine nicotinate **(535-83-1)**
Betaines, coco alkyldimethyl **(68424-94-2)**
Betain nicotinate **(535-83-1)**
Betalgil **(126-27-2)**
Betalin-12 **(68-19-9)**
Betalin S **(59-43-8)**
Betalin S **(67-03-8)**
Betalin 12 crystalline **(68-19-9)**
Betaline-12 **(68-19-9)**
Betamec **(741-58-2)**
Betameprodine **(468-50-8)**
Betametadol (Spanish) **(17199-55-2)**
Betametadolo **(17199-55-2)**
Betamethadol **(17199-55-2)**
Betamethadolum (Latin) **(17199-55-2)**
Betamethasone **(378-44-9)**
Betamethazone **(378-44-9)**
Betanal **(13684-63-4)**
Betanal AM **(13684-56-5)**
Betanaphthol Orange **(633-96-5)**
Betanex **(13684-56-5)**
Betapal **(120-23-0)**
Betapen-Vk **(132-98-9)**
d-Betaphedrine **(51-63-8)**
Betaphen **(60-13-9)**
Betaprene H **(9003-31-0)**
Betaprodina (Spanish) **(468-59-7)**
Betaprodine **(468-59-7)**
Betaprodinum (Latin) **(468-59-7)**
Betaprone **(57-57-8)**
Betapyrimidum **(59-26-7)**
Betasan **(741-58-2)**
Betaxin **(59-43-8)**
Betaxin **(67-03-8)**
Betaxina **(389-08-2)**
Betazed **(50-33-9)**
Bethabarra wood **(84-79-7)**
Bethiamin **(59-43-8)**
Bethiazine **(67-03-8)**
Bethrodine **(1861-40-1)**
Betnelan **(378-44-9)**
Betoxan **(581-96-4)**
Betoxon **(120-23-0)**
Betrachotoxinin A, 20-α-(2,4-dimethyl-1H-pyrrole-3-carboxylate) (9CI) **(23509-16-2)**
Betramin **(58-73-1)**
Betsolan **(378-44-9)**
Betula **(119-36-8)**
Betula Oil **(119-36-8)**
Beuion **(67-03-8)**
Bevatine-12 **(68-19-9)**
Bevidox **(68-19-9)**
Bevitex **(67-03-8)**
Bevitine **(67-03-8)**
Bewon **(59-43-8)**
Bewon **(67-03-8)**
Bexane **(94-81-5)**
Bexol **(58-89-9)**
Bexon **(443-48-1)**
Bexone **(94-81-5)**
Bexton **(1918-16-7)**
Bexton 4L **(1918-16-7)**
Bextrene XL 750 **(9003-53-6)**
Bezitramida (Spanish) **(15301-48-1)**
Bezitramide **(15301-48-1)**
Bezitramidum (Latin) **(15301-48-1)**
Bhang **(8063-14-7)**
Bhimsaim Camphor **(507-70-0)**
Bi 3411 **(302-17-0)**
Bi-Act **(21342-85-8)**
4',4'''-Biacetanilide **(613-35-4)**
4',4'''-Bi-o-acetoacetotoluidide (8CI) **(91-96-3)**
Biacetyl **(431-03-8)**
Biacetyl, dioxime **(95-45-4)**
Biacetylmonoxime **(57-71-6)**
Bialflavina **(8048-52-0)**
Bialminal **(50-06-6)**
Bialpirinia **(50-78-2)**
Bialzepam **(439-14-5)**
4,4'-Bianiline **(92-87-5)**
N,N'-Bianiline **(122-66-7)**
o,p'-Bianiline **(492-17-1)**
p,p-Bianiline **(92-87-5)**
Bianisidine **(119-93-7)**
(3,3'-Bianthra(1,9-cd)pyrazole)-6,6'(1H,1'H)-dione, 1,1'-diethyl **(4203-77-4)**
(1,1'-Bianthracene)-9,9',10,10'-tetrone, 4,4'-diamino- (9CI) **(4051-63-2)**
5,5'-Bianthranilic acid **(2130-56-5)**
5,5'-Bibarbituric acid, 5,5'-dihydroxy- (8CI) **(76-24-4)**
Bibenzal **(588-59-0)**
Bibenzene **(92-52-4)**
($\delta^{2,2'}$(3H,3'H)-Bibenzo(b)thiophene)-3,3'-dione, 6,6'-dichloro-4,4'-dimethyl- (8CI) **(2379-74-0)**
($\delta^{2,2'}$(3H,3'H)-Bibenzo(b)thiophene)-3,3'-dione, 5,5'-dichloro-7,7'-dimethyl- (8CI) **(5462-29-3)**
($\delta^{2,2'}$(3H,3'H)-Bibenzo(b)thiophene)-3,3'-dione, 6,6'-diethoxy **(3263-31-8)**
2,2'-Bibenzoic acid **(482-05-3)**
O,O'-Bibenzoic acid **(482-05-3)**
(8,8'-Bi-2H-1-benzopyran)-2,2'-dione, 7,7'-dihydroxy-4,4'-dimethoxy-5,5'-dimethyl- **(69975-77-5)**
Bibenzyl **(103-29-7)**
Bibenzyl, .α.-chloro- (6CI,7CI,8CI) **(4714-14-1)**
Bibenzyl, .α.,.α.'-dichloro- (6CI,7CI,8CI) **(5963-49-5)**
Bibenzyl, α,α'-diethyl-4,4'-dihydroxy- **(84-16-2)**
Bibenzyl, .α.,.α.'-dimethyl-, erythro- (8CI) **(4613-11-0)**
Bibenzyl, α,α'-epoxy- (8CI) **(17619-97-5)**
Bibenzylidene **(588-59-0)**
Bibenzylidine **(588-59-0)**
Bibenzyl, α-methyl- (8CI) **(5814-85-7)**
Bibenzyl, .α.,.α.,.α.',.α.'-tetrachloro- (6CI,7CI,8CI) **(13700-81-7)**
Bibenzyl, 3,3',4,4'-tetramethyl- (8CI) **(34101-86-5)**
Bibesol **(62-73-7)**
Bic **(5034-77-5)**
Bica-Penicillin **(1538-09-6)**
Bicalcium phosphate **(7757-93-9)**
Bicarbamamide **(110-21-4)**
Bicarbamimide (VAN) (8CI) **(3232-84-6)**
Bicarbamimidic acid (VAN) **(110-21-4)**
Bicarbonate of soda **(144-55-8)**
Bicarburet of hydrogen **(71-43-2)**
Bicarburretted hydrogen **(74-85-1)**
Bicep **(51218-45-2)**
Bichloracetic acid **(79-43-6)**
Bichlorendo **(2385-85-5)**
Bichlorhydrate d'histamine (French) **(56-92-8)**
Bichloride of mercury **(7487-94-7)**
1,2-Bichloroethane **(107-06-2)**
Bichlorure de mercure (French) **(7487-94-7)**
Bichlorure de propylene (French) **(78-87-5)**
Bichlorure d'ethylene (French) **(107-06-2)**
Bichromate d'ammonium (French) **(7789-09-5)**
Bichromate de sodium (French) **(10588-01-9)**
Bichromate of Potash **(7778-50-9)**
Bichromate of soda **(10588-01-9)**
Bicillin **(1538-09-6)**

Bickie-Mol **(103-90-2)**
Biclorhidrato de saxotoxina (Spanish) **(35554-08-6)**
Bicnu **(154-93-8)**
Bicol **(603-50-9)**
Bicolastic A 75 **(9003-53-6)**
Bicolene C **(9002-88-4)**
Bicolene H **(9003-53-6)**
Bicolene P **(9003-07-0)**
Bicortone **(53-03-2)**
Bicyclo(4.4.0)decane **(91-17-8)**
cis-Bicyclo(4.4.0)decane **(493-01-6)**
trans-Bicyclo(4.4.0)decane **(493-02-7)**
Bicyclo(0.3.5)deca-1,3,5,7,9-pentaene **(275-51-4)**
Bicyclo(5.3.0)-deca-2,4,6,8,10-pentaene **(275-51-4)**
Bicyclo(5.3.0)decapentaene **(275-51-4)**
Bicyclo(2,2,2)-1,4-diazaoctane **(280-57-9)**
Bicyclo(2.2.1)hept-5-ene-2,3-dicarboxylic acid, 1,4,5,6,7,7-hexachloro- (9CI) **(115-28-6)**
Bicyclo(2.2.1)hept-5-ene-2-methylol acrylate **(95-39-6)**
Bicyclo(2.2.1)heptadiene **(121-46-0)**
Bicyclo(2.2.1)hepta-2,5-diene, 1,2,3,4,7,7-hexachloro- (9CI) **(3389-71-7)**
Bicyclo(2.2.1)hepta-2,5-dien-7-ol (9CI) **(822-80-0)**
Bicyclo(2.2.1)heptan-2-amine, N-ethyl-3-phenyl- **(1209-98-9)**
cis-Bicyclo(4.1.0)heptane **(286-08-8)**
Bicyclo(4.1.0)heptane (9CI) **(286-08-8)**
Bicyclo(2.2.1)heptane-2-carbonitrile, 5-chloro-6-((((methylamino)-carbonyl)oxy)imino)- **(15271-41-7)**
Bicyclo(2.2.1)heptane-2-carbonitrile, 5-chloro-6-((((methylamino)carbonyl)oxy)imino)-, (1S-(1-α,2-β,4-α,5-α,6E))- **(15271-41-7)**
Bicyclo(3.1.1)heptane, 6,6-dimethyl-2-methylene **(127-91-3)**
Bicyclo(4.1.0)heptane, 7,7-dimethyl-3-methylene- **(554-60-9)**
Bicyclo(3.1.1)heptane, 6,6-dimethyl-2-methylene-, (1S)- (9CI) **(18172-67-3)**
Bicyclo(2.2.1)heptane-2,3-dione, 1,7,7-trimethyl-, (+-)- (9CI) **(10373-78-1)**
Bicyclo(2.2.1)heptane-2,5-dione, 1,7,7-trimethyl- (9CI) **(4230-32-4)**
Bicyclo(2.2.1)heptane-1-methanesulfonic acid, 7,7-dimethyl-2-oxo-, (1S)- (9CI) **(3144-16-9)**
Bicyclo(4.1.0)heptane, 7-(1-methylethylidene)- (9CI) **(53282-47-6)**
Bicyclo(2.2.1)heptane, 2-methyl-3-methylene-2-(4-methyl-3-pentenyl)-, (1S-exo)- (9CI) **(511-59-1)**
Bicyclo(2.2.1)heptane, 2,3,3,5,6-pentachloro-7,7-bis(chloromethyl)-(dichloromethyl)-, (2-endo,5-exo,6-exo)- (9CI) **(52819-39-3)**
Bicyclo(4.1.0)heptane, 7-pentyl- (9CI) **(41977-45-1)**
Bicyclo(2.2.1)heptane, 2,2,5,6-tetrachloro-1,7,7-tris(chloromethyl)-, (5-endo,6-exo)- **(51775-36-1)**
Bicyclo(3.1.1)heptane, 2,6,6-trimethyl- (9CI) **(473-55-2)**
Bicyclo(3.1.1)heptane, 2,6,6-trimethyl-, didehydro deriv. **(1330-16-1)**
Bicyclo(3.1.1)heptane, 2,6,6-trimethyl-, (1α,2β,5α)- (9CI) **(6876-13-7)**
Bicyclo(3.1.1)heptane, 2,6,6-trimethyl-3-(2-propenyl)-, (1.α.,2.β.,3.α.,5.α.)- (9CI) **(50746-55-9)**
Bicyclo(2.2.1)heptan-2-ol, 1,2,7,7-tetramethyl-, exo- (9CI) **(2371-42-8)**
Bicyclo(2.2.1)heptan-2-ol, 1,3,3-trimethyl- **(1632-73-1)**
Bicyclo(2.2.1)heptan-2-ol, 1,7,7-trimethyl-, endo- (9CI) **(507-70-0)**
Bicyclo(3.1.1)heptan-2-ol, 2,6,6-trimethyl- (9CI) **(473-54-1)**
Bicyclo(3.1.1)heptan-2-ol, 2,6,6-trimethyl-, (1α,2α,5α)- (9CI) **(4948-28-1)**
Bicyclo(2.2.1)heptan-2-ol, 1,7,7-trimethyl-, acetate, endo- (9CI) **(76-49-3)**
Bicyclo(2.2.1)heptan-2-ol, 1,7,7-trimethyl-, acetate, exo- (9CI) **(125-12-2)**
Bicyclo(2.2.1)heptan-2-ol, 1,3,3-trimethyl-, (1r-endo)- **(512-13-0)**
Bicyclo(2.2.1)heptan-2-ol, 1,7,7-trimethyl-, exo- **(124-76-5)**
Bicyclo(2.2.1)heptan-2-ol, 1,7,7-trimethyl-, propanoate, exo- (9CI) **(2756-56-1)**
Bicyclo(3.1.1)heptan-2-one, 6,6-dimethyl- (9CI) **(24903-95-5)**
Bicyclo(2.2.1)heptan-2-one, hydroxy-1,7,7-trimethyl- (9CI) **(12001-40-0)**
Bicyclo(2.2.1)heptan-2-one, 1,7,7-trimethyl- **(76-22-2)**
Bicyclo(2.2.1)heptan-2-one, 1,3,3-trimethyl- (9CI) **(1195-79-5)**
Bicyclo(3.1.1)heptan-3-one, 2,6,6-trimethyl-, (1α,2α,5α)- (9CI) **(547-60-4)**
Bicyclo(2.2.1)heptan-2-one, 1,7,7-trimethyl-, (1R)- **(464-49-3)**
Bicyclo(2.2.1)heptan-2-one, 1,3,3-trimethyl-, (1R)- (9CI) **(4695-62-9)**
Bicyclo(2.2.1)heptan-2-one, 4,7,7-trimethyl-, (1S)- (9CI) **(10292-98-5)**
Bicyclo(2.2.1)heptan-2-one, 1,7,7-trimethyl-3-(phenylmethylene)- (9CI) **(15087-24-8)**
Bicyclo(4.1.0)hept-3-ene (9CI) **(16554-83-9)**
Bicyclo(2.2.1)hept-5-ene-2,3-dicarboxylic acid, dimethyl ester, (endo,endo)- (9CI) **(39589-98-5)**
Bicyclo(2.2.1)heptene-2-dicarboxylic acid, 2-ethylhexylimide **(113-48-4)**
Bicyclo(2.2.1)hept-5-ene-2,3-dicarboxylic acid, 1,4,5,6,7,7-hexachloro-, bis(2-ethylhexyl) ester (9CI) **(4827-55-8)**
Bicyclo(2.2.1)hept-2-ene, 2,3-dimethyl- (9CI) **(529-16-8)**
Bicyclo(3.1.1)hept-2-ene, 6,6-dimethyl-2-(2-(phenylmethoxy)-ethyl)-, (1R)- (9CI) **(74851-17-5)**
Bicyclo(2.2.1)hept-2-ene, 5-ethenyl- **(3048-64-4)**
Bicyclo(2.2.1)heptene, heptachloro- **(28680-45-7)**
Bicyclo(2.2.1)hept-2-ene, heptachloro- (9CI) **(28680-45-7)**
Bicyclo(2.2.1)hept-5-ene-2-methanol (9CI) **(95-12-5)**
Bicyclo(3.1.1)hept-2-ene-2-methanol, 6,6-dimethyl- (9CI) **(515-00-4)**
Bicyclo(2.2.1)hept-2-ene, 1,2,3,4,7-pentachloro- (9CI) **(5825-64-9)**
Bicyclo(2.2.1)hept-2-ene, 1,2,3,4,7-pentachloro-, syn- (9CI) **(18317-90-3)**
Bicyclo(3.1.1)hept-2-ene, 2,6,6-trimethyl **(80-56-8)**
Bicyclo(2.2.1)hept-2-ene, 1,7,7-trimethyl- (9CI) **(464-17-5)**
Bicyclo(4.1.0)hept-3-ene, 3,7,7-trimethyl- (9CI) **(13466-78-9)**
Bicyclo(3.1.1)hept-2-ene, 2,6,6-trimethyl-, (1S)- (9CI) **(7785-26-4)**
Bicyclo(2.2.1)hept-2-en-7-ol (9CI) **(53783-87-2)**
Bicyclo(2.2.1)hept-5-en-2-ylmethyl monochloroacetate **(28693-00-7)**
Bicyclohexane **(92-51-3)**
Bicyclo(3.1.0)hexane, 4-methylene-1-(1-methylethyl)- (9CI) **(3387-41-5)**
Bicyclo(3.1.0)hexan-3-ol, 4-methyl-1-(1-methylethyl)- (9CI) **(513-23-5)**
Bicyclo(3.1.0)hexan-2-one, 1,5-bis(1,1-dimethylethyl)-3,3-dimethyl- (9CI) **(19377-95-8)**
Bicyclo(3.1.0)hexan-2-one, 1,5-di-tert-butyl-3,3-dimethyl- (8CI) **(19377-95-8)**
Bicyclo(3.1.0)hexan-3-one, 4-methyl-1-(1-methylethyl)-, (1S-1-α,4-α,5-α)- (9CI) **(546-80-5)**
Bicyclo(3.1.0)hex-2-ene, 2-methyl-5-(1-methylethyl)- (9CI) **(2867-05-2)**
1,1'-Bicyclohexyl (9CI) **(92-51-3)**
Bicyclohexyl (8CI) **(92-51-3)**
(Bicyclohexyl)-1-carboxylic acid, 2-(diethylamino)ethyl ester **(77-19-0)**
(1,1'-Bicyclohexyl)-1-carboxylic acid, 2-(diethylamino)ethyl ester (9CI) **(77-19-0)**
Bicyclohexyl, isopropyl- (6CI,7CI,8CI) **(31624-59-6)**
1,1'-Bicyclohexyl, (1-methylethyl)- (9CI) **(31624-59-6)**
(1,1'-Bicyclohexyl)-2-ol (9CI) **(6531-86-8)**
(Bicyclohexyl)-2-ol (8CI) **(6531-86-8)**
(1,1'-Bicyclohexyl)-2-one **(90-42-6)**
(1,1'-Bicyclohexyl)one **(56025-96-8)**
Bicyclohexyl, 4-phenyl- (8CI) **(20273-27-2)**
Bicyclo(4,3,0)nona-3,7-diene **(3048-65-5)**
Bicyclo(4.3.0)nonane **(496-10-6)**
Bicyclo(3.3.1)non-2-en-9-one (8CI,9CI) **(4844-11-5)**
Bicyclo(2.2.2)octane **(280-33-1)**
Bicyclo(3.2.1)octane, 2,3-bis(methylene)- (9CI) **(49826-54-2)**
Bicyclo(2.2.2)octane, 1-bromo-4-methyl- (7CI,8CI,9CI) **(697-40-5)**
Bicyclo(2.2.2)octane-1,4-diol, monoacetate (9CI) **(54774-94-6)**
Bicyclo(2.2.2)octane, 1-methoxy-4-methyl- (7CI,8CI,9CI) **(6555-95-9)**
2-exo-Bicyclo(3.2.1)octanol **(1965-38-4)**
exo-Bicyclo(3.2.1)octan-2-ol **(1965-38-4)**
Bicyclo(3.2.1)octan-2-ol, exo- (8CI,9CI) **(1965-38-4)**
Bicyclo(4.2.0)octa-1,3,5-triene (8CI,9CI) **(694-87-1)**
Bicyclo(2.2.2)oct-2-ene (8CI,9CI) **(931-64-6)**
Bicyclo(5.1.0)oct-3-ene (7CI,9CI) **(659-84-7)**
Bicyclo(3.2.1)oct-2-ene, 3-methyl-4-methylene- (9CI) **(49826-53-1)**
Bicyclopentadiene **(77-73-6)**
Bicyclopentadiene dioxide **(81-21-0)**
Bi-2,4-cyclopentadien-1-yl, decachloro **(2227-17-0)**
Bicyclo(2.1.0)pentane (6CI,7CI,8CI,9CI) **(185-94-4)**
1,1'-Bicyclopentyl (9CI) **(1636-39-1)**
Bicyclopentyl (8CI) **(1636-39-1)**
(1,1'-Bicyclopentyl)-2-one (9CI) **(4884-24-6)**

Bicyclo(7.2.0)undec-4-ene, 8-methylene-4,11,11-trimethyl-, (E)-(1R,9S)-(-) **(87-44-5)**
Bide **(102-79-4)**
Bideron **(34643-46-4)**
2,2'-Bi-1,3-dioxolane (8CI,9CI) **(6705-89-1)**
Bidiphen **(97-18-7)**
Bidirl **(141-66-2)**
Bidisin **(14437-17-3)**
Bidocef **(50370-12-2)**
Bidrin **(141-66-2)**
Biebrich Scarlet **(4196-99-0)**
Biebrich Scarlet BPC **(85-83-6)**
Biebrich Scarlet R Medicinal **(85-83-6)**
Biebrich Scarlet Red **(85-83-6)**
Biethylene **(106-99-0)**
1,1'-Bi(ethylene oxide) **(1464-53-5)**
Biethylxanthogentrisulfide **(502-55-6)**
Bifenox **(42576-02-3)**
4-Bifenylamin (Czech) **(92-67-1)**
Bifex **(114-26-1)**
Biflorine **(130-86-9)**
Bifluoriden (Dutch) **(7782-41-4)**
Bifluorure de potassium (French) **(7789-29-9)**
Bifluoruro potasico (Spanish) **(7789-29-9)**
Biformal **(107-22-2)**
Biformychlorazin **(26644-46-2)**
Biformyl **(107-22-2)**
Biformylchlorazin **(26644-46-2)**
Bifuron **(67-45-8)**
Big Dipper **(122-39-4)**
Biguanide, 1,1'-hexamethylenebis(5-(p-chlorophenyl) **(55-56-1)**
Biguanide, 1,1'-hexamethylenebis(5-(p-chlorophenyl)-, diacetate (8CI) **(56-95-1)**
Biguanide, 1,1'-hexamethylenebis(5-(p-chlorophenyl)-, digluconate **(18472-51-0)**
Biguanide, 1-phenethyl **(114-86-3)**
Biguanide, 1-phenethyl-, hydrochloride **(834-28-6)**
Biguanide, 1-phenethyl-, monohydrochloride **(834-28-6)**
Bihexyl **(112-40-3)**
1,1'-Biimidazole, 2,2'-bis(o-chlorophenyl)-4,4',5,5'-tetraphenyl- **(1707-68-2)**
1,1'-Bi-1H-imidazole, 2,2'-bis(2-chlorophenyl)-4,4',5,5'-tetraphenyl- (9CI) **(1707-68-2)**
(2,2'-Bi-1H-indole)-3,3'-diol (9CI) **(6537-68-4)**
(2,2'-Biindole)-3,3'-diol (8CI) **(6537-68-4)**
(2,2'-Biindoline)-3,3'-dione **(482-89-3)**
($\delta^{2,2'}$)-Biindoline)-3,3'-dione **(482-89-3)**
(δ(2,2')-Biindoline)-3,3'-dione, 5-bromo- (8CI) **(6492-73-5)**
(δ2,2'-Biindoline)-3,3'-dione, 5,5',7,7'-tetrabromo- (8CI) **(2475-31-2)**
(δ(2,2')-Biindoline)-5,5'-disulfonic acid, 3,3'-dioxo- (8CI) **(483-20-5)**
($\delta^{2,2'}$)-Biindoline)-5,5'-disulfonic acid, 3,3'-dioxo-, disodium salt **(860-22-0)**
Biisopropenyl **(513-81-5)**
Bilagen **(536-50-5)**
Bilarcil **(52-68-6)**
Bilcolic **(90-33-5)**
Biletan **(62-46-4)**
Bilevon **(70-30-4)**
Bilicante **(90-33-5)**
Bilijodon **(96-83-3)**
Biline-8,12-dipropionic acid, 1,10,19,22,23,24-hexahydro-2,7,13,17-tetramethyl-1,19-dioxo- 3,18-divinyl **(635-65-4)**
Bilineurine **(62-49-7)**
Bilirubin **(635-65-4)**
Bilirubin IX-α **(635-65-4)**
Biloborn **(6923-22-4)**
Bilobran **(6923-22-4)**
Bilopten **(5587-89-3)**
Biloxazol **(55179-31-2)**
Biltricide **(55268-74-1)**
Bilyn **(519-95-9)**
Bim **(41814-78-2)**
6,6'-Bimetanilic acid **(117-61-3)**
Bimethadol **(545-90-4)**
Bimethadolum **(545-90-4)**
Bimethyl **(74-84-0)**
Bimox M **(118-82-1)**
Binapacryl **(485-31-4)**
4',4'''-Bi-2-naphth-o-anisidide, 3,3''-dihydroxy- (8CI) **(91-92-9)**
1,1'-Binaphthalene (9CI) **(604-53-5)**
Binaphthalene (9CI) **(11068-27-2)**
(1,1'-Binaphthalene)-8,8'-dicarboxylic acid (9CI) **(29878-91-9)**
4',4'''-Bi-2-naphthanilide, 2',2'''-dichloro-4,4''-bis((o-chlorophenyl)azo)-3,3''-dihydroxy- (6CI,8CI) **(5280-74-0)**
1,1'-Binaphthyl (8CI) **(604-53-5)**
Binaphthyl (8CI) **(11068-27-2)**
1,1'-Binaphthyl-8,8'-dicarboxylic acid **(29878-91-9)**
2,3,1',8'-Binaphthylene **(207-08-9)**
β,β-Binaphthyleneethene **(213-46-7)**
Bindan **(83-12-5)**
Bineopentyl **(1071-81-4)**
Binitrobenzene **(99-65-0)**
Binnell **(1861-40-1)**
Binoctal **(57-43-2)**
Binotal **(69-53-4)**
Binox M **(118-82-1)**
Binox-M **(118-82-1)**
Bio 5,462 **(115-29-7)**
Bio-Beads S-S 2 **(9003-53-6)**
Bio-Clave **(120-32-1)**
Bio-DES **(56-53-1)**
Bio-Dac 50-22 **(7173-51-5)**
Bio-Luvil (9CI) **(57608-19-2)**
Bio-Perge **(55965-84-9)**
Bio-Quat 50-24 **(8001-54-5)**
Bio-Quat 50-25 **(8001-54-5)**
Bio-Quat 50-30 **(8001-54-5)**
Bio-Quat 50-40 **(8001-54-5)**
Bio-Quat 50-42 **(8001-54-5)**
Bio-Quat 50-60 **(8001-54-5)**
Bio-Quat 50-65 **(8001-54-5)**
Bio-Quat 80-24 **(8001-54-5)**
Bio-Quat 80-28 **(8001-54-5)**
Bio-Quat 80-40 **(8001-54-5)**
Bio-Quat 80-42 **(8001-54-5)**
Bio-Soft D-40 **(25155-30-0)**
Bio-Soft D-60 **(25155-30-0)**
Bio-Soft D-62 **(25155-30-0)**
Bio-Soft D-35X **(25155-30-0)**
Bio-Soft S 100 **(27176-87-0)**
Bio-Testiculina **(57-85-2)**
Bio-Tetra **(60-54-8)**
Bioacridin **(8048-52-0)**
Bioallethrin **(584-79-2)**
s-Bioallethrin **(28434-00-6)**
s-trans-Bioallethrin **(28434-00-6)**
Bioaltrina (Portuguese) **(584-79-2)**
Biobamat **(57-53-4)**
Bioban-C **(4075-81-4)**
Bioban P 1487 **(37304-88-4)**
Bioban P-1487 **(37304-88-4)**
Biobenzyfuroline **(28434-01-7)**
Biobor **(14697-50-8)**
Biobor JF **(2665-13-6)**
Biocetin **(56-75-7)**
Biocide **(107-02-8)**

Name	CAS No.
N-1386 Biocide	(3064-70-8)
Biocolina	(67-48-1)
Biocort acetate	(50-04-4)
Bioden	(9004-35-7)
Biodopa	(59-92-7)
Bioepiderm	(58-85-5)
Bioethanomethrin	(22431-62-5)
Biofanal	(1400-61-9)
Bioflavonoid	(153-18-4)
Biofuracina	(59-87-0)
Biofurea	(59-87-0)
Biogastrone	(5697-56-3)
Biogrisin-FP	(126-07-8)
Bioguard	(148-79-8)
Biokor	(86-87-3)
Biomet 14	(5035-58-5)
Biomet 204	(379-52-2)
Biomet TBTO	(56-35-9)
Biomitsin	(57-62-5)
Biomycin	(57-62-5)
Biomydrin	(61-76-7)
Bioneopynamin	(7696-12-0)
Bionex	(2642-71-9)
Bionic	(59-67-6)
Bionol	(8001-54-5)
Biopal CVL-10	(11096-42-7)
Biopal NR-20	(11096-42-7)
Biopal VRO 10	(11096-42-7)
Biopal VRO-20	(11096-42-7)
Biophedrin	(299-42-3)
Biophenicol	(56-75-7)
Bioprase	(9014-01-1)
Bioquin	(10380-28-6)
Bioquin	(148-24-3)
Bioquin 1	(10380-28-6)
Bioral	(5697-56-3)
Biorenine	(51-43-4)
Bioresmethrin	(28434-01-7)
Bioresmethrine	(28434-01-7)
Bioresmetrina (Portuguese)	(28434-01-7)
Bios II	(58-85-5)
Bioscleran	(637-07-0)
Biosechs	(54-47-7)
Biosedan	(57-33-0)
Biosept	(123-03-5)
Bioserpine	(50-55-5)
Biosol	(1405-10-3)
Biosol veterinary	(1405-10-3)
Biosolvomycin	(2058-46-0)
Biostat	(79-57-2)
Biostat PA	(79-57-2)
Biosterol	(68-26-8)
Biosupressin	(127-07-1)
Biothion	(3383-96-8)
(+)-Biotin	(58-85-5)
Biotin	(58-85-5)
d-(+)-Biotin	(58-85-5)
d-Biotin	(58-85-5)
Biotrol	(68038-71-1)
Bioxiran	(1464-53-5)
(R*,R*)-(+-)-2,2'-Bioxirane	(298-18-0)
(R*,S*)-2,2'-Bioxirane	(564-00-1)
(S-(R*,R*))-2,2'-Bioxirane	(30031-64-2)
2,2'-Bioxirane	(1464-53-5)
Bioxirane	(1464-53-5)
2,2'-Bioxirane, (R-(R*,R*))- (9CI)	(30419-67-1)
Bioxone	(20354-26-1)
Bioxyde d'azote (French)	(10102-43-9)
Bioxyde de plomb (French)	(1309-60-0)
2,2'-Biphenol	(1806-29-7)
4,4'-Biphenol	(92-88-6)
Biphenol	(26983-52-8)
o,o'-Biphenol	(1806-29-7)
p,p'-Biphenol	(92-88-6)
m,m'-Biphenol, 6,6'-diamino-	(2373-98-0)
1,1'-Biphenyl	(92-52-4)
Biphenyl (ACGIH,OSHA)	(92-52-4)
Biphenyl, Mixed with biphenyl oxide (3:7)	(8004-13-5)
1,1'-Biphenyl, Mixt. with 1,1'-Oxybis(benzene)	(8004-13-5)
4-Biphenylacetamide	(4075-79-0)
N-4-Biphenylacetamide	(4075-79-0)
(1,1'-Biphenyl)-4-acetic acid, 2-fluoroethyl ester	(4301-50-2)
4-Biphenylacetic acid 2-fluoroethyl ester	(4301-50-2)
4-Biphenylacetic acid, 2-fluoroethyl ester	(4301-50-2)
(1,1'-Biphenyl)-4-acetonitrile (9CI)	(31603-77-7)
Biphenyl-4-acetophenone	(92-91-1)
Biphenylacetylene	(501-65-5)
(1,1'-Biphenyl)-4-amine	(92-67-1)
2-Biphenylamine	(90-41-5)
4-Biphenylamine	(92-67-1)
Biphenylamine	(92-67-1)
o-Biphenylamine	(90-41-5)
p-Biphenylamine	(92-67-1)
(1,1'-Biphenyl)-2-amine (9CI)	(90-41-5)
4-Biphenylamine, 4,4'-dimethoxy	(101-70-2)
2-Biphenylamine, hydrochloride	(2185-92-4)
(1,1'-Biphenyl)-4,4'-bis(diazonium), 3,3'-dimethoxy- (9CI)	(20282-70-6)
1,1'-Biphenyl, ar,ar'-bis(1-methylethyl)- (9CI)	(36876-13-8)
1,1'-Biphenyl, bis(1-methylethyl)- (9CI)	(69009-90-1)
Biphenyl, 3-bromo	(2113-57-7)
Biphenyl, 4-bromo	(92-66-0)
1,1'-Biphenyl, 2-bromo- (9CI)	(2052-07-5)
1,1'-Biphenyl, 4'-bromo-3-methyl-	(56961-07-0)
1,1'-Biphenyl, butenylated	(81846-81-3)
1,1'-Biphenyl, butyl- (9CI)	(41638-55-5)
(1,1'-Biphenyl)-4-carbonitrile (9CI)	(2920-38-9)
(1,1'-Biphenyl)carbonitrile (9CI)	(28804-96-8)
Biphenylcarbonitrile (8CI)	(28804-96-8)
(1,1'-Biphenyl)-4-carbonitrile, 4'-hexyl- (9CI)	(41122-70-7)
(1,1'-Biphenyl)-3-carboxylic acid, 2-hydroxy-, copper(2+) salt (2:1) (9CI)	(5328-04-1)
1,1'-Biphenyl, chloro-	(27323-18-8)
Biphenyl, 2-chloro	(2051-60-7)
Biphenyl, 4-chloro	(2051-62-9)
Biphenyl, chloro	(27323-18-8)
1,1'-Biphenyl, chloro-ar,ar'-bis(trifluoromethyl)- (9CI)	(95998-64-4)
1,1'-Biphenyl, 2,2',3,3',4,4',5,5',6,6'-decabromo- (9CI)	(13654-09-6)
1,1'-Biphenyl, 2,2',3,3',4,4',5,5',6,6'-decachloro-	(2051-24-3)
(1,1'-Biphenyl)-2,4'-diamine	(492-17-1)
2,4'-Biphenyldiamine	(492-17-1)
4,4'-Biphenyldiamine	(92-87-5)
(1,1'-Biphenyl)-4,4'-diamine (9CI)	(92-87-5)
(1,1'-Biphenyl)-4,4'-diamine, N,N'-bis(2,4-dinitrophenyl)-3,3'-dimethoxy- (9CI)	(29398-96-7)
4,4'-Biphenyldiamine, N,N'-diacetoacetyl-3,3'-dimethyl-	(91-96-3)
(1,1'-Biphenyl)-4,4'-diamine, 3,3'-dichloro-	(91-94-1)
(1,1'-Biphenyl)-4,4'-diamine, 2,2'-dichloro- (9CI)	(84-68-4)
(1,1'-Biphenyl)-4,4'-diamine, 3,3'-dichloro-, dihydrochloride	(612-83-9)
(1,1'-Biphenyl)-4,4'-diamine, dihydrochloride	(531-85-1)
(1,1'-Biphenyl)-4,4'-diamine, 3,3-dimethoxy-, dihydrochloride (9CI)	(20325-40-0)
(1,1'-Biphenyl)-4,4'-diamine-3,3'-dimethyl-	(119-93-7)
(1,1'-Biphenyl)-4,4'-diamine, sulfate (1:1)	(531-86-2)
(1,1'-Biphenyl)-4,4'-diamine, 2,2',5,5'-tetrachloro- (9CI)	(15721-02-5)

Name	CAS No.
(1,1'-Biphenyl)-4,4'-diamine, N,N,N',N'-tetramethyl-	**(366-29-0)**
(1,1'-Biphenyl)-4,4-diamine, 3,3',5,5'-tetramethyl	**(54827-17-7)**
Biphenyl, 4,4'-diamino-	**(92-87-5)**
1,1'-Biphenyl, 4,4'-dibromo- (9CI)	**(92-86-4)**
Biphenyl, 4,4'-dibromo- (8CI)	**(92-86-4)**
2,2'-Biphenyldicarboxaldehyde	**(1210-05-5)**
2,2'-Biphenyldicarboxylic acid	**(482-05-3)**
(1,1'-Biphenyl)-2,2'-dicarboxylic acid (9CI)	**(482-05-3)**
3,3'-Biphenyldicarboxylic acid, 4,4'-diamino	**(2130-56-5)**
(1,1'-Biphenyl)-3,3'-dicarboxylic acid, 6,6'-dihydroxy-5,5'-dimethoxy- (9CI)	**(2134-90-9)**
3,3'-Biphenyldicarboxylic acid, 6,6'-dihydroxy-5,5'-dimethoxy- (8CI)	**(2134-90-9)**
4,4'-Biphenyldicarboxylic acid, dimethyl ester	**(792-74-5)**
(1,1'-Biphenyl)-4,4'-dicarboxylic acid, dimethyl ester (9CI)	**(792-74-5)**
Biphenyl, 2,2'-dichloro	**(13029-08-8)**
Biphenyl, 2,4'-dichloro	**(34883-43-7)**
Biphenyl, ar,ar'-dichloro-	**(33039-81-5)**
Biphenyl, dichloro	**(25512-42-9)**
1,1'-Biphenyl, 2,3'-dichloro- (9CI)	**(25569-80-6)**
1,1'-Biphenyl, 2,4'-dichloro- (9CI)	**(34883-43-7)**
1,1'-Biphenyl, 2,4-dichloro- (9CI)	**(33284-50-3)**
1,1'-Biphenyl, 3,4'-dichloro- (9CI)	**(2974-90-5)**
1,1'-Biphenyl, 3,4-dichloro- (9CI)	**(2974-92-7)**
1,1'-Biphenyl, 4,4'-dichloro- (9CI)	**(2050-68-2)**
1,1'-Biphenyl, ar,ar'-dichloro- (9CI)	**(33039-81-5)**
1,1'-Biphenyl, dichloro- (9CI)	**(25512-42-9)**
Biphenyl, 2,3'-dichloro- (8CI)	**(25569-80-6)**
Biphenyl, 2,4-dichloro- (8CI)	**(33284-50-3)**
Biphenyl, 3,4'-dichloro- (8CI)	**(2974-90-5)**
Biphenyl, 3,4-dichloro- (8CI)	**(2974-92-7)**
Biphenyl, 4,4'-dichloro- (8CI)	**(2050-68-2)**
Biphenyl, 3,3'-dichloro-4,4'-bis((2-hydroxy-3-(N-phenylcarbamyl)-1-naphthyl)azo)-	**(41709-76-6)**
1,1'-Biphenyl, 2,2'-dichloro-5,5'-bis(trifluoromethyl)- (9CI)	**(99686-52-9)**
1,1'-Biphenyl, 2,5-dichloro-4'-(1-methylethyl)- (9CI)	**(63028-27-3)**
Biphenyl, 2,2'-didehydro- (8CI)	**(28109-99-1)**
1,1'-Biphenyl, 3,4-diethyl- (9CI)	**(61141-66-0)**
1,1'-Biphenyl, 4,4'-difluoro- (9CI)	**(398-23-2)**
Biphenyl, 4,4'-difluoro- (8CI)	**(398-23-2)**
1,1'-Biphenyl, 4,4'-diiodo- (9CI)	**(3001-15-8)**
1,1'-Biphenyl, 4,4'-diisocyanato-3,3'-dimethoxy-	**(91-93-0)**
1,1'-Biphenyl, 4,4'-diisocyanato-3,3'-dimethyl	**(91-97-4)**
1,1'-Biphenyl, 4,4'-dimethoxy- (9CI)	**(2132-80-1)**
Biphenyl, 4,4'-dimethoxy- (8CI)	**(2132-80-1)**
1,1'-Biphenyl, 2,2'-dimethyl- (9CI)	**(605-39-0)**
1,1'-Biphenyl, 3,3'-dimethyl- (9CI)	**(612-75-9)**
1,1'-Biphenyl, 3,4'-dimethyl- (9CI)	**(7383-90-6)**
1,1'-Biphenyl, 3,4-dimethyl- (9CI)	**(4433-11-8)**
1,1'-Biphenyl, 4,4'-dimethyl- (9CI)	**(613-33-2)**
1,1'-Biphenyl, ar,ar'-dimethyl- (9CI)	**(28013-11-8)**
Biphenyl, 3,4-dimethyl- (8CI)	**(4433-11-8)**
Biphenyl, 4,4'-dinitro	**(1528-74-1)**
2,2'-Biphenyldiol	**(1806-29-7)**
2,5-Biphenyldiol	**(1079-21-6)**
4,4'-Biphenyldiol	**(92-88-6)**
(1,1'-Biphenyl)-2,3'-diol (9CI)	**(31835-45-7)**
(1,1'-Biphenyl)-2,3-diol (9CI)	**(1133-63-7)**
(1,1'-Biphenyl)-3,4'-diol (9CI)	**(18855-13-5)**
(1,1'-Biphenyl)-3,4-diol (9CI)	**(92-05-7)**
(1,1'-Biphenyl)-ar,ar'-diol (9CI)	**(26983-52-8)**
2,3-Biphenyldiol (8CI)	**(1133-63-7)**
3,4-Biphenyldiol (8CI)	**(92-05-7)**
ar,ar'-Biphenyldiol (8CI)	**(26983-52-8)**
(1,1'-Biphenyl)diol (VAN)(9CI)	**(68334-50-9)**
3,3'-Biphenyldiol, 4,4'-diamino	**(2373-98-0)**
(1,1'-Biphenyl)diol, nitro- (9CI)	**(116211-97-3)**
(1,1'-Biphenyl)-ar,ar'-diol, tetrabromo- (9CI)	**(36511-35-0)**
(1,1'-Biphenyl)-4,4'-diol, 2,2',6,6'-tetrachloro-	**(41363-16-0)**
(1,1'-Biphenyl)-2,2'-diol, 4,4',6,6'-tetrachloro- (9CI)	**(14477-61-3)**
2,2'-Biphenyldiol, 4,4',6,6'-tetrachloro- (8CI)	**(14477-61-3)**
1,1'-Biphenyl, 2,2'-diphenyl-	**(641-96-3)**
1,1'-Biphenyl, 4,4'-diphenyl-	**(135-70-6)**
Biphenyl-diphenyl ether mixture	**(8004-13-5)**
(1,1'-Biphenyl)-2,2'-disulfonic acid, 4,4'-diamino-	**(117-61-3)**
2,2'-Biphenyldisulfonic acid, 4,4'-diamino	**(117-61-3)**
4,4'-Biphenyldisulfonic acid, 2,2'-diamino-	**(117-61-3)**
(1,1'-Biphenyl)-2,2'-disulfonic acid, 4,4'-bis((4,5-dihydro-3-methyl-5-oxo-1-phenyl-1H- pyrazol-4-yl)azo)-, disodium salt	**(6375-55-9)**
(1,1'-Biphenyl)-2,2'-disulfonic acid, 4,4'-bis((2-hydroxy-1-naphthalenyl)azo)-, disodium salt	**(10169-02-5)**
N,N'-(1,1'-Biphenyl)-4,4'-diylbis-acetamide 4',4'''-biacetanilide	**(613-35-4)**
(1,1'-Biphenyl)-2,2'-diyl sulfide	**(132-65-0)**
1,1'-Biphenyl, dodecahydro-	**(92-51-3)**
Biphenylene (9CI)	**(259-79-0)**
4,4'-Biphenylene bis(thiourea)	**(1614-30-8)**
p,p'-Biphenylene-bis-1,1'-(2-thiourea)	**(1614-30-8)**
4,4'-Biphenylenediamine	**(92-87-5)**
o-Biphenylenemethane	**(86-73-7)**
2,2'-Biphenylene oxide	**(132-64-9)**
1,1'-Biphenyl, 4-ethenyl- (9CI)	**(2350-89-2)**
1,1'-Biphenyl, 2-ethyl- (9CI)	**(1812-51-7)**
1,1'-Biphenyl, ethyl- (9CI)	**(40529-66-6)**
Biphenyl, 2-ethyl- (8CI)	**(1812-51-7)**
Biphenyl, 2-fluoro-	**(321-60-8)**
1,1'-Biphenyl, 2-fluoro- (9CI)	**(321-60-8)**
1,1'-Biphenyl, 4-fluoro- (9CI)	**(324-74-3)**
Biphenyl, 4-fluoro- (8CI)	**(324-74-3)**
1,1'-Biphenyl, 2,2',3,4,4',5,5'-heptabromo- (9CI)	**(67733-52-2)**
1,1'-Biphenyl, 2,2',3,3',4,4',5-heptachloro- (9CI)	**(35065-30-6)**
1,1'-Biphenyl, 2,2',3,3',4,4',6-heptachloro- (9CI)	**(52663-71-5)**
1,1'-Biphenyl, 2,2',3,3',4,5',6'-heptachloro- (9CI)	**(52663-70-4)**
1,1'-Biphenyl, 2,2',3,3',4,5',6-heptachloro- (9CI)	**(40186-70-7)**
1,1'-Biphenyl, 2,2',3,3',4,5,5'-heptachloro- (9CI)	**(52663-74-8)**
1,1'-Biphenyl, 2,2',3,3',4,5,6'-heptachloro- (9CI)	**(38411-25-5)**
1,1'-Biphenyl, 2,2',3,3',4,5,6-heptachloro- (9CI)	**(68194-16-1)**
1,1'-Biphenyl, 2,2',3,3',5,5',6-heptachloro- (9CI)	**(52663-67-9)**
1,1'-Biphenyl, 2,2',3,3',5,6,6'-heptachloro- (9CI)	**(52663-64-6)**
1,1'-Biphenyl, 2,2',3,4',5,5',6-heptachloro- (9CI)	**(52663-68-0)**
1,1'-Biphenyl, 2,2',3,4,4',5',6-heptachloro- (9CI)	**(52663-69-1)**
1,1'-Biphenyl, 2,2',3,4,4',5,5'-heptachloro- (9CI)	**(35065-29-3)**
1,1'-Biphenyl, 2,2',3,4,4',5,6'-heptachloro- (9CI)	**(60145-23-5)**
1,1'-Biphenyl, 2,2',3,4,4',5,6-heptachloro- (9CI)	**(74472-47-2)**
1,1'-Biphenyl, 2,2',3,4,4',6,6'-heptachloro- (9CI)	**(74472-48-3)**
1,1'-Biphenyl, 2,2',3,4,5,5',6-heptachloro- (9CI)	**(52712-05-7)**
1,1'-Biphenyl, 2,3,3',4',5,5',6-heptachloro- (9CI)	**(69782-91-8)**
1,1'-Biphenyl, 2,3,3',4,4',5',6-heptachloro- (9CI)	**(74472-50-7)**
1,1'-Biphenyl, 2,3,3',4,4',5,6-heptachloro- (9CI)	**(41411-64-7)**
1,1'-Biphenyl, heptachloro- (9CI)	**(28655-71-2)**
Biphenyl, heptachloro- (8CI)	**(28655-71-2)**
1,1'-Biphenyl, 2,2',4,4',5,5'-hexabromo	**(59080-40-9)**
Biphenyl, hexabromo	**(36355-01-8)**
Biphenyl, hexabromo- (Technical grade)	**(59536-65-1)**
1,1'-Biphenyl, 2,2',3,3',4,4'-hexachloro	**(38380-07-3)**
1,1'-Biphenyl, 2,2',3,3',6,6'-hexachloro	**(38411-22-2)**
1,1'-Biphenyl, 2,3',4,4',5,5'-hexachloro-	**(52663-72-6)**
1,1'-Biphenyl, 2,3,3',4,4',5-hexachloro	**(38380-08-4)**
Biphenyl, 2,2',4,4',5,5'-hexachloro	**(35065-27-1)**
Biphenyl, 3,3',4,4',5,5'-hexachloro	**(32774-16-6)**
1,1'-Biphenyl, 2,2',3,3',4,5'-hexachloro- (9CI)	**(52663-66-8)**
1,1'-Biphenyl, 2,2',3,3',4,5-hexachloro- (9CI)	**(55215-18-4)**
1,1'-Biphenyl, 2,2',3,3',4,6'-hexachloro- (9CI)	**(38380-05-1)**
1,1'-Biphenyl, 2,2',3,3',5,6'-hexachloro- (9CI)	**(52744-13-5)**
1,1'-Biphenyl, 2,2',3,3',5,6-hexachloro- (9CI)	**(52704-70-8)**

1,1'-Biphenyl, 2,2',3,4',5',6-hexachloro- (9CI) **(38380-04-0)**
1,1'-Biphenyl, 2,2',3,4',5,5'-hexachloro- (9CI) **(51908-16-8)**
1,1'-Biphenyl, 2,2',3,4',5,6-hexachloro- (9CI) **(68194-13-8)**
1,1'-Biphenyl, 2,2',3,4',6,6'-hexachloro- (9CI) **(68194-08-1)**
1,1'-Biphenyl, 2,2',3,4,4',5-hexachloro- (9CI) **(35694-06-5)**
1,1'-Biphenyl, 2,2',3,4,4',6'-hexachloro- (9CI) **(59291-64-4)**
1,1'-Biphenyl, 2,2',3,4,5',6-hexachloro- (9CI) **(68194-14-9)**
1,1'-Biphenyl, 2,2',3,4,5,5'-hexachloro- (9CI) **(52712-04-6)**
1,1'-Biphenyl, 2,2',3,5,5',6-hexachloro- (9CI) **(52663-63-5)**
1,1'-Biphenyl, 2,3,3',4,4',5'-hexachloro- (9CI) **(69782-90-7)**
1,1'-Biphenyl, 2,3,3',4,4',6-hexachloro- (9CI) **(74472-42-7)**
1,1'-Biphenyl, 2,3,3',5,5',6-hexachloro- (9CI) **(74472-46-1)**
1,1'-Biphenyl, 2,3,4,4',5,6-hexachloro- (9CI) **(41411-63-6)**
1,1'-Biphenyl, 3,3',4,4',5,5'-hexachloro- (9CI) **(32774-16-6)**
1,1'-Biphenyl, hexachloro- (9CI) **(26601-64-9)**
Biphenyl, hexachloro- (8CI) **(26601-64-9)**
1,1'-Biphenyl, 3,3',4,4',5,5'-hexamethyl- (9CI) **(56667-01-7)**
Biphenyl, 2-hydroxy- **(90-43-7)**
1,1'-Biphenyl, 4-iodo- (9CI) **(1591-31-7)**
Biphenyl, 4-iodo- (8CI) **(1591-31-7)**
Biphenyl, isopropyl **(25640-78-2)**
o-Biphenylmethane **(86-73-7)**
4,4'-Biphenylmethanebismaleimide **(13676-54-5)**
(1,1'-Biphenyl)-4-methanol **(3597-91-9)**
4-Biphenylmethanol **(3597-91-9)**
1,1'-Biphenyl, 2-methoxy- (9CI) **(86-26-0)**
1,1'-Biphenyl, 4-methoxy- (9CI) **(613-37-6)**
Biphenyl, 4-methyl **(644-08-6)**
Biphenyl, methyl- **(28652-72-4)**
1,1'-Biphenyl, 2-methyl- (9CI) **(643-58-3)**
1,1'-Biphenyl, 3-methyl- (9CI) **(643-93-6)**
1,1'-Biphenyl, methyl- (9CI) **(28652-72-4)**
Biphenyl, 2-methyl- (8CI) **(643-58-3)**
1,1'-Biphenyl, (1-methylethyl)- (9CI) **(25640-78-2)**
Biphenyl, 2-nitro **(86-00-0)**
Biphenyl, 3-nitro **(2113-58-8)**
Biphenyl, 4-nitro **(92-93-3)**
1,1'-Biphenyl, 3-nitro- (9CI) **(2113-58-8)**
1,1-Biphenyl, nonabromo- **(27753-52-2)**
Biphenyl, nonabromo **(27753-52-2)**
1,1'-Biphenyl, 2,2',3,3',4,4',5,5',6-nonachloro- (9CI) **(40186-72-9)**
1,1'-Biphenyl, 2,2',3,3',4,5,5',6,6'-nonachloro- (9CI) **(52663-77-1)**
1,1'-Biphenyl, nonachloro- (9CI) **(53742-07-7)**
Biphenyl, octabromo **(27858-07-7)**
Biphenyl-, octachloro- **(55722-26-4)**
1,1'-Biphenyl, 2,2',3,3',4,4',5,5'-octachloro- (9CI) **(35694-08-7)**
1,1'-Biphenyl, 2,2',3,3',4,4',5,6'-octachloro- (9CI) **(42740-50-1)**
1,1'-Biphenyl, 2,2',3,3',4,4',5,6-octachloro- (9CI) **(52663-78-2)**
1,1'-Biphenyl, 2,2',3,3',4,5',6,6'-octachloro- (9CI) **(40186-71-8)**
1,1'-Biphenyl, 2,2',3,3',4,5,5',6'-octachloro- (9CI) **(52663-75-9)**
1,1'-Biphenyl, 2,2',3,3',4,5,5',6-octachloro- (9CI) **(68194-17-2)**
1,1'-Biphenyl, 2,2',3,3',4,5,6,6'-octachloro- (9CI) **(52663-73-7)**
1,1'-Biphenyl, 2,2',3,3',5,5',6,6'-octachloro- (9CI) **(2136-99-4)**
1,1'-Biphenyl, 2,2',3,4,4',5,5',6-octachloro- (9CI) **(52663-76-0)**
1,1'-Biphenyl, ar,ar,ar,ar,ar',ar',ar',ar'-octachloro- (9CI) **(31472-83-0)**
1,1'-Biphenyl, octachloro- (9CI) **(55722-26-4)**
Biphenyl, 2,2',3,3',5,5',6,6'-octachloro- (8CI) **(2136-99-4)**
(1,1'-Biphenyl)-2-ol **(90-43-7)**
2-Biphenylol **(90-43-7)**
4-Biphenylol **(92-69-3)**
o-Biphenylol **(90-43-7)**
(1,1'-Biphenyl)-3-ol (9CI) **(580-51-8)**
(1,1'-Biphenyl)ol (9CI) **(1322-20-9)**
3-Biphenylol (8CI) **(580-51-8)**
Biphenylol (8CI) **(1322-20-9)**
(1,1'-Biphenyl)-2-ol, 5-amino- (9CI) **(19434-42-5)**
(1,1'-Biphenyl)-4-ol, 3-amino- (9CI) **(1134-36-7)**
2-Biphenylol, 5-amino- (8CI) **(19434-42-5)**
(1,1'-Biphenyl)-4-ol, 3,5-bis(1,1-dimethylethyl)- (9CI) **(2668-47-5)**
4-Biphenylol, 3-chloro **(92-04-6)**
(1,1'-Biphenyl)-2-ol, 3-chloro- (9CI) **(85-97-2)**
(1,1'-Biphenyl)-2-ol, 5-chloro- (9CI) **(607-12-5)**
2-Biphenylol, 3-chloro- (8CI) **(85-97-2)**
2-Biphenylol, 5-chloro- (8CI) **(607-12-5)**
2-Biphenylol, 5-chloro-, sodium salt **(10605-10-4)**
4-Biphenylol, 2-chloro-, sodium salt (8CI) **(5578-88-1)**
(1,1'-Biphenyl)-2-ol, 3,5-dichloro- (9CI) **(5335-24-0)**
2-Biphenylol, 3,5-dichloro- (8CI) **(5335-24-0)**
2-Biphenylol, 3,5-dinitro **(731-92-0)**
4-Biphenylol, 4'-isopropyl **(22239-54-9)**
(1,1'-Biphenyl)-4-ol, 3-nitro- (9CI) **(885-82-5)**
4-Biphenylol, 3-nitro- (8CI) **(885-82-5)**
(1,1'-Biphenyl)-2-ol, potassium salt (9CI) **(13707-65-8)**
(1,1'-Biphenyl)-2-ol, sodium salt **(132-27-4)**
2-Biphenylol, sodium salt **(132-27-4)**
Biphenyl oxide **(101-84-8)**
1,1'-Biphenyl, 4,4''-oxybis- (9CI) **(58841-70-6)**
1,1'-Biphenyl, 2,2',4,5,5'-pentabromo- (9CI) **(67888-96-4)**
Biphenyl, 2,3,3',4,4'-pentachloro **(32598-14-4)**
Biphenyl, pentachloro **(25429-29-2)**
1,1'-Biphenyl, 2,2',3,3',4-pentachloro- (9CI) **(52663-62-4)**
1,1'-Biphenyl, 2,2',3,4',5'-pentachloro- (9CI) **(41464-51-1)**
1,1'-Biphenyl, 2,2',3,4',5-pentachloro- (9CI) **(68194-07-0)**
1,1'-Biphenyl, 2,2',3,4',6'-pentachloro- (9CI) **(60233-25-2)**
1,1'-Biphenyl, 2,2',3,4',6-pentachloro- (9CI) **(68194-05-8)**
1,1'-Biphenyl, 2,2',3,4,4'-pentachloro- (9CI) **(65510-45-4)**
1,1'-Biphenyl, 2,2',3,4,5'-pentachloro- (9CI) **(38380-02-8)**
1,1'-Biphenyl, 2,2',3,4,5-pentachloro- (9CI) **(55312-69-1)**
1,1'-Biphenyl, 2,2',3,4,6'-pentachloro- (9CI) **(73575-57-2)**
1,1'-Biphenyl, 2,2',3,4,6-pentachloro- (9CI) **(55215-17-3)**
1,1'-Biphenyl, 2,2',3,5,5'-pentachloro- (9CI) **(52663-61-3)**
1,1'-Biphenyl, 2,2',4,4',5-pentachloro- (9CI) **(38380-01-7)**
1,1'-Biphenyl, 2,2',4,4',6-pentachloro- (9CI) **(39485-83-1)**
1,1'-Biphenyl, 2,3',4,4',5'-pentachloro- (9CI) **(65510-44-3)**
1,1'-Biphenyl, 2,3',4,4',5-pentachloro- (9CI) **(31508-00-6)**
1,1'-Biphenyl, 2,3,3',4',5-pentachloro- (9CI) **(70424-68-9)**
1,1'-Biphenyl, 2,3,3',4,5-pentachloro- (9CI) **(70424-69-0)**
1,1'-Biphenyl, 2,3,3',5,6-pentachloro- (9CI) **(74472-36-9)**
1,1'-Biphenyl, 2,3,3'4,4'-pentachloro- (9CI) **(32598-14-4)**
1,1'-Biphenyl, 2,3,4,5,6-pentachloro- (9CI) **(18259-05-7)**
1,1'-Biphenyl, pentachloro- (9CI) **(25429-29-2)**
Biphenyl, 2,3',4,4',5-pentachloro- (8CI) **(31508-00-6)**
Biphenyl, 2,3,4,5,6-pentachloro- (8CI) **(18259-05-7)**
1,1'-Biphenyl, pentachloro-ar,ar'-bis(trifluoromethyl)- (9CI) **(95998-68-8)**
1,1'-Biphenyl, phenoxy- (9CI) **(28984-89-6)**
1,1'-Biphenyl, phenoxy-, Mixt. with 1,1'-oxybis(benzene) (9CI) **(62587-63-7)**
Biphenyl, 4-phenyl- **(92-94-4)**
Biphenyl, polychloro- **(1336-36-3)**
1,1'-Biphenyl, tetrabromo- (9CI) **(40088-45-7)**
1,1'-Biphenyl, 2,2',4,4'-tetrachloro- **(2437-79-8)**
1,1'-Biphenyl, 2,2',5,5'-tetrachloro **(35693-99-3)**
Biphenyl, 2,3',4,4'-tetrachloro **(32598-10-0)**
Biphenyl, 2,3,4,4'-tetrachloro- **(33025-41-1)**
Biphenyl, 2,3,4,5-tetrachloro- **(33284-53-6)**
Biphenyl, 3,3',4,4'-tetrachloro **(32598-13-3)**
1,1'-Biphenyl, 2,2',3,3'-tetrachloro- (9CI) **(38444-93-8)**
1,1'-Biphenyl, 2,2',3,4-tetrachloro- (9CI) **(52663-59-9)**
1,1'-Biphenyl, 2,2',3,5'-tetrachloro- (9CI) **(41464-39-5)**
1,1'-Biphenyl, 2,2',3,5-tetrachloro- (9CI) **(70362-46-8)**
1,1'-Biphenyl, 2,2',3,6'-tetrachloro- (9CI) **(41464-47-5)**
1,1'-Biphenyl, 2,2',4,5-tetrachloro- (9CI) **(70362-47-9)**
1,1'-Biphenyl, 2,2',4,6-tetrachloro- (9CI) **(62796-65-0)**
1,1'-Biphenyl, 2,2',5,6'-tetrachloro- (9CI) **(41464-41-9)**
1,1'-Biphenyl, 2,2',6,6'-tetrachloro- (9CI) **(15968-05-5)**

Name	CAS
1,1'-Biphenyl, 2,3',4',5'-tetrachloro- (9CI)	**(70362-48-0)**
1,1'-Biphenyl, 2,3',4',5-tetrachloro- (9CI)	**(32598-11-1)**
1,1'-Biphenyl, 2,3',4',6-tetrachloro- (9CI)	**(41464-46-4)**
1,1'-Biphenyl, 2,3',4,4'-tetrachloro- (9CI)	**(32598-10-0)**
1,1'-Biphenyl, 2,3',4,5'-tetrachloro- (9CI)	**(73575-52-7)**
1,1'-Biphenyl, 2,3',4,5-tetrachloro- (9CI)	**(73575-53-8)**
1,1'-Biphenyl, 2,3',5,5'-tetrachloro- (9CI)	**(41464-42-0)**
1,1'-Biphenyl, 2,3,3',4'-tetrachloro- (9CI)	**(41464-43-1)**
1,1'-Biphenyl, 2,3,3',6-tetrachloro- (9CI)	**(74472-33-6)**
1,1'-Biphenyl, 2,3,4',5-tetrachloro- (9CI)	**(74472-34-7)**
1,1'-Biphenyl, 2,3,4,4'-tetrachloro- (9CI)	**(33025-41-1)**
1,1'-Biphenyl, 2,3,4,5-tetrachloro- (9CI)	**(33284-53-6)**
1,1'-Biphenyl, 2,3,5,6-tetrachloro- (9CI)	**(33284-54-7)**
1,1'-Biphenyl, 2,4,4',5-tetrachloro- (9CI)	**(32690-93-0)**
1,1'-Biphenyl, 3,3',4,4'-tetrachloro- (9CI)	**(32598-13-3)**
1,1'-Biphenyl, 3,3',4,5-tetrachloro- (9CI)	**(70362-49-1)**
1,1'-Biphenyl, 3,3',5,5'-tetrachloro- (9CI)	**(33284-52-5)**
1,1'-Biphenyl, 3,4,4',5-tetrachloro- (9CI)	**(70362-50-4)**
1,1'-Biphenyl, tetrachloro- (9CI)	**(26914-33-0)**
Biphenyl, 2,3',4',5-tetrachloro- (8CI)	**(32598-11-1)**
Biphenyl, 2,3,5,6-tetrachloro- (8CI)	**(33284-54-7)**
Biphenyl, 2,4,4',5-tetrachloro- (8CI)	**(32690-93-0)**
Biphenyl, 3,3',5,5'-tetrachloro- (8CI)	**(33284-52-5)**
Biphenyl, tetrachloro- (8CI)	**(26914-33-0)**
1,1'-Biphenyl, tetrachloro-ar,ar'-bis(trifluoromethyl)- (9CI)	**(95998-67-7)**
1,1'-Biphenyl, 3,3',4,4'-tetramethyl- (9CI)	**(4920-95-0)**
Biphenyl, 3,3',4,4'-tetramethyl- (8CI)	**(4920-95-0)**
3,3'4,4'-Biphenyltetramine	**(91-95-2)**
3,3',4,4'-Biphenyltetramine, tetrahydrochloride	**(7411-49-6)**
(1,1'-Biphenyl)-2,2',4,4'-tetrol (9CI)	**(4371-31-7)**
2,2',4,4'-Biphenyltetrol (8CI)	**(4371-31-7)**
Biphenyl, 2,4',5-trichloro	**(16606-02-3)**
Biphenyl, 2,4,4'-trichloro-	**(7012-37-5)**
Biphenyl, 2,4,6-trichloro-	**(35693-92-6)**
Biphenyl, trichloro	**(25323-68-6)**
1,1'-Biphenyl, 2',3,4-trichloro- (9CI)	**(38444-86-9)**
1,1'-Biphenyl, 2,2',3-trichloro- (9CI)	**(38444-78-9)**
1,1'-Biphenyl, 2,2',4-trichloro- (9CI)	**(37680-66-3)**
1,1'-Biphenyl, 2,2',5-trichloro- (9CI)	**(37680-65-2)**
1,1'-Biphenyl, 2,2',6-trichloro- (9CI)	**(38444-73-4)**
1,1'-Biphenyl, 2,3',4-trichloro- (9CI)	**(55712-37-3)**
1,1'-Biphenyl, 2,3',5'-trichloro- (9CI)	**(37680-68-5)**
1,1'-Biphenyl, 2,3',5-trichloro- (9CI)	**(38444-81-4)**
1,1'-Biphenyl, 2,3',6-trichloro- (9CI)	**(38444-76-7)**
1,1'-Biphenyl, 2,3,3'-trichloro- (9CI)	**(38444-84-7)**
1,1'-Biphenyl, 2,3,4'-trichloro- (9CI)	**(38444-85-8)**
1,1'-Biphenyl, 2,3,4-trichloro- (9CI)	**(55702-46-0)**
1,1'-Biphenyl, 2,3,5-trichloro- (9CI)	**(55720-44-0)**
1,1'-Biphenyl, 2,3,6-trichloro- (9CI)	**(55702-45-9)**
1,1'-Biphenyl, 2,4',5-trichloro- (9CI)	**(16606-02-3)**
1,1'-Biphenyl, 2,4',6-trichloro- (9CI)	**(38444-77-8)**
1,1'-Biphenyl, 2,4,4'-trichloro- (9CI)	**(7012-37-5)**
1,1'-Biphenyl, 2,4,5-trichloro- (9CI)	**(15862-07-4)**
1,1'-Biphenyl, 2,4,6-trichloro- (9CI)	**(35693-92-6)**
1,1'-Biphenyl, 3,3',4-trichloro- (9CI)	**(37680-69-6)**
1,1'-Biphenyl, 3,3',5-trichloro- (9CI)	**(38444-87-0)**
1,1'-Biphenyl, 3,4',5-trichloro- (9CI)	**(38444-88-1)**
1,1'-Biphenyl, 3,4,4'-trichloro- (9CI)	**(38444-90-5)**
1,1'-Biphenyl, trichloro- (9CI)	**(25323-68-6)**
Biphenyl, 2,4',5-trichloro- (8CI)	**(16606-02-3)**
Biphenyl, 2,4,5-trichloro- (8CI)	**(15862-07-4)**
1,1'-Biphenyl, trichloro-ar,ar'-bis(trifluoromethyl)- (9CI)	**(95998-66-6)**
1,1'-Biphenyl, 2,4,6-trimethyl- (9CI)	**(3976-35-0)**
Biphenyl, 2,4,6-trimethyl- (6CI,7CI,8CI)	**(3976-35-0)**
2,3',4-Biphenyltriol	**(4190-05-0)**
(1,1'-Biphenyl)-2,3',4-triol (9CI)	**(4190-05-0)**
2,3',6-Biphenyltriol (8CI)	**(27949-30-0)**
N-(4-Biphenylyl)acetamide	**(4075-79-0)**
N,N'-4,4'-Biphenylylenebisacetamide	**(613-35-4)**
2,2'-Biphenylylene oxide	**(132-64-9)**
2,2'-Biphenylylene sulfide	**(132-65-0)**
1-(1,1'-Biphenyl)-4-ylethanone	**(92-91-1)**
4-Biphenylyl ether (6CI, 7CI)	**(58841-70-6)**
4-Biphenylyl methyl ketone	**(92-91-1)**
β-((1,1'-Biphenyl)-4-yloxy)-α-(1,1-dimethylethyl)-1H-1,2,4-triazole-1-ethanol	**(55179-31-2)**
(1,4'-Bipiperidine)-4'-carboxamide, 1'-(3-cyano-3,3-diphenylpropyl)-	**(302-41-0)**
Bipiquin	**(54-05-7)**
Bipotassium chromate	**(7789-00-6)**
$\delta^{2,2'}$-Bipseudoindoxyl	**(482-89-3)**
2,2'-Bipyridin	**(366-18-7)**
2,2'-Bipyridine	**(366-18-7)**
2,3'-Bipyridine	**(581-50-0)**
4,4'-Bipyridine	**(553-26-4)**
Bipyridine	**(366-18-7)**
α,α'-Bipyridine	**(366-18-7)**
α,β'-Bipyridine	**(581-50-0)**
Bipyridine (9CI)	**(37275-48-2)**
Bipyridine, methyl- (9CI)	**(64859-47-8)**
2,3'-Bipyridine, 1,2,3,6-tetrahydro-, (S)- (9CI)	**(581-49-7)**
2,3'-Bipyridine, 1,2,3,6-tetrahydro-1-nitroso	**(71267-22-6)**
1,4'-Bipyridinium, chloride	**(22752-98-3)**
4,4'-Bipyridinium, 1,1'-dimethyl	**(4685-14-7)**
4,4'-Bipyridinium, 1,1'-dimethyl-, bis(methyl sulfate)	**(2074-50-2)**
4,4'-Bipyridinium, 1,1'-dimethyl-, dichloride	**(1910-42-5)**
Bipyridinium, 1,1'-dimethyl-4,4'-, dichloride	**(1910-42-5)**
2,2'-Bipyridyl	**(366-18-7)**
2,3'-Bipyridyl	**(581-50-0)**
4,4'-Bipyridyl	**(553-26-4)**
4,4-Bipyridyl	**(553-26-4)**
α,α'-Bipyridyl	**(366-18-7)**
γ,γ'-Bipyridyl	**(553-26-4)**
(5,5'-Bipyrimidine)-2,2',4,4',6,6'(1H,1'H,3H,3'H,5H,5'H)-hexone, 5,5'-dihydroxy- (9CI)	**(76-24-4)**
Biquin durules	**(747-45-5)**
2,2'-Biquinoline (9CI)	**(119-91-5)**
Biquinoline (9CI)	**(51913-96-3)**
2,2'-Biquinolyl	**(119-91-5)**
Birch-Tar-Oil	**(8001-88-5)**
Birlane	**(470-90-6)**
Birlane 24	**(470-90-6)**
Birnenoel	**(628-63-7)**
Birutan	**(153-18-4)**
Bis Amine	**(101-14-4)**
Bis-Amine A	**(101-14-4)**
Bis-CME	**(542-88-1)**
Bis-GMA	**(1565-94-2)**
8000 Bis HC	**(152-20-5)**
β-Bisabolol	**(15352-77-9)**
4,4'-Bis(acetamido)azobenzene	**(15446-39-6)**
Bis(acetato)hydroxyaluminum	**(142-03-0)**
Bis(acetato-O)hydroxyaluminum	**(142-03-0)**
Bis(acetato)tetrahydroxytrilead	**(1335-32-6)**
Bis(acetato)trihydroxytrilead	**(6080-56-4)**
4,4'-Bis(acetoacetamido)-3,3'-dimethylbiphenyl	**(91-96-3)**
4,4'-Bis(o-acetoacetotoluidide)	**(91-96-3)**
p,p'-Bis(o-acetoacetotoluidide)	**(91-96-3)**
4,4'-Bis(acetoacetylamino)-3,3'-dimethylbiphenyl	**(91-96-3)**
N,N'-Bis(acetoacetyl)-3,3'-dimethylbenzidine	**(91-96-3)**
Bis(aceto)dihydroxytrilead	**(1335-32-6)**
Bis(acetoxy)cadmium	**(543-90-8)**
3-(N,N-Bisacetoxyethyl)aminoacetanilide	**(27059-08-1)**
N-(3-(Bis(2-acetoxyethyl)amino)-4-methoxyphenyl)acetamide	**(23128-51-0)**

N,N-Bis(2-acetoxyethyl)aniline **(19249-34-4)**
2,3-Bis(acetoxymethyl)quinoxaline di-N-oxide **(10103-89-6)**
Bis(p-acetoxyphenyl)-2-pyridylmethane **(603-50-9)**
Bis(acetylacetonato)cobalt(II) **(14024-48-7)**
Bis(acetylacetonato)copper **(13395-16-9)**
Bis(acetylacetonato)titanium oxide **(14024-64-7)**
Bis(acetylacetone)copper **(13395-16-9)**
Bis(acetylacetonyl)copper **(14024-48-7)**
1,1'-(3,17-Bis(acetyloxy)androstane-2,16-diyl)bis(1-methylpiperidinium) dibromide **(15500-66-0)**
Bis(acetyloxy)dibutylstannane **(1067-33-0)**
N,N-Bis(2-(acetyloxy)ethyl)-4-((2-cyano-4-nitrophenyl)azo)-benzeneamine **(30124-94-8)**
Bis(acetyloxy)mercury **(1600-27-7)**
Bis(acetyl-N-phenylcarbamylmethyl)-4,4'-disazo-3,3'-dichlorobiphenyl **(6358-85-6)**
Bisacodyl **(603-50-9)**
S-(1,2-Bis(aethoxy-carbonyl)-aethyl)-O,O-dimethyl-dithiophosphat (German) **(121-75-5)**
2,4-Bis(aethylamino)-6-chlor-1,3,5-triazin (German) **(122-34-9)**
Bisaklofen BP **(119-47-1)**
Bisalkofen MTSP **(77-62-3)**
2,2-Bis(4-allyloxy-3,5-dibromophenyl)propane **(25327-89-3)**
2,4-Bis(p-aminobenzyl)aniline **(25834-80-4)**
Bis(4-amino-3-chlorophenyl) ether **(28434-86-8)**
Bis(4-amino-3-chlorophenyl)methane **(101-14-4)**
Bis(4-aminocyclohexyl)methane **(1761-71-3)**
Bis(p-aminocyclohexyl)methane **(1761-71-3)**
Bis(2-aminoethyl)amine **(111-40-0)**
Bis(β-aminoethyl)amine **(111-40-0)**
N,N-Bis(2-aminoethyl)-1,2-diaminoethane **(112-24-3)**
N,N-Bis(2-aminoethyl)-1,2-ethanediamine **(4097-89-6)**
N,N'-Bis(2-aminoethyl)ethylenediamine **(112-24-3)**
Bis-p-aminofenylmethan (Czech) **(101-77-9)**
Bis(6-aminohexyl)amine **(143-23-7)**
4,4'-Bis(1-amino-8-hydroxy-2,4-disulfo-7-naphthylazo)-3,3'-bitolyl, tetrasodium salt **(314-13-6)**
4,4'-Bis(7-(1-amino-8-hydroxy-2,4-disulfo)naphthylazo)-3,3'-bitolyl, tetrasodium salt **(314-13-6)**
4,4'-Bis(1-amino-8-hydroxy-2,4-disulpho-7-naphthylazo)-3,3'-bitolyl, tetrasodium salt **(314-13-6)**
4,4'-Bis(7-(1-amino-8-hydroxy-2,4-disulpho)naphthylazo)-3,3'-bitolyl, tetrasodium salt **(314-13-6)**
1,3-Bis-aminomethylbenzen (Czech) **(1477-55-0)**
1,3-Bis(aminomethyl)benzene **(1477-55-0)**
Bis-4-amino-3-methylfenylmethan (Czech) **(838-88-0)**
Bis(4-amino-3-methylphenyl)methane **(838-88-0)**
1,7-Bis(p-aminophenoxy)heptane **(2519-77-9)**
Bis(p-aminophenyl)amine **(537-65-5)**
Bis(4-aminophenyl)ether **(101-80-4)**
Bis(p-aminophenyl)ether **(101-80-4)**
2',4-Bis(aminophenyl)methane **(1208-52-2)**
Bis(4-aminophenyl)methane **(101-77-9)**
Bis(p-aminophenyl)methane **(101-77-9)**
2,4-Bis((4-aminophenyl)methyl)benzenamine **(25834-80-4)**
Bis(4-aminophenyl) sulfide **(139-65-1)**
Bis(p-aminophenyl)sulfide **(139-65-1)**
Bis(4-aminophenyl) sulfone **(80-08-0)**
Bis(p-aminophenyl) sulfone **(80-08-0)**
Bis(4-aminophenyl)sulphide **(139-65-1)**
Bis(p-aminophenyl)sulphide **(139-65-1)**
Bis(4-aminophenyl)sulphone **(80-08-0)**
Bis(p-aminophenyl)sulphone **(80-08-0)**
Bis-(3-aminopropyl)amine **(56-18-8)**
1,4-Bis(aminopropyl) butanediamine **(71-44-3)**
N,N'-Bis(3-aminopropyl)-1,4-butanediamine **(71-44-3)**
N,N'-Bis(3-aminopropyl)-1,2-diaminoethane **(10563-26-5)**
N,N'-Bis(γ-aminopropyl)diaminoethane **(10563-26-5)**
N,N'-Bis(3-aminopropyl)ethylenediamine **(10563-26-5)**
Bis(3-aminopropyl)methylamine **(105-83-9)**
Bis(γ-aminopropyl)methylamine **(105-83-9)**
Bis(ω-aminopropyl)methylamine **(105-83-9)**
N,N-Bis(3-aminopropyl)methylamine **(105-83-9)**
N,N-Bis(γ-aminopropyl)methylamine **(105-83-9)**
4,4'-Bis((4-anilino-6-(bis(2-hydroxyethyl)amino)-1,3,5-triazin-2-yl)amino)stilbene-2,2'-disulfonic acid **(4404-43-7)**
4,4'-Bis((4-anilino-6-(bis(2-hydroxyethyl)amino)-s-triazin-2-yl)-amino)-2,2'-stilbene disulfonic acid **(4404-43-7)**
4,4'-Bis(4-anilino-6-chloro-s-triazin-2-yl)amino)-2,2'-stilbene-disulfonic acid, disodium salt **(37138-23-1)**
Bis(p-anisylamine) **(101-70-2)**
2,2-Bis(p-anisyl)-1,1,1-trichloroethane **(72-43-5)**
2,5-Bis(1-aziridinyl)-3,6-bis(2-methoxyethoxy)-p-benzoquinone **(800-24-8)**
2,5-Bis(1-aziridinyl)-3,6-bis(2-methoxyethoxy)-2,5-cyclo-hexadiene-1,4-dione **(800-24-8)**
Bis(1-aziridinyl)morpholinophosphine sulphide **(2168-68-5)**
Bis(2-benzamidophenyl) disulfide **(135-57-9)**
Bis(o-benzamidophenyl) disulfide **(135-57-9)**
Bis(benzhydryl) ether **(574-42-5)**
Bisbenzimidazo(2,1-b:2',1'-i)benzo(lmn)(3,8)phenanthroline-8,17-dione (9CI) **(4424-06-0)**
Bisbenzimidazo(2,1-b:1',2'-j)benzo(lmn)(3,8)phenanthroline-6,9-dione (9CI) **(4216-02-8)**
Bisbenzimidazole **(23491-45-4)**
Bisbenzimide **(23491-45-4)**
Bis(1,3)benzodioxolo(4,5-c:5',6'-g)azecin-13(5H)-one, 4,6,7,14-tetrahydro-5-methyl- **(130-86-9)**
Bis(1,3)benzodioxolo(4,5-c:5',6'-g)azecin-13(5H)-one, 4,6,7,14-tetrahydro-5-methyl-, hydrochloride (9CI) **(6164-47-2)**
Bis(benzothiazolyl) disulfide **(120-78-5)**
N,N'-Bis(2-benzothiazolylthiomethylene)urea **(95-35-2)**
1,3-Bis((2-benzothiazolylthio)methyl)urea **(95-35-2)**
Bis(2-benzothiazolylthio)zinc **(155-04-4)**
Bis(2-benzothiazyl) disulfide **(120-78-5)**
4,5'-Bis-benzoylamino-1,1'-dianthrimid (Czech) **(128-89-2)**
Bis-o-benzoylaminofenyl-disulfid (Czech) **(135-57-9)**
N,N-Bis(2-(bis(carboxymethyl)amino)ethyl)glycine calcium trisodium salt **(12111-24-9)**
N,N-Bis(2-(bis(carboxymethyl)amino)ethyl)glycine, pentasodium salt **(140-01-2)**
Bis((4-(bis(2-chloroethyl)amino)benzene)acetate)estra-1,3,5(10)-triene-3,17-diol(17β) **(22966-79-6)**
Bis((4-(bis(2-chloroethyl)amino)benzene)acetate)oestra-1,3,5(10)-triene-3,17-diol(17β) **(22966-79-6)**
Bis((p-(bis(2-chloroethyl)amino)phenyl)acetate)estradiol **(22966-79-6)**
Bis((p-(bis(2-chloroethyl)amino)phenyl)acetate)estra-1,3,5(10)-triene-3,17-β-diol **(22966-79-6)**
Bis((p-(bis(2-chloroethyl)amino)phenyl)acetate)oestradiol **(22966-79-6)**
Bis((p-bis(2-chloroethyl)aminophenyl)acetate)oestra-1,3,5(10)-triene-3,17-β-diol **(22966-79-6)**
Bis(bisdimethylaminophosphonous)anhydride **(152-16-9)**
Bis(bis(2-hydroxyethyl)aminoethyl) diisopropyl titanate **(36673-16-2)**
(T-4)-Bis(bis(phenylmethyl)carbamodithioato-S,S')zinc **(14726-36-4)**
1,4-Bis(bromoacetoxy)-2-butene **(20679-58-7)**
2,2-Bis(bromomethyl)-3-chloropropyl bis(2-chloro-1-(chloro-methyl)ethyl) phosphate **(66108-37-0)**
2,2-Bis(bromomethyl)-1,3-propanediol **(3296-90-0)**
Bis(p-bromophenyl) ether **(2050-47-7)**
1,4-Bis(3-bromopropionyl)-piperazine **(54-91-1)**
N,N-Bis-(3-bromopropionyl)-piperazine **(54-91-1)**
2,3,4,5-Bis(2-butenylene)tetrahydrofurfural **(126-15-8)**
2,3,4,5-Bis(δ^2-butenylene)tetrahydrofurfural **(126-15-8)**
Bisbutenylenetetrahydrofurfural **(126-15-8)**
1-Bis(2-(2-butoxyethoxy)ethoxy)methyl-3,4-methylenedioxy-

benzene (5281-13-0)
Bis(2-butoxyethyl) adipate (141-18-4)
Bis(2-butoxyethyl) decanedioate (141-19-5)
Bis(2-butoxyethyl) ether (112-73-2)
Bis(2-butoxyethyl)phthalate (117-83-9)
N,N'-Bis(butoxymethyl)urea (4981-47-9)
4,4'-Bis(sec-butylamino)diphenylmethane (5285-60-9)
Bis(butylcarbitol)formal (143-29-3)
Bis(4-tert-butylcyclohexyl)peroxydicarbonate (15520-11-3)
α,α'-Bis(tert-butyldioxy)-p-diisopropylbenzene (2781-00-2)
1,4-Bis(α-(t-butyldioxy)isopropyl)benzene (2781-00-2)
1,4-Bis(α-(tert-butyldioxy)isopropyl)benzene (2781-00-2)
2,3,4,5-Bis(2-butylene)tetrahydro-2-furaldehyde (126-15-8)
2,3,4,5-Bis(δ^2-butylene)tetrahydrofurfural (126-15-8)
2,3:4,5-Bis(2-butylene)tetrahydro-2-furfural (126-15-8)
Bis-δ^2-butylenetetrahydrofurfural (126-15-8)
Bis(2-butyl)ether (6863-58-7)
Bis(3-tert-butyl-4-hydroxy-6-methylphenyl) sulfide (96-69-5)
2,2'-Bis-6-terc.butyl-p-kresylmethan (Czech) (119-47-1)
2,2-Bis(tert-butylperoxy)butane (2167-23-9)
Bis(tert-butylperoxy)-2,2 butane (French) (2167-23-9)
Bis(tert-butylperoxy)-1,1 cyclohexane (French) (3006-86-8)
α,α'-Bis(tert-butylperoxy)-1,4-diisopropylbenzene (2781-00-2)
α,α'-Bis(tert-butylperoxy)-p-diisopropylbenzene (2781-00-2)
α,α'-Bis(tert-butylperoxy)diisopropylbenzene (25155-25-3)
Bis(tert-butylperoxy)-1,1 trimethyl-3,3,5 cyclohexane (French) (6731-36-8)
4,4-Bis(tert-butylperoxy)valeric acid butyl ester (995-33-5)
2,6-Bis(tert-butyl)phenol (128-39-2)
Bis(p-tert-butylphenyl) phenyl phosphate (115-87-7)
Bis(tert-butyl)phosphine oxide (684-19-5)
Bis(n-butyl)sebacate (109-43-3)
1,3-Bis(carbamoylthio)-2-(N,N-dimethylamino)propane hydrochloride (15263-52-2)
S-(1,2-Bis(carbethoxy)ethyl) O,O-dimethyl dithiophosphate (121-75-5)
Bis-β-carbobutoxyethyltin bisisooctylthioglycolate (63397-60-4)
Bis(carbonato(2-))dihydroxytriberyllium (66104-24-3)
Bis(carbonato(2-))dihydroxytrilead (1319-46-6)
1,5-Bis(2-carboxyanilino)anthraquinone (81-78-7)
1,3-Bis(2-carboxychromon-5-yloxy)-2-hydroxypropane (16110-51-3)
Bis(2-carboxyethyl) sulfide (111-17-1)
1,3-Bis-(carboxymethoxy)-1-propen-2-yl-dimethylfosfat (Czech) (122-10-1)
N-(2(Bis(carboxymethyl)amino)-ethyl)-N-(2-hydroxyethyl)-glycine trisodium salt (139-89-9)
N-(2-(Bis(carboxymethyl)amino)ethyl)-N-(2-hydroxyethyl)-glycine, trisodium salt (139-89-9)
Bis(carboxymethyl)ether (110-99-6)
(N,N-Bis(carboxymethyl)glycinato(3-)-N,O,O',O'')cuprate-(-1), ammonium (71484-80-5)
N,N-Bis(carboxymethyl)-glycine disodium salt (15467-20-6)
N,N-Bis(carboxymethyl)glycine disodium salt (15467-20-6)
N,N-Bis(carboxymethyl)glycine trisodium salt monohydrate (18662-53-8)
N,N-Bis(carboxymethyl)glysine (139-13-9)
Bis(3-carboxypropionyl) peroxide (123-23-9)
N,N-Bis(2-chloraethyl)methylamin-hydrochlorid (German) (55-86-7)
N,N-Bis-(β-chloraethyl)-N',O-propylen-phosphorsaeure-ester-diamid (German) (50-18-0)
Bis-(4-chlorbut-1-yl)-ether (German) (6334-96-9)
N,N-Bis-(2-chlorethyl)-2-naftylamin (Czech) (494-03-1)
Bis(5-chlor-2-hydroxyphenyl)-methan (German) (97-23-4)
4,4'-Bis(2-chloro-4-anilino-1,3,5-triazin-6-yl)aminostilbene-2,2'-disulfonic acid, disodium salt (37138-23-1)
Bis(p-chlorobenzoyl) peroxide (94-17-7)
Bis(p-chlorobenzyl) ether (56428-00-3)
Bis(4-chlorobutyl) ether (6334-96-9)
1,2-Bis(chlorocarbonyl)benzene (88-95-9)
Bis(2-chloro-1-(chloromethyl)ethyl) 2,3-dichloropropyl phosphate (68460-03-7)
1,2-Bis(2-chloroethoxy)ethane (112-26-5)
Bis(2-(2-chloroethoxy)ethyl) ether (638-56-2)
Bis(2-chloroethoxy)-methane (111-91-1)
4-(Bis(2-chloroethyl)amino)benzenebutanoic acid (305-03-3)
1,6-Bis(chloroethylamino)-1,6-bis-deoxy-D-mannitol (576-68-1)
1,6-Bis-(chloroethylamino)-1,6-desoxy-D-mannitol dihydrochloride (551-74-6)
1,6-Bis(chloroethylamino)-1,6-dideoxy-D-mannite (576-68-1)
1,6-Bis-(chloroethylamino)-1,6-dideoxy-D-mannite dihydrochloride (551-74-6)
1,6-Bis((2-chloroethyl)amino)-1,6-dideoxy-D-mannitol (576-68-1)
1,6-Bis((β-chloroethyl)amino)-1,6-dideoxy-D-mannitol (576-68-1)
1,6-Bis(β-chloroethylamino)-1,6-dideoxy-D-mannitol dihydrochloride (551-74-6)
3-(Bis(2-chloroethyl)amino)-4-methylbenzenesulfonylmorpholine (63905-03-3)
9-(4-Bis(2-chloroethyl)amino-1-methylbutylamino)-6-chloro-2-methoxyacridine dihydrochloride (4213-45-0)
2-Bis(2-chloroethyl)aminonaphthalene (494-03-1)
2-(Bis(2-chloroethyl)amino)-1-oxa-3-aza-2-phosphocyclohexane 2-oxide monohydrate (6055-19-2)
2-(Bis(2-chloroethyl)amino)-2H-1,3,2-oxazaphosphorine 2-oxide (50-18-0)
1-Bis(2-chloroethyl)amino-1-oxo-2-aza-5-oxaphosphoridine monohydrate (6055-19-2)
(p-(Bis(2-chloroethyl)amino)phenyl)acetic acid cholesterol ester (3546-10-9)
(+)-3-(p-(Bis(2-chloroethyl)amino)phenyl)alanine (13045-94-8)
3-(p-(Bis(2-chloroethyl)amino)phenyl)-L-alanine (148-82-3)
3-(p-(Bis(2-chloroethyl)amino)phenyl)alanine (531-76-0)
4-(Bis(2-chloroethyl)amino)-D-phenylalanine (13045-94-8)
4-(Bis(2-chloroethyl)amino)-DL-phenylalanine (531-76-0)
4-(Bis(2-chloroethyl)amino)-L-phenylalanine (148-82-3)
D-3-(p-(Bis(2-chloroethyl)amino)phenyl)alanine (13045-94-8)
DL-3-(p-(Bis(2-chloroethyl)amino)phenyl)alanine (531-76-0)
L-3-(p-(Bis(2-chloroethyl)amino)phenyl)alanine (148-82-3)
p-Bis(β-chloroethyl)aminophenylalanine (148-82-3)
p-N-Bis(2-chloroethyl)amino-L-phenylalanine (148-82-3)
4(p-Bis(β-chloroethyl)aminophenyl)butyric acid (305-03-3)
4-(p-(Bis(2-chloroethyl)amino)phenyl)butyric acid (305-03-3)
γ-(p-Bis(2-chloroethyl)aminophenyl)butyric acid (305-03-3)
N,N-Bis(β-chloroethyl)-amino-N',O-propylene-phosphoric acid ester diamide (3778-73-2)
5-(Bis(2-chloroethyl)amino)-2,4(1H,3H)pyrimidinedione (66-75-1)
(Bis(chloro-2-ethyl)amino)-2-tetrahydro-3,4,5,6-oxazaphosphorine-1,3,2-oxide-2 monohydrate (6055-19-2)
5-(Bis(2-chloroethyl)amino)uracil (66-75-1)
5-N,N-Bis(2-chloroethyl)aminouracil (66-75-1)
N,N-Bis(2-chloroethyl)aniline (553-27-5)
N,N-Bis(2-chloroethyl)benzenamine (553-27-5)
2,3-(N,N[1]-Bis(2-chloroethyl)diamido)-1,3,2-oxazaphosphor-idinoxyd (3778-73-2)
N,N-Bis(2-chloroethyl)-2,6-dinitro-p-toluidine (26389-78-6)
Bis(2-chloroethyl) ether (111-44-4)
Bis(α-chloroethyl) ether (6986-48-7)
Bis(β-chloroethyl) ether (111-44-4)
Bis(2-chloroethyl)ethylamine (538-07-8)
Bis(2-chloroethyl)formal (111-91-1)
Bis(β-chloroethyl)formal (111-91-1)
N,N-Bis(2-chloroethyl)-N'-(3-hydroxypropyl)phosphorodiamidate cyclohexylammonium salt (4465-94-5)
Bis(2-chloroethyl)methylamine (51-75-2)
Bis(β-chloroethyl)methylamine (51-75-2)
N,N-Bis(2-chloroethyl)methylamine (51-75-2)
Bis(2-chloroethyl)methylamine hydrochloride (55-86-7)
N,N-Bis(2-chloroethyl)methylamine hydrochloride (55-86-7)
Bis(2-chloroethyl)-β-naphthylamine (494-03-1)
N,N,-Bis(2-chloroethyl)-2-naphthylamine (494-03-1)
1,3-Bis(β-chloroethyl)-1-nitrosourea (154-93-8)
1,3-Bis-(2-chloroethyl)-1-nitrosourea (154-93-8)
Bis(2-chloroethyl)nitrosourea (154-93-8)
Bischloroethylnitrosourea (154-93-8)
N,N'-Bis(2-chloroethyl)-N-nitrosourea (154-93-8)

Bis(2-chloroethyl)phosphoramide cyclic propanolamide ester monohydrate (6055-19-2)
Bis(2-chloroethyl)phosphoramide-cyclic propanolamide ester (50-18-0)
N,N-Bis(2-chloroethyl)propylamine (621-68-1)
N,N-Bis(β-chloroethyl)-N',O-propylenephosphoric acid ester amide monohydrate (6055-19-2)
N,N-Bis(2-chloroethyl)-N',O-propylenephosphoric acid ester diamide (50-18-0)
Bis(2-chloroethyl)sulfide (505-60-2)
Bis(β-chloroethyl)sulfide (505-60-2)
Bis(2-chloroethyl)sulfone (471-03-4)
Bis(β-chloroethyl)sulfone (471-03-4)
1,2-Bis((2-chloroethyl)sulfonyl)ethane (3944-87-4)
Bis(β-chloroethyl)sulfoxide (5819-08-9)
Bis(2-chloroethyl)sulphide (505-60-2)
N,N-Bis(2-chloroethyl)tetrahydro-2H-1,3,2-oxaphosphorin-2-amine, 2-oxide monohydrate (6055-19-2)
N,3-Bis(2-chloroethyl)tetrahydro-2H-1,3,2-oxazaphosphorin-2-amine 2-oxide (3778-73-2)
N,N-Bis(2-chloroethyl)tetrahydro-2H-1,3,2-oxazaphosphorin-2-amine 2-oxide (50-18-0)
1,2-Bis(2-chloroethylthio)ethane (3563-36-8)
1,2-Bis(β-chloroethylthio)ethane (3563-36-8)
Bis(2-chloroethylthio)ethane (3563-36-8)
Bis(2-chloroethylthioethyl) ether (63918-89-8)
Bis(β-chloroethylthioethyl) ether (63918-89-8)
5-(3,3-Bis(2-chloroethyl)-1-triazeno)imidazole-4-carboxamide (5034-77-5)
N,N-Bis(β-chloroethyl)-N',O-trimethylenephosphoric acid ester diamide monohydrate (6055-19-2)
N,N-Bis(β-chloroethyl)-N',O-trimethylenephosphoric acid ester diamide (50-18-0)
Bis(5-chloro-2-hydroxyphenyl)methane (97-23-4)
Bis(2-chloroisopropyl) ether (108-60-1)
Bis(2-chloroisopropyl) ether (39638-32-9)
Bis(β-chloroisopropyl)ether (108-60-1)
Bis-1,2-(chloromethoxy)ethane (13483-18-6)
1,4-Bis(chloromethoxymethyl)benzene (56894-91-8)
Bis-1,4-(chloromethoxy)-p-xylene (56894-91-8)
1,2-Bis(chloromethyl)benzene (612-12-4)
1,3-Bis(chloromethyl)benzene (626-16-4)
Bis(chloromethyl)benzene (28347-13-9)
Bis(chloromethyl) ether (ACGIH,OSHA) (542-88-1)
Bis(2-chloro-1-methylethyl) ether (108-60-1)
Bis(2-chloro-1-methylethyl)-ether (39638-32-9)
Bis(chloromethyl)ketone (534-07-6)
3,3-Bis(chloromethyl)oxetane (78-71-7)
Bis(2-chlorooctyl) disulfide (70776-26-0)
Bis(p-chlorophenoxy)methane (555-89-5)
2,2-Bis(p-chlorophenyl)acetic acid (83-05-6)
Bis(4-chlorophenyl)acetic acid (83-05-6)
Bis(p-chlorophenyl)acetic acid (83-05-6)
O,O-Bis(p-chlorophenyl)acetimidophosphoroamidothioate (4104-14-7)
O,O-Bis(p-chlorophenyl) acetimidoylamidothiophosphat (4104-14-7)
O,O-Bis(p-chlorophenyl) acetimidoylphosphoramidothioate (4104-14-7)
O,O-Bis(p-chlorophenyl)acetimidoylphosphoramidothioate (4104-14-7)
1,6-Bis(5-(p-chlorophenyl)biguandino)hexane (55-56-1)
1,6-Bis(5-(p-chlorophenyl)biguandino)hexane diacetate (56-95-1)
1,6-Bis(5-(p-chlorophenyl)biguandino)hexane digluconate (18472-51-0)
1,6-Bis(p-chlorophenylbiguanido)hexane diacetate (56-95-1)
1,1-Bis(p-chlorophenyl)-2-chloroethylene (1022-22-6)
Bis(2-chlorophenyl)diazene (7334-33-0)
1,1-Bis(4-chlorophenyl)-2,2-dichloroethane (72-54-8)
1,1-Bis(p-chlorophenyl)-2,2-dichloroethane (72-54-8)
2,2-Bis(4-chlorophenyl)-1,1-dichloroethane (72-54-8)
2,2-Bis(p-chlorophenyl)-1,1-dichloroethane (72-54-8)
2,2-Bis(4-chlorophenyl)-1,1-dichloroethene (72-55-9)
1,1'-Bis(chlorophenyl)-2,2-dichloroethylene (68679-99-2)
2,2-Bis(p-chlorophenyl)-1,1-dichloroethylene (72-55-9)
1,6-Bis(p-chlorophenyldiguanido)hexane (55-56-1)
Bis(p-chlorophenyldiguanidohexane) diacetate (56-95-1)
N,N'-Bis(4-chlorophenyl)-3,12-diimino-2,4,11,13-tetraazatetradecanediimidamide, dihydrochloride (3697-42-5)
N,N'-Bis(4-chlorophenyl)-3,12-diimino-2,4,11,13-tetraazatetradecanediimidamide, diacetate (56-95-1)
Bis(4-chlorophenyl) disulfide (1142-19-4)
Bis(p-chlorophenyl)disulfide (1142-19-4)
Bis(4-chlorophenyl)ethanedione (3457-46-3)
1,1-Bis(4-chlorophenyl)ethanol (80-06-8)
1,1-Bis(p-chlorophenyl)ethanol (80-06-8)
1,2-Bis(p-chlorophenyl) ethanol (56960-97-5)
2,2-Bis(4-chlorophenyl)ethanol (2642-82-2)
2,2-Bis(p-chlorophenyl)ethanol (2642-82-2)
1,1-Bis(4-chlorophenyl)-2-ethoxyethanol (6012-83-5)
1,2-Bis(2-chlorophenyl)-hydrazine (782-74-1)
1,2-Bis(2-chlorophenyl)hydrazine (782-74-1)
2,2-Bis(4-chlorophenyl)-1-hydroxyethane (2642-82-2)
O,O-Bis(4-chlorophenyl) (1-iminoethyl)phosphoramidothioate (4104-14-7)
Bis(p-chlorophenyl)methane (101-76-8)
1,1-Bis(p-chlorophenyl)methylcarbinol (80-06-8)
Bis(p-chlorophenyl)methyl carbinol (80-06-8)
2,2-Bis(p-chlorophenyl)-1-monochloroethane (2642-80-0)
1,1-Bis(4-chlorophenyl)-2-nitrobutane (117-26-0)
1,1-Bis(p-chlorophenyl)-2-nitrobutane (117-26-0)
1,1-Bis(4-chlorophenyl)-2-nitropropane (117-27-1)
1,1-Bis(p-chlorophenyl)-2-nitropropane (117-27-1)
1,1-Bis(p-chlorophenyl)-2-nitropropane mixed with 1,1-bis-(p-chlorophenyl)-2-nitrobutane(1:2) (8027-00-7)
α,α-Bis(4-chlorophenyl)-3-pyridinemethanol (17781-31-6)
Bis(4-chlorophenyl)-3-pyridylmethanol (17781-31-6)
1,1'-Bis(chlorophenyl)-1,2,2,2-tetrachloroethylene (68631-02-7)
α,α-Bis(p-chlorophenyl)-β,β,β-trichlorethane (50-29-3)
1,1-Bis-(p-chlorophenyl)-2,2,2-trichloroethane (50-29-3)
2,2-Bis(o,p-chlorophenyl)-1,1,1-trichloroethane (789-02-6)
2,2-Bis(p-chlorophenyl)-1,1,1-trichloroethane (50-29-3)
1,1-Bis(4-chlorophenyl)-2,2,2-trichloroethanol (115-32-2)
1,1-Bis(chlorophenyl)-2,2,2-trichloroethanol (115-32-2)
1,1-Bis(p-chlorophenyl)-2,2,2-trichloroethanol (115-32-2)
Bis(1-chloro-2-propyl) ether (108-60-1)
Bis(4-chlorosulfonylphenyl) ether (121-63-1)
1,1-Bis(4-chlorphenyl)-aethanol (German) (80-06-8)
Bis(p-chlorphenyl)essigsaeure (German) (83-05-6)
N,N-Bis(β-cloraethyl) N'-O-propylenphosphorildiamid monohydratum (Romanian) (6055-19-2)
Biscomate (4044-65-9)
Bis(β-cyanoethyl)amine (111-94-4)
Bis-(2-cyanoethyl)amine (111-94-4)
N,N-Bis(2-cyanoethyl)amine (111-94-4)
N,N'-Bis(2-cyanoethyl)-1,2-ethanediamine (3217-00-3)
1,5-Bis(cyclohexylamino)-9,10-anthracenedione (15958-68-6)
Bis(cyclohexyl)disulfide (2550-40-5)
Biscyclopentadiene (77-73-6)
Bis(cyclopentadienyl)cobalt (1277-43-6)
Biscyclopentadienyliron (102-54-5)
1,4-Bis(dibrommethyl)benzen (Czech) (23488-38-2)
2,2-Bis(3,5-dibromo-4-(2-hydroxyethoxy)phenyl)propane (4162-45-2)
2,2-Bis(3,5-dibromo-4-hydroxyphenyl)propane (79-94-7)
Bis(dibutyldithiocarbamato)nickel (13927-77-0)
Bis(dibutyldithiocarbamato)zinc (136-23-2)
Bis(3,5-di-tert-butyl-4-hydroxyhydrocinnamoyl)hydrazine (32687-78-8)
Bis(di-n-butylthiocarbamoylthio)methane (10254-57-6)
2,2-Bis(4-(3,4-dicarboxyphenoxy)phenyl)propane dianhydride (38103-06-9)
Bis(2,4-dichlorobenzoyl)peroxide (133-14-2)
2,2-Bis(3,5-dichloro-4-hydroxyphenyl)propane (79-95-8)
Bis(3,5-dichloro-2-hydroxyphenyl) sulfide (97-18-7)

1,2-Bis(dichloromethyl)benzene (25641-99-0)
Bis(2,4-dichlorophenoxyethyl)phosphite mixed with tris(2,4-dichlorophenoxyethyl)phosphite (8005-49-0)
Bis(2,4-dichlorophenyl)ether (28076-73-5)
Bis(1,3-dichloro-2-propyl) 3-chloro-2,2-dibromomethyl-1-propyl phosphate (66108-37-0)
Bis(1,3-dichloro-2-propyl)-2,3-dichloro-1-propyl phosphate (68460-03-7)
2,6-Bis(diethanolamino)-4,8-dipiperidinopyrimido(5,4-d)pyrimidine (58-32-2)
Bis(S-(diethoxyphosphinothioyl)mercapto)methane (563-12-2)
4,4'-Bis(N,N-diethylamino)benzophenone (90-93-7)
p,p'-Bis(diethylamino)benzophenone (90-93-7)
2,4-Bis(diethylamino)-6-fluoro-s-triazine (1598-99-8)
2,4-Bis(diethylamino)-6-mercapto-s-triazine (4022-55-3)
Bis(4-(diethylamino)phenyl)methanone (90-93-7)
Bis((diethylamino)thioxomethyl)disulphide (97-77-8)
4,6-Bis(diethylamino)-s-triazine-2(1H)-thione (4022-55-3)
Bis(diethyldithiocarbamato)cadmium (14239-68-0)
Bis(diethyldithiocarbamato)zinc (14324-55-1)
Bis-O,O-diethylphosphoric anhydride (107-49-3)
Bis-O,O-diethylphosphorothionic anhydride (3689-24-5)
Bis(N,N-diethylthiocarbamoyl) disulfide (97-77-8)
Bis(diethylthiocarbamoyl) disulfide (97-77-8)
Bis(N,N-diethylthiocarbamoyl)disulphide* (97-77-8)
Bis(diethylthiocarbamoyl)disulphide (97-77-8)
Bis(2,4-dihydroxyl)benzoatolead(II) (41453-50-3)
2,2-Bisdihydroxymethyl-1,3-propanediol tetranitrate (78-11-5)
2,3-Bis(3,4-dihydroxyphenylmethyl)butane (500-38-9)
Bis(dimethylamido)fluorophosphate (115-26-4)
Bis(dimethylamido)-phosphoryl fluoride (115-26-4)
2,8-Bisdimethylaminoacridine (494-38-2)
3,6-Bis(dimethylamino)acridine (494-38-2)
3,6-Bis-(dimethylamino)akridin (Czech) (494-38-2)
Bis(dimethylamino)-3-amino-5-phenyltriazolyl phosphine oxide (1031-47-6)
1,4-Bis(dimethylamino)benzene (100-22-1)
p-Bis(dimethylamino)benzene (100-22-1)
4,4'-Bis(dimethylamino)-benzhydrylidenimine hydrochloride (2465-27-2)
4,4'-Bis(dimethylamino)benzohydrol (119-58-4)
4,4'-Bis(dimethylamino)benzophenone (90-94-8)
p,p'-Bis(N,N-dimethylamino)benzophenone (90-94-8)
4:4'-Bis(dimethylamino)benzophenone-imine hydrochloride (2465-27-2)
4,4'-Bis(N,N-dimethylamino)biphenyl (366-29-0)
Bis((dimethylamino)carbonothioyl) disulphide (137-26-8)
4,4'-Bis(dimethylamino)diphenylmethane (101-61-1)
p,p'-Bis(dimethylamino)diphenylmethane (101-61-1)
1,2-Bis-(dimethylamino)ethane (110-18-9)
Bis(2-dimethylaminoethyl)ether (3033-62-3)
Bis(dimethylamino)fluorophosphate (115-26-4)
Bisdimethylaminofluorophosphine oxide (115-26-4)
Bis(dimethylamino)methane (51-80-9)
3,7-Bis (dimethylamino) phenazathionium chloride (61-73-4)
3,7-Bis(dimethylamino)phenothiazin-5-ium chloride (61-73-4)
3,3-Bis(p-dimethylaminophenyl)-6-dimethylaminophthalate (1552-42-7)
3,3-Bis(4-dimethylaminophenyl)-6-dimethylaminophthalide (1552-42-7)
3,3-Bis(p-dimethylaminophenyl)-6-dimethylaminophthalide (1552-42-7)
Bis(p-(N,N-dimethylamino)phenyl)ketone (90-94-8)
Bis(4-(N,N-dimethylamino)phenyl)methane (101-61-1)
Bis(4-(dimethylamino)phenyl)methane (101-61-1)
Bis(p-(N,N-dimethylamino)phenyl)methane (101-61-1)
Bis(p-dimethylaminophenyl)methane (101-61-1)
p,p'-Bis(N,N-dimethylaminophenyl)methane (101-61-1)
α,α-Bis(p-dimethylaminophenyl)methanol (119-58-4)
Bis(4-(dimethylamino)phenyl)methanone (90-94-8)
Bis(p-dimethylaminophenyl)methyleneimine (492-80-8)
1,1-Bis(p-dimethylaminophenyl)methylenimine hydrochloride (2465-27-2)
Bis(dimethylamino)phosphonous anhydride (152-16-9)
Bis(dimethylamino)phosphoric anhydride (152-16-9)
Bis(3-dimethylaminopropyl)amine (6711-48-4)
N,N-Bis(3-(dimethylamino)propyl)-N',N'-dimethyl-1,3-propanediamine (33329-35-0)
N,N'-Bis(3-(dimethylamino)propyl)urea (52338-87-1)
Bis(α,α-dimethylbenzyl)peroxide (80-43-3)
O,O-Bis(1,3-dimethylbutyl)dithiophosphate zinc salt (2215-35-2)
Bis-(1,3-dimethylbutyl)ester kyseliny maleinove (Czech) (105-52-2)
1,7-Bis(1,3-dimethylbutylidene) diethylenetriamine (10595-60-5)
1,7-Bis(1,3-dimethylbutylidene)diethylenetriamine (10595-60-5)
Bis(1,3-dimethylbutyl) maleate (105-52-2)
O,O-Bis(1,3-dimethylbutyl) phosphorodithioate (6028-47-3)
Bis(dimethylcarbamodithioato-S,S')lead (19010-66-3)
Bis(dimethylcarbamodithioato-S,S')zinc (137-30-4)
Bis-dimethyldithiocarbamate de zinc (French) (137-30-4)
Bis(dimethyldithiocarbamato)lead (19010-66-3)
Bis(dimethyldithiocarbamato)zinc (137-30-4)
1,4-Bis(1,1-dimethylethyl)benzene (1012-72-2)
3,5-Bis(1,1-dimethylethyl)-(1,1'-biphenyl)-4-ol (2668-47-5)
N,N'-Bis(1,1-dimethylethyl)-1,2-ethanediamine (4062-60-6)
2,6-Bis(1,1-dimethylethyl)-4-ethylphenol (4130-42-1)
3,5-Bis(1,1-dimethylethyl)-4-hydroxybenzenepropanoic acid octadecyl ester (2082-79-3)
(((3,5-Bis(1,1-dimethylethyl)-4-hydroxyphenyl)methyl)thio)acetic acid 2-ethylhexyl ester (80387-97-9)
2,6-Bis(1,1-dimethylethyl)-4-methoxyphenol (489-01-0)
2,6-Bis(1,1-dimethylethyl)-4-methylphenol (128-37-0)
2,6-Bis(1,1-dimethylethyl)-4-nonylphenol (4306-88-1)
2,5-Bis(1,1-dimethylethyl)phenol (5875-45-6)
3,5-Bis(1,1-dimethylethyl)phenol (1138-52-9)
N,N'-Bis(1,4-dimethylpentyl)-p-phenylenediamine (3081-14-9)
2,9-Bis(3,5-dimethylphenyl)anthra(2,1,9-def:6,5,10-d',e',f')diisoquinoline-1,3,8,10(2H,9H)-tetrone (4948-15-6)
2,5-Bis(1,1-dimethylpropyl)hydroquinone (79-74-3)
N-(4-(2,4-Bis(1,1-dimethylpropyl)phenoxy)butyl)-1-hydroxy-2-naphthalenecarboxamide (32180-75-9)
Bis(dimethylsilyl) ether (3277-26-7)
Bis(dimethylsilyl) oxide (3277-26-7)
Bis(dimethyl-thiocarbamoyl)-disulfid (German) (137-26-8)
Bis(dimethylthiocarbamoyl) disulfide (137-26-8)
Bis(dimethylthiocarbamoyl) disulphide (137-26-8)
Bis(dimethylthiocarbamoyl)sulfide (97-74-5)
Bis(dimethylthiocarbamoyl) tetrasulfide (97-91-6)
Bis(dimethylthiocarbamoylthio)methyl-arsine (2445-07-0)
Bis(dimethylthiocarbamyl) disulfide (137-26-8)
Bis(dimethylthiocarbamyl) monosulfide (97-74-5)
Bis(O,O-dimethylthiophosphoryl)disulfide (5930-71-2)
Bis(N,N-dimetil-ditiocarbammato) di zinco (Italian) (137-30-4)
4,4'-Bis((2,4-dinitrophenyl)amino)-3,3'-dimethoxybiphenyl (29398-96-7)
(+,-)-1,2-Bis(3,5-dioxopiperazine-1-yl)propane (21416-87-5)
(+-)-1,2-Bis(3,5-dioxopiperazinyl)propane (21416-87-5)
(T-4)-Bis(dipentylcarbamodithioato-S,S')lead (36501-84-5)
(T-4)-Bis(dipentylcarbamodithioato-S,S')zinc (15337-18-5)
Bis(diphenylmethyl) ether (574-42-5)
Bis-O,O-di-n-propylphosphorothionic anhydride (3244-90-4)
Bis(dithiophosphate de O,O-diethyle) de S,S'-(1,4-dioxanne-2,3-diyle) (French) (78-34-2)
2,4-Bis(dodecylamino)-6-chloro-s-triazine (30355-03-4)
Bis(dodecyloxycarbonylethyl) sulfide (123-28-4)
Bis(dodecylthio)dimethylstannane (51287-84-4)
Bis((3,4-epoxycyclohexyl)methyl)adipate (3130-19-6)
Bis(2,3-epoxycyclopentyl) ether (2386-90-5)
Bis(2,3-epoxycyclopentyl)ether (2386-90-5)
Bis(2,3-epoxy-2-methylpropyl)ether (7487-28-7)
1,3-Bis(2,3-epoxypropoxy)benzene (101-90-6)
m-Bis(2,3-epoxypropoxy)benzene (101-90-6)
1,4-Bis(2,3-epoxypropoxy)butane (2425-79-8)
1,3-Bis(2,3-epoxypropoxy)-2,2-dimethylpropane (17557-23-2)
2,3-Bis(2,3-epoxypropoxy)-p-dioxan (10043-09-1)

Name	CAS
2,3-Bis(2,3-epoxypropoxy)-1,4-dioxane	(10043-09-1)
1,2-Bis(2-(2,3-epoxypropoxy)ethoxy)ethane	(1954-28-5)
1,4-Bis((2,3-epoxypropoxy)ethyl)cyclohexane	(14228-73-0)
2,2-Bis(p-(2,3-epoxypropoxy)phenyl)propane polymers	(25085-99-8)
1,3-Bis(3-(2,3-epoxypropoxy)propyl)tetramethyldisiloxane	(126-80-7)
Bis(2,3-epoxypropyl)aniline	(2095-06-9)
N,N-Bis(2,3-epoxypropyl)aniline	(2095-06-9)
Bis(2,3-epoxypropyl)-5-ethyl-5-methylhydantoin	(15336-82-0)
Bis(2,3-epoxypropyl)isophthalate	(7195-43-9)
2,2-Bis(4-(2,3-epoxypropyloxy)phenyl)propane	(1675-54-3)
Bis(epoxypropyl)phenylamine	(2095-06-9)
Bis(2,6-(2,3-epoxypropyl))phenyl glycidyl ether	(13561-08-5)
Biseptol	(8064-90-2)
S-(1,2-Bis(ethoxy-carbonyl)-ethyl)-O,O-dimethyl-dithio-fosfaat (Dutch)	(121-75-5)
S-(1,2-Bis(ethoxycarbonyl)ethyl) O,O-dimethyl phosphorodithioate	(121-75-5)
S-1,2-Bis(ethoxycarbonyl)ethyl-O,O-dimethyl thiophosphate	(121-75-5)
1,2-Bis(3-ethoxycarbonyl-2-thioureido) benzene	(23564-06-9)
1,2-Bis-(3-ethoxycarbonyl-thioureido)-benzene	(23564-06-9)
Bis(2-ethoxyethyl) ether	(112-36-7)
4,4'-Bis((p-ethoxyphenyl)azo)-2,2'-stilbenesulfonic acid disodium salt	(2870-32-8)
1,1-Bis(p-ethoxyphenyl)-2,2-dimethylpropane	(27955-87-9)
1,1-Bis(p-ethoxyphenyl)-2-nitrobutane	(26258-71-9)
1,1-Bis(p-ethoxyphenyl)-2-nitropropane	(26258-70-8)
2,2-Bis(p-ethoxyphenyl)-1,1,1-trichloroethane	(4329-03-7)
1,4-Bis(ethylamino)-9,10-anthracenedione	(6994-46-3)
2,4-Bis(ethylamino)-6-chloro-s-triazine	(122-34-9)
2,4-Bis(ethylamino)-6-methoxy-s-triazine	(673-04-1)
4,6-Bis(ethylamino)-2-methoxy-s-triazine	(673-04-1)
2,4-Bis(ethylamino)-6-methylmercapto-s-triazine	(1014-70-6)
2,4-Bis(ethylamino)-6-(methylthio)-s-triazine	(1014-70-6)
4,6-Bis(ethylamino)-2-methylthio-1,3,5-triazine	(1014-70-6)
4,6-Bis(ethylamino)-1,3,5-triazin-2(1H)-one	(2599-11-3)
(2-Bis(3-ethylbenzothiazolyl))pentamethine cyanine iodide	(514-73-8)
Bis-(2-ethylbutyl)ester kyseliny adipove (Czech)	(10022-60-3)
Bis(ethyl)-2-chloroethylamine hydrochloride	(869-24-9)
Bis(ethylene glycol) terephthalate	(959-26-2)
Bis-(N-ethyl-N-fenyl)mocovina (Czech)	(85-98-3)
Bis(2-ethylhexyl) adipate	(103-23-1)
Bis(2-ethylhexyl) adipate	(123-79-5)
Bis(2-ethylhexyl)azelate	(103-24-2)
Bis(2-ethylhexyl)-1,2-benzenedicarboxylate	(117-81-7)
Bis(2-ethylhexyl) diphosphorate	(26836-28-2)
Bis-(2-ethylhexyl)ester kyseliny adipove (Czech)	(103-23-1)
Bis-(2-ethylhexyl)ester kyseliny azelaove (Czech)	(103-24-2)
Bis-(2-ethylhexyl)ester kyseliny ftalove (Czech)	(117-81-7)
Bis-(2-ethylhexyl)ester kyseliny fumarove (Czech)	(141-02-6)
Bis-(2-ethylhexyl)ester kyseliny isoftalove (Czech)	(137-89-3)
Bis-(2-ethylhexyl)ester kyseliny maleinove (Czech)	(142-16-5)
Bis-(2-ethylhexyl)ester kyseliny sebakove (Czech)	(122-62-3)
Bis(ethylhexyl) ester of sodium sulfosuccinic acid	(577-11-7)
Bis-2-ethylhexylester sulfojantaranu sodneho (Czech)	(577-11-7)
Bis(2-ethylhexyl) fumarate	(141-02-6)
Bis(2-ethylhexyl)hydrogen phosphate	(298-07-7)
Bis(2-ethylhexyl) hydrogen phosphite	(3658-48-8)
Bis(2-ethylhexyl) isophthalate	(137-89-3)
Bis(2-ethylhexyl)maleate	(142-16-5)
Bis(2-ethylhexyloxycarbonylmethylthio)dibutylstannane	(25168-24-5)
Bis(2-ethylhexyl)phenyl phosphate	(16368-97-1)
Bis(2-ethylhexyl)phosphate	(298-07-7)
Bis(2-ethylhexyl)orthophosphoric acid	(298-07-7)
Bis(2-ethylhexyl)phosphoric acid	(298-07-7)
Bis(2-ethylhexyl)phthalate	(117-81-7)
Bis(2-ethylhexyl)sebacate	(122-62-3)
1,4-Bis(2-ethylhexyl) sodium sulfosuccinate	(577-11-7)
Bis(2-ethylhexyl) S-sodium sulfosuccinate	(577-11-7)
Bis(2-ethylhexyl)sodium sulfosuccinate	(577-11-7)
Bis(2-ethylhexyl) terephthalate	(6422-86-2)
N,N'-Bis(1-ethyl-3-methylpentyl)-p-phenylenediamine	(139-60-6)
Bis(N-ethyl-2-perfluorooctylsulfonaminoethyl)phosphate, ammonium salt	(30381-98-7)
1,1-Bis(p-ethylphenyl)-2,2-dichloroethane	(72-56-0)
2,2-Bis(p-ethylphenyl)-1,1-dichloroethane	(72-56-0)
Bis(N-ethyl-N-phenyl)urea	(85-98-3)
2,2-Bis(ethylsulfonyl)butane	(76-20-0)
2,2-Bis(ethylsulfonyl)propane	(115-24-2)
Bis(ethylthio)phenylarsine	(5582-58-1)
Bis(ethylxanthic)disulfide	(502-55-6)
Bis(ethylxanthogen)	(502-55-6)
Bisethylxanthogen disulfide	(502-55-6)
S-(1,2-Bis(etossi-carbonil)-etil)-O,O-dimetil-ditiofosfato (Italian)	(121-75-5)
Bisferol A (Czech)	(80-05-7)
1,1-Bis(p-fluorophenyl)-2,2,2-trichloroethane	(475-26-3)
2,2-Bis(4-fluorophenyl)-1,1,1-trichloroethane	(475-26-3)
2,2-Bis(p-fluorophenyl)-1,1,1-trichloroethane	(475-26-3)
1,4-Bis(1-formamido-2,2,2-trichloroethyl)piperazine	(26644-46-2)
N,N'-Bis(1-formamido-2,2,2-trichloroethyl)piperazine	(26644-46-2)
Bis(D-gluconato-O1,O2)copper	(527-09-3)
Bis(3-glycidoxypropyl)tetramethyldisiloxane	(126-80-7)
meta-Bis(glycidyloxy)benzene	(101-90-6)
1,4-Bis(glycidyloxy)butane	(2425-79-8)
2,2-Bis(p-(2-glycidyloxy-3-butoxypropyloxy)phenyl)propane	(71033-08-4)
2,3-Bis(glycidyloxy)-1,4-dioxane	(10043-09-1)
1,2-Bis(glycidyloxy)ethane	(2224-15-9)
Bis(4-glycidyloxyphenyl)dimethyamethane	(1675-54-3)
2,2-Bis(p-glycidyloxyphenyl)propane	(1675-54-3)
1,3-Bis(3-(glycidyloxy)propyl)-1,1,3,3-tetramethyldisiloxane	(126-80-7)
Bis(8-guanidino-octyl)amine	(13516-27-3)
Bis(hexamethylene)triamine	(143-23-7)
Bishexamethylenetriamine	(143-23-7)
Bishexamethylenetriamine, pentamethylenepentaphosphonic acid	(34690-00-1)
Bis(2-(hexyloxy)ethyl)adipate	(110-32-7)
Bis(hydrogenated tallow alkyl)benzylmethylammonium bentonite	(68153-30-0)
Bis(hydrogenated tallow alkyl)dimethylammonium bentonite	(68953-58-2)
Bis(hydroxyaethyl)-aether-dinitrat (German)	(693-21-0)
Bis(β-hydroxyaethyl)nitrosamin (German)	(1116-54-7)
(T-4)-Bis(2-hydroxybenzoato-O1,O2)magnesium	(18917-89-0)
Bis(2-hydroxy-3-tert-butyl-5-ethylphenyl)methane	(88-24-4)
Bis(4-hydroxy-5-tert-butyl-2-methylphenyl) sulfide	(96-69-5)
Bis-2-hydroxy-5-chlorfenylmethan (Czech)	(97-23-4)
Bis(2-hydroxy-5-chlorophenyl)methane	(97-23-4)
Bis(2-hydroxy-5-chlorophenyl)sulfide	(97-24-5)
Bishydroxycoumarin	(66-76-2)
Bis(4-hydroxycoumarin-3-yl)methane	(66-76-2)
Bis(1-hydroxycyclohexyl)peroxide	(2407-94-5)
2,2-Bis(4-hydroxycyclohexyl)propane	(80-04-6)
2,2-Bis(4-hydroxycyclohexyl)propane, epichlorohydrin polymer	(30583-72-3)
2,2-Bis(4-hydroxy-3,5-dibromophenyl)propane	(79-94-7)
2,2-Bis(4-hydroxy-3,5-dichlorophenyl)propane	(79-95-8)
Bis(2-hydroxy-3,5-dichlorophenyl) sulfide	(97-18-7)
1,4-Bis(2-hydroxyethoxy)benzene	(104-38-1)
1,4-Bis(β-hydroxyethoxy)benzene	(104-38-1)
1,2-Bis(2-hydroxyethoxy)ethane	(112-27-6)
Bis(2-hydroxyethyl)amine	(111-42-2)
N,N-Bis(2-hydroxyethyl)-4-aminoaniline sulfuric acid salt	(54381-16-7)
1,4-Bis((2-hydroxyethyl)amino)-9,10-anthracenedione	(4471-41-4)
5'-(Bis(2-hydroxyethyl)amino)-2'-((2-chloro-4-nitrophenyl)-azo)acetanilide, diacetate (ester)	(1533-78-4)
2-(Bis(2-hydroxyethyl)amino)ethanesulfonic acid	(10191-18-1)
N,N-Bis(2-hydroxyethyl)-2-aminoethanesulfonic acid	(10191-18-1)
N,N-Bis(2-hydroxyethyl)aminoethanesulfonic acid	(10191-18-1)
N,N-Bis(hydroxyethyl)-2-aminoethanesulfonic acid	(10191-18-1)
2-(Bis(2-hydroxyethyl)amino)-5-nitrophenol	(52551-67-4)

2-((4-((Bis(2-hydroxyethyl))amino)phenyl)azo)-5-nitro-benzonitrile, diacetate (ester) **(30124-94-8)**
Bis(2-hydroxyethyl)ammonium lauryl sulfate **(143-00-0)**
N,N-Bis(2-hydroxyethyl)aniline **(120-07-0)**
Bis(β-hydroxyethyl)butylamine **(102-79-4)**
Bis(2-hydroxyethyl)carbamodithioic acid, monopotassium salt **(23746-34-1)**
N,N-Bis(2-hydroxyethyl)chloroanilide **(92-00-2)**
Bis(2-hydroxyethyl)-2-(2-chloroethylthio)ethylsulfonium, chloride **(64036-91-5)**
Bis(hydroxyethyl)cocobenzylammonium chloride **(61789-68-2)**
N,N-Bis(2-hydroxyethyl)decanamide **(136-26-5)**
1,3-Bis(2-hydroxyethyl)-5,5-dimethyl-2,4-imidazolidinedione **(26850-24-8)**
Bis(2-hydroxyethyl)dithiocarbamic acid, monopotassium salt **(23746-34-1)**
Bis(2-hydroxyethyl)dithiocarbamic acid, potassium salt **(23746-34-1)**
N,N-Bis(2-hydroxyethyl)dodecanamide **(120-40-1)**
Bis(2-hydroxyethyl) ether **(111-46-6)**
N,N-Bis(2-hydroxyethyl)glycine, monosodium salt **(139-41-3)**
N,N-Bis(2-hydroxyethyl)glycine, sodium salt **(139-41-3)**
Bis(β-hydroxyethyl) hydroquinone ether **(104-38-1)**
Bis(2-hydroxyethyl)lauramide **(120-40-1)**
N,N-Bis(2-hydroxyethyl)lauramide **(120-40-1)**
N,N-Bis(β-hydroxyethyl)lauramide **(120-40-1)**
N,N-Bis(hydroxyethyl)lauramide **(120-40-1)**
Bis(2-hydroxyethyl)methylamine **(105-59-9)**
N,N-Bis(2-hydroxyethyl)-3-methylaniline **(91-99-6)**
N,N-Bis(β-hydroxyethyl)-3-methylaniline **(91-99-6)**
N',N'-Bis(2-hydroxyethyl)-N-methyl-2-nitro-p-phenylenediamine **(2784-94-3)**
N,N-Bis(2-hydroxyethyl)-N-methyl-N-tallow ammonium chloride **(67784-77-4)**
Bis(β-hydroxyethyl)nitrosamine **(1116-54-7)**
N,N-Bis(2-hydroxyethyl)nonanamide **(3077-37-0)**
N,N-Bis(2-hydroxyethyl)octadecanamide **(93-82-3)**
(Z)-N,N-Bis(2-hydroxyethyl)-9-octadecenamide **(93-83-4)**
N,N-Bis(2-hydroxyethyl)-9-octadecenamide **(93-83-4)**
N,N-Bis(2-hydroxyethyl)octanamide **(3077-30-3)**
N,N-Bis(2-hydroxyethyl)oleamide **(93-83-4)**
N,N-Bis(2-hydroxyethyl)-p-phenylenediamiamine sulfate **(54381-16-7)**
N,N-Bis(β-hydroxyethyl)-p-phenylenediamine sulfate **(54381-16-7)**
1,4-Bis(2-hydroxyethyl)piperazine **(122-96-3)**
N,N'-Bis(β-hydroxyethyl)piperazine **(122-96-3)**
N,N-Bis(2-hydroxyethyl)stearamide **(93-82-3)**
Bis(2-hydroxyethyl)sulfide **(111-48-8)**
Bis(β-hydroxyethyl)sulfide **(111-48-8)**
Bis(2-hydroxyethyl)sulfomethylamine, sodium salt **(25857-20-9)**
Bis(2-hydroxyethyl) terephthalate **(959-26-2)**
Bis(β-hydroxyethyl) terephthalate **(959-26-2)**
Bis(hydroxyethyl) terephthalate **(959-26-2)**
N,N-Bis(2-hydroxyethyl)-m-toluidine **(91-99-6)**
N,N-Bis(2-hydroxyethyl)-o-toluidine **(28005-74-5)**
2,2-Bis-4'-hydroxyfenylpropan (Czech) **(80-05-7)**
Bis(3-hydroxy-4-hydroxymethyl-2-methylpyridyl-(5)-methyl)-disulfide dihydrochloride monohydrate **(123-03-5)**
p-Bis(2-hydroxyisopropyl)benzene **(2948-46-1)**
p-Bis(α-hydroxyisopropyl)benzene **(2948-46-1)**
Bis(hydroxylamine) sulfate **(10039-54-0)**
(T-4)-Bis(hydroxymethanesulfinato-OS,O1)zinc **(24887-06-7)**
Bis(hydroxymethanesulfinato-O,O')zinc **(24887-06-7)**
Bis(8,8'-(7-hydroxy-4-methoxy-5-methylcoumarin)) **(69975-77-5)**
Bis(hydroxymethyl)acetylene **(110-65-6)**
3-Bis(hydroxymethyl)amino-6-(5-nitro-2-furylethenyl)-1,2,4-triazine **(794-93-4)**
5'-(Bis(2-hydroxymethyl)amino)-2'-((p-nitrophenyl)azo)-benzanilide, diacetate (ester) **(29765-00-2)**
1,4-Bis(hydroxymethyl)cyclohexane **(105-08-8)**
1,3-Bis(hydroxymethyl)-5,5-dimethylhydantoin **(6440-58-0)**
2,2-Bis(p-(2-hydroxy-2-methylethoxy)phenyl)propane **(116-37-0)**
N,N'-Bis(hydroxymethyl)ethyleneurea **(136-84-5)**
Bis(hydroxymethyl)furatrizine **(794-93-4)**
3,3-Bis(hydroxymethyl)heptane **(115-84-4)**
1,3-Bis(hydroxymethyl)-2-imidazolidinone **(136-84-5)**
3,3-Bis(hydroxymethyl)pentane **(115-76-4)**
2,2-Bis(hydroxymethyl)-1,3-propanediol **(115-77-5)**
2,2-Bis(hydroxymethyl)-1,3-propanediol tetranitrate **(78-11-5)**
d,N,N'-Bis(1-hydroxymethylpropyl)ethylenediamine **(74-55-5)**
2,3-Bis(hydroxymethyl)quinoxaline di-N-oxide **(17311-31-8)**
1,3-Bis(hydroxymethyl)urea **(140-95-4)**
Bis(hydroxymethyl)urea **(25155-29-7)**
N,N'-Bis(hydroxymethyl)urea **(140-95-4)**
N,N-Bis-(2-(2-hydroxy-5-octylbenzylamino)ethyl)-2-hydroxy-5-octylbenzylamine, calcium salt **(68568-82-1)**
Bis(4-hydroxyphenyl) dimethylmethane **(80-05-7)**
Bis(4-hydroxyphenyl)dimethylmethane diglycidyl ether **(1675-54-3)**
3,4-Bis(4-hydroxyphenyl)-2,4-hexadiene **(84-17-3)**
3,4-Bis(p-hydroxyphenyl)-2,4-hexadiene **(84-17-3)**
3,4-Bis(p-hydroxyphenyl)hexane **(84-16-2)**
meso-3,4-Bis(p-hydroxyphenyl)-n-hexane **(84-16-2)**
3,4-Bis(p-hydroxyphenyl)-3-hexene **(56-53-1)**
Bis(4-hydroxyphenyl)methane **(620-92-8)**
Bis(p-hydroxyphenyl)methane **(620-92-8)**
p,p'-Bis(hydroxyphenyl)methane **(620-92-8)**
4(Bis(p-hydroxyphenyl)methylene)-2,5-cyclohexadien-1-one **(603-45-2)**
4,4-Bis(4-hydroxyphenyl)pentanoic acid **(126-00-1)**
4,4-Bis(p-hydroxyphenyl)pentanoic acid **(126-00-1)**
3,3-Bis(p-hydroxyphenyl)phthalide **(77-09-8)**
2,2-Bis(4-hydroxyphenyl)propane **(80-05-7)**
2,2-Bis(p-hydroxyphenyl)propane **(80-05-7)**
Bis(4-hydroxyphenyl)propane **(80-05-7)**
2,2-Bis(4-hydroxyphenyl)propane, diglycidyl ether **(1675-54-3)**
2,2-Bis(p-hydroxyphenyl)propane, diglycidyl ether **(1675-54-3)**
4,4-Bis(4-hydroxyphenyl)valeric acid **(126-00-1)**
4,4-Bis(p-hydroxyphenyl)valeric acid **(126-00-1)**
γ,γ-Bis(p-hydroxyphenyl)valeric acid **(126-00-1)**
2,2-Bis(4-(2-hydroxypropoxy)phenyl)propane **(116-37-0)**
2,2-Bis(4-(β-hydroxypropoxy)phenyl)propane **(116-37-0)**
2,2-Bis(p-(2-hydroxypropoxy)phenyl)propane **(116-37-0)**
2,2-Bis(p-(β-hydroxypropoxy)phenyl)propane **(116-37-0)**
Bis(2-hydroxypropyl)amine **(110-97-4)**
4-(Bis(2-hydroxypropyl)amino)benzoic acid, ethyl ester **(58882-17-0)**
1,4-Bis(2-hydroxy-2-propyl)benzene **(2948-46-1)**
2,2'-Bishydroxypropylnitrosamine **(53609-64-6)**
N-Bis(2-hydroxypropyl)nitrosamine **(53609-64-6)**
N,N-Bis(2-hydroxypropyl)-p-toluidine **(38668-48-3)**
Bis(1-hydroxy-2(1H)-pyridinethionato)zinc **(13463-41-7)**
N,N'-Bis(1-hydroxy-3-sulfonaphthyl(6))urea **(134-47-4)**
1,3-Bis(1-hydroxy-2,2,2-trichloroethyl)urea **(116-52-9)**
Bis(2-hydroxy-3,5,6-trichlorophenyl)methane **(70-30-4)**
Bisina **(80-77-3)**
4,4'-Bis(iodooxy)-2,2'-dimethyl-5,5'-bis(1-methylethyl-1,1'-bi-phenyl **(552-22-7)**
Bis(isobutyl)aluminum chloride **(1779-25-5)**
Bis(isobutyl)hydroaluminum **(1191-15-7)**
Bis(4-isocyanatocyclohexyl)methane **(5124-30-1)**
Bis(isocyanatomethyl)benzene **(25854-16-4)**
1,3-Bis(1-isocyanato-1-methylethyl)benzene **(2778-42-9)**
Bis(1,4-isocyanatophenyl)methane **(101-68-8)**
Bis(4-isocyanatophenyl)methane **(101-68-8)**
Bis(p-isocyanatophenyl)methane **(101-68-8)**
Bis(isodecyl)phthalate **(26761-40-0)**
Bis(isopropylamido) fluorophosphate **(371-86-8)**
1,4-Bis(isopropylamino)-9,10-anthracenedione **(14233-37-5)**
1,4-Bis(N-isopropylamino)anthraquinone **(14233-37-5)**
1,4-Bis(isopropylamino)anthraquinone **(14233-37-5)**
2,4-Bis(isopropylamino)-6-chloro-s-triazine **(139-40-2)**
Bisisopropylaminofluorophosphine oxide **(371-86-8)**
2,4-Bis(isopropylamino)-6-methoxy-s-triazine **(1610-18-0)**
2,4-Bis(isopropylamino)-6-methyl mercapto-s-triazine **(7287-19-6)**
2,4-Bis(isopropylamino)-6-methylthio-1,3,5-triazine **(7287-19-6)**

2,4-Bis(isopropylamino)-6-methylthio-s-triazine	(7287-19-6)
Bis(isopropyl)naphthalene	(38640-62-9)
Bis-(2-kyanethyl)amin (Czech)	(111-94-4)
Bis(lauroyloxy)di(n-butyl)stannane	(77-58-7)
N,N'-Bismaleimido-4,4'-diphenylmethane	(13676-54-5)
Bismaleimide, 4,4'-diphenylmethane	(13676-54-5)
1,3-Bismaleimidobenzene	(3006-93-7)
4,4'-(N,N'-Bismaleimido)diphenylmethane	(13676-54-5)
4,4'-Bis(maleimido)diphenylmethane	(13676-54-5)
Bis(4-maleimidophenyl)methane	(13676-54-5)
Bismate	(21260-46-8)
Bis(mercaptobenzothiazolato)zinc	(155-04-4)
1,2-Bis(2-mercaptoethoxy)ethane	(14970-87-7)
Bis(2-mercaptoethyl)hexanedioate	(10194-00-0)
2,2-Bis(4-(2-(methacryloxy)ethoxy)phenyl) propane	(24448-20-2)
2,2-Bis-(4-(2-methacryloxyethoxy)phenyl)propane	(24448-20-2)
1,2-Bis(methacryloyloxy)ethane	(97-90-5)
1,4-Bis(methanesulfonoxy)butane	(55-98-1)
(1,4-Bis(methanesulfonyloxy)butane)	(55-98-1)
Bismethin	(390-64-7)
Bismethomyl thioether	(59669-26-0)
3,5-Bis(methoxycarbonyl)benzenesulfonic acid sodium salt	(3965-55-7)
N,N-(Bis(2-methoxycarbonyloxy)ethyl)-3-methylbenzenamine	(25790-28-7)
1,2-Bis(3-(methoxycarbonyl)-2-thioureido)benzene	(23564-05-8)
1,2-Bis(methoxycarbonylthioureido)benzene	(23564-05-8)
o-Bis(3-methoxycarbonyl-2-thioureido)benzene	(23564-05-8)
2,5-Bismethoxyethoxy-3,6-bisethyleneimino-1,4-benzoquinone	(800-24-8)
3,6-Bis(β-methoxyethoxy)-2,5-bis(ethyleneimino)-p-benzoquinone	(800-24-8)
3,6-Bis(β-methoxyethoxy)-2,5-bis(ethylenimino)-p-benzoquinone	(800-24-8)
Bis(2-(2-methoxyethoxy)ethyl) ether	(143-24-8)
Bis(2-methoxyethyl)amine	(111-95-5)
Bis(methoxyethyl)amine	(111-95-5)
Bis(2-methoxyethyl)ether	(111-96-6)
Bis(2-methoxyethyl)ether	(143-24-8)
Bis(2-methoxyethyl) phthalate	(117-82-8)
Bis(methoxyethyl) phthalate	(117-82-8)
3,5-Bis-(methoxykarbonyl)benzensulfonan sodny (Czech)	(3965-55-7)
1,3-Bis(methoxymethyl)-4,5-dihydroxy cyclic ethyleneurea	(3001-61-4)
Bis(methoxymethyl) ether	(628-90-0)
1,3-Bis(methoxymethyl)urea	(141-07-1)
Bis(methoxymethyl)urea	(141-07-1)
N,N'-Bis(methoxymethyl)urea	(141-07-1)
Bis(4-methoxyphenyl)amine	(101-70-2)
Bis(p-methoxyphenyl)amine	(101-70-2)
Bis(4-methoxyphenyl)diazene 1-oxide	(1562-94-3)
1,1-Bis(p-methoxyphenyl)-2,2-dimethylpropane	(4741-74-6)
1,1-Bis(p-methoxyphenyl)-2,2,2-trichloroethane	(72-43-5)
2,2-Bis(p-methoxyphenyl)-1,1,1-trichloroethane	(72-43-5)
2,4-Bis(3-methoxypropylamino)-6-methylthio-s-triazine	(845-52-3)
1,4-Bis(methylamino)anthraquinone	(2475-44-7)
4,6-Bis(methylamino)-2-chloro-s-triazine	(2911-36-6)
Bis(1-methylamyl) sodium sulfosuccinate	(3006-15-3)
Bis(N-methylaniline)methane	(1807-55-2)
Bis(N-methylanilino)methan (German)	(1807-55-2)
Bis(N-methylbenzamido)methylethoxysilane	(16230-35-6)
Bis(α-methylbenzyl)amine	(10024-74-5)
Bis(α-methylbenzyl) ether	(93-96-9)
Bis(methyl)-2-chloroethylamine hydrochloride	(4584-46-7)
Bis(methylcyclopentadiene)	(26472-00-4)
N,N'-Bis(1-methylethyl)-1,4-benzenediamine	(4251-01-8)
Bis(1-methylethyl)-1,1'-biphenyl	(69009-90-1)
ar,ar'-Bis(1-methylethyl)-1,1'-biphenyl	(36876-13-8)
Bis(1-methylethyl) carbamothioic acid, S-(2,3-dichloro-2-propenyl)ester	(2303-16-4)
2,3':4,6-Bis-O-(1-methylethylidene)-α-L-xylo-2-hexulofuranosonic acid sodium salt	(52508-35-7)
N,N'-Bis(1-methylethyl)-6-methyl-thio-1,3,5-triazine-2,4-diamine	(7287-19-6)
2,6-Bis(1-methylethyl)naphthalene	(24157-81-1)
Bis(1-methylethyl)naphthalene	(38640-62-9)
Bis(1-methylethyl)naphthalenesulfonic acid	(28757-00-8)
Bis(1-methylethyl)naphthalenesulfonic acid, cyclohexylamine salt	(68425-61-6)
2,6-Bis(1-methylethyl)phenol	(2078-54-8)
O,O-Bis(1-methylethyl) S-(phenylmethyl)phosphorothioate	(26087-47-8)
O,O-Bis(1-methylethyl) S-(2-((phenylsulfonyl)amino)ethyl)-pheosphorodithioate	(741-58-2)
N,N'-Bis(1-methylethyl)urea	(4128-37-4)
Bis(2-methylglycidyl) ether	(7487-28-7)
N,N'-Bis(1-methylheptyl)-1,4-benzenediamine	(103-96-8)
Bis(1-methylheptyl) hexanedioate	(108-63-4)
N,N'-Bis(1-methylheptyl)-p-phenylenediamine	(103-96-8)
N,N'-Bis(5-methyl-3-heptyl)-p-phenylenediamine	(139-60-6)
1,1-Bis(2-methyl-4-hydroxy-5-tert-butylphenyl)butane	(85-60-9)
3,5-Bis-methylkarboxy-benzensulfonan sodny (Czech)	(3965-55-7)
Bis-methylparathion (Czech)	(39004-94-9)
N,N-Bis(N-methyl-N-phenyl-tert-butylacetamido)-β-hydroxyethyl-amine	(126-27-2)
2,4-Bis(1-methyl-1-phenylethyl)phenol	(2772-45-4)
Bis(4-methylphenyl)mercury	(537-64-4)
Bis(methylphenyl) phenyl phosphate	(26446-73-1)
Bis(2-methylpropyl)carbamothioic acid S-ethyl ester	(2008-41-5)
Bis(2-methylpropyl)naphthalenesulfonic acid, sodium salt	(27213-90-7)
Bis(2-methylpropyl) nonanedioate	(105-80-6)
Bis(1-methylpropyl)phenol	(31291-60-8)
Bis(2-methylpropyl)phenol	(27515-66-8)
O,O-Bis(2-methylpropyl) phosphorodithioate	(2253-52-3)
1,4-Bis(2-methylpropyl)sulfobutanedioate, sodium salt	(127-39-9)
N-Bismethylpteroylglutamic acid	(59-05-2)
N,N-Bis(methylsulfonepropoxy)amine hydrochloride	(3458-22-8)
1,5-Bis(methylsulfonyl)naphthalene	(53135-94-7)
2,4-Bis(methylthio)-6-chloro-s-triazine	(4407-40-3)
Bis(methylthio)lead	(35029-96-0)
Bis(methylthio)methane	(1618-26-4)
1,5-Bis(methylthio)naphthalene	(10075-74-8)
2,2-Bis(p-(methylthio)phenyl)-1,1,1-trichloroethane	(19679-38-0)
Bis(monoisopropylamino)fluorophosphate	(371-86-8)
Bis(monoisopropylamino)fluorophosphine oxide	(371-86-8)
N,N'-Bismorpholine disulfide	(103-34-4)
Bismorpholino disulfide	(103-34-4)
Bismuth-209	(7440-69-9)
Bismuth-210	(14331-79-4)
Bismuth-214	(14733-03-0)
Bismuth	(7440-69-9)
Bismuth Magistery	(1304-85-4)
Bismuth Violet	(548-62-9)
Bismuth White	(1304-85-4)
Bismuth acetate	(22306-37-2)
Bismuth chromate	(61204-26-0)
Bismuth dimethyldithiocarbamate	(21260-46-8)
Bismuth-hydroxide-nitrate-oxide	(1304-85-4)
Bismuthiol I	(1072-71-5)
Bismuth, oxo(dihydrogen nitrilotriacetato)-, sodium salt, Compd. with disodium nitrilotriacetate (1:3)	(5798-43-6)
Bismuth, oxo(salicylato)	(14882-18-9)
Bismuth salicylate, Basic	(14882-18-9)
Bismuth sesquitelluride	(1304-82-1)
Bismuth sodium triglycollamate	(5798-43-6)
Bismuth subcarbonate	(5892-10-4)
Bismuth subnitrate	(1304-85-4)
Bismuth subnitricum	(1304-85-4)
Bismuth subsalicylate	(14882-18-9)
Bismuth-telluride	(1304-82-1)
Bismuth telluride (ACGIH,OSHA)	(1304-82-1)
Bismuth telluride, Undoped (OSHA)	(1304-82-1)
Bismuth triacetate	(22306-37-2)

Name	CAS
Bismuth, tris(dimethyldithiocarbamato)	(21260-46-8)
Bismuthyl nitrate	(1304-85-4)
N,N'-Bis-(2-naftyl)-p-fenylendiamin (Czech)	(93-46-9)
2,6-Bis(.α.-naphthylamino)-4-chloro-triazine	(30355-07-8)
N,N'-Bis(2-naphthyl)-p-phenylenediamine	(93-46-9)
O,O-Bis-p-nitrofenyl-O-ethylester kyseliny thiofosforecne (Czech)	(7508-73-8)
O,O-Bis-p-nitrofenyl-O-methylester kyseliny thiofosforecne (Czech)	(39004-94-9)
1,5-Bis(5-nitro-2-furanyl)-1,4-pentadien-3-one, (aminoimino-methyl)hydrazone	(804-36-4)
Bis(5-nitrofurfurylidene)acetone guanylhydrazone	(804-36-4)
sym-Bis(5-nitro-2-furfurylidene) acetone guanylhydrazone	(804-36-4)
1,5-Bis(5-nitro-2-furyl)-3-pentadienone amidinonhydrazone	(804-36-4)
1,5-Bis(5-nitro-2-furyl)-3-pentadienone guanylhydrazone	(804-36-4)
Bis(4-nitrophenyl)ether	(101-63-3)
Bis(p-nitrophenyl)ether	(101-63-3)
Bis(4-nitrophenyl)methyl phosphate	(799-87-1)
Bis(p-nitrophenyl)methyl phosphate	(799-87-1)
Bis(4-nitrophenyl) phosphate	(645-15-8)
Bis(4-nitrophenyl)phosphate	(645-15-8)
Bis(nonylphenyl) phenyl phosphate	(63302-94-3)
1,2-Bis(octadecanamido)ethane	(110-30-5)
Bis(octadecanoato)dioxodilead	(56189-09-4)
N,N'-Bis-n-octyl-3,3'-dithiopropionamide	(33312-01-5)
Bis(p-octylphenyl)amine	(101-67-7)
N,N'-Bis(2-octyl)-p-phenylenediamine	(103-96-8)
Bis(2-octyl)phthalate	(131-15-7)
Bisodium tartrate	(868-18-8)
Bisoflex 81	(117-81-7)
Bisoflex 91	(84-76-4)
Bisoflex DES	(110-40-7)
Bisoflex DNA	(151-32-6)
Bisoflex DOA	(103-23-1)
Bisoflex DOP	(117-81-7)
Bisoflex DOS	(122-62-3)
Bis-(2-oktyl)ester kyseliny ftalove (Czech)	(131-15-7)
Bisomel	(110-27-0)
Bisomer 2HEA	(818-61-1)
1,3-Bis(oxiranylmethoxy)-2-propanol	(3568-29-4)
1,3-Bis(1-oxiranylmethyl-5,5-dimethylhydantoin-3-yl)-2-oxiranylmethoxypropane	(38304-52-8)
1,3-Bis(oxiranylmethyl)-5-ethyl-5-(2-methylbutyl)hydantoin	(68444-05-3)
Bis(8-oxyquinoline)copper	(10380-28-6)
N,N'-Bis(1-oxy-3-sulfonaphthyl(6))urea	(134-47-4)
Bis-parathion (Czech)	(7508-73-8)
Bis(pentachlor-2,4-cyclopentadien-1-yl)	(2227-17-0)
Bis(pentachloro-2,4-cyclopentadien-1-yl)	(2227-17-0)
Bis(pentachlorocyclopentadienyl)	(2227-17-0)
4,6-Bis(pentachlorophenoxy)-2-chloro-s-triazine	(26396-34-9)
Bis(pentaerythritol)	(126-58-9)
Bis(pentamethylenethiuram)-tetrasulfide	(120-54-7)
(T-4)-Bis(2,4-pentanedionato-O,O')cobalt	(14024-48-7)
Bis(2,4-pentanedionato)cobalt	(14024-48-7)
Bis(2,4-pentanedionato)copper	(13395-16-9)
Bis(2,4-pentanedionato)titanium oxide	(14024-64-7)
2,2-Bis(p-phenetyl)-1,1,1-trichloroethane	(4329-03-7)
Bisphenol	(80-05-7)
4,4'-Bisphenol A	(80-05-7)
Bisphenol A	(80-05-7)
Bisphenol A bis(2-hydroxypropyl) ether	(116-37-0)
Bisphenol A bis(β-hydroxypropyl) ether	(116-37-0)
Bisphenol A diglycidyl ether	(1675-54-3)
Bisphenol A disodium salt	(2444-90-8)
Bisphenol A, disodium salt	(2444-90-8)
Bisphenol A, epichlorohydrin, tetrabromobisphenol A polymer	(26265-08-7)
Bisphenol A glycidylmethacrylate	(1565-94-2)
Bisphenol A-propylene oxide adduct (1:2)	(116-37-0)
Bisphenol A, propylene oxide, fumaric acid polymer	(38355-75-8)
Bis(phenoxarsin-10-yl) ether	(58-36-6)
Bis(10-phenoxarsyl) oxide	(58-36-6)
10,10'-Bis(phenoxyarsinyl) oxide	(58-36-6)
Bis(10-phenoxyarsinyl) oxide	(58-36-6)
1,4-Bis(phenylamino)benzene	(74-31-7)
Bisphenyl-(2-chlorphenyl)-1-imidazolyl-methan (German)	(23593-75-1)
Bis(α-phenylethyl) ether	(93-96-9)
2,6-Bis(1-phenylethyl)-4-methylphenol	(1817-68-1)
2,6-Bis(1-phenylethyl)-4-nonylphenol	(15860-96-5)
Bis(phenylmethyl) decanedioate	(140-24-9)
Bis(phenylmethyl) diazenedicarboxylate	(2449-05-0)
Bis(γ-phenylpropyl)ethylamine	(150-59-4)
N,N-Bis(phosphonomethyl)glycine	(2439-99-8)
Bis(piperidinothiocarbonyl) tetrasulfide	(120-54-7)
1,3-Bis(4-piperidyl)propane	(16898-52-5)
Bis(2-propanol)amine	(110-97-4)
2,4-Bis(propylamino)-6-chlor-1,3,5-triazin (German)	(139-40-2)
2,4-Bis(propylamino)-6-methylthio-1,3,5-triazin (German)	(7287-19-6)
N,N'-Bispropyleneisophthalamide	(7652-64-4)
Bis-(2-propylheptyl) phthalate	(53306-54-0)
trans-1,2-Bis(propylsulfonyl)ethene	(1113-14-0)
(E)-1,2-Bis(propylsulfonyl)ethylene	(1113-14-0)
trans-1,2-Bis(n-propylsulfonyl)ethylene	(1113-14-0)
trans-1,2-Bis(propylsulfonyl)ethylene	(1113-14-0)
Bis(8-quinolinato)copper	(10380-28-6)
Bis(8-quinolinolato)copper	(10380-28-6)
N,N-Bis(2-tallowamidoethyl)-N-(2-hydroxyethyl)-N-methyl-ammonium methylsulfate	(68153-35-5)
Bisteril	(136-40-3)
Bis(2,3,3,3-tetrachloropropyl) ether	(127-90-2)
Bis-N,N,N',N'-tetramethylphosphorodiamidic anhydride	(152-16-9)
Bisthiocarbamyl hydrazine	(142-46-1)
N,N'-Bisthiocarbamyl hydrazine	(142-46-1)
Bis(thiourea)	(142-46-1)
2,5-Bis(4-toluidino)terephthalic acid	(10291-28-8)
2,5-Bis(p-toluidino)terephthalic acid	(10291-28-8)
Bis-1,4-p-tolylaminoanthrchinon (Czech)	(128-80-3)
Bis(o-tolylcarbodiimide)	(1215-57-2)
Biston	(298-46-4)
Bis(2,4,6-tribromophenyl) carbonate	(67990-32-3)
Bis-(tri-n-butylcin)oxid (Czech)	(56-35-9)
Bis(tributyloxide) of tin	(56-35-9)
Bis(tributylstannium) oxide	(56-35-9)
Bis(tributylstannyl)oxide	(56-35-9)
Bis(tributyltin) adipate	(7437-35-6)
Bis(tri-n-butyltin)oxide	(56-35-9)
Bis(tributyltin)oxide	(56-35-9)
Bis(tributyltin) salicylate	(22330-14-9)
Bis(tributyltin) sulfosalicylate	(4419-22-1)
Bis(tri-n-butylzinn)-oxyd (German)	(56-35-9)
Bis-2,3,5-trichlor-6-hydroxyfenylmethan (Czech)	(70-30-4)
Bis(2,4,5-trichloro-6-carbopentoxyphenyl)oxalate	(30431-54-0)
1,3-Bis(2,2,2-trichloro-1-hydroxyethyl)urea	(116-52-9)
Bis(3,5,6-trichloro-2-hydroxyphenyl)methane	(70-30-4)
1,4-Bis-trichloromethyl benzene	(68-36-0)
Bis(trichloromethyl) sulfone	(3064-70-8)
Bis(trichloromethyl)sulfone	(3064-70-8)
2,4-Bis(trichloromethyl)-s-triazine	(3599-74-4)
4,6-Bis(trichloromethyl)-s-triazine	(3599-74-4)
Bis(tridecyl) sulfosuccinate, sodium salt	(2673-22-5)
Bis(trifluoroacetic) anhydride	(407-25-0)
3,5-Bis(trifluoromethyl)aniline	(328-74-5)
1,3-Bis(trifluoromethyl)benzene	(402-31-3)
m-Bis(trifluoromethyl)benzene	(402-31-3)
1,1-Bis(trifluoromethyl)ethene	(382-10-5)
Bistrimate	(5798-43-6)

Name	CAS
1,4-Bis(2,4,6-trimethylanilino)anthraquinone	(116-75-6)
Bis(trimethylsiloxy)methylsilane	(1873-88-7)
Bis(trimethylsilyl)amine	(999-97-3)
Bis(2,4,6-trinitro-phenyl)-amin (German)	(131-73-7)
Bis(2,4,6-trinitrophenyl)diazene	(19159-68-3)
N,N'-Bis(2,4,6-trinitrophenyl)ethanediamide	(29135-62-4)
Bis(triphenylsilyl)chromate	(1624-02-8)
Bis(tris(β,β-dimethylphenethyl)tin)oxide	(13356-08-6)
Bis(tris(2-methyl-2-phenylpropyl)tin)oxide	(13356-08-6)
Bisulfite	(7446-09-5)
Bisulfite de sodium (French)	(7631-90-5)
Bisulfite sodique de menadione (French)	(130-37-0)
Bisulfito sodico de menadiona (Spanish)	(130-37-0)
Bisvinyltetramethyldisiloxane	(2627-95-4)
N,N-Bis(2,4-xylyliminomethyl)methylamine	(33089-61-1)
Bitemol	(122-34-9)
Bitemol S 50	(122-34-9)
Bitertanol	(55179-31-2)
Bitesorb 40	(1343-88-0)
(4,4'-Bithiazole)-2,2'-diamine	(58139-59-6)
4,4'-Bithiazole, 2,2'-diamino- (7CI)	(58139-59-6)
Bithion	(3383-96-8)
Bithionol	(97-18-7)
Bithionol sulfide	(97-18-7)
Bithymol diiodide	(552-22-7)
Bitin	(97-18-7)
Bitoksybacillin	(68038-71-1)
4,4'-Bi-o-toluidine	(119-93-7)
Bitolyl	(28013-11-8)
m,m'-Bitolyl (8CI)	(612-75-9)
m,p'-Bitolyl (8CI)	(7383-90-6)
o,o'-Bitolyl (8CI)	(605-39-0)
p,p'-Bitolyl (8CI)	(613-33-2)
Bitoscanate	(4044-65-9)
Bitoscanato (Spanish)	(4044-65-9)
Bitoscanatum (Latin)	(4044-65-9)
Bitoxibacillin	(68038-71-1)
Bitrex	(3734-33-6)
Bitter Almond LS	(8001-21-6)
Bitter Almond Oil Camphor	(119-53-9)
Bitter Fennel Oil	(8006-84-6)
Bitter Root	(72968-42-4)
Bitumen	(8052-42-4)
Biuno	(67-03-8)
δ(1,1')-Biurea	(123-77-3)
Biurea (8CI)	(110-21-4)
Biurea, 1-amino	(4381-07-1)
Biurea, 1,6-dimethyl-1,6-dinitroso	(3844-60-8)
Biurea, 2,5-dithio	(142-46-1)
Biuret (8CI)	(108-19-0)
Biuret, 2,4-dithio	(541-53-7)
Biuret, 1-ethyl-1-nitroso	(32976-88-8)
5,5'-Bivanillic acid	(2134-90-9)
Bivatin	(67-03-8)
Biverm	(92-84-2)
Bivinyl	(106-99-0)
Bivita	(67-03-8)
Bizolin 200	(50-33-9)
B(j)F	(205-82-3)
γ-Bl	(96-48-0)
Bl P 152	(132-93-4)
1743 Black	(2519-30-4)
Black Copper Oxide	(1317-38-0)
Black 2EMBL	(1937-37-7)
Black 3EMBL	(2429-83-6)
Black 4EMBL	(1937-37-7)
Black Iron Sulfide	(1317-37-9)
Black Lead	(7440-44-0)
Black Lead	(7782-42-5)
Black Leaf	(54-11-5)
Black Leaf 40	(54-11-5)
Black Nickel Oxide	(1313-99-1)
Black PN	(2519-30-4)
Black Pearls	(1333-86-4)
Black-Pepper-Oil	(8006-82-4)
Black and White Bleaching Cream	(123-31-9)
Blackcurrant LS	(8001-21-6)
Black manganese oxide	(1313-13-9)
Black oxide of Iron	(1309-37-1)
Blacosolv	(79-01-6)
Bladafum	(3689-24-5)
Bladafume	(3689-24-5)
Bladafun	(3689-24-5)
Bladan	(107-49-3)
Bladan	(56-38-2)
Bladan	(563-12-2)
Bladan	(757-58-4)
Bladan Base	(757-58-4)
Bladan F	(56-38-2)
Bladan-M	(298-00-0)
Bladex	(21725-46-2)
Bladex-B	(1929-73-3)
Bladex G	(2008-39-1)
Bladex H	(2545-59-7)
Bladex 80WP	(21725-46-2)
Blanc Fixe	(7727-43-7)
Blanc de fard	(1304-85-4)
Blancol	(9084-06-4)
Blancol Dispersant	(9084-06-4)
Blancosolv	(79-01-6)
Blandlube	(8012-95-1)
Blankophor BBH	(16090-02-1)
Blankophor BHC	(37069-54-8)
Blankophor MBBH	(16090-02-1)
Blanose BS 190	(9004-32-4)
Blanose BWM	(9004-32-4)
Blascide	(41814-78-2)
Blast furnace slag	(65996-69-2)
Blasticidin	(2079-00-7)
Blasticidin-S	(2079-00-7)
Blastin	(16022-69-8)
Blasting Gelatin (DOT)	(55-63-0)
Blasting Oil	(55-63-0)
Blattanex	(114-26-1)
Blatteralkohol	(928-96-1)
Blaues Pyoktanin (German)	(548-62-9)
Blaud's Mass	(563-71-3)
Blausaeure (German)	(74-90-8)
Blauwzuur (Dutch)	(74-90-8)
Blazer	(62476-59-9)
Blazer 2S	(62476-59-9)
Blazing Red	(2814-77-9)
Bleaching Powder	(7778-54-3)
Bleaching Powder, Containing 39% or less chlorine (DOT)	(7778-54-3)
Bleaching clay	(70131-50-9)
Bleiacetat (German)	(301-04-2)
Bleiazetat (German)	(6080-56-4)
Bleiphosphat (German)	(7446-27-7)
Bleistearat (German)	(7428-48-0)
Bleisulfat (German)	(7446-14-2)
Blekit evansa (Polish)	(314-13-6)
Bleminol	(315-30-0)
Blended Red Oxides of Iron	(1309-37-1)
Blenoxane	(11056-06-7)

Blenoxane (9041-93-4)
Bleo (11056-06-7)
Bleocin (11056-06-7)
Bleomycin (11056-06-7)
Bleomycin chlorhydrate (67763-87-5)
Bleomycin, hydrochloride (67763-87-5)
Bleomycin, sulfate (9041-93-4)
Bleomycin, sulfate (Salt) (9CI) (9041-93-4)
Bleph-10 (144-80-9)
Bleph-10 Liquifilm (144-80-9)
Bleu Brilliant FCF (2650-18-2)
Bleu Diamine (72-57-1)
Bleu Diazole N 3B (72-57-1)
Bleu Directe 3B (72-57-1)
Bleu Patente V (129-17-9)
Bleu Solanthrene (French) (81-77-6)
Bleu Trypane N (72-57-1)
Bleue Diretto 3B (72-57-1)
Blex (29232-93-7)
Blexane (9041-93-4)
Blightox (12122-67-7)
Blitex (12122-67-7)
Blitox (1332-40-7)
Blitox 50 (1332-40-7)
Blizene (12122-67-7)
Blo (96-48-0)
Bloatguard (9003-11-6)
Bloc (60168-88-9)
Blocadren (63-92-3)
Blocan (155-41-9)
Blon (96-48-0)
Blood, Glyoxal-denatured, Dried (68911-49-9)
Blood Stone (1317-60-8)
Blotic (31218-83-4)
Blown asphalt (64742-93-4)
Bloxanth (315-30-0)
Blo-trol (77-90-7)
Blu-Phen (50-06-6)
1085 Blue (129-17-9)
1206 Blue (3844-45-9)
1311 Blue (860-22-0)
11388 Blue (2650-18-2)
11669 Blue (482-89-3)
12070 Blue (860-22-0)
Blue 1084 (129-17-9)
Blue 1206 (3844-45-9)
Blue Anthraquinone Pigment (81-77-6)
Blue 2B (2602-46-2)
Blue 3B (72-57-1)
Blue BH (2429-73-4)
Blue BN Base (119-90-4)
Blue BN Salt (119-90-4)
Blue 2B Salt (2602-46-2)
Blue Base Irga B (119-90-4)
Blue Base NB (119-90-4)
Blue 2B base (120-00-3)
Blue Black 12B (1064-48-8)
Blue Black BN (2519-30-4)
Blue Black SX (1064-48-8)
Blue Copper (1332-40-7)
Blue Copper (7758-98-7)
Blue Copper-50 (1332-40-7)
Blue Copper AS (7758-99-8)
Blue Cross (712-48-1)
Blue Dextran (87915-38-6)
Blue Dextran 2000 (87915-38-6)
Blue EMB (72-57-1)
Blue GLA (147-14-8)
Blue K (130-20-1)
Blue No. 201 (482-89-3)
Blue No. 404 (147-14-8)
Blue O (81-77-6)
Blue Oil (68308-34-9)
Blue-Ox (1314-84-7)
Blue Powder (7440-66-6)
Blue Salt NB (119-90-4)
Blue Salt NBB (120-00-3)
Blue Stone (7758-98-7)
Blue Toner GTNF (147-14-8)
Blue URS (129-17-9)
Blue VRS (129-17-9)
Blue Vitriol (7758-98-7)
Blue Vitriol (7758-99-8)
Blue X (483-20-5)
Blue asbestos [UN 2212] (12001-28-4)
Blue oil (62-53-3)
Bluestone (7758-99-8)
Blulan (1861-40-1)
Bluton (15687-27-1)
Bluzedrin (60-13-9)
Bo-Ana (52-85-7)
Bog manganese (1313-13-9)
Boiler slag (68476-96-0)
Boldo (28772-56-7)
Bolero (28249-77-6)
Boletic acid (110-17-8)
Bolinan (9003-39-8)
Bolls-Eye (124-65-2)
Bolls-Eye (75-60-5)
Bolstar (35400-43-2)
Bombita (300-42-5)
Bomyl (122-10-1)
Bon (92-70-6)
Bona (92-70-6)
Bonabol (61-68-7)
Bon acid (92-70-6)
Bonadettes (569-65-3)
Bonadoxin (569-65-3)
Bonalan (1861-40-1)
Bonamid (105-60-2)
Bonamid (25038-54-4)
Bonamine (569-65-3)
Bonapicillin (69-53-4)
Bonare (604-75-1)
Bonarka (7758-87-4)
Bonazen (7733-02-0)
Bonbrain (50-35-1)
Bond CH 3 (9003-20-7)
Bond CH 18 (9003-20-7)
Bond CH 1200 (9003-20-7)
Bondelane A (126-33-0)
Bonderlube 235 (822-16-2)
Bondolane A (126-33-0)
Bondolane M (9CI) (39457-26-6)
Bone Oil (8001-85-2)
Bonibal (97-77-8)
Bonide Blue Death Rat Killer (7723-14-0)
Bonide Krab Crabgrass Killer (590-28-3)
Bonide Ryatox (15662-33-6)
Bonine (569-65-3)
Boniton (58-61-7)
Bonoform (79-34-5)
Bonomold OE (120-47-8)
Bonomold OP (94-13-3)

Booksaver	**(9003-20-7)**
Boots BTS 27419	**(33089-61-1)**
Boracic acid	**(10043-35-3)**
Borane, Compd. with dimethylsulfide	**(13292-87-0)**
Borane, Compd. with trimethylamine (1:1)	**(75-22-9)**
Borane, Compd. with N,N-dimethylmethanamine (1:1)	**(75-22-9)**
Borane, tributoxy-	**(688-74-4)**
Borane, trichloro	**(10294-34-5)**
Borane, triethyl	**(97-94-9)**
Borane, trifluoro	**(7637-07-2)**
Borane, trifluoro-, Compd. with 1,1'-oxybis(ethane) (1:1)	**(109-63-7)**
Borane, triphenyl-	**(960-71-4)**
Borascu	**(1303-96-4)**
Borate (8CI,9CI)	**(14213-97-9)**
Borate(1-), (acetato-O)trifluoro-, hydrogen, (T-4)- (9CI)	**(7578-36-1)**
Borate(1-), bis(1,2-benzenediolato(2-)-O,O')-, (T-4)-, hydrogen, Compd. with N,N'-bis(2-methylphenyl)guanidine (1:1) (9CI)	**(16971-82-7)**
Borate de trimethyle (French)	**(121-43-7)**
Borate(1-), hydroxytriphenyl-, sodium, (T-4)- (9CI)	**(12113-07-4)**
Borates, tetra, sodium salt, Anhydrous (OSHA)	**(1330-43-4)**
Borates, tetra, sodium salt, decahydrate (ACGIH,OSHA)	**(1303-96-4)**
Borates, tetra, sodium salt, pentahydrate (ACGIH,OSHA)	**(11130-12-4)**
Borates, tetra, sodium salts, pentahydrate	**(12179-04-3)**
Borate(1-), tetrafluoro-, ammonium	**(13826-83-0)**
Borate(1-), tetrafluoro-, hydrogen	**(16872-11-0)**
Borate(1-), tetrafluoro-, lead (2+)	**(13814-96-5)**
Borate(1-), tetrafluoro-, nickel(2+) (2:1), hexahydrate (9CI)	**(15684-36-3)**
Borate(1-), tetrafluoro-, nickel(2+), hexahydrate (8CI)	**(15684-36-3)**
Borate(1-), tetrahydro-, aluminum (8CI) (VAN)	**(13771-22-7)**
Borate(1-), tetrahydro-, lithium (8CI,9CI)	**(16949-15-8)**
Borate(1-), tetrahydro-, potassium (8CI,9CI)	**(13762-51-1)**
Borate(1-), tetrahydro-, sodium	**(16940-66-2)**
Borate(1-), tetraphenyl-, sodium (9CI)	**(143-66-8)**
Borato de trimetilo (Spanish)	**(121-43-7)**
Borax 2335	**(1332-07-6)**
Borax (8CI)	**(1303-96-4)**
Borax Glass	**(1330-43-4)**
Borax decahydrate	**(1303-96-4)**
Bordeaux	**(915-67-3)**
Bordeaux EMBL	**(2610-11-9)**
Bordeaux RRN	**(4424-06-0)**
Bordeaux S	**(915-67-3)**
Bordeaux S Extra Conc. A.Export	**(915-67-3)**
Bordeaux S Extra Pure A	**(915-67-3)**
Borden 2123	**(9003-20-7)**
Bordermaster	**(94-74-6)**
Borea	**(314-40-9)**
Borer Sol	**(107-06-2)**
Borester 2	**(688-74-4)**
Borester O	**(121-43-7)**
Boric-acid	**(10043-35-3)**
Boric acid (VAN)(9CI)	**(11113-50-1)**
Boric acid, aluminum salt (VAN)(9CI)	**(11121-16-7)**
Boric acid, ammonium salt	**(12007-89-5)**
Boric acid, ammonium salt, octahydrate	**(12007-89-5)**
Boric acid, barium salt (8CI,9CI)	**(13701-59-2)**
Boric acid, calcium salt (9CI)	**(12040-58-3)**
Boric acid, dilithium salt	**(12007-60-2)**
Boric acid, disodium salt	**(1330-43-4)**
Boric acid, ethyl ester	**(34099-73-5)**
Boric acid, ion(3-)	**(14213-97-9)**
Boric acid, lead(2+) salt (8CI,9CI)	**(14720-53-7)**
Boric acid, lithium salt (9CI)	**(13453-69-5)**
Boric acid, monosodium salt	**(7775-19-1)**
Boric acid, phenylmercury deriv.	**(102-98-7)**
Boric acid, sodium salt (9CI)	**(1333-73-9)**
Boric acid, sodium salt, pentahydrate (9CI)	**(11130-12-4)**
Boric acid, tributyl ester	**(688-74-4)**
Boric acid, triethyl ester	**(150-46-9)**
Boric acid, triisobutyl ester (8CI)	**(13195-76-1)**
Boric acid, triisopropyl ester	**(5419-55-6)**
Boric acid, trimethyl ester	**(121-43-7)**
Boric acid, trimethyl ester (9CI)	**(121-43-7)**
Boric acid, tris(1-methylethyl) ester	**(5419-55-6)**
Boric acid, tris(2-methylpropyl) ester (9CI)	**(13195-76-1)**
Boric acid, zinc salt (9CI)	**(1332-07-6)**
Boric anhydride	**(1303-86-2)**
Boricin	**(1303-96-4)**
Borinic acid, diphenyl-	**(2622-89-1)**
2,3-Bornanedione, (+-)- (8CI)	**(10373-78-1)**
2,5-Bornanedione (8CI)	**(4230-32-4)**
Bornane, 2,2,5-endo,6-exo,8,9,10-heptachloro	**(51775-36-1)**
Bornane, 2-oxo-	**(76-22-2)**
10-Bornanesulfonic acid, 2-oxo-, (1S,4R)-(+)	**(3144-16-9)**
2-Bornanol, endo-	**(507-70-0)**
(+)-2-Bornanone	**(464-49-3)**
2-Bornanone	**(76-22-2)**
d-2-Bornanone	**(464-49-3)**
3-Bornanone, (1S,4S)-(-)- (8CI)	**(10292-98-5)**
2-Bornanone, hydroxy- (8CI)	**(12001-40-0)**
Bornate	**(115-31-1)**
2-Bornene (8CI)	**(464-17-5)**
Borneo Camphor	**(507-70-0)**
trans-Borneol	**(507-70-0)**
Borneol [UN 1312]	**(507-70-0)**
Borneol, acetate (8CI)	**(76-49-3)**
Bornyl acetate	**(76-49-3)**
Bornyl acetic ether	**(76-49-3)**
Bornyl alcohol	**(507-70-0)**
Bornylene	**(464-17-5)**
Boron-bromide	**(10294-33-4)**
Boroethane	**(19287-45-7)**
Borofax	**(10043-35-3)**
Borofluoric acid	**(16872-11-0)**
Borohidruro de litio (Spanish)	**(16949-15-8)**
Borohydrure de lithium (French)	**(16949-15-8)**
Borohydrure de potassium (French)	**(13762-51-1)**
Borohydrure de sodium (French)	**(16940-66-2)**
Borolin	**(1918-02-1)**
Boron	**(7440-42-8)**
Boron, (N,N-dimethylmethanamine)trihydro-, (t-4)- (9CI)	**(75-22-9)**
Boron carbide (9CI)	**(12069-32-8)**
Boron chloride	**(10294-34-5)**
Boron fluoride	**(7637-07-2)**
Boron fluoride, Compound with acetic acid	**(7578-36-1)**
Boron fluoride diethyl etherate	**(109-63-7)**
Boron fluoride etherate	**(109-63-7)**
Boron fluoride-ethyl etherate	**(109-63-7)**
Boron fluoride-ethyl ether complex	**(109-63-7)**
Boron fluoride monoetherate	**(109-63-7)**
Boron hydride	**(19287-45-7)**
Boron lithium oxide (9CI)	**(12007-60-2)**
Boron oxide (ACGIH,OSHA)	**(1303-86-2)**
Boron-oxide	**(1303-86-2)**
Boron sesquioxide	**(1303-86-2)**
Boron tribromide (ACGIH,OSHA) [UN 2692]	**(10294-33-4)**
Boron trichloride [UN 1741]	**(10294-34-5)**
Boron trifluoride (ACGIH,OSHA) [UN 1008]	**(7637-07-2)**
Boron trifluoride acetic acid complex	**(7578-36-1)**
Boron trifluoride-acetic acid complex [UN 1742]	**(7578-36-1)**
Boron trifluoride-diethyl etherate	**(109-63-7)**
Boron trifluoride diethyletherate [UN 2604]	**(109-63-7)**
Boron trifluoride-dimethyl ether	**(353-42-4)**

Name	CAS
Boron trifluoride dimethyl etherate [UN 2965]	(353-42-4)
Boron trifluoride etherate	(109-63-7)
Boron trifluoride-ether complex	(109-63-7)
Boron trifluoride-ethyl ether	(109-63-7)
Boron trifluoride-ethyl etherate	(109-63-7)
Borontrifluoride monoethylamine	(75-23-0)
Boron, trifluoro(1,1'-oxybis(ethane))-, (t-4)- (9CI)	(109-63-7)
Boron trioxide	(1303-86-2)
Borsaure (German)	(10043-35-3)
Bortran	(99-30-9)
Bortrifluorid-dimethylether (Czech)	(353-42-4)
Bortrysan	(101-05-3)
Boruta Black A	(1064-48-8)
Bosan Supra	(50-29-3)
Bosmin	(51-43-4)
Botran	(99-30-9)
Botrilex	(82-68-8)
Botrydienal	(97165-23-6)
Bourbonal	(121-32-4)
Bov	(7664-93-9)
Bovidermol	(50-29-3)
Bovine Lactogenic Hormone	(9002-62-4)
Bovine Prolactin	(9002-62-4)
Bovine serum albumin	(9048-46-8)
Bovinocidin	(504-88-1)
Bovinox	(52-68-6)
Bovizole	(148-79-8)
Bovoflavin	(8048-52-0)
Boy	(561-27-3)
Boygon	(114-26-1)
Bri-Nylon	(63428-84-2)
Bracken Fern Toxic Component	(138-59-0)
Bralen KB 2-11	(9002-88-4)
Bralen RB 03-23	(9002-88-4)
Brasilaca Red R	(1248-18-6)
Brasilamina Black GN	(1937-37-7)
Brasilamina Blue 2B	(2602-46-2)
Brasilamina Blue 3B	(72-57-1)
Brasilamina Congo 4B	(573-58-0)
Brasilamina Fast Brown 3RA	(2429-82-5)
Brasilamina Green B	(3626-28-6)
Brasilamina Red 4B	(992-59-6)
Brasilamina Sky Blue 6B	(2610-05-1)
Brasilamina Violet 3R	(2586-60-9)
Brasilan Azo Rubine 2NS	(3567-69-9)
Brasilan Black BS	(1064-48-8)
Brasilan Chrome Violet B	(2092-55-9)
Brasilan Fast Fuchsine G	(4197-07-3)
Brasilan Fuchsine D	(3567-66-6)
Brasilan Metanil Yellow	(587-98-4)
Brasilan Orange A	(633-96-5)
Brasilan Orange 2G	(1936-15-8)
Brasilan Red S	(1658-56-6)
Brasilazet Blue GR	(2475-45-8)
Brasilazina Blue 3B	(72-57-1)
Brasilazina Oil Orange	(842-07-9)
Brasilazina Oil Red B	(85-83-6)
Brasilazina Oil Scarlet	(85-86-9)
Brasilazina Oil Scarlet 6G	(3118-97-6)
Brasilazina Oil Yellow G	(60-09-3)
Brasilazina Oil Yellow O	(1689-82-3)
Brasilazina Oil Yellow R	(97-56-3)
Brasilazina Orange Y	(532-82-1)
Brasilazol Black BH	(2429-73-4)
Brasoran	(4658-28-0)
Brass, Dross	(69011-68-3)
Brass baghouse fume	(69012-56-2)

Name	CAS
Brassica Campestris Oil	(8002-13-9)
Brassicasterol	(474-67-9)
Brassicol	(82-68-8)
Brassoron	(4658-28-0)
Brass slag	(69012-26-6)
Brass slag (Secondary nonferrous plant)	(69012-26-6)
Brassylic acid	(505-52-2)
Braunstein (German)	(1313-13-9)
Bravo-W-75	(1897-45-6)
Bravo	(1897-45-6)
Bravo 6F	(1897-45-6)
Brazil Wax	(8015-86-9)
Brazil Yellow X 2866	(2512-29-0)
Break-Thru	(2536-31-4)
Brecolane NDG	(111-46-6)
Brefeldin A	(20350-15-6)
Brellin	(77-06-5)
Bremil	(58-93-5)
Brenol	(12192-57-3)
Brentamine Fast Blue BB Base	(120-00-3)
Brentamine Fast Blue 2B Base	(120-00-3)
Brentamine Fast Blue B Base	(119-90-4)
Brentamine Fast Blue B Salt	(119-90-4)
Brentamine Fast Orange GC Base	(141-85-5)
Brentamine Fast Orange GC Salt	(141-85-5)
Brentamine Fast Orange GR Base	(88-74-4)
Brentamine Fast Orange GR Salt	(88-74-4)
Brentamine Fast Red B Base	(97-52-9)
Brentamine Fast Red TR Base	(95-69-2)
Brentamine Fast Red TR Salt	(3165-93-3)
Brenthol AN	(132-68-3)
Brenthol AS	(92-77-3)
Brenthol MN	(135-65-9)
Brenthol OT	(135-61-5)
Brentosyn FR	(135-62-6)
Brentosyn OTN	(135-61-5)
Brenzcatechin	(931-17-9)
Brenzkatechin (German)	(931-17-9)
Breon 202	(9011-06-7)
Breon 351	(9003-22-9)
Breon 425	(9003-22-9)
Breon AS 60/41	(9003-22-9)
Breon CS 100/30	(9011-06-7)
Breox MPEG 550	(9004-74-4)
Breox 75W270	(9003-11-6)
Bresit	(637-07-0)
Brestan	(900-95-8)
Brestan 60	(900-95-8)
Brestanol	(639-58-7)
Bretol	(124-03-8)
Brevimytal	(309-36-4)
Brevinyl	(62-73-7)
Brevinyl E50	(62-73-7)
Brevirenin	(51-43-4)
Brevital	(309-36-4)
Brevital sodium	(309-36-4)
Brianil	(78-44-4)
Brican	(23031-25-6)
Bricanyl	(23031-25-6)
Bricar	(23031-25-6)
Bricaril	(23031-25-6)
Brick Oil	(8001-58-9)
Bricyn	(23031-25-6)
Bridal	(961-71-7)
Brietal sodium	(309-36-4)
Bright Red	(5160-02-1)
BRIJ 23	(9002-92-0)

BRIJ 30SP (9002-92-0)
BRIJ 30 (9002-92-0)
BRIJ 35 (9002-92-0)
BRIJ 72 (9005-00-9)
BRIJ 76 (9005-00-9)
BRIJ 78 (9005-00-9)
BRIJ 92 (9004-98-2)
BRIJ 92((2)-Oleyl) (9004-98-2)
BRIJ 93 (9004-98-2)
BRIJ 96 (9004-98-2)
BRIJ 96((10) Oleyl) (9004-98-2)
BRIJ 97 (9004-98-2)
BRIJ 98 (9004-98-2)
BRIJ 99 (9004-98-2)
BRIJ 721 (9005-00-9)
Brillant-Grun (German) (633-03-4)
Brillfast Violet (1325-82-2)
Brilliant Acid Black BNA Export (2519-30-4)
Brilliant Acid Black BN Extra Pure A (2519-30-4)
Brilliant Acid Blue A Export (129-17-9)
Brilliant Acid Blue V Extra (129-17-9)
Brilliant Acid Blue VS (129-17-9)
Brilliant Acid Red G (3734-67-6)
Brilliant Acid Rosamine 2G (3734-67-6)
Brilliant Acid Rubine M (3567-69-9)
Brilliant Acridine Orange E (494-38-2)
Brilliant Alizarine Cyanine R (4368-56-3)
Brilliant Alizarine Light Blue 3FR (4368-56-3)
Brilliant Benzo Blue 6BA-CF (2610-05-1)
Brilliant Black (2519-30-4)
Brilliant Black A (2519-30-4)
Brilliant Black BN (2519-30-4)
Brilliant Black NAF (2519-30-4)
Brilliant Black N.FQ (2519-30-4)
Brilliant Blue (2650-18-2)
Brilliant Blue (3844-45-9)
Brilliant Blue FCD No. 1 (3844-45-9)
Brilliant Blue FCF (2650-18-2)
Brilliant Blue FCF (3844-45-9)
Brilliant Blue GS (129-17-9)
Brilliant Carmoisine (3567-69-9)
Brilliant Chrome Leather Black H (1937-37-7)
Brilliant Colacid Red G (3734-67-6)
Brilliant Crimson 2R.FQ (3567-69-9)
Brilliant Crimson Red (3567-69-9)
Brilliant Fast Oil Yellow (60-11-7)
Brilliant Fast Spirit Yellow (60-11-7)
Brilliant Fast Yellow (60-11-7)
Brilliant Green (633-03-4)
Brilliant Green Aseptic (633-03-4)
Brilliant Green B (633-03-4)
Brilliant Green B.P. (633-03-4)
Brilliant Green BP Crystals (633-03-4)
Brilliant Green BPc (633-03-4)
Brilliant Green Crystals (633-03-4)
Brilliant Green Crystals H (633-03-4)
Brilliant Green DSC (633-03-4)
Brilliant Green 3EMBL (4680-78-8)
Brilliant Green G (633-03-4)
Brilliant Green GX (633-03-4)
Brilliant Green Lake (633-03-4)
Brilliant Green Oxalate (36351-18-5)
Brilliant Green P (633-03-4)
Brilliant Green Phthalocyanine (1328-53-6)
Brilliant Green S (128-58-5)
Brilliant Green Special (633-03-4)
Brilliant Green Sulfate (633-03-4)
Brilliant Green WP Crystals (633-03-4)
Brilliant Green Y (633-03-4)
Brilliant Green YN (633-03-4)
Brilliant Green YNS (633-03-4)
Brilliant Lake Green Y (633-03-4)
Brilliant Oil Blue BGS (14233-37-5)
Brilliant Oil Orange R (842-07-9)
Brilliant Oil Orange Y Base (532-82-1)
Brilliant Oil Scarlet B (3118-97-6)
Brilliant Oil Yellow (492-80-8)
Brilliant Oil Yellow (60-11-7)
Brilliant Orange (6CI) (1934-20-9)
Brilliant Orange G (1934-20-9)
Brilliant Orange GN (1934-20-9)
Brilliant Orange GN Type 8019 (1934-20-9)
Brilliant Orange GR (4424-06-0)
Brilliant Pink B (81-88-9)
Brilliant Ponceau G (3761-53-3)
Brilliant Ponceau 3R (2611-82-7)
Brilliant Ponceau 4R (2611-82-7)
Brilliant Ponceau 5R (2611-82-7)
Brilliant Ponceau 4RC (2611-82-7)
Brilliant Ponceau 4RC Specially Pure (2611-82-7)
Brilliant Ponceau 3RF (2611-82-7)
Brilliant Ponceau 3RF Extra (2302-96-7)
Brilliant Red (5160-02-1)
Brilliant Red 5SKH (17804-49-8)
Brilliant Safranine BR (477-73-6)
Brilliant Safranine G (477-73-6)
Brilliant Safranine GR (477-73-6)
Brilliant Scarlet (2611-82-7)
Brilliant Scarlet (5160-02-1)
Brilliant Scarlet 3R (2611-82-7)
Brilliant Scarlet 4R (2611-82-7)
Brilliant Tangerine 13030 (3468-63-1)
Brilliant Toner RB (1103-39-5)
Brilliant Toner Z (5160-02-1)
Brilliant Toning Red Amine (88-51-7)
Brilliant Tungstate Green Toner GT-288 (633-03-4)
Brilliant Violet 5B (548-62-9)
Brilliant Violet K (1324-55-6)
Brilliant Yellow (91-34-9)
Brilliant Yellow Slurry (6358-85-6)
Brilliantschwarz BN (German) (2519-30-4)
Brimstone (7704-34-9)
Brinderdin (50-55-5)
Briserine (50-55-5)
Bristab (135-09-1)
Bristaciclin α (60-54-8)
Bristaciclina (60-54-8)
Bristacycline (60-54-8)
Bristacycline (64-75-5)
Bristamox (26787-78-0)
Bristamycin (643-22-1)
Bristuric (73-48-3)
Bristurin (135-09-1)
Bristuron (73-48-3)
Britacil (69-53-4)
Britesorb (1343-88-0)
Britesorb 90 (1343-88-0)
Britesorb No 40 (1343-88-0)
British Aluminum AF 260 (21645-51-2)
British Antilewisite (59-52-9)
Briton (52-68-6)
Britone Red Y (1248-18-6)
Britten (52-68-6)
Brittox (1689-84-5)

Brom-o-Gas (74-83-9)
Brom-o-Gaz (74-83-9)
Brobamate (57-53-4)
Brocadisipal (83-98-7)
Brocadopa (59-92-7)
Brocide (107-06-2)
Brockmann, Aluminum oxide (1344-28-1)
Brocsil (132-93-4)
Brodan (2921-88-2)
Brodifacoum (56073-10-0)
Brodifakum (Czech) (56073-10-0)
Broenner's acid (93-00-5)
2-Brom-2-ethylbutyrylmocovina (Czech) (77-65-6)
Brofene (2104-96-3)
Brogdex 555 (128-04-1)
5-Brom-3-isopropyl-6-methyl-uracil (German) (314-42-1)
Brom (German) (7726-95-6)
Bromacetocarbamide (77-65-6)
Bromacil (ACGIH,OSHA) (314-40-9)
Bromacil, dimethylamine salt (69484-12-4)
Bromacil lithium salt (53404-19-6)
Bromacil, lithium salt (53404-19-6)
Bromadal (77-65-6)
Bromadel (77-65-6)
Bromadialone (28772-56-7)
Bromadiolone (28772-56-7)
Bromal (115-17-3)
Bromallylene (106-95-6)
Bromamine acid (116-81-4)
Bromaminic acid (116-81-4)
p-Bromanilid kyseliny 5-bromsalicylove (Czech) (87-12-7)
4-Bromanilinu (Czech) (106-40-1)
p-Bromanisole (104-92-7)
Bromat (57-09-0)
Bromate (15541-45-4)
Bromate d'ammonium (French) (13843-59-9)
Bromate de sodium (French) (7789-38-0)
Bromato (Spanish) (15541-45-4)
Bromato amonico (Spanish) (13843-59-9)
Bromax (314-40-9)
Bromazepam (1812-30-2)
Bromazepamum (Latin) (1812-30-2)
Bromazil (314-40-9)
3-Brombenzotrifluorid (Czech) (401-78-5)
Brombenzyl cyanide (5798-79-8)
α-Brombenzylkyanid (Czech) (5798-79-8)
Brombloom (109-84-2)
Bromchlophos (300-76-5)
Bromdian (79-94-7)
O-(4-Brom-2,5-dichlor-phenyl)-O,O-dimethyl-monothiophosphat (German) (2104-96-3)
Bromdiethylacetylurea (77-65-6)
Brome (French) (7726-95-6)
Bromek dwumetylolaurylobenzyloamoniowy (Polish) (7281-04-1)
Bromelia (93-18-5)
Bromeosin (15086-94-9)
Bromethalin (63333-35-7)
Bromethaline (French) (63333-35-7)
Brom-methan (German) (74-83-9)
Brometon (76-08-4)
Brometone (76-08-4)
Bromex (13360-45-7)
Bromex (300-76-5)
Bromex (97-17-6)
Bromfeniramina (Spanish) (86-22-6)
2-Bromfenol (Czech) (95-56-7)
Bromfenuron (3408-97-7)
Bromfenvinphos-methyl (13104-21-7)
3-Bromfluorbenzen (Czech) (1073-06-9)
Bromhexina (Spanish) (3572-43-8)
Bromhexine (3572-43-8)
Bromhexinum (Latin) (3572-43-8)
Bromic acid, ammonium salt (13843-59-9)
Bromic acid, potassium salt (7758-01-2)
Bromic acid, sodium salt (7789-38-0)
Bromide salt of potassium (7758-02-3)
Bromide salt of sodium (7647-15-6)
Bromid uhlicity (Czech) (558-13-4)
Brominal (1689-84-5)
Brominal M & Plus (94-74-6)
Brominated butyl rubber (68441-14-5)
Bromine (ACGIH,OSHA) [UN 1744] (7726-95-6)
Bromine, Solution (DOT) (7726-95-6)
Bromine azide (DOT) (13973-87-0)
Bromine chloride [UN 2901] (13863-41-7)
Bromine cyanide (506-68-3)
Bromine monochloride (13863-41-7)
Bromine nitride (13973-87-0)
Bromine pentafluoride (ACGIH,OSHA) [UN 1745] (7789-30-2)
Bromine trifluoride [UN 1746] (7787-71-5)
Brominex (1689-84-5)
Brominil (1689-84-5)
Bromkal 80 (27858-07-7)
Bromkal 80 (61288-13-9)
Bromkal 80-9D (27753-52-2)
Bromkal 82-ODE (1163-19-5)
Bromkal 83-10DE (1163-19-5)
Bromkal P 67-6HP (126-72-7)
Bromnatrium (German) (7647-15-6)
Bromo Acid (17372-87-1)
Bromo B (17372-87-1)
Bromo B (548-26-5)
Bromo 4D (548-26-5)
Bromo 4DC (17372-87-1)
Bromo 4DC (548-26-5)
Bromo 4DL (548-26-5)
4-Bromo-2,5-DMA hydrobromide (53581-53-6)
Bromo DNA (1817-73-8)
Bromo FL (17372-87-1)
Bromo FL (548-26-5)
Bromo (Italian) (7726-95-6)
Bromo JPS (17372-87-1)
Bromo JPS (548-26-5)
Bromo TS (17372-87-1)
Bromo TS (548-26-5)
Bromo X-100 (17372-87-1)
Bromo X-100 (548-26-5)
Bromo XX (17372-87-1)
Bromo XX (548-26-5)
N-Bromoacetamide (79-15-2)
2-Bromoacetamidopyridine (40086-66-6)
3'-Bromoacetanilide (621-38-5)
3-Bromoacetanilide (621-38-5)
4'-Bromoacetanilide (103-88-8)
4-Bromoacetanilide (103-88-8)
m-Bromoacetanilide (621-38-5)
p-Bromo-N-acetanilide (103-88-8)
p-Bromoacetanilide (103-88-8)
Bromoacetate ion (79-08-3)
Bromoacetic acid (79-08-3)
α-Bromoacetic acid (79-08-3)
Bromoacetic acid, Solid [UN 1938] (79-08-3)
Bromoacetic acid, Solution [UN 1938] (79-08-3)
Bromoacetic acid, ethyl ester (105-36-2)

Bromoacetic acid methyl ester (96-32-2)
Bromoacetone, Liquid (DOT) (598-31-2)
Bromoacetone [UN 1569] (598-31-2)
2-Bromoacetophenone (70-11-1)
4'-Bromoacetophenone (99-90-1)
4-Bromoacetophenone (99-90-1)
α-Bromoacetophenone (70-11-1)
m-Bromoacetophenone (2142-63-4)
ω-Bromoacetophenone (70-11-1)
Bromo acid (548-26-5)
γ-Bromoallylene (106-96-7)
Bromoanilide (103-88-8)
4-Bromoaniline (106-40-1)
m-Bromoaniline (591-19-5)
p-Bromoaniline (106-40-1)
2-Bromoanisole (578-57-4)
4-Bromoanisole (104-92-7)
m-Bromoanisole (2398-37-0)
o-Bromoanisole (578-57-4)
p-Bromoanisole (104-92-7)
Bromoantifebrin (103-88-8)
p-Bromoantipyrine (603-65-6)
3-Bromobenzaldehyde (3132-99-8)
m-Bromobenzaldehyde (3132-99-8)
2-Bromobenzenamine (615-36-1)
3-Bromobenzenamine (591-19-5)
4-Bromobenzenamine hydrochloride (624-19-1)
Bromobenzene [UN 2514] (108-86-1)
4-Bromobenzeneacetonitrile (16532-79-9)
3-Bromobenzenesulfonyl chloride (2905-24-0)
4-Bromobenzenesulfonyl chloride (98-58-8)
m-Bromobenzenesulfonyl chloride (2905-24-0)
p-Bromobenzenesulfonyl chloride (98-58-8)
6-Bromobenzo(a)pyrene (21248-00-0)
2-Bromobenzoate (88-65-3)
2-Bromobenzoic acid (88-65-3)
3-Bromobenzoic acid (585-76-2)
4-Bromobenzoic acid (586-76-5)
o-Bromobenzoic acid (88-65-3)
p-Bromobenzoic acid (586-76-5)
3-Bromobenzonitrile (6952-59-6)
4-Bromobenzonitrile (623-00-7)
m-Bromobenzonitrile (6952-59-6)
3-Bromobenzotrifluoride (401-78-5)
m-Bromobenzotrifluoride (401-78-5)
4-Bromobenzylcyanide (16532-79-9)
α-Bromobenzyl cyanide (5798-79-8)
Bromobenzyl cyanide (DOT) (5798-79-8)
Bromobenzyl cyanide, Liquid [UN 1694] (5798-79-8)
p-Bromobenzyl cyanide [UN 1694] (16532-79-9)
Bromobenzylnitrile (5798-79-8)
α-Bromobenzylnitrile (5798-79-8)
3-Bromobenzyltrifluoride (401-78-5)
10-Bromo-11b-(2-fluorophenyl)-2,3,7,11b-tetrahydrooxazolo-(3,2-d)(1,4)benzodiazepin-6(5H)-one (59128-97-1)
2-Bromo-1,1'-biphenyl (2052-07-5)
2-Bromobiphenyl (2052-07-5)
3-Bromobiphenyl (2113-57-7)
3-(3-(4'-Bromo(1,1'-biphenyl)-4-yl)-3-hydroxy-1-phenylpropyl)-4-hydroxy-2H-1-benzopyran-2-one (28772-56-7)
3-(3-(4'-Bromo(1,1'-biphenyl)-4-yl)3-hydroxy-1-phenylpropyl)-4-hydroxy-2H-1-benzopyran-2-one (28772-56-7)
3-(3-(4'-Bromo(1,1-biphenyl)-4-yl)-3-hydroxy-1-phenylpropyl)-4-hydroxy-2H-1-benzopyran-2-one (9CI) (28772-56-7)
3-(3-(4'-Bromo-(1,1'-biphenyl)-4-yl)-3-hydroxy-1-phenylpropyl)-4-hydroxy-2H-1-benzopyran-2-one (28772-56-7)
3-(3-(4'-Bromo-(1,1'-biphenyl)-4-yl)-3-hydroxy-1-phenylpropyl)-4-hydroxyc-oumarin (28772-56-7)
3-(3-(4'-Bromobiphenyl-4-yl)-3-hydroxy-1-phenylpropyl-4-hydroxycoumarin (28772-56-7)
3-(3-(4'-Bromo-1,1'-biphenyl-4-yl)-1,2,3,4-tetrahydro-1-naphthyl)-4-hydroxycoumarin (56073-10-0)
3-(3-(4'-Bromobiphenyl-4-yl)-1,2,3,4-tetrahydronaphth-1-yl)-4-hydroxycoumarin (56073-10-0)
2-Bromo-2-(bromomethyl)glutaronitrile (35691-65-7)
5-Bromo-N-(4-bromophenyl)-2-hydroxybenzamide (87-12-7)
4-Bromo-α-(4-bromophenyl)-α-hydroxybenzeneacetic acid 1-methyl ethyl ester (18181-80-1)
1-Bromobutane (109-65-9)
2-Bromobutane [UN 2339] (78-76-2)
2-Bromobutanoic acid (80-58-0)
Bromobutanol (76-08-4)
1-Bromo-2-butanone (816-40-0)
3-Bromo-2-butanone (814-75-5)
1-Bromo-1-butene (31844-98-1)
1-Bromo-2-butene (4784-77-4)
2-Bromo-2-butene (13294-71-8)
4-Bromo-1-butene (5162-44-7)
1-Bromo-3-buten-2-ol (64341-49-7)
5-Bromo-3-sec-butyl-6-methyluracil (314-40-9)
5-Bromo-3-tert-butyl-6-methyluracil (7286-76-2)
2-Bromobutyric acid (80-58-0)
α-Bromobutyric acid (80-58-0)
dl-2-Bromobutyric acid (80-58-0)
α-Bromo-n-caproic acid (616-05-7)
α-Bromocaproic acid (616-05-7)
Bromocet (140-72-7)
Bromochloride (13863-41-7)
Bromochloroacetonitrile (83463-62-1)
2-Bromo-3'-chloroacetophenone (41011-01-2)
α-Bromo-3-chloroacetophenone (41011-01-2)
1-Bromo-2-chlorobenzene (694-80-4)
1-Bromo-3-chlorobenzene (108-37-2)
3-Bromochlorobenzene (108-37-2)
4-Bromochlorobenzene (106-39-8)
m-Bromochlorobenzene (108-37-2)
p-Bromochlorobenzene (106-39-8)
trans-1-Bromo-2-chlorocyclopentane (14376-82-0)
Bromochlorodifluoromethane (353-59-3)
3-Bromo-1-chloro-5,5-dimethylhydantoin (126-06-7)
1-Bromo-3-chloro-5,5-dimethyl-2,4-imidazolidinedione (16079-88-2)
1-Bromo-2-chloroethane (107-04-0)
Bromochloromethane [UN 1887] (74-97-5)
Bromochloromethyl cyanide (83463-62-1)
2-Bromo-6-chloro-4-nitroaniline (99-29-6)
2-Bromo-6-chloro-4-nitrobenzenamine (99-29-6)
2,2'-((4-((2-Bromo-6-chloro-4-nitrophenyl)azo)-3-chloro-phenyl)imino)bis(ethanol) (17464-91-4)
1-Bromo-2-(4-chlorophenyl)ethane (6529-53-9)
2-Bromo-1-(3-chlorophenyl)ethanone (41011-01-2)
O-(4-Bromo-2-chlorophenyl)-O-ethyl-S-propyl phosphorothioate (41198-08-7)
3-(4-Bromo-3-chlorophenyl)-1-methoxy-1-methylurea (13360-45-7)
N-(4-Bromo-3-chlorophenyl)-N'-methoxy-N'-methylurea (13360-45-7)
1-Bromo-3-chloropropane (109-70-6)
2-Bromo-1-chloropropane (3017-95-6)
α-Bromo-o-chlorotoluene (611-17-6)
1-Bromo-1-chloro-2,2,2-trifluoroethane (151-67-7)
1-Bromo-2-chloro-1,1,2-trifluoroethane (354-06-3)
2-Bromo-2-chloro-1,1,1-trifluoroethane (151-67-7)
Bromochlorotrifluoroethane (151-67-7)
m-Bromocinnamic acid (32862-97-8)
Bromocyan (506-68-3)
Bromocyanide (506-68-3)
Bromocyanogen (506-68-3)

Bromocycloheptane **(2404-35-5)**
1-Bromocyclohexane **(108-85-0)**
Bromocyclohexane **(108-85-0)**
trans-4-Bromocyclohexanol **(32388-22-0)**
Bromocyclopentane **(137-43-9)**
(2-Bromocyclopropyl)benzene **(36617-02-4)**
1-Bromo-4-cyclopropylbenzene **(1124-14-7)**
Bromo-p-cymeme **(65724-11-0)**
1-Bromodecane **(112-29-8)**
Bromodeoxyglycerol **(4704-77-2)**
2-Bromo-1,5-diamino-4,8-dihydroxyanthraquinone **(27312-17-0)**
2-Bromodibenzofuran **(86-76-0)**
4-Bromo-3,5-dichloroaniline **(1940-29-0)**
1-Bromo-3,5-dichlorobenzene **(19752-55-7)**
2-Bromo-1,3-dichlorobenzene **(19393-92-1)**
4-Bromo-1,2-dichlorobenzene **(18282-59-2)**
Bromodichloromethane **(75-27-4)**
2-Bromo-4,6-dichlorophenol **(4524-77-0)**
4-Bromo-2,5-dichlorophenol **(1940-42-7)**
6-Bromo-2,4-dichlorophenol **(4524-77-0)**
O-(4-Bromo-2,5-dichlorophenyl) O,O-diethyl phosphorothioate **(4824-78-6)**
O-(4-Bromo-2,5 dichlorophenyl) O,O-diethylphosphorothionate **(4824-78-6)**
O-(4-Bromo-2,5-dichlorophenyl) O,O-dimethyl phosphorothioate **(2104-96-3)**
4-Bromo-2,5-dichlorophenyl dimethyl phosphorothionate **(2104-96-3)**
O-(4-Bromo-2,5-dichlorophenyl) O-methyl phenylphosphonothioate **(21609-90-5)**
O-(4-Bromo-2,5-dicloro-fenil)-O,O-dimetil-monotiofosfato (Italian) **(2104-96-3)**
Bromodiethylacetylcarbamide **(77-65-6)**
Bromodiethylacetylurea **(77-65-6)**
p-Bromo-N,N-diethylaniline **(2052-06-4)**
4-Bromo-N,N-diethylbenzenamine **(2052-06-4)**
Bromodiethylgold **(26645-10-3)**
7-Bromo-1,3-dihydro-5-(2-pyridyl)-2H-1,4-benzdiazepin-2-one **(1812-30-2)**
7-Bromo-1,3-dihydro-5-(2-pyridyl)-2H-1,4-benzodiazepin-2-one **(1812-30-2)**
4-Bromo-2,5-dimethoxyamphetamine hydrobromide **(53581-53-6)**
dl-4-Bromo-2,5-dimethoxyamphetamine hydrobromide **(53581-53-6)**
4-Bromo-2,5-dimethoxy-α-methylphenethyl-amine hydrobromide **(53581-53-6)**
DL-4-Bromo-2,5-dimethoxy-α-methylphenethylamine hydrobromide **(53581-53-6)**
2-(p-Bromo-α-(2-dimethylaminoethyl)benzyl)pyridine **(86-22-6)**
1-Bromo-2,5-dimethylbenzene **(553-94-6)**
2-Bromo-1,4-dimethylbenzene **(553-94-6)**
2-Bromo-3,3-dimethyl-N-N-(α-α-dimethylbenzyl) butyramide **(74712-19-9)**
1-Bromo-2,2-dimethylpropane **(630-17-1)**
2-Bromo-4,6-dinitroaniline **(1817-73-8)**
4-Bromo-1,2-dinitrobenzene **(610-38-8)**
5-((2-Bromo-4,6-dinitrophenyl)azo)-4-acetamido-2-((2-cyanoethyl-ethylamino)anisole **(22578-86-5)**
4-(2-Bromo-4,6-dinitrophenylazo)-5-acetylamino-2-ethoxy-N,N-bis(β-acetoxyethyl)aniline **(12239-34-8)**
4-(6-Bromo-2,4-dinitrophenylazo)-3-acetylamino-6-methoxy-N-bis(acetoxyethyl)aniline **(3618-72-2)**
2'-((2-Bromo-4,6-dinitrophenyl)azo)-5'-((2-cyanoethyl)ethyl-amino)-p-acetanisidide **(22578-86-5)**
4-Bromodiphenyl ether **(101-55-3)**
p-Bromodiphenyl ether **(101-55-3)**
1-Bromododecane **(143-15-7)**
1-Bromoeicosane **(4276-49-7)**
Bromoeosin **(15086-94-9)**
Bromoeosine **(17372-87-1)**
Bromoeosine **(548-26-5)**
3-Bromo-1,2-epoxypropane **(3132-64-7)**
Bromoethane **(74-96-4)**
2-Bromoethanesulfonic acid **(26978-65-4)**
Bromoethanoic acid **(79-08-3)**
α-Bromoethanoic acid **(79-08-3)**
2-Bromoethanol **(540-51-2)**
Bromoethanol **(540-51-2)**
Bromoethene (9CI) **(593-60-2)**
2-Bromoethylacetate **(927-68-4)**
Bromoethyl acetate **(927-68-4)**
2-Bromoethyl acrylate **(4823-47-6)**
(1-Bromoethyl)benzene **(585-71-7)**
(2-Bromoethyl)benzene **(103-63-9)**
(α-Bromoethyl)benzene **(585-71-7)**
1-Bromo-4-ethylbenzene **(1585-07-5)**
β-Bromoethylbenzene **(103-63-9)**
4-(2-Bromoethyl)benzenesulfonic acid **(54322-31-5)**
(α-Bromo-α-ethylbutyryl)carbamide **(77-65-6)**
(α-Bromo-α-ethylbutyryl)urea **(77-65-6)**
1-Bromo-ethyl-butyryl-urea **(77-65-6)**
2-Bromo-2-ethylbutyrylurea **(77-65-6)**
1-(2-Bromoethyl)-4-chlorobenzene **(6529-53-9)**
Bromoethylene **(593-60-2)**
2-(2-Bromoethyl)-1H-isoindole-1,3(2H)-dione **(574-98-1)**
2-Bromo-5-ethylnonane **(55162-38-4)**
2-(Bromoethyl)phthalimide **(574-98-1)**
N-(2-Bromoethyl)phthalimide **(574-98-1)**
β-Bromoethylphthalimide **(574-98-1)**
Bromofenoxim **(13181-17-4)**
p-Bromofenuron **(3408-97-7)**
Bromoflor **(16672-87-0)**
Bromofluoresceic acid **(15086-94-9)**
Bromofluoresceic acid **(17372-87-1)**
Bromofluoresceic acid **(548-26-5)**
Bromo fluorescein **(17372-87-1)**
Bromo fluorescein **(548-26-5)**
4'-Bromo-2-fluoroacetanilide **(351-05-3)**
1-Bromo-4-fluorobenzene **(460-00-4)**
3-Bromofluorobenzene **(1073-06-9)**
4-Bromofluorobenzene **(460-00-4)**
p-Bromofluorobenzene **(460-00-4)**
Bromofluoroform **(75-63-8)**
Bromofos-Ethyl **(4824-78-6)**
Bromofos-Methyl **(2104-96-3)**
Bromoform (ACGIH,DOT,OSHA) [UN 2515] **(75-25-2)**
Bromoforme (French) **(75-25-2)**
Bromoformio (Italian) **(75-25-2)**
Bromofos **(2104-96-3)**
Bromofume **(106-93-4)**
1-Bromoheptadecane **(3508-00-7)**
1-Bromoheptane **(629-04-9)**
2-Bromoheptane **(1974-04-5)**
3-Bromoheptane **(1974-05-6)**
Bromohexachlorodibenzofuran **(107207-47-6)**
1-Bromohexadecane **(112-82-3)**
2-Bromohexadecanoic acid **(18263-25-7)**
Bromo(hexahydro-2H-azepin-2-onato-N)magnesium **(17091-31-5)**
1-Bromohexane **(111-25-1)**
2-Bromohexane **(3377-86-4)**
3-Bromohexane **(3377-87-5)**
Bromohexane **(111-25-1)**
2-Bromohexanoic acid **(616-05-7)**
6-Bromohexanoic acid **(4224-70-8)**
α-Bromohydrin **(4704-77-2)**
2-Bromo-4'-hydroxyacetophenone **(2491-38-5)**
2-Bromo-4-hydroxyacetophenone **(2491-38-5)**
3-Bromo-6-hydroxybenz-p-bromanilide **(87-12-7)**
4-Bromo-3-hydroxy-2-(1,3-indandion-2-yl)quinoline **(10319-14-9)**
3-Bromo-4-hydroxy-5-methoxybenzaldehyde **(2973-76-4)**
1-Bromo-2-iodobenzene **(583-55-1)**
1-Bromo-3-iodobenzene **(591-18-4)**
1-Bromo-4-iodobenzene **(589-87-7)**

2-Bromoisobutane (507-19-7)
p-Bromoisopropylbenzene (586-61-8)
5-Bromo-3-isopropyl-6-methyl, 2,4-pyrimidinedione (French) (314-42-1)
5-Bromo-3-isopropyl-6-methyluracil (314-42-1)
5-Bromo-3-isopropyl-6-metil-uracil (Italian) (314-42-1)
2-Bromoisovaleric acid (565-74-2)
α-Bromoisovaleric acid (565-74-2)
Bromol (118-79-6)
1-Bromomercuri-2-hydroxypropane (18832-83-2)
Bromometano (Italian) (74-83-9)
Bromomethane (74-83-9)
1-Bromo-3-methoxybenzene (2398-37-0)
(Bromomethyl)benzene (100-39-0)
Bromomethylbenzene (28807-97-8)
4-(Bromomethyl)benzoic acid (6232-88-8)
4'-Bromo-3-methylbiphenyl (56961-07-0)
1-Bromo-2-methylbutane (10422-35-2)
1-Bromo-3-methylbutane [UN 2341] (107-82-4)
2-Bromo-3-methylbutanoic acid (565-74-2)
2-Bromo-3-methylbutyric acid (565-74-2)
1-Bromo-4-methylcyclohexane (6294-40-2)
2-Bromomethyl-4,6-diaminotriazine (4576-40-3)
1-Bromo-4-(1-methylethyl)benzene (586-61-8)
7-Bromomethyl-12-methylbenz(a)anthracene (16238-56-5)
1-(Bromomethyl)-2-methylbenzene (89-92-9)
1-(Bromomethyl)-3-methylbenzene (620-13-3)
1-(Bromomethyl)-4-methylbenzene (104-81-4)
5-Bromo-6-methyl-3-(1-methylethyl)-2,4-(1H,3H)-pyrimidinedione (314-42-1)
Bromomethyl methyl ketone (598-31-2)
5-Bromo-6-methyl-3-(1-methylpropyl)-2,4(1H,3H)-pyrimidinedione (314-40-9)
5-Bromo-6-methyl-3-(1-methylpropyl)uracil (314-40-9)
p-(Bromomethyl)nitrobenzene (100-39-0)
4-Bromo-2-methyl-5-nitroimidazole (18874-52-7)
5-Bromo-2-methyl-4-nitroimidazole (18874-52-7)
1-(Bromomethyl)-3-phenoxybenzene (51632-16-7)
Bromomethyl phenyl ketone (70-11-1)
1-Bromo-2-methylpropane [UN 2342] (78-77-3)
2-Bromo-2-methylpropane [UN 2342] (507-19-7)
2-(Bromomethyl)toluene (89-92-9)
4-(Bromomethyl)toluene (104-81-4)
p-(Bromomethyl)toluene (104-81-4)
1-Bromonaphthalene (90-11-9)
2-Bromonaphthalene (580-13-2)
α-Bromonaphthalene (90-11-9)
Bromone (28772-56-7)
4-Bromo-2-nitroaniline (875-51-4)
4-Bromo-2-nitrobenzenamine (875-51-4)
2-Bromonitrobenzene (577-19-5)
3-Bromonitrobenzene (585-79-5)
4-Bromonitrobenzene (586-78-7)
Bromonitrobenzene (61878-56-6)
m-Bromonitrobenzene (585-79-5)
o-Bromonitrobenzene (577-19-5)
p-Bromonitrobenzene (586-78-7)
1-Bromo-2-nitrobenzene (DOT) (577-19-5)
1-Bromo-3-nitrobenzene (DOT) (585-79-5)
4-Bromo-2-nitrobenzeneamine (875-51-4)
3,3'-((4-((2-Bromo-4-nitro-6-chlorophenyl)azo)phenyl)imino)bis-(propanoic acid), dimethyl ester (59709-38-5)
(2-Bromo-2-nitroethenyl)benzene (7166-19-0)
2-Bromo-2-nitropropan-1,3-diol (52-51-7)
2-Bromo-2-nitro-1,3-propanediol (52-51-7)
2-Bromo-2-nitropropane-1,3-diol (52-51-7)
β-Bromo-β-nitrostyrene (7166-19-0)
β-Bromo-β-nitrotrimethyleneglycol (52-51-7)
1-Bromononadecane (4434-66-6)
1-Bromononane (693-58-3)
1-Bromooctadecane (112-89-0)
2-Bromooctadecanoic acid (142-94-9)
1-Bromooctane (111-83-1)
2-Bromooctanoate (2623-82-7)
2-Bromopalmitate (18263-25-7)
2-Bromopalmitic acid (18263-25-7)
α-Bromopalmitic acid (18263-25-7)
1-Bromopentadecane (629-72-1)
Bromopentafluorobenzene (344-04-7)
1-Bromopentane (110-53-2)
3-Bromopentane (1809-10-5)
2-Bromopentane [UN 2343] (107-81-3)
5-Bromopentanoic acid (2067-33-6)
Bromoperfluorobenzene (344-04-7)
p-Bromophenacyl-8 (99-73-0)
4-Bromophenacyl bromide (99-73-0)
p-Bromophenacyl bromide (99-73-0)
Bromopheniramine (86-22-6)
2-Bromophenol (95-56-7)
3-Bromophenol (591-20-8)
4-Bromophenol (106-41-2)
Bromophenol (32762-51-9)
o-Bromophenol (95-56-7)
p-Bromophenol (106-41-2)
Bromophenol Blue (115-39-9)
Bromophenoxim (13181-17-4)
1-Bromo-4-phenoxy benzene (101-55-3)
1-Bromo-4-phenoxybenzene (101-55-3)
4-Bromophenoxybenzene (101-55-3)
p-Bromophenoxybenzene (101-55-3)
α-Bromo-3-phenoxytoluene (51632-16-7)
2-(4-Bromophenyl)acetonitrile (16532-79-9)
4-Bromophenylacetonitrile (16532-79-9)
α-Bromophenylacetonitrile (5798-79-8)
p-Bromophenylacetonitrile (16532-79-9)
p-Bromophenylamine (106-40-1)
p-Bromophenyl bromide ((106-37-6)
m-Bromophenyl chloride (108-37-2)
p-Bromophenyl chloride (106-39-8)
4-Bromophenylchloromethyl sulfone (54091-06-4)
(4-Bromophenyl)cyclopropane (1124-14-7)
(p-Bromophenyl)cyclopropane (1124-14-7)
1-Bromo-2-phenylcyclopropane (36617-02-4)
γ-(4-Bromophenyl)-N,N-dimethyl-2-pyridinepropanamine (86-22-6)
3-(4-Bromophenyl)-N,N-dimethyl-3-(3-pyridinyl)-2-propen-1-amine (56775-88-3)
1-(4-Bromophenyl)-3,3-dimethylurea (3408-97-7)
1-(p-Bromophenyl)-3,3-dimethylurea (3408-97-7)
N'-(4-Bromophenyl)-N,N-dimethylurea (3408-97-7)
N-(4-Bromophenyl)-N',N'-dimethylurea (3408-97-7)
1-Bromo-1-phenylethane (585-71-7)
1-Bromo-2-phenylethane (103-63-9)
1-(3-Bromophenyl)ethanone (2142-63-4)
1-(4-Bromophenyl)ethanone (99-90-1)
α-Bromo-β-phenylethylene (103-64-0)
3-(α-(p-(p-Bromophenyl)-β-hydroxyphenethyl)benzyl)-4-hydroxycoumarin (28772-56-7)
Bromophenylmethane (100-39-0)
3-(p-Bromophenyl)-1-methoxy-1-methylurea (3060-89-7)
N'-(4-Bromophenyl)-n-methoxy-n-methylurea (3060-89-7)
o-Bromophenyl methyl ether (578-57-4)
p-Bromophenyl methyl ether (104-92-7)
p-Bromophenyl methyl ketone (99-90-1)
3-(p-Bromophenyl)-1-methyl-1-methoxyurea (3060-89-7)
4-Bromophenyl phenyl ether (101-55-3)
p-Bromophenyl phenyl ether (101-55-3)
1-Bromo-3-phenylpropane (637-59-2)

3-(p-Bromophenyl)-3-(2-pyridyl)-N,N-dimethylpropylamine (86-22-6)
(Z)-3-(4'-Bromophenyl)-3-(3''-pyridyl)dimethylallylamine (56775-88-3)
1-(p-Bromophenyl)-1-(2-pyridyl)-3-dimethylaminopropane (86-22-6)
3-Bromophenyl radical (2973-44-6)
p-Bromophenylsulfonyl chloride (98-58-8)
Bromophos-Ethyl (4824-78-6)
Bromophos (2104-96-3)
1-Bromo-2-phthalimidoethane (574-98-1)
1-Bromopropane (DOT) (106-94-5)
2-Bromopropane [UN 2344] (75-26-3)
3-Bromo-1,2-propanediol (4704-77-2)
1-Bromo-2-propanol (19686-73-8)
3-Bromo-1-propanol (627-18-9)
3-Bromopropanol (627-18-9)
1-Bromo-2-propanone (598-31-2)
Bromo-2-propanone (598-31-2)
2-Bromopropanoyl bromide (563-76-8)
1-Bromo-2-propene (106-95-6)
2-Bromopropene (557-93-7)
3-Bromopropene (106-95-6)
3-Bromopropionic acid (590-92-1)
α-Bromopropionic acid (598-72-1)
β-Bromopropionic acid (590-92-1)
Bromopropylate (18181-80-1)
(3-Bromopropyl)benzene (637-59-2)
3-Bromopropyl chloride (109-70-6)
2-Bromopropylene (557-93-7)
3-Bromopropylene (106-95-6)
1-Bromo-2-propyne (106-96-7)
3-Bromo-1-propyne (106-96-7)
3-Bromopropyne [UN 2345] (106-96-7)
2-Bromopyridine (109-04-6)
3-Bromopyridine (626-55-1)
4-Bromopyridine (1120-87-2)
7-Bromo-5-(2-pyridyl)-3H-1,4-benzodiazepin-2(1H)-one (1812-30-2)
5-Bromopyrimidine (4595-59-9)
Bromopyrine (603-65-6)
Bromoquinine (549-49-5)
3-Bromoquinoline (5332-24-1)
Bromosalicylanilide (87-12-7)
5-Bromosalicyl-4-bromoanilide (87-12-7)
5-Bromosalicylhydroxamic acid (5798-94-7)
5-Bromosalicylic acid p-bromoanilide (87-12-7)
Bromo seltzer (62-44-2)
Bromosilane (13465-73-1)
Bromosilano (Spanish) (13465-73-1)
2-Bromostearic acid (142-94-9)
α-Bromostearic acid (142-94-9)
β-Bromostyrene (103-64-0)
ω-Bromostyrene (103-64-0)
Bromostyrol (103-64-0)
Bromostyrolene (103-64-0)
Bromosulfalein (71-67-0)
Bromosulfophthalein (71-67-0)
Bromosulphalein (71-67-0)
Bromosulphthalein (71-67-0)
Bromotaleina (71-67-0)
1-Bromotetradecane (112-71-0)
8-Bromotheophylline (10381-75-6)
Bromotheophylline (10381-75-6)
2-Bromothiophene (1003-09-4)
3-Bromothiophene (872-31-1)
Bromothymol Blue (76-59-5)
2-Bromotoluene (95-46-5)
3-Bromotoluene (591-17-3)
5-Bromotoluene (591-17-3)
α-Bromotoluene (100-39-0)
m-Bromotoluene (591-17-3)
o-Bromotoluene (95-46-5)
ω-Bromotoluene (100-39-0)
p-Bromotoluene (106-38-7)
Bromotoluene, α [UN 1737] (100-39-0)
α-Bromo-p-toluic acid (6232-88-8)
α-Bromo-α-tolunitrile (5798-79-8)
Bromotrichlorodibenzo(b,e)(1,4)dioxin (107227-75-8)
Bromotrichlorodibenzo-p-dioxin (107227-75-8)
Bromotrichlorodibenzofuran (107227-56-5)
Bromotrichloromethane (75-62-7)
1-Bromotridecane (765-09-3)
Bromotrifluoroethene (598-73-2)
Bromotrifluoroethylene [UN 2419] (598-73-2)
Bromotrifluoromethane [UN 1009] (75-63-8)
3-Bromotrifluoromethylbenzene (401-78-5)
m-Bromo(trifluoromethyl)benzene (401-78-5)
1-Bromoundecane (693-67-4)
5-Bromo-uracil (51-20-7)
5-Bromovaleric acid (2067-33-6)
α-Bromovaleric acid (584-93-0)
Bromovur (2104-96-3)
Bromowodor (Polish) (10035-10-6)
2-Bromo-1,4-xylene (553-94-6)
2-Bromo-p-xylene (553-94-6)
Bromo-p-xylene (553-94-6)
α-Bromo-m-xylene (620-13-3)
α-Bromo-o-xylene (89-92-9)
α-Bromo-ortho-xylene (89-92-9)
α-Bromo-p-xylene (104-81-4)
α-Bromoxylene (35884-77-6)
ω-Bromo-m-xylene (620-13-3)
ω-Bromo-p-xylene (104-81-4)
α-Bromo-p-xylol (104-81-4)
Bromoxynil (1689-84-5)
Bromoxynil butyrate (3861-41-4)
Bromoxynil octanoate (1689-99-2)
Brompheniramine (86-22-6)
Brompheniraminum (Latin) (86-22-6)
Bromphenol Blue (115-39-9)
3-(4-Bromphenyl)-1-methoxyharnstoff (German) (3060-89-7)
Bromphthal (632-79-1)
Bromsalicylanilide (87-12-7)
β-Bromstyrol (103-64-0)
Bromstyrole (103-64-0)
Bromsulfalein (71-67-0)
Bromsulfan (71-67-0)
Bromsulfophthalein (71-67-0)
Bromsulfthalein (71-67-0)
Bromsulphalein (71-67-0)
Bromsulphthalein (71-67-0)
Bromthalein (71-67-0)
Bromthymol Blue (76-59-5)
Bromure de cyanogen (French) (506-68-3)
Bromure de methyle (French) (74-83-9)
Bromure de tetraethylammonium (French) (71-91-0)
Bromure de vinyle (French) (593-60-2)
Bromure de xylyle (French) (35884-77-6)
Bromure d'ethyle (74-96-4)
Bromuro de xililo (Spanish) (35884-77-6)
Bromuro di etile (Italian) (106-93-4)
Bromuro di metile (Italian) (74-83-9)
Bromuron (3408-97-7)
Bromwasserstoff (German) (10035-10-6)
Bronate (1689-99-2)
Broncholysin (616-91-1)
Bronkaid mist (51-43-4)

Bronner acid (93-00-5)
Bronner's acid (93-00-5)
Bronocot (52-51-7)
Bronopol (52-51-7)
Bronopolu (Polish) (52-51-7)
Bronopolum (Latin) (52-51-7)
Bronosol (52-51-7)
Bronox (1912-26-1)
Bronze Bromo (17372-87-1)
Bronze Bromo (548-26-5)
Bronze Green Toner A-8002 (569-64-2)
Bronze Powder (7440-50-8)
Bronze Red RO (5160-02-1)
Bronze Red 16913 Yellowish (5160-02-1)
Bronze Scarlet (5160-02-1)
Bronze Scarlet C (5160-02-1)
Bronze Scarlet CA (5160-02-1)
Bronze Scarlet CBA (5160-02-1)
Bronze Scarlet CT (5160-02-1)
Bronze Scarlet CTA (5160-02-1)
Bronze Scarlet Toner (5160-02-1)
5-Broom-3-isopropyl-6-methyl-uracil-(Dutch) (314-42-1)
Brookswax D (67762-27-0)
Brookswax R (67762-27-0)
Broom (Dutch) (7726-95-6)
O-(4-Broom-2,5-dichloor-fenyl)-O,O-dimethyl-monothiofosfaat (Dutch) (2104-96-3)
Broommethaan (Dutch) (74-83-9)
Broomwaterstof (Dutch) (10035-10-6)
Broprodifacoum (28772-56-7)
1-Brom-2-propin (Czech) (106-96-7)
Broserpine (50-55-5)
Brom-tetragnost (71-67-0)
Brotopon (52-86-8)
11460 Brown (1300-73-8)
Brown Copper Oxide (1317-39-1)
Brown 4EMBL (16071-86-6)
Brown M (2429-82-5)
Brown No. 201 (Japan) (1320-07-6)
Brown SK (2475-33-4)
Brown acetate (62-54-4)
Brox S-2 (9005-00-9)
Brox S-20 (9005-00-9)
Brox S-30 (9005-00-9)
Brox HLB-13 (9005-00-9)
Brox OL-4 (9004-98-2)
Brox OL-5 (9004-98-2)
Brox OL-10 (9004-98-2)
Brox OL-20 (9004-98-2)
Broxil (132-93-4)
Broxynil (1689-84-5)
Brucin (German) (357-57-3)
Brucina (Italian) (357-57-3)
(-)-Brucine (357-57-3)
Brucine (357-57-3)
Brucine, Solid (DOT) (357-57-3)
Brucine [UN 1570] (357-57-3)
Brucine alkaloid (357-57-3)
Brufaneuxol (58-15-1)
Brufanic (15687-27-1)
Brufen (15687-27-1)
Bruinsteen (Dutch) (1313-13-9)
Brulan (34014-18-1)
Brumolin (81-81-2)
Bruomophos (Russian) (2104-96-3)
Brush-Off 445 low volatile brush killer (93-76-5)
Brush-Rhap (94-75-7)
Brush Buster (1918-00-9)
Brush Killer 64 (1929-73-3)
Brush Rhap (93-76-5)
Brushtox (93-76-5)
Brygou (114-26-1)
Cn-B_{12} (68-19-9)
Bud-Nip (101-21-3)
Bualta (31512-74-0)
Bubartal (143-81-7)
Bubartal TT (143-81-7)
Buburone (15687-27-1)
Bucacid Azo Rubine (3567-69-9)
Bucacid Azure Blue (2650-18-2)
Bucacid Blue Black (1064-48-8)
Bucacid Brilliant Scarlet 3R (2611-82-7)
Bucacid Fast Crimson (3734-67-6)
Bucacid Fast Orange G (1936-15-8)
Bucacid Fast Red A (1658-56-6)
Bucacid Fast Wool Blue R (3861-73-2)
Bucacid Guinea Green BA (4680-78-8)
Bucacid Indigotine B (860-22-0)
Bucacid Metanil Yellow (587-98-4)
Bucacid Orange A (633-96-5)
Bucacid Patent Blue VF (129-17-9)
Bucacid Tartrazine (1934-21-0)
Bucarban (339-43-5)
Buckwheat LS (8001-21-6)
Bucolan Red GRE (6408-31-7)
Bucril (1689-84-5)
Bucrol (339-43-5)
Buctril (1689-84-5)
Buctril (1689-99-2)
Buctril Industrial (1689-84-5)
Budium RK 622 (9003-17-2)
Budoform (130-26-7)
Budorm (77-28-1)
Bueno (2163-80-6)
Bueno 6 (2163-80-6)
Bufapto Methalose (9004-67-5)
Bufa-4,20,22-trienolide, 6-(acetyloxy)-3-(β-D-glucopyranosyloxy)-8,14-dihydroxy-, (3β,6β)- (507-60-8)
Bufen (62-38-4)
Bufencarb (2282-34-0)
Bufencarb (8065-36-9)
Bufenina (Spanish) (447-41-6)
Buff-A-Comp (62-44-2)
Bufon (56-53-1)
Bufopto zinc sulfate (7733-02-0)
Bufotenin (487-93-4)
Bufotenine (487-93-4)
Bug Master (63-25-2)
Buhach (8003-34-7)
Bukarban (339-43-5)
Bulan (117-26-0)
Bulana (25014-41-9)
Bulbosan (2631-68-7)
Bulen A (9002-88-4)
Bulen A 30 (9002-88-4)
Bulkaloid (9004-67-5)
Bulpur (590-28-3)
Bumetrizol (Spanish) (3896-11-5)
Bumetrizole (3896-11-5)
Bumetrizolum (Latin) (3896-11-5)
Buminafos (51249-05-9)
Bumyr (110-36-1)
Bunt-Cure (118-74-1)
Bunt-No-More (118-74-1)

Name	CAS No.
Bunema	**(51026-28-9)**
Bunker "C"	**(68476-32-4)**
Bunsenite	**(1313-99-1)**
Buphenin	**(447-41-6)**
Buphenine	**(447-41-6)**
Bupheninum (Latin)	**(447-41-6)**
Buprenorfina (Spanish)	**(52485-79-7)**
Buprenorphine	**(52485-79-7)**
Buprenorphinum (Latin)	**(52485-79-7)**
Buprofezin	**(69327-76-0)**
Buprofezine	**(69327-76-0)**
Buro-Sol Concentrate	**(139-12-8)**
Burcol	**(339-43-5)**
Burese	**(50-36-2)**
Burex (Czech)	**(1698-60-8)**
Burgodin	**(15301-48-1)**
Burma Green B	**(569-64-2)**
Burma Yellow X 1622	**(2512-29-0)**
Burnish Gold	**(7440-57-5)**
Burnol	**(8048-52-0)**
Burnol	**(86-40-8)**
Burnt Alum	**(10043-67-1)**
Burnt Ammonium Alum	**(7784-25-0)**
Burnt Island Red	**(1309-37-1)**
Burnt Lime	**(1305-78-8)**
Burnt Sienna	**(1309-37-1)**
Burnt Umber	**(1309-37-1)**
Buroflavin	**(8048-52-0)**
Bursine	**(123-41-1)**
Burtolin	**(123-33-1)**
Burtonite V-40-E	**(9000-07-1)**
Burtonite V-7-E	**(9000-30-0)**
Busan 72A	**(21564-17-0)**
Busan 77	**(31512-74-0)**
Busan 90	**(2491-38-5)**
Busone	**(50-33-9)**
Bustren	**(9003-53-6)**
Bustren K 500	**(9003-53-6)**
Bustren K 525-19	**(9003-53-6)**
Bustren U 825	**(9003-53-6)**
Bustren U 825E11	**(9003-53-6)**
Bustren Y 3532	**(9003-53-6)**
Bustren Y 825	**(9003-53-6)**
Busulfan	**(55-98-1)**
Busulphan	**(55-98-1)**
Busulphane	**(55-98-1)**
Butabarb	**(125-40-6)**
Butabarbital	**(125-40-6)**
Butabarbital	**(143-81-7)**
Butabarbital sodium	**(143-81-7)**
Butabarbitone	**(125-40-6)**
Butabarbitone sodium	**(143-81-7)**
Butabarpal	**(143-81-7)**
Butabarpal sodium	**(143-81-7)**
Butacarb	**(2655-19-8)**
Butacarbe (French)	**(2655-19-8)**
Butachlor	**(23184-66-9)**
Butacide	**(51-03-6)**
Butacompren	**(50-33-9)**
Butacote	**(50-33-9)**
Buta-1,3-dieen (Dutch)	**(106-99-0)**
Butadieen (Dutch)	**(106-99-0)**
Butadien-Furfural Copolymer	**(126-15-8)**
Buta-1,3-dien (German)	**(106-99-0)**
Butadien (Polish)	**(106-99-0)**
Butadiendioxyd (German)	**(1464-53-5)**
Buta-1,3-diene	**(106-99-0)**
Butadiene	**(106-99-0)**
Butadiene	**(25339-57-5)**
α,γ-Butadiene	**(106-99-0)**
1,3-Butadiene (ACGIH)	**(106-99-0)**
1,2-Butadiene (9CI)	**(590-19-2)**
1,3-Butadiene, Homopolymer (9CI)	**(9003-17-2)**
Butadiene, Inhibited [UN 1010]	**(25339-57-5)**
Butadiene (OSHA)	**(106-99-0)**
1,3-Butadiene, Polymers	**(9003-17-2)**
1,3-Butadiene, 1-chloro	**(627-22-5)**
1,3-Butadiene, 2-chloro	**(126-99-8)**
1,2-Butadiene, 3-chloro- (9CI)	**(34581-41-4)**
1,2-Butadiene, 4-chloro- (8CI,9CI)	**(25790-55-0)**
1,2-Butadiene, 1-chloro- (6CI,8CI,9CI)	**(627-23-6)**
1,3-Butadiene, 2-chloro-, Polymers	**(9010-98-4)**
1,3-Butadiene-1,4-dicarboxylic acid	**(505-70-4)**
1,3-Butadiene, 2,3-dichloro	**(1653-19-6)**
1,3-Butadiene, dichloro	**(28577-62-0)**
1,2-Butadiene, 1,4-dichloro- (9CI)	**(83682-44-4)**
1,2-Butadiene, 4,4-dichloro- (9CI)	**(83682-41-1)**
1,3-Butadiene, 1,1-dichloro- (8CI,9CI)	**(6061-06-9)**
1,3-Butadiene, 1,4-dichloro- (6CI,7CI,8CI,9CI)	**(2984-42-1)**
1,3-Butadiene diepoxide	**(1464-53-5)**
Butadiene diepoxide	**(1464-53-5)**
l-Butadiene diepoxide	**(30031-64-2)**
Butadiene dimer	**(100-40-3)**
1,3-Butadiene, 2,3-dimethoxy- (6CI,7CI,8CI,9CI)	**(3588-31-6)**
1,3-Butadiene, 2,3-dimethyl- (9CI)	**(513-81-5)**
Butadiene dioxide	**(1464-53-5)**
dl-Butadiene dioxide	**(298-18-0)**
1,3-Butadiene, 1,4-diphenyl- (8CI)	**(886-65-7)**
1,1'-(1,3-Butadiene-1,4-diyl)bisbenzene	**(886-65-7)**
1,3-Butadiene, hexachloro	**(87-68-3)**
1,2-Butadiene, 1,1,3,4,4,4-hexachloro- (9CI)	**(56827-79-3)**
Butadiene homopolymer	**(9003-17-2)**
1,3-Butadiene, 2-methyl-	**(78-79-5)**
1,2-Butadiene, 3-methyl- (9CI)	**(598-25-4)**
1,3-Butadiene, 2-methyl-, Homopolymer (9CI)	**(9003-31-0)**
1,3-Butadiene, 2-methyl-, Polymer with 2-methyl-1-propene (9CI)	**(9010-85-9)**
1,3-Butadiene, 2-methyl-, Polymer with 2-methyl-1-propene, brominated	**(68441-14-5)**
Butadiene monoxide	**(930-22-3)**
Butadiene oligomer	**(9003-17-2)**
1,3-Butadiene, pentachloro- (9CI)	**(55880-77-8)**
1,3-Butadiene, 1,1,2,3,4-pentachloro-4-(1-methylethoxy)- (9CI)	**(68334-67-8)**
Butadiene polymer	**(9003-17-2)**
Butadiene resin	**(9003-17-2)**
Butadiene-styrene Resin	**(9003-55-8)**
1,3-Butadiene-styrene copolymer	**(9003-55-8)**
1,3-Butadiene-styrene polymer	**(9003-55-8)**
Butadiene-styrene polymer	**(9003-55-8)**
Butadiene sulfone	**(77-79-2)**
1,3-Butadiene, 1,1,2,3-tetrachloro	**(921-09-5)**
1,3-Butadiene, tetrachloro-	**(58334-79-5)**
1,3-Butadiene, 1,1,4,4-tetrachloro- (9CI)	**(36038-53-6)**
1,3-Butadiene, 1,2,3,4-tetrachloro- (8CI,9CI)	**(1637-31-6)**
1,3-Butadiene-1,2,4-tricarboxylic acid (8CI,9CI)	**(2547-45-7)**
1,3-Butadiene, 1,1,2-trichloro	**(2852-07-5)**
1,2-Butadiene, 1,1,4-trichloro- (9CI)	**(58679-08-6)**
1,2-Butadiene, 4,4,4-trichloro- (9CI)	**(34819-62-0)**
1,3-Butadiene, 1,2,3-trichloro-, (E)- (9CI)	**(53978-04-4)**
1,3-Butadiene, trichloro- (9CI)	**(53317-48-9)**
1,3-Butadiene, 1,2,3-trichloro- (8CI,9CI)	**(1573-58-6)**
1,3-Butadiene, 1,2,3-trichloro-, (Z)- (9CI)	**(39083-26-6)**
1,3-Butadiene, 2-(4,8,12-trimethyltridecyl)-	**(504-96-1)**
Butadion	**(50-33-9)**
Butadiona	**(50-33-9)**

2,3-Butadione **(431-03-8)**
Butadione **(431-03-8)**
Butadione **(50-33-9)**
1,3-Butadiyne (9CI) **(460-12-8)**
Butaflogin **(129-20-4)**
Butafume **(13952-84-6)**
Butagesic **(50-33-9)**
Butak **(143-81-7)**
Butakon 85-71 **(9003-55-8)**
Butal **(123-72-8)**
Butalan **(50-33-9)**
Butalbarbital **(77-26-9)**
Butalbital **(115-44-6)**
Butalbital **(77-26-9)**
Butalbitale **(77-26-9)**
Butalbitalum (Latin) **(77-26-9)**
Butaldehyde **(123-72-8)**
Butalgina **(50-33-9)**
Butalidon **(50-33-9)**
Butalin **(33629-47-9)**
Butaluy **(50-33-9)**
Butalyde **(123-72-8)**
Butamben **(94-25-7)**
Butamid **(64-77-7)**
Butamifos **(36335-67-8)**
Butanal **(123-72-8)**
n-Butanal (Czech) **(123-72-8)**
Butanal, 2-hydroxy-3-methyl- (9CI) **(67755-97-9)**
1-Butanal, 3-methyl- **(590-86-3)**
Butanal, 4-(N-methyl-N-nitrosamino)-4-(3-pyridyl)- **(64091-90-3)**
Butanal oxime (9CI) **(110-69-0)**
Butanamide (9CI) **(541-35-5)**
Butanamide, N-acetyl- (9CI) **(22534-71-0)**
Butanamide, N-(4-amino-2-hydroxyphenyl)-2,2,3,3,4,4,4-hepta-fluoro- (9CI) **(847-51-8)**
Butanamide, 2-((5-(aminosulfonyl)-2-hydroxyphenyl)azo)-3-oxo-N-phenyl- (9CI) **(21811-92-7)**
Butanamide, N-(4-chloro-2,5-dimethoxyphenyl)-2-((2,5-dimethoxy-4-((phenylamino)sulfonyl)phenyl)azo)-3-oxo- (9CI) **(12225-18-2)**
Butanamide, N-(4-chloro-2,5-dimethoxyphenyl)-3-oxo- (9CI) **(4433-79-8)**
Butanamide, 2-chloro-N,N-dimethyl-3-oxo- (9CI) **(5810-11-7)**
Butanamide, N-(1-(2-chloro-4,6-dimethylphenyl)-4,5-dihydro-5-oxo-1H-pyrazol-3-yl)-2-(3-pentadecylphenoxy)- (9CI) **(33956-01-3)**
Butanamide, 2-chloro-N-methyl-3-oxo- (9CI) **(4116-10-3)**
Butanamide, N-(4-chloro-2-methylphenyl)-2-((4-chloro-2-nitro-phenyl)azo)-3-oxo- (9CI) **(32432-45-4)**
Butanamide, N-(4-chloro-2-methylphenyl)-3-oxo- **(20139-55-3)**
Butanamide, 2-((4-chloro-2-nitrophenyl)azo)-N-(2-chlorophenyl)-3-oxo- (9CI) **(6486-23-3)**
Butanamide, 2-((4-chloro-2-nitrophenyl)azo)-N-(2,3-dihydro-2-oxo-1H-benzimidazol-5-yl)-3-oxo- (9CI) **(12236-62-3)**
Butanamide, 2-((4-chloro-2-nitrophenyl)azo)-N-(2-methoxyphenyl)-3-oxo- (9CI) **(13515-40-7)**
Butanamide, N-(4-chlorophenyl)-3-oxo- (9CI) **(101-92-8)**
Butanamide, 2,2'-((3,3'-dichloro(1,1'-biphenyl)-4,4'-diyl)bis(azo))-bis(N-(4-chloro-2,5-dimethoxyphenyl)-3-oxo- (9CI) **(5567-15-7)**
Butanamide, 2,2'-((3,3'-dichloro(1,1'-biphenyl)-4,4'-diyl)bis(azo))-bis(N-(2,4-dimethylphenyl)-3-oxo- (9CI) **(5102-83-0)**
Butanamide, 2,2'-((3,3'-dichloro(1,1'-biphenyl)-4,4'-diyl)bis(azo))-bis(N-(4-methylphenyl)-3-oxo- (9CI) **(6358-37-8)**
Butanamide, 2,2'-((3,3'-dichloro(1,1'-biphenyl)-4,4'-diyl)bis(azo))-bis(N-(2-methylphenyl)-3-oxo- (9CI) **(5468-75-7)**
Butanamide, 2,2'-((3,3'-dichloro(1,1'-bipheny{l)-4,4'-diyl)bis(azo))-bis(N-(2-methoxyphenyl)-3-oxo- (9CI) **(4531-49-1)**
Butanamide, 2,2'-((3,3'-dichloro(1,1'-diphenyl)-4,4'-diyl)bis(azo))-bis(3-oxo-N-phenyl- (9CI) **(6358-85-6)**
Butanamide, N,N-diethyl-3-oxo- (9CI) **(2235-46-3)**
Butanamide, N-(2,3-dihydro-2-oxo-1H-benzimidazol-5-yl)-3-oxo-2-(phenylazo)- (9CI) **(51083-28-4)**
Butanamide, 2,4-dihydroxy-N-(3-hydroxypropyl)-3,3-di-methyl-, (+-)- (9CI) **(16485-10-2)**
Butanamide, 2,4-dihydroxy-N-(3-hydroxypropyl)-3,3-di-methyl-, (R)- (9CI) **(81-13-0)**
Butanamide, 2,2'-((3,3'-dimethoxy(1,1'-biphenyl)-4,4'-diyl)bis-(azo))bis(N-(2-methylphenyl)-3-oxo- (9CI) **(7147-42-4)**
Butanamide, 2,2'-((3,3'-dimethoxy(1,1'-biphenyl)-4,4'-diyl)bis-(azo))bis(3-oxo-N-phenyl- (9CI) **(6505-28-8)**
Butanamide, N-(2,5-dimethoxyphenyl)-3-oxo- (9CI) **(6375-27-5)**
Butanamide, N,N'-(3,3'-dimethyl(1,1'-biphenyl)-4,4'-diyl)bis-(2-((2,4-dichlorophenyl)azo)-3-oxo- (9CI) **(5979-28-2)**
Butanamide, N,N'-(3,3'-dimethyl(1,1'-biphenyl)-4,4'-diyl)-bis(3-oxo- (9CI) **(91-96-3)**
Butanamide, N,N-dimethyl-3-oxo- (9CI) **(2044-64-6)**
Butanamide, N-(2,6-dimethylphenyl)-2-(ethylpropylamino)-, (+-) **(36637-18-0)**
Butanamide, N-(2,4-dimethylphenyl)-3-oxo- (9CI) **(97-36-9)**
Butanamide, N-(4-ethoxyphenyl)-3-oxo- (9CI) **(122-82-7)**
Butanamide, 2,2,3,3,4,4,4-heptafluoro-N-(2-hydroxy-4-nitro-phenyl)- (9CI) **(2712-83-6)**
Butanamide, N-hexyl- (9CI) **(10264-17-2)**
Butanamide, 4-hydroxy- (9CI) **(927-60-6)**
Butanamide, 2-((2-methoxy-4-nitrophenyl)azo)-N-(2-methoxy-phenyl)-3-oxo- (9CI) **(6358-31-2)**
Butanamide, 2-((4-methoxy-2-nitrophenyl)azo)-N-(2-methoxy-phenyl)-3-oxo- (9CI) **(6528-34-3)**
Butanamide, N-(2-methoxyphenyl)-3-oxo- (9CI) **(92-15-9)**
Butanamide, 3-methyl- (9CI) **(541-46-8)**
Butanamide, 2-((4-methyl-2-nitrophenyl)azo)-3-oxo-N-phenyl-(9CI) **(2512-29-0)**
Butanamide, N-methyl-3-oxo- (9CI) **(20306-75-6)**
Butanamide, N-(2-methylphenyl)-3-oxo- (9CI) **(93-68-5)**
Butanamide, N-(4-methylphenyl)-3-oxo- (9CI) **(2415-85-2)**
Butanamide, 3-oxo-N-phenyl- (9CI) **(102-01-2)**
Butanamide, N-phenyl- (9CI) **(1129-50-6)**
Butanamide, 2,2'-((2,2',5,5'-tetrachloro(1,1'-biphenyl)-4,4'-diyl)bis(azo))bis(N-(2,4-dimethylphenyl)-3-oxo- (9CI) **(22094-93-5)**
1-Butanamine **(109-73-9)**
2-Butanamine **(13952-84-6)**
1-Butanamine, 4-(9-anthracenyloxy)- (9CI) **(96334-91-7)**
1-Butanamine, N,N-bis(3-aminopropyl)- **(1555-68-6)**
1-Butanamine, n-butyl- **(111-92-2)**
1-Butanamine, N-butyl-N-chloro- (9CI) **(999-33-7)**
1-Butanamine, n-butyl-N-nitroso **(924-16-3)**
1-Butanamine, N,N-diethyl- (9CI) **(4444-68-2)**
1-Butanamine, N,N-dimethyl- (9CI) **(927-62-8)**
2-Butanamine, 3,3-dimethyl- (9CI) **(3850-30-4)**
1-Butanamine, 2-ethyl- **(617-79-8)**
Butanamine, N-ethyl-N-nitroso- **(4549-44-4)**
t-Butanamine, N-ethyl-N-nitroso- **(3398-69-4)**
1-Butanamine, hydrochloride (9CI) **(3858-78-4)**
1-Butanamine, 2-methyl- (9CI) **(96-15-1)**
1-Butanamine, 3-methyl- (9CI) **(107-85-7)**
1-Butanamine, 3-methyl-N-(3-methylbutyl)- (9CI) **(544-00-3)**
Butanamine, N-methyl-N-nitroso- **(7068-83-9)**
1-Butanamine, N-(2-methylpropyl)- (9CI) **(20810-06-4)**
2-Butanamine, N-(1-methylpropyl)- (9CI) **(626-23-3)**
Butanamine, N-nitroso- **(56375-33-8)**
1-Butanamine, 1,1,2,2,3,3,4,4,4-nonafluoro-N,N-bis(nonafluoro-butyl)- (9CI) **(311-89-7)**
1-Butanaminium, N,N,N-tributyl-, Salt with 4-nitrophenol (1:1) (9CI) **(3002-48-0)**
1-Butanaminium, N,N,N-tributyl-, bromide (9CI) **(1643-19-2)**
1-Butanaminium, N,N,N-tributyl-, sulfate (1:1) (9CI) **(32503-27-8)**
1,3-Butandiol (German) **(107-88-0)**
n-Butane **(106-97-8)**

Butane (ACGIH,OSHA) [UN 1011] **(106-97-8)**
Butane, D-1,2:3,4-diepoxy- **(30419-67-1)**
Butaneamide, N-(2-chlorophenyl)-3-oxo- **(93-70-9)**
Butane, 1,1-bis(p-chlorophenyl)-2-nitro **(117-26-0)**
Butane, 1,4-bis(3,4-dihydroxyphenyl)-2,3-dimethyl- **(500-38-9)**
Butane, 1,4-bis(2,3-epoxypropoxy) **(2425-79-8)**
Butane, 1,1-bis(p-ethoxyphenyl)-2-nitro **(26258-71-9)**
Butane, 2,2-bis(ethylsulfonyl) **(76-20-0)**
Butaneboronic acid **(4426-47-5)**
Butane, 1-bromo **(109-65-9)**
Butane, 2-bromo **(78-76-2)**
Butane, 1-bromo-3-methyl **(107-82-4)**
Butane, 1-bromo-2-methyl- (9CI) **(10422-35-2)**
1-Butanecarboxylic acid **(109-52-4)**
Butanecarboxylic acid **(109-52-4)**
Butane, 1-chloro **(109-69-3)**
Butane, 2-chloro **(78-86-4)**
Butane, 1-(2-(2-chloroethoxy)ethoxy)- **(1120-23-6)**
Butane, 1-chloro-4-fluoro **(462-73-7)**
Butane, 1-chloro-2-methyl- (9CI) **(616-13-7)**
Butane, 1-chloro-3-methyl- (9CI) **(107-84-6)**
Butane, 2-chloro-2-methyl- (8CI,9CI) **(594-36-5)**
Butane, 2-chloro-3-nitroso-, dimer (8CI) **(6865-97-0)**
Butane, chlorooctafluoro- (9CI) **(71342-62-6)**
Butane, decachloro- (6CI,7CI,8CI,9CI) **(6820-74-2)**
Butane, decafluoro- (9CI) **(355-25-9)**
Butanediamide (9CI) **(110-14-5)**
1,4-Butanediamine **(110-60-1)**
1,3-Butanediamine (8CI,9CI) **(590-88-5)**
1,4-Butanediamine, N-(3-aminopropyl) **(124-20-9)**
1,4-Butanediamine, N,N'-bis(3-aminopropyl) **(71-44-3)**
1,4-Butanediamine, N-cyclohexyl- (9CI) **(79419-72-0)**
1,4-Butanediamine, N-cyclopentyl- (9CI) **(90853-11-5)**
1,3-Butanediamine, N,N,N',N'-tetramethyl **(97-84-7)**
Butane, 1,4-dibromo **(110-52-1)**
Butane, 1,2-dibromo- (9CI) **(533-98-2)**
Butane, 1,3-dibromo- (9CI) **(107-80-2)**
Butane, 2,3-dibromo- (9CI) **(5408-86-6)**
Butane, 1,4-dibromo-1,1,2,2-tetrafluoro- (9CI) **(18599-20-7)**
Butane, 1,1-dibutoxy- (9CI) **(5921-80-2)**
1,4-Butanedicarboxamide **(628-94-4)**
1,1-Butanedicarboxylic acid **(616-62-6)**
1,4-Butanedicarboxylic acid **(124-04-9)**
1,4-Butanedicarboxylic acid disodium salt **(7486-38-6)**
Butane, 1,1-dichloro **(541-33-3)**
Butane, 1,2-dichloro- (9CI) **(616-21-7)**
Butane, 1,3-dichloro- (9CI) **(1190-22-3)**
Butane, 1,4-dichloro- (9CI) **(110-56-5)**
Butane, 2,2-dichloro- (9CI) **(4279-22-5)**
Butane, 2,3-dichloro- (9CI) **(7581-97-7)**
Butane, 1,4-dichloro-2,3-epoxy **(3583-47-9)**
Butane, 1,3-dichloro-2-methyl- (8CI,9CI) **(23010-07-3)**
Butane, 2,2-dichloro-3-methyl- (8CI,9CI) **(17773-66-9)**
Butane, 2,3-dichloro-2-methyl- (8CI,9CI) **(507-45-9)**
Butane, 1,4-dicyclohexyl- (6CI,7CI,8CI) **(6165-44-2)**
Butane diepoxide **(1464-53-5)**
Butane, (+-)-1,2:3,4-diepoxy **(298-18-0)**
Butane, 1,2:3,4-diepoxy **(1464-53-5)**
Butane, 1,2:3,4-diepoxy-, meso **(564-00-1)**
Butane, 1,1-diethoxy-3-methyl- (9CI) **(3842-03-3)**
1,4-Butane diglycidyl ether **(2425-79-8)**
Butane, 2,2-dimethyl **(75-83-2)**
Butane, 2,3-dimethyl **(79-29-8)**
Butane, 2-(1,1-dimethylethoxy)- (9CI) **(32970-45-9)**
Butanedinitrile **(110-61-2)**
Butanedinitrile, ethyl- (9CI) **(17611-82-4)**
Butanedinitrile, hydroxy- (9CI) **(4341-85-9)**
Butanedinitrile, tetramethyl- (9CI) **(3333-52-6)**
Butanedioic acid **(110-15-6)**
Butanedioic acid, Compd. with N,N-dimethyl-2-(1-phenyl-1-(2-pyridinyl)ethoxy)ethanamine (1:1), Mixt. with 2-(diethylamino)ethyl(1,1'-bicyclohexyl)-1-carboxylate hydrochloride and 5- hydroxy-6-methyl-3,4-pyridinedimethanol hydrochloride **(8064-77-5)**
Butanedioic acid, Polymer with 1,2-ethanediol (9CI) **(25569-53-3)**
Butanedioic acid, acetyl-, diethyl ester (9CI) **(1115-30-6)**
Butanedioic acid, acetyl-, dimethyl ester (9CI) **(10420-33-4)**
Butanedioic acid, ((bis((2-ethylhexyl)oxy)phosphinothioyl)-thio)-, dibutyl ester (9CI) **(68413-48-9)**
Butanedioic acid, bis(phenylmethyl) ester (9CI) **(103-43-5)**
Butanedioic acid, (carboxymethoxy)- (9CI) **(38945-27-6)**
Butanedioic acid, (carboxymethoxy)-, trisodium salt (9CI) **(34128-01-3)**
Butanedioic acid, chloro- (9CI) **(16045-92-4)**
Butanedioic acid, cobalt(2+) salt (1:1) **(3267-76-3)**
Butanedioic acid dibutyl ester **(141-03-7)**
Butanedioic acid, 2,2-dichloro- (9CI) **(15519-38-7)**
Butanedioic acid, diethyl ester **(123-25-1)**
Butanedioic acid, 2,3-dihydroxy- **(87-69-4)**
Butanedioic acid, 2,3-dihydroxy-, (R*,R*)-(+-)- (9CI) **(133-37-9)**
Butanedioic acid, 2,3-dihydroxy-, (S-(R*,R*))- (9CI) **(147-71-7)**
Butanedioic acid, 2,3-dihydroxy- (R-(R*,R*))-, ammonium salt (9CI) **(14307-43-8)**
Butanedioic acid, 2,3-dihydroxy- (R-(R*,R*))-, copper(2+) salt (1:1) (9CI) **(815-82-7)**
Butanedioic acid, 2,3-dihydroxy-, (R-(R*,R*))-, diammonium salt **(3164-29-2)**
Butanedioic acid, 2,3-dihydroxy- (R-(R*,R*))-, dibutyl ester (9CI) **(87-92-3)**
Butanedioic acid, 2,3-dihydroxy- (R-(R*,R*))-, dimethyl ester (9CI) **(608-68-4)**
Butanedioic acid, 2,3-dihydroxy-(R-(R*,R*))-, disodium salt (9CI) **(868-18-8)**
Butanedioic acid, 2,3-dihydroxy- (R-(R*,R*))-, lead(2+) salt (1:1) (9CI) **(815-84-9)**
Butanedioic acid, 2,3-dihydroxy- (R-(R*,R*))-, monopotassium monosodium salt (9CI) **(304-59-6)**
Butanedioic acid, 2,3-dihydroxy-, (R-(R*,R*))-, monopotassium salt (9CI) **(868-14-4)**
Butanedioic acid, 2,3-dihydroxy-, (R-(R*,R*))-, monosodium salt **(526-94-3)**
Butanedioic acid, 2,3-dihydroxy- (R-(R*,R*))-, strontium salt (1:1) (9CI) **(868-19-9)**
Butanedioic acid, 2,3-dihydroxy-, dimethyl ester, (R*,S*)- (9CI) **(5057-96-5)**
Butanedioic acid, 2,3-dihydroxy-, monopotassium monosodium salt **(304-59-6)**
Butanedioic acid, ((dimethoxyphosphinothioyl)thio)- **(1190-28-9)**
Butanedioic acid, ((dimethoxyphosphinothioyl)thio)-, monoethyl ester (9CI) **(35884-76-5)**
Butanedioic acid, ((dimethoxyphosphinyl)thio)-, diethyl ester (9CI) **(1634-78-2)**
Butanedioic acid, 2,2-dimethyl- (9CI) **(597-43-3)**
Butanedioic acid, dimethyl ester (9CI) **(106-65-0)**
Butanedioic acid, dodecenyl- (9CI) **(29658-97-7)**
Butanedioic acid, ((dodecyloxy)sulfonyl)-, disodium salt **(26838-05-1)**
Butanedioic acid, ((dodecyloxy)sulfonyl)-, disodium salt (9CI) **(36409-57-1)**
Butanedioic acid, (ethoxymethylene)oxo-, diethyl ester **(52942-64-0)**
Butanedioic acid, hydroxy- (9CI) **(6915-15-7)**
Butanedioic acid, hydroxy-, (+-)- (9CI) **(617-48-1)**
Butanedioic acid, hydroxy-, diethyl ester **(7554-12-3)**
Butanedioic acid, ((methoxy(methylthio)phosphinothioyl)thio)-diethyl ester (9CI) **(3344-12-5)**
Butanedioic acid, methyl- (9CI) **(498-21-5)**
Butanedioic acid, methyl-, dimethyl ester **(1604-11-1)**
Butanedioic acid, methylene- (9CI) **(97-65-4)**
Butanedioic acid, mono(3,4-dihydro-2,5,7,8-tetramethyl-2-(4,8,12-trimethyltridecyl)-2H-1-benzopyran-6-yl) ester **(4345-03-3)**
Butanedioic acid, mono(3,4-dihydro-2,5,7,8-tetramethyl-2-(4,8,12-trimethyltridecyl)-2H-1-benzopyran-6-yl) ester, (2R-(2R*(4R*,8R*)))- (9CI) **(4345-03-3)**
Butanedioic acid mono(2,2-dimethylhydrazide) **(1596-84-5)**
Butanedioic acid, monomethyl ester (9CI) **(3878-55-5)**

Butanedioic acid, 4-(octadecylamino)-4-oxo-2-sulfo-, disodium salt **(14481-60-8)**
Butanedioic acid, oxo- (9CI) **(328-42-7)**
Butanedioic acid, 2,2'-oxybis- (9CI) **(7408-18-6)**
Butanedioic acid, sodium salt (9CI) **(14047-56-4)**
Butanedioic acid, sulfo- (9CI) **(5138-18-1)**
Butanedioic acid, sulfo-, C-dodecyl ester, disodium salt **(36409-57-1)**
Butanedioic acid, sulfo-, C-dodecyl ester, disodium salt (9CI) **(26838-05-1)**
Butanedioic acid, sulfo-, C-isodecyl ester, disodium salt (9CI) **(37294-49-8)**
Butanedioic acid, sulfo-, 1,4-bis(1,3-dimethylbutyl) ester, sodium salt (9CI) **(2373-38-8)**
Butanedioic acid, sulfo-, 1,4-bis(2-ethylhexyl) ester, sodium salt (9CI) **(577-11-7)**
Butanedioic acid, sulfo-, 1,4-bis(1-methylheptyl) ester, sodium salt (9CI) **(20727-33-7)**
Butanedioic acid, sulfo-, 1,4-bis(1-methylpentyl) ester, sodium salt (9CI) **(6001-97-4)**
Butanedioic acid, sulfo-, 1,4-bis(2-methylpropyl) ester, sodium salt **(127-39-9)**
Butanedioic acid, sulfo-, 1,4-dicyclohexyl ester, sodium salt (9CI) **(23386-52-9)**
Butanedioic acid, sulfo-, 1,4-didodecyl ester, sodium salt (9CI) **(4229-35-0)**
Butanedioic acid, sulfo-, 1,4-dihexyl ester, sodium salt **(3006-15-3)**
Butanedioic acid, sulfo-, 1,4-diisodecyl ester, sodium salt (9CI) **(29857-13-4)**
Butanedioic acid, sulfo-, 1,4-dioctyl ester (9CI) **(2373-23-1)**
Butanedioic acid, sulfo-, 1,4-dioctyl ester, sodium salt (9CI) **(1639-66-3)**
Butanedioic acid, sulfo-, 1,4-dipentyl ester, sodium salt **(922-80-5)**
Butanedioic acid, sulfo-, 1,4-ditridecyl ester, sodium salt (9CI) **(2673-22-5)**
Butanedioic acid, sulfo-, 4-isodecyl ester, disodium salt **(37294-49-8)**
Butanedioic acid, sulfo-, 4-(1-methyl-2-((1-oxo-9-octadecenyl)-amino)ethyl) ester, disodium salt (9CI) **(67815-88-7)**
Butanedioic acid, sulfo-, monooctadecyl ester, disodium salt **(14481-60-8)**
Butanedioic acid, sulfo-, trisodium salt (9CI) **(13419-59-5)**
Butanedioic acid, tetrahydroxy- (9CI) **(76-30-2)**
Butanedioic acid, (tetrapropenyl)- (9CI) **(27859-58-1)**
Butanedioic anhydride **(108-30-5)**
1,2-Butanediol **(584-03-2)**
1,3-Butanediol **(107-88-0)**
1,4-Butanediol **(110-63-4)**
2,3-Butanediol **(513-85-9)**
Butane-1,3-diol **(107-88-0)**
Butane-1,4-diol **(110-63-4)**
Butanediol **(110-63-4)**
2,3-Butanediol, (+-)- (8CI) **(6982-25-8)**
Butanediol (8CI,9CI) **(25265-75-2)**
2,3-Butanediol, L- (8CI) **(19132-06-0)**
2,3-Butanediol, (2R,3R)-(-)- (8CI) **(24347-58-8)**
2,3-Butanediol, (R*,R*)-(+-)- (9CI) **(6982-25-8)**
2,3-Butanediol, (R-(R*,R*))- (9CI) **(24347-58-8)**
2,3-Butanediol, (S-(R*,R*))- (9CI) **(19132-06-0)**
2,3-Butanediol, (R*,S*)- (9CI) **(5341-95-7)**
1,4-Butanediol, 1,4-benzenedicarboxylic acid, α-hydro-ω-hydroxy-poly(oxy-1,2-ethanediyl) polymer **(26062-94-2)**
1,3-Butanediol, cyclic ester with boric acid (H_3BO_3), 1-methyl-trimethylene ester (8CI) **(2665-13-6)**
1,3-Butanediol diacrylate **(19485-03-1)**
1,4-Butanediol diglycidyl ether **(2425-79-8)**
Butane-1:4-diol diglycidyl ether **(2425-79-8)**
Butanediol diglycidyl ether **(2425-79-8)**
1,4-Butanediol dimethacrylate **(2082-81-7)**
Butanediol dimethacrylate **(2082-81-7)**
1,4-Butanediol dimethanesulfonate **(55-98-1)**
1,4-Butanediol, dimethanesulfonate **(55-98-1)**
1,4-Butanediol dimethanesulphonate **(55-98-1)**
2,3-Butanediol, 2,3-dimethyl **(76-09-5)**
2,3-Butanediol, 2,3-dimethyl-, cyclic carbonate (8CI) **(19424-29-4)**
2,3-Butanediol, meso- (8CI) **(5341-95-7)**
1,4-Butanediol-oxalic acid copolymer **(34090-00-1)**
1,4-Butanediol-oxalic acid polymer **(34090-00-1)**
1,2-Butanediol, 1-phenyl- (6CI,8CI) **(22607-13-2)**
1,3-Butanediol, Polymer with α-butyl-ω-hydroxy-poly(oxy(methyl-1,2-ethanediyl)) and 1,3-diisocyanato-methylbenzene (9CI) **(68400-67-9)**
1,4-Butanediol, Polymer with ethanedioic acid (9CI) **(34090-00-1)**
1,4-Butanediol-terephthalic acid copolymer **(26062-94-2)**
Butanediol-terephthalic acid copolymer **(26062-94-2)**
1,4-Butanediol-terephthalic acid polymer **(26062-94-2)**
1,1-Butanediol, 2,2,3-trichloro- (8CI,9CI) **(76-40-4)**
2,3-Butanedione [UN 2346] **(431-03-8)**
2,3-Butanedione, dioxime **(95-45-4)**
2,3-Butanedione, monooxime **(57-71-6)**
2,3-Butanedione 2-oxime **(57-71-6)**
1,3-Butanedione, 1-phenyl **(93-91-4)**
1,3-Butanedione, 4,4,4-trifluoro-1-(2-thienyl)- (9CI) **(326-91-0)**
Butane, 1,3-diphenyl- **(1520-44-1)**
1,4-Butanediyl 2-methyl-2-propenoate **(2082-81-7)**
Butane, 1,2-epoxy **(106-88-7)**
Butane, 1,3-epoxy **(2167-39-7)**
Butane, 1,4-epoxy- **(109-99-9)**
Butane, 2,3-epoxy **(3266-23-7)**
Butane, 2,3-epoxy-, cis- (8CI) **(1758-33-4)**
Butane, 2,3-epoxy-, trans- (8CI) **(21490-63-1)**
Butane, 2,3-epoxy-2,3-dimethyl- (8CI) **(5076-20-0)**
Butane, 2,3-epoxy-2-methyl **(5076-19-7)**
Butane, 1,2-epoxy-4,4,4-trichloro **(3083-25-8)**
Butane, 1,1'-(1,2-ethanediylbis(oxy))bis- **(112-48-1)**
Butane, 1-(ethenyloxy)- **(111-34-2)**
Butane, 1-(2-(ethenyloxy)ethoxy)- **(4223-11-4)**
Butane, 2-ethoxy- (9CI) **(2679-87-0)**
Butane, 1-(2-(2-ethoxyethoxy)ethoxy)- (9CI) **(3895-17-8)**
Butane, 2-ethoxy-2-methyl- (9CI) **(919-94-8)**
Butane, 1-(ethylthio)- (9CI) **(638-46-0)**
Butane, 1-fluoro- (9CI) **(2366-52-1)**
Butane, 1,1,1,2,2,3,3-heptachloro- (9CI) **(83682-70-6)**
Butane, 1,1,2,2,3,4,4-heptachloro- (9CI) **(34973-41-6)**
Butane, 1,1,1,4,4,4-hexachloro- (9CI) **(79458-54-1)**
Butane, 1,1,2,2,3,3-hexachloro- (9CI) **(83682-69-3)**
Butane, 1,1,2,2,3,4-hexachloro- (6CI,7CI,9CI) **(2431-55-2)**
Butane, 1,2,2,3,3,4-hexachloro- (6CI,7CI,8CI,9CI) **(1573-57-5)**
Butane, 1-iodo **(542-69-8)**
Butane, 2-iodo **(513-48-4)**
Butane, iodo- (8CI,9CI) **(25267-27-0)**
Butane, 2-iodo-2-methyl- (8CI,9CI) **(594-38-7)**
Butane, l-1,2:3,4-diepoxy **(30031-64-2)**
Butane, 1-methoxy- (9CI) **(628-28-4)**
Butane, 2-methoxy- (9CI) **(6795-87-5)**
Butane, 2-methoxy-2-methyl- (9CI) **(994-05-8)**
Butane, 2-methyl **(78-78-4)**
Butane, 1,1'-(methylenebis(oxy))bis- (9CI) **(2568-90-3)**
Butane, 2,2'-(methylenebis(oxy))bis- (9CI) **(2568-92-5)**
Butane, 2-(1-methylethoxy)- (9CI) **(18641-81-1)**
Butane, 1-(methylthio)- (9CI) **(628-29-5)**
Butanen (Dutch) **(106-97-8)**
Butanenitrile **(109-74-0)**
n-Butanenitrile **(109-74-0)**
Butanenitrile, 2-amino-2,3-dimethyl- (9CI) **(13893-53-3)**
Butanenitrile, 2,2'-azobis(2-methyl- (9CI) **(13472-08-7)**
Butanenitrile, 2-hydroxy-4-(methylthio)- (9CI) **(17773-41-0)**
Butanenitrile, 2-methyl- (9CI) **(18936-17-9)**
Butane, 1-nitro **(627-05-4)**
Butane, 2-nitro **(600-24-8)**
Butane, 1,1,1,2,2,3,3,4,4-nonachloro- (8CI) **(21483-62-5)**
Butane, 1,1,2,2,3,3,4,4,4-nonachloro- (6CI) **(21483-62-5)**
Butane, 1,1,1,2,3,3,4,4-octachloro- (8CI,9CI) **(32694-76-1)**

Butane, 1,1,1,2,3,4,4,4-octachloro- (8CI,9CI) **(18791-19-0)**
Butane, 1,1,2,2,3,3,4,4-octachloro- (6CI,7CI,8CI) **(20338-26-5)**
Butane, 1,1'-oxybis **(142-96-1)**
Butane, 2,2'-oxybis- (9CI) **(6863-58-7)**
Butane, 1,1'-oxybis(4-chloro- (9CI) **(6334-96-9)**
Butane, 1,1'-(oxybis(2,1-ethanediyloxy))bis- **(112-73-2)**
Butane, 1,1'-oxybis(3-methyl- (9CI) **(544-01-4)**
Butane, 1,1,2,3,4-pentachloro- (9CI) **(77753-24-3)**
Butane, 1,2,2,3,3-pentachloro- (9CI) **(83293-82-7)**
1-Butanephosphonic acid **(3321-64-0)**
Butane-propane mixture **(68475-59-2)**
1-Butanestannonic acid **(2273-43-0)**
1-Butanesulfonamide, N-ethyl-1,1,2,2,3,3,4,4,4-nonafluoro-N-(2-hydroxyethyl)- (9CI) **(34449-89-3)**
1-Butanesulfonamide, 1,1,2,2,3,3,4,4,4-nonafluoro-N-(2-hydroxyethyl)-N-methyl- (9CI) **(34454-97-2)**
Butanesulfone **(1633-83-6)**
1-Butanesulfonic acid, 1,1,2,2,3,3,4,4,4-nonafluoro-, potassium salt (9CI) **(29420-49-3)**
1-Butanesulfonyl chloride (9CI) **(2386-60-9)**
1-Butanesulfonyl fluoride, 1,1,2,2,3,3,4,4,4-nonafluoro- (9CI) **(375-72-4)**
1,4-Butanesultone **(1633-83-6)**
Butane sultone **(1633-83-6)**
δ-Butane sultone **(1633-83-6)**
Butane, 1,2,3,4-tetrabromo **(1529-68-6)**
1,2,3,4-Butanetetracarboxylic acid **(1703-58-8)**
Butanetetracarboxylic acid **(1703-58-8)**
Butane, 1,2,3,4-tetrachloro **(3405-32-1)**
Butane, 1,1,4,4-tetrachloro- (6CI,8CI) **(33455-24-2)**
Butane, 1,2,3,3-tetrachloro- (7CI,8CI,9CI) **(13138-51-7)**
Butane, 2,2,3,3-tetrachloro- (6CI,8CI,9CI) **(14499-87-7)**
Butane, 2,2,3,3-tetramethyl- (9CI) **(594-82-1)**
5H,6H-6,5a,13a,14-(1,2,3,4)Butanetetraylcycloocta(1,2-b:5,6-b')dinaphthalene-5,8,13,16(14H)- tetrone, 1,4,7,9,12,15,17,20-octahydroxy-3,11-dimethyl **(21884-44-6)**
1,2,3,4-Butanetetrol **(149-32-6)**
Butane, 2,2'-thiobis- (9CI) **(626-26-6)**
Butane, 1,1'-thiobis(3-methyl- **(544-02-5)**
Butanethiol **(109-79-5)**
n-Butanethiol **(109-79-5)**
sec-Butanethiol **(513-53-1)**
tert-Butanethiol **(75-66-1)**
2-Butanethiol (9CI) **(513-53-1)**
Butanethiol (OSHA) **(109-79-5)**
1-Butanethiol, 2-methyl- (9CI) **(1878-18-8)**
1-Butanethiol, 3-methyl- (9CI) **(541-31-1)**
2-Butanethiol, 2-methyl- (9CI) **(1679-09-0)**
2-Butanethiol, 3-methyl- (8CI,9CI) **(2084-18-6)**
1,2,4-Butanetricarboxylic acid, 2-phosphono- (9CI) **(37971-36-1)**
Butane, 1,2,4-trichloro- (7CI,9CI) **(1790-22-3)**
Butane, 1,3,3-trichloro- (8CI,9CI) **(15187-71-0)**
Butane, trichloroheptafluoro- (8CI,9CI) **(28984-80-7)**
Butane, 2,2,3-trimethyl- (9CI) **(464-06-2)**
1,2,4-Butanetriol **(3068-00-6)**
1,3,4-Butanetriol **(3068-00-6)**
1,2,3-Butanetriol (8CI,9CI) **(4435-50-1)**
1,2,3-Butanetriol, trinitrate **(84002-64-2)**
1,2,4-Butanetriol trinitrate (DOT) **(6659-60-5)**
Butanex **(23184-66-9)**
Butani (Italian) **(106-97-8)**
Butanic acid **(107-92-6)**
Butanimide **(123-56-8)**
Butanimidic acid **(541-35-5)**
Butanoic acid **(107-92-6)**
Butanoic acid, amide **(541-35-5)**
Butanoic acid, 3-amino- **(541-48-0)**
Butanoic acid, 2-amino-, (+-)- (9CI) **(2835-81-6)**
Butanoic acid, 2-amino-4-(hydroxymethylphosphinyl)-, (S)- (9CI) **(35597-44-5)**
Butanoic acid, 2-amino-4-(hydroxymethylphosphinyl)-, monoammonium salt **(77182-82-2)**
Butanoic acid, 2-amino-4-(S-methylsulfonimidoyl)- (9CI) **(1982-67-8)**
Butanoic acid, 4-amino-4-oxo- (9CI) **(638-32-4)**
Butanoic acid, ammonium salt (9CI) **(14287-04-8)**
Butanoic acid, anhydride (9CI) **(106-31-0)**
Butanoic acid, 4-(bis(2-chloroethyl)amino)benzene- **(305-03-3)**
Butanoic acid, 3-bromo- (9CI) **(2623-86-1)**
Butanoic acid, 2-bromo-3-methyl- (9CI) **(565-74-2)**
Butanoic acid, 4-butoxy- (9CI) **(55724-73-7)**
Butanoic acid, 2-butoxyethyl ester (9CI) **(20442-06-2)**
Butanoic acid, 3-chloro- **(1951-12-8)**
Butanoic acid, 2-chloro- (9CI) **(4170-24-5)**
Butanoic acid, 4-chloro- (9CI) **(627-00-9)**
Butanoic acid, 4-chloro-, ethyl ester (9CI) **(3153-36-4)**
Butanoic acid, 4-(4-chloro-2-methylphenoxy)- **(94-81-5)**
Butanoic acid, 4-(4-chloro-2-methylphenoxy)-, sodium salt **(6062-26-6)**
Butanoic acid, 2-chloro-3-oxo-, ethyl ester (9CI) **(609-15-4)**
Butanoic acid, 2-chloro-3-oxo-, methyl ester (9CI) **(4755-81-1)**
Butanoic acid, 2-chloro-3-oxo-, 1-phenylethyl ester (9CI) **(68683-30-7)**
Butanoic acid, 4-(4-chlorophenoxy)- (9CI) **(3547-07-7)**
Butanoic acid, 4-cyano- (9CI) **(39201-33-7)**
Butanoic acid, 2,4-diamino-4-oxo-, (S)- **(70-47-3)**
Butanoic acid, 2,2-dichloro- (9CI) **(13023-00-2)**
Butanoic acid, 3-(3,5-dichlorophenoxy)- (9CI) **(67883-08-3)**
Butanoic acid, 4-(2,4-dichlorophenoxy)-, 2-butoxyethyl ester (9CI) **(32357-46-3)**
Butanoic acid, 4-(2,4-dichlorophenoxy)-, butyl ester (9CI) **(6753-24-8)**
Butanoic acid, 4-(2,4-dichlorophenoxy)-, isooctyl ester **(1320-15-6)**
Butanoic acid, 2,2-dichloro-, sodium salt (9CI) **(2517-16-0)**
Butanoic acid, 4-(diethylamino)-4-oxo- (9CI) **(1522-00-5)**
Butanoic acid, 2,3-dihydroxy-2-(1-methylethyl)-, (2,3,5,7a-tetrahydro-1-hydroxy-1H- pyrrolizin-7-yl)methyl ester, N-oxide, (1R-(1-α,7(2R*,3S),7-α,β)) **(41708-76-3)**
Butanoic acid, 2,3-dihydroxypropyl ester (9CI) **(557-25-5)**
Butanoic acid, 2,2-dimethyl- (9CI) **(595-37-9)**
Butanoic acid, 2,3-dimethyl- (9CI) **(14287-61-7)**
Butanoic acid, 3,3-dimethyl- (9CI) **(1070-83-3)**
Butanoic acid, 3,3-dimethyl-2-oxo- (9CI) **(815-17-8)**
Butanoic acid ethyl ester **(105-54-4)**
Butanoic acid, 2-ethyl-2-methyl- (9CI) **(19889-37-3)**
Butanoic acid, 2-ethyl-3-oxo- (9CI) **(4433-85-6)**
Butanoic acid, 2-hydroxy- (9CI) **(565-70-8)**
Butanoic acid, 3-hydroxy- (9CI) **(300-85-6)**
Butanoic acid, 2-hydroxy-3,3-dimethyl- (9CI) **(4026-20-4)**
Butanoic acid, 2-hydroxy-4-(methylthio)- (9CI) **(583-91-5)**
Butanoic acid, 2-(2-hydroxy-4-(methylthio)-1-oxobutoxy)-4-(methylthio)-, 1-carboxy-3-(methylthio)propyl ester (9CI) **(88083-39-0)**
Butanoic acid, 2-methyl- (9CI) **(116-53-0)**
Butanoic acid, 2-methyl-, (+-)- (9CI) **(600-07-7)**
Butanoic acid, methyl- (9CI) **(35915-22-1)**
Butanoic acid, 3-methyl-, butyl ester (9CI) **(109-19-3)**
Butanoic acid, 3-methylbutyl ester (9CI) **(106-27-4)**
Butanoic acid, 2-methylene- (9CI) **(3586-58-1)**
Butanoic acid, 3-methyl-, ethyl ester **(108-64-5)**
Butanoic acid, 2-methyl-, ethyl ester (9CI) **(7452-79-1)**
Butanoic acid, 2-methyl-, 1,2,3,7,8,8a-hexahydro-3,7-dimethyl-8-(2-(tetrahydro-4-hydroxy- 6-oxo-2H-pyran-2-yl)ethyl)-1-naphthalenyl ester, (1S-(1-α-(R*),3-α,7-β, 8-β-(2s*,4s*), 8a-β)) **(75330-75-5)**
Butanoic acid, 2-methyl-, 2-methylbutyl ester (9CI) **(2445-78-5)**
Butanoic acid, 2-methyl-, methyl ester **(868-57-5)**
Butanoic acid, 3-methyl-, methyl ester (9CI) **(556-24-1)**
Butanoic acid, 3-methyl-, 2-methylpropyl ester (9CI) **(589-59-3)**
Butanoic acid, 3-((3-(2-methyl-1-piperidinyl)propyl)amino)-(9CI) **(90853-24-0)**

Butanoic acid, 3-methyl-, 2-propenyl ester **(2835-39-4)**
Butanoic acid, 1-methylpropyl ester (9CI) **(819-97-6)**
Butanoic acid, 3-methyl-, propyl ester (9CI) **(557-00-6)**
Butanoic acid, 3-methyl-, sodium salt (9CI) **(539-66-2)**
Butanoic acid, 4-(octadecylamino)-4-oxo-2-sulfo-, disodium salt (9CI) **(14481-60-8)**
Butanoic acid, 4-oxo- (9CI) **(692-29-5)**
Butanoic acid, 3-oxo-, butyl ester **(591-60-6)**
Butanoic acid, 3-oxo-, 1,1-dimethylethyl ester (9CI) **(1694-31-1)**
Butanoic acid, 3-oxo-, ethyl ester **(141-97-9)**
Butanoic acid, 3-oxo-, methyl ester (9CI) **(105-45-3)**
Butanoic acid, 3-oxo-, 2-((2-methyl-1-oxo-2-propenyl)oxy)ethyl ester (9CI) **(21282-97-3)**
Butanoic acid, 3-oxo-, 1-phenylethyl ester (9CI) **(40552-84-9)**
Butanoic acid, 2-(3-pentadecylphenoxy)- (9CI) **(14230-52-5)**
Butanoic acid pentyl ester **(540-18-1)**
Butanoic acid, phenyl ester (9CI) **(4346-18-3)**
Butanoic acid, 1,2,3-propanetriyl ester **(60-01-5)**
Butanoic acid, propyl ester (9CI) **(105-66-8)**
Butanoic acid, sodium salt (9CI) **(156-54-7)**
Butanoic acid, 2,2,2-trichloro-1-(dimethoxyphosphinyl)ethyl este R **(126-22-7)**
Butanoic acid, 4,4,4-trifluoro-3-oxo-, ethyl ester (9CI) **(372-31-6)**
Butanoic anhydride **(106-31-0)**
1-Butanol **(71-36-3)**
2-Butanol **(78-92-2)**
Butan-1-ol **(71-36-3)**
Butan-2-ol **(78-92-2)**
Butanol **(71-36-3)**
Butanol-2 **(78-92-2)**
n-Butan-1-ol **(71-36-3)**
n-Butanol **(71-36-3)**
s-Butanol **(78-92-2)**
sec-Butanol **(78-92-2)**
t-Butanol **(75-65-0)**
Butanol (9CI) (VAN) **(35296-72-1)**
Butanol (French) **(71-36-3)**
Butanol [UN 1120] **(71-36-3)**
sec-Butanol [UN 1120] **(78-92-2)**
tert-Butanol [UN 1120] **(75-65-0)**
2-Butanol acetate **(105-46-4)**
3-Butanolal **(107-89-1)**
1-Butanol, aluminum salt (9CI) **(3085-30-1)**
2-Butanol, aluminum salt (9CI) **(2269-22-9)**
Butanol-2-amine **(96-20-8)**
1-Butanol, 2-amino **(96-20-8)**
2-Butanol, 1-amino- (9CI) **(13552-21-1)**
1-Butanol, 2-amino-, (S)- (9CI) **(5856-62-2)**
1-Butanol, 2-amino-, (S)-(+)- (8CI) **(5856-62-2)**
Butanol (4)-butyl-nitrosamine **(3817-11-6)**
1-Butanol, 4-(butylnitrosoamino) **(3817-11-6)**
1-Butanol, 4-chloro **(928-51-8)**
2-Butanol, 3-chloro- (8CI,9CI) **(563-84-8)**
Butanol, chloro- (VAN) (9CI) **(1320-66-7)**
2-Butanol-d1 **(4712-39-4)**
1-Butanol-d (9CI) **(4712-38-3)**
2-Butanol-d (9CI) **(4712-39-4)**
2-Butanol, 1,4-dichloro- (8CI,9CI) **(2419-74-1)**
1-Butanol, 2,2-diethyl-, acetate (6CI,8CI,9CI) **(10332-40-8)**
1-Butanol, 2,2-dimethyl **(1185-33-7)**
2-Butanol, 3,3-dimethyl **(464-07-3)**
1-Butanol, 3,3-dimethyl- (9CI) **(624-95-3)**
1-Butanol, dimethyl- (9CI) **(79956-98-2)**
2-Butanol, 2,3-dimethyl- (9CI) **(594-60-5)**
Butanol, 2,3-dimethyl- (9CI) **(54206-54-1)**
1-Butanol, 2,3-dimethyl- (8CI,9CI) **(19550-30-2)**
2-Butanol, 1-(dimethylamino)-2-methyl-, benzoate (ester) **(644-26-8)**
2-Butanol, 4-(dimethylamino)-3-methyl-1,2-diphenyl-, propionate, (+) **(469-62-5)**
2-Butanol, 4-(dimethylamino)-3-methyl-1,2-diphenyl-, propionate (Ester), (-)- **(2338-37-6)**
2-Butanol, 2-((1,1-dimethylethyl)azo)- (9CI) **(57910-79-9)**
2-Butanol, 3,3-dimethyl-, methylphosphonofluoridate **(96-64-0)**
Butanolen (Dutch) **(71-36-3)**
1-Butanol, 2,2'-(1,2-ethanediyldiimino)bis-, (R) **(74-55-5)**
1-Butanol, 2-ethyl **(97-95-0)**
2-Butanol, 2-ethynyl- **(77-75-8)**
1-Butanol, 2,2,3,3,4,4,4-heptafluoro **(375-01-9)**
1-Butanol, 4-(hexyloxy)- (7CI,8CI,9CI) **(4541-13-3)**
1,2-Butanolide **(96-48-0)**
1,4-Butanolide **(96-48-0)**
4-Butanolide **(96-48-0)**
2-Butanol, 1,1'-iminobis- (9CI) **(21838-75-5)**
2-Butanol, 1,1'-iminodi- (8CI) **(21838-75-5)**
1-Butanol, 4-methoxy **(111-32-0)**
1-Butanol, 3-methoxy- (9CI) **(2517-43-3)**
2-Butanol, 1-methoxy- (9CI) **(53778-73-7)**
1-Butanol, 3-methoxy-, acetate **(4435-53-4)**
1-Butanol, 3-methoxy-3-methyl- (9CI) **(56539-66-3)**
2-Butanol, 4-methoxy-3-(1-octenyl-ONN-azoxy)-, (E,Z)-(2S,3S) **(23315-05-1)**
1-Butanol, 2-methyl **(137-32-6)**
1-Butanol, 3-methyl **(123-51-3)**
2-Butanol, 2-methyl- **(75-85-4)**
2-Butanol, 3-methyl- (9CI) **(598-75-4)**
1-Butanol, 2-methyl-, acetate (9CI) **(624-41-9)**
1-Butanol, 3-methyl-, carbamate **(543-86-2)**
1-Butanol, 2-methylene-, acetate (6CI,9CI) **(55670-09-2)**
2-Butanol, 4-(1-methylethoxy)- (9CI) **(40091-57-4)**
1-Butanol, 3-methyl-, hydrogen phosphorodithioate (9CI) **(32650-55-8)**
2-Butanol, 2-methyl-4-phenyl **(103-05-9)**
2-Butanol, 1,1',1''-nitrilotris- (9CI) **(2421-02-5)**
2-Butanol, 1,1',1''-nitrilotri- (7CI,8CI) **(2421-02-5)**
1-Butanol, 2-nitro- (9CI) **(609-31-4)**
Butanolo (Italian) **(71-36-3)**
2,3-Butanolone **(513-86-0)**
2-Butanol-3-one **(513-86-0)**
1-Butanol, potassium salt (9CI) **(3999-70-0)**
Butanol secondaire (French) **(78-92-2)**
Butanol, sodium salt **(2372-45-4)**
1-Butanol, sodium salt (9CI) **(2372-45-4)**
Butanol tertiaire (French) **(75-65-0)**
Butanol, 2,4,4,4-tetrachloro- **(3290-70-8)**
1-Butanol, 2,4,4,4-tetrachloro- (9CI) **(3290-70-8)**
Butanol, tetrachloro- (9CI) **(75536-53-7)**
2-Butanol, 2,3,3-trimethyl- (9CI) **(594-83-2)**
2-Butanone **(78-93-3)**
3-Butanone **(78-93-3)**
Butanone **(78-93-3)**
Butanone 2 (French) **(78-93-3)**
2-Butanone (OSHA) **(78-93-3)**
2-Butanone, 1-bromo **(816-40-0)**
2-Butanone, 3-bromo- (9CI) **(814-75-5)**
1-Butanone, 2-bromo-1-phenyl-, (.+-.)- (9CI) **(73908-28-8)**
2-Butanone, 3-chloro- (9CI) **(4091-39-8)**
2-Butanone, 1-chloro-3,3-dimethyl- (9CI) **(13547-70-1)**
2-Butanone, 1-(4-chlorophenoxy)-3,3-dimethyl-1-(1,2,4-triazol-1-yl) **(43121-43-3)**
1-Butanone, 2-chloro-1-phenyl-, (.+-.)- (9CI) **(73908-29-9)**
1-Butanone, 4-(4-(4-chlorophenyl)-4-hydroxy-1-piperidinyl)-1-(4-fluorophenyl)- **(52-86-8)**
2-Butanone, cyclic 1,2-ethanediyl acetal **(126-39-6)**
2-Butanone, cyclic ethylene acetal **(126-39-6)**
2-Butanone, 4-(4,6-diamino-s-triazin-2-yl)- (8CI) **(30354-99-5)**
2-Butanone, 1,1-dichloro-3,3-dimethyl **(22591-21-5)**
2-Butanone, 3,3-dimethyl **(75-97-8)**

2-Butanone, 3,3-dimethyl-1-(methylthio)-, o-((methylamino)-carbonyl)oxime **(39196-18-4)**
2-Butanone, 3,4-epoxy- (6CI,7CI,8CI) **(4401-11-0)**
2-Butanone, 3-ethoxy-3-methyl- (6CI,7CI,9CI) **(36687-99-7)**
2-Butanone, 3-hydroxy **(513-86-0)**
2-Butanone, 4-hydroxy- (9CI) **(590-90-9)**
2-Butanone, 3-hydroxy-3-methoxy- (9CI) **(61996-25-6)**
2-Butanone, 4-(4-hydroxy-3-methoxyphenyl) **(122-48-5)**
2-Butanone, 3-hydroxy-3-methyl- (9CI) **(115-22-0)**
2-Butanone, 1-hydroxy-3-methyl- (6CI,9CI) **(36960-22-2)**
2-Butanone, 4-hydroxy-3-methyl- (8CI,9CI) **(3393-64-4)**
2-Butanone, 3-methoxy-3-methyl- (9CI) **(36687-98-6)**
2-Butanone, 3-methyl **(563-80-4)**
2-Butanone, 4-(3,4-methylenedioxyphenyl) **(3160-37-0)**
1-Butanone, 4-(methylnitrosoamino)-1-(3-pyridinyl)- **(64091-91-4)**
1-Butanone, 1-(10-(3-(4-methyl-1-piperazinyl)propyl)phenothiazin-2-yl) **(653-03-2)**
1-Butanone, 1-(10-(3-(4-methyl-1-piperazinyl)propyl)-10H-phenothiazin-2-yl)- (9CI) **(653-03-2)**
2-Butanone, O,O',O''-(methylsilylidyne)trioxime (9CI) **(22984-54-9)**
2-Butanone, 3-(methylsulfonyl)-, O-((methylamino)carbonyl)-oxime (10%) **(34681-23-7)**
2-Butanone, 3-methylsulfonyl-, O-(N-methylcarbamoyl)oxime **(34681-23-7)**
2-Butanone, 3-(methylthio)-, O-(N-methylcarbamoyl)oxime **(34681-10-2)**
2-Butanone, oxime **(96-29-7)**
2-Butanone, peroxide **(1338-23-4)**
2-Butanone peroxide, Containing more than 10% available oxygen **(1338-23-4)**
2-Butanone peroxide, Maximum concentration 60% **(1338-23-4)**
2-Butanone peroxide, With not more than 9% by weight active oxygen **(1338-23-4)**
2-Butanone, 4-phenyl **(2550-26-7)**
1-Butanone, 1-phenyl- (9CI) **(495-40-9)**
2-Butanone, 1-phenyl- (8CI,9CI) **(1007-32-5)**
Butanova **(129-20-4)**
Butanox LPT **(1338-23-4)**
Butanox M 105 **(1338-23-4)**
Butanox M 50 **(1338-23-4)**
Butanoyl chloride, 4-chloro- (9CI) **(4635-59-0)**
Butanoyl fluoride, heptafluoro- (9CI) **(335-42-2)**
1,2,4-Butantriol (German) **(3068-00-6)**
Butaperazine **(653-03-2)**
Butaphen **(50-33-9)**
Butaphene **(88-85-7)**
Butapirazol **(50-33-9)**
Butapirone **(129-20-4)**
Butapon **(94-80-4)**
Butapyrazole **(50-33-9)**
Butarecbon **(50-33-9)**
Butarez 15 **(9003-17-2)**
Butartril **(50-33-9)**
Butartrina **(50-33-9)**
Butasaron **(143-81-7)**
Butased **(143-81-7)**
Butatab **(125-40-6)**
Butatal **(125-40-6)**
Butatal sodium **(143-81-7)**
Butatensin **(64-55-1)**
Butazate **(136-23-2)**
Butazate 50-D **(136-23-2)**
Butazem **(143-81-7)**
Butazina **(50-33-9)**
Butazolidin **(50-33-9)**
Butazolidine **(50-33-9)**
Butazona **(50-33-9)**
Butazone **(50-33-9)**
Bute **(50-33-9)**
Butein **(487-52-5)**
2-Butenal, (E)- **(123-73-9)**
trans-2-Butenal **(123-73-9)**
2-Butenal (9CI) **(4170-30-3)**
2-Butenal, 2-ethenyl- (9CI) **(20521-42-0)**
2-Butenal, 2-methyl- (9CI) **(1115-11-3)**
2-Butenal, 2-methyl-, (E)- (9CI) **(497-03-0)**
3-Buten-2-amine, N,N-diethyl-4,4-di-2-thienyl- (9CI) **(86-14-6)**
3-Buten-2-amine, N,N-dimethyl-4,4-di-2-thienyl- (9CI) **(524-84-5)**
(E)-2-Butene **(624-64-6)**
(Z)-2-Butene **(590-18-1)**
2-Butene **(107-01-7)**
2-Butene-cis **(590-18-1)**
2-Butene-trans **(624-64-6)**
2-trans-Butene **(624-64-6)**
Butene-1 **(106-98-9)**
α-Butene **(106-98-9)**
cis-2-Butene **(590-18-1)**
cis-Butene **(590-18-1)**
n-Butene **(25167-67-3)**
trans-2-Butene **(624-64-6)**
trans-Butene **(624-64-6)**
1-Butene (9CI) **(106-98-9)**
2-Butene, (E)- (9CI) **(624-64-6)**
Butene (DOT) **(25167-67-3)**
Butene, Homopolymer (9CI) **(9003-29-6)**
Butene, Homopolymer (44% Formulation) **(9003-29-6)**
Butene, Homopolymer (n=ca. 5) **(9003-29-6)**
Butene, Homopolymer (n=ca. 8) **(9003-29-6)**
Butene, Homopolymer (n=ca. 9) **(9003-29-6)**
Butene, Homopolymer (n=ca. 12) **(9003-29-6)**
Butene, Homopolymer (n=ca. 13) **(9003-29-6)**
Butene, Homopolymer (n=ca. 14) **(9003-29-6)**
Butene, Homopolymer (n=ca. 15) **(9003-29-6)**
Butene, Homopolymer (n=ca. 20) **(9003-29-6)**
Butene, Homopolymer (n=ca. 32) **(9003-29-6)**
Butene, Homopolymer (n=ca. 34) **(9003-29-6)**
Butene, Polymers **(9003-29-6)**
2-Butene, (Z)- (9CI) **(590-18-1)**
2-Butene, 1-bromo- (86%) **(4784-77-4)**
1-Butene, 1-bromo- (9CI) **(31844-98-1)**
1-Butene, 4-bromo- (9CI) **(5162-44-7)**
2-Butene, 1-bromo- (9CI) **(4784-77-4)**
2-Butene, 2-bromo- (9CI) **(13294-71-8)**
2-Butene, 1-bromo-2-chloro- (9CI) **(54410-84-3)**
1-Butene, 1-chloro **(4461-42-1)**
1-Butene, 3-chloro **(563-52-0)**
2-Butene, 1-chloro **(591-97-9)**
2-Butene, 2-chloro- **(4461-41-0)**
1-Butene, 4-chloro- (9CI) **(927-73-1)**
Butene chlorohydrin **(1320-66-7)**
1-Butene, 3-chloro-2-methyl **(5166-35-8)**
2-Butene, 1-chloro-2-methyl **(13417-43-1)**
2-Butene, 1-chloro-3-methyl- (9CI) **(503-60-6)**
2-Butene, 2-chloro-3-methyl- (8CI,9CI) **(17773-65-8)**
2-Butene, 3-chloro-1-phenyl-, (Z)- (8CI) **(16608-68-7)**
2-Butene, 2-cyclopropyl-, (E)- (8CI) **(20479-71-4)**
2-Butene, 2-cyclopropyl-, (Z)- (8CI) **(20479-72-5)**
2-Butenedial, (Z)- (9CI) **(3675-13-6)**
1-Butene, 3,4-dichloro **(760-23-6)**
2-Butene, 1,3-dichloro **(926-57-8)**
2-Butene, 1,3-dichloro-, (E)- **(7415-31-8)**
2-Butene, 1,4-dichloro **(764-41-0)**
2-Butene, 1,4-dichloro-, (E) **(110-57-6)**
2-Butene, 1,4-dichloro-, cis- **(1476-11-5)**
Butene, dichloro- **(11069-19-5)**
Butene, 1,4-dichloro- (9CI) **(31423-92-4)**
1-Butene, 3,4-dichloro- (Racemic mixture) **(64037-54-3)**

Name	CAS
2-Butene, 1,4-dichloro-, (Z)- (9CI)	(1476-11-5)
2-Butene, 1,4-diethoxy- (7CI,8CI,9CI)	(7250-85-3)
2-Butene, 1,4-dihydroxy	(110-64-5)
1-Butene, 2,3-dimethyl- (9CI)	(563-78-0)
1-Butene, 3,3-dimethyl- (9CI)	(558-37-2)
2-Butene, 2,3-dimethyl- (9CI)	(563-79-1)
1-Butene, 3,3-dimethyl-1,1-diphenyl- (8CI)	(23586-64-3)
2-Butenedinitrile, 2,3-diamino-, (Z)- (9CI)	(1187-42-4)
Butenedioic acid, (E)-	(110-17-8)
Butenedioic acid, (Z)-	(110-16-7)
cis-Butenedioic acid	(110-16-7)
trans-Butenedioic acid	(110-17-8)
2-Butenedioic acid (9CI)	(6915-18-0)
Butenedioic acid (8CI)	(6915-18-0)
2-Butenedioic acid (Z)-, Polymer with 4,4'-(1-methylethylidene)-bis(phenol) and methyloxirane (9CI)	(38355-75-8)
2-Butenedioic acid, bis(1,3-dimethylbutyl) ester	(105-52-2)
2-Butenedioic acid, bis(2-ethylhexyl) ester	(141-02-6)
2-Butenedioic acid (Z)-, bis(2-methylpropyl) ester (9CI)	(14234-82-3)
2-Butenedioic acid, 2-chloro- (9CI)	(19071-21-7)
2-Butenedioic acid, dibutyl ester	(105-76-0)
2-Butenedioic acid, 2,3-dichloro- (9CI)	(43180-81-0)
2-Butenedioic acid, (Z)-, didodecyl ester (C10-C18)	(2915-52-8)
2-Butenedioic acid (Z)-, didodecyl ester (9CI)	(2915-52-8)
2-Butenedioic acid, diethyl ester, (E)	(623-91-6)
2-Butenedioic acid (Z)-, diisooctyl ester (9CI)	(1330-76-3)
2-Butenedioic acid, 2,3-dimethyl-, (Z)- (9CI)	(488-21-1)
2-Butenedioic acid, dimethyl ester, (Z)	(624-48-6)
2-Butenedioic acid, 2-ethyl-3-methyl-, (E)- (9CI)	(28098-80-8)
2-Butenedioic acid, 2-ethyl-3-methyl-, (Z)- (9CI)	(41654-09-5)
2-Butenedioic acid, 2-methyl-, (Z)- (9CI)	(498-23-7)
2-Butenedioic acid, (Z)-, 4,4'-(1-methylethylidene)bisphenol, methyloxirane polymer	(38355-75-8)
2-Butenedioic acid (Z)-, monobutyl ester (9CI)	(925-21-3)
2-Butenedioic acid (Z)-, monoethyl ester (9CI)	(3990-03-2)
2-Butenedioic acid (Z)-, mono(2-ethylhexyl) ester (9CI)	(7423-42-9)
2-Butenedioic acid (Z)-, mono(1-methylethyl) ester (9CI)	(924-83-4)
2-Butenedioic acid (z)-, diethyl ester	(141-05-9)
cis-Butenedioic anhydride	(108-31-6)
2-Butene-1,4-diol	(110-64-5)
3-Butene-1,2-diol	(497-06-3)
2-Butene-1,2-diol (9CI)	(50317-11-8)
2-Butene-1,4-diol, (E)- (8CI,9CI)	(821-11-4)
2-Butene-1,4-diol, (Z)- (8CI,9CI)	(6117-80-2)
2-Butene-1,4-diol bis(bromoacetate)	(20679-58-7)
2-Butene-1,4-diol, 2,3-dibromo	(3234-02-4)
1-Butene, 1,3-diphenyl- (7CI,8CI)	(7614-93-9)
2-Butene-1,1,1-d3, 2-methyl-, (E)- (9CI)	(69432-94-6)
1-Butene, 3,4-epoxy	(930-22-3)
1-Butene, 2-ethyl- (8CI)	(760-21-4)
2-Butene expoxide	(3266-23-7)
2-Butene, hexabromo-	(36678-45-2)
2-Butene, 1,1,2,3,4,4-hexabromo- (9CI)	(36678-45-2)
1-Butene, 2,3,3,4,4,4-hexachloro- (9CI)	(83682-71-7)
Butenemal	(125-42-8)
Butenemalum	(125-42-8)
1-Butene, 1-methoxy-2-methyl-, (E)- (9CI)	(69432-97-9)
1-Butene, 1-methoxy-2-methyl-, (Z)- (9CI)	(69432-98-0)
1-Butene, 2-methyl	(563-46-2)
1-Butene, 3-methyl	(563-45-1)
2-Butene, 2-methyl	(513-35-9)
Butene, 2-methyl	(26760-64-5)
2-Butene, 3-methyl-1-phenyl- (8CI)	(4489-84-3)
1-Butene-4-nitrile	(109-75-1)
2-Butenenitrile	(4786-20-3)
3-Butenenitrile	(109-75-1)
3-Butenenitrile, 2-methyl-	(16529-56-9)
3-Butenenitrile, 3-methyl	(4786-19-0)
2-Butenenitrile, 2-methyl- (9CI)	(4403-61-6)
2-Butenenitrile, 2-methyl-, (E)- (9CI)	(30574-97-1)
2-Butenenitrile, 2-methyl-, (Z)- (9CI)	(20068-02-4)
1-Butene, octachloro-	(2482-68-0)
1-Butene, 1,1,2,3,3,4,4,4-octachloro- (8CI,9CI)	(2482-68-0)
2-Butene, octachloro- (6CI,7CI,8CI)	(3050-42-8)
2-Butene, 1,1,1,2,3,4,4,4-octafluoro	(360-89-4)
2-Butene-1-ol	(6117-91-5)
3-Butene-2-one	(78-94-4)
3-Butene-2-one, 4-(2-furanyl)-	(623-15-4)
1,2-Butene oxide	(106-88-7)
1-Butene oxide	(106-88-7)
2-Butene oxide	(3266-23-7)
1-Butene, 4,4'-oxybis(3-chloro- (9CI)	(83682-66-0)
1-Butene, 1,1,4,4,4-pentachloro-2-(trichloromethyl)- (9CI)	(83682-40-0)
2-Buteneperoxoic acid, 1,1-dimethylethyl ester (9CI)	(23474-91-1)
1-Butene, 4-phenyl	(768-56-9)
2-Butene, 1-phenyl	(1560-06-1)
1-Butene, 1-phenyl-, (E)- (8CI)	(1005-64-7)
2-Butene, 1,1,4,4-tetrachloro-2,3-bis(dichloromethyl)- (9CI)	(83682-30-8)
2-Butene-1-thiol (8CI,9CI)	(5954-72-3)
1-Butene-1,2,4-tricarboxylic acid, 3-hydroxy-, γ-lactone	(16426-62-3)
1-Butene, 2,3,4-trichloro	(2431-50-7)
2-Butene, 1,2,4-trichloro	(2431-54-1)
Butene, trichloro	(51023-22-4)
2-Butene, 1,1,4-trichloro- (9CI)	(41601-59-6)
2-Butene, 1,1,4-trichloro-, (E)- (9CI)	(57808-36-3)
1-Butene, 1,1,3-trichloro- (8CI,9CI)	(13279-86-2)
(E)-2-Butenoic acid	(107-93-7)
α-Butenoic acid	(3724-65-0)
trans-2-Butenoic acid	(107-93-7)
2-Butenoic acid (9CI)	(3724-65-0)
2-Butenoic acid, (E)- (9CI)	(107-93-7)
3-Butenoic acid (9CI)	(625-38-7)
2-Butenoic acid, (Z)- (9CI)	(503-64-0)
2-Butenoic acid, 4-amino-4-oxo-, (E)- (9CI)	(2987-87-3)
2-Butenoic acid, 4-amino-4-oxo-, (Z)- (9CI)	(557-24-4)
2-Butenoic acid, anhydride (9CI)	(623-68-7)
2-Butenoic acid, 3-bromo-4-(4-methoxyphenyl)-4-oxo-, sodium salt, (E)- (9CI)	(21739-91-3)
2-Butenoic acid, butyl ester (9CI)	(7299-91-4)
2-Butenoic acid, 4-chloro- (9CI)	(16197-90-3)
2-Butenoic acid, 4,4'-((dibutylstannylene)bis(oxy))bis-(4-oxo-, diisooctyl ester, (Z,Z)- (9CI)	(25168-21-2)
2-Butenoic acid, 4,4-dichloro-, (E)- (9CI)	(16502-88-8)
2-Butenoic acid, 3-((dimethoxyphosphinyl)oxy)-, methyl ester (9CI)	(7786-34-7)
2-Butenoic acid, 3-((dimethoxyphosphinyl)oxy)-, methyl ester, (E)- (9CI)	(298-01-1)
2-Butenoic acid, 3-((dimethoxyphosphinyl)oxy)-, methyl ester, (Z)- (9CI)	(338-45-4)
2-Butenoic acid, 3-((dimethoxyphosphinyl)oxy)-, 1-phenylethyl ester, (E)- (9CI)	(7700-17-6)
2-Butenoic acid, 3-((((1,1-dimethylethyl)amino)carbonyl)amino)-, methyl ester (9CI)	(64346-47-0)
2-Butenoic acid, 2,4-dinitro-6-octylphenyl ester (9CI)	(49794-90-3)
2-Butenoic acid, 2,6-dinitro-4-octylphenyl ester (9CI)	(49794-91-4)
2-Butenoic acid, ethenyl ester (9CI)	(14861-06-4)
2-Butenoic acid, ethenyl ester, (E)- (9CI)	(3234-54-6)
2-Butenoic acid, 3-(((ethylamino)methoxyphosphinothioyl)oxy)-, isopropyl ester, (E)-	(31218-83-4)
2-Butenoic acid, 3-(((ethylamino)methoxyphosphinothioyl)oxy)-, 1-methylethyl ester, (E)-	(31218-83-4)
2-Butenoic acid, 3-(((ethylamino)methoxyphosphinothioyl)oxy)-, 1-methylethyl ester, (E)-, Mixt. with 2,2-Dichloroethenyl dimethyl phosphate	(77491-30-6)

Name	CAS No.
2-Butenoic acid, ethyl ester	**(10544-63-5)**
trans-2-Butenoic acid ethyl ester	**(623-70-1)**
2-Butenoic acid, ethyl ester, (E)- (9CI)	**(623-70-1)**
2-Butenoic acid, 3-formyl-2,4,4-trichloro-, (E)	**(115340-67-5)**
2-Butenoic acid, 4-hydroxy-, γ-lactone	**(497-23-4)**
2-Butenoic acid, 2-(or 4)-isooctyl-4,6(or 2,6)-dinitrophenyl ester (9CI)	**(39300-45-3)**
2-Butenoic acid γ-lactone	**(497-23-4)**
2-Butenoic acid, 2-methoxy-3-methyl- (9CI)	**(58973-18-5)**
2-Butenoic acid, 2-methyl-, (E)- (9CI)	**(80-59-1)**
2-Butenoic acid, 3-methyl- (9CI)	**(541-47-9)**
2-Butenoic acid, 2-methyl-, (Z)- (9CI)	**(565-63-9)**
2-Butenoic acid, 2-methyl-, 7-((2,3-dihydroxy-2-(1-hydroxyethyl)-3-methyl-1-oxobutoxy) methyl)-2,3,5,7a-tetrahydro-1H-pyrrolizin-1-yl ester, hydrochloride	**(520-68-3)**
Butenoic acid, 2-methyl-, 7-((2,3-dihydroxy-2-(1-methylethyl)-1-oxobutoxy)methyl)-2,3,5,7a- tetrahydro-1H-pyrrolizin-1-yl ester	**(22571-95-5)**
2-Butenoic acid, methyl ester (9CI)	**(18707-60-3)**
2-Butenoic acid, 2-methyl-, methyl ester, (E)	**(6622-76-0)**
2-Butenoic acid, 3-methyl-, methyl ester (9CI)	**(924-50-5)**
2-Butenoic acid, 3-methyl-, 2-(1-methylpropyl)-4,6-dinitrophenyl ester (9CI)	**(485-31-4)**
3-Butenoic acid, 2,2,3,4,4-pentachloro-, butyl ester (9CI)	**(75147-20-5)**
2-Butenoic acid, 4-phenyl- (9CI)	**(2243-52-9)**
(Z)-2-Buten-1-ol	**(4088-60-2)**
2-Buten-1-ol	**(6117-91-5)**
2-Butenol	**(6117-91-5)**
3-Buten-2-ol	**(598-32-3)**
2-Buten-1-ol, (Z)- (9CI)	**(4088-60-2)**
2-Buten-2-ol, acetate (8CI,9CI)	**(6203-88-9)**
3-Buteno-β-lactone	**(674-82-8)**
3-Buten-2-ol, 1-bromo- (7CI,9CI)	**(64341-49-7)**
3-Buten-1-ol, 2-chloro- (6CI,7CI,9CI)	**(75455-41-3)**
3-Buten-2-ol, 1-chloro- (6CI,7CI,8CI,9CI)	**(671-56-7)**
2-Buten-1-ol, 2-ethyl-4-(2,2,3-trimethyl-3-cyclopenten-1-yl)-(9CI)	**(28219-61-6)**
2-Buten-4-olide	**(497-23-4)**
δ,α,β-Butenolide	**(497-23-4)**
1-Buten-3-ol, 3-methyl	**(115-18-4)**
2-Buten-1-ol, 3-methyl	**(556-82-1)**
2-Buten-1-ol, 2-methyl- (9CI)	**(4675-87-0)**
3-Buten-1-ol, 3-methyl- (9CI)	**(763-32-6)**
3-Buten-1-ol, 2,2'-oxybis- (9CI)	**(83682-68-2)**
3-Buten-2-ol, 1,1'-oxybis- (9CI)	**(83682-67-1)**
3-Buten-2-one	**(78-94-4)**
Butenone	**(78-94-4)**
δ^3-2-Butenone	**(78-94-4)**
3-Buten-2-one, 3-chloro- (8CI,9CI)	**(683-70-5)**
3-Buten-2-one, 4-(2,2-dimethyl-6-methylenecyclohexyl)-3-methyl- (9CI)	**(7388-22-9)**
3-Buten-2-one, 4-(2-furanyl)- (9CI)	**(623-15-4)**
3-Buten-2-one, 4-(2-furyl)- (8CI)	**(623-15-4)**
3-Buten-2-one, 3-methyl	**(814-78-8)**
3-Buten-2-one, 3-methyl-, Polymer with ethenylbenzene (9CI)	**(25191-48-4)**
3-Buten-2-one, 3-methyl-, Polymer with methyl 2-methyl-2-propenoate (9CI)	**(51555-36-3)**
3-Buten-2-one, 3-methyl-, Polymer with styrene (8CI)	**(25191-48-4)**
3-Buten-2-one, 3-methyl-, dimer (6CI,9CI)	**(54789-11-6)**
3-Buten-2-one, 3-methyl-4-(2,6,6-trimethyl-2-cyclohexen-1-yl)- (9CI)	**(127-51-5)**
3-Buten-2-one, 4-phenyl	**(122-57-6)**
3-Buten-2-one, 1-phenyl- (9CI)	**(37442-55-0)**
3-Buten-2-one, 4-(2,5,6,6-tetramethyl-2-cyclohexen-1-yl)-, cis	**(79-69-6)**
3-Buten-2-one, 4-(2,5,6,6-tetramethyl-2-cyclohexen-1-yl)- (9CI)	**(79-69-6)**
3-Buten-2-one, 4-(2,6,6-trimethyl-1-cyclohexen-1-yl)	**(14901-07-6)**
3-Buten-2-one, 4-(2,6,6-trimethyl-1-cyclohexen-1-yl)-, (e)	**(79-77-6)**
3-Buten-2-one, 4-(2,6,6-trimethyl-2-cyclohexen-1-yl)	**(127-41-3)**
3-Buten-2-one, 1-(2,3,6-trimethylphenyl)- (9CI)	**(54789-45-6)**
β-Butenonitrile	**(109-75-1)**
2-Butenyl alcohol	**(6117-91-5)**
2-Butenyl chloride	**(591-97-9)**
3-Butenyl cyanide	**(592-51-8)**
1-Buten-3-yne	**(689-97-4)**
1-Buten-3-yne, 2-methyl- (8CI,9CI)	**(78-80-8)**
Butesin	**(94-25-7)**
Butethal	**(77-28-1)**
Butethanol	**(136-47-0)**
Buthidazole	**(55511-98-3)**
Buthiobate	**(51308-54-4)**
Buticaps	**(125-40-6)**
Buticaps	**(143-81-7)**
2-Butin-1,4-diol (Czech)	**(110-65-6)**
Butidiona	**(50-33-9)**
Butifos	**(78-48-8)**
n-Butilamina (Italian)	**(109-73-9)**
Butilate	**(2008-41-5)**
Butilchlorofos	**(126-22-7)**
Butilchlorofos	**(1689-84-5)**
O-(4-terz.-Butil-2-cloro-fenil)-O-metil-fosforammide (Italian)	**(299-86-5)**
Butile (acetati di) (Italian)	**(123-86-4)**
Butilene	**(129-20-4)**
Butil metacrilato (Italian)	**(97-88-1)**
Butinox	**(56-35-9)**
Butiphos	**(78-48-8)**
Butisan	**(21267-72-1)**
Butisan S	**(67129-08-2)**
Butisane	**(21267-72-1)**
Butiserpazide-25	**(50-55-5)**
Butiserpazide-50	**(50-55-5)**
Butiserpine	**(50-55-5)**
Butisol	**(125-40-6)**
Butisol	**(143-81-7)**
Butisol sodium	**(143-81-7)**
Butisulfina	**(339-43-5)**
Butiwas-simple	**(50-33-9)**
Butobarbital	**(77-28-1)**
Butobarbitone	**(77-28-1)**
Butobarbitural	**(77-28-1)**
Butoben	**(94-26-8)**
Butocarboxim (German)	**(34681-10-2)**
Butocarboxime	**(34681-10-2)**
Butocide	**(51-03-6)**
Butoflin	**(52918-63-5)**
Butoksyetylowy alkohol (Polish)	**(111-76-2)**
Butonat (German)	**(126-22-7)**
Butonate	**(126-22-7)**
Butone	**(50-33-9)**
Butophen	**(6365-83-9)**
Butopyronoxyl	**(532-34-3)**
(-)-Butorphanol	**(42408-82-2)**
Butorphanol	**(42408-82-2)**
2-Butossi-etanolo (Italian)	**(111-76-2)**
Butox	**(52918-63-5)**
Butoxicarboxim (German)	**(34681-23-7)**
Butoxide	**(51-03-6)**
Butoxon	**(94-82-6)**
Butoxone	**(94-82-6)**
Butoxone SB	**(10433-59-7)**
Butoxone amine	**(94-82-6)**
Butoxone ester	**(94-82-6)**
2-Butoxy-aethanol (German)	**(111-76-2)**
1-Butoxybutane	**(142-96-1)**
t-Butoxycarbonyl azide	**(1070-19-5)**

tert-Butoxycarbonyl azide (1070-19-5)
4-(Butoxycarbonyl)phenol (94-26-8)
Butoxycarboxim (34681-23-7)
Butoxycarboxime (French) (34681-23-7)
1-Butoxy-3-chloro-2-propanol (16224-33-2)
2-Butoxy-N-(2-(diethylamino)ethyl)cinchoninamide (85-79-0)
2-Butoxy-N-(β-diethylaminoethyl)cinchoninamide (85-79-0)
Butoxydiethylene glycol (112-34-5)
Butoxydiglycol (112-34-5)
2-Butoxy-1-ethanol (111-76-2)
2-sec-Butoxyethanol (7795-91-7)
2-tert-Butoxyethanol (7580-85-0)
Butoxyethanol (111-76-2)
n-Butoxyethanol (111-76-2)
2-Butoxyethanol (ACGIH,OSHA) (111-76-2)
2-Butoxyethanol acetate (112-07-2)
2-Butoxyethanol benzoate (5451-76-3)
Butoxyethanol 4-(2,4-dichlorophenoxy)butyrate (32357-46-3)
2-Butoxyethanol, phosphate (78-51-3)
2-Butoxyethanol sodium salt (52663-57-7)
Butoxyethene (111-34-2)
2-(2-Butoxyethoxy)ethanol (112-34-5)
2-(2-Butoxyethoxy)ethanol acetate (124-17-4)
2-(2-Butoxyethoxy)ethanol phosphate (3:1) (7332-46-9)
2-(2-(2-Butoxyethoxy)ethoxy)ethanol (143-22-6)
α-(2-(2-Butoxyethoxy)ethoxy)-4,5-methylenedioxy-2-propyltoluene (51-03-6)
α-(2-(2-n-Butoxyethoxy)-ethoxy)-4,5-methylenedioxy-2-propyl-toluene (51-03-6)
5-((2-(2-Butoxyethoxy)ethoxy)methyl)-6-propyl-1,3-benzodioxole (51-03-6)
2-(2-Butoxyethoxy)ethyl acetate (124-17-4)
2-β-Butoxyethoxyethyl chloride (1120-23-6)
2-(2-Butoxyethoxy)ethylester kyseliny octove (Czech) (124-17-4)
2-(2-Butoxyethoxy)ethyl thiocyanate (112-56-1)
2-(2-(Butoxy)ethoxy)ethyl thiocyanic acid ester (112-56-1)
2-(2-Butoxyethoxy)ethylthiokyanat (Czech) (112-56-1)
1-(2-Butoxyethoxy)-2-propanol (124-16-3)
1-Butoxyethoxy-2-propanol (124-16-3)
Butoxyethoxypropyl 2,4-dichlorophenoxyacetate (1928-57-0)
Butoxyethyl 2,4,5-T (2545-59-7)
2-Butoxyethyl acetate (112-07-2)
2-Butoxyethyl benzoate (5451-76-3)
2-Butoxyethyl butanoate (20442-06-2)
2-Butoxyethyl chloroacetate (5330-17-6)
Butoxyethyl 2,4-dichlorophenoxyacetate (1929-73-3)
Butoxyethyl 2-(2,4-dichlorophenoxy)propionate (53404-31-2)
2-Butoxyethylester kyseliny octove (Czech) (112-07-2)
Butoxyethyl isobutyrate (20442-06-2)
2-Butoxyethyl 2-methyl-4-chlorophenoxyacetate (19480-43-4)
2-Butoxyethyl octadecanoate (109-38-6)
β-Butoxyethyl phthalate (117-83-9)
2-Butoxyethyl vinyl ether (4223-11-4)
Butoxyl [UN 2708] (4435-53-4)
Butoxymethanol (3085-35-6)
1-tert-Butoxy-3-methoxybenzene (15359-99-6)
N-(Butoxymethyl)acrylamide (1852-16-0)
N-Butoxymethylakrylamid (Czech) (1852-16-0)
N-Butoxymethyl-2-chloro-2',6'-diethylacetanilide (23184-66-9)
N-(Butoxymethyl)-2-chloro-N-(2,6-diethylphenyl)acetamide (23184-66-9)
1-tert-Butoxy-2-methylcyclohexene (40648-26-8)
4-Butoxyphenol (122-94-1)
p-Butoxyphenol (122-94-1)
Butoxyphenyl (1126-79-0)
Butoxypolypropoxypolyethoxyethanol iodine complex (68610-00-4)
Butoxy polypropylene glycol (9003-13-8)
Butoxypolypropylene glycol (9003-13-8)
Butoxypropanediol polymer (9003-13-8)
1-Butoxy-2-propanol (5131-66-8)
2-Butoxy-1-propanol (29387-86-8)
Butoxypropanol (29387-86-8)
1-(3-Butoxypropoxy)-2-propanol (35075-24-2)
2-Butoxyquinoline-4-carboxylic acid diethylaminoethylamide (85-79-0)
Butoxyrhodanodiethyl ether (112-56-1)
Butoxyslovanik BK 61 (9038-95-3)
2-Butoxy-2'-thiocyanodiethyl ether (112-56-1)
β-Butoxy-β'-thiocyanodiethyl ether (112-56-1)
1-Butoxy-2-(2-thiocyanoethoxy)ethane (112-56-1)
1-Butoxy-α-(2-thiocyanoethoxy)ethane (112-56-1)
2-tert-Butoxythiophene (23290-55-3)
Butoxytriethylene glycol (143-22-6)
Butoxytriglycol (143-22-6)
Butoz (50-33-9)
Butralin (33629-47-9)
Butraline (33629-47-9)
Butrate (125-40-6)
Butrizol (16227-10-4)
Butryaldehydic acid (692-29-5)
Butter Yellow (60-11-7)
Butter Yellow (97-56-3)
Butter of Antimony (10025-91-9)
Butter of Antimony (7647-18-9)
Butter of Arsenic (7784-34-1)
Butter of Zinc (7646-85-7)
Buttersaeure (German) (107-92-6)
Buttercup Yellow (11103-86-9)
Buttercup Yellow (13530-65-9)
Buttercup Yellow (15930-94-6)
Buturon (3766-60-7)
Butvar (63148-65-2)
3-Butyn-2-amine, 2-methyl- (9CI) (2978-58-7)
Butyl 2,4-D (94-80-4)
Butyl Rubber (9010-85-9)
S-n-Butyl S'-p-tert-butylbenzyl N-3-pyridyldithiocarbonimidate (51308-54-4)
Butyl 2,4,5-T (93-79-8)
N-Butyl acetamide (1119-49-9)
N-Butylacetamide (1119-49-9)
N-tert-Butylacetamide (762-84-5)
Butylacetanilide (91-49-6)
N-Butylacetanilide (91-49-6)
Butylacetat (German) (123-86-4)
1-Butyl acetate (123-86-4)
2-Butyl acetate (105-46-4)
t-Butyl acetate (540-88-5)
sec-Butyl acetate (ACGIH,DOT,OSHA) [UN 1123] (105-46-4)
n-Butyl acetate (ACGIH,OSHA) (123-86-4)
tert-Butyl acetate (ACGIH,OSHA) [UN 1123] (540-88-5)
Butyl acetate [UN 1123] (123-86-4)
Butylacetaten (Dutch) (123-86-4)
Butylacetic acid (142-62-1)
Butyl acetoacetate (591-60-6)
4-tert-Butylacetophenone (943-27-1)
N-Butyl-N-(2-acetoxyethyl)-4-((4-nitro-2,6-dicyanophenyl)azo)-3-methylbenzeneamine (72828-64-9)
Butyl acetoxymethylnitrosamine (56986-36-8)
N-Butyl-N-(acetoxymethyl)nitrosamine (56986-36-8)
Butyl 12-acetoxyoleate (140-04-5)
Butyl 12-(acetyloxy)-9-octadecenoate (140-04-5)
Butyl acetyl ricinoleate (140-04-5)
Butyl acid phosphate (52933-01-4)
n-Butyl acid phosphate (12788-93-1)
3-Butylacrolein (18829-55-5)
β-Butylacrolein (18829-55-5)
N-tert-Butylacrylamide (107-58-4)
tert-Butyl acrylate (1663-39-4)
n-Butyl acrylate (ACGIH) (141-32-2)

Butylacrylate, Inhibited (DOT) (141-32-2)
Butylacrylate (OSHA) [UN 2348] (141-32-2)
Butyl acrylate, N-methylolacrylamide, vinyl acetate polymer (26428-41-1)
n-Butyl acrylate, N-methylolacrylamide, vinyl acetate polymer (26428-41-1)
Butyl acrylate, vinyl acetate, N-methylolacrylamide polymer (26428-41-1)
Butyl acrylate, vinyl acetate, methylolacrylamide polymer (26428-41-1)
Butyl adipate (105-99-7)
1-Butyl alcohol (71-36-3)
2-Butyl alcohol (78-92-2)
s-Butyl alcohol (78-92-2)
sec-Butyl alcohol (78-92-2)
tert-Butyl alcohol (75-65-0)
n-Butyl alcohol (ACGIH,OSHA) (71-36-3)
sec-Butyl alcohol (ACGIH,OSHA) [UN 1120] (78-92-2)
tert-Butyl alcohol (ACGIH,OSHA) [UN 1120] (75-65-0)
Butyl alcohol [UN 1120] (71-36-3)
sec-Butyl alcohol acetate (105-46-4)
Butyl alcohol-d (8CI) (4712-38-3)
sec-Butyl alcohol-d (6CI,7CI,8CI) (4712-39-4)
Butyl alcohol, hydrogen phosphite (1809-19-4)
Butyl alcohol, sodium salt (2372-45-4)
Butyl aldehyde (123-72-8)
n-Butyl aldehyde (123-72-8)
sec-Butyl allyl barbituric acid (115-44-6)
n-Butylamide (541-35-5)
n-Butylamin (German) (109-73-9)
(+)-2-Butylamine (513-49-5)
(RS)-sec-Butylamine (13952-84-6)
S-2-Butylamine (513-49-5)
sec-Butylamine (13952-84-6)
tert-Butylamine (75-64-9)
n-Butylamine (ACGIH) [UN 1125] (109-73-9)
tert-Butylamine, Compd. with Borane (1:1) (7337-45-3)
Butylamine (DOT,OSHA) (109-73-9)
sec-Butylamine, (S) (513-49-5)
Butyl amine, N-(1-acetoxymethyl)-N-nitroso- (56986-36-8)
Butylamine, N,N-bis(3-aminopropyl)- (8CI) (1555-68-6)
Butylamine, bis(1,3-dimethyl) (107-45-9)
tert-Butylamine borane (7337-45-3)
Butylamine, 4-(diethoxymethylsilyl) (3037-72-7)
Butylamine, N,N-diethyl- (6CI,7CI,8CI) (4444-68-2)
Butylamine, 1,3-dimethyl (108-09-8)
Butylamine, 2-ethyl (617-79-8)
Butylamine, N-ethyl (13360-63-9)
Butylamine, N-ethyl-1,3-dimethyl- (6CI,7CI) (42966-64-3)
Butylamine, N-ethyl-N-nitroso (4549-44-4)
tert-Butylamine, N-ethyl-N-nitroso (3398-69-4)
Butylamine, 3-methyl- (107-85-7)
Butylamine, N-methyl (110-68-9)
Butylamine, 1-methyl- (8CI) (625-30-9)
Butylamine, N-methyl-N-nitroso (7068-83-9)
Butylamine, N-nitroso (56375-33-8)
Butylamine, N-nitrosodi- (924-16-3)
Butylamine oleate (26094-13-3)
tert-Butylamine 2-pyridinethiol-1-oxide (33079-08-2)
Butylamine, tertiary (75-64-9)
2-tert.Butylamino-4-aethylamino-6-chlor-1,3,5-triazin (German) (5915-41-3)
2-tert.Butylamino-4-aethylamino-6-methylthio-1,3,5-triazin (German) (886-50-0)
Butyl 2-aminobenzoate (7756-96-9)
Butyl p-aminobenzoate (94-25-7)
p-(Butylamino)benzoic acid, 2-(dimethylamino)ethyl ester, hydrochloride (136-47-0)
p-Butylaminobenzoyl-2-dimethylaminoethanol hydrochloride (136-47-0)
4-(Butylamino)-N-butyl-1,8-naphthalimide (19125-99-6)
4-Butylamino-N-butyl-1,8-naphthalimide (19125-99-6)
2-tert-Butylamino-4-chloro-6-ethylamino-s-triazine (5915-41-3)
1-(Butylamino)cyclohexylphosphonic acid dibutyl ester (51249-05-9)
tert-Butylaminoethanol (4620-70-6)
2-sec-Butylamino-4-ethylamino-6-methoxy-1,3,5-triazine (26259-45-0)
2-sec-Butylamino-4-ethylamino-6-methoxy-s-triazine (26259-45-0)
2-tert-Butylamino-4-ethylamino-6-methoxy-1,3,5-triazine (33693-04-8)
2-tert-Butylamino-4-ethylamino-6-methoxy-s-triazine (33693-04-8)
2-tert-Butylamino-4-ethylamino-6-methylmercapto-s-triazine (886-50-0)
2-tert-Butylamino-4-ethylamino-6-methylthio-s-triazine (886-50-0)
2-(tert-Butylamino)ethyl methacrylate (3775-90-4)
tert-Butylaminoethyl methacrylate (3775-90-4)
p-(N-tert-Butylamino)nitrobenzene (4138-38-9)
N-Butyl-p-aminophenol (103-62-8)
N-n-Butyl-p-aminophenol (103-62-8)
p-(Butylamino)phenol (103-62-8)
1-(tert-Butylamino)-3-((5,6,7,8-tetrahydro-cis-6,7-dihydroxy-1-naphthyl)oxy)-2-propanol (42200-33-9)
Butylammonium oleate (26094-13-3)
N-(n-Butyl)aniline (1126-78-9)
N-n-Butylaniline (1126-78-9)
p-n-Butylaniline (104-13-2)
sec-Butylaniline (68400-78-2)
N-Butylaniline [UN 2738] (1126-78-9)
Butylate (2008-41-5)
Butylate 2,4,5-T (93-79-8)
N-Butylated-para-aminophenol (103-62-8)
2-tert-Butylated hydroxyanisole (121-00-6)
3-tert-Butylated hydroxyanisole (88-32-4)
Butylated hydroxyanisole (25013-16-5)
Butylated hydroxytoluene (128-37-0)
tert-Butyl azidoformate (1070-19-5)
1-Butylaziridine (1120-85-0)
n-Butylaziridine (1120-85-0)
2-tert-Butylazo-2-hydroxy-5-methylhexane (64819-51-8)
p-tert-Butylbenzoic acid, triethanolamine salt (59993-86-1)
4-Butylbenzaldehyde (1200-14-2)
n-Butylbenzene (104-51-8)
n-Butylbenzene [UN 2709] (104-51-8)
sec-Butylbenzene [UN 2709] (135-98-8)
tert-Butylbenzene [UN 2709] (98-06-6)
4-Butylbenzeneamine (104-13-2)
4-Butyl-1,2-benzenediamine (3663-23-8)
4-tert-Butyl-1,2-benzenediol (98-29-3)
n-Butylbenzenesulfonamide (3622-84-2)
5-Butyl-2-benzimidazolecarbamic acid methyl ester (14255-87-9)
n-(Butyl-5, benzimidazolyl)-2, carbamate de methyle (French) (14255-87-9)
(4-Butyl-1H-benzimidazol-2-yl)-carbamic acid methyl ester (14255-87-9)
Butyl benzoate (136-60-7)
n-Butyl benzoate (136-60-7)
4-Butylbenzoic acid (20651-71-2)
p-Butylbenzoic acid (20651-71-2)
p-tert-Butyl benzoic acid (98-73-7)
p-t-Butylbenzoic acid, triethanolamine salt (59993-86-1)
N-tert-Butyl-2-benzothiazolesulfenamide (95-31-8)
p-tert-Butylbenzyl alcohol (877-65-6)
Butyl benzyl phthalate (85-68-7)
n-Butyl benzyl phthalate (85-68-7)
Butyl-1,1'-biphenyl (41638-55-5)
Butylbis(2-hydroxyethyl)amine (102-79-4)
n-Butyl-N,N-bis(hydroxyethyl)amine (102-79-4)
Butyl borate (688-74-4)
n-Butyl borate (688-74-4)
Butyl borate, tri- (688-74-4)
n-Butylboronate (4426-47-5)
n-Butylboronic acid (4426-47-5)
sec-Butyl bromide (78-76-2)
tert-Butyl bromide (507-19-7)
n-Butyl bromide [UN 1126] (109-65-9)

Butyl bromide, normal (DOT) (109-65-9)
3-sek.Butyl-5-brom-6-methyluracil (German) (314-40-9)
n-Butyl-1-butanamine (111-92-2)
Butyl butanedioate (141-03-7)
n-Butyl n-butanoate (109-21-7)
Butyl-butanol(4)-nitrosamin (German) (3817-11-6)
Butyl-butanol-nitrosamine (3817-11-6)
Butyl 2-butenoate (7299-91-4)
Butyl butex (94-26-8)
2-Butyl-6-(butylamino)-1H-benz(de)isoquinoline-1,3(2H)-dione (19125-99-6)
Butyl 4-tert-butylbenzyl N-(3-pyridyl)dithiocarbonimidate (51308-54-4)
N-sec-Butyl-4-tert-butyl-2,6-dinitroaniline (33629-47-9)
Butyl butyrate (109-21-7)
n-Butyl butyrate (109-21-7)
n-Butyl n-butyrate (109-21-7)
γ-n-Butyl-γ-butyrolactone (104-50-7)
Butyl caproate (626-82-4)
Butyl caprylate (589-75-3)
n-Butylcaprylate (589-75-3)
Butyl carbamate (592-35-8)
tert-Butylcarbamic acid ester with 3-(m-hydroxyphenyl)-1,1-dimethylurea (4849-32-5)
1-(Butylcarbamoyl)-2-benzimidazolecarbamic acid, methyl ester (17804-35-2)
1-(Butylcarbamoyl)-2-benzimidazol-methylcarbamat (German) (17804-35-2)
1-(N-Butylcarbamoyl)-2-(methoxy-carboxamido)-benzimidazol (German) (17804-35-2)
N'-(Butylcarbamoyl)sulfanilamide (339-43-5)
N[1]-(Butylcarbamoyl)sulfanilamide (339-43-5)
dl-sec-Butyl carbinol (137-32-6)
n-Butylcarbinol (71-41-0)
tert-Butyl carbinol (75-84-3)
tert-Butylcarbinol (75-84-3)
Butyl carbitol (112-34-5)
Butyl carbitol acetate (124-17-4)
Butylcarbitol formal (143-29-3)
Butyl carbitol 6-propylpiperonyl ether (51-03-6)
Butyl carbitol rhodanate (112-56-1)
Butyl carbitol thiocyanate (112-56-1)
Butyl-carbityl (6-propylpiperonyl) ether (51-03-6)
Butyl carbobutoxymethyl phthalate (85-70-1)
5-Butyl-2-(carbomethoxyamino)benzimidazole (14255-87-9)
4-t-Butylcatechol (98-29-3)
tert-Butylcatechol (27213-78-1)
Butyl cellosolve (111-76-2)
Butyl cellosolve acetate (112-07-2)
Butyl "cellosolve" adipate (bca) (141-18-4)
Butyl "cellosolve" phthalate (117-83-9)
Butylcelosolv (Czech) (111-76-2)
Butylcelosolvacetat (Czech) (112-07-2)
Butyl chemosept (94-26-8)
O-(4-tert Butyl-2-chloor-fenyl)-O-methyl-fosforzuur-N-methyl-amide (Dutch) (299-86-5)
Butylchloral hydrate (76-40-4)
Butyl-chlorhydrinether (Czech) (16224-33-2)
sec-Butyl chloride (78-86-4)
tert-Butyl chloride (507-20-0)
Butyl chloride (DOT) (109-69-3)
n-Butyl chloride [UN 1127] (109-69-3)
3-tert.Butyl-5-chlor-6-methyluracil (German) (5902-51-2)
Butyl chloroacetate (590-02-3)
n-Butyl chloroacetate (590-02-3)
n-Butyl-chloroacetate (590-02-3)
2-t-Butyl-6-(5-chloro-2h-benzotriazol-2-yl)-p-cresol (3896-11-5)
2-tert-Butyl-6-(5-chloro-2H-benzotriazol-2-yl)-p-cresol (3896-11-5)
sec-Butyl chlorocarbonate (17462-58-7)
Butylchlorodihydroxystannane (13355-96-9)
sec-Butyl chloroformate (17462-58-7)
sec-Butylchloroformate (17462-58-7)
3-tert-Butyl-5-chloro-6-methyluracil (5902-51-2)
6-tert-Butyl-3-chloromethyl-2,4-xylenol (23500-79-0)
4-tert-Butyl-2-chlorophenol (98-28-2)
O-(4-tert-Butyl-2-chlorophenyl) O-methyl N-methylamido phosphate (299-86-5)
4-t-Butyl-2-chlorophenyl methyl methylphosphoramidate (299-86-5)
4-tert. Butyl 2-chlorophenyl methylphosphoramidate de methyle (French) (299-86-5)
Butylchlorotin dihydroxide (13355-96-9)
O-(4-tert-Butyl-2-chlor-phenyl)-O-methyl-phosphorsaeure-N-methylamid (German) (299-86-5)
t-Butyl chromate (1189-85-1)
tert-Butyl chromate (ACGIH,OSHA) (1189-85-1)
Butyl cinnamate (538-65-8)
n-Butyl cinnamate (538-65-8)
n-Butyl citrate (77-94-1)
Butyl citrate (VAN) (77-94-1)
2-t-Butyl-p-cresol (2409-55-4)
6-tert-Butyl-m-cresol (88-60-8)
tert-Butyl-m-cresol (1333-13-7)
Butyl crotonate (7299-91-4)
tert-Butyl cumene peroxide, Technically pure (DOT) (3457-61-2)
tert-Butyl cumyl peroxide (3457-61-2)
tert-Butyl cumyl peroxide, Technically pure (DOT) (3457-61-2)
Butyl cyanate (1768-24-7)
Butylcyclohexane (1678-93-9)
t-Butylcyclohexane (3178-22-1)
tert-Butylcyclohexane (3178-22-1)
4-tert-Butylcyclohexanol (98-52-2)
o-tert-Butylcyclohexanol (13491-79-7)
p-tert-Butylcyclohexanone (98-53-3)
4-tert-Butylcyclohexyl acetate (32210-23-4)
p-tert-Butylcyclohexyl acetate (32210-23-4)
N-Butylcyclohexylamine (10108-56-2)
Butyl cyclohexyl phthalate (84-64-0)
tert-Butyl-d9 alcohol-d (53001-22-2)
Butyl decanoate (30673-36-0)
Butyl decyl phthalate (89-19-0)
n-Butyl-4,4-di(tert-butylperoxy)valerate, Not more than 52% with inert solid (DOT) (995-33-5)
n-Butyl-4,4-di(tert-butylperoxy)valerate, Technically pure (DOT) (995-33-5)
2-tert-Butyl-4-(2,4-dichloro-5-isopropyloxyphenyl)-1,3,4-oxa-diazolin-5-one (19666-30-9)
Butyl (2,4-dichlorophenoxy)acetate (94-80-4)
Butyl dichlorophenoxyacetate (94-80-4)
Butyl 4-(2,4-dichlorophenoxy)butyrate (6753-24-8)
1-Butyl-3-(3,4-dichlorophenyl)-1-methylurea (555-37-3)
N-Butyl-N'-(3,4-dichlorophenyl)-N-methylurea (555-37-3)
Butyldiethanolamine (102-79-4)
N-Butyldiethanolamine (102-79-4)
tert-Butyldiethanolamine (2160-93-2)
O-Butyl diethylene glycol (112-34-5)
Butyl diglyme (112-73-2)
Butyl 3,4-dihydro-2,2-dimethyl-4-oxo-2h-pyran-6-carboxylate (532-34-3)
4'-tert-Butyl-2',6'-dimethyl-3',5'-dinitroacetophenone (81-14-1)
Butyl (4-(1,1-dimethylethyl)phenyl)methyl 3-pyridinyl carbon-imidodithioate (51308-54-4)
2-tert-Butyl-4,6-dimethylphenol (1879-09-0)
6-t-Butyl-2,4-dimethylphenol (1879-09-0)
2-tert-Butyl-4,6-dinitro-m-cresol (3996-59-6)
n-Butyl-2,6-dinitro-N-ethyl-4-trifluoromethylaniline (1861-40-1)
2-sek.Butyl-4,6-dinitrofenylester kyseliny 3-methylkrotonove (Czech) (485-31-4)
2-sek.Butyl-4,6-dinitrofenylester kyseliny octove (Czech) (2813-95-8)
(2-sek.Butyl-4,6-dinitrofenyl)-isopropylkarbonat (Czech) (973-21-7)
2-sec-Butyl-4,6-dinitrophenol (88-85-7)

2-tert-Butyl-4,6-dinitrophenol (1420-07-1)
2-sec-Butyl-4,6-dinitrophenol, ammonium salt (6365-83-9)
2-sec-Butyl-4,6-dinitrophenol 2,2',2''-nitrilotriethanol salt (6420-47-9)
o-sec-Butyl-4,6-dinitrophenol triethanolamine salt (6420-47-9)
2-sec-Butyl-4,6-dinitrophenylacetate (2813-95-8)
6-sec-Butyl-2,4-dinitrophenylacetate (2813-95-8)
2-sec-Butyl-4,6-dinitrophenyl-3,3-dimethylacrylate (485-31-4)
2-sec-Butyl-4,6-dinitrophenyl isopropyl carbonate (973-21-7)
2-sec-Butyl-4,6-dinitrophenyl 3-methyl-2-butenoate (485-31-4)
2-sec Butyl-4,6-dinitrophenyl 3-methylcrotonate (485-31-4)
2-sec-Butyl-4,6-dinitrophenyl senecioate (485-31-4)
Butyl dioxitol (112-34-5)
4-Butyl-1,2-diphenyl-3,5-dioxopyrazolidine (50-33-9)
.β.-tert-Butyl-1,1-diphenylethylene (23586-64-3)
Butyl diphenyl phosphate (2752-95-6)
4-Butyl-1,2-diphenyl-3,5-pyrazolidinedione (50-33-9)
4-Butyl-1,2-diphenylpyrazolidine-3,5-dione (50-33-9)
Butyl dipotassium phosphate (26290-70-0)
n-Butyl disulfide (629-45-8)
Butyl disulfide (8CI) (629-45-8)
Butyle (acetate de) (French) (123-86-4)
1-Butylene (106-98-9)
α-Butylene (106-98-9)
β-Butylene (107-01-7)
β-cis-Butylene (590-18-1)
β-trans-Butylene (624-64-6)
cis-Butylene (590-18-1)
γ-Butylene (115-11-7)
n-Butylene (25167-67-3)
Butylene (DOT) (25167-67-3)
Butylene [UN 1012] (25167-67-3)
Butylene chlorohydrin (1320-66-7)
1,3-Butylene diacrylate (19485-03-1)
1,4-Butylenediamine (110-60-1)
Butylenediamine (110-60-1)
α-Butylene dibromide (533-98-2)
2-Butylene dichloride (110-57-6)
1,4-Butylene dimethacrylate (2082-81-7)
Butylene dimethacrylate (2082-81-7)
1,2-Butylene glycol (584-03-2)
1,3-Butylene glycol (107-88-0)
1,4-Butylene glycol (110-63-4)
2,3-Butylene glycol (513-85-9)
Butylene glycol (25265-75-2)
β-Butylene glycol (107-88-0)
1,3-Butylene glycol diacrylate (19485-03-1)
1,4-Butylene glycol dimethacrylate (2082-81-7)
Butylene glycol methyl ether (111-32-0)
Butylene glycol monomethyl ether (111-32-0)
1,4-Butylene glycol-terephthalic acid copolymer (26062-94-2)
Butylene glycol-terephthalic acid copolymer (26062-94-2)
1,4-Butylene glycol-terephthalic acid polymer (26062-94-2)
Butylene glycol-terephthalic acid polymer (26062-94-2)
Butylene hydrate (78-92-2)
Butyleneimine (2549-67-9)
1,2-Butylene oxide (106-88-7)
2,3-Butylene oxide (3266-23-7)
Butylene oxide (106-88-7)
Butylene oxide (109-99-9)
β-Butylene oxide (3266-23-7)
1,4-Butylene sulfone (1633-83-6)
1,3-Butylenglykol (German) (107-88-0)
1,2-Butylenimine (2549-67-9)
Butylenin (15687-27-1)
Butyl 2,3-epoxypropyl fumarate (25876-07-7)
Butyl-9,10-epoxystearate (106-83-2)
Butyl ester 2,4-D (94-80-4)
n-Butylester kyselini 2,4,5-trichlorfenoxyoctove (Czech) (93-79-8)
Butylester kyseliny acetoctove (Czech) (591-60-6)
Butylester kyseliny akrylove (Czech) (141-32-2)
Butylester kyseliny p-aminobenzoove (Czech) (94-25-7)
Butylester kyseliny benzoove (Czech) (136-60-7)
n-Butylester kyseliny 2,4-dichlorfenoxyoctove (Czech) (94-80-4)
Butylester kyseliny methakrylove (Czech) (97-88-1)
Butylester kyseliny mlecne (Czech) (138-22-7)
Butylester kyseliny mravenci (Czech) (592-84-7)
Butylester kyseliny octove (Czech) (123-86-4)
terc.Butylester kyseliny peroxybenzoove (Czech) (614-45-9)
Butyl ester of 2,4-D (94-80-4)
n-Butyl ester of 3,4-dihydro-2,2-dimethyl-4-oxo-2H-pyran-6-carboxylic acid (532-34-3)
Butyl ethanoate (123-86-4)
tert-Butylethene (558-37-2)
n-Butyl ether (142-96-1)
sec-Butyl ether (DOT) (6863-58-7)
tert-Butyl ether (DOT) (6163-66-2)
Butyl ether [UN 1149] (142-96-1)
5-Butyl-5-ethyl-2,4,6(1H,3H,5H)-pyrimidinetrione (77-28-1)
Butyl ethyl acetaldehyde (123-05-7)
Butylethylacetic acid (149-57-5)
Butylethylamine (13360-63-9)
5-n-Butyl-2-ethylamino-4-hydroxy-6-methylpyrimidine (23947-60-6)
5-Butyl-2-(ethylamino)-6-methyl-4(1H)-pyrimidinone (23947-60-6)
5-Butyl-5-ethylbarbituric acid (77-28-1)
5-sec-Butyl-5-ethylbarbituric acid (125-40-6)
5-sec-Butyl-5-ethylbarbituric acid sodium salt (143-81-7)
n-Butyl-N-ethyl-2,6-dinitro-4-(trifluoromethyl)benzenamine (1861-40-1)
tert-Butylethylene (558-37-2)
O-Butyl ethylene glycol (111-76-2)
n-Butylethylenimine (1120-85-0)
Butyl ethyl ether (628-81-9)
Butyl 2-ethylhexyl phthalate (85-69-8)
Butyl ethyl ketone (106-35-4)
n-Butyl ethyl ketone (106-35-4)
5-sec-Butyl-5-ethylmalonyl urea (125-40-6)
Butylethylnitrosamin (Czech) (4549-44-4)
2-t-Butyl-4-ethylphenol (96-70-8)
2-tert-Butyl-4-ethylphenol (96-70-8)
2-tert-Butyl-5-ethylphenol (4237-25-6)
2-Butyl-2-ethyl-1,3-propanediol (115-84-4)
Butylethylthiocarbamic acid S-propyl ester (1114-71-2)
n-Butyl-N-ethyl-α,α,α-trifluoro-2,6-dinitro-p-toluidine (1861-40-1)
2-sec.Butylfenol (Czech) (89-72-5)
p-terc.Butylfenol (Czech) (98-54-4)
2-(p-terc.Butylfenoxy)isopropyl-2'-chlorethylester kyseliny siricite (Czech) (140-57-8)
2-sek.Butylfenylester kyseliny methylkarbaminove (Czech) (3766-81-2)
p-terc.Butylfenylester kyseliny salicylove (Czech) (87-18-3)
9-tert-Butylfluorene (17114-78-2)
tert-Butyl fluoride (353-61-7)
Butyl formal (110-62-3)
tert-Butylformamide (2425-74-3)
Butyl formate (DOT) (592-84-7)
n-Butyl formate [UN 1128] (592-84-7)
Butyl glycidyl ether (2426-08-6)
t-Butyl glycidyl ether (7665-72-7)
n-Butyl glycidyl ether (ACGIH,OSHA) (2426-08-6)
Butyl glycol (111-76-2)
Butylglycol (French,German) (111-76-2)
Butylglycol acetate (112-07-2)
Butyl glycol phthalate (117-83-9)
Butyl glycolyl butyl phthalate (85-70-1)
tert-Butylglyoxylic acid (815-17-8)
Butyl heptanoate (5454-28-4)

Butyl heptyl ketone (19780-10-0)
Butyl hexadecanoate (111-06-8)
n-Butyl hexadecanoate (111-06-8)
4-tert-Butylhexahydrophenyl acetate (32210-23-4)
Butyl hexanoate (626-82-4)
n-Butyl hexanoate (626-82-4)
n-Butylhydrazine hydrochloride (56795-65-4)
Butyl hydrogen phthalate (131-70-4)
terc. Butylhydroperoxid (Czech) (75-91-2)
tert-Butyl hydroperoxide (75-91-2)
tert-Butyl hydroperoxide, Maximum concentration 72% with water (DOT) (75-91-2)
tert-Butyl hydroperoxide, More than 72% but not more than 90% water (DOT) (75-91-2)
tert-Butyl hydroperoxide, More than 90% water (DOT) (75-91-2)
tert-Butylhydroquinone (1948-33-0)
Butyl hydroxide (71-36-3)
t-Butyl hydroxide (75-65-0)
Butyl hydroxyacetate (7397-62-8)
2(3)-tert-Butyl-4-hydroxyanisole (25013-16-5)
3-tert-Butyl-4-hydroxyanisole (121-00-6)
Butylhydroxyanisole (25013-16-5)
tert-Butyl-4-hydroxyanisole (25013-16-5)
tert-Butylhydroxyanisole (25013-16-5)
Butyl 4-hydroxybenzoate (94-26-8)
Butyl p-hydroxybenzoate (94-26-8)
n-Butyl-(4-hydroxybutyl)nitrosamine (3817-11-6)
n-Butyl-N-(4-hydroxybutyl)nitrosamine (3817-11-6)
2-(N-Butyl-N-2-hydroxyethylamino)ethanol (102-79-4)
4-Butyl-2-(4-hydroxyphenyl)-1-phenyl-3,5-dioxopyrazolidine (129-20-4)
4-Butyl-1-(4-hydroxyphenyl)-2-phenyl-3,5-pyrazolidinedione (129-20-4)
4-Butyl-1-(p-hydroxyphenyl)-2-phenyl-3,5-pyrazolidinedione (129-20-4)
4-Butyl-2-(p-hydroxyphenyl)-1-phenyl-3,5-pyrazolidinedione (129-20-4)
Butyl α-hydroxypropionate (138-22-7)
Butylhydroxytin oxide (2273-43-0)
Butylhydroxytoluene (128-37-0)
4,4'-Butylidenebis(6-tert-butyl-m-cresol) (85-60-9)
4,4'-Butylidenebis(6-tert-butyl-3-methylphenyl) (85-60-9)
4,4'-Butylidenebis(3-methyl-6-tert-butylphenol) (85-60-9)
Butylidene chloride (541-33-3)
1-Butylimidazole (4316-42-1)
Butylimidazole (4316-42-1)
N-(n-Butyl)imidazole (4316-42-1)
n-Butylimidazole (4316-42-1)
N-n-Butyl imidazole [UN 2690] (4316-42-1)
6-t-Butyl-3-(2-imidazolin-2-ylmethyl)-2,4-dimethylphenol (1491-59-4)
2-tert-Butylimino-3-isopropyl-5-phenylperhydro-1,3,5-thiadiazinan-4-one (69327-76-0)
2-tert-Butylimino-3-isopropyl-5-phenyl-3,4,5,6-tetrahydro-2H-1,3,5-thiadiazin-4-one (69327-76-0)
2-tert-Butylimino-3-isopropyl-5-phenyl-1,3,5-thiadiazinan-4-one (69327-76-0)
2,2'-(Butylimino)diethanol (102-79-4)
N-Butyl-2,2'-iminodiethanol (102-79-4)
Butyl iodide (542-69-8)
n-Butyl iodide (542-69-8)
sec-Butyl iodide (513-48-4)
iso-Butyl isobutyrate (97-87-0)
n-Butyl isocyanate [UN 2485] (111-36-4)
Butyl isocyanide, 1,1,3,3-tetramethyl- (8CI) (14542-93-9)
2-Butyl-1H-isoindole-1,3(2H)-dione (1515-72-6)
N-Butylisomaleimide (27396-39-0)
n-Butyl isopentanoate (109-19-3)
t-Butyl isopropyl benzene hydroperoxide (30026-92-7)
tert-Butyl isopropyl benzene hydroperoxide (30026-92-7)
2-tert-Butyl-4-isopropylphenol (7597-97-9)
1-Butyl isovalerate (109-19-3)
Butyl isovalerate (109-19-3)
n-Butyl isovalerate (109-19-3)
Butyl isovalerianate (109-19-3)
3-(5-tert-Butylisoxazol-3-yl)-1,1-dimethylurea (55861-78-4)
Butylkarbitolacetat (Czech) (124-17-4)
Butyl ketone (502-56-7)
2-terc.Butyl-p-kresol (Czech) (2409-55-4)
Butyl lactate (138-22-7)
n-Butyl lactate (ACGIH,OSHA) (138-22-7)
Butyl laevulinate (2052-15-5)
n-Butyl laevulinate (2052-15-5)
Butyl levulinate (2052-15-5)
n-Butyl levulinate (2052-15-5)
Butyllithium (109-72-8)
2-Butylmalonate (534-59-8)
2-n-Butylmalonate (534-59-8)
Butylmalonate (534-59-8)
2-Butylmalonic acid (534-59-8)
2-n-Butylmalonic acid (534-59-8)
n-Butylmalonic acid (534-59-8)
2-Butyl mercaptan (513-53-1)
sec-Butyl mercaptan (513-53-1)
secondary Butylmercaptan (513-53-1)
tert-Butyl mercaptan (75-66-1)
n-Butyl mercaptan (ACGIH) (109-79-5)
Butyl mercaptan (OSHA) [UN 2347] (109-79-5)
Butyl mesityl oxide (532-34-3)
n-Butyl mesityl oxide oxalate (532-34-3)
n-Butylmesityloxid oxalate (532-34-3)
Butylmethacrylaat (Dutch) (97-88-1)
Butyl 2-methacrylate (97-88-1)
Butyl methacrylate (97-88-1)
tert-Butyl methacrylate (585-07-9)
n-Butyl methacrylate [UN 2227] (97-88-1)
Butyl methacrylate, methyl methacrylate polymer (25608-33-7)
Butyl methoxydibenzoylmethane (70356-09-1)
2-terc.Butyl-4-methoxyfenol (Czech) (25013-16-5)
2-tert-Butyl-4-methoxyphenol (121-00-6)
2-tert-Butyl-4-methoxyphenol (25013-16-5)
3-tert-Butyl-4-methoxyphenol (88-32-4)
tert-Butyl-4-methoxyphenol (25013-16-5)
2-sec-Butyl-3-methoxypyrazine (24168-70-5)
N-tert-Butyl-3-methylbenzamide (42498-33-9)
Butyl 3-methylbutanoate (109-19-3)
n-Butyl 3-methylbutanoate (109-19-3)
Butyl 3-methylbutyrate (109-19-3)
tert-Butyl methyl carbinol (464-07-3)
2-tert-Butyl-6-methylchloroacetanilide (3785-20-4)
Butyl 2-methyl-4-chlorophenoxyacetate (1713-12-8)
6-tert-Butyl-3-methyl-2,4-dinitrophenol (3996-59-6)
2-tert-Butyl-2-methyl-1,3-dioxolane (6135-54-2)
tert-Butyl methyl ether (1634-04-4)
Butyl methyl ether [UN 2350] (628-28-4)
Butylmethyl ethyl malonyl urea sodium (57-33-0)
p-tert-Butyl-α-methylhydrocinnamaldehyde (80-54-6)
p-tert-Butyl-α-methylhydrocinnamic aldehyde (80-54-6)
Butyl methyl ketone (591-78-6)
n-Butyl methyl ketone (591-78-6)
sec-Butyl methyl ketone (565-61-7)
t-Butyl methyl ketone (75-97-8)
Butylmethylnitrosamine (7068-83-9)
2-sec-Butyl-2-methyloxirane (42328-43-8)
2-t-Butyl-4-methylphenol (2409-55-4)
2-tert-Butyl-5-methylphenol (88-60-8)
2-tert-Butyl-6-methylphenol (2219-82-1)
4-tert-Butyl-2-methylphenol (98-27-1)
6-tert-Butyl-3-methylphenol (88-60-8)
1-Butyl-3-(p-methylphenylsulfonyl)urea (64-77-7)

2-sec-Butyl-2-methyl-1,3-propanediol dicarbamate (64-55-1)
Butyl 2-methyl-2-propenoate (97-88-1)
N-Butyl-2-methyl-2-propyl-1,3-propanediol dicarbamate (4268-36-4)
N-n-Butyl-2-methyl-2-propyl-1,3-propanediol dicarbamate (4268-36-4)
3-Butyl-6-methyl-2,4-pyridinediol (6967-70-0)
2-sec-Butyl-2-methyltrimethylene dicarbamate (64-55-1)
3-sec-Butyl-6-methyluracil, sodium salt (65208-42-6)
3-tert-Butyl-6-methyluracil, sodium salt (65086-97-7)
tert-Butyl monopermaleate (1931-62-0)
tert-Butyl monoperoxymaleate (1931-62-0)
tert-Butyl monoperoxymaleate, Maximum concentration 55% as a paste (DOT) (1931-62-0)
tert-Butyl monoperoxymaleate, Maximum concentration 55% in solution (DOT) (1931-62-0)
tert-Butyl monoperoxymaleate, Technically pure (DOT) (1931-62-0)
Butyl monosulfide (544-40-1)
4-Butylmorpholine (1005-67-0)
Butyl myristate (110-36-1)
Butyl namate (136-30-1)
.α.-Butylnaphthalene (1634-09-9)
.β.-tert-Butylnaphthalene (2876-35-9)
1-Butylnaphthalene (1634-09-9)
2-tert-Butylnaphthalene (2876-35-9)
Butylnaphthalenesulfonate sodium salt (25638-17-9)
Butylnaphthalenesulfonic acid (26761-78-4)
Butylnaphthalenesulfonic acid, ammonium salt (27478-24-6)
Butyl nitrate (928-45-0)
n-Butyl nitrite (544-16-1)
sec-Butyl nitrite (924-43-6)
Butyl nitrite [UN 2351] (544-16-1)
N-tert-Butyl-4'-nitroaniline (4138-38-9)
N-tert-Butyl-4-nitroaniline (4138-38-9)
Butylnitrosamine (56375-33-8)
4-(Butylnitrosamino)-1-butanol (3817-11-6)
4-(n-Butylnitrosamino)-1-butanol (3817-11-6)
Butylnitrosaminomethyl acetate (56986-36-8)
n-Butyl-N-nitroso-1-butamine (924-16-3)
Butylnitrosoharnstoff (German) (869-01-2)
1-Butyl-1-nitrosourea (869-01-2)
n-Butyl-N-nitrosourea (869-01-2)
n-Butylnitrosourea (869-01-2)
Butylocaine (136-47-0)
Butyl octadecanoate (123-95-5)
n-Butyl octadecanoate (123-95-5)
Butyl octanoate (589-75-3)
2-Butyl-1-octanol (3913-02-8)
2-Butyloctyl alcohol (3913-02-8)
Butyl octyl 1,2-benzenedicarboxylate (84-78-6)
Butyl octyl phthalate (84-78-6)
Butylohydroksyanizol (Polish) (25013-16-5)
Butylohydroksytoluenu (Polish) (128-37-0)
Butyl oleate (142-77-8)
Butylone (57-33-0)
Butylowy alkohol (Polish) (71-36-3)
Butyl oxitol (111-76-2)
Butyl 4-oxopentanoate (2052-15-5)
Butyloxostannane (51590-67-1)
tert-Butyloxycarbonyl azide (1070-19-5)
α-Butyloxycinchoninic acid diethylethylenediamide (85-79-0)
Butyl palmitate (111-06-8)
n-Butyl palmitate (111-06-8)
Butyl paraben (94-26-8)
Butylparaben (94-26-8)
n-Butyl parahydroxybenzoate (94-26-8)
Butyl parasept (94-26-8)
t-Butyl peracetate (107-71-1)
tert-Butyl peracetate (107-71-1)
terc.Butylperbenzoan (Czech) (614-45-9)
t-Butyl perbenzoate (614-45-9)
tert-Butyl percrotonate (23474-91-1)
tert-Butyl perisobutyrate (109-13-7)
tert-Butyl perneodecanoate (26748-41-4)
tert-Butyl peroxide (DOT) (110-05-4)
t-Butyl peroxyacetate (107-71-1)
tert-Butyl peroxyacetate (107-71-1)
tert-Butyl peroxyacetate, More than 76% in solution (DOT) (107-71-1)
tert-Butyl peroxyacetate, More than 52% to a maximum concentration of 76% (UN 2095 DOT) (107-71-1)
tert-Butyl peroxyacetate, Not more than 52% in solution (UN 2096 DOT) (107-71-1)
t-Butyl peroxy benzoate (614-45-9)
tert-Butyl peroxybenzoate, Not more than 75% in solution (DOT) (614-45-9)
tert-Butyl peroxybenzoate, Not more than 50% with inert inorganic solid (DOT) (614-45-9)
tert-Butyl peroxybenzoate, Technically pure or in concentration of more than 75% (DOT) (614-45-9)
tert-Butyl peroxycrotonate (23474-91-1)
tert-Butyl peroxycrotonate (Not more than 76% in solutdion) (23474-91-1)
tert-Butyl peroxycrotonate, Not more than 76% in solution (23474-91-1)
Butyl peroxydicarbonate (16215-49-9)
sec-Butyl peroxydicarbonate (19910-65-7)
n-Butyl peroxydicarbonate, More than 52% in solution (16215-49-9)
n-Butyl peroxydicarbonate, More than 27% to a maximum concentration of 52% (16215-49-9)
n-Butyl peroxydicarbonate, Not more than 27% in solution (16215-49-9)
n-Butyl peroxydicarbonate, Not more than 52% in solution (16215-49-9)
tert-Butyl peroxy-2-ethylhexanoate (62695-55-0)
tert-Butyl peroxyethylhexanoate (62695-55-0)
tert-Butyl peroxy-2-ethylhexanoate (Not more than 12% with 2,2-di-(tert-butylperoxy)butane, not more than 14% with not less than 14% phlegmatizer and 60% inert inorganic solid) (62695-55-0)
tert-Butyl peroxy-2-ethylhexanoate (Not more than 30% with 2,2-di-(tert-butylperoxy)butane, not more than 35%, with not less than 35% phlegmatizer) (62695-55-0)
tert-Butyl peroxy-2-ethylhexanoate (Not more than 50% with phlegmatizer) (62695-55-0)
tert-Butyl peroxy-2-ethyl hexanoate, Technically pure (62695-55-0)
tert-Butyl peroxyisobutyrate (109-13-7)
tert-Butyl peroxyisobutyrate, Not more than 52% in solution (DOT) (109-13-7)
tert-Butyl peroxyisobutyrate, more than 52% but not more than 77% in solution (DOT) (109-13-7)
tert-Butyl peroxyisobutyrate, more than 77% in solution (DOT) (109-13-7)
tert-Butyl peroxy isopropyl carbonate, Technically pure (DOT) (2372-21-6)
tert-Butyl peroxymaleate, Not more than 55% as a paste (DOT) (1931-62-0)
tert-Butyl peroxymaleate, Not more than 55% in solution (DOT) (1931-62-0)
tert-Butyl peroxymaleate, Technically pure (DOT) (1931-62-0)
tert-Butyl peroxymaleic acid (1931-62-0)
tert-Butyl peroxy-2-methylbenzoate (22313-62-8)
tert-Butyl peroxy-o-methylbenzoate (22313-62-8)
tert-Butyl peroxyneodecanoate, Not more than 77% in solution (DOT) (26748-41-4)
tert-Butyl peroxyneodecanoate, Technically pure (DOT) (26748-41-4)
t-Butyl peroxypivalate (927-07-1)
tert-Butyl peroxypivalate, Not more than 77% in solution (DOT) (927-07-1)
tert-Butyl perpivalate (927-07-1)
Butylphen (98-54-4)
2-n-Butylphenol (3180-09-4)
2-t-Butylphenol (88-18-6)
4-n-Butylphenol (1638-22-8)
4-sec-Butylphenol (99-71-8)
4-t-Butylphenol (98-54-4)
p-(sec-Butyl)phenol (99-71-8)
p-tert-Butylphenol (98-54-4)
tert-Butylphenol (27178-34-3)

o-sec-Butylphenol (ACGIH,OSHA) **(89-72-5)**
o-Butylphenol, Liquid [UN 2228] **(3180-09-4)**
p-Butylphenol, Liquid [UN 2228] **(1638-22-8)**
o-Butylphenol, Solid [UN 2229] **(3180-09-4)**
p-Butylphenol, Solid [UN 2229] **(1638-22-8)**
Butyl phenol, formaldehyde resin in xylene **(9039-76-3)**
tert-Butylphenol glycidyl ether **(26447-45-0)**
tert-Butylphenol glycidyl ether **(3101-60-8)**
p-tert-Butylphenol, phosphate (3:1) **(78-33-1)**
p-tert-Butylphenol sodium salt **(5787-50-8)**
2-(p-t-Butylphenoxy)cyclohexanol **(1942-71-8)**
2-(p-t-Butylphenoxy)cyclohexyl propargyl sulfite **(2312-35-8)**
2-(p-tert-Butylphenoxy)cyclohexyl 2-propynyl sulfite **(2312-35-8)**
3-(p-tert-Butylphenoxy)-1,2-epoxypropane **(3101-60-8)**
2-(4-t-Butylphenoxy)isopropyl-2-chloroethyl sulfite **(140-57-8)**
2-(p-Butylphenoxy)isopropyl 2-chloroethyl sulfite **(140-57-8)**
Butylphenoxyisopropyl chloroethyl sulfite **(140-57-8)**
2-(p-t-Butylphenoxy)isopropyl 2'-chloroethyl sulphite **(140-57-8)**
2-(p-t-Butylphenoxy)-1-methylethyl 2-chloroethyl ester of sulphurous acid **(140-57-8)**
2-(p-Butylphenoxy)-1-methylethyl 2-chloroethyl sulfite **(140-57-8)**
2-(p-t-Butylphenoxy)-1-methylethyl-2-chloroethyl sulfite **(140-57-8)**
2-(p-t-Butylphenoxy)-1-methylethyl 2'-chloroethyl sulphite **(140-57-8)**
2-(p-t-Butylphenoxy)-1-methylethyl sulphite of 2-chloroethanol **(140-57-8)**
Butylphenyl **(41638-55-5)**
n-Butyl phenylacrylate **(538-65-8)**
2-(Butylphenylamino)ethanol **(3046-94-4)**
t-Butylphenyl diphenyl phosphate **(56803-37-3)**
tert-Butylphenyl diphenyl phosphate **(56803-37-3)**
4-n-Butyl-o-phenylenediamine **(3663-23-8)**
tert-Butylphenyl 2,3-epoxypropyl ether **(26447-45-0)**
Butyl phenyl ether **(1126-79-0)**
tert-Butylphenyl glycidyl ether **(26447-45-0)**
tert-Butylphenyl glycidyl ether **(3101-60-8)**
Butyl phenylmethyl 1,2-benzenedicarboxylate **(85-68-7)**
2-sec-Butylphenyl N-methylcarbamate **(3766-81-2)**
o-sec-Butylphenyl methylcarbamate **(3766-81-2)**
tert-Butylphenylphosphinoyl fluoride **(55236-56-1)**
p-tert-Butylphenyl salicylate **(87-18-3)**
Butyl phosphate, tri- **(126-73-8)**
Butyl phosphoric acid **(12788-93-1)**
Butyl phosphorotrithioate **(78-48-8)**
Butyl phthalate butyl glycolate **(85-70-1)**
N-Butylphthalimide **(1515-72-6)**
Butyl phthalyl butyl glycolate **(85-70-1)**
5-Butylpicolinic acid **(536-69-6)**
Butyl potassium phosphate (1:1:1) **(25238-99-7)**
Butyl propanoate **(590-01-2)**
Butyl 2-propenoate **(141-32-2)**
tert-Butyl propenoate **(1663-39-4)**
n-Butyl propionate **(590-01-2)**
Butylpropionate [UN 1914] **(590-01-2)**
3-Butylpyridine **(539-32-2)**
3-n-Butylpyridine **(539-32-2)**
5-Butyl-2-pyridinecarboxylic acid **(536-69-6)**
Butylpyrin **(50-33-9)**
4-t-Butylpyrocatechol **(98-29-3)**
p-t-Butylpyrocatechol **(98-29-3)**
4-tert-Butylpyrokatechin (Czech) **(98-29-3)**
Butyl rubber, brominated **(68441-14-5)**
Butylspiropentane **(6191-90-8)**
Butylstannium trichloride **(1118-46-3)**
Butylstannoic acid **(2273-43-0)**
Butyl stearate **(123-95-5)**
n-Butyl stearate **(123-95-5)**
1-Butyl-3-sulfanilylurea **(339-43-5)**
N-Butylsulfanilylurea **(339-43-5)**
Butyl-sulfide **(544-40-1)**
n-Butyl-sulfide **(544-40-1)**
sec-Butyl sulfide (8CI) **(626-26-6)**
tert-Butyl sulfide (8CI) **(107-47-1)**
tert-Butyl sulfone (8CI) **(1886-75-5)**
1-Butyl (sulfooxy)-9-octadecenoate **(38621-44-2)**
Butyl tegosept **(94-26-8)**
Butyl n-tetradecanoate **(110-36-1)**
Butyl tetradecanoate **(110-36-1)**
6-Butyltetrahydro-2H-pyran-2-one **(3301-94-8)**
tert-Butyl tetralin **(73090-68-3)**
1-(5-tert-Butyl-1,3,4-thiadiazol-2yl)-3-dimethylharnstoff (German) **(34014-18-1)**
1-(5-tert-Butyl-1,3,4-thiadiazol-2-yl)-1,3-dimethylurea **(34014-18-1)**
1-(5-tert-Butyl-1,3,4-thiadiazol-2-yl)-4-hydroxy-1-methyl-2-imidazolidinone **(55511-98-3)**
sec-Butyl thioalcohol **(513-53-1)**
Butylthiobutane **(544-40-1)**
2-(n-Butylthio)-4,6-dichloro-s-triazine **(30894-61-2)**
sec-Butyl thiol **(513-53-1)**
S-((tert-Butylthio)methyl)O,O-diethylphosphorodithioate **(13071-79-9)**
Butyltin hydroxide oxide **(2273-43-0)**
Butyltin tris(2-ethylhexoate) **(23850-94-4)**
Butyltintris(2-oleyoyloxyethylmercaptide) **(67361-76-6)**
Butyl titanate **(5593-70-4)**
Butyl toluene **(27458-20-4)**
p-tert-Butyltoluene (ACGIH,OSHA) **(98-51-1)**
N-Butyl-N'-p-toluenesulfonylurea **(64-77-7)**
N-Butyl-N'-toluene-p-sulfonylurea **(64-77-7)**
1-Butyl-3-(p-tolylsulfonyl)urea **(64-77-7)**
1-Butyl-3-tosylurea **(64-77-7)**
N-n-Butyl-N'-tosylurea **(64-77-7)**
4-n-Butyl-4H-1,2,4-triazole **(16227-10-4)**
Butyl 2,4,5-trichlorophenoxyacetate **(93-79-8)**
n-Butyl (2,4,5-trichlorophenoxy)acetate **(93-79-8)**
Butyltrichlorosilane [UN 1747] **(7521-80-4)**
Butyl 2-(4-(5-trifluoromethyl-2-pyridinyloxy)phenoxy)propanoate **(69806-50-4)**
5-tert-Butyl-1,2,3-trimethyl-4,6-dinitrobenzene **(145-39-1)**
tert-Butyl trimethylperoxyacetate **(927-07-1)**
5-tert-Butyl-2,4,6-trinitroxylene **(81-15-2)**
Butyltris(2-ethylhexyloxycarbonylmethylthio)stannane **(25852-70-4)**
Butyltris((2-ethyl-1-oxohexyl)oxy)stannane **(23850-94-4)**
n-Butylurea **(592-31-4)**
sec-Butylurea **(689-11-2)**
Butyl vinyl ether **(111-34-2)**
Butyl vinyl ether, Inhibited [UN 2352] **(111-34-2)**
Butylvinylether ethylenglykolu (Czech) **(4223-11-4)**
Butyl xanthate **(110-50-9)**
Butylxanthate **(110-50-9)**
6-t-Butyl-2,4-xylenol **(1879-09-0)**
Butyl zimate **(136-23-2)**
Butyl ziram **(136-23-2)**
1-Butyne **(107-00-6)**
2-Butyne (8CI,9CI) **(503-17-3)**
1-Butyne, 3-chloro- (9CI) **(21020-24-6)**
1-Butyne, 4-chloro- (9CI) **(51908-64-6)**
1-Butyne, 1-chloro- (7CI,9CI) **(62981-74-2)**
2-Butyne, 1,4-dichloro **(821-10-3)**
1-Butyne, 1,4-dichloro- (9CI) **(83682-45-5)**
1-Butyne, 4,4-dichloro- (9CI) **(83682-42-2)**
1-Butyne, 3,3-dimethyl- (8CI,9CI) **(917-92-0)**
2-Butynedioic acid **(142-45-0)**
Butynedioic acid **(142-45-0)**
2-Butynedioic acid, dimethyl ester **(762-42-5)**
2-Butyne-1,4-diol **(110-65-6)**
Butynediol **(110-65-6)**
1,4-Butynediol [UN 2716] **(110-65-6)**
1,4-Butynediol, ethoxylated **(32167-31-0)**

Name	CAS No.
2,2'-(2-Butyne-1,4-diylbis(oxy))bisethanol	**(1606-85-5)**
1-Butyne, 3-ethoxy-3-methyl- (9CI)	**(7740-69-4)**
1-Butyne, 1,3,3,4,4,4-hexachloro- (9CI)	**(83682-47-7)**
1-Butyne, 1,4,4-trichloro- (9CI)	**(83682-46-6)**
1-Butyne, 4,4,4-trichloro- (9CI)	**(83682-43-3)**
2-Butynoic acid, 4-cyclobutyl-4-oxo-, ethyl ester (9CI)	**(54966-51-7)**
Butynorate	**(77-58-7)**
3-Butynyl-m-chlorocarbanilate	**(1967-16-4)**
2-Butynyl-4-chloro-m-chlorocarbanilate	**(101-27-9)**
Butynyl-3N-3-chlorophenylcarbamate mixed with 3-cyclooctyl-1,1-dimethyl urea	**(8015-55-2)**
1-Butynyltrimethylsilane	**(62108-37-6)**
2-Butyn-1-ol	**(764-01-2)**
3-Butyn-2-ol, m-chlorocarbanilate	**(1967-16-4)**
2-Butyn-1-ol, 4-chloro-, m-chlorocarbanilate	**(101-27-9)**
Butyn-1-ol-3-ester of m-chlorophenylcarbamic acid	**(1967-16-4)**
1-Butyn-3-ol, 3-methyl-	**(115-19-5)**
3-Butyn-2-ol, 2-methyl	**(115-19-5)**
Butyrac	**(10433-59-7)**
Butyrac	**(94-82-6)**
Butyrac 118	**(94-82-6)**
Butyrac 200	**(94-82-6)**
Butyrac ester	**(94-82-6)**
Butyral	**(123-72-8)**
Butyraldehyd (German)	**(123-72-8)**
Butyraldehyde	**(123-72-8)**
n-Butyraldehyde	**(123-72-8)**
Butyraldehyde [UN 1129]	**(123-72-8)**
Butyraldehyde, dibutyl acetal (6CI,7CI,8CI)	**(5921-80-2)**
Butyraldehyde, 2-ethyl	**(97-96-1)**
Butyraldehyde, 3-hydroxy	**(107-89-1)**
Butyraldehyde, 2-methyl	**(96-17-3)**
Butyraldehyde, 3-methyl	**(590-86-3)**
Butyraldehyde, oxime (8CI)	**(110-69-0)**
Butyraldehyde, 4-phenyl- (8CI)	**(18328-11-5)**
Butyraldol	**(496-03-7)**
n-Butyraldoxime	**(110-69-0)**
Butyraldoxime [UN 2840]	**(110-69-0)**
n-Butyramide	**(541-35-5)**
Butyramide (8CI)	**(541-35-5)**
Butyramide, N-acetyl- (7CI,8CI)	**(22534-71-0)**
Butyramide, 2,4-dihydroxy-N-(3-hydroxypropyl)-3,3-dimethyl-, D-(+)	**(81-13-0)**
Butyramide, N-hexyl- (8CI)	**(10264-17-2)**
Butyramide, 4-hydroxy-	**(927-60-6)**
Butyranhydrid (Czech)	**(106-31-0)**
n-Butyranilide	**(1129-50-6)**
Butyranilide (8CI)	**(1129-50-6)**
Butyranilide, 4'-amino-2,2,3,3,4,4,4-heptafluoro-2'-hydroxy-	**(847-51-8)**
Butyranilide, 4'-chloro-2'-methyl-3-oxo-	**(20139-55-3)**
Butyranilide, 2,2,3,3,4,4,4-heptafluoro-2'-hydroxy-4'-nitro-	**(2712-83-6)**
Butyrate sodium	**(156-54-7)**
Butyric-acid	**(107-92-6)**
n-Butyric acid (DOT)	**(107-92-6)**
Butyric acid [UN 2820]	**(107-92-6)**
Butyric acid, 2-acetyl-	**(4433-85-6)**
Butyric acid, 4-amino	**(56-12-2)**
Butyric acid, 2-amino-, DL- (8CI)	**(2835-81-6)**
Butyric acid, 3-((2-amino-2-carboxyethyl)thio)-, Stereoisomer (8CI)	**(21861-11-0)**
Butyric acid, DL-2-amino-4-(ethylthio)-	**(67-21-0)**
Butyric acid, 2-amino-4-(ethylthio)-, DL	**(67-21-0)**
Butyric acid, 2-amino-4-(ethylthio)-, L	**(13073-35-3)**
Butyric acid, 2-amino-4-(ethylthio)-, d	**(535-32-0)**
Butyric acid, 2-amino-4-(guanidinooxy)-, L	**(543-38-4)**
Butyric acid, 2-amino-4-hydroxy-, L- (8CI)	**(672-15-1)**
Butyric acid, 2-amino-4-mercapto-, DL-	**(454-29-5)**
Butyric acid, 2-amino-4-(methylselenyl)	**(1464-42-2)**
Butyric acid, 2-amino-4-(methylthio)-	**(63-68-3)**
Butyric acid, 4-(p-aminophenyl)- (8CI)	**(15118-60-2)**
Butyric acid anhydride	**(106-31-0)**
n-Butyric acid anhydride	**(106-31-0)**
Butyric acid, benzyl ester	**(103-37-7)**
Butyric acid, 4-(p-bis(2-chloroethyl)aminophenyl)	**(305-03-3)**
Butyric acid, 2-bromo	**(80-58-0)**
Butyric acid, α-bromo-	**(80-58-0)**
Butyric acid, 3-bromo- (8CI)	**(2623-86-1)**
Butyric acid, 2-bromo-3-methyl	**(565-74-2)**
Butyric acid, 4-butoxy-, methyl ester (8CI)	**(29006-06-2)**
Butyric acid, butyl ester	**(109-21-7)**
Butyric acid, sec-butyl ester (8CI)	**(819-97-6)**
Butyric acid, 3-chloro	**(1951-12-8)**
Butyric acid, 2-chloro- (8CI)	**(4170-24-5)**
Butyric acid, 4-chloro- (8CI)	**(627-00-9)**
Butyric acid, 4-chloro-, ethyl ester (8CI)	**(3153-36-4)**
Butyric acid, 4-(4-chloro-2-methylphenoxy)-, sodium salt	**(6062-26-6)**
Butyric acid, 4-(p-chlorophenoxy)- (8CI)	**(3547-07-7)**
Butyric acid, 2-(p-chlorophenyl)-3-methyl	**(2012-74-0)**
Butyric acid, 4-((4-chloro-o-tolyl)oxy)	**(94-81-5)**
Butyric acid, 4-((4-chloro-o-tolyl)oxy)-, ethyl ester	**(10443-70-6)**
Butyric acid, 4-((4-chloro-o-tolyl)oxy)-, sodium salt	**(6062-26-6)**
Butyric acid, 4-cyano- (7CI)	**(39201-33-7)**
Butyric acid, 2,2-dichloro- (8CI)	**(13023-00-2)**
Butyric acid, 4-(2,4-dichlorophenoxy)	**(94-82-6)**
Butyric acid, 4-(2,4-dichlorophenoxy)-, butyl ester	**(6753-24-8)**
Butyric acid, 4-(2,4-dichlorophenoxy)-, isooctyl ester	**(1320-15-6)**
Butyric acid, 4-(2,4-dichlorophenoxy)-, sodium salt	**(10433-59-7)**
Butyric acid, 2,2-dichloro-, sodium salt (7CI,8CI)	**(2517-16-0)**
Butyric acid, 2,2-dimethyl- (8CI)	**(595-37-9)**
Butyric acid, 2,3-dimethyl- (8CI)	**(14287-61-7)**
Butyric acid, 3,3-dimethyl- (8CI)	**(1070-83-3)**
Butyric acid, 3,3-dimethyl-2-oxo-	**(815-17-8)**
Butyric acid, 2-(2,4-di-tert-pentylphenoxy)	**(13403-01-5)**
Butyric acid, 2,3-epoxy-, butyl ester	**(10138-34-8)**
Butyric acid, 2,3-epoxy-3-phenyl-	**(5669-15-8)**
Butyric acid, ester with dimethyl (2,2,2-trichloro-1-hydroxyethyl)phosphonate	**(126-22-7)**
Butyric acid, 1-ethoxybutyl ester (8CI)	**(33931-68-9)**
Butyric acid, 2-ethyl	**(88-09-5)**
Butyric acid, 2-ethyl-, diester with triethylene glycol	**(95-08-9)**
Butyric acid, ethyl ester	**(105-54-4)**
Butyric acid, heptafluoro	**(375-22-4)**
Butyric acid, hexyl ester	**(2639-63-6)**
Butyric acid, 2-hydroxy- (8CI)	**(565-70-8)**
Butyric acid, 3-hydroxy- (8CI)	**(300-85-6)**
Butyric acid, 4-hydroxy-, γ-lactone	**(96-48-0)**
Butyric acid, 2-hydroxy-4-(methylthio)	**(583-91-5)**
Butyric acid, 3-hydroxy-4-phenyl- (6CI,7CI,8CI)	**(6828-41-7)**
Butyric acid, 4-(indolyl)-	**(133-32-4)**
Butyric acid, isobutyl ester	**(539-90-2)**
Butyric acid, isopentyl ester	**(106-27-4)**
Butyric acid lactone	**(96-48-0)**
Butyric acid, 3-methyl-	**(503-74-2)**
Butyric acid, 2-methyl- (8CI)	**(116-53-0)**
Butyric acid, 3-methyl-, allyl ester	**(2835-39-4)**
Butyric acid, 2-methylene- (8CI)	**(3586-58-1)**
Butyric acid, methyl ester	**(623-42-7)**
Butyric acid, 3-methyl-, ethyl ester	**(108-64-5)**
Butyric acid, 2-methyl-, ethyl ester (8CI)	**(7452-79-1)**
Butyric acid, 2-methyl-, hexyl ester	**(10032-15-2)**
Butyric acid, 3-methyl-4-phenyl	**(7315-68-6)**
Butyric acid, 2-methyl-4-(2,5-xylyl)- (8CI)	**(30316-14-4)**
Butyric acid, 3-methyl-4-(2,5-xylyl)- (8CI)	**(30275-76-4)**
Butyric acid nitrile	**(109-74-0)**

Name	CAS No.
Butyric acid, 2-(p-(1-oxo-2-isoindolinyl)phenyl)-, (+-)	**(63610-08-2)**
Butyric acid, pentyl ester	**(540-18-1)**
Butyric acid, 4-phenoxy	**(6303-58-8)**
Butyric acid, 2-phenyl	**(90-27-7)**
Butyric acid, 4-phenyl- (8CI)	**(1821-12-1)**
Butyric acid, phenyl ester (8CI)	**(4346-18-3)**
Butyric acid, propyl ester	**(105-66-8)**
Butyric acid, sodium salt	**(156-54-7)**
Butyric acid, 4-p-tolyl-, methyl ester (8CI)	**(24306-23-8)**
Butyric acid, 4-(2,4,5-trichlorophenoxy)	**(93-80-1)**
Butyric acid triester with glycerin	**(60-01-5)**
Butyric acid, vinyl ester	**(123-20-6)**
Butyric acid, 4-(2,5-xylyl)- (8CI)	**(1453-06-1)**
Butyric aldehyde	**(123-72-8)**
Butyric-anhydride	**(106-31-0)**
n-Butyric anhydride	**(106-31-0)**
Butyric anhydride [UN 2739]	**(106-31-0)**
Butyric ether	**(105-54-4)**
Butyric or normal primary butyl alcohol	**(71-36-3)**
Butyrin, 1-mono- (8CI)	**(557-25-5)**
Butyrin, tri	**(60-01-5)**
Butyroguanamine	**(5962-23-2)**
Butyrolactam	**(616-45-5)**
γ-Butyrolactam	**(616-45-5)**
4-Butyrolactone	**(96-48-0)**
Butyrolactone	**(96-48-0)**
α-Butyrolactone	**(96-48-0)**
β-Butyrolactone	**(3068-88-0)**
γ-Butyrolactone	**(96-48-0)**
.β.-Butyrolactone homopolymer	**(36486-76-7)**
.β.-Butyrolactone polymer	**(36486-76-7)**
β-Butyrolakton (Czech)	**(3068-88-0)**
Butyron	**(3766-60-7)**
Butyrone	**(123-19-3)**
n-Butyronitrile	**(109-74-0)**
Butyronitrile [UN 2411]	**(109-74-0)**
Butyronitrile, 2-amino-2,3-dimethyl- (8CI)	**(13893-53-3)**
Butyronitrile, 4-(dichloromethylsilyl)	**(1190-16-5)**
Butyronitrile, 2-methyl	**(18936-17-9)**
n-Butyrophenone	**(495-40-9)**
Butyrophenone (8CI)	**(495-40-9)**
Butyrophenone, 4-(4-(p-chlorophenyl)-4-hydroxypiperidino)-4'-fluoro	**(52-86-8)**
Butyrophenone, 4'-fluoro-4-(4-(p-chlorophenyl)-4-hydroxy-piperidino)-	**(52-86-8)**
Butyrophenone, 2',4',5'-trihydroxy	**(1421-63-2)**
Butyrylactone	**(96-48-0)**
n-Butyryl chloride	**(141-75-3)**
Butyryl chloride [UN 2353]	**(141-75-3)**
Butyryl lactone	**(96-48-0)**
3-n-Butyryl-10-(3'-N-methyl-piperazino-N'-propyl)phenothiazin	**(653-03-2)**
Butyrylonitrile	**(109-74-0)**
Butyryl oxide	**(106-31-0)**
Butyrylperazine	**(653-03-2)**
Butyryltrichlorfon	**(126-22-7)**
Butyryl triglyceride	**(60-01-5)**
1-Butyn-3-yl-m-chlorophenylcarbamate	**(1967-16-4)**
Buvetzone	**(50-33-9)**
Buzon	**(50-33-9)**
Buzulfan	**(55-98-1)**
Bye Bugs	**(16919-19-0)**
Byladoce	**(68-19-9)**
Bypolet 34	**(25791-96-2)**
Bypolet 36	**(25791-96-2)**
153C51	**(2519-77-9)**
191C49	**(86-14-6)**
238 C	**(511-45-5)**
338C48	**(524-84-5)**
50-CS-46	**(2597-93-5)**
60-CS-16	**(999-81-5)**
C 2	**(104-74-5)**
C 4D	**(21645-51-2)**
C 31	**(21645-51-2)**
C 31F	**(21645-51-2)**
C 33	**(21645-51-2)**
C 178 (9CI)	**(102256-72-4)**
C 180 (9CI)	**(58033-85-5)**
C 570	**(13171-21-6)**
C 663	**(19395-62-1)**
C 709	**(141-66-2)**
C 1,006	**(80-33-1)**
C 1414	**(6923-22-4)**
C 1739	**(2801-68-5)**
C 1983	**(1982-47-4)**
C 2018	**(9004-70-0)**
C 2059	**(2164-17-2)**
C 2242	**(15545-48-9)**
C 2446	**(919-76-6)**
C 3172	**(57-39-6)**
C 3470	**(14214-32-5)**
C 5968	**(86-54-4)**
C 6379	**(300-42-5)**
C 6866	**(55-86-7)**
C 7019	**(4658-28-0)**
C 7337 Ciba	**(50-60-2)**
C 8514	**(6164-98-3)**
C 13437 SU	**(3771-19-5)**
C 18898	**(22936-75-0)**
C 19490	**(24151-93-7)**
C-31-F	**(21645-51-2)**
C-56	**(77-47-4)**
C-257	**(6237-24-7)**
C-272	**(1113-14-0)**
C-299	**(2481-94-9)**
C-847	**(101-27-9)**
C-854	**(80-33-1)**
C-1297	**(143-07-7)**
C-3126	**(3060-89-7)**
C-5068	**(86-54-4)**
C-5968	**(304-20-1)**
C-6313	**(13360-45-7)**
C-6989	**(15457-05-3)**
C-8353	**(6988-21-2)**
C-9491	**(18181-70-9)**
C-10015	**(470-90-6)**
C-12669	**(477-30-5)**
C1	**(50-76-0)**
C8949	**(470-90-6)**
C9122	**(13181-17-4)**
CA	**(5798-79-8)**
CA	**(69-74-9)**
CA 70203	**(26644-46-2)**
CA 80-15	**(9004-70-0)**
CAA	**(107-91-5)**
CAA	**(372-09-8)**
CA 24 (9CI)	**(39390-54-0)**
CACP	**(15663-27-1)**
CADG	**(61789-40-0)**
CAF	**(532-27-4)**
CAF	**(56-75-7)**
CAID	**(3691-35-8)**
CALX	**(1305-78-8)**
CAM	**(56-75-7)**
CAMA	**(5902-95-4)**

CAMP **(60-92-4)**
CAO 1 **(128-37-0)**
CAO 3 **(128-37-0)**
CAP **(302-22-7)**
CAP **(532-27-4)**
CAP **(56-75-7)**
C Acid **(131-27-1)**
C-8 Acid **(124-07-2)**
C 10 Alcohol **(112-30-1)**
C-9 Aldehyde **(124-19-6)**
C-10 Aldehyde **(112-31-2)**
C-16 Aldehyde **(77-83-8)**
C-14 Aldehyde, Myristic **(124-25-4)**
C-12 Aldehyde, lauric **(112-54-9)**
C-11 Aldehyde, undecylenic **(112-45-8)**
C-11 Aldehyde, undecylic **(112-44-7)**
C12-18 Alkyl alcohol ethoxylate **(68213-23-0)**
C12-18 Alkyl alcohol ethoxylate propoxylate **(69227-21-0)**
C(15)-Anteiso acid **(5502-94-3)**
2041 C.B. **(55-98-1)**
3025 C.B. **(148-82-3)**
3026 C.B. **(13045-94-8)**
4-CB **(32598-13-3)**
4261 CB **(10379-14-3)**
4361 CB **(10379-14-3)**
8075 C.B. **(53-39-4)**
8080 C. B. **(297-76-7)**
C.B. 2041 **(55-98-1)**
CB **(51-83-2)**
CB 10 **(9003-17-2)**
CB 11 **(467-84-5)**
CB 50 **(7782-42-5)**
CB 1348 **(305-03-3)**
CB 1639 **(52-52-8)**
CB 1678 **(362-29-8)**
CB 2487 **(7426-35-9)**
CB 2562 **(299-75-2)**
CB 3025 **(148-82-3)**
CB 4261 **(10379-14-3)**
CB 4564 **(50-18-0)**
CB 8000 **(120-97-8)**
CB 8019 **(78-44-4)**
CB-3026 **(13045-94-8)**
CB-3307 **(531-76-0)**
CB-4564 **(6055-19-2)**
CB-4835 **(66-75-1)**
2-CBA **(118-91-2)**
CBBP **(115-78-6)**
CBC 806495 **(52-24-4)**
CBD **(13956-29-1)**
CBD 90 **(87-90-1)**
CBN **(101-27-9)**
CBN **(521-35-7)**
p-CBP **(134-85-0)**
CBSP **(71-67-0)**
C-Blau 17 (Germany) **(14038-43-8)**
C8 Branched alkyl phenol ethoxylate sulfuric acid, sodium salt **(69011-84-3)**
C8 Branched alkylphenol ethoxylate sulfuric acid, sodium salt **(69011-84-3)**
C 31C **(21645-51-2)**
CC 11511 **(77-21-4)**
(C10-C16)Alcohol ethoxylate, sulfated, ammonium salt **(67762-19-0)**
(C10-C16) Alcohol ethoxylate, sulfated, sodium salt **(68585-34-2)**
(C10-C18) Alkanes **(73138-29-1)**
3-(C12-C15) Alkoxy-2-hydroxypropyltrimethylammonium chloride **(68187-63-3)**
(C6-C12) Alkyl alcohol **(68603-15-6)**
C10-C16 Alkyl alcohol **(67762-41-8)**
(C10-C16) Alkyl alcohol **(67762-41-8)**
(C16-C18) Alkyl alcohol **(67762-27-0)**
(C16-C18)-Alkyl alcohol **(67762-27-0)**
C6-C12 Alkyl alcohol ethoxylate **(68439-45-2)**
(C6-C12) Alkyl alcohol ethoxylate **(68439-45-2)**
(C12-C18) Alkyl alcohol ethoxylate **(68213-23-0)**
(C16-C18) Alkyl alcohol ethoxylate **(68439-49-6)**
C6-C12 (Alkyl) alcohol, ethoxylated **(68439-45-2)**
(C12-C18) Alkyl alcohol ethoxylate propoxylate **(69227-21-0)**
(C12-C18)Alkylalcohol ethoxylate propoxylate **(69227-21-0)**
(C10-C16)-Alkyl alcohol ethoxylate sulfuric acid ammonium salt **(67762-19-0)**
(C10-C16)Alkyl alcohol, ethoxylate, sulfuric acid, ammonium salt **(67762-19-0)**
(C10-C16)-Alkyl alcohol ethoxylate sulfuric acid magnesium salt **(67762-21-4)**
C10-C16 Alkyl (alcohol) ethoxylate sulfuric acid sodium salt **(68585-34-2)**
(C10-C16)Alkyl(alcohol)ethoxylate sulfuric acid, sodium salt **(68585-34-2)**
1-(2-(C14-C18)-Alkylamidoethyl)-2-nor(C14-C18)alkyl-3-methylimidazolinium methyl sulfate **(72623-82-6)**
(C8-C16) Alkylbenzene **(68648-87-3)**
(C10-C16)Alkylbenzene **(68890-99-3)**
(C10-C16) Alkylbenzenesulfonic acid **(68584-22-5)**
(C10-C16)Alkyl benzene sulfonic acid **(68584-22-5)**
C10-C16 Alkylbenzenesulfonic acid **(68584-22-5)**
(C6-C12) Alkylbenzenesulfonic acid, sodium salt **(68608-87-7)**
(C6-C12)Alkylbenzenesulfonic acid, sodium salt **(68608-87-7)**
(C10-C16)Alkylcarboxylic acid **(68002-90-4)**
(C6-C12) Alkylcarboxylic acid methyl ester **(67762-39-4)**
N-(C12-C18)Alkyldiethanolamine **(71786-60-2)**
(C10-C16)Alkyldimethylamine oxide **(70592-80-2)**
(C10-C16-Alkyl)dimethylamines, N-oxides **(70592-80-2)**
(C12-C18) Alkyl ethoxylate, sulfate, ammonium salt **(68610-22-0)**
(C10-C16) Alkylethoxylate sulfuric acid, ammonium salt **(67762-19-0)**
(C13-C16)Alkyl ethoxylate sulfuric acid, ammonium salt **(67762-19-0)**
C12-C18 Alkyl ethoxylate sulfuric acid ammonium salt **(68610-22-0)**
(C10-C16) Alkyl ethoxylate sulfuric acid, sodium salt **(68585-34-2)**
(C10-C16) Alkylethoxylate sulfuric acid, sodium salt **(68585-34-2)**
(C10-C16)Alkyl ethoxylate sulfuric acid, sodium salt **(68585-34-2)**
(C6-C12) Alkylglycidyl ether **(68987-80-4)**
α-(C10-C12)Alkyl-ω-hydroxypoly(oxy-1,2-ethanediyl), sulfate, ammonium salt **(68890-88-0)**
(C1-C3) Alkylphenol **(68555-24-8)**
(C12-C18) N-Alkylpropylenediamine **(68155-37-3)**
N-(C12-C18)Alkyl propylenediamine **(68155-37-3)**
(C10-C18)Alkylsulfonic acid, sodium salt **(68037-49-0)**
(C14-C18) Alkyltrimethylammonium bromide **(68424-92-0)**
((C12-C18)Alkyl)trimethylammonium chloride **(68391-03-7)**
(C12-C18)Alkyl trimethylammonium chloride **(68391-03-7)**
(C12-C18)Alkyltrimethylammonium chloride **(68391-03-7)**
(C12-C30) Aromatic Oil **(68602-80-2)**
(C9-C17) Branched alkylbenzenesulfonic acid, sodium salt **(68953-95-7)**
CCC **(156-62-7)**
CCC **(999-81-5)**
CCC Plant Growth Regulant **(999-81-5)**
(C10-C16) Carboxylic acid **(68002-90-4)**
(C14-C18) Dialkylmethylamine **(67700-99-6)**
(C16-C18) Fatty alcohol, ethylene oxide reaction product **(68439-49-6)**
CCH **(7778-54-3)**
CCHO **(286-20-4)**
CCL **(53994-73-3)**
(C5-C9) Monobasic acids **(68603-84-9)**
CCN52 **(52315-07-8)**
CCNU **(13010-47-4)**
(C18-C70) Paraffins **(70913-86-9)**
(C5-C18) Partial fraction (Petroleum) **(68477-58-7)**
CCS **(80-33-1)**
CCS 203 **(71-36-3)**
CCS 301 **(78-92-2)**
(C10-C16) Saturated alkylbenzenesulfonic acid **(68584-22-5)**

Synonym	CAS No.
(C14-C18) and (C16-C18) Alkylcarboxylic acid	**(67701-06-8)**
C14-C18 and C16-C18 Alkylcarboxylic acid	**(67701-06-8)**
(C14-C18) and (C16-C18)-Unsaturated alkyl carboxylic acid	**(67701-06-8)**
(C14-C18) and (C16-C18)-Unsaturated alkylcarboxylic acid	**(67701-06-8)**
(C16-C18) and (C18) Unsaturated alkylalcohol, ethoxylate	**(68920-66-1)**
(C14-C18) and C18 Unsaturated alkylcarboxylic acid	**(67701-06-8)**
"C" Carrie	**(50-36-2)**
CD	**(58-14-0)**
CD 2	**(58-25-3)**
CD 2	**(148-71-0)**
CD 3	**(25646-71-3)**
CD 4	**(25646-77-9)**
CD 9	**(13395-16-9)**
CD 68	**(57-74-9)**
CD 419	**(9003-07-0)**
CD 15006 A	**(2425-79-8)**
CDA	**(124-03-8)**
CDA 101	**(7440-50-8)**
CDA 102	**(7440-50-8)**
CDA 110	**(7440-50-8)**
CDA 122	**(7440-50-8)**
CDAA	**(93-71-0)**
CDAA-T	**(8005-43-4)**
CDAAT	**(93-71-0)**
CDAA + TCBC	**(8005-43-4)**
CDB 60	**(2782-57-2)**
CDB 63	**(2893-78-9)**
CDBM	**(124-48-1)**
CDC	**(474-25-9)**
CDCA	**(474-25-9)**
CDDP	**(15663-27-1)**
CDEA	**(2315-36-8)**
CDEA Br	**(13316-70-6)**
CDEC	**(95-06-7)**
CD III	**(25646-71-3)**
CDM	**(6164-98-3)**
CDNA	**(99-30-9)**
CDNB	**(97-00-7)**
3-CDO	**(58629-01-9)**
CDO	**(58-25-3)**
CDP	**(58-25-3)**
CDT	**(122-34-9)**
CDT	**(4904-61-4)**
CDTA	**(482-54-2)**
C21-Dicarboxylic acid	**(53980-88-4)**
CE	**(479-45-8)**
CE	**(9003-11-6)**
CE CE CE	**(999-81-5)**
CEP	**(16672-87-0)**
2-CEPA	**(16672-87-0)**
CEPHA	**(16672-87-0)**
CEPHA 10LS	**(16672-87-0)**
CER	**(50-59-9)**
CES	**(140-57-8)**
CET	**(122-34-9)**
CEX	**(15686-71-2)**
CEZ	**(25953-19-9)**
C-Ext. Blau 10 (Germany)	**(147-14-8)**
C-Ext. Braun 4 (Germany)	**(1320-07-6)**
C-Ext. Gelb 16 (Germany)	**(2321-07-5)**
CF 2	**(71-55-6)**
CF 8	**(7440-44-0)**
CF 8 (Carbon)	**(7440-44-0)**
CF 125	**(2536-31-4)**
CFC 22	**(75-45-6)**
CFC 31	**(593-70-4)**
CFC 133a	**(75-88-7)**
CFC 142b	**(75-68-3)**
CFNP	**(13738-63-1)**
21 CFR 182,5933	**(127-47-9)**
C.F.S.	**(7446-18-6)**
CFV	**(470-90-6)**
CFX	**(35607-66-0)**
CF 218 (polymer)	**(9002-98-6)**
CG	**(75-44-5)**
CG 117	**(57837-19-1)**
CG 3117	**(22204-53-1)**
CG-1283	**(2385-85-5)**
CG1	**(22232-54-8)**
CGA 10832	**(26399-36-0)**
CGA 14397	**(34128-99-9)**
CGA 15324	**(41198-08-7)**
CGA 16339	**(22175-22-0)**
CGA 16340	**(13142-64-8)**
CGA 26351	**(470-90-6)**
CGA 45156	**(59669-26-0)**
CGA 48988	**(57837-19-1)**
CGA 61837	**(4849-32-5)**
CGA 73102	**(65907-30-4)**
CGA 80000	**(67932-85-8)**
CGA-12223	**(42509-80-8)**
CGA-18731	**(34123-59-6)**
CGA-18762	**(32889-48-8)**
CGA-24705	**(51218-45-2)**
CGA-64250	**(60207-90-1)**
CGD 92710F	**(60207-90-1)**
CGP 2175	**(37350-58-6)**
CGP 9000	**(51762-05-1)**
CGT	**(516-21-2)**
CGTA	**(482-54-2)**
C-Green 10	**(128-80-3)**
CH	**(563-41-7)**
CH	**(693-30-1)**
CH 13-437	**(3771-19-5)**
CH 3565	**(3380-34-5)**
CHA	**(108-91-8)**
1,3-CHBP	**(109-70-6)**
CHE 1843	**(1113-14-0)**
CHEL 330	**(140-01-2)**
CHEL 330	**(67-43-6)**
CHEL 600	**(482-54-2)**
CHEL DM 41	**(139-89-9)**
CHEL DTPA	**(67-43-6)**
CHEL 330 acid	**(67-43-6)**
CHI 8	**(340-57-8)**
1,4-CHIDM	**(105-08-8)**
CHLZ	**(54749-90-5)**
CHP-Depot	**(101-40-6)**
Co-Hydeltra	**(50-24-8)**
C.I. 2	**(12108-13-3)**
C.I. 3/11855	**(2832-40-8)**
C.I. 23	**(85-86-9)**
C.I. 27	**(1936-15-8)**
C.I. 79	**(3761-53-3)**
C.I. 150	**(523-44-4)**
C.I. 184	**(915-67-3)**
C.I. 185	**(2611-82-7)**
C.I. 258	**(85-83-6)**
C.I. 337	**(115-02-6)**
C.I. 366	**(77-67-8)**
C.I. 395	**(956-90-1)**
C.I. 406	**(434-07-1)**
C.I. 473	**(61-68-7)**
C.I. 581	**(1867-66-9)**

C.I. 588	(366-18-7)	C.I. 15595	(5850-90-8)
C.I. 636	(17784-12-2)	C.I. 15620	(1658-56-6)
C.I. 640	(1934-21-0)	C.I. 15630	(1248-18-6)
C.I. 670	(5141-20-8)	C.I. 15670	(2092-55-9)
C.I. 671	(2650-18-2)	C.I. 15850	(5858-81-1)
C.I. 671	(3844-45-9)	C.I. 15970	(1934-20-9)
C.I. 712	(129-17-9)	C.I. 15980	(2347-72-0)
C.I. 749	(81-88-9)	C.I. 15985	(2783-94-0)
C.I. 766	(518-47-8)	C.I. 16035	(25956-17-6)
C.I. 773	(16423-68-0)	C.I. 16045	(2302-96-7)
C.I. 801	(8004-92-0)	C.I. 16050	(5858-93-5)
C.I. 918	(8004-92-0)	C.I. 16150	(3761-53-3)
C.I. 1037	(81-54-9)	C.I. 16155	(3564-09-8)
C.I. 1106	(81-77-6)	C.I. 16185	(915-67-3)
C.I. 10000	(131-91-9)	C.I. 16230	(1936-15-8)
C.I. 10305	(88-89-1)	C.I. 16255	(2611-82-7)
C.I. 10310	(534-52-1)	C.I. 16290	(5850-44-2)
C.I. 10315	(605-69-6)	C.I. 16570	(4197-07-3)
C.I. 10335	(119-75-5)	C.I. 18050	(3734-67-6)
C.I. 10345	(119-15-3)	C.I. 18800	(6408-31-7)
C.I. 10355	(122-39-4)	C.I. 19140	(1934-21-0)
C.I. 10360	(131-73-7)	C.I. 20170	(1320-07-6)
C.I. 10385	(6373-74-6)	C.I. 20470	(1064-48-8)
C.I. 11000	(60-09-3)	C.I. 21090	(6358-85-6)
C.I. 11020	(60-11-7)	C.I. 21100	(5102-83-0)
C.I. 11021	(2481-94-9)	C.I. 21110	(3520-72-7)
C.I. 11025	(539-17-3)	C.I. 22120	(573-58-0)
C.I. 11085	(14097-03-1)	C.I. 22245	(3567-65-5)
C.I. 11154	(12270-13-2)	C.I. 22311	(2429-82-5)
C.I. 11160	(97-56-3)	C.I. 22480	(6426-67-1)
C.I. 11270	(495-54-5)	C.I. 22570	(2586-60-9)
C.I. 11270	(532-82-1)	C.I. 22590	(2429-73-4)
C.I. 11285	(6416-57-5)	C.I. 22610	(2602-46-2)
C.I. 11350	(131-22-6)	C.I. 22890	(10169-02-5)
C.I. 11380	(85-84-7)	C.I. 22910	(6375-55-9)
C.I. 11390	(131-79-3)	C.I. 23050	(6548-29-4)
C.I. 11680	(2512-29-0)	C.I. 23060	(91-94-1)
C.I. 11710	(6486-23-3)	C.I. 23500	(992-59-6)
C.I. 11738	(13515-40-7)	C.I. 23630	(6358-29-8)
C.I. 11741	(6358-31-2)	C.I. 23635	(6459-94-5)
C.I. 11800	(1689-82-3)	C.I. 23850	(72-57-1)
C.I. 11850	(952-47-6)	C.I. 23860	(314-13-6)
C.I. 11855	(2832-40-8)	C.I. 24110	(119-90-4)
C.I. 11920	(2051-85-6)	C.I. 24400	(2429-74-5)
C.I. 12055	(842-07-9)	C.I. 24401	(28407-37-6)
C.I. 12075	(3468-63-1)	C.I. 24410	(2610-05-1)
C.I. 12085	(2814-77-9)	C.I. 24895	(2870-32-8)
C.I. 12100	(2646-17-5)	C.I. 26050	(6368-72-5)
C.I. 12120	(2425-85-6)	C.I. 26090	(6300-37-4)
C.I. 12140	(3118-97-6)	C.I. 26100	(85-86-9)
C.I. 12150	(1229-55-6)	C.I. 26105	(85-83-6)
C.I. 12156	(6358-53-8)	C.I. 26135	(119-28-8)
C.I. 12355	(6471-49-4)	C.I. 26690	(6406-45-7)
C.I. 12370	(6535-46-2)	C.I. 28160	(2610-11-9)
C.I. 13020	(493-52-7)	C.I. 28440	(2519-30-4)
C.I. 13025	(547-58-0)	C.I. 29160	(3441-14-3)
C.I. 13065	(587-98-4)	C.I. 30110	(2586-58-5)
C.I. 13080	(554-73-4)	C.I. 30145	(16071-86-6)
C.I. 13220	(10190-66-6)	C.I. 30235	(1937-37-7)
C.I. 13390	(3861-73-2)	C.I. 30245	(2429-83-6)
C.I. 13950	(10190-68-8)	C.I. 30280	(3626-28-6)
C.I. 14270	(547-57-9)	C.I. 31930	(7082-31-7)
C.I. 14600	(523-44-4)	C.I. 35255	(6428-31-5)
C.I. 14700	(4548-53-2)	C.I. 35570	(90-20-0)
C.I. 14720	(3567-69-9)	C.I. 35660	(2429-81-4)
C.I. 14815	(3257-28-1)	C.I. 35780	(2610-10-8)
C.I. 15510	(633-96-5)	C.I. 35811	(102-56-7)

C.I. 37005 (141-85-5)
C.I. 37010 (95-82-9)
C.I. 37020 (609-20-1)
C.I. 37025 (88-74-4)
C.I. 37030 (99-09-2)
C.I. 37035 (100-01-6)
C.I. 37040 (89-63-4)
C.I. 37077 (95-53-4)
C.I. 37085 (3165-93-3)
C.I. 37100 (99-52-5)
C.I. 37105 (99-55-8)
C.I. 37107 (106-49-0)
C.I. 37110 (89-62-3)
C.I. 37115 (134-29-2)
C.I. 37125 (97-52-9)
C.I. 37130 (99-59-2)
C.I. 37175 (120-00-3)
C.I. 37225 (92-87-5)
C.I. 37230 (119-93-7)
C.I. 37240 (101-54-2)
C.I. 37270 (91-59-8)
C.I. 37275 (82-45-1)
C.I. 37500 (135-19-3)
C.I. 37505 (92-77-3)
C.I. 37515 (135-65-9)
C.I. 37520 (135-61-5)
C.I. 37526 (135-63-7)
C.I. 37530 (135-62-6)
C.I. 37531 (137-52-0)
C.I. 37558 (92-74-0)
C.I. 37560 (132-68-3)
C.I. 37575 (91-92-9)
C.I. 37610 (91-96-3)
C.I. 40000 (1325-37-7)
C.I. 40645 (6416-68-8)
C.I. 41000 (2465-27-2)
C.I. 42000 (569-64-2)
C.I. 42040 (633-03-4)
C.I. 42045 (129-17-9)
C.I. 42053 (2353-45-9)
C.I. 42085 (4680-78-8)
C.I. 42090 (2650-18-2)
C.I. 42090 (3844-45-9)
C.I. 42095 (5141-20-8)
C.I. 42500 (569-61-9)
C.I. 42510 (632-99-5)
C.I. 42535 (8004-87-3)
C.I. 42555 (548-62-9)
C.I. 42581 (4129-84-4)
C.I. 42600 (2390-59-2)
C.I. 42640 (1694-09-3)
C.I. 42655 (6104-58-1)
C.I. 42760 (2152-64-9)
C.I. 42765 (1324-76-1)
C.I. 44040 (2185-86-6)
C.I. 44050 (90-30-2)
C.I. 45000 (2465-29-4)
C.I. 45010 (2150-48-3)
C.I. 45160 (989-38-8)
C.I. 45170 (81-88-9)
C.I. 45350 Disodium salt (518-47-8)
C.I. 45350 Sodium salt (518-47-8)
C.I. 45380 (548-26-5)
C.I. 45380 (17372-87-1)
C.I. 45405 (6441-77-6)
C.I. 45430 (16423-68-0)
C.I. 46000 (86-40-8)
C.I. 46005 (494-38-2)
C.I. 46005 (65-61-2)
C.I. 46500 (1047-16-1)
C.I. 47000 (8003-22-3)
C.I. 47005 (8004-92-0)
C.I. 47031 (947-02-4)
C.I. 48035 (3056-93-7)
C.I. 50240 (477-73-6)
C.I. 50415 (11099-03-9)
C.I. 50420 (8005-03-6)
C.I. 51015 (7199-02-2)
C.I. 53185 (1326-82-5)
C.I. 53185 (66241-11-0)
C.I. 53186 (1326-83-6)
C.I. 53228 (1326-96-1)
C.I. 53440 (1327-57-7)
C.I. 55005 (476-66-4)
C.I. 56200 (2478-20-8)
C.I. 57000 (528-21-2)
C.I. 58000 (72-48-0)
C.I. 58005 (130-22-3)
C.I. 58050 (81-64-1)
C.I. 58205 (81-54-9)
C.I. 58500 (81-61-8)
C.I. 59040 (6358-69-6)
C.I. 59100 (128-66-5)
C.I. 59700 (128-70-1)
C.I. 59815 (1324-54-5)
C.I. 59825 (128-58-5)
C.I. 60010 (1324-55-6)
C.I. 60505 (82-38-2)
C.I. 60700 (82-28-0)
C.I. 60710 (116-85-8)
C.I. 60725 (81-48-1)
C.I. 60756 (12222-78-5)
C.I. 60767 (15791-78-3)
C.I. 61100 (128-95-0)
C.I. 61105 (1220-94-6)
C.I. 61200 (2580-78-1)
C.I. 61500 (2475-44-7)
C.I. 61505 (2475-46-9)
C.I. 61505 (86722-66-9)
C.I. 61525 (128-85-8)
C.I. 61565 (128-80-3)
C.I. 61570 (4403-90-1)
C.I. 61720 (81-78-7)
C.I. 62015 (2872-48-2)
C.I. 62030 (82-33-7)
C.I. 62045 (4368-56-3)
C.I. 62130 (2666-17-3)
C.I. 63285 (12217-79-7)
C.I. 64500 (2475-45-8)
C.I. 67300 (129-09-9)
C.I. 69005 (2379-81-9)
C.I. 69500 (3271-76-9)
C.I. 69800 (81-77-6)
C.I. 70320 (4203-77-4)
C.I. 70800 (2475-33-4)
C.I. 70802 (2475-33-4)
C.I. 71105 (4424-06-0)
C.I. 71200 (6424-76-6)
C.I. 73000 (482-89-3)
C.I. 73015 (860-22-0)
C.I. 73335 (3263-31-8)
C.I. 73360 (2379-74-0)
C.I. 74160 (147-14-8)
C.I. 74260 (1328-53-6)

C.I. 74265 (14302-13-7)
C.I. 75130 (7235-40-7)
C.I. 75270 (476-66-4)
C.I. 75300 (458-37-7)
C.I. 75400 (481-74-3)
C.I. 75410 (81-54-9)
C.I. 75440 (518-82-1)
C.I. 75470 (1260-17-9)
C.I. 75480 (83-72-7)
C.I. 75490 (84-79-7)
C.I. 75500 (481-39-0)
C.I. 75640 (520-18-3)
C.I. 75670 (117-39-5)
C.I. 75720 (522-12-3)
C.I. 75730 (153-18-4)
C.I. 75781 (860-22-0)
C.I. 76000 (62-53-3)
C.I. 76001 (142-04-1)
C.I. 76005 (122-80-5)
C.I. 76010 (95-54-5)
C.I. 76015 (95-83-0)
C.I. 76020 (99-56-9)
C.I. 76025 (108-45-2)
C.I. 76027 (5131-60-2)
C.I. 76030 (5131-58-8)
C.I. 76035 (95-80-7)
C.I. 76042 (95-70-5)
C.I. 76043 (6369-59-1)
C.I. 76043 (615-50-9)
C.I. 76050 (615-05-4)
C.I. 76051 (39156-41-7)
C.I. 76060 (106-50-3)
C.I. 76061 (624-18-0)
C.I. 76065 (615-66-7)
C.I. 76066 (6219-71-2)
C.I. 76070 (5307-14-2)
C.I. 76075 (105-10-2)
C.I. 76085 (101-54-2)
C.I. 76120 (537-65-5)
C.I. 76500 (120-80-9)
C.I. 76505 (108-46-3)
C.I. 76515 (87-66-1)
C.I. 76520 (95-55-6)
C.I. 76535 (121-88-0)
C.I. 76545 (591-27-5)
C.I. 76555 (119-34-6)
C.I. 76605 (90-15-3)
C.I. 76625 (83-56-7)
C.I. 76630 (575-44-0)
C.I. 76645 (582-17-2)
C.I. 77000 (7429-90-5)
C.I. 77002 (21645-51-2)
C.I. 77004 (1327-36-2)
C.I. 77050 (7440-36-0)
C.I. 77052 (1309-64-4)
C.I. 77056 (10025-91-9)
C.I. 77060 (1345-04-6)
C.I. 77061 (1315-04-4)
C.I. 77086 (1303-33-9)
C.I. 77099 (513-77-9)
C.I. 77103 (10294-40-3)
C.I. 77120 (7727-43-7)
C.I. 77169 (1304-85-4)
C.I. 77180 (7440-43-9)
C.I. 77185 (543-90-8)
C.I. 77199 (1306-23-6)
C.I. 77201 (1345-09-1)
C.I. 77223 (13765-19-0)
C.I. 77265 (7440-44-0)
C.I. 77265 (7782-42-5)
C.I. 77266 (1333-86-4)
C.I. 77288 (1308-38-9)
C.I. 77295 (10025-73-7)
C.I. 77305 (10101-53-8)
C.I. 77320 (7440-48-4)
C.I. 77322 (1307-96-6)
C.I. 77400 (7440-50-8)
C.I. 77402 (1317-39-1)
C.I. 77403 (1317-38-0)
C.I. 77410 (12002-03-8)
C.I. 77450 (1317-40-4)
C.I. 77480 (7440-57-5)
C.I. 77489 (1345-25-1)
C.I. 77491 (1309-37-1)
C.I. 77510 (14038-43-8)
C.I. 77520 (14038-43-8)
C.I. 77540 (1317-37-9)
C.I. 77575 (7439-92-1)
C.I. 77577 (1317-36-8)
C.I. 77578 (1314-41-6)
C.I. 77580 (1309-60-0)
C.I. 77600 (1344-37-2)
C.I. 77600 (7758-97-6)
C.I. 77601 (18454-12-1)
C.I. 77603 (1344-37-2)
C.I. 77605 (12656-85-8)
C.I. 77610 (592-05-2)
C.I. 77622 (7446-27-7)
C.I. 77630 (7446-14-2)
C.I. 77640 (1314-87-0)
C.I. 77713 (546-93-0)
C.I. 77726 (1344-43-0)
C.I. 77727 (1317-34-6)
C.I. 77728 (1313-13-9)
C.I. 77755 (7722-64-7)
C.I. 77760 (21908-53-2)
C.I. 77775 (7440-02-0)
C.I. 77777 (1313-99-1)
C.I. 77779 (3333-67-3)
C.I. 77795 (7440-06-4)
C.I. 77805 (7782-49-2)
C.I. 77820 (7440-22-4)
C.I. 77837 (1633-05-2)
C.I. 77847 (1314-96-1)
C.I. 77864 (7772-99-8)
C.I. 77891 (13463-67-7)
C.I. 77901 (1314-35-8)
C.I. 77938 (1314-62-1)
C.I. 77940 (27774-13-6)
C.I. 77945 (7440-66-6)
C.I. 77947 (1314-13-2)
C.I. 77955 (13530-65-9)
C.I. 19140:1 (12225-21-7)
C.I. 42535:2 (1325-82-2)
C.I. 42595:2 (1325-87-7)
C.I. 45350:1 (2321-07-5)
C.I. 45380:2 (15086-94-9)
C.I. 45430:1 (12227-78-0)
C.I. 58055:1 (1328-04-7)
C.I. Acid Black 1 (7CI) (1064-48-8)
C.I. Acid Black 1, Disodium salt (8CI) (1064-48-8)
C.I. Acid Black 2 (9CI) (8005-03-6)
C.I. Acid Black 24, Disodium salt (8CI) (3071-73-6)
C.I. Acid Black 26:2 (6406-45-7)

Name	CAS
C.I. Acid Black 26B (8CI)	**(6406-45-7)**
C.I. Acid Black 84	**(6408-22-6)**
C.I. Acid Black 48 (9CI)	**(1328-24-1)**
C.I. Acid Black 52 (9CI)	**(5610-64-0)**
C.I. Acid Black 107 (9CI)	**(12218-96-1)**
C.I. Acid Blue 1	**(129-17-9)**
C.I. Acid Blue 1, Sodium salt	**(129-17-9)**
C.I. Acid Blue 3	**(129-17-9)**
C.I. Acid Blue 9	**(2650-18-2)**
C.I. Acid Blue 9, Diammonium salt	**(2650-18-2)**
C.I. Acid Blue 9, Disodium salt	**(3844-45-9)**
C.I. Acid Blue 41	**(2666-17-3)**
C.I. Acid Blue 62 (8CI)	**(4368-56-3)**
C.I. Acid Blue 71	**(2666-17-3)**
C.I. Acid Blue 74	**(860-22-0)**
C.I. Acid Blue 90, Monosodium salt (8CI)	**(6104-58-1)**
C.I. Acid Blue 92	**(3861-73-2)**
C.I. Acid Blue 92, Trisodium salt	**(3861-73-2)**
C.I. Acid Blue 113, Disodium salt (8CI)	**(3351-05-1)**
C.I. Acid Blue 277 (9CI)	**(61967-93-9)**
C.I. Acid Brown 14, Disodium salt (8CI)	**(5850-16-8)**
C.I. Acid Brown 100 (9CI)	**(61724-08-1)**
C.I. Acid Brown 159 (9CI)	**(61901-21-1)**
C.I. Acid Brown 311 (8CI,9CI)	**(12234-86-5)**
C.I. Acid Green 3	**(4680-78-8)**
C.I. Acid Green 3, Monosodium salt	**(4680-78-8)**
C.I. Acid Green 3, Sodium salt	**(4680-78-8)**
C.I. Acid Green 5	**(5141-20-8)**
C.I. Acid Green 5, Disodium salt	**(5141-20-8)**
C.I. Acid Green 25	**(4403-90-1)**
C.I. Acid Green 108 (9CI)	**(71872-22-5)**
C.I. Acid Orange 3	**(6373-74-6)**
C.I. Acid Orange 5	**(554-73-4)**
C.I. Acid Orange 5, Monosodium salt (8CI)	**(554-73-4)**
C.I. Acid Orange 6	**(547-57-9)**
C.I. Acid Orange 6, Monosodium salt (8CI)	**(547-57-9)**
C.I. Acid Orange 7	**(633-96-5)**
C.I. Acid Orange 7, Free acid	**(573-89-7)**
C.I. Acid Orange 7, Monosodium salt	**(633-96-5)**
C.I. Acid Orange 8, Monosodium salt (8CI)	**(5850-86-2)**
C.I. Acid Orange 10	**(1936-15-8)**
C.I. Acid Orange 10, Disodium salt	**(1936-15-8)**
C.I. Acid Orange 12 (7CI)	**(1934-20-9)**
C.I. Acid Orange 12, Monosodium salt (8CI)	**(1934-20-9)**
C.I. Acid Orange 14, Disodium salt (8CI)	**(5859-00-7)**
C.I. Acid Orange 20	**(523-44-4)**
C.I. Acid Orange 20, Monosodium salt	**(523-44-4)**
C.I. Acid Orange 52	**(547-58-0)**
C.I. Acid Orange 61 (9CI)	**(6408-33-9)**
C.I. Acid Red 1	**(3734-67-6)**
C.I. Acid Red 1, Disodium salt	**(3734-67-6)**
C.I. Acid Red 2	**(493-52-7)**
C.I. Acid Red 4, Monosodium salt (8CI)	**(5858-39-9)**
C.I. Acid Red 13 (7CI)	**(2302-96-7)**
C.I. Acid Red 13, Disodium salt (8CI)	**(2302-96-7)**
C.I. Acid Red 14, Disodium salt	**(3567-69-9)**
C.I. Acid Red 18	**(2611-82-7)**
C.I. Acid Red 18, Trisodium salt	**(2611-82-7)**
C.I. Acid Red 25	**(5858-93-5)**
C.I. Acid Red 25, Disodium salt	**(5858-93-5)**
C.I. Acid Red 26	**(3761-53-3)**
C.I. Acid Red 26, Disodium salt	**(3761-53-3)**
C.I. Acid Red 27	**(915-67-3)**
C.I. Acid Red 27, Trisodium salt	**(915-67-3)**
C.I. Acid Red 29 (7CI)	**(4197-07-3)**
C.I. Acid Red 29, Disodium salt (8CI)	**(4197-07-3)**
C.I. Acid Red 33	**(3567-66-6)**
C.I. Acid Red 41	**(5850-44-2)**
C.I. Acid Red 41, Tetrasodium salt (8CI)	**(5850-44-2)**
C.I. Acid Red 51	**(16423-68-0)**
C.I. Acid Red 66, Disodium salt (8CI)	**(4196-99-0)**
C.I. Acid Red 73, Disodium salt (8CI)	**(5413-75-2)**
C.I. Acid Red 85	**(3567-65-5)**
C.I. Acid Red 85, Disodium salt	**(3567-65-5)**
C.I. Acid Red 87	**(17372-87-1)**
C.I. Acid Red 87	**(548-26-5)**
C.I. Acid Red 88	**(1658-56-6)**
C.I. Acid Red 97	**(10169-02-5)**
C.I. Acid Red 97, Disodium salt (8CI)	**(10169-02-5)**
C.I. Acid Red 98	**(6441-77-6)**
C.I. Acid Red 114	**(6459-94-5)**
C.I. Acid Red 114, Disodium salt	**(6459-94-5)**
C.I. Acid Red 137, Monosodium salt (8CI)	**(6222-63-5)**
C.I. Acid Red 151, Monosodium salt (8CI)	**(6406-56-0)**
C.I. Acid Red 183 (8CI)	**(6408-31-7)**
C.I. Acid Red 348 (9CI)	**(61847-60-7)**
C.I. Acid Red 361 (9CI)	**(61931-22-4)**
C.I. Acid Red 413 (9CI)	**(102640-14-2)**
C.I. Acid Violet 12, Disodium salt (8CI)	**(6625-46-3)**
C.I. Acid Violet 17	**(4129-84-4)**
C.I. Acid Violet 17, Sodium salt (8CI)	**(4129-84-4)**
C.I. Acid Violet 19 (7CI)	**(3244-88-0)**
C.I. Acid Violet 19, Disodium salt (8CI)	**(3244-88-0)**
C.I. Acid Violet 49	**(1694-09-3)**
C.I. Acid Violet 49 (Sodium salt)	**(1694-09-3)**
C.I. Acid Yellow 3	**(8004-92-0)**
C.I. Acid Yellow 17, Disodium salt (VAN) (8CI)	**(6359-98-4)**
C.I. Acid Yellow 23	**(1934-21-0)**
C.I. Acid Yellow 23, Trisodium salt	**(1934-21-0)**
C.I. Acid Yellow 25, Monosodium salt (8CI)	**(6359-85-9)**
C.I. Acid Yellow 34, Monosodium salt (8CI)	**(6359-90-6)**
C.I. Acid Yellow 36	**(587-98-4)**
C.I. Acid Yellow 36 monosodium salt	**(587-98-4)**
C.I. Acid Yellow 42	**(6375-55-9)**
C.I. Acid Yellow 42 Disodium salt	**(6375-55-9)**
C.I. Acid Yellow 54 (8CI)	**(10127-05-6)**
C.I. Acid Yellow 61 (8CI)	**(12217-38-8)**
C.I. Acid Yellow 73	**(518-47-8)**
C.I. Acid Yellow 151	**(12715-61-6)**
C.I. Acid Yellow 176 (8CI,9CI)	**(12270-08-5)**
C.I. Acid Yellow 219 (9CI)	**(71819-57-3)**
C.I. Acid Yellow 237 (9CI)	**(77907-22-3)**
C.I. Azoic Coupling Component 1	**(135-19-3)**
C.I. Azoic Coupling Component 2	**(92-77-3)**
C.I. Azoic Coupling Component 3	**(91-92-9)**
C.I. Azoic Coupling Component 4	**(132-68-3)**
C.I. Azoic Coupling Component 5	**(91-96-3)**
C.I. Azoic Coupling Component 14	**(92-74-0)**
C.I. Azoic Coupling Component 17	**(135-65-9)**
C.I. Azoic Coupling Component 18	**(135-61-5)**
C.I. Azoic Coupling Component 20	**(135-62-6)**
C.I. Azoic Coupling Component 21	**(135-63-7)**
C.I. Azoic Coupling Component 34	**(137-52-0)**
C.I. Azoic Coupling Component 41	**(137-52-0)**
C.I. Azoic Coupling Component 107	**(106-49-0)**
C.I. Azoic Coupling Component 110	**(135-61-5)**
C.I. Azoic Diazo Component 2	**(141-85-5)**
C.I. Azoic Diazo Component 3	**(95-82-9)**
C.I. Azoic Diazo Component 5	**(97-52-9)**
C.I. Azoic Diazo Component 6	**(88-74-4)**
C.I. Azoic Diazo Component 7	**(99-09-2)**
C.I. Azoic Diazo Component 8	**(89-62-3)**
C.I. Azoic Diazo Component 9	**(89-63-4)**
C.I. Azoic Diazo Component 11	**(3165-93-3)**

Name	CAS No.
C.I. Azoic Diazo Component 12	(99-55-8)
C.I. Azoic Diazo Component 13	(99-59-2)
C.I. Azoic Diazo Component 20	(120-00-3)
C.I. Azoic Diazo Component 20, Base	(120-00-3)
C.I. Azoic Diazo Component 22	(101-54-2)
C.I. Azoic Diazo Component 37	(100-01-6)
C.I. Azoic Diazo Component 37	(117-27-1)
C.I. Azoic Diazo Component 48	(119-90-4)
C.I. Azoic Diazo Component 48, Fast Blue B Salt	(119-90-4)
C.I. Azoic Diazo Component 112	(92-87-5)
C.I. Azoic Diazo Component 113	(119-93-7)
C.I. Azoic Diazo Component 114	(134-32-7)
C.I. Azoic Red 83	(120-71-8)
C.I. 11160B	(97-56-3)
C.I. 41000B	(492-80-8)
C.I. 46005B	(494-38-2)
C.I. Basic Blue 4	(55840-82-9)
C.I. Basic Blue 1 (8CI)	(3521-06-0)
C.I. Basic Blue 3 (VAN) (9CI)	(55840-82-9)
C.I. Basic Blue 7 (8CI)	(2390-60-5)
C.I. Basic Blue 9	(61-73-4)
C.I. Basic Blue 22 (8CI,9CI)	(12217-41-3)
C.I. Basic Blue 26 (8CI)	(2580-56-5)
C.I. Basic Blue 41 (8CI)	(12270-13-2)
C.I. Basic Blue 69 (9CI)	(12235-47-1)
C.I. Basic Brown 4, Dihydrochloride (8CI)	(5421-66-9)
C.I. Basic Green 1	(633-03-4)
C.I. Basic Green 1, Sulfate (1:1)	(633-03-4)
C.I. Basic Green 4	(569-64-2)
C.I. Basic Orange 2	(532-82-1)
C.I. Basic Orange 2, Monohydrochloride	(532-82-1)
C.I. Basic Orange 3	(532-82-1)
C.I. Basic Orange 14	(494-38-2)
C.I. Basic Orange 14	(65-61-2)
C.I. Basic Orange 21	(3056-93-7)
C.I. Basic Orange 40 (8CI,9CI)	(12270-19-8)
C.I. Basic Red 1	(989-38-8)
C.I. Basic Red 1, Monohydrochloride	(989-38-8)
C.I. Basic Red 2	(477-73-6)
C.I. Basic Red 9, Monohydrochloride	(569-61-9)
C.I. Basic Red 18	(14097-03-1)
C.I. Basic Red 29	(42373-04-6)
C.I. Basic Red 22 (9CI)	(12221-52-2)
C.I. Basic Red 51 (8CI,9CI)	(12270-25-6)
C.I. Basic Violet	(8004-87-3)
C.I. Basic Violet 1	(8004-87-3)
C.I. Basic Violet 1, molybdatetungstatephosphate (9CI)	(1325-82-2)
C.I. Basic Violet 2, Monohydrochloride (8CI)	(3248-91-7)
C.I. Basic Violet 3	(548-62-9)
C.I. Basic Violet 4 (8CI)	(2390-59-2)
C.I. Basic Violet 10	(81-88-9)
C.I. Basic Violet 14	(632-99-5)
C.I. Basic Violet 14, Free Base	(3248-93-9)
C.I. Basic Violet 14, monohydrochloride (8CI)	(632-99-5)
C.I. Basic Yellow 2	(2465-27-2)
C.I. Basic Yellow 2, Free Base	(492-80-8)
C.I. Basic Yellow 2, Monohydrochloride	(2465-27-2)
C.I. 15850:1 (Ca Salt)	(5281-04-9)
C.I. 15880:1 (Ca Salt)	(6417-83-0)
C.I. 52015 (Czech)	(61-73-4)
C.I. Developer 1	(89-25-8)
C.I. Developer 3	(90-51-7)
C.I. Developer 4	(108-46-3)
C.I. Developer 5	(135-19-3)
C.I. Developer 8	(92-70-6)
C.I. Developer 13	(106-50-3)
C.I. Developer 15	(101-54-2)
C.I. Developer 17	(100-01-6)
C.I. Developer 21	(135-61-5)
C.I. Developer 22	(135-62-6)
C.I. Developer 20 (Obs.)	(92-70-6)
C.I. Direct Black 4	(2429-83-6)
C.I. Direct Black 4, Disodium salt (8CI)	(2429-83-6)
C.I. Direct Black 8 (9CI)	(61814-69-5)
C.I. Direct Black 19	(6428-31-5)
C.I. Direct Black 19, Disodium salt	(6428-31-5)
C.I. Direct Black 22	(6473-13-8)
C.I. Direct Black 38	(1937-37-7)
C.I. Direct Black 38, Disodium salt	(1937-37-7)
C.I. Direct Black 114 (9CI)	(61703-05-7)
C.I. Direct Blue 1	(2610-05-1)
C.I. Direct Blue 1, Tetrasodium salt	(2610-05-1)
C.I. Direct Blue 2	(2429-73-4)
C.I. Direct Blue 2, Trisodium salt	(2429-73-4)
C.I. Direct Blue 6	(2602-46-2)
C.I. Direct Blue 6, Tetrasodium salt	(2602-46-2)
C.I. Direct Blue 8	(2429-71-2)
C.I. Direct Blue 8, Disodium salt (8CI)	(2429-71-2)
C.I. Direct Blue 14	(72-57-1)
C.I. Direct Blue 14, Tetrasodium salt	(72-57-1)
C.I. Direct Blue 15	(2429-74-5)
C.I. Direct Blue 15, Tetrasodium salt	(2429-74-5)
C.I. Direct Blue 22, Disodium salt (8CI)	(2586-57-4)
C.I. Direct Blue 25, Tetrasodium salt (8CI)	(2150-54-1)
C.I. Direct Blue 26	(7082-31-7)
C.I. Direct Blue 26, trisodium salt (8CI)	(7082-31-7)
C.I. Direct Blue 53	(314-13-6)
C.I. Direct Blue 53, Tetrasodium salt	(314-13-6)
C.I. Direct Blue 106	(6527-70-4)
C.I. Direct Blue 148, Trisodium salt (8CI)	(3841-15-4)
C.I. Direct Blue 148 (VAN)	(3841-15-4)
C.I. Direct Blue 151, Disodium salt (8CI)	(6449-35-0)
C.I. Direct Blue 218	(28407-37-6)
C.I. Direct Brown	(16071-86-6)
C.I. Direct Brown 1:2	(2586-58-5)
C.I. Direct Brown 1A, Disodium salt	(2586-58-5)
C.I. Direct Brown 2	(2429-82-5)
C.I. Direct Brown 2, Disodium salt (8CI)	(2429-82-5)
C.I. Direct Brown 6, Disodium salt (8CI)	(2893-80-3)
C.I. Direct Brown 31	(2429-81-4)
C.I. Direct Brown 31, Tetrasodium salt	(2429-81-4)
C.I. Direct Brown 59 (VAN)	(3476-90-2)
C.I. Direct Brown 59, Disodium salt (8CI)	(3476-90-2)
C.I. Direct Brown 74 (8CI)	(8014-91-3)
C.I. Direct Brown 78	(2650-18-2)
C.I. Direct Brown 78, Diammonium salt	(2650-18-2)
C.I. Direct Brown 106 (8CI)	(6854-81-5)
C.I. Direct Brown 191	(10190-66-6)
C.I. Direct Green 1	(3626-28-6)
C.I. Direct Green 6, Disodium salt (8CI)	(4335-09-5)
C.I. Direct Green 8, Trisodium salt (8CI)	(5422-17-3)
C.I. Direct Orange 1	(54579-28-1)
C.I. Direct Orange 6, Disodium salt (8CI)	(6637-88-3)
C.I. Direct Orange 8	(2429-79-0)
C.I. Direct Orange 8, Disodium salt (8CI)	(2429-79-0)
C.I. Direct Orange 15 (9CI)	(1325-35-5)
C.I. Direct Orange 46 (9CI)	(50814-31-8)
C.I. Direct Orange 60 (8CI,9CI)	(12262-20-3)
C.I. Direct Red 2	(992-59-6)
C.I. Direct Red 7, Disodium salt (8CI)	(2868-75-9)
C.I. Direct Red 23, Disodium salt (8CI)	(3441-14-3)
C.I. Direct Red 24, Trisodium salt (8CI)	(6420-44-6)
C.I. Direct Red 28	(573-58-0)
C.I. Direct Red 28, Disodium salt	(573-58-0)

C.I. Direct Red 37, Disodium salt (8CI) **(3530-19-6)**
C.I. Direct Red 39 **(6358-29-8)**
C.I. Direct Red 46 **(6548-29-4)**
C.I. Direct Red 46, Tetrasodium salt **(6548-29-4)**
C.I. Direct Red 61 (VAN) **(6470-31-1)**
C.I. Direct Red 61, Disodium salt (8CI) **(6470-31-1)**
C.I. Direct Red 73, Trisodium salt (8CI) **(6460-01-1)**
C.I. Direct Red 80 **(2610-10-8)**
C.I. Direct Red 81 (7CI) **(2610-11-9)**
C.I. Direct Red 254 (9CI) **(101380-00-1)**
C.I. Direct Violet 1 **(2586-60-9)**
C.I. Direct Violet 1, Disodium salt **(2586-60-9)**
C.I. Direct Violet 22 **(6426-67-1)**
C.I. Direct Violet 22, trisodium salt (8CI) **(6426-67-1)**
C.I. Direct Violet 32 **(6428-94-0)**
C.I. Direct Violet 32, Disodium salt (8CI) **(6428-94-0)**
C.I. Direct Yellow 4 **(91-34-9)**
C.I. Direct Yellow 4, Disodium salt (8CI) **(3051-11-4)**
C.I. Direct Yellow 6 (9CI) **(1325-38-8)**
C.I. Direct Yellow 11 (9CI) **(1325-37-7)**
C.I. Direct Yellow 12 (8CI) **(2870-32-8)**
C.I. Direct Yellow 27 **(10190-68-8)**
C.I. Direct Yellow 27, Disodium salt (8CI) **(10190-68-8)**
C.I. Direct Yellow 28 (9CI) **(8005-72-9)**
C.I. Direct Yellow 50, Tetrasodium salt (8CI) **(3214-47-9)**
C.I. Direct Yellow 132 (9CI) **(61968-26-1)**
C.I. Direct Yellow 133 (9CI) **(60202-36-0)**
C.I. Disperse Black 3 **(539-17-3)**
C.I. Disperse Black 6 **(119-90-4)**
C.I. Disperse Black 6 Dihydrochloride **(20325-40-0)**
C.I. Disperse Blue 1 **(2475-45-8)**
C.I. Disperse Blue 3 **(2475-46-9)**
C.I. Disperse Blue 3 **(86722-66-9)**
C.I. Disperse Blue 14 **(2475-44-7)**
C.I. Disperse Blue 27 **(15791-78-3)**
C.I. Disperse Blue 41 **(86722-66-9)**
C.I. Disperse Blue 56 (8CI) **(12217-79-7)**
C.I. Disperse Blue 59 **(12217-79-7)**
C.I. Disperse Blue 71 **(12217-79-7)**
C.I. Disperse Blue 73 **(12222-78-5)**
C.I. Disperse Blue 81 (8CI,9CI) **(12222-79-6)**
C.I. Disperse Blue 110 **(2475-44-7)**
C.I. Disperse Blue 113 **(12222-78-5)**
C.I. Disperse Blue 134 **(14233-37-5)**
C.I. Disperse Blue 139 (9CI) **(50922-60-6)**
C.I. Disperse Brown 1 **(23355-64-8)**
C.I. Disperse Orange 11 **(82-28-0)**
C.I. Disperse Orange 21 (8CI,9CI) **(12217-83-3)**
C.I. Disperse Red 1 (8CI) **(2872-52-8)**
C.I. Disperse Red 8 **(62570-20-1)**
C.I. Disperse Red 9 **(82-38-2)**
C.I. Disperse Red 11 **(2872-48-2)**
C.I. Disperse Red 15 **(116-85-8)**
C.I. Disperse Red 50 (8CI,9CI) **(12223-35-7)**
C.I. Disperse Red 60 (8CI) **(17418-58-5)**
C.I. Disperse Red 71 **(17418-58-5)**
C.I. Disperse Red 83 **(17418-58-5)**
C.I. Disperse Red 135 (9CI) **(58051-96-0)**
C.I. Disperse Red 274 (9CI) **(83929-87-7)**
C.I. Disperse Red II **(129-44-2)**
C.I. Disperse Violet 1 **(128-95-0)**
C.I. Disperse Violet 4 **(1220-94-6)**
C.I. Disperse Violet 8 **(82-33-7)**
C.I. Disperse Yellow 1 **(119-15-3)**
C.I. Disperse Yellow 3 **(2832-40-8)**
C.I. Disperse Yellow 7 (7CI,8CI) **(6300-37-4)**
C.I. Disperse Yellow 23 (8CI) **(6250-23-3)**
C.I. Disperse Yellow 50 (9CI) **(61902-16-7)**
C.I. Disperse Yellow 64 **(10319-14-9)**
C.I. Disperse Yellow 86 (8CI,9CI) **(12223-97-1)**
C.I. Fluorescent Brightener 46 **(6416-68-8)**
C.I. Fluorescent Brightener 225 **(24019-80-5)**
C.I. Fluorescent Brightener 260 **(16090-02-1)**
C.I. Fluorescent Brightening Agent 28 **(4404-43-7)**
C.I. Fluorescent Brightening Agent 46, Sodium salt (8CI) **(6416-68-8)**
C.I. Food Black 1, Tetrasodium salt **(2519-30-4)**
C.I. Food Blue 1 **(860-22-0)**
C.I. Food Blue 1, Disodium salt **(860-22-0)**
C.I. Food Blue 2 **(2650-18-2)**
C.I. Food Blue 2 **(3844-45-9)**
C.I. Food Blue 3 **(129-17-9)**
C.I. Food Green 1 **(4680-78-8)**
C.I. Food Green 2 **(5141-20-8)**
C.I. Food Green 3 **(2353-45-9)**
C.I. Food Orange 1 **(1934-20-9)**
C.I. Food Orange 2 **(2347-72-0)**
C.I. Food Orange 3 **(2051-85-6)**
C.I. Food Orange 4 **(1936-15-8)**
C.I. Food Red 1 **(4548-53-2)**
C.I. Food Red 1, Disodium salt **(4548-53-2)**
C.I. Food Red 2 **(3257-28-1)**
C.I. Food Red 2, Disodium salt **(3257-28-1)**
C.I. Food Red 3 **(3567-69-9)**
C.I. Food Red 4 **(2302-96-7)**
C.I. Food Red 5 **(3761-53-3)**
C.I. Food Red 6 **(3564-09-8)**
C.I. Food Red 6, Disodium salt **(3564-09-8)**
C.I. Food Red 7 **(2611-82-7)**
C.I. Food Red 8 **(5850-44-2)**
C.I. Food Red 9 **(915-67-3)**
C.I. Food Red 10 **(3734-67-6)**
C.I. Food Red 12 **(3567-66-6)**
C.I. Food Red 14 **(16423-68-0)**
C.I. Food Red 15 **(81-88-9)**
C.I. Food Red 17 **(25956-17-6)**
C.I. Food Violet 2 **(1694-09-3)**
C.I. Food Violet 3 **(4129-84-4)**
C.I. Food Yellow 3 **(2783-94-0)**
C.I. Food Yellow 3 **(8004-92-0)**
C.I. Food Yellow 3, Disodium salt **(2783-94-0)**
C.I. Food Yellow 4 **(1934-21-0)**
C.I. Food Yellow 6 **(104-23-4)**
C.I. Food Yellow 6, Monosodium salt (8CI) **(2491-71-6)**
C.I. Food Yellow 8 **(547-57-9)**
C.I. Food Yellow 10 **(85-84-7)**
C.I. Food Yellow 11 **(131-79-3)**
C.I. Food Yellow 13 **(8004-92-0)**
C.I. 42535 Lake **(1325-82-2)**
C.I. Leuco Sulfur Black 1 **(66241-11-0)**
C.I. Leuco Sulfur Red 10 **(1326-96-1)**
C.I. Leuco Sulphur Black 1 (9CI) **(66241-11-0)**
C.I. Leuco Sulphur Black 2 (9CI) **(61814-27-5)**
C.I. Leuco Sulphur Blue 13 (9CI) **(12262-26-9)**
C.I. Leuco Sulphur Green 2 (9CI) **(12262-32-7)**
C.I. Mordant Black 1 (VAN) (8CI) **(3618-58-4)**
C.I. Mordant Black 11 **(1787-61-7)**
C.I. Mordant Black 11, Monosodium salt (8CI) **(1787-61-7)**
C.I. Mordant Red 3 **(130-22-3)**
C.I. Mordant Red 11 **(72-48-0)**
C.I. Mordant Violet 5 **(2092-55-9)**
C.I. Mordant Violet 5, Monosodium salt **(2092-55-9)**
C.I. Mordant Violet 26 **(81-61-8)**
C.I. Natural Black 1 (9CI) **(8005-33-2)**
C.I. Natural Blue 2 **(860-22-0)**

C.I. Natural Brown 7 (481-39-0)
C.I. Natural Brown 8 (1317-34-6)
C.I. Natural Orange 6 (83-72-7)
C.I. Natural Red 1 (117-39-5)
C.I. Natural Red 4 (1260-17-9)
C.I. Natural Yellow 3 (458-37-7)
C.I. Natural Yellow 10 (117-39-5)
C.I. Natural Yellow 10 & 13 (117-39-5)
C.I. Natural Yellow 14 (518-82-1)
C.I. Natural Yellow 16 (84-79-7)
C.I. Natural Yellow 23 (481-74-3)
C.I. No. 46005:1 (494-38-2)
C.I. No.77278 (1308-38-9)
C.I. Oxidation Base (95-80-7)
C.I. Oxidation base 1 (62-53-3)
C.I. Oxidation Base 2 (101-54-2)
C.I. Oxidation Base 4 (6369-59-1)
C.I. Oxidation Base 6A (123-30-8)
C.I. Oxidation Base 7 (591-27-5)
C.I. Oxidation Base 10 (106-50-3)
C.I. Oxidation Base 10A (624-18-0)
C.I. Oxidation Base 12 (615-05-4)
C.I. Oxidation Base 12A (39156-41-7)
C.I. Oxidation Base 13A (6219-71-2)
C.I. Oxidation Base 16 (95-54-5)
C.I. Oxidation Base 17 (95-55-6)
C.I. Oxidation Base 19 (122-80-5)
C.I. Oxidation Base 20 (95-80-7)
C.I. Oxidation Base 21 (96-91-3)
C.I. Oxidation Base 22 (5307-14-2)
C.I. Oxidation Base 26 (120-80-9)
C.I. Oxidation Base 31 (108-46-3)
C.I. Oxidation Base 32 (87-66-1)
C.I. Oxidation Base 33 (90-15-3)
C.I. Oxidation Base 35 (95-80-7)
C.I. Oxidation Base 200 (95-80-7)
C.I. Pigment Black 6 (1333-86-4)
C.I. Pigment Black 7 (1333-86-4)
C.I. Pigment Black 10 (7440-44-0)
C.I. Pigment Black 10 (7782-42-5)
C.I. Pigment Black 13 (1307-96-6)
C.I. Pigment Black 14 (1313-13-9)
C.I. Pigment Black 15 (1317-38-0)
C.I. Pigment Black 16 (7440-66-6)
C.I. Pigment Blue 1 (1325-87-7)
C.I. Pigment Blue 15 (147-14-8)
C.I. Pigment Blue 15:1 (147-14-8)
C.I. Pigment Blue 15:3 (147-14-8)
C.I. Pigment Blue 15:4 (147-14-8)
C.I. Pigment Blue 16 (574-93-6)
C.I. Pigment Blue 34 (1317-40-4)
C.I. Pigment Blue 60 (81-77-6)
C.I. Pigment Blue 61 (1324-76-1)
C.I. Pigment Blue 63 (9CI) (16521-38-3)
C.I. Pigment Brown 8 (1313-13-9)
C.I. Pigment Green 7 (1328-53-6)
C.I. Pigment Green 17 (1308-38-9)
C.I. Pigment Green 21 (9CI) (12002-03-8)
C.I. Pigment Green 36 (14302-13-7)
C.I. Pigment Green 38 (14302-13-7)
C.I. Pigment Green 41 (14302-13-7)
C.I. Pigment Green 42 (1328-53-6)
C.I. Pigment Metal 2 (7440-50-8)
C.I. Pigment Metal 3 (7440-57-5)
C.I. Pigment Metal 4 (7439-92-1)
C.I. Pigment Metal 6 (7440-66-6)
C.I. Pigment Orange (12236-62-3)
C.I. Pigment Orange 5 (3468-63-1)
C.I. Pigment Orange 7, Monosodium salt (8CI) (5850-81-7)
C.I. Pigment Orange 13 (3520-72-7)
C.I. Pigment Orange 20 (1306-23-6)
C.I. Pigment Oragne 21 (18454-12-1)
C.I. Pigment Orange 21 (9CI) (1344-38-3)
C.I. Pigment Orange 31 (5280-74-0)
C.I. Pigment Orange 36 (12236-62-3)
C.I. Pigment Orange 40 (128-70-1)
C.I. Pigment Orange 43 (4424-06-0)
C.I. Pigment Red (18454-12-1)
C.I. Pigment Red (5160-02-1)
C.I. Pigment Red 3 (2425-85-6)
C.I. Pigment Red 4 (2814-77-9)
C.I. Pigment Red 23 (6471-49-4)
C.I. Pigment Red 38 (VAN) (8CI) (6358-87-8)
C.I. Pigment Red 49 (1248-18-6)
C.I. Pigment Red 48:4 (9CI) (5280-66-0)
C.I. Pigment Red 49, Barium salt (2:1) (1103-38-4)
C.I. Pigment Red 49:1 (1103-38-4)
C.I. Pigment Red 49 Ca Salt (1103-39-5)
C.I. Pigment Red 49, Calcium salt (2:1) (8CI) (1103-39-5)
C.I. Pigment Red 49:2 (1103-39-5)
C.I. Pigment Red 52:2 (9CI) (12238-31-2)
C.I. Pigment Red 53, Barium salt (5160-02-1)
C.I. Pigment Red 53:1 (5160-02-1)
C.I. Pigment Red 57 (7CI) (5858-81-1)
C.I. Pigment Red 57, Disodium salt (8CI) (5858-81-1)
C.I. Pigment Red 69 (5850-90-8)
C.I. Pigment Red 69, monosodium salt (8CI) (5850-90-8)
C.I. Pigment Red 83 (72-48-0)
C.I. Pigment Red 49, Monosodium salt (1248-18-6)
C.I. Pigment Red 101 (1309-37-1)
C.I. Pigment Red 101 and 102 (1309-37-1)
C.I. Pigment Red 102 (1309-37-1)
C.I. Pigment Red 104 (12656-85-8)
C.I. Pigment Red 105 (1314-41-6)
C.I. Pigment Red 107 (1345-04-6)
C.I. Pigment Red 113 (1345-09-1)
C.I. Pigment Red 120 (2786-76-7)
C.I. Pigment Red 122 (1047-16-1)
C.I. Pigment Red 122 (980-26-7)
C.I. Pigment Red 123 (24108-89-2)
C.I. Pigment Red 169 (9CI) (12237-63-7)
C.I. Pigment Red 170 (2786-76-7)
C.I. Pigment Red 195 (4203-77-4)
C.I. Pigment Violet 3 (1325-82-2)
C.I. Pigment Violet 5:1 (9CI) (1328-04-7)
C.I. Pigment Violet 19 (1047-16-1)
C.I. Pigment Violet 23 (6358-30-1)
C.I. Pigment Violet 31 (1324-55-6)
C.I. Pigment White 1 (1319-46-6)
C.I. Pigment White 3 (7446-14-2)
C.I. Pigment White 4 (1314-13-2)
C.I. Pigment White 6 (13463-67-7)
C.I. Pigment White 7 (1314-98-3)
C.I. Pigment White 8 (1314-96-1)
C.I. Pigment White 10 (513-77-9)
C.I. Pigment White 11 (1309-64-4)
C.I. Pigment White 17 (1304-85-4)
C.I. Pigment White 21 (7727-43-7)
C.I. Pigment White 33 (11070-82-9)
C.I. Pigment Yellow (1303-33-9)
C.I. Pigment Yellow 1 (VAN) (8CI) (2512-29-0)
C.I. Pigment Yellow 3 (VAN) (8CI) (6486-23-3)
C.I. Pigment Yellow 12 (6358-85-6)
C.I. Pigment Yellow 13 (VAN) (8CI) (5102-83-0)

C.I. Pigment Yellow 14 (VAN) (8CI) (5468-75-7)
C.I. Pigment Yellow 16 (VAN) (8CI) (5979-28-2)
C.I. Pigment Yellow 31 (10294-40-3)
C.I. Pigment Yellow 34 (9CI) (1344-37-2)
C.I. Pigment Yellow 32 (7789-06-2)
C.I. Pigment Yellow 33 (13765-19-0)
C.I. Pigment Yellow 34 (1344-37-2)
C.I. Pigment Yellow 34 (7758-97-6)
C.I. Pigment Yellow 36 (13530-65-9)
C.I. Pigment Yellow 37 (1306-23-6)
C.I. Pigment Yellow 46 (1317-36-8)
C.I. Pigment Yellow 48 (592-05-2)
C.I. Pigment Yellow 53 (9CI) (8007-18-9)
C.I. Pigment Yellow 73 (13515-40-7)
C.I. Pigment Yellow 74 (6358-31-2)
C.I. Pigment Yellow 100 (9CI) (12225-21-7)
C.I. Pigment Yellow 104 (9CI) (15790-07-5)
C.I. Pigment Yellow 151 (31837-42-0)
C.I. Reactive Black 8 (12225-26-2)
C.I. Reactive Blue 19 (2580-78-1)
C.I. Reactive Blue 19, Disodium salt (2580-78-1)
C.I. Reactive Blue 21 (8CI,9CI) (12236-86-1)
C.I. Reactive Blue 49 (9CI) (12236-92-9)
C.I. Reactive Blue 170 (9CI) (71902-14-2)
C.I. Reactive Green 12 (8CI,9CI) (12225-80-8)
C.I. Reactive Orange 12 (9CI) (12225-84-2)
C.I. Reactive Red 1 (17752-85-1)
C.I. Reactive Red 1, Trisodium salt (8CI) (17752-85-1)
C.I. Reactive Red 2 (17804-49-8)
C.I. Reactive Red 16 (8CI,9CI) (12238-07-2)
C.I. Reactive Red 24 (8CI,9CI) (12238-00-5)
C.I. Reactive Red 31 (9CI) (12237-00-2)
C.I. Reactive Red 35 (8CI,9CI) (12226-12-9)
C.I. Reactive Red 133 (9CI) (72979-85-2)
C.I. Reactive Red 185 (9CI) (102640-17-5)
C.I. Reactive Violet 5 (9CI) (12226-38-9)
C.I. Reactive Yellow 17 (20317-19-5)
C.I. Reactive Yellow 17, Dipotassium salt (8CI) (20317-19-5)
C.I. Reactive Yellow 23 (8CI,9CI) (12226-50-5)
C.I. Reactive Yellow 37 (8CI,9CI) (12237-16-0)
C.I. Reactive Yellow 42 (8CI,9CI) (12226-63-0)
C.I. Reactive Yellow 155 (9CI) (102640-18-6)
C.I. Red 33, Disodium salt (3567-66-6)
C.I. Reducing Agent 6 (3567-66-6)
C.I. Solubilised Sulphur Black 1 (9CI) (1326-83-6)
C.I. Solubilised Sulphur Black 2 (9CI) (1326-86-9)
C.I. Solubilized Sulfur Black 1 (1326-83-6)
C.I. Solvent Black 3 (4197-25-5)
C.I. Solvent Black 5 (11099-03-9)
C.I. Solvent Black 7 (9CI) (8005-02-5)
C.I. Solvent Blue 7 (60-09-3)
C.I. Solvent Blue 11 (128-85-8)
C.I. Solvent Blue 18 (2475-45-8)
C.I. Solvent Blue 23 (2152-64-9)
C.I. Solvent Blue 23, Monohydrochloride (2152-64-9)
C.I. Solvent Blue 36 (14233-37-5)
C.I. Solvent Blue 38 (9CI) (1328-51-4)
C.I. Solvent Brown 1 (6416-57-5)
C.I. Solvent Brown PR (6416-57-5)
C.I. Solvent Green 3 (128-80-3)
C.I. Solvent Green 7 (6358-69-6)
C.I. Solvent Green 9 (6358-69-6)
C.I. Solvent Orange 1 (8CI) (2051-85-6)
C.I. Solvent Orange 2 (2646-17-5)
C.I. Solvent Orange 3 (495-54-5)
C.I. Solvent Orange 3 (532-82-1)
C.I. Solvent Orange 7 (3118-97-6)
C.I. Solvent Orange 15 (494-38-2)
C.I. Solvent Red (1229-55-6)
C.I. Solvent Red 17 (6410-20-4)
C.I. Solvent Red 19 (6368-72-5)
C.I. Solvent Red 23 (85-86-9)
C.I. Solvent Red 24 (85-83-6)
C.I. Solvent Red 41 (3248-93-9)
C.I. Solvent Red 43 (15086-94-9)
C.I. Solvent Red 49 (509-34-2)
C.I. Solvent Red 53 (116-85-8)
C.I. Solvent Red 80 (6358-53-8)
C.I. Solvent Red 111 (82-38-2)
C.I. Solvent Violet 11 (128-95-0)
C.I. Solvent Violet 12 (1220-94-6)
C.I. Solvent Violet 13 (81-48-1)
C.I. Solvent Violet 26 (2872-48-2)
C.I. Solvent Yellow 1 (60-09-3)
C.I. Solvent Yellow 2 (60-11-7)
C.I. Solvent Yellow 3 (97-56-3)
C.I. Solvent Yellow 4 (131-22-6)
C.I. Solvent Yellow 5 (85-84-7)
C.I. Solvent Yellow 6 (131-79-3)
C.I. Solvent Yellow 7 (1689-82-3)
C.I. Solvent Yellow 14 (842-07-9)
C.I. Solvent Yellow 33 (8003-22-3)
C.I. Solvent Yellow 34 (492-80-8)
C.I. Solvent Yellow 44 (2478-20-8)
C.I. Solvent Yellow 52 (119-15-3)
C.I. Solvent Yellow 56 (2481-94-9)
C.I. Solvent Yellow 58 (8CI) (2452-84-8)
C.I. Solvent Yellow 77 (2832-40-8)
C.I. Sulfur Black 11, C.I. 53290 (1327-14-6)
C.I. Sulfur Blue 7 (1327-57-7)
C.I. Sulfur Red 10 (1326-96-1)
C.I. Sulphur Black 1 (9CI) (1326-82-5)
C.I. Sulphur Black 2 (9CI) (1326-85-8)
C.I. Sulphur Black 11 (9CI) (1327-14-6)
C.I. Sulphur Blue 7 (9CI) (1327-57-7)
C.I. Sulphur Blue 13 (9CI) (1327-59-9)
C.I. Sulphur Red 10 (9CI) (1326-96-1)
C.I. Vat Black 27 (2379-81-9)
C.I. Vat Blue 1 (482-89-3)
C.I. Vat Blue 3 (6492-73-5)
C.I. Vat Blue 4 (81-77-6)
C.I. Vat Blue 6 (130-20-1)
C.I. Vat Blue 16 (6424-76-6)
C.I. Vat Blue 18 (1324-54-5)
C.I. Vat Blue 34 (6492-73-5)
C.I. Vat Blue 43 (9CI) (1327-79-3)
C.I. Vat Brown 1 (2475-33-4)
C.I. Vat Brown 44 (2475-33-4)
C.I. Vat Green 1 (128-58-5)
C.I. Vat Green 3 (3271-76-9)
C.I. Vat Orange 5 (3263-31-8)
C.I. Vat Orange 7 (4424-06-0)
C.I. Vat Orange 9 (128-70-1)
C.I. Vat Red 13 (4203-77-4)
C.I. Vat Red 14 (9CI) (8005-56-9)
C.I. Vat Violet 1 (8CI) (1324-55-6)
C.I. Vat Yellow (128-66-5)
C.I. Vat Yellow 2 (129-09-9)
C.I. Vat Yellow 4 (128-66-5)
C.I. Yellow 11 (1325-37-7)
CI 42595 Phosphotungstomolybdic acid salt (1325-87-7)
CICP (101-21-3)
CI-IPC (101-21-3)
CIM (303-49-1)

CIPC **(101-21-3)**
CK3 **(1333-86-4)**
CKB 1028A **(51249-05-9)**
CL 369 **(1867-66-9)**
CL 10304 **(60-32-2)**
CL 11344 **(55512-33-9)**
CL 12880 **(60-51-5)**
CL 13494 **(80-35-3)**
CL 18133 **(297-97-2)**
CL 36010 **(73-49-4)**
CL 39,148 **(1665-48-1)**
CL 47031 **(947-02-4)**
CL 47300 **(122-14-5)**
CL 52160 **(3383-96-8)**
CL 59806 **(10118-90-8)**
CL 64475 **(21548-32-3)**
CL 217300 **(67485-29-4)**
CL-14377 **(59-05-2)**
CL-38023 **(52-85-7)**
CL-47,470 **(950-10-7)**
CL-62362 **(1977-10-2)**
CL-71563 **(1977-10-2)**
CLF II **(7440-44-0)**
C-Level **(50-81-7)**
C-Long **(50-81-7)**
CM 1001 **(105-60-2)**
CM 1001 **(25038-54-4)**
CM 1011 **(105-60-2)**
CM 1011 **(25038-54-4)**
CM 1031 **(105-60-2)**
CM 1031 **(25038-54-4)**
CM 1041 **(105-60-2)**
CM 1041 **(25038-54-4)**
CM 6912 **(29177-84-2)**
CMA **(302-22-7)**
CMB 50 **(7440-44-0)**
CMB 200 **(7440-44-0)**
CMC **(9000-11-7)**
CMC **(9004-32-4)**
CMC 2 **(9004-32-4)**
CMC 41A **(9004-32-4)**
CMC 4H1 **(9004-32-4)**
CMC 7H **(9004-32-4)**
CMC 7H3SF **(9004-32-4)**
CMC 7L1 **(9004-32-4)**
CMC-4LF **(9000-11-7)**
CMC 4M6 **(9004-32-4)**
CMC 7M **(9004-32-4)**
CMC 3M5T **(9004-32-4)**
CMC 7MT **(9004-32-4)**
CMC Sodium Salt **(9004-32-4)**
CM-Cellulose **(9000-11-7)**
CM-Cellulose Na Salt **(9004-32-4)**
CM-Cellulose Sodium Salt **(9004-32-4)**
CMDP **(7786-34-7)**
CME **(8063-14-7)**
CME 74770 **(26644-46-2)**
CMH **(7791-18-6)**
CMHEC **(9004-30-2)**
CMME **(107-30-2)**
CMMP **(2307-68-8)**
CMPP **(7085-19-0)**
CMPP **(93-65-2)**
CM S 2957 **(21923-23-9)**
CM-S 2957 **(60238-56-4)**
CMU **(150-68-5)**
CMW Bone Cement **(9011-14-7)**
C-Methyl-1,3,5-triazine **(3599-87-9)**
C-Meton **(113-92-8)**
CN **(532-27-4)**
CN 38703 **(72-44-6)**
CN 8676 **(67-21-0)**
CN-15,757 **(115-02-6)**
CN-25,253-2 **(956-90-1)**
CN-35355 **(61-68-7)**
CN-52,372-2 **(1867-66-9)**
CNA **(99-30-9)**
CNC **(1338-02-9)**
CNP **(1836-77-7)**
CNP 1032 **(1836-77-7)**
CO 12 **(112-53-8)**
CO 12 **(136-52-7)**
CO 433 **(9014-90-8)**
CO 436 **(9051-57-4)**
CO 755 **(34681-10-2)**
CO 859 **(34681-23-7)**
CO-1214 **(112-53-8)**
CO-1670 **(36653-82-4)**
CO-1695 **(36653-82-4)**
CO-1895 **(112-92-5)**
CO-1897 **(112-92-5)**
COBH **(495-73-8)**
COD **(1552-12-1)**
CO-1214N **(112-53-8)**
CO-1214S **(112-53-8)**
Co-Op Hexa **(118-74-1)**
C-Orange 11 (Germany) **(7235-40-7)**
4-CP **(122-88-3)**
CP **(50-18-0)**
CP 10-40 **(14807-96-6)**
CP 34 **(110-02-1)**
CP 38-33 **(14807-96-6)**
CP 105 **(106-91-2)**
CP 2000 **(9004-74-4)**
CP 3438 **(97-18-7)**
CP 4572 **(95-06-7)**
CP 4,742 **(95-06-7)**
CP 6,343 **(93-71-0)**
CP 14,957 **(297-78-9)**
CP 15,336 **(2303-16-4)**
CP 16171 **(36322-90-4)**
CP 17029 **(845-52-3)**
CP 19699 **(13067-93-1)**
CP 23426 **(2303-17-5)**
CP 25017 **(24458-48-8)**
CP 31393 **(1918-16-7)**
CP 31675 **(3785-20-4)**
CP 40294 **(2665-30-7)**
CP 41845 **(2439-99-8)**
CP 45592 **(2163-81-7)**
CP 47114 **(122-14-5)**
CP 49674 **(13265-60-6)**
CP 50144 **(15972-60-8)**
CP 53619 **(23184-66-9)**
CP 53926 **(2540-82-1)**
CP-10,423-16 **(15686-83-6)**
CP-12,252-1 **(5591-45-7)**
CP-15467-61 **(554-13-2)**
CP-16533-1 **(52-53-9)**
4-CPA **(122-88-3)**
6-CPA **(4684-94-0)**
CPA **(122-88-3)**
CPA **(50-18-0)**
CPB **(80-38-6)**

CPB 5000 (7782-42-5)
CPBS (80-38-6)
CP Basic Copper TS-53 WP (1332-03-2)
CP Basic Sulfate (1332-03-2)
CP Basic Sulfate (7758-98-7)
CPC 3005 (9005-25-8)
CPC 6448 (9005-25-8)
CPCA (115-32-2)
CPCBS (80-33-1)
C.P. Chrome Light 2010 (18454-12-1)
C.P. Chrome Orange Dark 2030 (18454-12-1)
C.P. Chrome Orange Extra Dark 2040 (18454-12-1)
C.P. Chrome Orange Medium 2020 (18454-12-1)
C.P. Chrome Yellow Light (7758-97-6)
C.P. Chrome Yellow Light 1066 (1344-37-2)
C.P. Chrome Yellow Light 1074 (1344-37-2)
C.P. Chrome Yellow Medium (7758-97-6)
C.P. Chrome Yellow Medium 1074 (1344-37-2)
C.P. Chrome Yellow Medium 1085 (1344-37-2)
C.P. Chrome Yellow Medium 1298 (1344-37-2)
C.P. Chrome Yellow Primose (7758-97-6)
CPDC (15663-27-1)
CPDD (15663-27-1)
CPE (9002-88-4)
CPE 16 (9002-88-4)
CPE 25 (9002-88-4)
CPH (103-03-7)
CPH (56-75-7)
CPIB (637-07-0)
CPMC (3942-54-9)
2-p-CPP (3307-39-9)
CPPC (2150-32-5)
CPPO (30431-54-0)
CPP 25S (9003-07-0)
CPT (113-59-7)
CPT (95-74-9)
C.P. Titanium (7440-32-6)
C.P.Toluidine Toner A-2989 (2425-85-6)
C.P.Toluidine Toner A-2990 (2425-85-6)
C.P.Toluidine Toner Dark RS-3340 (2425-85-6)
C.P.Toluidine Toner Deep X-1865 (2425-85-6)
C.P.Toluidine Toner Light RS-3140 (2425-85-6)
C.P.Toluidine Toner RT-6101 (2425-85-6)
C.P.Toluidine Toner RT-6104 (2425-85-6)
CPX (113-59-7)
CPZ (50-53-3)
CPZ (69-09-0)
C.P. Zinc Yellow X-883 (13530-65-9)
C.P. Zinc Yellow X-2127 (13530-65-9)
C12-15 Pareth-2 (68131-39-5)
C12-15 Pareth-4 (68131-39-5)
C11-15 Pareth-3 (68131-40-8)
C12-15 Pareth-3 (68131-39-5)
C12-15 Pareth-5 (68131-39-5)
C11-15 Pareth-5 (68131-40-8)
C11-15 Pareth-7 (68131-40-8)
C12-15 Pareth-7 (68131-39-5)
C12-15 Pareth-9 (68131-39-5)
C11-15 Pareth-9 (68131-40-8)
C12-15 Pareth-12 (68131-39-5)
C11-15 Pareth-20 (68131-40-8)
C11-15 Pareth-30 (68131-40-8)
C11-15 Pareth-40 (68131-40-8)
CP 2000 (polyoxyalkylene) (9004-74-4)
3-CPs (20073-24-9)
C-Quens (302-22-7)
C-Quin (50-81-7)
CR (573-58-0)
CR 39 (142-22-3)
CR 305 (97-24-5)
CR 409 (115-26-4)
CR 1639 (39300-45-3)
CR 1693 (39300-45-3)
CR 3029 (12427-38-2)
CR 14658 (26225-79-6)
CR-144 (71878-19-8)
CRAG (533-74-4)
CRAG 974 (533-74-4)
CRAG Fungicide 974 (533-74-4)
CRAG Nemacide (533-74-4)
CRAG 85W (533-74-4)
CRILL 20,21,22,23 (9004-99-3)
C-Raban 1HB (9006-03-5)
C-Rot 14 (Germany) (6417-83-)
CS (2698-41-1)
CS 293 (2152-34-3)
CS 370 (24166-13-0)
CS 645A (117-27-1)
CS 674A (117-26-0)
CS 1483 (53-33-8)
CS 8890 (24201-58-9)
CS 12602 (321-64-2)
CS 58525 (57-47-6)
CS-430 (59128-97-1)
CS-847 (101-27-9)
CSAC (111-15-9)
C3S (Cement component) (12168-85-3)
CSF-Giftweizen (7446-18-6)
CS Lafarge (1344-95-2)
CSP (7758-99-8)
C-Sn-9 (56-35-9)
C-Span (50-81-7)
CT (153-61-7)
CT (58-94-6)
CTA (569-57-3)
CTAB (57-09-0)
CTC (57-62-5)
CTCP (25267-55-4)
CTFE (79-38-9)
CTH (7745-89-3)
CTMAB (57-09-0)
CTR 6669 (10605-21-7)
CTX (50-18-0)
CUZ 3 (7440-44-0)
CV 399 (1668-54-8)
CVMP (22248-79-9)
CVP (470-90-6)
C-Vimin (50-81-7)
CW 524 (26644-46-2)
CWN 2 (7440-44-0)
C-Weiss 7 (German) (13463-67-7)
CXD (51762-05-1)
CXM (55268-75-2)
CY (50-18-0)
CY 116 (60-32-2)
CY 39 (520-52-5)
CY-39 (520-52-5)
CYBIS (389-08-2)
CYDTA (482-54-2)
CY-L 500 (156-62-7)
CYP (13067-93-1)
CZT (54749-90-5)
Ca 33 (9005-35-0)
CaO 6 (90-66-4)

Caesium, Metal (DOT)	**(7440-46-2)**
Caesium, Powdered (DOT)	**(7440-46-2)**
Caesium [UN 1407]	**(7440-46-2)**
Caesium hydroxide, Solid [UN 2682]	**(21351-79-1)**
Caesium hydroxide, Solution [UN 2681]	**(21351-79-1)**
Caffearine	**(535-83-1)**
Caffeic acid	**(331-39-5)**
Caffein	**(58-08-2)**
Caffeine	**(58-08-2)**
3-Caffeoylquinic acid	**(327-97-9)**
3-o-Caffeoylquinic acid	**(327-97-9)**
Cairox	**(7722-64-7)**
Cajeputene	**(138-86-3)**
Cajeputol	**(470-82-6)**
Cake Alum	**(10043-01-3)**
Calo-Clor	**(8065-83-6)**
Calactin	**(20304-47-6)**
(+)-Calamenene	**(22339-23-7)**
(-)-Calamenene	**(483-77-2)**
Calamenene	**(483-77-2)**
Calamenene, (+)-	**(22339-23-7)**
Calamenene, (-)-	**(483-77-2)**
Calamine	**(1314-13-2)**
Calamine (Spray)	**(1314-13-2)**
Calar	**(5902-95-4)**
Calaroc EU	**(136-84-5)**
Calci-C	**(5743-27-1)**
Calcaloid Printing Orange RYW	**(3263-31-8)**
Calcamine	**(67-96-9)**
Calcia	**(1305-78-8)**
Calciate(3)-, (N,N-bis(2-(bis(carboxymethyl)amino)ethyl)-glycinato(5-))-, trisodium (9CI)	**(12111-24-9)**
Calciate(2-), ((ethylenedinitrilo)tetraacetato)-, disodium	**(62-33-9)**
Calcicat	**(7440-70-2)**
Calcifediol	**(19356-17-3)**
Calcifediolum (Latin)	**(19356-17-3)**
Calciferol	**(50-14-6)**
Calciferon 2	**(50-14-6)**
Calcigenol Simple	**(7758-87-4)**
Calcined Kentucky flint clay	**(66402-68-4)**
Calcined Missouri flint clay	**(66402-68-4)**
Calcined Soda	**(12401-86-4)**
Calcined baryta	**(1304-28-5)**
Calcined bastnasite	**(68909-13-7)**
Calcined bauxite	**(66402-68-4)**
Calcined brucite	**(1309-48-4)**
Calcined clay	**(66402-68-4)**
Calcined clays	**(66402-68-4)**
Calcined diatomite	**(14464-46-1)**
Calcined fireclay	**(66402-68-4)**
Calcined kaolin	**(66402-68-4)**
Calcined kaolin clay	**(66402-68-4)**
Calcined lightweight aggregate	**(66402-68-4)**
Calcined magnesia	**(1309-48-4)**
Calcined magnesite	**(1309-48-4)**
Calcined semi-flint clay (Blum)	**(66402-68-4)**
Calcined semi-flint clay (Harris)	**(66402-68-4)**
Calciopen K	**(132-98-9)**
Calcite (8CI,9CI)	**(13397-26-7)**
Calcitetracemate disodium	**(62-33-9)**
Calcitriol	**(32222-06-3)**
Calcium-DTPA	**(12111-24-9)**
Calcium-bisulfite	**(13780-03-5)**
Calcium-bromide	**(7789-41-5)**
Calcium-carbonate	**(1317-65-3)**
Calcium-chloride	**(10043-52-4)**
Calcium-fluoride	**(7789-75-5)**
Calcium-gluconate	**(299-28-5)**
Calcium-hydroxide	**(1305-62-0)**
Calcium (2:1)	**(814-71-1)**
Calcium, Acetate octanoate stearate complexes	**(68411-15-4)**
Calcium Chel-330	**(12111-24-9)**
Calcium Chrome Yellow	**(13765-19-0)**
Calcium D(+)-N-(α,γ-dihydroxy-β,β-dimethylbutyryl)-β-alaninate	**(137-08-6)**
Calcium Lithol Red	**(1103-39-5)**
Calcium, Metal, Crystalline (DOT)	**(7440-70-2)**
Calcium, Metal (DOT)	**(7440-70-2)**
Calcium, Non-pyrophoric (DOT)	**(7440-70-2)**
Calcium, Pyrophoric [UN 1855]	**(7440-70-2)**
Calcium Trisodium CHEL 330	**(12111-24-9)**
Calcium Trisodium DTPA	**(12111-24-9)**
Calcium [UN 1401]	**(7440-70-2)**
Calcium acetate	**(62-54-4)**
Calcium acid methanearsonate	**(5902-95-4)**
Calcium acid methyl arsonate	**(5902-95-4)**
Calcium acid phosphate	**(7757-93-9)**
Calcium alginate	**(9005-35-0)**
Calcium alkylaromatic sulfonate	**(26264-06-2)**
Calcium alkylbenzenesulfonate	**(26264-06-2)**
Calcium aluminum silicate	**(1318-02-1)**
Calcium ammonium nitrate	**(15245-12-2)**
Calciumarsenat	**(7778-44-1)**
Calcium arsenate	**(10103-62-5)**
Calcium arsenate	**(15194-99-7)**
Calcium orthoarsenate	**(7778-44-1)**
Calcium arsenate, Solid (DOT)	**(7778-44-1)**
Calcium arsenate [UN 1573]	**(7778-44-1)**
Calcium arsenite	**(27152-57-4)**
Calcium arsenite	**(52740-16-6)**
Calcium arsenite (1:1)	**(52740-16-6)**
Calcium arsenite (2:1)	**(15194-98-6)**
Calcium meta-arsenite	**(52740-16-6)**
Calcium arsenite, Solid [NA 1574]	**(27152-57-4)**
Calcium arsenite, Solid [NA 1574]	**(52740-16-6)**
Calcium arsentit (2:3)	**(27152-57-4)**
Calcium ascorbate	**(5743-27-1)**
Calcium benzoate	**(2090-05-3)**
Calcium o-benzosulfimide	**(6485-34-3)**
Calcium 2-benzosulphimide	**(6485-34-3)**
Calcium o-benzosulphimide	**(6485-34-3)**
Calcium, bis(nonylphenoxy)-	**(30977-64-1)**
Calcium bis(tetrahydroaluminate(1-))	**(16941-10-9)**
Calcium bis(tetrahydroaluminate)	**(16941-10-9)**
Calcium bisulfite, Solution (DOT)	**(13780-03-5)**
Calcium borate	**(12040-58-3)**
Calcium carbide [UN 1402]	**(75-20-7)**
Calcium carbimide	**(156-62-7)**
Calcium carbonate	**(471-34-1)**
Calcium carbonate (ACGIH,OSHA)	**(1317-65-3)**
Calcium chlorate	**(10137-74-3)**
Calcium chlorate, Aqueous solution [UN 2429]	**(10137-74-3)**
Calcium chlorate [UN 1452]	**(10137-74-3)**
Calcium chlorite	**(14674-72-7)**
Calcium chlorohydrochlorite	**(7778-54-3)**
Calcium chromate	**(13765-19-0)**
Calcium chromate	**(14307-33-6)**
Calcium chromate (VI)	**(13765-19-0)**
Calcium chromium oxide (CaCrO4)	**(13765-19-0)**
Calcium citrate	**(7693-13-2)**
Calcium citrate	**(813-94-5)**
Calcium cyanamid	**(156-62-7)**
Calcium cyanamide (ACGIH,OSHA)	**(156-62-7)**
Calcium cyanamide, Not hydrated (containing more than 0.1% calcium carbide) [UN 1403]	**(156-62-7)**

Name	CAS
Calcium cyanide, Solid (DOT)	(592-01-8)
Calcium cyanide [UN 1575]	(592-01-8)
Calcium cyclamate	(139-06-0)
Calcium cyclohexanesulfamate	(139-06-0)
Calcium cyclohexane sulphamate	(139-06-0)
Calcium cyclohexylsulfamate	(139-06-0)
Calcium cyclohexylsulphamate	(139-06-0)
Calcium diacetate	(62-54-4)
Calcium dibasic phosphate	(7757-93-9)
Calcium dibromide	(7789-41-5)
Calcium, dichloro(5-chloro-2-methyl-3(2H)-isothiazolone-O)-, Mixt. with dichloro(2-methyl-3(2H)-isothiazolone-O) calcium (9CI)	(50815-77-5)
Calcium dichromate(VI)	(14307-33-6)
Calcium difluoride	(7789-75-5)
Calcium dinonylnaphthalenesulfonate	(57855-77-3)
Calcium diphosphate	(7790-76-3)
Calcium dipropionate	(4075-81-4)
Calcium disodium edathamil	(62-33-9)
Calcium disodium edetate	(62-33-9)
Calcium disodium EDTA	(62-33-9)
Calcium disodium ethylenediaminetetraacetate	(62-33-9)
Calcium disodium (ethylenedinitrilo)tetraacetate	(62-33-9)
Calcium disodium versenate	(62-33-9)
Calcium distearate	(1592-23-0)
Calcium dodecylbenzene sulfonate	(26264-06-2)
Calcium dodecylbenzenesulfonate	(26264-06-2)
Calcium n-dodecylbenzenesulfonate	(26264-06-2)
Calcium dodecylbenzensulfonate	(26264-06-2)
Calcium 2-ethylhexanoate	(136-51-6)
Calcium formate	(544-17-2)
Calcium hexadecanoate	(542-42-7)
(E,E)-Calcium 2,4-hexadienoate	(7492-55-9)
Calcium hydrate	(1305-62-0)
Calcium hydride (9CI)	(7789-78-8)
Calcium hydrogen methanearsonate	(5902-95-4)
Calcium hydrogen orthophosphate	(7757-93-9)
Calcium hydrogen phosphate	(7757-93-9)
Calcium hydrogenphosphate	(7757-93-9)
Calcium hydrogen sulfite, Solution (DOT)	(13780-03-5)
Calcium hydrosilicate	(1344-95-2)
Calcium hydrosulphite [UN 1923]	(13780-03-5)
Calcium hydroxide (ACGIH,OSHA)	(1305-62-0)
Calcium hypochloride	(7778-54-3)
Calcium hypochlorite	(7778-54-3)
Calcium hypochlorite, Dry [UN 1748]	(7778-54-3)
Calcium hypochlorite mixture, Dry (containing more than 39% available chlorine) (DOT)	(7778-54-3)
Calcium hypochlorite mixtures, Dry, with 10% to 39% available chlorine [UN 2208]	(7778-54-3)
Calcium hypophosphite	(7789-79-9)
Calcium iodate	(7789-80-2)
Calcium iodobehenate	(1319-91-1)
Calcium 2-isovaleryl-1,3-indandione	(23710-76-1)
Calcium lactate	(814-80-2)
Calcium lignosulfonate	(8061-52-7)
Calcium lithol	(1103-39-5)
Calcium mercaptoacetate	(814-71-1)
Calcium metasilicate	(10101-39-0)
Calcium methanearsonate	(5902-95-4)
Calcium monochromate	(13765-19-0)
Calcium monohydrogen phosphate	(7757-93-9)
Calcium monohydrogen phosphate anhydrous	(7757-93-9)
Calcium monosilicate	(1344-95-2)
Calcium monosulfide	(20548-54-3)
Calcium neodecanoate	(27253-33-4)
Calcium(II) nitrate (1:2)	(10124-37-5)
Calcium nitrate [UN 1454]	(10124-37-5)
Calcium nitrite	(13780-06-8)
Calcium nonanoate	(29813-38-5)
Calcium orotate	(22454-86-0)
Calcium oxide (ACGIH,OSHA) [UN 1910]	(1305-78-8)
Calcium oxide silicate (9CI)	(12168-85-3)
Calcium oxychloride	(7778-54-3)
Calcium oxytetracycline	(7179-50-2)
Calcium panthothenate	(137-08-6)
Calcium d-pantothenate	(137-08-6)
Calcium pantothenate	(137-08-6)
d-Calcium pantothenate	(137-08-6)
Calcium pelargonate	(29813-38-5)
Calcium pentarate	(68568-63-8)
Calcium permanganate [UN 1456]	(10118-76-0)
Calcium peroxide [UN 1457]	(1305-79-9)
Calcium orthophosphate	(7758-87-4)
Calcium phosphate	(7758-87-4)
Calcium phosphate	(7790-76-3)
Calcium phosphate (1:1)	(7757-93-9)
Calcium phosphate (3:2)	(7758-87-4)
Calcium orthophosphate, basic	(1306-06-5)
Calcium phosphate dibasic anhydrous	(7757-93-9)
Calcium phosphate hydroxide	(1306-06-5)
Calcium phosphate, monobasic	(7758-23-8)
Calcium phosphate, tribasic	(7758-87-4)
Calcium orthophosphate, tri-(tert)	(7758-87-4)
Calcium phosphide [UN 1360]	(1305-99-3)
Calcium phosphinate	(7789-79-9)
Calcium phytate	(7776-28-5)
Calcium polysilicate	(1344-95-2)
Calcium polysulfide (9CI)	(1344-81-6)
Calcium polysulphide	(1344-81-6)
Calcium propanoate	(4075-81-4)
Calcium propionate	(4075-81-4)
Calcium pyrophosphate	(7790-76-3)
Calcium resinate, Fused [UN 1314]	(9007-13-0)
Calcium resinate, Technically pure (DOT)	(9007-13-0)
Calcium resinate [UN 1313]	(9007-13-0)
Calcium saccharin	(6485-34-3)
Calcium saccharina	(6485-34-3)
Calcium saccharinate	(6485-34-3)
Calcium salt of thiobis(C12-alkylated phenol)	(26998-97-0)
Calcium saltpeter	(10124-37-5)
Calcium secondary phosphate	(7757-93-9)
Calcium silicate	(10101-39-0)
Calcium silicate	(12168-85-3)
Calcium silicate (OSHA)	(1344-95-2)
Calciumsilicide [UN 1405]	(12737-18-7)
Calcium silicon (Powder)	(12737-18-7)
Calcium silicon, Powder	(12737-18-7)
Calcium silicon oxide	(12168-85-3)
Calcium sodium aluminosilicate	(1344-01-0)
Calcium sodium aluminosilicate hydrate	(1344-01-0)
Calcium sodium metaphosphate	(23209-59-8)
Calcium sodium nitrilotriacetic acid	(60034-45-9)
Calcium sorbate	(7492-55-9)
Calcium stearate (ACGIH)	(1592-23-0)
Calcium sulfate (ACGIH,OSHA)	(7778-18-9)
Calcium sulfide (9CI)	(20548-54-3)
Calcium tertiary phosphate	(7758-87-4)
Calcium tetradecanoate	(15284-51-2)
Calcium tetrahydroaluminate (7CI)	(16941-10-9)
Calcium thioglycolate	(814-71-1)
Calcium thioglycollate	(814-71-1)
Calcium thiosulfate	(10124-41-1)
Calcium titriplex	(62-33-9)

Calcium trisodium diethylenetriaminepentaacetate (12111-24-9)
Calcium trisodium pentetate (12111-24-9)
Calcium trisodium salt of diethylenetriaminepentaacetic acid (12111-24-9)
Calcium 10-undecenoate (1322-14-1)
Calco 2246 (119-47-1)
Calco Nigrosine O 2P (8005-03-6)
Calco Oil Orange 7078 (842-07-9)
Calco Oil Orange 7078-Y (842-07-9)
Calco Oil Orange Z-7078 (842-07-9)
Calco Oil Red D (85-83-6)
Calco Oil Red ZMQ (82-38-2)
Calco Oil Scarlet BL (3118-97-6)
Calco Oil Scarlet ZBL (3118-97-6)
Calcochrome Alizarine Red SC (130-22-3)
Calcocid Amaranth (915-67-3)
Calcocid Blue Black (1064-48-8)
Calcocid Blue Black 2R (1064-48-8)
Calcocid Blue EG (2650-18-2)
Calcocid Blue 2 G (2650-18-2)
Calcocid Brilliant Scarlet 3RN (2611-82-7)
Calcocid Erythrosine N (16423-68-0)
Calcocid Fast Blue SR (3861-73-2)
Calcocid Fast Light Orange 2G (1936-15-8)
Calcocid Fast Red A (1658-56-6)
Calcocid Green G (4680-78-8)
Calcocid Milling Red G (10169-02-5)
Calcocid Milling Yellow R (6375-55-9)
Calcocid Orange Y (633-96-5)
Calcocid Phloxine 2G (3734-67-6)
Calcocid 2RIL (3761-53-3)
Calcocid Rubine XX (3567-69-9)
Calcocid Scarlet 2R (3761-53-3)
Calcocid Scarlet 2RIL (3761-53-3)
Calcocid Uranine B4315 (518-47-8)
Calcocid Violet 4BNS (1694-09-3)
Calcocid Yellow MCG (1934-21-0)
Calcocid Yellow MXXX (587-98-4)
Calcocid Yellow XX (1934-21-0)
Calcodur Blue 6GFL (2610-05-1)
Calcodur Brown BRL (16071-86-6)
Calcodur Red 8BL (2610-11-9)
Calcodur Resin Fast Blue (2610-05-1)
Calcogas M (842-07-9)
Calcogas Orange NC (842-07-9)
Calcogene Black GX-CF (1326-82-5)
Calcogene Blue 2B-CF (1327-57-7)
Calcogene Blue 2R-CF (1327-57-7)
Calcogene Blue 6R-CF (1327-57-7)
Calcogene Navy Blue 2GS-CF (1327-57-7)
C12-15 alcohol (63393-82-8)
C9-11 alcohol (66455-17-2)
Calcolake Scarlet 2R (3761-53-3)
Calcoloid Blue BLC (130-20-1)
Calcoloid Blue BLR (130-20-1)
Calcoloid Blue RS (81-77-6)
Calcoloid Brown BR (2475-33-4)
Calcoloid Diazo Black BHL (2429-73-4)
Calcoloid Golden Orange GD (128-70-1)
Calcoloid Golden Orange GFD (128-70-1)
Calcoloid Golden Yellow (128-66-5)
Calcoloid Jade Green N (128-58-5)
Calcoloid Jade Green NC (128-58-5)
Calcoloid Jade Green NP (128-58-5)
Calcoloid Navy Blue (1324-54-5)
Calcoloid Navy Blue 2GC (6424-76-6)
Calcoloid Navy Blue NTC (1324-54-5)
Calcoloid Olive R (2379-81-9)
Calcoloid Olive RC (2379-81-9)
Calcoloid Olive RL (2379-81-9)
Calcoloid Pink FFC (2379-74-0)
Calcoloid Pink FFD (2379-74-0)
Calcoloid Pink FFRP (2379-74-0)
Calcoloid Printing Orange RE (3263-31-8)
Calcoloid Printing Pink FFE (2379-74-0)
Calcoloid Violet 4RD (1324-55-6)
Calcoloid Violet 4RP (1324-55-6)
Calcomine Black (1937-37-7)
Calcomine Black Exl (1937-37-7)
Calcomine Blue 2B (2602-46-2)
Calcomine Brown B (2429-81-4)
Calcomine Brown MCW (2429-82-5)
Calcomine Catechu 2B (2429-81-4)
Calcomine Dark Blue BN (7082-31-7)
Calcomine Dark Green BG (3626-28-6)
Calcomine Diazo Black BHD (2429-73-4)
Calcomine Diazo Black BTCW (2429-73-4)
Calcomine Red 4BX (992-59-6)
Calcomine Scarlet 3B (6358-29-8)
Calcomine Sky Blue FF (2610-05-1)
Calcomine Violet N (2586-60-9)
Calcomine Yellow 2G (1325-37-7)
Calcon (2538-85-4)
Calcophyl Red FF (2379-74-0)
Calcosyn Pink B (116-85-8)
Calcosyn Sapphire Blue 2GS (2475-46-9)
Calcosyn Sapphire Blue 2GS (86722-66-9)
Calcosyn Sapphire Blue R (2475-46-9)
Calcosyn Sapphire Blue R (86722-66-9)
Calcosyn Yellow GC (2832-40-8)
Calcosyn Yellow GCN (2832-40-8)
Calcotone Blue GP (147-14-8)
Calcotone Green G (1328-53-6)
Calcotone Hansa Yellow (2512-29-0)
Calcotone Orange 2R (3468-63-1)
Calcotone Orange R (3520-72-7)
Calcotone Red (1309-37-1)
Calcotone Red 2B (1103-39-5)
Calcotone Red 3B (6471-49-4)
Calcotone Red B (1103-38-4)
Calcotone Toluidine Red YP (2425-85-6)
Calcotone Violet RP (1325-82-2)
Calcotone White T (13463-67-7)
Calcozine Blue ZF (61-73-4)
Calcozine Brilliant Green G (633-03-4)
Calcozine Chrysoidine Y (532-82-1)
Calcozine Fuchsine HO (632-99-5)
Calcozine Green V (569-64-2)
Calcozine Magenta N (569-61-9)
Calcozine Magenta RIN (632-99-5)
Calcozine Magenta XX (632-99-5)
Calcozine Orange YS (532-82-1)
Calcozine Red BX (81-88-9)
Calcozine Red 6G (989-38-8)
Calcozine Red Y (477-73-6)
Calcozine Rhodamine BL (81-88-9)
Calcozine Rhodamine BX (81-88-9)
Calcozine Rhodamine BXP (81-88-9)
Calcozine Rhodamine 6GX (989-38-8)
Calcozine Violet 6BN (548-62-9)
Calcozine Violet C (548-62-9)
Calcozine Yellow OX (2465-27-2)
Calcspar (13397-26-7)
Calcyanide (592-01-8)
Caldan (15263-52-2)

Caldedon Navy Blue AR	(1324-54-5)
C-8 aldehyde	(124-13-0)
Caldon	(88-85-7)
Caledon Blue RN	(81-77-6)
Caledon Blue XRN	(81-77-6)
Caledon Brilliant Blue RN	(81-77-6)
Caledon Brilliant Purple 4R	(1324-55-6)
Caledon Brilliant Purple 4RP	(1324-55-6)
Caledon Dark Blue G	(6424-76-6)
Caledon Dark Brown 3R	(2475-33-4)
Caledon Gold Orange G	(128-70-1)
Caledon Gold Orange GN	(128-70-1)
Caledon Golden Yellow	(128-66-5)
Caledon Jade Green	(128-58-5)
Caledon Jade Green XBN	(128-58-5)
Caledon Jade Green XN	(128-58-5)
Caledon Navy Blue ART	(1324-54-5)
Caledon Navy Blue 2R	(1324-54-5)
Caledon Olive R	(2379-81-9)
Caledon Paper Blue RN	(81-77-6)
Caledon Paper Gold Orange G	(128-70-1)
Caledon Printing Blue RN	(81-77-6)
Caledon Printing Blue XRN	(81-77-6)
Caledon Printing Jade Green XBN	(128-58-5)
Caledon Printing Jade Green XN	(128-58-5)
Caledon Printing Navy G	(6424-76-6)
Caledon Printing Orange G	(128-70-1)
Caledon Printing Purple 4R	(1324-55-6)
Caledon Printing Yellow	(128-66-5)
Caledone Olive RP	(2379-81-9)
Calflo E	(1344-95-2)
Calfoam ES 30	(9004-82-4)
Calf serum ultralyzate	(9048-46-8)
Calginate	(9005-35-0)
Calgon	(10124-56-8)
Calidria RG 100	(12001-29-5)
Calidria RG 144	(12001-29-5)
Calidria RG 600	(12001-29-5)
Calioben	(1319-91-1)
Calirus	(15310-01-7)
Calixin	(24602-86-6)
Callatex	(9005-34-9)
Callitrisic acid	(5155-70-4)
Calmadin	(57-53-4)
Calmathion	(121-75-5)
Calmax	(57-53-4)
Calmday	(1088-11-5)
Calmetten	(50-06-6)
Calminal	(50-06-6)
Calmiren	(57-53-4)
Calmocitene	(439-14-5)
Calmodulin (ox brain) (9CI)	(73298-54-1)
Calmonal	(569-65-3)
Calmore	(50-35-1)
Calmorex	(50-35-1)
Calmosine	(15708-41-5)
Calmpose	(439-14-5)
Calocain	(113-45-1)
Calochlor	(7487-94-7)
Calogreen	(10112-91-1)
Calomel	(10112-91-1)
Calonat	(50-02-2)
Calophyl Pink ZFF	(2379-74-0)
Calorose	(8013-17-0)
Calotab	(10112-91-1)
Calotoxin	(20304-49-8)
Calotropin	(1986-70-5)
Calpanate	(137-08-6)
Calphosan	(814-80-2)
Calplus	(10043-52-4)
Calpol	(103-90-2)
Calsil	(1344-95-2)
Calsmin	(146-22-5)
Calsoft F-90	(25155-30-0)
Calsoft L-40	(25155-30-0)
Calsoft L-60	(25155-30-0)
Calsoft LAS 99	(27176-87-0)
Calsol	(64-02-8)
Calstar	(1592-23-0)
Caltac	(10043-52-4)
Calusterone	(17021-26-0)
Camazepam	(36104-80-0)
Camazepamum (Latin)	(36104-80-0)
Cambendichlor	(56141-00-5)
Cambendichlore	(56141-00-5)
Camcolit	(554-13-2)
Camite	(5798-79-8)
Campaprim A 1544	(61-82-5)
Camparol 3303	(8066-11-3)
Campbelline Oil Orange	(842-07-9)
Campesterin	(474-62-4)
Campesterol	(474-62-4)
Campestrol	(474-62-4)
Camphane, 2-hydroxy-	(507-70-0)
2-Camphanol	(507-70-0)
2-Camphanol acetate	(76-49-3)
2-Camphanone	(76-22-2)
d-2-Camphanone	(464-49-3)
endo-2-Camphanyl ethanoate	(76-49-3)
exo-2-Camphanyl β-hydroxyethyl ether	(7070-15-7)
Camphechlor	(8001-35-2)
Camphene (DOT)	(79-92-5)
Campherol	(520-18-3)
Camphersulfosaeure (German)	(3144-16-9)
Camphochlor	(8001-35-2)
Camphoclor	(8001-35-2)
Camphofene Huileux	(8001-35-2)
Camphogen	(99-87-6)
Camphol	(507-70-0)
(+)-Camphor	(464-49-3)
Camphor	(76-22-2)
Camphor, (+)-	(464-49-3)
D-(+)-Camphor	(464-49-3)
β-Camphor	(10292-98-5)
d-Camphor	(464-49-3)
epi-Camphor	(10292-98-5)
l-(-)-Camphor	(464-48-2)
l-Camphor	(464-48-2)
Camphor, Natural [UN 1130]	(76-22-2)
Camphor Oil	(8008-51-3)
Camphor Oil Brown	(8008-51-3)
Camphor Oil (Light)	(8008-51-3)
Camphor Oil, Rectified	(8008-51-3)
Camphor Oil White	(8008-51-3)
Camphor Oil Yellow	(8008-51-3)
Camphor, (1R,4R)-(+)	(464-49-3)
Camphor, Synthetic (ACGIH,OSHA) [UN 2717]	(76-22-2)
Camphor Tar	(91-20-3)
Camphor USP	(464-49-3)
Camphor White Oil	(8008-51-3)
Camphorated Opium Tincture USP XVI	(8029-99-0)
Camphoric acid (8CI)	(5394-83-2)
Camphor, l-, (-)	(464-48-2)
Camphor oil [UN 1130]	(76-22-2)

Camphorol (12001-40-0)
(+)-Camphorsulfonic acid (3144-16-9)
(+)-β-Camphorsulfonic acid (3144-16-9)
Camphorsulfonic acid (3144-16-9)
d-10-Camphorsulfonic acid (3144-16-9)
d-Camphorsulfonic acid (3144-16-9)
Camphozone (59-26-7)
Campilit (506-68-3)
Campogran (60568-05-0)
Camposan (16672-87-0)
Campoviton 6 (58-56-0)
Can (56-25-7)
Can-Trol (94-81-5)
Canacert Amaranth (915-67-3)
Canacert Brilliant Blue FCF (3844-45-9)
Canacert Erythrosine BS (16423-68-0)
Canacert Indigo Carmine (860-22-0)
Canacert Sunset Yellow FCF (2783-94-0)
Canacert Tartrazine (1934-21-0)
Canadian Balsam (8007-47-4)
Canadian Fir Needle Oil (8021-28-1)
Canadien 2000 (28772-56-7)
Canadol (8032-32-4)
Canary Chrome Yellow 40-2250 (7758-97-6)
Canavanin (543-38-4)
Canavanine (543-38-4)
L-Canavanine (543-38-4)
L-Canavanine sulfate (2219-31-0)
Cancarb (1333-86-4)
Candamide (554-13-2)
Candaseptic (59-50-7)
Candelilla (8006-44-8)
Candelilla Wax (8006-44-8)
Canderel (22839-47-0)
Candex (1400-61-9)
Candex (1912-24-9)
Candio-Hermal (1400-61-9)
Candle Scarlet 2B (85-83-6)
Candle Scarlet B (85-83-6)
Candle Scarlet G (85-83-6)
Canesten (23593-75-1)
Cane sugar (57-50-1)
Canlub (7782-42-5)
(-)-Cannabidiol (13956-29-1)
(-)-trans-Cannabidiol (13956-29-1)
Cannabidiol (13956-29-1)
Cannabinol (521-35-7)
Cannabinol, 1-trans-δ^9-tetrahydro- (1972-08-3)
Cannabis (8063-14-7)
Cannabis Resin (8063-14-7)
Canogard (62-73-7)
Canquil-400 (57-53-4)
Cantabilin (90-33-5)
Cantabiline (90-33-5)
Cantan (50-81-7)
Cantaxin (50-81-7)
Cantharides camphor (56-25-7)
Cantharidin (56-25-7)
Cantharidine (56-25-7)
Cantharone (56-25-7)
Cantrece (63428-84-2)
Cantrex (59-01-8)
Cantrol (6062-26-6)
Cao 14 (119-47-1)
Cao 5 (119-47-1)
Caocobre (1317-39-1)
Caparol (7287-19-6)
Capastat (11003-38-6)
Capathyn (135-23-9)
Capilan (456-59-7)
Capisten (22071-15-4)
Capitus (77-67-8)
Capla (64-55-1)
Caplenal (315-30-0)
Capmul (9005-64-5)
Capmul (9005-67-8)
Capmul POE-O (9005-65-6)
Caporit (7778-54-3)
Capostatin (11003-38-6)
Capoten (62571-86-2)
Capraldehyde (112-31-2)
Capralense (60-32-2)
Capramide DEA (136-26-5)
Capramol (60-32-2)
Capran 77C (105-60-2)
Capran 77C (25038-54-4)
Capran 80 (105-60-2)
Capran 80 (25038-54-4)
Caprane (39300-45-3)
Capreomycin (11003-38-6)
Capri Blue GON (7199-02-2)
Capric acid (334-48-5)
n-Capric acid (334-48-5)
Capric acid diethanolamide (136-26-5)
Capric acid ethyl ester (110-38-3)
Capric acid methyl ester (110-42-9)
Capric acid, potassium salt (13040-18-1)
Capric alcohol (112-30-1)
Capric aldehyde (112-31-2)
Caprin (50-78-2)
Caprinaldehyde (112-31-2)
Caprine (327-57-1)
Caprinic acid (334-48-5)
Caprinic acid sodium salt (1002-62-6)
Caprinic alcohol (112-30-1)
Caprinic aldehyde (112-31-2)
Caproaldehyde (66-25-1)
Caproamide (628-02-4)
Caproamide Polymer (25038-54-4)
Caprocid (60-32-2)
Caprodat (78-44-4)
Caproic acid (142-62-1)
n-Caproic acid (142-62-1)
Caproic aldehyde (66-25-1)
Caprokol (136-77-6)
6-Caprolactam (105-60-2)
Caprolactam (105-60-2)
ε-Caprolactam (105-60-2)
ω-Caprolactam (105-60-2)
ε-Caprolactam (ACGIH) (105-60-2)
Caprolactam (OSHA) (105-60-2)
Caprolactam Oligomer (25038-54-4)
Caprolactam Polymer (25038-54-4)
Caprolactam monomer (105-60-2)
ε-Caprolactam polymer (25038-54-4)
ε-Caprolactam polymere (German) (25038-54-4)
6-Caprolactone (695-06-7)
Caprolactone (502-44-3)
ε-Caprolactone (502-44-3)
γ-Caprolactone (695-06-7)
Caprolan (polyamide) (63428-84-2)
Caprolattame (French) (105-60-2)
Caprolin (63-25-2)
Caprolisin (60-32-2)

Caprolon	**(63428-83-1)**
Caprolon B	**(105-60-2)**
Caprolon B	**(25038-54-4)**
Caprolon V	**(105-60-2)**
Caprolon V	**(25038-54-4)**
Caprolyl peroxide	**(762-16-3)**
Capromycin	**(11003-38-6)**
Capron	**(105-60-2)**
Capron	**(25038-54-4)**
Capron	**(630-56-8)**
Capron 8250	**(105-60-2)**
Capron 8250	**(25038-54-4)**
Capron 8252	**(105-60-2)**
Capron 8252	**(25038-54-4)**
Capron 8253	**(105-60-2)**
Capron 8253	**(25038-54-4)**
Capron 8256	**(105-60-2)**
Capron 8256	**(25038-54-4)**
Capron 8257	**(105-60-2)**
Capron B	**(105-60-2)**
Capron B	**(25038-54-4)**
Capron GR 8256	**(105-60-2)**
Capron GR 8256	**(25038-54-4)**
Capron GR 8257	**(25038-54-4)**
Capron GR 8258	**(105-60-2)**
Capron GR 8258	**(25038-54-4)**
Capron PK 4	**(25038-54-4)**
Capron PK4	**(105-60-2)**
Capronaldehyde	**(66-25-1)**
Capronamide	**(628-02-4)**
Capronic acid	**(142-62-1)**
Capronitrile	**(628-73-9)**
n-Capronitrile	**(628-73-9)**
Caproyl alcohol	**(111-27-3)**
n-Caproylaldehyde	**(66-25-1)**
N-Capryl-N,N-dimethylbenzylammonium chloride	**(965-32-2)**
Capryl alcohol	**(123-96-6)**
Caprylamine	**(111-86-4)**
Caprylamine	**(693-16-3)**
Capryl chloride	**(111-85-3)**
2-Capryl-4,6-dinitrophenyl crotonate	**(39300-45-3)**
Capryldinitrophenyl crotonate	**(39300-45-3)**
1-Caprylene	**(111-66-0)**
Caprylene	**(111-66-0)**
Caprylic acid	**(124-07-2)**
n-Caprylic acid	**(124-07-2)**
Caprylic acid, aminoethylethanolamine amide-imidazoline	**(36060-61-4)**
Caprylic acid, diethanolamide	**(3077-30-3)**
Caprylic acid methyl ester	**(111-11-5)**
Caprylic acid sodium salt	**(1984-06-1)**
Caprylic acid triglyceride	**(538-23-8)**
Caprylic alcohol	**(111-87-5)**
Caprylic aldehyde	**(124-13-0)**
Caprylic diethanolamide	**(3077-30-3)**
Caprylic ether	**(629-82-3)**
Caprylnitrile	**(124-12-9)**
Caprylone	**(818-23-5)**
Caprylonitrile	**(124-12-9)**
Capryloyl peroxide (DOT)	**(762-16-3)**
Capryl peroxide	**(762-16-3)**
Capryl o-phthalate	**(131-15-7)**
Caprylyl acetate	**(112-14-1)**
Caprylylamine	**(111-86-4)**
Caprylyl chloride	**(111-64-8)**
Caprylyl hydroxyethyl imidazoline	**(36060-61-4)**
Caprylyl peroxide	**(19102-74-0)**
Caprylyl peroxide	**(762-16-3)**
Caprylyl peroxide, Solution (DOT)	**(762-16-3)**
Caprynic acid	**(334-48-5)**
Capsaicin	**(404-86-4)**
Capsaicine	**(404-86-4)**
Capsebon	**(1306-23-6)**
Capsine	**(534-52-1)**
Captab	**(133-06-2)**
Captaf	**(133-06-2)**
Captaf 85W	**(133-06-2)**
Captafol	**(2425-06-1)**
Captafol (ACGIH)	**(2939-80-2)**
Captafol (OSHA)	**(2425-06-1)**
Captan (OSHA)	**(133-06-2)**
Captan-Streptomycin 7.5-0.1 Potato seed piece protectant	**(133-06-2)**
Captan 50W	**(133-06-2)**
Captancapteneet 26,538	**(133-06-2)**
Captane	**(133-06-2)**
Captatol	**(2425-06-1)**
Captax	**(149-30-4)**
Captec	**(2463-84-5)**
Captex	**(133-06-2)**
Captopril	**(62571-86-2)**
Captopryl	**(62571-86-2)**
Caput Mortuum	**(1309-37-1)**
Capval	**(128-62-1)**
Caradate 30	**(101-68-8)**
Caradol 520 (9CI)	**(57219-31-5)**
Caradol 560 (9CI)	**(75881-81-1)**
Carafate	**(54182-58-0)**
Caragard	**(33693-04-8)**
Caragard	**(8072-81-9)**
Caramel	**(8028-89-5)**
Caramel (Color)	**(8028-89-5)**
Carastay	**(9000-07-1)**
Carastay C	**(9000-07-1)**
Caraway-Oil	**(8000-42-8)**
Carbo-Cort	**(8007-45-2)**
Carb-O-Sep	**(121-59-5)**
Carbachol	**(51-83-2)**
Carbachol chloride	**(51-83-2)**
Carbachol hydrochloride	**(51-83-2)**
Carbacholin	**(51-83-2)**
Carbacholine	**(51-83-2)**
Carbacholine Chloride	**(51-83-2)**
Carbacholine chloride	**(51-83-2)**
Carbacholini chloridum	**(51-83-2)**
Carbacholinium chloratum	**(51-83-2)**
Carbacholinum	**(51-83-2)**
Carbacholum (Latin)	**(51-83-2)**
Carbacholum chloratum	**(51-83-2)**
Carbacol (Spanish)	**(51-83-2)**
Carbacolina	**(51-83-2)**
Carbacolo	**(51-83-2)**
Carbacryl	**(107-13-1)**
Carbadine	**(12122-67-7)**
Carbadox	**(6804-07-5)**
Carbam	**(137-42-8)**
Carbamaldehyde	**(75-12-7)**
Carbamamidine	**(113-00-8)**
Carbamate	**(14484-64-1)**
Carbamate de l'ethinylcyclohexanol (French)	**(126-52-3)**
Carbamate, 4-dimethylamino-3,5-xylyl N-methyl-	**(315-18-4)**
Carbamate (isoamyl)	**(543-86-2)**
Carbamazepen	**(298-46-4)**
Carbamazepine	**(298-46-4)**
Carbamezepine	**(298-46-4)**
Carbamic acid (8CI,9CI)	**(463-77-4)**

Carbamic acid, Compd with 1,2-ethanediamine (109-58-0)
Carbamic acid, Compd with ethylenediamine (109-58-0)
Carbamic acid, allyl ester (2114-11-6)
Carbamic acid, (aminocarbonyl)-, ethyl ester (9CI) (626-36-8)
Carbamic acid, (2-aminoethyl)- (9CI) (109-58-0)
Carbamic acid, (6-aminohexyl)- (9CI) (143-06-6)
Carbamic acid, (aminoiminomethyl)methyl-, dimethyl deriv., ethyl ester, monohydrochloride (9CI) (65206-90-8)
Carbamic acid, ((4-aminophenyl)sulfonyl)-, methyl ester, monosodium salt (9CI) (2302-17-2)
Carbamic acid, ammonium salt (1111-78-0)
Carbamic acid, (1H-benzimidazol-2-yl)-, methyl ester, phosphate (1:1) (52316-55-9)
Carbamic acid, N-(5-benzoylbenzimidazol-2-yl)-, methyl ester (31431-39-7)
Carbamic acid, bis(2-hydroxyethyl)dithio-, monopotassium salt (23746-34-1)
Carbamic acid, bis(hydroxymethyl)-, (2-methoxy)ethyl ester (10143-22-3)
Carbamic acid, bis(hydroxymethyl)-, 2-(1-methylethoxy)ethyl ester (9CI) (68413-83-2)
Carbamic acid, bis(hydroxymethyl)-, 1-methylethyl ester (9CI) (4987-75-1)
Carbamic acid, bis(hydroxymethyl)-, 2-methylpropyl ester (9CI) (52304-17-3)
Carbamic acid, butyl-, 2-(((aminocarbonyl)oxy)methyl)-2-methylpentyl ester (4268-36-4)
Carbamic acid, (5-butyl-1H-benzimidazol-2-yl)-, methyl ester (14255-87-9)
Carbamic acid, tert-butyl-, (m-(3,3-dimethyureido)phenyl) ester (4849-32-5)
Carbamic acid, butyl ester (592-35-8)
Carbamic acid, butyl-, ester with 2-(hydroxymethyl)-2-methylpentyl carbamate (4268-36-4)
Carbamic acid, butylethylthio-, S-propyl ester (1114-71-2)
Carbamic acid, butyl-, 2-(hydroxymethyl)-2-methylpentyl ester, carbamate (4268-36-4)
Carbamic acid, butyl-, 3-iodo-2-propynyl ester (9CI) (55406-53-6)
Carbamic acid, 2-sec-butyl-, 2-methyltrimethylene ester (64-55-1)
Carbamic acid, 2-sec-butyl-2-methyltrimethylene ester (64-55-1)
Carbamic acid, (3-chloro-4-methoxyphenyl)-, 1-methylethyl ester (9CI) (94483-57-5)
Carbamic acid, (chloromethyl)phenyl-, ethyl ester (9CI) (35600-63-6)
Carbamic acid, 3-(p-chlorophenoxy)-2-hydroxypropyl ester (886-74-8)
Carbamic acid, (3-chlorophenyl)-, 2-chloroethyl ester (9CI) (587-56-4)
Carbamic acid, (3-chlorophenyl)-, 2-chloro-1-methylethyl ester (9CI) (2150-32-5)
Carbamic acid, (2-chlorophenyl)-, methyl ester (20668-13-7)
Carbamic acid, (o-chlorophenyl)-, methyl ester (20668-13-7)
Carbamic acid, (3-chlorophenyl)-, methyl ester (9CI) (2150-88-1)
Carbamic acid, (3-chlorophenyl)-, 1-methylpropyl ester (9CI) (2164-13-8)
Carbamic acid, (3-chlorophenyl)-, 1-methyl-2-propynyl ester (9CI) (1967-16-4)
Carbamic acid, cyano-, methyl ester (9CI) (21729-98-6)
Carbamic acid, cyanomethyl-, ethyl ester (9CI) (60754-24-7)
Carbamic acid, cyclohexylethylthio-, S-ethyl ester (1134-23-2)
Carbamic acid, ((dibutylamino)thio)methyl-, 2,2-dimethyl-2,3-dihydro-7-benzofuranyl ester (55285-14-8)
Carbamic acid, dibutyldithio-, nickel salt (13927-77-0)
Carbamic acid, dibutyldithio-, sodium salt (136-30-1)
Carbamic acid, dibutyldithio-, zinc complex (136-23-2)
Carbamic acid, di-sec-butylthio-, S-benzyl ester (36756-79-3)
Carbamic acid, dichloro-, ethyl ester (8CI,9CI) (13698-16-3)
Carbamic acid, (3,4-dichlorophenyl)-, methyl ester (1918-18-9)
Carbamic acid, (2,4-dichlorophenyl)-, 1-methylethyl ester (9CI) (2150-25-6)
Carbamic acid, (3,4-dichlorophenyl)-, 1-methylethyl ester (9CI) (2150-28-9)
Carbamic acid, (3,4-diethoxyphenyl)-, 1-methylethyl ester (9CI) (87130-20-9)
Carbamic acid, diethyldithio (147-84-2)
Carbamic acid, diethyldithio-, cadmium salt (14239-68-0)
Carbamic acid, diethyldithio-, 2-chloroallyl ester (95-06-7)
Carbamic acid, diethyldithio-, 4-chloro-6-methoxy-s-triazin-2-yl ester (7CI,8CI) (13733-96-5)
Carbamic acid, diethyldithio-, 4-chloro-6-phenyl-s-triazin-2-yl ester (7CI,8CI) (13733-97-6)
Carbamic acid, diethyldithio-, 6-chloro-s-triazine-2,4-diyl ester (6CI,8CI) (30863-06-0)
Carbamic acid, diethyldithio-, 4,6-dichloro-s-triazin-2-yl ester (7CI,8CI) (13733-95-4)
Carbamic acid, diethyldithio-, 6-(diethylamino)-s-triazine-2,4-diyl ester (6CI,8CI) (30863-11-7)
Carbamic acid, diethyldithio-, ester with 6-chloro-s-triazine-2,4-dithiol (7CI) (30863-06-0)
Carbamic acid, diethyldithio-, selenium(II) salt (136-92-5)
Carbamic acid, diethyldithio-, sodium salt (148-18-5)
Carbamic acid, diethyldithio-, sodium salt, trihydrate (20624-25-3)
Carbamic acid, diethyldithio-, s-triazine-2,4,6-triyl ester (8CI) (30863-12-8)
Carbamic acid, diethyldithio-, tris(anhydrosulfide) with trithiocyanuric acid (6CI) (30863-12-8)
Carbamic acid, diethylthio-, S-(o-chlorobenzyl) ester (34622-58-7)
Carbamic acid, diethylthio-, S-(p-chlorobenzyl) ester (28249-77-6)
Carbamic acid, diethylthio-, S-ethyl ester (2941-55-1)
Carbamic acid, diisobutylthio-, S-ethyl ester (2008-41-5)
Carbamic acid, diisopropylthio-, S-(2,3-dichloroallyl) ester (2303-16-4)
Carbamic acid, diisopropylthio-, S-(2,3-dichloroallyl)ester, (Z) (17708-57-5)
Carbamic acid, diisopropylthio-, S-(2,3,3-trichloroallyl) ester (2303-17-5)
Carbamic acid, dimethyl-, 3-aminophenyl ester (9CI) (19962-04-0)
Carbamic acid, dimethyl-, m-aminophenyl ester (8CI) (19962-04-0)
Carbamic acid, (3-(dimethylamino)propyl)-, propyl ester (24579-73-5)
Carbamic acid, (3-dimethylaminopropyl)thio-, S-ethyl ester, hydrochloride (19622-19-6)
Carbamic acid, dimethyl-, 7-chloro-2,3-dihydro-1-methyl-2-oxo-5-phenyl-1H-1,4-benzodiazepin-3-yl ester (36104-80-0)
Carbamic acid, dimethyl-, 1-((dimethylamino)carbonyl)-5-methyl-1H-pyrazol-3-yl ester (644-64-4)
Carbamic acid, dimethyl-, 2-(dimethylamino)-5,6-dimethyl-4-pyrimidyl ester (23103-98-2)
Carbamic acid, dimethyldithio (79-45-8)
Carbamic acid, dimethyldithio-, anhydrosulfide (97-74-5)
Carbamic acid, dimethyldithio-, bis(anhydrosulfide) with dithiomethanearsonous acid (2445-07-0)
Carbamic acid, dimethyldithio-, bismuth salt (21260-46-8)
Carbamic acid, dimethyldithio-, 6-chloro-s-triazine-2,4-diyl ester (6CI,8CI) (30863-05-9)
Carbamic acid, dimethyldithio-, copper(II) salt (137-29-1)
Carbamic acid, dimethyldithio-, 6-(dimethylamino)-s-triazine-2,4-diyl ester (6CI,8CI) (30863-10-6)
Carbamic acid, dimethyldithio-, iron salt (14484-64-1)
Carbamic acid, dimethyldithio-, lead salt (19010-66-3)
Carbamic acid, dimethyldithio-, potassium salt, hydrate (128-03-0)
Carbamic acid, dimethyldithio-, sodium salt (128-04-1)
Carbamic acid, dimethyldithio-, zinc salt (2:1) (137-30-4)
Carbamic acid, dimethyl-, ester with 3-hydroxy-5,5-dimethyl-2-cyclohexen-1-one (122-15-6)
Carbamic acid, dimethyl-, ester with (m-hydroxyphenyl)-trimethylammonium bromide (114-80-7)
Carbamic acid, dimethyl-, ethyl ester (687-48-9)
Carbamic acid, dimethyl-, 1-isopropyl-3-methylpyrazol-5-yl ester (119-38-0)
Carbamic acid, dimethyl-, 3-methyl-1-phenylpyrazol-5-yl ester (87-47-8)
Carbamic acid, dimethyl-, 5-methyl-1H-pyrazol-3-yl ester (644-64-4)
Carbamic acid, dimethyl-, 1-naphthalenyl ester (2619-00-3)
Carbamic acid, dimethyl-, 3-nitrophenyl ester (7304-99-6)
Carbamic acid, (2,6-dimethylphenyl)-, methyl ester (9CI) (20642-93-7)
Carbamic acid, N-(O,O-dimethylphosphorothioyl)-N-methyl-, 2,2-dimethyl-2,3-dihydrobenzofuran- 7-yl ester (28789-80-2)
Carbamic acid, dipropylthio-, S-ethyl ester (759-94-4)
Carbamic acid, dipropylthio-, S-propyl ester (1929-77-7)
Carbamic acid, dithio-, N,N-dimethyl-, dimethylaminomethyl ester (51-82-1)
Carbamic acid, dithio-, monoammonium salt (8CI) (513-74-6)
Carbamic acid, dithio-, sodium salt (4384-81-0)
Carbamic acid, ester with choline chloride (51-83-2)
Carbamic acid, ester with 2-(hydroxymethyl)-2-methylpentyl butylcarbamate (4268-36-4)

Carbamic acid, ester with 2-(hydroxymethyl)-2-methylpentyl isopropylcarbamate **(78-44-4)**
Carbamic acid, ester with 2-methyl-2-propyl-1,3-propanediol butylcarbamate **(4268-36-4)**
Carbamic acid, ester with 2-methyl-2-propyl-1,3-propanediol isopropylcarbamate **(78-44-4)**
Carbamic acid, ethylenebis(dithio- **(111-54-6)**
Carbamic acid, ethylenebis(dithio-, diammonium salt (8CI) **(3566-10-7)**
Carbamic acid, ethylenebis(dithio-, disodium salt **(142-59-6)**
Carbamic acid, ethylenebis(dithio-, manganese salt **(12427-38-2)**
Carbamic acid, ethylenebis(dithio-, manganese zinc complex (8CI) **(8018-01-7)**
Carbamic acid, ethyl ester **(51-79-6)**
Carbamic acid, ethyl-, ethyl ester **(623-78-9)**
Carbamic acid, ethylnitroso-, ethyl ester **(614-95-9)**
Carbamic acid, ethylphenyl-, ethyl ester (9CI) **(1013-75-8)**
Carbamic acid, ethylphenyl-, 1-methylethyl ester **(56961-11-6)**
Carbamic acid, 1-ethynylcyclohexyl ester **(126-52-3)**
Carbamic acid, hydrazide **(57-56-7)**
Carbamic acid, 2-hydroxyethyl ester **(5395-01-7)**
Carbamic acid, 2-hydroxy-3-(o-methoxyphenoxy)propyl ester **(532-03-6)**
Carbamic acid, N-hydroxymethyl-n-methyldithio-, potassium salt **(51026-28-9)**
Carbamic acid, (hydroxymethyl)-, 2-methylpropyl ester (9CI) **(67953-32-6)**
Carbamic acid, (hydroxymethyl)-, 2-propenyl ester (9CI) **(24935-97-5)**
Carbamic acid, (4-((4-((4-hydroxyphenyl)azo)-2-methylphenyl)azo)phenyl)-, methyl ester (9CI) **(6465-02-7)**
Carbamic acid, isobutyl ester (8CI) **(543-28-2)**
Carbamic acid, (3-isocyanatomethylphenyl)-, 1-methyl-1,3-propanediyl ester (VAN)(9CI) **(65105-00-2)**
Carbamic acid, isopentyl ester **(543-86-2)**
Carbamic acid, isopropyldithio-, 6-chloro-s-triazine-2,4-diyl ester (6CI,8CI) **(30863-07-1)**
Carbamic acid, isopropyl ester **(1746-77-6)**
Carbamic acid, isopropyl-, 2-(hydroxymethyl)-2-methylpentyl ester carbamate (ester) **(78-44-4)**
Carbamic acid, (mercaptoacetyl)methyl-, ethyl ester, S-ester with O,O- diethyl phosphorodithioate **(2595-54-2)**
Carbamic acid, 2-methoxyethyl ester **(1616-88-2)**
Carbamic acid, methyl-, benzo(b)thien-4-yl ester **(1079-33-0)**
Carbamic acid, methyl-, 1-(butylcarbamoyl)-2-benzimidazole ester **(17804-35-2)**
Carbamic acid, methyl-, o-sec-butylphenyl ester **(3766-81-2)**
Carbamic acid, methyl-, (2-chloro-4,5-dimethyl)phenyl ester **(671-04-5)**
Carbamic acid, methyl-, o-chlorophenyl ester **(3942-54-9)**
Carbamic acid, methyl-, 2-chloro-4,5-xylyl ester **(671-04-5)**
Carbamic acid, methyl-, 6-chloro-3,4-xylyl ester **(671-04-5)**
Carbamic acid, methyl-, o-cumenyl ester **(2631-40-5)**
Carbamic acid, methyl-, m-cumenyl ester (8CI) **(64-00-6)**
Carbamic acid, methyl-, m-cym-5-yl ester **(2631-37-0)**
Carbamic acid, methyl-, 3,5-di-t-butylphenyl ester **(2655-19-8)**
Carbamic acid, N-methyl-, 3,4-dichlorobenzyl ester **(1966-58-1)**
Carbamic acid, N-methyl-, 3,4-dichlorobenzyl ester, Mixed with carbamic acid, N-methyl-, 2,3-dichlorobenzyl ester (4:1) **(62046-37-1)**
Carbamic acid, methyl-, 2,3-dihydro-2,2-dimethyl-7-benzofuranyl ester **(1563-66-2)**
Carbamic acid, methyl-, 2,3-dihydro-2,2-dimethyl-3-hydroxy-7-benzofuranyl ester **(16655-82-6)**
Carbamic acid, methyl-, 4-(dimethylamino)-3,5-dimethylphenyl ester **(315-18-4)**
Carbamic acid, methyl-, m-(((dimethylamino)methylene)-amino)phenyl ester **(22259-30-9)**
Carbamic acid, methyl-, 4-(((dimethylamino)methylene) amino)-m-tolyl ester **(17702-57-7)**
Carbamic acid, methyl-, 4-(dimethylamino)-3-methylphenyl ester **(2032-59-9)**
Carbamic acid, methyl-, 4-dimethylamino-m-tolyl ester **(2032-59-9)**
Carbamic acid, methyl-, 4-dimethylamino-3,5-xylyl ester **(315-18-4)**
Carbamic acid, methyl-, 2,2-dimethyl-2,3-dihydrobenzofuran-7-yl ester **(1563-66-2)**
Carbamic acid, methyl-, O-(((2,4-dimethyl-1,3-dithiolan-2-yl)methylene)amino) deriv. **(26419-73-8)**
Carbamic acid, methyl-, O-(((2,4-dimethyl-1,3-dithiolan-2-yl)methylene)amino)- **(26419-73-8)**
Carbamic acid, methyl-, 2,3-(dimethylmethylenedioxy)phenyl ester **(22781-23-3)**
Carbamic acid, methyl-, 3,5-dimethyl-4-(methylthio)phenyl ester **(2032-65-7)**
Carbamic acid, N-methyl-, (3,4-dimethylphenyl) ester **(2425-10-7)**
Carbamic acid, methyl-, o-(1,3-dioxolan-2-yl)phenyl ester **(6988-21-2)**
Carbamic acid, N-methyldithio-, sodium salt **(137-42-8)**
Carbamic acid, methyldithio-, sodium salt, dihydrate **(6734-80-1)**
Carbamic acid, (methylenedi-4,1-phenylene)bis-, diphenyl ester (9CI) **(101-65-5)**
Carbamic acid, methyl ester **(598-55-0)**
Carbamic acid, methyl-, ester with 2,2-dimethyl-7-hydroxy-3(2H)-benzofuranone **(16709-30-1)**
Carbamic acid, methyl-, ester with eseroline **(57-47-6)**
Carbamic acid, methyl-, ester with N'-(m-hydroxyphenyl)-N,N-dimethylformamidine **(22259-30-9)**
Carbamic acid, methyl-, ester with N'-(m-hydroxyphenyl)-N,N-dimethylformamidine hydrochloride **(23422-53-9)**
Carbamic acid, 1-methylethyl ester **(1746-77-6)**
Carbamic acid, methyl-, ethyl ester **(105-40-8)**
Carbamic acid, methyl-, m-(1-ethylpropyl)phenyl ester **(672-04-8)**
Carbamic acid, methyl-, 2-(ethylthiomethyl)phenyl ester **(29973-13-5)**
Carbamic acid, methyl-, o-isopropoxyphenyl ester **(114-26-1)**
Carbamic acid, methyl-, 2,3-(isopropylidenedioxy)phenyl ester **(22781-23-3)**
Carbamic acid, N-methyl-, 3-isopropylphenyl ester **(64-00-6)**
Carbamic acid, methyl-, o-isopropylphenyl ester **(2631-40-5)**
Carbamic acid, methyl-, m-(1-methylbutyl)phenyl ester **(2282-34-0)**
Carbamic acid, methyl-, m-(1-methylbutyl)phenyl ester mixed with carbamic acid, methyl-, m-(1-ethylpropyl)phenyl ester **(8065-36-9)**
Carbamic acid, methyl-, 2-(1-methylethoxy)phenyl ester **(114-26-1)**
Carbamic acid, methyl-, 2-(1-methylethyl)phenyl ester **(2631-40-5)**
Carbamic acid, methyl-, 3-(1-methylethyl)phenyl ester **(64-00-6)**
Carbamic acid, N-methyl-, 3-methyl-5-isopropylphenyl ester **(2631-37-0)**
Carbamic acid, methyl-, 3-methyl-5-(1-methylethyl)phenyl ester **(2631-37-0)**
Carbamic acid, methyl-, 3-methyl-4-(methylthio)phenyl ester **(3566-00-5)**
Carbamic acid, methyl-, O-((2-methyl-2-(methylthio)propylidene)-amino) deriv. **(116-06-3)**
Carbamic acid, methyl((2-methylphenyl)thio)-, o-isopropoxyphenyl ester **(50539-85-0)**
Carbamic acid, methyl-, 4-methylthio-m-tolyl ester **(3566-00-5)**
Carbamic acid, N-methyl-, 4-(methylthio)-3,5-xylyl ester **(2032-65-7)**
Carbamic acid, N-methyl-N-(morpholinothio)-, 2,3-dihydro-2,2-dimethyl-7-benzofuranyl ester **(55285-05-7)**
Carbamic acid, methyl-, 1-naphthyl ester **(63-25-2)**
Carbamic acid, N-methyl-N-nitroso-, ethyl ester **(615-53-2)**
Carbamic acid, methyl-, phenyl ester **(1943-79-9)**
Carbamic acid, (3-methylphenyl)-, 3-((methoxycarbonyl)amino)-phenyl ester, Mixt. with 3-cyclohexyl-6,7-dihydro-1H-cyclopentapyrimidine-2,4(3H,5H)-dione (9CI) **(53028-35-6)**
Carbamic acid, (3-methylphenyl)-3-((methoxycarbonyl)amino)-phenyl ester (9CI) **(13684-63-4)**
Carbamic acid, methylphenyl-, 4-nitrophenyl ester (9CI) **(49839-35-2)**
Carbamic acid, methylphenyl-, phenyl ester (9CI) **(13599-69-4)**
Carbamic acid, 2-methylpropyl ester (9CI) **(543-28-2)**
Carbamic acid, (1-methylpropyl)-, methyl ester (9CI) **(39076-02-3)**
Carbamic acid, 2-methyl-2-propyltrimethylene ester **(57-53-4)**
Carbamic acid, methyl-, m-(2-propynyloxy)phenyl ester **(3692-90-8)**
Carbamic acid, methyl-, 2,6-pyridinediyldimethylene ester **(1882-26-4)**
Carbamic acid, methyl-, 3-tolyl ester **(1129-41-5)**
Carbamic acid, methyl-, m-tolyl ester **(1129-41-5)**
Carbamic acid, methyl-, 2,3,5(or 3,4,5)-trimethylphenyl ester **(12407-86-2)**
Carbamic acid, methyl-, 2,3,5-trimethylphenyl ester **(2655-15-4)**

Carbamic acid, methyl-, 3,4,5-trimethylphenyl ester **(2686-99-9)**
Carbamic acid, methyl-, 3,4-xylyl ester **(2425-10-7)**
Carbamic acid, methyl-, 3,5-xylyl ester **(2655-14-3)**
Carbamic acid, ((p-nitrophenyl)sulfonyl)-, methyl ester **(3337-70-0)**
Carbamic acid, octadecyl-, ethenyl ester, Homopolymer (9CI) **(36671-85-9)**
Carbamic acid, N,N-pentamethylene-, 4-dimethylaminothymyl ester, methochloride **(2438-53-1)**
Carbamic acid, (2-(4-phenoxyphenoxy)ethyl)-, ethyl ester **(72490-01-8)**
Carbamic acid, (2-(4-phenoxyphenoxy)ethyl)-, ethyl ester (9CI) **(72490-01-8)**
Carbamic acid, phenyl-, butyl ester (9CI) **(1538-74-5)**
Carbamic acid, phenylcarbamoyloxyphenyl-, ethyl ester **(13684-56-5)**
Carbamic acid, phenyl-, 3-chlorophenyl ester (9CI) **(16400-09-2)**
Carbamic acid, (1,2-phenylenebis(iminocarbonothioyl))bis-, diethyl ester (9CI) **(23564-06-9)**
Carbamic acid, (1,2-phenylenebis(iminocarbonothioyl))bis-, dimethyl ester **(23564-05-8)**
Carbamic acid, phenyl ester (9CI) **(622-46-8)**
Carbamic acid, 2-phenylhydrazide **(103-03-7)**
Carbamic acid, phenyl-, methyl ester (9CI) **(2603-10-3)**
Carbamic acid, phenyl-, 4-methylphenyl ester (9CI) **(16323-13-0)**
Carbamic acid, phenyl-, 3-nitrophenyl ester (9CI) **(35289-89-5)**
Carbamic acid, phenyl-, pentyl ester (9CI) **(63075-06-9)**
Carbamic acid, phenyl-, propyl ester (9CI) **(5532-90-1)**
Carbamic acid, (5-(phenylthio)-1H-benzimidazol-2-yl)-, methyl ester **(43210-67-9)**
Carbamic acid, propyl ester **(627-12-3)**
Carbamic acid, sulfanilyl-, methyl ester **(3337-71-1)**
Carbamic acid, sulfanilyl-, methyl ester, monosodium salt **(2302-17-2)**
Carbamic acid, (5-(2-thienylcarbonyl)-1H-benzimidazol-2-yl)-, methyl ester **(31430-18-9)**
Carbamic acid, thio-, S,S'-(2-(dimethylamino)trimethylene) ester, hydrochloride **(15263-52-2)**
Carbamic acid, (2,4,5-trichlorophenyl)-, 1-methylethyl ester (9CI) **(52819-37-1)**
Carbamic acid, (3-(triethoxysilyl)propyl)-, ethyl ester (8CI,9CI) **(17945-05-0)**
Carbamic anhydride, tetramethyltrithio- **(97-74-5)**
Carbamic chloride, diethyl- (9CI) **(88-10-8)**
Carbamic fluoride, (((5,5-dimethyl-1,3,2-dioxaphosphorinan-2-yl)(1,1-dimethylethyl)amino)thio)methyl-, P-sulfide (9CI) **(78051-43-1)**
Carbamidal **(59-26-7)**
Carbamide **(57-13-6)**
Carbamide, N-ethyl-N-nitroso- **(759-73-9)**
Carbamide, N-methyl-N-nitroso- **(684-93-5)**
Carbamide peroxide **(124-43-6)**
Carbamide peroxide, Solution **(124-43-6)**
Carbamide phenylacetate **(63-98-9)**
Carbamide resin **(57-13-6)**
Carbamidine **(113-00-8)**
p-Carbamidobenzenearsonic acid **(121-59-5)**
Carbamidoyl chloride, diethyl- **(88-10-8)**
Carbamidsaeure-aethylester (German) **(51-79-6)**
Carbamimidic acid **(57-13-6)**
Carbamimidothioic acid, 2-(dimethylamino)ethyl ester, dihydrochloride **(16111-27-6)**
Carbamimidothioic acid, methyl ester, sulfate (2:1) (9CI) **(867-44-7)**
Carbamine **(63-25-2)**
Carbaminocholine chloride **(51-83-2)**
Carbaminophenyl-p-arsonic acid **(121-59-5)**
p-Carbamino phenyl arsonic acid **(121-59-5)**
Carbaminoylcholine chloride **(51-83-2)**
Carbamiotin **(51-83-2)**
Carbamite **(85-98-3)**
Carbamodithioic acid, cyclohexylmethyl-, sodium salt (9CI) **(77100-49-3)**
Carbamodithioic acid, dibutyl-, methylene ester (9CI) **(10254-57-6)**
Carbamodithioic acid, diethyl-, 2-chloro-2-propenyl ester (9CI) **(95-06-7)**
Carbamodithioic acid, diethyl-, sodium salt (9CI) **(148-18-5)**
Carbamodithioic acid, diethyl-, tetrakis(anhydrosulfide) with thiotelluric acid (H4TES4) **(20941-65-5)**
Carbamodithioic acid, diethyl-, tetrakis(anhydrosulfide) with thiotelluric acid (H4TeS4) **(20941-65-5)**
Carbamodithioic acid, diethyl-, 1,3,5-triazine-2,4,6-triyl ester (9CI) **(30863-12-8)**
Carbamodithioic acid, dimethyl- (9CI) **(79-45-8)**
Carbamodithioic acid, dimethyl-, 6-chloro-1,3,5-triazine-2,4-diyl ester (9CI) **(30863-05-9)**
Carbamodithioic acid, dimethyl-, iron(2+) salt **(15339-38-5)**
Carbamodithioic acid, dimethyl-, sodium salt **(128-04-1)**
Carbamodithioic acid, 1,2-ethanediylbis- **(111-54-6)**
Carbamodithioic acid, 1,2-ethanediylbis-, diammonium salt (9CI) **(3566-10-7)**
Carbamodithioic acid, 1,2-ethanediylbis-, salts and esters **(111-54-6)**
Carbamodithioic acid, 1,2-ethanediylbis-, zinc salt **(12122-67-7)**
Carbamodithioic acid, monoammonium salt (9CI) **(513-74-6)**
Carbamohydroxamic acid **(127-07-1)**
Carbamohydroximic acid **(127-07-1)**
Carbamohydroxyamic acid **(127-07-1)**
Carbamol **(9011-05-6)**
Carbamol TsEM **(136-84-5)**
Carbamonitrile **(420-04-2)**
Carbamothioic acid, bis(1-methylethyl)-, S-(2,3-dichloro-2-propenyl) ester **(2303-16-4)**
Carbamothioic acid, bis(2-methylpropyl)-, S-ethyl ester **(2008-41-5)**
Carbamothioic acid, butylethyl-, S-propyl ester (9CI) **(1114-71-2)**
Carbamothioic acid, cyclohexylethyl-, S-ethyl ester **(1134-23-2)**
Carbamothioic acid, (4,6-dichloro-1,3,5-triazin-2-yl)-, O-methyl ester (9CI) **(5097-58-5)**
Carbamothioic acid, (4,6-dichloro-1,3,5-triazin-2-yl)-, O-(1-methylethyl) ester (9CI) **(30863-26-4)**
Carbamothioic acid, diethyl-, S-((4-chlorophenyl)methyl) ester **(28249-77-6)**
Carbamothioic acid, diethyl-, S-((2-chlorophenyl)methyl) ester (9CI) **(34622-58-7)**
Carbamothioic acid, S,S'-(2-(dimethylamino)-1,3-propanediyl) ester, monohydrochloride (9CI) **(15263-52-2)**
Carbamothioic acid, dimethyl-, O-(5-chloro-8-quinolinyl) ester (9CI) **(97073-94-4)**
Carbamothioic acid, dimethyl-, O-(5-nitro-8-quinolinyl) ester (9CI) **(97073-93-3)**
Carbamothioic acid, dimethyl-, S-(4-phenoxybutyl)ester **(62850-32-2)**
Carbamothioic acid, dimethyl-, O-8-quinolinyl ester (9CI) **(97073-95-5)**
Carbamothioic acid, dipropyl-, S-ethyl ester, Mixt. with 2-chloro-N-(2,6-diethylphenyl)-N-(methoxymethyl)acetamide (9CI) **(8074-35-9)**
Carbamothioic acid, ethyl-, O-(1-methylethyl) ester (9CI) **(141-98-0)**
Carbamothioic acid, methyl(3-methylphenyl)-, o-2-naphthalenyl ester (9CI) **(2398-96-1)**
Carbamothioic chloride, dimethyl- (9CI) **(16420-13-6)**
2-Carbamoylaniline **(88-68-6)**
N-Carbamoylarsanilic acid **(121-59-5)**
o-Carbamoylbenzoic acid **(88-97-1)**
Carbamoyl chloride, diethyl **(88-10-8)**
Carbamoyl chloride, dimethyl **(79-44-7)**
Carbamoylcholine chloride **(51-83-2)**
γ-Carbamoyl choline chloride **(51-83-2)**
Carbamoylcholine-hydrochloride **(51-83-2)**
5-Carbamoyl-5H-dibenz(b,f)azepine **(298-46-4)**
5-Carbamoyl-5H-dibenzo(b,f)azepine **(298-46-4)**
5-Carbamoyldibenzo(b,f)azepine **(298-46-4)**
Carbamoylglycine **(462-60-2)**
N-Carbamoylglycine **(462-60-2)**
N-Carbamoylhydroxylamine **(127-07-1)**
1-p-Carbamoylmethylphenoxy-3-isopropylamino-2-propanol **(29122-68-7)**
Carbamoyl oxime **(127-07-1)**
(2-Carbamoyloxyethyl)trimethylammonium chloride **(51-83-2)**
1-Carbamoyl-2-phenylhydrazine **(103-03-7)**
4-Carbamoylpyridine **(1453-82-3)**
Carbamoylurea **(108-19-0)**

Name	CAS No.
Carbamult	**(2631-37-0)**
4-Carbamylaminophenylarsonic acid	**(121-59-5)**
N-Carbamyl arsanilic acid	**(121-59-5)**
Carbamyl chloride, N,N-dimethyl-	**(79-44-7)**
Carbamylcholine chloride	**(51-83-2)**
5-Carbamyl-5H-dibenzo(b,f)azepine	**(298-46-4)**
5-Carbamyldibenzo(b,f)azepine	**(298-46-4)**
Carbamylhydrazine	**(57-56-7)**
Carbamylhydrazine hydrochloride	**(563-41-7)**
Carbamyl hydroxamate	**(127-07-1)**
4-((5-Carbamyl-2-methylphenyl)azo)-3-hydroxy-2-naphthanilide	**(16403-84-2)**
1-Carbamyl-2-phenylhydrazine	**(103-03-7)**
2-Carbamyl pyrazine	**(98-96-4)**
Carbamylurea	**(108-19-0)**
Carban T-10	**(28801-69-6)**
Carbanil	**(103-71-9)**
Carbanilaldehyde	**(103-70-8)**
Carbanilic acid	**(501-82-6)**
Carbanilic acid, butyl ester (8CI)	**(1538-74-5)**
Carbanilic acid, m-carbaniloyloxy-, ethyl ester	**(13684-56-5)**
Carbanilic acid, 1-carboxyethyl ester	**(73622-98-7)**
Carbanilic acid, m-chloro-, sec-butyl ester (8CI)	**(2164-13-8)**
Carbanilic acid, m-chloro-, 4-chloro-2-butynyl ester	**(101-27-9)**
Carbanilic acid, m-chloro-, 2-chloroethyl ester (8CI)	**(587-56-4)**
Carbanilic acid, m-chloro-, 2-chloro-1-methylethyl ester (8CI)	**(2150-32-5)**
Carbanilic acid, p-chlorodithio-, 6-chloro-s-triazine-2,4-diyl ester (6CI,8CI)	**(30863-09-3)**
Carbanilic acid, p-chloro-, .α.-ethoxy-4,6-dinitro-o-tolyl ester (7CI,8CI)	**(2542-29-2)**
Carbanilic acid, 2-chloroethyl ester	**(3747-48-6)**
Carbanilic acid, m-chloro-, isopropyl ester	**(101-21-3)**
Carbanilic acid, p-chloro-, isopropyl ester	**(2239-92-1)**
Carbanilic acid, o-chloro-, methyl ester	**(20668-13-7)**
Carbanilic acid, m-chloro-, methyl ester (8CI)	**(2150-88-1)**
Carbanilic acid, m-chloro-, 1-methyl-2-propynyl ester	**(1967-16-4)**
Carbanilic acid, m-chlorophenyl ester (7CI,8CI)	**(16400-09-2)**
Carbanilic acid, 3,4-dichloro-, isopropyl ester (8CI)	**(2150-28-9)**
Carbanilic acid, 2,4-dichloro-, isopropyl ester (7CI,8CI)	**(2150-25-6)**
Carbanilic acid, 3,4-dichloro-, methyl ester	**(1918-18-9)**
Carbanilic acid, 2,6-dimethyl-, methyl ester (8CI)	**(20642-93-7)**
Carbanilic acid, dithio-, 6-chloro-s-triazine-2,4-diyl ester (6CI,8CI)	**(30863-08-2)**
Carbanilic acid, (1-ethylcarbamoyl)ethyl ester, D-(-)	**(16118-49-3)**
Carbanilic acid, ethyl ester	**(101-99-5)**
Carbanilic acid, N-ethyl-, ethyl ester (8CI)	**(1013-75-8)**
Carbanilic acid, n-ethyl-isopropyl ester	**(56961-11-6)**
Carbanilic acid, m-hydroxy-, methyl ester, m-methylcarbanilate	**(13684-63-4)**
Carbanilic acid, isopropyl ester	**(122-42-9)**
Carbanilic acid, m,n-dimethylthio-, o-2-naphthyl ester	**(2398-96-1)**
Carbanilic acid, methyl ester (8CI)	**(2603-10-3)**
Carbanilic acid, N-methyl-, phenyl ester (6CI,7CI,8CI)	**(13599-69-4)**
Carbanilic acid, m-nitrophenyl ester	**(35289-89-5)**
Carbanilic acid, pentyl ester (6CI,7CI)	**(63075-06-9)**
Carbanilic acid, propyl ester (8CI)	**(5532-90-1)**
Carbanilic acid, p-tolyl ester (8CI)	**(16323-13-0)**
Carbanilide	**(102-07-8)**
Carbanilide, 4,4'-diacetyl-, 4,4'-bis(amidinohydrazone), dimethanesulfonate (8CI)	**(15427-93-7)**
Carbanilide, 4,4'-dichloro-3-(trifluoromethyl)- (8CI)	**(369-77-7)**
Carbanilide, N,N'-diethyl	**(85-98-3)**
Carbanilide, N,N'-dimethyl	**(611-92-7)**
Carbanilide, 2,2',4,4',6,6'-hexachloro	**(20632-35-3)**
Carbanilide, thio	**(102-08-9)**
Carbanilide, 3,4,4'-trichloro	**(101-20-2)**
Carbanmide	**(63-98-9)**
Carbanolate	**(671-04-5)**
Carbanthrene Blue 2R	**(81-77-6)**
Carbanthrene Blue RCS	**(130-20-1)**
Carbanthrene Blue RS	**(81-77-6)**
Carbanthrene Blue RSP	**(81-77-6)**
Carbanthrene Brilliant Green	**(128-58-5)**
Carbanthrene Brilliant Green G	**(128-58-5)**
Carbanthrene Brilliant Violet 4R	**(1324-55-6)**
Carbanthrene Brown BR	**(2475-33-4)**
Carbanthrene Golden Orange G	**(128-70-1)**
Carbanthrene Golden Orange GD	**(128-70-1)**
Carbanthrene Golden Orange GP	**(128-70-1)**
Carbanthrene Golden Yellow	**(128-66-5)**
Carbanthrene Navy Blue G	**(6424-76-6)**
Carbanthrene Navy Blue RA	**(1324-54-5)**
Carbanthrene Olive R	**(2379-81-9)**
Carbanthrene Printing Golden Orange G	**(128-70-1)**
Carbanthrene Red G 2B	**(4203-77-4)**
Carbanthrene Red G 2BP	**(4203-77-4)**
Carbanthrene Violet 2R	**(1324-55-6)**
Carbanthrene Violet 2RP	**(1324-55-6)**
Carbaril (Italian)	**(63-25-2)**
Carbarsone	**(121-59-5)**
Carbaryl (ACGIH,DOT,OSHA)	**(63-25-2)**
Carbasone	**(121-59-5)**
Carbasulam	**(1773-37-1)**
Carbatene	**(9006-42-2)**
Carbathiin	**(5234-68-4)**
Carbathione	**(137-42-8)**
Carbation	**(137-42-8)**
Carbatox	**(63-25-2)**
Carbatox-60	**(63-25-2)**
Carbatox-75	**(63-25-2)**
Carbavur	**(63-25-2)**
Carbax	**(115-32-2)**
Carbazaldehyde	**(624-84-0)**
Carbazamide	**(57-56-7)**
Carbazepine	**(298-46-4)**
Carbazic acid, dithio-, hydrazine (Salt)	**(20469-71-0)**
Carbazic acid, hydrazide	**(497-18-7)**
Carbazic acid, methyl ester (8CI)	**(6294-89-9)**
Carbazic acid, 3-(2-quinoxalinylmethylene)-, methyl ester, N^1,N^4-dioxide	**(6804-07-5)**
Carbazide (DOT)	**(497-18-7)**
Carbazinc	**(137-30-4)**
Carbazine	**(92-81-9)**
Carbazole	**(86-74-8)**
9H-Carbazole (9CI)	**(86-74-8)**
Carbazole, 3-amino-9-ethyl	**(132-32-1)**
Carbazole, 3-amino-9-ethyl-, hydrochloride	**(6109-97-3)**
9H-Carbazole, dichloro- (9CI)	**(28804-85-5)**
Carbazole, dichloro- (6CI,8CI)	**(28804-85-5)**
9H-Carbazole, 9-ethenyl- (9CI)	**(1484-13-5)**
9H-Carbazole, 9-ethenyl-, Homopolymer (9CI)	**(25067-59-8)**
9H-Carbazole, 9-ethyl- (9CI)	**(86-28-2)**
Carbazole, 9-ethyl- (8CI)	**(86-28-2)**
9H-Carbazole, 9-ethyl-3-nitro- (9CI)	**(86-20-4)**
Carbazole, 9-ethyl-3-nitro- (8CI)	**(86-20-4)**
Carbazole, 3-(p-hydroxyanilino)-	**(86-72-6)**
9H-Carbazole, 1-methyl- (9CI)	**(6510-65-2)**
9H-Carbazole, 2-methyl- (9CI)	**(3652-91-3)**
9H-Carbazole, 3-methyl- (9CI)	**(4630-20-0)**
9H-Carbazole, 4-methyl- (9CI)	**(3770-48-7)**
9H-Carbazole, 9-methyl- (9CI)	**(1484-12-4)**
9H-Carbazole, methyl- (9CI)	**(27323-29-1)**
Carbazole, 1-methyl- (8CI)	**(6510-65-2)**
Carbazole, 2-methyl- (8CI)	**(3652-91-3)**
Carbazole, 3-methyl- (8CI)	**(4630-20-0)**
Carbazole, 4-methyl- (8CI)	**(3770-48-7)**

Name	CAS No.
Carbazole, 9-methyl- (8CI)	**(1484-12-4)**
Carbazole, methyl- (8CI)	**(27323-29-1)**
9H-Carbazole, methylnitro- (9CI)	**(116232-63-4)**
Carbazole, nitro-	**(95273-11-3)**
9H-Carbazole, nitro- (9CI)	**(95273-11-3)**
Carbazole, 9-vinyl	**(1484-13-5)**
4-(3-Carbazolylamino)phenol	**(86-72-6)**
Carbazotic acid	**(88-89-1)**
Carbendazim	**(10605-21-7)**
Carbendazime	**(10605-21-7)**
Carbendazim phosphate	**(52316-55-9)**
Carbendazole	**(10605-21-7)**
Carbendazym	**(10605-21-7)**
Carbenicillin	**(4697-36-3)**
Carbenoxolone	**(5697-56-3)**
Carbethoxyacetic ester	**(105-53-3)**
Carbethoxy malaoxon	**(1634-78-2)**
Carbethoxy malathion	**(121-75-5)**
Carbethoxymethimazole	**(22232-54-8)**
Carbethoxymethyl ethyl phthalate	**(84-72-0)**
4-Carbethoxy-1-methyl-4-phenylazacycloheptane	**(77-15-6)**
4-Carbethoxy-1-methyl-4-phenylhexamethylenimine	**(77-15-6)**
p-Carbethoxyphenol	**(120-47-8)**
3-Carbethoxyphenyl N-phenylcarbamate	**(37070-83-0)**
N-Carbethoxyphthalimide	**(22509-74-6)**
3-Carbethoxypsoralen	**(20073-24-9)**
Carbetidine	**(469-82-9)**
Carbetovur	**(121-75-5)**
Carbetox	**(121-75-5)**
Carbicron	**(141-66-2)**
Carbide 6-12	**(94-96-2)**
Carbide Black E	**(1937-37-7)**
Carbide Black ER	**(2429-83-6)**
Carbimazol	**(22232-54-8)**
Carbimazole	**(22232-54-8)**
Carbimide	**(420-04-2)**
Carbin	**(101-27-9)**
Carbinamine	**(74-89-5)**
Carbinazole	**(22232-54-8)**
Carbinol	**(67-56-1)**
Carbinoxamide maleate	**(113-92-8)**
Carbitol	**(111-46-6)**
Carbitol	**(111-90-0)**
Carbitol acetate	**(112-15-2)**
Carbitol acrylate	**(7328-17-8)**
Carbitol cellosolve	**(111-90-0)**
Carbitol solvent	**(111-90-0)**
Carbon-12	**(7440-44-0)**
Carbon-Black	**(1333-86-4)**
Carbobenzoxy chloride	**(501-53-1)**
Carbobenzyloxy chloride	**(501-53-1)**
2-Carbo-n-butoxy-6,6-dimethyl-5,6-dihydro-1,4-pyrone	**(532-34-3)**
Carbochol	**(51-83-2)**
Carbocholin	**(51-83-2)**
Carbocholine	**(51-83-2)**
Carbocysteine	**(638-23-3)**
Carbodicyclohexylimide	**(538-75-0)**
Carbodihydrazide	**(497-18-7)**
Carbodiimide	**(420-04-2)**
Carbodiimide, dicyclohexyl- (8CI)	**(538-75-0)**
Carbodiimide, diisopropyl	**(693-13-0)**
Carbon-dioxide	**(124-38-9)**
Carbodis	**(1333-86-4)**
N-Carboethoxyphthalimide	**(22509-74-6)**
Carbofenothion (Dutch)	**(786-19-6)**
Carbofos	**(121-75-5)**
Carbofuran (ACGIH,DOT,OSHA)	**(1563-66-2)**
Carbofurane	**(1563-66-2)**
Carbofuran mixture, Liquid (DOT)	**(1563-66-2)**
Carbofuran 7-phenol	**(1563-38-8)**
Carbofuran phenol	**(1563-38-8)**
Carbogen (8CI)	**(8063-77-2)**
Carbohydrazide	**(497-18-7)**
Carbohydrazide, 1-sec-butylidene-3-thio- (8CI)	**(18801-52-0)**
Carbohydrazide-N-carboxamide	**(4381-07-1)**
Carbohydrazide, 1,5-diphenyl	**(140-22-7)**
Carbohydrazide, thio	**(2231-57-4)**
Carbolac	**(1333-86-4)**
Carbolac 1	**(1333-86-4)**
Carbolic acid (DOT)	**(108-95-2)**
Carbolic acid, Liquid (liquid tar acid containing over 50% phenol) [UN 2821]	**(108-95-2)**
Carbolith	**(554-13-2)**
Carbolon	**(409-21-2)**
Carbolsaure (German)	**(108-95-2)**
Carbomate	**(63-25-2)**
Carbomer 940	**(9003-01-4)**
Carbomer 934P	**(9003-01-4)**
Carbomet	**(1333-86-4)**
Carbomethene	**(463-51-4)**
2-Carbomethoxyaniline	**(134-20-3)**
o-Carbomethoxyaniline	**(134-20-3)**
4-Carbomethoxybenzaldehyde	**(1571-08-0)**
p-Carbomethoxybenzaldehyde	**(1571-08-0)**
4-(Carbomethoxy)benzoic acid	**(1679-64-7)**
2-β-Carbomethoxy-3-β-benzoxytropane	**(50-36-2)**
2-Carbomethoxy-1-methylvinyl dimethyl phosphate	**(7786-34-7)**
α-2-Carbomethoxy-1-methylvinyl dimethyl phosphate	**(7786-34-7)**
2-Carbomethoxy-1-propen-2-yl dimethyl phosphate	**(7786-34-7)**
p-Carbomethoxytoluene	**(99-75-2)**
Carbon-monoxide	**(630-08-0)**
Carbomonoxyhemoglobin	**(9061-29-4)**
Carbon	**(7440-44-0)**
Carbon, Activated [UN 1362]	**(7440-44-0)**
Carbon Black (ACGIH,OSHA)	**(1333-86-4)**
Carbon Black, Acetylene	**(1333-86-4)**
Carbon Black BV and V	**(1333-86-4)**
Carbon Black, Channel	**(1333-86-4)**
Carbon Black, Furnace	**(1333-86-4)**
Carbon Black, Lamp	**(1333-86-4)**
Carbon Black, Thermal	**(1333-86-4)**
Carbon D	**(142-59-6)**
Carbon, Isotope of mass 14 (8CI,9CI)	**(14762-75-5)**
Carbon Oil	**(71-43-2)**
Carbon S	**(128-04-1)**
Carbon Tet	**(56-23-5)**
Carbona	**(56-23-5)**
Carbonate (8CI,9CI)	**(3812-32-6)**
Carbonate d'ammoniaque (French)	**(506-87-6)**
Carbonate magnesium	**(546-93-0)**
(Carbonato)dihydroxydicopper	**(12069-69-1)**
(Carbonato(2-))tetrahydroxytrinickel	**(12607-70-4)**
Carbonazidic acid, 1,1-dimethylethyl ester (9CI)	**(1070-19-5)**
Carbonazidodithioic acid	**(4472-06-4)**
Carbon bichloride	**(127-18-4)**
Carbon bisulfide (DOT)	**(75-15-0)**
Carbon bisulphide	**(75-15-0)**
Carbon black oil (Petroleum)	**(64741-62-4)**
Carbon bromide	**(558-13-4)**
Carbon bromotrichloride	**(75-62-7)**
Carbon chloride (CCl_4)	**(56-23-5)**
Carbon chlorosulfide	**(463-71-8)**
Carbon dichloride	**(127-18-4)**
Carbon dichloride oxide	**(75-44-5)**

Carbon difluoride oxide **(353-50-4)**
Carbon dioxide (ACGIH,OSHA) [UN 1013] **(124-38-9)**
Carbon dioxide, Mixture with nitrogen oxide (9CI) **(53569-62-3)**
Carbon dioxide, Refrigerated liquid [UN 2187] **(124-38-9)**
Carbon dioxide, Solid [UN 1845] **(124-38-9)**
Carbon dioxide and nitrous oxide mixtures **(53569-62-3)**
Carbon dioxide mixed with oxygen **(8063-77-2)**
Carbon dioxide-nitrous oxide, Mixture **(53569-62-3)**
Carbon dioxide-oxygen, Mixture **(8063-77-2)**
Carbon disulfide (ACGIH,OSHA) [UN 1131] **(75-15-0)**
Carbon disulphide **(75-15-0)**
Carbone (oxychlorure de) (French) **(75-44-5)**
Carbone (oxyde de) (French) **(630-08-0)**
Carbone (sufure de) (French) **(75-15-0)**
Carbon ferrochromium **(11114-46-8)**
Carbon fluoride **(75-73-0)**
Carbon fluoride oxide **(353-50-4)**
Carbon hexachloride **(67-72-1)**
Carbon hydride nitride (CHN) **(74-90-8)**
Carbonic acid, Compd. with cyclohexanamine (9CI) **(20227-92-3)**
Carbonic acid, Compd. with cyclohexylamine (8CI) **(20227-92-3)**
Carbonic acid, allyl ester, diester with diethylene glycol **(142-22-3)**
Carbonic acid, ammonium copper salt (9CI) **(33113-08-5)**
Carbonic acid, ammonium salt **(506-87-6)**
Carbonic acid, barium salt (1:1) **(513-77-9)**
Carbonic acid beryllium salt (1:1) **(13106-47-3)**
Carbonic acid, 2-sec-butyl-4,6-dinitrophenyl isopropyl ester **(973-21-7)**
Carbonic acid, cadmium salt **(513-78-0)**
Carbonic acid, calcium salt (1:1) **(471-34-1)**
Carbonic acid, cerium(3+) salt (3:2) (9CI) **(537-01-9)**
Carbonic acid, chromium salt (9CI) **(29689-14-3)**
Carbonic acid, cobalt(2+) salt (1:1) **(513-79-1)**
Carbonic acid, Compd. with guanidine (1:2) **(593-85-1)**
Carbonic acid, copper(2+) salt (1:1) **(1184-64-1)**
Carbonic acid, copper(1+) salt (9CI) **(3444-14-2)**
Carbonic acid, copper salt (8CI,9CI) **(7492-68-4)**
Carbonic acid, cyclic ethylene ester **(96-49-1)**
Carbonic acid cyclic methylethylene ester **(108-32-7)**
Carbonic acid, cyclic propylene ester **(108-32-7)**
Carbonic acid, cyclic tetramethylethylene ester (6CI,8CI) **(19424-29-4)**
Carbonic acid, cyclic vinylene ester **(872-36-6)**
Carbonic acid, diallyl ester **(15022-08-9)**
Carbonic acid, diammonium salt (8CI,9CI) **(506-87-6)**
Carbonic acid, diethyl ester **(105-58-8)**
Carbonic acid, dihydrazide **(497-18-7)**
Carbonic acid, dilithium salt **(554-13-2)**
Carbonic acid, dimethyl ester **(616-38-6)**
Carbonic acid, diphenyl ester **(102-09-0)**
Carbonic acid, dipotassium salt **(584-08-7)**
Carbonic acid, di-2-propenyl ester (9CI) **(15022-08-9)**
Carbonic acid, disodium salt **(497-19-8)**
Carbonic acid, dithallium(1+) salt **(6533-73-9)**
Carbonic acid, dithio-, anhydrosulfide with O-ethyl thiocarbonate, O-ethyl ester (8CI) **(3278-35-1)**
Carbonic acid, dithio-, cyclic S,S-(6-methyl-2,3-quinoxalinediyl)-ester **(2439-01-2)**
Carbonic acid, dithio-, S,S-diethyl ester (8CI) **(623-80-3)**
Carbonic acid, dithio-, O-ethyl ester **(151-01-9)**
Carbonic acid, dithio-, O-ethyl ester, potassium salt **(140-89-6)**
Carbonic acid, dithio-, O-isopropyl ester, potassium salt **(140-92-1)**
Carbonic acid, dithio-, O-isopropyl ester, sodium salt **(140-93-2)**
Carbonic acid, dithio-, S-methyl O-(2-methylcyclohexyl) ester, cis- (8CI) **(15288-12-7)**
Carbonic acid, dithio-, O-pentyl ester, potassium salt **(2720-73-2)**
Carbonic acid gas **(124-38-9)**
Carbonic acid, ion(2-) **(3812-32-6)**
Carbonic acid, iron(2+) salt (1:1) (9CI) **(563-71-3)**
Carbonic acid, lanthanum(3+) salt (3:2) (9CI) **(587-26-8)**
Carbonic acid, lead(2+) salt (1:1) **(598-63-0)**
Carbonic acid, lead salt (8CI,9CI) **(13427-42-4)**
Carbonic acid, lead salt, basic **(1319-46-6)**
Carbonic acid lithium salt **(554-13-2)**
Carbonic acid, magnesium salt **(546-93-0)**
Carbonic acid, magnesium salt (2:1) (9CI) **(2090-64-4)**
Carbonic acid, magnesium salt (1:1), dihydrate (9CI) **(68973-26-2)**
Carbonic acid, manganese(2+) salt (1:1) (9CI) **(598-62-9)**
Carbonic acid, 1-methylethyl 2-(1-methylpropyl)-4,6-dinitrophenylester (9CI) **(973-21-7)**
Carbonic acid, monoammonium salt **(1066-33-7)**
Carbonic acid, monopotassium salt (9CI) **(298-14-6)**
Carbonic acid monosodium salt **(144-55-8)**
Carbonic acid, neodymium(3+) salt (3:2) (9CI) **(5895-46-5)**
Carbonic acid, nickel salt (1:1) **(3333-67-3)**
Carbonic acid, oxydiethylenedi-, diallyl ester **(142-22-3)**
Carbonic acid, praseodymium(3+) salt (3:2) (9CI) **(5895-45-4)**
Carbonic acid, sodium salt (2:3) (9CI) **(533-96-0)**
Carbonic acid, strontium salt (1:1) (9CI) **(1633-05-2)**
Carbonic acid, trithio-, cyclic ester with 2,3-quinoxalinedithiol **(93-75-4)**
Carbonic acid, zinc salt (1:1) **(3486-35-9)**
Carbonic anhydrase inhibitor No. 6063 **(59-66-5)**
Carbonic anhydride **(124-38-9)**
Carbonic chloride **(75-44-5)**
Carbonic difluoride (9CI) **(353-50-4)**
Carbonic dihydrazide (9CI) **(497-18-7)**
Carbonic dihydrazide, 2,2'-diphenyl- (9CI) **(140-22-7)**
Carbonice (DOT) **(124-38-9)**
Carbonic oxide **(630-08-0)**
Carbonimidic dichloride, (4,6-dichloro-1,3,5-triazin-2-yl)- (9CI) **(877-83-8)**
Carbonimidic dichloride, phenyl- (9CI) **(622-44-6)**
Carbonimidodithioic acid, 3-pyridinyl-, butyl (4-(1,1-dimethylethyl)phenyl)methyl ester **(51308-54-4)**
4,4'-Carbonimidoylbis(N,N-diethylbenzenamine) mononitrate **(43130-12-7)**
Carbon iodide **(507-25-5)**
Carbonio (ossicloruro di) (Italian) **(75-44-5)**
Carbonio (ossido di) (Italian) **(630-08-0)**
Carbonio (solfuro di) (Italian) **(75-15-0)**
Carbon monoxide (ACGIH,OSHA) [UN 1016] **(630-08-0)**
Carbon monoxide, Cryogenic liquid [NA 9202] **(630-08-0)**
Carbonmonoxyhemoglobin **(9061-29-4)**
Carbon nitride **(460-19-5)**
Carbon nitride ion (CN^{1-}) **(57-12-5)**
Carbonochloride acid, 1-methylethyl ester **(108-23-6)**
Carbonochloridic acid, 2-ethylhexyl ester **(24468-13-1)**
Carbonochloridic acid, hexyl ester (9CI) **(6092-54-2)**
Carbonochloridic acid, methyl ester **(79-22-1)**
Carbonochloridic acid, 1-methylpropyl ester (9CI) **(17462-58-7)**
Carbonochloridic acid, 2-methylpropyl ester (9CI) **(543-27-1)**
Carbonochloridic acid, oxydi-2,1-ethanediyl ester **(106-75-2)**
Carbonochloridic acid, phenyl ester **(1885-14-9)**
Carbonochloridic acid, propyl ester **(109-61-5)**
Carbonochloridic acid trichloromethyl ester **(503-38-8)**
Carbonochloridothioic acid, O-methyl ester (9CI) **(2812-72-8)**
Carbonochloridothioic acid, S-ethyl ester (9CI) **(2941-64-2)**
Carbonodithioic acid, O-butyl ester **(110-50-9)**
Carbonodithioic acid, O-ethyl ester, sodium salt (9CI) **(140-90-9)**
Carbonodithioic acid, O-(1-methylethyl) ester, potassium salt (9CI) **(140-92-1)**
Carbonodithioic acid, O-(1-methylethyl)ester, sodium salt (9CI) **(140-93-2)**
Carbonodithioic acid, O-(1-methylethyl) ester, zinc salt **(1000-90-4)**
Carbonodithioic acid, O-(2-methylpropyl) ester, sodium salt (9CI) **(25306-75-6)**
Carbonodithioic acid, O-pentyl S-2-propenyl ester (9CI) **(2956-12-9)**
Carbonodithioic acid, S,S-diethyl ester (9CI) **(623-80-3)**
Carbonodithioic acid, o-ethyl ester **(151-01-9)**
Carbonodithioic acid, o-ethyl ester, potassium salt (9CI) **(140-89-6)**
Carbonodithioic acid, S-methyl O-(1-methylethyl) ester (9CI) **(35200-02-3)**

Carbonodithioic acid, o-pentyl ester, potassium salt (9CI) (2720-73-2)
Carbonohydrazide (497-18-7)
Carbonothioic acid, O-(6-chloro-3-phenyl-4-pyridazinyl) S-octyl ester (55512-33-9)
Carbonothioic dichloride (9CI) (463-71-8)
Carbonothioic dihydrazide (9CI) (2231-57-4)
Carbonothioic dihydrazide, (1-methylheptylidene)- (9CI) (41361-12-0)
Carbonothioic dihydrazide, (1-methylhexylidene)- (9CI) (59653-29-1)
Carbonothioic dihydrazide, (1-methylpropylidene)- (9CI) (18801-52-0)
Carbon oxide (co) (630-08-0)
Carbon oxide sulfide (9CI) (463-58-1)
Carbon oxychloride (75-44-5)
Carbon oxyfluoride (353-50-4)
Carbon oxysulfide (463-58-1)
Carbon silicide (409-21-2)
Carbon sulfide (DOT) (75-15-0)
Carbon sulphide (DOT) (75-15-0)
Carbon tetrabromide (ACGIH,OSHA) [UN 2516] (558-13-4)
Carbon tetrachloride (ACGIH,OSHA) [UN 1846] (56-23-5)
Carbon trichlorobromide (75-62-7)
Carbon trifluoride (75-46-7)
Carbonyl J Acid (134-47-4)
3,3'-(Carbonylbis(imino(3-methoxy-4,1-phenylene)azo))bis(benzenesulfonic acid), disodium salt (10114-86-0)
5,5'-Carbonylbis-1,3-isobenzofurandione (2421-28-5)
Carbonylchlorid (German) (75-44-5)
Carbonyl chloride (DOT,OSHA) (75-44-5)
Carbonyl chloride, thio- (463-71-8)
Carbonyl diamide (57-13-6)
Carbonyldiamine (57-13-6)
Carbonyl dichloride (75-44-5)
Carbonyl difluoride (353-50-4)
7,7'-(Carbonyldiimino)bis(4-hydroxy-3-((6-sulfo-2-naphthalenyl)-azo)-2-naphthalenesulfonic acid), tetrasodium salt (28706-25-4)
Carbonyl fluoride (ACGIH,OSHA) [UN 2417] (353-50-4)
Carbonyl iron (7439-89-6)
Carbonyl nickel Powder (7440-02-0)
Carbonyl sulfide-^{32}S (463-58-1)
Carbonyl sulfide [UN 2204] (463-58-1)
Carbophenothion (786-19-6)
Carbophos (121-75-5)
Carbopol 934 (9003-01-4)
Carbopol 940 (9003-01-4)
Carbopol 941 (9003-01-4)
Carbopol 960 (9003-01-4)
Carbopol 961 (9003-01-4)
Carbopol Extra (7440-44-0)
Carbopol M (7440-44-0)
Carbopol 934P (9003-01-4)
Carbopol Z 4 (7440-44-0)
Carbopol Z Extra (7440-44-0)
Carborundeum (409-21-2)
Carborundum (409-21-2)
Carbose (9000-11-7)
Carbose 1M (9004-32-4)
Carboset (9003-01-4)
Carboset 511 (140-88-5)
Carboset 515 (9003-01-4)
Carboset Resin No. 515 (9003-01-4)
Carbosieve (7440-44-0)
Carbosorbit R (7440-44-0)
Carbospol (57-06-7)
Carbostyril (59-31-4)
Carbostyril, 1-methyl- (8CI) (606-43-9)
Carbosulfan (55285-14-8)
Carbocisteine (638-23-3)
Carbon-tetrabromide (558-13-4)
Carbon-tetrachloride (56-23-5)
Carbon-tetrafluoride (75-73-0)
Carbon-tetraiodide (507-25-5)
Carbothialdin (533-74-4)
Carbothialdine (533-74-4)
Carbothrone (90-44-8)
Carbowax (25322-68-3)
Carbowax 350 (9004-74-4)
Carbowax 550 (9004-74-4)
Carbowax 750 (9004-74-4)
Carbowax 1000 (25322-68-3)
Carbowax 1000 distearate (9005-08-7)
Carbowax 1000 monostearate (9004-99-3)
Carbowax 1500 (25322-68-3)
Carbowax 1540 (25322-68-3)
Carbowax 2000 (9004-74-4)
Carbowax 4000 (25322-68-3)
Carbowax 5000 (9004-74-4)
Carbowax 6000 (25322-68-3)
3-Carboxamido-4-hydroxy-α-((1-methyl-3-phenylpropylamino)-methyl)benzyl alcohol (36894-69-6)
5-Carboxanilido-2,3-dihydro-6-methyl-1,4-oxathiin (5234-68-4)
Carboxin (5234-68-4)
Carboxine (5234-68-4)
Carboxin sulfone (5259-88-1)
2-Carboxyacetanilide (89-52-1)
4'-Carboxyacetanilide (556-08-1)
4-Carboxyacetanilide (556-08-1)
Carboxyacetic acid (141-82-2)
Carboxyacetylene (471-25-0)
2-Carboxyaniline (118-92-3)
3-Carboxyaniline (99-05-8)
4-Carboxyaniline (150-13-0)
Carboxyaniline (118-92-3)
o-Carboxyaniline (118-92-3)
p-Carboxyaniline (150-13-0)
9-Carboxyanthracene (723-62-6)
3-Carboxybenzaldehyde (619-21-6)
4-Carboxybenzaldehyde methyl ester (1571-08-0)
Carboxybenzene (65-85-0)
1-(1'-Carboxy-2'-benzeneazo)-2-naphthol (29128-56-1)
p-Carboxybenzenesulfondichloroamide (80-13-7)
m-Carboxybenzenesulfonic acid (121-53-9)
3-Carboxy-1-benzylpyridinium chloride (16214-98-5)
p-Carboxybromobenzene (586-76-5)
α-Carboxycaproic acid (534-59-8)
p-Carboxychlorobenzene (74-11-3)
Carboxycyclohexane (98-89-5)
Carboxycyclophosphamide (22788-18-7)
Carboxy derivative of dimethoate (1113-01-5)
2-Carboxy-4'-(dimethylamino)azobenzene (493-52-7)
3'-Carboxy-4-dimethylaminoazobenzene (20691-84-3)
2-Carboxydiphenylamine (91-40-7)
Carboxyethane (79-09-4)
β-Carboxyethyl acrylate (24615-84-7)
N-(2-Carboxyethyl)-N-dodecyl-β-alanine, disodium salt (3655-00-3)
N-(2-Carboxyethyl)-N-dodecyl-β-alanine, monosodium salt (14960-06-6)
3-Carboxy-1-ethyl-7-methyl-1,8-naphthidin-4-one (389-08-2)
α-Carboxyethyl N-phenylcarbamate (73622-98-7)
2-Carboxyethyl 2-propenoate (24615-84-7)
2-Carboxyfuran (88-14-2)
Carboxyhemoglobin (9061-29-4)
Carboxyhemoglobin A (9061-29-4)
Carboxyhemoglobin C (9061-29-4)
5(6)-Carboxy-4-hexyl-2-cyclohexene-1-octanoic acid (53980-88-4)
5(or 6)-Carboxy-4-hexyl-2-cyclohexene-1-octanoic acid (53980-88-4)
5(6)-Carboxy-4-hexyl-2-cyclohexene-1-octanoic acid, potassium

salt (68127-33-3)
3-Carboxy-4-hydroxybenzenesulfonic acid (97-05-2)
3-Carboxy-2-hydroxy-N,N,N-trimethyl-1-propanaminium hydroxide, inner salt (541-15-1)
(3-Carboxy-2-hydroxypropyl)trimethyl-ammonium hydroxide, inner salt (541-15-1)
3-Carboxy-5-hydroxy-1-p-sulfophenyl-4-p-sulfophenylazopyrazole trisodium salt (1934-21-0)
5-Carboxyisophthalic acid (554-95-0)
Carboxylic acids, C1-5 (68937-68-8)
Carboxylic acids, C5-9 (68603-84-9)
Carboxylic acids, C6-18 and C5-15-di- (68937-69-9)
Carboxylic acids, fatty (67254-79-9)
Carboxylic acids, polyamides (63428-84-2)
Carboxymethanephosphonic acid (4408-78-0)
(Carboxymethoxy)malonic acid (55203-12-8)
(Carboxymethoxy)propanedioic acid (55203-12-8)
N-(Carboxymethyl)-N,N-dimethyl-1-hexadecanaminium hydroxide, inner salt (693-33-4)
N-(Carboxymethyl)-N,N-dimethyl-9-octadecen-1-aminium hydroxide, inner salt (871-37-4)
N-(Carboxymethyl)-N,N-dimethyl-3-((1-oxococonut)amino)-1-propanaminium hydroxide, inner salt (61789-40-0)
N-(Carboxymethyl)-N,N-dimethyl-3-((1-oxododecyl)amino)-1-propanaminium hydroxide, inner salt (4292-10-8)
4-Carboxymethylaniline (1197-55-3)
Carboxymethylated Cellulose Pulp (9000-11-7)
Carboxymethyl cellulose (9004-32-4)
Carboxymethylcellulose (9000-11-7)
Carboxymethyl cellulose ether (9000-11-7)
Carboxymethylcellulose sodium (9004-32-4)
Carboxymethylcellulose, sodium salt (9004-32-4)
l-Carboxymethylcysteine (638-23-3)
s-(Carboxymethyl)cysteine (638-23-3)
N-(Carboxymethyl)glycine (142-73-4)
N-(Carboxymethyl)-N'-(2-hydroxyethyl)-N,N'-ethylenediglycine (150-39-0)
N-(Carboxymethyl)-N'-(2-hydroxyethyl)-N,N'-ethylene-diglycine trisodium salt (139-89-9)
Carboxymethyl hydroxyethyl cellulose (9004-30-2)
Carboxymethyl hydroxyethylcellulose (9004-30-2)
(((Carboxymethyl)imino)bis(ethylenenitrilo))tetraacetic acid, pentasodium salt (140-01-2)
5-Carboxymethyl-3-methyl-2H-1,3,5-thiadiazine-2-thione (3655-88-7)
Carboxymethyloxysuccinate trisodium salt (34128-01-3)
Carboxymethyloxysuccinic acid, trisodium salt (34128-01-3)
3-Carboxy-1-methylpyridinium hydroxide inner salt (535-83-1)
Carboxymethyl starch (9057-06-1)
o-(Carboxymethyl)tartronic acid (55203-12-8)
(Carboxymethylthio)acetic acid (123-93-3)
3-(Carboxymethylthio)alanine (638-23-3)
l-3-((Carboxymethyl)thio)alanine (638-23-3)
(Carboxymethyl)trimethylammonium chloride (590-46-5)
(Carboxymethyl)trimethylammonium hydroxide, inner salt (107-43-7)
1-Carboxynaphthalene (86-55-5)
2-Carboxy-1-naphthol (86-48-6)
5-Carboxynicotinic acid (499-81-0)
1-Carboxy-4-nitrobenzene (62-23-7)
2-Carboxy-α-oxobenzeneacetic acid (528-46-1)
N-(3-Carboxy-1-oxosulfopropyl)-N-octadecyl-L-aspartic acid, tetrasodium salt (3401-73-8)
3-Carboxyphenol (99-06-9)
4-Carboxyphenol (99-96-7)
o-Carboxyphenyl acetate (50-78-2)
p-Carboxyphenylamine (150-13-0)
1-((2-Carboxyphenyl)azo)-2-naphthol. (29128-56-1)
1-(2'-Carboxyphenylazo)-2-naphthol (29128-56-1)
1-(3'-Carboxyphenylazo)-2-naphthol (32624-41-2)
9-o-Carboxyphenyl-6-diethylamino-3-ethylimino-3-isoxanthene, 3-ethochloride (81-88-9)
(9-(o-Carboxyphenyl)-6-(diethylamino)-3H-xanthen-3-ylidene) diethylammonium chloride (81-88-9)
9-(o-Carboxyphenyl)-6-hydroxy-3-isoxanthenone (2321-07-5)
9-o-Carboxyphenyl-6-hydroxy-3-isoxanthone, disodium salt (518-47-8)
9-(o-Carboxyphenyl)-6-hydroxy-2,4,5,7-tetraiodo-3-isoxanthone (16423-68-0)
9-(o-Carboxyphenyl)-6-hydroxy-3h-xanthen-3-one (2321-07-5)
3-Carboxy-1-(phenylmethyl)pyridinium chloride (16214-98-5)
((o-Carboxyphenyl)thio)ethylmercury sodium salt (54-64-8)
Carboxyphosphamide (22788-18-7)
4-Carboxyphthalic anhydride (552-30-7)
3-β-(3-Carboxypropionyloxy)-11-oxo-olean-12-en-30-oic acid (5697-56-3)
2-Carboxypyridine (98-98-6)
3-Carboxypyridine (59-67-6)
4-Carboxypyridine (55-22-1)
2-Carboxyresorcinol (303-07-1)
4-Carboxyresorcinol (89-86-1)
5-Carboxyresorcinol (99-10-5)
trans-β-Carboxystyrene (140-10-3)
p-Carboxy-α-toluic acid (1679-64-7)
6-Carboxyuracil (65-86-1)
Carbrital (57-33-0)
Carbromal (77-65-6)
Carbutamid (339-43-5)
Carbutamide (339-43-5)
Carbuten (64-55-1)
Carbyl (51-83-2)
Carbylamine, ethyl (624-79-3)
Carbyne (101-27-9)
Carcholin (51-83-2)
Cardamine (59-26-7)
Cardamist (55-63-0)
5-β-Cardanolide, 3-β-((O-2,6-dideoxy-β-D-ribo-hexopyranosyl-(1-4)-O-2,6-dideoxy- β-D-hexopyranosyl-(1-4)-2,6-dideoxy-β-d-ribo-hexopyranosyl)oxy)-14-hydroxy (3786-76-3)
Cardenal (50-06-6)
Card-20(22)-enolide, 16-(acetyloxy)-3-((2,6-dideoxy-3-O-methyl-l-arabino-hexopyranosyl)oxy)- 14-hydroxy-, (3-β,16-β) (465-16-7)
5-α-Card-20(22)-enolide, 3-β-((6-deoxy-β-d-hexopyranos-2-ulos-1-yl)oxy)- 2-α,14-dihydroxy-19-oxo (20304-49-8)
5-β-Card-20(22)-enolide, 19-oxo-3-β-((l-rhamnopyranosyl)oxy)-5,12-β,14-trihydroxy (639-13-4)
Cardio-Green (3599-32-4)
Cardiagen (59-26-7)
Cardiamid (59-26-7)
Cardiamina (59-26-7)
Cardiamine (59-26-7)
Cardiazol (54-95-5)
Cardiazole (54-95-5)
Cardidigin (71-63-6)
Cardifortan (54-95-5)
Cardigin (71-63-6)
Cardimon (59-26-7)
Cardiofilina (317-34-0)
Cardiol (54-95-5)
Cardiomin (317-34-0)
Cardiomone (61-19-8)
Cardioserpin (50-55-5)
Cardiotonicum (54-95-5)
Cardis (87-33-2)
Carditalin (71-63-6)
Carditin (390-64-7)
Carditivo (50-55-5)
Carditoxin (71-63-6)
Cardivix (68-90-6)
β-Cardone (3930-20-9)

Name	CAS No.
Cardophylin	(317-34-0)
Cardophyllin	(317-34-0)
Cardosal	(54-95-5)
Cardosan	(54-95-5)
Cardoverina	(61-25-6)
Cardoxin	(58-32-2)
Cardrase	(452-35-7)
Cardura E10	(71206-09-2)
Carena	(317-34-0)
3(10)-Carene	(554-60-9)
3-Carene	(13466-78-9)
S-3-Carene	(13466-78-9)
β-Carene	(554-60-9)
δ^3-Carene	(13466-78-9)
psi-Carene	(554-60-9)
Carfene	(86-50-0)
Carfentanil	(59708-52-0)
Carfentanila (Spanish)	(59708-52-0)
Carfentanilum (Latin)	(59708-52-0)
Carfentanyl	(59708-52-0)
Caricide	(1642-54-2)
Carinex GP	(9003-53-6)
Carinex HR	(9003-53-6)
Carinex HRM	(9003-53-6)
Carinex SB 59	(9003-53-6)
Carinex SB 61	(9003-53-6)
Carinex SL 273	(9003-53-6)
Carinex TGX/MF	(9003-53-6)
Cariomin	(317-34-0)
Carisol	(78-44-4)
Carisoma	(78-44-4)
Carisoprodate	(78-44-4)
Carisoprodatum	(78-44-4)
Carisoprodol	(78-44-4)
Caritrol	(1642-54-2)
Carlisle 280	(110-30-5)
Carlisle Wax 280	(110-30-5)
Carlisle metal deactivator	(94-91-7)
Carlona 58-030	(9002-88-4)
Carlona 900	(9002-88-4)
Carlona 18020 FA	(9002-88-4)
Carlona K 571	(9003-07-0)
Carlona KM 61	(9003-07-0)
Carlona P	(9003-07-0)
Carlona PM 61 Naturel	(9003-07-0)
Carlona PPLZ 074	(9003-07-0)
Carlona PXB	(9002-88-4)
Carlsoma	(78-44-4)
Carlsoprol	(78-44-4)
Carmazine	(8018-01-7)
Carmethose	(9004-32-4)
Carbetamex	(16118-49-3)
Carbetamid (German)	(16118-49-3)
Carbetamide	(16118-49-3)
Carmin Blue VS	(129-17-9)
Carminaph	(842-07-9)
Carmine	(1260-17-9)
Carmine	(1390-65-4)
Carmine Blue (Biological stain)	(860-22-0)
Carmine Blue VF	(129-17-9)
Carminic acid	(1260-17-9)
Carmoisin (German)	(3567-69-9)
Carmoisine	(3567-69-9)
Carmoisine Aluminum Lake	(3567-69-9)
Carmoisine BA	(3567-69-9)
Carmoisine BA-CF	(3567-69-9)
Carmoisine BSS	(3567-69-9)
Carmoisine FU	(3567-69-9)
Carmoisine GRN	(3567-69-9)
Carmoisine LAS	(3567-69-9)
Carmoisine S	(3567-69-9)
Carmoisine Supra	(3567-69-9)
Carmoisine W	(3567-69-9)
Carmoisine WS	(3567-69-9)
Carmubris	(154-93-8)
Carmustin	(154-93-8)
Carmustine	(154-93-8)
Carnation Red Toner B	(6471-49-4)
Carnauba	(8015-86-9)
Carnauba Wax	(8015-86-9)
Carnauba Waxes	(8015-86-9)
Carnelio Helio Red	(2425-85-6)
Carnelio Orange G	(3520-72-7)
Carnelio Pale Lithol Red	(1248-18-6)
Carnelio Red 2G	(3468-63-1)
Carnelio Red R	(2814-77-9)
Carnelio Rubine Lake	(130-22-3)
Carnelio Yellow G	(2512-29-0)
Carnelio Yellow GX	(6358-85-6)
(-)-Carnitine	(541-15-1)
(R)-Carnitine	(541-15-1)
Carnitine	(541-15-1)
l-Carnitine	(541-15-1)
Carob Bean Gum	(9000-40-2)
Carob Flour	(9000-40-2)
Carofam	(68-90-6)
Carolid AL	(92-52-4)
Carolysine	(55-86-7)
all-trans-β-Carotene	(7235-40-7)
β-Carotene	(7235-40-7)
β,β-Carotene (9CI)	(7235-40-7)
β-Carotene, all-trans (8CI)	(7235-40-7)
Carpen	(112-65-2)
Carpene	(2439-10-3)
Carbetapentane	(77-23-6)
Carpidor	(1861-40-1)
Carpol 2040	(9003-11-6)
Carpol 2050	(9003-11-6)
Carpolene	(9003-01-4)
Carpolin	(63-25-2)
Carragard	(8072-81-9)
Carrageen	(9000-07-1)
kappa-Carrageen	(11114-20-8)
Carrageenan	(9000-07-1)
kappa-Carrageenan	(11114-20-8)
lambda-Carrageenan	(9064-57-7)
iota-Carrageenan (9CI)	(9062-07-1)
Carrageenan Gum	(9000-07-1)
Carrageenin	(9000-07-1)
kappa-Carrageenin	(11114-20-8)
lambda-Carrageenin	(9064-57-7)
Carragheanin	(9000-07-1)
Carragheen	(9000-07-1)
Carragheenan	(9000-07-1)
Carrel-Dakin Solution	(7681-52-9)
Carrserp	(50-55-5)
Carrtime	(51-63-8)
Carsodal	(78-44-4)
Carsodol	(78-44-4)
Carsonol SLS	(151-21-3)
Carsonol SLS Paste B	(151-21-3)
Carsonol SLS Special	(151-21-3)
Carsonon N-9	(9016-45-9)
Carsonon PEG-4000	(25322-68-3)

Name	CAS No.
Carsoquat SDQ-25	(122-19-0)
Carsoquat SDQ-85	(122-19-0)
Carsoron	(1194-65-6)
Carstab DLTDP	(123-28-4)
Cartagyl	(637-07-0)
Cartap	(15263-52-2)
Cartap Hydrochloride	(15263-52-2)
Cartose	(50-99-7)
Cartwheels	(60-13-9)
Carvacrol	(499-75-2)
Carvacron	(133-67-5)
Carvacryl chloride	(4395-79-3)
Carvanil	(87-33-2)
Carvasin	(87-33-2)
l-Carveol	(99-48-9)
Carveol, dihydro-	(38049-26-2)
Carvil	(3766-81-2)
Carvol	(99-49-0)
3-Carvomenthenol	(491-04-3)
4-Carvomenthenol	(562-74-3)
3-Carvomenthenone	(89-81-6)
Carvomenthone	(499-70-7)
(+)-Carvone	(2244-16-8)
(-)-Carvone	(6485-40-1)
(R)-Carvone	(6485-40-1)
(S)-(+)-Carvone	(2244-16-8)
(S)-Carvone	(2244-16-8)
Carvone	(99-49-0)
Carvone, (+)-	(2244-16-8)
D(+)-Carvone	(2244-16-8)
L(-)-Carvone	(6485-40-1)
d-Carvone	(2244-16-8)
l-Carvone	(6485-40-1)
Carvone oxime	(31198-76-2)
Carvoxime	(31198-76-2)
l-Carvyl acetate	(97-42-7)
Carylderm	(63-25-2)
Caryne	(101-27-9)
Caryolysin	(51-75-2)
Caryolysine	(55-86-7)
Caryolysine hydrochloride	(55-86-7)
Caryophylene oxide	(1139-30-6)
Caryophyllene	(13877-93-5)
Caryophyllene	(87-44-5)
β-Caryophyllene	(87-44-5)
(-)-β-Caryophyllene epoxide	(1139-30-6)
Caryophyllene epoxide	(1139-30-6)
β-Caryophyllene epoxide	(1139-30-6)
(-)-Caryophyllene oxide	(1139-30-6)
Caryophyllene oxide	(1139-30-6)
β-Caryophyllene oxide	(1139-30-6)
Caryophyllic acid	(97-53-0)
Carzol	(23422-53-9)
Carzol	(6164-98-3)
Carzol SP	(23422-53-9)
Carzonal	(51-21-8)
Casalis Green	(1308-38-9)
Casaron	(1194-65-6)
Cascade Blue	(1325-87-7)
Cascamite	(9011-05-6)
Casco 5H	(9011-05-6)
Casco PR 335	(9011-05-6)
Casco Resin	(9011-05-6)
Casco UL 30	(9011-05-6)
Casco WS 114-79	(9011-05-6)
Casco WS 138-43	(9011-05-6)
Casco WS 138-44	(9011-05-6)
Cascorez	(9003-20-7)
Casein	(9000-71-9)
Caseins	(9000-71-9)
Casfen	(340-57-8)
Cashew Nut Shell Oil (Untreated)	(8007-24-7)
Cashew, Nutshell Liq.	(8007-24-7)
Cashew, Nutshell liq., Polymer with epichlorohydrin	(68413-24-1)
Cashew nutshell oil, Polymer with (chloromethyl)oxirane	(68413-24-1)
Casiflux	(13983-17-0)
Casiflux VP 413-004	(13983-17-0)
Casing Head Gasoline (DOT)	(8006-61-9)
Casoron	(1194-65-6)
Casoron 133	(1194-65-6)
Caspan	(115-09-3)
Cassaine	(468-76-8)
Cassappret SR	(25038-59-9)
Cassel Brown	(1317-34-6)
Cassel Green	(1344-43-0)
Cassella's Acid	(92-40-0)
Cassenne	(53-33-8)
Cassia-Oil	(8007-80-5)
Cassia aldehyde	(104-55-2)
Cassiar AK	(12001-29-5)
Cassic acid	(478-43-3)
Cassie ketone	(104-21-2)
Cassurit LR	(1854-26-8)
Cassurit RI	(136-84-5)
Castanea Sativa Mill Tannin	(1401-55-4)
Castor-Oil	(8001-79-4)
Castor Oil Acid, Methyl Ester	(141-24-2)
Castor Oil Aromatic	(8001-79-4)
Castor Oil, Hydrogenated	(8001-78-3)
Castor Oil, Sulfated	(8002-33-3)
Castor Oil, Sulfate, sodium salt	(68187-76-8)
Castor Oil, Sulfated, sodium salt	(68187-76-8)
Castorwax	(8001-78-3)
Castorwax MP-70	(8001-78-3)
Castorwax MP-80	(8001-78-3)
Castorwax NF	(8001-78-3)
Castrix	(535-89-7)
Casul 70HF	(26264-06-2)
Caswell No. 003B	(34256-82-1)
Caswell No. 003D	(81510-83-0)
Caswell No. 003F	(81334-34-1)
Caswell No. 005A	(34490-93-2)
Caswell No. 005AB	(77-90-7)
Caswell No. 005B	(40164-67-8)
Caswell No. 024	(97-59-6)
Caswell No. 029A	(1317-25-5)
Caswell No. 033BB	(141-94-6)
Caswell No. 033F	(1007-28-9)
Caswell No. 037A	(1031-47-6)
Caswell No. 037C	(1066-51-9)
Caswell No. 041	(7664-41-7)
Caswell No. 041B	(7784-25-0)
Caswell No. 043	(16919-19-0)
Caswell No. 044AA	(22228-82-6)
Caswell No. 045A	(9080-17-5)
Caswell No. 046C	(544-60-5)
Caswell No. 049	(7646-88-0)
Caswell No. 050B	(31366-95-7)
Caswell No. 050A	(53404-18-5)
Caswell No. 052	(65996-91-0)
Caswell No. 052B	(1397-94-0)
Caswell No. 054	(68477-31-6)
Caswell No. 055	(68602-80-2)
Caswell No. 056	(7778-39-4)

Caswell No. 061B (50-81-7)
Caswell No. 062B (2302-17-2)
Caswell No. 066 (68038-71-1)
Caswell No. 071 (13701-59-2)
Caswell No. 072 (1332-65-6)
Caswell No. 072A (74051-80-2)
Caswell No. 073 (1332-03-2)
Caswell No. 075C (5251-93-4)
Caswell No. 079A (2634-33-5)
Caswell No. 081EE (1214-39-7)
Caswell No. 083BB (3734-33-6)
Caswell No. 083B (3184-65-4)
Caswell No. 083A (35471-49-9)
Caswell No. 088A (20679-58-7)
Caswell No. 089A (26389-78-6)
Caswell No. 091A (4104-14-7)
Caswell No. 098AA (52508-35-7)
Caswell No. 099 (1113-14-0)
Caswell No. 104 (3064-70-8)
Caswell No. 106A (108-19-0)
Caswell No. 107 (8001-85-2)
Caswell No. 111A (53404-19-6)
Caswell No. 114A (16079-88-2)
Caswell No. 115 (2491-38-5)
Caswell No. 116A (52-51-7)
Caswell No. 116B (7166-19-0)
Caswell No. 120A (34681-23-7)
Caswell No. 122 (9003-13-8)
Caswell No. 125J (590-02-3)
Caswell No. 128BB (3996-59-6)
Caswell No. 129 (115-84-4)
Caswell No. 130A (94-26-8)
Caswell No. 130C (59756-60-4)
Caswell No. 130G (5787-50-8)
Caswell No. 146 (8061-52-7)
Caswell No. 147 (13780-06-8)
Caswell No. 148 (7758-87-4)
Caswell No. 150 (1344-81-6)
Caswell No. 156 (8008-51-3)
Caswell No. 159A (108-19-0)
Caswell No. 165F (78-21-7)
Caswell No. 167A (112-02-7)
Caswell No. 168B (1076-46-6)
Caswell No. 168E (7286-84-2)
Caswell No. 169 (127-52-6)
Caswell No. 186 (13347-42-7)
Caswell No. 192A (2464-37-1)
Caswell No. 192BB (2832-19-1)
Caswell No. 205A (5825-87-6)
Caswell No. 206 (101-10-0)
Caswell No. 210 (607-12-5)
Caswell No. 210B (10605-10-4)
Caswell No. 211 (85-97-2)
Caswell No. 211C (3691-35-8)
Caswell No. 216A (7745-89-3)
Caswell No. 217C (64628-44-0)
Caswell No. 221 (7738-94-5)
Caswell No. 225A (8002-29-7)
Caswell No. 229B (33113-08-5)
Caswell No. 235A (10125-13-0)
Caswell No. 235B (10402-15-0)
Caswell No. 239 (22221-10-9)
Caswell No. 240 (54453-03-1)
Caswell No. 248A (814-91-5)
Caswell No. 252 (10102-90-6)
Caswell No. 254 (61789-22-8)
Caswell No. 254A (9007-39-0)
Caswell No. 257 (10257-54-2)
Caswell No. 266A (1111-67-7)
Caswell No. 268C (31717-87-0)
Caswell No. 268E (31717-87-0)
Caswell No. 271 (6531-86-8)
Caswell No. 275AA (64726-91-6)
Caswell No. 276 (1446-61-3)
Caswell No. 276A (2026-24-6)
Caswell No. 277 (51344-62-8)
Caswell No. 286A (1249-84-9)
Caswell No. 286C (7783-28-0)
Caswell No. 287C (87-12-7)
Caswell No. 289 (2577-72-2)
Caswell No. 290 (4776-06-1)
Caswell No. 294 (1918-11-2)
Caswell No. 295A (25059-78-3)
Caswell No. 315AC (2569-01-9)
Caswell No. 315AL (94-80-4)
Caswell No. 315AS (1928-43-4)
Caswell No. 315B (3766-27-6)
Caswell No. 315K (5742-19-8)
Caswell No. 315M (20940-37-8)
Caswell No. 316C (2758-42-1)
Caswell No. 316A (32357-46-3)
Caswell No. 316B (6753-24-8)
Caswell No. 316D (1320-15-6)
Caswell No. 316E (10433-59-7)
Caswell No. 320A (53404-31-2)
Caswell No. 323E (5335-24-0)
Caswell No. 328B (40843-25-2)
Caswell No. 331C (68959-20-6)
Caswell No. 333AB (68476-31-3)
Caswell No. 333F (61228-92-0)
Caswell No. 353 (6659-45-6)
Caswell No. 353B (20018-09-1)
Caswell No. 355 (25155-18-4)
Caswell No. 359DD (17702-57-7)
Caswell No. 362A (3735-23-7)
Caswell No. 364B (26419-73-8)
Caswell No. 380A (34484-77-0)
Caswell No. 380AB (24307-26-4)
Caswell No. 390A (2312-76-7)
Caswell No. 392H (5538-94-3)
Caswell No. 394A (42721-99-3)
Caswell No. 399A (27236-65-3)
Caswell No. 399C (5035-58-5)
Caswell No. 400 (136-45-8)
Caswell No. 404 (135-37-5)
Caswell No. 406 (22232-25-3)
Caswell No. 411B (16974-11-1)
Caswell No. 413C (27176-87-0)
Caswell No. 413A (6843-97-6)
Caswell No. 413D (26545-53-9)
Caswell No. 414 (1330-85-4)
Caswell No. 419F (82558-50-7)
Caswell No. 419H (74223-64-6)
Caswell No. 424A (29804-22-6)
Caswell No. 427A (61790-81-6)
Caswell No. 427E (104-28-9)
Caswell No. 431C (66441-23-4)
Caswell No. 438A (7379-26-2)
Caswell No. 438C (17421-79-3)
Caswell No. 456DD (72269-48-8)
Caswell No. 456E (68603-15-6)
Caswell No. 456H (41096-46-2)
Caswell No. 458 (14484-64-1)
Caswell No. 459B (10045-89-3)

Caswell No. 462A	(4301-50-2)
Caswell No. 471D	(13516-27-3)
Caswell No. 472A	(64742-94-5)
Caswell No. 472C	(53939-28-9)
Caswell No. 481E	(56-95-1)
Caswell No. 481F	(3697-42-5)
Caswell No. 481G	(18472-51-0)
Caswell No. 482B	(118-56-9)
Caswell No. 486A	(8001-78-3)
Caswell No. 486AB	(28772-56-7)
Caswell No. 487C	(139-89-9)
Caswell No. 494C	(34375-28-5)
Caswell No. 495AB	(6542-37-6)
Caswell No. 499D	(24959-67-9)
Caswell No. 501A	(55406-53-6)
Caswell No. 505A	(64771-72-8)
Caswell No. 506	(78-59-1)
Caswell No. 508A	(4812-20-8)
Caswell No. 509D	(50723-80-3)
Caswell No. 512A	(64-00-6)
Caswell No. 513A	(2634-33-5)
Caswell No. 518	(8006-54-0)
Caswell No. 521A	(3772-94-9)
Caswell No. 525BB	(52316-55-9)
Caswell No. 528A	(13840-33-0)
Caswell No. 532	(16949-65-8)
Caswell No. 532A	(3097-08-3)
Caswell No. 532B	(12057-74-8)
Caswell No. 533	(1343-88-0)
Caswell No. 540	(1490-04-6)
Caswell No. 541C	(2492-26-4)
Caswell No. 541B	(7778-70-3)
Caswell No. 548	(108-62-3)
Caswell No. 549DD	(1771-07-9)
Caswell No. 557D	(19480-43-4)
Caswell No. 557E	(20405-19-0)
Caswell No. 557G	(2039-46-5)
Caswell No. 558A	(6062-26-6)
Caswell No. 559B	(32351-70-5)
Caswell No. 561BB	(63333-35-7)
Caswell No. 566A	(5736-15-2)
Caswell No. 568C	(67762-39-4)
Caswell No. 572A	(2682-20-4)
Caswell No. 573I	(1773-37-1)
Caswell No. 575	(3478-94-2)
Caswell No. 578	(4726-14-1)
Caswell No. 584C	(53905-38-7)
Caswell No. 589D	(61-31-4)
Caswell No. 589AA	(2122-70-5)
Caswell No. 589E	(61790-13-4)
Caswell No. 595A	(1405-10-3)
Caswell No. 601	(609-31-4)
Caswell No. 601A	(2224-44-4)
Caswell No. 604AA	(1124-33-0)
Caswell No. 609AB	(53120-27-7)
Caswell No. 612	(6379-37-9)
Caswell No. 613A	(32426-11-2)
Caswell No. 613B	(10361-16-7)
Caswell No. 613D	(27193-28-8)
Caswell No. 614	(9036-19-5)
Caswell No. 618	(8000-29-1)
Caswell No. 618A	(8000-48-4)
Caswell No. 618B	(8000-46-2)
Caswell No. 618D	(8007-02-1)
Caswell No. 625B	(2164-07-0)
Caswell No. 627	(14697-50-8)
Caswell No. 627B	(68607-28-3)
Caswell No. 632	(106-46-7)
Caswell No. 638B	(8007-44-1)
Caswell No. 641A	(7778-73-6)
Caswell No. 642B	(140-01-2)
Caswell No. 644	(79-21-0)
Caswell No. 645A	(8009-03-8)
Caswell No. 647	(64742-16-1)
Caswell No. 647B	(61789-85-3)
Caswell No. 652C	(72490-01-8)
Caswell No. 655D	(122-70-3)
Caswell No. 657	(32407-99-1)
Caswell No. 657E	(122-64-5)
Caswell No. 657F	(5822-97-9)
Caswell No. 657J	(104-60-9)
Caswell No. 657N	(23319-66-6)
Caswell No. 658D	(13707-65-8)
Caswell No. 663	(7723-14-0)
Caswell No. 663I	(10294-56-1)
Caswell No. 663L	(4948-28-1)
Caswell No. 666	(8011-48-1)
Caswell No. 676A	(9003-27-4)
Caswell No. 691A	(27177-77-1)
Caswell No. 692A	(125-67-7)
Caswell No. 696	(137-41-7)
Caswell No. 696A	(13429-27-1)
Caswell No. 699A	(10058-23-8)
Caswell No. 700A	(7778-53-2)
Caswell No. 701	(37199-66-9)
Caswell No. 701A	(7492-30-0)
Caswell No. 701B	(1312-76-1)
Caswell No. 702A	(13932-13-3)
Caswell No. 703B	(30526-22-8)
Caswell No. 704A	(30346-73-7)
Caswell No. 706	(869-29-4)
Caswell No. 714AA	(42588-37-4)
Caswell No. 719AA	(134-30-5)
Caswell No. 722	(507-60-8)
Caswell No. 728	(8051-02-3)
Caswell No. 733	(8008-74-0)
Caswell No. 734	(1343-98-2)
Caswell No. 735A	(7783-90-6)
Caswell No. 736	(7775-41-9)
Caswell No. 739D	(51170-59-3)
Caswell No. 739O	(37913-89-6)
Caswell No. 739L	(2818-16-8)
Caswell No. 741	(68952-95-4)
Caswell No. 743	(13464-38-5)
Caswell No. 744A	(26628-22-8)
Caswell No. 755	(7758-19-2)
Caswell No. 757A	(34689-46-8)
Caswell No. 778B	(137-16-6)
Caswell No. 780AA	(7346-80-7)
Caswell No. 789	(1344-08-7)
Caswell No. 790A	(15922-78-8)
Caswell No. 792	(1344-09-8)
Caswell No. 796	(25567-55-9)
Caswell No. 801C	(8002-24-2)
Caswell No. 811	(61790-19-0)
Caswell No. 819	(1401-55-4)
Caswell No. 821AA	(6515-38-4)
Caswell No. 823	(8000-41-7)
Caswell No. 829	(776-19-2)
Caswell No. 833A	(2136-79-0)
Caswell No. 850	(4418-66-0)
Caswell No. 859B	(1300-78-3)
Caswell No. 863	(87-10-5)
Caswell No. 867B	(56573-85-4)

Caswell No. 867D (53404-82-3)
Caswell No. 867E (24124-25-2)
Caswell No. 867EF (2155-70-6)
Caswell No. 867G (28801-69-6)
Caswell No. 881E (6369-97-7)
Caswell No. 881H (53404-86-7)
Caswell No. 881J (3813-14-7)
Caswell No. 881K (2008-46-0)
Caswell No. 881O (1928-48-9)
Caswell No. 881S (4938-72-1)
Caswell No. 881T (93-78-7)
Caswell No. 882FF (38827-35-9)
Caswell No. 882C (2439-00-1)
Caswell No. 883C (27519-02-4)
Caswell No. 884 (1330-78-5)
Caswell No. 887A (2224-49-9)
Caswell No. 887AA (27323-41-7)
Caswell No. 887D (2717-15-9)
Caswell No. 887C (41669-40-3)
Caswell No. 892B (27668-52-6)
Caswell No. 892A (3452-97-9)
Caswell No. 892AA (75673-43-7)
Caswell No. 899 (8002-33-3)
Caswell No. 900AA (59669-26-0)
Caswell No. 907B (60568-05-0)
Caswell No. 915 (557-41-5)
Caswell No. 928 (68813-94-5)
Caswell No. 930 (12122-67-7)
Caswell No. 931A (16509-79-8)
Caswell No. 934A (53939-27-8)
Cat (122-34-9)
Cat (Herbicide) (122-34-9)
Catalin CAO-3 (128-37-0)
Cataloid (7631-86-9)
Catalytic Coke (65996-77-2)
Catalytic-Dewaxed heavy naphthenic distillate (64742-68-3)
Catalytic-Dewaxed heavy paraffinic distillate (64742-70-7)
Catalytic-Dewaxed light naphthenic distillate (64742-69-4)
Catalytic-Dewaxed light paraffinic distillate (64742-71-8)
Catalytically cracked clarified oil (64741-62-4)
Catalytic reformed naphtha (Petroleum) (68955-35-1)
Catamin AB (8001-54-5)
Catamine AB (8001-54-5)
Catanil (94-20-2)
Catapal S (1344-28-1)
Catapal SB Alumina (1344-28-1)
Catavin C (50-81-7)
Cat cracker feed stock (68476-31-3)
(+)-Catechin (154-23-4)
Catechin (120-80-9)
Catechin (154-23-4)
D-(+)-Catechin (154-23-4)
d-Catechin (154-23-4)
Catechin (Flavan) (154-23-4)
Catechin hydrate (480-18-2)
Catechinic acid (154-23-4)
Catechins (1401-55-4)
(+)-Catechol (154-23-4)
Catechol (154-23-4)
D-Catechol (154-23-4)
Catechol (ACGIH,OSHA) (120-80-9)
Catechol (Flavan) (154-23-4)
Catecholcarboxylic acid (303-38-8)
Catechol, 5-chloro-3-methyl (31934-88-0)
Catechol monoethyl ether (94-71-3)
Catechuic acid (154-23-4)
Catenulin (7542-37-2)
Catergen (154-23-4)
Cateudyl (72-44-6)
Cathina (Spanish) (492-39-7)
Cathine (492-39-7)
Cathinum (Latin) (492-39-7)
Catilan (56-75-7)
Cation AB (112-03-8)
Cation BB (112-00-5)
Cation FB (112-00-5)
Cation PB 40 (112-02-7)
Cationic Orange ZH (3056-93-7)
Catolin 14 (119-47-1)
Caurite (140-95-4)
Causoin (57-41-0)
Caustic Alcohol (141-52-6)
Caustic Barley (8051-02-3)
Caustic Baryta (17194-00-2)
Caustic Magnesite (1309-42-8)
Caustic nickel skims (Secondary nonferrous plant) (69011-64-9)
Caustic potash, Dry, Solid, Flake, Bead, or Granular (DOT) (1310-58-3)
Caustic potash, Liquid or solution (DOT) (1310-58-3)
Caustic potash (1310-58-3)
Caustic soda (1310-73-2)
Caustic soda (DOT) (1310-73-2)
Caustic soda, Liquid (DOT) (1310-73-2)
Caustic soda, Solid (DOT) (1310-73-2)
Caustic soda, Solution (DOT) (1310-73-2)
Cavi-Trol (7681-49-4)
Cavalite Brilliant Blue R (2580-78-1)
Cavonyl (52-31-3)
Cbc 906288 (545-55-1)
Ccucol (446-86-6)
Ce Lent (50-81-7)
Ce-Mi-Lin (50-81-7)
Ce-Vi-Sol (50-81-7)
Cebetox (2587-90-8)
Cebicure (50-81-7)
Cebid (50-81-7)
Cebion (50-81-7)
Cebione (50-81-7)
Cebitate (134-03-2)
neo-Cebitate (6381-77-7)
Cebrogen (56-85-9)
Cecalgine TBV (9005-38-3)
Cecarbon (7440-44-0)
Cecenu (13010-47-4)
Cecil (50-36-2)
Cecolene (79-01-6)
Cecon (50-81-7)
Cedar Wood Oil (8000-27-9)
Cedarwood-Oil (8000-27-9)
Cedr-8-ene (469-61-4)
Cedin (54-85-3)
Cedocard (87-33-2)
8βH-Cedran-8-ol (8CI) (77-53-2)
8-β-H-Cedran-8-ol, acetate (77-54-3)
Cedranyl acetate (77-54-3)
α-Cedrene (469-61-4)
Cedro Oil (8008-56-8)
Cedrol (77-53-2)
α-Cedrol (77-53-2)
Cedrus Atlantica Oil (8000-27-9)
Cedryl acetate (77-54-3)
Cee-Caps TD (50-81-7)
Cee Dee (57-09-0)
Cee-Vite (50-81-7)
Ceenu (13010-47-4)

Ceeprin chloride	(123-03-5)
Ceepryn	(123-03-5)
Ceepryn chloride	(123-03-5)
Cefa-Iskia	(15686-71-2)
Cefacetrile sodium	(23239-41-0)
Cefaclor	(53994-73-3)
Cefadole	(34444-01-4)
Cefalexin	(15686-71-2)
Cefaloglycin	(3577-01-3)
Cefaloridin	(50-59-9)
Cefaloridine	(50-59-9)
Cefalorizin	(50-59-9)
Cefalotin	(153-61-7)
Cefaloto	(15686-71-2)
Cefamandol	(34444-01-4)
Cefamandole	(34444-01-4)
L-Cefamandole	(34444-01-4)
Cefamezin	(25953-19-9)
Cefapirin (German)	(21593-23-7)
Cefatin	(31566-31-1)
Cefazolin	(25953-19-9)
Cefazoline	(25953-19-9)
Ceflorin	(50-59-9)
Cefoxitin	(35607-66-0)
Cefracycline suspension	(60-54-8)
Cefracycline tablets	(64-75-5)
Cefradine	(38821-53-3)
Cefroxadin	(51762-05-1)
Cefroxadine	(51762-05-1)
Cefuroxim	(55268-75-2)
Cefuroxime	(55268-75-2)
Cegiolan	(50-81-7)
Ceglion	(50-81-7)
Ceglution	(554-13-2)
Cekiuron	(330-54-1)
Ceku C.B.	(118-74-1)
Cekubaryl	(63-25-2)
Cekudifol	(115-32-2)
Cekufon	(52-68-6)
Cekugib	(77-06-5)
Cekumeta	(108-62-3)
Cekumethion	(298-00-0)
Cekuquat	(1910-42-5)
Cekusan	(122-34-9)
Cekusan	(62-73-7)
Cekusil	(62-38-4)
Cekusil Universal A	(151-38-2)
Cekusil Universal C	(123-88-6)
Cekuthoate	(60-51-5)
Cekutrothion	(122-14-5)
Cekuzina-S	(122-34-9)
Cekuzina-T	(1912-24-9)
Cela 50	(26644-46-2)
Cela S-2225	(4824-78-6)
Cela S-2957	(21923-23-9)
Cela S 1942	(2104-96-3)
Cela W 524	(26644-46-2)
Celaburato (Spanish)	(9004-36-8)
Celacol EM	(9004-59-5)
Celacol M	(9004-67-5)
Celacol M20	(9004-67-5)
Celacol M450	(9004-67-5)
Celacol MM	(9004-67-5)
Celacol MM 10P	(9004-67-5)
Celacol M 20P	(9004-67-5)
Celamerck S-2957	(21923-23-9)
Celanar	(25038-59-9)
Celanex	(58-89-9)
Celanol 252	(9014-90-8)
Celanol DOS 75	(577-11-7)
Celanthrene Brilliant Blue	(2475-46-9)
Celanthrene Brilliant Blue	(86722-66-9)
Celanthrene Brilliant Blue FFS	(2475-46-9)
Celanthrene Brilliant Blue FFS	(86722-66-9)
Celanthrene Fast Pink 3B	(2872-48-2)
Celanthrene Pure Blue BRS	(2475-45-8)
Celanthrene Red 3BN	(116-85-8)
Celanthrene Red Violet R	(128-95-0)
Celanthrene Red Y	(82-38-2)
Celaskon	(50-81-7)
Celathion	(21923-23-9)
Celathion	(60238-56-4)
Celcot RF	(92-77-3)
Celcot RK	(135-62-6)
Celcot RM	(135-65-9)
Celcot RN	(132-68-3)
Celcot RTO	(135-61-5)
Celestolide	(13171-00-1)
Celestone	(378-44-9)
Celex	(9004-70-0)
Celfume	(74-83-9)
Celgard 2500	(9003-07-0)
Celgard 3501	(9003-07-0)
Celgard KKX 2	(9003-07-0)
Celgard 2400W	(9003-07-0)
Celin	(50-81-7)
Celinhol-A	(31566-31-1)
Celite	(14808-60-7)
Celkate T 21	(1343-88-0)
Cellaburato	(9004-36-8)
Cellaburatum (Latin)	(9004-36-8)
Cellapret	(9004-67-5)
Cellex MX	(9004-34-6)
Cellidor	(9004-35-7)
Cellidor A	(9004-35-7)
Cellidrin	(315-30-0)
Cellit K 700	(9004-35-7)
Cellit L 700	(9004-35-7)
Cellitazol B	(119-90-4)
Cellitazol BN	(119-90-4)
Cellitazol R	(60-09-3)
Celliton Blue BB-CF	(2475-45-8)
Celliton Blue Extra	(2475-45-8)
Celliton Blue FFR	(2475-46-9)
Celliton Blue FFR	(86722-66-9)
Celliton Blue G	(2475-45-8)
Celliton Blue GA-CF	(2475-45-8)
Celliton Blue RN	(81-77-6)
Celliton Discharge Yellow GL	(2832-40-8)
Celliton Discharge Yellow 5RL	(6300-37-4)
Celliton Fast Blue B	(2475-44-7)
Celliton Fast Blue FBBN	(2475-46-9)
Celliton Fast Blue FBBN	(86722-66-9)
Celliton Fast Blue FFR	(2475-46-9)
Celliton Fast Blue FFR	(86722-66-9)
Celliton Fast Blue FFRN	(2475-46-9)
Celliton Fast Blue FFRN	(86722-66-9)
Celliton Fast Blue FFRS	(2475-46-9)
Celliton Fast Blue FFRS	(86722-66-9)
Celliton Fast Pink BA-CF	(116-85-8)
Celliton Fast Pink BN	(116-85-8)
Celliton Fast Pink FF3B	(2872-48-2)
Celliton Fast Pink FF3BA-CF	(2872-48-2)
Celliton Fast Red Violet	(128-95-0)

Celliton Fast Red Violet R	(128-95-0)
Celliton Fast Red Violet RN	(128-95-0)
Celliton Fast Red Violet RNA-CF	(128-95-0)
Celliton Fast Violet 6B	(1220-94-6)
Celliton Fast Violet B	(82-33-7)
Celliton Fast Violet 6BA-CF	(1220-94-6)
Celliton Fast Violet BA-CF	(82-33-7)
Celliton Fast Yellow G	(2832-40-8)
Celliton Fast Yellow GA	(2832-40-8)
Celliton Fast Yellow GA-CF	(2832-40-8)
Celliton Fast Yellow 5R	(6300-37-4)
Celliton Fast Yellow RR	(119-15-3)
Celliton Orange R	(82-28-0)
Celliton Pink R	(82-38-2)
Celliton Rose FF3B	(2872-48-2)
Celliton Violet B	(82-33-7)
Celliton Yellow G	(2832-40-8)
Celliton Yellow 5R	(6300-37-4)
Cellmic S	(80-51-3)
Cellocol EM	(9004-59-5)
Cellofas	(9004-32-4)
Cellofas A	(9004-59-5)
Cellofas B	(9004-32-4)
Cellofas B5	(9004-32-4)
Cellofas B50	(9004-32-4)
Cellofas B6	(9004-32-4)
Cellofas C	(9004-32-4)
Cellofas WLD	(9004-59-5)
Cellofor (Czech)	(136-35-6)
Cellogel C	(9004-32-4)
Cellogen 3H	(9004-32-4)
Cellogen PR	(9004-32-4)
Cellogen WS-C	(9004-32-4)
Cellogran	(9004-67-5)
Celloidin	(9004-70-0)
Cellon	(79-34-5)
Cellophane	(9005-81-6)
Cellosize 4400H16	(9004-62-0)
Cellosize QP	(9004-62-0)
Cellosize QP 3	(9004-62-0)
Cellosize QP 1500	(9004-62-0)
Cellosize QP 30000	(9004-62-0)
Cellosize QP 4400	(9004-62-0)
Cellosize UT 40	(9004-62-0)
Cellosize WP	(9004-62-0)
Cellosize WP 300	(9004-62-0)
Cellosize WP 300H	(9004-62-0)
Cellosize WP 400H	(9004-62-0)
Cellosize WP 4400	(9004-62-0)
Cellosize WPO 9H17	(9004-62-0)
Cellosolve (DOT)	(110-80-5)
Cellosolve acetate (DOT,OSHA)	(111-15-9)
Cellosolve acrylate	(106-74-1)
Cellosolve, 2,4-dichlorophenyl-	(120-67-2)
Cellosolve, n-hexyl-	(112-25-4)
Cellosolve solvent	(110-80-5)
Cellothyl	(9004-67-5)
Cellpro	(9004-32-4)
Cellu-Quin	(10380-28-6)
Cellufix FF 100	(9004-32-4)
Celluflex	(115-96-8)
Celluflex 179C	(1330-78-5)
Celluflex DOP	(117-84-0)
Celluflex DPB	(84-74-2)
Celluflex FR-2	(78-43-3)
Celluflex tpp	(115-86-6)
Cellugel	(9004-32-4)
Cellulose 248	(9004-34-6)
α-Cellulose	(9004-34-6)
Cellulose (ACGIH,OSHA)	(9004-34-6)
Cellulose Crystalline	(9004-34-6)
Cellulose Gum 7H	(9000-11-7)
Cellulose, Polymer with 2-propenenitrile	(37243-36-0)
Cellulose 2,5-acetate	(9004-35-7)
Cellulose, acetate (9CI)	(9004-35-7)
Cellulose, acetate butanoate (9CI)	(9004-36-8)
Cellulose acetate butyrate	(9004-36-8)
Cellulose acetate-butyrate	(9004-36-8)
Cellulose carboxymethylate	(9000-11-7)
Cellulose, carboxymethyl ether	(9000-11-7)
Cellulose, carboxymethyl ether, sodium salt	(9004-32-4)
Cellulose, carboxymethyl 2-hydroxyethyl ether	(9004-30-2)
Cellulose, carboxymethyl hydroxyethyl ether	(9004-30-2)
Cellulose, carboxymethyl-2-hydroxyethyl ether, sodium salt	(9004-30-2)
Cellulose, carboxymethyl hydroxyethyl mixed ether	(9004-30-2)
Cellulose, carboxymethyl methyl ether (9CI)	(37206-01-2)
Cellulose, 2,5-diacetate	(9004-35-7)
Cellulose, diacetate	(9004-35-7)
Cellulose ethyl	(9004-57-3)
Cellulose ethylate	(9004-57-3)
Cellulose, ethyl ether	(9004-57-3)
Cellulose, ethyl methyl ether	(9004-59-5)
Celluloseglycolic acid	(9000-11-7)
Cellulose glycolic acid, sodium salt	(9004-32-4)
Cellulose hydroxyethylate	(9004-62-0)
Cellulose hydroxyethyl ether	(9004-62-0)
Cellulose, 2-hydroxyethyl ether	(9004-62-0)
Cellulose, hydroxymethyl ether (9CI)	(37353-59-6)
Cellulose methyl	(9004-67-5)
Cellulose methylate	(9004-67-5)
Cellulose, methyl ether (1/2%)	(9004-67-5)
Cellulose monoacetate	(9004-35-7)
Cellulose nitrate	(9004-70-0)
Cellulose, nitrate (9CI)	(9004-70-0)
Cellulose, octadecanoate (9CI)	(9085-22-7)
Cellulose-polyacrylonitrile copolymer	(37243-36-0)
Cellulose-polyacrylonitrile graft copolymer	(37243-36-0)
Cellulose sodium glycolate	(9004-32-4)
Cellulose tetranitrate	(9004-70-0)
Cellulose, triacetate	(9004-35-7)
Cellumeth	(9004-67-5)
Celluphos 4	(126-73-8)
Celmer	(123-88-6)
Celmer	(62-38-4)
Celmide	(106-93-4)
Celmone	(86-87-3)
Celogen BSH	(80-17-1)
Celogen OT	(80-51-3)
Celon	(63428-84-2)
Celon A	(60-00-4)
Celon ATH	(60-00-4)
Celon E	(64-02-8)
Celon H	(64-02-8)
Celon IS	(64-02-8)
Celontin	(77-41-8)
Celosen AZ	(123-77-3)
Celosolv (Czech)	(110-80-5)
Celosolvacetat (Czech)	(111-15-9)
Celospor	(23239-41-0)
Celphide	(20859-73-8)
Celphine	(20859-73-8)
Celphos	(20859-73-8)
Celphos	(7803-51-2)
Celthion	(121-75-5)

Celufi	(9004-34-6)
Celutate Blue BLT	(2475-46-9)
Celutate Blue BLT	(86722-66-9)
Celutate Blue RNH	(2475-46-9)
Celutate Blue RNH	(86722-66-9)
Celutate Brilliant Blue B	(2475-46-9)
Celutate Brilliant Blue B	(86722-66-9)
Celutate Pink B	(116-85-8)
Celutate Pink BN	(116-85-8)
Celutate Pink BY	(116-85-8)
Celutate Red Violet RH	(128-95-0)
Celutate Yellow GH	(2832-40-8)
Cemagyl	(50-81-7)
Cembrane	(1786-12-5)
Cembrane I	(1786-12-5)
Cemedine 196	(9003-20-7)
Cement Black	(1313-13-9)
Cemidon	(54-85-3)
Cemill	(50-81-7)
Cemiod	(66-02-4)
Cemulsol 1050	(9004-96-0)
Cemulsol A	(9004-96-0)
Cemulsol C 105	(9004-96-0)
Cemulsol D-8	(9004-96-0)
Cemulsol OP 16	(9036-19-5)
Cenalene-M	(54-95-5)
Cenazol	(54-95-5)
Cenetone	(50-81-7)
Cenitron OB	(80-51-3)
Cenol Garden Dust	(83-79-4)
Cenolate	(134-03-2)
Cenolate	(50-81-7)
Censtim	(50-49-7)
Censtin	(50-49-7)
Centedein	(113-45-1)
Centedrin	(298-59-9)
Centimide	(57-09-0)
Centralgin	(50-13-5)
Centraline Blue 3B	(72-57-1)
Centralite	(85-98-3)
Centralite-1	(85-98-3)
Centralite II	(611-92-7)
Centramin	(2152-34-3)
Centramina	(60-13-9)
Centrax	(2955-38-6)
Centrazole	(54-95-5)
Centredin	(113-45-1)
Century 1240	(57-11-4)
Century CD fatty acid	(112-80-1)
Centyl	(73-48-3)
Cenwax ME	(141-23-1)
Cepacilina	(1538-09-6)
Cepacillina	(1538-09-6)
Cepacol chloride	(123-03-5)
Cepaloridin	(50-59-9)
Cepalorin	(50-59-9)
Ceph 87/4	(50-59-9)
Cephacetrile sodium	(23239-41-0)
Cephadole	(34444-01-4)
(-)-Cephaeline dihydrochloride	(5853-29-2)
Cephaeline-hydrochloride	(5853-29-2)
Cephaeline methyl ether	(483-18-1)
Cephalexin	(15686-71-2)
Cephaloglycin	(3577-01-3)
Cephaloglycine	(3577-01-3)
D-Cephaloglycine	(3577-01-3)
Cephalon	(330-55-2)
Cephaloridin	(50-59-9)
Cephaloridine	(50-59-9)
Cephalothin	(153-61-7)
Cephalotin	(153-61-7)
Cephamandole	(34444-01-4)
Cephamezine	(25953-19-9)
Cephaoglycin acid	(3577-01-3)
Cephapirin	(21593-23-7)
Cephazolin	(25953-19-9)
Cephazoline	(25953-19-9)
Cephoxitin	(35607-66-0)
Cephradin	(38821-53-3)
Cephradine	(38821-53-3)
Cephrol	(106-22-9)
Cephuroxime	(55268-75-2)
Cepo	(9004-34-6)
Cepo CFM	(9004-34-6)
Cepo S 20	(9004-34-6)
Cepo S 40	(9004-34-6)
Ceporan	(50-59-9)
Ceporex	(15686-71-2)
Ceporexin	(15686-71-2)
Ceporexine	(15686-71-2)
Ceporin	(50-59-9)
Ceporine	(50-59-9)
Ceprim	(123-03-5)
Cequartyl	(8001-54-5)
Ceramic	(66402-68-4)
Ceramic Fibre	(1302-76-7)
Ceramic Materials and Wares, Chemicals	(66402-68-4)
Ceramic bonded alumina	(66402-68-4)
Ceramic bonded silicon carbide	(66402-68-4)
Ceraphyl 230	(6938-94-9)
Ceraphyl 368	(29806-73-3)
Cerasin Red	(85-86-9)
Cerasine Yellow GG	(60-11-7)
Cerasinrot	(85-86-9)
Cerasynt	(106-11-6)
Cerasynt 1000-D	(31566-31-1)
Cerasynt M	(9004-99-3)
Cerasynt MN	(9004-99-3)
Cerasynt PA	(1323-39-3)
Cerasynt PN	(1323-39-3)
Cerasynt S	(31566-31-1)
Cerasynt SD	(31566-31-1)
Cerasynt SE	(31566-31-1)
Cerasynt Special	(106-11-6)
Cerasynt WM	(31566-31-1)
Ceratinic acid	(506-46-7)
Cerazol (Suspension)	(72-14-0)
Cercine	(439-14-5)
Cercobin	(23564-06-9)
Cercobin M	(23564-05-8)
Cercobin methyl	(23564-05-8)
Cercosporin	(35082-49-6)
Cerebro-Nicin	(54-95-5)
Cerechlor 54	(63449-39-8)
Cereclor	(63449-39-8)
Cereclor 30	(63449-39-8)
Cereclor 42	(63449-39-8)
Cereclor 48	(63449-39-8)
Cereclor 50LV	(63449-39-8)
Cereclor 52	(63449-39-8)
Cereclor 54	(63449-39-8)
Cereclor 70	(63449-39-8)
Cereclor 511	(63449-39-8)
Cereclor 631	(63449-39-8)

Name	CAS No.
Cereclor 65l	(63449-39-8)
Cereclor 70l	(63449-39-8)
Cereclor S 42	(63449-39-8)
Cereclor S52	(63449-39-8)
Cereclor S70	(63449-39-8)
Cereden	(495-73-8)
Ceredon	(495-73-8)
Ceregulart	(439-14-5)
Cereline	(495-73-8)
Cerelose	(50-99-7)
Cerenox	(495-73-8)
Cereon	(50-81-7)
Cerepap	(59-92-7)
Cerepax	(846-50-4)
Ceres Blue BHR	(147-14-8)
Ceres Green 3B	(1328-53-6)
Ceres Orange G	(2051-85-6)
Ceres Orange GN	(2051-85-6)
Ceres Orange R	(842-07-9)
Ceres Oranges RR	(3118-97-6)
Ceres Red 7B	(6368-72-5)
Ceres Red BB	(85-83-6)
Ceres Yellow GGN	(2481-94-9)
Ceres Yellow R	(60-09-3)
Ceresan	(107-27-7)
Ceresan	(62-38-4)
Ceresan M	(517-16-8)
Ceresan M-DB	(517-16-8)
Ceresan M-2X	(517-16-8)
Ceresan Universal	(62-38-4)
Ceresan-Universal Nassbeize	(123-88-6)
Ceresan Universal Nazbeize	(123-88-6)
Ceresol	(62-38-4)
Cerespan	(61-25-6)
Cerex	(63428-84-2)
Cergona	(50-81-7)
Ceric acid	(506-46-7)
Ceric hydroxide	(12014-56-1)
Ceric oxide	(1306-38-3)
Cerinic acid	(506-46-7)
Cerise B	(632-99-5)
Cerise Toner X1127	(81-88-9)
Cerisol Scarlet G	(3118-97-6)
Cerisol Yellow AB	(85-84-7)
Cerisol Yellow GR	(2051-85-6)
Cerisol Yellow TB	(131-79-3)
Cerit Fac 3	(106-14-9)
Cerium-acetate	(537-00-8)
Cerium-fluoride	(7758-88-5)
Cerium	(7440-45-1)
Cerium, Crude, Powder (DOT)	(7440-45-1)
Cerium, Crude, Slabs or ingots [UN 1333]	(7440-45-1)
Cerium (IV) hydroxide	(12014-56-1)
Cerium, Isotope of mass 141 (8CI,9CI)	(13967-74-3)
Cerium, Isotope of mass 144 (8CI,9CI)	(14762-78-8)
Cerium(III) acetate	(537-00-8)
Cerium(III) carbonate	(537-01-9)
Cerium carbonate (VAN)	(537-01-9)
Cerium concentrate	(68909-12-6)
Cerium dioxide	(1306-38-3)
Cerium fluorure (French)	(7758-88-5)
Cerium hydrate	(12014-56-1)
Cerium hydroxide	(12014-56-1)
Cerium hydroxide, (T-4)- (9CI)	(12014-56-1)
Cerium nitrate	(10108-73-3)
Cerium(3+) nitrate	(10108-73-3)
Cerium(III) nitrate	(10108-73-3)
Cerium octadecanoate	(10119-53-6)
Cerium oxide (9CI)	(1306-38-3)
Cerium triacetate	(537-00-8)
Cerium tricarbonate	(537-01-9)
Cerium trifluoride	(7758-88-5)
Cerium trinitrate	(10108-73-3)
Cerium-trioxide	(1345-13-7)
Cern Brilantni PN (Czech)	(2519-30-4)
Cern Kypova 27 (Czech)	(2379-81-9)
Cern Kysela 1 (Czech)	(1064-48-8)
Cern Potravinarska 1 (Czech)	(2519-30-4)
Cern Prima 19 (Czech)	(6428-31-5)
Cern Prima 38 (Czech)	(1937-37-7)
Cern Reaktivni 8 (Czech)	(12225-26-2)
Cerobin	(23564-06-9)
Cerone	(16672-87-0)
Cerotic acid	(506-46-7)
Cerotine Ponceau 3B	(85-83-6)
Cerotinorange G	(842-07-9)
Cerotinscharlach G	(3118-97-6)
Cerotinscharlach R	(85-86-9)
Cerous acetate	(537-00-8)
Cerous carbonate	(537-01-9)
Cerous fluoride	(7758-88-5)
Cerous nitrate	(10108-73-3)
Ceroxin GL	(106-14-9)
Certicol Amaranth S	(915-67-3)
Certicol Black PNW	(2519-30-4)
Certicol Carmoisine S	(3567-69-9)
Certicol Fast Red E	(2302-96-7)
Certicol Orange GS	(1936-15-8)
Certicol Ponceau MXS	(3761-53-3)
Certicol Ponceau 4RS	(2611-82-7)
Certicol Ponceau SXS	(4548-53-2)
Certicol Red B	(3567-66-6)
Certicol Sunset Yellow CFS	(2783-94-0)
Certicol Tartrazol Yellow S	(1934-21-0)
Certinal	(123-30-8)
Certiqual Alizarine	(72-48-0)
Certiqual Eosine	(17372-87-1)
Certiqual Eosine	(548-26-5)
Certiqual Fluoresceine	(518-47-8)
Certiqual Lithol Red	(1248-18-6)
Certiqual Oil Red	(85-86-9)
Certiqual Orange I	(523-44-4)
Certiqual Orange II	(633-96-5)
Certiqual Rhodamine	(509-34-2)
Certiqual Rhodamine	(81-88-9)
Certol	(1689-83-4)
Certolake Sunset Yellow	(15790-07-5)
Certolake Sunset Yellow	(2783-94-0)
Certox	(57-24-9)
Certrol	(1689-83-4)
Cerubidin	(20830-81-3)
Cerulignol	(2785-87-7)
Ceruse	(1319-46-6)
Cerussa	(1319-46-6)
Cerussete	(598-63-0)
Cerven Brilantni Ostacetova F-LB (Czech)	(17418-58-5)
Cerven Disperzni 11 (Czech)	(2872-48-2)
Cerven Disperzni 15 (Czech)	(116-85-8)
Cerven Disperzni 60 (Czech)	(17418-58-5)
Cerven 2G (Czech)	(3734-67-6)
Cerven Kongo (Czech)	(573-58-0)
Cerven Kosenilova A (Czech)	(2611-82-7)
Cerven Kumidinova (Czech)	(3564-09-8)
Cerven Kypova 13 (Czech)	(4203-77-4)

Cerven Kysela 1 (Czech) (3734-67-6)
Cerven Kysela 2 (Czech) (493-52-7)
Cerven Kysela 14 (Czech) (3567-69-9)
Cerven Kysela 18 (Czech) (2611-82-7)
Cerven Kysela 26 (Czech) (3761-53-3)
Cerven Kysela 27 (Czech) (915-67-3)
Cerven Kysela 41 (Czech) (5850-44-2)
Cerven Kysela 51 (Czech) (16423-68-0)
Cerven Kysela 87 (Czech) (17372-87-1)
Cerven Kysela 114 (Czech) (6459-94-5)
Cerven Methylova (Czech) (493-52-7)
Cerven Pigment 3 (Czech) (2425-85-6)
Cerven Pigment 57 (Czech) (5858-81-1)
Cerven Potravinarska 1 (Czech) (4548-53-2)
Cerven Potravinarska 2 (Czech) (3257-28-1)
Cerven Potravinarska 3 (Czech) (3567-69-9)
Cerven Potravinarska 5 (Czech) (3761-53-3)
Cerven Potravinarska 6 (Czech) (3564-09-8)
Cerven Potravinarska 7 (Czech) (2611-82-7)
Cerven Potravinarska 8 (Czech) (5850-44-2)
Cerven Potravinarska 9 (Czech) (915-67-3)
Cerven Potravinarska 10 (Czech) (3734-67-6)
Cerven Potravinarska 14 (Czech) (16423-68-0)
Cerven Prima 2 (Czech) (992-59-6)
Cerven Prima 28 (Czech) (573-58-0)
Cerven Reaktivni 2 (Czech) (17804-49-8)
Cerven Rozpoustedlova 19 (Czech) (6368-72-5)
Cerven Rozpoustedlova 23 (Czech) (85-86-9)
Cerven Rozpoustedlova 24 (Czech) (85-83-6)
Cerven Rozpoustedlova 80 (Czech) (6358-53-8)
Cerven Zasadita 1 (Czech) (989-38-8)
Cerven Zasadita 2 (Czech) (477-73-6)
Cerven Zasadita 9 (Czech) (569-61-9)
Cervicundin (297-76-7)
Cervilaxin (9002-69-1)
Cesamet (51022-71-0)
Cescorbat (50-81-7)
Cesium-133 (7440-46-2)
Cesium-bromide (7787-69-1)
Cesium-chloride (7647-17-8)
Cesium-iodide (7789-17-5)
Cesium, Isotope of mass 134 (8CI,9CI) (13967-70-9)
Cesium, Isotope of mass 136 (8CI,9CI) (14234-29-8)
Cesium, Powdered (DOT) (7440-46-2)
Cesium [UN 1407] (7440-46-2)
Cesium chromate (13454-78-9)
Cesium hydrate (21351-79-1)
Cesium hydroxide (ACGIH,OSHA) (21351-79-1)
Cesium hydroxide, Solid (DOT) (21351-79-1)
Cesium hydroxide, Solution (DOT) (21351-79-1)
Cesium hydroxide dimer (21351-79-1)
Cesium, Isotope of mass 137 (8CI,9CI) (10045-97-3)
Cesium metal (DOT) (7440-46-2)
Cesium monochloride (7647-17-8)
Cesium monoiodide (7789-17-5)
Cesol (55268-74-1)
Cet (153-61-7)
Cetab (57-09-0)
Cetacort (50-23-7)
Cetadol (103-90-2)
Cetaffine (36653-82-4)
Cetain (51-05-8)
Cetal (36653-82-4)
Cetaldehyde methylformylhydrazone (16568-02-8)
Cetalol CA (36653-82-4)
Cetalox AT (9005-00-9)
Cetamid (50-81-7)
Cetamine oxide (7128-91-8)
Cetamium (123-03-5)
Cetane (50-81-7)
Cetane (544-76-3)
n-Cetane (544-76-3)
Cetane-Caps TC (50-81-7)
Cetane-Caps TD (50-81-7)
Cetapharm (140-72-7)
Cetarin (297-90-5)
Cetarol (57-09-0)
Cetasol (140-72-7)
Cetavlon (57-09-0)
Cetazol (140-72-7)
Cetearyl Alcohol (67762-27-0)
Cetemican (50-81-7)
1-Cetene (629-73-2)
Cetene (629-73-2)
Ceteth-1 (9004-95-9)
Ceteth-2 (9004-95-9)
Ceteth-4 (9004-95-9)
Ceteth-5 (9004-95-9)
Ceteth-6 (9004-95-9)
Ceteth-10 (9004-95-9)
Ceteth-12 (9004-95-9)
Ceteth-15 (9004-95-9)
Ceteth-16 (9004-95-9)
Ceteth-20 (9004-95-9)
Ceteth-24 (9004-95-9)
Ceteth-25 (9004-95-9)
Ceteth-30 (9004-95-9)
Ceteth-45 (9004-95-9)
Cetethyl morpholinium ethosulfate (78-21-7)
Cethylose (9004-67-5)
Cethytin (9004-67-5)
Cetic acid, cerium(3+) salt (8CI,9CI) (537-00-8)
Cetil Chromine Yellow GR (547-57-9)
Cetil Light Orange GG (1936-15-8)
Cetil Light Red GG (3734-67-6)
Cetin (540-10-3)
Cetobemidon (469-79-4)
Cetobemidona (Spanish) (469-79-4)
Cetobemidone (French) (469-79-4)
Cetobemidonum (Latin) (469-79-4)
Cetomacrogol (9004-95-9)
Cetomacrogol 1000 (9004-95-9)
Cetomacrogol 1000 BPC (9004-95-9)
Cetomacrogol Wax BP (67762-27-0)
Cetomacrogolum (Latin) (68439-49-6)
Cetone α (127-51-5)
α-Cetone (127-51-5)
Cetostearyl alcohol (67762-27-0)
Cetrimide (57-09-0)
Cetrimide bp (57-09-0)
Cetrimonium (6899-10-1)
Cetrimonium bromide (57-09-0)
Cetrimonium chloride (112-02-7)
Cetyl acetate (629-70-9)
Cetylacetic acid (57-11-4)
Cetyl acrylate (13402-02-3)
Cetyl alcohol (36653-82-4)
Cetyl alcohol, ethoxylated (9004-95-9)
Cetylamin (German) (143-27-1)
Cetylamine (143-27-1)
Cetylamine (57-09-0)
Cetyl Betaine (693-33-4)
Cetyl bromide (112-82-3)
Cetylcide (124-03-8)

Cetyldimethylamine (112-69-6)
Cetyl dimethyl amine oxide (7128-91-8)
Cetyldimethylethylammonium bromide (124-03-8)
Cetylene (629-73-2)
Cetylethyldimethylammonium bromide (124-03-8)
Cetyl ethyl morpholinium ethosulfate (78-21-7)
Cetylethylmorpholinium ethosulfate (78-21-7)
N-Cetyl-N-ethylmorpholinium ethosulfate (78-21-7)
N-Cetyl-N-ethyl morpholinium ethyl sulfate (78-21-7)
N-Cetyl-N-ethylmorpholinium ethyl sulfate (78-21-7)
N-Cetyl-N-ethylmorpholinium ethylsulfate (78-21-7)
Cetylic acid (57-10-3)
Cetylic alcohol (36653-82-4)
Cetyl methacrylate (2495-27-4)
αCetylmethadol (17199-58-5)
βCetylmethadol (17199-59-6)
Cetyl myristate (2599-01-1)
Cetylol (36653-82-4)
(Cetyloxymethyl)oxirane (15965-99-8)
Cetyl palmitate (540-10-3)
Cetyl phosphate (3539-43-3)
1-Cetylpyridinium bromide (140-72-7)
Cetylpyridinium bromide (140-72-7)
n-Cetylpyridinium bromide (140-72-7)
1-Cetylpyridinium chloride (123-03-5)
Cetylpyridinium chloride (123-03-5)
N-Cetylpyridinium chloride (123-03-5)
Cetyl sodium sulfate (1120-01-0)
Cetyl/stearyl alcohol (67762-27-0)
Cetyl sulfate (143-02-2)
Cetyl sulfate ammonium salt (4696-47-3)
Cetyl sulfate sodium salt (1120-01-0)
Cetyltriethylammonium bromide (13316-70-6)
Cetyltrimethylammonium (6899-10-1)
Cetyltrimethylammonium borohydride, hexadecyltrimethylammonium borohydride (19710-01-1)
Cetyltrimethylammonium bromide (57-09-0)
N-Cetyltrimethylammonium bromide (57-09-0)
Cetyltrimethylammonium cation (6899-10-1)
Cetyl trimethyl ammonium chloride (112-02-7)
Cetyltrimethylammonium chloride (112-02-7)
Cetyltrimethylammonium ion (6899-10-1)
Cetylureum (63-98-9)
Cevi-Bid (50-81-7)
Cevadene (62-59-9)
Cevadic acid (80-59-1)
Cevadilla (8051-02-3)
Cevadin (62-59-9)
Cevadine (62-59-9)
Cevadine (8051-02-3)
Cevalin (50-81-7)
Cevane-3-β,4-α,12,14,16-β,17,20-heptol, 4,9-epoxy-, 3-(3,4-dimethoxybenzoate) (71-62-5)
Cevane-3-β,4-β,12,14,16-β,17,20-heptol, 4,9-epoxy-, 3-((Z)-2-methylcrotonate), (Z) (62-59-9)
Cevatine (50-81-7)
Cevex (50-81-7)
Cevian 380 (9003-20-7)
Cevian A 678 (9003-20-7)
Cevian HL (9003-54-7)
Cevian N (9003-54-7)
Cevian NF (9003-54-7)
Cevimin (50-81-7)
Cevital (50-81-7)
Cevitamic acid (50-81-7)
Cevitamin (50-81-7)
Cevitan (50-81-7)
Cevitex (50-81-7)
Cewin (50-81-7)
Cextromaltose (69-79-4)
Ceylon (9002-18-0)
Ceylon Black Lead (7782-42-5)
Ceylon Isinglass (9002-18-0)
ChKhZ 18 (101-25-7)
ChKhZ 21 (123-77-3)
ChKhZ 21R (123-77-3)
ChS-RR2 (2425-79-8)
Chalcedony (14808-60-7)
Chalcone (94-41-7)
Chalk (1317-65-3)
Chalkone (94-41-7)
Chalothane (151-67-7)
Chaloxyd MEKP-HA 1 (1338-23-4)
Chaloxyd MEKP-LA 1 (1338-23-4)
Chameleon mineral (7722-64-7)
Chamigrene (18431-82-8)
β-Chamigrene (18431-82-8)
Channel Black (1333-86-4)
Channing's Solution (7783-33-7)
Chapco Cu-NAP (1338-02-9)
Charas (8063-14-7)
Charcoal (16291-96-6)
Charcoal Briquettes [NA 1361] (16291-96-6)
Charcoal Screenings, Made from "pinon" wood (DOT) (16291-96-6)
Charger E (9036-19-5)
Chavicol methyl ether (140-67-0)
Checkmate (74051-80-2)
Chee-o-gen (57-63-6)
Chee-o-genf (57-63-6)
Cheelox (60-00-4)
Cheelox BR-33 (64-02-8)
Cheelox BF (64-02-8)
Cheelox BF-12 (64-02-8)
Cheelox BF-13 (64-02-8)
Cheelox BF-78 (64-02-8)
Cheelox BF acid (60-00-4)
Cheelox NTA-14,-Na$_3$ (5064-31-3)
Chel 300 (139-13-9)
Chel DM Acid (150-39-0)
Cheladrate (139-33-3)
Chelafrin (51-43-4)
Chelaplex III (139-33-3)
Chelates of copper citrate (10402-15-0)
Chelates of copper gluconate (527-09-3)
Chelaton III (139-33-3)
Chelen (75-00-3)
Chelidamic acid (138-60-3)
Chelidonic acid (99-32-1)
Chelon 100 (64-02-8)
Chemagro 1,776 (78-48-8)
Chemagro 2353 (50-65-7)
Chemagro 9010 (114-26-1)
Chemagro 25141 (115-90-2)
Chemagro 37289 (327-98-0)
Chemagro B-1776 (150-50-5)
Chemagro B-1776 (78-48-8)
Chemagro B-1843 (1113-14-0)
Chemaid (124-65-2)
Chemal LA-4 (5274-68-0)
Chemal OA-2 (9004-98-2)
Chemal OA-4 (9004-98-2)
Chemal OA-5 (9004-98-2)
Chemal OA-10 (9004-98-2)
Chemal OA-20 (9004-98-2)

Chemal OA-23 (9004-98-2)
Chemanox 11 (128-37-0)
Chemanox 21 (119-47-1)
Chemathion (121-75-5)
Chemax DNP-150 (9014-93-1)
Chemax E-400-MS (9004-99-3)
Chemax E-400-MT (61791-00-2)
Chemax E-600-MS (9004-99-3)
Chemax E-1000-MS (9004-99-3)
Chemax NP-4 (7311-27-5)
Chemax NP Series (9016-45-9)
Chemax PEG-1000-DS (9005-08-7)
Chemax PEG-400-DS (9005-08-7)
Chemax TO-10 (61791-00-2)
Chem Bam (142-59-6)
Chembutazone (50-33-9)
Chemcarb (513-78-0)
Chemcolox 200 (64-02-8)
Chemcolox 240 Powder (64-02-8)
Chemcolox 340 (60-00-4)
Chemcolox 800 (139-89-9)
Chemcor (9002-88-4)
Chemeen C-2 (61791-14-8)
Chemeen C-5 (61791-14-8)
Chemeen C-10 (61791-14-8)
Chemeen C-15 (61791-14-8)
Chemeen C 12G (61791-14-8)
Chemester 300-OC (9004-96-0)
Chemetron 100 (110-30-5)
Chemetron Fire Shield (1309-64-4)
Chem Fish (83-79-4)
Chemform (123-33-1)
Chemform (134-62-3)
Chemform (57-92-1)
Chemform (72-43-5)
Chemform Brand Fixed Copper Fungicide (1332-03-2)
Chem-Hoe (122-42-9)
Chemiazid (54-85-3)
Chemical 109 (86-88-4)
Chemical Mace (532-27-4)
Chemical Oil (Coal) (65996-82-9)
Chemically neutralized kerosene (64742-31-0)
Chemically-neutralized light distillate (64742-31-0)
Chemically neutralized heavy naphthenic distillate (64742-34-3)
Chemicetin (56-75-7)
Chemicetina (56-75-7)
Chemi-Charl (10124-56-8)
Chemictive Brilliant Red 5b (17804-49-8)
Chemidon (54-85-3)
Chemifluor (7681-49-4)
Chemiochin (69-05-6)
Chemiofuran (67-20-9)
Chemipen (132-93-4)
Chemipen-C (132-93-4)
Chemithrene Brilliant Pink R (2379-74-0)
Chemithrene Brown BR (2475-33-4)
Chemitrim (8064-90-2)
Chemlon (105-60-2)
Chemlon (25038-54-4)
Chemlon 67/16 (63428-84-2)
Chemlon VPK (63428-84-2)
Chem-Mite (83-79-4)
Chem Neb (12427-38-2)
Chemochin (54-05-7)
Chemocide PK (94-13-3)
Chemocin (1332-40-7)
Chemofuran (59-87-0)
Chemosept (72-14-0)
Chemouag (127-69-5)
Chemox General (4097-36-3)
Chemox General (88-85-7)
Chemox P.E. (88-85-7)
Chemox PE (51-28-5)
Chemox Selective (6365-83-9)
Chempar (1332-40-7)
Chempar (1332-65-6)
Chem-Penta (87-86-5)
Chem Pels C (7784-46-5)
Chem-Phene (8001-35-2)
Chemplex 3006 (9002-88-4)
Chemquat 12-33 (112-00-5)
Chemquat 12-50 (112-00-5)
Chemquat 16-29 (112-02-7)
Chemquat 16-50 (112-02-7)
Chemrat (83-26-1)
Chem Rice (709-98-8)
Chemsect DNOC (534-52-1)
Chem-Sen 56 (7784-46-5)
Chem-Tol (87-86-5)
Chem zineb (12122-67-7)
Chendal (474-25-9)
Chendol (474-25-9)
Chenic acid (474-25-9)
Chenodeoxycholic acid (474-25-9)
Chenodesoxycholic acid (474-25-9)
Chenodesoxycholsaeure (German) (474-25-9)
Chenodiol (474-25-9)
Cheratina (Italian) (68238-35-7)
Cherry Red A Geigy (2302-96-7)
Cherts (14808-60-7)
Cheshunt Compound (12069-69-1)
Chestnut-Tannin (1401-55-4)
Chevron 6 (9003-29-6)
Chevron 12 (9003-29-6)
Chevron 16 (9003-29-6)
Chevron 18 (9003-29-6)
Chevron 100 (68602-80-2)
Chevron 9006 (10265-92-6)
Chevron Ortho 9006 (10265-92-6)
Chevron RE 12,420 (30560-19-1)
Chevron RE 5353 (2282-34-0)
Chevron acetone (67-64-1)
Chexmate (75-60-5)
Chip-Cal (7778-44-1)
Chip-Cal Granular (7778-44-1)
Chicago Acid (82-47-3)
Chicago Blue 6B (2610-05-1)
Chicago Sky Blue 6B (2610-05-1)
Chick Antidermatitis Factor (79-83-4)
Chiclida (569-65-3)
Chile Saltpeter (7631-99-4)
Chimassorb 944 (71878-19-8)
Chimipal AE 3 (9002-92-0)
China Green (569-64-2)
China Green (Biological stain) (569-64-2)
Chinacrin (69-05-6)
Chinacrin hydrochloride (69-05-6)
Chinaldine (91-63-4)
Chinalphos (13593-03-8)
Chinasaure (77-95-2)
China wood oil (8001-20-5)
Chinawood oil, glycerol, linseed oil, phthalic anhydride polymer (66071-18-9)
Chinese Blue (14038-43-8)
Chinese Isinglass (9002-18-0)

Chinese Red (18454-12-1)
Chinese White (1314-13-2)
Chinese seasoning (142-47-2)
Chinetazone (73-49-4)
Chinethazone (73-49-4)
Chinethazonum (73-49-4)
Chinetrin (52645-53-1)
Chingamin (54-05-7)
Chinhydron (Czech) (106-34-3)
Chinic acid (77-95-2)
Chinidin (German) (56-54-2)
Chinidin duriles (747-45-5)
Chinidine sulfate (50-54-4)
Chinidin hydrochlorid (German) (1668-99-1)
Chinidin vufb (747-45-5)
Chinimetten (130-89-2)
Chinin (German) (130-95-0)
Chinin hydrobromid (German) (549-49-5)
Chinizarin (81-64-1)
Chinofer (9004-66-4)
Chinoform (130-26-7)
Chinogelb Extra (German) (8004-92-0)
Chinogelb (German) (8004-92-0)
Chinogelb wasserloeslich (German) (8004-92-0)
Chinoin (13593-03-8)
Chinoin (389-08-2)
Chinoleine (91-22-5)
Chinolin (Czech) (91-22-5)
Chinoline (91-22-5)
Chinoline Yellow D Sol. In spirits (8003-22-3)
Chinoline Yellow ZSS (8003-22-3)
8-Chinolinol (Czech) (148-24-3)
Chinomethionat (2439-01-2)
Chinomethionate (2439-01-2)
Chinon (Dutch, German) (106-51-4)
p-Chinon (German) (106-51-4)
Chinone (106-51-4)
p-Chinonmonoxim (Czech) (104-91-6)
Chinonoxim-benzoylhydrazon (German) (495-73-8)
Chinonoxime-benzoylhydrazone (495-73-8)
Chinorta (311-45-5)
Chinosol (134-31-6)
Chinothionat (93-75-4)
Chinoxalin-2,3-dithiol-cyclo-thio-carbonat (German) (93-75-4)
Chinoxalin-2,3-diyl-trithiocarbonat (German) (93-75-4)
3-(2-Chinoxalinylmethylen-1,4-dioxid)methylkarbazat (Czech) (6804-07-5)
Chinoxidin (10103-89-6)
Chipco 26019 (36734-19-7)
Chipco Buctril (1689-84-5)
Chipco Crab Kleen (144-21-8)
Chipco Crab-Kleen (1689-84-5)
Chipco Turf Herbicide "D" (94-75-7)
Chipco Turf Herbicide MCPP (93-65-2)
Chipcote (2597-97-9)
Chipcote 75 (2597-97-9)
Chipco thiram 75 (137-26-8)
Chipman 11974 (2310-17-0)
Chipman 3,142 (117-18-0)
Chipman 6199 (3734-97-2)
Chipman 6200 (78-53-5)
Chipman R-6,199 (3734-97-2)
Chiptox (94-74-6)
Chisso Polypro 1014 (9003-07-0)
Chisso 507b (9003-07-0)
Chissonox 201 (141-37-7)
Chissonox 206 (106-87-6)
Chitin (1398-61-4)
β-Chitin (1398-61-4)
Chitina (Italian) (1398-61-4)
Chitosamine (3416-24-8)
Chixin (59-87-0)
Chkhz 9 (80-17-1)
Chlo-Amine (25523-97-1)
Chloditan (53-19-0)
Chlodithane (53-19-0)
Chlofenvinphos (470-90-6)
Chloflurecol-methyl (2536-31-4)
Chloflurecol-methyl ester (2536-31-4)
Chlomaphene (911-45-5)
Chlomethoxyfen (32861-85-1)
Chlomethoxynil (32861-85-1)
Chlomin (56-75-7)
Chlomizole (4897-31-8)
Chlomycol (56-75-7)
Chloor (Dutch) (7782-50-5)
3-Chlooranilinen (Dutch) (108-42-9)
2-Chloorbenzaldehyde (Dutch) (89-98-5)
o-Chloorbenzaldehyde (Dutch) (89-98-5)
Chloorbenzeen (Dutch) (108-90-7)
Chloorbenzide (Dutch) (103-17-3)
(4-Chloor-benzyl)-(4-chloor-fenyl)-sulfide (Dutch) (103-17-3)
2-Chloor-1,3-butadieen (Dutch) (126-99-8)
(4-Chloor-but-2-yn-yl)-N-(3-chloor-fenyl)-carbamaat (Dutch) (101-27-9)
Chloordaan (Dutch) (57-74-9)
O-2-Chloor-1-(2,4-dichloor-fenyl)-vinyl-O,O-diethylfosfaat (Dutch) (470-90-6)
(2-Chloor-3-diethylamino-1-methyl-3-oxo-prop-1-en-yl)-dimethyl-fosfaat (Dutch) (13171-21-6)
2-Chloor-4-dimethylamino-6-methyl-pyrimidine (dutch) (535-89-7)
1-Chloor-2,4-dinitrobenzeen (Dutch) (97-00-7)
1-Chloor-2,3-epoxy-propaan (Dutch) (106-89-8)
Chloorethaan (Dutch) (75-00-3)
2-Chloorethanol (Dutch) (107-07-3)
Chloorfacinon (Dutch) (3691-35-8)
3-(4-(4-Chloor-fenoxy)-fenyl)-1,1-dimethylureum (Dutch) (1982-47-4)
Chloorfenson (Dutch) (80-33-1)
(4-Chloor-fenyl)-benzeen-sulfonaat (Dutch) (80-38-6)
(4-Chloor-fenyl)-4-chloor-benzeen-sulfonaat (Dutch) (80-33-1)
3-(4-Chloor-fenyl)-1,1-dimethylureum (Dutch) (150-68-5)
2(2-(4-Chloor-fenyl-2-fenyl)-acetyl)-indaan-1,3-dion (Dutch) (3691-35-8)
N-(3-Chloor-fenyl)-isopropyl carbamaat (Dutch) (101-21-3)
3-(1-(4-Chloorfenyl)-3-oxo-butyl)-4-hydroxy-cumarine (Dutch) (81-82-3)
Chloor-methaan (Dutch) (74-87-3)
2-(4-Chloor-2-methyl-fenoxy)-propionzuur (Dutch) (93-65-2)
1-Chloor-4-nitrobenzeen (Dutch) (100-00-5)
O-(3-Chloor-4-nitro-fenyl)-O,O-dimethyl-monothiofosfaat (Dutch) (500-28-7)
O-(4-Chloor-3-nitro-fenyl)-O,O-dimethylmonothiofosfaat (Dutch) (2463-84-5)
Chloorpikrine (Dutch) (76-06-2)
Chloorthion (Dutch) (500-28-7)
Chloorwaterstof (Dutch) (7647-01-0)
Chlophen (1336-36-3)
β-Chlor (2218-68-0)
Chlor-Ethamine (333-18-6)
Chlor (German) (7782-50-5)
Chlor-IFC (101-21-3)
Chlor-IPC (101-21-3)
Chlor Kil (57-74-9)
Chlor-PZ (50-53-3)
Chlor-Trimeton (113-92-8)
Chlor-Trimeton (132-22-9)
Chlor-Trimeton Maleate (113-92-8)
Chlor-Tripolon (113-92-8)
Chlor-Tripolon (132-22-9)
Chloracetamid (German) (79-07-2)
Chloracetamide (79-07-2)

Name	CAS
Chloracetamide-N-metholol	(2832-19-1)
Chloracetic acid	(79-11-8)
Chloracetic anhydride	(541-88-8)
Chloracetone (French)	(78-95-5)
Chloracetonitrile	(107-14-2)
Chloracetophenone	(1341-24-8)
Chloracetophenone	(532-27-4)
Chloracetyl chloride	(79-04-9)
Chloractil	(69-09-0)
2-Chloraethanol (German)	(107-07-3)
N-(2-Chloraethyl)-N'-(2-chloraethyl)-N',O-propylen-phosphorsaureester-diamid (German)	(3778-73-2)
α-Chlor-6'-aethyl-n-(2-methoxy-1-methylaethyl)-acet-o-toluidin (German)	(51218-45-2)
4-Chlor-2-aethylphenol (German)	(18979-90-3)
2-Chloraethyl-phosphonsaeure (German)	(16672-87-0)
2-Chloraethyl-trimethylammoniumchlorid (German)	(999-81-5)
Chlorak	(52-68-6)
Chloral	(75-87-6)
Chloral, Anhydrous, Inhibited [UN 2075]	(75-87-6)
Chloraldehyde	(79-02-7)
Chloraldurat	(302-17-0)
Chloral-hydrate	(302-17-0)
Chloral betaine	(2218-68-0)
Chlorallyl diethyldithiocarbamate	(95-06-7)
Chlorallylene	(107-05-1)
Chloralone	(127-65-1)
Chloralosane	(15879-93-3)
Chloralose	(15879-93-3)
Chloralose, α	(15879-93-3)
α-Chloralose	(15879-93-3)
Chlorambed	(133-90-4)
Chloramben	(133-90-4)
Chloramben ammonium salt	(1076-46-6)
Chlorambene	(133-90-4)
Chloramben, methyl ester	(7286-84-2)
Chlorambucil	(305-03-3)
Chlorameisensaeure methylester (German)	(79-22-1)
Chloramex	(56-75-7)
Chloramfenikol (Czech)	(56-75-7)
Chloramficin	(56-75-7)
Chloramfilin	(56-75-7)
Chloramiblau 3B	(72-57-1)
Chloramide	(10599-90-3)
Chloramifene	(911-45-5)
Chloramin	(55-86-7)
Chloramin B	(127-52-6)
Chloramine	(10599-90-3)
Chloramine	(55-86-7)
Chloramine B	(127-52-6)
Chloramine-B	(127-52-6)
Chloramine Black BH	(2429-73-4)
Chloramine Black C	(1937-37-7)
Chloramine Black E2B	(2429-83-6)
Chloramine Black EC	(1937-37-7)
Chloramine Black ERT	(1937-37-7)
Chloramine Black EX	(1937-37-7)
Chloramine Black EXR	(1937-37-7)
Chloramine Black XO	(1937-37-7)
Chloramine Blue	(72-57-1)
Chloramine Blue 2B	(2602-46-2)
Chloramine Blue 3B	(72-57-1)
Chloramine Brilliant Red 8B	(6548-29-4)
Chloramine Brown M	(2429-82-5)
Chloramine Brown 2ME	(2429-82-5)
Chloramine Carbon Black S	(1937-37-7)
Chloramine Carbon Black SJ	(1937-37-7)
Chloramine Carbon Black SN	(1937-37-7)
Chloramine Fast Brown BRL	(16071-86-6)
Chloramine Fast Brown BRLL	(16071-86-6)
Chloramine Fast Cutch Brown PL	(16071-86-6)
Chloramine Fast Red 5BL	(2610-11-9)
Chloramine Fast Red K	(2610-11-9)
Chloramine (Inorganic compound)	(10599-90-3)
Chloramine Red 3B	(6358-29-8)
Chloramine Red 8B	(6548-29-4)
Chloramine Sky Blue A	(2429-74-5)
Chloramine Sky Blue 4B	(2429-74-5)
Chloramine Sky Blue FF	(2610-05-1)
Chloramine T	(127-65-1)
Chloramin hydrochloride	(55-86-7)
1-Chlor-5-aminoanthrachinon (Czech)	(117-11-3)
Chloraminophen	(305-03-3)
Chloraminophene	(305-03-3)
Chloramiphene	(50-41-9)
Chloramiphene	(911-45-5)
Chloramiphene citrate	(50-41-9)
Chloramizol	(35554-44-0)
Chloramp	(2545-60-0)
Chloramp (Russian)	(1918-02-1)
Chloramphenicol	(56-75-7)
D-(-)-threo-Chloramphenicol	(56-75-7)
D-Chloramphenicol	(56-75-7)
D-threo-Chloramphenicol	(56-75-7)
Chloramphenicol monosuccinate sodium salt	(982-57-0)
Chloramphenicol sodium monosuccinate	(982-57-0)
Chloramphenicol sodium succinate	(982-57-0)
Chloramphenicol succinate sodium	(982-57-0)
Chloramphenicol-sukzinat-natrium (German)	(982-57-0)
Chloramsaar	(56-75-7)
Chloran 542	(115-27-5)
Chloranautine	(523-87-5)
Chloraniformethan	(20856-57-9)
Chloraniformethane	(20856-57-9)
2-Chloranil	(2435-53-2)
Chloranil	(118-75-2)
o-Chloranil	(2435-53-2)
Chloranilic acid	(87-88-7)
4-Chloranilid kyseliny 2,2-dimethylvalerove (Czech)	(7287-36-7)
4-Chloranilin (Czech)	(106-47-8)
2-(2-Chloranilin)-4,6-dichlor-1,3,5-triazin (German)	(101-05-3)
m-Chloraniline	(108-42-9)
o-Chloraniline	(95-51-2)
p-Chloraniline	(106-47-8)
Chloranocryl	(2164-09-2)
1-Chloranthrachinon (Czech)	(82-44-0)
Chlorantine Fast Blue B5GL	(2610-05-1)
Chlorantine Fast Green BLL	(6388-26-7)
Chlorantine Fast Red	(2610-11-9)
Chlorantine Fast Red 5B (6CI)	(2610-11-9)
Chlorantine Fast Yellow 7GL	(10190-68-8)
Chloraquine	(54-05-7)
Chlorasan	(127-65-1)
Chloraseptine	(127-65-1)
Chlorasol	(56-75-7)
Chlora-tabs	(56-75-7)
Chlorate d'ammonium (French)	(10192-29-7)
Chlorate de calcium (French)	(10137-74-3)
Chlorate de potassium (French)	(3811-04-9)
Chlorate of potash (DOT)	(3811-04-9)
Chlorate of soda (DOT)	(7775-09-9)
Chlorate salt of magnesium	(10326-21-3)
Chlorate salt of sodium	(7775-09-9)
Chlorax	(7775-09-9)

Name	CAS
Chlorazan	**(127-65-1)**
Chlorazanil	**(500-42-5)**
Chlorazene	**(127-65-1)**
Chlorazin	**(69-09-0)**
Chlorazine	**(580-48-3)**
Chlorazinil	**(500-42-5)**
Chlorazodin	**(502-98-7)**
Chlorazodine (French)	**(502-98-7)**
Chlorazodinum (Latin)	**(502-98-7)**
Chlorazol Black BH	**(2429-73-4)**
Chlorazol Black E	**(1937-37-7)**
Chlorazol Black EA	**(1937-37-7)**
Chlorazol Black E (Biological stain)	**(1937-37-7)**
Chlorazol Black EN	**(1937-37-7)**
Chlorazol Black LF	**(2429-83-6)**
Chlorazol Blue 3B	**(72-57-1)**
Chlorazol Blue B	**(2602-46-2)**
Chlorazol Blue BP	**(2602-46-2)**
Chlorazol Brilliant Purpurine 8B	**(6548-29-4)**
Chlorazol Brown LF	**(2429-81-4)**
Chlorazol Brown M	**(2429-82-5)**
Chlorazol Burl Black E	**(1937-37-7)**
Chlorazol Dark Green PL	**(3626-28-6)**
Chlorazol Leather Black BH	**(2429-73-4)**
Chlorazol Leather Black ENP	**(1937-37-7)**
Chlorazol Orange Brown X	**(2586-58-5)**
Chlorazol Paper Yellow R	**(1325-37-7)**
Chlorazol Silk Black G	**(1937-37-7)**
Chlorazol Sky Blue FF	**(2610-05-1)**
Chlorazol Sky Blue ff	**(314-13-6)**
Chlorazol Steel Blue 6B	**(7082-31-7)**
Chlorazol Violet N	**(2586-60-9)**
Chlorazol Violet WB	**(6426-67-1)**
Chlorazol Violet WBS	**(6426-67-1)**
Chlorazol Viscose Black B	**(6428-31-5)**
Chlorazone	**(127-65-1)**
Chlorbensid (German)	**(103-17-3)**
Chlorbenside	**(103-17-3)**
Chlorbenxide	**(103-17-3)**
2-Chlorbenzaldehyd (German)	**(89-98-5)**
Chlorbenzene	**(108-90-7)**
p-Chlorbenzensulfochlorid (Czech)	**(98-60-2)**
p-Chlorbenzensulfonan sodny (Czech)	**(5138-90-9)**
Chlorbenzide	**(103-17-3)**
Chlorbenzilat	**(510-15-6)**
Chlorbenzilate	**(510-15-6)**
p-Chlorbenzoic acid	**(74-11-3)**
Chlorbenzol	**(108-90-7)**
o-Chlorbenzonitril (Czech)	**(873-32-5)**
N-p-Chlorbenzoyl-5-methoxy-2-methylindole-3-acetic acid	**(53-86-1)**
Chlorbenzylate	**(510-15-6)**
(4-Chlor-benzyl)-(4-chlor-phenyl)-sulfid (German)	**(103-17-3)**
4-Chlor-benzyl-cyanid (German)	**(140-53-4)**
Chlorbicyclene (French)	**(50-13-5)**
Chlorbisan	**(4418-66-0)**
Chlorbromuron	**(13360-45-7)**
Chlorbufam	**(1967-16-4)**
Chlorbufame	**(1967-16-4)**
Chlorbufan mixed with cyceuron	**(8015-55-2)**
Chlorbupham	**(1967-16-4)**
2-Chlor-1,3-butadien (German)	**(126-99-8)**
Chlorbutanol	**(57-15-8)**
4-Chlorbutan-1-ol (German)	**(928-51-8)**
(4-Chlor-but-2-in-yl)-N-(3-chlor-phenyl)-carbamat (German)	**(101-27-9)**
Chlorbutol	**(57-15-8)**
Chlorbycyclen	**(50-13-5)**
Chlorcholinchlorid (Czech,German)	**(999-81-5)**
Chlorcholine chloride	**(999-81-5)**
Chlorcosane	**(63449-39-8)**
p-Chlor-m-cresol	**(59-50-7)**
Chlorcyan	**(506-77-4)**
Chlorcycline	**(82-93-9)**
Chlorcyclizine	**(82-93-9)**
Chlorcyclizine hydrochloride	**(14362-31-3)**
Chlorcyclizinium chloride	**(14362-31-3)**
Chlorcylizine	**(14362-31-3)**
Chlordan	**(57-74-9)**
α-Chlordan	**(5103-71-9)**
β-Chlordan	**(5103-74-2)**
cis-Chlordan	**(5103-71-9)**
γ-Chlordan	**(5566-34-7)**
γ-Chlordan	**(57-74-9)**
trans-Chlordan	**(5103-74-2)**
trans-Chlordan	**(5566-34-7)**
Chlordane	**(12789-03-6)**
α(cis)-Chlordane	**(5103-71-9)**
α-Chlordane	**(5103-71-9)**
β-Chlordane	**(5103-74-2)**
cis-Chlordane	**(5103-71-9)**
γ(trans)-Chlordane	**(5566-34-7)**
trans-Chlordane	**(5103-74-2)**
Chlordane (ACGIH,OSHA)	**(57-74-9)**
Chlordane, Liquid (DOT)	**(57-74-9)**
Chlordane, Technical	**(12789-03-6)**
Chlordecone	**(143-50-0)**
Chlordecone alcohol	**(1034-41-9)**
Chlordene	**(3734-48-3)**
Chlordesmethyldiazepam	**(2894-67-9)**
(2-Chlor-3-diaethylamino-1-methyl-3-oxo-prop-1-en-yl)-dimethyl-phosphat (German)	**(13171-21-6)**
Chlordiazepoxide	**(58-25-3)**
O-2-Chlor-1-(2,4-dichlor-phenyl)-vinyl-O,O-diaethylphosphat (German)	**(470-90-6)**
O-2-Chlor-1-(2,5-dichlorfenyl)vinyl-O,O-diethylthiofosfat (Czech)	**(1757-18-2)**
Chlordiethylsulfid (Czech)	**(693-07-2)**
Chlor-difenylarsin (Czech)	**(712-48-1)**
4-Chlordifenylsulfon (Czech)	**(80-00-2)**
Chlordimeform	**(6164-98-3)**
Chlordimeform hydrochloride	**(19750-95-9)**
2-Chlor-4-dimethylamino-6-methylpyrimidin (German)	**(535-89-7)**
Chlordimethylether (Czech)	**(107-30-2)**
1-Chlor-2,4-dinitrobenzene	**(97-00-7)**
Chlore (French)	**(7782-50-5)**
Chlorefenizon (French)	**(80-33-1)**
Chlorendic acid	**(115-28-6)**
Chlorendic anhydride	**(115-27-5)**
Chlorene	**(75-00-3)**
Chlorepin	**(22316-47-8)**
1-Chlor-2,3-epoxy-propan (German)	**(106-89-8)**
Chloresene	**(58-89-9)**
Chloressigsaeure-N-isobutinylanilid (German)	**(21267-72-1)**
Chloressigsaeure-N-isopropylanilid (German)	**(1918-16-7)**
Chloressigsaeure-N-(methoxymethyl)-2,6-diaethylanilid (German)	**(15972-60-8)**
Chlorestrolo	**(569-57-3)**
Chlorethaminacil	**(66-75-1)**
Chlorethamine	**(55-86-7)**
2-Chlorethanol (German)	**(107-07-3)**
Chlorethazine	**(51-75-2)**
Chlorethazine	**(55-86-7)**
Chlorethene	**(75-01-4)**
Chlorethephon	**(16672-87-0)**
Chlorethiazol	**(533-45-9)**
Chlorethiazole	**(533-45-9)**
2-(2-Chlorethoxy)ethyl 2'-chlorethyl ether	**(112-26-5)**

Chlorethyl (75-00-3)
2-Chlorethylacetat (German) (542-58-5)
Chlorethylene (75-01-4)
2-Chlorethylester kyseliny akrylove (Czech) (2206-89-5)
2-Chlorethylester kyseliny chlormravenci (Czech) (627-11-2)
2-Chlorethyl vinyl ether (110-75-8)
Chloretin (21267-72-1)
Chloretone (57-15-8)
Chlorex (111-44-4)
Chlorextol (1336-36-3)
Chlorez 700 (63449-39-8)
Chlorez 700HMP (63449-39-8)
Chlorfacinon (German) (3691-35-8)
Chlorfenac (85-34-7)
Chlorfenac-sodium (2439-00-1)
Chlorfenamidine (6164-98-3)
Chlorfenethol (80-06-8)
Chlorfenidim (150-68-5)
p-Chlorfenol (Czech) (106-48-9)
Chlorfenprop-methyl (14437-17-3)
Chlorfenson (80-33-1)
Chlorfenvinfos (470-90-6)
Chlorfenvinphos (470-90-6)
Chlorfenwinfosem (Polish) (470-90-6)
p-Chlorfenylamid kyseliny mravenci (Czech) (2617-79-0)
1-p-Chlorfenyl-3,3-dimethyltriazen (Czech) (7203-90-9)
2-Chlor-p-fenylendiamin (Czech) (615-66-7)
3-Chlor-p-fenylendiamin (Czech) (615-66-7)
p-Chlor-m-fenylendiamin (Czech) (5131-60-2)
p-Chlorfenylester kyseliny p-chlorbenzensulfonove (Czech) (80-33-1)
p-Chlorfenylester kyseliny octove (Czech) (876-27-7)
m-Chlorfenylisokyanat (Czech) (2909-38-8)
p-Chlorfenylisokyanat (Czech) (104-12-1)
Chlorflurazole (3615-21-2)
Chlorflurecol (2464-37-1)
Chlorflurecol (2536-31-4)
Chlorflurenol (2464-37-1)
Chlorflurenol (2536-31-4)
Chlorflurenol-methyl ester (2536-31-4)
Chlorfonium (115-78-6)
Chlorfos (52-68-6)
Chlor-N-(2-furylmethyl)-5-sulfamylanthranilsaeure (German) (54-31-9)
Chlorguanide triazine (516-21-2)
Chlorhexadol (3563-58-4)
Chlorhexane (544-10-5)
Chlorhexidin (Czech) (55-56-1)
Chlorhexidine (55-56-1)
Chlorhexidine acetate (VAN) (56-95-1)
Chlorhexidine diacetate (56-95-1)
Chlorhexidine digluconate (18472-51-0)
Chlorhexidine dihydrochloride (3697-42-5)
Chlorhexidine gluconate (18472-51-0)
Chlorhexidine hydrochloride (3697-42-5)
Chlorhexidin glukonatu (Czech) (18472-51-0)
Chlorhydrate d'aniline (French) (142-04-1)
Chlorhydrate de 4-chloroorthotoluidine (French) (3165-93-3)
Chlorhydrate de 1-methyl 2 p (isopropylcarbamoyl)benzyl-hydrazine (French) (366-70-1)
Chlorhydrate de nicotine (French) (2820-51-1)
Chlorhydrate de papaverine (French) (61-25-6)
Chlorhydrate de tetracycline (French) (64-75-5)
Chlorhydrin (96-24-2)
α-Chlorhydrin (96-24-2)
Chlorhydrol (12042-91-0)
Chlorhydrol, Granular (12042-91-0)
Chlorhydrol, Impalpable (12042-91-0)
Chlorhydroxyaluminum allantoinate (1317-25-5)
3-Chlor-4-hydroxybifenyl (Czech) (92-04-6)
2-Chlor-9-hydroxyfluoren-carbonsaeure-(9)-methylester (German) (2536-31-4)
Chloriazid (58-94-6)
Chloric acid (DOT) (7790-93-4)
Chloric acid, aluminum salt (9CI) (15477-33-5)
Chloric acid, ammonium salt (10192-29-7)
Chloric acid, barium salt (13477-00-4)
Chloric acid, barium salt (Wet) (13477-00-4)
Chloric acid, calcium salt (10137-74-3)
Chloric acid, magnesium salt (10326-21-3)
Chloric acid, potassium salt (3811-04-9)
Chloric acid, sodium salt (7775-09-9)
Chloric acid solution, Containing not more than 10% acid [UN 2626] (7790-93-4)
Chloric acid, strontium salt (7791-10-8)
Chloricol (56-75-7)
Chlorid amonny (Czech) (12125-02-9)
Chlorid anilinu (Czech) (142-04-1)
Chlorid antimonity (Czech) (10025-91-9)
Chloridazon (1698-60-8)
Chloridazone (1698-60-8)
Chlorid n-butylcinicity (Czech) (1118-46-3)
Chlorid di-n-butylcinicity (Czech) (683-18-1)
Chlorid draselny (Czech) (7447-40-7)
Chloride (8CI,9CI) (16887-00-6)
Chloride de choline (French) (67-48-1)
Chloride of Lime (DOT) (7778-54-3)
Chloride of phosphorus (DOT) (7719-12-2)
Chloride of sulfur (DOT) (10545-99-0)
Chloride of sulfur [UN 1828] (10025-67-9)
Chlorid fenylrtutnaty (Czech) (100-56-1)
Chloridiazepide (58-25-3)
Chloridiazepoxide (58-25-3)
Chloridin (58-14-0)
Chloridine (58-14-0)
Chlorid kremicity (Czech) (10026-04-7)
Chlorid kyseliny akrylove (Czech) (814-68-6)
Chlorid kyseliny p-chlorbensulfonove (Czech) (98-60-2)
Chlorid kyseliny chloroctove (Czech) (79-04-9)
Chlorid kyseliny dichloroctove (Czech) (79-36-7)
Chlorid kyseliny dimethylkarbaminove (Czech) (79-44-7)
Chlorid kyseliny fluoroctove (Czech) (359-06-8)
Chlorid kyseliny m-nitrobenzoove (Czech) (121-90-4)
Chlorid medny (Czech) (7758-89-6)
Chlorid rtutnaty (Czech) (7487-94-7)
Chlorid tri-n-butylcinicity (Czech) (1461-22-9)
Chlorid uhlicity (Czech) (56-23-5)
Chloridum (75-00-3)
Chloridyn (58-14-0)
Chlorierte biphenyle, Chlorgehalt 42% (German) (53469-21-9)
Chlorierte biphenyle, Chlorgehalt 54% (German) (11097-69-1)
Chlorilen (79-01-6)
Chlorimide (8CI,9CI) (3400-09-7)
Chlorimipramine (303-49-1)
Chlorinat (101-27-9)
Chlorinated biphenyl (1336-36-3)
Chlorinated camphene 60% (ACGIH) (8001-35-2)
Chlorinated camphene (OSHA) (8001-35-2)
Chlorinated diphenyl (1336-36-3)
Chlorinated diphenylene (1336-36-3)
Chlorinated diphenyl oxide (55720-99-5)
Chlorinated diphenyl oxide (ACGIH,OSHA) (31242-93-0)
Chlorinated hydrochloric ether (75-34-3)
Chlorinated lime (DOT) (7778-54-3)
Chlorinated natural rubber (9006-03-5)
Chlorinated paraffins (106232-85-3)
Chlorinated paraffins (106232-86-4)

Chlorinated paraffins **(61788-76-9)**
Chlorinated paraffins **(63449-39-8)**
Chlorinated paraffins **(68920-70-7)**
Chlorinated paraffins **(71011-12-6)**
Chlorinated paraffins **(84082-38-2)**
Chlorinated paraffins **(84776-06-7)**
Chlorinated paraffins **(84776-07-8)**
Chlorinated paraffins **(85049-26-9)**
Chlorinated paraffins **(85535-84-8)**
Chlorinated paraffins **(85535-85-9)**
Chlorinated paraffins **(85535-86-0)**
Chlorinated paraffins **(85536-22-7)**
Chlorinated paraffins **(85681-73-8)**
Chlorinated paraffins **(97553-43-0)**
Chlorinated paraffins **(97659-46-6)**
Chlorinated n-paraffins (C6-C18) **(68920-70-7)**
Chlorinated paraffins (C12, 60% chlorine) **(63449-39-8)**
Chlorinated paraffins (C23, 43% chlorine) **(63449-39-8)**
Chlorinated rubber **(9006-03-5)**
Chlorinated trisodium phosphate **(56802-99-4)**
Chlorinated-trisodium-phosphate **(11084-85-8)**
Chlorindan **(57-74-9)**
Chlorindanol **(145-94-8)**
Chlorine (ACGIH,OSHA) [UN 1017] **(7782-50-5)**
Chlorine(IV) oxide **(10049-04-4)**
Chlorine Mol. **(7782-50-5)**
Chlorine azide **(13973-88-1)**
Chlorine control **(7772-98-7)**
Chlorine cure **(7772-98-7)**
Chlorine cyanide **(506-77-4)**
Chlorine dioxide **(10049-04-4)**
Chlorine dioxide (ACGIH,OSHA) **(10049-04-4)**
Chlorine dioxide, Not hydrated (DOT) **(10049-04-4)**
Chlorine dioxide hydrate, Frozen [NA 9191] **(10049-04-4)**
Chlorine fluoride **(7790-91-2)**
Chlorine fluoride oxide **(7616-94-6)**
Chlorine, ion (Cl(1-)) **(16887-00-6)**
Chlorine monoxide **(7791-21-1)**
Chlorine nitride **(10025-85-1)**
Chlorine-oxide **(10049-04-4)**
Chlorine oxide (9CI) **(7791-21-1)**
Chlorine oxyfluoride **(7616-94-6)**
Chlorine peroxide **(10049-04-4)**
Chlorine sulfide **(10545-99-0)**
Chlorine trifluoride (ACGIH,OSHA) [UN 1749] **(7790-91-2)**
Chlorite d'argent (French) **(7783-91-7)**
Chlorite de calcium (French) **(14674-72-7)**
Chlorite de sodium (French) **(7758-19-2)**
5-Chlor-7-jod-8-hydroxy-chinolin (German) **(130-26-7)**
Chlorku litu (Polish) **(7447-41-8)**
Chlormadinon acetate **(302-22-7)**
Chlormadinone acetate **(302-22-7)**
Chlormadinonu (Polish) **(302-22-7)**
Chlormefos **(24934-91-6)**
Chlormene **(113-92-8)**
Chlormephos **(24934-91-6)**
Chlormequat **(999-81-5)**
Chlormequat chloride **(999-81-5)**
Chlormerodrin **(62-37-3)**
Chlormerodrine **(62-37-3)**
Chlormeroprin **(62-37-3)**
Chlor-methan (German) **(74-87-3)**
Chlormethazanone **(80-77-3)**
Chlormethazone **(80-77-3)**
Chlormethiazol **(533-45-9)**
Chlormethiazole **(533-45-9)**
Chlormethine **(51-75-2)**
Chlormethine hydrochloride **(55-86-7)**
Chlormethine-N-oxide hydrochloride **(302-70-5)**
Chlormethinum **(55-86-7)**
3-(3-Chlor-4-methoxyphenyl)-1,1-dimethylharnstoff (German) **(19937-59-8)**
N'-(3-Chlor-4-methoxy-phenyl)-N,N-dimethylharnstoff (German) **(19937-59-8)**
α-(Chlormethyl)-2-methyl-5-nitro-imidazol-1-aethanol (German) **(16773-42-5)**
1-(Chlormethyl)naftalen (Czech) **(86-52-2)**
4-(4-Chlor-2-methyl-phenoxy)-buttersaeure (German) **(94-81-5)**
4-(4-Chlor-2-methylphenoxy)-buttersaeure (German) **(94-81-5)**
4-(4-Chlor-2-methyl-phenoxy)-buttersaeure natriumsalz (German) **(6062-26-6)**
2-(4-Chlor-2-methyl-phenoxy)-propionsaeure (German) **(93-65-2)**
3-(3-Chlor-4-methylphenyl)-1,1-dimethylharnstoff (German) **(15545-48-9)**
N-(3-Chlor-methylphenyl)-2-methylpentanamid (German) **(2307-68-8)**
3-Chlor-2-methyl-prop-1-en (German) **(563-47-3)**
Chlormezanone **(80-77-3)**
Chlormite **(5836-10-2)**
Chlormithiazole **(533-45-9)**
1-Chlornaftalen (Czech) **(90-13-1)**
2-Chlornaftalen (Czech) **(91-58-7)**
Chlornaftina **(494-03-1)**
Chlornaphazin **(494-03-1)**
Chlornaphazine **(494-03-1)**
α-Chlornaphthalene **(90-13-1)**
Chlornaphthin **(494-03-1)**
Chlornidine **(26389-78-6)**
1-Chlor-5-nitroanthrachinon (Czech) **(129-40-8)**
1-Chlor-4-nitrobenzol (German) **(100-00-5)**
Chlornitrofen **(1836-77-7)**
Chlornitromycin **(56-75-7)**
N-(2'-Chlor-4'-nitrophenyl)-5-chlorsalicylamid (German) **(50-65-7)**
O-(3-Chlor-4-nitro-phenyl)-O,O-dimethyl-monothiophosphat (German) **(500-28-7)**
O-(4-Chlor-3-nitro-phenyl)-O,O-dimethyl-monothiophosphat (German) **(2463-84-5)**
Chloro-IFK **(101-21-3)**
Chloro-IPC **(101-21-3)**
Chloro-PDMT **(7203-90-9)**
Chlor-o-Pic **(76-06-2)**
Chloro-S.C.T.Z. **(533-45-9)**
2-Chloroacetaldehyde **(107-20-0)**
Chloroacetaldehyde (ACGIH,OSHA) [UN 2232] **(107-20-0)**
2-Chloroacetaldehyde, dimethyl acetal **(97-97-2)**
Chloroacetaldehyde, dimethyl acetal **(97-97-2)**
Chloroacetaldehyde ethylene acetal **(2568-30-1)**
Chloroacetaldehyde monomer **(107-20-0)**
2-Chloroacetamide **(79-07-2)**
Chloroacetamide **(79-07-2)**
α-Chloroacetamide **(79-07-2)**
2'-Chloroacetanilide **(533-17-5)**
2-Chloroacetanilide **(587-65-5)**
4'-Chloroacetanilide **(539-03-7)**
4-Chloroacetanilide **(539-03-7)**
Chloroacetanilide **(587-65-5)**
α-Chloroacetanilide **(587-65-5)**
m-Chloroacetanilide **(588-07-8)**
o-Chloroacetanilide **(533-17-5)**
Chloroacetic acid **(79-11-8)**
α-Chloroacetic acid **(79-11-8)**
Chloroacetic acid, Liquid [UN 1750] **(79-11-8)**
Chloroacetic acid, Solid [UN 1751] **(79-11-8)**
Chloroacetic acid, Solution [UN 1750] **(79-11-8)**
Chloroacetic acid anhydride **(541-88-8)**
Chloroacetic acid, benzyl ester **(140-18-1)**
Chloroacetic acid chloride **(79-04-9)**
Chloroacetic acid, ethyl ester **(105-39-5)**
Chloroacetic acid isopropyl ester **(105-48-6)**
Chloroacetic acid methyl ester **(96-34-4)**

Chloroacetic acid sodium salt (3926-62-3)
2-Chloroacetic anhydride (541-88-8)
Chloroacetic anhydride (541-88-8)
Chloroacetic chloride (79-04-9)
2'-Chloroacetoacetanilide (93-70-9)
4'-Chloroacetoacetanilide (101-92-8)
o-Chloroacetoacetanilide (93-70-9)
p-Chloroacetoacetanilide (101-92-8)
Chloroacetoguanamine (10581-62-1)
Chloroacetone (78-95-5)
Chloroacetone, Stabilized [UN 1695] (78-95-5)
2-Chloroacetonitrile (107-14-2)
α-Chloroacetonitrile (107-14-2)
Chloroacetonitrile [UN 2668] (107-14-2)
1-Chloroacetophenone (532-27-4)
2-Chloroacetophenone (1341-24-8)
2-Chloroacetophenone (532-27-4)
4-Chloroacetophenone (99-91-2)
Chloroacetophenone (1341-24-8)
ω-Chloroacetophenone (532-27-4)
p-Chloroacetophenone (99-91-2)
α-Chloroacetophenone (ACGIH,OSHA) (532-27-4)
Chloroacetophenone (DOT) (532-27-4)
Chloroacetophenone, Gas, liquid, or solid [UN 1697] (532-27-4)
5-((Chloroacetoxy)methyl)-2-norbornene (28693-00-7)
6-Chloro-17-α-acetoxy-4,6-pregnadiene-3,20-dione (302-22-7)
6-Chloro-δ^6-(17-α)acetoxyprogesterone (302-22-7)
6-Chloro-δ^6-17-acetoxyprogesterone (302-22-7)
δ^6-6-Chloro-17-α-acetoxyprogesterone (302-22-7)
4'-Chloroacetyl (acetanilide) (140-49-8)
Chloroacetyl anhydride (541-88-8)
Chloroacetyl chloride (ACGIH,OSHA) [UN 1752] (79-04-9)
N-Chloroacetyldiethylamine (2315-36-8)
N-(Chloroacetyl)-N-(2,6-diethylphenyl)glycine ethyl ester (38727-55-8)
N-Chloroacetyl-N-(2,6-diethylphenyl)glycine ethyl ester (38727-55-8)
2-Chloroacrolein (683-51-2)
α-Chloroacrolein (683-51-2)
(E)-3-Chloroacrylic acid (2345-61-1)
2-Chloroacrylic acid (598-79-8)
Chloroacrylic acid (598-79-8)
α-Chloroacrylic acid (598-79-8)
trans-3-Chloroacrylic acid (2345-61-1)
trans-β-Chloroacrylic acid (2345-61-1)
2-Chloroacrylic acid, methyl ester (80-63-7)
Chloroaethan (German) (75-00-3)
Chloroalkylene-9 (9CI) (57308-11-9)
Chloroallene (3223-70-9)
β-Chloro allyl alcohol (5976-47-6)
pi-Chloro allyl alcohol (29560-84-7)
3-Chloroallyl chloride (542-75-6)
α-Chloroallyl chloride (542-75-6)
γ-Chloroallyl chloride (542-75-6)
2-Chloroallyl N,N-diethyldithiocarbamate (95-06-7)
2-Chloroallyl diethyldithiocarbamate (95-06-7)
Chloroallylene (107-05-1)
1-(3-Chloroallyl)-3,5,7-triaza-1-azoniaadamantane chloride (4080-31-3)
Chloroalonil (1897-45-6)
Chloroalosane (15879-93-3)
Chloroambucil (305-03-3)
Chloroamine (10599-90-3)
3-Chloro-4-aminoaniline (615-66-7)
3-Chloro-4-aminoaniline sulfate (6219-71-2)
5-Chloro-1-aminoanthraquinone (117-11-3)
2-Chloro-4-aminobenzoic acid (2457-76-3)
2-Chloro-3-amino-1,4-naphthoquinone (2797-51-5)
2-Chloro-4-aminotoluene (95-74-9)
3-Chloro-2-aminotoluene (87-63-8)
4-Chloro-2-aminotoluene (95-79-4)
5-Chloro-2-aminotoluene (95-69-2)
5-Chloro-2-aminotoluene hydrochloride (3165-93-3)
Chloroamphenicol (56-75-7)
Chloro-4-tert-amylphenol (73090-69-4)
2-Chloroaniline (95-51-2)
3-Chloroaniline (108-42-9)
4-Chloroaniline (106-47-8)
m-Chloroaniline (108-42-9)
o-Chloroaniline (95-51-2)
p-Chloroaniline (106-47-8)
m-Chloroaniline, Liquid [UN 2019] (108-42-9)
o-Chloroaniline, Liquid [UN 2019] (95-51-2)
p-Chloroaniline, Liquid [UN 2019] (106-47-8)
m-Chloroaniline, Solid [UN 2018] (108-42-9)
o-Chloroaniline, Solid [UN 2018] (95-51-2)
p-Chloroaniline, Solid [UN 2018] (106-47-8)
3-Chloroaniline hydrochloride (141-85-5)
4-Chloroaniline hydrochloride (20265-96-7)
m-Chloroaniline hydrochloride (141-85-5)
p-Chloroaniline hydrochloride (20265-96-7)
4-Chloroaniline-3-sulfonic acid (88-43-7)
p-Chloroaniline-m-sulfonic acid (88-43-7)
p-Chloroanilinium chloride (20265-96-7)
(o-Chloroanilino)dichlorotriazine (101-05-3)
3-Chloroanisidine (5345-54-0)
2-Chloro-9,10-anthracenedione (131-09-9)
1-Chloro-9,10-anthraquinone (82-44-0)
1-Chloroanthraquinone (82-44-0)
2-Chloroanthraquinone (131-09-9)
α-Chloroanthraquinone (82-44-0)
Chloroazinal (500-42-5)
Chloroazodin (502-98-7)
11-Chloro-8,12b-dihydro-2,8-dimethyl-12b-phenyl-4H-(1,3)-oxazino(3,2-d)-(1,4)benzodiazepine-4,7(6H)dione (27223-35-4)
11-Chloro-8,12b-dihydro-2,8-dimethyl-12b-phenyl-4H-(1,3)oxazino(3,2-d)(1,4)benzodiazepine-4,7(6H)-dione (27223-35-4)
Chloroben (95-50-1)
Chlorobenzal (98-87-3)
2-Chlorobenzaldehyde (89-98-5)
4-Chlorobenzaldehyde (104-88-1)
α-Chlorobenzaldehyde (98-88-4)
o-Chlorobenzaldehyde (89-98-5)
p-Chlorobenzaldehyde (104-88-1)
(o-Chlorobenzal)malononitrile (2698-41-1)
2-Chlorobenzalmalononitrile (2698-41-1)
2-Chlorobenzamide (609-66-5)
2'-Chlorobenzanilide (1020-39-9)
Chlorobenzen (Polish) (108-90-7)
3-Chlorobenzenamine (108-42-9)
4-Chlorobenzenamine (106-47-8)
2-Chlorobenzenamine hydrochloride (137-04-2)
3-Chlorobenzenamine hydrochloride (141-85-5)
Chlorobenzene (ACGIH,OSHA) [UN 1134] (108-90-7)
4-Chlorobenzeneacetic acid (1878-66-6)
4-Chlorobenzeneacetonitrile (140-53-4)
3-Chloro-benzenecarboperoxoic acid (937-14-4)
3-Chlorobenzenecarboperoxoic acid (937-14-4)
o-Chlorobenzenecarboxaldehyde (89-98-5)
p-Chlorobenzenecarboxaldehyde (104-88-1)
4-Chloro-1,3-benzenediamine (5131-60-2)
2-Chloro-1,4-benzenediamine dihydrochloride (615-46-3)
2-Chloro-1,4-benzenediamine sulfate (6219-71-2)
4-Chloro-1,2-benzenediamine sulfate (1:1) (68459-98-3)
4-Chloro-1,3-benzenediamine sulfate (68239-80-5)
4-Chloro-1,3-benzenediamine sulfate (1:1) (68239-80-5)
4-Chloro-1,2-benzenedicarboxylic acid (89-20-3)

p-Chlorobenzenesulfonamide **(98-64-6)**
(N-Chlorobenzenesulfonamide)sodium **(127-52-6)**
N-Chlorobenzenesulfonamide, sodium salt **(127-52-6)**
(N-Chlorobenzenesulfonamido)-sodium **(127-52-6)**
4-Chlorobenzenesulfonate de 4-chlorophenyle (French) **(80-33-1)**
4-Chlorobenzenesulfonic acid **(98-66-8)**
p-Chlorobenzenesulfonic acid, p-chlorophenyl ester **(80-33-1)**
p-Chlorobenzenesulfonyl chloride **(98-60-2)**
1-(p-Chlorobenzenesulfonyl)-3-propylurea **(94-20-2)**
N-(p-Chlorobenzenesulfonyl)-N'-propylurea **(94-20-2)**
3-Chlorobenzenethiol **(2037-31-2)**
4-Chlorobenzenethiol **(106-54-7)**
Chlorobenzenu (Czech) **(108-90-7)**
4-Chlorobenzhydrazide **(536-40-3)**
p-Chlorobenzhydrazide **(536-40-3)**
4-Chlorobenzhydrol **(119-56-2)**
Chlorobenzhydrol **(119-56-2)**
p-Chlorobenzhydrol **(119-56-2)**
4-Chlorobenzhydryl chloride **(134-83-8)**
1-(p-Chlorobenzhydryl)-4-(2-(2-hydroxyethoxy)ethyl)-diethylenediamine **(68-88-2)**
1-(p-Chlorobenzhydryl)-4-(2-(2-hydroxyethoxy)ethyl)piperazine **(68-88-2)**
N-(4-Chlorobenzhydryl)-N'-(hydroxyethoxyethyl)piperazine **(68-88-2)**
1-(p-Chlorobenzhydryl)-4-(m-methylbenzyl)diethylenediamine **(569-65-3)**
1-p-Chlorobenzhydryl-4-m-methylbenzylpiperazine **(569-65-3)**
1-(4-Chlorobenzhydryl)-4-methylpiperazine **(82-93-9)**
1-(p-Chlorobenzhydryl)-4-methylpiperazine hydrochloride **(14362-31-3)**
Chlorobenzilate **(510-15-6)**
4-Chlorobenzohydrazide **(536-40-3)**
p-Chlorobenzohydrazide **(536-40-3)**
2-Chlorobenzoic acid **(118-91-2)**
3-Chlorobenzoic acid **(535-80-8)**
4-Chlorobenzoic acid **(74-11-3)**
m-Chlorobenzoic acid **(535-80-8)**
o-Chlorobenzoic acid **(118-91-2)**
p-Chlorobenzoic acid **(74-11-3)**
4-Chlorobenzoic acid, hydrazide **(536-40-3)**
p-Chlorobenzoic acid, hydrazide **(536-40-3)**
m-Chlorobenzoic acid methyl ester **(2905-65-9)**
p-Chlorobenzoic hydrazide **(536-40-3)**
Chlorobenzol (DOT) **(108-90-7)**
o-Chlorobenzonitrile **(873-32-5)**
p-Chlorobenzonitrile **(623-03-0)**
4-Chlorobenzophenone **(134-85-0)**
p-Chlorobenzophenone **(134-85-0)**
para-Chlorobenzophenone **(134-85-0)**
6-Chloro-2H-1,2,4-benzothiadiazine-7-sulfonamide 1,1-dioxide **(58-94-6)**
5-Chloro-1H-benzotriazole **(94-97-3)**
5-Chlorobenzotriazole **(94-97-3)**
6-Chlorobenzotriazole **(94-97-3)**
2-(5-Chloro-2H-benzotriazol-2-yl)-6-(1,1-dimethylethyl)-4-methyl-phenol **(3896-11-5)**
4-Chlorobenzotrichloride **(5216-25-1)**
p-Chlorobenzotrichloride **(5216-25-1)**
2-Chlorobenzotrifluoride **(88-16-4)**
3-Chlorobenzotrifluoride **(98-15-7)**
o-Chlorobenzotrifluoride **(88-16-4)**
m-Chlorobenzotrifluoride [UN 2234] **(98-15-7)**
p-Chlorobenzotrifluoride [UN 2234] **(98-56-6)**
Chlorobenzotrifluorides [UN 2234] **(88-16-4)**
3-Chlorobenzoyl chloride **(618-46-2)**
4-Chlorobenzoyl chloride **(122-01-0)**
5-(4-Chlorobenzoyl)-1,4-dimethyl-1H-pyrrole-2-acetic acid sodium salt dihydrate **(64092-48-4)**
4-Chlorobenzoyl hydrazide **(536-40-3)**
p-Chlorobenzoyl hydrazide **(536-40-3)**
4-Chlorobenzoylhydrazine **(536-40-3)**
p-Chlorobenzoylhydrazine **(536-40-3)**
m-Chlorobenzoyl hydroperoxide **(937-14-4)**
1-(p-Chlorobenzoyl)-5-methoxy-2-methylindole **(6260-97-5)**
1-(p-Chlorobenzoyl)-5-methoxy-2-methylindole-3-acetic acid **(53-86-1)**
(1-p-Chlorobenzoyl-5-methoxy-2-methylindol-3-yl)acetic acid **(53-86-1)**
1-(p-Chlorobenzoyl)-2-methyl-5-methoxy-3-indole-acetic acid **(53-86-1)**
1-(p-Chlorobenzoyl)-2-methyl-5-methoxyindole-3-acetic acid **(53-86-1)**
α-(1-(p-Chlorobenzoyl)-2-methyl-5-methoxy-3-indolyl)acetic acid **(53-86-1)**
p-Chlorobenzoyl peroxide (DOT) **(94-17-7)**
p-Chlorobenzoyl peroxide, Not more than 52% as a paste (DOT) **(94-17-7)**
p-Chlorobenzoyl peroxide, Not more than 52% in solution (DOT) **(94-17-7)**
p-Chlorobenzoyl peroxide, Not more than 75% with water (DOT) **(94-17-7)**
1-(o-Chlorobenzoyl)-3-(p-(trifluoromethoxy)phenyl)urea (8CI) **(64628-44-0)**
Chlorobenzylate **(510-15-6)**
p-Chlorobenzylbenzene **(831-81-2)**
2-Chlorobenzyl bromide **(611-17-6)**
o-Chlorobenzyl bromide **(611-17-6)**
2-Chlorobenzyl chloride **(611-19-8)**
p-Chlorobenzyl chloride [UN 2235] **(104-83-6)**
p-Chlorobenzyl p-chlorophenyl sulfide **(103-17-3)**
4-Chlorobenzyl 4-chlorophenyl sulphide **(103-17-3)**
p-Chlorobenzyl p-chlorophenyl sulphide **(103-17-3)**
4-Chlorobenzyl cyanide **(140-53-4)**
p-Chlorobenzyl cyanide **(140-53-4)**
p-Chlorobenzyl 2,4-dichlorophenyl ether **(21571-58-4)**
S-(4-Chlorobenzyl) N,N-diethylthiocarbamate **(28249-77-6)**
S-(2-Chlorobenzyl)-N,N-diethylthiolcarbamate **(34622-58-7)**
2-(2-Chlorobenzyl)-4,4-dimethyl-1,2-oxazolidin-3-one **(81777-89-1)**
4-Chlorobenzyl 4'-fluorophenyl sulfide **(405-30-1)**
p-Chlorobenzyl p-fluorophenyl sulfide **(405-30-1)**
p-Chlorobenzyl p-fluorophenyl sulphide **(405-30-1)**
o-Chlorobenzylidene malonitrile **(2698-41-1)**
2-Chlorobenzylidene malononitrile **(2698-41-1)**
o-Chlorobenzylidene malononitrile (ACGIH,OSHA) **(2698-41-1)**
p-Chlorobenzyl phenyl ether **(19962-25-5)**
.α.-Chlorobibenzyl **(4714-14-1)**
2-Chloro-1,1'-biphenyl **(2051-60-7)**
2-Chlorobiphenyl **(2051-60-7)**
3-Chlorobiphenyl **(2051-61-8)**
4-Chloro-1,1'-biphenyl **(2051-62-9)**
4-Chlorobiphenyl **(2051-62-9)**
Chloro 1,1-biphenyl **(1336-36-3)**
Chloro biphenyl **(1336-36-3)**
Chlorobiphenyl **(27323-18-8)**
3-Chloro-(1,1'-biphenyl)-2-ol **(85-97-2)**
3-Chloro-2-biphenylol **(85-97-2)**
5-Chloro-(1,1'-biphenyl)-2-ol **(607-12-5)**
5-Chloro-2-biphenylol **(607-12-5)**
5-Chloro-(1,1'-biphenyl)-2-ol sodium salt **(10605-10-4)**
5-Chloro-2-biphenylol sodium salt **(10605-10-4)**
2-Chloro-4,6-bis(diethylamino)-s-triazine **(580-48-3)**
2-Chloro-4,6-bis(dodecylamino)-s-triazine **(30355-03-4)**
4-Chloro-2,6-bis(dodecylamino)-s-triazine **(30355-03-4)**
1-Chloro, 3,5-bisethylamino-2,4,6-triazine **(122-34-9)**
2-Chloro-4,6-bis(ethylamino)-1,3,5-triazine **(122-34-9)**
2-Chloro-4,6-bis(ethylamino)-s-triazine **(122-34-9)**
2-Chloro-4,6-bis(isopropylamino)-s-triazine **(139-40-2)**
4-Chloro-2,6-bis(laurylamino)-s-triazine **(30355-03-4)**
6-Chloro-N,N'-bis(1-methylethyl)-1,3,5-triazine-2,4-diamine **(139-40-2)**
2-Chloro-4,6-bis(methylthio)-s-triazine **(4407-40-3)**
Chloroble M **(12427-38-2)**
2-Chlorobmn **(2698-41-1)**
1-Chloro-3-bromobenzene **(108-37-2)**
1-Chloro-4-bromobenzene **(106-39-8)**
3-Chlorobromobenzene **(108-37-2)**
4-Chloro-1-bromobenzene **(106-39-8)**
4-Chlorobromobenzene **(106-39-8)**

Name	CAS
m-Chlorobromobenzene	(108-37-2)
p-Chlorobromobenzene	(106-39-8)
1-Chloro-2-bromoethane	(107-04-0)
sym-Chlorobromoethane	(107-04-0)
1-Chloro-4-(2-bromoethyl)benzene	(6529-53-9)
Chlorobromomethane (ACGIH,OSHA)	(74-97-5)
1-(3-Chloro-4-bromophenyl)-3-methyl-3-methoxyurea	(13360-45-7)
ω-Chlorobromopropane	(109-70-6)
1-Chloro-3-bromopropane [UN 2688]	(109-70-6)
Chlorobromuron	(13360-45-7)
Chlorobufam	(1967-16-4)
1-Chloro-1,3-butadiene	(627-22-5)
1-Chlorobutadiene	(627-22-5)
2-Chlorobuta-1,3-diene	(126-99-8)
3-Chloro-1,2-butadiene	(34581-41-4)
Chlorobutadiene	(126-99-8)
2-Chloro-1,3-butadiene (OSHA)	(126-99-8)
2-Chloro-1,3-butadiene homopolymer (9CI)	(9010-98-4)
2-Chloro-1,3-butadiene polymer	(9010-98-4)
Chlorobutadiene polymer	(9010-98-4)
1-Chlorobutane [UN 1127]	(109-69-3)
2-Chlorobutane[UN 1127]	(78-86-4)
4-Chloro-1-butane-ol	(928-51-8)
3-Chlorobutanoic acid	(1951-12-8)
4-Chlorobutanoic acid	(627-00-9)
4-Chloro-1-butanol	(928-51-8)
4-Chlorobutanol	(928-51-8)
Chlorobutanol	(1320-66-7)
Chlorobutanol	(57-15-8)
Chlorobutanol, Anhydrous	(57-15-8)
3-Chloro-2-butanone	(4091-39-8)
4-Chlorobutanoyl chloride	(4635-59-0)
1-Chloro-1-butene	(4461-42-1)
1-Chloro-2-butene	(591-97-9)
2-Chloro-2-butene	(4461-41-0)
2-Chlorobutene-2	(4461-41-0)
3-Chloro-1-butene	(563-52-0)
4-Chloro-1-butene	(927-73-1)
1-Chloro-3-buten-2-ol	(671-56-7)
Chlorobutin	(305-03-3)
Chlorobutine	(305-03-3)
(β-Chloro-tert-butyl)benzene	(515-40-2)
β-Chloro-tert-butylbenzene	(515-40-2)
5-Chloro-3-tert-butyl-6-methyluracil	(5902-51-2)
2-Chloro-4-tert-butylphenol	(98-28-2)
3-Chloro-1-butyne	(21020-24-6)
4-Chloro-1-butyne	(51908-64-6)
Chloro-2-butynyl m-chlorocarbamate	(101-27-9)
4-Chloro-2-butynyl-m-chlorocarbanilate	(101-27-9)
4-Chloro-2-butynyl N-(3-chlorophenyl)carbamate	(101-27-9)
3-Chlorobutyric acid	(1951-12-8)
β-Chlorobutyric acid	(1951-12-8)
γ-Chlorobutyryl chloride	(4635-59-0)
4-Chlorobut-2-ynyl-m-chlorocarbanilate	(101-27-9)
4-Chlorobut-2-ynyl 3-chlorophenylcarbamate	(101-27-9)
Chlorocaine	(51-05-8)
Chlorocamphene	(8001-35-2)
Chlorocaps	(56-75-7)
m-Chlorocarbanilic acid, 4-chloro-2-butynyl ester	(101-27-9)
3-Chlorocarbanilic acid, isopropyl ester	(101-21-3)
m-Chlorocarbanilic acid, isopropyl ester	(101-21-3)
Chlorocarbonate de methyle (French)	(79-22-1)
Chlorocarbonate D'ethyle (French)	(541-41-3)
Chlorocarbonic acid isobutyl ester	(543-27-1)
Chlorocarbonic acid methyl ester	(79-22-1)
3-Chlorocatechol	(4018-65-9)
4-Chlorocatechol	(2138-22-9)
Chlorochin	(54-05-7)
3-Chlorochlordene	(76-44-8)
Chloro(3-chloro-4-cyclohexylphenyl)acetic acid	(36616-52-1)
(3-Chloro-4-chlorodifluoromethylthiophenyl)-1,1-dimethylurea	(33439-45-1)
1-Chloro-2-(β-chloroethoxy)ethane	(111-44-4)
2-Chloro-N-(2-chloroethyl)-N-methylethanamine	(51-75-2)
2-Chloro-N-(2-chloroethyl)-N-methylethanamine hydrochloride	(55-86-7)
2-Chloro-N-(2-chloroethyl)-N-methylethanamine-N-oxide	(126-85-2)
2-Chloro-N-(2-chloroethyl)-N-methylethanamine-N-oxide hydrochloride	(302-70-5)
1-Chloro-2-(β-chloroethylthio)ethane	(505-60-2)
Chloro(chloromethoxy)methane	(542-88-1)
1-Chloro-2-(chloromethyl)benzene	(611-19-8)
1-Chloro-3-(chloromethyl)benzene	(620-20-2)
1-Chloro-4-chloromethylbenzene	(104-83-6)
3-Chloro-2-(chloromethyl)propene	(1871-57-4)
3-Chloro-4-(chloromethyl)-1-(3-(trifluoromethyl)phenyl)-2-pyrrolidinone	(61213-25-0)
N-(4-Chloro-2-((2-chloro-4-nitrophenyl)azo)-5-((2-hydroxypropyl)amino)phenyl)acetamide	(71617-28-2)
5-Chloro-N-(2-chloro-4-nitrophenyl)-2-hydroxybenzamide Compd. with 2-aminoethanol (1:1)	(1420-04-8)
3-Chloro-4-(3-chloro-2-nitrophenyl)pyrrole	(1018-71-9)
5-Chloro-2'-chloro-4'-nitrosalicylanilide	(50-65-7)
7-Chloro-5-(2-chlorophenyl)-1,3-dihydro-3-hydroxy-2H-1,4-benzodiazepin-2-one	(846-49-1)
7-Chloro-5-(o-chlorophenyl)-1,3-dihydro-3-hydroxy-2H-1,4-benzodiazepin-2-one	(846-49-1)
7-Chloro-5-(o-chlorophenyl)-1,3-dihydro-3-hydroxy-1-methyl-2H-1,4-benzodiazepin-2-one	(848-75-9)
7-Chloro-5-(2-chlorophenyl)-3-hydroxy-1H-1,4-benzodiazepin-2(3H)-one	(846-49-1)
7-Chloro-5-(2-chlorophenyl)-3-hydroxy-1-methyl-2,3-dihydro-1H-1,4-benzodiazepin-2-one	(848-75-9)
1-Chloro-4-(chlorophenylmethyl)benzene	(134-83-8)
2-Chloro-3-(4-chlorophenyl)-methylpropionate	(14437-17-3)
1-Chloro-4-(((4-chlorophenyl)methyl)thio)benzene	(103-17-3)
8-Chloro-6-(o-chlorophenyl)-1-methyl-4H-s-triazolo-(4,3-a)(1,4)benzodiazepine	(28911-01-5)
Chloro(p-chlorophenyl)phenylmethane	(134-83-8)
Chloro-(o-chlorophenyl)-phenylmethane	(56961-47-8)
2-Chloro-3-(4-chlorophenyl)propionic acid methyl ester	(14437-17-3)
4-Chloro-α-(4-chlorophenyl)-α-(trichloromethyl)benzenemethanol	(115-32-2)
Chlorocholine chloride	(999-81-5)
Chlorocid	(56-75-7)
Chlorocid S	(56-75-7)
Chlorocide	(103-17-3)
Chlorocide	(56-75-7)
Chlorocidin C	(56-75-7)
Chlorocidin c tetran	(56-75-7)
4-Chlorocinnamic acid	(1615-02-7)
p-Chlorocinnamic acid	(1615-02-7)
Chlorocol	(56-75-7)
4-Chloro-m-cresol	(59-50-7)
4-Chloro-o-cresol	(1570-64-5)
6-Chloro-m-cresol	(59-50-7)
6-Chloro-m-cresol	(615-74-7)
6-Chloro-o-cresol	(87-64-9)
Chlorocresol	(59-50-7)
p-Chloro-m-cresol	(59-50-7)
p-Chlorocresol	(59-50-7)
Chloro-m-cresol [UN 2669]	(54548-50-4)
4-Chloro-o-cresoxyacetic acid	(94-74-6)
.γ.-Chlorocrotonic acid	(16197-90-3)
Chlorocrotyl ester of 2,4-d	(2971-38-2)
Chloroctan sodny (Czech)	(3926-62-3)
o-Chlorocumene	(2077-13-6)

Chlorocyan (506-77-4)
Chlorocyanide (506-77-4)
Chlorocyanogen (506-77-4)
2-Chloro-4-(1-cyano-1-methylethylamino)-6-ethylamino-1,3,5-triazine (21725-46-2)
2-exo-Chloro-6-endo-cyano-2-norbornanone O-(methyl carbamoyl) oxime (15271-41-7)
3-Chloro-6-cyano-2-norbornanone-O-(methylcarbamoyl)oxime (15271-41-7)
2-exo-Chloro-6-endo-cyano-2-norbornanone-O-(methyl-carbamoyl)oxime (15271-41-7)
3-Chloro-6-cyanonorbornanone-2 oxime O,N-methylcarbamate (15271-41-7)
Chlorocycline (82-93-9)
Chlorocyclizine (82-93-9)
Chlorocyclizine hydrochloride (14362-31-3)
Chlorocyclohexane (542-18-7)
3-Chloro-1-cyclohexene (2441-97-6)
4-Chlorocyclohexene (930-65-4)
7-Chloro-5-(1-cyclohexen-1-yl)-1,3-dihydro-1-methyl-2H-1,4-benzodiazepin-2-one (10379-14-3)
7-Chloro-5-(cyclohexen-1-yl)-1,3-dihydro-1-methyl-2H-1,4-benzodiazepin-2-one (10379-14-3)
7-Chloro-5-(1-cyclohexenyl)-1-methyl-2-oxo-2,3-dihydro-1H-(1,4)-benzo(f)diazepine (10379-14-3)
Chlorocyclohexylmagnesium (931-51-1)
3-Chloro-2-cyclopenten-1-one (53102-14-0)
3-Chloro-2-cyclopentenone (53102-14-0)
trans-2-Chlorocyclopentyl bromide (14376-82-0)
4-Chloro-2-cyclopentyl phenol (13347-42-7)
4-Chloro-2-cyclopentylphenol (13347-42-7)
2-Chloro-4-cyclopropylamino-6-isopropylamino-1,3,5-triazine (22936-86-3)
2-Chloro-4-cyclopropylamino-6-isopropylamino-s-triazine (22936-86-3)
2-((4-Chloro-6-(cyclopropylamino)-1,3,5-triazin-2-yl)amino)-2-methylpropanenitrile (32889-48-8)
2-(4-Chloro-6-(cyclopropylamino)-s-triazin-2-yl)amino-2-methyl-propionitrile (32889-48-8)
7-Chloro-1-(cyclopropylmethyl)-1,3-dihydro-5-phenyl-2H-1,4-benzo-diazepin-2-one (2955-38-6)
7-Chloro-1-cyclopropylmethyl-5-phenyl-1H-1,4-benzodiazepin-2(3H)-one (2955-38-6)
Chlorodane (57-74-9)
1-Chlorodecane (1002-69-3)
3-Chlorodecane (1002-11-5)
6-Chloro-6-dehydro-17-α-acetoxyprogesterone (302-22-7)
6-Chloro-δ^{6}-dehydro-17-acetoxyprogesterone (302-22-7)
6-Chloro-6-dehydro-17-α-hydroxyprogesterone acetate (302-22-7)
7-Chloro-6-demethyltetracycline (127-33-3)
Chloroden (95-50-1)
Chlorodeoxyglycerol (96-24-2)
7(S)-Chloro-7-deoxylincomycin (18323-44-9)
5-Chloro-2'-deoxyuridine (50-90-8)
5-Chlorodeoxyuridine (50-90-8)
Chlorodiabina (94-20-2)
2-Chloro-N,N-diallylacetamide (93-71-0)
α-Chloro-N,N-diallylacetamide (93-71-0)
1-Chloro-2,4-diaminobenzene (5131-60-2)
4-Chloro-1,2-diaminobenzene (95-83-0)
2-Chloro-4,6-diamino-1,3,5-triazine (3397-62-4)
2-Chloro-4,6-di-tert-amylphenol (42350-99-2)
Chlorodiane blue (41709-76-6)
Chlorodiazepoxide (58-25-3)
2-Chlorodibenzo-para-dioxin (39227-54-8)
Chlorodibenzo(b,e)(1,4)dioxin (35656-51-0)
1-Chlorodibenzo-p-dioxin (39227-53-7)
1-Chlorodibenzodioxin (39227-53-7)
Chlorodibenzo-p-dioxin (35656-51-0)
1-Chlorodibenzofuran (84761-86-4)
2-Chlorodibenzofuran (51230-49-0)
3-Chlorodibenzofuran (25074-67-3)
4-Chlorodibenzofuran (74992-96-4)
1-Chloro-1,2-dibromoethane (598-20-9)
Chlorodibromomethane (124-48-1)
1-Chloro-2,3-dibromopropane (96-12-8)
3-Chloro-1,2-dibromopropane (96-12-8)
Chlorodibromotrifluoroethane (29256-79-9)
Chlorodibutylamine (999-33-7)
N-Chlorodibutylamine (999-33-7)
1-Chloro-2,2-dichloroethylene (79-01-6)
o-Chloro-α-(dichloromethyl)benzyl alcohol (27683-60-9)
6-Chloro-3-(dichloromethyl)-3,4-dihydro-2H-1,2,4-benzothia-diazine-7-sulfonamide 1,1-dioxide (133-67-5)
6-Chloro-3-(dichloromethyl)3,4-dihydro-7-sulfamyl-1,2,4-benzo-thiadiazine-1,1-dioxide (133-67-5)
3-Chloro-4-dichloromethyl-5-hydroxy-2(5H)-furanone (77439-76-0)
Chloro(dichloromethyl)-5-hydroxy-2(5H)-furanone (77439-76-0)
(E)-2-Chloro-3-(dichloromethyl)-4-oxo-butenoic acid (115340-67-5)
2,2'-((3-Chloro-4-((2,6-dichloro-4-nitrophenyl)azo)phenyl)-amino)bisethanol (23355-64-8)
O'-(2-Chloro-1-(2,4-dichlorophenyl)ethenyl) O,O-diethyl-phosphorothioate (1224-63-1)
2-Chloro-1-(2,4-dichlorophenyl)ethenyl dimethyl phosphate (2274-67-1)
2-Chloro-1-(2,4-dichlorophenyl)vinyl diethyl phosphate (470-90-6)
β-2-Chloro-1-(2',4'-dichlorophenyl) vinyl diethylphosphate (470-90-6)
O-(2-Chloro-1-(2,5-dichlorophenyl)vinyl) O,O-diethyl phosphorothioate (1757-18-2)
1-Chloro-2,5-diethoxybenzene (52196-74-4)
2-Chloro-1,4-diethoxybenzene (52196-74-4)
Chloro-2,5-diethoxybenzene (52196-74-4)
2-Chloro-1,1-diethoxyethane (621-62-5)
1-Chloro-2,5-diethoxy-4-nitrobenzene (91-43-0)
2-Chloro-α-((diethoxyphosphinothioyloxy)imino)benzeneaceto-nitrile (14816-20-7)
Chlorodiethoxysilane (6485-91-2)
Chlorodiethylaluminum (96-10-6)
2-Chloro-1-(p-(β-diethylaminoethoxy)phenyl)-1,2-diphenylethylene (50-41-9)
2-Chloro-4-(diethylamino)-6-(ethylamino)-s-triazine (1912-26-1)
7-Chloro-1-(2-(diethylamino)ethyl)-5-(2-fluorophenyl)-1H-1,4-benzodiazepin-2(3H)-one (17617-23-1)
2-Chloro-4-(diethylamino)-6-(isopropylamino)-s-triazine (1912-25-0)
6-Chloro-9-((4-(diethylamino)-1-methylbutyl)amino)-2-methoxy-acridine (83-89-6)
2-Chloro-5-(ω-diethylamino-α-methylbutylamino)-7-methoxy-acridine dihydrochloride (69-05-6)
3-Chloro-9-(4'-diethylamino-1'-methylbutylamino)-7-methoxy-acridine dihydrochloride (69-05-6)
6-Chloro-9-((4-(diethylamino)-1-methylbutyl)amino)-2-methoxy-acridine dihydrochloride (69-05-6)
7-Chloro-4-(4-diethylamino-1-methylbutylamino)quinoline (54-05-7)
2-Chloro-2',6'-diethyl-N-(butoxymethyl)acetanilide (23184-66-9)
2-Chloro-2-diethylcarbamoyl-1-methylvinyl dimethylphosphate (13171-21-6)
1-Chloro-diethylcarbamoyl-1-propen-2-yl dimethyl phosphate (13171-21-6)
2-Chloro-2',6'-diethyl-N-(methoxymethyl)acetanilide (15972-60-8)
2-Chloro-N-(2,6-diethyl)phenyl-N-methoxymethylacetamide (15972-60-8)
Chlorodiethylsilane (1609-19-4)
6-Chloro-N,N'-diethyl-1,3,5-triazine-2,4-diyldiamine (122-34-9)
Chlorodifluorethane (French) (25497-29-4)
Chlorodifluorobromomethane [UN 1974] (353-59-3)
Chlorodifluoroethane (25497-29-4)
1-Chloro-1,1-difluoroethane (DOT) (75-68-3)
Chlorodifluoroethane (DOT) (75-68-3)
Chlorodifluoroethanes (25497-29-4)
2-Chloro-1,1-difluoroethylene (359-10-4)
Chloro-1,1-difluoroethylene (359-10-4)
Chlorodifluoromethane (ACGIH,OSHA) [UN 1018] (75-45-6)
2-Chloro-1-(difluoromethoxy)-1,1,2-trifluoroethane (13838-16-9)

Chlorodifluoromonobromomethane **(353-59-3)**
10-Chloro-5,10-dihydroarsacridine **(578-94-9)**
6-Chloro-3,4-dihydro-2H-1,2,4-benzothiadiazine-7-sulfonamide 1,1-dioxide **(58-93-5)**
7-Chloro-1,3-dihydro-3-(N,N-dimethylcarbamoyl)-1-methyl-5-phenyl-2H-1,4-benzodiazepin-2-one **(36104-80-0)**
S-(2-Chloro-1-(1,3-dihydro-1,3-dioxo-2H-isoindol-2-yl)-ethyl)O,O-diethyl phosphorodithioate **(10311-84-9)**
7-Chloro-1,3-dihydro-3-hydroxy-1-methyl-5-phenyl-2H-1,4-benzodiazepin-2-one **(846-50-4)**
7-Chloro-1,3-dihydro-3-hydroxy-1-methyl-5-phenyl-1,4-benzodiazepin-2-one dimethylcarbamate **(36104-80-0)**
7-Chloro-1,3-dihydro-3-hydroxy-5-phenyl-2H-1,4-benzodiazepine-2-one **(604-75-1)**
7-Chloro-2,3-dihydro-1-methyl-5-phenyl-1H-1,4-benzodiazepine **(2898-12-6)**
7-Chloro-1,3-dihydro-1-methyl-5-phenyl-2H-1,4-benzodiazepin-2-one **(439-14-5)**
10-Chloro-5,10-dihydrophenarsazine **(578-94-9)**
7-Chloro-1,3-dihydro-5-phenyl-2H-1,4-benzodiazepin-2-one **(1088-11-5)**
7-Chloro-1,3-dihydro-5-phenyl-1-(2-propynyl)-2h-1,4-benzodiazepin-2-one **(52463-83-9)**
7-Chloro-1,3-dihydro-5-phenyl-1-(2,2,2-trifluoroethyl)-2H-1,4-benzodiazepin-2-one **(23092-17-3)**
6-Chloro-3,4-dihydro-7-sulfamoyl-2H-1,2,4-benzothiadiazine 1,1-dioxide **(58-93-5)**
1-Chloro-2,3-dihydroxypropane **(96-24-2)**
3-Chloro-1,2-dihydroxypropane **(96-24-2)**
Chlorodiisobutylaluminum **(1779-25-5)**
6-Chloro-N,N'-diisopropyl-1,3,5-triazine-2,4-diyldiamine **(139-40-2)**
1-Chloro-2,5-dimethoxybenzene **(2100-42-7)**
2-Chloro-1,4-dimethoxybenzene **(2100-42-7)**
2-Chloro-1,1-dimethoxyethane **(97-97-2)**
2-Chloro-N,N-dimethylacetoacetamide **(5810-11-7)**
2-(p-Chloro-α-(2-(dimethylamino)ethyl)benzyl)pyridine **(132-22-9)**
D-2-(p-Chloro-α-(2-dimethylaminoethyl)benzyl)pyridine **(25523-97-1)**
dl-2(-p-Chloro-α-2-(dimethylamino)ethylbenzyl)pyridine bimaleate **(113-92-8)**
2-Chloro-4-dimethylamino-6-methyl-pyrimidine **(535-89-7)**
(α-2-Chloro-9-ω-dimethylamino-propylamine)thioxanthene **(113-59-7)**
3-Chloro-5-(3-(dimethylamino)propyl)-10,11-dihydro-5H-dibenz(b,f)azepine **(303-49-1)**
2-Chloro-9-(3-(dimethylamino)propylidene)-thioxanthene **(113-59-7)**
2-Chloro-9-(ω-di-methylaminopropylidene)thioxanthene **(113-59-7)**
cis-2-Chloro-9-(3-dimethylaminopropylidene)thioxanthene **(113-59-7)**
2-Chloro-10-(3-(dimethylamino)propyl)phenothiazine **(50-53-3)**
Chloro-3 (dimethylamino-3 propyl)-10 phenothiazine (French) **(50-53-3)**
2-Chloro-10-(3-dimethylaminopropyl)phenothiazine monohydrochloride **(69-09-0)**
4-Chloro-5-(dimethylamino)-2-α,α,α- (trifluoro-m-tolyl)-3-(2H)-pyridazinone **(23576-23-0)**
2-Chloro-N,N-dimethylaniline **(698-01-1)**
3-Chloro-N,N-dimethylaniline **(6848-13-1)**
o-Chloro-N,N-dimethylaniline **(698-01-1)**
3-Chloro-N,N-dimethylbenzenamine **(6848-13-1)**
4-Chloro-N,N-dimethylbenzenesulfonamide **(7463-22-1)**
1-Chloro-3,3-dimethyl-2-butanone **(13547-70-1)**
2-Chloro-N,N-dimethylethylamine hydrochloride **(4584-46-7)**
(2-Chloro-1,1-dimethylethyl)benzene **(515-40-2)**
(β-Chloro-α,α-dimethyl)ethylbenzene **(515-40-2)**
5-Chloro-3-(1,1-dimethylethyl)-6-methyl-2,4(1H,3H)-pyrimidinedione **(5902-51-2)**
2-Chloro-4-(1,1-dimethylethyl)phenol **(98-28-2)**
3-Chloro-4,4-dimethyl-2-oxazolidinone **(58629-01-9)**
2-Chloro-N,N-dimethyl-3-oxobutanamide **(5810-11-7)**
o-Chloro-α,α-dimethylphenethylamine **(10389-73-8)**
p-Chloro-α,α-dimethylphenethylamine **(461-78-9)**
4-Chloro-3,5-dimethylphenol **(88-04-0)**
2-Chloro-N-(2,6-dimethylphenyl)-N-(2-methoxyethyl)acetamide **(50563-36-5)**
2-Chloro-4,5-dimethylphenyl methylcarbamate **(671-04-5)**
2-Chloro-N-(2,6-dimethylphenyl)-N-(1H-pyrazol-1-ylmethyl)-acetamide **(67129-08-2)**
Chlorodimethylphenylsilane **(768-33-2)**
Chlorodimethylphosphine sulfide **(993-12-4)**
1-Chloro-2,2-dimethylpropane **(753-89-9)**
1-Chloro-3,3-dimethylpropane **(107-84-6)**
3-Chloro-2,2-dimethylpropanoic acid **(13511-38-1)**
3-Chloro-2,2-dimethylpropionic acid **(13511-38-1)**
β-Chloro-α,α-dimethylpropionic acid **(13511-38-1)**
Chlorodimethylsilane **(1066-35-9)**
2-Chloro-N,N-dimethylthioxanthene-δ^9, γ-propylamine **(113-59-7)**
Chlorodimethylvinylsilane **(1719-58-0)**
4-Chloro-2,6-dinitroaniline **(5388-62-5)**
6-Chloro-2,4-dinitroaniline **(3531-19-9)**
4-Chloro-2,6-dinitrobenzenamine **(5388-62-5)**
1-Chloro-2,4-dinitrobenzene **(97-00-7)**
4-Chloro-1,3-dinitrobenzene **(97-00-7)**
6-Chloro-1,3-dinitrobenzene **(97-00-7)**
Chlorodinitrobenzene [UN 1577] **(25567-67-3)**
4-Chloro-3,5-dinitrobenzenesulfonic acid potassium salt **(38185-06-7)**
4-Chloro-3,5-dinitrobenzoate **(118-97-8)**
4-Chloro-3,5-dinitrobenzoic acid **(118-97-8)**
1-Chloro-2,4-dinitrobenzol (German) **(97-00-7)**
4-Chloro-3,5-dinitrobenzotrifluoride **(393-75-9)**
1-Chloro-2,4-dinitronaphthalene **(2401-85-6)**
2-Chloro-4,6-dinitrophenol **(946-31-6)**
4-Chloro-2,6-dinitrophenol **(88-87-9)**
4-Chloro-3,5-dinitro-α,α,α-trifluorotoluene **(393-75-9)**
p-Chloro-2,4-dioxa-5-methyl-p-thiono-3-phosphabicyclo-(4.4.0)decane **(2921-31-5)**
Chloro((2,5-dioxo-4-imidazolidinyl)ureato)tetrahydroxydialuminum **(1317-25-5)**
2-Chloro-4,6-di(pentachlorophenoxy)-2-triazine **(26396-34-9)**
N-(2-(Chloro-5-(4-(2,4-di-tert-pentylphenoxy)butyramido)-phenyl)-4,4-dimethyl-3-oxopentanamide **(26110-32-7)**
6-Chloro-2,4-diphenoxy-s-triazine **(2972-65-8)**
2-Chlorodiphenyl **(2051-60-7)**
4-Chlorodiphenyl **(2051-62-9)**
Chlorodiphenyl **(27323-18-8)**
o-Chlorodiphenyl **(2051-60-7)**
p-Chlorodiphenyl **(2051-62-9)**
Chlorodiphenyl (42% Chlorine) (ACGIH,OSHA) **(53469-21-9)**
Chlorodiphenyl (54% Chlorine) (ACGIH,OSHA) **(11097-69-1)**
Chlorodiphenyl (21% Cl) **(11104-28-2)**
Chlorodiphenyl (32% Cl) **(11141-16-5)**
Chlorodiphenyl (41% Cl) **(12674-11-2)**
Chlorodiphenyl (48% Cl) **(12672-29-6)**
Chlorodiphenyl (54% Cl) **(11097-69-1)**
Chlorodiphenyl (60% Cl) **(11096-82-5)**
Chlorodiphenyl (68% Cl) **(11100-14-4)**
Chlorodiphenylarsine **(712-48-1)**
1-(o-Chloro-α,α-diphenylbenzyl)imidazole **(23593-75-1)**
Chlorodiphenyl (42% cl) **(53469-21-9)**
1-Chloro-1,2-diphenylethane **(4714-14-1)**
1-Chloro-1,2-diphenylethene **(1460-06-6)**
2-(4-(2-Chloro-1,2-diphenylethenyl)phenoxy)-N,N-diethyl-ethanamine **(911-45-5)**
4-Chlorodiphenyl ether **(7005-72-3)**
Chlorodiphenylmethane **(90-99-3)**
p-Chlorodiphenylmethane **(831-81-2)**
1-(p-Chlorodiphenylmethyl)-4-(2-(2-hydroxyethoxy)ethyl)piperazine **(68-88-2)**
p-Chlorodiphenyl oxide **(7005-72-3)**
4-Chlorodiphenyl sulfone **(80-00-2)**
4-Chlorodiphenyl sulphone **(80-00-2)**
2-Chloro-4,6-diphenyl-s-triazine **(3842-55-5)**
6-Chloro-2,4-diphenyl-s-triazine **(3842-55-5)**

2-(p-(2-Chloro-1,2-diphenylvinyl)phenoxy)triethylamine citrate (50-41-9)
2-(p-(2-Chloro-1,2-diphenylvinyl)phenoxy)triethylamine citrate (1:1) (50-41-9)
2-Chloro-N,N-di-2-propenylacetamide (93-71-0)
4-Chloroditan (831-81-2)
α-Chloroditan (90-99-3)
1-Chlorododecane (112-52-7)
2-Chlorododecane (2350-11-0)
Chlorodracylic acid (74-11-3)
Chlorodwufenol (Polish) (27323-18-8)
1-Chloro-2,3-epoxypropane (106-89-8)
3-Chloro-1,2-epoxypropane (106-89-8)
1-Chloro-2,3-epoxypropane (OSHA) (106-89-8)
Chloroetene (71-55-6)
2-Chloro-1-ethanal (107-20-0)
2-Chloroethanal (107-20-0)
2-Chloroethanamide (79-07-2)
Chloroethane (75-00-3)
2-Chloroethanephosphonic acid (16672-87-0)
2-Chloroethane sulfochloride (1622-32-8)
2-Chloroethanesulfonyl chloride (1622-32-8)
β-Chloroethanesulfonyl chloride (1622-32-8)
Chloroethanoic acid (79-11-8)
Chloroethanoic anhydride (541-88-8)
2-Chloroethanol (107-07-3)
δ-Chloroethanol (107-07-3)
2-Chloroethanol (OSHA) (107-07-3)
2-Chloroethanol acetate (542-58-5)
2-Chloroethanol phosphate (115-96-8)
2-Chloroethanol phosphite (3:1) (140-08-9)
Chloroethene (71-55-6)
Chloroethene (75-01-4)
Chloroethene NU (71-55-6)
(2-Chloroethenyl)benzene (622-25-3)
1-Chloro-3-ethenylbenzene (2039-85-2)
1-Chloro-4-ethenylbenzene (1073-67-2)
Chloroethenyldimethylsilane (1719-58-0)
1,1',1''-(1-Chloro-1-ethenyl-2-ylidene)-tris(4-methoxybenzene) (569-57-3)
N-(2-Chloroethoxycarbonyl)aniline (3747-48-6)
(2-(((2-Chloroethoxy)(2-chloroethyl)phosphinyl)oxy)ethyl)-phosphonic acid, bis(2-chloroethyl) ester (58823-09-9)
3-(2-Chloroethoxy)-1,2-dichloropropene (84987-77-9)
1-Chloro-1-ethoxyethane (7081-78-9)
2-Chloroethoxyethane (628-34-2)
2-(2'-Chloroethoxy)ethanol (628-89-7)
2-(2-Chloroethoxy)ethanol (628-89-7)
(2-Chloroethoxy)ethene (110-75-8)
1-(2-(2-Chloroethoxy)ethoxy)butane (1120-23-6)
2-(2-Chloroethoxy)ethyl 2'-chloroethyl ether (112-26-5)
2-Chloro-N-(ethoxymethyl)-6'-ethylacet-o-toluidide (34256-82-1)
2-Chloro-N-(ethoxymethyl)-6'-ethyl-o-acetotoluidide (34256-82-1)
2-Chloro-N-(ethoxymethyl)-N-(2-ethyl-6-methylphenyl)acetamide (34256-82-1)
2-Chloro-1-(3-ethoxy-4-nitrophenoxy)-4-trifluoromethylbenzene (42874-03-3)
2-Chloroethyl acetate (542-58-5)
β-Chloroethyl acetate (542-58-5)
2-Chloroethyl acrylate (2206-89-5)
Chloroethyl acrylate (2206-89-5)
β-Chloroethyl acrylate (2206-89-5)
2-Chloroethyl alcohol (107-07-3)
β-Chloroethyl alcohol (107-07-3)
2-Chloro-4-ethylamineisopropylamine-s-triazine (1912-24-9)
1-Chloro-3-ethylamino-5-isopropylamino-2,4,6-triazine (1912-24-9)
1-Chloro-3-ethylamino-5-isopropylamino-s-triazine (1912-24-9)
2-Chloro-4-ethylamino-6-isopropylamino-1,3,5-triazine (1912-24-9)
2-Chloro-4-ethylamino-6-isopropylamino-s-triazine (1912-24-9)
2-(4-Chloro-6-ethylamino-1,3,5-triazine-2-ylamino)-2-methyl-propionitrile (21725-46-2)
2-(4-Chloro-6-ethylamino-s-triazine-2-ylamino)-2-methyl-propionitrile (21725-46-2)
2-((4-Chloro-6-(ethylamino)-s-triazin-2-yl)amino)-2-methylpropionitrile (21725-46-2)
(1-Chloroethyl)benzene (672-65-1)
(2-Chloroethyl)benzene (622-24-2)
β-Chloroethyl β-(bis(β-hydroxyethyl)sulfonium)ethyl sulfide chloride (64036-91-5)
β-Chloroethyl-β'-(p-t-butylphenoxy)-α'-methylethyl sulfite (140-57-8)
β-Chloroethyl-β-(p-t-butylphenoxy)-α-methylethyl sulphite (140-57-8)
3-(2-Chloroethyl)-2-((2-chloroethyl)amino)perhydro-2H-1,3,2-oxazaphosphorine oxide (3778-73-2)
6-Chloro-9-(3-(ethyl-2-chloroethyl)aminopropylamino)-2-methoxyacridine dihydrochloride (146-59-8)
3-(2-Chloroethyl)-2-((2-chloroethyl)amino)tetrahydro-2H-1,3,2-oxazaphosphorine 2-oxide (3778-73-2)
N-(2-Chloroethyl)-N'-(2-chloroethyl)-N',O-propylene-phosphoric acid diamide (3778-73-2)
N-(2-Chloroethyl)-N'-(2-chloroethyl)-N',O-propylene-phosphoric acid ester diamide (3778-73-2)
2-Chloroethyl chloroformate (627-11-2)
(Chloro-2-ethyl)-1-cyclohexyl-3-nitrosourea (13010-47-4)
1-(2-Chloroethyl)-3-cyclohexyl-1-nitrosourea (13010-47-4)
Chloroethylcyclohexylnitrosourea (13010-47-4)
N-(2-Chloroethyl)-N'-cyclohexyl-N-nitrosourea (13010-47-4)
N-(2-Chloroethyl)dibenzylamine (51-50-3)
(2-Chloroethyl)diethylamine hydrochloride (869-24-9)
β-Chloroethyldiethylamine hydrochloride (869-24-9)
N-(2-Chloroethyl)-2,6-dinitro-N-propyl-4-(trifluoromethyl)aniline (33245-39-5)
N-(2-Chloroethyl)-2,6-dinitro-N-propyl-4-(trifluoromethyl)-benzenamide (33245-39-5)
Chloroethylene (75-01-4)
Chloroethylene (OSHA) (75-01-4)
Chloroethylene-1,1-dichloroethylene polymer (9011-06-7)
Chloroethylene oxide (7763-77-1)
Chloroethylenevinyl acetate polymer (9003-22-9)
Chloroethyl ether (111-44-4)
N-(2-Chloroethyl)-N-ethylaniline (92-49-9)
2-Chloroethyl ethyl ether (628-34-2)
β-Chloroethyl ethyl ether (628-34-2)
2-Chloroethyl ethyl sulfide (693-07-2)
2-Chloroethyl ethyl thioether (693-07-2)
1-(2-Chloroethyl)-3-(D-glucopyranos-2-yl)-1-nitrosourea (54749-90-5)
1-Chloro-2-ethylhexane (123-04-6)
2-Chloroethyl 2-hydroxyethyl sulfide (693-30-1)
β-Chloroethyl β-hydroxyethyl sulfide (693-30-1)
Chloroethylidene fluoride (75-68-3)
α-Chloroethylidene fluoride (75-68-3)
2-Chloroethyl linoleate (25525-76-2)
Chloroethylmagnesium (2386-64-3)
Chloroethylmercury (107-27-7)
2-Chloro-6'-ethyl-N-(2-methoxy-1-methylethyl)acet-o-toluidide (51218-45-2)
2-Chloroethyl 1-methyl-2-(p-t-butylphenoxy)ethyl sulphate (140-57-8)
1-(2-Chloroethyl)-3-(4-methyl-cyclohexyl)-1-nitrosourea (13909-09-6)
1-(2-Chloroethyl)-3-(trans-4-methyl-cyclohexyl)-1-nitrosourea (13909-09-6)
N-(2-Chloroethyl)-N'-(trans-4-methylcyclohexyl)-N-nitrosourea (13909-09-6)
2-Chloroethyl methyl ether (627-42-9)
β-Chloroethyl methyl ether (627-42-9)
6-Chloro-N-ethyl-N'-(1-methylethyl)-1,3,5-triazine-2,4-diamine (1912-24-9)
α-Chloro-2'-ethyl-6'-methyl-N-(1-methyl-2-methoxyethyl)-acetanilide (51218-45-2)
N-(2-Chloroethyl)-N-(1-methyl-2-phenoxyethyl)benzene-methanamine (59-96-1)
N-(2-Chloroethyl)-N-(1-methyl-2-phenoxyethyl)benzene-methanamine hydrochloride (63-92-3)
N-(2-Chloroethyl)-N-(1-methyl-2-phenoxyethyl)benzylamine (59-96-1)
N-(2-Chloroethyl)-N-(1-methyl-2-phenoxyethyl)benzylamine, hydrochloride (63-92-3)

2-Chloro-N-(2-ethyl-6-methylphenyl)-N-(2-methoxy-1-methyl-ethyl)acetamide **(51218-45-2)**
2-Chloroethyl methyl sulfide **(542-81-4)**
5-(2-Chloroethyl)-4-methylthiazole **(533-45-9)**
2-((((2-Chloroethyl)nitrosoamino)carbonyl)amino)-2-deoxy-D-glucose **(54749-90-5)**
2-(3-(2-Chloroethyl)-3-nitrosoureido)-D-gluco-pyranose **(54749-90-5)**
2-(3-(2-Chloroethyl)-3-nitrosoureido)-2-deoxy-D-glucosopyranose **(54749-90-5)**
Chloroethylowy alkohol (Polish) **(107-07-3)**
2-Chloroethyl palmitate **(929-16-8)**
1-Chloro-3-ethyl-1-penten-4-yn-3-ol **(113-18-8)**
4-Chloro-2-ethylphenol **(18979-90-3)**
2-Chloroethyl N-phenylcarbamate **(3747-48-6)**
2-Chloroethyl phenylcarbamate **(3747-48-6)**
2-Chloroethylphosphonic acid **(16672-87-0)**
7-Chloro-2-ethyl-6-sulfamoyl-1,2,3,4-tetrahydro-4-quinazolinone **(73-49-4)**
2-Chloroethylsulfonyl chloride **(1622-32-8)**
2-Chloroethyl sulphite of 1-(p-t-butylphenoxy)-2-propanol **(140-57-8)**
7-Chloro-2-ethyl-1,2,3,4-tetrahydro-4-oxo-6-quinazolinesulfonamide **(73-49-4)**
7-Chloro-2-ethyl-1,2,3,4-tetrahydro-4-oxo-6-sulfamoylquinazoline **(73-49-4)**
1-Chloro-2-(ethylthio)ethane **(693-07-2)**
2-((2-Chloroethyl)thio)ethanol **(693-30-1)**
2-(2-Chloroethyl)thioethylbis(2-hydroxyethyl)-, chloride **(64036-91-5)**
N-(2-Chloroethyl)-α,α,α-trifluoro-2,6-dinitro-N-propyl-p-toluidine **(33245-39-5)**
(2-Chloroethyl)trimethylammonium chloride **(999-81-5)**
(β-Chloroethyl)trimethylammonium chloride **(999-81-5)**
2-Chloroethyl vinyl ether **(110-75-8)**
endo-3-Chloro-exo-6-cyano-2-norbornanone O-(methyl-carbamoyl)oxime **(15271-41-7)**
Chlorofenizon **(80-33-1)**
3-(4-(4-Chloro-fenossil)fenil)-1,1-dimetil-urea (Italian) **(1982-47-4)**
Chlorofensone **(80-33-1)**
Chlorofenvinphos **(470-90-6)**
p-Chlorofenylester kyseliny benzensulfonove (Czech) **(80-38-6)**
3-Chloro-4-fluoroaniline **(367-21-5)**
3-Chloro-4-fluorobenzenamine **(367-21-5)**
1-Chloro-2-fluorobenzene **(348-51-6)**
1-Chloro-3-fluorobenzene **(625-98-9)**
1-Chloro-4-fluorobenzene **(352-33-0)**
o-Chlorofluorobenzene **(348-51-6)**
p-Chlorofluorobenzene **(352-33-0)**
Chlorofluoromethane **(593-70-4)**
1-Chloro-3-fluoro-2-methylbenzene **(443-83-4)**
2-Chloro-1-fluoro-4-nitrobenzene **(350-30-1)**
3-Chloro-4-fluoronitrobenzene **(350-30-1)**
7-Chloro-5-(2-fluorophenyl)-1,3-dihydro-1-(2,2,2-trifluoroethyl)-2H-1,4-benzodiazepine-2-thione **(36735-22-5)**
7-Chloro-5-(o-fluorophenyl)-1,3-dihydro-1-(2,2,2-trifluoroethyl)-2H-1,4-benzodiazepine-2-thione **(36735-22-5)**
7-Chloro-5-(2-fluorophenyl)-1-methyl-1H-1,4-benzodiazepin-2(3H)-one **(3900-31-0)**
6-(3-(2-Chloro-6-fluorophenyl)-5-methyl-4-isoxazolecarboxamido)-3,3-dimethyl-7-oxo-4-thia-1-azabicyclo(3.2.0)heptane-2-carboxylic acid **(5250-39-5)**
3-(2-Chloro-6-fluorophenyl)-5-methyl-4-isoxazolylpenicillin **(5250-39-5)**
1-Chloro-4-(((4-fluorophenyl)thio)methyl)benzene **(405-30-1)**
2-Chloro-6-fluorotoluene **(443-83-4)**
α-Chloro-2-fluorotoluene **(345-35-7)**
α-Chloro-3-fluorotoluene **(456-42-8)**
α-Chloro-4-fluorotoluene **(352-11-4)**
α-Chloro-m-fluorotoluene **(456-42-8)**
α-Chloro-o-fluorotoluene **(345-35-7)**
α-Chloro-p-fluorotoluene **(352-11-4)**
Chloroflurazole **(3615-21-2)**
Chloroflurenol (French) **(2464-37-1)**
Chloroflurenol-methyl ester **(2536-31-4)**
Chloroform (ACGIH,OSHA) [UN 1888] **(67-66-3)**
4'-Chloroformanilide **(2617-79-0)**
p-Chloroformanilide **(2617-79-0)**
Chloroforme (French) **(67-66-3)**
Chloroformiate de methyle (French) **(79-22-1)**
Chloroformic acid allyl ester **(2937-50-0)**
Chloroformic acid benzyl ester **(501-53-1)**
Chloroformic acid 2-chloroethyl ester **(627-11-2)**
Chloroformic acid dimethylamide **(79-44-7)**
Chloroformic acid ethyl ester **(541-41-3)**
Chloroformic acid 2-fluoroethyl ester **(462-27-1)**
Chloroformic acid n-hexyl ester **(6092-54-2)**
Chloroformic acid isopropyl ester **(108-23-6)**
Chloroformic acid methyl ester **(79-22-1)**
Chloroformic acid phenyl ester **(1885-14-9)**
Chloroformic digitalin **(20830-75-5)**
Chloroform, methyl- **(71-55-6)**
Chloroform, nitro- **(76-06-2)**
Chloroformyl chloride **(75-44-5)**
Chlorofos **(52-68-6)**
Chloroftalm **(52-68-6)**
2-Chlorofumaric acid **(617-43-6)**
4-Chloro-N-furfuryl-5-sulfamoylanthranilic acid **(54-31-9)**
4-Chloro-N-(2-furylmethyl)-5-sulfamoylanthranilic acid **(54-31-9)**
Chlorogen **(127-52-6)**
Chlorogenic acid **(327-97-9)**
6-Chloroguanine **(10310-21-1)**
1-Chloroheptadecane **(62016-75-5)**
1-Chloroheptane **(629-06-1)**
Chloroheptane **(29756-37-4)**
1-Chlorohexadecane **(4860-03-1)**
1-Chloro-1,1,2,2,3,3-hexafluoropropane **(422-55-9)**
10-Chloro-2,3,5,6,7,11b-hexahydro-11b-(o-chlorophenyl)benzo-(6,7)-1,4-diazepino-(5,4-b)-oxazol-6-one **(24166-13-0)**
1-Chlorohexane **(544-10-5)**
2-Chlorohexane **(638-28-8)**
Chlorohexidine diacetate **(56-95-1)**
Chlorohydric acid **(7647-01-0)**
α-Chlorohydrin **(96-24-2)**
epi-Chlorohydrin **(106-89-8)**
Chlorohydrol **(12042-91-0)**
Chlorohydroquinone **(615-67-8)**
Chlorohydroquinone dimethyl ether **(2100-42-7)**
4-Chloro-α-hydroxybenzeneacetic acid **(492-86-4)**
3-Chloro-4-hydroxybenzoate **(3964-58-7)**
3-Chloro-6-hydroxybenzophenone **(85-19-8)**
5-Chloro-2-hydroxybenzophenone **(85-19-8)**
Chlorohydroxy benzophenone **(85-19-8)**
3-Chloro-4-hydroxybiphenyl **(92-04-6)**
4-Chloro-3-hydroxy-1-butene **(671-56-7)**
2-Chloro-5-hydroxy-1,3-dimethylbenzene **(88-04-0)**
4-Chloro-1-hydroxy-3,5-dimethylbenzene **(88-04-0)**
3-Chloro-4-hydroxydiphenyl **(92-04-6)**
5-Chloro-2-hydroxydiphenylmethane **(120-32-1)**
2-Chloro-10-3-(1-(2-hydroxyethyl)-4-piperazinyl)propyl phenothiazine **(58-39-9)**
2-Chloro-9-hydroxy-9H-fluorene-9-carboxylic acid **(2464-37-1)**
2-Chloro-9-hydroxyfluorene-9-carboxylic acid **(2464-37-1)**
5-Chloro-8-hydroxy-7-iodoquinoline **(130-26-7)**
2-Chloro-N-(hydroxymethyl)acetamide **(2832-19-1)**
2-Chloro-N-hydroxymethylacetamide **(2832-19-1)**
2-Chloro-9-hydroxy-9-methylcarboxylatefluorene **(2536-31-4)**
3-Chloro-7-hydroxy-4-methyl-coumarin O-ester with O,O-diethyl phosphorothioate **(56-72-4)**
3-Chloro-7-hydroxy-4-methyl-coumarin O,O-diethyl phosphorothioate **(56-72-4)**
(-)-N-((5-Chloro-8-hydroxy-3-methyl-1-oxo-7-isochromanyl)-carbonyl)-3-phenylalanine **(303-47-9)**

N-(((3R)-5-Chloro-8-hydroxy-3-methyl-1-oxo-7-isochromanyl)-carbonyl)-3-phenyl-l-alanine (303-47-9)
5'-Chloro-3-hydroxy-2-naphth-o-anisidide (137-52-0)
5-Chloro-2-((2-hydroxy-1-naphthalenyl)azo)-4-methylbenzene-sulfonic acid, barium salt (2:1) (5160-02-1)
5-Chloro-2-((2-hydroxy-1-naphthalenyl)azo)-4-methylbenzene-sulphonic acid, barium salt (5160-02-1)
5-Chloro-2-((2-hydroxy-1-naphthyl)azo)-p-toluenesulfonic acid, barium salt (5160-02-1)
N-(5-Chloro-2-hydroxy-4-nitrophenyl)benzamide (5099-06-9)
Chlorohydroxyoxozirconium (18428-88-1)
3-((5-Chloro-2-hydroxyphenyl)azo)-4,5-dihydroxy-2,7-naphthalene-disulfonic acid disodium salt (1058-92-0)
7-Chloro-3-hydroxy-5-phenyl-1,3-dihydro-2H-1,4-benzodiazepin-2-one (604-75-1)
(5-Chloro-2-hydroxyphenyl)phenylmethanone (85-19-8)
6-Chloro-17-α-hydroxypregna-4,6-diene-3,20-dione acetate (302-22-7)
6-Chloro-17-α-hydroxy-δ^6-progesterone acetate (302-22-7)
3-Chloro-2-hydroxy-1-propanesulfonic acid, sodium salt (126-83-0)
1-(3-Chloro-2-hydroxypropyl)-2-methyl-5-nitroimidazole (16773-42-5)
(3-Chloro-2-hydroxypropyl)trimethylammonium chloride (3327-22-8)
5-Chloro-8-hydroxyquinoline (130-16-5)
2-Chloro-hydroxytoluene (59-50-7)
6-Chloro-3-hydroxytoluene (59-50-7)
2-Chloro-5-hydroxy-m-xylene (88-04-0)
Chlorohyssopifolin C (41787-75-1)
4-Chloroimino-2,5-cyclohexadiene-1-one (637-61-6)
4-Chloroimino-2,6-dibromo-2,5-cyclohexadiene-1-one (537-45-1)
3-Chloroimipramine (303-49-1)
1-Chloro-2-iodobenzene (615-41-8)
1-Chloro-3-iodobenzene (625-99-0)
1-Chloro-4-iodobenzene (637-87-6)
5-Chloro-7-iodo-8-hydroxyquinoline (130-26-7)
Chloroiodoquine (130-26-7)
5-Chloro-7-iodo-8-quinolinol (130-26-7)
2-Chloroisobutane (507-20-0)
1-Chloroisobutylene (513-37-1)
α-Chloroisobutylene (513-37-1)
γ-Chloroisobutylene (563-47-3)
2-Chloroisobutyraldehyde (917-93-1)
Chloroisocyanuric acid (13057-78-8)
2-Chloro-N-isopropylacetanilide (1918-16-7)
α-Chloro-N-isopropylacetanilide (1918-16-7)
1-Chloroisopropyl alcohol (127-00-4)
2-Chloro-4-isopropylamino-6-methylamino-s-triazine (3004-71-59)
4-Chloro-4'-isopropylbiphenyl (17790-61-3)
2-Chloro-N-isopropyl-N-phenylacetamide (1918-16-7)
Chloroject L (56-75-7)
Chlorolignin (8068-02-8)
Chloromadinone acetate (302-22-7)
2-Chloromaleic acid (617-42-5)
4-Chloromandelic acid (492-86-4)
p-Chloromandelic acid (492-86-4)
Chloromax (56-75-7)
Chloromelamine (7673-09-8)
S-(6-Chloro-3-(mercaptomethyl)-2-benzoxazolinone) O,O-diethyl phosphorodithioate (2310-17-0)
(Chloromercuri)benzene (100-56-1)
p-Chloro-mercuric benzoic acid (59-85-8)
1-(3-(Chloromercuri)-2-methoxypropyl)urea (62-37-3)
Chloromercuriphenylsulfonate (554-77-8)
Chloromeridin (62-37-3)
Chloromerodrin (62-37-3)
4-Chlorometanilic acid (98-36-2)
6-Chlorometanilic acid (88-43-7)
Chloromethane (74-87-3)
Chloromethapyrilene (148-65-2)
1-Chloro-2-methoxybenzene (766-51-8)
1-Chloro-3-methoxybenzene (2845-89-8)
1-Chloro-4-methoxybenzene (623-12-1)
1-Chloro-2-methoxyethane (627-42-9)
Chloromethoxyethane (3188-13-4)
2-(2-Chloro-1-methoxyethoxy)phenol methylcarbamate (51487-69-5)
2-Chloro-N-(2-methoxyethyl)-o-acetotoluidide (50563-41-2)
2-Chloro-N-(2-methoxyethyl)acet-2',6'-xylidide (50563-36-5)
Chloro(2-methoxyethyl)mercury (123-88-6)
3-Chloro-7-methoxy-9-(1-methyl-4-diethylaminobutylamino)acridine (83-89-6)
3-Chloro-7-methoxy-9-(1-methyl-4-diethylaminobutylamino)-acridine dihydrochloride (69-05-6)
2-Chloro-N-((4-methoxy-6-methyl-1,3,5-triazin-2-yl)amino-carbonyl)-benzenesulfonamide (64902-72-3)
N-(3-Chloro-4-methoxyphenyl)-N',N'-dimethylurea (19937-59-8)
5-Chloro-2-methoxyprocainamide (364-62-5)
2-Chloro-6-methoxy-4-(trichloromethyl)pyridine (7159-34-4)
2-Chloro-N-methylacetoacetamide (4116-10-3)
1-Chloro-2-methylacetylene (7747-84-4)
Chloromethylacetylene (7747-84-4)
5-Chloro-6-((((methylamino)carbonyl)oxy)imino)bicyclo-(2.2.1)heptane-2-carbonitrile (15271-41-7)
7-Chloro-2-methylamino-5-phenyl-3H-1,4-benzodiazepine 4-oxide (58-25-3)
7-Chloro-2-methylamino-5-phenyl-3H-1,4-benzodiazepin-4-oxide (58-25-3)
4-Chloro-5-(methylamino)-2-(α,α,α-trifluoro-m-tolyl)-3(2H)-pyridazinone (27314-13-2)
2-Chloro-4-methylaniline (615-65-6)
3-Chloro-2-methylaniline (87-60-5)
3-Chloro-4-methylaniline (95-74-9)
3-Chloro-6-methylaniline (95-79-4)
4-Chloro-2-methylaniline (95-69-2)
4-Chloro-6-methylaniline (95-69-2)
5-Chloro-2-methylaniline (95-79-4)
6-Chloro-2-methylaniline (87-63-8)
4-Chloro-2-methylaniline hydrochloride (3165-93-3)
4-Chloro-6-methylaniline hydrochloride (3165-93-3)
1-Chloro-2-methyl-9,10-anthracenedione (129-35-1)
1-Chloro-2-methylanthraquinone (129-35-1)
4-Chloro-2-methylbenzenamine hydrochloride (3165-93-3)
5-Chloro-2-methylbenzenamine hydrochloride (6259-42-3)
2-Chloro-1-methylbenzene (95-49-8)
4-Chloro-1-methylbenzene (106-43-4)
Chloromethylbenzene (100-44-7)
Chloromethylbenzene (25168-05-2)
4-Chloro-2-methylbenzeneamine (95-69-2)
2-Chloro-α-methylbenzenemethanol (13524-04-4)
4-Chloro-α-methylbenzenemethanol (3391-10-4)
7-Chloro-1-methyl-5-3H-1,4-benzodiazepin-2(1H)-one (439-14-5)
5-Chloro-2-methylbenzoxazole (19219-99-9)
1-Chloro-2-methylbutane (616-13-7)
1-Chloro-3-methylbutane (107-84-6)
2-Chloro-2-methylbutane (594-36-5)
4-Chloro-2-methylbutane (107-84-6)
1-Chloro-2-methyl-2-butene (13417-43-1)
1-Chloro-3-methyl-2-butene (503-60-6)
3-Chloro-2-methyl-1-butene (5166-35-8)
2-Chloro-N-(2-methyl-6-tert-butylphenyl)acetamide (3785-20-4)
5-Chloro-3-methyl-catechol (31934-88-0)
3-Chloro-4-methyl-7-coumarinyl diethyl phosphorothioate (56-72-4)
O-3-Chloro-4-methyl-7-coumarinyl O,O-diethyl phosphorothioate (56-72-4)
Chloromethyl cyanide (107-14-2)
S-(Chloromethyl) O,O-diethylphosphorodithioate (24934-91-6)
S-Chloromethyl-O,O-diethylphosphorothiolothionate (24934-91-6)
2-Chloro-4-methyl-6-dimethylaminopyrimidine (535-89-7)
3-(Chloromethyl)-6-(1,1-dimethylethyl)-2,4-dimethylphenol (23500-79-0)
2-(Chloromethyl)-1,3-dioxolane (2568-30-1)
Chloromethyldiphenylsilane (144-79-6)

1,1'-(Chloromethylene)bisbenzene (90-99-3)
2-Chloro-6-methylergoline-8β-acetonitrile (36945-03-6)
(Chloromethyl)ethenylbenzene (30030-25-2)
Chloromethyl ether (542-88-1)
(Chloromethyl)ethylbenzene (26968-58-1)
1-Chloro-2-(1-methylethyl)benzene (2077-13-6)
1-Chloro-4-(1-methylethyl)benzene (2621-46-7)
(Chloromethyl)ethylene oxide (106-89-8)
(2-Chloro-1-methylethyl) ether (108-60-1)
Chloromethyl ethyl ether [UN 2354] (3188-13-4)
N-Chloro-N-(1-methylethyl)-2-propanamine (24948-81-0)
O-(5-Chloro-1-(1-methylethyl)-1H-1,2,4-triazol-3-yl) O,O-diethyl phosphorothioate (42509-80-8)
1-(Chloromethyl)-2-fluorobenzene (345-35-7)
1-(Chloromethyl)-3-fluorobenzene (456-42-8)
1-(Chloromethyl)-4-fluorobenzene (352-11-4)
3-Chloromethylheptane (123-04-6)
3-Chloro-4-methyl-7-hydroxycoumarin diethyl thiophosphoric acid ester (56-72-4)
2-(Chloromethyl)-1H-isoindole-1,3(2H)-dione (17564-64-6)
5-Chloro-2-methyl-4-isothiazolin-3-one (26172-55-4)
5-Chloro-2-methyl-4-isothiazolin-3-one calcium(II) chloride (57373-19-0)
3-Chloromethyl-4-keto-1,2,3-benzotriazine (24310-41-6)
Chloromethylmagnesium (676-58-4)
Chloromethylmercury (115-09-3)
1-(Chloromethyl)-2-methylbenzene (552-45-4)
1-(Chloromethyl)-4-methylbenzene (104-82-5)
Chloromethyl methyl ether (ACGIH,OSHA) (107-30-2)
α-(Chloromethyl)-2-methyl-5-nitro-1H-imidazole-1-ethanol (16773-42-5)
1-Chloromethyl naphthalene (86-52-2)
1-(Chloromethyl)-2-nitrobenzene (612-23-7)
1-Chloro-4-methyl-2-nitrobenzene (89-60-1)
2-Chloro-5-methylnitrobenzene (89-60-1)
4-(Chloromethyl)nitrobenzene (100-14-1)
4-Chloro-1-methyl-2-nitrobenzene (89-59-8)
p-(Chloromethyl)nitrobenzene (100-14-1)
5-Chloro-4-methyl-2-nitrobenzenesulfonic acid (6973-13-3)
2-Chloro-1-methyl-4-nitro-1H-imidazole (63634-21-9)
4-Chloro-1-methyl-5-nitroimidazole (4897-31-8)
5-Chloro-1-methyl-4-nitroimidazole (4897-25-0)
4-Chloro-5-methyl-2-nitrophenol (7147-89-9)
2-(Chloromethyl)oxirane (106-89-8)
Chloromethyloxirane (106-89-8)
Chloromethyloxirane, Polymer with 4,4'-(1-methylethylidene)-bis(2,6-dibromophenol) (40039-93-8)
2-Chloro-N-methyl-3-oxobutanamide (4116-10-3)
7-Chloro-1-methyl-2-oxo-5-phenyl-3H-1,4-benzodiazepine (439-14-5)
2-Chloro-4-methylphenol (6640-27-3)
2-Chloro-5-methylphenol (59-50-7)
2-Chloro-5-methylphenol (615-74-7)
2-Chloro-6-methylphenol (87-64-9)
4-Chloro-2-methylphenol (1570-64-5)
4-Chloro-3-methylphenol (59-50-7)
4-Chloro-2-methylphenol sodium salt (52106-86-2)
(4-Chloro-2-methylphenoxy)acetic acid (94-74-6)
4-Chloro-2-methylphenoxyacetic acid sodium salt (3653-48-3)
4-(4-Chloro-2-methylphenoxy)butyric acid (94-81-5)
γ-(4-Chloro-2-methylphenoxy)butyric acid (94-81-5)
4-(4-Chloro-2-methylphenoxy)butyric acid sodium salt (6062-26-6)
Chloromethylphenoxybutyric acid sodium salt (6062-26-6)
2-(4-Chloro-2-methylphenoxy)propanoic acid (93-65-2)
(+)-α-(4-Chloro-2-methylphenoxy) propionic acid (93-65-2)
2-(4-Chloro-2-methylphenoxy)propionic acid (93-65-2)
4-Chloro-2-methylphenoxy-α-propionic acid (93-65-2)
α-(4-Chloro-2-methylphenoxy)propionic acid (93-65-2)
5-((4-Chloro-6-(methylphenylamino)-1,3,5-triazin-2-yl)amino)-4-hydroxy-3-((4-methyl-2-sulfophenyl)azo)-2,7-naphthalene-disulfonic acid, trisodium salt (70210-46-7)
7-Chloro-1-methyl-5-phenyl-3H-1,4-benzodiazepin-2(1H)-one (439-14-5)
7-Chloro-1-methyl-5-phenyl-1H-1,5-benzodiazepine-2,4(3H,5H)-dione (22316-47-8)
7-Chloro-1-methyl-5-phenyl-2H-1,4-benzodiazepin-2-one (439-14-5)
7-Chloro-1-methyl-5-phenyl-1,3-dihydro-2H-1,4-benzodiazepin-2-one (439-14-5)
N'-(4-Chloro-2-methylphenyl)-N,N-dimethylmethanimidamide (6164-98-3)
3-(3-Chloro-4-methylphenyl)-1,1-dimethyl-urea (15545-48-9)
N-(3-Chloro-4-methylphenyl)-N',N'-dimethylurea (15545-48-9)
Chloromethyl phenyl ketone (532-27-4)
N-(3-Chloro-4-methylphenyl)-2-methylpentanamide (2307-68-8)
2-Chloro-2-methyl-1-phenyl-1-propanone (7473-99-6)
4-Chloro-2-methylphenylsulfonyl chloride (56157-92-7)
4-Chloro-2-methylphenylthioglycolic acid (94-76-8)
8-Chloro-1-methyl-6-phenyl-4H-s-triazolo(4,3-a)(1,4)benzo-diazepine (28981-97-7)
Chloromethylphthalimide (17564-64-6)
N-(Chloromethyl)phthalimide (17564-64-6)
2-Chloro-11-(4-methyl-1-piperazinyl)-dibenzo(b,f)(1,4)oxazepine (1977-10-2)
2-Chloro-11-(4-methyl-1-piperazinyl)-dibenzo(b,f)(1,4)oxoazepine (1977-10-2)
2-Chloro-10-(3-(1-methyl-4-piperazinyl)-propyl)-phenothiazine (58-38-8)
2-Chloro-10-(3-(4-methyl-1-piperazinyl)propyl)phenothiazine (58-38-8)
3-Chloro-10-(3-(1-methyl-4-piperazinyl)propyl)phenothiazine (58-38-8)
Chloro-3 (N-methylpiperazinyl-3 propyl)-10 phenothiazine (French) (58-38-8)
2-Chloro-2-methylpropanal (917-93-1)
1-Chloro-2-methylpropane (513-36-0)
2-Chloro-2-methylpropane (507-20-0)
1-Chloro-2-methyl-1-propene (513-37-1)
1-Chloro-2-methyl-2-propene (563-47-3)
1-Chloro-2-methylpropene (513-37-1)
3-Chloro-2-methyl-1-propene (563-47-3)
3-Chloro-2-methylpropene (563-47-3)
Chloro(1-methylpropyl)magnesium (15366-08-2)
2-Chloro-N-(1-methyl-2-propynyl)acetanilide (21267-72-1)
2-(Chloromethyl) pyridine, hydrochloride (6959-47-3)
3-(Chloromethyl) pyridine, hydrochloride (6959-48-4)
(6-Chloro-2-methyl-5-pyrimidyl)acetic acid, ethyl ester (14273-76-8)
Chloro methyl sulfone (124-63-0)
4-((5-Chloro-4-methyl-2-sulfophenyl)azo)-3-hydroxy-2-naphthalenecarboxylic acid, strontium salt (1:1) (15782-05-5)
2-Chloromethyltetrahydrofuran (3003-84-7)
1-Chloro-2-(methylthio)ethane (542-81-4)
(p-Chloromethyl)toluene (104-82-5)
2-(Chloromethyl)toluene (552-45-4)
4-(Chloromethyl)toluene (104-82-5)
p-Chloromethyltoluene (104-82-5)
(Chloromethyl)toluene (2-(chloromethyl) plus 4-(chloromethyl)) (104-82-5)
(Chloromethyl)trichlorosilane (1558-25-4)
N-Chloromethyltrimellitimide (17564-64-6)
3-Chloro-4-methylumbelliferone O-ester with O,O-diethyl phosphorothioate (56-72-4)
3'-Chloro-2-methyl-p-valerotoluidide (2307-68-8)
4-Chloromorpholine (23328-69-0)
Chloromycetin (56-75-7)
Chloromycetny (Polish) (56-75-7)
Chloronaftina (494-03-1)
1-Chloronaphthalene (90-13-1)
2-Chloronaphthalene (91-58-7)
Chloronaphthalene (25586-43-0)
α-Chloronaphthalene (90-13-1)
β-Chloronaphthalene (91-58-7)
4-Chloro-1-naphthalenol (604-44-4)
Chloronaphthine (494-03-1)
4-Chloro-1-naphthol (604-44-4)
Chloronase (94-20-2)
Chloroneb (2675-77-6)

Chloronebe (French)	(2675-77-6)
Chloronitrin	(56-75-7)
4-Chloro-3-nitro-N,N-dimethylbenzenesulfonamide	(137-47-3)
α-Chloro-m-nitro-acetophenone	(99-47-8)
6-Chloro-4-nitro-2-aminophenol	(6358-09-4)
2-Chloro-4-nitroaniline	(121-87-9)
2-Chloro-5-nitroaniline	(6283-25-6)
4-Chloro-2-nitroaniline	(89-63-4)
4-Chloro-3-nitroaniline	(635-22-3)
o-Chloro-p-nitroaniline	(121-87-9)
p-Chloro-o-nitroaniline	(89-63-4)
1-Chloro-5-nitroanthraquinone	(129-40-8)
5-Chloro-1-nitroanthraquinone	(129-40-8)
1-Chloro-2-nitrobenzene	(88-73-3)
1-Chloro-3-nitrobenzene	(121-73-3)
1-Chloro-4-nitrobenzene	(100-00-5)
2-Chloro-1-nitrobenzene	(88-73-3)
2-Chloronitrobenzene	(88-73-3)
4-Chloro-1-nitrobenzene	(100-00-5)
4-Chloronitrobenzene	(100-00-5)
Chloro-m-nitrobenzene	(121-73-3)
Chloro-o-nitrobenzene	(88-73-3)
Chloronitrobenzene	(25167-93-5)
o-Chloronitrobenzene (DOT)	(88-73-3)
p-Chloronitrobenzene (DOT)	(100-00-5)
m-Chloronitrobenzene, Solid [UN 1578]	(121-73-3)
5-Chloro-3-nitro-1,2-benzenediamine	(42389-30-0)
4-Chloro-2-nitrobenzenediazonium	(27165-22-6)
4-Chloro-2-nitrobenzenediazonium, chloride, zinc chloride	(14263-89-9)
4-Chloro-3-nitrobenzenesulfonamide	(97-09-6)
2-Chloro-5-nitrobenzenesulfonic acid	(96-73-1)
2-Chloro-5-nitrobenzenesulfonyl chloride	(4533-95-3)
4-Chloro-3-nitrobenzenesulfonyl chloride	(97-08-5)
2-Chloro-4-nitrobenzoic acid	(99-60-5)
2-Chloro-5-nitrobenzoic acid	(2516-96-3)
2-Chloro-5-nitrobenzotrifluoride	(777-37-7)
4-Chloro-3-nitrobenzotrifluoride	(121-17-5)
3-Chloro-4-(2'-nitro-3'-chlorophenyl)pyrrole	(1018-71-9)
5-Chloro-2-nitro-p-diethoxybenzene	(91-43-0)
3'-Chloro-4-nitrodiphenylamine	(15979-85-8)
1-Chloro-1-nitroethane	(598-92-5)
2-Chloro-4-nitrophenol	(619-08-9)
2-Chloro-6-nitrophenol	(603-86-1)
3-Chloro-4-nitrophenol	(491-11-2)
4-Chloro-2-nitrophenol	(89-64-5)
5-Chloro-2-nitrophenol	(611-07-4)
2-Chloro-4-nitrophenylamide-6-chlorosalicylic acid	(50-65-7)
4-((2-Chloro-4-nitrophenyl)azo)-N,N-biscyanoethylaniline	(4058-30-4)
2-((4-Chloro-2-nitrophenyl)azo)-N-(2-chlorophenyl)-3-oxobutanamide	(6486-23-3)
5-((4-Chloro-2-nitrophenyl)azo)-1-ethyl-1,2-dihydro-6-hydroxy-4-methyl-2-oxo-3-pyridinecarbonitrile	(70528-90-4)
4-(2-Chloro-4-nitrophenylazo)-N-ethyl-N-(β-succinimidoethyl)-3-acetamidoaniline	(29649-47-6)
3-(4-((2-Chloro-4-nitrophenyl)azo)-N-ethyl-m-toluidino)propionitrile	(16586-43-9)
2-((4-Chloro-2-nitrophenyl)azo)-N-(2-methoxyphenyl)-3-oxobutanamide	(13515-40-7)
4-(2-Chloro-4-nitrophenylazo)-3-methyl-N,N-bis(2-hydroxyethyl)-aniline	(3769-57-1)
1-((2-Chloro-4-nitrophenyl)azo)-2-naphthol	(2814-77-9)
3,3'-((4-((2-Chloro-4-nitrophenyl)azo)phenyl)imino)bispropanenitrile	(4058-30-4)
3,3'-((p-((2-Chloro-4-nitrophenyl)azo)phenylimino)bispropionitrile	(4058-30-4)
4-Chloro-2-nitrophenyl p-chlorophenyl ether	(135-12-6)
N-(2-Chloro-4-nitrophenyl)-5-chlorosalicylamide	(50-65-7)
O-(2-Chloro-4-nitrophenyl) O,O-dimethyl phosphorothioate	(2463-84-5)
O-(3-Chloro-4-nitrophenyl) O,O-dimethyl phosphorothioate	(500-28-7)
2-Chloro-1-(p-nitrophenyl)propane	(34197-98-3)
Chloronitropropan (Polish)	(2425-66-3)
1-Chloro-2-nitropropane	(2425-66-3)
2-Chloro-2-nitropropane	(594-71-8)
Chloronitropropane	(2425-66-3)
Chloronitropropane	(600-25-9)
1-Chloro-1-nitropropane (ACGIH,OSHA)	(600-25-9)
1-Chloro-4-nitrosobenzene	(932-98-9)
4-Chloronitrosobenzene	(932-98-9)
2-Chloro-3-nitrosobutane dimer	(6865-97-0)
2-Chloro-4-nitrotoluene	(121-86-8)
2-Chloro-6-nitrotoluene	(83-42-1)
4-Chloro-2-nitrotoluene	(89-59-8)
4-Chloro-3-nitrotoluene	(89-60-1)
6-Chloro-2-nitrotoluene	(83-42-1)
α-Chloro-m-nitrotoluene	(619-23-8)
α-Chloro-o-nitrotoluene	(612-23-7)
α-Chloro-p-nitrotoluene	(100-14-1)
p-Chloro-o-nitrotoluene	(89-59-8)
2-Chloro-5-nitro-1-toluene-4-sulfonic acid	(6973-13-3)
1-Chloro-4-nitro-2-(trifluoromethyl)benzene	(777-37-7)
4-Chloro-3-nitro-α,α,α-trifluorotoluene	(121-17-5)
1-Chlorononane	(2473-01-0)
2-Chloro-4-nonylphenol	(60044-33-9)
1-(Chloro-2-norbornyl)-3,3-dimethylurea	(1319-96-6)
1-Chlorooctane	(111-85-3)
2-Chlorooctane	(628-61-5)
2-Chloro-4-tert-octylphenol	(17199-24-5)
1-Chloro-8-(p-octylphenoxy)-3,6-dioxaoctane	(66028-01-1)
Chlorooxirane	(7763-77-1)
4-Chloro-2-oxobenzothiazol-3-ylacetic acid	(3813-05-6)
3-(6-Chloro-2-oxobenzoxazolin-3-yl)methyl O,O-diethyl phosphorothiolothionate	(2310-17-0)
S-((6-Chloro-2-oxo-3(2H)-benzoxazolyl)methyl) O,O-diethyl phosphorodithioate	(2310-17-0)
2-Chloro-3-oxobutanoic acid, ethyl ester	(609-15-4)
exo-5-Chloro-6-oxo-endo-2-norbornanecarbonitrile O-(methylcarbamoyl)oxime	(15271-41-7)
Chloroparacide	(103-17-3)
1-Chloropentadecane	(4862-03-7)
Chloropentafluoroacetone	(79-53-8)
Chloropentafluorobenzene	(344-07-0)
Chloropentafluoroethane (ACGIH,DOT,OSHA) [UN 1020]	(76-15-3)
Chloropentahydroxydialuminum	(12042-91-0)
1-Chloropentane	(543-59-9)
2-Chloropentane	(625-29-6)
3-Chloropentane	(616-20-6)
1-Chloro-1-pentene	(21450-13-5)
3-Chloroperbenzoic acid	(937-14-4)
m-Chloroperbenzoic acid	(937-14-4)
Chloroperfluorobenzene	(344-07-0)
3-Chloroperoxybenzoic acid	(937-14-4)
m-Chloroperoxybenzoic acid	(937-14-4)
3-Chloroperoxybenzoic acid, Maximum concentration 86%	(937-14-4)
3-Chloroperoxybenzoic acid (Not more than 86% with 3-chlorobenzoic acid)	(937-14-4)
Chloroperoxyl	(10049-04-4)
Chlorophacinone	(3691-35-8)
Chlorophen	(87-86-5)
m-Chlorophenacyl bromide	(41011-01-2)
Chlorophenamidin	(6164-98-3)
Chlorophenamidine	(6164-98-3)
Chlorophenamidine hydrochloride	(19750-95-9)
10-Chlorophenanthrene	(947-72-8)
9-Chlorophenanthrene	(947-72-8)
Chlorophene	(120-32-1)

4-Chlorophene-1,3-diamine (5131-60-2)
4-Chlorophenethyl bromide (6529-53-9)
p-Chloro-.β.-phenethyl bromide (6529-53-9)
p-Chlorophenethyl bromide (6529-53-9)
4-Chloropheniramine (132-22-9)
2-Chlorophenol (95-57-8)
3-Chlorophenol (108-43-0)
4-Chlorophenol (106-48-9)
m-Chlorophenol (108-43-0)
o-Chlorophenol (95-57-8)
p-Chlorophenol (106-48-9)
m-Chlorophenol, Liquid [UN 2021] (108-43-0)
o-Chlorophenol, Liquid [UN 2021] (95-57-8)
p-Chlorophenol, Liquid [UN 2021] (106-48-9)
Chlorophenol Red (4430-20-0)
m-Chlorophenol, Solid [UN 2020] (108-43-0)
o-Chlorophenol, Solid [UN 2020] (95-57-8)
p-Chlorophenol, Solid [UN 2020] (106-48-9)
Chlorophenothan (50-29-3)
Chlorophenothane (50-29-3)
4-(3-(2-Chlorophenothiazin-10-yl)propyl)-1-piperazineethanol (58-39-9)
4-(3-(2-Chlorophenothiazin-10-yl)propyl)-1-piperazineethanol acetate (84-06-0)
Chlorophenotoxum (50-29-3)
(2-Chlorophenoxy)acetic acid (614-61-9)
(4-Chlorophenoxy)acetic acid (122-88-3)
3-Chlorophenoxyacetic acid (588-32-9)
o-Chlorophenoxyacetic acid (614-61-9)
p-Chlorophenoxyacetic acid (122-88-3)
1-Chloro-4-phenoxybenzene (7005-72-3)
2-(4-Chlorophenoxy)-1-tert-butyl-2-(1H-1,2,4-triazole-1-yl)-ethanol (55219-65-3)
β-(4-Chlorophenoxy)-α-(1,1-dimethylethyl)-1H-1,2,4-triazole-1-ethanol (55219-65-3)
1-(4-Chlorophenoxy)-3,3-dimethyl-1-(1,2,4-triazol-1-yl)-butan-2-one (43121-43-3)
1-(4-Chlorophenoxy)-3,3-dimethyl-1-(1H-1,2,4-triazol-1-yl)-2-butanone (43121-43-3)
3-(p-Chlorophenoxy)-2-hydroxypropyl carbamate (886-74-8)
α-(p-Chlorophenoxy)isobutyric acid (882-09-7)
α-(p-Chlorophenoxy)isobutyric acid, ethyl ester (637-07-0)
α-p-Chlorophenoxyisobutyryl ethyl ester (637-07-0)
2-(4-Chlorophenoxy)-2-methylpropanoic acid (882-09-7)
2-(4-Chlorophenoxy)-2-methylpropanoic acid ethyl ester (637-07-0)
2-(4-Chlorophenoxy-2-methyl)propionic acid (93-65-2)
2-(m-Chlorophenoxy)-2-methylpropionic acid (17413-73-9)
2-(p-Chlorophenoxy)-2-methylpropionic acid (882-09-7)
2-(p-Chlorophenoxy)-2-methylpropionic acid ethyl ester (637-07-0)
3-(p-(p-Chlorophenoxy)phenyl)-1,1-dimethylurea (1982-47-4)
N'-4-(4-Chlorophenoxy)phenyl-N,N-dimethylurea (1982-47-4)
1-(4-(4-Chloro-phenoxy)phenyl)-3,3-d'methyluree (French) (1982-47-4)
2-(3-Chlorophenoxy)propanamide (5825-87-6)
3-(4-Chlorophenoxy)-1,2-propanediol-1-carbamate (886-74-8)
3-(p-Chlorophenoxy)-1,2-propanediol-1-carbamate (886-74-8)
2-(3-Chlorophenoxy)propanoic acid (101-10-0)
2-(m-Chlorophenoxy)propionamide (5825-87-6)
2-(4-Chlorophenoxy)propionic acid (3307-39-9)
2-(m-Chlorophenoxy)propionic acid (101-10-0)
2-(p-Chlorophenoxy)propionic acid (3307-39-9)
2-(m-Chlorophenoxy)propionic acid, sodium salt (53404-22-1)
Chlorophentermine (461-78-9)
N-(2-Chlorophenyl)acetamide (533-17-5)
N-(4-Chlorophenyl)acetamide (539-03-7)
N-(p-Chlorophenyl)acetamide (539-03-7)
4-Chlorophenyl acetate (876-27-7)
p-Chlorophenyl acetate (876-27-7)
(4-Chlorophenyl)acetic acid (1878-66-6)
(p-Chlorophenyl)acetic acid (1878-66-6)
2-(p-Chlorophenyl)acetic acid (1878-66-6)
N-(2-Chlorophenyl)acetoacetamide (93-70-9)
(4-Chlorophenyl)acetonitrile (140-53-4)
2-(4-Chlorophenyl)acetonitrile (140-53-4)
p-Chlorophenylacetonitrile (140-53-4)
3-(α-p-Chlorophenyl-β-acetylethyl)-4-hydroxycoumarin (81-82-3)
2-(α-p-Chlorophenylacetyl)indane-1,3-dione (3691-35-8)
3-(p-Chlorophenyl)acrylic acid (1615-02-7)
3-Chlorophenylamine (108-42-9)
4-Chlorophenylamine (106-47-8)
m-Chlorophenylamine (108-42-9)
p-Chlorophenylamine hydrochloride (20265-96-7)
2-(4-Chlorophenylamino)-4-amino-1,3,5-triazine (500-42-5)
N-(((4-Chlorophenyl)amino)carbonyl)-2,6-difluorobenzamide (35367-38-5)
2-Chloro-4-(phenylazo)phenol (6657-05-2)
N-(2-Chlorophenyl)benzamide (1020-39-9)
4-Chloro-α-phenylbenzenemethanol (119-56-2)
4-Chlorophenyl benzenesulfonate (80-38-6)
p-Chlorophenyl benzenesulfonate (80-38-6)
4-Chlorophenyl benzenesulphonate (80-38-6)
p-Chlorophenyl benzenesulphonate (80-38-6)
1-(α-(2-Chlorophenyl)benzhydryl)imidazole (23593-75-1)
α-(2-Chlorophenyl)benzyl chloride (56961-47-8)
1-(p-Chloro-α-phenylbenzyl)-4-(2-(2-hydroxyethoxy)ethyl)-piperazine (68-88-2)
1-(p-Chloro-α-phenylbenzyl)-4-(m-methylbenzyl)piperazine (569-65-3)
1-(p-Chloro-α-phenylbenzyl)-4-methylpiperazine (82-93-9)
1-(p-Chloro-α-phenylbenzyl)-4-methylpiperazine hydrochloride (14362-31-3)
2-(2-(4-(p-Chloro-α-phenylbenzyl)-1-piperazinyl)ethoxy)ethanol (68-88-2)
3-Chlorophenyl bromide (108-37-2)
4-Chlorophenyl bromide (106-39-8)
m-Chlorophenyl bromide (108-37-2)
p-Chlorophenyl bromide (106-39-8)
N-(3-Chloro phenyl) carbamate de 4-chloro 2-butynyle (French) (101-27-9)
N-(3-Chloro phenyl) carbamate d'isopropyle (French) (101-21-3)
(3-Chlorophenyl)carbamic acid 4-chloro-2-butynyl ester (101-27-9)
N-(3-Chlorophenyl)carbamic acid, isopropyl ester (101-21-3)
(3-Chlorophenyl)carbamic acid, 1-methylethyl ester (101-21-3)
3-Chlorophenylcarbamic acid 1-methylpropynyl ester (1967-16-4)
3-Chlorophenyl carbanilate (16400-09-2)
p-Chlorophenyl chloride (106-46-7)
4-Chlorophenyl 4-chlorobenzenesulfonate (80-33-1)
p-Chlorophenyl p-chlorobenzenesulfonate (80-33-1)
4-Chlorophenyl 4-chlorobenzenesulphonate (80-33-1)
p-Chlorophenyl p-chlorobenzenesulphonate (80-33-1)
4-Chlorophenyl 4'-chlorobenzyl sulfide (103-17-3)
2-(o-Chlorophenyl)-2-(p-chlorophenyl)-1,1-dichloroethane (53-19-0)
(2-Chlorophenyl)(4-chlorophenyl)methanone (85-29-0)
(2-Chlorophenyl)-α-(4-chlorophenyl)-5-pyrimidinemethanol (60168-88-9)
α-(2-Chlorophenyl)-α-(4-chlorophenyl)-5-pyrimidinemethanol (60168-88-9)
N-(4-Chlorophenyl)-N'-(4-chloro-3-(trifluoromethyl)phenyl)urea (369-77-7)
4-Chloro-α-phenyl-o-cresol (120-32-1)
1-(p-Chlorophenyl)-4,6-diamino-2,2-dimethyl-1,2-dihydro-s-triazine (516-21-2)
5-(4'-Chlorophenyl)-2,4-diamino-6-ethylpyrimidine (58-14-0)
N-(3-Chlorophenyl)diethanolamine (92-00-2)
N-(m-Chlorophenyl)diethanolamine (92-00-2)
2-(2-Chlorophenyl)-2-(diethoxyphosphinothioyloxyimino)acetonitrile (14816-20-7)
1-(4-Chlorophenyl)-3-(2,6-difluorobenzoyl)urea (35367-38-5)
1-p-Chlorophenyl-1,2-dihydro-2,2-dimethyl-4,6-diamino-s-triazine (516-21-2)
5-p-Chlorophenyl-2,3-dihydro-5H-imidazo(2,1-a)isoindol-5-ol (22232-71-9)
6-(2-Chlorophenyl)-2,4-dihydro-2-((4-methyl-1-piperazinyl)-methylene)-8-nitro-1H-imidazo(1,2-a) (1,4)benzodiazepin-1-one (61197-73-7)
5-(2-Chlorophenyl)-1,3-dihydro-7-nitro-2H-1,4-benzodiazepin-2-one (1622-61-3)
5-(o-Chlorophenyl)-1,3-dihydro-7-nitro-2H-1,4-benzodiazepin-2-one (1622-61-3)

4-(3-(4-Chlorophenyl)-4,5-dihydro-1H-pyrazol-1-yl)benzenesulfonamide (2744-49-2)
4-(4-Chlorophenyl)-N,N-dimethyl-α,α-diphenyl-4-hydroxy-1-piperidinebutanamide (53179-11-6)
β-(p-Chlorophenyl)-α,α-dimethylethylamine (461-78-9)
N-(4-Chlorophenyl)-2,2-dimethylpentanamide (7287-36-7)
γ-(4-Chlorophenyl)-N,N-dimethyl-2-pyridinepropanamine (25523-97-1)
1-(4-Chlorophenyl)-3,3-dimethyltriazene (7203-90-9)
1-(p-Chlorophenyl)-3,3-dimethyl-triazene (7203-90-9)
1-(m-Chlorophenyl)-3,3-dimethylurea (587-34-8)
1-(p-Chlorophenyl)-3,3-dimethylurea (150-68-5)
3-(4-Chlorophenyl)-1,1-dimethylurea (150-68-5)
3-(p-Chlorophenyl)-1,1-dimethylurea (150-68-5)
N'-(4-Chlorophenyl)-N,N-dimethylurea (150-68-5)
N-(p-Chlorophenyl)-N',N'-dimethylurea (150-68-5)
3-(p-Chlorophenyl)-1,1-dimethylurea, trichloroacetate (140-41-0)
1-(4-Chloro phenyl)-3,3-dimethyluree (French) (150-68-5)
N-(4-Chlorophenyl)-2,2-dimethylvaleroamide (7287-36-7)
2-(2-Chlorophenyl)-4,5-diphenyl-1H-imidazole (1707-67-1)
2-(o-Chlorophenyl)-4,5-diphenylimidazole (1707-67-1)
1-((2-Chlorophenyl)diphenylmethyl)-1H-imidazole (9CI) (23593-75-1)
p-Chlorophenyl disulfide (1142-19-4)
2-Chloro-p-phenylenediamine (615-66-7)
4-Chloro-1,2-phenylenediamine (95-83-0)
4-Chloro-1,3-phenylenediamine (5131-60-2)
4-Chloro-m-phenylenediamine (5131-60-2)
4-Chloro-o-phenylenediamine (95-83-0)
4-Chlorophenylene-1,3-diamine (5131-60-2)
o-Chloro-p-phenylenediamine (615-66-7)
p-Chloro-m-phenylenediamine (5131-60-2)
p-Chloro-o-phenylenediamine (95-83-0)
4-Chloro-o-phenylenediamine, monosulfate (68459-98-3)
2-Chloro-p-phenylenediamine sulfate (6219-71-2)
2-Chloro-1-phenylethanone (532-27-4)
7-(2-Chlorophenyl)-4-ethoxy-3,5-dioxa-6-aza-4-phosphaoct-6-ene-8-nitrile 4-sulfide (14816-20-7)
2-(p-Chlorophenyl)ethyl bromide (6529-53-9)
5-(o-Chlorophenyl)-7-ethyl-1,3-dihydro-1-methyl-2H-thieno(2,3-E)-1,4-diazepin-2-one (33671-46-4)
α-(2-(4-Chlorophenyl)ethyl)-α-(1,1-dimethylethyl)-1H-1,2,4-triazole-1-ethanol (107534-96-3)
S-(p-Chlorophenyl) O-ethyl ethanephosphonodithioate (2984-64-7)
S-(4-Chlorophenyl) O-ethyl ethylphosphonodithioate (2984-64-7)
5-(2-Chlorophenyl)-7-ethyl-1-methyl-1,3-dihydro-2H-thieno-(2,3-E)(1,4)diazepin-2-one (33671-46-4)
5-(4-Chlorophenyl)-6-ethyl-2,4-pyrimidinediamine (58-14-0)
9-m-Chlorophenylfluorene (32377-11-0)
(2-Chlorophenyl)(4-fluorophenyl)methanone (1806-23-1)
N-(4-Chlorophenyl)formamide (2617-79-0)
N-(p-Chlorophenyl)formamide (2617-79-0)
3-Chloro-7-D-(2-phenylglycinamido)-3-cephem-4-carboxylic acid (53994-73-3)
4-(2-Chlorophenylhydrazone)-3-methyl-5-isoxazolone (5707-69-7)
4-(2-Chlorophenylhydrazono)-3-methyl-5(4H)-isoxazolone (5707-69-7)
4-(4-(4-Chlorophenyl)-4-hydroxy-1-piperidinyl)-1-(4-fluorophenyl)-1-butanone (52-86-8)
N'-(4-Chlorophenyl)-N-isobutinyl-N-methylurea (3766-60-7)
m-Chlorophenyl isocyanate (2909-38-8)
p-Chlorophenyl isocyanate (104-12-1)
N-3-Chlorophenylisopropylcarbamate (101-21-3)
Chlorophenylmagnesium (100-59-4)
o-Chlorophenylmercaptoacetic acid (18619-18-6)
Chlorophenylmercury (100-56-1)
Chlorophenylmethane (100-44-7)
N-p-Chlorophenyl-m-methoxybenzohydroxamic acid (77915-81-2)
3-(4-Chlorophenyl)-1-methoxy-1-methylurea (1746-81-2)
N-(4-Chlorophenyl)-N'-methoxy-N-methylurea (1746-81-2)
3-(o-Chlorophenyl)-2-methyl-4(3H)-quinazolinone (340-57-8)
2-(o-Chlorophenyl)-2-(methylamino)cyclohexanone hydrochloride (1867-66-9)
1-(o-Chlorophenyl)-2-methyl-2-aminopropane (10389-73-8)
1-(p-Chlorophenyl)-2-methyl-2-aminopropane (461-78-9)
N-p-Chlorophenyl-m-methylbenzohydroxamic acid (36016-24-7)
2-(4-Chlorophenyl)-α-methyl-5-benzoxazoleacetic acid (51234-28-7)
2-(p-Chlorophenyl)-3-methylbutyric acid (2012-74-0)
2-Chlorophenyl-N-methylcarbamate (3942-54-9)
4-Chlorophenyl-N-methylcarbamate (2620-53-3)
o-Chlorophenyl methylcarbamate (3942-54-9)
p-Chlorophenyl-N-methylcarbamate (2620-53-3)
S-((4-Chlorophenyl)methyl)diethylcarbamothioate (28249-77-6)
2-((2-Chlorophenyl)methyl)-4,4-dimethyl-3-isoxazolidinone (81777-89-1)
2-(4-Chlorophenyl)-3-methyl-4-metathiazanone-1,1-dioxide (80-77-3)
3-(p-Chlorophenyl)-1-methyl-1-(1-methyl-2-propynyl)urea (3766-60-7)
N'-(4-Chlorophenyl)-N-methyl-N-(1-methyl-2-propynyl)-urea (3766-60-7)
3-(4-Chlorophenyl)-1-methyl-1-(1-methylprop-2-ynyl)urea (3766-60-7)
4-Chloro-2-(phenylmethyl)phenol sodium salt (3184-65-4)
1-(o-Chlorophenyl)-2-methyl-2-propylamine (10389-73-8)
3-(o-Chlorophenyl)-2-methyl-4-quinazolone (340-57-8)
3-(p-Chlorophenyl)-5-methyl rhodanine (6012-92-6)
p-Chlorophenyl methyl sulfide (123-09-1)
4-Chlorophenyl methyl sulfone (98-57-7)
p-Chlorophenyl methyl sulfone (98-57-7)
4-Chlorophenyl methyl sulfoxide (934-73-6)
p-Chlorophenyl methyl sulfoxide (934-73-6)
5-(o-Chlorophenyl)-7-nitro-1H-1,4-benzodiazepin-2(3H)-one (1622-61-3)
N-p-(Chlorophenyl)-m-nitrobenzohydroxamic acid (36016-30-5)
S-(((2-Chlorophenyl)(1-oxobutyl)amino)methyl) O,O-dimethyl-phosphorodithioate (83733-82-8)
3-(1-(4-Chlorophenyl)-3-oxo-butyl)-4-hydroxy-coumarine (French) (81-82-3)
2-Chloro-4-phenylphenol (92-04-6)
2-Chloro-6-phenylphenol (85-97-2)
4-Chloro-2-phenylphenol (607-12-5)
6-Chloro-2-phenylphenol (85-97-2)
4-Chloro-2-phenylphenol sodium salt (10605-10-4)
(o-Chlorophenyl)phenylacetylene (10271-57-5)
2(2-(4-Chlorophenyl)-2-phenylacetyl)indan-1,3-dione (3691-35-8)
2-(2-(4-Chlorophenyl)-2-phenylacetyl)indan-1,3-dione (3691-35-8)
2-((p-Chlorophenyl)phenylacetyl)-1,3-indandione (8CI) (3691-35-8)
2-(2-(4-Chlorophenyl)-2-phenylacetyl)indane-1,3-dione (3691-35-8)
2-(α-p-Chlorophenyl-α-phenylacetyl)indane-1,3-dione (3691-35-8)
2-((4-Chlorophenyl)phenylacetyl)-1H-indene-1,3(2H)-dione (9CI) (3691-35-8)
3-Chlorophenyl N-phenylcarbamate (16400-09-2)
3-Chlorophenyl phenylcarbamate (16400-09-2)
m-Chlorophenyl N-phenylcarbamate (16400-09-2)
4-Chlorophenyl phenyl ether (7005-72-3)
p-Chlorophenyl phenyl ether (7005-72-3)
(4-Chlorophenyl)phenylmethane (831-81-2)
(p-Chlorophenyl)phenylmethane (831-81-2)
(4-Chlorophenyl)phenylmethanol (119-56-2)
(4-Chlorophenyl)phenylmethanone (134-85-0)
p-Chlorophenyl phenyl sulphone (80-00-2)
2-Chlorophenylpropiolic acid (24654-08-8)
4-Chlorophenylpropiolic acid (3240-10-6)
o-Chlorophenylpropiolic acid (24654-08-8)
1-p-Chlorophenyl-3-(propylsulfonyl)urea (94-20-2)
2-(4-Chlorophenyl)-2-propynoic acid (3240-10-6)
3-(2-Chlorophenyl)-2-propynoic acid (24654-08-8)
3-(3-Chlorophenyl)-2-propynoic acid (7396-28-3)
Chlorophenylpyridamine (132-22-9)
O-(6-Chloro-3-phenyl-4-pyridazinyl) S-octyl carbonothioate (55512-33-9)
1-(p-Chlorophenyl)-1-(2-pyridyl)-3-dimethylaminopropane (132-22-9)
1-p-Chlorophenyl-1-(2-pyridyl)-3-dimethylaminopropane maleate (113-92-8)
1-(p-Chlorophenyl)-1-(2-pyridyl)-3-N,N-dimethylpropylamine (132-22-9)
2-(p-(β-Chloro-α-phenylstyryl)phenoxy)-triethylamine (911-45-5)
1-Chloro-4-(phenylsulfonyl)benzene (80-00-2)
1-((o-Chlorophenyl)sulfonyl)-3-(4-methoxy-6-methyl-s-triazin-

2-yl)urea (64902-72-3)
1-(p-Chlorophenylsulfonyl)-3-propylurea (94-20-2)
2-(p-Chlorophenyl)tetrahydro-3-methyl-4H-1,3-thiazin-4-one 1,1-dioxide (80-77-3)
10-(m-Chlorophenyl)-2,3,4,10-tetrahydropyrimido(1,2-a)indol-10-ol (37751-39-6)
o-Chlorophenyl thioacetic acid (18619-18-6)
S-((p-Chlorophenylthio)methyl) O,O-diethyl phosphorodithioate (786-19-6)
S-(4-Chlorophenylthiomethyl)diethyl phosphorothiolothionate (786-19-6)
S-(((4-Chlorophenyl)thio)methyl) O,O-dimethylphosphorodithioate (953-17-3)
S-(((p-Chlorophenyl)thio)methyl) O,O-dimethyl phosphorodithioate (953-17-3)
2-Chlorophenyl thiourea (5344-82-1)
N-(4-Chlorophenyl)-1,3,5-triazine-2,4-diamine (500-42-5)
8-Chloro-6-phenyl-4H-(1,2,4)triazolo-(4,3-a)(1,4)benzodiazepine (29975-16-4)
8-Chloro-6-phenyl-4H-(1,2,4)triazolo(4,3-a)(1,4)benzodiazepine (29975-16-4)
8-Chloro-6-phenyl-4H-s-triazolo(4,3-a)(1,4)benzodiazepine (29975-16-4)
p-Chlorophenyltrichloromethane (5216-25-1)
4-Chlorophenyl 2,4,5-trichlorophenyl sulfide (2227-13-6)
p-Chlorophenyl 2,4,5-trichlorophenyl sulfide (2227-13-6)
4-Chlorophenyl 2,4,5-trichlorophenyl sulfone (116-29-0)
p-Chlorophenyl 2,4,5-trichlorophenyl sulfone (116-29-0)
p-Chlorophenyl 2,4,5-trichlorophenyl sulphone (116-29-0)
Chlorophenyltrichlorosilane (26571-79-9)
(p-Chlorophenyl)trifluoromethane (98-56-6)
(o-Chlorophenyl)urea (114-38-5)
1-(p-Chlorophenyl)urea (140-38-5)
2-Chlorophenylurea (114-38-5)
3-Chlorophenylurea (1967-27-7)
4-Chlorophenylurea (140-38-5)
meta-Chlorophenylurea (1967-27-7)
Chlorophibrinic acid (882-09-7)
Chlorophos (52-68-6)
Chlorophose (52-68-6)
Chlorophosphoric acid, diethyl ester (814-49-3)
4-Chlorophthalic acid (89-20-3)
S-(2-Chloro-1-phthalimidoethyl) O,O-diethyl phosphorodithioate (10311-84-9)
Chlorophthalm (52-68-6)
(Chloro-29H,31H-phthalocyaninato(2-)-N29,N30,N31,N32)copper (12239-87-1)
Chlorophyll A (479-61-8)
Chlorophyll A2 (479-61-8)
6-Chloropicolinic acid (4684-94-0)
Chloropicrin (ACGIH,OSHA) [UN 1580] (76-06-2)
Chloropicrin, Absorbed (DOT) (76-06-2)
Chloropicrin, Liquid (DOT) (76-06-2)
Chloropicrin and methyl bromide, Mixture [UN 1581] (8004-09-9)
Chloropicrine (French) (76-06-2)
Chloropicrin-methyl bromide mixt. (8004-09-9)
Chloropicrin mixture, Flammable [UN 1582] (76-06-2)
Chloropicrin mixtures, N.O.S. [UN 1583] (76-06-2)
6-Chloropiperonyl (+-)-cis/trans-chrysanthemumate (70-43-9)
6-Chloropiperonyl chrysanthemumate (70-43-9)
6-Chloropiperonyl 2,2-dimethyl-3-(2-methylpropenyl)-cyclopropanecarboxylate (70-43-9)
6-Chloropiperonyl ester of chrysanthemummonocarboxylic acid (70-43-9)
Chloropiril (113-92-8)
Chloropiril (132-22-9)
3-Chloropivalic acid (13511-38-1)
Chloropivalic acid (13511-38-1)
β-Chloropivalic acid (13511-38-1)
Chloroplatinic (IV) acid (16941-12-1)
Chloroplatinic acid (16941-12-1)
Chloroplatinic acid, Solid [UN 2507] (16941-12-1)
Chloropon (3278-46-4)
Chloropotassuril (7447-40-7)
Chloropreen (Dutch) (126-99-8)
6-Chloro-$\delta^{4,6}$-pregnadiene-17-α-ol-3,20-dione 17-acetate (302-22-7)
6-Chloro-pregna-4,6-dien-17-α-ol-3,20-dione acetate (302-22-7)
Chloropren (German, Polish) (126-99-8)
3-Chloroprene (107-05-1)
Chloroprene (126-99-8)
β-Chloroprene (ACGIH,OSHA) (126-99-8)
Chloroprene, Inhibited [UN 1991] (126-99-8)
Chloroprene Polymer (9010-98-4)
Chloroprene, Uninhibited (DOT) (126-99-8)
Chloroprocaine (133-16-4)
Chloropromazine (50-53-3)
Chloropromazine hydrochloride (69-09-0)
Chloropromazine monohydrochloride (69-09-0)
Chloropropadiene (3223-70-9)
Chloropropamide (94-20-2)
N-Chloro-2-propanamine (26245-56-7)
1-Chloropropane (540-54-5)
2-Chloropropane [UN 2356] (75-29-6)
1-Chloro-2,3-propanediol (96-24-2)
1-Chloropropane-2,3-diol (96-24-2)
3-Chloro-1,2-propanediol (96-24-2)
3-Chloropropane-1,2-diol (96-24-2)
3-Chloropropanenitrile (542-76-7)
3-Chloropropanoic acid (107-94-8)
Chloro-propanoic acid (28554-00-9)
1-Chloro-2-propanol (127-00-4)
2-Chloro-1-propanol (78-89-7)
2-Chloropropanol (78-89-7)
3-Chloropropanol (627-30-5)
3-Chloropropanol-1 [UN 2849] (627-30-5)
2-Chloro-1-propanol phosphate (3:1) (6145-73-9)
1-Chloro-2-propanone (78-95-5)
Chloropropanone (78-95-5)
3-Chloropropanonitrile (542-76-7)
2-Chloropropanoyl chloride (7623-09-8)
7-Chloro-1-propargyl-5-phenyl-2H-1,4-benzodiazepin-2-one (52463-83-9)
2-Chloropropenaldehyde (683-51-2)
1-Chloro propene-2 (107-05-1)
1-Chloro-1-propene (590-21-6)
1-Chloro-2-propene (107-05-1)
1-Chloropropene (590-21-6)
2-Chloro-1-propene (557-98-2)
3-Chloropropene (107-05-1)
3-Chloropropene-1 (107-05-1)
2-Chloropropene [UN 2456] (557-98-2)
2-Chloro-2-propene-1-thiol diethyldithiocarbamate (95-06-7)
2-Chloro-2-propen-1-ol (5976-47-6)
3-Chloropropenyl chloride (542-75-6)
2-Chloro-2-propenyl diethylcarbamodithioate (95-06-7)
Chloropropham (101-21-3)
Chloroprophenpyridamine (132-22-9)
Chloroprophenpyridamine maleate (113-92-8)
4-Chloropropionanilide (2759-54-8)
3-Chloropropionic acid (107-94-8)
Chloropropionic acid (28554-00-9)
α-Chloropropionic acid (28554-00-9)
α-Chloropropionic acid (598-78-7)
β-Chloropropionic acid (107-94-8)
3-Chloropropionitrile (542-76-7)
β-Chloropropionitrile (542-76-7)
2-Chloropropionyl chloride (7623-09-8)
2-Chloropropyl alcohol (78-89-7)
4-Chloro-4-((propylamino)carbonyl)benzenesulfonamide (94-20-2)
2-Chloro-4-(2-propylamino)-6-ethylamino-s-triazine (1912-24-9)
Chloropropylate (5836-10-2)
3-Chloropropyl bromide (109-70-6)
2-Chloropropyl-dimethylamine hydrochloride (4584-49-0)
3-Chloro-1-propylene (107-05-1)
3-Chloropropylene (107-05-1)

α-Chloropropylene **(107-05-1)**
3-Chloropropylene glycol **(96-24-2)**
3-Chloro-1,2-propylene oxide **(106-89-8)**
Chloropropylene oxide **(106-89-8)**
γ-Chloropropylene oxide **(106-89-8)**
1,1'-(2-Chloropropylidene)bis(4-ethoxybenzene) **(56265-22-6)**
Chloropropylmercury **(2440-40-6)**
N-(3-Chloropropyl)-α-methylphenethylamine **(17243-57-1)**
3-Chloropropyl n-octylsulfoxide **(3569-57-1)**
3-Chloropropyl-n-octylsulfoxide **(3569-57-1)**
(3-Chloropropyl)trichlorosilane **(2550-06-3)**
Chloropropyltrichlorosilane **(2550-06-3)**
(3-Chloropropyl)trimethoxysilane **(2530-87-2)**
1-Chloro-1-propyne **(7747-84-4)**
1-Chloropropyne **(7747-84-4)**
3-Chloro-1-propyne **(624-65-7)**
Chloroprothixene **(113-59-7)**
Chloroptic **(56-75-7)**
1-Chloropyrene **(34244-14-9)**
3-Chloropyrene **(34244-14-9)**
3-Chloropyridine **(626-60-8)**
4-Chloropyridine **(626-61-9)**
4-Chloropyridine **(7379-35-3)**
α-Chloropyridine **(109-09-1)**
m-Chloropyridine **(626-60-8)**
o-Chloropyridine **(109-09-1)**
2-Chloropyridine [UN 2822] **(109-09-1)**
4-Chloropyridine hydrochloride **(7379-35-3)**
Chloropyrilene **(148-65-2)**
Chloropyriphos-methyl **(5598-13-0)**
4-Chloropyrocatechol **(2138-22-9)**
1-Chloro-2,5-pyrrolidinedione **(128-09-6)**
3-Chloro-4-(3-pyrrolin-1-yl)hydratropic acid **(31793-07-4)**
Chloroquina **(54-05-7)**
Chloroquine **(54-05-7)**
Chloroquinium **(54-05-7)**
5-Chloro-8-quinolinol **(130-16-5)**
N^{4}-(7-Chloro-4-quinolinyl)-N^{1},N^{1}-diethyl-1,4-pentanediamine **(54-05-7)**
2-(4-((6-Chloro-2-quinoxalinyl)oxy)phenoxy)propanoic acid ethyl ester **(76578-14-8)**
4-Chlororesorcinol **(95-88-5)**
Chloros **(7681-52-9)**
5-Chlorosalicyclanilide **(4638-48-6)**
5-Chlorosalicylic acid **(321-14-2)**
N-Chloro-N-sodiobenzenesulfonamide **(127-52-6)**
.α.-Chlorostilbene **(1460-06-6)**
2-Chlorostyrene **(2039-87-4)**
3-Chlorostyrene **(2039-85-2)**
4-Chlorostyrene **(1073-67-2)**
Chlorostyrene **(1331-28-8)**
m-Chlorostyrene **(2039-85-2)**
p-Chlorostyrene **(1073-67-2)**
o-Chlorostyrene (ACGIH,OSHA) **(2039-87-4)**
N-Chlorosuccinimide **(128-09-6)**
Chlorosulfacide **(103-17-3)**
6-Chloro-7-sulfamoyl-2H-1,2,4-benzothiadiazine 1,1-dioxide **(58-94-6)**
6-Chloro-7-sulfamoyl-3,4-dihydro-2H-1,2,4-benzothiadiazine 1,1-dioxide **(58-93-5)**
Chlorosulfona **(3064-70-8)**
Chlorosulfonic acid (DOT) **(7790-94-5)**
Chlorosulfonic acid (with or without sulfur trioxide) [UN 1754] **(7790-94-5)**
Chlorosulfonic anhydride **(7791-27-7)**
4'-(Chlorosulfonyl)acetanilide **(121-60-8)**
4-Chlorosulfonylacetanilide **(121-60-8)**
4-((5-Chloro-2-sulfo-p-tolyl)azo)-3-hydroxy-2-naphthalene-carboxylic acid barium salt (1:1) **(7585-41-3)**
4-((5-Chloro-2-sulfo-p-tolyl)azo)-3-hydroxy-2-naphthalene-carboxylic acid, calcium salt (1:1) **(7023-61-2)**
4-((5-Chloro-2-sulfo-p-tolyl)azo)-3-hydroxy-2-naphthalene-carboxylic acid, strontium salt **(15782-05-5)**
1-(4-Chloro-o-sulfo-5-tolylazo)-2-naphthol, barium salt **(5160-02-1)**
Chlorosulfuric-acid **(7790-94-5)**
Chlorosulfurous acid, (1-amino-9,10-dihydro-9,10-dioxo-2-anthracenyl)methyl ester (9CI) **(56594-22-0)**
Chlorosulfurous acid, 2-(p-t-butylphenoxy)cyclohexyl ester **(3021-31-6)**
Chlorosulfurous acid, 2-(p-tert-butylphenoxy)cyclohexyl ester (8CI) **(3021-31-6)**
Chlorosulphonic acid (DOT) **(7790-94-5)**
1-(4-Chloro-o-sulpho-5-tolylazo)-2-naphthol, barium salt **(5160-02-1)**
Chlorosulthiadil **(58-93-5)**
7-Chlorotetracycline **(57-62-5)**
1-Chlorotetradecane **(2425-54-9)**
6-Chloro-N,N,N',N'-tetraethyl-1,3,5-triazine-2,4-diamine **(580-48-3)**
Chlorotetrafluoroethane [UN 1021] **(63938-10-3)**
7-Chloro-1,2,3,4-tetrahydro-2-methyl-3-(2-methylphenyl)-4-oxo-6-quinazolinesulfonamide **(17560-51-9)**
7-Chloro-1,2,3,4-tetrahydro-2-methyl-4-oxo-3-o-tolyl-6-quinazolinesulfonamide **(17560-51-9)**
Chlorotetrahydroxy((2-hydroxy-5-oxo-2-imidazolin-4-yl)ureato)-dialuminum **(1317-25-5)**
2-Chloro-4-(1,1,3,3-tetramethylbutyl)phenol **(17199-24-5)**
Chlorothal **(1861-32-1)**
Chlorothalidone **(77-36-1)**
Chlorothalonil **(1897-45-6)**
Chlorothane NU **(71-55-6)**
Chlorothen **(148-65-2)**
Chlorothene **(71-55-6)**
Chlorothene NU **(71-55-6)**
Chlorothene SM **(71-55-6)**
Chlorothene VG **(71-55-6)**
Chlorothene(inhibited) **(71-55-6)**
2-((5-Chloro-2-thenyl)(2-dimethylaminoethyl)amino)pyridine **(148-65-2)**
Chlorothenylpyramine **(148-65-2)**
8-Chlorotheophylline, Compd. with 2-(diphenylmethoxy)-N,N-dimethylethylamine (1:1) **(523-87-5)**
Chlorothiamide **(1918-13-4)**
Chlorothiazid **(58-94-6)**
Chlorothiazide **(58-94-6)**
4-Chlorothioanisole **(123-09-1)**
p-Chlorothioanisole **(123-09-1)**
Chlorothion **(500-28-7)**
2,3,4,5-Chlorothiophene **(6012-97-1)**
2-Chlorothiophene **(96-43-5)**
4-Chlorothiophenol **(106-54-7)**
p-Chlorothiophenol **(106-54-7)**
4-Chloro-o-toloxyacetic acid **(94-74-6)**
2-Chlorotoluene **(95-49-8)**
3-Chlorotoluene **(108-41-8)**
4-Chlorotoluene **(106-43-4)**
Chlorotoluene **(25168-05-2)**
α-Chlorotoluene **(100-44-7)**
ar-Chlorotoluene **(25168-05-2)**
ω-Chlorotoluene **(100-44-7)**
o-Chlorotoluene (ACGIH,OSHA) [UN 2238] **(95-49-8)**
m-Chlorotoluene [UN 2238] **(108-41-8)**
p-Chlorotoluene [UN 2238] **(106-43-4)**
2-Chloro-p-toluidine **(615-65-6)**
3-Chloro-o-toluidine **(87-60-5)**
3-Chloro-p-toluidine **(95-74-9)**
4-Chloro-2-toluidine **(95-69-2)**
4-Chloro-o-toluidine **(95-69-2)**
5-Chloro-o-toluidine **(95-79-4)**
6-Chloro-2-toluidine **(87-63-8)**
6-Chloro-o-toluidine **(87-63-8)**

Name	CAS No.
3-Chloro-p-toluidine hydrochloride	**(7745-89-3)**
4-Chloro-2-toluidine hydrochloride	**(3165-93-3)**
4-Chloro-o-toluidine hydrochloride [UN 1579]	**(3165-93-3)**
Chlorotoluron	**(15545-48-9)**
N'-(4-Chloro-o-tolyl)-N,N-dimethylformamidine	**(6164-98-3)**
N'-(4-Chloro-o-tolyl)-N,N-dimethylformamidine, hydrochloride	**(19750-95-9)**
N^{2}-(4-Chloro-o-tolyl)-N^{1},N^{1}-dimethylformamidine	**(6164-98-3)**
((4-Chloro-o-tolyl)oxy)acetic acid	**(94-74-6)**
((4-Chloro-o-tolyl)oxy)acetic acid isooctyl ester	**(26544-20-7)**
(p-Chloro-o-tolyloxy)acetic acid sodium salt	**(3653-48-3)**
(4-Chloro-o-tolyloxy)butyric acid	**(94-81-5)**
4-((4-Chloro-o-tolyl)oxy)butyric acid	**(94-81-5)**
(4-Chloro-o-tolyloxy)butyric acid sodium salt	**(6062-26-6)**
2-(p-Chloro-o-tolyloxy)propionic acid	**(93-65-2)**
2-((4-Chloro-o-tolyl)oxy)propionic acid potassium salt	**(1929-86-8)**
Chlorotolylthioglycolic acid	**(94-76-8)**
Chlorotrianisene	**(569-57-3)**
Chlorotrianizen	**(569-57-3)**
Chlorotriazine	**(108-77-0)**
6-Chloro-1,3,5-triazine-2,4-diamine	**(3397-62-4)**
Chlorotributylstannane	**(1461-22-9)**
Chlorotributyltin	**(1461-22-9)**
1-Chloro-2-(trichloromethyl)benzene	**(2136-89-2)**
2-Chloro-6-(trichloromethyl)pyridine	**(1929-82-4)**
2-Chloro-6-trichloromethyl pyridine (OSHA)	**(1929-82-4)**
(Z)-2-Chloro-1-(2,4,5-trichlorophenyl)vinyl dimethyl phosphate	**(22248-79-9)**
2-Chloro-1-(2,4,5-trichlorophenyl)vinyl dimethyl phosphate	**(961-11-5)**
2-Chloro-1-(2,4,5-trichlorophenyl)vinyl phosphoric acid dimethyl ester	**(961-11-5)**
1-Chlorotridecane	**(822-13-9)**
Chlorotridecane	**(34214-84-1)**
2-Chlorotriethylamine hydrochloride	**(869-24-9)**
Chlorotriethylsilane	**(994-30-9)**
Chlorotriethylstannane	**(994-31-0)**
Chlorotriethyltin	**(994-31-0)**
6-Chloro-N,N,N'-triethyl-1,3,5-triazine-2,4-diamine	**(1912-26-1)**
Chlorotrifluoride	**(7790-91-2)**
1-Chloro-2,2,2-trifluoroethane	**(75-88-7)**
2-Chloro-1,1,1-trifluoroethane	**(75-88-7)**
2-Chloro-1,1,2-trifluoroethyl difluoromethyl ether	**(13838-16-9)**
1-Chloro-1,2,2-trifluoroethylene	**(79-38-9)**
2-Chloro-1,1,2-trifluoroethylene	**(79-38-9)**
Chlorotrifluoroethylene (DOT)	**(79-38-9)**
Chlorotrifluoromethane [UN 1022]	**(75-72-9)**
2-Chloro-N-(((4-trifluoromethoxy)phenyl)amino)carbonyl)benzamide	**(64628-44-0)**
4-Chloro-2-(trifluoromethyl)benzenamine	**(445-03-4)**
2-Chloro(trifluoromethyl)benzene	**(88-16-4)**
Chloro(trifluoromethyl)benzene	**(52181-51-8)**
p-Chlorotrifluoromethylbenzene	**(98-56-6)**
3-(2-Chloro-4-(trifluoromethyl)phenoxy)benzoic acid	**(63734-62-3)**
5-(2-Chloro-4-(trifluoromethyl)phenoxy)-2-nitrobenzoic acid sodium salt	**(62476-59-9)**
N-(2-Chloro-4-(trifluoromethyl)phenyl)-dl-valine cyano-(3-phenoxyphenyl)methyl ester	**(69409-94-5)**
2-(4-((3-Chloro-5-(trifluoromethyl)-2-pyridinyl)oxy)phenoxy)-propanoic acid methyl ester	**(69806-40-2)**
(RS)-2-(4-(3-Chloro-5-trifluoromethyl-2-pyridyloxy)phenoxy)-propionic acid	**(69806-34-4)**
1-Chloro-3,3,3-trifluoropropane	**(460-35-5)**
3-Chloro-1,1,1-trifluoropropane	**(460-35-5)**
2-Chloro-α,α,α-trifluoro-p-tolyl-3-ethoxy-4-nitrophenyl ether	**(42874-03-3)**
6-Chloro-11β,17,21-trihydroxypregna-1,4,6-triene-3,20-dione	**(5251-34-3)**
6-Chloro-11 β, 17 α, 21-trihydroxypregn-1,4,6-triene-3,20-dione	**(5251-34-3)**
7-Chloro-4,6,2'-trimethoxy-6'-methylgris-2'-en-3,4'-dione	**(126-07-8)**
Chlorotrimethylacetic acid	**(13511-38-1)**
2-Chloro-N,N,N-trimethylethanaminium chloride	**(999-81-5)**
Chlorotrimethylplumbane	**(1520-78-1)**
2-Chloro-N,N-6-trimethyl-4-pyrimidinamine	**(535-89-7)**
Chlorotrimethylsilane	**(75-77-4)**
Chlorotrimethylstannane	**(1066-45-1)**
Chlorotrimethyltin	**(1066-45-1)**
2-Chloro-1,3,5-trinitrobenzene	**(88-88-0)**
Chlorotrinitromethane	**(1943-16-4)**
Chlorotriphenylmethane	**(76-83-5)**
Chlorotriphenylstannane	**(639-58-7)**
Chlorotriphenyltin	**(639-58-7)**
Chlorotripropylstannane	**(2279-76-7)**
Chlorotrisin	**(569-57-3)**
Chlorotris(p-methoxyphenyl)ethylene	**(569-57-3)**
(Chlorotrityl)imidazole	**(23593-75-1)**
1-(o-Chlorotrityl)imidazole	**(23593-75-1)**
1-Chloroundecane	**(2473-03-2)**
5-Chlorouracil	**(1820-81-1)**
Chlorous acid, silver(1+) salt	**(7783-91-7)**
Chlorous acid, sodium salt (8CI,9CI)	**(7758-19-2)**
5-Chlorovanillin	**(19463-48-0)**
Chloro-25 vetag	**(56-75-7)**
Chlorovinylarsine dichloride	**(541-25-3)**
β-Chlorovinylbichloroarsine	**(541-25-3)**
2-Chlorovinyldichloroarsine	**(541-25-3)**
2-Chlorovinyl diethyl phosphate	**(311-47-7)**
β-Chlorovinyl ethyl ethynyl carbinol	**(113-18-8)**
3-(β-Chlorovinyl)-1-pentyn-3-ol	**(113-18-8)**
Chlorovules	**(56-75-7)**
Chlorowax	**(63449-39-8)**
Chlorowax 40	**(63449-39-8)**
Chlorowax 50	**(63449-39-8)**
Chlorowax 70	**(63449-39-8)**
Chlorowax 70S	**(63449-39-8)**
Chlorowax 500C	**(63449-39-8)**
Chlorowax S 70	**(63449-39-8)**
Chlorowodor (Polish)	**(7647-01-0)**
Chlorox	**(7681-52-9)**
8-Chloroxanthine	**(13548-68-0)**
Chloroxifenidim	**(1982-47-4)**
Chloroxone	**(94-75-7)**
Chloroxuron	**(1982-47-4)**
Chloroxylam	**(671-04-5)**
α-Chloro-o-xylene	**(552-45-4)**
α-Chloro-p-xylene	**(104-82-5)**
ω-Chloro-o-xylene	**(552-45-4)**
2-Chloro-m-xylenol	**(88-04-0)**
4-Chloro-3,5-xylenol	**(88-04-0)**
Chloro-xylenol	**(88-04-0)**
p-Chloro-m-xylenol	**(88-04-0)**
6-Chloro-3,4-xylenyl N-methylcarbamate	**(671-04-5)**
2-Chloro-4,5-xylyl N-methylcarbamate	**(671-04-5)**
6-Chloro-3,4-xylyl N-methylcarbamate	**(671-04-5)**
6-Chloro-3,4-xylyl methylcarbamate	**(671-04-5)**
Chloroxyphos	**(52-68-6)**
Chloroxyquinoline	**(130-16-5)**
Chlorozirconyl	**(7699-43-6)**
Chlorozone	**(127-65-1)**
Chlorozotocin	**(54749-90-5)**
Chlorparacide	**(103-17-3)**
Chlorparanitraniline Red	**(2814-77-9)**
Chlorperazine	**(58-38-8)**
Chlorphacinon (Italian)	**(3691-35-8)**
Chlorphenamidine	**(6164-98-3)**
Chlorphenamine	**(132-22-9)**
Chlorphenesin carbamate	**(886-74-8)**
(+)-Chlorpheniramine	**(25523-97-1)**
Chlorpheniramine	**(132-22-9)**

D-Chlorpheniramine	(25523-97-1)
Chlorpheniramine maleate	(113-92-8)
o-Chlorphenol (German)	(95-57-8)
3-(4-(4-Chlor-phenoxy)-phenyl)-1,1-dimethylharnstoff (German)	(1982-47-4)
Chlorphenprop-methyl	(14437-17-3)
Chlorphenteramine	(461-78-9)
Chlorphentermine	(461-78-9)
Chlorphenvinfos	(470-90-6)
Chlorphenvinphos	(470-90-6)
3-(α-(p-Chlorphenyl)-β-acetylaethyl)-4-hydroxycumarin (German)	(81-82-3)
(4-Chlor-phenyl)-benzolsulfonat (German)	(80-38-6)
3-Chlorphenyl-carbamidsaure-butin-(1)-yl(3)-ester (German)	(1967-16-4)
(4-Chlor-phenyl)-4-chlor-benzol-sulfonate (German)	(80-33-1)
4-Chlorphenyl-4'-chlorbenzolsulfonat (German)	(80-33-1)
3-(4-Chlorphenyl)-2-chlorpropionsaeuremethylester (German)	(14437-17-3)
3-(4-Chlor-phenyl)-1,1-dimethyl-harnstoff (German)	(150-68-5)
N-(4-Chlorphenyl)-2,2-dimethylpentamid (German)	(7287-36-7)
1-(p-Chlor-phenyl)-3,3-dimethyl-triazen (German)	(7203-90-9)
N-(4-Chlor-phenyl)-2,2-dimethyl-valeriansaeureamid (German)	(7287-36-7)
γ-(4-(p-Chlorphenyl)-4-hydroxpiperidino)-p-fluorbutyrophenone	(52-86-8)
N-(3-Chlor-phenyl)-isopropyl-carbamat (German)	(101-21-3)
4-Chlor-phenyl-isothiocyanat (German)	(2131-55-7)
3-(4-Chlorphenyl)-1-methoxy-1-methylharnstoff (German)	(1746-81-2)
3-(4-Chlorphenyl)-1-methyl-1-isobutinylharnstoff (German)	(3766-60-7)
N-(4-Chlorphenyl)-N'-methyl-N'-isobutinylharnstoff (German)	(3766-60-7)
2-(p-Chlorphenyl)-3-methyl-1,3-perhydrothiazin-4-on-1,1-dioxide	(80-77-3)
3-(1-(4-Chlor-phenyl)-3-oxo-butyl)-4-hydroxy-cumarin (German)	(81-82-3)
1-(4-Chlorphenyl)-1-phenyl-acetyl-indan-1,3-dion (German)	(3691-35-8)
2(2-(4-Chlor-phenyl-2-phenyl)acetyl)indan-1,3-dion (German)	(3691-35-8)
((4- Chlorphenyl)-1-phenyl)-acetyl-1,3-indandion (German)	(3691-35-8)
1-(p-Chlorphenyl)-1-(2-pyridyl)-3-dimethylaminopropan maleat (German)	(113-92-8)
Chlorphonium	(115-78-6)
Chlorphonium chloride	(115-78-6)
Chlorphoxim	(14816-20-7)
Chlorphoxime (French)	(14816-20-7)
Chlorphthalidolone	(77-36-1)
Chlorphthalidone	(77-36-1)
Chlorpikrin (German)	(76-06-2)
Chlorpromazin	(50-53-3)
Chlorpromazine	(50-53-3)
Chlorpromazine chloride	(69-09-0)
Chlorpromazine hydrochloride	(69-09-0)
Chlorpromazine monohydrochloride	(69-09-0)
Chlorpromazinium chloride	(69-09-0)
Chlorpropamid	(94-20-2)
Chlorpropamide	(94-20-2)
3-Chlorpropannitril (Czech)	(542-76-7)
3-Chlorpropan-1-ol (German)	(627-30-5)
3-Chlorpropen (German)	(107-05-1)
Chlorpropham	(101-21-3)
Chlorprophame (French)	(101-21-3)
Chlorprophenpyridamine	(132-22-9)
Chlorprophenpyridamine maleate	(113-92-8)
Chlorpropylat (Czech)	(5836-10-2)
Chlorprothixen	(113-59-7)
Chlorprothixene	(113-59-7)
α-Chlorprothixene	(113-59-7)
cis-Chlorprothixene	(113-59-7)
Chlorprotixen	(113-59-7)
Chlorprotixene	(113-59-7)
Chlorprotixine	(113-59-7)
Chlorpyrifos-ethyl	(2921-88-2)
Chlorpyrifos-methyl	(5598-13-0)
Chlorpyrifos-methyl oxon	(5598-52-7)
Chlorpyrifos (ACGIH,DOT,OSHA)	(2921-88-2)
Chlorpyriphos-ethyl	(2921-88-2)
Chlorpyriphos	(2921-88-2)
Chlorquin	(54-05-7)
Chlorsaure (German)	(7775-09-9)
Chlorseptol	(127-65-1)
Chlorsulfon	(64902-72-3)
Chlorsulfonamidodihydrobenzothiadiazine dioxide	(58-93-5)
Chlorsulfuron	(64902-72-3)
Chlorsulphacide	(103-17-3)
Chlortalidone	(77-36-1)
Chlorten	(71-55-6)
Chlorten (Czech)	(8001-50-1)
Chlortetracycline	(57-62-5)
Chlortetracycline, 6-demethyl-	(127-33-3)
Chlorthal	(2136-79-0)
Chlorthal-dimethyl	(1861-32-1)
Chlorthalidon	(77-36-1)
Chlorthalidone	(77-36-1)
Chlorthal-methyl	(1861-32-1)
Chlorthalonil (German)	(1897-45-6)
Chlorthiamid	(1918-13-4)
Chlorthiamide	(1918-13-4)
Chlorthiazide	(58-94-6)
Chlorthiepin	(115-29-7)
Chlorthioamide	(1918-13-4)
p-Chlorthiofenol (Czech)	(106-54-7)
Chlorthion	(2463-84-5)
Chlorthion	(500-28-7)
Chlorthion methyl	(500-28-7)
Chlorthiophos	(21923-23-9)
Chlorthiophos	(60238-56-4)
Chlortiamid	(1918-13-4)
Chlortion (Czech)	(500-28-7)
α-Chlortoluol (German)	(100-44-7)
Chlortoluron	(15545-48-9)
N'-(4-Chlor-o-tolyl)-N,N-dimethylformamidin (German)	(6164-98-3)
Chlortox	(57-74-9)
Chlortrianisen	(569-57-3)
Chlortrifluoraethylen (German)	(79-38-9)
Chlorure d'acriflavinium (French)	(8018-07-3)
Chlorure d'aluminium (French)	(7446-70-0)
Chlorure d'amyle (French)	(594-36-5)
Chlorure antimonieux (French)	(10025-91-9)
Chlorure d'arsenic (French)	(7784-34-1)
Chlorure arsenieux (French)	(7784-34-1)
Chlorure de benzenyle (French)	(98-07-7)
Chlorure de benzyle (French)	(100-44-7)
Chlorure de benzylidene (French)	(98-87-3)
Chlorure de bore (French)	(10294-34-5)
Chlorure de butyle (French)	(109-69-3)
Chlorure de chloracetyle (French)	(79-04-9)
Chlorure de chromyle (French)	(14977-61-8)
Chlorure de cyanogene (French)	(506-77-4)
Chlorure de dichloracetyle (French)	(79-36-7)
Chlorure de fumaryle (French)	(627-63-4)
Chlorure de lithium (French)	(7447-41-8)
Chlorure de magnesium hydrate (French)	(7791-18-6)
Chlorure de methallyle (French)	(563-47-3)
Chlorure de methylbenzethonium (French)	(25155-18-4)
Chlorure de methyle (French)	(74-87-3)
Chlorure de methylene (French)	(75-09-2)
Chlorure de propionyle (French)	(79-03-8)
Chlorure de valeryle (French)	(638-29-9)
Chlorure de vinyle (French)	(75-01-4)
Chlorure de vinylidene (French)	(75-35-4)
Chlorure de zinc (French)	(7646-85-7)
Chlorure d'ethyle (French)	(75-00-3)
Chlorure d'ethylene (French)	(107-06-2)

Name	CAS No.
Chlorure d'ethylidene (French)	**(75-34-3)**
Chlorure mercureux (French)	**(10112-91-1)**
Chlorure mercurique (French)	**(7487-94-7)**
Chlorure perrique (French)	**(7705-08-0)**
Chlorurit	**(58-94-6)**
Chlorvinphos	**(62-73-7)**
2-Chlorvinyl-diethylfosfat (Czech)	**(311-47-7)**
Chlorwasserstoff (German)	**(7647-01-0)**
Chlorxylam	**(62046-37-1)**
Chloryl	**(75-00-3)**
Chloryl Radical	**(10049-04-4)**
Chloryl anesthetic	**(75-00-3)**
Chlorylea	**(79-01-6)**
Chlorylen	**(79-01-6)**
Chlorzide	**(58-93-5)**
Chlothixen	**(113-59-7)**
2-Chlor-4-toluidin (Czech)	**(615-65-6)**
3-Chlor-2-toluidin (Czech)	**(87-60-5)**
Chlotride	**(58-94-6)**
Chlotrimazole	**(23593-75-1)**
Chlozolinate	**(72391-46-9)**
Chlrosal	**(58-94-6)**
Chocola A	**(68-26-8)**
Chocolate EMBL	**(2429-81-4)**
Choladine	**(96-83-3)**
Cholaic acid	**(81-24-3)**
Cholalin	**(81-25-4)**
5-β-Cholanic acid, 3-α-hydroxy-	**(434-13-9)**
5-β-Cholan-24-oic acid, 3-α,12-α-dihydroxy	**(83-44-3)**
5-β-Cholan-24-oic acid, 3-α,7-α-dihydroxy	**(474-25-9)**
Cholan-24-oic acid, 3,12-dihydroxy-, (3-α,5-β,12-α)- (9CI)	**(83-44-3)**
Cholan-24-oic acid, 3,7-dihydroxy-, (3-α,5-β,7-α)- (9CI)	**(474-25-9)**
5-β-Cholan-24-oic acid, 3-α-hydroxy	**(434-13-9)**
Cholan-24-oic acid, 3-hydroxy-, (3-α,5-β)- (9CI)	**(434-13-9)**
Cholan-24-oic acid, 3,7,12-trihydroxy-, (3-α,5-β,7-α,12-α)- (9CI)	**(81-25-4)**
Cholanthrene	**(479-23-2)**
Cholanthrene, 3-methyl	**(56-49-5)**
1-Cholanthrenol, 3-methyl	**(3342-98-1)**
Cholaxine	**(50-70-4)**
Cholebrine	**(16034-77-8)**
Cholecalciferol	**(67-97-0)**
Cholecalciferol, D3	**(67-97-0)**
Cholecalciferol, 1a,25-dihydroxy	**(32222-06-3)**
5.α.-Cholest-3-ene (8CI)	**(28338-69-4)**
Cholest-3-ene, (5.α.)- (9CI)	**(28338-69-4)**
Cholest-5-ene (9CI)	**(570-74-1)**
1'H-Cholest-2-eno(3,2-b)indole, 5'-methoxy-, (5.α.)- (9CI)	**(55493-86-2)**
Cholest-5-en-3-β-ol	**(57-88-5)**
Cholest-5-en-3-ol (3-β)- (9CI)	**(57-88-5)**
Cholest-5-en-3-ol (3β)-, acetate (9CI)	**(604-35-3)**
Cholest-5-en-3-ol, (3-β)-, 4-(bis(2-chloroethyl)amino)benzene-acetate	**(3546-10-9)**
Cholest-5-en-3-ol (3β)-, 4-methylbenzenesulfonate (9CI)	**(1182-65-6)**
Cholest-5-en-3-ol (3β)-, propanoate (9CI)	**(633-31-8)**
Cholest-5-en-3-one	**(601-54-7)**
Cholest-7-en-3-one, (5α)- (9CI)	**(15459-85-5)**
Choleic acid	**(83-44-3)**
Chol-3-ene, 23-methyl-, (5α)- (9CI)	**(119973-28-3)**
Cholerebic	**(83-44-3)**
Cholest-8(14)-en-3-ol, 4-methyl-, (3β,4α,5α)- (9CI)	**(62014-96-4)**
$\delta^{3,5}$-Cholestadien (German)	**(747-90-0)**
Cholesta-3,5-diene	**(747-90-0)**
$\delta^{3,5}$-Cholestadiene	**(747-90-0)**
δ-3,5-Cholestadiene	**(747-90-0)**
Cholestadiene (9CI) (VAN)	**(69760-73-2)**
Cholestadiene (9CI) (VAN)	**(81546-39-6)**
5,7-Cholestadien-3-β-ol	**(434-16-2)**
Cholesta-5,7-dien-3-β-ol	**(434-16-2)**
Cholesta-5,7-dien-3-ol, (3-β)- (9CI)	**(434-16-2)**
4,6-Cholestadien-3β-ol, benzoate	**(25485-34-1)**
Cholesta-4,6-dien-3-ol, benzoate, (3β)- (9CI)	**(25485-34-1)**
3,5-Cholestadien-7-one	**(567-72-6)**
δ3,5-Cholestadien-7-one	**(567-72-6)**
Cholesta-3,5-dien-7-one (9CI)	**(567-72-6)**
α-Cholestane	**(481-21-0)**
5α-Cholestane (8CI)	**(481-21-0)**
Cholestane (VAN)	**(481-21-0)**
5α-Cholestane-3β,5-diol (8CI)	**(3347-60-2)**
Cholestane-3,5-diol, (3β,5α)- (9CI)	**(3347-60-2)**
Cholestane, (5α)- (9CI)	**(481-21-0)**
3-β-Cholestanol	**(360-68-9)**
5-β-Cholestan-3-β-ol	**(360-68-9)**
Cholestan-3-ol	**(80-97-7)**
5α-Cholestan-3β-ol (8CI)	**(80-97-7)**
Cholestan-3-ol, (3-β,5-β)- (9CI)	**(360-68-9)**
5α-Cholestanol (VAN)	**(80-97-7)**
Cholestan-3β-ol (VAN)	**(80-97-7)**
Cholestanol (VAN)	**(80-97-7)**
5-α-Cholestan-3-β-ol, 5,6-α-epoxy	**(1250-95-9)**
Cholestan-3-ol, 5,6-epoxy-, (3-β,5-α,6-α)-	**(1250-95-9)**
Cholestan-3-ol, (3β,5α)- (9CI)	**(80-97-7)**
Cholestan-3-ol, 6-methyl-, (3.β.,5.α.,6.β.)- (9CI)	**(43217-65-8)**
Cholestan-3-ol, 4-methylbenzenesulfonate, (3.β.,5.α.)- (9CI)	**(3381-52-0)**
5.α.-Cholestan-3.β.-ol, p-toluenesulfonate (6CI,7CI,8CI)	**(3381-52-0)**
5.α.-Cholestan-3.β.-ol tosylate	**(3381-52-0)**
Cholestan-6-one, 3-(acetyloxy)-, (3β,5α)- (9CI)	**(1256-83-3)**
5α-Cholestan-6-one, 3β-hydroxy-, acetate (8CI)	**(1256-83-3)**
Cholestan-3.β.-yl toluene-p-sulfonate	**(3381-52-0)**
5-Cholesten-3-β-ol	**(57-88-5)**
5:6-Cholesten-3-β-ol	**(57-88-5)**
5:6-Cholesten-3-ol	**(57-88-5)**
δ^{5}-Cholesten-3-β-ol	**(57-88-5)**
5-Cholesten-3-β-ol 3-(p-(bis(2-chloroethyl)amino)phenyl)acetate	**(3546-10-9)**
5-Cholesten-3-one	**(601-54-7)**
Cholestenone	**(601-54-7)**
δ^{5}-Cholestenone	**(601-54-7)**
5α-Cholest-7-en-3-one (8CI)	**(15459-85-5)**
Cholesterilene	**(747-90-0)**
Cholesterin	**(57-88-5)**
Cholesterine	**(57-88-5)**
(-)-Cholesterol	**(57-88-5)**
Cholesterol	**(57-88-5)**
$\delta^{5,7}$-Cholesterol	**(434-16-2)**
δ^{7}-Cholesterol	**(434-16-2)**
Cholesterol acetate	**(604-35-3)**
Cholesterol, acetate (8CI)	**(604-35-3)**
Cholesterol base H	**(57-88-5)**
Cholesterol, (p-(bis(2-chloroethyl)amino)phenyl) acetate	**(3546-10-9)**
Cholesterol, 7-dehydro-	**(434-16-2)**
Cholesterol 5-α,6-α-epoxide	**(1250-95-9)**
Cholesterol-α-epoxide	**(1250-95-9)**
Cholesterol α-oxide	**(1250-95-9)**
Cholesterol oxide	**(1250-95-9)**
Cholesterol, p-toluenesulfonate (8CI)	**(1182-65-6)**
Cholesterone	**(601-54-7)**
Cholesteryl alcohol	**(57-88-5)**
Cholesteryl p-bis(2-chloroethyl)amino phenylacetate	**(3546-10-9)**
Cholesteryl n-propionate	**(633-31-8)**
Cholestrin	**(57-88-5)**
Cholestrol	**(57-88-5)**
Cholestyramine	**(11041-12-6)**
Cholestyramine chloride	**(11041-12-6)**
Cholestyramine resin	**(11041-12-6)**
Cholevid	**(96-83-3)**

Cholic-acid (81-25-4)
Cholic acid, deoxy- (83-44-3)
Cholic acid taurine conjugate (81-24-3)
Choliflavin (8048-52-0)
Cholimil (16034-77-8)
Choline (62-49-7)
Choline bicarbonate (78-73-9)
Choline bisulfite (28427-24-9)
Choline bitartrate (87-67-2)
Choline carbamate chloride (51-83-2)
Choline, carbonate (1:1) (Salt) (8CI) (78-73-9)
Choline chlorhydrate (67-48-1)
Choline chloride (67-48-1)
Choline chloride carbamate (51-83-2)
Choline chloride, carbamate (51-83-2)
Choline, chloride, carbamate (8CI) (51-83-2)
Choline, chloride carbamate(Ester) (51-83-2)
Choline, chloride, carbamate, hydrochloride (51-83-2)
Choline chloride, carbamoyl- (51-83-2)
Choline dichloride (999-81-5)
Choline hydrochloride (67-48-1)
Choline, hydrogen carbonate (78-73-9)
Choline hydrogen tartrate (87-67-2)
Choline-hydroxide (123-41-1)
Choline ion (62-49-7)
Choline, succinate (2:1) (ester) (306-40-1)
Choline, succinate (ester) (306-40-1)
Choline, succinyl- (306-40-1)
Choline, sulfite (1:1) (Salt) (8CI) (28427-24-9)
Choline tartrate (87-67-2)
Choline, tartrate (1:1) salt (87-67-2)
Cholinium chloride (67-48-1)
Cholly (50-36-2)
Cholorebic (83-44-3)
N-Choloyltaurine (81-24-3)
Cholsaeure (German) (81-25-4)
Cholyltaurine (81-24-3)
Chondrosamine (7535-00-4)
Chondrus (9000-07-1)
Chondrus Extract (9000-07-1)
Chooz (1343-88-0)
Chorigon (9002-61-3)
Choriogonadotropin (9002-61-3)
Choriogonin (9002-61-3)
Chorionic Gonadotrophin (9002-61-3)
Chorionic Gonadotropic Hormone (9002-61-3)
Chorionic Gonadotropin (9002-61-3)
Chorulon (9002-61-3)
Chorylen (79-01-6)
Chot (78-11-5)
Chromacid Violet R (2092-55-9)
Chromal Violet B (2092-55-9)
2-Chromanone (119-84-6)
Chroman, 2-oxo- (119-84-6)
Chromar (106-42-3)
Chromar (67-63-0)
Chromastral Green HM (1344-37-2)
Chromastral Green M (1344-37-2)
Chromastral Green Y (1344-37-2)
Chromate (9CI) (13907-45-4)
Chromate (8CI,9CI) (13907-47-6)
Chromate(1-), (6-amino-5-((2-hydroxy-4-nitrophenyl)azo)-1-naphthalenesulfonato(3-))hydroxy-, sodium (9CI) (10241-21-1)
Chromate(1-), bis(N-(8-((5-(aminosulfonyl)-2-hydroxyphenyl)-azo)-7-hydroxy-1-naphthalenyl)acetamidato(2-))-, sodium (9CI) (24305-97-3)
Chromate(1-), bis(1-((5-chloro-2-hydroxyphenyl)azo)-2-naphthalenolato(2-))-, hydrogen (9CI) (31714-55-3)
Chromate(5-), bis(2-((4,5-dihydro-3-methyl-1-(2-methyl-4-sulfophenyl)-5-oxo-1H-pyrazol-4-yl)azo)-4-sulfobenzoato (4-))-, tetrasodium hydrogen (9CI) (10127-05-6)
Chromate(1-), bis((1-(2-hydroxy-5-chlorophenyl)azo)-2-naphthalenolato)-, hydrogen (31714-55-3)
Chromate(1-), bis(N-(7-hydroxy-8-((2-hydroxy-5-((methylamino)sulfonyl)phenyl)azo)-1-naphthalenyl)-acetamidato(2-))-, hydrogen (9CI) (12218-95-0)
Chromate(3-), bis(3-hydroxy-4-((2-hydroxy-1-naphthalenyl)azo)-1-naphthalenesulfonato(3-))-, disodium hydrogen (9CI) (12392-64-2)
Chromate(3-), bis(3-hydroxy-4-((2-hydroxy-1-naphthalenyl)azo)-7-nitro-1-naphthalenesulfonato(3-))-, disodium hydrogen (5610-64-0)
Chromate(7-), bis(4-hydroxy-3-((2-hydroxy-5-nitro-3-sulfophenyl)azo)-6-((sulfomethyl)amino)-2-naphthalene-sulfonato(5-))-, hexasodium hydrogen (9CI) (6408-22-6)
Chromate(1-), bis(1-((2-hydroxy-4-nitrophenyl)azo)-2-naphthalenolato(2-))-, hydrogen, (OC-6-22')- (9CI) (32517-36-5)
Chromate(1-), bis(8-((2-hydroxy-5-sulfamoylphenyl)azo)-1-acetylamino-7-naphthalenolato(2-))-, sodium (24305-97-3)
Chromate(2-), (5-chloro-3-((4,5-dihydro-3-methyl-5-oxo-1-(3-sulfophenyl)-1H-pyrazol-4-yl)azo)-2-hydroxybenzene-sulfonato(4-))hydroxy-, disodium (9CI) (6408-31-7)
Chromated copper arsenate (37337-13-6)
Chromate de plomb (French) (7758-97-6)
Chromate(1-), (4-((4,5-dihydro-3-methyl-5-oxo-1-(3-sulfophenyl)-1H-pyrazol-4-yl)azo)-3-hydroxy-1-naphthalenesulfonato-(4-))-, sodium (9CI) (52677-44-8)
Chromate(3-)(1-(3,5-dinitro-2-hydroxyphenyl)azo)-2-naphthalenolato(2-))(3-hydroxy-4-((2-hydroxy-1-naphthyl)-azo)-7-nitro-1-naphthalenesulfonato(3-)), sodium dihydrogen (12218-96-1)
Chromate(2-), heptadecaoxotetrazincatetetra-, dipotassium (9CI) (12433-50-0)
Chromate(2-), hydroxy(3-hydroxy-4-((1-hydroxy-8-sulfo-2-naphthalenyl)azo)-1-naphthalenesulfonato(4-))-, disodium (9CI) (6370-08-7)
Chromate(2-), (3-hydroxy-4-((2-hydroxy-1-naphthalenyl)azo)-1-naphthalenesulfonato(3-))(1-((2-hydroxy-5-((2- methoxy-phenyl)azo)phenyl)azo)-2-naphthalenolato(2-))-, disodium (9CI) (27425-58-7)
Chromate(1-), hydroxy(4-hydroxy-3-nitro-5-((1,2,3,4-tetra-hydro-2,4-dioxo-3-quinolinyl)azo)benzenesulfonato(3-))-, sodium (9CI) (6656-02-6)
Chromate(1-), (3-hydroxy-4-((1-hydroxy-8-sulfo-2-naphthalenyl)-azo)-1-naphthalenesulfonato(4-))-, sodium (9CI) (70942-15-3)
Chromate(1-), hydroxyoctaoxodizincatedi-, potassium (11103-86-9)
Chromate of Potass (7789-00-6)
Chromate of soda (7775-11-3)
Chromatex Blue BN (147-14-8)
Chromatex Green G (1328-53-6)
Chromatex Orange R (3468-63-1)
Chromatex Red J (2425-85-6)
Chromatex Yellow J (2512-29-0)
Chromaven Violet B (2092-55-9)
Chrome (7440-47-3)
Chrome Alum (10141-00-1)
Chrome Alum (7788-99-0)
Chrome Alum (Dodecahydrate) (7788-99-0)
Chrome Fast Blue 2R (3567-69-9)
Chrome Fast Green CP (1344-37-2)
Chrome Fast Violet B (2092-55-9)
Chrome Ferroalloy (11114-46-8)
Chrome Green (1308-38-9)
Chrome Green (7758-97-6)
Chrome Green UC61 (7758-97-6)
Chrome Green UC74 (7758-97-6)
Chrome Green UC76 (7758-97-6)
Chrome Intra Red GRE (6408-31-7)
Chrome Leather Black BH (2429-73-4)
Chrome Leather Black CR (2429-73-4)

Chrome Leather Black DS	(2429-73-4)
Chrome Leather Black E	(1937-37-7)
Chrome Leather Black EC	(1937-37-7)
Chrome Leather Black EM	(1937-37-7)
Chrome Leather Black ER	(2429-83-6)
Chrome Leather Black G	(1937-37-7)
Chrome Leather Black GNA	(6428-31-5)
Chrome Leather Blue 2B	(2602-46-2)
Chrome Leather Blue 3B	(72-57-1)
Chrome Leather Brilliant Black ER	(1937-37-7)
Chrome Leather Brown BRLL	(16071-86-6)
Chrome Leather Brown BRSL	(16071-86-6)
Chrome Leather Brown BS	(2429-81-4)
Chrome Leather Brown M	(2429-82-5)
Chrome Leather Dark Blue BHM	(2429-73-4)
Chrome Leather Dark Green N	(3626-28-6)
Chrome Leather Dark Green S	(3626-28-6)
Chrome Leather Green B	(3626-28-6)
Chrome Leather Pure Blue	(2429-74-5)
Chrome Leather Red 4B	(992-59-6)
Chrome Leather Red 5B	(2610-11-9)
Chrome Leather Scarlet 3BS	(6358-29-8)
Chrome Leather Sky Blue	(2610-05-1)
Chrome Leather Violet BS	(6426-67-1)
Chrome Leather Yellow A	(1325-37-7)
Chrome Lemon	(7758-97-6)
Chrome Orange	(1344-37-2)
Chrome Orange	(18454-12-1)
Chrome Orange 54	(18454-12-1)
Chrome Orange 56	(18454-12-1)
Chrome Orange 57	(18454-12-1)
Chrome Orange 58	(18454-12-1)
Chrome Orange Dark	(18454-12-1)
Chrome Orange Extra Light	(18454-12-1)
Chrome Orange G	(18454-12-1)
Chrome Orange Medium	(18454-12-1)
Chrome Orange NC-22	(18454-12-1)
Chrome Orange 5R	(18454-12-1)
Chrome Orange R	(18454-12-1)
Chrome Orange RF	(18454-12-1)
Chrome Orange XL	(18454-12-1)
Chrome Oxide Green	(1308-38-9)
Chrome Oxide Green GN-M	(1308-38-9)
Chrome Oxide Pigment	(1308-38-9)
Chrome Oxide X1134	(1308-38-9)
Chrome Potash Alum	(10141-00-1)
Chrome Red	(18454-12-1)
Chrome Red Alizarine	(130-22-3)
Chrome Violet B	(2092-55-9)
Chrome Violet K	(2092-55-9)
Chrome Violet R	(2092-55-9)
Chrome Yellow	(1344-37-2)
Chrome Yellow	(7758-97-6)
Chrome Yellow A-241	(1344-37-2)
Chrome Yellow 62E	(1344-37-2)
Chrome Yellow 10G	(1344-37-2)
Chrome Yellow 4G	(1344-37-2)
Chrome Yellow 5G	(7758-97-6)
Chrome Yellow G	(7758-97-6)
Chrome Yellow 5GF	(1344-37-2)
Chrome Yellow GF	(7758-97-6)
Chrome Yellow 4GL Light	(1344-37-2)
Chrome Yellow GL Medium	(1344-37-2)
Chrome Yellow 6GL Primrose	(1344-37-2)
Chrome Yellow LF	(7758-97-6)
Chrome Yellow LF AA	(1344-37-2)
Chrome Yellow Lemon	(1344-37-2)
Chrome Yellow Light	(1344-37-2)
Chrome Yellow Light 1066	(7758-97-6)
Chrome Yellow Light 1075	(7758-97-6)
Chrome Yellow Light Y 434D	(1344-37-2)
Chrome Yellow Medium	(1344-37-2)
Chrome Yellow Medium 1074	(7758-97-6)
Chrome Yellow Medium 1085	(7758-97-6)
Chrome Yellow Medium 1295	(7758-97-6)
Chrome Yellow Medium 1298	(7758-97-6)
Chrome Yellow Medium Y 469D	(1344-37-2)
Chrome Yellow Middle	(1344-37-2)
Chrome Yellow Primrose	(1344-37-2)
Chrome Yellow Primrose 1010	(7758-97-6)
Chrome Yellow Primrose 1015	(7758-97-6)
Chrome(III) complex of 2-hydroxy-4,6-dipropoxybenzald1 (2'-oxy-5'-nitrophenyl)imine, 2-(3'-(2'',5''-dihydroxy-phenyl)propionyl)cyclopentanone, and water	(31303-42-1)
Chromedia CC 31	(9004-34-6)
Chromedia CF 11	(9004-34-6)
Chrome fluorure (French)	(7788-97-8)
Chromelin	(96-26-4)
Chrome ochre	(1308-38-9)
Chrome ore	(1308-31-2)
Chrome oxide	(1308-38-9)
Chrome (trioxyde de) (French)	(1333-82-0)
Chrome vermilion	(12656-85-8)
Chromia	(1308-38-9)
Chromic Acid Green	(1308-38-9)
Chromic (VI) acid	(1333-82-0)
Chromic(VI) acid	(7738-94-5)
Chromic acetate	(1066-30-4)
Chromic acetate (III)	(1066-30-4)
Chromic acid	(1333-82-0)
Chromic acid	(7738-94-5)
Chromic acid (9CI)	(13530-68-2)
Chromic acid, Compd. with cyclohexanamine (9CI)	(20736-64-5)
Chromic acid (H2CrO4) (9CI)	(7738-94-5)
Chromic acid (OSHA)	(7738-94-5)
Chromic acid, Solid	(7738-94-5)
Chromic acid, Solid [NA 1463]	(1333-82-0)
Chromic acid, Solid [NA 1463]	(13530-68-2)
Chromic acid Solution [UN 1755]	(1308-14-1)
Chromic acid, Solution [UN 1755]	(1333-82-0)
Chromic acid, barium salt (1:1)	(10294-40-3)
Chromic acid, bis(triphenylsilyl) ester	(1624-02-8)
Chromic acid, calcium salt (1:1)	(13765-19-0)
Chromic acid, calcium salt (1:1) (9CI)	(14307-33-6)
Chromic acid, chromium (3+) salt (3:2)	(24613-89-6)
Chromic acid, chromium salt	(41261-95-4)
Chromic acid, cobalt(2+) salt (1:1) (8CI,9CI)	(13455-25-9)
Chromic acid, copper(2+) salt (1:1) (8CI,9CI)	(13548-42-0)
Chromic acid, diammonium salt	(7788-98-9)
Chromic acid, di-t-butyl ester	(1189-85-1)
Chromic acid, dicesium salt (8CI,9CI)	(13454-78-9)
Chromic acid, dilithium salt (9CI)	(14307-35-8)
Chromic acid, dipotassium salt	(7778-50-9)
Chromic acid, dipotassium salt	(7789-00-6)
Chromic acid, dirubidium salt (8CI,9CI)	(13446-72-5)
Chromic acid, disilver(1+) salt (8CI,9CI)	(7784-01-2)
Chromic acid, disodium salt	(10588-01-9)
Chromic acid, disodium salt	(7775-11-3)
Chromic acid, disodium salt, decahydrate	(13517-17-4)
Chromic acid, disodium salt, tetrahydrate	(10034-82-9)
Chromic acid, ion(2-)	(13907-47-6)
Chromic acid, lead(2+) salt (1:1)	(7758-97-6)
Chromic acid, magnesium salt (1:1) (9CI)	(13423-61-5)
Chromic acid, potassium zinc salt (2:2:1)	(11103-86-9)

Chromic acid solution [UN 1755] (13530-68-2)
Chromic acid, strontium salt (1:1) (7789-06-2)
Chromic acid, zinc hydroxide hydrate (1:2:2:1) (15930-94-6)
Chromic acid, zinc salt (13530-65-9)
Chromic acid, zinc salt (1:2) (15930-94-6)
Chromic acid, zinc salt, Compd. with Zinc hydroxide and chromium oxide (9:1) (37300-23-5)
Chromic acid, zinc salt, basic (50922-29-7)
Chromic anhydride (DOT) (1333-82-0)
Chromic chloride (10025-73-7)
Chromic chloride stearate (15242-96-3)
Chromic chromate (24613-89-6)
Chromic fluoride (7788-97-8)
Chromic fluoride, Solid [UN 1756] (7788-97-8)
Chromic fluoride solution [UN 1757] (7788-97-8)
Chromic (III) hydroxide (1308-14-1)
Chromic ion (16065-83-1)
Chromic nitrate nonahydrate (7789-02-8)
Chromic oxide (1308-38-9)
Chromic oxychloride (14977-61-8)
Chromic phosphate (7789-04-0)
Chromic potassium sulfate (10141-00-1)
Chromic potassium sulphate (10141-00-1)
Chromic sulfate (10101-53-8)
Chromic sulphate (10101-53-8)
Chromic trifluoride (7788-97-8)
Chromic trioxide (DOT) (1333-82-0)
Chromitan B (10101-53-8)
Chromitan MS (10101-53-8)
Chromitan NA (10101-53-8)
Chromite (1308-31-2)
Chromite (Mineral) (1308-31-2)
Chromite ore (1308-31-2)
Chromium-carbonyl (13007-92-6)
Chromium-cobalt alloy (11114-92-4)
Chromium-hydroxide-sulfate (12336-95-7)
Chromium (3+) (16065-83-1)
Chromium (III) (16065-83-1)
Chromium(6+) (18540-29-9)
Chromium (ACGIH) (7440-47-3)
Chromium Alloy, Base, Cr,C,Fe,N,Si (Ferrochromium) (11114-46-8)
Chromium Alloy, Cr,C,Fe,N,Si (9CI) (11114-46-8)
Chromium (Cr +6) (18540-29-9)
Chromium (Cr^{6+}) (18540-29-9)
Chromium III oxide (1308-38-9)
Chromium III sulfate (10101-53-8)
Chromium (IV) oxide (12018-01-8)
Chromium, Isotope of mass 51 (8CI,9CI) (14392-02-0)
Chromium Orange (18454-12-1)
Chromium Oxide Green (1308-38-9)
Chromium Oxide Pigment (1308-38-9)
Chromium Oxide X1134 (1308-38-9)
Chromium(VI) (18540-29-9)
Chromium (VI) dioxychloride (14977-61-8)
Chromium (VI) oxide (1333-82-0)
Chromium(VI) oxide (1:3) (1333-82-0)
Chromium Yellow (7758-97-6)
Chromium acetate (1066-30-4)
Chromium(III) acetate (1066-30-4)
Chromium, aqua(2-(3-(2,5-dihydroxyphenyl)-1-oxopropyl)-cyclopentanonato)(2-(-((2-hydroxy-5-nitrophenyl)imino)-methyl)-3,5-dipropoxyphenolato(2-))- (9CI) (31303-42-1)
Chromium, bis(eta(5)-2,4-cyclopentadien-1-yl)- (1271-24-5)
Chromium carbonate (29689-14-3)
Chromium carbonyl (OC-6-11) (9CI) (13007-92-6)
Chromium chloride (10025-73-7)
Chromium chloride (10049-05-5)
Chromium(II) chloride (10049-05-5)
Chromium(II) chloride (1:2) (10049-05-5)
Chromium(III) chloride (1:3) (10025-73-7)
Chromium chloride, Anhydrous (10025-73-7)
Chromium chloride oxide (14977-61-8)
Chromium chromate (11118-57-3)
Chromium chromate (24613-89-6)
Chromium chromate (41261-95-4)
Chromium dichloride (10049-05-5)
Chromium dichloride dioxide (14977-61-8)
Chromium, dichlorodioxo (14977-61-8)
Chromium dioxide (12018-01-8)
Chromium dioxide dichloride (14977-61-8)
Chromium, di-pi-cyclopentadienyl- (8CI) (1271-24-5)
Chromium disodium oxide (7775-11-3)
Chromium(III) fluoride (7788-97-8)
Chromium fluoride (8CI,9CI) (7788-97-8)
Chromium hexacarbonyl (13007-92-6)
Chromium hexavalent ion (18540-29-9)
Chromium (III) ion (16065-83-1)
Chromium ion (3+) (16065-83-1)
Chromium(6+) ion (18540-29-9)
Chromium, ion (Cr 3+) (16065-83-1)
Chromium, ion (Cr 6+) (18540-29-9)
Chromium, ion (Cr^{6+}) (8CI,9CI) (18540-29-9)
Chromium iron lignosulfonate (8075-74-9)
Chromium lead oxide (18454-12-1)
Chromium lead silicate (11113-70-5)
Chromium lignosulfonate (9066-50-6)
Chromium lithium oxide (14307-35-8)
Chromium metal (OSHA) (7440-47-3)
Chromium nitrate nonahydrate (7789-02-8)
Chromium(III) nitrate, nonahydrate (1:3:9) (7789-02-8)
Chromium oxide (1308-38-9)
Chromium oxide (1333-82-0)
Chromium oxide (7738-94-5)
Chromium(3+) oxide (1308-38-9)
Chromium(III) oxide (2:3) (1308-38-9)
Chromium oxide (9CI) (12018-01-8)
Chromium oxide (VAN)(9CI) (11118-57-3)
Chromium oxychloride [UN 1758] (14977-61-8)
Chromium orthophosphate (7789-04-0)
Chromium potassium sulfate (10141-00-1)
Chromium potassium sulfate (1:1:2) (10141-00-1)
Chromium(III) potassium sulfate (1:1:2), dodecahydrate (7788-99-0)
Chromium potassium sulphate (10141-00-1)
Chromium potassium zinc oxide (9CI) (37224-57-0)
Chromium salt of oxidized lignosulfonate (9066-50-6)
Chromium salt of spent sulfite liquor (9066-50-6)
Chromium sesquioxide (1308-38-9)
Chromium sodium oxide (10588-01-9)
Chromium sodium oxide (7775-11-3)
Chromium (III) sulfate (2:3) (10101-53-8)
Chromium sulfate (2:3) (10101-53-8)
Chromium sulphate (10101-53-8)
Chromium sulphate (2:3) (10101-53-8)
Chromium, tetrachloro-mu-hydroxy(mu-(2-methyl-2-propenoato-O:O'))di- (9CI) (15096-41-0)
Chromium, tetrachloro-mu-hydroxy(mu-(octadecanoato-O:O'))di (15242-96-3)
Chromium, tetrachloro-mu-hydroxy(mu-stearato)di- (15242-96-3)
Chromium, tetrachloro-mu-hydroxy(mu-(tetradecanoato-O:O'))di- (9CI) (15659-56-0)
Chromium, tetrachloro-u-hydroxy(u-(2-methyl-2-propenoato)-O:O'))di- (15096-41-0)
Chromium triacetate (1066-30-4)
Chromium trichloride (10025-73-7)
Chromium trifluoride (7788-97-8)

Chromium trinitrate nonahydrate (7789-02-8)
Chromium trioxide (1333-82-0)
Chromium(3+) trioxide (1308-38-9)
Chromium(6+) trioxide (1333-82-0)
Chromium trioxide, Anhydrous [UN 1463] (1333-82-0)
Chromium zinc oxide (13530-65-9)
Chromium(6+) zinc oxide hydrate (1:2:6:1) (15930-94-6)
Chromium-phosphate (7789-04-0)
Chromium-zinc-oxide (12018-19-8)
Chromocene (9CI) (1271-24-5)
Chromoflavine (8048-52-0)
Chromoflavine (86-40-8)
Chromolan Red GRE (6408-31-7)
Chromone, 5-hydroxy-7-methoxy-2-methyl- (8CI) (480-34-2)
Chromosmon (61-73-4)
Chromotrichia Factor (150-13-0)
anti-Chromotrichia Factor (150-13-0)
Chromotrop FB (3567-69-9)
Chromotrope FB (3567-69-9)
Chromotrope 2R (6CI) (4197-07-3)
Chromotrope Red 2R (4197-07-3)
Chromotropic acid disodium salt (129-96-4)
Chromous chloride (10049-05-5)
Chromous oxalate (814-90-4)
Chromous(II) sulfate (13825-86-0)
Chromoxychlorid (German) (14977-61-8)
Chromsaeureanhydrid (German) (1333-82-0)
Chromtrioxid (German) (1333-82-0)
Chromylchlorid (German) (14977-61-8)
Chromyl chloride (ACGIH,DOT) (14977-61-8)
Chroomoxychloride (Dutch) (14977-61-8)
Chroomtrioxyde (Dutch) (1333-82-0)
Chroomzuuranhydride (Dutch) (1333-82-0)
Chrysammic acid (517-92-0)
Chrysamminic acid (517-92-0)
(+)-trans-Chrysanthemic acid-carboxy-14C (32511-06-1)
Chrysanthemic acid (10453-89-1)
Chrysanthemum cinerareaefolium (8003-34-7)
Chrysanthemumdicarboxylic acid monomethyl ester pyrethrolone ester (121-29-9)
Chrysanthemumic acid (10453-89-1)
Chrysanthemumic acid 6-chloropiperonyl ester (70-43-9)
Chrysanthemummonocarboxylic acid (10453-89-1)
Chrysanthemummonocarboxylic acid 6-chloropiperonyl ester (70-43-9)
Chrysanthemum monocarboxylic acid pyrethrolone ester (121-21-1)
Chrysarobin (491-59-8)
Chrysazin (117-10-2)
Chrysazin-3-carboxylic acid (478-43-3)
6-Chrysenamine (2642-98-0)
Chrysene (OSHA) (218-01-9)
Chrysene, 1,2-dihydro (41593-31-1)
Chrysene, dihydro- (9CI) (41593-31-1)
Chrysene, 1,2-dimethyl (15914-23-5)
Chrysene, 5,6-dimethyl (3697-27-6)
Chrysene, dimethyl- (9CI) (41637-92-7)
Chrysene, 1-methyl (3351-28-8)
Chrysene, 2-methyl (3351-32-4)
Chrysene, 3-methyl (3351-31-3)
Chrysene, 4-methyl (3351-30-2)
Chrysene, 5-methyl (3697-24-3)
Chrysene, 6-methyl (1705-85-7)
Chrysene, methyl- (9CI) (41637-90-5)
Chrysene, methylnitro- (9CI) (80182-33-8)
Chrysene, 6-nitro (7496-02-8)
Chrysene, nitro- (9CI) (63021-85-2)
Chrysene, octahydro- (9CI) (91741-91-2)
Chrysene, 1,2,3,4-tetrahydro-3,3,7-trimethyl- (9CI) (65755-17-1)
Chrysene, 1,2,3,4-tetrahydro-3,4,7-trimethyl- (9CI) (74229-83-7)
Chrysene, tetramethyl- (9CI) (71277-90-2)
Chrysenex (2642-98-0)
Chrysenol, nitro- (9CI) (116212-01-2)
α-Chrysidine (225-51-4)
Chrysodermol (1143-38-0)
Chrysofluorene (238-84-6)
Chrysogen (92-24-0)
Chrysoidin (532-82-1)
Chrysoidin A (495-54-5)
Chrysoidin FB (532-82-1)
Chrysoidin Y (532-82-1)
Chrysoidin YN (532-82-1)
Chrysoidine (532-82-1)
Chrysoidine (II) (532-82-1)
Chrysoidine A (532-82-1)
Chrysoidine B (532-82-1)
Chrysoidine Base (495-54-5)
Chrysoidine Base A (495-54-5)
Chrysoidine Base B (495-54-5)
Chrysoidine C Crystals (532-82-1)
Chrysoidine Crystals (532-82-1)
Chrysoidine G (532-82-1)
Chrysoidine G Base (495-54-5)
Chrysoidine GN (532-82-1)
Chrysoidine GS (532-82-1)
Chrysoidine HR (532-82-1)
Chrysoidine J (532-82-1)
Chrysoidine J Base (495-54-5)
Chrysoidine M (532-82-1)
Chrysoidine Orange (532-82-1)
Chrysoidine PRL (532-82-1)
Chrysoidine PRR (532-82-1)
Chrysoidine SL (532-82-1)
Chrysoidine SS (532-82-1)
Chrysoidine Special (Biological stain and indicator) (532-82-1)
Chrysoidine Y (532-82-1)
Chrysoidine Y Base (495-54-5)
Chrysoidine Y Base New (495-54-5)
Chrysoidine Y Base New (532-82-1)
Chrysoidine Y Crystals (532-82-1)
Chrysoidine YD Base (495-54-5)
Chrysoidine Y EX (532-82-1)
Chrysoidine YGH (532-82-1)
Chrysoidine YL (532-82-1)
Chrysoidine YN (532-82-1)
Chrysoidine Y Special (532-82-1)
Chrysoin G (547-57-9)
Chrysoin S (547-57-9)
Chrysoin S Specially Pure (547-57-9)
Chrysoine (547-57-9)
Chrysoine Extra (547-57-9)
Chrysoine Extra Pure A (547-57-9)
Chrysoine N (547-57-9)
Chrysoine S (547-57-9)
Chrysoine S Extra Pure (547-57-9)
allo-Chrysoketone (6051-98-5)
Chrysomykine (57-62-5)
Chryson (10453-86-8)
Chrysonex (2642-98-0)
Chrysonine S (547-57-9)
Chrysonis S (547-57-9)
Chrysophanic acid (481-74-3)
Chrysophanic acid anthranol (491-59-8)
Chrysophanol (481-74-3)
Chrysophenine (2870-32-8)
Chrysotile [UN 2590] (12001-29-5)

Chrysotile asbestos (12001-29-5)
Chrysron (10453-86-8)
Chryzoidyna F.B. (Polish) (532-82-1)
Chwastoks (3653-48-3)
Chwastox (3653-48-3)
Chwastox (94-74-6)
Chwastox 80 (3653-48-3)
Chymar (9004-07-3)
α Chymar (9004-07-3)
α-Chymar Ophth (9004-07-3)
Chymase (9001-98-3)
Chymosin (9001-98-3)
Chymotest (9004-07-3)
Chymotrypsin (9004-07-3)
α-Chymotrypsin (9004-07-3)
Chymotrypsin A (9004-07-3)
Chymotrypsin B (9004-07-3)
Ci-705 (72-44-6)
Ciafos (2636-26-2)
Ciamin (50-81-7)
Cianazil (140-87-4)
Cianidanol (154-23-4)
Cianuro de plomo (Spanish) (592-05-2)
Cianuro di sodio (Italian) (143-33-9)
Cianuro di vinile (Italian) (107-13-1)
Ciba 570 (13171-21-6)
Ciba 709 (141-66-2)
Ciba 1414 (6923-22-4)
Ciba 1983 (1982-47-4)
Ciba 2059 (2164-17-2)
Ciba 2446 (919-76-6)
Ciba-3126 (3060-89-7)
Ciba 5968 (304-20-1)
Ciba 5968 (86-54-4)
Ciba 6313 (13360-45-7)
Ciba 7115 (469-79-4)
Ciba 8353 (6988-21-2)
Ciba 8514 (6164-98-3)
Ciba 8514 (143-50-0)
Ciba 9491 (18181-70-9)
Ciba 12223 (42509-80-8)
Ciba 12669A (477-30-5)
Ciba 17309 BA (72-63-9)
Ciba 20-684BA (911-65-9)
Ciba 32644 (61-57-4)
Ciba 32644-BA (61-57-4)
Ciba 36278-BA (23239-41-0)
Ciba Brilliant Pink FR (2379-74-0)
Ciba C 7019 (4658-28-0)
Ciba C-9491 (18181-70-9)
Ciba-C8514 (6164-98-3)
Ciba-Geigy C-9491 (18181-70-9)
Ciba-Geigy GS 13005 (950-37-8)
Ciba-Geigy GS 19851 (18181-80-1)
Ciba Orange R (3263-31-8)
Ciba Orange RDL (3263-31-8)
Ciba Orange RP (3263-31-8)
Ciba Pink FF (2379-74-0)
Ciba 13437 SU (3771-19-5)
Ciba Thiocron (919-76-6)
Cibacet Blue BNG (2475-46-9)
Cibacet Blue BNG (86722-66-9)
Cibacet Blue BR (2475-44-7)
Cibacet Blue F3R (2475-46-9)
Cibacet Blue F3R (86722-66-9)
Cibacet Brilliant Blue BG New (2475-46-9)
Cibacet Brilliant Blue BG new (86722-66-9)
Cibacet Brilliant Pink 4BN (2872-48-2)
Cibacet Brilliant Violet 3B (82-33-7)
Cibacet Red 3B (116-85-8)
Cibacet Red E3B (116-85-8)
Cibacet Sapphire Blue G (2475-45-8)
Cibacet Violet E2R (128-95-0)
Cibacet Violet 2R (128-95-0)
Cibacet Yellow GBA (2832-40-8)
Cibacet Yellow 2GC (2832-40-8)
Cibacete Brilliant Blue BG New (2475-46-9)
Cibacete Brilliant Pink 4BN (2872-48-2)
Cibacete Diazo Navy Blue 2B (119-90-4)
Cibacete Red 3B (116-85-8)
Cibacete Violet 2R (128-95-0)
Cibacete Yellow GBA (2832-40-8)
Cibacron Black B-D (12225-26-2)
Cibacron Brilliant Green T 3G-E (12225-80-8)
Cibacron Brilliant Red B-D (12238-00-5)
Cibacron Scarlet 4G-P (12238-07-2)
Cibanaphthol AG (91-96-3)
Cibanaphthol RCA (137-52-0)
Cibanaphthol RF (92-77-3)
Cibanaphthol RK (135-62-6)
Cibanaphthol RM (135-65-9)
Cibanaphthol RN (132-68-3)
Cibanaphthol RTO (135-61-5)
Cibanone Blue FG (130-20-1)
Cibanone Blue FRS (81-77-6)
Cibanone Blue FRSN (81-77-6)
Cibanone Blue GF (130-20-1)
Cibanone Blue RS (81-77-6)
Cibanone Brilliant Blue FR (81-77-6)
Cibanone Brilliant Green 2BF (128-58-5)
Cibanone Brilliant Green FBF (128-58-5)
Cibanone Brilliant Orange GR (4424-06-0)
Cibanone Brown BR (2475-33-4)
Cibanone Brown FBR (2475-33-4)
Cibanone Golden Orange FG (128-70-1)
Cibanone Golden Orange G (128-70-1)
Cibanone Golden Yellow (128-66-5)
Cibanone Navy Blue FRA (1324-54-5)
Cibanone Navy Blue RA (1324-54-5)
Cibanone Olive F2R (2379-81-9)
Cibanone Olive 2R (2379-81-9)
Cibanone Red 6B (4203-77-4)
Cibanone Red F 6B (4203-77-4)
Cibanone Violet F 4R (1324-55-6)
Cibanone Violet F 2RB (1324-55-6)
Cibanone Violet 2R (1324-55-6)
Cibanone Violet 4R (1324-55-6)
Ciba 2020/r (835-31-4)
Cibazol (72-14-0)
Cichorigenin (305-01-1)
Cichoriin aglycon (305-01-1)
Ciclazindol (37751-39-6)
Ciclazindolum (Latin) (37751-39-6)
Ciclizina (82-92-8)
Cicloesano (Italian) (110-82-7)
Cicloesanolo (Italian) (108-93-0)
Cicloesanone (Italian) (108-94-1)
6-Cicloesil-2,4-dinitr-fenolo (Italian) (131-89-5)
Ciclohexenitriclorosilano (Spanish) (10137-69-6)
Ciclohexiltriclorosilano (Spanish) (98-12-4)
Cicloral (339-43-5)
Cicloserina (Italian) (68-41-7)
Ciclosom (52-68-6)
Ciclospasmol (456-59-7)

Ciclosporin (59865-13-3)
Cicutin (458-88-8)
Cicutine (458-88-8)
Cicutoxin (505-75-9)
Cidal (65-45-2)
Cidalon (115-31-1)
Cidamex (59-66-5)
Cidanchin (54-05-7)
Cidandopa (59-92-7)
Cidemul (2597-03-7)
Cidex (111-30-8)
Cidial (2597-03-7)
Cidocetine (56-75-7)
Cidrex (58-93-5)
Cignolin (1143-38-0)
Cignolin (480-22-8)
Cigthranol (1143-38-0)
Cigthranol (480-22-8)
Cilefa Black B (2519-30-4)
Cilefa Blue R (860-22-0)
Cilefa Orange S (2783-94-0)
Cilefa Pink B (16423-68-0)
Cilefa Ponceau 4R (2611-82-7)
Cilefa Red E (2302-96-7)
Cilefa Red G (3257-28-1)
Cilefa Rubine 2B (915-67-3)
Cilefa Rubine R (3567-69-9)
Cilefa Yellow T (1934-21-0)
Cilicaine (6130-64-9)
Cilifor (50-59-9)
Cilla Blue Extra (2475-45-8)
Cilla Fast Blue B (2475-44-7)
Cilla Fast Blue FFR (2475-46-9)
Cilla Fast Blue FFR (86722-66-9)
Cilla Fast Pink BN (116-85-8)
Cilla Fast Pink FF3B (2872-48-2)
Cilla Fast Red Violet RN (128-95-0)
Cilla Fast Violet 6B (1220-94-6)
Cilla Fast Violet B (82-33-7)
Cilla Fast Yellow G (2832-40-8)
Cilla Fast Yellow 5R (6300-37-4)
Cilla Fast Yellow RR (119-15-3)
Cilla Orange R (82-28-0)
Cillenta (1538-09-6)
Cilloral (113-98-4)
Cilloral (61-33-6)
Cilopen (61-33-6)
Cimagel (9002-92-0)
Cimcool Wafers (126-11-4)
Cimetidine (51481-61-9)
Cimexan (121-75-5)
Cin-Quin (56-54-2)
Cinamine (24815-24-5)
Cinatabs (24815-24-5)
Cinchocaine (85-79-0)
Cinchol (83-46-5)
Cincholepidine (491-35-0)
Cinchomeronic acid (490-11-9)
Cinchonan-9-ol, (8-α,9R)- (9CI) (485-71-2)
Cinchonan-9-ol, 10,11-dihydro-6'-methoxy-, (8α,9R)- (9CI) (522-66-7)
Cinchonan-9-ol, 6'-methoxy-, (8-α,9R)- (9CI) (130-95-0)
Cinchonan-9-ol, 6'-methoxy-, (9S)-, sulfate (1:1) (Salt) (9CI) (747-45-5)
Cinchonan-9-ol, 6'-methoxy-, (9S)-, sulfate (2:1) (Salt) (9CI) (50-54-4)
Cinchonan-9-ol, 6'-methoxy-, (8α,9R)-, mono(2-hydroxypropanoate) (Salt) (9CI) (749-49-5)
Cinchonan-9-ol, 6'-methoxy-, (8α,9R)-, monophosphinate (Salt) (9CI) (6119-53-5)
Cinchonan-9-ol, 6'-methoxy-, (8α,9R)-, sulfate (1:1) (Salt) (9CI) (549-56-4)
Cinchonan-9-ol, 6'-methoxy-, monohydrochloride, (8-α,9R)- (9CI) (130-89-2)
Cinchonan-9-ol, 6'-methoxy-, monohydrochloride, (9S)- (9CI) (1668-99-1)
Cinchonan-9-ol, 6'-methoxy-, (9s)- (56-54-2)
Cinchonan-9-ol, (9s)- (9CI) (118-10-5)
Cinchonan-9-one, 6'-methoxy-, (8α)- (9CI) (84-31-1)
(-)-Cinchonidine (485-71-2)
Cinchonidine (485-71-2)
Cinchoninamide, 2-butoxy-N-(2-(diethylamino)ethyl) (85-79-0)
Cinchonine (118-10-5)
d-Cinchonine (118-10-5)
Cinchoninic acid, 2-phenyl (132-60-5)
Cinchophen (132-60-5)
Cinchophene (132-60-5)
Cinchophenic acid (132-60-5)
Cinchovatine (485-71-2)
Cinconal (132-60-5)
Cincophen (132-60-5)
Cincosal (132-60-5)
Cineb (12122-67-7)
Cinene (138-86-3)
1,4-Cineol (470-67-7)
1,8-Cineol (470-82-6)
1,4-Cineole (470-67-7)
1,8-Cineole (470-82-6)
Cineole (470-82-6)
Cinerin I Allyl Homolog (584-79-2)
Cinerin I or II (8003-34-7)
Cinmethylin (87818-31-3)
Cinnamal (104-55-2)
(E)-Cinnamaldehyde (14371-10-9)
Cinnamaldehyde (104-55-2)
Cinnamaldehyde, (E) (14371-10-9)
trans-Cinnamaldehyde (14371-10-9)
Cinnamaldehyde, p-tert-butyl-α-methyl- (13586-68-0)
Cinnamaldehyde, α-hexyl (8CI) (101-86-0)
Cinnamaldehyde, 4-hydroxy-3-methoxy- (8CI) (458-36-6)
Cinnamaldehyde, o-methoxy (1504-74-1)
Cinnamaldehyde, α-methyl (101-39-3)
Cinnamaldehyde, α-pentyl- (6CI,7CI,8CI) (122-40-7)
Cinnamate de n-butyle (French) (538-65-8)
Cinnamein (103-41-3)
Cinnamene (100-42-5)
Cinnamenol (100-42-5)
(E)-Cinnamic acid (140-10-3)
Cinnamic acid, (E) (140-10-3)
Cinnamic-acid (621-82-9)
trans-Cinnamic acid (140-10-3)
Cinnamic acid, benzyl ester (103-41-3)
trans-Cinnamic acid benzyl ester (103-41-3)
Cinnamic acid, m-bromo (32862-97-8)
Cinnamic acid n-butyl ester (538-65-8)
Cinnamic acid, butyl ester (538-65-8)
Cinnamic acid chloride (102-92-1)
Cinnamic acid, p-chloro (1615-02-7)
Cinnamic acid, 3-chloro-4-hydroxy-5-methoxy- (8CI) (5438-40-4)
Cinnamic acid, 2,4-dichloro (1201-99-6)
Cinnamic acid, 3,4-dihydroxy (331-39-5)
Cinnamic acid, 3,4-dimethoxy- (8CI) (2316-26-9)
Cinnamic acid, ethyl ester (103-36-6)
Cinnamic acid, o-hydroxy-, (E) (614-60-8)
Cinnamic acid, p-hydroxy (7400-08-0)
Cinnamic acid, p-hydroxy-, (E) (501-98-4)
Cinnamic acid, ar-hydroxy- (8CI) (25429-38-3)
Cinnamic acid, o-hydroxy-, δ-lactone (91-64-5)
Cinnamic acid, 4-hydroxy-3-methoxy-, (E) (537-98-4)
Cinnamic acid, 4-hydroxy-3-methoxy-, trans- (537-98-4)

Name	CAS
Cinnamic acid, 4-hydroxy-3-methoxy- (8CI)	**(1135-24-6)**
Cinnamic acid, 4-hydroxy-3-methoxy-, (Z)- (8CI)	**(1014-83-1)**
Cinnamic acid, α-methyl- (8CI)	**(1199-77-5)**
Cinnamic acid, p-methyl- (6CI,7CI,8CI)	**(1866-39-3)**
Cinnamic acid, 3,4-(methylenebis(oxy))-	**(2373-80-0)**
Cinnamic acid, 3,4-(methylenedioxy)- (8CI)	**(2373-80-0)**
Cinnamic acid, methyl ester	**(103-26-4)**
Cinnamic acid, propyl ester	**(7778-83-8)**
Cinnamic acid, 3,4,5-trimethoxy	**(90-50-6)**
Cinnamic alcohol	**(104-54-1)**
Cinnamic aldehyde	**(104-55-2)**
trans-Cinnamic aldehyde	**(14371-10-9)**
Cinnamic chloride	**(102-92-1)**
Cinnamol	**(100-42-5)**
Cinnamon Bark Oil	**(8007-80-5)**
Cinnamon Oil	**(8007-80-5)**
Cinnamophenone	**(94-41-7)**
Cinnamoylchloride	**(102-92-1)**
Cinnamoyl chloride (8CI)	**(102-92-1)**
Cinnamyl-alcohol	**(104-54-1)**
Cinnamyl alcohol, anthranilate	**(87-29-6)**
Cinnamyl aldehyde	**(104-55-2)**
trans-Cinnamylaldehyde	**(14371-10-9)**
Cinnamyl 2-aminobenzoate	**(87-29-6)**
Cinnamyl o-aminobenzoate	**(87-29-6)**
Cinnamyl anthranilate	**(87-29-6)**
Cinnamylester kyseliny anthranilove (Czech)	**(87-29-6)**
Cinnarizin	**(637-07-0)**
Cinnarizine	**(637-07-0)**
Cinnoline (8CI,9CI)	**(253-66-7)**
Cinnoline, nitro- (9CI)	**(116211-87-1)**
Cinobac	**(28657-80-9)**
Cinobufotenine	**(487-93-4)**
Cinoxacin	**(28657-80-9)**
Cinoxate	**(104-28-9)**
Cinoxato (Spanish)	**(104-28-9)**
Cinoxatum (Latin)	**(104-28-9)**
Cinquasia Red	**(1047-16-1)**
Cinquasia Red B	**(1047-16-1)**
Cinquasia Red Y	**(1047-16-1)**
Cinquasia Violet	**(1047-16-1)**
Cinquasia Violet R	**(1047-16-1)**
Cin-quin	**(50-54-4)**
Cinu	**(13010-47-4)**
Cinx	**(28657-80-9)**
Ciodrin	**(7700-17-6)**
Ciovap	**(62-73-7)**
Ciovap	**(7700-17-6)**
Cipca	**(50-81-7)**
Cipe	**(9002-88-4)**
Ciplamin H	**(13422-51-0)**
Ciplamycetin	**(56-75-7)**
Cipoviol W 72	**(9002-89-5)**
Ciprenorfina (Spanish)	**(4406-22-8)**
Cipromid	**(2759-71-9)**
Ciram	**(137-30-4)**
Circain	**(479-18-5)**
Circair	**(479-18-5)**
Circosolv	**(79-01-6)**
Cirene Brilliant Blue R	**(3861-73-2)**
Cirpon	**(57-53-4)**
Cirponyl	**(57-53-4)**
Cirrasol 185A	**(112-05-0)**
Cirrasol LN-GS	**(31556-45-3)**
Cirrasol-OD	**(57-09-0)**
Cisanilide	**(34484-77-0)**
Cisclomiphene	**(911-45-5)**
Cisplatin	**(15663-27-1)**
Cisplatino (Spanish)	**(15663-27-1)**
Cisplatyl	**(15663-27-1)**
Cisteamina (Italian)	**(60-23-1)**
Cistobil	**(96-83-3)**
Cistoplex	**(519-95-9)**
Citanest	**(721-50-6)**
Citco Tri-basic Copper Sulfate-50XF	**(1332-03-2)**
Citex BCL 462	**(3322-93-8)**
Citexal	**(72-44-6)**
Cithrol 4MS	**(9004-99-3)**
Cithrol 10DS	**(9005-08-7)**
Cithrol 10MS	**(9004-99-3)**
Cithrol 60DS	**(9005-08-7)**
Cithrol PO	**(9004-96-0)**
Cithrol PS	**(9004-99-3)**
Citiflus	**(637-07-0)**
Citilat	**(21829-25-4)**
Citiolase	**(1195-16-0)**
Citiolone	**(1195-16-0)**
Citnatin	**(68-04-2)**
Citobaryum	**(7727-43-7)**
Citocor	**(59-26-7)**
Citodon	**(56-29-1)**
Citodorm	**(77-75-8)**
Citol	**(123-30-8)**
Citomulgan M	**(31566-31-1)**
Citopan	**(56-29-1)**
Citosulfan	**(55-98-1)**
Citox	**(50-29-3)**
Citraconic-acid	**(498-23-7)**
Citraconic acid anhydride	**(616-02-4)**
Citraconic-anhydride	**(616-02-4)**
Citra-fort	**(62-44-2)**
(E)-Citral	**(141-27-5)**
Citral	**(5392-40-5)**
Citral α	**(141-27-5)**
α-Citral	**(141-27-5)**
trans-Citral	**(141-27-5)**
Citral diethyl acetal	**(7492-66-2)**
Citral dimethyl acetal	**(7549-37-3)**
Citralka	**(144-33-2)**
Citral oxime	**(13372-77-5)**
Citral terpenes	**(8007-02-1)**
Citram	**(3734-97-2)**
Citram	**(78-53-5)**
Citramon	**(8003-03-0)**
Citrazinic acid	**(99-11-6)**
Citreme	**(68-04-2)**
Citretten	**(77-92-9)**
Citric-acid	**(77-92-9)**
Citric acid, Anhydrous	**(77-92-9)**
Citric acid, acetyl triethyl ester	**(77-89-4)**
Citric acid, ammonium iron(3+) salt	**(1185-57-5)**
Citric acid, calcium salt	**(7693-13-2)**
Citric acid, calcium salt (2:3)	**(813-94-5)**
Citric acid, cobalt(2+) salt (2:3)	**(866-81-9)**
Citric acid, copper(2+) salt (8CI)	**(866-82-0)**
Citric acid, diammonium salt	**(3012-65-5)**
Citric acid, diethanolamine salt	**(23349-61-3)**
Citric acid, disodium salt	**(144-33-2)**
Citric acid, ion(3-) (8CI)	**(126-44-3)**
Citric acid, iron(2+) salt	**(23383-11-1)**
Citric acid, lead(2+) salt (2:3) (8CI)	**(512-26-5)**
Citric acid, manganese(3+) salt (1:1) (8CI)	**(5968-88-7)**
Citric acid, monosodium salt (8CI)	**(18996-35-5)**
Citric acid, sodium salt	**(18996-35-5)**

Citric acid, tributyl ester (8CI)	(77-94-1)	Clark I	(712-48-1)
Citric acid, tributyl ester, acetate (8CI)	(77-90-7)	Claro 5591	(9005-25-8)
Citric acid, triethyl ester	(77-93-0)	Clarosan	(886-50-0)
Citric acid, triethyl ester, acetate	(77-89-4)	Claudelite	(1327-53-3)
Citric acid, trioctadecyl ester	(7775-50-0)	Claudetite	(1327-53-3)
Citric acid, tripotassium salt	(866-84-2)	Clavacin	(149-29-1)
Citric acid, trisodium salt	(68-04-2)	Clavatin	(149-29-1)
Citric acid, trisodium salt, monohydrate (8CI)	(19287-96-8)	Claviform	(149-29-1)
Citric acid, zinc salt (2:3)	(546-46-3)	Claviformin	(149-29-1)
Citrical	(813-94-5)	Clay (Kaolin)	(1332-58-7)
Citridic acid	(499-12-7)	Clay adsorbent	(70131-50-9)
Citrinin	(518-75-2)	Clayamine #4	(68953-58-2)
Citriscorb	(50-81-7)	Clayamine ARO	(68953-58-2)
Citro	(77-92-9)	Clayamine EP	(68953-58-2)
Citroflex 2	(77-93-0)	Clayamine EPA	(68953-58-2)
Citroflex 4	(77-94-1)	Clay bonded mordenite	(66402-68-4)
Citroflex A	(77-90-7)	Clay bonded natural zeolite	(66402-68-4)
Citroflex A 2	(77-89-4)	Cleansweep	(85-00-7)
Citroflex A 4	(77-90-7)	Clearasil BP Acne Treatment	(94-36-0)
Citrofluyl	(18996-35-5)	Clearasil Benzoyl Peroxide Lotion	(94-36-0)
Citron Yellow	(11103-86-9)	Clearate G	(9004-99-3)
Citron Yellow	(13530-65-9)	Clearcide	(33439-45-1)
Citronellal	(106-23-0)	Clearjel	(9005-25-8)
β-Citronellal	(106-23-0)	Clearjrel	(9005-25-8)
Citronellal hydrate	(107-75-5)	Cleartuf	(25038-59-9)
Citronellal, hydroxy-	(107-75-5)	Cleary 3336	(23564-06-9)
Citronella oil	(8000-29-1)	Cleocin	(18323-44-9)
Citronellel	(106-23-0)	Clera	(550-99-2)
Citronellene	(2436-90-0)	Cleridium 150	(58-32-2)
Citronellol	(106-22-9)	Clertuf	(25038-59-9)
Citronellol,(d)	(106-23-0)	Clestol	(577-11-7)
Citronellol, dihydro-	(106-21-8)	Cleve's Acid-1,6	(119-79-9)
Citronellol, hydroxy-	(107-74-4)	1,6-Cleve's Acid	(119-79-9)
Citronellyl acetate	(150-84-5)	1,7-Cleve's Acid	(119-28-8)
Citrosodine	(68-04-2)	Cleve's Acid	(119-28-8)
Citrosodna	(68-04-2)	Cleve's β-Acid	(119-79-9)
Citroviol	(128-51-8)	Cleve's Theta-Acid	(119-28-8)
Citrullamon	(57-41-0)	1,6-Clev's Acid	(93-00-5)
Citrullamon	(630-93-3)	Clhorameisensaeureaethylester (German)	(541-41-3)
Citrulliamon	(57-41-0)	Cliacil	(132-98-9)
Citrus Red 2	(6358-53-8)	Climaterine	(56-53-1)
Citrus Red No. 2	(6358-53-8)	Clin	(54-21-7)
Civettal	(91-61-2)	Clindamycin	(18323-44-9)
Cizaron	(91-81-6)	Clindamycine (French)	(18323-44-9)
Cl 12503	(119-12-0)	Clindrol 101CG	(120-40-1)
Cl 18706	(116-01-8)	Clindrol 203CG	(120-40-1)
Cl 26691	(115-93-5)	Clindrol 200CGN	(68603-42-9)
Cl-395	(77-10-1)	Clindrol 202CGN	(68603-42-9)
Cl-MDA	(101-14-4)	Clindrol 210CGN	(120-40-1)
2-Cl-p-PD	(6219-71-2)	Clindrol 200L	(120-40-1)
4-Cl-m-PD	(5131-60-2)	Clindrol 200-MS	(111-57-9)
4-Cl-o-PD	(95-83-0)	Clindrol SDG	(106-11-6)
ClUDR	(50-90-8)	Clindrol Seg	(111-60-4)
Cladosporin	(35818-31-6)	Clindrol Superamide 100CG	(68603-42-9)
Clafen	(50-18-0)	Clindrol Superamide 100L	(120-40-1)
Clafen	(6055-19-2)	Clinestrol	(130-80-3)
Clairformin	(149-29-1)	Clinoptilolite	(1318-02-1)
Clairsit	(594-42-3)	Clinoril	(38194-50-2)
Clam Poison Dihydrochloride	(35554-08-6)	Clinoxan	(10379-14-3)
Clamoxyl	(26787-78-0)	Clionasterol	(83-47-6)
Clandilon	(456-59-7)	Clioquinol	(130-26-7)
Claodical	(87-33-2)	Cliquinol	(130-26-7)
Claphene	(50-18-0)	Cliradon	(469-79-4)
Clarified oils, Petroleum, Catalytic cracked	(64741-62-4)	Cliradone	(469-79-4)
Clarified slurry oil	(64741-62-4)	Clistanol	(62-44-2)
Claripex	(637-07-0)	Clixodyne	(103-90-2)
Claripex CPIB	(637-07-0)	Cloazepam	(1622-61-3)

Clobazam (22316-47-8)
Clobber (2759-71-9)
Cloberat (637-07-0)
Clobrat (637-07-0)
Clobren-SF (637-07-0)
Cloethocarb (51487-69-5)
Cloetocarb (51487-69-5)
Clofar (637-07-0)
Clofenotane (50-29-3)
Clofenvinfos (470-90-6)
Clofibram (637-07-0)
Clofibrat (637-07-0)
Clofibrate (637-07-0)
Clofibrato (Spanish) (637-07-0)
Clofibric acid (882-09-7)
Clofibrinic acid (882-09-7)
Clofibrinsaeure (German) (882-09-7)
Clofinit (637-07-0)
Clofipront (637-07-0)
Cloflucarban (369-77-7)
Cloflucarbon (369-77-7)
Clomethiazole (533-45-9)
Clomethiazolum (533-45-9)
Clomid (50-41-9)
Clomifen citrate (50-41-9)
Clomifene (911-45-5)
Clomifeno (50-41-9)
Clomiphene (911-45-5)
Clomiphene B (911-45-5)
Clomiphene citrate (50-41-9)
Clomiphene dihydrogen citrate (50-41-9)
Clomiphene-r (50-41-9)
Clomiphine (50-41-9)
Clomipramine (303-49-1)
Clomivid (50-41-9)
Clomphid (50-41-9)
Clonazepam (1622-61-3)
Clonidin (4205-90-7)
Clonidine (4205-90-7)
Clonitralid (1420-04-8)
Clonitralid (50-65-7)
Clont (443-48-1)
Clophen (1336-36-3)
Clophen A 50 (8068-44-8)
Clophen A-30 (55600-34-5)
Clophen A60 (11096-82-5)
Clopidol (ACGIH,OSHA) (2971-90-6)
Clopoxide (58-25-3)
Cloprednol (5251-34-3)
Cloprednolum (Latin) (5251-34-3)
Cloprop (101-10-0)
Clopyralid (1702-17-6)
Clor Chem T-590 (8001-35-2)
Clori-Clean (8007-32-7)
Clorafin (63449-39-8)
Cloral betaina (Spanish) (2218-68-0)
Cloral betaine (2218-68-0)
Cloralio (Italian) (75-87-6)
Cloralum betainum (Latin) (2218-68-0)
Cloramficin (56-75-7)
Cloramicol (56-75-7)
Cloramidina (56-75-7)
Cloramin (51-75-2)
Clorato amonico (Spanish) (10192-29-7)
Clorazepate (23887-31-2)
Clorazepic Acid (23887-31-2)
Clorazodina (Spanish) (502-98-7)
Clordan (Italian) (57-74-9)
Clordano (57-74-9)
Clordesmetildiazepam (Spanish) (2894-67-9)
Clordiazepossido (Italian) (58-25-3)
Clordion (302-22-7)
Clorepin (22316-47-8)
Clorestrolo (569-57-3)
Cloretilo (75-00-3)
Clorex (111-44-4)
Clorfeniramina (Italian) (132-22-9)
Clorilax (80-77-3)
Clorina (127-65-1)
Clorindanol (145-94-8)
Clorito calcico (Spanish) (14674-72-7)
Clorito de plata (Spanish) (7783-91-7)
Clorito sodico (Spanish) (7758-19-2)
Clormetazanone (80-77-3)
Clormetazon (80-77-3)
Cloro (Italian) (7782-50-5)
Cloroamfenicolo (Italian) (56-75-7)
3-Cloroaniline (Italian) (108-42-9)
Cloroben (95-50-1)
2-Clorobenzaldeide (Italian) (89-98-5)
Clorobenzene (Italian) (108-90-7)
(4-Cloro-benzil)-(4-cloro-fenil)-solfuro (Italian) (103-17-3)
1-p-Cloro-benzoil-5-metoxi-2-metilindol-3-acido acetico (Spanish) (53-86-1)
(4-Cloro-but-2-in-il)-N-(3-cloro-fenil)-carbammato (Italian) (101-27-9)
2-Cloro-1,3-butadiene (Italian) (126-99-8)
Clorochina (54-05-7)
Clorocyn (56-75-7)
O-2-Cloro-1-(2,4-dicloro-fenil)-vinil-O,O-dietilfosfato (Italian) (470-90-6)
(2-Cloro-3-dietilamino-1-metil-3-oxo-prop-1-en-il)-dimetil-fosfato (Italian) (13171-21-6)
Clorodifenili, Cloro 54% (Italian) (11097-69-1)
Clorodifenili, cloro 42% (Italian) (53469-21-9)
Clorodifluoretano (Spanish) (25497-29-4)
2-Cloro-4-dimetilamino-6-metil-pirimidina (Italian) (535-89-7)
2-Cloro-10 (3-dimetilaminopropil)fenotiazina (Italian) (50-53-3)
1-Cloro-2,4-dinitrobenzene (Italian) (97-00-7)
1-Cloro-2,3-epossipropano (Italian) (106-89-8)
Cloroetano (Italian) (75-00-3)
2-Cloroetanolo (Italian) (107-07-3)
(Cloro-2-etil)-1-cicloesil-3-nitrosourea (Italian) (13010-47-4)
Clorofene (120-32-1)
(4-Cloro-fenil)-benzol-solfonato (Italian) (80-38-6)
(4-Cloro-fenil)-4-cloro-venzol-solfonato (Italian) (80-33-1)
3-(4-Cloro-fenil)-1,1-dimetil-urea (Italian) (150-68-5)
2(2-(4- Cloro-fenil-2fenil)-acetil)indan-1,3-dione (Italian) (3691-35-8)
N-(3-Cloro-fenil)-isopropil-carbammato (Italian) (101-21-3)
3-(1-(4-Cloro-fenil)-3-oxo-butil)-4-idrossicumarina (Italian) (81-82-3)
Clorofeniltriclorosilano (Spanish) (26571-79-9)
Cloroformio (Italian) (67-66-3)
Clorofos (Russian) (52-68-6)
Clorometano (Italian) (74-87-3)
7-Cloro-2-metilamino-5-fenil-3H-1,4-benzodiazepina 4-ossido (Italian) (58-25-3)
3-Cloro-2-metil-prop-1-ene (Italian) (563-47-3)
Cloromisan (56-75-7)
1-Cloro-4-nitrobenzene (Italian) (100-00-5)
O-(3-Cloro-4-nitro-fenil)-O,O-dimetil-monotiofosfato (Italian) (500-28-7)
O-(4-Cloro-3-nitro-fenil)-O,O-dimetil-monotiofosfato (Italian) (2463-84-5)
Clorophene (120-32-1)
Cloropicrina (Italian) (76-06-2)
Cloropiril (113-92-8)
Cloropiril (132-22-9)
Cloroprene (Italian) (126-99-8)
Clorosan (127-65-1)

Synonym	CAS No.
Clorosintex	**(56-75-7)**
Clorotrisin	**(569-57-3)**
Clorox	**(7681-52-9)**
Clorpromazina (Italian)	**(50-53-3)**
Clorpropamide (Italian)	**(94-20-2)**
Clortermina (Spanish)	**(10389-73-8)**
Clortermine	**(10389-73-8)**
Clorterminum (Latin)	**(10389-73-8)**
Clortokem	**(15545-48-9)**
Clortran	**(57-15-8)**
Cloruro de acriflavinio (Spanish)	**(8018-07-3)**
Cloruro de amilo (Spanish)	**(594-36-5)**
Cloruro de metilbenzetonio (Spanish)	**(25155-18-4)**
Cloruro de propionilo (Spanish)	**(79-03-8)**
Cloruro de valerilo (Spanish)	**(638-29-9)**
Cloruro di ethene (Italian)	**(107-06-2)**
Cloruro di etile (Italian)	**(75-00-3)**
Cloruro di etilidene (Italian)	**(75-34-3)**
Cloruro di mercurio (Italian)	**(7487-94-7)**
Cloruro di metallile (Italian)	**(563-47-3)**
Cloruro di metile (Italian)	**(74-87-3)**
Cloruro di vinile (Italian)	**(75-01-4)**
Clotiamina	**(67-03-8)**
Clotiazepam	**(33671-46-4)**
Clotiazepamum (Latin)	**(33671-46-4)**
Clotride	**(58-94-6)**
Clotrimazol	**(23593-75-1)**
Clotrimazole	**(23593-75-1)**
Clout	**(144-21-8)**
Clout	**(55635-13-7)**
Clovene	**(469-92-1)**
Cloxacillin	**(61-72-3)**
Cloxazepine	**(1977-10-2)**
Cloxazolam	**(24166-13-0)**
Cloxazolamum (Latin)	**(24166-13-0)**
Cloxazolazepam	**(24166-13-0)**
Cluytyl alcohol	**(557-61-9)**
Clysar	**(9003-07-0)**
Clyzerin, Wasserfrei (German)	**(56-81-5)**
Co-Ral	**(56-72-4)**
Co-Rax	**(81-81-2)**
Co-Trimoxazole	**(723-46-6)**
Co-Trimoxazole	**(8064-90-2)**
Coad	**(557-05-1)**
Coal Ash	**(68131-74-8)**
Coal-Fly-Ash	**(68131-74-8)**
Coal Liquid	**(8002-05-9)**
Coal Oil	**(8002-05-9)**
Coal Oil	**(8008-20-6)**
Coal Oil (Export shipment only) (DOT)	**(8008-20-6)**
Coal-Tar	**(65996-89-6)**
Coal-Tar	**(8007-45-2)**
Coal Tar, Aerosol	**(8007-45-2)**
Coal-Tar-Creosote	**(8001-58-9)**
Coal Tar Distillate	**(65996-92-1)**
Coal Tar Distillates	**(65996-92-1)**
Coal Tar Hydrocarbons	**(8002-29-7)**
Coal Tar Light Oil	**(65996-91-0)**
Coal Tar Naphtha (DOT)	**(8030-30-6)**
Coal Tar Neutral Oils	**(65996-82-9)**
Coal Tar Neutral Oils	**(8002-29-7)**
Coal Tar Oil (DOT)	**(8001-58-9)**
Coal-Tar Pitch	**(65996-93-2)**
Coal Tar Pitch Volatiles (ACGIH,OSHA)	**(65996-93-2)**
Coal-Tar-Solution-USP	**(8007-45-2)**
Coalite NTP	**(25155-23-1)**
Coal naphtha	**(71-43-2)**
Coal slag	**(68476-96-0)**
Coal tar pitch volatiles: phenanthrene	**(85-01-8)**
Coal-tars, Low temperature	**(65996-90-9)**
Coapt	**(137-05-3)**
Coathylene HA 1671	**(9002-88-4)**
Coathylene PF 0548	**(9003-07-0)**
Cobadex	**(50-23-7)**
Cobadoce forte	**(68-19-9)**
Cobalt-59	**(7440-48-4)**
Cobalamin, hydroxo-	**(13422-51-0)**
Cobalt-beryllium copper	**(55158-44-6)**
Cobalt-chromium alloy	**(11114-92-4)**
Cobalex	**(13422-51-0)**
Cobalin	**(68-19-9)**
Cobalt (ACGIH,OSHA)	**(7440-48-4)**
Cobalt Alloy, Co,Cr	**(11114-92-4)**
Cobalt Black	**(1307-96-6)**
Cobalt (2+) NTA	**(53108-50-2)**
Cobalt Resinate, Precipitated [UN 1318]	**(68956-82-1)**
Cobalt Rosinate	**(68956-82-1)**
Cobalt acetate	**(71-48-7)**
Cobalt(2+) acetate	**(71-48-7)**
Cobalt(II) acetate	**(71-48-7)**
Cobalt acetate (co(oac)2)	**(71-48-7)**
Cobalt acetate tetrahydrate	**(6147-53-1)**
Cobalt(II) acetate tetrahydrate	**(6147-53-1)**
Cobalt(II) acetylacetonate	**(14024-48-7)**
Cobalt arsenate	**(24719-19-5)**
Cobaltate(1-), bis(2-((5-(aminosulfonyl)-2-hydroxyphenyl)azo)-3-oxo-N-phenylbutanamidato- (2-))-, sodium (9CI)	**(72496-88-9)**
Cobaltate(2-), bis(2-((5-(aminosulfonyl)-2-hydroxyphenyl)azo)-3-oxo-N-phenylbutanamidato- (2-))-, dihydrogen (9CI)	**(12715-61-6)**
Cobaltate(1-), (N,N-bis(carboxymethyl)glycinato(3-)-N,O,O',O'')-, hydrogen (9CI)	**(53108-50-2)**
Cobaltate(1-), (N,N-bis(carboxymethyl)glycinato(3-)-N,O,O',O'')-, hydrogen, (T-4)-	**(53108-50-2)**
Cobaltate(1-), bis(3-((4,5-dihydro-3-methyl-5-oxo-1-phenyl-1H-pyrazol-4-yl)azo)-4-hydroxybenzenesulfonamidato(2-))-, sodium, (OC-6-22')- (9CI)	**(34664-47-6)**
Cobaltate(2-), (29H,31H-phthalocyaninedisulfonato(4-)-N29,N30,N31,N32)-, dihydrogen (9CI)	**(29383-29-7)**
Cobaltate(1-), (29H,31H-phthalocyaninesulfonato(3-)-N29,N30,N31,N32)-, hydrogen (9CI)	**(30638-08-5)**
Cobaltate(4-), (29H,31H-phthalocyanine-2,9,16,23-tetrasulfonato-(6-)-N29,N30,N31,N32)-, tetrahydrogen, (SP-4-1)- (9CI)	**(14285-59-7)**
Cobalt 1,4-benzenedicarboxylate	**(34262-88-9)**
Cobalt bis(acetylacetonate)	**(14024-48-7)**
Cobalt(II) bis(acetylacetonate)	**(14024-48-7)**
Cobalt bis(2-ethylhexanoate)	**(136-52-7)**
Cobalt, bis(3-fluorosalicylaldehyde)-ethylenediimine-	**(62207-76-5)**
Cobalt, bis(3-fluorosalicylaldehyde)ethylenediimine-	**(62207-76-5)**
Cobalt, bis(2,4-pentanedionato)- (8CI)	**(14024-48-7)**
Cobalt, bis(2,4-pentanedionato-O,O')-, (T-4)- (9CI)	**(14024-48-7)**
Cobalt(II) bromide	**(7789-43-7)**
Cobalt carbonate, cobalt dihydroxide (2:3)	**(12602-23-2)**
Cobalt carbonyl (ACGIH,OSHA)	**(10210-68-1)**
Cobalt chloride	**(7646-79-9)**
Cobalt(II) chloride	**(7646-79-9)**
Cobalt chromate	**(13455-25-9)**
Cobalt citrate	**(866-81-9)**
Cobalt diacetate	**(71-48-7)**
Cobalt diacetate tetrahydrate	**(6147-53-1)**
Cobalt diacetylacetonate	**(14024-48-7)**
Cobalt dibromide	**(7789-43-7)**
Cobalt, di-mu-carbonylhexacarbonyldi-, (Co-Co)	**(10210-68-1)**
Cobalt dichloride	**(7646-79-9)**
Cobalt difluoride	**(10026-17-2)**

Cobalt diformate	(544-18-3)
Cobalt dinitrate	(10141-05-6)
Cobalt, di-pi-cyclopentadienyl	(1277-43-6)
Cobalt, ((2,2'-(1,2-ethanediylbis(nitrilomethylidyne))bis-(6-fluorophenolato))(2-)-N,N',O,O')-	(62207-76-5)
Cobalt, ((2,2'-(1,2-ethanediylbis(nitrilomethylidyne))bis-(6-fluorophenolato))(2-)-N,N',O,O')-, (SP-4-2)- (9CI)	(62207-76-5)
Cobalt, ((2,2'-(1,2-ethanediylbis(nitrilomethylidyne))bis-(phenolato))(2-)-N,N',O,O')-	(14167-18-1)
Cobalt, ((2,2'-(1,2-ethanediylbis(nitrilomethylidyne))bis-(phenolato))(2-)-N,N',O,O')-, (SP-4-2)- (9CI)	(14167-18-1)
Cobalt(II), N,N'-ethylenebis(3-fluorosalicylideneiminato)-	(62207-76-5)
Cobalt, N,N'-ethylenebis(salicylideneiminato)-	(14167-18-1)
Cobalt, ((α,α'-(ethylenedinitrilo)di-o-cresolato)(2-))- (8CI)	(14167-18-1)
Cobalt 2-ethylhexanoate	(13586-82-8)
Cobalt(2+) 2-ethylhexanoate	(136-52-7)
Cobalt(II) 2-ethylhexanoate	(136-52-7)
Cobalt 2-ethylhexoate	(136-52-7)
Cobalt(II) fluoride	(10026-17-2)
Cobalt(2+) formate	(544-18-3)
Cobalt formate (VAN)	(544-18-3)
Cobalt hydrocarbonyl (ACGIH,OSHA)	(16842-03-8)
Cobalt, (hydrogen phthalocyaninesulfonato(2-))-	(30638-08-5)
Cobalt, Isotope of mass 60 (8CI,9CI)	(10198-40-0)
Cobalt monooxide	(1307-96-6)
Cobalt monosulfate heptahydrate	(10026-24-1)
Cobalt monosulfide	(1317-42-6)
Cobalt monoxide	(1307-96-6)
Cobalt muriate	(7646-79-9)
Cobalt naphthenate	(61789-51-3)
Cobalt naphthenate, Powder [UN 2001]	(61789-51-3)
Cobalt neodecanoate	(27253-31-2)
Cobalt nitrate	(10141-05-6)
Cobalt(2+) nitrate	(10141-05-6)
Cobalt(II) nitrate	(10141-05-6)
Cobalt(II) nitrate (1:2)	(10141-05-6)
Cobalt nitrilotriacetic acid	(53108-50-2)
Cobaltocene	(1277-43-6)
Cobalt octacarbonyl	(10210-68-1)
Cobalt octadecanoate	(13586-84-0)
Cobalt octoate	(136-52-7)
Cobaltous acetate	(71-48-7)
Cobaltous acetate tetrahydrate	(6147-53-1)
Cobaltous acetylacetonate	(14024-48-7)
Cobaltous ammonium sulfate	(13596-46-8)
Cobaltous bromide	(7789-43-7)
Cobaltous carbonate	(513-79-1)
Cobaltous chloride	(7646-79-9)
Cobaltous citrate	(866-81-9)
Cobaltous diacetate	(71-48-7)
Cobaltous dichloride	(7646-79-9)
Cobaltous 2-ethylhexanoate	(136-52-7)
Cobaltous fluoride	(10026-17-2)
Cobaltous formate	(544-18-3)
Cobaltous nitrate	(10141-05-6)
Cobaltous octoate	(136-52-7)
Cobaltous oxide	(1307-96-6)
Cobaltous succinate	(3267-76-3)
Cobaltous sulfamate	(14017-41-5)
Cobaltous sulfate	(10124-43-3)
Cobaltous sulfate heptahydrate	(10026-24-1)
Cobaltous sulfide	(1317-42-6)
Cobalt oxide	(1307-96-6)
Cobalt(2+) oxide	(1307-96-6)
Cobalt(II) oxide	(1307-96-6)
Cobalt oxide (9CI)	(1308-06-1)
Cobalt, (phthalocyaninato(2-))- (8CI)	(3317-67-7)
Cobalt, (29H,31H-phthalocyaninato(2-)-N29,N30,N31,N32)-, (SP-4-1)- (9CI)	(3317-67-7)
Cobalt 4,4',4'',4'''-phthalocyaninetetrasulfonate	(14285-59-7)
Cobalt succinate	(3267-76-3)
Cobalt (2+) sulfate	(10124-43-3)
Cobalt sulfate	(10124-43-3)
Cobalt sulfate (1:1)	(10124-43-3)
Cobalt(II) sulfate	(10124-43-3)
Cobalt(II) sulfate (1:1)	(10124-43-3)
Cobalt(II) sulfate (1:1), heptahydrate	(10026-24-1)
Cobalt sulfide	(1317-42-6)
Cobalt(2+) sulfide	(1317-42-6)
Cobalt sulfide (Amorphous)	(1317-42-6)
Cobalt(II) sulphate	(10124-43-3)
Cobalt(II) sulphate heptahydrate	(10026-24-1)
Cobalt terephthalate	(34262-88-9)
Cobalt tetracarbonyl	(10210-68-1)
Cobalt tetracarbonyl dimer	(10210-68-1)
Cobalt, tetracarbonylhydro- (8CI,9CI)	(16842-03-8)
Cobalt thiocyanate	(3017-60-5)
Cobalt(2+) thiocyanate	(3017-60-5)
Cobalt titanium oxide (9CI)	(12017-38-8)
Cobamin	(68-19-9)
Coban	(22373-78-0)
Coban 45	(22373-78-0)
Cobex	(29091-05-2)
Cobex (Herbicide)	(29091-05-2)
Cobexo	(29091-05-2)
Cobinamide, cyanide phosphate 3'-ester with 5,6-dimethyl-1-α-D- ribofuranosylbenzimidazole, inner salt	(68-19-9)
Cobinamide dihydroxide dihydrogen phosphate (ester), mono(inner salt), 3'-ester with 5,6-dimethyl-1-α-D-ribofuranosylbenzimidazole	(13422-51-0)
Cobinamide, dihydroxide, dihydrogen phosphate (ester), mono-(inner salt), 3'-ester with 5,6-dimethyl-1-α-D-ribofuranosyl-1H-benzimidazole (9CI)	(13422-51-0)
Cobinamide hydroxide phosphate 3'-ester with 5,6-dimethyl-1-α-D-ribofuranosylbenzimidazole inner salt	(13422-51-0)
Cobione	(68-19-9)
α Cobione	(13422-51-0)
Cobox	(1332-40-7)
Cobox Blue	(1332-40-7)
Cobratec #99	(95-14-7)
Cobratec 99	(95-14-7)
Cobratec TT 100	(29385-43-1)
Coco-Diazine	(68-35-9)
Cocafurin	(59-87-0)
Cocain-chlorhydrat (German)	(53-21-4)
(-)-Cocaine	(50-36-2)
1-Cocaine	(50-36-2)
Cocaine	(50-36-2)
β-Cocaine	(50-36-2)
Cocaine chloride	(53-21-4)
(-)-Cocaine hydrochloride	(53-21-4)
Cocaine hydrochloride	(53-21-4)
l-Cocaine hydrochloride	(53-21-4)
Cocaine muriate	(53-21-4)
Cocamidopropyl dimethyl glycine	(61789-40-0)
N-Cocamidopropyl-N,N-dimethylglycine, hydroxide, inner salt	(61789-40-0)
Cocamidopropyl betaine	(61789-40-0)
Cocao Bean Oil	(8002-31-1)
Cocartrit	(54-05-7)
Coccidine A	(148-01-6)
Coccidiostat C	(2971-90-6)
Coccidot	(148-01-6)
Coccin Red	(2611-82-7)
Coccine	(2611-82-7)

Coccine Nouvelle (French) (2611-82-7)
Coccoclase (144-83-2)
Cocculin (124-87-8)
Cocculus, Solid (DOT) (124-87-8)
Cocculus, Solid (Fishberry) (DOT) (124-87-8)
Cocculus [UN 1584] (124-87-8)
N-Coco-1,3-diaminopropane acetate (61791-64-8)
Cochenillerot A (German) (2611-82-7)
Cochineal Red A (2611-82-7)
Cochineal Red A Specially Pure (2611-82-7)
Cochineal Red 4R (2611-82-7)
Cochineal Tincture (1260-17-9)
Cocoa Absolute (8002-31-1)
Cocoa Bean Extract (8002-31-1)
Cocoa Beans Absolute, Colourless MD (8002-31-1)
Cocoa Beans, Methanol Extract (8002-31-1)
Cocoa Butter (8002-31-1)
Cocoa Essence, Dark (8002-31-1)
Cocoa Essence, White (8002-31-1)
Cocoa Oil (8002-31-1)
Cocoa Oil Absolute (8002-31-1)
Cocoa Shell Extract (8002-31-1)
(Coco alkyl)amine monobenzoate (68526-65-8)
1-((Coco alkyl)amino)-3-aminopropane (61791-63-7)
1-((Coco alkyl)amino)-3-aminopropane acetates (61791-64-8)
1-((Coco alkyl)amino)-3-aminopropane adipate (68155-42-0)
1-((Coco alkyl)amino)-3-aminopropane benzoate (68188-29-4)
1-((Coco alkyl)amino)-3-aminopropane hydroxyacetate (68155-43-1)
2-(Cocoalkyl)-1-benzyl-1-(2-hydroxyethyl)-2-imidazolinium chloride (61791-52-4)
(Coco alkyl) bis(2-hydroxyethyl) methyl ammonium chloride (70750-47-9)
N-Cocoalkyltrimethylendiamine benzoate (68188-29-4)
N-(3-Cocoamidopropyl)-N,N-dimethyl-N-carboxymethyl-ammonium hydroxide, inner salt (61789-40-0)
N-(Cocoamidopropyl)-N,N-dimethyl-N-carboxymethyl ammonium, betaine (61789-40-0)
N-(3-Cocoamidopropyl)-N,N-dimethyl-N-carboxymethyl betaine (61789-40-0)
Cocoamidopropylbetaine (61789-40-0)
Cocoamine, benzoic acid salt (68526-65-8)
Cocoamine, ethoxylated (61791-14-8)
Cocoanut Oil (8001-31-8)
Cocobis(hydroxyethyl)benzylammonium, chloride (61789-68-2)
Coco diethanolamide (120-40-1)
Coconut Butter (8001-31-8)
Coconut Diethanolamide (68603-42-9)
Coconut Meal Pellets, Containing 6% to 13% moisture and no more than 10% residual fat (DOT) (8001-31-8)
Coconut-Oil (8001-31-8)
Coconut-Oil-Acid-Diethanolamine (68603-42-9)
Coconut Palm Oil (8001-31-8)
Coconut aldehyde (104-61-0)
Coconut fatty acid, aminoethylethanolamine amide-imidazoline, carboxymethylated, sodium salt (68647-53-0)
Coconut fatty acid, aminoethylethanolamine amide-imidazoline, dicarboxymethylated, disodium salt (68647-53-0)
Coconut fatty acid, aminoethylethanolamine imidazoline, carboxymethylated, sodium salt (68647-53-0)
Coconut fatty acid, aminoethylethanolamine reaction product, dicarboxymethylated, disodium salt (68647-53-0)
Coconut oil, Polymer with glycerol and phthalic anhydride (66070-87-9)
(Coconut oil alkyl)amine, ethoxylated (61791-14-8)
1,1'-((Coconut oil alkyl)imino)di-2-propanol (68516-06-3)
N-(Coconut oil alkyl)-trimethylenediamine, benzoic acid salt (68188-29-4)
N-(Coconut oil alkyl)trimethylenediamine, hydroxyacetate (68155-43-1)
Coconut oil amide of diethanolamine (120-40-1)
Coconut oil amidopropyl betaine (61789-40-0)
Coconut oil fatty acids, phthalic anhydride, glycerin polymer (66070-87-9)
Coconut oil, glycerine, phthalic anhydride polymer (66070-87-9)
Coconut oil, glycerin, phthalic anhydride polymer (66070-87-9)
Coconut oil, glycerin, phthalic anhydride resin (66070-87-9)
Coconut oil, glycerol, phthalic anhydride alkyd resin (66070-87-9)
Coconut oil, glycerol, phthalic anhydride polymer (66070-87-9)
Coconut oil, glycerol, phthalic anhydride resin (66070-87-9)
Coconut oil, phthalic anhydride, glycerin polymer (66070-87-9)
Coconut oil, phthalic anhydride, glycerol polyester (66070-87-9)
Coconut oil, phthalic anhydride, glycerol polymer (66070-87-9)
Coconut oil, phthalic anhydride, glycerol resin (66070-87-9)
Cocopropylenediamine (61791-63-7)
Cocoyl amide propylbetaine (61789-40-0)
N-Coco-1,3-propylenediamine (61791-63-7)
Codal (51218-45-2)
Code H 133 (1194-65-6)
Codecarboxylase (54-47-7)
Codechine (58-89-9)
Codeigene (3688-65-1)
Codeine (63905-03-3)
Codeine (76-57-3)
Codeine, acetyldihydro- (3861-72-1)
Codeine bromomethylate (125-27-9)
Codeine, dihydro (125-28-0)
Codeine hydrochloride (1422-07-7)
Codeine methylbromide (125-27-9)
Codeine, nicotinate (ester) (3688-66-2)
Codeine N-oxide (3688-65-1)
Codeine-N-oxide (3688-65-1)
Codeine phosphate (52-28-8)
Codeine sulfate (1420-53-7)
Codeinone, 7,8-dihydro-14-hydroxy- (76-42-6)
Codeinone, dihydro-14-hydroxy- (76-42-6)
Codelcortone (50-24-8)
Codempiral (62-44-2)
Codethyline (76-58-4)
Codethyline hydrochloride (125-30-4)
Codhydrine (125-28-0)
Codiazine (68-35-9)
Codibarbita (50-06-6)
Codroxomin (13422-51-0)
Codylin (509-67-1)
Coe 536 (3658-77-3)
Coenzyme R (58-85-5)
Coerulignol (2785-87-7)
Coffearin (535-83-1)
Coffearine (535-83-1)
Coffein (German) (58-08-2)
Coffeine (58-08-2)
Cogentine (86-13-5)
Cogilor Blue 512.12 (3844-45-9)
Cogilor Red 321.10 (81-88-9)
Cogilor Violet 411.12 (1694-09-3)
Cognac Oil (106-30-9)
Cohasal-IH (9005-38-3)
Cohoba (487-93-4)
Cohydrin (125-28-0)
Coir Deep Black C (1937-37-7)
Coir Deep Black R (2429-83-6)
Coke (50-36-2)
Coke (65996-77-2)
Coke (Chaud) (French) (65996-77-2)
Coke, Coal (65996-77-2)
Coke, Hot (65996-77-2)
Coke Powder (7440-44-0)
Para-Col (1910-42-5)
γ-Col (58-89-9)
Col-Evac (144-55-8)

Colace	(577-11-7)
Colacid Black 10A	(1064-48-8)
Colacid Blue A	(3861-73-2)
Colacid Orange	(633-96-5)
Colacid Orange G	(1936-15-8)
Colacid Ponceau 4R	(2611-82-7)
Colacid Ponceau Special	(3761-53-3)
Colacid Red 2A	(3567-66-6)
Colacid Red AV	(1658-56-6)
Colalin	(81-25-4)
Colamine	(141-43-5)
Colamine hydrochloride	(2002-24-6)
Colanyl Green GG	(1328-53-6)
Colascor	(50-81-7)
Colcemid	(477-30-5)
Colcemide	(477-30-5)
Colchamin	(477-30-5)
Colchamine	(477-30-5)
Colchicin (German)	(64-86-8)
Colchicina (Italian)	(64-86-8)
7-α-H-Colchicine	(64-86-8)
Colchicine	(64-86-8)
Colchicine, 7-deacetamido-7-(methylamino)-	(477-30-5)
Colchicine, N-deacetyl-N-methyl	(477-30-5)
Colchicine, deacetyl-N-methyl-	(477-30-5)
Colchine, N-deacetyl-N-methyl	(477-30-5)
Colchineos	(64-86-8)
Colchisol	(64-86-8)
Colcin	(64-86-8)
Colcothar	(1309-37-1)
Coldan	(550-99-2)
Colebenz	(120-51-4)
Colecalciferol	(67-97-0)
Colemid	(477-30-5)
Colep	(2665-30-7)
Colepax	(96-83-3)
Colestyramin	(11041-12-6)
Coletyl	(51-83-2)
Colfarit	(50-78-2)
Colibil	(41826-92-0)
Colimycin	(1066-17-7)
Colisone	(53-03-2)
Colisticina	(1066-17-7)
Colistin	(1066-17-7)
Colistinase	(9014-01-1)
Collatex Arm Extra	(9005-34-9)
Colletotrichin	(61235-00-5)
2,4,6-Collidine	(108-75-8)
Collidine	(29611-84-5)
α,γ,α'-Collidine	(108-75-8)
γ-Collidine	(108-75-8)
s-Collidine	(108-75-8)
sym-Collidine	(108-75-8)
Collidine, aldehydecollidine	(104-90-5)
Colliron I.V.	(8047-67-4)
Collocarb	(1333-86-4)
Collodial Lead Phosphate	(7446-27-7)
Collodion Cotton	(9004-70-0)
Collodion (DOT)	(9004-70-0)
Collodion Wool	(9004-70-0)
Colloid 775	(9000-07-1)
Colloidal Arsenic	(7440-38-2)
Colloidal Cadmium	(7440-43-9)
Colloidal Ferric Oxide	(1309-37-1)
Colloidal Gold	(7440-57-5)
Colloidal Manganese	(7439-96-5)
Colloidal Mercury	(7439-97-6)
Colloidal-S	(7704-34-9)
Colloidal Selenium	(7782-49-2)
Colloidal Silica	(7631-86-9)
Colloidal Silicon Dioxide	(7631-86-9)
Colloidal Silver Iodide	(7783-96-2)
Colloidal Sulfur	(7704-34-9)
Colloidox	(1332-40-7)
Collokit	(7704-34-9)
Collomide	(63-74-1)
Colloresine	(9000-11-7)
Collowel	(9004-32-4)
Colloxylin	(9004-70-0)
Collu Hextril	(141-94-6)
Collunosol	(95-95-4)
Color-Set	(93-72-1)
Cologne Earth	(1317-34-6)
Cologne Spirit	(64-17-5)
Cologne Umber	(1317-34-6)
Cologne Yellow	(7758-97-6)
Colonatrast	(7727-43-7)
Colonial Spirit	(67-56-1)
Colpovister	(50-27-1)
Colsaloid	(64-86-8)
Colsul	(7704-34-9)
Colsulanyde	(63-74-1)
Coltsfoot LS	(8001-21-6)
Columbia Black EP	(1937-37-7)
Columbia Brown M	(2429-82-5)
Columbia Fast Black G	(6428-31-5)
Columbia Fast Black GB	(6428-31-5)
Columbia LCK	(7440-44-0)
Columbia carbon	(1333-86-4)
Columbian spirit	(67-56-1)
Columbian spirits (DOT)	(67-56-1)
Columbium pentachloride	(10026-12-7)
Coly-Mycin	(1066-17-7)
Colyer Pectin	(9000-69-5)
Colymysin S	(1066-17-7)
Comac	(20427-59-2)
Comacid Blue Black B	(1064-48-8)
Comacid Scarlet 2R	(3761-53-3)
Combat	(67485-29-4)
Combat White Fly Insecticide	(28434-01-7)
Combinace	(9005-35-0)
Combinal K1	(84-80-0)
Combi-schutz	(67-63-0)
Combot	(52-68-6)
Combot Equine	(52-68-6)
Combustion Improver-2	(12108-13-3)
Comesa	(77-75-8)
Comestrol	(56-53-1)
Comestrol estrobene	(56-53-1)
Comital	(57-41-0)
Comite	(2312-35-8)
Comitiadone	(63-98-9)
Comitoina	(57-41-0)
Command	(81777-89-1)
Common Bean LS	(8001-21-6)
Common Fumitory LS	(8001-21-6)
Common Salt	(7647-14-5)
Common sense cockroach and rat preparations	(7723-14-0)
Commotional	(62-44-2)
Compactin	(73573-88-3)
Compalox	(1344-28-1)
Compazine	(58-38-8)
Compendium	(1812-30-2)
Comperlan HS	(111-57-9)

Comperlan LD	(120-40-1)
Comperlan LM	(142-78-9)
Comperlan MM	(142-58-5)
Compitox	(7085-19-0)
Compitox	(93-65-2)
Compitox Plus	(7085-19-0)
Complemix	(577-11-7)
Complexon I	(139-13-9)
Complexon II	(60-00-4)
Complexon III	(139-33-3)
Complexon IV	(482-54-2)
Complexone	(64-02-8)
Compocillin G	(61-33-6)
Compocillin-VK	(132-98-9)
(Component of) Dri-Die	(16919-19-0)
(Component of) Drianone	(16919-19-0)
Compound-4018	(137-29-1)
Compound-666	(608-73-1)
Compound-1452-F	(517-16-8)
Compound 3-120	(59-40-5)
Compound 22/190	(500-28-7)
Compound 42	(81-81-2)
Compound 47-83	(82-92-8)
Compound 105	(137-16-6)
Compound 118	(309-00-2)
Compound 269	(72-20-8)
Compound 338	(510-15-6)
Compound 347	(13838-16-9)
Compound 497	(60-57-1)
Compound 604	(117-80-6)
Compound 711	(465-73-6)
Compound 732	(5902-51-2)
Compound 733	(7286-76-2)
Compound 864	(3458-22-8)
Compound 889	(117-81-7)
Compound 923	(97-16-5)
Compound 01013	(135-23-9)
Compound 1081	(640-19-7)
Compound 1189	(143-50-0)
Compound 01748	(514-73-8)
Compound 1836	(311-47-7)
Compound 2046	(7786-34-7)
Compound 3422	(56-38-2)
Compound 3956	(8001-35-2)
Compound 3961	(8001-50-1)
Compound 4049	(121-75-5)
Compound 4072	(470-90-6)
Compound 7744	(63-25-2)
Compound 10854	(64-00-6)
Compound 17309	(72-63-9)
Compound 33355	(72-33-3)
Compound 42339	(7008-42-6)
Compound 64716	(28657-80-9)
Compound 90459	(51234-28-7)
Compound 121607	(82558-50-7)
Compound R-242	(80-00-2)
Compound R-25788	(37764-25-3)
Compound S-6,999	(991-42-4)
Compound B	(50-22-6)
Compound B Dicamba	(1918-00-9)
Compound C-9491	(18181-70-9)
Compound E	(53-06-5)
Compound E acetate	(50-04-4)
Compound F	(50-23-7)
Compound F-2	(17924-92-4)
Compound F (Kendall)	(50-23-7)
Compound G-11	(70-30-4)
Compound M-81	(640-15-3)
Compound No. 1080	(62-74-8)
Compound 88R	(140-57-8)
Compound 2339 RP	(2045-52-5)
Compound Tincture of Camphor	(8029-99-0)
Compound UC-20047 A	(15271-41-7)
Compound W	(26644-46-2)
Compound XI*	(99-32-1)
Compound 6-12 insect repellent	(94-96-2)
Comprox	(50642-03-0)
Comycetin	(56-75-7)
d-Con	(81-81-2)
Conc Blue B	(1325-87-7)
Conc. Violet R	(1325-82-2)
Concemin	(50-81-7)
Conchinin	(56-54-2)
Conclyte calcium	(814-80-2)
Conco AAS-35	(25155-30-0)
Conco AAS-40	(25155-30-0)
Conco AAS-65	(25155-30-0)
Conco AAS-90	(25155-30-0)
Conco NI-90	(9016-45-9)
Conco NIX-100	(9002-93-1)
Conco SXS	(1300-72-7)
Conco Sulfate A	(2235-54-3)
Conco Sulfate C	(1120-01-0)
Conco Sulfate WA	(151-21-3)
Conco Sulfate WA-1200	(151-21-3)
Conco Sulfate WA-1245	(151-21-3)
Conco Sulfate Wag	(151-21-3)
Conco Sulfate Wan	(151-21-3)
Conco Sulfate Was	(151-21-3)
Conco Sulfate WE	(9004-82-4)
Conco Sulfate WN	(151-21-3)
Conco XAL	(1643-20-5)
Concogel 2 Conc.	(137-20-2)
Concord	(67375-30-8)
Condacaps	(50-14-6)
Condanol DLS	(143-00-0)
Condensate PL	(120-40-1)
Condensates (Petroleum), Vacuum tower (9CI)	(64741-49-7)
Condensation products, epoxy	(61788-97-4)
Condensation products, polyamides (VAN)	(63428-84-2)
Condensed Phosphoric Acid	(8017-16-1)
Condition	(439-14-5)
Conditioner 1	(143-28-2)
Condocaps	(50-14-6)
Condol	(50-14-6)
Conductex	(1333-86-4)
Conductex	(7440-44-0)
Condylon	(64-86-8)
Condy's Crystals	(7722-64-7)
Conestoral	(438-67-5)
Confectioner's sugar	(57-50-1)
Confortid	(53-86-1)
Congo Blue	(72-57-1)
Congo Blue 3B	(72-57-1)
Congo Red	(573-58-0)
Congo Red R-138	(6471-49-4)
Congo Red 4B	(573-58-0)
Congo Red 4BX	(573-58-0)
Congo Red CR	(573-58-0)
Congo Red H	(573-58-0)
Congo Red ICI	(573-58-0)
Congo Red L	(573-58-0)
Congo Red M	(573-58-0)
Congo Red N	(573-58-0)

Name	CAS No.
Congo Red R	(573-58-0)
Congo Red RS	(573-58-0)
Congo Red W	(573-58-0)
Congoblau 3B	(72-57-1)
δ-Coniceine	(13618-93-4)
d-Conicine	(458-88-8)
.α.-Conidendrol	(2316-10-1)
Coniferaldehyde	(458-36-6)
p-Coniferaldehyde	(458-36-6)
Coniferyl alcohol	(458-35-5)
Coniferyl aldehyde	(458-36-6)
Conigon BC	(64-02-8)
Coniin	(458-88-8)
(+)-Coniine	(458-88-8)
Coniine	(458-88-8)
Conine	(458-88-8)
α-Conine	(458-88-8)
Conjugated estrogen	(16680-47-0)
Conjugated estrogen	(7280-37-7)
Conoco C 650	(26248-24-8)
Conoco C-50	(25155-30-0)
Conoco C-60	(25155-30-0)
Conoco SD 40	(25155-30-0)
Conopal	(1344-28-1)
Conquinine	(56-54-2)
Consdrin	(61-76-7)
Consdrin hydrochloride	(61-76-7)
Consol Violet	(1325-82-2)
Constimol	(2152-34-3)
Constonate	(577-11-7)
Cont	(443-48-1)
Contaverm	(92-84-2)
Contergan	(50-35-1)
Contimet 30	(7440-32-6)
Continal	(57-33-0)
Continental	(1332-58-7)
Continental	(1333-86-4)
Continex	(1333-86-4)
Contomin	(69-09-0)
Contra Creme	(62-38-4)
Contrac	(28772-56-7)
Contradol	(62-44-2)
Contradouleur	(62-44-2)
Contralin	(97-77-8)
Contrapot	(97-77-8)
Cóntrheuma retard	(50-78-2)
Conturex	(126-31-8)
Convallaotoxin	(508-75-8)
Convallaton	(508-75-8)
Convallatoxin	(508-75-8)
Convallatoxoside	(508-75-8)
Convallotoxin	(508-75-8)
Convul	(57-41-0)
Convulex	(1069-66-5)
Coolspan	(15676-16-1)
Coomassie Blue	(3861-73-2)
Coomassie Blue G	(6104-58-1)
Coomassie Blue Medicinal	(3861-73-2)
Coomassie Blue RL	(3861-73-2)
Coomassie Brilliant Blue	(6104-58-1)
Coomassie Brilliant Blue G250	(6104-58-1)
Coomassie Milling Scarlet G	(10169-02-5)
Coomassie Milling Scarlet GP	(10169-02-5)
Coomassie Red PG	(3567-65-5)
Coomassie Red PGP	(3567-65-5)
Coomassie Violet	(1694-09-3)
Coomassie Yellow R	(6375-55-9)
Coomassie Yellow RP	(6375-55-9)
Coopex	(52645-53-1)
Cop Tox	(1332-65-6)
Cop-Tox	(1332-40-7)
Copaene	(3856-25-5)
α-Copaene	(3856-25-5)
Copagel PB 25	(9004-32-4)
Copal Z	(9003-53-6)
Copanoic	(96-83-3)
Copar 100	(64742-16-1)
Copellidin	(104-89-2)
Copharcilin	(69-53-4)
Copox	(1317-39-1)
Copper-8	(10380-28-6)
Copper-Airborne	(7440-50-8)
Copper-Milled	(7440-50-8)
Copper-Sandoz	(1317-39-1)
Copper-hydroxide	(20427-59-2)
Copper-oxide	(1317-38-0)
Copper (ACGIH,OSHA)	(7440-50-8)
Copper Blue	(1317-40-4)
Copper Bronze	(7440-50-8)
Copper Brown	(1317-38-0)
Copper D-gluconate (1:2)	(527-09-3)
Copper Dross (Lead refinery)	(69029-52-3)
Copper EDTA complex	(54453-03-1)
Copper(I) bromide	(7787-70-4)
Copper(I) chloride	(7758-89-6)
Copper(I) citrate	(866-82-0)
Copper(I) cyanide	(544-92-3)
Copper(I) iodide	(7681-65-4)
Copper Inhibitor 50	(94-91-7)
Copper(I) oxide	(1317-39-1)
Copper(I) sulfide	(22205-45-4)
Copper(3+), (N,N,N,N',N',N',N'',N'',N''-nonamethyl-29H,31H-phthalocyaninetrimethanaminiumato(2-)-N29,N30,N31,N32)-, trichloride (9CI)	(26854-10-4)
Copper (2+) NTA	(34831-02-2)
Copper OC Fungicide	(1332-40-7)
Copper Phthalocyanine Blue	(147-14-8)
Copper Phthalocyanine Green	(1328-53-6)
Copper Pride	(1332-03-2)
Copper Sardez	(1317-39-1)
Copper Slag-Airborne	(7440-50-8)
Copper Slag-Milled	(7440-50-8)
Copper Uversol	(1338-02-9)
Copper aceto-arsenite	(12002-03-8)
Copper(2+) acetate	(142-71-2)
Copper(II) acetate	(142-71-2)
Copper acetate arsenite	(12002-03-8)
Copper acetate (cu(c2H3o2)2)	(142-71-2)
Copper acetoarsenite, Solid (DOT)	(12002-03-8)
Copper acetoarsenite [UN 1585]	(12002-03-8)
Copper acetylacetonate	(13395-16-9)
Copper(II) acetylacetonate	(13395-16-9)
Copper acetylide	(12540-13-5)
Copper alloy, Cu,Be	(11133-98-5)
Copper alloy, Cu,Be,Co	(55158-44-6)
Copper alloy, nonbase, Ni,Cu (9CI)	(11102-90-2)
Copper ammonium acetate	(23087-46-9)
Copper ammonium carbonate	(33113-08-5)
Copper orthoarsenite	(10290-12-7)
Copper arsenite, Solid (DOT)	(10290-12-7)
Copper arsenite [UN 1586]	(10290-12-7)
Copperas	(7720-78-7)
Copperas	(7782-63-0)
Copper bichloride	(7447-39-4)

Name	CAS
Copper, bis(D-gluconato)-	**(527-09-3)**
Copper, bis(D-gluconato-O1,O2)- (9CI)	**(527-09-3)**
Copper, bis(acetato)hexametaarsenitotetra	**(12002-03-8)**
Copper bis(acetylacetonate)	**(13395-16-9)**
Copper bis(acetylacetone)	**(13395-16-9)**
Copper(1+), bis(2,9-dichloro-1,10-phenanthroline-N1,N10)-, (T-4)- (9CI)	**(69742-55-8)**
Copper, bis(dimethyldithiocarbamato)-	**(137-29-1)**
Copper bis(2,4-pentanedionate)	**(13395-16-9)**
Copper, bis(2,4-pentanedionato)	**(13395-16-9)**
Copper, bis(2,4-pentanedionato-O,O')-	**(13395-16-9)**
Copper, bis(2,4-pentanedionato-O,O')-, (SP-4-1)- (9CI)	**(13395-16-9)**
Copper(II), bis(3-phenylsalicylato)	**(5328-04-1)**
Copper, bis(8-quinolinolato-N^1,O^8)-	**(10380-28-6)**
Copper(1+) bromide	**(7787-70-4)**
Copper(II) bromide	**(7789-45-9)**
Copper bromide (9CI)	**(7787-70-4)**
Copper bromide (9CI)	**(7789-45-9)**
Copper carbide	**(12540-13-5)**
Copper (II) carbonate	**(1184-64-1)**
Copper carbonate (1:1)	**(1184-64-1)**
Copper carbonate, basic	**(12069-69-1)**
Copper carbonate hydroxide	**(12069-69-1)**
Copper(II) carbonate hydroxide (2:1:2)	**(12069-69-1)**
Copper, (carbonato)dihydroxydi-	**(12069-69-1)**
Copper, (carbonato(2-))dihydroxydi- (9CI)	**(12069-69-1)**
Copper chloride	**(7758-89-6)**
Copper(1+) chloride	**(7758-89-6)**
Copper(2+) chloride	**(7447-39-4)**
Copper(II) chloride	**(7447-39-4)**
Copper(II) chloride (1:2)	**(7447-39-4)**
Copper chloride, Basic	**(1332-40-7)**
Copper chloride, Mixed with copper oxide, hydrate	**(1332-40-7)**
Copper chloride [UN 2802]	**(1344-67-8)**
Copper chloride dihydrate	**(10125-13-0)**
Copper(2+) chloride dihydrate	**(10125-13-0)**
Copper chloride hydroxide (9CI)	**(1332-65-6)**
Copper chloride oxide	**(1332-40-7)**
Copper chloride oxide, hydrate (9CI)	**(1332-40-7)**
Copper, (chloro-29H,31H-phthalocyaninato(2-)-N29,N30, N31,N32)- (9CI)	**(12239-87-1)**
Copper chromate	**(13548-42-0)**
Copper chromium oxide	**(13548-42-0)**
Copper citrate	**(10402-15-0)**
Copper citrate	**(866-82-0)**
Copper, Compd. with beryllium (9CI)	**(64535-95-1)**
Copper(II) cyanide	**(14763-77-0)**
Copper cyanide [UN 1587]	**(14763-77-0)**
Copper cyanide [UN 1587]	**(544-92-3)**
Copper cynanamide	**(14763-77-0)**
Copper dehydroabietyl ammonium 2-ethylhexanoate	**(53404-24-3)**
Copper diacetate	**(142-71-2)**
Copper(2+) diacetate	**(142-71-2)**
Copper diacetylacetonate	**(13395-16-9)**
Copper dibromide	**(7789-45-9)**
Copper(dihydrogen phthalocyaninedisulfonato(2)) disodium salt	**(1330-38-7)**
Copper dihydroxide	**(20427-59-2)**
Copper, (5-((4'-((2,5-dihydroxy-4-((2-hydroxy-5-sulfophenyl) azo)phenyl)azo)(1,1'-biphenyl)- 4-yl)azo)-2-hydroxybenzoato-(2-))-, disodium salt	**(16071-86-6)**
Copper dimethyldithiocarbamate	**(137-29-1)**
Copper dinitrate	**(3251-23-8)**
Copper diphosphorate	**(10102-90-6)**
Copper ethylenediaminetetraacetate	**(54453-03-1)**
Copper 2-ethylhexanoate	**(22221-10-9)**
Copper etidronic acid complex	**(50376-91-5)**
Copperfine-Zinc	**(7758-99-8)**
Copper(2+) formate	**(544-19-4)**
Copper formate (VAN)	**(544-19-4)**
Copper gluconate	**(527-09-3)**
Copper heptanoate	**(5128-10-9)**
Copper(2+) heptanoate	**(5128-10-9)**
Copper, (1,3,8,16,18,24-hexabromo-2,4,9,10,11,15,17,22,23, 25-decachlorophthalocyaninato(2-))	**(14302-13-7)**
Copper, (1,2,3,4,8,9,10,11,15,16,17,18,22,23,24,25-hexadeca-chloro-29H,31H-phthalocyaninato(2-)-N29,N30,N31,N32)-, (SP-4-1)- (9CI)	**(14832-14-5)**
Copper(2+) hydroxide	**(20427-59-2)**
Copper(II) hydroxide	**(20427-59-2)**
Copper hydroxide chloride	**(1332-65-6)**
Copper (II) hydroxide sulfate	**(1332-03-2)**
Copper(2+) hydroxide sulfate, monohydrate	**(1332-03-2)**
Copper hydroxide sulfate, monohydrate (8CI,9CI)	**(1332-03-2)**
Copper 8-hydroxyquinolate	**(10380-28-6)**
Copper 8-hydroxyquinolinate	**(10380-28-6)**
Copper hydroxyquinolinate	**(10380-28-6)**
Copper 8-hydroxyquinoline	**(10380-28-6)**
Copper in the form of an ammonia complex	**(16828-95-8)**
Copper iodide (9CI)	**(7681-65-4)**
Copper, ion (Cu(2+)) (8CI,9CI)	**(15158-11-9)**
Copper linoleate	**(7721-15-5)**
Copper(2+) 4-methylbenzenesulfonate	**(7144-37-8)**
Copper monobromide	**(7787-70-4)**
Copper monochloride	**(7758-89-6)**
Copper monoiodide	**(7681-65-4)**
Copper monooxide	**(1317-38-0)**
Copper monosulfate	**(7758-98-7)**
Copper monosulfide	**(1317-40-4)**
Copper monoxide	**(1317-38-0)**
Copper naphthenate	**(1338-02-9)**
Copper neodecanoate	**(32276-75-8)**
Copper, nickel base	**(11102-90-2)**
Copper (II) nitrate	**(3251-23-8)**
Copper(2+) nitrate	**(3251-23-8)**
Copper nitrilotriacetic acid	**(34831-02-2)**
Copper nordox	**(1317-39-1)**
Copper, (octachloro-29H,31H-phthalocyaninato(2-)-N29,N30, N31,N32)- (9CI)	**(1330-37-6)**
Copper(2+) octadecanoate	**(660-60-6)**
Copper oxalate	**(814-91-5)**
Copper(II) oxalate	**(814-91-5)**
Copper (1+) oxide	**(1317-39-1)**
Copper(2+) oxide	**(1317-38-0)**
Copper(II) oxide	**(1317-38-0)**
Copper oxide (8CI,9CI)	**(1317-39-1)**
Copper (2+) oxinate	**(10380-28-6)**
Copper oxinate	**(10380-28-6)**
Copper oxine	**(10380-28-6)**
Copper oxychloride	**(1332-40-7)**
Copper oxychloride	**(1332-65-6)**
Copper(II) oxychloride	**(1332-65-6)**
Copper oxychloride sulfate	**(8012-69-9)**
Copper oxyquinolate	**(10380-28-6)**
Copper oxyquinoline	**(10380-28-6)**
Copper 3-phenylsalicylate	**(5328-04-1)**
Copper orthophosphate	**(10103-48-7)**
Copper phosphate	**(10103-48-7)**
Copper phosphate	**(7798-23-4)**
Copper phosphate (3:2)	**(7798-23-4)**
Copper(II) phosphate	**(7798-23-4)**
Copper phthalocyanin	**(147-14-8)**
Copper, (phthalocyaninato(2-))- (8CI)	**(147-14-8)**
Copper, (29H,31H-phthalocyaninato(2-)-N(29),N(30), N(31),N(32))-, (SP-4-1)- (9CI)	**(147-14-8)**

Copper phthalocyanine (147-14-8)
Copper,(29H,31H-phthalocyaninetrisulfonato(2-)-N29,N30, N31,N32)- (30638-09-6)
Copper, (29H,31H-phthalocyanine-ar,ar',ar''-trisulfonato-(2-)-N29,N30,N31,N32)-, triammonium salt (25512-11-2)
Copper, (29H,31H-phthalocyanine-ar,ar',ar''-trisulfonato-(2-)-N29,N30,N31,N32)-, trisodium salt (1330-39-8)
Copper phthalocyanine trisulfonic acid (1330-39-8)
Copper phthalocyaninetrisulfonic acid, sodium salt (1330-39-8)
Copper, (29H,31H-phthalocyaninetrisulfonyl trichloridato-(2-)-N29,N30,N31,N32)- (9CI) (27121-30-8)
Copper (II) pyrophosphate (10102-90-6)
Copper pyrophosphate (10102-90-6)
Copper 8-quinolate (10380-28-6)
Copper quinolate (10380-28-6)
Copper 8-quinolinol (10380-28-6)
Copper 8-quinolinolate (10380-28-6)
Copper quinolinolate (10380-28-6)
Copper resinate (9007-39-0)
Copper salts of fatty and rosin acids (9007-39-0)
Copper salts of the acids of tall oil (61789-22-8)
Coppersan (1332-40-7)
Copper sodium cyanide (14264-31-4)
Copper suboxide (1317-39-1)
Copper (II) sulfate (1:1) (7758-98-7)
Copper sulfate (7758-98-7)
Copper sulfate (7758-99-8)
Copper sulfate (1:1) (7758-98-7)
Copper(2+) sulfate (7758-98-7)
Copper(2+) sulfate (1:1) (7758-98-7)
Copper(II) sulfate (7758-98-7)
Copper sulfate basic (1344-73-6)
Copper sulfate basic (7758-98-7)
Copper sulfate monohydrate (10257-54-2)
Copper sulfate, monohydrate (10257-54-2)
Copper(2+) sulfate pentahydrate (7758-99-8)
Copper(II) sulfate pentahydrate (7758-99-8)
Copper(II) sulfate, pentahydrate (1:1:5) (7758-99-8)
Copper sulfate, tribasic (1332-03-2)
Copper (II) sulfide (1317-40-4)
Copper sulfide (1317-40-4)
Copper sulfide (22205-45-4)
Copper(2+) sulfide (1317-40-4)
Copper sulphate (7758-99-8)
Copper tallate (61789-22-8)
Copper, (mu-((tetrahydrogen 3,3'-((3,3'-dihydroxy-4,4'-bi-phenylene)bis(azo))bis(5-amino-4-hydroxy-2,7-naphthalenedi-sulfonato))(4-)))di-, tetrasodium salt (28407-37-6)
Copper, (tetraphenyl-29H,31H-phthalocyaninato(2-)-N29,N30, N31,N32)- (9CI) (1330-40-1)
Copper(1+) thiocyanate (1111-67-7)
Coppertone (118-56-9)
Coppertrace (10125-13-0)
Copper 2,4,5-trichlorophenolate (25267-55-4)
Copper trichlorophenolate (25267-55-4)
Copper triethanolamine complex (82027-59-6)
Copper, (trihydrogen 7-amino-3-((3,3'-dihydroxy-4'-((2-hydroxy-3,6-disulfonato(4-)-1-naphthyl)azo)-4-biphenyl)azo)-4-hydroxy-2-naphthalenesulfonato(2-))di-, trisodium salt (66418-17-5)
Copper zinc hydroxide sulfate (55072-57-6)
Coppesan (1332-40-7)
Coppesan Blue (1332-40-7)
Copra Oil (8001-31-8)
Copra Pellets (DOT) (8001-31-8)
Copra [UN 1363] (8001-31-8)
Copramyl (142-78-9)
Coprantol (1332-40-7)
Copren (70-18-8)
Coprex (1332-40-7)
Coprol (577-11-7)
Coprosan Blue (1332-40-7)
Coprostan-3-β-ol (360-68-9)
Coprostanol (360-68-9)
Coprosterol (360-68-9)
Copsamine (91-84-9)
Copticide (63-74-1)
Coque Caliente (Spanish) (65996-77-2)
Coques du levant (French) (124-87-8)
Cort-Dome (50-23-7)
Corn-Starch (9005-25-8)
Coracon (59-26-7)
Coradon (91-84-9)
Coraethamide (59-26-7)
Coraethamidum (59-26-7)
Coralept (59-26-7)
Coramine (59-26-7)
Coranil Direct Black B (6428-31-5)
Coranormal (54-95-5)
Coranormol (54-95-5)
Corasol (54-95-5)
Coratoline (54-95-5)
Coravita (59-26-7)
Corax (1333-86-4)
Corax P (1333-86-4)
Corazol (54-95-5)
Corazole (54-95-5)
Corazole (analeptic) (54-95-5)
Corazone (59-26-7)
Corbella (55-48-1)
Corbit (84-65-1)
Corcat P 18 (9002-98-6)
Corcat P 100 (9002-98-6)
Corcat P 145 (9002-98-6)
Corcat P 200 (9002-98-6)
Corcat P 600 (9002-98-6)
Cordan (1524-88-5)
Cordiamid (59-26-7)
Cordiamin (59-26-7)
Cordiamine (59-26-7)
Cordianine (97-59-6)
Cordierite (66402-68-4)
Cordilox (52-53-9)
Cordipin (21829-25-4)
Corditon (59-26-7)
Cordran (1524-88-5)
Cordulan (57-88-5)
Cordynil (59-26-7)
Corediol (59-26-7)
Coreine (9000-07-1)
Corespin (59-26-7)
Coretal (6452-71-7)
Corethamide (59-26-7)
Coretone (59-26-7)
Corexit 9527 (60617-06-3)
Corexit (9CI) (39362-29-3)
Corexit 8666 (9CI) (51158-21-5)
Corflex 880 (27554-26-3)
Corgard (42200-33-9)
Corglycon (508-75-8)
Corglycone (508-75-8)
Corglykon (508-75-8)
Coriacid Scarlet R (10169-02-5)
Corial EM Finish F (9004-70-0)
Coriantin (9002-61-3)

Coricidin	(62-44-2)
Coriforte	(62-44-2)
Corine	(50-36-2)
Corisan	(54-95-5)
Corisol	(51-43-4)
Corizium	(67-45-8)
Corlin	(53-06-5)
Corlutin	(57-83-0)
Corlutin L.A.	(630-56-8)
Corlutina	(57-83-0)
Corluvite	(57-83-0)
Cormed	(59-26-7)
Cormid	(59-26-7)
Cormotyl	(59-26-7)
Corn Products	(9005-25-8)
Corn Sugar Gum	(11138-66-2)
Corn Sugar Syrup	(8029-43-4)
Corn Syrup	(8029-43-4)
Cornflower LS	(8001-21-6)
Cornocentin	(60-79-7)
Cornotone	(59-26-7)
Iso-Cornox	(93-65-2)
Iso-Cornox 57	(7085-19-0)
Cornox CWK	(3813-05-6)
Cornox-M	(94-74-6)
Cornox RD	(120-36-5)
Cornox RK	(120-36-5)
Corn sugar	(50-99-7)
Corodane	(57-74-9)
Corodinoc	(2312-76-7)
Corn-oil	(8001-30-7)
Corona corozate	(137-30-4)
Coronal	(479-18-5)
Coronal-Crinos	(68-90-6)
Coronaletta	(50-06-6)
Coronarin	(479-18-5)
Coronarine	(58-32-2)
Coronatine	(62251-96-1)
Coronene	(191-07-1)
Corontin	(390-64-7)
Corosorbide	(87-33-2)
Corosul D and S	(7704-34-9)
Corothion	(56-38-2)
Corotonin	(59-26-7)
Corotran	(80-33-1)
Corovit	(59-26-7)
Corozate	(137-30-4)
Corpax	(390-64-7)
Corphyllin	(479-18-5)
Corporin	(57-83-0)
Corps Praline	(118-71-8)
Corps 2339 R P (French)	(2045-52-5)
Corpus luteum hormone	(57-83-0)
Correx	(52316-55-9)
Corrigen	(465-16-7)
Corronarobetin	(50-35-1)
Corrosive mercury chloride	(7487-94-7)
Corrosive sublimate	(7487-94-7)
Corry's Slug Death	(108-62-3)
Corsair	(52645-53-1)
Corsedrol	(54-95-5)
Corsodyl	(18472-51-0)
Corsone	(50-02-2)
Cortadren	(50-04-4)
Cortadren	(53-06-5)
Cortan	(53-03-2)
Cortancyl	(53-03-2)
Cortef	(50-23-7)
δ-Cortef	(50-24-8)
Cortelan	(50-04-4)
δ Cortelan	(53-03-2)
δ-Cortelan	(53-03-2)
Cortenema	(50-23-7)
Corthion	(56-38-2)
Corthione	(56-38-2)
Corticosteron	(50-22-6)
Corticosterone	(50-22-6)
Corticosterone, 1-dehydro-6α-fluoro-	(53-34-9)
β-1,24-Corticotrophin	(16960-16-0)
Corticotropin-(1-24)	(16960-16-0)
α^{1-24}-Corticotropin	(16960-16-0)
β^{1-24}-Corticotropin	(16960-16-0)
α(1-39)-Corticotropin (Ox)	(39319-42-1)
α(1-39)-Corticotropin (Pig), 31-L-serine-33-L-glutamine- (9CI)	(39319-42-1)
α(1-39)-Corticotropin (Sheep)	(39319-42-1)
Cortidelt	(53-03-2)
Cortiden	(53-33-8)
Cortifan	(50-23-7)
Cortilan-Neu	(57-74-9)
Cortinazine	(54-85-3)
Cortis	(54-95-5)
Cortisal	(50-04-4)
Cortisal	(53-06-5)
Cortisate	(50-04-4)
Cortisate	(53-06-5)
Cortisol	(50-23-7)
δ^1-Cortisol	(50-24-8)
Cortisol alcohol	(50-23-7)
Cortisone	(53-06-5)
δ^1-Cortisone	(53-03-2)
δ-1-Cortisone	(53-03-2)
δ-Cortisone	(53-03-2)
Cortisone 21-acetate	(50-04-4)
Cortisone acetate	(50-04-4)
Cortisone monoacetate	(50-04-4)
Cortispray	(50-23-7)
Cortistab	(50-04-4)
Cortistal	(53-06-5)
Cortisyl	(50-04-4)
Cortivite	(50-04-4)
Cortivite	(53-06-5)
Cortogen	(50-04-4)
Cortogen	(53-06-5)
Cortogen acetate	(50-04-4)
Cortone	(50-04-4)
Cortone	(53-06-5)
δ-Cortone	(53-03-2)
Cortone acetate	(50-04-4)
Cortonema	(50-23-7)
Cortril	(50-23-7)
Cortrophin S	(16960-16-0)
Cortrosinta	(16960-16-0)
Cortrosyn	(16960-16-0)
Cortussin	(93-14-1)
Corundum	(1302-74-5)
α-Corundum	(1302-74-5)
Corvasol	(54-95-5)
Corvic 236581	(9003-22-9)
Corvic 51/83	(9003-22-9)
Corvic R 46/88	(9003-22-9)
Corvis	(54-95-5)
Corvitan	(59-26-7)
Corvitin	(300-42-5)
Corvitol	(59-26-7)

Corvotone	(59-26-7)
Coryban-D	(62-44-2)
Corydinine	(130-86-9)
Corylon	(80-71-7)
Corylone	(80-71-7)
Corynine	(146-48-5)
Coryvet	(54-95-5)
Corywas	(59-26-7)
Coryzol	(135-23-9)
Cosan	(7704-34-9)
Cosan 80	(7704-34-9)
Cosban	(2655-14-3)
Cosbiol	(111-01-3)
Coscopin	(128-62-1)
Coscopin	(6035-40-1)
Coscotabs	(128-62-1)
Cosden 550	(9003-53-6)
Cosden 945E	(9003-53-6)
Coslan	(61-68-7)
Cosmegen	(50-76-0)
Cosmelan	(8006-54-0)
Cosmetic Blue Lake	(3844-45-9)
Cosmetic Brilliant Pink Bluish D Conc	(81-88-9)
Cosmetic Coral Red KO Bluish	(5160-02-1)
Cosmetic DVR	(5160-02-1)
Cosmetic Green Blue R25396	(129-17-9)
Cosmetic Lanolin	(8006-54-0)
Cosmetic Pigment Yellow Red DVR	(5160-02-1)
Cosmetic White	(1304-85-4)
Cosmetic White C47-5175	(13463-67-7)
Cosmetic White C47-9623	(13463-67-7)
Cosmetol	(8001-79-4)
Cosmoline	(8009-03-8)
Cosmopen	(113-98-4)
Cosmopen	(61-33-6)
Cosmowax S	(67762-27-0)
Cosyntropin	(16960-16-0)
Cotel	(68-19-9)
Cor-theophylline	(479-18-5)
Cotinazin	(54-85-3)
(-)-Cotinine	(486-56-6)
(S)-Cotinine	(486-56-6)
Cotinine	(486-56-6)
Cotinizin	(54-85-3)
Cotneon	(86-50-0)
Cotnion-Ethyl	(2642-71-9)
Cotnion methyl	(86-50-0)
Cotofilm	(70-30-4)
Cotofor	(4147-51-7)
Cotone	(53-03-2)
Cotoran	(2164-17-2)
Cotoran Multi	(51218-45-2)
Cotoran Multi 50WP	(2164-17-2)
Cotton Black MT	(2429-83-6)
Cotton Red 4B	(992-59-6)
Cotton Red 5B	(573-58-0)
Cotton Red 4BC	(573-58-0)
Cotton Red L	(573-58-0)
Cotton Violet R	(2586-60-9)
Cottonex	(2164-17-2)
Cottonseed Oil (Deodorized winterized)	(8001-29-4)
Coumachlor	(81-82-3)
Coumachlore (French)	(81-82-3)
Coumadin	(81-81-2)
Coumadin sodium	(129-06-6)
Coumafen	(81-81-2)
Coumafene	(81-81-2)
Coumafene sodium	(129-06-6)
Coumafos	(56-72-4)
Coumafuryl	(117-52-2)
Coumalin	(504-31-4)
Coumaphos	(56-72-4)
Coumaphos [UN 2783]	(56-72-4)
Coumaphos mixture, Liquid [UN 2783]	(56-72-4)
Coumaran	(496-16-2)
Coumaranone	(7169-34-8)
(E)-p-Coumaric acid	(501-98-4)
3-Coumaric acid	(588-30-7)
4-Coumaric acid	(7400-08-0)
p-Coumaric acid	(7400-08-0)
trans-o-Coumaric acid	(614-60-8)
trans-p-Coumaric acid	(501-98-4)
Coumarin	(91-64-5)
Coumarin 1	(91-44-1)
Coumarin 4	(90-33-5)
Coumarin 311	(87-01-4)
Coumarin, 3-(α-acetonylbenzyl)-4-hydroxy	(81-81-2)
Coumarin, 3-(α-acetonylbenzyl)-4-hydroxy-, sodium salt	(129-06-6)
Coumarin, 3-(α-acetonyl-p-chlorobenzyl)-4-hydroxy	(81-82-3)
Coumarin, 3-(α-acetonylfurfuryl)-4-hydroxy	(117-52-2)
Coumarin, 3-(α-acetonylfurfuryl)-4-hydroxy-, sodium salt	(34490-93-2)
Coumarin, 3-(α-acetonyl-p-nitrobenzyl)-4-hydroxy	(152-72-7)
Coumarin, 3-(3-(4'-bromo-1,1'-biphenyl-4-yl)-3-hydroxy-1-phenylpropyl)-4-hydroxy-	(28772-56-7)
Coumarin, 3-(3-(4'-bromo-1,1'-biphenyl-4-yl)-1,2,3,4-tetrahydro-1-naphthyl)-4-hydroxy	(56073-10-0)
Coumarin, 3-(α-(p-(p-bromophenyl)-β-hydroxyphenethyl)benzyl)-4-hydroxy-	(28772-56-7)
Coumarin, 3-chloro-7-hydroxy-4-methyl-, O-ester with O,O-diethyl phosphorothioate	(56-72-4)
Coumarin, 7-diethylamino-4-methyl	(91-44-1)
Coumarin, 3,4-dihydro-	(119-84-6)
Coumarin, 6,7-dihydroxy	(305-01-1)
Coumarin, 5,7-dihydroxy-4-methyl-	(2107-76-8)
Coumarin, 6,7-dihydroxy-4-methyl	(529-84-0)
Coumarin, 7,8-dihydroxy-4-methyl	(2107-77-9)
Coumarin, 6,7-dimethoxy	(120-08-1)
Coumarin, 7-dimethylamino-4-methyl	(87-01-4)
Coumarin, 3-(α-ethylbenzyl)-4-hydroxy	(435-97-2)
Coumarin, 7-hydroxy	(93-35-6)
Coumarin, 7-hydroxy-4-methyl	(90-33-5)
Coumarin, 7-hydroxy-4-methyl-, O-ester with O,O-diethyl phosphorothioate	(299-45-6)
Coumarin, 4-hydroxy-3-(1,2,3,4-tetrahydro-1-naphthyl)	(5836-29-3)
trans-p-Coumarinic acid	(501-98-4)
cis-o-Coumarinic acid lactone	(91-64-5)
Coumarinic anhydride	(91-64-5)
Coumarin, 6-methyl	(92-48-8)
Coumarin, 3,3'-methylenebis(4-hydroxy	(66-76-2)
Coumarins	(81-81-2)
Coumarone	(271-89-6)
Coumatetralyl	(5836-29-3)
Coumefene	(81-81-2)
Counter	(13071-79-9)
Counter 15G Soil Insecticide	(13071-79-9)
Counter 15G Soil Insecticide-Nematicide	(13071-79-9)
Courlene PY	(9003-07-0)
Courlene-X3	(9002-88-4)
Courlose A 590	(9004-32-4)
Courlose A 610	(9004-32-4)
Courlose A 650	(9004-32-4)
Courlose F 4	(9004-32-4)
Courlose F 8	(9004-32-4)
Courlose F 20	(9004-32-4)

Courlose F 370 (9004-32-4)
Courlose F 1000G (9004-32-4)
Covi-Ox (59-02-9)
Covit (68-19-9)
Covol (9002-89-5)
Covol 971 (9002-89-5)
Cov-R-Tox (81-81-2)
Coxigon (51234-28-7)
Coxistat (59-87-0)
Coxysan (1332-40-7)
Coyden (2971-90-6)
Coyden 25 (2971-90-6)
Cozyme (81-13-0)
Crab-E-Rad (144-21-8)
Crab-3-Rad 100 (144-21-8)
Cradex (51-63-8)
Crag 2 (116-52-9)
Crag 341 (556-22-9)
Crag DCU-73W (116-52-9)
Crag Experimental Herbicide 2 (116-52-9)
Crag Fly Repellent (9003-13-8)
Crag Fruit Fungicide 341 (556-22-9)
Crag Fungicide 341 (556-22-9)
Crag Herbicide 2 (116-52-9)
Crag Herbicide I (136-78-7)
Crag Herbicide (OSHA) (136-78-7)
Crag Sesone (136-78-7)
Crag Sevin (63-25-2)
Cralo-E-Rad (144-21-8)
Crastin S 330 (25038-59-9)
Crastin S 350 (25038-59-9)
Crastin S 440 (25038-59-9)
Cratecil (50-06-6)
Crategine (123-11-5)
Crawhaspol (79-01-6)
Cream White (8009-03-8)
Cream of Tartar (868-14-4)
Creatin (57-00-1)
Creatine (8CI) (57-00-1)
Creatine, hydrate (57-00-1)
Creatinine (60-27-5)
Creatinine (VAN) (8CI) (60-27-5)
Crechlor S 45 (63449-39-8)
Credazine (14491-59-9)
Credo (7681-49-4)
Crein (72-63-9)
Crellate (9004-35-7)
Cremart (36335-67-8)
Cremodiazine (68-35-9)
Cremomethazine (57-68-1)
Cremophor A (9004-99-3)
Cremor Tartari (868-14-4)
p-Creosol (93-51-6)
Creosote (8001-58-9)
Creosote, From coal tar (8001-58-9)
Creosote Oil (8001-58-9)
Creosote Oil (Derived from any source) (70321-79-8)
Creosote Oil, High-boiling Distillate (70321-79-8)
Creosote P1 (8001-58-9)
Creosotum (8001-58-9)
Cresidine (120-71-8)
m-Cresidine (102-50-1)
o-Cresidine (16452-01-0)
p-Cresidine (120-71-8)
Cresoate, Wood (8021-39-4)
2-Cresol (95-48-7)
3-Cresol (108-39-4)
4-Cresol (106-44-5)
para-Cresol (106-44-5)
Cresol (ACGIH,OSHA) [UN 2076] (1319-77-3)
m-Cresol (OSHA) [UN 2076] (108-39-4)
o-Cresol (OSHA) [UN 2076] (95-48-7)
p-Cresol (OSHA) [UN 2076] (106-44-5)
p-Cresol acetate (140-39-6)
p-Cresol, 3-amino- (8CI) (2836-00-2)
p-Cresol, 2-amino-3-nitro- (8CI) (6265-05-0)
p-Cresol, 2-amino-5-nitro- (8CI) (6265-06-1)
p-Cresol, 2-(2H-benzotriazol-2-yl) (2440-22-4)
p-Cresol, 2,6-bis(α-methylbenzyl) (1817-68-1)
p-Cresol, 2-bromo-6-nitro- (8CI) (20039-91-2)
p-Cresol, 2-tert-butyl (2409-55-4)
m-Cresol, 6-tert-butyl- (8CI) (88-60-8)
m-Cresol, tert-butyl- (8CI) (1333-13-7)
m-Cresol, 4,4'-butylidenebis(6-tert-butyl (85-60-9)
m-Cresol, 4-chloro (59-50-7)
m-Cresol, chloro (54548-50-4)
o-Cresol, 4-chloro (1570-64-5)
o-Cresol, 6-chloro (87-64-9)
Cresol, chloro- (8CI) (1321-10-4)
p-Cresol, 2-chloro- (8CI) (6640-27-3)
o-Cresol, 4-chloro-α-phenyl (120-32-1)
o-Cresol, 4,6-diamino- (15872-73-8)
m-Cresol, 4,6-di-tert-butyl (497-39-2)
p-Cresol, 2,6-di-tert-butyl (128-37-0)
o-Cresol, 4,6-dichloro- (8CI) (1570-65-6)
p-Cresol, 2,6-dichloro- (8CI) (2432-12-4)
p-Cresol, 5-(diethylamino)-2-nitroso- (8CI) (6265-09-4)
p-Cresol, 3-(dimethylamino)- (6CI,8CI) (119-31-3)
m-Cresol, 4-(dimethylamino)-, methylcarbamate (ester) (2032-59-9)
p-Cresol, 5-(dimethylamino)-2-nitroso- (8CI) (6265-11-8)
m-Cresol, 4,6-dinitro (616-73-9)
o-Cresol, 3,5-dinitro (497-56-3)
o-Cresol, 4,6-dinitro (534-52-1)
o-Cresol, dinitro (1335-85-9)
p-Cresol, 2,6-dinitro (609-93-8)
p-Cresol, 3,5-dinitro (63989-82-2)
o-Cresol, 4,6-dinitro-, Sodium deriv. (2312-76-7)
m-Cresol, 2,4-dinitro-6-isobropyl- (303-21-9)
o-Cresol, 4,6-dinitro-, sodium salt (2312-76-7)
Cresol diphenyl phosphate (26444-49-5)
m-Cresole (108-39-4)
Cresol, ar-ethyl- (8CI) (30230-52-5)
m-Cresol, 5-ethyl- (8CI) (698-71-5)
Cresol glycidyl ether (26447-14-3)
p-Cresol glycidyl ether (2186-24-5)
Cresoli (Italian) (1319-77-3)
m-Cresol, 6-isopropyl- (89-83-8)
o-Cresol, 5-isopropyl- (499-75-2)
p-Cresol, 2-methoxy (93-51-6)
p-Cresol, 3-(methylamino)- (8CI) (6265-13-0)
p-Cresol, 2,2'-methylenebis(6-tert-butyl- (119-47-1)
p-Cresol, 2,2'-methylenebis(6-(1-methylcyclohexyl)- (77-62-3)
p-Cresol, 2,2'-methylenebis(6-nonyl (7786-17-6)
m-Cresol methyl ether (100-84-5)
p-Cresol methyl ether (104-93-8)
m-Cresol, 4-(methylthio) (3120-74-9)
m-Cresol, 4-(methylthio)-, O-ester with O,O-dimethyl phosphorothioate (55-38-9)
Cresol, nitro (12167-20-3)
m-Cresol, 4-nitro (2581-34-2)
o-Cresol, 4-nitro (99-53-6)
p-Cresol, 2-nitro (119-33-5)
p-Cresol, 3-nitro- (8CI) (2042-14-0)
m-Cresol, 4-nitro-, dimethyl phosphate (2255-17-6)

Name	CAS
m-Cresol, 4-nitro-, O-ester with O,O-dimethyl phosphorothioate	(122-14-5)
m-Cresol, 4-nitro-α,α,α-trifluoro-	(88-30-2)
p-Cresol, 2-nitro-α,α,α-trifluoro	(400-99-7)
p-Cresol, α-phenyl- (8CI)	(101-53-1)
p-Cresol, 2-(phenylazo)- (6CI,7CI,8CI)	(952-47-6)
p-Cresol, α-phenyl-, carbamate (8CI)	(101-71-3)
o-Cresol, α,α'-(propylenedinitrilo)di	(94-91-7)
p-Cresol, sodium salt	(1121-70-6)
o-Cresol, 3,4,5,6-tetrabromo-	(576-55-6)
o-Cresol, tetrabromo- (8CI)	(576-55-6)
m-Cresol, 4,4'-thiobis(6-tert-butyl	(96-69-5)
p-Cresol, 2,2'-thiobis(6-tert-butyl- (8CI)	(90-66-4)
o-Cresol, 6,6'-thiobis(4-chloro-	(4418-66-0)
m-Cresol, 2,4,6-trichloro- (8CI)	(551-76-8)
m-Cresol, α,α,α-trifluoro	(98-17-9)
m-Cresol, α,α,α-trifluoro-4-nitro	(88-30-2)
m-Cresol, 2,4,6-trinitro	(602-99-3)
Creson	(93-14-1)
Cresopur	(3813-05-6)
Cresorcinol diisocyanate	(584-84-9)
2,3-Cresotic acid	(83-40-9)
2,4-Cresotic acid	(50-85-1)
2,5-Cresotic acid	(89-56-5)
2,6-Cresotic acid	(567-61-3)
Cresotic acid	(83-40-9)
γ-Cresotic acid	(50-85-1)
m-Cresotic acid	(50-85-1)
o-Cresotic acid	(83-40-9)
p-Cresotic acid	(89-56-5)
2,3-Cresotic acid, 5-(p-aminobenzamido)- (6CI,8CI)	(6265-15-2)
4,2-Cresotic acid, 6-methoxy-, bimol. ester, methyl ester, 4,6-dimethoxy-o-toluate (8CI)	(19314-74-0)
4,2-Cresotic acid, 6-methoxy-, methyl ester, ester with 6-methoxy-4,2-cresotic acid 4,6-dimethoxy-o-toluate (7CI)	(19314-74-0)
2,3-Cresotic acid, methyl ester (8CI)	(23287-26-5)
Cresotine Blue 2B	(2602-46-2)
Cresotine Blue 3B	(72-57-1)
Cresotine Blue 6B	(2610-05-1)
Cresotine Brown RC	(2429-82-5)
Cresotine Dark Green B	(3626-28-6)
Cresotine Pure Blue	(2429-74-5)
Cresotine Yellow A	(1325-37-7)
2,3-Cresotinic acid	(83-40-9)
Cresotinic acid	(83-40-9)
α-Cresotinic acid	(89-56-5)
β-Cresotinic acid	(83-40-9)
m-Cresotinic acid	(50-85-1)
o-Cresotinic acid	(83-40-9)
p-Cresotinic acid	(89-56-5)
Cresotol	(2312-76-7)
Crestanil	(57-53-4)
Crestomycin	(7542-37-2)
Crestoxo	(8001-35-2)
Cresyl acetate	(140-39-6)
o-Cresyl acetate	(533-18-6)
p-Cresyl acetate	(140-39-6)
Cresyl aerofloat	(27157-94-4)
Cresylate Spent Caustic	(68815-21-4)
Cresylate Spent Caustic Solution	(68815-21-4)
Cresyl diphenyl phosphate	(26444-49-5)
p-Cresyl diphenyl phosphate	(78-31-9)
m-Cresyl ester of N-methylcarbamic acid	(1129-41-5)
Cresylglycide ether	(26447-14-3)
Cresyl glycidyl ether	(26447-14-3)
p-Cresyl glycidyl ether	(2186-24-5)
Cresylic Acid Tar	(68555-24-8)
Cresylic Acids Residue (Coal)	(68555-24-8)
o-Cresylic acetate	(533-18-6)
m-Cresylic acid	(108-39-4)
o-Cresylic acid	(95-48-7)
p-Cresylic acid	(106-44-5)
Cresylic acid [UN 2022]	(1319-77-3)
Cresylic acid, potassium salt	(12002-51-6)
Cresylic acid, sodium salt	(34689-46-8)
Cresylic acid, sodium salt solution	(34689-46-8)
Cresylic creosote	(8001-58-9)
Cresylite	(602-99-3)
m-Cresyl methylcarbamate	(1129-41-5)
p-Cresyl methyl ether	(104-93-8)
Cresyl phosphate	(1330-78-5)
o-Cresyl phosphate	(78-30-8)
Criasazin	(117-10-2)
Crill 2	(26266-57-9)
Crill 3	(1338-41-6)
Crill 5	(26266-58-0)
Crill 10	(9005-65-6)
Crill 11	(9005-65-6)
Crill 16	(8007-43-0)
Crill 26	(1323-39-3)
Crill K 3	(1338-41-6)
Crill K 16	(8007-43-0)
Crill S 10	(9005-65-6)
Crillet 4	(9005-65-6)
Crillon LME	(142-78-9)
Crillon l.D.E.	(120-40-1)
Crimidin (German)	(535-89-7)
Crimidina (Italian)	(535-89-7)
Crimidine	(535-89-7)
Crimson Antimony	(1345-04-6)
Crimson EMBL	(3567-69-9)
Crimson 4R	(4548-53-2)
Crimson SX	(2611-82-7)
Crimson 2embl	(3567-69-9)
Crinodora	(69-05-6)
Crinothene	(9011-14-7)
Crinovaryl	(53-16-7)
Crinuryl	(58-54-8)
Crisalin	(1582-09-8)
Crisapon	(75-99-0)
Crisatrina	(1912-24-9)
Crisatrine	(834-12-8)
Crisazine	(1912-24-9)
Criscobre	(1332-65-6)
Criscobre	(20427-59-2)
Criseociclina	(60-54-8)
Crisfolatan	(2939-80-2)
Crisfuran	(1563-66-2)
Crisodin	(6923-22-4)
Crisodrin	(6923-22-4)
Crisonar	(846-50-4)
Crispin Fast Red E	(2302-96-7)
Crispin Red GM	(10169-02-5)
Crisquat	(1910-42-5)
Cristallose	(128-44-9)
Cristallovar	(53-16-7)
Cristapen	(113-98-4)
Cristapurat	(71-63-6)
Cristerone T	(58-22-0)
Cristobalite	(66402-68-4)
Cristobalite (ACGIH)	(14464-46-1)
Cristoxo	(8001-35-2)
Cristoxo 90	(8001-35-2)
Crisulfan	(115-29-7)
Crisuron	(330-54-1)

Crittox **(12122-67-7)**
Crocein Orange **(1934-20-9)**
Crocein Orange G **(1934-20-9)**
Croceine 3BX **(5858-93-5)**
Croceine Orange **(1934-20-9)**
Croceine Orange EN **(1934-20-9)**
Croceine Orange 2G **(1934-20-9)**
Croceine Orange Y **(1934-20-9)**
Croceine Scarlet 3BX **(5858-93-5)**
Crocidolite (9CI) **(61105-31-5)**
Crocidolite [UN 2212] **(12001-28-4)**
Crocidolite asbestos **(12001-28-4)**
Crociodolite **(12001-28-4)**
Crocoite **(7758-97-6)**
Crocoite (7CI,8CI) **(14654-05-8)**
Crocoite (Pb(CrO4)) (9CI) **(14654-05-8)**
Crocus martis adstringens **(1309-37-1)**
Crodacid **(544-63-8)**
Crodacol A.10 **(143-28-2)**
Crodacol CS-50 **(67762-27-0)**
Crodacol-CAS **(36653-82-4)**
Crodacol-CAT **(36653-82-4)**
Crodacol-O **(143-28-2)**
Crodacol-S **(112-92-5)**
Crodamide O **(301-02-0)**
Crodamide OR **(301-02-0)**
Crodamol DA **(6938-94-9)**
Crodamol IPM **(110-27-0)**
Crodamol IPP **(142-91-6)**
Crodet O 6 **(9004-96-0)**
Croflex **(1333-86-4)**
Crolac **(1333-86-4)**
Crolean **(107-02-8)**
Cromile, cloruro di (Italian) **(14977-61-8)**
Cromoglicic acid **(16110-51-3)**
Cromoglycate **(15826-37-6)**
Cromoglycate disodium **(15826-37-6)**
Cromoglycic acid **(16110-51-3)**
Cromolyn **(16110-51-3)**
Cromolyn sodium **(15826-37-6)**
Cromolyn sodium salt **(15826-37-6)**
Cromo, ossicloruro di (Italian) **(14977-61-8)**
Cromophtal Blue A 3R **(81-77-6)**
Cromophtal Blue 4G **(147-14-8)**
Cromophtal Blue GF **(147-14-8)**
Cromophtal Blue 4GN **(147-14-8)**
Cromophtal Green GF **(1328-53-6)**
Cromophtal Orange 4R **(5280-74-0)**
Cromosan **(142-72-3)**
Cromo solfato basificato (Spanish) **(12336-95-7)**
Cromo(triossido di) (Italian) **(1333-82-0)**
Cronetal **(97-77-8)**
Croneton **(29973-13-5)**
Crop Rider **(94-75-7)**
Crotaline **(315-22-0)**
Crotenaldehyde **(123-73-9)**
Crotilin **(2971-38-2)**
Crotilin **(94-75-7)**
Crotiline **(2971-38-2)**
Croton-Oil **(8001-28-3)**
γ-Crotolactone **(497-23-4)**
Croton Factor F1 **(53202-98-5)**
Croton Resin **(8001-28-3)**
Croton Tiglium L. Oil **(8001-28-3)**
Crotonal **(123-73-9)**
Crotonaldehyde, (E) **(123-73-9)**
Crotonaldehyde (ACGIH,OSHA) **(123-73-9)**
Crotonaldehyde (DOT,OSHA) **(4170-30-3)**
Crotonaldehyde, 2-methyl- (8CI) **(1115-11-3)**
Crotonaldehyde, 2-methyl-, (E)- (8CI) **(497-03-0)**
Crotonaldehyde, Stabilized [UN 1143] **(4170-30-3)**
Crotonaldehyde, 2-vinyl- (8CI) **(20521-42-0)**
Crotonamide, 2-chloro-N,N-diethyl-3-hydroxy-, dimethyl phosphate **(13171-21-6)**
Crotonamide, 3-hydroxy-N,N-dimethyl-, cis-, dimethyl phosphate **(141-66-2)**
Crotonamide, 3-hydroxy-N-N-dimethyl-, dimethyl phosphate, (E)- **(141-66-2)**
Crotonamide, 3-hydroxy-N-N-dimethyl-, dimethyl phosphate, cis- **(141-66-2)**
Crotonamide, 3-hydroxy-N-methyl-, dimethylphosphate, (E)- **(6923-22-4)**
Crotonamide, 3-hydroxy-N-methyl-, dimethylphosphate, cis- **(6923-22-4)**
Crotonate de 2,4-dinitro 6-(1-methyl-heptyl)-phenyle (French) **(39300-45-3)**
Crotonate d'ethyle (French) **(623-70-1)**
(E)-Crotonic acid **(107-93-7)**
Crotonic acid **(107-93-7)**
α-Crotonic acid **(3724-65-0)**
trans-Crotonic acid **(107-93-7)**
Crotonic acid, (E)- (8CI) **(107-93-7)**
Crotonic acid, Solid [UN 2823] **(3724-65-0)**
Crotonic acid, (Z)- (8CI) **(503-64-0)**
Crotonic acid anhydride **(623-68-7)**
Crotonic acid, butyl ester **(7299-91-4)**
Crotonic acid, 4-chloro- (6CI,7CI,8CI) **(16197-90-3)**
Crotonic acid, 4,4-dichloro-, (E)- (8CI) **(16502-88-8)**
Crotonic acid 2,4-dinitro-6-(1-methylheptyl)phenyl ester **(39300-45-3)**
Crotonic acid 2,4-dinitro-6-(2-octyl)phenyl ester **(39300-45-3)**
Crotonic acid, ethyl ester **(10544-63-5)**
Crotonic acid, ethyl ester, (E) **(623-70-1)**
α-Crotonic acid ethyl ester **(623-70-1)**
Crotonic acid, 3-hydroxy-, isopropyl ester, O-ester with O-methyl ethylphosphoramidothioate, (E) **(31218-83-4)**
Crotonic acid, 4-hydroxy-, γ-lactone (6CI,7CI) **(497-23-4)**
Crotonic acid, 3-hydroxy-, α-methylbenzyl ester, dimethyl phosphate, (E) **(7700-17-6)**
Crotonic acid, 3-hydroxy-, methyl ester, dimethyl phosphate **(7786-34-7)**
Crotonic acid, 3-hydroxy-, methyl ester, dimethyl phosphate, (E) **(298-01-1)**
Crotonic acid, 3-hydroxy-, methyl ester, dimethyl phosphate, (Z) **(338-45-4)**
Crotonic acid, 2-methyl-, (E) **(80-59-1)**
Crotonic acid, 3-methyl **(541-47-9)**
Crotonic acid, 2-methyl-, (Z)- (8CI) **(565-63-9)**
Crotonic acid, 3-methyl-, 2-sec-butyl-4,6-dinitrophenyl ester **(485-31-4)**
Crotonic acid, methyl ester (8CI) **(18707-60-3)**
Crotonic acid, 2-methyl-, methyl ester, (E)- (8CI) **(6622-76-0)**
Crotonic acid, 2(or 4)-(1-methylheptyl)-4,6(or 2,6)-dinitrophenyl ester **(39300-45-3)**
Crotonic acid, 4-phenyl- (6CI,8CI) **(2243-52-9)**
Crotonic acid, vinyl ester (8CI) **(14861-06-4)**
Crotonic aldehyde **(123-73-9)**
Crotonic aldehyde **(4170-30-3)**
Crotonic-anhydride **(623-68-7)**
Crotonic nitrile **(4786-20-3)**
Crotonique nitrile (French) **(4786-20-3)**
Crotonitrile **(4786-20-3)**
Crotonoel (German) **(8001-28-3)**
α,β-Crotonolactone **(497-23-4)**
γ-Crotonolactone **(497-23-4)**
Crotononitrile **(4786-20-3)**
Crotononitrile, 2-methyl- (8CI) **(4403-61-6)**
Crotononitrile, 2-methyl-, (Z)- **(20068-02-4)**
Crotonyl alcohol **(6117-91-5)**
Crotonylene [UN 1144] **(503-17-3)**
Crotothane **(39300-45-3)**
Crotoxyfos **(7700-17-6)**
Crotoxyphos **(7700-17-6)**
Crotyl alcohol **(6117-91-5)**
1-Crotyl bromide **(4784-77-4)**

Crotyl bromide	(4784-77-4)	Crystalline	(71-63-6)
Crotyl chloride	(591-97-9)	Crystalline digitalin	(71-63-6)
Crotylin	(2971-38-2)	Crystallized Verdigris	(142-71-2)
Crovaril	(129-20-4)	Crystallose	(128-44-9)
12-Crown-4	(294-93-9)	Crystals of Venus	(142-71-2)
18-Crown-6	(17455-13-9)	Crystamet	(6834-92-0)
Croysulfone	(80-08-0)	Crystamin	(68-19-9)
Croysulphone	(80-08-0)	Crystapen	(113-98-4)
Crtron	(50-14-6)	Crystapen	(69-57-8)
Crude Arsenic	(1327-53-3)	Crystar	(50-78-2)
Crude Butadiene (Petroleum)	(64742-83-2)	Crystex	(7704-34-9)
Crude Coal Tar	(8007-45-2)	Crysthion	(2642-71-9)
Crude Light Oil (Coal)	(65996-78-3)	Crysthion 2L	(86-50-0)
Crude Oil	(8002-05-9)	Crysthyon	(86-50-0)
Crude Oil, Petroleum (DOT)	(8002-05-9)	Crysticillin	(6130-64-9)
Crude Opium	(8008-60-4)	Crystodigin	(71-63-6)
Crude Petroleum	(8002-05-9)	Crystogen	(53-16-7)
Crude Rice Bran Oil	(68553-81-1)	Crystoids	(136-77-6)
Crude Rice Oil	(68553-81-1)	Crystol 325	(8012-95-1)
Crude Shale Oils	(68308-34-9)	Crystol carbonate	(497-19-8)
Crude Tar Acids (Coal)	(65996-85-2)	Crystoserpine	(50-55-5)
Crude Tar Bases (Coal)	(65996-84-1)	Crystosol	(8012-95-1)
Crufomat	(299-86-5)	Crystwel	(68-19-9)
Crufomate A	(299-86-5)	Csi Paste	(140-95-4)
Crufomate (ACGIH,OSHA)	(299-86-5)	Cu-56	(1332-40-7)
Crunch	(63-25-2)	Cu-Be25	(56700-77-7)
Cry-O-Vac L	(9002-88-4)	Cu-NTA	(15844-52-7)
Cryoflex	(143-29-3)	Cube	(83-79-4)
Cryofluoran	(76-14-2)	Cube Extract	(83-79-4)
Cryofluorane	(76-14-2)	Cube-Pulver	(83-79-4)
Cryogenine	(103-03-7)	Cube Root	(83-79-4)
Cryolite	(15096-52-3)	Cubes	(50-37-3)
Cryopolythene	(9002-88-4)	Cubic Niter	(7631-99-4)
Cryptogil ol	(87-86-5)	Cubor	(83-79-4)
Cryptonol	(134-31-6)	Cucumber Dust	(7778-44-1)
Cryptopimaric acid	(471-74-9)	Cucurbitacin B	(6199-67-3)
Crysalba	(7778-18-9)	Cucurbitacine (C)	(5988-76-1)
Crystal Chrome Alum	(10141-00-1)	Cucurbitacine b	(6199-67-3)
Crystal O	(8001-79-4)	Cudrox	(1332-65-6)
Crystal Orange 2G	(1936-15-8)	Cuemid	(11041-12-6)
Crystal Propanil-4	(709-98-8)	Cuidrox	(1332-65-6)
Crystal Violet	(548-62-9)	Cuidrox	(20427-59-2)
Crystal Violet AO	(548-62-9)	Cullen Earth	(1317-34-6)
Crystal Violet AON	(548-62-9)	Culminal K 42	(9004-67-5)
Crystal Violet 6B	(548-62-9)	Cuma	(66-76-2)
Crystal Violet 10B	(548-62-9)	Cumachloor (Dutch)	(81-82-3)
Crystal Violet 5BO	(548-62-9)	Cumachlor (German)	(81-82-3)
Crystal Violet 6BO	(548-62-9)	Cumadin	(129-06-6)
Crystal Violet BP	(548-62-9)	Cumafos (Dutch)	(56-72-4)
Crystal Violet BPC	(548-62-9)	Cumafuryl (German)	(117-52-2)
Crystal Violet Base	(548-62-9)	Cumaldehyde	(122-03-2)
Crystal Violet Chloride	(548-62-9)	Cuman	(137-30-4)
Crystal Violet Extra Pure	(548-62-9)	Cuman L	(137-30-4)
Crystal Violet Extra Pure APN	(548-62-9)	p-Cumaric acid	(7400-08-0)
Crystal Violet Extra Pure APNX	(548-62-9)	Cumarin	(91-64-5)
Crystal Violet FN	(548-62-9)	Cumate	(137-29-1)
Crystal Violet HL2	(548-62-9)	Cumatetralyl (German, Dutch)	(5836-29-3)
Crystal Violet Lactone	(1552-42-7)	Cumeen (Dutch)	(98-82-8)
Crystal Violet O	(548-62-9)	Cumeenhydroperoxyde (Dutch)	(80-15-9)
Crystal Violet Pure DSC	(548-62-9)	psi-Cumene	(95-63-6)
Crystal Violet Pure DSC Brilliant	(548-62-9)	Cumene (ACGIH,OSHA)	(98-82-8)
Crystal Violet SS	(548-62-9)	Cumene aldehyde	(93-53-8)
Crystal Violet Technical	(548-62-9)	Cumene, p-amino-	(99-88-7)
Crystal Violet USP	(548-62-9)	Cumene, o-bromo- (8CI)	(7073-94-1)
Crystalets	(127-47-9)	Cumene, p-bromo- (8CI)	(586-61-8)
Crystalite CRS 6002	(14807-96-6)	Cumene, 3,5-dimethyl- (8CI)	(4706-90-5)
Crystallina	(50-14-6)	Cumene, 2,4-dimethyl- (6CI,7CI,8CI)	(4706-89-2)

Cumene, α,β-epoxy- (6CI,7CI,8CI) **(2085-88-3)**
Cumene, m-ethyl- (8CI) **(4920-99-4)**
Cumene, p-ethyl- (8CI) **(4218-48-8)**
Cumene hydroperoxide (DOT) **(80-15-9)**
Cumene hydroperoxide, Technically pure (DOT) **(80-15-9)**
Cumene, p-isopropenyl- **(2388-14-9)**
Cumene, p-methyl- **(99-87-6)**
Cumene, p-nitro- (8CI) **(1817-47-6)**
Cumene peroxide **(80-43-3)**
Cumene, trichloro- (7CI) **(61465-79-0)**
m-Cumenol **(618-45-1)**
p-Cumenol **(99-89-8)**
m-Cumenol methylcarbamate **(64-00-6)**
Cument hydroperoxide **(80-15-9)**
Cumenyl hydroperoxide **(80-15-9)**
m-Cumenyl methylcarbamate **(64-00-6)**
p-Cumenyl phenyl phosphate **(55864-04-5)**
m-Cumenyl phosphate **(72668-27-0)**
p-Cumenyl phosphate **(2502-15-0)**
Cumic alcohol **(536-60-7)**
Cumic aldehyde **(122-03-2)**
p-Cumic aldehyde **(122-03-2)**
Cumid **(66-76-2)**
o-Cumidine **(643-28-7)**
p-Cumidine **(99-88-7)**
psi-Cumidine **(137-17-7)**
Cumidine (8CI) **(99-88-7)**
Cumidine, 2,6-dinitro-N,N-dipropyl- **(33820-53-0)**
psi-Cumidine hydrochloride **(21436-97-5)**
Cumidine, N-phenyl- (8CI) **(5650-10-2)**
Cumin LS **(8001-21-6)**
Cumin-Oil **(8014-13-9)**
Cuminal **(122-03-2)**
Cuminaldehyde **(122-03-2)**
Cuminic acid **(536-66-3)**
Cuminic alcohol **(536-60-7)**
Cuminic aldehyde **(122-03-2)**
Cuminol **(536-60-7)**
Cuminyl alcohol **(536-60-7)**
Cuminyl aldehyde **(122-03-2)**
psi-Cumohydroquinone **(700-13-0)**
Cumol **(98-82-8)**
Cumolhydroperoxid (German) **(80-15-9)**
Cumyl alcohol **(536-60-7)**
α-Cumyl alcohol **(617-94-7)**
Cumyl tert-butyl peroxide **(3457-61-2)**
Cumyl hydroperoxide **(80-15-9)**
α-Cumyl hydroperoxide **(80-15-9)**
Cumyl hydroperoxide, Technically pure (DOT) **(80-15-9)**
Cumyl peroxide **(80-43-3)**
α-Cumyl peroxyneodecanoate **(26748-47-0)**
p-(α-Cumyl)phenol **(599-64-4)**
p-Cumylphenol **(599-64-4)**
Cumylphenyl diphenyl phosphate **(66594-31-8)**
p-Cumylphenyl glycidyl ether **(61578-04-9)**
Cunapsol **(1338-02-9)**
Cunilate **(10380-28-6)**
Cunilate 2472 **(10380-28-6)**
Cupferon (Czech) **(135-20-6)**
Cupferron **(135-20-6)**
Cupola slag (Secondary nonferrous plant) **(69012-26-6)**
Cupper oxide (Russian) **(1317-39-1)**
Cupral **(148-18-5)**
Cupral 45 **(1332-40-7)**
Cupramar **(1317-39-1)**
Cupramar **(1332-40-7)**
Cupramer **(1332-40-7)**
Cupranil Brown BCW **(2429-81-4)**
Cupranil Brown BCWR **(2429-81-4)**
Cuprantol **(1332-40-7)**
Cuprasulfide **(22205-45-4)**
Cuprate(3-), (mu-(4-((4'-((6-amino-1-hydroxy-3-sulfo-2-naphthalenyl)azo)-3,3'-dihydroxy(1,1'-biphenyl)-4-yl)azo)-3-hydroxy-2,7-naphthalenedisulfonato(7-)))di-, trisodium (9CI) **(66418-17-5)**
Cuprate(1-), (N,N-bis(carboxymethyl)glycinato(3-)-N,O,O',O'')-, ammonium, (T-4)- (9CI) **(71484-80-5)**
Cuprate(1-), (N,N-bis(carboxymethyl)glycinato(3-)-N,O,O',O'')-, hydrogen, (T-4)- **(34831-02-2)**
Cuprate(2-), (bis(chlorosulfonyl)-29H,31H-phthalocyaninedi sulfonato(4-)-N29,N30,N31,N32)-, dihydrogen (9CI) **(31361-57-6)**
Cuprate(4-), (mu-((7,7'-(carbonyldiimino)bis(4-hydroxy-3-((2-hydroxy-5-sulfophenyl)azo)-2-naphthalenesulfonato))-(8-)))di-, tetrasodium (9CI) **(15418-16-3)**
Cuprate(4-), (mu-((6,6'-((3,3'-dihydroxy(1,1'-biphenyl)-4,4'-diyl)bis(azo))bis(4-amino-5-hydroxy-1,3-naphthalenedi-sulfonato))(8-)))di-, tetrasodium (9CI) **(16143-79-6)**
Cuprate(4-), (mu-((3,3'-((3,3'-dihydroxy(1,1'-biphenyl)-4,4'-diyl)bis(azo))bis(5-amino- 4-hydroxy-2,7-naphthalenedi-sulfonato))(8-)))di-, tetrasodium **(28407-37-6)**
Cuprate(4-), (mu-((3,3'-((3,3'-dihydroxy(1,1'-biphenyl)-4,4'-diyl)bis(azo))bis(5-amino-4-hydroxy- 2,7-naphthalenedi-sulfonato))(8-)))di-, tetrasodium (9CI) **(28407-37-6)**
Cuprate(4-), (mu-((4,4'-((3,3'-dihydroxy(1,1'-biphenyl)-4,4'-diyl)bis(azo))bis(3-hydroxy-2,7-naphthalenedisulfonato))-(8-)))di-, tetrasodium (9CI) **(12222-00-3)**
Cuprate(3-), (mu-(7-((3,3'-dihydroxy-4'-((1-hydroxy-6-(phenyl-amino)-3-sulfo-2-naphthalenyl)azo)(1,1'-biphenyl)-4-yl)azo)-8-hydroxy-1,6-naphthalenedisulfonato(7-)))di-, trisodium (9CI) **(6656-03-7)**
Cuprate(2-), ((N,N'-1,2-ethanediylbis(N-(carboxymethyl)-glycinato))(4-)-N,N',O,O',ON,ON')-, dihydrogen, (OC-6-21)-(9CI) **(54453-03-1)**
Cuprate(2-), ((ethylenedinitrilo)tetraacetato)-, disodium **(14025-15-1)**
Cuprate(2-), (29H,31H-phthalocyaninedisulfonato(4-)-N29,N30,N31,N32)-, dihydrogen (9CI) **(29188-28-1)**
Cuprate(2-), (29H,31H-phthalocyaninedisulfonato(4-)-N29,N30,N31,N32)-, disodium (9CI) **(1330-38-7)**
Cuprate(1-), (29H,31H-phthalocyaninesulfonato(3-)-N29,N30,N31,N32)-, hydrogen (9CI) **(28901-96-4)**
Cuprate(3-), (29H,31H-phthalocyaninetrisulfonato(5-)-N29,N30,N31,N32)-, triammonium (9CI) **(25512-11-2)**
Cuprate(3-), (29H,31H-phthalocyaninetrisulfonato(5-)-N29,N30,N31,N32)-, trihydrogen (9CI) **(30638-09-6)**
Cuprate(3-), (29H,31H-phthalocyaninetrisulfonato(5-)-N29,N30,N31,N32)-, trisodium (9CI) **(1330-39-8)**
Cuprate(1-), (sulfato(2-)-O)- (9CI) **(12400-75-8)**
Cuprate(3-), tetrakis(cyano-C)-, tripotassium, (T-4)- (9CI) **(14263-73-1)**
Cuprate(2-), tris(cyano-C)-, disodium **(14264-31-4)**
Cupravet **(1332-40-7)**
Cupravit **(1332-40-7)**
Cupravit **(1332-65-6)**
Cupravit Blau **(20427-59-2)**
Cupravit Blue **(20427-59-2)**
Cupravit Forte **(1332-40-7)**
Cupravit Green **(1332-40-7)**
Cuprenil **(52-67-5)**
Cupreol **(83-46-5)**
Cupric Green **(10290-12-7)**
Cupric acetate **(142-71-2)**
Cupric acetoarsenite **(12002-03-8)**
Cupric acetylacetonate **(13395-16-9)**
Cupric arsenite **(10290-12-7)**
Cupric bromide **(7789-45-9)**
Cupric carbonate **(12069-69-1)**
Cupric carbonate (1:1) **(1184-64-1)**

Cupric carbonate, basic	(12069-69-1)
Cupric chloride	(10125-13-0)
Cupric chloride	(7447-39-4)
Cupric citrate	(866-82-0)
Cupric cyanide (DOT)	(14763-77-0)
Cupric diacetate	(142-71-2)
Cupric dichloride	(7447-39-4)
Cupric diformate	(544-19-4)
Cupric dinitrate	(3251-23-8)
Cupricellulose	(9004-34-6)
Cupric ferric subsulfate complex	(12168-20-6)
Cupric formate	(544-19-4)
Cupric gluconate	(527-09-3)
Cupric hydroxide	(20427-59-2)
Cupric 8-hydroxyquinolate	(10380-28-6)
Cupricin	(544-92-3)
Cupric lactate	(814-81-3)
Cupric neodecanoate	(32276-75-8)
Cupric nitrate (DOT)	(3251-23-8)
Cupric nitrilotriacetate	(15844-52-7)
Cupricol	(1332-40-7)
Cupric oxalate	(5893-66-3)
Cupric oxalate	(814-91-5)
Cupric oxide	(1317-38-0)
Cupric oxide chloride	(1332-40-7)
Cupric phosphate	(7798-23-4)
Cupric 8-quinolinolate	(10380-28-6)
Cupric sulfate	(7758-98-7)
Cupric sulfate, ammoniated	(10380-29-7)
Cupric sulfate anhydrous	(7758-98-7)
Cupric sulfate pentahydrate	(7758-99-8)
Cupric sulfide	(1317-40-4)
Cupric sulphate	(7758-98-7)
Cupric tartrate	(815-82-7)
Cupriethylene diamine	(13426-91-0)
Cupriethylenediamine, Solution [UN 1761]	(13426-91-0)
Cuprimine	(52-67-5)
Cuprinol	(1338-02-9)
Cupritox	(1332-40-7)
Cuprocaffaro	(1332-65-6)
Cuprocide	(1317-39-1)
Cuprocitrol	(866-82-0)
Cuprofix Brown GL	(16071-86-6)
Cuproin	(119-91-5)
Cuproine	(119-91-5)
Cuprokylt	(1332-40-7)
Cuprokylt	(1332-65-6)
Cuprol	(1332-40-7)
Cupron (Czech)	(441-38-3)
Cuprone	(441-38-3)
Cuprosan	(1332-40-7)
Cuprosan Blue	(1332-40-7)
Cuprosana	(1332-40-7)
Cuprosana	(1332-65-6)
Cuprous and cupric oxide, mixed	(82010-82-0)
Cuprous bromide	(7787-70-4)
Cuprous chloride	(7758-89-6)
Cuprous cyanide	(544-92-3)
Cuprous dichloride	(7758-89-6)
Cuprous iodide	(7681-65-4)
Cuprous oxide	(1317-39-1)
Cuprous sulfide	(22205-45-4)
Cuprous thiocyanate	(1111-67-7)
Cuprovinol	(1332-40-7)
Cuprovinol	(1332-65-6)
Cuprox	(1332-40-7)
Cuproxol	(1332-40-7)
Curacron	(41198-08-7)
Curafume	(74-83-9)
Curalin M	(101-14-4)
Curamil	(13457-18-6)
Curantyl	(58-32-2)
Curaterr	(1563-66-2)
Curb	(7784-25-0)
Curbiset	(2536-31-4)
Curcuma	(458-37-7)
Curcuma Oil	(8024-37-1)
Curcumin	(458-37-7)
Curcumin	(8024-37-1)
Curcumine	(8024-37-1)
Cure-Rite 18	(13752-51-7)
Curene	(9009-54-5)
Curene 442	(101-14-4)
Curesan	(123-88-6)
Curetard A	(86-30-6)
Curex Flea Duster	(83-79-4)
Curitan	(2439-10-3)
Curithane	(101-77-9)
Curithane 103	(96-33-3)
Curithane C126	(91-94-1)
Curling Factor	(126-07-8)
Curol Bright Red 4R	(2611-82-7)
Curol Orange	(633-96-5)
Curol Orange G	(547-57-9)
Curon Fast Yellow 5G	(1934-21-0)
Curral	(52-43-7)
Curtacain	(136-47-0)
Curzate	(57966-95-7)
Curzate M	(12427-38-2)
Cusiter	(101-20-2)
Cut-Back, Asphalt or Bitumen (DOT)	(8052-42-4)
Cutamin Black CG	(6428-31-5)
Cutamin Brilliant Red CG	(3567-65-5)
Cutamin Dark Blue CB	(2429-73-4)
Cutamine Brown CM	(2429-82-5)
Cuticura Acne Cream	(94-36-0)
Cutisan	(101-20-2)
Cutlass	(52508-35-7)
Cuvan 80	(94-91-7)
Cyaanwaterstof (Dutch)	(74-90-8)
Cyacetacid	(140-87-4)
Cyacetacide	(140-87-4)
Cyacetazid	(140-87-4)
Cyacetazide	(140-87-4)
Cyagard RF-1	(10310-38-0)
Cyalane	(947-02-4)
Cyamopsis Gum	(9000-30-0)
Cyan Blue BNC 55-3745	(147-14-8)
Cyan Blue BNF 55-3753	(147-14-8)
Cyan Blue GT 55-3295	(147-14-8)
Cyan Blue GTNF	(147-14-8)
Cyan Green 15-3100	(1328-53-6)
Cyan Peacock Blue G	(147-14-8)
Cyanacetamide	(107-91-5)
Cyanacetate ethyle (German)	(105-56-6)
Cyanacethydrazide	(140-87-4)
Cyanacetic acid hydrazide	(140-87-4)
Cyanacetohydrazide	(140-87-4)
Cyanacetylhydrazide	(140-87-4)
Cyanaein	(20350-15-6)
Cyanamid	(156-62-7)
Cyanamid Granular	(156-62-7)
Cyanamid Special Grade	(156-62-7)
Cyanamide	(156-62-7)

Cyanamide (ACGIH,OSHA) **(420-04-2)**
Cyanamide, (4,6-bis((1-methylethyl)amino)-1,3,5-triazin-2-yl)-methyl **(67704-68-1)**
Cyanamide calcique (French) **(156-62-7)**
Cyanamide, calcium salt (1:1) **(156-62-7)**
Cyanamide, cyano- (9CI) **(504-66-5)**
Cyanamide, diallyl **(538-08-9)**
Cyanamide, dimethyl **(1467-79-4)**
Cyanamide, di-2-propenyl- (9CI) **(538-08-9)**
Cyanamide, lead(2+) salt (1:1) (8CI,9CI) **(20837-86-9)**
Cyanamide, sodium salt (8CI,9CI) **(19981-17-0)**
Cyanaset **(101-14-4)**
Cyanatryn **(21689-84-9)**
Cyanazide **(140-87-4)**
Cyanazine **(21725-46-2)**
Cyanessigsaeure (German) **(372-09-8)**
Cyanhydrine d'acetone (French) **(75-86-5)**
Cyanic acid (8CI,9CI) **(420-05-3)**
Cyanic acid, butyl ester (7CI,8CI,9CI) **(1768-24-7)**
Cyanic acid, methylenebis(2,6-dimethyl-4,1-phenylene) ester (9CI) **(101657-77-6)**
Cyanic acid, potassium salt **(590-28-3)**
Cyanic acid, sodium salt **(917-61-3)**
Cyanide **(57-12-5)**
Cyanide(1-) **(57-12-5)**
Cyanide (CN^{1-}) **(57-12-5)**
Cyanide anion **(57-12-5)**
Cyanide ion **(57-12-5)**
Cyanide(1-) ion **(57-12-5)**
Cyanidelonon 1522 **(117-39-5)**
Cyanide of potassium **(151-50-8)**
Cyanide of sodium **(143-33-9)**
Cyanides (OSHA) **(151-50-8)**
Cyanidine **(290-87-9)**
Cyanine Acid Blue R **(3861-73-2)**
Cyanine Acid Blue R New **(3861-73-2)**
Cyanine Blue BB **(147-14-8)**
Cyanine Blue C **(147-14-8)**
Cyanine Blue LBG **(147-14-8)**
Cyanine Blue LC **(147-14-8)**
Cyanine Blue PRPD **(147-14-8)**
Cyanine Blue S 2100 **(147-14-8)**
Cyanine Blue SR 150A **(147-14-8)**
Cyanine Fast Scarlet G **(10169-02-5)**
Cyanine Fast Yellow M **(6375-55-9)**
Cyanine Fast Yellow R New **(5858-93-5)**
Cyanine Green G Base **(128-80-3)**
Cyanine Green GP **(1328-53-6)**
Cyanine Green NB **(1328-53-6)**
Cyanine Green T **(1328-53-6)**
Cyanine Green Toner **(1328-53-6)**
Cyanite **(1302-76-7)**
Cyanizide **(140-87-4)**
Cyano-B12 **(68-19-9)**
2-Cyanoacetamide **(107-91-5)**
Cyanoacetamide **(107-91-5)**
α-Cyanoacetamide **(107-91-5)**
Cyanoacethydrazide **(140-87-4)**
Cyanoacetic acid **(372-09-8)**
Cyanoacetic acid ethyl ester **(105-56-6)**
Cyanoacetic acid, 2-ethylhexyl ester **(13361-34-7)**
Cyanoacetic acid hydrazide **(140-87-4)**
Cyanoacetic acid methyl ester **(105-34-0)**
Cyanoacetic ester **(105-56-6)**
Cyanoacetohydrazide **(140-87-4)**
α-Cyanoacetohydrazide **(140-87-4)**
Cyanoacetonitrile **(109-77-3)**
Cyanoacetylhydrazide **(140-87-4)**
2-Cyanoacrylic acid, methyl ester **(137-05-3)**
α-Cyanoacrylic acid methyl ester **(137-05-3)**
Cyanoamine **(420-04-2)**
N-Cyanoamine **(420-04-2)**
2-Cyanoaniline **(1885-29-6)**
3-Cyanoaniline **(2237-30-1)**
4-Cyanoaniline **(873-74-5)**
m-Cyanoaniline **(2237-30-1)**
o-Cyanoaniline **(1885-29-6)**
p-Cyanoaniline **(873-74-5)**
Cyanobenzene **(100-47-0)**
4-Cyanobenzenesulfonyl chloride **(49584-26-1)**
p-Cyanobenzenesulfonyl chloride **(49584-26-1)**
4-Cyanobenzoic acid **(619-65-8)**
Cyanobrik **(143-33-9)**
Cyanobromide **(506-68-3)**
1-Cyanobutane **(110-59-8)**
4-Cyanobutanoic acid **(39201-33-7)**
4-Cyano-1-butene **(592-51-8)**
.γ.-Cyanobutyric acid **(39201-33-7)**
4-Cyanobutyric acid **(39201-33-7)**
Cyanocobalamin **(68-19-9)**
Cyanocobalamine **(68-19-9)**
Cyanocyanamide **(504-66-5)**
1-(1-Cyanocyclohexyl)piperidine **(3867-15-0)**
4-Cyano-2,6-diiodophenol **(1689-83-4)**
4-Cyano-2,6-dijodphenol (German) **(1689-83-4)**
4-Cyano-2,6-dijodphenol caprysaeureester (German) **(3861-47-0)**
α-Cyanodiphenylmethane **(86-29-3)**
1'-(3-Cyano-3,3-diphenylpropyl)(1,4'-bipiperidine)-4'-carboxamide **(302-41-0)**
1-(3-Cyano-3,3-diphenylpropyl)-4-(2-oxo-3-propionyl-1-benzimidazolinyl)piperidine **(15301-48-1)**
1-(3-Cyano-3,3-diphenylpropyl)-4-phenylisonipecotic acid **(28782-42-5)**
1-(3-Cyano-3,3-diphenylpropyl)-4-phenyl-isonipecotic acid ethyl ester **(915-30-0)**
1-(1-(3-Cyano-3,3-diphenylpropyl)-4-piperidyl)-3-propionyl-2-benzimidazolinone **(15301-48-1)**
Cyanoethane **(107-12-0)**
2-Cyanoethanol **(109-78-4)**
Cyanoethydrazide **(140-87-4)**
2-Cyanoethyl acrylate **(106-71-8)**
Cyanoethyl acrylate **(106-71-8)**
2-Cyanoethyl alcohol **(109-78-4)**
β-Cyanoethylamine **(151-18-8)**
2-Cyano-N-((ethylamino)carbonyl)-2-(methoxyimino)-acetamide **(57966-95-7)**
3-(N-Cyanoethyl)amino-4-methoxyacetanilide **(26408-28-6)**
N-(2-Cyanoethyl)cyclohexylamine **(702-03-4)**
Cyanoethylene **(107-13-1)**
3-(N-Cyanoethyl-N-ethyl)amino-4-methoxyacetanilide **(19433-94-4)**
3-(N-Cyanoethyl-N-ethylamino)-4-methoxyacetanilide **(19433-94-4)**
3-(N-Cyanoethyl-N-ethylamino)-4-methoxy-6-((2-bromo-4,6-dinitrophenyl)azo)acetanilide **(22578-86-5)**
N-(3-((2-Cyanoethyl)ethylamino)-4-methoxyphenyl)acetamide **(19433-94-4)**
N-(2-Cyanoethyl)-N-ethyl-m-toluidine **(148-69-6)**
N-β-Cyanoethyl-N-β-hydroxyethylaniline **(92-64-8)**
N-β-Cyanoethyl-N-methylaniline **(94-34-8)**
Cyanoethylmorpholine **(4542-47-6)**
2-Cyanoethyl propenoate **(106-71-8)**
2-Cyanoethyltrichlorosilane **(1071-22-3)**
β-Cyanoethyltrichlorosilane **(1071-22-3)**
(2-Cyanoethyl)triethoxysilane **(919-31-3)**
β-Cyanoethyltriethoxysilane **(919-31-3)**
Cyanofenphos **(13067-93-1)**
Cyanogas **(592-01-8)**
Cyanogen (ACGIH,DOT,OSHA) **(460-19-5)**
Cyanogen, Liquefied [UN 1026] **(460-19-5)**

Cyanogenamide (420-04-2)
Cyanogen-bromide (506-68-3)
Cyanogen bromide [UN 1889] (506-68-3)
Cyanogen-chloride (506-77-4)
Cyanogen chloride (ACGIH,OSHA) (506-77-4)
Cyanogen chloride, Containing less than 0.9% water (DOT) (506-77-4)
Cyanogen chloride, Inhibited [UN 1589] (506-77-4)
Cyanogen chloride, tetramer (6CI) (877-83-8)
Cyanogene (French) (460-19-5)
Cyanogen gas (DOT) (460-19-5)
Cyanogen iodide (506-78-5)
Cyanogen monobromide (506-68-3)
Cyanogen nitride (420-04-2)
Cyanogran (143-33-9)
Cyanoguanidine (461-58-5)
Cyanoiminoacetic acid (107-91-5)
Cyanol (62-53-3)
Cyanolit (137-05-3)
Cyanomethane (75-05-8)
Cyanomethanol (107-16-4)
Cyanomethyl acetate (1001-55-4)
(Cyanomethyl)benzene (140-29-4)
N-(Cyanomethyl)dimethylamine (926-64-7)
2-(1-Cyano-1-methylethylamino)-4-ethylamino-6-methylthio-1,3,5-triazine (21689-84-9)
Cyano(methylmercuri)guanidine (502-39-6)
1-Cyano-2-methyl-3-(2-(((5-methyl-4-imidazolyl)methyl)-thio)ethyl)guanidine (51481-61-9)
2-Cyano-1-methyl-3-(2-(((5-methylimidazol-4-yl)methyl)thio)-ethyl)guanidine (51481-61-9)
N-Cyanomethylmorpholine (5807-02-3)
3-Cyano-1-methyl-4-pyridone (767-98-6)
1-Cyanonaphthalene (86-53-3)
2-Cyanonaphthalene (613-46-7)
α-Cyanonaphthalene (86-53-3)
β-Cyanonaphthalene (613-46-7)
2-Cyano-4-nitroaniline (17420-30-3)
3-Cyanonitrobenzene (619-24-9)
4-Cyanonitrobenzene (619-72-7)
m-Cyanonitrobenzene (619-24-9)
p-Cyanonitrobenzene (619-72-7)
1-Cyanooctadecane (28623-46-3)
1-Cyanooctane (2243-27-8)
o-Cyanophenol (611-20-1)
α-Cyano-3-phenoxybenzyl 2-(4-chlorophenyl)isovalerate (51630-58-1)
(S)-α-Cyano-3-phenoxybenzyl(S)-2-(4-chlorophenyl)-3-methyl-butyrate (66230-04-4)
α-Cyano-3-phenoxybenzyl-2-(4-chlorophenyl)-3-methylbutyrate (51630-58-1)
(+-)-α-Cyano-3-phenoxybenzyl 2,2-dimethyl-3-(2,2-dichloro-vinyl)cyclopropane carboxylate (52315-07-8)
α-Cyano-3-phenoxybenzyl 2,2,3,3-tetramethyl-1-cyclo-propanecarboxylate (39515-41-8)
Cyano(3-phenoxyphenyl)methyl 4-chloro-α-(1-methylethyl)-benzeneacetate (51630-58-1)
(+-)-Cyano(3-phenoxyphenyl)methyl(+)-4-(difluoromethoxy)-α-(1-methylethyl)benzeneacetate (70124-77-5)
Cyano(3-phenoxyphenyl)methyl 2,2-dimethyl-3-(1,2,2,2-tetra-bromoethyl)cyclopropanecarboxylate (66841-25-6)
Cyanophenphos (13067-93-1)
p-Cyanophenyl dimethyl phosphate (61090-94-6)
O-(4-Cyanophenyl) O,O-dimethyl phosphorothioate (2636-26-2)
O-p-Cyanophenyl O,O-dimethyl phosphorothioate (2636-26-2)
O-(4-Cyanophenyl) O-ethyl phenylphosphonothioate (13067-93-1)
O-p-Cyanophenyl O-ethyl phenyl-phosphonothioate (13067-93-1)
4-Cyanophenyl N-phenylcarbamate (37070-85-2)
4-Cyanophenyl phenylcarbamate (37070-85-2)
p-Cyanophenylsulfonyl chloride (49584-26-1)
Cyanophos (2636-26-2)
1-Cyanopropane (109-74-0)
1-Cyanopropene (4786-20-3)
2-Cyanopropene-1 (126-98-7)
3-Cyanopropyldichloromethylsilane (1190-16-5)
2-Cyanopyridine (100-70-9)
3-Cyanopyridine (100-54-9)
4-Cyanopyridine (100-48-1)
γ-Cyanopyridine (100-48-1)
.α.-Cyanostyrene (495-10-3)
4-Cyanosuberonitrile (1772-25-4)
4-Cyanothiazole (1452-15-9)
N-Cyano-N'-methyl-N''-(2-(((5-methyl-1H-imidazol-4-yl)methyl)thio)ethyl)guanidine (51481-61-9)
2-Cyanotoluene (529-19-1)
4-Cyanotoluene (104-85-8)
α-Cyanotoluene (140-29-4)
o-Cyanotoluene (529-19-1)
ω-Cyanotoluene (140-29-4)
p-Cyanotoluene (104-85-8)
Cyanotrichloromethane (545-06-2)
Cyanotubericidin (606-58-6)
Cyanox (2636-26-2)
Cyanox 425 (88-24-4)
Cyanox LTDP (123-28-4)
Cyanox STDP (693-36-7)
4-Cyano-4'-hexylbiphenyl (41122-70-7)
N-Cyano-N'-(hydroxymethyl)guanidine (13101-26-3)
Cyansan (917-61-3)
Cyantin (67-20-9)
Cyanuramide (108-78-1)
Cyanurchloride (108-77-0)
Cyanure (French) (57-12-5)
Cyanure d'argent (French) (506-64-9)
Cyanure de calcium (French) (592-01-8)
Cyanure de cuivre (French) (14763-77-0)
Cyanure de mercure (French) (592-04-1)
Cyanure de methyl (French) (75-05-8)
Cyanure de plomb (French) (592-05-2)
Cyanure de potassium (French) (151-50-8)
Cyanure de sodium (French) (143-33-9)
Cyanure de vinyle (French) (107-13-1)
Cyanure de zinc (French) (557-21-1)
Cyanuric acid (108-80-5)
Cyanuric acid chloride (108-77-0)
Cyanuric acid, trithio- (638-16-4)
Cyanuric bromide (6CI) (14921-00-7)
Cyanuric chloride [UN 2670] (108-77-0)
Cyanuric fluoride (675-14-9)
Cyanuric triamide (108-78-1)
Cyanuric triazide (DOT) (5637-83-2)
Cyanuric trichloride (108-77-0)
Cyanurotriamide (108-78-1)
Cyanurotriamine (108-78-1)
Cyanurtriamide (108-78-1)
Cyanuryl chloride (108-77-0)
Cyanwasserstoff (German) (74-90-8)
Cyap (2636-26-2)
Cyaphenine (493-77-6)
Cyasorb 5411 (3147-75-9)
Cyasorb UV 9 (131-57-7)
Cyasorb UV 12 (131-54-4)
Cyasorb UV 24 (131-53-3)
Cyasorb UV 24 Light Absorber (131-53-3)
Cyazid (140-87-4)
Cyazide (140-87-4)
Cyazin (1912-24-9)

Cybolt (70124-77-5)
Cycas Revoluta Glucoside (14901-08-7)
Cycasin (14901-08-7)
Cycasin acetate (592-62-1)
Cyceuron plus chlorbufan (8015-55-2)
Cyclo-D-serine (68-41-7)
Cycladiene (84-17-3)
Cyclal cetyl alcohol (36653-82-4)
Cyclalia (107-75-5)
Cyclamal (103-95-7)
Cyclamate (100-88-9)
Cyclamate (139-05-9)
Cyclamate calcium (139-06-0)
Cyclamate, calcium salt (139-06-0)
Cyclamate sodium (139-05-9)
Cyclamate, sodium salt (139-05-9)
Cyclamen aldehyde (103-95-7)
Cyclamic acid (100-88-9)
Cyclamic acid sodium salt (139-05-9)
Cyclamide (968-81-0)
Cyclan (139-06-0)
Cyclandelate (456-59-7)
Cyclaprop (17511-60-3)
Cycle (32889-48-8)
Cyclic 3',5'-AMP (60-92-4)
Cyclic AMP (60-92-4)
Cyclic PNCl2 (25034-79-1)
Cyclic adenosine 3',5'-phosphate (60-92-4)
Cyclic 3',5'-adenylic acid (60-92-4)
Cyclic dimethylolethyleneurea (136-84-5)
Cyclic ethylene carbonate (96-49-1)
Cyclic ethylene(diethoxyphosphinothioyl)dithioimidocarbonate (947-02-4)
Cyclic ethylene ethylboronate ((C2H4O2)EtB) (10173-38-3)
Cyclic ethylene p,p-diethyl phosphonodithioimidocarbonate (947-02-4)
Cyclic(glycylglycyl) (106-57-0)
Cyclic methylethylene carbonate (108-32-7)
Cyclic S,S-(6-methyl-2,3-quinoxalinediyl) dithiocarbonate (2439-01-2)
Cyclic 1,2-propylene carbonate (108-32-7)
Cyclic propylene carbonate (108-32-7)
Cyclic propylene (diethoxyphosphinyl)dithioimidocarbonate (950-10-7)
Cyclic N',O-propylene ester of N,N-bis(2-chloroethyl)-phosphorodiamidic acid monohydrate (6055-19-2)
Cyclic tetramethylene sulfone (126-33-0)
Cyclizine (82-92-8)
Cycloamide SM (111-57-9)
Cycloate (1134-23-2)
Cyclobarbital (52-31-3)
Cyclobarbitol (52-31-3)
Cyclobarbiton (52-31-3)
Cyclobarbitone (52-31-3)
Cyclobutadibenzene (259-79-0)
Cyclobutane [UN 2601] (287-23-0)
Cyclobutaneacetaldehyde, 3-acetyl-2,2-dimethyl- (8CI,9CI) (2704-78-1)
Cyclobutaneacetic acid, 3-acetyl-2,2-dimethyl- (8CI,9CI) (473-72-3)
Cyclobutane, 1,2-bis(1,2-dichloroethyl)- (9CI) (83682-62-6)
Cyclobutanecarbonitrile, 3,3-dimethyl- (9CI) (53783-86-1)
Cyclobutanecarboxylic-acid (3721-95-7)
Cyclobutane, 1,2-dichloro 3,4-bis(dichloromethylene)- (9CI) (55044-46-7)
Cyclobutane, 1,2-dichloro-1,2-diethenyl- (9CI) (14112-00-6)
Cyclobutane, 1,2-dichloro-1,2-divinyl- (8CI) (14112-00-6)
Cyclobutane, 1,2-diethenyl- (9CI) (2422-85-7)
Cyclobutane, 1,2-diethyl- (6CI,9CI) (61141-83-1)
1,3-Cyclobutanedione (7CI,8CI,9CI) (15506-53-3)
1,3-Cyclobutanedione, 2,2,4,4-tetramethyl (933-52-8)
Cyclobutane, 1,2-divinyl- (6CI,7CI,8CI) (2422-85-7)
Cyclobutane, methyl- (8CI,9CI) (598-61-8)
Cyclobutane, methylene- (9CI) (1120-56-5)
Cyclobutane, octafluoro (115-25-3)
Cyclobutanol, 2-ethyl- (9CI) (35301-43-0)
Cyclobutanone (9CI) (1191-95-3)
Cyclobutanone, 2-tert-butyl- (8CI) (4579-31-1)
Cyclobutanone, 2-(1,1-dimethylethyl)- (9CI) (4579-31-1)
Cyclobutanone, 2-methyl- (6CI,7CI,8CI,9CI) (1517-15-3)
Cyclobutanone, 2,2,3-trimethyl- (7CI,8CI,9CI) (1449-49-6)
3-Cyclobutene-1,2-dione, 3-hydroxy-, potassium salt, hydrate (52591-22-7)
Cyclobutene, hexachloro- (8CI,9CI) (6130-82-1)
2-Cyclobuten-1-one, dichloro- (9CI) (103339-60-2)
2-Cyclobuten-1-one, trichloro- (9CI) (78099-58-8)
Cyclobutylcarboxylic acid (3721-95-7)
17-(Cyclobutylmethyl)morphinan-3,14-diol (42408-82-2)
Cyclocel (999-81-5)
Cyclochem GMS (31566-31-1)
Cyclochlorotine (12663-46-6)
α-Cyclocitrylideneacetone (127-41-3)
β-Cyclocitrylideneacetone (14901-07-6)
Cyclodan (115-29-7)
Cyclodecane (9CI) (293-96-9)
Cyclodecanol (9CI) (1502-05-2)
1H-Cyclodecapyrazole, 4,5,6,7,8,9,10,11-octahydro- (8CI) (34176-71-1)
Cyclodecene, (E)- (8CI,9CI) (2198-20-1)
Cyclodecene, (Z)- (8CI,9CI) (935-31-9)
Cyclodiglycine (106-57-0)
Cyclodisilaselenane (8CI,9CI) (287-68-3)
Cyclododecalactam (947-04-6)
Cyclododecane (9CI) (294-62-2)
Cyclododecane, 1,2-epoxy- (286-99-7)
Cyclododecane, ethanol deriv. (9CI) (32399-56-7)
Cyclododecane, 1,2,5,6,9,10-hexabromo- (9CI) (3194-55-6)
Cyclododecane, hexabromo- (9CI) (25637-99-4)
Cyclododecanol (9CI) (1724-39-6)
Cyclododecanone (9CI) (830-13-7)
Cyclododecatriene (9CI) (27070-59-3)
1,5,9-Cyclododecatriene, (E,E,E)- (8CI,9CI) (676-22-2)
1,5,9-Cyclododecatriene [UN 2518] (4904-61-4)
Cyclododecene (8CI,9CI) (1501-82-2)
Cyclododecene epoxide (286-99-7)
N-Cyclododecyl-2,6-dimethylmorpholinacetat (German) (31717-87-0)
4-Cyclododecyl-2,6-dimethylmorpholine (1593-77-7)
4-Cyclododecyl-2,6-dimethylmorpholine acetate (31717-87-0)
Cyclododecyl-2,6-dimethylmorpholine acetate (31717-87-0)
4-Cyclododecyl-2,6-dimethylmorpholine benzoate (59145-63-0)
4-Cyclododecyl-2,6-dimethylmorpholinium acetate (31717-87-0)
N-Cyclododecyl-2,6-dimethylmorpholinium acetate (31717-87-0)
2-(Cyclododecyloxy)ethanol (32399-56-7)
Cyclodorm (52-31-3)
Cycloestrol (84-16-2)
Cyclogallipharaol (501-24-6)
Cyclogest (57-83-0)
Cycloglycylglycine (106-57-0)
Cyclo(glycylglycyl) (106-57-0)
Cycloguanil (516-21-2)
Cycloguanyl (516-21-2)
4H-Cyclohepta(def)phenanthrene (8CI,9CI) (19561-31-0)
1,3-Cycloheptadiene (8CI,9CI) (4054-38-0)
1,4-Cycloheptadiene, 6-(1-butenyl)-, (S-(Z))- (9CI) (33156-92-2)
1,4-Cycloheptadiene, 6-(1-butenyl)-, (Z)-(S)-(+)- (8CI) (33156-92-2)
Cycloheptanamine (9CI) (5452-35-7)
Cycloheptane [UN 2241] (291-64-5)
Cycloheptane, bromo- (9CI) (2404-35-5)
Cycloheptanecarboxylic acid (8CI,9CI) (1460-16-8)
Cycloheptanecarboxylic acid, 1-amino- (9CI) (6949-77-5)
Cycloheptane, methyl- (8CI,9CI) (4126-78-7)
Cycloheptane, 1-methyl-4-methylene- (8CI,9CI) (23799-25-9)
Cycloheptanol (8CI,9CI) (502-41-0)

Cycloheptanone (502-42-1)
Cycloheptasiloxane, tetradecamethyl- (8CI,9CI) (107-50-6)
1,3,5-Cycloheptatriene (544-25-2)
Cycloheptatriene [UN 2603] (544-25-2)
2,4,6-Cycloheptatrien-1-one (539-80-0)
2,4,6-Cycloheptatrien-1-one, 2-hydroxy (533-75-5)
2,4,6-Cycloheptatrien-1-one, 2-hydroxy-5-isopropyl- (8CI) (672-76-4)
2,4,6-Cycloheptatrien-1-one, 2-hydroxy-5-(1-methylethyl) (672-76-4)
Cycloheptene [UN 2242] (628-92-2)
Cycloheptene, 5-ethylidene-1-methyl- (8CI,9CI) (15402-94-5)
5-(1-Cyclohepten-1-yl)-5-ethyl-2,4,6(1H,3H,5H)-pyrimidinetrione (509-86-4)
5-(1-Cyclohepten-1-yl)-5-ethylbarbituric acid (509-86-4)
Cycloheptenylethylbarbituric acid (509-86-4)
Cycloheptenylethylmalonylurea (509-86-4)
1-Cyclohepten-1-yl methyl ketone (14377-11-8)
Cycloheptylamine (5452-35-7)
Cycloheptyl bromide (2404-35-5)
1-Cycloheptyl-3-phenylurea (19095-79-5)
Cyclohexaan (Dutch) (110-82-7)
2,5-Cyclohexadien-1,4-dione, 2,5-bis(1,1-dimethylethyl)- (2460-77-7)
1,3-Cyclohexadiene (9CI) (592-57-4)
1,4-Cyclohexadiene (9CI) (628-41-1)
Cyclohexadiene (8CI,9CI) (29797-09-9)
1,4-Cyclohexadienedione (106-51-4)
2,5-Cyclohexadiene-1,4-dione (106-51-4)
Cyclohexadienedione (106-51-4)
3,5-Cyclohexadiene-1,2-dione (9CI) (583-63-1)
2,5-Cyclohexadiene-1,4-dione, 2,5-bis(1,1-dimethylethyl)- (9CI) (2460-77-7)
2,5-Cyclohexadiene-1,4-dione, Compd. with 1,4-benzenediol (1:1) (106-34-3)
2,5-Cyclohexadiene-1,4-dione, Compd. with 2-methyl-1,4-benzenediol (1:1) (9CI) (55836-33-4)
2,5-Cyclohexadiene-1,4-dione, 2,5-dichloro-3,6-dihydroxy- (9CI) (87-88-7)
2,5-Cyclohexadiene-1,4-dione, dioxime (105-11-3)
2,5-Cyclohexadiene-1,4-dione, monooxime (9CI) (637-62-7)
2,5-Cyclohexadiene-1,4-dione, 2,3,5,6-tetraazido- (22826-61-5)
3,5-Cyclohexadiene-1,2-dione, 3,4,5,6-tetrachloro- (9CI) (2435-53-2)
2,5-Cyclohexadiene-1,4-dione, 2,3,5-trimethyl- (9CI) (935-92-2)
1,4-Cyclohexadiene dioxide (106-51-4)
1,3-Cyclohexadiene, 1-methyl-4-isopropyl- (99-86-5)
1,4-Cyclohexadiene, 1-methyl-4-isopropyl- (99-85-4)
1,3-Cyclohexadiene, 2-methyl-5-(1-methylethyl)- (99-83-2)
1,3-Cyclohexadiene, 2-methyl-5-(1-methylethyl)-, (R)- (9CI) (4221-98-1)
1,3-Cyclohexadiene, 1,5,5,6-tetramethyl- (7CI,8CI,9CI) (514-94-3)
2,4-Cyclohexadien-1-one, acetyldichloro-2,6-dimethoxy- (9CI) (108673-04-7)
2,4-Cyclohexadien-1-one, acetyl-2,6-dimethoxy-, trichloro deriv. (9CI) (108548-70-5)
2,5-Cyclohexadien-1-one, 4-acetylimino- (50700-49-7)
2,5-Cyclohexadien-1-one, 4-((4-amino-3-methylphenyl)imino)- (9CI) (101-15-5)
2,5-Cyclohexadien-1-one, 4-(bis(4-hydroxyphenyl)methylene)- (9CI) (603-45-2)
2,5-Cyclohexadien-1-one, 4-(bis(p-hydroxyphenyl)methylene)- (8CI) (603-45-2)
2,5-Cyclohexadien-1-one, 4-chloroimino-2,6-dibromo (537-45-1)
2,5-Cyclohexadien-1-one, 4-chloroimino (637-61-6)
2,4-Cyclohexadien-1-one, 6-diazo-2,4-dinitro- (9CI) (4682-03-5)
2,5-Cyclohexadien-1-one, 2,6-dibromo-4-(chloroimino)- (8CI,9CI) (537-45-1)
2,5-Cyclohexadien-1-one, 4-hydroxy- (8CI,9CI) (4323-21-1)
2,5-Cyclohexadien-1-one, 4-hydroxyimino (637-62-7)
2,5-Cyclohexadien-1-one, 4-((4-(phenylamino)phenyl)imino)- (9CI) (6201-64-5)
2,5-Cyclohexadien-1-one, 4-(phenylimino)-, oxime, sodium salt (9CI) (63451-40-1)
Cyclohexamethylenimine (111-49-9)
Cyclohexamine (2201-15-2)
Cyclohexan (German) (110-82-7)
Cyclohexanaecarboxaldehyde (2043-61-0)
Cyclohexanamine, 2-((4-aminocyclohexyl)methyl)- (9CI) (24650-10-0)
Cyclohexanamine, N-(2-aminoethyl)- (5700-53-8)
Cyclohexanamine, N-cyclohexyl-, nitrite (9CI) (3129-91-7)
Cyclohexanamine, N,N-diethyl- (9CI) (91-65-6)
Cyclohexanamine, N,N-dimethyl- (9CI) (98-94-2)
Cyclohexanamine, N-(2-ethylhexyl)- (9CI) (5432-61-1)
Cyclohexanamine, N-ethyl-1-phenyl (2201-15-2)
Cyclohexanamine, N,N'-methanetetraylbis- (9CI) (538-75-0)
Cyclohexanamine, 2-methyl- (9CI) (7003-32-9)
Cyclohexanamine, N-methyl- (9CI) (100-60-7)
Cyclohexanamine, 4,4'-methylenebis- (9CI) (1761-71-3)
Cyclohexanamine, 4,4'-methylenebis-, (cis(cis))- (9CI) (6693-31-8)
Cyclohexanamine, 4,4'-methylenebis-, (trans(cis))- (9CI) (6693-30-7)
Cyclohexanamine, 4,4'-methylenebis-, (trans(trans))- (9CI) (6693-29-4)
Cyclohexanamine, N-(1-methylethyl)- (9CI) (1195-42-2)
Cyclohexanamine, 1-phenyl- (2201-24-3)
Cyclohexanamine, N-propyl- (9CI) (3592-81-2)
Cyclohexanaminium, N-heptyl-2-hydroxy-N,N,2-trimethyl-5-(1-methylethyl)- (20091-61-6)
1-Cyclohexanaminoanthracene-9,10-dione (1096-48-6)
1,3-Cyclohexandione (504-02-9)
Cyclohexane (ACGIH,OSHA) [UN 1145] (110-82-7)
Cyclohexaneacetic-acid (5292-21-7)
Cyclohexane, 1,4-bis(methylene)- (9CI) (4982-20-1)
Cyclohexane, bromo- (9CI) (108-85-0)
Cyclohexane, 1-bromo-2-chloro-, cis- (9CI) (51422-75-4)
Cyclohexane, (2-bromoethyl)- (9CI) (1647-26-3)
Cyclohexane, 1-bromo-4-methyl- (9CI) (6294-40-2)
Cyclohexane, 1,1'-(1,4-butanediyl)bis- (9CI) (6165-44-2)
Cyclohexane, sec-butyl- (7058-01-7)
Cyclohexane, butyl- (9CI) (1678-93-9)
Cyclohexane, tert-butyl- (8CI) (3178-22-1)
Cyclohexanecarbamic acid, N-ethylthio-, S-ethyl ester (1134-23-2)
Cyclohexanecarbinol (100-49-2)
Cyclohexanecarbonitrile, 1,1'-azobis (2094-98-6)
Cyclohexanecarbonitrile, 1-hydroxy (931-97-5)
Cyclohexanecarbonitrile, 1-piperidino- (3867-15-0)
Cyclohexanecarbonitrile, 1-(1-piperidinyl)- (9CI) (3867-15-0)
Cyclohexanecarboxaldehyde (9CI) (2043-61-0)
Cyclohexanecarboxylic-acid (98-89-5)
Cyclohexanecarboxylic acid, 3-(1-(allyloxyamino)butylidene)-6,6-dimethyl-2,4-dioxo-, methyl ester, sodium salt (55635-13-7)
Cyclohexanecarboxylic acid, 3-((3-(3,4-dihydroxyphenyl)-1-oxo-2-propenyl)oxy)- 1,4,5-trihydroxy-, (1s-(1-α,3-β,4-α,5-α)) (327-97-9)
Cyclohexanecarboxylic acid, 4-(1,5-dimethyl-3-oxohexyl)-, (4(R)-cis)- (9CI) (38963-91-6)
Cyclohexanecarboxylic acid, ethenyl ester (9CI) (4840-76-0)
Cyclohexanecarboxylic acid, 2-ethyl-2-(((1-oxononyl)oxy)-methyl)-1,3-propanediyl ester (9CI) (99554-33-3)
Cyclohexanecarboxylic acid, lead salt (61790-14-5)
Cyclohexanecarboxylic acid, lead salt (9CI) (50825-29-1)
Cyclohexanecarboxylic acid, 4-methyl- (8CI,9CI) (4331-54-8)
Cyclohexanecarboxylic acid, 4-nitrophenyl ester (9CI) (13551-17-2)
Cyclohexanecarboxylic acid, p-nitrophenyl ester (7CI,8CI) (13551-17-2)
Cyclohexanecarboxylic acid, 1,3,4,5-tetrahydroxy-, (1R-(1-α,3-α,4-α,5-β))- (77-95-2)
Cyclohexanecarboxylic acid, 1,3,4,5-tetrahydroxy-, (-)- (8CI) (77-95-2)
Cyclohexanecarboxylic acid, 1,3,4,5-tetrahydroxy- (8CI,9CI) (562-73-2)
Cyclohexanecarboxylic acid, 1,3,4,5-tetrahydroxy-, D-(-) (77-95-2)
Cyclohexanecarboxylic acid, 1,3,4,5-tetrahydroxy-, (1α,3α,4α,5β)- (9CI) (36413-60-2)
Cyclohexanecarboxylic acid, vinyl ester (8CI) (4840-76-0)
Cyclohexane, chloro (542-18-7)
Cyclohexane, 1-chloro-2,3,4,5,6-pentabromo- (87-84-3)
Cyclohexane, (3-chloro-1-propynyl)- (9CI) (55723-99-4)
Cyclohexane, cyclohexyl- (92-51-3)
Cyclohexane, 1-(cyclohexylmethyl)-2-ethyl-, trans- (9CI) (54934-92-8)

Name	CAS No.
Cyclohexane, 1-(cyclohexylmethyl)-2-methyl-, trans- (9CI)	**(54823-94-8)**
Cyclohexane, 1-(cyclohexylmethyl)-3-methyl-, trans- (9CI)	**(54823-95-9)**
Cyclohexane, 1-(cyclohexylmethyl)-4-methyl-, trans- (9CI)	**(54823-98-2)**
Cyclohexane, 1-(cyclohexylmethyl)-4-(1-methylethyl)- (9CI)	**(54965-61-6)**
Cyclohexane, (cyclopentylmethyl)- (9CI)	**(4431-89-4)**
Cyclohexane, decyl- (9CI)	**(1795-16-0)**
1,2-Cyclohexanediamine (9CI)	**(694-83-7)**
1,4-Cyclohexanediamine (8CI,9CI)	**(3114-70-3)**
1,2-Cyclohexanediamine-N,N,N',N'-tetraacetic acid	**(482-54-2)**
1,2-Cyclohexanediaminetetraacetic acid	**(482-54-2)**
Cyclohexane, 1,4-dibromo- (9CI)	**(35076-92-7)**
Cyclohexane, 1,2-dibromo-4-(1,2-dibromoethyl)	**(3322-93-8)**
Cyclohexane, 1,1'-(1,2-dibromo-1,2-ethanediyl)bis(3,4-dibromo- (9CI)	**(18122-77-5)**
(1-Cyclohexane-1,2-dicarboximido)methyl chrysanthemumate	**(7696-12-0)**
1,2-Cyclohexanedicarboxylic acid (9CI)	**(1687-30-5)**
1,2-Cyclohexanedicarboxylic acid anhydride	**(85-42-7)**
1,2-Cyclohexanedicarboxylic acid, bis(2,3-epoxypropyl) ester	**(5493-45-8)**
1,2-Cyclohexanedicarboxylic acid, bis(2-ethylhexyl) ester (8CI,9CI)	**(84-71-9)**
1,2-Cyclohexanedicarboxylic acid, bis(oxiranylmethyl)ester (9CI)	**(5493-45-8)**
1,4-Cyclohexanedicarboxylic acid, dimethyl ester	**(94-60-0)**
1,4-Cyclohexanedicarboxylic acid, 2,5-dioxo-, dimethyl ester (9CI)	**(6289-46-9)**
1,2-Cyclohexanedicarboxylic acid, 3,6-endo-epoxy-	**(145-73-3)**
1,2-Cyclohexanedicarboxylic anhydride (8CI)	**(85-42-7)**
1,4-Cyclohexanedicarboxylic dimethyl ester	**(94-60-0)**
Cyclohexane, 1,2-dichloro- (9CI)	**(1121-21-7)**
Cyclohexane, 1,1-dichloro- (8CI,9CI)	**(2108-92-1)**
Cyclohexane, 1,2-dichloro-, trans- (8CI,9CI)	**(822-86-6)**
Cyclohexane, 1,3-dichloro-, cis- (8CI,9CI)	**(24955-63-3)**
Cyclohexane, 1,2-dichloro-4-(1,2-dichloroethyl)- (9CI)	**(51962-63-1)**
Cyclohexane, dichloro(dichloropropoxy)- (9CI)	**(99308-23-3)**
Cyclohexane, 1,2-dichloro-4-ethenyl- (9CI)	**(45803-84-7)**
Cyclohexane, diethyl-	**(1331-43-7)**
Cyclohexane, diethyl- (Mixed isomers)	**(1331-43-7)**
Cyclohexane diisocyanate	**(2556-36-7)**
Cyclohexane, 1,4-diisocyanato- (9CI)	**(2556-36-7)**
1,4-Cyclohexanedimethanol	**(105-08-8)**
1,4-Cyclohexanedimethanol, hexahydro-2-oxo-	**(105-08-8)**
Cyclohexane, 1,1-dimethoxy- (9CI)	**(933-40-4)**
Cyclohexane, 1,2-dimethyl	**(583-57-3)**
Cyclohexane, 1,3-dimethyl	**(591-21-9)**
Cyclohexane, 1,4-dimethyl	**(589-90-2)**
Cyclohexane, 1,1-dimethyl- (9CI)	**(590-66-9)**
Cyclohexane, 1,2-dimethyl-, cis- (9CI)	**(2207-01-4)**
Cyclohexane, 1,2-dimethyl-, trans- (9CI)	**(6876-23-9)**
Cyclohexane, 1,3-dimethyl-, cis- (9CI)	**(638-04-0)**
Cyclohexane, 1,4-dimethyl-, trans- (9CI)	**(2207-04-7)**
Cyclohexane, 1,3-dimethyl-, trans- (8CI,9CI)	**(2207-03-6)**
Cyclohexane, 1,4-dimethylene- (7CI,8CI)	**(4982-20-1)**
Cyclohexane, (1,1-dimethylethyl)- (9CI)	**(3178-22-1)**
Cyclohexane, 1-(1,5-dimethylhexyl)-4-(4-methylpentyl)- (9CI)	**(56009-20-2)**
1,2-Cyclohexanediol	**(931-17-9)**
Cyclohexane-1,3-diol	**(504-01-8)**
1,3-Cyclohexanediol (9CI)	**(504-01-8)**
1,4-Cyclohexanediol (8CI,9CI)	**(556-48-9)**
1,2-Cyclohexanediol, 1-methyl-4-(1-methylethyl)- (9CI)	**(33669-76-0)**
1,2-Cyclohexanedione	**(765-87-7)**
1,3 Cyclohexanedione	**(504-02-9)**
Cyclohexane-1,3-dione	**(504-02-9)**
1,3-Cyclohexanedione (9CI)	**(504-02-9)**
1,4-Cyclohexanedione (9CI)	**(637-88-7)**
1,3-Cyclohexanedione, 5,5-dimethyl- (9CI)	**(126-81-8)**
1,3-Cyclohexanedione, 5-(2-(ethylthio)propyl)-2-(1-oxopropyl)	**(99422-01-2)**
1,4-Cyclohexanedione, 2,2,6-trimethyl- (6CI,7CI,8CI,9CI)	**(20547-99-3)**
Cyclohexane, 1,1-diphenyl- (7CI,8CI)	**(21113-55-3)**
Cyclohexane, dodecafluoro- (9CI)	**(355-68-0)**
Cyclohexane, dodecyl- (9CI)	**(1795-17-1)**
Cyclohexane, eicosyl- (9CI)	**(4443-55-4)**
Cyclohexane, 3,6-endoxo-	**(279-49-2)**
Cyclohexane, 1,2-epithio-	**(286-28-2)**
Cyclohexane, 1,2-epoxy-	**(286-20-4)**
Cyclohexane, 1,4-epoxy-	**(279-49-2)**
Cyclohexaneethanamine, N,α-dimethyl-, (+-)	**(101-40-6)**
Cyclohexaneethanamine, N,α-dimethyl- (9CI)	**(101-40-6)**
Cyclohexaneethanethiol, 3-mercapto-β,4-dimethyl- (9CI)	**(4802-20-4)**
Cyclohexane, 1-ethenyl-1-methyl-2,4-bis(1-methylethenyl)-, (1α,2β,4β)- (9CI)	**(33880-83-0)**
Cyclohexane, ethyl- (9CI)	**(1678-91-7)**
(Cyclohexaneethyl)amine, N-α-dimethyl-	**(101-40-6)**
Cyclohexaneethylamine, α,n-dimethyl-	**(101-40-6)**
Cyclohexaneethylamine, N-α-dimethyl- (8CI)	**(101-40-6)**
Cyclohexane, ethylmethyl- (9CI)	**(30677-34-0)**
Cyclohexane, 1-ethyl-3-methyl-, cis- (8CI,9CI)	**(19489-10-2)**
Cyclohexane, 1-ethyl-4-methyl- (8CI,9CI)	**(3728-56-1)**
Cyclohexaneglycolic acid, α-phenyl-,4-(diethylamino)-2-butynyl ester	**(5633-20-5)**
Cyclohexane, heptadecyl- (9CI)	**(19781-73-8)**
Cyclohexane, heptyl- (9CI)	**(5617-41-4)**
Cyclohexane, 1,1'-heptylidenebis- (9CI)	**(2090-15-5)**
Cyclohexane, 1,2,3,4,5,6-hexabromo- (8CI)	**(1837-91-8)**
Cyclohexane, 1,2,3,4,5,6-hexachloro	**(608-73-1)**
Cyclohexane, 1,2,3,4,5,6-hexachloro-, (1α,2α,3α,4α,5α,6α)- (9CI)	**(6108-11-8)**
Cyclohexane, 1,2,3,4,5,6-hexachloro-, (1α,2α,3α,4α,5α,6β)- (9CI)	**(6108-13-0)**
Cyclohexane, 1,2,3,4,5,6-hexachloro-, (1α,2α,3α,4α,5β,6β)- (8CI,9CI)	**(6108-12-9)**
Cyclohexane, 1,2,3,4,5,6-hexachloro-, (1α,2α,3α,4β,5β,6β)-	**(6108-10-7)**
Cyclohexane, 1,2,3,4,5,6-hexachloro-, α-	**(319-84-6)**
Cyclohexane, 1,2,3,4,5,6-hexachloro-, β-	**(319-85-7)**
Cyclohexane, 1,2,3,4,5,6-hexachloro-, trans-	**(319-85-7)**
Cyclohexane, α-1,2,3,4,5,6-hexachloro-	**(319-84-6)**
Cyclohexane, β-1,2,3,4,5,6-hexachloro-	**(319-85-7)**
Cyclohexane, δ-1,2,3,4,5,6-hexachloro-	**(319-86-8)**
Cyclohexane, hexachloro- (9CI)	**(27154-44-5)**
Cyclohexane, 1,2,3,4,5,6-hexachloro-, α-isomer	**(319-84-6)**
Cyclohexane, 1,2,3,4,5,6-hexachloro-, β-isomer	**(319-85-7)**
Cyclohexane, 1,2,3,4,5,6-hexachloro-, δ-isomer	**(319-86-8)**
Cyclohexane, 1,2,3,4,5,6-hexachloro-, γ-isomer	**(58-89-9)**
Cyclohexane, hexadecyl- (9CI)	**(6812-38-0)**
Cyclohexanehexanol (6CI,8CI,9CI)	**(4354-58-9)**
1,2,3,5/4,6-Cyclohexanehexol	**(87-89-8)**
cis-1,2,3,5-trans-4,6-Cyclohexanehexol	**(87-89-8)**
Cyclohexane, hexyl- (9CI)	**(4292-75-5)**
Cyclohexane, iodo- (9CI)	**(626-62-0)**
Cyclohexane, isobutyl- (8CI)	**(1678-98-4)**
Cyclohexane, isocyanato- (9CI)	**(3173-53-3)**
Cyclohexane, 5-isocyanato-1-(isocyanatomethyl)-1,3,3-trimethyl- (9CI)	**(4098-71-9)**
Cyclohexane, isopropyl- (8CI)	**(696-29-7)**
Cyclohexanemethanamine, 3-amino-α,α,4-trimethyl- (9CI)	**(90853-12-6)**
Cyclohexanemethanamine, 5-amino-1,3,3-trimethyl- (9CI)	**(2855-13-2)**
Cyclohexanemethanol	**(100-49-2)**
Cyclohexanemethanol, 4-hydroxy-α,α,4-trimethyl- (9CI)	**(80-53-5)**
Cyclohexanemethanol, 4-hydroxy-α,α,4-trimethyl-, monohydrate	**(2451-01-6)**
Cyclohexanemethanol, 4-hydroxy-α,α,4-trimethyl-, monohydrate, cis-	**(2451-01-6)**
Cyclohexanemethanol, α-methyl- (9CI)	**(1193-81-3)**
Cyclohexanemethanol, 2-methyl- (8CI,9CI)	**(2105-40-0)**
Cyclohexanemethanol, 4-methylene- (7CI,8CI,9CI)	**(1004-24-6)**
Cyclohexanemethanol, 4-(1-methylethyl)-, cis- (9CI)	**(13828-37-0)**
Cyclohexanemethanol, α,α,4-trimethyl- (9CI)	**(498-81-7)**
Cyclohexane, methoxy- (9CI)	**(931-56-6)**
Cyclohexane, methyl	**(108-87-2)**
Cyclohexanemethylamine, 5-amino-1,3,3-trimethyl	**(2855-13-2)**
Cyclohexane, methylene- (9CI)	**(1192-37-6)**

Cyclohexane, 1,1'-methylenebis(isocyanato- (9CI) **(28605-81-4)**
Cyclohexane, 1-methylene-3-(1-methylethenyl)-, (R)- (9CI) **(13837-95-1)**
Cyclohexane, 1-methylene-3-(1-methylethyl)-, (R)- (9CI) **(13837-71-3)**
Cyclohexane, (1-methylethyl)- (9CI) **(696-29-7)**
Cyclohexane, 1-methyl-4-(1-methylethyl)- **(99-82-1)**
Cyclohexane, 1-methyl-3-(1-methylethyl)- (9CI) **(16580-24-8)**
Cyclohexane, 1-methyl-4-(1-methylethyl)-, cis- (9CI) **(6069-98-3)**
Cyclohexane, 1-methyl-4-(1-methylethyl)-, trans- (9CI) **(1678-82-6)**
Cyclohexane, 1-methyl-4-(1-methylethylidene)- (9CI) **(1124-27-2)**
Cyclohexane, 1-methyl-4-(1-methylethyl)-, monohydroperoxy deriv. (9CI) **(26762-92-5)**
Cyclohexane, (1-methylpropyl)- (9CI) **(7058-01-7)**
Cyclohexane, (2-methylpropyl)- (9CI) **(1678-98-4)**
Cyclohexane, nitro **(1122-60-7)**
Cyclohexane, nonyl- (9CI) **(2883-02-5)**
Cyclohexane, octadecyl- (9CI) **(4445-06-1)**
Cyclohexane, octyl- (9CI) **(1795-15-9)**
Cyclohexane, (1-octyldodecyl)- (9CI) **(4443-61-2)**
Cyclohexane oxide **(286-20-4)**
Cyclohexane, 1,1'-oxybis- (9CI) **(4645-15-2)**
Cyclohexane, 1,2,3,4,5-pentabromo-6-chloro- (9CI) **(87-84-3)**
Cyclohexane, pentadecyl- (9CI) **(6006-95-7)**
Cyclohexane, pentyl- (9CI) **(4292-92-6)**
Cyclohexane, (2-(pentyloxy)ethyl)- (9CI) **(54852-75-4)**
Cyclohexane, 1,2-propadienyl- (9CI) **(5664-17-5)**
Cyclohexane, propadienyl- (8CI) **(5664-17-5)**
Cyclohexanepropionic acid, allyl ester **(2705-87-5)**
Cyclohexane, propyl- (9CI) **(1678-92-8)**
Cyclohexane, 2-propynylidene- (7CI,8CI,9CI) **(2806-45-3)**
Cyclohexanesulfamic-acid **(100-88-9)**
Cyclohexanesulfamic acid, calcium salt **(139-06-0)**
Cyclohexanesulfamic acid, calcium salt (2:1) **(139-06-0)**
Cyclohexanesulfamic acid, monopotassium salt **(7758-04-5)**
Cyclohexanesulfamic acid, monosodium salt **(139-05-9)**
Cyclohexanesulfenyl chloride (9CI) **(17797-03-4)**
Cyclohexanesulphamic acid **(100-88-9)**
Cyclohexanesulphamic acid, monosodium salt **(139-05-9)**
Cyclohexane, tetrabromodichloro- (9CI) **(30554-72-4)**
Cyclohexane, tetradecyl- (9CI) **(1795-18-2)**
Cyclohexane, 1,1,4,4-tetramethyl-2,6-bis(methylene)- (9CI) **(40482-18-6)**
Cyclohexane, 1,1'-thiobis- (9CI) **(7133-46-2)**
Cyclohexanethiol **(1569-69-3)**
Cyclohexane, tribromotrichloro- (9CI) **(30554-73-5)**
Cyclohexane, 1-(trichlorosilyl)- **(98-12-4)**
Cyclohexane, 1,2,4-trichloro-4-(1,1,2-trichloroethyl)- (9CI) **(83682-64-8)**
Cyclohexane, tridecyl- (9CI) **(6006-33-3)**
Cyclohexane, 1-trifluoromethyl-1,2,2,3,3,4,4,5,5,6,6-undecafluoro **(355-02-2)**
Cyclohexane, 1,1,3-trimethyl **(3073-66-3)**
Cyclohexane, 1,2,3-trimethyl-, (1.α.,2.α.,3.β.)- (9CI) **(7667-55-2)**
Cyclohexane, 1,2,3-trimethyl-, cis-1,2,trans-1,3- (8CI) **(7667-55-2)**
Cyclohexane, 1,3,5-trimethyl-, cis- (8CI) **(1795-27-3)**
Cyclohexane, trimethyl- (9CI) **(30498-63-6)**
Cyclohexane, 1,2,3-trimethyl- (8CI,9CI) **(1678-97-3)**
Cyclohexane, 1,3,5-trimethyl-, (1α,3α,5α)- (9CI) **(1795-27-3)**
Cyclohexane, 1,3,5-trimethyl-2-octadecyl- (9CI) **(55282-34-3)**
1,3,5-Cyclohexanetriol, cyclic orthoformate **(281-32-3)**
Cyclohexane, undecyl- (9CI) **(54105-66-7)**
Cyclohexanoic acid **(98-89-5)**
Cyclohexanol (ACGIH,OSHA) **(108-93-0)**
Cyclohexanol, acetate **(622-45-7)**
Cyclohexanol, 3-allyl-, propionate **(2705-87-5)**
Cyclohexanol, 2-amino- (8CI,9CI) **(6850-38-0)**
Cyclohexanolazetat (German) **(622-45-7)**
Cyclohexanol, 4-bromo-, trans- **(32388-22-0)**
Cyclohexanol, 4-tert-butyl **(98-52-2)**
Cyclohexanol, 4-tert-butyl-, acetate **(32210-23-4)**
Cyclohexanol, 2-(p-tert-butylphenoxy)- **(1942-71-8)**
Cyclohexanol, 2-chloro- **(1561-86-0)**
Cyclohexanol, 2,6-dimethyl- (8CI,9CI) **(5337-72-4)**
Cyclohexanol, 3,5-dimethyl- (8CI,9CI) **(5441-52-1)**
Cyclohexanol, 2-(1,1-dimethylethyl)- (9CI) **(13491-79-7)**
Cyclohexanol, 2-(1,1-dimethylethyl)-, trans- (9CI) **(5448-22-6)**
Cyclohexanol, 2-(1,1-dimethylethyl)-, acetate (9CI) **(88-41-5)**
Cyclohexanol, 1-((1,1-dimethylethyl)azo)- (9CI) **(54043-65-1)**
Cyclohexanol, 2-(4-(1,1-dimethylethyl)phenoxy)- (9CI) **(1942-71-8)**
Cyclohexanol, 1,1'-dioxybis- (9CI) **(2407-94-5)**
Cyclohexanol, 1,1'-dioxydi- **(2407-94-5)**
Cyclohexanol, 1-ethyl **(1940-18-7)**
Cyclohexanol, 2-ethyl- (8CI,9CI) **(3760-20-1)**
Cyclohexanol, 1-ethynyl-, carbamate **(126-52-3)**
Cyclohexanol, 1-((1-hydroperoxycyclohexyl)dioxy) **(78-18-2)**
Cyclohexanol, p-isopropyl **(4621-04-9)**
Cyclohexanol, 4,4'-isopropylidenedi- (8CI) **(80-04-6)**
Cyclohexanol, 2-isopropyl-5-methyl- **(89-78-1)**
Cyclohexanol, m-methyl **(591-23-1)**
Cyclohexanol, methyl **(25639-42-3)**
Cyclohexanol, o-methyl **(583-59-5)**
Cyclohexanol, 1-methyl- (9CI) **(590-67-0)**
Cyclohexanol, 3-methyl-, cis- (9CI) **(5454-79-5)**
Cyclohexanol, 4-methyl- (9CI) **(589-91-3)**
Cyclohexanol, 2-methyl-, cis- (8CI,9CI) **(7443-70-1)**
Cyclohexanol, 2-methyl-, trans- (8CI,9CI) **(7443-52-9)**
Cyclohexanol, 3-methyl-, trans- (8CI,9CI) **(7443-55-2)**
Cyclohexanol, 4-methyl-, cis- (8CI,9CI) **(7731-28-4)**
Cyclohexanol, 4-methyl-, trans- (8CI,9CI) **(7731-29-5)**
Cyclohexanol, 4,4'-(1-methylethylidene)bis- (9CI) **(80-04-6)**
Cyclohexanol, 4,4'-(1-methylethylidene)bis-, Polymer with (chloromethyl)oxirane (9CI) **(30583-72-3)**
Cyclohexanol, 1-methyl-4-(1-methylethenyl)- (9CI) **(138-87-4)**
Cyclohexanol, 2-methyl-5-(1-methylethenyl)-, (1α,2β,5α)- (9CI) **(38049-26-2)**
Cyclohexanol, 5-methyl-2-(1-methylethyl)- **(1490-04-6)**
Cyclohexanol, 5-methyl-2-(1-methylethyl)-, (1-α,2-β,5-α) **(15356-70-4)**
Cyclohexanol, 5-methyl-2-(1-methylethyl)-, acetate **(16409-45-3)**
Cyclohexanol, 5-methyl-2-(1-methylethyl)-, (1R-(1α,2β,5α))- (9CI) **(2216-51-5)**
Cyclohexanol, 2-phenyl **(1444-64-0)**
Cyclohexanol, 3,3,5-trimethyl **(116-02-9)**
Cyclohexanol, trimethyl- (8CI,9CI) **(1321-60-4)**
Cyclohexanon (Dutch) **(108-94-1)**
1,3-Cyclohexanone **(504-02-9)**
Cyclohexanone (ACGIH,OSHA) [UN 1915] **(108-94-1)**
Cyclohexanone, 2-acetyl- (9CI) **(874-23-7)**
Cyclohexanone, p-tert-butyl **(98-53-3)**
Cyclohexanone, 2-(o-chlorophenyl)-2-(methylamino)-, hydrochloride **(1867-66-9)**
Cyclohexanone, 2-cyclohexylidene- (9CI) **(1011-12-7)**
Cyclohexanone, 2,6-dimethyl- (9CI) **(2816-57-1)**
Cyclohexanone, dimethyl acetal (8CI) **(933-40-4)**
Cyclohexanone, 4-(1,1-dimethylpropyl)- **(16587-71-6)**
Cyclohexanone, 2-(ethoxymethylene)-4-methyl- (7CI,8CI,9CI) **(704-76-7)**
Cyclohexanone, 2-ethyl- (8CI,9CI) **(4423-94-3)**
Cyclohexanone, 2-hydroxy- (9CI) **(533-60-8)**
Cyclohexanone, 2-(hydroxymethyl)- (6CI,7CI,8CI,9CI) **(5331-08-8)**
Cyclohexanone, 2-isopropylidene- (6CI,7CI,8CI) **(13747-73-4)**
Cyclohexanone, 2-methyl **(583-60-8)**
Cyclohexanone, 3-methyl **(591-24-2)**
Cyclohexanone, 4-methyl **(589-92-4)**
Cyclohexanone, methyl **(1331-22-2)**
Cyclohexanone, 2-(1-methylethylidene)- (9CI) **(13747-73-4)**
Cyclohexanone, 2-methyl-5-(1-methylethenyl)-, (2R-trans)- (9CI) **(5524-05-0)**
Cyclohexanone, 5-methyl-2-(1-methylethyl)-, trans- **(89-80-5)**
Cyclohexanone, 2-methyl-5-(1-methylethyl)-, trans- (9CI) **(499-70-7)**
Cyclohexanone, 5-methyl-2-(1-methylethyl)- (9CI) **(10458-14-7)**
Cyclohexanone, 5-methyl-2-(1-methylethyl)-, (Z)- **(491-07-6)**

Cyclohexanone iso-oxime (105-60-2)
Cyclohexanone, oxime (100-64-1)
Cyclohexanone, 4-tert-pentyl (16587-71-6)
Cyclohexanone peroxide (78-18-2)
Cyclohexanone peroxide (As a paste with not more than 9% by weight active oxygen) (4904-55-6)
Cyclohexanone peroxide (In solution with not more than 9% by weight active oxygen) (4904-55-6)
Cyclohexanone peroxide (Not over 50% peroxide) (4904-55-6)
Cyclohexanone peroxide and di-(1-hydroxycyclohexyl)peroxide mixture (4904-55-6)
Cyclohexanone peroxide (50 to 85% Peroxide) (4904-55-6)
Cyclohexanone, 2-propyl- (9CI) (94-65-5)
Cyclohexanone, 3,3,5-trimethyl- (873-94-9)
Cyclohexanone, 2,2,6-trimethyl- (9CI) (2408-37-9)
Cyclohexanone, 2,6-vanillylidene (579-23-7)
Cyclohexanyl acetate (622-45-7)
Cyclohexasiloxane, dodecamethyl- (9CI) (540-97-6)
Cyclohexasiloxane, 2,2,4,6,8,8,10,12-octamethyl-4,6,10,12-tetraphenyl- (9CI) (60573-48-0)
Cyclohexatriene (71-43-2)
Cyclohexene (ACGIH,OSHA) [UN 2256] (110-83-8)
3-Cyclohexene-1-acetaldehyde (8CI,9CI) (24480-99-7)
1-Cyclohexene-1-acetonitrile (9CI) (6975-71-9)
Cyclohexene, 3-(bromomethyl)- (9CI) (34825-93-9)
Cyclohexene, 4-tert-butyl- (8CI) (2228-98-0)
4-Cyclohexene-1-carboxaldehyde (100-50-5)
Cyclohexenecarboxaldehyde (1321-16-0)
3-Cyclohexene-1-carboxaldehyde (8CI,9CI) (100-50-5)
3-Cyclohexene-1-carboxaldehyde, dimethyl- (VAN)(8CI) (27939-60-2)
3-Cyclohexene-1-carboxaldehyde, 4-(4-hydroxy-4-methylpentyl)-(9CI) (31906-04-4)
1-Cyclohexene-1-carboxaldehyde, 4-isopropenyl (2111-75-3)
1-Cyclohexene-1-carboxaldehyde, 4-(1-methylethenyl)- (9CI) (2111-75-3)
3-Cyclohexene-1-carboxaldehyde, 4-(5-methyl-3-penten-1-yl)- (37677-14-8)
3-Cyclohexene-1-carboxaldehyde, 4-(4-methyl-3-pentenyl)- (9CI) (37677-14-8)
1-Cyclohexene-1-carboxaldehyde, 2,6,6-trimethyl- (432-25-7)
3-Cyclohexene-1-carboxaldehyde, 1,3,4-trimethyl- (9CI) (40702-26-9)
3-Cyclohexene-1-carboxaldehyde, 2,4,6-trimethyl- (8CI) (1423-46-7)
3-Cyclohexene-1-carboxylic acid (4771-80-6)
3-Cyclohexene-1-carboxylic acid, 3-cyclohexen-1-yl methyl ester (2611-00-9)
3-Cyclohexene-1-carboxylic acid, 3-cyclohexen-1-ylmethyl ester (9CI) (2611-00-9)
1-Cyclohexene-1-carboxylic acid, 4-(1,5-dimethyl-3-oxohexyl)-(8CI,9CI) (VAN) (17844-07-4)
3-Cyclohexene-1-carboxylic acid, 4-methyl- (8CI,9CI) (4342-60-3)
1-Cyclohexene-1-carboxylic acid, 3,4,5-trihydroxy (138-59-0)
Cyclohexene, 3-chloro (2441-97-6)
Cyclohexene, 1-chloro- (9CI) (930-66-5)
Cyclohexene, 4-chloro- (6CI,7CI,8CI,9CI) (930-65-4)
Cyclohexene, 1-chloro-4-(1-chloroethenyl)- (9CI) (13547-06-3)
Cyclohexene, 1-chloro-5-(1-chloroethenyl)- (9CI) (13547-07-4)
Cyclohexene, 1-chloro-4-(1-chlorovinyl)- (8CI) (13547-06-3)
Cyclohexene, 1-chloro-5-(1-chlorovinyl)- (8CI) (13547-07-4)
4-Cyclohexene-1,2-dicarboximide (85-40-5)
4-Cyclohexene-1,2-dicarboximide, N-propyl- (8CI) (2021-20-7)
4-Cyclohexene-1,2-dicarboximide, N-((1,1,2,2-tetrachloroethyl)-thio) (2425-06-1)
4-Cyclohexene-1,2-dicarboximide, N-((1,1,2,2-tetrachloroethyl)-thio)-, cis (2939-80-2)
4-Cyclohexene-1,2-dicarboximide, N-(trichloromethyl)thio (133-06-2)
cis-4-Cyclohexene-1,2-dicarboxylic acid (2305-26-2)
4-Cyclohexene-1,2-dicarboxylic acid (9CI) (88-98-2)
4-Cyclohexene-1,2-dicarboxylic acid, cis- (9CI) (2305-26-2)
1-Cyclohexene-1,2-dicarboxylic acid (8CI,9CI) (635-08-5)
4-Cyclohexene-1,2-dicarboxylic anhydride (85-43-8)
Cyclohexene, 1,4-dichloro-4-(1-chloroethenyl)- (9CI) (83682-63-7)
Cyclohexene, 1,4-dichloro-4-ethenyl- (9CI) (65122-21-6)
Cyclohexene, 1,4-dichloro-4-vinyl- (6CI) (65122-21-6)
Cyclohexene, 1,2-dimethyl- (9CI) (1674-10-8)
Cyclohexene, 4,4-dimethyl- (8CI,9CI) (14072-86-7)
Cyclohexene, 1-(1,1-dimethylethoxy)-2-methyl- (9CI) (40648-26-8)
Cyclohexene, 4-(1,1-dimethylethyl)- (9CI) (2228-98-0)
2-Cyclohexene-1,4-dione, 2-hydroxy-3,5,5-trimethyl- (9CI) (35692-98-9)
2-Cyclohexene-1,4-dione, 2-methoxy-3,5,5-trimethyl- (9CI) (41654-27-7)
Cyclohexene episulfide (286-28-2)
Cyclohexene epoxide (286-20-4)
1-Cyclohexene-1-ethanamine (9CI) (3399-73-3)
Cyclohexene, 3,6-endo-(1,2-ethanediyl)- (931-64-6)
Cyclohexene, 4-ethenyl- (100-40-3)
Cyclohexene, ethenyl- (9CI) (25168-07-4)
Cyclohexene, 3-isobutyl- (6CI,7CI,8CI) (4104-56-7)
3-Cyclohexene-1-methanol (9CI) (1679-51-2)
Cyclohex-1-ene-1-methanol, 4-(1-methylethenyl)- (536-59-4)
3-Cyclohexene-1-methanol, α,α,4-trimethyl- (9CI) (98-55-5)
3-Cyclohexene-1-methanol, α,α,4-trimethyl-, (R)- (9CI) (7785-53-7)
3-Cyclohexene-1-methanol, α,α,4-trimethyl-, (S)- (9CI) (10482-56-1)
Cyclohexene, 1-methyl- (9CI) (591-49-1)
Cyclohexene, 3-methyl- (9CI) (591-48-0)
Cyclohexene, 4-methyl- (9CI) (591-47-9)
Cyclohexene, methyl- (9CI) (1335-86-0)
Cyclohexene, 3-(3-methyl-1-butenyl)-, (E)- (9CI) (56030-49-0)
Cyclohexene, 3-methylene-6-(1-methylethyl)- (9CI) (555-10-2)
Cyclohexene, 1-methyl-3-(1-methylethenyl)-, (+-)- (9CI) (499-03-6)
Cyclohexene, 1-methyl-4-(1-methylethenyl)-, (+-)- (9CI) (7705-14-8)
Cyclohexene, 5-methyl-3-(1-methylethenyl)- (9CI) (86853-03-4)
Cyclohexene, 1-methyl-4-(1-methylethenyl)-, (R) (5989-27-5)
Cyclohexene, 1-methyl-4-(1-methylethenyl)-, (S)- (5989-54-8)
Cyclohexene, 1-methyl-4-(1-methylethenyl)-, Homopolymer (9CI) (9003-73-0)
Cyclohexene, 1-methyl-4-(1-methylethyl)- (9CI) (5502-88-5)
Cyclohexene, 1-methyl-4-(1-methylethylidene)- (586-62-9)
Cyclohexene, 3-(2-methylpropyl)- (9CI) (4104-56-7)
2-Cyclohexene-1-octanoic acid, 5(or 6)-carboxy-4-hexyl- (9CI) (53980-88-4)
2-Cyclohexene-1-octanoic acid, 5(or 6)-carboxy-4-hexyl-, potassium salt (9CI) (68127-33-3)
1,2-Cyclohexene oxide (286-20-4)
Cyclohexene oxide (286-20-4)
Cyclohexene-1-oxide (286-20-4)
Cyclohexene, 1,3,4,5,6-pentachloro-, γ (319-94-8)
Cyclohexene, 1,3,4,5,6-pentachloro-, (3-α,4-β,5-β,6-α)- (9CI) (319-94-8)
3-Cyclohexene-1-propanol, 4-methyl-γ-methylene-, acetate (9CI) (6819-19-8)
Cyclohexene, 3-(2-propynyl)- (9CI) (55956-43-9)
Cyclohexene sulfide (286-28-2)
Cyclohexene, sulfide (286-28-2)
Cyclohexene, 3,4,5,6-tetrachloro- (8CI,9CI) (1782-00-9)
Cyclohexene, 1,1'-(3,7,12,16-tetramethyl-1,3,5,7,9,11,13,15,17-octadecanonaene-1,18-diyl)bis(2,6,6-trimethyl-, (all-E)- (7235-40-7)
Cyclohexene, 4-(trichlorosilyl)- (10137-69-6)
1-Cyclohexene, 4-vinyl (100-40-3)
Cyclohexene, 1-vinyl (2622-21-1)
2-Cyclohexen-1-ol (9CI) (822-67-3)
3-Cyclohexen-1-ol, 1-(1,5-dimethyl-4-hexenyl)-4-methyl-(8CI,9CI) (VAN) (15352-77-9)
2-Cyclohexen-1-ol, 1-methyl- (8CI,9CI) (23758-27-2)
3-Cyclohexen-1-ol, 1-methyl- (7CI,9CI) (33061-16-4)
2-Cyclohexen-1-ol, 2-methyl-5-(1-methylethenyl)-, acetate (97-42-7)
2-Cyclohexen-1-ol, 3-methyl-6-(1-methylethyl)- (9CI) (491-04-3)
3-Cyclohexen-1-ol, 1-methyl-4-(1-methylethyl)- (9CI) (586-82-3)
3-Cyclohexen-1-ol, 4-methyl-1-(1-methylethyl)- (9CI) (562-74-3)
3-Cyclohexen-1-ol, 4-methyl-1-(1-methylethyl)-, acetate (9CI) (4821-04-9)
2-Cyclohexen-1-one (930-68-7)
2-Cyclohexenone (930-68-7)
Cyclohexenone (930-68-7)
2-Cyclohexen-1-one, 3,5-dimethyl- (8CI,9CI) (1123-09-7)

Name	CAS No.
2-Cyclohexen-1-one, 2-(1-(ethoxyimino)butyl)-5-(2-(ethylthio)-propyl)-3-hydroxy	(74051-80-2)
2-Cyclohexen-1-one, 2-(1-(ethoxyimino)butyl)-5-(2-(ethylthio)-propyl)-3-hydroxy- (9CI)	(74051-80-2)
2-Cyclohexen-1-one, 5-(2-(ethylthio)propyl)-3-hydroxy-2-(1-oxobutyl)- (9CI)	(79419-43-5)
2-Cyclohexen-1-one, 3-hexyl-5-(3,4-(methylenedioxy)phenyl)	(119-89-1)
2-Cyclohexen-1-one, 3-hydroxy-5,5-dimethyl-, dimethylcarbamate	(122-15-6)
2-Cyclohexen-1-one, 3-methyl	(1193-18-6)
2-Cyclohexen-1-one, 2-methyl-5-(1-methylethenyl)-, (R)- (9CI)	(6485-40-1)
2-Cyclohexen-1-one, 2-methyl-5-(1-methylethenyl)-, (S)-	(2244-16-8)
2-Cyclohexen-1-one, 2-methyl-5-(1-methylethenyl)-, oxime (9CI)	(31198-76-2)
2-Cyclohexen-1-one, 3-methyl-6-(1-methylethyl)-	(89-81-6)
2-Cyclohexen-1-one, 6-methyl-3-(1-methylethyl)-	(499-74-1)
2-Cyclohexen-1-one, 3-methyl-6-(1-methylethyl)-, (R)- (9CI)	(4573-50-6)
2-Cyclohexen-1-one, 2,3,4,4,5,5,6,6-octachloro- (8CI,9CI)	(4024-81-1)
2-Cyclohexen-1-one, octachloro- (6CI,7CI)	(4024-81-1)
2-Cyclohexen-1-one, 2(or 4)-chloro-3,5,5-trimethyl- (9CI)	(72175-27-0)
2-Cyclohexen-1-one, 2,5,6-trimethyl- (9CI)	(20030-30-2)
2-Cyclohexen-1-one, 3,5,5-trimethyl- (9CI)	(78-59-1)
2-Cyclohexen-1-one, 4,4,5-trimethyl- (8CI,9CI)	(17429-29-7)
3-Cyclohexen-1-one, 3,5,5-trimethyl- (8CI,9CI)	(471-01-2)
2-Cyclohexen-1-one, 2,4,4-trimethyl-3-(3-oxo-1-butenyl)-(8CI,9CI)	(27185-77-9)
Cyclohexenylacetonitrile	(6975-71-9)
3-Cyclohexenyl chloride	(2441-97-6)
3-Cyclohexenyl 3-cyclohexene 1-carboxylate	(2611-00-9)
5-(1-Cyclohexen-1-yl)-1,5-dimethyl-2,4,6(1H,3H,5H)-pyrimidine-trione	(56-29-1)
5-(1-Cyclohexen-1-yl)-1,5-dimethylbarbituric acid	(56-29-1)
Cyclohexenylethylamine	(3399-73-3)
5-(1-Cyclohexen-1-yl)-5-ethylbarbituric acid	(52-31-3)
5-(1-Cyclohexenyl)-5-ethylbarbituric acid	(52-31-3)
Cyclohexenyl-ethyl barbituric acid	(52-31-3)
Cyclohexenylethylene	(100-40-3)
5-(1-Cyclohexen-1-yl)-5-ethyl-2,4,6(1H,3H,5H)-pyrimidinetrione	(52-31-3)
3-Cyclohexenylmethyl 3-cyclohexenecarboxylate	(2611-00-9)
5-(1-Cyclohexenyl-1)-1-methyl-5-methylbarbituric acid	(56-29-1)
5-(δ-1,2-Cyclohexenyl)-5-methyl-N-methyl-barbitursaeure (German)	(56-29-1)
1-(1-Cyclohexen-1-yl)-2-propanone	(768-50-3)
3-Cyclohexenyltrichlorosilane	(10137-69-6)
Cyclohexenyl trichlorosilane	(10137-69-6)
Cyclohexenyltrichlorosilane	(10137-69-6)
Cycloheximide	(66-81-9)
N-Cyclohexylacetamide	(1124-53-4)
Cyclohexyl acetate [UN 2243]	(622-45-7)
Cyclohexylacetic acid	(5292-21-7)
Cyclohexyl alcohol	(108-93-0)
Cyclohexylallene	(5664-17-5)
Cyclohexylamidosulphuric acid	(100-88-9)
Cyclohexylamine (ACGIH,OSHA) [UN 2357]	(108-91-8)
Cyclohexylamine acetate	(58695-41-3)
Cyclohexylamine, 1-aminomethyl	(5062-67-9)
Cyclohexylamine, N-(3-aminopropyl)-	(3312-60-5)
Cyclohexylamine, N-butyl	(10108-56-2)
Cyclohexylamine, N,N-diethyl- (8CI)	(91-65-6)
Cyclohexylamine, N,N-dimethyl	(98-94-2)
Cyclohexylamine, N-ethyl	(5459-93-8)
Cyclohexylamine, N-(2-ethylhexyl)-	(5432-61-1)
Cyclohexylamine, N-ethyl-1-phenyl- (8CI)	(2201-15-2)
Cyclohexylamine, N-methyl	(100-60-7)
Cyclohexylamine, 2,4'-methylenebis-	(24650-10-0)
Cyclohexylamine, 4,4'-methylenebis	(1761-71-3)
Cyclohexylamine, N-propyl- (8CI)	(3592-81-2)
Cyclohexylaminesulphonic acid	(100-88-9)
1-(Cyclohexylamino)-9,10-anthracenedione	(1096-48-6)
(2-Cyclohexylamino)ethanesulfonic acid, sodium salt	(3076-05-9)
2-(Cyclohexylamino)ethanol	(2842-38-8)
2-(Cyclohexylamino)ethanol Aerosol	(2842-38-8)
N-Cyclohexyl-N'-(3-amino-4-methylbenzenesulfonyl)urea	(565-33-3)
1-Cyclohexylamino-2-propanol	(103-00-4)
Cyclohexylammonium benzoate	(3129-92-8)
N-Cyclohexyl-2-benzothiazolesulfenamide	(95-33-0)
Cyclohexyl bromide	(108-85-0)
1-Cyclohexylbutane	(1678-93-9)
Cyclohexyl butyl phthalate	(84-64-0)
Cyclohexylcarbinol	(100-49-2)
2-Cyclohexylcarbonyl-1,2,3,6,7,11b-hexahydro-4H-pyrazino-(2,1-a)isoquinolin-4-one	(55268-74-1)
Cyclohexylcarboxylic acid	(98-89-5)
Cyclohexyl chloride	(542-18-7)
N-Cyclohexylcyclohexanamine	(101-83-7)
Cyclohexylcyclohexane	(92-51-3)
2-Cyclohexylcyclohexanol	(6531-86-8)
2-Cyclohexylcyclohexanone	(90-42-6)
1-Cyclohexyldecane	(1795-16-0)
Cyclohexyldiethylamine	(91-65-6)
Cyclohexyldimethylamine	(98-94-2)
N-Cyclohexyldimethylamine	(98-94-2)
3-Cyclohexyl-6-(dimethylamino)-1-methyl-1,3,5-triazine-2,4(1H,3H)-dione	(51235-04-2)
3-Cyclohexyl-6-(dimethylamino)-1-methyl-s-triazine-2,4(1H,3H)-dione	(51235-04-2)
2-Cyclohexyl-4,6-dinitrofenol (Dutch)	(131-89-5)
2-Cyclohexyl-4,6-dinitrophenol	(131-89-5)
6-Cyclohexyl-2,4-dinitrophenol	(131-89-5)
Cyclohexyl disulfide	(2550-40-5)
1-Cyclohexyldodecane	(1795-17-1)
1,2-Cyclohexylenediaminetetraacetic acid	(482-54-2)
(1,2-Cyclohexylenedinitrilo)tetraacetic acid	(482-54-2)
Cyclohexylene oxide	(286-20-4)
1,2-Cyclohexylene sulfide	(286-28-2)
Cyclohexylester kyseliny octove (Czech)	(622-45-7)
2-Cyclohexylethanol	(4442-79-9)
Cyclohexyl ether (8CI)	(4645-15-2)
Cyclohexylethyl alcohol	(4442-79-9)
1-Cyclohexylheptane	(5617-41-4)
1-Cyclohexylhexadecane	(6812-38-0)
1-Cyclohexylhexane	(4292-75-5)
2-Cyclohexylidenecyclohexanone	(1011-12-7)
2,2'-Cyclohexyliminodiethanol	(4500-29-2)
2,2'-Cyclohexyliminodiethanol aerosol	(4500-29-2)
Cyclohexyliodide	(626-62-0)
Cyclohexyl isocyanate [UN 2488]	(3173-53-3)
Cyclohexyl-isothiocyanat (German)	(1122-82-3)
Cyclohexylmagnesium chloride	(931-51-1)
Cyclohexyl methacrylate	(101-43-9)
Cyclohexylmethane	(108-87-2)
Cyclohexylmethanol	(100-49-2)
N-Cyclohexyl-N-methoxy-2,5-dimethyl-3-furancarboxamide	(60568-05-0)
Cyclohexylmethylamine	(100-60-7)
1-Cyclohexyl-2-methylaminopropan (German)	(101-40-6)
1-Cyclohexyl-2-(methylamino)propane	(101-40-6)
N-Cyclohexyl-4-methylbenzenesulfonamide	(80-30-8)
Cyclohexyl 4-methylbenzenesulfonate	(953-91-3)
Cyclohexylmethylcarbinol	(1193-81-3)
Cyclohexyl methyl ether	(931-56-6)
1-Cyclohexyl-N-methyl-2-propanamine	(101-40-6)
Cyclohexyl 2-methyl-2-propenoate	(101-43-9)
1-Cyclohexylnonane	(2883-02-5)
1-Cyclohexyloctadecane	(4445-06-1)
2-(2-(Cyclohexyloxy)ethoxy)ethanol	(25961-84-6)
1-Cyclohexylpentadecane	(6006-95-7)
1-Cyclohexylpentane	(4292-92-6)

2-Cyclohexylphenol	**(119-42-6)**
3-Cyclohexylphenol	**(1943-95-9)**
4-Cyclohexylphenol	**(1131-60-8)**
Cyclohexylphenol	**(26570-85-4)**
m-Cyclohexylphenol	**(1943-95-9)**
o-Cyclohexylphenol	**(119-42-6)**
N-Cyclohexyl-N'-phenyl-p-phenylendiamine	**(101-87-1)**
1-Cyclohexyl-1,2-propadiene	**(5664-17-5)**
N-Cyclohexyl-1,3-propanediamine	**(3312-60-5)**
N-Cyclohexylpropylene 1,3-diamine	**(3312-60-5)**
N-Cyclohexylsulfenylphthalimide	**(17796-82-6)**
Cyclohexyl sulfide (8CI)	**(7133-46-2)**
Cyclohexylsulphamate sodium	**(139-05-9)**
Cyclohexylsulphamic acid	**(100-88-9)**
N-Cyclohexylsulphamic acid	**(100-88-9)**
Cyclohexylsulphamic acid, calcium salt	**(139-06-0)**
Cyclohexylsulphamic acid, monosodium salt	**(139-05-9)**
1-Cyclohexyltetradecane	**(1795-18-2)**
N-(Cyclohexylthio)phthalimide	**(17796-82-6)**
N-Cyclohexyl-p-toluenesulfonamide	**(80-30-8)**
Cyclohexyl tosylate	**(953-91-3)**
Cyclohexyl trichlorosilane	**(98-12-4)**
Cyclohexyltrichlorosilane	**(98-12-4)**
1-Cyclohexyltridecane	**(6006-33-3)**
3-Cyclohexyl-5,6-trimethyleneuracil	**(2164-08-1)**
3-Cyclohexyl-5,6-trimethylenuracil (German)	**(2164-08-1)**
Cyclol	**(95-12-5)**
Cyclol acrylate	**(95-39-6)**
Cycloleucin	**(52-52-8)**
Cycloleucine	**(52-52-8)**
Cycloleucyl-N-methylalanylglycyl-N-methyl dehydrophenyl-alanine	**(28540-82-1)**
Cyclolyt	**(456-59-7)**
Cyclomandol	**(456-59-7)**
Cyclomethicone	**(69430-24-6)**
Cyclomethone	**(126-81-8)**
Cyclomide LM	**(142-78-9)**
Cyclomorph	**(31717-87-0)**
Cyclomycin	**(60-54-8)**
Cyclomycin	**(68-41-7)**
Cyclon	**(74-90-8)**
Cyclonal	**(56-29-1)**
Cyclone B	**(74-90-8)**
Cyclonite (ACGIH,OSHA)	**(121-82-4)**
Cyclonol	**(116-02-9)**
1H-Cyclonona(1,2-c:5,6-c')difuran-1,3,6,8(4H)-tetrone, 10-((3,6-dihydro-6-oxo-2H-pyran-2-yl) hydroxymethyl)-5,9,10,11-tetrahydro-4-hydroxy-5-(1-hydroxyheptyl)	**(21794-01-4)**
1H-Cyclonona(c)furan-6,7-dicarboxylic anhydride, 9-((3,6-dihydro-6-oxo-2H-pyran-2-yl) hydroxymethyl)-3,4,5,8,9,10-hexahydro-1,5-dihydroxy-4-(1-hydroxyheptyl)-3-oxo	**(22467-31-8)**
Cyclononasiloxane, octadecamethyl- (8CI,9CI)	**(556-71-8)**
1,5-Cyclooctadiene	**(1552-12-1)**
Cycloocta-1,5-diene	**(111-78-4)**
cis,cis-Cycloocta-1,5-diene	**(1552-12-1)**
1,3-Cyclooctadiene (9CI)	**(1700-10-3)**
1,5-Cyclooctadiene (9CI)	**(111-78-4)**
1,5-Cyclooctadiene (Z,Z)	**(1552-12-1)**
1,4-Cyclooctadiene, 1,5-dichloro- (9CI)	**(83682-65-9)**
Cyclooctafluorobutane	**(115-25-3)**
Cyclooctane (9CI)	**(292-64-8)**
Cyclooctane, ethyl- (6CI,7CI,8CI,9CI)	**(13152-02-8)**
Cyclooctanemethanol, .α.,.α.-dimethyl- (8CI)	**(16624-06-9)**
Cyclooctane, methyl- (6CI,7CI,8CI,9CI)	**(1502-38-1)**
Cyclooctanepropanol (8CI,9CI)	**(16782-30-2)**
Cyclooctanol (9CI)	**(696-71-9)**
Cyclooctanone	**(502-49-8)**
Cyclooctasiloxane, hexadecamethyl- (8CI,9CI)	**(556-68-3)**
1,3,5,7-Cyclooctatetraene	**(629-20-9)**
Cyclooctatetraene [UN 2358]	**(629-20-9)**
Cyclooctatriene (6CI,8CI,9CI)	**(29759-77-1)**
5H-Cyclooct(b)indole, 5-(3-(dimethylamino)propyl)-6,7,8,9,10,11-hexahydro	**(5560-72-5)**
Cyclooctene (9CI)	**(931-88-4)**
3-Cyclooctyl-1,1-dimethylharnstoff (German)	**(2163-69-1)**
3-Cyclooctyl-1,1-dimethylurea	**(2163-69-1)**
N-Cyclooctyl-N',N'-dimethylurea	**(2163-69-1)**
3-Cyclooctyl-1,1-dimethyl urea mixed with butynyl-3n-3-chlorophenylcarbamate	**(8015-55-2)**
Cyclooxabutane	**(503-30-0)**
Cyclopamine	**(4449-51-8)**
Cyclopan	**(56-29-1)**
Cyclopar	**(64-75-5)**
(-)-Cyclopenin	**(20007-87-8)**
Cyclopenin	**(20007-87-8)**
(-)-Cyclopenine	**(20007-87-8)**
Cyclopenine (8CI)	**(20007-87-8)**
Cyclopenta(cd)perylenone (9CI)	**(84665-39-4)**
Cyclopenta(cd)pyrene	**(27208-37-3)**
Cyclopenta(c)furo(3',2':4,5)furo(2,3-h)(1)benzopyran-1,11-dione, 2,3,6a,8,9,9a-hexahydro- 9a-hydroxy-4-methoxy	**(6885-57-0)**
Cyclopenta(c)furo(3',2':4,5)furo(2,3-h)(1)benzopyran-1,11-dione, 2,3,6a-α,8,9,9a- α-hexahydro-4-methoxy	**(7220-81-7)**
Cyclopenta(c)furo(3',2':4,5)furo(2,3-h)(1)benzopyran-1,11-dione, 2,3,6a,9a-tetrahydro- 9a-hydroxy-4-methoxy	**(6795-23-9)**
Cyclopenta(c)furo(3',2':4,5)furo(2,3-h)(1)benzopyran-1,11-dione, 2,3,6a,9a-tetrahydro- 4-methoxy	**(1162-65-8)**
Cyclopenta(c)furo(3',2':4,5)furo(2,3-h)(1)benzopyran-11(1H)-one, 2,3,6a,9a-tetrahydro-1,9a- dihydroxy-4-methoxy-, (6ar-cis)	**(64330-03-6)**
Cyclopentacycloheptene	**(275-51-4)**
2H-Cyclopentacycloocten-2-one, decahydro-3a-methyl-, trans-(9CI)	**(55103-65-6)**
Cyclopentadecanolide	**(106-02-5)**
Cyclopentadecanone	**(502-72-7)**
Cyclopentadecanone, 3-methyl	**(541-91-3)**
4H-Cyclopenta(def)chrysen-4-one	**(86853-91-0)**
4H-Cyclopenta(def)phenanthrene	**(203-64-5)**
4H-Cyclopenta(def)phenanthren-4-one (8CI,9CI)	**(5737-13-3)**
4H-Cyclopenta(def)triphenylen-4-one (9CI)	**(86853-90-9)**
Cyclopenta(de)naphthalene	**(208-96-8)**
1,3-Cyclopentadiene	**(542-92-7)**
Cyclopentadiene (ACGIH,OSHA)	**(542-92-7)**
1,3-Cyclopentadiene-1-carboxylic acid, 6-methylheptyl ester (8CI)	**(16219-25-3)**
1,3-Cyclopentadiene, 5-chloro- (9CI)	**(41851-50-7)**
1,3-Cyclopentadiene, dimer	**(77-73-6)**
1,3-Cyclopentadiene, 1,2,3,4,5,5-hexachloro	**(77-47-4)**
Cyclopentadiene, hexachloro-, dimer	**(2385-85-5)**
1,3-Cyclopentadiene, methyl	**(26519-91-5)**
1,3-Cyclopentadiene, 1,2,3,4,5-pentachloro	**(25329-35-5)**
Cyclopentadiene polymers	**(25568-84-7)**
1,3-Cyclopentadiene, 1,2,3,4-tetrachloro	**(695-77-2)**
pi-Cyclopentadienyl Compd. with nickel	**(1271-28-9)**
Cyclopentadienylmanganese tricarbonyl	**(12079-65-1)**
Cyclopenta(g)-2-benzopyran, 1,3,4,6,7,8-hexahydro-4,6,6,7,8,8-hexamethyl	**(1222-05-5)**
1H-Cyclopenta(ghi)perylen-1-one (9CI)	**(83622-91-7)**
6H-Cyclopenta(ghi)picen-6-one (9CI)	**(83484-79-1)**
α,β-Cyclopentamethylenetetrazole	**(54-95-5)**
Cyclopentanamine (9CI)	**(1003-03-8)**
Cyclopenta(5,6)naphtho(2,1-b)carbazole, 1'H-cholest-2-eno (3,2-b)indole deriv. (9CI)	**(55493-86-2)**
Cyclopentane (ACGIH,OSHA) [UN 1146]	**(287-92-3)**

Cyclopentaneacetic acid, ethenyl ester (9CI) **(45955-66-6)**
Cyclopentaneacetic acid, 3-oxo-2-pentyl-, methyl ester (9CI) **(24851-98-7)**
Cyclopentane, bromo- (9CI) **(137-43-9)**
Cyclopentane, 1-bromo-2-chloro-, trans- (8CI,9CI) **(14376-82-0)**
Cyclopentane, butyl- (8CI,9CI) **(2040-95-1)**
Cyclopentane, 1-butyl-2-propyl- (9CI) **(62199-50-2)**
Cyclopentanecarboxylic acid (8CI,9CI) **(3400-45-1)**
Cyclopentanecarboxylic acid, 1-amino **(52-52-8)**
Cyclopentanecarboxylic acid, 1-amino-, l- **(52-52-8)**
Cyclopentanecarboxylic acid, ethenyl ester (9CI) **(16523-06-1)**
Cyclopentanecarboxylic acid, 1-phenyl-, 2-(2-(diethylamino)-ethoxy)ethyl ester **(77-23-6)**
Cyclopentanecarboxylic acid, vinyl ester (8CI) **(16523-06-1)**
Cyclopentane, cyclopentyl- **(1636-39-1)**
Cyclopentane, 1,1'-(3-(2-cyclopentylethylidene)-1,5-pentanediyl) bis- (9CI) **(54934-71-3)**
Cyclopentane, (1-decylundecyl)- (9CI) **(6703-81-7)**
1,3-Cyclopentanedicarboxylic acid, 1,2,2-trimethyl-, (1R-cis)- **(124-83-4)**
1,3-Cyclopentanedicarboxylic acid, 1,2,2-trimethyl-, cis- (9CI) **(5394-83-2)**
Cyclopentane, 1,1-dimethyl- (9CI) **(1638-26-2)**
Cyclopentane, 1,2-dimethyl-, cis- (9CI) **(1192-18-3)**
Cyclopentane, 1,3-dimethyl-, cis- (9CI) **(2532-58-3)**
Cyclopentane, 1,3-dimethyl-, trans- (9CI) **(1759-58-6)**
Cyclopentane, dimethyl- (9CI) **(28729-52-4)**
Cyclopentane, 1,2-dimethyl- (8CI,9CI) **(2452-99-5)**
Cyclopentane, 1,2-dimethyl-, trans- (8CI,9CI) **(822-50-4)**
Cyclopentane, 1,3-dimethyl- (8CI,9CI) **(2453-00-1)**
1,2-Cyclopentanediol, 3-methyl- (8CI,9CI) **(27583-37-5)**
1,3-Cyclopentanedione, 4-butyl- (9CI) **(54244-72-3)**
1,3-Cyclopentanedione, 2,2-dimethyl- (7CI,8CI,9CI) **(3883-58-7)**
1,3-Cyclopentanedione, 2-methyl- (8CI,9CI) **(765-69-5)**
Cyclopentane, 1,2-epoxy- **(285-67-6)**
Cyclopentane, ethyl **(1640-89-7)**
Cyclopentane, 2-ethyl-1,1-dimethyl- (9CI) **(54549-80-3)**
Cyclopentane, ethylmethyl- (9CI) **(61593-45-1)**
Cyclopentane, 1-ethyl-1-methyl- (8CI,9CI) **(16747-50-5)**
Cyclopentane, 1-ethyl-2-methyl-, trans- (8CI,9CI) **(930-90-5)**
Cyclopentane, 1-ethyl-3-methyl- (8CI,9CI) **(3726-47-4)**
Cyclopentane, heneicosyl- (9CI) **(6703-82-8)**
Cyclopentane, hexachloro- (9CI) **(68258-91-3)**
Cyclopentane, isopropyl- (8CI) **(3875-51-2)**
Cyclopentanemethanamine, 2-amino- (9CI) **(21544-02-5)**
Cyclopentane, methyl [UN 2298] **(96-37-7)**
Cyclopentanemethylamine, 2-amino- (8CI) **(21544-02-5)**
Cyclopentanemethylamine, 2-isopropylidene-N,N,5-trimethyl-, (1R,5R)-(-)- (8CI) **(17943-83-8)**
Cyclopentane, (1-methylethyl)- (9CI) **(3875-51-2)**
Cyclopentane oxide **(285-67-6)**
Cyclopentane, pentyl- (9CI) **(3741-00-2)**
Cyclopentanepropanoic acid (9CI) **(140-77-2)**
Cyclopentane, 1-propenyl- (9CI) **(5623-78-9)**
Cyclopentane, propenyl- (8CI) **(5623-78-9)**
Cyclopentanepropionic acid (8CI) **(140-77-2)**
Cyclopentane, propyl **(2040-96-2)**
1,2,3,4-Cyclopentanetetracarboxylic acid, all-cis- (8CI) **(3786-91-2)**
1,2,3,4-Cyclopentanetetracarboxylic acid, (1α,2α,3α,4α)- (9CI) **(3786-91-2)**
Cyclopentane, tetrachloro- (9CI) **(59808-78-5)**
Cyclopentanethiol **(1679-07-8)**
Cyclopentane, 1,1,3-trimethyl- (9CI) **(4516-69-2)**
Cyclopentane, 1,2,3-trimethyl-, cis-1,2,trans-1,3- (8CI) **(15890-40-1)**
Cyclopentane, trimethyl- (9CI) **(30498-64-7)**
Cyclopentane, 1,2,4-trimethyl- (8CI,9CI) **(2815-58-9)**
Cyclopentane, 1,2,3-trimethyl-, (1α,2α,3β)- (9CI) **(15890-40-1)**
1,2,4-Cyclopentanetrione (6CI,8CI,9CI) **(15849-14-6)**
1,2,4-Cyclopentanetrione, 3-butyl- (9CI) **(46005-09-8)**
1,2,4-Cyclopentanetrione, 3-methyl- (8CI,9CI) **(4505-54-8)**
Cyclopentanol [UN 2244] **(96-41-3)**
Cyclopentanol, 1,2-dimethyl-3-(1-methylethenyl)-, (1R-(1α,2β,3α))- (9CI) **(4099-07-4)**
Cyclopentanol, 3-isopropenyl-1,2-dimethyl-, (1R,2R,3S)-(-)- (8CI) **(4099-07-4)**
Cyclopentanol, 2-methyl- (8CI,9CI) **(24070-77-7)**
Cyclopentanol, 2-methyl-, cis- (8CI,9CI) **(25144-05-2)**
Cyclopentanol, 2-methyl-, trans- (8CI,9CI) **(25144-04-1)**
Cyclopentanol, 2-methyl-, acetate, cis- (9CI) **(40991-93-3)**
Cyclopentanone [UN 2245] **(120-92-3)**
Cyclopentanone, dichloro- (9CI) **(103339-61-3)**
Cyclopentanone, 2-n-heptyl **(137-03-1)**
Cyclopentanone, 2-methyl- (9CI) **(1120-72-5)**
Cyclopentanone, 3-methyl- (9CI) **(1757-42-2)**
Cyclopentanone, 2-methyl-3-(1-methylethyl)- (9CI) **(54549-81-4)**
Cyclopentanone, oxime (8CI,9CI) **(1192-28-5)**
Cyclopentanone, 2-pentyl- (9CI) **(4819-67-4)**
Cyclopentaphenanthrene **(203-64-5)**
Cyclopentaphenanthrene (9CI) **(80455-52-3)**
4H-Cyclopenta(pqr)picen-4-one (9CI) **(86854-16-2)**
5H-Cyclopentapyrazine, 6,7-dihydro-5-methyl- (9CI) **(23747-48-0)**
Cyclopentapyrene (9CI) **(83381-96-8)**
1H-Cyclopentapyrimidine-2,4(3H,5H)-dione, 6,7-dihydro-3-cyclohexyl **(2164-08-1)**
13H-Cyclopenta(rst)pentaphen-13-one (9CI) **(86854-15-1)**
Cyclopentasiloxane, decamethyl **(541-02-6)**
Cyclopentasiloxane, 2,2,4,4,6,8,10-heptamethyl-6,8,10-tris-(3,3,3-trifluoropropyl)- (8CI,9CI) **(22474-57-3)**
Cyclopentene [UN] **(142-29-0)**
3-Cyclopentene-1-acetaldehyde, 2,2,3-trimethyl- **(4501-58-0)**
3-Cyclopentene-1-acetaldehyde, 2,2,3-trimethyl-, (R)- (9CI) **(4501-58-0)**
Cyclopentene, 3-chloro- (8CI,9CI) **(96-40-2)**
3-Cyclopentene-1,2-dione, chloro- (9CI) **(103339-62-4)**
3-Cyclopentene-1,2-dione, dichloro- (9CI) **(94650-97-2)**
4-Cyclopentene-1,3-dione, 4-methoxy-5-methyl- (7CI,8CI,9CI) **(7180-62-3)**
3-Cyclopentene-1,2-dione, trichloro- (9CI) **(103354-08-1)**
Cyclopentene epoxide **(285-67-6)**
Cyclopentene, 3-ethylidene-1-methyl- (9CI) **(62338-00-5)**
Cyclopentene, hexachloro- (9CI) **(72030-26-3)**
Cyclopentene, 1-methyl- (9CI) **(693-89-0)**
Cyclopentene, 3-methyl- (8CI,9CI) **(1120-62-3)**
Cyclopentene, 4-methyl- (8CI,9CI) **(1759-81-5)**
Cyclopentene, octachloro- (9CI) **(706-78-5)**
Cyclopenteneoxide **(285-67-6)**
Cyclopentene, 3-propyl- (8CI,9CI) **(34067-75-9)**
Cyclopenteno(c,d)pyrene **(27208-37-3)**
2-Cyclopenten-1-ol, 1-phenyl- (7CI,9CI) **(56667-10-8)**
Cyclopentenone **(930-30-3)**
2-Cyclopenten-1-one (9CI) **(930-30-3)**
2-Cyclopenten-1-one, 2-allyl-4-hydroxy-3-methyl- (VAN) (8CI) **(551-45-1)**
2-Cyclopenten-1-one, 4-butyl-3-methoxy- (9CI) **(53690-92-9)**
2-Cyclopenten-1-one, 3-chloro- (9CI) **(53102-14-0)**
2-Cyclopenten-1-one, 2,3-dimethyl- (8CI,9CI) **(1121-05-7)**
2-Cyclopenten-1-one, 2,5-dimethyl- (8CI,9CI) **(4041-11-6)**
2-Cyclopenten-1-one, 2-hydroxy-3-methyl **(80-71-7)**
2-Cyclopenten-1-one, 4-hydroxy-3-methyl-2-(2-propenyl)- (9CI) **(551-45-1)**
2-Cyclopenten-1-one, 3-methyl- (9CI) **(2758-18-1)**
2-Cyclopenten-1-one, 2-methyl- (8CI,9CI) **(1120-73-6)**
2-Cyclopenten-1-one, 3-methyl-2-(2-pentenyl)-, (Z) **(488-10-8)**
2-Cyclopenten-1-one, 2,3,4,5-tetramethyl- (9CI) **(54458-61-6)**
2-Cyclopenten-1-one, tetramethyl- (9CI) **(92366-34-2)**
2-Cyclopenten-1-one, 2,3,5-trimethyl- (9CI) **(54562-24-2)**
2-Cyclopenten-1-one, 3,4,5-trimethyl- (9CI) **(55683-21-1)**
2-Cyclopenten-1-one, 2,3,4-trimethyl- (8CI,9CI) **(28790-86-5)**
2-Cyclopenten-1-one, 3,5,5-trimethyl- (6CI,8CI,9CI) **(24156-95-4)**
4H-Cyclopent(f)oxacyclotridecin-4-one, 1,6,7,8,9,11a-β,12,13,14,14a-α-decahydro-1- β-13-α-dihydroxy-6-β-methyl **(20350-15-6)**
Cyclopentimine **(110-89-4)**
Cyclopentylamine (8CI) **(1003-03-8)**

Cyclopentylamine, 2-(aminomethyl)- **(21544-02-5)**
Cyclopentyl bromide **(137-43-9)**
2-Cyclopentyl-p-chlorophenol **(13347-42-7)**
Cyclopentylcyclohexylmethane **(4431-89-4)**
2-Cyclopentylcyclopentanone **(4884-24-6)**
Cyclopentyl ether **(10137-73-2)**
1-Cyclopentylheneicosane **(6703-82-8)**
Cyclopentyl mercaptan **(1679-07-8)**
3-Cyclopentylpropionic acid **(140-77-2)**
Cyclopentylpropionic acid **(140-77-2)**
Cyclophil SXS30 **(1300-72-7)**
Cyclophos **(13396-41-3)**
Cyclophosphamid **(50-18-0)**
(-)-Cyclophosphamide **(50-18-0)**
Cyclophosphamide **(50-18-0)**
Cyclophosphamide hydrate **(6055-19-2)**
Cyclophosphamide monohydrate **(6055-19-2)**
Cyclophosphamidum **(50-18-0)**
Cyclophosphamidum **(6055-19-2)**
Cyclophosphan **(50-18-0)**
Cyclophosphan **(6055-19-2)**
Cyclophosphane **(50-18-0)**
Cyclophosphane **(6055-19-2)**
Cyclophosphanum **(6055-19-2)**
Cyclophosphoramide **(50-18-0)**
Cyclopolydimethylsiloxane **(69430-24-6)**
Cycloposine **(23185-94-6)**
Cycloprate **(54460-46-7)**
5H-Cyclopropa(3,4)benz(1,2-e)azulen-5-one, 1,1a-β,1b-α,4,4a,7a-β,7b,8,9,9a- decahydro-4a-α,7b-α,9-β,9a-α-tetrahydroxy-3-(hydroxymethyl)-1,1,6,8-α- tetramethyl **(17673-25-5)**
Cyclopropanamine (9CI) **(765-30-0)**
Cyclopropane (DOT) **(75-19-4)**
Cyclopropaneacrylic acid, 3-carboxy-α,2,2-trimethyl-, 1-methyl ester, ester with 4- hydroxy-3-methyl-2-(2,4-pentadienyl)-2-cyclopenten-1-one **(121-29-9)**
Cyclopropanealanine, 2-methylene, l- **(156-56-9)**
Cyclopropane, 1-butyl-1-methyl-2-propyl- (9CI) **(41977-34-8)**
Cyclopropanecarbonitrile (9CI) **(5500-21-0)**
Cyclopropanecarbonyl chloride, 3-(2,2-dichloroethenyl)-2,2-dimethyl- (9CI) **(52314-67-7)**
Cyclopropanecarboxamide (9CI) **(6228-73-5)**
Cyclopropanecarboxanilide, 3',4'-dichloro **(2759-71-9)**
Cyclopropanecarboxylic-14C acid, 2,2-dimethyl-3-(2-methyl-1-propenyl)-, (1R-trans)- (9CI) **(32511-06-1)**
Cyclopropanecarboxylic-acid **(1759-53-1)**
Cyclopropanecarboxylic acid, 3-(2-chloro-3,3,3-trifluoro-1-propenyl)-2,2-dimethyl-, (2,3,5,6-tetrafluoro-4-methylphenyl)-methyl ester, (1-α,3-α(Z))-(+-) **(79538-32-2)**
Cyclopropanecarboxylic acid, 3-(2-chloro-3,3,3-trifluoro-1-propenyl)-2,2-dimethyl-, (3-phenoxyphenyl)methyl ester (9CI) **(71698-60-7)**
Cyclopropanecarboxylic acid, 3-(2-chloro-3,3,3-trifluoro-1-propenyl)-2,2-dimethyl-, cyano(3-phenoxyphenyl)methyl ester **(68085-85-8)**
Cyclopropanecarboxylic acid, 3-(cyclopentylidenemethyl)-2,2-dimethyl-, (5-benzyl-3-furyl) methyl ester, (1R,3R) **(22431-62-5)**
Cyclopropanecarboxylic acid, 3-(2,2-dibromoethenyl)-2,2-dimethyl-, cyano(3-phenoxyphenyl) methyl ester **(52820-00-5)**
Cyclopropanecarboxylic acid, 3-(2,2-dibromoethenyl)-2 ,2-dimethyl-, cyano(3-phenoxyphenyl) methyl ester, (1r-(1-α(S*),3-α)) **(52918-63-5)**
Cyclopropanecarboxylic acid, 3-(2,2-dibromoethenyl)-2,2-dimethyl-, cyano(3-phenoxyphenyl)methyl ester, (1α(R*),3α)-(+-)- (9CI) **(120710-23-8)**
Cyclopropanecarboxylic acid, 3-(2,2-dibromoethenyl)-2,2-dimethyl-, cyano(3-phenoxyphenyl)methyl ester, (1α(R*),3β)-(+-)- (9CI) **(120710-25-0)**
Cyclopropanecarboxylic acid, 3-(2,2-dibromoethenyl)-2,2-dimethyl-, cyano(3-phenoxyphenyl)methyl ester, (1α(S*),3α)-(+-)- (9CI) **(80845-12-1)**
Cyclopropanecarboxylic acid, 3-(2,2-dibromoethenyl)-2,2-dimethyl-, cyano(3-phenoxyphenyl)methyl ester, (1α(S*),3β)-(+-)- (9CI) **(120710-24-9)**
Cyclopropanecarboxylic acid, 3-(2,2-dichloroethenyl)-2,2-dimethyl-, cyano(3-phenoxyphenyl) methyl ester, (+-) **(52315-07-8)**
Cyclopropanecarboxylic acid, 3-(2,2-dichloroethenyl)-2,2-dimethyl-, cyano(3-phenoxyphenyl) methyl ester, (1-α(S*),3-β)-(+-) **(71697-59-1)**
Cyclopropanecarboxylic acid, 3-(2,2-dichloroethenyl)-2,2-dimethyl-, cyano(3-phenoxyphenyl) methyl ester, (1-α(S*),3-α)-(+-) **(67375-30-8)**
Cyclopropanecarboxylic acid, 3-(2,2-dichloroethenyl)-2,2-dimethyl-, cyano(3-phenoxyphenyl)methyl ester, Mixt. with 2-chloro-1-(2,4-dichlorophenyl)ethenyl diethyl phosphate (9CI) **(85682-59-3)**
Cyclopropanecarboxylic acid, 3-(2,2-dichloroethenyl)-2,2-dimethyl-, methyl ester (9CI) **(61898-95-1)**
Cyclopropanecarboxylic acid, 3-(2,2-dichloroethenyl)-2,2-dimethyl-, (3-phenoxyphenyl) methyl ester, cis **(61949-76-6)**
Cyclopropanecarboxylic acid, 3-(2,2-dichloroethenyl)-2,2-dimethyl-, (3-phenoxyphenyl) methyl ester, cis-(+-) **(52341-33-0)**
Cyclopropanecarboxylic acid, 3-(2,2-dichloroethenyl)-2,2-dimethyl-, (3-phenoxyphenyl)methyl ester, trans- (9CI) **(61949-77-7)**
Cyclopropanecarboxylic acid, 3-(2,2-dichloroethenyl)-2,2-dimethyl-, (3-phenoxyphenyl)methyl ester, trans-(+-)- (9CI) **(52341-32-9)**
Cyclopropanecarboxylic acid, 2-(2,2-dichlorovinyl)-3,3-dimethyl-, ester with (4-fluoro-3-phenoxyphenyl)-hydroxyacetonitrile **(68359-37-5)**
Cyclopropanecarboxylic acid, 3-(2,2-dichlorovinyl)-2,2-dimethyl-, 3-phenoxybenzyl ester, (+-)-, (cis,trans) **(52645-53-1)**
Cyclopropanecarboxylic acid, 3-((dihydro-2-oxo-3(2H)-thienylidene)methyl)-2,2-dimethyl-, (5-(phenylmethyl)-3-furanyl)methyl ester, (1R-(1-α,3-α(E))) **(58769-20-3)**
Cyclopropanecarboxylic acid, 2,2-dimethyl-3-(2-methylpropenyl) **(10453-89-1)**
Cyclopropanecarboxylic acid, 2,2-dimethyl-3-(2-methyl-propenyl)-, (5-benzyl-3-furyl) methyl ester **(10453-86-8)**
Cyclopropanecarboxylic acid, 2,2-dimethyl-3-(2-methyl-propenyl)-, (5-benzyl-3-furyl) methyl ester, d-trans **(28434-01-7)**
Cyclopropanecarboxylic acid, 2,2-dimethyl-3-(2-methyl-propenyl)-, 6-chloropiperonyl ester, (+-)-cis, trans **(70-43-9)**
Cyclopropanecarboxylic acid, 2,2-dimethyl-3-(2-methyl-propenyl)-, 2,4-dimethylbenzyl ester **(70-38-2)**
Cyclopropanecarboxylic acid, 2,2-dimethyl-3-(2-methyl-propenyl)-, (+)-(E)-, ester with (+)- 2-allyl-4-hydroxy-3-methyl-2-cyclopenten-1-one **(28434-00-6)**
Cyclopropanecarboxylic acid, 2,2-dimethyl-3-(2-methylpropenyl)-, ester with (+-)-2-furfuryl- 4-hydroxy-3-methyl-2-cyclopenten-1-one, (+-)-cis,trans **(17080-02-3)**
Cyclopropanecarboxylic acid, 2,2-dimethyl-3-(2-methylpropenyl)-, ester with 4-hydroxy- 3-methyl-2-(2,4-pentadienyl)-2-cyclopenten-1-one **(121-21-1)**
Cyclopropanecarboxylic acid, 2,2-dimethyl-3-(2-methyl-1-propenyl)-, ethyl ester **(97-41-6)**
Cyclopropanecarboxylic acid, 2,2-dimethyl-3-(2-methyl-1-propenyl)-, ethyl ester (9CI) **(97-41-6)**
Cyclopropanecarboxylic acid, 2,2-dimethyl-3-(2-methylpropenyl)-, ethyl ester (8CI) **(97-41-6)**
Cyclopropanecarboxylic acid, 2,2-dimethyl-3-(2-methyl-1-propenyl)-, (1,3,4,5,6,7-hexahydro-1,3-dioxo-2H-isoindol-2-yl)methyl ester **(7696-12-0)**
Cyclopropanecarboxylic acid, 2,2-dimethyl-3-(2-methyl-1-propenyl)-, 2-methyl-4-oxo- 3-(2-propenyl)-2-cyclopenten-1-yl ester **(584-79-2)**

Cyclopropanecarboxylic acid, 2,2-dimethyl-3-(2-methylpropenyl)-, (2-methyl-5-(2-propynyl)- 3-furyl)methyl ester **(27223-49-0)**
Cyclopropanecarboxylic acid, 2,2-dimethyl-3-(2-methylpropenyl)-, m-phenoxybenzyl ester **(26002-80-2)**
Cyclopropanecarboxylic acid, 2,2-dimethyl-3-(2-methylpropenyl)-, 5-(2-propynyl)furfuryl ester **(23031-38-1)**
Cyclopropanecarboxylic acid, 2,2-dimethyl-3-(1,2,2,2-tetrabromo-ethyl)-, cyano(3-phenoxy phenyl)methyl ester **(66841-25-6)**
Cyclopropanecarboxylic acid, 2-ethyl-1-(((6-ethyl-2,3,3a,6,7,7a-hexahydro-1-oxo-1H-inden-4-yl)carbonyl)amino)-, (3aS-(3aα,4(1R*,2R*),6β,7aα))- (9CI) **(62251-96-1)**
Cyclopropanecarboxylic acid, hexadecyl ester **(54460-46-7)**
Cyclopropanecarboxylic acid, 2,2,3,3-tetramethyl-, cyano-(3-phenoxyphenyl)methyl ester **(39515-41-8)**
Cyclopropanecarboxylic-carboxy-14C acid, 2,2-dimethyl-3-(2-methylpropenyl)-, trans-(+)- (8CI) **(32511-06-1)**
1,2-Cyclopropanedicarboximide, N-(3,5-dichlorophenyl)-1,2-dimethyl **(32809-16-8)**
Cyclopropane, 1-ethenyl-2-hexenyl-, (1.α.,2.β.(E))-(.+-.)- (9CI) **(22822-99-7)**
Cyclopropane, ethoxy- (9CI) **(5614-38-0)**
Cyclopropane, Liquefied [UN 1027] **(75-19-4)**
Cyclopropanemethanamine, N-propyl- (9CI) **(26389-60-6)**
Cyclopropane, (2-methoxy-1-methylethenyl)-, (E)- (9CI) **(35200-80-7)**
Cyclopropane, (2-methoxy-1-methylethenyl)-, (Z)- (9CI) **(35200-79-4)**
Cyclopropane, methyl- (9CI) **(594-11-6)**
Cyclopropane, 1-methyl-2-(1-methylethyl)-3-(1-methylethylidene)-, cis- (9CI) **(24524-52-5)**
Cyclopropane, 1-methyl-1-(1-methylethyl)-2-nonyl- (9CI) **(41977-40-6)**
Cyclopropane, 1-methyl-1-(2-methylpropyl)-2-nonyl- (9CI) **(41977-41-7)**
Cyclopropane, (1-methyl-1-propenyl)-, (E)- (9CI) **(20479-71-4)**
Cyclopropane, (1-methyl-1-propenyl)-, (Z)- (9CI) **(20479-72-5)**
Cyclopropane, pentachloro **(6262-51-7)**
Cyclopropanepropanoic acid, α-amino-2-methylene- (9CI) **(156-56-9)**
Cyclopropanepropionic acid, α-amino-2-methylene-, l-(+) **(156-56-9)**
Cyclopropane, propyl- (9CI) **(2415-72-7)**
1H-Cycloprop(e)azulene, decahydro-1,1,7-trimethyl-4-methylene-, (1aR,4aR,7R,7aR,7bS)-(+)- (8CI) **(489-39-4)**
1H-Cycloprop(e)azulene, decahydro-1,1,7-trimethyl-4-methylene-, (1aR-(1aα,4aα,7α,7aβ,7bα))- (9CI) **(489-39-4)**
1-Cyclopropene-1-heptanoic acid, 2-octyl- **(503-05-9)**
1-Cyclopropene-1-octanoic acid, 2-octyl- **(738-87-4)**
Cyclopropylamine (8CI) **(765-30-0)**
Cyclopropylamine, 2-phenyl-, trans-(+-) **(155-09-9)**
4-Cyclopropylbromobenzene **(1124-14-7)**
Cyclopropylcarboxamide **(6228-73-5)**
Cyclopropyl cyanide **(5500-21-0)**
α-Cyclopropyl-α-(4-methoxyphenyl)-5-pyrimidinemethanol **(12771-68-5)**
α-Cyclopropyl-α-(p-methoxyphenyl)-5-pyrimidinemethanol **(12771-68-5)**
α-Cyclopropyl-4-methoxy-α-(pyrimidin-5-yl)benzyl alcohol **(12771-68-5)**
α-Cyclopropyl-4-methoxy-α-(pyrimidin-5-yl)-benzylalkohol (German) **(12771-68-5)**
N-Cyclopropylmethyl-6,14-endoetheno-7 α- (1-hydroxy-1-methylethyl)-6,7,8,14-tetrahydronororipavine **(4406-22-8)**
N-(Cyclopropylmethyl)-α,α,α-trifluoro-2,6-dinitro-N-propyl-p-toluidine **(26399-36-0)**
4-Cyclopropylphenyl bromide **(1124-14-7)**
Cycloryl 21 **(151-21-3)**
Cycloryl 31 **(151-21-3)**
Cycloryl 580 **(151-21-3)**
Cycloryl 585N **(151-21-3)**
Cycloryl NA **(9004-82-4)**
Cycloryl OS **(142-31-4)**
Cycloryl TAWF **(139-96-8)**
Cycloryl WAT **(139-96-8)**
Cyclosan **(10112-91-1)**
(+)-Cycloserine **(68-41-7)**
Cycloserine **(68-41-7)**
D-Cycloserine **(68-41-7)**
Cyclosia **(107-75-5)**
Cyclosiloxanes, di-Me **(69430-24-6)**
Cyclosol 63 (9CI) **(89072-60-6)**
Cyclospasmol **(456-59-7)**
Cyclosporin **(59865-13-3)**
Cyclosporin-A **(59865-13-3)**
Cyclosporine **(59865-13-3)**
Cyclosporine A **(59865-13-3)**
Cyclostin **(50-18-0)**
Cycloten **(80-71-7)**
Cyclotene **(80-71-7)**
Cyclotetramethylene oxide **(109-99-9)**
Cyclotetramethylene sulfone **(126-33-0)**
Cyclotetramethylenetetranitramine **(2691-41-0)**
Cyclotetramethylenetetranitramine, Containing at least 10%-25% water (DOT) **(2691-41-0)**
Cyclotetramethylenetetranitramine, Desensitized with not less than 10% phlegmatizer (DOT) **(2691-41-0)**
Cyclotetramethylene tetranitramine, Dry (DOT) **(2691-41-0)**
Cyclotetramethylenetetranitramine (HMX; Octogen), Wetted with not less than 15 per cent water, by mass [UN 0226] **(2691-41-0)**
Cyclotetrasiloxane, 2,6-diphenyl-2,4,4,6,8,8-hexamethyl **(4657-20-9)**
Cyclotetrasiloxane, 2,4-diphenyl-2,4,6,6,8,8-hexamethyl-, racemic mixture **(18604-02-9)**
Cyclotetrasiloxane, heptamethylphenyl **(10448-09-6)**
Cyclotetrasiloxane, hexamethyldiphenyl- (8CI,9CI) **(30026-85-8)**
Cyclotetrasiloxane, octamethyl **(556-67-2)**
Cyclotetrasiloxane, octaphenyl- (9CI) **(546-56-5)**
Cyclotetrasiloxane, 2,2,4,6,8-pentamethyl-4,6,8-triphenyl-(6CI,7CI,8CI,9CI) **(10448-10-9)**
Cyclotetrasiloxane, 2,4,6,8-tetraethenyl-2,4,6,8-tetramethyl- (9CI) **(2554-06-5)**
Cyclotetrasiloxane, 2,4,6,8-tetramethyl-2,4,6,8-tetraphenyl- (9CI) **(77-63-4)**
Cyclotetrasiloxane, 2,4,6,8-tetramethyl-2,4,6,8-tetravinyl- **(2554-06-5)**
Cyclothiazide **(2259-96-3)**
Cycloton V **(57-09-0)**
Cyclotrimethylenenitramine **(121-82-4)**
Cyclotrimethylenetrinitramine **(121-82-4)**
Cyclotrimethylenetrinitramine, Desensitized (DOT) **(121-82-4)**
Cyclotrimethylenetrinitramine, Wetted with not less than 15 per cent water, by mass [UN 0072] **(121-82-4)**
Cyclotrisiloxane, hexamethyl- **(541-05-9)**
Cyclotrisiloxane, 2,4,6-trimethyl-2,4,6-triphenyl-, cis,trans-(8CI) **(6138-53-0)**
Cyclotrisiloxane, 2,4,6-trimethyl-2,4,6-triphenyl-, (Z) **(3424-57-5)**
Cyclotrisiloxane, 2,4,6-trimethyl-2,4,6-triphenyl-, (2α,4α,6β)- (9CI) **(6138-53-0)**
Cyclotrisiloxane, 2,4,6-trimethyl-2,4,6-tris(3,3,3-trifluoropropyl)-(9CI) **(2374-14-3)**
Cycloundecanone (8CI,9CI) **(878-13-7)**
1,4,8-Cycloundecatriene, 2,6,6,9-tetramethyl-, (E,E,E)- (9CI) **(6753-98-6)**
1H-Cycloundec(d)isoindole-1,11(2H)-dione, 3-benzyl-3,3-α,4,5,6,6-α,9,10,12,15- decahydro-6,12,15-trihydroxy-4,10,12-trimethyl-5-methylene-, 15-acetate **(22144-77-0)**
Cyclouron **(2163-69-1)**
Cycluron-chlorbufam mixture **(8015-55-2)**
Cycluron-chlorbuyam mixture **(8015-55-2)**
Cycluron **(2163-69-1)**
Cycocel **(999-81-5)**
Cycocel-Extra **(999-81-5)**
Cycogan **(999-81-5)**
Cycogan Extra **(999-81-5)**
Cycolamin **(68-19-9)**
Cycosin **(23564-05-8)**
Cydeal **(2597-03-7)**
Cyethoxydim **(74051-80-2)**
Cyfen **(122-14-5)**
Cyflee **(115-93-5)**

Name	CAS No.
l-Cystine (9CI)	(56-89-3)
Cystine, L	(56-89-3)
Cystine acid	(56-89-3)
Cystisine	(485-35-8)
Cystoceva	(482-89-3)
Cystogen	(100-97-0)
Cystoids anthelmintic	(136-77-6)
Cystopyrin	(136-40-3)
Cystural	(136-40-3)
Cytacon	(68-19-9)
Cytamen	(68-19-9)
Cytarabin	(147-94-4)
Cytarabina	(147-94-4)
Cytarabine	(147-94-4)
Cytarabine hydrochloride	(69-74-9)
Cytarabinoside	(147-94-4)
Cytel	(122-14-5)
Cytembena	(21739-91-3)
Cyten	(122-14-5)
Cythioate	(115-93-5)
Cythion	(121-75-5)
Cythrin	(70124-77-5)
Cytidine	(65-46-3)
Cytidine, 2'-deoxy-5-fluoro	(10356-76-0)
Cytidine, 2',3'-dideoxy	(7481-89-2)
5'-Cytidylic acid, adenylyl-(5'-3')- (9CI)	(15648-73-4)
Cytisine	(485-35-8)
Cytiton	(485-35-8)
Cytitone	(485-35-8)
Cytobion	(68-19-9)
Cytochalasin B	(14930-96-2)
Cytochalasin D	(22144-77-0)
Cytochalasin-E	(36011-19-5)
Cytomycin (9CI)	(2005-98-3)
Cytophosphan	(50-18-0)
Cytophosphan	(6055-19-2)
Cytosar-U	(147-94-4)
Cytosar	(147-94-4)
Cytosar hydrochloride	(69-74-9)
Cytosine (8CI)	(71-30-7)
Cytosine, 1-β-D-arabinofuranosyl-, hydrochloride	(69-74-9)
Cytosine, 1-β-D-arabinofuranosyl-, monohydrochloride	(69-74-9)
Cytosine β-D-arabinoside	(147-94-4)
Cytosine, 1-β-D-arabinosyl-	(147-94-4)
Cytosine, 1-β-d-arabinofuranosyl	(147-94-4)
Cytosine-β-arabinoside	(147-94-4)
Cytosinearabinoside	(147-94-4)
Cytosine arabinoside hydrochloride	(69-74-9)
Cytosine, 5-fluoro	(2022-85-7)
Cytosine riboside	(65-46-3)
Cytosinimine	(71-30-7)
Cytovirin	(2079-00-7)
Cytoxal alcohol	(4465-94-5)
Cytoxan	(50-18-0)
Cytoxan	(6055-19-2)
Cytoxyl alcohol cyclohexylammonium salt	(4465-94-5)
Cytrol	(61-82-5)
Cytrol Amitrole-T	(61-82-5)
Cytrolane	(950-10-7)
Cytrole	(61-82-5)
Cyuram DS	(137-26-8)
m-Cym-5-yl methylcarbamate	(2631-37-0)
Cyzone	(101-72-4)
Czterochlorek wegla (Polish)	(56-23-5)
2,3,7,8-Czterochlorodwubenzo-p-dwuoksyny (Polish)	(1746-01-6)
1,1,2,2-Czterochloroetan (Polish)	(79-34-5)
Czterochloroetylen (Polish)	(127-18-4)
Czteroetylek olowiu (Polish)	(78-00-2)
2,6-D	(575-90-6)
3,4-D	(588-22-7)
D 25	(97-24-5)
D 50	(94-75-7)
D 50	(9003-20-7)
D 151	(9003-07-0)
D 209	(67465-67-2)
D 735	(5234-68-4)
D 854	(80-33-1)
D 860	(64-77-7)
D 1221	(1563-66-2)
D 1991	(17804-35-2)
D 3520	(3655-88-7)
D 31,717	(3004-70-4)
D-50	(98-96-4)
D-365	(52-53-9)
D-1410	(23135-22-0)
D & C 22	(17372-87-1)
D & C Blue No. 1	(61-73-4)
D & C Blue No. 4	(2650-18-2)
D & C Blue No. 4	(3844-45-9)
D & C Blue No. 6	(482-89-3)
D & C Blue No. 9	(130-20-1)
D & C Brown No. 1	(1320-07-6)
D & C Green No. 4	(5141-20-8)
D & C Green No. 6	(128-80-3)
D & C Green No. 8	(6358-69-6)
D & C Orange No. 2	(2646-17-5)
D & C Orange No. 3	(1936-15-8)
D & C Orange No. 3	(523-44-4)
D & C Orange No. 4	(633-96-5)
D & C Orange No. 17	(3468-63-1)
D & C Orange No. 15	(72-48-0)
D & C Red 2	(915-67-3)
D & C Red No. 3	(16423-68-0)
D & C Red No. 5	(3761-53-3)
D & C Red No. 6, Barium Lake	(17852-98-1)
D & C Red No. 6	(5858-81-1)
D & C Red No. 7	(5281-04-9)
D & C Red No. 9	(5160-02-1)
D & C Red No. 10	(1248-18-6)
D & C Red No. 11	(1103-39-5)
D & C Red No. 12	(1103-38-4)
D & C Red No. 17	(85-86-9)
D & C Red No. 19	(81-88-9)
D & C Red No. 21	(15086-94-9)
D & C Red No. 22	(17372-87-1)
D & C Red No. 30	(2379-74-0)
D & C Red No. 33	(3567-66-6)
D & C Red No. 34	(6417-83-0)
D & C Red No. 35	(2425-85-6)
D & C Red No. 36	(2814-77-9)
D & C Violet No. 1	(1694-09-3)
D & C Violet No. 2	(81-48-1)
D & C Yellow No. 7	(2321-07-5)
D & C Yellow No. 8	(518-47-8)
D & C Yellow No. 5	(1934-21-0)
D & C Yellow No. 10	(8004-92-0)
D & C Yellow No. 11	(8003-22-3)
D & D	(14808-60-7)
D & P Double O Crabgrass Killer	(590-28-3)
2,4-D Acid	(94-75-7)
2,4-D n-Butyl ester Mixed with 2,4,5-T n-butyl ester (1:1)	(39277-47-9)
2,4-D Ethyl ester	(533-23-3)
2,4-D 2-Ethylhexyl ester	(1928-43-4)
2,4-D PGBE	(1320-18-9)

Name	CAS
2,4-D PGBE	**(1928-45-6)**
2,4-D, Propylene glycol butyl ether ester	**(1320-18-9)**
2,4-D amine	**(2008-39-1)**
2,4-D amine salt	**(2008-39-1)**
2,4-D ammonium salt	**(2307-55-3)**
2,4-D butoxyethanol ester	**(1929-73-3)**
2,4-D 2-butoxyethyl ester	**(1929-73-3)**
2,4-D butoxyethyl ester	**(1929-73-3)**
2,4-D-butyl	**(94-80-4)**
2,4-D butyl ester	**(94-80-4)**
2,4-D n-butyl ester	**(94-80-4)**
2,4-D butyric	**(94-82-6)**
2,4-D, α-chlorocrotyl ester	**(2971-38-2)**
2,4-D diethylamine salt	**(20940-37-8)**
2,4-D dimethylamine salt	**(2008-39-1)**
2,4-D isooctyl ester	**(25168-26-7)**
2,4-D, isopropyl ester	**(94-11-1)**
2,4-D propylene glycol butyl ether ester	**(1928-45-6)**
2,4-D, propylene glycol butyl ether ester	**(1320-18-9)**
2,4-D sodium salt	**(2702-72-9)**
2,4-D-tris(2-hydroxyethyl)ammonium	**(2569-01-9)**
3,4-DA	**(588-22-7)**
D-90-A	**(7287-36-7)**
DA	**(59-92-7)**
DA	**(712-48-1)**
DA-192	**(50-33-9)**
2,4 DAA	**(615-05-4)**
DAAB	**(136-35-6)**
1,5-DAA (Russian)	**(129-44-2)**
2,4-DAA sulfate	**(39156-41-7)**
DAB	**(60-11-7)**
DAB (Carcinogen)	**(60-11-7)**
DAC 2787	**(1897-45-6)**
DAC 893	**(1861-32-1)**
2,4-D (ACGIH,DOT,OSHA)	**(94-75-7)**
4,4-DADPE	**(101-80-4)**
DADPE	**(101-80-4)**
DADPM	**(101-77-9)**
DAEP	**(13265-60-6)**
DAF 68	**(117-81-7)**
DAFF	**(16368-97-1)**
DAG 325	**(1317-33-5)**
DAGC	**(142-22-3)**
D-P-A Injection	**(81-13-0)**
DAM	**(57-71-6)**
DAMC	**(87-01-4)**
1,4-DA-2-MOA (Russian)	**(2872-48-2)**
DAMS	**(51-63-8)**
DAP	**(439-14-5)**
DAP	**(58-15-1)**
DAP	**(7783-28-0)**
DAPA	**(140-56-7)**
DAPM	**(101-77-9)**
DAPT	**(490-55-1)**
DAS	**(140-56-7)**
DAS	**(51-63-8)**
DASC 2	**(59453-69-9)**
DASC 3	**(56776-30-8)**
DASC 4	**(56776-29-5)**
DASD	**(81-11-8)**
DATB	**(1630-08-6)**
DATC	**(2303-16-4)**
2,4-DB	**(94-82-6)**
4(2,4-DB)	**(94-82-6)**
DB 905	**(117-18-0)**
DB-905	**(879-39-0)**
1,2,5,6-DBA	**(53-70-3)**
DBA	**(53-70-3)**
DBA	**(57-97-6)**
DBAE	**(102-81-8)**
DBB	**(110-52-1)**
DBCP	**(96-12-8)**
DBD	**(10318-26-0)**
DBD	**(86-50-0)**
1,2-DBDCE	**(683-68-1)**
DBDPO	**(1163-19-5)**
2,4-DB-Dimethylammonium	**(2758-42-1)**
2,4-DBE	**(94-80-4)**
DBE	**(106-93-4)**
DBED diacetate	**(122-75-8)**
DBF	**(761-65-9)**
DBH	**(58-89-9)**
DBH	**(608-73-1)**
DBHMD	**(4835-11-4)**
DBI	**(114-86-3)**
DBI-TD	**(834-28-6)**
DBM	**(105-76-0)**
DBM	**(488-41-5)**
DBMC	**(497-39-2)**
DBMP	**(128-37-0)**
2,6-DBN	**(1194-65-6)**
DBN	**(924-16-3)**
DBNA	**(924-16-3)**
DBNPA	**(10222-01-2)**
DBN (The herbicide)	**(1194-65-6)**
DBOT	**(818-08-6)**
DBP	**(84-74-2)**
DBP	**(90-98-2)**
DBPC	**(128-37-0)**
DBPC (Technical grade)	**(128-37-0)**
DBQ	**(719-22-2)**
DBS	**(27176-87-0)**
DBS	**(523-88-6)**
D.B.T.C.	**(683-18-1)**
DBTL	**(77-58-7)**
DB(a,c)A	**(215-58-7)**
DB(a,e)P	**(192-65-4)**
DB(a,h)A	**(53-70-3)**
DB(a,h)AC	**(226-36-8)**
DB(a,h)P	**(189-64-0)**
DB(a,i)P	**(189-55-9)**
DB(a,j)AC	**(224-42-0)**
DB(a,l)P	**(191-30-0)**
2,4-DB butyl ester	**(6753-24-8)**
7H-DB(c,g)C	**(194-59-2)**
2,4-D-Bee	**(1929-73-3)**
2,4-DB isoocytl ester	**(1320-15-6)**
2,4-DB-sodium	**(10433-59-7)**
2,4-DB sodium salt	**(10433-59-7)**
DC 0572	**(539-21-9)**
DC 2	**(7782-42-5)**
DC 360	**(63148-62-9)**
3,4-DCA	**(95-76-1)**
DCA	**(112-69-6)**
DCA	**(118-52-5)**
DCA	**(79-43-6)**
DCA	**(95-76-1)**
DCA 70	**(9003-20-7)**
DCAB	**(1602-00-2)**
DCAOB	**(614-26-6)**
DC Antifoam A	**(8050-81-5)**
1,4-DCB	**(764-41-0)**
DCB	**(1194-65-6)**
DCB	**(764-41-0)**

DCB	(90-98-2)	DD-Mencs	(8066-01-1)
DCB	(91-94-1)	D-D Methylisothiocyanate	(8066-01-1)
DCB	(95-50-1)	DD-Methyl isothiocyanate mixt.	(8066-01-1)
DCBD	(28577-62-0)	DD Mixture	(8003-19-8)
DCBN	(1918-13-4)	DDNO	(1643-20-5)
DCBZ	(3978-67-4)	DDNU (VAN)	(530-48-3)
DCC	(538-75-0)	DDOA	(828-00-2)
DCCD	(538-75-0)	DDOD	(24201-58-9)
DCCI	(538-75-0)	DDOH	(2642-82-2)
DCDD	(33857-26-0)	p,p'-DDOH	(2642-82-2)
2,3-DCDT	(2303-16-4)	DDOM	(2642-82-2)
1,1-DCE	(75-35-4)	DDP	(15663-27-1)
1,2-DCE	(107-06-2)	cis-DDP	(15663-27-1)
DCEE	(111-44-4)	DDS	(25377-73-5)
DC 200 Fluid	(107-46-0)	DDS	(80-08-0)
DCHA	(101-83-7)	DDS A	(25377-73-5)
DCI Light Magnesium carbonate	(546-93-0)	DD Soil Fumigant	(8003-19-8)
DCIP	(108-60-1)	DDS (Pharmaceutical)	(80-08-0)
3,4-DCIPC	(2150-28-9)	DDS(Van)	(80-08-0)
DCIP (Nematocide)	(108-60-1)	o,p'-DDT	(789-02-6)
DCM	(116-52-9)	p,p'-DDT	(50-29-3)
DCM	(2164-09-2)	DDT (ACGIH,OSHA) [UN 2761]	(50-29-3)
DCM	(528-74-5)	DDT-Technical	(8017-34-3)
DCM	(75-09-2)	DDT dehydrochloride	(72-55-9)
DCMA	(2164-09-2)	DDUG	(15427-93-7)
DCMO	(5234-68-4)	DDUG diMS	(15427-93-7)
DCMOD	(5259-88-1)	DDVF	(62-73-7)
DCMU	(330-54-1)	DDVP	(62-73-7)
DCMX	(133-53-9)	DDVP (Insecticide)	(62-73-7)
DCNA	(99-30-9)	2,4-D-Dicamba mixt.	(8068-77-7)
DCNA (Fungicide)	(99-30-9)	DN-Dry Mix No. 2	(534-52-1)
DCNB	(99-54-7)	DEA	(111-42-2)
DCNU	(54749-90-5)	DEA	(20940-37-8)
2,4-DCP	(120-83-2)	DEA	(91-66-7)
DCP	(120-83-2)	DEA-Dodecylbenzenesulfonate	(26545-53-9)
DCP	(513-88-2)	DEAE	(100-37-8)
DCP	(542-75-6)	DEA No. 1230	(492-39-7)
DCP 340	(7757-93-9)	DEA No. 1475	(457-87-4)
DCPA	(1861-32-1)	DEA No. 1480	(49681-82-5)
DCPA	(50976-02-8)	DEA No. 1485	(3563-49-3)
DCPA	(709-98-8)	DEA No. 1503	(3736-08-1)
DCPC	(80-06-8)	DEA No. 1575	(15686-61-0)
DCPE	(80-06-8)	DEA No. 1635	(14148-99-3)
DCPM	(555-89-5)	DEA No. 1647	(10389-73-8)
DCPMU	(3567-62-2)	DEA No. 1750	(467-60-7)
D.C.S.	(132-27-4)	DEA No. 1760	(1209-98-9)
DCTA	(482-54-2)	DEA No. 2100	(76-75-5)
DCU	(116-52-9)	DEA No. 2100	(77-26-9)
D-D	(8003-19-8)	DEA No. 2264	(151-83-7)
D-D92	(542-75-6)	DEA No. 2271	(115-58-2)
DDA	(112-18-5)	DEA No. 2316	(115-58-2)
DDA	(83-05-6)	DEA No. 2460	(2218-68-0)
p,p'-DDA	(83-05-6)	DEA No. 2510	(3563-58-4)
DDAB	(2390-68-3)	DEA No. 2572	(340-57-8)
DDC	(148-18-5)	DEA No. 2591	(78-12-6)
DDC	(79-44-7)	DEA No. 2748	(1812-30-2)
2,4'-DDD	(53-19-0)	DEA No. 2749	(36104-80-0)
DDD	(72-54-8)	DEA No. 2752	(33671-46-4)
o,p'-DDD	(53-19-0)	DEA No. 2753	(24166-13-0)
p,p'-DDD	(72-54-8)	DEA No. 2754	(2894-67-9)
DDDM	(97-23-4)	DEA No. 2756	(29975-16-4)
DDDS (Pesticide)	(1142-19-4)	DEA No. 2758	(29177-84-2)
DDE	(72-55-9)	DEA No. 2762	(23092-17-3)
o,p'-DDE	(3424-82-6)	DEA No. 2763	(1622-62-4)
p,p'-DDE	(72-55-9)	DEA No. 2764	(2955-38-6)
DDM	(101-77-9)	DEA No. 2768	(23887-31-2)
DDM	(97-23-4)	DEA No. 2771	(59128-97-1)

DEA No. 2772 (27223-35-4)
DEA No. 2773 (61197-73-7)
DEA No. 2774 (848-75-9)
DEA No. 2800 (64-55-1)
DEA No. 2836 (2898-12-6)
DEA No. 2837 (2011-67-8)
DEA No. 2839 (24143-17-7)
DEA No. 2881 (36735-22-5)
DEA No. 2882 (28981-97-7)
DEA No. 2883 (52463-83-9)
DEA No. 2884 (59467-70-8)
DEA No. 2886 (10379-14-3)
DEA No. 2887 (28911-01-5)
DEA No. 7260 (83-74-9)
DEA No. 7300 (82-58-6)
DEA No. 7310 (478-94-4)
DEA No. 7360 (8063-14-7)
DEA No. 7374 (117-51-1)
DEA No. 7379 (51022-71-0)
DEA No. 7390 (1082-88-8)
DEA No. 7391 (53581-53-6)
DEA No. 7395 (15588-95-1)
DEA No. 7396 (24973-25-9)
DEA No. 7401 (13674-05-0)
DEA No. 7405 (54946-52-0)
DEA No. 7411 (23239-32-9)
DEA No. 7415 (11006-96-5)
DEA No. 7433 (487-93-4)
DEA No. 7434 (61-51-8)
DEA No. 7437 (520-52-5)
DEA No. 7438 (520-52-5)
DEA No. 7458 (2201-39-0)
DEA No. 7460 (2201-24-3)
DEA No. 7470 (21500-98-1)
DEA No. 7482 (3567-12-2)
DEA No. 7484 (3321-80-0)
DEA No. 8161 (101-40-6)
DEA No. 8603 (3867-15-0)
DEA No. 9020 (144-14-9)
DEA No. 9041 (50-36-2)
DEA No. 9050 (63905-03-3)
DEA No. 9051 (3861-72-1)
DEA No. 9052 (14297-87-1)
DEA No. 9053 (3688-65-1)
DEA No. 9054 (4406-22-8)
DEA No. 9055 (427-00-9)
DEA No. 9056 (14521-96-1)
DEA No. 9058 (14357-78-9)
DEA No. 9059 (13764-49-3)
DEA No. 9064 (52485-79-7)
DEA No. 9070 (125-27-9)
DEA No. 9145 (509-60-4)
DEA No. 9168 (28782-42-5)
DEA No. 9180 (481-37-8)
DEA No. 9180 (519-09-5)
DEA No. 9210 (125-70-2)
DEA No. 9226 (466-40-0)
DEA No. 9233 (77-17-8)
DEA No. 9240 (3734-52-9)
DEA No. 9260 (143-52-2)
DEA No. 9301 (2183-56-4)
DEA No. 9302 (16008-36-9)
DEA No. 9304 (7732-92-5)
DEA No. 9305 (125-23-5)
DEA No. 9307 (639-46-3)
DEA No. 9308 (467-18-5)
DEA No. 9309 (3688-66-2)
DEA No. 9312 (639-48-5)
DEA No. 9313 (466-97-7)
DEA No. 9314 (509-67-1)
DEA No. 9315 (466-90-0)
DEA No. 9319 (25333-77-1)
DEA No. 9335 (3176-03-2)
DEA No. 9601 (509-74-0)
DEA No. 9603 (17199-58-5)
DEA No. 9604 (468-51-9)
DEA No. 9605 (17199-54-1)
DEA No. 9606 (3691-78-9)
DEA No. 9607 (17199-59-6)
DEA No. 9608 (468-50-8)
DEA No. 9609 (17199-55-2)
DEA No. 9611 (468-59-7)
DEA No. 9615 (552-25-0)
DEA No. 9616 (86-14-6)
DEA No. 9617 (509-78-4)
DEA No. 9618 (545-90-4)
DEA No. 9619 (524-84-5)
DEA No. 9621 (467-86-7)
DEA No. 9622 (467-83-4)
DEA No. 9623 (441-61-2)
DEA No. 9624 (911-65-9)
DEA No. 9625 (469-82-9)
DEA No. 9626 (2385-81-1)
DEA No. 9627 (468-56-4)
DEA No. 9628 (469-79-4)
DEA No. 9629 (5666-11-5)
DEA No. 9631 (10061-32-2)
DEA No. 9633 (1477-39-0)
DEA No. 9634 (1531-12-0)
DEA No. 9635 (467-85-6)
DEA No. 9636 (561-48-8)
DEA No. 9637 (467-84-5)
DEA No. 9638 (129-83-9)
DEA No. 9639 (8008-60-4)
DEA No. 9641 (562-26-5)
DEA No. 9642 (302-41-0)
DEA No. 9643 (77-14-5)
DEA No. 9644 (561-76-2)
DEA No. 9645 (545-59-5)
DEA No. 9646 (64-39-1)
DEA No. 9647 (468-07-5)
DEA No. 9649 (15686-91-6)
DEA No. 9652 (76-41-5)
DEA No. 9661 (13147-09-6)
DEA No. 9663 (64-52-8)
DEA No. 9715 (127-35-5)
DEA No. 9730 (13495-09-5)
DEA No. 9732 (510-53-2)
DEA No. 9733 (297-90-5)
DEA No. 9737 (71195-58-9)
DEA No. 9740 (56030-54-7)
DEA No. 9743 (59708-52-0)
DEA No. 9750 (20380-58-9)
DEA No. 9800 (15301-48-1)
DEA No. 9812 (90736-23-5)
DEA No. 9813 (42045-86-3)
DEA No. 9814 (79704-88-4)
DEA No. 9830 (78995-10-5)
DEA No. 9831 (78995-14-9)
DEA-lauryl sulfate (143-00-0)
2,4-DEB (94-83-7)
DEB (1464-53-5)
DEB (56-53-1)
DEBA (57-44-3)

DEC **(105-58-8)**
DECHAN **(3129-91-7)**
DEDC **(148-18-5)**
DEDK **(148-18-5)**
DEDM hydantoin **(26850-24-8)**
DEF **(78-48-8)**
DEF Defoliant **(78-48-8)**
DEG **(111-46-6)**
D.E.H. 20 **(111-40-0)**
D.E.H. 26 **(112-57-2)**
DEH 24 **(112-24-3)**
DEHA **(103-23-1)**
DEHA **(3710-84-7)**
DEHB **(13898-68-5)**
DEHP **(117-81-7)**
DEHPA Extractant **(298-07-7)**
DEK **(96-22-0)**
DEL **(364-62-5)**
DEL 1267 **(364-62-5)**
DEMA **(55-86-7)**
DEN **(109-89-7)**
DEN **(55-18-5)**
DES-N **(4035-89-6)**
DENA **(55-18-5)**
2,4-DEP **(8005-49-0)**
2,4-DEP **(94-84-8)**
DEP **(115-76-4)**
DEP **(52-68-6)**
DEPD **(2497-07-6)**
DEPEG **(9002-92-0)**
DEPIL **(814-71-1)**
DEPP **(16368-97-1)**
DEPPT **(32345-29-2)**
DEP (Pesticide) **(52-68-6)**
D.E.R. 332 **(1675-54-3)**
DE 83R **(1163-19-5)**
DES **(56-53-1)**
DESD **(130-80-3)**
DESMA **(56-53-1)**
2,4-DES-Na **(136-78-7)**
2,4-DES-Natrium (German) **(136-78-7)**
DESdp **(522-40-7)**
DES (Synthetic estrogen) **(56-53-1)**
D.E.T. **(61-51-8)**
DET **(134-62-3)**
DET **(61-51-8)**
m-DET **(134-62-3)**
DETA **(111-40-0)**
DETA **(134-62-3)**
DETA-20 **(134-62-3)**
m-DETA **(134-62-3)**
DETF **(52-68-6)**
DETP **(2465-65-8)**
DEX **(502-55-6)**
DF **(26007-63-6)**
DF (Formaldehyde polymer) **(26007-63-6)**
DF 118 **(125-28-0)**
DF 125 **(57520-17-9)**
DF 468 **(56-12-2)**
DFA **(122-39-4)**
DF B **(70-51-9)**
DFD 0173 **(9002-88-4)**
DFD 0188 **(9002-88-4)**
DFD 2005 **(9002-88-4)**
DFD 6005 **(9002-88-4)**
DFD 6032 **(9002-88-4)**
DFD 6040 **(9002-88-4)**
DFDJ 5505 **(9002-88-4)**
DFDT **(475-26-3)**
DFO **(70-51-9)**
DFOA **(70-51-9)**
DFOM **(70-51-9)**
DFP **(55-91-4)**
DFS **(2664-63-3)**
DFT **(102-08-9)**
2 DG **(154-17-6)**
DGE (OSHA) **(2238-07-5)**
DGF 25 **(37099-12-0)**
DGM 2 **(2358-84-1)**
DGNB 3825 **(9002-88-4)**
DHA **(1740-19-8)**
DHA **(520-45-6)**
DHA-245 **(490-55-1)**
DHA-S **(4418-26-2)**
DHA-Sodium **(4418-26-2)**
2,3 DHB **(303-38-8)**
2,4-DHBA **(89-86-1)**
2,5-DHBA **(490-79-9)**
DHBP **(548-73-2)**
β-DHC **(14168-01-5)**
DHIC **(795-38-0)**
DHMS **(128-46-1)**
DHN **(4418-26-2)**
DHNT **(794-93-4)**
DHT2 **(67-96-9)**
4,5-DHPA **(63958-66-7)**
DHPN **(53609-64-6)**
DHPT **(92-36-4)**
DHPTA **(3148-72-9)**
DHS **(520-45-6)**
DIAK 1 **(143-06-6)**
DIANA **(3129-91-7)**
DIAPP **(121-54-0)**
DIBA **(141-04-8)**
DIBP **(84-69-5)**
DIC **(4342-03-4)**
DIC 1468 **(21087-64-9)**
DICHA **(101-83-7)**
DICHAN (Czech) **(3129-91-7)**
DID 47 **(58-36-6)**
DIDP (Plasticizer) **(26761-40-0)**
DIF 4 **(957-51-7)**
DIMID **(957-51-7)**
DIMO **(466-99-9)**
DIMP **(1445-75-6)**
DIN 2.4602 **(11114-92-4)**
DIN 2.4964 **(11114-92-4)**
DINEX **(131-89-5)**
DINOC **(2312-76-7)**
DINOC **(534-52-1)**
DINOK **(534-52-1)**
DINP **(28553-12-0)**
DIOA **(1330-86-5)**
DIOP **(27554-26-3)**
DIPA **(108-18-9)**
DIPA **(110-97-4)**
DIPAR **(834-28-6)**
DIPN **(53609-64-6)**
DIPRAM **(709-98-8)**
DIPS **(2973-10-6)**
DISTAQUAINE V-K **(132-98-9)**
DIT (VAN) **(66-02-4)**
2,4-D Isobutyl ester **(1713-15-1)**
DKC 1347 **(95-74-9)**

DKP	(7758-11-4)
DKhM	(116-52-9)
DLP	(9003-07-0)
DLP 787	(53558-25-1)
DLP-87	(53558-25-1)
DLT	(123-28-4)
DLTDP	(123-28-4)
DLTP	(123-28-4)
D 33LV	(280-57-9)
2,4-DM	(94-82-6)
D 50 M	(9003-20-7)
DM	(20830-81-3)
DM	(578-94-9)
DMS-70	(67-68-5)
DMS-90	(67-68-5)
DMA	(124-40-3)
DMA	(127-19-5)
DMA	(144-21-8)
DMA 100	(144-21-8)
DMA-4	(94-75-7)
DMAA	(75-60-5)
DMAB	(60-11-7)
DMAC	(127-19-5)
DMAE	(108-01-0)
DMAM	(105-52-2)
DMAMP	(7005-47-2)
DMAP	(595-33-5)
DMASA	(1596-84-5)
DMA (VAN)	(103-49-1)
2,5-DMA hydrochloride	(24973-25-9)
DMA-sol	(93746-34-0)
DMB	(150-78-7)
7,12-DMBA	(57-97-6)
DMBA	(57-97-6)
DMBC	(100-86-7)
DMBCA	(151-05-3)
DMC	(80-06-8)
DMCC	(79-44-7)
DMCT	(127-33-3)
DMD	(94-91-7)
DMDJ 4309	(9002-88-4)
DMDJ 5140	(9002-88-4)
DMDJ 7008	(9002-88-4)
DMDK	(128-04-1)
DMDMH	(6440-58-0)
DMDMH 55	(6440-58-0)
DMDM Hydantoin	(6440-58-0)
DMDT	(72-43-5)
p,p'-DMDT	(72-43-5)
DMDZ	(1088-11-5)
DMEP	(117-82-8)
DMF	(115-26-4)
DMF	(68-12-2)
DMFA	(68-12-2)
DMH	(306-37-6)
DMH	(540-73-8)
DMH	(56400-60-3)
DMH	(57-14-7)
DMH	(77-71-4)
DMHP	(32904-22-6)
DMI	(50-47-5)
DMI 50475	(50-47-5)
DMMP	(756-79-6)
DMMPA	(597-25-1)
DMN	(62-75-9)
DMNA	(62-75-9)
DMNM	(1456-28-6)
DMNM	(4164-28-7)
DMNO	(4164-28-7)
DMNP	(1630-17-7)
DMN-OAC	(56856-83-8)
DMOC	(5234-68-4)
2,5-DMP	(95-87-4)
2,6-DMP	(576-26-1)
3,4-DMP	(95-65-8)
3,5-DMP	(108-68-9)
DMP	(131-11-3)
DMP	(90-72-2)
DMP-30	(90-72-2)
DMPA	(299-85-4)
DMPA	(71-58-9)
DMPD	(105-10-2)
DMPT	(7227-91-0)
DMPTP	(123-28-4)
DMS	(75-18-3)
DMS	(77-78-1)
DMSA	(1596-84-5)
DMSO	(67-68-5)
DMSP	(115-90-2)
DMS(Methyl sulfate)	(77-78-1)
D 65MT	(28981-97-7)
DMT	(120-61-6)
DMT	(61-50-7)
DMTP	(55-38-9)
DMTP(Japan)	(950-37-8)
DMTT	(533-74-4)
DMU	(140-95-4)
DMU	(330-54-1)
DN	(131-89-5)
DN	(534-52-1)
DN 1	(131-89-5)
DN 289	(88-85-7)
DNA	(97-02-9)
DNAP	(4097-36-3)
DNB	(117-26-0)
DNBA	(99-34-3)
DNBP	(88-85-7)
DNBP Ammonium salt	(6365-83-9)
DNBS	(141-03-7)
DNC	(534-52-1)
DNCB	(97-00-7)
DNCDE	(20115-34-8)
DN Dry Mix No. 1	(131-89-5)
DN Dust No. 12	(131-89-5)
2,4-DNFB	(70-34-8)
1,3-DNG	(623-87-0)
DNOC	(2312-76-7)
DNOC	(534-52-1)
DNOCHP	(131-89-5)
DNOCP	(39300-45-3)
DNOC sodium salt	(2312-76-7)
DNOK (Czech)	(534-52-1)
DNOP	(117-84-0)
DNOPC	(39300-45-3)
DNOSAP	(4097-36-3)
DNOSBP	(88-85-7)
2,4-DNP	(51-28-5)
2,5-DNP	(329-71-5)
DNP	(117-27-1)
DNP	(305-85-1)
DNPC	(609-93-8)
DNPD	(93-46-9)
DNPDA	(93-46-9)
DNPMT	(101-25-7)

DNPT	(101-25-7)
DNPZ	(140-79-4)
DNRB	(1594-56-5)
DNSBP	(88-85-7)
2,3-DNT	(602-01-7)
2,4-DNT	(121-14-2)
2,5-DNT	(619-15-8)
2,6-DNT	(606-20-2)
3,4-DNT	(610-39-9)
3,5-DNT	(618-85-9)
DNT	(121-14-2)
DNTB	(1594-56-5)
DNTBP	(1420-07-1)
DNTP	(56-38-2)
DO 9	(9002-92-0)
DO 14	(2312-35-8)
DOA	(103-23-1)
DOA	(956-90-1)
DOBK	(303-38-8)
DOBO	(317-34-0)
DOB hydrobromide	(53581-53-6)
1,4-DOEA-5,8-DAPFA (Russian)	(3179-90-6)
DOF	(141-02-6)
DOIP	(137-89-3)
DOM	(142-16-5)
DOM	(15588-95-1)
"DOM" and "STP"	(15588-95-1)
DOP	(117-81-7)
DOPAC	(102-32-9)
DOS	(122-62-3)
D.O.T.	(148-01-6)
DOTC	(3542-36-7)
DOTG	(26401-97-8)
DOTG Accelerator	(97-39-2)
DOWCO 118	(299-85-4)
DOWCO 132	(299-86-5)
DOWCO 139	(315-18-4)
DOWCO-163	(1929-82-4)
DOWCO 169	(1754-58-1)
DOWCO 179	(2921-88-2)
DOWCO 184	(4080-31-3)
DOWCO 186	(76-87-9)
DOWCO-199	(5131-24-8)
DOWCO-213	(13121-70-5)
DOWCO 214	(5598-13-0)
DOWCO 217	(5598-52-7)
DOWCO 233	(55335-06-3)
DOWCO 269	(7159-34-4)
DOWCO 275	(39624-86-7)
DOWCO 290	(1702-17-6)
DOWCO 453	(69806-40-2)
DOWCO 453	(87237-48-7)
DOWCO 453ME	(69806-40-2)
DOW HCB	(85-19-8)
D-OX	(7775-14-6)
D-3,17-β-Oestradiol	(50-28-2)
D-Oxamicina (Italian)	(68-41-7)
D-Oxamycin	(68-41-7)
2,4-DP	(120-36-5)
2-(2,4-DP)	(120-36-5)
3,4-DP	(3307-41-3)
DP 02	(9004-35-7)
DP 06	(9004-35-7)
2,2-DPA	(127-20-8)
DPA	(117-34-0)
DPA	(122-39-4)
DPA	(131-73-7)
DPA	(709-98-8)
n-DPA	(99-66-1)
DPA (VAN)	(126-00-1)
DPA sodium	(1069-66-5)
DPBS	(97-16-5)
DPC	(104-74-5)
DPC	(140-22-7)
DPC	(39300-45-3)
DPD	(93-05-0)
DPD 63760M	(74223-64-6)
DPG	(102-06-7)
DPG Accelerator	(102-06-7)
DPGME	(34590-94-8)
DPH	(57-41-0)
DPH	(630-93-3)
DPID	(50-49-7)
DPN	(621-64-7)
DPNA	(621-64-7)
DPP	(131-18-0)
DPP	(56-38-2)
DPP	(94-78-0)
DPPD	(74-31-7)
DPS	(127-63-9)
DPT	(365-07-1)
DPTA	(3148-72-9)
DPX 1108	(25954-13-6)
DPX 1410	(23135-22-0)
DPX 3217	(57966-95-7)
DPX 3674	(51235-04-2)
DPX 4189	(64902-72-3)
DPX 6376	(74223-64-6)
DPX-F 5384	(83055-99-6)
DPX 3217M	(57966-95-7)
DPX T6376	(74223-64-6)
DPX-T 6376	(74223-64-6)
DPX-Y 6202	(76578-14-8)
D 50 (Polymer)	(9003-20-7)
DQ12	(14808-60-7)
DQDA 1868	(9002-88-4)
DQV-K	(132-98-9)
DQWA 0355	(9002-88-4)
D-Quinic acid	(77-95-2)
DRB	(1594-56-5)
DRC 1201	(16034-77-8)
DRC 1339	(95-74-9)
DRC 3340	(2655-14-3)
DRC 3341	(1129-41-5)
DRC 714	(4104-14-7)
DRC-1,339	(7745-89-3)
DRC-1339	(7745-89-3)
DRC-714	(4104-14-7)
DRP 859025	(51-18-3)
DRW 1139	(41394-05-2)
DS	(3696-28-4)
DS	(64-67-5)
DS 8620	(9003-07-0)
DS-15647	(39196-18-4)
DS 18302	(973-21-7)
DSDP	(78-53-5)
DSE	(142-59-6)
DSMA	(144-21-8)
DSMA Liquid	(144-21-8)
DSP	(7558-79-4)
D-S-S	(577-11-7)
DSS	(80-08-0)
DST	(128-46-1)
DST 50	(9003-55-8)

DST 75 (9003-55-8)
DSTDP (693-36-7)
DT (2971-22-4)
DT (479-18-5)
DT (50-89-5)
D-40TA (29975-16-4)
DTA (3148-72-9)
DTA (3347-22-6)
D'e.D.T.A. Disodique (French) (139-33-3)
DTB (541-53-7)
DTBP (110-05-4)
DTDP (119-06-2)
DTHYD (50-89-5)
DTIC (4342-03-4)
DTIC-Dome (4342-03-4)
DTM (73928-09-3)
DTMC (115-32-2)
5'-DTMP (365-07-1)
DTMP (365-07-1)
DTPA (67-43-6)
DTPA Calcium trisodium salt (12111-24-9)
DTPA pentasodium salt (140-01-2)
DU 112307 (35367-38-5)
DV (84-17-3)
DV 400 (9011-14-7)
D 43 (VAN) (142-30-3)
DW 62 (1165-48-6)
DW3418 (21725-46-2)
DX (23214-92-8)
DXM 100 (9002-88-4)
DXMS (50-02-2)
DYMID (957-51-7)
DYP-97F (105-74-8)
DZhp-4K (6300-37-4)
Dai-Ei Roccelline (1658-56-6)
Dab-O-Lite P 4 (13983-17-0)
Dabco (280-57-9)
Dabco Crystal (280-57-9)
Dabco EG (280-57-9)
Dabco 33LV (280-57-9)
Dabco R-8020 (280-57-9)
Dabco S-25 (280-57-9)
Dabeersen 503 (67-43-6)
Dabrosin (315-30-0)
Dabylen (147-24-0)
Dabylen (58-73-1)
Dacamine (93-76-5)
Dacamine (94-75-7)
Dacamox (39196-18-4)
Dacarbazine (4342-03-4)
Dacartil (467-85-6)
2,4-D acetate (2008-39-1)
D'acide tannique (French) (1401-55-4)
Daconate (2163-80-6)
Daconate 6 (2163-80-6)
Daconil (1897-45-6)
Daconil 2787 (1897-45-6)
Daconil 2787 Flowable Fungicide (1897-45-6)
Dacortin (53-03-2)
Dacosoil (1897-45-6)
Dacpm (101-14-4)
Dacthal (1861-32-1)
Dacthalor (1861-32-1)
Dactin (118-52-5)
Dactinol (83-79-4)
Dactinomycin (50-76-0)
Dactinomycin D (50-76-0)
Dadex (51-63-8)
Dadibutol (74-55-5)
Dadox D-Citramine (51-63-8)
Dadps (80-08-0)
Daftazol (490-55-1)
Dagadip (786-19-6)
Dagenan (144-83-2)
Dagenan chloride (121-60-8)
Dagutan (128-44-9)
Dai Cari XBN (136-60-7)
Daicel 1150 (9004-32-4)
Daicel 1180 (9004-32-4)
Daiflon (79-38-9)
Daiflon 22 (75-45-6)
Daiflon S 3 (76-13-1)
Dailon (330-54-1)
Dainichi Benzidine Yellow G (6358-85-6)
Dainichi Benzidine Yellow GRT (6358-85-6)
Dainichi Benzidine Yellow GT (6358-85-6)
Dainichi Benzidine Yellow GY (6358-85-6)
Dainichi Benzidine Yellow GYT (6358-85-6)
Dainichi Chrome Orange 5R (18454-12-1)
Dainichi Chrome Orange R (18454-12-1)
Dainichi Chrome Yellow 5G (1344-37-2)
Dainichi Chrome Yellow 10G (1344-37-2)
Dainichi Chrome Yellow G (7758-97-6)
Dainichi Cyanine Blue B (147-14-8)
Dainichi Cyanine Blue FPG (147-14-8)
Dainichi Cyanine Green FG (1328-53-6)
Dainichi Cyanine Green FGH (1328-53-6)
Dainichi Fast Blue EX (1325-87-7)
Dainichi Fast Blue Toner (1325-87-7)
Dainichi Fast Orange RR (3520-72-7)
Dainichi Fast Red B Base (97-52-9)
Dainichi Fast Scarlet G Base (99-55-8)
Dainichi Fast Violet MX (1325-82-2)
Dainichi Fast Violet M toner (1325-82-2)
Dainichi Fast Yellow G (2512-29-0)
Dainichi Lake Red C (5160-02-1)
Dainichi Lithol Red R (1103-38-4)
Dainichi Permanent Red GG (3468-63-1)
Dainichi Permanent Red 4 R (2425-85-6)
Dainichi Permanent Red RX (2814-77-9)
Dairylide Yellow YT 553D (6358-85-6)
Daisen (12122-67-7)
Daishiki Amaranth (915-67-3)
Daishiki Brilliant Scarlet 3R (2611-82-7)
Daisolac (9002-88-4)
Daito Blue Base BB (120-00-3)
Daito Brown Salt RR (609-20-1)
Daito Grounder BS (135-65-9)
Daito Grounder D (135-61-5)
Daito Grounder G (91-96-3)
Daito Grounder OL (135-62-6)
Daito Grounder Phenyl (92-74-0)
Daito Orange Base GC (141-85-5)
Daito Orange Base R (99-09-2)
Daito Orange Salt GC (141-85-5)
Daito Red Base B (97-52-9)
Daito Red Base 3GL (89-63-4)
Daito Red Base RL (99-52-5)
Daito Red Base TR (95-69-2)
Daito Red Salt 3GL (89-63-4)
Daito Red Salt TR (3165-93-3)
Daito Scarlet Base G (99-55-8)
Daito Scarlet Base GG (95-82-9)
Daiya Foil (25038-59-9)

Dakins Solution (7681-52-9)
Daktal (Czech) (1861-32-1)
Daktin (118-52-5)
Daktose B (105-16-8)
Dakuron (2307-68-8)
Dal-E-Rad (2163-80-6)
Dal-E-Rad 100 (144-21-8)
Dal-E-Rad 120 (2163-80-6)
Dalamar Yellow (6358-31-2)
Dalapon (127-20-8)
Dalapon (75-99-0)
Dalapon 85 (75-99-0)
Dalapon sodium (127-20-8)
Dalapon sodium salt (127-20-8)
Dalf (298-00-0)
Dalf Fast Red (2302-96-7)
Dalgol (77-75-8)
Dalkon 11 (9CI) (96320-70-6)
Dalmadorm (1172-18-5)
Dalmadorm hydrochloride (1172-18-5)
Dalmane (1172-18-5)
Dalmate (1172-18-5)
Dalmation Insect Flowers (8003-34-7)
Dalpac (128-37-0)
Daltogen (102-71-6)
Daltolite Fast Blue B (147-14-8)
Daltolite Fast Green GN (1328-53-6)
Daltolite Fast Orange G (3520-72-7)
Daltolite Fast Yellow GT (6358-85-6)
Daltolite Pink FF (2379-74-0)
Dalzic (6673-35-4)
Damar (9000-16-2)
Damar Gum (9000-16-2)
Damar Resin (9000-16-2)
Dambose (87-89-8)
Damilan (50-48-6)
Daminozide (1596-84-5)
Dammar (9000-16-2)
Damoral (50-06-6)
Dana (55-18-5)
Danabol (72-63-9)
Danamid (105-60-2)
Danamid (25038-54-4)
Danamine (59-26-7)
Danantizol (60-56-0)
Danazol (17230-88-59)
2,4-D and 2,4,5-T (2:1) (8015-35-8)
Dane Salt (961-69-3)
Daneral (132-20-7)
Danex (52-68-6)
Danfirm (9003-20-7)
Danilone (83-12-5)
Danitol (39515-41-8)
Danitrol (39515-41-8)
Danizol (443-48-1)
Danocrine (17230-88-5)
Danol (17230-88-5)
Dantafur (67-20-9)
Danten (57-41-0)
Danten (630-93-3)
Danthion (56-38-2)
Danthron (117-10-2)
Dantinal (57-41-0)
Dantoin (118-52-5)
Dantoin (630-93-3)
Dantoin DMDMH 55 (6440-58-0)
Dantoinal (57-41-0)
Dantoinal klinos (57-41-0)
Dantoine (57-41-0)
Dantrolene (7261-97-4)
Dantromin (2152-34-3)
Dantron (117-10-2)
Danylen (90-84-6)
Dapacryl (485-31-4)
Dapaz (57-53-4)
Daphene (60-51-5)
Daphnetoxin, 12-((1-oxo-5-phenyl-2,4-pentadienyl)oxy)-, (12-β(E,E)) (34807-41-5)
Daplen (9002-88-4)
Daplen AD (9003-07-0)
Daplen APP (9003-07-0)
Daplen AS 50 (9003-07-0)
Daplen AT 10 (9003-07-0)
Daplen ATK 92 (9003-07-0)
Daplen DM 55U (9003-07-0)
Daplen 1810 H (9002-88-4)
Dapon 35 (131-17-9)
Dapon R (131-17-9)
Daprisal (62-44-2)
Dapson (80-08-0)
Dapsone (80-08-0)
Dapsonum (80-08-0)
Daptazile (490-55-1)
Daptazole (490-55-1)
Daquin (500-42-5)
Dar-Chem 14 (57-11-4)
Daraclor (58-14-0)
Daral (50-14-6)
Daramin (6381-61-9)
Daramin (6485-34-3)
Darammon (12125-02-9)
Daran (9011-06-7)
Daran CR 6795H (9011-06-7)
Daranide (120-97-8)
Darapram (58-14-0)
Daraprim (58-14-0)
Daraprime (58-14-0)
Daratak (9003-20-7)
Darcil (132-93-4)
Darco (7440-44-0)
Darebon (50-55-5)
Darid QH (9005-38-3)
Dariloid QH (9005-38-3)
Dark Blue Z (1327-57-7)
Dark Cocoa Essence (8002-31-1)
Dark Green EMBL (3626-28-6)
Daropervamin (300-42-5)
Dartal (84-06-0)
Dartalan (84-06-0)
Darvon (469-62-5)
Darvon compound (62-44-2)
Darvyl (657-27-2)
Dasanide (120-97-8)
Dasanit (OSHA) (115-90-2)
Dasanit sulfone (14255-72-2)
Dasanit sulphone (14255-72-2)
Dasikon (62-44-2)
Dasin (62-44-2)
Dasin CH (62-44-2)
Daskil (59-67-6)
Datril (103-90-2)
Daturine (101-31-5)
Daunamycin (20830-81-3)
Daunoblastina (20830-81-3)

Name	CAS No.
Daunomycin	(20830-81-3)
Daunorubicin	(20830-81-3)
Daunorubicine	(20830-81-3)
Dauran	(357-56-2)
Davison SG-67	(7631-86-9)
Davitamon C	(50-81-7)
Davitamon PP	(59-67-6)
Davitamon d	(50-14-6)
Davitin	(50-14-6)
Davosin	(80-35-3)
Davurtrop	(55-48-1)
Dawe's Destrol	(56-53-1)
Dawson 100	(74-83-9)
Daxtron	(1970-40-7)
Dazomet-Powder BASF	(533-74-4)
Dazomet	(533-74-4)
Dazzel	(333-41-5)
Dbed Dipenicillin G	(1538-09-6)
Dbed Penicillin	(1538-09-6)
2,4-D-bis(2-hydroxyethyl)ammonium	(5742-19-8)
De-Cut	(123-33-1)
De-Fend	(60-51-5)
De-Fol-Ate	(10326-21-3)
De-Fol-Ate	(7775-09-9)
De-Green	(78-48-8)
De-Kalin	(91-17-8)
De-Sprout	(123-33-1)
Dea Oxo-5	(300-42-5)
Deacetyl-N-methylcolchicine	(477-30-5)
Deacetylmethylcolchicine	(477-30-5)
N-Deacetyl-N-methylcolchicine	(477-30-5)
Deactivator E	(111-46-6)
Deactivator H	(111-46-6)
Deadopa	(59-92-7)
Dealca TP1	(9000-30-0)
Dealca TP2	(9000-30-0)
Dealkylprazepam	(1088-11-5)
Deaminated Sencor	(35045-02-4)
Deaminoisocytosine	(4562-27-0)
Deaminometribuzin	(35045-02-4)
Deamocard	(54-95-5)
Deanol	(108-01-0)
Deanox	(1309-37-1)
Deanox DNX Pigments	(1309-37-1)
Deapasil	(65-49-6)
Debantic	(22248-79-9)
Debecillin	(1538-09-6)
Debecylina	(1538-09-6)
Debenal	(68-35-9)
Debendox	(8064-77-5)
Debendrin	(58-73-1)
Debroussaillant 600	(94-75-7)
Debroussaillant Concentre	(93-76-5)
Debroussaillant Super Concentre	(93-76-5)
Debroxide	(94-36-0)
Dec	(91-17-8)
Decabane	(1194-65-6)
Decaborane(14)	(17702-41-9)
Decaborane (ACGIH,OSHA) [UN 1868]	(17702-41-9)
2,2',3,3',4,4',5,5',6,6'-Decabromo-1,1'-biphenyl	(13654-09-6)
Decabromobiphenyl ether	(1163-19-5)
Decabromobiphenyl oxide	(1163-19-5)
Decabromodiphenyl oxide	(1163-19-5)
Decabromophenyl ether	(1163-19-5)
Decachlor	(2227-17-0)
1,1',2,2',3,3',4,4',5,5'-Decachlorobi-2,4-cyclopentadien-1-yl	(2227-17-0)
Decachlorobi-2,4-cyclopentadien-1-yl	(2227-17-0)
Decachlorobiphenyl	(2051-24-3)
Decachlorobutane	(6820-74-2)
1,2,3,5,6,7,8,9,10,10-Decachloro(5.2.1.$0^{2,6}$.$0^{3,9}$.$0^{5,8}$)decano-4-one	(143-50-0)
Decachloroketone	(143-50-0)
Decachloro-1,3,4-metheno-2H-cyclobuta(cd)pentalen-2-one	(143-50-0)
1,1a,3,3a,4,5,5,5a,5b,6-Decachlorooctahydro-1,3,4-metheno-2H-cyclobuta(cd)pentalen-2-one	(143-50-0)
Decachlorooctahydro-1,3,4-metheno-2H-cyclobuta(cd)pentalen-2-one	(143-50-0)
Decachloropentacyclo(5.2.1.$0^{2,6}$.$0^{3,9}$.$0^{5,8}$)Decan-4-one	(143-50-0)
Decachloropentacyclo(5.3.0.$0^{2,6}$.$0^{4,10}$.$0^{5,9}$)Decan-3-one	(143-50-0)
Decachlorotetracyclodecanone	(143-50-0)
Decachlorotetrahydro-4,7-methanoindeneone	(143-50-0)
Decacil	(58-25-3)
Decaderm	(50-02-2)
(E,E)-2,4-Decadienal	(25152-84-5)
2,4-trans,trans-Decadienal	(25152-84-5)
2,4-Decadienal (9CI)	(2363-88-4)
2,4-Decadienal, (E,E)- (9CI)	(25152-84-5)
1,9-Decadiene (9CI)	(1647-16-1)
3,5-Decadiene, 2,2-dimethyl-, (Z,Z)- (9CI)	(55638-50-1)
2,8-Decadiyne (7CI,8CI,9CI)	(4116-93-2)
3,5-Decadiyne, 2,2-dimethyl- (9CI)	(55682-73-0)
Decadron	(50-02-2)
Decaethoxy oleyl ether	(9004-98-2)
Decafluorobutane	(355-25-9)
Decaglycerol (9CI)	(9041-07-0)
Decahydroazecine	(4396-27-4)
1,2,3,4,4a,4b,7,9,10,10a-Decahydro-2-hydroxy-2,4b-dime thyl-7-oxo-1-phenanthrenepropionic acid δ-lactone	(968-93-4)
Decahydronaphthalene, trans-	(493-02-7)
cis-Decahydronaphthalene	(493-01-6)
trans-Decahydronaphthalene	(493-02-7)
Decahydronaphthalene [UN 1147]	(91-17-8)
Decahydro-β-naphthol	(825-51-4)
Decahydronaphthol-2	(825-51-4)
trans-Decahydro-β-naphthol	(825-51-4)
δ-Decalactone	(705-86-2)
γ-n-Decalactone	(706-14-9)
Decaldehyde	(112-31-2)
n-Decaldehyde	(112-31-2)
cis-Decalin	(493-01-6)
trans-Decalin	(493-02-7)
Decalin (DOT)	(91-17-8)
Decalin Solvent	(91-17-8)
cis-Decalin-9-carboxylic acid	(3021-73-6)
2-Decalinol	(825-51-4)
2-Decalol	(825-51-4)
Decalso	(1344-00-9)
Decalso F	(1344-00-9)
Decamethonium	(156-74-1)
Decamethonum	(156-74-1)
Decamethrin	(52918-63-5)
Decamethrine	(52918-63-5)
Decamethylcyclopentasiloxane	(541-02-6)
Decamethylene dibromide	(4101-68-2)
Decamethylenedicarboxylic acid	(693-23-2)
Decamethylene glycol-decamethylene dicarboxylic acid copolymer	(35464-94-9)
Decamethylene glycol-dodecanedioic acid copolymer	(35464-94-9)
Decamethylene glycol-.α.,.ω.-dodecanedioic acid polymer	(35464-94-9)
Decamethyltetrasiloxane	(141-62-8)
Decamine	(94-75-7)
Decamine 4T	(93-76-5)
1-Decaminium, N-octyl-N,N-dimethyl-, chloride	(32426-11-2)
1-Decanal	(112-31-2)
Decanal	(112-31-2)

Name	CAS No.
n-Decanal	**(112-31-2)**
Decanaldehyde	**(112-31-2)**
Decanamide, N,N-bis(2-hydroxyethyl)- (9CI)	**(136-26-5)**
Decanamide, N,N-dimethyl	**(14433-76-2)**
1-Decanamine	**(2016-57-1)**
1-Decanamine, N-decyl- (9CI)	**(1120-49-6)**
1-Decanamine, N-decyl-N-methyl- (9CI)	**(7396-58-9)**
1-Decanamine, N,N-dimethyl- (9CI)	**(1120-24-7)**
1-Decanamine, hydrochloride (9CI)	**(143-09-9)**
1-Decanamine, N-methyl-N-octyl- (9CI)	**(22020-14-0)**
1-Decanaminium, N-decyl-N,N-dimethyl-, bromide (9CI)	**(2390-68-3)**
1-Decanaminium, N-decyl-N,N-dimethyl-, chloride (9CI)	**(7173-51-5)**
1-Decanaminium, N-decyl-N-methyl-N-(3-(trimethoxysilyl)-propyl)-, chloride (9CI)	**(68959-20-6)**
1-Decanaminium, N,N-dimethyl-N-octyl-, chloride (9CI)	**(32426-11-2)**
1-Decanaminium, N,N,N-trimethyl-, chloride (9CI)	**(10108-87-9)**
Decane	**(124-18-5)**
n-Decane [UN 2247]	**(124-18-5)**
Decane, 2,9-bis(chlorodifluoromethyl)-1,1,1,2,3,3,4,4,5,5,6,6,7,7,8,8,9,10,10,10-eicosafluoro- (9CI)	**(103188-54-1)**
Decane, 1-bromo	**(112-29-8)**
1-Decanecarboxylic acid	**(112-37-8)**
Decane, chloro	**(28519-06-4)**
Decane, 1-chloro- (9CI)	**(1002-69-3)**
Decane, 3-chloro- (6CI,7CI,8CI,9CI)	**(1002-11-5)**
Decane, 1-cyclohexyl- (8CI)	**(1795-16-0)**
1,10-Decanediaminium, N,N,N,N',N',N'-hexamethyl-	**(156-74-1)**
Decane, 1,10-dibromo- (9CI)	**(4101-68-2)**
1,10-Decanedicarboxylic acid	**(693-23-2)**
Decanedinitrile	**(1871-96-1)**
Decanedioic acid	**(111-20-6)**
Decanedioic acid, Compd. with 1,6-Hexanediamine (1:1) (9CI)	**(6422-99-7)**
Decanedioic acid, bis(2-butoxyethyl) ester	**(141-19-5)**
Decanedioic acid, bis(2-ethylhexyl) ester	**(122-62-3)**
Decanedioic acid, bis(phenylmethyl) ester (9CI)	**(140-24-9)**
Decanedioic acid, dibutyl ester	**(109-43-3)**
Decanedioic acid, didecyl ester (9CI)	**(2432-89-5)**
Decanedioic acid, dimethyl ester (9CI)	**(106-79-6)**
Decanedioic acid, dinonyl ester (9CI)	**(4121-16-8)**
Decanedioic acid, lead(2+) salt (1:1) (9CI)	**(29473-77-6)**
Decanedioic acid, Polymer with 1,6-hexanediol (9CI)	**(26745-88-0)**
1,10-Decanediol-dodecanedioic acid copolymer	**(35464-94-9)**
1,10-Decanediol-1,12-dodecanedioic acid polymer	**(35464-94-9)**
1,10-Decanediol-dodecanedioic acid polymer	**(35464-94-9)**
1,10-Decanediol, Polymer with dodecanedioic acid (9CI)	**(35464-94-9)**
Decanedioyl chloride	**(111-19-3)**
Decanedioyl dichloride (9CI)	**(111-19-3)**
Decane, 1,1,1,2,3,3,4,4,5,5,6,6,7,7,8,8,9,10,10,10-eicosa-fluoro-2,9-bis(trifluoromethyl)- (9CI)	**(103188-55-2)**
Decane, 1-(ethenyloxy)- (9CI)	**(765-05-9)**
Decane, 6-ethyl-2-methyl- (9CI)	**(62108-21-8)**
Decane, 1-fluoro	**(334-56-5)**
Decane, 1-iodo	**(2050-77-3)**
Decane, 2-methyl- (9CI)	**(6975-98-0)**
Decane, 4-methyl- (8CI,9CI)	**(2847-72-5)**
Decanenitrile (9CI)	**(1975-78-6)**
Decane, 1,1'-oxybis- (9CI)	**(2456-28-2)**
Decane,perchloropentacyclo-	**(2385-85-5)**
Decane, 1-phenyl- (8CI)	**(104-72-3)**
Decane, 2-phenyl- (8CI)	**(4537-13-7)**
Decane, 3-phenyl- (8CI)	**(4621-36-7)**
Decane, 4-phenyl- (8CI)	**(4537-12-6)**
Decane, 5-phenyl- (8CI)	**(4537-11-5)**
1-Decanethiol (9CI)	**(143-10-2)**
tert-Decanethiol (9CI)	**(30174-58-4)**
Decane, 2,2,7-trimethyl- (9CI)	**(62237-99-4)**
Decane, 2,6,7-trimethyl- (9CI)	**(62108-25-2)**
Decane, 3,3,4-trimethyl- (9CI)	**(49622-18-6)**
Decane, 3,3,8-trimethyl- (9CI)	**(62338-16-3)**
Decanoic-acid	**(334-48-5)**
n-Decanoic acid	**(334-48-5)**
Decanoic acid, barium salt (9CI)	**(13098-41-4)**
Decanoic acid, butyl ester	**(30673-36-0)**
Decanoic acid, cadmium salt (9CI)	**(2847-16-7)**
Decanoic acid, 2-chloroethyl ester (8CI,9CI)	**(15175-04-9)**
Decanoic acid, decyl ester (9CI)	**(1654-86-0)**
Decanoic acid, ethyl ester	**(110-38-3)**
Decanoic acid, 2-hydroxy- (9CI)	**(5393-81-7)**
Decanoic acid, 5-hydroxy- (8CI,9CI)	**(624-00-0)**
Decanoic acid, 3-hydroxy- (8CI,9CI) (VAN)	**(14292-26-3)**
Decanoic acid, isopropyl ester (8CI)	**(2311-59-3)**
Decanoic acid, 8-methyl- (8CI,9CI)	**(5601-60-5)**
Decanoic acid methyl ester	**(110-42-9)**
Decanoic acid, methyl ester (9CI)	**(110-42-9)**
Decanoic acid, 1-methylethyl ester (9CI)	**(2311-59-3)**
Decanoic acid, octyl ester (9CI)	**(2306-92-5)**
tert-Decanoic acid oxiranylmethyl ester	**(71206-09-2)**
Decanoic acid, 9-oxo- (8CI,9CI)	**(1422-26-0)**
Decanoic acid, 9-oxo-, methyl ester (6CI,7CI,9CI)	**(2575-07-7)**
Decanoic acid, 10-phenyl- (8CI)	**(18017-73-7)**
Decanoic acid, potassium salt (9CI)	**(13040-18-1)**
Decanoic acid, sodium salt	**(1002-62-6)**
1-Decanol	**(112-30-1)**
Decanol	**(112-30-1)**
n-Decanol	**(112-30-1)**
2-Decanol (9CI)	**(1120-06-5)**
4-Decanol (8CI,9CI)	**(2051-31-2)**
1-Decanol acetate	**(112-17-4)**
1-Decanol, aluminum salt (9CI)	**(26303-54-8)**
1-Decanol, 3,3,4,4,5,5,6,6,7,7,8,8,9,9,10,10,10-heptadecafluoro-(9CI)	**(678-39-7)**
Decanolide-1,4	**(706-14-9)**
Decanolide-1,5	**(705-86-2)**
1-Decanol, 1-phenyl- (8CI)	**(21078-95-5)**
2-Decanone	**(693-54-9)**
3-Decanone (9CI)	**(928-80-3)**
5-Decanone (8CI,9CI)	**(820-29-1)**
4-Decanone (6CI,7CI,8CI,9CI)	**(624-16-8)**
1-Decanone, 1-phenyl- (9CI)	**(6048-82-4)**
Decanophenone	**(6048-82-4)**
Decanox	**(762-12-9)**
Decanoyl chloride (9CI)	**(112-13-0)**
Decanoyl-peroxide	**(762-12-9)**
Decanoyl peroxide, Technically pure (DOT)	**(762-12-9)**
3,6,9,12,15,18,21,24,27,30-Decaoxatriacontan-1-ol, 30-(p-(1,1,3,3-tetramethylbutyl)phenyl)-	**(9002-93-1)**
Decapol A 33 (9CI)	**(63439-57-6)**
Decapryn	**(562-10-7)**
Decapryn succinate	**(562-10-7)**
Decaps	**(50-14-6)**
Decarboxycysteine	**(60-23-1)**
Decarpyn succinate (1:1)	**(562-10-7)**
Decasept	**(133-53-9)**
Decasiloxane, 1,1,1,3,3,5,7,7,9,11,11,13,15,15,17,19,19,19-octadecamethyl-5,9,13,1 7-tetraphenyl- (9CI)	**(60573-46-8)**
Decasone	**(50-02-2)**
Decaspray	**(50-02-2)**
n-Decatyl alcohol	**(112-30-1)**
Decco Salt No 5	**(7673-09-8)**
Deccoscald 282	**(122-39-4)**
Deccotane	**(13952-84-6)**
Decemthion	**(732-11-6)**
Decemthion p-6	**(732-11-6)**
2-Decen-1-al	**(3913-71-1)**

2-Decenal (3913-71-1)
2-Decenal, (E)- (8CI,9CI) (3913-81-3)
Decenaldehyde (3913-71-1)
1-n-Decene (872-05-9)
Decene (25339-53-1)
α-Decene (872-05-9)
n-1-Decene (872-05-9)
1-Decene (9CI) (872-05-9)
5-Decene, 4-ethynyl-, (E)- (9CI) (55976-10-8)
Decene, hydroformylation products (68516-18-7)
4-Decenoic acid, ethyl ester, cis- (8CI) (7367-84-2)
4-Decenoic acid, ethyl ester, (Z)- (9CI) (7367-84-2)
4-Decenoic acid, methyl ester, (Z)- (8CI,9CI) (7367-83-1)
5-Decen-1-ol, acetate, (E)- (9CI) (38421-90-8)
3-Decen-1-ol, acetate, (Z)- (9CI) (81634-99-3)
4-Decen-1-ol, acetate, (Z)- (9CI) (67452-27-1)
5-Decen-1-ol, acetate, (Z)- (9CI) (67446-07-5)
6-Decen-1-ol, acetate, (Z)- (9CI) (68760-70-3)
3-Decen-2-one, 3-methyl- (9CI) (54411-03-9)
Decentan (58-39-9)
(E)-5-Decen-1-yl acetate (38421-90-8)
(E)-5-Decenyl acetate (38421-90-8)
(R,Z)-5-(1-Decenyl)dihydro-2(3H)-furanone (64726-91-6)
(R-(Z))-5-(1-Decenyl)dihydro-2(3H)-furanone (64726-91-6)
(R-(Z))-5-(1-Decenyl)dihydro-2-(3H)-furanone (64726-91-6)
Deceth-4 phosphate (9004-80-2)
Dechlorane (2385-85-5)
Dechlorane 605 (13560-89-9)
Dechlorane 4070 (2385-85-5)
Dechlorane A-O (1309-64-4)
Dechlorane 604 (9CI) (71245-27-7)
Dechlorane Plus (13560-89-9)
Dechlorane Plus 515 (13560-89-9)
Dechlorane Plus 2520 (13560-89-9)
2,4-Dechlorophenyl p-nitrophenyl ether (1836-75-5)
Decicain (136-47-0)
Decicaine (136-47-0)
Deciquam (2390-68-3)
Deciquam 222 (2390-68-3)
Decis (52918-63-5)
Declid (112-38-9)
Declor-It (7772-98-7)
Declomycin (127-33-3)
Decofol (115-32-2)
n-Decoic acid (334-48-5)
Decopperizing Dross (Secondary nonferrous plant) (69029-52-3)
Decorpa (9000-30-0)
Decortancyl (53-03-2)
Decortin (53-03-2)
Decortin H (50-24-8)
Decortisyl (53-03-2)
3-Decoxypropane-1-amine (7617-78-9)
3-Decoxypropanenitrile (16728-51-1)
Decrotox (7700-17-6)
Dectancyl (50-02-2)
Decumbin (20350-15-6)
Decurvon (9004-10-8)
2-Decyl-1-Tetradecanol (58670-89-6)
Decyl acetate (112-17-4)
n-Decyl acetate (112-17-4)
Decyl acrylate (2156-96-9)
n-Decyl acrylate (2156-96-9)
Decyl-alcohol (112-30-1)
n-Decyl alcohol (112-30-1)
Decyl alcohol, ethoxylated (26183-52-8)
Decyl alcohol, ethoxylated, dihydrogen phosphate (9004-80-2)
1-Decyl aldehyde (112-31-2)
1-Decyl aldehyde (112-44-7)
Decyl aldehyde (112-31-2)
n-Decyl aldehyde (112-31-2)
Decylamine (2016-57-1)
Decylamine, N,N-dimethyl- (8CI) (1120-24-7)
Decylamine, hydrochloride (8CI) (143-09-9)
Decyl benzene, n- (104-72-3)
Decylbenzene (104-72-3)
n-Decylbenzene (104-72-3)
Decyl benzene sodium sulfonate (1322-98-1)
p-n-Decylbenzenesulfonate (140-60-3)
4-Decylbenzenesulfonic acid (140-60-3)
p-Decylbenzenesulfonic acid (140-60-3)
4-Decylbenzenesulfonic acid, sodium salt (2627-06-7)
1-Decyl bromide (112-29-8)
Decyl bromide (112-29-8)
n-Decyl bromide (112-29-8)
Decyl butyl phthalate (89-19-0)
Decyl chloride (Mixed isomers) (28519-06-4)
Decylcyclohexane (1795-16-0)
N-Decyl-1-decanamine (1120-49-6)
Decyl decanoate (1654-86-0)
Decyl dimethyl octyl ammonium chloride (32426-11-2)
Decylene (872-05-9)
n-Decylene (872-05-9)
n-Decyl ethanoate (112-17-4)
Decyl ether (8CI) (2456-28-2)
Decyl hexyl hexanedioate (22707-35-3)
Decylic acid (334-48-5)
n-Decylic acid (334-48-5)
Decylic alcohol (112-30-1)
Decylic aldehyde (112-31-2)
1-Decyl iodide (2050-77-3)
Decyl iodide (2050-77-3)
Decyl mercaptan (143-10-2)
tert-Decylmercaptan (30174-58-4)
n-Decyl methacrylate (3179-47-3)
N-Decyl-N-methyl-1-decanamine (7396-58-9)
cis-2-Decyl-3-(5-methylhexyl)oxirane (29804-22-6)
Decyl 2-methyl-2-propenoate (3179-47-3)
Decyl 9-octadecenoate (3687-46-5)
Decyl octanoate (2306-89-0)
Decyl octyl adipate (110-29-2)
n-Decyl n-octyl adipate (110-29-2)
Decyl octyl alcohol (112-92-5)
Decyloctyldimethylammonium chloride (32426-11-2)
Decyl octyl hexanedioate (110-29-2)
Decyl octyl phthalate (119-07-3)
n-Decyl n-octyl phthalate (119-07-3)
Decyl oleate (3687-46-5)
Decyl oxirane (2855-19-8)
3-(Decyloxy)-1-propanamine (7617-78-9)
3-(Decyloxy)propanenitrile (16728-51-1)
3-Decyloxysulfolane (18760-44-6)
3-(Decyloxy)tetrahydrothiophene 1,1-dioxide (18760-44-6)
Decylphenol (27157-66-0)
n-Decyl phosphoric acid (3921-30-0)
Decylpolyethyleneglycol 300 (26183-52-8)
1-Decylpyridinium chloride (1609-21-8)
Decylpyridinium chloride (1609-21-8)
Decyl(sulfophenoxy)benzenesulfonic acid, disodium salt (36445-71-3)
Decyltetradecanol (58670-89-6)
Decyltrimethylammonium bromide (2082-84-0)
Decyl vinyl ether (765-05-9)
1-Decyne (9CI) (764-93-2)
5-Decyne (9CI) (1942-46-7)
4-Decyne (8CI,9CI) (2384-86-3)

5-Decyne-4,7-diol, 2,4,7,9-tetramethyl- (9CI) **(126-86-3)**
Decynediol, tetramethyl- (8CI,9CI) **(1333-17-1)**
Ded-Weed **(75-99-0)**
Ded-Weed **(93-72-1)**
Ded-Weed **(94-74-6)**
Ded-Weed **(94-75-7)**
Ded-Weed Brush Killer **(93-76-5)**
Ded-Weed Crabgrass Killer **(590-28-3)**
Ded-Weed LV-69 **(94-75-7)**
Ded-Weed LV-6 Brush Kil and T-5 Brush Kil **(93-76-5)**
Dedelo **(50-29-3)**
Dederon **(63428-84-2)**
Dedevap **(62-73-7)**
Dedotex **(63428-84-2)**
Dee-Ron **(50-14-6)**
Dee-Ronal **(50-14-6)**
Dee-Roual **(50-14-6)**
Deenax **(128-37-0)**
Dee-osterol **(50-14-6)**
Deep Crimson Madder 10821 **(72-48-0)**
Deep Fastona Red **(2425-85-6)**
Deep Lemon Yellow **(7789-06-2)**
Deer's Tongue **(8024-14-4)**
Deertongue-Incolore **(8024-14-4)**
Deet **(134-62-3)**
4-Deethylatrazine **(6190-65-4)**
Deethylatrazine **(6190-65-4)**
Deferoxamine **(70-51-9)**
Deferoxaminum **(70-51-9)**
Deferrioxamine **(70-51-9)**
Deferrioxamine B **(70-51-9)**
Defilin **(577-11-7)**
Defiltran **(59-66-5)**
Deflamon-wirkstoff **(443-48-1)**
Deflogin **(129-20-4)**
Deflorin **(50-59-9)**
Defoamer S-10 **(7440-21-3)**
Defolit **(51707-55-2)**
Defonin **(54-85-3)**
Deftor **(19937-59-8)**
Defy **(2008-39-1)**
Degalan LP 59/03 **(9011-14-7)**
Degalan S 85 **(9011-14-7)**
Degalol **(83-44-3)**
De graafina **(50-50-0)**
Degranol **(551-74-6)**
Degranol **(576-68-1)**
Degranol chinoin **(551-74-6)**
Degrassan **(534-52-1)**
Degreaser P (9CI) **(83589-41-7)**
Degummed Soybean Oil **(8001-22-7)**
Degussa **(1333-86-4)**
Degussa P820 **(1344-00-9)**
Dehacodin **(125-28-0)**
Dehidrobenzperidol **(548-73-2)**
Dehistin **(154-69-8)**
Dehistin **(91-81-6)**
Dehistin monohydrochloride **(154-69-8)**
Dehydag Sulfate GL Emulsion **(151-21-3)**
Dehydag Sulphate GL Emulsion **(151-21-3)**
Dehydol 100 **(9004-98-2)**
Dehydol LS 4 **(9002-92-0)**
Dehydracetic acid **(520-45-6)**
Dehydratin **(59-66-5)**
Dehydrite **(10034-81-8)**
Dehydroabietic acid **(1740-19-8)**
Dehydroabietylamine **(1446-61-3)**
Dehydroabietylamine acetate **(2026-24-6)**
Dehydroabietylamine, ethylene oxide adduct **(51344-62-8)**
Dehydroabietylamine- ethylene oxide condensate **(51344-62-8)**
Dehydroabietylamine-ethylene oxide condensate **(51344-62-8)**
Dehydroacetic acid **(520-45-6)**
Dehydroacetic acid, sodium salt **(4418-26-2)**
Dehydrobenzperidol **(548-73-2)**
6-Dehydro-6-chloro-17-α-acetoxyprogesterone **(302-22-7)**
7-Dehydrocholesterin **(434-16-2)**
Dehydrocholesterin (German) **(434-16-2)**
7-Dehydrocholesterol **(434-16-2)**
Dehydrocholesterol **(434-16-2)**
7-Dehydrocholestrol, activated **(67-97-0)**
δ^1-Dehydrocortisol **(50-24-8)**
1,2-Dehydrocortisone **(53-03-2)**
1-Dehydrocortisone **(53-03-2)**
δ-1-Dehydrocortisone **(53-03-2)**
Dehydrodivanillic acid **(2134-90-9)**
2-Dehydroemetine **(4914-30-1)**
Dehydroemetine **(4914-30-1)**
Dehydro-4-epiabietal **(24035-50-5)**
Dehydro-4-epiabietic acid **(5155-70-4)**
Dehydroheliotridine **(26400-24-8)**
1-Dehydrohydrocortisone **(50-24-8)**
δ^1-Dehydrohydrocortisone **(50-24-8)**
11-Dehydro-17-hydroxycorticosterone **(53-06-5)**
11-Dehydro-17-hydroxycorticosterone acetate **(50-04-4)**
11-Dehydro-17-hydroxycorticosterone-21-acetate **(50-04-4)**
Dehydroisophytol **(29171-23-1)**
6-Dehydro-6-methyl-17-α-acetoxyprogesterone **(595-33-5)**
1-Dehydro-16-α-methyl-9-α-fluorohydrocortisone **(50-02-2)**
Dehydromethyltesterone **(72-63-9)**
a1-Dehydromethyltesterone **(72-63-9)**
1-Dehydro-17-α-methyltestosterone **(72-63-9)**
Dehydronivalenol **(51481-10-8)**
6-Dehydro-retro-progesterone **(152-62-5)**
Dehydrostilbestrol **(84-17-3)**
Dehydrostilboestrol **(84-17-3)**
δ(1)-Dehydrotestolactone **(968-93-4)**
1,2-Dehydrotestololactone **(968-93-4)**
1-Dehydrotestololactone **(968-93-4)**
δ(1)-Dehydrotestololactone **(968-93-4)**
Dehydro-p-toluidine **(92-36-4)**
Dehyquart A **(112-02-7)**
Dehyquart C **(104-74-5)**
Dehyquart LT **(112-00-5)**
Dehyquart STC-25 **(122-19-0)**
Deidrobenzperidolo **(548-73-2)**
Deinait **(68-88-2)**
Deiquat **(85-00-7)**
6-Deisopropylatrazine **(1007-28-9)**
Dejo **(514-73-8)**
Dekacort **(50-02-2)**
Dekalina (Polish) **(91-17-8)**
Dekamethylcyklopentasiloxan (Czech) **(541-02-6)**
Dekametrin (Hungarian) **(52918-63-5)**
Dekarbon T (9CI) **(83589-42-8)**
Dekortin **(53-03-2)**
Dekrysil **(534-52-1)**
Deksonal **(140-56-7)**
Delac J **(86-30-6)**
Delacillin **(26787-78-0)**
Deladiol **(979-32-8)**
Delagil **(54-05-7)**
Delahormone Unimatic **(979-32-8)**
Delalande 69276 **(29218-27-7)**
Delalutin **(630-56-8)**

Delan (3347-22-6)
Delan-Col (3347-22-6)
Delatestryl (315-37-7)
Delatestryl (58-18-4)
Delcortol (50-24-8)
Deleaf Defoliant (150-50-5)
Delestrogen (979-32-8)
Delestrogen 4X (979-32-8)
Delgesic (50-78-2)
Delicia (20859-73-8)
Delicia (7803-51-2)
Delicia gastoxin (20859-73-8)
Deliva (637-07-0)
Dellipsoids (51-63-8)
Delmofulvina (126-07-8)
Delmoneurina (51-12-7)
Delnatex (78-34-2)
Delnav (78-34-2)
Delnav (OSHA) (78-34-2)
Delonin Amide (98-92-0)
Delorazepam (2894-67-9)
Delorazepamum (Latin) (2894-67-9)
Delowas S (26140-60-3)
Delowax OM (26140-60-3)
Delpet 50M (9011-14-7)
Delpet 60N (9011-14-7)
Delpet 80N (9011-14-7)
Delphene (134-62-3)
m-Delphene (134-62-3)
Delphinic acid (503-74-2)
Delsene (10605-21-7)
Delsene M (12427-38-2)
Delsterol (67-97-0)
Delta (3691-35-8)
Deltacortenol (50-24-8)
Deltacortisone (53-03-2)
Deltacortone (53-03-2)
Deltacortril (50-24-8)
Deltafluorene (50-02-2)
Deltagluconolactone (90-80-2)
Deltalin (50-14-6)
Deltamethrin (52918-63-5)
Deltamethrine (52918-63-5)
Deltamine (2152-34-3)
Deltan (67-68-5)
Deltanet (65907-30-4)
Deltasone (53-03-2)
Deltathione (70-18-8)
Deltazina (68-35-9)
Deltic (78-34-2)
Deltison (53-03-2)
Deltisone (53-03-2)
Deltra (53-03-2)
Deltrate-20 (78-11-5)
Deltyl (142-91-6)
Deltyl Prime (142-91-6)
Deltylextra (110-27-0)
Delussa Black FW (1333-86-4)
Delvex (514-73-8)
Delvinal (125-42-8)
Delysid (50-37-3)
Delzol-W (54-95-5)
Demal 14 (9007-48-1)
Demarol (57-42-1)
Demasan (2675-77-6)
Demasorb (67-68-5)
Demavet (67-68-5)
Demecarium bromide (56-94-0)
Demeclocycline (127-33-3)
Demecolcin (477-30-5)
Demecolcine (477-30-5)
Demelverine (552-82-9)
Demephion-S (2587-90-8)
Demephion (2587-90-8)
Demephion (8065-62-1)
Demerol (50-13-5)
Demerol (57-42-1)
Demerol hydrochloride (50-13-5)
Demeso (67-68-5)
Demethylamitriptylene (72-69-5)
Demethylamitriptyline (72-69-5)
Demethylamitryptyline (72-69-5)
5-O-Demethylavermectin A1A (65195-55-3)
6-Demethyl-7-chlorotetracycline (127-33-3)
6-Demethylchlorotetracycline (127-33-3)
Demethylchlorotetracycline (127-33-3)
Demethylchlortetracyclin (127-33-3)
6-Demethyl-7-chlortetracycline (127-33-3)
6-Demethylchlortetracycline (127-33-3)
Demethylchlortetracycline (127-33-3)
Demethylchlortetracycline Base (127-33-3)
1-Demethyldiazepam (1088-11-5)
Demethyldiazepam (1088-11-5)
N-Demethyldiazepam (1088-11-5)
Demethyldihydrothebaine, acetate (466-90-0)
Demethyldihydrothebanine acetate (466-90-0)
Demethyldopan (66-75-1)
Demethylimipramine (50-47-5)
Demethylmisonidazole (13551-92-3)
Demethylmorphine (466-97-7)
Demeton (298-03-3)
Demeton-O (298-03-3)
Demeton-S (126-75-0)
Demeton (ACGIH,OSHA) (8065-48-3)
Demeton-O + Demeton-S (8065-48-3)
Demeton-O-Methyl (867-27-6)
Demeton-O-Metile (Italian) (867-27-6)
Demeton-Sulfone (4891-54-7)
Demeton methyl (8022-00-2)
Demeton-S-methyl (919-86-8)
Demeton-S-methylsulfon (German) (17040-19-6)
Demeton-S-methyl-sulfoxid (German) (301-12-2)
Demeton-S-methyl sulfoxide (301-12-2)
Demeton-S-methyl-sulphone (17040-19-6)
Demeton-methyl sulphoxide (301-12-2)
Demeton-S-metile (Italian) (919-86-8)
Demetrin (2955-38-6)
Demise (2008-39-1)
Democracin (60-54-8)
Demos-l40 (60-51-5)
Demosan (2675-77-6)
Demox (8065-48-3)
Demsodrox (67-68-5)
Denacol EX 314 (31305-91-6)
Denacol EX 721 (37099-12-0)
Denamone (59-02-9)
Denapon (63-25-2)
Denatonii benzoas (Latin) (3734-33-6)
Denatonium benzoate (3734-33-6)
Dendritis (7647-14-5)
Denka AC 50 (9003-22-9)
Denka AS-CY (9003-54-7)
Denka QP3 (9003-53-6)
Denka Vinyl MM 90 (9003-22-9)

Denkalac 61 (9003-22-9)
Denkalac 41M (9003-22-9)
Denmert (51308-54-4)
Densinfluat (79-01-6)
Denydryl (147-24-0)
Denyl (57-41-0)
Denyl (630-93-3)
Denyl sodium (630-93-3)
Deobase (8008-20-6)
Deobase (8020-83-5)
Deodorized Kerosene (8020-83-5)
Deodorized Kerosine (8020-83-5)
Deodorized Winterized Cottonseed Oil (8001-29-4)
Deofed (300-42-5)
Deorlene Blue 5G (55840-82-9)
Deorlene Fast Blue BL (12217-41-3)
Deorlene Fast Blue RL (12270-13-2)
Deorlene Green JJO (633-03-4)
Deoval (50-29-3)
11-Deoxojervine (4449-51-8)
2'Deoxy-5'-AMP (653-63-4)
Deoxy TMP (365-07-1)
2'-Deoxy-5'-adenosine monophosphate (653-63-4)
Deoxyadenosine monophosphate (653-63-4)
Deoxyadenylate (653-63-4)
2'-Deoxy-5'-adenylic acid (653-63-4)
2-Deoxy-D-arabino-hexose (154-17-6)
Deoxybenzoin (451-40-1)
Deoxycholatic acid (83-44-3)
7-α-Deoxycholic acid (83-44-3)
Deoxycholic acid (83-44-3)
17-Deoxycortisol (50-22-6)
6-Deoxy-6-demethyl-6-methylene-5-oxytetracycline (914-00-1)
N-Deoxydemoxapam (1088-11-5)
Deoxyephedrine (300-42-5)
2-Deoxyerythritol (3068-00-6)
2'-Deoxy-5-fluorouridine (50-91-9)
Deoxyfluorouridine (50-91-9)
6-Deoxy-L-galactose (2438-80-4)
2-Deoxy-D-glucose (154-17-6)
2-Deoxyglucose (154-17-6)
D-2-Deoxyglucose (154-17-6)
2-Deoxyglycerol (504-63-2)
2'-Deoxyguanosine (961-07-9)
Deoxyguanosine (961-07-9)
Deoxyhemoglobin (9008-02-0)
α-6-Deoxy-5-hydroxytetracycline (564-25-0)
6-Deoxy-L-mannose (3615-41-6)
1-Deoxy-1-(methylamino)-D-glucitol (6284-40-8)
2-Deoxy-2-(((methylnitrosoamino)carbonyl)amino)-D-glucopyranose (18883-66-4)
2-Deoxy-2-(3-methyl-3-nitrosoureido)-D-glucopyranose (18883-66-4)
2-Deoxy-2-(3-methyl-3-nitrosoureido)-α(and β)-D-glucopyranose (18883-66-4)
4-Deoxynivalenol (51481-10-8)
Deoxynivalenol (51481-10-8)
Deoxynorephedrine (60-13-9)
6-α-Deoxy-5-oxytetracycline (564-25-0)
α-6-Deoxyoxytetracycline (564-25-0)
2-Deoxy-D-erythro-pentose (533-67-5)
2-Deoxyphenobarbital (125-33-7)
1-β-D-2'-Deoxyribofuranosyl-5-flurouracil (50-91-9)
Deoxyribose (533-67-5)
Deoxyribosylthymine monophosphate (365-07-1)
Deoxytetraric acid (6915-15-7)
4-Deoxytetronic acid (96-48-0)
β-2'-Deoxythioguanosine (789-61-7)
2'-Deoxythymidine (50-89-5)
Deoxythymidine (50-89-5)
Deoxythymidine 5'-monophosphate (365-07-1)
Deoxythymidine monophosphate (365-07-1)
Deoxythymidine 5'-phosphate (365-07-1)
Deoxythymydilic acid (365-07-1)
2'-Deoxy-5-(trifluoromethyl)uridine (70-00-8)
Depakene (1069-66-5)
Depakene (99-66-1)
Depakine (1069-66-5)
Depakine (99-66-1)
Depallethrin (584-79-2)
Deparal (67-97-0)
Depekane (1069-66-5)
Depen (52-67-5)
Dephadren (51-63-8)
Dephadren (51-64-9)
Depigman (103-16-2)
Depinar (68-19-9)
Depo-MPA (71-58-9)
Depo-Medrate (53-36-1)
Depo-Medrol (53-36-1)
Depo-Medrone (53-36-1)
Depo-Methylprednisolone (53-36-1)
Depo-Methylprednisolone acetate (53-36-1)
Depo-Penicillin (6130-64-9)
Depo-Proluton (630-56-8)
Depo-Provera (57-83-0)
Depo-Provera (71-58-9)
Depogamma (13422-51-0)
Depo-medroxyprogesterone acetate (71-58-9)
Depot-Medrol (53-36-1)
Depocid (526-08-9)
Depocillin (54-35-3)
Depotsulfonamide (526-08-9)
Depovernil (80-35-3)
Depoxin (300-42-5)
Deprelin (110-61-2)
Depremol G (1854-26-8)
Depremol M (9011-05-6)
Depthon (52-68-6)
Dequest 2000 (6419-19-8)
Dequest 2006 (2235-43-0)
Dequest 2010 (2809-21-4)
Dequest 2015 (2809-21-4)
Dequest 2041 (1429-50-1)
Dequest Z 010 (2809-21-4)
Deracil (141-90-2)
Deratol (50-14-6)
Dereuma (58-15-1)
Derfon (90-84-6)
Dergramin (50-02-2)
Deriban (62-73-7)
Deril (83-79-4)
Deriphat 160 (3655-00-3)
Derizene (61-76-7)
Derizene (630-93-3)
Derma Fast Brown W-GL (16071-86-6)
Derma Yellow P (6373-74-6)
Dermacaine (85-79-0)
Dermacort (50-23-7)
Dermadex (70-30-4)
Dermaffine (143-28-2)
Dermafix Brown PL (16071-86-6)
Dermafosu (Polish) (299-84-3)
Dermagan (83-63-6)
Dermagen (83-63-6)
Dermagine (31566-31-1)

Dermalar (67-73-2)
Dermaline (480-22-8)
Dermaphos (299-84-3)
Dermarone (557-05-1)
Dermasorb (67-68-5)
Dermaton (470-90-6)
Dermina Yellow G (547-57-9)
Dermistina (58-73-1)
Dermodrin (58-73-1)
Dermofural (59-87-0)
Dermoxin (2398-96-1)
Derobin (480-22-8)
Deronil (50-02-2)
Derosal (10605-21-7)
Derribante (62-73-7)
Derrin (83-79-4)
Derris (83-79-4)
Des-I-Cate (129-67-9)
Des-Oxa-D (300-42-5)
2,4-Des Sodium (136-78-7)
Desacetylmethylcolchicine (477-30-5)
N-Desacetyl-N-methylcolchicine (477-30-5)
N-Desacetylmethylcolchicine (477-30-5)
Desadrene (50-02-2)
Desalkylprazepam (1088-11-5)
Desametasone (50-02-2)
Desamethasone (50-02-2)
Desameton (50-02-2)
Desamine (300-42-5)
Desapon (333-41-5)
Deschlorobiomycin (60-54-8)
N-Descyclopropylmethylprazepam (1088-11-5)
Desdemin (54-31-9)
Desenex (112-38-9)
Desenex solution (112-38-9)
Desentol (58-73-1)
Deseptyl (63-74-1)
Deseril (361-37-5)
Desernyl (361-37-5)
Deserpine (50-55-5)
Desert Red (5160-02-1)
Desertalc 57 (14807-96-6)
Deseryl (361-37-5)
Desethyl atrazine (6190-65-4)
Desfedran (300-42-5)
Desfedrin (300-42-5)
Desferal (70-51-9)
Desferral (70-51-9)
Desferrin (70-51-9)
Desferrioxamine (70-51-9)
Desferrioxamine B (70-51-9)
Desiccant L-10 (7778-39-4)
Desimipramine (50-47-5)
Desimpramine (50-47-5)
Desinfect (127-65-1)
Desipramin (50-47-5)
Desipramina (Spanish) (50-47-5)
Desipramine (50-47-5)
Desipraminum (Latin) (50-47-5)
Desmecolchine (477-30-5)
Desmecolcine (477-30-5)
Desmedipham (13684-56-5)
Desmel (60207-90-1)
Desmethryn (1014-69-3)
Desmethylamitriptyline (72-69-5)
N-Desmethylcodeine (467-15-2)
Desmethyldiazepam (1088-11-5)
N-Desmethyldiazepam (1088-11-5)
Desmethyldopan (66-75-1)
Desmethylimipramine (50-47-5)
Desmethylmethadone (467-85-6)
Desmethylmisonidazole (13551-92-3)
Desmethylmorphine (466-97-7)
Desmethylnitrophos (950-35-6)
Desmetryn (German, Dutch) (1014-69-3)
Desmetryne (1014-69-3)
Desmodur 44 (101-68-8)
Desmodur H (822-06-0)
Desmodur N (822-06-0)
Desmodur T80 (584-84-9)
Desmodur T100 (26471-62-5)
Desmophen 7100 (9003-11-6)
Desolet (7775-09-9)
Desomorfin (Czech) (427-00-9)
Desomorfina (Spanish) (427-00-9)
Desomorphine (427-00-9)
Desomorphinum (Latin) (427-00-9)
Desormone (120-36-5)
Desormone (94-75-7)
Desossiefedrina (300-42-5)
Desoxedrine (300-42-5)
Desoxin (300-42-5)
Desoxo-5 (300-42-5)
Desoxon 1 (79-21-0)
Desoxybenzoin (451-40-1)
Desoxycholic acid (83-44-3)
Desoxycholsaeure (German) (83-44-3)
Desoxyephedrine hydrochloride (300-42-5)
Desoxyfed (300-42-5)
2-Desoxy-D-glucose (French) (154-17-6)
Desoxyn (300-42-5)
Desoxyn (51-63-8)
Desoxynivalenol (51481-10-8)
(+-)-Desoxynorephedrine (300-62-9)
Desoxynorephedrine (60-13-9)
Desoxynorephedrine (60-15-1)
racemic-Desoxynor-ephedrine (300-62-9)
3-Desoxynorlutin (52-76-6)
Desoxyphed (300-42-5)
2-Desoxyphenobarbital (125-33-7)
Desoxyphenobarbitone (125-33-7)
Despacilina (6130-64-9)
Desphen (56-75-7)
Despirol (4234-79-1)
Dessicant L-10 (7778-39-4)
Dessicated Thyroid (8028-36-2)
Dessin (973-21-7)
Desson (88-04-0)
Destilados de Alquitran de Hulla (Spanish) (65996-92-1)
Destim (300-42-5)
Destriol (50-27-1)
Destrol (56-53-1)
Destrone (53-16-7)
Destruxol Applex (16893-85-9)
Destruxol Borer-Sol (107-06-2)
Destruxol orchid spray (54-11-5)
Destun (37924-13-3)
Desulfonated Spent Pulping Liquor (8061-51-6)
Desyphed (300-42-5)
Desyphen (300-42-5)
Detal (534-52-1)
Detalup (50-14-6)
Detamide (134-62-3)
Detapac (140-01-2)

Detapac	(67-43-6)
Detarex	(67-43-6)
Detarex PY	(140-01-2)
Detaril	(467-60-7)
Detarol trisodium salt	(139-89-9)
Detergent 66	(151-21-3)
Detergent Alkylate	(123-01-3)
Detergent HD-90	(25155-30-0)
Dethmor	(81-81-2)
Dethnel	(81-81-2)
Dethylandiamine	(958-93-0)
Detia	(20859-73-8)
Detia	(7803-51-2)
Detia-Ex-B	(20859-73-8)
Detia Gas EX-B	(7803-51-2)
Detia Gas EX-M	(74-83-9)
Detia Gas Ex-B	(20859-73-8)
Deticene	(4342-03-4)
Detmol-Extrakt	(58-89-9)
Detmol MA	(121-75-5)
Detmol MA 96%	(121-75-5)
Detmol U.A.	(2921-88-2)
Detol	(534-52-1)
Detox	(50-29-3)
Detox 25	(58-89-9)
Detoxan	(50-29-3)
Detoxargin	(1119-34-2)
Detoxin	(68238-35-7)
1,2,4,8-Detrachlorodibenzofuran	(64126-87-0)
Detreomycine	(56-75-7)
Detrex	(300-42-5)
Dettol	(88-04-0)
Deumacard	(54-95-5)
Deuslon-A	(50-27-1)
Deuterium [UN 1957]	(7782-39-0)
Deuterium oxide	(7789-20-0)
Deval Red K	(95-69-2)
Deval Red TR	(95-69-2)
Developer 11	(108-45-2)
Developer 13	(106-50-3)
Developer 14	(95-80-7)
Developer A	(135-19-3)
Developer AMS	(135-19-3)
Developer B	(95-80-7)
Developer BN	(135-19-3)
Developer Bon	(92-70-6)
Developer C	(108-45-2)
Developer DB	(95-80-7)
Developer DBJ	(95-80-7)
Developer H	(108-45-2)
Developer H	(95-80-7)
Developer M	(108-45-2)
Developer MC	(95-80-7)
Developer MT	(95-80-7)
Developer MT-CF	(95-80-7)
Developer MTD	(95-80-7)
Developer O	(108-46-3)
Developer P	(100-01-6)
Developer PF	(106-50-3)
Developer R	(108-46-3)
Developer RS	(108-46-3)
Developer Sodium	(135-19-3)
Developer T	(95-80-7)
Developer Z	(89-25-8)
Devicarb	(63-25-2)
Devicopper	(1332-40-7)
Devicopper	(1332-65-6)
Devicoran	(652-67-5)
Devigon	(60-51-5)
Devikol	(62-73-7)
Devinal	(125-42-8)
Devipon	(75-99-0)
Devisulphan	(115-29-7)
Devithion	(298-00-0)
Devol Orange B	(88-74-4)
Devol Orange C	(141-85-5)
Devol Orange GC	(141-85-5)
Devol Orange GC Salt	(141-85-5)
Devol Orange R	(99-09-2)
Devol Orange Salt B	(88-74-4)
Devol Red E	(97-52-9)
Devol Red F	(89-63-4)
Devol Red G	(89-62-3)
Devol Red GG	(100-01-6)
Devol Red K	(3165-93-3)
Devol Red Salt E	(99-52-5)
Devol Red Salt F	(89-63-4)
Devol Red Salt G	(89-62-3)
Devol Red TA Salt	(3165-93-3)
Devol Red TR	(3165-93-3)
Devol Scarlet A (Free Base)	(95-82-9)
Devol Scarlet B	(99-55-8)
Devol Scarlet 2GS Base	(95-82-9)
Devol Scarlet G Salt	(99-55-8)
Devoran	(58-89-9)
Devoton	(79-20-9)
Devrinol	(15299-99-7)
Dewaxed Lanolin	(8006-54-0)
Dex OB	(51-63-8)
Dex-Sule	(51-63-8)
Dexa	(50-02-2)
Dexa-Cortidelt	(50-02-2)
Dexa-Cortidelt Hostacortin H	(50-24-8)
Dexa-Cortisyl	(50-02-2)
Dexa-Scheroson	(50-02-2)
Dexacort	(50-02-2)
Dexadeltone	(50-02-2)
Dexaime	(51-63-8)
Dexaline	(51-63-8)
Dexalme	(51-63-8)
Dexalona	(50-02-2)
Dexalone	(51-63-8)
Dexamed	(51-63-8)
Dexameth	(50-02-2)
Dexamethasone	(50-02-2)
Dexamethasone alcohol	(50-02-2)
Dexamethazone	(50-02-2)
Dexamine	(51-63-8)
Dexamphamine	(51-63-8)
Dexamphetamine	(51-63-8)
Dexamphetamine	(51-64-9)
Dexamphetamine sulfate	(51-63-8)
Dexamyl	(51-63-8)
Dexapolcort	(50-02-2)
Dexaprol	(50-02-2)
Dexason	(50-02-2)
Dexchlorpheniramine	(25523-97-1)
Dexchlorpheniraminum (Latin)	(25523-97-1)
Dexclorfeniramina (Spanish)	(25523-97-1)
Dexedrina	(51-63-8)
Dexedrine	(51-63-8)
Dexedrine	(51-64-9)
Dexedrine sulfate	(51-63-8)
Dexies	(51-63-8)

Dexon	(140-56-7)
Dexon	(547-58-0)
Dexon E 117	(9003-07-0)
Dexone	(50-02-2)
Dexophrine	(300-42-5)
Dexosyn	(300-42-5)
Dexoval	(300-42-5)
Dexoval	(51-63-8)
Dexoxon	(140-56-7)
Dexpanthenol	(81-13-0)
Dexstim	(300-42-5)
Dextelan	(50-02-2)
Dexten	(51-63-8)
Dextenal	(51-63-8)
Dextim	(300-42-5)
Dextran Blue	(87915-38-6)
Dextran Iron Complex	(9004-66-4)
Dextriferron	(9004-51-7)
Dextriferron Injection	(9004-51-7)
Dextrin, hydrogen 1-octenylbutanedioate (9CI)	(68070-94-0)
Dextrins	(9004-53-9)
Dextroamphetamine	(51-63-8)
Dextroamphetamine sulfate	(51-63-8)
Dextroanfetamina	(51-63-8)
Dextro calcium pantothenate	(137-08-6)
Dextrofer 75	(9004-66-4)
Dextromethorfan (Czech)	(125-71-3)
Dextromethorphan	(125-71-3)
Dextromoramide	(357-56-2)
Dextromycetin	(56-75-7)
Dextrone	(1910-42-5)
Dextrone	(85-00-7)
Dextrone-X	(1910-42-5)
Dextronic acid	(526-95-4)
Dextropimaric acid	(127-27-5)
Dextropropoxyphene	(469-62-5)
Dextroproxifeno (Spanish)	(469-62-5)
Dextropur	(50-99-7)
Dextrose	(50-99-7)
Dextrose, Anhydrous	(50-99-7)
Dextrosol	(50-99-7)
Dextrosule	(51-63-8)
Dextro-profetamine	(51-63-8)
Dexuron	(1910-42-5)
Dezibarbitur	(50-06-6)
Dezone	(50-02-2)
Di-Ademil	(135-09-1)
Di-Adreson	(53-03-2)
Di-Adreson F	(50-24-8)
Di-Aethyl-propanediol (German)	(115-76-4)
Di-Allate	(2303-16-4)
Di-Azo	(136-40-3)
2,6-Di-terc.Butyl-p-kresol (Czech)	(128-37-0)
Di(C6-C10)alkyl phthalate	(68515-51-5)
Di-(C9-branched alkyl) phthalate	(71549-78-5)
Di-Chlor-Mulsion	(107-06-2)
Di-Chloricide	(106-46-7)
Di(2-Chloroethyl)methylamine hydrochloride	(55-86-7)
Di-α-Cumyl peroxide	(80-43-3)
Di-Cup	(80-43-3)
Di-Cup 40 KE	(80-43-3)
Di-Cupr	(80-43-3)
Di-Estryl	(56-53-1)
Di-Halo	(16079-88-2)
Di-Hydan	(57-41-0)
Di-Hydan	(630-93-3)
Di-β-Hydroxyethoxyethane	(112-27-6)
α-Di(p-Hydroxyphenyl)phthalide	(77-09-8)
Di-Isobutylcetone (French)	(108-83-8)
Di-Lan	(57-41-0)
Di-Len	(630-93-3)
Di-On	(330-54-1)
Di-PIP	(16898-52-5)
Di-Paralen	(82-93-9)
Di-Phetine	(57-41-0)
Di-Phetine	(630-93-3)
Di-Quinol	(83-73-8)
Di-Septon	(298-03-3)
Di-Sipidin	(50-56-6)
Di-Syston	(298-04-4)
Di-Tac	(144-21-8)
Di-m-Tolyl ketone	(2852-68-8)
Di-Trapex	(8066-01-1)
Dia-Tuss	(509-67-1)
Diabaril	(94-20-2)
Diabase Orange GC Base	(141-85-5)
Diabase Red B	(97-52-9)
Diabase Red RL	(99-52-5)
Diabase Scarlet G	(99-55-8)
Diabasic Magenta	(632-99-5)
Diabasic Malachite Green	(569-64-2)
Diabasic Rhodamine B	(81-88-9)
Diabet-Pages	(94-20-2)
Diabechlor	(94-20-2)
Diaben	(64-77-7)
Diabenal	(94-20-2)
Diabenese	(94-20-2)
Diabeneza	(94-20-2)
Diabenyl	(58-73-1)
Diabetol	(64-77-7)
Diabetoral	(94-20-2)
Diabinese	(94-20-2)
Diaboral	(339-43-5)
Diabutal	(57-33-0)
Diabuton	(64-77-7)
Diabylen	(58-73-1)
Diacarb	(59-66-5)
Diacel Navy DC	(119-90-4)
Diacelliton Fast Blue R	(2475-45-8)
Diacelliton Fast Brilliant Blue B	(2475-46-9)
Diacelliton Fast Brilliant Blue B	(86722-66-9)
Diacelliton Fast Brilliant Blue BF	(2475-46-9)
Diacelliton Fast Brilliant Blue BF	(86722-66-9)
Diacelliton Fast Grey G	(119-90-4)
Diacelliton Fast Pink B	(116-85-8)
Diacelliton Fast Pink R	(82-38-2)
Diacelliton Fast Violet B	(82-33-7)
Diacelliton Fast Violet BF	(1220-94-6)
Diacelliton Fast Violet 5R	(128-95-0)
Diacelliton Fast Yellow G	(2832-40-8)
Diacepan	(439-14-5)
Diacephin	(561-27-3)
Diacesal	(552-94-3)
Diacetamate	(2623-33-8)
Diacetamato (Spanish)	(2623-33-8)
Diacetamatum (Latin)	(2623-33-8)
Diacetamide (8CI)	(625-77-4)
Diacetamide, N-hexyl- (7CI,8CI)	(25457-47-0)
Diacetamide, N-phenethyl- (8CI)	(27179-64-2)
Diacetamide, N-propyl- (8CI)	(1563-84-4)
4,4'-Diacetamidobiphenyl	(613-35-4)
3,5-Diacetamido-2,4,6-triiodobenzoic acid	(117-96-4)
α,5-Diacetamido-2,4,6-triiodo-m-toluic acid	(440-58-4)
10,040 Diacetate	(56-95-1)

Diacetazotol **(83-63-6)**
Diacetic ether **(141-97-9)**
Diacetin **(25395-31-7)**
Diacetonalcohol (Dutch) **(123-42-2)**
Diacetonalcool (Italian) **(123-42-2)**
Diacetonalkohol (German) **(123-42-2)**
Diacetone acrylamide **(2873-97-4)**
Diacetone alcohol (ACGIH,OSHA) [UN 1148] **(123-42-2)**
Diacetone alcohol peroxide **(54693-46-8)**
Diacetone alcohol peroxide, More than 57% in solution **(54693-46-8)**
Diacetone alcohol peroxide, Not more than 57% in solution **(54693-46-8)**
Diacetone alcohol peroxides (More than 57% in solution with more than 9% hydrogen peroxide, less than 26% diacetone alcohol and less than 9% water; total active oxygen content more than 9% by weight) **(54693-46-8)**
Diacetone alcohol peroxides (Not more than 57% in solution with not more than 9% hydrogen peroxide, not less than 26% diacetone alcohol and not less than 9% water, total active oxygen content not more than 9% by weight) **(54693-46-8)**
Diacetone-alcool (French) **(123-42-2)**
Diacetonyl **(110-13-4)**
Diacetotoluide **(83-63-6)**
o-Diacetotoluidide, 4''-(o-tolylazo)- (8CI) **(83-63-6)**
Diacetoxybutyltin **(1067-33-0)**
Diacetoxydibutylstannane **(1067-33-0)**
Diacetoxydibutyltin **(1067-33-0)**
2-(4,4'-Diacetoxydiphenylmethyl)pyridine **(603-50-9)**
(4,4'-Diacetoxydiphenyl)(2-pyridyl)methane **(603-50-9)**
4,4'-Diacetoxydiphenylpyrid-2-ylmethane **(603-50-9)**
3-α,17-β-Diacetoxy-2-β,16-β-dipiperidino-5-α-androstane dimethobromide **(15500-66-0)**
1,1-Diacetoxyethane **(542-10-9)**
3-Diacetoxyethylamino-4-methoxyacetanilide **(23128-51-0)**
3-β, 17-β-Diacetoxy-17-α-ethynyl-4-oestrene **(297-76-7)**
Diacetoxymercury **(1600-27-7)**
4,15-Diacetoxy-8-(3-methylbutyryloxy)-12,13-epoxy-δ-9-trichothecen-3-ol **(21259-20-1)**
4-β,15-Diacetoxy-8-α-(3-methylbutyryloxy)-3-α-hydroxy-12,13-epoxytrichothec-9-ene **(21259-20-1)**
4,15-Diacetoxy-8-(3-methylbutyryloxy)scirp-9-en-3-ol **(21259-20-1)**
3-β,17-β-Diacetoxy-19-nor-17-α-pregn-4-en-20-yne **(297-76-7)**
Di-(4-acetoxyphenyl)-2-pyridylmethane **(603-50-9)**
Di-(p-acetoxyphenyl)-2-pyridylmethane **(603-50-9)**
1,1-Diacetoxypropene-2 **(869-29-4)**
3,3-Diacetoxypropene **(869-29-4)**
Diacetoxypropene **(869-29-4)**
Diacetyl (DOT) **(431-03-8)**
Diacetylacetotolidide **(91-96-3)**
Diacetylaminoazotoluene **(83-63-6)**
4,4'-Diacetylaminobiphenyl **(613-35-4)**
4,4'-Diacetylbenzidine **(613-35-4)**
Diacetylbenzidine **(613-35-4)**
N,N'-Diacetyl benzidine **(613-35-4)**
Diacetylcholine **(306-40-1)**
3,4-Diacetyl-1,2-5,6-dianhydro-dulcitol **(57230-48-5)**
3,4-Diacetyldianhydrogalactitol **(57230-48-5)**
2,6-Diacetyl-7,9-dihydroxy-8,9b-dimethyl-1,3(2H,9bh)-dibenzofurandione **(125-46-2)**
Diacetyldioxime **(95-45-4)**
1,2-Diacetylethane **(110-13-4)**
α,β-Diacetylethane **(110-13-4)**
Diacetylglycerol **(25395-31-7)**
Diacetylmanganese **(638-38-0)**
Diacetylmethane **(123-54-6)**
Diacetylmonooxime **(57-71-6)**
Diacetylmonoxime **(57-71-6)**
Diacetylmorfin **(561-27-3)**
Diacetylmorphine **(561-27-3)**
Diacetylmorphine hydrochloride **(1502-95-0)**
N,O-Diacetylmuramidase **(9001-63-2)**
3-((N,N-Diacetyloxyethyl)amino)-6-((2'-chloro-4'-nitrophenyl)-azo)acetanilide **(1533-78-4)**
Diacetyl peroxide **(110-22-5)**
Diacetyl peroxide (Solution) **(110-22-5)**
N,N-Diacetyl-o-tolylazo-o-toluidine **(83-63-6)**
Diacid **(77-65-6)**
Diacid Blue Black 10B **(1064-48-8)**
Diacid Light Blue BR **(2666-17-3)**
Diacid Metanil Yellow **(587-98-4)**
Diacid Orange II **(633-96-5)**
Diacid Red A **(1658-56-6)**
α,γ-Diacipiperazine **(106-57-0)**
Diacon **(40596-69-8)**
Diacotton Benzopurpurine 4B **(992-59-6)**
Diacotton Black BH **(2429-73-4)**
Diacotton Blue BB **(2602-46-2)**
Diacotton Brown M **(2429-82-5)**
Diacotton Congo Red **(573-58-0)**
Diacotton Dark Green **(3626-28-6)**
Diacotton Deep Black **(1937-37-7)**
Diacotton Deep Black RX **(1937-37-7)**
Diacotton Navy Blue BS **(7082-31-7)**
Diacotton Sky Blue 5B **(2429-74-5)**
Diacotton Sky Blue 6B **(2610-05-1)**
Diacrid **(8048-52-0)**
Diacromo Violet N **(2092-55-9)**
Diacryl Supra Red GTL **(14097-03-1)**
Diactol **(50-14-6)**
Diacycine **(64-75-5)**
Diadem Chrome Blue G **(3567-69-9)**
Diadem Chrome Blue R **(3567-69-9)**
Diadilan **(83-12-5)**
Diadol **(52-43-7)**
Diadur **(1344-28-1)**
Diaethanolamin (German) **(111-42-2)**
Diaethanolnitrosamin (German) **(1116-54-7)**
1,1-Diaethoxy-aethan (German) **(105-57-7)**
2-Diaethoxyphosphinyl-thioaethyl-trimethyl-ammonium-jodid (German) **(513-10-0)**
Diaethylacetal (German) **(105-57-7)**
Diaethylaether (German) **(60-29-7)**
O,O-Diaethyl-S-(2-aethylthio-aethyl)-dithiophosphat (German) **(298-04-4)**
O,O-Diaethyl-S(2-aethylthio-aethyl)-monothiophosphat (German) **(126-75-0)**
O,O-Diaethyl-S-(aethylthio-methyl)-dithiophosphat (German) **(298-02-2)**
Diaethylamin (German) **(109-89-7)**
Diaethylaminoaethanol (German) **(100-37-8)**
Diaethylanilin (German) **(91-66-7)**
O,O-Diaethyl-O-(4-brom-2,5-dichlor)-phenyl-monothiophosphat (German) **(4824-78-6)**
Diaethylcarbonat (German) **(105-58-8)**
O,O-Diaethyl-O-(chinoxalyl-(2))-monothiophosphat (German) **(13593-03-8)**
O,O-Diaethyl-O-(3-chlor-4-methyl-cumarin-7-yl)-monothiophosphat (German) **(56-72-4)**
O,O-Diaethyl-S-(6-chlor-benzoxazolon-2-on-3-yl)-dithiophosphat (German) **(2310-17-0)**
O,O-Diaethyl-S-(6-chlor-2-oxo-ben(b)-1,3-oxalin-3-yl)-methyl-dithiophosphat (German) **(2310-17-0)**
O,O-Diaethyl-S-((4-chlor-phenyl-thio)-methyl)dithiophosphat (German) **(786-19-6)**
O,O-Diaethyl-S-2-chlor-1-(phthalimido)-aethyl-dithiophosphat (German) **(10311-84-9)**
O,O-Diaethyl-O-(α-cyanbenzyliden-amino)-thionphosphat (German) **(14816-18-3)**
O,O-Diaethyl-O-(α-cyano-benzylidenamino)-monothiophosphat

(German) **(14816-18-3)**
O,O-Diaethyl-O-(2,5-dichlor-4-bromphenyl)-thionophosphat (German) **(4824-78-6)**
O,O-Diaethyl-O-1-(4,5-dichlorphenyl)-2-chlor-vinyl-phosphat (German) **(470-90-6)**
O,O-Diaethyl-O-2,4-dichlor-phenyl-monothiophosphat (German) **(97-17-6)**
O,O-Diaethyl-S((2,5-dichlor-phenyl-thio)-methyl)-dithiophosphat (German) **(2275-14-1)**
O,O-Diaethyl-O-2,4-dichlorphenyl-thionophosphat (German) **(97-17-6)**
Diaethyl-dichlorvinyl-phosphat (German) **(72-00-4)**
1,2-Diaethylhydrazin (German) **(1615-80-1)**
O,O-Diaethyl-O-(2-isopropyl-4-methyl-pyrimidin-6-yl)-monothiophosphat (German) **(333-41-5)**
O,O-Diaethyl-O-(2-isopropyl-4-methyl)-6-pyrimidyl-thionophosphat (German) **(333-41-5)**
O,O-Diaethyl-O-(4-methyl-coumarin-7-yl)-monothiophosphat (German) **(299-45-6)**
O,O-Diaethyl-S-(3-methyl-2,4-dioxo-5-oxa-3-aza-heptyl)-dithiophosphat (German) **(2595-54-2)**
O,O-Diaethyl-O-(3-methyl-1H-pyrazol-5-yl)-phosphat (German) **(108-34-9)**
O,O-Diaethyl-O-4-methylsulfinyl-phenyl-monothiophosphat (German) **(115-90-2)**
O,O-Diaethyl-O-(4-methylsulfinyl-phenyl)-thionophosphat (German) **(115-90-2)**
Diaethyl-nicotinamid (German) **(59-26-7)**
O,O-Diaethyl-O-(4-nitro-phenyl)-monothiophosphat (German) **(56-38-2)**
O,O'Diaethyl-p-nitrophenylphosphat (German) **(311-45-5)**
O,O-Diaethyl-S-(p-nitrophenyl)-phosphat (German) **(3270-86-8)**
O,S-Diaethyl-O-(p-nitrophenyl)-phosphat (German) **(597-88-6)**
Diaethyl-p-nitrophenylphosphorsaeureester (German) **(311-45-5)**
Diaethylnitrosamin (German) **(55-18-5)**
O,O-Diaethyl-S-(4-oxobenzotriazin-3-methyl)-dithiophosphat (German) **(2642-71-9)**
O,O-Diaethyl-S-((4-oxo-3H-1,2,3-benzotriazin-3-yl)-methyl)-dithiophosphat (German) **(2642-71-9)**
O,O-Diaethyl-N-phtalimido-thiophosphat (German) **(5131-24-8)**
O,O-Diaethyl-O-(pyrazin-2yl)-monothiophosphat (German) **(297-97-2)**
O,O-Diaethyl-O-(2-pyrazinyl)-thionophosphat (German) **(297-97-2)**
Diaethylsulfat (German) **(64-67-5)**
Diaethylthiambutenum **(86-14-6)**
O,O-Diaethyl-S-(3-thia-pentyl)-dithiophosphat (German) **(298-04-4)**
Diaethylthiophosphorsaeureester des aethylthioglykol (German) **(126-75-0)**
Diaethylthiophosphorsaeureester des aethylthioglykol (German) **(298-03-3)**
O,O-Diaethyl-O-3,5,6-trichlor-2-pyridylmonothiophosphat (German) **(2921-88-2)**
Diafen 13 **(793-24-8)**
Diaform UR **(9011-05-6)**
Diafuron **(67-45-8)**
Diagnorenol **(126-31-8)**
Diagran **(333-41-5)**
Diakarb **(59-66-5)**
Diakarmon **(50-70-4)**
Diakol DM **(9011-05-6)**
Diakol F **(9011-05-6)**
Diakol M **(9011-05-6)**
Diakon **(80-62-6)**
Diakon **(9011-14-7)**
Diakon LO 951 **(9011-14-7)**
Diakon MG **(9011-14-7)**
Diakon MG 101 **(9011-14-7)**
Dial **(52-43-7)**
Dial **(69-05-6)**
Dial-A-Gesic **(103-90-2)**
Dial (barbiturate) **(52-43-7)**
Dialifor **(10311-84-9)**
Dialifos **(10311-84-9)**
Dialin **(447-53-0)**
Dialkyl dimethyl ammonium chloride **(68514-95-4)**
Diallaat (Dutch) **(2303-16-4)**
Diallat (German) **(2303-16-4)**
Diallate **(2303-16-4)**
cis-Diallate **(17708-57-5)**
Diallyl adipate **(2998-04-1)**
Diallylamid kyseliny chloroctove (Czech) **(93-71-0)**
Diallylamine [UN 2359] **(124-02-7)**
Diallylamine, N-nitroso **(16338-97-9)**
Diallylbarbital **(52-43-7)**
5,5-Diallylbarbituric acid **(52-43-7)**
Diallylbarbituric acid **(52-43-7)**
Diallyl carbonate **(15022-08-9)**
Diallylchloroacetamide **(93-71-0)**
N,N-Diallyl-2-chloroacetamide **(93-71-0)**
N,N-Diallyl-α-chloroacetamide **(93-71-0)**
N,N-Diallylchloroacetamide **(93-71-0)**
Diallylcyanamide **(538-08-9)**
N,N-Diallyl-2,2-dichloroacetamide **(37764-25-3)**
N,N-Diallyldichloroacetamide **(37764-25-3)**
Diallyldichlorosilane **(3651-23-8)**
Diallyl diglycol carbonate **(142-22-3)**
Diallyldimethylammonium chloride **(7398-69-8)**
Diallyl disulfide **(2179-57-9)**
Diallyl disulphide **(2179-57-9)**
O,O-Diallyl dithiophosphate **(5851-14-9)**
Diallylester kyseliny adipove (Czech) **(2998-04-1)**
Diallylester kyseliny ftalove (Czech) **(131-17-9)**
Diallylester kyseliny maleinove (Czech) **(999-21-3)**
Diallylether [UN 2360] **(557-40-4)**
Diallylkyanamid (Czech) **(538-08-9)**
Diallyl maleate **(999-21-3)**
N,N'-Diallylmelamine **(30360-15-7)**
Diallyl monosulfide **(592-88-1)**
Diallylnitrosamin (German) **(16338-97-9)**
Diallylnitrosamine **(16338-97-9)**
2,6-Diallylphenyl 2,3-epoxypropyl ether **(40693-04-7)**
O,O-Diallyl phosphorodithioate **(5851-14-9)**
Diallyl phthalate **(131-17-9)**
Diallyl phthalate, ethyl acrylate, methacrylic acid polymer **(28411-49-6)**
Diallyl sulfide **(592-88-1)**
Diallyl thioether **(592-88-1)**
Diallymal **(52-43-7)**
Dialuminous Brown BRS **(16071-86-6)**
Dialuminous Red 4B **(2610-11-9)**
Dialuminum dipotassium sulfate **(10043-67-1)**
Dialuminum sulphate **(10043-01-3)**
Dialuminum tetrahydroxychloro allantoinate **(1317-25-5)**
Dialuminum trioxide **(1344-28-1)**
Dialuminum trisulfate **(10043-01-3)**
Dialux **(9003-54-7)**
Diamarin **(523-87-5)**
Diamazo **(83-63-6)**
Diamet **(3653-48-3)**
Diabetamid **(64-77-7)**
Diamidafos **(1754-58-1)**
Diamide **(302-01-2)**
Diamide **(957-51-7)**
Diamidfos **(1754-58-1)**
4,4'-Diamidino-α,ω-diphenoxypentane isethionate **(140-64-70)**
Diamidine **(140-64-70)**
4,4'-Diamidinodiphenoxypentane di(β-hydroxyethanesulfonate) **(140-64-70)**
p,p'-Diamidinodiphenylmethane **(63690-09-5)**
Diaminblau 3B **(72-57-1)**
Diamine **(302-01-2)**
Diamine Black BH **(2429-73-4)**
Diamine Black BHM **(2429-73-4)**

Name	CAS
Diamine Blue 2B	**(2602-46-2)**
Diamine Blue 3B	**(72-57-1)**
Diamine Blue BB	**(2602-46-2)**
Diamine Brown M	**(2429-82-5)**
Diamine Brown MBA-CF	**(2429-82-5)**
Diamine Dark Green B	**(3626-28-6)**
Diamine Dark Green N	**(3626-28-6)**
Diamine Deep Black EC	**(1937-37-7)**
Diamine Deep Black RW	**(2429-83-6)**
Diamine Direct Black E	**(1937-37-7)**
Diamine Fast Black B	**(6428-31-5)**
Diamine Fast Yellow A	**(1325-37-7)**
Diamine Penicillin	**(1538-09-6)**
Diamine Purpurine 4B	**(992-59-6)**
Diamine Scarlet 3BA-CF	**(6358-29-8)**
Diamine Sky Blue CI	**(2429-74-5)**
Diamine Sky Blue FF	**(314-13-6)**
Diamine Violet N	**(2586-60-9)**
2,4-Diamineanisole	**(615-05-4)**
Diamineblue	**(72-57-1)**
cis-Diaminedichloroplatinum	**(15663-27-1)**
Diaminide maleate	**(59-33-6)**
3,6-Diaminoacridine	**(92-62-6)**
3,6-Diaminoacridine bisulphate	**(553-30-0)**
3,6-Diaminoacridine dihydrochloride	**(531-73-7)**
2,8-Diaminoacridine (European)	**(92-62-6)**
3,6-Diaminoacridine hemisulfate	**(1811-28-5)**
3,6-Diaminoacridine mixture with 3,6-diamino-10-methyl-acridinium chloride	**(8048-52-0)**
3,6-Diaminoacridine monohydrochloride	**(952-23-8)**
3,6-Diaminoacridine sulphate (1:1)	**(553-30-0)**
2,8-Diaminoacridinium	**(92-62-6)**
3,6-Diaminoacridinium	**(92-62-6)**
3,6-Diaminoacridinium chloride	**(952-23-8)**
2,8-Diaminoacridinium chloride hydrochloride	**(531-73-7)**
3,6-Diaminoacridinium chloride hydrochloride	**(531-73-7)**
3,6-Diaminoacridinium chloride hydrochloride	**(952-23-8)**
2,8-Diaminoacridinium chloride monohydrochloride	**(952-23-8)**
3,6-Diaminoacridinium monohydrogen sulphate	**(553-30-0)**
2,8-Diaminoacridinium sulphate	**(553-30-0)**
1,2-Diaminoaethan (German)	**(107-15-3)**
2,4-Diaminoanisol	**(615-05-4)**
2,4-Diaminoanisole	**(615-05-4)**
2,5-Diaminoanisole	**(5307-02-8)**
2,4-Diaminoanisole Base	**(615-05-4)**
m-Diaminoanisole 1,3-diamino-4-methoxybenzene	**(615-05-4)**
2,4-Diaminoanisole dihydrochloride	**(614-94-8)**
2,4-Diaminoanisole sulfate	**(39156-41-7)**
2,4-Diaminoanisole sulphate	**(39156-41-7)**
2,4-Diamino-anisol sulphate	**(39156-41-7)**
1,4-Diaminoanthrachinon (Czech)	**(128-95-0)**
1,5-Diaminoanthrachinon (Czech)	**(129-44-2)**
1,8-Diaminoanthrachinon (Czech)	**(129-42-0)**
1,4-Diaminoanthraquinon-N-γ-methoxypropyl-2,3-dicarboximide	**(12217-80-0)**
1,4-Diaminoanthraquinone	**(128-95-0)**
1,5-Diamino-9,10-anthraquinone	**(129-44-2)**
1,5-Diaminoanthraquinone	**(129-42-0)**
1,5-Diaminoanthraquinone	**(129-44-2)**
1,8-Diaminoanthraquinone	**(129-42-0)**
1,4-Diamino-2,3-anthraquinonedicarboximide	**(128-81-4)**
1,4-Diaminoanthraquinone-2,3-dicarboximide	**(128-81-4)**
1,5-Diaminoanthrarufin	**(145-49-3)**
4,8-Diaminoanthrarufin	**(145-49-3)**
Diaminoanthrarufin	**(145-49-3)**
3,7-Diamino-5-azaanthracene	**(92-62-6)**
1,13-Diamino-7-azatridecane	**(143-23-7)**
2,4-Diaminoazobenzen (Czech)	**(495-54-5)**
2,4-Diaminoazobenzene	**(495-54-5)**
Diaminoazobenzene	**(495-54-5)**
2,4-Diaminoazobenzene hydrochloride	**(532-82-1)**
4,4'-Diaminobenzanilide	**(785-30-8)**
1,2-Diaminobenzene	**(95-54-5)**
1,3-Diaminobenzene	**(108-45-2)**
1,4-Diaminobenzene	**(106-50-3)**
m-Diaminobenzene	**(108-45-2)**
o-Diaminobenzene	**(95-54-5)**
p-Diaminobenzene	**(106-50-3)**
1,3-Diaminobenzene dihydrochloride	**(541-69-5)**
1,4-Diaminobenzene dihydrochloride	**(624-18-0)**
m-Diaminobenzene dihydrochloride	**(541-69-5)**
p-Diaminobenzene dihydrochloride	**(624-18-0)**
4,6-Diamino-1,3-benzenedisulfonic acid	**(137-50-8)**
1,3-Diaminobenzene-4-sulfonic acid	**(88-63-1)**
1,3-Diaminobenzene-6-sulfonic acid	**(88-63-1)**
1,3-Diaminobenzenesulfonic acid	**(88-63-1)**
2,4-Diaminobenzenesulfonic acid	**(88-63-1)**
3,3'-Diaminobenzidene	**(91-95-2)**
3,3'-Diaminobenzidine tetrahydrochloride	**(7411-49-6)**
2,4'-Diaminobifenyl (Czech)	**(492-17-1)**
2,4'-Diaminobiphenyl	**(492-17-1)**
4,4'-Diamino-1,1'-biphenyl	**(92-87-5)**
4,4'-Diaminobiphenyl	**(92-87-5)**
o,p'-Diaminobiphenyl	**(492-17-1)**
p,p'-Diaminobiphenyl	**(92-87-5)**
4,4'-Diamino-3,3'-biphenyldicarboxylic acid	**(2130-56-5)**
4,4'-Diaminobiphenyl-3,3'-dicarboxylic acid	**(2130-56-5)**
4,4'-Diamino-3,3'-biphenyldiol	**(2373-98-0)**
4,4'-Diamino-2,2'-biphenyldisulfonic acid	**(117-61-3)**
4,4'-Diaminobiphenyl-2,2'-disulfonic acid	**(117-61-3)**
4,4'-Diaminobiphenyloxide	**(101-80-4)**
1,3-Diaminobutane	**(590-88-5)**
1,4-Diaminobutane	**(110-60-1)**
1,4-Diaminobutane, N-(3-aminopropyl)-	**(124-20-9)**
1,4-Diaminobutane, N,N'-bis(3-aminopropyl)-	**(71-44-3)**
1,3-Diaminobutane, N,N,N',N'-tetramethyl-	**(97-84-7)**
(Z)-2,3-Diamino-2-butenedinitrile	**(1187-42-4)**
α,ε-Diaminocaproic acid	**(56-87-1)**
1,2-Diamino-4-chlorobenzene	**(95-83-0)**
3,4-Diamino-1-chlorobenzene	**(95-83-0)**
3,4-Diaminochlorobenzene	**(95-83-0)**
2,4-Diamino-5-(4-chlorophenyl)-6-ethylpyrimidine	**(58-14-0)**
2,4-Diamino-5-p-chlorophenyl-6-ethylpyrimidine	**(58-14-0)**
Di(-4-amino-3-chlorophenyl)methane	**(101-14-4)**
2,4-Diamino-6-chloro-1,3,5-triazine	**(3397-62-4)**
4,5-Diaminochrysazin	**(128-94-9)**
1,8-Diaminochrysazine	**(128-94-9)**
Diaminocillina	**(1538-09-6)**
Di-(4-amino-3-clorofenil)metano (Italian)	**(101-14-4)**
2,4-Diamino-o-cresol	**(15872-73-8)**
2,4-Diamino-6-cyanomethyl-1,3,5-triazine	**(13301-35-4)**
1,2-Diaminocyclohexane	**(694-83-7)**
1,2-Diaminocyclohexane-n,N'-tetraacetic acid	**(482-54-2)**
1,2-Diaminocyclohexanetetraacetic acid	**(482-54-2)**
Di(p-aminocyclohexyl)methane	**(1761-71-3)**
1,12'-Diaminodecane	**(2783-17-7)**
1,4-Diamino-2,3-dichloro-9,10-anthracenedione	**(81-42-5)**
1,4-Diamino-2,3-dichloroanthraquinone	**(81-42-5)**
1,4-Diamino-2,6-dichlorobenzene	**(609-20-1)**
1,4-Diamino-3,6-dichlorobenzene	**(20103-09-7)**
2,5-Diamino-1,3-dichlorobenzene	**(609-20-1)**
4,4'-Diamino-3,3'-dichlorobiphenyl	**(91-94-1)**
4,4'-Diamino-3,3'-dichlorodiphenyl	**(91-94-1)**
4,4'-Diamino-3,3'-dichlorodiphenylmethane	**(101-14-4)**

2,4-Diamino-6-(3,4-dichlorophenyl)-s-triazine	**(30354-89-3)**
cis-Diaminodichloroplatinum(II)	**(15663-27-1)**
4,4'-Diaminodicyclohexylmethane	**(1761-71-3)**
p,p'-Diaminodicyclohexylmethane	**(1761-71-3)**
2,2'-Diaminodiethylamine	**(111-40-0)**
Diaminodifenilsulfona (Spanish)	**(80-08-0)**
p,p'-Diaminodifenylmethan (Czech)	**(101-77-9)**
4,4-Diaminodifenylsulfon (Czech)	**(80-08-0)**
6,6'-Diaminodihexylamine	**(143-23-7)**
1,4-Diamino-2,3-dihydroanthraquinone	**(81-63-0)**
1,4-Diamino-9,10-dihydro-9,10-dioxo-2,3-anthracenedicarboximide	**(128-81-4)**
1,4-Diamino-9,10-dihydro-N-(3-methoxypropyl)-9,10-dioxo-2,3-anthracenedicarboximide	**(12217-80-0)**
1,5-Diamino-4,8-dihydroxyanthrachinon (Czech)	**(145-49-3)**
1,8-Diamino-4,5-dihydroxyanthrachinon (Czech)	**(128-94-9)**
1,5-Diamino-4,8-dihydroxyanthraquinone	**(145-49-3)**
1,8-Diamino-4,5-dihydroxy-9,10-anthraquinone	**(128-94-9)**
1,8-Diamino-4,5-dihydroxyanthraquinone	**(128-94-9)**
4,5-Diamino-1,8-dihydroxyanthraquinone	**(128-94-9)**
4,8-Diamino-1,5-dihydroxyanthraquinone	**(145-49-3)**
leuco-1,5-Diamino-4,8-dihydroxyanthraquinone	**(145-49-3)**
4,8-Diamino-1,5-dihydroxy-9,10-dihydro-9,10-dioxo-2,6-anthracene-disulfonic acid, disodium salt	**(128-86-9)**
1,5-Diamino-4,8-dihydroxy-3-(4-hydroxy-3-methylphenyl)-anthraquinone	**(4702-65-2)**
4,8-Diamino-1,5-dihydroxy-2-(p-hydroxyphenyl)anthracen-9,10-dione	**(7098-08-0)**
4,8-Diamino-1,5-dihydroxy-2-(4-hydroxyphenyl)-9,10-anthracene-dione	**(7098-08-0)**
1,5-Diamino-4,8-dihydroxy(p-hydroxyphenyl)anthraquinone	**(31529-83-6)**
1,5-Diamino-4,8-dihydroxy-3-(4-hydroxyphenyl)anthraquinone	**(7098-08-0)**
1,5-Diamino-4,8-dihydroxy(p-methoxyphenyl)anthraquinone	**(31288-44-5)**
1,5-Diamino-4,8-dihydroxy-3-(p-methoxyphenyl)anthraquinone	**(4702-64-1)**
4,4'-Diamino-3,3'-dimethoxybiphenyl	**(119-90-4)**
2,4-Diamino-6-dimethoxyphosphinothionylthiomethyl-s-triazine	**(78-57-9)**
4,4'-Diamino-3,3'-dimethylbiphenyl	**(119-93-7)**
4,4'-Diamino-3,3'-dimethyldiphenyl	**(119-93-7)**
2,4'-Diaminodiphenyl	**(492-17-1)**
4,4'-Diaminodiphenyl	**(92-87-5)**
p-Diaminodiphenyl	**(92-87-5)**
4,4'-Diaminodiphenylamine	**(537-65-5)**
p,p'-Diaminodiphenylamine	**(537-65-5)**
4,4'-Diaminodiphenylamine-2'-sulfonic acid	**(119-70-0)**
4,4'-Diaminodiphenyl-2,2'-disulfonic acid	**(117-61-3)**
4,4-Diaminodiphenyl ether	**(101-80-4)**
Diaminodiphenyl ether	**(101-80-4)**
p,p'-Diaminodiphenyl ether	**(101-80-4)**
2,4'-Diaminodiphenylmethan (German)	**(1208-52-2)**
4,4'-Diaminodiphenylmethan (German)	**(101-77-9)**
2,4'-Diaminodiphenylmethane	**(1208-52-2)**
4,4'-Diaminodiphenylmethane	**(101-77-9)**
Diaminodiphenylmethane	**(101-77-9)**
o,p'-Diaminodiphenylmethane	**(1208-52-2)**
p,p'-Diaminodiphenylmethane	**(101-77-9)**
4,4'-Diaminodiphenyl methane [UN 2651]	**(101-77-9)**
N,N'-4,4'-Diaminodiphenylmethanebismaleimide	**(13676-54-5)**
4,4'-Diaminodiphenyl oxide	**(101-80-4)**
4,4'-Diaminodiphenyl sulfide	**(139-65-1)**
p,p'-Diaminodiphenyl sulfide	**(139-65-1)**
4,4'-Diaminodiphenyl sulfone	**(80-08-0)**
Diamino-4,4'-diphenyl sulfone	**(80-08-0)**
p,p'-Diaminodiphenyl sulfone	**(80-08-0)**
4,4'-Diaminodiphenylsulphide	**(139-65-1)**
4,4-Diaminodiphenyl sulphide	**(139-65-1)**
p,p'-Diaminodiphenyl sulphide	**(139-65-1)**
4,4'-Diaminodiphenyl sulphone	**(80-08-0)**
Diamino-4,4'-diphenyl sulphone	**(80-08-0)**
p,p-Diaminodiphenyl sulphone	**(80-08-0)**
3,3'-Diaminodipropylamine	**(56-18-8)**
3,3-Diaminodipropylamine	**(56-18-8)**
Diaminoditolyl	**(119-93-7)**
1,2-Diamino-ethaan (Dutch)	**(107-15-3)**
1,2-Diaminoethane	**(107-15-3)**
1,2-Diaminoethane (OSHA)	**(107-15-3)**
1,2-Diamino-ethano (Italian)	**(107-15-3)**
2,4-Diaminoethoxybenzene dihydrochloride	**(67801-06-3)**
2,7-Diamino-10-ethyl-9-phenylphenanthridinium bromide	**(1239-45-8)**
3,8-Diamino-5-ethyl-6-phenylphenanthridinium bromide	**(1239-45-8)**
2,4-Diamino-6-fluoro-s-triazine	**(823-95-0)**
2,4-Diamino-6-(2-furyl)-s-triazine	**(4685-18-1)**
Diaminogene Velour Black B	**(2429-73-4)**
1,6-Diaminohexane	**(124-09-4)**
1,6-Diaminohexane dihydrochloride	**(6055-52-3)**
2,6-Diaminohexanoic acid	**(56-87-1)**
4,6-Diamino-2-hydroxy-1,3-cyclohexane 3,6'diamino-3,6'-dideoxy-di-α-D-glucoside	**(59-01-8)**
1,3-Diamino-2-hydroxypropane	**(616-29-5)**
Diaminomaleonitrile	**(1187-42-4)**
1,8-Diamino-p-menthane	**(80-52-4)**
1,4-Diamino-2-methoxyanthraquinone	**(2872-48-2)**
2,4-Diamino-1-methoxybenzene	**(39156-41-7)**
2,4-Diamino-1-methoxybenzene	**(615-05-4)**
1,3-Diamino-4-methoxybenzene sulphate	**(39156-41-7)**
2,4-Diamino-1-methoxybenzene sulphate	**(39156-41-7)**
1,4-Diamino-N-(3-methoxypropyl)anthraquinone-2,3-dicarboximide	**(12217-80-0)**
2,8-Diamino-10-methylacridinium chloride	**(86-40-8)**
3,6-Diamino-10-methylacridinium chloride	**(86-40-8)**
3,6-Diamino-10-methylacridinium chloride mixture with 3,6-acridinediamine	**(8048-52-0)**
2,8-Diamino-10-methylacridinium chloride mixture with 2,8-diaminoacridine	**(8048-52-0)**
1,3-Diamino-4-methylbenzene	**(95-80-7)**
1,3-Diamino-5-methylbenzene	**(108-71-4)**
2,4-Diamino-1-methylbenzene	**(95-80-7)**
3,7'-Diamino-N-methyldipropylamine	**(105-83-9)**
N^1-(Diaminomethylene)sulfanilamide	**(57-67-0)**
2,4-Diamino-6-methylphenol	**(15872-73-8)**
2,4-Diamino-6-methylphenol hydrochloride	**(65879-44-9)**
4,6-Diamino-2-methylphenol, hydrochloride	**(65879-44-9)**
2,4-Diamino-6-methyl-1,3,5-triazine	**(542-02-9)**
1,3-Diaminomocovina (Czech)	**(497-18-7)**
Diaminon	**(76-99-3)**
1,5-Diaminonaphthalene	**(2243-62-1)**
1,8-Diaminonaphthalene	**(479-27-6)**
1,4-Diamino-5-nitroanthraquinone	**(82-33-7)**
1,2-Diamino-4-nitrobenzene	**(99-56-9)**
1,4-Diamino-2-nitrobenzene	**(5307-14-2)**
2,4-Diaminonitrobenzene	**(5131-58-8)**
3,4-Diaminonitrobenzene	**(99-56-9)**
1,8-Diaminooktan hydrochlorid (Czech)	**(7613-16-3)**
1,5-Diaminopentane	**(462-94-2)**
(S)-2,5-Diaminopentanoic acid	**(70-26-8)**
2,3-Diaminophenazine	**(655-86-7)**
2,4-Diaminophenol	**(95-86-3)**
2,4-Diaminophenol hydrochloride	**(137-09-7)**
2-(2,4-Diaminophenoxy)ethanol dihydrochloride	**(66422-95-5)**
Di(p-aminophenyl)amine	**(537-65-5)**
2,6-Diamino-3-phenylazopyridine	**(94-78-0)**
2,6-Diamino-3-phenylazopyridine hydrochloride	**(136-40-3)**
2,6-Diamino-3-(phenylazo)pyridine monohydrochloride	**(136-40-3)**
4,4'-Diaminophenyl ether	**(101-80-4)**
2,7-Diamino-9-phenyl-10-ethylphenanthridinium bromide	**(1239-45-8)**
Di-(4-aminophenyl)methane	**(101-77-9)**
4,4'-Diaminophenyl oxide	**(101-80-4)**

2,7-Diamino-9-phenylphenanthridine ethobromide (1239-45-8)
Di(p-aminophenyl) sulfide (139-65-1)
Di(4-aminophenyl)sulfone (80-08-0)
Di(p-aminophenyl) sulfone (80-08-0)
Di(p-aminophenyl)sulphide (139-65-1)
Di(4-aminophenyl)sulphone (80-08-0)
Di(p-aminophenyl)sulphone (80-08-0)
2,4-Diamino-5-phenylthiazole (490-55-1)
2,4-Diamino-6-phenyl-s-triazine (91-76-9)
4,6-Diamino-2-phenyl-s-triazine (91-76-9)
1,2-Diaminopropane (78-90-0)
1,3-Diaminopropane (109-76-2)
1,3-Diamino-2-propanol (616-29-5)
Diaminopropanol tetra acetic acid (3148-72-9)
Di(3-aminopropyl) ether of diethylene glycol (4246-51-9)
N,N'-Di(3-aminopropyl)-1,2-ethylenediamine (10563-26-5)
Diaminopropyltetramethylenediamine (71-44-3)
N-(4-(((2,4-Diamino-6-pteridinyl)methyl)amino)benzoyl)-l-glutamic acid (54-62-6)
l-(+)-N-(p-(((2,4-Diamino-6-pteridinyl)methyl)methylamino)-benzoyl)glutamic acid (59-05-2)
N-(p-(((2,4-Diamino-6-pteridinyl)methyl)methylamino)benzoyl)-l-(+)-glutamic acid (59-05-2)
N-(p-(((2,4-Diamino-6-pteridyl)methyl)methylamino)benzoyl)-glutamic acid (59-05-2)
2,3-Diaminopyridine (452-58-4)
2,5-Diaminopyridine (4318-76-7)
2,6-Diaminopyridine (141-86-6)
3,4-Diaminopyridine (54-96-6)
Diamino-3,4 pyridine (54-96-6)
Diaminopyritamin (58-14-0)
2,4-Diaminosole sulphate (39156-41-7)
4,4'-Diamino-2,2'-stilbenedisulfonic acid (81-11-8)
4,4'-Diaminostilbene-2,2'-disulfonic acid, sodium salt (25394-13-2)
4,4'-Diamino-2-sulfodiphenylamine (119-70-0)
2,4-Diaminotoluen (Czech) (95-80-7)
2,3-Diaminotoluene (2687-25-4)
2,4-Diamino-1-toluene (95-80-7)
2,4-Diaminotoluene (95-80-7)
2,5-Diaminotoluene (95-70-5)
2,6-Diaminotoluene (823-40-5)
3,4-Diaminotoluene (496-72-0)
3,5-Diaminotoluene (108-71-4)
Diaminotoluene (25376-45-8)
Diaminotoluene (95-80-7)
2,4-Diaminotoluene dihydrochloride (636-23-7)
2,5-Diaminotoluene dihydrochloride (615-45-2)
2,6-Diaminotoluene dihydrochloride (15481-70-6)
2,5-Diaminotoluene sulfate (615-50-9)
p-Diaminotoluene sulfate (615-50-9)
2,5-Diaminotoluene sulphate (615-50-9)
2,4-Diaminotoluol (95-80-7)
4,6-Diamino-1,3,5-triazin-2(1H)-one (645-92-1)
2,6-Diamino-s-triazine (504-08-5)
4,6-Diamino-s-triazine (504-08-5)
Diamino-s-triazine (504-08-5)
4,6-Diamino-1,3,5-triazine 2-thione (767-17-9)
S-((4,6-Diamino-1,3,5-triazin-2-yl)-methyl)-O,O-dimethyl-dithio-fosfaat (Dutch) (78-57-9)
S-((4,6-Diamino-1,3,5-triazin-2-yl)-methyl)-O,O-dimethyl-dithio-phosphat (German) (78-57-9)
4,6-Diamino-1,3,5-triazin-2-ylmethyl O,O-dimethyl phosphoro-dithioate (78-57-9)
S-((4,6-Diamino-s-triazin-2-yl)methyl)-O,O-dimethyl phosphoro-dithioate (78-57-9)
S-(4,6-Diamino-1,3,5-triazin-2-ylmethyl) O,O-dimethyl phosphoro-dithioate (78-57-9)
S-(4,6-Diamino-1,3,5-triazin-2-ylmethyl) dimethyl phosphoro-thiolothionate (78-57-9)
S-((4,6-Diamino-1,3,5-triazin-2-yl)methyl)phosphorodithioic acid O,O-dimethyl ester (78-57-9)
2,4-Diamino-5-(3,4,5-trimethoxybenzyl)pyrimidine (738-70-5)
1,3-Diaminourea (497-18-7)
(S)-α,δ-Diaminovaleric acid (70-26-8)
Diamira Golden Yellow G (20317-19-5)
Diamminedichloroplatinum (15663-27-1)
cis-Diamminedichloroplatinum (15663-27-1)
trans-Diamminedichloroplatinum (II) (14913-33-8)
Diammonium acid phosphate (7783-28-0)
Diammonium arsenate (7784-44-3)
Diammonium carbonate (506-87-6)
Diammonium chromate (7788-98-9)
Diammonium citrate (3012-65-5)
Diammonium citrate (Secondary) (3012-65-5)
Diammonium dithiocarbazate (20469-71-0)
Diammonium fluorosilicate (16919-19-0)
Diammonium fluosilicate (16919-19-0)
Diammonium hexachloroplatinate(2-) (16919-58-7)
Diammonium hexafluorosilicate (16919-19-0)
Diammonium hexafluorosilicate(2-) (16919-19-0)
Diammonium hydrogen arsenate (7784-44-3)
Diammonium hydrogen citrate (3012-65-5)
Diammonium hydrogen orthophosphate (7783-28-0)
Diammonium hydrogen phosphate (7783-28-0)
Diammonium molybdate (13106-76-8)
Diammonium monohydrogen arsenate (7784-44-3)
Diammonium monohydrogen phosphate (7783-28-0)
Diammonium oxalate monohydrate (6009-70-7)
Diammonium peroxydisulfate (7727-54-0)
Diammonium peroxydisulphate (7727-54-0)
Diammonium persulfate (7727-54-0)
Diammonium orthophosphate (7783-28-0)
Diammonium phosphate (7783-28-0)
Diammonium silicon hexafluoride (16919-19-0)
Diammonium sulfate (7783-20-2)
Diammonium sulfide (12135-76-1)
Diammonium sulfite (10196-04-0)
Diammonium sulphate (7783-20-2)
Diammonium tartrate (3164-29-2)
Diammonium thiosulfate (7783-18-8)
Diammonium zinc disulfate hexahydrate (7783-24-6)
Diamond (9CI) (7782-40-3)
Diamond Corinth N (2092-55-9)
Diamond Fuchsine (632-99-5)
Diamond Green B (569-64-2)
Diamond Green B Extra (569-64-2)
Diamond Green BX (569-64-2)
Diamond Green G (633-03-4)
Diamond Green P Extra (569-64-2)
Diamond Red W (130-22-3)
Diamond Shamrock 744 (9003-22-9)
Diamond Shamrock 7401 (9003-22-9)
Diamond Shamrock DS-15647 (39196-18-4)
Diamorfina (561-27-3)
Diamorphine (561-27-3)
Diamorphine hydrochloride (1502-95-0)
4-Diamox (59-66-5)
Diamox (59-66-5)
Diampromid (552-25-0)
Diampromida (Spanish) (552-25-0)
Diampromide (552-25-0)
Diampromidum (Latin) (552-25-0)
Diamyl acid phosphate (3138-42-9)
Diamyl amine (2050-92-2)

2,5-Di-tert-amylbenzene-1,4-diol **(79-74-3)**
Diamylester kyseliny maleinove (Czech) **(10099-71-5)**
Di-iso-amyl ether **(544-01-4)**
Diamyl ether **(693-65-2)**
2,5-Di-t-amylhydroquinone **(79-74-3)**
2,6-Di-tert-amylhydroquinone **(2349-85-1)**
Diamyl ketone **(927-49-1)**
Diamyl maleate **(10099-71-5)**
Diamylnitrosamin (German) **(13256-06-9)**
2,4-Di-tert-amylphenol **(120-95-6)**
2,4-Diamylphenol **(138-00-1)**
Di-tert-amylphenol **(120-95-6)**
Diamylphenol **(28652-04-2)**
Diamyl phthalate **(131-18-0)**
Diamyl sodium sulfosuccinate **(922-80-5)**
Diamyl sulfide **(872-10-6)**
Dian **(80-05-7)**
Dianabol **(58-18-4)**
Dianabol **(72-63-9)**
Dianabole **(72-63-9)**
Dianat (Russian) **(1918-00-9)**
Dianate **(1918-00-9)**
Dianate **(2300-66-5)**
Dian-bis-glycidylether (Czech) **(1675-54-3)**
Dianex **(40596-69-8)**
1,2:3,4-Dianhydroerythritol **(564-00-1)**
D-1,4:3,6-Dianhydroglucitol **(652-67-5)**
1,4:3,6-Dianhydro-D-glucitol, di(Z)-9-octadecenoate **(4252-85-1)**
1,4:3,6-Dianhydrosorbitol **(652-67-5)**
1,4:3,6-Dianhydrosorbitol 2,5-dinitrate **(87-33-2)**
1,2:3,4-Dianhydro-DL-threitol **(298-18-0)**
Dianil Blue **(72-57-1)**
Dianil Blue H3G **(72-57-1)**
Dianil Dark Blue H **(2429-73-4)**
Dianilblau **(72-57-1)**
Dianilblau H3G **(72-57-1)**
2,4-Dianilino-6-hydroxy-s-triazine **(30303-58-3)**
o,p'-Dianiline **(492-17-1)**
p,p'-Dianiline **(92-87-5)**
Dianilinemethane **(101-77-9)**
Dianilinomethane **(101-77-9)**
2,5-Dianilinoterephthalic acid **(10109-95-2)**
2,4-Dianilino-s-triazine **(13107-54-5)**
o-Dianisidin (Czech, German) **(119-90-4)**
o-Dianisidina (Italian) **(119-90-4)**
3,3'-Dianisidine **(119-90-4)**
Dianisidine **(119-90-4)**
o,o'-Dianisidine **(119-90-4)**
o-Dianisidine **(119-90-4)**
o-Dianisidine dihydrochloride **(20325-40-0)**
Dianisidine diisocyanate **(91-93-0)**
Di-p-anisylamine **(101-70-2)**
Dianisylneopentane **(4741-74-6)**
Dianisyltrichlorethane **(72-43-5)**
2,2-Di-p-anisyl-1,1,1-trichloroethane **(72-43-5)**
Dianix Blue BG-FS **(12222-78-5)**
Dianix Developer ND **(135-61-5)**
Dianix Fast Blue BG-FS **(12222-78-5)**
Dianix Fast Violet B **(82-33-7)**
Dianix Fast Yellow 5R **(6300-37-4)**
Dianix Yellow 5R-E **(6300-37-4)**
Dianol 33 **(116-37-0)**
Dianon **(333-41-5)**
Diantimony pentaoxide **(1314-60-9)**
Diantimony pentoxide **(1314-60-9)**
Diantimony trioxide **(1309-64-4)**
Diapadrin **(141-66-2)**
Diapam **(439-14-5)**
Diaphenylsulfon **(80-08-0)**
Diaphenylsulfone **(80-08-0)**
Diaphenylsulphon **(80-08-0)**
Diaphenylsulphone **(80-08-0)**
Diaphorm **(561-27-3)**
Diaphtamine Black BH **(2429-73-4)**
Diaphtamine Black MT **(2429-83-6)**
Diaphtamine Black V **(1937-37-7)**
Diaphtamine Blue BB **(2602-46-2)**
Diaphtamine Blue BS **(2610-05-1)**
Diaphtamine Blue TH **(72-57-1)**
Diaphtamine Brown M **(2429-82-5)**
Diaphtamine Fast Black KG **(6428-31-5)**
Diaphtamine Fast Brown TB **(2429-81-4)**
Diaphtamine Fast Purpurine 8B **(6548-29-4)**
Diaphtamine Fast Scarlet **(6358-29-8)**
Diaphtamine Light Brown BRLL **(16071-86-6)**
Diaphtamine Light Red 4B **(2610-11-9)**
Diaphtamine Pure Blue **(2429-74-5)**
Diaphtamine Purpurine **(992-59-6)**
Diaphtamine Violet N **(2586-60-9)**
Diaphthamine Fast Black FE **(3626-28-6)**
Diaphylline **(317-34-0)**
Diaquone **(117-10-2)**
Diarex 600 **(9003-55-8)**
Diarex 43G **(9003-53-6)**
Diarex HF 55 **(9003-53-6)**
Diarex HF 55-247 **(9003-53-6)**
Diarex HF 77 **(100-42-5)**
Diarex HF 77 **(9003-53-6)**
Diarex HS 77 **(9003-53-6)**
Diarex HT 88 **(9003-53-6)**
Diarex HT 88A **(9003-53-6)**
Diarex HT 90 **(9003-53-6)**
Diarex HT 190 **(9003-53-6)**
Diarex HT 500 **(9003-53-6)**
Diarex YH 476 **(9003-53-6)**
Diarsen **(144-21-8)**
Diarsenic pentoxide **(1303-28-2)**
Diarsenic trioxide **(1327-53-3)**
Diarsenic trisulfide **(1303-33-9)**
Diarsenic trisulphide **(1303-33-9)**
Diarylanilide Yellow **(6358-85-6)**
Diarylide Orange **(3520-72-7)**
Diarylide Yellow AAA **(6358-85-6)**
Diasetielmorfien **(561-27-3)**
Diasetilmorfin **(561-27-3)**
Diasetylmorfiimi **(561-27-3)**
Diastatin **(1400-61-9)**
Diastyl **(56-53-1)**
Diat (German) **(117-96-4)**
Diater **(330-54-1)**
Diaterr-Fos **(333-41-5)**
Diathal **(5490-27-7)**
Diathesin **(90-01-7)**
Diathol AS **(92-77-3)**
Diathol ASF **(92-77-3)**
Diathol BO **(135-62-6)**
Diathol BS **(135-65-9)**
Diathol D **(135-61-5)**
Diathol OL **(135-62-6)**
Diato Blue Base B **(119-90-4)**
Diato Blue Salt B **(119-90-4)**
Diatol **(105-58-8)**
Diatomaceous earth, Flux-calcined **(68855-54-9)**
Diatomaceous earth, Natural **(61790-53-2)**

Name	CAS
Diatomaceous earth	**(61790-53-2)**
Diatomaceous silica, Calcined	**(61790-53-2)**
Diatomaceous silica, Flux-calcined	**(66402-68-4)**
Diatomite	**(61790-53-2)**
Diatomite	**(68855-54-9)**
Diatrizoesaure (German)	**(117-96-4)**
Diatrizoic acid	**(117-96-4)**
p-Diazo-N,N'-diethylaniline chloride zinc chloride	**(5149-85-9)**
1,2-Diazabenzene	**(289-80-5)**
1,3-Diazabenzene	**(289-95-2)**
1,4-Diazabenzene	**(290-37-9)**
1,4-Diazabicyclo(2.2.2)octane	**(280-57-9)**
1,2-Diazabicyclo(2.2.2)octan-3-one (7CI,8CI,9CI)	**(1632-26-4)**
1,5-Diazabicyclo(5.4.0)undec-5-ene	**(6674-22-2)**
1,8-Diazabicyclo(5.4.0)undec-7-ene	**(6674-22-2)**
20,25-Diazachlolestenol dihydrochloride	**(1249-84-9)**
20,25-Diazacholesterol dihydrochloride	**(1249-84-9)**
Diazacholesterol dihydrochloride	**(1249-84-9)**
Diazacosterol hydrochloride	**(1249-84-9)**
1,3-Diaza-2,4-cyclopentadiene	**(288-32-4)**
4,7-Diazadecane-1,10-dinitrile	**(3217-00-3)**
4,9-Diaza-1,12-dodecanediamine	**(71-44-3)**
1,3-Diazaindene	**(51-17-2)**
2,3-Diazaindole	**(95-14-7)**
3,5-Diazaindole	**(272-97-9)**
Diazajet	**(333-41-5)**
Diazald	**(80-11-5)**
Diazale	**(80-11-5)**
Diazamine Purpurine 4B	**(992-59-6)**
1,2-Diazanaphthalene	**(253-66-7)**
1,3-Diazanaphthalene	**(253-82-7)**
1,4-Diazanaphthalene	**(91-19-0)**
1,7-Diazanaphthalene	**(253-69-0)**
2,3-Diazanaphthalene	**(253-52-1)**
3,6-Diazaoctane-1,8-diamine	**(112-24-3)**
3,6-Diazaoctanedioic acid, 3,6-bis(carboxymethyl)-	**(60-00-4)**
4,5-Diazaphenanthrene	**(66-71-7)**
5,6-Diazaphenanthrene	**(230-17-1)**
9,10-Diazaphenanthrene	**(230-17-1)**
Diazatol	**(333-41-5)**
N-(4-Diazo-2,5-diethoxyphenyl)morpholine, chloride, zinc chloride (2:1)	**(6023-29-6)**
2-Diazo-1,2-dihydro-1-oxo-5-naphthalenesulfonyl chloride	**(3770-97-6)**
6-Diazo-5,6-dihydro-5-oxo-1-naphthalenesulfonyl chloride	**(3770-97-6)**
6-Diazo-2,4-dinitro-2,4-cyclohexadien-1-one	**(4682-03-5)**
Diazemuls	**(439-14-5)**
Diazene 42	**(30812-87-4)**
Diazene, (1,1'-biphenyl)-4-ylphenyl- (9CI)	**(7466-42-4)**
Diazene, bis(4-chlorophenyl)-	**(1602-00-2)**
Diazene, bis(2-chlorophenyl)- (9CI)	**(7334-33-0)**
Diazene, bis(3-chlorophenyl)- (9CI)	**(15426-14-9)**
Diazene, bis(4-chlorphenyl)-, 1-oxide (9CI)	**(614-26-6)**
Diazene, bis(3,4-dichlorophenyl)- (9CI)	**(14047-09-7)**
Diazene, bis(3,4-dichlorophenyl)-, 1-oxide (9CI)	**(21232-47-3)**
Diazene, bis(2,4-dimethylphenyl)- (9CI)	**(29418-25-5)**
Diazene, bis(2,5-dimethylphenyl)- (9CI)	**(6311-44-0)**
Diazene, bis(4-ethylphenyl)- (9CI)	**(61653-33-6)**
Diazene, bis(2-methoxyphenyl)- (9CI)	**(613-55-8)**
Diazene, bis(3-methoxyphenyl)- (9CI)	**(6319-23-9)**
Diazene, bis(4-methoxyphenyl)- (9CI)	**(501-58-6)**
Diazene, bis(4-methoxyphenyl)-, 1-oxide (9CI)	**(1562-94-3)**
Diazene, bis(2-methylphenyl)- (9CI)	**(584-90-7)**
Diazene, bis(3-methylphenyl)- (9CI)	**(588-04-5)**
Diazene, bis(4-methylphenyl)- (9CI)	**(501-60-0)**
Diazene, bis(methylphenyl)- (9CI)	**(26444-20-2)**
Diazene, bis(2,4,6-trimethylphenyl)- (9CI)	**(5692-66-0)**
Diazene, bis(2,4,6-trinitrophenyl)- (9CI)	**(19159-68-3)**
Diazene, (4-chlorophenyl)phenyl- (9CI)	**(4340-77-6)**
Diazenedicarboxamide	**(123-77-3)**
Diazenedicarboximidamide, N,N''-dichloro	**(502-98-7)**
Diazenedicarboxylic acid, bis(phenylmethyl) ester (9CI)	**(2449-05-0)**
Diazene, diethyl-, 1-oxide	**(16301-26-1)**
Diazene, (2,6-dimethylphenyl)(2-methylphenyl)- (9CI)	**(6319-26-2)**
Diazene, (2,6-dimethylphenyl)phenyl- (9CI)	**(17590-87-3)**
Diazene, diphenyl-, 1-oxide (9CI)	**(495-48-7)**
Diazene, (2-methoxyphenyl)phenyl- (9CI)	**(6319-21-7)**
Diazene, (4-methoxyphenyl)phenyl- (9CI)	**(2396-60-3)**
Diazene, (methoxyphenyl)phenyl- (9CI)	**(63460-08-2)**
Diazene, (2-methylphenyl)(4-methylphenyl)- (9CI)	**(29418-22-2)**
Diazene, (2-methylphenyl)(4-nitrophenyl)- (9CI)	**(7030-18-4)**
Diazene, (2-methylphenyl)phenyl- (9CI)	**(6676-90-0)**
Diazene, (4-methylphenyl)phenyl- (9CI)	**(949-87-1)**
Diazene, (4-nitrophenyl)phenyl- (9CI)	**(2491-52-3)**
Diazenesulfonic acid, (4-(dimethylamino)phenyl)-, sodium salt	**(140-56-7)**
Diazepam	**(439-14-5)**
Diazepamu (Polish)	**(439-14-5)**
Diazepan	**(439-14-5)**
Diazetard	**(439-14-5)**
1,3-Diazetidine-2,4-dione, 1,3-bis(3-isocyanatomethylphenyl)- (9CI)	**(26747-90-0)**
Diazetylmorphine	**(561-27-3)**
Diazide	**(333-41-5)**
Diazido-1,2 ethane (French)	**(629-13-0)**
p-Diazidobenceno (Spanish)	**(2294-47-5)**
1,4-Diazidobenzene	**(2294-47-5)**
p-Diazidobenzene	**(2294-47-5)**
1,2-Diazidoetano (Spanish)	**(629-13-0)**
1,2-Diazidoethane	**(629-13-0)**
Diazil	**(302-40-9)**
Diazil	**(57-68-1)**
1,2-Diazine	**(289-80-5)**
1,4-Diazine	**(290-37-9)**
m-Diazine	**(289-95-2)**
p-Diazine	**(290-37-9)**
Diazine Black BHC	**(2429-73-4)**
Diazine Black E	**(1937-37-7)**
Diazine Black H	**(2429-73-4)**
Diazine Black HDW	**(2429-73-4)**
Diazine Black HNJ	**(2429-73-4)**
Diazine Blue 2B	**(2602-46-2)**
Diazine Blue 3B	**(72-57-1)**
Diazine Brown M	**(2429-82-5)**
Diazine Dark Green BO	**(3626-28-6)**
Diazine Dark Green P	**(3626-28-6)**
Diazine Direct Black E	**(1937-37-7)**
Diazine Direct Black G	**(1937-37-7)**
Diazine Direct Black R	**(2429-83-6)**
Diazine Fast Brown RSL	**(16071-86-6)**
Diazine Fast Red 8BK	**(2610-11-9)**
Diazine Red 4B	**(992-59-6)**
Diazine Sky Blue FF	**(2610-05-1)**
Diazine Violet N	**(2586-60-9)**
Diazine Yellow R	**(1325-37-7)**
Diazinon (ACGIH,OSHA) [UN 2783]	**(333-41-5)**
Diazinone	**(333-41-5)**
Diazirine	**(334-88-3)**
Diazitol	**(333-41-5)**
2-Diazo-1-naphthol-5-sulfonic acid sodium salt	**(2657-00-3)**
2-Diazo-1-naphthol-5-sulfonyl chloride	**(3770-97-6)**
2-Diazo-1-naphthone-5-sulfonic acid chloride	**(3770-97-6)**
2-Diazo-1-naphthone-5-sulfonic acid, sodium salt	**(2657-00-3)**
2-Diazo-1-naphthone-5-sulfonyl chloride	**(3770-97-6)**
Diazo Black BH	**(2429-73-4)**
Diazo Black BHN-CF	**(2429-73-4)**

Diazo Black BHSW (2429-73-4)
Diazo Black BHSWK (2429-73-4)
Diazo Black CR (2429-73-4)
Diazo Black RW (2429-83-6)
Diazo Brown MC (2429-82-5)
Diazo Direct Black N (2429-73-4)
Diazo Fast Black BH (2429-73-4)
Diazo Fast Black MBH (2429-73-4)
Diazo Fast Blue B (119-90-4)
Diazo Fast Blue BB (120-00-3)
Diazo Fast Orange GC (141-85-5)
Diazo Fast Orange GR (88-74-4)
Diazo Fast Orange R (99-09-2)
Diazo Fast Red AL (82-45-1)
Diazo Fast Red B (97-52-9)
Diazo Fast Red GG (100-01-6)
Diazo Fast Red 3GL (89-63-4)
Diazo Fast Red GL (89-62-3)
Diazo Fast Red RL (99-52-5)
Diazo Fast Red TR (3165-93-3)
Diazo Fast Red TRA (3165-93-3)
Diazo Fast Red TRA (95-69-2)
Diazo Fast Scarlet G (99-55-8)
Diazo Fast Scarlet GG (95-82-9)
Diazo Light Red 8B (2610-11-9)
Diazo Light Red 8BD (2610-11-9)
Diazo Navy Blue BH (2429-73-4)
Diazo Nero Microsetile G (539-17-3)
Diazoacetate (ester) l-serine (115-02-6)
l-Diazoacetate (ester) serine (115-02-6)
o-Diazoacetyl-l-serine (115-02-6)
Diazoaminobenzen (Czech) (136-35-6)
Diazoaminobenzene (136-35-6)
p-Diazoaminobenzene (136-35-6)
Diazoaminobenzol (German) (136-35-6)
Diazoben (140-56-7)
Diazoben (547-58-0)
Diazobenzene (103-33-3)
Diazobleu (314-13-6)
Diazocard Chrysoidine G (532-82-1)
Diazodifenilmetano (Spanish) (883-40-9)
Diazodinitrofenol (Spanish) (4682-03-5)
Diazodinitrophenol (Dry) (4682-03-5)
Diazodiphenylmethane (883-40-9)
Diazodiphenylmethane (French) (883-40-9)
Diazoimide (7782-79-8)
Diazol (333-41-5)
Diazol Black BH (2429-73-4)
Diazol Black C (537-65-5)
Diazol Black ER (2429-83-6)
Diazol Black 2V (1937-37-7)
Diazol Blue 2B (2602-46-2)
Diazol Blue 3B (72-57-1)
Diazol Brown M (2429-82-5)
Diazol Cutch F (2429-81-4)
Diazol Cutch FB (2429-81-4)
Diazol Fast Purpurine 8B (6548-29-4)
Diazol Fast Yellow A (1325-37-7)
Diazol Green Black N (3626-28-6)
Diazol Light Brown BRN (16071-86-6)
Diazol Light Yellow 7JL (10190-68-8)
Diazol Pure Blue 4B (2429-74-5)
Diazol Pure Blue 6B (2610-05-1)
Diazol Pure Blue BF (314-13-6)
Diazol Purpurine 4B (992-59-6)
Diazol Scarlet 3B (6358-29-8)
Diazol Violet N (2586-60-9)
1,2-Diazole (288-13-1)
1,3-Diazole (288-32-4)
Diazolone (68-35-9)
Diazomethane (ACGIH,OSHA) (334-88-3)
p-Diazonium N,N-diethylaniline chloride, zinc chloride salt (2:1) (5149-85-9)
p-Diazonium-N,N-diethylaniline, zinc chloride (2:1) (5149-85-9)
Diazophenyl Black BH (2429-73-4)
1,3-Diazopropane (5239-06-5)
Diazoresorcinol (550-82-3)
Diazotizing salts (7632-00-0)
p-Diazoviolet (74-39-5)
Diazoxon (962-58-3)
Diazyl (57-68-1)
Diazyl (68-35-9)
Dibam (128-04-1)
Dibam A (128-04-1)
Dibar (101-42-8)
Dibasic ammonium arsenate (7784-44-3)
Dibasic ammonium citrate (3012-65-5)
Dibasic ammonium phosphate (7783-28-0)
Dibasic lead acetate (301-04-2)
Dibasic lead arsenate (7784-40-9)
Dibasic lead carbonate (598-63-0)
Dibasic potassium phosphate (7758-11-4)
Dibasic sodium arsenate (10048-95-0)
Dibasic sodium phosphate (7558-79-4)
Dibasic zinc stearate (557-05-1)
Dibask (87-12-7)
Dibenamine (51-50-3)
Dibencil (1538-09-6)
Dibencillin (1538-09-6)
Dibendrin (58-73-1)
Dibenylin (59-96-1)
Dibenylin (63-92-3)
Dibenyline (59-96-1)
Dibenyline (63-92-3)
1,2:5,6-Dibenz(a)anthracene (53-70-3)
Dibenz(a,c)acridine (215-62-3)
Dibenz(a,c)anthracene (215-58-7)
Dibenzacepin (1977-10-2)
1,2,5,6-Dibenzacridine (226-36-8)
1,2,7,8-Dibenzacridine (224-42-0)
3,4,5,6-Dibenzacridine (224-42-0)
3,4:5,6-Dibenzacridine (224-53-3)
1,2,7,8-Dibenzacridine (French) (224-53-3)
Dibenz(a,d)acridine (226-36-8)
13H-Dibenz(a,de)anthracen-13-one (9CI) (86854-06-0)
8H-Dibenz(a,de)anthracen-8-one (8CI,9CI) (28609-66-7)
Dibenz(a,e)aceanthrylene (5385-75-1)
Dibenz(a,e)acephanthrylene (2997-45-7)
Dibenz(a,f)acridine (224-42-0)
Dibenz(a,h)acridine (226-36-8)
Dibenz(a,h)acridine, 1-ethyl (63021-33-0)
Dibenz(a,h)anthracene (53-70-3)
Dibenz(a,i)pyrene (189-55-9)
Dibenz(a,j)aceanthrylene (203-20-3)
Dibenz(a,j)acridine (224-42-0)
Dibenz(a,j)acridine, 1-ethyl (63021-35-2)
Dibenz(a,j)acridine, 14-methyl (59652-20-9)
Dibenz(a,j)anthracene (224-41-9)
7H-Dibenz(a,kl)anthracen-7-one (9CI) (60848-01-3)
o,o'-Dibenzamidodiphenyl disulfide (135-57-9)
Di-o-benzamidophenyl disulphide (135-57-9)
1,2,5,6-Dibenzanthraceen (Dutch) (53-70-3)
1,2:3,4-Dibenzanthracene (215-58-7)
1,2:5,6-Dibenzanthracene (53-70-3)
1,2:7,8-Dibenzanthracene (224-41-9)

2,3:6,7-Dibenzanthracene **(135-48-8)**
3,4,5,6-Dibenzanthracene **(224-41-9)**
Dibenzanthracene **(414-29-9)**
Dibenzo-1,2,7,8-anthracene **(224-41-9)**
lin-Dibenzanthracene **(135-48-8)**
Dibenzanthrone **(116-71-2)**
2,3,6,7-Dibenzazepine **(256-96-2)**
13H-Dibenz(bc,j)aceanthrylen-13-one (9CI) **(86854-17-3)**
13H-Dibenz(bc,l)aceanthrylen-13-one (9CI) **(86854-18-4)**
Dibenz(b,e)fluoranthene **(2997-45-7)**
Dibenz(b,e)oxepin-$\delta^{11(6H)}$,γ-propylamine, N,N-dimethyl **(1668-19-5)**
Dibenz(b,e)oxepin-δ(11(6H),γ)-propylamine, N-methyl-, hydrochloride (8CI) **(2887-91-4)**
Dibenz(b,f)azepine **(256-96-2)**
5H-Dibenz(b,f)azepine (9CI) **(256-96-2)**
5H-Dibenz(b,f)azepine, 5-(2-aminoethyl)- (7CI,8CI) **(2064-28-0)**
5H-Dibenz(b,f)azepine-5-carboxamide **(298-46-4)**
5H-Dibenz(b,f)azepine, 3-chloro-5-(3-(dimethylamino)propyl)-10,11-dihydro- **(303-49-1)**
5H-Dibenz(b,f)azepine, 10,11-dihydro **(494-19-9)**
5H-Dibenz(b,f)azepine, 10,11-dihydro-3-chloro-5-(3-(dimethylamino)propyl) **(303-49-1)**
5H-Dibenz(b,f)azepine, 10,11-dihydro-5-(3-(dimethylamino)-2-methylpropyl)- **(739-71-9)**
5H-Dibenz(b,f)azepine, 10,11-dihydro-5-(3-(dimethylamino)propyl)- **(50-49-7)**
5H-Dibenz(b,f)azepine, 10,11-dihydro-5-(3-(methylamino)propyl) (8CI) **(50-47-5)**
5H-Dibenz(b,f)azepine, 5-(3-(dimethylamino)-2-methylpropyl)-10,11-dihydro **(739-71-9)**
5H-Dibenz(b,f)azepine, 5-(3-(dimethylamino)propyl)-10,11-dihydro **(50-49-7)**
5H-Dibenz(b,f)azepine-5-ethanamine (9CI) **(2064-28-0)**
5H-Dibenz(b,f)azepine-5-propanamine, 10,11-dihydro-N-methyl- (9CI) **(50-47-5)**
5H-Dibenz(b,f)azepine-5-propanamine, 10,11-dihydro-N,N,β-trimethyl- **(739-71-9)**
Dibenz(b,f)(1,4)oxazepine, 2-chloro-11-(4-methyl-1-piperazinyl) **(1977-10-2)**
Dibenz(b,f)(1,4)oxazepin-8-ol, 2-chloro-11-(4-methyl-1-piperazinyl)- **(61443-77-4)**
Dibenz(b,f)(1,4)oxazepin-8-ol, 2-chloro-11-(1-piperazinyl)- **(61443-78-5)**
3,4,5,6-Dibenzcarbazol **(194-59-2)**
1,2,7,8-Dibenzcarbazole **(239-64-5)**
3,4,5,6-Dibenzcarbazole **(194-59-2)**
Dibenz(c,e)azocine, 5,6,7,8-tetrahydro- (8CI,9CI) **(6196-54-9)**
6H-Dibenz(c,e)(1,2)oxaphosphorin, 6-oxide (9CI) **(35948-25-5)**
Dibenz(c,h)acridine **(224-53-3)**
Dibenz(c,h)acridine, 7-methyl **(59652-21-0)**
Dibenz(de,kl)anthracene **(198-55-0)**
2,4-Dibenzeneazoresorcinol **(2051-85-6)**
Dibenzhydryl ether **(574-42-5)**
1,2,3,4-Dibenznaphthalene **(217-59-4)**
Dibenzo PQD **(105-11-3)**
Dibenzo(a,b)pyrene-7,14-dione **(128-66-5)**
Dibenzo(a,c)anthracene **(215-58-7)**
6H-Dibenzo(a,c)cyclohepten-6-one, 5,7-dihydro- (7CI,8CI,9CI) **(1139-82-8)**
Dibenzo(a,c)naphthacene (8CI,9CI) **(216-00-2)**
1,2,5,6-Dibenzoacridine **(226-36-8)**
5H-Dibenzo(a,d)cycloheptene-δ^5,γ-Propylamine, 10,11-dihydro-N-methyl **(72-69-5)**
5H-Dibenzo(a,d)cycloheptene-δ^5,γ-propylamine, 10,11-dihydro-N,N-dimethyl **(50-48-6)**
5H-Dibenzo(a,d)cycloheptene-5-propylamine, N-methyl **(438-60-8)**
5H-Dibenzo(a,d)cyclohepten-5-one **(2222-33-5)**
5H-Dibenzo(a,d)cyclohepten-5-one, 10,11-dihydro- (9CI) **(1210-35-1)**
4-(5H-Dibenzo(a,d)cyclohepten-5-ylidene)-1-methylpiperidine **(129-03-3)**
4-(5-Dibenzo(a,d)cyclohepten-5-ylidine)-1-methylpiperidine **(129-03-3)**
N-3-(5H-Dibenzo(a,d)cyclohepten-5-yl)propyl-N-methylamine **(438-60-8)**
Dibenzo(a,d)pyrene **(191-30-0)**
Dibenzo(a,e)fluoranthene **(5385-75-1)**
Dibenzo(a,e)pyrene **(192-65-4)**
13H-Dibenzo(a,g)fluorene **(207-83-0)**
13H-Dibenzo(a,g)fluoren-13-one (9CI) **(63041-47-4)**
Dibenzo(a,h)anthracene **(53-70-3)**
1H-Dibenzo(a,h)fluorene (8CI,9CI) **(23143-01-3)**
13H-Dibenzo(a,h)fluoren-13-one (8CI,9CI) **(4599-94-4)**
Dibenzo(a,h)pyrene **(189-64-0)**
7H-Dibenzo(a,i)carbazole **(239-64-5)**
13H-Dibenzo(a,i)fluoren-13-one (9CI) **(86854-01-5)**
Dibenzo(a,i)phenanthrene **(213-46-7)**
Dibenzo(a,i)pyrene **(189-55-9)**
Dibenzo(a,j)acridine **(224-42-0)**
Dibenzo(a,j)anthracene **(224-41-9)**
Dibenzo(a,jk)fluorene **(205-82-3)**
Dibenzo(a,j)naphthacene (8CI,9CI) **(227-04-3)**
Dibenzo(a,l)naphthacene (8CI,9CI) **(226-86-8)**
Dibenzo(a,l)pyrene **(191-30-0)**
1,2:3,4-Dibenzoanthracene **(215-58-7)**
1,2:5,6-Dibenzoanthracene **(53-70-3)**
Dibenzo(a,o)naphtho(1,2,3,4-rst)pentaphene (9CI) **(72382-90-2)**
Dibenzoazepine **(1977-10-2)**
Dibenzo(b,def)chrysene **(189-64-0)**
Dibenzo(b,def)chrysene-7,14-dione **(128-66-5)**
Dibenzo(b,d)furan **(132-64-9)**
6H-Dibenzo(b,d)pyran-1-ol, 3-(1',2'-dimethylheptyl)-7,8,9,10-tetra-hydro-6,6,9-trimethyl **(32904-22-6)**
6H-Dibenzo(b,d)pyran-1-ol, 3-hexyl-7,8,9,10-tetrahydro-6,6,9-trimethyl- **(117-51-1)**
6H-Dibenzo(b,d)pyran-1-ol, 6a,7,8,10a-tetrahydro-6,6,9-tri-methyl-3-pentyl **(1972-08-3)**
6H-Dibenzo(b,d)pyran-1-ol, 6,6,9-trimethyl-3-pentyl **(521-35-7)**
9H-Dibenzo(b,d)pyran-9-one, 3-(1,1-dimethylheptyl)-6,6a,7,8,10,10a-hexahydro-1-hydroxy-6,6-dimethyl-, trans-(+-)- **(51022-71-0)**
Dibenzo(b,d)pyrrole **(86-74-8)**
Dibenzo(b,d)thiophene **(132-65-0)**
Dibenzo(b,e)(1,4)dioxin **(262-12-4)**
Dibenzo(b,e)(1,4)dioxin, bromotrichloro- (9CI) **(107227-75-8)**
Dibenzo(b,e)(1,4)dioxin, 1-chloro- **(39227-53-7)**
Dibenzo(b,e)(1,4)dioxin, 2-chloro- **(39227-54-8)**
Dibenzo(b,e)(1,4)dioxin, chloro- (9CI) **(35656-51-0)**
Dibenzo(b,e)(1,4)dioxin, 1,6-dibromo- (9CI) **(91371-14-1)**
Dibenzo(b,e)(1,4)dioxin, 2,3-dibromo- (9CI) **(50585-37-0)**
Dibenzo(b,e)(1,4)dioxin, 2,7-dibromo- (9CI) **(39073-07-9)**
Dibenzo(b,e)(1,4)dioxin, 2,3-dibromo-7,8-dichloro- (9CI) **(50585-40-5)**
Dibenzo(b,e)(1,4)dioxin, 2,3-dibromo-7,8-difluoro- (9CI) **(50585-43-8)**
Dibenzo(b,e)(1,4)dioxin, 1,3-dichloro- **(50585-39-2)**
Dibenzo(b,e)(1,4)dioxin, 1,6-dichloro- **(38178-38-0)**
Dibenzo(b,e)(1,4)dioxin, 2,3-dichloro- **(29446-15-9)**
Dibenzo(b,e)(1,4)dioxin, 2,7-dichloro- **(33857-26-0)**
Dibenzo(b,e)(1,4)dioxin, 2,8-dichloro- **(38964-22-6)**
Dibenzo(b,e)(1,4)dioxin, 1,2-dichloro- (9CI) **(54536-18-4)**
Dibenzo(b,e)(1,4)dioxin, 1,4-dichloro- (9CI) **(54536-19-5)**
Dibenzo(b,e)(1,4)dioxin, dichloro- (9CI) **(64501-00-4)**
Dibenzo(b,e)(1,4)dioxin, 2,3-dichloro-7,8-difluoro- (9CI) **(50585-42-7)**
Dibenzo(b,e)(1,4)dioxin, 2,3-difluoro- (9CI) **(50585-38-1)**
Dibenzo(b,e)(1,4)dioxin, heptabromo- (9CI) **(103456-43-5)**
Dibenzo(b,e)(1,4)dioxin, 1,2,3,4,6,7,8-heptachloro- **(35822-46-9)**
Dibenzo(b,e)(1,4)dioxin, 1,2,3,4,6,7,9-heptachloro- **(58200-70-7)**
Dibenzo(b,e)(1,4)dioxin, 1,2,3,4,6,8-hexabromo- (9CI) **(116490-11-0)**
Dibenzo(b,e)(1,4)dioxin, hexabromo- (9CI) **(103456-42-4)**
Dibenzo(b,e)(1,4)dioxin, 1,2,4,6,7,9-hexabromo-3,8-dichloro- (9CI) **(2170-44-7)**
Dibenzo(b,e)(1,4)dioxin, 1,2,3,6,7,8-hexachloro- **(57653-85-7)**
Dibenzo(b,e)(1,4)dioxin, 1,2,3,6,7,9-hexachloro- **(64461-98-9)**
Dibenzo(b,e)(1,4)dioxin, 1,2,4,6,7,9-hexachloro- **(39227-62-8)**
Dibenzo(b,e)(1,4)dioxin, 1,2,3,4,6,7-hexachloro- (9CI) **(58200-66-1)**

Dibenzo(b,e)(1,4)dioxin, 1,2,3,4,6,8-hexachloro- (9CI) **(58200-67-2)**
Dibenzo(b,e)(1,4)dioxin, 1,2,3,4,6,9-hexachloro- (9CI) **(58200-68-3)**
Dibenzo(b,e)(1,4)dioxin, 1,2,3,6,8,9-hexachloro- (9CI) **(58200-69-4)**
Dibenzo(b,e)(1,4)dioxin, 1,2,4,6,8,9-hexachloro- (9CI) **(58802-09-8)**
Dibenzo(b,e)(1,4)dioxin, 2-iodo- (9CI) **(101714-96-9)**
Dibenzo(b,e)(1,4)dioxin, octabromo- (9CI) **(2170-45-8)**
Dibenzo(b,e)(1,4)dioxin, pentabromo- (9CI) **(103456-36-6)**
Dibenzo(b,e)(1,4)dioxin, 1,2,3,4,6-pentachloro- (9CI) **(67028-19-7)**
Dibenzo(b,e)(1,4)dioxin, 1,2,3,6,7-pentachloro- (9CI) **(71925-15-0)**
Dibenzo(b,e)(1,4)dioxin, 1,2,3,6,8-pentachloro- (9CI) **(71925-16-1)**
Dibenzo(b,e)(1,4)dioxin, 1,2,3,6,9-pentachloro- (9CI) **(82291-34-7)**
Dibenzo(b,e)(1,4)dioxin, 1,2,3,7,9-pentachloro- (9CI) **(71925-17-2)**
Dibenzo(b,e)(1,4)dioxin, 1,2,3,8,9-pentachloro- (9CI) **(71925-18-3)**
Dibenzo(b,e)(1,4)dioxin, 1,2,4,6,8-pentachloro- (9CI) **(71998-76-0)**
Dibenzo(b,e)(1,4)dioxin, 1,2,4,6,9-pentachloro- (9CI) **(82291-36-9)**
Dibenzo(b,e)(1,4)dioxin, 1,2,4,7,9-pentachloro- (9CI) **(82291-37-0)**
Dibenzo(b,e)(1,4)dioxin, 1,2,4,8,9-pentachloro- (9CI) **(82291-38-1)**
Dibenzo(b,e)(1,4)dioxin, pentachloro- (9CI) **(36088-22-9)**
Dibenzo(b,e)(1,4)dioxin, 1,3,6,8-tetrabromo- (9CI) **(76584-71-9)**
Dibenzo(b,e)(1,4)dioxin, 2,3,7,8-tetrabromo- (9CI) **(50585-41-6)**
Dibenzo(b,e)(1,4)dioxin, tetrabromo- (9CI) **(103456-39-9)**
Dibenzo(b,e)(1,4)dioxin, 1,2,3,4-tetrachloro- **(30746-58-8)**
Dibenzo(b,e)(1,4)dioxin, 1,2,3,8-tetrachloro- **(53555-02-5)**
Dibenzo(b,e)(1,4)dioxin, 1,2,7,8-tetrachloro- **(34816-53-0)**
Dibenzo(b,e)(1,4)dioxin, 1,3,6,8-tetrachloro- **(33423-92-6)**
Dibenzo(b,e)(1,4)dioxin, 1,3,7,8-tetrachloro- **(50585-46-1)**
Dibenzo(b,e)(1,4)dioxin, 2,3,7,8-tetrachloro- **(1746-01-6)**
Dibenzo(b,e)(1,4)dioxin, 1,2,3,6-tetrachloro- (9CI) **(71669-25-5)**
Dibenzo(b,e)(1,4)dioxin, 1,2,3,7-tetrachloro- (9CI) **(67028-18-6)**
Dibenzo(b,e)(1,4)dioxin, 1,2,3,9-tetrachloro- (9CI) **(71669-26-6)**
Dibenzo(b,e)(1,4)dioxin, 1,2,4,6-tetrachloro- (9CI) **(71669-27-7)**
Dibenzo(b,e)(1,4)dioxin, 1,2,4,7-tetrachloro- (9CI) **(71669-28-8)**
Dibenzo(b,e)(1,4)dioxin, 1,2,4,8-tetrachloro- (9CI) **(71669-29-9)**
Dibenzo(b,e)(1,4)dioxin, 1,2,4,9-tetrachloro- (9CI) **(71665-99-1)**
Dibenzo(b,e)(1,4)dioxin, 1,2,6,7-tetrachloro- (9CI) **(40581-90-6)**
Dibenzo(b,e)(1,4)dioxin, 1,2,6,8-tetrachloro- (9CI) **(67323-56-2)**
Dibenzo(b,e)(1,4)dioxin, 1,2,6,9-tetrachloro- (9CI) **(40581-91-7)**
Dibenzo(b,e)(1,4)dioxin, 1,2,7,9-tetrachloro- (9CI) **(71669-23-3)**
Dibenzo(b,e)(1,4)dioxin, 1,2,8,9-tetrachloro- (9CI) **(62470-54-6)**
Dibenzo(b,e)(1,4)dioxin, 1,3,6,9-tetrachloro- (9CI) **(71669-24-4)**
Dibenzo(b,e)(1,4)dioxin, 1,4,6,9-tetrachloro- (9CI) **(40581-93-9)**
Dibenzo(b,e)(1,4)dioxin, 1,4,7,8-tetrachloro- (9CI) **(40581-94-0)**
Dibenzo(b,e)(1,4)dioxin, 1,3,8-tribromo- (9CI) **(80246-33-9)**
Dibenzo(b,e)(1,4)dioxin, 1,2,4-trichloro- **(39227-58-2)**
Dibenzo(b,e)(1,4)dioxin, 1,3,7-trichloro- (9CI) **(67028-17-5)**
Dibenzo(b,e)(1,4)dioxin, trichloro- (9CI) **(69760-96-9)**
Dibenzo(b,e)pyridine **(260-94-6)**
3-Dibenzo(b,e)thiepin-11(6H)-ylidene-N,N-dimethyl-1-propamine **(113-53-1)**
5H-Dibenzo(b,f)azepine **(256-96-2)**
7H-Dibenzo(b,g)fluoren-7-one (9CI) **(86854-02-6)**
12H-Dibenzo(b,h)fluoren-12-one (9CI) **(53223-75-9)**
Dibenzo(b,h)phenanthrene **(222-93-5)**
Dibenzo(b,h)pyrene **(189-55-9)**
Dibenzo(b,jk)fluorene **(207-08-9)**
Dibenzo(b,k)chrysene **(217-54-9)**
Dibenzo(b,k)fluoranthene (8CI) **(205-97-0)**
3,4,5,6-Dibenzocarbazole **(194-59-2)**
Dibenzocarbazole (9CI) **(71012-25-4)**
Dibenzo(c,d,e)quinoline **(194-03-6)**
Dibenzo(cd,jk)pyrene **(191-26-4)**
7H-Dibenzo(c,g)carbazole **(194-59-2)**
7H-Dibenzo(c,g)fluoren-7-one (9CI) **(86853-97-6)**
Dibenzo(c,lm)fluorene **(203-33-8)**
Dibenzocycloheptanone **(1210-35-1)**
Dibenzo(def,mno)chrysene **(191-26-4)**
Dibenzo(def,mno)chrysene-6,12-dione, 4,10-dibromo- (9CI) **(4378-61-4)**
Dibenzo(def,p)chrysene **(191-30-0)**
13H-Dibenzo(def,qr)chrysen-13-one (9CI) **(86854-24-2)**
4H-Dibenzo(de,g)quinoline-10,11-diol, 5,6,6a,7-tetrahydro-6-methyl- **(58-00-4)**
4H-Dibenzo(de,g)quinoline-10,11-diol, 5,6,6a,7-tetrahydro-6-methyl-, hydrochloride, (R) **(314-19-2)**
Dibenzo(de,qr)naphthacene (8CI,9CI) **(193-09-9)**
Dibenzo(de,uv)pentacene (7CI,8CI,9CI) **(193-11-3)**
Dibenzo(de,yz)hexacene (7CI,8CI,9CI) **(313-71-3)**
Dibenzo(de,yz)naphtho(8,1,2-hij)hexaphene (7CI,8CI,9CI) **(435-02-9)**
Dibenzo(1,4)dioxin **(262-12-4)**
Dibenzo-p-dioxin **(262-12-4)**
Dibenzodioxin **(262-12-4)**
Dibenzo-p-dioxin, 1-chloro **(39227-53-7)**
Dibenzo-p-dioxin, 2-chloro **(39227-54-8)**
Dibenzo-p-dioxin, 1,6-dibromo- (9) **(91371-14-1)**
Dibenzo-p-dioxin, 2,7-dibromo- **(39073-07-9)**
Dibenzo-p-dioxin, 1,3-dichloro **(50585-39-2)**
Dibenzo-p-dioxin, 1,6-dichloro **(38178-38-0)**
Dibenzo-p-dioxin, 2,3-dichloro **(29446-15-9)**
Dibenzo-p-dioxin, 2,7-dichloro **(33857-26-0)**
Dibenzo-p-dioxin, 2,8-dichloro **(38964-22-6)**
Dibenzo-p-dioxin, 1,2,3,4,6,7,8-heptachloro **(35822-46-9)**
Dibenzo-p-dioxin, 1,2,3,4,6,7,9-heptachloro **(58200-70-7)**
Dibenzo-p-dioxin, 1,2,4,6,7,9-hexabromo-3,8-dichloro- (8CI) **(2170-44-7)**
Dibenzo-p-dioxin, 1,2,3,6,7,8-hexachloro **(57653-85-7)**
Dibenzo-p-dioxin, 1,2,3,6,7,9-hexachloro **(64461-98-9)**
Dibenzo-p-dioxin, 1,2,3,7,8,9-hexachloro **(19408-74-3)**
Dibenzo-p-dioxin, 1,2,4,6,7,9-hexachloro **(39227-62-8)**
Dibenzo-p-dioxin, hexachloro **(39227-28-6)**
Dibenzo-p-dioxin, 2-iodo- **(101714-96-9)**
Dibenzo-p-dioxin, octabromo- (8CI) **(2170-45-8)**
Dibenzo-p-dioxin, 1,2,3,4,6,7,8,9-octachloro **(3268-87-9)**
Dibenzo-p-dioxin, 1,2,3,4,7-pentachloro **(39227-61-7)**
Dibenzo-p-dioxin, 1,2,3,7,8-pentachloro **(40321-76-4)**
Dibenzo-p-dioxin, 1,2,4,7,8-pentachloro **(58802-08-7)**
Dibenzo-p-dioxin, 1,2,4,6,7-pentachloro- (9CI) **(82291-35-8)**
Dibenzo-p-dioxin, 2,3,7,8-tetrabromo- **(50585-41-6)**
Dibenzo-p-dioxin, 1,2,3,4-tetrachloro **(30746-58-8)**
Dibenzo-p-dioxin, 1,2,3,8-tetrachloro **(53555-02-5)**
Dibenzo-p-dioxin, 1,2,7,8-tetrachloro **(34816-53-0)**
Dibenzo-p-dioxin, 1,3,6,8-tetrachloro **(33423-92-6)**
Dibenzo-p-dioxin, 1,3,7,8-tetrachloro **(50585-46-1)**
Dibenzo-p-dioxin, 2,3,7,8-tetrachloro **(1746-01-6)**
Dibenzo-p-dioxin, 1,2,4-trichloro **(39227-58-2)**
Dibenzo-p-dioxin, 2,3,7-trichloro **(33857-28-2)**
Dibenzo(e,l)pyrene **(192-51-8)**
Dibenzo(fg,op)naphthacene (9CI) **(192-51-8)**
Dibenzo(fg,st)pentacene (7CI,8CI,9CI) **(192-59-6)**
Dibenzo(fg,wx)hexacene (7CI,8CI,9CI) **(192-60-9)**
2,3,5,6-Dibenzofluoranthene **(5385-75-1)**
Dibenzofluoranthene (9CI) **(60382-88-9)**
1,2,5,6-Dibenzofluorene **(207-83-0)**
Dibenzofluorene (9CI) **(73560-78-8)**
Dibenzofluorenone (9CI) **(83589-46-2)**
Dibenzofuran **(132-64-9)**
Dibenzofuran, 2-bromo- (9CI) **(86-76-0)**
Dibenzofuran, bromohexachloro- (9CI) **(107207-47-6)**
Dibenzofuran, bromotrichloro- (9CI) **(107227-56-5)**
4a(4H)-Dibenzofurancarboxaldehyde, 1,5a,6,9,9a,9b-hexahydro **(126-15-8)**
Dibenzofuran, 2,7-dibromo- (9CI) **(65489-80-7)**
Dibenzofuran, dichloro **(43047-99-0)**
Dibenzofuran, 1,2-dichloro- (9CI) **(64126-85-8)**
Dibenzofuran, 1,3-dichloro- (9CI) **(94538-00-8)**
Dibenzofuran, 1,4-dichloro- (9CI) **(94538-01-9)**
Dibenzofuran, 1,6-dichloro- (9CI) **(74992-97-5)**
Dibenzofuran, 1,7-dichloro- (9CI) **(94538-02-0)**
Dibenzofuran, 1,8-dichloro- (9CI) **(81638-37-1)**

Name	CAS
Dibenzofuran, 1,9-dichloro- (9CI)	**(70648-14-5)**
Dibenzofuran, 2,3-dichloro- (9CI)	**(64126-86-9)**
Dibenzofuran, 2,6-dichloro- (9CI)	**(60390-27-4)**
Dibenzofuran, 2,7-dichloro- (9CI)	**(74992-98-6)**
Dibenzofuran, 2,8-dichloro- (9CI)	**(5409-83-6)**
Dibenzofuran, 3,4-dichloro- (9CI)	**(94570-83-9)**
Dibenzofuran, 3,6-dichloro- (9CI)	**(74918-40-4)**
Dibenzofuran, 3,7-dichloro- (9CI)	**(58802-21-4)**
Dibenzofuran, 4,6-dichloro- (9CI)	**(64560-13-0)**
Dibenzofuran, 2,4-dichloro- (8CI,9CI)	**(24478-74-8)**
1,3(2H,9bH)-Dibenzofurandione, 2,6-diacetyl-7,9-dihydroxy-8,9b-dimethyl	**(125-46-2)**
Dibenzofuran, heptabromo- (9CI)	**(62994-32-5)**
Dibenzofuran, 1,2,3,4,6,7,8-heptachloro- (9CI)	**(67562-39-4)**
Dibenzofuran, 1,2,3,4,6,7,9-heptachloro- (9CI)	**(70648-25-8)**
Dibenzofuran, 1,2,3,4,6,8,9-heptachloro- (9CI)	**(69698-58-4)**
Dibenzofuran, 1,2,3,4,7,8,9-heptachloro- (9CI)	**(55673-89-7)**
Dibenzofuran, heptachloro- (9CI)	**(38998-75-3)**
Dibenzofuran, hexabromo- (9CI)	**(103456-33-3)**
Dibenzofuran, 1,2,3,4,7,8-hexachloro	**(70648-26-9)**
Dibenzofuran, 2,3,4,6,7,8-hexachloro	**(60851-34-5)**
Dibenzofuran, 1,2,3,4,6,7-hexachloro- (9CI)	**(79060-60-9)**
Dibenzofuran, 1,2,3,4,6,8-hexachloro- (9CI)	**(69698-60-8)**
Dibenzofuran, 1,2,3,4,6,9-hexachloro- (9CI)	**(91538-83-9)**
Dibenzofuran, 1,2,3,4,7,9-hexachloro- (9CI)	**(91538-84-0)**
Dibenzofuran, 1,2,3,4,8,9-hexachloro- (9CI)	**(92341-07-6)**
Dibenzofuran, 1,2,3,6,7,8-hexachloro- (9CI)	**(57117-44-9)**
Dibenzofuran, 1,2,3,6,7,9-hexachloro- (9CI)	**(92341-06-5)**
Dibenzofuran, 1,2,3,6,8,9-hexachloro- (9CI)	**(75198-38-8)**
Dibenzofuran, 1,2,3,7,8,9-hexachloro- (9CI)	**(72918-21-9)**
Dibenzofuran, 1,2,4,6,7,8-hexachloro- (9CI)	**(67562-40-7)**
Dibenzofuran, 1,2,4,6,7,9-hexachloro- (9CI)	**(75627-02-0)**
Dibenzofuran, 1,2,4,6,8,9-hexachloro- (9CI)	**(69698-59-5)**
Dibenzofuran, 1,3,4,6,7,8-hexachloro- (9CI)	**(71998-75-9)**
Dibenzofuran, 1,3,4,6,7,9-hexachloro- (9CI)	**(92341-05-4)**
Dibenzofuran, hexachloro- (9CI)	**(55684-94-1)**
Dibenzofuran, methyl- (9CI)	**(60826-62-2)**
Dibenzofuran, octabromo- (9CI)	**(103582-29-2)**
Dibenzofuran, pentabromo- (9CI)	**(68795-14-2)**
Dibenzofuran, 1,2,3,7,8-pentachloro	**(57117-41-6)**
Dibenzofuran, 2,3,4,7,8-pentachloro	**(57117-31-4)**
Dibenzofuran, 1,2,3,4,6-pentachloro- (9CI)	**(83704-47-6)**
Dibenzofuran, 1,2,3,4,7-pentachloro- (9CI)	**(83704-48-7)**
Dibenzofuran, 1,2,3,4,8-pentachloro- (9CI)	**(67517-48-0)**
Dibenzofuran, 1,2,3,4,9-pentachloro- (9CI)	**(83704-49-8)**
Dibenzofuran, 1,2,3,6,7-pentachloro- (9CI)	**(57117-42-7)**
Dibenzofuran, 1,2,3,6,8-pentachloro- (9CI)	**(83704-51-2)**
Dibenzofuran, 1,2,3,6,9-pentachloro- (9CI)	**(83704-52-3)**
Dibenzofuran, 1,2,3,7,9-pentachloro- (9CI)	**(83704-53-4)**
Dibenzofuran, 1,2,3,8,9-pentachloro- (9CI)	**(83704-54-5)**
Dibenzofuran, 1,2,4,6,7-pentachloro- (9CI)	**(83704-50-1)**
Dibenzofuran, 1,2,4,6,8-pentachloro- (9CI)	**(69698-57-3)**
Dibenzofuran, 1,2,4,6,9-pentachloro- (9CI)	**(70648-24-7)**
Dibenzofuran, 1,2,4,7,8-pentachloro- (9CI)	**(58802-15-6)**
Dibenzofuran, 1,2,4,7,9-pentachloro- (9CI)	**(71998-74-8)**
Dibenzofuran, 1,2,4,8,9-pentachloro- (9CI)	**(70648-23-6)**
Dibenzofuran, 1,2,6,7,8-pentachloro- (9CI)	**(69433-00-7)**
Dibenzofuran, 1,2,6,7,9-pentachloro- (9CI)	**(70872-82-1)**
Dibenzofuran, 1,3,4,6,7-pentachloro- (9CI)	**(83704-36-3)**
Dibenzofuran, 1,3,4,6,8-pentachloro- (9CI)	**(83704-55-6)**
Dibenzofuran, 1,3,4,6,9-pentachloro- (9CI)	**(70648-15-6)**
Dibenzofuran, 1,3,4,7,8-pentachloro- (9CI)	**(58802-16-7)**
Dibenzofuran, 1,3,4,7,9-pentachloro- (9CI)	**(70648-20-3)**
Dibenzofuran, 1,3,6,7,8-pentachloro- (9CI)	**(70648-21-4)**
Dibenzofuran, 1,4,6,7,8-pentachloro- (9CI)	**(83704-35-2)**
Dibenzofuran, 2,3,4,6,7-pentachloro- (9CI)	**(57117-43-8)**
Dibenzofuran, 2,3,4,6,8-pentachloro- (9CI)	**(67481-22-5)**
Dibenzofuran, pentachloro- (8CI,9CI)	**(30402-15-4)**
Dibenzofuran, 2,3,7,8-tetrabromo	**(67733-57-7)**
Dibenzofuran, 1,2,7,8-tetrabromo- (9CI)	**(84761-80-8)**
Dibenzofuran, 2,3,7,8-tetrachloro	**(51207-31-9)**
Dibenzofuran, 1,2,3,6-tetrachloro- (9CI)	**(83704-21-6)**
Dibenzofuran, 1,2,3,7-tetrachloro- (9CI)	**(83704-22-7)**
Dibenzofuran, 1,2,3,8-tetrachloro- (9CI)	**(62615-08-1)**
Dibenzofuran, 1,2,3,9-tetrachloro- (9CI)	**(83704-23-8)**
Dibenzofuran, 1,2,4,6-tetrachloro- (9CI)	**(71998-73-7)**
Dibenzofuran, 1,2,4,7-tetrachloro- (9CI)	**(83719-40-8)**
Dibenzofuran, 1,2,4,8-tetrachloro- (9CI)	**(64126-87-0)**
Dibenzofuran, 1,2,4,9-tetrachloro- (9CI)	**(83704-24-9)**
Dibenzofuran, 1,2,6,7-tetrachloro- (9CI)	**(83704-25-0)**
Dibenzofuran, 1,2,6,8-tetrachloro- (9CI)	**(83710-07-0)**
Dibenzofuran, 1,2,6,9-tetrachloro- (9CI)	**(70648-18-9)**
Dibenzofuran, 1,2,7,8-tetrachloro- (9CI)	**(58802-20-3)**
Dibenzofuran, 1,2,7,9-tetrachloro- (9CI)	**(83704-26-1)**
Dibenzofuran, 1,2,8,9-tetrachloro- (9CI)	**(70648-22-5)**
Dibenzofuran, 1,3,4,6-tetrachloro- (9CI)	**(83704-27-2)**
Dibenzofuran, 1,3,4,7-tetrachloro- (9CI)	**(70648-16-7)**
Dibenzofuran, 1,3,4,8-tetrachloro- (9CI)	**(92341-04-3)**
Dibenzofuran, 1,3,4,9-tetrachloro- (9CI)	**(83704-28-3)**
Dibenzofuran, 1,3,6,7-tetrachloro- (9CI)	**(57117-36-9)**
Dibenzofuran, 1,3,6,9-tetrachloro- (9CI)	**(83690-98-6)**
Dibenzofuran, 1,3,7,8-tetrachloro- (9CI)	**(57117-35-8)**
Dibenzofuran, 1,3,7,9-tetrachloro- (9CI)	**(64560-17-4)**
Dibenzofuran, 1,4,6,7-tetrachloro- (9CI)	**(66794-59-0)**
Dibenzofuran, 1,4,6,8-tetrachloro- (9CI)	**(82911-58-8)**
Dibenzofuran, 1,4,6,9-tetrachloro- (9CI)	**(70648-19-0)**
Dibenzofuran, 1,4,7,8-tetrachloro- (9CI)	**(83704-29-4)**
Dibenzofuran, 1,6,7,8-tetrachloro- (9CI)	**(83704-33-0)**
Dibenzofuran, 2,3,4,6-tetrachloro- (9CI)	**(83704-30-7)**
Dibenzofuran, 2,3,4,7-tetrachloro- (9CI)	**(83704-31-8)**
Dibenzofuran, 2,3,4,8-tetrachloro- (9CI)	**(83704-32-9)**
Dibenzofuran, 2,3,6,7-tetrachloro- (9CI)	**(57117-39-2)**
Dibenzofuran, 2,3,6,8-tetrachloro- (9CI)	**(57117-37-0)**
Dibenzofuran, 2,4,6,7-tetrachloro- (9CI)	**(57117-38-1)**
Dibenzofuran, 3,4,6,7-tetrachloro- (9CI)	**(57117-40-5)**
Dibenzofuran, ar,ar,ar',ar'-tetrachloro- (8CI,9CI)	**(30402-14-3)**
Dibenzofuran, 1,2,8-tribromo- (9CI)	**(84761-81-9)**
Dibenzofuran, 2,3,8-tribromo- (9CI)	**(84761-82-0)**
Dibenzofuran, 1,2,3-trichloro- (9CI)	**(83636-47-9)**
Dibenzofuran, 1,2,6-trichloro- (9CI)	**(64560-15-2)**
Dibenzofuran, 1,2,7-trichloro- (9CI)	**(83704-37-4)**
Dibenzofuran, 1,2,8-trichloro- (9CI)	**(83704-34-1)**
Dibenzofuran, 1,2,9-trichloro- (9CI)	**(83704-38-5)**
Dibenzofuran, 1,3,4-trichloro- (9CI)	**(82911-61-3)**
Dibenzofuran, 1,3,6-trichloro- (9CI)	**(83704-39-6)**
Dibenzofuran, 1,3,7-trichloro- (9CI)	**(64560-16-3)**
Dibenzofuran, 1,3,8-trichloro- (9CI)	**(76621-12-0)**
Dibenzofuran, 1,3,9-trichloro- (9CI)	**(83704-40-9)**
Dibenzofuran, 1,4,6-trichloro- (9CI)	**(82911-60-2)**
Dibenzofuran, 1,4,7-trichloro- (9CI)	**(83704-41-0)**
Dibenzofuran, 1,4,9-trichloro- (9CI)	**(70648-13-4)**
Dibenzofuran, 1,6,7-trichloro- (9CI)	**(83704-46-5)**
Dibenzofuran, 1,6,8-trichloro- (9CI)	**(82911-59-9)**
Dibenzofuran, 1,7,8-trichloro- (9CI)	**(58802-18-9)**
Dibenzofuran, 2,3,4-trichloro- (9CI)	**(57117-34-7)**
Dibenzofuran, 2,3,6-trichloro- (9CI)	**(57117-33-6)**
Dibenzofuran, 2,3,7-trichloro- (9CI)	**(58802-17-8)**
Dibenzofuran, 2,3,8-trichloro- (9CI)	**(57117-32-5)**
Dibenzofuran, 2,4,6-trichloro- (9CI)	**(58802-14-5)**
Dibenzofuran, 2,4,7-trichloro- (9CI)	**(83704-42-1)**
Dibenzofuran, 2,4,8-trichloro- (9CI)	**(54589-71-8)**
Dibenzofuran, 2,6,7-trichloro- (9CI)	**(83704-45-4)**
Dibenzofuran, 3,4,6-trichloro- (9CI)	**(83704-43-2)**
Dibenzofuran, 3,4,7-trichloro- (9CI)	**(83704-44-3)**

Name	CAS No.
Dibenzofuran, trichloro- (9CI)	**(43048-00-6)**
Dibenzofuran, 1,2,4-trichloro- (8CI,9CI)	**(24478-73-7)**
Dibenzo(hi,uv)hexacene (7CI,8CI,9CI)	**(192-54-1)**
Dibenzohydryl ether	**(574-42-5)**
Dibenzo(j,xyz)heptaphene (7CI,8CI,9CI)	**(192-46-1)**
1,2,5,6-Dibenzonaphthalene	**(218-01-9)**
Dibenzoparadiazine	**(92-82-0)**
Dibenzoparathiazine	**(92-84-2)**
Dibenzoperylene (9CI)	**(65256-40-8)**
1,2,3,4-Dibenzophenanthrene	**(196-78-1)**
1,2:6,7-Dibenzophenanthrene	**(214-17-5)**
1,2:7,8-Dibenzophenanthrene	**(213-46-7)**
2,3:6,7-Dibenzophenanthrene	**(222-93-5)**
2,3:7,8-Dibenzophenanthrene	**(214-17-5)**
Dibenzo-2,3,7,8-phenanthrene	**(214-17-5)**
Dibenzopyrazine	**(92-82-0)**
1,2,4,5-Dibenzopyrene	**(192-65-4)**
1,2,6,7-Dibenzopyrene	**(189-64-0)**
1,2,7,8-Dibenzopyrene	**(189-55-9)**
1,2,9,10-Dibenzopyrene	**(191-30-0)**
1,2:3,4-Dibenzopyrene	**(191-30-0)**
2,3:4,5-Dibenzopyrene	**(191-30-0)**
3,4,8,9-Dibenzopyrene	**(189-64-0)**
3,4:9,10-Dibenzopyrene	**(189-55-9)**
Dibenzopyrene (9CI)	**(58615-36-4)**
2,3,7,8-Dibenzopyrene-1,6-quinone	**(128-66-5)**
Dibenzopyrrole	**(86-74-8)**
Dibenzo(h,rst)pentaphene	**(192-47-2)**
Dibenzosuberone	**(1210-35-1)**
1,2:3,4-Dibenzotetracene	**(216-00-2)**
1,2:7,8-Dibenzotetracene	**(227-04-3)**
Dibenzo-1,4-thiazine	**(92-84-2)**
Dibenzothiazine	**(92-84-2)**
Di-2-benzothiazolyl disulfide	**(120-78-5)**
Dibenzothiazolyl disulphide	**(120-78-5)**
2,2'-Dibenzothiazyl disulfide	**(120-78-5)**
Dibenzothiazyl disulfide	**(120-78-5)**
Dibenzothiophene (9CI)	**(132-65-0)**
Dibenzothiophene, dimethyl- (9CI)	**(70021-47-5)**
Dibenzothiophene, 5,5-dioxide	**(1016-05-3)**
Dibenzothiophene, methyl- (9CI)	**(30995-64-3)**
Dibenzothiophene, 1-methyl- (8CI,9CI)	**(31317-07-4)**
Dibenzothiophene, 2-methyl- (8CI,9CI)	**(20928-02-3)**
Dibenzothiophene, 3-methyl- (8CI,9CI)	**(16587-52-3)**
Dibenzothiophene, 4-methyl- (8CI,9CI)	**(7372-88-5)**
Dibenzothiophene, 5-oxide (9CI)	**(1013-23-6)**
Dibenzothiophene sulfone	**(1013-23-6)**
Dibenzothiophene, trimethyl- (9CI)	**(70021-48-6)**
Dibenzothioxin	**(262-20-4)**
1,4-Dibenzothioxine	**(262-20-4)**
Dibenzoyl	**(134-81-6)**
2,2'-Dibenzoylaminodiphenyl disulfide	**(135-57-9)**
Dibenzoyldiethyleneglycol ester	**(120-55-8)**
Dibenzoyl-methane	**(120-46-7)**
Dibenzoylmethane	**(120-46-7)**
Dibenzoylperoxid (German)	**(94-36-0)**
Dibenzoyl peroxide	**(94-36-0)**
Dibenzoylperoxyde (Dutch)	**(94-36-0)**
Dibenzoylthiazyl disulfide	**(120-78-5)**
1,2,3,4-Dibenzphenanthrene	**(196-78-1)**
1,2,5,6-Dibenzphenanthrene	**(194-69-4)**
2,3:6,7-Dibenzphenanthrene	**(222-93-5)**
1,2,3,4-Dibenzpyrene	**(191-30-0)**
1,2:7,8-Dibenzpyrene	**(189-55-9)**
3,4,8,9-Dibenzpyrene	**(189-64-0)**
3,4:9,10-Dibenzpyrene	**(189-55-9)**
4,5,6,7-Dibenzpyrene	**(191-30-0)**
1,2:6,7-Dibenzpyrene (VAN)	**(192-51-8)**
1',2',6',7'-Dibenzpyrene-7,14-quinone	**(128-66-5)**
Dibenzsuberone	**(1210-35-1)**
Dibenzthiazyl disulfide	**(120-78-5)**
Dibenzyl	**(103-29-7)**
Dibenzylamine	**(103-49-1)**
Dibenzylamine (8CI)	**(103-49-1)**
Dibenzylamine, N-(2-chloroethyl)	**(51-50-3)**
Dibenzylamine, α,α'-dimethyl	**(10024-74-5)**
Dibenzylamine, hydrochloride	**(20455-68-9)**
Dibenzylammonium chloride	**(20455-68-9)**
Dibenzyl azodicarboxylate	**(2449-05-0)**
Dibenzyl chlorethylamine	**(51-50-3)**
N,N-Dibenzyl-β-chloroethylamine	**(51-50-3)**
Dibenzyldichlorosilane	**(18414-36-3)**
Dibenzyldithiocarbamic acid, zinc salt	**(14726-36-4)**
Dibenzylene	**(63-92-3)**
Dibenzyl ether	**(103-50-4)**
N,N'-Dibenzylethylenediamine, Compounded with penicillin G (1:2)	**(1538-09-6)**
N,N'-Dibenzylethylenediamine bis(benzyl penicillin)	**(1538-09-6)**
N,N'-Dibenzylethylenediamine diacetate	**(122-75-8)**
Dibenzylethylenediamine-di-penicillin G	**(1538-09-6)**
N,N'-Dibenzylethylenediamine salt of benzylpenicillin	**(1538-09-6)**
Dibenzyline	**(59-96-1)**
Dibenzyline	**(63-92-3)**
Dibenzyline hydrochloride	**(63-92-3)**
Dibenzyl ketone	**(102-04-5)**
Dibenzylmethylbenzene	**(26898-17-9)**
Dibenzyl monosulfide	**(538-74-9)**
Dibenzyl peroxydicarbonate	**(2144-45-8)**
Dibenzyl peroxydicarbonate, Maximum concentration 87% with water	**(2144-45-8)**
Dibenzyl peroxydicarbonate (More than 87% with water)	**(2144-45-8)**
Dibenzyl peroxydicarbonate, More than 87% with water	**(2144-45-8)**
Dibenzyl peroxydicarbonate (Not more than 87% with water)	**(2144-45-8)**
Dibenzyl sebacate	**(140-24-9)**
Dibenzyl sulfide	**(538-74-9)**
Dibenzyltoluene	**(26898-17-9)**
Dibenzyran	**(63-92-3)**
Dibestil	**(130-80-3)**
Dibestrol	**(56-53-1)**
Dibestrol (2) premix	**(56-53-1)**
Dibondrin	**(58-73-1)**
Diborane(6)	**(19287-45-7)**
Diborane (ACGIH,DOT,OSHA)	**(19287-45-7)**
Diborane (ACGIH,OSHA) [UN 1911]	**(19287-45-7)**
Diborane(6) (9CI)	**(19287-45-7)**
Diborane Mixture (DOT)	**(19287-45-7)**
Diborane mixture	**(19287-45-7)**
Diborane or diborane mixture	**(19287-45-7)**
Diborane(6), tris(methyl-d3)- (8CI)	**(23797-84-4)**
Diborano (Spanish)	**(19287-45-7)**
Diboron hexahydride	**(19287-45-7)**
Dibovan	**(50-29-3)**
1,2-Dibrom-3-chlor-propan (German)	**(96-12-8)**
1,6-Dibrom-1,6-didesoxy-d-mannit (German)	**(488-41-5)**
5,5'-Dibrom-2,2'-dioxybenzil (German)	**(523-88-6)**
3,5-Dibrom-4-hydroxylbenzaldoxim-O-(2',4'-dinitrophenyl)-aether (German)	**(13181-17-4)**
Dibrom	**(300-76-5)**
1,2-Dibromaethan (German)	**(106-93-4)**
Dibromannit	**(488-41-5)**
Dibromantin	**(77-48-5)**
Dibromantine	**(77-48-5)**
1,4-Dibrombutan (German)	**(110-52-1)**
Dibromchlorpropan (German)	**(96-12-8)**

O-(1,2-Dibrom-2,2-dichlor-aethyl)-O,O-dimethyl-phosphat (German) **(300-76-5)**
Dibromdulcit **(10318-26-0)**
Dibromdulcitol **(10318-26-0)**
Dibromoacetic acid **(631-64-1)**
Dibromoacetonitrile **(3252-43-5)**
2,4'-Dibromoacetophenone **(99-73-0)**
α,p-Dibromoacetophenone **(99-73-0)**
Dibromoacetylene **(624-61-3)**
2,4-Dibromoaniline **(615-57-6)**
2,4-Dibromo-1-anthraquinonylamine **(81-49-2)**
2,4-Dibromobenzenamine **(615-57-6)**
1,2-Dibromobenzene **(583-53-9)**
1,3-Dibromobenzene **(108-36-1)**
1,4-Dibromobenzene **(106-37-6)**
m-Dibromobenzene **(108-36-1)**
p-Dibromobenzene **(106-37-6)**
Dibromobenzene [UN 2711] **(26249-12-7)**
4,4'-Dibromobenzilic acid isopropyl ester **(18181-80-1)**
4,4'-Dibromo-1,1'-biphenyl **(92-86-4)**
4,4'-Dibromobiphenyl **(92-86-4)**
p,p'-Dibromobiphenyl **(92-86-4)**
3,5-Dibromo-N-(4-bromophenyl)-2-hydroxybenzamide **(87-10-5)**
1,2-Dibromobutane **(533-98-2)**
1,3-Dibromobutane **(107-80-2)**
1,4-Dibromobutane **(110-52-1)**
2,3-Dibromobutane **(5408-86-6)**
2,6-Dibromocaproic acid **(13137-43-4)**
Dibromochloromethane **(124-48-1)**
1,2-Dibromo-3-chloro-2-methylpropane **(10474-14-3)**
1,1-Dibromo-2-chloropropane **(55162-35-1)**
Dibromochloropropane **(96-12-8)**
1,2-Dibromo-3-chloropropane (DOT,OSHA) **(96-12-8)**
1,2-Dibromo-3-cloro-propano (Italian) **(96-12-8)**
Dibromocyanoacetamide **(10222-01-2)**
α,α-Dibromo-α-cyanoacetamide **(10222-01-2)**
2,6-Dibromo-4-cyanophenol **(1689-84-5)**
2,6-Dibromo-4-cyanophenyl octanoate **(1689-99-2)**
1,4-Dibromocyclohexane **(35076-92-7)**
1,10-Dibromodecane **(4101-68-2)**
2,3-Dibromodibenzo(b,e)(1,4)dioxin **(50585-37-0)**
1,6-Dibromodibenzo-p-dioxin **(91371-14-1)**
2,3-Dibromodibenzo-p-dioxin **(50585-37-0)**
2,7-Dibromodibenzofuran **(65489-80-7)**
1,2-Dibromo-4-(1,2-dibromoethyl)cyclohexane **(3322-93-8)**
2,3-Dibromo-7,8-dichlorobenzo(b,e)(1,4)dioxin **(50585-40-5)**
2,3-Dibromo-7,8-dichlorodibenzo-p-dioxin **(50585-40-5)**
1,2-Dibromo-1,1-dichloroethane **(75-81-0)**
1,2-Dibromo-1,2-dichloroethane **(683-68-1)**
1,2-Dibromo-2,2-dichloroethane **(75-81-0)**
1,2-Dibromo-2,2-dichloroethyl dimethyl phosphate **(300-76-5)**
O-(1,2-Dibromo-2,2-dicloro-etil)-O,O-dimetil-fostato (Italian) **(300-76-5)**
1,2-Dibromo-2,4-dicyanobutane **(35691-65-7)**
1,6-Dibromo-1,6-dideoxydulcitol **(10318-26-0)**
1,6-Dibromodideoxydulcitol **(10318-26-0)**
1,6-Dibromo-1,6-dideoxy-D-galactitol **(10318-26-0)**
1,6-Dibromo-1,6-dideoxygalactitol **(10318-26-0)**
1,6-Dibromo-1,6-dideoxy-d-mannitol **(488-41-5)**
1,6-Dibromo-1,6-d-didesoxymannitol **(488-41-5)**
β,β'-Dibromodiethyl ether **(5414-19-7)**
2,3-Dibromo-7,8-difluorodibenzo(b,e)(1,4)dioxin **(50585-43-8)**
2,3-Dibromo-7,8-difluorodibenzo-p-dioxin **(50585-43-8)**
Dibromodifluoromethane [UN 1941] **(75-61-6)**
5,5'-Dibromo-2,2'-dihydroxybenzil **(523-88-6)**
5,5'-Dibromo-2,2'-dihydroxybibenzoyl **(523-88-6)**
N,N'-Dibromodimethylhydantoin **(77-48-5)**
1,3-Dibromo-2,2-dimethylpropane **(5434-27-5)**
4,4'-Dibromodiphenyl **(92-86-4)**
1,6-Dibromodulcitol **(10318-26-0)**
Dibromodulcitol **(10318-26-0)**
1,2-Dibromoetano (Italian) **(106-93-4)**
1,1-Dibromoethane **(557-91-5)**
Dibromoethane **(106-93-4)**
α,β-Dibromoethane **(106-93-4)**
sym-Dibromoethane **(106-93-4)**
1,2-Dibromoethane (DOT) **(106-93-4)**
1,1'-(1,2-Dibromo-1,2-ethanediyl)bis(3,4-dibromocyclohexane) **(18122-77-5)**
1,2-Dibromoethanol acetate **(24442-57-7)**
Dibromoethene **(25429-23-6)**
1,2-Dibromoethyl acetate **(24442-57-7)**
α,β-Dibromoethyl acetate **(24442-57-7)**
(1,2-Dibromoethyl)benzene **(93-52-7)**
α,β-Dibromoethylbenzene **(93-52-7)**
1-(1,2-Dibromoethyl)-3,4-dibromocyclohexane **(3322-93-8)**
1,1-Dibromoethylene **(593-92-0)**
1,2-Dibromoethylene **(540-49-8)**
Dibromoethylene **(624-61-3)**
2,6-Dibromohexanoic acid **(13137-43-4)**
3,5-Dibromo-4-hydroxybenzaldehyde-O-(2',4'-dinitrophenyl)-oxime **(13181-17-4)**
3,5-Dibromo-4-hydroxybenzaldehyde (2',4'-dinitrophenyl)oxime **(13181-17-4)**
3,5-Dibromo-4-hydroxybenzaldehyde 2,4-dinitrophenyl oxime **(13181-17-4)**
3,5-Dibromo-4-hydroxybenzonitrile **(1689-84-5)**
3,5-Dibromo-4-hydroxyphenylcyanide **(1689-84-5)**
3,5-Dibromo-2-hydroxy-N-(3-(trifluoromethyl)phenyl)benzamide **(4776-06-1)**
1,6-Dibromomannitol **(488-41-5)**
D-Dibromomannitol **(488-41-5)**
Dibromomannitol **(488-41-5)**
Dibromomethane [UN 2664] **(74-95-3)**
1,4-Dibromo-2-methylbenzene **(615-59-8)**
2,4-Dibromo-1-methylbenzene **(31543-75-6)**
2,4-Dibromo-6-methylphenol **(609-22-3)**
((2,4-Dibromo-6-methylphenoxy)methyl)oxirane **(75150-13-9)**
((2,6-Dibromo-4-methylphenoxy)methyl)oxirane **(22421-59-6)**
1,2-Dibromo-2-methylpropane **(594-34-3)**
Dibromoneopentyl glycol **(3296-90-0)**
2,2-Dibromo-3-nitrilopropionamide **(10222-01-2)**
2,6-Dibromo-4-nitroaniline **(827-94-1)**
2,6-Dibromo-4-nitrobenzenamine **(827-94-1)**
2,6-Dibromo-4-nitrophenol **(99-28-5)**
1,1-Dibromononane **(62168-27-8)**
1,9-Dibromononane **(4549-33-1)**
1,1-Dibromooctane **(62168-26-7)**
3,5-Dibromo-4-octanoyloxy-benzonitrile **(1689-99-2)**
Dibromopentaerythritol **(3296-90-0)**
1,5-Dibromopentane **(111-24-0)**
2,4-Dibromopentane **(19398-53-9)**
1,2-Dibromoperfluoroethane **(124-73-2)**
2,3-Dibromophenol **(57383-80-9)**
2,4-Dibromophenol **(615-58-7)**
2,6-Dibromophenol **(608-33-3)**
3,4-Dibromophenol **(615-56-5)**
2,4(or 2,6)-Dibromophenol homopolymer **(69882-11-7)**
(3,5-Dibromophenoxy) acetic acid **(7507-35-9)**
((2,4-Dibromophenoxy)methyl)oxirane **(20217-01-0)**
1,2-Dibromo-1-phenylethane **(93-52-7)**
2,4-Dibromophenyl glycidyl ether **(20217-01-0)**
1,2-Dibromopropane **(78-75-1)**
1,3-Dibromopropane **(109-64-8)**
2,2-Dibromopropane **(594-16-1)**
α,γ-Dibromopropane **(109-64-8)**
ω,ω'-Dibromopropane **(109-64-8)**
2,3-Dibromopropanoic acid **(600-05-5)**
1,3-Dibromo-2-propanol **(96-21-9)**

2,3-Dibromo-1-propanol (96-13-9)
2,3-Dibromopropanol (96-13-9)
2,3-Dibromo-1-propanol dihydrogen phosphate (5324-12-9)
2,3-Dibromo-1-propanol phosphate (126-72-7)
2,3-Dibromopropene (513-31-5)
2,3-Dibromopropionic acid (600-05-5)
α,β-Dibromopropionic acid (600-05-5)
((1,2-Dibromopropoxy)methyl)oxirane (35243-89-1)
2,3-Dibromopropyl acrylate (19660-16-3)
2,3-Dibromopropyl methacrylate (3066-70-4)
(2,3-Dibromopropyl) phosphate (126-72-7)
2,3-Dibromopropylphosphate (5324-12-9)
Dibromo-8,16-pyranthrenedione (1324-35-2)
2,6-Dibromoquinone chlorimide (537-45-1)
2,6-Dibromoquinone chloroimide (537-45-1)
2,6-Dibromoquinone chloroimine (537-45-1)
5,5'-Dibromosalicil (523-88-6)
Dibromosalicil (523-88-6)
Dibromosalicyl (523-88-6)
3,5-Dibromosalicylanilide (2577-72-2)
4',5-Dibromosalicylanilide (87-12-7)
3,5-Dibromosalicylic acid p-bromoanilide (87-10-5)
1,4-Dibromo-1,1,2,2-tetrafluorobutane (18599-20-7)
1,2-Dibromotetrafluoroethane (124-73-2)
sym-Dibromotetrafluoroethane (124-73-2)
2,5-Dibromothiophene (3141-27-3)
3,3'-Dibromothymolsulfonphthalein (76-59-5)
Dibromothymolsulfophthalein (76-59-5)
3,5-Dibromo-3'-(trifluoromethyl)salicylanilide (4776-06-1)
3,5-Dibromo-3'-trifluoromethylsalicylanilide (4776-06-1)
Dibromotrifluoromonochloroethane (29256-79-9)
3,5-Dibromo-α,α,α-trifluoro-m-salicylotoluidide (4776-06-1)
Dibromsalan (87-12-7)
Dibromsalanum (Latin) (87-12-7)
Dibromsalen (87-12-7)
Dibromure d'ethylene (French) (106-93-4)
Dibronsalan (Spanish) (87-12-7)
1,2-Dibroom-3-chloorpropaan (Dutch) (96-12-8)
O-(1,2-Dibroom-2,2-dichloor-ethyl)-O,O-dimethyl-fosfaat (Dutch) (300-76-5)
1,2-Dibroomethaan (Dutch) (106-93-4)
Dibrosal (523-88-6)
Dibucain (85-79-0)
Dibucaine (85-79-0)
Dibucaine base (85-79-0)
Dibunol (128-37-0)
Dibutalin (33629-47-9)
2,3:4,5-Di(2-butenyl)tetrahydrofurfural (126-15-8)
Dibutilamina (Romanian) (111-92-2)
2,2-Di-(terc-butilperoxi) butano (Spanish) (2167-23-9)
1,1-Di-(terc-butilperoxi) ciclohexano (Spanish) (3006-86-8)
Di-(terc-butilperoxi)ftalato (Spanish) (2155-71-7)
1,1-Di-(terc-butilperoxi)-3,3,5-trimetilciclohexano (Spanish) (6731-36-8)
Dibutin (54-85-3)
1,1-Dibutoxybutane (5921-80-2)
1,2-Dibutoxyethane (112-48-1)
Dibutoxyethoxyethyl adipate (141-17-3)
Di(2-butoxyethyl) adipate (141-18-4)
Dibutoxyethyl adipate (141-18-4)
Di-(2-butoxyethyl)ester kyseliny ftalove (Czech) (117-83-9)
2,2'-Dibutoxyethyl ether (112-73-2)
Di(butoxyethyl)phthalate (117-83-9)
Dibutyl acid phosphate (107-66-4)
Di-n-butyl adipate (105-99-7)
Dibutyl adipate (105-99-7)
Dibutyl adipinate (105-99-7)
Dibutylamid kyseliny mravenci (Czech) (761-65-9)
Di-sec-butylamine (626-23-3)
Dibutylamine (111-92-2)
n-Dibutylamine (111-92-2)
Di-sec-butylamine (8CI) (626-23-3)
Di(n-butyl)amine (DOT) (111-92-2)
Di-n-butylamine [UN 2248] (111-92-2)
Dibutylamine, N-chloro- (6CI,7CI,8CI) (999-33-7)
Dibutylamine, 4-hydroxy-N-nitroso- (3817-11-6)
Dibutylamine, N-nitroso- (6CI) (924-16-3)
2-Di-n-butylaminoethanol (102-81-8)
2-Dibutylaminoethanol (102-81-8)
2-N-Dibutylaminoethanol (ACGIH,OSHA) (102-81-8)
N,N-Di-n-butylaminoethanol (DOT) (102-81-8)
Dibutylaminoethanol [UN 2873] (102-81-8)
β-n-Dibutylaminoethyl alcohol (102-81-8)
((Dibutylamino)thio)methylcarbamic acid, 2,2-dimethyl-2,3-dihydro-7-benzofuranyl ester (55285-14-8)
N,N-Dibutylaniline (613-29-6)
Dibutylated hydroxytoluene (128-37-0)
Dibutyl azelate (2917-73-9)
N,N-Dibutylbenzamide (25033-65-2)
N,N-Dibutylbenzenamine (613-29-6)
1,4-Di-tert-butylbenzene (1012-72-2)
p-Di-tert-butylbenzene (1012-72-2)
Dibutyl 1,2-benzenedicarboxylate (84-74-2)
2,5-Di-tert-butylbenzene-1,4-diol (88-58-4)
Di-N-n-butylbenzoic acid amide (25033-65-2)
2,5-Di-t-butyl-p-benzoquinone (2460-77-7)
2,5-Di-tert-butyl-1,4-benzoquinone (2460-77-7)
2,5-Di-tert-butylbenzoquinone (2460-77-7)
2,6-Di-tert-butyl-p-benzoquinone (719-22-2)
N,N-Di-sec-butyl-S-benzylthiocarbamate (36756-79-3)
Dibutylbis(dodecylthio)stannane (1185-81-5)
Dibutylbis((β-hydroxyethyl)thio)tin (3026-81-1)
Dibutylbis(lauroyloxy)tin (77-58-7)
Dibutylbis(stearoyloxy)stannane (5847-55-2)
Dibutyl butanephosphonate (78-46-6)
O,O-Dibutyl 1-butylamino-cyclohexylphosphonate (51249-05-9)
2,6-Di-tert-butyl-4-sec-butylphenol (17540-75-9)
Dibutyl butylphosphonate (78-46-6)
Dibutyl carbitol (112-73-2)
Dibutylcarbitolformal (143-29-3)
Dibutyl o-(o-carboxybenzoyl) glycolate (85-70-1)
Dibutyl o-carboxybenzoyloxyacetate (85-70-1)
Dibutyl cellosolve (112-48-1)
Dibutyl cellosolve adipate (141-18-4)
Dibutylcellosolve ftalat (Czech) (117-83-9)
Dibutyl cellosolve phthalate (117-83-9)
Dibutylchloramine (999-33-7)
4,6-Di-t-butyl-m-cresol (497-39-2)
2,6-Di-tert-butyl-p-cresol (ACGIH,OSHA) (128-37-0)
Di-(4-tert-butylcyclohexyl)peroxydicarbonate (15520-11-3)
Di-(4-tert-butylcyclohexyl)peroxydicarbonate (Not more than 42% stable dispersion, in water) (15520-11-3)
Di-(4-tert-butylcyclohexyl)peroxydicarbonate (Technically pure) (15520-11-3)
Dibutyldichlorostannane (683-18-1)
Dibutyldichlorotin (683-18-1)
Dibutyldifluorostannane (563-25-7)
2,2-Dibutyl-1,3,2-dioxastannepin-4,7-dione (78-04-6)
Dibutyldithiocarbamic acid, nickel salt (13927-77-0)
Dibutyldithiocarbamic acid sodium salt (136-30-1)
Dibutyldithiocarbamic acid zinc salt (136-23-2)
Dibutyldithiokarbaman sodny (Czech) (136-30-1)
Dibutyldithioxodistannathiane (15666-29-2)
1,3-Dibutyl-1,3-dithioxodistannthiane (15666-29-2)
Dibutylester kyseliny adipove (Czech) (105-99-7)
Di-n-butylester kyseliny ftalove (Czech) (84-74-2)
Dibutylester kyseliny fumarove (Czech) (105-75-9)

Di-n-butylester kyseliny jantarove (Czech) **(141-03-7)**
Dibutylester kyseliny maleinove (Czech) **(105-76-0)**
Dibutylester kyseliny sebakove (Czech) **(109-43-3)**
Dibutyl ethanedioate **(2050-60-4)**
N,N-Dibutylethanolamine **(102-81-8)**
Di-sec-butyl ether **(6863-58-7)**
Dibutyl ether **(142-96-1)**
n-Dibutyl ether **(142-96-1)**
Di-n-butyl ether [UN 1149] **(142-96-1)**
Di-sec-butyl ether [UN 1149] **(6863-58-7)**
Di-tert-butyl ether [UN 1149] **(6163-66-2)**
Dibutylether diethylenglykolu (Czech) **(112-73-2)**
Dibutylether ethylenglykolu (Czech) **(112-48-1)**
Dibutylether tetraethylenglykolu (Czech) **(112-98-1)**
N,N'-Di-tert-butylethylenediamine **(4062-60-6)**
N,N'-Di-sek.butyl-p-fenylendiamin (Czech) **(101-96-2)**
N,N-Di-n-butylformamide **(761-65-9)**
Dibutylfosfit (Czech) **(1809-19-4)**
Dibutyl fumarate **(105-75-9)**
Dibutylglycol phthalate **(117-83-9)**
Dibutylhexamethylenediamine **(4835-11-4)**
N,N'-Dibutylhexamethylenediamine **(4835-11-4)**
N,N'-Dibutyl-1,6-hexanediamine **(4835-11-4)**
Dibutyl hexanedioate **(105-99-7)**
Dibutyl hydrogen phosphate **(107-66-4)**
Dibutyl hydrogen phosphite **(1809-19-4)**
2,5-Di-t-butylhydroquinone **(88-58-4)**
3,5-Di-tert-butyl-4-hydroxybenzoic acid **(1421-49-4)**
3,5-Di-tert-butyl-4-hydroxybenzoic acid, (2,4-di-tert-butylphenyl) ester **(4221-80-1)**
3,5-Di-tert-butyl-4-hydroxybenzyl alcohol **(88-26-6)**
3,5-Ditert-butyl-4-hydroxy-(2,4-ditert-butylphenyl)benzoate **(4221-80-1)**
N,N-Dibutyl-N-(2-hydroxyethyl)amine **(102-81-8)**
3,5-Di-tert-butyl-4-hydroxyhydrocinnamic acid, hydrazide **(32687-77-7)**
3,5-Di-tert-butyl-4-hydroxyhydrocinnamic acid, 1,3,5-tris-(2-hydroxyethyl)-s-triazine-2,4,6(1H,3H,5H)trione triester **(34137-09-2)**
2,6-Di-tert-butyl-1-hydroxy-4-methylbenzene **(128-37-0)**
2,6-Di-tert-butyl-4-hydroxymethylphenol **(88-26-6)**
2-(3',5'-Di-tert-butyl-2'-hydroxyphenyl)-5-chloro-2H-benzotriazole **(3864-99-1)**
N,N-Dibutyl(2-hydroxypropyl)amine **(2109-64-0)**
3,5-Di-tert-butyl-4-hydroxytoluene **(128-37-0)**
Dibutylisopropanolamine **(2109-64-0)**
Dibutyl ketone **(502-56-7)**
Dibutyl maleate **(105-76-0)**
Dibutyl(maleoyldioxy)tin **(78-04-6)**
2,6-Di-tert-butyl-4-methoxyphenol **(489-01-0)**
2,4-Di-t-butyl-5-methylphenol **(497-39-2)**
2,6-Di-tert-butyl-4-methylphenol **(128-37-0)**
2,6-Di-tert-butyl-p-methylphenol **(128-37-0)**
2,6-Di-t-butyl-4-methylphenyl-N-methylcarbamate **(1918-11-2)**
2,7-Di-tert-butylnaphthalene **(10275-58-8)**
Dibutyl-naphthalene sulfate, sodium salt **(25417-20-3)**
2,6-Di-tert-butyl-4-nitroaniline **(5180-59-6)**
Di-n-butylnitrosamin (German) **(924-16-3)**
Di-n-butylnitrosamine **(924-16-3)**
Dibutylnitrosamine **(924-16-3)**
N,N-Di-n-butylnitrosamine **(924-16-3)**
N,N-Dibutylnitrosoamine **(924-16-3)**
N,N'-Dibutyl-N-nitrosourea **(56654-52-5)**
Dibutyl nonanedioate **(2917-73-9)**
2,2-Dibutyl-1-oxa-2-stanna-3-thiacyclohexan-6-one **(78-06-8)**
Dibutyl oxide **(142-96-1)**
Dibutyloxide of tin **(818-08-6)**
Dibutyloxostannane **(818-08-6)**
Dibutyloxotin **(818-08-6)**
Di-tert-butylperoxid (German) **(110-05-4)**
Di-tert-butyl peroxide **(110-05-4)**
Di-tert-butyl peroxide, Technically pure (DOT) **(110-05-4)**
2,2-Di-(tert-butylperoxy)butane **(2167-23-9)**
2,2-Di-(tert-butylperoxy)butane (More than 55% in solution) **(2167-23-9)**
2,2-Di-(tert-butylperoxy)butane (Not more than 55% in solution) **(2167-23-9)**
1,1-Di-(tert-butylperoxy)cyclohexane **(3006-86-8)**
1,1-Di-(tert-butylperoxy)cyclohexane (Not more than 77% in solution) **(3006-86-8)**
1,1-Di-(tert-butylperoxy)cyclohexane (Not more than 40% with inert inorganic solid with less than 13% in phlegmatizer) **(3006-86-8)**
1,1-Di-(tert-butylperoxy)cyclohexane (Not more than 50% with phlegmatizer) **(3006-86-8)**
1,1-Di-(tert-butylperoxy)cyclohexane (Technically pure) **(3006-86-8)**
Di-tert-butyl peroxyde (Dutch) **(110-05-4)**
Di-n-butyl peroxydicarbonate **(16215-49-9)**
Di-sec-butyl peroxydicarbonate **(19910-65-7)**
Dibutyl peroxydicarbonate **(16215-49-9)**
Di-sec-butyl peroxydicarbonate, Not more than 52% in solution (DOT) **(19910-65-7)**
Di-sec-butyl peroxydicarbonate, Technically pure (DOT) **(19910-65-7)**
Di-(tert-butylperoxy) phthalate **(2155-71-7)**
Di-(tert-butylperoxy) phthalate (More than 55% in solution) **(2155-71-7)**
Di-(tert-butylperoxy) phthalate (Not more than 55% as a paste) **(2155-71-7)**
Di-(tert-butylperoxy) phthalate (Not more than 55% in solution) **(2155-71-7)**
Di-(tert-butylperoxy) phthalate (Technically pure) **(2155-71-7)**
1,1-Di-(tert-butylperoxy)-3,3,5-trimethylcyclohexane **(6731-36-8)**
1,1-Di-(tert-butylperoxy)-3,3,5-trimethylcyclohexane, Maximum 57% in solution (DOT) **(6731-36-8)**
1,1-Di-(tert-butylperoxy)-3,3,5-trimethylcyclohexane, Not > 58% inert solid (DOT) **(6731-36-8)**
1,1-Di-(tert-butylperoxy)-3,3,5-trimethylcyclohexane, Technically pure (DOT) **(6731-36-8)**
2,4-Di-sec-butylphenol **(1849-18-9)**
2,4-Di-tert-butylphenol **(96-76-4)**
2,6-Di-tert-butylphenol **(128-39-2)**
Di-sec-butylphenol **(31291-60-8)**
N,N'-Di-s-butyl-p-phenylenediamine **(101-96-2)**
3,5-Di-t-butylphenylmethylcarbamate **(2655-19-8)**
Di(tert-butylphenyl) phenyl phosphate **(65652-41-7)**
Di-tert-butylphenyl phenyl phosphate **(65652-41-7)**
Dibutyl phenyl phosphate (ACGIH) **(2528-36-1)**
Di-n-butyl phosphate **(107-66-4)**
Dibutyl phosphate (ACGIH,OSHA) **(107-66-4)**
Di-tert-butylphosphine oxide **(684-19-5)**
Dibutyl-phosphite **(1809-19-4)**
Di-n-butyl phthalate **(84-74-2)**
Dibutyl phthalate (ACGIH,OSHA) **(84-74-2)**
Di-n-butyl sebacate **(109-43-3)**
Dibutyl sebacate **(109-43-3)**
Dibutylstannane **(1002-53-5)**
Dibutylstannane oxide **(818-08-6)**
Dibutylstannium diacetate **(1067-33-0)**
Dibutylstannium dichloride **(683-18-1)**
Dibutylstannium dilaurate **(77-58-7)**
Dibutylstannium oxide **(818-08-6)**
2,2'-((Dibutylstannylene)bis(thio))bisethanol **(3026-81-1)**
Dibutylstannylene maleate **(78-04-6)**
Di-n-butylsuccinate **(141-03-7)**
Dibutyl succinate **(141-03-7)**
Di-n-butylsulfide **(544-40-1)**
Di-tert-butyl sulfide **(107-47-1)**
n-Dibutyl sulfide **(544-40-1)**
Dibutyl sulphide **(544-40-1)**
Dibutyl terephthalate **(1962-75-0)**
2,6-Di-tert-butyltetralin **(42981-76-0)**
Dibutyl thioether **(544-40-1)**
1,3-Di-n-butyl-2-thiourea **(109-46-6)**
1,3-Dibutylthiourea **(109-46-6)**

N,N'-Dibutylthiourea (109-46-6)
Dibutyltin (1002-53-5)
Dibutyltin bis(2-ethylhexanoate) (2781-10-4)
Dibutyltin bis(α-ethylhexanoate) (2781-10-4)
Dibutyltin bis(2-hydroxyethylmercaptide) (3026-81-1)
Dibutyltin bis(2-mercaptoethyl dodecanoate) (28570-24-3)
Dibutyltin bis(methyl maleate) (15546-11-9)
Dibutyltin bis(monomethyl maleate) (15546-11-9)
Dibutyltinbis(2-oleoyloxyethylmercaptide) (67361-77-7)
Dibutyltin chloride (683-18-1)
Dibutyl tin diacetate (1067-33-0)
Di-n-butyltin dichloride (683-18-1)
Dibutyltin dichloride (683-18-1)
Di-n-butyltin di-2-ethylhexanoate (2781-10-4)
Dibutyltin di(2-ethylhexanoate) (2781-10-4)
Dibutyltin di(2-ethylhexoate) (2781-10-4)
Dibutyltin difluoride (563-25-7)
Dibutyltin dihydride (6CI) (1002-53-5)
Dibutyltin dilaurate (77-58-7)
Di-n-butyltin di(monobutyl)maleate (15546-16-4)
Dibutyltin distearate (5847-55-2)
Dibutyltin hydride (1002-53-5)
Dibutyltin laurate (77-58-7)
Dibutyltin maleate (78-04-6)
Dibutyltin S,O-3 mercaptopropionate (78-06-8)
Dibutyltin S,O-β-mercaptopropionate (78-06-8)
Dibutyltin mercaptopropionate (78-06-8)
Dibutyltin, O,S-mercaptopropionate (78-06-8)
Dibutyltin methyl maleate (15546-11-9)
Di-n-butyltin oxide (818-08-6)
Dibutyltin oxide (818-08-6)
Dibutyltin stearate (5847-55-2)
Dibutyltin sulfide (4253-22-9)
2,6-Di-tert-butyl-p-tolyl methylcarbamate (1918-11-2)
1,3-Di-sec-butylurea (869-79-4)
Di-n-butyl-zinn-dichlorid (German) (683-18-1)
Di-n-butyl-zinn-di(monobutyl)maleinat (German) (15546-16-4)
Di-n-butylzinn-dimonomethylmaleinat (German) (15546-11-9)
Dibutyl-zinn (German) (1002-53-5)
Dibutylzinn-S,S'-bis(isooctylthioglycolat) (German) (25168-24-5)
Dibutyl-zinn-dilaurat (German) (77-58-7)
Di-n-butyl-zinn-oxyd (German) (818-08-6)
Dicaine hydrochloride (136-47-0)
Dicainum (136-47-0)
Dicalcium orthophosphate (7757-93-9)
Dicalcium phosphate (7757-93-9)
Dicalcium-O-phosphate (7757-93-9)
Dicalcium phosphate anhydrous (7757-93-9)
Dicalcium pyrophosphate (7790-76-3)
Dicalite (7631-86-9)
Dicamba (1918-00-9)
Dicamba-2,4-D mixt. (8068-77-7)
Dicamba amine (2300-66-5)
Dicamba diethanolamine salt (25059-78-3)
Dicamba dimethylamine salt (2300-66-5)
Dicamba-sodium (1982-69-0)
Dicamba sodium salt (1982-69-0)
Dicambe (1918-00-9)
Dicamoylmethtane (64-55-1)
Dicandiol (57-53-4)
Dicapryl adipate (105-97-5)
Dicapryl 1,2-benzenedicarboxylate (131-15-7)
Dicapryl phthalate (131-15-7)
Dicaprylyl adipate (123-79-5)
Dicaprylyl peroxide (762-16-3)
Dicaptan (2463-84-5)
Dicapthon (2463-84-5)
Dicaptol (59-52-9)
Dicapton (2463-84-5)
Dicarbam (63-25-2)
2,2-Di(carbamoyloxymethyl)pentane (57-53-4)
1,3-Dicarbamoylthio-2-(N,N-dimethylamino)propane hydrochloride (15263-52-2)
Dicarbamylamine (108-19-0)
2,2-Dicarbamyloxymethyl-3-methylpentane (64-55-1)
Dicarbasulf (59669-26-0)
S-(1,2-Dicarbethoxyethyl) O,O-dimethyldithiophosphate (121-75-5)
Dicarbethoxymethane (105-53-3)
Dicarbocalm (14987-04-3)
Dicarboethoxyethyl O,O-dimethyl phosphorodithioate (121-75-5)
Dicarbomethoxyzinc (557-34-6)
2,3-Dicarbonitrilo-1,4-diathiaanthrachinon (German) (3347-22-6)
Di-mu-carbonylhexacarbonyldicobalt (10210-68-1)
o-Dicarboxybenzene (88-99-3)
3,5-Dicarboxybenzenesulfonic acid, sodium salt (6362-79-4)
3,3'-Dicarboxybenzidine (2130-56-5)
2,2'-Dicarboxybiphenyl (482-05-3)
1,10-Dicarboxydecane (693-23-2)
Dicarboxymethane (141-82-2)
1,3-Di(2-carboxy-4-oxochromen-5-yloxy)propan-2-ol (16110-51-3)
Dicarocide (1642-54-2)
Dicarzol (22259-30-9)
Dicarzol (23422-53-9)
Dicerium trioxide (1345-13-7)
Dicesium dichloride (7647-17-8)
Dicesium diiodide (7789-17-5)
Dicestal (97-23-4)
Dicetyl adipate (26720-21-8)
Dicetyldimethylammonium chloride (1812-53-9)
Dicetyl peroxydicarbonate (26322-14-5)
Dicetyl peroxydicarbonate, Not more than 42% in water (26322-14-5)
Dicetyl peroxydicarbonate (Not more than 42% stable dispersion, in water) (26322-14-5)
Dicetyl peroxydicarbonate, Technically pure (26322-14-5)
Dicetyl phosphate (2197-63-9)
Dicetylphosphate (2197-63-9)
Dichinalex (54-05-7)
Dichlobenil (1194-65-6)
Dichlofenamide (120-97-8)
Dichlofenthion (97-17-6)
Dichlofention (97-17-6)
Dichlofluanid (1085-98-9)
Dichlofluanide (1085-98-9)
Dichlofluanid-methyl (731-27-1)
Dichlone (117-80-6)
1,4-Dichloorbenzeen (Dutch) (106-46-7)
p-Dichloorbenzeen (Dutch) (106-46-7)
1,1-Dichloor-2,2-bis(4-chloor fenyl)-ethaan (Dutch) (72-54-8)
1,1-Dichloorethaan (Dutch) (75-34-3)
1,2-Dichloorethaan (Dutch) (107-06-2)
2,2'-Dichloorethylether (Dutch) (111-44-4)
Dichloorfeen (Dutch) (97-23-4)
(2,4-Dichloor-fenoxy)-azijnzuur (Dutch) (94-75-7)
2-(2,4-Dichloor-fenoxy)-propionzuur (Dutch) (120-36-5)
3-(3,4-Dichloor-fenyl)-1,1-dimethylureum (Dutch) (330-54-1)
3-(3,4-Dichloor-fenyl)-1-methoxy-1-methylureum (Dutch) (330-55-2)
3,6-Dichloor-2-methoxy-benzoeizuur (Dutch) (1918-00-9)
1,1-Dichloor-1-nitroethaan (Dutch) (594-72-9)
(2,2-Dichloor-vinyl)-dimethyl-fosfaat (Dutch) (62-73-7)
Dichloorvo (Dutch) (62-73-7)
Dichlor (563-54-2)
Dichloracetic acid (79-43-6)
Dichloracetyl chloride (79-36-7)
1,1-Dichloraethan (German) (75-34-3)

1,2-Dichlor-aethan (German) **(107-06-2)**
1,2-Dichlor-aethen (German) **(540-59-0)**
p-Di-(2-chloraethyl)-amino-DL-phenyl-alanin (German) **(531-76-0)**
S-(2,3-Dichlor-allyl)-N,N-diisopropyl-monothiocarbamaat (Dutch) **(2303-16-4)**
2,3-Dichlorallyl-N,N-(diisopropyl)-thiocarbamat (German) **(2303-16-4)**
Dichloral urea **(116-52-9)**
Dichloraluree **(116-52-9)**
Dichlor amine **(51-75-2)**
Dichloramine **(3400-09-7)**
Dichloran **(102-30-7)**
Dichloran **(99-30-9)**
Dichloran (amine fungicide) **(99-30-9)**
3,4-Dichloranilid kyseliny cyklopropankarboxylove (Czech) **(2759-71-9)**
3,4-Dichloranilid kyseliny methakrylove (Czech) **(2164-09-2)**
3,4-Dichloranilid kyseliny α-methylvalerove (Czech) **(2533-89-3)**
3,4-Dichloranilid kyseliny propionove (Czech) **(709-98-8)**
3,4-Dichloranilin **(95-76-1)**
2,5-Dichloranilin (Czech) **(95-82-9)**
2,4-Dichloranilin (German) **(554-00-7)**
3,4-Dichloraniline **(95-76-1)**
1-(3,4-Dichloranilino)-1-formylamino-2,2,2-trichloraethan (German) **(20856-57-9)**
1,5-Dichloranthrachinon (Czech) **(82-46-2)**
1,8-Dichloranthrachinon (Czech) **(82-43-9)**
Dichlorantin **(118-52-5)**
o-Dichlorbenzene **(95-50-1)**
3,3'-Dichlorbenzidin (Czech) **(91-94-1)**
4,4'-Dichlorbenzilsaeureaethylester (German) **(510-15-6)**
o-Dichlor benzol **(95-50-1)**
1,4-Dichlor-benzol (German) **(106-46-7)**
p-Dichlorbenzol (German) **(106-46-7)**
2,6-Dichlorbenzonitril (German) **(1194-65-6)**
3,4-Dichlorbenzylester kyseliny methylkarbaminove (Czech) **(1966-58-1)**
2,2'-Dichlorbiphenyl (German) **(13029-08-8)**
1,1-Dichlor-2,2-bis(4-chlor-phenyl)-aethan (German) **(72-54-8)**
2,3-Dichlor-1,3-butadien (Czech) **(1653-19-6)**
2,2'-Dichlor-diaethylaether (German) **(111-44-4)**
3,3'-Dichlor-4,4'-diamino-diphenylaether (German) **(28434-86-8)**
3,3'-Dichlor-4,4'-diaminodiphenylmethan (German) **(101-14-4)**
Dichlor-difenylsilan (Czech) **(80-10-4)**
Dichlordimethylaether (German) **(542-88-1)**
Dichloremulsion **(107-06-2)**
Dichloren **(55-86-7)**
Dichloren (German) **(51-75-2)**
Dichloren hydrochloride **(55-86-7)**
1,1-Dichlorethane **(75-34-3)**
1,2-Dichlorethane **(107-06-2)**
Dichlorethanoic acid **(79-43-6)**
1,2-Dichlorethylester kyseliny octove (Czech) **(10140-87-1)**
2,2'-Dichlorethyl ether **(111-44-4)**
β,β-Dichlor-ethyl-sulphide **(505-60-2)**
Dichlorfenidim **(330-54-1)**
2,6-Dichlorfenol (Czech) **(87-65-0)**
2-(2,4-Dichlorfenoxy)ethylsiran sodny (Czech) **(136-78-7)**
Dichlorfenthion **(97-17-6)**
Dichlor-fenylarsin (Czech) **(696-28-6)**
2,5-Dichlor-1,4-fenylendiamin (Czech) **(20103-09-7)**
2,4-Dichlorfenylester kyseliny benzensulfonove (Czech) **(97-16-5)**
3,4-Dichlorfenylisokyanat (Czech) **(102-36-3)**
Dichlor-fenyl-methylsilane (Czech) **(149-74-6)**
Dichlorfluanid (Czech) **(1085-98-9)**
N-Dichlorfluormethylthio-N',N'-dimethylaminosulfonsaeure-anilid (German) **(1085-98-9)**
N-(Dichlor-fluor-methyl-thio)-N',N'-dimethyl-N-phenyl-schwefel-saeurediamid (German) **(1085-98-9)**
Dichlorfos (Polish) **(62-73-7)**
Dichloricide aerosol **(8001-50-1)**
Dichloricide mothproofer **(8001-50-1)**
Dichlorid dimethylcinicity (Czech) **(753-73-1)**
Dichlorid kyseliny fumarove (Czech) **(627-63-4)**
Dichlorid kyseliny isoftalove (Czech) **(99-63-8)**
Dichlor-3-kyanpropyl-methylsilan (Czech) **(1190-16-5)**
Dichlorman **(62-73-7)**
3,6-Dichlor-3-methoxy-benzoesaeure (German) **(1918-00-9)**
Dichlormid **(37764-25-3)**
2,3-Dichlor-1,4-naftochinon (Czech) **(117-80-6)**
2,3-Dichlor-1,4-naphthochinon (German) **(117-80-6)**
1,1-Dichlor-1-nitroaethan (German) **(594-72-9)**
2,6-Dichlor-4-nitroanilin (Czech) **(99-30-9)**
2,5-Dichlornitrobenzen (Czech) **(89-61-2)**
3,4-Dichlornitrobenzen (Czech) **(99-54-7)**
4,4'-Dichlor-2-nitrodifenylether (Czech) **(135-12-6)**
2,4-Dichlor-6-nitrofenol (Czech) **(609-89-2)**
2',5-Dichlor-4'-nitro-salizylsaeureanilid (German) **(50-65-7)**
2,2-Dichloroacetaldehyde **(79-02-7)**
α,α-Dichloroacetaldehyde **(79-02-7)**
D-(-)-threo-2-Dichloroacetamido-1-p-nitrophenyl-1,3-propanediol **(56-75-7)**
2,2-Dichloroacetic acid **(79-43-6)**
Dichloroacetic acid **(79-43-6)**
Dichloroacetic acid [UN 1764] **(79-43-6)**
Dichloroacetic acid, ethyl ester **(535-15-9)**
Dichloroacetic acid methyl ester **(116-54-1)**
α,α'-Dichloroacetic anhydride **(541-88-8)**
sym-Dichloroacetic anhydride **(541-88-8)**
1,1-Dichloroacetone **(513-88-2)**
α,α'-Dichloroacetone **(534-07-6)**
α,α-Dichloroacetone **(513-88-2)**
α,γ-Dichloroacetone **(534-07-6)**
sym-Dichloroacetone **(534-07-6)**
1,3-Dichloroacetone [UN 2649] **(534-07-6)**
Dichloroacetonitrile **(3018-12-0)**
2',4'-Dichloroacetophenone **(2234-16-4)**
2,2-Dichloroacetyl chloride **(79-36-7)**
α,α-Dichloroacetyl chloride **(79-36-7)**
Dichloroacetyl chloride [UN 1765] **(79-36-7)**
Dichloroacetylene (ACGIH,DOT,OSHA) **(7572-29-4)**
D-threo-N-Dichloroacetyl-1-p-nitrophenyl-2-amino-1,3-propanediol **(56-75-7)**
S-(2,3-Dichloro-allil)-N,N-diisopropil-monotiocarbammato (Italian) **(2303-16-4)**
Dichloroallyl diisopropylthiocarbamate **(2303-16-4)**
S-2,3-Dichloroallyl diisopropylthiocarbamate **(2303-16-4)**
2,3-Dichloroallyl N,N-diisopropylthiolcarbamate **(2303-16-4)**
3',5'-Dichloroamethopterin **(528-74-5)**
Dichloroamethopterin **(528-74-5)**
2,5-Dichloro-3-aminobenzoic acid **(133-90-4)**
3',5'-Dichloro-4-amino-4-deoxy-N^{10}-methylpteroglutamic acid **(528-74-5)**
2,5-Dichloro-4-((5-amino-3-methyl-1-phenyl-1H-pyrazol-4-yl)-azo)benzenesulfonic acid **(12239-15-5)**
2,3-Dichloroaniline **(608-27-5)**
2,4-Dichloroaniline **(554-00-7)**
2,5-Dichloroaniline **(95-82-9)**
2,6-Dichloroaniline **(608-31-1)**
3,4-Dichloroaniline **(95-76-1)**
3,5-Dichloroaniline **(626-43-7)**
4,5-Dichloroaniline **(95-76-1)**
o-Dichloroaniline **(27134-27-6)**
Dichloroanilines [UN 1590] **(27134-27-6)**
2,5-Dichloroaniline-4-sulfonic acid **(88-50-6)**
(o-(2,6-Dichloroanilino)phenyl)acetic acid monosodium salt **(15307-79-6)**
(o-(2,6-Dichloroanilino)phenyl)acetic acid sodium salt **(15307-79-6)**
4-(3,4-Dichloroanilino)-3,3',4'-trichloroazobenzene **(27125-68-4)**
3,6-Dichloro-o-anisic acid **(1918-00-9)**
3,6-Dichloro-o-anisic acid, sodium salt **(1982-69-0)**
2,4-Dichloroanisole **(553-82-2)**

2,5-Dichloroanisole	(1984-58-3)
2,6-Dichloroanisole	(1984-65-2)
9,10-Dichloroanthracene	(605-48-1)
1,5-Dichloro-9,10-anthraquinone	(82-46-2)
1,5-Dichloroanthraquinone	(82-46-2)
1,8-Dichloro-9,10-anthraquinone	(82-43-9)
1,8-Dichloroanthraquinone	(82-43-9)
2,2'-Dichloroazobenzene	(7334-33-0)
4,4'-Dichloroazobenzene	(1602-00-2)
p,p'-Dichloroazobenzene	(1602-00-2)
N,N'-Dichloroazodicarbonamidine (Salts of) (Dry)	(502-98-7)
4,4'-Dichloroazoxybenzene	(614-26-6)
p,p'-Dichloroazoxybenzene	(614-26-6)
2,6-Dichlorobenzal chloride	(81-19-6)
2,4-Dichlorobenzaldehyde	(874-42-0)
2,6-Dichlorobenzaldehyde	(83-38-5)
3,4-Dichlorobenzaldehyde	(6287-38-3)
Dichlorobenzaldehyde	(31155-09-6)
Dichlorobenzalkonium chloride	(8023-53-8)
2,6-Dichlorobenzamide	(2008-58-4)
2,3-Dichlorobenzenamine	(608-27-5)
2,6-Dichlorobenzenamine	(608-31-1)
3,4-Dichlorobenzenamine	(95-76-1)
3,5-Dichlorobenzenamine	(626-43-7)
1,2-Dichlorobenzene	(95-50-1)
1,3-Dichlorobenzene	(541-73-1)
1,4-Dichlorobenzene	(106-46-7)
Dichlorobenzene	(25321-22-6)
m-Dichlorobenzene	(541-73-1)
o-Dichlorobenzene (ACGIH,OSHA) [UN 1591]	(95-50-1)
p-Dichlorobenzene (ACGIH,OSHA) [UN 1592]	(106-46-7)
Dichlorobenzene, ortho, Liquid [UN 1591]	(95-50-1)
2,6-Dichlorobenzenecarbothioamide	(1918-13-4)
2,6-Dichloro-1,4-benzenediamine	(609-20-1)
2,5-Dichloro-1,4-benzenediol	(824-69-1)
3,4-Dichloro-1,2-benzenediol	(3978-67-4)
4,5-Dichloro-m-benzenedisulfonamide	(120-97-8)
Dichlorobenzene, para, Solid (DOT)	(106-46-7)
3,4-Dichlorobenzenesulfonyl chloride	(98-31-7)
4,4'-Dichlorobenzhydrol	(90-97-1)
3,3'-Dichlorobenzidina (Spanish)	(91-94-1)
2,2'-Dichlorobenzidine	(84-68-4)
3,3'-Dichlorobenzidine	(91-94-1)
Dichlorobenzidine	(91-94-1)
o,o'-Dichlorobenzidine	(91-94-1)
3',3'-Dichlorobenzidine (ACGIH)	(91-94-1)
Dichlorobenzidine Base	(91-94-1)
3,3'-Dichlorobenzidine (OSHA)	(91-94-1)
3,3'-Dichlorobenzidine dihydrochloride	(612-83-9)
4,4'-Dichlorobenzil	(3457-46-3)
4,4'-Dichlorobenzilate	(510-15-6)
4,4'-Dichlorobenzilic acid	(23851-46-9)
4,4'-Dichlorobenzilic acid ethyl ester	(510-15-6)
1,2-Dichlorobenzo(b,e)(1,4)dioxin	(54536-18-4)
2,3-Dichlorobenzoic acid	(50-45-3)
2,4(or 2,5)-Dichlorobenzoic acid	(35915-19-6)
2,4-Dichlorobenzoic acid	(50-84-0)
2,5-Dichlorobenzoic acid	(50-79-3)
2,6-Dichlorobenzoic acid	(50-30-6)
3,4-Dichlorobenzoic acid	(51-44-5)
3,5-Dichlorobenzoic acid	(51-36-5)
Dichlorobenzoic acid	(35915-19-6)
m-Dichlorobenzol	(541-73-1)
p-Dichlorobenzol	(106-46-7)
2,6-Dichlorobenzonitrile	(1194-65-6)
2,4'-Dichlorobenzophenone	(85-29-0)
4,4'-Dichlorobenzophenone	(90-98-2)
p,p'-Dichlorobenzophenone	(90-98-2)
5,6-Dichloro-2-benzothiazolamine	(24072-75-1)
2,4-Dichlorobenzotrichloride	(13014-18-1)
3,4-Dichlorobenzotrichloride	(13014-24-9)
2,4-Dichlorobenzotrifluoride	(320-60-5)
3,4-Dichlorobenzotrifluoride	(328-84-7)
2,3-Dichlorobenzoyl chloride	(2905-60-4)
2,4-Dichlorobenzoyl chloride	(89-75-8)
2,5-Dichlorobenzoyl chloride	(2905-61-5)
3,5-Dichlorobenzoyl chloride	(2905-62-6)
Dichlorobenzoyl chloride	(25134-08-1)
4-(2,4-Dichlorobenzoyl)-1,3-dimethyl-5-pyrazolyl p-toluene-sulfonate	(58011-68-0)
p,p'-Dichlorobenzoyl peroxide	(94-17-7)
2,4-Dichlorobenzoyl peroxide, More than 75% with water	(96881-25-3)
2,4-Dichlorobenzoylperoxide, More than 75% with water (DOT)	(133-14-2)
2,4-Dichlorobenzoyl peroxide, Not more than 52% as a paste (DOT)	(133-14-2)
Di-(4-chlorobenzoyl)peroxide, Not more than 52% as a paste (DOT)	(94-17-7)
2,4-Dichlorobenzoyl peroxide, Not more than 52% in solution (DOT)	(133-14-2)
Di-(4-chlorobenzoyl)peroxide, Not more than 52% in solution (DOT)	(94-17-7)
2,4-Dichlorobenzoyl peroxide, Not more than 75% with water (DOT)	(133-14-2)
Di-(4-chlorobenzoyl) peroxide, Not more than 75% with water (DOT)	(94-17-7)
2,4-Dichlorobenzyl alcohol	(1777-82-8)
2,6-Dichlorobenzyl alcohol	(15258-73-8)
2,4-Dichlorobenzyl chloride	(94-99-5)
3,4-Dichlorobenzyl chloride	(102-47-6)
Dichlorobenzyl chloride	(38721-71-0)
(3,4-Dichlorobenzyl)dodecyldimethylammonium chloride	(102-30-7)
((2,6-Dichlorobenzylidene)amino)guanidine	(5051-62-7)
2,6-Dichlorobenzylidene chloride	(81-19-6)
3,4-Dichlorobenzyl-lauryl-dimethylammonium chloride	(102-30-7)
3,4-Dichlorobenzyl methylcarbamate	(1966-58-1)
3,4-Dichlorobenzyl methylcarbamate mixed with 2,3-dichloro-benzyl methylcarbamate (80%:20%)	(62046-37-1)
2,4-Dichlorobenzyltributylphosphonium chloride	(115-78-6)
2,2'-Dichloro-1,1'-biphenyl	(13029-08-8)
2,2'-Dichlorobiphenyl	(13029-08-8)
2,3'-Dichlorobiphenyl	(25569-80-6)
2,3-Dichlorobiphenyl	(16605-91-7)
2,4'-Dichlorobiphenyl	(34883-43-7)
2,4-Dichlorobiphenyl	(33284-50-3)
2,5-Dichloro-1,1'-biphenyl	(34883-39-1)
2,5-Dichlorobiphenyl	(34883-39-1)
2,6-Dichloro-1,1'-biphenyl	(33146-45-1)
2,6-Dichlorobiphenyl	(33146-45-1)
3,3'-Dichloro-1,1'-biphenyl	(2050-67-1)
3,3'-Dichlorobiphenyl	(2050-67-1)
3,4'-Dichlorobiphenyl	(2974-90-5)
3,5-Dichloro-1,1'-biphenyl	(34883-41-5)
3,5-Dichlorobiphenyl	(34883-41-5)
4,4'-Dichlorobiphenyl	(2050-68-2)
Dichlorobiphenyl	(25512-42-9)
ar,ar'-Dichlorobiphenyl	(33039-81-5)
3,3'-Dichloro-4,4'-biphenyldiamine	(91-94-1)
3,3'-Dichlorobiphenyl-4,4'-diamine	(91-94-1)
2,2'-((3,3'-Dichloro(1,1'-biphenyl)-4,4'-diyl)bis(azo))bis-(N-(2,4-dimethylphenyl)-3-oxobutanamide)	(5102-83-0)
2,2'-((3,3'-Dichloro(1,1'-biphenyl)-4,4'-diyl)-bis(azo))bis-(3-oxo-N-phenyl)-butanamide)	(6358-85-6)
4,4'-(3,3'-Dichloro-4,4'-biphenylene)bis(azo))bis(1-(4-methyl-phenyl)-3-methyl-5-pyrazolone)	(15793-73-4)
3.5-Dichloro-(1,1'-biphenyl)-2-ol	(5335-24-0)

Name	CAS No.
3.5-Dichloro-2-biphenylol	**(5335-24-0)**
1,1-Dichloro-2,2-bis(p-chlorophenyl)ethane (DOT)	**(72-54-8)**
1,1-Dichloro-2,2-bis(4-chlorophenyl)-ethane (French)	**(72-54-8)**
1,1-Dichloro-2,2-bis(p-chlorophenyl)ethylene	**(72-55-9)**
1,1-Dichloro-2,2-bis(2,4'-dichlorophenyl)ethane	**(53-19-0)**
Dichlorobis(eta^5-2,4-cyclopentadien-1-yltitanium)	**(1271-19-8)**
1,1-Dichloro-2,2-bis(4-ethylphenyl)ethane	**(72-56-0)**
1,1-Dichloro-2,2-bis(p-ethylphenyl)ethane	**(72-56-0)**
2,2-Dichloro-1,1-bis(p-ethylphenyl)ethane	**(72-56-0)**
a,a-Dichloro-2,2-bis(p-ethylphenyl)ethane	**(72-56-0)**
1,1-Dichloro-2,2-bis(parachlorophenyl)ethane	**(72-54-8)**
1,3-Dichloro-2-bromobenzene	**(19393-92-1)**
2,6-Dichlorobromobenzene	**(19393-92-1)**
Dichlorobromomethane	**(75-27-4)**
2,4-Dichloro-6-bromophenol	**(4524-77-0)**
O-(2,5-Dichloro-4-bromophenyl) O-methyl phenylthiophosphonate	**(21609-90-5)**
1,4-Dichloro-1,3-butadiene	**(2984-42-1)**
2,3-Dichloro-1,3-butadiene	**(1653-19-6)**
Dichlorobutadiene	**(28577-62-0)**
1,1-Dichlorobutane	**(541-33-3)**
1,2-Dichlorobutane	**(616-21-7)**
1,3-Dichlorobutane	**(1190-22-3)**
1,4-Dichlorobutane	**(110-56-5)**
1,4-Dichlorobutane	**(110-87-2)**
2,2-Dichlorobutane	**(4279-22-5)**
2,3-Dichlorobutane	**(7581-97-7)**
(Z)-1,4-Dichloro-2-butene	**(1476-11-5)**
1,3-Dichloro-2-butene	**(7415-31-8)**
1,3-Dichloro-2-butene	**(926-57-8)**
1,4-Dichloro-2-butene	**(110-57-6)**
1,4-Dichloro-2-butene	**(764-41-0)**
1,4-Dichloro-cis-2-butene	**(1476-11-5)**
1,4-Dichlorobutene	**(31423-92-4)**
1,4-Dichlorobutene-2	**(764-41-0)**
1,4-Dichlorobutene-2 (trans)	**(110-57-6)**
3,4-Dichloro-1-butene	**(760-23-6)**
3,4-Dichlorobutene-1	**(64037-54-3)**
Dichlorobutene	**(11069-19-5)**
cis-1,4-Dichloro-2-butene	**(1476-11-5)**
trans-1,3-Dichlorobutene-2	**(7415-31-8)**
1,2-Dichloro-3-butene (Racemic mixture)	**(64037-54-3)**
3,4-Dichlorobutene-1 (Racemic mixture)	**(64037-54-3)**
1,4-Dichlorobutene-2,3-epoxide	**(3583-47-9)**
Dichlorobutylene	**(11069-19-5)**
1,4-Dichloro-2-butyne	**(821-10-3)**
1,4-Dichlorobutyne	**(821-10-3)**
2,2-Dichlorobutyric acid, sodium salt	**(2517-16-0)**
3,4-Dichlorocarbanilic acid methyl ester	**(1918-18-9)**
3,4-Dichlorocatechol	**(3978-67-4)**
4,5-Dichlorocatechol	**(3428-24-8)**
2,4-Dichloro-6-o-chloranilino-s-triazine	**(101-05-3)**
Dichlorochlordene	**(57-74-9)**
2,4-Dichloro-6-(2-chloroanilino)-1,3,5-triazine	**(101-05-3)**
2,4-Dichloro-6-(o-chloroanilino)-s-triazine	**(101-05-3)**
1,1-Dichloro-2-chloroethylene	**(79-01-6)**
1,2-Dichloro-4-(chloromethyl)benzene	**(102-47-6)**
2,4-Dichloro-1-(chloromethyl)benzene	**(94-99-5)**
Dichloro(chloromethyl)benzene	**(38721-71-0)**
Dichloro(chloromethyl)methylsilane	**(1558-33-4)**
1,3-Dichloro-2-chloromethyl-1-propene	**(13245-65-3)**
1,1-Dichloro-2-(o-chlorophenyl)-2-(p-chlorophenyl)ethane	**(53-19-0)**
1,1-Dichloro-2-(p-chlorophenyl)-2-(o-chlorophenyl)ethane	**(53-19-0)**
2,2-Dichloro-1-(2-chlorophenyl) ethanol	**(27683-60-9)**
4,6-Dichloro-N-(2-chlorophenyl)-1,3,5-triazin-2-amine	**(101-05-3)**
Dichloro(2-chlorovinyl)arsine	**(541-25-3)**
2,4-Dichlorocinnamic acid	**(1201-99-6)**
Dichlorocyanuric acid	**(2782-57-2)**
1,2-Dichlorocyclohexane	**(1121-21-7)**
cis-1,3-Dichlorocyclohexane	**(24955-63-3)**
α,3-Dichloro-4-cyclohexylbenzeneacetic acid	**(36616-52-1)**
4,5-Dichloro-2-cyclohexyl-4-isothiazolin-3-one	**(57063-29-3)**
3',4'-Dichlorocyclopropanecarboxanilide	**(2759-71-9)**
Dichloro-p-cymene	**(65724-12-1)**
3,3'-Dichloro-4,4'-diamino(1,1-biphenyl)	**(91-94-1)**
3,3'-Dichloro-4,4'-diaminobiphenyl	**(91-94-1)**
3,3'-Dichloro-4,4'-diaminodifenilmetano (Italian)	**(101-14-4)**
3,3'-Dichloro-4,4'-diaminodiphenyl ether	**(28434-86-8)**
3,3'-Dichloro-4,4'-diaminodiphenylmethane	**(101-14-4)**
N-(3,5-Dichloro-4-((2,4-diamino-6-pteridinylmethyl)methyl-amino)benzoyl)glutamic acid	**(528-74-5)**
cis-Dichlorodiammine platinum(II)	**(15663-27-1)**
cis-Dichlorodiammineplatinum	**(15663-27-1)**
trans-Dichlorodiammineplatinum (II)	**(14913-33-8)**
cis-Dichlorodiamminoplatinum(II)	**(15663-27-1)**
1,4-Dichlorodibenzo(b,e)(1,4)dioxin	**(54536-19-5)**
Dichloro dibenzo(b,e)(1,4)dioxin	**(64501-00-4)**
1,2-Dichlorodibenzo-p-dioxin	**(54536-18-4)**
1,3-Dichlorodibenzo-para-dioxin	**(50585-39-2)**
1,4-Dichlorodibenzo-p-dioxin	**(54536-19-5)**
1,6-Dichlorodibenzo-para-dioxin	**(38178-38-0)**
2,3-Dichlorodibenzo-para-dioxin	**(29446-15-9)**
2,3-Dichlorodibenzodioxin	**(29446-15-9)**
2,7-Dichlorodibenzo-p-dioxin	**(33857-26-0)**
2,7-Dichlorodibenzodioxin	**(33857-26-0)**
2,8-Dichlorodibenzo-para-dioxin	**(38964-22-6)**
2,8-Dichlorodibenzodioxin	**(38964-22-6)**
Dichloro dibenzo-p-dioxin	**(64501-00-4)**
1,2-Dichlorodibenzofuran	**(64126-85-8)**
1,4-Dichlorodibenzofuran	**(94538-01-9)**
1,6-Dichlorodibenzofuran	**(74992-97-5)**
1,7-Dichlorodibenzofuran	**(94538-02-0)**
1,9-Dichlorodibenzofuran	**(70648-14-5)**
2,3-Dichlorodibenzofuran	**(64126-86-9)**
2,6-Dichlorodibenzofuran	**(60390-27-4)**
2,7-Dichlorodibenzofuran	**(74992-98-6)**
2,8-Dichlorodibenzofuran	**(5409-83-6)**
3,6-Dichlorodibenzofuran	**(74918-40-4)**
3,7-Dichlorodibenzofuran	**(58802-21-4)**
4,4'-Dichlorodibutyl ether	**(6334-96-9)**
Dichlorodibutylstannane	**(683-18-1)**
Dichlorodibutyltin	**(683-18-1)**
1,1-Dichloro-2,2-dichloroethane	**(79-34-5)**
1,3-Dichloro-2-(dichloromethyl)benzene	**(81-19-6)**
2,4-Dichloro-1-(dichloromethyl)benzene	**(134-25-8)**
3,4-Dichloro-5-(dichloromethyl)-5-hydroxy-2-furanone	**(108082-06-0)**
1,1-Dichloro-2,2-di(4-chlorophenyl)ethane	**(72-54-8)**
Dichlorodi-pi-cyclopentadienyltitanium	**(1271-19-8)**
Dichlorodicyclopentadienyltitanium	**(1271-19-8)**
β,β'-Dichlorodiethylaniline	**(553-27-5)**
β,β-Dichlorodiethyl ether	**(111-44-4)**
2,2-Dichlorodiethyl ether [UN 1916]	**(111-44-4)**
2,2'-Dichlorodiethyl-methylamine	**(51-75-2)**
β,β'-Dichlorodiethyl-N-methylamine	**(51-75-2)**
β,β'-Dichlorodiethyl-N-methylamine hydrochloride	**(55-86-7)**
2,2'-Dichlorodiethylmethylamine oxide	**(302-70-5)**
Dichlorodiethylplumbane	**(13231-90-8)**
2,2'-Dichlorodiethyl sulfide	**(505-60-2)**
2,3-Dichloro-7,8-difluorodibenzo(b,e)(1,4)dioxin	**(50585-42-7)**
2,3-Dichloro-7,8-difluorodibenzo-p-dioxin	**(50585-42-7)**
1,2-Dichloro-1,1-difluoroethane	**(1649-08-7)**
1,1-Dichloro-2,2-difluoroethylene	**(79-35-6)**
Dichlorodifluoroethylene	**(27156-03-2)**
2,2-Dichloro-1,1-difluoroethyl methyl ether	**(76-3[illegible])**

Dichlorodifluoromethane (ACGIH,OSHA) [UN 1028]	**(75-71-8)**
5,5'-Dichloro-2,2'-dihydroxydiphenylmethane	**(97-23-4)**
Dichlorodiisopropyl ether	**(108-60-1)**
β,β'-Dichlorodiisopropyl ether	**(108-60-1)**
1,4-Dichloro-2,5-dimethoxybenzene	**(2675-77-6)**
3,5-Dichloro-2,6-dimethoxyphenol	**(78782-46-4)**
1,1-Dichloro-N-((dimethylamino)sulfonyl)-1-fluoro-N-phenyl-methane sulfenamide	**(1085-98-9)**
2,4-Dichloro-6-dimethylamino-s-triazine	**(2401-64-1)**
2,6-Dichloro-N,N-dimethylaniline	**(56961-05-8)**
1,4-Dichloro-2,5-dimethylbenzene	**(1124-05-6)**
1,1-Dichloro-3,3-dimethyl-2-butanone	**(22591-21-5)**
1,1'-Dichlorodimethyl ether	**(542-88-1)**
sym-Dichloro-dimethyl ether	**(542-88-1)**
Dichlorodimethyl ether, symmetrical [UN 2249]	**(542-88-1)**
1,3-Dichloro-5,5-dimethylhydantoin	**(118-52-5)**
1,3-Dichloro-5,5-dimethyl hydantoin (ACGIH,OSHA)	**(118-52-5)**
2,4-Dichloro-3,5-dimethylphenol	**(133-53-9)**
3,5-Dichloro-N-(1,1-dimethyl-2-propynyl)benzamide	**(23950-58-5)**
3,5-Dichloro-2,6-dimethyl-4-pyridinol	**(2971-90-6)**
Dichlorodimethylsilane [UN 1162]	**(75-78-5)**
Dichlorodimethylstannane	**(753-73-1)**
5,5'-Dichloro-3,3'-dimethyl-thioindigo	**(2379-74-0)**
Dichlorodimethyltin	**(753-73-1)**
Dichlorodioctylstannane	**(3542-36-7)**
2,3-Dichloro-1,4-dioxane	**(95-59-0)**
2,3-Dichloro-p-dioxane	**(95-59-0)**
trans-2,3-Dichloro-1,4-dioxane	**(3883-43-0)**
trans-2,3-Dichloro-p-dioxane	**(3883-43-0)**
Dichlorodioxochromium	**(14977-61-8)**
Di-p-chlorodiphenoxymethane	**(555-89-5)**
Dichlorodiphenyl	**(25512-42-9)**
Dichlorodiphenylacetic acid	**(83-05-6)**
p,p'-Dichlorodiphenylacetic acid	**(83-05-6)**
3,4'-Dichlorodiphenylamine	**(15979-79-0)**
2,4-Dichloro-6-(diphenylamino)-s-triazine	**(16033-74-2)**
2,4-Dichloro-6-diphenylamino-1,3,5-triazine	**(16033-74-2)**
Dichlorodiphenyl dichloroethane	**(72-54-8)**
o,p'-Dichlorodiphenyldichloroethane	**(53-19-0)**
p,p'-Dichlorodiphenyldichloroethane	**(72-54-8)**
Dichlorodiphenyldichloroethylene	**(72-55-9)**
p,p'-Dichlorodiphenyl dichloroethylene	**(72-55-9)**
4,4'-Dichlorodiphenyl disulfide	**(1142-19-4)**
p,p'-Dichlorodiphenyl disulfide	**(1142-19-4)**
1,2-Dichloro-1,2-diphenylethane	**(5963-49-5)**
Dichlorodiphenylethanol	**(80-06-8)**
(E)-1,2-Dichloro-1,2-diphenylethene	**(951-86-0)**
Dichloro diphenyl ether	**(28675-08-3)**
trans-1,2-Dichlorodiphenylethylene	**(951-86-0)**
p,p'-Dichlorodiphenylmethylcarbinol	**(80-06-8)**
Dichloro diphenyl oxide	**(28675-08-3)**
Dichlorodiphenylsilane	**(80-10-4)**
p,p'-Dichlorodiphenyl sulfide	**(103-17-3)**
4,4'-Dichlorodiphenyltrichloroethane	**(50-29-3)**
p,p'-Dichlorodiphenyltrichloroethane	**(50-29-3)**
Dichlorodiphenyltrichloroethane (OSHA) [UN 2761]	**(50-29-3)**
Dichlorodi-2-propenylsilane	**(3651-23-8)**
4,5-Dichloro-1,3-disulfamoylbenzene	**(120-97-8)**
1,2-Dichloro-1,2-divinylcyclobutane	**(14112-00-6)**
4,6-Dichloro-2-(dodecylamino)-s-triazine	**(26113-25-7)**
1,4-Dichloro-2,3-epoxybutane	**(3583-47-9)**
1,2-Dichloroethane	**(107-06-2)**
Dichloroethane	**(1300-21-6)**
α,β-Dichloroethane	**(107-06-2)**
sym-Dichloroethane	**(107-06-2)**
1,1-Dichloroethane (ACGIH,DOT,OSHA) [UN 2362]	**(75-34-3)**
Dichloro-1,2-ethane (French)	**(107-06-2)**
Dichloroethanoic acid	**(79-43-6)**
2,2-Dichloroethanol	**(598-38-9)**
Dichloroethanoyl chloride	**(79-36-7)**
1,2-Dichloroethene	**(540-59-0)**
1,1-Dichloroethene (9CI)	**(75-35-4)**
1,1-Dichloroethene polymer with chloroethene	**(9011-06-7)**
2,2-Dichloroethenyl diethyl phosphate	**(72-00-4)**
3-(2,2-Dichloroethenyl)-2,2-dimethylcyclopropanecarbonyl chloride	**(52314-67-7)**
2,2-Dichloroethenyl dimethyl phosphate	**(62-73-7)**
1,1'-(Dichloroethenylidene)bis(4-chlorobenzene)	**(72-55-9)**
2,2-Dichloroethenyl phosphoric acid dimethyl ester	**(62-73-7)**
Dichloroether	**(111-44-4)**
4,4'-Dichloro-α-(ethoxymethyl)benzhydrol	**(6012-83-5)**
1,2-Dichloroethyl acetate	**(10140-87-1)**
Dichloroethylaluminum	**(563-43-9)**
p-Di-(2-chloroethyl)-amino-D-phenylalanine	**(13045-94-8)**
p-Di-(2-chloroethyl)amino-L-phenylalanine	**(148-82-3)**
2-(Di(2-chloroethyl)amino)-1-oxa-3-aza-2-phosphacyclohexane-2-oxide monohydrate	**(6055-19-2)**
3-p-(Di(2-chloroethyl)amino)-phenyl-L-alanine	**(148-82-3)**
p-Di(2-chloroethyl)amino-DL-phenylalanine	**(531-76-0)**
p-N-Di(chloroethyl)aminophenylalanine	**(148-82-3)**
p-Di(2-chloroethyl)amino-d-phenylalanine d	**(13045-94-8)**
N,N-Di-2-chloroethyl-γ-p-aminophenylbutyric acid	**(305-03-3)**
γ-(p-Di(2-chloroethyl)aminophenyl)butyric acid	**(305-03-3)**
p-(N,N-Di-2-chloroethyl)aminophenyl butyric acid	**(305-03-3)**
p-N,N-Di-(β-chloroethyl)aminophenyl butyric acid	**(305-03-3)**
N,N-Di(2-chloroethyl)amino-n,o-propylene phosphoric acid ester diamide monohydrate	**(6055-19-2)**
5-(Di-(β-chloroethyl)amino)uracil	**(66-75-1)**
5-(Di-2-chloroethyl)aminouracil	**(66-75-1)**
N,N,-Di(2-chloroethyl)aniline	**(553-27-5)**
Dichloroethylarsine	**(598-14-1)**
1,1-Dichloroethylene	**(75-35-4)**
Dichloroethylene	**(107-06-2)**
cis-1,2-Dichloroethylene	**(156-59-2)**
cis-Dichloroethylene	**(156-59-2)**
sym-Dichloroethylene	**(540-59-0)**
trans-1,2-Dichloroethylene	**(156-60-5)**
trans-Dichloroethylene	**(156-60-5)**
1,2-Dichloroethylene (ACGIH,OSHA)	**(540-59-0)**
Dichloro-1,2-ethylene (French)	**(540-59-0)**
Dichloroethylene [UN 1150]	**(25323-30-2)**
1,1-Dichloroethylene-monochloroethylene polymer	**(9011-06-7)**
1,1-Dichloroethylene polymer with chloroethylene	**(9011-06-7)**
2,2'-Dichloroethyl ether	**(111-44-4)**
2,2'-Dichloroethyl ether	**(1191-17-9)**
Di(2-chloroethyl) ether	**(111-44-4)**
Di(β-chloroethyl)ether	**(111-44-4)**
Dichloroethyl ether	**(111-44-4)**
β,β'-Dichloroethyl ether	**(111-44-4)**
sym-Dichloroethyl ether	**(111-44-4)**
Dichloroethyl ether (ACGIH,DOT,OSHA)	**(111-44-4)**
Di-2-chloroethyl formal	**(111-91-1)**
Dichloroethyl formal	**(111-91-1)**
1,1'-(2,2-Dichloroethylidene)bis(4-ethoxybenzene)	**(7388-32-1)**
Di(2-chloroethyl)methylamine	**(51-75-2)**
N,N-Di(chloroethyl)methylamine	**(51-75-2)**
2-N,N-Di(2-chloroethyl)naphthylamine	**(494-03-1)**
Di(2-chloroethyl)-β-naphthylamine	**(494-03-1)**
Dichloroethyl-β-naphthylamine	**(494-03-1)**
NN-Di(2-chloroethyl)-β-naphthylamine	**(494-03-1)**
Dichloroethyl oxide	**(111-44-4)**
2,2-Dichloroethyl N-phenylcarbamate	**(35661-56-4)**
Dichloroethylphenylsilane	**(1125-27-5)**
Dichloroethylphosphine sulfide	**(993-43-1)**

N,N-Di(2-chloroethyl)-N,O-propylene-phosphoric acid ester diamide **(50-18-0)**
Di-2-chloroethyl sulfide **(505-60-2)**
β,β'-Dichloroethyl sulfide **(505-60-2)**
2,2'-Dichloroethyl sulphide **(505-60-2)**
2-2'-Di(3-chloroethylthio)-diethyl ether **(63918-89-8)**
Dichloroethylvinylsilane **(10138-21-3)**
Dichloroethyne **(7572-29-4)**
Dichlorofen (Czech) **(97-23-4)**
3-(3,4-Dichloro-fenil)-1-metossi-1-metil-urea (Italian) **(330-55-2)**
Dichlorofenthion **(97-17-6)**
Dichlorofluoromethane (ACGIH) **(75-43-4)**
N'-Dichlorofluoromethylthio-N,N-dimethyl-N'-(4-tolyl)sulfamide **(731-27-1)**
N-(Dichlorofluoromethylthio)-N',N'-dimethyl-N-phenylsulfamide **(1085-98-9)**
N-(Dichlorofluoromethylthio)-N-(dimethylsulfamoyl)-aniline **(1085-98-9)**
2,4-Dichloro-6-fluorophenyl p-nitrophenyl ether **(13738-63-1)**
2,4-Dichloro-6-fluorophenyl-4'-nitrophenyl ether **(13738-63-1)**
.α.,.α.-Dichloroglutaric acid **(50901-13-8)**
2,2-Dichloroglutaric acid **(50901-13-8)**
4,5-Dichloroguaiacol **(2460-49-3)**
1,1-Dichloroheptane **(821-25-0)**
1,6-Dichloro-1,5-hexadiene **(67546-51-4)**
1,2-Dichlorohexane **(2162-92-7)**
1,6-Dichlorohexane **(2163-00-0)**
2,5-Dichlorohexane **(13275-18-8)**
4,4-Dichloro-3-hexanone **(2648-60-4)**
2,2'-Dichlorohydrazobenzene **(782-74-1)**
Dichlorohydrin **(96-23-1)**
α-Dichlorohydrin **(96-23-1)**
2,3-Dichlorohydroquinone monomethylether **(39542-65-9)**
3,5-Dichloro-4-hydroxybenzoic acid **(3336-41-2)**
D-(-)-threo-2,2-Dichloro-N-(β-hydroxy-α-(hydroxymethyl))-p-nitrophenethylacetamide **(56-75-7)**
D-(-)-2,2-Dichloro-N-(β-hydroxy-α-(hydroxymethyl)-p-nitrophenylethyl)acetamide **(56-75-7)**
Di-(5-chloro-2-hydroxyphenyl)methane **(97-23-4)**
7:16-Dichloro-6:15-indanthrone **(130-20-1)**
Dichloroindanthrone **(130-20-1)**
O-(2,5-Dichloro-4-iodophenyl) O,O-dimethyl phosphorothioate **(18181-70-9)**
1,3-Dichloroisobutylene **(3375-22-2)**
2,3-Dichloroisobutyrate **(10411-52-6)**
2,3-Dichloroisobutyric acid **(10411-52-6)**
Dichloroisocyanurate **(2782-57-2)**
Dichloroisocyanuric acid **(2782-57-2)**
Dichloroisocyanuric acid, Dry [UN 2465] **(2782-57-2)**
Dichloroisocyanuric acid potassium salt [UN 2465] **(2244-21-5)**
Dichloroisocyanuric acid sodium salt [UN 2465] **(2893-78-9)**
sym-Dichloroisopropyl alcohol **(96-23-1)**
2,2'-Dichloroisopropyl ether **(108-60-1)**
Dichloroisopropyl ether **(108-60-1)**
Dichloroisopropyl ether **(39638-32-9)**
Dichloroisopropyl ether **(63283-80-7)**
Dichloroisopropyl ether [UN 2490] **(108-60-1)**
3-(2,4-Dichloro-5-isopropyloxy-phenyl)-δ^4-5-(tert-butyl)-1,3,4-oxadiazoline-2-one **(19666-30-9)**
Dichloroisoviolanthrone **(1324-55-6)**
Dichlorokelthane **(115-32-2)**
Dichloromercury **(7487-94-7)**
Dichlorometaxylenol **(133-53-9)**
3',4'-Dichloro-2-methacrylanilide **(2164-09-2)**
Dichloromethane (DOT,OSHA) **(75-09-2)**
Dichloromethazanone **(80-77-3)**
3',5'-Dichloromethotrexate **(528-74-5)**
Dichloromethotrexate **(528-74-5)**
1,2-Dichloro-3-methoxybenzene **(1984-59-4)**
1,3-Dichloro-2-methoxybenzene **(1984-65-2)**
1,4-Dichloro-2-methoxybenzene **(1984-58-3)**
1,5-Dichloro-2-methoxybenzene **(553-82-2)**
2,4-Dichloro-1-methoxybenzene **(553-82-2)**
2,5-Dichloro-6-methoxybenzoic acid **(1918-00-9)**
3,6-Dichloro-2-methoxybenzoic acid **(1918-00-9)**
3,6-Dichloro-2-methoxybenzoic acid, Compd. with N-methylmethanamine (1:1) **(2300-66-5)**
3,6-Dichloro-2-methoxybenzoic acid, sodium salt **(1982-69-0)**
2,3-Dichloro-4-methoxyphenol **(39542-65-9)**
2,5-Dichloro-4-methoxyphenol **(18113-14-9)**
2,6-Dichloro-4-methoxyphenol **(2423-72-5)**
4,5-Dichloro-2-methoxyphenol **(2460-49-3)**
3',4'-Dichloro-2-methylacrylanilide **(2164-09-2)**
Dichloromethylarsine **(593-89-5)**
1,5-Dichloro-3-methyl-3-azapentane hydrochloride **(55-86-7)**
(Dichloromethyl)benzene **(98-87-3)**
1,2-Dichloro-3-methylbenzene **(32768-54-0)**
1,2-Dichloro-4-methylbenzene **(95-75-0)**
1,3-Dichloro-2-methylbenzene **(118-69-4)**
1,4-Dichloro-2-methylbenzene **(19398-61-9)**
Dichloromethylbenzene **(29797-40-8)**
4,4'-Dichloro-(methyl benzhydrol) **(80-06-8)**
4,4'-Dichloro-α-methylbenzhydrol **(80-06-8)**
4,4'-Dichloro-α-methylbenzohydrol **(80-06-8)**
1,3-Dichloro-2-methylbutane **(23010-07-3)**
2,2-Dichloro-3-methylbutane **(17773-66-9)**
Dichloromethyl tert-butyl ketone **(22591-21-5)**
3-Dichloromethyl-6-chloro-7-sulfamoyl-3,4-dihydro-1,2,4-benzothiadiazine-1,1-dioxide **(133-67-5)**
3-Dichloromethyl-6-chloro-7-sulfamyl-3,4-dihydro-1,2,4-benzothiadiazine 1,1-dioxide **(133-67-5)**
Dichloromethyl cyanide **(3018-12-0)**
2,2'-Dichloro-N-methyldiethylamine **(51-75-2)**
2,2'-Dichloro-N-methyldiethylamine hydrochloride **(55-86-7)**
2,2'-Dichloro-N-methyldiethylamine-N-oxide **(126-85-2)**
2,2'-Dichloro-N-methyldiethylamine N-oxide hydrochloride **(302-70-5)**
Dichloromethyl O,O-diphenyl phosphonate **(40911-36-2)**
N-(Dichloromethylene)aniline **(622-44-6)**
1,1'-(Dichloromethylene)bisbenzene **(2051-90-3)**
(2,3-Dichloro-4-(2-methylenebutyryl)phenoxy)-acetic acid **(58-54-8)**
2,3-Dichloro-4-(2-methylenebutyryl)phenoxy acetic acid **(58-54-8)**
4,4'-Dichloro-2,2'-methylenediphenol **(97-23-4)**
(2,3-Dichloro-4-(2-methylene-1-oxobutyl)phenoxy)-acetic acid **(58-54-8)**
α,α-Dichloromethyl ether **(4885-02-3)**
sym-Dichloromethyl ether **(542-88-1)**
1,3-Dichloro-5,5'-methylhydantoin **(118-52-5)**
α,α-Dichloromethyl methyl ether **(4885-02-3)**
Dichloromethyl methyl ketone **(513-88-2)**
2,5-Dichloro-4-(3-methyl-5-oxo-2-pyrazolin-1-yl)-benzenesulfonic acid **(84-57-1)**
3,3-Dichloromethyloxycyclobutane **(78-71-7)**
1,1-Dichloro-4-methyl-1,3-pentadiene **(55667-43-1)**
2,4-Dichloro-6-methylphenol **(1570-65-6)**
2,6-Dichloro-4-methylphenol **(2432-12-4)**
Dichloromethyl(2-phenylethyl)silane **(772-65-6)**
Dichloromethylphenylsilane **(149-74-6)**
(Dichloromethyl)phosphonic acid, diphenyl **(40911-36-2)**
1,2-Dichloro-2-methylpropane **(594-37-6)**
1,3-Dichloro-2-methyl-1-propene **(3375-22-2)**
1,3-Dichloro-2-methylpropene **(3375-22-2)**
2,3-Dichloro-2-methylpropionaldehyde **(10141-22-7)**
Dichloromethylsilane **(75-54-7)**
O-(Dichloro(methylthio)phenyl) O,O-diethyl phosphorothioate (3 isomers) **(21923-23-9)**

2,4-Dichloro-6-(methylthio)-s-triazine **(13705-05-0)**
Dichloromethyl-3,3,3-trifluoropropylsilane **(675-62-7)**
Dichloromethylvinylsilane **(124-70-9)**
Dichloromonobromomethane **(75-27-4)**
Dichloromonoethylaluminum **(563-43-9)**
Dichloromonofluoromethane (OSHA) [UN 1029] **(75-43-4)**
2,4-Dichloro-m,5-xylenol **(133-53-9)**
1,2-Dichloronaphthalene **(2050-69-3)**
1,3-Dichloronaphthalene **(2198-75-6)**
1,4-Dichloronaphthalene **(1825-31-6)**
1,5-Dichloronaphthalene **(1825-30-5)**
1,6-Dichloronaphthalene **(2050-72-8)**
1,7-Dichloronaphthalene **(2050-73-9)**
1,8-Dichloronaphthalene **(2050-74-0)**
2,3-Dichloronaphthalene **(2050-75-1)**
2,6-Dichloronaphthalene **(2065-70-5)**
2,7-Dichloronaphthalene **(2198-77-8)**
2,3-Dichloro-1,4-naphthalenedione **(117-80-6)**
2,4-Dichloro-1-naphthalenol **(2050-76-2)**
2,3-Dichloro-1,4-naphthaquinone **(117-80-6)**
2,3-Dichloro-1,4-naphthoquinone **(117-80-6)**
2,3-Dichloro-α-naphthoquinone **(117-80-6)**
2,3-Dichloronaphthoquinone **(117-80-6)**
2,3-Dichloronaphthoquinone-1,4 **(117-80-6)**
Dichloronaphthoquinone **(117-80-6)**
2,4-Dichloro-6-(1-naphthylamino)-s-triazine **(30369-88-1)**
2,5-Dichloro-4-nitroaniline **(6627-34-5)**
2,6-Dichloro-4-nitroaniline **(99-30-9)**
3,4-Dichloronitrobenzen (Czech) **(99-54-7)**
2,6-Dichloro-4-nitrobenzenamine **(99-30-9)**
1,2-Dichloro-4-nitrobenzene **(99-54-7)**
1,3-Dichloro-5-nitrobenzene **(618-62-2)**
1,4-Dichloro-2-nitrobenzene **(89-61-2)**
2,3-Dichloronitrobenzene **(3209-22-1)**
2,4-Dichloronitrobenzene **(611-06-3)**
2,5-Dichloronitrobenzene **(89-61-2)**
3,4-Dichloronitrobenzene **(99-54-7)**
2,5-Dichloro-3-nitrobenzoic acid **(88-86-8)**
2',4'-Dichloro-4-nitrobiphenyl ether **(1836-75-5)**
2,4-Dichloro-4'-nitrodiphenyl ether **(1836-75-5)**
Dichloronitroethane **(594-72-9)**
1,1-Dichloro-1-nitroethane (ACGIH,OSHA) [UN 2650] **(594-72-9)**
2,4-Dichloro-6-nitrophenol **(609-89-2)**
2,6-Dichloro-4-nitrophenol **(618-80-4)**
2,4-Dichloro-1-(4-nitrophenoxy)benzene **(1836-75-5)**
4-(2,6-Dichloro-4-nitrophenylazo)-N-(β-acetoxy-ethyl)-N-(β-cyanoethyl)aniline **(5261-31-4)**
4-((2,6-Dichloro-4-nitrophenyl)azo)-N-(cyanoethyl)-N-(acetoxyethyl)aniline **(5261-31-4)**
3-(p-((2,6-Dichloro-4-nitrophenyl)azo)-N-ethyl-anilino)propionitrile **(13301-61-6)**
3-(4-((2,6-Dichloro-4-nitrophenyl)azo)-N-(2-hydroxy-ethyl)anilino)propionitrile, acetate (ester) **(5261-31-4)**
3-((4-((2,6-Dichloro-4-nitrophenyl)azo)phenyl)ethylamino)-propanenitrile **(13301-61-6)**
1,1-Dichloro-1-nitropropane **(595-44-8)**
2',5-Dichloro-4'-nitrosalicylanilide **(50-65-7)**
2',5-Dichloro-4'-nitrosalicylanilide, 2-aminoethanol salt **(1420-04-8)**
5,2'-Dichloro-4'-nitrosalicylanilide, ethanolamine salt **(1420-04-8)**
5,2-Dichloro-4-nitrosalicylic anilide 2-aminoethanol salt **(1420-04-8)**
2',5-Dichloro-4'-nitrosalicyloylanilide ethanolamine salt **(1420-04-8)**
1,1-Dichlorononane **(821-88-5)**
1,9-Dichlorononane **(821-99-8)**
1,1-Dichlorooctane **(20395-24-8)**
2,6-Dichloro-4-octylphenol- **(73986-52-4)**
Dichlorooxozirconium **(7699-43-6)**

1,2-Dichloropentane **(1674-33-5)**
1,5-Dichloropentane **(628-76-2)**
2,4-Dichloropentane **(625-67-2)**
Dichloropentane [UN 1152] **(30586-10-8)**
Dichlorophen **(97-23-4)**
Dichlorophen B **(97-23-4)**
Dichlorophenamide **(120-97-8)**
Dichlorophene **(97-23-4)**
2,4-Dichlorophenol **(120-83-2)**
2,5-Dichlorophenol **(583-78-8)**
2,6-Dichlorophenol **(87-65-0)**
3,4-Dichlorophenol **(95-77-2)**
3,5-Dichlorophenol **(591-35-5)**
Dichlorophenol **(25167-81-1)**
2,4-Dichlorophenol, benzenesulfonate **(97-16-5)**
3-(3,4-Dichlorophenol)-1,1-dimethylurea **(330-54-1)**
2,5-Dichlorophenol potassium salt **(68938-81-8)**
2,4-Dichlorophenol sodium salt **(3757-76-4)**
2,5-Dichlorophenol, sodium salt **(52166-72-0)**
3',3''-Dichlorophenolsulfonphthalein **(4430-20-0)**
2,6-Dichlorophenoxyacetate **(575-90-6)**
(2,6-Dichlorophenoxy)acetic acid **(575-90-6)**
3,4-Dichlorophenoxyacetic acid **(588-22-7)**
Dichlorophenoxyacetic acid **(94-75-7)**
2,4-Dichlorophenoxyacetic acid (DOT) **(94-75-7)**
Dichlorophenoxyacetic acid (OSHA) **(94-75-7)**
(2,4-Dichlorophenoxy)acetic acid ammonium salt **(2307-55-3)**
2,4-Dichlorophenoxyacetic acid butoxyethanol ester **(1929-73-3)**
(2,4-Dichlorophenoxy)acetic acid butoxyethyl ester **(1929-73-3)**
(2,4-Dichlorophenoxy)acetic acid, butyl ester **(94-80-4)**
2,4-Dichlorophenoxyacetic acid butyl ester **(94-80-4)**
2,4-Dichlorophenoxyacetic acid, 4-chlorocrotonyl ester **(2971-38-2)**
2,4-Dichlorophenoxyacetic acid diethanolamine salt **(5742-19-8)**
2,4-Dichlorophenoxyacetic acid, diethanolamine salt solution **(20940-37-8)**
2,4-Dichlorophenoxyacetic acid, diethylamine salt **(20940-37-8)**
(2,4-Dichlorophenoxy)acetic acid dimethylamine **(2008-39-1)**
(2,4-Dichlorophenoxy)acetic acid ethyl ester **(533-23-3)**
(2,4-Dichlorophenoxy)acetic acid 2-ethylhexyl ester **(1928-43-4)**
(2,4-Dichlorophenoxy)acetic acid isobutyl ester **(1713-15-1)**
2,4-Dichlorophenoxyacetic acid, isooctyl ester **(25168-26-7)**
2,4-Dichlorophenoxyacetic acid, isopropyl ester **(94-11-1)**
2,4-Dichlorophenoxyacetic acid lithium salt **(3766-27-6)**
2,4-Dichlorophenoxyacetic acid propylene glycol butyl ether ester **(1320-18-9)**
2,4-Dichlorophenoxyacetic acid, propylene glycol butyl ether ester **(1928-45-6)**
2,4-Dichlorophenoxyacetic acid, sodium salt **(2702-72-9)**
2,4-Dichlorophenoxyacetic acid triethanolamine salt **(2569-01-9)**
2,4-Dichlorophenoxy), 2-butoxymethylethyl ester **(1320-18-9)**
4-(2,4-Dichlorophenoxy)butyric acid **(94-82-6)**
γ-(2,4-Dichlorophenoxy)butyric acid **(94-82-6)**
4-(2,4-Dichlorophenoxy)butyric acid butoxyethanol ester **(32357-46-3)**
2,4-Dichlorophenoxybutyric acid, butoxyethyl ester **(32357-46-3)**
4-(2,4-Dichlorophenoxy)butyric acid butyl ester **(6753-24-8)**
4-(2,4-Dichlorophenoxy)butyric acid dimethylamine salt **(2758-42-1)**
4-(2,4-Dichlorophenoxy)butyric acid isooctyl ester **(1320-15-6)**
2,4-Dichlorophenoxybutyric acid, sodium salt **(10433-59-7)**
4-(2,4-Dichlorophenoxy)butyric acid sodium salt **(10433-59-7)**
γ-(2,4-Dichlorophenoxy)butyric acid, sodium salt **(10433-59-7)**
5,6-Dichloro-1-phenoxycarbonyl-2-trifluoromethylbenzimidazole **(14255-88-0)**
2-(2,4-Dichlorophenoxy)ethanol **(120-67-2)**
2-(2,4-Dichlorophenoxy)ethanol hydrogen sulfate sodium salt **(136-78-7)**
2-(2,4-Dichlorophenoxy)ethyl benzoate **(94-83-7)**
2,4-Dichlorophenoxyethyl sulfate, sodium salt **(136-78-7)**

Di-(4-chlorophenoxy)methane **(555-89-5)**
Di-(p-chlorophenoxy)methane **(555-89-5)**
4-(2,4-Dichlorophenoxy)-2-methoxy-1-nitrobenzene **(32861-85-1)**
5-(2,4-Dichlorophenoxy)-2-nitroanisole **(32861-85-1)**
4-(2,4-Dichlorophenoxy)nitrobenzene **(1836-75-5)**
2-(4-(2,4-Dichlorophenoxy)phenoxy)-methyl-propionate **(51338-27-3)**
2-(4-(2,4-Dichlorophenoxy)phenoxy)propanoate **(40843-25-2)**
(RS)-2-(4-(2,4-Dichlorophenoxy)phenoxy)propanoic acid **(40843-25-2)**
2-(4-(2,4-Dichlorophenoxy)phenoxy)propanoic acid **(40843-25-2)**
2-(2,4-Dichlorophenoxy) propionic acid **(120-36-5)**
α-(2,4-Dichlorophenoxy) propionic acid **(120-36-5)**
2-(2,4-Dichlorophenoxy)propionic acid 2-butoxyethanol ester **(53404-31-2)**
2-(2,4-Dichlorophenoxy)propionic acid, 2-butoxyethyl ester **(53404-31-2)**
2,2-Dichloro-2-phenylacetic acid **(61031-72-9)**
Di(p-chlorophenyl)acetic acid **(83-05-6)**
2-((2,6-Dichlorophenyl)amino)benzeneacetic acid monosodium salt **(15307-79-6)**
6-(((2,3-Dichlorophenyl)amino)carbonyl)-2,3,4,5-tetrachloro-benzoic acid **(76280-91-6)**
2-(2,6-Dichlorophenylamino)-2-imidazoline **(4205-90-7)**
Dichlorophenylarsine **(696-28-6)**
2,4-Dichlorophenyl benzenesulfonate **(97-16-5)**
2,4-Dichlorophenyl benzenesulphonate **(97-16-5)**
O-1-(2,4-Dichlorophenyl)-2-bromovinyl-O,O-dimethyl phosphate **(13104-21-7)**
N-(3,4-Dichlorophenyl)-N'-(4-chlorophenyl)urea **(101-20-2)**
N-3,4-Dichlorophenyl N-5-chloro-2-(2-sodium sulfonyl-4-chlorophenoxy)phenyl urea **(3567-25-7)**
N-(3,4-Dichlorophenyl)cyclopropanecarboxamide **(2759-71-9)**
2,4'-Dichlorophenyldichlorethane **(53-19-0)**
O-2,4-Dichlorophenyl O,O-diethyl phosphorothioate **(97-17-6)**
2,4-Dichloro-phenyl diethyl phosphorothionate **(97-17-6)**
1,6-Di(4'-chlorophenyldiguanidino)hexane diacetate **(56-95-1)**
1,6-Di(4'-chlorophenyldiguanido)hexane **(55-56-1)**
N-(3,4-Dichlorophenyl)-N-((dimethylamino)carbonyl)benzamide **(3134-12-1)**
N-(3',5'-Dichlorophenyl)-1,2-dimethylcyclopropane-1,2-dicarboximide **(32809-16-8)**
3-(3,5-Dichlorophenyl)-5,5-dimethyl oxazoline-dione-2,4 **(24201-58-9)**
1-(2,4-Dichlorophenyl)-4,4-dimethyl-2-(1,2,4-triazol-1-yl)pentan-3-ol **(75736-33-3)**
3-(3,4-Dichlorophenyl)-1,1-dimethylurea **(330-54-1)**
N'-(3,4-Dichlorophenyl)-N,N-dimethylurea **(330-54-1)**
1-(3,4-Dichlorophenyl)-3,3-dimethyluree (French) **(330-54-1)**
1-((2-(2,4-Dichlorophenyl)-1,3-dioxolan-2-yl)methyl)-1H-1,2,4-triazole **(60207-31-0)**
Di(p-chlorophenyl) disulfide **(1142-19-4)**
2,5-Dichloro-p-phenylenediamine **(20103-09-7)**
2,6-Dichloro-1,4-phenylenediamine **(609-20-1)**
2,6-Dichloro-p-phenylenediamine **(609-20-1)**
2,4-Dichlorophenyl ester of benzenesulfonic acid **(97-16-5)**
Di-(p-chlorophenyl)-ethanol **(80-06-8)**
1-(2,4-Dichlorophenyl)ethanone **(2234-16-4)**
3-(3,5-Dichlorophenyl)-5-ethenyl-5-methyl-2,4-oxazolidinedione **(50471-44-8)**
O-(2,4-Dichlorophenyl) O-ethyl S-propylphosphorodithioate **(34643-46-4)**
3,4-Dichlorophenyl isocyanate **(102-36-3)**
N-(3,4-Dichlorophenyl)methacrylamide **(2164-09-2)**
Di-(4-chlorophenyl)methane **(101-76-8)**
Di-(p-chlorophenyl)methane **(101-76-8)**
3-(3,4-Dichlorophenyl)-1-methoxy-1-methylurea **(330-55-2)**
3-(3,4-Dichlorophenyl)-1-methoxymethylurea **(330-55-2)**
N'-(3,4-Dichlorophenyl)-N-methoxy-N-methylurea **(330-55-2)**
1-(3,4-Dichlorophenyl)3-methoxy-3-methyluree (French) **(330-55-2)**
2,4-Dichlorophenyl 3'-methoxy-4'-nitrophenyl ether **(32861-85-1)**
3-(3,4-Dichlorophenyl)-1-methyl-1-butylurea **(555-37-3)**
Di(p-chlorophenyl) methylcarbinol **(80-06-8)**
3-(3,5-Dichlorophenyl)-N-(1-methylethyl)-2,4-dioxo-1-imidazolidinecarboxamide **(36734-19-7)**
O-(2,4-Dichlorophenyl) O-methyl N-isopropylphosphoramidothioate **(299-85-4)**
O-(2,4-Dichlorophenyl) O-methyl isopropylphosphoramidothioate **(299-85-4)**
6-(3-(2,6-Dichlorophenyl)-5-methyl-4-isoxazolecarboxamido)-3,3-dimethyl-7-oxo-4-thia-1-azabicyclo(3.2.0)heptane-2-carboxylic acid **(3116-76-5)**
6-(3-(2,6-Dichlorophenyl)-5-methyl-4-isoxazolecarboxamido penicillanic acid **(3116-76-5)**
3-(2,6-Dichlorophenyl)-5-methyl-4-isoxazolylpenicillin **(3116-76-5)**
N-(3,4-Dichlorophenyl)-N'-methyl-N'-methoxyurea **(330-55-2)**
2-(3,4-Dichlorophenyl)-4-methyl-1,2,4-oxadiazolidine-3,5-dione **(20354-26-1)**
N-(3,4-Dichlorophenyl)-2-methylpentamide **(2533-89-3)**
N-(2,4-Dichlorophenyl)-2-methylpentanamide **(2533-89-3)**
N-(3,4-Dichlorophenyl)-2-methyl-2-propenamide **(2164-09-2)**
2,4-Dichlorophenyl 4-nitrophenyl ether **(1836-75-5)**
2,4-Dichlorophenyl p-nitrophenyl ether **(1836-75-5)**
4,6-Dichloro-2-phenylphenol **(5335-24-0)**
α-(2,4-Dichlorophenyl)-α-phenyl-5-pyrimidinemethanol **(26766-27-8)**
Dichlorophenylphosphine **(644-97-3)**
Dichlorophenylphosphine sulfide **(3497-00-5)**
N-(3,4-Dichlorophenyl)propanamide **(709-98-8)**
1-(2-(2,4-Dichlorophenyl)-2-(2-propenyloxy)ethyl)-1H-imidazole **(35554-44-0)**
N-(3,4-Dichlorophenyl)propionamide **(709-98-8)**
3',4'-Dichlorophenylpropionanilide **(709-98-8)**
1-(2-(2,4-Dichlorophenyl)-4-propyl-1,3-dioxolan-2-ylmethyl)-1H-1,2,4-triazole **(60207-90-1)**
N-(3,4-Dichlorophenyl)-N'-2-(2-sulfo-4-chlorophenoxy)-5-chlorophenyl urea sodium salt **(3567-25-7)**
3,4-Dichlorophenylsulfonyl chloride **(98-31-7)**
N-(2,3-Dichlorophenyl)-3,4,5,6-tetrachlorophthalamic acid **(76280-91-6)**
((2,5-Dichlorophenyl)thio)acetic acid **(6274-27-7)**
2,5-Dichlorophenylthiomethyl O,O-diethyl phosphorodithioate **(2275-14-1)**
S-(2,5-Dichlorophenylthiomethyl) O,O-diethyl phosphorodithioate **(2275-14-1)**
S-(2,5-Dichlorophenylthiomethyl) diethyl phosphorothiolothionate **(2275-14-1)**
S-(((2,5-Dichlorophenyl)thio)methyl) O,O-dimethyl phosphorodithioate **(3735-23-7)**
Di-4-chlorophenyl thiourea **(1220-00-4)**
Di-p-chlorophenylthiourea **(1220-00-4)**
2,4-Dichlorophenyltrichloromethane **(13014-18-1)**
3,4-Dichlorophenyltrichloromethane **(13014-24-9)**
Di-(p-chlorophenyl)trichloromethylcarbinol **(115-32-2)**
Dichlorophenyltrichlorosilane [UN 1766] **(27137-85-5)**
3,4-Dichlorophenyltrifluoromethane **(328-84-7)**
Dichlorophos **(62-73-7)**
Dichlorophosphoric acid, ethyl ester **(1498-51-7)**
3,6-Dichloropicolinic acid **(1702-17-6)**
Dichloropinacolin **(22591-21-5)**
α,α-Dichloropinacolin **(22591-21-5)**
Dichloropinakolin **(22591-21-5)**
Dichloroprop **(120-36-5)**
1,1-Dichloropropane **(78-99-9)**
1,2-Dichloropropane **(78-87-5)**
1,3-Dichloropropane **(142-28-9)**
2,2-Dichloropropane **(594-20-7)**
α,β-Dichloropropane **(78-87-5)**
Dichloropropane (DOT) **(26638-19-7)**
1,2-Dichloropropane (OSHA) **(78-87-5)**
Dichloropropane-dichloropropene mixture **(8003-19-8)**
1,2-Dichloro-3-propanol **(616-23-9)**
1,2-Dichloropropanol-3 **(616-23-9)**
1,3-Dichloro-2-propanol **(96-23-1)**
2,3-Dichloro-1-propanol **(616-23-9)**
2,3-Dichloropropanol **(616-23-9)**
Dichloropropanol **(26545-73-3)**
1,3-Dichloropropanol-2 [UN 2750] **(96-23-1)**
1,3-Dichloro-2-propanol phosphate (3:1) **(13674-87-8)**
1,1-Dichloropropanone **(513-88-2)**
1,3-Dichloro-2-propanone **(534-07-6)**

(E)-1,3-Dichloropropene (10061-02-6)
(Z)-1,3-Dichloropropene (10061-01-5)
1,1-Dichloropropene (563-58-6)
1,2-Dichloro-2-propene (78-88-6)
1,2-Dichloropropene (563-54-2)
1,3-Dichloro-2-propene (542-75-6)
1,3-Dichloropropene-1 (542-75-6)
2,3-Dichloro-1-propene (78-88-6)
2,3-Dichloropropene (78-88-6)
Dichloropropene (542-75-6)
cis-1,3-Dichloropropene (10061-01-5)
trans-1,3-Dichloropropene (10061-02-6)
1,3-Dichloropropene (ACGIH,OSHA) (542-75-6)
Dichloropropene [UN 2047] (26952-23-8)
1,3-Dichloropropene and 1,2-dichloropropane mixture (8003-19-8)
2,3-Dichloro-2-propene-1-thiol, diisopropylcarbamate (2303-16-4)
3',4'-Dichloropropionanilide (709-98-8)
3,4-Dichloropropionanilide (709-98-8)
Dichloropropionanilide (709-98-8)
Dichloropropionate (127-20-8)
2,3-Dichloropropionic acid (565-64-0)
α,α-Dichloropropionic acid (75-99-0)
α-Dichloropropionic acid (75-99-0)
2,2-Dichloropropionic acid (ACGIH,DOT,OSHA) (75-99-0)
2,2-Dichloropropionic acid, magnesium salt (29110-22-3)
2,2-Dichloropropionic acid, sodium salt (127-20-8)
α-α-Dichloropropionic acid sodium salt (127-20-8)
2,2-Dichloropropionic acid, 2-(2,4,5-trichlorophenoxy)ethyl ester (136-25-4)
1,1-Dichloropropylene (563-58-6)
1,2-Dichloropropylene (563-54-2)
1,3-Dichloropropylene (542-75-6)
2,3-Dichloropropylene (78-88-6)
Dichloropropylene (26952-23-8)
α,γ-Dichloropropylene (542-75-6)
cis-1,3-Dichloropropylene (10061-01-5)
trans-1,3-Dichloropropylene (10061-02-6)
1,1'-(2,2-Dichloropropylidene)bis(4-ethoxybenzene) (56265-23-7)
2,3-Dichloro-4-(propylsulfonyl)pyridine (85847-73-0)
3,6-Dichloro-2-pyridinecarboxylic acid (1702-17-6)
4,6-Dichloro-2-pyrimidinamine (56-05-3)
4,5-Dichloropyrocatechol (3428-24-8)
4,7-Dichloroquinoline (86-98-6)
5,7-Dichloro-8-quinolinol, 5-chloro-8-quinolinol, and 7-chloro-8-quinolinol in proportions resulting naturally from chlorination of 8-quinolinol (8067-69-4)
5,7-Dichloro-8-quinolinol mixt. with 5-chloro-8-quinolinol and 7-chloro-8-quinolinol (8067-69-4)
2,3-Dichloroquinoxaline (2213-63-0)
4-(((2,3-Dichloro-6-quinoxalinyl)carbonyl)amino)-5-hydroxy-6-((2-sulfophenyl)azo)-2,7-naphthalenedisulfonic acid, trisodium salt (2407-13-8)
5-(((2,3-Dichloroquinoxalin-6-yl)carbonyl)amino)-4-hydroxy-3-((2-sulfophenyl)azo)-2,7-naphthalenedisulfonic acid, trisodium salt (2407-13-8)
Dichlorosal (58-93-5)
3,6-Dichlorosalicylate, sodium, potassium salt (68938-79-4)
3,5-Dichlorosalicylic acid (320-72-9)
3,6-Dichlorosalicylic acid (3401-80-7)
Dichlorosilane [UN 2189] (4109-96-0)
trans-.α.,.α.'-Dichlorostilbene (951-86-0)
trans-Dichlorostilbene (951-86-0)
2,3-Dichlorostyrene (2123-28-6)
Dichlorostyrene (6607-45-0)
p-Dichlorosulfamoylbenzoic acid (80-13-7)
3,4-Dichloro-5-sulfamylbenzenesulfonamide (120-97-8)
p-(N,N-Dichlorosulfamyl)benzoic acid (80-13-7)
Dichlorosulfane (10545-99-0)
2,5-Dichlorosulfanilic acid (88-50-6)
3,5-Dichlorosyringol (78782-46-4)
4,6-Dichloro-2',4',5',7'-tetrabromofluorescein dipotassium salt (6441-77-6)
Di-mu-chlorotetrachlorodialuminum (13845-12-0)
1,1-Dichloro-1,2,2,2-tetrafluoroethane (374-07-2)
1,1-Dichlorotetrafluoroethane (374-07-2)
1,2-Dichloro-1,1,2,2-tetrafluoroethane (76-14-2)
Dichlorotetrafluoroethane (374-07-2)
sym-Dichlorotetrafluoroethane (76-14-2)
Dichlorotetrafluoroethane (ACGIH,OSHA) (76-14-2)
Dichlorotetrafluoroethane [UN 1958] (1320-37-2)
2,3-Dichlorotetrahydrofuran (3511-19-1)
1,2-Dichloro-1,1,2,2-tetramethyldisilane (4342-61-4)
8,19-Dichlorotetrasulfodiphenaleno(1,9-ab:1',9'-lm)triphenodioxazine, tetrasodium salt (33700-25-3)
2,6-Dichlorothiobenzamide (1918-13-4)
Dichlorothiocarbonyl (463-71-8)
2,5-Dichlorothiophene (3172-52-9)
Dichlorotitanocene (1271-19-8)
2,4-Dichlorotoluene (95-73-8)
2,5-Dichlorotoluene (19398-61-9)
2,6-Dichlorotoluene (118-69-4)
3,4-Dichlorotoluene (95-75-0)
Dichlorotoluene (29797-40-8)
α,α-Dichlorotoluene (98-87-3)
Dichloro-s-triazine-2,4,6(1H,3H,5H)-trione potassium deriv (2244-21-5)
1,3-Dichloro-s-triazine-2,4,6(1H,3H,5H)trione potassium salt (2244-21-5)
4-((4,6-Dichloro-1,3,5-triazin-2-yl)amino)benzenesulfonic acid (16110-89-7)
4-(2,4-Dichloro-1,3,5-triazinylamino)benzenesulfonic acid (16110-89-7)
2-((6-((4,6-Dichloro-1,3,5-triazin-2-yl)methylamino)-1-hydroxy-3-sulfo-2-naphthalenyl)azo)-1,5-naphthalenedisulfonic acid, trisodium salt (70616-90-9)
1,2-Dichloro-4-(trichloromethyl)benzene (13014-24-9)
1,2-Dichloro-4-trichloromethylbenzene (13014-24-9)
1,3-Dichloro-4-(trichloromethyl)benzene (13014-18-1)
2,4-Dichloro-1-(trichloromethyl)benzene (13014-18-1)
4,4'-Dichloro-α-(trichloromethyl)benzhydrol (115-32-2)
2,5-Dichloro-6-(trichloromethyl)pyridine (1817-13-6)
3,5-Dichloro-2-(trichloromethyl)pyridine (1128-16-1)
3,6-Dichloro-2-(trichloromethyl)pyridine (1817-13-6)
4,5-Dichloro-2-(2,4,5-trichlorophenoxy)phenol (61639-90-5)
2,2-Dichloro-1-(2,4,5-trichlorophenyl)ethanone (1203-86-7)
2,2'-Dichlorotriethylamine (538-07-8)
1,2-Dichloro-1,1,2-trifluoroethane (354-23-4)
2,2-Dichloro-1,1,1-trifluoroethane (306-83-2)
1,2-Dichloro-4-(trifluoromethyl)benzene (328-84-7)
2,4-Dichloro-1-(trifluoromethyl)benzene (320-60-5)
4,5-Dichloro-2-trifluoromethylbenzimidazole (3615-21-2)
5,6-Dichloro-2-trifluoromethylbenzimidazole (2338-25-2)
5,6-Dichloro-2-trifluoromethylbenzimidazole-1-carboxylate (14255-88-0)
4,4'-Dichloro-3-(trifluoromethyl)-carbanilide (369-77-7)
4,4'-Dichloro-3-(trifluoromethyl)carbanilide (369-77-7)
3,4-Dichloro-α,α,α-trifluorotoluene (328-84-7)
1,3-Dichloro-4,5,6-trimethoxybenzene (99849-00-0)
4,6-Dichloro-1,2,3-trimethoxybenzene (99849-00-0)
1,11-Dichloro-3,6,9-trioxaundecane (638-56-2)
Dichlorovanadocene (12083-48-6)
Dichlorovas (62-73-7)
(2,2-Dichloro-vinil)dimetil-fosfato (Italian) (62-73-7)
2,2-Dichlorovinyl diethyl phosphate (72-00-4)
3-(2,2-Dichlorovinyl)-2,2-dimethylcyclopropanecarbonyl chloride (52314-67-7)
2,2-Dichlorovinyl dimethyl phosphate (62-73-7)
2,2-Dichlorovinyl dimethyl phosphoric acid ester (62-73-7)
Dichlorovos (62-73-7)
Dichlorovos mixture, Dry (DOT) (62-73-7)
2,5-Dichloro-p-xylene (1124-05-6)
α,α'-Dichloro-m-xylene (626-16-4)

α,α'-Dichloro-o-xylene **(612-12-4)**
α,α'-Dichloroxylene **(28347-13-9)**
2,4-Dichloro-3,5-xylenol **(133-53-9)**
Dichloro-m-xylenol **(133-53-9)**
Dichloroxylenol **(133-53-9)**
Dichloroxylenolum (Latin) **(133-53-9)**
Dichloroxylylene **(28347-13-9)**
Dichlorphen **(97-23-4)**
Dichlorphenamide **(120-97-8)**
2,4-Dichlorphenoxyacetic acid **(94-75-7)**
2,4-Dichlorphenoxyacetic acid octyl ester **(1928-44-5)**
(2,4-Dichlor-phenoxy)-essigsaeure (German) **(94-75-7)**
2-(2,4-Dichlor-phenoxy)-propionsaeure (German) **(120-36-5)**
3-(3,4-Dichlorphenyl)-1-n-butyl-harnstoff (German) **(555-37-3)**
2,4-Dichlorphenyl cellosolve **(120-67-2)**
3-(3,4-Dichlor-phenyl)-1,1-dimethyl-harnstoff (German) **(330-54-1)**
3-(3,4-Dichlor-phenyl)-1-methoxy-1-methyl-harnstoff (German) **(330-55-2)**
3-(4,5-Dichlorphenyl)-1-methoxy-1-methylharnstoff (German) **(330-55-2)**
3-(3,5-Dichlorphenyl)-5-methyl-5-vinyl-1,3-oxazolidin-2,4-dion (German) **(50471-44-8)**
2,4-Dichlorphenyl-4-nitrophenylaether (German) **(1836-75-5)**
1-(2-(2,4-Dichlorphenyl)-2-(2-propenyloxy)aethyl)-1H-imidazol (German) **(35554-44-0)**
Dichlorphos **(62-73-7)**
ω,ω-Dichlorpinakolin (German) **(22591-21-5)**
Dichlorprop **(120-36-5)**
Dichlorpropan **(26638-19-7)**
Dichlorpropan-dichlorpropengemisch (German) **(8003-19-8)**
Dichlorpropaphos **(34643-46-4)**
Dichlorpropen-gemisch (German) **(563-54-2)**
2,2-Dichlorpropionsaeure natrium (German) **(127-20-8)**
Dichlorsulfofenyl-methylpyrazolon (Czech) **(84-57-1)**
2,6-Dichlor-thiobenzamid (German) **(1918-13-4)**
Dichlor-s-triazin-2,4,6(1H,3H,5H)trione potassium **(2244-21-5)**
O-(2,2-Dichlorvinyl)-O,O-diethylphosphat (German) **(72-00-4)**
O-(2,2-Dichlorvinyl)-O,O-dimethylphosphat (German) **(62-73-7)**
Dichlorvos-ethyl **(72-00-4)**
Dichlorvos (ACGIH,DOT,OSHA) **(62-73-7)**
Dichlosale **(50-65-7)**
Dichlotiazid **(58-93-5)**
Dichlotride **(58-93-5)**
(2,2-Dichlor-vinyl)-dimethyl-phosphat (German) **(62-73-7)**
Dichlozolinate **(72391-46-9)**
Dichlozoline **(24201-58-9)**
Dicholine succinate **(306-40-1)**
Dichromic acid **(13530-68-2)**
Dichromic acid, calcium salt (1:1) **(14307-33-6)**
Dichromic acid, diammonium salt **(7789-09-5)**
Dichromic acid, dipotassium salt **(7778-50-9)**
Dichromic acid, disodium salt **(10588-01-9)**
Dichromic acid, zinc salt (1:1) **(14018-95-2)**
Dichromium sulfate **(10101-53-8)**
Dichromium sulphate **(10101-53-8)**
Dichromium trioxide **(1308-38-9)**
Dichromium trisulfate **(10101-53-8)**
Dichromium trisulphate **(10101-53-8)**
Dichronic **(15307-79-6)**
Dichystrolum **(67-96-9)**
Di(2-cianoetil)ammina (Italian) **(111-94-4)**
Dicicloverina (Spanish) **(77-19-0)**
Dick (German) **(598-14-1)**
Diclobutrazol **(75736-33-3)**
Diclofenac sodium **(15307-79-6)**
Diclofop-Methyl **(51338-27-3)**
Diclofop **(40843-25-2)**
N,N-Diclohexylamine **(101-83-7)**
Diclophenac sodium **(15307-79-6)**
Dicloralurea **(116-52-9)**
Dicloran **(99-30-9)**
N,N'-Dicloroazodicarbonamidina (Spanish) **(502-98-7)**
p-Diclorobenceno (Spanish) **(106-46-7)**
1,4-Diclorobenzene (Italian) **(106-46-7)**
p-Diclorobenzene (Italian) **(106-46-7)**
1,1-Dicloro-2,2-bis(4-cloro-fenil)-etano (Italian) **(72-54-8)**
1,1-Dicloroetano (Italian) **(75-34-3)**
1,2-Dicloroetano (Italian) **(107-06-2)**
2,2'-Dicloroetiletere (Italian) **(111-44-4)**
3-(3,4-Dicloro-fenyl)-1,1-dimetil-urea (Italian) **(330-54-1)**
1,1-Dicloro-1-nitroetano (Italian) **(594-72-9)**
Dicloroxilenol (Spanish) **(133-53-9)**
Dicloroxilenolo **(133-53-9)**
Diclossacillina **(3116-76-5)**
Diclotride **(58-93-5)**
Dicloxacilin **(3116-76-5)**
Dicloxacilina (Spanish) **(3116-76-5)**
Dicloxacillin **(3116-76-5)**
Dicloxacilline (French) **(3116-76-5)**
Dicloxacillinum (Latin) **(3116-76-5)**
Dicloxacycline **(3116-76-5)**
Dico **(125-29-1)**
Dicobalt carbonyl **(10210-68-1)**
Dicobalt octacarbonyl **(10210-68-1)**
Di(coco alkyl) dimethyl ammonium chloride **(61789-77-3)**
Dicoco dimethyl ammonium chloride **(61789-77-3)**
Dicocodimethylammonium chloride **(61789-77-3)**
Dicocodimonium Chloride **(61789-77-3)**
Dicodid **(125-29-1)**
Dicofol **(115-32-2)**
Dicol **(111-46-6)**
Diconirt **(2702-72-9)**
Diconirt D **(2702-72-9)**
Dicophane **(50-29-3)**
Dicopper chloride trihydroxide **(1332-65-6)**
Dicopper dichloride **(7758-89-6)**
Dicopper dihydroxycarbonate **(12069-69-1)**
Dicopper monosulfide **(22205-45-4)**
Dicopper monoxide **(1317-39-1)**
Dicopper sulfide **(22205-45-4)**
Dicopur **(94-75-7)**
Dicopur-M **(94-74-6)**
Dicorel Brown LMR **(16071-86-6)**
Dicortol **(50-24-8)**
Dicorvin **(56-53-1)**
Dicotex **(3653-48-3)**
Dicotex **(94-74-6)**
Dicotex 80 **(3653-48-3)**
Dicotox **(533-23-3)**
Dicotox **(94-75-7)**
Dicoumarin **(66-76-2)**
Dicoumarol **(66-76-2)**
Dicresyl **(1129-41-5)**
Dicresyl N-methylcarbamate **(1129-41-5)**
Dicrotofos (Dutch) **(141-66-2)**
Dicrotophos (ACGIH,OSHA) **(141-66-2)**
Dicryl **(2164-09-2)**
Dictycide **(140-87-4)**
Dictyopteren D' **(33156-92-2)**
Dictyopterene D' **(33156-92-2)**
(.+-.)-Dictyopterene A **(22822-99-7)**
Dictyzide **(140-87-4)**
Dicuman **(66-76-2)**
Dicumaol R **(66-76-2)**
Dicumarine **(66-76-2)**
Dicumarol **(66-76-2)**

Dicumol **(66-76-2)**
Dicumyl **(36876-13-8)**
Dicumyl peroxide **(80-43-3)**
Dicumyl peroxide, Dry (DOT) **(80-43-3)**
Dicumyl peroxide, Technically pure or with inert solid (DOT) **(80-43-3)**
Dicupral **(97-77-8)**
Dicuran **(15545-48-9)**
Dicyan **(460-19-5)**
Dicyanamide **(504-66-5)**
Dicyandiamide **(461-58-5)**
Dicyandiamin (German) **(461-58-5)**
Dicyanimide **(504-66-5)**
2,2'-Dicyano-2,2'-azopropane **(78-67-1)**
1,2-Dicyanobenzene **(91-15-6)**
1,3-Dicyanobenzene **(626-17-5)**
m-Dicyanobenzene **(626-17-5)**
o-Dicyanobenzene **(91-15-6)**
1,4-Dicyanobutane **(111-69-3)**
1,4-Dicyano-2-butene **(1119-85-3)**
β,β-Dicyano-o-chlorostyrene **(2698-41-1)**
β,β'-Dicyanodiethyl ether **(1656-48-0)**
β,β'-Dicyanodiethyl sulfide **(111-97-7)**
2,3-Dicyano-1,4-dithia-anthraquinone **(3347-22-6)**
s-Dicyanoethane **(110-61-2)**
Di-(2-cyanoethyl)amine **(111-94-4)**
N,N'-Di(2-cyanoethyl)-1,2-ethylenediamine **(3217-00-3)**
Di(2-cyanoethyl)sulfide **(111-97-7)**
Dicyanogen **(460-19-5)**
1,6-Dicyanohexane **(629-40-3)**
Dicyanomethane **(109-77-3)**
N-(2-((2,6-Dicyano-4-nitrophenyl)azo)-5-(diethylamino)phenyl)-acetamide **(41642-51-7)**
1,8-Dicyanooctane **(1871-96-1)**
1,5-Dicyanopentane **(646-20-8)**
1,3-Dicyanopropane **(544-13-8)**
2,2-Dicyanopropane **(7321-55-3)**
1,3-Dicyanotetrachlorobenzene **(1897-45-6)**
Dicyclohexane **(92-51-3)**
3,9-Di-(3-cyclohexenyl)-2,4,8,10-tetraoxaspiro(5,5)undecane **(6600-31-3)**
Dicyclohexyl **(92-51-3)**
Dicyclohexyl adipate **(849-99-0)**
Dicyclohexylamine [UN 2565] **(101-83-7)**
Dicyclohexylamine, N-methyl **(7560-83-0)**
Dicyclohexylamine, nitrite **(3129-91-7)**
Dicyclohexylamine, N-nitroso **(947-92-2)**
1,5-Di(cyclohexylamino)anthracene-9,10-dione **(15958-68-6)**
Dicyclohexylaminonitrite **(3129-91-7)**
Dicyclohexylammonium nitrite **(3129-91-7)**
1,4-Dicyclohexylbutane **(6165-44-2)**
1,3-Dicyclohexylcarbodiimide **(538-75-0)**
Dicyclohexylcarbodiimide **(538-75-0)**
N,N'-Dicyclohexylcarbodiimide **(538-75-0)**
Dicyclohexyl disulfide **(2550-40-5)**
Dicyclohexylnitrosamin (German) **(947-92-2)**
Dicyclohexylnitrosamine **(947-92-2)**
Dicyclohexyl peroxide carbonate **(1561-49-5)**
Dicyclohexyl peroxydicarbonate **(1561-49-5)**
Dicyclohexyl peroxydicarbonate (Not more than 91% with water) **(1561-49-5)**
Dicyclohexyl peroxydicarbonate, Not more than 91% with water **(1561-49-5)**
Dicyclohexyl peroxydicarbonate, Technically pure **(1561-49-5)**
N,N'-Dicyclohexyl-p-phenylenediamine **(4175-38-6)**
Dicyclohexylphosphine **(829-84-5)**
Dicyclohexyl phthalate **(84-61-7)**
Dicyclohexyl sodium sulfosuccinate **(23386-52-9)**
Dicyclohexyl thiourea **(1212-29-9)**
N,N'-Dicyclohexylthiourea **(1212-29-9)**
Dicyclohexyltin oxide **(22771-17-1)**
Dicyclomine **(77-19-0)**
Dicyclopentadiene (ACGIH,OSHA) [UN 2048] **(77-73-6)**
Dicyclopentadiene acrylate **(33791-58-1)**
Dicyclopentadiene diepoxide **(81-21-0)**
Dicyclopentadiene dioxide **(81-21-0)**
Dicyclopentadiene, 3,4,5,6,7,8,8a-heptachloro- **(76-44-8)**
Dicyclopentadienyl acrylate **(50976-02-8)**
Dicyclopentadienylcobalt **(1277-43-6)**
Dicyclopentadienyldichlorotitanium **(1271-19-8)**
Di-2,4-cyclopentadien-1-yliron **(102-54-5)**
Dicyclopentadienyl iron (OSHA) **(102-54-5)**
Dicyclopentadienyltitanium dichloride **(1271-19-8)**
Dicyclopentenyl acrylate **(50976-02-8)**
Dicyclopentenyloxyethyl methacrylate **(68586-19-6)**
Dicyclopentyloxyethyl acrylate **(65983-31-5)**
Dicycloverin **(77-19-0)**
Dicycloverine **(77-19-0)**
Dicycloverinum (Latin) **(77-19-0)**
Dicyklohexylamin (Czech) **(101-83-7)**
Dicyklohexylamin nitrit (Czech) **(3129-91-7)**
N,N-Dicyklohexylbenzthiazolsulfenamid (Czech) **(4979-32-2)**
Dicyklopentadien (Czech) **(77-73-6)**
Dicylcohexylcarbodiimide **(538-75-0)**
Dicynit (Czech) **(3129-91-7)**
Dicysteine **(56-89-3)**
Didakene **(127-18-4)**
Didan-TDC-250 **(57-41-0)**
Didandin **(82-66-6)**
Didecanoyl peroxide **(762-12-9)**
Didecanoyl peroxide, Technically pure (DOT) **(762-12-9)**
Di-n-decyl adipate **(105-97-5)**
Didecyl adipate **(105-97-5)**
Didecyl decanedioate **(2432-89-5)**
Didecyl dimethyl ammonium chloride **(7173-51-5)**
Didecyl hexanedioate **(105-97-5)**
Di(decyl)methylamine **(7396-58-9)**
N,N-Didecyl-N-methyl-3-(trimethoxysilyl)propanediol **(68959-20-6)**
Di-N-decylmethyl(3-trimethoxysilylpropyl)ammonium chloride **(68959-20-6)**
Didecylmethyl(3-(trimethoxysilyl)propyl)ammonium chloride. **(68959-20-6)**
Didecyl phenyl phosphite **(1254-78-0)**
Di-n-decyl phthalate **(84-77-5)**
Didecyl phthalate **(84-77-5)**
9,10-Didehydro-N,N-diethyl-6-methyl-ergoline-8-β-carboxamide **(50-37-3)**
7,8-Didehydrocholesterol **(434-16-2)**
2,3-Didehydroemetine **(4914-30-1)**
7,8-Didehydro-4,5-α-epoxy-3,6-α-dihydroxy-17,17-dimethy morphinanium bromide **(125-23-5)**
7,8-Didehydro-4,5-α-epoxy-3-ethoxy-17-methylmorphinan-6-α-ol hydrochloride dihydrate **(6746-59-4)**
(5 α,6 α)-7,8-Didehydro-4,5-epoxy-3-methoxy-17-methyl-morphinan-6-ol, 3-pyridinecarboxylate (Ester) **(3688-66-2)**
7,8-Didehydro-4,5-α-epoxy-3-methoxymorphinan-6-α-ol **(467-15-2)**
7,8-Didehydro-4,5-α-epoxy-17-methyl-3-(2-morpholino-ethoxy)morphinan-6-α-ol **(509-67-1)**
9,10-Didehydro-N-(α-(hydroxymethyl)ethyl)-6-methylergoline-8-β-carboxamide **(60-79-7)**
9,10-Didehydro-N-(α-(hydroxymethyl)propyl)-6-methyl-ergoline-8-β-carboxamide **(113-42-8)**
9,10-Didehydro-6-methyl-ergoline-8-β-carboxamide **(478-94-4)**
1,2-Didehydrotestololactone **(968-93-4)**
2',3'-Dideoxycytidine **(7481-89-2)**
1,6-Dideoxy-1,6-di(2-chloroethylamino)-D-mannitol dihydro-chloride **(551-74-6)**
Dideuterium oxide **(7789-20-0)**
Di-2,4-dichlorobenzoyl peroxide, Maximum concentration 52% as a paste or in solution (DOT) **(133-14-2)**
Di-2,4-dichlorobenzoyl peroxide, Not more than 75% with water

(DOT) **(133-14-2)**
4,4'-Di(diethylamino)-4',6'-disulphotriphenylmethanol anhydride, sodium salt **(129-17-9)**
Didigam **(50-29-3)**
Didimac **(50-29-3)**
3,6-Di(dimethylamino)acridine **(494-38-2)**
1,2-Di-(dimethylamino)ethane [UN 2372] **(110-18-9)**
N,N-Di(1,4-dimethylpentyl)-p-phenylenediamine **(3081-14-9)**
1,5-Di(2,4-dimethylphenyl)-3-methyl-1,3,5-triazapenta-1,4-diene **(33089-61-1)**
Di(2,5-dimethylphenyl) phenyl phosphate **(72121-83-6)**
Di(2,6-dimethylphenyl) phenyl phosphate **(23666-93-5)**
Di(2,6-dimethylphenyl) 2,4,6-trimethylphenyl phosphate **(73179-37-0)**
Didoc **(59-66-5)**
Didodecylamine (8CI) **(3007-31-6)**
2,4-Didodecylamino-6-chloro-1,3,5-triazine **(30355-03-4)**
Di-tert-dodecyl disulfide **(27458-90-8)**
N,N-Didodecyl-1-dodecanamine **(102-87-4)**
Didodecyl hydrogen phosphate **(7057-92-3)**
Didodecylphenol **(25482-47-7)**
Di-n-dodecyl phosphate **(7057-92-3)**
Didodecyl phosphate **(7057-92-3)**
Di-n-dodecyl phosphite **(21302-09-0)**
Didodecyl phosphite **(21302-09-0)**
Didodecyl phosphonate **(21302-09-0)**
Di-n-dodecyl phthalate **(2432-90-8)**
Didodecyl phthalate **(2432-90-8)**
Didodecyl 3,3'-thiodipropionate **(123-28-4)**
Didrate **(125-28-0)**
Didrex **(5411-22-3)**
Didrex **(60-13-9)**
Didromycin **(5490-27-7)**
Didronel R **(7414-83-7)**
Didroxan **(97-23-4)**
Didroxane **(97-23-4)**
Dieca **(147-84-2)**
Dieldrex **(60-57-1)**
Dieldrin (ACGIH,OSHA) [NA 2761] **(60-57-1)**
Dieldrine (French) **(60-57-1)**
Dieldrin, photo- **(13366-73-9)**
Dieldrite **(60-57-1)**
Dieltamid **(134-62-3)**
Diemal **(57-44-3)**
Para-Dien **(84-17-3)**
Diene 221 **(2611-00-9)**
Diene 35 NF **(9003-17-2)**
Dienestrol **(84-17-3)**
Dienite 556 **(9003-17-2)**
Dienite 643 **(9003-17-2)**
Dienite X 555 **(9003-17-2)**
Dienite X 644 **(9003-17-2)**
Dienochlor **(2227-17-0)**
Dienochlore **(2227-17-0)**
Dienoestrol **(84-17-3)**
β-Dienoestrol **(84-17-3)**
Dienol **(84-17-3)**
Dienol S **(9003-55-8)**
Dienpax **(439-14-5)**
Diepin **(2898-12-6)**
(+-)-1,2:3,4-Diepoxybutane **(298-18-0)**
(2S,3S)-1,2:3,4-Diepoxybutane **(30031-64-2)**
(2S,3S)-Diepoxybutane **(30031-64-2)**
(R*,S*)-Diepoxybutane **(564-00-1)**
1,2:3,4-Diepoxybutane **(1464-53-5)**
2,4-Diepoxybutane **(1464-53-5)**
2R:3R-Diepoxybutane **(30419-67-1)**
D,L-Diepoxybutane **(298-18-0)**
D-Diepoxybutane **(30419-67-1)**
Diepoxybutane **(1464-53-5)**
dl-1,2:3,4-Diepoxybutane **(298-18-0)**
l-1,2:3,4-Diepoxybutane **(30031-64-2)**
l-Diepoxybutane **(30031-64-2)**
meso-1,2,3,4-Diepoxybutane **(564-00-1)**
meso-Diepoxybutane **(564-00-1)**
(1Z,5E)-1,10(14)-Diepoxy-4(15),5-germacradiene-9-one **(61228-92-0)**
1,10(14)-Diepoxy-4(15),5-germecradiene-9-one **(61228-92-0)**
1,2:5,6-Diepoxy-3a,4,5,6,7,7a-hexahydro-4,7-methanoindan **(81-21-0)**
1,2:5,6-Diepoxyhexahydro-4,7-methanoindan **(81-21-0)**
1,2:5,6-Diepoxyhexane **(1888-89-7)**
1,2,8,9-Diepoxylimonene **(96-08-2)**
1,2:8,9-Diepoxy-p-menthane **(96-08-2)**
1,2:7,8-Diepoxyoctane **(2426-07-5)**
Di(2,3-epoxy)propyl ether **(2238-07-5)**
2,6-Di(2,3-epoxypropyl)phenyl 2,3-epoxypropyl **(13561-08-5)**
1,2:15,16-Diepoxy-4,7,10,13-tetraoxahexadecane **(1954-28-5)**
Diesel Fuel [NA 1993] **(68334-30-5)**
Diesel Fuel No.2 **(68476-34-6)**
Diesel Oil (Petroleum) **(68334-30-5)**
Diesel Test Fuel **(68334-30-5)**
Diesel Fuel Oil #2 **(68476-31-3)**
Dietadione (Italian) **(702-54-5)**
Dietamine **(60-13-9)**
Diethadion **(702-54-5)**
Diethadione **(702-54-5)**
Diethamine **(109-89-7)**
Diethanolamin (Czech) **(111-42-2)**
Diethanolamine (ACGIH,OSHA) **(111-42-2)**
Diethanolamine 4-chlorophenoxyacetate **(53404-23-2)**
Diethanolamine 4-chloro-2-phenylphenate **(53537-63-6)**
Diethanolamine dicamba **(25059-78-3)**
Diethanolamine 2,4-dichlorophenoxyacetate **(5742-19-8)**
Diethanolamine dodecylbenzene sulfonate **(26545-53-9)**
Diethanolamine dodecylbenzenesulfonate **(26545-53-9)**
Diethanolamine hydrochloride **(14426-21-2)**
Diethanolamine lauryl sulfate **(143-00-0)**
Diethanolamine mefluidide **(53780-36-2)**
Diethanolamine 2-methyl-4-chlorophenoxyacetate **(20405-19-0)**
Diethanolamine 2-(2-methyl-4-chlorophenoxy)propionate **(1432-14-0)**
Diethanolamine monolaurate **(7487-79-8)**
Diethanolamine oleate **(13961-86-9)**
Diethanolamine oleic acid amide **(93-83-4)**
Diethanolamine 4(or 6)-chloro-2-phenylphenate **(53537-63-6)**
Diethanolamine stearic acid amide **(93-82-3)**
Diethanolaminobenzene **(120-07-0)**
Diethanolaminochlorobenzene **(92-00-2)**
Diethanolammonium chloride **(14426-21-2)**
Diethanolammonium maleic hydrazide **(5716-15-4)**
Diethanolammonium oleate **(13961-86-9)**
N,N-Diethanolanilide, 3-chloro- **(92-00-2)**
Diethanolaniline **(120-07-0)**
N,N-Diethanolaniline **(120-07-0)**
Diethanolchloroanilide **(92-00-2)**
Diethanolethylamine **(139-87-7)**
Diethanolisopropylamine **(121-93-7)**
Diethanollauramide **(120-40-1)**
N,N-Diethanollauramide **(120-40-1)**
N,N-Diethanollauric acid amide **(120-40-1)**
Diethanolmethylamine **(105-59-9)**
Diethanolnitrosoamine **(1116-54-7)**
Diethanol sulfone **(2580-77-0)**
Diethanol-m-toluidine **(91-99-6)**
Diethatyl ethyl **(38727-55-8)**
1,4-Diethenylbenzene **(105-06-6)**
1,3-Diethenyl-1,1,3,3-tetramethyl disiloxane **(2627-95-4)**
1,3-Diethenyl-1,1,3,3-tetramethyldisiloxane **(2627-95-4)**

Diethibutin **(86-14-6)**
Diethion **(563-12-2)**
2,2-Diethoxyacetophenone **(6175-45-7)**
α,α-Diethoxyacetophenone **(6175-45-7)**
2,5-Diethoxy-4-benzamidoaniline **(120-00-3)**
1,4-Diethoxy-2-butene **(7250-85-3)**
1,2-Di(ethoxycarbonyl)ethyl O,O-dimethyl phosphorodithioate **(121-75-5)**
S-(1,2-Diethoxycarbonyl)ethyl O,O-dimethyl phosphorothioate **(1634-78-2)**
S-(1,2-Di(ethoxycarbonyl)ethyl) dimethyl phosphorothiolothionate **(121-75-5)**
2,5-Diethoxy-1-chlorobenzene **(52196-74-4)**
2,5-Diethoxychlorobenzene **(52196-74-4)**
Diethoxychlorosilane **(6485-91-2)**
1,1-Diethoxy-3,7-dimethyl-2,6-octadiene **(7492-66-2)**
1,1-Diethoxy-ethaan (Dutch) **(105-57-7)**
1,1-Diethoxyethane **(105-57-7)**
1,2-Diethoxyethane **(629-14-1)**
Diethoxymethane [UN 2373] **(462-95-3)**
1,1-Diethoxy-3-methylbutane **(3842-03-3)**
1,1-Diethoxy-2-methylpropane **(1741-41-9)**
2,5-Diethoxy-4-morpholinobenzenediazonium chloride, zinc chloride double salt **(6023-29-6)**
2,5-Diethoxy-4-nitrochlorobenzene **(91-43-0)**
Diethoxyphenylarsine **(3141-11-5)**
Diethoxyphenylarsine oxide **(3141-11-5)**
α-(((Diethoxyphosphinothioyl)oxy)imino)benzeneacetonitrile **(14816-18-3)**
(Diethoxyphosphinyl)dithioimidocarbonic acid cyclic ethylene ester **(947-02-4)**
Diethoxyphosphinylimino-2 dithietanne-1,3 (French) **(21548-32-3)**
(Diethoxyphosphinylimino)-1,3-dithietane **(21548-32-3)**
2-(Diethoxyphosphinylimino)-1,3-dithiolane **(947-02-4)**
2-(Diethoxyphosphinylimino)-4-methyl-1,3-dithiolane **(950-10-7)**
2-Diethoxy-phosphinylthioethyl-trimethylammonium iodide **(513-10-0)**
N-(2-(Diethoxyphosphinylthio)ethyl)trimethylammonium iodide **(513-10-0)**
Diethoxyphosphorus oxychloride **(814-49-3)**
Diethoxyphosphoryl-thiocholine iodide **(513-10-0)**
1,1-Diethoxypropane **(4744-08-5)**
2,2-Diethoxypropane **(126-84-1)**
2,4-Diethoxypyrimidine **(20461-60-3)**
Diethoxytetraethylene glycol **(4353-28-0)**
6,6'-Diethoxythioindigo **(3263-31-8)**
Di-ethoxythiokarbonyl-disulfid (Czech) **(502-55-6)**
Diethoxy thiophosphoric acid ester of 2-ethylmercaptoethanol **(8065-48-3)**
Diethoxy thiophosphoric acid ester of 7-hydroxy-4-methyl coumarin **(299-45-6)**
(Diethoxy-thiophosphoryloxyimino)-phenyl acetonitrile **(14816-18-3)**
Diethquinalphion **(13593-03-8)**
Diethquinalphione **(13593-03-8)**
Diethyl **(106-97-8)**
Diethyl Yellow **(2481-94-9)**
Diethyl acetal **(105-57-7)**
Diethyl acetaldehyde **(97-96-1)**
N,N-Diethylacetamide **(685-91-6)**
Diethylacetic acid **(88-09-5)**
Diethylacetoacetamide **(2235-46-3)**
N,N-Diethylacetoacetamide **(2235-46-3)**
Diethyl acetoacetate **(1619-57-4)**
Diethyl acetylbutanedioate **(1115-30-6)**
Diethyl adipate **(141-28-6)**
Diethylaluminium chloride (DOT) **(96-10-6)**
Diethylaluminum chloride **(96-10-6)**
Diethylaluminum ethoxide **(1586-92-1)**
Diethylaluminum hydride **(871-27-2)**
Diethylaluminum monochloride **(96-10-6)**
Diethylamide of 3-ethoxy-4-hydroxy-benzoic acid **(13898-68-5)**
Diethylamid kyseliny acetoctove (Czech) **(2235-46-3)**
Diethylamid kyseliny chlormravenci (Czech) **(88-10-8)**
Diethylamid kyseliny chloroctove (Czech) **(2315-36-8)**
Diethylamid kyseliny lysergove (Czech) **(50-37-3)**
Diethylamid kyseliny mravenci (Czech) **(617-84-5)**
Diethylamid kyseliny nikotinove (Czech) **(59-26-7)**
N,N-Diethylamine **(109-89-7)**
Diethylamine (ACGIH,OSHA) [UN 1154] **(109-89-7)**
Diethylamine, 2,2'-diamino- **(111-40-0)**
Diethylamine, 2,2'-dichloro-N-methyl **(51-75-2)**
Diethylamine, 2,2'-dichloro-N-methyl, hydrochloride **(55-86-7)**
Diethylamine, 2,2'-dichloro-N-methyl-, N-oxide **(126-85-2)**
Diethylamine, 2,2'-dichloro-N-methyl-, oxide **(126-85-2)**
Diethylamine, 2,2'-dichloro-N-methyl-, N-oxide, hydrochloride **(302-70-5)**
Diethylamine 2,4-dichlorophenoxyacetate **(20940-37-8)**
Diethylamine, 2,2'-dicyano- **(111-94-4)**
Diethylamine, 2,2'-dihydroxy- **(111-42-2)**
Diethylamine, 2,2'-dihydroxy-N-nitroso- **(1116-54-7)**
Diethylamine, 2,2'-dimethoxy- (8CI) **(111-95-5)**
Diethylamine, hydriodide (8CI) **(19833-78-4)**
Diethylamine, hydrobromide (8CI) **(6274-12-0)**
Diethylamine, hydrochloride **(660-68-4)**
Diethylamine, 1-methyl-N-nitroso **(16339-04-1)**
Diethylamine, N-nitroso **(55-18-5)**
2-(Diethylamino)-2',6'-acetoxylidide **(137-58-6)**
α-Diethylamino-2,6-acetoxylidide **(137-58-6)**
Diethylaminoaceto-2,6-xylidide **(137-58-6)**
α-Diethylaminoaceto-2,6-xylidide **(137-58-6)**
Diethylaminoacet-2,6-xylidide **(137-58-6)**
4-(Diethylamino)aniline **(93-05-0)**
p-(Diethylamino)aniline **(93-05-0)**
4-(Diethylamino)azobenzene **(2481-94-9)**
N,N-Diethyl-4-aminoazobenzene **(2481-94-9)**
p-(Diethylamino)azobenzene **(2481-94-9)**
4-(Diethylamino)benzaldehyde **(120-21-8)**
p-(Diethylamino)benzaldehyde **(120-21-8)**
p-Diethylaminobenzaldehyde, 1,1-diphenylhydrazone **(68189-23-1)**
N,N-Diethylaminobenzene **(91-66-7)**
4-Diethylaminobenzenediazonium chloride zinc chloride (2:1) **(5149-85-9)**
3-(Diethylamino)benzenesulfonic acid sodium salt **(5123-63-7)**
3-Diethylamino-1,1-bis(2-thienyl)-1-butene **(86-14-6)**
4-Diethylamino-2-butynyl α-phenylcyclohexaneglycolate **(5633-20-5)**
α-Diethylamino-2,6-dimethylacetanilide **(137-58-6)**
ω-Diethylamino-2,6-dimethylacetanilide **(137-58-6)**
3-Diethylamino-2,4-dinitro-6-trifluoromethylaniline **(29091-05-2)**
3-Diethylamino-1,1-di(2'-thienyl)but-1-ene **(86-14-6)**
3-Diethylamino-1,1-dithienylbut-1-ene **(86-14-6)**
3-Diethylamino-1,1-di(2'-thienyl)-1-butene **(86-14-6)**
(Diethylamino)ethane **(121-44-8)**
2-(Diethylamino)ethanol **(100-37-8)**
2-N-Diethylaminoethanol **(100-37-8)**
Diethylaminoethanol **(100-37-8)**
N-Diethylaminoethanol **(100-37-8)**
β-Diethylaminoethanol **(100-37-8)**
2-Diethylaminoethanol (ACGIH,OSHA) **(100-37-8)**
Diethylaminoethanol [UN 2686] **(100-37-8)**
Diethylaminoethanol 4-aminobenzoate hydrochloride **(51-05-8)**
2-(2-(Diethylamino)ethoxy)ethanol **(140-82-9)**
2-β-Diethylaminoethoxyethanol **(140-82-9)**
Diethylaminoethoxyethanol **(140-82-9)**
1-(p-(β-Diethylaminoethoxy)-phenyl)-1,2-diphenylchloroethylene **(911-45-5)**
1-(p-(β-Diethylaminoethoxy)phenyl)-1,2-diphenyl-2-chloroethylene citrate **(50-41-9)**
S-2-Diethylaminoethyl-O-ethylester kyseliny methylthiofosfonove (Czech) **(21770-86-5)**
2-(Diethylamino)ethyl acrylate **(2426-54-2)**
Diethylaminoethyl acrylate **(2426-54-2)**
N,N-Diethylaminoethyl acrylate **(2426-54-2)**
β-Diethylaminoethyl acrylate **(2426-54-2)**
β-Diethylaminoethyl alcohol **(100-37-8)**

2-Diethylaminoethylamid kyseliny p-aminobenzoove (Czech) **(51-06-9)**
2-(Diethylamino)ethylamine **(100-36-7)**
2-(N,N-Diethylamino)ethylamine **(100-36-7)**
N,N-(Diethylamino)ethylamine **(100-36-7)**
N-(2-(Diethylamino)ethyl)amine **(100-36-7)**
(Diethylamino)ethylamino **(100-36-7)**
2-Diethylaminoethyl p-aminobenzoate **(59-46-1)**
Diethylaminoethyl p-aminobenzoate **(59-46-1)**
β-Diethylaminoethyl 4-aminobenzoate **(59-46-1)**
2-Diethylaminoethyl p-aminobenzoate hydrochloride **(51-05-8)**
2-(Diethylamino)ethyl 4-amino-2-chlorobenzoate **(133-16-4)**
1-((2-(Diethylamino)ethyl)amino)-4-(hydroxymethyl)-9H-thioxanthen-9-one **(3105-97-3)**
1-((2-(Diethylamino)ethyl)amino)-4-(hydroxymethyl)thioxanthen-9-one **(3105-97-3)**
2-(Diethylamino)ethyl benzilate **(302-40-9)**
Diethylaminoethyl benzilate **(302-40-9)**
β-Diethylaminoethyl benzilate **(302-40-9)**
N-(2-(Diethylamino)ethyl)-2-butoxycinchoninamide **(85-79-0)**
Diethylaminoethyl chloride hydrochloride **(869-24-9)**
β-Diethylamino-ethyl chloride hydrochloride **(869-24-9)**
S-(Diethylaminoethyl) O,O-diethyl phosphorothioate **(78-53-5)**
S-(2-Diethylaminoethyl) O,O-diethylphosphorothioate hydrogenoxalate **(3734-97-2)**
2-Diethylaminoethyl diphenylacetate **(64-95-9)**
2-(Diethylamino)ethyl diphenylglycolate **(302-40-9)**
2-Diethylaminoethylester kyseliny akrylove (Czech) **(2426-54-2)**
2-Diethylaminoethylester kyseliny p-aminobenzoove (Czech) **(59-46-1)**
2-Diethylaminoethylester kyseliny difenyloctove (Czech) **(64-95-9)**
2-Diethylaminoethylester kyseliny methakrylove (Czech) **(105-16-8)**
1-(2-(Diethylamino)ethyl)-2-(p-ethoxybenzyl)-5-nitrobenzimidazole **(911-65-9)**
2-(Diethylamino)ethyl methacrylate **(105-16-8)**
2-(N,N-Diethylamino)ethyl methacrylate **(105-16-8)**
Diethylaminoethyl methacrylate **(105-16-8)**
β-(Diethylamino)ethyl methacrylate **(105-16-8)**
N-(Diethylaminoethyl)-2-methoxy-4-amino-5-chlorobenzamide **(364-62-5)**
N-(2-(Diethylamino)ethyl)octadecanamide **(16889-14-8)**
N-(2-Diethylamino)ethyl)octadecanamide **(16889-14-8)**
(2-Diethylamino)ethylphosphorothioic acid O,O-diethyl ester **(78-53-5)**
S-(2-(Diethylamino)ethyl)phosphorothioic acid O,O-diethyl ester **(78-53-5)**
Diethylaminoethyl stearamide **(16889-14-8)**
N-(2-(Diethylamino)ethyl)stearamide **(16889-14-8)**
β-Diethylaminoethyl 9-xanthenecarboxylate methobromide **(53-46-3)**
β-Diethylaminoethyl xanthene-9-carboxylate methobromide **(53-46-3)**
4-(Diethylamino)-2-hydroxybenzaldehyde **(17754-90-4)**
2-Diethylamino-4-isopropylamino-6-methoxy-s-triazine **(3004-70-4)**
N-(3-(Diethylamino)-4-methoxyphenyl)acetamide **(19433-93-3)**
1-(Diethylamino)-3-methylbenzene **(91-67-8)**
3-(Diethylamino)-1-methylbenzene **(91-67-8)**
7-Diethylamino-4-methylcoumarin **(91-44-1)**
O-(2-(Diethylamino)-6-methyl-4-pyrimidinyl)O,O-diethyl phosphorothioate **(23505-41-1)**
2-Diethylamino-6-methylpyrimidin-4-yl diethylphosphorothionate **(23505-41-1)**
O-(2-(Diethylamino)-6-methyl-4-pyrimidinyl)O,O-dimethyl phosphorothioate **(29232-93-7)**
O-(2-Diethylamino-6-methylpyrimidin-4-yl) O,O-dimethyl phosphorothioate **(29232-93-7)**
2-Diethylamino-6-methylpyrimidin-4-yl dimethyl phosphorothionate **(29232-93-7)**
5-(Diethylamino)-2-nitrosophenol hydrochloride **(25953-06-4)**
3-(Diethylamino)phenol **(91-68-9)**
m-(Diethylamino)phenol **(91-68-9)**
Diethyl-m-amino-phenolphthalein hydrochloride **(81-88-9)**
N-(3-(Diethylamino)phenyl)acetamide **(6375-46-8)**
O,O-Diethyl O-(4-aminophenyl) phosphorothioate **(3735-01-1)**
2-(Diethylamino)propiophenone **(90-84-6)**
α-Diethylaminopropiophenone **(90-84-6)**
N,N-Diethylaminopropylamine **(104-78-9)**
N-(3-Diethylaminopropyl)amine **(104-78-9)**
3-(Diethylamino)propylamine [UN 2684] **(104-78-9)**
4-N,N-Diethylaminosalicylic aldehyde **(17754-90-4)**
3-(Diethylamino)toluene **(91-67-8)**
3-(N,N-Diethylamino)toluene **(91-67-8)**
Diethylaminotrimethylenamine **(104-78-9)**
(6-(Diethylamino)-3H-xanthen-3-ylidine)diethylammonium chloride **(2150-48-3)**
N-(6-(Diethylamino)-3H-xanthen-3-ylidine)-N-ethylethanaminium chloride **(2150-48-3)**
Diethylammonium chloride **(660-68-4)**
Diethylandiamine **(91-79-2)**
N,N-Diethylanilin (Czech) **(91-66-7)**
2,6-Diethylaniline **(579-66-8)**
3,4-Diethylaniline **(54675-14-8)**
Diethylaniline **(91-66-7)**
N,N-Diethylaniline [UN 2432] **(91-66-7)**
Diethylarsine **(692-42-2)**
Diethylbarbitone **(57-44-3)**
5,5-Diethylbarbituric acid **(57-44-3)**
Diethyl-barbituric acid **(57-44-3)**
6,8-Diethylbenz(a)anthracene **(36911-94-1)**
8,12-Diethylbenz(a)anthracene **(36911-95-2)**
2,6-Diethylbenzamide **(89151-70-2)**
N,N-Diethylbenzamide **(1696-17-9)**
9,10-Diethyl-1,2-benzanthracene **(16354-52-2)**
1,4-Diethylbenzene **(105-05-5)**
Diethyl benzene **(25340-17-4)**
m-Diethylbenzene **(141-93-5)**
o-Diethylbenzene **(135-01-3)**
p-Diethyl benzene **(105-05-5)**
p-Diethylbenzene **(105-05-5)**
Diethylbenzene [UN 2049] **(25340-17-4)**
O,O-Diethyl S-benzyl thiophosphate **(13286-32-3)**
Diethyl (N,N-bis(2-hydroxyethyl)amino)methanephosphonate **(2781-11-5)**
Diethyl ((bis(2-hydroxyethyl)amino)methyl)phosphonate **(2781-11-5)**
O,O-Diethyl-O-(4-broom-2,5-dichloor-fenyl)-monothiofosfaat (Dutch) **(4824-78-6)**
N,N-Diethyl-1,3-butanediamine **(32280-46-9)**
Di-2-(2-ethylbutoxy)ethyl adipate **(7790-07-0)**
Di-(2-(2-ethylbutoxy)ethyl)ester kyseliny adipove (Czech) **(7790-07-0)**
Di-2-ethyl-butyl-acetate **(10332-40-8)**
Di(2-ethylbutyl) adipate **(10022-60-3)**
N,N-Diethylbutylamine **(4444-68-2)**
Diethyl butylmethylmalonate **(55114-29-9)**
Diethyl tert-butylmethylmalonate **(53268-44-3)**
Diethylcarbamazane citrate **(1642-54-2)**
Diethylcarbamazine acid citrate **(1642-54-2)**
Diethylcarbamazine citrate **(1642-54-2)**
Diethylcarbamazine hydrogen citrate **(1642-54-2)**
Diethylcarbamodithioic acid 2-chloro-2-propenyl ester **(95-06-7)**
Diethylcarbamodithioic acid, sodium salt **(148-18-5)**
Diethylcarbamoyl chloride **(88-10-8)**
N,N-Diethylcarbamoyl chloride **(88-10-8)**
1-Diethylcarbamoyl-4-methylpiperazine dihydrogen citrate **(1642-54-2)**
Diethylcarbamyl chloride **(88-10-8)**
N,N-Diethylcarbanilide **(85-98-3)**
O,O-Diethyl S-(carbethoxy)methyl phosphorothiolate **(2425-25-4)**
Diethyl carbinol **(584-02-1)**
Diethylcarbinol (DOT) **(584-02-1)**
Diethyl carbitol **(112-36-7)**
O,O-Diethyl S-carboethoxymethyl dithiophosphate **(919-54-0)**
O,O-Diethyl S-carboethoxymethyl phosphorodithioate **(919-54-0)**
O,O-Diethyl S-carboethoxymethyl phosphorothioate **(2425-25-4)**
O,O-Diethyl S-carboethoxymethyl thiophosphate **(2425-25-4)**
Diethyl carbonate [UN 2366] **(105-58-8)**

Diethyl o-carboxybenzoyloxyacetate **(84-72-0)**
Diethyl cellosolve (DOT) **(629-14-1)**
Diethylcetone (French) **(96-22-0)**
O,O-Diethyl O-(2-chinoxalyl)-phosphorothioate **(13593-03-8)**
O,O-Diethyl-S-((4-chloor-fenyl-thio)-methyl)-dithiofosfaat (Dutch) **(786-19-6)**
O,O-Diethyl-O-(3-chloor-4-methyl-cumarin-7-yl)monothiofosfaat (Dutch) **(56-72-4)**
O,O-Diethyl-S-((6-chloor-2-oxo-benzoxazolin-3-yl)-methyl)-dithiofosfaat (Dutch) **(2310-17-0)**
O,O-Diethyl-S-p-chlorfenylthiomethylester kyseliny dithiofosforecne (Czech) **(786-19-6)**
Diethylchlorfosfat (Czech) **(814-49-3)**
N,N-Diethylchloroacetamide **(2315-36-8)**
Diethylchloroaluminum **(96-10-6)**
O,O-Diethyl S-(6-chlorobenzoxazolinyl-3-methyl) dithiophosphate **(2310-17-0)**
O,O-Diethyl 2-chloro-α-cyanobenzylideneamino-oxyphosphonothioate **(14816-20-7)**
O,O-Diethyl O-(2-chloro-1-(2',4'-dichlorophenyl)vinyl) phosphate **(470-90-6)**
O,O-Diethyl O-(2-chloro-1-2,5-dichlorophenylvinyl) phosphorothioate **(1757-18-2)**
Diethyl-β-chloroethylamine hydrochloride **(869-24-9)**
O,O-Diethyl O-(3-chloro-4-methyl-7-coumarinyl)phosphorothioate **(56-72-4)**
O,O-Diethyl O-(3-chloro-4-methylcoumarinyl-7) thiophosphate **(56-72-4)**
O,O-Diethyl O-(3-chloro-4-methyl-2-oxo-2H-benzopyran-7-yl)phosphorothioate **(56-72-4)**
O,O-Diethyl 3-chloro-4-methyl-7-umbelliferone thiophosphate **(56-72-4)**
O,O-Diethyl O-(3-chloro-4-methylumbelliferyl)phosphorothioate **(56-72-4)**
Diethyl 3-chloro-4-methylumbelliferyl thionophosphate **(56-72-4)**
O,O-Diethyl S-((6-chloro-2-oxobenzoxazolin-3-yl)methyl) phosphorodithioate **(2310-17-0)**
O,O-Diethyl-S-(6-chloro-2-oxo-benzoxazolin-3-yl)methyl-phosphorothiolothionate **(2310-17-0)**
O,O-Diethyl p-chlorophenylmercaptomethyl dithiophosphate **(786-19-6)**
O,O-Diethyl S-(4-chlorophenylthiomethyl) dithiophosphate **(786-19-6)**
O,O-Diethyl S-p-chlorophenylthiomethyl dithiophosphate **(786-19-6)**
O,O-Diethyl S-(p-chlorophenylthiomethyl) phosphorodithioate **(786-19-6)**
Diethyl chlorophosphate **(814-49-3)**
O,O-Diethyl S-(2-chloro-1-phthalimidoethyl)phosphorodithioate **(10311-84-9)**
Diethylchlorosilane **(1609-19-4)**
Diethylchlorothiophosphate **(2524-04-1)**
Diethyl 2-chlorovinyl phosphate **(311-47-7)**
O,O-Diethyl O-(2-chlorovinyl) phosphate **(311-47-7)**
Diethylchlorthiofosfat (Czech) **(2524-04-1)**
O,O-Diethyl-α-cyanobenzylidineaminooxyphosphonothiate **(14816-18-3)**
p,p-Diethyl cyclic ethylene ester of phosphonodithioimidocarbonic acid **(947-02-4)**
p,p-Diethyl cyclic propylene ester of phosphonodithioimidocarbonic acid **(950-10-7)**
N,N-Diethylcyclohexanamine **(91-65-6)**
Diethylcyclohexane **(1331-43-7)**
Diethylcyclohexane (Mixed isomers) **(1331-43-7)**
Diethylcyclohexylamine **(91-65-6)**
N,N-Diethylcyclohexylamine **(91-65-6)**
Diethyl 1,10-decanedioate **(110-40-7)**
Diethyl decanedioate **(110-40-7)**
N,N-Diethyl-1,2-diaminoethane **(100-36-7)**
N,N-Diethyl-1,3-diaminopropane **(104-78-9)**
O,O-Diethyl-O-(2,4-dichloor-fenyl)-monothiofosfaat (Dutch) **(97-17-6)**
O,O-Diethyl O-2,5-dichloro-4-bromophenyl-phosphorothioate **(4824-78-6)**
O,O-Diethyl O-(2,5-dichloro-4-bromophenyl) thiophosphate **(4824-78-6)**
O,O-Diethyl O-(dichloro(methylthio)phenyl) phosphorothioate **(60238-56-4)**
O,O-Diethyl-O-2,4,5-dichloro-(methylthio)phenyl thionophosphate **(21923-23-9)**
O,O-Diethyl O-dichloro(methylthio)phenyl thiophosphate **(60238-56-4)**
Diethyl 1-(2,4-dichlorophenyl)-2-chlorovinyl phosphate **(470-90-6)**
O,O-Diethyl O-(2,4-dichlorophenyl) phosphorothioate **(97-17-6)**
Diethyl 2,4-dichlorophenyl phosphorothionate **(97-17-6)**
O,O-Diethyl S-(2,5-dichlorophenylthiomethyl) dithiophosphate **(2275-14-1)**
O,O-Diethyl-S-(2,5-dichlorophenylthiomethyl) dithiophosphoran **(2275-14-1)**
O,O-Diethyl S-(2,5-dichlorophenylthiomethyl) phosphorodithioate **(2275-14-1)**
O,O-Diethyl S-(2,5-dichlorophenylthiomethyl) phosphorothiolothionate **(2275-14-1)**
O,O-Diethyl O-2,4-dichlorophenyl thiophosphate **(97-17-6)**
Diethyldichlorosilane [UN 1767] **(1719-53-5)**
Diethyl ((diethanolamino)methyl)phosphonate **(2781-11-5)**
O,O-Diethyl-S-2-(diethylamino)ethylester kyseliny thiofosforecne (Czech) **(78-53-5)**
Diethyl S-2-diethylaminoethyl phosphorothioate **(78-53-5)**
O,O-Diethyl S-2-diethylaminoethyl phosphorothioate **(78-53-5)**
O,O-Diethyl-S-(2-diethylamino)ethylphosphorothioate hydrogen oxalate **(3734-97-2)**
O,O-Diethyl S-(β-diethylamino)ethyl phosphorothiolate **(78-53-5)**
O,O-Diethyl S-2-diethylaminoethyl phosphorothiolate **(78-53-5)**
O,O-Diethyl S-diethylaminoethyl phosphorothiolate **(78-53-5)**
O,O-Diethyl S-(β-diethylamino)ethyl phosphorothiolate hydrogen oxalate **(3734-97-2)**
O,O-Diethyl S-(2-diethylaminoethyl) thiophosphate **(78-53-5)**
O,O-Diethyl O-(2-diethylamino-6-methyl-4-pyrimidinyl)phosphorothioate **(23505-41-1)**
1,3-Diethyl-1,3-difenylmocovina (Czech) **(85-98-3)**
2,2'-Diethyldihexylamine **(106-20-7)**
O,O-Diethyl (1,2-dihydro-1,3-dioxo-2H-isoindol-2-yl)phosphonothioate **(5131-24-8)**
5,5-Diethyldihydro-2H-1,3-oxazine-2,4(3H)-dione **(702-54-5)**
O,O-Diethyl O-(2,3-dihydro-3-oxo-2-phenyl-6-pyridazinyl)-phosphorothioate **(119-12-0)**
Diethyl (dimethoxyphosphinothioylthio) butanedioate **(121-75-5)**
Diethyl (dimethoxyphosphinothioylthio)succinate **(121-75-5)**
Diethyldimethyllead **(1762-27-2)**
Diethyldimethylplumbane **(1762-27-2)**
N^{3},N^{3}-Diethyl-2,4-dinitro-6-(trifluoromethyl)-1,3-benzenediamine **(29091-05-2)**
3,3-Diethyl-2,4-dioxo-5-methylpiperidine **(125-64-4)**
Diethyl-diphenyl dichloroethane **(72-56-0)**
4,4'-Diethyldiphenylethane **(57364-79-1)**
Diethyl diphenylethane **(68398-19-6)**
N,N'-Diethyl-N,N'-diphenylurea **(85-98-3)**
sym-Diethyldiphenylurea **(85-98-3)**
Diethyldisulfid (Czech) **(110-81-6)**
Diethyl disulfide **(110-81-6)**
3,3'-Diethyldithiacarbodicyanine iodide **(514-73-8)**
N,N-Diethyl-4,4-di-2-thienyl-3-buten-2-amine **(86-14-6)**
Diethyldithio bis(thionoformate) **(502-55-6)**
Diethyldithiocarbamate sodium **(148-18-5)**
Diethyldithiocarbamate sodium trihydrate **(20624-25-3)**
Diethyldithiocarbamic acid **(147-84-2)**
Diethyldithiocarbamic acid 2-chloroallyl ester **(95-06-7)**
Diethyldithiocarbamic acid sodium **(148-18-5)**
Diethyldithiocarbamic acid, sodium salt **(148-18-5)**
Diethyldithiocarbamic acid sodium salt trihydrate **(20624-25-3)**
Diethyldithio carbamic acid tellurium salt **(20941-65-5)**
Diethyldithiocarbamic acid zinc salt **(14324-55-1)**
Diethyldithiocarbaminic acid **(147-84-2)**
O,O-Diethyldithiofosforecnan sodny (Czech) **(3338-24-7)**
Diethyldithiokarbaman sodny trihydrat (Czech) **(20624-25-3)**
Diethyldithione **(147-84-2)**
O,O-Diethyl dithiophosphoric acid p-chlorophenylthiomethyl ester **(786-19-6)**
O,O-Diethyldithiophosphorylacetic acid, N-monoisopropylamide **(2275-18-5)**
3-Diethyldithiophosphorylmethyl-6-chlorobenzoxazolone-2 **(2310-17-0)**
Diethyl dixanthogen **(502-55-6)**
N,N-Diethyldodecanamide **(3352-87-2)**
1,4-Diethylenediamine **(110-85-0)**
1,4-Diethylene dioxide **(123-91-1)**
Diethylene dioxide (OSHA) **(123-91-1)**
Diethylene ether **(123-91-1)**
Diethylene-glycol **(111-46-6)**

Diethylene glycol, bis(allyl carbonate)- (142-22-3)
Diethylene glycol, bischloroformate (106-75-2)
Diethylene glycol bis(methacrylate) (2358-84-1)
Diethylene glycol bisphthalate (13988-26-6)
Diethylene glycol n-butyl ether (112-34-5)
Diethylene glycol butyl ether acetate (124-17-4)
Diethylene glycol diacrylate (4074-88-8)
Diethylene glycol, di(3-aminopropyl) ether (4246-51-9)
Diethylene glycol, dibenzoate (120-55-8)
Diethyleneglycoldibutyl ether (112-73-2)
Diethylene glycol, dicarbamate (8CI) (5952-26-1)
Diethylene glycol diethyl ether (112-36-7)
Diethylene glycol diglycidyl ether (4206-61-5)
Diethylene glycol dimethacrylate (2358-84-1)
Diethylene glycol dimethacrylate monomer (2358-84-1)
Diethylene glycol dimethyl ether (111-96-6)
Diethylene glycol di-n-butyl ether (112-73-2)
Diethylene glycol dinitrate (DOT) (693-21-0)
Diethyleneglycol dinitrate, desensitized [UN 0075] (693-21-0)
Diethylene glycol distearate (109-30-8)
Diethylene glycol ethyl ether (111-90-0)
Diethylene glycol ethyl methyl ether (1002-67-1)
Diethylene glycol formal (1779-19-7)
Diethylene glycol n-hexyl ether (112-59-4)
Diethylene glycol laurate (141-20-8)
Diethylene glycol lauric acid monoester (141-20-8)
Diethylene glycol methyl ether (111-77-3)
Diethylene glycol monobutyl ether (112-34-5)
Diethylene glycol, monobutyl ether, acetate (124-17-4)
Diethylene glycol monocyclohexyl ether (25961-84-6)
Diethylene glycol monododecyl ether sodium sulfate (3088-31-1)
Diethylene glycol monododecyl ether sulfate sodium salt (3088-31-1)
Diethylene glycol, monoester with stearic acid (106-11-6)
Diethylene glycol monoethyl ether (111-90-0)
Diethylene glycol monoethyl ether acetate (112-15-2)
Diethylene glycol monohexyl ether (112-59-4)
Diethylene glycol monolaurate (141-20-8)
Diethylene glycol monolauryl ether sodium sulfate (3088-31-1)
Diethylene glycol monolauryl ether sulfate sodium salt (3088-31-1)
Diethylene glycol monomethyl ether (111-77-3)
Diethylene glycol monomethyl ether acetate (629-38-9)
Diethylene glycol monooleate (106-12-7)
Diethylene glycol monophenyl ether (104-68-7)
Diethylene glycol monostearate (106-11-6)
Diethylene glycolphenyl ether (104-68-7)
Diethylene glycol phthalate (7447-67-8)
Diethylene glycol sesquilaurate (141-20-8)
Diethylene glycol stearate (106-11-6)
Diethyleneimide oxide (110-91-8)
Diethylene imidoxide (110-91-8)
N,N'-Diethylenemorpholinophosphinothioic diamide (2168-68-5)
N,N'-Diethylene-N'-(3-oxapentamethylene)phosphorothioic triamide (2168-68-5)
Diethylene oxapentamethylenethiophosphoramide (2168-68-5)
Di(ethylene oxide) (123-91-1)
Diethylene oxide (109-99-9)
Diethylene oximide (110-91-8)
Diethylenetriamine (ACGIH,OSHA) [UN 2079] (111-40-0)
Diethylenetriamine, cyanoguanidine, epichlorohydrin polymer (67953-54-2)
Diethylenetriamine, ethylenedichloride, oleylamine polymer (67905-86-6)
Diethylenetriaminepentaacetate, pentasodium salt (140-01-2)
1,1,4,7,7-Diethylenetriaminepentaacetic acid (67-43-6)
Diethylenetriaminepentaacetic acid (67-43-6)
Diethylenetriamine pentaacetic acid, calcium trisodium salt (12111-24-9)
Diethylenetriaminepentaacetic acid pentasodium salt (140-01-2)
Diethylenetriaminepentaacetic acid sodium salt (140-01-2)
Diethylenetriamine, 1,1,4,7,7-pentamethyl (3030-47-5)
Diethylenetriamine, pentamethylenepentaphosphonic acid (15827-60-8)
Diethylenetriaminepenta(methylenephosphonic) acid (15827-60-8)
(Diethylenetrinitrilo)pentaacetic acid (67-43-6)
Diethylenglykol (Czech) (111-46-6)
Diethylenglykoldinitrate (Czech) (693-21-0)
Diethylenimide oxide (110-91-8)
Diethylester kyseliny acetylaminomalonove (Czech) (1068-90-2)
Diethylester kyseliny adipove (Czech) (141-28-6)
Diethylester kyseliny ftalove (Czech) (84-66-2)
Diethylester kyseliny fumarove (Czech) (623-91-6)
Diethylester kyseliny jantarove (Czech) (123-25-1)
Diethylester kyseliny maleinove (Czech) (141-05-9)
Diethylester kyseliny sirove (Czech) (64-67-5)
Diethylester kyseliny stavelove (Czech) (95-92-1)
Diethylester kyseliny uhlicite (Czech) (105-58-8)
N,N-Diethylethanamine (121-44-8)
N,N-Diethylethanamine, Compd. with ((3,5,6-trichloro-2-pyridinyl) oxy)acetic acid (1:1) (57213-69-1)
N,N-Diethylethanamine hydrobromide (636-70-4)
N,N-Diethylethanamine phosphate (10138-93-9)
N,N-Diethyl-1,2-ethanediamine (100-36-7)
N,N-Diethylethanediamine (100-36-7)
Diethyl ethanediimidate (13534-15-1)
Diethyl ethanedioate (95-92-1)
Diethylethanolamine (100-37-8)
N,N-Diethylethanolamine (100-37-8)
4,4'-(1,2-Diethyl-1,2-ethenediyl)bis-phenol (56-53-1)
trans-4,4'-(1,2-Diethyl-1,2-ethenediyl)bisphenol (56-53-1)
trans-4,4'-(1,2-Diethyl-1,2-ethenediyl)bisphenol dipropionate (130-80-3)
Diethyl ether (60-29-7)
Diethyl ether (OSHA) [UN 1155] (60-29-7)
Diethylether diethylenglykolu (Czech) (112-36-7)
Diethylether ethylenglykolu (Czech) (629-14-1)
Diethylether tetraethylenglykolu (Czech) (4353-28-0)
Diethyl S-(2-ethioethyl)thiophosphate (126-75-0)
O,O-Diethyl S-(N-ethoxycarbonyl-N-methylcarbamoylmethyl) phosphorodithioate (2595-54-2)
O,O-Diethyl S-(N-ethoxycarbonyl-N-methylcarbamoylmethyl) phosphorothiolothionate (2595-54-2)
Diethylethoxymethyleneoxalacetate (52942-64-0)
O,O-Diethyl S-(2-ethsulfonylethyl) phosphorodithioate (2497-06-5)
O,O-Diethyl S-(2-ethsulfonylethyl)phosphorothioate (2496-91-5)
Diethyl S-(2-ethsulfonylethyl) thiophosphate (2496-91-5)
O,O-Diethyl S-(2-ethsulfonylethyl) thiothionophosphate (2497-06-5)
O,O-Diethyl S-ethsulfonylmethyl thiothionophosphate (2588-04-7)
O,O-Diethyl S-(2-eththioethyl) phosphorodithioate (298-04-4)
O,O-Diethyl O-(2-eththioethyl)phosphorothioate (298-03-3)
O,O-Diethyl S-(2-eththioethyl)phosphorothioate (126-75-0)
Diethyl 2-eththioethyl thionophosphate (298-03-3)
O,O-Diethyl S-(2-eththioethyl) thiothionophosphate (298-04-4)
O,O-Diethyl S-eththionylmethyl phosphorothioate (2588-05-8)
Diethyl S-eththionylmethyl thiophosphate (2588-05-8)
O,O-Diethyl S-eththionylmethyl thiothionophosphate (2588-03-6)
O,O-Diethyl S-(2-ethyl-N,N-diethylamino)phosphorothioate hydrogen oxalate (3734-97-2)
N,N-(Diethylethyl)diamine (100-36-7)
N,N-(Diethylethylene)diamine (100-36-7)
N,N-Diethylethylene diamine (100-36-7)
N,N-Diethylethylenediamine (100-36-7)
N,N-Diethylethylenediamine (French) (100-36-7)
4,4'-(1,2-Diethylethylene)diphenol (84-16-2)
O,O-Diethyl S-ethyl-2-ethylmercaptophosphorothiolate (126-75-0)
O,O-Diethyl S-ethyl 2-ethylmercaptophosphorothiolate sulfone (2496-91-5)
Diethyl 2-ethylmalonate (133-13-1)
Diethyl ethylmalonate (133-13-1)
O,O-Diethyl S-(2-ethylmercaptoethyl) dithiophosphate (298-04-4)
O,O-Diethyl 2-ethylmercaptoethyl thiophosphate (8065-48-3)

O,O-Diethyl 2-ethylmercaptoethyl thiophosphate, thiolo isomer (2496-91-5)
O,O-Diethyl 2-ethylmercaptoethyl thiophosphate, thiono isomer (4891-54-7)
O,O-Diethyl S-ethylmercaptomethyl dithiophosphonate (298-02-2)
Diethyl ethylphosphonate (78-38-6)
O,O-Diethyl S-ethyl phosphorothioate (1186-09-0)
O,O-Diethyl O(and S)-2-(ethylthio)ethyl phosphorothioate Mixture (8065-48-3)
Diethyl ethylpropanedioate (133-13-1)
O,O-Diethyl S-(2-(ethylsulfinyl)ethyl) phosphorodithioate (2497-07-6)
O,O-Diethyl-S-((ethylsulfinyl)ethyl)phosphorodithioate (2497-07-6)
O,O-Diethyl-S-(2-ethylsulfonylethyl)phosphorodithioate (2497-06-5)
O,O-Diethyl-O-(2-ethylsulfonylethyl)phosphorothioate (4891-54-7)
Diethyl 2-ethylsulfonylethyl thionophosphate (4891-54-7)
O,O-Diethyl S-(2-ethylsulfonylethyl)thionophosphate (2497-06-5)
O,O-Diethyl-S-ethylsulfonyl methylphosphorodithioate (2588-04-7)
O,O-Diethyl S-ethylsulfonylmethyl thionophosphate (2588-04-7)
O,O-Diethyl-S-(2-ethylthio-ethyl)-dithiofosfaat (Dutch) (298-04-4)
O,O-Diethyl-S-(2-ethylthio-ethyl)-monothiofosfaat (Dutch) (126-75-0)
O,O-Diethyl 2-ethylthioethyl phosphorodithioate (298-04-4)
O,O-Diethyl S-2-(ethylthio)ethyl phosphorodithioate (298-04-4)
O,O-Diethyl O-2-(ethylthio)ethyl phosphorothioate (298-03-3)
O,O-Diethyl S-2-(ethylthio)ethyl phosphorothioate (126-75-0)
O,O-Diethyl-2-ethylthio ethyl phosphorothioate (298-03-3)
O,O-Diethyl S-(2-(ethylthio)ethyl) phosphorothiolate (126-75-0)
Diethyl 2-(ethylthio)ethyl phosphorothionate (298-03-3)
O,O-Diethyl-S-(ethylthio-methyl)-dithiofosfaat (Dutch) (298-02-2)
O,O-Diethyl S-ethylthiomethyl dithiophosphonate (298-02-2)
O,O-Diethyl S-(ethylthio)methyl phosphorodithioate (298-02-2)
O,O-Diethyl ethylthiomethyl phosphorodithioate (298-02-2)
O,O-Diethyl S-ethylthiomethyl thiothionophosphate (298-02-2)
O,O-Diethyl-S-ethylthionylmethylphosphorodithioate (2588-03-6)
O,O-Diethyl-S-ethylthionylmethyl thionophosphate (2588-03-6)
N,N'-Diethyl-p-fenylendiamin (Czech) (93-05-0)
O,O-Diethyl O-(6-fluoro-2-pyridyl ester phosphorothioic) acid (39624-86-7)
Diethylformal (462-95-3)
Diethyl formamide (617-84-5)
Diethylfosfit (Czech) (762-04-9)
Diethyl fumarate (623-91-6)
Diethyl glutarate (818-38-2)
Diethyl glycol dimethyl ether (111-96-6)
Diethylgold bromide (26645-10-3)
N,N-Diethyl-1,3,4,6,7,11b-hexahydro-2-hydroxy-9,10-dimethoxy-2H-benzo(a)quinolizine-3-carboxamide acetate (63-12-7)
N,N-Diethyl-1,3,4,6,7,11b-hexahydro-2-hydroxy-9,10-dimethoxy-2H-benzo(a)quinolizine-3-carboxamide acetate (ester) (63-12-7)
Diethyl hexanedioate (141-28-6)
Di(2-ethylhexyl) adipate (123-79-5)
Di-2-ethylhexyl adipate (103-23-1)
Di(2-ethylhexyl)amine (106-20-7)
Di-2-ethylhexyl chlorendate (4827-55-8)
O,O'-Di(2-ethylhexyl) dithiophosphoric acid (5810-88-8)
Di-(2-ethylhexyl) ether (10143-60-9)
Di(2-ethylhexyl) fumarate (141-02-6)
Di-2-ethylhexyl isophthalate (137-89-3)
Di-(2-ethylhexyl)maleate (142-16-5)
Di(2-ethylhexyl)peroxydicarbonate (16111-62-9)
Di-(2-ethylhexyl)peroxydicarbonate, 77% In solution (DOT) (16111-62-9)
Di-(2-ethylhexyl)peroxydicarbonate, Maximum concentration 32% (DOT) (16111-62-9)
Di-(2-ethylhexyl)peroxydicarbonate, Technically pure (DOT) (16111-62-9)
Di(2-ethylhexyl)phenyl phosphate (16368-97-1)
Di(2-ethylhexyl)phosphate (298-07-7)
Di(2-ethylhexyl)orthophosphoric acid (298-07-7)
Di-2(ethylhexyl)phosphoric acid (298-07-7)
Di(2-ethylhexyl)phosphoric acid (DOT) (298-07-7)
Di(2-ethylhexyl)orthophthalate (117-81-7)
Di(2-ethylhexyl)phthalate (117-81-7)
Di-2-ethylhexylphthalate (OSHA) (117-81-7)
Di(2-ethylhexyl)sebacate (122-62-3)
Di-(2-ethylhexyl) sodium sulfosuccinate (577-11-7)
5,5-Diethylhydantoin (5455-34-5)
1,2-Diethylhydrazine (1615-80-1)
N-N'-Diethylhydrazine (1615-80-1)
sym-Diethylhydrazine (1615-80-1)
1,2-Diethylhydrazine dihydrochloride (7699-31-2)
Diethylhydroaluminum (871-27-2)
Diethyl hydrogen phosphite (762-04-9)
N,N-Diethylhydroxyamine (3710-84-7)
Diethyl hydroxybutanedioate (7554-12-3)
N,N-Diethyl-N-(β-hydroxyethyl)amine (100-37-8)
Diethyl ((N,N-bis(2-hydroxyethyl)amino)methyl)phosphonate (2781-11-5)
Diethyl(2-hydroxyethyl)methylammoniumbromide xanthene-9-carboxylate (53-46-3)
Diethylhydroxylamine (3710-84-7)
N,N-Diethylhydroxylamine (3710-84-7)
4,4'-(1,2-Diethylidene-1,2-ethanediyl)bisphenol (84-17-3)
4,4'-(Diethylideneethylene)diphenol (84-17-3)
p,p'-(Diethylideneethylene)diphenol (84-17-3)
Diethyliodoaluminum (2040-00-8)
Diethyl isobutylmalonate (10203-58-4)
Diethyl isophthalate (636-53-3)
O,O-Diethyl S-(N-isopropylcarbamoylmethyl) dithiophosphate (2275-18-5)
O,O-Diethyl S-(N-isopropylcarbamoylmethyl) phosphorodithioate (2275-18-5)
O,O-Diethyl S-isopropylcarbamoylmethyl phosphorodithioate (2275-18-5)
O,O-Diethyl-S-2-isopropylmercaptomethyl dithiophosphate (78-52-4)
O,O-Diethyl S-(isopropylmercaptomethyl) phosphorodithioate (78-52-4)
O,O-Diethyl-o-(2-isopropyl-4-methyl-pyrimidin-6-yl)-monothiofosfaat (Dutch) (333-41-5)
O,O-Diethyl O-(2-isopropyl-6-methyl-4-pyrimidinyl) phosphorothioate (333-41-5)
O,O-Diethyl-O-(2-isopropyl-4-methyl-6-pyrimidinyl)-phosphorothioate (333-41-5)
Diethyl 4-(2-isopropyl-6-methylpyrimidinyl)phosphorothionate (333-41-5)
O,O-Diethyl-O-(2-isopropyl-4-methyl-6-pyrimidyl)phosphorothioate (333-41-5)
O,O-Diethyl O-(2-isopropyl-4-methyl-6-pyrimidyl) thionophosphate (333-41-5)
O,O-Diethyl 2-isopropyl-4-methylpyrimidyl-6-thiophosphate (333-41-5)
O,O-Diethyl O-(2-isopropyl-4-methylpyrimid-6-yl)phosphate (962-58-3)
O,O-Diethyl-S-(isopropylthio)methylester kyseliny dithiofosforecne (Czech) (78-52-4)
O,O-Diethyl S-(isopropylthiomethyl) phosphorodithioate (78-52-4)
Diethylkarbonat (Czech) (105-58-8)
O,O-Diethyl O-(2-keto-4-methyl-7-α',β'-benzo-α'-pyranyl) thiophosphate (299-45-6)
Diethyl ketone (ACGIH,OSHA) [UN 1156] (96-22-0)
Diethyllauramide (3352-87-2)
N,N-Diethyllauramide (3352-87-2)
N,N-Diethyllaurylamide (3352-87-2)
Diethyllead (24952-65-6)
Diethyllead dichloride (13231-90-8)
N,N-Diethyllysergamide (50-37-3)
Diethyl malate (7554-12-3)
Diethyl maleate (141-05-9)
Diethyl malonate (105-53-3)
Diethylmalonylurea (57-44-3)
Diethyl mercaptosuccinate, O,O-dimethyl dithiophosphate, S-ester (121-75-5)
Diethyl mercaptosuccinate, O,O-dimethyl phosphorodithioate (121-75-5)
Diethyl mercaptosuccinate, O,O-dimethyl thiophosphate (121-75-5)
Diethyl mercaptosuccinate S-ester with O,O-dimethylphosphorodithioate (121-75-5)
Diethyl mercaptosuccinic acid O,O-dimethyl phosphorodithioate (121-75-5)
Diethyl mercury (627-44-1)
N,N-Diethylmetanilan sodny (Czech) (5123-63-7)
N,N-Diethyl-3-methylaniline (91-67-8)
5,5-Diethyl-1-methylbarbituric acid (50-11-3)
N,N-Diethyl-3-methylbenzamide (134-62-3)

N,N-Diethyl-3-methylbenzenamine **(91-67-8)**
1,3-Diethyl-5-methylbenzene **(2050-24-0)**
N4,N4-Diethyl-2-methyl-1,4-benzenediamine **(148-71-0)**
ar,ar-Diethyl-ar-methylbenzenediamine **(68479-98-1)**
O,O-Diethyl S-(N-methyl-N-carboethoxycarbamoylmethyl) dithiophosphate **(2595-54-2)**
O,O-Diethyl-O-(4-methylcoumarin-7-yl)-monothiofosfaat (Dutch) **(299-45-6)**
O,O-Diethyl O-(4-methyl-7-coumarinyl) phosphorothioate **(299-45-6)**
O,O-Diethyl O-(4-methyl-7-coumarinyl) thionophosphate **(299-45-6)**
Diethyl methylcoumarinyl thiophosphate **(299-45-6)**
O,O-Diethyl O-(4-methylcoumarinyl-7) thiophosphate **(299-45-6)**
O,O-Diethyl-S-(3-methyl-2,4-dioxo-5-oxa-3-aza-heptyl)-dithio-fosfaat (Dutch) **(2595-54-2)**
N,N-Diethyl-1-methyl-3,3-di-2-thienylallylamine **(86-14-6)**
Diethyl (4-methyl-1,3-dithiolan-2-ylidene)phosphoroamidate **(950-10-7)**
O,O-Diethyl-S-methyl-dithiophosphate **(3288-58-2)**
2,6-Diethyl-N-methylenebenzenamine **(35203-08-8)**
O,O-Diethyl-O-(5-methyl-6-ethoxy-carbonyl-pyrazolo-(1.5-a)pyrimid-2-yl)-thionophosphate **(13457-18-6)**
O,O-Diethyl O-6-methyl-2-isopropyl-4-pyrimidinyl phosphorothioate **(333-41-5)**
O,O-Diethyl-O-(4-methyl-7-kumarinyl) ester kyseliny thiofosforescne (Czech) **(299-45-6)**
1,1-Diethyl-3-methyl-3-nitrosourea **(50285-72-8)**
N,N-Diethyl-N-methyl-2-((1-oxo-2-propenyl)oxy)-ethanaminium, methyl sulfate **(21810-39-9)**
N,N-Diethyl-N-methyl-2-(1-oxo-2-propenyloxy)ethanaminium methyl sulfate **(21810-39-9)**
N,N'-Di(1-ethyl-3-methylpentyl)-p-phenylenediamine **(139-60-6)**
Diethyl methylphosphonate **(683-08-9)**
O,S-Diethyl methylphosphonothioate **(2511-10-6)**
O,O-Diethyl S-methyl phosphorothioate **(2404-05-9)**
N,N-Diethyl-4-methyl-1-piperazine carboxamide citrate **(1642-54-2)**
N,N-Diethyl-4-methyl-1-piperazinecarboxamide dihydrogen citrate **(1642-54-2)**
3,3-Diethyl-5-methyl-2,4-piperidinedione **(125-64-4)**
3,3-Diethyl-5-methylpiperidine-2,4-dione **(125-64-4)**
Diethyl methylpropanedioate **(609-08-5)**
Diethyl methylpropylmalonate **(55898-43-6)**
2,3-Diethyl-5-methylpyrazine **(18138-04-0)**
O,O-Diethyl-O-(3-methyl-1H-pyrazol-5-yl)-fosfaat (Dutch) **(108-34-9)**
Diethyl 3-methyl-5-pyrazolyl phosphate **(108-34-9)**
O,O-Diethyl O-(3-methyl-5-pyrazolyl) phosphate **(108-34-9)**
O,O-Diethyl O-(3-methyl-5-pyrazolyl) phosphorothionate **(108-35-0)**
Diethyl 3-methyl-5-pyrazolyl thionophosphate **(108-35-0)**
O,O-Diethyl O-(p-(methylsulfinyl)phenyl) phosphorothioate **(115-90-2)**
O,O-Diethyl O-p-(methylsulfinyl)phenyl thiophosphate **(115-90-2)**
O,O-Diethyl O-(p-methylsulfonyl)phenyl phosphorothioate **(14255-72-2)**
O,S-Diethyl methylthiophosphonate **(2511-10-6)**
N,N'-Diethyl-6-(methylthio)-1,3,5-triazine-2,4-diamine **(1014-70-6)**
O,O-Diethyl O-(4-methylumbelliferone) ester of thiophosphoric acid **(299-45-6)**
O,O-Diethyl O-(4-methylumbelliferone) phosphorothioate **(299-45-6)**
Diethyl (4-methylumbelliferyl) thionophosphate **(299-45-6)**
Diethylmethylvinylsilane **(18292-29-0)**
N,N-Diethyl-2-(1-naphthalenyloxy)propionamide **(15299-99-7)**
N,N-Diethyl-2-(1-naphthyloxy)propionamide **(15299-99-7)**
Diethyl-nicotamide **(59-26-7)**
N,N-Diethylnicotinamide **(59-26-7)**
O,O-Diethyl-O-(4-nitro-fenil)-monothiofosfaat (Dutch) **(56-38-2)**
Diethyl-p-nitrofenyl ester kyseliny fosforecne (Czech) **(311-45-5)**
O,O-Diethyl-O-p-nitrofenylester kyseliny thiofosforecne (Czech) **(56-38-2)**
O,O-Diethyl-S-p-nitrofenylester kyseliny thiofosforecne (Czech) **(3270-86-8)**
O,S-Diethyl-O-p-nitrofenyl ester kyseliny thiofosforecne (Czech) **(597-88-6)**
O,O-Diethyl-S-p-nitrofenylthiofosfat (Czech) **(3270-86-8)**
O,S-Diethyl-O-p-nitrofenylthiofosfat (Czech) **(597-88-6)**
Diethyl para-nitrophenol thiophosphate **(56-38-2)**
Diethyl p-nitrophenyl phosphate **(311-45-5)**
O,O-Diethyl O-p-nitrophenyl phosphate **(311-45-5)**
O,O-Diethyl O-(p-nitrophenyl) phosphorothioate **(56-38-2)**
O,O-Diethyl S-(4-nitrophenyl) phosphorothioate **(3270-86-8)**
O,O-Diethyl-O-(4-nitrophenyl) phosphorothioate **(56-38-2)**
O,S-Diethyl O-(4-nitrophenyl)phosphorothioate **(597-88-6)**
O,S-Diethyl O-(p-nitrophenyl) phosphorothioate **(597-88-6)**
Diethyl 4-nitrophenyl phosphorothionate **(56-38-2)**
Diethyl p-nitrophenyl phosphorothionate **(56-38-2)**
Diethyl p-nitrophenyl thionophosphate **(56-38-2)**
O,O-Diethyl-O-(p-nitrophenyl)thionophosphate **(56-38-2)**
Diethyl p-nitrophenyl thiophosphate **(56-38-2)**
O,O-Diethyl O-4-nitrophenyl thiophosphate **(56-38-2)**
O,O-Diethyl O-p-nitrophenyl thiophosphate **(56-38-2)**
O,O-Diethyl S-(4-nitrophenyl)thiophosphate **(3270-86-8)**
O,S-Diethyl O-(4-nitrophenyl)thiophosphate **(597-88-6)**
Diethylnitrosamide **(55-18-5)**
Diethylnitrosamine **(55-18-5)**
N,N-Diethylnitrosamine **(55-18-5)**
Diethylnitrosoamine **(55-18-5)**
Diethyl nonanedioate **(624-17-9)**
cis,trans-4,5-Diethyl-3,5-octadiene **(21293-02-7)**
Diethylolamine **(111-42-2)**
Diethylol dimethyl hydantoin **(26850-24-8)**
Diethyl oxalate **(95-92-1)**
5,5-Diethyl-1,3-oxazin-2,4-dione **(702-54-5)**
5,5-Diethyl-1,3-oxazine-2,4-dione **(702-54-5)**
Diethyl oxide **(60-29-7)**
O,O-Diethyl S-(4-oxobenzotriazino-3-methyl)phosphorodithioate **(2642-71-9)**
O,O-Diethyl-S-(4-oxo-3H-1,2,3-benzotriazine-3-yl)-methyl-dithio-phosphate **(2642-71-9)**
O,O-Diethyl-S-((4-oxo-3H-1,2,3-benzotriazin-3-yl)-methyl)-dithio-fosfaat (Dutch) **(2642-71-9)**
Diethyl (3-oxobutyl)phosphonate **(1067-90-9)**
Diethyl paraoxon **(311-45-5)**
Diethylparathion **(56-38-2)**
3,3'-Diethylpentamethinethiacyanine iodide **(514-73-8)**
3,3-Diethylpentane **(1067-20-5)**
Diethyl pentanedioate **(818-38-2)**
2,2-Diethyl-1-pentanol **(14202-62-1)**
Diethyl peroxide **(628-37-5)**
Diethyl peroxydicarbonate **(14666-78-5)**
Diethyl peroxydicarbonate (More than 27% in solution) **(14666-78-5)**
Diethyl peroxydicarbonate, Not more than 27% in solution **(14666-78-5)**
Diethyl peroxydiformate **(14666-78-5)**
Diethylphenylamine **(91-66-7)**
N,N-Diethyl-p-(phenylazo)aniline **(2481-94-9)**
N,N-Diethyl-4-(phenylazo)benzenamine **(2481-94-9)**
Di(p-ethylphenyl)dichloroethane **(72-56-0)**
Diethyl 4,4'-o-phenylenebis(3-thioallophanate) **(23564-06-9)**
Diethyl-para-phenylenediamine **(93-05-0)**
N,N'-Diethyl-p-phenylenediamine **(93-05-0)**
N,N-Diethyl-p-phenylenediamine hydrochloride **(2198-58-5)**
O,O-Diethyl O-(5-phenyl-3-isoxazolyl) phosphorothioate **(18854-01-8)**
O,O-Diethyl O-(3-(5-phenyl)-1,2-isoxazolyl)phosphorothionate **(18854-01-8)**
O,O-Diethyl-O-phenylphosphorothioate **(32345-29-2)**
O,O-Diethyl O-(1-phenyl-1H-1,2,4-triazol-3-yl)phosphorothioate **(24017-47-8)**
Diethyl phosphate **(598-02-7)**
Diethyl phosphite **(762-04-9)**
O,O-Diethyl phosphoric acid O-p-nitrophenyl ester **(311-45-5)**
O,O-Diethylphosphorochloridothioate **(2524-04-1)**
O,O-Diethyl phosphorodithioate, S-ester with 6-chloro-3-(mercapto-methyl)-2-benzoxazolinone **(2310-17-0)**
O,O-Diethyl phosphorodithioate s-ester with 3-(mercaptomethyl)-1,2,3-benzotriazin-4(3H)-one **(2642-71-9)**
O,O-Diethylphosphorothioate, O-ester with 6-hydroxy-2-phenyl-3(2H)-pyridazinone **(119-12-0)**
O,O-Diethyl phosphorothioate, O-ester with phenylglyoxylo-nitrile oxime **(14816-18-3)**

O,O-Diethyl phosphorothionate **(2465-65-8)**
Diethyl phthalate (ACGIH,OSHA) **(84-66-2)**
O,O-Diethyl phthalimidothiophosphate **(5131-24-8)**
O,O-Diethyl phthalimido-phosphonothioate **(5131-24-8)**
Diethylplumbium dichloride **(13231-90-8)**
Diethyl propanedioate **(105-53-3)**
2,2-Diethyl-1,3-propanediol **(115-76-4)**
2,2-Diethylpropane-1,3-diol **(115-76-4)**
2,2-Diethylpropanediol-1,3 **(115-76-4)**
Diethylpropion **(90-84-6)**
Diethylpropione **(90-84-6)**
O,O-Diethyl-O-(2-propyl-4-methylpyrimidinyl-6) phosphorothioate **(5826-91-5)**
O,O-Diethyl-O-(2-n-propyl-4-methyl-pyrimidyl-6)phosphorothioate **(5826-91-5)**
Diethyl propylmethylpyrimidyl thiophosphate **(5826-91-5)**
2,3-Diethylpyrazine **(15707-24-1)**
O,O-Diethyl O,2-pyrazinyl phosphorothioate **(297-97-2)**
Diethyl O-2-pyrazinyl phosphorothionate **(297-97-2)**
O,O-Diethyl O-2-pyrazinyl phosphothionate **(297-97-2)**
O,O-Diethyl O-pyrazinyl thiophosphate **(297-97-2)**
N,N-Diethyl-3-pyridinecarboxamide **(59-26-7)**
Diethylquinoline **(68228-10-4)**
O,O-Diethyl O-quinoxalin-2-yl phosphorothioate **(13593-03-8)**
O,O-Diethyl-O-(2-quinoxalinyl) phosphorothioate **(13593-03-8)**
O,O-Diethyl-O-(2-quinoxalyl) phosphorothioate **(13593-03-8)**
O,O-Diethyl-O-2-quinoxalyl-thiophosphate **(13593-03-8)**
Diethyl sebacate **(110-40-7)**
Diethyl selenide **(627-53-2)**
Diethylselenide **(627-53-2)**
Diethylselenium **(627-53-2)**
Diethyl sodium dithiocarbamate **(148-18-5)**
α,α'-Diethyl-(E)-4,4'-stilbenediol **(56-53-1)**
α,α'-Diethyl-4,4'-stilbenediol **(56-53-1)**
α,α'-Diethylstilbenediol **(56-53-1)**
trans-α,α'-Diethyl-4,4'-stilbenediol **(56-53-1)**
α,α'-Diethyl-(E)-4,4'-stilbenediol bis(dihydrogen phosphate) **(522-40-7)**
α,α'-Diethyl-4,4'-stilbenediol trans-dipropionate **(130-80-3)**
α,α'-Diethyl-4,4'-stilbenediol, dipropionate **(130-80-3)**
trans-α,α'-Diethyl-4,4'-stilbenediol dipropionate **(130-80-3)**
α,α'-Diethyl-4,4'-stilbenediol dipropionyl ester **(130-80-3)**
Diethylstilbene dipropionate **(130-80-3)**
Diethylstilbesterol **(56-53-1)**
trans-Diethylstilbesterol **(56-53-1)**
Diethylstilbesterol diphosphate **(522-40-7)**
Diethylstilbesterol dipropionate **(130-80-3)**
Diethylstilbestrol **(56-53-1)**
trans-Diethylstilbestrol **(56-53-1)**
Diethylstilbestrol diphosphate **(522-40-7)**
Diethylstilbestrol dipropionate **(130-80-3)**
Diethylstilbestrol phosphate **(522-40-7)**
Diethylstilbestrol propionate **(130-80-3)**
Diethylstilbestryl diphosphate **(522-40-7)**
Diethylstilboesterol **(56-53-1)**
trans-Diethylstilboesterol **(56-53-1)**
N,N-Diethylsuccinamic acid **(1522-00-5)**
Diethyl succinate **(123-25-1)**
N,N-Diethylsuccinic acid monoamide **(1522-00-5)**
Diethyl sulfate **(64-67-5)**
Diethylsulfid (Czech) **(352-93-2)**
Diethyl sulfide [UN 2375] **(352-93-2)**
Diethyl sulfide 2,2'-dicarboxylic acid **(111-17-1)**
Diethylsulfonate **(1912-30-7)**
Diethylsulfondimethylmethane **(115-24-2)**
Diethylsulfonmethylethylmethane **(76-20-0)**
Diethyl sulphate [UN 1594] **(64-67-5)**
Diethyl terephthalate **(636-09-9)**
Diethyl terephthaloylbisacetate **(93-94-7)**
5,10-Diethyl-7-tetradecyn-6,9-diol **(25430-52-8)**
5,10-Diethyl-7-tetradecyne-6,9-diol **(25430-52-8)**
5,5-Diethyltetrahydro-2H-1,3-oxazine-2,4(3H)-dione **(702-54-5)**
3,3'-Diethyl-2,2'-thiadicarbocyanine iodide **(514-73-8)**
3,3'-Diethylthiadicarbocyanine iodide **(514-73-8)**
3,3-Diethylthiadicarbocyanine iodide **(514-73-8)**
Diethylthiadicarbocyanine iodide **(514-73-8)**
Diethylthiambutene **(86-14-6)**
Diethylthiambutenum (Latin) **(86-14-6)**
Diethylthiocarbamic acid S-(o-chlorobenzyl) ester **(34622-58-7)**
N,N'-Diethylthiocarbamide **(105-55-5)**
Diethylthioether **(352-93-2)**
2-(O,O-Diethyl-thionophosphoryl)-5-methyl-6-carbethoxy-pyrazolo-(1,5a)pyrimidine **(13457-18-6)**
Diethyl thiophosphoric acid ester of 3-chloro-4-methyl-7-hydroxycoumarin **(56-72-4)**
Diethylthiophosphoryl chloride [UN 2751] **(2524-04-1)**
1,3-Diethyl-2-thiourea **(105-55-5)**
1,3-Diethylthiourea **(105-55-5)**
N,N'-Diethylthiourea **(105-55-5)**
Diethyl-m-toluamide **(134-62-3)**
Diethyltoluamide **(134-62-3)**
N,N-Diethyl-m-toluamide **(134-62-3)**
3,5-Diethyltoluene **(2050-24-0)**
Diethyltoluenediamine **(68479-98-1)**
N,N-Diethyl-m-toluidine **(91-67-8)**
N,N-Diethyl-m-toluidinium ion **(91-67-8)**
O,O-Diethyl O-3,5,6-trichloro-2-pyridyl phosphorothioate **(2921-88-2)**
3,9-Diethyl-tridecyl-6-sulfate **(3282-85-7)**
N^4,N^4-Diethyl-α,α,α-trifluoro-3,5-dinitrotoluene, 2,4-diamine **(29091-05-2)**
O,O-Diethyl S-2-trimethylammonium ethylphosphonothiolate iodide **(513-10-0)**
Diethyltryptamine **(61-51-8)**
N,N-Di-ethyltryptamine **(61-51-8)**
N,N-Diethyltryptamine **(61-51-8)**
1,1-Diethylurea **(634-95-7)**
Diethylurea **(50816-31-4)**
N,N-Diethylurea **(634-95-7)**
asym-Diethylurea **(634-95-7)**
Diethyl xanthogenate **(502-55-6)**
Diethylxanthogen disulfide **(502-55-6)**
Diethyl zinc **(557-20-0)**
Diethylzinc [UN 1366] **(557-20-0)**
Dietil **(139-06-0)**
Dietilamide-carbopiridina **(59-26-7)**
Dietilamina (Italian) **(109-89-7)**
alfa-Dietilamino-2,6-dimetilacetanilide (Italian) **(137-58-6)**
O,O-Dietil-O-(4-bromo-2,5 dicloro-fenil)-monotiofosfato (Italian) **(4824-78-6)**
O,O-Dietil-S-((4-cloro-fenil-tio)-metile)-ditiofosfato (Italian) **(786-19-6)**
O,O-Dietil-O-(3-cloro-4-metil-cumarin-7-il-monotiofosfato) (Italian) **(56-72-4)**
O,O-Dietil-S-((6-cloro-2-oxo-benzossazolin-3-il)-metil)-ditiofosfato (Italian) **(2310-17-0)**
O,O-Dietil-O-(2,4-dicloro-fenil)-monotiofosfato (Italian) **(97-17-6)**
5,5-Dietildiidro-1,3-ossazin-2,4-dione (Italian) **(702-54-5)**
Dietilestilbestrol (Spanish) **(56-53-1)**
N,N-Dietiletilendiamina (Spanish) **(100-36-7)**
O,O-Dietil-S-(2-etiltio-etil)-ditiofosfato (Italian) **(298-04-4)**
O,O-Dietil-S-(2-etiltio-etil)-monotiofosfato (Italian) **(126-75-0)**
O,O-Dietil-S-(etiltio-metil)-ditiofosfato (Italian) **(298-02-2)**
O,O-Dietil-S-(N-etossi-carbonil-N-metil-carbamoil-metil)-ditiofosfato (Italian) **(2595-54-2)**
O,O-Dietil-O-(2-isopropil-4-metil-pirimidin-6-il)-monotiofosfato (Italian) **(333-41-5)**
O,O-Dietil-O-(4-metilcumarin-7-il)-monotiofosfato (Italian) **(299-45-6)**
O,O-Dietil-O-(3-metil-1H-pirazol-5-il)-fosfato (Italian) **(108-34-9)**
O,O-Dietil-O-(4-nitro-fenil)-monotiofosfato (Italian) **(56-38-2)**
O,O-Dietil-S-((4-oxo-3H-1,2,3-benzotriazin-3-il)-metil)-ditiofosfato (Italian) **(2642-71-9)**

Dietilpropandiolo (115-76-4)
Dietiltiambutene (86-14-6)
Dietiltiambuteno (Spanish) (86-14-6)
Dietil tiofosfato de p-nitrofenila (Portuguese) (56-38-2)
1,1-Dietossietano (Italian) (105-57-7)
Dietreen (22248-79-9)
Dietroxine (702-54-5)
O,O-Dietyl-S-2-etylmerkaptoetyltiofosfat (Czech) (126-75-0)
O,O-Dietyl-O-4-methylkumarinyl(7)tiofosfat (Czech) (299-45-6)
O,O-Dietyl-O-p-nitrofenylfosfat (Czech) (311-45-5)
O,O-Dietyl-O-p-nitrofenyltiofosfat (Czech) (56-38-2)
Difacil (64-95-9)
Difedryl (58-73-1)
Difenamid (Czech) (957-51-7)
Difenano (Spanish) (101-71-3)
Difenhydramin (58-73-1)
Difenhydramine hydrochloride (147-24-0)
Difenidol (972-02-1)
Difenidramina (Italian) (58-73-1)
Difenilhidantoina (Spanish) (57-41-0)
Difenil-metan-diisocianato (Italian) (101-68-8)
Difenin (57-41-0)
Difenin (630-93-3)
Difenossina (28782-42-5)
Difenoxin (28782-42-5)
Difenoxina (Spanish) (28782-42-5)
Difenoxine (French) (28782-42-5)
Difenoxinum (Latin) (28782-42-5)
Difenoxuron (14214-32-5)
1,5-Difenoxyanthrachinon (Czech) (82-21-3)
Difenson (80-33-1)
Difenthos (3383-96-8)
Difenylacetonitril (Czech) (86-29-3)
Difenylamin (Czech) (122-39-4)
Difenylchlorarsin (Czech) (712-48-1)
N,N'-Difenyl-p-fenylendiamin (Czech) (74-31-7)
1,3-Difenylguanid (Czech) (102-06-7)
Difenylin (492-17-1)
Difenylmethaan-dissocyanaat (Dutch) (101-68-8)
N,N'-Difenylmocovina (Czech) (102-07-8)
Difenylnitrosamin (Czech) (86-30-6)
Difenylrtut (Czech) (587-85-9)
Difenylstanniumdichlorid (Czech) (1135-99-5)
Difenylsulfon (Czech) (127-63-9)
1,3-Difenylthiomocovina (Czech) (102-08-9)
Difenzoquat (49866-87-7)
Difenzoquat methyl sulfate (43222-48-6)
Diferuloylmethane (458-37-7)
Difetoin (57-41-0)
Difetoin (630-93-3)
Diffollisterol (50-50-0)
Diffusil H (52645-53-1)
Difhydan (57-41-0)
Difhydan (630-93-3)
Diflubenzuron (35367-38-5)
Difluoramine (10405-27-3)
Difluoretano (Spanish) (25497-28-3)
Difluorethane (French) (25497-28-3)
Difluoroacetic acid (381-73-7)
2,4-Difluoroaniline (367-25-9)
m-Difluorobenzene (372-18-9)
o-Difluorobenzene (367-11-3)
para-Difluorobenzene (540-36-3)
2,6-Difluorobenzoic acid (385-00-2)
1,1-Difluoro-1-chloroethane (75-68-3)
1,1-Difluorochloroethylene (359-10-4)
Difluorochloromethane (75-45-6)
2,3-Difluorodibenzo(b,e)(1,4)dioxin (50585-38-1)
2,3-Difluorodibenzo-p-dioxin (50585-38-1)
Difluorodibromomethane (ACGIH,OSHA) (75-61-6)
1,1-Difluoro-2,2-dichloroethylene (79-35-6)
Difluorodichloromethane (75-71-8)
Difluorodimethylsilane (353-66-2)
Difluorodimethylstannane (3582-17-0)
4,4'-Difluorodiphenyl (398-23-2)
Difluorodiphenyltrichloroethane (475-26-3)
p,p'-Difluorodiphenyltrichloroethane (475-26-3)
1,2-Difluoroethane (624-72-6)
Difluoroethane (25497-28-3)
Difluoroethane (75-37-6)
1,1-Difluoroethane (DOT) (75-37-6)
1,1-Difluoroethene (75-38-7)
1,1-Difluoroethene homopolymer (24937-79-9)
2,4-Difluoro-6-(ethylamino)-s-triazine (30369-76-7)
1,1-Difluoroethylene [UN 1959] (75-38-7)
Difluoro(fluorosulfonyl)acetyl fluoride (677-67-8)
Difluoroformaldehyde (353-50-4)
1,1-Difluoroheptane (407-96-5)
6-α,9-α-Difluoro-16-α-hydroxyprednisolone 16,17-acetonide (67-73-2)
Difluoromethane (75-10-5)
Difluoromethylphenylsilane (328-57-4)
Difluoromonochloroethane (25497-29-4)
Difluoromonochloroethane (DOT) (75-68-3)
Difluoromonochloromethane (75-45-6)
1,1-Difluorooctane (61350-03-6)
2,4-Difluoro-6-(phenylamino)-s-triazine (717-90-8)
Difluorophosphoric acid (13779-41-4)
Difluorophosphoric acid, Anhydrous [UN 1768] (13779-41-4)
1,2-Difluoro-1,1,2,2-tetrachloroethane (76-12-0)
Diflupyl (55-91-4)
Difluron (35367-38-5)
Diflurophate (55-91-4)
Difo (115-26-4)
Difolatan (2939-80-2)
Difolatan (OSHA) (2425-06-1)
Difolliculine (50-50-0)
Difonate (944-22-9)
Diforin (54-85-3)
Diformal (107-22-2)
Diformatozinc dihydrate (5970-62-7)
Diformyl (107-22-2)
1,4-Diformylbenzene (623-27-8)
2,2'-Diformylbiphenyl (1210-05-5)
1,2-Diformylhydrazin (German) (628-36-4)
1,2-Diformylhydrazine (628-36-4)
Difosan (2425-06-1)
Difosgen (Czech) (503-38-8)
Difuran (804-36-4)
1,2-Di-2-furanyl-2-hydroxyethanone (552-86-3)
1,5-Di-2-furanyl-1,4-pentadien-3-one (886-77-1)
Difurazone (804-36-4)
Difurfurylideneacetone (886-77-1)
1,5-Difuryl-1,4-pentadien-3-one (886-77-1)
1,5-Di-2-furyl-3-pentanone (6075-11-2)
Digacin (20830-75-5)
Digadolinium trioxide (12064-62-9)
Digallium trioxide (12024-21-4)
Digamon (9005-34-9)
Digenea Simplex Mucilage (9002-18-0)
Digermin (1582-09-8)
Digibutina (50-33-9)
Digilong (71-63-6)
Digimed (71-63-6)
Digimerck (71-63-6)

Digisidin **(71-63-6)**
Digitalin **(71-63-6)**
Digitaline (French) **(71-63-6)**
Digitaline cristallisee **(71-63-6)**
Digitaline nativelle **(71-63-6)**
Digitalinum verum **(71-63-6)**
Digitalis glycoside **(20830-75-5)**
Digitin **(11024-24-1)**
Digitonin **(11024-24-1)**
Digitophyllin **(71-63-6)**
Digitoxigenin-tridigitoxosid (German) **(71-63-6)**
Digitoxigenin tridigitoxoside **(71-63-6)**
Digitoxin **(71-63-6)**
Digitoxinum **(71-63-6)**
Digitoxoside **(71-63-6)**
Digitrin **(71-63-6)**
Diglycerol **(59113-36-9)**
Diglycerol tetranitrate **(20600-96-8)**
1,4-Diglycidloxybutane **(2425-79-8)**
N,N-Diglycidylanilin (Czech) **(2095-06-9)**
N-N-Diglycidylaniline **(2095-06-9)**
Diglycidyl bisphenol A ether **(1675-54-3)**
Diglycidylester kyseliny ftalove (Czech) **(7195-45-1)**
Diglycidylester kyseliny hexahydroftalove (Czech) **(5493-45-8)**
Diglycidyl ether (ACGIH,OSHA) **(2238-07-5)**
Diglycidyl ether of 2,2-bis(4-hydroxyphenyl)propane **(1675-54-3)**
Diglycidyl ether of 2,2-bis(p-hydroxyphenyl)propane **(1675-54-3)**
Diglycidyl ether of bisphenol A **(1675-54-3)**
Diglycidyl ether of 4,4'-isopropylidenediphenol **(1675-54-3)**
Diglycidyl ether of neopentyl gylcol **(17557-23-2)**
Diglycidyl ether of resorcinol **(101-90-6)**
Diglycidylethylene glycol **(2224-15-9)**
Diglycidyl hexahydrophthalate **(5493-45-8)**
Diglycidyl isophthalate **(7195-43-9)**
1,3-Diglycidyloxybenzene **(101-90-6)**
1,2-Diglycidyloxyethane **(2224-15-9)**
N-N-Diglycidylphenylamine **(2095-06-9)**
Diglycidyl phthalate **(7195-45-1)**
Diglycidyl resorcinol ether **(101-90-6)**
Diglycidyltriethylene glycol **(1954-28-5)**
Diglycin **(142-73-4)**
Diglycine **(142-73-4)**
Diglycol **(111-46-6)**
Diglycolamine **(929-06-6)**
Diglycol chlorhydrin **(628-89-7)**
Diglycol dimethacrylate **(97-90-5)**
Diglycoldinitraat (Dutch) **(693-21-0)**
Diglycol (dinitrate de) (French) **(693-21-0)**
Diglycol formal **(1779-19-7)**
Diglycolic acid (6CI) **(110-99-6)**
Diglycol laurate **(141-20-8)**
Diglycol monobutyl ether **(112-34-5)**
Diglycol monobutyl ether acetate **(124-17-4)**
Diglycol monoethyl ether **(111-90-0)**
Diglycol monoethyl ether acetate **(112-15-2)**
Diglycol monolaurate **(141-20-8)**
Diglycol monomethyl ether **(111-77-3)**
Diglycol monostearate **(106-11-6)**
Diglycol oleate **(106-12-7)**
Diglycol stearate **(106-11-6)**
Diglycolurethane **(5952-26-1)**
Diglycolyl diamide **(106-57-0)**
Diglycylglycine **(556-33-2)**
Diglykokoll **(142-73-4)**
Diglykoldinitrat (German) **(693-21-0)**
Diglyme **(111-96-6)**
Digoksyna (Polish) **(20830-75-5)**
Digoxigenin-tridigitoxosid (German) **(20830-75-5)**
Digoxin **(20830-75-5)**
Digoxine **(20830-75-5)**
Diheptadecyldimethylammonium chloride (6CI,7CI) **(1118-41-8)**
Di-n-heptadecyl ketone **(504-53-0)**
Diheptadecyl ketone **(504-53-0)**
Diheptyl hexanedioate **(14697-48-4)**
Diheptyl ketone **(818-23-5)**
Di-n-heptyl phthalate **(3648-21-3)**
Diheptyl phthalate **(3648-21-3)**
Di-n-hexadecyl adipate **(26720-21-8)**
Dihexadecyl hexanedioate **(26720-21-8)**
Dihexadecyl peroxydicarbonate **(26322-14-5)**
Dihexadecyl phosphate **(2197-63-9)**
Dihexyl **(112-40-3)**
Di-n-hexyl adipate **(110-33-8)**
Dihexyl adipate **(110-33-8)**
Di-n-hexylamine **(143-16-8)**
Dihexylamine **(143-16-8)**
Dihexylamine, 6,6'-diamino- (8CI) **(143-23-7)**
Dihexylamine, 2,2'-diethyl **(106-20-7)**
Di-n-hexyl azelate **(109-31-9)**
Dihexylenetriamine **(143-23-7)**
Dihexylenetriaminepentakismethylenephosphonic acid, sodium salt **(35657-77-3)**
Di-n-hexylester kyseliny azelaove (Czech) **(109-31-9)**
Dihexylester kyseliny ftalove (Czech) **(84-75-3)**
Di-n-hexyl ether **(112-58-3)**
Dihexyl ether **(112-58-3)**
N,N-Dihexyl-1-hexanamine **(102-86-3)**
Dihexyl hexanedioate **(110-33-8)**
Di-n-hexyl ketone **(462-18-0)**
Dihexyl ketone **(462-18-0)**
Dihexyl maleate **(105-52-2)**
1,1-Di(hexyloxy)ethane **(5405-58-3)**
Dihexyloxyethyl adipate **(110-32-7)**
Di-(2-(2-hexyloxy)ethyl)ester kyseliny adipove (Czech) **(110-32-7)**
Di-n-hexyl phosphorodithioate, zinc salt **(7282-28-2)**
Di-n-hexyl phthalate **(84-75-3)**
Dihexyl phthalate **(84-75-3)**
Dihexyl sodium sulfosuccinate **(3006-15-3)**
Dihidral **(58-73-1)**
Dihıdrobenzperidol **(548-73-2)**
Dihidrocloruro de benzidina (Spanish) **(531-85-1)**
2,3-Dihidropirano (Spanish) **(110-87-2)**
1,8-Dihidroxi-2,4,5,7-tetranitroantraquinona (Spanish) **(517-92-0)**
Dihomo-γ-linolenic acid **(1783-84-2)**
Dihycon **(57-41-0)**
Dihydantoin **(57-41-0)**
Dihydantoin **(630-93-3)**
Dihydrin **(125-28-0)**
Dihydro-2(3H)-furanone **(96-48-0)**
1,2-Dihydro-5-acenaphthylenamine **(4657-93-6)**
Dihydroactinidiolide **(17092-92-1)**
Dihydroaflatoxin B1 **(7220-81-7)**
9,10-Dihydro-1-amino-4-(cyclohexylamino)-9,10-dioxo-2-anthracenesulfonic acid sodium salt **(4368-56-3)**
5,6-Dihydro-4-amino-1-β-d-ribofuranosyl-s-triazin-2(1H)-one **(62488-57-7)**
Dihydroanethole **(104-45-0)**
1,4-Dihydro-9,10-anthracenediol disodium salt **(73347-80-5)**
N,N-Dihydro-1,1,1',2'-anthraquinone-azine **(81-77-6)**
22,23-Dihydroavermectin B1 **(70288-86-7)**
5,6-Dihydro-5-azacytidine **(62488-57-7)**
Dihydroazirene **(151-56-4)**
Dihydro-1H-azirine **(151-56-4)**
Dihydrobaikiane **(535-75-1)**
4,5-Dihydrobenz(k)acephenanthrylene **(5779-79-3)**

2,3-Dihydrobenzofuran **(496-16-2)**
1,4-Dihydrobiphenyl **(4794-05-2)**
Dihydrobis(2-methoxyethoxy)aluminate **(22722-98-1)**
1,3-Dihydro-7-bromo-5-(2-pyridyl)-2H-1,4-benzodiazepin-2-one **(1812-30-2)**
Dihydrobutadiene sulphone **(126-33-0)**
Dihydrocaffeic acid **(1078-61-1)**
2,3-Dihydro-5-carboxanilido-6-methyl-1,4-oxathiin **(5234-68-4)**
2,3-Dihydro-5-carboxanilido-6-methyl-1,4-oxathiin, 4,4-dioxide **(5259-88-1)**
Dihydrocarvone **(7764-50-3)**
d-Dihydrocarvone **(5524-05-0)**
Dihydrochloride salt of diethylenediamine **(142-64-3)**
1,3-Dihydro-7-chloro-5-(o-chlorophenyl)-2H-1,4-benzodiazepin-2-one **(2894-67-9)**
1,3-Dihydro-7-chloro-3-hydroxy-1-methyl-5-phenyl-2H-1,4-benzodiazepin-2-one **(846-50-4)**
2,3-Dihydro-7-chloro-1-methyl-5-phenyl-1H-1,4-benzodiazepine **(2898-12-6)**
3,4-Dihydro-6-chloro-7-sulfamyl-1,2,4-benzothiadiazine-1,1-dioxide **(58-93-5)**
Dihydrochlorothiazid **(58-93-5)**
3,4-Dihydrochlorothiazide **(58-93-5)**
Dihydrochlorothiazide **(58-93-5)**
Dihydrocholesterin **(80-97-7)**
5α-Dihydrocholesterol **(80-97-7)**
Dihydrocholesterol **(360-68-9)**
Dihydrocholesterol **(80-97-7)**
1,2-Dihydrochrysene **(41593-31-1)**
Dihydrochrysene **(41593-31-1)**
Dihydrocinnamic acid **(501-52-0)**
2,3-Dihydrocitral **(106-23-0)**
Dihydrocitronellol **(106-21-8)**
7,8-Dihydrocodeine **(125-28-0)**
Dihydrocodeine **(125-28-0)**
Dihydrocodeinone **(125-29-1)**
Dihydrocodeinone enol acetate **(466-90-0)**
Dihydro-p-coumaric acid **(501-97-3)**
3,4-Dihydrocoumarin **(119-84-6)**
Dihydrocoumarin **(119-84-6)**
Dihydrocuminyl alcohol **(536-59-4)**
Dihydrocuminyl aldehyde **(2111-75-3)**
Dihydrodeoxymorphine **(427-00-9)**
Dihydrodesoxymorphine-d **(427-00-9)**
9,10-Dihydro-8a,10,-diazoniaphenanthrene dibromide **(85-00-7)**
9,10-Dihydro-8a,10a-diazoniaphenanthrene(1,1'-ethylene-2,2'-bipyridylium)dibromide **(85-00-7)**
10,11-Dihydro-5-dibenz(b,f)azepine **(494-19-9)**
10,11-Dihydro-5H-dibenzo(a,d)cyclohepten-5-one **(1210-35-1)**
3-(10,11-Dihydro-5H-dibenzo(a,d)cyclohepten-5-ylidene)-N,N-dimethyl-1-propanamine **(50-48-6)**
Dihydro-5,5-diethyl-2H-1,3-oxazine-2,4(3H)-dione **(702-54-5)**
Dihydrodiethylstilbestrol **(84-16-2)**
20,22-Dihydrodigitoxin **(3786-76-3)**
Dihydrodigitoxin **(3786-76-3)**
2,3-Dihydro-9,10-dihydroxy-1,4-anthracenedione **(17648-03-2)**
1,4-Dihydro-9,10-dihydroxy anthracene, disodium salt solution **(73347-80-5)**
9,10-Dihydro-4,5-dihydroxy-9,10-dioxo-2-anthracenecarboxylic acid **(478-43-3)**
9,10-Dihydro-3,4-dihydroxy-9,10-dioxo-2-anthracenesulfonic acid monosodium salt **(130-22-3)**
(13-α,14-α)-14,19-Dihydro-12,13-dihydroxy-20-norcrotalanan-11,15-dione **(315-22-0)**
1,4-Dihydro-1,4-diketonaphthalene **(130-15-4)**
2,5-Dihydro-2,5-dimethoxyfuran **(332-77-4)**
Dihydro-4,4-dimethyl-2(3H)-furanone **(13861-97-7)**
9,10-Dihydro-9,9-dimethylacridine **(6267-02-3)**
10,11-Dihydro-5-(3-dimethylamino-2-methylpropyl)-5H-dibenz-(b,f)azepine **(739-71-9)**
5,6-Dihydro-N-(3-(dimethylamino)propyl)-11H-dibenz(b,e)azepine **(50-49-7)**
10,11-Dihydro-5-(γ-dimethylaminopropylidene)-5H-dibenzo-(a,d)cycloheptene **(50-48-6)**
2,3-Dihydro-2,2-dimethyl-7-benzofuranamine **(68298-46-4)**
2,3-Dihydro-2,2-dimethyl-7-benzofurandiazonium, sulfate (2:1) **(68298-47-5)**
2,3-Dihydro-2,2-dimethyl-7-benzofuranol **(1563-38-8)**
2,3-Dihydro-2,2-dimethyl-7-benzofuranyl((dibutylamino)thio) methyl carbamate **(55285-14-8)**
2,3-Dihydro-2,2-dimethyl-7-benzofuranyl (di-n-butylamino-sulfenyl)methylcarbamate **(55285-14-8)**
2,3-Dihydro-2,2-dimethyl-7-benzofuranyl methylcarbamate **(1563-66-2)**
2,3-Dihydro-2,2-dimethylbenzofuran-7-yl methylcarbamate **(1563-66-2)**
2,3-Dihydro-2,2-dimethylbenzofuranyl-7-N-methylcarbamate **(1563-66-2)**
10,11-Dihydro-N,N-dimethyl-5H-dibenzo(a,d)heptalene-δ^5,γ-propylamine **(50-48-6)**
2,3-Dihydro-5,6-dimethyl-1,4-dithiin 1,1,4,4-tetroxide **(55290-64-7)**
2,3-Dihydro-2,5-dimethyl-2-formyl-1,4-pyran **(1920-21-4)**
2,3-Dihydro-2,2-dimethyl-7-hydroxybenzofuran **(1563-38-8)**
2,3-Dihydro-2,2-dimethyl-7-nitrobenzofuran **(13414-55-6)**
1,4-Dihydro-2,6-dimethyl-4-(2-nitrophenyl)-3,5-pyridinedicarboxylic acid dimethyl ester **(21829-25-4)**
3,4-Dihydro-2,2-dimethyl-4-oxo-2h-pyran-6-carboxylic acid, n-butyl ester **(532-34-3)**
2,3-Dihydro-2,2-dimethyl-6-((4-(phenylazo)-1-naphthyl)azo)-perimidine **(4197-25-5)**
1,2-Dihydro-1,5-dimethyl-2-phenyl-3H-pyrazol-3-one **(60-80-0)**
3,7-Dihydro-1,3-dimethyl-1H-purine-2,6-dione **(58-55-9)**
3,7-Dihydro-3,7-dimethyl-1H-purine-2,6-dione **(83-67-0)**
3,7-Dihydro-1,3-dimethyl-1H-purine-2,6-dione, Compd. with 1,2-ethanediamine (2:1) **(317-34-0)**
endo,endo-Dihydrodi(norbornadiene) **(66289-74-5)**
9,10-Dihydro-9,10-dioxo-1-anthracenesulfonic acid sodium salt **(128-56-3)**
1,3-Dihydro-1,3-dioxo-5-isobenzofurancarboxylic acid **(552-30-7)**
5,10-Dihydro-5,10-dioxonaphtho(2,3-b)-p-dithiin-2,3-dicarbonitrile **(3347-22-6)**
5,6-Dihydro-dipyrido(1,2a;2,1c)pyrazinium dibromide **(85-00-7)**
3,4-Dihydro-3,4-epoxy-1,2-benzanthracene **(962-32-3)**
7,8-Dihydro-4,5-α-epoxy-17-methylmorphinan-3,6-α-diol diacetate **(561-27-3)**
α-Dihydroequilenin **(6639-99-2)**
alfa-Dihydroequilenina (Spanish) **(6639-99-2)**
.β.-Dihydroequilin **(3563-27-7)**
17 α-Dihydroequilin **(651-55-8)**
17.β.-Dihydroequilin **(3563-27-7)**
Dihydroequilin **(651-55-8)**
α-Dihydroequilin **(651-55-8)**
β-Dihydroequilin **(651-55-8)**
Dihydroestrin benzoate **(50-50-0)**
3-(9,10-Dihydro-9,10-ethanoanthracen-9-yl)propylmethylamine **(10262-69-8)**
1,2-Dihydro-6-ethoxy-2,2,4-trimethylquinoline **(91-53-2)**
1,4-Dihydro-1-ethyl-7-methyl-4-oxo-1,8-naphthyridine-3-carboxylic acid **(389-08-2)**
Dihydroeugenol **(2785-87-7)**
Dihydrofarnesol **(1335-48-4)**
Dihydroflumethazide **(135-09-1)**
Dihydroflumethiazide **(135-09-1)**
1,3-Dihydro-5-(o-fluorophenyl)-1-methyl-7-nitro-2H-1,4-benzodiazepin-2-one **(1622-62-4)**
Dihydrofollicular hormone **(50-28-2)**
Dihydrofolliculin **(50-28-2)**
Dihydrofolliculin benzoate **(50-50-0)**
α-Dihydrofucosterol **(83-46-5)**
β-Dihydrofucosterol **(83-47-6)**
2,5-Dihydrofuran **(1708-29-8)**
Dihydro-2,5-furandione **(108-30-5)**
Dihydrogen ammonium phosphate **(7722-76-1)**
(SP-4-2)-Dihydrogen (7,12-diethenyl-3,8,13,17-tetramethyl-21H,23H-porphine-2,18-dipropanoato(4-)-N(21),N(22),N(23),N(24))zincate(2-) **(15442-64-5)**
Dihydrogen dioxide **(7722-84-1)**
Dihydrogen hexachloroplatinate **(16941-12-1)**

Dihydrogen hexachloroplatinate(2-) **(16941-12-1)**
Dihydrogen hexafluorosilicate **(16961-83-4)**
Dihydrogen hexafluorosilicate (2-) **(16961-83-4)**
Dihydrogen methyl phosphate **(812-00-0)**
Dihydrogen oxide **(7732-18-5)**
β-Dihydroheptachlor **(14168-01-5)**
Dihydroheterocodeine **(7732-92-5)**
Dihydro-14-hydroxycodeinone **(76-42-6)**
Dihydrohydroxycodeinone **(76-42-6)**
3a,12c-Dihydro-8-hydroxy-6-methoxy-7H-furo(3',2':4,5)furo-(2,3-c)xanthen-7-one **(10048-13-2)**
7,8-Dihydro-7-α-(1-(R)-hydroxy-1-methylbutyl)-O^{6}-methyl-6,14-endo-ethenomorphine **(14521-96-1)**
4,5-Dihydro-2-hydroxymethylene-17-α-methyltestosterone **(434-07-1)**
2,3-Dihydro-3-α-hydroxy-2-β-methyl-7-propenyl-4H,5H-pyrano-(4,3-b)pyran-4,5-dione **(10088-95-6)**
Dihydro-14-hydroxy-4-o-methyl-6-β-thebainol **(3176-03-2)**
7,8-Dihydro-14-hydroxymorphine **(2183-56-4)**
7,8-Dihydro-14-hydroxymorphinone **(76-41-5)**
Dihydro-14-hydroxymorphinone **(76-41-5)**
Dihydrohydroxymorphinone **(76-41-5)**
Dihydro-14-hydroxy-6-β-thebainol 4 methyl ester **(3176-03-2)**
(3R,4S)-4,6-Dihydro-8-hydroxy-3,4,5-trimethyl-6-oxo-3H-2-benzopyran-7-carboxylic acid **(518-75-2)**
5,6-Dihydro-3H-imidazo(2,1-c)-1,2,4-dithiazole-3-thione **(33813-20-6)**
4,5-Dihydroimidazole-2(3H)-thione **(96-45-7)**
1,6-Dihydro-6-iminopurine **(73-24-5)**
3,6-Dihydro-6-iminopurine **(73-24-5)**
2,3-Dihydroindene **(496-11-7)**
2,3-Dihydro-1H-inden-1-ol **(6351-10-6)**
1,3-Dihydro-2H-inden-2-one **(615-13-4)**
2,3-Dihydro-1H-inden-1-one **(83-33-0)**
2,3-Dihydroindole **(496-15-1)**
Dihydroisocodeine **(795-38-0)**
Dihydro-isophorone **(873-94-9)**
1,2-Dihydro-2-ketobenzisosulfonazole **(81-07-2)**
1,2-Dihydro-2-ketobenzisosulphonazole **(81-07-2)**
Dihydrokodein (Czech) **(125-28-0)**
Dihydrolinalool **(18479-51-1)**
2,3-Dihydro-3α-hydroxytropidine **(120-29-6)**
Dihydromenformon **(50-28-2)**
3,4-Dihydro-6-methoxy-1(2H)-naphthalenone **(1078-19-9)**
1,2-Dihydro-4-methoxy-1-methyl-2-oxonicotinonitrile **(524-40-3)**
S-(2,3-Dihydro-5-methoxy-2-oxo-1,3,4-thiadiazol-3-methyl) dimethyl phosphorothiolothionate **(950-37-8)**
S-2,3-Dihydro-5-methoxy-2-oxo-1,3,4-thiadiazol-3-ylmethyl O,O-dimethylphosphorodithioate **(950-37-8)**
3,4-Dihydro-2-methoxy-2H-pyran **(4454-05-1)**
3,12-Dihydro-6-methoxy-3,3,12-trimethyl-7H-pyrano(2,3-c)-acridin-7-one **(7008-42-6)**
10,11-Dihydro-5-(3-methylaminopropyl)-5H-dibenz(b,f)azepine **(50-47-5)**
Dihydro(2-methyl-1,3-butadienyl)aluminum **(24683-32-7)**
5,6-Dihydro-2-methyl-3-carboxanilido-1,4-oxathiin-4,4-dioxid (German) **(5259-88-1)**
5,6-Dihydro-2-methyl-3-carboxanilido-1,4-oxathiin (German) **(5234-68-4)**
6,7-Dihydro-5-methyl-5H-cyclopentapyrazine **(23747-48-0)**
10,11-Dihydro-N-methyl-5H-dibenzo(a,d)cycloheptane-δ,γ-propylamine **(72-69-5)**
4,5-Dihydro-2-(1-methylethenyl)oxazole **(10471-78-0)**
Dihydro-3-methyl-2,5-furandione **(4100-80-5)**
4,5-Dihydro-2-methyl-1H-imidazole **(534-26-9)**
2-(4,5-Dihydro-4-methyl-4-(1-methylethyl)-5-oxo-1H-imidazol-2-yl)-5-ethyl-3-pyridinecarboxylic acid **(81335-77-5)**
2-(4,5-Dihydro-4-methyl-4-(1-methylethyl)-5-oxo-1H-imidazol-2-yl)-3-pyridinecarboxylic acid **(81334-34-1)**
2-(4,5-Dihydro-4-methyl-4-(1-methylethyl)-5-oxo-1H-imidazol-2-yl)-3-pyridnecarboxylic acid, monoisopropylamine **(81510-83-0)**
2-(4,5-Dihydro-4-methyl-4-(1-methylethyl)-5-oxo-1H-imidazol-2-yl)-3-quinolinecarboxylic acid **(81335-37-7)**
Dihydro-6-methylmorphinone **(143-52-2)**
1,3-Dihydro-1-methyl-7-nitro-5-phenyl-2H-1,4-benzodiazepin-2-one **(2011-67-8)**
3,4-Dihydro-6-methyl-1,2,3-oxathiazin-4-one-2,2-dioxide potassium salt **(33665-90-6)**
2,3-Dihydro-6-methyl-1,4-oxathiin-5-carboxanilide **(5234-68-4)**
5,6-Dihydro-2-methyl-1,4-oxathiin-3-carboxanilide **(5234-68-4)**
5,6-Dihydro-2-methyl-1,4-oxathiin-3-carboxanilide 4,4-dioxide **(5259-88-1)**
3,4-Dihydro-2-methyl-4-oxo-3-o-tolylquinazoline **(72-44-6)**
5-6-Dihydro-2-methyl-3-(phenylcarbamoyl)-4H-pyrane **(24691-76-7)**
5,6-Dihydro-2-methyl-N-phenyl-1,4-oxathiin-3-carboxamide **(5234-68-4)**
5,6-Dihydro-2-methyl-N-phenyl-1,4-oxathiin-3-carboxamide 4,4-dioxide **(5259-88-1)**
3,4-Dihydro-6-methyl-N-phenyl-2H-pyran-5-carboxamide **(24691-76-7)**
3,4-Dihydro-6-methyl-2H-pyran-5-carboxanilide **(24691-76-7)**
(S)-(+)-5,6-Dihydro-6-methyl-2H-pyran-2-one **(10048-32-5)**
Dihydro-2-α-methyltestosterone **(58-19-5)**
2,3-Dihydro-6-methyl-2-thioxo-4(1H)-pyrimidinone **(56-04-2)**
Dihydromorfin (Czech) **(509-60-4)**
Dihydromorfinon (Czech) **(466-99-9)**
Dihydromorphine **(509-60-4)**
Dihydromorphinone **(466-99-9)**
Dihydromucodinitrile **(13042-02-9)**
Dihydromyrcene **(2436-90-0)**
Dihydromyrcenol **(18479-58-8)**
1,2-Dihydronaphthalene **(447-53-0)**
Dihydronaringenin **(60-82-2)**
Dihydrone **(76-42-6)**
Dihydroneopine **(125-28-0)**
Dihydronereistoxin dicarbamate **(15263-52-2)**
1,2-Dihydro-5-nitro-acenaphthylene **(602-87-9)**
1,3-Dihydro-7-nitro-5-(2-chlorophenyl)-2H-1,4.Benzodiazepin-2-one **(1622-61-3)**
1,3-Dihydro-7-nitro-5-phenyl-2H-1,4-benzodiazepin-2-one **(146-22-5)**
Dihydro-nordicyclopentadienyl acetate **(5413-60-5)**
Dihydronorguaiaretic acid **(500-38-9)**
4,5-Dihydro-7-nortall oil-1H-imidazole-1-ethanol **(61791-39-7)**
Dihydro-3-(octadecenyl)-2,5-furandione **(28777-98-2)**
Dihydro-3-(octenyl)-2,5-furandione **(26680-54-6)**
9,10-Dihydro-9-oxa-10-phosphaphenanthrene 10-oxide **(35948-25-5)**
Dihydrooxirene **(75-21-8)**
9,10-Dihydro-9-oxoanthracene **(90-44-8)**
2,3-Dihydro-3-oxobenzisosulfonazole **(81-07-2)**
2,3-Dihydro-3-oxobenzisosulphonazole **(81-07-2)**
3,4-Dihydro-4-oxo-3-benzotriazinylmethyl O,O-diethyl phosphorodithioate **(2642-71-9)**
S-(3,4-Dihydro-4-oxo-1,2,3-benzotriazin-3-ylmethyl) O,O-diethyl phosphorodithioate **(2642-71-9)**
S-(3,4-Dihydro-4-oxo-1,2,3-benzotriazin-3-ylmethyl) O,O-dimethyl phosphorodithioate **(86-50-0)**
S-(3,4-Dihydro-4-oxo-benzo(α)(1,2,3)triazin-3-ylmethyl) O,O-dimethyl phosphorodithioate **(86-50-0)**
4-(1,3-Dihydro-1-oxo-2H-isoindol-2-yl)-α-ethyl-benzeneacetic acid **(63610-08-2)**
O-(1,6-Dihydro-6-oxo-1-phenylpyridazin-3-yl) O,O-diethyl phosphorothioate **(119-12-0)**
Dihydrooxophorone **(20547-99-3)**
2,3-Dihydro-1,1,2,3,3-pentamethyl-1H-indene **(1203-17-4)**
2,3-Dihydro-1,1,3,3,5-pentamethyl-1H-indene **(81-03-8)**
1,1-Dihydroperfluorobutanol **(375-01-9)**
α,α-Dihydroperfluorobutanol **(375-01-9)**
2,2-Dihydroperoxy propane (Not more than 25% with inert organic solid) **(2614-76-8)**
1,2-Dihydro-4-phenylnaphthalene **(7469-40-1)**
3,4-Dihydro-1-phenylnaphthalene **(7469-40-1)**

Dihydropinene (473-55-2)
2,3-Dihydro-6-propyl-2-thioxo-4(1H)-pyrimidinone (51-52-5)
3,7-Dihydro-1H-purine-2,6-dione (69-89-6)
1,7-Dihydro-6H-purin-6-one (68-94-0)
1,7-Dihydro-6H-purin-6-thion, monohydrat (Czech) (6112-76-1)
1,2-Dihydro-3,6-pyradizinedione (123-33-1)
2,3-Dihydro-4H-pyran (110-87-2)
2,3-Dihydropyran (110-87-2)
2H-3,4-Dihydropyran (110-87-2)
3,4-Dihydro-2H-pyran (110-87-2)
5,6-Dihydro-4H-pyran (110-87-2)
δ2-Dihydropyran (110-87-2)
3,4-Dihydropyran [UN 2376] (110-87-2)
Dihydropyran (VAN) (110-87-2)
3,4-Dihydro-2H-pyran-2-carboxaldehyde (100-73-2)
2,3-Dihydro-1,4-pyran-2-karboxaldehyd (Czech) (100-73-2)
Dihydro-2,3 pyranne (French) (110-87-2)
1,5-Dihydro-4H-pyrazolo(3,4-d)pyrimidin-4-one (315-30-0)
1,2-Dihydro-3,6-pyridazinedione (123-33-1)
1,2-Dihydropyridazine-3,6-dione (123-33-1)
1,2-Dihydro-3,6-pyridazinedione monopotassium salt (28382-15-2)
6,7-Dihydropyrido(1,2-a;2',1'-c)pyrazinedium dibromide (85-00-7)
Dihydropyrone (532-34-3)
3,4-Dihydropyrrole-2,5-dione (123-56-8)
3,4-Dihydropyrrolidine (123-56-8)
Dihydro-3-pyrroline-2,5-dione (123-56-8)
2,3-Dihydro-1H-pyrrolo(2,3-b)pyridine (10592-27-5)
(+)-Dihydroquercetin (480-18-2)
(2R,3R)-Dihydroquercetin (480-18-2)
2,3-Dihydroquercetin (480-18-2)
Dihydroquercetin (480-18-2)
6,13-Dihydroquinacridone (5862-38-4)
Dihydroquinine (522-66-7)
5,12-Dihydroquino(2,3-b)acridine-7,14-dione (1047-16-1)
Dihydroresorcinol (504-02-9)
Dihydroretenone (6659-45-6)
Dihydrorotenone (6659-45-6)
Dihydrosafrole (94-58-6)
22,23-Dihydrostigmasterol (83-46-5)
Dihydrostilbestrol (84-16-2)
Dihydrostreptomycin (128-46-1)
Dihydrostreptomycin Sulfate (5490-27-7)
Dihydrostreptomycin-kaolinite-pectin-aluminum hydroxide mixture (8047-42-5)
Dihydrostreptomycin sulfate (5490-27-7)
2,3-Dihydrosuccinic acid (87-69-4)
3,4-Dihydro-7-sulfamyl-6-trifluoromethyl-2H-1,2,4-benzothiadiazine 1,1-dioxide (135-09-1)
Dihydrotachysterol (67-96-9)
Dihydrotachysterol2 (67-96-9)
Dihydro-α-terpineol (498-81-7)
Dihydro-3-(tetradecenyl)-2,5-furandione (33806-58-5)
6,7-Dihydro-1,2,3,10-tetramethoxy-7-(methylamino)-benzo-(α)heptalen-9(5H)-one (477-30-5)
1,3-Dihydrotetramethyldisiloxane (3277-26-7)
Dihydrotheelin (50-28-2)
2,3-Dihydrothiirene (420-12-2)
2,5-Dihydrothiophene 1,1-dioxide (77-79-2)
2,5-Dihydrothiophene dioxide (77-79-2)
2,5-Dihydrothiophene sulfone (77-79-2)
2,3-Dihydro-2-thioxo-4(1H)-pyrimidinone (141-90-2)
cis-Dihydrotodomatuic acid (38963-91-6)
Dihydro-1,3,5-triazine-2,4(1H,3H)-dione (27032-78-6)
3,4-Dihydro-6-trifluoromethyl-2H-1,2,4-benzothiadiazine-7-sulfonamide 1,1-dioxide (135-09-1)
3,4-Dihydro-6-trifluoromethyl-7-sulfamoylbenzo-1,2,4-thiadiazine 1,1-dioxide (135-09-1)
3,4-Dihydro-3,7,8-trihydroxy-3-methyl-10-oxo-1H, 10H-pyrano-(4,3-b)(1)benzopyran-9-carboxylic acid (479-66-3)
1,2-Dihydro-2,2,4-trimethyl-6-ethoxyquinoline (91-53-2)
2,3-Dihydro-1,4,7-trimethyl-1H-indene (54340-87-3)
1,2-Dihydro-1,5,8-trimethylnaphthalene (4506-36-9)
2,3-Dihydro-1,1,3-trimethyl-3-phenyl-1H-indene (3910-35-8)
3,7-Dihydro-1,3,7-trimethyl-1H-purine-2,6-dione (58-08-2)
1,2-Dihydro-2,2,4-trimethylquinoline (147-47-7)
4,5-Dihydro-2-undecyl-1H-imidazole-1-ethanol (136-99-2)
5,6-Dihydrouridine (18771-50-1)
1,3-Dihydroxyacetone (96-26-4)
Dihydroxyacetone (96-26-4)
2',4'-Dihydroxyacetophenone (89-84-9)
2',5'-Dihydroxyacetophenone (490-78-8)
2,4-Dihydroxyacetophenone (89-84-9)
2,5-Dihydroxyacetophenone (490-78-8)
3,4-Dihydroxyacetophenone (1197-09-7)
Dihydroxyaluminum aminoacetate (13682-92-3)
4,6-Dihydroxy-2-aminopyrimidine (56-09-7)
3,5-Dihydroxyamylbenzene (500-66-3)
1,8-Dihydroxy-9,10-anthracenedione (117-10-2)
1,2-Dihydroxyanthrachinon (Czech) (72-48-0)
1,4-Dihydroxyanthrachinon (Czech) (81-64-1)
1,5-Dihydroxyanthrachinon (Czech) (117-12-4)
1,8-Dihydroxyanthrachinon (Czech) (117-10-2)
1,8-Dihydroxy-9-anthranol (480-22-8)
1,8-Dihydroxyanthranol (480-22-8)
Dihydroxy-anthranol (480-22-8)
1,2-Dihydroxy-9,10-anthraquinone (72-48-0)
1,2-Dihydroxyanthraquinone (72-48-0)
1,4-Dihydroxy-9,10-anthraquinone (81-64-1)
1,4-Dihydroxyanthraquinone (81-64-1)
1,5-Dihydroxy-9,10-anthraquinone (117-12-4)
1,5-Dihydroxyanthraquinone (117-12-4)
1,8-Dihydroxyanthraquinone (117-10-2)
2,6-Dihydroxyanthraquinone (84-60-6)
1,8-Dihydroxyanthraquinone-3-carboxylic acid (478-43-3)
4,5-Dihydroxy-2-anthraquinonecarboxylic acid (478-43-3)
1,8-Dihydroxy-9-anthrone (1143-38-0)
1,8-Dihydroxy-9-anthrone (480-22-8)
1,8-Dihydroxyanthrone (1143-38-0)
1,8-Dihydroxyanthrone (480-22-8)
2,4-Dihydroxyazobenzene (2051-85-6)
2,4-Dihydroxyazobenzene-4'-sulfonate sodium salt (547-57-9)
2,4-Dihydroxybenzaldehyde (95-01-2)
3,4-Dihydroxybenzaldehyde (139-85-5)
3,4-Dihydroxybenzaldehyde methylene ketal (120-57-0)
1,4-Dihydroxy-benzeen (Dutch) (123-31-9)
1,4-Dihydroxybenzen (Czech) (123-31-9)
1,2-Dihydroxybenzene (120-80-9)
1,3-Dihydroxybenzene (108-46-3)
1,4-Dihydroxybenzene (123-31-9)
m-Dihydroxybenzene (108-46-3)
o-Dihydroxybenzene (120-80-9)
p-Dihydroxybenzene (123-31-9)
Dihydroxybenzene (OSHA) (123-31-9)
2,5-Dihydroxybenzeneacetic acid (451-13-8)
3,4-Dihydroxybenzeneacetic acid (102-32-9)
3,4-Dihydroxybenzeneacrylic acid (331-39-5)
2,4-Dihydroxybenzenecarbonal (95-01-2)
3,4-Dihydroxybenzenecarbonal (139-85-5)
2,5-Dihydroxy-p-benzenedisulfonic acid, dipotassium salt (15763-57-2)
2,5-Dihydroxy-para-benzenedisulfonic acid dipotassium salt (15763-57-2)
2,5-Dihydroxybenzenesulfonic acid, monopotassium salt (21799-87-1)
2,5-Dihydroxybenzenesulfonic acid, monosodium salt (10021-55-3)
3,3'-Dihydroxybenzidine (2373-98-0)
2,4-Dihydroxybenzofenon (Czech) (131-56-6)
2,4-Dihydroxybenzoic acid (89-86-1)

2,5-Dihydroxybenzoic acid **(490-79-9)**
2,6-Dihydroxybenzoic acid **(303-07-1)**
3,4-Dihydroxybenzoic acid **(99-50-3)**
3,5-Dihydroxybenzoic acid **(99-10-5)**
1,4-Dihydroxy-benzol (German) **(123-31-9)**
2,4-Dihydroxybenzophenone **(131-56-6)**
3,4-Dihydroxybenzylamine **(37491-68-2)**
5,5'-Dihydroxy-5,5'-bibarbituric acid **(76-24-4)**
2,2'-Dihydroxybiphenyl **(1806-29-7)**
2,3'-Dihydroxybiphenyl **(31835-45-7)**
2,5-Dihydroxybiphenyl **(1079-21-6)**
3,4-Dihydroxybiphenyl **(92-05-7)**
(3,3'-((3,3'-Dihydroxy-1,1'-biphenyl-4,4'-diyl)bis(azo)bis-(5-amino-2,7-naphthalenedisulfonato-(O4,O3)))dicopper, tetrasodium salt **(28407-37-6)**
2,6-Dihydroxy-5-bis(2-chloroethyl)aminopyrimidine **(66-75-1)**
4,5-Dihydroxy-1,3-bis(hydroxymethyl)-2-imidazolidinone **(1854-26-8)**
4,5-Dihydroxy-1,3-bis(hydroxymethyl)-2-imidazolidinone, methylated **(68411-81-4)**
4,5-Dihydroxy-1,3-bis(methoxymethyl)-2-imidazolidinone **(3001-61-4)**
1,3-Dihydroxybutane **(107-88-0)**
1,4-Dihydroxybutane **(110-63-4)**
2,3-Dihydroxybutane **(513-85-9)**
(R*,R*)-(+-)-2,3-Dihydroxybutanedioic acid **(133-37-9)**
(S-(R*,R*))-2,3-Dihydroxybutanedioic acid **(147-71-7)**
2,3-Dihydroxybutanedioic acid **(87-69-4)**
2,3-Dihydroxybutanedioic acid copper salt **(815-82-7)**
2,3-Dihydroxy-butanedioic acid, diammonium salt (9CI) **(3164-29-2)**
2,3-Dihydroxybutanedioic acid, monopotassium monosodium salt **(304-59-6)**
1,4-Dihydroxy-2-butene **(110-64-5)**
2,6-Dihydroxy-4-carboxypyridine **(99-11-6)**
Dihydroxychlorothiazidum **(58-93-5)**
3,12-Dihydroxycholanic acid **(83-44-3)**
3-α,12-α-Dihydroxycholanic acid **(83-44-3)**
3-α,7-α-Dihydroxycholanic acid **(474-25-9)**
3-α,12-α-Dihydroxy-5-β-cholanoic acid **(83-44-3)**
3-α,7-α-Dihydroxy-5-β-cholan-24-oic acid **(474-25-9)**
3-α,12-α-Dihydroxycholansaeure (German) **(83-44-3)**
3-α,7-α-Dihydroxycholansaeure (German) **(474-25-9)**
1,25-Dihydroxycholecalciferol **(32222-06-3)**
1-α,25-Dihydroxycholecalciferol **(32222-06-3)**
1a,25-Dihydroxycholecalciferol **(32222-06-3)**
3,4-Dihydroxycinnamic acid **(331-39-5)**
6,7-Dihydroxycoumarin **(305-01-1)**
Di-(4-hydroxy-3-coumarinyl)methane **(66-76-2)**
1,3-Dihydroxycyclohexane **(504-01-8)**
Di-(1-hydroxycyclohexyl) peroxide **(2407-94-5)**
Dihydroxycyclohexylperoxide **(2407-94-5)**
Di-(1-hydroxycyclohexyl) peroxide (Technically pure) **(2407-94-5)**
Di-(1-hydroxycyclohexyl)peroxide, Technically pure **(2407-94-5)**
1,5-Dihydroxy-4,8-diaminoanthrachinon (Czech) **(145-49-3)**
1,8-Dihydroxy-4,5-diaminoanthrachinon (Czech) **(128-94-9)**
1,5-Dihydroxy-4,8-diaminoanthraquinone **(145-49-3)**
Di(4-hydroxy-3,5-di-tert-butylphenyl)methane **(118-82-1)**
2,5-Dihydroxy-3,6-dichlorobenzoquinone **(87-88-7)**
2,2'-Dihydroxy-5,5'-dichlorodiphenylmethane **(97-23-4)**
2,2'-Dihydroxy-5,5'-dichlorodiphenyl sulfide **(97-24-5)**
2,2'-Dihydroxydiethylamine **(111-42-2)**
4,4'-Dihydroxy-α,β-diethyldiphenylethane **(84-16-2)**
Dihydroxydiethyl ether **(111-46-6)**
β,β'-Dihydroxydiethyl ether **(111-46-6)**
4,4'-Dihydroxy-α,β-diethylstilbene **(56-53-1)**
4,4'-Dihydroxydiethylstilbene **(56-53-1)**
4,4'-Dihydroxy-α,β-diethylstilbene dipropionate **(130-80-3)**
Dihydroxydiethylstilbene dipropionate **(130-80-3)**
β,β'-Dihydroxydiethyl sulfide **(111-48-8)**
5,8-Dihydroxy-1,4-dihydroxyethylaminoanthraquinone **(3179-90-6)**
α,α'-Dihydroxy-p-diisopropylbenzene **(2948-46-1)**
Dihydroxydi(methoxymethyl)ethyleneurea **(3001-61-4)**
6-β,14-Dihydroxy-3,4-dimethoxy-N-methylmorphinan **(3176-03-2)**
1,3-Dihydroxy-4,5-dimethylbenzene **(527-55-9)**
3,5-Dihydroxy-1,2-dimethylbenzene **(527-55-9)**
2,4-Dihydroxy-3,6-dimethylbenzoic acid, methyl ester **(4707-47-5)**
N-(2,4-Dihydroxy-3,3-dimethylbutyryl)-β-alanine calcium **(137-08-6)**
2,2'-Dihydroxy-3,3'-dimethyl-5,5'-dichlorodiphenyl sulfide **(4418-66-0)**
Dihydroxydimethylolethyleneurea, methylated **(68411-81-4)**
2,2'-Dihydroxy-6,6'-dinaphthyldisulfide **(6088-51-3)**
1,5-Dihydroxy-4,8-dinitro-9,10-anthracenedione **(128-91-6)**
1,5-Dihydroxy-4,8-dinitroanthraquinone **(128-91-6)**
1,8-Dihydroxy-4,5-dinitroanthraquinone **(81-55-0)**
4,8-Dihydroxy-1,5-dinitroanthraquinone **(128-91-6)**
3,4'-Dihydroxydiphenyl **(18855-13-5)**
4,4'-Dihydroxydiphenyldimethylmethane **(80-05-7)**
p,p'-Dihydroxydiphenyldimethylmethane **(80-05-7)**
4,4'-Dihydroxydiphenyldimethylmethane diglycidyl ether **(1675-54-3)**
p,p'-Dihydroxydiphenyldimethylmethane diglycidyl ether **(1675-54-3)**
4,4'-Dihydroxy-γ,δ-diphenylhexane **(84-16-2)**
2,2-(4,4'-Dihydroxydiphenyl)propane **(80-05-7)**
4,4'-Dihydroxy-2,2-diphenylpropane **(80-05-7)**
4,4'-Dihydroxydiphenyl-2,2-propane **(80-05-7)**
4,4'-Dihydroxydiphenylpropane **(80-05-7)**
p,p'-Dihydroxydiphenylpropane **(80-05-7)**
4,4'-Dihydroxydiphenyl sulfide **(2664-63-3)**
2,2'-Dihydroxydipropyl ether **(110-98-5)**
2,2'-Dihydroxy-di-n-propylnitrosoamine **(53609-64-6)**
3,14-Dihydroxy-4,5-α-epoxy-17-methylmorphinan-6-one **(76-41-5)**
3,17-Dihydroxyestratriene **(57-91-0)**
3,17-β-Dihydroxy-1,3,5(10)-estratriene **(50-28-2)**
3,17-β-Dihydroxyestra-1,3,5(10)-triene **(50-28-2)**
Dihydroxyestrin **(50-28-2)**
1,2-Dihydroxyethane **(107-21-1)**
Di(2-hydroxyethyl)amine **(111-42-2)**
Dihydroxyethylaniline **(120-07-0)**
N,N-Di(2-hydroxyethyl)aniline **(120-07-0)**
N,N-Di(β-hydroxyethyl)aniline **(120-07-0)**
1,2-Dihydroxyethylbenzene **(93-56-1)**
2,4-Dihydroxy-1-ethylbenzene **(2896-60-8)**
α,β-Dihydroxyethylbenzene **(93-56-1)**
N,N-Dihydroxyethyl-3-chloroaniline **(92-00-2)**
N,N-Dihydroxyethyl-m-chloroaniline **(92-00-2)**
N,N-Di(2-hydroxyethyl)-N-coco-N-methylammonium chloride **(70750-47-9)**
N,N-Di(2-hydroxyethyl)cyclohexylamine **(4500-29-2)**
1,3-Di(hydroxyethyl)-5,5-dimethylhydantoin **(26850-24-8)**
2,2'-Dihydroxyethyl ether **(111-46-6)**
Di(hydroxyethyl) ether dinitrate **(693-21-0)**
2,2'-Dihydroxyethyl ether monododecanoate **(141-20-8)**
1,4-Di(2-hydroxyethyl)piperazine **(122-96-3)**
N,N'-Di(2-hydroxyethyl)piperazine **(122-96-3)**
β,β'-Dihydroxyethyl sulfide **(111-48-8)**
N,N-Dihydroxyethyl-m-toluidine **(91-99-6)**
Di-(hydroxyethyl)-o-tolylamine **(28005-74-5)**
3,17-β-Dihydroxy-17-α-ethynyl-1,3,5(10)-estratriene **(57-63-6)**
3,17-β-Dihydroxy-17-α-ethynyl-1,3,5(10)-estratriene-3-methyl ether **(72-33-3)**
3,17-β-Dihydroxy-17-α-ethynyl-1,3,5(10)-oestratriene **(57-63-6)**
3',6'-Dihydroxyfluoran **(2321-07-5)**
11-β,17-β-Dihydroxy-9-α-fluoro-17-α-methyl-4-androster-3-one **(76-43-7)**
2,2'-Dihydroxy-3,3',5,5',6,6'-hexachlorodiphenylmethane **(70-30-4)**
2,2'-Dihydroxy-3,5,6,3',5',6'-hexachlorodiphenylmethane **(70-30-4)**
3,4-Dihydroxyhydrocinnamic acid **(1078-61-1)**
1,8-Dihydroxy-4-(4'-β-hydroxyethyl)anilino-6-nitroanthroquinone **(15791-78-3)**
1,8-Dihydroxy-4-(p-(2-hydroxyethyl)anilino)-5-nitroanthraquinone **(15791-78-3)**
1,8-Dihydroxy-3-hydroxymethyl-10-(6-hydroxymethyl-3,4,5-trihydroxy-2-pyranyl)anthrone **(1415-73-2)**
7-(3,5-Dihydroxy-2-(3-hydroxy-1-octenyl)cyclopentyl)-

5-heptenoic acid **(551-11-1)**
(+-)-2,4-Dihydroxy-N-(3-hydroxypropyl)-3,3-dimethylbutyramide **(16485-10-2)**
D-(+)-2,4-Dihydroxy-N-(3-hydroxypropyl)-3,3-dimethylbutyramide **(81-13-0)**
4,5-Dihydroxy-2-imidazolidinone **(3720-97-6)**
4,5-Dihydroxyimidazolidone-2 **(3720-97-6)**
2,2-Dihydroxy-1,3-indandione **(485-47-2)**
2,2-Dihydroxy-1H-indene-1,3(2H)-dione **(485-47-2)**
2,6-Dihydroxyisonicotinic acid **(99-11-6)**
3,4-Dihydroxy-α-((isopropylamino)methyl)benzyl alcohol hydrochloride **(51-30-9)**
β,β'-Dihydroxyisopropyl chloride **(96-24-2)**
2,2'-Dihydroxyisopropyl ether **(110-98-5)**
3,4-Dihydroxy-l-phenylalanine **(59-92-7)**
Dihydroxy-l-phenylalanine **(59-92-7)**
2,3-Dihydroxy-4-methoxyacetophenone **(708-53-2)**
α,4-Dihydroxy-3-methoxybenzeneacetic acid **(55-10-7)**
2,2'-Dihydroxy-4-methoxybenzophenone **(131-53-3)**
1,2-Dihydroxy-3-(2-methoxyphenoxy)propane **(93-14-1)**
3,4-Dihydroxy-α-((methylamino)methyl)benzyl alcohol **(51-43-4)**
(+-)-3,4-Dihydroxy-α-((methylamino)methyl)benzyl alcohol hydrochloride **(329-63-5)**
(-)-3,4-Dihydroxy-α-((methylamino)methyl)benzyl alcohol (+)-tartrate (1:1) salt **(51-42-3)**
3-Di(hydroxymethyl)amino-6-(2-(5-nitro-2-furyl)vinyl)-1,2,4-triazine **(794-93-4)**
1,8-Dihydroxy-3-methylanthraquinone **(481-74-3)**
1,3-Dihydroxy-5-methylbenzene **(504-15-4)**
2,5-Dihydroxy-3-methylbenzoic acid **(5981-39-5)**
5,7-Dihydroxy-4-methyl-2H-1-benzopyran-2-one **(2107-76-8)**
5,7-Dihydroxy-4-methylcoumarin **(2107-76-8)**
6,7-Dihydroxy-4-methylcoumarin **(529-84-0)**
7,8-Dihydroxy-4-methylcoumarin **(2107-77-9)**
2,12-Dihydroxy-4-methyl-11,16-dioxosenecionanium **(2318-18-5)**
Di-4-hydroxy-3,3'-methylenedicoumarin **(66-76-2)**
2,3-Dihydroxy-2-(1-methylethyl)-butanoic acid) 2,3,5,7a-tetrahydro-1-hydroxy-1H-pyrrolizin-7-yl) methyl ester **(480-82-0)**
Dihydroxymethyl furatrizine **(794-93-4)**
1,3-Dihydroxymethyl-2-imidazolidone **(136-84-5)**
2,4-Dihydroxy-2-methylpentane **(107-41-5)**
N,N'-Dihydroxymethylurea **(140-95-4)**
1,2-Dihydroxynaphthalene **(574-00-5)**
1,5-Dihydroxynaphthalene **(83-56-7)**
1,6-Dihydroxynaphthalene **(575-44-0)**
2,7-Dihydroxynaphthalene **(582-17-2)**
6,7-Dihydroxy-2-naphthalenesulfonic acid **(92-27-3)**
2,3-Dihydroxynaphthoic acid **(16715-77-8)**
1,5-Dihydroxynapthalene **(83-56-7)**
2,4-Dihydroxy-4'-nitroazobenzene **(74-39-5)**
4-((2,4-Dihydroxy-3-((4-nitrophenyl)azo)-5-((3,5-dinitro-2-hydroxyphenyl)azo)phenyl)azo)-5-hydroxy-2,7-naphthalenedisulfonic acid, disodium salt **(6637-87-2)**
D-threo-N-(1,1'-Dihydroxy-1-p-nitrophenylisopropyl)dichloroacetamide **(56-75-7)**
3,17-α-Dihydroxyoestra-1,3,5(10)-triene **(57-91-0)**
3,17-β-Dihydroxy-1,3,5(10)-oestratriene **(50-28-2)**
3,17-β-Dihydroxyoestra-1,3,5-triene **(50-28-2)**
Dihydroxyoestrin **(50-28-2)**
1,5-Dihydroxypentane **(111-29-5)**
3,5-Dihydroxyphenol **(108-73-6)**
2,5-Dihydroxyphenylacetic acid **(451-13-8)**
3,4-Dihydroxy-phenylacetic acid **(102-32-9)**
Dihydroxyphenylacetic acid **(102-32-9)**
(-)-3,4-Dihydroxyphenylalanine **(59-92-7)**
3,4-Dihydroxyphenylalanine **(59-92-7)**
β-(3,4-Dihydroxyphenyl)-α-alanine **(59-92-7)**
l-3,4-Dihydroxyphenyl-α-alanine **(59-92-7)**
l-3,4-Dihydroxyphenylalanine **(59-92-7)**
l-Dihydroxyphenylalanine **(59-92-7)**
l-α-Dihydroxyphenylalanine **(59-92-7)**
l-β-(3,4-Dihydroxyphenyl)alanine **(59-92-7)**
l-1-(3,4-Dihydroxyphenyl)-2-aminoethanol **(51-41-2)**
p-((2,4-Dihydroxyphenyl)azo)benzenesulfonic acid, sodium salt **(547-57-9)**
p-(2,4-Dihydroxyphenylazo)benzenesulfonic acid, sodium salt **(547-57-9)**
l-3,4-Dihydroxyphenylethanolamine **(51-41-2)**
γ,δ-Di(p-hydroxyphenyl)-hexane **(84-16-2)**
meso-3,4-Di(p-hydroxyphenyl)-n-hexane **(84-16-2)**
3,4'(4,4'-Dihydroxyphenyl)hex-3-ene **(56-53-1)**
(-)-3-(3,4-Dihydroxyphenyl)-l-alanine **(59-92-7)**
3,4-Dihydroxyphenyl-l-alanine **(59-92-7)**
3-(3,4-Dihydroxyphenyl)-l-alanine **(59-92-7)**
β-(3,4-Dihydroxyphenyl)-l-alanine **(59-92-7)**
L(-)-β-(3,4-Dihydroxyphenyl)-α-methylalanine **(555-30-6)**
L-(-)-3-(3,4-Dihydroxyphenyl)-2-methylalanine **(555-30-6)**
L-3-(3,4-Dihydroxyphenyl)-2-methylalanine **(555-30-6)**
l-1-(3,4-Dihydroxyphenyl)-2-methylaminoethanol **(51-43-4)**
l-1-(3,4-Dihydroxyphenyl)-2-methylamino-1-ethanol hydrochloride **(55-31-2)**
1-(2,5-Dihydroxyphenyl)-1-octanone **(4693-19-0)**
1,2-Dihydroxy-1-phenylpropane **(1855-09-0)**
2,2-Di(4-hydroxyphenyl)propane **(80-05-7)**
β-Di-p-hydroxyphenylpropane **(80-05-7)**
4,5-Dihydroxyphthalic acid **(63958-66-7)**
Dihydroxyphthalophenone **(77-09-8)**
11-β,21-Dihydroxypregn-3,20-dione **(50-22-6)**
17,21-Dihydroxypregna-1,4-diene-3,11,20-trione **(53-03-2)**
17α,21-Dihydroxy-4-pregnene-3,11,20-trione **(53-06-5)**
17α,21-Dihydroxypregn-4-ene-3,11,20-trione **(53-06-5)**
17,21-Dihydroxypregn-4-ene-3,11,20-trione acetate **(50-04-4)**
11,12-Dihydroxyprogesterone **(50-22-6)**
11-β,21-Dihydroxyprogesterone **(50-22-6)**
1,2-Dihydroxypropane **(57-55-6)**
1,3-Dihydroxypropane **(504-63-2)**
1,3-Dihydroxypropanone **(96-26-4)**
1-(2,3-Dihydroxypropoxy)naphthalene **(36112-95-5)**
2,3-Dihydroxypropylamine **(616-30-8)**
2,3-Dihydroxypropyl butanoate **(557-25-5)**
2,3-Dihydroxypropyl chloride **(96-24-2)**
7-(2,3-Dihydroxypropyl)-3,7-dihydro-1,3-dimethyl-1H-purine-2,6-dione **(479-18-5)**
N-(2,3-Dihydroxypropyl)dodecylamine **(821-91-0)**
Di(2-hydroxypropyl)nitrosamine **(53609-64-6)**
N,N-Di-(2-hydroxypropyl)nitrosamine **(53609-64-6)**
2,3-Dihydroxypropyl octadecanoate **(123-94-4)**
4-(o-(2',3'-Dihydroxypropyloxycarbonyl)phenyl)-amino-8-trifluoromethylquinoline **(23779-99-9)**
7-(2,3-Dihydroxypropyl)theophylline **(479-18-5)**
Dihydroxypropyl theophylline **(479-18-5)**
Dihydroxypropyl theopylin (German) **(479-18-5)**
(1,2-Dihydroxy-3-propyl)thiophyllin **(479-18-5)**
2,3-Dihydroxypropyl N-(8-(trifluoromethyl)-4-quinolyl) anthranilate **(23779-99-9)**
2,4-Dihydroxy-2H-pyran-δ-3(6H),α-acetic acid-3,4-lactone **(149-29-1)**
(2,4-Dihydroxy-2H-pyran-3(6H)-ylidene)acetic acid-3,4-lactone **(149-29-1)**
2,3-Dihydroxypyridine **(16867-04-2)**
2,4-Dihydroxypyrimidine **(66-22-8)**
8,8'-Dihydroxy-rugulosin **(21884-44-6)**
12,18-Dihydroxy-senecionan-11,16-dione **(480-54-6)**
3',6'-Dihydroxyspiro(isobenzofuran-1(3h),9'-(9h)xanthen)-3-one **(2321-07-5)**
3,6-Dihydroxyspiro(xanthene-9,3'-phthalide) **(2321-07-5)**
Dihydroxytartaric acid **(76-30-2)**
2,2'-Dihydroxy-3,3',5,5'-tetrachlorodiphenylsulfide **(97-18-7)**
4,5-Dihydroxytetrahydroimidazol-2-one **(3720-97-6)**
1,8-Dihydroxy-2,4,5,7-tetranitroanthraquinone **(517-92-0)**
Dihydroxy-1,8 tetranitro-2,4,5,7 anthraquinone (French) **(517-92-0)**
2,3-Dihydroxytoluene **(488-17-5)**
2,5-Dihydroxytoluene **(95-71-6)**

2,6-Dihydroxytoluene (608-25-3)
3,4-Dihydroxytoluene (452-86-8)
3,5-Dihydroxytoluene (504-15-4)
α,2-Dihydroxytoluene (90-01-7)
2,5-Dihydroxy-α-toluic acid (451-13-8)
2,5-Dihydroxy-m-toluic acid (5981-39-5)
1,3-Dihydroxy-2,4,6-trinitrobenzene (82-71-3)
2,4-Dihydroxy-1,3,5-trinitrobenzene (82-71-3)
Dihydroxy(2,4,6-trinitro-1,3-benzenediolato(2-)dilead (12403-82-6)
6-(6,10-Dihydroxyundecyl)-β-resorcylic acid, mu-lactone (26538-44-3)
1-α,25-Dihydroxyvitamin D3 (32222-06-3)
Dihydroxyvitamin D3 (32222-06-3)
Dihyxal (96-26-4)
Diidro-5,5-dietil-2H-1,3-ossazin-2,4(3H)-dione (Italian) (702-54-5)
1,4-Diidrobenzene (Italian) (123-31-9)
5,28:14,19-Diimino-7,12:26,21-dinitrilotetrabenzo-(c,h,m,r)(1,6,11,16) tetraazacycloeicosine (574-93-6)
1,3-Diiminoisoindolin (Czech) (3468-11-9)
1,3-Diiminoisoindoline (3468-11-9)
Diindogen (482-89-3)
Diindolo(3,2-b:3',2'-m)triphenodioxazine, 8,18-dichloro-5,15-diethyl-5,15-dihydro- (9CI) (6358-30-1)
Diindolo(3,2-b:3',2'-m)triphenodioxazinetrisulfonic acid, 8,18-dichloro-5,15-diethyl-5,15-dihydro-, trisodium salt (9CI) (1324-58-9)
Diioacetylene (624-74-8)
Diiodoacetylene (624-74-8)
1,2-Diiodobenzene (615-42-9)
1,3-Diiodobenzene (626-00-6)
1,4-Diiodobenzene (624-38-4)
4,4'-Diiodo-1,1'-biphenyl (3001-15-8)
Diiododithymol (552-22-7)
1,2-Diiodoethane (624-73-7)
cis-1,2-Diiodoethene (590-26-1)
trans-1,2-Diiodoethene (590-27-2)
trans-Diiodoethene (590-27-2)
(E)-1,2-Diiodoethylene (590-27-2)
1,2-Diiodoethylene (20244-70-6)
cis-1,2-Diiodoethylene (590-26-1)
cis-Diiodoethylene (590-26-1)
trans-1,2-Diiodoethylene (590-27-2)
trans-Diiodoethylene (590-27-2)
3,5-Diiodo-2-hydroxybenzoic acid (133-91-5)
3,5-Diiodo-4-hydroxybenzonitrile (1689-83-4)
3,5-Diiodo-4-hydroxybenzonitrile octanoate (3861-47-0)
3,5-Diiodo-4-hydroxyphenyl 2-ethyl-3-benzofuranyl ketone (68-90-6)
Diiodohydroxyquin (83-73-8)
5,7-Diiodo-8-hydroxyquinoline (83-73-8)
Diiodohydroxyquinoline (83-73-8)
Diiodo-l-tryosine (66-02-4)
Diiodomethane (75-11-6)
1-((Diiodomethyl)sulfonyl)-4-methylbenzene (20018-09-1)
Diiodomethyl p-tolyl sulfone (20018-09-1)
2,6-Diiodo-4-nitrophenol (305-85-1)
3,5-Diiodo-4-octanoyloxybenzonitrile (3861-47-0)
5,7-Diiodo-oxine (83-73-8)
Diiodoquin (83-73-8)
5,7-Diiodo-8-quinolinol (83-73-8)
3,5-Diiodotyrocine (300-39-0)
3,5-Diiodo-L-tyrosine (300-39-0)
3,5-L-Diiodotyrosine (300-39-0)
Diiodotyrosine (66-02-4)
3,5-Diiodotyrosine (VAN) (66-02-4)
Diiodozinc (10139-47-6)
Diiron monophosphide (1310-43-6)
Di-iron phosphide (1310-43-6)
Diiron trisulfate (10028-22-5)
Diisoamyl (1072-16-8)
Diisoamylamine (544-00-3)
Diisoamyl ether (544-01-4)
Di(isobornylphenyl)amine (68586-20-9)
Diisobutene (25167-70-8)
Diisobutilchetone (Italian) (108-83-8)
Diisobutyl adipate (141-04-8)
Diisobutylaluminium hydride (1191-15-7)
Diisobutylaluminum (1191-15-7)
Diisobutylaluminum chloride (1779-25-5)
Diisobutylaluminum hydride (1191-15-7)
Diisobutylaluminum monochloride (1779-25-5)
Diisobutylamine [UN 2361] (110-96-3)
Diisobutyl carbinol (108-82-7)
Diisobutylchloroaluminum (1779-25-5)
2-(2-(p-(Diisobutyl)cresoxy)ethoxy)ethyl dimethyl benzyl ammonium chloride (25155-18-4)
Diisobutyl cresoxy ethoxy ethyl dimethyl benzyl ammonium chloride (25155-18-4)
Diisobutylcresoxyethoxyethyl dimethyl benzyl ammonium 8-93chloride (25155-18-4)
1,4-Diisobutyl-1,4-dimethylbutynediol (126-86-3)
Diisobutylene (107-40-4)
β-Diisobutylene (107-40-4)
Diisobutylene (DOT) (25167-70-8)
Diisobutylene, isomeric compounds [UN 2050] (25167-70-8)
Diisobutylester kyseliny ftalove (Czech) (84-69-5)
Diisobutylhydroaluminum (1191-15-7)
Diisobutylketon (Dutch, German) (108-83-8)
Diisobutyl ketone (ACGIH,DOT,OSHA) [UN 1157] (108-83-8)
Diisobutylnaphthalenesulfonic acid, sodium salt (27213-90-7)
Diisobutylphenoxyethoxyethyl dimethyl benzyl ammonium chloride (121-54-0)
Diisobutyl phthalate (84-69-5)
Diisobutyl sodium sulfosuccinate (127-39-9)
Diisobutylthiocarbamic acid S-ethyl ester (2008-41-5)
Diisobutyryl peroxide (3437-84-1)
Diisobutyryl peroxide, Not more than 52% in solution (3437-84-1)
Diisocarb (2008-41-5)
O,O'-Diisoctylphosphorodithioic acid (26999-29-1)
4-4'-Diisocyanate de diphenylmethane (French) (101-68-8)
Di-isocyanate de toluylene (French) (584-84-9)
1,4-Diisocyanatocyclohexane (2556-36-7)
4,4'-Diisocyanato-3,3'-dimethoxy-1,1'-biphenyl (91-93-0)
4,4'-Diisocyanatodiphenylmethane (101-68-8)
1,6-Diisocyanatohexane (822-06-0)
Di-iso-cyanatoluene (584-84-9)
Diisocyanat-toluol (German) (584-84-9)
2,6-Diisocyanato-1-methylbenzene (91-08-7)
Diisocyanatomethylbenzene (26471-62-5)
2,4-Diisocyanato-1-methylbenzene (9CI) (584-84-9)
2,4-Diisocyanatotoluene (584-84-9)
2,6-Diisocyanatotoluene (91-08-7)
Diisocyanatotoluene (26471-62-5)
1,6-Diisocyanato-2,2,4-trimethylhexane (16938-22-0)
1,6-Diisocyanato-2,4,4-trimethylhexane (15646-96-5)
Diisodecyl adipate (27178-16-1)
Diisodecyl glutarate (29733-18-4)
N,N-Diisodecylisodecanamine (35723-89-8)
Diisodecyl nonanedioate (28472-97-1)
Diisodecyl pentanedioate (29733-18-4)
Diisodecyl phenyl phosphate (51363-64-5)
O,O-Diisodecyl phosphorodithioate (28631-44-9)
Diisodecyl phthalate (26761-40-0)
Diisodecyl sulfosuccinate, sodium salt (29857-13-4)
2,3-Diisonitrosobutane (95-45-4)
Diisononyl adipate (33703-08-1)
Diisononyl 1,2-benzenedicarboxylate (28553-12-0)

Name	CAS No.
Diisononyl hexanedioate	(33703-08-1)
Diisononylnaphthalene	(63512-64-1)
Diisononyl phthalate	(28553-12-0)
Diisooctyl acid phosphate [UN 1902]	(27215-10-7)
Di-iso-octyl adipate	(1330-86-5)
Diisooctyl adipate	(1330-86-5)
Diisooctyl 1,2-benzenedicarboxylate	(27554-26-3)
Diisooctyl ((dioctylstannylene)dithio)diacetate	(26401-97-8)
Diisooctyl hexanedioate	(1330-86-5)
N,N-Diisooctylisooctanamine	(25549-16-0)
Diisooctyl maleate	(1330-76-3)
Diisooctyl nonanedioate	(26544-17-2)
Diisooctyl phosphate	(27215-10-7)
O,O-Diisooctyl phosphorodithioate	(26999-29-1)
O,O'-Diisooctyl phosphorodithioic acid	(26999-29-1)
Di-iso-octyl phthalate	(27554-26-3)
Diisooctyl phthalate	(1330-86-5)
Diisooctyl phthalate	(27554-26-3)
Diisooctyl phthalate	(27554-26-3)
Diisopentylamine (8CI)	(544-00-3)
Diisopentyl ether	(544-01-4)
Diisophenol	(305-85-1)
Diisopropanolamine	(110-97-4)
Diisopropanolnitrosamine	(53609-64-6)
Diisopropenyl	(513-81-5)
N,N'-Diisopropil-fosforodiammido-fluoruro (Italian)	(371-86-8)
Diisopropoxybis(2-(bis(2-hydroxyethyl)amino)ethoxy)titanium	(36673-16-2)
Diisopropoxydi(ethoxyacetoacetyl)titanate	(27858-32-8)
Diisopropoxyphosphoryl fluoride	(55-91-4)
s-Diisopropylacetone	(108-83-8)
sym-Diisopropylacetone	(108-83-8)
Diisopropyl adipate	(6938-94-9)
Diisopropyl aluminum ethyl acetoacetate	(14782-75-3)
Di(isopropylamido)phosphoryl fluoride	(371-86-8)
Diisopropylamine (ACGIH,OSHA) [UN 1158]	(108-18-9)
Diisopropylamine, nitrate (8CI)	(6143-52-8)
Diisopropylamine, N-nitroso	(601-77-4)
2-Diisopropylaminoethanol	(96-80-0)
S-(2-Diisopropylaminoethyl) O-ethyl methyl phosphonothiolate	(50782-69-9)
2-(Diisopropylamino)ethyl methacrylate	(16715-83-6)
Diisopropylaminoethyl methacrylate	(16715-83-6)
N,N-Diisopropylaminoethyl methacrylate	(16715-83-6)
2,6-Diisopropylamino-4-methoxytriazine	(1610-18-0)
2,6-Diisopropyl aniline	(24544-04-5)
1,3-Diisopropylbenzene	(99-62-7)
1,4-Diisopropylbenzene	(100-18-5)
Diisopropylbenzene	(25321-09-9)
o-Diisopropylbenzene	(577-55-9)
p-Diisopropylbenzene	(100-18-5)
Diisopropylbenzene hydroperoxide (DOT)	(26762-93-6)
Diisopropylbenzene peroxide	(80-43-3)
N,N-Diisopropyl-2-benzothiazolesulfenamide	(95-29-4)
O,O-Diisopropyl-S-benzylester kyseliny thiofosforecne (Czech)	(26087-47-8)
O,O-Diisopropyl S-benzyl phosphorothiolate	(26087-47-8)
O,O-Diisopropyl S-benzyl thiophosphate	(26087-47-8)
Diisopropyl biphenyl	(36876-13-8)
Diisopropylbiphenyl	(36876-13-8)
Diisopropylbiphenyl	(69009-90-1)
Diisopropylcarbodiimide	(693-13-0)
N,N'-Diisopropyl-diamido-fosforzuur-fluoride (Dutch)	(371-86-8)
N,N'-Diisopropyl-diamido-phosphorsaeure-fluorid (German)	(371-86-8)
N,N'-Diisopropyldiamidophosphoryl fluoride	(371-86-8)
4,6-Diisopropyl-1,3-dimethylbenzene	(5186-68-5)
Di-isopropyl 1,3-dithiolane-2-ylidenemalonate	(50512-35-1)
O,O-Diisopropyl dithiophosphate	(107-56-2)
O,O-Diisopropyl dithiophosphoric acid	(107-56-2)
N-(2-(O,O-Diisopropyldithiophosphoryl)ethyl)benzenesulfonamide	(741-58-2)
N-(β-O,O-Diisopropyldithiophosphorylethyl)benzenesulfonamide	(741-58-2)
1,1-Diisopropylethanol	(3054-92-0)
Diisopropyl ethanolamine	(96-80-0)
N,N-Diisopropylethanolamine	(96-80-0)
N,N-Diisopropyl ethanolamine (DOT)	(96-80-0)
Diisopropyl ether	(108-20-3)
Diisopropyl ether [UN 1159]	(108-20-3)
N,N-Diisopropylethylamine	(7087-68-5)
O,O-Diisopropyl-S-ethylsulfinylmethyldithiophosphate	(5827-05-4)
O,O-Diisopropyl-S-ethylsulfinylmethyl phosphorodithioate	(5827-05-4)
Diisopropylfluorfosfat (Czech)	(55-91-4)
Diisopropyl fluorophosphate	(55-91-4)
O,O-Diisopropyl fluorophosphate	(55-91-4)
Diisopropyl fluorophosphonate	(55-91-4)
Diisopropylfluorophosphoric acid ester	(55-91-4)
Diisopropylfluorphosphorsaeureester (German)	(55-91-4)
O,O-Diisopropyl hydrogen dithiophosphate	(107-56-2)
Diisopropyl hydrogen phosphite	(1809-20-7)
O,O-Diisopropyl hydrogen phosphorodithioate	(107-56-2)
Diisopropyl(2-hydroxyethyl)methylammonium bromide xanthene-9-carboxylate	(50-34-0)
Diisopropylidene acetone	(504-20-1)
sym-Diisopropylidene acetone	(504-20-1)
Diisopropyl ketone	(565-80-0)
Diisopropyl methanephosphonate	(1445-75-6)
N,N'-Diisopropyl-6-methoxy-1,3,5-triazine-2,4-diyldiamine	(1610-18-0)
Diisopropylmethylcarbinol	(3054-92-0)
Diisopropyl methylphosphonate	(1445-75-6)
N,N'-Diisopropyl-6-methylthio-1,3,5-triazine-2,4-diyldiamine	(7287-19-6)
2,6-Diisopropylnaphthalene	(24157-81-1)
Diisopropyl naphthalene	(38640-62-9)
Diisopropylnaphthalene	(38640-62-9)
Diisopropylnitrosamin (German)	(601-77-4)
Diisopropyl oxide	(108-20-3)
Diisopropyl perdicarbonate	(105-64-6)
Diisopropyl peroxydicarbonate	(105-64-6)
Diisopropyl peroxydicarbonate, Maximum concentration 52% in solution (DOT)	(105-64-6)
Diisopropyl peroxydicarbonate, Technically pure (DOT)	(105-64-6)
2,4-Diisopropylphenol	(2934-05-6)
2,6-Diisopropylphenol	(2078-54-8)
2,4-Diisopropyl phenylglycidic acid ethyl ester	(1334-99-2)
2,6-Diisopropylphenyl isocyanate	(28178-42-9)
Di(isopropylphenyl) phenyl phosphate	(28109-00-4)
Diisopropyl phosphite	(1809-20-7)
Diisopropyl phosphofluoridate	(55-91-4)
Diisopropylphosphonate	(1809-20-7)
O,O-Diisopropyl phosphonate	(1809-20-7)
N,N'-Diisopropylphosphorodiamidic fluoride	(371-86-8)
Diisopropyl phosphorodithioate	(107-56-2)
O,O-Diisopropyl phosphorodithioate	(107-56-2)
S-(O,O-Diisopropyl phosphorodithioate) ester of N-(2-mercaptoethyl)benzenesulfonamide	(741-58-2)
O,O-Diisopropylphosphorodithioic acid	(107-56-2)
Diisopropyl phosphorofluoridate	(55-91-4)
O,O'-Diisopropyl phosphoryl fluoride	(55-91-4)
Diisopropyl phthalate	(605-45-8)
Di-isopropylsulfat (German)	(2973-10-6)
Di-isopropylsulfate	(2973-10-6)
Diisopropyl sulfide	(625-80-9)
Diisopropylthiocarbamic acid, S-(2,3-dichloroallyl) ester	(2303-16-4)
N-Diisopropylthiocarbamic acid S-2,3,3-trichloro-2-propenyl ester	(2303-17-5)
Di-isopropylthiolocarbamate de S-(2,3-dichloro allyle) (French)	(2303-16-4)
.α.,.α.-Diisopropyltoluene	(21777-84-4)
N,N-Diisopropyl-2,3,3-trichlorallyl-thiolcarbamat (German)	(2303-17-5)
Diisopropyltrichloroallylthiocarbamate	(2303-17-5)
4,6-Diisopropyl-m-xylene	(5186-68-5)

1,4-Diisothiocyanatobenzene **(4044-65-9)**
Diisotridecyl peroxydicarbonate (Technically pure) **(82065-80-3)**
Diisotridecyl phosphate **(27073-01-4)**
3,5-Dijod-4-hydroxy-benzonitril (German) **(1689-83-4)**
3,5-Dijod-4-hydroxy-benzonitril caprysaeureester (German) **(3861-47-0)**
Dijodmethan (Czech) **(75-11-6)**
Dikain hydrochloride **(136-47-0)**
Dikalium phosphate **(7758-11-4)**
Dikar (8CI) **(8064-42-4)**
3,5-Dikarboxybenzensulfonan sodny (Czech) **(6362-79-4)**
Dikaril **(9003-54-7)**
Dikegulac **(52508-35-7)**
Dikegulac (German) **(52508-35-7)**
Dikegulac sodium **(52508-35-7)**
Dikegulac-sodium **(52508-35-7)**
Diketene **(674-82-8)**
Diketene, Inhibited [UN 2521] **(674-82-8)**
2,3-Diketobutane **(431-03-8)**
2,5-Diketohexane **(110-13-4)**
1,3-Diketohydrindene **(606-23-5)**
2,3-Diketoindoline **(91-56-5)**
Diketone alcohol **(123-42-2)**
2,5-Diketopiperazine **(106-57-0)**
Diketopiperazine **(106-57-0)**
2,5-Diketopyrrolidine **(123-56-8)**
2,5-Diketotetrahydrofuran **(108-30-5)**
Dikonirt **(2702-72-9)**
Dikonirt D **(2702-72-9)**
Dikonit **(2893-78-9)**
Dikoteks **(3653-48-3)**
Dikoteks 40 **(19480-39-8)**
Dikoteks AM **(19480-39-8)**
Dikotes **(94-74-6)**
Dikotex **(3653-48-3)**
Dikotex **(94-74-6)**
Dikotex 30 **(3653-48-3)**
Dikotex 40 **(19480-39-8)**
Dikumarol **(66-76-2)**
Dilabid **(57-41-0)**
Dilacoran **(52-53-9)**
Dilactone actinomycin D acid **(50-76-0)**
Dilactone actinomycindioic D acid **(50-76-0)**
Dilafurane **(68-90-6)**
Dilan **(8027-00-7)**
Dilangil **(15825-70-4)**
Dilantin **(57-41-0)**
Dilantin **(630-93-3)**
Dilantin DB **(95-50-1)**
Dilantine **(57-41-0)**
Dilantin sodium **(630-93-3)**
Dilar **(53-33-8)**
Dilatan **(456-59-7)**
Dilatin DB **(95-50-1)**
Dilatin DBI **(25321-22-6)**
Dilatrate-SR **(87-33-2)**
Dilaudid **(466-99-9)**
Dilauroyl peroxide **(105-74-8)**
Dilauroyl peroxide, Not more than 42% (DOT) **(105-74-8)**
Dilauroyl peroxide, Technically pure (DOT) **(105-74-8)**
Dilauryl acid phosphate **(7057-92-3)**
Dilauryl dimethyl ammonium chloride **(3401-74-9)**
Dilauryldimonium chloride **(3401-74-9)**
Dilaurylester kyseliny β',β'-thiodipropionove (Czech) **(123-28-4)**
Dilauryl hydrogen phosphite **(21302-09-0)**
Dilauryl peroxide **(105-74-8)**
Dilauryl phosphate **(7057-92-3)**
Dilauryl phosphite **(21302-09-0)**
Dilauryl phthalate **(2432-90-8)**
Dilauryl 3,3'-thiodipropionate **(123-28-4)**
Dilauryl β',β'-thiodipropionate **(123-28-4)**
Dilauryl β-thiodipropionate **(123-28-4)**
Dilauryl thiodipropionate **(123-28-4)**
Dila-vasal **(68-90-6)**
Dilavase **(395-28-8)**
Dilene **(72-54-8)**
Dilic **(75-60-5)**
Dilinoleic acid, ditridecyl ester **(16958-92-2)**
Dilithium carbonate **(554-13-2)**
Dilithium chromate **(14307-35-8)**
Dillantin **(57-41-0)**
Dillar **(53-33-8)**
Dilombrin **(514-73-8)**
Dilombrine **(514-73-8)**
Dilor **(14168-01-5)**
Dilor **(479-18-5)**
Dilospan S **(108-73-6)**
Dilosyn **(1229-35-2)**
Dilosyn **(1982-37-2)**
Diluex **(12174-11-7)**
Diluran **(59-66-5)**
Dilyn **(93-14-1)**
1,3-Dimaleimidobenzene **(3006-93-7)**
4,4'-Dimaleimidodiphenylmethane **(13676-54-5)**
p,p'-Dimaleimidodiphenylmethane **(13676-54-5)**
Dimanganese trioxide **(1317-34-6)**
Dimanin A **(8001-54-5)**
Dimanin C **(2893-78-9)**
Dimapp **(60-87-7)**
Dimapyrin **(58-15-1)**
Dimas **(1596-84-5)**
Dimate 267 **(60-51-5)**
Dimaz **(298-04-4)**
Dimazin **(57-14-7)**
Dimazine **(57-14-7)**
Dimazon **(83-63-6)**
Dimecron **(13171-21-6)**
Dimecron 100 **(13171-21-6)**
Dimedon **(126-81-8)**
Dimedone **(126-81-8)**
Dimedrol **(147-24-0)**
Dimedrol **(58-73-1)**
Dimedryl **(58-73-1)**
Dimefenthoat **(2597-03-7)**
Dimefeptanol (Spanish) **(545-90-4)**
Dimefeptanolo **(545-90-4)**
Dimefline **(1165-48-6)**
Dimefox **(115-26-4)**
Dimelin **(968-81-0)**
Dimelone **(39589-98-5)**
Dimelor **(968-81-0)**
Dimenformon **(50-28-2)**
Dimenformon benzoate **(50-50-0)**
Dimenformon dipropionate **(113-38-2)**
Dimenformone **(50-50-0)**
Dimenformon prolongatum **(50-28-2)**
Dimenhydrinate **(523-87-5)**
Dimenossadolo **(509-78-4)**
Dimenoxadol **(509-78-4)**
Dimenoxadolum (Latin) **(509-78-4)**
Dimephenthioate **(2597-03-7)**
Dimephenthoate **(2597-03-7)**
Dimepheptanol **(17199-55-2)**
Dimepheptanol **(545-90-4)**
Dimepheptanolum (Latin) **(545-90-4)**

Dimer acid, ditridecyl ester **(16958-92-2)**
Dimercaprol **(59-52-9)**
Dimercaprol propanol **(59-52-9)**
1,2-Dimercaptoethane **(540-63-6)**
Dimercaptol **(59-52-9)**
2,3-Dimercaptol-1-propanol **(59-52-9)**
1,5-Dimercaptonaphthalene **(5325-88-2)**
1,3-Dimercaptopropane **(109-80-8)**
2,3-Dimercapto-1-propanesulfonic acid **(74-61-3)**
2,3-Dimercaptopropanesulfonic acid **(74-61-3)**
2,3-Dimercaptopropan-1-ol **(59-52-9)**
2,3-Dimercaptopropanol **(59-52-9)**
Dimercaptopropanol **(59-52-9)**
2,5-Dimercapto-1,3,4-thiadiazole **(1072-71-5)**
2,5-Dimercaptothiadiazole **(1072-71-5)**
Dimercury dichloride **(10112-91-1)**
Dimercury diiodide **(15385-57-6)**
Dimer cyklopentadienu (Czech) **(77-73-6)**
Dimerin **(125-64-4)**
Dimerkaprol (Czech) **(59-52-9)**
Dimero de la acroleina (Spanish) **(100-73-2)**
1,4-Dimesidinoanthraquinone **(116-75-6)**
Dimesylmannitol **(551-74-6)**
1,4-Dimesyloxybutane **(55-98-1)**
Dimet **(144-21-8)**
Dimet **(60-51-5)**
Dimetan **(122-15-6)**
Dimetate **(60-51-5)**
Dimethachlor **(50563-36-5)**
Dimethachlore **(50563-36-5)**
β,β-Dimethacrylic acid **(541-47-9)**
Dimethametryn **(22936-75-0)**
Dimethametryne **(22936-75-0)**
1,4-Dimethanesulfonoxybutane **(55-98-1)**
1,4-Dimethanesulfonoxylbutane **(55-98-1)**
1,4-Dimethanesulfonyloxybutane **(55-98-1)**
1,4-Dimethanesulphonyloxybutane **(55-98-1)**
1,4:7,10-Dimethanodibenzo(a,e)cyclooctane, 1,2,3,4,7,8,9,10, 13,13,14,14-dodecachloro- 1,4,4a,5,6,6a,7,10,10a,11, 12,12a-dodecahydro **(13560-89-9)**
1,4:6,9-Dimethanodibenzofuran, 1,2,3,4,6,7,8,9,10,10, 11,11-dodecachloro-1,4,4a,5a,6,9,9a,9b-octahydro- **(31107-44-5)**
5,9:7,10a-Dimethano-10ah-(1,3)dioxocino(6,5-d)pyrimidine- 4,7,10,11,12-pentol, octahydro- 12-(hydroxymethyl)-2-imino **(4368-28-9)**
Dimethanol urea **(140-95-4)**
1,4:5,8-Dimethanonaphthalene, 1,2,3,4,10,10-hexachloro- 6,7-epoxy-1,4,4a,5,6,7,8,8a-octahydro- (8CI) **(128-10-9)**
1,4:5,8-Dimethanonaphthalene, 1,2,3,4,10,10-hexachloro- 6,7-epoxy-1,4,4a,5,6,7,8,8a- octahydro-, endo,endo **(72-20-8)**
1,4:5,8-Dimethanonaphthalene, 1,2,3,4,10,10-hexachloro- 6,7-epoxy-1,4,4a,5,6,7,8,8a- octahydro, endo,exo **(60-57-1)**
1,4:5,8-Dimethanonaphthalene, 1,2,3,4,10,10-hexachloro- 1,4,4a,5,8,8a-hexahydro- (8CI,9CI) **(124-96-9)**
1,4:5,8-Dimethanonaphthalene, 1,2,3,4,10,10-hexachloro- 1,4,4a,5,8,8a-hexahydro-, endo, endo **(465-73-6)**
1,4:5,8-Dimethanonaphthalene, 1,2,3,4,10,10-hexachloro- 1,4,4a,5,8,8a-hexahydro-, endo,exo **(309-00-2)**
1,4:5,8-Dimethanonaphthalene, 1,4,4a,5,8,8a-hexahydro- 1,2,3,4,10,10-hexachloro-, endo, exo mixture (65% or less aldrin) **(309-00-2)**
1,4:5,8-Dimethanonaphthalene, 1,4,4a,5,8,8a-hexahydro-1,2,3,4, 10,10-hexachloro-, endo, exo mixture (More than 60% aldrin) **(309-00-2)**
2,7:3,6-Dimethanonaphth(2,3-b)oxirene-4-d, 3,5,6,9,9-pentachloro- 1a,2,2a,3,6,6a,7,7a-octahydro-, (1aα,2β,2aα,3β,6β,6aα,7β,7aα)- (9CI) **(60468-28-2)**
2,7:3,6-Dimethanonaphth(2,3-b)oxirene, 3,4,5,6,9,9-hexachloro- 1a,2,2a,3,6,6a,7,7a-octahydro- (9CI) **(128-10-9)**
Dimethazone **(81777-89-1)**
Dimethenthoate **(2597-03-7)**
Dimethesterone **(79-64-1)**
Dimethibutin **(524-84-5)**
Dimethicone **(9006-65-9)**
Dimethicone 350 **(9016-00-6)**
Dimethicones **(9006-65-9)**
Dimethipin **(55290-64-7)**
Dimethisteron **(79-64-1)**
Dimethisterone **(79-64-1)**
Dimethoaat (Dutch) **(60-51-5)**
Dimethoat (German) **(60-51-5)**
Dimethoate **(60-51-5)**
Dimethoate-267 **(60-51-5)**
Dimethoate O-Analog **(1113-02-6)**
Dimethoate Oxygen Analog **(1113-02-6)**
Dimethoate Po Isologue **(1113-02-6)**
Dimethoate-ethyl **(116-01-8)**
Dimethoat technisch 95% **(60-51-5)**
Dimethogen **(60-51-5)**
Dimethoxane **(828-00-2)**
Dimethoxon **(1113-02-6)**
Dimethoxy-DDT **(72-43-5)**
Dimethoxy-DT **(72-43-5)**
2,6-Dimethoxyacetophenone **(2040-04-2)**
1,2-Dimethoxy-4-allylbenzene **(93-15-2)**
2,5-Dimethoxyamphetamine hydrochloride **(24973-25-9)**
2,4-Dimethoxyaniline **(2735-04-8)**
2,5-Dimethoxyaniline **(102-56-7)**
2,4-Dimethoxyaniline hydrochloride **(54150-69-5)**
1,5-Dimethoxyanthrachinon (Czech) **(6448-90-4)**
1,5-Dimethoxyanthraquinone **(6448-90-4)**
2,2'-Dimethoxyazobenzene **(613-55-8)**
3,3'-Dimethoxyazobenzene **(6319-23-9)**
4,4'-Dimethoxyazobenzene **(501-58-6)**
2,5-Dimethoxybenzaldehyde **(93-02-7)**
3,4-Dimethoxybenzaldehyde **(120-14-9)**
2,4-Dimethoxybenzenamine hydrochloride **(54150-69-5)**
1,2-Dimethoxybenzene **(91-16-7)**
1,3-Dimethoxybenzene **(151-10-0)**
1,4-Dimethoxybenzene **(150-78-7)**
m-Dimethoxybenzene **(151-10-0)**
o-Dimethoxybenzene **(91-16-7)**
p-Dimethoxybenzene **(150-78-7)**
3,4-Dimethoxybenzeneacetic acid **(93-40-3)**
2,5-Dimethoxybenzeneazo-β-naphthol **(6358-53-8)**
3,4-Dimethoxybenzenecarbonal **(120-14-9)**
3,4-Dimethoxybenzenemethanol **(93-03-8)**
3,3'-Dimethoxybenzidin (Czech) **(119-90-4)**
3,3'-Dimethoxybenzidine **(119-90-4)**
3,3'-Dimethoxybenzidine dihydrochloride **(20325-40-0)**
3,3'-Dimethoxybenzidine-4,4'-diisocyanate **(91-93-0)**
2,4-Dimethoxybenzoic acid **(91-52-1)**
2,6-Dimethoxybenzoic acid **(1466-76-8)**
3,4-Dimethoxybenzoic acid **(93-07-2)**
3,5-Dimethoxybenzoic acid **(1132-21-4)**
3,3'-Dimethoxy-(1,1'-biphenyl)-4,4'-bis(diazonium) **(20282-70-6)**
3,3'-Dimethoxybiphenyl-4,4'-bisdiazonium **(20282-70-6)**
3,3'-((3,3'-Dimethoxy(1,1'-biphenyl)-4,4'-diyl)bis(azo))bis (4,5-dihydroxy-2,7-naphthalenedisulfonic acid), tetrasodium salt **(4198-19-0)**
3,3'-Dimethoxy-4,4'-biphenylene diisocyanate **(91-93-0)**
3,3'-Dimethoxy-4,4'-biphenylylene isocyanate **(91-93-0)**
3,3'-Dimethoxy-4,4'-biphenylylene isocyanic acid ester **(91-93-0)**
4,5-Dimethoxy-1,3-bis(methoxymethyl)-2-imidazolidinone **(4356-60-9)**
dl-2,5-Dimethoxy-4-bromoamphetamine hydrobromide **(53581-53-6)**
2,3-Dimethoxybuta-1,3-diene **(3588-31-6)**
2,3-Dimethoxybutadiene **(3588-31-6)**

Dimethoxycarbenium hexafluorophosphate **(50318-32-6)**
3,3'-Dimethoxy-4-((8-((2-carboxy-4-nitrophenyl)azo)-7-amino-4-hydroxy-2-sulfonaphth-3-yl)azo)-4'-((2-hydroxy-6-sulfonaphth-1-yl)azo)-1,1'-biphenyl, trisodium salt **(71566-41-1)**
2,5-Dimethoxy-4-chloroacetoacetanilide **(4433-79-8)**
2,5-Dimethoxychlorobenzene **(2100-42-7)**
N-(3',4'-Dimethoxycinnamoyl)anthranilic acid **(53902-12-8)**
N-(3,4-Dimethoxycinnamoyl)anthranilic acid **(53902-12-8)**
6,7-Dimethoxycoumarin **(120-08-1)**
2,2'-Dimethoxydiethylamine **(111-95-5)**
2,5-Dimethoxy-2,5-dihydrofuran **(332-77-4)**
.α.,.α.'-Dimethoxydimethyl ether **(628-90-0)**
1,1-Dimethoxy-3,7-dimethyl-2,6-octadiene **(7549-37-3)**
Dimethoxydimethylolurea **(141-07-1)**
4,4'-Dimethoxydiphenylamine **(101-70-2)**
p,p'-Dimethoxydiphenylamine **(101-70-2)**
2,2-Dimethoxy-1,2-diphenylethanone **(24650-42-8)**
Dimethoxydiphenylsilane **(6843-66-9)**
p,p'-Dimethoxydiphenyltrichloroethane **(72-43-5)**
Dimethoxyethane **(110-71-4)**
α,β-Dimethoxyethane **(110-71-4)**
1,1-Dimethoxyethane [UN 2377] **(534-15-6)**
1,2-Dimethoxyethane [UN 2252] **(110-71-4)**
2,2-Dimethoxyethylamine **(22483-09-6)**
Dimethoxyethylamine **(111-95-5)**
Di-(2-methoxyethyl)ester kyseliny ftalove (Czech) **(117-82-8)**
2,2-Dimethoxyethyl(methyl)amine **(122-07-6)**
Di(2-methoxyethyl)phthalate **(117-82-8)**
Dimethoxy ethyl phthalate **(117-82-8)**
2,5-Dimethoxyfuran **(332-77-4)**
3,5-Dimethoxy-4-hydroxyacetophenone **(2478-38-8)**
3,5-Dimethoxy-4-hydroxybenzaldehyde **(134-96-3)**
4,6-Dimethoxy-2-mercapto-1,3,5-triazine **(30886-14-7)**
Dimethoxymethane **(109-87-5)**
Dimethoxymethane (OSHA) **(109-87-5)**
2,4-Dimethoxy-6-(methylamino)-1,3,5-triazine **(30357-98-3)**
2,5-Dimethoxy-4-methylamphetamine **(15588-95-1)**
1,2-Dimethoxy-4-methylbenzene **(494-99-5)**
2,2-Dimethoxy-N-methylethanamine **(122-07-6)**
Dimethoxymethylium hexafluorophosphate(1-) **(50318-32-6)**
1-(Dimethoxymethyl)-4-methoxybenzene **(2186-92-7)**
3,4-Dimethoxy-17-methylmorphinan-6 β,14-diol **(3176-03-2)**
2,5-Dimethoxy-α-methylphenethylamine hydrochloride **(24973-25-9)**
1,1-Dimethoxy-2-methylpropane **(41632-89-7)**
1,3-Dimethoxymethylurea **(141-07-1)**
N,N'-Dimethoxymethylurea **(141-07-1)**
Dimethoxymethyluron **(7388-44-5)**
1,5-Dimethoxypentane **(111-89-7)**
5-((3,4-Dimethoxyphenethyl)methylamino)-2-(3,4-dimethoxyphenyl)-2-isopropylvaleronitrile **(52-53-9)**
2,6-Dimethoxyphenol **(91-10-1)**
3,4-Dimethoxyphenol **(2033-89-8)**
3,5-Dimethoxyphenol **(500-99-2)**
N-(2,5-Dimethoxyphenyl)acetamide **(3467-59-2)**
3,4-Dimethoxyphenyl acetic acid **(93-40-3)**
α,α-Dimethoxy-α-phenylacetophenone **(24650-42-8)**
Di-p-methoxyphenylamine **(101-70-2)**
1-(2,5-Dimethoxyphenyl)-2-aminopropanehydrochloride **(24973-25-9)**
1-(2,5-Dimethoxyphenyl)-2-aminopropanol **(61-16-5)**
Dimethoxyphenylarsine **(24582-52-3)**
1-((2,5-Dimethoxyphenyl)azo)-2-naphthalenol **(6358-53-8)**
1-(1-(2,5-Dimethoxyphenyl)azo)-2-naphthol **(6358-53-8)**
1-(2,5-Dimethoxyphenylazo)-2-naphthol **(6358-53-8)**
2,5-Dimethoxy-1-(phenylazo)-2-naphthol **(6358-53-8)**
1-(3,4-Dimethoxyphenyl)ethanone **(1131-62-0)**
3,4-Dimethoxyphenyl ethyl ketone **(1835-04-7)**
β-(2,5-Dimethoxyphenyl)-β-hydroxyisopropylamine hydrochloride **(61-16-5)**
2-(2,5-Dimethoxyphenyl)isopropylamine hydrochloride **(2801-68-5)**
3,4-Dimethoxyphenylmethyl alcohol **(93-03-8)**
1-((3,4-Dimethoxyphenyl)methyl)-6,7-dimethoxyisoquinoline **(58-74-2)**
N-(2,5-Dimethoxyphenyl)-3-oxobutanamide **(6375-27-5)**
Dimethoxyphenylpenicillin **(61-32-5)**
1-(3,4-Dimethoxyphenyl)-2-phenoxy-1,3-propanediol **(75217-43-5)**
1-(3,4-Dimethoxyphenyl)-1-propanone **(1835-04-7)**
1-(3,4-Dimethoxyphenyl)-2-propene **(93-15-2)**
2,2-Di-(p-methoxyphenyl)-1,1,1-trichloroethane **(72-43-5)**
Di(p-methoxyphenyl)-trichloromethyl methane **(72-43-5)**
((Dimethoxyphosphinothioyl)thio)butanedioic acid diethyl ester **(121-75-5)**
2-Dimethoxyphosphinothioylthiomethyl-4,6-diamino-s-triazine **(78-57-9)**
3-((Dimethoxyphosphinyl)oxy)-2-butenoic acid methyl ester **(7786-34-7)**
3-(Dimethoxyphosphinyloxy)-N,N-dimethyl-cis-crotonamide **(141-66-2)**
3-(Dimethoxyphosphinyloxy)-N,N dimethylisocrotonamide **(141-66-2)**
3-(Dimethoxyphosphinyloxy)n-methyl-cis-crotonamide **(6923-22-4)**
Dimethoxy polyethylene glycol **(24991-55-7)**
2,2-Dimethoxypropane **(77-76-9)**
1,1-Dimethoxy-2-propanone **(6342-56-9)**
3',4'-Dimethoxypropiophenone **(1835-04-7)**
3,4-Dimethoxypropiophenone **(1835-04-7)**
2-((4-((2,3,-Dimethoxypropyl)methylamino)phenyl)azo)-6-methoxy-3-methylbenzothiazolium chloride **(12270-13-2)**
2,6-Dimethoxy-4-pyrimidinamine **(3289-50-7)**
2,3-Dimethoxy-strychnine **(357-57-3)**
Dimethoxy strychnine **(357-57-3)**
Dimethoxy strychnine (DOT) **(357-57-3)**
Dimethoxytetraethylene glycol **(143-24-8)**
Dimethoxytetraglycol **(143-24-8)**
3,4-Dimethoxytoluene **(494-99-5)**
4,6-Dimethoxy-1,3,5-triazine-2-thiol **(30886-14-7)**
Dimethoxy-2,2,2-trichloro-1-hydroxy-ethyl-phosphine oxide **(52-68-6)**
Dimethoxy-2,2,2-trichloro-l-n-butyryloxy-ethylphosphine oxide **(126-22-7)**
3,3'-Dimethoxytriphenylmethane-4,4'-bis(1''-azo-2''-naphthol) **(6483-64-3)**
6,7-Dimethoxy-1-veratrylisoquinoline **(58-74-2)**
16,17-Dimethoxyviolanthrone **(128-58-5)**
Dimethoxyviolanthrone **(128-58-5)**
Dimethrin **(70-38-2)**
Dimethyl **(74-84-0)**
Dimethyl Yellow **(60-11-7)**
Dimethyl Yellow Analar **(60-11-7)**
Dimethyl Yellow N,N-dimethylaniline **(60-11-7)**
Dimethylacetal **(534-15-6)**
Dimethylacetamide **(127-19-5)**
N,N-Dimethylacetamide **(127-19-5)**
Dimethyl acetamide (ACGIH,OSHA) **(127-19-5)**
2',4'-Dimethylacetanilide **(2050-43-3)**
2,4-Dimethylacetanilide **(2050-43-3)**
Dimethylacetic acid **(79-31-2)**
N,N-Dimethylacetoacetamide **(2044-64-6)**
2',4'-Dimethylacetoacetanilide **(97-36-9)**
Dimethylacetone **(96-22-0)**
Dimethylacetone amide **(127-19-5)**
2',4'-Dimethylacetophenone **(89-74-7)**
2,3-Dimethylacetophenone **(2142-71-4)**
2,4-Dimethylacetophenone **(89-74-7)**
O,O-Dimethyl-S-(2-(acetylamino)ethyl) dithiophosphate **(13265-60-6)**
O,O-Dimethyl S-(2-acetylaminoethyl) phosphorodithioate **(13265-60-6)**
Dimethyl acetylbutanedioate **(10420-33-4)**
Dimethylacetylcarbinol **(115-22-0)**
Dimethylacetylene **(503-17-3)**
Dimethylacetylenecarbinol **(115-19-5)**
Dimethyl acetylenedicarboxylic acid **(762-42-5)**
Dimethylacetylenylcarbinol **(115-19-5)**
O,S-Dimethylacetylphosphoroamidothioate **(30560-19-1)**
N,N-Dimethylacrylamide **(2680-03-7)**
3,3 Dimethyl-acrylate de 2,4-dinitro-6-(1-methylpropyle)

phenyle (French) (485-31-4)
(E)-2,3-Dimethylacrylic acid (80-59-1)
3,3-Dimethylacrylic acid (541-47-9)
β,β-Dimethylacrylic acid (541-47-9)
trans-2,3-Dimethylacrylic acid (80-59-1)
trans-α,β-Dimethylacrylic acid (80-59-1)
3,3-Dimethylacrylic acid 2-sec-butyl-4,6-dinitrophenyl ester (485-31-4)
Dimethyl adipate (627-93-0)
Dimethylaethanolamin (German) (108-01-0)
O,O-Dimethyl-S-(2-aethylsulfinyl-aethyl)-thiolphosphat (German) (301-12-2)
O,O-Dimethyl-S-(2-aethylsulfonyl-aethyl)-thiolphosphat (German) (17040-19-6)
N-(5-(1,1-Dimethylaethyl)-1,3,4-thiadiazol-2-yl)-N,N'-dimethyl-harnstoff (German) (34014-18-1)
O,O-Dimethyl-S-(2-aethylthio-aethyl)-dithio phosphat (German) (640-15-3)
O,O-Dimethyl-O-(2-aethylthio-aethyl)monothiophosphat (German) (867-27-6)
O,O-Dimethyl-S-(2-aethylthio-aethyl)-monothiophosphat (German) (919-86-8)
Dimethyl aldehyde (534-15-6)
3,3-Dimethylallyl alcohol (556-82-1)
Dimethylallyl alcohol (556-82-1)
γ,γ-Dimethylallyl alcohol (556-82-1)
3,3-Dimethylallyl chloride (503-60-6)
γ,γ-Dimethylallyl chloride (503-60-6)
2-(3,3-Dimethylallyl)cyclazocine (359-83-1)
2-Dimethylallyl-5,9-dimethyl-2'-hydroxybenzomorphan (359-83-1)
2-(3,3-Dimethylallyl)-2',2'-hydroxy-5,9-dimethyl-6,7-benzomorphan (359-83-1)
N-Dimethyl amino-β-carbamyl propionic acid (1596-84-5)
Dimethylamide acetate (127-19-5)
Dimethylamid kyseliny acetoctove (Czech) (2044-64-6)
Dimethylamid kyseliny akrylove (Czech) (2680-03-7)
Dimethylamid kyseliny chlormravenci (Czech) (79-44-7)
Dimethylamid kyseliny difenyloctove (Czech) (957-51-7)
Dimethylamid kyseliny mravenci (Czech) (68-12-2)
Dimethylamid kyseliny octove (Czech) (127-19-5)
Dimethylamidoethoxyphosphoryl cyanide (77-81-6)
N-(3-Dimethylamidopropyl)stearamide (7651-02-7)
5-(γ-Dimethylamino-β-methylpropyl)-10,11-dihydro-5H-dibenzo-(b,f)azepine (739-71-9)
Dimethylamine (ACGIH,OSHA) (124-40-3)
Dimethylamine, Anhydrous [UN 1032] (124-40-3)
Dimethylamine, Solution [UN 1160] (124-40-3)
Dimethylamine, 1-acetoxy-N-nitroso- (56856-83-8)
Dimethylamine benzhydryl ester hydrochloride (147-24-0)
4-Dimethylamine m-cresyl methylcarbamate (2032-59-9)
Dimethylamine, (2,4-dichlorophenoxy)acetate (2008-39-1)
Dimethylamine 4-(2,4-dichlorophenoxy)butyrate (2758-42-1)
Dimethylamine 2-(2,4-dichlorophenoxy)propionate (53404-32-3)
Dimethylamine, hydrochloride (506-59-2)
Dimethylamine 2-methyl-4-chlorophenoxyacetate (2039-46-5)
Dimethylamine 2-(2-methyl-4-chlorophenoxy)propionate (32351-70-5)
Dimethylamine, N-nitro (4164-28-7)
Dimethylamine, N-nitroso (62-75-9)
4-Dimethylaminepyridine (1122-58-3)
Dimethylamine salt of 2,4-D (2008-39-1)
Dimethylamine salt of dicamba (2300-66-5)
Dimethylamine salts of mixed polychlorobenzoic acids (1338-32-5)
Dimethylamine, 2,3,6-trichlorobenzoate (3426-62-8)
Dimethylamine 2,3,6-trichlorophenylacetate (69462-13-1)
4-(Dimethylamine)-3,5-xylyl N-methylcarbamate (315-18-4)
Dimethylaminoacetonitrile (926-64-7)
2-Dimethylaminoacetonitrile [UN 2378] (926-64-7)
Dimethylaminoaethanol (German) (108-01-0)
β-Dimethylamino-aethyl-benzhydryl-aether (German) (58-73-1)
N,N-Dimethyl-β-aminoaethyl-isothiuronium dihydrochlorid (German) (16111-27-6)
Dimethylamino-analgesine (58-15-1)
4-Dimethylaminoaniline hydrochloride (540-24-9)
4-(Dimethylamino)antipyrine (58-15-1)
Dimethylaminoantipyrine (58-15-1)
p-Dimethylaminoazobenzen (Czech) (60-11-7)
2',3-Dimethyl-4-aminoazobenzene (97-56-3)
4-(N,N-Dimethylamino)azobenzene (60-11-7)
4-Dimethylaminoazobenzene (539-17-3)
4-Dimethylaminoazobenzene (60-11-7)
Dimethylaminoazobenzene (60-11-7)
N,N-Dimethyl-4-aminoazobenzene (60-11-7)
N,N-Dimethyl-p-aminoazobenzene (60-11-7)
p-Dimethylaminoazobenzene (60-11-7)
4-Dimethylaminoazobenzene (OSHA) (60-11-7)
4'-Dimethylaminoazobenzene-2-carboxylic acid (493-52-7)
p-(Dimethylamino)azobenzene-o-carboxylic acid (493-52-7)
4-Dimethylaminoazobenzene-4'-sulphonic acid sodium salt (547-58-0)
4-Dimethylaminoazobenzol (60-11-7)
Dimethylaminoazobenzol (60-11-7)
p-Dimethylamino-azobenzol (German) (60-11-7)
Dimethylaminoazophene (58-15-1)
(Dimethylamino)benzaldehyde (28602-27-9)
4-(Dimethylamino)benzaldehyde (100-10-7)
p-(Dimethylamino)benzaldehyde (100-10-7)
(Dimethylamino)benzene (121-69-7)
3,4-Dimethylaminobenzene (95-64-7)
Dimethylaminobenzene (OSHA) (1300-73-8)
4-Dimethylaminobenzenecarbonal (100-10-7)
p-Dimethylaminobenzene diazo sodium sulfonate (140-56-7)
p-Dimethylaminobenzenediazosodium sulphonate (140-56-7)
p-(Dimethylamino)benzenediazosulfonate (140-56-7)
4-Dimethylaminobenzenediazosulfonic acid, sodium salt (140-56-7)
p-Dimethylaminobenzenediazosulfonic acid, sodium salt (140-56-7)
p-(Dimethylamino)benzenediazosulphonate (140-56-7)
4-Dimethylaminobenzenediazosulphonic acid, sodium salt (140-56-7)
p-(Dimethylamino)benzenediazosulphonic acid, sodium salt (140-56-7)
2,3-Dimethyl-5-aminobenzenesulfethanolamide (25797-78-8)
4-Dimethylaminobenzenethiol (4946-22-9)
p-Dimethylaminobenzenethiol (4946-22-9)
3-(Dimethylamino)benzoic acid (99-64-9)
4-(Dimethylamino)benzoic acid (619-84-1)
N,N-Dimethyl-4-aminobenzoic acid (619-84-1)
N,N-Dimethyl-m-aminobenzoic acid (99-64-9)
m-(Dimethylamino)benzoic acid (99-64-9)
p-Dimethylaminobenzoic acid (619-84-1)
4-(Dimethylamino)benzoic acid, 2-ethylhexyl ester (21245-02-3)
p-Dimethylaminobenzoic acid, pentyl ester (14779-78-3)
p-Dimethylaminobenzoldiazosulfonat (natriumsalz) (German) (140-56-7)
4-(Dimethylamino)benzophenone (530-44-9)
4-Dimethylaminobenzophenone (530-44-9)
4-N,N-Dimethylaminobenzophenone (530-44-9)
p-Dimethylaminobenzophenone (530-44-9)
4,4'-Dimethylaminobenzophenonimide (492-80-8)
2-(4-(Dimethylamino)benzoyl)benzoic acid (21528-31-4)
6-(Dimethylamino)-3,3-bis(4-(dimethylamino)phenyl)phthalide (1552-42-7)
3-Dimethylamino-1,1-bis(2-thienyl)-1-butene (524-84-5)
3-(((Dimethylamino)carbonyl)amino)phenyl 1,1-dimethylethyl-carbamate (4849-32-5)
(Dimethylamino)carbonyl chloride (79-44-7)
N,N-Dimethylaminocarbonyl chloride (79-44-7)
1-(N,N-Dimethylamino)-3-(p-chlorophenyl-3-α-pyridyl)-propane maleate (113-92-8)
3-(Dimethylamino)-p-cresol (119-31-3)
4-Dimethylamino-3-cresyl methylcarbamate (2032-59-9)
Dimethylaminocyanphosphorsaeureaethylester (German) (77-81-6)
(Dimethylamino)cyclohexane (98-94-2)
N,N-Dimethylaminocyclohexane (98-94-2)
7-Dimethylamino-6-demethyl-6-deoxytetracycline (10118-90-8)
4-(Dimethylamino)-3,5-dimethylphenol methylcarbamate (ester) (315-18-4)
4-(Dimethylamino)-3,5-dimethylphenyl N-methylcarbamate (315-18-4)

4-Dimethylamino-2,3-dimethyl-1-phenyl-3-pyrazolin-5-one (58-15-1)
4-Dimethylamino-2,3-dimethyl-1-phenyl-5-pyrazolone (58-15-1)
6-(Dimethylamino)-2-(2-(2,5-dimethyl-1-phenyl-1H-pyrrol-3-yl)-ethenyl)-1-methylquinolinium (7187-62-4)
2-(Dimethylamino)-5,6-dimethyl-4-pyrimidinyldimethylcarbamate (23103-98-2)
(-)-1-Dimethylamino-1,2-diphenyl-ethane (14148-99-3)
6-Dimethylamino-4,4-diphenyl-3-heptanol (17199-55-2)
6-(Dimethylamino)-4,4-diphenyl-3-heptanol acetate (ester) (509-74-0)
p,p-Dimethylaminodiphenylmethane (101-61-1)
α-(+)-4-Dimethylamino-1,2-diphenyl-3-methyl-2-butanol propionate ester (469-62-5)
3-Dimethylamino-1,1-di-(2'-thienyl)but-1-ene (524-84-5)
3-Dimethylamino-1,1-di-(2'-thienyl)-1-butene (524-84-5)
2-(Dimethylamino)ethanol (108-01-0)
2-Dimethylaminoethanol (108-01-0)
Dimethylaminoethanol (108-01-0)
N,N-Dimethylaminoethanol (108-01-0)
N-Dimethylaminoethanol (108-01-0)
β-Dimethylaminoethanol (108-01-0)
2-Dimethylaminoethanol 4-N-butylaminobenzoate hydrochloride (136-47-0)
β-Dimethylaminoethanol diphenylmethyl ether (58-73-1)
N-((2-Dimethylaminoethoxy)benzyl)-3,4,5-trimethoxybenzamide (138-56-7)
α-(2-Dimethylaminoethoxy)diphenylmethane (58-73-1)
2-(2-Dimethylaminoethoxy)ethanol (1704-62-7)
2-(α-(2-(Dimethylamino)ethoxy)-α-methylbenzyl)pyridine (469-21-6)
2-(α-(2-Dimethylaminoethoxy)-α-methylbenzyl)pyridine succinate (562-10-7)
Dimethylaminoethoxy-methyl-benzyl-pyridine succinate (562-10-7)
2-Dimethylaminoethoxyphenylmethyl-2-picoline (469-21-6)
2-Dimethylaminoethoxyphenylmethyl-2-picoline succinate (562-10-7)
4-(2-Dimethylaminoethoxy)-N-(3,4,5-trimethoxybenzoyl)benzyl-amine (138-56-7)
Dimethylaminoethyl acrylate (2439-35-2)
β-Dimethylaminoethyl alcohol (108-01-0)
β-(Dimethylamino)ethylamine (100-36-7)
β-Dimethylaminoethylbenzhydryl ether (58-73-1)
β-Dimethylaminoethyl benzhydryl ether hydrochloride (147-24-0)
N-Dimethylaminoethylbenzylaniline hydrochloride (2045-52-5)
2-(α-(2-Dimethylaminoethyl)benzyl)pyridine (86-21-5)
2-(α-(2-(Dimethylamino)ethyl)benzyl)pyridine, bimaleate (132-20-7)
2-(α-(2-(Dimethylamino)ethyl)benzyl)pyridine, maleate (132-20-7)
2-(Dimethylamino)ethyl p-(butylamino)benzoate hydrochloride (136-47-0)
Dimethylaminoethyl-p-N-butylaminobenzoate hydrochloride (136-47-0)
2-(Dimethylamino)ethyl p-(butylamino)benzoate monohydrochloride (136-47-0)
S-(2-Dimethylaminoethyl)-O.O-diethylphosphorothioate methiodide (513-10-0)
2-Dimethylaminoethylester kyseliny methakrylove (Czech) (2867-47-2)
2-(Dimethylamino)ethyl ethoxydiphenylacetate (509-78-4)
3-((β-Dimethylamino)ethyl)-5-hydroxyindole (487-93-4)
3-(β-Dimethylaminoethyl)-5-hydroxyindole (487-93-4)
3-((2-Dimethylamino)ethyl)-5-indolol (487-93-4)
3-(2-Dimethylaminoethyl)-5-indolol (487-93-4)
3-(2-Dimethylaminoethyl)indol-5-ol (487-93-4)
3-(2-(Dimethylamino)ethyl)-1H-indol-4-ol dihydrogen phosphate ester (520-52-5)
3-2'-Dimethylaminoethylindol-4-phosphate (520-52-5)
3-(2-Dimethylaminoethyl)indol-4-yl dihydrogen phosphate (520-52-5)
2-Dimethylaminoethylisothiuronium chloride hydrochloride (16111-27-6)
2-(Dimethylamino)ethyl methacrylate (2867-47-2)
2-Dimethylaminoethyl methacrylate (2867-47-2)
Dimethylaminoethyl methacrylate (2867-47-2)
N,N-Dimethylaminoethyl methacrylate (2867-47-2)
β-Dimethylaminoethyl methacrylate (2867-47-2)
Dimethylaminoethyl methacrylate [UN 2522] (2867-47-2)
Dimethylaminoethyl methacrylate homopolymer (25154-86-3)
N,N-Dimethylaminoethyl methacrylate polymer (25154-86-3)
N-Dimethylaminoethyl-N-p-methoxy-α-aminopyridine maleate (59-33-6)
2-((2-(Dimethylamino)ethyl)-(p-methoxybenzyl)amino)pyridine (91-84-9)
2-((2-(Dimethylamino)ethyl)(p-methoxybenzyl)amino)pyridine bimaleate (59-33-6)
2-((2-(Dimethylamino)ethyl)(p-methoxybenzyl)amino)pyridine maleate (59-33-6)
2-((2-(Dimethylamino)ethyl)(p-methoxybenzyl)amino)pyrimidine (91-85-0)
2-((2-(Dimethylamino)ethyl)(p-methoxy-benzyl)amino)-pyrimidine hydrochloride (63-56-9)
6-((6-(2-(Dimethylamino)ethyl)-2-methoxy-3,4-(methylenedioxy)-phenyl)acetyl)-o-veratric acid (131-28-2)
2-((2-(Dimethylamino)ethyl)methylamino)-ethanol (2212-32-0)
β-Dimethylaminoethyl-2-methylbenzhydryl ether (83-98-7)
N-(2-(Dimethylamino)ethyl)-N-methylethanolamine (2212-32-0)
2-(Dimethylamino)ethyl 2-propenoate (2439-35-2)
S-(2-(Dimethylamino)ethyl)pseudothiourea dihydrochloride (16111-27-6)
β-Dimethylaminoethyl-2-pyridylaminotoluene (91-81-6)
β-Dimethylaminoethyl-2-pyridylbenzylamine (91-81-6)
N-(2-Dimethylaminoethyl)-N-2-pyridyl-3-thenylamine (91-79-2)
2-((2-(Dimethylamino)ethyl)-2-thenylamino)pyridine (91-80-5)
2-((2-Dimethylaminoethyl)-3-thenylamino)pyridine (91-79-2)
2-((2-(Dimethylamino)ethyl)-2-thenyl-amino)pyridine hydrochloride (135-23-9)
2-((2-(Dimethylamino)ethyl)-3-thenyl-amino)-pyridine hydrochloride (958-93-0)
2-(2-(Dimethylamino)ethyl)-2-thiopseudourea dihydrochloride (16111-27-6)
2-Dimethylamino-4-hydroxytoluene (119-31-3)
10-(3-Dimethylaminoisopropyl)phenothiazine hydrochloride (58-33-3)
Dimethylamino-isopropyl-phenthiazin (German) (60-87-7)
10-Dimethylaminoisopropyl-2-propionylphenothiazine (362-29-8)
(Dimethylamino)methanol (14002-21-2)
2-(Dimethylamino)-N-(((methylamino)carbonyl)oxy)-2-oxoethan-imidothioic acid methyl ester (23135-22-0)
4-(N,N-Dimethylamino)-3'-methylazobenzene (55-80-1)
7-Dimethylamino-4-methylcoumarin (87-01-4)
α-L-4-Dimethylamino-3-methyl-1,2-diphenyl-2-butanol propionate (2338-37-6)
6-(Dimethylamino)-5-methyl-4,4-diphenyl-3-hexanone (466-40-0)
6-Dimethylamino-5-methyl-4,4-diphenyl-3-hexanone (466-40-0)
m-(((Dimethylamino)methylene)amino)phenyl methyl carbamate, hydrochloride (23422-53-9)
4-(((Dimethylamino)methylene)amino)-m-tolylmethylcarbamate (17702-57-7)
3-Dimethylaminomethyleneiminophenyl-N-methylcarbamate, hydrochloride (23422-53-9)
(2-Dimethylamino-2-methyl)ethyl-N-dibenzoparathiazine (60-87-7)
10-(2-(Dimethylamino)-2-methylethyl)phenothiazine (60-87-7)
N-(2'-Dimethylamino-2'-methyl)ethylphenothiazine (60-87-7)
N-(2'-Dimethylamino-2'-methyl)ethylphenothiazine hydrochloride (58-33-3)
(R)-α-(2-(Dimethylamino)-1-methylethyl)-α-phenylbenzene-ethanol propanoate (2338-37-6)
N-Dimethylamino-2-methylethyl thiodiphenylamine (60-87-7)
trans-2-((Dimethylamino)methylimino)-5-(2-(5-nitro-2-furyl)-vinyl)-1,3,4-oxadiazole (55738-54-0)
8-(Dimethylaminomethyl)-7-methoxy-3-methylflavone (1165-48-6)
8-((Dimethylamino)methyl)-7-methoxy-3-methyl-2-phenylflavone (1165-48-6)
3-(Dimethylamino)-1-methyl-3-oxo-1-propenyl dimethyl phosphate (141-66-2)
((Dimethylamino)methyl)phenol (25338-55-0)
3-Dimethylamino-4-methylphenol (119-31-3)
4-(Dimethylamino)-3-methylphenol methyl carbamate (ester) (2032-59-9)
(4-Dimethylamino-3-methyl-phenyl)N-methyl-carbamaat (Dutch) (2032-59-9)
(4-Dimethylamino-3-methyl-phenyl)N-methyl-carbamat (German) (2032-59-9)
(4-Dimethylamino-3-methyl-phenyl)N-methyl-carbamate (2032-59-9)
2-Dimethylamino-2-methyl-1-propanol (7005-47-2)
1-(3-Dimethylamino-2-methylpropyl)-4,5-dihydro-2,3:6,7-di-benzazepine (739-71-9)
5-(3-(Dimethylamino)-2-methylpropyl)-10,11-dihydro-5H-dibenz-(b,f)azepine (739-71-9)
2-Dimethylamino-1-(methylthio)glyoxal O-methylcarbamoyl-monoxime (23135-22-0)
1-Dimethylaminonaphthalene (86-56-6)
1-(Dimethylamino)-4-nitrobenzene (100-23-2)

4-(Dimethylamino)nitrobenzene (100-23-2)
p-(Dimethylamino)nitrobenzene (100-23-2)
4-(Dimethylamino)nitrosobenzene (138-89-6)
p-(Dimethylamino)nitrosobenzene (138-89-6)
Dimethylaminophenazon (German) (58-15-1)
4-Dimethylaminophenazone (58-15-1)
Dimethylaminophenazone (58-15-1)
3-(Dimethylamino)phenol (99-07-0)
m-(Dimethylamino)phenol (99-07-0)
p-Dimethylaminophenylamine (105-10-2)
4-Dimethylaminophenylazobenzene (60-11-7)
p-((p-(Dimethylamino)phenyl)azo)benzenesulfonic acid sodium salt (547-58-0)
2-((4-Dimethylamino)phenylazo)benzoic acid (493-52-7)
3-((p-(Dimethylamino)phenyl)azo)benzoic acid (20691-84-3)
o-((p-(Dimethylamino)phenyl)azo)benzoic acid (493-52-7)
5-((4-(Dimethylamino)phenyl)azo)-1,4-dimethyl-1H-1,2,4-triazolium (12221-52-2)
(4-(Dimethylamino)phenyl)diazenesulfonic acid, sodium salt (140-56-7)
4-((Dimethylamino)phenyl)diazenesulfonic acid, sodium salt (140-56-7)
p-(Dimethylamino)-phenyldiazo-natriumsulfonat (German) (140-56-7)
2-(p-(Dimethylamino)phenyl)-3,6-dimethylbenzothiazolium chloride (2390-54-7)
Dimethylaminophenyldimethylpyrazolin (58-15-1)
4-Dimethylamino-1-phenyl-2,3-dimethylpyrazolone (58-15-1)
1-Dimethylamino-1-phenylethane (2449-49-2)
N-(4-((4-(Dimethylamino)phenyl)(4-(phenylamino)-1-naphthalenyl)methylene)-2,5-cyclohexadien-1-ylidene)-N-methylmethanaminium, chloride (2580-56-5)
(4-(Dimethylamino)phenyl)phenylmethanone (530-44-9)
2-(3-Dimethylamino-1-phenylpropyl)pyridine (86-21-5)
1-(N,N-Dimethylamino)-3-(phenyl-3-α-pyridyl)propane maleate (132-20-7)
3-Dimethylaminopropannitril (Czech) (1738-25-6)
1,1-Dimethylaminopropan-2-ol (108-16-7)
1,1-Dimethylaminopropanol-2 (108-16-7)
3-(Dimethylamino)propionitrile (1738-25-6)
β-Dimethylaminopropionitrile (1738-25-6)
3-(Dimethylamino)propylamine (109-55-7)
N,N-Dimethyl-N-(3-aminopropyl)amine (109-55-7)
10-(3-Dimethylaminopropyl)-2-chlorophenothiazine monohydrochloride (69-09-0)
1-(3-Dimethylaminopropyl)-4,5-dihydro-2,3,6,7-dibenzazepine (50-49-7)
5-(3-(Dimethylamino)propyl)-10,11-dihydro-5H-dibenz(b,f)azepine (50-49-7)
5-(3-Dimethylaminopropyl)-10,11-dihydro-5H-dibenzo(b,f)azepine (50-49-7)
N-(3-(Dimethylamino)propyl)dodecanamide (3179-80-4)
N-(3-Dimethylaminopropyl)dodecanamide (3179-80-4)
β-(2-(Dimethylamino)propyl)-α-ethyl-β-phenylbenzeneethanol acetate hydrochloride (509-74-0)
N-(3-(Dimethylamino)propyl)-1,1,2,2,3,3,4,4,5,5,6,6,7,7,8,8,8-heptadecafluoro-1-octanesulfonamide (13417-01-1)
5-(3-(Dimethylamino)propyl)-6,7,8,9,10,11-hexahydro-5H-cyclooct(b)indole (5560-72-5)
5-(3'-Dimethylaminopropylidene)-dibenzo-(a,d)(1,4)-cycloheptadiene (50-48-6)
5-(γ-Dimethylaminopropylidene)-5H-dibenzo(a,d)-10,11-dihydrocycloheptene (50-48-6)
11-(3-Dimethylaminopropylidene)-6,11-dihydrodibenz(b,e)oxipin (1668-19-5)
5-(3-Dimethylaminopropylidene)-10,11-dihydro-5H-dibenzo-(a,d)cycloheptatriene (50-48-6)
5-(3-Dimethylaminopropylidene)-10,11-dihydro-5H-dibenzo-(a,d)cycloheptene (50-48-6)
5-(γ-Dimethylaminopropylidene)-10,11-dihydro-5H-dibenzo-(a,d)cycloheptene (50-48-6)
11-(3-Dimethylaminopropylidene)-6,11-dihydrodibenzo(b,e)thiepin (113-53-1)
2,2'-(3-Dimethylaminopropylimino)bibenzyl (50-49-7)
2,2'-(3-Dimethylaminopropylimino)dibenzyl (50-49-7)
N-(γ-Dimethylaminopropyl)iminodibenzyl (50-49-7)
Dimethylaminopropyl lauramide (3179-80-4)
N-(3-Dimethylaminopropyl)lauramide (3179-80-4)
17-β-((3-(Dimethylamino)-propyl)methylamino)androst-5-en-3-β-ol dihydrochloride (1249-84-9)
17β-((3-(Dimethylamino)-propyl)methylamino)androst-5-en-3β-ol dihydrochloride (1249-84-9)
N-(3-Dimethylaminopropyl)octadecamide (7651-02-7)
N-(3-(Dimethylamino)propyl)octadecanamide (7651-02-7)
10-(2-(Dimethylamino)propyl)phenothiazine (60-87-7)
10-(3-(Dimethylamino)propyl)phenothiazine (58-40-2)
10-(2-Dimethylamino-1-propyl)phenothiazine hydrochloride (58-33-3)
10-(2-Dimethylaminopropyl)phenothiazine hydrochloride (58-33-3)
10-(3-(Dimethylamino)propyl)phenothiazine hydrochloride (53-60-1)
10-(γ-Dimethylamino-n-propyl)phenothiazine hydrochloride (53-60-1)
N-(2-Dimethylaminopropyl-1)phenothiazine hydrochloride (58-33-3)
10-(2-(Dimethylamino)propyl)phenothiazine monohydrochloride (58-33-3)
10-(3-(Dimethylamino)propyl)phenothiazine monohydrochloride (53-60-1)
1-(10-(2-(Dimethylamino)propyl)phenothiazin-2-yl)-1-propanone (362-29-8)
10-(2-Dimethylaminopropyl)-2-propionylphenothiazine (362-29-8)
Dimethylaminopropyl stearamide (7651-02-7)
N-Dimethylaminopropylstearamide (7651-02-7)
N-(3-Dimethylaminopropyl)-thiocarbaminsaeure-S-aethylester-hydrochlorid (German) (19622-19-6)
4-Dimethylaminopyridine (1122-58-3)
γ-(Dimethylamino)pyridine (1122-58-3)
p-Dimethylaminopyridine (1122-58-3)
2-(4-Dimethylaminostyryl)benzothiazole (1628-58-6)
2-(p-(Dimethylamino)styryl)benzothiazole (1628-58-6)
2-(4-N,N-Dimethylaminostyryl)quinoline (897-55-2)
4-(4-Dimethylaminostyryl)quinoline (897-55-2)
4-(p-(Dimethylamino)styryl)quinoline (897-55-2)
Dimethylaminosuccinamic acid (1596-84-5)
N-(Dimethylamino)succinamic acid (1596-84-5)
N-Dimethylamino-succinamidsaeure (German) (1596-84-5)
O-(4-((Dimethylamino)sulfonyl)phenyl) O,O-dimethyl phosphorothioate (52-85-7)
3-((Dimethylamino)thioxomethyl)thio)-1-propanesulfonic acid, sodium salt (18880-36-9)
4-(Dimethylamino)-m-tolyl methylcarbamate (2032-59-9)
S,S'-(2-(Dimethylamino)trimethylene)bis(thiocarbamate) hydrochloride (15263-52-2)
5-Dimethylamino-1,2,3-trithiane hydrogeneoxalate (31895-22-4)
4-(Dimethylamino)-3,5-xylenol, methylcarbamate (ester) (315-18-4)
4-(N,N-Dimethylamino)-3,5-xylyl N-methylcarbamate (315-18-4)
4-Dimethylamino-3,5-xylyl N-methylcarbamate (315-18-4)
4-Dimethylamino-3,5-xylyl methylcarbamate (315-18-4)
Dimethylammonium chloride (506-59-2)
Dimethylammonium 2,4-dichlorophenoxyacetate (2008-39-1)
N,N-Dimethylamphetamine (49681-82-5)
Di(4-methyl-2-amyl) maleate (105-52-2)
3,7-Dimethyl-2,6(and 2,7)-octadienoic acid (459-80-3)
4',4-Dimethylangelicin (4063-41-6)
4,5'-Dimethyl angelicin (4063-41-6)
4,4'-Dimethylangelicin plus ultraviolet a radiation (22975-76-4)
2,3-Dimethylaniline (87-59-2)
2,4-Dimethylaniline (95-68-1)
2,5-Dimethylaniline (95-78-3)
2,6-Dimethylaniline (87-62-7)
3,4-Dimethylaniline (95-64-7)
3,5-Dimethylaniline (108-69-0)
Dimethylaniline (1300-73-8)
Dimethylaniline (ACGIH,OSHA) (121-69-7)
N-Dimethyl-aniline (OSHA) (121-69-7)
N,N-Dimethylaniline [UN 2253] (121-69-7)
N,N-Dimethyl-p-anilinediazosulfonic acid sodium salt (140-56-7)
2,4-Dimethylaniline hydrochloride (21436-96-4)
2,5-Dimethyl aniline hydrochloride (51786-53-9)
2,5-Dimethylaniline hydrochloride (51786-53-9)

9,10-Dimethylanthracene (781-43-1)
Dimethylanthracene (29063-00-1)
Dimethyl anthranilate (85-91-6)
Dimethylarsenic acid (75-60-5)
Dimethylarsinat sodny (Czech) (124-65-2)
Dimethylarsinic acid (75-60-5)
((Dimethylarsino)oxy)sodium-as-oxide (124-65-2)
2,2-Dimethylaziridine (2658-24-4)
N,N-Dimethyl-p-azoaniline (60-11-7)
2,6-Dimethylazobenzene (17590-87-3)
4,4'-Dimethylazobenzene (501-60-0)
8,12-Dimethylbenz(a)acridine (3518-05-6)
9,12-Dimethylbenz(a)acridine (17401-48-8)
4,5-Dimethylbenz(a)anthracene (18429-70-4)
6,12-Dimethylbenz(a)anthracene (568-81-0)
7,11-Dimethylbenz(a)anthracene (35187-28-1)
7,12-Dimethylbenz(a)anthracene (57-97-6)
8,12-Dimethylbenz(a)anthracene (20627-31-0)
9,10-Dimethylbenz(a)anthracene (57-97-6)
9,10-Dimethylbenz(a)anthracene (58429-99-5)
Dimethylbenz(a)anthracene (57-97-6)
7,12-Dimethylbenz(a)anthracene-d16 (32976-87-7)
7,12-Dimethylbenz(a)anthracene, deuterated (32976-87-7)
1,10-Dimethyl-5,6-benzacridine (3518-05-6)
2,10-Dimethyl-5,6-benzacridine (17401-48-8)
5,7-Dimethyl-1,2-benzacridine (53-69-0)
6,9-Dimethyl-1,2-benzacridine (2381-40-0)
1,10-Dimethyl-7,8-benzacridine (French) (32740-01-5)
2,10-Dimethyl-7,8-benzacridine (French) (2381-40-0)
3,10-Dimethyl-7,8-benzacridine (French) (963-89-3)
Dimethylbenzamide (611-74-5)
N,N-Dimethylbenzamide (611-74-5)
3,4'-Dimethyl-1,2-benzanthracene (18429-70-4)
4,9-Dimethyl-1,2-benzanthracene (568-81-0)
5,9-Dimethyl-1,2-benzanthracene (20627-31-0)
6,7-Dimethyl-1,2-benzanthracene (58429-99-5)
8,10-Dimethyl-1,2-benzanthracene (35187-28-1)
9,10-Dimethyl-1,2-benzanthracene (57-97-6)
9,10-Dimethyl-benzanthracene (57-97-6)
Dimethylbenzanthracene (57-97-6)
7,12-Dimethylbenzanthrancene (57-97-6)
9,10-Dimethyl-1,2-benzanthrazen (German) (57-97-6)
Dimethylbenzanthrene (57-97-6)
O,O-Dimethyl-S-(benzaziminomethyl) dithiophosphate (86-50-0)
7,10-Dimethylbenz(c)acridine (2381-40-0)
7,11-Dimethylbenz(c)acridine (32740-01-5)
7,9-Dimethylbenz(c)acridine (963-89-3)
2,2-Dimethyl-1,3-benzdioxol-4-yl N-methylcarbamate (22781-23-3)
2,2-Dimethylbenzo-1,3-dioxol-4-yl methylcarbamate (22781-23-3)
2,3-Dimethylbenzenamine (87-59-2)
2,4-Dimethylbenzenamine (95-68-1)
2,5-Dimethylbenzenamine (95-78-3)
2,6-Dimethylbenzenamine (87-62-7)
3,5-Dimethylbenzenamine (108-69-0)
N,4-Dimethylbenzenamine (623-08-5)
2,4-Dimethylbenzenamine hydrochloride (21436-96-4)
2,5-Dimethylbenzenamine hydrochloride (51786-53-9)
1,2-Dimethylbenzene (95-47-6)
1,3-Dimethylbenzene (108-38-3)
1,4-Dimethylbenzene (106-42-3)
m-Dimethylbenzene (108-38-3)
o-Dimethylbenzene (95-47-6)
p-Dimethylbenzene (106-42-3)
Dimethylbenzene (OSHA) (1330-20-7)
N,N-Dimethylbenzeneamine (121-69-7)
Dimethyl benzenearsonite (24582-52-3)
Dimethyl 1,2-benzenedicarboxylate (131-11-3)
Dimethyl 1,4-benzenedicarboxylate (120-61-6)
Dimethyl benzeneorthodicarboxylate (131-11-3)
3,5-Dimethyl-1,2-benzenediol (2785-75-3)
N,N-Dimethylbenzenemethanamine (103-83-3)
α,α-Dimethylbenzenemethanol (617-94-7)
α,4-Dimethylbenzenemethanol (9CI) (536-50-5)
N,4-Dimethylbenzenesulfonamide (640-61-9)
2,4-Dimethylbenzenesulfonic acid (88-61-9)
2,5-Dimethylbenzenesulfonic acid (609-54-1)
2,4-Dimethylbenzenesulfonyl chloride (609-60-9)
Dimethylbenzenethiol (25550-52-1)
3,3'-Dimethylbenzidin (119-93-7)
3,3'-Dimethylbenzidine (119-93-7)
3,3'-Dimethylbenzidine dihydrochloride (612-82-8)
5,6-Dimethylbenzimidazole (582-60-5)
5,6-Dimethylbenzimidazolylcobamide cyanide (68-19-9)
α-(5,6-Dimethylbenzimidazolyl)hydroxocobamide (13422-51-0)
Dimethylbenzimidazoylcobamide (68-19-9)
7,12-Dimethylbenzo(a)anthracene (57-97-6)
1,2-Dimethylbenzo(a)pyrene (16757-85-0)
1,3-Dimethylbenzo(a)pyrene (16757-86-1)
1,4-Dimethylbenzo(a)pyrene (16757-88-3)
1,6-Dimethylbenzo(a)pyrene (16757-90-7)
2,3-Dimethylbenzo(a)pyrene (16757-87-2)
3,12-Dimethylbenzo(a)pyrene (16757-84-9)
3,6-Dimethylbenzo(a)pyrene (16757-91-8)
4,5-Dimethylbenzo(a)pyrene (16757-89-4)
2,2-Dimethyl-1,3-benzodioxol-4-ol methylcarbamate (22781-23-3)
2,3-Dimethylbenzofuran (3782-00-1)
2,3-Dimethylbenzoic acid (603-79-2)
2,4-Dimethylbenzoic acid (611-01-8)
2,5-Dimethylbenzoic acid (610-72-0)
2,6-Dimethylbenzoic acid (632-46-2)
3,4-Dimethylbenzoic acid (619-04-5)
3,5-Dimethylbenzoic acid (499-06-9)
3,3'-Dimethylbenzophenone (2852-68-8)
1,3-Dimethyl-3-(2-benzothiazolyl)urea (18691-97-9)
O,O-Dimethyl-S-(1,2,3-benzotriazinyl-4-keto)methyl phosphorodithioate (86-50-0)
α,β-Di(5-methylbenzoxazol-2-yl)ethene (1041-00-5)
Di-(2-methylbenzoyl) peroxide (Not more than 85% with water) (895-85-2)
1,4-Dimethyl-2,3-benzphenanthrene (57-97-6)
1,3-Dimethyl-3-(2-benzthiazolyl)-harnstoff (German) (18691-97-9)
α,α-Dimethylbenzyl alcohol (617-94-7)
p,α-Dimethylbenzyl alcohol (536-50-5)
Dimethylbenzylamine (103-83-3)
N,N-Dimethylbenzylamine (103-83-3)
Dimethylbenzylamine hydrochloride (1875-92-9)
Dimethylbenzylammonium chloride (1875-92-9)
Dimethylbenzylcarbinol (100-86-7)
Dimethylbenzyl carbinol acetate (151-05-3)
2,4-Dimethylbenzyl-(i)-cis-trans-chrysanthemumate (70-38-2)
2,4-Dimethylbenzylchrysanthemumate (70-38-2)
2,4-Dimethylbenzyl 2,2-dimethyl-3-(2-methylpropenyl)-cyclopropanecarboxylate (70-38-2)
2,4-Dimethylbenzylester kyseliny chrysanthemove (Czech) (70-38-2)
2,4-Dimethylbenzyl ester of cis,trans-chrysanthemumic acid (70-38-2)
Dimethyl benzyl hydrogenated tallow ammonium chloride, Reaction product with hectorite (71011-26-2)
α,α-Dimethylbenzyl hydroperoxide (80-15-9)
1-(α,α-Dimethylbenzyl)-3-methyl-3-phenylurea (42609-52-9)
Dimethylbenzyloctadecylammonium chloride (122-19-0)
p-(α,α-Dimethylbenzyl)phenol (599-64-4)
α,α-Dimethylbenzyl propyl ether (24142-77-6)
N,N-Dimethyl-N'-benzyl-N'-(2-pyridyl)ethylenediamine (91-81-6)
N,N-Dimethyl-N'-benzyl-N'-(α-pyridyl)ethylenediamine (91-81-6)
6,6-Dimethylbicyclo(3.1.1)-2-heptene-2-ethyl acetate (128-51-8)

6,6-Dimethylbicyclo(3.1.1)heptan-2-one **(24903-95-5)**
3,4'-Dimethylbiphenyl **(7383-90-6)**
3,4-Dimethylbiphenyl **(4433-11-8)**
ar,ar'-Dimethylbiphenyl **(28013-11-8)**
3,3'-Dimethyl-4,4'-biphenyldiamine **(119-93-7)**
3,3'-Dimethylbiphenyl-4,4'-diamine **(119-93-7)**
Dimethyl 4,4'-biphenyldicarboxylate **(792-74-5)**
3,3'-Dimethyl-4,4'-biphenylene diisocyanate **(91-97-4)**
N,N'-Dimethyl-4,4'-bipyridinium dichloride **(1910-42-5)**
N,N'-Dimethyl-4,4'-bipyridylium dichloride **(1910-42-5)**
1,1'-Dimethyl-4,4'-bipyridynium dichloride **(1910-42-5)**
1,1'-Dimethyl-4,4'-bipyridynium dimethylsulfate **(2074-50-2)**
3,3'-Dimethyl-4,4'-bis(acetoacetylamino)biphenyl **(91-96-3)**
Dimethyl-2,5 bis(benzoylperoxy)-2,5 hexane (French) **(2618-77-1)**
2,5-Dimethyl-2,5-bis-(tert-butylperoxy)hexyne-3, Max. concent. 52% with inert solid (DOT) **(1068-27-5)**
Dimethyl 1,3-bis(carbomethoxy)-1-propen-2-yl phosphate **(122-10-1)**
N,N'-Dimethyl-2,2-bis(4-(3,4-dicarboxyphenoxy)phenyl)-propane diimide **(54395-52-7)**
β,γ-Dimethyl-α,δ-bis(3,4-dihydroxyphenyl)butane **(500-38-9)**
O,O-Dimethyl S-(1,2-bis(ethoxycarbonyl)ethyl)dithiophosphate **(121-75-5)**
O,O-Dimethyl-S-1,2-bis(ethoxycarbonyl)ethyl phosphorothioate **(1634-78-2)**
Dimethyl bis(p-hydroxyphenyl)methane **(80-05-7)**
1,5-Dimethyl-2,4-bis(1-methylethyl)benzene **(5186-68-5)**
Dimethylbis((1-oxoneodecyl)oxy)stannane **(68928-76-7)**
Dimethyl brassylate **(1472-87-3)**
2,5-Dimethylbromobenzene **(553-94-6)**
O,O-Dimethyl O-(4-bromo-2,5-dichlorophenyl) phosphorothioate **(2104-96-3)**
2,3-Dimethylbut-2-ene **(563-79-1)**
1,2-Dimethyl-1,3-butadiene **(4549-74-0)**
2,3-Dimethyl-1,3-butadiene **(513-81-5)**
2,3-Dimethylbuta-1,3-diene **(513-81-5)**
2,3-Dimethylbutadiene **(513-81-5)**
2,4-Dimethyl-1,3-butadiene **(1118-58-7)**
3,4-Dimethylbutadiene **(4549-74-0)**
3,3-Dimethyl-2-butanamine **(3850-30-4)**
N,N-Dimethyl-1-butanamine **(927-62-8)**
2,3-Dimethylbutane [UN 2457] **(79-29-8)**
Dimethyl butanedioate **(106-65-0)**
2,3-Dimethyl-2,3-butanediol **(76-09-5)**
3,3-Dimethylbutanoic acid **(1070-83-3)**
1,3-Dimethyl butanol **(105-30-6)**
2,2-Dimethylbutanol **(1185-33-7)**
3,3-Dimethyl-1-butanol **(624-95-3)**
3,3-Dimethyl-2-butanol **(464-07-3)**
3,3-Dimethylbutan-1-ol **(624-95-3)**
3,3-Dimethyl-2-butanone **(75-97-8)**
2,2-Dimethyl-3-butene **(558-37-2)**
2,3-Dimethyl-1-butene **(563-78-0)**
2,3-Dimethyl-2-butene **(563-79-1)**
2,3-Dimethylbutene-2 **(563-79-1)**
3,3-Dimethyl-1-butene **(558-37-2)**
3,3-Dimethylbutene **(558-37-2)**
((1,3-Dimethylbutoxy)methyl)oxirane **(68134-06-5)**
1,3-Dimethylbutyl acetate **(108-84-9)**
1,3-Dimethyl butylamine **(108-09-8)**
Dimethylbutylamine **(927-62-8)**
1,3-Dimethylbutylamine [UN 2379] **(108-09-8)**
(1,1-Dimethylbutyl)benzene **(1985-57-5)**
(2,2-Dimethylbutyl)benzene **(28080-86-6)**
1,3-Dimethyl-5-tert-butylbenzene **(98-19-1)**
1,1-Dimethyl-3-(3-(N-tert-butylcarbamyloxy)phenyl)urea **(4849-32-5)**
1,3-Dimethylbutylester kyseliny octove (Czech) **(108-84-9)**
3,3-Dimethyl-2-butyl methylphosphonofluoridate **(96-64-0)**
N-(1,3-Dimethylbutyl)-N'-phenyl-p-phenylenediamine **(793-24-8)**
β,β-Dimethylbutylrolacton (German) **(13861-97-7)**
4,4-Dimethylbutyrolactone **(13861-97-7)**
O,O-Dimethyl-(1-n-butyryloxy-2,2,2-trichloraethyl)-phosphonsaeureester **(126-22-7)**
O,O-Dimethyl-(1-butyryloxy-2,2,2-trichloroethyl) phosphonate **(126-22-7)**
3,3-Dimethyl-n-but-2-yl methylphosphonofluoridate **(96-64-0)**
Dimethylcadmium **(506-82-1)**
N,N-Dimethylcapramide **(14433-76-2)**
N,N-Dimethylcaprylamide **(1118-92-9)**
Dimethylcarbamate de 5,5-dimethyl dihydroresorcinol (French) **(122-15-6)**
Dimethylcarbamate d'l-isopropyl 3-methyl 5-pyrazolyle (French) **(119-38-0)**
Dimethylcarbamic acid chloride **(79-44-7)**
N,N-Dimethylcarbamic acid chloride **(79-44-7)**
Dimethylcarbamic acid 1-((dimethylamino)carbonyl)-5-methyl-1H-pyrazol-3-yl ester **(644-64-4)**
Dimethylcarbamic acid 2-(dimethylamino)-5,6-dimethyl-4-pyrimidinyl ester **(23103-98-2)**
Dimethylcarbamic acid ester of 3-hydroxy-1-methylpyridinium bromide **(101-26-8)**
Dimethylcarbamic acid ester with 3-hydroxy-N,N,5-trimethyl-pyrazole-1-carboxamide **(644-64-4)**
Dimethylcarbamic acid 3-methyl-1-(1-methylethyl)-1H-pyrazol-5-yl ester **(119-38-0)**
Dimethylcarbamic acid 3-methyl-1-phenyl-1H-pyrazol-5-yl ester **(87-47-8)**
Dimethylcarbamic acid, 3-methyl-1-phenylpyrazol-5-yl ester **(87-47-8)**
Dimethylcarbamic chloride **(79-44-7)**
N,N-Dimethylcarbamidoyl chloride **(79-44-7)**
Dimethylcarbamidoyl chloride [UN 2262] **(79-44-7)**
Dimethylcarbamodithioic acid **(79-45-8)**
Dimethylcarbamodithioic acid, iron complex **(14484-64-1)**
Dimethylcarbamodithioic acid, iron(3+) salt **(14484-64-1)**
Dimethylcarbamodithioic acid, zinc complex **(137-30-4)**
Dimethylcarbamodithioic acid, zinc salt **(137-30-4)**
Dimethylcarbamothioic acid S-(4-phenoxybutyl)ester **(62850-32-2)**
Dimethylcarbamothioic chloride **(16420-13-6)**
3-Dimethylcarbamoxyphenyl trimethyl ammonium bromide **(114-80-7)**
Dimethylcarbamoyl chloride **(79-44-7)**
N,N-Dimethylcarbamoyl chloride **(79-44-7)**
Dimethyl carbamoyl chloride (ACGIH) **(79-44-7)**
2-Dimethylcarbamoyl-3-methylpyrazolyl-(5)-N,N-dimethyl-carbamat (German) **(644-64-4)**
1-Dimethylcarbamoyl-5-methyl-3-pyrazolyl dimethylcarbamate **(644-64-4)**
2-Dimethylcarbamoyl-3-methyl-5-pyrazolyl dimethylcarbamate **(644-64-4)**
cis-2-Dimethylcarbamoyl-1-methylvinyl dimethylphosphate **(141-66-2)**
3-N,N-Dimethylcarbamoyloxy-7-chloro-5-phenyl-1-methyl-1,3-dihydro-2H-1,4-benzodiazepin-2-one **(36104-80-0)**
Dimethylcarbamyl chloride **(79-44-7)**
N,N-Dimethylcarbamyl chloride **(79-44-7)**
2-(N,N-Dimethylcarbamyl)-3-methylpyrazolyl-5 N,N-dimethylcarbamate **(644-64-4)**
Dimethyl 2-carbamyl-3-methylpyrazolyldimethylcarbamate **(644-64-4)**
N,N'-Dimethyl carbanilide **(611-92-7)**
Dimethyl carbate **(39589-98-5)**
9,9-Dimethylcarbazine **(6267-02-3)**
O,O-Dimethyl S-(carbethoxy)methyl phosphorothiolate **(2088-72-4)**
Dimethylcarbinol **(67-63-0)**
2,2-Dimethyl-6-carbobutoxy-2,3-dihydro-4-pyrone **(532-34-3)**
α,α-Dimethyl-α'-carbobutoxy-dihydro-γ-pyrone **(532-34-3)**
O,O-Dimethyl S-(1-carboethoxybenzyl) dithiophosphate **(2597-03-7)**
O,O-Dimethyl S-carboethoxymethyl thiophosphate **(2088-72-4)**
O,O-Dimethyl-O-(2-carbomethoxy-1-methylvinyl) phosphate **(7786-34-7)**
Dimethyl-1-carbomethoxy-1-propen-2-yl phosphate **(7786-34-7)**
Dimethyl carbonate [UN 1161] **(616-38-6)**
O,O-Dimethyl S-carboxymethyl phosphorodithioate **(1113-01-5)**
3,4-Dimethylcatechol **(2785-76-4)**
3,5-Dimethylcatechol **(2785-75-3)**
Dimethylcellosolve **(110-71-4)**
Dimethylcetylamine **(112-69-6)**
N,N-Dimethylcetylamine **(112-69-6)**

Dimethyl chloracetal (97-97-2)
O,O-Dimethyl-O-3-chlor-4-nitrofenylester kyseliny thiofosforecne (Czech) (500-28-7)
O,O-Dimethyl-O-3-chlor-4-nitrofenyltiofosfat (Czech) (500-28-7)
O,O-Dimethyl-O-(3-chlor-4-nitrophenyl)-monothiophosphat (German) (500-28-7)
N,N-Dimethyl-3-chloroaniline (6848-13-1)
O,O-Dimethyl O-(2-chloro-2-(N,N-diethylcarbamoyl)-1-methylvinyl) phosphate (13171-21-6)
Dimethyl 2-chloro-2-diethylcarbamoyl-1-methylvinyl phosphate (13171-21-6)
Dimethylchloroether (107-30-2)
Dimethyl(2-chloroethyl)amine hydrochloride (4584-46-7)
Dimethyl-β-chloroethylamine hydrochloride (4584-46-7)
Dimethylchloroformamide (79-44-7)
2,4-Dimethyl-3-(chloromethyl)-6-tert-butylphenol (23500-79-0)
O,O-Dimethyl O-(3-chloro-4-nitrophenyl) phosphorothioate (500-28-7)
O,O-Dimethyl O-2-chloro-4-nitrophenyl phosphorothioate (2463-84-5)
Dimethyl 3-chloro-4-nitrophenyl thionophosphate (500-28-7)
O,O-Dimethyl-O-(2-chloro-4-nitrophenyl)thionophosphate (2463-84-5)
Dimethyl 2-chloronitrophenyl thiophosphate (2463-84-5)
O,O-Dimethyl O-(3-chloro-4-nitrophenyl) thiophosphate (500-28-7)
Dimethyl p-chlorophenylthiomethyl dithiophosphate (953-17-3)
O,O-Dimethyl-S-(p-chlorophenylthiomethyl)phosphorodithioate (953-17-3)
1,1-Dimethyl-3-(p-chlorophenyl)urea (150-68-5)
N,N-Dimethyl-N'-(3-chlorophenyl)-urea (587-34-8)
N,N-Dimethyl-N'-(4-chlorophenyl)urea (150-68-5)
Dimethylchlorosilane (1066-35-9)
Dimethyl chlorothiophosphate (DOT) (2524-03-0)
Dimethylchlorovinylsilane (1719-58-0)
Dimethylchlorthiofosfat (Czech) (2524-03-0)
O,O-Dimethyl-O-2-chlor-1-(2,4,5-trichlorphenyl)-vinyl-phosphat (German) (961-11-5)
1:2-Dimethylchrysene (15914-23-5)
5,6-Dimethylchrysene (3697-27-6)
Dimethylcocoamine, bis(chloroethyl) ether, diquaternary ammonium salt (68607-28-3)
2,2-Dimethyl-7-coumaranyl N-methylcarbamate (1563-66-2)
Dimethylcyanamide (1467-79-4)
O,O-Dimethyl-O-(4-cyano-phenyl)-monothiophosphat (German) (2636-26-2)
O,O-Dimethyl O-4-cyanophenyl phosphorothioate (2636-26-2)
O,O-Dimethyl-O-p-cyanophenyl phosphorothioate (2636-26-2)
O,O-Dimethyl O-4-cyanophenyl thiophosphate (2636-26-2)
N,N-Dimethylcyclohexanamine (98-94-2)
1,4-Dimethylcyclohexane-cis (624-29-3)
1,4-Dimethylcyclohexane-trans (2207-04-7)
Dimethylcyclohexane (27195-67-1)
cis-1,2-Dimethylcyclohexane (2207-01-4)
m-Dimethylcyclohexane (591-21-9)
o-Dimethylcyclohexane (583-57-3)
trans-1,2-Dimethylcyclohexane (6876-23-9)
1,2-Dimethylcyclohexane [UN 2263] (583-57-3)
1,3-Dimethylcyclohexane [UN 2263] (591-21-9)
1,4-Dimethylcyclohexane [UN 2263] (589-90-2)
Dimethyl 1,4-cyclohexanedicarboxylate (94-60-0)
1,1-Dimethyl-3,5-cyclohexanedione (126-81-8)
5,5-Dimethyl-1,3-cyclohexanedione (126-81-8)
5,5-Dimethylcyclohexane-1,3-dione (126-81-8)
(+-)-N,α-Dimethylcyclohexaneethylamine (101-40-6)
N,α-Dimethylcyclohexaneethylamine (101-40-6)
N-α-(Dimethylcyclohexane)ethylamine (101-40-6)
α,N-Dimethylcyclohexaneethylamine (101-40-6)
Dimethyl-3-cyclohexene-1-carboxaldehyde (27939-60-2)
1,5-Dimethyl-5-(1-cyclohexenyl)barbituric acid (56-29-1)
Dimethylcyclohexylamine (98-94-2)
N,N-Dimethylcyclohexylamine [UN 2264] (98-94-2)
cis-3,3-Dimethylcyclohexylideneethanal (26532-24-1)
1,3-Dimethylcyclopentane (2453-00-1)
cis-1,3-Dimethylcyclopentane (2532-58-3)
2,2-Dimethyl-1,3-cyclopentanedione (3883-58-7)
Dimethylcyclopolysiloxane (69430-24-6)
D-β,β-Dimethylcysteine (52-67-5)
Dimethylcysteine (52-67-5)
β,β-Dimethylcysteine (52-67-5)
N,N-Dimethyldecanamide (14433-76-2)
N,N-Dimethyl-1-decanamine (1120-24-7)
Dimethyl decanedioate (106-79-6)
N,N-Dimethyldecylamine (1120-24-7)
Dimethyldiallylammonium chloride (7398-69-8)
3,3'-Dimethyl-4,4'-diaminodiphenylmethane (838-88-0)
Dimethyldiaminodiphenylmethane (1807-55-2)
N,N-Dimethyl-1,3-diaminopropane (109-55-7)
O,O-Dimethyl S-(4,6-diamino-1,3,5-triazinyl-2-methyl) dithiophosphate (78-57-9)
O,O-Dimethyl S-(4,6-diamino-1,3,5-triazin-2-yl)methyl phosphorodithioate (78-57-9)
O,O-Dimethyl S-(4,6-diamino-s-triazin-2-ylmethyl)-phosphorodithioate (78-57-9)
O,O-Dimethyl S-(4,6-diamino-1,3,5-triazin-2-yl)methyl phosphorothiolothionate (78-57-9)
Dimethyldiaminoxanthenyl chloride (2465-29-4)
N,N-Dimethyldibenzo(b,e)thiepin-δ-sup(11(6H),γ)propylamine (113-53-1)
2,5-Dimethyl-2,5-di-(benzoylperoxy) hexane (2618-77-1)
2,5-Dimethyl-2,5-di-(benzoylperoxy)hexane, Not more than 82% with inert solid (DOT) (2618-77-1)
2,5-Dimethyl-2,5-di-(benzoylperoxy)hexane, Not more than 82% with water (DOT) (2618-77-1)
2,5-Dimethyl-2,5-di-(benzoylperoxy)hexane, Technically pure (DOT) (2618-77-1)
O,O-Dimethyl-O-(1,2-dibrom-2,2-dichlor-aethyl)-phosphat (German) (300-76-5)
O,O-Dimethyl-O-(1,2-dibromo-2,2-dichloroethyl)phosphate (300-76-5)
Dimethyl-1,2-dibromo-2,2-dichloroethyl phosphate (OSHA) (300-76-5)
2,5-Dimethyl-2,5-di(t-butylperoxy)hexane (78-63-7)
2,5-Dimethyl-2,5-di-(tert-butylperoxy)hexane, Technically pure (DOT) (78-63-7)
2,5-Dimethyl-2,5-di(t-butylperoxy)hexyne-3 (1068-27-5)
2,5-Dimethyl-2,5-di-(tert-butylperoxy)hexyne-3, Technically pure (DOT) (1068-27-5)
O,O-Dimethyl-S-1,2-(dicarbaethoxyaethyl)-dithiophosphat (German) (121-75-5)
O,O-Dimethyl-S-(1,2-dicarbethoxyethyl) dithiophosphate (121-75-5)
O,O-Dimethyl S-(1,2-dicarbethoxyethyl)phosphorodithioate (121-75-5)
O,O-Dimethyl-S-(1,2-dicarbethoxy)ethyl phosphorothioate (1634-78-2)
O,O-Dimethyl S-(1,2-dicarbethoxyethyl) thiothionophosphate (121-75-5)
Dimethyl-1,3-di(carbomethoxy)-1-propen-2-yl phosphate (122-10-1)
Dimethyldicetylammonium chloride (1812-53-9)
O,O-Dimethyl-O-(2,5-dichlor-4-bromphenyl)-thionophosphat (German) (2104-96-3)
O,O-Dimethyl-O-(2,5-dichlor-4-jodphenyl)-monothiophosphat (German) (18181-70-9)
O,O-Dimethyl-O-(2,5-dichlor-4-jodphenyl)-thionophosphat (German) (18181-70-9)
O,O-Dimethyl-O-(2,5-dichloro-4-bromophenyl)phosphorothioate (2104-96-3)
O,O-Dimethyl O-(2,5-dichloro-4-bromophenyl) thiophosphate (2104-96-3)
O,O-Dimethyl O-2,2-dichloro-1,2-dibromoethyl phosphate (300-76-5)
Dimethyl 2,2-dichloroethenyl phosphate (62-73-7)
Dimethyl-1,1'-dichloroether (542-88-1)
O,O-Dimethyl-O-2,5-dichloro-4-iodophenyl thiophosphate (18181-70-9)
3,5-Dimethyl-2,4-dichlorophenol (133-53-9)
O,O-Dimethyl S-(((2,5-dichlorophenyl)thio)methyl) phosphorodithioate (3735-23-7)
O,O-Dimethyl S-(2,5-dichlorophenylthio)methyl phosphorodithioate (3735-23-7)
1,1-Dimethyl-3-(3,4-dichlorophenyl)urea (330-54-1)

Dimethyldichlorostannane (753-73-1)
Dimethyldichlorotin (753-73-1)
Dimethyl 2,2-dichlorovinyl phosphate (62-73-7)
Dimethyl dichlorovinyl phosphate (62-73-7)
O,O-Dimethyl 2,2-dichlorovinyl phosphate (62-73-7)
O,O-Dimethyl O-2,2-dichlorovinyl phosphate (62-73-7)
O,O-Dimethyl dichlorovinyl phosphate (62-73-7)
Dimethyl-dichlorsilan (Czech) (75-78-5)
O,O-Dimethyl-O-(2,2-dichlor-vinyl)-phosphat (German) (62-73-7)
Dimethyldicocoammonium chloride (61789-77-3)
Dimethyldidecylammonium chloride (7173-51-5)
O,O-Dimethyl S-1,2-di(ethoxycarbamyl)ethyl phosphorodithioate (121-75-5)
Dimethyl diethylamido-1-chlorocrotonyl (2) phosphate (13171-21-6)
2,5-Dimethyl-2,5-di-(2-ethylhexanoylperoxy)hexane, Technically pure (DOT) (13052-09-0)
Dimethyldiethyllead (1762-27-2)
Dimethyldifluorosilane (353-66-2)
2,2-Dimethyl-2,3-dihydrobenzofuranyl-7-N-(O,O-dimethyl-phosphorothioyl)-N-methylcarbamate (28789-80-2)
2,2-Dimethyl-2,3-dihydro-7-benzofuranyl N-methylcarbamate (1563-66-2)
Dimethyl 1,4-dihydro-2,6-dimethyl-4-(2'-nitrophenyl)-3,5-pyridine-dicarboxylate (21829-25-4)
Dimethyl dihydrogenated tallow ammonium chloride, Reaction product with bentonite (68953-58-2)
Dimethyl dihydrogenated tallow ammonium chloride, Reaction product with hectorite (71011-27-3)
O,O-Dimethyl S-(3,4-dihydro-4-keto-1,2,3-benzotriazinyl-3-methyl) dithiophosphate (86-50-0)
5,5-Dimethyldihydroresorcinol (126-81-8)
5,5-Dimethyl-dihydroresorcinol-N,N-dimethylcarbamat (German) (122-15-6)
5,5-Dimethyldihydroresorcinol dimethylcarbamate (122-15-6)
5,5-Dimethyl-4,5-dihydro-3-resorcyl-dimethyl-carbamat (German) (122-15-6)
1,2-Dimethyl-3,5-dihydroxybenzene (527-55-9)
1,3-Dimethyl-7-(2,3-dihydroxypropyl)xanthine (479-18-5)
1,3-Dimethyl-4,6-diisopropylbenzene (5186-68-5)
O,O-Dimethyl-S-1,2-dikarbetoxylethylditiofosfat (Czech) (121-75-5)
1,1-Dimethyl-3,5-diketocyclohexane (126-81-8)
Dimethyl diketone (431-03-8)
Dimethyl 3-(dimethoxyphosphinyloxy)glutaconate (122-10-1)
Dimethyl 2,2-dimethyladipate (17219-21-5)
Dimethyl((dimethylamino)methyl)amine (51-80-9)
1,5-Dimethyl-4-dimethylamino-2-phenyl-3-pyrazolone (58-15-1)
2,3-Dimethyl-4-dimethylamino-1-phenyl-5-pyrazolone (58-15-1)
5,6-Dimethyl-2-dimethylamino-4-pyrimidinyldimethylcarbamate (23103-98-2)
O,O-Dimethyl-O-(2-dimethyl-carbamoyl-1-methyl-vinyl)phosphat (German) (141-66-2)
2-Dimethyl cis-2-dimethyl-carbamoyl-1-methylvinyl phosphate (141-66-2)
O,O-Dimethyl O-(N,N-dimethylcarbamoyl-1-methylvinyl) phosphate (141-66-2)
N,N-Dimethyl-N'-(5-(1,1-dimethylethyl)-3-isoxazolyl)urea (55861-78-4)
O,O-Dimethyl-O-(1,4-dimethyl-3-oxo-4-aza-pent-1-enyl)fosfaat (Dutch) (141-66-2)
O,O-Dimethyl-O-(1,4-dimethyl-3-oxo-4-aza-pent-1-enyl)phosphate (141-66-2)
O,O-Dimethyl O-(p-(N,N-dimethylsulfamoyl)phenyl) phosphorothioate (52-85-7)
1,3-Dimethyl-4,6-dinitrobenzene (616-72-8)
1,5-Dimethyl-2,4-dinitrobenzene (616-72-8)
1,2-Dimethyl-4,5-dinitroimidazole (19183-17-6)
1,6-Dimethyl-1,6-dinitrosobiurea (3844-60-8)
Dimethyldioctadecylammonium chloride (107-64-2)
1,4-Dimethyl-3,6-dioxa-1-heptanol (34590-94-8)
4,4-Dimethyl-1,3-dioxane (766-15-4)
4,4-Dimethyl-m-dioxane (766-15-4)
4,4-Dimethyldioxane-1,3 (766-15-4)
Dimethyl dioxane [UN 2707] (25136-55-4)
Dimethyl-p-dioxane [UN 2702] (25136-55-4)
2,6-Dimethyl-m-dioxan-4-ol acetate (828-00-2)
2,6-Dimethyl-m-dioxan-4-yl acetate (828-00-2)
Dimethyl 2,5-dioxo-1,4-cyclohexanedicarboxylate (6289-46-9)
2,2-Dimethyl-1,3-dioxolane-4-methanol (100-79-8)
N,N-Dimethyl-2,2-diphenylacetamide (957-51-7)
N,N-Dimethyl-α,α-diphenylacetamide (957-51-7)
N,N-Dimethyldiphenylacetamide (957-51-7)
3,3'-Dimethyl-4,4'-diphenyldiamine (119-93-7)
3,3'-Dimethyldiphenyl-4,4'-diamine (119-93-7)
(-)-N,N-Dimethyl-1,2-diphenylethylamine hydrochloride (14148-99-3)
1,2-Dimethyl-3,5-diphenyl-1H-pyrazolium (49866-87-7)
1,2-Dimethyl-3,5-diphenylpyrazolium ion (49866-87-7)
1,2-Dimethyl-3,5-diphenyl-1-h-pyrazolium methyl sulfate (43222-48-6)
1,1'-Dimethyl-4,4'-dipyridinium-dichlorid (German) (1910-42-5)
1',1'-Dimethyl-4,4'-dipyridinium di(methyl sulfate) (2074-50-2)
4,4'-Dimethyldipyridyl dichloride (1910-42-5)
1,1'-Dimethyl-4,4'-dipyridylium chloride (1910-42-5)
N,N'-Dimethyl-4,4'-dipyridylium dichloride (1910-42-5)
Dimethyldiselenide (7101-31-7)
Dimethyldisulfide (624-92-0)
Dimethyl disulfide [UN 2381] (624-92-0)
Dimethyl ditallow ammonium chloride (68783-78-8)
Dimethylditallowammonium chloride (68783-78-8)
N,N-Dimethyl-4,4-di-2-thienyl-3-buten-2-amine (524-84-5)
Dimethyl 3,3'-dithiobispropanoate (15441-06-2)
Dimethyldithiocarbamate (79-45-8)
Dimethyldithiocarbamate zinc salt (137-30-4)
Dimethyldithiocarbamic acid (79-45-8)
N,N-Dimethyldithiocarbamic acid (79-45-8)
Dimethyldithiocarbamic acid copper salt (137-29-1)
N,N-Dimethyldithiocarbamic acid dimethylaminomethyl ester (51-82-1)
Dimethyldithiocarbamic acid, iron salt (14484-64-1)
Dimethyldithiocarbamic acid iron(3+) salt (14484-64-1)
Dimethyldithiocarbamic acid, iron(3+) salt (14484-64-1)
Dimethyldithiocarbamic acid, sodium salt (128-04-1)
Dimethyldithiocarbamic acid, zinc salt (137-30-4)
N,N-Dimethyl-dithiocarbaminsaeure-dimethylaminomethyl-ester (German) (51-82-1)
Dimethyl 3,3'-dithiodipropionate (15441-06-2)
Dimethyl dithiodipropionate (15441-06-2)
O,O-Dimethyldithiofosforecnan sodny (Czech) (26377-29-7)
2,4-Dimethyl-1,3-dithiolane-2-carboxaldehyde O-(methyl-carbamoyl)oxime (26419-73-8)
O,O-Dimethyldithiophosphate diethylmercaptosuccinate (121-75-5)
Dimethyldithiophosphoric acid N-methylbenzazimide ester (86-50-0)
O,O,-Dimethyldithiophosphorylacetic acid (1113-01-5)
O,O-Dimethyl dithiophosphorylacetic acid N-methyl-N-formyl amide (2540-82-1)
O,O-Dimethyldithiophosphorylacetic acid, N-monomethylamide salt (60-51-5)
O,O-Dimethyl-dithiophosphorylessigsaeure monomethylamid (German) (60-51-5)
Dimethyldithioxodistannathiane (33397-79-4)
N,N-Dimethyl-N-(3-dodecanamidopropyl)amine (3179-80-4)
2,6-Dimethyldodeca-2,6,8-trien-10-one (26651-96-7)
7,11-Dimethyl-4,6,10-dodecatrien-3-one (26651-96-7)
N,N-Dimethyl-(3-dodecylamidopropyl)amine (3179-80-4)
N,N-Dimethyldodecylamine (112-18-5)
Dimethyldodecylamine chlorhydrate (French) (2016-48-0)
N,N-Dimethyldodecylamine hydrochloride (2016-48-0)
Dimethyldodecylamine N-oxide (1643-20-5)
N,N-Dimethyldodecylamine oxide (1643-20-5)
N,N-Dimethyl-dodecylaminoxid (Czech) (1643-20-5)
7,8-Dimethylenebenz(a)anthracene (479-23-2)
3:4-Dimethylene-1:2-benzanthracene (5779-79-3)
2,3-Dimethylenebutane (513-81-5)
1,4-Dimethylenecyclohexane (4982-20-1)
Dimethylenediamine (107-15-3)

Dimethylenedioxy benzphenanthridine **(2447-54-3)**
Dimethylene glycol **(513-85-9)**
Dimethyleneimine **(151-56-4)**
Dimethylenemethane **(463-49-0)**
Dimethylene oxide **(75-21-8)**
1,3-Dimethylene-2,2,5,5-tetramethylcyclohexane **(40482-18-6)**
Dimethylenimine **(151-56-4)**
exo-1,2-cis-Dimethyl-3,6-epoxyhexahydrophthalic anhydride **(56-25-7)**
6,7-Dimethylesculetin **(120-08-1)**
O,S-Dimethyl ester amide of amidothioate **(10265-92-6)**
O,O-Dimethylester kyseliny chlorthiofosforecne (Czech) **(2524-03-0)**
Dimethylester kyseliny fosforite (Czech) **(868-85-9)**
Dimethylester kyseliny ftalove (Czech) **(131-11-3)**
Dimethylester kyseliny isoftalove (Czech) **(1459-93-4)**
Dimethylester kyseliny maleinove (Czech) **(624-48-6)**
Dimethylester kyseliny sirove (Czech) **(77-78-1)**
Dimethylester kyseliny tereftalove (Czech) **(120-61-6)**
Dimethylester kyseliny tetrachlortereftalove (Czech) **(1861-32-1)**
O,S-Dimethylester kyseliny tetrachlorthiotereftalove (Czech) **(3765-57-9)**
N,N-Dimethylethanamine **(598-56-1)**
Dimethyl ethanedioate **(553-90-2)**
1,1-Dimethylethanol **(75-65-0)**
N,N-Dimethylethanolamine **(108-01-0)**
Dimethylethanolamine [UN 2051] **(108-01-0)**
Dimethyl ether [UN 1033] **(115-10-6)**
Dimethylether hydrochinonu (Czech) **(150-78-7)**
Dimethyl ether hydroquinone **(150-78-7)**
Dimethylether pyrokatechinu (Czech) **(91-16-7)**
Dimethylether resorcinolu (Czech) **(151-10-0)**
6-α,21-Dimethyl-17-α-ethinyltestosterone **(79-64-1)**
6-α,21-Dimethylethisterone **(79-64-1)**
O,O-Dimethyl S-α-ethoxycarbonylbenzyl phosphorodithioate **(2597-03-7)**
O,O-Dimethyl S-(5-ethoxy-1,3,4-thiadiazol-2(3H)-onyl-(3)-methyl)dithiophosphate **(2669-32-1)**
O,O-Dimethyl-S-(5-ethoxy-1,3,4-thiadiazol-2(3H)-onyl-(3)-methyl)-phosphorodithioate **(2669-32-1)**
O,O-Dimethyl-S-(5-ethoxy-1,3,4-thiadiazolinyl-3-methyl)dithiophosphate **(2669-32-1)**
2-(1,1-Dimethylethoxy)thiophene **(23290-55-3)**
O,O-Dimethyl S-(2-ethsulfonylethyl)phosphorothioate **(17040-19-6)**
Dimethyl S-(2-ethsulfonylethyl)thiophosphate **(17040-19-6)**
O,O-Dimethyl S-(2-eththioethyl)phosphorothioate **(919-86-8)**
Dimethyl S-(2-eththioethyl)thiophosphate **(919-86-8)**
O,O-Dimethyl S-(2-eththionylethyl) phosphorothioate **(301-12-2)**
Dimethyl S-(2-eththionylethyl) thiophosphate **(301-12-2)**
N-(1,1-Dimethylethyl)acetamide **(762-84-5)**
1,1-Dimethylethylamine **(75-64-9)**
Dimethylethylamine **(598-56-1)**
N,N-Dimethylethylamine **(598-56-1)**
2-((1,1-Dimethylethyl)amino)ethanol **(4620-70-6)**
5-(2-((1,1-Dimethylethyl)amino)-1-hydroxyethyl)-1,3-benzenediol **(23031-25-6)**
5-(3-((1,1-Dimethylethyl)amino)-2-hydroxypropoxy)-1,2,3,4-tetrahydro-2,3-naphthalenediol **(42200-33-9)**
2-((1,1-Dimethylethyl)azo)-2-butanol **(57910-79-9)**
1-((1,1-Dimethylethyl)azo)cyclohexanol **(54043-65-1)**
2-((1,1-Dimethylethyl)azo)-5-methyl-2-hexanol **(64819-51-8)**
1,3-Dimethyl-5-ethylbarbituric acid **(7391-61-9)**
4-(1,1-Dimethylethyl)benzaldehyde **(939-97-9)**
(1,1-Dimethylethyl)-1,2-benzenediol **(27213-78-1)**
4-(1,1-Dimethylethyl)benzenemethanol **(877-65-6)**
(1,1-Dimethylethyl)benzoic acid **(1320-16-7)**
1,1-Dimethylethyl 2-buteneperoxoate **(23474-91-1)**
Di-N-methyl ethyl carbamate **(687-48-9)**
O,O-Dimethyl S-(N-ethylcarbamoylmethyl) dithiophosphate **(116-01-8)**
O,O-Dimethyl S-(N-ethylcarbamoylmethyl) phosphorodithioate **(116-01-8)**
Dimethylethylcarbinol **(75-85-4)**
1,1-Dimethylethyl carbonazidate **(1070-19-5)**
2-(1,1-Dimethylethyl)cyclohexanol **(13491-79-7)**
trans-2-(1,1-Dimethylethyl)cyclohexanol **(5448-22-6)**
2-(1,1-Dimethylethyl)cyclohexanol acetate **(88-41-5)**
(1,1-Dimethylethyl)dimethylphenol **(36812-13-2)**
4-(1,1-Dimethylethyl)-2,5-dimethylphenol **(17696-37-6)**
4-(1,1-Dimethylethyl)-2,6-dimethylphenol **(879-97-0)**
2-(1,1-Dimethylethyl)-4,6-dinitrophenol **(1420-07-1)**
cis-1,2-Dimethylethylene **(590-18-1)**
trans-1,2-Dimethylethylene **(624-64-6)**
2,2-Dimethylethylenimine **(2658-24-4)**
1,1-Dimethylethyl ethaneperoxoate **(107-71-1)**
1-(1,1-Dimethylethyl)-4-ethenylbenzene **(1746-23-2)**
2-(1,1-Dimethylethyl)-4-ethylphenol **(96-70-8)**
2-(1,1-Dimethylethyl)-5-ethylphenol **(4237-25-6)**
1,1-Dimethylethyl glycidyl ether **(7665-72-7)**
Dimethylethylhexadecylammonium bromide **(124-03-8)**
1,1-Dimethylethyl hydroperoxide **(75-91-2)**
2,2'-((1,1-Dimethylethyl)imino)bisethanol **(2160-93-2)**
2-((1,1-Dimethylethyl)imino)tetrahydro-3-(1-methylethyl)-5-phenyl-4H-1,3,5-thiadiazin-4-one **(69327-76-0)**
N'-(5-(1,1-Dimethylethyl)-3-isoxazolyl)-N,N-dimethylurea **(55861-78-4)**
O,O-Dimethyl-S-(2-ethylmercaptoethyl) dithiophosphate **(640-15-3)**
O,O-Dimethyl 2-ethylmercaptoethyl thiophosphate **(2633-54-7)**
O,O-Dimethyl O-ethylmercaptoethyl thiophosphate **(867-27-6)**
O,O-Dimethyl S-ethylmercaptoethyl thiophosphate **(919-86-8)**
O,O-Dimethyl 2-ethylmercaptoethyl thiophosphate, thiolo isomer **(919-86-8)**
O,O-Dimethyl 2-ethylmercaptoethyl thiophosphate, thiono isomer **(867-27-6)**
O,O-Dimethyl-S-2-ethylmerkaptoethylester kyseliny dithiofosforecne (Czech) **(640-15-3)**
(1,1-Dimethylethyl)-4-methoxyphenol **(25013-16-5)**
(1,1-Dimethylethyl)-3-methylphenol **(1333-13-7)**
(1,1-Dimethylethyl)-4-methylphenol **(25567-40-2)**
2-(1,1-Dimethylethyl)-5-methylphenol **(88-60-8)**
4-(1,1-Dimethylethyl)-3-methylphenol **(2219-72-9)**
4-(1,1-Dimethylethyl)-N-(1-methylpropyl)-2,6-dinitrobenzenamine **(33629-47-9)**
6-(1,1-Dimethylethyl)-3-(methylthio)-1,2,4-triazin-5(2H)-one **(35045-02-4)**
1,1-Dimethyl-3-ethyl-3-nitrosourea **(50285-71-7)**
3,5-Dimethyl-5-ethyloxazolidine-2,4-dione **(115-67-3)**
2,4-Dimethyl-3-ethyl pentane **(1068-87-7)**
(1,1-Dimethylethyl)phenol **(27178-34-3)**
3-(1,1-Dimethylethyl)phenol **(585-34-2)**
4-(1,1-Dimethylethyl)phenol **(98-54-4)**
4-(1,1-Dimethylethyl)phenol phosphate (3:1) **(78-33-1)**
4-(1,1-Dimethylethyl)phenol sodium salt **(5787-50-8)**
2-(4-(1,1-Dimethylethyl)phenoxy)cyclohexanol **(1942-71-8)**
2-(4-(1,1-Dimethylethyl)phenoxy)cyclohexyl 2-propynyl sulfite **(2312-35-8)**
1-(4-(1,1-Dimethylethyl)phenyl)ethanone **(943-27-1)**
1-(4-(1,1-Dimethylethyl)phenyl)-3-(4-methoxyphenyl)-1,3-propanedione **(70356-09-1)**
1,1-Dimethylethyl 2-propenoate **(1663-39-4)**
2,6-Dimethyl-4-ethylpyridine **(36917-36-9)**
Dimethylethylsoyaammonium ethosulfate **(68308-67-8)**
O,O-Dimethyl-S-(2-ethylsulfinyl-ethyl)-monothiofosfaat (Dutch) **(301-12-2)**
O,O-Dimethyl S-(2-(ethylsulfinyl)ethyl) phosphorothioate **(301-12-2)**
O,O-Dimethyl S-(2-ethylsulfinyl)ethyl thiophosphate **(301-12-2)**
O,O-Dimethyl S-ethyl-2-sulfonylethyl phosphorothiolate **(17040-19-6)**
O,O-Dimethyl S-ethylsulphinylethyl phosphorothiolate **(301-12-2)**
O,O-Dimethyl S-ethylsulphonylethyl phosphorothiolate **(17040-19-6)**
O,O-Dimethyl-S-(2-ethylthio-ethyl)-dithiofosfaat (Dutch) **(640-15-3)**
O,O-Dimethyl-O-(2-ethyl-thio-ethyl)-monothiofosfaat (Dutch) **(867-27-6)**
O,O-Dimethyl-S-(2-ethylthio-ethyl)-monothiofosfaat (Dutch) **(919-86-8)**
O,O-Dimethyl S-(2-(ethylthio)ethyl) phosphorodithioate **(640-15-3)**
O,O-Dimethyl O-2-(ethylthio)ethyl phosphorothioate **(867-27-6)**
O,O-Dimethyl S-(2-(ethylthio)ethyl)phosphorothioate **(919-86-8)**
S-(((1,1-Dimethylethyl)thio)methyl)-O,O-diethyl phosphorodithioate **(13071-79-9)**
5-(1,1-Dimethylethyl)-1,2,3-trimethylbenzene **(98-23-7)**

Dimethylethynylcarbinol **(115-19-5)**
Dimethylethynylmethanol **(115-19-5)**
N,N'-Dimethyl-p-fenylendiamin (Czech) **(105-10-2)**
N,N-Dimethyl-p-fenylendiamin (Czech) **(99-98-9)**
1,1-Dimethyl-3-(p-fluorophenyl)urea **(332-33-2)**
Dimethyl formal **(109-87-5)**
Dimethylformaldehyde **(67-64-1)**
Dimethylformamid (German) **(68-12-2)**
Dimethyl formamide **(68-12-2)**
N,N-Dimethyl formamide **(68-12-2)**
Dimethylformamide (ACGIH,OSHA) **(68-12-2)**
N,N-Dimethylformamide [UN 2265] **(68-12-2)**
Dimethylformocarbothialdine **(533-74-4)**
2,4-Dimethyl-2-formyl-1,3-dithiolane oxime methylcarbamate **(26419-73-8)**
2,4-Dimethyl-2-formyl-1,3-dithiolane-oxime-methylcarbamate **(26419-73-8)**
2,4-Dimethyl-2-formyl-1,3-dithiolanoximmethylkarbamat (Czech) **(26419-73-8)**
O,O-Dimethyl S-(N-formyl-N-methylcarbamoylmethyl) phosphorodithioate **(2540-82-1)**
Dimethylfosfit (Czech) **(868-85-9)**
Dimethylfosfonat (Czech) **(868-85-9)**
2,5-Dimethylfuran **(625-86-5)**
Dimethyl furane **(28802-49-5)**
4,8-Dimethyl-2H-furo(2,3-h)-1-benzopyran-2-one **(4063-41-6)**
4,9-Dimethyl-2H-furo(2,3-h)-1-benzopyran-2-one plus ultraviolet a radiation **(22975-76-4)**
Dimethyl germanium sulfide **(16090-49-6)**
3,3-Dimethylglutarate **(4839-46-7)**
Dimethyl glutarate **(1119-40-0)**
Dimethylglyoxal **(431-03-8)**
Dimethylglyoxime **(95-45-4)**
Dimethyl-guanidin (German) **(3324-71-8)**
Dimethylguanidine **(3324-71-8)**
as-Dimethylguanidine hydrochloride **(22583-29-5)**
N,N'-Dimethylharnstoff (German) **(96-31-1)**
m-(3,3-Dimethylharnstoff)-phenyl-tert-butylcarbamat (German) **(4849-32-5)**
2,6-Dimethyl-2,5-heptadien-4-one **(504-20-1)**
3,3-Dimethylheptane **(4032-86-4)**
3,5-Dimethylheptane **(926-82-9)**
(R)-(-)-2,6-Dimethylheptanoic acid **(60148-94-9)**
2,6-Dimethyl heptanol-4 **(108-82-7)**
2,6-Dimethyl-2-heptanol **(13254-34-7)**
2,6-Dimethyl-4-heptanol **(108-82-7)**
2,6-Dimethylheptan-2-ol **(13254-34-7)**
2,6-Dimethyl-heptan-4-on (Dutch, German) **(108-83-8)**
2,6-Dimethyl-4-heptanone **(108-83-8)**
2,6-Dimethylheptan-4-one **(108-83-8)**
4,6-Dimethyl-2-heptanone **(19549-80-5)**
2,6-Dimethyl-4-heptanone (OSHA) **(108-83-8)**
Dimethyl heptyladipate **(1330-86-5)**
1,4-Dimethylheptylamine **(67953-04-2)**
(+)-trans-3-(1,1-Dimethylheptyl)-6,6a,7,8,10,10a-hexahydro-1-hydroxy-6,6-dimethyl-9H-dibenzo(b,d)pyran-9-one) **(51022-71-0)**
(+-)-3-(1,1-Dimethylheptyl-6,6aβ,7,8,10,10aα-hexahydro-1-hydroxy-6,6-dimethyl-9H-dibenzo(b,d)pyran-9-one **(51022-71-0)**
Dimethylheptyl(1-hydroxy-p-menth-2-yl)ammonium.Br **(20091-61-6)**
N,N'-Di(1-methylheptyl)-p-phenylenediamine **(103-96-8)**
Dimethylheptylpyran **(32904-22-6)**
3-(1,2-Dimethylheptyl)-7,8,9,10-tetrahydrocannabinol **(32904-22-6)**
5,5-Dimethyl-3-heptyne **(23097-98-5)**
N,N-Dimethyl-1-hexadecanamine **(112-69-6)**
N,N-Dimethyl-1-hexadecanamine N-oxide **(7128-91-8)**
Dimethyl-n-hexadecylamine **(112-69-6)**
Dimethylhexadecylamine **(112-69-6)**
N,N-Dimethyl-n-hexadecylamine **(112-69-6)**
N,N-Dimethylhexadecylamine **(112-69-6)**
2,3-Dimethyl-1,4-hexadiene **(18669-52-8)**
2,5-Dimethyl-1,5-hexadiene **(627-58-7)**
2,5-Dimethyl-2,4-hexadiene **(764-13-6)**
2,5-Dimethylhexa-1,5-diene **(627-58-7)**
6,11-Dimethyl-1,2,3,4,5,6-hexahydro-8-hydroxy-3-phenethyl-2,6-methano-3-benzazocine **(127-35-5)**
Dimethyl hexahydroterephthalate **(94-60-0)**
4,4-Dimethylhexanal **(5932-91-2)**
2,3-Dimethylhexane **(584-94-1)**
2,4-Dimethylhexane **(589-43-5)**
3,3-Dimethylhexane **(563-16-6)**
3,4-Dimethylhexane **(583-48-2)**
Dimethylhexane dihydroperoxide, Dry (DOT) **(3025-88-5)**
Dimethylhexane dihydroperoxide, (With 18% or more water) (DOT) **(3025-88-5)**
Dimethyl hexanedioate **(627-93-0)**
2,5-Dimethyl-2,5-hexanediol **(110-03-2)**
2,5-Dimethylhexane-2,5-diol **(110-03-2)**
Dimethylhexanediol **(110-03-2)**
3,5-Dimethyl-3-hexanol **(4209-91-0)**
3,3-Dimethyl-2-hexanone **(26118-38-7)**
2,5-Dimethyl-3-hexene **(15910-22-2)**
3,4-Dimethyl-3-hexene **(30951-95-2)**
3,4-Dimethylhexene-3 **(30951-95-2)**
5,5-Dimethyl-1-hexene **(7116-86-1)**
5,5-Dimethylhexene-1 **(7116-86-1)**
1,5-Dimethylhexyl acetate **(67952-57-2)**
N-(5,5-Dimethylhexyl)acrylamide **(4223-03-4)**
2,5-Dimethyl-3-hexyne-2,5-diol **(142-30-3)**
Dimethylhexynediol **(142-30-3)**
3,5-Dimethyl-1-hexyn-3-ol **(107-54-0)**
5,5-Dimethylhydantoin **(77-71-4)**
2,2-Dimethylhydrazid kyseliny jantarove (Czech) **(1596-84-5)**
1,1-Dimethylhydrazin (German) **(57-14-7)**
1,2-Dimethylhydrazin (German) **(540-73-8)**
1,2-Dimethyl-hydrazine **(540-73-8)**
Dimethylhydrazine **(57-14-7)**
N,N'-Dimethylhydrazine **(540-73-8)**
N,N-Dimethylhydrazine **(57-14-7)**
as-Dimethyl hydrazine **(57-14-7)**
asymmetric Dimethylhydrazine **(57-14-7)**
sym-Dimethylhydrazine **(540-73-8)**
u-Dimethylhydrazine **(57-14-7)**
uns-Dimethylhydrazine **(57-14-7)**
unsym-Dimethylhydrazine **(57-14-7)**
1,1-Dimethylhydrazine (ACGIH,OSHA) **(57-14-7)**
1,2-Dimethylhydrazine dihydrochloride **(306-37-6)**
N,N'-Dimethylhydrazine dihydrochloride **(306-37-6)**
sym-Dimethylhydrazine dihydrochloride **(306-37-6)**
1,1-Dimethylhydrazine hydrochloride **(593-82-8)**
1,2-Dimethylhydrazine hydrochloride **(56400-60-3)**
sym-Dimethylhydrazine hydrochloride **(56400-60-3)**
Dimethylhydrazine, symmetrical [UN 2382] **(540-73-8)**
Dimethylhydrazine, unsymmetrical [UN 1163] **(57-14-7)**
β,β-Dimethylhydrocinnamic acid **(1010-48-6)**
Dimethyl hydrogen phosphate **(813-78-5)**
O,O-Dimethyl hydrogen phosphate **(813-78-5)**
Dimethylhydrogenphosphite **(868-85-9)**
Dimethylhydroquinone **(150-78-7)**
Dimethylhydroquinone ether **(150-78-7)**
5,5-Dimethylhydroresorcinol **(126-81-8)**
2,5-Dimethyl-4-hydroxy-3(2H)-furanone **(3658-77-3)**
3,3-Dimethyl-2-hydroxybutyric acid **(4026-20-4)**
N,N-Dimethyl-2-hydroxyethylamine **(108-01-0)**
N,N-Dimethyl-N-(2-hydroxyethyl)amine **(108-01-0)**
Dimethyl 3-hydroxyglutaconate dimethyl phosphate **(122-10-1)**
N,O-Dimethylhydroxylamine **(1117-97-1)**
O,N-Dimethylhydroxylamine **(1117-97-1)**
2,2-Dimethyl-5-hydroxymethyl-1,3-dioxolane **(100-79-8)**

2',3-Dimethyl-4-(2-hydroxynaphthylazo)azobenzene **(85-83-6)**
3,7-Dimethyl-7-hydroxyoctanal **(107-75-5)**
6-α,21-Dimethyl-17-β-hydroxy-17-α-pregn-4-en-20-yn-3-one **(79-64-1)**
2,2-Dimethyl-3-hydroxypropanal **(597-31-9)**
2,2-Dimethyl-β-hydroxypropionaldehyde **(597-31-9)**
α,α-Dimethyl-β-hydroxypropionaldehyde **(597-31-9)**
Dimethyl(2-hydroxypropyl)amine **(108-16-7)**
O,O-Dimethyl-(1-hydroxy-2,2,2-trichloraethyl)phosphonsaeure ester (German) **(52-68-6)**
O,O-Dimethyl-(1-hydroxy-2,2,2-trichlorathyl)-phosphat (German) **(52-68-6)**
O,O-Dimethyl-(1-hydroxy-2,2,2-trichloro)ethyl phosphate **(52-68-6)**
Dimethyl 1-hydroxy-2,2,2-trichloroethyl phosphonate **(52-68-6)**
O,O-Dimethyl (1-hydroxy-2,2,2-trichloroethyl)phosphonate **(52-68-6)**
N,N-Dimethyl-5-hydroxytryptamine **(487-93-4)**
3,4-Dimethyl-3H-imidazo(4,5-f)quinolin-2-amine **(77094-11-2)**
3,8-Dimethyl-3H-imidazo(4,5-f)quinoxalin-2-amine **(77500-04-0)**
1,2-Dimethyl-1H-imidazole **(1739-84-0)**
1,2-Dimethylimidazole **(1739-84-0)**
Dimethylimipramine **(50-47-5)**
1,2-Dimethylindan **(17057-82-8)**
4,6-Dimethylindan **(1685-82-1)**
1,2-Dimethylindane **(17057-82-8)**
3,3-Dimethyl-1-indanol **(38393-92-9)**
(+)-1,2-Dimethylindene **(53204-57-2)**
1,1-Dimethylindene **(18636-55-0)**
1,4-Dimethyl-2-isobutylbenzene **(55669-88-0)**
Dimethyl isophthalate **(1459-93-4)**
Dimethylisopropanolamine **(108-16-7)**
1,3-Dimethyl-4-isopropylbenzene **(4706-89-2)**
O,O-Dimethyl-S-2-(isopropylthio)ethylphosphorodithioate **(36614-38-7)**
4,8-Dimethylisopsoralen **(4063-41-6)**
4,4'-Dimethylisopsoralen plus ultraviolet a radiation **(22975-76-4)**
1,3-Dimethylisothiourea **(534-13-4)**
3,4-Dimethylisoxale-5-sulfanilamide **(127-69-5)**
3,4-Dimethylisoxazole-5-sulphanilamide **(127-69-5)**
N^1-(3,4-Dimethyl-5-isoxazolyl)sulfanilamide **(127-69-5)**
N'-(3,4)Dimethylisoxazol-5-yl-sulphanilamide **(127-69-5)**
N^1-(3,4-Dimethyl-5-isoxazolyl)sulphanilamide **(127-69-5)**
Dimethylkarbamoylchlorid (Czech) **(79-44-7)**
Dimethylketal **(67-64-1)**
Dimethyl .α.-ketoglutarate **(13192-04-6)**
Dimethylketol **(513-86-0)**
Dimethyl ketone **(67-64-1)**
Dimethylkyanamid (Czech) **(1467-79-4)**
N,N-Dimethyllaurylamine **(112-18-5)**
Dimethyl laurylbenzene ammonium bromide **(7281-04-1)**
Dimethyl-5-(l-isopropyl-3-methyl-pyrazolyl)-carbamate **(119-38-0)**
Dimethylmagnesium (DOT) **(2999-74-8)**
Dimethyl maleate **(624-48-6)**
α,β-Dimethylmaleic anhydride **(766-39-2)**
Dimethyl malonate **(108-59-8)**
Dimethylmalonic acid **(595-46-0)**
Dimethylmalononitrile **(7321-55-3)**
Dimethyl mercury **(593-74-8)**
Dimethyl meso-tartrate **(5057-96-5)**
N,N-Dimethylmethanamine N-oxide **(1184-78-7)**
Dimethylmethane **(74-98-6)**
O,O-Dimethyl-S-((2-methoxy-1,3,4 (4H)-thiodiazol-5-on-4-yl)-methyl)-dithiofosfaat (Dutch) **(950-37-8)**
N,N-Dimethyl-N'-(4-methoxybenzyl)-N'-(2-pyridyl)ethylene-diamine maleate **(59-33-6)**
N,N-Dimethyl-N'-(p-methoxybenzyl)-N'-(2-pyrimidyl)ethylene-diamine **(91-85-0)**
N,N-Dimethyl-N'-(4-methoxybenzyl)-N'-(2-pyrimidyl)ethylene-diamine hydrochloride **(63-56-9)**
O,O-Dimethyl-O-2-methoxycarbonyl-1-methyl-vinyl-phosphat (German) **(7786-34-7)**
Dimethyl 2-methoxycarbonyl-1-methylvinyl phosphate **(7786-34-7)**
Dimethyl methoxycarbonylpropenyl phosphate **(7786-34-7)**
Dimethyl (1-methoxycarboxypropen-2-yl)phosphate **(7786-34-7)**
N,N-Dimethyl-N'-(4-methoxy-3-chlorophenyl)urea **(19937-59-8)**
O,O-Dimethyl-S-(2-methoxyethylcarbamoylmethyl)dithiophosphate **(919-76-6)**
O,O-Dimethyl S-(2-methoxyethylcarbamoyl methyl) phosphoro-dithioate **(919-76-6)**
2,6-Dimethyl-N-(2-methoxyethyl)chloroacetanilide **(50563-36-5)**
O,O-Dimethyl S-(5-methoxy-4-oxo-4H-pyran-2-yl)phosphoro-thioate **(2778-04-3)**
O,O-Dimethyl S-(5-methoxypyronyl-2-methyl) thiophosphate **(2778-04-3)**
O,O-Dimethyl-S-((5-methoxy-pyron-2-yl)-methyl)-thiol-phosphat (German) **(2778-04-3)**
O,O-Dimethyl-S-(2-methoxy-1,3,4-thiadiazol-5-(4H)-onyl-(4)-methyl)-dithiophosphat (German) **(950-37-8)**
O,O-Dimethyl-S-(2-methoxy-1,3,4-thiadiazol-5(4H)-onyl-(4)-methyl) phosphorodithioate **(950-37-8)**
(O,O-Dimethyl)-S-(-2-methoxy-δ^2-1,3,4-thiadiazolin-5-on-4-ylmethyl)dithiophosphate **(950-37-8)**
O,O-Dimethyl S-(5-methoxy-1,3,4-thiadiazolinyl-3-methyl) dithiophosphate **(950-37-8)**
O,O-Dimethyl-S-(2-methoxy-1,3,4-thiadiazol-5-on-4-ly)-methyl-dithiophosphat (German) **(950-37-8)**
N,N-Dimethyl-N'-(((methylamino)carbonyl)oxy)phenylmethan-imidamide monohydrochloride **(23422-53-9)**
O,O-Dimethyl S-(2-(methylamino)-2-oxoethyl) phosphorodithioate **(60-51-5)**
O,O-Dimethyl S-(2-(methylamino)-2-oxoethyl)phosphorothioate **(1113-02-6)**
O,O-Dimethyl-O-(3-methyl-4-aminophenyl) phosphorothioate **(13306-69-9)**
Dimethyl 2-methyl-1,4-benzenedicarboxylate **(14186-60-8)**
Dimethyl 4-methyl-1,2-benzenedicarboxylate **(20116-65-8)**
N,N-Dimethyl-α-methylbenzylamine **(2449-49-2)**
Dimethyl methylbutanedioate **(1604-11-1)**
O,O-Dimethyl S-2-(1-N-methylcarbamoylethylmercapto)ethyl thiophosphate **(2275-23-2)**
Dimethyl S-(2-(1-methylcarbamoylethylthio)ethyl) phosphoro-thiolate **(2275-23-2)**
O,O-Dimethyl S-(2-(1-methylcarbamoylethylthio)ethyl) phosphorothioate **(2275-23-2)**
O,O-Dimethyl-S-(N-methyl-carbamoyl)-methyl-dithiofosfaat (Dutch) **(60-51-5)**
(O,O-Dimethyl-S-(N-methyl-carbamoyl-methyl)-dithiophosphat) (German) **(60-51-5)**
O,O-Dimethyl S-(N-methylcarbamoylmethyl) dithiophosphate **(60-51-5)**
O,O-Dimethyl-S-((N-methyl-carbamoyl)-methyl)-monothio-fosfaat (Dutch) **(1113-02-6)**
O,O-Dimethyl-S-(N-methyl-carbamoyl)-methyl-monothio-phosphat (German) **(1113-02-6)**
O,O-Dimethyl S-(N-methylcarbamoylmethyl) phosphorodithioate **(60-51-5)**
O,O-Dimethyl methylcarbamoylmethyl phosphorodithioate **(60-51-5)**
O,O-Dimethyl S-((methylcarbamoyl)methyl)phosphorothioate **(1113-02-6)**
O,O-Dimethyl-S-(N-methylcarbamoylmethyl)phosphorothioate **(1113-02-6)**
Dimethyl-S-(N-methyl-carbamoyl-methyl)phosphorothiolate **(1113-02-6)**
O,O-Dimethyl S-(N-methylcarbamoylmethyl) phosphorothiolate **(1113-02-6)**
O,O-Dimethyl-S-(N-methyl-carbamoyl-methyl)-thiolphosphat (German) **(1113-02-6)**
O,O-Dimethyl S-(N-methylcarbamoylmethyl) thiophosphate **(1113-02-6)**
O,O-Dimethyl-O-(2-N-methylcarbamoyl-1-methyl-vinyl)-fosfaat (Dutch) **(6923-22-4)**
O,O-Dimethyl-O-(2-N-methylcarbamoyl-1-methyl)-vinyl-phosphat (German) **(6923-22-4)**
O,O-Dimethyl-O-(2-N-methylcarbamoyl-1-methyl-vinyl) phosphate **(6923-22-4)**
N,N-Dimethyl-α-methylcarbamoyloxyimino-α-(methylthio)-acetamide **(23135-22-0)**
N',N'-Dimethyl-N-((methylcarbamoyl)oxy)-1-thiooxamimidic acid methyl ester **(23135-22-0)**
O,O-Dimethyl S-(N-methylcarbamylmethyl) thiothionophosphate **(60-51-5)**
O,O-Dimethyl O-(1-methyl-2-carboxy-α-phenylethyl)vinyl phosphate **(7700-17-6)**

O,O-Dimethyl O-(1-methyl-2-carboxyvinyl) phosphate **(7786-34-7)**
O,O-Dimethyl-O-(1-methyl-2-chlor-2-N,N-diaethyl-carbamoyl)-vinyl-phosphat (German) **(13171-21-6)**
(O,O-Dimethyl-O-(1-methyl-2-chloro-2-diethylcarbamoyl-vinyl) phosphate) **(13171-21-6)**
N,N-Dimethyl-N'-(2-methyl-4-chlorophenyl)formamidine **(6164-98-3)**
N,N-Dimethyl-N'-(2-methyl-4-chlorophenyl)-formamidine, hydrochloride **(19750-95-9)**
N,N-Dimethyl-N'-(2-methyl-4-chlorphenyl)-formadin (German) **(6164-98-3)**
2,4-Dimethyl-6-(1-methylcyclohexyl)phenol **(77-61-2)**
O,O-Dimethyl-O-(1-methyl-2-N,N-dimethyl-carbamoyl)-vinyl-phosphat (German) **(141-66-2)**
O,O-Dimethyl-S-(3-methyl-2,4-dioxo-3-aza-butyl)-dithiofosfaat (Dutch) **(2540-82-1)**
O,O-Dimethyl-S-(3-methyl-2,4-dioxo-3-aza-butyl)-dithiophosphat (German) **(2540-82-1)**
N,N'-Dimethyl-4,4'-methylenedianiline **(1807-55-2)**
Dimethylmethylene-p,p'-diphenol **(80-05-7)**
2,6-Dimethyl-10-methylene-2,6,11-dodecatrienal **(60066-88-8)**
O,O-Dimethyl-s-(N-methyl-N-formyl-carbamoylmethyl)-dithio-phosphat (German) **(2540-82-1)**
O,O-Dimethyl S-(N-methyl-N-formylcarbamoylmethyl)phosphoro-dithioate **(2540-82-1)**
O,O-Dimethyl O-4-(methylmercapto)-3-methylphenyl phosphoro-thioate **(55-38-9)**
O,O-Dimethyl-O-4-(methylmercapto)-3-methylphenyl thiophosphate **(55-38-9)**
O,O-Dimethyl O-(4-methylmercaptophenyl) phosphate **(3254-63-5)**
N,N-Dimethyl-N'-(2-methyl-4-(((methylamino)carbonyl)oxy)-phenyl)methanimidamide **(17702-57-7)**
(E)-Dimethyl 1-methyl-3-(methylamino)-3-oxo-1-propenyl phosphate **(6923-22-4)**
O,O-Dimethyl-O-(1-methyl-2-N-methyl-carbamoyl)-vinyl-phosphat (German) **(6923-22-4)**
Dimethyl 1-methyl-2-(methylcarbamoyl)vinyl phosphate, cis **(6923-22-4)**
O,O-Dimethyl O-(3-methyl-4-methylmercaptophenyl)phosphoro-thioate **(55-38-9)**
O,O-Dimethyl-O-(3-methyl-4-methylthio-fenyl)-monothiofosfaat (Dutch) **(55-38-9)**
O,O-Dimethyl-O-(3-methyl-4-methylthiophenyl)-monothiophosphat (German) **(55-38-9)**
O,O-Dimethyl O-3-methyl-4-methylthiophenyl phosphorothioate **(55-38-9)**
O,O-Dimethyl-O-(3-methyl-4-methylthio-phenyl)-thionophosphat (German) **(55-38-9)**
O,O-Dimethyl-O-(3-methyl-4-nitrofenyl)-monothiofosfaat (Dutch) **(122-14-5)**
O,S-Dimethyl-O-(3-methyl-4-nitrofenyl)thiofosfat (Czech) **(3344-14-7)**
O,O-Dimethyl-O-(3-methyl-4-nitro-phenyl)-monothiophosphat (German) **(122-14-5)**
O,O-Dimethyl O-(3-methyl-4-nitrophenyl)phosphorate **(2255-17-6)**
O,O-Dimethyl O-(3-methyl-4-nitrophenyl) phosphorothioate **(122-14-5)**
Dimethyl 3-methyl-4-nitrophenyl phosphorothionate **(122-14-5)**
O,O-Dimethyl O-(3-methyl-4-nitrophenyl) thiophosphate **(122-14-5)**
Dimethyl-cis-1-methyl-2-(1-phenylethoxycarbonyl)vinyl phosphate **(7700-17-6)**
Dimethyl 5-(3-methyl-1-phenylpyrazolyl) carbamate **(87-47-8)**
Dimethyl methylphosphonate **(756-79-6)**
O,O-Dimethyl O-(3-methyl) phosphorothioate **(122-14-5)**
N,N-Dimethyl-9-(3-(4-methyl-1-piperazinyl)propylidene)-thiaxanthene-2-sulfonamide **(5591-45-7)**
N,N-Dimethyl-9-(3-(4-methyl-1-piperazinyl)propylidene)-thioxanthene-2-sulfonamide **(5591-45-7)**
Dimethyl 3-(2-methyl-1-propenyl)cyclopropanecarboxylate **(10453-86-8)**
2,2-Dimethyl-3-(2-methylpropenyl)cyclopropanecarboxylic acid ethyl ester **(97-41-6)**
Dimethyl methylsuccinate **(1604-11-1)**
O,O-Dimethyl O-(4-(methylsulfinyl)-m-tolyl) phosphorothioate **(3761-41-9)**
O,O-Dimethyl O-(4-(methylsulfonyl)-m-tolyl) phosphorothioate **(3761-42-0)**
Dimethyl 3-methylterephthalate **(14186-60-8)**
N,N-Dimethyl-N-(2-(2-(methyl-4-(1,1,3,3-tetramethylbutyl)-phenoxy)ethoxy)ethyl)benzenemethanaminium chloride **(25155-18-4)**
3,3-Dimethyl-1-(methylthio)-2-butanone-O-((methylamino)-carbonyl)oxime **(39196-18-4)**
O,O-Dimethyl S-(2-(methylthio)ethyl) phosphorothioate **(2587-90-8)**
Dimethyl-p-(methylthiofenyl)fosfat (Czech) **(3254-63-5)**
O,O-Dimethyl O-(4-methylthio-3-methylphenyl) phosphorothioate **(55-38-9)**
3,5-Dimethyl-4-(methylthio)phenol **(7379-51-3)**
3,5-Dimethyl-4-(methylthio)phenol methylcarbamate **(2032-65-7)**
3,5-Dimethyl-4-methyl-thiophenyl-N-carbamat (German) **(2032-65-7)**
3,5-Dimethyl-4-methylthiophenyl N-methylcarbamate **(2032-65-7)**
Dimethyl p-(methylthio)phenyl phosphate **(3254-63-5)**
O,O-Dimethyl O-(4-(methylthio)-m-tolyl) phosphorothioate **(55-38-9)**
O,O-Dimethyl O-((4-methylthio)-m-tolyl)phosphorothioate sulfone **(3761-42-0)**
O,O-Dimethyl O-((4-methylthio)-m-tolyl)phosphorothioate sulf-oxide **(3761-41-9)**
O,O-Dimethyl-S-(N-monomethyl)-carbamyl methyl dithiophosphate **(60-51-5)**
Dimethyl monosulfate **(77-78-1)**
O,O-Dimethyl-S-((morfolino-carbonyl)-methyl)-dithiofosfaat (Dutch) **(144-41-2)**
2,6-Dimethylmorfolin (Czech) **(141-91-3)**
O,O-Dimethyl-S-((morpholino-carbonyl)-methyl)-dithio-phosphat (German) **(144-41-2)**
2,6-Dimethylmorpholine **(141-91-3)**
N,N-Dimethyl-4-morpholineethanamine **(4385-05-1)**
O,O-Dimethyl S-(morpholinocarbamoylmethyl) dithiophosphate **(144-41-2)**
O,O-Dimethyl S-(morpholinocarbonylmethyl) phosphorodithioate **(144-41-2)**
O,O-Dimethyl morpholinocarbonylmethyl phosphorodithioate **(144-41-2)**
Dimethyl S-(morpholinocarbonylmethyl) phosphorothiolothionate **(144-41-2)**
Dimethylmorpholinophosphonate **(597-25-1)**
Dimethyl morpholinophosphoramidate **(597-25-1)**
Dimethyl myristamine **(112-75-4)**
Dimethylmyristamine **(112-75-4)**
Dimethyl myristylamine **(112-75-4)**
Dimethylmyristylamine **(112-75-4)**
N,N-Dimethylmyristylamine **(112-75-4)**
N,N-Dimethyl-1-naftylamin (Czech) **(86-56-6)**
1,2-Dimethylnaphthalene **(573-98-8)**
1,3-Dimethylnaphthalene **(575-41-7)**
1,4-Dimethylnaphthalene **(571-58-4)**
1,5-Dimethylnaphthalene **(571-61-9)**
1,6-Dimethylnaphthalene **(575-43-9)**
1,7-Dimethylnaphthalene **(575-37-1)**
2,6-Dimethylnaphthalene **(581-42-0)**
Dimethylnaphthalene **(28804-88-8)**
Dimethylnaphthalene **(575-43-9)**
Dimethyl 2,6-naphthalenedicarboxylate **(840-65-3)**
Dimethylnaphthalene, formaldehyde polymer **(52613-22-6)**
Dimethyl-α-naphthylamine **(86-56-6)**
N,N-Dimethyl-1-naphthylamine **(86-56-6)**
N,N-Dimethyl-α-naphthylamine **(86-56-6)**
α-Dimethylnaphthylamine **(86-56-6)**
1,4-Dimethylnapthalene **(571-58-4)**
Dimethylnitramin (German) **(4164-28-7)**
Dimethylnitramine **(4164-28-7)**
Dimethylnitroamine **(4164-28-7)**
3,5-Dimethyl-4-nitroaniline **(34761-82-5)**
N,N-Dimethyl-4-nitroaniline **(100-23-2)**
N,N-Dimethyl-p-nitroaniline **(100-23-2)**
N,N-Dimethyl-4-nitrobenzenamine **(100-23-2)**
1,2-Dimethyl-3-nitrobenzene **(83-41-0)**
1,2-Dimethyl-4-nitrobenzene **(99-51-4)**
1,3-Dimethyl-4-nitrobenzene **(89-87-2)**
1,3-Dimethyl-5-nitrobenzene **(99-12-7)**
1,4-Dimethyl-2-nitrobenzene **(89-58-7)**
2,3-Dimethylnitrobenzene **(83-41-0)**
2,4-Dimethyl-1-nitrobenzene **(89-87-2)**
2,6-Dimethylnitrobenzene **(81-20-9)**

3,4-Dimethyl-1-nitrobenzene **(99-51-4)**
3,5-Dimethyl-1-nitrobenzene **(99-12-7)**
3,5-Dimethylnitrobenzene **(99-12-7)**
2,3-Dimethyl-5-nitrobenzenesulfethanolamide **(25959-70-0)**
O,O-Dimethyl O-4-nitro-3-chlorophenyl thiophosphate **(500-28-7)**
O,O-Dimethyl p-nitro-m-chlorophenyl thiophosphate **(500-28-7)**
O,O-Dimethyl-O-(4-nitro-5-chlorphenyl)-thionophosphat (German) **(500-28-7)**
Dimethyl-p-nitrofenylester kyseliny fosforecne (Czech) **(950-35-6)**
O,O-Dimethyl-O-p-nitrofenylester kyseliny thiofosforecne (Czech) **(298-00-0)**
O,O-Dimethyl-O-(4-nitro-fenyl)-monothiofosfaat (Dutch) **(298-00-0)**
O,O-Dimethyl-O-p-nitrofenylthiofosfat (Czech) **(298-00-0)**
1,2-Dimethyl-5-nitroimidazole **(551-92-8)**
1,2-Dimethyl-4-nitro-1H-imidazole (9CI) **(13230-04-1)**
1,5-Dimethyl-4-nitro-1H-imidazole (9CI) **(7464-68-8)**
Dimethyl 5-nitroisophthalate **(13290-96-5)**
Dimethylnitromethane **(79-46-9)**
O,O-Dimethyl-O-(4-nitro-5-methylphenyl)-thionophosphat (German) **(122-14-5)**
O,O-Dimethyl O-(4-nitro-3-methylphenyl)thiophosphate **(122-14-5)**
N,N-Dimethyl-p-((o-nitrophenyl)azo)aniline **(3010-38-6)**
N,N-Dimethyl-4-((4-nitrophenyl)azo)benzenamine **(2491-74-9)**
O,O-Dimethyl-O-(4-nitro-phenyl)-monothiophosphat (German) **(298-00-0)**
Dimethyl p-nitrophenyl monothiophosphate **(298-00-0)**
Dimethyl-4-nitrophenyl phosphate **(950-35-6)**
Dimethyl-p-nitrophenyl phosphate **(950-35-6)**
O,O-Dimethyl O-(p-nitrophenyl) phosphorothioate **(298-00-0)**
O,O-Dimethyl-O-(4-nitrophenyl) phosphorothioate **(298-00-0)**
O,S-Dimethyl O-(p-nitrophenyl) phosphorothioate **(597-89-7)**
Dimethyl 4-nitrophenyl phosphorothionate **(298-00-0)**
O,O-Dimethyl-O-(4-nitrophenyl)-thionophosphat (German) **(298-00-0)**
O,O-Dimethyl-O-(p-nitrophenyl)-thionophosphat (German) **(298-00-0)**
O,O-Dimethyl O-(p-nitrophenyl) thionophosphate **(298-00-0)**
Dimethyl-p-nitrophenyl thionphosphate **(298-00-0)**
Dimethyl p-nitrophenyl thiophosphate **(298-00-0)**
O,O-Dimethyl O-p-nitrophenyl thiophosphate **(298-00-0)**
O,S-Dimethyl O-(4-nitrophenyl)thiophosphate **(597-89-7)**
Dimethylnitrosamin (German) **(62-75-9)**
Dimethylnitrosamine **(62-75-9)**
N,N-Dimethylnitrosamine **(62-75-9)**
Dimethylnitrosoamine **(62-75-9)**
N,N-Dimethyl-p-nitrosoaniline **(138-89-6)**
Dimethyl-p-nitrosoaniline (DOT) **(138-89-6)**
2,6-Dimethylnitrosomorpholine **(1456-28-6)**
Dimethylnitrosomorpholine **(1456-28-6)**
Dimethyl(p-nitrosophenyl)amine **(138-89-6)**
3,5-Dimethyl-4-nitroso-1-phenylpyrazole **(715-99-1)**
O,O-Dimethyl O-4-nitro-m-tolyl phosphorothioate **(122-14-5)**
Dimethyl 4-nitro-m-tolyl phosphorothionate **(122-14-5)**
Dimethyl nonanedioate **(1732-10-1)**
17,17-Dimethyl-18-norandrost-4,13-dien-3-one **(18869-73-3)**
6,6-Dimethyl-2-norpinene-2-ethanol, acetate **(128-51-8)**
N,N-Dimethyl-1-octadecanamine acetate **(19855-61-9)**
N,N-Dimethyl-1-octadecanamine N-oxide **(2571-88-2)**
(Z)-N,N-Dimethyl-9-octadecenamide **(2664-42-8)**
(Z)-N,N-Dimethyl-9-octadecen-1-amine N-oxide **(14351-50-9)**
N,N-Dimethyl-9-octadecen-1-amine-N-oxide **(14351-50-9)**
N,N-Dimethyloctadecenylamine **(28061-69-0)**
N,N-Dimethyl-N-9-octadecenylbenzenemethanaminium chloride **(37139-99-4)**
N,N-Dimethyloctadecylamine **(124-28-7)**
Dimethyloctadecylbenzylammonium chloride **(122-19-0)**
Dimethyl octadecylphosphonate **(25371-54-4)**
Dimethyloctadecylphosphonate **(25371-54-4)**
(Z)-3,7-Dimethyl-2,6-octadienal **(106-26-3)**
3,7-Dimethyl-2,6-octadienal **(5392-40-5)**
trans-3,7-Dimethyl-2,6-octadienal **(141-27-5)**
3,7-Dimethyl-2,6-octadienal diethyl acetal **(7492-66-2)**
3,7-Dimethyl-2,6-octadienal oxime **(13372-77-5)**
3,7-Dimethyl-1,6-octadiene **(2436-90-0)**
(E)-3,7-Dimethyl-2,6-octadienenitrile **(5585-39-7)**
(Z)-3,7-Dimethyl-2,6-octadienenitrile **(31983-27-4)**
3,7-Dimethyl-2,6-octadienenitrile **(5146-66-7)**
cis-3,7-Dimethyl-2,6-octadienenitrile **(31983-27-4)**
2,6-Dimethyl-2,7-octadiene-6-ol **(78-70-6)**
2,6-Dimethyl-trans-2,6-octadien-8-ol **(106-24-1)**
2,6-Dimethylocta-2,7-dien-6-ol **(78-70-6)**
2-cis-3,7-Dimethyl-2,6-octadien-1-ol **(106-25-2)**
3,7-Dimethyl-1,6-octadien-3-ol **(78-70-6)**
3,7-Dimethyl-2,6-octadien-1-ol **(624-15-7)**
3,7-Dimethyl-4,6-octadien-3-ol **(18479-54-4)**
3,7-Dimethyl-trans-2,6-octadien-1-ol **(106-24-1)**
3,7-Dimethylocta-1,6-dien-3-ol **(78-70-6)**
trans-3,7-Dimethyl-2,6-octadien-1-ol **(624-15-7)**
3,7-Dimethyl-1,6-octadien-3-ol acetate **(115-95-7)**
trans-3,7-Dimethyl-2,6-octadien-1-ol, acetate **(105-87-3)**
trans-3,7-Dimethyl-2,6-octadien-1-ol formate **(105-86-2)**
3,7-Dimethyl-1,6-octadien-3-yl acetate **(115-95-7)**
3,7-Dimethyl-2-trans, 6-octadienyl acetate **(105-87-3)**
trans-3,7-Dimethyl-2,6-octadien-1-yl acetate **(105-87-3)**
3,7-Dimethyl-1,6-octadien-3-yl o-aminobenzoate **(7149-26-0)**
trans-2,6-Dimethyl-2,6-octadien-8-yl ethanoate **(105-87-3)**
trans-3,7-Dimethyl-2,6-octadien-1-yl formate **(105-86-2)**
2,6-Dimethyl octanal **(1321-89-7)**
3,7-Dimethyloctanal **(5988-91-0)**
N,N-Dimethyloctanamide **(1118-92-9)**
N,N-Dimethyl-1-octanamine **(7378-99-6)**
Dimethyl octane-1,8-dicarboxylate **(106-79-6)**
Dimethyl octanedioate **(1732-09-8)**
3,7-Dimethyl-1,7-octanediol **(107-74-4)**
Dimethyloctanoic acid **(29662-90-6)**
2,6-Dimethyl octanoic aldehyde **(1321-89-7)**
2,6-Dimethyl-2-octanol **(18479-57-7)**
2,6-Dimethyl-8-octanol **(106-21-8)**
3,6-Dimethyl-3-octanol **(151-19-9)**
3,7-Dimethyl-1-octanol **(106-21-8)**
3,7-Dimethyloctanol-3 **(78-69-3)**
Dimethyloctanol **(106-21-8)**
(Z)-3,7-Dimethyl-1,3,6-octatriene **(3338-55-4)**
2,6-Dimethyl-2,4,6-octatriene **(673-84-7)**
3,7-Dimethyl-1,3,6-octatriene **(13877-91-3)**
Dimethyloctatriene (Mixed isomer) **(29714-87-2)**
(S)-3,7-Dimethyl-6-octenal **(5949-05-3)**
3,7-Dimethyl-6-octen-1-al **(106-23-0)**
3,7-Dimethyl-6-octenal **(106-23-0)**
(+-)-3,7-Dimethyl-6-octen-1-ol **(26489-01-0)**
2,6-Dimethyl-2-octen-8-ol **(106-22-9)**
2,6-Dimethyl-6-octen-2-ol **(30385-25-2)**
2,6-Dimethyl-7-octen-2-ol **(18479-58-8)**
3,7-Dimethyl-6-octen-1-ol **(106-22-9)**
3,7-Dimethyl-6-octen-3-ol **(18479-51-1)**
2,6-Dimethyl-7-octen-2-ol formate **(25279-09-8)**
3,7-Dimethyl-6-octen-1-ol propanoate **(141-14-0)**
3,7-Dimethyl-6-octen-1-yl acetate **(150-84-5)**
2,6-Dimethyl-7-octen-2-yl formate **(25279-09-8)**
3,6-Dimethyl-octin-4-diol-(3,6) (German) **(78-66-0)**
N,N-Dimethyl-N-octyl-1-decanaminium chloride **(32426-11-2)**
N,N-Dimethyl-N-octyl-1-octanaminium chloride **(5538-94-3)**
Dimethyloctynediol **(1321-87-5)**
N,N-Dimethyloktadecylamin (Czech) **(124-28-7)**
Dimethylol cyclic ethyleneurea **(136-84-5)**
Dimethylolcycloethyleneurea **(136-84-5)**
Dimethyloldihydroxyethyleneurea **(1854-26-8)**
Dimethylol-5,5-dimethylhydantoin **(6440-58-0)**
N,N-Dimethyloleamide **(2664-42-8)**
1,3-Dimethylolethyleneurea **(136-84-5)**
Dimethylolethyleneurea **(136-84-5)**

N,N'-Dimethylol-N,N'-ethyleneurea	(136-84-5)
N,N'-Dimethylolethyleneurea	(136-84-5)
N,N-Dimethyloleylamine 2,4-dichlorophenoxyacetate	(53535-36-7)
Dimethyloleylamine oxide	(14351-50-9)
N,N-Dimethyl oleyl-linoleyl amine 2,4-dichlorophenoxyacetate	(55256-32-1)
Dimethylolglyoxalurea	(1854-26-8)
1,3-Dimethylol-5-β-hydroxyethylhexahydrotriazinone-2	(1852-21-7)
Dimethylol hydroxyethyltriazone	(35503-54-9)
Dimethylol(hydroxyethyl)triazone	(1852-21-7)
1,3-Dimethylol-2-imidazolidinone	(136-84-5)
Dimethylol isobutyl carbamate	(52304-17-3)
N,N-Dimethylol isopropoxyethyl carbamate	(68413-83-2)
Dimethylol isopropyl carbamate	(4987-75-1)
N,N-Dimethylol-2-methoxyethyl carbamate	(10143-22-3)
1,1-Dimethylol-1-nitroethane	(77-49-6)
N,N'-Dimethylolpiperazine	(3312-58-1)
Dimethylolpropane	(126-30-7)
Dimethylolpropane diacrylate	(2223-82-7)
1,3-Dimethylolurea	(140-95-4)
Dimethylolurea	(140-95-4)
N,N'-Dimethylolurea	(140-95-4)
Dimethylol urea dibutyl ether	(4981-47-9)
Dimethylolurea dimethyl ether	(141-07-1)
N,N'-Dimethylolurea dimethyl ether	(141-07-1)
3,5-Dimethyloluron	(7327-69-7)
2,3-Dimethyl-7-oxabicyclo(2.2.1)heptane-2,3-dicarboxylic anhydride	(56-25-7)
Dimethyl oxalate	(553-90-2)
4,4-Dimethyloxazolidine	(51200-87-4)
Dimethyl oxazolidine	(51200-87-4)
3,3-Dimethyloxetane	(6921-35-3)
3,3-Dimethyl-2-oxetanone	(1955-45-9)
3,3-Dimethyl-2-oxethanone	(1955-45-9)
2,3-Dimethyloxirane	(3266-23-7)
O,O-Dimethyl-S-(2-oxo-3-aza-butyl)-dithiophosphat (German)	(60-51-5)
O,O-Dimethyl-S-(2-oxo-3-azabutyl)-monothiophosphate	(1113-02-6)
O,O-Dimethyl S-(4-oxobenzotriazino-3-methyl)phosphorodithioate	(86-50-0)
O,O-Dimethyl S-4-oxo-1,2,3-benzotriazin-3(4H)-ylmethyl phosphorodithioate	(86-50-0)
O,O-Dimethyl S-(4-oxo-3H-1,2,3-benzotriazine-3-methyl)-phosphorodithioate	(86-50-0)
O,O-Dimethyl-S-(4-oxobenzotriazin-3-methyl)-dithiophosphat (German)	(86-50-0)
O,O-Dimethyl S-(4-oxo-1,2,3-benzotriazino(3)-methyl) thiothionophosphate	(86-50-0)
O,O-Dimethyl-S-((4-oxo-3H-1,2,3-benzotriazin-3-yl)-methyl)-di-thiofosfaat (Dutch)	(86-50-0)
O,O-Dimethyl-S-((4-oxo-3H-1,2,3-benzotriazin-3-yl)-methyl)-dithio-phosphat (German)	(86-50-0)
3,3-Dimethyl-2-oxobutanoic acid	(815-17-8)
N-(1,1-Dimethyl-3-oxobutyl)acrylamide	(2873-97-4)
N-(1,1-Dimethyl-3-oxobutyl)-2-propenamide	(2873-97-4)
3,3-Dimethyl-2-oxobutyric acid	(815-17-8)
(5,5-Dimethyl-3-oxo-cyclohex-1-en-yl)-N,N-dimethyl-carbamaat (Dutch)	(122-15-6)
(5,5-Dimethyl-3-oxo-cyclohex-1-en-yl)-N,N-dimethyl-carbamat (German)	(122-15-6)
5,5-Dimethyl-3-oxo-1-cyclohexen-1-yl dimethylcarbamate	(122-15-6)
5,5-Dimethyl-3-oxocyclohex-1-enyl dimethylcarbamate	(122-15-6)
3-(2-(3,5-Dimethyl-2-oxocyclohexyl)-2-hydroxyethyl)glutarimide	(66-81-9)
β-(2-(3,5-Dimethyl-2-oxocyclohexyl)-2-hydroxyethyl)glutarimide	(66-81-9)
5,5-Dimethyl-3-oxo-1-cyklohexenylester kyseliny dimethyl-karbaminove (Czech)	(122-15-6)
Dimethyl .α.-oxoglutarate	(13192-04-6)
Dimethyl 2-oxoglutarate	(13192-04-6)
2-(2,2-Dimethyl-1-oxopropyl)-1H-indene-1,3(2H)-dione	(83-26-1)
O,O-Dimethyl-S-(3-oxo-3-thia-pentyl)-monothiophosphat (German)	(301-12-2)
Dimethyloxychinizin	(60-80-0)
2,2-Dimethyl-4-oxymethyl-1,3-dioxolane	(100-79-8)
1-(2,5-Dimethyloxyphenylazo)-2-naphthol	(6358-53-8)
Dimethyloxyquinazine	(60-80-0)
O,O-Dimethyl-1-oxy-2,2,2-trichloroethyl phosphonate	(52-68-6)
Dimethyl palmitamine	(112-69-6)
Dimethyl palmitylamine	(112-69-6)
Dimethyl paraoxon	(950-35-6)
Dimethyl-paraphenylenediamine	(99-98-9)
Dimethyl parathion	(298-00-0)
2,3-Dimethyl pentaldehyde	(32749-94-3)
2,3-Dimethylpentaldehyde	(32749-94-3)
2,3-Dimethyl-pentanal	(32749-94-3)
2,3-Dimethylpentanal	(32749-94-3)
2,2-Dimethylpentane	(590-35-2)
2,3-Dimethylpentane	(565-59-3)
2,4-Dimethylpentane	(108-08-7)
3,3-Dimethylpentane	(562-49-2)
3,4-Dimethylpentane	(565-59-3)
Dimethyl pentanedioate	(1119-40-0)
3,3-Dimethylpentanedioic acid	(4839-46-7)
2,2-Dimethylpentanoic acid	(1185-39-3)
4,4-Dimethylpentanoic acid	(1118-47-4)
2,3-Dimethyl-1-pentanol	(10143-23-4)
2,3-Dimethylpentanol	(10143-23-4)
2,4-Dimethyl-3-pentanol	(600-36-2)
2,4-Dimethyl-3-pentanone	(565-80-0)
2,4-Dimethylpentan-3-one	(565-80-0)
3,3-Dimethyl-4-pentenoic acid, methyl ester	(63721-05-1)
Di(4-methyl-2-pentyl) maleate	(105-52-2)
3,5-Dimethylperhydro-1,3,5-thiadiazin-2-thion (Czech, German)	(533-74-4)
1,4-Dimethylphenanthrene	(22349-59-3)
α,α-Dimethylphenethyl acetate	(151-05-3)
α,α-Dimethylphenethyl alcohol	(100-86-7)
α,α-Dimethylphenethyl alcohol, acetate	(151-05-3)
α,α-Dimethylphenethylamine	(122-09-8)
N,α-Dimethylphenethylamine hydrochloride	(300-42-5)
β,β-Dimethylphenethyl chloride	(515-40-2)
2,3-Dimethylphenol	(526-75-0)
2,4-Dimethylphenol	(105-67-9)
2,5-Dimethylphenol	(95-87-4)
2,6-Dimethylphenol	(576-26-1)
3,4-Dimethylphenol	(95-65-8)
3,5-Dimethylphenol	(108-68-9)
3,6-Dimethylphenol	(95-87-4)
4,5-Dimethylphenol	(95-65-8)
4,6-Dimethylphenol	(105-67-9)
Dimethylphenol	(1300-71-6)
2,6-Dimethylphenol homopolymer	(25134-01-4)
2,4-Dimethylphenol phosphate (3:1)	(3862-12-2)
2,6-Dimethylphenol phosphate (3:1)	(121-06-2)
Dimethylphenol phosphate (3:1)	(25155-23-1)
2,8-Dimethylphenosafranine	(477-73-6)
5-((3,5-Dimethylphenoxy)methyl)-2-oxazolidinone	(1665-48-1)
O,O-Dimethyl-S-(phenylacetic acid ethyl ester) phosphorodithioate	(2597-03-7)
2,3-Dimethylphenylamine	(87-59-2)
2,4-Dimethylphenylamine	(95-68-1)
2,5-Dimethylphenylamine	(95-78-3)
3,4-Dimethylphenylamine	(95-64-7)
3,5-Dimethylphenylamine	(108-69-0)
Dimethylphenylamine	(121-69-7)
Dimethylphenylamine	(1300-73-8)
N,N-Dimethylphenylamine	(121-69-7)
2-((2,3-Dimethylphenyl)amino)benzoic acid	(61-68-7)
N-(2-((2,6-Dimethylphenyl)amino)-2-oxoethyl)-N,N-diethyl-benzenemethanaminium benzoate	(3734-33-6)
N-(2,3-Dimethylphenyl)anthranilic acid	(61-68-7)
Dimethyl phenylarsenite	(24582-52-3)

N,N-Dimethyl-p-phenylazoaniline **(60-11-7)**
N,N-Dimethyl-4-(phenylazo)benzamine **(60-11-7)**
N,N-Dimethyl-4-(phenylazo)benzenamine **(60-11-7)**
4-((3-((2,4-Dimethylphenyl)azo)-2,4-dihydroxyphenyl)azo)-benzenesulfonic acid, monosodium salt **(1320-07-6)**
4-((2,4-Dimethylphenyl)azo)-3-hydroxy-2,7-naphthalene-disulfonic acid, disodium salt **(3761-53-3)**
4-((2,4-Dimethylphenyl)azo)-3-hydroxy-2,7-naphthalene-disulphonic acid, disodium salt **(3761-53-3)**
1-((2,4-Dimethylphenyl)azo)-2-naphthalenol **(3118-97-6)**
N,N-Dimethyl-α-phenylbenzeneacetamide **(957-51-7)**
2,5-Dimethylphenyl bromide **(553-94-6)**
2-Di(N-methyl-N-phenyl-tert-butyl-carbamoylmethyl)aminoethanol **(126-27-2)**
Dimethylphenylcarbinol **(617-94-7)**
O,O-Dimethyl S-(phenyl)(carboethoxy)methyl phosphorodithioate **(2597-03-7)**
N'-(2,4-Dimethylphenyl)-N-(((2,4-dimethylphenyl)imino)-methyl)-N-methylmethanimidamide **(33089-61-1)**
1,3-Dimethyl-3-phenyl-2,5-dioxopyrrolidine **(77-41-8)**
Dimethyl ((1,2-phenylene)bis-(iminocarbonothioyl))bis-(carbamate) **(23564-05-8)**
Dimethyl-4,4'-o-phenylene-bis-(3-thioallophanate) **(23564-05-8)**
Dimethyl-p-phenylenediamine **(105-10-2)**
N,N'-Dimethyl-p-phenylenediamine **(105-10-2)**
N,N-Dimethyl-p-phenylenediamine **(99-98-9)**
1,1-Dimethyl-2-phenylethanol **(100-86-7)**
1-(2,4-Dimethylphenyl)ethanone **(89-74-7)**
1-(Dimethylphenyl)ethanone **(1335-42-8)**
(Dimethyl S-(phenylethoxycarbonylmethyl)phosphorothiolo-thionate) **(2597-03-7)**
α,α-Dimethyl-β-phenylethylamine **(122-09-8)**
1,2-Dimethyl-4-(1-phenylethyl)benzene **(6196-95-8)**
Dimethylphenylethyl carbinol **(103-05-9)**
N-(2,6-Dimethylphenyl)-2-(ethylpropylamino)butanamide **(36637-18-0)**
N,N-Dimethyl-N'-phenyl-N'-fluorodichloromethylthiosulfamide **(1085-98-9)**
Dimethylphenylmethanol **(617-94-7)**
N-(2,6-Dimethylphenyl)-N-(methoxyacetyl)-alanine methyl ester **(57837-19-1)**
N-(2,6-Dimethylphenyl)-N-(methoxyacetyl)-dl-alanine methyl ester **(57837-19-1)**
N-(2,6-Dimethylphenyl)-2-methoxy-N-(2-oxo-3-oxazolidinyl)-acetamide **(77732-09-3)**
3,4-Dimethylphenyl-N-methylcarbamate **(2425-10-7)**
3,5-Dimethylphenyl N-methylcarbamate **(2655-14-3)**
(2,4-Dimethylphenyl)methyl 2,2-dimethyl-3-(2-methyl-1-propenyl)cyclopropanecarboxylate **(70-38-2)**
4-(Dimethylphenylmethyl)phenol **(599-64-4)**
1-(3,4-Dimethylphenyl)-1-phenylethane **(6196-95-8)**
1,3-Dimethyl-4-phenyl-4-piperidinol propionate (ester) **(77-20-3)**
α-1,3-Dimethyl-4-phenyl-4-piperidinyl propionate **(77-20-3)**
1,1-Dimethyl-3-phenyl-1-propanol **(103-05-9)**
1,1-Dimethyl-3-phenylpropanol **(103-05-9)**
1,3-Dimethyl-4-phenyl-4-propionoxypiperidine **(77-20-3)**
α-1,3-Dimethyl-4-phenyl-4-propionoxypiperidine **(77-20-3)**
α,α-Dimethyl-δ-phenylpropyl alcohol **(103-05-9)**
4,4-Dimethyl-1-phenyl-3-pyrazolidinone **(2654-58-2)**
2,3-Dimethyl-1-phenyl-3-pyrazolin-5-one **(60-80-0)**
2,3-Dimethyl-1-phenyl-5-pyrazolone **(60-80-0)**
N,N-Dimethyl-3-phenyl-3-(2-pyridyl)propylamine **(86-21-5)**
1,3-Dimethyl-3-phenyl-pyrrolidin-2,5-dione **(77-41-8)**
cis-2,5-Dimethyl-N-phenyl-1-pyrrolidinecarboxamide **(34484-77-0)**
Dimethylphenylsilane **(766-77-8)**
N,2-Dimethyl-2-phenylsuccinimide **(77-41-8)**
3,3-Dimethyl-1-phenyltriazene **(7227-91-0)**
1,1-Dimethyl-3-phenylurea **(101-42-8)**
N,N-Dimethyl-N'-phenylurea **(101-42-8)**
1,1-Dimethyl-3-phenylurea trichloroacetate **(4482-55-7)**
N,N-Dimethyl-N'-phenyluronium trichloracetate **(4482-55-7)**
Dimethyl phosphate **(813-78-5)**
Dimethyl phosphate ester of 3-hydroxy-N-methyl-cis-crotonamide **(6923-22-4)**
Dimethyl phosphate ester with 3-hydroxy-N,N-dimethyl-cis-crotonamide **(141-66-2)**
Dimethyl phosphate ester with (E)-3-hydroxy-N-methylcroton-amide **(6923-22-4)**
Dimethyl phosphate of 2-chloro-N,N-diethyl-3-hydroxycroton-amide **(13171-21-6)**
Dimethyl phosphate of 3-hydroxy-N,N-dimethyl-cis-crotonamide **(141-66-2)**
Dimethyl phosphate of 3-hydroxy-N-methyl-cis-crotonamine **(6923-22-4)**
Dimethyl phosphate of α-methylbenzyl 3-hydroxy-cis-crotonate **(7700-17-6)**
Dimethylphosphinothioic chloride **(993-12-4)**
Dimethylphosphinothioyl chloride **(993-12-4)**
Dimethyl phosphite **(868-85-9)**
Dimethyl phosphonate **(868-85-9)**
Dimethylphosphoramidocyanidic acid, ethyl ester **(77-81-6)**
O,S-Dimethyl phosphoramidothioate **(10265-92-6)**
Dimethyl phosphorochloridothioate **(993-12-4)**
O,O-Dimethylphosphorochloridothioate **(2524-03-0)**
Dimethyl phosphorochloridothioate (DOT) **(2524-03-0)**
O,O-Dimethylphosphorodithioate **(756-80-9)**
O,O-Dimethyl phosphorodithioate N-formyl-2-mercapto-N-methylacetamide s-ester **(2540-82-1)**
N-((O,O-Dimethylphosphorodithioyl)ethyl)acetamide **(13265-60-6)**
O,O-Dimethyl phosphorothioate O,O-diester with 4,4'-thiodi-phenol **(3383-96-8)**
O,O-Dimethyl phosphorothionate **(1112-38-5)**
5-(O,O-Dimethylphosphoryl)-6-chlorobicyclo(3.2.0)-hepta-1,5-diene **(34783-40-9)**
Dimethyl p-phthalate **(120-61-6)**
Dimethyl phthalate (ACGIH,OSHA) **(131-11-3)**
(O,O-Dimethyl-phthalimidiomethyl-dithiophosphate) **(732-11-6)**
O,O-Dimethyl 5-(phthalimidomethyl)dithiophosphate **(732-11-6)**
O,O-Dimethyl S-(N-phthalimidomethyl) dithiophosphate **(732-11-6)**
O,O-Dimethyl S-phthalimidomethyl phosphorodithioate **(732-11-6)**
1,4-Dimethylpiperazine **(106-58-1)**
2,5-Dimethylpiperazine **(106-55-8)**
Dimethyl piperazine-cis **(6284-84-0)**
N,N'-Dimethylpiperazine **(106-58-1)**
1,1-Dimethylpiperidinium chloride **(24307-26-4)**
N,N-Dimethyl-piperidinium chloride **(24307-26-4)**
N,N-Dimethylpiperidinium chloride **(24307-26-4)**
Dimethylpolysiloxane **(9006-65-9)**
Dimethylpolysiloxane hydrolyzate **(63148-62-9)**
2,2-Dimethylpropanal **(630-19-3)**
2,2-Dimethylpropanamide **(754-10-9)**
N,2-Dimethyl-1-propanamine **(625-43-4)**
2,2-Dimethylpropane [UN 2044] **(463-82-1)**
2,2-Dimethyl-1,3-propanediamine **(7328-91-8)**
N,N-Dimethyl-1,3-propanediamine **(109-55-7)**
2,2-Dimethylpropane dimethacrylate **(1985-51-9)**
Dimethyl propanedioate **(108-59-8)**
Dimethylpropanedioic acid **(595-46-0)**
2,2-Dimethyl-1,3-propanediol **(126-30-7)**
2,2-Dimethyl-1,3-propanediol dibenzoate **(4196-89-8)**
2,2'-((2,2-Dimethyl-1,3-propanediyl)bis(oxymethylene))bis-oxirane **(17557-23-2)**
2,2-Dimethyl-1,3-propanediyl nonanoate **(15834-05-6)**
2,2-Dimethylpropanenitrile **(630-18-2)**
2,2-Dimethylpropanoic acid **(75-98-9)**
2,2-Dimethylpropanoic acid anhydride **(1538-75-6)**
2,2-Dimethyl-1-propanol **(75-84-3)**
2,2-Dimethyl-1-propanol tribromo deriv. **(36483-57-5)**
2,2-Dimethylpropanoyl chloride **(3282-30-2)**
1,1-Dimethylpropargyl alcohol **(115-19-5)**
α,α-Dimethylpropargyl alcohol **(115-19-5)**
3,3-Dimethyl-β-propiolactone **(1955-45-9)**
Dimethyl propiolactone **(1955-45-9)**

2,2-Dimethylpropionic acid (75-98-9)
α,α-Dimethylpropionic acid (75-98-9)
2,2-Dimethylpropionyl chloride (3282-30-2)
Dimethylpropiothetin (7314-30-9)
2,2-Dimethylpropyl alcohol (75-84-3)
N,2-Dimethylpropylamine (625-43-4)
4-(1,2-Dimethyl-N-propylamino)-2-ethylamino-6-methylthio-s-triazine (22936-75-0)
4-(1,1-Dimethylpropyl)cyclohexanone (16587-71-6)
N,N-Dimethyl-1,3-propylenediamine (109-55-7)
1,1-Dimethylpropyl hydroperoxide (3425-61-4)
5-(3-Dimethylpropylidene)dibenzo(a,d)(1,4)cycloheptadiene (50-48-6)
1,1-Dimethylpropyl neodecaneperoxoate (68299-16-1)
2-(1,1-Dimethylpropyl)phenol (3279-27-4)
p-(1,1-Dimethylpropyl)phenol (80-46-6)
p-(α,α-Dimethylpropyl)phenol (80-46-6)
4-(1,1-Dimethylpropyl)phenol potassium salt (53404-18-5)
4-(1,1-Dimethylpropyl)phenol sodium salt (31366-95-7)
2,5-Dimethyl-3-propylpyrazine (18433-97-1)
3,6-Dimethyl-2-propylpyrazine (18433-97-1)
1,1-Dimethylpropynol (115-19-5)
N-(1,1-Dimethylpropynyl)-3,5-dichlorobenzamide (23950-58-5)
3,4-Dimethylprotocatechuic acid (93-07-2)
2,3-Dimethylpyrazine (5910-89-4)
2,5-Dimethylpyrazine (123-32-0)
2,6-Dimethylpyrazine (108-50-9)
2,7-Dimethylpyrene (15679-24-0)
2,3-Dimethylpyridine (583-61-9)
2,4-Dimethylpyridine (108-47-4)
2,5-Dimethylpyridine (589-93-5)
2,6-Dimethylpyridine (108-48-5)
3,4-Dimethylpyridine (583-58-4)
3,5-Dimethylpyridine (591-22-0)
α,α'-Dimethylpyridine (108-48-5)
α,γ-Dimethylpyridine (108-47-4)
2,6-Dimethylpyridine 1-oxide (1073-23-0)
N,N-Dimethyl-N'-2-pyridinyl-N'-(2-thienylmethyl)-1,2-ethanediamide (91-80-5)
N,N-Dimethyl-N'-(2-pyridyl)-N'-benzylethylenediamine hydrochloride (154-69-8)
N,N-Dimethyl-N'-(2-pyridyl)-N'-(5-chloro-2-thenyl)ethylenediamine (148-65-2)
N,N-Dimethyl-N'-(2-pyridyl)-N'-(2-thenyl)ethylenediamine hydrochloride (135-23-9)
4,6-Dimethyl-2-pyrimidinamine (767-15-7)
4,6-Dimethylpyrimidine (1558-17-4)
2-(((((4,6-Dimethyl-2-pyrimidinyl)amino)carbonyl)amino)-sulfonyl)benzoic acid (74223-56-6)
N[1]-(4,6-Dimethyl-2-pyrimidinyl)sulfanilamide (57-68-1)
2-(3-(4,6-Dimethylpyrimidin-2-yl)ureidosulfonyl)benzoic acid (74223-56-6)
2-(3-(4,6-Dimethylpyrimidin-2-yl)ureidosulphonyl)benzoic acid (74223-56-6)
N[1]-(4,6-Dimethyl-2-pyrimidyl)sulfanilamide (57-68-1)
N-(4,6-Dimethyl-2-pyrimidyl)sulfanilamide (57-68-1)
N,N-Dimethyl-N'-pyrid-2-yl-N'-2-thenylethylenediamine (91-80-5)
3,5-Dimethylpyrocatechol (2785-75-3)
1,3-Dimethyl pyrogallate (91-10-1)
2,5-Dimethyl-1-pyrrolidinecarboxanilide (34484-77-0)
2,4-Dimethylquinoline (1198-37-4)
2,6-Dimethylquinoline (877-43-0)
2,7-Dimethylquinoline (93-37-8)
3,4-Dimethylquinoline (2436-92-2)
6,7-Dimethyl-9-D-ribitylisoalloxazine (83-88-5)
Dimethyl sebacate (106-79-6)
Dimethyl selenide (593-79-3)
Dimethylselenium (593-79-3)
N,N-Dimethylserotonin (487-93-4)
Dimethylsilanediol (1066-42-8)
Dimethyl silicone (9006-65-9)
Dimethylsilyl ether (3277-26-7)
Dimethylstannylene (23120-99-2)
N,N-Dimethyl-N-(3-stearamidopropyl)amine (7651-02-7)
2,4-Dimethylstyrene (2234-20-0)
2,5-Dimethylstyrene (2039-89-6)
3,5-Dimethylstyrene (5379-20-4)
2,4-Di(α-methylstyryl)phenol (2772-45-4)
Dimethyl suberate (1732-09-8)
Dimethyl succinate (106-65-0)
Dimethyl succinylsuccinate (6289-46-9)
Dimethylsulfaat (Dutch) (77-78-1)
2-(Dimethylsulfamoyl)-(9-(4-methyl-1-piperazinyl)propylidene)-thioxanthene (5591-45-7)
O-4-Dimethylsulfamoylphenyl O,O-dimethyl phosphorothioate (52-85-7)
O,O-Dimethyl O-p-sulfamoylphenyl phosphorothioate (115-93-5)
3,4-Dimethyl-5-sulfanilamidoisoxazole (127-69-5)
4,6-Dimethyl-2-sulfanilamidopyrimidine (57-68-1)
Dimethylsulfat (Czech) (77-78-1)
Dimethyl sulfate (ACGIH,OSHA) [UN 1595] (77-78-1)
1-(3,4-Dimethyl-5-sulfethanolamide anilino)-4-amino-3-anthraquinonesulfonic acid (36897-88-8)
Dimethylsulfid (Czech) (75-18-3)
Dimethyl sulfide [UN 1164] (75-18-3)
Di-methylsulfide borane (13292-87-0)
Dimethylsulfide-α-α'-dicarboxylic acid (123-93-3)
Dimethyl sulfone (67-71-0)
1,4-Dimethylsulfonoxybutane (55-98-1)
1,4-Dimethylsulfonyloxybutane (55-98-1)
3-((2,4-Dimethyl-5-sulfophenyl)azo)-4-hydroxy-1-naphthalenesulfonic acid, disodium salt (4548-53-2)
Dimethyl sulfoxide (67-68-5)
3,4-Dimethyl-5-sulphanilamidoisoxazole (127-69-5)
Dimethyl sulphate (77-78-1)
Dimethyl sulphide [UN 1164] (75-18-3)
as-Dimethyl sulphite (66-27-3)
3,4-Dimethyl-5-sulphonamidoisoxazole (127-69-5)
3-((2,4-Dimethyl-5-sulphophenyl)azo)-4-hydroxy-1-naphthalenesulphonic acid, disodium salt (4548-53-2)
Dimethyl sulphoxide (67-68-5)
Dimethyl terephthalate (120-61-6)
Dimethyl terephthalate, polyethylene glycol polyester (58782-15-3)
7-β,17-Dimethyltestosterone (17021-26-0)
7-β,17-α-Dimethyl testosterone (17021-26-0)
1,2-Dimethyltetrabromobenzene (2810-69-7)
1,2-Dimethyltetrachlorodisilane (4518-98-3)
Dimethyl 2,3,5,6-tetrachloroterephthalate (1861-32-1)
Dimethyl tetrachloroterephthalate (1861-32-1)
O,S-Dimethyltetrachlorothioterephthalate (3765-57-9)
N,N-Dimethyl-1-tetradecanamine (112-75-4)
N,N-Dimethyltetradecanamine (112-75-4)
N,N-Dimethyl-1-tetradecanamine N-oxide (3332-27-2)
Dimethyl-n-tetradecylamine (112-75-4)
Dimethyltetradecylamine (112-75-4)
N,N-Dimethyl-N-tetradecylamine (112-75-4)
N,N-Dimethyltetradecylamine (112-75-4)
1,1-Dimethyl-3-tetrahydrodicyclopentadienylurea (2163-79-3)
2,5-Dimethyltetrahydrofuran (1003-38-9)
1,3,3-Dimethyl-2-(2-(1,2,3,4-tetrahydro-6-methoxy-1-quinolyl)-vinyl)-3H-indolium chloride (27326-17-6)
1,8-Dimethyl-1,2,3,4-tetrahydronaphthalene (25419-33-4)
2,6-Dimethyl-2,3,5,6-tetrahydro-4H-1,4-oxazine (141-91-3)
3,5-Dimethyl-1,2,3,5-tetrahydro-1,3,5-thiadiazinethione-2 (533-74-4)
3,5-Dimethyl-1,3,5-2h-tetrahydrothiadiazine-2-thione (533-74-4)
3,5-Dimethyltetrahydro-1,3,5-2h-thiadiazine-2-thione (533-74-4)
3,5-Dimethyltetrahydro-1,3,5-thiadiazine-2-thione (533-74-4)
3,5-Dimethyltetrahydro-2h-1,3,5-thiadiazine-2-thione (533-74-4)
1,1-Dimethyltetralin (1985-59-7)

1,8-Dimethyltetralin **(25419-33-4)**
2,7-Dimethyltetralin **(13065-07-1)**
4,4'-(2,3-Dimethyltetramethylene)dipyrocatechol **(500-38-9)**
N,N-Dimethyl-N'-(2-thenyl)-N'-(2-pyridyl-ethylene-diamine hydrochloride) **(135-23-9)**
N,N-Dimethyl-N'-(3-thenyl)-N'-(2-pyridyl) ethylenediamine hydrochloride **(958-93-0)**
O,O-Dimethyl S-9-thiabicyclo(4.2.1)nonenyl phosphorodithioate (Isomeric mixture) **(39624-86-7)**
Dimethylthiambutene **(524-84-5)**
Dimethylthiambutenum (Latin) **(524-84-5)**
O,O-Dimethyl-S-(3-thia-pentyl)-monothiophosphat (German) **(919-86-8)**
2,4-Dimethylthiazole **(541-58-2)**
Dimethyl N,N'-(thiobis((methylimino)carbonyloxy))bis(ethan-imidothioate) **(59669-26-0)**
Dimethyl-N,N'-(thiobis(((methylimino)carbonyl)oxy))bis(ethan-imidothioate) **(59669-26-0)**
Dimethyl 3,3'-thiobispropanoate **(4131-74-2)**
Dimethylthiocarbamide **(534-13-4)**
m,N-Dimethylthiocarbanilic acid o-2-naphthyl ester **(2398-96-1)**
Dimethylthiomethylphosphate **(152-20-5)**
3,5-Dimethyl-2-thionotetrahydro-1,3,5-thiadiazine **(533-74-4)**
O,O-Dimethylthiophosphoric acid p-chlorophenyl ester **(953-17-3)**
Dimethyl thiophosphoryl chloride **(993-12-4)**
Dimethylthiophosphoryl chloride **(993-12-4)**
1,3-Dimethylthiourea **(534-13-4)**
N,N'-Dimethylthiourea **(534-13-4)**
sym-Dimethylthiourea **(534-13-4)**
Dimethylthioxostannane **(13269-74-4)**
Dimethyltin **(23120-99-2)**
Dimethyltinbis(2-linoleoyloxyethylmercaptide) **(67859-64-7)**
Dimethyltinbis(2-oleoyloxyethylmercaptide) **(67859-63-6)**
Dimethyltin dichloride **(753-73-1)**
Dimethyltin difluoride **(3582-17-0)**
Dimethyltindineodecanoate **(68928-76-7)**
Dimethyltin fluoride **(3582-17-0)**
Dimethyltin sulfide **(13269-74-4)**
N,N-Dimethyl p-toluenesulfonamide **(599-69-9)**
Dimethyl-o-toluidine **(609-72-3)**
Dimethyl-p-toluidine **(99-97-8)**
N,N-Dimethyl-o-toluidine **(609-72-3)**
N,N-Dimethyl-p-((o-tolyl)azo)aniline **(3731-39-3)**
N,N-Dimethyl-p-(m-tolylazo)aniline **(55-80-1)**
N,N-Dimethyl-N'-(4-tolyl)-N'-(dichlorfluormethylthio)-sulfamid (German) **(731-27-1)**
N,N-Dimethyl-N-(4-tolyl)-N-(dichlorofluor-methylthio)-sulfamide **(731-27-1)**
N,N-Dimethyl-N'-(o-tolyl)formamidine **(10278-71-4)**
4-(4-(3,3-Dimethyl-1-triazene)-phenylsulfamide)-5,6-dimethoxy-pyrimidine **(103947-07-5)**
(Dimethyltriazeno)imidazolecarboxamide **(4342-03-4)**
4-(3,3-Dimethyl-1-triazeno)imidazole-5-carboxamide **(4342-03-4)**
4-(5)-(3,3-Dimethyl-1-triazeno)imidazole-5(4)-carboxamide **(4342-03-4)**
4-(Dimethyltriazeno)imidazole-5-carboxamide **(4342-03-4)**
5-(3,3-Dimethyl-1-triazeno)imidazole-4-carboxamide **(4342-03-4)**
5-(3,3-Dimethyltriazeno)imidazole-4-carboxamide **(4342-03-4)**
5-(Dimethyltriazeno)imidazole-4-carboxamide **(4342-03-4)**
N,N-Dimethyl-1,3,5-triazin-2-amine **(4040-00-0)**
2,4-Dimethyl-1,3,5-triazine **(1722-15-2)**
Dimethyl-s-triazine **(1722-15-2)**
O,O-Dimethyl-(2,2,2-trichloor-1-hydroxy-ethyl)-fosfonaat (Dutch) **(52-68-6)**
O,O-Dimethyl-(2,2,2-trichlor-1-hydroxy-aethyl)phosphonat (German) **(52-68-6)**
O,O-Dimethyl 2,2,2-trichloro-1-(n-butyryloxy)ethylphosphonate **(126-22-7)**
Dimethyl 2,2,2-trichloro-1-hydroxyethylphosphonate **(52-68-6)**
Dimethyltrichlorohydroxyethyl phosphonate **(52-68-6)**
O,O-Dimethyl-2,2,2-trichloro-1-hydroxyethyl phosphonate **(52-68-6)**
O,O-Dimethyl O-2,4,5-trichlorophenyl phosphorothioate **(299-84-3)**
Dimethyl trichlorophenyl thiophosphate **(299-84-3)**
O,O-Dimethyl O-(2,4,5-trichlorophenyl)thiophosphate **(299-84-3)**
Dimethyl 3,5,6-trichloro-2-pyridyl phosphate **(5598-52-7)**
O,O-Dimethyl O-(3,5,6-trichloro-2-pyridyl)phosphorothioate **(5598-13-0)**
O,O-Dimethyl-O-(2,4,5-trichlorphenyl)-thionophosphat (German) **(299-84-3)**
Dimethyl-3,5,6-trichlor-2-pyridylfosfat (Czech) **(5598-52-7)**
Dimethyl tridecanedioate **(1472-87-3)**
2,6-Dimethyl-4-tridecylmorpholine **(24602-86-6)**
1,1-Dimethyl-3-(3-trifluoromethylphenyl)urea **(2164-17-2)**
N-(2,4-Dimethyl-5-(((trifluoromethyl)sulfonyl)amino)phenyl)-acetamide **(53780-34-0)**
1,1-Dimethyl-3-(α,α,α-trifluoro-m-tolyl) urea **(2164-17-2)**
3,7-Dimethyl-9-(2,6,6-trimethyl-1-cyclohexen-1-yl)-2,4,6,8-nona-tetraenoic acid **(302-79-4)**
3,7-Dimethyl-9-(2,6,6-trimethyl-1-cyclohexen-1-yl)-2,4,6,8-nona-tetraen-1-ol **(68-26-8)**
2,2-Dimethyltrimethylene acrylate **(2223-82-7)**
Dimethyltrimethylene glycol **(126-30-7)**
3,3-Dimethyltrimethylene oxide **(6921-35-3)**
β,β-Dimethyltrimethylene oxide **(6921-35-3)**
Dimethyl trisulfide **(3658-80-8)**
N,N-Dimethyl-1,2,3-trithian-5-amine, ethanedioate (1:1) **(31895-22-4)**
N,N-Dimethyl-1,2,3-trithian-5-amine hydrogenoxalate **(31895-22-4)**
N,N-Dimethyl-1,2,3-trithian-5-ylammonium hydrogen oxalate **(31895-22-4)**
3,5-Dimethyl-1,2,4-trithiolane **(23654-92-4)**
N,N-Dimethyltryptamine **(61-50-7)**
(E)-6,10-Dimethyl-5,9-undecadien-2-one **(3796-70-1)**
6,10-Dimethyl-2-undecanone **(1604-34-8)**
6,10-Dimethyl-3,5,9-undecatrien-2-one **(141-10-6)**
1,1-Dimethylurea **(598-94-7)**
1,3-Dimethylurea **(96-31-1)**
N,N'-Dimethylurea **(96-31-1)**
sym-Dimethylurea **(96-31-1)**
m-(3,3-Dimethylureido)phenyl-tert-butyl carbamate **(4849-32-5)**
2,3-Dimethylvaleraldehyde **(32749-94-3)**
2,2-Dimethylvaleric acid **(1185-39-3)**
2,2-Dimethylvinyl chloride **(513-37-1)**
Dimethylvinylchloride **(513-37-1)**
β, β-Dimethylvinyl chloride **(513-37-1)**
Dimethylvinylchlorosilane **(1719-58-0)**
1,5-Dimethyl-1-vinyl-4-hexen-1-yl o-aminobenzoate **(7149-26-0)**
Dimethylvinylsilyl chloride **(1719-58-0)**
Dimethyl viologen **(4685-14-7)**
Dimethyl viologen chloride **(1910-42-5)**
1,3-Dimethylxanthine **(58-55-9)**
1,7-Dimethylxanthine **(611-59-6)**
3,7-Dimethylxanthine **(83-67-0)**
Dimethylzinn-S,S'-bis(isooctylthioglycolat) (German) **(26636-01-1)**
Dimethyl zinc **(544-97-8)**
Dimethylzinc [UN 1370] **(544-97-8)**
2,6-Dimethypyridine **(108-48-5)**
10,11-Dimethystrychnine **(357-57-3)**
Dimeticona (Spanish) **(9006-65-9)**
Dimeticone **(9006-65-9)**
Dimeticonum (Latin) **(9006-65-9)**
(4-Dimetilamino-3-metil-fenil)-N-metil-carbammato (Italian) **(2032-59-9)**
Dimetilan **(644-64-4)**
Dimetilane **(644-64-4)**
Dimetilciclohexano (Spanish) **(27195-67-1)**
2,5-Dimetil-2,5-di-(benzoilperoxi)hexano (Spanish) **(2618-77-1)**
O,O-Dimetil-O-(1,4-dimetil-3-oxo-4-aza-pent-1-enil)-fosfato (Italian) **(141-66-2)**
2,6-Dimetil-eptan-4-one (Italian) **(108-83-8)**
O,O-Dimetil-S-(2-etil-solfinil-etil)-monotiofosfato (Italian) **(301-12-2)**
O,O-Dimetil-S-(etiltio-etil)-ditiofosfato (Italian) **(640-15-3)**
O,O-Dimetil-O-(2-etiltio-etil)-monotiofosfato (Italian) **(867-27-6)**
O,O-Dimetil-S-(2-etiltio-etil)-monotiofosfato (Italian) **(919-86-8)**

Dimetilformamide (Italian) (68-12-2)
O,O-Dimetil-S-(N-formil-N-metil-carbamoil-metil)-di-tiofosfato (Italian) (2540-82-1)
O,O-Dimetil-S-(N-metil-carbamoil-metil)-ditiofosfato (Italian) (60-51-5)
O,O-Dimetil-S-(N-metil-carbamoil)-metil-monotiofosfato (Italian) (1113-02-6)
O,O-Dimetil-O-(2-N-metilcarbamoil-1-metil-vinil)-fosfato (Italian) (6923-22-4)
O,O-Dimetil-O-(3-metil-4-metiltio-fenil)-monotiofosfato (Italian) (55-38-9)
O,O-Dimetil-O-(3-metil-4-nitrofenil) fosforotioato (Portugese) (122-14-5)
O,O-Dimetil-O-(3-metil-4-nitro-fenil)-monotiofosfato (Italian) (122-14-5)
O,O-Dimetil-S-((2-metossi-1,3,4-(4H)-tiadiazol-5-on-4-il)-metil)-di-tifosfato (Italian) (950-37-8)
O,O-Dimetil-S-((morfolino-carbonil)-metil)-ditiofosfato (Italian) (144-41-2)
O,O-Dimetil-O-(4-nitro-fenil)-monotiofosfato (Italian) (298-00-0)
O,O-Dimetil-S-((4-oxo-3H-1,2,3-benzotriazin-3-il)-metil)-di-tiofosfato (Italian) (86-50-0)
(5,5-Dimetil-3-oxo-cicloes-1-en-il)-N,N-dimetil-carbammato (Italian) (122-15-6)
3,5-Dimetil-peridro-1,3,5-tiadiazin-2-tione (Italian) (533-74-4)
Dimetilsolfato (Italian) (77-78-1)
Dimetiltiambutene (524-84-5)
Dimetiltiambuteno (Spanish) (524-84-5)
Dimetil-p-toluidina (Italian) (99-97-8)
O,O Dimetil 2,2,2-trichloro 1 hidroxietil fosfonato (portugese) (52-68-6)
O,O-Dimetil-(2,2,2-tricloro-1-idrossi-etil)-fosfonato (Italian) (52-68-6)
Dimetina (961-71-7)
Dimeton (60-51-5)
3,3'-Dimetossibenzodina (Italian) (119-90-4)
Dimetox (52-68-6)
Dimetridazole (551-92-8)
Dimetylformamidu (Czech) (68-12-2)
O,O-Dimetyl-O-p-nitrofenylfosfat (Czech) (950-35-6)
Dimevur (60-51-5)
Dimexide (67-68-5)
Dimezathine (57-68-1)
Dimilin (35367-38-5)
Dimipressin (50-49-7)
Dimit (80-06-8)
Dimitan (55-86-7)
Dimite (80-06-8)
Dimorlin (357-56-2)
2,4-Dimorpholino-6-methyltriazine (26234-42-4)
Dimorpholine disulfide (103-34-4)
Dimorpholinodiethyl ether (6425-39-4)
Dimorpholino disulfide (103-34-4)
Dimpylate (333-41-5)
Dimyristyl peroxydicarbonate (53220-22-7)
Dimyristyl peroxydicarbonate, Not more than 22% in water (53220-22-7)
Dimyristyl peroxydicarbonate (Not more than 22% stable dispersion, in water) (53220-22-7)
Dimyristyl peroxydicarbonate, Technically pure (53220-22-7)
Dimyristyl thiodipropionate (16545-54-3)
Dina acid (29878-91-9)
Dinacoryl (59-26-7)
Dinacrin (54-85-3)
Di-n-amylamine [UN 2841] (2050-92-2)
Di-n-amyl ether (693-65-2)
Di-n-amylnitrosamine (13256-06-9)
Dinapacryl (485-31-4)
3,4,5,6-Dinaphthacarbazole (194-59-2)
1,2,5,6-Dinaphthacridine (226-36-8)
3,4,6,7-Dinaphthacridine (224-42-0)
Dinaphtho(2,3-a:2',3'-h)pyrene (189-45-7)
16H-Dinaphtho(2,3-a:2',3'-i)carbazole-5,10,15,17-tetraone, 6,9-dibenzamido (2379-81-9)
Dinaphtho(2,3-a:2',3'-i)naphth(2',3':6,7)indolo(2,3-c)carbazole-5,10,15,17,22,24-hexaone, 16,23-dihydro (2475-33-4)
Dinaphtho(1,2,3-cd:3',2',1'-lm)perylene-5,10-dione (116-71-2)
Dinaphtho(1,2,3-cd:3',2',1'-lm)perylene-5,10-dione, 16,17-dimethoxy (128-58-5)
Dinaphtho(1,2,3-cd:3',2',1'-lm)perylene-5,10-dione, nitro derivs. (68478-60-4)
Dinaphtho(2,1,8-c1d1a:2',1',8'-nop)heptacene (6CI,8CI,9CI) (190-09-0)
Dinaphtho(3,2,1-fg:1',2',3'-st)pentacene (8CI,9CI) (31541-10-3)
Dinaphtho(2,3-i:2',3'-i')benzo(1,2-a:4,5-a')dicarbazole-5,7,12,17,19,24-hexone, 6,18-dihydro- (9CI) (2172-33-0)
Di-2-naphthol, 1,1'-(benzylidene)bis((2-methoxy-p-phenylene)-(azo)) (6483-64-3)
Dinaphtho(1',2',3':3,4;3'',2'',1'':9,10)perylo(1,12-efg)(1,4)dioxocin-5,10-dione, 17,18- dihydro (6424-76-6)
Dinaphtho(2',3',3,4)(2'',3'',9,10)pyrene (189-43-5)
Di-(1-naphthoyl)peroxide (38512-20-8)
Di-β-naphthyl-p-phenyldiamine (93-46-9)
Di-β-naphthyl-p-phenylenediamine (93-46-9)
N,N'-Di-2-naphthyl-p-phenylenediamine (93-46-9)
N,N'-Di-β-naphthyl-p-phenylenediamine (93-46-9)
sym-Di-β-naphthyl-p-phenylenediamine (93-46-9)
Dinate (144-21-8)
Di-natrium-aethylenbisdithiocarbamat (German) (142-59-6)
Dinatrium-(N,N'-aethylen-bis(dithiocarbamat)) (German) (142-59-6)
Dinatrium-(3,6-epoxy-cyclohexaan-1,2-dicarboxylaat) (Dutch) (129-67-9)
Dinatrium-(3,6-epoxy-cyclohexan-1,2-dicarboxylat) (German) (129-67-9)
Dinatrium-(N,N'-ethyleen-bis(dithiocarbamaat)) (Dutch) (142-59-6)
Dinatrium ethylendiamintetraacetat (Czech) (139-33-3)
Dinatriumpyrophosphat (German) (7758-16-9)
Dindevan (83-12-5)
Dineval (83-12-5)
Dinickel trioxide (1314-06-3)
Dinicotinic acid (499-81-0)
Dinil (8004-13-5)
Dinile (110-61-2)
Dinitolmid (148-01-6)
Dinitolmide (ACGIH,OSHA) (148-01-6)
Dinitramine (29091-05-2)
2,4-Dinitraniline (97-02-9)
Dinitraniline Orange (3468-63-1)
Dinitrate de diethylene-glycol (French) (693-21-0)
Dinitrate de methylene glycol (French) (38483-28-2)
1,3-Dinitrato-2,2-bis(nitratomethyl)propane (78-11-5)
Dinitrato de metilenglicol (Spanish) (38483-28-2)
Dinitrile of isophthalic acid (626-17-5)
2,3-Dinitrilo-1,4-dithia-anthraquinone (3347-22-6)
2,3-Dinitrilo-1,4-dithioanthrachinon (German) (3347-22-6)
Dinitro (88-85-7)
Dinitro-3 (88-85-7)
Dinitroamine (29091-05-2)
4,6-Dinitro-2-aminophenol (96-91-3)
4,6-Dinitro-o-sec-amylphenol (4097-36-3)
2,4-Dinitroanilin (German) (97-02-9)
2,4-Dinitroanilina (Italian) (97-02-9)
2,4-Dinitroaniline (97-02-9)
2,6-Dinitroaniline (606-22-4)
3,5-Dinitroaniline (618-87-1)
Dinitroaniline Orange ND-204 (3468-63-1)
Dinitroaniline Red (3468-63-1)
Dinitroanilines [UN 1596] (26471-56-7)
2-(2,4-Dinitroanilino)phenol (6358-23-2)
2,4-Dinitroanisol (119-27-7)
2,4-Dinitroanisole (119-27-7)
Dinitroanisole (119-27-7)
α-Dinitroanisole (119-27-7)
4,4'-Dinitrobenzanilide (6333-15-9)
1,3-Dinitrobenzene (99-65-0)
2,4-Dinitrobenzene (99-65-0)
m-Dinitrobenzene (ACGIH,DOT,OSHA) (99-65-0)
Dinitrobenzene (ACGIH) (25154-54-5)

o-Dinitrobenzene (ACGIH,OSHA) [UN 1597]	**(528-29-0)**
p-Dinitrobenzene (ACGIH,OSHA) [UN 1597]	**(100-25-4)**
Dinitrobenzene, Solid [UN 1597]	**(25154-54-5)**
Dinitrobenzene, Solution (DOT)	**(25154-54-5)**
2,4-Dinitrobenzeneacetic acid	**(643-43-6)**
Dinitrobenzenes, Liquid [UN 1597]	**(99-65-0)**
2,4-Dinitrobenzoic acid	**(610-30-0)**
2,5-Dinitrobenzoic acid	**(610-28-6)**
3,4-Dinitrobenzoic acid	**(528-45-0)**
3,5-Dinitrobenzoic acid	**(99-34-3)**
1,3-Dinitrobenzol	**(99-65-0)**
Dinitrobenzol, Solid (DOT)	**(25154-54-5)**
4,4'-Dinitrobifenyl (Czech)	**(1528-74-1)**
4,4'-Dinitrobiphenyl	**(1528-74-1)**
2,4-Dinitro-6-bromanilin (Czech)	**(1817-73-8)**
3-(2,4-Dinitro-6-bromophenylazo)-4-acetamido-6-(N,N-bis-(acetoxyethyl)amino)anisole	**(3618-72-2)**
2,4-Dinitro-6-s-butylfenylester kyseliny octove (Czech)	**(2813-95-8)**
2,4-Dinitro-6-sec-butylphenol	**(88-85-7)**
2,4-Dinitro-6-tert-butylphenol	**(1420-07-1)**
4,6-Dinitro-2-sec-butylphenol	**(88-85-7)**
4,6-Dinitro-o-sec-butylphenol	**(88-85-7)**
Dinitrobutylphenol	**(88-85-7)**
4,6-Dinitro-2-sec-butylphenol ammonium salt	**(6365-83-9)**
4,6-Dinitro-o-sec-butylphenol ammonium salt	**(6365-83-9)**
Dinitrobutylphenol 2,2',2''-nitrilotriethanol salt	**(6420-47-9)**
4,6-Dinitro-2-s-butylphenyl acetate	**(2813-95-8)**
4,6-Dinitro-2-sec-butylphenyl β,β-dimethylacrylate	**(485-31-4)**
2,4-Dinitro-6-sec-butylphenyl isopropyl carbonate	**(973-21-7)**
2,4-Dinitro-6-sec-butylphenyl 2-methylcrotonate	**(485-31-4)**
4,6-Dinitro-2-(2-capryl)phenyl crotonate	**(39300-45-3)**
4,6-Dinitro-2-caprylphenyl crotonate	**(39300-45-3)**
2,6-Dinitro-4-chloroaniline	**(5388-62-5)**
1,3-Dinitro-4-chlorobenzene	**(97-00-7)**
2,4-Dinitro-1-chlorobenzene	**(97-00-7)**
2,4-Dinitrochlorobenzene	**(97-00-7)**
2,6-Dinitrochlorobenzene	**(606-21-3)**
Dinitrochlorobenzene (DOT)	**(25567-67-3)**
3,5-Dinitro-4-chlorobenzoic acid	**(118-97-8)**
Dinitrochlorobenzol	**(97-00-7)**
2,4-Dinitro-1-chloro-naphthalene	**(2401-85-6)**
2,6-Dinitro-4-chlorophenol	**(88-87-9)**
3,5-Dinitro-4-chloro-α,α,α-trifluorotoluene	**(393-75-9)**
4,5-Dinitrochrysazin	**(81-55-0)**
Dinitro-o-cresate de sodium (French)	**(2312-76-7)**
2,4-Dinitro-o-cresol	**(534-52-1)**
3,5-Dinitro-o-cresol	**(497-56-3)**
3,5-Dinitro-p-cresol	**(63989-82-2)**
4,6-Dinitro-m-cresol	**(616-73-9)**
4,6-Dinitro-o-cresol	**(534-52-1)**
Dinitro-o-cresol	**(1335-85-9)**
Dinitro-p-cresol	**(609-93-8)**
Dinitrocresol	**(534-52-1)**
Dinitro-o-cresol (ACGIH,OSHA)	**(534-52-1)**
Dinitro-o-cresol, Liquid (DOT)	**(1335-85-9)**
Dinitro-o-cresol, Solid [UN 1598]	**(1335-85-9)**
Dinitro-o-cresolato sodico (Spanish)	**(2312-76-7)**
4,6-Dinitro-o-cresolo (Italian)	**(534-52-1)**
3,5-Dinitro-o-cresol sodium salt	**(2312-76-7)**
4,6-Dinitro-o-cresol sodium salt	**(2312-76-7)**
Dinitro-o-cresol sodium salt	**(2312-76-7)**
2,4-Dinitro-6-cyclohexylphenol	**(131-89-5)**
4,6-Dinitro-o-cyclohexylphenol	**(131-89-5)**
Dinitro-o-cyclohexylphenol	**(131-89-5)**
Dinitrocyclohexylphenol (DOT)	**(131-89-5)**
Dinitrodendtroxal	**(534-52-1)**
Dinitrodiglicol (Italian)	**(693-21-0)**
Dinitrodiglykol (Czech)	**(693-21-0)**
1,5-Dinitro-4,8-dihydroxyanthraquinone	**(128-91-6)**
4,8-Dinitro-1,5-dihydroxyanthraquinone	**(128-91-6)**
2,4-Dinitrodiphenylamine	**(961-68-2)**
4,4'-Dinitrodiphenyl ether	**(101-63-3)**
p,p'-Dinitrodiphenyl ether	**(101-63-3)**
4,4'-Dinitrodiphenyl oxide	**(101-63-3)**
2,6-Dinitro-N,N-dipropylcumidine	**(33820-53-0)**
2,6-Dinitro-N,N-dipropyl-4-isopropylaniline	**(33820-53-0)**
3,5-Dinitro-N^4,N^4-dipropylsulfanilamide	**(19044-88-3)**
2,6-Dinitro-N,N-di-n-propyl-p-toluidine	**(1918-08-7)**
2,6-Dinitro-N,N-dipropyl-p-toluidine	**(1918-08-7)**
2,6-Dinitro-N,N-dipropyl-4-(trifluoromethyl)benzenamine	**(1582-09-8)**
2,6-Dinitro-N,N-di-n-propyl-α,α,α-trifluoro-p-toluidine	**(1582-09-8)**
2,2-Dinitroestilbeno (Spanish)	**(6275-02-1)**
1,1-Dinitroetano (Spanish)	**(600-40-8)**
1,2'-Dinitroetano (Spanish)	**(7570-26-5)**
1,1-Dinitroethane	**(600-40-8)**
1,2-Dinitroethane	**(7570-26-5)**
1,1-Dinitroethane (Dry)	**(600-40-8)**
Dinitro-1,1 ethane (French)	**(600-40-8)**
Dinitro-1,2 ethane (French)	**(7570-26-5)**
2,4-Dinitrofenetol (Czech)	**(610-54-8)**
2,3-Dinitrofenol (Czech)	**(66-56-8)**
2,6-Dinitrofenol (Czech)	**(573-56-8)**
3,4-Dinitrofenol (Czech)	**(577-71-9)**
3,5-Dinitrofenol (Czech)	**(586-11-8)**
2,4-Dinitrofenol (Dutch)	**(51-28-5)**
2,5-Dinitrofenol (Czech)	**(329-71-5)**
Dinitrofenolo (Italian)	**(51-28-5)**
2,4-Dinitrofenylthiokyanat (Czech)	**(1594-56-5)**
3,7-Dinitrofluoranthene	**(105735-71-5)**
3,9-Dinitrofluoranthene	**(22506-53-2)**
4,12-Dinitrofluoranthene	**(22506-53-2)**
2,5-Dinitrofluorene	**(15110-74-4)**
2,7-Dinitrofluorene	**(5405-53-8)**
2,7-Dinitro-9-fluorenone	**(31551-45-8)**
2,7-Dinitrofluorenone	**(31551-45-8)**
2,4-Dinitro-1-fluorobenzene	**(70-34-8)**
2,4-Dinitrofluorobenzene	**(70-34-8)**
Dinitrogen monoxide	**(10024-97-2)**
Dinitrogen tetroxide (DOT)	**(10544-72-6)**
Dinitrogen trioxide	**(10544-73-7)**
Dinitroglicol (Italian)	**(628-96-6)**
1,3-Dinitroglycerin	**(623-87-0)**
Dinitroglycol	**(628-96-6)**
Dinitroglycoluril	**(55510-04-8)**
2,4-Dinitro-p-hydroxydiphenylamine	**(119-15-3)**
3,5-Dinitro-2-hydroxytoluene	**(534-52-1)**
2,4-Dinitroimidazole	**(5213-49-0)**
2,5-Dinitroimidazole	**(5213-49-0)**
Dinitroisooctylphenol	**(37224-61-6)**
2,6-Dinitro-4-isopropylphenol	**(4097-47-6)**
4,6-Dinitro-o-kresol (Czech)	**(534-52-1)**
4,6-Dinitrokresol (Dutch)	**(534-52-1)**
Dinitrol	**(534-52-1)**
2,4-Dinitromesitylene	**(608-50-4)**
Dinitrometano (Spanish)	**(625-76-3)**
Dinitromethane	**(625-76-3)**
2,6-Dinitro-3-methoxy-4-tert-butyltoluene	**(83-66-9)**
2,6-Dinitro-4-methylaniline	**(6393-42-6)**
Dinitromethyl cyclohexyltrienol	**(534-52-1)**
2,4-Dinitro-6-(1-methylheptyl)-phenylcrotonat (German)	**(39300-45-3)**
2,4-Dinitro-6-(1-methylheptyl)phenyl crotonate	**(39300-45-3)**
4,6-Dinitro-2-(1-methylheptyl)phenyl crotonate	**(39300-45-3)**
Dinitro(1-methylheptyl)phenyl crotonate	**(39300-45-3)**
2,4-Dinitro-6-methylphenol	**(534-52-1)**

2,4-Dinitro-6-methylphenol sodium salt	**(2312-76-7)**
4,6-Dinitro-2-(1-methyl-n-propyl)phenol	**(88-85-7)**
2,4-Dinitro-6-(1-methyl-propyl)phenol (French)	**(88-85-7)**
2,6-Dinitro-4-methylsulfonyl-N,N-dipropylaniline	**(4726-14-1)**
4,6-Dinitro-N-methyl-N-(2,4,6-tribromophenyl)-α,α,α-trifluoro-o-toluidine	**(63333-35-7)**
2,4-Dinitro-1-naftol (Czech)	**(605-69-6)**
1,3-Dinitronaphthalene	**(606-37-1)**
1,5-Dinitronaphthalene	**(605-71-0)**
1,8-Dinitronaphthalene	**(602-38-0)**
Dinitronaphthalene	**(27478-34-8)**
2,4-Dinitro-1-naphthol	**(605-69-6)**
2-4 Dinitro-α-naphtol (French)	**(605-69-6)**
2,6-Dinitro-4-octylphenol	**(4097-33-0)**
2,4-Dinitro-6-(2-octyl)phenyl crotonate	**(39300-45-3)**
2,4-Dinitro-6-octylphenyl crotonate	**(49794-90-3)**
2,6-Dinitro-4-octylphenyl crotonate	**(49794-91-4)**
2,4-Dinitro-4'-(perfluoromethyl)diphenylamine	**(13744-79-1)**
2,4-Dinitrophenetole	**(610-54-8)**
2,3-Dinitrophenol	**(66-56-8)**
2,4-Dinitrophenol	**(51-28-5)**
2,5-Dinitrophenol	**(329-71-5)**
2,6-Dinitrophenol	**(573-56-8)**
3,4-Dinitrophenol	**(577-71-9)**
3,5-Dinitrophenol	**(586-11-8)**
Dinitrophenol	**(25550-58-7)**
α-Dinitrophenol	**(51-28-5)**
β-Dinitrophenol	**(573-56-8)**
γ-Dinitrophenol	**(329-71-5)**
Dinitrophenol, Dry or containing, by weight, less than 15% water [UN 0076]	**(25550-58-7)**
Dinitrophenol, Solution in water or flammable liquid [UN 1599]	**(25550-58-7)**
Dinitrophenol, Wetted with, by weight, at least 15% water [UN 1320]	**(25550-58-7)**
2,4-Dinitrophenol acetate (ester)	**(4232-27-3)**
2,4-Dinitrophenoxyethanol	**(2831-60-9)**
2-(2,4-Dinitrophenoxy)ethanol	**(2831-60-9)**
2,4-Dinitrophenylacetic acid	**(643-43-6)**
N-(2,4-Dinitrophenyl)-1,4-benzenediamine	**(6373-73-5)**
4,6-Dinitrophenyl-2-sec-butyl-3-methyl-2-butenonate	**(485-31-4)**
Di-4-nitrophenyl ether	**(101-63-3)**
Dinitrophenylmethane	**(25321-14-6)**
2,4-Dinitrophenylmethyl ether	**(119-27-7)**
2,4-Dinitro-6-phenylphenol	**(731-92-0)**
2,4-Dinitrophenyl thiocyanate	**(1594-56-5)**
2,2-Dinitropropyl acrylate	**(17977-09-2)**
2,2-Dinitropropyl 2-propenoate	**(17977-09-2)**
1,3-Dinitropyrene	**(75321-20-9)**
1,6-Dinitropyrene	**(42397-64-8)**
1,8-Dinitropyrene	**(42397-65-9)**
Dinitropyrene	**(78432-19-6)**
Dinitro-2,4 resorcinol (French)	**(519-44-8)**
2,4-Dinitroresorcinol (Heavy metal salts of) (Dry)	**(519-44-8)**
4,6-Dinitroresorcinol (Heavy metal salts of) (Dry)	**(616-74-0)**
2,4-Dinitroresorcinol (Spanish)	**(519-44-8)**
2,4-Dinitro-rhodanbenzol (German)	**(1594-56-5)**
3,5-Dinitrosalicylic acid	**(609-99-4)**
3,5-Dinitrosalicylic acid, methyl ester	**(22633-33-6)**
4,6-Dinitro-2-sec.butylfenol (Czech)	**(88-85-7)**
4,6-Dinitro-2-sec.butylfenolate ammony (Czech)	**(6365-83-9)**
2,4-Dinitro-6-sek.butyl-isopropylphenylcarbonat (German)	**(973-21-7)**
2,4-Dinitro-6-sek.butyl-phenylacetat (German)	**(2813-95-8)**
1,4-Dinitrosobenzene	**(105-12-4)**
p-Dinitrosobenzene	**(105-12-4)**
N,4-Dinitroso-N-methylaniline	**(99-80-9)**
3,4-Di-N-nitrosopentamethylenetetramine	**(101-25-7)**
3,7-Di-N-nitrosopentamethylenetetramine	**(101-25-7)**
Dinitrosopentamethylenetetramine	**(101-25-7)**
N[1],N[3]-Dinitrosopentamethylenetetramine	**(101-25-7)**
N,N'-Dinitrosopentamethylenetetramine	**(101-25-7)**
N,N-Dinitrosopentamethylenetetramine	**(101-25-7)**
Dinitrosopiperazin (German)	**(140-79-4)**
1,4-Dinitrosopiperazine	**(140-79-4)**
Dinitrosopiperazine	**(140-79-4)**
N,N'-Dinitrosopiperazine	**(140-79-4)**
Dinitrosorbide	**(87-33-2)**
3,7-Dinitroso-1,3,5,7-tetraazabicyclo-(3,3,1)-nonane	**(101-25-7)**
2,2'-Dinitrostilbene	**(6275-02-1)**
2,2-Dinitrostilbene	**(6275-02-1)**
Dinitro-2,2 stilbene (French)	**(6275-02-1)**
4,4'-Dinitro-2,2'-stilbenedisulfonic acid	**(128-42-7)**
Dinitrostilbenedisulfonic acid	**(128-42-7)**
4,4'-Dinitrostilbene-2,2'-disulfonic acid, disodium salt	**(3709-43-1)**
2,4-Dinitrothiocyanatobenzene	**(1594-56-5)**
2,4-Dinitro-1-thiocyanobenzene	**(1594-56-5)**
2,4-Dinitrothiocyanobenzene	**(1594-56-5)**
2,6-Dinitrothymol	**(303-21-9)**
Dinitrothymol 1-2-4 (French)	**(303-21-9)**
3,5-Dinitro-o-toluamide (OSHA)	**(148-01-6)**
2,3-Dinitrotoluene	**(602-01-7)**
2,4-Dinitrotoluene	**(121-14-2)**
2,5-Dinitrotoluene	**(619-15-8)**
2,6-Dinitrotoluene	**(606-20-2)**
3,4-Dinitrotoluene	**(610-39-9)**
3,5-Dinitrotoluene	**(618-85-9)**
Dinitrotoluene	**(25321-14-6)**
Dinitrotoluene (ACGIH,OSHA)	**(121-14-2)**
Dinitrotoluene, Liquid [UN 2038]	**(25321-14-6)**
Dinitrotoluene, Molten [UN 1600]	**(25321-14-6)**
Dinitrotoluene, Solid (DOT)	**(25321-14-6)**
3,5-Dinitro-o-toluic acid	**(28169-46-2)**
3,5-Dinitrotoluic acid	**(28169-46-2)**
2,4-Dinitro-m-toluidine	**(19406-51-0)**
2,6-Dinitro-m-toluidine	**(70343-06-5)**
2,6-Dinitro-p-toluidine	**(6393-42-6)**
3,5-Dinitro-o-toluidine	**(35572-78-2)**
3,6-Dinitro-o-toluidine	**(56207-39-7)**
4,6-Dinitro-m-toluidine	**(5267-27-6)**
2,4-Dinitrotoluol	**(121-14-2)**
4,6-Dinitro-1,2,3-trichlorobenzene	**(6379-46-0)**
Dinitrotrichlorobenzene	**(8003-46-1)**
2,6-Dinitro-4-trifluormethyl-N,N-dipropylanilin (German)	**(1582-09-8)**
2,4'-Dinitro-4-trifluoromethyl-diphenyl ether	**(15457-05-3)**
2,4-Dinitro-1,3,5-trimethylbenzene	**(608-50-4)**
4,6-Dinitro-1,3-xylene	**(616-72-8)**
4,6-Dinitro-m-xylene	**(616-72-8)**
Dinkum Oil	**(8000-48-4)**
Dinoben	**(88-86-8)**
Dinobuton	**(973-21-7)**
Dinocap	**(131-72-6)**
Dinocap	**(39300-45-3)**
Dinofan	**(51-28-5)**
Dinofen	**(973-21-7)**
Dinokap	**(39300-45-3)**
Dinoleine	**(83-73-8)**
Di-n-nonanoyl peroxide, Technically pure	**(762-13-0)**
Di-n-nonyl adipate	**(151-32-6)**
Dinonyl adipate	**(151-32-6)**
Dinonyl 1,2-benzenedicarboxylate	**(84-76-4)**
Dinonyl decanedioate	**(4121-16-8)**
Dinonyl hexanedioate	**(151-32-6)**
Di-n-nonyl ketone	**(504-57-4)**
Dinonyl ketone	**(504-57-4)**
Dinonylnaphthalenesulfonic acid	**(25322-17-2)**

Dinonylnaphthalene sulfonic acid barium salt	**(25619-56-1)**
Dinonylnaphthalenesulfonic acid, calcium salt	**(57855-77-3)**
Dinonylnaphthalenesulfonic acid, lithium salt	**(28214-91-7)**
Dinonylnaphthalenesulfonic acid, sodium salt	**(26834-28-6)**
2,4-Dinonylphenol	**(137-99-5)**
Dinonyl phenol	**(1323-65-5)**
Dinonylphenol	**(1323-65-5)**
Dinonylphenol, ethoxylated, phosphated	**(39464-64-7)**
Dinonylphenol, ethoxylate, phosphate	**(39464-64-7)**
Dinonylphenol phosphite (3:1)	**(1333-21-7)**
Di-n-nonyl phthalate	**(84-76-4)**
Dinonyl phthalate	**(84-76-4)**
Dinopol 235	**(119-07-3)**
Dinopol IDO	**(1330-96-7)**
Dinopol NOP	**(117-84-0)**
Dinoprost	**(551-11-1)**
18,19-Dinorcholestane, 5,14-dimethyl-, (5β,8α,9β,10α,14β)- (9CI)	**(57030-15-6)**
18,19-Dinorcholestane, 5,14-dimethyl-, (5β,8α,9β,10α,14β,20S)- (9CI)	**(56975-84-9)**
26,27-Dinorergosta-3,5-diene (9CI)	**(119973-29-4)**
26,27-Dinorergosta-3,5,22-triene, (22E)- (9CI)	**(119973-31-8)**
26,27-Dinorergost-3-ene, (5α)- (9CI)	**(119973-30-7)**
26,27-Dinorergost-5-en-3-ol, benzoate, (3.β.)- (9CI)	**(58003-48-8)**
Dinorguaiaretic acid, dihydro-	**(500-38-9)**
18,19-Dinor-17-α-pregn-4-en-20-yn-3-one, 13-ethyl-17-hydroxy-, (+)	**(797-63-7)**
18,19-Dinor-17-α-pregn-4-en-20-yn-3-one, 13-ethyl-17-hydroxy-, (+-)	**(6533-00-2)**
18,19-Dinorpregn-4-en-20-yn-3-one, 13-ethyl-17-hydroxy-, (17-α)-(+-)- (9CI)	**(6533-00-2)**
18,19-Dinorstigmastane, 5,14-dimethyl-, (5β,8α,9β,10α,14β,20S,24xi)- (9CI)	**(67597-34-6)**
18,19-Dinorstigmastane, 5,14-dimethyl-, (5β,8α,9β,10α,14β,24xi)- (9CI)	**(67597-35-7)**
Dinosam	**(4097-36-3)**
Dinosame (French)	**(4097-36-3)**
Dinoseb	**(88-85-7)**
Dinoseb (Amine)	**(6365-83-9)**
Dinoseb-acetate	**(2813-95-8)**
Dinoseb, 3,3-dimethylacryl ester	**(485-31-4)**
Dinosebe (French)	**(88-85-7)**
Dinosebe acetate	**(2813-95-8)**
Dinoseb methacrylate	**(485-31-4)**
Dinoterb	**(1420-07-1)**
Dinoterbe	**(1420-07-1)**
Dinovex	**(84-17-3)**
Dinoxol	**(93-76-5)**
Dinoxol	**(94-75-7)**
Dintoin	**(57-41-0)**
Dintoina	**(57-41-0)**
Dintoina	**(630-93-3)**
Dinurania	**(534-52-1)**
Dinyl	**(8004-13-5)**
Dioctadecenyl phosphate	**(14450-07-8)**
Dioctadecyl 1,2-benzenedicarboxylate	**(14117-96-5)**
Dioctadecyl phosphate	**(3037-89-6)**
Dioctadecyl 3,3'-thiobispropanoate	**(693-36-7)**
3,3'-Dioctadecyl thiodipropionate	**(693-36-7)**
Dioctadecyl 3,3'-thiodipropionate	**(693-36-7)**
Dioctadecyl thiodipropionate	**(693-36-7)**
Dioctanol-2-phthalate	**(131-15-7)**
Dioctanoyl peroxide	**(762-16-3)**
Di-n-octanoyl peroxide, Technically pure (DOT)	**(762-16-3)**
Dioctlyn	**(577-11-7)**
Dioctyl-Medo Forte	**(577-11-7)**
Dioctyl acid pyrophosphate	**(26658-09-3)**
Di-n-octyl adipate	**(123-79-5)**
Dioctyl adipate	**(103-23-1)**
Dioctyl adipate	**(123-79-5)**
Dioctylal	**(577-11-7)**
Dioctylamine (8CI)	**(1120-48-5)**
Dioctyl azelate	**(103-24-2)**
Dioctyl o-benzenedicarboxylate	**(117-84-0)**
Dioctyl dimethyl ammonium chloride	**(5538-94-3)**
Dioctyldimethylammonium chloride	**(5538-94-3)**
2,2-Dioctyl-1,3,2-dioxastannepin-4,7-dione	**(16091-18-2)**
4,4'-Dioctyldiphenylamine	**(101-67-7)**
Di-N-octyl diphenylamine	**(101-67-7)**
p,p'-Dioctyldiphenylamine	**(101-67-7)**
p,p-Dioctyldiphenylamine	**(101-67-7)**
Dioctyl diphosphorate	**(26658-09-3)**
Dioctyl ester hexanedioic acid	**(123-79-5)**
Dioctyl ester of sodium sulfosuccinate	**(577-11-7)**
Dioctyl ester of sodium sulfosuccinic acid	**(577-11-7)**
Dioctyl ether	**(629-82-3)**
Dioctyl fumarate	**(141-02-6)**
Dioctyl hexanedioate	**(123-79-5)**
Di(octyl) hydrogen phosphate	**(3115-39-7)**
Dioctyl isophthalate	**(137-89-3)**
Di-n-octyl maleate	**(2915-53-9)**
Dioctyl maleate	**(142-16-5)**
Dioctyl maleate	**(2915-53-9)**
Di(octyl)methylamine	**(4455-26-9)**
Dioctyloxostannane	**(870-08-6)**
Dioctyl peroxide	**(19102-74-0)**
Dioctylphenol	**(29988-16-7)**
4,4'-Dioctylphenylamine	**(101-67-7)**
Di-2-octyl-p-phenylenediamine	**(103-96-8)**
N,N'-Di(2-octyl)-p-phenylenediamine	**(103-96-8)**
N,N'-Di(2-octyl)-para-phenylenediamine	**(103-96-8)**
Dioctyl phosphate	**(3115-39-7)**
O,O-Dioctyl phosphorodithioate	**(2253-57-8)**
Di-n-octyl phthalate	**(117-84-0)**
Dioctyl phthalate	**(117-81-7)**
Dioctyl phthalate	**(117-84-0)**
Di-sec-octyl phthalate (ACGIH,OSHA)	**(117-81-7)**
Dioctyl pyrophosphate	**(26658-09-3)**
Dioctyl sebacate	**(122-62-3)**
Di-n-octyl sodium sulfosuccinate	**(1639-66-3)**
Dioctyl sodium sulfosuccinate	**(577-11-7)**
Dioctylstannium dichloride	**(3542-36-7)**
Dioctylstannylene maleate	**(16091-18-2)**
Dioctyl sulfosuccinate	**(2373-23-1)**
Dioctyl sulfosuccinate sodium	**(577-11-7)**
Dioctyl sulfosuccinate sodium salt	**(577-11-7)**
Di-n-octyltin bis(2-ethylhexyl mercaptoacetate)	**(15571-58-1)**
Di-(n-octyl)tin-S,S'-bis(isooctylmercaptoacetate)	**(26401-97-8)**
Dioctyltin S,S'-bis(isooctyl mercaptoacetate)	**(26401-97-8)**
Dioctyltin bis(isooctyl mercaptoacetate)	**(26401-97-8)**
Dioctyltin bis(isooctyl thioglycolate)	**(26401-97-8)**
Di-n-octyltindichloride	**(3542-36-7)**
Dioctyltin dichloride	**(3542-36-7)**
Di-n-octyltin diisooctyl thioglycolate	**(26401-97-8)**
Di-n-octyltin dilaurate	**(3648-18-8)**
Dioctyltin dilaurate	**(3648-18-8)**
Di-n-octyltin-dithioglycolic acid 2-ethylhexyl ester	**(15571-58-1)**
Di-n-octyltin-2-ethylhexyl-dimercaptoethanoate	**(15571-58-1)**
Di-n-octyltin maleate	**(16091-18-2)**
Dioctyltin maleate	**(16091-18-2)**
Di-n-octyltin β-mercaptopropionate	**(3033-29-2)**
Di-n-octyltin oxide	**(870-08-6)**
Dioctyltin oxide	**(870-08-6)**
Di-n-octyl-zinn dichlorid (German)	**(3542-36-7)**

Di-n-octyl-zinn-di-isooctylthioglykolat (German) **(26401-97-8)**
Di-n-octyl-zinn dilaurat (German) **(3648-18-8)**
Di-n-octylzinn maleinat (German) **(16091-18-2)**
Di-n-octyl-zinn β-mercaptopropionat (German) **(3033-29-2)**
Di-n-octyl-zinn oxyd (German) **(870-08-6)**
Diocurb **(51-63-8)**
Diocyl **(77-19-0)**
Diodohydroxyquin **(83-73-8)**
Diodoquin **(83-73-8)**
Diodoxylin **(83-73-8)**
Dioform **(540-59-0)**
Diognat-E **(57-63-6)**
Diogyn-E **(57-63-6)**
Diogyn **(50-28-2)**
Diogyn B **(50-50-0)**
Diogynets **(50-28-2)**
Diohin **(76-58-4)**
Diokan **(123-91-1)**
Dioksan (Polish) **(123-91-1)**
Dioksyny (Polish) **(1746-01-6)**
Dioktylester kyseliny ftalove (Czech) **(117-84-0)**
Dioktylester sulfojantaranu sodneho (Czech) **(1639-66-3)**
Diol 14B **(110-63-4)**
Diolamine **(111-42-2)**
Diolane **(107-41-5)**
Diolein **(25637-84-7)**
Diolene **(78-44-4)**
N,N'-Dioleoylethylenediamine **(110-31-6)**
Dioleoyl phosphate **(14450-07-8)**
Dioleyl hydrogen phosphite **(25088-57-7)**
Dioleyl phosphate **(14450-07-8)**
Dioleyl phosphite **(25088-57-7)**
Dioleyl phosphonate **(25088-57-7)**
Diolice **(56-72-4)**
Diom **(1330-76-3)**
Diomedicone **(577-11-7)**
Dione 21-Acetate **(13292-46-1)**
Dionin **(6746-59-4)**
Dionine **(76-58-4)**
Dionine hydrochloride **(125-30-4)**
Dionin hydrochloride **(125-30-4)**
Dionone **(117-10-2)**
Diopal **(155-41-9)**
Diophyllin **(317-34-0)**
Dioscorine (8CI) **(3329-91-7)**
Diose **(141-46-8)**
Diossano-1,4 (Italian) **(123-91-1)**
1,4-Diossan-2,3-diyl-bis(O,O-dietil-ditiofosfato) (Italian) **(78-34-2)**
1,4-Diossibenzene (Italian) **(106-51-4)**
Diossidone **(50-33-9)**
Diosuccin **(577-11-7)**
Diothene **(9002-88-4)**
Diothyl **(5221-49-8)**
Diotilan **(577-11-7)**
Diovac **(577-11-7)**
Diovascole **(54-95-5)**
Diovocyclin **(113-38-2)**
Diovocylin **(113-38-2)**
Dioxaan-1,4 (Dutch) **(123-91-1)**
1,4-Dioxaan-2,3-diyl-bis(O,O-diethyl-dithiofosfaat) (Dutch) **(78-34-2)**
3,5-Dioxa-6-aza-4-phosphaoct-6-ene-8-nitrile, 7-(2-chlorophenyl)-4-ethoxy-, 4-sulfide **(14816-20-7)**
Dioxabenzofos **(3811-49-2)**
2,3-Dioxabicyclo(2.2.2)oct-5-ene, 1-isopropyl-4-methyl- **(512-85-6)**
1,3,2-Dioxaborinane, 2,2'-((1-methyl-1,3-propanediyl)bis(oxy))bis(4-methyl- (9CI) **(2665-13-6)**
1,3,2-Dioxaborinane, 2,2'-oxybis(4,4,6-trimethyl- **(14697-50-8)**
1,3,2-Dioxaborolane, 2-ethyl- (9CI) **(10173-38-3)**
Dioxacarb **(6988-21-2)**
Dioxacarbe **(6988-21-2)**
1,6-Dioxacyclododecane-7,12-dione (9CI) **(777-95-7)**
1,3-Dioxacyclohexane **(505-22-6)**
1,4-Dioxacyclohexane **(123-91-1)**
1,3-Dioxacyclopentane **(646-06-0)**
2,4-Dioxa-1,5-dibismapentane, 1,3,5-trioxo- (9CI) **(5892-10-4)**
3,6-Dioxa-1-dodecanol **(112-59-4)**
3,6-Dioxadodecanol-1 **(112-59-4)**
2,5-Dioxahexane **(110-71-4)**
Dioxan **(123-91-1)**
p-Dioxan (Czech) **(123-91-1)**
Dioxan-1,4 (German) **(123-91-1)**
2,3-p-Dioxan-S,S'-bis(O,O-diaethyldithiophosphat) (German) **(78-34-2)**
2,3-p-Dioxandithiol S,S-bis(O,O-diethyl phosphorodithioate) **(78-34-2)**
1,4-Dioxan-2,3-diyl-bis(O,O-diaethyl-dithiophosphat) (German) **(78-34-2)**
1,4-Dioxan-2,3-diyl bis(O,O-diethylphosphorothiolothionate) **(78-34-2)**
1,4-Dioxan-2,3-diyl O,O,O',O'-tetraethyl di(phosphoromithioate) **(78-34-2)**
1,3-Dioxane **(505-22-6)**
1,4-Dioxane **(123-91-1)**
Dioxane-1,4 **(123-91-1)**
m-Dioxane **(505-22-6)**
p-Dioxane **(123-91-1)**
Dioxane (ACGIH,OSHA) [UN 1165] **(123-91-1)**
2,3-p-Dioxane S,S-bis(O,O-diethylphosphoroithioate) **(78-34-2)**
p-Dioxane, 2,3-bis(2,3-epoxypropoxy) **(10043-09-1)**
p-Dioxane, 2,3-bis(glycidyloxy)- **(10043-09-1)**
1,4-Dioxane, trans-2,3-dichloro- **(3883-43-0)**
p-Dioxane, 2,3-dichloro-, trans **(3883-43-0)**
1,4-Dioxane, 2,3-dichloro- (9CI) **(95-59-0)**
p-Dioxane, 2,3-dichloro- (8CI) **(95-59-0)**
m-Dioxane, 4,4-dimethyl **(766-15-4)**
p-Dioxane, dimethyl **(25136-55-4)**
p-Dioxane-2,3-dithiol, S,S-diester with O,O-diethyl phosphorodithioate **(78-34-2)**
p-Dioxane-2,3-diyl ethyl phosphorodithioate **(78-34-2)**
1,4-Dioxane, 2-ethyl-5-methyl- (9CI) **(53907-91-8)**
1,3-Dioxane, 4-methyl **(1120-97-4)**
Dioxane phosphate **(78-34-2)**
Dioxanne (French) **(123-91-1)**
1,3-Dioxan-4-ol, 2,6-dimethyl-, acetate **(828-00-2)**
m-Dioxan-4-ol, 2,6-dimethyl-, acetate **(828-00-2)**
1,3-Dioxan-5-ol, 4,4,5-trimethyl- (9CI) **(54063-14-8)**
3,6-Dioxaoctane-1,8-diol **(112-27-6)**
3,6-Dioxaoctane-1,8-diyl dibenzoate **(120-56-9)**
3,6-Dioxa-1-octanol **(111-90-0)**
3,6-Dioxa-1-oktanol (Czech) **(111-90-0)**
Dioxaphetyl butyrate **(467-86-7)**
2-(1,3,2-Dioxaphospholan-2-yloxy)ethanol **(1073-75-2)**
1,4-Dioxaspiro(4.5)decane **(177-10-6)**
1,6-Dioxaspiro(4.5)decane-7-butyric acid, 2-(5-ethyltetrahydro-5-(tetrahydro-3-methyl- 5-(tetrahydro-6-hydroxy-6-(hydroxymethyl)-3,5-dimethyl-2H-pyran-2-yl)-2-furyl)-2-furyl)-9-hydroxy-β-methoxy-α,γ,2,8-tetramethyl **(17090-79-8)**
1,6-Dioxaspiro(4.5)decane-7-butyric acid, 2-(5-ethyltetrahydro-5-(tetrahydro-3-methyl-5- (tetrahydro-6-hydroxy-6-(hydroxymethyl)-3,5-dimethyl-2H-pyran-2-yl)-2-furyl)-2-furyl)-9- hydroxy-β-methoxy-α,γ,2,8-tetramethyl-, monosodium salt **(22373-78-0)**
1,3,2-Dioxastannepin-4,7-dione, 2,2-dibutyl **(78-04-6)**
1,3,2-Dioxastannepin-4,7-dione, 2,2-dioctyl **(16091-18-2)**
Dioxathion **(78-34-2)**
Dioxathion (ACGIH,OSHA) **(78-34-2)**
Dioxation **(78-34-2)**
3,5-Dioxe-4 buty-1, diphenyl-pyrazolidine **(50-33-9)**
1,3-Dioxepane, 2-methylene-, Polymer with ethene (9CI) **(90588-08-2)**
1,4-Di-N-oxide 2,3-bis(oxymethyl)quinoxline **(17311-31-8)**

Dioxide of Vitavax (5259-88-1)
1,4-Di-N-oxide of dihydroxymethylquinoxaline (17311-31-8)
1,1-Dioxidetetrahydrothiofuran (126-33-0)
1,1-Dioxidetetrahydrothiophene (126-33-0)
Dioxidin (17311-31-8)
Dioxidine (17311-31-8)
Dioxime p-benzoquinone (105-11-3)
Dioxime 1,4-cyclohexadienedione (105-11-3)
Dioxime 2,5-cyclohexadiene-1,4-dione (105-11-3)
Dioxin (Bactericide) (Obs.) (828-00-2)
1,4-Dioxin, 2,3-dihydro-5,6-dimethyl- (9CI) (25465-18-3)
p-Dioxin, 2,3-dihydro-5,6-dimethyl- (8CI) (25465-18-3)
Dioxine (1746-01-6)
Dioxin (herbicide contaminant) (1746-01-6)
(1,4)Dioxino(2,3-b)-1,4-dioxin, hexahydro-, cis- (9CI) (13405-83-9)
p-Dioxino(2,3-b)-p-dioxin, hexahydro-, cis- (8CI) (13405-83-9)
p-Dioxin, tetrahydro- (123-91-1)
Dioxitol (111-90-0)
9,10-Dioxoanthracene (84-65-1)
p-Dioxobenzene (123-31-9)
Dioxocarb (6988-21-2)
1,1',1''-(3,6-Dioxo-1,4-cyclohexadiene-1,2,4-triyl)trisaziridine (68-76-8)
1,2-Dioxocyclohexane (765-87-7)
Dioxodichlorochromium (14977-61-8)
3,5-Dioxo-1,2-diphenyl-4-n-butylpyrazolidene (50-33-9)
3,5-Dioxo-1,2-diphenyl-4-n-butyl-pyrazolidin (50-33-9)
3,5-Dioxo-1,2-diphenyl-4-n-butylpyrazolidine (50-33-9)
Dioxohexahydrotriazine (27032-78-6)
(2,5-Dioxo-4-imidazolidinyl)urea (97-59-6)
2-(2-(1,3-Dioxo-2-indanyl)-6-quinolyl)-6-methyl-7-benzothiazole-sulfonic acid, sodium salt (4121-67-9)
2,3-Dioxoindoline (91-56-5)
1,3-Dioxolan (646-06-0)
Dioxolan (100-79-8)
Dioxolan (Czech) (646-06-0)
1,3-Dioxolane (646-06-0)
Dioxolane [UN 1166] (100-79-8)
1,3-Dioxolane, 2-tert-butyl-2-methyl- (7CI,8CI) (6135-54-2)
1,3-Dioxolane, 2-(chloromethyl)- (9CI) (2568-30-1)
1,3-Dioxolane, 2-(dichloromethyl)- (8CI,9CI) (2612-35-3)
1,3-Dioxolane, 2-(1,1-dimethylethyl)-2-methyl- (9CI) (6135-54-2)
1,3-Dioxolane, 2-ethyl- (8CI,9CI) (2568-96-9)
1,3-Dioxolane, 4-ethyl-5-hexyl-2,2-bis(trifluoromethyl)-, trans- (9CI) (38274-67-8)
1,3-Dioxolane, 2-ethyl-2-isobutyl- (7CI,8CI) (935-45-5)
1,3-Dioxolane, 2-ethyl-2-methyl- (8CI,9CI) (126-39-6)
1,3-Dioxolane, 2-isopropyl- (8CI) (822-83-3)
1,3-Dioxolane, 2-isopropyl-2-methyl- (7CI,8CI) (4405-16-7)
1,3-Dioxolane-4-methanol, 2,2-dimethyl (100-79-8)
1,3-Dioxolane-4-methanol, 2-(1-iodoethyl) (5634-39-9)
1,3-Dioxolane, 2-methoxy- (9CI) (19693-75-5)
1,3-Dioxolane, 2-methoxy-2-(1-methylethyl)- (9CI) (66822-98-8)
1,3-Dioxolane, 2-methyl (497-26-7)
Dioxolane, methyl (1331-09-5)
1,3-Dioxolane, 2-(1-methylethyl)- (9CI) (822-83-3)
1,3-Dioxolane, 2-methyl-2-(1-methylethyl)- (9CI) (4405-16-7)
1,3-Dioxolane, 2-methyl-2-propyl- (7CI,8CI,9CI) (4352-98-1)
1,3-Dioxolane, 2-(2-phenylethenyl)- (9CI) (5660-60-6)
1,3-Dioxolane, 2-(1-propenyl)- (9CI) (4528-26-1)
1,3-Dioxolane, 2-propenyl- (8CI) (4528-26-1)
1,3-Dioxolane, 2-styryl- (8CI) (5660-60-6)
2-(1,3-Dioxolane-2-yl)phenyl N-methylcarbamate (6988-21-2)
1,3-Dioxolan-2-one (96-49-1)
1,3-Dioxolan-2-one, 4-methyl- (108-32-7)
1,3-Dioxolan-2-one, 4,4,5,5-tetramethyl- (9CI) (19424-29-4)
2-(1,3-Dioxolan-2-yl)phenyl-N-methylcarbamat (German) (6988-21-2)
o-(1,3-Dioxolan-2-yl)phenyl methylcarbamate (6988-21-2)
(1,3)Dioxolo(4,5-g)cinnoline-3-carboxylic acid, 1,4-dihydro-1-ethyl-4-oxo (28657-80-9)
1,3-Dioxolo(4,5-g)quinoline-7-carboxylic acid, 5-ethyl-5,8-dihydro-8-oxo (14698-29-4)
Dioxolone-2 (96-49-1)
1,3-Dioxol-2-one (9CI) (872-36-6)
Dioxone (702-54-5)
3,5-Dioxo-1-phenyl-2-(p-hydroxyphenyl)-4-n-butylpyrazolidene (129-20-4)
1,3-Dioxophthalan (85-44-9)
1,3-Dioxo-5-phthalancarboxylic acid (552-30-7)
2,6-Dioxo-3-phthalimidopiperidine (50-35-1)
2,5-Dioxopiperazine (106-57-0)
N-(2,6-Dioxo-3-piperidyl)phthalimide (50-35-1)
2,6-Dioxopurine (69-89-6)
2,4-Dioxopyrimidine (66-22-8)
2,5-Dioxopyrrolidine (123-56-8)
Dioxo(sulfato(2-)-O)uranium (1314-64-3)
3,5-Dioxo-2,3,4,5-tetrahydro-1,2,4-triazine riboside (54-25-1)
1,1-Dioxothiolan (126-33-0)
Dioxothiolan (126-33-0)
Dioxothion (78-34-2)
2,4-Dioxovalerate (5699-58-1)
Dioxyanthranol (480-22-8)
1,4-Dioxyanthraquinone (Russian) (81-64-1)
2,6-Dioxy-8-azapurine (1468-26-4)
1,4-Dioxybenzene (106-51-4)
m-Dioxybenzene (108-46-3)
o-Dioxybenzene (120-80-9)
3,3'-Dioxybenzidine (2373-98-0)
1,4-Dioxy-benzol (German) (106-51-4)
Dioxybenzon (131-53-3)
Dioxybenzone (131-53-3)
1,1'-Dioxybis-cyclohexanol (2407-94-5)
1,1'-Dioxybiscyclohexanol (2407-94-5)
Dioxybutadiene (1464-53-5)
Dioxyde de baryum (French) (1304-29-6)
Dioxyde de carbone et oxygene en melange (French) (8063-77-2)
Dioxyde de carbone et protoxyde d'azote en melange (French) (53569-62-3)
Dioxydemeton-S-methyl (17040-19-6)
2,4-Dioxy-3,3-diethyl-5-methylpiperidine (125-64-4)
Dioxydine (17311-31-8)
4,4'-Dioxydiphenylsulfide (2664-63-3)
1,4-Dioxyethylamino-5,8-dioxyanthraquinone (Russian) (3179-90-6)
N,N-Dioxyethylaniline (120-07-0)
Dioxyethylene ether (123-91-1)
Dioxymethylene-protocatechuic aldehyde (120-57-0)
Di(p-oxyphenyl)-2,4-hexadiene (84-17-3)
7-(2,3-Dioxypropyl)theophylline (479-18-5)
Diozol (50-33-9)
Dipac (95-29-4)
Dipalmityl adipate (26720-21-8)
Dipam (439-14-5)
Dipan (86-29-3)
Dipane (91-84-9)
Dipanol (138-86-3)
Di-paralene (14362-31-3)
Diparalene (82-93-9)
Diparalene hydrochloride (14362-31-3)
Di-paralene monohydrochloride (14362-31-3)
Dipaxin (82-66-6)
Di-pi-cyclopentadienylnickel (1271-28-9)
Dipegyl (98-92-0)
Dipel (68038-71-1)
Dipelargonyl peroxide (762-13-0)
Dipentaerythritol (8CI) (126-58-9)
Dipentamethylenethiuram tetrasulfide (120-54-7)
Dipentek (126-58-9)

Dipentene [UN 2052] **(138-86-3)**
Dipentene dioxide **(96-08-2)**
Dipentene polymer **(9003-73-0)**
Dipentylamine **(2050-92-2)**
Dipentylamine, N-nitroso **(13256-06-9)**
2,5-Di-tert-pentylbenzene-1,4-diol **(79-74-3)**
Dipentyl ether **(693-65-2)**
2,5-Di-t-pentylhydroquinone **(79-74-3)**
Dipentyl ketone **(927-49-1)**
Dipentyl maleate **(10099-71-5)**
Di-n-pentylnitrosamine **(13256-06-9)**
Dipentylnitrosamine **(13256-06-9)**
N,N-Dipentyl-1-pentanamine **(621-77-2)**
2,4-Di-tert-pentylphenol **(120-95-6)**
2-(2,4-Di-tert-pentylphenoxy)butyric acid **(13403-01-5)**
Dipentyl phosphate **(3138-42-9)**
Di-n-pentylphthalate **(131-18-0)**
Dipentyl phthalate **(131-18-0)**
1,4-Dipentylsulfobutanedioic acid, sodium salt **(922-80-5)**
Dipeptide sweetener **(22839-47-0)**
1,9:5,10-Di(perinaphthylene)anthracene **(191-79-7)**
Diperoxyphtalate de tert-butyle (French) **(2155-71-7)**
Dipezona **(439-14-5)**
Diphacil **(64-95-9)**
Diphacin **(82-66-6)**
Diphacinone **(82-66-6)**
Diphacinone, monosodium salt **(42721-99-3)**
Diphacinone sodium salt **(42721-99-3)**
Diphacyl **(64-95-9)**
Diphantine **(58-73-1)**
Diphantoin **(57-41-0)**
Diphantoine sodium **(630-93-3)**
Diphaston **(152-62-5)**
Diphebuzol **(50-33-9)**
Diphedal **(57-41-0)**
Diphedan **(57-41-0)**
Diphedan **(630-93-3)**
Diphenacin **(82-66-6)**
Diphenadione **(82-66-6)**
Diphenaldehyde **(1210-05-5)**
Diphenaleno(1,9-ab:1',9'-lm)triphenodioxazinetetrasulfonic acid, 8,19-dichloro-, tetrasodium salt (9CI) **(33700-25-3)**
Diphenamid **(957-51-7)**
Diphenamide **(957-51-7)**
Diphenan (VAN) **(101-71-3)**
Diphenane (French) **(101-71-3)**
Diphenanum (Latin) **(101-71-3)**
Diphenasone **(80-08-0)**
Diphenate **(630-93-3)**
Diphenatrile **(86-29-3)**
Diphenex **(32861-85-1)**
Diphenhydramine **(58-73-1)**
Diphenhydramine 8-chlorotheophylline **(523-87-5)**
Diphenhydramine hydrochloride **(147-24-0)**
Diphenhydrinate **(523-87-5)**
2,2'-Diphenic acid **(482-05-3)**
Diphenic acid **(482-05-3)**
O,O'-Diphenic acid **(482-05-3)**
Diphenidol **(972-02-1)**
Diphenin **(57-41-0)**
Diphenin **(630-93-3)**
Diphenine **(57-41-0)**
Diphenine sodium **(630-93-3)**
o-Diphenol **(120-80-9)**
Diphenolic acid **(126-00-1)**
Diphenoxin **(28782-42-5)**
1,8-Diphenoxy-9,10-anthracenedione **(82-17-7)**
1,5-Diphenoxyanthraquinone **(82-21-3)**
Diphenoxychloro-s-triazine **(2972-65-8)**
1,2-Diphenoxyethane **(104-66-5)**
Diphenoxylate **(915-30-0)**
Diphenoxylic acid **(28782-42-5)**
Diphenthane 70 **(97-23-4)**
Diphentoin **(57-41-0)**
Diphentoin **(630-93-3)**
Diphentyn **(57-41-0)**
Diphenydramine hydrochloride **(147-24-0)**
Diphenyl **(92-52-4)**
Diphenyl (OSHA) **(92-52-4)**
Diphenyl Black **(101-54-2)**
Diphenyl Blue 2B **(2602-46-2)**
Diphenyl Blue 3B **(72-57-1)**
Diphenyl Blue Black GHS **(2429-73-4)**
Diphenyl Blue Black MBH **(2429-73-4)**
Diphenyl Blue KF **(2602-46-2)**
Diphenyl Blue M2B **(2602-46-2)**
Diphenyl Brilliant Blue **(2429-74-5)**
Diphenyl Brilliant Blue FF **(2610-05-1)**
Diphenyl Brown BS **(2429-81-4)**
Diphenyl Brown PT **(2586-58-5)**
Diphenyl Brown TB **(2429-81-4)**
Diphenyl Brown V **(2429-82-5)**
Diphenyl Chrome Black Blue 2B **(7082-31-7)**
Diphenyl Chrome Blue Black 2B **(7082-31-7)**
Diphenyl Dark Green B **(3626-28-6)**
Diphenyl Dark Green BN **(3626-28-6)**
Diphenyl Deep Black G **(1937-37-7)**
Diphenyl Deep Black VN **(2429-83-6)**
Diphenyl Fast 5BL Supra I Red **(2610-11-9)**
Diphenyl Fast Brilliant Yellow 8GL **(10190-68-8)**
Diphenyl Fast Brown BRL **(16071-86-6)**
Diphenyl Fast Brown F **(2429-81-4)**
Diphenyl Fast Brown MD **(2429-82-5)**
Diphenyl Fast Red 5BL **(2610-11-9)**
Diphenyl Fast Red 5BLN **(2610-11-9)**
Diphenyl Fast Yellow FA **(1325-37-7)**
Diphenyl Red 4B **(992-59-6)**
Diphenyl Red 8B **(6548-29-4)**
Diphenyl Red 3BS **(6358-29-8)**
Diphenyl Red 4BS **(992-59-6)**
Diphenyl Scarlet 3BS **(6358-29-8)**
Diphenyl Sky Blue 6B **(2429-74-5)**
Diphenyl Violet TS **(6426-67-1)**
Diphenyl Yellow G **(1325-37-7)**
2,2-Diphenylacetaldehyde **(947-91-1)**
Diphenylacetaldehyde **(947-91-1)**
N,N'-Diphenylacetamidine **(621-09-0)**
Diphenylacetic acid **(117-34-0)**
α,α-Diphenylacetic acid **(117-34-0)**
Diphenylacetic acid diethylaminoethyl ester **(64-95-9)**
Diphenylacetic acid, 2-(diethylamino)ethyl ester **(64-95-9)**
1,3-Diphenylacetone **(102-04-5)**
α,α'-Diphenylacetone **(102-04-5)**
Diphenylacetonitrile **(86-29-3)**
Diphenylacetyldiethylaminoethanol **(64-95-9)**
2-Diphenylacetyl-1,3-diketohydrindene **(82-66-6)**
1,2-Diphenylacetylene **(501-65-5)**
2-Diphenyl-acetyl-indan-1,3-dion (German) **(82-66-6)**
2-(Diphenylacetyl)indan-1,3-dione **(82-66-6)**
2-Diphenylacetyl-1,3-indandione **(82-66-6)**
2-(Diphenylacetyl)-1,3-indandione sodium salt **(42721-99-3)**
2-(Diphenylacetyl)-1H-indene-1,3(2H)-dione **(82-66-6)**
Diphenylamide **(957-51-7)**
Diphenylamine **(122-39-4)**

Name	CAS No.
N,N-Diphenylamine	**(122-39-4)**
Diphenylamine (ACGIH,OSHA)	**(122-39-4)**
Diphenylamine Orange	**(554-73-4)**
Diphenylamine, 4-amino-	**(101-54-2)**
Diphenylamine, p-amino-	**(101-54-2)**
Diphenylamine, 4,4'-bis(dimethylamino)- (8CI)	**(637-31-0)**
Diphenylamine-2-carboxylic acid	**(91-40-7)**
2-Diphenylaminecarboxylic acid, 2',3'-dimethyl-	**(61-68-7)**
Diphenylaminechlorarsine	**(578-94-9)**
Diphenylamine, 3-chloro- (8CI)	**(101-17-7)**
Diphenylamine, 4-chloro- (8CI)	**(1205-71-6)**
Diphenylamine chloroarsine [UN 1698]	**(578-94-9)**
Diphenylamine, 3'-chloro-2,4-dinitro- (8CI)	**(16220-58-9)**
Diphenylamine, 3-chloro-4'-nitro- (8CI)	**(15979-85-8)**
Diphenylamine, 4,4'-diamino- (8CI)	**(537-65-5)**
Diphenylamine, 3,4'-dichloro- (8CI)	**(15979-79-0)**
Diphenylamine, 2,4-dinitro	**(961-68-2)**
Diphenylamine, 3,4'-dinitro- (7CI,8CI)	**(15979-87-0)**
Diphenylamine, 4,4'-dinonyl- (8CI)	**(24925-59-5)**
Diphenylamine, 4,4'-dioctyl-	**(101-67-7)**
Diphenylamine, 2,2',4,4',6,6'-hexanitro	**(131-73-7)**
Diphenylamine, hexanitro-	**(131-73-7)**
Diphenylamine, hydrogen sulfate	**(587-84-8)**
Diphenylamine, 4-hydroxy-	**(122-37-2)**
Diphenylamine, 4-isopropoxy	**(101-73-5)**
Diphenylamine, N-methyl- (8CI)	**(552-82-9)**
Diphenylamine, 4-(methylsulfonyl)- (8CI)	**(15979-81-4)**
Diphenylamine, 4-nitro	**(836-30-6)**
Diphenylamine, 2-nitro- (8CI)	**(119-75-5)**
Diphenylamine, 3-nitro- (8CI)	**(4531-79-7)**
Diphenylamine, 4-((p-nitrophenyl)azo)	**(2581-69-3)**
Diphenylamine, 4-nitroso	**(156-10-5)**
Diphenylamine, N-nitroso	**(86-30-6)**
Diphenylamine sulfate	**(587-84-8)**
Diphenylamine-2-sulfonic acid, 4,4'-diamino-	**(119-70-0)**
Diphenylamine, 2,4,4'-trinitro- (8CI)	**(970-76-3)**
Diphenylamine, 2,3',4-trinitro- (7CI,8CI)	**(970-91-2)**
(Diphenylamino)cyanuric chloride	**(16033-74-2)**
2-Diphenylamino-4,6-dichloro-1,3,5-triazine	**(16033-74-2)**
2-Diphenylamino-4,6-dichloro-s-triazine	**(16033-74-2)**
Diphenylamino-4,6-dichloro-s-triazine	**(16033-74-2)**
Diphenylan	**(57-41-0)**
N,N-Diphenylaniline	**(603-34-9)**
Diphenylan sodium	**(630-93-3)**
9,10-Diphenylanthracene	**(1499-10-1)**
1,2-Diphenylbenzene	**(84-15-1)**
1,4-Diphenylbenzene	**(92-94-4)**
Diphenylbenzene	**(26140-60-3)**
m-Diphenylbenzene	**(92-06-8)**
p-Diphenylbenzene	**(92-94-4)**
Diphenyl 1,3-benzenedicarboxylate	**(744-45-6)**
Diphenyl 1,4-benzenedicarboxylate	**(1539-04-4)**
α,α-Diphenylbenzenemethanol	**(76-84-6)**
2,2'-Diphenylbiphenyl	**(641-96-3)**
4,4'-Diphenylbiphenyl	**(135-70-6)**
Diphenylborinic acid	**(2622-89-1)**
1,4-Diphenylbutadiene	**(886-65-7)**
1,3-Diphenylbutane	**(1520-44-1)**
erythro-2,3-Diphenylbutane	**(4613-11-0)**
meso-2,3-Diphenylbutane	**(4613-11-0)**
Diphenylbutazone	**(50-33-9)**
1,3-Diphenyl-1-butene	**(7614-93-9)**
1,2-Diphenyl-4-butyl-3,5-dioxopyrazolidine	**(50-33-9)**
1,2-Diphenyl-4-butyl-3,5-pyrazolidinedione	**(50-33-9)**
1,5-Diphenylcarbazide	**(140-22-7)**
2,2'-Diphenylcarbazide	**(140-22-7)**
Diphenylcarbazide	**(140-22-7)**
sym-Diphenylcarbazide	**(140-22-7)**
Diphenyl carbinol	**(91-01-0)**
1,5-Diphenylcarbohydrazide	**(140-22-7)**
Diphenyl carbonate	**(102-09-0)**
Diphenylchloorarsine (Dutch)	**(712-48-1)**
Diphenylchloride	**(27323-18-8)**
Diphenylchloroarsine, Solid or liquid [UN 1699]	**(712-48-1)**
Diphenylchloromethane	**(90-99-3)**
Diphenyl-(2-chlorophenyl)-1-imidazolylmethane	**(23593-75-1)**
Diphenyl cresol phosphate	**(26444-49-5)**
Diphenyl cresyl phosphate	**(26444-49-5)**
Diphenyl-α-cyanomethane	**(86-29-3)**
2,4'-Diphenyldiamine	**(492-17-1)**
1,2-Diphenyldiazene	**(103-33-3)**
Diphenyldiazene	**(103-33-3)**
1,1'-Diphenyldiazomethane	**(883-40-9)**
Diphenyl dichlorosilane	**(80-10-4)**
Diphenyldichlorosilane [UN 1769]	**(80-10-4)**
1,1'-Diphenyldiethyl ether	**(93-96-9)**
Diphenyldiimide	**(103-33-3)**
Diphenyl-α,β-diketone	**(134-81-6)**
Diphenyldimethoxysilane	**(6843-66-9)**
2,2-Diphenyl-N,N-dimethylacetamide	**(957-51-7)**
l-1,2-Diphenyl-1-dimethylaminoethane hydrochloride	**(14148-99-3)**
1,1-Diphenyl-1-(dimethylaminoisopropyl)butanone-2	**(466-40-0)**
Diphenyl 2,6-dimethylphenyl phosphate	**(23666-94-6)**
1,2-Diphenyl-3,5-dioxo-4-butylpyrazolidine	**(50-33-9)**
1,2-Diphenyl-2,3-dioxo-4-N-butylpyrazoline	**(50-33-9)**
1,2-Diphenyl-3,5-dioxo-4-(2'-phenyl-sulfinyl-aethyl)-pyrazolidin (German)	**(57-96-5)**
4,4'-Diphenyldiphenyl ether	**(58841-70-6)**
Diphenyl, diphenyl ether	**(58841-70-6)**
Diphenyl disulfide	**(882-33-7)**
Diphenyle chlore, 54% De Chlore (French)	**(11097-69-1)**
Diphenyle chlore, 42% De Chlore (French)	**(53469-21-9)**
Diphenylenazone	**(230-17-1)**
Diphenylene	**(259-79-0)**
4,4'-Diphenylenediamine	**(92-87-5)**
Diphenylene dioxide	**(262-12-4)**
Diphenyleneimine	**(86-74-8)**
Diphenylenemethane	**(86-73-7)**
Diphenylene oxide	**(132-64-9)**
Diphenylene sulfide	**(132-65-0)**
Diphenylenimide	**(86-74-8)**
Diphenylenimine	**(86-74-8)**
1,1-Diphenylethane	**(612-00-0)**
1,2-Diphenylethane	**(103-29-7)**
1,2-Diphenylethanedione	**(134-81-6)**
1,2-Diphenylethane-1,2-dione, dimethyl ketal	**(24650-42-8)**
1,2-Diphenylethanone	**(451-40-1)**
1,1-Diphenylethene	**(530-48-3)**
cis-Diphenylethene	**(645-49-8)**
trans-1,2-Diphenylethene	**(103-30-0)**
trans-Diphenylethene	**(103-30-0)**
Diphenyl ether	**(101-84-8)**
Diphenyl ether, 4-bromo-	**(101-55-3)**
Diphenyl ether 4,4'-disulfonyl chloride	**(121-63-1)**
Diphenyl ether-formaldehyde copolymer	**(26007-63-6)**
Diphenyl ether-formaldehyde polymer	**(26007-63-6)**
Diphenyl ether-formaldehyde resin	**(26007-63-6)**
Diphenylethoxyarsine	**(24582-55-6)**
1,2-Diphenylethyl chloride	**(4714-14-1)**
(E)-1,2-Diphenylethylene	**(103-30-0)**
1,1-Diphenylethylene	**(530-48-3)**
1,2-Diphenylethylene	**(588-59-0)**
Diphenylethylene	**(530-48-3)**
α,α-Diphenylethylene	**(530-48-3)**

Name	CAS No.
α,β-Diphenylethylene	**(588-59-0)**
as-Diphenylethylene	**(530-48-3)**
cis-1,2-Diphenylethylene	**(645-49-8)**
trans-1,2-Diphenylethylene	**(103-30-0)**
trans-α,β-Diphenylethylene	**(103-30-0)**
Diphenyl 2-ethylhexyl phosphate	**(1241-94-7)**
Diphenyl 2-ethylhexyl phosphite	**(15647-08-2)**
Diphenylethyne	**(501-65-5)**
Diphenylglycolic acid	**(76-93-7)**
α,α-Diphenylglycolic acid	**(76-93-7)**
Diphenylglycolic acid 2-(diethylamino)ethyl ester	**(302-40-9)**
Diphenylglyoxal	**(134-81-6)**
Diphenylglyoxal peroxide	**(94-36-0)**
1,3-Diphenylguanidine	**(102-06-7)**
Diphenylguanidine	**(102-06-7)**
N,N'-Diphenylguanidine	**(102-06-7)**
2,6-Diphenyl-2,4,6,6,8,8-hexamethylcyclotetrasiloxane	**(4657-20-9)**
racemic-2,4-Diphenyl-2,4,6,6,8,8-hexamethylcyclotetrasiloxane	**(18604-02-9)**
5,5-Diphenylhydantoin	**(57-41-0)**
Diphenylhydantoin	**(57-41-0)**
Diphenylhydantoine (French)	**(57-41-0)**
5,5-Diphenylhydantoin sodium	**(630-93-3)**
Diphenylhydantoin sodium	**(630-93-3)**
Diphenylhydramine	**(58-73-1)**
Diphenylhydramine hydrochloride	**(147-24-0)**
(sym)-Diphenylhydrazine	**(122-66-7)**
1,1-Diphenylhydrazine	**(530-50-7)**
N,N'-Diphenylhydrazine	**(122-66-7)**
N,N-Diphenylhydrazine	**(530-50-7)**
α,α-Diphenylhydrazine	**(530-50-7)**
1,2-Diphenylhydrazine (9CI)	**(122-66-7)**
N,N-Diphenylhydrazine hydrochloride	**(530-47-2)**
1,1-Diphenylhydrazine hydrochloride (VAN)	**(530-47-2)**
1,1-Diphenylhydrazine monohydrochloride	**(530-47-2)**
Diphenylhydroxyacetic acid	**(76-93-7)**
2,4-Diphenyl-6-(2-hydroxyphenyl)-s-triazine	**(3202-86-6)**
5,5-Diphenylimidazolidin-2,4-dione	**(57-41-0)**
5,5-Diphenyl-2,4-imidazolidinedione	**(57-41-0)**
5,5-Diphenyl-2,4-imidazolidine-dione, monosodium salt	**(630-93-3)**
Diphenyline	**(492-17-1)**
Diphenyliodonium hexafluoroarsenate	**(62613-15-4)**
Diphenyliodonium hexafluoroarsenate(1-)	**(62613-15-4)**
Diphenyliodonium hexafluorophosphate	**(58109-40-3)**
Diphenyliodonium hexafluorophosphate(1-)	**(58109-40-3)**
1,3-Diphenylisobenzofuran	**(5471-63-6)**
Diphenylketen	**(947-91-1)**
Diphenyl ketone	**(119-61-9)**
8,10-Diphenyllobelionol	**(90-69-7)**
Diphenylmercury	**(587-85-9)**
Di(phenylmercury) dodecenylsuccinate	**(27236-65-3)**
Diphenylmethan-4,4'-diisocyanat (German)	**(101-68-8)**
Diphenylmethane	**(101-81-5)**
4,4'-Diphenylmethanebismaleimide	**(13676-54-5)**
Diphenylmethanebismaleimide	**(13676-54-5)**
N,N'-4,4'-Diphenylmethanebismaleimide	**(13676-54-5)**
N,N'-p,p'-Diphenylmethanebismaleimide	**(13676-54-5)**
2,4'-Diphenylmethanediamine	**(1208-52-2)**
4,4'-Diphenylmethanediamine	**(101-77-9)**
4,4'-Diphenylmethane diisocyanate	**(101-68-8)**
Diphenyl methane diisocyanate	**(101-68-8)**
p,p'-Diphenylmethane diisocyanate	**(101-68-8)**
Diphenylmethane diisocyanate (OSHA)	**(101-68-8)**
Diphenylmethane 4,4'-diisocyanate [UN 2489]	**(101-68-8)**
Diphenylmethane-4,4'-diisocyanate-trimellic anhydride-ethomid HT polymer	**(552-30-7)**
4,4'-Diphenylmethanedimaleimide	**(13676-54-5)**
Diphenylmethane, tetramethyldiamino-	**(101-61-1)**
Diphenylmethanol	**(91-01-0)**
Diphenylmethanone	**(119-61-9)**
Diphenylmethoxyarsine	**(24582-54-5)**
2-(Diphenylmethoxy)-N,N-dimethylethylamine	**(58-73-1)**
2-Diphenylmethoxy-N,N-dimethylethylamine hydrochloride	**(147-24-0)**
Diphenylmethyl alcohol	**(91-01-0)**
Diphenylmethylamine	**(552-82-9)**
N,N-Diphenylmethylamine	**(552-82-9)**
Diphenyl methyl bromide, Solid (DOT)	**(776-74-9)**
Diphenyl methyl bromide, Solution (DOT)	**(776-74-9)**
Diphenylmethyl bromide [UN 1770]	**(776-74-9)**
Diphenylmethyl chloride	**(90-99-3)**
Diphenylmethylcyanide	**(86-29-3)**
Diphenyl 4,4'-methylenebis(phenylcarbamate)	**(101-65-5)**
Diphenylmethyl ether	**(574-42-5)**
1-Diphenylmethyl-4-methylpiperazine	**(82-92-8)**
(+)-2,2-Diphenyl-3-methyl-4-morpholinobutyrylpyrrolidine	**(357-56-2)**
(-)-2,2-Diphenyl-3-methyl-4-morpholinobutyrylpyrrolidine	**(5666-11-5)**
Diphenyl methyl phosphate	**(115-89-9)**
Diphenylmethylsilanol	**(778-25-6)**
N,N-Diphenyl-N'-methylurea	**(13114-72-2)**
Diphenyl mixed with diphenyl oxide	**(8004-13-5)**
4,4-Diphenyl-6-morpholino-3-heptanone	**(467-84-5)**
Diphenylnitrosamin (German)	**(86-30-6)**
Diphenylnitrosamine	**(86-30-6)**
N,N-Diphenylnitrosamine	**(86-30-6)**
Diphenyl N-nitrosoamine	**(86-30-6)**
1,3-Diphenyloctane	**(79606-18-1)**
Diphenyl octyl phosphate	**(115-88-8)**
o-Diphenylol	**(90-43-7)**
2,2-Di(4-phenylol)propane	**(80-05-7)**
Diphenylolpropane	**(80-05-7)**
Diphenyl oxide	**(101-84-8)**
Diphenyl oxide-formaldehyde copolymer	**(26007-63-6)**
Diphenyl oxide-formaldehyde polymer	**(26007-63-6)**
Diphenyl pentachloride	**(25429-29-2)**
Diphenyl-p-phenylenediamine	**(74-31-7)**
N,N'-Diphenyl-p-phenylenediamine	**(74-31-7)**
1,2-Diphenyl-4-(2'-phenylsulfinethyl)-3,5-pyrazolidinedione	**(57-96-5)**
Diphenyl phosphate	**(838-85-7)**
Diphenylphosphinous chloride	**(1079-66-9)**
Diphenyl phosphonate	**(4712-55-4)**
Diphenyl phthalate	**(84-62-8)**
Diphenylpicrylhydrazyl free radical	**(1898-66-4)**
α,α-Diphenyl-1-piperidinebutanol	**(972-02-1)**
α,α-Diphenyl-2-piperidinemethanol	**(467-60-7)**
α,α-Diphenyl-1-piperidinepropanol	**(511-45-5)**
2,2-Diphenyl-4-(4-piperidino-4-carbamoylpiperidino)butyronitrile	**(302-41-0)**
N,N'-Diphenyl-1,3-propanediamine	**(104-69-8)**
1,3-Diphenyl-1,3-propanedione	**(120-46-7)**
1,3-Diphenyl-2-propanone	**(102-04-5)**
1,3-Diphenylpropanone	**(102-04-5)**
(E)-1,2-Diphenyl-1-propene	**(833-81-8)**
trans-1,2-Diphenyl-1-propene	**(833-81-8)**
trans-1,2-Diphenylpropene	**(833-81-8)**
1,3-Diphenyl-1-propen-3-one	**(94-41-7)**
1,3-Diphenylpropenone	**(94-41-7)**
3,4-Di-β-phenylpropionyl-1,2-5,6-dianhydro-dulcitol	**(57230-49-6)**
Di(phenylpropyl)ethylamine	**(150-59-4)**
N-(3,3-Diphenylpropyl)-α-methylphenaethylamin (German)	**(390-64-7)**
N-(3,3-Diphenylpropyl)-α-methylphenethylamine	**(390-64-7)**
Diphenylpyrazone	**(57-96-5)**
Diphenylstannium dichloride	**(1135-99-5)**
Diphenylstibene 2-ethylhexanoate	**(5035-58-5)**
Diphenylstibine 2-ethylhexanoate	**(5035-58-5)**
Diphenyl sulfide	**(139-66-2)**
Diphenyl sulfone	**(127-63-9)**

Diphenyl sulphide **(139-66-2)**
Diphenyl sulphone **(127-63-9)**
Diphenyl terephthalate **(1539-04-4)**
1,2-Diphenyltetrachloroethane **(13700-81-7)**
1,3-Diphenyl-1,1,3,3-tetramethyldisiloxane **(56-33-7)**
3,3',4,4'-Diphenyltetramine **(91-95-2)**
1,3-Diphenyl-2-thiapropane **(538-74-9)**
N,N'-Diphenylthiocarbamide **(102-08-9)**
s-Diphenylthiocarbamide **(102-08-9)**
Diphenyl thioether **(139-66-2)**
1,3-Diphenyl-2-thiourea **(102-08-9)**
1,3-Diphenylthiourea **(102-08-9)**
Diphenylthiourea **(102-08-9)**
N,N'-Diphenylthiourea **(102-08-9)**
sym-Diphenylthiourea **(102-08-9)**
Diphenyltin dichloride **(1135-99-5)**
Diphenyl p-tolyl phosphate **(78-31-9)**
Diphenyl tolyl phosphate **(26444-49-5)**
1,3-Diphenyltriazene **(136-35-6)**
2,2-Diphenyl-1,1,1-trichloroethane **(2971-22-4)**
Diphenyltrichloroethane **(2971-22-4)**
Diphenyltrichloroethane **(50-29-3)**
Diphenyl 2,4,6-trimethylphenyl phosphate **(73179-43-8)**
2,2-Diphenyl-1-(2,4,6-trinitrophenyl)hydrazyl **(1898-66-4)**
1,3-Diphenylurea **(102-07-8)**
N,N'-Diphenylurea **(102-07-8)**
N,N-Diphenylurea **(603-54-3)**
s-Diphenylurea **(102-07-8)**
sym-Diphenylurea **(102-07-8)**
Dipher **(12122-67-7)**
Diphergan **(58-33-3)**
Diphone **(80-08-0)**
Diphosgen **(503-38-8)**
Diphosgene [UN 1076] **(503-38-8)**
Diphosphonic acid, dimethyl-, diethyl ester (9CI) **(32288-17-8)**
Diphosphonic acid, (1-hydroxyethylidene)-, tetrasodium salt **(3794-83-0)**
Diphosphonic acid, (1-hydroxyethylidene)-, trisodium salt **(2666-14-0)**
Diphosphoramide, octamethyl- (9CI) **(152-16-9)**
Diphosphoric acid, bis(2-ethylhexyl) ester (9CI) **(26836-28-2)**
Diphosphoric acid, calcium salt (1:2) (9CI) **(7790-76-3)**
Diphosphoric acid, copper salt (9CI) **(10102-90-6)**
Diphosphoric acid, dioctyl ester **(26658-09-3)**
Diphosphoric acid, disodium salt **(7758-16-9)**
Diphosphoric acid, iron(3+) salt (3:4) (9CI) **(10058-44-3)**
Diphosphoric acid, iron(3+) sodium salt (1:1:1) **(10045-87-1)**
Diphosphoric acid, lead(2+) salt (1:2) (9CI) **(13453-66-2)**
Diphosphoric acid tetraethyl ester **(107-49-3)**
Diphosphoric acid, tetrapotassium salt (9CI) **(7320-34-5)**
Diphosphorus pentoxide **(1314-56-3)**
Diphosphorus trioxide **(1314-24-5)**
Diphyl **(8004-13-5)**
Diphylets **(51-63-8)**
Diphyllin **(479-18-5)**
Dipicolinate **(499-83-2)**
2,6-Dipicolinic acid **(499-83-2)**
Dipicolinic acid **(499-83-2)**
Dipicrylamine **(131-73-7)**
Dipicryloxamide **(29135-62-4)**
Dipidolor **(302-41-0)**
Dipigyl **(98-92-0)**
Dipikrylamin (Czech) **(131-73-7)**
Dipipanona (Spanish) **(467-83-4)**
Dipipanone **(467-83-4)**
Dipipanonum (Latin) **(467-83-4)**
2,4-Dipiperidino-6-methyltriazine **(26234-41-3)**
2,2',2'',2'''-(4,8-Dipiperidinopyrimido(5,4-d)pyrimidine-2,6-diyldinitrilo)tetraethanol **(58-32-2)**
1,3-Di-4-piperidylpropane **(16898-52-5)**
Dipirartril-Tropico **(67-68-5)**
Dipirin **(58-15-1)**
Dipiritramide **(302-41-0)**
Diploptene **(1615-91-4)**
Diplosal **(552-94-3)**
Dipo-Saft **(1538-09-6)**
Dipotassium chromate **(7789-00-6)**
Dipotassium dichloride **(7447-40-7)**
Dipotassium dichromate **(7778-50-9)**
Dipotassium endothall **(2164-07-0)**
Dipotassium hexafluorozirconate **(16923-95-8)**
Dipotassium hydrogen phosphate **(7758-11-4)**
Dipotassium monochromate **(7789-00-6)**
Dipotassium monohydrogen phosphate **(7758-11-4)**
Dipotassium monophosphate **(7758-11-4)**
Dipotassium persulfate **(7727-21-1)**
Dipotassium orthophosphate **(7758-11-4)**
Dipotassium phosphate **(7758-11-4)**
Dipotassium-O-phosphate **(7758-11-4)**
Dipotassium phosphate, dibasic **(7758-11-4)**
Dipotassium sulfite **(10117-38-1)**
Dipotassium zirconium hexafluoride **(16923-95-8)**
Dippel's Oil **(8001-85-2)**
Dipping Acid **(7664-93-9)**
Diprazine **(60-87-7)**
Diprenorfina (Spanish) **(14357-78-9)**
Diprenorphine **(14357-78-9)**
Diprenorphinum (Latin) **(14357-78-9)**
Diprivan **(2078-54-8)**
Diprofillin **(479-18-5)**
Diprofilline **(479-18-5)**
Diprol **(57-63-6)**
Dipron **(63-74-1)**
Dipropalin **(1918-08-7)**
Di-2-propenylamine **(124-02-7)**
Di-2-propenyl 1,3-benzenedicarboxylate **(1087-21-4)**
Di-2-propenyl carbonate **(15022-08-9)**
5,5-Di-2-propenyl-2,4,6(1H,3H,5H)-pyrimidinetrione **(52-43-7)**
Dipropetryn **(4147-51-7)**
Dipropetryne **(4147-51-7)**
Diprophyllin **(479-18-5)**
Diprophylline **(479-18-5)**
Dipropionate d'oestradiol (French) **(113-38-2)**
Dipropionato de estilbene (Spanish) **(130-80-3)**
p,p'-Dipropionoxy-trans-α,β-diethylstilbene **(130-80-3)**
Dipropionyl peroxide **(3248-28-0)**
Dipropionyl peroxide, Maximum conc. 28% in solution **(3248-28-0)**
1-(Di-N-propoxyphosphinothioylthiomethylcarbonyl-2-methyl-piperidine) **(24151-93-7)**
Dipropylacetate sodium **(1069-66-5)**
Di-n-propylacetic acid **(99-66-1)**
Dipropylacetic acid **(99-66-1)**
n-Dipropylacetic acid **(99-66-1)**
Di-n-propyl adipate **(106-19-4)**
Dipropyl adipate **(106-19-4)**
Dipropylaluminum hydride **(2036-15-9)**
Di-n-propylamine **(142-84-7)**
n-Dipropylamine **(142-84-7)**
Dipropylamine [UN 2383] **(142-84-7)**
Dipropylamine, 3,3'-diamino **(56-18-8)**
Dipropylamine, 3,3'-diamino-N-methyl **(105-83-9)**
Dipropylamine, 2,2'-dihydroxy-N-nitroso **(53609-64-6)**
Dipropylamine, N-ethyl-3,3'-diphenyl- **(150-59-4)**
Dipropylamine, N-nitroso **(621-64-7)**
4-(Dipropylamino)-3,5-dinitrobenzenesulfonamide **(19044-88-3)**
4-(Di-n-propylamino)-3,5-dinitro-1-trifluoromethylbenzene **(1582-09-8)**

4-((Dipropylamino)sulfonyl)benzoic acid	**(57-66-9)**
Dipropylcarbamothioic acid S-ethyl ester	**(759-94-4)**
Dipropyl-2,2'-dihydroxy-amine	**(110-97-4)**
N,N-Di-n-propyl-2,6-dinitro-4-methylaniline	**(1918-08-7)**
N,N-Dipropyl-2,6-dinitro-p-toluidine	**(1918-08-7)**
N,N-Dipropyl-2,6-dinitro-4-trifluormethylanilin (German)	**(1582-09-8)**
N,N-Di-n-propyl-2,6-dinitro-4-trifluoromethylaniline	**(1582-09-8)**
N,N-Dipropyl-2,6-dinitro-4-trifluoromethylaniline	**(1582-09-8)**
Dipropyl disulfide	**(629-19-6)**
Dipropyl dithiophosphate	**(2253-43-2)**
O,O-Dipropyl dithiophosphate	**(2253-43-2)**
O,O-Dipropyldithiophosphoric acid	**(2253-43-2)**
Dipropylene carbonate	**(108-32-7)**
α,α'-Dipropylenedinitrilodi-o-cresol	**(94-91-7)**
Dipropylene glycol	**(110-98-5)**
Dipropylene glycol	**(25265-71-8)**
Dipropylene glycol, butyl ether	**(29911-28-2)**
Dipropylene glycol diacrylate	**(57472-68-1)**
Dipropylene glycol dibenzoate	**(27138-31-4)**
Dipropylene glycol diglycidyl ether	**(41638-13-5)**
Dipropylene glycol methyl ether (ACGIH,OSHA)	**(34590-94-8)**
Dipropylene glycol monomethyl ether	**(34590-94-8)**
Dipropylene glycol, monomethyl ether	**(34590-94-8)**
Dipropylenetriamine	**(56-18-8)**
Dipropylenetriamine, N,N,N',N'-tetramethyl	**(6711-48-4)**
Dipropylenglykol (Czech)	**(110-98-5)**
Dipropylentriamin (German)	**(56-18-8)**
Di-n-propylessigsaure (German)	**(99-66-1)**
Dipropylester kyseliny pyridin-2,5-dikarboxylove (Czech)	**(136-45-8)**
Dipropyl ether [UN 2384]	**(111-43-3)**
Dipropylin	**(150-59-4)**
Dipropyline	**(150-59-4)**
Di-n-propyl isocinchomeronate	**(136-45-8)**
Di-propylisocinchomeronate	**(136-45-8)**
Dipropyl isocinchomeronate	**(136-45-8)**
Di-n-propyl-isocinchomeronate (German)	**(136-45-8)**
Dipropylketone (ACGIH,OSHA) [UN 2710]	**(123-19-3)**
Di-n-propyl maleate	**(83-59-0)**
Di-n-propyl maleate-isosafrole condensate	**(83-59-0)**
Dipropyl methane	**(142-82-5)**
Di-n-propyl 6,7-methylenedioxy-3-methyl-1,2,3,4-tetra-hydronaphthalene	**(83-59-0)**
Di-n-propyl-3-methyl-6,7-methylenedioxy-1,2,3,4-tetra-hydronaphthalene-1,2-dicarboxylate	**(83-59-0)**
O,O-Dipropyl S-2-methyl-piperidinocarbonyl-methyl phosphorodithioate	**(24151-93-7)**
O,O-Di-n-propyl-O-(4-methylthiophenyl)phosphate	**(7292-16-2)**
Di-n-propylnitrosamine	**(621-64-7)**
Dipropylnitrosamine	**(621-64-7)**
Dipropyl oxide	**(111-43-3)**
Di-n-propyl peroxydicarbonate	**(16066-38-9)**
Di-n-propyl peroxydicarbonate, Technically pure (DOT)	**(16066-38-9)**
Dipropyl phosphate	**(1804-93-9)**
Dipropyl phosphorodithioate	**(2253-43-2)**
O,O-Dipropyl phosphorodithioate	**(2253-43-2)**
Di-n-propylphosphorodithioic acid	**(2253-43-2)**
O,O-Dipropyl phosphorodithiotic acid	**(2253-43-2)**
Di-n-propyl phthalate	**(131-16-8)**
Dipropyl phthalate	**(131-16-8)**
Di-n-propyl 2,5-pyridinedicarboxylate	**(136-45-8)**
Dipropyl 2,5-pyridinedicarboxylate	**(136-45-8)**
Dipropyl pyridine-2,5-dicarboxylate	**(136-45-8)**
4-(Dipropylsulfamoyl)benzoic acid	**(57-66-9)**
p-(Dipropylsulfamoyl)benzoic acid	**(57-66-9)**
p-(Dipropylsulfamyl)benzoic acid	**(57-66-9)**
Di-n-propyl sulfide	**(111-47-7)**
Dipropyl sulfide	**(111-47-7)**
Dipropyl-5,6,7,8-tetrahydro-7-methylnaphtho(2,3-d)-1,3-dioxole-5,6-dicarboxylate	**(83-59-0)**
N,N-Dipropylthiocarbamic acid S-ethyl ester	**(759-94-4)**
Dipropylthiocarbamic acid S-propyl ester	**(1929-77-7)**
Dipropyl thioether	**(111-47-7)**
N,N-Dipropyl-4-trifluoromethyl-2,6-dinitroaniline	**(1582-09-8)**
Diprostron	**(113-38-2)**
Diprozin	**(60-87-7)**
Dipterax	**(52-68-6)**
Dipterex	**(52-68-6)**
Dipterex 50	**(52-68-6)**
Diptevur	**(52-68-6)**
Dipthal	**(2303-17-5)**
Dipyridamine	**(58-32-2)**
Dipyridamol	**(58-32-2)**
Dipyridamole	**(58-32-2)**
Dipyridan	**(58-32-2)**
2,3'-Dipyridine	**(581-50-0)**
4,4'-Dipyridine	**(553-26-4)**
Dipyrido(1,2-a:2',1'-c)pyrazinediium, 6,7-dihydro	**(2764-72-9)**
Dipyrido(1,2-a;2',1'-c)pyrazinediium, 6,7-dihydro-, dibromide	**(85-00-7)**
Dipyrido(1,2-a:2',1'-c)pyrazinediium, 6,7-dihydro-, dibromide, Mixt. with Cutrine Plus (9CI)	**(66630-68-0)**
Dipyrido(1,2-a;2',1'-c)pyrazinediium, 6,7-dihydro-, dichloride	**(4032-26-2)**
Dipyrido(1,2-a:3',2'-d)imidazol-2-amine	**(67730-10-3)**
Dipyrido(1,2-a:3',2'-d)imidazol-2-amine, 6-methyl-	**(67730-11-4)**
Dipyrido(1,2-a:3',2'-d)imidazole, 2-amino	**(67730-10-3)**
Dipyrido(1,2-a:3',2'-d)imidazole, 2-amino-6-methyl	**(67730-11-4)**
2,2'-Dipyridyl	**(366-18-7)**
2,3'-Dipyridyl	**(581-50-0)**
4,4'-Dipyridyl	**(553-26-4)**
4,4-Dipyridyl	**(553-26-4)**
α,α'-Dipyridyl	**(366-18-7)**
α,β-Dipyridyl	**(581-50-0)**
γ,γ'-Dipyridyl	**(553-26-4)**
2,2'-Dipyridylamine	**(1202-34-2)**
1,2-Di-3-pyridyl-2-methyl-1-propanone	**(54-36-4)**
Dipyrin	**(58-15-1)**
Dipyrine	**(58-15-1)**
Dipyudamine	**(58-32-2)**
Diquat	**(2764-72-9)**
Diquat (ACGIH,OSHA)	**(85-00-7)**
Diquat dibromide	**(85-00-7)**
Diquat dichloride	**(4032-26-2)**
2,2'-Diquinolyl	**(119-91-5)**
Diralgan	**(23779-99-9)**
Dirax	**(86-88-4)**
Dirbomoacetileno (Spanish)	**(624-61-3)**
Direct Artificial Silk Black G	**(6428-31-5)**
Direct Black (9CI)	**(56449-31-1)**
Direct Black 3	**(1937-37-7)**
Direct Black 19	**(6428-31-5)**
Direct Black 38	**(1937-37-7)**
Direct Black A	**(1937-37-7)**
Direct Black BH	**(2429-73-4)**
Direct Black BRN	**(1937-37-7)**
Direct Black CX	**(1937-37-7)**
Direct Black CXR	**(1937-37-7)**
Direct Black E	**(1937-37-7)**
Direct Black EW	**(1937-37-7)**
Direct Black EX	**(1937-37-7)**
Direct Black FR	**(1937-37-7)**
Direct Black GAC	**(1937-37-7)**
Direct Black GW	**(1937-37-7)**
Direct Black GX	**(1937-37-7)**
Direct Black GXR	**(1937-37-7)**
Direct Black Green	**(3626-28-6)**

Direct Black JET (1937-37-7)
Direct Black K (2429-83-6)
Direct Black Meta (1937-37-7)
Direct Black Methyl (1937-37-7)
Direct Black N (1937-37-7)
Direct Black R (2429-83-6)
Direct Black RX (1937-37-7)
Direct Black SD (1937-37-7)
Direct Black WS (1937-37-7)
Direct Black Z (1937-37-7)
Direct Blue 1 (2610-05-1)
Direct Blue 2 (2429-73-4)
Direct Blue 6 (2602-46-2)
Direct Blue 14 (72-57-1)
Direct Blue 15 (2429-74-5)
Direct Blue 26 (7082-31-7)
Direct Blue 53 (314-13-6)
Direct Blue 218 (28407-37-6)
Direct Blue A (2602-46-2)
Direct Blue 2B (2602-46-2)
Direct Blue 3B (72-57-1)
Direct Blue 6B (2610-05-1)
Direct Blue BB (2602-46-2)
Direct Blue 6BS (2610-05-1)
Direct Blue 3BX (72-57-1)
Direct Blue Black BH (2429-73-4)
Direct Blue D3B (72-57-1)
Direct Blue FF (2610-05-1)
Direct Blue FFN (2610-05-1)
Direct Blue FFN (72-57-1)
Direct Blue 10G (2429-74-5)
Direct Blue GS (2602-46-2)
Direct Blue H3G (72-57-1)
Direct Blue HH (2429-74-5)
Direct Blue K (2602-46-2)
Direct Blue M3B (72-57-1)
Direct Bright Blue (2610-05-1)
Direct Brilliant Blue FF (2610-05-1)
Direct Brilliant Blue MFF (2610-05-1)
Direct Brilliant Sky Blue 6B (2610-05-1)
Direct Brilliant Violet 2R (2586-60-9)
Direct Brown 1:2 (2586-58-5)
Direct Brown 1A (2586-58-5)
Direct Brown 2 (2429-82-5)
Direct Brown 31 (2429-81-4)
Direct Brown 95 (16071-86-6)
Direct Brown 3B (2429-81-4)
Direct Brown B (2429-81-4)
Direct Brown BR (108-45-2)
Direct Brown BRL (16071-86-6)
Direct Brown BS (2429-81-4)
Direct Brown BSB (2429-81-4)
Direct Brown 5C (2586-58-5)
Direct Brown CGN (2586-58-5)
Direct Brown FS (2429-81-4)
Direct Brown 5G (2586-58-5)
Direct Brown GG (108-45-2)
Direct Brown 2GS (2586-58-5)
Direct Brown KX (2429-82-5)
Direct Brown 3RB (2429-82-5)
Direct Brown TRB (2429-81-4)
Direct Chrome Black Blue 2B (7082-31-7)
Direct Chrome Black Blue B (7082-31-7)
Direct Chrome Dark Blue 2B (7082-31-7)
Direct Dark Blue BH (2429-73-4)
Direct Dark Green A (3626-28-6)
Direct Dark Green B (3626-28-6)
Direct Dark Green BF (3626-28-6)
Direct Dark Green BG (3626-28-6)
Direct Dark Green MB (3626-28-6)
Direct Dark Green S (3626-28-6)
Direct Dark Green Supra (3626-28-6)
Direct Dark Green WS (3626-28-6)
Direct Deep Black E (1937-37-7)
Direct Deep Black EAC (1937-37-7)
Direct Deep Black EA-CF (1937-37-7)
Direct Deep Black E Extra (1937-37-7)
Direct Deep Black EW (1937-37-7)
Direct Deep Black EX (1937-37-7)
Direct Deep Black RW (2429-83-6)
Direct Deep Green A (3626-28-6)
Direct Diazo Black (2429-73-4)
Direct Diazo Black C (2429-73-4)
Direct Diazo Black N (2429-73-4)
Direct Diazo Black RW (2429-83-6)
Direct Diazo Black S (2429-73-4)
Direct Fast Black G (6428-31-5)
Direct Fast Black GU (6428-31-5)
Direct Fast Black SA (6428-31-5)
Direct Fast Blue Black MB (7082-31-7)
Direct Fast Brown BP (2429-81-4)
Direct Fast Brown B (Polish) (3476-90-2)
Direct Fast Brown BRL (16071-86-6)
Direct Fast Brown LMR (16071-86-6)
Direct Fast Brown M (2429-82-5)
Direct Fast Brown TSN (2429-81-4)
Direct Fast Brown TWC (2429-81-4)
Direct Fast Purpurine 8B (6548-29-4)
Direct Fast Red 5B (2610-11-9)
Direct Fast Red 8BL (2610-11-9)
Direct Fast Red 2S (2610-11-9)
Direct Fast Scarlet 3B (6358-29-8)
Direct Fast Violet MN (2586-60-9)
Direct Fast Violet N (2586-60-9)
Direct Green WAC (3626-28-6)
Direct Light Brown BRS (16071-86-6)
Direct Light Red 4B (2610-11-9)
Direct Light Red 8B (2610-11-9)
Direct Light Red M 8BL (2610-11-9)
Direct Lightfast Red 2S (2610-11-9)
Direct Navy Blue BH (2429-73-4)
Direct Orange G (1325-37-7)
Direct Pure Blue (2429-74-5)
Direct Pure Blue 6B (2610-05-1)
Direct Pure Blue FF (2610-05-1)
Direct Pure Blue M (2429-74-5)
Direct Purpurine 4B (992-59-6)
Direct Purpurine M4B (992-59-6)
Direct Rayon Black KSG (6428-31-5)
Direct Red 2 (992-59-6)
Direct Red 23 (3441-14-3)
Direct Red 28 (573-58-0)
Direct Red 39 (6358-29-8)
Direct Red 4A (992-59-6)
Direct Red 4B (992-59-6)
Direct Red 8BS (6548-29-4)
Direct Red 80 (2610-10-8)
Direct Red 81 (2610-11-9)
Direct Red C (573-58-0)
Direct Red DCB (992-59-6)
Direct Red DC-CF (573-58-0)
Direct Red K (573-58-0)
Direct Scarlet 3BS (6358-29-8)
Direct Sky Blue A (2429-74-5)

Direct Sky Blue 6B **(2610-05-1)**
Direct Sky Blue 6BS **(2610-05-1)**
Direct Sky Blue FF **(2610-05-1)**
Direct Sky Blue GS **(2610-05-1)**
Direct Sky Blue Green Shade **(2610-05-1)**
Direct Supra Light Brown ML **(16071-86-6)**
Direct Violet 22 **(6426-67-1)**
Direct Violet BS **(6426-67-1)**
Direct Violet C **(2586-60-9)**
Direct Violet FR **(2586-60-9)**
Direct Violet N **(2586-60-9)**
Direct Violet R **(2586-60-9)**
Direct Yellow 12 **(2870-32-8)**
Direct Yellow 27 **(10190-68-8)**
Direct Yellow 4 Dye **(91-34-9)**
Direct Yellow F **(1325-37-7)**
Directakol Blue 3BL **(72-57-1)**
Directblau 3B **(72-57-1)**
Direktan **(59-67-6)**
Direma **(58-93-5)**
Diren **(396-01-0)**
Diresul Black P **(1326-82-5)**
Diresul Blue 8RS **(1327-57-7)**
Diresul Blue 9RS **(1327-57-7)**
Diresul Navy Blue GIS **(1327-57-7)**
Direx 4L **(330-54-1)**
Direxiode **(83-73-8)**
Direz **(101-05-3)**
Diridone **(136-40-3)**
Diridone **(94-78-0)**
Dirimal **(19044-88-3)**
Dirochrome Dark Blue B **(7082-31-7)**
Dirox **(103-90-2)**
Disalcid **(552-94-3)**
Disalicylalpropylenediimine **(94-91-7)**
Disalicylic acid **(552-94-3)**
N,N'-Disalicylidene-1,2-diaminopropane **(94-91-7)**
N,N'-Disalicylidene-1,2-propanediamine **(94-91-7)**
Disalunil **(58-93-5)**
Disalyl **(552-94-3)**
Disan **(741-58-2)**
Disanyl **(87-12-7)**
Disapol M **(9011-14-7)**
Disatabs Tabs **(68-26-8)**
Discolite **(149-44-0)**
Discon **(12122-67-7)**
Diselenide, diethyl **(628-39-7)**
Disetil **(97-77-8)**
Disflamoll DPK **(26444-49-5)**
Disflamoll DPO **(115-88-8)**
Disflamoll TKP **(1330-78-5)**
Disflamoll TOF **(78-42-2)**
Disilane, 1,2-dichloro-1,1,2,2-tetramethyl- (9CI) **(4342-61-4)**
Disilane, 1,1,2,2-tetrachloro-1,2-dimethyl- (9CI) **(4518-98-3)**
Disilane, 1,1,2-trichloro-1,2,2-trimethyl- (9CI) **(13528-88-6)**
Disilazane, 1,1,1,3,3,3-hexamethyl **(999-97-3)**
Disiloxane, 1,3-bis(3-(2,3-epoxypropoxy)propyl)-1,1,3,3-tetramethyl- (8CI) **(126-80-7)**
Disiloxane, 1-chloro-1,3-dimethyl-1,3,3-triphenyl- (9CI) **(53634-34-7)**
Disiloxane, 1,3-diethenyl-1,1,3,3-tetramethyl- (9CI) **(2627-95-4)**
Disiloxane, 1,3-dimethyl-1,1,3,3-tetraphenyl- (9CI) **(807-28-3)**
Disiloxane, 1,3-diphenyl-1,1,3,3-tetramethyl **(56-33-7)**
Disiloxane, hexamethyl **(107-46-0)**
Disiloxane, pentamethylphenyl- (8CI,9CI) **(14920-92-4)**
Disiloxane, 1,1,3,3-tetramethyl **(3277-26-7)**
Disiloxane, 1,1,3,3-tetramethyl-1,3-bis(3-(oxiranylmethoxy)-propyl)- (9CI) **(126-80-7)**
Disiloxane, 1,1,3-trimethyl-1,3,3-triphenyl- (8CI,9CI) **(14920-93-5)**
Disilver oxalate **(533-51-7)**
Disilyn **(121-54-0)**
Disiquonium Chloride **(68959-20-6)**
Disodium-1,8-dihydroxynaphthalene-3,6-disulfonate **(129-96-4)**
Disodium cupric EDTA **(14025-15-1)**
Disodium EDTA **(139-33-3)**
Disodium Eosin **(17372-87-1)**
Disodium adipate **(7486-38-6)**
Disodium anthraquinone-1,5-disulfonate **(853-35-0)**
Disodium arsenate **(7778-43-0)**
Disodium arsenate, heptahydrate **(10048-95-0)**
Disodium arsenic acid **(7778-43-0)**
Disodium 4,4'-bis((4-anilino-6-morpholino-1,3,5-triazin-2-yl)amino)stilbene-2,2'-disulfonate **(16090-02-1)**
Disodium 4,4'-bis(2-sulfostyryl)biphenyl **(27344-41-8)**
Disodium bromosulfophthalein **(71-67-0)**
Disodium calcium ethylenediaminetetraacetate **(62-33-9)**
Disodium carbonate **(497-19-8)**
Disodium chromate **(7775-11-3)**
Disodium chromoglycate **(15826-37-6)**
Disodium chromotropate **(129-96-4)**
Disodium citrate **(144-33-2)**
Disodium cromoglicate **(15826-37-6)**
Disodium cromoglycate **(15826-37-6)**
Disodium cyanodithioimidocarbonate **(138-93-2)**
Disodium cyanurate **(36452-21-8)**
Disodium diacid ethylenediaminetetraacetate **(139-33-3)**
Disodium dichromate **(10588-01-9)**
Disodium difluoride **(7681-49-4)**
Disodium dihydrogen(ethylenedinitrilo)tetraacetate **(139-33-3)**
Disodium dihydrogen ethylenediaminetetraacetate **(139-33-3)**
Disodium dihydrogen (1-hydroxyethylidene)diphosphonate **(7414-83-7)**
Disodium dihydrogen hypophosphate **(7782-95-8)**
Disodium dihydrogen pyrophosphate **(7758-16-9)**
Disodium dihydrogen subphosphate **(7782-95-8)**
Disodium dihydroxyethyl ethylenediaminediacetate **(38011-25-5)**
Disodium 1,8-dihydroxylnaphthalene-3,6-disulfonate **(129-96-4)**
Disodium (2,4-dimethylphenylazo)-2-hydroxynaphthalene-3,6-disulfonate **(3761-53-3)**
Disodium (2,4-dimethylphenylazo)-2-hydroxynaphthalene-3,6-disulphonate **(3761-53-3)**
Disodium dioxide **(1313-60-6)**
Disodium diphosphate **(7758-16-9)**
Disodium 3,3'-(dodecylimino)bis(propionate) **(3655-00-3)**
Disodium 4-dodecyl-2,4'-oxydibenzenesulfonate **(7575-62-4)**
Disodium edathamil **(139-33-3)**
Disodium edetate **(139-33-3)**
Disodium 3,6-endoxohexahydrophthalate **(129-67-9)**
Disodium 3,6-epoxycyclohexane-1,2-dicarboxylate **(129-67-9)**
Disodium ethanol-1,1-diphosphonate **(7414-83-7)**
Disodium 2,2'-(1,2-ethenediyl)bis(5-nitrobenzenesulfonate) **(3709-43-1)**
Disodium ethydronate **(7414-83-7)**
Disodium-2,2'-methylenebis(4-chlorophenate) **(22232-25-3)**
Disodium ethylene-1,2-bisdithiocarbamate **(142-59-6)**
Disodium ethylenebis(dithiocarbamate) **(142-59-6)**
Disodium ethylenediaminetetraacetate **(139-33-3)**
Disodium ethylenediaminetetraacetic acid **(139-33-3)**
Disodium (ethylenedinitrilo)tetraacetate **(139-33-3)**
Disodium ((ethylenedinitrilo)tetraacetato)manganese **(15375-84-5)**
Disodium (ethylenedinitrilo)tetraacetic acid **(139-33-3)**
Disodium etidronate **(7414-83-7)**
Disodium fluorophosphate **(10163-15-2)**
Disodium hexafluorosilicate **(16893-85-9)**
Disodium hexafluorosilicate (2-) **(16893-85-9)**
Disodium hydrogen arsenate **(7778-43-0)**
Disodium hydrogen orthoarsenate **(7778-43-0)**

Disodium hydrogen citrate (144-33-2)
Disodium hydrogen phosphate (7558-79-4)
Disodium 1-hydroxyethylidene phosphonate (7414-83-7)
Disodium N-(2-hydroxyethyl)iminodiacetate (135-37-5)
Disodium 6-hydroxy-3-oxo-9-xanthene-o-benzoate (518-47-8)
Disodium 5,5'-((2-hydroxytrimethylene)dioxy)-bis(4-oxo-4H-1-benzopyran-2-carboxylate) (15826-37-6)
Disodium 3-hydroxy-4-((2,4,5-trimethylphenyl)azo)-2,7-naphthalenedisulfonate (3564-09-8)
Disodium 3-hydroxy-4-((2,4,5-trimethylphenyl)azo)-2,7-naphthalenedisulfonic acid (3564-09-8)
Disodium 3-hydroxy-4-((2,4,5-trimethylphenyl)azo)-2,7-naphthalenedisulphonate (3564-09-8)
Disodium 3-hydroxy-4-((2,4,5-trimethylphenyl)azo)-2,7-naphthalenedisulphonic acid (3564-09-8)
Disodium hypophosphate (7782-95-8)
Disodium hypophosphorate (7782-95-8)
Disodium iminodiacetate (928-72-3)
Disodium indigo-5,5-disulfonate (860-22-0)
Disodium isodecyl sulfosuccinate (37294-49-8)
Disodium lauriminodipropionate (3655-00-3)
Disodium N-lauryl-β,β'-iminodipropionate (3655-00-3)
Disodium N-lauryl-β-iminodipropionate (3655-00-3)
Disodium β,β'-(laurylimino)dipropionate (3655-00-3)
Disodium lauryl sulfosuccinate (26838-05-1)
Disodium lauryl sulfosuccinate (36409-57-1)
Disodium magnesium ethylenediaminetetraacetate (14402-88-1)
Disodium manganese EDTA (15375-84-5)
Disodium manganese ethylenediaminetetraacetate (15375-84-5)
Disodium metasilicate (6834-92-0)
Disodium methanearsenate (144-21-8)
Disodium methanearsonate (144-21-8)
Disodium methoxyimidodisulfurate (63450-73-7)
Disodium methylarsenate (144-21-8)
Disodium methylarsonate (144-21-8)
Disodium O-methylhydroxylamine-N,N-disulfonate (63450-73-7)
Disodium molybdate (7631-95-0)
Disodium monofluorophosphate (10163-15-2)
Disodium monohydrogen arsenate (7778-43-0)
Disodium monohydrogen citrate (144-33-2)
Disodium monohydrogen phosphate (7558-79-4)
Disodium monomethylarsonate (144-21-8)
Disodium monosilicate (6834-92-0)
Disodium monoxide (12401-86-4)
Disodium naphthalene 1,8-dihydroxy-3,6-disulfonate (129-96-4)
Disodium nitrilotriacetate (15467-20-6)
Disodium nitrilotriacetic acid monohydrate (23255-03-0)
Disodium octaborate tetrahydrate (12280-03-4)
Disodium (2-oleoylamido-1-methylethyl)sulfosuccinate (67815-88-7)
Disodium 7-oxabicyclo(2.2.1)heptane-2,3-dicarboxylate (129-67-9)
Disodium oxide (12401-86-4)
Disodium 2,2'-oxybis(4-dodecylbenzenesulfonate) (5136-51-6)
Disodium peroxide (1313-60-6)
Disodium orthophosphate (7558-79-4)
Disodium phosphate (7558-79-4)
Disodium phosphoric acid (7558-79-4)
Disodium phosphorofluoridate (10163-15-2)
Disodium pyrophosphate (7758-16-9)
Disodium pyrosulfite (7681-57-4)
Disodium salt of EDTA (139-33-3)
Disodium salt of endothall (129-67-9)
Disodium salt of 1-indigotin-S,S'-disulphonic acid (860-22-0)
Disodium salt of 7-oxabicyclo(2.2.1)heptane-2,3-dicarboxylic acid (129-67-9)
Disodium salt of 2-(4-sulpho-1-naphthylazo)-1-naphthol-4-sulphonic acid (3567-69-9)
Disodium salt of 1-p-sulphophenylazo-2-naphthol-6-sulphonic acid (2783-94-0)
Disodium salt of 1-(2,4-xylylazo)-2-naphthol-3,6-disulfonic acid (3761-53-3)
Disodium salt of 1-(2,4-xylylazo)-2-naphthol-3,6-disulphonic acid (3761-53-3)
Disodium selenate (13410-01-0)
Disodium selenite (10102-18-8)
Disodium sequestrene (139-33-3)
Disodium silicofluoride (16893-85-9)
Disodium stearyl sulfosuccinamate (14481-60-8)
Disodium sulfate (7757-82-6)
Disodium sulfite (7757-83-7)
Disodium 2-(4-sulfo-1-naphthylazo)-1-naphthol-4-sulfonate (3567-69-9)
Disodium 2-(4-sulpho-1-naphthylazo)-1-naphthol-4-sulphonate (3567-69-9)
Disodium L-(+)-tartrate (868-18-8)
Disodium tartrate (868-18-8)
Disodium tetraborate (1330-43-4)
Disodium tetracemate (139-33-3)
Disodium tetracyanozincate (15333-24-1)
Disodium 1-tetradecenedisulfonate (68003-17-8)
Disodium thiosulfate (7772-98-7)
Disodium 1,3,5-triazine-2,4,6(1H,3H,5H)-trione (36452-21-8)
Disodium versenate (139-33-3)
Disodium versene (139-33-3)
Disofen (305-85-1)
Disolfuro di tetrametiltiourame (Italian) (137-26-8)
Disomar (144-21-8)
Disomear (144-21-8)
Disophenol (305-85-1)
Disoquin (83-73-8)
Dispadol (50-13-5)
Dispal (1344-28-1)
Dispal Alumina (1344-28-1)
Dispal M (1344-28-1)
Dispamil (61-25-6)
Disparlure (29804-22-6)
Dispermine (110-85-0)
Disperse Blue 1 (2475-45-8)
Disperse Blue 3 (2475-46-9)
Disperse Blue 3 (86722-66-9)
Disperse Blue 14 (2475-44-7)
Disperse Blue 56 (12217-79-7)
Disperse Blue 72 (81-48-1)
Disperse Blue 73 (12222-78-5)
Disperse Blue K (2475-46-9)
Disperse Blue K (86722-66-9)
Disperse Blue No 1 (2475-45-8)
Disperse Brilliant Pink (2872-48-2)
Disperse Brilliant Rose (2872-48-2)
Disperse Dye Fast Yellow 4K (6300-37-4)
Disperse Fast Pink B (116-85-8)
Disperse Fast Violet B (1220-94-6)
Disperse Fast Yellow G (2832-40-8)
Disperse Fast Yellow 2K (119-15-3)
Disperse Fast Yellow 4K (6300-37-4)
Disperse MB-61 (96-69-5)
Disperse-Oil (9CI) (53763-23-8)
Disperse Orange (82-28-0)
Disperse Orange 3 (730-40-5)
Disperse Polyester Pink 2S (17418-58-5)
Disperse Red 9 (82-38-2)
Disperse Red 11 (2872-48-2)
Disperse Red 15 (116-85-8)
Disperse Red 25 (116-85-8)
Disperse Red 60 (17418-58-5)
Disperse Violet 4K (41541-13-3)
Disperse Violet K (128-95-0)
Disperse Violet 2S (82-33-7)
Disperse Violet 4S (1220-94-6)
Disperse Yellow 3 (2832-40-8)
Disperse Yellow 7 (6300-37-4)

Disperse Yellow G (2832-40-8)
Disperse Yellow R (119-15-3)
Disperse Yellow Stable 2K (119-15-3)
Disperse Yellow Z (2832-40-8)
Dispersed Blue 12195 (3844-45-9)
Dispersed Orange 11348 (2783-94-0)
Dispersed Violet 12197 (1694-09-3)
Dispersed Yellow 12116 (2783-94-0)
Dispersive Blue K (2475-46-9)
Dispersive Blue K (86722-66-9)
Dispersive Rubin Polyether (16889-10-4)
Dispersive Violet K (128-95-0)
Dispersol Blue B-R (12217-79-7)
Dispersol Fast Yellow A (119-15-3)
Dispersol Fast Yellow G (2832-40-8)
Dispersol Orange D-G (116-85-8)
Dispersol Printing Yellow A (119-15-3)
Dispersol Printing Yellow G (2832-40-8)
Dispersol Red B 2B (17418-58-5)
Dispersol Red B 3B (2872-48-2)
Dispersol Red PP (85-83-6)
Dispersol Violet B (1220-94-6)
Dispersol Yellow A-G (2832-40-8)
Dispersol Yellow B-A (119-15-3)
Dispersol Yellow PP (842-07-9)
Dispex C40 (9003-01-4)
Disposlips (34149-92-3)
Disproportionated tall oil fatty acid (61790-12-3)
Disrupt (16974-11-1)
Dissolvan 4411 (9003-11-6)
Dissolvant APV (111-46-6)
Distakaps V-K (132-98-9)
Distannathiane, dibutyldithioxo- (9CI) (15666-29-2)
Distannathiane, dimethyldithioxo- (9CI) (33397-79-4)
Distannoxane, hexabutyl (56-35-9)
Distannoxane, hexaethyl (1112-63-6)
Distannoxane, hexakis(β,β-dimethylphenethyl) (13356-08-6)
Distannthiane, 1,3-dibutyl-1,3-dithioxo- (15666-29-2)
Distaquaine (6130-64-9)
Distaquaine V (87-08-1)
Distaval (50-35-1)
Distaxal (50-35-1)
Distearin (31566-31-1)
N,N'-Distearoylethylenediamine (110-30-5)
Distearyl acid phosphate (3037-89-6)
Distearyl dimethylammonium chloride (107-64-2)
Distearyl peroxydicarbonate (52326-66-6)
Distearyl peroxydicarbonate (Not more than 85% with stearyl alcohol) (52326-66-6)
Distearylperoxydicarbonate, With 15% stearyl alcohol (52326-66-6)
Distearyl phthalate (14117-96-5)
Distearyl 3,3'-thiodipropionate (693-36-7)
Distearyl β,β'-thiodipropionate (693-36-7)
Distearyl β-thiodipropionate (693-36-7)
Distearyl thiodipropionate (693-36-7)
Disteryl (18472-51-0)
Disthene (1302-76-7)
Distilbene (130-80-3)
Distilbene (56-53-1)
Distillate Fuel Oils, Light (64742-31-0)
Distillates (Coal), Solvent-refining (SRC), Heavy (68410-07-1)
Distillates (Coal), Solvent-refining (SRC), Recycle (68410-08-2)
Distillates, Coal, Solvent-refining (SRC), Wash (68410-09-3)
Distillates (Coal), Solvent-refining (SRC), Middle (68911-57-9)
Distillates, Coal Tar (65996-92-1)
Distillates, Coal Tar, Upper (65996-91-0)
Distillates (Petroleum), Acid-treated heavy naphthenic (9CI) (64742-18-3)
Distillates (Petroleum), Acid-treated heavy paraffinic (9CI) (64742-20-7)
Distillates (Petroleum), Acid-treated light naphthenic (9CI) (64742-19-4)
Distillates (Petroleum), Acid-treated light paraffinic (9CI) (64742-21-8)
Distillates, Petroleum, C12-30-arom. (68602-80-2)
Distillates, Petroleum, Catalytic reformer fractionator residue, Low-boiling (68477-31-6)
Distillates, Petroleum, Chemically neutralized light (64742-31-0)
Distillates, Petroleum, Chemically neutralized heavy naphthenic (64742-34-3)
Distillates, Petroleum, Chemically neutralized heavy paraffinic (64742-27-4)
Distillates, Petroleum, Chemically neutralized light naphthenic (64742-35-4)
Distillates, Petroleum, Chemically neutralized light paraffinic (64742-28-5)
Distillates, Petroleum, Chemically neutralized middle (64742-30-9)
Distillates, Petroleum, Clay-treated heavy paraffinic (64742-36-5)
Distillates, Petroleum, Clay-treated light paraffinic (64742-37-6)
Distillates, Petroleum, Crude oil (68410-00-4)
Distillates (Petroleum), Heavy catalytic cracked (64741-61-3)
Distillates (Petroleum), Heavy naphthenic (9CI) (64741-53-3)
Distillates (Petroleum), Heavy paraffinic (9CI) (64741-51-1)
Distillates, Petroleum, Hydrodesulfurized middle (64742-80-9)
Distillates (Petroleum), Hydrotreated heavy naphthenic (9CI) (64742-52-5)
Distillates (Petroleum), Hydrotreated heavy paraffinic (9CI) (64742-54-7)
Distillates (Petroleum), Hydrotreated light naphthenic (9CI) (64742-53-6)
Distillates (Petroleum), Hydrotreated light paraffinic (9CI) (64742-55-8)
Distillates (Petroleum), Hydrotreated middle (64742-46-7)
Distillates, Petroleum, Intermediate catalytic cracked (64741-60-2)
Distillates, Petroleum, Light thermal cracked (64741-82-8)
Distillates (Petroleum), Light catalytic cracked (64741-59-9)
Distillates (Petroleum), Light naphthenic (99CI) (64741-52-2)
Distillates (Petroleum), Light paraffinic (9CI) (64741-50-0)
Distillates, Petroleum, Naphtha-Raffinate Pyrolyzate-derived, Gasoline-blending (68425-29-6)
Distillates (Petroleum), Solvent-dewaxed heavy naphthenic (9CI) (64742-63-8)
Distillates (Petroleum), Solvent-dewaxed heavy paraffinic (9CI) (64742-65-0)
Distillates (Petroleum), Solvent-dewaxed light naphthenic (9CI) (64742-64-9)
Distillates (Petroleum), Solvent-dewaxed light paraffinic (9CI) (64742-56-9)
Distillates (Petroleum), Solvent-refined heavy naphthenic (9CI) (64741-96-4)
Distillates (Petroleum), Solvent-refined heavy paraffinic (9CI) (64741-88-4)
Distillates (Petroleum), Solvent-refined light naphthenic (9CI) (64741-97-5)
Distillates (Petroleum), Solvent-refined light paraffinic (9CI) (64741-89-5)
Distillates, Petroleum, Solvent-refined middle (64741-91-9)
Distillates, Petroleum, Steam-cracked petroleum distillates, C5-18 fraction (68477-58-7)
Distillates, Petroleum, Straight-run light (68410-05-9)
Distilled Mustard (505-60-2)
Distivit (B12 peptide) (68-19-9)
Distobram (32986-56-4)
Distokal (67-72-1)
Distol (64-02-8)
Distol 8 (64-02-8)
Distopan (67-72-1)
Distopin (67-72-1)
Distoval (50-35-1)
Distraneurin (533-45-9)
Distylin (480-18-2)
Disuccinic acid peroxide, Maximum concentration 72% (123-23-9)
Disuccinic acid peroxide, Technically pure (DOT) (123-23-9)
Disul (136-78-7)
Disul-Na (136-78-7)
Disul-Sodium (136-78-7)
1,3-Disulfamyl-4,5-dichlorobenzene (120-97-8)
Disulfan (97-77-8)
Disulfaton (298-04-4)
Disulfatozirconic acid (14644-61-2)
Disulfide, allyl propyl (2179-59-1)
Disulfide, bis(2-chlorooctyl) (9CI) (70776-26-0)
Disulfide, bis(4-chlorophenyl) (9CI) (1142-19-4)
Disulfide, bis(p-chlorophenyl) (8CI) (1142-19-4)

Disulfide, bis(dibutylthiocarbamoyl) (1634-02-2)
Disulfide, bis(diethylthiocarbamoyl) (97-77-8)
Disulfide, bis(1,1-dimethylethyl) (110-06-5)
Disulfide, bis(dimethylthiocarbamoyl) (137-26-8)
Disulfide, bis(dodecylphenyl) (8CI) (28986-55-2)
Disulfide, bis(2-nitrophenyl) (9CI) (1155-00-6)
Disulfide, bis(o-nitrophenyl) (8CI) (1155-00-6)
Disulfide, bis(thiocarbamoyl) (504-90-5)
Disulfide, dibutyl (9CI) (629-45-8)
Disulfide, dicyclohexyl (9CI) (2550-40-5)
Disulfide, di-tert-dodecyl (9CI) (27458-90-8)
Disulfide, diethyl (110-81-6)
Disulfide, dimethyl (624-92-0)
Disulfide, dimorpholino- (103-34-4)
Disulfide, dipentyl (112-51-6)
Disulfide diphenyl (882-33-7)
Disulfide, di-2-propenyl (9CI) (2179-57-9)
Disulfide, dipropyl (9CI) (629-19-6)
Disulfide, ethenyl ethyl (9CI) (24298-49-5)
Disulfide, ethyl propyl (8CI,9CI) (30453-31-7)
Disulfide, ethyl vinyl (8CI) (24298-49-5)
Disulfide, methyl 2-methyl-1-(methylthio)butyl (9CI) (69078-83-7)
Disulfide, methyl (methylthio)methyl (9CI) (42474-44-2)
Disulfide, methyl propyl (2179-60-4)
Disulfine Blue VN (129-17-9)
Disulfiram (ACGIH,OSHA) (97-77-8)
4,8-Disulfo-2-naphthalamine (131-27-1)
1,8-Disulfoanthracene (61736-92-3)
1,5-Disulfoanthraquinone (117-14-6)
1,8-Disulfoanthraquinone (82-48-4)
2,2'-Disulfobenzidine (117-61-3)
3,5-Disulfocatechol disodium salt (149-45-1)
Disulfoton (ACGIH,DOT,OSHA) (298-04-4)
Disulfoton disulide (2497-07-6)
Disulfoton mixture, Dry (DOT) (298-04-4)
Disulfoton mixture, Liquid (DOT) (298-04-4)
Disulfoton sulfoxide (2497-07-6)
2,2'-Disulfo-4,4'-stilbenetetrazonium dichloride (13954-62-6)
Disulfuram (97-77-8)
Disulfur decafluoride (5714-22-7)
Disulfur dichloride (10025-67-9)
Disulfure de tetramethylthiourame (French) (137-26-8)
Disulfur pentoxydichloride (7791-27-7)
Disulfuryl chloride (7791-27-7)
Disulone (80-08-0)
Disulphine Blue VN 150 (129-17-9)
Disulphine Lake Blue EG (2650-18-2)
Disulphine VN (129-17-9)
Disulphuram (97-77-8)
Disulphuric acid (8014-95-7)
Disyncram (1982-37-2)
Disyncran (1229-35-2)
Disyncran (1982-37-2)
Disynformon (53-16-7)
Disyston S (2497-07-6)
Disyston sulfone (2497-06-5)
Disyston sulfoxide (2497-07-6)
Disyston sulphoxide (2497-07-6)
Disystox (298-04-4)
Ditab (51-63-8)
Ditak (396-01-0)
Ditalimfos (5131-24-8)
Ditalimphos (5131-24-8)
Di(tallow alkyl) dimethyl ammonium bentonite (68953-58-2)
N,N-Di(2-tallow amidoethyl)-N-(2-hydroxyethyl)-N-methyl ammonium methyl sulfate (68153-35-5)
N,N-Di(2-tallowamidoethyl)-N-(2-hydroxyethyl)-N-methyl-ammonium methylsulfate (68153-35-5)
Ditallow dimethyl ammonium chloride (68783-78-8)
Ditallowdimonium chloride (68783-78-8)
Ditan (101-81-5)
Ditane (101-81-5)
Ditaven (71-63-6)
Ditek (23564-05-8)
Ditetradecyl peroxydicarbonate (53220-22-7)
Ditetradecyl 3,3'-thiobispropanoate (16545-54-3)
Dithallium carbonate (6533-73-9)
Dithallium propanedioate (2757-18-8)
Dithallium sulfate (7446-18-6)
Dithallium(1+) sulfate (7446-18-6)
Dithallium trioxide (1314-32-5)
Dithane (12656-69-8)
Dithane 65 (12122-67-7)
Dithane A-4 (100-25-4)
Dithane A-40 (142-59-6)
Dithane D-14 (142-59-6)
Dithane M 22 (12427-38-2)
Dithane M 45 (8018-01-7)
Dithane M 22 Special (12427-38-2)
Dithane R-24 (16227-10-4)
Dithane S 60 (8018-01-7)
Dithane SPC (8018-01-7)
Dithane Stainless (3566-10-7)
Dithane Ultra (8018-01-7)
Dithane Z (12122-67-7)
Dithane Z 78 (12122-67-7)
Dithane Z-78 (12122-67-7)
9,10-Dithiaanthracene (92-85-3)
1,4-Dithiaanthraquinone-2,3-dicarbonitrile (3347-22-6)
1,4-Dithiaanthraquinone-2,3-dinitrile (3347-22-6)
1,4-Dithiacyclohexane (505-29-3)
3,4-Dithiaheptane (30453-31-7)
1,4-Dithiane (505-29-3)
p-Dithiane (505-29-3)
Dithiane Z-78 (12122-67-7)
p-Dithiane, 2,3-dehydro-2,3-dimethyl-, tetroxide (55290-64-7)
Dithianon (3347-22-6)
Dithianone (3347-22-6)
4,5-Dithia-1,7-octadiene (2179-57-9)
3,6-Dithia-1,8-octanediol (5244-34-8)
Dithiazanine iodide (514-73-8)
Dithiazanini iodidum (Latin) (514-73-8)
Dithiazanin iodide (514-73-8)
Dithiazine (514-73-8)
4H-1,3,5-Dithiazine, dihydro-2,4,6-trimethyl-, (2-α,4-α,6-α) (638-17-5)
Dithiazinine (514-73-8)
1,4-Dithiin, 2,3-dihydro-5,6-dimethyl-, 1,1,4,4-tetraoxide (55290-64-7)
Dithio (3689-24-5)
2',2'''-Dithiobisbenzanilide (135-57-9)
2,2'-Dithiobis(benzothiazole) (120-78-5)
1,1'-Dithiobis(N,N-diethylthioformamide) (97-77-8)
α,α'-Dithiobis(dimethylthio)formamide (137-26-8)
2,2'-(Dithiobis(methylene))bisfuran (4437-20-1)
4,4'-Dithiobis(morpholine) (103-34-4)
Dithiobismorpholine (103-34-4)
6,6'-Dithiobis-2-naphthalenol (6088-51-3)
3,3'-Dithiobis(N-octylpropionamide) (33312-01-5)
3,3'-Dithiobis(1-propanesulfonic acid), disodium salt (27206-35-5)
Dithiobis(thioformic acid) O,O-diethyl ester (502-55-6)
2,5-Dithiobiurea (142-46-1)
Dithiobiuret (541-53-7)
Dithiocarb (148-18-5)
Dithiocarb (20624-25-3)
Dithiocarbamate (148-18-5)

Dithiocarbamic acid monoammonium salt **(513-74-6)**
Dithiocarbamoyl disulfide **(504-90-5)**
Dithiocarbazic acid hydrazine (Salt) **(20469-71-0)**
Dithiocarbonic acid O-isopropyl ester potassium salt **(140-92-1)**
Dithiocarbonic acid o-pentyl ester potassium salt **(2720-73-2)**
Dithiocarbonic anhydride **(75-15-0)**
1,6-Dithiocyanatomannitol **(73928-09-3)**
1,6-Dithiocyano-1,6-dideoxy-D-mannitol **(73928-09-3)**
β,β'-Dithiocyano diethyl ether **(4617-17-8)**
Dithiocyanomannitol **(73928-09-3)**
Dithiodemeton **(298-04-4)**
β,β'-Dithiodialanine **(56-89-3)**
2',2'''-Dithiodibenzanilide **(135-57-9)**
N,N'-(Dithiodicarbonothioyl)bis(N-methylmethanamine) **(137-26-8)**
2,2'-(Dithiodimethylene)difuran **(4437-20-1)**
N,N'-Dithiodimorfolin (Czech) **(103-34-4)**
4,4'-Dithiodimorpholine **(103-34-4)**
N,N-Dithiodimorpholine **(103-34-4)**
N,N'-(Dithiodi-2,1-phenylene)bisbenzamide **(135-57-9)**
Dithiodiphosphoric acid, tetraethyl ester **(3689-24-5)**
2,2'-Dithiodipyridine-1,1'-dioxide **(3696-28-4)**
Dithioethyleneglycol **(540-63-6)**
Dithiofos **(3689-24-5)**
1,2-Dithioglycerol **(59-52-9)**
Dithioglycerol **(59-52-9)**
Dithioglycol **(540-63-6)**
1,3-Dithiolane-2-carboxaldehyde, 2,4-dimethyl-, O-(methy-carbamoyl)oxime **(26419-73-8)**
1,3-Dithiolane-2-carboxaldehyde, 2,4-dimethyl-, O-((methyl-amino)carbonyl)oxime **(26419-73-8)**
1,3-Dithiolane, 2-(diethoxyphosphinylimino)-4-methyl **(950-10-7)**
1,2-Dithiolane-3-valeric acid **(62-46-4)**
5-(1,2-Dithiolan-3-yl)valeric acid **(62-46-4)**
1,3-Dithiolo(4,5-b)quinoxaline-2-thione **(93-75-4)**
Dithiolo(4,5-b)quinoxalin-2-one, 6-methyl- (9CI) **(2439-01-2)**
Dithiometasystox **(640-15-3)**
Dithiomethon **(640-15-3)**
Dithiometon (French) **(640-15-3)**
4,4'-Dithiomorpholine **(103-34-4)**
Dithion **(3689-24-5)**
Dithione **(3689-24-5)**
Dithionic acid **(8014-95-7)**
Dithionic acid, disodium salt **(14970-71-9)**
Dithionous acid, disodium salt **(7775-14-6)**
Dithionous acid, zinc salt (1:1) **(7779-86-4)**
6,8-Dithiooctanoic acid **(62-46-4)**
Dithiooxamide **(79-40-3)**
Dithiophos **(3689-24-5)**
Dithiophosphate de O,O-diethyle et de S(2,5-dichlorophenyl) thiomethyle (French) **(2275-14-1)**
Dithiophosphate de O,O-diethyle et de (4-chloro-phenyl) thiomethyle (French) **(786-19-6)**
Dithiophosphate de O,O-diethyle et de S-(2-ethylthio-ethyle) (French) **(298-04-4)**
Dithiophosphate de O,O-diethyle et de S-N-methyl N-carboethoxy carbamoylmethyle (French) **(2595-54-2)**
Dithiophosphate de O,O-diethyle et d'ethylthiomethyle (French) **(298-02-2)**
Dithiophosphate de O,O-dimethyle et de S(-N-methyl-carbamoyl-methyle) (French) **(60-51-5)**
Dithiophosphate de O,O-dimethyle et de S-((4,6-diamino-1,3,5-triazine-2-yl)-methyle) (French) **(78-57-9)**
Dithiophosphate de O,O-dimethyle et de S-(1,2-dicarboethoxy-ethyle) (French) **(121-75-5)**
Dithiophosphate de O,O-dimethyle et de S-(2-ethylthio-ethyle) (French) **(640-15-3)**
Dithiophosphate de O,O-dimethyle et de S-((morpholino-carbonyle)-methyle) (French) **(144-41-2)**
Dithiophosphoric acid, O,O-di(C1-C14)alkyl ester **(68187-41-7)**
Dithiophosphoric acid, O,O'-diisooctyl ester, zinc salt **(28629-66-5)**
Dithiophosphoric acid, O,O'-isobutyl amyl esters **(68516-01-8)**
Dithiophosphoric acid, O,O'-isobutyl amyl ester, zinc salt **(68457-79-4)**
Di(thiophosphoric) acid, tetraethyl ester **(3689-24-5)**
Dithiophosphorsaeure-O-aethyl-S,S-diphenylester (German) **(17109-49-8)**
Di(thiopropane sodium sulfonate) **(27206-35-5)**
2,3-Dithiopropanol **(59-52-9)**
Dithiopyrophosphate de tetraethyle (French) **(3689-24-5)**
Dithioquinox **(2439-01-2)**
Dithiosystox **(298-04-4)**
Dithiotep **(3689-24-5)**
Dithiotrimethyleneglycol **(109-80-8)**
Dithioxamide **(79-40-3)**
Dithranol **(1143-38-0)**
Dithranol **(480-22-8)**
Dithrocream **(480-22-8)**
Dithymol diiodide **(552-22-7)**
Ditiamina **(12122-67-7)**
Ditilin **(306-40-1)**
Ditiline **(306-40-1)**
Ditoin **(630-93-3)**
Ditoinate **(57-41-0)**
4,4'-Di-o-toluidine **(119-93-7)**
2,5-Di-p-toluidinoterephthalic acid **(10291-28-8)**
Ditolylbis(azonaphthionic acid) **(992-59-6)**
1,3-Di-o-tolylguanidine **(97-39-2)**
Di-o-tolylguanidine **(97-39-2)**
Diorthotolylguanidine **(97-39-2)**
Di-p-tolyl mercury **(537-64-4)**
O,O-Ditolyl phosphorodithioate **(27157-94-4)**
Ditranil **(99-30-9)**
Ditrazin **(1642-54-2)**
Ditrazin citrate **(1642-54-2)**
Ditrazine **(1642-54-2)**
Ditrazine citrate **(1642-54-2)**
Ditridecyl acid phosphate **(5116-95-0)**
Ditridecylamine **(5910-75-8)**
Ditridecyl azelate **(26719-40-4)**
Ditridecyl dilinoleate **(16958-92-2)**
Ditridecyl dimerate **(16958-92-2)**
Ditridecyl hexanedioate **(16958-92-2)**
Ditridecyl nonanedioate **(26719-40-4)**
Ditridecyl phthalate **(119-06-2)**
Ditridecyl sodium sulfosuccinate **(2673-22-5)**
Ditridecyl 3,3'-thiobispropanoate **(10595-72-9)**
Di(tridecyl) thiodipropionate **(10595-72-9)**
Ditridecyl thiodipropionate **(10595-72-9)**
Di(tri-(2,2-dimethyl-2-phenylethyl)tin)oxide **(13356-08-6)**
Ditrifon **(52-68-6)**
Di(2,4,6-trimethylphenyl) 2,6-dimethylphenyl phosphate **(73195-13-8)**
Di(2,4,6-trimethylphenyl) phenyl phosphate **(73179-44-9)**
1,3-Di(2,4,6-trioxohexahydro-5-pyrimidinylidene)isoindole **(36888-99-0)**
Ditripentat **(12111-24-9)**
Ditrosol **(534-52-1)**
Ditsianamid **(504-66-5)**
Ditubin **(54-85-3)**
Dityrin **(66-02-4)**
Diucardin **(135-09-1)**
Diulo **(17560-51-9)**
Diumate **(330-54-1)**
Di-n-undecyl ketone **(540-09-0)**
Diundecyl phthalate **(3648-20-2)**
Diupres **(50-55-5)**
Diural **(54-31-9)**
Diuramid **(59-66-5)**
Diurazine **(500-42-5)**

Diurea glyoxalate	**(496-46-8)**
1,1-Diureidisobutane	**(6104-30-9)**
Diureidoisobutane	**(6104-30-9)**
Diuresal	**(58-94-6)**
Diurese	**(133-67-5)**
Diuretic C	**(452-35-7)**
Diuretic salt	**(127-08-2)**
Diureticum-holzinger	**(59-66-5)**
Diuretin	**(54-21-7)**
Diurex	**(330-54-1)**
Diuril	**(58-94-6)**
Diurilix	**(58-94-6)**
Diurite	**(58-94-6)**
Diuriwas	**(59-66-5)**
Diurnal	**(57-53-4)**
Diurnal-Penicillin	**(6130-64-9)**
Diurobromine	**(83-67-0)**
Diurol	**(330-54-1)**
Diurol	**(61-82-5)**
Diurol 5030	**(61-82-5)**
Diuron 4L	**(330-54-1)**
Diuron (OSHA)	**(330-54-1)**
Diurone	**(62-37-3)**
Diutazol	**(59-66-5)**
Diutensen-R	**(50-55-5)**
Diutrid	**(58-94-6)**
Diuxanthine	**(317-34-0)**
2,6-Divanillylidenecyclohexanone	**(579-23-7)**
Divaric acid	**(4707-50-0)**
Divercillin	**(69-53-4)**
Divercillin	**(7177-48-2)**
Diveron	**(57-53-4)**
Divinyl	**(106-99-0)**
Divinyl acetylene	**(821-08-9)**
m-Divinylbenzen (Czech)	**(108-57-6)**
Divinylbenzene	**(1321-74-0)**
m-Divinylbenzene	**(108-57-6)**
Divinylbenzene (ACGIH,OSHA)	**(108-57-6)**
1,2-Divinylcyclobutane	**(2422-85-7)**
Divinylene oxide	**(110-00-9)**
Divinylene sulfide	**(110-02-1)**
Divinylenimine	**(109-97-7)**
Divinyl ether (DOT)	**(109-93-3)**
Divinyl ether, Inhibited [UN 1167]	**(109-93-3)**
Divinylethylene	**(2235-12-3)**
Divinyl sulfone	**(77-77-0)**
Divinyl sulfoxide	**(1115-15-7)**
1,3-Divinyltetramethyldisiloxane	**(2627-95-4)**
Divinyltetramethyldisiloxane	**(2627-95-4)**
Divipan	**(62-73-7)**
Divit Urto	**(50-14-6)**
Divulsan	**(630-93-3)**
Divynyl oxide	**(109-93-3)**
Dixanthogen	**(502-55-6)**
Dixiben	**(389-08-2)**
Dixie	**(1332-58-7)**
Dixie	**(1333-86-4)**
Dixiecell	**(1333-86-4)**
Dixiedensed	**(1333-86-4)**
Dixitherm	**(1333-86-4)**
Dixol	**(133-53-9)**
Dixon	**(13171-21-6)**
Dixopak	**(9002-88-4)**
N,N-Di-(2,4-xylyliminomethyl)methylamine	**(33089-61-1)**
Di-p-xylyl phenyl phosphate	**(72121-83-6)**
Diyodoacetileno (Spanish)	**(624-74-8)**
Dizan	**(514-73-8)**
Dizene	**(95-50-1)**
Dizinon	**(333-41-5)**
Dizol	**(490-55-1)**
Dizzitol	**(333-41-5)**
Dmdheu	**(1854-26-8)**
2,4-D methyl ester	**(1928-38-7)**
DnHA	**(109-31-9)**
1,4-Doa (Russian)	**(81-64-1)**
Dobane 055 (9CI)	**(37203-41-1)**
Dobane 83 (9CI)	**(59763-33-6)**
Dobane IN (9CI)	**(60529-17-1)**
Dobane INQ (9CI)	**(60529-18-2)**
Dobanic Acid 83	**(27176-87-0)**
Dobanic Acid JN	**(27176-87-0)**
Dobanol 911	**(66455-17-2)**
Dobanol 45-7, acetate (9CI)	**(59536-56-0)**
Dobendan	**(123-03-5)**
Dobesin	**(90-84-6)**
Dobetin	**(68-19-9)**
Dobren	**(15676-16-1)**
Doburil	**(2259-96-3)**
Docclan	**(13422-51-0)**
Docemine	**(68-19-9)**
Docevita	**(13422-51-0)**
Docibin	**(68-19-9)**
Docigram	**(68-19-9)**
Dociton	**(525-66-6)**
Docosanamide (9CI)	**(3061-75-4)**
1-Docosanamine (9CI)	**(14130-??-?)**
Docosane (9CI)	**(629-97-0)**
Docosane, 11-decyl- (6CI,7CI,9CI)	**(55401-55-3)**
1-Docosanoic acid	**(112-85-6)**
n-Docosanoic acid	**(112-85-6)**
Docosanoic acid (9CI)	**(112-85-6)**
Docosanoic acid, ethyl ester (8CI,9CI)	**(5908-87-2)**
Docosanoic acid, iodo-, calcium salt	**(1319-91-1)**
Docosanoic acid, lithium salt (9CI)	**(4499-91-6)**
Docosanoic acid, methyl ester (9CI)	**(929-77-1)**
Docosanoic acid, silver salt	**(2489-05-6)**
Docosanoic acid, silver(1+) salt (9CI)	**(2489-05-6)**
1-Docosanol (9CI)	**(661-19-8)**
1-Docosanol, aluminum salt (9CI)	**(67905-30-0)**
(Z)-13-Docosenamide	**(112-84-5)**
13-Docosenamide	**(112-84-5)**
13-Docosenamide, cis-	**(112-84-5)**
13-Docosenamide, (Z)- (9CI)	**(112-84-5)**
13-Docosenamide, N-octadecyl-	**(10094-45-8)**
13-Docosenamide, N-octadecyl-, (Z)- (9CI)	**(10094-45-8)**
1-Docosene (9CI)	**(1599-67-3)**
(Z)-13-Docosenoic acid	**(112-86-7)**
13-Docosenoic acid (cis)	**(112-86-7)**
cis-13-Docosenoic acid	**(112-86-7)**
δ 13:14-Docosenoic acid	**(112-86-7)**
δ(13)-cis-Docosenoic acid	**(112-86-7)**
Docosenoic acid (8CI,9CI) (VAN)	**(25378-26-1)**
13-Docosenoic acid, (Z)- (9CI)	**(112-86-7)**
Docosoic acid	**(112-85-6)**
Docosyl alcohol	**(661-19-8)**
N-Docosyl-1,3-propanediamine	**(15268-40-3)**
Doctamicina	**(56-75-7)**
2,4-D-octyl ester	**(1928-44-5)**
Docusate sodium	**(577-11-7)**
Dodat	**(50-29-3)**
Dodecabee	**(68-19-9)**
Dodecachlorodicyclopentadiene	**(14979-34-1)**
1,1a,2,2,3,3a,4,5,5,5a,5b,6-Dodecachlorooctahydro-1,3,4-metheno-1H-cyclobuta(cd)pentalene	**(2385-85-5)**

Dodecachlorooctahydro-1,3,4-metheno-2H-cyclobuta(c,d)pentalene **(2385-85-5)**
Dodecachloropentacyclo(3.2.2.$0^{2,6}$,$0^{3,9}$,$0^{5,10}$Decane **(2385-85-5)**
Dodecachloropentacyclodecane **(2385-85-5)**
(E,E)-2,4-Dodecadien-1-al **(21662-16-8)**
(E,E)-2,4-Dodecadienal **(21662-16-8)**
(E,Z)-2,4-Dodecadienal **(21662-15-7)**
(trans,cis)-2,4-Dodecadienal **(21662-15-7)**
(trans,trans)-2,4-Dodecadienal **(21662-16-8)**
2,4-Dodecadienal, (E,E)- (9CI) **(21662-16-8)**
2,4-Dodecadienal, (E,Z)- (9CI) **(21662-15-7)**
1,11-Dodecadiene (8CI,9CI) **(5876-87-9)**
2,4-Dodecadienoic acid, 11-methoxy-3,7,11-trimethyl-, 1-methylethyl ester, (E,E) **(40596-69-8)**
Dodeca-2,4-dienoic acid, 3,7,11-trimethyl-, ethyl ester, (2E,4E)- **(41096-46-2)**
2,4-Dodecadienoic acid, 3,7,11-trimethyl-, ethyl ester, (E,E)- (9CI) **(41096-46-2)**
2,4-Dodecadienoic acid, 3,7,11-trimethyl-, 2-propynyl ester, (E,E)- (9CI) **(42588-37-4)**
Dodecadien-1-ol, 3,7,11-trimethyl- (8CI,9CI) **(1335-48-4)**
Dodecafluorocyclohexane **(355-68-0)**
Dodecafluoropentane **(678-26-2)**
Dodecahydrobiphenyl **(92-51-3)**
Dodecahydrobisphenol A **(80-04-6)**
Dodecahydrodiphenylamine **(101-83-7)**
(Dodecahydro-7β-hydroxy-1α,4bβ,8,8-tetramethyl-10-oxo-2-(1H)-phenanthrenylidene)acetic acid 2-(dimethylamino)-ethyl ester **(468-76-8)**
Dodecahydrophenylamine nitrite **(3129-91-7)**
δ-Dodecalactone **(713-95-1)**
γ-Dodecalactone **(2305-05-7)**
Dodecamethylcyclohexasiloxane **(540-97-6)**
1,12'-Dodecamethylenediamine **(2783-17-7)**
Dodecamethylenediamine **(2783-17-7)**
Dodecamethylpentasiloxane **(141-63-9)**
n-Dodecan (German) **(112-40-3)**
1-Dodecanal **(112-54-9)**
Dodecanamide (9CI) **(1120-16-7)**
Dodecanamide, N,N-bis(2-hydroxyethyl) **(120-40-1)**
Dodecanamide, N,N-diethyl- **(3352-87-2)**
Dodecanamide, N-(3-(dimethylamino)propyl)- (9CI) **(3179-80-4)**
Dodecanamide, N-(3-(dimethylamino)propyl)-, monohydro-chloride (9CI) **(71732-95-1)**
Dodecanamide, N-(2-hydroxyethyl)- (9CI) **(142-78-9)**
Dodecanamide, N-(2-hydroxypropyl)- (9CI) **(142-54-1)**
2-Dodecanamidoethanol **(142-78-9)**
Dodecanamine acetate **(2016-56-0)**
1-Dodecanamine, N,N-didodecyl- (9CI) **(102-87-4)**
1-Dodecanamine, N,N-dimethyl- **(112-18-5)**
1-Dodecanamine, N-dodecyl- (9CI) **(3007-31-6)**
Dodecanamine hydrochloride **(929-73-7)**
1-Dodecanamine, hydrochloride (9CI) **(929-73-7)**
1-Dodecanamine, (2,4,5-trichlorophenoxy)acetate (9CI) **(53404-84-5)**
1-Dodecanaminium, N,N-bis(2-hydroxyethyl)-N-((octylthio)-methyl)-, chloride (9CI) **(78865-89-1)**
1-Dodecanaminium, N-(carboxymethyl)-N,N-dimethyl-, hydroxide, inner salt (9CI) **(683-10-3)**
1-Dodecanaminium, N-((decylthio)methyl)-N,N-bis(2-hydroxyethyl), chloride (9CI) **(78865-90-4)**
1-Dodecanaminium, N,N-dimethyl-N-(3-sulfopropyl)-, hydroxide, inner salt (9CI) **(14933-08-5)**
1-Dodecanaminium, N-dodecyl-N,N-dimethyl-, bromide (9CI) **(3282-73-3)**
1-Dodecanaminium, N-dodecyl-N,N-dimethyl-, chloride (9CI) **(3401-74-9)**
1-Dodecanaminium, N,N,N-trimethyl-, chloride (9CI) **(112-00-5)**
1-Dodecanaminium, N,N,N-trimethyl-, methyl sulfate (9CI) **(13623-06-8)**
Dodecane **(112-40-3)**
Dodecane, 1-bromo- (9CI) **(143-15-7)**
Dodecane, 1-chloro- (9CI) **(112-52-7)**
Dodecane, 2-chloro- (9CI) **(2350-11-0)**
Dodecane, chloro- (8CI,9CI) **(28519-07-5)**
1,12'-Dodecanediamine **(2783-17-7)**
1,12-Dodecanediamine **(2783-17-7)**
Dodecane, 2,5-dimethyl- (7CI,9CI) **(56292-65-0)**
1,12-Dodecanedioic acid **(693-23-2)**
Dodecanedioic acid (9CI) **(693-23-2)**
Dodecanedioic acid, Compd. with 1,6-hexanediamine (1:1) (9CI) **(13188-60-8)**
Dodecanedioic acid, Polymer with 1,10-decanediol (9CI) **(35464-94-9)**
1,1-Dodecanediol, diacetate (9CI) **(56438-07-4)**
6,7-Dodecanedione (8CI,9CI) **(13757-90-9)**
Dodecane, 1,2-epoxy **(2855-19-8)**
Dodecane, 1-fluoro **(334-68-9)**
Dodecane, 1-iodo- (9CI) **(4292-19-7)**
Dodecanenitrile **(2437-25-4)**
Dodecane, 1,1'-oxybis- (9CI) **(4542-57-8)**
Dodecaneperoxoic acid (9CI) **(2388-12-7)**
Dodecane, 2-phenyl- (8CI) **(2719-61-1)**
Dodecane, 4-phenyl- (8CI) **(2719-64-4)**
Dodecane, 5-phenyl- (8CI) **(2719-63-3)**
Dodecane, 6-phenyl- (8CI) **(2719-62-2)**
1-Dodecanephosphonic acid **(5137-70-2)**
n-Dodecanephosphonic acid **(5137-70-2)**
Dodecane, 1,1'-selenobis- (9CI) **(5819-01-2)**
1-Dodecanesulfonic acid, sodium salt (9CI) **(2386-53-0)**
1-Dodecanethiol **(112-55-0)**
t-Dodecanethiol **(25103-58-6)**
Dodecane, 2,6,10-trimethyl- (8CI,9CI) **(3891-98-3)**
Dodecane, 2,6,11-trimethyl- (8CI,9CI) **(31295-56-4)**
Dodecanoic acid **(143-07-7)**
n-Dodecanoic acid **(143-07-7)**
Dodecanoic acid, Compd. with 2,2'-iminobis(ethanol) (1:1) (9CI) **(7487-79-8)**
Dodecanoic acid, Compd. with 2,2',2''-nitrilotris(ethanol) (1:1) (9CI) **(2224-49-9)**
Dodecanoic acid, ammonium salt **(2437-23-2)**
Dodecanoic acid, barium salt (9CI) **(4696-57-5)**
Dodecanoic acid, cadmium salt (9CI) **(2605-44-9)**
Dodecanoic acid, chloride **(112-16-3)**
Dodecanoic acid, 2-chloroethyl ester (9CI) **(64919-15-9)**
Dodecanoic acid, (dibutylstannylene)bis(thio-2,1-ethanediyl) ester (9CI) **(28570-24-3)**
Dodecanoic acid, 2-ethoxyethyl ester (9CI) **(106-13-8)**
Dodecanoic acid, ethyl ester (9CI) **(106-33-2)**
Dodecanoic acid, 3-((2-ethyl-1-oxohexyl)oxy)-2,2-dimethylpropyl ester (9CI) **(99562-17-1)**
Dodecanoic acid, glycidyl ester **(63978-73-4)**
Dodecanoic acid, 6-hydroxy- (9CI) **(35875-13-9)**
Dodecanoic acid, 5-hydroxy- (8CI,9CI) **(7779-95-5)**
Dodecanoic acid, 2-(2-hydroxyethoxy)ethyl ester (9CI) **(141-20-8)**
Dodecanoic acid, methyl ester (9CI) **(111-82-0)**
Dodecanoic acid, 1-methylethyl ester (9CI) **(10233-13-3)**
Dodecanoic acid, 1-methylethyl ester (Crude) **(10233-13-3)**
Dodecanoic acid, pentachlorophenyl ester (9CI) **(3772-94-9)**
Dodecanoic acid, potassium salt **(10124-65-9)**
Dodecanoic acid, 1,2,3-propanetriyl ester (9CI) **(538-24-9)**
Dodecanoic acid, 1,2,3-propantriyl ester **(538-24-9)**
Dodecanoic acid, 2-sulfo- (6CI,7CI,8CI,9CI) **(3054-88-4)**
Dodecanoic acid, 2-sulfo-, 1-(2-(2-aminoethoxy)ethyl) ester (9CI) **(59997-79-4)**
Dodecanoic acid, 2-sulfo-, Compd. with 2-(2-aminoethoxy)-ethanol (1:1) (9CI) **(65520-66-3)**
Dodecanoic acid, 2-sulfo-, 1-(2-(dimethylamino)ethyl) ester (9CI) **(59997-83-0)**
Dodecanoic acid, 2-sulfo-, 1-(2-((2-hydroxyethyl)amino)ethyl) ester (9CI) **(65520-65-2)**
Dodecanoic acid, 2-sulfo-, 1-(2-(methylamino)ethyl) ester (9CI) **(59997-81-8)**
Dodecanoic acid, 2-sulfo-, 1-methyl ester, sodium salt (9CI) **(4016-21-1)**
Dodecanoic acid 2-thiocyanatoethyl ester **(301-11-1)**
Dodecanoic acid, 2,2,2-trichloro-1-(dimethoxyphosphinyl)ethyl

Name	CAS No.
ester (9CI)	**(4414-15-7)**
Dodecanoic acid, zinc salt (9CI)	**(2452-01-9)**
1-Dodecanol	**(112-53-8)**
n-Dodecanol	**(112-53-8)**
n-Dodecanol	**(27342-88-7)**
2-Dodecanol (8CI,9CI)	**(10203-28-8)**
5-Dodecanol (8CI,9CI)	**(10203-33-5)**
6-Dodecanol (8CI,9CI)	**(6836-38-0)**
Dodecanol (VAN)(9CI)	**(27342-88-7)**
1-Dodecanol acetate	**(112-66-3)**
Dodecanol acetate	**(112-66-3)**
1-Dodecanol, aluminum salt (9CI)	**(14624-15-8)**
1-Dodecanol, ethoxy-	**(29718-44-3)**
Dodecanol, ethoxylate	**(9002-92-0)**
Dodecanol-ethylene oxide (9.5 moles) condensate	**(9002-92-0)**
1-Dodecanol, 3,3,4,4,5,5,6,6,7,7,8,8,9,9,10,10,11,11, 12,12,12-heneicosafluoro- (9CI)	**(865-86-1)**
Dodecanolide-1,4	**(2305-05-7)**
1-Dodecanol, 2-methyl-, (S)- (9CI)	**(57289-26-6)**
1-Dodecanol, 2-octyl	**(5333-42-6)**
Dodecanol, polyethoxylated	**(9002-92-0)**
3-Dodecanol, 3,7,11-trimethyl- (VAN)(9CI)	**(7278-65-1)**
2-Dodecanone	**(6175-49-1)**
3-Dodecanone (8CI,9CI)	**(1534-27-6)**
Dodecanoylamidopropyldimethylamine	**(3179-80-4)**
Dodecanoyl chloride	**(112-16-3)**
n-Dodecanoyl chloride	**(112-16-3)**
N-Dodecanoyl-N-methylglycine, sodium salt	**(137-16-6)**
Dodecanoyl peroxide	**(105-74-8)**
Dodecansulfonic acid, hydroxy-	**(151-41-7)**
Dodecan-1-yl acetate	**(112-66-3)**
2,6,9,11-Dodecatetraenal, 2,6,10-trimethyl- (9CI)	**(4955-32-2)**
2,6,9,11-Dodecatetraenal, 2,6,10-trimethyl-, (E,E,E)- (9CI)	**(17909-77-2)**
1,3,6,10-Dodecatetraene, 3,7,11-trimethyl-, (E,E)- (9CI)	**(502-61-4)**
2,6,11-Dodecatrienal, 2,6-dimethyl-10-methylene- (9CI)	**(60066-88-8)**
1,6,10-Dodecatrien-3-ol, 3,7,11-trimethyl-	**(7212-44-4)**
2,6,10-Dodecatrien-1-ol, 3,7,11-trimethyl	**(4602-84-0)**
1,6,10-Dodecatrien-3-ol, 3,7,11-trimethyl-, (S-(Z))- (9CI)	**(142-50-7)**
2,6,10-Dodecatrien-1-ol, 3,7,11-trimethyl-, (Z,E)- (8CI,9CI)	**(3790-71-4)**
1,6,10-Dodecatrien-3-ol, 3,7,11-trimethyl-, (Z)-(S)-(+)- (8CI)	**(142-50-7)**
4,6,10-Dodecatrien-3-one, 7,11-dimethyl	**(26651-96-7)**
5,7,11-Dodecatriyn-1-ol	**(76379-66-3)**
Dodecavite	**(68-19-9)**
2-Dodecenal	**(4826-62-4)**
1-Dodecene	**(25378-22-7)**
Dodecene	**(6842-15-5)**
α-Dodecene	**(112-41-4)**
n-Dodec-1-ene	**(112-41-4)**
1-Dodecene (9CI)	**(112-41-4)**
Dodecene (9CI)	**(25378-22-7)**
Dodecene epoxide	**(2855-19-8)**
Dodecenesulfonic acid, sodium salt (9CI)	**(99744-82-8)**
9-Dodecenoic acid (8CI,9CI)	**(2382-40-3)**
7-Dodecen-1-ol, (Z)	**(20056-92-2)**
(E)-9-Dodecen-1-ol acetate	**(35148-19-7)**
(Z)-7-Dodecen-1-ol acetate	**(14959-86-5)**
(Z)-9-Dodecen-1-ol acetate	**(16974-11-1)**
9-Dodecen-1-ol, acetate, (E)-	**(35148-19-7)**
cis-7-Dodecen-1-ol acetate	**(14959-86-5)**
1-Dodecen-1-ol, acetate (6CI,9CI)	**(56438-08-5)**
7-Dodecen-1-ol, acetate, (Z)-	**(14959-86-5)**
9-Dodecen-1-ol, acetate, (Z)-	**(16974-11-1)**
11-Dodecen-2-one, 7,7-dimethyl- (9CI)	**(35194-22-0)**
(E)-9-Dodecenyl acetate	**(29868-16-4)**
(E)-9-Dodecenyl acetate	**(35148-19-7)**
(Z)-9-Dodecenyl acetate	**(16974-11-1)**
(Z)-9-Dodecenyl acetate	**(16974-12-2)**
Dodecenylbutanedioic acid	**(29658-97-7)**
3-(2-Dodecenyl)dihydro-2,5-furandione	**(19780-11-1)**
Dodecenylsuccinic acid	**(29658-97-7)**
Dodecenylsuccinic anhydride	**(25377-73-5)**
cis-5,7-Dodecenyne	**(16336-83-7)**
5-Dodecen-7-yne, (Z)- (8CI,9CI)	**(16336-83-7)**
Dodecoic acid	**(143-07-7)**
Dodecto	**(76379-66-3)**
Dodecyl acetate	**(112-66-3)**
n-Dodecyl acetate	**(112-66-3)**
N-Dodecyl-β-alanine	**(1462-54-0)**
Dodecyl alcohol	**(27342-88-7)**
Dodecyl-alcohol	**(112-53-8)**
n-Dodecyl alcohol	**(112-53-8)**
Dodecyl alcohol acetate	**(112-66-3)**
Dodecyl alcohol (ethoxylated)	**(29718-44-3)**
Dodecyl alcohol, ethoxylated	**(29718-44-3)**
Dodecyl alcohol, ethoxylated	**(9002-92-0)**
Dodecyl alcohol, hydrogen sulfate, sodium salt	**(151-21-3)**
Dodecyl alcohol, phosphate ester	**(12751-23-4)**
1-Dodecyl aldehyde	**(112-54-9)**
Dodecylamine	**(124-22-1)**
n-Dodecylamine	**(124-22-1)**
1-Dodecylamine acetate	**(2016-56-0)**
Dodecylamine, acetate	**(2016-56-0)**
Dodecylamine, N,N-dimethyl	**(112-18-5)**
Dodecylamine, N,N-dimethyl-, hydrochloride	**(2016-48-0)**
Dodecylamine, N,N-dimethyl-, N-oxide	**(1643-20-5)**
Dodecylamine, hydrochloride	**(929-73-7)**
n-Dodecylamine hydrochloride	**(929-73-7)**
Dodecylamine, N-methyl-N-nitroso	**(55090-44-3)**
N-(2-((-(Dodecylamino)ethyl)amino)ethyl)glycine	**(6843-97-6)**
N-(2-((2-(Dodecylamino)ethyl)amino)ethyl)glycine	**(6843-97-6)**
3-(Dodecylamino)-1,2-propanediol	**(821-91-0)**
Dodecylammonium chloride	**(929-73-7)**
n-Dodecylammonium chloride	**(929-73-7)**
Dodecylammonium methanearsonate	**(53404-47-0)**
Dodecyl ammonium sulfate	**(2235-54-3)**
p-Dodecylaniline	**(104-42-7)**
4-Dodecylbenzenamine	**(104-42-7)**
Dodecylbenzene	**(123-01-3)**
Dodecyl benzene sodium sulfonate	**(25155-30-0)**
4-Dodecylbenzenesulfonic acid	**(121-65-3)**
Dodecyl benzene sulfonic acid	**(27176-87-0)**
Dodecylbenzene sulfonic acid	**(27176-87-0)**
n-Dodecylbenzenesulfonic acid	**(27176-87-0)**
Dodecylbenzenesulfonic acid, Compd. with 1-Amino-2-propanol (1:1)	**(42504-46-1)**
Dodecylbenzenesulfonic acid, Compd. with 2,2',2''-nitrilotris-(ethanol) (1:1)	**(27323-41-7)**
Dodecylbenzenesulfonic acid, Compound with 2-aminoethanol (1:1)	**(26836-07-7)**
Dodecylbenzenesulfonic acid, Comp. with 2-Propanamine (1:1)	**(26264-05-1)**
Dodecylbenzenesulfonic acid [NA 2584]	**(27176-87-0)**
Dodecylbenzenesulfonic acid, 2-aminoethanol salt	**(26836-07-7)**
Dodecylbenzenesulfonic acid, ammonium salt	**(1331-61-9)**
Dodecylbenzenesulfonic acid calcium salt	**(26264-06-2)**
Dodecylbenzenesulfonic acid diethanolamine salt	**(26545-53-9)**
Dodecylbenzenesulfonic acid, ethanolamine salt	**(26836-07-7)**
Dodecylbenzene sulfonic acid, monoethanolamine salt	**(26836-07-7)**
Dodecylbenzenesulfonic acid monoethanolamine salt	**(26836-07-7)**
Dodecylbenzenesulfonic acid, monoethanolamine salt	**(26836-07-7)**
Dodecylbenzenesulfonic acid monoisopropanolamine salt	**(26264-05-1)**
Dodecyl benzene sulfonic acid, potassium salt	**(27177-77-1)**
Dodecylbenzenesulfonic acid, potassium salt	**(27177-77-1)**
Dodecylbenzenesulfonic acid sodium salt	**(25155-30-0)**
Dodecylbenzenesulfonic acid triethanolamine salt	**(27323-41-7)**

Dodecylbenzenesulphonate, sodium salt	**(25155-30-0)**
Dodecylbenzenesulphonic acid	**(27176-87-0)**
Dodecylbenzensulfonan sodny (Czech)	**(25155-30-0)**
p-Dodecylbenzensulfonan sodny (Czech)	**(25155-30-0)**
Dodecylbenzensulfonic acid calcium salt	**(26264-06-2)**
(Dodecylbenzyl)trimethylammonium chloride	**(1330-85-4)**
Dodecylbenzyl trimethyl ammonium chloride	**(1330-85-4)**
Dodecylbenzyltrimonium chloride	**(1330-85-4)**
Dodecylbis(aminoethyl)glycine	**(6843-97-6)**
Dodecylbis(2-hydroxypropyl)amine	**(1541-66-8)**
Dodecyl bromide	**(143-15-7)**
n-Dodecyl bromide	**(143-15-7)**
Dodecyl chloride	**(112-52-7)**
n-Dodecyl chloride	**(112-52-7)**
Dodecylcyclohexane	**(1795-17-1)**
Dodecyldiamine	**(2783-17-7)**
2-Dodecyl-4,5-dihydro-1H-imidazole-1-ethanol	**(16058-17-6)**
Dodecyldimethylamine	**(112-18-5)**
N-Dodecyldimethylamine	**(112-18-5)**
Dodecyldimethylamine oxide	**(1643-20-5)**
n-Dodecyldimethylamine oxide	**(1643-20-5)**
Dodecyl-dimethyl-benzylammonium chloride	**(139-07-1)**
Dodecyl dimethyl 3,4-dichlorobenzyl ammonium chloride	**(102-30-7)**
Dodecyldimethyl(3,4-dichlorobenzyl)ammonium chloride	**(102-30-7)**
N-Dodecyl-N,N-dimethyl-1-dodecanaminium chloride	**(3401-74-9)**
Dodecyl dimethyl ethylbenzyl ammonium chloride	**(27479-28-3)**
Dodecyldimethyl(ethylbenzyl)ammonium chloride	**(27479-28-3)**
N-Dodecyl-N,N-dimethyl-N-(1-naphthylmethyl)ammonium chloride	**(1733-96-6)**
Dodecyl disodium sulfosuccinate	**(26838-05-1)**
N-Dodecyl-1-dodecanamine	**(3007-31-6)**
Dodecylene	**(25378-22-7)**
Dodecylene	**(6842-15-5)**
Dodecylene α-	**(112-41-4)**
α-Dodecylene	**(112-41-4)**
α-Dodecylene	**(25378-22-7)**
1,12'-Dodecylenediamine	**(2783-17-7)**
Dodecyl ether (8CI)	**(4542-57-8)**
N-Dodecyl-ar-ethyl-N,N-dimethylbenzenemethanaminium chloride	**(27479-28-3)**
Dodecyl glycidyl ether	**(2461-18-9)**
n-Dodecyl glycidyl ether	**(2461-18-9)**
n-Dodecylguanidinacetat (German)	**(2439-10-3)**
Dodecylguanidine acetate	**(112-65-2)**
Dodecylguanidine acetate	**(2439-10-3)**
N-Dodecylguanidine acetate	**(112-65-2)**
n-Dodecylguanidine acetate	**(2439-10-3)**
Dodecylguanidine hydrochloride	**(13590-97-1)**
N-Dodecylguanidine terephthalate	**(19727-17-4)**
Dodecyl hexaethylene glycol	**(3055-96-7)**
Dodecyl hexaoxyethylene monoether	**(3055-96-7)**
α-Dodecyl-ω-hydroxy-polyoxyethylene	**(9002-92-0)**
tert-Dodecyl 2-hydroxypropyl sulfide	**(67124-09-8)**
2,2'-(Dodecylimino)bisethanol	**(1541-67-9)**
1,1'-(Dodecylimino)bis-2-propanol	**(1541-66-8)**
n-Dodecyl iodide	**(4292-19-7)**
1-Dodecyl mercaptan	**(112-55-0)**
Dodecyl mercaptan	**(112-55-0)**
m-Dodecyl mercaptan	**(112-55-0)**
tert-Dodecylmercaptan	**(25103-58-6)**
Dodecyl mercaptoacetate	**(3746-39-2)**
terc.Dodecylmerkaptan (Czech)	**(25103-58-6)**
Dodecyl methacrylate	**(142-90-5)**
1-Dodecyl-4-methylbenzene	**(104-41-6)**
Dodecyl 2-methyl-2-propenoate	**(142-90-5)**
N-Dodecylmorpholine	**(1541-81-7)**
Dodecylnaphthalene	**(38641-16-6)**
Dodecylnitrobenzene	**(58353-63-2)**
Dodecyloxirane	**(3234-28-4)**
2-(Dodecyloxy)ethanol	**(4536-30-5)**
2-(Dodecyloxy)ethanol hydrogen sulfate sodium salt	**(15826-16-1)**
2-(2-(Dodecyloxy)ethoxy)ethanol	**(3055-93-4)**
2-(2-(2-(Dodecyloxy)ethoxy)ethoxy)ethanol	**(3055-94-5)**
2-(2-(2-(Dodecyloxy)ethoxy)ethoxy)ethanol hydrogen sulfate sodium salt	**(13150-00-0)**
2-(2-Dodecyloxyethoxy)ethyl sodium sulfate	**(3088-31-1)**
((Dodecyloxy)methyl)oxirane	**(2461-18-9)**
Dodecylphenol (Mixed isomers)	**(27193-86-8)**
Dodecylphenol hydrogen phosphorodithioate	**(30304-41-7)**
(Dodecylphenoxy)benzene	**(25619-63-0)**
Dodecylphenoxybenzene	**(25619-63-0)**
Dodecyl phosphate	**(12751-23-4)**
Dodecyl-polyaethylenoxyd-aether (German)	**(9002-92-0)**
Dodecyl poly(oxyethylene)ether	**(9002-92-0)**
Dodecyl 2-propenoate	**(2156-97-0)**
1-Dodecylpyridinium chloride	**(104-74-5)**
Dodecylpyridinium chloride	**(104-74-5)**
N-Dodecylpyridinium chloride	**(104-74-5)**
1-Dodecylpyridinium iodide	**(3026-66-2)**
Dodecylpyridinium iodide	**(3026-66-2)**
n-Dodecylpyridinium iodide	**(3026-66-2)**
N-(n-Dodecyl)-pyridinium-jodid (German)	**(3026-66-2)**
Dodecyl selenide	**(5819-01-2)**
Dodecyl sodium ethoxysulfate	**(15826-16-1)**
Dodecyl sodium sulfate	**(151-21-3)**
Dodecyl sodium sulfoacetate	**(1847-58-1)**
Dodecyl sulfate	**(151-41-7)**
Dodecyl sulfate diethanolamine salt	**(143-00-0)**
Dodecyl sulfate sodium	**(151-21-3)**
n-Dodecyl sulfate sodium	**(151-21-3)**
Dodecyl sulfate, sodium salt	**(151-21-3)**
Dodecylsulfonic acid	**(1510-16-3)**
Dodecyl(sulfophenoxy)benzenesulfonic acid	**(30260-72-1)**
Dodecyl(sulfophenoxy)benzenesulfonic acid, disodium salt	**(28519-02-0)**
Dodecylsulfuric acid	**(151-41-7)**
Dodecyltetraethylene glycol monoether	**(5274-68-0)**
Dodecyl thioglycolate	**(3746-39-2)**
tert-Dodecylthiol	**(25103-58-6)**
1-((Dodecylthio)methyl)-3-methylpyridinium chloride	**(70700-60-6)**
1-(tert-Dodecylthio)-2-propanol	**(67124-09-8)**
Dodecyl p-tolyl trimethyl ammonium chloride	**(1399-80-0)**
8-Dodecyl-2,5,8-triazaoctane-1-carboxylic acid	**(6843-97-6)**
Dodecyl trichlorosilane	**(4484-72-4)**
Dodecyltrichlorosilane [UN 1771]	**(4484-72-4)**
Dodecyl triethylene glycol ether	**(3055-94-5)**
Dodecyltrimethylammonium	**(10182-91-9)**
Dodecyltrimethylammonium bromide	**(10182-91-9)**
Dodecyltrimethylammonium chloride	**(112-00-5)**
n-Dodecyltrimethylammonium chloride	**(112-00-5)**
Dodecyltrimethylammonium methosulfate	**(13623-06-8)**
Dodecyltrimethylammonium methyl sulfate	**(13623-06-8)**
4-Dodecyl-N,N,N-trimethylbenzenemethanaminium chloride	**(1330-85-4)**
6-Dodecyl-2,2,4-trimethyl-1,2-dihydroquinoline	**(89-28-1)**
Dodecylurea	**(2158-09-0)**
1-Dodecyn-3-ol, 3,7,11-trimethyl- (8CI,9CI)	**(1604-35-9)**
Dodemorfe (French)	**(31717-87-0)**
Dodemorph acetate	**(31717-87-0)**
Dodex	**(68-19-9)**
Dodicin	**(6843-97-6)**
Dodigen 226	**(8001-54-5)**
Dodin	**(2439-10-3)**
Dodine	**(112-65-2)**
Dodine	**(2439-10-3)**
Dodine, Mixture with glyodin	**(2439-10-3)**

Dodine acetate	**(2439-10-3)**
Doe	**(300-42-5)**
Dofsol	**(68-26-8)**
Dogmatil	**(15676-16-1)**
Dogmatyl	**(15676-16-1)**
Doguadine	**(2439-10-3)**
Dojyopicrin	**(76-06-2)**
Dokirin	**(10380-28-6)**
Doktacillin	**(69-53-4)**
Doladene	**(53-46-3)**
Dolamin	**(7783-20-2)**
Dolantal	**(50-13-5)**
Dolantin	**(50-13-5)**
Dolantin	**(57-42-1)**
Dolantin hydrochloride	**(50-13-5)**
Dolantol	**(50-13-5)**
Dolaren	**(50-13-5)**
Dolargan	**(50-13-5)**
Dolco mouse cereal	**(57-24-9)**
Dolcontral	**(50-13-5)**
Dolcontral	**(57-42-1)**
Dolcymene	**(99-87-6)**
Dolean pH 8	**(50-78-2)**
Dolen-Pur	**(87-68-3)**
Dolenal	**(50-13-5)**
Dolene	**(469-62-5)**
Dolenol	**(50-13-5)**
Dolestan	**(147-24-0)**
Dolestine	**(50-13-5)**
Dolgin	**(15687-27-1)**
Dol granule	**(58-89-9)**
Dolicur	**(67-68-5)**
Doligur	**(67-68-5)**
Dolin	**(50-13-5)**
Dolipol	**(64-77-7)**
Doliprane	**(103-90-2)**
Dolkwal Amaranth	**(915-67-3)**
Dolkwal Brilliant Blue	**(3844-45-9)**
Dolkwal Erythrosine	**(16423-68-0)**
Dolkwal Indigo Carmine	**(860-22-0)**
Dolkwal Orange SS	**(2646-17-5)**
Dolkwal Ponceau 3R	**(3564-09-8)**
Dolkwal Sunset Yellow	**(2783-94-0)**
Dolkwal Tartrazine	**(1934-21-0)**
Dolkwal Yellow AB	**(85-84-7)**
Dolkwal Yellow OB	**(131-79-3)**
Dolochlor	**(76-06-2)**
Dologal	**(50-13-5)**
Dolomide	**(65-45-2)**
Dolomite (9CI)	**(16389-88-1)**
Doloneurine	**(50-13-5)**
Dolonil	**(136-40-3)**
Dolopethin	**(50-13-5)**
Dolophin	**(76-99-3)**
Dolophine	**(76-99-3)**
Dolosal	**(50-13-5)**
Dolosal	**(57-42-1)**
Dolostop	**(62-44-2)**
Dolovin	**(53-86-1)**
Doloxene	**(469-62-5)**
Dolsin	**(57-42-1)**
Dolvanol	**(50-13-5)**
Dolviran	**(62-44-2)**
Domafate	**(51-63-8)**
Domalium	**(439-14-5)**
Domar	**(52463-83-9)**
Domarax	**(78-44-4)**
Domatol	**(61-82-5)**
Domatol 88	**(61-82-5)**
δ-Dome	**(53-03-2)**
Domeboro	**(139-12-8)**
Domestrol	**(56-53-1)**
Domolite	**(1317-65-3)**
Domoso	**(67-68-5)**
Don	**(51481-10-8)**
Donmox	**(59-66-5)**
Dooje	**(561-27-3)**
(-)-Dopa	**(59-92-7)**
Dopa	**(59-92-7)**
l-Dopa	**(59-92-7)**
Dopacetic acid	**(102-32-9)**
Dopaflex	**(59-92-7)**
Dopaidan	**(59-92-7)**
Dopal	**(59-92-7)**
Dopal-Fher	**(59-92-7)**
Dopalina	**(59-92-7)**
Dopamet	**(41372-08-1)**
Dopamet	**(555-30-6)**
Dopamine	**(51-61-6)**
Dopar	**(59-92-7)**
Doparkine	**(59-92-7)**
Doparl	**(59-92-7)**
Dopasol	**(59-92-7)**
Dopaston	**(59-92-7)**
Dopastral	**(59-92-7)**
Dopatec	**(41372-08-1)**
Dopatec	**(555-30-6)**
Dopegyt	**(41372-08-1)**
Dopegyt	**(555-30-6)**
Dopidrin	**(300-42-5)**
Dopram	**(309-29-5)**
Doprin	**(59-92-7)**
Doquadine	**(2439-10-3)**
Dora-C-500	**(50-81-7)**
Doral	**(50-14-6)**
Dorantamin	**(91-84-9)**
Dorbane	**(117-10-2)**
Dorbanex	**(117-10-2)**
Dorico	**(56-29-1)**
Doriden	**(77-21-4)**
Doriden-Sed	**(77-21-4)**
Dorison	**(77-75-8)**
Dorlone II	**(542-75-6)**
Dorlotyn	**(57-43-2)**
Dorm	**(52-43-7)**
Dormabrol	**(57-53-4)**
Dormal	**(302-17-0)**
Dormalest	**(77-75-8)**
Dormalin	**(36735-22-5)**
Dormallyl	**(52-43-7)**
Dormate	**(64-55-1)**
Dorme	**(58-33-3)**
Dormidin	**(77-75-8)**
Dormigen	**(77-75-8)**
Dormigoa	**(72-44-6)**
Dormin	**(91-80-5)**
Dormin-5	**(146-22-5)**
Dormina	**(50-06-6)**
Dormiphen	**(77-75-8)**
Dormiral	**(50-06-6)**
Dormison	**(77-75-8)**
Dormiturin	**(77-65-6)**
Dormocit	**(77-75-8)**
Dormodor	**(1172-18-5)**

Name	CAS No.
Dormogen	**(72-44-6)**
Dormonal	**(57-44-3)**
Dormone	**(94-75-7)**
Dormosan	**(77-75-8)**
Dormutil	**(72-44-6)**
Dormytal	**(57-43-2)**
Dorsallin A.R.	**(6130-64-9)**
Dorsedin	**(72-44-6)**
Dorsital	**(76-74-4)**
Dorsulfan	**(127-69-5)**
Dorvicide A	**(132-27-4)**
Dorvon	**(9003-53-6)**
Dorvon FR 100	**(9003-53-6)**
Doryl (Pharmaceutical)	**(51-83-2)**
Doryl V 505-50	**(26007-63-6)**
Doryl (VAN)	**(51-83-2)**
Dosaflo	**(19937-59-8)**
Dosagran	**(19937-59-8)**
Dosanex	**(19937-59-8)**
Dosanex FL	**(19937-59-8)**
Dosanex MG	**(19937-59-8)**
Doscalun	**(50-06-6)**
Dosulepin	**(113-53-1)**
Dotan	**(24934-91-6)**
Dothiepin	**(113-53-1)**
Dotite Kalibor	**(143-66-8)**
Dotment 324	**(1344-28-1)**
Dotment 358	**(1344-28-1)**
Dotriacontane	**(544-85-4)**
Dotriacontane, 2-methyl- (8CI,9CI)	**(1720-11-2)**
Dotriacontane, 3-methyl- (8CI,9CI)	**(20129-49-1)**
Dotriacontanoic acid	**(3625-52-3)**
Dotycin	**(114-07-8)**
Double-M EC	**(72-43-5)**
Double Strength	**(93-72-1)**
Dovip	**(52-85-7)**
Dow 209	**(9003-55-8)**
Dow 234	**(9003-55-8)**
Dow 360	**(9003-53-6)**
Dow 456	**(9003-53-6)**
Dow 460	**(9003-55-8)**
Dow 620	**(9003-55-8)**
Dow 665	**(9003-53-6)**
Dow 680	**(9003-55-8)**
Dow 816	**(9003-55-8)**
Dow 860	**(9003-53-6)**
Dow 874	**(9011-06-7)**
Dow 1329	**(299-85-4)**
Dow 1683	**(9003-53-6)**
Dow Corning 93-120	**(63148-62-9)**
Dow Corning 200	**(107-46-0)**
Dow Corning 200	**(63148-62-9)**
Dow Corning 346	**(9016-00-6)**
Dow Corning 561	**(63148-62-9)**
Dow Defoliant	**(3926-62-3)**
Dow Dormant Fungicide	**(131-52-2)**
Dow ET 14	**(299-84-3)**
Dow ET 57	**(299-84-3)**
Dow General	**(88-85-7)**
Dow General Weed Killer	**(88-85-7)**
Dow Latex 612	**(9003-55-8)**
Dow Latex 874	**(9011-06-7)**
Dow MCP Amine Weed Killer	**(94-74-6)**
Dow MX 5514	**(9003-53-6)**
Dow MX 5516	**(9003-53-6)**
Dow Pentachlorophenol DP-2 Antimicrobial	**(87-86-5)**
Dow-Per	**(127-18-4)**
Dow Resin 565	**(116-37-0)**
Dow Seed Disinfectant No. 5	**(118-75-2)**
Dow Selective	**(6365-83-9)**
Dow Selective Weed Killer	**(88-85-7)**
Dow Sodium TCA Inhibited	**(650-51-1)**
Dow-Tri	**(79-01-6)**
Dow Z-200	**(141-98-0)**
Dowanol	**(10213-77-1)**
Dowanol	**(111-90-0)**
Dowanol 33B	**(107-98-2)**
Dowanol 50B	**(34590-94-8)**
Dowanol 62B	**(10213-77-1)**
Dowanol 62B	**(20324-33-8)**
Dowanol BM	**(111-32-0)**
Dowanol DB	**(112-34-5)**
Dowanol DE	**(111-90-0)**
Dowanol DM	**(111-77-3)**
Dowanol DPM	**(34590-94-8)**
Dowanol EB	**(111-76-2)**
Dowanol EE	**(110-80-5)**
Dowanol EM	**(109-86-4)**
Dowanol EP	**(122-99-6)**
Dowanol EPH	**(122-99-6)**
Dowanol Eipat	**(109-59-1)**
Dowanol PIB-T	**(23436-19-3)**
Dowanol PM	**(107-98-2)**
Dowanol PM Glycol Ether	**(107-98-2)**
Dowanol PPH Glycol Ether	**(4169-04-4)**
Dowanol TE	**(112-50-5)**
Dowanol TMAT	**(112-35-6)**
Dowanol TPM	**(25498-49-1)**
Dowanol TPM glycol ether	**(20324-33-8)**
Dowathurm A	**(58841-70-6)**
Dowchlor	**(57-74-9)**
Dowcide 1	**(90-43-7)**
Dowcide 2	**(95-95-4)**
Dowcide 31	**(85-97-2)**
Dowcide 32	**(85-97-2)**
Dowcide 7	**(87-86-5)**
Dowcide 2S	**(88-06-2)**
Dowco 453EE	**(87237-48-7)**
Dowell L 37	**(6419-19-8)**
Dowflake	**(10043-52-4)**
Dowfrost	**(57-55-6)**
Dowfroth 250	**(37286-64-9)**
Dowfume	**(106-93-4)**
Dowfume	**(74-83-9)**
Dowfume 40	**(106-93-4)**
Dowfume EDB	**(106-93-4)**
Dowfume MC 2	**(8004-09-9)**
Dowfume MC 33	**(8004-09-9)**
Dowfume MC-2	**(74-83-9)**
Dowfume MC-33	**(74-83-9)**
Dowfume MC-2 soil fumigant	**(74-83-9)**
Dowfume N	**(8003-19-8)**
Dowfume W-100	**(106-93-4)**
Dowfume W-8	**(106-93-4)**
Dowfume W-85	**(106-93-4)**
Dowfume W-90	**(106-93-4)**
Dowicide	**(132-27-4)**
Dowicide 1	**(90-43-7)**
Dowicide 2	**(95-95-4)**
Dowicide 4	**(92-04-6)**
Dowicide 6	**(58-90-2)**
Dowicide 7	**(87-86-5)**
Dowicide 9	**(13347-42-7)**
Dowicide 31	**(10605-10-4)**

Dowicide 31	**(85-97-2)**
Dowicide 32	**(85-97-2)**
Dowicide A	**(132-27-4)**
Dowicide A & A Flakes	**(132-27-4)**
Dowicide 1 Antimicrobial	**(90-43-7)**
Dowicide B	**(136-32-3)**
Dowicide B	**(95-95-4)**
Dowicide EC-7	**(87-86-5)**
Dowicide G	**(131-52-2)**
Dowicide G	**(87-86-5)**
Dowicide G-ST	**(131-52-2)**
Dowicide Q	**(4080-31-3)**
Dowicide 2S	**(88-06-2)**
Dowicil 75	**(4080-31-3)**
Dowicil 100	**(4080-31-3)**
Dowicil A-40	**(38827-35-9)**
Dowizid A	**(132-27-4)**
Dowlap F	**(88-30-2)**
Dowlex	**(25038-59-9)**
Dowlex Film	**(9002-88-4)**
Dowmycin E	**(643-22-1)**
Dowpen V-K	**(132-98-9)**
Dowpon	**(127-20-8)**
Dowpon	**(75-99-0)**
Dowpon M	**(75-99-0)**
Dowspray 9	**(93-52-7)**
Dowspray 17	**(131-89-5)**
Dowtex TL 612	**(9003-55-8)**
Dowtherm	**(8004-13-5)**
Dowtherm 209	**(107-98-2)**
Dowtherm A	**(8004-13-5)**
Dowtherm E	**(95-50-1)**
Dowtherm G	**(62587-63-7)**
Dowtherm SR 1	**(107-21-1)**
Dowzene DHC	**(142-64-3)**
Dox	**(25316-40-9)**
Doxapram	**(309-29-5)**
Doxcide 50	**(10049-04-4)**
Doxephin	**(300-42-5)**
Doxephrin	**(300-42-5)**
Doxepin	**(1668-19-5)**
Dox hydrochloride	**(25316-40-9)**
Doxiciclina (Italian)	**(564-25-0)**
Doxinate	**(577-11-7)**
Doxol	**(577-11-7)**
Doxorubicin	**(23214-92-8)**
Doxorubicin	**(25316-40-9)**
Doxorubicin hydrochloride	**(25316-40-9)**
Doxycycline	**(564-25-0)**
Doxyfed	**(300-42-5)**
Doxylamine	**(469-21-6)**
Doxylamine succinate	**(562-10-7)**
Doxylamine succinate (1:1)	**(562-10-7)**
No-Doz	**(58-08-2)**
Dozar	**(135-23-9)**
Dozer	**(4482-55-7)**
Dqoci	**(514-73-8)**
Dri-Die Insecticide 67	**(7631-86-9)**
Dri-Kil	**(83-79-4)**
Dri-Tri	**(7601-54-9)**
Drabet	**(64-77-7)**
Draconic acid	**(100-09-4)**
Dracylic acid	**(65-85-0)**
Dragil-P	**(1323-39-3)**
Dragnet	**(52645-53-1)**
Dragonthol A	**(92-77-3)**
Dragonthol BO	**(132-68-3)**
Dragonthol BS	**(135-65-9)**
Dragonthol D	**(135-61-5)**
Dragonthol OL	**(135-62-6)**
Drakeol	**(8012-95-1)**
Dralon T	**(25014-41-9)**
Dralzine	**(304-20-1)**
Dramamin	**(523-87-5)**
Dramamine	**(523-87-5)**
Dramarin	**(523-87-5)**
Dramcillin-S	**(132-93-4)**
Dramyl	**(523-87-5)**
Drapex 3.2	**(106-84-3)**
Drapex 4.4	**(106-84-3)**
Drapex 4.4	**(61788-72-5)**
Draplex 3.2	**(106-84-3)**
Drapolene	**(8001-54-5)**
Drapolex	**(8001-54-5)**
Drat	**(3691-35-8)**
Drawin 755	**(34681-10-2)**
Drawinol	**(973-21-7)**
Draza	**(2032-65-7)**
Drazoxolon	**(5707-69-7)**
Drazoxolone	**(5707-69-7)**
Dreft	**(151-21-3)**
Drenamist	**(51-43-4)**
Drene	**(139-96-8)**
Drenison	**(1524-88-5)**
Drenol	**(58-93-5)**
Drenusil	**(346-18-9)**
Drenusil-R	**(50-55-5)**
Drepamon	**(36756-79-3)**
Drewamine	**(110-91-8)**
Drewmulse POE-SMO	**(9005-65-6)**
Drewmulse TP	**(31566-31-1)**
Drewmulse V	**(31566-31-1)**
Drewsorb 60	**(1338-41-6)**
Drexel	**(330-54-1)**
Drexel DSMA Liquid	**(144-21-8)**
Drexel Defol	**(7775-09-9)**
Drexel Diuron 4L	**(330-54-1)**
Drexel Methyl Parathion 4E	**(298-00-0)**
Drexel-Super P	**(123-33-1)**
Drexel Parathion 8E	**(56-38-2)**
Dridol	**(548-73-2)**
Dried Blood	**(68911-49-9)**
Drierite	**(7778-18-9)**
Drill tox-spezial aglukon	**(58-89-9)**
Drillzid	**(2104-96-3)**
Drimarene Brilliant Green X 3G	**(12225-80-8)**
Drimarene Turquoise X-2G	**(1658-56-6)**
Drinalfa	**(300-42-5)**
Drinox	**(309-00-2)**
Drinox	**(76-44-8)**
Drinox H-34	**(76-44-8)**
Drisdol	**(50-14-6)**
Dristan	**(101-40-6)**
Dristan Inhaler	**(101-40-6)**
Drocode	**(125-28-0)**
Drocort	**(1524-88-5)**
Droleptan	**(548-73-2)**
Drometrizole	**(2440-22-4)**
Dromilac	**(1239-45-8)**
Dromisol	**(67-68-5)**
Dromoran	**(297-90-5)**
levo-Dromoran	**(77-07-6)**
Dromostanolone	**(58-19-5)**
Dromyl	**(523-87-5)**

Dronactin	(129-03-3)
Droncit	(55268-74-1)
Drop Leaf	(7775-09-9)
Dropcillin	(61-33-6)
Droperidol	(548-73-2)
Dropp	(51707-55-2)
Dropsprin	(65-45-2)
Drostanolone	(58-19-5)
Drotebanol	(3176-03-2)
Drotebanolum (Latin)	(3176-03-2)
Droxolan	(83-44-3)
Drumulse AA	(31566-31-1)
Drupina 90	(137-30-4)
Dry Ice [UN 1845]	(124-38-9)
Dry Mix No. 1	(131-89-5)
Dry and Clear	(94-36-0)
Dryistan	(58-73-1)
Drylin	(8064-90-2)
Drylistan	(58-73-1)
Dryobalanops camphor	(507-70-0)
Dryptal	(54-31-9)
Du Pont 326	(330-55-2)
Du Pont 634	(2164-08-1)
Du Pont 732	(5902-51-2)
Du Pont 1179	(16752-77-5)
Du Pont 1318	(1982-49-6)
Du Pont 1519	(6988-21-2)
Du Pont 1991	(17804-35-2)
Du Pont Gasoline Antioxidant No. 5	(103-62-8)
Du Pont Herbicide 82	(314-42-1)
Du Pont Herbicide 732	(5902-51-2)
Du Pont Herbicide 976	(314-40-9)
Du Pont Herbicide 1,318	(1982-49-6)
Du Pont Insecticide 1179	(16752-77-5)
Du Pont Insecticide 1519	(6988-21-2)
Du Pont Metal Deactivator	(94-91-7)
Du Pont WK	(9002-92-0)
Du-Sprex	(1194-65-6)
Du-Ter	(76-87-9)
Du-Ter W-50	(76-87-9)
Dus-Top	(7786-30-3)
Dual	(51218-45-2)
Duatok	(72-14-0)
Dublofix	(75-00-3)
Duboisine	(101-31-5)
Dubronax	(80-08-0)
Duckalgin	(9005-38-3)
Ducobee	(68-19-9)
Ducobee-Hy	(13422-51-0)
Dufalone	(66-76-2)
Dufaston	(152-62-5)
Duiramid	(59-66-5)
Dukeron	(79-01-6)
Duksen	(439-14-5)
Dulcidor	(15879-93-3)
Dulcin	(150-69-6)
Dulcine	(150-69-6)
Dulcitol	(608-66-2)
Dulcolan	(603-50-9)
Dulcolax	(603-50-9)
Dull 704	(105-60-2)
Dull 704	(25038-54-4)
Dulsivac	(577-11-7)
Dulzor-Etas	(139-05-9)
Dumasin	(120-92-3)
Dumitone	(80-08-0)
Dumocycin	(64-75-5)
Dumogran	(58-18-4)
Duncaine	(137-58-6)
Duneryl	(50-06-6)
Dunkelgelb	(842-07-9)
Duo-Kill	(62-73-7)
Duo-Kill	(7700-17-6)
Duodecane	(112-40-3)
Duodecibin	(68-19-9)
Duodecyl alcohol	(112-53-8)
Duodecylic acid	(143-07-7)
Duodecylic aldehyde	(112-54-9)
Duodex	(2492-26-4)
Duolax	(117-10-2)
Duomycin	(57-62-5)
Duoscorb	(50-81-7)
Duosol	(577-11-7)
Duotrate	(78-11-5)
Duphapen	(54-35-3)
Duphar	(116-29-0)
Duphaston	(152-62-5)
Duplex Permaton Red L 20-7022	(2814-77-9)
Duplex Red Lake C20-5925	(5160-02-1)
Duplex Toluidine Red L 20-3140	(2425-85-6)
Dupon 4472	(97-77-8)
Duponal	(151-21-3)
Duponal Waqe	(151-21-3)
Duponol	(151-21-3)
Duponol 80	(142-31-4)
Duponol C	(151-21-3)
Duponol ME	(151-21-3)
Duponol Methyl	(151-21-3)
Duponol QX	(151-21-3)
Duponol WA	(151-21-3)
Duponol WA Dry	(151-21-3)
Duponol Waq	(151-21-3)
Duponol Waqa	(151-21-3)
Duponol Waqe	(151-21-3)
Duponol Waqm	(151-21-3)
Dupont Fungicide 4472	(97-77-8)
Dupont Herbicide 326	(330-55-2)
Dupont PC Crabgrass Killer	(590-28-3)
Duprene	(9010-98-4)
Dur-EM 204	(25496-72-4)
Dura Clofibrat	(637-07-0)
Dura Dex	(51-63-8)
Dura-Tab S.M. Aminophylline	(317-34-0)
Durabiotic	(1538-09-6)
Durabolin-O	(965-90-2)
Duraboral	(965-90-2)
Duracillin	(6130-64-9)
Durad	(1330-78-5)
Duradoce	(13422-51-0)
Dura-estradiol	(979-32-8)
Durafur Black R	(106-50-3)
Durafur Black RC	(624-18-0)
Durafur Brown	(5307-14-2)
Durafur Brown MN	(39156-41-7)
Durafur Brown 2R	(5307-14-2)
Durafur Brown RB	(123-30-8)
Durafur Developer C	(120-80-9)
Durafur Developer D	(90-15-3)
Durafur Developer E	(83-56-7)
Durafur Developer G	(108-46-3)
Dural	(1344-28-1)
Duralta-12	(13422-51-0)
Duraluton	(630-56-8)
Duramax	(50-78-2)

Name	CAS No.
Duran	**(330-54-1)**
Duranest	**(36637-18-0)**
Duranil Aerosol	**(141-94-6)**
Duranit	**(9003-55-8)**
Duranit 40	**(9003-55-8)**
Duranol Blue PP	**(14233-37-5)**
Duranol Blue TR	**(12217-79-7)**
Duranol Brilliant Blue B	**(2475-46-9)**
Duranol Brilliant Blue B	**(86722-66-9)**
Duranol Brilliant Blue BN	**(2475-46-9)**
Duranol Brilliant Blue BN	**(86722-66-9)**
Duranol Brilliant Blue CB	**(2475-45-8)**
Duranol Brilliant Blue G	**(2475-44-7)**
Duranol Brilliant Blue Violet BR	**(82-33-7)**
Duranol Brilliant Red T 2B	**(17418-58-5)**
Duranol Brilliant Violet B	**(1220-94-6)**
Duranol Brilliant Violet BR	**(82-33-7)**
Duranol Orange G	**(82-28-0)**
Duranol Printing Blue B	**(2475-46-9)**
Duranol Printing Blue B	**(86722-66-9)**
Duranol Red 2B	**(116-85-8)**
Duranol Red GN	**(82-38-2)**
Duranol Red X3B	**(2872-48-2)**
Duranol Violet 2R	**(128-95-0)**
Durapatite	**(1306-06-5)**
Dura-penita	**(1538-09-6)**
Duraphos	**(7786-34-7)**
Duraset	**(85-72-3)**
Duraset 20W	**(85-72-3)**
Durasorb	**(67-68-5)**
Duratint Blue 1001	**(147-14-8)**
Duratint Green 1001	**(1328-53-6)**
Duratox	**(8022-00-2)**
Duratox	**(919-86-8)**
Duravos	**(62-73-7)**
Duravos	**(7700-17-6)**
Durax	**(95-33-0)**
Durazol Brillant Red BS	**(2610-10-8)**
Durazol Brown BR	**(16071-86-6)**
Durazol Red 2B	**(2610-11-9)**
Durazol Red 2BP	**(2610-11-9)**
Durene	**(95-93-2)**
Durenol	**(527-35-5)**
Durethan BK	**(105-60-2)**
Durethan BK	**(25038-54-4)**
Durethan BK 30S	**(105-60-2)**
Durethan BK 30S	**(25038-54-4)**
Durethan BKV 30H	**(105-60-2)**
Durethan BKV 30H	**(25038-54-4)**
Durethan BKV 55H	**(105-60-2)**
Durethan BKV 55H	**(25038-54-4)**
Durette	**(63428-84-2)**
Duretter	**(7720-78-7)**
Durex	**(1333-86-4)**
Durfax 80	**(9005-65-6)**
Durgacet Yellow G	**(2832-40-8)**
Duricef	**(50370-12-2)**
Durindone Orange R	**(3263-31-8)**
Durindone Orange RP	**(3263-31-8)**
Durindone Pink FF	**(2379-74-0)**
Durindone Pink FF-FA	**(2379-74-0)**
Durindone Printing Orange R	**(3263-31-8)**
Durindone Printing Pink FF	**(2379-74-0)**
Durochrome Violet B	**(2092-55-9)**
Durofast Brown BRL	**(16071-86-6)**
Duroferon	**(7720-78-7)**
Duromine	**(122-09-8)**
Duropenin	**(1538-09-6)**
Durophet	**(60-13-9)**
Durosperse Yellow G	**(2832-40-8)**
Durotox	**(87-86-5)**
Durox	**(80-35-3)**
Dursban	**(2921-88-2)**
Dursban F	**(2921-88-2)**
Dursban Methyl	**(5598-13-0)**
Durtan 60	**(1338-41-6)**
Durylaldehyde	**(5779-72-6)**
Durylic acid	**(528-90-5)**
Dusicnan barnaty (Czech)	**(10022-31-8)**
Dusicnan cerity (Czech)	**(10108-73-3)**
Dusicnan zinecnaty (Czech)	**(10196-18-6)**
Dusicnan zirkonicity (Czech)	**(13746-89-9)**
Dusitan dicyklohexylaminu (Czech)	**(3129-91-7)**
Dusitan sodny (Czech)	**(7632-00-0)**
Dusoline	**(57-88-5)**
Dusoran	**(57-88-5)**
Dust, Ferrous metal, Blast furnace	**(65996-70-5)**
Dust M	**(22248-79-9)**
Dust, Steelmaking	**(65996-72-7)**
Dutch-Treat	**(124-65-2)**
Dutch Liquid	**(107-06-2)**
Dutch Oil	**(107-06-2)**
Duter	**(76-87-9)**
Dutom	**(2307-68-8)**
Duvadilan	**(395-28-8)**
Duvaron	**(152-62-5)**
Duvilax	**(9003-20-7)**
Duvilax BD 20	**(9003-20-7)**
Duvilax HN	**(9003-20-7)**
Duvilax LM 52	**(9003-20-7)**
Duxen	**(439-14-5)**
Duxon	**(40704-75-4)**
Dwell	**(2593-15-9)**
Dwubromoetan (Polish)	**(106-93-4)**
Dwuchlorantyny (Polish)	**(118-52-5)**
Dwuchloroczterofluoroetan (Polish)	**(1320-37-2)**
Dwuchlorodwuetylowy eter (Polish)	**(111-44-4)**
Dwuchlorodwufluorometan (Polish)	**(75-71-8)**
1,3-Dwuchloro-5,5-dwumetylohydantoina (Polish)	**(118-52-5)**
2,4-Dwuchlorofenoksyoctowy kwas (Polish)	**(94-75-7)**
N-(3,4-Dwuchlorofenylo)N'-metoksy-N'-metylomocznik (Polish)	**(330-55-2)**
Dwuchlorofluorometan (Polish)	**(75-43-4)**
Dwuchloropropan (Polish)	**(78-87-5)**
Dwuchlorostyren (Polish)	**(6607-45-0)**
O,O-Dwuetylo-O-1-(2,4-dwuchlorofenylo)-2-chloro-winylofosforan (Polish)	**(470-90-6)**
Dwuetyloamina (Polish)	**(109-89-7)**
Dwuetylowy eter (Polish)	**(60-29-7)**
Dwufenyloguanidyna (Polish)	**(102-06-7)**
5,5-Dwufenylohydantoina (Polish)	**(57-41-0)**
p,p'-Dwumetoksydwufenylotrojchloroetan (Polish)	**(72-43-5)**
2,6-Dwumetoksyfenol (Polish)	**(91-10-1)**
O,O-Dwumetylo-O-1-(2,4-dwuchlorofenylo)-2-bromo-winylofosforan (Polish)	**(13104-21-7)**
O,O-Dwumetylo-O-dwuchlorowinylofosforan (Polish)	**(62-73-7)**
Dwumetyloanilina (Polish)	**(121-69-7)**
O,O-Dwumetylo-S-1,2-bis(karboetoksyetylo)-dwutio-fosforan (Polish)	**(121-75-5)**
Dwumetyloformamid (Polish)	**(68-12-2)**
Dwumetylosulfotlenku (Polish)	**(109-77-3)**
Dwumetylowy siarczan (Polish)	**(77-78-1)**
Dwu-β-naftylo-p-fenylodwuamina (Polish)	**(93-46-9)**
Dwunitrobenzen (Polish)	**(99-65-0)**
Dwunitro-o-krezol (Polish)	**(534-52-1)**

Name	CAS No.
3,3'-Dwuoksybenzydyna (Polish)	**(2373-98-0)**
Dwusiarczek dwubenzotiazylu (Polish)	**(120-78-5)**
Dwutiofosforan S-N-etylokarbamylometylo-O,O-dwumetylowy (Polish)	**(116-01-8)**
Dxewmulse POE-SML	**(9005-64-5)**
Dyall	**(9002-88-4)**
Dyanacide	**(62-38-4)**
Dyanap	**(132-66-1)**
Dyazide	**(58-93-5)**
Dybar	**(101-42-8)**
Dybar	**(122-14-5)**
Dybenal	**(1777-82-8)**
Dycarb	**(22781-23-3)**
Dydeltrone	**(50-24-8)**
Dydrogesterone	**(152-62-5)**
Dye, Benzanthrone	**(82-05-3)**
Dye Evans Blue	**(314-13-6)**
Dye FD & C Red 2	**(915-67-3)**
Dye FD & C Red No. 3	**(16423-68-0)**
Dye FD & C Red No. 4	**(4548-53-2)**
Dye FD & C Yellow Lake 6	**(2783-94-0)**
Dye FD & C Yellow No. 5	**(1934-21-0)**
Dye FD & C Yellow No. 6	**(2783-94-0)**
Dye FDC Yellow No. 6	**(2783-94-0)**
Dye GS	**(5307-14-2)**
Dye Orange No. 1	**(523-44-4)**
Dye Quinoline Yellow	**(8004-92-0)**
Dye Red Raspberry	**(915-67-3)**
Dye Sunset Yellow	**(2783-94-0)**
Dyestrol	**(56-53-1)**
Dyetone	**(7789-38-0)**
Dyflos	**(55-91-4)**
Dyfonate	**(944-22-9)**
Dygratyl	**(67-96-9)**
Dykanol	**(1336-36-3)**
Dykol	**(50-29-3)**
Dykon	**(126-96-5)**
Dylamon	**(58-73-1)**
Dylan	**(9002-88-4)**
Dylan Super	**(9002-88-4)**
Dylan WPD 205	**(9002-88-4)**
Dylark 250	**(9003-53-6)**
Dylene	**(9003-53-6)**
Dylene 8	**(9003-53-6)**
Dylene 9	**(9003-53-6)**
Dylene 8G	**(9003-53-6)**
Dylephrin	**(51-43-4)**
Dylite F 40	**(9003-53-6)**
Dylite F 40L	**(9003-53-6)**
Dyloform	**(57-63-6)**
Dylox	**(52-68-6)**
Dylox-Metasystox-R	**(52-68-6)**
Dymadon	**(103-90-2)**
Dymel 22	**(75-45-6)**
Dymelor	**(968-81-0)**
Dymex	**(98-86-2)**
Dyna-Zina	**(50-49-7)**
Dynacoryl	**(59-26-7)**
Dynalione	**(102-30-7)**
Dynaltone	**(102-30-7)**
Dynamicarde	**(59-26-7)**
Dynaphenyl	**(51-63-8)**
Dynaprin	**(50-49-7)**
Dynasan 114	**(555-45-3)**
Dynasan 118	**(555-43-1)**
Dynasten	**(434-07-1)**
Dynazone	**(59-87-0)**
Dyneric	**(50-41-9)**
Dynex	**(330-54-1)**
Dynh	**(9002-88-4)**
Dynium chloride	**(102-30-7)**
Dynk 2	**(9002-88-4)**
Dynomin UI 16	**(9011-05-6)**
Dynomin UM 15	**(9011-05-6)**
Dynone	**(19622-19-6)**
Dynosol	**(2312-76-7)**
Dyodin	**(83-73-8)**
Dypertane Compound	**(50-55-5)**
Dyphonate	**(944-22-9)**
Dyphylline	**(479-18-5)**
Dyprin	**(59-51-8)**
Dyrene	**(101-05-3)**
Dyrene 50W	**(101-05-3)**
Dyrenium	**(396-01-0)**
Dyrex	**(52-68-6)**
Dyspne-inhal	**(51-43-4)**
Dysprosium (8CI,9CI)	**(7429-91-6)**
Dytac	**(396-01-0)**
Dythol	**(57-88-5)**
Dytol E-46	**(112-92-5)**
Dytol F-11	**(36653-82-4)**
Dytol J-68	**(112-53-8)**
Dytol M-83	**(111-87-5)**
Dytol R-52	**(112-72-1)**
Dytol S-91	**(112-30-1)**
Dyvon	**(52-68-6)**
Dyzol	**(333-41-5)**
686E	**(9003-53-6)**
E 39 Soluble	**(800-24-8)**
E 102	**(1934-21-0)**
E 104	**(8004-92-0)**
E 110	**(2783-94-0)**
E 111	**(2347-72-0)**
E 120	**(1260-17-9)**
E 123	**(915-67-3)**
E 124	**(2611-82-7)**
E 125	**(3257-28-1)**
E 126	**(5850-44-2)**
E 127	**(16423-68-0)**
E 130	**(81-77-6)**
E 131	**(129-17-9)**
E 132	**(860-22-0)**
E 151	**(2519-30-4)**
E 158	**(17040-19-6)**
E 376-40	**(9004-35-7)**
E 383-40	**(9004-35-7)**
E 394-30	**(9004-35-7)**
E 394-40	**(9004-35-7)**
E 394-45	**(9004-35-7)**
E 394-60	**(9004-35-7)**
E 398-10	**(9004-35-7)**
E 400-25	**(9004-35-7)**
E 600	**(311-45-5)**
E 601	**(298-00-0)**
E 605	**(56-38-2)**
E 605 F	**(56-38-2)**
E 605 forte	**(56-38-2)**
E 605 Reduced	**(3735-01-1)**
E 702	**(118-82-1)**
E 828	**(25068-38-6)**
E 834	**(299-45-6)**
E 838	**(299-45-6)**
E 8573	**(3244-90-4)**
E 1001	**(25068-38-6)**

Name	CAS No.
E 1004	**(25068-38-6)**
E 1059	**(8065-48-3)**
E 1059	**(298-03-3)**
E 1440	**(9004-70-0)**
E 3314	**(76-44-8)**
E 7256	**(27176-87-0)**
E-48	**(18854-01-8)**
E-103	**(34014-18-1)**
E-111	**(2347-72-0)**
E-212	**(606-58-6)**
E-212-1	**(606-58-6)**
E-217	**(50-35-1)**
E-236	**(24602-86-6)**
E-400-25	**(9004-35-7)**
E122	**(3567-69-9)**
E2	**(9004-96-0)**
E393	**(3689-24-5)**
δ E	**(53-03-2)**
E-733-A	**(68-41-7)**
EA 1152	**(55-91-4)**
EA 1205	**(77-81-6)**
EA 1208	**(107-44-8)**
EA 1210	**(96-64-0)**
EA 1285	**(107-49-3)**
EA 1701	**(50782-69-9)**
EAA	**(141-97-9)**
EAK	**(106-68-3)**
EB	**(100-41-4)**
EB	**(314-13-6)**
EBDC	**(111-54-6)**
EBHP	**(3071-32-7)**
EBNA	**(3398-69-4)**
EBS	**(16423-68-0)**
EBZ	**(50-50-0)**
E.C. 3.2.1.17	**(9001-63-2)**
E.C. 3.4.21.1	**(9004-07-3)**
E.C. 3.4.21.14	**(9014-01-1)**
E.C. 3.4.21.7	**(9001-90-5)**
E.C. 3.4.4.14	**(9001-90-5)**
E.C. 3.4.4.16	**(9014-01-1)**
E.C. 3.4.4.5	**(9004-07-3)**
E.C. 3.4.4.6	**(9004-07-3)**
EC 3443	**(9001-98-3)**
EC 34234	**(9001-98-3)**
ECF	**(541-41-3)**
ECH	**(106-89-8)**
ECM	**(50-78-2)**
ECP	**(97-17-6)**
ED	**(598-14-1)**
ED 5661	**(37099-12-0)**
EDB	**(106-93-4)**
EDB-85	**(106-93-4)**
E-D-Bee	**(106-93-4)**
EDC	**(107-06-2)**
ED 5662 (9CI)	**(57425-31-7)**
EDDHI	**(5700-49-2)**
EDDI	**(5700-49-2)**
EDDP	**(17109-49-8)**
EDEMO	**(21770-86-5)**
EDEMO 3	**(21770-86-5)**
EDTA	**(60-00-4)**
EDTA disodium copper salt	**(14025-15-1)**
EDTA acid	**(60-00-4)**
EDTA calcium disodium salt	**(62-33-9)**
EDTA (chelating agent)	**(60-00-4)**
EDTA disodium	**(139-33-3)**
EDTA disodium manganese salt	**(15375-84-5)**
EDTA, disodium salt	**(139-33-3)**
EDTA, sodium salt	**(64-02-8)**
EDTA tetrasodium salt	**(64-02-8)**
EDTA trisodium salt	**(150-38-9)**
EDTPO	**(1429-50-1)**
EE	**(57-63-6)**
EEC No. 123	**(915-67-3)**
EEC No. E320	**(25013-16-5)**
EEC No. E924	**(7758-01-2)**
EED	**(57-63-6)**
EEM	**(5248-48-6)**
EE3ME	**(72-33-3)**
EENA	**(13147-25-6)**
EES	**(1912-30-7)**
EE_2	**(57-63-6)**
EF Corlin	**(50-23-7)**
EFED	**(101-02-0)**
EFV 250/400	**(7439-89-6)**
EG 0	**(7782-42-5)**
EGDME	**(110-71-4)**
EGDN	**(628-96-6)**
EGM	**(109-86-4)**
EGME	**(109-86-4)**
EH 121	**(73-22-3)**
EH2	**(116-52-9)**
EHB-M	**(25104-37-4)**
EHDP	**(2809-21-4)**
EHEN	**(13147-25-6)**
EI	**(151-56-4)**
EI 18682	**(2275-18-5)**
EI 38,555	**(999-81-5)**
EI 47031	**(947-02-4)**
EI 47300	**(122-14-5)**
EI 52160	**(3383-96-8)**
EI-103	**(34014-18-1)**
EI-12880	**(60-51-5)**
EI-18706	**(116-01-8)**
EI-47470	**(950-10-7)**
EI3911	**(298-02-2)**
EK 1700	**(95-54-5)**
EK 54	**(2312-76-7)**
EK 7011	**(123-39-7)**
EKKO capsules	**(57-41-0)**
EL 171	**(59756-60-4)**
EL 211	**(3478-94-2)**
EL 222	**(60168-88-9)**
EL 241	**(17781-31-6)**
EL 400	**(2104-96-3)**
EL 402	**(9002-98-6)**
EL 420	**(9002-98-6)**
EL 614	**(63333-35-7)**
EL 4049	**(121-75-5)**
EL-103	**(34014-18-1)**
EL-110	**(1861-40-1)**
EL-119	**(19044-88-3)**
EL-161	**(55283-68-6)**
EL-179	**(33820-53-0)**
EL-273	**(26766-27-8)**
EL-291	**(41814-78-2)**
EL-531	**(12771-68-5)**
EL-620	**(9004-98-2)**
EL-719	**(9004-98-2)**
ELA	**(7057-92-3)**
ELCO 106	**(25103-54-2)**
ELPI	**(637-07-0)**
EM	**(114-07-8)**
EM 490	**(9003-07-0)**

EM 923	(97-16-5)	ENT 9,735	(8001-35-2)
EMB	(74-55-5)	ENT 9,932	(57-74-9)
EMC	(107-27-7)	ENT 14,250	(51-03-6)
EMMI	(2597-93-5)	ENT 14,611	(103-33-3)
EMP	(2235-25-8)	ENT 14,689	(14484-64-1)
EMPG	(77-83-8)	ENT 14,874	(12122-67-7)
EMQ	(91-53-2)	ENT 14,875	(12427-38-2)
EMS	(62-50-0)	ENT 15,108	(56-38-2)
EMT 25,299	(59-05-2)	ENT 15,152	(76-44-8)
EMTAL 500	(14807-96-6)	ENT 15,208	(555-89-5)
EMTAL 549	(14807-96-6)	ENT 15,266	(83-59-0)
EMTAL 596	(14807-96-6)	ENT 15,349	(106-93-4)
EMTAL 599	(14807-96-6)	ENT 15,406	(78-87-5)
EMTS	(517-16-8)	ENT 15,949	(309-00-2)
EN	(2894-67-9)	ENT 16,087	(311-45-5)
EN 18133	(297-97-2)	ENT 16,225	(60-57-1)
EN 237	(1302-74-5)	ENT 16,273	(3689-24-5)
EN-1733A	(7416-34-4)	ENT 16,358	(80-33-1)
ENBU	(32976-88-8)	ENT 16,391	(143-50-0)
EN-COR	(9003-20-7)	ENT 16,436	(2439-10-3)
ENE 11183 B	(5836-29-3)	ENT 16,519	(140-57-8)
ENIDE	(957-51-7)	ENT 16,634	(120-62-7)
ENIDE 50	(957-51-7)	ENT 16,894	(3244-90-4)
ENJ 2065	(28553-12-0)	ENT 17,034	(121-75-5)
17-ENT	(51-98-9)	ENT 17,035	(2463-84-5)
ENT 4.9	(1646-88-4)	ENT 17,251	(72-20-8)
ENT 5	(301-11-1)	ENT 17,291	(152-16-9)
ENT 6	(112-56-1)	ENT 17,292	(298-00-0)
ENT 9	(532-34-3)	ENT 17,295	(8065-48-3)
ENT 38	(92-84-2)	ENT 17,296	(299-45-6)
ENT 54	(107-13-1)	ENT 17,470	(97-17-6)
ENT 92	(115-31-1)	ENT 17,510	(584-79-2)
ENT 123	(8051-02-3)	ENT 17,588	(87-47-8)
ENT 133	(83-79-4)	ENT 17,591	(136-45-8)
ENT 154	(534-52-1)	ENT 17,596	(126-15-8)
ENT 157	(131-89-5)	ENT 17,798	(2104-64-5)
ENT 262	(131-11-3)	ENT 17,941	(80-00-2)
ENT 375	(94-96-2)	ENT 17,957	(56-72-4)
ENT 666	(141-03-7)	ENT 18,060	(101-21-3)
ENT 884	(12002-03-8)	ENT 18,065	(117-26-0)
ENT 988	(137-30-4)	ENT 18,066	(8027-00-7)
ENT 1,025	(629-70-9)	ENT 18,544	(122-70-3)
ENT 1,122	(88-85-7)	ENT 18,596	(510-15-6)
ENT 1,501	(16893-85-9)	ENT 18,771	(107-49-3)
ENT 1,506	(50-29-3)	ENT 18,861	(500-28-7)
ENT 1,656	(107-06-2)	ENT 18,862	(867-27-6)
ENT 1,716	(72-43-5)	ENT 18,862	(8022-00-2)
ENT 1,860	(127-18-4)	ENT 18,870	(123-33-1)
ENT 2,435	(65-30-5)	ENT 19,060	(119-38-0)
ENT 3,424	(54-11-5)	ENT 19,109	(115-26-4)
ENT 3,776	(117-80-6)	ENT 19,244	(465-73-6)
ENT 3,797	(118-75-2)	ENT 19,442	(8001-50-1)
ENT 4,225	(72-54-8)	ENT 19,507	(333-41-5)
ENT 4,504	(111-44-4)	ENT 19,763	(52-68-6)
ENT 4,585	(80-38-6)	ENT 20,218	(134-62-3)
ENT 4,705	(56-23-5)	ENT 20,696	(103-17-3)
ENT 7,543	(121-29-9)	ENT 20,738	(62-73-7)
ENT 7,796	(58-89-9)	ENT 20,852	(1689-84-5)
ENT 8,184	(113-48-4)	ENT 20,852	(126-22-7)
ENT 8,286	(9003-13-8)	ENT 20,871	(51-14-9)
ENT 8,420	(8003-19-8)	ENT 20,993	(3734-97-2)
ENT 8,538	(94-75-7)	ENT 21,040	(93-15-2)
ENT 8,601	(608-73-1)	ENT 21,170	(70-38-2)
ENT 9,232	(319-84-6)	ENT 21,557	(70-43-9)
ENT 9,233	(319-85-7)	ENT 22,014	(2642-71-9)
ENT 9,234	(319-86-8)	ENT 22,335	(2438-88-2)
ENT 9,624	(80-06-8)	ENT 22,374	(7786-34-7)

ENT 22,542	**(134-62-3)**	ENT 25,671	**(114-26-1)**
ENT 22,784	**(117-27-1)**	ENT 25,675	**(2636-26-2)**
ENT 22,865	**(78-52-4)**	ENT 25,705	**(732-11-6)**
ENT 22,897	**(78-34-2)**	ENT 25,712	**(327-98-0)**
ENT 23,233	**(86-50-0)**	ENT 25,713	**(333-43-7)**
ENT 23,284	**(299-84-3)**	ENT 25,715	**(122-14-5)**
ENT 23,437	**(298-04-4)**	ENT 25,718	**(2227-17-0)**
ENT 23,438	**(2597-03-7)**	ENT 25,719	**(2385-85-5)**
ENT 23,648	**(115-32-2)**	ENT 25,723	**(2984-64-7)**
ENT 23,708	**(786-19-6)**	ENT 25,726	**(2032-65-7)**
ENT 23,737	**(116-29-0)**	ENT 25,732	**(3692-90-8)**
ENT 23,968	**(119-12-0)**	ENT 25,734	**(3254-63-5)**
ENT 23,969	**(63-25-2)**	ENT 25,736	**(62046-37-1)**
ENT 23,970	**(2921-31-5)**	ENT 25,760	**(78-57-9)**
ENT 23,979	**(115-29-7)**	ENT 25,764	**(6012-97-1)**
ENT 24,042	**(298-02-2)**	ENT 25,766	**(315-18-4)**
ENT 24,105	**(563-12-2)**	ENT 25,776	**(1113-02-6)**
ENT 24,482	**(141-66-2)**	ENT 25,777	**(3566-00-5)**
ENT 24,650	**(60-51-5)**	ENT 25,784	**(2032-59-9)**
ENT 24,652	**(2275-18-5)**	ENT 25,787	**(2665-30-7)**
ENT 24,653	**(2778-04-3)**	ENT 25,793	**(485-31-4)**
ENT 24,717	**(7700-17-6)**	ENT 25,796	**(944-22-9)**
ENT 24,723	**(108-34-9)**	ENT 25,818	**(2274-67-1)**
ENT 24,727	**(39300-45-3)**	ENT 25,823	**(50-65-7)**
ENT 24,738	**(122-15-6)**	ENT 25,830	**(947-02-4)**
ENT 24,833	**(122-10-1)**	ENT 25,832	**(13067-93-1)**
ENT 24,915	**(545-55-1)**	ENT 25,832-A	**(13067-93-1)**
ENT 24,945	**(115-90-2)**	ENT 25,841	**(22248-79-9)**
ENT 24,964	**(301-12-2)**	ENT 25,843	**(2686-99-9)**
ENT 24,969	**(470-90-6)**	ENT 25,922	**(644-64-4)**
ENT 24,979	**(56-35-9)**	ENT 25,962	**(15271-41-7)**
ENT 24,980-X	**(78-53-5)**	ENT 25,991	**(950-10-7)**
ENT 24,984	**(15096-52-3)**	ENT 26,058	**(101-05-3)**
ENT 24,988	**(300-76-5)**	ENT 26,079	**(54-62-6)**
ENT 25,208	**(900-95-8)**	ENT 26,263	**(75-21-8)**
ENT 25,294	**(51-75-2)**	ENT 26,316	**(52-46-0)**
ENT 25,296	**(51-18-3)**	ENT 26,396	**(62-50-0)**
ENT 25,419	**(142-09-6)**	ENT 26,538	**(133-06-2)**
ENT 25,445	**(61-82-5)**	ENT 26,592	**(1464-53-5)**
ENT 25,456	**(127-90-2)**	ENT 26,613	**(2275-23-2)**
ENT 25,500	**(64-00-6)**	ENT 26,925	**(101-20-2)**
ENT 25,506	**(116-01-8)**	ENT 26,999	**(5836-10-2)**
ENT 25,515	**(13171-21-6)**	ENT 27,041	**(1079-33-0)**
ENT 25,516	**(87-10-5)**	ENT 27,043	**(13104-21-7)**
ENT 25,540	**(55-38-9)**	ENT 27,093	**(116-06-3)**
ENT 25,543	**(64-00-6)**	ENT 27,097	**(1224-63-1)**
ENT 25,545	**(297-78-9)**	ENT 27,102	**(1757-18-2)**
ENT 25,545-X	**(297-78-9)**	ENT 27,115	**(2227-13-6)**
ENT 25,550	**(7631-86-9)**	ENT 27,127	**(2282-34-0)**
ENT 25,552-X	**(57-74-9)**	ENT 27,129	**(6923-22-4)**
ENT 25,554-X	**(3735-23-7)**	ENT 27,160	**(919-76-6)**
ENT 25,557-X	**(108-35-0)**	ENT 27,162	**(2104-96-3)**
ENT 25,579	**(93-75-4)**	ENT 27,163	**(2310-17-0)**
ENT 25,580	**(297-97-2)**	ENT 27,164	**(1563-66-2)**
ENT 25,584	**(1024-57-3)**	ENT 27,164	**(56-23-5)**
ENT 25,585	**(2275-14-1)**	ENT 27,165	**(3383-96-8)**
ENT 25,586	**(953-17-3)**	ENT 27,193	**(950-37-8)**
ENT 25,595-X	**(644-64-4)**	ENT 27,226	**(2312-35-8)**
ENT 25,599	**(953-17-3)**	ENT 27,238	**(2669-32-1)**
ENT 25,602-X	**(299-86-5)**	ENT 27,244	**(973-21-7)**
ENT 25,606	**(2439-01-2)**	ENT 27,257	**(2540-82-1)**
ENT 25,612	**(2703-13-1)**	ENT 27,258	**(4824-78-6)**
ENT 25,640	**(115-93-5)**	ENT 27,300	**(2631-37-0)**
ENT 25,644	**(52-85-7)**	ENT 27,300-A	**(2631-37-0)**
ENT 25,647	**(299-85-4)**	ENT 27,305	**(17702-57-7)**
ENT 25,650	**(919-54-0)**	ENT 27,311	**(2921-88-2)**
ENT 25,670	**(2631-40-5)**	ENT 27,318	**(13194-48-4)**

ENT 27,320	(10311-84-9)
ENT 27,335	(6164-98-3)
ENT 27,339	(7696-12-0)
ENT 27,340	(97-77-8)
ENT 27,341	(16752-77-5)
ENT 27,346	(13265-60-6)
ENT 27,386	(2597-03-7)
ENT 27,386GC	(2597-03-7)
ENT 27,389	(6988-21-2)
ENT 27,394	(13593-03-8)
ENT 27,395-X	(13121-70-5)
ENT 27,395	(13121-70-5)
ENT 27,396	(10265-92-6)
ENT 27,408	(18181-70-9)
ENT 27,438	(14255-88-0)
ENT 27,439	(5918-93-4)
ENT 27,474	(10453-86-8)
ENT 27,488	(14816-18-3)
ENT 27,520	(5598-13-0)
ENT 27,521	(5598-52-7)
ENT 27,552	(18181-80-1)
ENT 27,566	(23422-53-9)
ENT 27,567	(19750-95-9)
ENT 27,567	(6164-98-3)
ENT 27,572	(22224-92-6)
ENT 27,635	(60238-56-4)
ENT 27,635	(21923-23-9)
ENT 27,696	(26419-73-8)
ENT 27,699GC	(29232-93-7)
ENT 27,738	(13356-08-6)
ENT 27,766	(23103-98-2)
ENT 27,822	(30560-19-1)
ENT 27,851	(39196-18-4)
ENT 27,920	(13071-79-9)
ENT 27,967	(33089-61-1)
ENT 27,989	(31218-83-4)
ENT 28,009	(76-87-9)
ENT 28,344	(5281-13-0)
ENT 29,054	(35367-38-5)
ENT 29,117	(58769-20-3)
ENT 29,126	(38260-54-7)
ENT 50,003	(57-39-6)
ENT 50,146	(50-55-5)
ENT 50,324	(151-56-4)
ENT 50,434	(28300-74-5)
ENT 50,439	(66-75-1)
ENT 50,715	(542-02-9)
ENT 50,852	(645-05-6)
ENT 50,882	(680-31-9)
ENT 51,762	(991-42-4)
ENT 70,459	(41096-46-2)
ENT 70,460	(40596-69-8)
ENT 70,531	(42588-37-4)
ENT AI3-29261	(1646-88-4)
ENTA	(51-98-9)
ENU	(614-95-9)
ENU	(759-73-9)
E.O.	(75-21-8)
EO	(57-63-6)
EO 5A	(7439-89-)
EOCT	(294-93-9)
EP 30	87-86-5)
EP 160	9002-89-5)
EP 201	141-37-7)
EP 207	81-21-0)
EP 316	2631-37-0)
EP 333	19750-95-9)

EP 587	55838-67-0)
EP 1208	9003-20-7)
EP 1436	9003-20-7)
EP 1437	9003-20-7)
EP 1463	9003-20-7)
EP-101	(106-86-5)
EP-145	(123-36-4)
EP-162	(8066-01-1)
EP-205	(2386-90-5)
EP-206	(106-87-6)
EP-332	(23422-53-9)
EP-333	(6164-98-3)
EP-411	(156-51-4)
EP-452	(13684-63-4)
EP-475	(13684-56-5)
EP-1086	(68955-06-6)
EP-2017	(7487-28-7)
EPA Pesticide Chemical Code 000174	(137-16-6)
EPA Pesticide Chemical Code 002201	(8051-02-3)
EPA Pesticide Chemical Code 004201	(2026-24-6)
EPA Pesticide Chemical Code 004203	(51344-62-8)
EPA Pesticide Chemical Code 004206	(1446-61-3)
EPA Pesticide Chemical Code 005302	(7664-41-7)
EPA Pesticide Chemical Code 006101	(65996-91-0)
EPA Pesticide Chemical Code 006202	(5035-58-5)
EPA Pesticide Chemical Code 006313	(1405-10-3)
EPA Pesticide Chemical Code 006314	(1397-94-0)
EPA Pesticide Chemical Code 006315	(16079-88-2)
EPA Pesticide Chemical Code 006401	(68038-71-1)
EPA Pesticide Chemical Code 006501	(68477-31-6)
EPA Pesticide Chemical Code 006601	(68602-80-2)
EPA Pesticide Chemical Code 006602	(64742-94-5)
EPA Pesticide Chemical Code 006801	(7778-39-4)
EPA Pesticide Chemical Code 006902	(1344-08-7)
EPA Pesticide Chemical Code 008001	(1332-65-6)
EPA Pesticide Chemical Code 008101	(1332-03-2)
EPA Pesticide Chemical Code 008707	(2491-38-5)
EPA Pesticide Chemical Code 009106	(3734-33-6)
EPA Pesticide Chemical Code 009901	(7745-89-3)
EPA Pesticide Chemical Code 010801	(8001-85-2)
EPA Pesticide Chemical Code 011101	(13701-59-2)
EPA Pesticide Chemical Code 011401	(64742-16-1)
EPA Pesticide Chemical Code 011403	(9003-27-4)
EPA Pesticide Chemical Code 011701	(1113-14-0)
EPA Pesticide Chemical Code 011901	(9003-13-8)
EPA Pesticide Chemical Code 012302	(53404-19-6)
EPA Pesticide Chemical Code 012402	(14697-50-8)
EPA Pesticide Chemical Code 013505	(13464-38-5)
EPA Pesticide Chemical Code 013804	(6379-37-9)
EPA Pesticide Chemical Code 014506	(12122-67-7)
EPA Pesticide Chemical Code 014702	(13840-33-0)
EPA Pesticide Chemical Code 019202	(6062-26-6)
EPA Pesticide Chemical Code 020502	(7758-19-2)
EPA Pesticide Chemical Code 021101	(7738-94-5)
EPA Pesticide Chemical Code 021201	(101-10-0)
EPA Pesticide Chemical Code 021203	(5825-87-6)
EPA Pesticide Chemical Code 021901	(8000-29-1)
EPA Pesticide Chemical Code 022703	(33113-08-5)
EPA Pesticide Chemical Code 023103	(61789-22-8)
EPA Pesticide Chemical Code 023104	(9007-39-0)
EPA Pesticide Chemical Code 023305	(814-91-5)
EPA Pesticide Chemical Code 023701	(10125-13-0)
EPA Pesticide Chemical Code 024402	(10257-54-2)
EPA Pesticide Chemical Code 025001	(8002-29-7)
EPA Pesticide Chemical Code 025602	(1111-67-7)
EPA Pesticide Chemical Code 027601	(4776-06-1)
EPA Pesticide Chemical Code 029803	(25059-78-3)

EPA Pesticide Chemical Code 029902 **(1076-46-6)**
EPA Pesticide Chemical Code 029905 **(7286-84-2)**
EPA Pesticide Chemical Code 030002 **(3766-27-6)**
EPA Pesticide Chemical Code 030016 **(5742-19-8)**
EPA Pesticide Chemical Code 030017 **(20940-37-8)**
EPA Pesticide Chemical Code 030033 **(2569-01-9)**
EPA Pesticide Chemical Code 030056 **(94-80-4)**
EPA Pesticide Chemical Code 030063 **(1928-43-4)**
EPA Pesticide Chemical Code 030511 **(20405-19-0)**
EPA Pesticide Chemical Code 030516 **(2039-46-5)**
EPA Pesticide Chemical Code 030553 **(19480-43-4)**
EPA Pesticide Chemical Code 030804 **(10433-59-7)**
EPA Pesticide Chemical Code 030819 **(2758-42-1)**
EPA Pesticide Chemical Code 030853 **(32357-46-3)**
EPA Pesticide Chemical Code 030856 **(6753-24-8)**
EPA Pesticide Chemical Code 030863 **(1320-15-6)**
EPA Pesticide Chemical Code 031453 **(53404-31-2)**
EPA Pesticide Chemical Code 031519 **(32351-70-5)**
EPA Pesticide Chemical Code 031601 **(8006-54-0)**
EPA Pesticide Chemical Code 031604 **(8001-78-3)**
EPA Pesticide Chemical Code 031607 **(61790-81-6)**
EPA Pesticide Chemical Code 034801 **(14484-64-1)**
EPA Pesticide Chemical Code 034806 **(16509-79-8)**
EPA Pesticide Chemical Code 035601 **(3064-70-8)**
EPA Pesticide Chemical Code 037508 **(2312-76-7)**
EPA Pesticide Chemical Code 037601 **(4726-14-1)**
EPA Pesticide Chemical Code 038801 **(1918-11-2)**
EPA Pesticide Chemical Code 038904 **(2164-07-0)**
EPA Pesticide Chemical Code 039002 **(137-41-7)**
EPA Pesticide Chemical Code 039102 **(135-37-5)**
EPA Pesticide Chemical Code 039103 **(17421-79-3)**
EPA Pesticide Chemical Code 039105 **(54453-03-1)**
EPA Pesticide Chemical Code 039109 **(139-89-9)**
EPA Pesticide Chemical Code 039117 **(7379-26-2)**
EPA Pesticide Chemical Code 039120 **(140-01-2)**
EPA Pesticide Chemical Code 040502 **(8007-02-1)**
EPA Pesticide Chemical Code 040503 **(8000-48-4)**
EPA Pesticide Chemical Code 040509 **(8007-44-1)**
EPA Pesticide Chemical Code 041003 **(115-84-4)**
EPA Pesticide Chemical Code 041201 **(22221-10-9)**
EPA Pesticide Chemical Code 043802 **(125-67-7)**
EPA Pesticide Chemical Code 044005 **(10402-15-0)**
EPA Pesticide Chemical Code 044902 **(5736-15-2)**
EPA Pesticide Chemical Code 045502 **(56-95-1)**
EPA Pesticide Chemical Code 047201 **(136-45-8)**
EPA Pesticide Chemical Code 047401 **(78-59-1)**
EPA Pesticide Chemical Code 047801 **(64-00-6)**
EPA Pesticide Chemical Code 050506 **(10045-89-3)**
EPA Pesticide Chemical Code 051601 **(1490-04-6)**
EPA Pesticide Chemical Code 051704 **(2492-26-4)**
EPA Pesticide Chemical Code 051707 **(7778-70-3)**
EPA Pesticide Chemical Code 053001 **(108-62-3)**
EPA Pesticide Chemical Code 053200 **(24959-67-9)**
EPA Pesticide Chemical Code 055003 **(22232-25-3)**
EPA Pesticide Chemical Code 056007 **(61-31-4)**
EPA Pesticide Chemical Code 056008 **(2122-70-5)**
EPA Pesticide Chemical Code 056901 **(609-31-4)**
EPA Pesticide Chemical Code 059802 **(134-30-5)**
EPA Pesticide Chemical Code 061205 **(94-26-8)**
EPA Pesticide Chemical Code 061501 **(106-46-7)**
EPA Pesticide Chemical Code 062202 **(35471-49-9)**
EPA Pesticide Chemical Code 062203 **(3184-65-4)**
EPA Pesticide Chemical Code 062208 **(607-12-5)**
EPA Pesticide Chemical Code 062210 **(85-97-2)**
EPA Pesticide Chemical Code 062212 **(10605-10-4)**
EPA Pesticide Chemical Code 063002 **(7778-73-6)**
EPA Pesticide Chemical Code 063005 **(25567-55-9)**
EPA Pesticide Chemical Code 063010 **(3772-94-9)**
EPA Pesticide Chemical Code 063201 **(79-21-0)**
EPA Pesticide Chemical Code 063514 **(68476-31-3)**
EPA Pesticide Chemical Code 063604 **(10058-23-8)**
EPA Pesticide Chemical Code 064108 **(13707-65-8)**
EPA Pesticide Chemical Code 064111 **(53404-18-5)**
EPA Pesticide Chemical Code 064112 **(31366-95-7)**
EPA Pesticide Chemical Code 064115 **(5787-50-8)**
EPA Pesticide Chemical Code 064118 **(27193-28-8)**
EPA Pesticide Chemical Code 064202 **(13347-42-7)**
EPA Pesticide Chemical Code 064208 **(4418-66-0)**
EPA Pesticide Chemical Code 064216 **(5335-24-0)**
EPA Pesticide Chemical Code 065001 **(6531-86-8)**
EPA Pesticide Chemical Code 066001 **(27236-65-3)**
EPA Pesticide Chemical Code 066008 **(32407-99-1)**
EPA Pesticide Chemical Code 066012 **(122-64-5)**
EPA Pesticide Chemical Code 066013 **(5822-97-9)**
EPA Pesticide Chemical Code 066021 **(23319-66-6)**
EPA Pesticide Chemical Code 066022 **(104-60-9)**
EPA Pesticide Chemical Code 066502 **(7723-14-0)**
EPA Pesticide Chemical Code 067005 **(8000-41-7)**
EPA Pesticide Chemical Code 067204 **(8011-48-1)**
EPA Pesticide Chemical Code 067705 **(42721-99-3)**
EPA Pesticide Chemical Code 067707 **(3691-35-8)**
EPA Pesticide Chemical Code 068402 **(869-29-4)**
EPA Pesticide Chemical Code 069125 **(1330-85-4)**
EPA Pesticide Chemical Code 069133 **(112-02-7)**
EPA Pesticide Chemical Code 069134 **(25155-18-4)**
EPA Pesticide Chemical Code 069165 **(32426-11-2)**
EPA Pesticide Chemical Code 069166 **(5538-94-3)**
EPA Pesticide Chemical Code 069173 **(68607-28-3)**
EPA Pesticide Chemical Code 069187 **(78-21-7)**
EPA Pesticide Chemical Code 069190 **(10361-16-7)**
EPA Pesticide Chemical Code 069701 **(10102-90-6)**
EPA Pesticide Chemical Code 070801 **(507-60-8)**
EPA Pesticide Chemical Code 071002 **(6659-45-6)**
EPA Pesticide Chemical Code 072401 **(8008-74-0)**
EPA Pesticide Chemical Code 072502 **(7775-41-9)**
EPA Pesticide Chemical Code 072506 **(7783-90-6)**
EPA Pesticide Chemical Code 072601 **(1343-88-0)**
EPA Pesticide Chemical Code 072602 **(1343-98-2)**
EPA Pesticide Chemical Code 072603 **(1344-09-8)**
EPA Pesticide Chemical Code 072606 **(1312-76-1)**
EPA Pesticide Chemical Code 075301 **(16919-19-0)**
EPA Pesticide Chemical Code 075304 **(16949-65-8)**
EPA Pesticide Chemical Code 075903 **(13932-13-3)**
EPA Pesticide Chemical Code 076002 **(10294-56-1)**
EPA Pesticide Chemical Code 076202 **(13780-06-8)**
EPA Pesticide Chemical Code 076401 **(7758-87-4)**
EPA Pesticide Chemical Code 076407 **(7778-53-2)**
EPA Pesticide Chemical Code 076501 **(127-52-6)**
EPA Pesticide Chemical Code 076603 **(118-56-9)**
EPA Pesticide Chemical Code 076604 **(104-28-9)**
EPA Pesticide Chemical Code 076701 **(9080-17-5)**
EPA Pesticide Chemical Code 076702 **(1344-81-6)**
EPA Pesticide Chemical Code 076703 **(37199-66-9)**
EPA Pesticide Chemical Code 076801 **(7646-88-0)**
EPA Pesticide Chemical Code 077402 **(87-12-7)**
EPA Pesticide Chemical Code 077404 **(87-10-5)**
EPA Pesticide Chemical Code 077405 **(2577-72-2)**
EPA Pesticide Chemical Code 078401 **(776-19-2)**
EPA Pesticide Chemical Code 078502 **(1401-55-4)**
EPA Pesticide Chemical Code 079008 **(27177-77-1)**
EPA Pesticide Chemical Code 079009 **(68952-95-4)**
EPA Pesticide Chemical Code 079013 **(61790-19-0)**
EPA Pesticide Chemical Code 079014 **(8002-33-3)**
EPA Pesticide Chemical Code 079015 **(26545-53-9)**

EPA Pesticide Chemical Code 079017 (3097-08-3)
EPA Pesticide Chemical Code 079020 (27323-41-7)
EPA Pesticide Chemical Code 079022 (13429-27-1)
EPA Pesticide Chemical Code 079023 (7492-30-0)
EPA Pesticide Chemical Code 079024 (30346-73-7)
EPA Pesticide Chemical Code 079025 (2717-15-9)
EPA Pesticide Chemical Code 079029 (68603-15-6)
EPA Pesticide Chemical Code 079031 (30526-22-8)
EPA Pesticide Chemical Code 079034 (67762-39-4)
EPA Pesticide Chemical Code 079043 (2224-49-9)
EPA Pesticide Chemical Code 079044 (41669-40-3)
EPA Pesticide Chemical Code 079068 (7346-80-7)
EPA Pesticide Chemical Code 079100 (9036-19-5)
EPA Pesticide Chemical Code 082019 (6369-97-7)
EPA Pesticide Chemical Code 082033 (3813-14-7)
EPA Pesticide Chemical Code 082034 (2008-46-0)
EPA Pesticide Chemical Code 082038 (53404-86-7)
EPA Pesticide Chemical Code 082055 (1928-48-9)
EPA Pesticide Chemical Code 082062 (4938-72-1)
EPA Pesticide Chemical Code 082066 (93-78-7)
EPA Pesticide Chemical Code 082503 (2818-16-8)
EPA Pesticide Chemical Code 082504 (37913-89-6)
EPA Pesticide Chemical Code 082516 (51170-59-3)
EPA Pesticide Chemical Code 082602 (2439-00-1)
EPA Pesticide Chemical Code 083108 (56573-85-4)
EPA Pesticide Chemical Code 083109 (24124-25-2)
EPA Pesticide Chemical Code 083111 (28801-69-6)
EPA Pesticide Chemical Code 083115 (53404-82-3)
EPA Pesticide Chemical Code 083120 (2155-70-6)
EPA Pesticide Chemical Code 083401 (1330-78-5)
EPA Pesticide Chemical Code 085701 (97-59-6)
EPA Pesticide Chemical Code 086004 (34490-93-2)
EPA Pesticide Chemical Code 087802 (557-41-5)
EPA Pesticide Chemical Code 088004 (15922-78-8)
EPA Pesticide Chemical Code 089101 (68813-94-5)
EPA Pesticide Chemical Code 097003 (3478-94-2)
EPA Pesticide Chemical Code 098002 (27176-87-0)
EPA Pesticide Chemical Code 098101 (1249-84-9)
EPA Pesticide Chemical Code 098501 (7784-25-0)
EPA Pesticide Chemical Code 098901 (2634-33-5)
EPA Pesticide Chemical Code 099001 (34375-28-5)
EPA Pesticide Chemical Code 099102 (52316-55-9)
EPA Pesticide Chemical Code 099701 (8002-24-2)
EPA Pesticide Chemical Code 100801 (2224-44-4)
EPA Pesticide Chemical Code 101002 (20018-09-1)
EPA Pesticide Chemical Code 101401 (7166-19-0)
EPA Pesticide Chemical Code 101501 (22228-82-6)
EPA Pesticide Chemical Code 102601 (122-70-3)
EPA Pesticide Chemical Code 102801 (38827-35-9)
EPA Pesticide Chemical Code 103201 (27519-02-4)
EPA Pesticide Chemical Code 103901 (50723-80-3)
EPA Pesticide Chemical Code 107002 (6542-37-6)
EPA Pesticide Chemical Code 107104 (2682-20-4)
EPA Pesticide Chemical Code 107401 (27668-52-6)
EPA Pesticide Chemical Code 107501 (42588-37-4)
EPA Pesticide Chemical Code 107701 (26628-22-8)
EPA Pesticide Chemical Code 107801 (55406-53-6)
EPA Pesticide Chemical Code 109101 (24307-26-4)
EPA Pesticide Chemical Code 109501 (2832-19-1)
EPA Pesticide Chemical Code 109601 (52508-35-7)
EPA Pesticide Chemical Code 110401 (31717-87-0)
EPA Pesticide Chemical Code 112001 (28772-56-7)
EPA Pesticide Chemical Code 112802 (63333-35-7)
EPA Pesticide Chemical Code 112900 (59756-60-4)
EPA Pesticide Chemical Code 113001 (34681-23-7)
EPA Pesticide Chemical Code 114301 (29804-22-6)
EPA Pesticide Chemical Code 114501 (59669-26-0)
EPA Pesticide Chemical Code 114802 (75673-43-7)
EPA Pesticide Chemical Code 115101 (8061-52-7)
EPA Pesticide Chemical Code 116501 (64726-91-6)
EPA Pesticide Chemical Code 116901 (1214-39-7)
EPA Pesticide Chemical Code 117201 (53120-27-7)
EPA Pesticide Chemical Code 117701 (16974-11-1)
EPA Pesticide Chemical Code 118201 (64628-44-0)
EPA Pesticide Chemical Code 119501 (53905-38-7)
EPA Pesticide Chemical Code 120001 (53939-28-9)
EPA Pesticide Chemical Code 121001 (74051-80-2)
EPA Pesticide Chemical Code 121601 (34256-82-1)
EPA Pesticide Chemical Code 121902 (72269-48-8)
EPA Pesticide Chemical Code 122010 (74223-64-6)
EPA Pesticide Chemical Code 122601 (60568-05-0)
EPA Pesticide Chemical Code 124801 (61228-92-0)
EPA Pesticide Chemical Code 125301 (72490-01-8)
EPA Pesticide Chemical Code 125851 (82558-50-7)
EPA Pesticide Chemical Code 128001 (40164-67-8)
EPA Pesticide Chemical Code 128701 (66441-23-4)
EPA Pesticide Chemical Code 128821 (81334-34-1)
EPA Pesticide Chemical Code 128829 (81510-83-0)
EPA Pesticide Chemical Code 169160 (68959-20-6)
EPA Pesticide Chemical Code 205400 (4812-20-8)
EPA Pesticide Chemical Code 206200 (108-19-0)
EPA Pesticide Chemical Code 206900 (6515-38-4)
EPA Pesticide Chemical Code 207800 (1066-51-9)
EPA Pesticide Chemical Code 213600 (31717-87-0)
EPA Pesticide Chemical Code 216400 (52-51-7)
EPA Pesticide Chemical Code 218100 (12057-74-8)
EPA Pesticide Chemical Code 228300 (3996-59-6)
EPA Pesticide Chemical Code 237200 (1031-47-6)
EPA Pesticide Chemical Code 275400 (5251-93-4)
EPA Pesticide Chemical Code 292200 (2464-37-1)
EPA Pesticide Chemical Code 357200 (34689-46-8)
EPA Pesticide Chemical Code 359700 (17702-57-7)
EPA Pesticide Chemical Code 362200 (3735-23-7)
EPA Pesticide Chemical Code 364300 (26419-73-8)
EPA Pesticide Chemical Code 375300 (34484-77-0)
EPA Pesticide Chemical Code 413300 (6843-97-6)
EPA Pesticide Chemical Code 459300 (1300-78-3)
EPA Pesticide Chemical Code 462200 (4301-50-2)
EPA Pesticide Chemical Code 481700 (3697-42-5)
EPA Pesticide Chemical Code 486300 (41096-46-2)
EPA Pesticide Chemical Code 492200 (3452-97-9)
EPA Pesticide Chemical Code 498200 (13516-27-3)
EPA Pesticide Chemical Code 505200 (64771-72-8)
EPA Pesticide Chemical Code 549300 (1771-07-9)
EPA Pesticide Chemical Code 573800 (1773-37-1)
EPA Pesticide Chemical Code 589600 (61790-13-4)
EPA Pesticide Chemical Code 597300 (1124-33-0)
EPA Pesticide Chemical Code 597500 (8000-46-2)
EPA Pesticide Chemical Code 598400 (8009-03-8)
EPA Pesticide Chemical Code 598500 (61789-85-3)
EPC (The plant regulator) (101-99-5)
EP-161E (556-61-6)
EPIB (637-07-0)
EPN 300 (2104-64-5)
EPN (ACGIH,OSHA) (2104-64-5)
EPON 820 (25068-38-6)
EPON 828 (25068-38-6)
EPON 1001 (25068-38-6)
EPON 1007 (25068-38-6)
EPTAM (759-94-4)
EPTC (759-94-4)
EPTC Plus Inert Herbicide Safener (51990-04-6)
EPTC Plus R-25788 (51990-04-6)
E-Pam (439-14-5)

EQ	(91-53-2)
ER 115	(846-50-4)
ER5461	(26399-36-0)
ERBN	(136-25-4)
ERE 1359	(101-90-6)
ERL 4206	(25086-25-3)
ERL-2774	(1675-54-3)
ERL-2795	(25068-38-6)
ERL-4206	(25086-25-3)
ERL-4221	(2386-87-0)
ERLA-2270	(106-87-6)
ERLA-2271	(106-87-6)
ERR 4205	(2386-90-5)
ES685	(126-72-7)
ESBP	(21722-85-0)
ESK 1 (demulsifier)	(9005-00-9)
ESNN	(1444-64-0)
ESZ	(1338-16-5)
ET 14	(299-84-3)
ET 57	(299-84-3)
ET 67	(9011-06-7)
ET-394	(87-10-5)
ETEM	(33813-20-6)
ETH	(536-33-4)
ETH	(67-21-0)
ETM	(33813-20-6)
ETM (Heterocycle)	(33813-20-6)
ETMT	(2593-15-9)
ETO	(75-21-8)
ETP	(536-33-4)
ETS	(9004-57-3)
ETS (Cyanuric acid derivative)	(28825-96-9)
ETS (Polysaccharide)	(9004-57-3)
ETU	(96-45-7)
EVE	(109-92-2)
EX 10781	(3813-05-6)
EXD	(502-55-6)
EX-IT	(14807-96-6)
EXP 5598	(52315-07-8)
EXP-105-1	(768-94-5)
E-Z-Off	(10326-21-3)
E-Z-Off D	(78-48-8)
E-Z-Paque	(7727-43-7)
Eaca	(60-32-2)
Eaca Kabi	(60-32-2)
Eacs	(60-32-2)
Eagle Germantown	(1333-86-4)
α-Earleine	(107-43-7)
Earthcide	(82-68-8)
Earthnut Oil	(8002-03-7)
Easeptol	(120-47-8)
East Indian Copaiba	(8030-55-5)
East Indian Sandalwood Oil	(8006-87-9)
Eastern States Duocide	(81-81-2)
Eastman 298-10	(9004-35-7)
Eastman 7663	(514-73-8)
Eastman 910	(137-05-3)
Eastman 910 Adhesive	(137-05-3)
Eastman Blue BNN	(2475-46-9)
Eastman Blue BNN	(86722-66-9)
Eastman Blue GBN	(2475-46-9)
Eastman Blue GBN	(86722-66-9)
Eastman Fast Blue B-GLF	(15791-78-3)
Eastman Fast Yellow 2R-GLF	(12223-97-1)
Eastman Inhibitor DHPB	(131-56-6)
Eastman Inhibitor HPT	(680-31-9)
Eastman Inhibitor RMB	(136-36-7)
Eastman 910 Monomer	(137-05-3)
Eastman Polyester Yellow 2R	(12223-97-1)
Eastobond L 8080-270A	(9003-07-0)
Eastobond M 3	(9003-07-0)
Eastobond M 5	(9003-07-0)
Eastobond M 5H	(9003-07-0)
Eastone Yellow GN	(2832-40-8)
Eastozone	(3081-14-9)
Eastozone 31	(139-60-6)
Eastozone 33	(3081-14-9)
Easy Off-D	(150-50-5)
Eatan	(146-22-5)
Eau Grison	(1344-81-6)
Ebacream	(814-71-1)
Eberpine	(50-55-5)
Eberspine	(50-55-5)
Ebidene	(54-85-3)
Ebserpine	(50-55-5)
Ecarlate GN (French)	(3257-28-1)
Ecatox	(56-38-2)
Eccothal	(7446-18-6)
Ecgonine	(481-37-8)
Ecgonine, methyl ester, benzoate (ester)	(50-36-2)
Echimidine hydrochloride	(520-68-3)
Echlomezol	(2593-15-9)
Echlomezole	(2593-15-9)
Echodide	(513-10-0)
Echothiophate	(513-10-0)
Echothiophate iodide	(513-10-0)
Eciphin	(299-42-3)
Eclipse Black BG	(1326-82-5)
Eclipse Dark Blue B	(1327-57-7)
Eclipse Deep Black S	(1326-82-5)
Eclipse Red	(992-59-6)
Ecloril	(305-03-3)
Eclorion	(561-27-3)
Ecobutazone	(50-33-9)
Ecolyte PS 102	(25191-48-4)
Ecolyte PS 108	(25191-48-4)
Econazole nitrate	(24169-02-6)
Econochlor	(56-75-7)
Ecopro	(3383-96-8)
Ecothiopate	(513-10-0)
Ecothiopate iodide	(513-10-0)
Ecothiophate iodide	(513-10-0)
Ecotrin	(50-78-2)
Ecsumin	(52645-53-1)
Ectiban	(52645-53-1)
Ectiluran	(50-35-1)
(+)-(6S)-Ectocarpene	(33156-92-2)
S-(+)-Ectocarpene	(33156-92-2)
Ectodex	(33089-61-1)
Ectoral	(299-84-3)
Ectrin	(51630-58-1)
Ecuanil	(57-53-4)
Eczecidin	(130-26-7)
Edathamil	(60-00-4)
Edathamil calcium disodium	(62-33-9)
Edathamil disodium	(139-33-3)
Edathamil monosodium ferric salt	(15708-41-5)
Edathanil tetrasodium	(64-02-8)
Edco	(74-83-9)
Edecril	(58-54-8)
Edecrin	(58-54-8)
Edecrina	(58-54-8)
Edemex	(91-33-8)
Edemox	(59-66-5)

Eden (58-25-3)
Edenal (57-53-4)
Edestin (9007-57-2)
Edetamin (62-33-9)
Edetamine (62-33-9)
Edetate calcium (62-33-9)
Edetate disodium (139-33-3)
Edetate sodium (64-02-8)
Edetate trisodium (150-38-9)
Edetic (60-00-4)
Edetic acid (60-00-4)
Edetic acid calcium disodium salt (62-33-9)
Edetic acid disodium salt (139-33-3)
Edetic acid tetrasodium salt (64-02-8)
Edible Vegetable Oil (68956-68-3)
Edicol Amaranth (915-67-3)
Edicol Blue CL 2 (3844-45-9)
Edicol Ponceau RS (3761-53-3)
Edicol Suppa Rose BS (81-88-9)
Edicol Supra Amaranth A (915-67-3)
Edicol Supra 10B (3567-66-6)
Edicol Supra Black BN (2519-30-4)
Edicol Supra Blue E6 (2650-18-2)
Edicol Supra Blue VR (129-17-9)
Edicol Supra Blue X (860-22-0)
Edicol Supra Carmoisine W (3567-69-9)
Edicol Supra Carmoisine WS (3567-69-9)
Edicol Supra Erythrosine A (16423-68-0)
Edicol Supra Geranine 2G (3734-67-6)
Edicol Supra Geranine 2GS (3734-67-6)
Edicol Supra Ponceau 4R (2611-82-7)
Edicol Supra Ponceau R (3761-53-3)
Edicol Supra Ponceau SX (4548-53-2)
Edicol Supra Rose B (81-88-9)
Edicol Supra Tartrazine N (1934-21-0)
Edicol Supra Yellow FC (2783-94-0)
Edicol Supra 10bS (3567-66-6)
Edifas A (9004-59-5)
Edifas B (9004-32-4)
Edifas Grade "A" (9004-59-5)
Edifenphos (17109-49-8)
Ediphenphos (17109-49-8)
Edisol M (9004-67-5)
Edistir RB (9003-53-6)
Edistir RB 268 (9003-55-8)
Editempa (1429-50-1)
Editempa acid (1429-50-1)
Edrisal (62-44-2)
Edtacal (62-33-9)
Eerex Granular Weed Killer (314-40-9)
Eerex Water soluble concentrate weed killer (314-40-9)
Efo-Dine (25655-41-8)
Efacin (59-67-6)
Efcorbin (50-23-7)
Efcortelin (50-23-7)
Efdolan Red GRN (6408-31-7)
Efedrin (299-42-3)
Eferon (63-98-9)
Effemoll DOA (103-23-1)
Effisax (4268-36-4)
Effluderm (Free Base) (51-21-8)
Effomoll DOA (103-23-1)
Effroxine (300-42-5)
Effusan (534-52-1)
Effusan 3436 (534-52-1)
Efiran 99 (2000-43-3)
Efloran (443-48-1)
Efmethrin (52645-53-1)
Efosite Aluminum (39148-24-8)
Efpenix (26787-78-0)
Efricel (61-76-7)
Efroxine (300-42-5)
Eftolon (526-08-9)
Efudex (51-21-8)
Efudix (51-21-8)
Efurix (51-21-8)
Egacid Orange GG (523-44-4)
Egacid Red G (3734-67-6)
Egg Yellow A (1934-21-0)
Egitol (67-72-1)
Eglonyl (15676-16-1)
Egomaketone (59204-74-9)
Ehrlich's Reagent (100-10-7)
11,14-Eicosadienoic acid, methyl ester (8CI,9CI) (2463-02-7)
Eicosamethylnonasiloxane (2652-13-3)
n-Eicosane (112-95-8)
Eicosane (9CI) (112-95-8)
Eicosane, 1-bromo- (9CI) (4276-49-7)
Eicosane, 1-chloro- (9CI) (42217-02-7)
Eicosane, 1-cyclohexyl- (8CI) (4443-55-4)
Eicosane, 9-cyclohexyl- (6CI,8CI) (4443-61-2)
Eicosane, 10-methyl- (9CI) (54833-23-7)
Eicosane, 3-phenyl- (8CI) (2400-02-4)
Eicosane, 2-phenyl- (7CI,8CI) (2398-66-5)
Eicosane, 4-phenyl- (7CI,8CI) (2400-03-5)
Eicosane, 5-phenyl- (7CI,8CI) (2400-04-6)
Eicosane, 7-phenyl- (7CI,8CI) (2398-64-3)
Eicosane, 1-phenyl- (6CI,7CI,8CI) (2398-68-7)
Eicosane, 9-phenyl- (6CI,7CI,8CI) (2398-65-4)
Eicosanoic-acid (506-30-9)
Eicosanoic acid, ethyl ester (8CI,9CI) (18281-05-5)
Eicosanoic acid, methyl ester (1120-28-1)
Eicosanoic acid, silver salt (8CI) (24687-57-8)
n-1-Eicosanol (629-96-9)
n-Eicosanol (629-96-9)
1-Eicosanol (9CI) (629-96-9)
1-Eicosanol, aluminum salt (9CI) (67905-31-1)
3-Eicosanone (7CI,8CI,9CI) (2955-56-8)
(all-Z)-5,8,11,14-Eicosatetraenoic acid (506-32-1)
5,8,11,14-Eicosatetraenoic acid, ethyl ester (1808-26-0)
8,11,14-Eicosatrienoic acid, (Z,Z,Z)- (8CI,9CI) (1783-84-2)
11,14,17-Eicosatrienoic acid, methyl ester (9CI) (55682-88-7)
1-Eicosene (9CI) (3452-07-1)
Eicosyl alcohol (629-96-9)
Eicosylbenzene (2398-68-7)
Eicosyl methacrylate (45294-18-6)
Eicosyl 2-methyl-2-propenoate (45294-18-6)
Eicosyl octadecanoate (22413-02-1)
9-Eicosyl-9-phosphabicyclo(3.3.1)nonane (13887-00-8)
9-Eicosyl-9-phosphabicyclo(4.2.1)nonane (13886-99-2)
Eicosyl stearate (22413-02-1)
1-Eicosyne (8CI,9CI) (765-27-5)
Einalon S (52-86-8)
Eisendextran (German) (9004-66-4)
Eisendimethyldithiocarbamat (14484-64-1)
Eisendimethyldithiocarbamat (German) (14484-64-1)
Eisenoxyd (1309-37-1)
Eisen-sorbitol-zitrat (German) (1338-16-5)
Eisen(III)-tris(N,N-dimethyldithiocarbamat) (14484-64-1)
Eisen(III)-tris(N,N-dimethyldithiocarbamat) (German) (14484-64-1)
Ejibil (93-54-9)
Ekagom TB (137-26-8)
Ekagom Teds (97-77-8)
Ekaline G 80 (9005-00-9)

Ekalux	**(13593-03-8)**
Ekalux 25EC	**(13593-03-8)**
Ekamet	**(38260-54-7)**
Ekamet G	**(38260-54-7)**
Ekamet ULV	**(38260-54-7)**
Ekatin	**(640-15-3)**
Ekatin F	**(144-41-2)**
Ekatin M	**(144-41-2)**
Ekatin ULV	**(640-15-3)**
Ekatin WF & WF ULV	**(56-38-2)**
Ekatin aerosol	**(640-15-3)**
Ekatine-25	**(640-15-3)**
Ekatox	**(56-38-2)**
Ekkusagoni	**(32861-85-1)**
Ektafos	**(141-66-2)**
Ektasolve DB	**(112-34-5)**
Ektasolve DB acetate	**(124-17-4)**
Ektasolve DE Acetate	**(112-15-2)**
Ektasolve EB	**(111-76-2)**
Ektasolve EB Acetate	**(112-07-2)**
Ektasolve EE	**(110-80-5)**
Ektasolve EE Acetate Solvent	**(111-15-9)**
Ektasolve EIB	**(4439-24-1)**
Ektasolve EP	**(2807-30-9)**
El 60	**(95-35-2)**
El Petn	**(78-11-5)**
El Rexene PP 115	**(9003-07-0)**
Elagostasine	**(476-66-4)**
Elaic acid	**(112-80-1)**
Elaidic-acid	**(112-79-8)**
Elaiomycin	**(23315-05-1)**
Elaiomycin	**(499-48-9)**
Elaldehyde	**(123-63-7)**
Elancoban	**(17090-79-8)**
Elancolan	**(1582-09-8)**
Elanil	**(50-48-6)**
Elaol	**(84-74-2)**
Elastik (Fiber)	**(63428-84-2)**
Elastil	**(63428-84-2)**
Elastonin	**(60-13-9)**
Elastonon	**(300-62-9)**
Elastonon	**(51-63-8)**
Elastonon	**(60-13-9)**
Elastonon	**(60-15-1)**
Elastopar	**(99-80-9)**
Elastopax	**(99-80-9)**
Elastozone 30	**(103-96-8)**
Elastozone 31	**(139-60-6)**
Elastozone 33	**(3081-14-9)**
Elastozone 34	**(101-72-4)**
Elavil	**(50-48-6)**
Elayl	**(74-85-1)**
Elbanil	**(101-21-3)**
Elbrus	**(2898-12-6)**
Elcema F 150	**(9004-34-6)**
Elcema G 250	**(9004-34-6)**
Elcema P 050	**(9004-34-6)**
Elcema P 100	**(9004-34-6)**
Elcide 75	**(54-64-8)**
Elcoril	**(305-03-3)**
Elcozine Chrysoidine Y	**(532-82-1)**
Elcozine Rhodamine B	**(81-88-9)**
Elcozine Rhodamine 6GDN	**(989-38-8)**
Elcron	**(6988-21-2)**
Eldadryl	**(147-24-0)**
Eldecort	**(50-23-7)**
Elder (Fiber)	**(63428-84-2)**
Eldezol	**(59-87-0)**
Eldezol F-6	**(59-87-0)**
Eldiatric C	**(58-08-2)**
Eldodram	**(523-87-5)**
Eldopal	**(59-92-7)**
Eldopaque	**(123-31-9)**
Eldopar	**(59-92-7)**
Eldopatec	**(59-92-7)**
Eldoquin	**(123-31-9)**
Eldrin	**(153-18-4)**
Eleagic acid	**(476-66-4)**
Elecor	**(390-64-7)**
Elecron 50	**(6988-21-2)**
Electro-CF 11	**(75-69-4)**
Electro-CF 12	**(75-71-8)**
Electro-CF 22	**(75-45-6)**
Electric utility boiler slag (Coal)	**(68476-96-0)**
Electrocorundum	**(1302-74-5)**
Electrographite	**(7782-42-5)**
Electronic E-2	**(7783-07-5)**
Eledtrox 2500	**(1314-13-2)**
β-Elemene	**(33880-83-0)**
Elemental Selenium	**(7782-49-2)**
Elenium	**(58-25-3)**
Elephant tranquilizer	**(956-90-1)**
Elepsindon	**(57-41-0)**
Elestol	**(54-05-7)**
Eleudron	**(72-14-0)**
Elf	**(1333-86-4)**
Elfan 242	**(9004-82-4)**
Elfan NS 242	**(9004-82-4)**
Elfan NS 243	**(9004-82-4)**
Elfan NS 243S	**(9004-82-4)**
Elfan 4240 T	**(139-96-8)**
Elfan WA Sulphonic Acid	**(27176-87-0)**
Elfanex	**(50-55-5)**
Elftex	**(1333-86-4)**
Elgacid Orange 2G	**(523-44-4)**
Elgetol	**(2312-76-7)**
Elgetol	**(534-52-1)**
Elgetol	**(88-85-7)**
Elgetol 30	**(534-52-1)**
Elgetol 318	**(88-85-7)**
Eliamina Light Brown BRL	**(16071-86-6)**
Eliamina Red 8BL	**(2610-11-9)**
Elicide	**(54-64-8)**
Elimin	**(2011-67-8)**
Eliminoxy	**(538-65-8)**
Elipol	**(534-52-1)**
Elite Fast Red G	**(3567-65-5)**
Elite Fast Red GRS	**(3567-65-5)**
Elitone	**(59-26-7)**
Elixicon	**(58-55-9)**
Elixophyllin	**(58-55-9)**
Elixophylline	**(58-55-9)**
Eljon Bordeaux	**(1103-39-5)**
Eljon Fast Orange G	**(3520-72-7)**
Eljon Fast Scarlet PV Extra	**(2425-85-6)**
Eljon Fast Scarlet RN	**(2425-85-6)**
Eljon Fast Yellow GN-GX	**(2512-29-0)**
Eljon Fast Yellow PV Extra	**(2512-29-0)**
Eljon Lake Red C	**(5160-02-1)**
Eljon Lithol Red BS	**(1103-39-5)**
Eljon Lithol Red MS	**(1103-38-4)**
Eljon Lithol Red No. 10	**(1248-18-6)**
Eljon Madder	**(72-48-0)**
Eljon Magenta Toner	**(509-34-2)**

Name	CAS No.
Eljon Pink Toner	(989-38-8)
Eljon Violet Toner	(1325-82-2)
Eljon Yellow BG	(6358-85-6)
Ellagic acid	(476-66-4)
Elmasil	(61-82-5)
Elmedal	(50-33-9)
Elmer's Glue All	(9003-20-7)
Elobromol	(10318-26-0)
Elocron	(6988-21-2)
Elocron 8353	(6988-21-2)
Elocron 50WP	(6988-21-2)
Elodrin	(135-09-1)
Elon	(55-55-0)
Elon (Developer)	(55-55-0)
Elosal	(7704-34-9)
Elpon	(9003-07-0)
Elrodorm	(77-21-4)
Elsan	(2597-03-7)
Elserpine	(50-55-5)
Elsix	(141-94-6)
Eltesol SX 30	(1300-72-7)
Eltex	(9002-88-4)
Eltex 6037	(9002-88-4)
Eltex A 1050	(9002-88-4)
Eltren	(104-74-5)
Eltrianyl	(8064-90-2)
Elvacet 81-900	(9003-20-7)
Elvacite	(9011-14-7)
Elvacite 2008	(9011-14-7)
Elvacite 2009	(9011-14-7)
Elvacite 2010	(9011-14-7)
Elvacite 2013	(25608-33-7)
Elvacite 2021	(9011-14-7)
Elvacite 2041	(9011-14-7)
Elvacite 6011	(9011-14-7)
Elvacite 6012	(9011-14-7)
Elvacite 6016	(25608-33-7)
Elvanol	(9002-89-5)
Elvanol 50-42	(9002-89-5)
Elvanol 52-22	(9002-89-5)
Elvanol 70-05	(9002-89-5)
Elvanol 71-30	(9002-89-5)
Elvanol 90-50	(9002-89-5)
Elvanol 522-22	(9002-89-5)
Elvanol 73125G	(9002-89-5)
Elvaron	(1085-98-9)
Elysion	(956-90-1)
Elyzol	(443-48-1)
Elzogram	(25953-19-9)
Emo-Nik	(54-11-5)
Emaform	(130-26-7)
Emagrin	(61-76-7)
Emal 10	(151-21-3)
Emal 108	(9002-92-0)
Emal O	(151-21-3)
Emal T	(139-96-8)
Emalex 515	(9004-98-2)
Emalex 715	(9002-92-0)
Emanay Atomized Aluminum Powder	(7429-90-5)
Emanay Zinc Dust	(7440-66-6)
Emanay Zinc Oxide	(1314-13-2)
Emandione	(83-12-5)
Emanon 3113	(9004-99-3)
Emanon 3115	(9004-99-3)
Emanon 3199	(9004-99-3)
Emanon 4115	(9004-96-0)
Emar	(1314-13-2)
Emasol 430	(26266-58-0)
Emasol 41S	(8007-43-0)
Emathlite	(1332-58-7)
Embacetin	(56-75-7)
Embafume	(74-83-9)
Embanox	(25013-16-5)
Embarin	(315-30-0)
Embark	(53780-34-0)
Embark Plant Growth Regulator	(53780-34-0)
Embathion	(563-12-2)
Embay 8440	(55268-74-1)
Embechine	(55-86-7)
Embequin	(83-73-8)
Embichin	(51-75-2)
Embichin	(55-86-7)
Embichin hydrochloride	(55-86-7)
Embikhine	(55-86-7)
Embiol	(68-19-9)
Emblem	(1861-40-1)
Embutal	(57-33-0)
Embutox	(10433-59-7)
Embutox	(94-82-6)
Embutox E	(94-82-6)
Embutox Klean-Up	(94-82-6)
Emcepan	(94-74-6)
Emcol 12-14-18	(9007-48-1)
Emcol H-2A	(9004-96-0)
Emcol CA	(31566-31-1)
Emcol DS-50 CAD	(106-11-6)
Emcol E-607	(6272-74-8)
Emcol ETS	(106-11-6)
Emcol H 31A	(9004-96-0)
Emcol-IM	(110-27-0)
Emcol-IP	(142-91-6)
Emcol MSK	(31566-31-1)
Emcol NA-30	(61789-40-0)
Emcol O	(25496-72-4)
Emcol PS-50 RHP	(1323-39-3)
Emcol Q	(26006-22-4)
Emcol RDC-D	(141-20-8)
Emedan	(339-43-5)
Emerald Green	(12002-03-8)
Emerald Green	(633-03-4)
Emeressence 1160	(122-99-6)
Emerest 2301	(112-62-9)
Emerest 2314	(110-27-0)
Emerest 2316	(142-91-6)
Emerest 2325	(123-95-5)
Emerest 2350	(111-60-4)
Emerest 2381	(1323-39-3)
Emerest 2400	(31566-31-1)
Emerest 2401	(31566-31-1)
Emerest 2407	(123-94-4)
Emerest 2640	(9004-99-3)
Emerest 2642	(9005-08-7)
Emerest 2646	(9004-96-0)
Emerest 2660	(9004-96-0)
Emerest 2801	(112-62-9)
Emerex	(653-03-2)
Emerlube 6717	(9002-98-6)
Emerox 1110	(123-99-9)
Emerox 1144	(123-99-9)
Emersal 6400	(151-21-3)
Emersal 6434	(139-96-8)
Emersal 6465	(126-92-1)
Emersol 120	(57-11-4)
Emersol 132	(57-11-4)

Emersol 140	**(57-10-3)**
Emersol 143	**(57-10-3)**
Emersol 150	**(57-11-4)**
Emersol 210	**(112-80-1)**
Emersol 213	**(112-80-1)**
Emersol 310	**(60-33-3)**
Emersol 315	**(60-33-3)**
Emersol 6321	**(112-80-1)**
Emersol 221 Low titer white oleic acid	**(112-80-1)**
Emersol 220 White oleic acid	**(112-80-1)**
Emersol 233ll	**(112-80-1)**
Emery 655	**(544-63-8)**
Emery 2218	**(112-61-8)**
Emery 2219	**(112-62-9)**
Emery 2302	**(111-59-1)**
Emery 2310	**(112-62-9)**
Emery 2423	**(122-32-7)**
Emery 5700	**(122-98-5)**
Emery 5703	**(120-07-0)**
Emery 5709	**(91-99-6)**
Emery 5711	**(136-80-1)**
Emery 5712	**(28005-74-5)**
Emery 5714	**(91-88-3)**
Emery 5715	**(92-00-2)**
Emery 5717	**(92-00-2)**
Emery 5724	**(92-64-8)**
Emery 5770	**(92-49-9)**
Emery 5791	**(60-24-2)**
Emery 6705	**(122-99-6)**
Emery 6717	**(9002-98-6)**
Emery 15393	**(9004-99-3)**
Emery (ACGIH,OSHA)	**(112-62-9)**
Emery Oleic Acid Ester 2221	**(25496-72-4)**
Emery Oleic Acid Ester 2230	**(122-32-7)**
Emery Oleic Acid Ester 2301	**(112-62-9)**
Emery Oleic Acid Ester 2302	**(111-59-1)**
Emeside	**(77-67-8)**
Emetan, 2,3-didehydro-6',7',10,11-tetramethoxy- (9CI)	**(4914-30-1)**
Emetan-6'-ol, 7',10,11-trimethoxy-, dihydrochloride (9CI)	**(5853-29-2)**
Emete-Con	**(63-12-7)**
Emeticon	**(63-12-7)**
(-)-Emetine	**(483-18-1)**
Emetine	**(483-18-1)**
Emetine, 2,3-didehydro	**(4914-30-1)**
(-)-Emetine dihydrochloride	**(316-42-7)**
Emetine, dihydrochloride	**(316-42-7)**
l-Emetine dihydrochloride	**(316-42-7)**
Emetine, hydrochloride	**(316-42-7)**
Emetique (French)	**(28300-74-5)**
Emetren	**(56-75-7)**
Emfac 1202	**(112-05-0)**
Emid 6511	**(120-40-1)**
Emid 6541	**(120-40-1)**
Emineurina	**(533-45-9)**
Emipherol	**(59-02-9)**
Emisan 6	**(123-88-6)**
Emisan 6	**(7487-94-7)**
Emisol	**(61-82-5)**
Emisol 50	**(61-82-5)**
Emisol F	**(61-82-5)**
Emkalyx EP 64	**(9003-11-6)**
Emkalyx L 101	**(9003-11-6)**
Emkapyl 1839 (9CI)	**(58969-15-6)**
Emmatos	**(121-75-5)**
Emmatos Extra	**(121-75-5)**
Emociclina	**(68-19-9)**
Emodin	**(15687-27-1)**
Emodin	**(518-82-1)**
Emodol	**(518-82-1)**
Emoquil	**(82-92-8)**
Emoren	**(126-27-2)**
Emotival	**(846-49-1)**
Empal	**(94-74-6)**
Empecid	**(23593-75-1)**
Empicol ESB 3	**(9004-82-4)**
Empicol ESB 30	**(9004-82-4)**
Empicol LPZ	**(151-21-3)**
Empicol LS 30	**(151-21-3)**
Empicol LX 28	**(151-21-3)**
Empilan 2848	**(111-60-4)**
Empilan BP 100	**(9004-96-0)**
Empilan BQ 100	**(9004-96-0)**
Empimin KSN	**(9004-82-4)**
Empimin KSN 27	**(9004-82-4)**
Empimin KSN 60	**(9004-82-4)**
Empimin KSN 70	**(9004-82-4)**
Empimin OT	**(2373-23-1)**
Empiphos STP-D	**(7758-29-4)**
Empiral	**(62-44-2)**
Empirin	**(50-78-2)**
Empirin Compound	**(62-44-2)**
Empirin Compound	**(8003-03-0)**
Emplets potassium chloride	**(7447-40-7)**
Emprazil	**(62-44-2)**
Emprazil-C	**(62-44-2)**
Emrite 6009	**(25496-72-4)**
Emsorb 2500	**(1338-43-8)**
Emsorb 2502	**(8007-43-0)**
Emsorb 2503	**(26266-58-0)**
Emsorb 2505	**(1338-41-6)**
Emsorb 2510	**(26266-57-9)**
Emsorb 2515	**(1338-39-2)**
Emsorb 6900	**(9005-65-6)**
Emsorb 6915	**(9005-64-5)**
Emtexate	**(59-05-2)**
Emtryl	**(551-92-8)**
Emtrylvet	**(551-92-8)**
Emtrymix	**(551-92-8)**
Emul P.7	**(31566-31-1)**
Emulgator 8972	**(8007-43-0)**
Emulgen 100	**(9002-92-0)**
Emulgen 105	**(9002-92-0)**
Emulgen 108	**(9002-92-0)**
Emulgen 109	**(9002-92-0)**
Emulgen 120	**(9002-92-0)**
Emulgen 147	**(9002-92-0)**
Emulgen 404	**(9004-98-2)**
Emulgen 408	**(9004-98-2)**
Emulgen 420	**(9004-98-2)**
Emulgen 430	**(9004-98-2)**
Emulgen 808	**(9036-19-5)**
Emulgen 810	**(9036-19-5)**
Emulgen 306P	**(9005-00-9)**
Emulgen 320P	**(9005-00-9)**
Emulgen 409P	**(9004-98-2)**
Emulgen PP	**(9003-11-6)**
Emulgen PP 150	**(9003-11-6)**
Emulgen PP 250	**(9003-11-6)**
Emulgen PP 290	**(9003-11-6)**
Emulgin O 10	**(9004-98-2)**
Emulgin O 5	**(9004-98-2)**
Emulmin L 500	**(9002-92-0)**
Emulphor	**(9004-98-2)**
Emulphor A	**(9004-96-0)**

Emulphor ON 870	(9004-98-2)	Endosulphan	(115-29-7)
Emulphor Surfactants	(9004-98-2)	Endotal	(129-67-9)
Emulphor UN-430	(9004-96-0)	Endothal	(129-67-9)
Emulphor VN 430	(9004-96-0)	Endothal	(145-73-3)
Emulphor VT-650	(9004-99-3)	Endothal Technical	(145-73-3)
Emulsamine BK	(94-75-7)	Endothal Weed Killer	(129-67-9)
Emulsamine E-3	(94-75-7)	Endothal Weed Killer	(145-73-3)
Emulsans	(80450-55-1)	Endothall	(129-67-9)
Emulsifier No. 104	(151-21-3)	Endothall	(145-73-3)
Emulsion 212	(13674-87-8)	7-Endothall, dipotassium salt	(2164-07-0)
Emulsiphos 440/660	(7601-54-9)	Endothall dipotassium salt	(2164-07-0)
Emulsogen MS 12	(9004-98-2)	Endothal-natrium (Dutch)	(129-67-9)
Emultex F	(9003-20-7)	Endothal-sodium	(129-67-9)
Emulthin M-35	(8002-43-5)	Endothion	(2778-04-3)
Emyrenil	(14698-29-4)	Endox	(5836-29-3)
Ens-Zem Weevil Bait	(16893-85-9)	Endoxan	(50-18-0)
Enadel	(24166-13-0)	Endoxan-ASTA	(6055-19-2)
Enallynymal sodium	(309-36-4)	Endoxan R	(50-18-0)
Enamel White	(7727-43-7)	Endoxan R	(6055-19-2)
Enanthal	(111-71-7)	Endoxana	(50-18-0)
Enanthaldehyde	(111-71-7)	Endoxana	(6055-19-2)
Enanthic acid	(111-14-8)	Endoxanal	(50-18-0)
Enanthic alcohol	(111-70-6)	Endoxan-asta	(50-18-0)
Enanthole	(111-71-7)	Endoxane	(50-18-0)
Enanthone	(462-18-0)	Endoxan monohydrate	(6055-19-2)
Enanthylic acid	(111-14-8)	3,6-Endoxohexahydrophthalic acid	(145-73-3)
Enanthylic ether	(106-30-9)	3,6-Endoxohexahydrophthalic acid disodium salt	(129-67-9)
Enarmon	(57-85-2)	Endrate	(60-00-4)
Enaven	(22212-55-1)	Endrate disodium	(139-33-3)
Encapla	(64-55-1)	Endrate tetrasodium	(64-02-8)
Encephabol	(123-03-5)	Endrex	(72-20-8)
Encorton	(53-03-2)	Endrin (ACGIH,DOT,OSHA)	(72-20-8)
Encortone	(53-03-2)	Endrin Ketone	(53494-70-5)
Endecril	(58-54-8)	Endrin alcohol	(33058-12-7)
Endiemalum	(50-11-3)	Endrin aldehyde	(7421-93-4)
Endo E	(59-02-9)	Endrine (French)	(72-20-8)
Endobion	(98-92-0)	Endrin mixture (DOT)	(72-20-8)
Endocel	(115-29-7)	Endrocid	(5836-29-3)
Endocid	(2778-04-3)	Endrocide	(5836-29-3)
Endocide	(115-29-7)	Endurol Golden Orange G	(128-70-1)
Endocide	(2778-04-3)	Enduron	(135-07-9)
Endodan	(33813-20-6)	Enduxan	(50-18-0)
1,4-Endoethylenecyclohexane	(280-33-1)	Enduxan	(6055-19-2)
3,6-Endoethylenecyclohexene	(931-64-6)	Endydol	(50-78-2)
Endofollicolina D.P.	(113-38-2)	Endyl	(786-19-6)
Endofolliculina	(53-16-7)	Enelfa	(103-90-2)
Endolat	(50-13-5)	Eneril	(103-90-2)
Endolin	(2152-34-3)	Enfenemal	(115-38-8)
1,5-Endomethylene-3,7-dinitroso-1,3,5,7-tetraazacyclooctane	(101-25-7)	Enflurane (ACGIH)	(13838-16-9)
Endomethylenetetrahydrophthalic acid, N-2-ethylhexyl imide	(113-48-4)	Engemycin	(2058-46-0)
3,6-Endooxohexahydrophthalic acid	(145-73-3)	English Red	(1309-37-1)
3,6-Endooxycyclohexane	(279-49-2)	Enheptin	(121-66-4)
Endopancrine	(9004-10-8)	Enheptin Premix	(121-66-4)
Endopituitrina	(50-56-6)	Enheptin-T	(121-66-4)
Endosan	(485-31-4)	Enhexymal	(56-29-1)
Endosol	(115-29-7)	Eniacid Black IVS	(1064-48-8)
Endosulfan 1	(959-98-8)	Eniacid Black SH	(1064-48-8)
Endosulfan 2	(33213-65-9)	Eniacid Brilliant Rubine 3B	(3567-69-9)
a-Endosulfan-α	(959-98-8)	Eniacid Fast Red A	(1658-56-6)
α-Endosulfan	(959-98-8)	Eniacid Fuchsine BN	(3567-66-6)
b-Endosulfan-β	(33213-65-9)	Eniacid Light Orange G	(1936-15-8)
β-Endosulfan	(33213-65-9)	Eniacid Light Red 3G	(3734-67-6)
Endosulfan A	(959-98-8)	Eniacid Metanil Yellow GN	(587-98-4)
Endosulfan (ACGIH,DOT,OSHA)	(115-29-7)	Eniacid Orange I	(523-44-4)
Endosulfan B	(33213-65-9)	Eniacid Sunset Yellow	(2783-94-0)
Endosulfan mixture, Liquid (DOT)	(115-29-7)	Eniacid Yellow RS	(547-57-9)
Endosulfan sulfate	(1031-07-8)	Enial Orange I	(842-07-9)

Enial Red IV (85-83-6)
Enial Yellow 2G (60-11-7)
Enialit Light Red RL (2425-85-6)
Eniamethyl Orange (547-58-0)
Enianil Black CN (1937-37-7)
Enianil Black RCN (2429-83-6)
Enianil Blue 2BN (2602-46-2)
Enianil Brilliant Blue FF (2610-05-1)
Enianil Brown 2GS (2586-58-5)
Enianil Dark Green BG (3626-28-6)
Enianil Fast Brown M (2429-82-5)
Enianil Light Brown BRL (16071-86-6)
Enianil Pure Blue AN (2429-74-5)
Eniazol Blue Black BHN (2429-73-4)
Enicol (56-75-7)
Enidrel (604-75-1)
Eniloconazol (SP) (35554-44-0)
Enipresser (50-55-5)
Enjay CD 392 (9003-07-0)
Enjay CD 460 (9003-07-0)
Enjay CD 490 (9003-07-0)
Enjay E 117 (9003-07-0)
Enjay E 11S (9003-07-0)
Enkalon (63428-83-1)
Enkefal (630-93-3)
Enkelfel (57-41-0)
Enkorton (53-03-2)
Enodrin (96-23-1)
Enolofos (13104-21-7)
Enorden (57-53-4)
Enovit (23564-06-9)
Enovit M (23564-05-8)
Enovit Methyl (23564-05-8)
Enovit-Supper (23564-05-8)
Enphenemal (115-38-8)
Enrumay (91-84-9)
Enseal (7447-40-7)
Ensobarb (50-06-6)
Ensodorm (50-06-6)
Enstar (42588-37-4)
Enstar IGR (42588-37-4)
Ensure (115-29-7)
Ent (68-22-4)
Ent 2,818 (119-89-1)
Ent 25,515 (297-99-4)
Entepas (65-49-6)
Entero-Bio Form (130-26-7)
Entero-Vioform (130-26-7)
Enteramine (50-67-9)
Entericin (50-78-2)
Enteromycetin (56-75-7)
Enterophen (50-78-2)
Enteroquinol (130-26-7)
Enterosalicyl (54-21-7)
Enterosalil (54-21-7)
Enterosarine (50-78-2)
Enterosediv (50-35-1)
Enterosept (83-73-8)
Enteroseptol (130-26-7)
Enterotoxon (67-45-8)
Enterozol (130-26-7)
Enterum Locorten (130-26-7)
Entex (55-38-9)
Entizol (443-48-1)
Entol (140-07-8)
Entomoxan (58-89-9)
Entprol (102-60-3)
Entramin (121-66-4)
Entrokin (130-26-7)
Entrophen (50-78-2)
Entsufon (118-96-7)
Entsufon sodium (2917-94-4)
Enturen (57-96-5)
Entusil (127-69-5)
Entusul (127-69-5)
Enuclen (8001-54-5)
Envert-T (93-76-5)
Envert 171 (94-75-7)
Envert DT (94-75-7)
Enzactin (102-76-1)
Enzaprost (551-11-1)
Enzaprost F (551-11-1)
Enzeon (9004-07-3)
Enzianwurzel (72968-42-4)
Eosin (15086-94-9)
Eosin (17372-87-1)
Eosin (548-26-5)
Eosin G (17372-87-1)
Eosin Gelblich (German) (17372-87-1)
Eosin Y (17372-87-1)
Eosin YS (17372-87-1)
Eosin YS (548-26-5)
Eosine (15086-94-9)
Eosine (17372-87-1)
Eosine B (17372-87-1)
Eosine B (548-26-5)
Eosine BPC (17372-87-1)
Eosine BPC (548-26-5)
Eosine BS (17372-87-1)
Eosine BS (548-26-5)
Eosine BS-SF (17372-87-1)
Eosine BS-SF (548-26-5)
Eosine DA (17372-87-1)
Eosine DA (548-26-5)
Eosine DWC 73 (17372-87-1)
Eosine DWC 73 (548-26-5)
Eosine Extra Conc. A. Export (17372-87-1)
Eosine Extra conc. A. export (548-26-5)
Eosine FA (17372-87-1)
Eosine FA (548-26-5)
Eosine 3G (17372-87-1)
Eosine 3G (548-26-5)
Eosine G (17372-87-1)
Eosine GF (17372-87-1)
Eosine GF (548-26-5)
Eosine J (17372-87-1)
Eosine J (548-26-5)
Eosine K Salt Free (548-26-5)
Eosine Lake Red Y (548-26-5)
Eosine OJ (17372-87-1)
Eosine S13 (Bluish) (17372-87-1)
Eosine Salt Free (548-26-5)
Eosine Sodium Salt (17372-87-1)
Eosine W/S (17372-87-1)
Eosine W/S (548-26-5)
Eosine 3Y (17372-87-1)
Eosine Y (17372-87-1)
Eosine YB (17372-87-1)
Eosine YB (548-26-5)
Eosine YS (17372-87-1)
Eosine YS (548-26-5)
Eosine Yellowish-(YS) (17372-87-1)
Eosine Yellowish (17372-87-1)
Epi-Clear (94-36-0)

Epi-Rez 505 (9CI) **(53664-71-4)**
Epi-Rez 508 **(1675-54-3)**
Epi-Rez 510 **(1675-54-3)**
Epi-Rez 5011 (9CI) **(85255-90-9)**
Epi-Rez 5014 **(29298-03-1)**
Epi-Rez 5071 (9CI) **(67725-14-8)**
Epi-Rez 5077 (9CI) **(86903-93-7)**
Epal **(39148-24-8)**
Epal 6 **(111-27-3)**
Epal 8 **(111-87-5)**
Epal 10 **(112-30-1)**
Epal 12 **(112-53-8)**
Epal 16NF **(36653-82-4)**
Epamin **(57-41-0)**
Epamin **(630-93-3)**
Epan 420 **(9003-11-6)**
Epan 450 **(9003-11-6)**
Epan 485 **(9003-11-6)**
Epan 710 **(9003-11-6)**
Epan 720 **(9003-11-6)**
Epan 740 **(9003-11-6)**
Epan 742 **(9003-11-6)**
Epan 750 **(9003-11-6)**
Epan U 102 **(9003-11-6)**
Epan U 103 **(9003-11-6)**
Epan U 105 **(9003-11-6)**
Epan U 108 **(9003-11-6)**
Epanal **(50-06-6)**
Epanutin **(57-41-0)**
Epanutin **(630-93-3)**
Eparen **(1085-98-9)**
Epasmir "5" **(57-41-0)**
Epdantoine simple **(57-41-0)**
Epelin **(57-41-0)**
Epelin **(630-93-3)**
Ephadren **(51-63-8)**
Ephedral **(299-42-3)**
Ephedrate **(299-42-3)**
Ephedremal **(299-42-3)**
Ephedrin **(299-42-3)**
Ephedrine **(299-42-3)**
L(-)-Ephedrine **(299-42-3)**
d-psi-Ephedrine **(90-82-4)**
l-Ephedrine **(299-42-3)**
psi-Ephedrine **(90-82-4)**
Ephedrine, L-(-) **(299-42-3)**
l-Ephedrine sulfate **(134-72-5)**
Ephedrine sulfate (2:1) (Salt), (-) **(134-72-5)**
Ephedrital **(299-42-3)**
Ephedrol **(299-42-3)**
Ephedrosan **(299-42-3)**
Ephedrotal **(299-42-3)**
Ephedsol **(299-42-3)**
Ephendronal **(299-42-3)**
Epheron **(63-98-9)**
Ephirsulphonate **(80-33-1)**
Ephorran **(97-77-8)**
Ephoxamin **(299-42-3)**
Ephynal **(59-02-9)**
4-Epiabietal, dehydro- **(24035-50-5)**
4-Epiabietic acid, dehydro- **(5155-70-4)**
Epibenzalin **(146-22-5)**
Epibloc **(96-24-2)**
Epibromhydrin **(3132-64-7)**
Epibromohydrin [UN 2558] **(3132-64-7)**
Epibromohydrine **(3132-64-7)**
Epic **(60568-05-0)**
Epicamphor **(10292-98-5)**
Epichloorhydrine (Dutch) **(106-89-8)**
Epichlorhydrin (German) **(106-89-8)**
Epichlorhydrine (French) **(106-89-8)**
(DL)-α-Epichlorohydrin **(106-89-8)**
α-Epichlorohydrin **(106-89-8)**
Epichlorohydrin (ACGIH,OSHA) [UN 2023] **(106-89-8)**
Epichlorohydrin-glycerine copolymer **(25038-04-4)**
Epichlorohydryna (Polish) **(106-89-8)**
Epichlorophydrin **(106-89-8)**
Epiclase **(63-98-9)**
Epicloridrina (Italian) **(106-89-8)**
Epicur **(57-53-4)**
Epicure DDM **(101-77-9)**
Epidermol **(83-63-6)**
Epidian 5 **(25068-38-6)**
3,17-Epidihydroxyestratriene **(50-28-2)**
3,17-Epidihydroxyoestratriene **(50-28-2)**
10H-3,10a-Epidithiopyrazino(1,2-a)indole-1,4-dione,2,3,5a,6-tetra-hydro-6-hydroxy-3- (hydroxymethyl)-2-methyl **(67-99-2)**
Epidorm **(50-06-6)**
Epidropal **(315-30-0)**
Epifenyl **(57-41-0)**
Epifenyl **(630-93-3)**
Epifluorohydrin **(503-09-3)**
Epifrin **(51-43-4)**
Epihydan **(57-41-0)**
Epihydan **(630-93-3)**
Epihydrin alcohol **(556-52-5)**
Epihydrinaldehyde **(765-34-4)**
Epihydrine aldehyde **(765-34-4)**
11-Epiisoeusantona-1,4-dienic acid, 6α-hydroxy-3-oxo-, γ-lactone **(481-06-1)**
Epikote 812 **(31305-91-6)**
Epikote 828 **(25068-38-6)**
Epikote 1001 **(25068-38-6)**
Epikote 1004 **(25068-38-6)**
Epikote RXE 15 **(28825-96-9)**
Epikur **(57-53-4)**
Epikure DDM **(101-77-9)**
Epilan **(50-12-4)**
Epilan **(57-41-0)**
Epilan-D **(630-93-3)**
Epilan-d **(57-41-0)**
Epilantin **(57-41-0)**
Epilantin **(630-93-3)**
Epileo petit mal **(77-67-8)**
Epilim **(1069-66-5)**
Epilim **(99-66-1)**
Epilol **(50-06-6)**
Epimid **(86-34-0)**
Epinat **(57-41-0)**
Epinat **(630-93-3)**
Epinefrin (Czech) **(51-43-4)**
Epinefrina **(51-43-4)**
Epinelbon **(146-22-5)**
Epinephran **(51-43-4)**
(-)-Epinephrine **(51-43-4)**
(R)-Epinephrine **(51-43-4)**
Epinephrine **(51-43-4)**
dl-Epinephrine **(329-65-7)**
l-Epinephrine **(51-43-4)**
Epinephrine Racemic **(329-65-7)**
(-)-Epinephrine bitartrate **(51-42-3)**
Epinephrine bitartrate **(51-42-3)**
Epinephrine d-bitartrate **(51-42-3)**
l-Epinephrine bitartrate **(51-42-3)**
l-Epinephrine d-bitartrate **(51-42-3)**

Epinephrine chloride **(55-31-2)**
l-Epinephrine chloride **(55-31-2)**
(+-)-Epinephrine hydrochloride **(329-63-5)**
(-)-Epinephrine hydrochloride **(55-31-2)**
Epinephrine, hydrochloride **(55-31-2)**
dl-Epinephrine hydrochloride **(329-63-5)**
l-Epinephrine hydrochloride **(55-31-2)**
Epinephrine hydrogen tartrate **(51-42-3)**
l-Epinephrine (synthetic) **(51-43-4)**
Epinephrine tartrate **(51-42-3)**
l-Epinephrine tartrate **(51-42-3)**
Epirenamine **(51-43-4)**
l-Epirenamine **(51-43-4)**
Epirenan **(51-43-4)**
Epirenin **(51-43-4)**
Epised **(57-41-0)**
Episedal **(50-06-6)**
Epiteliol **(68-26-8)**
Epithelone **(83-63-6)**
1,12-Epithiotriphenylene **(68558-73-6)**
Epitrate **(51-43-4)**
Epobron **(15687-27-1)**
Epodyl **(1954-28-5)**
Epok U 9048 **(9011-05-6)**
Epolene C **(9002-88-4)**
Epolene C 10 **(9002-88-4)**
Epolene C 11 **(9002-88-4)**
Epolene E **(9002-88-4)**
Epolene E 10 **(9002-88-4)**
Epolene E 12 **(9002-88-4)**
Epolene M 5H **(9003-07-0)**
Epolene M 5K **(9003-07-0)**
Epolene M 5W **(9003-07-0)**
Epolene N **(9002-88-4)**
Epomine 1000 **(9002-98-6)**
Epomine D 3000 **(9002-98-6)**
Epomine P 1000 **(9002-98-6)**
Epomine P 1500 **(9002-98-6)**
Epomine P 500 **(9002-98-6)**
Epomine SP 003 **(9002-98-6)**
Epomine SP 006 **(9002-98-6)**
Epomine SP 012 **(9002-98-6)**
Epomine SP 018 **(9002-98-6)**
Epomine SP 103 **(9002-98-6)**
Epomine SP 110 **(9002-98-6)**
Epomine SP 200 **(9002-98-6)**
Epomine 150T **(9002-98-6)**
Epon 812 **(31305-91-6)**
Epon 815 **(29407-84-9)**
Epon 828 **(1675-54-3)**
Eporal **(80-08-0)**
(3,6-Epossi-cicloesan-1,2-dicarbossilato) disodico (Italian) **(129-67-9)**
Epoxide 101 **(106-86-5)**
Epoxide 207 **(81-21-0)**
Epoxide 269 **(96-08-2)**
Epoxide 7 **(55838-67-0)**
Epoxide 8 **(39390-62-0)**
Epoxide-201 **(141-37-7)**
Epoxide A **(1675-54-3)**
Epoxides, Polymers, epoxy resins **(61788-97-4)**
Epoxidized Soybean Oil **(8013-07-8)**
Epoxy Resin ERL-2795 **(25068-38-6)**
1,2-Epoxyaethan (German) **(75-21-8)**
1,2-Epoxybutane **(106-88-7)**
2,3-Epoxybutane **(3266-23-7)**
Epoxybutane **(106-88-7)**
1,2-Epoxy-3-butanone **(4401-11-0)**
3,4-Epoxybutanone **(4401-11-0)**
1,2-Epoxybutene-3 **(930-22-3)**
3,4-Epoxy-1-butene **(930-22-3)**
2,3-Epoxybutyric acid, butyl ester **(10138-34-8)**
Epoxycaryophyllene **(1139-30-6)**
4,9-Epoxycevane-3,4,12,14,16,17,20-heptol 3-(3,4-dimethoxybenzoate) **(71-62-5)**
(Z)-4-α,9-Epoxycevane-3-β,4,12,14,16-β,17,20-heptol 3-(2-methyl-2-butenoate) **(62-59-9)**
1,2-Epoxy-3-chloropropane **(106-89-8)**
5-α,6-α-Epoxycholestanol **(1250-95-9)**
Epoxycholesterol **(1250-95-9)**
Epoxy compounds (VAN) **(61788-97-4)**
Epoxycyclododecane **(286-99-7)**
1,2-Epoxycyclohexane **(286-20-4)**
1,4-Epoxycyclohexane **(279-49-2)**
3,6-Epoxy-cyclohexane 1,2-carboxylate disodique (French) **(129-67-9)**
Epoxycyclohexylethyl trimethoxy silane **(3388-04-3)**
β-(3,4-Epoxycyclohexyl)ethyltrimethoxysilane **(3388-04-3)**
3,4-Epoxycyclohexylmethyl 3,4-epoxycyclohexane carboxylate **(2386-87-0)**
1,2-Epoxycyclopentane **(285-67-6)**
1,2-Epoxydecane **(2404-44-6)**
5,6-Epoxy-5,6-dihydrobenz(a)anthracene **(962-32-3)**
(-)-Epoxydihydrocaryophyllene **(1139-30-6)**
15,20-Epoxy-15,20-dihydro-12-hydroxysenecionan-11,16-dione **(6870-67-3)**
4α,5-Epoxy-3,17β-dihydroxy-5α-androst-2-ene-2-carbonitrile **(13647-35-3)**
4,5-Epoxy-3,6-dihydroxymorphin-7-ene **(466-97-7)**
2,3-Epoxy-2,3-dimethylbutane **(5076-20-0)**
6,7-Epoxy-3,7-dimethyl-1,3-octadiene **(69103-20-4)**
1,3-Epoxy-2,2-dimethylpropane **(6921-35-3)**
1,2-Epoxydodecane **(2855-19-8)**
1,2-Epoxydodekan (Czech) **(2855-19-8)**
1,2-Epoxy-4-(epoxyethyl)cyclohexane **(106-87-6)**
1,2-Epoxy-7,8-epoxyoctane **(2426-07-5)**
1,2-Epoxyethane **(75-21-8)**
Epoxyethane **(75-21-8)**
1,2-Epoxy-3-ethoxypropane **(4016-11-9)**
1,2-Epoxyethylbenzene **(96-09-3)**
Epoxyethylbenzene (8CI) **(96-09-3)**
1-Epoxyethyl-3,4-epoxycyclohexane **(106-87-6)**
2,3-Epoxy-2-ethyl hexanol **(78-72-8)**
α-Epoxyethyl-1,2-(methylenedioxy)benzyl alcohol acetate **(59901-90-5)**
3-(1,2-Epoxyethyl)-7-oxabicyclo(4.1.0)heptane **(106-87-6)**
3-(Epoxyethyl)-7-oxabicyclo(4.1.0)heptane **(106-87-6)**
4-(1,2-Epoxyethyl)-7-oxabicyclo(4.1.0)heptane **(106-87-6)**
4-(Epoxyethyl)-7-oxabicyclo(4.1.0)heptane **(106-87-6)**
1,2-Epoxy-3-ethyloxy propane [UN 2752] **(4016-11-9)**
6,7-Epoxy-1-(p-ethylphenoxy)-3-ethyl-7-methylnonane **(57342-02-6)**
1,2-Epoxy-3-fluoropropane **(503-09-3)**
Epoxyheptachlor **(1024-57-3)**
1,2-Epoxyheptadecane **(22092-38-2)**
2,3-Epoxyheptane **(14925-96-3)**
trans-3,4-Epoxyheptane **(56052-95-0)**
1,2-Epoxyhexadecane **(7320-37-8)**
cis-2,3-Epoxyhexane **(6124-90-9)**
4,5-α-Epoxy-3-hydroxy-5,17-diethylmorphinan-6-one **(143-52-2)**
4,5-Epoxy-3-hydroxy-N-methylmorphinan **(427-00-9)**
(2-α,4-α,5-α,17-β)-4,5-Epoxy-17-hydroxy-3-oxoandrostane-2-carbonitrile **(13647-35-3)**
4-α-5-Epoxy-17-β-hydroxy-3-oxo-5-α-androstane-2-α-carbonitrile **(13647-35-3)**
1,4-Epoxy-p-menthane **(470-67-7)**
1,8-Epoxy-p-menthane **(470-82-6)**
Epoxymethamine bromide **(155-41-9)**
5,8-Epoxy-1,4-methanonaphthalene, 1,2,3,4,10,10-hexachloro-1,4,4a,5,6,7,8,8a-octahydro- **(61167-23-5)**
5,8-Epoxy-1,4-methanonaphthalene, 1,2,3,4,10,10-hexachloro-1,4,4a,5,6,7,8,8a-octahydro-, (1α,4α,4aβ,5α,8α,8aβ)- (9CI) **(61167-23-5)**

Name	CAS No.
2,7-Epoxy-3,6-methanonaphth(2,3-b)oxirene, 3,4,5,6,9,9-hexa-chloro-1a,2,2a,3,6,6a,7,7a-octahydro-, (1aα,2β,2aα,3β,6β,6aα,7β,7aα)- (9CI)	**(61217-08-1)**
4,5-α-Epoxy-3-methoxy-17-methylmorphinan-6-one	**(125-29-1)**
1,2-Epoxy-3-methoxypropane	**(930-37-0)**
3,4-Epoxy-6-methylcyclohexenecarboxylic acid (3,4-epoxy-6-methylcyclohexylmethyl) ester	**(141-37-7)**
3,4-Epoxy-6-methylcyclohexylmethyl 3,4-epoxy-6-methyl-cyclohexanecarboxylate	**(141-37-7)**
3,4-Epoxy-6-methylcyclohexylmethyl-3',4'-epoxy-6'-methyl-cyclohexane carboxylate	**(141-37-7)**
4,5-Epoxy-2-methylcyclohexylmethyl-4,5-epoxy-2-methyl-cyclohexanecarboxylate	**(141-37-7)**
4-(1,2-Epoxy-1-methylethyl)-1-methyl-7-oxabicyclo(4.1.0)heptane	**(96-08-2)**
α,β-Epoxy-β-methylhydrocinnamic acid, ethyl ester	**(77-83-8)**
cis-7,8-Epoxy-2-methyloctadecane	**(29804-22-6)**
2,3-Epoxy-2-methylpentane	**(1192-22-9)**
1,2-Epoxy-3(p-nitrophenoxy)-propane	**(5255-75-4)**
1,2-Epoxynonadecane	**(67860-04-2)**
exo-2,3-Epoxynorbornane	**(3146-39-2)**
1,2-Epoxyoctadecane	**(7390-81-0)**
cis-9,10-Epoxyoctadecanoate	**(2443-39-2)**
cis-9,10-Epoxyoctadecanoic acid	**(2443-39-2)**
9,10-Epoxyoctadecanoic acid, butyl ester	**(106-83-2)**
9,10-Epoxyoctadecanoic acid, 2-ethylhexyl ester	**(141-38-8)**
1,2-Epoxyoctane	**(2984-50-1)**
1,2-Epoxyoktan (Czech)	**(2984-50-1)**
Epoxyoleic acid	**(2443-39-2)**
1,2-Epoxypentane	**(1003-14-1)**
Epoxyperchlorovinyl	**(16650-10-5)**
1,2-Epoxy-3-phenoxypropane	**(122-60-1)**
1,2-Epoxy-1-phenylethane	**(96-09-3)**
2,3-Epoxypinane	**(1686-14-2)**
2,3-Epoxy-1-propanal	**(765-34-4)**
2,3-Epoxypropanal	**(765-34-4)**
1,2-Epoxypropane	**(75-56-9)**
1,3-Epoxypropane	**(503-30-0)**
2,3-Epoxypropane	**(75-56-9)**
Epoxypropane	**(75-56-9)**
1,2-Epoxypropane (OSHA)	**(75-56-9)**
2,3-Epoxypropanol	**(556-52-5)**
2,3-Epoxy-1-propanol (OSHA)	**(556-52-5)**
2,3-Epoxy-1-propanol oleate	**(5431-33-4)**
2,3-Epoxy-1-propanol stearate	**(7460-84-6)**
2,3-Epoxypropionaldehyde	**(765-34-4)**
1,2-Epoxy-3-propoxypropane	**(3126-95-2)**
3-(2,3-Epoxypropoxy)propyltrimethoxysilane	**(2530-83-8)**
2,3-Epoxypropyl acrylate	**(106-90-1)**
2,3-Epoxypropyl butyl ether	**(2426-08-6)**
2,3-Epoxypropyl chloride	**(106-89-8)**
2,3-Epoxypropyl ester of oleic acid	**(5431-33-4)**
2,3-Epoxypropyl ester of stearic acid	**(7460-84-6)**
2,3-Epoxypropylhexyl ether	**(5926-90-9)**
2,3-Epoxypropyl methacrylate	**(106-91-2)**
2,3-Epoxypropyl-4-methoxyphenyl ether	**(2211-94-1)**
2,3-Epoxypropyl oleate	**(5431-33-4)**
2,3-Epoxypropylphenyl ether	**(122-60-1)**
2,3-Epoxypropyl-phenyl glycidyl ether	**(63919-02-8)**
N-(2,3-Epoxypropyl)-phthalimide	**(5455-98-1)**
2,3-Epoxypropyl stearate	**(7460-84-6)**
Epoxy resins	**(61788-97-4)**
9,10-Epoxystearic acid	**(2443-39-2)**
cis-9,10-Epoxystearic acid	**(2443-39-2)**
9,10-Epoxystearic acid, allyl ester	**(123-36-4)**
9,10-Epoxystearic acid, 2-ethylhexyl ester	**(141-38-8)**
Epoxystyrene	**(96-09-3)**
α,β-Epoxystyrene	**(96-09-3)**
1,2-Epoxytetradecane	**(3234-28-4)**
12,13-Epoxy-3,4,7,15-tetrahydroxytrichothec-9-en-8-one	**(23282-20-4)**
12,13-Epoxy-3-α,4-β,7-β,15-tetrahydroxytrichothec-9-en-8-one 4-acetate	**(23255-69-8)**
1,2-Epoxy-3-(o-tolyloxy)propane	**(2210-79-9)**
1,2-Epoxy-3-(p-tolyloxy)propane	**(2186-24-5)**
1,2-Epoxy-3-(tolyloxy)propane	**(26447-14-3)**
1,2-Epoxy-4,4,4-trichlorobutane	**(3083-25-8)**
Epoxy-1,1,2-trichloroethane	**(16967-79-6)**
1,2-Epoxy-3,3,3-trichloropropane	**(3083-23-6)**
2,3-Epoxy-3,5,5-trimethyl-1-cyclohexanone	**(10276-21-8)**
2,3-Epoxy-3,5,5-trimethylcyclohexanone	**(10276-21-8)**
6-β,7-β-Epoxy-3-α-tropanyl S-(-)-tropate	**(51-34-3)**
Epoxytropine tropate	**(51-34-3)**
Epoxytropine tropate methylbromide	**(155-41-9)**
Epoxytropine tropate methylnitrate	**(6106-46-3)**
17,23 β-Epoxyveratraman-3 β-yl-β-D-glucopyranoside	**(23185-94-6)**
1,2-Epoxy-4-vinylcyclohexane	**(106-86-5)**
Epragen	**(62-44-2)**
Eprazin	**(98-96-4)**
Eprofil	**(148-79-8)**
Eprolin	**(59-02-9)**
Epsamon	**(60-32-2)**
Epsicapron	**(60-32-2)**
Epsikapron	**(60-32-2)**
Epsilan	**(59-02-9)**
Epsilcapramin	**(60-32-2)**
Epsilon S	**(60-32-2)**
Epsom Salts	**(7487-88-9)**
Epsylon Kaprolaktam (Polish)	**(105-60-2)**
Epsylone	**(50-06-6)**
Eptac 1	**(137-30-4)**
Eptacloro (Italian)	**(76-44-8)**
1,4,5,6,7,8,8-Eptacloro-3a,4,7,7a-tetraidro-4,7-endo-metano-indene (Italian)	**(76-44-8)**
Eptal	**(57-41-0)**
Eptani (Italian)	**(142-82-5)**
Eptan-3-one (Italian)	**(106-35-4)**
Eptapur	**(3766-60-7)**
Eptoin	**(57-41-0)**
Eptoin	**(630-93-3)**
Equal	**(22839-47-0)**
Equanil	**(57-53-4)**
Equanil suspension	**(57-53-4)**
Equatrate	**(57-53-4)**
Equi Bute	**(50-33-9)**
Equigard	**(62-73-7)**
Equigel	**(62-73-7)**
Equilin	**(474-86-2)**
Equilium	**(57-53-4)**
Equinil	**(57-53-4)**
Equino-Acid	**(52-68-6)**
Equino-Aid	**(52-68-6)**
Equipoise	**(68-88-2)**
Equiproxen	**(22204-53-1)**
Equisetic acid	**(499-12-7)**
Equitar	**(57-53-4)**
Equizole	**(148-79-8)**
Erade	**(2439-01-2)**
Eradex	**(2921-88-2)**
Eradex	**(93-75-4)**
Eradicane	**(51990-04-6)**
Eradicane	**(759-94-4)**
Eraditon	**(93-75-4)**
Eraldin	**(6673-35-4)**
Eralon	**(54-85-3)**
Eramide	**(14362-31-3)**

Eramin	(51-45-6)	Ergotrate	(60-79-7)
Erase	(75-60-5)	Ergot sugar	(99-20-7)
Erasol	(55-86-7)	Eributazone	(50-33-9)
Erasol-Ido	(55-86-7)	Eridan	(439-14-5)
Erasol hydrochloride	(55-86-7)	Erie Benzo 4BP	(992-59-6)
Erazidon	(2439-01-2)	Erie Black B	(1937-37-7)
Erazidon	(93-75-4)	Erie Black BF	(1937-37-7)
Erbaplast	(56-75-7)	Erie Black GAC	(1937-37-7)
Erbaprelina	(58-14-0)	Erie Black GXOO	(1937-37-7)
Erbium (9CI)	(7440-52-0)	Erie Black Jet	(1937-37-7)
Erbon	(136-25-4)	Erie Black Nug	(1937-37-7)
Ercorax	(50-34-0)	Erie Black RB	(2429-83-6)
Ercotina	(50-34-0)	Erie Black RF	(2429-83-6)
4βH,5α-Eremophilane (8CI)	(15404-63-4)	Erie Black RRAC	(2429-83-6)
4βH,5α-Eremorphila-1(10)11-dien-2-one	(4674-50-4)	Erie Black RW	(2429-83-6)
Ergadenylic acid	(61-19-8)	Erie Black RX	(2429-83-6)
Ergam	(379-79-3)	Erie Black RXOO	(1937-37-7)
Ergamine	(51-45-6)	Erie Brilliant Black S	(1937-37-7)
Ergaseptine	(63-74-1)	Erie Congo 4B	(573-58-0)
Ergate	(379-79-3)	Erie Fast Brown B	(2429-81-4)
Ergenyl	(1069-66-5)	Erie Fast Brown 3RB	(2429-82-5)
Ergine	(478-94-4)	Erie Fibre Black VP	(1937-37-7)
Ergoatetrine	(60-79-7)	Erie Green WT	(3626-28-6)
Ergobasine	(60-79-7)	Erie Red 4B	(992-59-6)
Ergocalciferol	(50-14-6)	Erie Scarlet 3B	(6358-29-8)
Ergoklinine	(60-79-7)	Erie Violet BW	(6426-67-1)
Ergoline-8-acetonitrile, 2-chloro-6-methyl-, (8β)-	(36945-03-6)	Erie Violet 3R	(2586-60-9)
Ergoline-8-β-carboxamide, 9,10-didehydro-N,N-diethyl-6-methyl	(50-37-3)	Erie Yellow FP	(1325-37-7)
Ergoline-8-β-carboxamide, 9,10-didehydro-N-((S)-1-(hydroxymethyl)propyl)-6-methyl	(113-42-8)	Erie Yellow SR	(1325-37-7)
Ergoline-8-β-carboxamide, 9,10-didehydro-N-(1-(hydroxymethyl)propyl)-1,6-dimethyl	(361-37-5)	Erigeron Oil	(8007-27-0)
Ergoline-8-β-carboxamide, 9,10-didehydro-N-((S)-2-hydroxy-1-methylethyl)-6-methyl	(60-79-7)	Erina	(57-53-4)
Ergomar	(379-79-3)	Erinit	(78-11-5)
Ergometrine	(60-79-7)	Erinitrit	(7632-00-0)
Ergon	(64-02-8)	Erio Anthracene Brilliant Blue RFF	(4368-56-3)
Ergonovine	(60-79-7)	Erio Blue BGL	(61967-93-9)
Ergoplast ADC	(849-99-0)	Erio Brilliant Blue V	(129-17-9)
Ergoplast AZDB	(2917-73-9)	Erio Chinoline Yellow 4G	(83-08-9)
Ergoplast ADDO	(103-23-1)	Erio Chrome Violet BA	(2092-55-9)
Ergoplast.FDC	(84-61-7)	Erio Chrome Violet BR	(2092-55-9)
Ergoplast FDO	(117-81-7)	Erio Fast Blue BRL	(2666-17-3)
Ergorone	(50-14-6)	Erio Fast Orange AS	(1936-15-8)
Ergosta-3,5-diene, (24xi)- (9CI)	(77327-07-2)	Erio Fast Yellow AE	(6373-74-6)
Ergosta-5,22-dien-3-ol, (3β,22E)-	(474-67-9)	Erio Fast Yellow AEN	(6373-74-6)
Ergostat	(379-79-3)	Erio Fast Yellow RL	(6375-55-9)
Ergosta-5,7,22-trien-3β-ol	(57-87-4)	Erio Floxine 2G	(3734-67-6)
Ergosta-5:6,7:8,22:23-trien-3-ol	(57-87-4)	Erio Floxine 2GN	(3734-67-6)
δ-5,7,22-Ergostatrien-3β-ol	(57-87-4)	Erio Orange II	(633-96-5)
Ergosta-5,7,22-trien-3-ol, (3β,22E)-	(57-87-4)	Erio Rubine B	(3567-69-9)
(24R)-5-Ergosten-3β-ol	(474-62-4)	Erio Tartrazine	(1934-21-0)
24xi-Ergost-7-en-3β-ol (8CI)	(17105-75-8)	Eriochrome Black T	(1787-61-7)
Ergost-5-en-3-ol, (3β,24R)- (9CI)	(474-62-4)	Eriochrome Blue SE	(1058-92-0)
Ergost-5-en-3β-ol, (24R)- (8CI)	(474-62-4)	Eriochrome Violet B	(2092-55-9)
Ergost-7-en-3-ol, (3β,24xi)- (9CI)	(17105-75-8)	Erioglaucine	(129-17-9)
Ergosterin	(57-87-4)	Erioglaucine	(2650-18-2)
Ergosterol	(57-87-4)	Erioglaucine A	(2650-18-2)
Ergosterol activated	(50-14-6)	Erioglaucine E	(2650-18-2)
Ergosterol, irradiated	(50-14-6)	Erioglaucine G	(3844-45-9)
Ergot alkaloid	(12126-57-7)	Erioglaucine Supra	(129-17-9)
Ergotamine	(113-15-5)	Erion	(69-05-6)
Ergotamine bitartrate	(379-79-3)	Erion	(83-89-6)
Ergotamine-tartrate	(379-79-3)	Erionite	(12510-42-8)
Ergotartrate	(379-79-3)	Erionite	(1318-02-1)
Ergotidine	(51-45-6)	Erionite (Cakna (Al2Si7O18)2.14H2O)	(66733-21-9)
Ergotocine	(60-79-7)	Erionyl Blue E-RFF	(4368-56-3)
		Erionyl Red G	(3567-65-5)
		Erionyl Red RS	(6459-94-5)
		Erionyl Yellow E-AEN	(6373-74-6)

Eriosin Blue Black B (1064-48-8)
Eriosin Fast Blue RFF (4368-56-3)
Eriosin Rhodamine B (81-88-9)
Eriosin Roccelline (1658-56-6)
Eriosin Roccelline SS (1658-56-6)
Eriosin Violet 3B (1694-09-3)
Eriosky Blue (2650-18-2)
Erispan (3900-31-0)
Eritrone (68-19-9)
Eritroxilina (50-36-2)
Ermetrine (60-79-7)
Eroina (561-27-3)
Errolon (54-31-9)
Erserine (57-47-6)
Ertalon 6SA (105-60-2)
Ertalon 6SA (25038-54-4)
Ertilen (56-75-7)
Ertron (50-14-6)
Ertuban (54-85-3)
Erucamide (112-84-5)
Erucic acid (112-86-7)
Erucic acid amide (112-84-5)
Erucyl amide (112-84-5)
Erycin (114-07-8)
Erycorbin (89-65-6)
Erycytol (68-19-9)
Erypar (643-22-1)
Erysan (494-03-1)
Erysipan (63-74-1)
D-Erythorbic acid (89-65-6)
Erythorbic acid (89-65-6)
Erythrene (106-99-0)
Erythrite (149-32-6)
Erythritol (149-32-6)
L-Erythritol (149-32-6)
meso-Erythritol (149-32-6)
Erythritol anhydride (1464-53-5)
Erythritol anhydride (564-00-1)
Erythritol, 1,2:3,4-dianhydro- (564-00-1)
Erythritol, meso- (149-32-6)
Erythrocin (114-07-8)
Erythrocin ethyl succinate (1264-62-6)
Erythrocin stearate (643-22-1)
Erythroglucin (149-32-6)
Erythrogran (114-07-8)
Erythroguent (114-07-8)
Erythrol (149-32-6)
Erythromycin (114-07-8)
Erythromycin A (114-07-8)
Erythromycin, ethyl succinate (1264-62-6)
Erythromycin, mono(ethyl succinate) (1264-62-6)
Erythromycin, octadecanoate (Salt) (643-22-1)
Erythromycin 2'-propanoate (134-36-1)
Erythromycin 2'-propionate (134-36-1)
Erythromycin propionate (134-36-1)
Erythromycin, stearate (Salt) (643-22-1)
Erythromycin stearic acid salt (643-22-1)
Erythrophlamine hydrochloride (23451-24-3)
Erythrophleine (36150-73-9)
Erythrosin (16423-68-0)
Erythrosin B (16423-68-0)
Erythrosine (16423-68-0)
Erythrosine 3B (16423-68-0)
Erythrosine B (16423-68-0)
Erythrosine B (Biological stain) (16423-68-0)
Erythrosine B-FO (Biological stain) (16423-68-0)
Erythrosine BS (16423-68-0)
Erythrosine Bluish (16423-68-0)
Erythrosine Bluish (Biological stain) (16423-68-0)
Erythrosine Extra Bluish (16423-68-0)
Erythrosine Extra Conc. A Export (16423-68-0)
Erythrosine Extra Pure A (16423-68-0)
Erythrosine (Indicator) (16423-68-0)
Erythrosine K-FO (Biological stain) (16423-68-0)
Erythrosine Lake (16423-68-0)
Erythrosine Sodium (16423-68-0)
Erythrosine TB (16423-68-0)
Erythrosine TB Extra (16423-68-0)
Erythrotin (68-19-9)
Erytroxylin (50-36-2)
Esaclorobenzene (Italian) (118-74-1)
Esaidro-1,3,5-trinitro-1,3,5-triazina (Italian) (121-82-4)
Esametilentetramina (Italian) (100-97-0)
Esani (Italian) (110-54-3)
Esanitrodifenilamina (Italian) (131-73-7)
Esasorb (50-70-4)
Esbecythrin (52918-63-5)
Esbioallethrin (28434-00-6)
Esbiol (28434-00-6)
Esbiol Concentrate 90% (28434-00-6)
Esbiothrin (28434-00-6)
Esbrite (9003-53-6)
Esbrite 2 (9003-53-6)
Esbrite 4 (9003-53-6)
Esbrite 4-62 (9003-53-6)
Esbrite 8 (9003-53-6)
Esbrite G 10 (9003-53-6)
Esbrite G-P 2 (9003-53-6)
Esbrite 500HM (9003-53-6)
Esbrite LBL (9003-53-6)
Escalol 106 (136-44-7)
Escalol 507 (21245-02-3)
Escaspere (50-55-5)
Escon 622 (9003-07-0)
Escon CD 44A (9003-07-0)
Escon EX 375 (9003-07-0)
Escoparone (120-08-1)
Escorez 100 (64742-16-1)
Escorez 1304 (64742-16-1)
Escorez 1310 (64742-16-1)
Escorez 1401 (64742-16-1)
Escorez 2101 (64742-16-1)
Escorez 2203 (64742-16-1)
Escorez 5280 (64742-16-1)
Escorez 7404 (9003-53-6)
Esculetin (305-01-1)
Esculetin dimethyl ether (120-08-1)
Esculetol (305-01-1)
Esculin aglycon (305-01-1)
Esdepallethrine (28434-00-6)
Esdragol (140-67-0)
Esdragon (140-67-0)
Esen (85-44-9)
Eserine (57-47-6)
Eserine salicylate (57-64-7)
Eserine sulfate (64-47-1)
Eserine sulphate (64-47-1)
Eserolein, methylcarbamate (ester) (57-47-6)
Eserpine (50-55-5)
Esfenvalerate (66230-04-4)
Esgram (1910-42-5)
Estercide T-2 and T-245 (93-76-5)
Esidrex (58-93-5)
Esidrix (58-93-5)

Esilgan (29975-16-4)
Eskabarb (50-06-6)
Eskacillian V (87-08-1)
Eskacillin (113-98-4)
Eskacillin (6130-64-9)
Eskadiazine (68-35-9)
Eskalith (554-13-2)
Eskapen (67-03-8)
Eskaphen (67-03-8)
Eskaserp (50-55-5)
Eskimon 11 (75-69-4)
Eskimon 12 (75-71-8)
Eskimon 22 (75-45-6)
Eslec C (9003-22-9)
Esmail (2898-12-6)
Esmarin (133-67-5)
Esnil P 18 (9003-20-7)
Esobarbitale (Italian) (56-29-1)
Esophotrast (7727-43-7)
Esorb (59-02-9)
Esorben (87-79-6)
Espadol (88-04-0)
Esparin (58-40-2)
Esparinal (53-60-1)
Espenal (97-77-8)
Esperal (97-77-8)
Esperfoam FR (1338-23-4)
Esperox (26748-41-4)
Esperox 10 (614-45-9)
Esperox 25 (1931-62-0)
Esperox 41-40 (1931-62-0)
Esperox 24M (109-13-7)
Esperox 31M (927-07-1)
Esperox 33M (26748-41-4)
Esphygmogenina (51-43-4)
Espril (51-12-7)
Essence de terebenthine (French) (9005-90-7)
Essence de terebenthine, succedane d' (French) (9005-90-7)
Essence of Mirbane (98-95-3)
Essence of Myrbane (98-95-3)
Essence of Niobe (93-58-3)
Essence of Niobe (93-89-0)
Essential Oil of Cymbopogon Nardus (8000-29-1)
Essential Oils (8014-17-3)
Essex (1333-86-4)
Essigester (German) (141-78-6)
Essigsaeure (German) (64-19-7)
Essigsaeureanhydrid (German) (108-24-7)
Esso Corexit 8666 (51158-21-5)
Esso Herbicide 10 (94-80-4)
Essofungicide 406 (133-06-2)
Estabex U 18 (16091-18-2)
Estanozolol (Spanish) (10418-03-8)
Estar (25038-59-9)
Estar (8007-45-2)
Estasil (57-53-4)
Estazol (10418-03-8)
Estazolam (29975-16-4)
Estazolamum (Latin) (29975-16-4)
Esteed (57-63-6)
Estr-4-en-3-one, 17-β-hydroxy (434-22-0)
Estr-4-en-3-one, 17-hydroxy-, (17-β)- (9CI) (434-22-0)
Ester 25 (311-45-5)
Esterase, choline (9CI) (9001-08-5)
Esterdiol 204 (1115-20-4)
Ester dwuetyloaminoetylowy kwasu dwufenylooctowego (Polish) (64-95-9)
Estere cianoacetico (Italian) (105-56-6)
Ester etylowykwasu p-hydroksybenzoesowego (Polish) (120-47-8)
S-Ester of (2-mercaptoethyl)trimethylammonium iodide with O,O-diethyl phosphorothioate (513-10-0)
Esteron (93-76-5)
Esteron (94-75-7)
Esteron 44 (94-11-1)
Esteron 99 (3966-11-8)
Esteron 99 (94-75-7)
Esteron 245 (93-76-5)
Esteron 245 BE (93-76-5)
Esteron 76 BE (94-75-7)
Esteron Brush Killer (93-76-5)
Esteron Brush Killer (94-75-7)
Esteron 99 Concentrate (94-75-7)
Esteron 44 Weed Killer (94-75-7)
Esterone (53-16-7)
Esterone Four (94-75-7)
Esteroquinone Light Yellow 4JL (2832-40-8)
Ester sulfonate (80-33-1)
'Esteve' (50-33-9)
Estifnato de plomo (Spanish) (15245-44-0)
Estigyn (57-63-6)
Estilben (130-80-3)
Estilben (56-53-1)
Estilbin (130-80-3)
Estilbin "MCO" (56-53-1)
Estimulex (300-42-5)
Estinerval (156-51-4)
Estinyl (57-63-6)
Eston-B (50-50-0)
Eston-E (57-63-6)
Estol 103 (142-91-6)
Estol 1550 (84-66-2)
Estol 603 (31566-31-1)
Estomycin (7542-37-2)
Estonate (50-29-3)
Estone (94-75-7)
Estone Yellow GN (2832-40-8)
Estonmite (80-33-1)
Estonox (8001-35-2)
Estoral (57-63-6)
Estoral (orion) (57-63-6)
Estorals (57-63-6)
Estosteril (79-21-0)
Estracyt (4891-15-0)
17-α-Estradiol (57-91-0)
17-β-Estradiol (50-28-2)
17-β-OH-Estradiol (50-28-2)
3,17-β-Estradiol (50-28-2)
D-3,17-β-Estradiol (50-28-2)
Estradiol (50-28-2)
Estradiol-17-β (50-28-2)
α-Estradiol (50-28-2)
β-Estradiol (50-28-2)
cis-Estradiol (50-28-2)
d-Estradiol (50-28-2)
Estradiol Mustard (22966-79-6)
Estradiol, Polyester with phosphoric acid (28014-46-2)
17-β-Estradiol 3-benzoate (50-50-0)
17-β-Estradiol benzoate (50-50-0)
Estradiol benzoate (50-50-0)
Estradiol, 3-benzoate (50-50-0)
Estradiol-17-β-3-benzoate (50-50-0)
Estradiol-17-β-benzoate (50-50-0)
β-Estradiol 3-benzoate (50-50-0)
β-Estradiol benzoate (50-50-0)
Estradiol, 3-(bis(2-chloroethyl)carbamate) dihydrogen phosphate (4891-15-0)

17-β-Estradiol dipropionate (113-38-2)
3,17-β-Estradiol dipropionate (113-38-2)
Estradiol 3,17-dipropionate (113-38-2)
Estradiol, dipropionate (113-38-2)
β-Estradiol 3,17-dipropionate (113-38-2)
β-Estradiol dipropionate (113-38-2)
Estradiol, 17-ethynyl- (57-63-6)
17-β-Estradiol monobenzoate (50-50-0)
Estradiol monobenzoate (50-50-0)
Estradiol phosphate, Polymer (28014-46-2)
Estradiol 17-β-valerate (979-32-8)
Estradiol valerate (979-32-8)
Estradiol, 17-valerate (979-32-8)
Estradiol valerianate (979-32-8)
Estradurin (28014-46-2)
Estragard (84-17-3)
Estragole (140-67-0)
Estragole, 1'-hydroxy- (51410-44-7)
Estraldine (50-28-2)
Estralutin (630-56-8)
Estramustine phosphate (4891-15-0)
α-Estra-1,3,5,7,9-pentaen-3,17-diol (6639-99-2)
Estra-1,3,5(10),7-tetraene-3,17-diol, (17.β.)- (9CI) (3563-27-7)
Estra-1,3,5(10),7-tetraene-3,17.β.-diol (6CI, 7CI, 8CI) (3563-27-7)
1,3,5,7-Estratetraen-3-ol-17-one (474-86-2)
Estra-1,3,5(10),7-tetraen-17-one, 3-hydroxy (474-86-2)
1,3,5-Estratriene-3,17-α-diol (57-91-0)
1,3,5-Estratriene-3,17-β-diol (50-28-2)
17-β-Estra-1,3,5(10)-triene-3,17-diol (50-28-2)
Estra-1,3,5(10)-triene-3,17-α-diol (57-91-0)
Estra-1,3,5(10)-triene-3,17-β-diol (50-28-2)
(17-β)-Estra-1,3,5(10)-triene-3,17-diol, Polymer with phosphoric acid (28014-46-2)
Estra-1,3,5(10)-triene-3,17 diol (17-β)-, Polymer with phosphoric acid (28014-46-2)
1,3,5(10)-Estratriene-3,17-β-diol 3-benzoate (50-50-0)
Estra-1,3,5(10)-triene-3,17-β-diol, 3-benzoate (50-50-0)
Estra-1,3,5(10)-triene-3,17-diol (17-β)-3-benzoate (50-50-0)
Estra-1,3,5(10)-triene-3,17-β-diol, bis(p-(bis(2-chloroethylamino)phenyl)acetate) (22966-79-6)
1,3,5(10)-Estratriene-3,17β-diol dipropionate (113-38-2)
Estra-1,3,5(10)-triene-3,17-diol (17-β)-dipropionate (113-38-2)
Estra-1,3,5(10)-triene-3,17-β-diol, 17-α-ethynyl- (57-63-6)
Estra-1,3,5(10)-triene-3,17-diol (17-β)-, 17-pentanoate (9CI) (979-32-8)
Estra-1,3,5(10)-triene-3,16-α, 17-β-triol (50-27-1)
(16-α,17-β)-Estra-1,3,5(10)-triene-3,16,17-triol (50-27-1)
1,3,5-Estratriene-3-β,16-α,17-β-triol (50-27-1)
Estra-1,3,5(10)-trien-17-β-ol, 17-α-ethynyl-3-methoxy- (72-33-3)
1,3,5(10)-Estratrien-3-ol-17-one (53-16-7)
1,3,5-Estratrien-3-ol-17-one (53-16-7)
δ-1,3,5-Estratrien-3-β-ol-17-one (53-16-7)
Estra-1,3,5(10)-trien-17-one, 3-hydroxy- (53-16-7)
Estra-1,3,5(10)-trien-17-one, 3-(sulfooxy)-, Compd. with piperazine (1:1) (7280-37-7)
Estra-1,3,5(10)-trien-17-one, 3-(sulfooxy)-, sodium salt (9CI) (438-67-5)
Estratriol (50-27-1)
Estraval (979-32-8)
Estreptocida (63-74-1)
Estril (56-53-1)
Estrin (53-16-7)
16-α,17-β-Estriol (50-27-1)
3,16-α,17-β-Estriol (50-27-1)
Estriol (50-27-1)
Estriolo (Italian) (50-27-1)
Estroben (130-80-3)
Estroben DF (130-80-3)
Estrobene (130-80-3)
Estrobene (56-53-1)
Estrobene DP (130-80-3)
Estrodienol (84-17-3)
Estrofol (25038-59-9)
Estrofol B (25038-59-9)
Estrofol Ow (25038-59-9)
Estrogen (56-53-1)
Estrogen (57-63-6)
Estrogenin (130-80-3)
Estroici (113-38-2)
Estrol (53-16-7)
Estromenin (56-53-1)
Estron (53-16-7)
Estrona (Spanish) (53-16-7)
Estrone (53-16-7)
Estrone-A (53-16-7)
Estrone, benzoate (2393-53-5)
Estrone hydrogen sulfate (481-97-0)
Estrone hydrogen sulfate compound with piperazine (1:1) (7280-37-7)
Estrone, hydrogen sulfate, sodium salt (438-67-5)
Estrone sodium sulfate (438-67-5)
Estrone sulfate (481-97-0)
Estrone sulfate sodium (438-67-5)
Estrone sulfate sodium salt (438-67-5)
Estrone-3-sulfate sodium salt (438-67-5)
Estronex (113-38-2)
Estropipate (7280-37-7)
Estroral (84-17-3)
Estrosel (62-73-7)
Estrosol (62-73-7)
Estrostilben (130-80-3)
Estrosyn (56-53-1)
Estrovite (50-28-2)
Estrugenone (53-16-7)
Estrusol (53-16-7)
Estynox 330 (106-81-0)
Estyrene 4-62 (9003-53-6)
Estyrene AS (9003-54-7)
Estyrene G 15 (9003-53-6)
Estyrene G 20 (9003-53-6)
Estyrene G-P 4 (9003-53-6)
Estyrene H 61 (9003-53-6)
Estyrene 500SH (9003-53-6)
E_1 (53-16-7)
E_2 (50-28-2)
Etabus (97-77-8)
Etacrinic acid (58-54-8)
Etadrol (53-34-9)
Etain (tetrachlorure d') (French) (7646-78-8)
Etakrinic acid (58-54-8)
Etambro (71-91-0)
Etamican (59-02-9)
Etamide (938-73-8)
Etaminal sodium (57-33-0)
Etamon chloride (56-34-8)
Etamucine (9004-61-9)
Etanautine (58-73-1)
Etanolamina (Italian) (141-43-5)
Etanolo (Italian) (64-17-5)
Etantiolo (Italian) (75-08-1)
Etaperazin (58-39-9)
Etaperazine (58-39-9)
Etatryn (3440-19-5)
Etavit (59-02-9)
Etazine (26259-45-0)
Etazol (94-19-9)
Etazole (94-19-9)

Etchlorvinolo **(113-18-8)**
Etcmtb **(2593-15-9)**
Eter dicloroisopropilico (Spanish) **(63283-80-7)**
Etere etilico (Italian) **(60-29-7)**
Ethaanthiol (Dutch) **(75-08-1)**
Ethacrylic acid **(3586-58-1)**
Ethacrynic acid **(58-54-8)**
Ethal **(36653-82-4)**
Ethal LA-X **(9002-92-0)**
Ethalfluralin **(55283-68-6)**
Ethalflurlin **(55283-68-6)**
Ethambutol **(74-55-5)**
Ethamide **(57-66-9)**
Ethaminal **(57-33-0)**
Ethaminal **(76-74-4)**
Ethaminal sodium **(57-33-0)**
Ethaminium, N-(9-(2-carboxyphenyl)-6-(diethylamino)-3H-xanthen-3-ylidene)-N-ethyl)-, hydroxide, inner salt **(3375-25-5)**
Ethana NU **(71-55-6)**
Ethanal **(75-07-0)**
Ethanal oxime **(107-29-9)**
Ethanamide **(60-35-5)**
Ethanamine, (Aqueous solution) **(75-04-7)**
Ethanamine, N-butyl-N-ethyl- **(4444-68-2)**
Ethanamine, 2-chloro-N-(2-chloroethyl)-N-methyl-, hydrochloride **(55-86-7)**
Ethanamine, 2-chloro-N-(2-chloroethyl)-N-methyl-, N-oxide, hydrochloride (9CI) **(302-70-5)**
Ethanamine, 2-cyano-N-(2-cyanoethyl)- **(111-94-4)**
Ethanamine, 2-(3,5-dichlorophenoxy)- (9CI) **(67883-07-2)**
Ethanamine, N,N-diethyl- **(121-44-8)**
Ethanamine, N,N-diethyl-, hydrobromide (9CI) **(636-70-4)**
Ethanamine, N,N-diethyl-, phosphate (9CI) **(10138-93-9)**
Ethanamine, N,N-diethyl-, 2-(2,4,5-trichlorophenoxy)propanoate **(53404-74-3)**
Ethanamine, 2,2-dimethoxy- (9CI) **(22483-09-6)**
Ethanamine, 2,2-dimethoxy-N-methyl- (9CI) **(122-07-6)**
Ethanamine, N,N-dimethyl- **(598-56-1)**
Ethanamine, 1,1-dimethyl-2-phenyl- **(122-09-8)**
Ethanamine, N,N-dimethyl-2-(1-phenyl-1-(2-pyridinyl)ethoxy)- (9CI) **(469-21-6)**
Ethanamine, 2-(diphenylmethoxy)-N,N-dimethyl-, hydrochloride **(147-24-0)**
Ethanamine, N-ethyl- **(109-89-7)**
Ethanamine, N-ethyl-, hydriodide (9CI) **(19833-78-4)**
Ethanamine, N-ethyl-, hydrobromide (9CI) **(6274-12-0)**
Ethanamine, N-ethyl-, hydrochloride (9CI) **(660-68-4)**
Ethanamine, N-ethyl-N-methyl- (9CI) **(616-39-7)**
Ethanamine, N-hydroxy-2-methoxy-N-(2-methoxyethyl)- (9CI) **(5815-11-2)**
Ethanamine, 2-methoxy-N-(2-methoxyethyl)- (9CI) **(111-95-5)**
Ethanamine, N-methyl- (9CI) **(624-78-2)**
Ethanamine, N-methyl-, hydrochloride (9CI) **(624-60-2)**
Ethanaminium, 2-amino-N-(2-aminoethyl)-N-(2-hydroxyethyl)-N-methyl-, N,N'-ditallow acyl derivs., Me sulfates (Salts) **(68153-35-5)**
Ethanaminium, 2-(aminocarbonyl)oxy-N,N,N-trimethyl-, chloride **(51-83-2)**
Ethanaminium, 2-((aminocarbonyl)oxy)-N,N,N-trimethyl-, chloride (9CI) **(51-83-2)**
Ethanaminium, N-(4-(bis(4-(diethylamino)phenyl)methylene)-2,5-cyclohexadien-1-ylidene)-N- ethyl-, chloride **(2390-59-2)**
Ethanaminium, 2-((4-((2-chloro-4-nitrophenyl)azo)phenyl)-ethylamino)-N,N,N-trimethyl **(14097-03-1)**
Ethanaminium, 2-chloro-N,N,N-trimethyl-, chloride (9CI) **(999-81-5)**
Ethanaminium, N-(9-(2,4-dicarboxyphenyl)-6-(diethylamino)-3H-xanthen-3-ylidene)-N-ethyl-, chloride, disodium salt **(37299-86-8)**
Ethanaminium, 2-((diethoxyphosphinyl)thio)-N,N,N-trimethyl-, iodide **(513-10-0)**
Ethanaminium, N-(4-((4-(diethylamino)phenyl)(4-(ethylamino)-1-naphthalenyl)methylene)-2,5-cyclohexadien-1-ylidene)-N-ethyl-, chloride (9CI) **(2390-60-5)**
Ethanaminium, N-(4-((4-(diethylamino)phenyl)(4-(ethylamino)-1-naphthalenyl)methylene)-2,5-cyclohexadien-1-ylidene)-N-ethyl-, molybdatetungstatephosphate (9CI) **(1325-87-7)**
Ethanaminium, N-(4-((4-(diethylamino)phenyl)phenyl-methylene)-2,5-cyclohexadien-1-y lidene)-N-ethyl-, ethane-dioate, ethanedioate (2:1:2) (9CI) **(36351-18-5)**
Ethanaminium, N-(6-(diethylamino)-3H-xanthen-3-ylidene)-N-ethyl-, chloride (9CI) **(2150-48-3)**
Ethanaminium, N,N-diethyl-N-methyl-2-((1-oxo-2-propenyl)oxy)-, methyl sulfate (9CI) **(21810-39-9)**
Ethanaminium, N,N-diethyl-N-methyl-2-((9H-xanthen-9-yl-carbonyl)oxy)-, bromide (9CI) **(53-46-3)**
Ethanaminium, 2-((((2,3-dihydro-2,2-dimethyl-7-benzofuranyl)-oxy)carbonyl)methylamino)-N,N,N-trimethyl-2-oxo-, chloride **(75096-86-5)**
Ethanaminium, 2,2'-((1,4-dioxo-1,4-butanediyl)bis(oxy))bis-(N,N,N-trimethyl- **(306-40-1)**
Ethanaminium, N-ethyl-2-hydroxy-N,N-bis(2-hydroxyethyl)-, ethylsulfate (Salt) (9CI) **(31774-90-0)**
Ethanaminium, 2-hydroxy-N,N-bis(2-hydroxyethyl)-N-methyl-, salt with silicic acid (9CI) **(12687-85-3)**
Ethanaminium, 2-hydroxy-N,N,N-trimethyl- (9CI) **(62-49-7)**
Ethanaminium, 2-hydroxy-N,N,N-trimethyl-, Salt with (R-(R*,R*))-2,3-dihydroxybutanedioic acid (1:1) (9CI) **(87-67-2)**
Ethanaminium, 2-hydroxy-N,N,N-trimethyl-, Salt with (R-(R+,R+))-2,3-dihydroxybutanedioic acid (1:1) **(87-67-2)**
Ethanaminium, 2-hydroxy-N,N,N-trimethyl-, carbonate (1:1) (Salt) (9CI) **(78-73-9)**
Ethanaminium, 2-hydroxy-N,N,N-trimethyl-, chloride **(67-48-1)**
Ethanaminium, N,N,N-triethyl-, chloride (9CI) **(56-34-8)**
Ethanaminium, N,N,N-triethyl-, iodide (9CI) **(68-05-3)**
Ethanaminium, N,N,N-triethyl-, perchlorate (9CI) **(2567-83-1)**
Ethanaminium, N,N,N-trimethyl-, chloride (9CI) **(27697-51-4)**
Ethanaminium, N,N,N-trimethyl-2-((2-methyl-1-oxo-2-propenyl)-oxy)-, chloride (9CI) **(5039-78-1)**
Ethanaminium, N,N,N-trimethyl-2-((2-methyl-1-oxo-2-propenyl)-oxy)-, methyl sulfate (9CI) **(6891-44-7)**
Ethanaminium, N,N,N-trimethyl-2-((2-methyl-1-oxo-2-propenyl)-oxy)-, methyl sulfate, Homopolymer (9CI) **(27103-90-8)**
Ethanaminium, N,N,N-trimethyl-2-((2-methyl-1-oxo-2-propenyl)-oxy)-, methyl sulfate, Polymer with 2-propenamide (9CI) **(26006-22-4)**
Ethanaminium, N,N,N-trimethyl-2-((2-methyl-1-oxo-2-propenyl)-oxy)-, methylsulfate, Polymer with 2-propenamine **(26006-22-4)**
Ethanaminium, N,N,N-trimethyl-2-((1-oxo-2-propenyl)oxy)-, chloride (9CI) **(44992-01-0)**
Ethanaminium, N,N,N-trimethyl-2-((1-oxo-2-sulfododecyl)oxy)-, hydroxide, inner salt (9CI) **(65520-64-1)**
Ethanaminium, N,N,N-trimethyl-2-((1-oxo-2-sulfohexadecyl)-oxy)-, hydroxide, inner salt (9CI) **(59997-85-2)**
Ethandial **(107-22-2)**
1,2-Ethandiol **(107-21-1)**
Ethane **(74-84-0)**
Ethane [UN 1035] **(74-84-0)**
Ethane, 1-amino-2-(3,4,5-trimethoxyphenyl)- **(54-04-6)**
Ethane, azoxy- **(16301-26-1)**
Ethane, 2,2-bis(p-anisyl)-1,1,1-trichloro- **(72-43-5)**
Ethane, 1,2-bis(2-chloroethoxy) **(112-26-5)**
Ethane, 1,2-bis(2-chloroethylmercapto)- **(3563-36-8)**
Ethane, 1,2-bis((2-chloroethyl)sulfonyl) **(3944-87-4)**
Ethane, 1,2-bis(2-chloroethylthio) **(3563-36-8)**
Ethane, 1,2-bis(chloromethoxy) **(13483-18-6)**
Ethane, 2,2-bis(p-chlorophenyl)-1-chloro **(2642-80-0)**
Ethane, 1,2-bis(2,3-epoxy-2-methylpropoxy) **(3775-85-7)**
Ethane, 1,2-bis(2,3-epoxypropoxy) **(2224-15-9)**
Ethane, 2,2-bis(p-ethoxyphenyl)-1,1,1-trichloro **(4329-03-7)**
Ethane, 1,1-bis(p-ethylphenyl)- (8CI) **(10224-91-6)**
Ethane, 2,2-bis(p-ethylphenyl)-1,1-dichloro **(72-56-0)**
Ethane, 2,2-bis(p-fluorophenyl)-1,1,1-trichloro- **(475-26-3)**
Ethane, 1,2-bis(methylthio)- (9CI) **(6628-18-8)**

Name	CAS No.
Ethane, 2,2-bis(p-(methylthio)phenyl)-1,1,1-trichloro	**(19679-38-0)**
Ethane, 1,2-bis(propylsulfonyl)-, trans- (8CI)	**(3563-34-6)**
Ethane, 1,1-bis(p-tolyl)-2,2,2-trichloro	**(4413-31-4)**
Ethaneboronic acid, cyclic ethylene ester (7CI,8CI)	**(10173-38-3)**
Ethane, bromo	**(74-96-4)**
Ethane, 1-bromo-2-chloro	**(107-04-0)**
Ethane, 1-bromo-1-chloro- (8CI,9CI)	**(593-96-4)**
Ethane, 1-bromo-1-chloro-2,2,2-trifluoro-	**(151-67-7)**
Ethane, 1-bromo-2-chloro-1,1,2-trifluoro	**(354-06-3)**
Ethane, 2-bromo-2-chloro-1,1,1-trifluoro	**(151-67-7)**
Ethane, 2-bromo-1,1-dichloro- (8CI,9CI)	**(683-53-4)**
Ethane, bromopentachloro- (7CI,9CI)	**(79504-02-2)**
Ethane, 1-butoxy-2-(2-chloroethoxy)-	**(1120-23-6)**
Ethane, 1-butoxy-2-(2,3-epoxypropoxy)- (8CI)	**(13483-47-1)**
Ethane, 1-butoxy-2-(2-ethoxyethoxy)- (8CI)	**(3895-17-8)**
Ethane, 1-butoxy-2-(2-thiocyanatoethoxy)-	**(112-56-1)**
Ethane, 1-butoxy-2-(vinyloxy)	**(4223-11-4)**
Ethanecarboxylic acid	**(79-09-4)**
Ethane, chloro	**(75-00-3)**
Ethane, chlorodibromotrifluoro	**(29256-79-9)**
Ethane, 2-chloro-1,1-diethoxy- (9CI)	**(621-62-5)**
Ethane, 1-chloro-1,1-difluoro	**(75-68-3)**
Ethane, chlorodifluoro-	**(25497-29-4)**
Ethane, 2-chloro-1-(difluoromethoxy)-1,1,2-trifluoro-	**(13838-16-9)**
Ethane, 2-chloro-2-(difluoromethoxy)-1,1,1-trifluoro- (9CI)	**(26675-46-7)**
Ethane, 2-chloro-1,1-dimethoxy- (9CI)	**(97-97-2)**
Ethane, chloroepoxy	**(7763-77-1)**
Ethane, 1-chloro-1-ethoxy-	**(7081-78-9)**
Ethane, 1-chloro-1-ethoxy- (9CI)	**(7081-78-9)**
Ethane, 1-chloro-2-ethoxy- (9CI)	**(628-34-2)**
Ethane, 1-chloro-2-(ethylthio)-	**(693-07-2)**
Ethane, 1-chloro-2-methoxy	**(627-42-9)**
Ethane, (chloromethoxy)- (9CI)	**(3188-13-4)**
Ethane, 1-chloro-2-(methylthio)- (9CI)	**(542-81-4)**
Ethane, 1-chloro-1-nitro- (8CI,9CI)	**(598-92-5)**
Ethane, chloropentafluoro	**(76-15-3)**
Ethane, 2-(o-chlorophenyl)-2-(p-chlorophenyl)-1,1-dichloro	**(53-19-0)**
Ethane, 2-(o-chlorophenyl)-2-(p-chlorophenyl)-1,1,1-trichloro	**(789-02-6)**
Ethane, chlorotetrafluoro	**(63938-10-3)**
Ethane, 2-chloro-1,1,1,2-tetrafluoro- (8CI,9CI)	**(2837-89-0)**
Ethane, 2-chloro-1,1,1-trifluoro	**(75-88-7)**
Ethane, compressed [UN 1035]	**(74-84-0)**
Ethanedial	**(107-22-2)**
Ethanedial, cyclic 1,1:2,2-bis(1,2-ethanediyl acetal)	**(6705-89-1)**
Ethanediamide	**(471-46-5)**
Ethanediamide, N,N'-bis(2,4,6-trinitrophenyl)- (9CI)	**(29135-62-4)**
1,2-Ethanediamine	**(107-15-3)**
1,2-Ethanediamine, N-(2-aminoethyl)-, Polymer with 1,2-dichloroethane and (Z)-9-octadecen-1-amine (9CI)	**(67905-86-6)**
1,2-Ethanediamine, N-(2-aminoethyl)-N'-(2-((2-aminoethyl)amino)ethyl)	**(112-57-2)**
1,2-Ethanediamine, N-(2-aminoethyl)-N'-(2-(1-piperazinyl)ethyl)- (9CI)	**(31295-49-5)**
1,2-Ethanediamine, N-(2-aminoethyl)-N'-(3-(trimethoxysilyl)propyl)- (9CI)	**(35141-30-1)**
1,2-Ethanediamine, N,N'-bis(2-aminoethyl)-	**(112-24-3)**
1,2-Ethanediamine, N,N-bis(2-aminoethyl)- (9CI)	**(4097-89-6)**
1,2-Ethanediamine, N,N'-bis(1,3-dimethylbutylidene)- (9CI)	**(25707-70-4)**
1,2-Ethanediamine, N,N'-bis(1,1-dimethylethyl)- (9CI)	**(4062-60-6)**
1,2-Ethanediamine, N-cyclohexyl- (9CI)	**(5700-53-8)**
1,2-Ethanediamine, N,N-diethyl- (9CI)	**(100-36-7)**
1,2-Ethanediamine, dihydriodide (9CI)	**(5700-49-2)**
1,2-Ethanediamine, N-(2-(dimethylamino)ethyl)-N,N',N'-trimethyl- (9CI)	**(3030-47-5)**
1,2-Ethanediamine, N-(1,3-dimethylbutylidene)-N'-(2-((1,3-dimethylbutylidene)amino)ethyl)- (9CI)	**(10595-60-5)**
1,2-Ethanediamine, N,N-dimethyl-N'-(phenylmethyl)-N'-2-pyridinyl-, monohydrochloride	**(154-69-8)**
1,2-Ethanediamine, N,N-dimethyl-N'-phenyl-N'-(phenylmethyl)- (9CI)	**(961-71-7)**
1,2-Ethanediamine, N,N-dimethyl-N'-2-pyridinyl-N'-(2-thienylmethyl)-, monohydrochloride	**(135-23-9)**
1,2-Ethanediamine, dinitrate (9CI)	**(20829-66-7)**
1,2-Ethanediamine, N-((4-ethenylphenyl)methyl)-N'-(3-(trimethoxysilyl)propyl)-, monohydrochloride (9CI)	**(33401-49-9)**
1,2-Ethanediamine, N-ethyl-N-(3-methylphenyl)- (9CI)	**(19248-13-6)**
1,2-Ethanediamine, N-(2-furanylmethyl)-N',N'-dimethyl-N-2-pyridinyl-, (E)-2-butenedioate (1:1) (9CI)	**(5429-41-4)**
1,2-Ethanediamine, hydrochloride (9CI)	**(15467-15-9)**
1,2-Ethanediamine, N,N,N',N'-tetramethyl- (9CI)	**(110-18-9)**
1,2-Ethanediamine, N,N,N'-tris(2-aminoethyl)- (9CI)	**(31295-46-2)**
Ethane, 1,2-diamino-, copper complex	**(13426-91-0)**
Ethane, 1,2-diazido-	**(629-13-0)**
Ethane, 1,1-dibromo	**(557-91-5)**
Ethane, 1,2-dibromo	**(106-93-4)**
Ethane, dibromo- (8CI,9CI)	**(25620-62-6)**
Ethane, 1,2-dibromo-1,2-bis(3,4-dibromocyclohexyl)- (8CI)	**(18122-77-5)**
Ethane, 1,2-dibromo-1-chloro-	**(598-20-9)**
Ethane, 1,1-dibromo-2-chloro- (8CI)	**(27949-36-6)**
Ethane, 1,2-dibromo-1,1-dichloro- (9CI)	**(75-81-0)**
Ethane, 1,2-dibromo-1,2-dichloro- (9CI)	**(683-68-1)**
Ethane, 1,1-dibromo-1,2,2,2-tetrachloro- (9CI)	**(630-24-0)**
Ethane, 1,2-dibromotetrafluoro	**(124-73-2)**
Ethane, 1,2-dibromo-1,1,2,2-tetrafluoro- (9CI)	**(124-73-2)**
Ethane, 1,2-dibutoxy	**(112-48-1)**
1,2-Ethanedicarbamic acid, tetrathio-	**(111-54-6)**
1,2-Ethanedicarbamic acid, tetrathio-	**(12656-69-8)**
1,2-Ethanedicarboxylic acid	**(110-15-6)**
Ethane dichloride	**(107-06-2)**
Ethane, 1,1-dichloro	**(75-34-3)**
Ethane, 1,2-dichloro	**(107-06-2)**
Ethane, dichloro	**(1300-21-6)**
Ethane, 1,1-dichloro-2,2-bis(p-chlorophenyl)	**(72-54-8)**
Ethane, 1,1-dichloro-2,2-bis(p-ethylphenyl)-	**(72-56-0)**
Ethane, 1,1-dichloro-2,2-diethoxy- (9CI)	**(619-33-0)**
Ethane, 1,2-dichloro-1,1-difluoro	**(1649-08-7)**
Ethane, dichlorodifluoro- (8CI,9CI)	**(25915-78-0)**
Ethane, 2,2-dichloro-1,1-difluoro-1-methoxy-	**(76-38-0)**
Ethane, 1,1-dichloro-1-nitro	**(594-72-9)**
Ethane, 1,1-dichlorotetrafluoro-	**(374-07-2)**
Ethane, 1,2-dichloro-1,1,2,2-tetrafluoro	**(76-14-2)**
Ethane, 1,2-dichlorotetrafluoro-	**(76-14-2)**
Ethane, dichlorotetrafluoro	**(1320-37-2)**
Ethane, 1,1-dichloro-1,2,2,2-tetrafluoro- (9CI)	**(374-07-2)**
Ethane, 1,2-dichloro-1,1,2-trifluoro	**(354-23-4)**
Ethane, 2,2-dichloro-1,1,1-trifluoro	**(306-83-2)**
Ethane, 1,2-dicyano-	**(110-61-2)**
Ethane, 1,1-diethoxy-	**(105-57-7)**
Ethane, 1,2-diethoxy	**(629-14-1)**
Ethane, 1,1-diethoxy-2-(2-methoxyethoxy)- (9CI)	**(62005-54-3)**
Ethane, 1,1-difluoro	**(75-37-6)**
Ethane, 1,2-difluoro	**(624-72-6)**
Ethane, difluoro-	**(25497-28-3)**
Ethane, 1,2-difluoro-1,1,2,2-tetrachloro	**(76-12-0)**
Ethane, 2,2-difluoro-1,1,1,2-tetrachloro	**(76-11-9)**
Ethanediimidic acid, diethyl ester (9CI)	**(13534-15-1)**
Ethane, 1,2-diiodo- (9CI)	**(624-73-7)**
Ethane, 1,1-dimethoxy-	**(534-15-6)**
Ethane, 1,2-dimethoxy	**(110-71-4)**
Ethane, 1,1-dimethoxy-2-phenyl-	**(101-48-4)**
Ethanedinitrile	**(460-19-5)**
Ethane, 1,2-dinitro-	**(7570-26-5)**
Ethane, 1,1-dinitro- (8CI,9CI)	**(600-40-8)**
Ethanedioic acid	**(144-62-7)**

Ethanolamine O-sulfate **(926-39-6)**
Ethanol, 1-amino- **(75-39-8)**
Ethanol, 2-amino **(141-43-5)**
Ethanol, 2-amino-, dihydrogen phosphate (ester) **(1071-23-4)**
Ethanol, 2-((4-amino-2,5-dimethoxyphenyl)sulfonyl)-, hydrogen sulfate (ester) (9CI) **(26672-24-2)**
Ethanol, 2-(2-aminoethoxy) **(929-06-6)**
Ethanol, 2-(2-aminoethoxy)-, 2-sulfododecanoate (Salt) (9CI) **(65520-66-3)**
Ethanol, 2-((2-aminoethyl)amino) **(111-41-1)**
Ethanolaminoethyl stearamide **(141-21-9)**
Ethanol, 2-amino-, hydrochloride **(2002-24-6)**
Ethanol, 2-amino-, hydrogen sulfate (ester) (9CI) **(926-39-6)**
Ethanol, 2-((4-amino-5-methoxy-2-methylphenyl)sulfonyl)-, hydrogen sulfate (ester) (9CI) **(21635-69-8)**
Ethanol, 2-((2-amino-1-methylethyl)amino)- **(10138-74-6)**
Ethanol, 2-((4-amino-3-methylphenyl)ethylamino)-, sulfate (1:1) (Salt) **(25646-77-9)**
Ethanol, 2-amino-, nitrate (Salt) (9CI) **(20748-72-5)**
Ethanol, 2-(4-amino-2-nitroanilino) **(2871-01-4)**
Ethanol, 2-((4-amino-2-nitrophenyl)amino)- **(2871-01-4)**
Ethanol, 2-amino-1-phenyl- **(7568-93-6)**
Ethanol, 2,2'-((4-((4-aminophenyl)azo)phenyl)imino)bis- (9CI) **(20721-50-0)**
Ethanol, 2,2'-((4-aminophenyl)imino)bis-, sulfate (Salt) **(54381-16-7)**
Ethanol, 2,2'-((4-aminophenyl)imino)bis-, sulfate (1:1) (Salt) (9CI) **(54381-16-7)**
Ethanol, 2-((3-aminophenyl)sulfonyl)- (9CI) **(5246-57-1)**
Ethanol, 2-((4-aminophenyl)sulfonyl)-, hydrogen sulfate (ester) (9CI) **(2494-89-5)**
Ethanol, 2-amino-, phosphate **(1071-23-4)**
Ethanol, 2-amino-, phosphate (Salt) (9CI) **(29868-05-1)**
Ethanol, 2-((3-aminopropyl)butylamino)- (9CI) **(90853-15-9)**
Ethanol, 2,2'-(aminopropylimino) **(4985-85-7)**
Ethanol, 2-amino-, sulfite (2:1) (Salt) (9CI) **(15535-29-2)**
Ethanol, 1,1'-((6-amino-1,3,5-triazine-2,4-diyl)diimino)bis-(2,2,2-trichloro- (9CI) **(2797-59-3)**
Ethanol, 1,1'-((6-amino-s-triazine-2,4-diyl)diimino)bis-(2,2,2-trichloro- (8CI) **(2797-59-3)**
Ethanol, 2-amino-, 2-(2,4,5-trichlorophenoxy)propionate (Salt) (8CI) **(7374-47-2)**
Ethanol, 2-anilino **(122-98-5)**
Ethanol, 2-azido-, nitrate (Ester) **(53422-49-4)**
Ethanol, 2-(benzylamino)- (8CI) **(104-63-2)**
Ethanol, 2,2'-(benzylimino)di- (8CI) **(101-32-6)**
Ethanol, 2-(benzyloxy) **(622-08-2)**
Ethanol, 1,1-bis(p-chlorophenyl)- **(80-06-8)**
Ethanol, 1,2-bis(p-chlorophenyl)- **(56960-97-5)**
Ethanol, 2,2-bis(p-chlorophenyl)- (8CI) **(2642-82-2)**
Ethanol, 2-((4,6-bis(trichloromethyl)-1,3,5-triazin-2-yl)amino)- (9CI) **(24803-11-0)**
Ethanol, 2-((4,6-bis(trichloromethyl)-s-triazin-2-yl)amino)-(8CI) **(24803-11-0)**
Ethanol, 2-bromo **(540-51-2)**
Ethanol, 2-bromo-, acetate **(927-68-4)**
Ethanol, 2,2'-((4-((2-bromo-6-chloro-4-nitrophenyl)azo)-3-chlorophenyl)imino)bis- (9CI) **(17464-91-4)**
Ethanol, 2-((4-bromophenyl)butylamino)- (9CI) **(63455-63-0)**
Ethanol, 2-butoxy **(111-76-2)**
Ethanol, 2-sec-butoxy **(7795-91-7)**
Ethanol, 2-tert-butoxy **(7580-85-0)**
Ethanol, 2-butoxy-, acetate **(112-07-2)**
Ethanol, 2-butoxy-, benzoate (9CI) **(5451-76-3)**
Ethanol, 2-(2-butoxyethoxy) **(112-34-5)**
Ethanol, 1-(2-butoxyethoxy)- (9CI) **(54446-78-5)**
Ethanol, 2-(2-butoxyethoxy)-, acetate **(124-17-4)**
Ethanol, 2-(2-(2-butoxyethoxy)ethoxy) **(143-22-6)**
Ethanol, 2-(2-butoxyethoxy)-, phosphate (3:1) (9CI) **(7332-46-9)**
Ethanol, 2-(2-butoxyethoxy)-, thiocyanate **(112-56-1)**
Ethanol, 2-butoxy-, phosphate (3:1) **(78-51-3)**
Ethanol, 2-butoxy-, phthalate (2:1) **(117-83-9)**
Ethanol, 2-butoxy-, sodium salt (9CI) **(52663-57-7)**
Ethanol, 2-(tert-butylamino)-, methacrylate (ester) **(3775-90-4)**
Ethanol, 2,2'-(butylimino)bis- (9CI) **(102-79-4)**
Ethanol, 2,2'-(butylimino)di **(102-79-4)**
Ethanol, 2-(p-tert-butylphenoxy)- (8CI) **(713-46-2)**
Ethanol, 2-(butylphenylamino)- (9CI) **(3046-94-4)**
Ethanol, 2-(2-(butylphenylamino)ethoxy)- (9CI) **(63455-64-1)**
Ethanol, 2,2'-(2-butyne-1,4-diylbis(oxy))bis- (9CI) **(1606-85-5)**
Ethanol, 2,2'-(2-butynylenedioxy)di- (8CI) **(1606-85-5)**
Ethanol, 2-chloro **(107-07-3)**
Ethanol, 2-chloro-, acetate **(542-58-5)**
Ethanol, 2-chloro-, acrylate **(2206-89-5)**
Ethanol, 2-chloro-, 2-(p-t-butylphenoxy)-1-methylethyl sulfite **(140-57-8)**
Ethanol, 2-chloro-, carbanilate (8CI) **(3747-48-6)**
Ethanol, 2,2'-((3-chloro-4-((2,6-dichloro-4-nitrophenyl)azo)-phenyl)imino)bis- (9CI) **(23355-64-8)**
Ethanol, 2,2'-((4-((2-chloro-4,6-dinitrophenyl)azo)-3-methyl-phenyl)imino)bis- (9CI) **(65125-87-3)**
Ethanol, 2-chloro-, ester with 2-(p-tert-butylphenoxy)-1-methyl-ethyl sulfite **(140-57-8)**
Ethanol, 2-(2-chloroethoxy) **(628-89-7)**
Ethanol, 2-((2-chloroethyl)thio) **(693-30-1)**
Ethanol, 2-(4-chloro-2-methylphenoxy)- (9CI) **(36220-29-8)**
Ethanol, 2,2'-((4-((2-chloro-4-nitrophenyl)azo)-3-methylphenyl)-imino)bis- (9CI) **(3769-57-1)**
Ethanol, 2-((4-((2-chloro-4-nitrophenyl)azo)phenyl)ethylamino)-(9CI) **(3180-81-2)**
Ethanol, 2,2'-((3-chloro-4-((4-nitrophenyl)azo)phenyl)imino)bis-(9CI) **(4540-00-5)**
Ethanol, 2-(4-chlorophenoxy)-1-tert-butyl-2-(1H-1,2,4-triazole-1-yl) **(55219-65-3)**
Ethanol, 2-(2-(4-(p-chloro-α-phenylbenzyl)-1-piperazinyl)ethoxy) **(68-88-2)**
Ethanol, 2-chloro-, phenylcarbamate (9CI) **(3747-48-6)**
Ethanol, 2,2'-((3-chlorophenyl)imino)bis **(92-00-2)**
Ethanol, 2-chloro-, phosphate (3:1) **(115-96-8)**
Ethanol, 2-chloro-, phosphite (3:1) **(140-08-9)**
Ethanol, 2-(cyclododecyloxy)- (8CI,9CI) **(32399-56-7)**
Ethanol, 2-cyclohexyl **(4442-79-9)**
Ethanol, 2-(cyclohexylamino) **(2842-38-8)**
Ethanol, 2-(cyclohexylamino)-, (Aerosol) **(2842-38-8)**
Ethanol, 2,2'-cyclohexyliminodi **(4500-29-2)**
Ethanol, 2,2'-cyclohexyliminodi-, (aerosol) **(4500-29-2)**
Ethanol, 2-(2-(cyclohexyloxy)ethoxy)- (8CI,9CI) **(25961-84-6)**
Ethanol-d6 (9CI) **(1516-08-1)**
Ethanol, 2-((1-decylcyclohexyl)oxy)- (9CI) **(70710-00-8)**
Ethanol, 2-(2,4-diaminophenoxy)-, dihydrochloride **(66422-95-5)**
Ethanol, 1,2-dibromo-, acetate (9CI) **(24442-57-7)**
Ethanol, 1,2-dibromo-2,2-dichloro-, dimethyl phosphate **(300-76-5)**
Ethanol, 2-(dibutylamino) **(102-81-8)**
Ethanol, 2-((1,1-dibutylpentyl)oxy)- (9CI) **(54661-98-2)**
Ethanol, 2,2'-((dibutylstannylene)bis(thio))bis- (9CI) **(3026-81-1)**
Ethanol, 2,2-dichloro **(598-38-9)**
Ethanol, 1,2-dichloro-, acetate **(10140-87-1)**
Ethanol, 2,2'-((4-((2,6-dichloro-4-nitrophenyl)azo)-3-methyl-phenyl)imino)bis- (9CI) **(58528-60-2)**
Ethanol, 2-((4-((2,6-dichloro-4-nitrophenyl)azo)phenyl)methyl-amino)- (9CI) **(6232-56-0)**
Ethanol, 2,2'-(4-(2,6-dichloro-4-nitrophenylazo)-m-tolylimino)di- **(58528-60-2)**
Ethanol, 2-(2,4-dichlorophenoxy) **(120-67-2)**
Ethanol, 2-(2,4-dichlorophenoxy)-, benzoate **(94-83-7)**
Ethanol, 2-(2,4-dichlorophenoxy)-, hydrogen sulfate, sodium salt **(136-78-7)**
Ethanol, 2-(2,4-dichlorophenoxy)-, phosphite (3:1) **(94-84-8)**
Ethanol, 2,2-dichloro-, phenylcarbamate (9CI) **(35661-56-4)**
Ethanol, 2,2-diethoxy- (9CI) **(621-63-6)**
Ethanol, 2-(diethylamino) **(100-37-8)**
Ethanol, 2-(2-(diethylamino)ethoxy)- (9CI) **(140-82-9)**

Ethanol, 2-(diisopropylamino) **(96-80-0)**
Ethanol, 2-dimethylamino **(108-01-0)**
Ethanol, 2-((2-(dimethylamino)-1,1-dimethylethyl)amino)- (9CI) **(90853-17-1)**
Ethanol, 2-(2-dimethylaminoethoxy) **(1704-62-7)**
Ethanol, 2-((2-(dimethylamino)ethyl)methylamino)- (9CI) **(2212-32-0)**
Ethanol, 2-(dimethylamino)-, methacrylate **(2867-47-2)**
Ethanol, 2-(1,1-dimethylethoxy)- (9CI) **(7580-85-0)**
Ethanol, 2-((1,1-dimethylethyl)amino)- (9CI) **(4620-70-6)**
Ethanol, 2,2'-((1,1-dimethylethyl)imino)bis- (9CI) **(2160-93-2)**
Ethanol, 2-(4-(1,1-dimethylethyl)phenoxy)- (9CI) **(713-46-2)**
Ethanol, 2-((1,1-dimethyl-2-((1-methylethyl)amino)ethyl)-amino)- (9CI) **(90853-16-0)**
Ethanol, 2-(2,4-dinitrophenoxy)- (9CI) **(2831-60-9)**
Ethanol, 2-(p-(2,4-dinitrophenylazo)-N-ethylanilino)- **(62570-20-1)**
Ethanol, 2,2'-((4-((2,4-dinitrophenyl)azo)-3-methylphenyl)-imino)bis- (9CI) **(41541-13-3)**
Ethanol, 2-((4-((2,4-dinitrophenyl)azo)phenyl)ethylamino)- (9CI) **(62570-20-1)**
Ethanol, 2,2'-((4-((2,4-dinitrophenyl)azo)phenyl)imino)bis- (9CI) **(60129-67-1)**
Ethanol, 2,2'-(p-(2,4-dinitrophenylazo)phenylimino)di- **(60129-67-1)**
Ethanol, 2,2'-(4-(2,4-dinitrophenylazo)-m-tolylimino)di- **(41541-13-3)**
Ethanol, 2-(1,3,2-dioxaphospholan-2-yloxy)- (9CI) **(1073-75-2)**
Ethanol, 2-((1,1-dipentylhexyl)oxy)- (9CI) **(70709-98-7)**
Ethanol, 2,2',2'',2'''-((4,8-dipiperidinopyrimido(5,4-d)pyrimi-dine-2,6-diyl)dinitrilo)tetra **(58-32-2)**
Ethanol, 2-(1,1-dipropylbutoxy)- (9CI) **(70709-97-6)**
Ethanol, 2,2'-(dodecylimino)bis- (9CI) **(1541-67-9)**
Ethanol, 2-(dodecyloxy) **(4536-30-5)**
Ethanol, 2-(2-(dodecyloxy)ethoxy)- (9CI) **(3055-93-4)**
Ethanol, 2-(2-(2-(dodecyloxy)ethoxy)ethoxy) **(3055-94-5)**
Ethanol, 2-(2-(2-(dodecyloxy)ethoxy)ethoxy)-, hydrogen sulfate sodium salt **(13150-00-0)**
Ethanol, 2-(2-(2-(dodecyloxy)ethoxy)ethoxy)-, sulfate ester, monosodium salt **(13150-00-0)**
Ethanol, 2-(2-(dodecyloxy)ethoxy)-, hydrogen sulfate, sodium salt (9CI) **(3088-31-1)**
Ethanol, 2-(dodecyloxy)-, hydrogen sulfate, sodium salt **(15826-16-1)**
Ethanol, 2,2'-(1,2-ethanediylbis(oxy))bis-, dibenzoate (9CI) **(120-56-9)**
Ethanol, 2,2'-(1,2-ethanediylbis(thio))bis- (9CI) **(5244-34-8)**
Ethanol, 2,2'-(1,2-ethanediyldiimino)bis- (9CI) **(4439-20-7)**
Ethanol, 2,2',2'',2'''-(1,2-ethanediyldinitrilo)tetrakis- (9CI) **(140-07-8)**
Ethanol, 2-ethoxy **(110-80-5)**
Ethanol, 2-ethoxy-, acetate **(111-15-9)**
Ethanol, 2-ethoxy-, acrylate **(106-74-1)**
Ethanol, 2-(2-ethoxyethoxy) **(111-90-0)**
Ethanol, 2-(2-ethoxyethoxy)-, acetate **(112-15-2)**
Ethanol, 2-(2-(2-ethoxyethoxy)ethoxy) **(112-50-5)**
Ethanol, 2-ethoxy-, magnesium salt (9CI) **(14064-03-0)**
Ethanol, 2-(ethylamino) **(110-73-6)**
Ethanol, 2-(N-ethylanilino)- (8CI) **(92-50-2)**
Ethanolethylene diamine **(111-41-1)**
Ethanol, 2,2'-(ethylenediimino)di- (8CI) **(4439-20-7)**
Ethanol, 2,2',2'',2'''-(ethylenedinitrilo)tetra- (8CI) **(140-07-8)**
Ethanol, 2,2'-(ethylenedioxy)di- **(112-27-6)**
Ethanol, 2,2'-ethylenedioxydi-, diacetate **(111-21-7)**
Ethanol, 2,2'-(ethylimino)bis- (9CI) **(139-87-7)**
Ethanol, 2,2'-(ethylimino)di **(139-87-7)**
Ethanol, 2-(ethyl(4-((6-methoxy-2-benzothiazolyl)azo)phenyl)-amino)- (9CI) **(13486-43-6)**
Ethanol, 2-(ethyl(4-((6-methoxy-3-methyl-2-benzothiazolium)-azo)phenyl)amino)-, methyl sulfate **(12270-13-2)**
Ethanol, 2-(ethyl(4-((4-nitrophenyl)azo)phenyl)amino)- (9CI) **(2872-52-8)**
Ethanol, 2-(ethylnitrosamino) **(13147-25-6)**
Ethanol, 2-((1-ethylpentyl)oxy) **(10138-47-3)**
Ethanol, 2-(ethylphenylamino)- (9CI) **(92-50-2)**
Ethanol, 2-(ethylsulfonyl)-, O-ester with O,O-diethyl phosphoro-thioate **(4891-54-7)**
Ethanol, 2-(ethylthio) **(110-77-0)**
Ethanol, 2-(ethylthio)-, O-ester with O,O-diethyl phosphorothioate **(298-03-3)**
Ethanol, 2-(ethylthio)-, O-ester with O,O-dimethyl phosphorothioate **(867-27-6)**
Ethanol, 2-(N-ethyl-m-toluidino) **(91-88-3)**
Ethanol, 2-fluoro **(371-62-0)**
Ethanol, 2-((1-heptyloctyl)oxy)- (9CI) **(70709-96-5)**
Ethanol, 2-(hexadecyloxy)- (8CI,9CI) **(2136-71-2)**
Ethanol, 2-(hexadecyloxy)-, hydrogen sulfate, sodium salt (9CI) **(14858-54-9)**
Ethanol, 2-((1-hexylcyclohexyl)oxy)- (9CI) **(70709-99-8)**
Ethanol, 2-((1-hexylheptyl)oxy)- (7CI,8CI,9CI) **(924-06-1)**
Ethanol, 2-(hexyloxy) **(112-25-4)**
Ethanol, 2-((2-hexyloxy)ethoxy) **(112-59-4)**
Ethanol, 2-hydrazino **(109-84-2)**
Ethanol, 2-(2-hydroxyethoxy)-, laurate **(141-20-8)**
Ethanol, 2,2'-((4-((2-hydroxyethyl)amino)-3-nitrophenyl)imino)-bis- (9CI) **(33229-34-4)**
Ethanol, 2,2'-((4-((2-hydroxyethyl)amino)-3-nitrophenyl)imino)di **(33229-34-4)**
Ethanol, 2-(hydroxymethylamino)- (9CI) **(34375-28-5)**
Ethanol, 2,2'-iminobis-, N-C12-18-alkyl derivs. **(71786-60-2)**
Ethanol, 2,2'-iminobis-, acetate (Salt) (9CI) **(23251-72-1)**
Ethanol, 2,2'-iminobis-, 3,6-dichloro-2-methoxybenzoate (Salt) **(25059-78-3)**
Ethanol, 2,2'-iminobis-, hydrochloride **(14426-21-2)**
Ethanol, 2,2'-iminobis-, 2-hydroxy-1,2,3-propanetricarboxylate (Salt) (9CI) **(23349-61-3)**
Ethanol, 2,2'-iminobis-, N-tallow alkyl derivs. **(61791-44-4)**
Ethanol, 2,2'-iminodi **(111-42-2)**
Ethanol, 2,2-iminodi-, Compd. with 2-((4-chloro-o-tolyl)-oxy)propionic acid (1:1) **(1432-14-0)**
Ethanol, 2,2'-iminodi-, Compd. with 1,2-dihydro-3,6-pyri-dazinedione (1:1) **(5716-15-4)**
Ethanol, 2,2'-iminodi-, 3,6-dichloro-o-anisate (Salt) **(25059-78-3)**
Ethanol, 2,2'-iminodi-, hydrochloride **(14426-21-2)**
Ethanol, 2-iodo **(624-76-0)**
Ethanol, 2-isobutoxy **(4439-24-1)**
Ethanol, 2-isopropoxy **(109-59-1)**
Ethanolisopropylamine **(109-56-8)**
Ethanol, 2-(isopropylamino) **(109-56-8)**
Ethanol, 2,2'-(isopropylimino)di- (8CI) **(121-93-7)**
Ethanol, magnesium salt (9CI) **(2414-98-4)**
Ethanol, 2-mercapto **(60-24-2)**
Ethanol, 2-mercapto-, monosodium salt (9CI) **(37482-11-4)**
Ethanol, 2-methoxy **(109-86-4)**
Ethanol, 2-methoxy-, acetate **(110-49-6)**
Ethanol, 2-methoxy-, acrylate **(3121-61-7)**
Ethanol, 2-methoxy-, carbamate **(1616-88-2)**
Ethanol, 2-(2-methoxyethoxy) **(111-77-3)**
Ethanol, 2-(2-methoxyethoxy)-, acetate **(629-38-9)**
Ethanol, 2-(2-(2-methoxyethoxy)ethoxy) **(112-35-6)**
Ethanol, 2-methoxy-, formate (7CI,8CI,9CI) **(628-82-0)**
Ethanol, 2,2'-((4-((2-methoxy-4-nitrophenyl)azo)-3-methylphenyl)-zimino)bis- (9CI) **(41541-11-1)**
Ethanol, 2,2'-((4-((2-methoxy-4-nitrophenyl)azo)phenyl)imino)-bis- (9CI) **(41541-14-4)**
Ethanol, 2,2'-(p-(2-methoxy-4-nitrophenylazo)phenylimino)di- **(41541-14-4)**
Ethanol, 2-(methylamino) **(109-83-1)**
Ethanol, 2,2'-((4-(methylamino)-3-nitrophenyl)imino)di **(2784-94-3)**
Ethanol, 2-(N-methylanilino)- (8CI) **(93-90-3)**
Ethanol, 2-((1-methylethyl)amino)- (9CI) **(109-56-8)**
Ethanol, 2,2'-((1-methylethylidene)bis((2,6-dibromo-4,1-phenylene)-oxy))bis- (9CI) **(4162-45-2)**
Ethanol, 2,2'-((1-methylethylidene)bis(4,1-phenyleneoxy))bis- (9CI) **(901-44-0)**
Ethanol, 2,2'-((1-methylethyl)imino)bis- (9CI) **(121-93-7)**
Ethanol, 2,2'-(methylimino)bis- **(105-59-9)**
Ethanol, 2,2'-(methylimino)di **(105-59-9)**
Ethanol, 2,2'-((3-methyl-4-((2-(methylsulfonyl)-4-nitrophenyl)-azo)phenyl)imino)bis-, diacetate (ester) (9CI) **(29426-52-6)**
Ethanol, 2,2'-((3-methyl-4-((4-nitrophenyl)azo)phenyl)imino)-bis- (9CI) **(3179-89-3)**

Name	CAS No.
Ethanone, 1-(4-aminophenyl)- (9CI)	**(99-92-3)**
Ethanone, 1-(2-aminophenyl)-2-nitro- (9CI)	**(63892-06-8)**
Ethanone, 1-(1,1'-biphenyl)-4-yl- (9CI)	**(92-91-1)**
Ethanone, 2-bromo-1-(3-chlorophenyl)-	**(41011-01-2)**
Ethanone, 1-(3-bromophenyl)- (9CI)	**(2142-63-4)**
Ethanone, 1-(4-bromophenyl)- (9CI)	**(99-90-1)**
Ethanone, 2-bromo-1-phenyl- (9CI)	**(70-11-1)**
Ethanone, 2-bromo-1-(2-pyridinyl)-	**(40086-66-6)**
Ethanone, 2-chloro-1-(4-methylphenyl)-2-phenyl-, (.+-.)- (9CI)	**(73908-26-6)**
Ethanone, 1-(4-chlorophenyl)	**(99-91-2)**
Ethanone, 2-chloro-1-phenyl-	**(532-27-4)**
Ethanone, 1-(1-cyclohepten-1-yl)- (9CI)	**(14377-11-8)**
Ethanone, 1-(1-cyclohexen-1-yl)- (9CI)	**(932-66-1)**
Ethanone, 1-cyclohexyl- (9CI)	**(823-76-7)**
Ethanone, 1-cyclopentyl- (9CI)	**(6004-60-0)**
Ethanone, 1-cyclopropyl- (9CI)	**(765-43-5)**
Ethanone, 1-(dichlorohydroxydimethoxyphenyl)- (9CI)	**(108548-73-8)**
Ethanone, 1-(2,4-dichlorophenyl)- (9CI)	**(2234-16-4)**
Ethanone, 2,2-dichloro-1-(2,4,5-trichlorophenyl)- (9CI)	**(1203-86-7)**
Ethanone, 1-(2,6-dichloro-3,4,5-trimethoxyphenyl)- (9CI)	**(75315-44-5)**
Ethanone, 2,2-diethoxy-1-phenyl-	**(6175-45-7)**
Ethanone, 1-(4,5-diethyl-2-methyl-1-cyclopenten-1-yl)- (9CI)	**(62338-24-3)**
Ethanone, 1,2-di-2-furanyl-2-hydroxy- (9CI)	**(552-86-3)**
Ethanone, 1,2-di-2-furyl-2-hydroxy-	**(552-86-3)**
Ethanone, 1-(2,3-dihydro-4-methyl-2-thioxo-5-thiazolyl)- (9CI)	**(7725-93-1)**
Ethanone, 1-(2,3-dihydroxy-4-methoxyphenyl)- (9CI)	**(708-53-2)**
Ethanone, 1-(2,6-dihydroxy-4-methoxyphenyl)- (9CI)	**(7507-89-3)**
Ethanone, 1-(2,4-dihydroxyphenyl)- (9CI)	**(89-84-9)**
Ethanone, 1-(2,5-dihydroxyphenyl)- (9CI)	**(490-78-8)**
Ethanone, 2,2-dimethoxy-1,2-diphenyl- (9CI)	**(24650-42-8)**
Ethanone, 1-(2,6-dimethoxyphenyl)- (9CI)	**(2040-04-2)**
Ethanone, 1-(3,4-dimethoxyphenyl)- (9CI)	**(1131-62-0)**
Ethanone, 1-(6-(1,1-dimethylethyl)-2,3-dihydro-1,1-dimethyl-1H-inden-4-yl)-	**(13171-00-1)**
Ethanone, 1-(4-(1,1-dimethylethyl)-2,6-dimethyl-3,5-dinitrophenyl)- (9CI)	**(81-14-1)**
Ethanone, 1-(4-(1,1-dimethylethyl)phenyl)- (9CI)	**(943-27-1)**
Ethanone, 1-(2,3-dimethylphenyl)- (9CI)	**(2142-71-4)**
Ethanone, 1-(2,4-dimethylphenyl)- (9CI)	**(89-74-7)**
Ethanone, 1-(2,6-dimethylphenyl)- (9CI)	**(2142-76-9)**
Ethanone, 1-(dimethylphenyl)- (9CI)	**(1335-42-8)**
Ethanone, 1,2-diphenyl- (9CI)	**(451-40-1)**
Ethanone, 1-(2-ethylphenyl)- (9CI)	**(2142-64-5)**
Ethanone, 1-(3-ethylphenyl)- (9CI)	**(22699-70-3)**
Ethanone, 1-(4-ethylphenyl)- (9CI)	**(937-30-4)**
Ethanone, 1-(3-fluorophenyl)- (9CI)	**(455-36-7)**
Ethanone, 1-(4-fluorophenyl)- (9CI)	**(403-42-9)**
Ethanone, 1-(2-furanyl)- (9CI)	**(1192-62-7)**
Ethanone, 1-(2,3,4,7,8,8a-hexahydro-3,6,8,8-tetramethyl-1H-3a,7-methanoazulen-5-yl)-, (3R-(3α,3aβ,7β,8aα))- (9CI)	**(32388-55-9)**
Ethanone, 1-(4-hydroxy-3,5-dimethoxyphenyl)- (9CI)	**(2478-38-8)**
Ethanone, 2-hydroxy-1,2-diphenyl-	**(119-53-9)**
Ethanone, 1-(2-hydroxy-5-methoxy-4-methylphenyl)- (9CI)	**(4223-84-1)**
Ethanone, 1-(4-hydroxy-3-methoxyphenyl)- (9CI)	**(498-02-2)**
Ethanone, 1-(4-(1-hydroxy-1-methylethyl)phenyl)- (9CI)	**(54549-72-3)**
Ethanone, 1-(2-hydroxy-5-methylphenyl)- (9CI)	**(1450-72-2)**
Ethanone, 1-(2-hydroxyphenyl)- (9CI)	**(118-93-4)**
Ethanone, 1-(3-hydroxyphenyl)- (9CI)	**(121-71-1)**
Ethanone, 1-(4-hydroxyphenyl)- (9CI)	**(99-93-4)**
Ethanone, 1-(4-hydroxy-3-thienyl)- (9CI)	**(5556-16-1)**
Ethanone, 1-(3-methoxyphenyl)- (9CI)	**(586-37-8)**
Ethanone, 1-(4-methoxyphenyl)- (9CI)	**(100-06-1)**
Ethanone, 1-(2-(1-methylethyl)phenyl)- (9CI)	**(2142-65-6)**
Ethanone, 1-(4-(1-methylethyl)phenyl)- (9CI)	**(645-13-6)**
Ethanone, 1-(2-methylphenyl)- (9CI)	**(577-16-2)**
Ethanone, 1-(3-methylphenyl)- (9CI)	**(585-74-0)**
Ethanone, 1-(4-methylphenyl)- (9CI)	**(122-00-9)**
Ethanone, 1-(methylphenyl)- (9CI)	**(26444-19-9)**
Ethanone, 2-(2-methylpropoxy)-1,2-diphenyl- (9CI)	**(22499-12-3)**
Ethanone, 1-(3-methylpyrazinyl)- (9CI)	**(23787-80-6)**
Ethanone, 1-(5-methylpyrazinyl)- (9CI)	**(22047-27-4)**
Ethanone, 1-(1-naphthalenyl)- (9CI)	**(941-98-0)**
Ethanone, 1-(2-naphthalenyl)- (9CI)	**(93-08-3)**
Ethanone, 1-(naphthalenyl)- (9CI)	**(1333-52-4)**
Ethanone, 1-(4'-nitro(1,1'-biphenyl)-4-yl)- (9CI)	**(135-69-3)**
Ethanone, 1-(4-nitrophenyl)-	**(100-19-6)**
Ethanone, 1-(2-nitrophenyl)- (9CI)	**(577-59-3)**
Ethanone, 1-(3-nitrophenyl)- (9CI)	**(121-89-1)**
Ethanone, 1-oxiranyl- (9CI)	**(4401-11-0)**
Ethanone, 1-phenyl- (9CI)	**(98-86-2)**
Ethanone, 1-(4-(((phenylamino)carbonyl)oxy)phenyl)- (9CI)	**(37070-86-3)**
Ethanone, 1,1'-(1,3-phenylene)bis- (9CI)	**(6781-42-6)**
Ethanone, 1,1'-(1,4-phenylene)bis- (9CI)	**(1009-61-6)**
Ethanone, 1,1'-(phenylene)bis- (9CI)	**(30773-71-8)**
Ethanone, 1-phenyl-2-(3-phenyl-1H-1,2,4-triazol-1-yl)- (9CI)	**(79746-01-3)**
Ethanone, 1-pyrazinyl- (9CI)	**(22047-25-2)**
Ethanone, 1-(2-pyridinyl)- (9CI)	**(1122-62-9)**
Ethanone, 1-(1H-pyrrol-2-yl)- (9CI)	**(1072-83-9)**
Ethanone, 1-(5,6,7,8-tetrahydro-3,5,5,6,8,8-hexamethyl-2-naphthalenyl)-	**(21145-77-7)**
Ethanone, 1-(5,6,7,8-tetrahydro-3,5,5,6,8,8-hexamethyl-2-naphthalenyl)- (VAN) (9CI)	**(1506-02-1)**
Ethanone, 1-(3-thienyl)- (9CI)	**(1468-83-3)**
Ethanone, 2,2,2-trifluoro-1-phenyl- (9CI)	**(434-45-7)**
Ethanone, 1-(2,3,4-trihydroxyphenyl)- (9CI)	**(528-21-2)**
Ethanone, 1-(3,4,5-trimethoxyphenyl)- (9CI)	**(1136-86-3)**
Ethanone, 1-(trimethyloxiranyl)- (9CI)	**(15120-99-7)**
Ethanone, 1-(2,4,6-trimethylphenyl)- (9CI)	**(1667-01-2)**
1H-2,10a-Ethanophenanthrene, kauran-13-ol deriv.	**(5749-44-0)**
Ethanox	**(563-12-2)**
Ethanox 330	**(1709-70-2)**
Ethanox 701	**(128-39-2)**
Ethanoylaminoethanoic acid	**(543-24-8)**
Ethanoyl chloride	**(75-36-5)**
Ethaperazine	**(58-39-9)**
Ethavan	**(121-32-4)**
Ethazate	**(14324-55-1)**
Ethazol	**(2593-15-9)**
Ethazole	**(94-19-9)**
Ethazole (Fungicide)	**(2593-15-9)**
Ethazole (Pharmaceutical)	**(94-19-9)**
Ethbenzamide	**(938-73-8)**
Ethchlorovynol	**(113-18-8)**
Ethchlorvinyl	**(113-18-8)**
Ethclorvynol	**(113-18-8)**
Ethefon	**(16672-87-0)**
Ethel	**(16672-87-0)**
Ethenamine, N-ethyl-N-nitroso- (9CI)	**(13256-13-8)**
Ethenamine, N-methylene- (9CI)	**(38239-27-9)**
Ethenaminium, N,N,N-trimethyl-, hydroxide (9CI)	**(463-88-7)**
Ethene	**(74-85-1)**
Ethene, Homopolymer, chlorinated, chlorosulfonated	**(68037-39-8)**
Ethene Polymer	**(9002-88-4)**
Ethene, Polymer with 2-methylene-1,3-dioxepane (9CI)	**(90588-08-2)**
Ethene, bromo-	**(593-60-2)**
Ethene, bromotrifluoro- (9CI)	**(598-73-2)**
Ethene, chloro-	**(75-01-4)**
Ethene, 2-chloro-1,1-difluoro-	**(359-10-4)**
Ethene, 2-chloroethoxy-	**(110-75-8)**
1,1-Ethenediamine, N-(2-(((5-((dimethylamino)methyl)-2-furanyl)methyl)thio)ethyl)-N'-methyl- 2-nitro-, hydrochloride	**(66357-35-5)**
Ethene, dibromo- (9CI)	**(25429-23-6)**
1,2-Ethenedicarboxylic acid	**(6915-18-0)**
1,2-Ethenedicarboxylic acid, trans-	**(110-17-8)**

Ether, bis(2-(2,3-epoxypropoxy)ethyl)	**(4206-61-5)**
Ether, bis(2,3-epoxypropyl)	**(2238-07-5)**
Ether, 2,6-bis(2,3-epoxypropyl)phenyl 2,3-epoxypropyl	**(13561-08-5)**
Ether, bis(2-(2-ethoxyethoxy)ethyl)	**(4353-28-0)**
Ether, bis(2-ethoxyethyl)	**(112-36-7)**
Ether, bis(2-ethylhexyl)	**(10143-60-9)**
Ether, bis(2-(2-methoxyethoxy)ethyl)	**(143-24-8)**
Ether, bis(2-methoxyethyl)	**(111-96-6)**
Ether, bis(methoxymethyl) (6CI,7CI,8CI)	**(628-90-0)**
Ether, bis(α-methylbenzyl) (8CI)	**(93-96-9)**
Ether, bis(2-methylglycidyl)	**(7487-28-7)**
Ether, bis(p-nitrophenyl)	**(101-63-3)**
Ether, bis(pentabromophenyl)	**(1163-19-5)**
Ether, bis(pentachlorophenyl) (8CI)	**(31710-30-2)**
Ether, bis(2,3,3,3-tetrachloropropyl)	**(127-90-2)**
Ether, 4-bromophenyl phenyl	**(101-55-3)**
Ether, bromophenyl phenyl	**(36563-47-0)**
Ether, p-bromophenyl phenyl (8CI)	**(101-55-3)**
Ether, butyl (3-chloro-2-hydroxypropyl)	**(16224-33-2)**
Ether, butyl 2,3-epoxypropyl	**(2426-08-6)**
Ether, butyl ethyl	**(628-81-9)**
Ether, sec-butyl ethyl	**(2679-87-0)**
Ether, tert-butyl ethyl (8CI)	**(637-92-3)**
Ether, butyl glycidyl	**(2426-08-6)**
Ether butylique (French)	**(142-96-1)**
Ether, sec-butyl isopropyl (8CI)	**(18641-81-1)**
Ether, butyl methyl	**(628-28-4)**
Ether, tert-butyl methyl	**(1634-04-4)**
Ether, sec-butyl methyl (8CI)	**(6795-87-5)**
Ether, butyl pentyl (6CI,8CI)	**(18636-66-3)**
Ether, butyl phenyl	**(1126-79-0)**
Ether, butyl tolyl (8CI)	**(29225-54-5)**
Ether, butyl vinyl	**(111-34-2)**
Ether chloratus	**(75-00-3)**
Ether, 2-chloroallyl phenyl (6CI)	**(53299-53-9)**
Ether, p-chlorobenzyl 2,4-dichlorophenyl	**(21571-58-4)**
Ether, p-chlorobenzyl phenyl	**(19962-25-5)**
Ether, 2-chloroethyl 2,3-dichloroallyl	**(84987-77-9)**
Ether, 2-chloroethyl ethyl	**(628-34-2)**
Ether, 2-chloroethyl methyl	**(627-42-9)**
Ether, 2-chloroethyl vinyl	**(110-75-8)**
Ether, chloromethyl ethyl	**(3188-13-4)**
Ether, chloromethyl methyl	**(107-30-2)**
Ether, 4-chloro-2-nitrophenyl p-nitrophenyl (8CI)	**(20115-34-8)**
Ether, 4-chlorophenyl (4'-chloro-2'-nitro)phenyl	**(135-12-6)**
Ether, o-chlorophenyl p-chlorophenyl (8CI)	**(6903-65-7)**
Ether, p-chlorophenyl phenyl	**(7005-72-3)**
Ether, p-chlorophenyl phenyl (8CI)	**(7005-72-3)**
Ether, 1-chloro-2,2,2-trifluoroethyl difluoromethyl	**(26675-46-7)**
Ether, 2-chloro-1,1,2-trifluoroethyl difluoromethyl	**(13838-16-9)**
Ether, 2-chloro-α,α,α-trifluoro-p-tolyl 3-ethoxy-4-nitrophenyl	**(42874-03-3)**
Ether, cinnamyl methyl (8CI)	**(16277-67-1)**
Ether cyanatus	**(107-12-0)**
Ether, cyclohexyl methyl (6CI,7CI,8CI)	**(931-56-6)**
Ether, cyclopropyl ethyl (8CI)	**(5614-38-0)**
Ether, decabromodiphenyl	**(1163-19-5)**
Ether, diallyl	**(557-40-4)**
Ether, 2,6-diallylphenyl 2,3-epoxypropyl	**(40693-04-7)**
Ether, 4,4'-diaminodiphenyl	**(101-80-4)**
Ether dichlore (French)	**(111-44-4)**
Ether, 2,2-dichloro-1,1-difluoroethyl methyl	**(76-38-0)**
Ether, 2,4-dichloro-6-fluorophenyl p-nitrophenyl	**(13738-63-1)**
Ether dichloroisopropylique (French)	**(63283-80-7)**
Ether, dichloromethyl methyl	**(4885-02-3)**
Ether, dichlorophenyl	**(28675-08-3)**
Ether, (2,4-dichlorophenyl) (3-methoxy-4-nitrophenyl)	**(32861-85-1)**
Ether, 2,4-dichlorophenyl p-nitrophenyl	**(1836-75-5)**

Ether, dicyclopentyl	**(10137-73-2)**
Ether, diglycidyl	**(2238-07-5)**
Ether, dihexyl	**(112-58-3)**
Ether, dimethyl	**(115-10-6)**
Ether, dimethyl chloro	**(107-30-2)**
Ether, 4,4-dimethyl-2-pentenyl ethyl (6CI)	**(55702-60-8)**
Ether, 1,1-dimethyl-2-propynyl ethyl (6CI,7CI,8CI)	**(7740-69-4)**
Ether, di-n-octyl-	**(629-82-3)**
Ether, di-n-pentyl-	**(693-65-2)**
Ether, di-n-propyl-	**(111-43-3)**
Ether, diphenyl	**(101-84-8)**
Ether, diphenyl, 4-bromo-	**(101-55-3)**
Ether, divinyl	**(109-93-3)**
Ether, dodecyl 2,3-epoxypropyl	**(2461-18-9)**
Ether, 2,3-epoxypropyl (2,3-epoxypropyl)phenyl	**(63919-02-8)**
Ether, 2,3-epoxypropyl phenyl	**(122-60-1)**
Ether, 2,3-epoxypropyl propyl	**(3126-95-2)**
Ether, 2-ethoxyethyl 2-methoxyethyl	**(1002-67-1)**
Ether ethylbutylique (French)	**(628-81-9)**
Ether, ethyl 1-ethylpropyl	**(36749-13-0)**
Ether, ethyl ethynyl	**(927-80-0)**
Ether, 2-ethylhexyl vinyl (8CI)	**(103-44-6)**
Ether ethylique (French)	**(60-29-7)**
Ether, ethyl methyl	**(540-67-0)**
Ether, ethyl α-methylbenzyl (8CI)	**(3299-05-6)**
Ether, ethyl phenyl	**(103-73-1)**
Ether, ethyl propenyl	**(928-55-2)**
Ether, ethyl propyl	**(628-32-0)**
Ether, 1-ethylpropyl methyl (6CI,7CI)	**(36839-67-5)**
Ether, ethyl vinyl	**(109-92-2)**
Ether, heptachlorodiphenyl	**(55684-92-9)**
Ether, hexachlorophenyl	**(31242-93-0)**
Ether, hexyl vinyl (8CI)	**(5363-64-4)**
Ether hydrochloric	**(75-00-3)**
Ether, 2'-hydroxy-2,4,4'-trichlorodiphenyl	**(3380-34-5)**
Etherin	**(9002-88-4)**
Ether, 3-iodo-2-propynyl 2,4,5-trichlorophenyl	**(777-11-7)**
Ether, isobutyl methyl (8CI)	**(625-44-5)**
Ether, isobutyl vinyl	**(109-53-5)**
Ether, isopropyl	**(108-20-3)**
Ether isopropylique (French)	**(108-20-3)**
Ether, isopropyl methyl (8CI)	**(598-53-8)**
Ether, 2-methoxyethyl vinyl	**(1663-35-0)**
Ether, methyl	**(115-10-6)**
Ether methylique monochlore (French)	**(107-30-2)**
Ether, methyl tert-pentyl (8CI)	**(994-05-8)**
Ether, methyl phenyl	**(100-66-3)**
Ether, methyl propyl	**(557-17-5)**
Ether, methyl vinyl	**(107-25-5)**
Ether monoethylique de l'ethylene-glycol (French)	**(110-80-5)**
Ether monoethylique de l'hydroquinone (French)	**(622-62-8)**
Ether monomethylique de l'ethylene-glycol (French)	**(109-86-4)**
Ether muriatic	**(75-00-3)**
Ether, 4-nitrophenyl phenyl	**(620-88-2)**
Ether, p-nitrophenyl o-tolyl	**(2444-29-3)**
Ether, p-nitrophenyl m-tolyl (8CI)	**(2303-25-5)**
Ether, p-nitrophenyl 2,4,6-trichlorophenyl	**(1836-77-7)**
Ether, (p-nitrophenyl) (α,α,α-trifluoro-2-nitro-p-tolyl)	**(15457-05-3)**
Ether, p-nitrophenyl 3,5-xylyl (8CI)	**(1630-17-7)**
Ether, octadecyl vinyl (8CI)	**(930-02-9)**
Etherol E	**(9002-88-4)**
Etheron	**(9009-54-5)**
Etheron Sponge	**(9009-54-5)**
Etherophenol	**(2883-98-9)**
Ether, pentachlorophenyl	**(42279-29-8)**
Ether, 3-pentenyl pentyl (8CI)	**(34061-80-8)**
Ether, pentyl vinyl (6CI,7CI,8CI)	**(5363-63-3)**

Ether, pentyl xylyl (8CI) **(1320-21-4)**
Ether, phenylglycidyl **(122-60-1)**
Ether, phenyl m-tolyl- **(3586-14-9)**
Ether, poly(allyl-glycidyl) **(25639-25-2)**
Ether, propenyl **(557-40-4)**
Ether, propyl phenyl **(622-85-5)**
Ethers, cyclic, epoxides, Polymers **(61788-97-4)**
Ethersulfonate **(80-33-1)**
Ether, tetrachlorophenyl **(31242-94-1)**
Ether, trichlorophenyl **(31242-93-0)**
Ether, 2,2,2-trifluoroethyl vinyl **(406-90-6)**
Ether, trifluoromethyl trifluorovinyl **(1187-93-5)**
Ether, vinyl ethyl **(109-92-2)**
Etheverse **(16672-87-0)**
Ethide **(594-72-9)**
Ethidium bromide **(1239-45-8)**
Ethidol **(57-63-6)**
Ethienocarb **(58270-08-9)**
Ethilon **(63428-84-2)**
Ethimide **(536-33-4)**
Ethina **(536-33-4)**
Ethinamate **(126-52-3)**
Ethinamide **(536-33-4)**
Ethine **(74-86-2)**
Ethinodiol diacetate **(297-76-7)**
Ethinoral **(57-63-6)**
Ethinylbenzene **(536-74-3)**
2-Ethinylbutanol-2 **(77-75-8)**
1-Ethinylcyclohexyl carbamate **(126-52-3)**
1-Ethinylcyclohexyl carbonate **(126-52-3)**
17-α-Ethinyl-3,17-dihydroxy-δ^{1,3,5}Oestratriene **(57-63-6)**
17-α-Ethinyl-3,17-dihydroxy-δ^{1,3,5}-estratriene **(57-63-6)**
17-Ethinyl-3,17-estradiol **(57-63-6)**
17-Ethinylestradiol **(57-63-6)**
17-α-Ethinyl-17-β-estradiol **(57-63-6)**
17-α-Ethinylestradiol **(57-63-6)**
Ethinyl estradiol **(57-63-6)**
17-α-Ethinyl estradiol 3-methyl ether **(72-33-3)**
Ethinylestradiol 3-methyl ether **(72-33-3)**
17-Ethinyl-5(10)-estraeneolone **(68-23-5)**
17-α-Ethinyl-estra(5,10)eneolone **(68-23-5)**
17-α-Ethinylestra-4-en-17-β-ol-3-one **(68-22-4)**
17-α-Ethinylestra-1,3,5(10)-triene-3,17-β-diol **(57-63-6)**
Ethinylestrenol **(52-76-6)**
δ^{4}-17-α-Ethinylestren-17-β-ol **(52-76-6)**
17-α-Ethinyl-5,10-estrenolone **(68-23-5)**
Ethinylestriol **(57-63-6)**
17-α-Ethinyl-13-β-ethyl-17-β-hydroxy-4-estren-3-one **(797-63-7)**
17-α-Ethinyl-17-β-hydroxyestr-4-ene **(52-76-6)**
17-α-Ethinyl-17-β-hydroxy-δ$^{5(10)}$-estren-3-one **(68-23-5)**
17-α-Ethinyl-17-β-hydroxy-δ^{4}-estren-3-one **(68-22-4)**
17-α-Ethinyl-17-β-hydroxyoestr-4-ene **(52-76-6)**
Ethinylmethylethylcarbinol **(77-75-8)**
17-Ethinyl-19-nortestosterone **(68-22-4)**
17-α-Ethinyl-19-nortestosterone **(68-22-4)**
Ethinyl-19-nortestosterone **(68-22-4)**
Ethinylnortestosterone **(68-22-4)**
17-α-Ethinyl-19-nortestosterone acetate **(51-98-9)**
17-α-Ethinyl-19-nortestosterone-17-β-acetate **(51-98-9)**
17-Ethinyl-3,17-oestradiol **(57-63-6)**
Ethinyloestradiol **(57-63-6)**
17-α-Ethinyl oestradiol 3-methyl ether **(72-33-3)**
Ethinyloestradiol 3-methyl ether **(72-33-3)**
Ethinyl-oestranol **(57-63-6)**
Ethinyloestranol **(52-76-6)**
17-α-Ethinyl-δ$^{1,3,5(10)}$oestratriene-3,17-β-diol **(57-63-6)**
17-α-Ethinyloestra-1,3,5(10)-triene-3,17-β-diol **(57-63-6)**
Ethinyl oestrenol **(52-76-6)**
δ^{4}-17-α-Ethinyloestren-17-β-ol **(52-76-6)**
Ethinyloestriol **(57-63-6)**
17-α-Ethinyl-δ$^{5,10-19}$-nortestosterone **(68-23-5)**
Ethinyl trichloride **(79-01-6)**
Ethiodan **(99-79-6)**
Ethiofencarb **(29973-13-5)**
Ethiol **(563-12-2)**
Ethiolacar **(121-75-5)**
Ethiolate **(2941-55-1)**
Ethion (ACGIH,DOT,OSHA) [NA 2783] **(563-12-2)**
Ethion, Dry **(2921-88-2)**
Ethionamide **(536-33-4)**
Ethioniamide **(536-33-4)**
Ethionin **(67-21-0)**
(+-)-Ethionine **(67-21-0)**
D-Ethionine **(535-32-0)**
DL-Ethionine **(67-21-0)**
Ethionine **(13073-35-3)**
Ethionine **(67-21-0)**
Ethionine, DL- **(67-21-0)**
L-Ethionine **(13073-35-3)**
Ethion mixture, Dry (DOT) **(563-12-2)**
Ethiophencarbe **(29973-13-5)**
Ethiophencarp **(29973-13-5)**
Ethirimol **(23947-60-6)**
Ethlon **(56-38-2)**
Ethnine **(509-67-1)**
Ethnine Simplex **(509-67-1)**
Ethoate methyl **(116-01-8)**
Ethocaine **(51-05-8)**
Ethocel **(9004-32-4)**
Ethocel **(9004-57-3)**
Ethocel 150 **(9004-57-3)**
Ethocel 890 **(9004-57-3)**
Ethocel E7 **(9004-57-3)**
Ethocel E50 **(9004-57-3)**
Ethocel MED **(9004-57-3)**
Ethocel N10 **(9004-57-3)**
Ethocel N200 **(9004-57-3)**
Ethocel N7 **(9004-57-3)**
Ethocel STD **(9004-57-3)**
Ethochlorvynol **(113-18-8)**
Ethodan **(563-12-2)**
Ethodryl citrate **(1642-54-2)**
Ethofat 60/15 **(9004-99-3)**
Ethofat 60/20 **(9004-99-3)**
Ethofat 60/25 **(9004-99-3)**
Ethofat O **(9004-96-0)**
Ethofat O 15 **(9004-96-0)**
Ethofumesate **(26225-79-6)**
Ethoglucid **(1954-28-5)**
Ethoglucide **(1954-28-5)**
Ethoheptazine **(77-15-6)**
Ethohexadiol **(94-96-2)**
Ethol **(36653-82-4)**
Ethomeen C **(61791-14-8)**
Ethomeen S/12 **(61791-24-0)**
Ethomeen S/15 **(61791-24-0)**
Ethone **(122-51-0)**
Ethophylline **(317-34-0)**
Ethoprop **(13194-48-4)**
Ethoprophos **(13194-48-4)**
Ethosalicyl **(938-73-8)**
Ethosperse LA 12 **(9002-92-0)**
Ethosperse LA 23 **(9002-92-0)**
Ethosperse LA-4 **(9002-92-0)**

Ethosuccimide **(77-67-8)**
Ethosuccinimide **(77-67-8)**
Ethosuxide **(77-67-8)**
Ethosuximide **(77-67-8)**
Ethovan **(121-32-4)**
Ethox **(75-21-8)**
Ethoxazolamide **(452-35-7)**
Ethoxose **(9004-32-4)**
3'-Ethoxyacetanilide **(591-33-3)**
3-Ethoxyacetanilide **(591-33-3)**
4'-Ethoxyacetanilide **(62-44-2)**
4-Ethoxyacetanilide **(62-44-2)**
m-Ethoxyacetanilide **(591-33-3)**
p-Ethoxyacetanilide **(62-44-2)**
Ethoxy acetate **(111-15-9)**
2-Ethoxyacetic acid **(627-03-2)**
Ethoxyacetic acid **(627-03-2)**
4'-Ethoxyacetoacetanilide **(122-82-7)**
4-Ethoxyacetoacetanilide **(122-82-7)**
Ethoxyacetylene **(927-80-0)**
6-Ethoxy-2-aminobenzothiazole **(94-45-1)**
N-(Ethoxyaminoethyl)stearamide **(141-21-9)**
p-Ethoxyanilid kyseliny octove (Czech) **(62-44-2)**
3-Ethoxyaniline **(621-33-0)**
4-Ethoxyaniline **(156-43-4)**
m-Ethoxyaniline **(621-33-0)**
p-Ethoxyaniline **(156-43-4)**
2-Ethoxybenzaldehyde **(613-69-4)**
4-Ethoxybenzaldehyde **(10031-82-0)**
Ethoxybenzaldehyde **(10031-82-0)**
p-Ethoxybenzaldehyde **(10031-82-0)**
2-Ethoxybenzamide **(938-73-8)**
o-Ethoxybenzamide **(938-73-8)**
3-Ethoxybenzenamine **(621-33-0)**
Ethoxybenzene **(103-73-1)**
4-Ethoxy-1,2-benzenediamine **(1197-37-1)**
4-Ethoxy-1,3-benzenediamine sulfate **(68015-98-5)**
4-Ethoxy-1,3-benzenediamine sulfate (1:1) **(68015-98-5)**
4-Ethoxybenzoic acid **(619-86-3)**
6-Ethoxy-2-benzothiazolesulfonamide **(452-35-7)**
2-Ethoxybutane **(2679-87-0)**
3-Ethoxycarbonylaminophenyl-N-phenylcarbamate **(13684-56-5)**
S-α-Ethoxycarbonylbenzyl O,O-dimethyl phosphorodithioate **(2597-03-7)**
S-α-Ethoxycarbonylbenzyl dimethyl phosphorothiolothionate **(2597-03-7)**
Ethoxycarbonylethylene **(140-88-5)**
Ethoxycarbonylmethyl bromide **(105-36-2)**
N-Ethoxycarbonyl-N-methylcarbamoylmethyl O,O-diethyl phosphorodithioate **(2595-54-2)**
S-((Ethoxycarbonyl)methylcarbamoyl)methyl O,O-diethyl phosphorodithioate **(2595-54-2)**
S-(N-Ethoxycarbonyl-N-methylcarbamoylmethyl)-diethyl phosphorodithioate **(2595-54-2)**
3-Ethoxycarbonylphenyl phenylcarbamate **(37070-83-0)**
3-(Ethoxycarbonyl)-1-phenyl-5-pyrazolone **(89-33-8)**
N-(Ethoxycarbonyl)phthalimide **(22509-74-6)**
3-Ethoxycarbonylpsoralen **(20073-24-9)**
Ethoxychin (Czech) **(91-53-2)**
Ethoxychloromethane **(3188-13-4)**
7-Ethoxycoumarin **(31005-02-4)**
cis-9-Ethoxydecalin **(51953-10-7)**
Ethoxydiethylaluminum **(1586-92-1)**
3-Ethoxy-N,N-diethylbenzenamine **(1864-92-2)**
3-Ethoxy-N,N-diethyl-4-hydroxybenzamide **(13898-68-5)**
Ethoxydiethylphosphine oxide **(4775-09-1)**
Ethoxy diglycol **(111-90-0)**
(+-)-2-Ethoxy-2,3-dihydro-3,3-dimethyl-5-benzofuranol methanesulfonate **(26225-79-6)**
2-Ethoxy-2,3-dihydro-γ-pyran **(103-75-3)**
2-Ethoxy-3,4-dihydro-1,2-pyran **(103-75-3)**
2-Ethoxy-3,4-dihydro-2H-pyran **(103-75-3)**
2-Ethoxydihydropyran, In pregnancy diagnosis **(103-75-3)**
6-Ethoxy-1,2-dihydro-2,2,4-trimethylquinoline **(91-53-2)**
Ethoxydiphenylarsine **(24582-55-6)**
Ethoxy-1-dodecanol **(29718-44-3)**
Ethoxyethane **(60-29-7)**
2-Ethoxyethanol (ACGIH,OSHA) **(110-80-5)**
2-Ethoxyethanol acetate **(111-15-9)**
2-Ethoxyethanol, ester with acetic acid **(111-15-9)**
2-Ethoxyethanol magnesium salt **(14064-03-0)**
2-Ethoxyethanol, phosphated **(68554-00-7)**
(1-Ethoxyethoxy)benzene **(5426-78-8)**
2-(2-Ethoxyethoxy)ethanol **(111-90-0)**
2-(2-Ethoxyethoxy)ethanol acetate **(112-15-2)**
1-Ethoxy-2-(β-ethoxyethoxy)ethane **(112-36-7)**
2-(2-(2-Ethoxyethoxy)ethoxy)ethanol **(112-50-5)**
5-(1-(2-(2-Ethoxyethoxy)ethoxy)ethoxy)-1,3-benzodioxole **(51-14-9)**
2-(2-Ethoxyethoxy)ethyl acetate **(112-15-2)**
2-(2-Ethoxyethoxy)ethyl 2-benzimidazole carbamate **(62732-91-6)**
2-(2-Ethoxyethoxy)ethylester kyseliny octove (Czech) **(112-15-2)**
2-(2-Ethoxyethoxy)ethyl-3,4-(methylenedioxy)phenyl acetal of acetaldehyde **(51-14-9)**
2-(2-Ethoxyethoxy)ethyl 2-propenoate **(7328-17-8)**
2-(2-Ethoxyethoxy)-2-methylpropane **(51422-54-9)**
1-(1-Ethoxyethoxy)propane **(20680-10-8)**
2-Ethoxy-ethylacetaat (Dutch) **(111-15-9)**
2-Ethoxyethylacetate **(111-15-9)**
Ethoxyethyl acetate **(111-15-9)**
β-Ethoxyethyl acetate **(111-15-9)**
2-Ethoxyethyl acetate (ACGIH,OSHA) **(111-15-9)**
2-Ethoxyethyl acrylate **(106-74-1)**
Ethoxyethyl acrylate **(106-74-1)**
2-Ethoxyethyl chloride **(628-34-2)**
2-Ethoxyethyl dodecanoate **(106-13-8)**
2-Ethoxyethyle, acetate de (French) **(111-15-9)**
2-Ethoxyethylester kyseliny akrylove (Czech) **(106-74-1)**
2-Ethoxyethylester kyseliny octove (Czech) **(111-15-9)**
(1-Ethoxyethylidene)propanedinitrile **(5417-82-3)**
2-Ethoxy ethyl methacrylate **(2370-63-0)**
2-Ethoxyethyl methacrylate **(2370-63-0)**
2-Ethoxyethyl p-methoxycinnamate **(104-28-9)**
2-Ethoxyethyl 2-methoxyethyl ether **(1002-67-1)**
2-Ethoxyethyl 2-methyl-2-propenoate **(2370-63-0)**
2-Ethoxyethyl-2-propenoate **(106-74-1)**
O-(6-Ethoxy-2-ethyl-4-pyrimidinyl) O,O-dimethyl phosphorothioate **(38260-54-7)**
O-6-Ethoxy-2-ethylpyrimidin-4-yl O,O-dimethyl phosphorothioate **(38260-54-7)**
Ethoxy-ethyne **(927-80-0)**
Ethoxyethyne **(927-80-0)**
p-Ethoxyfenylmocovina (Czech) **(150-69-6)**
Ethoxyformic anhydride **(105-58-8)**
3-Ethoxy-4-hydroxybenzaldehyde **(121-32-4)**
3-Ethoxy-4-hydroxy-diethylbenzamide **(13898-68-5)**
2'-Ethoxy-3-hydroxy-2-naphthanilide **(92-74-0)**
(+-)-(ZE)-2-(1-Ethoxyiminobutyl)-5-(2-(ethylthio)propyl)-3-hydroxycyclohex-2-enone **(74051-80-2)**
(+-)-2-(1-(Ethoxyimino)butyl)-5-(2-(ethylthio)propyl)-3-hydroxy-2-cyclohexen-1-one **(74051-80-2)**
2-((1-Ethoxyimino)butyl)-5-((ethylthio)propyl)-3-hydroxy-2-cyclohexen-1-one **(74051-80-2)**
2-(1-(Ethoxyimino)butyl)-5-(2-(ethylthio)propyl)-3-hydroxy-2-cyclohexen-1-one **(74051-80-2)**
Ethoxykarbonylmethyl-methylester kyseliny ftalove (Czech) **(85-71-2)**
Ethoxylated cetyl alcohol **(9004-95-9)**

Ethoxylated glycerin **(31694-55-0)**
Ethoxylated glycerine **(31694-55-0)**
Ethoxylated lanolin **(61790-81-6)**
Ethoxylated lauryl alcohol **(9002-92-0)**
Ethoxylated nonylphenol phosphate **(51811-79-1)**
Ethoxylated nonyl phenol, polyphosphates, sodium salt **(37340-60-6)**
Ethoxylated octyl phenol **(9036-19-5)**
Ethoxylated oleic monoethanolamide **(26027-37-2)**
Ethoxylated oleoamide **(26027-37-2)**
Ethoxylated oleylamine **(26635-93-8)**
Ethoxylated sorbitan monooleate **(9005-65-6)**
Ethoxylated stearic acid **(9004-99-3)**
Ethoxylated stearylamine **(26635-92-7)**
Ethoxylated tall oil fatty acid **(61791-00-2)**
Ethoxylated tall oil fatty acids **(61791-00-2)**
Ethoxylated tetrabromobisphenol A **(4162-45-2)**
1-Ethoxy-1-methoxyethane **(10471-14-4)**
(Ethoxymethyl)benzene **(539-30-0)**
1-Ethoxy-3-methylbenzene **(621-32-9)**
Ethoxymethylbis(p-chlorophenyl)carbinol **(6012-83-5)**
2-Ethoxy-2-methylbutane **(919-94-8)**
Ethoxymethyl chloride **(3188-13-4)**
1-Ethoxymethyl-1,1-di-(p-chlorophenyl)carbinol **(6012-83-5)**
Ethoxymethyl-di-(p-chlorophenyl)carbinol **(6012-83-5)**
Ethoxymethyldichlorophenylcarbinol **(6012-83-5)**
2-((Ethoxy((1-methylethyl)amino)phosphinothioyl)oxy)benzoic acid 1-methylethyl ester **(25311-71-1)**
(Ethoxymethyl)oxirane **(4016-11-9)**
(+-)-2-Ethoxy-1-methyl-2-oxoethyl 5-(2-chloro-4-(trifluoromethyl)phenoxy)-2 nitrobenzoate **(77501-63-4)**
2-Ethoxy-2-methylpropane **(637-92-3)**
2-Ethoxynaphthalene **(93-18-5)**
2-Ethoxy-1-naphthalenecarboxylic acid **(2224-00-2)**
2-Ethoxy-1-naphthoic acid **(2224-00-2)**
4-Ethoxy-3-nitroanilid kyseliny octove (Czech) **(1777-84-0)**
4-Ethoxynitrobenzene **(100-29-8)**
p-Ethoxynitrobenzene **(100-29-8)**
Ethoxy-4-nitrophenoxyphenylphosphine sulfide **(2104-64-5)**
N-(4-Ethoxy-3-nitro)phenylacetamide **(1777-84-0)**
S-((5-Ethoxy-2-oxo-1,3,4-thiadiazol-3(2H)-yl)methyl) O,O-dimethyl phosphorodithioate **(2669-32-1)**
Ethoxyphas **(919-54-0)**
2-Ethoxyphenol **(94-71-3)**
4-Ethoxyphenol **(622-62-8)**
o-Ethoxyphenol **(94-71-3)**
p-Ethoxyphenol **(622-62-8)**
2-((2-Ethoxyphenoxy)methyl)morpholine **(46817-91-8)**
2-(2-Ethoxyphenoxymethyl)tetrahydro-1,4-oxazine **(46817-91-8)**
N-(4-Ethoxyphenyl)acetamide **(62-44-2)**
N-para-Ethoxyphenylacetamide **(62-44-2)**
4-(p-Ethoxyphenylamino)-3-nitrobenzenesulfonamide **(22025-44-1)**
4-Ethoxy-7-phenyl-3,5-dioxa-6-aza-4-phosphaoct-6-ene-8-nitrile 4-sulfide **(14816-18-3)**
4-Ethoxy-m-phenylenediamine sulfate **(68015-98-5)**
N-(4-Ethoxyphenyl)-3'-nitroacetamide **(1777-84-0)**
N4-(p-Ethoxyphenyl)-3-nitrosulfanilamide **(22025-44-1)**
2-(p-Ethoxyphenyl)-2-p-tolyl-1,1,1-trichloroethane **(34197-05-2)**
4-Ethoxyphenylurea **(150-69-6)**
N-(4-Ethoxyphenyl)urea **(150-69-6)**
p-Ethoxyphenylurea **(150-69-6)**
Ethoxyphos **(919-54-0)**
3-Ethoxypropanal **(2806-85-1)**
1-Ethoxypropane **(628-32-0)**
2-Ethoxypropane **(625-54-7)**
3-Ethoxypropanenitrile **(2141-62-0)**
1-Ethoxy-2-propanol **(1569-02-4)**
2-Ethoxy-1-propanol **(19089-47-5)**
3-Ethoxy-1-propanol **(111-35-3)**
3-Ethoxy-1-propene **(557-31-3)**
3-Ethoxypropionaldehyde **(2806-85-1)**
β-Ethoxypropionaldehyde **(2806-85-1)**
3-Ethoxypropionic acid **(1331-11-9)**
Ethoxypropionic acid **(1331-11-9)**
Ethoxypropionic acid, ethyl ester **(763-69-9)**
1-Ethoxy-1-propoxyethane **(20680-10-8)**
Ethoxyquin **(91-53-2)**
Ethoxyquine **(91-53-2)**
Ethoxysodium **(141-52-6)**
5-Ethoxy-3-trichloromethyl-1,2,4-thiadiazole **(2593-15-9)**
Ethoxytriethylene glycol **(112-50-5)**
Ethoxytriglycol **(112-50-5)**
6-Ethoxy-2,2,4-trimethyl-1,2-dihydroquinoline **(91-53-2)**
Ethoxytrimethylsilane **(1825-62-3)**
Ethoxyzolamide **(452-35-7)**
Ethoxzolamide **(452-35-7)**
Ethrane **(13838-16-9)**
Ethrel **(16672-87-0)**
Ethril **(643-22-1)**
Ethriol **(77-99-6)**
Ethychlorvynol **(113-18-8)**
Ethyl 702 **(118-82-1)**
Ethyl 733 (9CI) **(60083-44-5)**
Ethyl Cadmate **(14239-68-0)**
Ethyl Cetab **(124-03-8)**
Ethyl cymate **(14324-55-1)**
p,p'-Ethyl-DDD **(72-56-0)**
p,p-Ethyl DDD **(72-56-0)**
Ethyl Etrinoate **(54350-48-0)**
Ethyl Glutamate **(1119-33-1)**
Ethyl Green **(633-03-4)**
Ethyl Hydrate **(64-17-5)**
Ethyl Mustard Oil **(542-85-8)**
Ethyl PTS **(80-40-0)**
Ethyl-S **(538-07-8)**
Ethyl Toluenesulfonamide **(80-39-7)**
Ethyl Tuads **(14239-68-0)**
Ethyl Violet **(2390-59-2)**
Ethyl Violet AX **(2390-59-2)**
Ethyl Violet GGA **(2390-59-2)**
Ethyl zimate **(14324-55-1)**
Ethyl ziram **(14324-55-1)**
Ethylacetaat (Dutch) **(141-78-6)**
Ethylacetacetat (Czech) **(141-97-9)**
Ethylacetamide **(625-50-3)**
N-Ethylacetamide **(625-50-3)**
D-N-Ethylacetamide carbanilate **(16118-49-3)**
Ethyl orthoacetate **(78-39-7)**
Ethyl acetate (ACGIH,OSHA) [UN 1173] **(141-78-6)**
Ethylacetic acid **(107-92-6)**
Ethyl acetic ester **(141-78-6)**
Ethyl acetoacetate **(141-97-9)**
Ethyl acetoacetate aluminum diidopropylate di isopropyl aluminium ethyl acetoacetate **(14782-75-3)**
.α.-Ethylacetoacetic acid **(4433-85-6)**
Ethyl acetone **(107-87-9)**
Ethyl acetyl acetate **(141-97-9)**
Ethyl acetylacetonate **(141-97-9)**
Ethylacetylene **(107-00-6)**
Ethyl acetylene, Inhibited [UN 2452] **(107-00-6)**
Ethyl 3-acetylpropionate **(539-88-8)**
Ethylacrylaat (Dutch) **(140-88-5)**
Ethyl acrylate (ACGIH,OSHA) **(140-88-5)**
Ethyl acrylate, Inhibited [UN 1917] **(140-88-5)**
Ethyl acrylate, Polymer with methyl methacrylate **(9010-88-2)**

Ethyl acrylate-acrylamide-N-methylolacrylamide-methylmethacrylate polymer **(30394-81-1)**
Ethyl acrylate, methacrylic acid, 1,2-benzenedicarboxylic acid, di-2-propenyl ester polymer **(28411-49-6)**
Ethyl acrylate, methyl methacrylate, acrylamide, methylol-acrylamide polymer **(30394-81-1)**
Ethyl acrylate, methyl methacrylate polymer **(9010-88-2)**
cis-3-Ethylacrylic acid **(16666-42-5)**
Ethyl adipate **(141-28-6)**
Ethylakrylat (Czech) **(140-88-5)**
Ethylal **(462-95-3)**
Ethyl alcohol (ACGIH,OSHA) [UN 1170] **(64-17-5)**
Ethylalcohol (Dutch) **(64-17-5)**
Ethyl alcohol, aluminum salt **(555-75-9)**
Ethyl alcohol anhydrous **(64-17-5)**
Ethyl alcohol, sodium salt **(141-52-6)**
Ethyl aldehyde (DOT) **(75-07-0)**
Ethylallene **(591-95-7)**
Ethyl allyl ether **(557-31-3)**
Ethyl aluminum dichloride (DOT) **(563-43-9)**
Ethylaluminum sesquichloride **(12075-68-2)**
Ethyl aluminum sesquichloride (DOT) **(12075-68-2)**
Ethylamine (ACGIH,DOT,OSHA) [UN 1036] **(75-04-7)**
Ethylamine, Aqueous solution with not less than 50 per cent but not more than 70 per cent ethylamine [UN 2270] **(75-04-7)**
Ethylamine, 2-chloro-N,N-dimethyl-, hydrochloride **(4584-46-7)**
Ethylamine, Compd. with boron fluoride (1:1) **(75-23-0)**
Ethylamine, N,N-dimethyl- **(598-56-1)**
Ethylamine, N,N-dimethyl-1,2-diphenyl-, hydrochloride, (R) (-)- **(14148-99-3)**
Ethylamine, N,N-dimethyl-2-(diphenylmethoxy)- **(58-73-1)**
Ethylamine, N,N-dimethyl-2-(diphenylmethoxy)-, Compd. with 8 chlorotheophylline **(523-87-5)**
Ethylamine, N,N-dimethyl-2-(diphenylmethoxy)-, hydrochloride **(147-24-0)**
Ethylamine, N,N-dimethyl-2-((o-methyl-α-phenylbenzyl)oxy) **(83-98-7)**
Ethylamine, 2-(diphenylmethoxy)-N,N-dimethyl **(58-73-1)**
Ethylamine, 2-(diphenylmethoxy)-N,N-dimethyl-, hydrochloride **(147-24-0)**
Ethylamine, hydrochloride **(557-66-4)**
Ethylamine, β-hydroxy-β-phenyl- **(7568-93-6)**
Ethylamine, 2-imidazol-4-yl- **(51-45-6)**
Ethylamine, 2,2'-iminobis- **(111-40-0)**
Ethylamine, 2-methoxy **(109-85-3)**
Ethylamine, N-methyl-N-nitroso **(10595-95-6)**
Ethylamine, N-nitrosodi- **(55-18-5)**
Ethylamine, 2,2'-oxybis(N,N-dimethyl **(3033-62-3)**
Ethylamine, 1-phenyl- **(98-84-0)**
Ethylamine, 2-phenyl- **(64-04-0)**
N-Ethylaminobenzene **(103-69-5)**
Ethyl 3-aminobenzoate **(582-33-2)**
Ethyl 4-aminobenzoate **(94-09-7)**
Ethyl aminobenzoate **(94-09-7)**
Ethyl o-aminobenzoate **(87-25-2)**
Ethyl p-aminobenzoate **(94-09-7)**
Ethyl (aminocarbonyl)carbamate **(626-36-8)**
2-Ethylamino-4-diethylamino-6-chloro-s-triazine **(1912-26-1)**
2-Ethylaminoethanol **(110-73-6)**
N-Ethyl-N-(β-aminoethyl)-m-toluidine **(19248-13-6)**
2-Ethylamino-4-isopropylamino-6-methoxy-s-triazine **(1610-17-9)**
4-Ethylamino-6-isopropylamino-2-methoxy-s-triazine **(1610-17-9)**
6-Ethylamino-4-isopropylamino-2-methoxy-1,3,5-triazine **(1610-17-9)**
2-Ethylamino-4-isopropylamino-6-methylmercapto-s-triazine **(834-12-8)**
2-Ethylamino-4-isopropylamino-6-methylthio-1,3,5-triazine **(834-12-8)**
2-Ethylamino-4-isopropylamino-6-methylthio-s-triazine **(834-12-8)**
2-Ethylamino-4-methyl-5-n-butyl-6-hydroxypyrimidine **(23947-60-6)**
3-Ethylamino-4-methylphenol **(120-37-6)**
2-(Ethylamino)naphthalene **(118-44-5)**
S-(2-(Ethylamino)-2-oxoethyl) O,O-dimethyl phosphorodithioate **(116-01-8)**
Ethyl 1-(4-aminophenethyl)-4-phenylisonipecotate **(144-14-9)**
Ethyl 1-(p-aminophenethyl)-4-phenylisonipecotate **(144-14-9)**
3-(Ethylamino)phenol **(621-31-8)**
Ethyl p-aminophenyl ketone **(70-69-9)**
2-Ethylamino-3-phenylnorbornane **(1209-98-9)**
2-Ethyl-2-aminopropanediol **(115-70-8)**
2-(Ethylamino)toluene **(94-68-8)**
2-Ethyl-3-(3-amino-2,4,6-triiodophenyl)propionic acid **(96-83-3)**
Ethyl ammonium chloride **(557-66-4)**
Ethylamphetamine **(457-87-4)**
N-Ethylamphetamine **(457-87-4)**
Ethyl-n-amylcarbinol **(589-98-0)**
Ethylamylcarbinol **(589-98-0)**
Ethyl tert-amyl ether **(919-94-8)**
Ethyl iso-amyl ketone **(624-42-0)**
Ethyl sec-amyl ketone **(541-85-5)**
Ethyl amyl ketone (ACGIH) **(541-85-5)**
Ethyl amyl ketone (OSHA) [UN 2271] **(106-68-3)**
2-(1-Ethylamyloxy)ethanol **(10138-47-3)**
Ethylan **(72-56-0)**
Ethylan A3 **(9004-96-0)**
Ethylan A6 **(9004-96-0)**
Ethylan CP **(9036-19-5)**
Ethylan CPX **(9036-19-5)**
Ethylan HB 4 **(9004-78-8)**
Ethylan LD **(68603-42-9)**
Ethylan MLD **(120-40-1)**
2-Ethyl aniline **(578-54-1)**
3-Ethylaniline **(587-02-0)**
Ethylaniline **(103-69-5)**
m-Ethylaniline **(587-02-0)**
o-Ethylaniline **(578-54-1)**
p-Ethylaniline **(589-16-2)**
2-Ethylaniline [UN 2273] **(578-54-1)**
N-Ethylaniline [UN 2272] **(103-69-5)**
2-(N-Ethylanilino)ethanol **(92-50-2)**
N-Ethylanilinoethanol **(92-50-2)**
β-(Ethylanilino)ethyl alcohol **(92-50-2)**
Ethyl anthranilate **(87-25-2)**
2-Ethyl-9,10-anthraquinone **(84-51-5)**
2-Ethylanthraquinone **(84-51-5)**
Ethyl arachidonate **(1808-26-0)**
Ethyl auramine nitrate **(43130-12-7)**
Ethyl azelaaldehydate **(3433-16-7)**
2-Ethylaziridine **(2549-67-9)**
Ethyl azobenzene-2-carboxylate **(18277-91-3)**
Ethylbarbital **(57-44-3)**
12-Ethylbenz(a)anthracene **(18868-66-1)**
7-Ethylbenz(a)anthracene **(3697-30-1)**
4-Ethylbenzal chloride **(54789-29-6)**
4-Ethylbenzaldehyde **(4748-78-1)**
Ethylbenzaldehyde **(53951-50-1)**
10-Ethyl-1,2-benzanthracene **(3697-30-1)**
5-Ethyl-1,2-benzanthracene **(56961-62-7)**
Ethylbenzeen (Dutch) **(100-41-4)**
2-Ethylbenzenamine **(578-54-1)**
N-Ethylbenzenamine **(103-69-5)**
N-Ethylbenzenamino **(103-69-5)**
Ethyl benzene (ACGIH,OSHA) [UN 1175] **(100-41-4)**
Ethyl benzeneacetate **(101-97-3)**
4-Ethyl-1,3-benzenediol **(2896-60-8)**
Ethylbenzene hydroperoxide **(3071-32-7)**
α-Ethylbenzenemethanol **(93-54-9)**
Ethyl benzenepropanoate **(2021-28-5)**
4-Ethylbenzenesulfonic acid **(98-69-1)**
Ethylbenzenesulfonic acid **(57352-34-8)**
p-Ethylbenzenesulfonic acid **(98-69-1)**
p-Ethylbenzenesulfonic acid, sodium salt **(14995-38-1)**

2-Ethylbenzenesulfonyl fluoride **(34586-49-7)**
4-Ethylbenzenesulfonyl fluoride **(455-20-9)**
4-Ethylbenzoate **(619-64-7)**
Ethyl benzoate **(93-89-0)**
4-Ethylbenzoic acid **(619-64-7)**
Ethylbenzoic acid **(28134-31-8)**
Ethylbenzol **(100-41-4)**
Ethyl benzoyl acetate **(94-02-0)**
Ethyl N-benzoyl-N-(3,4-dichlorophenyl)-2-aminopropionate **(22212-55-1)**
α-Ethylbenzyl alcohol **(93-54-9)**
Ethylbenzylaniline **(92-59-1)**
N-Ethyl-N-benzylaniline [UN 2274] **(92-59-1)**
Ethyl 1-(2-benzyloxyethyl)-4-phenylpiperidine-4-carboxylate **(3691-78-9)**
Ethyl-1,1'-biphenyl **(40529-66-6)**
Ethylbiphenyl **(40529-66-6)**
Ethylbis(2-chloroethyl)amine **(538-07-8)**
Ethylbis(β-chloroethyl)amine **(538-07-8)**
Ethylbis(2-hydroxyethyl)amine **(139-87-7)**
S-Ethyl bis(2-methylpropyl)carbamothioate **(2008-41-5)**
O-Ethyl-O,O-bis-(p-nitrofenyl)thiofosfat (Czech) **(7508-73-8)**
Ethyl borate **(34099-73-5)**
Ethyl borate **(51845-86-4)**
Ethyl bromacetate **(105-36-2)**
Ethyl bromide (ACGIH,OSHA) [UN 1891] **(74-96-4)**
Ethyl α-bromoacetate **(105-36-2)**
Ethyl bromoacetate [UN 1603] **(105-36-2)**
Ethyl 4-bromobenzoate **(5798-75-4)**
Ethyl α-bromoisobutyrate **(600-00-0)**
Ethyl 2-bromo-2-methylpropanoate **(600-00-0)**
Ethyl bromophos **(4824-78-6)**
2-Ethylbutanal **(97-96-1)**
Ethylbutanedinitrile **(17611-82-4)**
Ethyl butanoate **(105-54-4)**
2-Ethyl butanoic acid **(88-09-5)**
2-Ethyl-1-butanol **(97-95-0)**
2-Ethylbutanol-1 **(97-95-0)**
2-Ethylbutanol [UN 2275] **(97-95-0)**
2-Ethyl-1-butanol, silicate **(78-13-7)**
2-Ethyl-1-butene **(760-21-4)**
3-Ethylbutinol **(77-75-8)**
3-(2-Ethylbutoxy) propionic acid **(10213-74-8)**
3-(2-Ethylbutoxy)propionic acid **(10213-74-8)**
Ethylbutylacetaldehyde **(123-05-7)**
Ethylbutyl acetate **(40780-64-1)**
Ethylbutyl acetate [UN 1177] **(123-66-0)**
2-Ethylbutylacrylate **(3953-10-4)**
2-Ethylbutyl alcohol **(97-95-0)**
2-Ethylbutylamine **(617-79-8)**
N-Ethylbutylamine **(13360-63-9)**
5-Ethyl-5-n-butylbarbituric acid **(77-28-1)**
Ethyl butyl carbonate **(30714-78-4)**
Ethylbutylcetone (French) **(106-35-4)**
2-Ethylbutylester kyseliny akrylove (Czech) **(3953-10-4)**
Ethyl butyl ether [UN 1179] **(628-81-9)**
Ethylbutylketon (Dutch) **(106-35-4)**
Ethyl butyl ketone (ACGIH,OSHA) **(106-35-4)**
Ethyl-n-butylnitrosamine **(4549-44-4)**
Ethyl-tert-butylnitrosamine **(3398-69-4)**
Ethyl-t-butylnitrosoamine **(3398-69-4)**
4-Ethyl-2-tert-butylphenol **(96-70-8)**
2-Ethyl-2-butyl-1,3-propanediol **(115-84-4)**
2-Ethyl-2-butylpropanediol-1,3 **(115-84-4)**
3-Ethylbutynol **(77-75-8)**
α-Ethylbutyraldehyde **(97-96-1)**
Ethyl butyraldehyde (DOT) **(97-96-1)**
2-Ethylbutyraldehyde [UN 1178] **(97-96-1)**
Ethyl n-butyrate **(105-54-4)**
Ethyl butyrate [UN 1180] **(105-54-4)**
2-Ethylbutyric acid **(88-09-5)**
α-Ethylbutyric acid **(88-09-5)**
2-Ethylbutyric acid, diester with triethylene glycol **(95-08-9)**
2-Ethylbutyric acid, triethylene glycol diester **(95-08-9)**
2-Ethylbutyric aldehyde **(97-96-1)**
γ-Ethyl-n-butyrolactone **(695-06-7)**
γ-Ethylbutyrolactone **(695-06-7)**
Ethyl caprate **(110-38-3)**
Ethyl caprinate **(110-38-3)**
α-Ethylcaproaldehyde **(123-05-7)**
Ethyl caproate **(123-66-0)**
α-Ethylcaproic acid **(149-57-5)**
2-Ethylcaproyl chloride **(760-67-8)**
Ethyl caprylate **(106-32-1)**
Ethyl carbamate **(51-79-6)**
Ethylcarbamic acid, ethyl ester **(623-78-9)**
Ethylcarbamothioic acid, O-(1-methylethyl) ester **(141-98-0)**
D-(-)-1-(Ethylcarbamoyl)ethyl phenylcarbamate **(16118-49-3)**
S-(N-Ethylcarbamoylmethyl) dimethyl phosphorodithioate **(116-01-8)**
Ethyl carbanilate **(101-99-5)**
9-Ethyl-9H-carbazole **(86-28-2)**
9-Ethylcarbazole **(86-28-2)**
N-Ethylcarbazole **(86-28-2)**
Ethyl carbethoxymethyl phthalate **(84-72-0)**
Ethyl carbinol **(71-23-8)**
Ethyl carbitol **(111-90-0)**
Ethyl δ-carboethoxyvalerate **(141-28-6)**
Ethyl carbonate **(105-58-8)**
S-Ethyl carbonochloridothioate **(2941-64-2)**
O-Ethyl carbonodithioate sodium salt **(140-90-9)**
Ethylcarbylamine **(624-79-3)**
3-Ethylcatechol **(933-99-3)**
Ethyl cellosolve **(110-80-5)**
Ethyl cellosolve acetaat (Dutch) **(111-15-9)**
Ethylcellulose **(9004-57-3)**
Ethyl centralite **(85-98-3)**
Ethylchloorformiaat (Dutch) **(541-41-3)**
Ethyl chloracetate **(105-39-5)**
Ethyl chloride (ACGIH,OSHA) [UN 1037] **(75-00-3)**
Ethyl α-chloroacetate **(105-39-5)**
Ethyl chloroacetate [UN 1181] **(105-39-5)**
Ethyl 2-chloroacetoacetate **(609-15-4)**
Ethyl 2-(4-((6-chloro-2-benzothiazolyl)oxy)phenoxy)propanoate **(66441-11-0)**
(+-)-Ethyl 2-(4-((6-chloro-2-benzoxazolyl)oxy)phenoxy)propanoate **(66441-23-4)**
Ethyl-2-((4-(6-chloro-2-benzoxazolyloxy))-phenoxy)propionate **(66441-23-4)**
Ethyl chlorocarbonate (DOT) **(541-41-3)**
Ethyl 4-chloro-α-(4-chlorophenyl)-α-hydroxybenzeneacetate **(510-15-6)**
Ethyl chloroethanoate **(105-39-5)**
9-(3-(Ethyl(2-chloroethyl)amino)propylamino)-6-chloro-2-methoxyacridine dihydrochloride **(146-59-8)**
Ethyl(chloroethyl)aniline **(92-49-9)**
Ethyl β-chloroethyl ether **(628-34-2)**
Ethyl 2-chloroethyl sulfide **(693-07-2)**
Ethyl β-chloroethyl sulfide **(693-07-2)**
Ethyl 7-chloro-5-(o-fluorophenyl)-2,3-dihydro-2-oxo-1H-1,4-benzodiazepine-3-carboxylate **(29177-84-2)**
Ethyl chloroformate [UN 1182] **(541-41-3)**
Ethyl chloromethyl ether **(3188-13-4)**
Ethyl 4-(4-chloro-2-methylphenoxy)butylate **(10443-70-6)**
Ethyl 4-(4-chloro-2-methylphenoxy)butyrate **(10443-70-6)**
Ethyl chlorooxoacetate **(4755-77-5)**
Ethyl 2-chloro-3-oxobutanoate **(609-15-4)**
Ethyl 2-(p-chlorophenoxy)isobutyrate **(637-07-0)**
Ethyl α-(4-chlorophenoxy)isobutyrate **(637-07-0)**
Ethyl chlorophenoxyisobutyrate **(637-07-0)**

Ethyl para-chlorophenoxyisobutyrate	**(637-07-0)**
Ethyl-α-p-chlorophenoxy-isobutyrate	**(637-07-0)**
Ethyl 2-(4-chlorophenoxy)-2-methylpropionate	**(637-07-0)**
Ethyl 2-(p-chlorophenoxy)-2-methylpropionate	**(637-07-0)**
Ethyl α-(4-chlorophenoxy)-α-methylpropionate	**(637-07-0)**
Ethyl α-(p-chlorophenoxy)-α-methylpropionate	**(637-07-0)**
O-Ethyl S-4-chlorophenyl ethylphosphonodithioate	**(2984-64-7)**
Ethyl-2-chloropropionate [UN 2935]	**(535-13-7)**
Ethyl 2-(4-(6-chloro-2-quinoxalinyloxy)phenoxy)propanoate	**(76578-14-8)**
Ethyl chlorothioformate	**(2941-64-2)**
S-Ethyl chlorothioformate	**(2941-64-2)**
Ethyl chlorothioformate [UN 2826]	**(2812-73-9)**
S-Ethyl chlorothiolformate	**(2941-64-2)**
Ethyl chlorothiolformate [UN 2826]	**(2941-64-2)**
Ethyl chlorothioloformate	**(2941-64-2)**
Ethyl β-chlorovinyl ethynyl carbinol	**(113-18-8)**
24-α-Ethylcholesterol	**(83-46-5)**
Ethyl chrysanthemate	**(97-41-6)**
Ethyl chrysanthemumate	**(97-41-6)**
Ethyl trans-cinnamate	**(103-36-6)**
Ethylcinnamate	**(103-36-6)**
Ethyl citrate	**(77-93-0)**
Ethyl clofibrate	**(637-07-0)**
Ethyl (E)-crotonate	**(623-70-1)**
Ethyl crotonate	**(10544-63-5)**
Ethyl trans-crotonate	**(623-70-1)**
Ethylcrotonate	**(623-70-1)**
Ethyl crotonate [UN 1862]	**(623-70-1)**
Ethyl cyanide	**(107-12-0)**
Ethyl cyanoacetate [UN 2666]	**(105-56-6)**
Ethyl 2-cyanoacrylate	**(7085-85-0)**
Ethyl cyanoacrylate	**(7085-85-0)**
Ethyl cyanoethanoate	**(105-56-6)**
N-Ethyl-N-(2-cyanoethyl)-5-acetamido-4-((6-bromo-2,4-dinitrophenyl)azo)-2-methoxyaniline	**(22578-86-5)**
N-Ethyl-N-(2-cyanoethyl)aniline	**(148-87-8)**
Ethyl cyanomethylcarbamate	**(60754-24-7)**
O-Ethyl O-4-cyanophenyl phenylphosphorothioate	**(13067-93-1)**
Ethyl 2-cyano-2-propenoate	**(7085-85-0)**
Ethylcyclobutane	**(4806-61-5)**
5-Ethyl-5-(1'-cycloheptenyl)-barbituric acid	**(509-86-4)**
5-Ethyl-5-cycloheptenylbarbituric acid	**(509-86-4)**
Ethyl cyclohexane	**(1678-91-7)**
Ethylcyclohexane	**(1678-91-7)**
1-Ethylcyclohexanol	**(1940-18-7)**
5-Ethyl-5-cyclohexenylbarbituric acid	**(52-31-3)**
N-Ethyl(cyclohexyl)amine	**(5459-93-8)**
S-Ethyl cyclohexylethylthiocarbamate	**(1134-23-2)**
Ethylcyclooctane	**(13152-02-8)**
Ethylcyclopentane	**(1640-89-7)**
Ethyl-d5 alcohol-d (8CI)	**(1516-08-1)**
Ethyl decanoate	**(110-38-3)**
Ethyl decylate	**(110-38-3)**
(1-Ethyldecyl)benzene	**(2400-00-2)**
4-(1-Ethyldecyl)benzenesulfonic acid	**(18777-54-3)**
1'-Ethyl-1,2,5,6-dibenzacridine (French)	**(63021-33-0)**
1'-Ethyl-3,4,5,6-dibenzacridine (French)	**(63021-35-2)**
1-Ethyl-dibenz(a,j)acridine	**(63021-35-2)**
Ethyl dibromobenzene	**(30812-87-4)**
Ethyldichlorarsine	**(598-14-1)**
Ethyl 2,2-dichloroacetate	**(535-15-9)**
Ethyl dichloroacetate	**(535-15-9)**
Ethyldichloroaluminum	**(563-43-9)**
Ethyldichloroarsine [UN 1892]	**(598-14-1)**
Ethyl 4,4'-dichlorobenzilate	**(510-15-6)**
Ethyl p,p'-dichlorobenzilate	**(510-15-6)**
Ethyl 4,4'-dichlorodiphenyl glycollate	**(510-15-6)**
Ethyl dichloroethanoate	**(535-15-9)**
Ethyl (2,4-dichlorophenoxy)acetate	**(533-23-3)**
Ethyl 4,4'-dichlorophenyl glycollate	**(510-15-6)**
Ethyl 3-(3,5-dichlorophenyl)-5-methyl-2,4-dioxo-5-oxazolidine carboxylate	**(72391-46-9)**
O-Ethyl-O-(2,4-dichlorophenyl)-S-n-propyl-dithiophosphate	**(34643-46-4)**
Ethyl dichlorosilane	**(1789-58-8)**
Ethyldichlorosilane [UN 1183]	**(1789-58-8)**
Ethyldiethanolamine	**(139-87-7)**
N-Ethyldiethanolamine	**(139-87-7)**
Ethyl 2-((diethoxyphosphinothioyl)oxy)-5-methyl-pyrazolo(1,5-a)pyrimidine-6-carboxylate	**(13457-18-6)**
Ethyl ((diethoxyphosphinothioyl)thio)acetate	**(919-54-0)**
Ethyl (diethoxyphosphinyl)acetate	**(867-13-0)**
Ethyl diethoxyphosphoryl acetate	**(867-13-0)**
S-Ethyl diethylcarbamothioate	**(2941-55-1)**
Ethyl diethylene glycol	**(111-90-0)**
Ethyl diethylphosphinate	**(4775-09-1)**
Ethyl (diethylphosphono)acetate	**(867-13-0)**
S-Ethyl diethylthiocarbamate	**(2941-55-1)**
Ethyl difluoroacetate	**(454-31-9)**
5-Ethyl-1,3-diglycidyl-5-methylhydantoin	**(15336-82-0)**
Ethyl diglyme	**(112-36-7)**
Ethyl-9,10-dihydro-9,10-anthracenediol	**(67923-88-0)**
1-Ethyl-1,2-dihydro-6-hydroxy-4-methyl-2-oxo-3-pyridinecarbonitrile	**(28141-13-1)**
1-Ethyl-1,4-dihydro-6,7-methylenedioxy-4-oxo-3-quinolinecarboxylic acid	**(14698-29-4)**
(3S-cis)-3-Ethyldihydro-4-((1-methyl-1H-imidazol-5-yl)methyl)-2(3H)-furanone	**(92-13-7)**
1-Ethyl-1,4-dihydro-7-methyl-4-oxo-1,8-naphthyridine-3-carboxilic acid	**(389-08-2)**
2-Ethyl-4,5-dihydrooxazole	**(10431-98-8)**
1-Ethyl-1,4-dihydro-4-oxo(1,3)dioxolo(4,5-g)cinnoline-3-carboxylic acid	**(28657-80-9)**
5-Ethyl-5,8-dihydro-8-oxo-1,3-dioxolo(4,5-g)quinoline-7-carboxylic acid	**(14698-29-4)**
N(sup1)-(1-Ethyl-1,2-dihydro-2-oxo-4-pyrimidinyl)-sulfanilamide	**(17784-12-2)**
N1-(1-Ethyl-1,2-dihydro-2-oxo-4-pyrimidinyl)sulfanil-amide	**(17784-12-2)**
5-Ethyldihydro-5-phenyl-4,6(1H,5H)-pyrimidinedione	**(125-33-7)**
Ethyl dihydroxypropyl PABA	**(58882-17-0)**
Ethyl dihydroxypropyl p-aminobenzoate	**(58882-17-0)**
2-Ethyl-3-(3',5'-diiodo-4'-hydroxybenzoyl)-cumarone	**(68-90-6)**
Ethyl N,N-diisobutylthiocarbamate	**(2008-41-5)**
S-Ethyl N,N-diisobutylthiocarbamate	**(2008-41-5)**
S-Ethyldiisobutyl thiocarbamate	**(2008-41-5)**
Ethyl-N,N-diisobutyl thiolcarbamate	**(2008-41-5)**
O-Ethyl-S-2-diisopropylaminoethylester kyseliny methylthiofosfonove (Czech)	**(50782-69-9)**
Ethyl S-2-diisopropylaminoethyl methylphosphonothiolate	**(50782-69-9)**
O-Ethyl S-2-diisopropylaminoethyl methylphosphonothiote	**(50782-69-9)**
Ethyl-S-diisopropylaminoethyl methylthiophosphonate	**(50782-69-9)**
Ethyl α-((dimethoxyphosphenothioyl)thio)benzene-acetate	**(2597-03-7)**
Ethyl (dimethylamidino)methylcarbamate, hydrochloride	**(65206-90-8)**
Ethyl dimethylamidocyanophosphate	**(77-81-6)**
N-Ethyldimethylamine	**(598-56-1)**
Ethyl 4-(dimethylamino)benzoate	**(10287-53-3)**
Ethyl N,N-dimethylamino cyanophosphate	**(77-81-6)**
Ethyl S-dimethylaminoethyl methylphosphono-	

thiolate (50782-69-9)
S-Ethyl N-(3-dimethylaminopropyl)thiol carbamate hydrochloride (19622-19-6)
1-Ethyl-2,3-dimethylbenzene (933-98-2)
1-Ethyl-2,4-dimethylbenzene (874-41-9)
1-Ethyl-3,5-dimethylbenzene (934-74-7)
2-Ethyl-1,3-dimethylbenzene (2870-04-4)
2-Ethyl-1,4-dimethylbenzene (1758-88-9)
4-Ethyl-1,2-dimethylbenzene (934-80-5)
N-Ethyl-1,3-dimethylbutylamine (42966-64-3)
Ethyl N,N-dimethyl carbamate (687-48-9)
Ethyl dimethyl carbinol (75-85-4)
N-Ethyl-N,N-dimethyl-1-hexadecanaminium bromide (124-03-8)
Ethyldimethylmethane (78-78-4)
5-Ethyl-3,5-dimethyloxazolidine-2,4-dione (115-67-3)
Ethyl N,N-dimethylphosphoramidocyanidate (77-81-6)
Ethyl dimethylphosphoramidocyanidate (77-81-6)
Ethyl O,O-dimethyl phosphorodithioylphenyl acetate (2597-03-7)
Ethyl 2,2-dimethylpropanoate (3938-95-2)
2-Ethyl-3,5-dimethyl pyrazine (13925-07-0)
3-Ethyl-2,5-dimethylpyrazine (13360-65-1)
4-Ethyl-2,6-dimethylpyridine (36917-36-9)
Ethyldiol metacrylate (97-90-5)
Ethyl diphenylarsinite (24582-55-6)
N-Ethyl-3,3'-diphenyldipropylamine (150-59-4)
O-Ethyl S,S-diphenyl dithiophosphate (17109-49-8)
Ethyl diphenylethane (64800-83-5)
Ethyldiphenylmethane (42504-54-1)
O-Ethyl-S,S-diphenyl phosphorodithioate (17109-49-8)
O-Ethyl S,S-dipropylphosphorodithioate (13194-48-4)
S-Ethyl N,N-di-n-propylthiocarbamate (759-94-4)
S-Ethyl-N,N-dipropylthiocarbamate (759-94-4)
Ethyl N,N-di-n-propylthiolcarbamate (759-94-4)
Ethyl di-n-propylthiolcarbamate (759-94-4)
Ethyl-N,N-dipropylthiolcarbamate (759-94-4)
o-Ethyl dithiocarbamate (151-01-9)
(o-Ethyl dithiocarbonato)potassium (140-89-6)
Ethyldithiourame (97-77-8)
Ethyldithiurame (97-77-8)
Ethyl dodecanoate (106-33-2)
Ethyl dodecylate (106-33-2)
Ethyle (acetate d') (French) (141-78-6)
Ethyle, chloroformiat d' (French) (541-41-3)
Ethyleen-chloorhydrine (Dutch) (107-07-3)
Ethyleendiamine (Dutch) (107-15-3)
Ethyleendichloride (Dutch) (107-06-2)
Ethyleenimine (Dutch) (151-56-4)
Ethyleenoxide (Dutch) (75-21-8)
Ethyle (formiate d') (French) (109-94-4)
Ethyl enanthate (106-30-9)
Ethylen-bis-dithiokarbaman sodny (Czech) (142-59-6)
Ethylendiamine (107-15-3)
Ethylene (74-85-1)
Ethylene Homopolymer (9002-88-4)
Ethylene Polymer (9002-88-4)
Ethylene Polymers (8CI) (9002-88-4)
Ethylene [UN 1962] (74-85-1)
Ethylene/Vinyl Acetate Copolymer (24937-78-8)
Ethylene acetate (111-55-7)
Ethyleneacetic acid (126-39-6)
Ethylene alcohol (107-21-1)
Ethylene aldehyde (107-02-8)
1,1'-Ethylene-2,2'-bipyridylium dibromide (85-00-7)
1,1'-Ethylene-2,2'-bipyridylium ion (2764-72-9)
Ethylene, 1,1-bis(p-chlorophenyl)-2-chloro (1022-22-6)
Ethylenebis(5,6-dibromonorbornane-2,3-dicarboximide) (52907-07-0)
N,N'-Ethylene bis(dithiocarbamate de sodium) (French
Ethylenebis(dithiocarbamate), disodium salt (142-59-6)
Ethylenebisdithiocarbamate manganese (12427-38-2)
N,N'-Ethylene bis(dithiocarbamate manganeux) (French) (12427-38-2)
Ethylenebis(dithiocarbamato), manganese (12427-38-2)
Ethylenebis(dithiocarbamato)zinc (12122-67-7)
(Ethylenebis(dithiocarbamato))zinc (8CI) (12122-67-7)
Ethylenebis(dithiocarbamic acid) (111-54-6)
Ethylenebisdithiocarbamic acid (111-54-6)
Ethylenebis(dithiocarbamic acid) disodium salt (142-59-6)
Ethylenebis(dithiocarbamic acid), manganese salt (12427-38-2)
Ethylenebis(dithiocarbamic acid) manganous salt (12427-38-2)
Ethylenebisdithiocarbamic acid, salts & esters (111-54-6)
Ethylenebis(dithiocarbamic acid), zinc salt (12122-67-7)
N,N'-Ethylenebis(3-fluorosalicylideneiminato)cobalt (II) (62207-76-5)
Ethylenebis(iminodiacetic acid) disodium salt (139-33-3)
Ethylenebis(iminodiacetic acid) tetrasodium salt (64-02-8)
Ethylene bis(meraptoacetate) (123-81-9)
Ethylene bis(mercaptoacetate) (123-81-9)
trans-2,2'-Ethylenebis(5-methylbenzoxazole) (17233-65-7)
Ethylene bis(nitrilodimethylene)tetraphosphonic acid, hexasodium salt (15142-96-8)
Ethylene bis(nitrilodimethylene)tetraphosphonic acid, tetrapotassium salt (68188-96-5)
Ethylene bis(oleamide) (110-31-6)
N,N'-Ethylenebisoleamide (110-31-6)
Ethylene, 1,2-bis(propylsulfonyl)-, (E)- (8CI) (1113-14-0)
(N,N'-Ethylenebis(salicylaldehyde iminato))cobalt(II) (14167-18-1)
N,N'-Ethylenebis(salicylideneiminato)cobalt (II) (14167-18-1)
N,N-Ethylenebis(salicylideneiminato)cobalt(II) (14167-18-1)
N,N-Ethylenebis(salicylidene iminato)cobalt II (14167-18-1)
(N,N'-Ethylenebis(salicylideniminato))cobalt (14167-18-1)
N,N'-Ethylenebis(salicylideniminato)cobalt(II) (14167-18-1)
Ethylenebis(stearamide) (110-30-5)
Ethylenebisstearamide (110-30-5)
N,N'-Ethylene bisstearamide (110-30-5)
N,N'-Ethylenebis(stearamide) (110-30-5)
N,N'-Ethylenebisstearamide (110-30-5)
Ethylenebisstearoamide (110-30-5)
Ethylenebis(stearylamide) (110-30-5)
N,N'-Ethylenebis(1,2,3,6-tetrahydro-3,6-endo-methylenephthalimide) (25502-52-7)
Ethylene bis(thioglycolate) (123-81-9)
Ethylenebis(thioglycolate) (123-81-9)
Ethylene bisthiuram monosulfide (33813-20-6)
Ethylenebis(tris(2-cyanoethyl)phosphonium bromide) (10310-38-0)
1,1'-Ethylenebisurea (1852-14-8)
Ethylene bromide (106-93-4)
Ethylene, bromo (593-60-2)
Ethylenebromohydrin (540-51-2)
Ethylene, bromo-, Polymer (25951-54-6)
Ethylene, bromotrifluoro (598-73-2)
Ethylene carbonate (96-49-1)
Ethylene carbonic acid (96-49-1)
Ethylenecarboxamide (79-06-1)
Ethylenecarboxylic acid (79-10-7)
Ethylene chlorhydrin (107-07-3)
Ethylene chloride (107-06-2)
Ethylene, chloro (75-01-4)
Ethylene chlorobromide (107-04-0)
Ethylene, 2-chloro-1,1-difluoro (359-10-4)
Ethylene chlorohydrin (ACGIH,OSHA) [UN 1135] (107-07-3)
Ethylene, 1-(o-chlorophenyl)-1-(p-chlorophenyl)-2,2-dichloro- (3424-82-6)
Ethylene, chlorotrifluoro (79-38-9)

Ethylene, chlorotris(p-methoxyphenyl) **(569-57-3)**
Ethylene, compressed [UN 1962] **(74-85-1)**
Ethylene cyanide **(110-61-2)**
Ethylene cyanohydrin **(109-78-4)**
Ethylene cyanohydrin **(109-78-4)**
1,2-Ethylenediamine **(107-15-3)**
Ethylenediamine (ACGIH,OSHA) [UN 1604] **(107-15-3)**
Ethylenediamine, Compd. with theophylline (1:2) **(317-34-0)**
Ethylene-diamine (French) **(107-15-3)**
Ethylenediamineacetic acid trisodium salt **(150-38-9)**
Ethylenediamine, N-(2-aminoethyl)- **(111-40-0)**
Ethylenediamine, N-benzyl-N',N'-dimethyl-N-phenyl **(961-71-7)**
Ethylenediamine, N-benzyl-N',N'-dimethyl-N-phenyl-, hydrochloride **(2045-52-5)**
Ethylenediamine, N-benzyl-N',N'-dimethyl-N-(2-pyridyl)- **(91-81-6)**
Ethylenediamine, N-benzyl-N',N'-dimethyl-N-(2-pyridyl)-, hydrochloride **(154-69-8)**
Ethylenediamine, N,N'-bis(2-aminoethyl)- **(112-24-3)**
Ethylenediamine, N,N'-bis(1,3-dimethylbutylidene)- **(25707-70-4)**
Ethylenediamine bisstearamide **(110-30-5)**
Ethylenediamine carbamate **(109-58-0)**
Ethylenediamine, N-(5-chloro-2-thenyl)-N',N'-dimethyl-N-2-pyridyl- **(148-65-2)**
Ethylenediamine, N-cyclohexyl- (8CI) **(5700-53-8)**
N,N'-Ethylenediaminediacetic acid tetrasodium salt **(64-02-8)**
Ethylenediamine, N,N'-dibenzyl-, Compd. with penicillin G (1:2) **(1538-09-6)**
Ethylenediamine, N,N'-dibenzyl-, diacetate **(122-75-8)**
Ethylenediamine, N,N-diethyl- (8CI) **(100-36-7)**
Ethylenediamine, dihydriodide (8CI) **(5700-49-2)**
Ethylenediamine, dihydrochloride **(333-18-6)**
Ethylenediamine dihydroiodide **(5700-49-2)**
Ethylenediamine, N,N-dimethyl-N'-(2-pyridyl)-N'-(2-thenyl)-, hydrochloride **(135-23-9)**
Ethylenediamine dinitrate **(20829-66-7)**
Ethylenediamine di(tetrabromophthalic acid) **(66046-78-4)**
Ethylenediamine, N-ethyl-N-m-tolyl- (8CI) **(19248-13-6)**
Ethylenediamine hydrochloride **(333-18-6)**
Ethylenediamine, hydrochloride (8CI) (VAN) **(15467-15-9)**
Ethylenediamine, N-(p-methoxybenzyl)-N',N'-dimethyl-N-2-pyrimidinyl- **(91-85-0)**
Ethylenediamine, N-(1-naphthyl)-, dihydrochloride **(1465-25-4)**
Ethylenediamine steardiamide **(110-30-5)**
Ethylenediaminetetraacetate **(60-00-4)**
Ethylenediaminetetraacetate, disodium salt **(139-33-3)**
Ethylenediamine-N,N,N',N'-tetraacetic acid **(60-00-4)**
Ethylenediaminetetraacetic acid **(60-00-4)**
Ethylenediaminetetraacetic acid, calcium disodium chelate **(62-33-9)**
Ethylenediaminetetraacetic acid, disodium manganese salt **(15375-84-5)**
Ethylenediaminetetraacetic acid, disodium salt **(139-33-3)**
Ethylenediaminetetraacetic acid, sodium salt **(17421-79-3)**
Ethylenediaminetetraacetic acid, tetrasodium salt **(64-02-8)**
Ethylenediaminetetraacetic acid, trisodium salt **(150-38-9)**
Ethylenediaminetetracetic acid, ferric ammonium salt **(21265-50-9)**
Ethylenediamine tetraethanol **(140-07-8)**
Ethylenediamine, N,N,N',N'-tetrakis(2-hydroxyethyl)- **(140-07-8)**
Ethylenediamine, N,N,N',N'-tetramethyl **(110-18-9)**
Ethylenediaminetetra(methylenephosphonic)acid **(1429-50-1)**
N,N,N',N' Ethylenediamine tetra(methylenephosphonic acid) **(1429-50-1)**
Ethylenediamine, N-(3-(trimethoxysilyl)propyl) **(1760-24-3)**
1,2-Ethylene dibromide **(106-93-4)**
Ethylene dibromide (ACGIH,OSHA) [UN 1605] **(106-93-4)**
Ethylene, 1,2-dibromo **(540-49-8)**
1,2-Ethylenedicarboxylic acid, (E) **(110-17-8)**
cis-1,2-Ethylenedicarboxylic acid **(110-16-7)**
trans-1,2-Ethylenedicarboxylic acid **(110-17-8)**
1,2-Ethylenedicarboxylic acid, (Z) **(110-16-7)**
1,2-Ethylene dichloride **(107-06-2)**
Ethylene dichloride (ACGIH,OSHA) [UN 1184] **(107-06-2)**
Ethylene, 1,2-dichloro **(540-59-0)**
Ethylene, 1,2-dichloro-, (E) **(156-60-5)**
Ethylene, dichloro **(25323-30-2)**
Ethylene, 1,1-dichloro- (8CI) **(75-35-4)**
Ethylene, 1,1-dichloro-, Polymer with chloroethylene **(9011-06-7)**
Ethylene, 1,2-dichloro-, (Z) **(156-59-2)**
Ethylene, 1,1-dichloro-2,2-bis(p-chlorophenyl) **(72-55-9)**
Ethylene, 1,1-dichloro-2-(o-chlorophenyl)-2-(p-chlorophenyl) **(3424-82-6)**
Ethylene, 1,1-dichloro-2,2-difluoro **(79-35-6)**
Ethylene, dichlorodifluoro- **(27156-03-2)**
Ethylene, 1,2-dichloro-1,2-difluoro- (8CI) **(598-88-9)**
Ethylene dicyanide **(110-61-2)**
Ethylene, 1,1-difluoro **(75-38-7)**
Ethylene, 1,2-difluoro- (8CI) **(1691-13-0)**
Ethylene, 1,2-difluoro-, (E)- (8CI) **(1630-78-0)**
Ethylene, difluoro- (8CI) **(1320-41-8)**
Ethylene, 1,2-difluoro-, (Z)- (8CI) **(1630-77-9)**
Ethylene diglycidyl ether **(2224-15-9)**
Ethylene diglycol **(111-46-6)**
Ethylene diglycol monoethyl ether **(111-90-0)**
Ethylene diglycol monomethyl ether **(111-77-3)**
Ethylene dihydrate **(107-21-1)**
(+)-2,2'-(Ethylenediimino)di-1-butanol **(74-55-5)**
2,2',2'',2'''-(Ethylenediimino)tetraethanol **(140-07-8)**
Ethylene, 1,2-diiodo-, (E)- (8CI) **(590-27-2)**
Ethylene, 1,2-diiodo- (6CI,7CI,8CI) **(20244-70-6)**
Ethylene, 1,2-diiodo-, (Z)- (8CI) **(590-26-1)**
Ethylene dimercaptan **(540-63-6)**
α-Ethylene dimercaptan **(540-63-6)**
Ethylene dimethyl ether **(110-71-4)**
Ethylene dinitrate **(628-96-6)**
((Ethylenedinitrilo)tetraacetato)cuprate(2-) disodium **(14025-15-1)**
(Ethylenedinitrilo)tetraacetato zincate(2-), diammonium salt **(67859-51-2)**
Ethylenedinitrilotetraacetic acid **(60-00-4)**
(Ethylenedinitrilo)-tetraacetic acid disodium salt **(139-33-3)**
(Ethylenedinitrilo)tetraacetic acid, disodium zinc salt **(14025-21-9)**
Ethylenedinitrilotetraethanol **(140-07-8)**
(Ethylenedinitrilo)-tetramethylenephosphonic acid **(1429-50-1)**
1,1',1'',1'''-(Ethylenedinitrilo)tetra-2-propanol **(102-60-3)**
Ethylene dioleamide **(110-31-6)**
N,N'-Ethylenedioleamide **(110-31-6)**
1,2-(Ethylenedioxy)benzene **(493-09-4)**
2,2'-Ethylenedioxybis(ethanol) **(112-27-6)**
2,2'-(Ethylenedioxy)diethanethiol **(14970-87-7)**
2,2'-Ethylenedioxydiethanol **(112-27-6)**
2,2'-(Ethylenedioxy)di(ethyl acetate) **(111-21-7)**
2,2'-(Ethylenedioxy)di(ethyl 2-ethylbutyrate) **(95-08-9)**
2,2'-Ethylenedioxyethanol **(112-27-6)**
16,17-Ethylenedioxyviolanthrone **(6424-76-6)**
Ethylene, 1,1-diphenyl- (8CI) **(530-48-3)**
Ethylene dipropionate **(123-73-9)**
1,1'-Ethylene-2,2'-dipyridinium dichloride **(4032-26-2)**
1,1'-Ethylene-2,2'-dipyridylium dibromide **(85-00-7)**
1,1-Ethylene 2,2-dipyridylium dibromide **(85-00-7)**
Ethylene dipyridylium dibromide **(85-00-7)**
Ethylene distearamide **(110-30-5)**
Ethylenedistearamide **(110-30-5)**
N,N'-Ethylenedistearamide **(110-30-5)**
N,N'-Ethylene distearylamide **(110-30-5)**
Ethylenedithioethanol **(5244-34-8)**
Ethylene dithioglycol **(540-63-6)**
Ethylenedithiol **(540-63-6)**
1,1'-Ethylenediurea **(1852-14-8)**
Ethylenediurea **(1852-14-8)**
2,2'-(1,2-Ethylenediyl)bis(5-aminobenzenesulfonic acid) **(81-11-8)**
1,2-Ethylenediylbis(carbamodithioato)manganese **(12427-38-2)**

Ethylene episulfide (420-12-2)
Ethylene episulphide (420-12-2)
Ethylene fluoride (75-37-6)
Ethylene, fluoro- (8CI) (75-02-5)
Ethylene-glycol (107-21-1)
Ethylene glycol (ACGIH,OSHA) (107-21-1)
Ethylene glycol acetate (111-55-7)
Ethylene glycol acetate (542-59-6)
Ethylene glycol, acrylate (818-61-1)
Ethylene glycol bis(chloromethyl)ether (13483-18-6)
Ethylene glycol bis(2,3-epoxy-2-methylpropyl) ether (3775-85-7)
Ethylene glycol-bis-(2-hydroxyethyl ether) (112-27-6)
Ethylene glycol bis(mercaptoacetate) (123-81-9)
Ethylene glycol bis(methacrylate) (97-90-5)
Ethylene glycol bis(thioglycolate) (123-81-9)
Ethylene glycol bis(thioglycolic ester) (123-81-9)
Ethylene glycol n-butyl ether (111-76-2)
Ethylene glycol butyl vinyl ether (4223-11-4)
Ethylene glycol, carbanilate (7CI) (709-93-3)
Ethylene glycol carbonate (96-49-1)
Ethylene glycol cetyl ether (9004-95-9)
Ethylene glycol, chlorohydrin (107-07-3)
Ethylene glycol, cyclic carbonate (96-49-1)
Ethylene glycol, cyclic ethaneboronate (10173-38-3)
Ethylene glycol, diacetate (111-55-7)
Ethylene glycol dibutyl ether (112-48-1)
Ethylene glycol di(2,3-epoxy-2-methylpropyl) ether (3775-85-7)
Ethylene glycol diethyl ether [UN 1153] (629-14-1)
Ethylene glycol diglycidyl ether (2224-15-9)
Ethylene glycol dihydroxydiethyl ether (112-27-6)
Ethylene glycol dimethacrylate (97-90-5)
Ethylene glycol dimethyl ether (110-71-4)
Ethylene glycol dinitrate (ACGIH,DOT,OSHA) (628-96-6)
Ethylene glycol diphenyl ether (104-66-5)
Ethylene glycol, dipropionate (8CI) (123-73-9)
Ethylene glycol distearate (627-83-8)
Ethylene glycol ethyl ether (110-80-5)
Ethylene glycol ethyl ether acetate (111-15-9)
Ethylene glycol formal (646-06-0)
Ethylene glycol n-hexyl ether (112-25-4)
Ethylene glycolide, (2,3-epoxy-2-methylpropyl)ether (3775-85-7)
Ethylene glycol isopropyl ether (109-59-1)
Ethylene glycol methacrylate (868-77-9)
Ethylene glycol methyl acetate (OSHA) (110-49-6)
Ethylene glycol methyl ether (109-86-4)
Ethylene glycol methyl ether acetate (110-49-6)
Ethylene glycol, monoacetate (542-59-6)
Ethylene glycol, monoacrylate (818-61-1)
Ethylene glycol monobenzyl ether (622-08-2)
Ethylene glycol mono-sec-butyl ether (7795-91-7)
Ethylene glycol, mono-tert-butyl ether (7580-85-0)
Ethylene glycol, monobutyl ether (111-76-2)
Ethylene glycol monobutyl ether [UN 2369] (111-76-2)
Ethylene glycol monobutyl ether acetate (112-07-2)
Ethylene glycol, monocarbanilate (8CI) (709-93-3)
Ethylene glycol monocetyl ether (9004-95-9)
Ethylene glycol monocyclododecyl ether (32399-56-7)
Ethylene glycol mono(1-decyl-1-cyclohexyl) ether (70710-00-8)
Ethylene glycol mono(1,1-dibutylpentyl) ether (54661-98-2)
Ethylene glycol mono(di-n-hexylcarbonyl) ether (924-06-1)
Ethylene glycol mono(1,1-dipentylhexyl) ether (70709-98-7)
Ethylene glycol mono(1,1-dipropylbutyl) ether (70709-97-6)
Ethylene glycol monododecyl ether (4536-30-5)
Ethylene glycol monoethyl ether [UN 1171] (110-80-5)
Ethylene glycol monoethyl ether acetate [UN 1172] (111-15-9)
Ethylene glycol monoethyl ether acrylate (106-74-1)
Ethylene glycol monoethyl ether propenoate (106-74-1)
Ethylene glycol mono(1-hexyl-1-cyclohexyl) ether (70709-99-8)
Ethylene glycol monohexyl ether (112-25-4)
Ethylene glycol monoisobutyl ether (4439-24-1)
Ethylene glycol, monoisopropyl ether (109-59-1)
Ethylene glycol monolauryl ether (4536-30-5)
Ethylene glycol, monomethacrylate (868-77-9)
Ethylene glycol monomethyl ether [UN 1188] (109-86-4)
Ethylene glycol monomethyl ether acetate [UN 1189] (110-49-6)
Ethylene glycol monomethyl ether acetylricinoleate (140-05-6)
Ethylene glycol monomethyl ether acrylate (3121-61-7)
Ethylene glycol mono-8-pentadecyl ether (70709-96-5)
Ethylene glycol monophenyl ether (122-99-6)
Ethylene glycol mono propyl ether (2807-30-9)
Ethylene glycol mono-n-propyl ether (2807-30-9)
Ethylene glycol, monostearate (111-60-4)
Ethylene glycol, monothio- (60-24-2)
Ethylene glycol mono-7-tridecyl ether (924-06-1)
Ethylene glycol mono-6-undecyl ether (70709-95-4)
Ethylene glycol phenyl ether (122-99-6)
Ethylene glycol polyethylene-polypropylene glycol ether (1:2) (9003-11-6)
Ethylene glycol-propylene glycol copolymer (9003-11-6)
Ethylene glycol-propylene glycol polymer (9003-11-6)
Ethylene glycol stearate (111-60-4)
Ethylene glykol methyl vinyl ether (1663-35-0)
Ethylene hexachloride (67-72-1)
Ethyleneimine (ACGIH,OSHA) (151-56-4)
Ethyleneimine, Homopolymer (9002-98-6)
Ethylene imine, Inhibited [UN 1185] (151-56-4)
Ethylene iodohydrin (624-76-0)
Ethylene mercaptoacetate (123-81-9)
Ethylene methacrylate (97-90-5)
Ethylene-2-methylene-1,3-dioxepane copolymer (90588-08-2)
Ethylene monochloride (75-01-4)
1,8-Ethylenenaphthalene (83-32-9)
Ethylene nitrate (628-96-6)
Ethylene oxide (ACGIH,OSHA) [UN 1040] (75-21-8)
Ethylene oxide, Containing not more than 0.2% nitrogen [UN 1040] (75-21-8)
Ethylene oxide cyclic tetramer (294-93-9)
Ethylene oxide, ethyl- (106-88-7)
Ethylene oxide, lanolin adduct (61790-81-6)
Ethylene oxide-methanol adduct (9004-74-4)
Ethylene oxide, methyl- (75-56-9)
Ethylene oxide-propylene oxide copolymer (9003-11-6)
Ethylene oxide-propylene oxide copolymer ethylene glycol ether (9003-11-6)
Ethylene (oxyde d') (French) (75-21-8)
1-Ethyleneoxy-3,4-epoxycyclohexane (106-87-6)
Ethylene, phenyl- (100-42-5)
Ethylene o-phenylene dioxide (493-09-4)
1,4-Ethylenepiperazine (280-57-9)
Ethylene propionate (123-73-9)
Ethylene, refrigerated liquid [UN 1038] (74-85-1)
Ethylenester kyseliny uhlicite (Czech) (96-49-1)
Ethylenesuccinic acid (110-15-6)
Ethylene-sulfide (420-12-2)
Ethylene sulphide (420-12-2)
Ethylene terephthalate polymer (25038-59-9)
Ethylene, tetrabromo- (8CI) (79-28-7)
Ethylene tetrachloride (127-18-4)
Ethylene, tetrachloro (127-18-4)
Ethylene, tetrafluoro (116-14-3)
Ethylene, tetraphenyl- (8CI) (632-51-9)
Ethylenethiocarbamyl sulfide (33813-20-6)
1,3-Ethylene-2-thiourea (96-45-7)
Ethylene thiourea (96-45-7)
N,N'-Ethylenethiourea (96-45-7)
Ethylene thiuram monosulfide (33813-20-6)
Ethylene thiuram monosulphide (33813-20-6)

Ethylenethiuram sulfide **(33813-20-6)**
Ethylene, tribromo **(598-16-3)**
Ethylene trichloride **(79-01-6)**
Ethylene, trichloro **(79-01-6)**
Ethylene, trifluoro- **(359-11-5)**
Ethylene, trifluorochloro- **(79-38-9)**
Ethylene, trimethyl- **(513-35-9)**
Ethylene, triphenyl **(58-72-0)**
Ethylene urea **(120-93-4)**
Ethylenevinylacetate copolymer **(24937-78-8)**
Ethylene-vinyl acetate copolymer emulsion **(24937-78-8)**
Ethylene-vinylacetate resin **(24937-78-8)**
Ethylenglykoldiglycidylether (Czech) **(2224-15-9)**
Ethylenglykoldinitrat (Czech) **(628-96-6)**
Ethylenimine **(151-56-4)**
Ethylenimine, Polymers (8CI) **(9002-98-6)**
2,3,5-Ethylenimine-1,4-benzoquinone **(68-76-8)**
Ethyl α,β-epoxyhydrocinnamate **(121-39-1)**
Ethyl α,β-epoxy-β-methylhydrocinnamate **(77-83-8)**
Ethyl 2,3-epoxy-3-methyl-3-phenylpropionate **(77-83-8)**
Ethyl α,β-epoxy-α-phenylpropionate **(121-39-1)**
Ethylester-dimethylamid kyseliny kyanfosfonove (Czech) **(77-81-6)**
Ethylester kyseliny acetoctove (Czech) **(141-97-9)**
Ethylester kyseliny akrylove (Czech) **(140-88-5)**
Ethylester kyseliny p-aminobenzoove (Czech) **(94-09-7)**
Ethylester kyseliny benzoove (Czech) **(93-89-0)**
Ethylester kyseliny chlormravenci (Czech) **(541-41-3)**
Ethylester kyseliny chloroctove (Czech) **(105-39-5)**
Ethylester kyseliny 4-chlor-2-tolyloxythiooctove (Czech) **(25319-90-8)**
Ethylester kyseliny 4,4-dichlorbenzilove (Czech) **(510-15-6)**
Ethylester kyseliny dusicne (Czech) **(625-58-1)**
Ethylester kyseliny dusite (Czech) **(109-95-5)**
Ethylester kyseliny 3-ethoxypropionove (Czech) **(763-69-9)**
Ethylester kyseliny N-ethyl-N-nitrosokarbaminove (Czech) **(614-95-9)**
Ethylester kyseliny fluoroctove (Czech) **(459-72-3)**
Ethylester kyseliny gallove (Czech) **(831-61-8)**
Ethylester kyseliny p-hydroxybenzoove (Czech) **(120-47-8)**
Ethylester kyseliny jodoctove (Czech) **(623-48-3)**
Ethylester kyseliny karbaminove (Czech) **(51-79-6)**
Ethylester kyseliny karbanilove (Czech) **(101-99-5)**
Ethylester kyseliny krotonove (Czech) **(10544-63-5)**
Ethylester kyseliny kyanoctove (Czech) **(105-56-6)**
Ethylester kyseliny methakrylove (Czech) **(97-63-2)**
Ethylester kyseliny methansulfonove (Czech) **(62-50-0)**
Ethylester kyseliny methylkarbaminove (Czech) **(105-40-8)**
Ethylester kyseliny N-methyl-N-nitrosokarbaminove (Czech) **(615-53-2)**
Ethylester kyseliny mlecne (Czech) **(97-64-3)**
Ethylester kyseliny mravenci (Czech) **(109-94-4)**
Ethylester kyseliny orthomravenci (Czech) **(122-51-0)**
Ethylester kyseliny octove (Czech) **(141-78-6)**
Ethylester kyseliny propionove (Czech) **(105-37-3)**
Ethylester kyseliny p-toluensulfonove (Czech) **(80-40-0)**
Ethyl ester of 4,4'-dichlorobenzilic acid **(510-15-6)**
Ethyl ester of O,O-dimethyldithiophosphoryl α-phenyl acetate acid **(2597-03-7)**
Ethyl ester of 2,3-epoxy-3-phenylbutanoic acid **(77-83-8)**
Ethyl ester of methanesulfonic acid **(62-50-0)**
Ethyl ester of methanesulphonic acid **(62-50-0)**
Ethyl ester of methylnitroso-carbamic acid **(615-53-2)**
Ethyl ester of methylsulfonic acid **(62-50-0)**
Ethyl ester of methylsulphonic acid **(62-50-0)**
Ethyl ester of monoacetic acid **(79-20-9)**
Ethylestrenol **(965-90-2)**
O-Ethyl ethanephosphonic acid **(7305-61-5)**
O-Ethyl ethanephosphonochloridothioate **(1497-68-3)**
O-Ethyl ethanephosphonothionochloridate **(1497-68-3)**
Ethyl ethane sulfonate **(1912-30-7)**
S-Ethyl ethanethioate **(625-60-5)**
Ethyl ethanoate **(141-78-6)**
Ethyl ether (ACGIH,OSHA) [UN 1155] **(60-29-7)**
Ethyl ether, Compd. with boron fluoride (BF3) (1:1) **(109-63-7)**
Ethyl 3-ethoxy-3-iminopropanoate **(27317-59-5)**
Ethyl β-ethoxypropionate **(763-69-9)**
2-Ethyl-4-ethylamino-6-methoxy-s-triazine **(5248-48-6)**
p-Ethylethylbenzene **(105-05-5)**
3-Ethyl-2-(5-(3-ethyl-2-benzothiazolinylidene)-1,3-pentadienyl)-benzothiazolium iodide **(514-73-8)**
Ethyl-N-ethyl carbamate **(623-78-9)**
S-Ethyl N-ethyl N-cyclohexylthiolcarbamate **(1134-23-2)**
Ethylethylene **(106-98-9)**
Ethyl ethylene oxide **(106-88-7)**
2-Ethylethylenimine **(2549-67-9)**
Ethylethylenimine **(2549-67-9)**
Ethyl N-ethyl-N-((heptadecafluorooctyl)sulfonyl)glycinate **(1869-77-8)**
4-Ethyl-1-(3-ethylpentyl)oktylsiran sodny (Czech) **(3282-85-7)**
Ethyl N-ethyl-N-phenylcarbamate **(1013-75-8)**
O-Ethyl ethylphosphonate **(7305-61-5)**
O-Ethyl ethylphosphonochloridothioate **(1497-68-3)**
O,O-Ethyl S-2(ethylthio)ethyl phosphorodithioate **(298-04-4)**
O-Ethyl ethylthiophosphonyl chloride **(1497-68-3)**
Ethylethyne **(107-00-6)**
Ethyl ethynyl ether **(927-80-0)**
13-Ethyl-17-α-ethynylgon-4-en-17-β-ol-3-one **(797-63-7)**
dl-13-β-Ethyl-17-α-ethynyl-17-β-hydroxygon-4-en-3-one **(6533-00-2)**
13-Ethyl-17-α-ethynyl-17-β-hydroxy-4-gonen-3-one **(797-63-7)**
Ethyl ethynyl methyl carbinol **(77-75-8)**
dl-13-β-Ethyl-17-α-ethynyl-19-nortestosterone **(6533-00-2)**
5-Ethyl-5-fenyl-3-methylhydantoin (Czech) **(50-12-4)**
Ethyl fluclozepate **(29177-84-2)**
Ethyl fluoride [UN 2453] **(353-36-6)**
Ethyl fluoroacetate **(459-72-3)**
Ethylformamide **(627-45-2)**
N-Ethylformamide **(627-45-2)**
Ethyl formate (ACGIH,OSHA) [UN 1190] **(109-94-4)**
Ethylformiaat (Dutch) **(109-94-4)**
Ethylformic acid **(79-09-4)**
Ethyl formic ester **(109-94-4)**
Ethyl fumarate **(623-91-6)**
2-Ethylfuran **(3208-16-0)**
Ethyl furoate **(614-99-3)**
Ethyl gallate **(831-61-8)**
5-Ethyl-L-glutamate **(1119-33-1)**
γ-Ethylglutamate **(1119-33-1)**
Ethyl glycidate **(4660-80-4)**
Ethyl glycidyl ether **(4016-11-9)**
Ethylglycol acetate **(111-15-9)**
Ethylglykolacetat (German) **(111-15-9)**
Ethyl glyme **(629-14-1)**
Ethyl gusathion **(2642-71-9)**
Ethyl guthion **(2642-71-9)**
Ethyl 10-hendecenoate **(692-86-4)**
N-Ethyl-1,1,2,2,3,3,4,4,5,5,6,6,7,7,8,8,8-heptadecafluoro-N-(2-hydroxyethyl)-1-octanesulfonamide **(1691-99-2)**
2-(Ethyl((heptadecafluorooctyl)sulfonyl)amino)ethyl ester (9CI) **(376-14-7)**
N-Ethyl-N-((heptadecafluorooctyl)sulfonyl)glycine, sodium salt **(3871-50-9)**
1-Ethyl-2-(heptadecenyl)-1-(2-hydroxyethyl)-2-imidazolinium ethyl sulfate **(26266-76-2)**
Ethyl heptadecyl ketone **(2955-56-8)**
Ethyl heptanoate **(106-30-9)**
Ethyl n-heptanoate **(106-30-9)**
2-Ethylheptanoic acid **(3274-29-1)**
Ethyl heptazine **(77-15-6)**
Ethyl heptyl ketone **(928-80-3)**
(3-Ethyl-n-heptyl)methylcarbinol **(103-08-2)**
Ethylhexabital **(52-31-3)**

Ethyl hexadecanoate	**(628-97-7)**
Ethylhexadecyldimethylammonium bromide	**(124-03-8)**
Ethyl-N-hexadecylmorpholinium ethosulfate	**(78-21-7)**
4-Ethyl-4-hexadecyl morpholinium ethyl sulfate	**(78-21-7)**
4-Ethyl-4-hexadecylmorpholinium ethyl sulfate	**(78-21-7)**
(E,E)-Ethyl 2,4-hexadienoate	**(2396-84-1)**
S-Ethyl hexahydro-1H-azepine-1-carbothioate	**(2212-67-1)**
5-Ethylhexahydro-4,6-dioxo-5-phenylphrimidine	**(125-33-7)**
Ethyl hexahydro-1-methyl-4-phenyl-azepine-4-carboxylate	**(77-15-6)**
5-Ethylhexahydro-5-phenylpyrimidine-4,6-dione	**(125-33-7)**
2-Ethylhexaldehyde	**(123-05-7)**
Ethylhexaldehyde [UN 1191]	**(123-05-7)**
Ethyl 1-hexamethyleneiminecarbothiolate	**(2212-67-1)**
S-Ethyl 1-hexamethyleneiminothiocarbamate	**(2212-67-1)**
S-Ethyl-N-hexamethylenethiocarbamate	**(2212-67-1)**
2-Ethylhexanal	**(123-05-7)**
3-Ethylhexane	**(619-99-8)**
2-Ethyl-1,3-hexanediol	**(94-96-2)**
2-Ethylhexane-1,3-diol	**(94-96-2)**
2-Ethylhexanediol-1,3	**(94-96-2)**
Ethyl hexanediol	**(94-96-2)**
Ethyl hexanoate	**(123-66-0)**
2-Ethylhexanoic acid	**(149-57-5)**
2-Ethylhexanoic acid chloride	**(760-67-8)**
2-Ethylhexanoic acid, 2-ethylhexyl ester	**(7425-14-1)**
2-Ethylhexanoic acid, iron salt	**(19583-54-1)**
2-Ethylhexanoic acid, potassium salt	**(3164-85-0)**
2-Ethylhexanoic acid stannous salt	**(301-10-0)**
2-Ethylhexanoic acid, vinyl ester	**(94-04-2)**
2-Ethyl-1-hexanol	**(104-76-7)**
2-Ethylhexanol	**(104-76-7)**
Ethylhexanol	**(104-76-7)**
2-Ethyl-1-hexanol hydrogen phosphate	**(298-07-7)**
2-Ethyl-1-hexanol hydrogen sulfate sodium salt	**(126-92-1)**
2-Ethyl-1-hexanol phosphate	**(78-42-2)**
2-Ethylhexanol phosphate	**(12645-31-7)**
2-Ethylhexanol, phosphate	**(12645-31-7)**
2-Ethyl-1-hexanol sulfate sodium salt	**(126-92-1)**
2-Ethyl-1-hexanol titanium(4+) salt	**(1070-10-6)**
2-Ethylhexanoyl chloride	**(760-67-8)**
2-Ethylhexanyl acetate	**(103-09-3)**
2-Ethyl-2-hexenal	**(645-62-5)**
2-Ethylhexenal	**(645-62-5)**
2-Ethyl hexene-1	**(1632-16-2)**
2-Ethyl-1-hexene	**(1632-16-2)**
2-Ethyl-2-hexenoic acid	**(5309-52-4)**
2-Ethylhexenoic acid	**(5309-52-4)**
N-(2-Ethyl-2-hexenylidene)aniline	**(35331-89-6)**
N-(2-Ethyl-2-hexenylidene)benzenamine	**(35331-89-6)**
2-Ethylhexoic acid	**(149-57-5)**
2-Ethylhexoic acid, vinyl ester	**(94-04-2)**
2-Ethylhexyl acetate	**(103-09-3)**
β-Ethylhexyl acetate	**(103-09-3)**
2-Ethylhexyl acrylate	**(103-11-7)**
2-Ethylhexyl alcohol	**(104-76-7)**
Bis-2-ethylhexylamin	**(106-20-7)**
2-Ethyl hexylamine	**(104-75-6)**
2-Ethylhexylamine [UN 2276]	**(104-75-6)**
2-Ethylhexyl 3-aminopropyl ether	**(5397-31-9)**
N-(2-Ethylhexyl)aniline	**(10137-80-1)**
N-2-(Ethylhexyl) aniline	**(10137-80-1)**
N-(2-Ethylhexyl)-benzenamine	**(10137-80-1)**
2-Ethylhexyl benzoate	**(5444-75-7)**
N-(2-Ethylhexyl)bicyclo-(2,2,1)-hept-5-ene-2,3-dicarboximide	**(113-48-4)**
2-Ethylhexyl butyl phthalate	**(85-69-8)**
2-Ethylhexyl chloride	**(123-04-6)**
Ethylhexyl chloroformate	**(24468-13-1)**
2-Ethylhexylchloroformate [UN 2748]	**(24468-13-1)**
2-Ethylhexyl cyanoacetate	**(13361-34-7)**
2-Ethylhexyl 2-cyano-3,3-diphenylacrylate	**(6197-30-4)**
2-Ethylhexyl 2-cyano-3,3-diphenyl-2-propenoate	**(6197-30-4)**
N-(2-Ethylhexyl)cyclohexylamine	**(5432-61-1)**
N-2-(Ethylhexyl)-cyclohexylamine	**(5432-61-1)**
2-Ethylhexyl (2,4-dichlorophenoxy)acetate	**(1928-43-4)**
2-Ethylhexyl 2,4-dichlorophenoxyacetate	**(1928-43-4)**
(2-Ethylhexyl)-difenylfosfat (Czech)	**(1241-94-7)**
2-Ethylhexyl dihydrogen phosphate	**(1070-03-7)**
2-Ethylhexyl p-(dimethylamino)benzoate	**(21245-02-3)**
2-Ethylhexyl p-dimethylaminobenzoate	**(21245-02-3)**
2-Ethylhexyl diphenylphosphate	**(1241-94-7)**
2-Ethylhexyl diphenyl phosphite	**(15647-08-2)**
Ethyl hexylene glycol	**(94-96-2)**
2-Ethylhexyl 9,10-epoxyoctadecanoate	**(141-38-8)**
2-Ethylhexyl epoxystearate	**(141-38-8)**
2-Ethylhexylester kyseliny akrylove (Czech)	**(103-11-7)**
2-Ethylhexylester kyseliny 2-ethylkapronove (Czech)	**(7425-14-1)**
2-Ethylhexylester kyseliny octove (Czech)	**(103-09-3)**
2-Ethylhexyl ester of tall oil acids	**(68334-13-4)**
2-Ethylhexyl ethanoate	**(103-09-3)**
2-Ethylhexyl-2-ethylhexanoate	**(7425-14-1)**
2-Ethylhexyl fumarate	**(141-02-6)**
2-Ethylhexyl glycidyl ether	**(2461-15-6)**
N-2-Ethylhexylimide endomethylenetetrahydrophthalic acid	**(113-48-4)**
N-2-Ethylhexylimid kyseliny bicyklo-(2,2,1)-5-hepten-2,3-dikarboxylove (Czech)	**(113-48-4)**
2-Ethylhexyl isodecyl adipate	**(68052-04-0)**
2-Ethylhexyl isodecyl hexanedioate	**(68052-04-0)**
2-Ethylhexyl mercaptoacetate	**(7659-86-1)**
2-Ethylhexyl methacryate	**(688-84-6)**
2-Ethyl-1-hexyl methacrylate	**(688-84-6)**
2-Ethylhexyl methoxycinnamate	**(5466-77-3)**
2-Ethylhexyl-4-methoxycinnamate	**(5466-77-3)**
2-Ethylhexyl nitrate	**(27247-96-7)**
N-(2-Ethylhexyl)-5-norbornene-2,3-dicarboximide	**(113-48-4)**
2-Ethylhexyl octadecanoate	**(22047-49-0)**
(((2-Ethylhexyl)oxy)methyl)oxirane	**(2461-15-6)**
3-(2-Ethylhexyloxy)propannitril (Czech)	**(10213-75-9)**
3-(2-Ethylhexyloxy)propionitrile	**(10213-75-9)**
2-Ethylhexyloxypropylamine	**(5397-31-9)**
3-(2-Ethylhexyloxy)propylamine	**(5397-31-9)**
2-Ethylhexyl palmitate	**(29806-73-3)**
2-Ethylhexyl phosphate	**(12645-31-7)**
2-Ethylhexyl phthalate	**(117-81-7)**
Ethylhexyl phthalate	**(117-81-7)**
2-Ethylhexyl 2-propenoate	**(103-11-7)**
2-Ethylhexyl sebacate	**(122-62-3)**
2-Ethylhexylsiran sodny (Czech)	**(126-92-1)**
2-Ethylhexyl sodium phosphate	**(31044-12-9)**
2-Ethylhexyl sodium sulfate	**(126-92-1)**
2-Ethylhexyl stearate	**(22047-49-0)**
2-Ethylhexylsulfate sodium	**(126-92-1)**
2-Ethylhexyl sulfosuccinate sodium	**(577-11-7)**
2-Ethylhexyl tallate	**(68334-13-4)**
Ethyl hexyl tallate	**(68334-13-4)**
2-(2-Ethylhexyl)-3a,4,7,7a-tetrahydro-4,7-methano-1H-isoindole-1,3(2H)-dione	**(113-48-4)**
2-Ethylhexyl thioglycolate	**(7659-86-1)**
2-Ethylhexyl vinyl ether	**(103-44-6)**
Ethyl hydride	**(74-84-0)**
β-Ethylhydrocinnamic acid	**(5669-17-0)**
O-Ethyl hydrogen ethylphosphonate	**(7305-61-5)**
Ethyl(hydrogen p-mercaptobenzenesulfonato)mercury sodium salt	**(5964-24-9)**
Ethyl hydrogen peroxide	**(3031-74-1)**
Ethyl hydrogen phosphonate (9CI)	**(15845-66-6)**

Ethyl hydrogen 1-propylphosphonate	**(21921-96-0)**
Ethyl hydroperoxide	**(3031-74-1)**
Ethyl hydropersulfide	**(540-63-6)**
Ethyl hydrosulfide	**(75-08-1)**
Ethyl hydroxide	**(64-17-5)**
1-Ethyl-3-hydroxybenzene	**(620-17-7)**
Ethyl 4-hydroxybenzeneacetate	**(17138-28-2)**
Ethyl o-hydroxybenzoate	**(118-61-6)**
Ethyl p-hydroxybenzoate	**(120-47-8)**
Ethyl-2-hydroxy-2,2-bis(4-chlorophenyl)acetate	**(510-15-6)**
Ethyl m-hydroxycarbanilate carbanilate (Ester)	**(13684-56-5)**
(+-)-13-Ethyl-17-hydroxy-18,19-dinor-17-α-pregn-4-en-20-yn-3-one	**(6533-00-2)**
2-(N-Ethyl-N-2-hydroxyethylamino)ethanol	**(139-87-7)**
Ethyl(β-hydroxyethyl)aniline	**(92-50-2)**
N-Ethyl(β-hydroxyethyl)aniline	**(92-50-2)**
N-Ethyl-N-(2-hydroxyethyl)aniline	**(92-50-2)**
N-Ethyl-N-(β-hydroxyethyl)aniline	**(92-50-2)**
N-Ethyl-N-(hydroxyethyl)aniline	**(92-50-2)**
Ethyl-2-hydroxyethylnitrosamine	**(13147-25-6)**
N-Ethyl-N-hydroxyethylnitrosamine	**(13147-25-6)**
Ethyl 2-hydroxyethyl sulfide	**(110-77-0)**
Ethyl 2-hydroxyethyl thioether	**(110-77-0)**
N-Ethyl-N-(2-hydroxyethyl)-m-toluidine	**(91-88-3)**
2-Ethyl-3-hydroxyhexanal	**(496-03-7)**
2-Ethyl-2-(hydroxymethyl)-1,3-propanediol, cyclic phosphate (1:1)	**(1005-93-2)**
2-Ethyl-2-(hydroxymethyl)-1,3-propanediol triacrylate	**(15625-89-5)**
5-Ethyl-5-(4-hydroxyphenyl)-2,4,6(1H,3H,5H)-pyrimidinetrione	**(379-34-0)**
Ethyl 4-hydroxyphenylacetate	**(17138-28-2)**
Ethyl p-hydroxyphenyl ketone	**(70-70-2)**
Ethyl (4-(m-hydroxyphenyl)-1-methyl)-4-piperidyl ketone	**(469-79-4)**
Ethyl 2-hydroxypropionate	**(97-64-3)**
Ethyl α-hydroxypropionate	**(97-64-3)**
Ethylic acid	**(64-19-7)**
Ethylidene acetate	**(542-10-9)**
Ethylidene acetone	**(625-33-2)**
α-Ethylidenebenzeneacetaldehyde	**(4411-89-6)**
5-Ethylidenebicyclo(2.2.1)hept-2-ene	**(16219-75-3)**
1,1'-(Ethylidenebis(oxy))bishexane	**(5405-58-3)**
Ethylidene bromide	**(557-91-5)**
Ethylidene chloride	**(75-34-3)**
Ethylidene chloride (OSHA)	**(75-34-3)**
Ethylidene diacetate	**(542-10-9)**
Ethylidene dibromide	**(557-91-5)**
Ethylidene dichloride	**(75-34-3)**
Ethylidene diethyl ether	**(105-57-7)**
Ethylidene difluoride	**(75-37-6)**
Ethylidene dimethyl ether	**(534-15-6)**
Ethylidene fluoride	**(75-37-6)**
Ethylidene gyromitrin	**(16568-02-8)**
Ethylidenehydroxylamine	**(107-29-9)**
trans-15-Ethylidene-12-β-hydroxy-4,12-α,13-β-trimethyl 8-oxo-4,8 secosenec-1-enine	**(2318-18-5)**
Ethylidenelactic acid	**(50-21-5)**
Ethylidenelactic acid	**(598-82-3)**
5-Ethylidene-2-norbornene	**(16219-75-3)**
Ethylidene norbornene (ACGIH,OSHA)	**(16219-75-3)**
Ethylidenimine, N-chloro-1-methyl-	**(34508-68-4)**
Ethylimine	**(151-56-4)**
2,2'-(Ethylimino)diethanol	**(139-87-7)**
N-Ethyl-2,2'-iminodiethanol	**(139-87-7)**
Ethyl iodide	**(75-03-6)**
Ethyl iodoacetate	**(623-48-3)**
Ethyl 10-(p-iodophenyl)undecanoate	**(99-79-6)**
Ethyl 10-(p-iodophenyl)undecylate	**(99-79-6)**
Ethylis loflazepas (Latin)	**(29177-84-2)**
5-Ethyl-5-isoamylbarbituric acid	**(57-43-2)**
Ethyl isoamyl ketone	**(624-42-0)**
5-Ethyl-5-isoamylmalonyl urea	**(57-43-2)**
Ethyl isobutanoate	**(97-62-1)**
4-Ethyl-1-isobutyloktylsiran sodny (Czech)	**(139-88-8)**
Ethylisobutyrate	**(97-62-1)**
Ethyl isobutyrate [UN 2385]	**(97-62-1)**
Ethyl isocyanate [UN 2481]	**(109-90-0)**
Ethyl isocyanide (8CI)	**(624-79-3)**
2-Ethylisohexanol	**(106-67-2)**
Ethylisokyanid (Czech)	**(624-79-3)**
Ethyl isonicotinate	**(1570-45-2)**
2-Ethylisonicotinic acid thioamide	**(536-33-4)**
α-Ethylisonicotinic acid thioamide	**(536-33-4)**
2-Ethylisonicotinic thioamide	**(536-33-4)**
α-Ethylisonicotinoylthioamide	**(536-33-4)**
Ethyl isonitrile	**(624-79-3)**
5-Ethyl-5-isopentylbarbituric acid	**(57-43-2)**
Ethylisopentylbarbituric acid	**(57-43-2)**
O-Ethyl-O-(2-isopropoxy-carbonyl)-phenyl isopropyl-phosphoramidothioate	**(25311-71-1)**
Ethyl isopropyl ether	**(625-54-7)**
(RS)-5-Ethyl-2-(4-isopropyl-4-methyl-5-oxo-2-imidazolin-2-yl)nicotinic acid	**(81335-77-5)**
5-Ethyl-2-(4-isopropyl-4-methyl-5-oxo-2-imidazolin-2-yl) nicotinic acid	**(81335-77-5)**
N-Ethyl-N'-isopropyl-6-methylthio-1,3,5-triazine-2,4-diyldiamine	**(834-12-8)**
Ethylisopropylnitrosoamine	**(16339-04-1)**
N-Ethyl O-isopropyl thiocarbamate	**(141-98-0)**
Ethylisothiamide	**(536-33-4)**
Ethyl isothiocyanate	**(542-85-8)**
2-Ethylisothionicotinamide	**(536-33-4)**
α-Ethylisothionicotinamide	**(536-33-4)**
Ethyl isovalerate	**(108-64-5)**
Ethyljodid (Czech)	**(75-03-6)**
Ethylkarbylamin (Czech)	**(624-79-3)**
Ethyl 4-ketovalerate	**(539-88-8)**
Ethyl ketovalerate	**(539-88-8)**
N-Ethyl-N-2-kyanethylanilin (Czech)	**(148-87-8)**
Ethylkyanid (Czech)	**(107-12-0)**
D-N-Ethyllactamide carbanilate (ester)	**(16118-49-3)**
Ethyl lactate [UN 1192]	**(97-64-3)**
Ethyl laevulinate	**(539-88-8)**
Ethyl laurate	**(106-33-2)**
Ethyl laurinate	**(106-33-2)**
Ethyl levulate	**(539-88-8)**
Ethyl loflazepate	**(29177-84-2)**
Ethylloflazepate	**(29177-84-2)**
Ethyl maleate	**(141-05-9)**
Ethyl malonate	**(105-53-3)**
Ethylmalonate	**(601-75-2)**
Ethylmalonic acid	**(601-75-2)**
Ethyl mandelate	**(774-40-3)**
Ethylmercaptaan (Dutch)	**(75-08-1)**
Ethyl mercaptan (ACGIH,DOT,OSHA) [UN 2363]	**(75-08-1)**
Ethyl 2-mercaptoacetate	**(623-51-8)**
Ethyl α-mercaptoacetate	**(623-51-8)**
Ethyl mercaptoacetate	**(623-51-8)**
Ethyl mercaptoacetic acid	**(623-51-8)**
Ethyl(2-mercaptobenzoato-s)mercury sodium salt	**(54-64-8)**
β-Ethylmercaptoethyl dimethyl thionophosphate	**(867-27-6)**
2-Ethyl-mercaptomethyl-phenyl-N-methylcarbamate	**(29973-13-5)**
Ethyl mercaptophenylacetate-O,O-dimethylphosphorodithioate	**(2597-03-7)**
Ethyl 3-mercaptopropanoate	**(5466-06-8)**
N-Ethylmercuri-3,4,5,6,7,7-hexachloro-3,6-endomethylene-1,2,3,6-tetrahydrophthalimide	**(2597-93-5)**
Ethylmercuric acetate	**(109-62-6)**

Ethylmercuric chloride **(107-27-7)**
Ethylmercurichlorendimide **(2597-93-5)**
Ethylmercuric phosphate **(2235-25-8)**
N-(Ethylmercuri)-1,4,5,6,7,7-hexachlorobicyclo(2.2.1)hept-5-ene-2,3-dicarboximide **(2597-93-5)**
o-(Ethylmercurithio)benzoic acid sodium salt **(54-64-8)**
Ethylmercurithiosalicylic acid sodium salt **(54-64-8)**
N-(Ethylmercuri)-p-toluenesulfonanilide **(517-16-8)**
N-(Ethylmercuri)-p-toluenesulphonanilide **(517-16-8)**
N-Ethylmercuri-N-phenyl-p-toluenesulfonamide **(517-16-8)**
N-Ethylmercuri-1,2,3,6-tetrahydro-3,6-endomethano-3,4,5,6,7,7-hexachlorophthalimide **(2597-93-5)**
Ethylmercury chloride **(107-27-7)**
Ethylmercury phosphate **(2235-25-8)**
Ethylmercury p-toluenesulfanilide **(517-16-8)**
Ethylmercury p-toluene sulfonamide **(517-16-8)**
Ethylmercury p-toluenesulfonanilide **(517-16-8)**
Ethylmerkaptan (Czech) **(75-08-1)**
β-Ethylmerkaptoethanol (Czech) **(110-77-0)**
β-Ethylmerkaptoethylchlorid (Czech) **(693-07-2)**
Ethylmerkuriacetat (Czech) **(109-62-6)**
Ethylmerkurichlorid (Czech) **(107-27-7)**
Ethylmerkuridihydrogenfosfat (Czech) **(2235-25-8)**
Ethylmerkurithiosalicilan sodny (Czech) **(54-64-8)**
N-Ethylmerkuri-p-toluensulfoanilid (Czech) **(517-16-8)**
Ethyl methacrylate, Inhibited (DOT) **(97-63-2)**
Ethyl methacrylate [UN 2277] **(97-63-2)**
Ethyl methanesulfonate **(62-50-0)**
Ethyl methanesulphonate **(62-50-0)**
Ethyl methanoate **(109-94-4)**
Ethyl methansulfonate **(62-50-0)**
Ethyl methansulphonate **(62-50-0)**
2-(Ethyl(4-((6-methoxy-2-benzothiazolyl)azo)phenyl)amino)-ethanol **(13486-43-6)**
Ethyl o-(o-(methoxycarbonyl)benzoyl)glycolate **(85-71-2)**
Ethyl o-(methoxycarbonyl)benzoyloxyacetate **(85-71-2)**
N-Ethyl-6-methoxy-N'-(1-methylethyl)-1,3,5-triazine-2,4-diamine **(1610-17-9)**
4-Ethyl-2-methoxyphenol **(2785-89-9)**
2-Ethyl-3-methoxypyrazine **(25680-58-4)**
Ethyl (all-E)-9-(4-methoxy-2,3,6-trimethylphenyl)-3,7-dimethyl-2,4,6,8-nonatetraenoate **(54350-48-0)**
Ethyl all-trans-9-(4-methoxy-2,3,6-trimethylphenyl)-3,7-dimethyl-2,4,6,8-nonatetraenoate **(54350-48-0)**
Ethylmethylacetic acid **(116-53-0)**
Ethyl 2-methylacrylate **(97-63-2)**
Ethyl α-methyl acrylate **(97-63-2)**
N-Ethyl-2-methylallylamine **(18328-90-0)**
α-Ethyl-β- (2-(methylamino)propyl)-β-phenylbenzeneethanol, acetate (ester) **(1477-39-0)**
2-Ethyl-6-methylaniline **(24549-06-2)**
N-Ethyl-3-methylaniline **(102-27-2)**
N-Ethyl-N-methylaniline **(613-97-8)**
N-Ethyl-N-methylbenzenamine **(613-97-8)**
1-Ethyl-2-methylbenzene **(611-14-3)**
1-Ethyl-4-methylbenzene **(622-96-8)**
Ethylmethylbenzene **(25550-14-5)**
o-Ethyl methylbenzene **(611-14-3)**
p-Ethylmethylbenzene **(622-96-8)**
ar-Ethyl-ar-methyl-1,3-benzenediamine **(68966-84-7)**
N-Ethyl-2-methylbenzenesulfonamide **(1077-56-1)**
N-Ethyl-4-methylbenzenesulfonamide **(80-39-7)**
N-Ethyl-p-methylbenzenesulfonamide **(80-39-7)**
Ethyl p-methyl benzenesulfonate **(80-40-0)**
Ethyl 2-methylbutanoate **(7452-79-1)**
2-Ethyl-2-methylbutanoic acid **(19889-37-3)**
5-Ethyl-5-(1-methyl-1-butenyl)barbiturate **(125-42-8)**
5-Ethyl-5-(1-methyl-1-butenyl)barbituric acid **(125-42-8)**
5-Ethyl-5-(1-methylbutyl)barbituric acid **(76-74-4)**
5-Ethyl-5-(3-methylbutyl)barbituric acid **(57-43-2)**
5-Ethyl-5-(1-methylbutyl)barbituric acid sodium salt **(57-33-0)**
5-Ethyl-5-(1-methylbutyl)malonylurea **(76-74-4)**
5-Ethyl-5-(1-methylbutyl)-2,4,6(1H,3H,5H)-pyrimidinetrione monosodium salt **(57-33-0)**
5-Ethyl-5-(1-methylbutyl)-2-thiobarbituric acid **(76-75-5)**
Ethyl methylcarbamate **(105-40-8)**
Ethylmethyl carbinol **(78-92-2)**
Ethyl methyl cellulose **(9004-59-5)**
Ethyl methyl cetone (French) **(78-93-3)**
1-Ethyl-7-methyl-1,4-dihydro-1,8-naphthyridin-4-one-3-carboxylic acid **(389-08-2)**
2-Ethyl-2-methyl-1,3-dioxolane **(126-39-6)**
2-Ethyl-2-methyldioxolane **(126-39-6)**
1-Ethyl-6,7-methylenedioxy-4(1H)-oxocinnoline-3-carboxylic acid **(28657-80-9)**
1-Ethyl-6,7-methylenedioxy-4-quinolone-3-carboxylic acid **(14698-29-4)**
Ethyl methylene phosphorodithioate **(563-12-2)**
N-Ethyl-N-methylethanamine **(616-39-7)**
Ethyl methyl ether [UN 1039] **(540-67-0)**
Ethylmethylether diethylenglykolu (Czech) **(1002-67-1)**
N-Ethyl-N-(1-methylethyl)-2-propanamine **(7087-68-5)**
2-Ethyl-3-methylfumaric acid **(28098-80-8)**
7-Ethyl-2-methyl-4-hendecanol sulfate sodium salt **(1191-50-0)**
7-Ethyl-2-methyl-4-hendecanol sulfate sodium salt **(139-88-8)**
5-Ethyl-5-methylhydantoin **(5394-36-5)**
2-Ethyl-4-methyl-1H-imidazole **(931-36-2)**
2-Ethyl-4-methylimidazole **(931-36-2)**
2-Ethyl-5-methyl-2-imidazoline **(931-35-1)**
1-Ethyl-1-methylindan **(56298-75-0)**
Ethylmethylketon (Dutch) **(78-93-3)**
Ethyl methyl ketone [UN 1193] **(78-93-3)**
Ethyl methyl ketone oxime **(96-29-7)**
Ethyl methyl ketone peroxide **(1338-23-4)**
Ethyl methyl ketone peroxide, Max. conc. 50% containing more than 10% available oxygen (UN 2550 DOT) **(1338-23-4)**
Ethyl methyl ketone peroxide, Maximum concentration 60% **(1338-23-4)**
Ethyl methyl ketone peroxide, Maximum concentration 50% (DOT) **(1338-23-4)**
Ethyl-methylketonoxim (Czech) **(96-29-7)**
Ethyl methyl ketoxime **(96-29-7)**
2-Ethyl-3-methylmaleimide **(20189-42-8)**
Ethylmethylmaleimide **(20189-42-8)**
O-Ethyl O-(4-(methylmercapto)phenyl)-S-n-propylphosphorothionothiolate **(35400-43-2)**
2-Ethyl-6-methyl-1-N-(2-methoxy-1-methylethyl)chloroacetanilide **(51218-45-2)**
2-Ethyl-6-methyl-N-methylenebenzenamine **(35203-06-6)**
2-Ethyl-6-methyl-N-(1'-methyl-2-methoxyethyl)aniline **(51219-00-2)**
Ethyl 3-methyl-4-(methylthio)phenyl(1-methylethyl)phosphoramidate **(22224-92-6)**
Ethyl 7-methylmyristates **(17670-75-6)**
O-Ethyl O-(3-methyl-6-nitrophenyl) N-sec-butylphosphorothioamidate **(36335-67-8)**
O-Ethyl-O-(5-methyl-2-nitrophenyl)(1-methylpropyl)phosphoramidothioate **(36335-67-8)**
Ethylmethylnitrosamine **(10595-95-6)**
N-(2-(Ethyl(3-methyl-4-nitrosophenyl)amino)ethyl)-methanesulfonamide **(56046-62-9)**
1-Ethyl-7-methyl-4-oxo-1,4-dihydro-1,8-naphthyridine-3-carboxylic acid **(389-08-2)**
2-Ethyl-4-methyl-1-pentanol **(106-67-2)**
2-Ethyl-4-methylpentanol **(106-67-2)**
2-Ethyl-3-methyl-1-pentene **(3404-67-9)**
2-Ethyl-3-methylpentene-1 **(3404-67-9)**
2-Ethyl-4-methylpentyl 2,4-dichlorophenoxyacetate **(53404-37-8)**

3-(Ethyl(3-methylphenyl)amino)-1,2-propanediol	(92-11-5)
3-(Ethyl(3-methylphenyl)amino)propanenitrile	(148-69-6)
5-Ethyl-1-methyl-5-phenylbarbituric acid	(115-38-8)
5-Ethyl-N-methyl-5-phenylbarbituric acid	(115-38-8)
N-Ethylmethylphenylbarbituric acid	(115-38-8)
N-Ethyl-N-(3-methylphenyl)-1,2-ethanediamine	(19248-13-6)
O-Ethyl S-(4-methylphenyl) ethylphosphonodithioate	(333-43-7)
Ethyl methylphenylglycidate	(77-83-8)
trans-Ethyl 3-methyl-3-phenylglycidate	(19464-92-7)
5-Ethyl-3-methyl-5-phenylhydantoin	(50-12-4)
5-Ethyl-3-methyl-5-phenyl-2,4(3H,5H)-imidazoledione	(50-12-4)
5-Ethyl-3-methyl-5-phenylimidazolidin-2,4-dione	(50-12-4)
5-Ethyl-3-methyl-5-phenyl-2,4-imidazolidinedione	(50-12-4)
Ethyl 1-methyl-4-phenylisonipecotate	(57-42-1)
Ethyl-1-methyl-4-phenylisonipecotate hydrochloride	(50-13-5)
Ethyl 1-methyl-4-phenylpiperidine-4-carboxylate	(57-42-1)
Ethyl 1-methyl-4-phenylpiperidine-4-carboxylate hydrochloride	(50-13-5)
Ethyl 1-methyl-4-phenylpiperidyl-4-carboxylate hydrochloride	(50-13-5)
3-Ethyl-6-methylpiperidine	(104-89-2)
5-Ethyl-2-methylpiperidine	(104-89-2)
Ethyl 2-methylpropanoate	(97-62-1)
N-Ethyl-2-methyl-2-propen-1-amine	(18328-90-0)
Ethyl 2-methyl-2-propenoate	(97-63-2)
N-Ethyl-N-(2-methyl-2-propenyl)-2,6-dinitro-4-(trifluoromethyl)-benzenamine	(55283-68-6)
Ethyl 2-methylpropionate	(97-62-1)
5-Ethyl-5-(1-methylpropyl)barbiturate	(125-40-6)
5-Ethyl-5-(1-methylpropyl)barbituric acid	(125-40-6)
5-Ethyl-5-(1-methylpropyl)barbituric acid sodium salt	(143-81-7)
N-(3-(1-Ethyl-1-methylpropyl)isoxazol-5-yl)-2,6-dimethoxy-benzamide	(82558-50-7)
N-(3-(1-Ethyl-methylpropyl)-5-isoxazolyl)-2,6-dimethoxybenz-amide	(82558-50-7)
2-Ethyl-3-methyl pyrazine	(15707-23-0)
2-Ethyl-5-methyl pyrazine	(13360-64-0)
3-Ethyl-6-methylpyridine	(104-90-5)
5-Ethyl-2-methylpyridine	(104-90-5)
3-Ethyl-3-methyl-2,5-pyrrolidine-dione	(77-67-8)
3-Ethyl-3-methylpyrrolidine-2,5-dione	(77-67-8)
2-Ethyl-2-methylsuccinimide	(77-67-8)
α-Ethyl-α-methylsuccinimide	(77-67-8)
Ethylmethylthiambutene	(441-61-2)
2-Ethyl-4-methylthiazole	(15679-12-6)
O-Ethyl O-(4-(methylthio)phenyl)phosphorodithioic acid S-propyl ester	(35400-43-2)
O-Ethyl O-(4-(methylthio)phenyl) S-propyl phosphorodithioate	(35400-43-2)
Ethyl 4-(methylthio)-m-tolyl isopropylphosphoramidate	(22224-92-6)
N-Ethyl-α-methyl-3-trifluoromethylphenethylamine	(458-24-2)
7-Ethyl-2-methyl-4-undecanol sulfate sodium salt	(139-88-8)
Ethyl monobromoacetate	(105-36-2)
Ethyl monochloracetate	(105-39-5)
Ethyl monochloroacetate	(105-39-5)
Ethyl monoiodoacetate	(623-48-3)
Ethyl monosulfide	(352-93-2)
N-Ethylmorfolin (Czech)	(100-74-3)
3-O-Ethylmorphine	(76-58-4)
Ethylmorphine	(76-58-4)
Ethylmorphine hydrochloride	(125-30-4)
Ethylmorphine hydrochloride	(6746-59-4)
o-Ethylmorphine hydrochloride	(125-30-4)
Ethylmorphine hydrochloride dihydrate	(6746-59-4)
4-Ethylmorpholine	(100-74-3)
N-Ethylmorpholine (ACGIH,OSHA)	(100-74-3)
1-Ethyl-4-(2-morpholinoethyl)-3,3-diphenyl-2-pyrrolidinone	(309-29-5)
Ethyl myristate	(124-06-1)
Ethylnandrol	(965-90-2)
N-Ethyl-1-naphthalenamine	(118-44-5)
1-Ethylnaphthalene	(1127-76-0)
2-Ethylnaphthalene	(939-27-5)
Ethyl 1-naphthaleneacetate	(2122-70-5)
Ethyl β-naphtholate	(93-18-5)
Ethyl 1-naphthylacetate	(2122-70-5)
Ethyl-α-naphthylamine	(118-44-5)
N-Ethyl-1-naphthylamine	(118-44-5)
N-Ethyl-α-naphthylamine	(118-44-5)
Ethyl 2-naphthyl ether	(93-18-5)
Ethyl β-naphthyl ether	(93-18-5)
3-Ethylnirvanol	(50-12-4)
Ethyl nitrate (DOT)	(625-58-1)
Ethyl nitrile	(75-05-8)
Ethyl nitrite (DOT)	(109-95-5)
Ethyl nitrite, Solution [UN 1194]	(109-95-5)
1-Ethyl-4-nitrobenzene	(100-12-9)
4-Ethylnitrobenzene	(100-12-9)
p-Ethylnitrobenzene	(100-12-9)
Ethyl p-nitrobenzoate	(99-77-4)
Ethyl nitrobenzoate, para ester	(99-77-4)
3-(N-Ethyl-4-((6-nitro-2-benzothiazolyl)azo)-m-toluidino)propio-nitrile	(16586-42-8)
9-Ethyl-3-nitro-9H-carbazole	(86-20-4)
9-Ethyl-3-nitrocarbazole	(86-20-4)
O-Ethyl-O-p-nitrofenylester kyseliny fenylthiofosfonove (Czech)	(2104-64-5)
O-Ethyl-O-((4-nitro-fenyl)-fenyl)-monothiofosfonaat (Dutch)	(2104-64-5)
3-(Ethyl(4-((4-nitrophenyl)azo)phenyl)amino)propanenitrile	(31482-56-1)
Ethyl p-nitrophenyl benzenethionophosphonate	(2104-64-5)
O-Ethyl O-(4-nitrophenyl)benzenethionophosphonate	(2104-64-5)
Ethyl p-nitrophenyl benzenethiophosphate	(2104-64-5)
Ethyl p-nitrophenyl benzenethiophosphonate	(2104-64-5)
Ethyl p-nitrophenyl ether	(100-29-8)
Ethyl p-nitrophenyl ethylphosphate	(311-45-5)
Ethyl p-nitrophenyl phenylphosphonothioate	(2104-64-5)
O-Ethyl O-(4-nitrophenyl) phenylphosphonothioate	(2104-64-5)
O-Ethyl O-p-nitrophenyl phenylphosphonothiolate	(2104-64-5)
O-Ethyl O-p-nitrophenyl phenylphosphorothioate	(2104-64-5)
Ethyl p-nitrophenyl thionobenzenephosphate	(2104-64-5)
Ethyl p-nitrophenyl thionobenzenephosphonate	(2104-64-5)
2-Ethyl-2-nitro-1,3-propanediol	(597-09-1)
2-(Ethylnitrosamino)ethanol	(13147-25-6)
Ethylnitrosoaniline	(612-64-6)
Ethylnitrosobiuret	(32976-88-8)
N-Ethyl-N-nitrosobiuret	(32976-88-8)
N-Ethyl-N-nitrosobutylamine	(4549-44-4)
Ethylnitrosocarbamic acid, ethyl ester	(614-95-9)
N-Ethyl-N-nitrosocarbamic acid ethyl ester	(614-95-9)
N-Ethyl-N-nitroso-ethanamine	(55-18-5)
1-Ethyl-1-nitrosomocovina (Czech)	(759-73-9)
1-Ethyl-1-nitrosourea	(759-73-9)
Ethylnitrosourea	(759-73-9)
n-Ethyl-N-nitroso-urea	(759-73-9)
n-Ethylnitrosourea	(759-73-9)
N-Ethyl-N-nitrosourethan	(614-95-9)
N-Ethyl-N-nitrosourethane	(614-95-9)
N-Ethyl-N-nitrosovinylamine	(13256-13-8)
1-Ethyl-3-(5-nitro-2-thiazolyl) urea	(139-94-6)
N-Ethyl-N'-(5-nitro-2-thiazolyl)urea	(139-94-6)
4,4'-(2-Ethyl-2-nitrotrimethylene)dimorpholine	(1854-23-5)
N-Ethylnonafluoro-N-(2-hydroxymethyl)-1-butanesulfonamide	(34449-89-3)
Ethyl nonanoate	(123-29-5)
5-Ethyl-2-nonanol	(103-08-2)
Ethyl 8-nonenoate	(35194-39-9)
Ethyl nonylate	(123-29-5)
Ethyl n-octadecanoate	(111-61-5)
Ethyl octadecanoate	(111-61-5)
Ethyl cis-9-octadecenoate	(111-62-6)

7-Ethyl-6,6β,7,8,9,10,12,13-octahydro-2-methoxy-6,9-methano-5H-pyrido(1',2':1,2)azepino(5,4-b)indole **(83-74-9)**
Ethyl octanoate **(106-32-1)**
Ethyl octylate **(106-32-1)**
4-Ethyl-1-octyn-3-ol **(5877-42-9)**
Ethyl oenanthate **(106-30-9)**
Ethyl oenanthylate **(106-30-9)**
Ethylolamine **(141-43-5)**
Ethyl oleate **(111-62-6)**
1-(β-Ethylol)-2-methyl-5-nitro-3-azapyrrole **(443-48-1)**
5-Ethyl-5-(3 or 6-oxo-1-cyclohexen-1-yl)barbituric acid **(25104-37-4)**
Ethyl orthoformate [UN 2524] **(122-51-0)**
Ethyl orthosilicate **(78-10-4)**
Ethyl oxalate [UN 2525] **(95-92-1)**
2-Ethyl-2-oxazoline **(10431-98-8)**
2-Ethyloxazoline **(10431-98-8)**
Ethyloxirane **(106-88-7)**
Ethyl α-oxobenzeneacetate **(1603-79-8)**
Ethyl β-oxobenzenepropanoate **(94-02-0)**
Ethyl 3-oxobutanoate **(141-97-9)**
2-Ethyl-3-oxobutanoic acid **(4433-85-6)**
Ethyl 3-oxobutyrate **(141-97-9)**
Ethyl 7-oxo-7H-furo(3,2-g)(1)benzopyran-6-carboxylate **(20073-24-9)**
Ethyl 9-oxononanoate **(3433-16-7)**
Ethyl 4-oxopentanoate **(539-88-8)**
Ethyl 2-oxopropanoate **(617-35-6)**
Ethyl 4-oxovalerate **(539-88-8)**
Ethyl p-oxybenzoate **(120-47-8)**
Ethyl-oxyhydrate **(8030-89-5)**
2-Ethyloxyphenol **(94-71-3)**
4-Ethyloxyphenol **(622-62-8)**
Ethyl palmitate **(628-97-7)**
Ethyl paraben **(120-47-8)**
Ethyl paraoxon **(311-45-5)**
Ethyl parasept **(120-47-8)**
Ethyl parathion **(56-38-2)**
S-Ethyl parathion **(597-88-6)**
Ethyl pelargonate **(123-29-5)**
Ethyl pentanoate **(539-82-2)**
3-Ethyl-3-pentanol **(597-49-9)**
2-Ethyl-4-pentenal **(5204-80-8)**
3-Ethyl-2-pentene **(816-79-5)**
5-Ethyl-5-pentylbarbituric acid **(115-58-2)**
2-((1-Ethylpentyl)oxy)ethanol **(10138-47-3)**
Ethyl perchlorate **(22750-93-2)**
Ethyl peroxycarbonate **(14666-78-5)**
Ethyl-2 peroxyhexanoate de tert-butyle (French) **(62695-55-0)**
2-Ethylperoxyhexanoic acid tert-pentyl ester **(686-31-7)**
Ethyl phenacetate **(101-97-3)**
Ethylphencyclidine **(2201-15-2)**
2-Ethylphenol **(90-00-6)**
3-Ethylphenol **(620-17-7)**
4-Ethylphenol **(123-07-9)**
m-Ethylphenol **(620-17-7)**
meta-Ethylphenol **(620-17-7)**
o-Ethylphenol **(90-00-6)**
4-Ethylphenol sodium salt **(19277-91-9)**
p-Ethylphenol, sodium salt **(19277-91-9)**
Ethyl phenoxyacetate **(2555-49-9)**
Ethyl (2-(4-phenoxyphenoxy)ethyl)carbamate **(72490-01-8)**
Ethyl 2-(4-phenoxyphenoxy)ethylcarbamate **(72490-01-8)**
Ethyl(2-(p-phenoxyphenoxy)ethyl)carbamate **(72490-01-8)**
Ethyl phenylacetate **(101-97-3)**
Ethyl β-phenylacrylate **(103-36-6)**
Ethylphenylamine **(103-69-5)**
N-Ethyl-N-phenyl-o-aminobenzenesulfonamide **(81-10-7)**
(R)-N-Ethyl-2-(((phenylamino)carbonyl)oxy)propanamide **(16118-49-3)**
2-(Ethylphenylamino)ethanol **(92-50-2)**
2-(N-Ethyl-N-phenylamino)ethanol **(92-50-2)**
N-Ethyl-N-phenylaminoethanol **(92-50-2)**
N-Ethyl-1-((p-(phenylazo)phenyl)azo)-2-naphthalenamine **(6368-72-5)**
N-Ethyl-1-((4-(phenylazo)phenyl)azo)-2-naphthylamine **(6368-72-5)**
5-Ethyl-5-phenylbarbituric acid **(50-06-6)**
5-Ethyl-5-phenylbarbituric acid sodium **(57-30-7)**
5-Ethyl-5-phenylbarbituric acid sodium salt **(57-30-7)**
Ethyl 4-(4-phenylbenzalamino)benzoate **(3782-80-7)**
Ethyl-N-phenylcarbamate **(101-99-5)**
Ethyl phenylcarbamoyloxyphenylcarbamate **(13684-56-5)**
Ethyl phenyl carbinol **(93-54-9)**
N-Ethyl-1-phenylcyclohexylamine **(2201-15-2)**
Ethyl phenyl dichlorosilane **(1125-27-5)**
Ethylphenyldichlorosilane [UN 2435] **(1125-27-5)**
3-Ethyl-3-phenyl-2,6-diketopiperidine **(77-21-4)**
3-Ethyl-3-phenyl-2,6-dioxopiperidine **(77-21-4)**
Ethyl 2-phenylethanoate **(101-97-3)**
Ethylphenylethanolamine **(92-50-2)**
N-Ethyl-N-phenylethanolamine **(92-50-2)**
1-(4-Ethylphenyl)ethanone **(937-30-4)**
Ethyl phenyl ether **(103-73-1)**
N-(2-Ethylphenyl)-2-ethylbenzenamine, 1-propene trimer reaction product **(68608-77-5)**
Ethyl(phenylethyl)benzene **(64800-83-5)**
O-Ethyl S-phenyl ethyldithiophosphonate **(944-22-9)**
O-Ethyl-S-phenyl ethylphosphonodithioate **(944-22-9)**
2-Ethyl-2-phenylglutarimide **(77-21-4)**
α-Ethyl-α-phenylglutarimide **(77-21-4)**
Ethyl 3-phenylglycidate **(121-39-1)**
Ethyl phenylglycidate **(121-39-1)**
Ethyl phenylglyoxylate **(1603-79-8)**
5-Ethyl-5-phenylhexahydropyrimidine-4,6-dione **(125-33-7)**
Ethyl phenyl ketone **(93-55-0)**
5-Ethyl-5-phenyl-N-methyl-bartituric acid **(115-38-8)**
O-Ethyl phenyl p-nitrophenyl thiophosphonate **(2104-64-5)**
Ethyl p-phenylphosphonochloridothioate **(5075-13-8)**
O-Ethyl phenylphosphonothioate, O-ester with p-hydroxybenzonitrile **(13067-93-1)**
Ethyl-4-phenyl-piperidine-4-carboxylate **(77-17-8)**
3-Ethyl-3-phenyl-2,6-piperidinedione **(77-21-4)**
2-Ethyl-2-phenylpropanediamide **(7206-76-0)**
Ethyl 3-phenylpropenoate **(103-36-6)**
N-Ethyl-N-(3-phenylpropyl)benzenepropanamine **(150-59-4)**
5-Ethyl-5-phenyl-2,4,6-(1H,3H,5H)pyrimidinetrione **(50-06-6)**
5-Ethyl-5-phenyl-2,4,6-(1H,3H,5H)pyrimidinetrione monosodium salt **(57-30-7)**
Ethyl 4-phenyl-1-(2-tetrahydrofurfuryloxyethyl)piperidine-4-carboxylate **(2385-81-1)**
Ethyl phenylurethane **(1013-75-8)**
N-Ethyl-N-phenylurethane **(1013-75-8)**
Ethyl phosphate **(78-40-0)**
4-Ethyl-1-phospha-2,6,7-trioxabicyclo(2.2.2)octane **(824-11-3)**
4-Ethyl-1-phospha-2,6,7-trioxabicyclo(2.2.2)octane-1-oxide **(1005-93-2)**
Ethylphosphonic acid ethyl ester **(7305-61-5)**
Ethylphosphonothioic dichloride **(993-43-1)**
Ethyl phosphonothioic dichloride, Anhydrous (DOT) **(993-43-1)**
Ethylphosphonothionic dichloride **(993-43-1)**
Ethyl phosphonothioyl dichloride **(993-43-1)**
Ethyl phosphorodichloridate (DOT) **(1498-51-7)**
Ethyl phthalate **(84-66-2)**
Ethyl N-phthaloylcarbamate **(22509-74-6)**
Ethyl phthalyl ethyl glycolate **(84-72-0)**
Ethyl picolinate **(2524-52-9)**
5-Ethyl-2-picoline **(104-90-5)**
5-Ethyl-α-picoline **(104-90-5)**
1-Ethylpiperidine **(766-09-6)**

1-Ethyl piperidine [UN 2386] **(766-09-6)**
N-Ethyl-3-piperidyl benzilate **(3567-12-2)**
Ethyl pirimiphos **(23505-41-1)**
Ethyl pivalate **(3938-95-2)**
o-Ethyl potassium dithiocarbonate **(140-89-6)**
Ethyl potassium xanthate **(140-89-6)**
Ethyl potassium xanthogenate **(140-89-6)**
Ethyl 2-propenoate **(140-88-5)**
Ethyl propenoate **(140-88-5)**
Ethyl 2-propenoate, methyl 2-methyl-2-propenoate, N-(hydroxymethyl)-2-propenamide, 2-propenamide polymer **(30394-81-1)**
Ethyl 1-propenyl ether **(928-55-2)**
Ethyl propenyl ether **(928-55-2)**
Ethyl orthopropionate **(115-80-0)**
Ethyl propionate **(105-73-7)**
Ethyl propionate [UN 1195] **(105-37-3)**
2-Ethyl-3-propyl acrolein **(645-62-5)**
α-Ethyl-β-n-propylacrolein **(645-62-5)**
2-Ethyl-3-propylacrylic acid **(5309-52-4)**
2-(Ethylpropylamino)-2',6'-butyroxylidide **(36637-18-0)**
N-(1-Ethylpropyl)-3,4-dimethylaniline **(56038-89-2)**
N-(1-Ethylpropyl)-3,4-dimethyl-2,6-dinitrobenzenamine **(40487-42-1)**
Ethyl n-propyl ether **(628-32-0)**
Ethyl propyl ether [UN 2615] **(628-32-0)**
Ethyl propyl ketone **(589-38-8)**
1-(1-Ethylpropyl)-2-methylbenzene **(54410-74-1)**
Ethyl-propylmethylcarbinylbarbituric acid **(76-74-4)**
1-Ethylpropyl methyl ether **(36839-67-5)**
m-(1-Ethylpropyl)phenyl methylcarbamate **(672-04-8)**
2-Ethyl-3-propyl-1,3-propanediol **(94-96-2)**
3-Ethyl-2-propylquinoline **(3290-24-2)**
N-(1-Ethylpropyl)-3,4-xylidine **(56038-89-2)**
Ethylprotal **(121-32-4)**
Ethyl 3-psoralencarboxylate **(20073-24-9)**
2-Ethylpyrazine **(13925-00-3)**
Ethylpyrazine **(13925-00-3)**
Ethyl pyrazinyl phosphorothioate **(297-97-2)**
2-Ethylpyridine **(100-71-0)**
3-Ethylpyridine **(536-78-7)**
4-Ethylpyridine **(536-75-4)**
α-Ethylpyridine **(100-71-0)**
γ-Ethylpyridine **(536-75-4)**
2-Ethyl-4-pyridinecarbothioamide **(536-33-4)**
Ethyl 2-pyridinecarboxylate **(2524-52-9)**
1-Ethylpyridinium bromide **(1906-79-2)**
Ethyl pyrophosphate, tetra- **(107-49-3)**
1-Ethyl-2-pyrrolidinone **(2687-91-4)**
N-Ethylpyrrolidinone **(2687-91-4)**
N-((1-Ethyl-2-pyrrolidinyl)methyl)-2-methoxy-5-sulfamoyl-benzamide **(15676-16-1)**
N-((1-Ethyl-2-pyrrolidinyl)methyl)-5-sulfamoyl-o-anisamide **(15676-16-1)**
N-Ethylpyrrolidone **(2687-91-4)**
Ethyl pyruvate **(617-35-6)**
7-Ethylquinoline **(7661-47-4)**
4-Ethylresorcinol **(2896-60-8)**
Ethyl rhodanate **(542-90-5)**
Ethyl salicylate **(118-61-6)**
Ethyl sebacate **(110-40-7)**
Ethyl selenac **(136-92-5)**
Ethyl selenac **(5456-28-0)**
Ethyl seleram **(5456-28-0)**
Ethylsilanetriol triacetate **(17689-77-9)**
Ethyl silicate **(11099-06-2)**
Ethyl silicate (ACGIH,DOT,OSHA) **(78-10-4)**
Ethyl silicon trichloride **(115-21-9)**
1-N-Ethylsisomicin **(56391-56-1)**
Ethyl (sodium o-mercaptobenzoato)mercury **(54-64-8)**
Ethyl stearate **(111-61-5)**
m-Ethylstyrene **(7525-62-4)**
Ethyl succinate **(123-25-1)**
Ethylsuccinonitrile **(17611-82-4)**
1-Ethyl-N-sulfanilylcytosine **(17784-12-2)**
Ethyl sulfate **(64-67-5)**
Ethyl sulfhydrate **(75-08-1)**
Ethyl-sulfide **(352-93-2)**
S-(2-(Ethylsulfinyl)ethyl) O,O-dimethyl phosphorothioate **(301-12-2)**
S-Ethylsulfinylmethyl-O,O-diisopropyldithiofosfat (Czech) **(5827-05-4)**
S-(Ethylsulfinyl)methyl O,O-diisopropyl phosphorodithioate **(5827-05-4)**
Ethyl sulfocyanate **(542-90-5)**
Ethylsulfonal **(76-20-0)**
Ethyl ((p-sulfophenyl)thio)mercury, sodium salt **(5964-24-9)**
Ethyl tellurac **(20941-65-5)**
Ethyl tellurac **(30145-38-1)**
Ethyl tetradecanoate **(124-06-1)**
6-Ethyl-1,2,3,4-tetrahydro-9,10-anthracenediol **(68279-54-9)**
2-Ethyl-5,6,7,8-tetrahydroanthraquinone **(15547-17-8)**
2-Ethyltetralin **(32367-54-7)**
6-Ethyltetralin **(22531-20-0)**
Ethyl tetraphosphate **(757-58-4)**
Ethyl tetraphosphate, hexa- **(757-58-4)**
N-(5-Ethyl-1,3,4-thiadiazol-2-yl)sulfanilamide **(94-19-9)**
Ethyl thioalcohol **(75-08-1)**
2-Ethyl-4-thioamidylpyridine **(536-33-4)**
2-Ethylthio-4,6-bis(isopropylamino)-s-triazine **(4147-51-7)**
6-(Ethylthio)-N,N'-bis(1-methylethyl)-1,3,5-triazine-2,4-diamine **(4147-51-7)**
2-Ethyl-4-thiocarbamoylpyridine **(536-33-4)**
2-(Ethylthio)chloroethane **(693-07-2)**
Ethyl thiochloroformate **(2941-64-2)**
S-Ethyl thiochloroformate **(2941-64-2)**
Ethyl thiocyanate **(542-90-5)**
Ethylthiocyanate **(542-90-5)**
Ethylthiodemeton **(298-04-4)**
Ethylthioethane **(352-93-2)**
2-(Ethylthio)ethanol **(110-77-0)**
Ethyl thioether **(352-93-2)**
2-Ethylthioethyl chloride **(693-07-2)**
S-2-(Ethylthio)ethyl O,O-diethyl ester of phosphorodithioic acid **(298-04-4)**
2-Ethylthioethyl O,O-dimethyl phosphorodithioate **(640-15-3)**
S-(2-(Ethylthio)ethyl) O,O-dimethylphosphorodithionate **(640-15-3)**
O-(2-(Ethylthio)ethyl) O,O-dimethyl phosphorothioate **(867-27-6)**
S(and O)-2-(Ethylthio)ethyl O,O-dimethyl phosphorothioate **(8022-00-2)**
S-(2-(Ethylthio)ethyl) O,O-dimethyl phosphorothioate **(919-86-8)**
S-(2-(Ethylthio)ethyl) dimethyl phosphorothiolate **(919-86-8)**
S-(2-(Ethylthio)ethyl)dimethyl phosphorothiolothionate **(640-15-3)**
2-(Ethylthio)ethyl dimethyl phosphorothionate **(867-27-6)**
S-(2-(Ethylthio)ethyl) O,O-dimethyl thiophosphate **(919-86-8)**
Ethyl thioglycolate **(623-51-8)**
2-Ethylthioisonicotinamide **(536-33-4)**
α-Ethylthioisonicotinamide **(536-33-4)**
Ethylthiokyanat (Czech) **(542-90-5)**
2-((Ethylthio)methyl)phenol methylcarbamate **(29973-13-5)**
(2-Ethylthiomethyl-phenyl)-N-methylcarbamate **(29973-13-5)**
2-((Ethylthio)methyl)phenyl methylcarbamate **(29973-13-5)**
Ethylthiometon **(298-03-3)**
Ethylthiometon sulfoxide **(2497-07-6)**
Ethylthionophosphonic acid O-ethyl ester chloride **(1497-68-3)**
Ethylthionophosphonyl dichloride **(993-43-1)**
Ethylthioperazine **(1420-55-9)**
Ethyl thiophanate **(23564-06-9)**
Ethylthiophosphonic dichloride **(993-43-1)**
5-((2-Ethylthio)propyl)-2-(1-oxopropyl)-1,3-cyclohexanedione **(99422-01-2)**
Ethyl thiopyrophosphate **(3689-24-5)**
5-(Ethylthio)-1,3,4-thiadiazol-2-amine **(25660-70-2)**

α-Ethylthio-o-tolyl methylcarbamate **(29973-13-5)**
1-Ethylthiourea **(625-53-6)**
Ethyl thiourea **(625-53-6)**
Ethyl thiram **(97-77-8)**
Ethyl thiudad **(97-77-8)**
Ethyl thiurad **(97-77-8)**
Ethyl titanate **(3087-36-3)**
Ethyl α-toluate **(101-97-3)**
2-Ethyltoluene **(611-14-3)**
3-Ethyltoluene **(620-14-4)**
4-Ethyltoluene **(622-96-8)**
Ethyl toluene **(25550-14-5)**
Ethyltoluene **(25550-14-5)**
o-Ethyltoluene **(611-14-3)**
p-Ethyltoluene **(622-96-8)**
N-Ethyl-o-toluenesulfonamide **(1077-56-1)**
N-Ethyl-p-toluenesulfonamide **(80-39-7)**
Ethyl-p-toluenesulfonate **(80-40-0)**
6-Ethyl-o-toluidine **(24549-06-2)**
N-Ethyltoluidine **(102-27-2)**
N-Ethyl-m-toluidine [UN 2754] **(102-27-2)**
N-Ethyl-o-toluidine [UN 2754] **(94-68-8)**
2-(N-Ethyl-m-toluidino)ethanol **(91-88-3)**
3-(N-Ethyl-m-toluidino)propionitrite **(148-69-6)**
O-Ethyl-S-p-tolylester kyseliny ethyldithiofosfonove (Czech) **(333-43-7)**
O-Ethyl S-p-tolyl ethylphosphonodithioate **(333-43-7)**
N-Ethyl-p-tolylsulfonamide **(80-39-7)**
Ethyl p-tosylate **(80-40-0)**
Ethyl tosylate **(80-40-0)**
Ethyltriacetoxysilane **(17689-77-9)**
Ethyl trichloroacetate **(515-84-4)**
Ethyl trichlorophenylethylphosphonothioate **(327-98-0)**
O-Ethyl O-2,4,5-trichlorophenyl ethylphosphonothioate **(327-98-0)**
Ethyl trichlorosilane [UN 1196] **(115-21-9)**
Ethyl tricosanoate **(18281-07-7)**
N-Ethyltridecafluoro-N-(2-hydroxyethyl)-1-hexanesulfonamide **(34455-03-3)**
Ethyltriethoxysilane **(78-07-9)**
Ethyl trifluoroacetate **(383-63-1)**
Ethyl 4,4,4-trifluoroacetoacetate **(372-31-6)**
Ethyl trifluoroacetoacetate **(372-31-6)**
Ethyl (trifluoroacetyl)acetate **(372-31-6)**
Ethyl trifluoroethanoate **(383-63-1)**
Ethyl 3,4,5-trihydroxybenzoate **(831-61-8)**
Ethyltrimethylammonium chloride **(27697-51-4)**
2-Ethyl-4-(2,2,3-trimethyl-3-cyclopenten-1-yl)-2-buten-1-ol **(28219-61-6)**
(E,E)-Ethyl 3,7,11-trimethyl-2,4-dodecadienoate **(41096-46-2)**
Ethyl (2E,4E)-3,7,11-trimethyl-dodeca-2-4-dienoate **(41096-46-2)**
Ethyl (E,E)-3,7,11-trimethyl-2,4-dodecadienoate **(41096-46-2)**
Ethyl (E,E)-3,7,11-trimethyldodeca-2,4-dienoate **(41096-46-2)**
Ethyl 3,7,11-trimethyldodeca-2,4-dienoate **(41096-46-2)**
Ethyl(2E,4E)-3,7,11-trimethyl-2,4-dodecadienoate **(41096-46-2)**
Ethyltrimethyllead **(1762-26-1)**
Ethyltrimethylolmethane **(77-99-6)**
Ethyltrimethylplumbane **(1762-26-1)**
4-Ethyl-2,6,7-trioxa-1-phosphabicyclo(2.2.2)octane **(824-11-3)**
4-Ethyl-2,6,7-trioxa-1-phosphabicyclo(2.2.2)octane-1-oxide **(1005-93-2)**
Ethyltriphenylphosphonium acetate **(35835-94-0)**
Ethyltriphenylphosphonium iodide **(4736-60-1)**
Ethyltris(2-hydroxyethyl)ammonium ethyl sulfate **(31774-90-0)**
Ethyl tuads **(97-77-8)**
Ethyl tuex **(97-77-8)**
Ethyl undecanoate **(627-90-7)**
Ethyl 10-undecenoate **(692-86-4)**
Ethyl undecenoate **(692-86-4)**
Ethyl undecylenate **(692-86-4)**
1-Ethylurea **(625-52-5)**
Ethylurea **(625-52-5)**
n-Ethylurea **(625-52-5)**
Ethylurethan **(51-79-6)**
Ethyl urethane **(51-79-6)**
o-Ethylurethane **(51-79-6)**
Ethyl n-valerate **(539-82-2)**
Ethyl valerate **(539-82-2)**
Ethyl vanillin **(121-32-4)**
m-Ethyl vinylbenzen (Czech) **(7525-62-4)**
m-Ethylvinylbenzene **(7525-62-4)**
Ethylvinyldichlorosilane **(10138-21-3)**
Ethyl vinyl ether **(109-92-2)**
Ethyl vinyl ketone **(1629-58-9)**
Ethylvinylnitrosamine **(13256-13-8)**
2-Ethyl-6-vinylpyrazine **(32736-90-6)**
3-Ethyl-6-vinylpyridine **(5408-74-2)**
5-Ethyl-2-vinylpyridine **(5408-74-2)**
Ethyl xanthate **(151-01-9)**
Ethylxanthic acid **(151-01-9)**
Ethylxanthic acid potassium salt **(140-89-6)**
Ethylxanthic disulfide **(502-55-6)**
Ethyl xanthogenate **(151-01-9)**
Ethyl xanthogen disulfide **(502-55-6)**
Ethyl xanthogen ethyl formate **(3278-35-1)**
Ethyl zimate **(12122-67-7)**
Ethymal **(77-67-8)**
Ethyne **(74-86-2)**
Ethyne, dichloro- (9CI) **(7572-29-4)**
Ethyne, diiodo- **(624-74-8)**
Ethyne, diphenyl- **(501-65-5)**
1,1'-(1,2-Ethynediyl)bisbenzene **(501-65-5)**
Ethyne, ethoxy- (9CI) **(927-80-0)**
17-α-Ethynil-δ-4-estrene-17-β-ol **(52-76-6)**
Ethynloestrenol **(52-76-6)**
Ethynodiol acetate **(297-76-7)**
Ethynodiol diacetate **(297-76-7)**
β-Ethynodiol diacetate **(297-76-7)**
17-α-Ethynyl-17-β-acetoxy-19-norandrost-4-en-3-one **(51-98-9)**
Ethynylbenzene **(536-74-3)**
2-Ethynyl-2-butanol **(77-75-8)**
Ethynylcarbinol **(107-19-7)**
1-Ethynylcyclohexanol carbamate **(126-52-3)**
1-Ethynylcyclohexyl carbamate **(126-52-3)**
17-α-Ethynyl-3,17-dihydroxy-4-estrene diacetate **(297-76-7)**
17-Ethynyl-3,17-dihydroxy-1,3,5-oestratriene **(57-63-6)**
Ethynyldimethylcarbinol **(115-19-5)**
17-α-Ethynyl-6-α,21-dimethyltestosterone **(79-64-1)**
17-α-Ethynylestr-4-ene-3-β,17-β-diol acetate **(297-76-7)**
17-α-Ethynylestradiol **(57-63-6)**
Ethynylestradiol **(57-63-6)**
17-α-Ethynylestradiol-l7-β **(57-63-6)**
17-Ethynylestradiol 3-methyl ether **(72-33-3)**
17-α-Ethynylestradiol 3-methyl ether **(72-33-3)**
Ethynylestradiol 3-methyl ether **(72-33-3)**
17-α-Ethynyl-1,3,5(10)-estratriene-3,17-β-diol **(57-63-6)**
17-α-Ethynylestra-1,3,5(10)-triene-3,17-β-diol **(57-63-6)**
17-α-Ethynyl-4-estrene-3-β,17-β-diol diacetate **(297-76-7)**
17-α-Ethynyl-4-estrene-3-β,17-diol diacetate **(297-76-7)**
17-α-Ethynylestr-4-en-17-β-ol **(52-76-6)**
17-α-Ethynylestrenol **(52-76-6)**
Ethynylestrenol **(52-76-6)**
17-α-Ethynyl-4-estren-17-ol-3-one **(68-22-4)**
17-α-Ethynyl-5(10)-estren-17-ol-3-one **(68-23-5)**
17-α-Ethynylestr-5(10)-en-17-β-ol-3-one **(68-23-5)**
17-α-Ethynyl-estr-5(10)-en-3-on-17-β-ol **(68-23-5)**
17-α-Ethynyl-17-hydroxy-6-α,21-dimethylandrost-4-en-3-one **(79-64-1)**
17-α-Ethynyl-17-β-hydroxy-4-estren-3-one **(68-22-4)**
17-α-Ethynyl-17-β-hydroxy-5(10)-estren-3-one **(68-23-5)**

17-α-Ethynyl-17-β-hydroxy-δ$^{5(10)}$-estren-3-one **(68-23-5)**
17-α-Ethynyl-17-β-hydroxyestr-4-en-3-one **(68-22-4)**
17-α-Ethynyl-17-β-hydroxyestr-5(10)-en-3-one **(68-23-5)**
17-α-Ethynyl-17-hydroxy-4-estren-3-one **(68-22-4)**
17-α-Ethynyl-17-hydroxy-5(10)-estren-3-one **(68-23-5)**
17-α-Ethynyl-17-hydroxyestr-5(10)-en-3-one **(68-23-5)**
17-α-Ethynyl-17-hydroxyestr-4-en-3-one acetate **(51-98-9)**
(+)-17-α-Ethynyl-17-β-hydroxy-3-methoxy-1,3,5(10)-estratriene **(72-33-3)**
(+)-17-α-Ethynyl-17-β-hydroxy-3-methoxy-1,3,5(10)-oestratriene **(72-33-3)**
17-α-Ethynyl-17-β-hydroxy-19-norandrost-4-en-3-one **(68-22-4)**
17-α-Ethynyl-17-β-hydroxy-3-oxo-δ$^{5(10)}$-estrene **(68-23-5)**
17-Ethynyl-3-methoxy-1,3,5(10)-estratrien-17-β-ol **(72-33-3)**
17-α-Ethynyl-3-methoxy-1,3,5(10)-estratrien-17-β-ol **(72-33-3)**
17-α-Ethynyl-3-methoxy-17-β-hydroxy-δ-1,3,5(10)-estratriene **(72-33-3)**
17-α-Ethynyl-3-methoxy-17-β-hydroxy-δ-1,3,5(10)-oestratriene **(72-33-3)**
17-Ethynyl-3-methoxy-1,3,5(10)-oestratien-17-β-ol **(72-33-3)**
17-Ethynyl-18-methyl-19-nortestosterone **(797-63-7)**
17-α-Ethynyl-19-norandrost-4-ene-3-β,17-β-diol diacetate **(297-76-7)**
17-α-Ethynyl-19-nor-4-androsten-17-β-ol-3-one **(68-22-4)**
17-α-Ethynyl-19-nor-5(10)-androsten-17-β-ol-3-one **(68-23-5)**
17-α-Ethynyl-19-norandrost-4-en-17-β-ol-3-one **(68-22-4)**
17-α-Ethynyl-19-nortestosterone **(68-22-4)**
17-α-Ethynyl-19-nortestosterone acetate **(51-98-9)**
17-α-Ethynyloestr-4-en-17-β-ol **(52-76-6)**
17-Ethynyloestradiol **(57-63-6)**
17-α-Ethynyl-17-β-oestradiol **(57-63-6)**
17-α-Ethynyloestradiol **(57-63-6)**
17-α-Ethynyloestradiol-17-β **(57-63-6)**
Ethynyloestradiol **(57-63-6)**
17-Ethynyloestradiol 3-methyl ether **(72-33-3)**
17-α-Ethynyloestradiol 3-methyl ether **(72-33-3)**
17-α-Ethynyloestradiol methyl ether **(72-33-3)**
Ethynyloestradiol 3-methyl ether **(72-33-3)**
Ethynyloestradiol methyl ether **(72-33-3)**
17-Ethynyloestra-1,3,5(10)-triene-3,17-β-diol **(57-63-6)**
17-α-Ethynyl-1,3,5(10)-oestratriene-3,17-β-diol **(57-63-6)**
17-α-Ethynyl-1,3,5-oestratriene-3,17-β-diol **(57-63-6)**
17-α-Ethynyloestra-1,3,5(10)-triene-3,17-β-diol **(57-63-6)**
17-α-Ethynyloestrenol **(52-76-6)**
Ethyonomide **(536-33-4)**
Ethyrimol **(23947-60-6)**
Eticol **(311-45-5)**
Eticyclin **(57-63-6)**
Eticyclol **(57-63-6)**
Etidocaine **(36637-18-0)**
Etidronate disodium **(7414-83-7)**
Etidronic acid **(2809-21-4)**
Etil 702 **(118-82-1)**
Etil acrilato (Italian) **(140-88-5)**
Etilacrilatului (Romanian) **(140-88-5)**
Etilamfetamina (Spanish) **(457-87-4)**
Etilamfetamine **(457-87-4)**
Etilamfetaminum (Latin) **(457-87-4)**
Etilamina (Italian) **(75-04-7)**
Etilbenzene (Italian) **(100-41-4)**
Etilbutilchetone (Italian) **(106-35-4)**
Etil clorocarbonato (Italian) **(541-41-3)**
Etil cloroformiato (Italian) **(541-41-3)**
Etile (acetato di) (Italian) **(141-78-6)**
Etile (formiato di) (Italian) **(109-94-4)**
Etilen-Xantisan Tabl. **(317-34-0)**
N,N'-Etilen-bis(ditiocarbammato) di manganese (Italian) **(12427-38-2)**
N,N'-Etilen-bis(ditiocarbammato) di sodio (Italian) **(142-59-6)**
Etilene (ossido di) (Italian) **(75-21-8)**
Etilenimina (Italian) **(151-56-4)**
Etilfen **(50-06-6)**
Etilmercaptano (Italian) **(75-08-1)**
O-Etil-O-((4-nitro-fenil)-fenil)-monotiofosfonato (Italian) **(2104-64-5)**
Etilon **(56-38-2)**
Etimid **(536-33-4)**
Etin **(379-79-3)**
Etinamate **(126-52-3)**
Etinestrol **(57-63-6)**
Etinestryl **(57-63-6)**
Etinoestryl **(57-63-6)**
EtioCIdan **(536-33-4)**
Etiol **(121-75-5)**
Etionamid **(536-33-4)**
Etionid **(536-33-4)**
Etionizina **(536-33-4)**
Etionizine **(536-33-4)**
Etistradiol **(57-63-6)**
Etoat metylowy (Polish) **(116-01-8)**
Etobedolum **(911-65-9)**
Etocil **(938-73-8)**
Etoglucid **(1954-28-5)**
Etoglucide **(1954-28-5)**
Etoksyetylowy alkohol (Polish) **(110-80-5)**
Etomal **(77-67-8)**
Etonitazene **(911-65-9)**
Etonitazeno (Spanish) **(911-65-9)**
Etonitazenum (Latin) **(911-65-9)**
Etonitazine **(911-65-9)**
Etonitazinum **(911-65-9)**
Etorfina **(14521-96-1)**
(-)-Etorphine **(14521-96-1)**
7-α-Etorphine **(14521-96-1)**
Etorphine **(14521-96-1)**
Etorphine (Except hydrochloride salt) **(14521-96-1)**
Etorphine hydrochloride **(13764-49-3)**
Etorphinum (Latin) **(14521-96-1)**
Etosalicil **(938-73-8)**
Etosalicyl **(938-73-8)**
Etosseridina **(469-82-9)**
2-Etossietil-acetato (Italian) **(111-15-9)**
Etosuximida **(77-67-8)**
Etoval **(77-28-1)**
Etoxeridina (Spanish) **(469-82-9)**
Etoxeridine **(469-82-9)**
Etoxeridinum (Latin) **(469-82-9)**
Etoxinol **(6012-83-5)**
Etoxon AF5 **(9014-90-8)**
Etoxzolamide **(452-35-7)**
Etrenol **(23255-93-8)**
Etretinate **(54350-48-0)**
Etretinato (Spanish) **(54350-48-0)**
Etretinatum (Latin) **(54350-48-0)**
Etridiazol **(2593-15-9)**
Etridiazole **(2593-15-9)**
Etrimfos **(38260-54-7)**
Etriol **(77-99-6)**
Etroflex **(532-03-6)**
Etrofol **(3942-54-9)**
Etrofolan **(2631-40-5)**
Etrol OEM **(9004-35-7)**
Etrolene **(299-84-3)**
Etrozolidina **(129-20-4)**
Etsan **(1134-23-2)**
Ettriol **(77-99-6)**
Etylenu tlenek (Polish) **(75-21-8)**
2-Etylo-6-metylo-N-(1'-metylo-2'-metoksyetylo)chloro-acetanilid (Polish) **(51218-45-2)**
Etyloamina (Polish) **(75-04-7)**
Etylobenzen (Polish) **(100-41-4)**

Etylon	**(71-91-0)**
Etylowy alkohol (Polish)	**(64-17-5)**
Etylparation (Czech)	**(56-38-2)**
Etylu bromek (Polish)	**(74-96-4)**
Etylu chlorek (Polish)	**(75-00-3)**
Etylu krzemian (Polish)	**(78-10-4)**
Eubasin	**(144-83-2)**
Eubasinum	**(144-83-2)**
β-Eucaine	**(500-34-5)**
Eucaine B	**(500-34-5)**
Eucalyptol	**(470-82-6)**
Eucalyptole	**(470-82-6)**
Eucalyptus Citriodora	**(8000-48-4)**
Eucalyptus Citriodora Distillate	**(8000-48-4)**
Eucalyptus Citriodora Oil	**(8000-48-4)**
Eucalyptus Globulus Distillate	**(8000-48-4)**
Eucalyptus Oil	**(8000-48-4)**
Eucalyptus Oil Citriodora	**(8000-48-4)**
Eucanine GB	**(95-80-7)**
Euchessina	**(77-09-8)**
Eucheuma Spinosum Gum	**(9000-07-1)**
Eucheuma Spinosum Gum	**(9062-07-1)**
Euchrysine	**(494-38-2)**
Euclorina	**(127-65-1)**
Eucodin	**(125-27-9)**
Eucopon	**(467-85-6)**
Eucoran	**(59-26-7)**
Eudalene (6CI)	**(490-65-3)**
Eudalin	**(490-65-3)**
Eudesma-3,11-diene (8CI)	**(473-13-2)**
Eudesma-4(14),11-diene (8CI)	**(17066-67-0)**
Eudesma-1,4-dien-12-oic acid, 6-α-hydroxy-3-oxo-, γ-lactone, (11S)	**(481-06-1)**
Eudesma-1,4-dien-12-oic acid, 6-α-hydroxy-3-oxo-, γ-lactone, (11S)-(-)-	**(481-06-1)**
4αH-Eudesmane (8CI)	**(30824-81-8)**
Eudesmic acid	**(118-41-2)**
Eudesmol	**(77-53-2)**
Eufilina (Polish)	**(317-34-0)**
Eufin	**(105-58-8)**
Euflavin	**(8048-52-0)**
Euflavine	**(8048-52-0)**
Euflavine	**(86-40-8)**
Eufodrin	**(300-42-5)**
Eufodrinal	**(300-42-5)**
Eugenic acid	**(97-53-0)**
Eugenin	**(480-34-2)**
1,3,4-Eugenol	**(97-53-0)**
Eugenol	**(97-53-0)**
p-Eugenol	**(97-53-0)**
1,3,4-Eugenol acetate	**(93-28-7)**
Eugenol acetate	**(93-28-7)**
Eugenol formate	**(10031-96-6)**
1,3,4-Eugenol methyl ether	**(93-15-2)**
Eugenyl acetate	**(93-28-7)**
Eugenyl formate	**(10031-96-6)**
Eugenyl methyl ether	**(93-15-2)**
Eugerase	**(51-05-8)**
Euglycin	**(565-33-3)**
Euhaemon	**(68-19-9)**
Euhypnos	**(846-50-4)**
Eukain B	**(500-34-5)**
Eukalyptol (Czech)	**(470-82-6)**
Eukalyptus Oel (German)	**(8000-48-4)**
Eukystol	**(52-86-8)**
Eulan SP	**(68359-37-5)**
Eulan WA New (9CI)	**(55069-01-7)**
Eumicton	**(59-66-5)**
Eumin	**(443-48-1)**
Eunasin	**(3813-05-6)**
Eunatrol	**(143-19-1)**
Euneryl	**(50-06-6)**
Eunoctal	**(57-43-2)**
Eunoctin	**(146-22-5)**
Eunoktin	**(146-22-5)**
Euparen	**(1085-98-9)**
Euparen M	**(731-27-1)**
Euparene	**(1085-98-9)**
Euphobine	**(60-13-9)**
Euphodine	**(60-13-9)**
Euphodrin	**(300-42-5)**
Euphodrinal	**(300-42-5)**
Euphodyn	**(60-13-9)**
Euphorin	**(101-99-5)**
Euphyllin	**(317-34-0)**
Euphylline	**(317-34-0)**
Euphylline	**(58-55-9)**
Eupirina	**(636-46-4)**
Eupramin	**(50-49-7)**
Eureceptor	**(51481-61-9)**
Eurekene	**(1069-66-5)**
Euresol	**(102-29-4)**
Eurex	**(1134-23-2)**
Eurinol	**(133-67-5)**
Eurocert Amaranth	**(915-67-3)**
Eurocert Azorubine	**(3567-69-9)**
Eurocert Chrysoine S	**(547-57-9)**
Eurocert Cochineal Red A	**(2611-82-7)**
Eurocert Orange FCF	**(2783-94-0)**
Eurocert Ponceau 6R	**(5850-44-2)**
Eurocert Scarlet GN	**(3257-28-1)**
Eurocert Tartrazine	**(1934-21-0)**
Eurodin	**(29975-16-4)**
Eurodopa	**(59-92-7)**
Eurogale	**(90-33-5)**
Europium (9CI)	**(7440-53-1)**
Eurphyllin	**(317-34-0)**
Eusal	**(938-73-8)**
Eusaprim	**(723-46-6)**
Eusaprim	**(8064-90-2)**
Euscopol	**(114-49-8)**
Euspiran	**(51-30-9)**
Eustigmin	**(59-99-4)**
Eustigmin bromide	**(114-80-7)**
Eustigmine	**(59-99-4)**
Eutanol G	**(5333-42-6)**
Eucisten	**(389-08-2)**
Eutensin	**(54-31-9)**
Euthatal	**(57-33-0)**
Eucistin	**(136-40-3)**
Eutizon	**(54-85-3)**
Euufilin	**(317-34-0)**
Euvestin	**(130-80-3)**
Euvinyl Blue 702	**(147-14-8)**
Euvitol	**(1209-98-9)**
Euxyl K 100	**(50815-77-5)**
Evablin	**(314-13-6)**
Evac-Q-Kit	**(77-09-8)**
Evac-Q-Kwik	**(77-09-8)**
Evac-U-Gen	**(77-09-8)**
Evalon	**(54-85-3)**
Evans Blue	**(314-13-6)**
Evans Blue Dye	**(314-13-6)**
Evans Blue, Sodium salt	**(314-13-6)**

FD & C Red No. 19 **(81-88-9)**
FD & C Red No. 40 **(25956-17-6)**
FD & C Violet 1 **(1694-09-3)**
FD & C Violet No. 1 **(1694-09-3)**
FD & C Yellow 3 **(85-84-7)**
FD & C Yellow 4 **(131-79-3)**
FD & C Yellow 5 **(1934-21-0)**
FD & C Yellow 6 **(2783-94-0)**
FD & C Yellow Lake No. 6 **(2783-94-0)**
FD & C Yellow No. 3 **(85-84-7)**
FD & C Yellow No. 4 **(131-79-3)**
FD & C Yellow No. 5 Aluminum Lake **(12225-21-7)**
FD & C Yellow No. 5 Tartrazine **(1934-21-0)**
FD & C Yellow No. 6 **(2783-94-0)**
FD & C Yellow No. 6 Aluminium Lake **(2783-94-0)**
FD & C Yellow No. 10 **(8004-92-0)**
FDA **(696-28-6)**
FDA 0101 **(7681-49-4)**
FDA 0345 **(154-93-8)**
FDA 1446 **(584-79-2)**
FDA 1541 **(759-94-4)**
FDN **(957-51-7)**
F3DThd **(70-00-8)**
FDUR **(50-91-9)**
FE3 **(8075-80-7)**
FEMA 3150 **(13925-07-0)**
FEMA 3186 **(644-08-6)**
FES **(17924-92-4)**
FF **(3688-53-7)**
FFB 32 **(7784-30-7)**
FG 10 **(63148-62-9)**
FG 2000 **(79-94-7)**
FG 4000 **(632-79-1)**
FG 834 **(9003-53-6)**
FH 099 **(126-27-2)**
FH 122-A **(6533-00-2)**
F. I. 6150 **(53-34-9)**
F.I 106 **(23214-92-8)**
F.I. 58-30 **(536-33-4)**
FI 106 **(25316-40-9)**
FI 6120 **(739-71-9)**
FI 6804 **(25316-40-9)**
FI6339 **(20830-81-3)**
FI Clor 71 **(2782-57-2)**
FI Clor 91 **(87-90-1)**
FI Clor 60S **(2893-78-9)**
FIO **(2152-34-3)**
FIP **(60-51-5)**
6FK **(684-16-2)**
FKS **(16961-83-4)**
FL **(19937-59-8)**
FM 510 **(9002-88-4)**
FMA **(3570-80-7)**
FMA **(62-38-4)**
FMA (Analytical reagent) **(3570-80-7)**
FMB CAP B **(61789-40-0)**
(+-)-cis-FMC 33297 **(52341-33-0)**
FMC 249 **(584-79-2)**
FMC 2995 **(1918-18-9)**
FMC 4512 **(2307-68-8)**
FMC 4556 **(2164-09-2)**
FMC 5273 **(51-03-6)**
FMC 5462 **(115-29-7)**
FMC 5767 **(2778-04-3)**
FMC 9044 **(485-31-4)**
FMC 10242 **(1563-66-2)**
FMC 11092 **(4849-32-5)**
FMC 17370 **(10453-86-8)**
FMC 18739 **(28434-01-7)**
FMC 30980 **(52315-07-8)**
FMC 33297 **(52645-53-1)**
FMC 35001 **(55285-14-8)**
FMC 35171 **(52341-33-0)**
FMC 41655 **(52645-53-1)**
FMC 45497 **(52315-07-8)**
FMC 45498 **(52918-63-5)**
FMC 45806 **(52315-07-8)**
FMC 57020 **(81777-89-1)**
FMC-1240 **(563-12-2)**
FMC-9102 **(9006-42-2)**
FMC-9260 **(7696-12-0)**
FM-NTS **(9004-70-0)**
FN 20 **(9003-54-7)**
FN 25 **(9003-54-7)**
4-F-3NA **(364-76-1)**
FNT **(3570-75-0)**
FP 4 **(9002-88-4)**
FPA **(51-65-0)**
FPF 1002 **(51481-61-9)**
p-FPHE **(60-17-3)**
FPL 670 **(15826-37-6)**
FR 2 **(78-43-3)**
FR 28 **(1330-43-4)**
FR 40 (Petroleum resin) **(64742-16-1)**
FR 80 (Petroleum resin) **(64742-16-1)**
FR 100 (Petroleum resin) **(64742-16-1)**
FR 214 **(9004-78-8)**
FR 222 **(1338-23-4)**
FR 300 **(1163-19-5)**
FR 300BA **(1163-19-5)**
FR 1138 **(3296-90-0)**
FRP 53 **(1163-19-5)**
FSH **(9002-68-0)**
FSH-P **(9002-68-0)**
FSR 3 **(54-85-3)**
F3T **(70-00-8)**
FT **(794-93-4)**
FT 8 **(73-48-3)**
FT 257 **(9003-11-6)**
FT-207 **(51-21-8)**
F3TDR **(70-00-8)**
FTR 6100 **(64742-16-1)**
F1-Tabs **(7681-49-4)**
F-2 Toxin **(17924-92-4)**
5-FU **(51-21-8)**
FU **(51-21-8)**
5-FUDR **(50-91-9)**
FUDR **(50-91-9)**
5-FUR **(316-46-1)**
FUR **(316-46-1)**
FW 50 **(13983-17-0)**
FW 200 (Mineral) **(13983-17-0)**
FW 293 **(115-32-2)**
FW 325 **(13983-17-0)**
FW 734 **(709-98-8)**
FW 925 **(1836-75-5)**
FWH-352 **(2152-34-3)**
FW-XO **(14807-96-6)**
FZ 132 **(63148-62-9)**
Fabedrine **(60-13-9)**
Fabritone PE **(9002-88-4)**
Faccla **(868-14-4)**
Faccula **(868-14-4)**
Factitious Air **(10024-97-2)**

Factor A1	(16561-29-8)
Factor A1 (Croton oil)	(16561-29-8)
Factor A1 (VAN)	(16561-29-8)
Factor II	(68-19-9)
Factor II (vitamin)	(68-19-9)
Factor PP	(98-92-0)
Factor S	(58-85-5)
Factor S (vitamin)	(58-85-5)
Fadormir	(72-44-6)
Faecla	(868-14-4)
Faecula	(868-14-4)
Fagine	(123-41-1)
Fair 30	(123-33-1)
Fair PS	(123-33-1)
Falapen	(113-98-4)
Faligruen	(1332-40-7)
Falisan	(123-88-6)
Falithion	(122-14-5)
Falithrom	(435-97-2)
Falitiram	(137-26-8)
Falkitol	(67-72-1)
Fall	(7775-09-9)
Falliocor	(390-64-7)
Falodin	(8005-49-0)
Falone	(8005-49-0)
Falone E 44	(94-84-8)
Falone-44-E	(94-84-8)
Famfos	(13171-21-6)
Famfos	(52-85-7)
Famid	(6988-21-2)
Famofos	(52-85-7)
Famophos	(52-85-7)
Famophos warbex	(52-85-7)
Famosept	(102-98-7)
Famphos	(52-85-7)
Famphur	(52-85-7)
Fanal Blue BG Supra Powder	(1325-87-7)
Fanal Blue B Supra	(1325-87-7)
Fanal Pink GFK	(989-38-8)
Fanal Red 25532	(989-38-8)
Fanal Violet RA Supra	(1325-82-2)
Fanal Violet R Supra	(1325-82-2)
Fanatone Blue B	(1325-87-7)
Fanatone Violet	(1325-82-2)
Fanchon Yellow G-YH 1	(2512-29-0)
Fanchon Yellow WD 259	(2512-29-0)
Faneron	(13181-17-4)
Fanfos	(52-85-7)
Fannoform	(50-00-0)
Fanodormo	(52-31-3)
Farbenfabriken Bayer	(299-45-6)
Farbruss	(1333-86-4)
Faredina	(50-59-9)
Fargan	(58-33-3)
Fargan	(60-87-7)
Far-Go	(2303-17-5)
Farinex 100	(9005-25-8)
Faringosept	(539-21-9)
Farlutal	(520-85-4)
Farlutin	(71-58-9)
Farmaid	(1630-17-7)
Farmco	(94-75-7)
Farmco Diuron	(330-54-1)
Farmco Fence Rider	(93-76-5)
Farmco atrazine	(1912-24-9)
Farmco propanil	(709-98-8)
Farmicetina	(56-75-7)
Farmiserine	(68-41-7)
Farmotal	(76-75-5)
Farnesane	(3891-98-3)
Farnesol	(4602-84-0)
Farnesyl alcohol	(4602-84-0)
Fartox	(82-68-8)
Fas-Cile	(57-53-4)
Fasciolin	(56-23-5)
Fasciolin	(67-72-1)
Fasco Fascrat Powder	(81-81-2)
Fasco-Terpene	(8001-35-2)
Fasco Wy-HOE	(101-21-3)
Faserton	(1344-28-1)
Fasertonerde	(1344-28-1)
Fast Acid Blue RL	(3861-73-2)
Fast Acid Green N	(5141-20-8)
Fast Acid Magenta	(3567-66-6)
Fast Acid Magenta B	(3567-66-6)
Fast Acid Red G	(1658-56-6)
Fast Acid Violet 5BN	(1694-09-3)
Fast Benzidene Orange YB 3	(3520-72-7)
Fast Blue B	(20282-70-6)
Fast Blue BB Base	(120-00-3)
Fast Blue BBN	(120-00-3)
Fast Blue B Base	(119-90-4)
Fast Blue BN Salt	(119-90-4)
Fast Blue B Supra	(1325-87-7)
Fast Blue Base BB	(120-00-3)
Fast Blue DSC Base	(119-90-4)
Fast Blue DS Salt	(119-90-4)
Fast Blue EB Base	(120-00-3)
Fast Blue Lake	(1325-87-7)
Fast Blue R Salt	(101-54-2)
Fast Blue Salt B	(119-90-4)
Fast Blue Salt BN	(119-90-4)
Fast Blue Toner B	(1325-87-7)
Fast Bronze Violet	(1325-82-2)
Fast Brown RR Salt	(609-20-1)
Fast Corinth Base B	(92-87-5)
Fast Crimson GR	(3734-67-6)
Fast Dark Blue Base R	(119-93-7)
Fast Disperse Yellow 2K	(119-15-3)
Fast Drimson GR	(3734-67-6)
Fast Fuchsin G	(4197-07-3)
Fast Fuchsine G	(4197-07-3)
Fast Garnet Base B	(134-32-7)
Fast Garnet GBC Base	(97-56-3)
Fast Green	(569-64-2)
Fast Green FCF	(2353-45-9)
Fast Green J	(633-03-4)
Fast Green JJO	(633-03-4)
Fast Green O	(569-64-2)
Fast Light Orange GA	(1936-15-8)
Fast Light Orange GA-CF	(1936-15-8)
Fast Light Yellow E	(6373-74-6)
Fast Oil Orange	(842-07-9)
Fast Oil Orange I	(842-07-9)
Fast Oil Orange II	(3118-97-6)
Fast Oil Orange T	(2051-85-6)
Fast Oil Pink B	(509-34-2)
Fast Oil Red B	(85-83-6)
Fast Oil Scarlet III	(85-86-9)
Fast Oil Yellow	(97-56-3)
Fast Oil Yellow 64403	(2481-94-9)
Fast Oil Yellow B	(60-11-7)
Fast Oil Yellow 2G	(1689-82-3)
Fast Oil Yellow 2G	(2051-85-6)

Name	CAS No.
Fast Oil Yellow G	(2051-85-6)
Fast Orange	(842-07-9)
Fast Orange Base GC	(141-85-5)
Fast Orange Base GR	(88-74-4)
Fast Orange Base JR	(88-74-4)
Fast Orange Base JS	(141-85-5)
Fast Orange Base R	(99-09-2)
Fast Orange G	(1936-15-8)
Fast Orange G	(3520-72-7)
Fast Orange G Base	(141-85-5)
Fast Orange GC Base	(108-42-9)
Fast Orange GC New Salt	(141-85-5)
Fast Orange GR Base	(88-74-4)
Fast Orange GR Salt	(88-74-4)
Fast Orange JS Salt	(141-85-5)
Fast Orange M Base	(99-09-2)
Fast Orange MC Base	(141-85-5)
Fast Orange MC Salt	(141-85-5)
Fast Orange MM Base	(99-09-2)
Fast Orange O Base	(88-74-4)
Fast Orange O Salt	(88-74-4)
Fast Orange 3R	(2814-77-9)
Fast Orange R Base	(99-09-2)
Fast Orange 3RJ	(2814-77-9)
Fast Orange R Salt	(99-09-2)
Fast Orange Salt GC	(141-85-5)
Fast Orange Salt GCS	(141-85-5)
Fast Orange Salt GR	(88-74-4)
Fast Orange Salt JR	(88-74-4)
Fast Pink Y	(2379-74-0)
Fast Purpurine	(6548-29-4)
Fast Red	(915-67-3)
Fast Red A	(1658-56-6)
Fast Red A	(2425-85-6)
Fast Red AB	(2425-85-6)
Fast Red AE	(1658-56-6)
Fast Red AG	(1658-56-6)
Fast Red ALS	(1658-56-6)
Fast Red A (Pigment)	(2425-85-6)
Fast Red AV	(1658-56-6)
Fast Red B	(97-52-9)
Fast Red BB	(85-83-6)
Fast Red BB Base	(134-29-2)
Fast Red B Base	(97-52-9)
Fast Red Base B	(97-52-9)
Fast Red Base GG	(100-01-6)
Fast Red Base GL	(89-62-3)
Fast Red Base 3GL Special	(89-63-4)
Fast Red Base 2J	(100-01-6)
Fast Red Base 3JL	(89-63-4)
Fast Red Base JL	(89-62-3)
Fast Red Base RL	(99-52-5)
Fast Red Base TR	(95-69-2)
Fast Red 5CT Base	(95-69-2)
Fast Red 5CT Salt	(3165-93-3)
Fast Red E (6CI)	(2302-96-7)
Fast Red 2G Base	(100-01-6)
Fast Red G Base	(89-62-3)
Fast Red GG Base	(100-01-6)
Fast Red GG Salt	(100-01-6)
Fast Red GL	(89-62-3)
Fast Red 3GL Base	(89-63-4)
Fast Red GL Base	(89-62-3)
Fast Red 3GL Salt	(89-63-4)
Fast Red 3GL special Base	(89-63-4)
Fast Red 3GL special Salt	(89-63-4)
Fast Red 2G Salt	(100-01-6)
Fast Red J	(2425-85-6)
Fast Red JE	(2425-85-6)
Fast Red KB Amine	(95-79-4)
Fast Red KB Base	(95-79-4)
Fast Red KBS Salt	(95-79-4)
Fast Red KB Salt	(95-79-4)
Fast Red KB Salt Supra	(95-79-4)
Fast Red MA	(1658-56-6)
Fast Red MGL Base	(89-62-3)
Fast Red MP Base	(100-01-6)
Fast Red 5NA Base	(97-52-9)
Fast Red 2NC Base	(89-63-4)
Fast Red 2NC Salt	(89-63-4)
Fast Red 5NT	(99-52-5)
Fast Red 3NT Base	(89-62-3)
Fast Red 3NT Salt	(89-62-3)
Fast Red 5NT Salt	(99-52-5)
Fast Red P Base	(100-01-6)
Fast Red P Salt	(100-01-6)
Fast Red R	(2425-85-6)
Fast Red R	(85-86-9)
Fast Red RL Base	(99-52-5)
Fast Red S	(1658-56-6)
Fast Red SG Base	(99-55-8)
Fast Red SGG Base	(95-82-9)
Fast Red Salt GG	(100-01-6)
Fast Red Salt 3GL	(89-63-4)
Fast Red Salt 2J	(100-01-6)
Fast Red Salt 3JL	(89-63-4)
Fast Red Salt RL	(99-52-5)
Fast Red Salt TR	(3165-93-3)
Fast Red Salt TRA	(3165-93-3)
Fast Red Salt TRN	(3165-93-3)
Fast Red Specially Pure	(2302-96-7)
Fast Red TR	(95-69-2)
Fast Red TR11	(95-69-2)
Fast Red TR Base	(95-69-2)
Fast Red TRO Base	(95-69-2)
Fast Red TR Salt	(3165-93-3)
Fast Scarlet	(992-59-6)
Fast Scarlet 4BSA	(3441-14-3)
Fast Scarlet Base B	(91-59-8)
Fast Scarlet Base G	(99-55-8)
Fast Scarlet Base GGT	(95-82-9)
Fast Scarlet Base 2J	(95-82-9)
Fast Scarlet Base J	(99-55-8)
Fast Scarlet Base 2JS	(95-82-9)
Fast Scarlet DS Base	(95-82-9)
Fast Scarlet 2G	(95-82-9)
Fast Scarlet G	(99-55-8)
Fast Scarlet 2G Base	(95-82-9)
Fast Scarlet G Base	(99-55-8)
Fast Scarlet GC Base	(99-55-8)
Fast Scarlet GG Base	(95-82-9)
Fast Scarlet GGS Base	(95-82-9)
Fast Scarlet G Salt	(99-55-8)
Fast Scarlet J Salt	(99-55-8)
Fast Scarlet MDC Base	(95-82-9)
Fast Scarlet M4NT Base	(99-55-8)
Fast Scarlet R	(99-59-2)
Fast Scarlet S	(3567-65-5)
Fast Scarlet T Base	(99-55-8)
Fast Scarlet TR Base	(87-60-5)
Fast Silk Yellow SH	(6375-55-9)
Fast Spirit Yellow	(60-09-3)
Fast Spirit Yellow AAB	(60-09-3)
Fast Sulon Black BN	(1064-48-8)

Name	CAS No.
Fast Violet Toner R	**(1325-82-2)**
Fast White	**(7446-14-2)**
Fast Wool Blue R	**(3861-73-2)**
Fast Yellow	**(60-11-7)**
Fast Yellow AT	**(97-56-3)**
Fast Yellow G	**(1325-37-7)**
Fast Yellow GC Base	**(95-51-2)**
Fast Yellow J	**(2512-29-0)**
Fast Yellow JT	**(2512-29-0)**
Fastac	**(67375-30-8)**
Fastballs	**(51-63-8)**
Fastel Violet R	**(1325-82-2)**
Fastel Violet R Supra	**(1325-82-2)**
Fast Garnet B Base	**(134-32-7)**
Fastogen Blue 5007	**(147-14-8)**
Fastogen Blue 5110	**(147-14-8)**
Fastogen Blue B	**(147-14-8)**
Fastogen Blue FGF	**(147-14-8)**
Fastogen Blue FP-3100	**(3468-11-9)**
Fastogen Blue FSN	**(147-14-8)**
Fastogen Blue SH-100	**(3468-11-9)**
Fastogen Blue TGR	**(147-14-8)**
Fastogen Green 5005	**(1328-53-6)**
Fastogen Green B	**(1328-53-6)**
Fastogen Green Y	**(14302-13-7)**
Fastogen Green 2YK	**(14302-13-7)**
Fastolite Brown BRL	**(16071-86-6)**
Fastolite Red 8BL	**(2610-11-9)**
Fastolux Blue	**(147-14-8)**
Fastolux Green	**(1328-53-6)**
Fastona Orange G	**(3520-72-7)**
Fastona Red B	**(2425-85-6)**
Fastona Red 2G	**(3468-63-1)**
Fastona Red R	**(2814-77-9)**
Fastona Scarlet RL	**(2425-85-6)**
Fastona Scarlet YS	**(2425-85-6)**
Fastona Yellow G	**(2512-29-0)**
Fastona Yellow G Transparent	**(2512-29-0)**
Fastum	**(22071-15-4)**
Fastusol Blue 9GLP	**(28407-37-6)**
Fastusol Brown LBRSA	**(16071-86-6)**
Fastusol Brown LBRSN	**(16071-86-6)**
Fastusol Red 4BA-CF	**(2610-11-9)**
Fastusol Yellow L5GA	**(10190-68-8)**
neo-Fat 8	**(124-07-2)**
Fat Brown 2G	**(6416-57-5)**
Fat Brown GG	**(495-54-5)**
Fat Brown 2R	**(6416-57-5)**
Fat Brown RR	**(6416-57-5)**
Fat Orange 4A	**(842-07-9)**
Fat Orange A	**(2051-85-6)**
Fat Orange G	**(2051-85-6)**
Fat Orange G	**(842-07-9)**
Fat Orange GS	**(2051-85-6)**
Fat Orange I	**(842-07-9)**
Fat Orange II	**(2646-17-5)**
Fat Orange R	**(842-07-9)**
Fat Orange RG	**(2051-85-6)**
Fat Orange RR	**(2646-17-5)**
Fat Orange RS	**(842-07-9)**
Fat Ponceau R	**(85-83-6)**
Fat Red 2B	**(85-83-6)**
Fat Red 7B	**(6368-72-5)**
Fat Red B	**(85-83-6)**
Fat Red BB	**(85-83-6)**
Fat Red BS	**(85-83-6)**
Fat Red G	**(85-86-9)**
Fat Red HRR	**(85-86-9)**
Fat Red R	**(85-86-9)**
Fat Red RS	**(85-86-9)**
Fat Red TS	**(85-83-6)**
Fat Red (bluish)	**(85-86-9)**
Fat Scarlet 2G	**(3118-97-6)**
Fat Scarlet LB	**(85-86-9)**
Fat Soluble Anthraquinone Green	**(128-80-3)**
Fat Soluble Green Anthraquinone	**(128-80-3)**
Fat Soluble Red ZH	**(85-86-9)**
Fat Victoria Yellow D	**(2051-85-6)**
Fat Yellow	**(60-11-7)**
Fat Yellow A	**(60-11-7)**
Fat Yellow AAB	**(60-09-3)**
Fat Yellow AD OO	**(60-11-7)**
Fat Yellow B	**(97-56-3)**
Fat Yellow ES	**(60-11-7)**
Fat Yellow ES Extra	**(60-11-7)**
Fat Yellow Extra Conc.	**(60-11-7)**
Fat Yellow GGN	**(2481-94-9)**
Fat Yellow R	**(60-11-7)**
Fat Yellow R (8186)	**(60-11-7)**
Fatal	**(1861-32-1)**
Fatoliamid	**(536-33-4)**
Fat red (Yellowish)	**(3118-97-6)**
Fatsco Ant Poison	**(7631-89-2)**
Fatty (C8-C18) alcohol	**(68551-07-5)**
Fatty acid, Soybean Oil, Epoxidized	**(8013-07-8)**
Fatty acid, Tall Oil, Epoxidized, Octyl ester	**(61788-72-5)**
Fatty acids	**(67254-79-9)**
Fatty acids, C10-16	**(68002-90-4)**
Fatty acids, C6-12, Me esters	**(67762-39-4)**
Fatty acids, C14-18 and C16-18-unsatd.	**(67701-06-8)**
Fatty acids, C9-11-branched, glycidyl esters	**(81412-56-8)**
Fatty acids, C9-11-branched, glycidyl esters, Polymers with castor oil, formaldehyde, 6-phenyl-1,3,5-triazine-2,4-diamine and phthalic anhydride	**(68459-31-4)**
Fatty acids, dehydrated castor-oil, Polymers with glycerol, linseed oil, phthalic anhydride and rosin	**(66071-28-1)**
Fatty acids, lanolin, iso-Pr esters	**(63393-93-1)**
Fatty acids, lanolin, isopropyl esters	**(63393-93-1)**
Fatty acids, methyl esters	**(67762-39-4)**
Fatty acids, tall oil	**(61790-12-3)**
Fatty acids, tall-oil, Copper salts	**(61789-22-8)**
Fatty acids, tall oil, monoesters with polyethylene glycol	**(61791-00-2)**
Fatty acids, tallow	**(61790-37-2)**
Fatty acids, tall-oil, Polymers with benzoic acid, pentaerythritol and phthalic anhydride	**(66070-84-6)**
Fatty acids, tall-oil, Polymers with glycerol and phthalic anhydride	**(66070-71-1)**
Fatty acids, tall-oil, ethoxylated	**(61791-00-2)**
Fatty acids, tall-oil, 2-ethylhexyl esters	**(68334-13-4)**
Fatty alcohols	**(68603-15-6)**
Faustan,	**(439-14-5)**
Favistan	**(60-56-0)**
Fe-Dextran	**(9004-66-4)**
Febrilix	**(103-90-2)**
Febrinina	**(58-15-1)**
Febro-Gesic	**(103-90-2)**
Febrolin	**(103-90-2)**
Febron	**(58-15-1)**
Febuzina	**(50-33-9)**
Fecama	**(62-73-7)**
Fecto	**(1333-86-4)**
Fectrim	**(723-46-6)**
Fectrim	**(8064-90-2)**
Fedacin	**(59-87-0)**

Fedal-UN (127-18-4)
Federal Fast Violet 7001 (1325-82-2)
Fedrin (299-42-3)
Feen-A-Mint Gum (77-09-8)
Feeno (92-84-2)
Fekabit (3811-04-9)
Fekama (62-73-7)
Felan (2212-67-1)
Felazine (156-51-4)
Felben (147-24-0)
Feldene (36322-90-4)
Felicur (93-54-9)
Felison (1172-18-5)
Felitrope (93-54-9)
α-Fellandrene (99-83-2)
Felling zinc oxide (1314-13-2)
Fellozine (58-33-3)
Felsamid-OI (111-05-7)
Felsamid-OPD (93-83-4)
Felsamid-SM (111-57-9)
Felsamid-SPD (93-82-3)
Felsapon-GDS (627-83-8)
Felsules (302-17-0)
Felurea (63-98-9)
Fema 3174 (3658-77-3)
Fema No. 2034 (57-06-7)
Fema No. 2045 (2835-39-4)
Fema No. 2433 (75-21-8)
Fema No. 2467 (97-53-0)
Femanthren Golden Yellow (128-66-5)
Femergin (379-79-3)
Femestral (50-28-2)
Femestrone (50-50-0)
Femestrone injection (53-16-7)
Femidyn (53-16-7)
Feminone (57-63-6)
Femma (62-38-4)
Femogen (50-28-2)
Femogex (979-32-8)
Femproporex (Spanish) (15686-61-0)
Femulen (297-76-7)
Fen-All (50-31-7)
Fenac (85-34-7)
Fenacamfamin (1209-98-9)
Fenacemid (63-98-9)
Fenacemida (Spanish) (63-98-9)
Fenacemide (63-98-9)
Fenacet Blue G (2475-45-8)
Fenacet Fast Blue FF (2475-46-9)
Fenacet Fast Blue FF (86722-66-9)
Fenacet Fast Blue FFN (2475-46-9)
Fenacet Fast Blue FFN (86722-66-9)
Fenacet Fast Pink B (116-85-8)
Fenacet Fast Pink 3BE (2872-48-2)
Fenacet Fast Violet 6B (1220-94-6)
Fenacet Fast Violet B (82-33-7)
Fenacet Fast Violet 5R (128-95-0)
Fenacet Fast Yellow G (2832-40-8)
Fenacet Fast Yellow 2R (119-15-3)
Fenacet Yellow G (2832-40-8)
Fenacetamide (63-98-9)
Fenacetil-Karbamide (63-98-9)
Fenacetin (Czech) (62-44-2)
Fenacetina (62-44-2)
Fenacilin (87-08-1)
Fenactil (50-53-3)
Fenadossone (467-84-5)
Fenadoxona (Spanish) (467-84-5)
Fenae (85-34-7)
Fenafor Red PB (6459-94-5)
Fenafor Red PG (3567-65-5)
Fenakrom Red W (130-22-3)
Fenaktyl (50-53-3)
Fenalac Blue B Disp (147-14-8)
Fenalac Green G (1328-53-6)
Fenalac Green G Disp (1328-53-6)
Fenalac Red FKB Extra (6471-49-4)
Fenalan Blue B (2666-17-3)
Fenalan Yellow E (6373-74-6)
Fenaluz Brown BRL (16071-86-6)
Fenaluz Red 4B (2610-11-9)
Fenaluz Yellow 4G (10190-68-8)
Fenam (957-51-7)
Fenamic acid (91-40-7)
Fenamin (1912-24-9)
Fenamin (60-13-9)
Fenamin Black E (1937-37-7)
Fenamin Black GR (6428-31-5)
Fenamin Black RW (2429-83-6)
Fenamin Blue 2B (2602-46-2)
Fenamin Brown M (2429-82-5)
Fenamin Brown PBL (2429-81-4)
Fenamin Green M (3626-28-6)
Fenamin Navy Blue H (2429-73-4)
Fenamin Scarlet 3B (6358-29-8)
Fenamin Sky Blue (2429-74-5)
Fenamin Sky Blue 3F (2610-05-1)
Fenamin Yellow TP (1325-37-7)
Fenamine (1912-24-9)
Fenamine (61-82-5)
Fenaminosulf (140-56-7)
Fenamiphos (ACGIH,OSHA) (22224-92-6)
Fenamizol (490-55-1)
Fenampromida (Spanish) (129-83-9)
Fenampromide (129-83-9)
Fenan Blue BCS (130-20-1)
Fenan Blue RSN (81-77-6)
Fenanthren Blue BC (130-20-1)
Fenanthren Blue RS (81-77-6)
Fenanthren Brilliant Green B (128-58-5)
Fenanthren Brilliant Orange GR (4424-06-0)
Fenanthren Brilliant Pink R (2379-74-0)
Fenanthren Brilliant Violet 2R (1324-55-6)
Fenanthren Brilliant Violet 4R (1324-55-6)
Fenanthren Brown BR (2475-33-4)
Fenanthren Golden Orange G (128-70-1)
Fenanthren Olive R (2379-81-9)
Fenanthren Pink R Spura (2379-74-0)
Fenanthren Rubine R (4203-77-4)
1,10-Fenanthrolin (Czech) (66-71-7)
Fenantoin (57-41-0)
Fenantoin (630-93-3)
Fenara (60-13-9)
Fenarimol (60168-88-9)
Fenarol (68-88-2)
Fenarol (80-77-3)
Fenarsazinchlorid (Czech) (578-94-9)
Fenarsone (121-59-5)
Fenartil (50-33-9)
Fenasal (50-65-7)
Fenate (9004-66-4)
Fenatrol (1912-24-9)
Fenatrol (85-34-7)
Fenavar (61-82-5)

Name	CAS
Fenazaflor	(14255-88-0)
Fenazil	(58-33-3)
Fenazil	(60-87-7)
Fenazo Blue Black	(1064-48-8)
Fenazo Blue SR	(3861-73-2)
Fenazo Blue XF	(129-17-9)
Fenazo Blue XI	(3844-45-9)
Fenazo Blue XR	(2650-18-2)
Fenazo Blue XV	(129-17-9)
Fenazo Eosine XG	(17372-87-1)
Fenazo Eosine XG	(548-26-5)
Fenazo Green 7G	(5141-20-8)
Fenazo Green L	(4680-78-8)
Fenazo Light Blue RA	(4368-56-3)
Fenazo Orange	(633-96-5)
Fenazo Red B	(3734-67-6)
Fenazo Red C	(3567-69-9)
Fenazo Red FG	(10169-02-5)
Fenazo Red M	(1658-56-6)
Fenazo Scarlet 2R	(3761-53-3)
Fenazo Scarlet 3R	(2611-82-7)
Fenazo Yellow M	(587-98-4)
Fenazo Yellow T	(1934-21-0)
Fenazocina (Spanish)	(127-35-5)
Fenazon (Czech)	(60-80-0)
Fenazone	(60-80-0)
Fenazox (German)	(495-48-7)
Fenazoxine	(13669-70-0)
Fenbendazol	(43210-67-9)
Fenbendazole	(43210-67-9)
Fenbital	(50-06-6)
Fenbutatin oxide	(13356-08-6)
Fenbutatin-oxyde	(13356-08-6)
Fencal	(7778-44-1)
Fencamfamin	(1209-98-9)
Fencamfamine	(1209-98-9)
Fencamfamine (French)	(1209-98-9)
Fencamfaminum (Latin)	(1209-98-9)
Fencanfamina (Spanish)	(1209-98-9)
Fence Rider	(93-76-5)
Fenchel Oel (German)	(8006-84-6)
Fenchloorfos (Dutch)	(299-84-3)
Fenchlorfos	(299-84-3)
Fenchlorfosu (Polish)	(299-84-3)
Fenchlorophos	(299-84-3)
Fenchlorphos	(299-84-3)
Fenchol	(1632-73-1)
α-Fenchol	(512-13-0)
endo-Fenchol	(512-13-0)
Fenchon	(1195-79-5)
Fenchon (German)	(1195-79-5)
(+)-Fenchone	(4695-62-9)
D(+)-Fenchone	(4695-62-9)
Fenchone	(1195-79-5)
d-Fenchone	(4695-62-9)
α-Fenchyl alcohol	(512-13-0)
Fenclor	(1336-36-3)
Fenclorac	(36616-52-1)
Fencumar	(435-97-2)
Fendilina (Spanish)	(13042-18-7)
Fendiline	(13042-18-7)
Fendilinum (Latin)	(13042-18-7)
Fendon	(103-90-2)
Fendona	(52315-07-8)
Fendona	(67375-30-8)
Fenedrin	(60-13-9)
Fenelsin	(156-51-4)
Fenelzin	(156-51-4)
Fenelzina	(156-51-4)
Fenelzine	(156-51-4)
Fenelzyna (Polish)	(51-71-8)
Fenelzyne (Polish)	(51-71-8)
Fenemal	(50-06-6)
Fenergan	(58-33-3)
Fenergan	(60-87-7)
Fenesterin	(3546-10-9)
Fenestrin	(3546-10-9)
Fenetazina	(60-87-7)
β-Fenethylalkohol (Czech)	(60-12-8)
Fenethylline	(3736-08-1)
p-Fenetidin (Czech)	(156-43-4)
1-Fenetilbiguanide cloridrato (Italian)	(834-28-6)
N'-β-Fenetilformamidiniliminourea (Italian)	(114-86-3)
Fenetilina (Spanish)	(3736-08-1)
Fenetillina	(3736-08-1)
Fenetsin	(156-51-4)
Fenetylinum	(3736-08-1)
Fenetyllinum (Latin)	(3736-08-1)
Fenfluramine	(458-24-2)
Fenformina	(114-86-3)
Fenhydren	(83-12-5)
Fenia	(62-44-2)
Fenibutasan	(50-33-9)
Fenibutazona	(50-33-9)
Fenibutol	(50-33-9)
Fenicol	(56-75-7)
Fenicol	(93-54-9)
Fenidantoin "S"	(57-41-0)
Fenidin	(101-42-8)
Fenidina	(62-44-2)
Fenidon Pink R	(2379-74-0)
Fenilalanina (Spanish)	(63-91-2)
Fenilbutazona	(50-33-9)
Fenilbutine	(50-33-9)
Fenildicloroarsina (Italian)	(696-28-6)
Fenilep	(63-98-9)
Fenilfar	(61-76-7)
Fenilidina	(50-33-9)
Fenilidrazina (Italian)	(100-63-0)
Fenilin	(83-12-5)
2-Fenilpropano (Italian)	(98-82-8)
Fenina	(62-44-2)
Fenised	(63-98-9)
Fenitoin	(630-93-3)
Fenitoina	(57-41-0)
Fenitox	(122-14-5)
Fenitrooxon	(2255-17-6)
Fenitrooxone	(2255-17-6)
Fenitrothion	(122-14-5)
Fenitrothion Amino Analog	(13306-69-9)
Fenitrothion oxon	(2255-17-6)
Fenitrotion (Hungarian)	(122-14-5)
Fenitroxon	(2255-17-6)
Fenizin	(156-51-4)
Fenizon (French)	(80-38-6)
Fenkapton (Dutch)	(2275-14-1)
Fenmedifam	(13684-63-4)
Fennel-Oil	(8006-84-6)
Fennosan	(148-24-3)
Fennosan B 100	(533-74-4)
Fenobarbital	(50-06-6)
Fenobcarb	(3766-81-2)
Fenobucarb	(3766-81-2)
Fenoflurazole	(14255-88-0)

Name	CAS No.
Fenol (Dutch, Polish)	(108-95-2)
Fenolftalein (Czech)	(77-09-8)
Fenolipuna	(143-74-8)
Fenolo (Italian)	(108-95-2)
Fenolovo	(76-87-9)
Fenolovo acetate	(900-95-8)
Fenomorfano (Spanish)	(468-07-5)
Fenoperidina (Spanish)	(562-26-5)
Fenophosphon	(327-98-0)
Fenopon CO 436	(9051-57-4)
Fenopon CO 433N	(9014-90-8)
Fenopon EP 110	(9051-57-4)
Fenopon EP 120	(9051-57-4)
Fenoprofen	(31879-05-7)
Fenopromin	(60-13-9)
Fenopromin	(60-15-1)
Fenoprop	(93-72-1)
Fenormone	(93-72-1)
Fenosed	(50-06-6)
Fenosept	(102-98-7)
Fenospen	(87-08-1)
Fenostenyl	(63-98-9)
Fenosuccimide	(86-34-0)
Fenothiazine (Dutch)	(92-84-2)
Fenothiocarb	(62850-32-2)
Fenotiazina (Italian)	(92-84-2)
Fenotone	(50-33-9)
Fenoverm	(92-84-2)
Fenox	(61-76-7)
Fenoxaprop-Ethyl	(66441-23-4)
Fenoxaprop-ethyl	(66441-23-4)
Fenoxaprop ethyl ester	(66441-23-4)
Fenoxazol	(2152-34-3)
Fenoxybenzamin	(63-92-3)
Fenoxycarb	(72490-01-8)
2-Fenoxyethanol (Czech)	(122-99-6)
Fenoxyl Black RD	(1326-82-5)
Fenoxyl Blue L	(1327-57-7)
Fenoxyl Blue LC	(1327-57-7)
Fenoxyl Blue LCR	(1327-57-7)
Fenoxyl Blue 9R	(1327-57-7)
Fenoxyl Carbon N	(51-28-5)
Fenoxypen	(132-98-9)
Fenoxypen	(87-08-1)
1-Fenoxy-2,3-propandiol (Czech)	(538-43-2)
Fenozaflor	(14255-88-0)
Fenpropanate	(39515-41-8)
Fenpropathrin	(39515-41-8)
Fenproporex	(15686-61-0)
Fenproporexum (Latin)	(15686-61-0)
Fenpyrate	(55512-33-9)
Fenretinide	(65646-68-6)
Fenson	(80-38-6)
Fensone	(80-38-6)
Fensulfothion (ACGIH,OSHA)	(115-90-2)
Fensulfothion sulfone	(14255-72-2)
Fensulfoxide	(3761-41-9)
Fentanest	(437-38-7)
Fentanil	(437-38-7)
Fentanyl	(437-38-7)
Fentazin	(58-39-9)
Fenthiaprop-ethyl	(66441-11-0)
Fenthiaprop-ethyl	(93921-16-5)
Fenthion (ACGIH,OSHA)	(55-38-9)
Fenthion 4E	(55-38-9)
Fenthione sulfone	(3761-42-0)
Fenthion sulfoxide	(3761-41-9)
Fenthoate	(2597-03-7)
Fentiazin	(92-84-2)
Fentin acetaat (Dutch)	(900-95-8)
Fentin acetat (German)	(900-95-8)
Fentin acetate	(900-95-8)
Fentin azetat (German)	(900-95-8)
Fentin chloride	(639-58-7)
Fentine acetate (French)	(900-95-8)
Fentin hydroxide	(76-87-9)
Fentoin	(57-41-0)
Fenulon	(101-42-8)
Fenural	(63-98-9)
Fenurea	(63-98-9)
Fenuron	(101-42-8)
Fenuron TCA	(4482-55-7)
Fenuron TCA Salt	(4482-55-7)
Fenurone	(63-98-9)
Fenuron trichloroacetate	(4482-55-7)
Fenvalerate	(51630-58-1)
Fenvalerate α	(66230-04-4)
Fenvalerate β	(66267-77-4)
Fenvalerate A α	(66230-04-4)
4'-Fenylacetanilid (Czech)	(4075-79-0)
Fenylacetylmocovina (Czech)	(63-98-9)
1-Fenyl-4-amino-5-chlor-6-pyridazinon (Czech)	(1698-60-8)
N-Fenyl-1-aminonaftalen (Czech)	(90-30-2)
N-Fenyl-2-aminonaftalen (Czech)	(135-88-6)
N-Fenylanilin (Czech)	(122-39-4)
Fenylbutatin oxide	(13356-08-6)
Fenylbutazon	(50-33-9)
Fenylbutylstannium oxide (Czech)	(13356-08-6)
Fenyl-cellosolve (Czech)	(122-99-6)
Fenylcelosolv (Czech)	(122-99-6)
2-Fenylcyklohexanol (Czech)	(1444-64-0)
N-Fenyl-N'-cyklohexyl-p-fenylendiamin (Czech)	(101-87-1)
Fenyldichlorarsin (Czech)	(696-28-6)
Fenyl-(1-diethylaminoethyl)keton (Czech)	(90-84-6)
1-Fenyl-3,3-dimethyltriazin	(7227-91-0)
1,2-Fenylendiamin (Czech)	(95-54-5)
m-Fenylendiamin (Czech)	(108-45-2)
o-Fenylendiamin (Czech)	(95-54-5)
p-Fenylendiamin (Czech)	(106-50-3)
Fenylenodwuamina (Polish)	(106-50-3)
Fenylepsin	(57-41-0)
Fenylester kyseliny chlormravenci (Czech)	(1885-14-9)
Fenylester kyseliny octove (Czech)	(122-79-2)
Fenylester kyseliny salicylove (Czech)	(118-55-8)
1-Fenyl-1,2-ethandiol (Czech)	(93-56-1)
1-Fenylethanol (Czech)	(98-85-1)
β-Fenylethanol (Czech)	(60-12-8)
1-Fenylethylamin (Czech)	(98-84-0)
2-Fenylethylamin (Czech)	(64-04-0)
Fenylettae	(50-06-6)
2-Fenylfenol (Czech)	(90-43-7)
N-Fenyl-p-fenylendiamin (Czech)	(101-54-2)
Fenylfosfin (Czech)	(638-21-1)
Fenyl-glycidylether (Czech)	(122-60-1)
Fenylglycol (Czech)	(93-56-1)
Fenylhist	(147-24-0)
Fenylhydrazid kyseliny octove (Czech)	(114-83-0)
Fenylhydrazine (Dutch)	(100-63-0)
Fenyl-α-hydroxybenzylketon (Czech)	(119-53-9)
N-Fenylimid kyseliny maleinove (Czech)	(941-69-5)
Fenylisokyanat (Czech)	(103-71-9)
N-Fenyl-N'-isopropyl-p-fenylendiamin (Czech)	(101-72-4)
Fenylisothiokyanat (Czech)	(103-72-0)
Fenylo-izopropylaminyl (Polish)	(300-62-9)

O-(N-Fenylkarbamoyl)propanonoxim (Czech) **(2828-42-4)**
Fenylkarbitol (Czech) **(104-68-7)**
Fenylkyanid (Czech) **(100-47-0)**
Fenylmercurichlorid (Czech) **(100-56-1)**
Fenylmerkuriacetat (Czech) **(62-38-4)**
Fenylmerkurinitrat (Czech) **(55-68-5)**
Fenylmerkuri-tris-(2-hydroxyethyl)ammoniumlaktat (Czech) **(23319-66-6)**
2-(N-Fenyl-N-methylamino)ethanol (Czech) **(93-90-3)**
Fenyl-methylkarbinol (Czech) **(98-85-1)**
2-Fenyl-3-methylmorfolin (Czech) **(134-49-6)**
1-Fenyl-3-methyl-2-pyrazolin-5-on (Czech) **(89-25-8)**
1-Fenyl-3-methyl-5-pyrazolylester kyseliny dimethylkarbaminove (Czech) **(87-47-8)**
Fenyl-α-naftylamin (Czech) **(90-30-2)**
Fenyl-β-naftylamin (Czech) **(135-88-6)**
O-Fenyl-O-p-nitrofenylester kyseliny methylthiofosfonove (Czech) **(2665-30-7)**
1-Fenylosilatranu (Polish) **(2097-19-0)**
2-Fenylotiomocznik (Polish) **(102-08-9)**
Fenyloxiran (Czech) **(96-09-3)**
1-Fenylpiperazin (Czech) **(92-54-6)**
2-Fenyl-propaan (Dutch) **(98-82-8)**
2-Fenyl-1-propanal (Czech) **(93-53-8)**
3-Fenylpropenal (Czech) **(104-55-2)**
3-Fenyl-2-propen-1-ol (Czech) **(104-54-1)**
Fenylsemikarbazid (Czech) **(103-03-7)**
Fenylsilatran (Czech) **(2097-19-0)**
Fenylthiomocovina (Czech) **(103-85-5)**
Fenyprin **(300-42-5)**
Fenytan **(63-98-9)**
Fenytoine **(57-41-0)**
Fenytoine **(630-93-3)**
Fenzaflor **(14255-88-0)**
Fenzen (Czech) **(71-43-2)**
Feojectin **(8047-67-4)**
Feosol **(7720-78-7)**
Feosol **(7782-63-0)**
Feospan **(7720-78-7)**
Fero-Gradumet **(7720-78-7)**
Fero-Gradumet **(7782-63-0)**
Ferbam **(14484-64-1)**
Ferbam 50 **(14484-64-1)**
Ferbam (ACGIH,OSHA) **(14484-64-1)**
Ferbam, Iron Salt **(14484-64-1)**
Ferbam, Iron salt **(14484-64-1)**
Ferbame **(14484-64-1)**
Ferbame (French) **(14484-64-1)**
Ferbeck **(14484-64-1)**
Ferberk **(14484-64-1)**
Ferdex 100 **(9004-66-4)**
Fergon **(299-29-6)**
Fergon preparations **(299-29-6)**
Fer-In-Sol **(7720-78-7)**
Fer-In-Sol **(7782-63-0)**
Ferisan **(15708-41-5)**
Ferkethion **(60-51-5)**
Ferlucon **(299-29-6)**
Fermate **(14484-64-1)**
Fermate Ferbam Fungicide **(14484-64-1)**
Fermenicide Liquid **(7446-09-5)**
Fermenicide Powder **(7446-09-5)**
Fermentation Alcohol **(64-17-5)**
Fermentation amyl alcohol **(123-51-3)**
Fermentation butyl alcohol **(78-83-1)**
Fermi Strept **(5490-27-7)**
Fermide **(137-26-8)**
Fermine **(131-11-3)**
Fermocide **(14484-64-1)**
Fernacol **(137-26-8)**
Fernacot **(1332-65-6)**
Fernasan **(137-26-8)**
Fernasan A **(137-26-8)**
Fernesta **(94-75-7)**
Fernesta **(94-80-4)**
Fernex **(23505-41-1)**
Fernide **(137-26-8)**
Fernimine **(94-75-7)**
Fernisolone **(50-24-8)**
Fernos **(23103-98-2)**
Fernoxene **(2702-72-9)**
Fernoxone **(94-75-7)**
Fer pentacarbonyle (French) **(13463-40-6)**
Ferradow **(14484-64-1)**
Ferralyn **(7720-78-7)**
Ferrate(2-), (N,N-bis(2-(bis(carboxymethyl)amino)ethyl)glycinato-(5-))-, sodium hydrogen, (PB-7-13-12564)- (9CI) **(12389-75-2)**
Ferrate(1-), (N-(2-(bis(carboxymethyl)amino)ethyl)-N-(2-hydroxyethyl)glycinato(3-))-, sodium (9CI) **(16485-47-5)**
Ferrate(-1), ((N,N'-1,2-ethanediylbis(N-(carboxymethyl)glycinato))(4-)-N,N',O,O',ON,ON')-, ammonium **(21265-50-9)**
Ferrate(1-), ((N,N'-1,2-ethanediylbis(N-(carboxymethyl)-glycinato))(4-)-N,N',O,O',ON,ON')-, ammonium, (OC-6-21)- (9CI) **(21265-50-9)**
Ferrate(1-), ((ethylenedinitrilo)tetraacetato)-, sodium (8CI) **(15708-41-5)**
Ferrate(4-), hexacyano-, iron(3+) (3:4) (8CI) **(14038-43-8)**
Ferrate(4-), hexacyano-, tetraammonium **(14481-29-9)**
Ferrate(4-), hexacyano-, tetrasodium **(13601-19-9)**
Ferrate(4-), hexacyano-, tetrasodium, decahydrate (8CI) **(14434-22-1)**
Ferrate(4-), hexakis(cyano-C)-, ammonium iron(3+) (1:1:1), (OC-6-11)- (9CI) **(25869-00-5)**
Ferrate(1-), hexakis(cyano-C)di- **(14038-43-8)**
Ferrate(4-), hexakis(cyano-C)-, iron(3+) (3:4), (OC-6-11)- (9CI) **(14038-43-8)**
Ferrate(4-), hexakis(cyano-C)-, tetraammonium, (OC-6-11)- (9CI) **(14481-29-9)**
Ferrate(4-), hexakis(cyano-C)-, tetrasodium, (OC-6-11)- (9CI) **(13601-19-9)**
Ferrate(4-), hexakis(cyano-C)-, tetrasodium, decahydrate, (OC-6-11)- (9CI) **(14434-22-1)**
Ferrate, strontium (1:1) (9CI) **(12023-91-5)**
Ferrate(3-), tris(ethanedioato(2-)-O,O')-, triammonium, (OC-6-11)-(9CI) **(14221-47-7)**
Ferrate(3-), tris(oxalato)-, triammonium **(14221-47-7)**
Ferriamicide **(2385-85-5)**
Ferric HEDTA **(17084-02-5)**
Ferric ammonium citrate **(1185-57-5)**
Ferric ammonium edetate **(21265-50-9)**
Ferric ammonium oxalate **(14221-47-7)**
Ferric ammonium oxalate **(2944-67-4)**
Ferric ammonium oxalate **(55488-87-4)**
Ferric arsenate, Solid (DOT) **(10102-49-5)**
Ferric arsenate [UN 1606] **(10102-49-5)**
Ferric arsenite **(63989-69-5)**
Ferric arsenite, Basic **(63989-69-5)**
Ferric arsenite, Solid **(63989-69-5)**
Ferric cacodylate **(5968-84-3)**
Ferric chloride, Anhydrous (DOT) **(7705-08-0)**
Ferric chloride, Solid, Anhydrous (DOT) **(7705-08-0)**
Ferric chloride, Solid (DOT) **(7705-08-0)**
Ferric chloride, Solution [UN 2582] **(7705-08-0)**
Ferric chloride [UN 1773] **(7705-08-0)**
Ferric chloride, hexahydrate **(10025-77-1)**
Ferric dextran **(9004-66-4)**
Ferric dimethylarsinate **(5968-84-3)**
Ferric dimethyl dithiocarbamate **(14484-64-1)**
Ferric dimethyldithiocarbamate **(14484-64-1)**
Ferric ferrocyanide **(14038-43-8)**
Ferric fluoride **(7783-50-8)**

Ferric hexacyanoferrate (II) (14038-43-8)
Ferric-methanearsonate (6585-53-1)
Ferric monomethylarsonate (6585-53-1)
Ferric nitrate [UN 1466] (10421-48-4)
Ferric nitrilotriacetate (16448-54-7)
Ferricon (135-51-3)
Ferric oxide (1309-37-1)
Ferric oxide (Colloidal) (1309-37-1)
Ferric oxide, saccharated (8047-67-4)
Ferric orthophosphate (10045-86-0)
Ferric phosphate (10045-86-0)
Ferric pyrophosphate (10058-44-3)
Ferric saccharate ...Iron oxide (Mix.) (8047-67-4)
Ferric sodium EDTA (15708-41-5)
Ferric sodium edetate (15708-41-5)
Ferric sodium pyrophosphate (10045-87-1)
Ferric sulfate (10028-22-5)
Ferric trichloride hexahydrate (10025-77-1)
Ferridextran (9004-66-4)
Ferrigen (9004-51-7)
Ferrioxamine B, N-benzoyl- (70-51-9)
Ferritin (9007-73-2)
Ferrivenin (8047-67-4)
Ferro-Gradumet (7720-78-7)
Ferro Lemon Yellow (1306-23-6)
Ferro Orange Yellow (1306-23-6)
Ferro-Theron (7720-78-7)
Ferro Yellow (1306-23-6)
Ferroactinolite (12172-67-7)
Ferroactinolyte (12172-67-7)
Ferroanthophyllite (77536-67-5)
Ferrocene (102-54-5)
Ferrocene, acetyl- (1271-55-2)
Ferrochrome (11114-46-8)
Ferrochrome (Exothermic) (11114-46-8)
Ferrochrome, Exothermic (DOT) (11114-46-8)
Ferrochrome lignosulfonate (8075-74-9)
Ferrochromium (11114-46-8)
Ferrocyanide (13408-63-4)
Ferrodextran (9004-66-4)
Ferrofos 509 (6419-19-8)
Ferrofos 510 (2809-21-4)
Ferroglucin (9004-66-4)
Ferroglukin 75 (9004-66-4)
Ferromanganese, Exothermic (DOT) (12604-53-4)
Ferronicum (299-29-6)
Ferrophosphorus (8049-19-2)
Ferrosilicon (8049-17-0)
Ferrosilicon, Containing more than 30% but less than 90% silicon [UN 1408] (8049-17-0)
Ferrosoferric oxide (1317-61-9)
Ferrosulfat (German) (7720-78-7)
Ferrosulfate (7720-78-7)
Ferrous ammonium sulfate (10045-89-3)
Ferrous ammonium sulfate hexahydrate (10045-89-3)
Ferrous arsenate, Solid (DOT) (10102-50-8)
Ferrous arsenate [UN 1608] (10102-50-8)
Ferrous carbonate (563-71-3)
Ferrous chloride (7758-94-3)
Ferrous chloride, Solid [NA 1759] (7758-94-3)
Ferrous chloride, Solution (DOT) (7758-94-3)
Ferrous chloride, tetrahydrate (13478-10-9)
Ferrous citrate (23383-11-1)
Ferrous diammonium disulfate (10045-89-3)
Ferrous ferrite (1332-37-2)
Ferrous gluconate (299-29-6)
Ferrous lactate (5905-52-2)
Ferrous monosulfide (1317-37-9)
Ferrous oxalate (516-03-0)
Ferrous oxide (1345-25-1)
Ferrous phosphide (1310-43-6)
Ferrous sodium HEDTA (16485-47-5)
Ferrous sulfate (7720-78-7)
Ferrous sulfate (1:1) (7720-78-7)
Ferrous sulfate heptahydrate (7782-63-0)
Ferrous sulfate monohydrate (17375-41-6)
Ferrous sulfide (1317-37-9)
Ferrous sulphate (7720-78-7)
Ferrovac E (7439-89-6)
Ferrovanadium-Dust (12604-58-9)
Ferrovanadium (ACGIH,OSHA) (12604-58-9)
Ferrox (516-03-0)
Ferrugo (1309-37-1)
Fersolate (7720-78-7)
Fertene (9002-88-4)
Ferulaldehyde (458-36-6)
Ferulic acid (1135-24-6)
Ferulic acid (537-98-4)
Ferulic acid, trans- (537-98-4)
Ferulic acid, α,β-dihydro- (1135-23-5)
Fervin (55635-13-7)
Fervinal (74051-80-2)
Ferxone (94-75-7)
Fesia-sept (101-53-1)
Fesiasept (101-53-1)
Fesofor (7782-63-0)
Fesotyme (7782-63-0)
Fettorange 4A (842-07-9)
Fettorange B (3118-97-6)
Fettorange LG (842-07-9)
Fettorange R (842-07-9)
Fettponceau G (85-86-9)
Fettrot (85-86-9)
Fettscharlach (85-86-9)
Fettscharlach LB (85-86-9)
Fezudin (333-41-5)
Fiber A (37243-36-0)
Fiber V (25038-59-9)
Fibralem (637-07-0)
Fibraset TC (9011-05-6)
Fibre Black VF (1937-37-7)
Fibrene C 400 (14807-96-6)
Fibrinase (9001-90-5)
Fibrinolysin (9001-90-5)
Fibrinolysin (Human) (9001-90-5)
Fibrinolysin (Human) (9004-09-5)
Fibrous Grunerite (1332-21-4)
Fibrous crocidolite asbestos (12001-28-4)
Fibrous tremolite (77536-68-6)
Ficam (22781-23-3)
Ficam D (22781-23-3)
Ficam ULV (22781-23-3)
Ficam 80W (22781-23-3)
Ficam W (22781-23-3)
Fichlor 91 (87-90-1)
Ficortril (50-23-7)
Ficusin (66-97-7)
Field ethane (68475-58-1)
Fig Tree LS (8001-21-6)
Filariol (4824-78-6)
Film (DOT) (9004-70-0)
Filmerine (7632-00-0)
Filtersol "A" (118-56-9)
Filtrasorb (7440-44-0)

Filtrasorb 200	(7440-44-0)
Filtrasorb 400	(7440-44-0)
Fimalene	(54-85-3)
Finasol OSR$_2$	(151-21-3)
Finaven	(43222-48-6)
Fine Gum HES	(9004-32-4)
Finemeal	(7727-43-7)
Finimal	(103-90-2)
Finish EN	(140-95-4)
Finlepsin	(298-46-4)
Finntalc C10	(14807-96-6)
Finntalc M05	(14807-96-6)
Finntalc M15	(14807-96-6)
Finntalc P40	(14807-96-6)
Finntalc PF	(14807-96-6)
Fintin acetato (Italian)	(900-95-8)
Fintine hydroxyde (French)	(76-87-9)
Fintin hydroxid (German)	(76-87-9)
Fintin hydroxyde (Dutch)	(76-87-9)
Fintin idrossido (Italian)	(76-87-9)
Finuret	(135-09-1)
Fiocortril	(50-23-7)
Fiorinal	(62-44-2)
Fir Balsam	(8021-28-1)
Fir Balsam Absolute	(8007-47-4)
Fir Balsam Resinoid	(8021-28-1)
Fir Needle Oil, Balsam	(8021-28-1)
Fir Needle Oil, Siberian	(8021-28-1)
Fire Damp	(74-82-8)
Fire Guard 2000	(79-94-7)
Fire Guard 5000	(1529-68-6)
Fire Master PHT 4	(632-79-1)
Fire Master TSA	(74082-93-2)
Fire Master TSA-PO 64P	(74082-93-2)
Fireclay, calcined	(66402-68-4)
Fired clay	(66402-68-4)
Firemaster BP-6	(59536-65-1)
Firemaster BP4A	(79-94-7)
Firemaster LV-T 23P	(126-72-7)
Firemaster T23P	(126-72-7)
Firemaster T23P-LV	(126-72-7)
Firemaster FF-1	(67774-32-7)
Firmatex RK	(1854-26-8)
Firmazolo	(526-08-9)
Firmotox	(8003-34-7)
Fish-Tox	(83-79-4)
Fish Berry	(124-87-8)
Fish Oil	(8016-13-5)
Fisher's Aldehyde	(84-83-3)
Fisons B25	(101-27-9)
Fisons NC 2964	(950-37-8)
Fisons NC 5016	(14255-88-0)
Fitios	(116-01-8)
Fitios B/77	(116-01-8)
Fitrol	(1332-58-7)
Fitrol desiccite 25	(1332-58-7)
Fixanol Black E	(1937-37-7)
Fixanol Blue 2B	(2602-46-2)
Fixanol Blue BH	(2429-73-4)
Fixanol Brown LF	(2429-81-4)
Fixanol C	(140-72-7)
Fixanol Orange Brown X	(2586-58-5)
Fixanol Sky Blue FF	(2610-05-1)
Fixanol Violet N	(2586-60-9)
Fixanol Yellow GS	(1325-37-7)
Fixapret AH	(136-84-5)
Fixapret CP	(1854-26-8)
Fixapret CP 40	(1854-26-8)
Fixapret CPK	(1854-26-8)
Fixapret CPN	(1854-26-8)
Fixapret CPNS	(1854-26-8)
Fixierer P	(959-52-4)
Fixol	(107-75-5)
Flac	(7778-44-1)
Flacavon R	(126-72-7)
Flacedil	(65-29-2)
Flagecidin	(22862-76-6)
Flagemona	(443-48-1)
Flagesol	(443-48-1)
Flagil	(443-48-1)
Flagyl	(443-48-1)
Flaianina	(66-02-4)
Flake Lead	(1319-46-6)
Flake White	(1304-85-4)
Flamaril	(129-20-4)
Flame Tones	(2814-77-9)
Flamenco	(13463-67-7)
Flamex T 23P	(126-72-7)
Flaming Red	(2814-77-9)
Flamithin	(138-15-8)
Flammex AP	(126-72-7)
Flammex B 10	(13654-09-6)
Flammex 5BT	(87-83-2)
Flammex LV-T 23P	(126-72-7)
Flammex T 23P	(126-72-7)
Flamolin MF 15711	(9002-88-4)
Flamprop-Isopropyl	(52756-22-6)
Flamprop-Methyl	(52756-25-9)
Flamprop	(58667-63-3)
Flamruss	(1333-86-4)
Flamycin	(57-62-5)
Flanaril	(129-20-4)
Flanithin	(138-15-8)
Flanogen Ela	(9000-07-1)
Flavacridinum hydrochloricum	(8048-52-0)
Flavanone, 3,3',4',5,7-pentahydroxy	(480-18-2)
Flavaxin	(83-88-5)
Flavazone	(59-87-0)
Flavin	(86-40-8)
Flavine	(553-30-0)
Flavine	(8048-52-0)
Flavine	(86-40-8)
Flavinetten	(8048-52-0)
Flavin sulphate	(553-30-0)
Flavioform	(8048-52-0)
Flavipin	(8048-52-0)
Flavisept	(8048-52-0)
Flavolutan	(57-83-0)
Flavomycelin	(21884-44-6)
Flavone, 2,3-dihydro-3,3',4',5,7-pentahydroxy-	(480-18-2)
Flavone, 8-(dimethylaminomethyl)-7-methoxy-3-methyl	(1165-48-6)
Flavone, 3,3',4',5,7-pentahydroxy	(117-39-5)
Flavone, 3,3',4',5,7-pentahydroxy-, 3-(O-rhamnosylglucoside)	(153-18-4)
Flavone, 3,3',4',5,7-pentahydroxy-, 3-(6-deoxy-α-l-mannopyranoside)	(522-12-3)
Flavone, 3,3',4',5,7-pentahydroxy-, 3-rhamnoside	(522-12-3)
Flavone, 3,4',5,7-tetrahydroxy	(520-18-3)
Flavonic acid	(81-11-8)
Flavosan	(86-40-8)
Flawil	(63428-84-2)
Flaxain	(83-88-5)
Flaxedil	(65-29-2)
Fleabane Oil	(8007-27-0)
Fleck-Flip	(79-01-6)

Flectol A (147-47-7)
Flectol H (147-47-7)
Flectol Pastilles (147-47-7)
Flectol H, Polymer (26780-96-1)
Flectron (52315-07-8)
Fleet-X (108-67-8)
Flegyl (443-48-1)
Flexal (78-44-4)
Flexamine G (74-31-7)
Flexan 500 (9080-79-9)
Flexartal (78-44-4)
Flexartel (78-44-4)
Flexazone (50-33-9)
Flexchlor (63449-39-8)
Flexible Collodion (9004-70-0)
Flexichem (1592-23-0)
Flexichem B (822-16-2)
Flexichem CS (1592-23-0)
Fleximel (117-81-7)
Flexol A 26 (103-23-1)
Flexol DOP (117-81-7)
Flexol-Epo (8013-07-8)
Flexol 4GO (18268-70-7)
Flexol Plasticizer 380 (137-89-3)
Flexol Plasticizer 10-A (103-23-1)
Flexol Plasticizer A-26 (103-23-1)
Flexol Plasticizer DIOP (27554-26-3)
Flexol Plasticizer DOP (117-81-7)
Flexol Plasticizer 3GH (95-08-9)
Flexol Plasticizer 3GO (94-28-0)
Flexol Plasticizer TCP (1330-78-5)
Flexol TOF (78-42-2)
Flexricin P 6 (140-04-5)
Flexricin P-1 (141-24-2)
Flexricin P-4 (140-03-4)
Flexzone 3C (101-72-4)
Flexzone 6H (101-87-1)
Flibol E (52-68-6)
Fliegenteller (52-68-6)
Flindix (87-33-2)
Flint (14808-60-7)
Flintshot (14808-60-7)
Flit 406 (133-06-2)
Flo-Cillin (6130-64-9)
Flo-Cillin Aqueous (6130-64-9)
Flo-Mor (30525-89-4)
Flo Pro Mcseed Protectant (72-43-5)
Flo Pro T Seed Protectant (137-26-8)
Flo Pro V Seed Protectant (5234-68-4)
Flo-Tin 4l (76-87-9)
Flock Flip (79-01-6)
Flocool 180 (1313-60-6)
Floctafenic acid (36783-34-3)
Floctafenine (23779-99-9)
Flogal (129-20-4)
Floghene (129-20-4)
Floginax (22204-53-1)
Flogistin (129-20-4)
Flogitolo (129-20-4)
Flogodin (129-20-4)
Flogoril (129-20-4)
Flogostop (129-20-4)
Flomine (122-40-7)
Flomore (93-79-8)
Flopirina (129-20-4)
Floraltone (77-06-5)
Floraltone (88-82-4)
Florantirona (Spanish) (519-95-9)
Florantyron (519-95-9)
Florantyrone (519-95-9)
Florantyronum (Latin) (519-95-9)
Floraquin (83-73-8)
Flordimex (16672-87-0)
Florel (16672-87-0)
Flores martis (7705-08-0)
Floridin (50-59-9)
Floridine (7681-49-4)
Florinef (127-31-1)
Florisil (1343-88-0)
Florite (1318-16-7)
Florite R (1344-95-2)
Florocid (7681-49-4)
Floroglucin (Czech) (108-73-6)
Floroglucinol (Czech) (108-73-6)
Florol (Czech) (90-00-6)
Floropryl (55-91-4)
Floroxene (406-90-6)
Flothene (9002-88-4)
Flour Sulphur (7704-34-9)
Flovic (9003-22-9)
Flower of Paradise (83-72-7)
Flowers of Antimony (1309-64-4)
Flowers of Sulfur (DOT) (7704-34-9)
Flowers of Sulphur (7704-34-9)
Flowers of Zinc (1314-13-2)
Floxacillin (5250-39-5)
Floxapen (5250-39-5)
Floxiridina (Spanish) (50-91-9)
Floxuridin (50-91-9)
Floxuridine (50-91-9)
Floxuridinum (Latin) (50-91-9)
Flozenges (7681-49-4)
Fluate (79-01-6)
Fluazifop-Butyl (69806-50-4)
Fluazifop-p-butyl (79241-46-6)
Fluchloralin (33245-39-5)
Flucinar (67-73-2)
Flucloxacilina (Spanish) (5250-39-5)
Flucloxacillin (5250-39-5)
Flucloxacilline (French) (5250-39-5)
Flucloxacillinum (Latin) (5250-39-5)
Flucort (53-34-9)
Flucort (67-73-2)
Flucythrinate (70124-77-5)
Flucytosine (2022-85-7)
Fludestrin (968-93-4)
Fludiazepam (3900-31-0)
Fludrocortisone (127-31-1)
Fludrocortone (127-31-1)
Flue Dust, Arsenic-contg. (8028-73-7)
Flue Dust, Brass-manufg. (69012-56-2)
Flue Dust, Copper alloy-manufg., Zinc oxide-contg. (69012-58-4)
Flue Gas (630-08-0)
Flue Gases, Ferrous metal, Blast furnace (65996-68-1)
Fluenethyl (4301-50-2)
Fluenetil (4301-50-2)
Fluenthyl (4301-50-2)
Fluenyl (4301-50-2)
Fluflon (63428-84-2)
Flugene 22 (75-45-6)
Flugex 12B1 (353-59-3)
Fluid-AP (9003-13-8)
Fluimucetin (616-91-1)
Fluimucil (616-91-1)

Fluitran **(133-67-5)**
Flukoids **(56-23-5)**
Flumen **(58-94-6)**
Flumesil **(73-48-3)**
Flumethiazide **(148-56-1)**
Flumethone **(53-33-8)**
Flumicil **(616-91-1)**
Flunitrazepam **(1622-62-4)**
Flunitrazepamum (Latin) **(1622-62-4)**
Fluoboric acid [UN 1775] **(16872-11-0)**
Fluobrene **(124-73-2)**
Fluochol **(519-95-9)**
Fluocinolone 16,17-acetonide **(67-73-2)**
Fluocinolone acetonide **(67-73-2)**
Fluodrocortisone **(127-31-1)**
Fluohydric acid gas **(74-90-8)**
Fluohydrisone **(127-31-1)**
Fluohydrocortisone **(127-31-1)**
Fluometuron **(2164-17-2)**
Fluomine **(62207-76-5)**
Fluomine Dust **(62207-76-5)**
Fluooxene **(406-90-6)**
Fluoperazine **(117-89-5)**
Fluophosgene **(353-50-4)**
Fluophosphoric acid di(dimethylamide) **(115-26-4)**
Fluophosphoric acid, diisopropyl ester **(55-91-4)**
Fluor (Dutch, French, German, Polish) **(7782-41-4)**
Fluor-I-Strip A.T. **(518-47-8)**
Fluor-O-Kote **(7681-49-4)**
5-Fluoracil (German) **(51-21-8)**
Fluorad FC 128 **(2991-51-7)**
Fluorakil 100 **(640-19-7)**
Fluoral **(7681-49-4)**
Fluoran, 3',6'-bis(diethylamino)- (8CI) **(509-34-2)**
3',6'-Fluorandiol **(2321-07-5)**
Fluorandrenolone **(1524-88-5)**
Fluorandrenolone acetonide **(1524-88-5)**
Fluorane 114 **(76-14-2)**
3-Fluoranilin (Czech) **(372-19-0)**
4-Fluoranilin (Czech) **(371-40-4)**
Fluoranthene **(206-44-0)**
8-Fluoranthenebutyric acid, γ-oxo- **(519-95-9)**
Fluoranthene, 3-chloro- (8CI,9CI) **(25911-51-7)**
Fluoranthene, dichloro- (9CI) **(86329-60-4)**
Fluoranthene, dihydro- (9CI) **(41593-24-2)**
Fluoranthene, 3,7-dinitro- **(105735-71-5)**
Fluoranthene, 3,9-dinitro- **(22506-53-2)**
Fluoranthene, 2-methyl **(33543-31-6)**
Fluoranthene, 3-methyl **(1706-01-0)**
Fluoranthene, methyl- (9CI) **(30997-39-8)**
Fluoranthene, 1-nitro **(13177-28-1)**
Fluoranthene, 2-nitro **(13177-29-2)**
Fluoranthene, 3-nitro **(892-21-7)**
Fluoranthene, nitro- (9CI) **(77468-36-1)**
4-(8-Fluoranthenyl)-4-oxobutyric acid **(519-95-9)**
β-(8-Fluoranthoyl)propionic acid **(519-95-9)**
β-(8-Fluoranthyloyl)propionic acid **(519-95-9)**
Fluorantyrone **(519-95-9)**
Fluoraquin **(83-73-8)**
Fluorbenside **(405-30-1)**
o-Fluorbenzoesaeure (German) **(445-29-4)**
5-Fluor-1-(β-2'-deoxyribofuranosyl)pyrimidin-2,4(1H,3H)-dion (Czech) **(50-91-9)**
5-Fluor-desoxycytidin (German) **(10356-76-0)**
5-Fluor-2,4-dihydroxypyrimidin (Czech) **(51-21-8)**
2-Fluorenamine **(153-78-6)**
Fluoren-2-amine **(153-78-6)**
9H-Fluorene **(86-73-7)**
Fluorene **(86-73-7)**
1H-Fluorene (8CI,9CI) **(244-36-0)**
2-Fluoreneamine **(153-78-6)**
Fluorene, 2-amino- **(153-78-6)**
Fluorene, 9-benzyl- (6CI,7CI,8CI) **(1572-46-9)**
9H-Fluorene, 9-(1,1'-biphenyl)-4-yl- (9CI) **(17165-86-5)**
Fluorene, 9-(4-biphenylyl)- (8CI) **(17165-86-5)**
9H-Fluorene, 2-bromo- (9CI) **(1133-80-8)**
Fluorene, 2-bromo- (8CI) **(1133-80-8)**
Fluorene, 9-tert-butyl- (6CI,8CI) **(17114-78-2)**
9H-Fluorene-2-carbonitrile (9CI) **(2523-48-0)**
Fluorene-2-carbonitrile (8CI) **(2523-48-0)**
9H-Fluorene-4-carbonitrile, 9-phenyl- (9CI) **(32377-09-6)**
Fluorene-4-carbonitrile, 9-phenyl- (8CI) **(32377-09-6)**
9H-Fluorene-9-carboxylic acid, 2-chloro-9-hydroxy- (9CI) **(2464-37-1)**
Fluorene-9-carboxylic acid, 2-chloro-9-hydroxy- (8CI) **(2464-37-1)**
Fluorene-9-carboxylic acid, 2-chloro-9-hydroxy-, methyl ester **(2536-31-4)**
9H-Fluorene-9-carboxylic acid, 9-hydroxy-, butyl ester, Mixt. with isooctyl (4-chloro-2-methylphenoxy)acetate (9CI) **(53568-85-7)**
9H-Fluorene, 2-chloro- (9CI) **(2523-44-6)**
Fluorene, 2-chloro- (8CI) **(2523-44-6)**
9H-Fluorene, 9-(3-chlorophenyl)- (9CI) **(32377-11-0)**
9H-Fluorene, 9-(4-chlorophenyl)- (9CI) **(21846-07-1)**
Fluorene, 9-(m-chlorophenyl)- (8CI) **(32377-11-0)**
Fluorene, 9-(p-chlorophenyl)- (8CI) **(21846-07-1)**
9H-Fluorene, 2,3-dimethyl- (9CI) **(4612-63-9)**
Fluorene, 2,3-dimethyl- (8CI) **(4612-63-9)**
9H-Fluorene, 9-(1,1-dimethylethyl)- (9CI) **(17114-78-2)**
Fluorene, 2,5-dinitro **(15110-74-4)**
Fluorene, 2,7-dinitro **(5405-53-8)**
9H-Fluorene, 2,5-dinitro- (9CI) **(15110-74-4)**
1H-Fluorene, dodecahydro- (9CI) **(5744-03-6)**
Fluorene, dodecahydro- (8CI) **(5744-03-6)**
9H-Fluorene, 9-ethyl- (9CI) **(2294-82-8)**
Fluorene, 9-ethyl- (8CI) **(2294-82-8)**
Fluorene, 9-isopropyl- (8CI) **(3299-99-8)**
9H-Fluorene, 2-methoxy- (9CI) **(2523-46-8)**
Fluorene, 2-methoxy- (6CI,7CI,8CI) **(2523-46-8)**
9H-Fluorene, 9-(3-methoxyphenyl)- (9CI) **(32377-13-2)**
9H-Fluorene, 9-(4-methoxyphenyl)- (9CI) **(21846-08-2)**
Fluorene, 9-(m-methoxyphenyl)- (8CI) **(32377-13-2)**
Fluorene, 9-(p-methoxyphenyl)- (8CI) **(21846-08-2)**
9H-Fluorene, 9-methyl- **(2523-37-7)**
Fluorene, 9-methyl **(2523-37-7)**
Fluorene, methyl- **(26914-17-0)**
9H-Fluorene, 1-methyl- (9CI) **(1730-37-6)**
9H-Fluorene, 2-methyl- (9CI) **(1430-97-3)**
9H-Fluorene, 4-methyl- (9CI) **(1556-99-6)**
9H-Fluorene, methyl- (9CI) **(26914-17-0)**
Fluorene, 1-methyl- (8CI) **(1730-37-6)**
Fluorene, 4-methyl- (8CI) **(1556-99-6)**
9H-Fluorene, 9-(1-methylethyl)- (9CI) **(3299-99-8)**
9H-Fluorene, 9-(3-methylphenyl)- (9CI) **(18153-42-9)**
9H-Fluorene, 9-(4-methylphenyl)- (9CI) **(18153-43-0)**
9H-Fluorene, 2-nitro **(607-57-8)**
Fluorene, 2-nitro- **(607-57-8)**
Fluorene, nitro **(55345-04-5)**
9H-Fluorene, nitro- (9CI) **(55345-04-5)**
Fluorene, 2-nitroso **(2508-20-5)**
9H-Fluorene, 9-phenyl- (9CI) **(789-24-2)**
Fluorene, 9-phenyl- (8CI) **(789-24-2)**
9H-Fluorene, 9-(phenylmethyl)- (9CI) **(1572-46-9)**
Fluorene, 9-m-tolyl- (8CI) **(18153-42-9)**
Fluorene, 9-p-tolyl- (8CI) **(18153-43-0)**
9H-Fluorene, 9-(3-(trifluoromethyl)phenyl)- (9CI) **(32377-12-1)**
Fluorene, 9-(.α.,.α.,.α.-trifluoro-m-tolyl)- (8CI) **(32377-12-1)**

9H-Fluoren-9-imine (9CI) (4440-33-9)
Fluoren-9-imine (8CI) (4440-33-9)
9-Fluorenone (486-25-9)
9H-Fluoren-9-one (486-25-9)
Fluoren-9-one (486-25-9)
9H-Fluoren-9-one, 2-chloro- (9CI) (3096-47-7)
9H-Fluoren-9-one, chloro- (9CI) (85897-29-6)
Fluoren-9-one, 2-chloro- (8CI) (3096-47-7)
9H-Fluoren-9-one, dichloro- (9CI) (90077-74-0)
9-Fluorenone, 2,7-dinitro (31551-45-8)
9H-Fluoren-9-one, hydroxy- (9CI) (119620-42-7)
9H-Fluoren-9-one, 2-methyl- (9CI) (2840-51-9)
9H-Fluoren-9-one, methyl- (9CI) (77468-39-4)
Fluoren-9-one, 2-methyl- (8CI) (2840-51-9)
Fluorenone, methyl- (9CI) (79147-47-0)
9H-Fluoren-9-one, 3-nitro (42135-22-8)
9H-Fluoren-9-one, pentachloro- (9CI) (90077-77-3)
9H-Fluoren-9-one, tetrachloro- (9CI) (90077-76-2)
9H-Fluoren-9-one, 2,4,5,7-tetranitro- (9CI) (746-53-2)
Fluoren-9-one, 2,4,5,7-tetranitro- (8CI) (746-53-2)
9H-Fluoren-9-one, trichloro- (9CI) (90077-75-1)
Fluoren-9-one, 2,4,7-trinitro (129-79-3)
2-Fluorenylacetamide (53-96-3)
N-2-Fluorenylacetamide (53-96-3)
N-4-Fluorenylacetamide (28322-02-3)
N-Fluoren-2-ylacetamide (53-96-3)
N-Fluoren-4-ylacetamide (28322-02-3)
Fluorenyl-2-acethydroxamic acid (53-95-2)
N-2-Fluorenylacetohydroxamic acid (53-95-2)
N-Fluoren-2-ylacetohydroxamic acid (53-95-2)
2-Fluorenylhydroxylamine (53-94-1)
Fluoren-2-yl methyl ether (2523-46-8)
Fluorescein (8CI) (2321-07-5)
Fluorescein, Soluble (518-47-8)
Fluorescein, 4,7-dichloro-2',4',5',7'-tetrabromo-, dipotassium salt (6441-77-6)
Fluorescein, disodium salt (518-47-8)
Fluoresceine (2321-07-5)
Fluorescein mercuriacetate (3570-80-7)
Fluorescein mercuric acetate (3570-80-7)
Fluorescein mercury acetate (3570-80-7)
Fluorescein sodium (518-47-8)
Fluorescein sodium B.P (518-47-8)
Fluorescein, 2',4',5',7'-tetrabromo (15086-94-9)
Fluorescein, 2',4',5',7'-tetrabromo-4,7-dichloro-, dipotassium salt (6441-77-6)
Fluorescein, 2',4',5',7'-tetrabromo-, disodium salt (17372-87-1)
Fluorescein, 2',4',5',7'-tetrabromo-, disodium salt (548-26-5)
Fluorescein, 2',4',5',7'-tetraiodo (15905-32-5)
Fluorescein, 2',4',5',7'-tetraiodo-, disodium salt (16423-68-0)
Fluorescent Brightener 46 (6416-68-8)
Fluoressigaeure (German) (62-74-8)
2-Fluorethylester kyseliny chlormravenci (Czech) (462-27-1)
2-Fluorethylester kyseliny fluoroctove (Czech) (459-99-4)
2-Fluorethylester kyseliny xenyloctove (Czech) (4301-50-2)
Fluoric acid (DOT) (7664-39-3)
Fluoridamid (47000-92-0)
Fluorid bority-dimethylether (1:1) (Czech) (353-42-4)
Fluoride (16984-48-8)
Fluoride(1-) (16984-48-8)
Fluoride, sodium (7681-49-4)
Fluoride ion(1-) (16984-48-8)
Fluoride ion (16984-48-8)
Fluorident (7681-49-4)
Fluorid hlinity (Czech) (7784-18-1)
Fluorid sodny (Czech) (7681-49-4)
Fluorigard (7681-49-4)
Fluorimide (8CI,9CI) (10405-27-3)
Fluorine (ACGIH,DOT,OSHA) (7782-41-4)
Fluorine, Compressed [UN 1045] (7782-41-4)
Fluorineed (7681-49-4)
Fluorine monoxide (7783-41-7)
Fluorine oxide (7783-41-7)
Fluorinert FC 43 (311-89-7)
Fluorinert FC 70 (338-84-1)
Fluorinse (7681-49-4)
Fluoristan (7783-47-3)
Fluoritab (7681-49-4)
Fluorite (9CI) (14542-23-5)
Fluorite (9CI) (7789-75-5)
1-Fluornaftalen (Czech) (321-38-0)
Fluoro-DDT (475-26-3)
Fluoro (Italian) (7782-41-4)
2-Fluoroacetamide (640-19-7)
Fluoroacetamide (640-19-7)
4'-Fluoroacetanilide (351-83-7)
4-Fluoroacetanilide (351-83-7)
p-Fluoroacetanilide (351-83-7)
Fluoroacetate (144-49-0)
2-Fluoroacetic acid (144-49-0)
Fluoroacetic acid [UN 2642] (144-49-0)
Fluoroacetic acid amide (640-19-7)
Fluoroacetic acid, sodium salt (62-74-8)
Fluoroaceto-p-bromoanilide (351-05-3)
4'-Fluoroacetophenone (403-42-9)
4-Fluoroacetophenone (403-42-9)
p-Fluoroacetophenone (403-42-9)
Fluoroacetyl chloride (359-06-8)
Fluoroandrenolone acetonide (1524-88-5)
3-Fluoroaniline (372-19-0)
o-Fluoroaniline (348-54-9)
p-Fluoroaniline (371-40-4)
4-Fluoroaniline (DOT) (371-40-4)
2-Fluoroaniline [UN 2941] (348-54-9)
4-Fluoroanisole (459-60-9)
p-Fluoroanisole (459-60-9)
2-Fluorobenzenamine (348-54-9)
4-Fluorobenzenamine (371-40-4)
Fluorobenzene [UN 2387] (462-06-6)
4-Fluorobenzeneacetic acid (405-50-5)
4-Fluorobenzenesulfonyl chloride (349-88-2)
p-Fluorobenzenesulfonyl chloride (349-88-2)
3-Fluorobenzoic acid (455-38-9)
4-Fluorobenzoic acid (456-22-4)
m-Fluorobenzoic acid (455-38-9)
o-Fluorobenzoic acid (445-29-4)
p-Fluorobenzoic acid (456-22-4)
2-Fluorobenzotrifluoride (392-85-8)
o-Fluorobenzotrifluoride (392-85-8)
2-Fluorobenzoyl chloride (393-52-2)
o-Fluorobenzoyl chloride (393-52-2)
1-(3-p-Fluorobenzoylpropyl)-4-p-chlorophenyl-4-hydroxypiperidine (52-86-8)
1-(1-(3-(p-Fluorobenzoyl)propyl)-1,2,3,6-tetrahydro-4-pyridyl)-2-benzimidazolinone (548-73-2)
2-Fluorobenzyl chloride (345-35-7)
3-Fluorobenzyl chloride (456-42-8)
4-Fluorobenzyl chloride (352-11-4)
m-Fluorobenzyl chloride (456-42-8)
o-Fluorobenzyl chloride (345-35-7)
p-Fluorobenzyl chloride (352-11-4)
2-Fluoro-1,1'-biphenyl (321-60-8)
2-Fluorobiphenyl (321-60-8)
4-Fluorobiphenyl (324-74-3)
Fluorobisisopropylaminophosphine oxide (371-86-8)
Fluoroblastin (51-21-8)
1-Fluoro-4-bromobenzene (460-00-4)

4-Fluoro-1-bromobenzene **(460-00-4)**
4-Fluorobromobenzene **(460-00-4)**
p-Fluorobromobenzene **(460-00-4)**
1-Fluorobutane **(2366-52-1)**
4-Fluorobutyl chloride **(462-73-7)**
Fluorocarbon-12 **(75-71-8)**
Fluorocarbon-22 **(75-45-6)**
Fluorocarbon 113 **(76-13-1)**
Fluorocarbon 114 **(76-14-2)**
Fluorocarbon-115 **(76-15-3)**
Fluorocarbon 1211 **(353-59-3)**
Fluorocarbon FC 43 **(311-89-7)**
Fluorocarbon FC 70 **(338-84-1)**
Fluorocarbon FC142b **(75-68-3)**
Fluorocarbon FC143 **(624-72-6)**
Fluorocarbon FC143a **(420-46-2)**
Fluorocarbon No. 11 **(75-69-4)**
Fluorochloridone **(61213-25-0)**
1,4-Fluorochlorobenzene **(352-33-0)**
1-Fluoro-2-chlorobenzene **(348-51-6)**
1-Fluoro-4-chlorobenzene **(352-33-0)**
o-Fluorochlorobenzene **(348-51-6)**
p-Fluorochlorobenzene **(352-33-0)**
4'-Fluoro-4-(4-(p-chlorophenyl)-4-hydroxypiperidinyl)butyrophenone **(52-86-8)**
Fluorocid **(7681-49-4)**
Fluorocort **(50-02-2)**
9-α-Fluorocortisol **(127-31-1)**
Fluorocortisone **(127-31-1)**
Fluoroctan sodny (Czech) **(62-74-8)**
5-Fluorocystosine **(2022-85-7)**
5-Fluorocytosine **(2022-85-7)**
1-Fluorodecane **(334-56-5)**
6α-Fluoro-1-dehydrohydrocortisone **(53-34-9)**
5-Fluoro-2'-deoxycytidine **(10356-76-0)**
5-Fluorodeoxycytidine **(10356-76-0)**
5-Fluoro-2'-deoxyuridine **(50-91-9)**
5-Fluoro-2-deoxyuridine **(50-91-9)**
5-Fluorodeoxyuridine **(50-91-9)**
Fluorodeoxyuridine **(50-91-9)**
β-5-Fluoro-2'-deoxyuridine **(50-91-9)**
Fluorodichloromethane **(75-43-4)**
Fluorodifen **(15457-05-3)**
Fluorodifene **(15457-05-3)**
9-α-Fluoro-11-β,21-dihydroxy-16-α-isopropylidenedioxy-1,4-pregnadiene,3,20-dione **(76-25-5)**
9-α-Fluoro-11-β,17-β-dihydroxy-17-α-methyl-4-androstene-3-one **(76-43-7)**
Fluoro-9-α dihydroxy-11-β,17-β methyl-17-α androstene-4 one-3 (French) **(76-43-7)**
9-Fluoro-11-β,17-β-dihydroxy-17-methylandrost-4-en-3-one **(76-43-7)**
Fluorodiisopropyl phosphate **(55-91-4)**
Fluorodimethylphenylsilane **(454-57-9)**
1,2,4-Fluorodinitrobenzene **(70-34-8)**
1-Fluoro-2,4-dinitrobenzene **(70-34-8)**
o-Fluorodiphenyl **(321-60-8)**
p-Fluorodiphenyl **(324-74-3)**
1-Fluorododecane **(334-68-9)**
Fluoroethane **(353-36-6)**
Fluoroethanoic acid **(144-49-0)**
2-Fluoroethanol **(371-62-0)**
β-Fluoroethanol **(371-62-0)**
Fluoroethene **(75-02-5)**
Fluoroethene homopolymer **(24981-14-4)**
2-Fluoroethyl chloroformate **(462-27-1)**
Fluoroethylene **(75-02-5)**
2-Fluoroethyl fluoroacetate **(459-99-4)**
β-Fluoroethyl fluoroacetate **(459-99-4)**
para-Fluorofentanyl **(90736-23-5)**
Fluoroform **(75-46-7)**
Fluoroformyl fluoride **(353-50-4)**
Fluorogesarol **(475-26-3)**
1-Fluoroheptane **(661-11-0)**
1-Fluorohexane **(373-14-8)**
Fluorohexane **(373-14-8)**
9-α-Fluorohydrocortisone **(127-31-1)**
4'-Fluoro-4-(4-hydroxy-4-(4'-chlorophenyl)piperidino)butyrophenone **(52-86-8)**
9-α-Fluoro-17-hydroxycorticosterone **(127-31-1)**
9-α-Fluoro-11-β-hydroxy-17-methyltestosterone **(76-43-7)**
9-α-Fluoro-16-α-hydroxyprednisolone **(124-94-7)**
9-α-Fluoro-16-α-hydroxyprednisolone 16-α,17-α-acetonide **(76-25-5)**
9-α-Fluoro-16-hydroxyprednisolone acetonide **(76-25-5)**
1-Fluoro-4-iodobenzene **(352-34-1)**
p-Fluoroiodobenzene **(352-34-1)**
9-α-Fluoro-16-α-17-α-isopropyledenedioxyprednisolone **(76-25-5)**
9-α-Fluoro-16-α-17-α-isopropylidenedioxy-δ-1-hydrocortisone **(76-25-5)**
Fluorol **(7681-49-4)**
Fluoromar **(406-90-6)**
Fluoromethane **(593-53-3)**
1-Fluoro-4-methoxybenzene **(459-60-9)**
p-Fluoromethoxybenzene **(459-60-9)**
6-Fluoro-7-methylbenz(a)anthracene **(2541-68-6)**
(Fluoromethyl)benzene **(350-50-5)**
1-Fluoro-3-methylbenzene **(352-70-5)**
1-Fluoro-4-methylbenzene **(352-32-9)**
δ^{1}-9-α-Fluoro-16-α-methylcortisol **(50-02-2)**
9-α-Fluoro-17-α-methyl-11-β,17-dihydroxy-4-androsten-3-one **(76-43-7)**
(Z)-5-Fluoro-2-methyl-1-((p-(methylsulfinyl)phenyl)methylene)-1H-indene-3-acetic acid **(38194-50-2)**
cis-5-Fluoro-2-methyl-1-((4-(methylsulfinyl)phenyl)methylene)-1H-indene-3-acetic acid **(38194-50-2)**
6α-Fluoro-16α-methylprednisolone **(53-33-8)**
9-α-Fluoro-16-α-methylprednisolone **(50-02-2)**
9-α-Fluoro-16-β-methylprednisolone **(378-44-9)**
9-α-Fluoro-16-α-methyl-1,4-pregnadiene-11-β,17-α,21-triol-3,20-dione **(50-02-2)**
9-α-Fluoro-16-β-methyl-1,4-pregnadiene-11-β,17-α,21-triol-3,20-dione **(378-44-9)**
2-Fluoro-2-methylpropane **(353-61-7)**
4-α-Fluoro-16-α-methyl-11-β,17,21-trihydroxypregna-1,4-diene-3,20-dione **(50-02-2)**
α-Fluoronaphthalene **(321-38-0)**
4-Fluoro-3-nitroaniline **(364-76-1)**
1-Fluoro-3-nitrobenzene **(402-67-5)**
3-Fluoronitrobenzene **(402-67-5)**
4-Fluoronitrobenzene **(350-46-9)**
m-Fluoronitrobenzene **(402-67-5)**
p-Fluoronitrobenzene **(350-46-9)**
1-Fluorononane **(463-18-3)**
1-Fluorooctane **(463-11-6)**
Fluoroparacide **(405-30-1)**
1-Fluoropentane **(592-50-7)**
Fluorophene **(4776-06-1)**
2-Fluorophenol **(367-12-4)**
3-Fluorophenol **(372-20-3)**
4-Fluorophenol **(371-41-5)**
m-Fluorophenol **(372-20-3)**
o-Fluorophenol **(367-12-4)**
(4-Fluorophenoxy)acetic acid 3,3,5-trimethylcyclohexyl ester **(58327-09-6)**
(4-Fluorophenyl)acetic acid **(405-50-5)**
(p-Fluorophenyl)acetic acid **(405-50-5)**
4-Fluorophenylalanine **(60-17-3)**
D,L-Fluorophenylalanine **(51-65-0)**
D,L-p-Fluorophenylalanine **(51-65-0)**

DL-4-Fluorophenylalanine (51-65-0)
Fluorophenylalanine (51-65-0)
p-Fluorophenylalanine (51-65-0)
p-Fluorophenylalanine (60-17-3)
p-Fluorophenylamine (371-40-4)
4-Fluorophenyl bromide (460-00-4)
p-Fluorophenyl bromide (460-00-4)
5-(o-Fluorophenyl)-1,3-dihydro-1-methyl-7-nitro-2H-1,4-benzodiazepin-2-one (1622-62-4)
N'-(4-Fluorophenyl)-N,N-dimethylurea (332-33-2)
1-(4-Fluorophenyl)ethanone (403-42-9)
p-Fluorophenyl methyl ether (459-60-9)
1-(1-(4-(p-Fluorophenyl)-4-oxobutyl)-1,2,3,6-tetrahydro-4-pyridyl)-2-benzimidazolinone (548-73-2)
N-(4-Fluorophenyl)-N-(1-(2-phenethyl)-4-piperidinyl)-propanamide (90736-23-5)
4-Fluorophenylsulfonyl chloride (349-88-2)
(4-Fluorophenyl)urea (659-30-3)
1-(p-Fluorophenyl)urea (659-30-3)
Fluorophlogopite (12003-38-2)
Fluorophosgene (353-50-4)
Fluorophosphoric acid, Anhydrous [UN 1776] (13537-32-1)
Fluoropicrin (335-02-4)
Fluoroplast 3 (79-38-9)
Fluoroplast 4 (116-14-3)
Fluoroplex (51-21-8)
6α-Fluoroprednisolone (53-34-9)
6α-Fluoro-1,4-pregnadiene-11β,17α,21-triol-3,20-dione (53-34-9)
1-Fluoropropane (460-13-9)
n-Fluoropropane (460-13-9)
3-Fluoropropionic acid (461-56-3)
ω-Fluoropropionic acid (461-56-3)
Fluoropryl (55-91-4)
5-Fluoro-2,4(1H,3H)-pyrimidinedione (51-21-8)
5-Fluoro-2,4-pyrimidinedione (51-21-8)
5-Fluoropyrimidine-2,4-dione (51-21-8)
Fluorosalan (4776-06-1)
Fluorosilicic acid [UN 1778] (16961-83-4)
Fluorosilicic acid, ammonium salt (16919-19-0)
Fluorosilicone trimer (2374-14-3)
Fluorosulfacide (405-30-1)
Fluorosulfonic acid [UN 1777] (7789-21-1)
Fluorosulfuric acid, methyl ester (421-20-5)
Fluorosulphacide (405-30-1)
Fluorosulphonic acid (DOT) (7789-21-1)
Fluorotane (151-67-7)
9-Fluoro-11-β,16-α,17,21-tetrahydroxypregna-1,4-diene-3,20-dione (124-94-7)
9-α-Fluoro-11-β,16-α,17,21-tetrahydroxy-1,4-pregnadiene-3,20-dione (124-94-7)
9-α-Fluoro-11-β,16-α,17,21-tetrahydroxypregna-1,4-diene-3,20-dione (124-94-7)
9-α-Fluoro-11-β,16-α,17-α,21-tetrahydroxypregna-1,4-diene-3,20-dione (124-94-7)
2-Fluorotoluene (95-52-3)
3-Fluorotoluene (352-70-5)
4-Fluorotoluene (352-32-9)
α-Fluorotoluene (350-50-5)
m-Fluorotoluene [UN 2388] (352-70-5)
o-Fluorotoluene [UN 2388] (95-52-3)
p-Fluorotoluene [UN 2388] (352-32-9)
Fluorotributylstannane (1983-10-4)
1-Fluoro-1,2,2-trichloroethane (359-28-4)
2-Fluoro-1,1,2-trichloroethane (359-28-4)
Fluorotrichloromethane (75-69-4)
Fluorotrichloromethane (OSHA) (75-69-4)
1-Fluoro-2-(trifluoromethyl)benzene (392-85-8)
9-Fluoro-11-β,17,21-trihydroxy-16-α-methylpregna-1,4-diene-3,20-dione (50-02-2)
9-Fluoro-11-β,17,21-trihydroxy-16-β-methylpregna-1,4-diene-3,20-dione (378-44-9)
9-α-Fluoro-11-β,17,21-trihydroxy-16-β-methylpregna-1,4-diene-3,20-dione (378-44-9)
9-α-Fluoro-11-β,17-α,21-trihydroxy-16-α-methylpregna-1,4-diene-3,20-dione (50-02-2)
9-α-Fluoro-11-β,17-α,21-trihydroxy-16-β-methylpregna-1,4-diene-3,20-dione (378-44-9)
6α-Fluoro-11β,17,21-trihydroxypregna-1,4-diene-3,20-dione (53-34-9)
6α-Fluoro-11β,17α,21-trihydroxypregna-1,4-diene-3,20-dione (53-34-9)
9-Fluoro-11-β,17,21-trihydroxypregn-4-ene-3,20-dione (127-31-1)
9-α-Fluoro-11-β,17-α,21-trihydroxy-4-pregnene-3,20-dione (127-31-1)
Fluorotriphenylstannane (379-52-2)
Fluorotrojchlorometan (Polish) (75-69-4)
1-Fluoroundecane (506-05-8)
5-Fluorouracil (51-21-8)
Fluorouracil (51-21-8)
5-Fluorouracil 2'-deoxyriboside (50-91-9)
5-Fluorouracil deoxyriboside (50-91-9)
5-Fluorouridine (316-46-1)
Fluorowodor (Polish) (7664-39-3)
Fluoroxene (406-90-6)
Fluorparacide (405-30-1)
para-Fluorphenylalanine (60-17-3)
Fluorphlogopite (9CI) (12003-38-2)
5-Fluor-2,4-pyrimidindiol (Czech) (51-21-8)
5-Fluor-2,4(1H,3H)-pyrimidindion (Czech) (51-21-8)
Fluorspar (7789-75-5)
5-Fluoruracil (German) (51-21-8)
Fluorure acide de potassium (French) (7789-29-9)
Fluorure de bore (French) (7637-07-2)
Fluorure de chrome III (French) (7788-97-8)
Fluorure de N,N'-diisopropyle phosphorodiamide (French) (371-86-8)
Fluorure de potassium (French) (7789-23-3)
Fluorure de sodium (French) (7681-49-4)
Fluorure de sulfuryle (French) (2699-79-8)
Fluorure de N,N,N',N'-tetramethyle phosphoro-diamide (French) (115-26-4)
Fluorure de thionyle (French) (7783-42-8)
Fluorures acide (French) (7782-41-4)
Fluoruri acidi (Italian) (7782-41-4)
Fluoruridine deoxyribose (50-91-9)
Fluoruro cromico (Spanish) (7788-97-8)
Fluorwasserstoff (German) (7664-39-3)
Fluorwaterstof (Dutch) (7664-39-3)
Fluorxene (406-90-6)
Fluoryl (75-46-7)
Fluosilicate de sodium (16893-85-9)
Fluosilicic acid (DOT) (16961-83-4)
Fluosol 43 (311-89-7)
Fluostigmine (55-91-4)
Fluosulfonic acid (DOT) (7789-21-1)
Fluotestin (76-43-7)
Fluothane (151-67-7)
Fluothiuron (33439-45-1)
Fluovitif (67-73-2)
Fluoximesterone (76-43-7)
Fluoxymesterone (76-43-7)
Fluoxymestrone (76-43-7)
Fluoxyprednisolone (124-94-7)
α-Flupenthixol (53772-82-0)
Fluphenazine (69-23-8)
Fluprednisolone (53-34-9)
Fluprednisolonum (Latin) (53-34-9)
Flura (7681-49-4)
Flura Drops (7681-49-4)
Flura-Gel (7681-49-4)

Flura-Loz (7681-49-4)
Fluracil (51-21-8)
Fluracilum (51-21-8)
Flurandrenolide (1524-88-5)
Flurandrenolone (1524-88-5)
Flurandrenolone acetonide (1524-88-5)
Flurazepam (17617-23-1)
Flurazepam dihydrochloride (1172-18-5)
Flurazepam hydrochloride (1172-18-5)
Flurcare (7681-49-4)
Fluri (51-21-8)
Fluridone (59756-60-4)
Fluril (51-21-8)
Flurochloridone (61213-25-0)
Fluroxene (406-90-6)
Flursol (7681-49-4)
Flusalan (4776-06-1)
Flusalanum (Latin) (4776-06-1)
Flusteron (76-43-7)
Flutestos (76-43-7)
Flutolanil (66332-96-5)
Flutone (76-25-5)
Flutra (133-67-5)
Fluvalinate (69409-94-5)
Fluvin (58-93-5)
Flux Maag (54-11-5)
Flux-calcined diatomaceous earth (68855-54-9)
Fly Bait Grits (122-10-1)
Fly-Die (62-73-7)
Fly Fighter (62-73-7)
Flypel (134-62-3)
Fmc 5488 (116-29-0)
Fos-Fall "A" (78-48-8)
For-Syn (10453-86-8)
Foamkill 8D (63148-62-9)
Focusan (2398-96-1)
Folacin (59-30-3)
Folan Red B (6459-94-5)
Folan Yellow G (3567-65-5)
Folate (59-30-3)
Folbex (510-15-6)
Folbex Smoke-Strips (510-15-6)
Folcid (2425-06-1)
Folcid (2939-80-2)
Folcodina (Spanish) (509-67-1)
Folcodine (509-67-1)
Folcord (52315-07-8)
Folcysteine (59-30-3)
Folex (150-50-5)
Foliandrin (465-16-7)
Folic-acid (59-30-3)
Folic acid, 4-amino- (54-62-6)
Folidol (56-38-2)
Folidol-80 (298-00-0)
Folidol E (56-38-2)
Folidol E605 (56-38-2)
Folidol E & E 605 (56-38-2)
Folidol M (298-00-0)
Folidol Oil (56-38-2)
Foligan (315-30-0)
Folikrin (53-16-7)
Folimat (1113-02-6)
Folinerin (465-16-7)
Folinevin (465-16-7)
Folipex (53-16-7)
Folisan (53-16-7)
Folithion (122-14-5)
Folithion EC 50 (122-14-5)
Folkodin (Czech) (509-67-1)
Follestrine (53-16-7)
Follicle-Stimulating Hormone (9002-68-0)
Follicormon (50-50-0)
Follicular hormone (53-16-7)
Follicular hormone hydrate (50-27-1)
Folliculin (53-16-7)
Folliculine (53-16-7)
Folliculine benzoate (53-16-7)
Follicunodis (53-16-7)
Follicyclin P (113-38-2)
Follidiene (56-53-1)
Follidiene (84-17-3)
Follidrin (50-50-0)
Follidrin (53-16-7)
Follitropin (9002-68-0)
Follormon (84-17-3)
Follutein (9002-61-3)
Fologenon (57-83-0)
Folosan (117-18-0)
Folosan (82-68-8)
Folosan DB-905 Fumite (879-39-0)
Folpan (133-07-3)
Folpel (133-07-3)
Folpet (133-07-3)
Folsan (879-39-0)
Fomac (70-30-4)
Fomac 2 (82-68-8)
Fomrez Sul-3 (1067-33-0)
Fomrez sul-4 (77-58-7)
Fonatol (56-53-1)
Fonofos (ACGIH,OSHA) (944-22-9)
Fonoline (8012-95-1)
Fonoline, White (8009-03-8)
Fonoline, Yellow (8009-03-8)
Fonophos (944-22-9)
Fonurit (59-66-5)
Food Blue 1 (2650-18-2)
Food Blue 2 (3844-45-9)
Food Blue 2 (860-22-0)
Food Blue 3 (129-17-9)
Food Blue 4 (81-77-6)
Food Blue Dye No. 1 (3844-45-9)
Food Dye Red No. 104 (6441-77-6)
Food Green 1 (4680-78-8)
Food Green 2 (5141-20-8)
Food Green 3 (2353-45-9)
Food Orange No. 1 (1934-20-9)
Food Red 2 (915-67-3)
Food Red 4 (4548-53-2)
Food Red 5 (3567-69-9)
Food Red 6 (2611-82-7)
Food Red 7 (2611-82-7)
Food Red 9 (915-67-3)
Food Red 14 (16423-68-0)
Food Red 15 (81-88-9)
Food Red No. 101 (3761-53-3)
Food Red No. 102 (2611-82-7)
Food Violet 2 (1694-09-3)
Food Yellow 3 (2783-94-0)
Food Yellow 4 (1934-21-0)
Food Yellow 5 (1934-21-0)
Food Yellow 6 (2783-94-0)
Food Yellow No. 4 (1934-21-0)
Foodcol Sunset Yellow FCF (2783-94-0)
Foraat (Dutch) (298-02-2)

Forane	**(26675-46-7)**
Forane	**(76-13-1)**
Forane 22	**(75-45-6)**
Force	**(79538-32-2)**
Fore	**(8018-01-7)**
Foredex 75	**(94-75-7)**
Forlamin	**(5429-41-4)**
Forlex	**(8066-01-1)**
Forlien	**(63428-84-2)**
Forlin	**(58-89-9)**
Formagene	**(30525-89-4)**
Formal	**(109-87-5)**
Formal	**(121-75-5)**
Formal Fast Black 2B	**(6428-31-5)**
Formaldehyd (Czech, Polish)	**(50-00-0)**
Formaldehyde (ACGIH,OSHA)	**(50-00-0)**
Formaldehyde, Gas	**(50-00-0)**
Formaldehyde, Phenol polymer	**(9003-35-4)**
Formaldehyde, Polymer with benzenamine	**(25214-70-4)**
Formaldehyde, Polymer with benzenamine, hydrochloride (VAN) (9CI)	**(57138-85-9)**
Formaldehyde, Polymer with (chloromethyl)oxirane and 4,4'-(1-methylethylidene)bis(phenol) (9CI)	**(28906-96-9)**
Formaldehyde, Polymer with (chloromethyl)oxirane, 4,4'-(1-methylethylidene)bis(phenol) and phenol (9CI)	**(40216-08-8)**
Formaldehyde, Polymer with dimethylnaphthalene (9CI)	**(52613-22-6)**
Formaldehyde, Polymer with 1,1'-oxybis(benzene) (9CI)	**(26007-63-6)**
Formaldehyde, Polymer with paraformaldehyde and phenol	**(9003-35-4)**
Formaldehyde, Solutions, flammable [UN 1198]	**(50-00-0)**
Formaldehyde, Urea adduct	**(68611-64-3)**
Formaldehyde-aniline Copolymer	**(25214-70-4)**
Formaldehyde bis(β-chloroethyl) acetal	**(111-91-1)**
Formaldehyde bisulfite	**(75-92-3)**
Formaldehyde copolymer with urea	**(9011-05-6)**
Formaldehyde cyanohydrin	**(107-16-4)**
Formaldehyde dimethylacetal	**(109-87-5)**
Formaldehyde hydrosulfite	**(149-44-0)**
Formaldehyde sodium bisulfite	**(75-92-3)**
Formaldehyde sodium bisulfite adduct	**(149-44-0)**
Formaldehyde sodium sulfoxylate	**(149-44-0)**
Formaldehyde solution [UN 2209]	**(50-00-0)**
Formaldehydesulfoxylic acid, sodium salt	**(149-44-0)**
Formaldehyde, thio-, trimer	**(291-21-4)**
Formaldehyde-urea condensate	**(9011-05-6)**
Formaldehyde-urea copolymer	**(9011-05-6)**
Formaldehyde-urea polymer	**(9011-05-6)**
Formaldehyde-urea precondensate	**(9011-05-6)**
Formaldehyde-urea prepolymer	**(9011-05-6)**
Formaldehyde-urea resin	**(9011-05-6)**
Formal glycol	**(646-06-0)**
Formal hydrazine	**(624-84-0)**
Formalin 40	**(50-00-0)**
Formalin [UN 2209]	**(50-00-0)**
Formalina (Italian)	**(50-00-0)**
Formaline Black C	**(1937-37-7)**
Formaline (German)	**(50-00-0)**
Formalin-loesungen (German)	**(50-00-0)**
Formalin-urea copolymer	**(9011-05-6)**
Formalith	**(50-00-0)**
Formamide (ACGIH,OSHA)	**(75-12-7)**
Formamide, 1,1'-azobis	**(123-77-3)**
Formamide, N-(tert-butyl)	**(2425-74-3)**
Formamide, N-(4-chlorophenyl)- (9CI)	**(2617-79-0)**
Formamide, N-(4,6-diamino-1,3,5-triazin-2-yl)- (9CI)	**(13236-84-5)**
Formamide, N-(4,6-diamino-s-triazin-2-yl)- (8CI)	**(13236-84-5)**
Formamide, N,N-di-n-butyl	**(761-65-9)**
Formamide, N-(1-(3,4-dichloroanilino)-2,2,2-trichloroethyl)	**(20856-57-9)**
Formamide, N,N-diethyl	**(617-84-5)**
Formamide, N,N-dimethyl	**(68-12-2)**
Formamide, N,N-dimethyl-1-((4-phenoxybutyl)sulfinyl)- (9CI)	**(103614-75-1)**
Formamide, 1,1'-dithiobis(N,N-diethylthio-	**(97-77-8)**
Formamide, 1,1'-dithiobis(N,N-dimethylthio-	**(137-26-8)**
Formamide, N-ethyl	**(627-45-2)**
Formamide, 1,1'-hydrazobis-	**(110-21-4)**
Formamide, N-(hydroxymethyl)- (9CI)	**(13052-19-2)**
Formamide, N-methyl	**(123-39-7)**
Formamide, N-(2-methylphenyl)- (9CI)	**(94-69-9)**
Formamide, N-(4-(5-nitro-2-furyl)-2-thiazolyl)	**(24554-26-5)**
Formamide, N-phenyl- (9CI)	**(103-70-8)**
Formamide, N,N'-(1,4-piperazinediylbis(2,2,2-trichloro-ethylidene))bis- (8ci,9CI)	**(26644-46-2)**
Formamide, N-(1,2,2,2-tetrachloroethyl)- (8CI,9CI)	**(3659-66-3)**
Formamide, 1,1'-thiobis(N,N-dimethylthio-	**(97-74-5)**
Formamide, N-(2,2,2-trichloro-1-(morpholinyl)ethyl)	**(60029-23-4)**
Formamidine, N'-(4-chloro-o-tolyl)-N,N-dimethyl	**(6164-98-3)**
Formamidine, N'-(4-chloro-o-tolyl)-N,N-dimethyl-, hydrochloride	**(19750-95-9)**
Formamidine, N,N-dimethyl-N'-(5-(2-(5-nitro-2-furyl)vinyl)-1,3,4-oxadiazol-2-yl)	**(25962-77-0)**
Formamidine, N,N-dimethyl-N'-(o-tolyl)	**(10278-71-4)**
Formamidine sulfinic acid	**(1758-73-2)**
Formamidine, N,N'-vinylene-	**(288-32-4)**
Formamidobenzene	**(103-70-8)**
N-(1-Formamido-2,2,2-trichloretyl)morfolin (Czech)	**(60029-23-4)**
Formamine	**(100-97-0)**
Formanilide	**(103-70-8)**
Formanilide, 4'-chloro	**(2617-79-0)**
Formasept	**(102-98-7)**
Formetanat	**(22259-30-9)**
Formetanate	**(22259-30-9)**
Formetanate hydrochloride	**(23422-53-9)**
Formhydrazid (German)	**(624-84-0)**
Formhydrazide	**(624-84-0)**
Formiate de methyle (French)	**(107-31-3)**
Formiate de propyle (French)	**(110-74-7)**
Formic Black BA	**(1937-37-7)**
Formic Black C	**(1937-37-7)**
Formic Black CW	**(1937-37-7)**
Formic Black MTG	**(1937-37-7)**
Formic Black MTR	**(2429-83-6)**
Formic Black TG	**(1937-37-7)**
Formic-acid	**(64-18-6)**
Formic acid (ACGIH,OSHA) [UN 1779]	**(64-18-6)**
Formic acid, Solution [UN 1779]	**(64-18-6)**
Formic acid, aluminum salt (9CI)	**(7360-53-4)**
Formic acid, amino-	**(463-77-4)**
Formic acid, (aminocarbonyl)-	**(471-47-6)**
Formic acid, ammonium salt	**(540-69-2)**
Formic acid, azido-, tert-butyl ester (8CI)	**(1070-19-5)**
Formic acid, azidodithio-	**(4472-06-4)**
Formic acid, benzyl ester	**(104-57-4)**
Formic acid, butyl ester	**(592-84-7)**
Formic acid, calcium salt	**(544-17-2)**
Formic acid, carbamoyl-	**(471-47-6)**
Formic acid, chloro-, allyl ester	**(2937-50-0)**
Formic acid, chloro-, benzyl ester	**(501-53-1)**
Formic acid, chloro-, 2-chloroethyl ester	**(627-11-2)**
Formic acid, chloro-, ethyl ester	**(541-41-3)**
Formic acid, chloro-, 2-fluoroethyl ester	**(462-27-1)**
Formic acid, chloro-, hexyl ester	**(6092-54-2)**
Formic acid, chloro-, isobutyl ester (8CI)	**(543-27-1)**
Formic acid, chloro-, isopropyl ester	**(108-23-6)**
Formic acid, chloro-, methyl ester	**(79-22-1)**
Formic acid, chloro-, oxydiethylene ester	**(106-75-2)**
Formic acid p-chlorophenylamide	**(2617-79-0)**

Formic acid, chloro-, phenyl ester	**(1885-14-9)**
Formic acid, chloro-, propyl ester	**(109-61-5)**
Formic acid, chlorothio-, O-methyl ester (6CI,7CI)	**(2812-72-8)**
Formic acid, chlorothio-, S-ethyl ester	**(2941-64-2)**
Formic acid, chlorothio-, ethyl ester	**(2812-73-9)**
Formic acid, chlorothio-, S-propyl ester	**(13889-92-4)**
Formic acid, chloro-, trichloromethyl ester	**(503-38-8)**
Formic acid, cobalt(2+) salt (9CI)	**(544-18-3)**
Formic acid, Compd. with 2,2',2''-nitrilotris(ethanol) (1:1) (9CI)	**(24794-58-9)**
Formic acid, copper(2+) salt (1:1)	**(544-19-4)**
Formic acid, copper(2+) salt (9CI)	**(544-19-4)**
Formic acid, cyclohexyl ester (9CI)	**(4351-54-6)**
Formic acid, 3,7-dimethyl-2,6-octadienyl ester, (E)-	**(105-86-2)**
Formic acid, dithiobis(thio-, O,O-diethyl ester	**(502-55-6)**
Formic acid, ethyl ester	**(109-94-4)**
Formic acid, ethylidenemethylhydrazide	**(16568-02-8)**
Formic acid, geraniol ester	**(105-86-2)**
Formic acid, hydrazide	**(624-84-0)**
Formic acid, hydrazodi	**(628-36-4)**
Formic acid, ion(1-) (8CI,9CI) (VAN)	**(71-47-6)**
Formic acid, isobutyl ester	**(542-55-2)**
Formic acid, isopentyl ester	**(110-45-2)**
Formic acid, isopropyl ester	**(625-55-8)**
Formic acid, lead(2+) salt (8CI,9CI)	**(811-54-1)**
Formic acid, lithium salt (9CI)	**(556-63-8)**
Formic acid, methyl ester	**(107-31-3)**
Formic acid, methylhydrazide	**(758-17-8)**
Formic acid, nickel salt	**(15843-02-4)**
Formic acid, nickel(2+) salt	**(15843-02-4)**
Formic acid, nickel(2+) salt, dihydrate (8CI,9CI)	**(15694-70-9)**
Formic acid, 2-(4-(5-nitro-2-furyl)-2-thiazolyl)hydrazide	**(3570-75-0)**
Formic acid, pentyl ester	**(638-49-3)**
Formic acid, potassium salt	**(590-29-4)**
Formic acid, propyl ester	**(110-74-7)**
Formic acid, sodium salt	**(141-53-7)**
Formic acid, vinyl ester	**(692-45-5)**
Formic acid, zinc salt (9CI)	**(557-41-5)**
Formic acid, zinc salt, dihydrate (8CI,9CI)	**(5970-62-7)**
Formic aldehyde	**(50-00-0)**
Formic anammonide	**(74-90-8)**
Formic ether	**(109-94-4)**
Formic hydrazide	**(624-84-0)**
Formic 2-(4-(5-nitrofuryl)-2-thiazolyl)hydrazide	**(3570-75-0)**
Formimidic acid, 1-carbamoyl-	**(471-46-5)**
Formimidic acid, 1-semicarbazido-	**(110-21-4)**
Formin	**(100-97-0)**
Formison	**(77-75-8)**
Formoguanamine	**(504-08-5)**
Formohydrazide	**(624-84-0)**
Formol	**(50-00-0)**
Formomalenic thallium	**(2757-18-8)**
Formonitrile	**(74-90-8)**
Formopan	**(149-44-0)**
Formosa Camphor	**(76-22-2)**
Formosa Camphor Oil	**(8008-51-3)**
Formose Oil of Camphor	**(8008-51-3)**
Formosulfacetamide	**(144-80-9)**
Formosulfathiazole	**(72-14-0)**
Formothion	**(2540-82-1)**
o-Formotoluidide (8CI)	**(94-69-9)**
Formparanate	**(17702-57-7)**
Formula 40	**(2008-39-1)**
Formula 40	**(94-75-7)**
Formula 300	**(57-11-4)**
Formvar 1285	**(9003-20-7)**
p-Formyl-N,N-diethylaniline	**(120-21-8)**
Formyl Violet S4BN	**(1694-09-3)**
2-Formylamino-4-(5-nitro-2-furyl)thiazole	**(24554-26-5)**
Formylaniline	**(103-70-8)**
N-Formylaniline	**(103-70-8)**
4-Formylbenzaldehyde	**(623-27-8)**
p-Formylbenzaldehyde	**(623-27-8)**
2-Formylbenzenesulfonic acid	**(91-25-8)**
3-Formylbenzoic acid	**(619-21-6)**
4-Formylbenzoic acid	**(619-66-9)**
p-Formylbenzoic acid methyl ester	**(1571-08-0)**
4a-Formyl-1,4,4a,5a,6,9,9a,9b-octahydrodibenzofuran	**(126-15-8)**
4-Formylcyclohexene	**(100-50-5)**
N-Formyl-N'-(3',4'-dichlorphenyl)-2,2,2-trichloracetaldehydam (German)	**(20856-57-9)**
2-Formyl-3,4-dihydro-2H-pyran	**(100-73-2)**
N-Formyldimethylamine	**(68-12-2)**
p-Formyldimethylaniline	**(100-10-7)**
N-Formylethylamine	**(627-45-2)**
α-Formylethylbenzene	**(93-53-8)**
Formyl fluoride (8CI,9CI)	**(1493-02-3)**
2-Formylfuran	**(98-01-1)**
5-Formylguaiacol	**(621-59-0)**
6-Formylguaiacol	**(148-53-8)**
Formylhydrazide	**(624-84-0)**
Formylhydrazine	**(624-84-0)**
N-Formylhydrazine	**(624-84-0)**
2-(2-Formylhydrazino)-4-(5-nitro-2-furyl)thiazole	**(3570-75-0)**
Formylic acid	**(64-18-6)**
2-Formylimidazole	**(10111-08-7)**
1-Formyl-4-isohexenyl-4-cyclohexene	**(37677-14-8)**
1-Formyl-4-ixohexenyl-4-cyclohexene	**(37677-14-8)**
S-(2-(Formylmethylamino)-2-oxoethyl) O,O-dimethyl-phosphorodithioate	**(2540-82-1)**
N-Formyl-N-methylcarbamoylmethyl O,O-dimethyl phosphoro-dithioate	**(2540-82-1)**
S-(N-Formyl-N-methylcarbamoylmethyl) O,O-dimethyl phosphoro-dithioate	**(2540-82-1)**
S-(N-Formyl-N-methylcarbamoylmethyl) dimethyl phosphoro-thiolothionate	**(2540-82-1)**
2-(Formylmethylene)-1,3,3-trimethylindoline	**(84-83-3)**
2-Formyl-5-methylfuran	**(620-02-0)**
N-Formyl-N-methylhydrazine	**(758-17-8)**
α-Formyl methyl phenyl acetate	**(5894-79-1)**
N-Formylmorfolin (Czech)	**(4394-85-8)**
4-Formylmorpholine	**(4394-85-8)**
N-Formylmorpholine	**(4394-85-8)**
1-Formylnaphthalene	**(66-77-3)**
2-Formylnaphthalene	**(66-99-9)**
β-Formylnaphthalene	**(66-99-9)**
3-Formylnitrobenzene	**(99-61-6)**
p-Formylnitrobenzene	**(555-16-8)**
Formyloxiran	**(765-34-4)**
2-Formylphenol	**(90-02-8)**
3-Formylphenol	**(100-83-4)**
4-Formylphenol	**(123-08-0)**
m-Formylphenol	**(100-83-4)**
o-Formylphenol	**(90-02-8)**
p-Formylphenol	**(123-08-0)**
3-Formylphenyl N-phenylcarbamate	**(37070-87-4)**
3-Formylphenyl phenylcarbamate	**(37070-87-4)**
4-Formylphenyl N-phenylcarbamate	**(37076-88-3)**
4-Formylphenyl phenylcarbamate	**(37076-88-3)**
n-Formylpiperidin (German)	**(2591-86-8)**
1-Formylpiperidine	**(2591-86-8)**
3-Formylpropanoic acid	**(692-29-5)**
3-Formylpropionic acid	**(692-29-5)**
β-Formylpropionic acid	**(692-29-5)**
3-Formylpyridine	**(500-22-1)**

Name	CAS No.
4-Formylpyridine	**(872-85-5)**
2-Formylpyridine ketoxime	**(873-69-8)**
2-Formylpyridine oxime	**(873-69-8)**
2-Formylquinoxaline 1,4-dioxide carbomethoxyhydrazone	**(6804-07-5)**
4-Formylresorcinol	**(95-01-2)**
2-Formylthiophene	**(98-03-3)**
α-Formylthiophene	**(98-03-3)**
2-Formyltoluene	**(529-20-4)**
p-Formyltoluene	**(104-87-0)**
Formyl trichloride	**(67-66-3)**
(E)-3-Formyl-2,4,4-trichloro-2-butenoic acid	**(115340-67-5)**
Foron Blue SBGL	**(12222-78-5)**
Foron Brilliant Red E 2BL	**(17418-58-5)**
Foron Orange E-RFL	**(12217-83-3)**
Foron Scarlet E 2GFL	**(12223-35-7)**
Foron Yellow SE-2GL	**(61902-16-7)**
Forotox	**(52-68-6)**
Forpen	**(113-98-4)**
Forron	**(93-76-5)**
Forst-nexen	**(58-89-9)**
Forst U 46	**(93-76-5)**
Forstan	**(2439-01-2)**
Fortacyl	**(62-44-2)**
Fortafil 5Y	**(7782-42-5)**
Fortalgesic	**(359-83-1)**
Fortalin	**(359-83-1)**
Fortamine	**(25523-97-1)**
Fortecortin	**(50-02-2)**
Fortex	**(93-76-5)**
Forthion	**(121-75-5)**
Fortiflex 6015	**(9002-88-4)**
Fortiflex A 60/500	**(9002-88-4)**
Fortigro	**(6804-07-5)**
Fortion NM	**(60-51-5)**
Fortodyl	**(50-14-6)**
Fortracin	**(1405-87-4)**
Fortral	**(359-83-1)**
Fortrol	**(21725-46-2)**
Forturf	**(1897-45-6)**
Forza	**(79538-32-2)**
Fosalon	**(2310-17-0)**
Fosamine ammonium	**(25954-13-6)**
Foschlor	**(52-68-6)**
Foschlor 25	**(52-68-6)**
Foschlor R-50	**(52-68-6)**
Foschlor R	**(52-68-6)**
Foschlorem (Polish)	**(52-68-6)**
Fosdrin	**(7786-34-7)**
Fosetyl Aluminum	**(39148-24-8)**
Fosfakol	**(311-45-5)**
Fosfamid	**(60-51-5)**
Fosfamidon (Dutch)	**(13171-21-6)**
Fosfamidone (Italian)	**(13171-21-6)**
Fosfato de tricresilo (Spanish)	**(1330-78-5)**
Fosfermo	**(56-38-2)**
Fosferno	**(56-38-2)**
Fosferno M 50	**(298-00-0)**
Fosfestrol	**(522-40-7)**
Fosfex	**(56-38-2)**
Fosfive	**(56-38-2)**
Fosfon D	**(115-78-6)**
Fosfonet	**(4408-78-0)**
Fosfono 50	**(563-12-2)**
Fosforan O-1,2-dwubromo-2,2-dwuchloroetylo-O,O-dwumetylowy (Polish)	**(300-76-5)**
Fosforan troj-(1,3-dwuchloroizopropylowy) (Polish)	**(13674-87-8)**
Fosforo (Spanish)	**(7723-14-0)**
Fosforo bianco (Italian)	**(7723-14-0)**
Fosforo blanco (Spanish)	**(7723-14-0)**
Fosforo(pentacloruro di) (Italian)	**(10026-13-8)**
Fosforo(tricloruro di) (Italian)	**(7719-12-2)**
Fosforowodor (Polish)	**(7803-51-2)**
Fosforoxychlorid (Czech)	**(10025-87-3)**
Fosforpentachloride (Dutch)	**(10026-13-8)**
Fosforthiochlorid (Czech)	**(3982-91-0)**
Fosfortrichloride (Dutch)	**(7719-12-2)**
Fosforyn trojetylowy (Czech)	**(122-52-1)**
Fosforyn trojmetylowy (Czech)	**(121-45-9)**
Fosforzuuroplossingen (Dutch)	**(7664-38-2)**
Fosfothion	**(121-75-5)**
Fosfotion	**(121-75-5)**
Fosfotion 550	**(121-75-5)**
Fosfotox	**(60-51-5)**
Fosfotox R	**(60-51-5)**
Fosfotox R 35	**(60-51-5)**
Fosfuri di alluminio (Italian)	**(20859-73-8)**
Fosfuri di magnesio (Italian)	**(12057-74-8)**
Fosfuro magnesico (Spanish)	**(12057-74-8)**
Fosgeen (Dutch)	**(75-44-5)**
Fosgen (Polish)	**(75-44-5)**
Fosgene (Italian)	**(75-44-5)**
Fosmet	**(732-11-6)**
Fosmethilan	**(83733-82-8)**
Fosmetilan	**(83733-82-8)**
Fosova	**(56-38-2)**
Fospirat	**(5598-52-7)**
Fospirate	**(5598-52-7)**
Fospirate methyl	**(5598-52-7)**
Fossil Flour	**(7631-86-9)**
Foster Grant 834	**(9003-53-6)**
Fostern	**(56-38-2)**
Fostex	**(94-36-0)**
Fosthietan	**(21548-32-3)**
Fostion	**(2275-18-5)**
Fostion MM	**(60-51-5)**
Fostox	**(56-38-2)**
Fostril	**(70-30-4)**
Fosvel	**(21609-90-5)**
Fosvex	**(107-49-3)**
Foszfamidon (Hungarian)	**(13171-21-6)**
Foumarin	**(117-52-2)**
Four Thousand Forty-nine	**(121-75-5)**
Fouramine	**(95-80-7)**
Fouramine BA	**(39156-41-7)**
Fouramine Brown AP	**(87-66-1)**
Fouramine D	**(106-50-3)**
Fouramine EG	**(591-27-5)**
Fouramine ERN	**(90-15-3)**
Fouramine J	**(95-80-7)**
Fouramine OP	**(95-55-6)**
Fouramine P	**(123-30-8)**
Fouramine PCH	**(120-80-9)**
Fouramine 2R	**(5307-14-2)**
Fouramine RS	**(108-46-3)**
Fouramine STD	**(6369-59-1)**
Fouramine Standard	**(615-50-9)**
Fourneau 1162	**(63-74-1)**
Fourneau 2559	**(65-29-2)**
Fourrine 1	**(106-50-3)**
Fourrine 36	**(5307-14-2)**
Fourrine 57	**(119-34-6)**
Fourrine 64	**(624-18-0)**
Fourrine 65	**(591-27-5)**
Fourrine 68	**(120-80-9)**

Fourrine 76 **(39156-41-7)**
Fourrine 79 **(108-46-3)**
Fourrine 81 **(6219-71-2)**
Fourrine 84 **(123-30-8)**
Fourrine 85 **(87-66-1)**
Fourrine 88 **(122-80-5)**
Fourrine 93 **(96-91-3)**
Fourrine 94 **(95-80-7)**
Fourrine 99 **(90-15-3)**
Fourrine A **(122-80-5)**
Fourrine Brown PR **(119-34-6)**
Fourrine Brown 2R **(5307-14-2)**
Fourrine Brown propyl **(119-34-6)**
Fourrine D **(106-50-3)**
Fourrine DS **(624-18-0)**
Fourrine EG **(591-27-5)**
Fourrine ERN **(90-15-3)**
Fourrine EW **(108-46-3)**
Fourrine 4G **(6219-77-8)**
Fourrine M **(95-80-7)**
Fourrine P Base **(123-30-8)**
Fourrine PG **(87-66-1)**
Fourrine 4R **(96-91-3)**
Fourrine SLA **(39156-41-7)**
Fourrine SO **(6219-71-2)**
Fovane **(91-33-8)**
Fozalon **(2310-17-0)**
Frabel **(129-20-4)**
Fracine **(59-87-0)**
Fraction AB **(50-53-3)**
Fradiomycin sulfate **(1405-10-3)**
Fraeseol **(77-83-8)**
Framed **(122-34-9)**
Frangula emodin **(518-82-1)**
Franklin **(1317-65-3)**
Franocide **(1642-54-2)**
Franozan **(1642-54-2)**
Fratol **(62-74-8)**
Free Benzylpenicillin **(61-33-6)**
Free Coconut Oil **(8001-31-8)**
Free Histamine **(51-45-6)**
Free Penicillin G **(61-33-6)**
Free Penicillin II **(61-33-6)**
Freemans White Lead **(7446-14-2)**
Freeuril **(91-33-8)**
Frenantol **(70-70-2)**
Frenasma **(15826-37-6)**
French Green **(12002-03-8)**
Frenogastrico **(53-46-3)**
Frenohypon **(70-70-2)**
Frentirox **(60-56-0)**
Freon **(75-45-6)**
Freon 11 **(75-69-4)**
Freon 11A **(75-69-4)**
Freon 11B **(75-69-4)**
Freon 113TR-T **(76-13-1)**
Freon 12 **(75-71-8)**
Freon 12B1 **(353-59-3)**
Freon 12-B2 **(75-61-6)**
Freon 13 **(75-72-9)**
Freon 13B1 **(75-63-8)**
Freon 14 **(75-73-0)**
Freon 21 **(75-43-4)**
Freon 22 **(75-45-6)**
Freon 23 **(75-46-7)**
Freon 31 **(593-70-4)**
Freon 41 **(593-53-3)**
Freon 112 **(76-12-0)**
Freon 113 **(76-13-1)**
Freon 114 **(76-14-2)**
Freon 115 **(76-15-3)**
Freon 116 **(76-16-4)**
Freon 124 **(2837-89-0)**
Freon 133a **(75-88-7)**
Freon 134 **(359-35-3)**
Freon 142 **(75-68-3)**
Freon 142b **(75-68-3)**
Freon 143 **(430-66-0)**
Freon 152 **(75-37-6)**
Freon 152 **(624-72-6)**
Freon 218 **(76-19-7)**
Freon 253 **(460-35-5)**
Freon 114B2 **(124-73-2)**
Freon C-318 **(115-25-3)**
Freon F-12 **(75-71-8)**
Freon F-23 **(75-46-7)**
Freon F113 **(76-13-1)**
Freon FT **(354-58-5)**
Freon HE **(75-69-4)**
Freon MF **(75-69-4)**
Freon R 112 **(76-12-0)**
Freon TF **(76-13-1)**
Frescon **(1420-06-0)**
Fresenius D 6 **(9004-34-6)**
Fresmin **(68-19-9)**
Freudal **(439-14-5)**
Frideron **(26538-44-3)**
Frigen **(75-45-6)**
Frigen 11 **(75-69-4)**
Frigen 12 **(75-71-8)**
Frigen 22 **(75-45-6)**
Frigen 113 **(76-13-1)**
Frigen 113a **(76-13-1)**
Frigen 113TR **(76-13-1)**
Frigen 113TR-N **(76-13-1)**
Frigen 113 TR-T **(76-13-1)**
Frigen 114 **(76-14-2)**
Frigen 114a **(374-07-2)**
Frigiderm **(76-14-2)**
Frisium **(22316-47-8)**
Frucote **(13952-84-6)**
Fructofuranoside, α-D-glucopyranosyl, β-d **(57-50-1)**
β-D-Fructofuranoside, α-D-glucopyranosyl **(57-50-1)**
β-D-Fructofuranosyl-α-D-glucopyranoside benzoate **(12738-64-6)**
β-D-Fructofuranosyl-α-D-glucopyranoside, monododecanoate **(25339-99-5)**
β-D-Fructofuranosyl-α-D-glucopyranoside, monooctadecanoate **(25168-73-4)**
β-D-Fructofuranosyl-α-D-glucopyranoside octakis-(hydrogen sulfate) aluminum complex **(54182-58-0)**
Fructopyranose, β-D **(7660-25-5)**
Fructopyranose, 1,2:4,5-di-O-isopropylidene-, .α.-D- (8CI) **(20880-93-7)**
D-Fructose **(57-48-7)**
Fructose **(57-48-7)**
Fructose **(7660-25-5)**
Fructose solution **(57-48-7)**
Fruit Red A Extra Yellowish Geigy **(3567-69-9)**
Fruit Red A Geigy **(915-67-3)**
Fruit Sugar **(7660-25-5)**
Fruitdo **(10380-28-6)**
Fruitofix **(86-87-3)**
Fruitone **(86-86-2)**
Fruitone **(86-87-3)**
Fruitone A **(93-76-5)**
Fruitone N **(86-87-3)**
Fruitone T **(93-72-1)**

Frumidor **(23564-05-8)**
Frumin AL **(298-04-4)**
Frumin G **(298-04-4)**
Frusemide **(54-31-9)**
Frusemin **(54-31-9)**
Frusid **(54-31-9)**
Frustan **(439-14-5)**
Frutabs **(7660-25-5)**
Ftaalzuuranhydride (Dutch) **(85-44-9)**
Ftaflex DIBA **(141-04-8)**
Ftalan **(133-07-3)**
Ftalanhydrid (Czech) **(85-44-9)**
Ftalimmide (Italian) **(85-41-6)**
Ftalodinitril (Czech) **(91-15-6)**
Ftalonitril (Czech) **(91-15-6)**
Ftalophos **(732-11-6)**
Ftalowy bezwodnik (Polish) **(85-44-9)**
Fthalide **(27355-22-2)**
Ftorotan (Russian) **(151-67-7)**
Ftoruracil **(51-21-8)**
Fuberidatol **(3878-19-1)**
Fuberidazole **(3878-19-1)**
Fuberisazol **(3878-19-1)**
Fubol **(75701-74-5)**
Fubridazole **(3878-19-1)**
Fuchsin(E) Acid **(3244-88-0)**
p-Fuchsin **(569-61-9)**
Fuchsin (Basic) **(3248-93-9)**
Fuchsin NB **(3248-91-7)**
Fuchsin SP **(569-61-9)**
Fuchsin dye base **(467-62-9)**
Fuchsine **(632-99-5)**
Fuchsine A **(632-99-5)**
Fuchsine Acid **(3244-88-0)**
p-Fuchsine Acid **(3244-88-0)**
Fuchsine Base **(3248-93-9)**
Fuchsine CS **(632-99-5)**
Fuchsine DR-001 **(569-61-9)**
Fuchsine G **(632-99-5)**
Fuchsine HF Base **(3248-93-9)**
Fuchsine HO **(632-99-5)**
Fuchsine N **(632-99-5)**
Fuchsine RTN **(632-99-5)**
Fuchsine SPC **(569-61-9)**
Fuchsine Y **(632-99-5)**
Fuchsin nitrate **(61467-64-9)**
Fuclasin **(137-30-4)**
Fuclasin Ultra **(137-30-4)**
Fucosterol, β-dihydro- **(83-47-6)**
Fudiolan **(50512-35-1)**
Fuel Gases, Low and medium B.T.U. **(8006-20-0)**
Fuel Oil, Diesel (DOT) **(68334-30-5)**
Fuel Oil, No. 2 **(68476-30-2)**
Fuel Oil, No. 4 **(68476-31-3)**
Fuel Oil, No. 6 **(68553-00-4)**
Fuel Oil, Pyrolysis **(69013-21-4)**
Fuel Oil, Residual **(68476-33-5)**
Fuel Oil, Residues-straight-run gas oils, high-sulfur **(68476-32-4)**
Fuels, Diesel **(68334-30-5)**
Fuels, Diesel, No. 2 **(68476-34-6)**
Fuels, Gasoline **(86290-81-5)**
Fugu Poison **(4368-28-9)**
Fuji 1 **(50512-35-1)**
Fuji HEC-BL 20 **(9004-62-0)**
Fujione **(50512-35-1)**
Fuji-one **(50512-35-1)**
Fukinan **(31230-13-4)**
Fukinane **(31230-13-4)**
Fukinotoxin **(60102-37-6)**
Fukinotoxin (Neutral) **(60102-37-6)**
Fuklasin **(137-30-4)**
Fuklasin **(14484-64-1)**
Fuklasin Ultra **(14484-64-1)**
Fuklazin **(14484-64-1)**
Ful-Glo **(518-47-8)**
Fulcin **(126-07-8)**
Fulcine **(126-07-8)**
Full-range alkylate naphtha **(64741-64-6)**
Full-range reformed naphtha **(68919-37-9)**
Full range reformed naphtha (Petroleum) **(68919-37-9)**
Fulminate d'argent (French) **(5610-59-3)**
Fulminate of mercury, Dry (DOT) **(628-86-4)**
Fulminate of mercury, Wet **(628-86-4)**
Fulminate of mercury, Wet (DOT) **(628-86-4)**
Fulminato de plata (Spanish) **(5610-59-3)**
Fulminene **(217-37-8)**
Fulminic acid **(506-85-4)**
Fulminic acid, silver(1+) salt **(5610-59-3)**
Fulsix **(54-31-9)**
Fuluvamide **(54-31-9)**
Fulvic acid **(479-66-3)**
Fulvican Grisactin **(126-07-8)**
Fulvicin **(126-07-8)**
Fulvicin-P/G **(126-07-8)**
Fulvicin-U/F **(126-07-8)**
Fulvina **(126-07-8)**
Fulvistatin **(126-07-8)**
Fumo-Gas **(106-93-4)**
Fumagon **(96-12-8)**
Fumaramic acid (8CI) **(2987-87-3)**
Fumaric-acid **(110-17-8)**
Fumaric acid, bis(2-ethylhexyl) ester **(141-02-6)**
Fumaric acid, butyl 2,3-epoxypropyl ester **(25876-07-7)**
Fumaric acid, dibutyl ester **(105-75-9)**
Fumaric acid, diethyl ester **(623-91-6)**
Fumaric acid, ethyl 2,3-epoxypropyl ester **(25876-47-5)**
Fumaric acid, ethylmethyl- (8CI) **(28098-80-8)**
Fumarin **(117-52-2)**
Fumarine **(130-86-9)**
Fumaronitrile **(764-42-1)**
Fumaroyl chloride **(627-63-4)**
Fumaroyl dichloride **(627-63-4)**
Fumarylchlorid (Czech) **(627-63-4)**
Fumaryl chloride [UN 1780] **(627-63-4)**
Fumasol **(117-52-2)**
Fumasol **(34490-93-2)**
Fumazone **(96-12-8)**
Fumazone 86 **(96-12-8)**
Fumazone 86E **(96-12-8)**
Fumed Silica **(7631-86-9)**
Fumed Silicon Dioxide **(7631-86-9)**
Fumes, Zinc **(69012-65-3)**
Fumetobac **(54-11-5)**
Fumette **(558-25-8)**
Fumigant-1 (Obs.) **(74-83-9)**
Fumigrain **(107-13-1)**
Fuming Liquid Arsenic **(7784-34-1)**
Fuming Sulfuric acid (DOT) **(8014-95-7)**
Fumitoxin **(20859-73-8)**
Fundal **(6164-98-3)**
Fundal 500 **(6164-98-3)**
Fundal SP **(19750-95-9)**
Fundasol **(17804-35-2)**
Fundazol **(17804-35-2)**

Fundex (6164-98-3)
Funduscein (518-47-8)
Fungo-Pulvit (12122-67-7)
Fungacetin (102-76-1)
Fungacide 337 (95-19-2)
Fungaflor (35554-44-0)
Fungchex (7487-94-7)
Fungicide 1991 (17804-35-2)
Fungicide FX (97-23-4)
Fungiclor (82-68-8)
Fungifen (87-86-5)
Fungilin (1397-89-3)
Fungimar (1317-39-1)
Funginex (26644-46-2)
Fungisone (1397-89-3)
Fungitox (23564-05-8)
Fungitox OR (62-38-4)
Fungivin (126-07-8)
Fungizone (1397-89-3)
Fungo (23564-05-8)
Fungo 50 (23564-05-8)
Fungol (16871-71-9)
Fungol B (7681-49-4)
Fungonit gf 2 (16871-71-9)
Fungostop (137-30-4)
Fungus Ban Type II (133-06-2)
Funjeb (12122-67-7)
Fur Black 41867 (106-50-3)
Fur Brown 41866 (106-50-3)
Fur Yellow (106-50-3)
Furacilin (59-87-0)
Furacillin (59-87-0)
Furacin (59-87-0)
Furacin (61-76-7)
Furacin-E (59-87-0)
Furacin-HC (59-87-0)
Furacine (59-87-0)
Furacinetten (59-87-0)
Furacoccid (59-87-0)
Furacort (59-87-0)
Furacycline (59-87-0)
Furadan (OSHA) (1563-66-2)
Furadane (1563-66-2)
Furadantin (67-20-9)
Furadantine (67-20-9)
Furadantoin (67-20-9)
Furadonin (67-20-9)
Furadonine (67-20-9)
Fural (98-01-1)
2-Furaldehyde (98-01-1)
3-Furaldehyde (8CI) (498-60-2)
2-Furaldehyde, 2,3:4,5-bis(2-butenylene)tetrahydro- (126-15-8)
2-Furaldehyde, 5-(hydroxymethyl) (67-47-0)
2-Furaldehyde, 5-methyl (620-02-0)
2-Furaldehyde, 5-nitro (698-63-5)
2-Furaldehyde, 5-nitro-, (4,6-bis(ethylamino)-s-triazin-2-yl)-hydrazone (8CI) (10422-01-2)
2-Furaldehyde, 5-nitro-, (4,6-dimethoxy-s-triazin-2-yl)hydrazone (8CI) (1037-57-6)
2-Furaldehyde, 5-nitro-, (4-(dimethylamino)-6-(trichloromethyl)-s-triazin-2-yl)hydrazone (8CI) (30576-26-2)
2-Furaldehyde, 5-nitro-, (4-methyl-6-(methylamino)-s-triazin-2-yl)-hydrazone (8CI) (30355-55-6)
2-Furaldehyde, 5-nitro-, semicarbazone (59-87-0)
Furaldon (59-87-0)
Furale (98-01-1)
Furalone (59-87-0)
Furaltadone (139-91-3)
Furaltadone (3031-51-4)
Furamethrin (23031-38-1)
Furametral (59-87-0)
Furan [UN 2389] (110-00-9)
2-Furanacetamide, α-((5-nitro-2-furanyl)methylene)- (9CI) (3688-53-7)
2-Furanacetic acid, 3-carboxy-2,5-dihydro-5-oxo- (8CI,9CI) (16426-62-3)
2-Furanacetic acid, 2,5-dihydro-5-oxo-, (+)- (8CI,9CI) (1124-48-7)
Furan, 2-acetyl- (1192-62-7)
Furan, 2-acetyl-5-methyl (1193-79-9)
2-Furanacrolein (8CI) (623-30-3)
2-Furanacrylamide, α-2-furyl-5-nitro- (8CI) (3688-53-7)
2-Furanaldehyde (98-01-1)
Furan, 2-butyl- (8CI,9CI) (4466-24-4)
Furan, 2-butyltetrahydro- (8CI,9CI) (1004-29-1)
Furan, 3-butyltetrahydro-2-methyl-, trans- (9CI) (36712-20-6)
2-Furancarbinol (98-00-0)
2-Furancarbonal (98-01-1)
2-Furancarboxaldehyde (98-01-1)
3-Furancarboxaldehyde (9CI) (498-60-2)
2-Furancarboxaldehyde, 5-(hydroxymethyl)- (9CI) (67-47-0)
5-Furancarboxaldehyde, 5-nitro- (9CI) (698-63-5)
3-Furancarboxamide, N-cyclohexyl-N-methoxy-2,5-dimethyl- (60568-05-0)
α-Furancarboxylic acid (88-14-2)
2-Furancarboxylic acid (9CI) (88-14-2)
3-Furancarboxylic acid (9CI) (488-93-7)
Furancarboxylic acid, ethyl ester (9CI) (1335-40-6)
3-Furancarboxylic acid, 5-(methoxymethyl)-2-methyl-, methyl ester (9CI) (35340-00-2)
Furancarboxylic acid, methyl ester (9CI) (1334-76-5)
2-Furancarboxylic acid, 5-nitro (9CI) (645-12-5)
Furan, 2-(chloromethyl)tetrahydro (3003-84-7)
2,5-Furandicarboxylic acid, tetrahydro-, trans- (8CI) (2044-00-0)
Furan, 2,3-dichlorotetrahydro (3511-19-1)
Furan, 4,5-diethyl-2,3-dihydro-2,3-dimethyl- (9CI) (54244-89-2)
Furan, 2,5-dihydro- (9CI) (1708-29-8)
Furan, 2,5-dihydro-2,5-dimethoxy- (332-77-4)
Furan, 2,3-dihydro-4-methyl- (9CI) (34314-83-5)
Furan, 2,3-dihydro-5-methyl- (8CI,9CI) (1487-15-6)
Furan, 2,3-dihydro-4-(1-methylethyl)- (9CI) (34314-84-6)
Furan, 2,3-dihydro-2-(methylphenyl)- (9CI) (103433-72-3)
Furan, 2,5-dimethoxy- (332-77-4)
Furan, 2,5-dimethyl (625-86-5)
Furan, dimethyl (28802-49-5)
Furan, 2,4-dimethyl- (8CI,9CI) (3710-43-8)
2,5-Furandione (108-31-6)
2,5-Furandione, dihydro- (108-30-5)
2,5-Furandione, dihydro-3-methyl- (9CI) (4100-80-5)
2,5-Furandione, dihydro-3-(nonenyl)- (9CI) (28928-97-4)
2,5-Furandione, dihydro-3-(octadecenyl)- (9CI) (28777-98-2)
2,5-Furandione, dihydro-3-(1-octenyl)- (9CI) (7757-96-2)
2,5-Furandione, dihydro-3-(octenyl)- (9CI) (26680-54-6)
2,5-Furandione, dihydro-3-(tetradecenyl)- (9CI) (33806-58-5)
2,5-Furandione, dihydro-3-(tetrapropenyl)- (9CI) (26544-38-7)
2,5-Furandione, 3-(dodecenyl)dihydro- (25377-73-5)
2,5-Furandione, 3-(2-dodecenyl)dihydro- (9CI) (19780-11-1)
2,5-Furandione, 3-(hexadecenyl)dihydro- (9CI) (32072-96-1)
2,5-Furandione, 3-methyl- (616-02-4)
Furan, 2,2'-(dithiobis(methylene))bis- (9CI) (4437-20-1)
Furaneol (3658-77-3)
Furan, 2-ethenyl- (9CI) (1487-18-9)
Furan, 2-ethyl- (9CI) (3208-16-0)
Furan, 2-heptyl- (9CI) (3777-71-7)
Furanidine (109-99-9)
Furanium (518-47-8)
2-Furankarbaldehyd (Czech) (98-01-1)
2-Furanmethanamine, N-(2,6-dinitro-4-(trifluoromethyl)-

phenyl)-N-ethyltetrahydro- (9CI)	**(34129-07-2)**
2-Furanmethanamine, N-(2,6-dinitro-4-(trifluoromethyl)-phenyl)tetrahydro-N-propyl- (9CI)	**(34128-99-9)**
2-Furanmethanol	**(98-00-0)**
3-Furanmethanol (8CI,9CI)	**(4412-91-3)**
2-Furanmethanol, acetate (9CI)	**(623-17-6)**
2-Furanmethanol, 5-ethenyltetrahydro-α,α,5-trimethyl-, cis- (VAN)(9CI)	**(5989-33-3)**
2-Furanmethanol, tetrahydro	**(97-99-4)**
2-Furanmethanol, tetrahydro-, acetate (9CI)	**(637-64-9)**
2-Furanmethanol, tetrahydro-5-methyl-, trans- (9CI)	**(54774-28-6)**
Furan, 2-methyl	**(534-22-5)**
Furan, 3-methyl	**(930-27-8)**
2-Furanmethyl acetate	**(623-17-6)**
2-Furanmethylamine	**(617-89-0)**
Furan, 2-methyl-tetrahydro	**(96-47-9)**
Furan, 2-((methylthio)methyl)- (9CI)	**(1438-91-1)**
Furan, 2-nitro	**(609-39-2)**
Furan-ofteno	**(59-87-0)**
2-Furanol, tetrahydro-2,3-dimethyl-, trans- (9CI)	**(61142-77-6)**
2-Furanol, tetrahydro-2-methyl- (8CI,9CI)	**(7326-46-7)**
2(5H)-Furanone	**(497-23-4)**
2(3H)-Furanone, 3-acetyldihydro- (9CI)	**(517-23-7)**
3(2H)-Furanone, 5-tert-butyldihydro- (8CI)	**(34003-77-5)**
2(3H)-Furanone, 5-butyldihydro-4-methyl-, cis- (9CI)	**(55013-32-6)**
2(5H)-Furanone, 5-(butylimino)- (9CI)	**(27396-39-0)**
2(5H)-Furanone, 3-chloro-4-dichloromethyl-5-hydroxy	**(77439-76-0)**
2(3H)-Furanone, 5-(1-decenyl)dihydro-, (R-(Z))-	**(64726-91-6)**
2(3H)-Furanone, 5-(1-decenyl)dihydro-, (Z)-(R)-(-)-	**(64726-91-6)**
2(5H)-Furanone, 3,4-dichloro-5-(dichloromethyl)-5-hydroxy	**(108082-06-0)**
2(3H)-Furanone, dihydro	**(96-48-0)**
2(3H)-Furanone, dihydro-5-butyl	**(104-50-7)**
2(3H)-Furanone, dihydro-4,4-dimethyl	**(13861-97-7)**
2(3H)-Furanone, dihydro-3,5-dimethyl- (8CI,9CI)	**(5145-01-7)**
2(3H)-Furanone, dihydro-5,5-dimethyl- (8CI,9CI)	**(3123-97-5)**
2(3H)-Furanone, dihydro-5-methyl	**(108-29-2)**
3(2H)-Furanone, dihydro-2-methyl	**(3188-00-9)**
2(3H)-Furanone, dihydro-5-octyl	**(2305-05-7)**
2(3H)-Furanone, dihydro-5-pentyl	**(104-61-0)**
2(3H)-Furanone, dihydro-5-propyl	**(105-21-5)**
2(5H)-Furanone, 5,5-dimethyl- (8CI,9CI)	**(20019-64-1)**
3(2H)-Furanone, 2,5-dimethyl-4-hydroxy	**(3658-77-3)**
2(3H)-Furanone, 5-ethyldihydro	**(695-06-7)**
2(3H)-Furanone, 5-heptyldihydro-	**(104-67-6)**
2(3H)-Furanone, 5-hexyldihydro	**(706-14-9)**
2(?H)-Furanone, 5-methyl- (9CI)	**(1333-38-6)**
2-Furanone, 5-methyl- (8CI)	**(1333-38-6)**
2(5H)-Furanone, 3-methyl- (8CI,9CI)	**(22122-36-7)**
2(5H)-Furanone, 3,5,5-trimethyl- (9CI)	**(50598-50-0)**
Furan, 2-pentyl	**(3777-69-3)**
Furan, pentyl- (9CI)	**(64079-01-2)**
Furan, tetrahydro	**(109-99-9)**
Furan, tetrahydro-2,5-dimethyl- (9CI)	**(1003-38-9)**
Furan, tetrahydro-2,5-dimethyl-, cis- (8CI,9CI)	**(2144-41-4)**
Furan, tetrahydro-2,2,5,5-tetramethyl- (8CI,9CI)	**(15045-43-9)**
Furanthril	**(54-31-9)**
Furanthryl	**(54-31-9)**
Furantoin	**(67-20-9)**
Furantril	**(54-31-9)**
Furan, 2,3,4-trimethyl- (7CI,8CI,9CI)	**(10599-57-2)**
Furan, 2-vinyl- (8CI)	**(1487-18-9)**
4-(2-Furanyl)-3-buten-2-one	**(623-15-4)**
1-(2-Furanyl)ethanone	**(1192-62-7)**
3-(2-Furanyl)-2-propenal	**(623-30-3)**
Furaplast	**(59-87-0)**
Furaseptyl	**(59-87-0)**
Furaskin	**(59-87-0)**
Furathiazole	**(531-82-8)**
Furathiocarb	**(65907-30-4)**
Furatoin	**(67-20-9)**
Furatol	**(62-74-8)**
Furatone	**(794-93-4)**
Furatone-S	**(794-93-4)**
Furaxon	**(67-45-8)**
Furaxone	**(67-45-8)**
Furaziline	**(59-87-0)**
Furazin	**(59-87-0)**
Furazina	**(59-87-0)**
Furazol	**(67-45-8)**
Furazol W	**(59-87-0)**
Furazolidon	**(67-45-8)**
Furazolidone	**(67-45-8)**
Furazolin	**(139-91-3)**
Furazoline	**(139-91-3)**
Furazon	**(67-45-8)**
Furazone	**(59-87-0)**
Furazosin	**(19216-56-9)**
Furesis	**(54-31-9)**
Furesol	**(59-87-0)**
Furethidine	**(2385-81-1)**
Furethidinum (Latin)	**(2385-81-1)**
(+-)-Furethionyl (+-)-cis, trans-chrysanthemate	**(17080-02-3)**
Furethrin	**(17080-02-3)**
Furetidina	**(2385-81-1)**
2-Furfural	**(98-01-1)**
Furfural (ACGIH,OSHA) [UN 1199]	**(98-01-1)**
Furfural alcohol	**(98-00-0)**
Furfuralcohol	**(98-00-0)**
Furfuraldehyde	**(98-01-1)**
Furfurale (Italian)	**(98-01-1)**
Furfuralu (Polish)	**(98-01-1)**
Furfuran	**(110-00-9)**
Furfurin	**(59-87-0)**
Furfurol	**(98-01-1)**
Furfurole	**(98-01-1)**
Furfuryl acetate	**(623-17-6)**
6-Furfuryladenine	**(525-79-1)**
N^6-Furfuryladenine	**(525-79-1)**
N-Furfuryladenine	**(525-79-1)**
Furfuryl-alcohol	**(98-00-0)**
Furfuryl alcohol (ACGIH,OSHA) [UN 2874]	**(98-00-0)**
Furfuryl alcohol, acetate	**(623-17-6)**
Furfuryl alcohol, tetrahydro-	**(97-99-4)**
Furfuryl alcohol, tetrahydro-, acetate (8CI)	**(637-64-9)**
2-Furfurylalkohol (Czech)	**(98-00-0)**
Furfurylamine [UN 2526]	**(617-89-0)**
Furfurylamine, N-ethyltetrahydro-N-(α,α,α-trifluoro-2,6-dinitro-p-tolyl)- (8CI)	**(34129-07-2)**
Furfurylamine, tetrahydro-N-propyl-N-(α,α,α-trifluoro-2,6-dinitro-p-tolyl)- (8CI)	**(34128-99-9)**
6-(Furfurylamino)purine	**(525-79-1)**
N^6-(Furfurylamino)purine	**(525-79-1)**
1-Furfurylpyrrole	**(1438-94-4)**
N-(2-Furfuryl)pyrrole	**(1438-94-4)**
N-Furfuryl pyrrole	**(1438-94-4)**
DL-Furfurylrethronyl DL-cis-trans-chrysanthemate	**(17080-02-3)**
Furidazina	**(712-68-5)**
Furidazol	**(3878-19-1)**
Furidazole	**(3878-19-1)**
Furidiazina	**(712-68-5)**
Furidiazine	**(712-68-5)**
Furidon	**(67-45-8)**
2-Furil-metanale (Italian)	**(98-01-1)**
Furium	**(531-82-8)**

Furloe (101-21-3)
Furloe 4EC (101-21-3)
Furmarin (117-52-2)
Furmecyclox (60568-05-0)
Furmetamide (60568-05-0)
Furmethanol (139-91-3)
Furmethonol (139-91-3)
Furmethonol (3031-51-4)
Furmetonol (139-91-3)
Furnal (1333-86-4)
Furnex (1333-86-4)
Furnex N 765 (1333-86-4)
Furobactina (67-20-9)
5H-Furo(3',2':6,7)(1)benzopyrano(3,4-c)pyridin-5-one (85878-62-2)
5H-Furo(3',2':6,7)(1)benzopyrano(3,4-c)pyridin-5-one, 7-methyl- (85878-63-3)
5H-Furo(3,2-c)(2)benzopyran-5-one, 2,3,3a,9b-tetrahydro-6-hydroxy-7,8-dimethoxy-2-propyl- (8CI) (VAN) (30270-60-1)
5H-Furo(3,2-c)(2)benzopyran-5-one, 2,3,3a,9b-tetrahydro-6-hydroxy-7,8-dimethoxy-2-propyl-, (2S-(2α,3aβ,9bβ))- (9CI) (30270-60-1)
Furo(2',3',7,6)coumarin (66-97-7)
Furo(4',5',6,7)coumarin (66-97-7)
Furo(5',4',7,8)coumarin (523-50-2)
Furocoumarin (66-97-7)
4H-Furo(3,2-c)pyran-2(6H)-one, 4-hydroxy (149-29-1)
Furo(2',3':5,6)cyclohepta(1,2-c)pyran-2(3H)-one, decahydro-5,7-dihydroxy-4a,9-dimethyl-3- methylene-, (3ar-(3a-α,4a-β,5-α,7-β,8a-α,9-α,10a-α)) (57377-32-9)
Furodan (1563-66-2)
Furo(2,3-d)-1,3-dioxole, .β.-D-talofuranose deriv. (9CI) (23262-79-5)
7H-Furo(3',2':4,5)furo(2,3-c)xanthen-7-one, 3a,12c-dihydro-8-hydroxy-6-methoxy (10048-13-2)
1H,12H-Furo(3',2':4,5)furo(2,3-h)pyrano(3,4-c)(1)benzopyran-1,12-dione, 3,4,7a-α,9, 10,10a-α-hexahydro-5-methoxy (7241-98-7)
1H,12H-Furo(3',2':4,5)furo(2,3-h)pyrano(3,4-c)(1)benzopyran-1,12-dione, 3,4,7a,10a- tetrahydro-5-methoxy (1165-39-5)
7H-Furo(3,2-g)(1)benzopyran-6-carboxylic acid, 7-oxo-, ethyl ester (20073-24-9)
7H-Furo(3,2-g)(1)benzopyran-7-one (66-97-7)
5H-Furo(3,2-g)(1)benzopyran-5-one, 4-hydroxy-9-methoxy-7-methyl- (6CI,7CI,8CI,9CI) (478-42-2)
7H-Furo(3,2-g)(1)benzopyran-7-one, 4-methoxy (484-20-8)
7H-Furo(3,2-g)(1)benzopyran-7-one, 9-methoxy (298-81-7)
7H-Furo(3,2-g)(1)benzopyran-7-one, 9-((3-methyl-2-butenyl)oxy) (482-44-0)
7H-Furo(3,2-g)(1)benzopyran-7-one, 2,5,9-trimethyl (3902-71-4)
2H-Furo(2,3-h)(1)benzopyran-2-one (523-50-2)
7H-Furo(3,2-h)(2)benzopyran-3(2H)-one, 6,9-dihydro-7-hydroxy-7-methyl-2-(1-methylethylidene)-, (R)- (9CI) (74798-20-2)
2H-Furo(2,3-h)(1)benzopyran-2-one, 4,8-dimethyl (4063-41-6)
2H-Furo(2,3-h)(1)benzopyran-2-one, 4,9-dimethyl (22975-76-4)
2H-Furo(2,3-h)-1-benzopyran-2-one, 5-methyl (73459-03-7)
2H-Furo(2,3-h)(1)benzopyran-2-one, 4,6,9-trimethyl (90370-29-9)
Furo(2,3-h)coumarin (523-50-2)
2-Furoic acid (88-14-2)
Furoic-acid (26447-28-9)
α-Furoic acid (88-14-2)
3-Furoic acid (8CI) (488-93-7)
2-Furoic acid, ethyl ester (614-99-3)
Furoic acid, ethyl ester (8CI) (1335-40-6)
2-Furoic acid, 5-nitro (645-12-5)
Furoin (552-86-3)
α-Furoin (552-86-3)
Furole (98-01-1)
α-Furole (98-01-1)
Furo(3',4':6,7)naphtho(2,3-d)-1,3-dioxol-6(5ah)-one, 5,8,8a,9-tetrahydro-9-hydroxy-5- (3,4,5-trimethoxyphenyl) (518-28-5)
Furore (66441-23-4)
Furose (53973-98-1)
Furosedon (54-31-9)
Furosem (59-87-0)
Furosemid (54-31-9)
Furosemide (54-31-9)
Furosemide "Mita" (54-31-9)
Furosemidu (Polish) (54-31-9)
Furothiazole (531-82-8)
Furovag (67-45-8)
Furox (67-45-8)
Furoxal (67-45-8)
Furoxane (67-45-8)
Furoxone (67-45-8)
Furoxone Swine Mix (67-45-8)
Furoylfurylcarbinol (552-86-3)
2-(4-(2-Furoyl)piperazin-1-yl)-4-amino-6,7-dimethoxyquinazoline (19216-56-9)
Furozolidine (67-45-8)
Furro D (106-50-3)
Furro EG (591-27-5)
Furro ER (90-15-3)
Furro L (615-05-4)
Furro P Base (123-30-8)
Furro 4R (96-91-3)
Furro SLA (39156-41-7)
Fursemid (54-31-9)
Fursemide (54-31-9)
3-(α-Furyl-β-acetylaethyl)-4-hydroxycumarin (German) (117-52-2)
3-(1-Furyl-3-acetylethyl)-4-hydroxycoumarin (117-52-2)
3-(1-Furyl-2-acetylethyl)-4-hydroxycoumarin, sodium salt (34490-93-2)
Furyl alcohol (98-00-0)
2-Furylaldehyde (98-01-1)
Furylamide (3688-53-7)
2-(2'-Furyl)-benzimidazole (3878-19-1)
2-(2-Furyl)benzimidazole (3878-19-1)
2-Furylcarbinol (98-00-0)
α-Furylcarbinol (98-00-0)
Furylfuramide (3688-53-7)
4-(2-Furyl)-4-(4-hydroxy-3-coumarinyl)-2-butanone (117-52-2)
4-(2-Furyl)-4-(4-hydroxy-3-kumarinyl)-2-butanon (Czech) (117-52-2)
1-(3-Furyl)-4-hydroxypentanone (32954-58-8)
1-(β-Furyl)-4-hydroxypentanone (32954-58-8)
2-Furyl-methanal (98-01-1)
2-Furylmethanol (98-00-0)
(2-Furylmethyl)amine (617-89-0)
2-Furyl methyl ketone (1192-62-7)
α-2-Furyl-5-nitro-2-furanacrylamide (3688-53-7)
2-(2-Furyl)-3-(5-nitro-2-furyl)acrylamide (3688-53-7)
2-(2-Furyl)-3-(5-nitro-2-furyl)acrylic acid amide (3688-53-7)
α-(Furyl)-β-(5-nitro-2-furyl)acrylic amide (3688-53-7)
Fusarenon (23255-69-8)
Fusarenon X (23255-69-8)
Fusarenone X (23255-69-8)
Fusarex (117-18-0)
Fusarex (879-39-0)
Fusaric acid (536-69-6)
Fusarinic acid (536-69-6)
Fusariotoxin T 2 (21259-20-1)
Fusariotoxine T2 (21259-20-1)
Fusarium Toxin (17924-92-4)
Fused Borax (1330-43-4)
Fused Boric Acid (1303-86-2)
Fused Quartz (60676-86-0)
Fused Silica (60676-86-0)
Fusel Oil (8013-75-0)
Fusel Oil, Sugar Beet (8013-75-0)
Fuseloel (German) (8013-75-0)
Fusid (54-31-9)
Fusilade (69806-50-4)

Fusilade 2000 (79241-46-6)
Fusilade 5 (79241-46-6)
Fusilade Super (79241-46-6)
Fussol (640-19-7)
Futramine D (106-50-3)
Futramine EG (591-27-5)
Fuvacillin (59-87-0)
Fybrene (8009-03-8)
Fycol 8 (1332-40-7)
Fydalin (77-65-6)
Fyde (50-00-0)
Fyfanon (121-75-5)
Fypro (63428-84-2)
Fyrex (7783-28-0)
Fyrol 6 (2781-11-5)
Fyrol 76 (53529-45-6)
Fyrol CEF (115-96-8)
Fyrol FR 2 (13674-87-8)
Fyrol HB32 (126-72-7)
Fyrol PCF (6145-73-9)
Fyrquel 150 (1330-78-5)
Fyrquel 220 (9CI) (55957-10-3)
Fysostigmin (Czech) (57-47-6)
Fytic acid (83-86-3)
Fytolan (1332-40-7)
2-G (93-14-1)
2100 GP (9002-88-4)
G 0 (111-49-9)
G 0 (Oxide) (1344-28-1)
G 1 (103-90-2)
G 2 (Oxide) (1344-28-1)
G 4 (97-23-4)
G 25 (76-06-2)
G 50 (9004-57-3)
G 50 (Polysaccharide) (9004-57-3)
G 87 (93-14-1)
G 200 (9004-57-3)
G 251 (VAN) (78-21-7)
G 263 (78-21-7)
G 301 (333-41-5)
G 338 (510-15-6)
G 339 S (1592-23-0)
G 444E (565-33-3)
G 475 (509-86-4)
G 996 (16672-87-0)
G 2124 (141-20-8)
G 3707 (9002-92-0)
G 3710 (9005-00-9)
G 3711 (9002-92-0)
G 3720 (9005-00-9)
G 3910 (9004-98-2)
G 3920 (9004-98-2)
G 13,871 (50-33-9)
G 19258 (122-15-6)
G 23611 (119-38-0)
G 23992 (510-15-6)
G 24,163 (5836-10-2)
G 25,804 (565-33-3)
G 27202 (129-20-4)
G 27550 (2814-20-2)
G 27692 (122-34-9)
G 27901 (1912-26-1)
G 28029 (2275-14-1)
G 28315 (57-96-5)
G 28509 (3084-92-2)
G 30027 (1912-24-9)
G 30031 (1912-25-0)
G 30033 (6190-65-4)
G 30494 (3735-23-7)
G 30888 (2630-10-6)
G 31432 (13532-26-8)
G 31709 (30360-56-6)
G 32292 (3035-45-8)
G 32883 (298-46-4)
G 32911 (1014-70-6)
G 33182 (77-36-1)
G 34161 (7287-19-6)
G 34360 (1014-69-3)
G 34698 (1824-09-5)
G 36393 (841-06-5)
G-11 (70-30-4)
G-444E (118-75-2)
G-2130A (9002-92-0)
G-22008 (87-47-8)
G-22355 (50-49-7)
G-22870 (644-64-4)
G-23133 (81-82-3)
G-23350 (152-72-7)
G-23645 (6012-83-5)
G-24027 (108-35-0)
G-24480 (333-41-5)
G-24483 (108-34-9)
G-24622 (5826-91-5)
G-25804 (118-75-2)
G-28279 (1007-28-9)
G-29288 (953-17-3)
G-30,026 (3004-71-5)
G-30028 (139-40-2)
G-30044 (673-04-1)
G-31,717 (3004-70-4)
G-31435 (1610-18-0)
G-32293 (1610-17-9)
G-34162 (834-12-8)
GA (77-06-5)
GA (77-81-6)
GA-10832 (26399-36-0)
GA3 (77-06-5)
G Acid (118-32-1)
Gp-Amin (Czech) (96-96-8)
GB (107-44-8)
GBH (495-73-8)
GBL (123-19-3)
GBS (7681-38-1)
GC 1189 (143-50-0)
GC 1283 (2385-85-5)
GC 3707 (122-10-1)
GC 3944-3-4 (82-68-8)
GC 4072 (470-90-6)
GC 6506 (3254-63-5)
GC 6936 (900-95-8)
GC 7787 (34202-69-2)
GC 8993 (639-58-7)
GC-928 (80-38-6)
GC-1106 (116-16-5)
GC-2603 (4482-55-7)
GC-2996 (140-41-0)
GC-9160 (4234-79-1)
G-Cure (9003-01-4)
GD (96-64-0)
GDMA (123-81-9)
G-Eleven (70-30-4)
GF 58 (1067-53-4)
GGBGE (7382-59-4)
GGE (93-14-1)

Name	CAS No.
GGG	(93-14-1)
GH	(9002-72-6)
GH	(97-23-4)
GH 20	(9002-89-5)
GH 44	(27955-87-9)
GH 74	(26258-70-8)
GHA 331	(21645-51-2)
GHA 332	(21645-51-2)
G.H. Woods Degreaser-Formula 11470 (9CI)	(53988-70-8)
G-I	(4720-09-6)
G-II	(70-30-4)
GINK	(54-85-3)
GIX	(475-26-3)
GK 2	(7782-42-5)
GK 3	(7782-42-5)
GK (Oxide)	(1344-28-1)
GL 02	(9002-89-5)
GL 03	(9002-89-5)
GLC 935P	(74082-93-2)
GLO 5	(9002-89-5)
GLU-P-1	(67730-11-4)
GLU-P-2	(67730-10-3)
GM 14	(9002-89-5)
GR-M	(9010-98-4)
G-M-F	(105-11-3)
GMO 8903	(25496-72-4)
5'-GMP	(85-32-5)
GMP	(85-32-5)
GM sulfate	(1405-41-0)
GN 8384	(9004-74-4)
GP 60	(7782-42-5)
GP 63	(7782-42-5)
GP 3000	(25791-96-2)
GP 45840	(15307-79-6)
GP-121	(956-90-1)
GP-40-66:120	(87-68-3)
GPKh	(76-44-8)
GP 60S	(7782-42-5)
GR	(118-00-3)
GR-9234	(22228-82-6)
G-Resins	(9002-88-4)
GS 2	(7782-42-5)
GS 6	(7439-89-6)
GS-2876	(914-00-1)
GS-3065	(564-25-0)
GS 6244	(6804-07-5)
GS 11526	(7374-53-0)
GS 12515	(49624-61-5)
GS-12968	(2669-32-1)
GS 13005	(950-37-8)
GS-13332	(644-64-4)
GS 13528	(7286-69-3)
GS 13529	(5915-41-3)
GS 14253	(836-24-8)
GS 14254	(26259-45-0)
GS 14259	(33693-04-8)
GS 14260	(886-50-0)
GS-16068	(4147-51-7)
GS 18182	(33692-99-8)
GS 18183	(34333-27-2)
GS-19851	(18181-80-1)
GS 34360	(1014-69-3)
GS 38946	(34129-07-2)
GS 39985	(34128-99-9)
GSH	(70-18-8)
G-Strophanthin	(630-60-4)
GT	(9000-70-8)
GT 41	(55-98-1)
GT 2041	(55-98-1)
GTG	(12192-57-3)
GTN	(55-63-0)
G1V Gard DXN	(828-00-2)
GY 70	(7782-42-5)
Gas-EX-B	(7803-51-2)
Gas-Furnace Black	(1333-86-4)
Gaba	(56-12-2)
Gabbromycin	(7542-37-2)
Gabbropas	(65-49-6)
Gabrite	(9011-05-6)
Gadexyl	(57-53-4)
Gadolinia	(12064-62-9)
Gadolinium	(7440-54-2)
Gadolinium oxide (9CI)	(12064-62-9)
Gafcol EB	(111-76-2)
Gagexyl	(57-53-4)
Gagro	(16672-87-0)
Gaiamar	(93-14-1)
Galactaric-acid	(526-99-8)
Galactasol	(9000-30-0)
Galactasol A	(9005-25-8)
Galacticol	(10318-26-0)
Galactitol (9CI)	(608-66-2)
Galactitol, 1,2:5,6-dianhydro-, bis(benzenepropionate)	(57230-49-6)
Galactitol, 1,2:5,6-dianhydro-, diacetate	(57230-48-5)
Galactitol, 1,6-dibromo-1,6-dideoxy	(10318-26-0)
D-erythro-D-Galacto-octopyranoside, methyl-6,8-dideoxy-6-(1-methyl-4-propyl-l-2- pyrrolidinecarboxamido)-1-thio-, trans-α	(154-21-2)
α-D-Galactometasaccharinic acid	(18521-63-6)
L-threo-D-Galacto-octopyranoside, methyl-7-chloro-6,7,8-trideoxy-6-(1-methyl-4-propyl-L-2- pyrrolidine-carboxamido)-1-thio-, trans-α	(18323-44-9)
α-D-Galactopyranose, 1,2:3,4-bis-O-(1-methylethylidene)- (9CI)	(4064-06-6)
Galactopyranose, 1,2:3,4-di-O-isopropylidene-, α-D- (8CI)	(4064-06-6)
Galactosaccharic acid	(526-99-8)
D-(+)-Galactosamine	(7535-00-4)
D-Galactosamine	(7535-00-4)
Galactosamine	(7535-00-4)
Galactose	(59-23-4)
Galactose, D	(59-23-4)
Galactose, 2-amino-2-deoxy-, D	(7535-00-4)
L-Galactose, 6-deoxy- (VAN)(9CI)	(2438-80-4)
4-(β-D-Galactosido)-D-glucose	(63-42-3)
Galatur	(5560-72-5)
Galben	(71626-11-4)
Galctin	(9002-62-4)
Galecron	(6164-98-3)
Galecron SP	(19750-95-9)
Galena	(1314-87-0)
Galipan	(3813-05-6)
Gallacetophenone	(528-21-2)
Gallacetophenone 4-O-methyl ether	(708-53-2)
Gallaldehyde 3,5-dimethyl ether	(134-96-3)
Gallamin	(65-29-2)
Gallamine	(65-29-2)
Gallamine-3eti	(65-29-2)
Gallamine iodide	(65-29-2)
Gallamine triethiodide	(65-29-2)
Gallamine triiodoethylate	(65-29-2)
Gallamin triethiodide	(65-29-2)
Gallant	(69806-40-2)
Gallant	(87237-48-7)
Gallia	(12024-21-4)
Gallic-acid	(149-91-7)

Gallic acid, ethyl ester **(831-61-8)**
Gallic acid, methyl ester **(99-24-1)**
Gallic acid, propyl ester **(121-79-9)**
Gallic acid trimethyl ether **(118-41-2)**
Gallimycin **(643-22-1)**
Gallium-arsenide **(1303-00-0)**
Gallium [UN 2803] **(7440-55-3)**
Gallium(3+) chloride **(13450-90-3)**
Gallium chloride (8CI,9CI) **(13450-90-3)**
Gallium, metal (DOT) **(7440-55-3)**
Gallium metal, Liquid (DOT) **(7440-55-3)**
Gallium metal, Solid (DOT) **(7440-55-3)**
Gallium monoarsenide **(1303-00-0)**
Gallium sesquioxide **(12024-21-4)**
Gallium trichloride **(13450-90-3)**
Gallium, trimethyl- (9CI) **(1445-79-0)**
Gallium trioxide **(12024-21-4)**
Gallium-oxide **(12024-21-4)**
Galloacetophenone **(528-21-2)**
Gallodesoxycholic acid **(474-25-9)**
Gallogama **(58-89-9)**
Gallogen **(476-66-4)**
Gallogen (Astringent) **(476-66-4)**
Gallotannic acid **(1401-55-4)**
Gallotannic acids **(1401-55-4)**
Gallotannin **(1401-55-4)**
Gallotannins **(1401-55-4)**
Gallotox **(62-38-4)**
Galofak **(61-33-6)**
Galoperidol **(52-86-8)**
Galoxolide **(1222-05-5)**
Galozone **(9000-07-1)**
Galvatol 1-60 **(9002-89-5)**
Gamacarbatox **(58-89-9)**
Gamacid **(58-89-9)**
Gamaphex **(58-89-9)**
Gamarex **(56-12-2)**
Gamaserpin **(50-55-5)**
Gamasol 90 **(67-68-5)**
Gamene **(58-89-9)**
Gamiso **(58-89-9)**
Gamit **(81777-89-1)**
Gamlen Sea Clean (9CI) **(53988-69-5)**
Gamma-6480 **(96-48-0)**
Gammacorten **(50-02-2)**
Gammahexa **(58-89-9)**
Gammahexane **(58-89-9)**
Gammalin **(58-89-9)**
Gammalin 20 **(58-89-9)**
Gammalon **(56-12-2)**
Gammaserpine **(50-55-5)**
Gammaterr **(58-89-9)**
Gammex **(58-89-9)**
Gammexane **(58-89-9)**
Gammexane **(608-73-1)**
Gammopaz **(58-89-9)**
Gamonil **(63-25-2)**
Gamophen **(70-30-4)**
Gamophene **(70-30-4)**
Ganda **(127-69-5)**
Ganex P 804 **(9003-39-8)**
Gangesol **(133-67-5)**
Ganidan **(57-67-0)**
Ganja **(8063-14-7)**
Ganocide **(5707-69-7)**
Ganozan **(107-27-7)**
Ganphen **(58-33-3)**
Gansil **(127-65-1)**
Gantanol **(723-46-6)**
Gantaprin **(8064-90-2)**
Gantrim **(8064-90-2)**
Gantrisin **(127-69-5)**
Gantrisine **(127-69-5)**
Gantrisona **(127-69-5)**
Gantrosan **(127-69-5)**
Garamycin **(1403-66-3)**
Garamycin **(1405-41-0)**
Garantose **(81-07-2)**
Gardcide **(22248-79-9)**
Gardenal **(50-06-6)**
Gardenal sodium **(57-30-7)**
Gardeniol II **(93-92-5)**
Gardenol **(93-92-5)**
Gardentox **(333-41-5)**
Gardepanyl **(50-06-6)**
Gardinol **(151-21-3)**
Gardiquart SV480 **(8001-54-5)**
Gardiquat 1450 **(8001-54-5)**
Gardol **(137-16-6)**
Gardona **(22248-79-9)**
Gardoprim **(5915-41-3)**
Garlic Oil **(8000-78-0)**
Garlon **(55335-06-3)**
Garlon 3A **(57213-69-1)**
Garlon 4 **(64700-56-7)**
Garlon 4E **(64700-56-7)**
Garnitan **(330-55-2)**
Garox **(94-36-0)**
Garrathion **(786-19-6)**
Garvox **(22781-23-3)**
Gas, Blast furnace **(65996-68-1)**
Gas, Natural **(8006-14-2)**
Gas Oil **(68476-30-2)**
Gas Oil, Blend **(64741-44-2)**
Gas Oils, Petroleum, Heavy Vacuum **(64741-57-7)**
Gas Oils, Petroleum, Light Vacuum **(64741-58-8)**
Gas Oils, Petroleum, Straight-run **(64741-43-1)**
Gas, Producer **(8006-20-0)**
Gascon **(63148-62-9)**
Gases, Petroleum, C3-5 Olefinic-Paraffinic Alkylation Feed **(68477-83-8)**
Gases, Petroleum, isomerized naphtha fractionater, C4-rich, hydrogen sulfide-free **(68477-99-6)**
Gases de Petroleo (Spanish) **(68476-85-7)**
Gas oils, Petroleum, Straight-run, high-boiling **(68915-97-9)**
Gasoline **(86290-81-5)**
Gasoline (ACGIH) [UN 1203] **(8006-61-9)**
Gasoline Blend Stock **(68425-29-6)**
Gasoline Blending Stock, Alkylates **(68425-29-6)**
Gasoline Blending Stock, Reformates **(68425-29-6)**
Gasoline, Straight-run, Topping-plant **(68606-11-1)**
Gastex **(1333-86-4)**
Gastomag **(1343-88-0)**
Gastracid **(94-78-0)**
Gastrinide **(50-35-1)**
Gastromet **(51481-61-9)**
Gastron **(53-46-3)**
Gastrosedan **(53-46-3)**
Gastrotest **(94-78-0)**
Gastuloric **(138-15-8)**
Gatinon **(1929-88-0)**
Gatnon **(1929-88-0)**
Gaultheria Oil **(119-36-8)**
Gaultheria Oil, Artificial **(119-36-8)**
Gaultherin (8CI) **(490-67-5)**

Gazarin	(4460-86-0)
Gaz de Petrole (French)	(68476-85-7)
Geno-Cristaux Gremy	(58-22-0)
Geabol	(72-63-9)
Gearphos	(298-00-0)
Gearphos	(56-38-2)
Gebutox	(88-85-7)
Gechloreerdedifenyl (Dutch)	(53469-21-9)
Gedex	(9003-53-6)
Geigy	(1912-25-0)
Geigy 337	(6012-83-5)
Geigy 338	(510-15-6)
Geigy 12968	(2669-32-1)
Geigy 13005	(950-37-8)
Geigy 19258	(122-15-6)
Geigy 22870	(644-64-4)
Geigy 24163	(5836-10-2)
Geigy 24480	(333-41-5)
Geigy 27,692	(122-34-9)
Geigy 28029	(2275-14-1)
Geigy 30026	(3004-71-5)
Geigy 30,027	(1912-24-9)
Geigy 30,028	(139-40-2)
Geigy 30,044	(673-04-1)
Geigy 30494	(3735-23-7)
Geigy 32,293	(1610-17-9)
Geigy 32883	(298-46-4)
Geigy-Blau 536	(314-13-6)
Geigy Blue 536, Med	(314-13-6)
Geigy 444E	(565-33-3)
Geigy-444E	(118-75-2)
Geigy G-22008	(87-47-8)
Geigy G-23611	(119-38-0)
Geigy G-24027	(108-35-0)
Geigy G-24483	(108-34-9)
Geigy G-28029	(2275-14-1)
Geigy G-29288	(953-17-3)
Geigy G-30494	(3735-23-7)
Geigy GS-12968	(2669-32-1)
Geigy GS-13005	(950-37-8)
Geigy GS-13332	(644-64-4)
Geigy GS-19851	(18181-80-1)
Geigy G.S. 14254	(26259-45-0)
Geigy Herbicide 444E	(565-33-3)
Geigy Rodenticide Exp. 332	(81-82-3)
Gel II	(7681-49-4)
Gelacillin	(61-33-6)
Gelan I	(77-81-6)
Gelatin	(9000-70-8)
Gelatin Foam	(9000-70-8)
Gelatine	(9000-70-8)
Gelatin-epinephrine	(55-31-2)
Gelatins	(9000-70-8)
L-Gelb 2 (German)	(1934-21-0)
L-Gelb 3 (German)	(8004-92-0)
Gelber phosphor (German)	(7723-14-0)
Gelbin	(13765-19-0)
Gelbkreuz (Czech)	(505-60-2)
Gelborange-S (German)	(2783-94-0)
Gelcarin	(9000-07-1)
Gelcarin HMR	(9000-07-1)
Gelcarin SI	(9062-07-1)
Gelfoam	(9000-70-8)
Gelocatil	(103-90-2)
Gelonida	(62-44-2)
Gelose	(9002-18-0)
Gelovermin	(136-77-6)
Gelozone	(9000-07-1)
Gelsemin	(509-15-9)
Gelsemine	(509-15-9)
Gelsemine, monohydrochloride	(35306-33-3)
Geltabs	(50-14-6)
Gelution	(7681-49-4)
Gelva	(9003-20-7)
Gelva 25	(9003-20-7)
Gelva CSV 16	(9003-20-7)
Gelva GP 702	(9003-20-7)
Gelva S 55H	(9003-20-7)
Gelva TS 22	(9003-20-7)
Gelva TS 23	(9003-20-7)
Gelva TS 30	(9003-20-7)
Gelva TS 85	(9003-20-7)
Gelva V 15	(9003-20-7)
Gelva V 25	(9003-20-7)
Gelva V 100	(9003-20-7)
Gelva V 800	(9003-20-7)
Gelvatol	(9002-89-5)
Gelvatol 1-30	(9002-89-5)
Gelvatol 1-90	(9002-89-5)
Gelvatol 20-30	(9002-89-5)
Gelvatol 2090	(9002-89-5)
Gelvatol 3-91	(9002-89-5)
Gemalgene	(79-01-6)
Gemonal	(50-11-3)
Gemonil	(50-11-3)
Gemonit	(50-11-3)
Genacort	(50-23-7)
Genacron Yellow G	(2832-40-8)
Genacryl Orange G	(3056-93-7)
Genamin C	(61791-14-8)
Genamin CTAC	(112-02-7)
Genamin DSAC	(107-64-2)
Genapol LRO	(9004-82-4)
Genapol O	(9004-98-2)
Genapol PF	(9003-11-6)
Genapol PF 10	(9003-11-6)
Genapol PF 20	(9003-11-6)
Genapol PF 40	(9003-11-6)
Genapol PF 80	(9003-11-6)
Genapol S	(9005-00-9)
Genazo Red KB Soln	(95-79-4)
Gendriv 162	(9000-30-0)
R-Gene	(1119-34-2)
Genep EPTC	(759-94-4)
General Chemical 9160	(4234-79-1)
General Chemicals 1189	(143-50-0)
General Chemicals 3707	(122-10-1)
General Chemicals 8993	(639-58-7)
General Chemicals 9160	(4234-79-1)
General Weed Killer	(33213-65-9)
Genetron-23	(75-46-7)
Genetron 11	(75-69-4)
Genetron 12	(75-71-8)
Genetron 13	(75-72-9)
Genetron 21	(75-43-4)
Genetron 22	(75-45-6)
Genetron 100	(75-37-6)
Genetron 101	(75-68-3)
Genetron 112	(76-12-0)
Genetron 113	(76-13-1)
Genetron 114	(76-14-2)
Genetron 115	(76-15-3)
Genetron 133a	(75-88-7)
Genetron 142b	(75-68-3)

Genetron 152a	**(75-37-6)**
Genetron 218	**(76-19-7)**
Genetron 316	**(76-14-2)**
Genetron 1112A	**(79-35-6)**
Genetron 1113	**(79-38-9)**
Genetrone 1112A	**(79-35-6)**
Geniphene	**(8001-35-2)**
Genite	**(97-16-5)**
Genite 883	**(80-33-1)**
Genite 923	**(97-16-5)**
Genite EM-923	**(97-16-5)**
Genite-R99	**(97-16-5)**
Genithion	**(56-38-2)**
Genitol	**(97-16-5)**
Genitol 923	**(97-16-5)**
Genitox	**(50-29-3)**
Genitron AC	**(123-77-3)**
Genitron AC 2	**(123-77-3)**
Genitron AC 4	**(123-77-3)**
Genitron BSH	**(80-17-1)**
Genocodein	**(3688-65-1)**
Genocodeine	**(3688-65-1)**
Genoface	**(101-20-2)**
Genol	**(55-55-0)**
Genomorphin	**(639-46-3)**
Genomorphine	**(639-46-3)**
Genophyllin	**(317-34-0)**
Genoptic	**(1405-41-0)**
Genoptic S.O.P.	**(1405-41-0)**
Genoxal	**(50-18-0)**
Genoxal	**(6055-19-2)**
Genozym	**(50-41-9)**
Gentalline	**(1405-41-0)**
Gentamicin	**(1403-66-3)**
Gentamicin, sulfate	**(1405-41-0)**
Gentamycin	**(1403-66-3)**
Gentamycin-Creme (German)	**(1403-66-3)**
Gentamycin sulfate	**(1405-41-0)**
Gentersal	**(548-62-9)**
Gentian Extract	**(72968-42-4)**
Gentian LS	**(8001-21-6)**
Gentian Violet	**(548-62-9)**
Gentianaviolett (German)	**(548-62-9)**
Gentiaverm	**(548-62-9)**
Genticid	**(548-62-9)**
Gentimon	**(115-69-5)**
Gentioletten	**(548-62-9)**
Gentisate	**(490-79-9)**
Gentisic-acid	**(490-79-9)**
Gentisic acid, 3-methyl- (6CI,8CI)	**(5981-39-5)**
Gentron 142b	**(75-68-3)**
Genu	**(9000-07-1)**
Genu	**(9000-69-5)**
Genugel	**(9000-07-1)**
Genugel CJ	**(9000-07-1)**
Genugol RLV	**(9000-07-1)**
Genuine Acetate Chrome Orange	**(18454-12-1)**
Genuine Orange Chrome	**(18454-12-1)**
Genuine Paris Green	**(12002-03-8)**
Genuvisco J	**(9000-07-1)**
Genvis	**(9005-25-8)**
Geobilan	**(58-89-9)**
Geofos	**(21548-32-3)**
Geolin G 3	**(58-89-9)**
Geon	**(9016-00-6)**
Geon D 100	**(64742-16-1)**
Geon 100X150	**(9003-22-9)**
Geon 103EP-J	**(9003-22-9)**
Geon 103 ZX	**(9003-22-9)**
Geon 130X10	**(9003-22-9)**
Geon 135	**(9003-22-9)**
Geon 150XML	**(9003-22-9)**
Geon 222	**(9011-06-7)**
Geon 351	**(9003-22-9)**
Geon 400X47	**(9003-22-9)**
Geon 421	**(9003-22-9)**
Geon 427	**(9003-22-9)**
Geon 434	**(9003-22-9)**
Geon 440L2	**(9003-22-9)**
Geon 450X150PN	**(9003-22-9)**
Geon 652	**(9011-06-7)**
Geon 1032X	**(9003-22-9)**
Geosmin; (-)-geosmin; 4a-α-(2H)-naphthol, octahydro-4-α, 8a-β-dimethyl- (8CI)	**(19700-21-1)**
Gepiron	**(2163-80-6)**
Geranaldehyde	**(141-27-5)**
Geranial	**(141-27-5)**
Geranic acid	**(459-80-3)**
Geranine 2GS	**(3734-67-6)**
Geraniol	**(106-24-1)**
Geraniol	**(624-15-7)**
Geraniol acetate	**(105-87-3)**
Geraniol alcohol	**(106-24-1)**
Geraniol extra	**(106-24-1)**
Geraniol formate	**(105-86-2)**
Geraniol, perhydro-	**(106-21-8)**
Geraniol tetrahydride	**(106-21-8)**
Geraniol, tetrahydro-	**(106-21-8)**
Geranium Absolute	**(8000-46-2)**
Geranium Concrete	**(8000-46-2)**
Geranium Lake N	**(81-88-9)**
Geranium Oil	**(8000-46-2)**
Geranium Oil Algerian	**(8000-46-2)**
Geranium Oil, Algerian	**(8000-46-2)**
Geranium Oil, Grasse	**(8000-46-2)**
Geranium Oil, Morocco	**(8000-46-2)**
Geranium Sur Roses	**(8000-46-2)**
Geranium crystals	**(101-84-8)**
Geranonitrile	**(5585-39-7)**
Geranyl acetate	**(105-87-3)**
Geranyl alcohol	**(106-24-1)**
Geranyl formate	**(105-86-2)**
Geranyl nitrile	**(5585-39-7)**
Gerastop	**(637-07-0)**
Gerfil	**(9003-07-0)**
Gerison	**(63-74-1)**
Gerlach 1396	**(120-23-0)**
Germa-Medica	**(70-30-4)**
Germain's	**(63-25-2)**
Germalgene	**(79-01-6)**
Germane [UN 2192]	**(7782-65-2)**
Germane, dimethylthioxo-	**(16090-49-6)**
Germania	**(1310-53-8)**
Germanic acid	**(1310-53-8)**
Germanium-bromide	**(13450-92-5)**
Germanium-dioxide	**(1310-53-8)**
Germanium (9CI)	**(7440-56-4)**
Germanium hydride	**(7782-65-2)**
Germanium oxide	**(1310-53-8)**
Germanium oxide (geo2)	**(1310-53-8)**
Germanium sulfide, dimethyl-	**(16090-49-6)**
Germanium tetrabromide	**(13450-92-5)**
Germanium, tetrabromo-	**(13450-92-5)**
Germanium tetrahydride (ACGIH,OSHA)	**(7782-65-2)**

Name	CAS
Germany: C-Blau 17	(14038-43-8)
Germany: C-Ext. Blau 10	(147-14-8)
Germany: C-Ext. Braun 4	(1320-07-6)
Germany: C-Ext. Gelb 16	(2321-07-5)
Germany: C-Orange 11	(7235-40-7)
Germany: C-Rot 14	(6417-83-0)
Germicin	(8001-54-5)
Germidine	(123-03-5)
Germinol	(8001-54-5)
Germitol	(8001-54-5)
Gernebcin	(32986-56-4)
Gerobit	(300-42-5)
Gerodyl	(467-60-7)
Gerot-epilan	(50-12-4)
Gerot-epilan-D	(57-41-0)
Gerontine	(71-44-3)
Gerovit	(300-42-5)
Gerovital	(59-46-1)
Gerox	(57-92-1)
Gertley borate	(1303-96-4)
Gesabal	(1912-25-0)
Gesadural	(673-04-1)
Gesafid	(50-29-3)
Gesafloc	(1912-26-1)
Gesafram	(1610-18-0)
Gesafram 50	(1610-18-0)
Gesagard	(7287-19-6)
Gesagram	(1610-18-0)
Gesakur	(5836-10-2)
Gesamil	(139-40-2)
Gesapax	(834-12-8)
Gesapon	(50-29-3)
Gesaprim	(1912-24-9)
Gesaprim 1798	(8073-77-6)
Gesaprim Multy	(8073-77-6)
Gesaran	(122-34-9)
Gesaran	(841-06-5)
Gesarex	(50-29-3)
Gesarol	(50-29-3)
Gesatamin	(1610-17-9)
Gesatop	(122-34-9)
Gesatop 50	(122-34-9)
Gesemine hydrochloride	(35306-33-3)
Gesfid	(7786-34-7)
Gesoprim	(1912-24-9)
Gestageno gador	(68-96-2)
Gestatron	(152-62-5)
Gesterol 100	(57-83-0)
Gesterol L.A.	(630-56-8)
Gestid	(7786-34-7)
Gestone	(57-83-0)
Gestormone	(57-83-0)
Gettysolve-B	(110-54-3)
Gettysolve-C	(142-82-5)
Gewazol	(54-95-5)
Gewodin	(62-44-2)
Gexane	(58-89-9)
Giallo Cromo (Italian)	(7758-97-6)
Giap 10	(1314-13-2)
Giardil	(67-45-8)
Giarlam	(67-45-8)
Giatricol	(443-48-1)
Gib-Sol	(77-06-5)
Gib-Tabs	(77-06-5)
Gibberellic-acid	(77-06-5)
Gibberellic acid, monopotassium salt	(125-67-7)
Gibberellic acid potassium salt	(125-67-7)
Gibberellin	(77-06-5)
Gibberellin A3	(77-06-5)
Gibberellin A3 potassium salt	(125-67-7)
Gibberellin X	(77-06-5)
Gibbrel	(77-06-5)
Gibrel	(125-67-7)
Gibrofit	(125-67-7)
Gibs	(7778-18-9)
Gichtex	(315-30-0)
Gie	(100-79-8)
Gifblaar Poison	(144-49-0)
Gigantin	(149-29-1)
Gihitan	(439-14-5)
Gillebachite	(13983-17-0)
Gilotherm OM 2	(26140-60-3)
Gilsonite	(12002-43-6)
Gilucard	(50-55-5)
Gilucor nitro	(55-63-0)
Gimid	(77-21-4)
Gineflavir	(443-48-1)
Ginestrene	(57-63-6)
Gingerone	(122-48-5)
Gingicain M	(136-47-0)
Gingilli Oil	(8008-74-0)
Ginseng LS	(8001-21-6)
Gips	(13397-24-5)
Giracid	(136-40-3)
Girl	(50-36-2)
Girostan	(52-24-4)
Giv-Cide BNS	(7166-19-0)
Giv-Gard	(7166-19-0)
Giv-Tan F	(104-28-9)
Gl-Amin (Czech)	(119-32-4)
Glacial acetic acid	(64-19-7)
Glacial acrylic acid	(79-10-7)
Glacidet K (9CI)	(83589-63-3)
Glandubolin	(53-16-7)
Glanducorpin	(57-83-0)
Glass	(65997-17-3)
Glass, Oxide, Chemicals	(65997-17-3)
Glass enamel 19 E 110	(65997-17-3)
Glassy sodium phosphate	(65997-17-3)
Glasure 40X	(25608-33-7)
Glaucosan	(51-43-4)
Glaupax	(59-66-5)
Glauramine	(492-80-8)
Glaurin	(141-20-8)
Glaxoridin	(50-59-9)
Glazd Penta	(87-86-5)
Glean	(64902-72-3)
Glean 20DF	(64902-72-3)
Gleem	(7681-49-4)
Glenbar	(3765-57-9)
Glicol monocloridrina (Italian)	(107-07-3)
Gliddex	(8002-43-5)
Glikocel TA	(9004-32-4)
Glimid	(77-21-4)
Gliotoxin	(67-99-2)
Glisema	(94-20-2)
Glob-P-1	(68006-83-7)
Glob-P-2	(26148-68-5)
Globalin	(7097-60-1)
Globenicol	(56-75-7)
Globoid	(50-78-2)
Globol	(106-46-7)
Globucid	(94-19-9)
Globucin	(94-19-9)

Globulariacitrin **(153-18-4)**
Globularicitrin **(153-18-4)**
Globulin G **(9001-63-2)**
Globulin G1 **(9001-63-2)**
Globulus **(8000-48-4)**
Globuzid (German) **(94-19-9)**
Glomax **(1332-58-7)**
Glonoin **(55-63-0)**
Glonsen **(527-07-1)**
Glore Phos 36 **(6923-22-4)**
Glorous **(56-75-7)**
Glosanto **(526-95-4)**
Glosso sterandryl **(58-18-4)**
Glover **(7439-92-1)**
Glucagon **(16941-32-5)**
Glucagon (Human) **(16941-32-5)**
Glucagon, Porcine, For bioassay **(16941-32-5)**
Glucagon, Porcine, For immunoassay **(16941-32-5)**
Glucagone **(16941-32-5)**
Glucagon (Pig) **(16941-32-5)**
Glucagonum (Latin) **(16941-32-5)**
Gluceptate sodium **(13007-85-7)**
Gluco-ferrum **(299-29-6)**
D-Gluco-heptonic acid, monosodium salt, (2.xi.)- (9CI) **(31138-65-5)**
D-Gluco-heptonic acid, (2.XI.)- (9CI) **(23351-51-1)**
Glucid **(81-07-2)**
Glucidoral **(339-43-5)**
Glucinium **(7440-41-7)**
Glucinum **(7440-41-7)**
Glucitol **(50-70-4)**
D-Glucitol, Anhydro-, monooctadecanoate **(1338-41-6)**
D-Glucitol (9CI) **(50-70-4)**
Glucitol, D- (8CI) **(50-70-4)**
D-Glucitol, Polymer with (chloromethyl)oxirane (9CI) **(29994-68-1)**
D-Glucitol, 1,4-anhydro-, 6-dodecanoate (9CI) **(5959-89-7)**
D-Glucitol, 1-chloro-2,3-epoxy propane polymer **(29994-68-1)**
D-Glucitol, 1-deoxy-1-(methylamino)- (9CI) **(6284-40-8)**
Glucitol, 1-deoxy-1-(methylamino)-, D- (8CI) **(6284-40-8)**
Glucitol, 1,4:3,6-dianhydro-, D **(652-67-5)**
Glucitol, 1,4:3,6-dianhydro-, dinitrate, D **(87-33-2)**
D-Glucitol, 1,4:3,6-dianhydro-, di-9-octadecenoate, (Z,Z)- (9CI) **(4252-85-1)**
D-Glucitol, hexaacetate (9CI) **(7208-47-1)**
Glucitol, hexaacetate, D- (8CI) **(7208-47-1)**
Glucono-δ-lactone **(90-80-2)**
Glucobasin **(56-65-5)**
Glucochloral **(15879-93-3)**
Glucochloralose **(15879-93-3)**
α-D-Glucochloralose **(15879-93-3)**
Glucocorticoid **(1524-88-5)**
Glucocorticoid **(53-34-9)**
Glucodigin **(71-63-6)**
Glucofren **(339-43-5)**
α-D-Glucofuranose, 1,2-O-(2,2,2-trichloroethylidene)-, (R)- (9CI) **(15879-93-3)**
(2.xi.)-D-Gluco-heptonic acid **(23351-51-1)**
α-D-Glucoisosaccharinic acid **(1518-54-3)**
β-D-Glucoisosaccharinic acid **(1518-56-5)**
Glucolin **(50-99-7)**
α-D-Glucometasaccharinic acid **(498-43-1)**
β-D-Glucometasaccharinic acid **(1518-59-8)**
Glucometasaccharinic acid, α-D- (8CI) **(498-43-1)**
Gluconate de calcium (French) **(299-28-5)**
Gluconato di sodio (Italian) **(527-07-1)**
Gluconic acid **(526-95-4)**
D-Gluconic acid (9CI) **(526-95-4)**
Gluconic acid, Compd. with 1,1'-Hexamethylene bis-(5-(p-chlorophenyl)biguanide) (2:1), D- (8CI) **(18472-51-0)**
Gluconic acid, D- (8CI) **(526-95-4)**
D-Gluconic acid, ammonium salt (9CI) **(10361-31-6)**
Gluconic acid, ammonium salt, D- **(10361-31-6)**
D-Gluconic acid, Compd. with N,N''-bis(4-chlorophenyl)-3,12-diimino-2,4,11,13- tetraazatetradecane diimidamide (2:1) **(18472-51-0)**
Gluconic acid, copper(2+) salt **(527-09-3)**
Gluconic acid, copper(2+) salt (2:1), D- **(527-09-3)**
Gluconic acid, iron(2+) salt (2:1) **(299-29-6)**
Gluconic acid, δ-lactone, D **(90-80-2)**
D-Gluconic acid, manganese salt (2:1) **(6485-39-8)**
Gluconic acid, manganese salt (2:1) **(6485-39-8)**
D-Gluconic acid, monopotassium salt (9CI) **(299-27-4)**
Gluconic acid, monopotassium salt, D **(299-27-4)**
Gluconic acid, monosodium salt, D **(527-07-1)**
Gluconic acid potassium salt **(299-27-4)**
D-Gluconic acid, potassium salt (9CI) **(35087-77-5)**
Gluconic acid sodium salt **(527-07-1)**
D-Gluconic acid, sodium salt (9CI) **(14906-97-9)**
D-Gluconic acid, strontium salt (2:1) (9CI) **(10101-21-0)**
Gluconic acid, strontium salt (2:1) (8CI) **(10101-21-0)**
D-Gluconic δ-lactone **(90-80-2)**
Gluconolactone **(90-80-2)**
Gluconsan K **(299-27-4)**
D-Glucopyranose, 2-((((2chloroethyl)nitrosoamino)carbonyl)-amino)-2-deoxy- **(54749-90-5)**
Glucopyranose, 2-deoxy-2-(3-(2-chloroethyl)-3-nitrosoureido)-, D **(54749-90-5)**
D-Glucopyranose, 2-deoxy-2-(((methylnitrosoamino)carbonyl)-amino)- **(18883-66-4)**
Glucopyranose, 2-deoxy-2-(3-methyl-3-nitrosoureido)-, D **(18883-66-4)**
α-D-Glucopyranose, 1-(dihydrogen phosphate) (9CI) **(59-56-3)**
Glucopyranoside, β-D-fructofuranosyl, α-d **(57-50-1)**
Glucopyranoside, 4,6-diamino-2-hydroxy-1,3-cyclohexylene 3,6'-diamino-3,6'-dideoxydi-, D- **(59-01-8)**
Glucopyranoside, 3,5-dihydroxy-4-(3-hydroxy-4-methoxyhy-drocinnamoyl)phenyl- 2-O-(6-deoxy-α-l-mannopyranosyl)-, β-d **(20702-77-6)**
β-D-Glucopyranoside, (3β,23β)-17,23-epoxyveratraman-3-yl **(23185-94-6)**
α-D-Glucopyranoside, β-D-fructofuranosyl, Polymer with methyloxirane and oxirane (9CI) **(26301-10-0)**
α-D-Glucopyranoside, β-D-fructofuranosyl, benzoate (9CI) **(12738-64-6)**
α-D-Glucopyranoside, β-D-fructofuranosyl O-α-D-galacto-pyranosyl (1 to 6)-, hydrate **(512-69-6)**
α-D-Glucopyranoside, β-D-fructofuranosyl O-α-D-galacto-pyranosyl-(1-6)- (9CI) **(512-69-6)**
α-D-Glucopyranoside, β-D-fructofuranosyl, monododecanoate **(25339-99-5)**
α-D-Glucopyranoside, β-D-fructofuranosyl, monohexadecanoate (9CI) **(26446-38-8)**
α-D-Glucopyranoside, β-D-fructofuranosyl, monooctadecanoate (9CI) **(25168-73-4)**
α-D-Glucopyranoside, β-D-fructofuranosyl, mono-2-propenyl ether (9CI) **(12002-22-1)**
α-D-Glucopyranoside, β-D-fructofuranosyl-, octakis(hydrogen sulfate), aluminum complex **(54182-58-0)**
α-D-Glucopyranoside, α-D-glucopyranosyl (9CI) **(99-20-7)**
ß-D-Glucopyranoside, 2-(hydroxymethyl)phenyl **(138-52-3)**
β-D-Glucopyranoside, 2-(hydroxymethyl)phenyl (9CI) **(138-52-3)**
α-D-Glucopyranoside, methyl (9CI) **(97-30-3)**
Glucopyranoside, methyl, α-D- (8CI) **(97-30-3)**
β-D-Glucopyranoside, (methyl-onn-azoxy)methyl **(14901-08-7)**
Glucopyranoside, methyl tetra-O-methyl-, β-D- (8CI) **(3149-65-3)**
β-D-Glucopyranoside, methyl 2,3,4,6-tetra-O-methyl- (9CI) **(3149-65-3)**
α-D-Glucopyranoside, methyl-, tetranitrate **(13225-10-0)**
Glucopyranoside, quercetin-3 6-O-(6-deoxy-α-l-manno-pyranosyl)-, β-D- **(153-18-4)**
Glucopyranoside, quercetin-3 6-O-α-l-rhamnopyranosyl-, β-D **(153-18-4)**
α-D-Glucopyranoside, 1,3,4,6-tetra-O-acetyl-β-D-fructofuranosyl-, tetraacetate (9CI) **(126-14-7)**
4-(α-D-Glucopyranosido)-α-glucopyranose **(69-79-4)**
β-D-Glucopyranosiduronic acid, (3β,4α,16α)-17-carboxy-

16-hydroxy-23-oxo-28-norolean-12-en-3-yl (9CI) **(1393-03-9)**
7-α-D-Glucopyranosyl-9,10-dihydro-3,5,6,8-tetrahydr oxy-1-methyl-9,10-dioxo-2-anthracene-carboxylic acid **(1260-17-9)**
10-Glucopyranosyl-1,8-dihydroxy-3-(hydroxymethyl)-9(10H)-anthracenone **(1415-73-2)**
10-β-D-Glucopyranosyl-1,8-dihydroxy-3-(hydroxymethyl)-, 9(10H)-anthracenone-, (+) **(1415-73-2)**
α-D-Glucopyranosyl β-D-fructofuranoside **(57-50-1)**
(R)-α-((6-O-β-D-Glucopyranosyl-β-D-glucopyranosyl)oxy)-benzeneacetonitrile **(29883-15-6)**
4-O-β-D-Glucopyranosyl-D-glucose **(528-50-7)**
2-(β-D-Glucopyranosyloxy)isobutyronitrile **(554-35-8)**
2-(β-D-Glucopyranosyloxy)-2-methylpropanenitrile **(554-35-8)**
(D-Glucopyranosylthio)gold **(12192-57-3)**
3β-(β-D-Glucopyropyanosyloxy)-6β,8,14-trihydroxybufa-4,20,22-trienolide 6-accetate **(507-60-8)**
Glucosaccharonic acid **(89-65-6)**
D-Glucosamine **(3416-24-8)**
Glucosamine **(3416-24-8)**
D-Glucose **(50-99-7)**
Glucose **(50-99-7)**
D-Glucose, Anhydrous **(50-99-7)**
Glucose, Anhydrous **(50-99-7)**
D-Glucose-6,6-C-d2 (9CI) **(18991-62-3)**
Glucose, 2-Deoxy- **(154-17-6)**
D-Glucose, 2-amino-2-deoxy **(3416-24-8)**
D-Glucose, 2-((((2-chloroethyl)nitrosoamino)carbonyl)amino)-2-deoxy- (9CI) **(54749-90-5)**
D-Glucose-6,6-d2 (8CI) **(18991-62-3)**
D-Glucose, 2-deoxy- **(154-17-6)**
D-Glucose, 2-deoxy-2-(((methylnitrosoamino)carbonyl)amino)-(9CI) **(18883-66-4)**
D-Glucose, 2-deoxy-2-(3-methyl-3-nitrosoureido)- **(18883-66-4)**
D-Glucose, 6-(dihydrogen phosphate) (9CI) **(56-73-5)**
D-Glucose, 4-o-β-D-galactopyranosyl- **(63-42-3)**
D-Glucose, 4-O-β-D-glucopyranosyl- **(528-50-7)**
D-Glucose, 4-o-α-D-glucopyranosyl- **(69-79-4)**
Glucose liquid **(50-99-7)**
Glucose-1-phosphate **(59-56-3)**
Glucose-6-phosphate **(56-73-5)**
α-Glucose-1-phosphate **(59-56-3)**
β-Glucose-1-phosphate **(59-56-3)**
Glucose syrup **(8029-43-4)**
(α-D-Glucosido)-β-D-fructofuranoside **(57-50-1)**
4-(α-D-Glucosido)-D-glucose **(69-79-4)**
3-Glucosyl-11-deoxojervine **(23185-94-6)**
N-d-Glucosyl(2)-N'-nitrosomethylharnstoff (German) **(18883-66-4)**
N-D-Glucosyl-(2)-N'-nitrosomethylurea **(18883-66-4)**
β-D-Glucosyloxyazoxymethane **(14901-08-7)**
β-D-Glucosyloxyazoxymethase **(14901-08-7)**
(1-D-Glucosylthio)gold **(12192-57-3)**
Glufosinate-Ammonium **(77182-82-2)**
Glukagon novo **(16941-32-5)**
Glumin **(56-85-9)**
Glupan **(50-35-1)**
Glupax **(59-66-5)**
Glusate **(56-86-0)**
Glusatin **(138-15-8)**
Gluside **(81-07-2)**
Glutacid **(56-86-0)**
Glutaconaldehydedianil hydrochloride **(1497-49-0)**
Glutaconaldehyde dianil monohydrochloride **(1497-49-0)**
Glutaconic acid, 3-hydroxy-, dimethyl ester, dimethyl phosphate **(122-10-1)**
Glutacyl **(142-47-2)**
Glutamat sodny (Czech) **(142-47-2)**
(+-)-Glutamic acid **(617-65-2)**
Glutamic acid **(56-86-0)**
α-Glutamic acid **(56-86-0)**
l-Glutamic acid **(56-86-0)**
DL-Glutamic acid (9CI) **(617-65-2)**
L-Glutamic acid, Compd. with L-arginine (1:1) (9CI) **(4320-30-3)**
Glutamic acid, DL **(617-65-2)**
Glutamic acid, L **(56-86-0)**
Glutamic acid, L-, Compd. with L-arginine (1:1) (8CI) **(4320-30-3)**
Glutamic acid 5-amide **(56-85-9)**
Glutamic acid amide **(56-85-9)**
Glutamic acid, N-(p-(((2-amino-4-hydroxy-6-pteridinyl)-methyl)amino)benzoyl)-, l- **(59-30-3)**
Glutamic acid, N-(p-(((2,4-diamino-6-pteridinyl)methyl)-amino)benzoyl)-, L **(54-62-6)**
Glutamic acid, N-(p-(((2,4-diamino-6-pteridinyl)methyl)-methylamino)benzoyl)-, L **(59-05-2)**
Glutamic acid, N-(p-(((2,4-diamino-6-pteridinyl)methyl)-methylamino)benzoyl)-, l-(+)- (8CI) **(59-05-2)**
Glutamic acid, N-(p-(((2,4-diamino-6-pteridinyl)methyl)-methylamino)benzoyl)-, sodium salt **(15475-56-6)**
l-Glutamic acid, N-(4-(((2,4-diamino-6-pteridinyl)methyl)-methylamino)benzoyl)- (9CI) **(59-05-2)**
Glutamic acid, N-(3,5-dichloro-4-(((2,4-diamino-6-pteridinyl)-methyl)methylamino)benzoyl) **(528-74-5)**
L-Glutamic acid, 5-ethyl ester (9CI) **(1119-33-1)**
Glutamic acid, 5-ethyl ester, L- (8CI) **(1119-33-1)**
Glutamic acid hydrochloride **(138-15-8)**
L-Glutamic acid, hydrochloride (9CI) **(138-15-8)**
Glutamic acid, hydrochloride, L- (8CI) **(138-15-8)**
Glutamic acid hydrogen chloride **(138-15-8)**
L-Glutamic acid, 5-(2-(4-(hydroxymethyl)phenyl)hydrazide) (9CI) **(2757-90-6)**
L-Glutamic acid, 5-(2-(α-hydroxy-para-tolyl)hydrazide) **(2757-90-6)**
Glutamic acid, 5-(2-(α-hydroxy-p-tolyl)hydrazide), L **(2757-90-6)**
L-Glutamic acid, ion(1-) (9CI) **(11070-68-1)**
Glutamic acid, ion(1-), L- (8CI) **(11070-68-1)**
Glutamic acid, monoammonium salt, L **(7558-63-6)**
L-Glutamic acid monohydrochloride **(138-15-8)**
Glutamic acid, monosodium salt, l-(+) **(142-47-2)**
Glutamic acid, N-nitroso-N-pteroyl-, L **(29291-35-8)**
Glutamic acid, pteroyl-, l- **(59-30-3)**
Glutamic acid, sodium salt **(142-47-2)**
Glutamidin **(138-15-8)**
D-Glutamiensuur **(56-86-0)**
Glutamine **(56-85-9)**
γ-Glutamine **(56-85-9)**
l-Glutamine (9CI) **(56-85-9)**
Glutamine, L **(56-85-9)**
Glutaminic acid **(56-86-0)**
l-Glutaminic acid **(56-86-0)**
Glutaminol **(56-86-0)**
Glutammato monosodico (Italian) **(142-47-2)**
β-N-(γ-L(+)Glutamyl)4-hydroxymethylphenylhydrazine **(2757-90-6)**
Glutan HCl **(138-15-8)**
Glutan hydrochloric **(138-15-8)**
Glutan hydrochloride **(138-15-8)**
Glutanon **(50-35-1)**
Glutaral **(111-30-8)**
Glutaraldehyd (Czech) **(111-30-8)**
Glutaraldehyde (ACGIH,OSHA) **(111-30-8)**
Glutardialdehyde **(111-30-8)**
Glutaric-acid **(110-94-1)**
Glutaric acid, 2,2-dichloro- (6CI) **(50901-13-8)**
Glutaric acid, diethyl ester (8CI) **(818-38-2)**
Glutaric acid, 2,2-dimethyl- (8CI) **(681-57-2)**
Glutaric acid, 3,3-dimethyl- (8CI) **(4839-46-7)**
Glutaric acid, dimethyl ester (8CI) **(1119-40-0)**
Glutaric acid dinitrile **(544-13-8)**
Glutaric acid, α keto **(328-50-7)**

Glutaric acid, 2-methyl- (8CI) **(617-62-9)**
Glutaric acid, 3-methyl-, dibutyl ester (6CI) **(56051-60-6)**
Glutaric acid, monomethyl ester (8CI) **(1501-27-5)**
Glutaric acid, 2-oxo- (8CI) **(328-50-7)**
Glutaric acid, 2-oxo-, dimethyl ester (6CI,7CI,8CI) **(13192-04-6)**
Glutaric-anhydride **(108-55-4)**
Glutaric dialdehyde **(111-30-8)**
Glutarimide, 3-(2-(3,5-dimethyl-2-oxocyclohexyl)-2-hydroxyethyl) **(66-81-9)**
Glutarimide, 2-ethyl-2-phenyl **(77-21-4)**
Glutarimide, 2-phthalimido- **(50-35-1)**
Glutarodinitrile **(544-13-8)**
Glutaronitrile (8CI) **(544-13-8)**
Glutaronitrile, 2-bromo-2-(bromomethyl) **(35691-65-7)**
Glutaryl chloride **(2873-74-7)**
Glutasin **(138-15-8)**
Glutathimid **(77-21-4)**
Glutathion **(70-18-8)**
Glutathione **(70-18-8)**
Glutathione SH **(70-18-8)**
Glutathione (reduced) **(70-18-8)**
Glutatiol **(70-18-8)**
Glutatione **(70-18-8)**
l-Glutatione **(70-18-8)**
Glutaton **(56-86-0)**
Glutavene **(142-47-2)**
Glutethimid **(77-21-4)**
Glutethimide **(77-21-4)**
Glutetimide **(77-21-4)**
Glutetimidu (Polish) **(77-21-4)**
Glutide **(70-18-8)**
Glutinal **(70-18-8)**
Glutofix 600 **(9004-62-0)**
Gly-Cel A 100 **(37099-12-0)**
Gly-Cel C 200 (9CI) **(61840-33-3)**
Glybutamide **(339-43-5)**
Glycel **(38641-94-0)**
Glyceraldehyde, (+-)- **(56-82-6)**
Glyceraldehyde, DL **(56-82-6)**
Glycereth-12 **(31694-55-0)**
Glycereth-26 **(31694-55-0)**
Glyceric acid **(473-81-4)**
DL-Glyceric aldehyde **(56-82-6)**
per-Glycerin **(72-17-3)**
Glycerin (ACGIH,OSHA) **(56-81-5)**
Glycerin, Anhydrous **(56-81-5)**
Glycerin, Synthetic **(56-81-5)**
Glycerin diacetate **(25395-31-7)**
Glycerine **(56-81-5)**
Glycerine diacetate **(25395-31-7)**
Glycerine, 1,3-diglycidyl ether **(3568-29-4)**
Glycerine ethoxylate **(31694-55-0)**
Glycerine, ethoxylated **(31694-55-0)**
Glycerine monooleate **(25496-72-4)**
Glycerine, phthalic anhydride, coconut oil fatty acid polymer **(66070-87-9)**
Glycerine, soybean oil, phthalic anhydride, litharge polymer **(66070-61-9)**
Glycerin, ethylene oxide condensate **(31694-55-0)**
Glycerine triglycidyl ether **(13236-02-7)**
Glycerine tripropionate **(139-45-7)**
Glycerin guaiacolate **(93-14-1)**
Glycerin monoallyl ether **(25136-53-2)**
Glycerin α-monochlorhydrin **(96-24-2)**
Glycerinmonoguaiacol ether **(93-14-1)**
Glycerin 1-monooleate **(111-03-5)**
Glycerin monooleate **(25496-72-4)**
Glycerin 1-monostearate **(123-94-4)**
Glycerin monostearate **(31566-31-1)**
Glycerin, pentaerythritol, phthalic anhydride, soybean oil polymer **(66070-93-7)**
Glycerin 1-stearate **(123-94-4)**
Glycerin triacetate **(102-76-1)**
Glycerin trilaurate **(538-24-9)**
Glycerintrinitrate (Czech) **(55-63-0)**
Glycerin trioleate **(122-32-7)**
Glycerite **(1401-55-4)**
Glyceritol **(56-81-5)**
Glycero-guaiacol ether **(93-14-1)**
D-Glycero-D-gulo-heptonic acid, monosodium salt (9CI) **(13007-85-7)**
D-Glycero-D-ido-heptonic acid, monosodium salt (9CI) **(30080-50-3)**
Glycerol **(56-81-5)**
Glycerol, Polymers **(25618-55-7)**
Glycerolacetone **(100-79-8)**
Glycerol allyl ether **(25136-53-2)**
Glycerol α-allyl ether **(123-34-2)**
Glycerol, 1-p-aminobenzoate **(136-44-7)**
Glycerol α-chlorohydrin **(96-24-2)**
Glycerol β-chlorohydrin **(497-04-1)**
Glycerol chlorohydrin **(96-24-2)**
Glycerol diacetate **(25395-31-7)**
Glycerol-α,γ-dibromohydrine **(96-21-9)**
Glycerol α,β-dichlorohydrin **(616-23-9)**
Glycerol α,γ-dichlorohydrin **(96-23-1)**
sym-Glycerol dichlorohydrin **(96-23-1)**
Glycerol 1,3-diglycidyl ether **(3568-29-4)**
Glycerol, 1-(dihydrogen phosphate) (8CI) **(57-03-4)**
Glycerol dimethylketal **(100-79-8)**
Glycerol, 1,3-dinitrate **(623-87-0)**
Glycerol-1,3-dinitrate **(623-87-0)**
Glycerol, 1,2-dithio- **(59-52-9)**
Glycerol epichlorhydrin **(106-89-8)**
Glycerol, ethoxylated **(31694-55-0)**
Glycerol, ethylene oxide, propylene oxide polymer **(9082-00-2)**
Glycerol guaiacolate **(93-14-1)**
Glycerol-α-guajakolether (Czech) **(93-14-1)**
Glycerol, 1,2-O-isopropylidene **(100-79-8)**
Glycerol α-(o-methoxyphenyl)ether **(93-14-1)**
Glycerol monoallyl ether **(25136-53-2)**
Glycerol-α-mono-n-butyrate **(557-25-5)**
Glycerol-α-monochlorohydrin [UN 2689] **(96-24-2)**
Glycerol, mono(dihydrogen phosphate), potassium salt **(1335-34-8)**
Glycerol-α-monoguaiacol ether **(93-14-1)**
Glycerol mono(2-methoxyphenyl)ether **(93-14-1)**
Glycerol 1-mononitrate **(624-43-1)**
Glycerol 1-monooleate **(111-03-5)**
Glycerol α-monooleate **(111-03-5)**
Glycerol monooleate **(25496-72-4)**
Glycerol α-monophenyl ether **(538-43-2)**
Glycerol 1-monostearate **(123-94-4)**
Glycerol α-monostearate **(123-94-4)**
Glycerol monostearate **(31566-31-1)**
Glycerol-1-nitrat (German) **(624-43-1)**
Glycerol, 1-nitrate **(624-43-1)**
Glycerol, nitric acid triester **(55-63-0)**
Glycerol α-cis-9-octadecenate **(111-03-5)**
Glycerol oleate **(25496-72-4)**
Glycerol, pentaerythritol, phthalic anhydride, soybean oil polymer **(66070-93-7)**
Glycerol α-phosphate **(57-03-4)**
β-Glycerol phosphate **(17181-54-3)**
Glycerol, phthalic anhydride, soybean oil polymer **(66070-61-9)**
Glycerol, phthalic anhydride, tall oil acids polymer **(66070-71-1)**
Glycerol, phthalic anhydride, tung oil, linseed oil polymer **(66071-18-9)**
Glycerol poly(oxyethylene) ether **(31694-55-0)**
Glycerol poly(oxyethylene, oxypropylene) ether **(9082-00-2)**
Glycerol poly(oxypropylene)triol **(25791-96-2)**

Glycerol, propylene oxide, ethylene oxide polymer	(9082-00-2)
Glycerol-propylen oxide polymer	(25791-96-2)
Glycerol 1-stearate	(123-94-4)
Glycerol, 1-thio-	(96-27-5)
Glycerol triacetate	(102-76-1)
Glycerol, tribenzoate (8CI)	(614-33-5)
Glycerol tribromohydrin	(96-11-7)
Glycerol tributyrate	(60-01-5)
Glycerol tricaprylate	(538-23-8)
Glycerol trichlorohydrin	(96-18-4)
Glycerol(tri(chloromethyl))ether	(38571-73-2)
Glycerol triglycidyl ether	(13236-02-7)
Glycerol-trihexanoate	(621-70-5)
Glycerol trilaurate	(538-24-9)
Glycerol trimyristate	(555-45-3)
Glyceroltrinitraat (Dutch)	(55-63-0)
Glycerol trinitrate	(55-63-0)
Glycerol(trinitrate de) (French)	(55-63-0)
Glycerol trioctadecanoate	(555-43-1)
Glycerol, trioctadecanoate	(555-43-1)
Glycerol, tri(cis-9-octadecenoate)	(122-32-7)
Glycerol trioctanoate	(538-23-8)
Glycerol trioleate	(122-32-7)
Glycerol triolein	(122-32-7)
Glycerol tripropionate	(139-45-7)
Glycerol tris(2,3-epoxypropyl) ether	(13236-02-7)
Glycerol tristearate	(555-43-1)
1-Glycerophosphate	(57-03-4)
3-Glycerophosphate	(57-03-4)
Glycerophosphate	(57-03-4)
β-Glycerophosphate	(17181-54-3)
1-Glycerophosphoric acid	(57-03-4)
2-Glycerophosphoric acid	(17181-54-3)
Glycerophosphoric acid	(57-03-4)
α-Glycerophosphoric acid	(57-03-4)
β-Glycerophosphoric acid	(17181-54-3)
Glycerophosphoric acid, potassium salt	(1335-34-8)
Glyceryl para-aminobenzoate	(136-44-7)
Glyceryl-α-chlorohydrin	(96-24-2)
Glyceryl diacetate	(25395-31-7)
1,3-Glyceryl dinitrate	(623-87-0)
Glyceryl dioleate	(25637-84-7)
Glyceryl guaiacolate	(93-14-1)
Glycerylguaiacolate carbamate	(532-03-6)
α-Glyceryl guaiacolate ether	(93-14-1)
Glyceryl guaiacyl ether	(93-14-1)
Glycerylguajacol-carbamat	(532-03-6)
Glyceryl 1-mononitrate	(624-43-1)
Glyceryl monooleate	(25496-72-4)
Glyceryl monooleate (VAN)	(111-03-5)
Glyceryl 1-monostearate	(123-94-4)
Glyceryl monostearate	(123-94-4)
Glyceryl monostearate	(31566-31-1)
Glyceryl nitrate	(55-63-0)
1-Glyceryl oleate	(111-03-5)
Glyceryl oleate	(111-03-5)
Glyceryl oleate	(25496-72-4)
Glyceryl polyethylene glycol ether	(31694-55-0)
1-Glyceryl stearate	(123-94-4)
Glyceryl stearate	(123-94-4)
Glyceryl stearate (ACGIH)	(31566-31-1)
Glyceryl triacetate	(102-76-1)
Glyceryl tribromohydrin	(96-11-7)
Glyceryl trichlorohydrin	(96-18-4)
Glyceryl tridodecanoate	(538-24-9)
Glyceryl trilaurate	(538-24-9)
Glyceryl trimyristate	(555-45-3)
Glyceryl trinitrate	(55-63-0)
Glyceryl trioctanoate	(538-23-8)
Glyceryl trioleate	(122-32-7)
Glyceryl-1,2,3-trioleate	(122-32-7)
Glyceryl tripropionate	(139-45-7)
Glyceryl tristearate	(555-43-1)
Glycidal	(765-34-4)
Glycidaldehyde [UN 2622]	(765-34-4)
Glycide	(556-52-5)
Glycidic acid, (2,4-diisopropylphenyl)-, ethyl ester (8CI)	(1334-99-2)
Glycidic acid ethyl ester	(4660-80-4)
Glycidic acid, 3-methyl-3-phenyl-	(5669-15-8)
Glycidic acid, 3-phenyl-, ethyl ester	(121-39-1)
Glycidol (ACGIH,OSHA)	(556-52-5)
Glycidol acetate	(6387-89-9)
Glycidol methyl ether	(930-37-0)
Glycidol oleate	(5431-33-4)
Glycidol stearate	(7460-84-6)
(3-Glycidoxypropyl)bis(trimethylsiloxy)methylsilane	(7422-52-8)
(3-Glycidoxypropyl)(3-chloropropyl)dimethoxysilane	(71808-64-5)
(3-Glycidoxypropyl)dimethylethoxysilane	(17963-04-1)
(γ-Glycidoxypropyl)methyldiethoxysilane	(2897-60-1)
γ-Glycidoxypropyltrimethoxysilane	(2530-83-8)
Glycidyl acrylate	(106-90-1)
Glycidyl alcohol	(556-52-5)
Glycidylaldehyde	(765-34-4)
Glycidyl butyl ether	(2426-08-6)
Glycidylester kyseliny akrylove (Czech)	(106-90-1)
Glycidylester kyseliny olejove (Czech)	(5431-33-4)
Glycidyl ester of dodecanoic acid	(63978-73-4)
Glycidyl ester of hexanoic acid	(17526-74-8)
Glycidyl 2-ethylhexyl ether	(2461-15-6)
Glycidyl hexyl ether	(5926-90-9)
Glycidyl isopropyl ether	(4016-14-2)
Glycidyl laurate	(1984-77-6)
Glycidyl methacrylate	(106-91-2)
Glycidyl 4-methoxyphenyl ether	(2211-94-1)
Glycidyl p-methoxyphenyl ether	(2211-94-1)
Glycidyl α-methyl acrylate	(106-91-2)
Glycidyl methyl ether	(930-37-0)
Glycidyl 2-methylphenyl ether	(2210-79-9)
Glycidyl 4-methylphenyl ether	(2186-24-5)
Glycidyl methylphenyl ether	(26447-14-3)
Glycidyl o-methylphenyl ether	(2210-79-9)
Glycidyl neodecanoate	(26761-45-5)
Glycidyl 4-nitrophenyl ether	(5255-75-4)
Glycidyl p-nitrophenyl ether	(5255-75-4)
Glycidyl octadecanoate	(7460-84-6)
Glycidyl octadecenoate	(5431-33-4)
Glycidyl oleate	(5431-33-4)
Glycidyl phenyl ether	(122-60-1)
N-Glycidylphthalimide	(5455-98-1)
Glycidyl propenate	(106-90-1)
Glycidyl propyl ether	(3126-95-2)
Glycidyl sorbate dimer	(63915-78-6)
Glycidyl stearate	(7460-84-6)
Glycidyl o-tolyl ether	(2210-79-9)
Glycidyl p-tolyl ether	(2186-24-5)
Glycidyl-trimethyl-ammonium chloride	(3033-77-0)
Glycinanilide, hydrochloride	(4801-39-2)
(Glycinato)dihydroxyaluminum	(13682-92-3)
(Glycinato-N,O)dihydroxyaluminum	(13682-92-3)
(T-4)-(Glycinato-N,O)dihydroxyaluminum	(13682-92-3)
Glycine	(56-40-6)
Glycine, Bimol. cyclic peptide	(106-57-0)
Glycine, Polymer with hexahydro-2H-azepin-2-one (9CI)	(54590-59-9)
Glycine, N-acetyl- (9CI)	(543-24-8)

Glycine, N-(4-aminobenzoyl)- (9CI) **(61-78-9)**
Glycine, N-(aminocarbonyl)- (9CI) **(462-60-2)**
Glycine, N-(aminoiminomethyl)-N-methyl- (9CI) **(57-00-1)**
Glycine, anhydride (9CI) **(4202-74-8)**
Glycine anilide **(555-48-6)**
Glycine, N-benzoyl- **(495-69-2)**
Glycine, N-benzoyl-, methyl ester (9CI) **(1205-08-9)**
Glycine, N,N-bis(2-(bis(carboxymethyl)amino)ethyl) **(67-43-6)**
Glycine, N,N-bis(2-(bis(carboxymethyl)amino)ethyl)-, calcium trisodium salt **(12111-24-9)**
Glycine, N,N-bis(2-(bis(carboxymethyl)amino)ethyl)-, pentasodium salt (9CI) **(140-01-2)**
Glycine, N,N-bis(carboxymethyl)- (9CI) **(139-13-9)**
Glycine, N-(2-(bis(carboxymethyl)amino)ethyl)-N-(2-hydroxyethyl)-, trisodium salt (9CI) **(139-89-9)**
Glycine, N,N-bis(carboxymethyl)-, calcium sodium salt (1:1:1) **(60034-45-9)**
Glycine, N,N-bis(carboxymethyl)-, disodium salt (9CI) **(15467-20-6)**
Glycine, N,N-bis(carboxymethyl)-, disodium salt, monohydrate (9CI) **(23255-03-0)**
Glycine, N,N-bis(carboxymethyl)-, sodium salt (9CI) **(10042-84-9)**
Glycine, N,N-bis(carboxymethyl)-, tripotassium salt (9CI) **(2399-85-1)**
Glycine, N,N-bis(carboxymethyl)-, trisodium salt **(5064-31-3)**
Glycine, N,N-bis(carboxymethyl)-, trisodium salt, monohydrate **(18662-53-8)**
Glycine, N,N-bis(2-hydroxyethyl)-, monosodium salt (9CI) **(139-41-3)**
Glycine, N,N-bis(phosphonomethyl) **(2439-99-8)**
Glycine, N-carbamoyl- (8CI) **(462-60-2)**
Glycine, N-(carboxymethyl)- **(142-73-4)**
Glycine, N-(carboxymethyl)-N-(2-hydroxyethyl)- (8CI) **(93-62-9)**
Glycine, N-(carboxymethyl)-N'-(2-hydroxyethyl)-N,N'-ethylenedi **(150-39-0)**
Glycine, N-(carboxymethyl)-N'-(2-hydroxyethyl)-N,N'-ethylenedi-, trisodium salt (8CI) **(139-89-9)**
Glycine, N-(carboxymethyl)-N-(2-hydroxyethyl)-, disodium salt (9CI) **(135-37-5)**
Glycine, N-(carboxymethyl)-N-methyl- (9CI) **(4408-64-4)**
Glycine, N-(carboxymethyl)-N-(phosphonomethyl)- (9CI) **(5994-61-6)**
Glycine, N-(chloroacetyl)-N-(2,6-diethylphenyl)-, ethyl ester **(38727-55-8)**
Glycine, N,N[1]-1,2-cyclohexanediylbis(N-(carboxymethyl)- (9CI) **(482-54-2)**
Glycine dipeptide **(556-50-3)**
Glycine, N-(2-((2-(dodecylamino)ethyl)amino)ethyl)- (9CI) **(6843-97-6)**
Glycine, N,N'-1,2-ethanediylbis(N-(carboxymethyl)- (9CI) **(60-00-4)**
Glycine, N,N'-1,2-ethanediylbis(N-(carboxymethyl)-, ammonium salt (9CI) **(7379-26-2)**
Glycine, N,N'-1,2-ethanediylbis(N-(carboxymethyl)-, diammonium salt (9CI) **(20824-56-0)**
Glycine, N,N'-1,2-ethanediylbis(N-carboxymethyl)-, diammonium zinc salt **(67859-51-2)**
Glycine, N,N'-1,2-ethanediylbis(N-(carboxymethyl)-, disodium salt (9CI) **(139-33-3)**
Glycine, N,N'-1,2-ethanediylbis(N-(carboxymethyl)-, manganese disodium salt **(15375-84-5)**
Glycine, N,N'-1,2-ethanediylbis(N-(carboxymethyl)-, monosodium salt (9CI) **(17421-79-3)**
Glycine, N,N'-1,2-ethanediylbis(N-(carboxymethyl)-, potassium salt (9CI) **(7379-27-3)**
Glycine, N,N'-1,2-ethanediylbis(N-(carboxymethyl)-, sodium salt (9CI) **(7379-28-4)**
Glycine, N,N'-1,2-ethanediylbis(N-(carboxymethyl)-, tetraammonium salt (9CI) **(22473-78-5)**
Glycine, N,N'-1,2-ethanediylbis(N-(carboxymethyl)-, tetrapotassium salt (9CI) **(5964-35-2)**
Glycine, N,N'-1,2-ethanediylbis(N-(carboxymethyl)-, tetrasodium salt (9CI) **(64-02-8)**
Glycine, N,N'-1,2-ethanediylbis(N-(carboxymethyl)-, tripotassium salt (9CI) **(17572-97-3)**
Glycine, N,N'-1,2-ethanediylbis(N-(carboxymethyl)-, trisodium salt (9CI) **(150-38-9)**
Glycine, ethyl ester, hydrochloride **(623-33-6)**
Glycine, N-ethyl-N-((heptadecafluorooctyl)sulfonyl)-, ethyl ester (8CI) **(1869-77-8)**
Glycine, N-ethyl-N-((heptadecafluorooctyl)sulfonyl)-, potassium salt **(2991-51-7)**
Glycine, N-ethyl-N-((heptadecafluorooctyl)sulfonyl)-, sodium salt (9CI) **(3871-50-9)**
Glycine, N-glycyl- (9CI) **(556-50-3)**
Glycine, N-glycyl-, cyclic peptide **(106-57-0)**
Glycine, N-(N-glycylglycyl)- (9CI) **(556-33-2)**
Glycine, hydrochloride **(6000-43-7)**
Glycine, N-(2-hydroxyethyl)-N-(2-((1-oxooctadecyl)amino)-ethyl)- (9CI) **(139-92-4)**
Glycine, N,N'-(2-hydroxy-1,3-propanediyl)bis(N-(carboxymethyl)- (9CI) **(3148-72-9)**
Glycine betaine **(107-43-7)**
Glycine, N-(N-l-γ-glutamyl-l-cysteinyl) **(70-18-8)**
Glycine, N-methyl- **(107-97-1)**
Glycine, N-methyl-N-nitroso- **(13256-22-9)**
Glycine, N-methyl-N-(1-oxododecyl)-, sodium salt (9CI) **(137-16-6)**
Glycine, N-methyl-N-(1-oxo-9-octadecenyl)- **(110-25-8)**
Glycine, N-methyl-N-(1-oxo-9-octadecenyl)-, (Z)- (9CI) **(110-25-8)**
Glycine, N-methyl-N-(1-oxo-9-octadecenyl)-, sodium salt (9CI) **(3624-77-9)**
Glycine, N-nicotinoyl- (8CI) **(583-08-4)**
Glycine, 2-oxo- **(471-47-6)**
Glycine, 2-phenyl-, DL- (8CI) **(2835-06-5)**
Glycine, N-phenyl-, monopotassium salt (9CI) **(19525-59-8)**
Glycine, N-phenyl-, monosodium salt (9CI) **(10265-69-7)**
Glycine, N-(phosphonomethyl) **(1071-83-6)**
Glycine, N-(phosphonomethyl)-, Compd. with 2-propanamine (1:1) **(38641-94-0)**
Glycine, N-(phosphonomethyl)-, monosodium salt (9CI) **(34494-03-6)**
Glycine, N-(3-pyridinylcarbonyl)- (9CI) **(583-08-4)**
Glycinol **(141-43-5)**
Glycinonitrile, N,N-dimethyl- **(926-64-7)**
Glycirenan **(51-43-4)**
Glyco-Flavine **(8048-52-0)**
Glyco-Hydrochloride **(6000-43-7)**
Glycocel TA **(9000-11-7)**
Glycocollanilide **(555-48-6)**
Glycocoll hydrochloride **(6000-43-7)**
Glycocoll betaine **(107-43-7)**
Glycodine **(509-67-1)**
Glycogen (9CI) **(9005-79-2)**
Glycogenic acid **(526-95-4)**
Glycol **(107-21-1)**
Glycol Ether DB **(112-34-5)**
Glycol Ether DB Aceatate **(124-17-4)**
Glycol Ether EB **(111-76-2)**
Glycol Ether EB Acetate **(111-76-2)**
Glycol Ether EE **(110-80-5)**
Glycol Ether EE Acetate **(111-15-9)**
Glycol Ether EM **(109-86-4)**
Glycol Ether PM **(107-98-2)**
Glycol Ether TPM **(10213-77-1)**
Glycol alcohol **(107-21-1)**
Glycolaldehyde **(141-46-8)**
Glycolaldehyde, diethyl acetal (8CI) **(621-63-6)**
Glycolanilide **(4746-61-6)**
Glycol bis(hydroxyethyl) ether **(112-27-6)**
Glycol bis(mercaptoacetate) **(123-81-9)**
Glycol bromide **(106-93-4)**
Glycol bromohydrin **(540-51-2)**
Glycol butyl ether **(111-76-2)**
Glycol carbonate **(96-49-1)**
Glycol chlorohydrin **(107-07-3)**
Glycol cyanohydrin **(109-78-4)**
Glycol diacetate **(111-55-7)**

Glycol dibromide (106-93-4)
Glycol dichloride (107-06-2)
Glycol, diethoxytetraethylene (4353-28-0)
Glycol diglycidyl ether (2224-15-9)
Glycol dimercaptoacetate (123-81-9)
Glycol dimethacrylate (97-90-5)
Glycol dimethyl ether (110-71-4)
Glycoldinitraat (Dutch) (628-96-6)
Glycol dinitrate (628-96-6)
Glycol (dinitrate de) (French) (628-96-6)
Glycol distearate (627-83-8)
Glycol ether (111-46-6)
Glycol ether de acetate (112-15-2)
Glycol ether em acetate (110-49-6)
Glycol ethylene ether (123-91-1)
Glycol ethyl ether (110-80-5)
Glycol ethyl ether (111-46-6)
Glycoleucine (327-57-1)
Glycol formal (646-06-0)
Glycolic-acid (79-14-1)
Glycolic acid, butyl ester, butyl phthalate (85-70-1)
Glycolic acid cellulose ether (9000-11-7)
Glycolic acid, ethyl ester, methyl phthalate (85-71-2)
Glycolic acid, monosodium salt (2836-32-0)
Glycolic acid, phenyl- (90-64-2)
Glycolic acid phenyl ether (122-59-8)
Glycolic acid, phthalate, dibutyl ester (85-70-1)
Glycolic acid, 2-thio- (68-11-1)
Glycolic acid, thio- (68-11-1)
Glycolic aldehyde (141-46-8)
Glycolic nitrile (107-16-4)
Glycolixir (56-40-6)
Glycollic acid phenyl ether (122-59-8)
Glycol methacrylate (868-77-9)
Glycolmethyl ether (109-86-4)
Glycol monoacetate (542-59-6)
Glycol-monoacetin (542-59-6)
Glycol monobutyl ether (111-76-2)
Glycol monobutyl ether acetate (112-07-2)
Glycolmonochloorhydrine (Dutch) (107-07-3)
Glycol monochlorohydrin (107-07-3)
Glycol monoethyl ether (110-80-5)
Glycol monoethyl ether (OSHA) (110-80-5)
Glycol monoethyl ether acetate (111-15-9)
Glycol monohexyl ether (112-25-4)
Glycol monomethacrylate (868-77-9)
Glycol monomethyl ether (109-86-4)
Glycol monomethyl ether acetate (110-49-6)
Glycol monomethyl ether acetylricinoleate (140-05-6)
Glycol monomethyl ether acrylate (3121-61-7)
Glycol monophenyl ether (122-99-6)
Glycol monostearate (111-60-4)
Glycolmul SOC (8007-43-0)
Glycolonitrile (8CI) (107-16-4)
Glycolonitrile, acetate (1001-55-4)
Glycolonitrile, phenyl- (532-28-5)
Glycol polyethylene monostearate #200 (9004-99-3)
Glycol, polyethylene monostearate #6000 (9004-99-3)
Glycols, polyethylene, dibenzoate (9004-86-8)
Glycols, polyethylene, dimethyl ether (24991-55-7)
Glycols, polyethylene, distearate (8CI) (9005-08-7)
Glycols, polyethylene, dodecyl ether, monosulfonate, ammonium salt (32612-48-9)
Glycols, polyethylene, mono(2-(dodecylamino)ethyl) ether (8CI) (25190-01-6)
Glycols, polyethylene, monododecyl ether (9002-92-0)
Glycols, polyethylene, mono(2,3-epoxy-2-methylpropyl) ether (8CI) (32196-63-7)
Glycols, polyethylene, mono(hydrogen sulfate), dodecyl ether, sodium salt (8CI) (9004-82-4)
Glycols, polyethylene, monomethyl ether (8CI) (9004-74-4)
Glycols, polyethylene, mono(nonylphenyl) ether (9016-45-9)
Glycols, polyethylene, mono(p-nonylphenyl) ether (26027-38-3)
Glycols, polyethylene, mono-9-octadecenyl ether, (Z)- (8CI) (9004-98-2)
Glycols, polyethylene, monooctadecyl ether (8CI) (9005-00-9)
Glycols, polyethylene, monophenyl ether (9004-78-8)
Glycols, polyethylene, monostearate (8CI) (9004-99-3)
Glycols, polyethylene, mono((1,1,3,3-tetramethylbutyl)phenyl) ether (8CI) (9036-19-5)
Glycols, polyethylene, mono(p-(1,1,3,3-tetramethylbutyl)-phenyl) ether (9002-93-1)
Glycols, polyethylene-polypropylene (8CI) (9003-11-6)
Glycols, polyethylene-polypropylene, monobutyl ether (8CI) (9038-95-3)
Glycols, polyethylenepolypropylene, monobutyl ether (Nonionic) (9038-95-3)
Glycols, polypropylene, 1,2,3-propanetriyl ether (25791-96-2)
Glycol stearate (111-60-4)
Glycoluric acid (462-60-2)
Glycoluril (8CI) (496-46-8)
Glycolylurea (461-72-3)
Glycomonochlorhydrin (107-07-3)
Glycomul O (1338-43-8)
Glycomul P (26266-57-9)
Glycomul S (1338-41-6)
Glycomul SOC (8007-43-0)
Glycomul SOC Special (8007-43-0)
Glycomul TO (26266-58-0)
Glycon S-70 (57-11-4)
Glycon S-80 (57-11-4)
Glycon S-90 (57-11-4)
Glycon B-70 (112-85-6)
Glycon DP (57-11-4)
Glycon RO (112-80-1)
Glycon TP (57-11-4)
Glycon WO (112-80-1)
Glyconic acid (526-95-4)
Glyconitrile (107-16-4)
Glycophen (36734-19-7)
Glycophene (36734-19-7)
Glycoproteins, emulsans (80450-55-1)
Glycoproteins, specific or class, emulsans (80450-55-1)
Glycosperse S-20 (9005-67-8)
Glycosperse L-20 (9005-64-5)
Glycosperse L-20X (9005-64-5)
Glycosperse O-20 (9005-65-6)
Glycosperse O-20 VEG (9005-65-6)
Glycosperse O-20X (9005-65-6)
Glyco stearin (106-11-6)
Glycotuss (93-14-1)
Glycowax S 932 (555-43-1)
Glycyl alcohol (56-81-5)
N-Glycylaniline (555-48-6)
Glycylglycine (556-50-3)
N-Glycylglycine (556-50-3)
α-Glycylglycine (556-50-3)
Glycylglycine lactam (106-57-0)
Glycyl-glycyl-glycine (556-33-2)
Glycylglycylglycine (556-33-2)
N-(N-Glycylglycyl)glycine (556-33-2)
Glycylbetaine (107-43-7)
Glycyrrhiza (68916-91-6)
Glycyrrhizae (Latin) (68916-91-6)
Glycyrrhiza extract (68916-91-6)
Glycyrrhiza-extract (68916-91-6)
Glycyrrhizinic-acid (1405-86-3)
Glydant (6440-58-0)

Name	CAS No.
Glyecine A	**(111-48-8)**
Glyfyllin	**(479-18-5)**
Glyhexylamide	**(565-33-3)**
Glyhexylamine isodiane	**(565-33-3)**
Glykokolan sodny (Czech)	**(2836-32-0)**
Glykokollbetain (German)	**(107-43-7)**
Glykokollbetain-chlorid (German)	**(590-46-5)**
Glykoldinitrat (German)	**(628-96-6)**
Glykolonitril (Czech)	**(107-16-4)**
Glyme	**(110-71-4)**
Glyme-23	**(24991-55-7)**
Glyme-3	**(112-49-2)**
Glymol	**(8012-95-1)**
Glyodex 3722	**(133-06-2)**
Glyodin acetate	**(556-22-9)**
Glyoxal	**(107-22-2)**
Glyoxal, 29.2%	**(107-22-2)**
Glyoxal, 40%	**(107-22-2)**
Glyoxalbiuret	**(496-46-8)**
Glyoxal, dimethyl-	**(431-03-8)**
Glyoxal, diphenyl-	**(134-81-6)**
Glyoxaldiureine	**(496-46-8)**
Glyoxaldiurene	**(496-46-8)**
Glyoxalin	**(288-32-4)**
Glyoxaline	**(288-32-4)**
Glyoxaline-5-alanine	**(71-00-1)**
Glyoxal, methyl	**(78-98-8)**
Glyoxal, phenyl-, 2-(diethyl acetal)	**(6175-45-7)**
Gly-oxide	**(124-43-6)**
Glyoxide	**(124-43-6)**
Glyoxide Dry	**(556-22-9)**
Glyoxime, dimethyl-	**(95-45-4)**
Glyoxylaldehyde	**(107-22-2)**
Glyoxyldiureid	**(97-59-6)**
Glyoxyldiureide	**(97-59-6)**
Glyoxylic-acid	**(298-12-4)**
Glyoxylic acid, amino-	**(471-47-6)**
Glyoxylic acid, methyl ester, 2-(methyl phenyl acetal) (8CI)	**(24607-12-3)**
Glyoxylic acid, phenyl-	**(611-73-4)**
Glyoxylic acid, phenyl-, ethyl ester (8CI)	**(1603-79-8)**
Glyoxylonitrile, phenyl- (8CI)	**(613-90-1)**
Glyoxylonitrile, phenyl-, oxime, O,O-diethyl phosphorothioate	**(14816-18-3)**
Glyped	**(102-76-1)**
Glypesin	**(141-94-6)**
Glyphosate	**(1071-83-6)**
Glyphosate isopropylamine salt	**(38641-94-0)**
Glyphosine	**(2439-99-8)**
Glyphyllin	**(479-18-5)**
Glyphylline	**(479-18-5)**
Glysanol B	**(12192-57-3)**
Glysoletten	**(50-06-6)**
Gnaifenesin	**(93-14-1)**
Gnoscopine	**(6035-40-1)**
Goal	**(42874-03-3)**
Gofron	**(63428-84-2)**
Gohensil E 50Y	**(9003-20-7)**
Gohsenol	**(9002-89-5)**
Gohsenol AH 22	**(9002-89-5)**
Gohsenol GH	**(9002-89-5)**
Gohsenol GH 17	**(9002-89-5)**
Gohsenol GH 20	**(9002-89-5)**
Gohsenol GH 23	**(9002-89-5)**
Gohsenol GL 03	**(9002-89-5)**
Gohsenol GL 05	**(9002-89-5)**
Gohsenol GL 08	**(9002-89-5)**
Gohsenol GM 14	**(9002-89-5)**
Gohsenol GM 94	**(9002-89-5)**
Gohsenol GM 141	**(9002-89-5)**
Gohsenol KH 17	**(9002-89-5)**
Gohsenol NH 05	**(9002-89-5)**
Gohsenol NH 17	**(9002-89-5)**
Gohsenol NH 18	**(9002-89-5)**
Gohsenol NH 20	**(9002-89-5)**
Gohsenol NH 26	**(9002-89-5)**
Gohsenol NK 114	**(9002-89-5)**
Gohsenol NL 05	**(9002-89-5)**
Gohsenol NM 14	**(9002-89-5)**
Gohsenyl E 50 Y	**(9003-20-7)**
Gold-chloride	**(13453-07-1)**
1721 Gold	**(7440-50-8)**
Gold	**(7440-57-5)**
Gold Bond	**(8001-79-4)**
Gold Bond R	**(14808-60-7)**
Gold Bronze	**(7440-50-8)**
Gold, (D-glucopyranosylthio)-	**(12192-57-3)**
Gold Flake	**(7440-57-5)**
Gold Leaf	**(7440-57-5)**
Gold Orange	**(547-58-0)**
Gold Orange MP	**(140-56-7)**
Gold Powder	**(7440-57-5)**
Gold Salt	**(128-56-3)**
Gold Satinobre	**(1314-41-6)**
Gold Yellow	**(547-57-9)**
Gold, bromodiethyl-	**(26645-10-3)**
Gold(III) chloride	**(13453-07-1)**
Gold dust	**(50-36-2)**
Golden Antimony Sulfide	**(1315-04-4)**
Golden Salt	**(128-56-3)**
Golden Seal	**(458-37-7)**
Golden Yellow	**(128-66-5)**
Golden Yellow	**(605-69-6)**
Goldenleaf tobacco spray	**(115-29-7)**
Goldquat 276	**(1910-42-5)**
Gold, (1-thio-D-glucopyranosato)	**(12192-57-3)**
Gold thioglucose	**(12192-57-3)**
Gold trichloride	**(13453-07-1)**
Goltix	**(41394-05-2)**
Gombardol	**(63-74-1)**
Gonabion	**(9002-61-3)**
Gonacin	**(8048-52-0)**
Gonacrine	**(8048-52-0)**
Gonacrine	**(86-40-8)**
Gonad Stimulating Factor	**(9002-68-0)**
Gonadex	**(9002-61-3)**
Gonadotropin, chorionic	**(9002-61-3)**
Gonocrin	**(86-40-8)**
Gontochin	**(54-05-7)**
Gonyaulax Catenella Poison Dihydrochloride	**(35554-08-6)**
Gonyaulax Toxic Dihydrochloride	**(35554-08-6)**
Good-Rite	**(9016-00-6)**
Good-Rite GP 264	**(117-81-7)**
Good-Rite K 37	**(9003-01-4)**
Good-Rite K 702	**(9003-01-4)**
Good-Rite K-700	**(9003-01-4)**
Good-Rite K727	**(9003-01-4)**
Good-Rite Nix	**(140-93-2)**
Good-Rite WS 801	**(9003-01-4)**
Goodrite 1800X73	**(9003-55-8)**
Gophacide	**(4104-14-7)**
Gordona	**(22248-79-9)**
Gordon's Mecomec	**(1929-86-8)**
Gorgoic acid, diiodo-	**(66-02-4)**
Gorgonic acid, diiodo-	**(66-02-4)**
Gossypimine	**(477-73-6)**

Gossypine (123-41-1)
Gossyplure (50933-33-0)
Gotamine tartrate (379-79-3)
Gothnion (86-50-0)
Goudron de Houille, Distillats de (French) (65996-92-1)
Goulard's extract (8006-24-4)
Gpcd 398 (9003-07-0)
Graafina (50-50-0)
Gracet Violet 2R (128-95-0)
Grafestrol (56-53-1)
Grafoil (7782-42-5)
Grafoil GTA (7782-42-5)
Graham's Salt (10361-03-2)
Grain Alcohol (64-17-5)
Grain Sorghum Harvest-Aid (7775-09-9)
Gramevin (127-20-8)
Gramevin (75-99-0)
Gramicidin D (8CI,9CI) (1393-88-0)
Gramicidin Dubos (1393-88-0)
Graminon-Plus (120-36-5)
Graminon (34123-59-6)
Gramisan (123-88-6)
Gramixel (1910-42-5)
Gramonol (1910-42-5)
Gramoxon (1910-42-5)
Gramoxone (1910-42-5)
Gramoxone D (1910-42-5)
Gramoxone S (1910-42-5)
Gramoxone S (4685-14-7)
Gramoxone W (1910-42-5)
Gramoxone dichloride (1910-42-5)
Gramoxone methyl sulfate (2074-50-2)
Grampenil (69-53-4)
Gramuron (1910-42-5)
Granada Green Lake GL (1328-53-6)
Granex O (7775-09-9)
Granmag (1309-48-4)
Granosan (107-27-7)
Granosan (517-16-8)
Granosan M (2235-25-8)
Granosan M (517-16-8)
Granosan MDB (517-16-8)
Granox NM (118-74-1)
Granox PFM (133-06-2)
Granular zinc (7440-66-6)
Granulated sugar (57-50-1)
Granulestin (8002-43-5)
Granulin (83-63-6)
Granurex (555-37-3)
Granutox (298-02-2)
Grape Blue A Geigy (860-22-0)
Grape Oil (106-30-9)
Grapefruit Oil (8016-20-4)
Grape sugar (50-99-7)
Graphic Red Y (1248-18-6)
Graphite (7782-42-5)
Graphite, Natural (ACGIH,OSHA) (7782-42-5)
Graphite Synthetic (ACGIH,OSHA) (7440-44-0)
Graphlox (77-47-4)
Graphnol N 3M (7782-42-5)
Graphol (55-55-0)
Graphtal Red RL (2814-77-9)
Graphtol Blue BL (147-14-8)
Graphtol Blue BLF (147-14-8)
Graphtol Blue 2GLS (147-14-8)
Graphtol Blue RL (81-77-6)
Graphtol Green 2GLS (1328-53-6)
Graphtol Orange GP (3520-72-7)
Graphtol Red A-4RL (2425-85-6)
Graphtol Red 2GL (3468-63-1)
Graphtol Red RL (2814-77-9)
Graphtol Yellow 4813-0 (2512-29-0)
Graphtol Yellow A-HG (6358-85-6)
Graphtol Yellow GL (2512-29-0)
Grasal Brilliant Red B (85-83-6)
Grasal Brilliant Red G (85-86-9)
Grasal Brilliant Yellow (60-11-7)
Grasal Orange (842-07-9)
Grasal Yellow (85-84-7)
Grasan Brilliant Red B (85-83-6)
Grasan Brown DT New (6416-57-5)
Grasan Chrysoidine (495-54-5)
Grasan Orange 3R (3118-97-6)
Grasan Orange R (842-07-9)
Grascide (709-98-8)
Grasex (75-87-6)
Grasip (55635-13-7)
Grasipan (55635-13-7)
Graslan (34014-18-1)
Grasol Blue 2GS (2475-45-8)
Grasol Violet R (128-95-0)
Grasol Yellow RSF (2051-85-6)
Graspaz (55635-13-7)
Grassland Weedkiller (3813-05-6)
Gratibain (630-60-4)
Gratus strophanthin (630-60-4)
Gravidox (65-23-6)
Gravinol (523-87-5)
Gravocain (137-58-6)
Gravol (523-87-5)
Gray acetate (62-54-4)
Grayanotoxane-3,5,6,10,14,16-hexol 14-acetate (4720-09-6)
Grayanotoxane-3,5,6,10,14,16-hexol, 14-acetate, (3-β,6-β,14R)- (4720-09-6)
Grayanotoxin I (4720-09-6)
Grease (68153-81-1)
Grease (Animal) (68153-81-1)
Great Burdock LS (8001-21-6)
1724 Green (2353-45-9)
11091 Green (128-80-3)
11389 Green (6358-69-6)
11661 Green (1308-38-9)
12415 Green (633-03-4)
Green Chrome Oxide (1308-38-9)
Green Chromic Oxide (1308-38-9)
Green Chromium Oxide (1308-38-9)
Green Cinnabar (1308-38-9)
Green Cross Couch Grass Killer (650-51-1)
Green Cross Crabgrass Killer (590-28-3)
Green Cross Warble Powder (83-79-4)
Green-Daisen M (8018-01-7)
Green EN (633-03-4)
Green Hellebore (65072-04-0)
Green Hydroquinone (106-34-3)
Green MX (569-64-2)
Green Nickel Oxide (1313-99-1)
Green No. 202 (128-80-3)
Green No. 203 (5141-20-8)
Green No. 204 (6358-69-6)
Green Oil (120-12-7)
Green Oil (68308-34-9)
Green Oxide of Chromium (1308-38-9)
Green Oxide of Chromium OC-31 (1308-38-9)
Green Rouge (1308-38-9)
Green Seal-8 (1314-13-2)

Green Vitriol	(7720-78-7)
Greenharten	(84-79-7)
Greenockite	(1306-23-6)
Green vitrol	(7782-63-0)
Grenade	(68085-85-8)
Grenoble Green	(569-64-2)
Greosin	(126-07-8)
Gresfeed	(126-07-8)
Grex	(9002-88-4)
Grex PP 60-002	(9002-88-4)
Grey arsenic	(7440-38-2)
Gricin	(126-07-8)
Griffex	(1912-24-9)
Griffin Manex	(12427-38-2)
Griffin Super Cu	(1332-03-2)
Griffin super Cu	(7758-98-7)
Grifomin	(317-34-0)
Grifulvin	(126-07-8)
Grifulvin V	(126-07-8)
Grilon	(105-60-2)
Grilon	(25038-54-4)
Grilon	(63428-83-1)
Grilonit RV 1806	(2425-79-8)
Grinding dust	(65996-72-7)
Grippex	(50-35-1)
Gris-Peg	(126-07-8)
Grisactin	(126-07-8)
Griscofulvin	(126-07-8)
Grisefuline	(126-07-8)
Grisemin	(1967-16-4)
Griseo	(126-07-8)
(+)-Griseofulvin	(126-07-8)
Griseofulvin	(126-07-8)
Griseofulvin-Forte	(126-07-8)
Griseofulvinum	(126-07-8)
Grisetin	(126-07-8)
Grisin	(1967-16-4)
Grisofulvin	(126-07-8)
Grisol	(107-49-3)
Grisolen	(9002-88-4)
Grisovin	(126-07-8)
Grocel	(77-06-5)
Groco	(8001-26-1)
Groco 2	(112-80-1)
Groco 4	(112-80-1)
Groco 6	(112-80-1)
Groco 5l	(112-80-1)
Groco 54	(57-11-4)
Groco 55	(57-11-4)
Groco 58	(57-11-4)
Groco 59	(57-11-4)
Groco 55l	(57-11-4)
Groco 5810	(123-95-5)
Grocolene	(56-81-5)
Grocor 5500	(31566-31-1)
Grocor 6000	(31566-31-1)
Grosafe	(7440-44-0)
Grota	(2832-19-1)
Grotan B	(4719-04-4)
Grotan BK	(4719-04-4)
Grotan DF-35	(2832-19-1)
Ground Ryania Specisa(vahl) Stemwood (Alkoloid Ryanodine)	(15662-33-6)
Ground Vocle Sulphur	(7704-34-9)
Groundnut Oil	(8002-03-7)
Growth Hormone	(9002-72-6)
Gruenau S	(78-23-9)
Grundier arbezol	(87-86-5)
Grysio	(126-07-8)
Gryzbol	(1594-56-5)
Grzybol	(1594-56-5)
Guar-Gum	(9000-30-0)
Gum-Tara	(39300-88-4)
Guaethol	(94-71-3)
Guaiac-Resin	(9000-29-7)
Guaiac-Wood-Oil	(8016-23-7)
Guaiacol	(90-05-1)
m-Guaiacol	(150-19-6)
Guaiacolglicerinetere	(93-14-1)
Guaiacol glyceryl ether	(93-14-1)
Guaiacol glyceryl ether carbamate	(532-03-6)
Guaiacurane	(93-14-1)
Guaiacylglycerol	(1208-42-0)
Guaiacylglycerol-β-O-4-guaiacyl ether	(7382-59-4)
Guaiacylglycerol-β-guaiacyl ether	(7382-59-4)
Guaiacyl glyceryl ether	(93-14-1)
Guaiacylpropane	(2785-87-7)
Guaia-1(5),7(11)-diene	(88-84-6)
Guaiamar	(93-14-1)
Guaianesin	(93-14-1)
Guaicol	(90-05-1)
Guaiene	(88-84-6)
β-Guaiene	(88-84-6)
Guaiethol	(94-71-3)
Guaifenesin	(93-14-1)
Guaiphenesine	(93-14-1)
Guaiphesin	(93-14-1)
Guajacol-glycerinaether (German)	(93-14-1)
Guajacol-α-glycerinether	(93-14-1)
Guajacuran	(93-14-1)
Guajakol (Czech)	(90-05-1)
Guajamar	(93-14-1)
Guamide	(57-67-0)
Guanabenz	(5051-62-7)
Guanabenzo (Spanish)	(5051-62-7)
Guanabenzum (Latin)	(5051-62-7)
Guanamine	(504-08-5)
Guanar	(93-14-1)
Guanazol	(134-58-7)
Guanazole	(1455-77-2)
Guanazolo	(134-58-7)
Guanethidine	(55-65-2)
Guanicil	(57-67-0)
Guanidan	(57-67-0)
Guanidine	(113-00-8)
Guanidine, amino	(79-17-4)
Guanidine, amino-, hydrochloride	(1937-19-5)
Guanidine, 2-benzyl-1,3-dimethyl	(55-73-2)
Guanidine carbonate	(593-85-1)
Guanidine carboxamide	(141-83-3)
Guanidine chloride	(50-01-1)
Guanidine, conjugate monoacid (9CI)	(25215-10-5)
Guanidine, cyano	(461-58-5)
Guanidine, N-cyano-N'-(hydroxymethyl)- (9CI)	(13101-26-3)
Guanidine, N-cyano-N'-methyl-,	(1609-07-0)
Guanidine, cyano-, methylmercury deriv.	(502-39-6)
Guanidine, N-cyano-N'-methyl-N''-(2-(((5-methyl-1H-imidazol-4-yl)methyl)thio)ethyl)	(51481-61-9)
Guanidine, cyano-, Polymer with N-(2-aminoethyl)-1,2-ethanediamine and (chloromethyl)oxirane (9CI)	(67953-54-2)
Guanidine, ((2,6-dichlorobenzylidene)amino)- (8CI)	(5051-62-7)
Guanidine, dimethyl	(3324-71-8)
Guanidine, 1,1-dimethyl-, monohydrochloride	(22583-29-5)
Guanidine, N,N'-dimethyl-N''-(phenylmethyl)-	(55-73-2)
Guanidine, 1,3-diphenyl	(102-06-7)

Guanidine, 1,3-di-o-tolyl	(97-39-2)
Guanidine, dodecyl-, acetate	(2439-10-3)
Guanidine, dodecyl-, monohydrochloride	(13590-97-1)
Guanidine, (2-(hexahydro-1(2H)-azocinyl)ethyl)	(55-65-2)
Guanidine hydrochloride	(50-01-1)
Guanidine, 1,1'-(iminobis(octamethylene))di-, triacetate	(57520-17-9)
Guanidine, N,N'''-(iminodi-8,1-octanediyl)bis-, triacetate	(57520-17-9)
Guanidine, methyl	(471-29-4)
Guanidine, 1-methyl-3-nitro-1-nitroso	(70-25-7)
Guanidine, N-methyl-N'-nitro-N-nitroso- (9CI)	(70-25-7)
Guanidine, monohydrochloride	(50-01-1)
Guanidine, mononitrate	(506-93-4)
Guanidine, mononitrate (8CI,9CI)	(506-93-4)
Guanidine monophosphate	(85-32-5)
Guanidine nitrate	(506-93-4)
Guanidine nitrate (1:1)	(506-93-4)
Guanidine, nitrate (9CI) (VAN)	(52470-25-4)
Guanidine nitrate [UN 1467]	(506-93-4)
Guanidine, 1-nitro	(556-88-7)
Guanidine, nitro-	(556-88-7)
Guanidine, ((3-(5-nitro-2-furyl)-1-(2-(5-nitro-2-furyl)vinyl)-allylidene)amino)	(804-36-4)
Guanidine, nitroso	(674-81-7)
Guanidine, ((4-oxo-2,5-cyclohexadien-1-ylidene)amino)-, thio-semicarbazone	(539-21-9)
Guanidine, sulfanilyl-	(57-67-0)
Guanidine, 1,1,3,3-tetramethyl- (8CI)	(80-70-6)
Guanidine, N,N,N',N'-tetramethyl- (9CI)	(80-70-6)
Guanidinium chloride	(50-01-1)
Guanidinium hydrochloride	(50-01-1)
Guanidinium nitrate	(506-93-4)
Guanine	(73-40-5)
Guanine deoxyriboside	(961-07-9)
Guanine, hydrochloride	(635-39-2)
Guanine, 3-hydroxy-7-methyl	(30345-27-8)
Guanine, 3-hydroxy-9-methyl	(30345-28-9)
Guanine, 9-β-D-ribofuranosyl-	(118-00-3)
Guanine riboside	(118-00-3)
Guanine, thio-	(154-42-7)
Guaniol	(106-24-1)
Guanoctine triacetate	(57520-17-9)
Guanosine	(118-00-3)
Guanosine, 2'-deoxy	(961-07-9)
Guanosine, 2'-deoxy-6-thio- (9CI)	(789-61-7)
Guanosine 5'-monophosphate	(85-32-5)
Guanosine monophosphate	(85-32-5)
Guanosine 5'-monophosphoric acid	(85-32-5)
Guanosine 5'-phosphate	(85-32-5)
Guanothiazon	(539-21-9)
Guanyl hydrazine	(79-17-4)
Guanylhydrazine hydrochloride	(1937-19-5)
5'-Guanylic acid	(85-32-5)
Guanylic acid	(85-32-5)
1-Guanyl-4-nitrosaminoguanyltetrazene	(109-27-3)
Guanyl nitrosamino guanyl tetrazene	(109-27-3)
Guanyl nitrosamino guanyl tetrazene (DOT)	(109-27-3)
Guanyl nitrosaminoguanyltetrazene (Tetrazene), Wetted with not less than 30 per cent water or mixture of alcohol and water, by mass [UN 0114]	(109-27-3)
N^1-Guanylsulfanilamide	(57-67-0)
Guanylurea sulfate	(591-01-5)
Guar	(9000-30-0)
Guar Flour	(9000-30-0)
Guaran	(9000-30-0)
Guaranine	(58-08-2)
Guastil	(15676-16-1)
Guayanesin	(93-14-1)
Guazatine	(13516-27-3)
Guazatine triacetate	(57520-17-9)
Guazatin triacetate	(57520-17-9)
Guesapon	(50-29-3)
Guesarol	(50-29-3)
Guethol	(94-71-3)
Guiacol-Glicerileteré Monocarbammato	(532-03-6)
Guiaphenesin	(93-14-1)
Guicitrina	(69-53-4)
Guicitrine	(69-53-4)
Guignet's Green	(1308-38-9)
Guinea Green	(4680-78-8)
Guinea Green B	(4680-78-8)
Guinea Green BA	(4680-78-8)
Gulitol	(50-70-4)
L-Gulitol	(50-70-4)
Gulliostin	(58-32-2)
Gum	(9016-00-6)
Gum Acacia	(9000-01-5)
Gum Arabic	(9000-01-5)
Gum Benjamin	(9000-05-9)
Gum Benzoin	(9000-05-9)
Gum Benzoin Siam	(9000-05-9)
Gum Camphor	(76-22-2)
Gum Carrageenan	(9000-07-1)
Gum Chon 2	(9000-07-1)
Gum Chond	(9000-07-1)
Gum Chrond	(9000-07-1)
Gum Cyamopsis	(9000-30-0)
Gum Damar	(9000-16-2)
Gum Guaiac	(9000-29-7)
Gum Guar	(9000-30-0)
Gum Ovaline	(9000-01-5)
Gum Senegal	(9000-01-5)
Gum Sterculia	(9000-36-6)
Gum Sumatra	(9000-05-9)
Gum Tragacanth	(9000-65-1)
Gum Turpentine	(9005-90-7)
Gum rosin modified phenol formaldehyde polymer	(67700-45-2)
Guncotton (DOT)	(9004-70-0)
Gurjun-Balsam	(8030-55-5)
Gus 215	(60568-05-0)
Gusathion-20	(86-50-0)
Gusathion	(86-50-0)
Gusathion 25	(86-50-0)
Gusathion A	(2642-71-9)
Gusathion H	(2642-71-9)
Gusathion K	(2642-71-9)
Gusathion K	(86-50-0)
Gusathion M	(86-50-0)
Gusathion methyl	(86-50-0)
Guservin	(126-07-8)
Gustafson Captan 30-DD	(133-06-2)
Guthion (DOT,OSHA)	(86-50-0)
Guthion (Ethyl)	(2642-71-9)
Guthion mixture, Liquid (DOT)	(86-50-0)
Gy-Bon	(1014-70-6)
Gy-Phene	(8001-35-2)
Gycolan Red GRL	(6408-31-7)
Gynaesan	(50-27-1)
Gyne-Lotrimin	(23593-75-1)
Gyneclorina	(127-65-1)
Gynecormone	(50-50-0)
Gynefollin	(84-17-3)
Gynergen	(379-79-3)
Gynergon	(50-28-2)
Gynesine	(535-83-1)

Gynestrel	**(50-28-2)**
Gynformone	**(50-50-0)**
Gynoestryl	**(50-28-2)**
Gynolett	**(130-80-3)**
Gyno-Pevaryl	**(24169-02-6)**
Gyno-Pevaryl 150	**(24169-02-6)**
Gynopharm	**(56-53-1)**
Gynorest	**(152-62-5)**
Gypsine	**(3687-31-8)**
Gypsine	**(7784-40-9)**
Gypsite	**(13397-24-5)**
Gypsum (OSHA)	**(13397-24-5)**
Gyromitrin	**(16568-02-8)**
Gyron	**(50-29-3)**
"H"	**(561-27-3)**
052 H	**(1982-55-4)**
15H	**(9003-29-6)**
2000H	**(9003-29-6)**
2903 H	**(21267-72-1)**
300H	**(9003-29-6)**
454H	**(9003-53-6)**
75H380000	**(9003-11-6)**
75H90000	**(9003-11-6)**
7H	**(9000-11-7)**
H	**(76-44-8)**
H 46	**(21645-51-2)**
H 56/28	**(13655-52-2)**
H 93/26	**(37350-58-6)**
H 95	**(3766-60-7)**
H 95-1	**(3766-60-7)**
H 100	**(9003-29-6)**
H 119	**(1698-60-8)**
H 133	**(1194-65-6)**
H 300	**(9003-29-6)**
H 321	**(2032-65-7)**
H 451	**(7782-42-5)**
H 520	**(50-55-5)**
H 940	**(77-67-8)**
H 1313	**(1194-65-6)**
H 1318	**(1982-49-6)**
H 1500	**(9003-29-6)**
H 1803	**(122-34-9)**
H 1900	**(9003-29-6)**
H 3104	**(2152-34-3)**
H 5727	**(64-00-6)**
H 8717	**(3692-90-8)**
H 8757	**(64-00-6)**
H 9789	**(27314-13-2)**
H 22234	**(38727-55-8)**
H 22949	**(2150-28-9)**
H 26905	**(36335-67-8)**
H 33258	**(23491-45-4)**
H 52143	**(27314-13-2)**
H-34	**(76-44-8)**
H-60	**(76-44-8)**
H-69	**(2655-14-3)**
H-82	**(314-42-1)**
H-365	**(70-70-2)**
H-490	**(77-67-8)**
H-722	**(14491-59-9)**
H-4723	**(22316-47-8)**
H4 099	**(126-27-2)**
cis-H 102/09	**(56775-88-3)**
HA-1200	**(3254-63-5)**
HAG 107	**(66841-25-6)**
4HAQO	**(4637-56-3)**
H Acid	**(90-20-0)**

3 HBA	**(300-85-6)**
m-HBA	**(99-06-9)**
HBB	**(36355-01-8)**
HBBN	**(3817-11-6)**
HBF 386	**(50-76-0)**
HBPA	**(80-04-6)**
7900-HC	**(152-18-1)**
8056HC	**(298-00-0)**
8057HC	**(122-14-5)**
8062 HC	**(3344-14-7)**
8063HC	**(3344-12-5)**
HC	**(50-23-7)**
HC 1281	**(50-31-7)**
HC 1717	**(597-25-1)**
HC 2072	**(311-45-5)**
HC 7900	**(152-18-1)**
HC 7901	**(152-20-5)**
HCA	**(116-16-5)**
HCA Weedkiller	**(116-16-5)**
HCB	**(118-74-1)**
HCB	**(38380-07-3)**
HCBD	**(87-68-3)**
HC Blue 1	**(2784-94-3)**
HC Blue No. 2	**(33229-34-4)**
HCC	**(58-89-9)**
HCCH	**(58-89-9)**
HCCH	**(608-73-1)**
HCCPD	**(77-47-4)**
HCDD	**(39227-28-6)**
HCE	**(1024-57-3)**
HCFC-132b	**(1649-08-7)**
HCG	**(9002-61-3)**
ε-HCH	**(6108-10-7)**
HCH	**(58-89-9)**
HCH	**(608-73-1)**
α-HCH	**(319-84-6)**
β-HCH	**(319-85-7)**
δ-HCH	**(319-86-8)**
γ-HCH	**(58-89-9)**
HCN	**(74-90-8)**
HCP	**(57-15-8)**
HCP	**(70-30-4)**
HC Red No. 3	**(2871-01-4)**
HCS 3260	**(57-74-9)**
HCS 3510	**(51461-71-3)**
HCT	**(23255-93-8)**
HCTZ	**(58-93-5)**
1,2,4,5,7,8-HCX	**(38178-99-3)**
HC Yellow No.4	**(52551-67-4)**
HCZ	**(58-93-5)**
HD	**(505-60-2)**
HD Amaranth B	**(915-67-3)**
HD Amaranth Supra	**(915-67-3)**
HD Carmoisine	**(3567-69-9)**
HD Carmoisine Supra	**(3567-69-9)**
HDEHP	**(298-07-7)**
H.D. Eutanol	**(143-28-2)**
HD Fast Red E	**(2302-96-7)**
HD Fast Red E Supra	**(2302-96-7)**
HD Fast Red GL Base	**(89-62-3)**
HD Indigo Carmine	**(860-22-0)**
HD Indigo Carmine Supra	**(860-22-0)**
HDMTX	**(59-05-2)**
HDO	**(629-11-8)**
HD Oleyl alcohol 70/75	**(143-28-2)**
HD Oleyl alcohol 80/85	**(143-28-2)**
HD Oleyl alcohol 90/95	**(143-28-2)**

HD Oleyl alcohol CG (143-28-2)
HD Ponceau 4R (2611-82-7)
HD Ponceau 4R Supra (2611-82-7)
HD Sunset Yellow FCF (2783-94-0)
HD Sunset Yellow FCF Supra (2783-94-0)
HD Tartrazine (1934-21-0)
HD Tartrazine Supra (1934-21-0)
HE 1975 (9003-29-6)
HE 314 (2303-25-5)
HE 375 (9003-29-6)
HEC (9004-62-0)
HEC-AL 5000 (9004-62-0)
HEDP (2809-21-4)
HEDTA (150-39-0)
HEEDTA (150-39-0)
HEIDA (93-62-9)
HER. 5727 (64-00-6)
HET (757-58-4)
HETP (757-58-4)
HET anhydride (115-27-5)
HEV-4 (11114-92-4)
HF 10 (9003-53-6)
HF 20 (9003-07-0)
HF 55 (9003-53-6)
HF 77 (9003-53-6)
HF 191 (84-61-7)
HF 3170 (1977-10-2)
HFA (144-49-0)
HFA (34202-69-2)
H-35-F 87 (BVM) (122-14-5)
HFDb 4201 (9002-88-4)
HFIP (920-66-1)
H.G. Blending (7647-14-5)
HGI (58-89-9)
HG-203 chlormerodrin (62-37-3)
HH 102 (9003-53-6)
HH 212 (511-45-5)
HHDN (309-00-2)
HHI 11 (9003-53-6)
HHPA (85-42-7)
HIA (54-85-3)
HICO CCC (999-81-5)
HI-Enterol (130-26-7)
HIP (64-00-6)
4-HIPA (636-46-4)
H.K. Formula No. K. 7117 (2650-18-2)
HL 2447 (50-65-7)
HL 8700 (58-33-3)
HL-331 (62-38-4)
HL-Dex (50-02-2)
HLR 4219 (1761-71-3)
HLR 4448 (1761-71-3)
HLS 831 (64-77-7)
4HMB (3597-91-9)
HMD (434-07-1)
HMDA (124-09-4)
HMDI (822-06-0)
HMDS (999-97-3)
HMF (67-47-0)
H-MG (693-07-2)
HMM (645-05-6)
HMP (10124-56-8)
HMPA (306-08-1)
HMPA (680-31-9)
HMPT (680-31-9)
HMT (100-97-0)
HMX (DOT) (2691-41-0)
β HMY (2691-41-0)
HN1 (538-07-8)
HN2 (51-75-2)
HN3 (555-77-1)
HNAB (19159-68-3)
HN2 Amine Oxide (126-85-2)
HNED Kypova 1 (Czech) (2475-33-4)
HNED Prima 95 (Czech) (16071-86-6)
HNED Rozpoustedlova 1 (Czech) (6416-57-5)
HNED Sudan RR (Czech) (6416-57-5)
HN2.HCl (55-86-7)
HN3 HCl (817-09-4)
HN3 Hydrochloride (817-09-4)
HN2 Oxide Hydrochloride (302-70-5)
HN2 Oxide Mustard (126-85-2)
HN2 Hydrochloride (55-86-7)
HO 11513 (132-20-7)
HO 50 (9003-07-0)
HO-2,474 (475-26-3)
HOE 35 609 (66441-11-0)
HOE 00661 (77182-82-2)
HOE 881 (43210-67-9)
HOE 2,671 (115-29-7)
HOE 2747 (1746-81-2)
HOE 2784 OA (485-31-4)
HOE 2784 (485-31-4)
HOE 2810 (330-55-2)
HOE-2824 (900-95-8)
HOE 2872 (639-58-7)
HOE 2873 (13457-18-6)
HOE 2874 (9CI) (39456-75-2)
HOE 2904 (2813-95-8)
HOE 2960 OJ (24017-47-8)
HOE 2960 (24017-47-8)
HOE 2982 (34783-40-9)
HOE 2989 (24691-76-7)
HOE 6052 (24691-76-7)
HOE 6053 (24691-76-7)
HOE 13764 OF (24691-76-7)
HOE 13764 (24691-76-7)
HOE 16410 (34123-59-6)
HOE 17411 (10605-21-7)
HOE 23408 (51338-27-3)
HOE 026014 (82-68-8)
HOE 33058 (23491-45-4)
HOE 33171 (66441-23-4)
HOE 33258 (23491-45-4)
HOE 39866 (77182-82-2)
HOE-A 25-01 (66441-23-4)
HOG (956-90-1)
H.P. 209 (938-73-8)
H.P. 34 (65-45-2)
HPA (999-61-1)
HPC (630-56-8)
mHPHA (3247-75-4)
β-HPN (109-78-4)
HPP (315-30-0)
HPT (680-31-9)
2-(1H)-Pyridinethione, 1-hydroxy-, sodium salt (15922-78-8)
HR 376 (22316-47-8)
HRS-16 (2227-17-0)
HRS 16A (2227-17-0)
HRS 587 (1861-44-5)
HRS 860 (103-17-3)
HRS 942 (405-30-1)
HRS 1276 (2385-85-5)
HRS 1654 (2227-17-0)

HRS 1655 (77-47-4)
HRW 13 (9005-25-8)
HS (10034-93-2)
HS (2517-16-0)
HS (302-17-0)
HS 21 (11114-92-4)
HS 25 (11114-92-4)
HS 55 (8015-55-2)
HS 61 (2163-69-1)
HS 95 (3766-60-7)
HS-119-1 (11694-09-3)
HS-14260 (886-50-0)
7H3SF (9004-32-4)
H Sulfone (471-03-4)
H Sulfoxide (5819-08-9)
5-HT (50-67-9)
HT 88 (9003-53-6)
HT 88A (9003-53-6)
HT 91-1 (9003-53-6)
HT 972 (101-77-9)
5-HTA (50-67-9)
HTAC (112-02-7)
HT-F 76 (9003-53-6)
H-1TG (64036-91-5)
HTH (7778-54-3)
5-HTP (56-69-9)
HTP (757-58-4)
HT 4 (VAN) (63428-84-2)
HU (127-07-1)
HV 490 (63148-62-9)
HV 1900 (9003-29-6)
HV-1900 (9003-29-6)
HVA (306-08-1)
HVA 2 (3006-93-7)
HVA-2 Curing Agent (3006-93-7)
HW 4 (2691-41-0)
HW 920 (330-54-1)
HX (68-94-0)
HX 3/5 (9004-70-0)
HX 752 (7652-64-4)
HX-868 (7722-73-8)
HXR (58-63-9)
HY 951 (112-24-3)
HYVIS 10 (9003-29-6)
HYVIS 7000/45 (9003-29-6)
Hachi-Sugar (139-05-9)
Haelan (1524-88-5)
Haematite (1317-60-8)
Haemofort (7782-63-0)
Haemostasin (51-43-4)
Haemostatin (51-43-4)
Haffkinine (83-89-6)
Hafnium (ACGIH,OSHA) (7440-58-6)
Hafnium metal (DOT) (7440-58-6)
Hafnium metal, Wet (DOT) (7440-58-6)
Haiari (83-79-4)
Haidr (458-37-7)
Haimased (540-72-7)
Haipen (2939-80-2)
Hairoxal (54-47-7)
Hairy (561-27-3)
Haitin (76-87-9)
Hal-Lub-N (7428-48-0)
Halad (458-37-7)
Halamid (127-65-1)
Halane (118-52-5)
Halazepam (23092-17-3)
Halazepamum (Latin) (23092-17-3)
Halazone (80-13-7)
Halbmond (147-24-0)
Halcion (28911-01-5)
Haldar (458-37-7)
Haldol (52-86-8)
Haldrate (53-33-8)
Haldrone-F (1524-88-5)
Half-Cysteine (52-90-4)
Half-Cystine (52-90-4)
Half Mustard Gas (693-30-1)
Half-Mustard Gas (693-07-2)
Half-Myleran (62-50-0)
Half Sulfur Mustard (693-30-1)
Halidol (52-86-8)
Halite (7647-14-5)
Halizan (108-62-3)
Halkan (548-73-2)
Halloo-Wach (60-13-9)
Halltex (9005-38-3)
Halmark (66230-04-4)
Halocarban (369-77-7)
Halocarbano (Spanish) (369-77-7)
Halocarbanum (Latin) (369-77-7)
Halocarbon 11 (75-69-4)
Halocarbon 14 (75-73-0)
Halocarbon 23 (75-46-7)
Halocarbon 112 (76-12-0)
Halocarbon 112a (76-11-9)
Halocarbon 113 (76-13-1)
Halocarbon 13/Ucon 13 (75-72-9)
Halocarbon 114 (76-14-2)
Halocarbon 115 (76-15-3)
Halocarbon 152A (75-37-6)
Halocarbon 1132A (75-38-7)
Halocarbon C-138 (115-25-3)
Haloflex 202 (9003-01-4)
Haloflex 208 (9003-01-4)
Halol (52-86-8)
Halomycetin (56-75-7)
Halon (75-71-8)
Halon 14 (75-73-0)
Halon 1001 (74-83-9)
Halon 1011 (74-97-5)
Halon 1202 (75-61-6)
Halon 1211 (353-59-3)
Halon 1301 (75-63-8)
Halon 2001 (74-96-4)
Halon 2312 (29256-79-9)
Halon 2402 (124-73-2)
Halon 10001 (74-88-4)
Halopal (52-86-8)
Haloperidol (52-86-8)
Halopidol (52-86-8)
Halopoidol (52-86-8)
Halopont Blue BGM (1325-87-7)
Halopont Violet NM (1325-82-2)
Haloprogin (777-11-7)
Halosten (52-86-8)
Halotan (151-67-7)
Halotestin (76-43-7)
Halotex (777-11-7)
Halothane (ACGIH) (151-67-7)
Halowax (1321-65-9)
Halowax (1335-88-2)
Halowax 1000 (9CI) (58718-66-4)
Halowax 1001 (58718-67-5)

Halowax 1013 (9CI)	(12616-35-2)
Halowax 1014	(12616-36-3)
Halowax 1014	(1335-87-1)
Halowax 1031	(25586-43-0)
Halowax 1099 (9CI)	(39450-05-0)
Haloxazolam	(59128-97-1)
Haloxyfop-Methyl	(69806-40-2)
Haloxyfop-(2-ethoxyethyl)	(87237-48-7)
Haloxyfop	(69806-34-4)
Halquinol	(8067-69-4)
Halquinols	(8067-69-4)
Hals 3	(71878-19-8)
Halsan	(151-67-7)
Halso 99	(95-49-8)
Haltox	(74-83-9)
Haltron 22	(75-45-6)
Halud	(458-37-7)
Hamamelis	(68916-39-2)
Hamidop	(10265-92-6)
Hamilton Red	(5160-02-1)
Hamp-EX 80	(140-01-2)
Hamp-Ene acid	(60-00-4)
Hamp-Ex Acid	(67-43-6)
Hamp-ene 100	(64-02-8)
Hamp-ene 215	(64-02-8)
Hamp-ene 220	(64-02-8)
Hamp-ene Na4	(64-02-8)
Hamp-ol 120	(139-89-9)
Hamp-ol Crystals	(139-89-9)
Hamp-ol acid	(150-39-0)
Hamposyl L-30	(137-16-6)
Hamposyl O	(110-25-8)
Hampshire	(142-73-4)
Hampshire NTA	(5064-31-3)
Hampshire NTA acid	(139-13-9)
Hampshire glycine	(56-40-6)
Hana	(83-72-7)
Hanane	(115-26-4)
Hancock Yellow 1008	(2512-29-0)
Hancock Yellow 10010	(6358-85-6)
Hansa Brilliant Yellow 5GX	(6358-31-2)
Hansa Orange RN	(3468-63-1)
Hansa Red B	(2425-85-6)
Hansa Red G	(2425-85-6)
Hansa Scarlet RB	(2425-85-6)
Hansa Scarlet RN	(2425-85-6)
Hansa Scarlet RNC	(2425-85-6)
Hansa Yellow	(2512-29-0)
Hansa Yellow G	(2512-29-0)
Hansa Yellow G 45-4050	(2512-29-0)
Hansa Yellow GAD	(2512-29-0)
Hansa Yellow G Extra	(2512-29-0)
Hansa Yellow GT	(2512-29-0)
Hansa Yellow G Toner	(2512-29-0)
Hansa Yellow S 3155	(2512-29-0)
Hansa Yellow Toner G-YA 8365	(2512-29-0)
Hansacor	(59-26-7)
Hansamid	(98-92-0)
Haplopan	(50-06-6)
Haplos	(50-06-6)
Happy	(134-31-6)
Happy dust	(50-36-2)
Haptocil	(144-83-2)
Hardened Oil	(555-43-1)
Hardlen 15L	(9006-03-5)
Harflex 321	(27941-08-8)
Harflex 330	(27941-09-9)
Harman	(486-84-0)
Harmane	(486-84-0)
Harmone B 79	(130-20-1)
Harmonin	(57-53-4)
Harness	(1689-84-5)
Harowax L 9	(25496-72-4)
Harrical	(87-33-2)
Harry	(561-27-3)
Hartol	(57-53-4)
Hartosol	(67-63-0)
Harvade	(55290-64-7)
Harvamine	(91-84-9)
Harven	(4418-26-2)
Harvest-Aid	(7775-09-9)
Harwax A	(106-14-9)
Hasach	(8063-14-7)
Hasethrol	(78-11-5)
Hashish	(8063-14-7)
Hastelloy C	(11114-92-4)
Hastings Carmine 2G	(3734-67-6)
Hatcol DOP	(117-81-7)
Havero-extra	(50-29-3)
Havidote	(60-00-4)
Havoc	(56073-10-0)
Haydite	(68334-37-2)
Haynes 25	(11114-92-4)
Haynes Alloy Number 25	(11114-92-4)
Haynes Stellite 21	(11114-92-4)
Haynon	(132-22-9)
Hazodrin	(6923-22-4)
Hb(T)	(9008-02-0)
Healon	(9004-61-9)
Hearts	(51-63-8)
Hearts	(60-13-9)
Heat Pre	(99-80-9)
Heavenly blue	(50-37-3)
Heavy Oil	(8001-58-9)
Heavy Vacuum Distillate (Heavy Vacuum Gas Oil)	(64741-57-7)
Heavy Water-D2	(7789-20-0)
Heavy Water	(7789-20-0)
Heavy aromatic naphtha	(64742-94-5)
Heavy catalytically cracked distillate	(64741-61-3)
Heavy catalytically cracked naphtha	(64741-54-4)
Heavy naphthenic distillate	(64741-53-3)
Heavy naphthenic distillates (Petroleum)	(64741-53-3)
Heavy naphthenic distillate solvent extract	(64742-11-6)
Heavy normal paraffins (Petroleum)	(64771-72-8)
Heavy normal paraffins concentrate (Petroleum)	(64771-72-8)
Heavy paraffinic distillate	(64741-51-1)
Heavy paraffinic distillate, Solvent extract	(64742-04-7)
Heavy reformed naphtha	(64741-68-0)
Heavy straight-run naphtha	(64741-41-9)
Heavy straight run naphtha (Petroleum)	(64741-41-9)
Heavy thermally cracked naphtha	(64741-83-9)
Heazlewoodite	(12035-72-2)
Heb-Cort	(50-23-7)
Hebanil	(69-09-0)
Hebin	(9002-68-0)
Heclotox	(58-89-9)
Hecto Violet R	(548-62-9)
Hectograph Violet SR	(548-62-9)
Hedapur M 52	(94-74-6)
Hedarex M	(94-74-6)
Hedex	(103-90-2)
Hedolit	(534-52-1)
Hedolite	(534-52-1)
Hedonal	(120-36-5)

Hedonal (94-75-7)
Hedonal DP (120-36-5)
Hedonal M (94-74-6)
Hedonal MCPP (1929-86-8)
Hedonal MCPP (93-65-2)
Hedonal (The herbicide) (94-75-7)
Hedulin (83-12-5)
Hefti CE-55-2 (9004-95-9)
Hefti CE-55-20 (9004-95-9)
Hefti DMO-33 (106-12-7)
Hefti GMO-33 (111-03-5)
Hefti GMS-33 (123-94-4)
Hefti GMS-66 (123-94-4)
Hefti GMS-99 (123-94-4)
Hefti LA-55-4 (5274-68-0)
Hefti MYM-33 (3234-85-3)
Hefti NP-55-40 (7311-27-5)
Hefti OL-55-F-2 (9004-98-2)
Hefti OL-55-F-10 (9004-98-2)
Hefti OL-55-F-20 (9004-98-2)
Hefti PGE-400-DO (9005-07-6)
Hefti PGE-400-DS (9005-08-7)
Hefti PGE-600-DS (9005-08-7)
Hefti PGE-400-MS (9004-99-3)
Hefti PGE-1000-MS (9004-99-3)
Hefti RS-55-9 (9004-99-3)
Hefti RS-55-40 (9004-99-3)
Hefti RS-55-100 (9004-99-3)
Hefti ST-55-2 (9005-00-9)
Hefti ST-55-20 (9005-00-9)
Hefti Sorbex-R (50-70-4)
Hefti Sorbex-RP (50-70-4)
Hefti TS-33-F (26658-19-5)
Heksan (Polish) (110-54-3)
Heksogen (Polish) (121-82-4)
Hektalin (51-43-4)
Hel-Fire (88-85-7)
Helaktyn Black DN (12225-26-2)
Helanca (63428-84-2)
Helanthrene Blue BC (130-20-1)
Helanthrene Brilliant Pink R (2379-74-0)
Helanthrene Brown GR (2475-33-4)
Helanthrene Green B (128-58-5)
Helanthrene Pink R (2379-74-0)
Helanthrene Yellow (128-66-5)
Helenalin (6754-13-8)
Helfo Dopa (59-92-7)
Helfoserpin (50-55-5)
Heliane Orange RF (3263-31-8)
Helianthine (547-58-0)
Helianthine B (547-58-0)
Helianthrene Blue RS (81-77-6)
Helic Yellow GW (6358-85-6)
Helicon (50-78-2)
Helindon Orange R (3263-31-8)
Helindon Pink R (2379-74-0)
Helindone Pink CN (2379-74-0)
Helio Blue B (147-14-8)
Helio Fast Blue BRN (147-14-8)
Helio Fast Blue GO (147-14-8)
Helio Fast Blue HG (147-14-8)
Helio Fast Green GN (14302-13-7)
Helio Fast Green GT (14302-13-7)
Helio Fast Orange 3RN (3468-63-1)
Helio Fast Orange RN (3468-63-1)
Helio Fast Orange 3RT (3468-63-1)
Helio Fast Orange RT (3468-63-1)
Helio Fast Red BN (2425-85-6)
Helio Fast Red RL (2425-85-6)
Helio Fast Red RN (2425-85-6)
Helio Fast Yellow GN (2512-29-0)
Helio Fast Yellow GNS (2512-29-0)
Helio Fast Yellow GT (2512-29-0)
Helio Orange CAG (1934-20-9)
Helio Red RL (2425-85-6)
Helio Red Toner (2425-85-6)
Helio Red Toner LCLL (5160-02-1)
Helio Yellow GWN (6358-85-6)
Heliogen (127-65-1)
Heliogen Blue 6470 (81-77-6)
Heliogen Blue 6840 (147-14-8)
Heliogen Blue 6960 (147-14-8)
Heliogen Blue 7080 (147-14-8)
Heliogen Blue 7100 (147-14-8)
Heliogen Blue 7560 (574-93-6)
Heliogen Blue 7800 (574-93-6)
Heliogen Blue A (147-14-8)
Heliogen Blue B (147-14-8)
Heliogen Blue BG (147-14-8)
Heliogen Blue 6902K (147-14-8)
Heliogen Blue K (147-14-8)
Heliogen Blue LBG (147-14-8)
Heliogen Blue LBGN (147-14-8)
Heliogen Blue SBL (1330-38-7)
Heliogen Blue 7044T (147-14-8)
Heliogen Blue WX (147-14-8)
Heliogen Blue g (574-93-6)
Heliogen Green 8680 (1328-53-6)
Heliogen Green 8730 (1328-53-6)
Heliogen Green 9360 (14302-13-7)
Heliogen Green A (1328-53-6)
Heliogen Green 6G (14302-13-7)
Heliogen Green G (1328-53-6)
Heliogen Green 6GA (14302-13-7)
Heliogen Green 8GA (14302-13-7)
Heliogen Green GA (1328-53-6)
Heliogen Green GN (1328-53-6)
Heliogen Green GNA (1328-53-6)
Heliogen Green GTA (1328-53-6)
Heliogen Green GV (1328-53-6)
Heliogen Green GWS (1328-53-6)
Heliogen Green 8681K (1328-53-6)
Heliogen Green 8682T (1328-53-6)
Helion Brown BRSL (16071-86-6)
Helion Red 8B (2610-11-9)
Helion Yellow 5G (10190-68-8)
Helional (50-06-6)
Heliopan (118-56-9)
Heliopar (54-05-7)
Heliophan (118-56-9)
Heliostable Blue B (1325-87-7)
Heliostable Brilliant Blue B Extra (1325-87-7)
Heliostable Brilliant Pink B Extra (989-38-8)
Heliothis pheromone (53939-28-9)
Heliothis virescens (53939-28-9)
Heliotridine (520-63-8)
Heliotridine, 3,8-didehydro- (26400-24-8)
Heliotridine ester with lasiocarpum and angelic acid (303-34-4)
Heliotrine (303-33-3)
Heliotron (303-33-3)
Heliotropic acid (94-53-1)
Heliotropin (120-57-0)
Heliotropin acetal (5281-13-0)
Heliotropine (120-57-0)

Heliotropyl acetate (326-61-4)
Heliotropyl acetone (3160-37-0)
Heliox (58933-55-4)
Helioyellow GW (6358-85-6)
Helium, Compressed [UN 1046] (7440-59-7)
Helium (DOT) (7440-59-7)
Helium, Isotope of mass 3 (8CI,9CI) (14762-55-1)
Helium, Refrigerated liquid [UN 1963] (7440-59-7)
Helium-oxygen (Mixture) (58933-55-4)
Hellipidyl (65-49-6)
Helmatac (14255-87-9)
Helmerco Blue M 4G (1325-87-7)
Helmerco Violet MR (1325-82-2)
Helmetina (92-84-2)
Helmox (140-87-4)
Helothion (35400-43-2)
Heloxy WC68 (17557-23-2)
Helvagit (62-44-2)
Hemo-B-Doze (68-19-9)
Hematite (1317-60-8)
Hematoidin (635-65-4)
Hematoxylin (517-28-2)
Hemel (645-05-6)
Hemellitic acid (603-79-2)
Hemetoidin (635-65-4)
Hemimellitene (526-73-8)
3,4,5-Hemimellitenol (527-54-8)
Hemimellitic acid (569-51-7)
Hemineurine (137-00-8)
Heminevrin (533-45-9)
Hemisine (51-43-4)
Hemisulfur mustard (542-81-4)
Hemlock Oil (8021-28-1)
Hemocaprol (60-32-2)
Hemodal (58-27-5)
Hemodesis (9003-39-8)
Hemodez (9003-39-8)
Hemofuran (59-87-0)
Hemoglobins (9008-02-0)
Hemomin (68-19-9)
Hemopar (60-32-2)
Hemostasin (51-43-4)
Hemostatin (51-43-4)
Hemostyp (71-36-3)
Hemostyptanon (50-27-1)
Hempa (680-31-9)
Hendecanal (112-44-7)
Hendecanaldehyde (112-44-7)
Hendecane (1120-21-4)
Hendecanoic acid (112-37-8)
Hendecanoic alcohol (112-42-5)
1-Hendecanol (112-42-5)
2-Hendecanone (112-12-9)
Hendecenal (112-44-7)
Hendecenal (112-45-8)
1-Hendecene (821-95-4)
10-Hendecenoic (112-38-9)
Hendecyl alcohol (112-42-5)
n-Hendecylenic alcohol (112-42-5)
10-Henedecenoic acid (112-38-9)
Heneicosane (629-94-7)
Heneicosane, 1-cyclopentyl- (8CI) (6703-82-8)
Heneicosane, 11-cyclopentyl- (8CI) (6703-81-7)
Heneicosanoic acid (8CI,9CI) (2363-71-5)
Heneicosanoic acid, methyl ester (8CI,9CI) (6064-90-0)
11-Heneicosanone (8CI,9CI) (19781-72-7)
Henna (83-72-7)
Hennoletten (50-06-6)
1-Hentetracontanol (9CI) (40710-42-7)
Hentriacontane (630-04-6)
Hentriacontanoic acid (9CI) (38232-01-8)
Hentriacontanoic acid, methyl ester (9CI) (77630-51-4)
Heod (60-57-1)
Hepacholine (67-48-1)
Hepagon (68-19-9)
Heparin (9005-49-6)
α-Heparin (9005-49-6)
Heparinate (9005-49-6)
Heparinic acid (9005-49-6)
Heparin sulfate (9005-49-6)
Heparlipon (62-46-4)
Hepartest (71-67-0)
Hepartestabrome (71-67-0)
Hepatosulfalein (71-67-0)
Hepavis (68-19-9)
Hepcovite (68-19-9)
Hepin (60-32-2)
Hept (107-49-3)
Hepta (76-44-8)
Heptabarb (509-86-4)
Heptabarbital (509-86-4)
Heptabarbitone (509-86-4)
Heptabarbum (509-86-4)
2,2',3,4,4',5,5'-Heptabromobiphenyl (67733-52-2)
Heptabromodibenzo(b,e)(1,4)dioxin (103456-43-5)
Heptabromodibenzo-p-dioxin (103456-43-5)
Heptabromodibenzofuran (62994-32-5)
Heptabromodiphenyl ether (68928-80-3)
Heptabromodiphenyl oxide (68928-80-3)
Heptabromophenoxybenzene (68928-80-3)
Heptachloor (Dutch) (76-44-8)
1,4,5,6,7,8,8-Heptachloor-3a,4,7,7a-tetrahydro-4,7-endo-methano-indeen (Dutch) (76-44-8)
Heptachlor (ACGIH,OSHA) (76-44-8)
Heptachlorane (76-44-8)
Heptachlore (French) (76-44-8)
Heptachlor epoxide (1024-57-3)
Heptachlorobenzo-p-dioxin (37871-00-4)
2,3,4,5,6,7,7-Heptachloro-1a,1b,5,5a,6,6a-hexahydro-2,5-methano-2H-indeno(1,2-b)oxirene (1024-57-3)
Heptachlorobicyclo(2.2.1)hept-2-ene (28680-45-7)
2,2',3,3',4,4',5-Heptachlorobiphenyl (35065-30-6)
2,2',3,3',4,4',6-Heptachlorobiphenyl (52663-71-5)
2,2',3,3',4,5',6'-Heptachlorobiphenyl (52663-70-4)
2,2',3,3',4,5',6-Heptachlorobiphenyl (40186-70-7)
2,2',3,3',4,5,5'-Heptachlorobiphenyl (52663-74-8)
2,2',3,3',4,5,6'-Heptachlorobiphenyl (38411-25-5)
2,2',3,3',4,5,6-Heptachlorobiphenyl (68194-16-1)
2,2',3,3',5,5',6-Heptachlorobiphenyl (52663-67-9)
2,2',3,3',5,6,6'-Heptachlorobiphenyl (52663-64-6)
2,2',3,4',5,5',6-Heptachlorobiphenyl (52663-68-0)
2,2',3,4,4',5',6-Heptachlorobiphenyl (52663-69-1)
2,2',3,4,4',5,5'-Heptachlorobiphenyl (35065-29-3)
2,2',3,4,4',5,6'-Heptachlorobiphenyl (60145-23-5)
2,2',3,4,4',5,6-Heptachlorobiphenyl (74472-47-2)
2,2',3,4,4',6,6'-Heptachlorobiphenyl (74472-48-3)
2,2',3,4,5,5',6-Heptachlorobiphenyl (52712-05-7)
2,3,3',4',5,5',6-Heptachlorobiphenyl (69782-91-8)
2,3,3',4,4',5',6-Heptachlorobiphenyl (74472-50-7)
2,3,3',4,4',5,6-Heptachlorobiphenyl (41411-64-7)
2,3,4,5,3',4',5'-Heptachlorobiphenyl (39635-31-9)
Heptachlorobiphenyl (28655-71-2)
2,2,5-endo,6-exo,8,9,10-Heptachlorobornane (51775-36-1)
1,2,3,4,6,7,8-Heptachlorodibenzo-para-dioxin (35822-46-9)

1,2,3,4,6,7,8-Heptachlorodibenzodioxin (35822-46-9)
1,2,3,4,6,7,9-Heptachlorodibenzo-para-dioxin (58200-70-7)
1,2,3,4,6,7,9-Heptachlorodibenzodioxin (58200-70-7)
Heptachlorodibenzo-p-dioxin (35822-46-9)
1,2,3,4,6,7,8-Heptachlorodibenzofuran (67562-39-4)
1,2,3,4,6,7,9-Heptachlorodibenzofuran (70648-25-8)
1,2,3,4,6,8,9-Heptachlorodibenzofuran (69698-58-4)
3,4,5,6,7,8,8-Heptachlorodicyclopentadiene (76-44-8)
3,4,5,6,7,8,8a-Heptachlorodicyclopentadiene (76-44-8)
Heptachloro diphenyl ether (55684-92-9)
1,4,5,6,7,8,8-Heptachloro-2,3-epoxy-2,3,3a,4,7,7a-hexahydro-4,7-methanoindene (1024-57-3)
1,4,5,6,7,8,8-Heptachloro-2,3-epoxy-3a,4,7,7a-tetrahydro-4,7-methanoindan (1024-57-3)
Heptachloronaphthalene (32241-08-0)
Heptachloronorbornene (28680-45-7)
Heptachloro-2-picoline (1134-04-9)
1,4,5,6,7,8,8-Heptachloro-3a,4,7,7a-tetrahydro-4,7-endomethano-indene (76-44-8)
1,4,5,6,7,10,10-Heptachloro-4,7,8,9-tetrahydro-4,7-endomethylene-indene (76-44-8)
1,4,5,6,7,8,8a-Heptachloro-3a,4,7,7a-tetrahydro-4,7-methanoindane (76-44-8)
1(3a),4,5,6,7,8,8-Heptachloro-3a(1),4,7,7a-tetrahydro-4,7-methano-indene (76-44-8)
1,4,5,6,7,8,8-Heptachloro-3a,4,7,7a-tetrahydro-4,7-methanoindene (76-44-8)
1,4,5,6,7,8,8-Heptachloro-3a,4,7,7a-tetrahydro-4,7-methanol-1H-indene (76-44-8)
1,4,5,6,7,10,10-Heptachloro-4,7,8,9-tetrahydro-4,7-methyleneindene (76-44-8)
1,4,5,6,7,8,8-Heptachloro-3a,4,7,7,7a-tetrahydro-4,7-methylene indene (76-44-8)
1,4,5,6,7,8,8-Heptachlor-3a,4,7,7,7a-tetrahydro-4,7-endo-methano-inden (German) (76-44-8)
Heptacosafluorotributylamine (311-89-7)
Heptacosane (593-49-7)
Heptacosane, 3-methyl- (8CI,9CI) (14167-66-9)
Heptacosanoic acid (8CI,9CI) (7138-40-1)
Heptacosanoic acid, methyl ester (9CI) (55682-91-2)
14-Heptacosanone (8CI,9CI) (542-50-7)
1,1,2,2,3,3,4,4,5,5,6,6,7,7,8,8,8-Heptadecafluoro-N-(2-hydroxy-ethyl)-N-methyl-1-octanesulfonamide (24448-09-7)
Heptadecafluoro-N-methyl-1-octanesulfonamide (31506-32-8)
1,1,2,2,3,3,4,4,5,5,6,6,7,7,8,8,8-Heptadecafluoro-1-octane-sulfonic acid, potassium salt (2795-39-3)
1,1,2,2,3,3,4,4,5,5,6,6,7,7,8,8,8-Heptadecafluoro-1-octane-sulfonyl fluoride (307-35-7)
3-(((Heptadecafluorooctyl)sulfonyl)amino)-N,N,N-trimethyl-1-propanaminium chloride (38006-74-5)
3-(((Heptadecafluorooctyl)sulfonyl)amino)-N,N,N-trimethyl-1-propanaminium iodide (1652-63-7)
1-Heptadecanaminium, 1-carboxy-N,N,N-trimethyl-, hydroxide, inner salt (9CI) (96-56-0)
1-Heptadecanaminium, N-heptadecyl-N,N-dimethyl-, chloride (9CI) (1118-41-8)
Heptadecane (629-78-7)
n-Heptadecane (629-78-7)
Heptadecane, 9-aza-1,17-diguanidino-, triacetate (57520-17-9)
Heptadecane, 1-bromo- (9CI) (3508-00-7)
1-Heptadecanecarboxylic acid (57-11-4)
Heptadecane, 1-chloro- (9CI) (62016-75-5)
Heptadecane, 1-cyclohexyl- (8CI) (19781-73-8)
Heptadecanedinitrile (5399-02-0)
Heptadecane, 2-methyl- (9CI) (1560-89-0)
Heptadecane, 7-methyl- (8CI,9CI) (20959-33-5)
Heptadecanenitrile (9CI) (5399-02-0)
1,2-Heptadecane oxide (22092-38-2)
Heptadecane, 2,6,10,15-tetramethyl- (9CI) (54833-48-6)
Heptadecane, 2,6,10,14-tetramethyl- (8CI,9CI) (18344-37-1)
Heptadecanoic-acid (506-12-7)
Heptadecanoic acid, ethyl ester (8CI,9CI) (14010-23-2)
Heptadecanoic acid, 2-methyl- (9CI) (5638-12-0)
Heptadecanoic acid, methyl ester (9CI) (1731-92-6)
Heptadecanoic acid, 15-methyl-, ethyl ester (9CI) (57274-46-1)
1-Heptadecanol (9CI) (1454-85-9)
7-Heptadecanol, 7-methyl- (9CI) (55723-93-8)
2-Heptadecanone (8CI,9CI) (2922-51-2)
8,10,12-Heptadecatriene-4,6-diyne-1,14-diol, (E,E,E)-(-) (505-75-9)
5,8,11-Heptadecatrien-1-ol (8CI,9CI) (22117-09-5)
1-Heptadecene (9CI) (6765-39-5)
Heptadecene (8CI,9CI) (26266-05-7)
2-Heptadecenyl-2-imidazoline (30968-43-5)
2-(8-Heptadecenyl)-2-imidazoline-1-ethanol (95-38-5)
Heptadecenyl imidazolinium chloride (82078-98-6)
n-Heptadecoic acid (506-12-7)
Heptadecylbenzene (14752-75-1)
Heptadecylbenzenesulfonic acid (39735-13-2)
Heptadecyl bromide (3508-00-7)
Heptadecyl cyanide (638-65-3)
2-Heptadecyl-4,5-dihydro-1H-imidazolyl monoacetate (556-22-9)
2-Heptadecyl glyoxalidine acetate (556-22-9)
2-Heptadecyl-1-hydroxyethylimidazoline (95-19-2)
Heptadecyl hydroxyethyl imidazoline (53466-91-4)
Heptadecyl hydroxyethylimidazolinium chloride (53466-92-5)
n-Heptadecylic acid (506-12-7)
2-Heptadecyl-2-imidazoline acetate (556-22-9)
Heptadecyl ketone (504-53-0)
2-Heptadecyl-1-methyl-1-(2-(stearoylamido)ethyl)-2-imidazolinium methyl sulfate (13470-50-3)
Heptadecyloxirane (67860-04-2)
(E,E)-2,4-Heptadien-1-al (4313-03-5)
(E,E)-2,4-Heptadienal (4313-03-5)
trans,trans-2,4-Heptadienal (4313-03-5)
trans-2-trans-4-Heptadienal (4313-03-5)
2,4-Heptadienal, (E,E)- (9CI) (4313-03-5)
1,6-Heptadiene (9CI) (3070-53-9)
1,5-Heptadiene-3,4-diol (9CI) (51945-98-3)
1,6-Heptadiene-3,5-dione, 1,7-bis(4-hydroxy-3-methoxyphenyl) (458-37-7)
1,4-Heptadiene, 3-methyl- (8CI,9CI) (1603-01-6)
2,5-Heptadien-4-one, 2,6-dimethyl (504-20-1)
3,5-Heptadien-2-one, 6-methyl-, (E)- (8CI,9CI) (16647-04-4)
1,6-Heptadien-3-one, 1-(2,6,6-trimethyl-2-cyclohexen-1-yl) (79-78-7)
1,6-Heptadiyne (8CI,(CI) (2396-63-6)
Heptadone (76-99-3)
Heptadorm (509-86-4)
Heptafluorjodpropan (Czech) (27636-85-7)
1,1-H,H-Heptafluorobutanol (375-01-9)
2,2,3,3,4,4,4-Heptafluorobutanol (375-01-9)
Heptafluorobutanoyl fluoride (335-42-2)
Heptafluorobutyric acid (375-22-4)
1,1,1,2,2,3,3-Heptafluoro-7,7-dimethyl-4,6-octanedione (17587-22-3)
6,6,7,7,8,8,8-Heptafluoro-2,2-dimethyl-3,5-octanedione (17587-22-3)
2,2,3,3,4,4,4-Heptafluoro-2'-hydroxy-4'-nitrobutyranilide (2712-83-6)
Heptafluoroiodopropane (27636-85-7)
Heptagran (76-44-8)
γ-Heptalactone (105-21-5)
Heptaldehyde (111-71-7)
Heptamal (509-86-4)
Heptamethylenimine, 1-(2-guanidinoethyl)- (55-65-2)
2,2,4,4,6,8,8-Heptamethylnonane (4390-04-9)
Heptamethylphenylcyclotetrasiloxane (10448-09-6)
1,1,1,3,5,5,5-Heptamethyltrisiloxane (1873-88-7)
Heptamine (123-82-0)
Heptamul (76-44-8)
Heptan (Polish) (142-82-5)
Heptanal (111-71-7)

Heptanal, 2-benzylidene- **(122-40-7)**
Heptanal, 2-(phenylmethylene)- (9CI) **(122-40-7)**
Heptanamide, N,N-dimethyl- (8CI,9CI) **(1115-96-4)**
1-Heptanamine **(111-68-2)**
2-Heptanamine **(123-82-0)**
1-Heptanamine, N-ethyl- (9CI) **(66793-76-8)**
Heptandioic acid **(111-16-0)**
Heptane (ACGIH,OSHA) [UN 1206] **(142-82-5)**
n-Heptane (OSHA) **(142-82-5)**
Heptane, 1-bromo **(629-04-9)**
Heptane, 2-bromo- (9CI) **(1974-04-5)**
Heptane, 3-bromo- (9CI) **(1974-05-6)**
1-Heptanecarboxylic acid **(124-07-2)**
Heptane, 1-chloro- (9CI) **(629-06-1)**
Heptane, chloro- (8CI,9CI) **(29756-37-4)**
Heptane, 3-chloromethyl **(123-04-6)**
Heptane, 1-cyclohexyl- (8CI) **(5617-41-4)**
Heptane, 1,1-dibromo- (9CI) **(59104-79-9)**
1,7-Heptanedicarboxylic acid **(123-99-9)**
Heptanedicarboxylic acid **(123-99-9)**
Heptane, 1,1-dichloro- (6CI,7CI,8CI,9CI) **(821-25-0)**
Heptane, 1,1-dicyclohexyl- (6CI,7CI,8CI) **(2090-15-5)**
Heptane, 1,1-difluoro- (6CI,7CI,8CI,9CI) **(407-96-5)**
Heptane, 2,2-dimethyl- (9CI) **(1071-26-7)**
Heptane, 3,4-dimethyl- (9CI) **(922-28-1)**
Heptane, 3,5-dimethyl- (9CI) **(926-82-9)**
Heptane, dimethyl- (9CI) **(30498-66-9)**
Heptane, 2,4-dimethyl- (8CI,9CI) **(2213-23-2)**
Heptane, 2,5-dimethyl- (8CI,9CI) **(2216-30-0)**
Heptane, 2,6-dimethyl- (8CI,9CI) **(1072-05-5)**
Heptanedinitrile **(646-20-8)**
1,7-Heptanedioic acid **(111-16-0)**
Heptane-1,7-dioic acid **(111-16-0)**
Heptanedioic acid **(111-16-0)**
Heptanedioic acid, diethyl ester (9CI) **(2050-20-6)**
1,7-Heptanediol (9CI) **(629-30-1)**
2,5-Heptanedione **(1703-51-1)**
2,4-Heptanedione (8CI,9CI) **(7307-02-0)**
Heptane, 3,4-epoxy- **(53897-32-8)**
Heptane, 2,3-epoxy- (7CI,8CI) **(14925-96-3)**
Heptane, 3-((ethenyloxy)methyl)- (9CI) **(103-44-6)**
Heptane, ethyl- (9CI) **(73507-01-4)**
Heptane, 4-ethyl- (8CI,9CI) **(2216-32-2)**
Heptane, 5-ethyl-2-methyl- (8CI,9CI) **(13475-78-0)**
Heptane, 4-ethyl-2,2,6,6-tetramethyl- (9CI) **(62108-31-0)**
Heptane, 1-fluoro **(661-11-0)**
Heptane, 2,2,3,3,5,6,6-heptamethyl- (8CI,9CI) **(7225-67-4)**
Heptane, hexadecafluoro **(335-57-9)**
Heptane, 1-iodo- (9CI) **(4282-40-0)**
Heptane, 2-methyl- (9CI) **(592-27-8)**
Heptane, 3-methyl- (9CI) **(589-81-1)**
Heptane, 4-methyl- (9CI) **(589-53-7)**
Heptane, 3-methylene- **(1632-16-2)**
Heptanen (Dutch) **(142-82-5)**
Heptanenitrile (9CI) **(629-08-3)**
Heptane, 1,1,1,2,2,3,3,4,4,5,5,6,6,7,7-pentadecafluoro- (6CI,7CI,8CI,9CI) **(375-83-7)**
Heptane, 2,2,4,6,6-pentamethyl- (9CI) **(13475-82-6)**
Heptane, 1-phenyl- (8CI) **(1078-71-3)**
Heptane, 4-phenyl- (6CI,7CI,8CI) **(2132-86-7)**
Heptane, 1-(2-propenyloxy)- (9CI) **(16519-24-7)**
1-Heptanesulfonyl fluoride, 1,1,2,2,3,3,4,4,5,5,6,6,7,7,7-pentadecafluoro- (9CI) **(335-71-7)**
1-Heptanethiol **(1639-09-4)**
Heptane, 3,3,5-trimethyl- (9CI) **(7154-80-5)**
Heptane, 2,2,4-trimethyl- (8CI,9CI) **(14720-74-2)**
Heptane, 2,3,6-trimethyl- (8CI,9CI) **(4032-93-3)**
Heptane, 2,5,5-trimethyl- (8CI,9CI) **(1189-99-7)**
Heptanoic-acid **(111-14-8)**
Heptanoic acid, 2-bromo- (8CI,9CI) **(2624-01-3)**
Heptanoic acid, butyl ester (9CI) **(5454-28-4)**
Heptanoic acid, copper(2+) salt (9CI) **(5128-10-9)**
Heptanoic acid, cupric salt **(5128-10-9)**
Heptanoic acid, 2,6-dimethyl-, (R)- (9CI) **(60148-94-9)**
Heptanoic acid, ester with testosterone **(315-37-7)**
Heptanoic acid, 2-ethyl- (9CI) **(3274-29-1)**
Heptanoic acid, ethyl ester (9CI) **(106-30-9)**
Heptanoic acid, 2-ethyl-2-(((1-oxoheptyl)oxy)methyl)-1,3-propanediyl ester (9CI) **(78-16-0)**
Heptanoic acid, 3-hexenyl ester, (Z)- (9CI) **(61444-39-1)**
Heptanoic acid, 2-methyl- (9CI) **(1188-02-9)**
Heptanoic acid, methyl ester **(106-73-0)**
Heptanoic acid, oxybis(2,1-ethanediyloxy-2,1-ethanediyl) ester **(70729-68-9)**
Heptanoic acid, potassium salt (9CI) **(16761-12-9)**
Heptanoic acid, 2-propyl **(31080-39-4)**
1-Heptanol **(111-70-6)**
2-Heptanol **(543-49-7)**
3-Heptanol **(589-82-2)**
Heptanol-2 **(543-49-7)**
n-Heptanol **(111-70-6)**
4-Heptanol (9CI) **(589-55-9)**
n-Heptanol-1 (French) **(111-70-6)**
Heptanol (VAN)(9CI) **(53535-33-4)**
γ-Heptanolactone **(105-21-5)**
4-Heptanol, 2-6-dimethyl **(108-82-7)**
2-Heptanol, 2,6-dimethyl- (9CI) **(13254-34-7)**
3-Heptanol, 3,6-dimethyl- (9CI) **(1573-28-0)**
3-Heptanol, 2,6-dimethyl- (8CI,9CI) **(19549-73-6)**
4-Heptanol, 2,4-dimethyl- (8CI,9CI) **(19549-77-0)**
4-Heptanol, 3,5-dimethyl- (8CI,9CI) **(19549-79-2)**
3-Heptanol, 6-(dimethylamino)-4,4-diphenyl- **(17199-55-2)**
3-Heptanol, 6-(dimethylamino)-4,4-diphenyl- (8CI) **(545-90-4)**
3-Heptanol, 6-(dimethylamino)-4,4-diphenyl-, acetate **(509-74-0)**
3-Heptanol, 6-(dimethylamino)-4,4-diphenyl-, acetate (ester) **(509-74-0)**
2-Heptanol, 5-ethyl- (8CI,9CI) **(19780-40-6)**
Heptanolide-4,1 **(105-21-5)**
3-Heptanol, 4-methyl **(14979-39-6)**
2-Heptanol, 2-methyl- (9CI) **(625-25-2)**
2-Heptanol, 3-methyl- (9CI) **(31367-46-1)**
3-Heptanol, 3-methyl- (9CI) **(5582-82-1)**
1-Heptanol, 6-methyl- (8CI,9CI) **(1653-40-3)**
2-Heptanol, 5-methyl- (7CI,9CI) **(54630-50-1)**
2-Heptanol, 6-methyl- (8CI,9CI) **(4730-22-7)**
3-Heptanol, 3-methyl-, (.+-.)- (8CI,9CI) **(598-06-1)**
4-Heptanol, 4-methyl- (6CI,8CI,9CI) **(598-01-6)**
2-Heptanol, 6-methyl-, acetate (9CI) **(67952-57-2)**
1-Heptanol, 2-propyl **(10042-59-8)**
Heptanon **(76-99-3)**
Heptan-3-on (Dutch, German) **(106-35-4)**
2-Heptanone **(110-43-0)**
3-Heptanone **(106-35-4)**
4-Heptanone **(123-19-3)**
Heptan-3-one **(106-35-4)**
Heptan-4-one **(123-19-3)**
3-Heptanone (OSHA) **(106-35-4)**
4-Heptanone, 5,5-diethyl-2,2,3,3-tetramethyl- (8CI,9CI) **(16424-67-2)**
4-Heptanone, 2,6-dimethyl **(108-83-8)**
2-Heptanone, 4,6-dimethyl- (9CI) **(19549-80-5)**
3-Heptanone, 2,4-dimethyl- (8CI,9CI) **(18641-71-9)**
3-Heptanone, 6-(dimethylamino)-4,4-diphenyl **(76-99-3)**
3-Heptanone, 4,4-diphenyl-6-morpholino- **(467-84-5)**
2-Heptanone, 1-ethoxy- (9CI) **(51149-70-3)**
3-Heptanone, 4-methyl- **(6137-11-7)**
3-Heptanone, 5-methyl **(541-85-5)**

Name	CAS No.
3-Heptanone, 6-methyl-	**(624-42-0)**
2-Heptanone, 6-methyl- (9CI)	**(928-68-7)**
3-Heptanone, 2-methyl- (9CI)	**(13019-20-0)**
2-Heptanone, 5-methyl- (8CI,9CI)	**(18217-12-4)**
4-Heptanone, 3-methyl- (8CI,9CI)	**(15726-15-5)**
3-Heptanone, 6-(4-morpholinyl)-4,4-diphenyl- (9CI)	**(467-84-5)**
2-Heptanone, 3-propylidene- (8CI,9CI)	**(32064-70-3)**
Heptanoyl chloride (9CI)	**(2528-61-2)**
Heptanyl acetate	**(112-06-1)**
3,6,9,12,15,18,21-Heptaoxatritriacontan-1-ol (9CI)	**(3055-97-8)**
Heptasiloxane, hexadecamethyl- (9CI)	**(541-01-5)**
Heptatriacontane (9CI)	**(7194-84-5)**
Heptazine	**(1912-25-0)**
Heptbarbital	**(509-86-4)**
Heptedrine	**(123-82-0)**
(E)-2-Hepten-1-al	**(18829-55-5)**
2-Heptenal	**(2463-63-0)**
2-Heptenal, (E)	**(18829-55-5)**
3-Heptenal (9CI)	**(89896-73-1)**
(E)-2-Heptene	**(14686-13-6)**
(E)-3-Heptene	**(14686-14-7)**
(Z)-3-Heptene	**(7642-10-6)**
1-n-Heptene	**(592-76-7)**
2-Heptene, trans-	**(14686-13-6)**
Heptene	**(592-76-7)**
cis-3-Heptene	**(7642-10-6)**
n-Hept-1-ene	**(592-76-7)**
n-Heptene	**(592-76-7)**
trans-2-Heptene	**(14686-13-6)**
trans-3-Heptene	**(14686-14-7)**
1-Heptene (9CI)	**(592-76-7)**
2-Heptene, (E)- (9CI)	**(14686-13-6)**
3-Heptene, (E)- (9CI)	**(14686-14-7)**
2-Heptene (8CI,9CI)	**(592-77-8)**
3-Heptene (8CI,9CI)	**(592-78-9)**
Heptene (VAN)(9CI)	**(25339-56-4)**
2-Heptene, (Z)- (8CI,9CI)	**(6443-92-1)**
3-Heptene, (Z)- (8CI,9CI)	**(7642-10-6)**
1-Heptene, 1-bromo- (7CI)	**(89942-12-1)**
3-Heptene, 4-ethyl- (8CI,9CI)	**(33933-74-3)**
1-Heptene, 6-methyl-	**(5026-76-6)**
1-Heptene, 2-methyl- (9CI)	**(15870-10-7)**
1-Heptene, 3-methyl- (8CI,9CI)	**(4810-09-7)**
2-Heptene, 2-methyl- (8CI,9CI)	**(627-97-4)**
2-Heptene, 2-nitro	**(6065-14-1)**
2-Heptene, 3-nitro	**(6065-13-0)**
3-Heptene, 3-nitro	**(6187-24-2)**
Hept-2-ene oxide	**(14925-96-3)**
1-Heptene, 2,4,4,6,6-pentamethyl- (9CI)	**(14031-86-8)**
3-Heptene, 2,2,4,6,6-pentamethyl- (9CI)	**(123-48-8)**
2-Heptene, 1-phenyl- (8CI)	**(26447-63-2)**
1-Heptene, 1-phenyl- (7CI,8CI)	**(829-99-2)**
n-Hepteno (Spanish)	**(592-76-7)**
3-Heptenoic acid (8CI,9CI)	**(29901-85-7)**
5-Heptenoic acid, 7-(3,5-dihydroxy-2-(3-hydroxy-1-octenyl)-cyclopentyl)-	**(551-11-1)**
5-Heptenoic acid, 7-(3,5-dihydroxy-2-(3-hydroxy-1-octenyl)-cyclopentyl)-, L-	**(551-11-1)**
(E)-2-Hepten-1-ol	**(33467-76-4)**
trans-2-Hepten-1-ol	**(33467-76-4)**
1-Hepten-3-ol (9CI)	**(4938-52-7)**
1-Hepten-4-ol (9CI)	**(3521-91-3)**
2-Hepten-1-ol, (E)- (9CI)	**(33467-76-4)**
2-Hepten-3-ol, 4,5-dimethyl- (9CI)	**(55956-37-1)**
6-Hepten-2-ol, 6-methyl	**(1335-09-7)**
5-Hepten-2-ol, 6-methyl-, (+-)- (8CI,9CI)	**(4630-06-2)**
3-Hepten-2-one	**(1119-44-4)**
2-Hepten-4-one (9CI)	**(4643-25-8)**
5-Hepten-2-one (9CI)	**(6714-00-7)**
2-Hepten-4-one, 2,6-dimethyl- (8CI,9CI)	**(16525-05-6)**
4-Hepten-3-one, 5-ethyl-2-methyl- (9CI)	**(49833-96-7)**
5-Hepten-2-one, 6-methyl	**(409-02-9)**
5-Hepten-2-one, 6-methyl- (9CI)	**(110-93-0)**
2-Hepten-4-one, 2-methyl- (8CI,9CI)	**(22319-24-0)**
3-Hepten-2-one, 4-methyl- (8CI,9CI)	**(22319-25-1)**
4-Hepten-3-one, 4-methyl- (8CI,9CI)	**(22319-31-9)**
Heptenophos	**(34783-40-9)**
5-Hepten-1-yne, 1-chloro- (9CI)	**(83682-36-4)**
n-Heptoic acid	**(111-14-8)**
Heptox	**(76-44-8)**
1-Heptyl acetate	**(112-06-1)**
Heptyl acetate	**(112-06-1)**
n-Heptyl acetate	**(112-06-1)**
Heptyl-alcohol	**(111-70-6)**
Heptyl aldehyde	**(111-71-7)**
1-Heptylamine	**(111-68-2)**
2-Heptylamine	**(123-82-0)**
Heptylamine	**(111-68-2)**
n-Heptylamine	**(111-68-2)**
Heptylamine 2,4-dichlorophenoxyacetate	**(37102-63-9)**
Heptylamine, 1-methyl	**(693-16-3)**
(4-Heptyl)benzene	**(2132-86-7)**
Heptylbenzene	**(1078-71-3)**
Heptylbenzenesulfonic acid sodium salt	**(33660-91-2)**
Heptyl bromide	**(629-04-9)**
n-Heptyl bromide	**(629-04-9)**
γ-Heptylbutyrolactone	**(104-67-6)**
γ-n-Heptylbutyrolactone	**(104-67-6)**
Heptyl carbinol	**(111-87-5)**
Heptylcyclohexane	**(5617-41-4)**
2-n-Heptyl cyclopentanone	**(137-03-1)**
α-Heptyl cyclopentanone	**(137-03-1)**
2-Heptyl-4,5-dihydro-1H-imidazole-1-ethanol	**(36060-61-4)**
Heptylene	**(25339-56-4)**
Heptylene-2, trans-	**(14686-13-6)**
Heptylene-2-trans	**(14686-13-6)**
α-Heptylene	**(592-76-7)**
2-Heptylfuran	**(3777-71-7)**
Heptyl hydride	**(142-82-5)**
Heptylic acid	**(111-14-8)**
n-Heptylic acid	**(111-14-8)**
2-Heptyl-2-imidazoline-1-ethanol	**(36060-61-4)**
2-(2-Heptylimidazolin-1-yl)ethanol	**(36060-61-4)**
Heptyl iodide	**(4282-40-0)**
Heptyl ketone	**(818-23-5)**
Heptyl mercaptan	**(1639-09-4)**
n-Heptylmercaptan	**(1639-09-4)**
Heptyl-2-naphthalenol	**(31215-04-0)**
x-Heptyl-2-naphthol	**(31215-04-0)**
2-(3-Heptyloxy)ethanol	**(10138-47-3)**
4-(Heptyloxy)phenol	**(13037-86-0)**
4-Heptylphenol	**(1987-50-4)**
Heptyl phthalate	**(3648-21-3)**
n-Heptyl-δ-valerolactone	**(713-95-1)**
1-Heptyne (9CI)	**(628-71-7)**
3-Heptyne, 5,5-dimethyl- (8CI,9CI)	**(23097-98-5)**
Hepzide	**(139-94-6)**
Hepzide 30	**(139-94-6)**
Heraclin	**(484-20-8)**
Herald	**(39515-41-8)**
Herb-All	**(2163-80-6)**
Herbadox	**(40487-42-1)**
Herban	**(18530-56-8)**
Herban M	**(2163-80-6)**

Herbatim (6734-80-1)
Herbatox (120-36-5)
Herbatox (330-54-1)
Herbax Technical (709-98-8)
Herbaxon (1910-42-5)
Herbazin (122-34-9)
Herbazin 50 (122-34-9)
Herbazolin (3813-05-6)
Herbex (122-34-9)
Herbicide 82 (314-42-1)
Herbicide 273 (129-67-9)
Herbicide 326 (330-55-2)
Herbicide 976 (314-40-9)
Herbicide 6602 (19937-59-8)
Herbicide C-2059 (2164-17-2)
Herbicide ES (86290-81-5)
Herbicide M (94-74-6)
Herbicides, Monuron (150-68-5)
Herbicides, Silvex (93-72-1)
Herbidal (94-75-7)
Herbidal Total (61-82-5)
Herbisan (502-55-6)
Herbit (25319-90-8)
Herbitox (3813-05-6)
Herbixol (330-54-1)
Herbizole (61-82-5)
Herbogil (1420-07-1)
Herbon 2,4-des-Sodium (136-78-7)
Herboxone (1910-42-5)
Herboxy (122-34-9)
Herbrak (41394-05-2)
Herco Prills (6484-52-2)
Hercoflat 135 (9003-07-0)
Hercoflex 260 (117-81-7)
Hercoflex 290 (110-29-2)
Hercofloc 1021 (9CI) (106946-84-3)
Hercon disrupt (53939-28-9)
Hercotuf 110A (9003-07-0)
Hercotuf PB 681 (9003-07-0)
Hercules 528 (78-34-2)
Hercules 3956 (8001-35-2)
Hercules 4580 (919-54-0)
Hercules 5727 (64-00-6)
Hercules 6523 (9003-07-0)
Hercules 7175 (1319-96-6)
Hercules 7531 (18530-56-8)
Hercules 8717 (3692-90-8)
Hercules 14503 (10311-84-9)
Hercules 22234 (38727-55-8)
Hercules 26905 (36335-67-8)
Hercules AC 5727 (64-00-6)
Hercules AC528 (78-34-2)
Hercules N 100 (9004-62-0)
Hercules P6 (115-77-5)
Hercules toxaphene (8001-35-2)
Herculon (9003-07-0)
Hercurez A 100 (64742-16-1)
Hercurez AR 100 (64742-16-1)
Herkal (62-73-7)
Herkol (62-73-7)
Hermal (137-26-8)
Hermat TMT (137-26-8)
Hermat ZDM (137-30-4)
Hermat Zn-MBT (155-04-4)
Hermesetas (81-07-2)
Heroien (561-27-3)
Heroiin (561-27-3)
Heroin (561-27-3)
Heroine hydrochloride (1502-95-0)
Heroin hydrochloride (1502-95-0)
Herolan (561-27-3)
Heropon (300-42-5)
Heryl (137-26-8)
Hesofen (77-75-8)
Hespan (9004-62-0)
Het Acid (115-28-6)
Hetamide ML (120-40-1)
Hetamine (51-63-8)
Hetastarch (9004-62-0)
Heteroauxin (87-51-4)
Heteroxanthin (552-62-5)
Hetrazan (1642-54-2)
Heweten 10 (9004-34-6)
Hewsol (133-53-9)
Hexa (58-89-9)
Hexa (608-73-1)
Hexa C.B. (118-74-1)
Hexa-Nema (97-17-6)
1,3,5,7,9,11-Hexaaza-2,4,6,8,10,12-hexaphosphacyclododeca-1,3,5,7,9,11-hexaene, 2,2,4,4,6,6,8,8,10,10,12,12-dodecachloro-2,2,4,4,6,6,8,8,10,10,12,12-dodecahydro- (8CI,9CI) (2851-52-7)
1,3,5,7,9,11-Hexaaza-2,4,6,8,10,12-hexaphosphacyclododeca-hexaene, 2,2,4,4,6,6,8,8,10,10,12,12-dodecachloride (2851-52-7)
Hexa(1-aziridinyl)triphosphotriazine (52-46-0)
Hexabalm (70-30-4)
Hexabarbital (56-29-1)
Hexabenzobenzene (191-07-1)
Hexabione hydrochloride (58-56-0)
Hexabromobenzene (87-82-1)
2,4,5,2',4',5'-Hexabromobiphenyl (59080-40-9)
2,4,5,2',4',5'-Hexabromobiphenyl (67774-32-7)
Hexabromobiphenyl (36355-01-8)
Hexabromobiphenyl (Technical grade) (59536-65-1)
1,1,2,3,4,4-Hexabromo-2-butene (36678-45-2)
Hexabromo-2-butene (36678-45-2)
1,2,5,6,9,10-Hexabromocyclododecane (3194-55-6)
Hexabromocyclododecane (25637-99-4)
Hexabromocyclododecane (3194-55-6)
1,2,3,4,5,6-Hexabromocyclohexane (1837-91-8)
1,2,3,4,6,8-Hexabromodibenzo(b,e)(1,4)dioxin (116490-11-0)
Hexabromodibenzo(b,e)(1,4)dioxin (103456-42-4)
1,2,3,4,6,8-Hexabromodibenzo-p-dioxin (116490-11-0)
Hexabromodibenzo-p-dioxin (103456-42-4)
Hexabromodibenzofuran (103456-33-3)
1,2,4,6,7,9-Hexabromo-3,8-dichlorodibenzo-p-dioxin, (2170-44-7)
Hexabromodiphenyl ether (36483-60-0)
Hexabromodiphenyl oxide (36483-60-0)
Hexabromonaphthalene (56480-06-9)
Hexabromophenoxybenzene (36483-60-0)
Hexabutyldistannioxan (Czech) (56-35-9)
Hexabutyldistannoxane (56-35-9)
Hexabutylditin (56-35-9)
Hexacalcium phytate (7776-28-5)
Hexacap (133-06-2)
Hexacarbonyl chromium (13007-92-6)
Hexacarboxybenzene (517-60-2)
Hexacert Blue No. 2 (860-22-0)
Hexacert Red No. 2 (915-67-3)
Hexacert Red No. 3 (16423-68-0)
Hexacert Yellow No. 5 (1934-21-0)
Hexachlor-aethan (German) (67-72-1)
Hexachlor-1,3-butadien (Czech) (87-68-3)
Hexachlor (608-73-1)
γ-Hexachlor (58-89-9)

Name	CAS No.
1,2,3,6,7,8-Hexachloro-p-dioxin	**(57653-85-7)**
Hexachlorodiphenyl ether	**(31242-93-0)**
Hexachloro diphenyl oxide	**(31242-93-0)**
Hexachlorodiphenyl oxide	**(55720-99-5)**
Hexachloroendomethylene tetrahydrophthalic anhydride	**(115-27-5)**
Hexachloroepoxyoctahydro-endo,endo-dimethanonaphthalene	**(72-20-8)**
Hexachloroepoxyoctahydro-endo,exo-dimethanonaphthalene	**(60-57-1)**
1,1,1,2,2,2-Hexachloroethane	**(67-72-1)**
Hexachloroethane (ACGIH,DOT,OSHA)	**(67-72-1)**
Hexachloroethylene	**(67-72-1)**
Hexachlorofen (Czech)	**(70-30-4)**
1,2,3,4,10,10-Hexachloro-1,4,4a,5,8,8a-hexahydro-1,4,5,8-dimethanonaphthalene	**(309-00-2)**
1,2,3,4,10,10-Hexachloro-1,4,4a,5,8,8a-hexahydro-1,4:5,8-endo-endo-dimethanonaphthalene	**(465-73-6)**
1,2,3,4,10,10-Hexachloro-1,4,4a,5,8,8a-hexahydro-1,4-endo, endo-5,8-dimethanonaphthalene	**(465-73-6)**
1,2,3,4,10,10-Hexachloro-1,4,4a,5,8,8a-hexahydro-1,4-endo-exo-5,8-dimethanonaphthalene	**(309-00-2)**
1,2,3,4,10,10-Hexachloro-1,4,4a,5,8,8a-hexahydro-exo-1,4-endo-5,8-dimethanonaphthalene	**(309-00-2)**
Hexachlorohexahydro-endo-exo-dimethanonaphthalene	**(309-00-2)**
6,7,8,9,10,10-Hexachloro-1,5,5a,6,9,9a-hexahydro-6,9-methano-2,4,3-benzodioxathiepin-3-oxide	**(115-29-7)**
Hexachlorohexahydromethano 2,4,3-benzodioxathiepin-3-oxide	**(115-29-7)**
1,2,3,4,5,6-Hexachloro-3-hexene	**(1725-74-2)**
Hexachloromelamine	**(2428-04-8)**
Hexachloronaphthalene (ACGIH,OSHA)	**(1335-87-1)**
1,2,3,4,7,7-Hexachloro-2,5-norbornadiene	**(3389-71-7)**
1,2,3,4,7,7-Hexachloronorbornadiene	**(3389-71-7)**
Hexachloronorbornadiene	**(3389-71-7)**
Hexachloro-5-norbornene-2,3-dicarboxylic anhydride	**(115-27-5)**
1,4,5,6,7,7-Hexachloro-5-norbornene-2,3-dimethanol cyclic sulfite	**(115-29-7)**
3,4,5,6,9,9-Hexachloro-1a,2,2a,3,6,6a,7,7a-octahydro-2,7:3,6-dimethanonaphth(2,3-b)oxirene	**(60-57-1)**
3,4,5,6,9,9-Hexachloro-1a,2,2a,3,6,6a,7,7a-octahydro-2,7:3,6-dimethanonaphth(2,3-b)oxirene	**(72-20-8)**
Hexachlorophane	**(70-30-4)**
Hexachlorophen	**(70-30-4)**
Hexachlorophene	**(5736-15-2)**
Hexachlorophene [UN 2875]	**(70-30-4)**
Hexachloroplatinic (IV) acid	**(16941-12-1)**
Hexachloroplatinic acid	**(16941-12-1)**
Hexachloroplatinic(4+) acid, hydrogen-	**(16941-12-1)**
1,1,1,3,3,3-Hexachloropropane	**(3607-78-1)**
1,1,1,3,3,3-Hexachloro-2-propanone	**(116-16-5)**
Hexachloro-2-propanone	**(116-16-5)**
Hexachloropropene	**(1888-71-7)**
Hexachloropropylene	**(1888-71-7)**
4,5,6,7,8,8-Hexachloro-3a,4,7,7a-tetrahydro-4,7-methanoindene	**(3734-48-3)**
α,α'-Hexachloroxylene	**(68-36-0)**
α,α,α,α',α',α'-Hexachloro-p-xylene	**(68-36-0)**
4,5,6,7,8,8-Hexachlor-$\delta^{1,5}$-tetrahydro-4,7-methanoinden	**(3734-48-3)**
Hexacid 698	**(142-62-1)**
Hexacid 898	**(124-07-2)**
Hexacid 1095	**(334-48-5)**
Hexacid C-7	**(111-14-8)**
Hexacid C-9	**(112-05-0)**
Hexaco Blue VRS	**(129-17-9)**
Hexacol Amaranth B Extra	**(915-67-3)**
Hexacol Black PN	**(2519-30-4)**
Hexacol Blue VRS	**(129-17-9)**
Hexacol Brilliant Blue A	**(3844-45-9)**
Hexacol Carmoisine	**(3567-69-9)**
Hexacol Carmoisine Conc.	**(3567-69-9)**
Hexacol Erythrosine BS	**(16423-68-0)**
Hexacol Fast Red E	**(2302-96-7)**
Hexacol Indigo Carmine Supra	**(860-22-0)**
Hexacol Oil Orange SS	**(2646-17-5)**
Hexacol Oil Yellow GG	**(2051-85-6)**
Hexacol Orange GG Crystals	**(1936-15-8)**
Hexacol Orange RN	**(1934-20-9)**
Hexacol Ponceau MX	**(3761-53-3)**
Hexacol Ponceau 2R	**(3761-53-3)**
Hexacol Ponceau 4R	**(2611-82-7)**
Hexacol Ponceau SX	**(4548-53-2)**
Hexacol Red 10B	**(3567-66-6)**
Hexacol Red 2G	**(3734-67-6)**
Hexacol Rhodamine B Extra	**(81-88-9)**
Hexacol Sunset Yellow F & F Supra	**(2783-94-0)**
Hexacol Sunset Yellow FCF	**(2783-94-0)**
Hexacol Sunset Yellow FCF Supra	**(2783-94-0)**
Hexacol Sunset Yellow FCP	**(2783-94-0)**
Hexacol Tartrazine	**(1934-21-0)**
Hexacontane (8CI,9CI)	**(7667-80-3)**
Hexacosane (9CI)	**(630-01-3)**
Hexacosanic acid	**(506-46-7)**
Hexacosanoic acid (9CI)	**(506-46-7)**
Hexacosanoic acid, methyl ester (8CI,9CI)	**(5802-82-4)**
n-Hexacosanol	**(506-52-5)**
1-Hexacosanol (9CI)	**(506-52-5)**
1-Hexacosanol, aluminum salt (9CI)	**(67905-28-6)**
Hexacose	**(111-28-4)**
1-Hexacosene	**(18835-33-1)**
Hexacosyl alcohol	**(506-52-5)**
Hexacyanoferrate II	**(13408-63-4)**
1,2-Hexadecadiene, 3,7,11,15-tetramethyl- (9CI)	**(2437-92-5)**
(E,Z)-7,11-Hexadecadien-1-ol, acetate	**(53042-79-8)**
(Z,E)-7,11-Hexadecadien-1-ol, acetate	**(50933-33-0)**
(Z,Z)-7,11-Hexadecadien-1-ol, acetate	**(52207-99-5)**
7,11-Hexadecadien-1-ol, acetate (9CI)	**(50933-33-0)**
7,11-Hexadecadien-1-ol, acetate (Stereoisomer unspec.)	**(51607-94-4)**
7,11-Hexadecadien-1-ol, acetate, (Z,E)-	**(51607-94-4)**
7,11-Hexadecadien-1-ol, acetate, (Z,Z)-	**(52207-99-5)**
Hexadecadrol	**(50-02-2)**
Hexadecafluoroheptane	**(335-57-9)**
1,2,3,3a,3b,4,5,5a,6,8,10,10a,10b,11,12,12a-Hexadecahydro-1,10a,12a-trimethylcyclopenta(7,8)-phenanthro(2,3-c)-pyrazol-1-ol	**(10418-03-8)**
Hexadeca-mu-hydroxytetracosahydroxy-(u8-(1,3,4,6-tetra-O-sulfo-β-D-fructofuranosyl-α-D-glucopyranoside tetrakis-(hydrogen sulfato)(8-))) hexadecaaluminum	**(54182-58-0)**
Hexadecamethylheptasiloxane	**(541-01-5)**
Hexadecanal	**(629-80-1)**
Hexadecanamide, N-acetyl- (9CI)	**(65882-22-6)**
Hexadecanamide, N-9-octadecenyl-	**(16260-09-6)**
Hexadecanamide, N-9-octadecenyl-, (Z)- (9CI)	**(16260-09-6)**
1-Hexadecanamine	**(143-27-1)**
1-Hexadecanamine, acetate (9CI)	**(2016-52-6)**
1-Hexadecanamine, N,N-dimethyl- (9CI)	**(112-69-6)**
1-Hexadecanamine, N,N-dimethyl-, N-oxide (9CI)	**(7128-91-8)**
1-Hexadecanamine, N,N-dimethyl-, N-oxide (40% solution)	**(7128-91-8)**
1-Hexadecanamine, N-hexadecyl-, acetate (9CI)	**(71764-17-5)**
1-Hexadecanaminium, N-(carboxymethyl)-N,N-dimethyl-, hydroxide, inner salt (9CI)	**(693-33-4)**
1-Hexadecanaminium, N-ethyl-N,N-dimethyl-, bromide	**(16919-58-7)**
1-Hexadecanaminium, N-hexadecyl-N,N-dimethyl-, chloride (9CI)	**(1812-53-9)**
1-Hexadecanaminium, N,N,N-triethyl-, bromide (9CI)	**(13316-70-6)**
1-Hexadecanaminium, N,N,N-trimethyl- (9CI)	**(6899-10-1)**
1-Hexadecanaminium, N,N,N-trimethyl-, bromide	**(57-09-0)**
1-Hexadecanaminium, N,N,N-trimethyl-, chloride (9CI)	**(112-02-7)**
1-Hexadecanaminium, N,N,N-trimethyl-, chloride (50% in 2-propanol)	**(112-02-7)**
Hexadecane	**(544-76-3)**

n-Hexadecane

n-Hexadecane (544-76-3)
Hexadecane, 1,1-bis(dodecyloxy)- (9CI) (56554-64-4)
Hexadecane, 1-bromo- (9CI) (112-82-3)
Hexadecane, 1-chloro- (9CI) (4860-03-1)
Hexadecane, chloro- (8CI,9CI) (34214-79-4)
Hexadecane, 1-cyclohexyl- (6CI,8CI) (6812-38-0)
1,16-Hexadecanedioic acid (505-54-4)
Hexadecane-1,16-dioic acid (505-54-4)
Hexadecanedioic acid (9CI) (505-54-4)
Hexadecane, 1,2-epoxy (7320-37-8)
Hexadecane, 1-iodo- (9CI) (544-77-4)
Hexadecane, 2-methyl- (8CI,9CI) (1560-92-5)
Hexadecanenitrile (9CI) (629-79-8)
1-Hexadecanesulfonyl chloride (9CI) (38775-38-1)
1-Hexadecanethiol (9CI) (2917-26-2)
1-Hexadecanethiol (50% concentrate) (2917-26-2)
Hexadecane, 2,6,10-trimethyl- (9CI) (55000-52-7)
Hexadecanoic acid (57-10-3)
n-Hexadecanoic acid (57-10-3)
Hexadecanoic acid, aluminum salt (3:1) (555-35-1)
Hexadecanoic acid, aluminum salt (9CI) (555-35-1)
Hexadecanoic acid, ammonium salt (9CI) (593-26-0)
Hexadecanoic acid, 2-bromo- (9CI) (18263-25-7)
Hexadecanoic acid, butyl ester (9CI) (111-06-8)
Hexadecanoic acid, calcium salt (9CI) (542-42-7)
Hexadecanoic acid, chloride (112-67-4)
Hexadecanoic acid, 2-chloroethyl ester (929-16-8)
Hexadecanoic acid, ethyl ester (9CI) (628-97-7)
Hexadecanoic acid, 2-ethylhexyl ester (9CI) (29806-73-3)
Hexadecanoic acid, hexadecyl ester (540-10-3)
Hexadecanoic acid, 16-hydroxy- (9CI) (506-13-8)
Hexadecanoic acid, isopropyl ester (142-91-6)
Hexadecanoic acid, 2-methoxyethyl ester (9CI) (111-07-9)
n-Hexadecanoic acid methyl ester (112-39-0)
Hexadecanoic acid, methyl ester (9CI) (112-39-0)
Hexadecanoic acid, 14-methyl-, methyl ester (8CI,9CI) (2490-49-5)
Hexadecanoic acid, 2-methylpropyl ester (110-34-9)
Hexadecanoic acid, propyl ester (2239-78-3)
Hexadecanoic acid, sodium salt (9CI) (408-35-5)
Hexadecanoic acid, 2-sulfo-, 1-(2-(2-aminoethoxy)ethyl) ester (9CI) (59997-80-7)
Hexadecanoic acid, 2-sulfo-, 1-(2-aminoethyl) ester (9CI) (59997-74-9)
Hexadecanoic acid, 2-sulfo-, 1-methyl ester (9CI) (58849-75-5)
Hexadecanoic acid, 2-sulfo-, 1-methyl ester, sodium salt (9CI) (4016-24-4)
Hexadecanoic acid, zinc salt (9CI) (4991-47-3)
1-Hexadecanol (36653-82-4)
Hexadecan-1-ol (36653-82-4)
Hexadecanol (36653-82-4)
n-Hexadecanol (36653-82-4)
Hexadecanol (8CI,9CI) (VAN) (29354-98-1)
1-Hexadecanol, acetate (629-70-9)
1-Hexadecanol, aluminum salt (9CI) (19141-82-3)
2-Hexadecanol, 1-(bis(2-hydroxyethyl)amino)-, N-oxide (8CI) (28865-36-3)
1-Hexadecanol, dihydrogen phosphate (9CI) (3539-43-3)
1-Hexadecanol, hydrogen phosphate (9CI) (2197-63-9)
1-Hexadecanol, hydrogen sulfate (9CI) (143-02-2)
1-Hexadecanol, hydrogen sulfate, ammonium salt (9CI) (4696-47-3)
1-Hexadecanol, hydrogen sulfate, sodium salt (1120-01-0)
1-Hexadecanol, 2-methyl- (8CI,9CI) (2490-48-4)
Hexadecanoyl chloride (9CI) (112-67-4)
(Z)-11-Hexadecenal (53939-28-9)
11-Hexadecenal (53939-28-9)
11-Hexadecenal, (Z)- (53939-28-9)
α-Hexadecene (629-73-2)
n-Hexadec-1-ene (629-73-2)
1-Hexadecene (9CI) (629-73-2)
Hexadecene epoxide (7320-37-8)
1-Hexadecene, 7,11,15-trimethyl-3-methylene- (504-96-1)
(E)-9-Hexadecenoic acid (10030-73-6)
9-Hexadecenoic acid, (E)- (10030-73-6)
9-cis-Hexadecenoic acid (373-49-9)
cis-9-Hexadecenoic acid (373-49-9)
9-Hexadecenoic acid (9CI) (2091-29-4)
9-Hexadecenoic acid, (Z)- (9CI) (373-49-9)
9-Hexadecenoic acid, methyl ester, (Z)- (1120-25-8)
(Z)-11-Hexadecen-1-ol (56683-54-6)
(Z)-9-Hexadecen-1-ol (10378-01-5)
11-Hexadecen-1-ol, (Z)- (56683-54-6)
9-Hexadecen-1-ol, (Z)- (9CI) (10378-01-5)
1-Hexadecen-3-ol, 3,7,11,15-tetramethyl- (6CI,7CI,9CI) (60046-87-9)
2-Hexadecen-1-ol, 3,7,11,15-tetramethyl-, (R-(R*,R*-(E)))- (9CI) (150-86-7)
1-Hexadecen-3-ol, 3,7,11,15-tetramethyl- (VAN) (9CI) (505-32-8)
3-(Hexadecenyl)dihydro-2,5-furandione (32072-96-1)
n-Hexadecoic acid (57-10-3)
Hexadecyl acetate (629-70-9)
Hexadecyl acrylate (13402-02-3)
Hexadecyl alcohol (36653-82-4)
n-Hexadecyl alcohol (36653-82-4)
n-Hexadecylamine (143-27-1)
Hexadecylamine, acetate (8CI) (2016-52-6)
Hexadecylamine, N,N-dimethyl- (8CI) (112-69-6)
Hexadecylbenzene (1459-09-2)
4-Hexadecylbenzenesulfonic acid (16722-32-0)
p-Hexadecylbenzenesulfonic acid (16722-32-0)
Hexadecylbis(2-hydroxypropyl) amine (2269-21-8)
1-Hexadecyl bromide (112-82-3)
Hexadecyl bromide (112-82-3)
n-Hexadecyl bromide (112-82-3)
n-Hexadecyl-1-bromide (112-82-3)
Hexadecylcyclohexane (6812-38-0)
Hexadecyl cyclopropanecarboxylate (54460-46-7)
Hexadecyldimethylamine (112-69-6)
Hexadecyldimethyl(ethylbenzyl)ammonium chloride (29656-52-8)
α-Hexadecylene (629-73-2)
N-Hexadecyl-N-ethylmorpholinium ethyl sulfate (78-21-7)
Hexadecylic acid (57-10-3)
1,1'-(Hexadecylimino)bis-2-propanol (2269-21-8)
Hexadecyl methacrylate (2495-27-4)
Hexadecyl 2-methyl-2-propenoate (2495-27-4)
Hexadecyloxirane (7390-81-0)
2-(Hexadecyloxy)ethyl sodium sulfate (14858-54-9)
((Hexadecyloxy)methyl)oxirane (15965-99-8)
Hexadecyl palmitate (540-10-3)
2-Hexadecylphenol (25401-86-9)
4-Hexadecylphenol (2589-78-8)
o-Hexadecylphenol (25401-86-9)
Hexadecyl phosphate (3539-43-3)
Hexadecyl phosphate (1:1) (3539-43-3)
Hexadecyl 2-propenoate (13402-02-3)
Hexadecylpyridine bromide (140-72-7)
1-Hexadecylpyridinium bromide (140-72-7)
Hexadecylpyridinium bromide (140-72-7)
n-Hexadecylpyridinium bromide (140-72-7)
1-Hexadecylpyridinium chloride (123-03-5)
Hexadecylpyridinium chloride (123-03-5)
n-Hexadecylpyridinium chloride (123-03-5)
5-Hexadecylsalicylic acid, calcium salt (68540-40-9)
Hexadecyl sodium sulfate (1120-01-0)
Hexadecylsulfonyl chloride (38775-38-1)
Hexadecyl tetradecanoate (2599-01-1)
Hexadecyltrichlorosilane [UN 1781] (5894-60-0)
Hexadecyltriethylammonium bromide (13316-70-6)
Hexadecyltrimethylammonium (6899-10-1)
(1-Hexadecyl)trimethylammonium bromide (57-09-0)

Hexadecyltrimethylammonium bromide **(57-09-0)**
n-Hexadecyl-N,N,N-trimethylammonium bromide **(57-09-0)**
n-Hexadecyltrimethylammonium bromide **(57-09-0)**
Hexadecyl trimethyl ammonium chloride **(112-02-7)**
Hexadecyltrimethylammonium chloride **(112-02-7)**
n-Hexadecyltrimethylammonium chloride **(112-02-7)**
Hexadecyltrimethylammonium ion **(6899-10-1)**
1-Hexadecyne (9CI) **(629-74-3)**
1-Hexadecyn-3-ol, 3,7,11,15-tetramethyl- (8CI,9CI) **(29171-23-1)**
Hexadenol **(111-28-4)**
Hexaderm Red MRG **(10169-02-5)**
Hexa-2,4-dienal **(142-83-6)**
trans,trans-2,4-Hexadienal **(142-83-6)**
2,4-Hexadienal, (E,E)- **(142-83-6)**
1,4-Hexadiene (9CI) **(592-45-0)**
1,5-Hexadiene (9CI) **(592-42-7)**
2,4-Hexadiene (9CI) **(592-46-1)**
1,3-Hexadiene (8CI,9CI) **(592-48-3)**
1,3-Hexadiene, (E)- (8CI,9CI) **(20237-34-7)**
2,4-Hexadiene, (E,E)- (8CI,9CI) **(5194-51-4)**
2,4-Hexadiene, (E,Z)- (8CI,9CI) **(5194-50-3)**
Hexadiene [UN 2458] **(42296-74-2)**
1,5-Hexadiene, 2,5-bis(chloromethyl)- (9CI) **(83682-51-3)**
1,5-Hexadiene, 3,4-dichloro- (9CI) **(83682-33-1)**
1,5-Hexadiene, 1,6-dichloro- (7CI,9CI) **(67546-51-4)**
1,5-Hexadiene, 3,4-dichloro-2,5-bis(chloromethyl)- (9CI) **(83682-52-4)**
1,5-Hexadiene, 1,6-dichloro-2,5-dimethyl- (9CI) **(83682-48-8)**
1,5-Hexadiene, 3,4-dichloro-2,5-dimethyl- (6CI,9CI) **(83682-49-9)**
1,5-Hexadiene, 2,5-dimethyl- (9CI) **(627-58-7)**
2,4-Hexadiene, 2,5-dimethyl- (9CI) **(764-13-6)**
1,4-Hexadiene, 2,3-dimethyl- (8CI,9CI) **(18669-52-8)**
2,4-Hexadienedioic acid (9CI) **(505-70-4)**
2,4-Hexadienedioic acid, (Z,Z)- (9CI) **(1119-72-8)**
2,4-Hexadienedioic acid, 3-carboxy- **(2547-45-7)**
1,4-Hexadiene, 3-ethyl- (8CI,9CI) **(2080-89-9)**
1,3-Hexadiene, 3-ethyl-2-methyl- (9CI) **(61142-36-7)**
1,5-Hexadiene, 2-methyl- (8CI,9CI) **(4049-81-4)**
1,5-Hexadiene, 3,3,4,4-tetrachloro-2,5-bis(chloromethyl)- (9CI) **(83682-53-5)**
1,5-Hexadiene, 3,3,4,4-tetrachloro-2,5-dimethyl- (9CI) **(83682-50-2)**
Hexadienic acid **(110-44-1)**
2,4-Hexadienoic acid **(110-44-1)**
Hexadienoic acid **(110-44-1)**
trans-trans-2,4-Hexadienoic acid **(110-44-1)**
2,4-Hexadienoic acid, calcium salt **(7492-55-9)**
2,4-Hexadienoic acid, calcium salt, (E,E)- (9CI) **(7492-55-9)**
2,4-Hexadienoic acid, ethyl ester **(2396-84-1)**
2,4-Hexadienoic acid, ethyl ester, (E,E)- **(2396-84-1)**
2,5-Hexadienoic acid, 3-methoxy-5-methyl-4-oxo **(90-65-3)**
2,4-Hexadienoic acid, methyl ester, (E,E)- **(689-89-4)**
2,4-Hexadienoic acid potassium salt **(590-00-1)**
2,4-Hexadienoic acid, potassium salt, (E,E)- (9CI) **(24634-61-5)**
2,4-Hexadien-1-ol **(111-28-4)**
2,4-Hexadienol **(111-28-4)**
Hexadiona **(125-33-7)**
2,4-Hexadiyn-1,6-bis-p-toluenesulfonate **(32527-15-4)**
2,4-Hexadiyne-1,6-diol, di-p-toluenesulfonate **(32527-15-4)**
Hexadrin **(50-18-0)**
Hexadrin **(72-20-8)**
Hexadrol **(50-02-2)**
Hexaethoxydisiloxane **(2157-42-8)**
Hexaethylbenzene **(604-88-6)**
1,1,1,3,3,3-Hexaethyldistannoxane **(1112-63-6)**
Hexaethyldistannoxane **(1112-63-6)**
Hexaethylene-glycol **(2615-15-8)**
Hexaethylene glycol dodecyl ether **(3055-96-7)**
Hexaethylene glycol lauryl ether **(3055-96-7)**
Hexaethyltetrafosfat (Czech) **(757-58-4)**
Hexaethyl tetraphosphate (DOT) **(757-58-4)**
Hexaethyl tetraphosphate, Liquid [UN 1611] **(757-58-4)**
Hexaethyl tetraphosphate, Liquid, containing not more than 25% hexaethyl tetraphosphate (DOT) **(757-58-4)**
Hexaethyl tetraphosphate Mixture, Dry containing more than 2% hexaethyl tetraphosphate (DOT) **(757-58-4)**
Hexaethyl tetraphosphate Mixture, Dry containing not more than 2% hexaethyl tetraphosphate (DOT) **(757-58-4)**
Hexaethyl tetraphosphate Mixture, Liquid containing > 25% hexaethyl tetraphosphate (DOT) **(757-58-4)**
Hexaethyl tetraphosphate and compressed gas mixture [UN 1612] **(757-58-4)**
Hexafen **(70-30-4)**
Hexaferb **(14484-64-1)**
Hexafluorenium **(4844-10-4)**
Hexafluoroacetic anhydride **(407-25-0)**
Hexafluoroacetone (ACGIH,OSHA) [UN 2420] **(684-16-2)**
Hexafluoroacetone sesquihydrate **(13098-39-0)**
Hexafluoroacetone trihydrate **(34202-69-2)**
Hexafluorobenzene **(392-56-3)**
Hexafluoro-1,2-epoxypropane **(428-59-1)**
Hexafluoroepoxypropane **(428-59-1)**
Hexafluoroethane (DOT) **(76-16-4)**
3,3,3,4,4,4-Hexafluoroisobutylene **(382-10-5)**
Hexafluoroisobutylene **(382-10-5)**
Hexafluoroisopropanol **(920-66-1)**
Hexafluorokieselsaiure (German) **(16961-83-4)**
Hexafluorokiezelzuur (Dutch) **(16961-83-4)**
Hexafluorophosphoric acid [UN 1782] **(16940-81-1)**
1,1,1,3,3,3-Hexafluoro-2-propanol **(920-66-1)**
Hexafluoro-2-propanone sesquihydrate **(13098-39-0)**
Hexafluoropropene **(116-15-4)**
Hexafluoropropene epoxide **(428-59-1)**
Hexafluoropropene oxide **(428-59-1)**
Hexafluoropropylene [UN 1858] **(116-15-4)**
Hexafluoropropylene oxide [NA 1956] **(428-59-1)**
Hexafluorosilicate(2-) diammonium **(16919-19-0)**
Hexafluorosilicate(2-) lead(2+) (1:1) **(25808-74-6)**
Hexafluorosilicate(2-) magnesium (1:1) **(16949-65-8)**
Hexafluoro-m-xylene **(402-31-3)**
α,α,α,α',α',α'-Hexafluoro-m-xylene **(402-31-3)**
Hexafluorure de soufre (French) **(2551-62-4)**
Hexafluorure de tungstene (French) **(7783-82-6)**
Hexafluoruro de tungsteno (Spanish) **(7783-82-6)**
Hexafluosilicic acid **(16961-83-4)**
Hexaform **(100-97-0)**
Hexaglycerine **(77-99-6)**
Hexaglycerol (9CI) **(36675-34-0)**
Hexahydroaniline **(108-91-8)**
Hexahydro-1H-azepine **(111-49-9)**
Hexahydroazepine **(111-49-9)**
Hexahydro-2-azepinone **(105-60-2)**
Hexahydro-2H-azepin-2-one (9CI) **(105-60-2)**
Hexahydro-2H-azepin-2-one homopolymer **(25038-54-4)**
1-(Hexahydro-1-azepinyl)-3-p-tolylsulfonylurea **(1156-19-0)**
1-(Hexahydro-1H-azepin-1-yl)-3-(p-tolylsulfonyl)urea **(1156-19-0)**
Hexahydrobenzenamine **(108-91-8)**
Hexahydrobenzene **(110-82-7)**
Hexahydrobenzoic acid **(98-89-5)**
Hexahydrobenzyl alcohol **(100-49-2)**
Hexahydrocresol **(25639-42-3)**
Hexahydrodesoxyephedrine **(101-40-6)**
Hexahydro-1,4-diazine **(110-85-0)**
1,5a,6,9,9a,9b-Hexahydro-4a(4H)-dibenzofurancarboxaldehyde **(126-15-8)**
Hexahydro-3a,7a-dimethyl-4,7-epoxyisobenzofuran-1,3-dione **(56-25-7)**
4,4a,5,6,7,8-Hexahydro-4,4a-dimethyl-6-(1-methylethenyl)-2(3H)-naphthalenone, (4R-(4α,4aα,6β))- **(4674-50-4)**
1,2,3,4,5,6-Hexahydro-6,11-dimethyl-3-phenethyl-2,6-methano-

3-benzazocin-8-ol (127-35-5)
Hexahydrofarnesylacetone (502-69-2)
2,2,4,4,6,6-Hexahydro-2,2,4,4,6,6-hexakis(1-aziridinyl)-1,3,5,2,4,6-triazatriphosphorine (52-46-0)
1,3,4,6,7,8-Hexahydro-4,6,6,7,8,8-hexamethyl-cyclopenta-γ-2-benzopyran (1222-05-5)
1,2,3,4,5,6-Hexahydro-8-hydroxy-6,11-dimethyl-3-phenethyl-2,6-methano-3-benzazocine (127-35-5)
Hexahydro-1,3-isobenzofurandione (85-42-7)
exo-Hexahydro-4,7-methanoindan (2825-82-3)
1-(3a,4,5,6,7,7a-Hexahydro-4,7-methano-5-indanyl)-3,3-dimethyl-urea (18530-56-8)
1-(5-(3a,4,5,6,7,7a-Hexahydro-4,7-methanoindanyl))-3,3-dimethyl-urea (18530-56-8)
3-(5-(3a,4,5,6,7,7a-Hexahydro-4,6-methanoindanyl))-1,1-dimethyl-urea (18530-56-8)
3-(Hexahydro-4,7-methanoindan-5-yl)-1,1-dimethylurea (18530-56-8)
3-(Hexahydro-4,7-methanoindan-5-yl)-1,1-dimethylurea (2163-79-3)
3a,4,5,6,7,7a-Hexahydro-4,7-methano-1H-inden-5-yl acetate (2500-83-6)
3a,4,5,6,7,7a-Hexahydro-4,7-methanoindenyl acrylate (33791-58-1)
1,3,4,9,10,10a-Hexahydro-11-methyl-2H-10,4a-iminoethano-phenanthren-6-ol, dl-Mixture (297-90-5)
1,3,4,9,10,10a-Hexahydro-11-methyl-2H-10,4a-iminoethano-phenanthren-6-ol, l- (77-07-6)
Hexahydromethyl-1,3-isobenzofurandione (25550-51-0)
7-(1,2,6,7,8,8a-Hexahydro-2-methyl-8- (2-methylbutyryloxy)-naphthyl)-3-hydroxyheptan-5-olide (73573-88-3)
Hexahydromethylphenol (25639-42-3)
Hexahydro-1-methyl-4-phenyl-4-azepinecarboxylic acid ethyl ester (77-15-6)
Hexahydro-1-nitroso-1H-azepine (932-83-2)
Hexahydrophenol (108-93-0)
Hexahydrophenylethyl alcohol (4442-79-9)
Hexahydrophthalic acid (1687-30-5)
Hexahydrophthalic acid anhydride (85-42-7)
Hexahydrophthalic anhydride (85-42-7)
Hexahydrophthalic anhydride, (chloromethyl)oxirane polymer (31095-87-1)
Hexahydropicolinic acid (535-75-1)
Hexahydropyrazine (110-85-0)
Hexahydropyridine (110-89-4)
Hexahydropyridine hydrochloride (6091-44-7)
Hexahydropyrrolizine (643-20-9)
Hexahydrothymol (89-78-1)
Hexahydrotoluene (108-87-2)
Hexahydro-1,3,5-triethyl-s-triazine (7779-27-3)
Hexahydro-1,3,5-trimethyl-s-triazine (108-74-7)
Hexahydro-1,3,5-trinitro-1,3,5-triazin (German) (121-82-4)
Hexahydro-1,3,5-trinitro-1,3,5-triazine (121-82-4)
Hexahydro-1,3,5-trinitro-s-triazine (121-82-4)
Hexahydro-1,3,5-triphenyl-1,3,5-triazine (91-78-1)
Hexahydro-1,3,5-tris(hydroxyethyl)triazine (4719-04-4)
Hexaiodobenzene (608-74-2)
Hexakis-(1-aziridinyl)phosphonitrile (52-46-0)
Hexakis-methoxymethylmelamin (Czech) (3089-11-0)
Hexakis(mu-(acetato-O:O'))-mu^4-oxotetraberyllium (19049-40-2)
Hexakis(mu-acetato)-mu^4-oxotetraberyllium (19049-40-2)
2,2,4,4,6,6-Hexakis(1-aziridinyl)cyclotriphosphaza-1,3,5-triene (52-46-0)
2,2,4,4,6,6-Hexakis(1-aziridinyl)-2,2,4,4,6,6-hexahydro-1,3,5,2,4,6-triazatriphosphorine (52-46-0)
Hexakis(aziridinyl)phosphotriazine (52-46-0)
(OC-6-11)-Hexakis(cyano-C)ferrate(4-) (13408-63-4)
Hexakis(dihydrogen phosphate) myo-inositol (9CI) (83-86-3)
Hexakis(β,β-dimethylphenethyl)-distannoxane (13356-08-6)
Hexakis(methoxymethyl)melamine (3089-11-0)
Hexakis(methoxymethyl)-s-triazine-2,4,6-triamine (3089-11-0)
Hexakis (2-methyl-2-phenylpropyl)-distannoxane (13356-08-6)
Hexakis(2-methyl-2-phenylpropyl)distannoxane (13356-08-6)
Hexaklor (608-73-1)
Hexakose (111-28-4)
γ-Hexalactone (695-06-7)
Hexalan Fast Red E (2302-96-7)
Hexalan Red B (3567-66-6)
Hexalan Red 2G (3734-67-6)
Hexaldehyde [UN 1207] (66-25-1)
Hexabetalin (58-56-0)
Hexalin (108-93-0)
Hexametaphosphate, sodium salt (10124-56-8)
Hexametapol (680-31-9)
Hexa(methoxymethyl)melamine (3089-11-0)
Hexamethyl Violet (548-62-9)
2,2',4,4',6,6'-Hexamethylazobenzene (5692-66-0)
Hexamethylbenzene (87-85-4)
3,4,5,3',4',5'-Hexamethylbiphenyl (56667-01-7)
Hexamethylcyclotrisiloxane (541-05-9)
Hexamethyldisilazane (999-97-3)
Hexamethyldisiloxane (107-46-0)
Hexamethylenamine (100-97-0)
Hexamethylendiisokyanat (Czech) (822-06-0)
Hexamethylene (110-82-7)
Hexamethyleneamine (100-97-0)
1,1'-Hexamethylenebis(5-(p-chlorophenyl)biguanide) (55-56-1)
1,1'-Hexamethylene bis(5-(p-chlorophenyl)biguanide) diacetate (56-95-1)
1,1'-Hexamethylene bis(5-(p-chlorophenyl)biguanide), diacetate (56-95-1)
1,1'-Hexamethylenebis(5-(p-chlorophenyl)biguanide) diacetate (56-95-1)
1,1'-Hexamethylenebis(5-(p-chlorophenyl)biguanide)diacetate (56-95-1)
1,1'-Hexamethylenebis(5-(p-chlorophenyl)biguanide)digluconate (18472-51-0)
1,1'-Hexamethylene bis(5-(p-chlorophenyl)biguanide), dihydrochloride (3697-42-5)
1,1'-Hexamethylenebis(5-(p-chlorophenyl)biguanide) dihydrochloride (3697-42-5)
1,6-Hexamethylenediamine (124-09-4)
Hexamethylenediamine (124-09-4)
Hexamethylenediamine, Solid [UN 2280] (124-09-4)
Hexamethylenediamine, Solution [UN 1783] (124-09-4)
Hexamethylenediamine adipate (3323-53-3)
Hexamethylenediamine adipate (1:1) (3323-53-3)
Hexamethylenediamine carbamate (143-06-6)
1,6-Hexamethylenediamine dihydrochloride (6055-52-3)
Hexamethylenediamine dihydrochloride (6055-52-3)
1,6-Hexamethylenediamine-1,6-hexamethylenediisocyanate copolymer, SRU (25035-67-0)
Hexamethylenediamine monoadipate (3323-53-3)
Hexamethylenediamine sebacate (6422-99-7)
Hexamethylenediammonium adipate (3323-53-3)
Hexamethylenediammonium sebacate (6422-99-7)
Hexamethylenedicarboxylic acid (505-48-6)
1,6-Hexamethylene diisocyanate (822-06-0)
Hexamethylene diisocyanate (822-06-0)
Hexamethylene-1,6-diisocyanate (822-06-0)
Hexamethylenediisocyanate [UN 2281] (822-06-0)
Hexamethylene glycol (629-11-8)
Hexamethyleneimine [UN 2493] (111-49-9)
Hexamethylenetetraamine (100-97-0)
Hexamethylenetetramine (100-97-0)
Hexamethylene triperoxidediamine (283-66-9)
Hexamethylenetriperoxidediamine (283-66-9)
Hexamethylene triperoxide diamine, Dry (French) (283-66-9)
Hexamethylenimine (111-49-9)
Hexamethylenimine, 1-acetyl- (5809-41-6)
Hexamethylenimine, 1-(2-aminoethyl)- (51388-00-2)
Hexamethylenimine, 1-(3-aminopropyl)- (3437-33-0)
Hexamethylenimine, 2-oxo- (105-60-2)
Hexamethylentetramin (German) (100-97-0)
Hexamethylentetramine (100-97-0)
Hexamethylfosforamid (Czech) (680-31-9)

2,3,3,4,4,5-Hexamethyl-2-hexanethiol **(25103-58-6)**
Hexamethylmelamine **(645-05-6)**
Hexamethyl methylolmelamine **(3089-11-0)**
Hexamethylol benzene hexanitrate **(105554-30-1)**
Hexamethylol-melamin-hexa-methylaether (German) **(3089-11-0)**
Hexamethylparaosaniline chloride **(548-62-9)**
Hexamethyl phosphoramide (ACGIH) **(680-31-9)**
Hexamethylphosphoric acid triamide **(680-31-9)**
Hexamethylphosphoric triamide **(680-31-9)**
N,N,N,N,N,N-Hexamethylphosphoric triamide **(680-31-9)**
Hexamethylphosphorotriamide **(680-31-9)**
Hexamethylphosphotriamide **(680-31-9)**
Hexamethyl-p-rosaniline chloride **(548-62-9)**
Hexamethyl p-rosaniline hydrochloride **(548-62-9)**
Hexamethyl(silaacenaphthenyl)cyclotetrasiloxane **(13093-12-4)**
Hexamethyl(silacyclohexyl)cyclotetrasiloxane **(35331-58-9)**
N,N,N',N',N'',N''-Hexamethylsilanetriamine **(15112-89-7)**
Hexamethylsilazane **(999-97-3)**
2,6,10,15,19,23-Hexamethyltetracosane **(111-01-3)**
3,3,6,6,9,9-Hexamethyl-1,2,4,5-tetraoxacyclononane (Not more than 52% in solution) **(22397-33-7)**
3,3,6,6,9,9-Hexamethyl-1,2,4,5-tetraoxacyclononane (Not more than 52% with inert solid) **(22397-33-7)**
3,3,6,6,9,9-Hexamethyl-1,2,4,5-tetraoxacyclononane (Technically pure) **(22397-33-7)**
Hexamethyl-3,3,6,6,9,9 tetraoxanonane-1,2,4,5 (French) **(22397-33-7)**
3,3,6,6,9.9-Hexamethyl-1,2,4,5-tetraoxocyclononane **(22397-33-7)**
3,3,6,6,9,9-Hexamethyl-1,2,4,5-tetroxonane **(22397-33-7)**
Hexametilenotriperoxidiamina (Spanish) **(283-66-9)**
3,3,6,6,9,9-Hexametil-1,2,4,5-tetraoxaciclononano (Spanish) **(22397-33-7)**
Hexamic acid **(100-88-9)**
Hexamidine **(125-33-7)**
Hexamidine (The antispasmodic) **(125-33-7)**
Hexamine [UN 1328] **(100-97-0)**
Hexamite **(107-49-3)**
Hexamol SLS **(151-21-3)**
1-Hexanal **(66-25-1)**
Hexanal **(66-25-1)**
Hexanal, 5,5-dimethyl- (9CI) **(55320-58-6)**
Hexanal, 4,4-dimethyl- (7CI,8CI,9CI) **(5932-91-2)**
Hexanal, 2-ethyl **(123-05-7)**
Hexanal, 2-ethyl-3-hydroxy- **(496-03-7)**
Hexanal, 4-methyl- (9CI) **(41065-97-8)**
Hexanal, 5-methyl- (8CI,9CI) **(1860-39-5)**
Hexanamide **(628-02-4)**
Hexanamide, N-acetyl- (7CI,8CI,9CI) **(10601-70-4)**
Hexanamide, 2-(2,4-bis(1,1-dimethylpropyl)phenoxy)-N-(4-((2,2,3,3,4,4,4-heptafluoro-1-oxobutyl)amino)-3-hydroxyphenyl)-(9CI) **(2923-93-5)**
Hexanamide, N-hexyl- (7CI,8CI,9CI) **(10264-29-6)**
1-Hexanamine **(111-26-2)**
2-Hexanamine (9CI) **(5329-79-3)**
1-Hexanamine, N,N-dihexyl- (9CI) **(102-86-3)**
1-Hexanamine, 2-ethyl-N-(2-ethylhexyl)- **(106-20-7)**
1-Hexanamine, n-hexyl- **(143-16-8)**
1-Hexanamine, N-propyl- (9CI) **(20193-23-1)**
1-Hexanamine, 1,1,2,2,3,3,4,4,5,5,6,6,6-tridecafluoro-N,N-bis-(tridecafluorohexyl)- (9CI) **(432-08-6)**
1-Hexanaminium, N,N-bis(2-hydroxyethyl)-N-((octylthio)methyl)-, chloride (9CI) **(78865-85-7)**
1-Hexanaminium, N-((dodecylthio)methyl)-N,N-bis(2-hydroxyethyl)-, chloride (9CI) **(78865-87-9)**
Hexanaphthene **(110-82-7)**
Hexanaphthylene **(110-83-8)**
Hexanastab oral **(56-29-1)**
1,6-Hexandiamine, Vegetable oil fatty acids diamide **(73398-58-0)**
Hexane **(110-54-3)**
n-Hexane (ACGIH,OSHA) **(110-54-3)**
Hexane [UN 1208] **(110-54-3)**
Hexane, 3,4-bis(p-hydroxyphenyl)- **(84-16-2)**
Hexane, 1-bromo **(111-25-1)**
Hexane, 2-bromo- (9CI) **(3377-86-4)**
Hexane, 3-bromo- (9CI) **(3377-87-5)**
1-Hexanecarboxylic acid **(111-14-8)**
Hexane, 1-chloro- **(544-10-5)**
Hexane, 2-chloro- (9CI) **(638-28-8)**
Hexane, chloro- (8CI,9CI) **(25495-90-3)**
Hexane, 1-cyclohexyl- (8CI) **(4292-75-5)**
Hexanediamide (9CI) **(628-94-4)**
1,6-Hexanediamine **(124-09-4)**
1,6-Hexanediamine, N-(6-aminohexyl)- (9CI) **(143-23-7)**
1,6-Hexanediamine, N,N'-dibutyl **(4835-11-4)**
1,6-Hexanediamine, dihydrochloride **(6055-52-3)**
Hexanediamine dihydrochloride **(6055-52-3)**
1,6-Hexanediamine, N,N,N',N'-tetrabutyl- (9CI) **(27090-63-7)**
1,6-Hexanediamine, N,N,N',N'-tetramethyl **(111-18-2)**
1,6-Hexanediamine, trimethyl **(25620-58-0)**
1,6-Hexanediaminium,N,N'-di-9H-fluoren-9-yl-N,N,N',N'-tetramethyl **(4844-10-4)**
1,1-Hexanedicarboxylic acid **(616-88-6)**
1,6-Hexanedicarboxylic acid **(505-48-6)**
Hexane, 1,2-dichloro- (9CI) **(2162-92-7)**
Hexane, 1,6-dichloro- (9CI) **(2163-00-0)**
Hexane, 2,5-dichloro- (6CI,7CI,8CI,9CI) **(13275-18-8)**
Hexane, dichloro-1,6-bis(dichloropropoxy)- (9CI) **(99308-27-7)**
Hexane, 1,2:5,6-diepoxy **(1888-89-7)**
Hexane, 1,6-diisocyanato **(822-06-0)**
Hexane, 1,6-diisocyanato-2,2,4-trimethyl- (9CI) **(16938-22-0)**
Hexane, 1,6-diisocyanato-2,4,4-trimethyl- (9CI) **(15646-96-5)**
Hexane, 2,2-dimethyl- (9CI) **(590-73-8)**
Hexane, 2,3-dimethyl- (9CI) **(584-94-1)**
Hexane, 2,4-dimethyl- (9CI) **(589-43-5)**
Hexane, 2,5-dimethyl- (9CI) **(592-13-2)**
Hexane, 3,3-dimethyl- (9CI) **(563-16-6)**
Hexane, 3,4-dimethyl- (9CI) **(583-48-2)**
Hexane, dimethyl- (8CI,9CI) **(28777-67-5)**
Hexane, 2,5-dimethyl-2,5-di(t-butylperoxy) **(78-63-7)**
Hexane, 2,5-dimethyl-2,5-di(t-butylperoxy)-, Maximum concentration 52% with inert solid **(78-63-7)**
Hexane, dimethyl-, dihydroperoxide **(3025-88-5)**
Hexanedinitrile **(111-69-3)**
1,6-Hexanedioic acid **(124-04-9)**
Hexanedioic acid **(124-04-9)**
Hexanedioic acid, Compd. with 1,6-hexanediamine (1:1) (9CI) **(3323-53-3)**
Hexanedioic acid, Compd. with 1,6-hexanediamine (9CI) **(15511-81-6)**
Hexanedioic acid, Polymer with 1,4-butanediol (9CI) **(25103-87-1)**
Hexanedioic acid, Polymer with 1,2-ethanediol (9CI) **(24938-37-2)**
Hexanedioic acid, Polymer with 1,2-propanediol (9CI) **(25101-03-5)**
Hexanedioic acid, bis(2-(2-butoxyethoxy)ethyl) ester (9CI) **(141-17-3)**
Hexanedioic acid, bis(2-butoxyethyl) ester (9CI) **(141-18-4)**
Hexanedioic acid, bis(2-((4-(2,2-dicyanoethenyl)-3-methylphenyl)ethylamino)ethyl) ester (9CI) **(25857-05-0)**
Hexanedioic acid, bis(2-ethylhexyl) ester (9CI) **(103-23-1)**
Hexanedioic acid, bis(2-(hexyloxy)ethyl)ester (9CI) **(110-32-7)**
Hexanedioic acid, bis(2-mercaptoethyl)ester) **(10194-00-0)**
Hexanedioic acid, bis(2-mercaptoethyl) ester (9CI) **(10194-00-0)**
Hexanedioic acid, bis(1-methylethyl) ester (9CI) **(6938-94-9)**
Hexanedioic acid, bis(1-methylheptyl) ester (9CI) **(108-63-4)**
Hexanedioic acid, bis(2-methylpropyl) ester (9CI) **(141-04-8)**
Hexanedioic acid, bis(7-oxabicyclo(4.1.0)hept-3-ylmethyl ester **(3130-19-6)**
Hexanedioic acid, bis(oxiranylmethyl) ester (9CI) **(2754-17-8)**
Hexanedioic acid, decyl hexyl ester (9CI) **(22707-35-3)**
Hexanedioic acid, decyl octyl ester (9CI) **(110-29-2)**
Hexanedioic acid, dibutyl ester **(105-99-7)**

Hexanedioic acid, dicyclohexyl ester (9CI)	**(849-99-0)**
Hexanedioic acid, didecyl ester (9CI)	**(105-97-5)**
Hexanedioic acid, diethyl ester (9CI)	**(141-28-6)**
Hexanedioic acid, diheptyl ester (9CI)	**(14697-48-4)**
Hexanedioic acid, dihexadecyl ester (9CI)	**(26720-21-8)**
Hexanedioic acid, dihexyl ester	**(110-33-8)**
Hexanedioic acid, diisodecyl ester (9CI)	**(27178-16-1)**
Hexanedioic acid, diisononyl ester (9CI)	**(33703-08-1)**
Hexanedioic acid, diisooctyl ester (9CI)	**(1330-86-5)**
Hexanedioic acid, 2,2-dimethyl-, dimethyl ester (6CI,7CI,8CI, 9CI)	**(17219-21-5)**
Hexanedioic acid, dimethyl ester (9CI)	**(627-93-0)**
Hexanedioic acid, dinitrile	**(111-69-3)**
Hexanedioic acid, dinonyl ester (9CI)	**(151-32-6)**
Hexanedioic acid, dioctyl ester	**(103-23-1)**
Hexanedioic acid, dioctyl ester (9CI)	**(123-79-5)**
Hexanedioic acid, dipentyl ester (9CI)	**(14027-78-2)**
Hexanedioic acid, di-2-propenyl ester (9CI)	**(2998-04-1)**
Hexanedioic acid, dipropyl ester (9CI)	**(106-19-4)**
Hexanedioic acid, disodium salt (9CI)	**(7486-38-6)**
Hexanedioic acid, ditridecyl ester (9CI)	**(16958-92-2)**
Hexanedioic acid, ester with 2,2'-oxybis(ethanol) (9CI)	**(58984-19-3)**
Hexanedioic acid, 2-ethylhexyl isodecyl ester (9CI)	**(68052-04-0)**
Hexanedioic acid, isodecyl isooctyl ester (9CI)	**(31474-57-4)**
Hexanedioic acid, 2-methyl- (8CI,9CI)	**(626-70-0)**
Hexanedioic acid, 3-methyl-, (S)- (9CI)	**(61898-58-6)**
Hexanedioic acid, mono(2-ethylhexyl) ester (9CI)	**(4337-65-9)**
Hexanedioic acid, monomethyl ester (9CI)	**(627-91-8)**
Hexanedioic acid, sodium salt (9CI)	**(23311-84-4)**
1,2-Hexanediol	**(107-41-5)**
1,6-Hexanediol	**(629-11-8)**
2,5-Hexanediol	**(2935-44-6)**
1,6-Hexanediol diacrylate	**(13048-33-4)**
1,6-Hexanediol diisocyanate	**(822-06-0)**
1,6-Hexanediol dimethacrylate	**(6606-59-3)**
2,5-Hexanediol, 2,5-dimethyl- (9CI)	**(110-03-2)**
1,3-Hexanediol, 2-ethyl	**(94-96-2)**
2,5-Hexanedione	**(110-13-4)**
Hexanedioyl dichloride	**(111-50-2)**
1,6-Hexanediyl 2-methyl-2-propenoate	**(6606-59-3)**
Hexane, 2,3-epoxy-, cis- (8CI)	**(6124-90-9)**
Hexane, 2,3-epoxy-2-methyl- (8CI)	**(17612-35-0)**
Hexane, 1-(ethenyloxy)- (9CI)	**(5363-64-4)**
Hexane, 3-ethyl- (9CI)	**(619-99-8)**
Hexane, 1,1'-(ethylidenebis(oxy))bis- (9CI)	**(5405-58-3)**
Hexane, 3-ethyl-3-methyl- (8CI,9CI)	**(3074-76-8)**
Hexane, 1-fluoro	**(373-14-8)**
Hexane, 1,2,2,5,5,6-hexachloro- (9CI)	**(83682-29-5)**
Hexane, 1,2,3,4,5,6-hexachloro- (8CI)	**(18585-38-1)**
Hexane, 2,2,3,4,5,5-hexachloro- (9CI)	**(83682-28-4)**
1,2,3,4,5,6-Hexanehexol	**(69-65-8)**
Hexane, 1-iodo- (9CI)	**(638-45-9)**
Hexane, iodo- (8CI,9CI)	**(25495-92-5)**
Hexane, 1-(isopropylidenecyclopropyl)- (8CI)	**(24524-53-6)**
6-Hexanelactam	**(105-60-2)**
Hexane, 2-methyl- (9CI)	**(591-76-4)**
Hexane, 3-methyl- (9CI)	**(589-34-4)**
Hexane, 3-methyl-4-methylene- (9CI)	**(3404-67-9)**
Hexanen (Dutch)	**(110-54-3)**
Hexanenitrile	**(628-73-9)**
Hexanenitrile, 6-amino- (9CI)	**(2432-74-8)**
Hexanenitrile, 5-methyl- (6CI,7CI,8CI,9CI)	**(19424-34-1)**
Hexane, 1-nitro- (8CI,9CI)	**(646-14-0)**
Hexane, 1,1,1,2,2,3,3,4,4-nonafluoro-6-iodo- (9CI)	**(2043-55-2)**
Hexane, 1,1'-oxybis-	**(112-58-3)**
Hexane, 1,1'-oxybis(2-ethyl-	**(10143-60-9)**
Hexane, 1-(pentyloxy)- (9CI)	**(32357-83-8)**
Hexaneperoxoic acid, 2-ethyl-, 1,1-dimethylethyl ester (9CI)	**(3006-82-4)**
Hexaneperoxoic acid, 2-ethyl-, 1,1,4,4-tetramethyl-1,4-butanediyl ester	**(13052-09-0)**
Hexane, 1-phenyl- (8CI)	**(1077-16-3)**
Hexane, 2-phenyl- (6CI,7CI,8CI)	**(6031-02-3)**
Hexane, 3-phenyl- (6CI,7CI,8CI)	**(4468-42-2)**
1-Hexanephosphonic acid	**(4721-24-8)**
1-Hexanesulfonamide, N-ethyl-1,1,2,2,3,3,4,4,5,5,6,6,6-tridecafluoro-N-(2-hydroxyethyl)- (9CI)	**(34455-03-3)**
1-Hexanesulfonyl fluoride, 1,1,2,2,3,3,4,4,5,5,6,6,6-tridecafluoro- (9CI)	**(423-50-7)**
Hexane, tetrachloro-1-(dichloropropoxy)- (9CI)	**(99308-25-5)**
Hexane, tetradecachloro- (9CI)	**(83682-34-2)**
Hexane, tetradecafluoro- (9CI)	**(355-42-0)**
Hexane, 2,2,3,3-tetramethyl- (8CI,9CI)	**(13475-81-5)**
Hexane, 2,2,4,5-tetramethyl- (8CI,9CI)	**(16747-42-5)**
Hexane, 2,2,5,5-tetramethyl- (8CI,9CI)	**(1071-81-4)**
Hexanethioic acid, S-butyl ester (7CI,9CI)	**(2432-79-3)**
1-Hexanethiol	**(111-31-9)**
1,3,6-Hexanetricarbonitrile (9CI)	**(1772-25-4)**
Hexane, trichloro-1,6-bis(dichloropropoxy)- (9CI)	**(99308-28-8)**
Hexane, trichloro-1-(dichloropropoxy)- (9CI)	**(99308-24-4)**
Hexane, 2,2,4-trimethyl-	**(16747-26-5)**
Hexane, 2,2,5-trimethyl- (9CI)	**(3522-94-9)**
Hexane, 3,3,4-trimethyl- (9CI)	**(16747-31-2)**
Hexane, 2,2,3-trimethyl- (8CI,9CI)	**(16747-25-4)**
Hexane, 2,3,3-trimethyl- (8CI,9CI)	**(16747-28-7)**
Hexane, 2,3,4-trimethyl- (8CI,9CI)	**(921-47-1)**
Hexane, 2,3,5-trimethyl- (8CI,9CI)	**(1069-53-0)**
Hexane, 2,4,4-trimethyl- (8CI,9CI)	**(16747-30-1)**
1,2,6-Hexanetriol	**(106-69-4)**
Hexane-1,2,6-triol	**(106-69-4)**
Hexanetriol-1,2,6	**(106-69-4)**
1,2,6-Hexanetriol tris(glycidyl) ether	**(68959-23-9)**
Hexanitrate de mannitol (French)	**(15825-70-4)**
Hexanitrato de manitol (Spanish)	**(15825-70-4)**
2,2',4,4',6,6'-Hexanitroazobenzene	**(19159-68-3)**
Hexanitroazoxi benceno (Spanish)	**(19159-68-3)**
Hexanitroazoxybenzene	**(19159-68-3)**
Hexanitroazoxy benzene (French)	**(19159-68-3)**
2,2',4,4',6,6'-Hexanitrodifenylamin (Czech)	**(131-73-7)**
Hexanitrodifenylamine (Dutch)	**(131-73-7)**
2,2',4,4',6,6'-Hexanitrodiphenylamine	**(131-73-7)**
2,4,6,2',4',6'-Hexanitrodiphenylamine	**(131-73-7)**
Hexanitrodiphenylamine (French)	**(131-73-7)**
Hexanitroetano (Spanish)	**(918-37-6)**
Hexanitroethane	**(918-37-6)**
Hexanitroethane (French)	**(918-37-6)**
Hexanitrol	**(15825-70-4)**
Hexanitrooxanilida (Spanish)	**(29135-62-4)**
2,2',4,4',6,6'-Hexanitrooxanilide	**(29135-62-4)**
Hexanitrooxanilide	**(29135-62-4)**
Hexanitrooxanilide (French)	**(29135-62-4)**
Hexanoestrol	**(84-16-2)**
Hexanoic-acid	**(142-62-1)**
n-Hexanoic acid	**(142-62-1)**
Hexanoic acid (DOT)	**(142-62-1)**
Hexanoic acid, allyl ester	**(123-68-2)**
Hexanoic acid, 6-amino	**(60-32-2)**
Hexanoic acid, 6-amino-, cyclic lactam	**(105-60-2)**
Hexanoic acid, 6-amino-, lactam	**(105-60-2)**
Hexanoic acid, benzyl ester (8CI)	**(6938-45-0)**
Hexanoic acid, 2-bromo	**(616-05-7)**
Hexanoic acid, 6-bromo- (9CI)	**(4224-70-8)**
Hexanoic acid, butyl ester	**(626-82-4)**
Hexanoic acid, 2,6-dibromo- (7CI,8CI,9CI)	**(13137-43-4)**
Hexanoic acid, 2,4-dimethyl-, (R-(R*,S*))- (9CI)	**(42329-90-8)**

Hexanoic acid, 2-((1,1-dimethylethyl)dioxy)ethyl ester **(62695-55-0)**
Hexanoic acid, 3,3-dimethyl-4,6,6,6-tetrachloro, methyl ester **(64667-33-0)**
Hexanoic acid, 2,3-epoxypropyl ester **(17526-74-8)**
Hexanoic acid, 2-ethyl **(149-57-5)**
Hexanoic acid, 3-ethyl- (9CI) **(41065-91-2)**
Hexanoic acid, 4-ethyl- (8CI,9CI) **(6299-66-7)**
Hexanoic acid, 2-ethyl-, Compd. with 2,4,6-tris((dimethylamino)-methyl)phenol (9CI) **(51365-70-9)**
Hexanoic acid, 2-ethyl-, barium salt (9CI) **(2457-01-4)**
Hexanoic acid, 2-ethyl-, cadmium salt (9CI) **(2420-98-6)**
Hexanoic acid, 2-ethyl-, calcium salt (9CI) **(136-51-6)**
Hexanoic acid, 2-ethyl-, cobalt salt (9CI) **(13586-82-8)**
Hexanoic acid, 2-ethyl-, cobalt(2+) salt (9CI) **(136-52-7)**
Hexanoic acid, 2-ethyl-, copper salt (9CI) **(22221-10-9)**
Hexanoic acid, 2-ethyl-, diester with tetraethylene glycol **(18268-70-7)**
Hexanoic acid, ethyl-, diester with triethylene glycol **(94-28-0)**
Hexanoic acid, ethyl ester **(123-66-0)**
Hexanoic acid, 2-ethyl-, ethenyl ester (9CI) **(94-04-2)**
Hexanoic acid, 2-ethyl-, 2-ethylhexyl ester **(7425-14-1)**
Hexanoic acid, 2-ethyl-, iron salt (9CI) **(19583-54-1)**
Hexanoic acid, 2-ethyl-, lead salt (9CI) **(16996-40-0)**
Hexanoic acid, 2-ethyl-, lead(2+) salt (9CI) **(301-08-6)**
Hexanoic acid, 2-ethyl-, manganese salt (9CI) **(15956-58-8)**
Hexanoic acid, 2-ethyl-, nickel salt (9CI) **(7580-31-6)**
Hexanoic acid, 2-ethyl-, nickel(2+) salt (9CI) **(4454-16-4)**
Hexanoic acid, 2-ethyl-, oxybis(2,1-ethyldiyloxy-2,1-ethanediyl) ester (9CI) **(18268-70-7)**
Hexanoic acid, 2-ethyl-, potassium salt (9CI) **(3164-85-0)**
Hexanoic acid, 2-ethyl-, sodium salt (9CI) **(19766-89-3)**
Hexanoic acid, 2-ethyl-, tin(2+) salt (9CI) **(301-10-0)**
Hexanoic acid, 2-ethyl-, vinyl ester (8CI) **(94-04-2)**
Hexanoic acid, 2-ethyl-, zinc salt (9CI) **(136-53-8)**
Hexanoic acid, 2-ethyl-, zirconium salt (9CI) **(22464-99-9)**
Hexanoic acid, hexyl ester **(6378-65-0)**
Hexanoic acid, 5-hydroxy-3,5-dimethyl-, δ-lactone **(20628-36-8)**
Hexanoic acid, 5-hydroxy-, δ-lactone **(823-22-3)**
Hexanoic acid, 5-hydroxy-, lactone **(823-22-3)**
Hexanoic acid, 5-hydroxy-2-methyl-, .δ.-lactone (7CI) **(3720-22-7)**
Hexanoic acid, isobutyl ester **(105-79-3)**
Hexanoic acid, isopentyl ester **(2198-61-0)**
Hexanoic acid, isopropyl ester (8CI) **(2311-46-8)**
Hexanoic acid, ε-lactone **(502-44-3)**
Hexanoic acid, 2-methyl- (9CI) **(4536-23-6)**
Hexanoic acid, 4-methyl- (9CI) **(1561-11-1)**
Hexanoic acid, 5-methyl- (8CI,9CI) **(628-46-6)**
Hexanoic acid, methyl ester **(106-70-7)**
Hexanoic acid, 1-methylethyl ester (9CI) **(2311-46-8)**
Hexanoic acid, 5-methyl-4-oxo- (9CI) **(41654-04-0)**
Hexanoic acid, 2-methyl-3-oxo-, ethyl ester (8CI,9CI) **(29304-40-3)**
Hexanoic acid, 6-(methyl(phenylsulfonyl)amino)-, Compd. with 2,2',2''-nitrilotris(ethanol) (1:1) (9CI) **(26919-50-6)**
Hexanoic acid, 2-methyl-2-propyl- (8CI,9CI) **(31080-37-2)**
Hexanoic acid, 2-methylpropyl ester **(105-79-3)**
Hexanoic acid, 6-(nitrooxy)- (9CI) **(74754-55-5)**
Hexanoic acid, octyl ester (8CI,9CI) **(4887-30-3)**
Hexanoic acid, 5-oxo- (9CI) **(3128-06-1)**
Hexanoic acid, pentyl ester (9CI) **(540-07-8)**
Hexanoic acid, phenylmethyl ester (9CI) **(6938-45-0)**
Hexanoic acid, propyl ester (9CI) **(626-77-7)**
Hexanoic acid, sodium salt (9CI) **(10051-44-2)**
Hexanoic acid, 4,6,6,6-tetrachloro-3,3-dimethyl-, methyl ester (9CI) **(64667-33-0)**
Hexanoic acid, thio-, S-butyl ester (8CI) **(2432-79-3)**
1-Hexanol **(111-27-3)**
Hexanol **(111-27-3)**
2-Hexanol (9CI) **(626-93-7)**
3-Hexanol (9CI) **(623-37-0)**
n-Hexanol [UN 2282] **(111-27-3)**
sec-Hexanol [UN 2282] **(97-95-0)**
Hexanol (VAN)(9CI) **(25917-35-5)**
Hexanolactam **(105-60-2)**
6-Hexanolactone **(502-44-3)**
γ-Hexanolactone **(695-06-7)**
1-Hexanol, aluminum salt (9CI) **(23275-26-5)**
3-Hexanol, 3,5-dimethyl- (9CI) **(4209-91-0)**
2-Hexanol, 2,3-dimethyl- (8CI,9CI) **(19550-03-9)**
2-Hexanol, 2,5-dimethyl- (8CI,9CI) **(3730-60-7)**
3-Hexanol, 2,4-dimethyl- (8CI,9CI) **(13432-25-2)**
2-Hexanol-2-((1,1-dimethylethyl)azo)-5-methyl- **(64819-51-8)**
2-Hexanol, 2-((1,1-dimethylethyl)azo)-5-methyl- (9CI) **(64819-51-8)**
1-Hexanol, 2,3-epoxy-2-ethyl **(78-72-8)**
1-Hexanol, 2-ethyl **(104-76-7)**
3-Hexanol, 3-ethyl- (9CI) **(597-76-2)**
1-Hexanol, 2-ethyl-, acrylate **(103-11-7)**
1-Hexanol, 2-ethyl-, ester with diphenyl phosphate **(1241-94-7)**
1-Hexanol, 2-ethyl-, hydrogen sulfate, sodium salt **(126-92-1)**
1-Hexanol, 2-ethyl-, phosphate **(78-42-2)**
1-Hexanol, 2-ethyl-, sebacate **(122-62-3)**
1-Hexanol, 2-ethyl-, titanium(4+) salt (9CI) **(1070-10-6)**
1,6-Hexanolide **(502-44-3)**
Hexanolide-1,4 **(695-06-7)**
3-Hexanol, 3-methyl **(597-96-6)**
1-Hexanol, methyl- (9CI) **(61949-26-6)**
2-Hexanol, 2-methyl- (9CI) **(625-23-0)**
1-Hexanol, 3-methyl- (8CI,9CI) **(13231-81-7)**
1-Hexanol, 5-methyl- (8CI,9CI) **(627-98-5)**
2-Hexanol, 3-methyl- (8CI,9CI) **(2313-65-7)**
2-Hexanol, 5-methyl- (8CI,9CI) **(627-59-8)**
3-Hexanol, 2-methyl- (8CI,9CI) **(617-29-8)**
3-Hexanol, 4-methyl- (8CI,9CI) **(615-29-2)**
3-Hexanol, 5-methyl- (8CI,9CI) **(623-55-2)**
1-Hexanol, 3,3,4,4,5,5,6,6,6-nonafluoro- (9CI) **(2043-47-2)**
1-Hexanol, 3,5,5-trimethyl- (9CI) **(3452-97-9)**
Hexanon **(108-94-1)**
3-Hexanone **(589-38-8)**
Hexanone-2 **(591-78-6)**
2-Hexanone (OSHA) **(591-78-6)**
2-Hexanone, 6-(acetyloxy)- (9CI) **(4305-26-4)**
2-Hexanone, 6-bromo- (8CI,9CI) **(10226-29-6)**
2-Hexanone, 1,1-dichloro- (7CI,8CI,9CI) **(2648-59-1)**
3-Hexanone, 4,4-dichloro- (7CI,8CI,9CI) **(2648-60-4)**
3-Hexanone, 2,5-dimethyl- (9CI) **(1888-57-9)**
2-Hexanone, 3,3-dimethyl- (6CI,8CI,9CI) **(26118-38-7)**
3-Hexanone, 6-(dimethylamino)-4,4-diphenyl- (8CI,9CI) **(467-85-6)**
3-Hexanone, 6-(dimethylamino)-4,4-diphenyl-5-methyl- **(466-40-0)**
3-Hexanone, 4,4-diphenyl-6-piperidino- (8CI) **(561-48-8)**
3-Hexanone, 4,4-diphenyl-6-(1-piperidinyl)- (9CI) **(561-48-8)**
2-Hexanone, 6-hydroxy-, acetate (7CI,8CI) **(4305-26-4)**
Hexanone isoxime **(105-60-2)**
2-Hexanone, 6-methoxy- (8CI,9CI) **(29006-00-6)**
2-Hexanone, 5-methyl **(110-12-3)**
3-Hexanone, 2-methyl **(7379-12-6)**
2-Hexanone, 4-methyl- (9CI) **(105-42-0)**
3-Hexanone, 5-methyl- (8CI,9CI) **(623-56-3)**
2-Hexanone, 5-methyl-3-methylene- (6CI,7CI,8CI,9CI) **(1187-87-7)**
2-Hexanone, 5-methyl-5-phenyl- (8CI,9CI) **(14128-61-1)**
2-Hexanone, 1-phenyl- (8CI,9CI) **(25870-62-6)**
2-Hexanone, 6-phenyl- (8CI,9CI) **(14171-89-2)**
Hexanonisoxim (German) **(105-60-2)**
1,4,7,10,13,16-Hexanoxacyclooctadecane **(17455-13-9)**
Hexanoyl chloride, 2-ethyl **(760-67-8)**
Hexanoyl chloride, 3,5,5-trimethyl- (9CI) **(36727-29-4)**
17-α-Hexanoyloxypregn-4-ene-3,20-dione **(630-56-8)**
3,4,8,9,12,13-Hexaoxa-1,6-diazabicyclo(4.4.4)tetradecane, Dry **(283-66-9)**

2,5,7,10,11,14-Hexaoxa-1,6-distibabicyclo(4.4.4)tetradecane (9CI) **(29736-75-2)**
3,6,9,12,15,18-Hexaoxaeicosane (9CI) **(23601-39-0)**
3,6,9,12,15,18-Hexaoxatetracosan-1-ol, 21,23-dimethyl-19-(2-methylpropyl)- (9CI) **(14149-99-6)**
3,6,9,12,15,18-Hexaoxatetracosan-1-ol, 19-isobutyl-21,23-dimethyl- (8CI) **(14149-99-6)**
3,6,9,12,15,18-Hexaoxatriacontan-1-ol **(3055-96-7)**
5,8,11,13,16,19-Hexaoxatricosane (9CI) **(143-29-3)**
Hexa(oxydiethanol) monodecyl ether **(3055-96-7)**
Hexaoxyethylene dodecyl monoether **(3055-96-7)**
Hexaoxyethylene glycol **(2615-15-8)**
Hexaplas BUT **(27941-09-9)**
Hexaplas DIOP **(27554-26-3)**
Hexaplas M/1B **(84-69-5)**
Hexaplas M/B **(84-74-2)**
Hexaplas M/O **(27554-26-3)**
Hexaplas PPA **(27941-08-8)**
Hexaplin **(50-55-5)**
Hexapon (9CI) **(39425-24-6)**
threo-Hexaric acid, 2,5-anhydro-3,4-dideoxy- (9CI) **(2044-00-0)**
Hexasiloxane, 1,1,1,3,5,7,9,11,11,11-decamethyl-3,5,7,9-tetraphenyl- (7CI,8CI,9CI) **(13271-58-4)**
Hexasiloxane, 1,1,1,3,3,5,7,7,9,11,11,11-dodecamethyl-5,9-diphenyl- (9CI) **(60573-45-7)**
Hexasiloxane, tetradecamethyl- (9CI) **(107-52-8)**
Hexastat **(645-05-6)**
Hexasul **(7704-34-9)**
Hexathane **(12122-67-7)**
1,1'-(Hexathiodicarbonothioyl)bispiperidine **(971-15-3)**
Hexathir **(137-26-8)**
Hexatox **(58-89-9)**
Hexatriacontane (9CI) **(630-06-8)**
2,6,10,14,18,22,26,30,34-Hexatriacontanonaen-1-ol, 3,7,11,15,19,23,27,31,35-nonamethyl-, (Z,Z,Z,Z,Z,Z,E,E)- (9CI) **(13190-97-1)**
1,3,5-Hexatriene **(2235-12-3)**
1,3,5-Hexatriene, 2,5-bis(chloromethyl)- (9CI) **(83682-56-8)**
1,2,5-Hexatriene, 3-chloro- (9CI) **(83682-35-3)**
1,3,5-Hexatriene, 1,6-dichloro- (9CI) **(69645-07-4)**
1,3,5-Hexatriene, 3,4-dichloro- (6CI,9CI) **(83682-37-5)**
1,3,5-Hexatriene, 3,4-dichloro-2,5-bis(chloromethyl)- (9CI) **(83682-55-7)**
1,3,5-Hexatriene, 3,4-dichloro-2,5-bis(dichloromethyl)- (9CI) **(83682-57-9)**
1,3,5-Hexatriene, 3,4-dichloro-2,5-bis(trichloromethyl)- (9CI) **(83682-58-0)**
1,3,5-Hexatriene, 1,6-dichloro-2,5-dimethyl- (9CI) **(83682-54-6)**
1,3,5-Hexatriene, 2,5-dimethyl- (6CI,7CI,8CI,9CI) **(4916-63-6)**
1,3,4-Hexatriene, 3-methoxy- (9CI) **(53783-88-3)**
1,3,5-Hexatriyne (9CI) **(3161-99-7)**
Hexatriyne (6CI,7CI,8CI) **(3161-99-7)**
Hexatype Brown N **(6416-57-5)**
Hexatype Carmine B **(6368-72-5)**
Hexaverm **(58-89-9)**
Hexavibex **(58-56-0)**
Hexavin **(63-25-2)**
Hexazane **(110-89-4)**
Hexazinone **(51235-04-2)**
Hexazir **(137-30-4)**
Hexemal **(52-31-3)**
2-Hexenal **(505-57-7)**
2-Hexenal, (E) **(6728-26-3)**
2-trans-Hexenal **(6728-26-3)**
3-(Z)-Hexenal **(6789-80-6)**
3-Hexenal, (E) **(69112-21-6)**
Hex-2-en-1-al **(505-57-7)**
Hex-2-enal **(505-57-7)**
Hexenal **(56-29-1)**
cis-3-Hexenal **(6789-80-6)**
trans-2-Hexenal **(6728-26-3)**
trans-3-Hexenal **(69112-21-6)**
Hexenal (VAN)(9CI) **(1335-39-3)**
3-Hexenal, (Z) **(6789-80-6)**
2-Hexenal, (Z)- (8CI,9CI) **(16635-54-4)**
Hexenal (barbiturate) **(56-29-1)**
5-Hexenal, 2,6-epoxy- **(100-73-2)**
2-Hexenal, 2-ethyl **(645-62-5)**
Hexenal, 2-ethyl **(26266-68-2)**
(E)-3-Hexene **(13269-52-8)**
2-Hexene **(592-43-8)**
2-Hexene-cis **(7688-21-3)**
Hexene **(25264-93-1)**
tert-Hexene **(558-37-2)**
trans-3-Hexene **(13269-52-8)**
2-Hexene, (E)- (9CI) **(4050-45-7)**
3-Hexene, (E)- (9CI) **(13269-52-8)**
3-Hexene (8CI,9CI) **(592-47-2)**
Hex-1-ene (DOT) **(592-41-6)**
2-Hexene (Mixed cis & trans) **(592-43-8)**
1-Hexene [UN 2370] **(592-41-6)**
2-Hexene, (Z)- (9CI) **(7688-21-3)**
3-Hexene, (Z)- (9CI) **(7642-09-3)**
3-Hexene,3,4-bis(p-hydroxyphenyl)- **(56-53-1)**
1-Hexene, 5,5-dimethyl- (9CI) **(7116-86-1)**
3-Hexene, 3,4-dimethyl- (9CI) **(30951-95-2)**
1-Hexene, 4,5-dimethyl- (8CI,9CI) **(16106-59-5)**
2-Hexene, 2,3-dimethyl- (8CI,9CI) **(7145-20-2)**
2-Hexene, 2,5-dimethyl- (8CI,9CI) **(3404-78-2)**
Hexene, 2,5-dimethyl- (8CI,9CI) **(26266-69-3)**
3-Hexene, 2,5-dimethyl- (6CI,7CI,8CI,9CI) **(15910-22-2)**
2-Hexenedinitrile **(13042-02-9)**
3-Hexenedinitrile **(1119-85-3)**
3-Hexenedinitrile (8CI,9CI) **(1119-85-3)**
3-Hexene-2,5-diol (8CI,9CI) **(7319-23-5)**
3-Hexene-2,5-dione (9CI) **(4436-75-3)**
3-Hexene-2,5-dione, (E)- (8CI,9CI) **(820-69-9)**
3-Hexene-2,5-dione, (Z)- (8CI,9CI) **(17559-81-8)**
1-Hexene, 2-ethyl **(1632-16-2)**
3-Hexene, 2,2,3,4,5,5-hexachloro- (9CI) **(83682-31-9)**
3-Hexene, 1,2,3,4,5,6-hexachloro- (6CI,7CI,9CI) **(1725-74-2)**
2-Hexene, 1-methoxy-, (E)- (9CI) **(56052-83-6)**
1-Hexene, 3-methyl- (9CI) **(3404-61-3)**
2-Hexene, 5-methyl- (9CI) **(3404-62-4)**
3-Hexene, 2-methyl- (9CI) **(42154-69-8)**
1-Hexene, 2-methyl- (8CI,9CI) **(6094-02-6)**
1-Hexene, 4-methyl- (8CI,9CI) **(3769-23-1)**
2-Hexene, 2-methyl- (8CI,9CI) **(2738-19-4)**
2-Hexene, 3-methyl- (6CI,7CI,8CI,9CI) **(17618-77-8)**
3-Hexene, 3-methyl- (6CI,7CI,8CI,9CI) **(3404-65-7)**
2-Hexene, 3-methyl-, (Z)- (9CI) **(10574-36-4)**
2-Hexene, 2-nitro **(6065-17-4)**
1-Hexene, 3,3,4,4,5,5,6,6,6-nonafluoro- (9CI) **(19430-93-4)**
Hexene-ol **(111-28-4)**
2-Hexene-4-one **(2497-21-4)**
3-Hexene, 2,2,5,5-tetramethyl- (8CI,9CI) **(22808-06-6)**
2-Hexene, 1,?,?-trichloro- (9CI) **(99308-22-2)**
1-Hexene, 3,3,5-trimethyl- (9CI) **(13427-43-5)**
1-Hexene, 3,4,5-trimethyl- (9CI) **(56728-10-0)**
Hexene, trimethyl- (9CI) **(95461-54-4)**
1-Hexene, 1-(2-vinylcyclopropyl)-, (E)-trans-(.+-.)- (8CI) **(22822-99-7)**
2-Hexenoic acid (9CI) **(1191-04-4)**
5-Hexenoic acid (8CI,9CI) **(1577-22-6)**
4-Hexenoic acid, 2-acetyl-5-hydroxy-3-oxo-, δ-lactone **(520-45-6)**
4-Hexenoic acid, 2-acetyl-5-hydroxy-3-oxo, δ-lactone, sodium deriv. **(4418-26-2)**
2-Hexenoic acid, 2-ethyl- **(5309-52-4)**
2-Hexenoic acid, 5-hydroxy-, δ-lactone **(10048-32-5)**

3-Hexenoic acid, 2-methyl-, (E)- (9CI) **(62243-57-6)**
4-Hexenoic acid, 3-methyl-2,6-dioxo- (9CI) **(56771-77-8)**
2-Hexenoic acid, 3,4,4-trimethyl-5-oxo-, (E)- (8CI) **(6994-95-2)**
(E)-3-Hexen-1-ol **(928-97-2)**
2-Hexen-1-ol, (E) **(928-95-0)**
2-Hexen-1-ol, trans- **(928-95-0)**
2-Hexenol **(928-95-0)**
3-Hexen-1-ol **(544-12-7)**
3-Hexen-1-ol, (E)- **(928-97-2)**
3-Hexen-1-ol, cis- **(928-96-1)**
Hexenol **(111-28-4)**
β-γ-Hexenol **(928-96-1)**
cis-3-Hexen-1-ol **(928-96-1)**
cis-3-Hexenol **(928-96-1)**
trans-2-Hexenol **(928-95-0)**
trans-3-Hexen-1-ol **(928-97-2)**
trans-3-Hexenol **(928-97-2)**
1-Hexen-3-ol (9CI) **(4798-44-1)**
2-Hexen-1-ol (9CI) **(2305-21-7)**
5-Hexen-2-ol, (.+-.)- (9CI) **(54774-27-5)**
4-Hexen-3-ol (8CI,9CI) **(4798-58-7)**
3-Hexen-1-ol, (Z) **(928-96-1)**
2-Hexen-1-ol acetate **(2497-18-9)**
3-Hexen-1-ol, acetate (9CI) **(1708-82-3)**
3-Hexen-1-ol, acetate, (Z)- **(3681-71-8)**
2-Hexen-1-ol, acetate, (Z)- (9CI) **(56922-75-9)**
γ-Hexenolactone **(10048-32-5)**
4-Hexen-1-ol, 1,5-dimethyl-1-vinyl-, o-aminobenzoate **(7149-26-0)**
2-Hexen-1-ol, 2-ethyl- (9CI) **(50639-00-4)**
2-Hexen-5,1-olide **(10048-32-5)**
D''-Hexenollactone **(10048-32-5)**
2-Hexen-1-ol, 3-methyl-, (E)- (8CI,9CI) **(30801-96-8)**
1-Hexen-5-one **(109-49-9)**
2-Hexen-4-one **(2497-21-4)**
4-Hexen-2-one **(25659-22-7)**
5-Hexen-2-one (9CI) **(109-49-9)**
1-Hexen-3-one (8CI,9CI) **(1629-60-3)**
3-Hexen-2-one (8CI,9CI) **(763-93-9)**
3-Hexen-2-one, 3,4-dimethyl-, (Z)- (8CI) **(20685-45-4)**
3-Hexen-2-one, 5-methyl **(5166-53-0)**
D-erythro-Hex-2-enonic acid, γ-lactone **(89-65-6)**
D-erythro-Hex-2-enonic acid, γ-lactone, monosodium salt (9CI) **(6381-77-7)**
D-erythro-Hex-2-enonic acid, γ-lactone, sodium salt (9CI) **(7378-23-6)**
(E)-2-Hexenyl acetate **(2497-18-9)**
2-Hexen-1-yl-acetate **(2497-18-9)**
2-Hexenyl acetate **(2497-18-9)**
Hex-2-enyl acetate **(2497-18-9)**
trans-2-Hexenyl acetate **(2497-18-9)**
Hexermin **(58-56-0)**
Hexermin P **(54-47-7)**
Hexestrol **(84-16-2)**
meso-Hexestrol **(84-16-2)**
Hexetidina (Spanish) **(141-94-6)**
Hexetidine **(141-94-6)**
Hexetidinum (Latin) **(141-94-6)**
Hexicide **(58-89-9)**
Hexide **(70-30-4)**
Hexilmethylenamine **(100-97-0)**
Hexmethylphosphoramide **(680-31-9)**
Hexobarbital **(56-29-1)**
Hexobarbitone **(56-29-1)**
Hexobion **(58-56-0)**
Hexoestrol **(84-16-2)**
Hexofen **(77-75-8)**
Hexogeen (Dutch) **(121-82-4)**
Hexogen (Explosive), Wetted with not less than 15 per cent water, by mass [UN 0072] **(121-82-4)**
Hexogen 5W **(121-82-4)**
n-Hexoic acid **(142-62-1)**
1,6-Hexolactam **(105-60-2)**
Hexolite, Dry or wetted with less than 15 per cent water, by mass [UN 0118] **(121-82-4)**
Hexon (Czech) **(108-10-1)**
Hexone **(108-10-1)**
Hexone (OSHA) **(108-10-1)**
Hexophene **(70-30-4)**
Hexopyrimidine **(141-94-6)**
Hexoral **(141-94-6)**
Hexosan **(70-30-4)**
n-Hexoxyethoxyethanol **(112-59-4)**
Hexteroauxin **(87-51-4)**
Hextril **(141-94-6)**
Hexyclan **(58-89-9)**
Hexyl (German, Dutch) **(131-73-7)**
N-Hexylacetamide **(7501-79-3)**
Hexyl acetate **(142-92-7)**
l-Hexyl acetate **(142-92-7)**
n-Hexyl acetate **(142-92-7)**
sec-Hexyl acetate (ACGIH,OSHA) **(108-84-9)**
Hexylacetylene **(629-05-0)**
β-Hexylacrolein **(2463-53-8)**
Hexyl acrylate **(2499-95-8)**
n-Hexyl acrylate **(2499-95-8)**
Hexyl-alcohol **(111-27-3)**
n-Hexyl alcohol **(111-27-3)**
sec-Hexyl alcohol **(97-95-0)**
Hexyl alcohol, acetate **(142-92-7)**
Hexyl alcohol, ethoxylated **(31726-34-8)**
Hexylamine **(111-26-2)**
n-Hexylamine **(111-26-2)**
Hexylamine, N-cyclohexyl-2-ethyl- (8CI) **(5432-61-1)**
Hexylamine, 2-ethyl **(104-75-6)**
Hexylamine, 2-ethyl-n-phenyl- **(10137-80-1)**
Hexylamine, 1-methyl **(123-82-0)**
Hexylamine, N-propyl- (8CI) **(20193-23-1)**
Hexylan **(608-73-1)**
p-Hexylaniline **(33228-45-4)**
(2-Hexyl)benzene **(6031-02-3)**
(3-Hexyl)benzene **(4468-42-2)**
Hexylbenzene **(1077-16-3)**
n-Hexylbenzene **(1077-16-3)**
tert-Hexylbenzene **(1985-57-5)**
4-Hexyl-1,3-benzenediol **(136-77-6)**
Hexylbenzenesulfonic acid **(58425-67-5)**
Hexyl benzoate **(6789-88-4)**
n-Hexylbenzoate **(6789-88-4)**
4'-Hexyl-(1,1'-biphenyl)-4-carbonitrile **(41122-70-7)**
1-Hexyl bromide **(111-25-1)**
Hexyl bromide **(111-25-1)**
n-Hexyl bromide **(111-25-1)**
Hexyl butanoate **(2639-63-6)**
n-Hexyl butanoate **(2639-63-6)**
n-Hexyl n-butanoate **(2639-63-6)**
1-Hexyl butyrate **(2639-63-6)**
Hexyl butyrate **(2639-63-6)**
n-Hexyl butyrate **(2639-63-6)**
γ-n-Hexyl-γ-butyrolactone **(706-14-9)**
Hexyl caproate **(6378-65-0)**
Hexyl carbitol **(112-59-4)**
n-Hexyl carbitol **(112-59-4)**
Hexyl carbonochloridate **(6092-54-2)**
Hexyl cellosolve **(112-25-4)**
n-Hexyl cellosolve **(112-25-4)**
Hexyl chlorocarbonate **(6092-54-2)**

Name	CAS No.
Hexyl chloroformate	**(6092-54-2)**
n-Hexyl chloroformate	**(6092-54-2)**
Hexyl cinnamaldehyde	**(101-86-0)**
α-Hexylcinnamaldehyde	**(101-86-0)**
Hexyl cinnamic aldehyde	**(101-86-0)**
α-Hexylcinnamic aldehyde	**(101-86-0)**
Hexylcyclohexane	**(4292-75-5)**
4-Hexyl-1,3-dihydroxybenzene	**(136-77-6)**
Hexylene	**(25264-93-1)**
Hexylene glycol (ACGIH,OSHA)	**(107-41-5)**
Hexylenic aldehyde	**(505-57-7)**
cis-β,γ-Hexylenic aldehyde	**(6789-80-6)**
Hexylester kyseliny benzoove (Czech)	**(6789-88-4)**
Hexylester kyseliny octove (Czech)	**(142-92-7)**
Hexyl ethanoate	**(142-92-7)**
Hexyl ether	**(112-58-3)**
n-Hexyl ether	**(112-58-3)**
Hexyl ethyl carbinol	**(624-51-1)**
Hexyl glycidyl ether	**(5926-90-9)**
N-Hexylhexanamide	**(10264-29-6)**
n-Hexyl hexanoate	**(6378-65-0)**
Hexyl hexoate	**(6378-65-0)**
Hexyl 2-hydroxybenzoate	**(6259-76-3)**
α-Hexyl,ω-hydroxypoly(oxy-1,2-ethanediyl)	**(31726-34-8)**
3-Hexyl-1-hydroxy-7,8,9,10-tetrahydro-6,6,9-trimethyl-6H-dibenzo-(b,d)pyran	**(117-51-1)**
Hexyl isobutanoate	**(2349-07-7)**
n-Hexyl isobutanoate	**(2349-07-7)**
1-Hexyl isobutyrate	**(2349-07-7)**
Hexyl isobutyrate	**(2349-07-7)**
n-Hexyl isobutyrate	**(2349-07-7)**
Hexylkarbitol (Czech)	**(112-59-4)**
Hexyl ketone	**(462-18-0)**
Hexyl mercaptan	**(111-31-9)**
Hexyl methacrylate	**(142-09-6)**
n-Hexyl methacrylate	**(142-09-6)**
Hexyl 2-methylbutyrate	**(10032-15-2)**
3-Hexyl-5-(3,4-methylenedioxyphenyl)-2-cyclohexen-1-one	**(119-89-1)**
Hexyl 2-methyl-2-propenoate	**(142-09-6)**
Hexyl nitrate	**(20633-11-8)**
2-(Hexyloxy)ethanol	**(112-25-4)**
2-((2-Hexyloxy)ethoxy)ethanol	**(112-59-4)**
4-(Hexyloxy)phenol	**(18979-55-0)**
4-Hexylphenol	**(2446-69-7)**
α-n-Hexyl-β-phenylacrolein	**(101-86-0)**
Hexyl phosphate	**(3900-04-7)**
Hexyl poly(oxyethylene) ether	**(31726-34-8)**
Hexyl propanoate	**(2445-76-3)**
Hexyl 2-propenoate	**(2499-95-8)**
Hexyl propyl ketone	**(624-16-8)**
Hexylresorcin (German)	**(136-77-6)**
4-Hexylresorcine	**(136-77-6)**
4-Hexylresorcinol	**(136-77-6)**
4-n-Hexylresorcinol	**(136-77-6)**
Hexylresorcinol	**(136-77-6)**
p-Hexylresorcinol	**(136-77-6)**
3-Hexyl-7,8,9,10-tetrahydro-6,6,9-trimethyl-6H-dibenzo-(b,d)pyran-1-ol	**(117-51-1)**
Hexylthiocarbam	**(1134-23-2)**
Hexyltrichlorosilane [UN 1784]	**(928-65-4)**
1-Hexyne (9CI)	**(693-02-7)**
Hexyne (9CI)	**(26856-30-4)**
3-Hexyne, 2,5-dimethyl-2,5-di(t-butylperoxy)	**(1068-27-5)**
3-Hexyne, 2,5-dimethyl-2,5-di(t-butylperoxy)-, Maximum concentration 52% with inert solid	**(1068-27-5)**
3-Hexyne-2,5-diol	**(3031-66-1)**
Hexyne-3-diol-2,5	**(3031-66-1)**
3-Hexyne-2,5-diol, 2,5-dimethyl- (9CI)	**(142-30-3)**
Hexynol	**(105-31-7)**
1-Hexyn-3-ol	**(105-31-7)**
3-Hexyn-2-ol (9CI)	**(109-50-2)**
1-Hexyn-3-ol, 3,5-dimethyl- (9CI)	**(107-54-0)**
Hi-Alazin	**(2398-96-1)**
Hi-A-vita	**(68-26-8)**
Hi-Boot	**(5700-49-2)**
Hi-Deratol	**(50-14-6)**
Hi-Dry	**(112-60-7)**
Hi-Ester 2,4-D	**(94-80-4)**
Hi-Fax	**(9002-88-4)**
Hi-Fax 1900	**(9002-88-4)**
Hi-Fax 4601	**(9002-88-4)**
Hi-Fax 4401	**(9002-88-4)**
Hi-Flash Naphtha	**(8030-30-6)**
Hi-Jel	**(1302-78-9)**
Hi-O-Dide	**(5700-49-2)**
Hi-Point 90	**(1338-23-4)**
Hi-Point 180	**(1338-23-4)**
Hi-Point PD-1	**(1338-23-4)**
Hi-Pyridoxin	**(54-47-7)**
Hi-Styrol	**(9003-53-6)**
Hi-Yield Desiccant H-10	**(7778-39-4)**
Hiadelon	**(54-47-7)**
Hiamine	**(5700-49-2)**
Hibanil	**(50-53-3)**
Hibanil	**(69-09-0)**
Hiberna	**(60-87-7)**
Hibernal	**(50-53-3)**
Hibernal	**(69-09-0)**
Hibernyl	**(509-67-1)**
Hibestrol	**(56-53-1)**
Hibiclens	**(18472-51-0)**
Hibiscrub	**(18472-51-0)**
Hibitane	**(18472-51-0)**
Hibitane	**(55-56-1)**
Hibitane diacetate	**(56-95-1)**
Hibrom	**(300-76-5)**
Hicee	**(50-81-7)**
Hico DCPAS	**(127-20-8)**
Hicophor PR	**(108-78-1)**
Hidacian	**(140-87-4)**
Hidaciann	**(140-87-4)**
Hidacid Amaranth	**(915-67-3)**
Hidacid Azo Rubine	**(3567-69-9)**
Hidacid Azure Blue	**(2650-18-2)**
Hidacid Blue V	**(129-17-9)**
Hidacid Boiling Bromo	**(17372-87-1)**
Hidacid Boiling Bromo	**(548-26-5)**
Hidacid Bromo Acid Regular	**(17372-87-1)**
Hidacid Bromo Acid Regular	**(548-26-5)**
Hidacid Dibromo Fluorescein	**(17372-87-1)**
Hidacid Dibromo Fluorescein	**(548-26-5)**
Hidacid Emerald Green	**(4680-78-8)**
Hidacid Eosine Soda Salt	**(17372-87-1)**
Hidacid Eosine Soda Salt	**(548-26-5)**
Hidacid Fast Crimson	**(3734-67-6)**
Hidacid Fast Orange 2G	**(1936-15-8)**
Hidacid Fast Orange G	**(1936-15-8)**
Hidacid Fast Red A	**(1658-56-6)**
Hidacid Fast Scarlet 3R	**(2611-82-7)**
Hidacid Fluorescein	**(2321-07-5)**
Hidacid Metanil Yellow	**(587-98-4)**
Hidacid Orange II	**(633-96-5)**
Hidacid Scarlet 2R	**(3761-53-3)**
Hidacid Uranine	**(518-47-8)**

Hidacid White Bromo (17372-87-1)
Hidacid White Bromo (548-26-5)
Hidacid Wool Violet 5B (1694-09-3)
Hidaco Brilliant Crystal Violet (548-62-9)
Hidaco Brilliant Green (633-03-4)
Hidaco Crystal Violet (548-62-9)
Hidaco Malachite Green Base (569-64-2)
Hidaco Malachite Green LC (569-64-2)
Hidaco Malachite Green SC (569-64-2)
Hidaco Methylene Blue Salt Free (61-73-4)
Hidaco Oil Orange (842-07-9)
Hidaco Oil Red (85-83-6)
Hidaco Oil Yellow (97-56-3)
Hidaco Safranine (477-73-6)
Hidaco Victoria Blue R (2185-86-6)
Hidan (57-41-0)
Hidantal (57-41-0)
Hidantal sodium (630-93-3)
Hidantilo (57-41-0)
Hidantina (57-41-0)
Hidantina senosian (57-41-0)
Hidantina vitoria (57-41-0)
Hidantomin (57-41-0)
Hidazid Tartrazine (1934-21-0)
Hidro-Colisona (50-23-7)
Hidralazin (304-20-1)
Hidralazin (86-54-4)
Hidralazina (Spanish) (86-54-4)
Hidranizil (54-85-3)
Hidrasonil (54-85-3)
Hidril (58-93-5)
Hidrix (127-07-1)
Hidrochlortiazid (58-93-5)
Hidroestron (50-50-0)
Hidromedin (58-54-8)
Hidromorfinol (Spanish) (2183-56-4)
Hidroperoxido de etilo (Spanish) (3031-74-1)
Hidroperoxido de p-mentilo (Spanish) (3031-73-0)
Hidroperoxido de tetrahidronaftilo (Spanish) (771-29-9)
Hidroronol (58-93-5)
Hidrotiazida (58-93-5)
Hidroxipetidina (Spanish) (468-56-4)
Hidroxiteofillina (479-18-5)
Hidroxocobalamina (Spanish) (13422-51-0)
Hidrulta (54-85-3)
Hidrun (54-85-3)
Hiestrone (53-16-7)
Hifix SL (1401-55-4)
Hifol (115-32-2)
Higalcoton (2164-17-2)
High-boiling Butene-2 (590-18-1)
High Flash Acid Fraction (65996-82-9)
Highflex 443 (9003-11-6)
High sulfur fuel oil (68476-32-4)
Higilite (21645-51-2)
Higilite H 32 (21645-51-2)
Higilite H 42 (21645-51-2)
Higilite H 31S (21645-51-2)
Higosan (123-88-6)
Hilbeech (58-89-9)
Hildan (115-29-7)
Hildit (50-29-3)
Hilfol 18.5 EC (115-32-2)
Hilite 60 (2782-57-2)
Hilong (604-75-1)
Hilthion (121-75-5)
Hilthion 25WDP (121-75-5)
Hiltonaphthol AS-BO (132-68-3)
Hiltonaphthol AS-BR (91-92-9)
Hiltonaphthol AS-BS (135-65-9)
Hiltonaphthol AS-D (135-61-5)
Hiltonaphthol AS-EL (137-52-0)
Hiltonaphthol AS-G (91-96-3)
Hiltonaphthol AS-KB (135-63-7)
Hiltonaphthol AS-OL (135-62-6)
Hiltonaphthol AS (92-77-3)
Hiltonaphthol AS-phenyl (92-74-0)
Hiltonil Fast Blue BB Base (120-00-3)
Hiltonil Fast Blue B Base (119-90-4)
Hiltonil Fast Orange GC Base (141-85-5)
Hiltonil Fast Orange GR Base (88-74-4)
Hiltonil Fast Orange R Base (99-09-2)
Hiltonil Fast Red B Base (97-52-9)
Hiltonil Fast Red 3GL Base (89-63-4)
Hiltonil Fast Red GL Base (89-62-3)
Hiltonil Fast Red KB Base (95-79-4)
Hiltonil Fast Red RL Base (99-52-5)
Hiltonil Fast Scarlet 2G Base (95-82-9)
Hiltonil Fast Scarlet G Base (99-55-8)
Hiltonil Fast Scarlet GC Base (99-55-8)
Hiltonil Fast Scarlet 2GS Base (95-82-9)
Hiltonil Fast Scarlet G Salt (99-55-8)
Hiltosal Fast Blue B Salt (119-90-4)
Hiltosal Fast Orange GC Salt (141-85-5)
Hiltosal Fast Orange GR Salt (88-74-4)
Hiltosal Fast Red 3GL Salt (89-63-4)
Hiltosal Fast Red GL Salt (89-62-3)
Hiltosal Fast Red RL Salt (99-52-5)
β-Himachalene (1461-03-6)
Hindamine Scarlet GG (95-82-9)
Hindasol Blue B Salt (119-90-4)
Hindasol Orange GC Salt (141-85-5)
Hindasol Orange GR Salt (88-74-4)
Hindasol Red TR Salt (3165-93-3)
Hindatal (57-41-0)
Hinosan (17109-49-8)
Hioxyl (7722-84-1)
Hipercilina (113-98-4)
Hiphyllin (479-18-5)
Hipnax (146-22-5)
Hipoclorito de litio (Spanish) (13840-33-0)
Hipoftalin (304-20-1)
Hipoftalin (86-54-4)
Hiposerpil (50-55-5)
Hippophain (50-67-9)
Hippuric-acid (495-69-2)
Hippuric acid, p-amino- (8CI) (61-78-9)
Hippuric acid, methyl ester (8CI) (1205-08-9)
Hippuzon (50-35-1)
Hipsal (146-22-5)
Hiptachlor epoxide (1024-57-3)
Hiptagenic acid (504-88-1)
Hiresin (Petroleum resin) (64742-16-1)
Hi-rez (64742-16-1)
Hiropon (300-42-5)
Hiserpia (50-55-5)
Hispacet Fast Yellow G (2832-40-8)
Hispacid Brilliant Scarlet (5850-44-2)
Hispacid Brilliant Scarlet 3RF (2611-82-7)
Hispacid Fast Blue R (3861-73-2)
Hispacid Fast Carmoisine G (3734-67-6)
Hispacid Fast Orange 2G (1936-15-8)
Hispacid Fast Red A (1658-56-6)
Hispacid Fast Red E (2302-96-7)

Hispacid Fast Yellow T	(1934-21-0)
Hispacid Fuchsin G	(4197-07-3)
Hispacid Green GB	(4680-78-8)
Hispacid Orange 1	(523-44-4)
Hispacid Orange AF	(633-96-5)
Hispacid Orange CG	(1934-20-9)
Hispacid Orange IV	(554-73-4)
Hispacid Ponceau R	(3761-53-3)
Hispacid Red AM	(915-67-3)
Hispacid Rubine F	(3567-69-9)
Hispacid Yellow CG	(547-57-9)
Hispacid Yellow MG	(587-98-4)
Hispacrom Violet B	(2092-55-9)
Hispalit Fast Scarlet RN	(2425-85-6)
Hispaluz Brown BRL	(16071-86-6)
Hispaluz Red 8BL	(2610-11-9)
Hispaluz Yellow 5G	(10190-68-8)
Hispamin Black EF	(1937-37-7)
Hispamin Black 3RX	(2429-83-6)
Hispamin Blue 2B	(2602-46-2)
Hispamin Blue 3BX	(72-57-1)
Hispamin Congo 4B	(573-58-0)
Hispamin Fast Black CG	(6428-31-5)
Hispamin Fast Brown NZ	(2429-81-4)
Hispamin Fast Brown 3R2B	(2429-82-5)
Hispamin Green WT	(3626-28-6)
Hispamin Red 4B	(992-59-6)
Hispamin Sky Blue 3B	(2429-74-5)
Hispamin Sky Blue 6B	(2610-05-1)
Hispamin Violet 3R	(2586-60-9)
Hispamin Yellow F	(1325-37-7)
Hisperse Yellow G	(2832-40-8)
Histabid	(61-76-7)
Histacap	(91-84-9)
Histadur	(113-92-8)
Histadur	(132-22-9)
Histadur Dura-Tabs	(113-92-8)
Histadyl	(135-23-9)
Histadyl	(91-80-5)
Histadyl hydrochloride	(135-23-9)
Histafed	(135-23-9)
Histalen	(113-92-8)
Histalon	(91-84-9)
Histamethine	(569-65-3)
Histamethizine	(569-65-3)
Histametizine	(569-65-3)
Histametizyne	(569-65-3)
Histamine	(51-45-6)
Histamine dichloride	(56-92-8)
Histamine, dihydrochloride	(56-92-8)
Histan	(91-84-9)
Histantin	(14362-31-3)
Histantin	(82-93-9)
Histantine	(82-93-9)
Histapan	(113-92-8)
Histapyran	(91-84-9)
Histargan	(60-87-7)
Histasan	(91-84-9)
Histaspan	(113-92-8)
Histatex	(59-33-6)
Histaxin	(58-73-1)
Histidine	(71-00-1)
L-Histidine	(71-00-1)
DL-Histidine (9CI)	(4998-57-6)
Histidine, l	(71-00-1)
Histidyl	(135-23-9)
Histocarb	(121-59-5)
Histocillin	(26787-78-0)
Histyrene S 6F	(9003-55-8)
Hitco HMG 50	(7782-42-5)
Hitenol N 093	(9051-57-4)
Hizarocin	(66-81-9)
Hizex	(9002-88-4)
Hizex 1091J	(9002-88-4)
Hizex 1291J	(9002-88-4)
Hizex 1300J	(9002-88-4)
Hizex 2100LP	(9002-88-4)
Hizex 2100J	(9002-88-4)
Hizex 2200J	(9002-88-4)
Hizex 3000S	(9002-88-4)
Hizex 3000B	(9002-88-4)
Hizex 3300S	(9002-88-4)
Hizex 3300F	(9002-88-4)
Hizex 5000	(9002-88-4)
Hizex 5000S	(9002-88-4)
Hizex 5100LP	(9002-88-4)
Hizex 5100	(9002-88-4)
Hizex 6100P	(9002-88-4)
Hizex 7000F	(9002-88-4)
Hizex 7300F	(9002-88-4)
Hjorton's powder	(62-44-2)
Hned Ostanthrenova BR (Czech)	(2475-33-4)
Hoca	(97-77-8)
Hoch	(50-00-0)
Hocophen	(62-44-2)
Hodag GMS	(31566-31-1)
Hodag PSML-20	(9005-64-5)
Hodag SMS	(1338-41-6)
Hodag SSO	(8007-43-0)
Hodag SVO 9	(9005-65-6)
Hodson	(36614-38-7)
33258 Hoechst	(23491-45-4)
Hoechst 10495	(561-48-8)
Hoechst 10582	(467-85-6)
Hoechst 10720	(469-79-4)
Hoechst 33258	(23491-45-4)
Hoechst Dye 33258	(23491-45-4)
Hoechst PA 190	(9002-88-4)
Hoechst Wax PA 520	(9002-88-4)
Hoegrass	(51338-27-3)
Hoelite	(84-65-1)
Hoelon	(51338-27-3)
Hoelon 3EC	(51338-27-3)
Hoffman Bonded E-D-D Iodine Compound	(5700-49-2)
Hog	(77-10-1)
Hoggar	(77-65-6)
Hoggar N	(562-10-7)
Hok 7501	(25319-90-8)
Hokko-Mycin	(57-92-1)
Hokmate	(14484-64-1)
Hokupanon	(34863-74-6)
Hol 1302	(18024-00-5)
Holbamate	(57-53-4)
Holin	(50-27-1)
Holmium (9CI)	(7440-60-0)
Holodorm	(72-44-6)
Holopan	(155-41-9)
Holoxan	(3778-73-2)
Holoxan 1000	(3778-73-2)
Homandren	(58-18-4)
Homandren (amps)	(57-85-2)
Homatropin	(87-00-3)
Homatropine	(87-00-3)
(+-)-Homatropine bromide	(51-56-9)

Homatropine hydrobromide (51-56-9)
Hombitan (13463-67-7)
Home Heating Oil No.2 (68476-30-2)
#2 Home Heating Oils (68476-30-2)
Homidium bromide (1239-45-8)
1-Homoadamantanol (31061-64-0)
Homoanisic acid (104-01-8)
Homoatropine (87-00-3)
Homocatechol (452-86-8)
Homochroman (6169-78-4)
Homocodeine (509-67-1)
Homocysteine, DL- (454-29-5)
DL-Homocysteine (Free base) (454-29-5)
DL-Homocysteine, S-ethyl- (67-21-0)
Homocysteine, S-ethyl- (67-21-0)
L-Homocysteine, S-ethyl- (13073-35-3)
Homogentisate acid (451-13-8)
Homogentisic acid (451-13-8)
Homogentisinic acid (451-13-8)
Homoguaiacol (93-51-6)
Homohopane (53584-62-6)
Homolinalyl acetate (6819-19-8)
Homolle's Digitalin (20830-75-5)
Homomenthol (116-02-9)
Homomenthyl salicylate (118-56-9)
m-Homomenthyl salicylate (118-56-9)
Homoneurine chloride (1516-27-4)
Homoolan (103-90-2)
Homophthalic acid anhydride (703-59-3)
Homopiperidine (111-49-9)
Homoproline (535-75-1)
Homoprotocatechuic acid (102-32-9)
Homopyrocatechol (452-86-8)
Homosalate (118-56-9)
Homosalato (Spanish) (118-56-9)
Homosalatum (Latin) (118-56-9)
Homosalicylic acid (83-40-9)
m-Homosalicylic acid (50-85-1)
p-Homosalicylic acid (89-56-5)
L-Homoserine (9CI) (672-15-1)
L-Homoserine, O-((aminoiminomethyl)amino)- (9CI) (543-38-4)
L-Homoserine, O-((aminoiminomethyl)amino)-, sulfate (1:1) (2219-31-0)
Homosteron (58-22-0)
Homosterone (58-22-0)
Homoterephthalic acid (1679-64-7)
3-Homotetrahydrocannibinol (117-51-1)
Homotropine (87-00-3)
Homovanillic acid (306-08-1)
Homoveratric acid (93-40-3)
Homoveratrole (494-99-5)
Honey Yellow 3GNT (2586-58-5)
Honey diazine (68-35-9)
Hong Nien (62-38-4)
Honvan (522-40-7)
Hooker HRS-16 (2227-17-0)
Hooker HRS 1654 (2227-17-0)
Hooker No. 1 Chrysotile Asbestos (12001-29-5)
Hopane (471-62-5)
Hopcide (3942-54-9)
Hopcin (3766-81-2)
22(29)-Hopene (1615-91-4)
Horace Vernet's Blue (1317-40-4)
Horbadox (40487-42-1)
Horfemine (130-80-3)
Horizon (439-14-5)
Hormale (58-18-4)
Hormatox (120-36-5)
Hormex Rooting Powder (133-32-4)
Hormin (2008-39-1)
Hormit (2702-72-9)
Hormocel-2CCC (999-81-5)
Hormodin (133-32-4)
Hormoestrol (84-16-2)
Hormofemin (84-17-3)
Hormoflaveine (57-83-0)
Hormofollin (53-16-7)
Hormofort (630-56-8)
Hormogynon (50-50-0)
Hormoluton (57-83-0)
Hormomed (50-27-1)
Hormone Somatotrope (French) (9002-72-6)
Hormonin (50-27-1)
Hormonisene (569-57-3)
Hormoslyr 64 (8015-35-8)
Hormoslyr 500t (2545-59-7)
Hormoteston (57-85-2)
Hormotuho (94-74-6)
Hormovarine (53-16-7)
Hornotuho (94-74-6)
Horse (561-27-3)
Horse Head R-710 (13463-67-7)
Horse Head A-410 (13463-67-7)
Horse Head A-420 (13463-67-7)
Hortex (58-89-9)
Hortfenicol (56-75-7)
Hortocritt (33813-20-6)
Hosalon (36614-38-7)
Hosdon (36614-38-7)
Hosdon Granule (36614-38-7)
Hostacillin (54-35-3)
Hostacortin (50-24-8)
Hostacortin (53-03-2)
Hostacyclin (60-54-8)
Hostadur (25038-59-9)
Hostadur A (25038-59-9)
Hostadur K (25038-59-9)
Hostadur K-VP 4022 (25038-59-9)
Hostaflex VP 150 (9003-22-9)
Hostaginan (390-64-7)
Hostalen (9002-88-4)
Hostalen GD 620 (9002-88-4)
Hostalen GD 6250 (9002-88-4)
Hostalen GF 4760 (9002-88-4)
Hostalen GF 5750 (9002-88-4)
Hostalen GM 5010 (9002-88-4)
Hostalen GUR (9002-88-4)
Hostalen HDPE (9002-88-4)
Hostalen N 1060 (9003-07-0)
Hostalen PP (9003-07-0)
Hostalen PPH 1050 (9003-07-0)
Hostalen PPN (9003-07-0)
Hostalen PPN 1060 (9003-07-0)
Hostalen PPN 1075 F (9003-07-0)
Hostalen PPN 1076 F (9003-07-0)
Hostalen PPR 1042 (9003-07-0)
Hostalen PPT VP 7090A (9003-07-0)
Hostalen PP-U (9003-07-0)
Hostalit PVP (9003-22-9)
Hostaperm Blue AFN (147-14-8)
Hostaperm Blue A 2R (147-14-8)
Hostaperm Blue A 3R (147-14-8)
Hostaperm Blue B 2G (147-14-8)
Hostaperm Blue B 3G (147-14-8)
Hostaperm Blue BG (147-14-8)

Name	CAS No.
Hostaperm Green 8G	(14302-13-7)
Hostaperm Green GG	(1328-53-6)
Hostaperm Orange GR	(4424-06-0)
Hostaperm Red E 3B	(1047-16-1)
Hostaperm Red E 5B	(1047-16-1)
Hostaperm Red Violet ER	(1047-16-1)
Hostaperm Vat Orange GR	(4424-06-0)
Hostaperm Yellow H 4G	(31837-42-0)
Hostaphan	(25038-59-9)
Hostaphan BNH	(25038-59-9)
Hostaphan RN	(25038-59-9)
Hostapon T	(137-20-2)
Hostaquick	(34783-40-9)
Hostaquick	(62-38-4)
Hostaquik	(62-38-4)
Hostatherm Pink FBL	(17418-58-5)
Hostathion	(24017-47-8)
Hostavat Brilliant Orange GR	(4424-06-0)
Hostavat Brilliant Pink R	(2379-74-0)
Hostavat Golden Yellow	(128-66-5)
Hostavat Orange R	(3263-31-8)
Hostetex l-Pec	(120-82-1)
Hostyren N	(9003-53-6)
Hostyren N 4000	(9003-53-6)
Hostyren N 7001	(9003-53-6)
Hostyren N 4000V	(9003-53-6)
Hostyren S	(9003-53-6)
Hotrienol	(20053-88-7)
Hotrienol, trans-(-)-	(20053-88-7)
trans-(-)-Hotrienol	(20053-88-7)
Housane	(185-94-4)
Howflex GBP	(13988-26-6)
Hox 1901	(29973-13-5)
H 130 (Resin)	(64742-16-1)
Hubbuck's White	(1314-13-2)
Huber	(1333-86-4)
Huile H50	(110-02-1)
Huile HSO	(110-02-1)
Huile d'aniline (French)	(62-53-3)
Huile de camphre (French)	(76-22-2)
Huile de camphre (French)	(8008-51-3)
Huile de colophane (French)	(8002-16-2)
Huile de fusel (French)	(8013-75-0)
Huls P 6500	(9003-07-0)
Human Chorionic Gonadotropin	(9002-61-3)
Humatin	(7542-37-2)
Humenegro	(1333-86-4)
Humic acids	(1415-93-6)
Humifen WT 27G	(577-11-7)
Humorsol	(56-94-0)
Humulene	(6753-98-6)
α-Humulene	(6753-98-6)
Humycin	(7542-37-2)
Hungaria L7	(58-89-9)
Hungazin	(1912-24-9)
Hungazin DT	(122-34-9)
Hungazin PK	(1912-24-9)
Husept extra	(88-04-0)
Hustodil	(93-14-1)
Hustosil	(93-14-1)
Hy-Chlor	(7778-54-3)
Hy-Odide	(5700-49-2)
Hy-Phi 1055	(112-80-1)
Hy-Phi 1088	(112-80-1)
Hy-Phi 1199	(57-11-4)
Hy-Phi 1205	(57-11-4)
Hy-Phi 1303	(57-11-4)
Hy-Phi 1401	(57-11-4)
Hy-Phi 2066	(112-80-1)
Hy-Phi 2088	(112-80-1)
Hy-Phi 2102	(112-80-1)
Hyacinth-Absolute	(8023-94-7)
Hyacinth Base	(103-64-0)
Hyacinthal	(93-53-8)
Hyacinthin	(122-78-1)
Hyadrine	(58-73-1)
Hyadur	(67-68-5)
Hyaluronan	(9004-61-9)
Hyaluronate, sodium	(9004-61-9)
Hyaluronic acid (9CI)	(9004-61-9)
Hyamine	(121-54-0)
Hyamine 10	(25155-18-4)
Hyamine 1622	(121-54-0)
Hyamine 2389	(1399-80-0)
Hyamine 3500	(68391-01-5)
Hyamine 3500	(8001-54-5)
Hyasorb	(113-98-4)
Hybar X	(66-22-8)
Hybernal	(69-09-0)
Hybrin	(50-81-7)
Hycanthon	(3105-97-3)
Hycanthone	(3105-97-3)
Hycanthone mesylate	(23255-93-8)
Hycanthone methanesulfonate	(23255-93-8)
Hycanthone methanesulphonate	(23255-93-8)
Hycanthone monomethanesulphonate	(23255-93-8)
Hycar	(9016-00-6)
Hycar LX 407	(9003-55-8)
Hychol 705	(21645-51-2)
Hychotine	(68-88-2)
Hyclorate	(637-07-0)
Hyclorite	(7681-52-9)
Hycort	(50-23-7)
Hycortol	(50-23-7)
Hycortole	(50-23-7)
Hycozid	(54-85-3)
Hydan	(118-52-5)
Hydan (antiseptic)	(118-52-5)
Hydantal	(57-41-0)
Hydantin	(57-41-0)
Hydantin sodium	(630-93-3)
Hydantoic acid	(462-60-2)
Hydantoin	(461-72-3)
Hydantoin	(57-41-0)
Hydantoinal	(57-41-0)
Hydantoinal sodium	(630-93-3)
Hydantoin, bis(2,3-epoxypropyl)-5-ethyl-5-methyl	(15336-82-0)
Hydantoin, 1,3-bis(hydroxymethyl)-5,5-dimethyl	(6440-58-0)
Hydantoin, 3-bromo-1-chloro-5,5-dimethyl	(126-06-7)
Hydantoin, 1,3-dibromo-5,5-dimethyl	(77-48-5)
Hydantoin, 1,3-dichloro-5,5-dimethyl	(118-52-5)
Hydantoin, dichlorodimethyl-	(118-52-5)
Hydantoin, 5,5-diethyl- (8CI)	(5455-34-5)
Hydantoin, 5,5-dimethyl	(77-71-4)
Hydantoin, 5,5-diphenyl	(57-41-0)
Hydantoin, 5,5-diphenyl-, monosodium salt	(630-93-3)
Hydantoin, 5,5-diphenyl-, sodium salt	(630-93-3)
Hydantoin, 5-ethyl-5-methyl	(5394-36-5)
Hydantoin, 5-ethyl-3-methyl-5-phenyl	(50-12-4)
Hydantoin, 3-(2-hydroxyethyl)-5,5-dimethyl- (8CI)	(29071-93-0)
Hydantoin, 1-(hydroxymethyl)-5,5-dimethyl- (8CI)	(116-25-6)
Hydantoin, 1-nitro-	(2825-15-2)
Hydantoin, 1-((5-nitrofurfurylidene)amino)	(67-20-9)
Hydantoin, 1-((5-(p-nitrophenyl)furfurylidene)amino)	(7261-97-4)

Hydantoin sodium (630-93-3)
Hydantoin, 5-ureido- (97-59-6)
Hydeltra (50-24-8)
Hydeltrone (50-24-8)
Hydense (557-05-1)
Hydol (135-09-1)
Hydout (129-67-9)
Hydout (145-73-3)
Hydoxin (65-23-6)
Hydracetin (114-83-0)
Hydracillin (54-35-3)
Hydracillin (6130-64-9)
Hydracrylaldehyde, 2,2-dimethyl- (597-31-9)
Hydracrylic acid, N,N-bis(2-chloroethyl)phosphorodiamidate (22788-18-7)
Hydracrylic acid, 2,2-dimethyl-, 3-hydroxy-2,2-dimethylpropyl ester (1115-20-4)
Hydracrylic acid β-lactone (57-57-8)
Hydracrylic acid, 2-phenyl- (529-64-6)
Hydracrylonitrile (109-78-4)
Hydracrylonitrile, acrylate (106-71-8)
Hydracrylonitrile, 2-methyl- (8CI) (2567-01-3)
Hydral (302-17-0)
Hydral 705 (21645-51-2)
Hydral 710 (21645-51-2)
Hydralazin (86-54-4)
Hydralazine (86-54-4)
Hydralazine HCl (304-20-1)
Hydralazine chloride (304-20-1)
Hydralazine hydrochloride (304-20-1)
Hydralazine monohydrochloride (304-20-1)
Hydralin (108-93-0)
Hydrallazine (86-54-4)
Hydrallazine hydrochloride (304-20-1)
Hydram (2212-67-1)
Hydramethylnon (67485-29-4)
Hydrangin (93-35-6)
Hydrangine (93-35-6)
Hydrapress (304-20-1)
Hydrargyrum bijodatum (German) (7774-29-0)
Hydrastine (118-08-1)
β-Hydrastine (118-08-1)
Hydrastis (458-37-7)
Hydrated alkali-aluminum silicate (1318-02-1)
Hydrated alumina (21645-51-2)
Hydrate de chloral (302-17-0)
Hydrated lime (1305-62-0)
Hydrated silica (1343-98-2)
Hydratropa aldehyde (93-53-8)
Hydratropaldehyde (93-53-8)
Hydratropic acid (492-37-5)
Hydratropic acid, 3-chloro-4-(3-pyrrolin-1-yl) (31793-07-4)
Hydratropic acid, β-hydroxy- (529-64-6)
Hydratropic acid, p-isobutyl (15687-27-1)
Hydratropic acid, α-methyl- (8CI) (826-55-1)
Hydratropic acid, m-phenoxy-, (+-) (31879-05-7)
Hydratropic alcohol (1123-85-9)
Hydratropic aldehyde (93-53-8)
Hydratropyl alcohol (1123-85-9)
Hydrazid (54-85-3)
Hydrazide (54-85-3)
Hydrazide BSG (80-17-1)
Hydrazid kyseliny benzoove (Czech) (613-94-5)
Hydrazid kyseliny maleinove (Czech) (123-33-1)
Hydrazine (ACGIH,OSHA) (302-01-2)
Hydrazine, Anhydrous [UN 2029] (302-01-2)
Hydrazine, Aqueous solution [UN 2030] (302-01-2)
Hydrazine, Aqueous solution containing more than 64% hydrazine [UN 2029] (302-01-2)
Hydrazine Base (302-01-2)
Hydrazine Yellow (1934-21-0)
Hydrazine, 1-acetyl-2-picolinoyl (17433-31-7)
Hydrazine, 1-((4-aminopyrazolo(5,1-c)-as-triazin-3-yl)-carbonyl)-2-formyl- (8CI) (16111-79-8)
Hydrazine, azido- (14546-44-2)
Hydrazine azide (14546-44-2)
Hydrazine-benzene (100-63-0)
Hydrazine, benzoyl- (613-94-5)
Hydrazine, 2-benzyl-1-methyl (10309-79-2)
Hydrazine, 1,2-bis(aminocarbonyl)- (110-21-4)
Hydrazine, 1,2-bis(2-chlorophenyl)- (9CI) (782-74-1)
Hydrazine, 1,2-bis(3,4-dichlorophenyl)- (9CI) (71753-42-9)
Hydrazine, butyl-, hydrochloride (56795-65-4)
Hydrazine, tert-butyl-, monohydrochloride (8CI) (7400-27-3)
Hydrazine, carbamoyl- (57-56-7)
Hydrazine, 1-carbamoyl-2-phenyl- (103-03-7)
Hydrazinecarbodithioic acid, Compd. with hydrazine (1:1) (20469-71-0)
Hydrazinecarbohydrazonothioic acid (2231-57-4)
Hydrazine, carbonyldi- (497-18-7)
Hydrazinecarbothioamide (79-19-6)
Hydrazinecarbothioamide, 2-(4-((aminoiminomethyl)hydrazono)-2,5-cyclohexadien-1-ylidene)- (539-21-9)
Hydrazinecarbothioamide, N-methyl- (9CI) (6610-29-3)
Hydrazinecarbothioamide, 2-(1-methylethylidene)- (9CI) (1752-30-3)
Hydrazinecarboxaldehyde (9CI) (624-84-0)
Hydrazine carboxaldehyde, ethylidenemethyl- (16568-02-8)
Hydrazinecarboxaldehyde, 2-(4-(5-nitro-2-furanyl)-2-thiazolyl)- (3570-75-0)
Hydrazinecarboxamide (57-56-7)
Hydrazinecarboxamide, hydrochloride (563-41-7)
Hydrazinecarboxamide monohydrochloride (563-41-7)
Hydrazinecarboxamide, monohydrochloride (9CI) (563-41-7)
Hydrazinecarboxamide, 2-((5-nitro-2-furanyl)methylene)- (59-87-0)
Hydrazinecarboxamide, 2-phenyl- (103-03-7)
Hydrazinecarboxamide, N-phenyl- (9CI) (537-47-3)
Hydrazinecarboximidamide (79-17-4)
Hydrazinecarboximidamide, 2,2'-(carbonylbis(imino-4,1-phenylene-ethylidyne))bis-, dimethanesulfonate (9CI) (15427-93-7)
Hydrazinecarboximidamide, 2-((2,6-dichlorophenyl)methylene)-(9CI) (5051-62-7)
Hydrazinecarboximidamide hydrochloride (1937-19-5)
Hydrazinecarboximidic acid, 2-(aminocarbonyl)- (110-21-4)
Hydrazinecarboxylic acid, methyl ester (9CI) (6294-89-9)
Hydrazinecarboxylic acid, (2-quinoxalinylmethylene)-, methyl ester, N,N'-dioxide (9CI) (6804-07-5)
Hydrazine, 1,2-dicarbamoyl- (110-21-4)
Hydrazine dicarbonic acid diazide (67880-17-5)
1,2-Hydrazinedicarbonyl diazide (67880-17-5)
1,2-Hydrazinedicarbothioamide (142-46-1)
1,2-Hydrazinedicarboxamide (9CI) (110-21-4)
1,2-Hydrazinedicarboxamide, N,N'-dimethyl-N,N'-dinitroso- (9CI) (3844-60-8)
1,2-Hydrazinedicarboxylic acid, monoethyl ester (9CI) (60913-86-2)
Hydrazine, 1,2-diethyl (1615-80-1)
Hydrazine, 1,2-diethyl-, dihydrochloride (7699-31-2)
Hydrazine, 1,2-diformyl- (628-36-4)
Hydrazine, 1,1-dimethyl (57-14-7)
Hydrazine, 1,2-dimethyl (540-73-8)
Hydrazine, 1,2-dimethyl-, dihydrochloride (306-37-6)
Hydrazine, (1,1-dimethylethyl)-, monohydrochloride (9CI) (7400-27-3)
Hydrazine, 1,1-dimethyl-, hydrochloride (593-82-8)
Hydrazine, 1,2-dimethyl-, hydrochloride (56400-60-3)
Hydrazine, 1,1-diphenyl- (530-50-7)
Hydrazine, 1,2-diphenyl- (122-66-7)
Hydrazine, diphenyl- (9CI) (38622-18-3)
Hydrazine, 1,1-diphenyl-, monohydrochloride (9CI) (530-47-2)
Hydrazine, formyl- (624-84-0)

Hydrazine, 1-formyl-1-methyl- **(758-17-8)**
Hydrazine hydrate **(7803-57-8)**
Hydrazine, hydrochloride **(2644-70-4)**
Hydrazine, methyl **(60-34-4)**
Hydrazine, methyl-, hydrochloride **(7339-53-9)**
Hydrazine, 1-methyl-2-(phenylmethyl)- **(10309-79-2)**
Hydrazine, monoacetate (9CI) **(7335-65-1)**
Hydrazine monochloride **(2644-70-4)**
Hydrazine, monohydrate **(7803-57-8)**
Hydrazine monosulfate **(10034-93-2)**
Hydrazine perchlorate **(27978-54-7)**
Hydrazine, perchlorate **(27978-54-7)**
Hydrazine, phenethyl **(51-71-8)**
Hydrazine, phenethyl-, sulfate (1:1) **(156-51-4)**
Hydrazine, phenyl **(100-63-0)**
Hydrazine, (2-phenylethyl)- **(51-71-8)**
Hydrazine, (2-phenylethyl)-, sulfate (1:1) **(156-51-4)**
Hydrazine, phenyl-, hydrochloride **(59-88-1)**
Hydrazine, phenyl-, monohydrochloride (8CI,9CI) **(59-88-1)**
Hydrazine selenate **(73506-32-8)**
Hydrazine, selenate **(73506-32-8)**
Hydrazine, sulfate (1:1) **(10034-93-2)**
Hydrazine, sulfate (2:1) (9CI) **(13464-80-7)**
Hydrazine sulphate **(10034-93-2)**
Hydrazine, tetramethyl-, monohydrochloride **(61556-82-9)**
Hydrazine, p-tolylsulfonyl **(1576-35-8)**
Hydrazine, (2,4,6-trichlorophenyl)- (9CI) **(5329-12-4)**
Hydrazine, trimethyl-, hydrochloride **(60597-20-8)**
Hydrazinium chloride **(2644-70-4)**
Hydrazinium, 1,1-dimethyl-1-(2-hydroxypropyl)-2-(2-methyl-1-oxo-2-propenyl)-, hydroxide, inner salt **(17341-40-1)**
Hydrazinium, 1-(2-hydroxypropyl)-1,1-dimethyl-2-(1-oxotetradecyl)-, hydroxide, inner salt (9CI) **(38848-76-9)**
Hydrazinium, 1-(2-hydroxypropyl)-2-methacryloyl-1,1-dimethyl-, hydroxide, inner salt (8CI) **(17341-40-1)**
Hydrazinium monochloride **(2644-70-4)**
Hydrazinium sulfate **(10034-93-2)**
Hydrazinobenzene **(100-63-0)**
2-Hydrazinobenzothiazole **(615-21-4)**
2-Hydrazinoethanol **(109-84-2)**
1-Hydrazino-2-phenylethane **(51-71-8)**
1-Hydrazino-2-phenylethane hydrogen sulphate **(156-51-4)**
1-Hydrazinophthalazine **(86-54-4)**
Hydrazinophthalazine **(86-54-4)**
1-Hydrazinophthalazine hydrochloride **(304-20-1)**
1-Hydrazinophthalazine monohydrochloride **(304-20-1)**
Hydrazobenzen (Czech) **(122-66-7)**
Hydrazobenzene **(122-66-7)**
Hydrazobenzene, 2,2'-dichloro- **(782-74-1)**
Hydrazocarbonamide **(110-21-4)**
Hydrazodicarbonamide **(110-21-4)**
Hydrazodicarbonsaeureabis(methylnitrosamid) (German) **(3844-60-8)**
Hydrazodicarboxamide **(110-21-4)**
Hydrazodicarboxylic acid bis(methylnitrosamide) **(3844-60-8)**
Hydrazoethane **(1615-80-1)**
Hydrazoic-acid **(7782-79-8)**
Hydrazoic acid (ACGIH) **(7782-79-8)**
Hydrazoic acid, sodium salt. **(26628-22-8)**
Hydrazomethane **(540-73-8)**
Hydrazomethane **(60-34-4)**
Hydrazonium sulfate **(10034-93-2)**
Hydrazyl, 2,2-diphenyl-1-(2,4,6-trinitrophenyl) **(1898-66-4)**
Hydrazyna (Polish) **(302-01-2)**
Hydrea **(127-07-1)**
Hydreia **(127-07-1)**
Hydrenox **(135-09-1)**
Hydridotetracarbonylcobalt **(16842-03-8)**
Hydril **(58-93-5)**
S-Hydril **(7772-98-7)**
s-Hydrindacene, 1,1,4,5,5,8-hexamethyl- (8CI) **(17465-59-7)**
s-Hydrindacene, 1,1,4,7,7,8-hexamethyl- (8CI) **(17465-58-6)**
Hydrindan **(496-10-6)**
1,2-Hydrindene **(496-11-7)**
Hydrindene **(496-11-7)**
Hydrindonaphthene **(496-11-7)**
α-Hydrindone **(83-33-0)**
Hydriodic acid (DOT) **(10034-85-2)**
Hydriodic acid, Solution [UN 1787] **(10034-85-2)**
Hydriodic acid, ion(1-) **(20461-54-5)**
Hydriodic ether **(75-03-6)**
Hydriodide-Boot **(5700-49-2)**
Hydriodide-Enterol **(130-26-7)**
Hydriodide-O-Dide **(5700-49-2)**
Hydrionic **(138-15-8)**
Hydrite **(1332-58-7)**
Hydro-Adreson **(50-23-7)**
Hydro-Aquil **(58-93-5)**
Hydro-Colisona **(50-23-7)**
Hydro-Giene **(7681-65-4)**
Hydro-Rapid **(54-31-9)**
Hydroazodicarboxybis(methylnitrosamide) **(3844-60-8)**
Hydroazoethane **(1615-80-1)**
Hydrobromic-acid **(10035-10-6)**
Hydrobromic acid, Anhydrous [UN 1048] **(10035-10-6)**
Hydrobromic acid, More than 49% strength (DOT) **(10035-10-6)**
Hydrobromic acid, Not more than 49% strength (DOT) **(10035-10-6)**
Hydrobromic acid, Solution [UN 1788] **(10035-10-6)**
Hydrobromic acid monoammoniate **(12124-97-9)**
Hydrocaffeic acid **(1078-61-1)**
Hydrocarb 90 **(13397-26-7)**
Hydrocarbons, C3-5, Polymn. Unit Feed **(68476-54-0)**
Hydrocarbons, petroleum resins **(64742-16-1)**
Hydrocardanol **(501-24-6)**
Hydrocerin **(57-88-5)**
Hydrochinon (Czech, Polish) **(123-31-9)**
Hydrochinon monobenzylether (Czech) **(103-16-2)**
Hydrochloric acid, Anhydrous [UN 1050] **(7647-01-0)**
Hydrochloric acid (DOT) **(7647-01-0)**
Hydrochloric acid, Mixed with nitric acid (3:1) **(8007-56-5)**
Hydrochloric acid, Solution, Inhibited (DOT) **(7647-01-0)**
Hydrochloric acid, Solution [UN 1789] **(7647-01-0)**
Hydrochloric acid dimethylamine **(506-59-2)**
Hydrochloric acid, ion(1-) **(16887-00-6)**
Hydrochloric ether **(75-00-3)**
Hydrochloride **(7647-01-0)**
Hydrochloride benzenamide **(142-04-1)**
Hydrochlorofluorocarbon 142b **(75-68-3)**
Hydrochlorothiazid **(58-93-5)**
Hydrochlorothiazide **(58-93-5)**
Hydrochlorthiazide **(58-93-5)**
Hydrocinnamaldehyde, p-tert-butyl-α-methyl **(80-54-6)**
Hydrocinnamaldehyde, p-isopropyl-α-methyl **(103-95-7)**
Hydrocinnamamide, o-hydroxy- (6CI,7CI,8CI) **(22367-76-6)**
Hydrocinnamic acid (8CI) **(501-52-0)**
Hydrocinnamic acid, α-amino- **(63-91-2)**
Hydrocinnamic acid, 3-amino-α-ethyl-2,4,6-triiodo **(96-83-3)**
Hydrocinnamic acid, 3,5-di-tert-butyl-4-hydroxy-, methyl ester **(6386-38-5)**
Hydrocinnamic acid, 3,5-di-t-butyl-4-hydroxy-, octadecyl ester **(2082-79-3)**
Hydrocinnamic acid, 3,5-di-tert-butyl-4-hydroxy-, octadecyl ester **(2082-79-3)**
Hydrocinnamic acid, 3,4-dihydroxy **(1078-61-1)**
Hydrocinnamic acid, β,β-dimethyl **(1010-48-6)**
Hydrocinnamic acid, 3-(((dimethylamino)methylene)amino)-2,4,6-triiodo **(5587-89-3)**
Hydrocinnamic acid, .α.,.β.-epoxy-.β.-methyl- (6CI,7CI,8CI) **(5669-15-8)**

Name	CAS No.
Hydrocinnamic acid, α,β-epoxy-β-methyl-, ethyl ester	**(77-83-8)**
Hydrocinnamic acid, .α.,.β.-epoxy-.β.-methyl-, ethyl ester, trans- (8CI)	**(19464-92-7)**
Hydrocinnamic acid, β-ethyl	**(5669-17-0)**
Hydrocinnamic acid, ethyl ester (8CI)	**(2021-28-5)**
Hydrocinnamic acid, p-hydroxy	**(501-97-3)**
Hydrocinnamic acid, o-hydroxy-, δ-lactone	**(119-84-6)**
Hydrocinnamic acid, 4-hydroxy-3-methoxy- (8CI)	**(1135-23-5)**
Hydrocinnamic acid, o-methyl- (7CI,8CI)	**(22084-89-5)**
Hydrocinnamic acid, methyl ester (8CI)	**(103-25-3)**
Hydrocinnamic acid, p,α-dichloro- (8CI)	**(14437-20-8)**
Hydrocinnamic alcohol	**(122-97-4)**
Hydrocinnamonitrile, p,α-dichloro- (8CI)	**(17849-64-8)**
Hydrocinnamyl alcohol	**(122-97-4)**
Hydrocobalamin	**(13422-51-0)**
Hydrocobalt tetracarbonyl	**(16842-03-8)**
Hydrocodin	**(125-28-0)**
Hydrocodone	**(125-29-1)**
11-β-Hydrocortisone	**(50-23-7)**
Hydrocortisone	**(50-23-7)**
δ^1-Hydrocortisone	**(50-24-8)**
Hydrocortisone free alcohol	**(50-23-7)**
Hydrocortistab	**(50-23-7)**
Hydrocortisyl	**(50-23-7)**
Hydrocortone	**(50-23-7)**
Hydro-p-coumaric acid	**(501-97-3)**
Hydrocoumarin	**(119-84-6)**
Hydrocyanic-acid	**(74-90-8)**
Hydrocyanic acid, Aqueous solutions not more than 20% hydrocyanic acid [UN 1613]	**(74-90-8)**
Hydrocyanic acid, Liquefied (DOT)	**(74-90-8)**
Hydrocyanic acid (Prussic), Unstabilized (DOT)	**(74-90-8)**
Hydrocyanic acid, ion(1-)	**(57-12-5)**
Hydrocyanic acid, potassium salt	**(151-50-8)**
Hydrocyanic acid, sodium salt	**(143-33-9)**
Hydrocyanic ether	**(107-12-0)**
Hydrocyclin	**(2058-46-0)**
Hydrodarco	**(7440-44-0)**
Hydrodeltisone	**(50-24-8)**
Hydrodesulfurized Heavy Naphtha	**(64742-82-1)**
Hydrodesulfurized Kerosene	**(64742-81-0)**
Hydrodesulfurized Middle Distillate	**(64742-80-9)**
Hydrodiisobutylaluminum	**(1191-15-7)**
Hydrodine	**(5700-49-2)**
Hydrodiuretic	**(58-93-5)**
Hydro-diuril	**(58-93-5)**
17-β-Hydroestr-4-en-3-one	**(434-22-0)**
Hydroferulic acid	**(1135-23-5)**
Hydrofining (VAN)	**(86290-81-5)**
Hydroflumethiazide	**(135-09-1)**
Hydrofluoboric acid	**(16872-11-0)**
Hydrofluoric-acid	**(7664-39-3)**
Hydrofluoric acid, Anhydrous (DOT)	**(7664-39-3)**
Hydrofluoric acid, Mixt. with sodium fluoride (9CI)	**(51273-71-3)**
Hydrofluoric acid, Solution (DOT)	**(7664-39-3)**
Hydrofluoride	**(7664-39-3)**
Hydrofluoroboric acid (DOT)	**(16872-11-0)**
Hydrofluorosilicic acid	**(56977-47-0)**
Hydrofluorosilicic acid (DOT)	**(16961-83-4)**
Hydrofluosilicic acid (DOT)	**(16961-83-4)**
Hydrofol	**(57-10-3)**
Hydrofol 1895	**(57-11-4)**
Hydrofol 2022-55	**(112-85-6)**
Hydrofol Acid 200	**(106-14-9)**
Hydrofol Acid 560	**(112-85-6)**
Hydrofol Acid 1255	**(143-07-7)**
Hydrofol Acid 1295	**(143-07-7)**
Hydrofol Acid 1495	**(544-63-8)**
Hydrofol Acid 1655	**(57-11-4)**
Hydrofol Acid 1855	**(57-11-4)**
Hydrofuran	**(109-99-9)**
Hydrogel	**(25852-47-5)**
Hydrogen, Compressed [UN 1049]	**(1333-74-0)**
Hydrogen Cyanamide	**(420-04-2)**
Hydrogen (DOT)	**(1333-74-0)**
Hydrogen Oxalate of Amiton	**(3734-97-2)**
Hydrogen, Refrigerated liquid [UN 1966]	**(1333-74-0)**
Hydrogen antimonide	**(7803-52-3)**
Hydrogen arsenide	**(7784-42-1)**
Hydrogenated Castor Oil	**(8001-78-3)**
Hydrogenated bisphenol A	**(80-04-6)**
Hydrogenated terphenyls (ACGIH,OSHA)	**(61788-32-7)**
Hydrogen azide	**(7782-79-8)**
Hydrogen bromide (ACGIH,DOT,OSHA)	**(10035-10-6)**
Hydrogen bromide, Anhydrous (DOT)	**(10035-10-6)**
Hydrogen bromide (9CI)	**(11071-85-5)**
Hydrogen carboxylic acid	**(64-18-6)**
Hydrogen chloride (ACGIH,DOT,OSHA)	**(7647-01-0)**
Hydrogen chloride (Aerosol)	**(7647-01-0)**
Hydrogen chloride, Anhydrous (DOT)	**(7647-01-0)**
Hydrogen chloride, Refrigerated liquid [UN 2186]	**(7647-01-0)**
Hydrogen cyanide (ACGIH,OSHA)	**(74-90-8)**
Hydrogen cyanide, Anhydrous, Stabilized [UN 1051]	**(74-90-8)**
Hydrogen cyanide, Anhydrous, Stabilized, absorbed in a porous inert material [UN 1614]	**(74-90-8)**
Hydrogen dibromide (8CI)	**(11071-85-5)**
Hydrogen dioxide	**(7722-84-1)**
Hydrogen dipotassium phosphate	**(7758-11-4)**
Hydrogene sulfure (French)	**(7783-06-4)**
Hydrogen fluoride (ACGIH,OSHA) [UN 1052]	**(7664-39-3)**
Hydrogen hexachloroplatinate(4+)	**(16941-12-1)**
Hydrogen hexachloroplatinate(IV)	**(16941-12-1)**
Hydrogen hexafluorophosphate	**(16940-81-1)**
Hydrogen hexafluorosilicate	**(16961-83-4)**
Hydrogen iodide	**(10034-85-2)**
Hydrogen iodide, Anhydrous [UN 2197]	**(10034-85-2)**
Hydrogen iodide, Solution (DOT)	**(10034-85-2)**
Hydrogen isothiocyanate	**(3129-90-6)**
Hydrogen methyl terephthalate	**(1679-64-7)**
Hydrogen nitrate	**(7697-37-2)**
Hydrogen oxide	**(3352-57-6)**
Hydrogen peroxide, 30%	**(7722-84-1)**
Hydrogen peroxide, 90%	**(7722-84-1)**
Hydrogen peroxide (ACGIH,OSHA)	**(7722-84-1)**
Hydrogen peroxide, Compd. with urea (1:1)	**(124-43-6)**
Hydrogen peroxide, Solution (over 52% peroxide) (DOT)	**(7722-84-1)**
Hydrogen peroxide, Solution (8% to 40% Peroxide) (DOT)	**(7722-84-1)**
Hydrogen peroxide, Stabilized (over 60% peroxide) (DOT)	**(7722-84-1)**
Hydrogen peroxide carbamide	**(124-43-6)**
Hydrogen phosphide	**(7803-51-2)**
Hydrogen potassium fluoride	**(7789-29-9)**
Hydrogen-selenide	**(7783-07-5)**
Hydrogen selenide (ACGIH,DOT,OSHA)	**(7783-07-5)**
Hydrogen selenide, Anhydrous [UN 2202]	**(7783-07-5)**
Hydrogen sodium selenite	**(7782-82-3)**
Hydrogen sodium selenium oxide	**(7782-82-3)**
Hydrogen sulfate (DOT)	**(7664-93-9)**
Hydrogen sulfide (ACGIH,DOT,OSHA)	**(7783-06-4)**
Hydrogen sulfide, Liquefied [UN 1053]	**(7783-06-4)**
Hydrogen sulfite sodium	**(7631-90-5)**
Hydrogen sulfuric acid	**(7783-06-4)**
Hydrogen sulphide (DOT)	**(7783-06-4)**
Hydrogen tetrafluoroborate	**(16872-11-0)**
Hydrogestrone	**(152-62-5)**

Hydroginkgol	(501-24-6)
Hydrogrisevit	(13422-51-0)
4-Hydroksyacetofenol (Polish)	(99-93-4)
4-Hydroksy-3',5'-dwumetoksyacetofenon (Polish)	(2478-38-8)
4-Hydroksy-3,5-dwumetoksybenzaldehyd (Polish)	(134-96-3)
Hydroksyzyny (Polish)	(68-88-2)
Hydrol SW	(9002-93-1)
Hydrolin	(7775-14-6)
Hydrolit	(149-44-0)
Hydrodeltalone	(50-24-8)
Hydrolose	(9004-67-5)
Hydrolyzed protein	(73049-73-7)
Hydromagnesite	(546-93-0)
Hydromedin	(58-54-8)
Hydromethamphetamine	(101-40-6)
Hydromirex	(39801-14-4)
Hydromorphine	(509-60-4)
Hydromorphinol	(2183-56-4)
Hydromorphinolum (Latin)	(2183-56-4)
Hydromorphone	(466-99-9)
Hydromox	(73-49-4)
Hydromox R	(50-55-5)
Hydronitric acid	(7782-79-8)
Hydronol	(652-67-5)
1-Hydroperfluoroheptane	(375-83-7)
Hydroperit	(124-43-6)
Hydroperite	(124-43-6)
Hydroperoxide	(7722-84-1)
Hydroperoxide, acetyl	(79-21-0)
Hydroperoxide, bis(1-methylethyl)phenyl (9CI)	(26762-93-6)
Hydroperoxide, tert-butyl	(75-91-2)
Hydroperoxide, t-butyl-α,α-dimethylbenzyl-	(30026-92-7)
Hydroperoxide, diisopropylphenyl (8CI)	(26762-93-6)
Hydroperoxide, α,α-dimethylbenzyl	(80-15-9)
Hydroperoxide, 1,1-dimethylethyl- (9CI)	(75-91-2)
Hydroperoxide, 1,1-dimethylpropyl (9CI)	(3425-61-4)
Hydroperoxide, ethyl	(3031-74-1)
Hydroperoxide, 1-methylethyl (9CI)	(3031-75-2)
Hydroperoxide, (1-methylethylidene)bis-	(2614-76-8)
Hydroperoxide, 1-methylhexyl (8CI,9CI)	(762-46-9)
Hydroperoxide, 1-phenylethyl (9CI)	(3071-32-7)
Hydroperoxide, 1,2,3,4-tetrahydro-1-naphthalenyl	(771-29-9)
Hydroperoxide, 1,2,3,4-tetrahydro-1-naphthyl-	(771-29-9)
Hydroperoxide, 2,6,6-trimethylbicyclo(3.1.1)heptyl	(28324-52-9)
Hydroperoxide, 2,6,6-trimethylbicyclo(3.1.1)heptyl-, Not over 45% peroxide	(28324-52-9)
1-Hydroperoxycyclohexyl-1-hydroxycyclohexyl peroxide	(78-18-2)
Hydroperoxyde de butyle tertiaire (French)	(75-91-2)
Hydroperoxyde de cumene (French)	(80-15-9)
Hydroperoxyde de cumyle (French)	(80-15-9)
Hydroperoxyde de p-menthyle (French)	(3031-73-0)
Hydroperoxyde de tetrahydronaphtyle (French)	(771-29-9)
Hydroperoxyde de tetraline (French)	(771-29-9)
Hydroperoxyde d'ethyle (French)	(3031-74-1)
7-Hydroperoxykumen (Czech)	(80-15-9)
Hydropetidine	(468-56-4)
Hydrophenol	(108-93-0)
Hydroprene	(41096-46-2)
Hydropres	(50-55-5)
Hydropres KA	(50-55-5)
Hydroquinine (8CI)	(522-66-7)
Hydroquinol	(123-31-9)
Hydroquinole	(123-31-9)
α-Hydroquinone	(123-31-9)
m-Hydroquinone	(108-46-3)
o-Hydroquinone	(120-80-9)
p-Hydroquinone	(123-31-9)
Hydroquinone, Compd. with p-benzoquinone	(106-34-3)
Hydroquinone, Solid or liquid (ACGIH,OSHA) [UN 2662]	(123-31-9)
Hydroquinone benzyl ether	(103-16-2)
Hydroquinone bis(2-hydroxyethyl) ether	(104-38-1)
Hydroquinone bis(β-hydroxyethyl) ether	(104-38-1)
Hydroquinone, tert-butyl	(1948-33-0)
Hydroquinonecarboxylic acid	(490-79-9)
Hydroquinone, chloro	(615-67-8)
Hydroquinone, 2,5-di-tert-amyl-	(79-74-3)
Hydroquinone, 2,5-di-tert-butyl	(88-58-4)
Hydroquinone, 2,3-dichloro- (8CI)	(608-44-6)
Hydroquinone, 2,5-dichloro- (8CI)	(824-69-1)
Hydroquinone diethylol ether	(104-38-1)
Hydroquinone 1,4-diglycidyl ether	(2425-01-6)
Hydroquinone di(2-hydroxyethyl) ether	(104-38-1)
Hydroquinone di(β-hydroxyethyl) ether	(104-38-1)
Hydroquinone, di(β-hydroxyethyl) ether	(104-38-1)
Hydroquinone, dimethyl ether	(150-78-7)
Hydroquinone, 2,5-di-t-pentyl	(79-74-3)
Hydroquinone, 2,6-di-tert-pentyl- (6CI,7CI,8CI)	(2349-85-1)
Hydroquinone, hydroxy-	(533-73-3)
Hydroquinone, methyl	(95-71-6)
Hydroquinone monobenzyl ether	(103-16-2)
Hydroquinone monoethyl ether	(622-62-8)
Hydroquinone monomethyl ether	(150-76-5)
Hydroquinone, phenyl-	(1079-21-6)
Hydroquinone, tetrachloro	(87-87-6)
Hydroquinone, trichloro- (8CI)	(608-94-6)
Hydroquinone, trimethyl	(700-13-0)
Hydroresorcinol	(504-02-9)
Hydroretrocortin	(50-24-8)
Hydrorubeanic acid	(79-40-3)
Hydrosaluric	(58-93-5)
Hydroscine hydrobromide	(114-49-8)
Hydrosilicofluoric acid (DOT)	(16961-83-4)
Hydrosulfite AWC	(149-44-0)
2-Hydrosy-1-((4-carboxamidophenyl)azo)-N-(2-methoxyphenyl)-3-naphthalenecarboxamide	(36968-27-1)
Hydrothal-191	(145-73-3)
Hydrothal-47	(145-73-3)
Hydrothide	(58-93-5)
Hydrothol	(129-67-9)
Hydrothol	(145-73-3)
Hydrothol-191 (9CI)	(66330-88-9)
Hydrotreated heavy naphthenic distillate	(64742-52-5)
Hydrotreated heavy naphthenic distillates (Petroleum)	(64742-52-5)
Hydrotreated heavy paraffinic distillate	(64742-54-7)
Hydrotreated kerosene	(64742-47-8)
Hydrotreated light naphthenic distillate	(64742-53-6)
Hydrotreated light naphthenic distillates (Petroleum)	(64742-53-6)
Hydrotreated light paraffinic distillate	(64742-55-8)
Hydrotrichlorothiazide	(133-67-5)
Hydrotrope	(1300-72-7)
Hydrouracil, 6-hydroxy-	(67-52-7)
Hydrovit	(13422-51-0)
Hydroxine	(68-88-2)
Hydroxine Yellow L	(1934-21-0)
Hydroxocobalamin	(13422-51-0)
Hydroxocobalamine	(13422-51-0)
Hydroxocobalamine (French)	(13422-51-0)
Hydroxocobalaminum (Latin)	(13422-51-0)
Hydroxocobemine	(13422-51-0)
N-Hydroxy-AAF	(53-95-2)
N-Hydroxy-2-FAA	(53-95-2)
Hydroxy Vitamin B12	(13422-51-0)
Hydroxyacetaldehyde	(141-46-8)
2-(N-Hydroxyacetamido)fluorene	(53-95-2)

N-Hydroxy-2-acetamidofluorene (53-95-2)
1-Hydroxy-2-(β-(2'-acetamidophenyl)-ethyl)-naphthamide (5254-41-1)
2'-Hydroxyacetanilide (614-80-2)
3'-Hydroxyacetanilide (621-42-1)
3-Hydroxyacetanilide (621-42-1)
4'-Hydroxyacetanilide (103-90-2)
4-Hydroxyacetanilide (103-90-2)
m-Hydroxyacetanilide (621-42-1)
o-Hydroxyacetanilide (614-80-2)
p-Hydroxyacetanilide (103-90-2)
Hydroxyacetic acid (79-14-1)
Hydroxyacetic acid, cocopropylenediamine salt (68155-43-1)
Hydroxyacetic acid homopolymer (26124-68-5)
Hydroxyacetic acid, potassium salt (1932-50-9)
Hydroxyacetone (116-09-6)
2-Hydroxyacetonitrile (107-16-4)
Hydroxyacetonitrile (107-16-4)
2'-Hydroxyacetophenone (118-93-4)
3'-Hydroxyacetophenone (121-71-1)
4'-Hydroxyacetophenone (99-93-4)
m-Hydroxyacetophenone (121-71-1)
o-Hydroxyacetophenone (118-93-4)
p-Hydroxyacetophenone (99-93-4)
N-Hydroxy-2-acetylaminofluorene (53-95-2)
N-Hydroxy-N-acetyl-2-aminofluorene (53-95-2)
1-Hydroxy-2-acetyl-4-methylbenzene (1450-72-2)
4-Hydroxyaflatoxin B1 (6795-23-9)
4-Hydroxyaflatoxin B2 (6885-57-0)
β-Hydroxyalanine (56-45-1)
(-)-3-Hydroxy-N-allylmorphinan (152-02-3)
levo-3-Hydroxy-N-allyl morphinan (152-02-3)
Hydroxyamine hydrochloride (5470-11-1)
1-Hydroxy-4-aminoanthraquinone (116-85-8)
4-Hydroxy-3-aminobenzenesulfonic acid (98-37-3)
2-Hydroxy-4-aminobenzoic acid (65-49-6)
5-Hydroxy-3-(β-aminoethyl)indole (50-67-9)
N-Hydroxy-2-aminofluorene (53-94-1)
2-Hydroxy-4-amino-5-fluoropyrimidine (2022-85-7)
3-Hydroxy-5-aminomethylisoxazole (2763-96-4)
3-Hydroxy-5-aminomethylisoxazole-agarin (2763-96-4)
N-Hydroxy-1-aminonaphthalene (607-30-7)
6-(p-Hydroxy-α-aminophenylacetamido)penicillanic acid (26787-78-0)
dl-α-Hydroxy-β-aminopropylbenzene (14838-15-4)
α-Hydroxy-β-aminopropylbenzene hydrochloride (154-41-6)
4-(Hydroxyamino)quinoline 1-oxide (4637-56-3)
Hydroxyammonium chloride (5470-11-1)
8-Hydroxyamoxapine (61443-78-5)
p-Hydroxyampicillin (26787-78-0)
2-Hydroxyanaline (95-55-6)
17-β-Hydroxy-4-androsten-3-one (58-22-0)
17-β-Hydroxy-δ^4-androsten-3-one (58-22-0)
17-β-Hydroxyandrost-4-en-3-one (58-22-0)
7-β-Hydroxyandrost-4-en-3-one (58-22-0)
2-Hydroxyanilid kyseliny octove (Czech) (614-80-2)
4-Hydroxyanilid kyseliny octove (Czech) (103-90-2)
3-Hydroxyaniline (591-27-5)
4-Hydroxyaniline (123-30-8)
o-Hydroxyaniline (95-55-6)
p-Hydroxyaniline (123-30-8)
3-Hydroxy-m-anisic acid (877-22-5)
3-Hydroxyanisic acid (645-08-9)
2-Hydroxyanisole (90-05-1)
3-Hydroxyanisole (150-19-6)
m-Hydroxyanisole (150-19-6)
o-Hydroxyanisole (90-05-1)
1-Hydroxyanthrachinon (Czech) (129-43-1)
3-Hydroxyanthranilic acid (548-93-6)
3-Hydroxy-anthranilsaeure (German) (548-93-6)
1-Hydroxy-9,10-anthraquinone (129-43-1)
1-Hydroxyanthraquinone (129-43-1)
2-Hydroxyanthraquinone (605-32-3)
β-Hydroxyanthraquinone (605-32-3)
4-Hydroxy-1-anthraquinonylamine (116-85-8)
N-(4-Hydroxy-1-anthraquinonyl)-4-methylaniline (81-48-1)
N-(4-Hydroxy-1-anthraquinonyl)-p-toluidine (81-48-1)
Hydroxyapatite (1306-06-5)
2-Hydroxyatrazine (2163-68-0)
4-Hydroxyazobenzene (1689-82-3)
p-Hydroxyazobenzene (1689-82-3)
4-Hydroxy-3,4'-azodi-1-naphthalenesulfonic acid, disodium salt (3567-69-9)
4-Hydroxy-3,4'-azodi-1-naphthalenesulphonic acid, disodium salt (3567-69-9)
2-Hydroxy-1,1'-azonaphthalene-3,6,4'-trisulfonic acid trisodium salt (915-67-3)
2-Hydroxybenzaldehyde (90-02-8)
3-Hydroxybenzaldehyde (100-83-4)
4-Hydroxybenzaldehyde (123-08-0)
m-Hydroxybenzaldehyde (100-83-4)
meta-Hydroxybenzaldehyde (100-83-4)
o-Hydroxybenzaldehyde (90-02-8)
p-Hydroxybenzaldehyde (123-08-0)
2-Hydroxybenzamide (65-45-2)
o-Hydroxybenzamide (65-45-2)
Hydroxybenzene (108-95-2)
(+-)-α-Hydroxybenzeneacetic acid (611-72-3)
2-Hydroxybenzeneacetic acid (614-75-5)
3-Hydroxybenzeneacetic acid (621-37-4)
4-Hydroxybenzeneacetic acid (156-38-7)
α-Hydroxybenzeneacetic acid, (+-)- (611-72-3)
α-Hydroxybenzeneacetic acid 3,3,5-trimethylcyclohexyl ester (456-59-7)
4-Hydroxy-1,2-benzenedicarboxylic acid (610-35-5)
4-Hydroxybenzeneethanol (501-94-0)
2-Hydroxybenzenemethanol (90-01-7)
4-Hydroxybenzenemethanol (623-05-2)
4-Hydroxybenzenepropanoic acid (501-97-3)
Hydroxybenzenesulfonic acid (1333-39-7)
Hydroxybenzenesulfonic acid, monosodium salt (1300-51-2)
p-Hydroxybenzenesulfonic acid zinc salt (127-82-2)
3-Hydroxybenzisothiazole-S,S-dioxide (81-07-2)
2-Hydroxybenzoate (63-36-5)
o-Hydroxybenzoate (63-36-5)
o-Hydroxybenzoate anion (63-36-5)
4-Hydroxy-5-benzofuranacrylic acid γ-lactone (523-50-2)
((6-Hydroxy-5-benzofuranyl)methylene)malonic acid, γ-lactone, ethyl ester (20073-24-9)
3-(4-Hydroxy-5-benzofuranyl)-2-propenoic acid γ-lactone (523-50-2)
2-Hydroxybenzoic acid (69-72-7)
3-Hydroxybenzoic acid (99-06-9)
4-Hydroxybenzoic acid (99-96-7)
m-Hydroxybenzoic acid (99-06-9)
o-Hydroxybenzoic acid (69-72-7)
p-Hydroxybenzoic acid (99-96-7)
p-Hydroxybenzoic acid benzyl ester (94-18-8)
4-Hydroxybenzoic acid butyl ester (94-26-8)
p-Hydroxybenzoic acid butyl ester (94-26-8)
p-Hydroxybenzoic acid ethyl ester (120-47-8)
2-Hydroxybenzoic acid methyl ester (119-36-8)
o-Hydroxybenzoic acid, methyl ester (119-36-8)
p-Hydroxybenzoic acid methyl ester (99-76-3)
2-Hydroxybenzoic acid monosodium salt (54-21-7)
4-Hydroxybenzoic acid propyl ester (94-13-3)
p-Hydroxybenzoic acid propyl ester (94-13-3)
2-Hydroxybenzoic acid strontium salt (2:1) (526-26-1)
p-Hydroxybenzoic ethyl ester (120-47-8)
o-Hydroxybenzoic sodium salt (54-21-7)

2-Hydroxybenzoic-5-sulfonic acid **(97-05-2)**
2-Hydroxybenzonitrile **(611-20-1)**
4-Hydroxybenzonitrile **(767-00-0)**
o-Hydroxybenzonitrile **(611-20-1)**
2-Hydroxybenzophenone **(117-99-7)**
4-Hydroxybenzophenone **(1137-42-4)**
o-Hydroxybenzophenone **(117-99-7)**
Hydroxybenzopyridine **(148-24-3)**
3-Hydroxybenzotrifluoride **(98-17-9)**
m-Hydroxybenzotrifluoride **(98-17-9)**
2-Hydroxybenzoxazole **(59-49-4)**
2-Hydroxybenzyl alcohol **(90-01-7)**
3-Hydroxybenzyl alcohol **(620-24-6)**
4-Hydroxybenzyl alcohol **(623-05-2)**
o-Hydroxybenzyl alcohol **(90-01-7)**
p-Hydroxybenzyl alcohol **(623-05-2)**
4-Hydroxybenzyl cyanide **(14191-95-8)**
p-Hydroxybenzyl cyanide **(14191-95-8)**
α-Hydroxybenzyl phenyl ketone **(119-53-9)**
exo-2-Hydroxybicyclo(3.2.1)octane **(1965-38-4)**
2-Hydroxybifenyl (Czech) **(90-43-7)**
2-Hydroxybiphenyl **(90-43-7)**
3-Hydroxybiphenyl **(580-51-8)**
4-Hydroxybiphenyl **(92-69-3)**
o-Hydroxybiphenyl **(90-43-7)**
p-Hydroxybiphenyl **(92-69-3)**
2-Hydroxybiphenyl sodium salt **(132-27-4)**
Hydroxybis(salicylic acid acetato)aluminum **(23413-80-1)**
3-Hydroxybutanal **(107-89-1)**
4-Hydroxybutanamide **(927-60-6)**
1-Hydroxybutane **(71-36-3)**
2-Hydroxybutane **(78-92-2)**
(+-)-Hydroxybutanedioic acid **(617-48-1)**
2-Hydroxybutanoic acid **(565-70-8)**
3-Hydroxybutanoic acid **(300-85-6)**
3-Hydroxybutanoic acid, β-lactone **(3068-88-0)**
4-Hydroxybutanoic acid lactone **(96-48-0)**
4-Hydroxybutanoic acid, γ-lactone **(96-48-0)**
3-Hydroxy-2-butanone **(513-86-0)**
4-Hydroxy-2-butanone **(590-90-9)**
4-Hydroxy-2-butenoic acid γ-lactone **(497-23-4)**
4-Hydroxy-2-butenoic acid lactone **(497-23-4)**
1-Hydroxy-4-tert-butylbenzene **(98-54-4)**
4-Hydroxybutylbutylnitrosamine **(3817-11-6)**
2-(2'-Hydroxy-3'-t-butyl-5'-methylphenyl)-5-chlorobenzotriazole **(3896-11-5)**
2-(2'-Hydroxy-3'-tert-butyl-5'-methylphenyl)-5-chlorobenzotriazole **(3896-11-5)**
3-Hydroxybutyraldehyde **(107-89-1)**
β-Hydroxybutyraldehyde **(107-89-1)**
4-Hydroxybutyramide **(927-60-6)**
Hydroxybutyramide **(927-60-6)**
γ-Hydroxybutyramide **(927-60-6)**
2-Hydroxybutyric acid **(565-70-8)**
3-Hydroxybutyric acid **(300-85-6)**
α-Hydroxy-n-butyric acid **(565-70-8)**
α-Hydroxybutyric acid **(565-70-8)**
β-Hydroxy-n-butyric acid **(300-85-6)**
4-Hydroxybutyric acid amide **(927-60-6)**
γ-Hydroxybutyric acid cyclic ester **(96-48-0)**
3-Hydroxybutyric acid lactone **(3068-88-0)**
3-Hydroxybutyric acid, β-lactone **(3068-88-0)**
4-Hydroxybutyric acid lactone **(96-48-0)**
4-Hydroxybutyric acid, γ-lactone **(96-48-0)**
Hydroxybutyric acid lactone **(3068-88-0)**
γ-Hydroxybutyric acid lactone **(96-48-0)**
γ-Hydroxybutyrolactone **(96-48-0)**
2-Hydroxycamphane **(507-70-0)**
ε-Hydroxycaproic acid **(1191-25-9)**
Hydroxycarbamide **(127-07-1)**
Hydroxycarbamine **(127-07-1)**
m-Hydroxycarbanilic acid methyl ester m-methylcarbanilate **(13684-63-4)**
3-Hydroxycarbofuran **(16655-82-6)**
3-Hydroxy-4-carboxyaniline **(65-49-6)**
Hydroxycellulose **(9004-34-6)**
8-Hydroxy-chinolin (German) **(148-24-3)**
8-Hydroxy-chinolin-sulfat (German) **(134-31-6)**
1-exo-Hydroxychlordene **(24009-05-0)**
Hydroxychlordene **(24009-05-0)**
2-Hydroxy-5-chlorobenzophenone **(85-19-8)**
2-Hydroxy-3-chlorobiphenyl **(85-97-2)**
2-Hydroxy-5-chloro-N-(2-chloro-4-nitrophenyl)benzamide **(50-65-7)**
4-(4-Hydroxy-4'-chloro-4-phenylpiperidino)-4'-fluorobutyrophenone **(52-86-8)**
1-Hydroxy-5-chlorosulfonyl-2-naphthalenediazonium hydroxide, inner salt **(3770-97-6)**
3-α-Hydroxy-5-β-cholanic acid **(434-13-9)**
3-α-Hydroxycholanic acid **(434-13-9)**
3-β-Hydroxycholest-5-ene **(57-88-5)**
3-β-Hydroxycholestane **(360-68-9)**
3β-Hydroxycholestane **(80-97-7)**
Hydroxycine **(68-88-2)**
(E)-p-Hydroxycinnamic acid **(501-98-4)**
4'-Hydroxycinnamic acid **(7400-08-0)**
4-Hydroxycinnamic acid **(7400-08-0)**
p-Hydroxycinnamic acid **(7400-08-0)**
trans-2-Hydroxycinnamic acid **(614-60-8)**
trans-4-Hydroxycinnamic acid **(501-98-4)**
trans-o-Hydroxycinnamic acid **(614-60-8)**
trans-p-Hydroxycinnamic acid **(501-98-4)**
o-Hydroxycinnamic acid lactone **(91-64-5)**
o-Hydroxycinnamic lactone **(91-64-5)**
7-Hydroxycitronellal **(107-75-5)**
Hydroxycitronellal **(107-75-5)**
Hydroxycitronellal methyl ether **(3613-30-7)**
Hydroxycitronellol **(107-74-4)**
Hydroxycobalamin **(13422-51-0)**
Hydroxycobalamine **(13422-51-0)**
11-Hydroxycorticoaldosterone **(50-22-6)**
17-Hydroxycorticosterone **(50-23-7)**
11-β-Hydroxycortisone **(50-23-7)**
Hydroxycortisone **(50-23-7)**
7-Hydroxycoumarin **(93-35-6)**
(Hydroxy-4 coumarinyl 3)-3 phenyl-3 (bromo-4 biphenylyl-4)-1 propanol-1 (French) **(28772-56-7)**
1-(4'-Hydroxy-3'-coumarinyl)-1-phenyl-3-butanone **(81-81-2)**
5-Hydroxy-2,3-cresotic acid **(5981-39-5)**
γ-Hydroxycrotonic acid lactone **(497-23-4)**
3-Hydroxycrotonic acid methyl ester dimethyl phosphate **(7786-34-7)**
1-Hydroxycumene **(617-94-7)**
1-Hydroxycyclobut-1-ene-3,4-dione potassium salt hydrate **(52591-22-7)**
1-Hydroxycycloheptanecarbonitrile **(931-97-5)**
3-Hydroxycyclohexadien-1-one **(108-46-3)**
Hydroxycyclohexane **(108-93-0)**
1-Hydroxycyclohexanecarbonitrile **(931-97-5)**
2-Hydroxy-1-cyclohexanone **(533-60-8)**
2-Hydroxycyclohexanone **(533-60-8)**
1-Hydroxycyclohexyl phenyl ketone **(947-19-3)**
(1-Hydroxycyclohexyl)phenylmethanone **(947-19-3)**
2-Hydroxy-p-cymene **(499-75-2)**
3-Hydroxy-p-cymene **(89-83-8)**
14'-Hydroxydaunomycin **(23214-92-8)**
14-Hydroxydaunomycin **(23214-92-8)**
14-Hydroxydaunorubicine **(23214-92-8)**
Hydroxydaunorubicin hydrochloride **(25316-40-9)**
2-Hydroxydecalin **(825-51-4)**
2-Hydroxydecanoate **(5393-81-7)**

5-Hydroxydecanoic acid **(624-00-0)**
Hydroxydecanoic acid γ-lactone **(706-14-9)**
Hydroxyde de potassium (French) **(1310-58-3)**
Hydroxyde de sodium (French) **(1310-73-2)**
Hydroxyde de tetramethylammonium (French) **(75-59-2)**
Hydroxyde de triphenyl-etain (French) **(76-87-9)**
17-Hydroxy-11-dehydrocorticosterone **(53-06-5)**
17-α-Hydroxy-11-dehydrocorticosterone **(53-06-5)**
5-Hydroxy-α-6-deoxytetracycline **(564-25-0)**
3-Hydroxy-4,15-diacetoxy-8-(3-methylbutyryloxy)-12,13-epoxy-δ^9-trichothecene **(21259-20-1)**
3-Hydroxydiazepam **(846-50-4)**
Hydroxydiazepam **(846-50-4)**
4-Hydroxy-3,5-dibromobenzonitrile **(1689-84-5)**
4-Hydroxy-3,5-di-tert-butyltoluene **(128-37-0)**
2-Hydroxy-3,5-dichlorophenyl sulphide **(97-18-7)**
N-Hydroxydiethylamine **(3710-84-7)**
2-Hydroxy-4-diethylaminobenzaldehyde **(17754-90-4)**
2-Hydroxy-3-diethylcarbamyl-9,10-dimethoxy-1,2,3,4,6,7-hexahydro-11bh-benzoquinolizine acetate **(63-12-7)**
β-Hydroxydiethyl sulfide **(110-77-0)**
p-Hydroxydifenylamin (Czech) **(122-37-2)**
para-Hydroxydifenylamin (Czech) **(122-37-2)**
14-Hydroxydihydrocodeinone **(76-42-6)**
(14S)-14-Hydroxydihydromorphinone **(76-41-5)**
14-Hydroxydihydromorphinone **(76-41-5)**
14-Hydroxydihydro-6-β-thebainol 4-methyl ether **(3176-03-2)**
4-Hydroxy-3,5-diiodobenzonitrile **(1689-83-4)**
β-(4-Hydroxy-3,5-diiodophenyl)alanine **(66-02-4)**
8-Hydroxy-5,7-diiodoquinoline **(83-73-8)**
1-Hydroxy-3,5-dimethoxybenzene **(500-99-2)**
4-Hydroxy-3,5-dimethoxybenzoic acid **(530-57-4)**
β-Hydroxy-β-(2,5-dimethoxyphenyl)-isopropylamine hydrochloride **(61-16-5)**
17-β-Hydroxy-7-β,17-α-dimethylandrost-4-ene-3-one **(17021-26-0)**
Hydroxydimethylarsine oxide **(75-60-5)**
Hydroxydimethylarsine oxide, sodium salt **(124-65-2)**
1-Hydroxy-2,4-dimethylbenzene **(105-67-9)**
2-Hydroxy-3,3-dimethylbutanoic acid **(4026-20-4)**
3-Hydroxy-N,N-dimethyl-cis-crotonamide dimethyl phosphate **(141-66-2)**
3-Hydroxydimethyl crotonamide dimethyl phosphate **(141-66-2)**
2'-Hydroxy-5,9-dimethyl-2-(3,3-dimethylallyl)-6,7-benzomorphan **(359-83-1)**
dl-2'-Hydroxy-5,9-dimethyl-2-(3,3-dimethylallyl)-6,7-benzomorphan **(359-83-1)**
4-Hydroxy-2,5-dimethyl-3(2H)furanone **(3658-77-3)**
7-Hydroxy-3,7-dimethyloctan-1-al **(107-75-5)**
7-Hydroxy-3,7-dimethyloctan-1-ol **(107-74-4)**
3-Hydroxy-4,5-dimethylol-α-picoline **(65-23-6)**
3-Hydroxy-4,5-dimethylol-α-picoline hydrochloride **(58-56-0)**
(+)2'-Hydroxy-5,9-dimethyl-2-phenethyl-6,7-benzomorphan **(127-35-5)**
2'-Hydroxy-5,9-dimethyl-2-phenethyl-6,7-benzomorphan **(127-35-5)**
3-Hydroxy-2,2-dimethylpropanal **(597-31-9)**
3-Hydroxy-2,2-dimethylpropanoic acid **(4835-90-9)**
3-Hydroxy-2,2-dimethylpropionaldehyde **(597-31-9)**
3-Hydroxy-2,2-dimethylpropyl 2,2-dimethylhydracrylate **(1115-20-4)**
3-Hydroxy-2,2-dimethylpropyl hydroxypivalate **(1115-20-4)**
5-Hydroxy-N,N-dimethyltryptamine **(487-93-4)**
1-Hydroxy-2,4-dinitrobenzene **(51-28-5)**
β-(2-Hydroxy-3,5-dinitrophenyl)butane acetate **(2813-95-8)**
2-Hydroxydiphenyl **(90-43-7)**
4-Hydroxydiphenyl **(92-69-3)**
o-Hydroxydiphenyl **(90-43-7)**
p-Hydroxydiphenyl **(92-69-3)**
Hydroxydiphenylacetic acid **(76-93-7)**
3-Hydroxydiphenylamine **(101-18-8)**
4-Hydroxydiphenylamine **(122-37-2)**
m-Hydroxydiphenylamine **(101-18-8)**
p-Hydroxydiphenylamine **(122-37-2)**
p-Hydroxydiphenylamine isopropyl ether **(101-73-5)**
4-Hydroxydiphenyldimethylmethane **(599-64-4)**
4,4'-Hydroxy-γ,δ-diphenyl-β,δ-hexadiene **(84-17-3)**
Hydroxydiphenylmethane **(91-01-0)**
p-Hydroxydiphenyl methane **(101-53-1)**
α-Hydroxydiphenylmethane-β-dimethylaminoethyl ether hydrochloride **(147-24-0)**
2-Hydroxydiphenyl sodium **(132-27-4)**
2-Hydroxydiphenyl, sodium salt **(132-27-4)**
1-Hydroxy-1,1-diphosphonoethane **(2809-21-4)**
4-Hydroxyditane **(101-53-1)**
6-Hydroxydodecanoic acid **(35875-13-9)**
5-Hydroxydodecanoic acid δ-lactone **(713-95-1)**
5-Hydroxydodecanoic acid lactone **(713-95-1)**
3-Hydroxy-1,2-epoxypropane **(556-52-5)**
16-α-Hydroxyestradiol **(50-27-1)**
1'-Hydroxyestragole **(51410-44-7)**
3-Hydroxyestra-1,3,5(10),7-tetraen-17-one **(474-86-2)**
3-Hydroxyestra-1,3,5(10)-trien-17-one **(53-16-7)**
3-Hydroxyestra-1,3,5(10)-trien-17-one benzoate **(2393-53-5)**
Hydroxyestrin benzoate **(50-50-0)**
1-Hydroxyethanecarboxylic acid **(50-21-5)**
1-Hydroxyethanediphosphonic acid **(2809-21-4)**
Hydroxyethanediphosphonic acid **(2809-21-4)**
1-Hydroxyethanediphosphonic acid, sodium salt **(29329-71-3)**
(1-Hydroxyethane-1,1-diyl)diphosphonic acid disodium salt **(7414-83-7)**
2-Hydroxyethanesulfonic acid, sodium salt **(1562-00-1)**
2-Hydroxy-1-ethanethiol **(60-24-2)**
Hydroxyethanoic acid **(79-14-1)**
Hydroxy ether **(110-80-5)**
17-β-Hydroxy-17-α-ethinyl-5(10)-estren-3-one **(68-23-5)**
(2-Hydroxyethoxy)acetic acid **(13382-47-3)**
β-Hydroxyethoxyacetic acid **(13382-47-3)**
4-Hydroxy-3-ethoxybenzaldehyde **(121-32-4)**
4-Hydroxy-3-ethoxy-benzoic acid diethylamide **(13898-68-5)**
2-(2-Hydroxyethoxy)ethyl dodecanoate **(141-20-8)**
2-(2-Hydroxyethoxy)ethyl laurate **(141-20-8)**
Hydroxyethyl acetamide **(142-26-7)**
N-(2-Hydroxyethyl)acetamide **(142-26-7)**
N-β-Hydroxyethylacetamide **(142-26-7)**
β-Hydroxyethylacetamide **(142-26-7)**
2-Hydroxyethyl acetate **(542-59-6)**
α-(2-Hydroxyethyl)acetoacetic acid γ-lactone **(517-23-7)**
2-Hydroxyethyl acrylate **(818-61-1)**
Hydroxyethyl acrylate **(818-61-1)**
β-Hydroxyethyl acrylate **(818-61-1)**
2-Hydroxyethylamine **(141-43-5)**
β-Hydroxyethylamine **(141-43-5)**
1-((2-Hydroxyethyl)amino)-9,10-anthracenedione **(4465-58-1)**
1-(N-(2-Hydroxyethyl)amino)anthraquinone **(4465-58-1)**
2-Hydroxyethylaminodiacetic acid **(93-62-9)**
2-(2-Hydroxyethylamino)ethanol **(111-42-2)**
(-)-α-Hydroxy-β-(methylamino)ethyl-α-(3-hydroxybenzene) hydrochloride **(61-76-7)**
N-(2-((2-Hydroxyethyl)amino)ethyl)octadecanamide **(141-21-9)**
N-(2-((2-Hydroxyethyl)amino)ethyl)stearamide **(141-21-9)**
2-Hydroxy-4-(ethylamino)-6-(isopropylamino)-s-triazine **(2163-68-0)**
1-(β-Hydroxyethylamino)-4-(methylamino)anthraquinone **(86722-66-9)**
4-((Hydroxyethyl)amino)-1-(methylamino)anthraquinone **(86722-66-9)**
4-Hydroxyethylamino-1-methylaminoanthraquinone **(2475-46-9)**
3-((2-Hydroxyethyl)amino)propionamide **(27076-30-8)**
4-(2-Hydroxyethyl)aniline **(104-10-9)**
p-(2-Hydroxyethyl)aniline **(104-10-9)**
2-Hydroxy-1-ethylaziridine **(1072-52-2)**
N-(2-Hydroxyethyl)aziridine **(1072-52-2)**
N-(β-Hydroxyethyl)aziridine **(1072-52-2)**
β-Hydroxy-1-ethylaziridine **(1072-52-2)**
β-Hydroxyethylbenzene **(60-12-8)**

5-Hydroxyethyl-1,3-bis(hydroxymethyl)hexahydro-s-triazin-2-one (1852-21-7)
2-Hydroxyethylcarbamate (5395-01-7)
β-Hydroxyethylcarbamate (5395-01-7)
1-(((2-Hydroxyethyl)carbamoyl)methyl)pyridinium chloride laurate (ester) (6272-74-8)
2-Hydroxyethyl cellulose (9004-62-0)
Hydroxyethyl cellulose (9004-62-0)
2-Hydroxyethyl cellulose ether (9004-62-0)
Hydroxyethyl cellulose ether (9004-62-0)
2-Hydroxyethyl 2-chloroethyl sulfide (693-30-1)
1-(2-Hydroxyethyl)-4-(3-(2-chloro-10-phenothiazinyl)propyl)-piperazine (58-39-9)
2-Hydroxyethyl coco amine, ethoxylated (61791-14-8)
N-(2-Hydroxyethyl)cyclohexylamine (2842-38-8)
β-Hydroxyethyldimethylamine (108-01-0)
N-(2-Hydroxyethyl)dodecanamide (142-78-9)
N-Hydroxyethylenediaminetriacetic acid (150-39-0)
2-Hydroxyethylester kyseliny akrylove (Czech) (818-61-1)
2-Hydroxyethylester kyseliny octove (Czech) (542-59-6)
N-Hydroxyethyl-1,2-ethanediamine (111-41-1)
Hydroxyethyl ether cellulose (9004-62-0)
Hydroxyethylethylaniline (92-50-2)
N-(2-Hydroxyethyl)-N-ethylaniline (92-50-2)
N-(2-Hydroxyethyl)ethylenediamine (111-41-1)
N-(β-Hydroxyethyl)ethylenediamine (111-41-1)
N-(β-Hydroxyethylethylenediamine)-N,N',N'-triacetic acid (150-39-0)
N-(2-Hydroxyethyl)ethylenediaminetriacetic acid (150-39-0)
N-(2-Hydroxyethyl)-N,N',N'-ethylenediaminetriacetic acid tri-sodium salt (139-89-9)
N-(Hydroxyethyl)-N,N',N'-ethylenediaminetriacetic acid trisodium (139-89-9)
N-(Hydroxyethyl)-N,N',N'-ethylenediaminetriacetic acid trisodium salt (139-89-9)
(2-Hydroxyethyl)ethylenediaminetriacetic acid, trisodium salt (139-89-9)
(Hydroxyethyl)ethylenediaminetriacetic acid, trisodium salt (139-89-9)
N-Hydroxyethylethylenediaminetriacetic acid trisodium salt (139-89-9)
(N-Hydroxyethylethylenedinitrilo)triacetic acid (150-39-0)
N-Hydroxyethyl ethylene imine (1072-52-2)
Hydroxyethylethyleneurea (3699-54-5)
1-(2-Hydroxyethyl)ethylenimine (1072-52-2)
N-(2-Hydroxyethyl)ethylenimine (1072-52-2)
N-Hydroxyethyl-N-ethyl-m-toluidine (91-88-3)
1-(2-Hydroxyethyl)-2-heptadecenylglyoxalidine (95-38-5)
1-Hydroxyethyl-2-heptadecenylglyoxalidine (95-38-5)
1-(2-Hydroxyethyl)-2-(8-heptadecenyl)-2-imidazoline (95-38-5)
1-(2-Hydroxyethyl)-2-heptadecenyl-2-imidazoline (27136-73-8)
1-(2-Hydroxyethyl)-2-heptadecenyl-2-imidazoline (95-38-5)
1-(2-Hydroxyethyl)-2-n-heptadecenyl-2-imidazoline (95-38-5)
1-(Hydroxyethyl)-2-(8-heptadecenyl)imidazoline acetate (3388-72-5)
3-Hydroxy-2-ethylhexanal (496-03-7)
Hydroxyethyl hydrazine (109-84-2)
N-(2-Hydroxyethyl)hydrazine (109-84-2)
β-Hydroxyethylhydrazine (109-84-2)
(1-Hydroxyethylidene)bisphosphonic acid, potassium salt (67953-76-8)
1-Hydroxyethylidene-1,1'-diphosphonate, dipotassium salt (21089-06-5)
(1-Hydroxyethylidene)diphosphonic acid, disodium salt (7414-83-7)
1-Hydroxyethylidene-1,1-diphosphonic acid disodium salt (7414-83-7)
Hydroxyethylidene diphosphonic acid, potassium salt (67953-76-8)
Hydroxyethylidene diphosphonic acid, sodium salt (29329-71-3)
(1-Hydroxyethylidene)diphosphonic acid, tetrasodium salt (3794-83-0)
(1-Hydroxyethylidene)diphosphonic acid, trisodium salt (2666-14-0)
3-(1-Hydroxyethylidene)-6-methyl-2H-pyran-2,4(3H)-dione, sodium salt (4418-26-2)
1-(2-Hydroxyethyl)-2-imidazolidinone (3699-54-5)
2,2'-((2-Hydroxyethyl)imino)bis(N-(α,α-dimethylphenethyl)-N-methylacetamide) (126-27-2)
2,2'-((2-Hydroxyethyl)imino)bis(N-(1,1-dimethyl-2-phenyl-ethyl)-N-methylacetamide) (126-27-2)
2-Hydroxyethyliminobis(methylene phosphonic acid) (5995-42-6)
(2-Hydroxyethyl)iminodiacetic acid (93-62-9)
β-Hydroxyethyl isobornyl ether (7070-15-7)
(N-Hydroxyethyl)isopropylamine (109-56-8)
β-Hydroxyethyl isopropyl ether (109-59-1)
N-2-Hydroxyethyl-N-2-kyanethylanilin (Czech) (92-64-8)
N-β-Hydroxyethyl-N-β-kyanethylanilin (Czech) (92-64-8)
2-Hydroxyethylkyanid (Czech) (109-78-4)
N-(2-Hydroxyethyl)lauramide (142-78-9)
2-Hydroxyethyl mercaptan (60-24-2)
2-Hydroxyethyl methacrylate (868-77-9)
Hydroxyethyl methacrylate (868-77-9)
β-Hydroxyethyl methacrylate (868-77-9)
2-(N-2-Hydroxyethyl-N-methylamino)ethanol (105-59-9)
1-Hydroxyethyl methyl ketone (513-86-0)
1-(2-Hydroxy-1-ethyl)-2-methyl-5-nitroimidazole (443-48-1)
1-(2-Hydroxyethyl)-2-methyl-5-nitroimidazole (443-48-1)
1-(β-Hydroxyethyl)-2-methyl-5-nitroimidazole (443-48-1)
1-Hydroxyethyl-2-methyl-5-nitroimidazole (443-48-1)
5-(Hydroxyethyl)-4-methylthiazole (137-00-8)
N-(2-Hydroxyethyl)morfolin (Czech) (622-40-2)
N-(2-Hydroxyethyl)morpholine (622-40-2)
N-β-Hydroxyethylmorpholine (622-40-2)
β-Hydroxyethyl p-nitrophenyl ether (16365-27-8)
N-(2-Hydroxyethyl)octadecanamide (111-57-9)
N-(2-Hydroxyethyl)-N'-octadecanoylethylenediamine (141-21-9)
N-(2-Hydroxyethyl)octylamine (32582-63-1)
2-Hydroxyethyl-n-octyl sulfide (3547-33-9)
N-(2-Hydroxyethyl)phenylamine (122-98-5)
2-Hydroxyethyl N-phenylcarbamate (709-93-3)
β-Hydroxyethyl phenyl ether (122-99-6)
1-(2-Hydroxyethyl)piperazine (103-76-4)
Hydroxyethylpiperazine (25154-38-5)
N-(β-Hydroxyethyl)piperazine (103-76-4)
10-(3-(2-Hydroxyethyl)piperazinopropyl)-2-(trifluoromethyl)-phenothiazine (69-23-8)
γ-(4-(β-Hydroxyethyl)piperazin-1-yl)propyl-2-chlorophenothiazine (58-39-9)
N-(2-Hydroxyethyl)propanamide (18266-55-2)
N-(β-Hydroxyethyl)propionamide (18266-55-2)
β-Hydroxyethylpropionamide (18266-55-2)
Hydroxyethylpropylenediamine (10138-74-6)
N-(2-Hydroxyethyl) propylene diamine (10138-74-6)
2-(2-Hydroxyethyl)pyridine (103-74-2)
1-(2-Hydroxyethyl)-2-pyrrolidinone (3445-11-2)
n-(2-Hydroxyethyl)-2-pyrrolidone (3445-11-2)
2-Hydroxyethyl sodium sulfide (37482-11-4)
Hydroxyethyl starch (9004-62-0)
N-(2-Hydroxyethyl)stearamide (111-57-9)
N-(Hydroxyethyl)stearamide (111-57-9)
N-(2-Hydroxyethyl)-N'-stearoylethylenediamine (141-21-9)
β-Hydroxyethyl sulfide (111-48-8)
α-Hydroxyethylsulfide, monosodium salt (37482-11-4)
4'-(2-Hydroxyethylsulfonyl)acetanilide (27375-52-6)
1-(2-Hydroxyethyl)-2-(tall oil alkyl)-2-imidazoline (61791-39-7)
N-(2-Hydroxyethyl)tetradecanamide (142-58-5)
4-(α-Hydroxyethyl)toluene (536-50-5)
N-β-Hydroxyethyl-o-toluidino- (136-80-1)
(2-Hydroxyethyl)trimethylammonium (62-49-7)
(2-Hydroxyethyl)trimethylammonium bicarbonate (78-73-9)
Hydroxyethyltrimethylammonium bicarbonate (78-73-9)
(2-Hydroxyethyl)trimethylammonium bitartrate (87-67-2)
(2-Hydroxyethyl)trimethylammonium chloride (67-48-1)
(2-Hydroxyethyl)trimethyl ammonium chloride carbamate (51-83-2)
1-(2-Hydroxyethyl)-2-undecylimidazoline (136-99-2)
1-Hydroxyethyl-2-undecylimidazoline (136-99-2)
N-(2-Hydroxyethyl)-2,3-xylenesulfonamide (25959-70-0)
7-Hydroxy-3H-fenoxazin-3-on-10-oxid (Czech) (550-82-3)

β-Hydroxyfentanyl **(78995-10-5)**
3-(4'-Hydroxyfenyl)aminokarbazol (Czech) **(86-72-6)**
N-Hydroxy-N-2-fluorenylacetamide **(53-95-2)**
4-Hydroxy-4H-furo(3,2-c)pyran-2(6H)-one **(149-29-1)**
3-Hydroxyglutaconic acid, dimethyl ester, dimethyl phosphate **(122-10-1)**
1-Hydroxyheptane **(111-70-6)**
2-Hydroxyheptane **(543-49-7)**
3-Hydroxyheptane **(589-82-2)**
4-Hydroxyheptanoic acid lactone **(105-21-5)**
4-Hydroxyheptanoic acid, γ-lactone **(105-21-5)**
16-Hydroxyhexadecanoic acid **(506-13-8)**
N-(2-Hydroxyhexadecyl)diethanolamine oxide **(28865-36-3)**
1-Hydroxy-2,4-hexadiene **(111-28-4)**
1-Hydroxyhexane **(111-27-3)**
6-Hydroxyhexanoate **(1191-25-9)**
6-Hydroxyhexanoic acid **(1191-25-9)**
4-Hydroxyhexanoic acid lactone **(695-06-7)**
6-Hydroxyhexanoic acid lactone **(502-44-3)**
5-Hydroxy-2-hexenoic acid lactone **(10048-32-5)**
1-Hydroxy-3-n-hexyl-6,6,9-trimethyl-7,8,9,10-tetrahydro-6-dibenzopyran **(117-51-1)**
1-Hydroxyhomoadamantane **(31061-64-0)**
5-Hydroxyhydrindene **(1470-94-6)**
p-Hydroxyhydrocinnamic acid **(501-97-3)**
1-Hydroxy-1'-hydroperoxydicyclohexyl peroxide **(78-18-2)**
1-Hydroxy-1-hydroperoxydicyclohexyl peroxide **(78-18-2)**
Hydroxyhydroquinone **(533-73-3)**
Hydroxy(2-hydroxybenzoato-O1,O2)lead **(87903-39-7)**
1-Hydroxy-4-(1-(4-hydroxy-3-(methoxycarbonyl)-1-naphthalenyl)-3-oxo-1H,3H-naphtho(1,8-cd)pyran-1-yl)-6-(octadecyloxy)-2-naphthalenecarboxylic acid **(38122-80-4)**
17-β-Hydroxy-2-(hydroxymethylene)-17-α-methyl-5-α-androstan-3-one **(434-07-1)**
17-β-Hydroxy-2-(hydroxymethylene)-17-methyl-5-α-androstan-3-one **(434-07-1)**
17-β-Hydroxy-2-hydroxymethylene-17-α-methyl-3-androstanone **(434-07-1)**
17-Hydroxy-2-(hydroxymethylene)-17-methyl-5-α-17-β-androst-3-one **(434-07-1)**
3-Hydroxy-5-(hydroxymethyl)-2-methylisonicotinaldehyde, hydrochloride **(65-22-5)**
3-Hydroxy-5-(hydroxymethyl)-2-methylisonicotinaldehyde 5-phosphate monohydrate **(41468-25-1)**
3-Hydroxy-4-((1-hydroxy-2-naphthalenyl)azo)-7-nitro-1-naphthalenesulfonic acid sodium salt **(1787-61-7)**
3-Hydroxy-4-((2-hydroxy-1-naphthalenyl)azo)-7-nitro-1-naphthalenesulfonic acid, monosodium salt, chromium complex **(5610-64-0)**
3-Hydroxy-4-(2-hydroxy-1-naphthylazo)-1-naphthalenesulfonic acid, sodium salt **(2538-85-4)**
4-Hydroxy-γ-(4-hydroxyphenyl)-γ-methylbenzenebutanoic acid **(126-00-1)**
4-Hydroxyimino-2,5-cyclohexadiene-1-one **(637-62-7)**
2-Hydroxyimidazole **(5918-93-4)**
2-(Hydroxyimino)-N,N-dimethyl-2-(methylmercapto)acetamide **(30558-43-1)**
3-Hydroxyiminoacetone **(306-44-5)**
α-(Hydroxyimino)benzeneacetonitrile **(825-52-5)**
(Hydroxyimino)cyclohexane **(100-64-1)**
2-Hydroxyiminomethyl pyridine **(873-69-8)**
4-(Hydroxyiminomethyl)pyridine **(696-54-8)**
3-Hydroxy-2-(1,3-indandione-2-yl)quinoline **(7576-65-0)**
m-Hydroxyiodobenzene **(626-02-8)**
L-3-(4-(4-Hydroxy-3-iodophenoxy)-3,5-diiodophenyl)alanine **(6893-02-3)**
o-(4-Hydroxy-3-iodophenyl)-3,5-diiodo-l-tyrosine **(6893-02-3)**
2-Hydroxyisobutyric acid **(594-61-6)**
α-Hydroxyisobutyronitrile **(75-86-5)**
4-Hydroxyisophthalic acid **(636-46-4)**
4'-(2-Hydroxy-3-(isopropylamino)propoxy)acetanilide **(6673-35-4)**
2-(p-(2-Hydroxy-3-(isopropylamino)propoxy)phenyl)acetamide **(29122-68-7)**
3-Hydroxy-4-isopropyl-1-methylcyclohexene **(491-04-3)**
1-Hydroxyisoquinoline **(491-30-5)**
Hydroxyisoxazole **(10004-44-1)**
3-Hydroxy-17-keto-estra-1,3,5-triene **(53-16-7)**
4-Hydroxy-2-keto-4-methylpentane **(123-42-2)**
3-Hydroxy-17-keto-oestra-1,3,5-triene **(53-16-7)**
Hydroxyl **(3352-57-6)**
Hydroxyl No. 253 **(25154-52-3)**
Hydroxylamine **(7803-49-8)**
Hydroxylamine, N-acetyl- **(546-88-3)**
Hydroxylamine, N-(aminocarbonyl)- **(127-07-1)**
Hydroxylamine, N-(4,6-bis(trichloromethyl)-s-triazin-2-yl)- (6CI) **(30357-61-0)**
Hydroxylamine, N-carbamoyl- **(127-07-1)**
Hydroxylamine chloride **(5470-11-1)**
Hydroxylamine chloride (1:1) **(5470-11-1)**
Hydroxylamine, N-decyl- (8CI) **(26228-72-8)**
Hydroxylamine, O-decyl- (8CI,9CI) **(29812-79-1)**
Hydroxylamine, N,N-diethyl **(3710-84-7)**
Hydroxylamine, N,O-dimethyl- **(1117-97-1)**
Hydroxylamine, N-fluoren-2-yl **(53-94-1)**
Hydroxylamine, hydriodide **(59917-23-6)**
Hydroxylamine, hydrochloride **(5470-11-1)**
Hydroxyl amine iodide **(59917-23-6)**
Hydroxylamine iodide **(59917-23-6)**
Hydroxylamine, N-methyl **(593-77-1)**
Hydroxylamine, o-methyl **(67-62-9)**
Hydroxylamine, o-methyl-, hydrochloride **(593-56-6)**
Hydroxylamine, N-1-naphthyl **(607-30-7)**
Hydroxylamine neutral sulfate **(10039-54-0)**
Hydroxylamine nitrate **(13465-08-2)**
Hydroxylamine, nitrate (Salt) (9CI) **(13465-08-2)**
Hydroxylamine, N-nitroso-N-phenyl-, ammonium salt **(135-20-6)**
Hydroxylamine, N-phenyl **(100-65-2)**
Hydroxylamine, N-(4-quinolyl)-, 1'-oxide **(4637-56-3)**
Hydroxylamine, sulfate (2:1) **(10039-54-0)**
Hydroxylamine, sulfate (1:1) (Salt) (9CI) **(10046-00-1)**
Hydroxylamine, sulfate [UN 2865] **(10039-54-0)**
Hydroxylamine sulphate (DOT) **(10039-54-0)**
Hydroxylammonium acid sulfate **(10046-00-1)**
Hydroxylammonium chloride **(5470-11-1)**
Hydroxylammonium sulfate **(10039-54-0)**
6-Hydroxylauric acid **(35875-13-9)**
1-Hydroxylethanediphosphonic acid, potassium salt **(67953-76-8)**
Hydroxylizaric acid **(81-54-9)**
8-Hydroxyloxapine **(61443-77-4)**
3β-Hydroxy-5α-cholestane **(80-97-7)**
17β-Hydroxy-17α-methyl-androstano(3,2-c)pyrazole **(10418-03-8)**
(E)-3β-Hydroxy-14α-methyl-7-oxopodocarpane-δ13α-acetic acid 2-(dimethylamino)-ethyl ester **(468-76-8)**
Hydroxyl radical **(3352-57-6)**
5-Hydroxy-l-tryptophan **(4350-09-8)**
3-Hydroxy-l-tyrosine **(59-92-7)**
Hydroxylurea **(127-07-1)**
6-Hydroxy-2-mercaptopyrimidine **(141-90-2)**
2-Hydroxymesitylene **(527-60-6)**
4-Hydroxymetanilamide **(98-32-8)**
4-Hydroxymetanilic acid **(98-37-3)**
Hydroxymethanesulfinic acid, sodium salt **(149-44-0)**
Hydroxymethanesulfonate, sodium salt **(75-92-3)**
4'-Hydroxy-3'-methoxyacetophenone **(498-02-2)**
1-Hydroxy-2-methoxy-4-allylbenzene **(97-53-0)**
4-Hydroxy-3-methoxyallylbenzene **(97-53-0)**
4-Hydroxy-3-methoxybenzaldehyde **(121-33-5)**
p-Hydroxy-m-methoxybenzaldehyde **(121-33-5)**
1-Hydroxy-2-methoxybenzene **(90-05-1)**
4-Hydroxy-3-methoxybenzeneacetic acid **(306-08-1)**
6-Hydroxy-4-methoxy-5-benzofuranacrylic acid, γ-lactone **(484-20-8)**
6-Hydroxy-7-methoxy-5-benzofuranacrylic acid δ-lactone **(298-81-7)**

Name	CAS No.
2-Hydroxy-3-methoxybenzoic acid	**(877-22-5)**
3-Hydroxy-4-methoxybenzoic acid	**(645-08-9)**
4-Hydroxy-3-methoxybenzoic acid	**(121-34-6)**
2-Hydroxy-4-methoxybenzophenone	**(131-57-7)**
2-Hydroxy-4-methoxybenzophenone-5-sulfonic acid	**(4065-45-6)**
4-Hydroxy-3-methoxycinnamic acid	**(537-98-4)**
5-Hydroxymethoxymethyl-1-aza-3,7-dioxabicyclo(3.3.0)octane	**(59720-42-2)**
4-Hydroxy-3-methoxy-1-methylbenzene	**(93-51-6)**
6-Hydroxy-3-methoxy-N-methyl-4,5-epoxymorphinan	**(125-28-0)**
3-Hydroxy-2'-methoxy-2-naphthanilide	**(135-62-6)**
4-Hydroxy-3-methoxy-α-oxobenzenepropanoic acid	**(1081-71-6)**
2-Hydroxy-3-(o-methoxyphenoxy)propyl 1-carbamate	**(532-03-6)**
(4-Hydroxy-3-methoxyphenyl)acetic acid	**(306-08-1)**
4-(4-Hydroxy-3-methoxyphenyl)-2-butanone	**(122-48-5)**
(4-Hydroxy-3-methoxyphenyl)ethyl methyl ketone	**(122-48-5)**
N-((4-Hydroxy-3-methoxyphenyl)methyl)-8-methyl-6-nonenamide	**(404-86-4)**
trans-N-((4-Hydroxy-3-methoxyphenyl)methyl)-8-methyl-6-nonenamide	**(404-86-4)**
3-Hydroxy-N-(2-methoxyphenyl)-2-naphthalenecarboxamide	**(135-62-6)**
(2-Hydroxy-4-methoxyphenyl)phenylmethanone	**(131-57-7)**
3-(4-Hydroxy-3-methoxyphenyl)-2-propenoic acid	**(1135-24-6)**
3-(4-Hydroxy-3-methoxyphenyl)propenoic acid	**(537-98-4)**
3-(4-Hydroxy-3-methoxyphenyl)-2-propen-1-ol	**(458-35-5)**
1-Hydroxy-2-methoxy-4-prop-2-enylbenzene	**(97-53-0)**
1-Hydroxy-2-methoxy-4-propenylbenzene	**(97-54-1)**
4-Hydroxy-3-methoxy-1-propenylbenzene	**(97-54-1)**
4-Hydroxy-3-methoxypropylbenzene	**(2785-87-7)**
1-(2-Hydroxy-3-methoxypropyl)-2-nitroimidazole	**(13551-87-6)**
4-Hydroxy-3-methoxystyrene polymer	**(31853-85-7)**
4-Hydroxy-3-methoxytoluene	**(93-51-6)**
N-(Hydroxymethyl)acrylamide	**(924-42-5)**
α-Hydroxy-β-methyl amine propylbenzene	**(299-42-3)**
2-((Hydroxymethyl)amino)ethanol	**(34375-28-5)**
2-(Hydroxymethylamino)ethanol	**(34375-28-5)**
l-α-Hydroxy-β-methylamino-3-hydroxy-l-ethylbenzene	**(59-42-7)**
(R)-3-Hydroxy-α-((methylamino)methyl)benzenemethanol hydrochloride	**(61-76-7)**
(-)-m-Hydroxy-α-(methylaminomethyl)benzyl alcohol	**(59-42-7)**
l-m-Hydroxy-α-((methylamino)methyl)benzyl alcohol	**(59-42-7)**
1-m-Hydroxy-α-(methylaminomethyl)benzyl alcohol hydrochloride	**(61-76-7)**
2-((Hydroxymethyl)amino)-2-methyl-1-propanal	**(52299-20-4)**
2-((Hydroxymethyl)amino)-2-methyl-1-propanol	**(52299-20-4)**
1-Hydroxy-2-methylamino-1-phenylpropane	**(299-42-3)**
17-β-Hydroxy-2-α-methyl-5-α-androstan-3-one	**(58-19-5)**
17-β-Hydroxy-17-methylandrost-4-en-3-one	**(58-18-4)**
17-β-Hydroxy-17-α-methylandrostra-1,4-dien-3-one	**(72-63-9)**
5-(Hydroxymethyl)-1-aza-3,7-dioxabicyclo(3.3.0)octane	**(6542-37-6)**
5-Hydroxymethyl-1-aza-3,7-dioxabicyclo(3.3.0)octane	**(6542-37-6)**
2-Hydroxy-5-methylazobenzene	**(952-47-6)**
1-Hydroxy-2-methylbenzene	**(95-48-7)**
1-Hydroxy-3-methylbenzene	**(108-39-4)**
1-Hydroxy-4-methylbenzene	**(106-44-5)**
α-(Hydroxymethyl)benzeneacetic acid	**(529-64-6)**
α-Hydroxy-α-methylbenzeneacetic acid	**(515-30-0)**
α-(Hydroxymethyl)benzeneacetic acid 8-methyl-8-azabicyclo(3.2.1)-oct-3-yl ester	**(51-55-8)**
5-Hydroxymethyl-1,3-benzodioxole	**(495-76-1)**
4-Hydroxy-6-methyl-5-benzofuranacrylic acid γ-lactone	**(73459-03-7)**
2-Hydroxymethylbenzoic acid, γ-lactone	**(87-41-2)**
2-Hydroxy-3-methylbenzoic acid methyl ester	**(23287-26-5)**
2-(Hydroxymethyl)bicyclo(2.2.1)hept-5-ene	**(95-12-5)**
5-Hydroxymethylbicyclo(2.2.1)hept-2-ene	**(95-12-5)**
4-(Hydroxymethyl)biphenyl	**(3597-91-9)**
2-Hydroxy-3-methylbutanal	**(67755-97-9)**
1-Hydroxy-3-methyl-2-butanone	**(36960-22-2)**
2-Hydroxy-2-methyl-3-butanone	**(115-22-0)**
3-Hydroxy-3-methyl-2-butanone	**(115-22-0)**
2-Hydroxy-3-(3-methyl-2-butenyl)-1,4-naphthalenedione	**(84-79-7)**
2-Hydroxy-3-(3-methyl-2-butenyl)-1,4-naphthoquinone	**(84-79-7)**
7-α-(1-(R)-Hydroxy-1-methylbutyl)-6,14-endoethenotetrahydro-oripavine hydrochloride	**(13764-49-3)**
7-α-(1-(R)-Hydroxy-1-methylbutyl)-6,14-endo-ethenotetrahydro-oripavine	**(14521-96-1)**
2-Hydroxy-2-methyl-3-butyne	**(115-19-5)**
(Hydroxymethyl)carbamic acid, isobutyl ester	**(67953-32-6)**
1-Hydroxy-3-methylcholanthrene	**(3342-98-1)**
15-Hydroxy-20-methylcholanthrene	**(3342-98-1)**
7-Hydroxy-4-methylcoumarin	**(90-33-5)**
7-Hydroxy-4-methylcoumarin, O,O-diethyl thiophosphoric acid ester	**(299-45-6)**
7-Hydroxy-4-methylcoumarin, O-ester with O,O-diethyl phosphorothioate	**(299-45-6)**
3-Hydroxy-N-methyl-cis-crotonamide dimethyl phosphate	**(6923-22-4)**
3-Hydroxy-N-methylcrotonamide dimethyl phosphate	**(6923-22-4)**
Hydroxymethylcyclohexane	**(100-49-2)**
1-Hydroxymethyl-2-cyclohexanone	**(5331-08-8)**
2-(Hydroxymethyl)cyclohexanone	**(5331-08-8)**
2-Hydroxy-3-methyl-2-cyclopenten-1-one	**(80-71-7)**
N-Hydroxy-methyl-2,6-dichlorothiobenzamide	**(1953-89-5)**
4-Hydroxymethyl-2,2-dimethyl-1,3-dioxolane	**(100-79-8)**
1-(Hydroxymethyl)-5,5-dimethyl hydantoin	**(116-25-6)**
1-(Hydroxymethyl)-5,5-dimethylhydantoin	**(116-25-6)**
1-(Hydroxymethyl)-5,5-dimethyl-2,4-imidazolidinedione	**(116-25-6)**
3-(Hydroxymethyl)diphenyl ether	**(13826-35-2)**
2-Hydroxymethylene-17-α-methyl-5-α-androstan-17-β-ol-3-one	**(434-07-1)**
2-(Hydroxymethylene)-17-α-methyldihydrotestosterone	**(434-07-1)**
2-Hydroxymethylene-17-α-methyl-dihydrotestosterone	**(434-07-1)**
2-Hydroxymethylene-17-α-methyl-17-β-hydroxy-3-androstanone	**(434-07-1)**
2-(Hydroxymethyl)ethanol	**(504-63-2)**
4-(2-Hydroxy-3-((1-methylethyl)amino)propoxy)benzene-acetamide	**(29122-68-7)**
N-(4-(2-Hydroxy-3-((1-methylethyl)amino)propoxy)phenyl)-acetamide	**(6673-35-4)**
N-(1-(Hydroxymethyl)ethyl)-D-lysergomide	**(60-79-7)**
N-(α-(Hydroxymethyl)ethyl)-D-lysergomide	**(60-79-7)**
1-Hydroxy-1-methylethyl methyl ketone	**(115-22-0)**
2-(Hydroxymethyl)-2-ethyl-1,3-propanediol, cyclic phosphate (1:1)	**(1005-93-2)**
2-(Hydroxymethyl)-2-ethyl-1,3-propanediol, cyclic phosphite (1:1)	**(824-11-3)**
β-Hydroxy-3-methylfentanyl	**(78995-14-9)**
(Hydroxymethyl)formamide	**(13052-19-2)**
N-(Hydroxymethyl)formamide	**(13052-19-2)**
N-Hydroxymethylformamide	**(13052-19-2)**
5-Hydroxymethylfuraldehyde	**(67-47-0)**
2-Hydroxymethylfuran	**(98-00-0)**
5-(Hydroxymethyl)furfural	**(67-47-0)**
Hydroxymethylfurfurole	**(67-47-0)**
3-Hydroxy-7-methylguanine	**(30345-27-8)**
3-Hydroxy-9-methylguanine	**(30345-28-9)**
4-Hydroxy-4-methylheptane	**(598-01-6)**
3-Hydroxymethyl-n-heptan-4-ol	**(94-96-2)**
3-Hydroxy-1-methyl-4-isopropylbenzene	**(89-83-8)**
3-Hydroxy-5-methylisoxazole	**(10004-44-1)**
Hydroxymethylkyanid (Czech)	**(107-16-4)**
2-Hydroxy-4-(methylmercapto)butyronitrile	**(17773-41-0)**
1-Hydroxymethyl-2-methylditmide-2-oxide	**(590-96-5)**
1-Hydroxymethyl-3,4-methylenedioxybenzene	**(495-76-1)**
2-(Hydroxymethyl)-2-(methylpentyl) butylcarbamate carbamate	**(4268-36-4)**
5-(Hydroxymethyl)-3-(3-methylphenyl)-2-oxazolidinone	**(29218-27-7)**
4-Hydroxymethyl-4-methyl-1-phenyl-3-pyrazolidone	**(13047-13-7)**
2-(Hydroxymethyl)-2-methylpropanal	**(597-31-9)**
2-(Hydroxymethyl)-2-methyl-1,3-propanediol	**(77-85-0)**
N-(Hydroxymethyl)-2-methyl-2-propenamide	**(923-02-4)**

(+-)-3-Hydroxy-N-methylmorphinan **(297-90-5)**
(-)-3-Hydroxy-N-methylmorphinan **(77-07-6)**
DL-3-Hydroxy-N-methylmorphinan **(297-90-5)**
2-Hydroxy-3-methyl-1,4-naphthoquinone **(483-55-6)**
3-Hydroxy-2-methyl-1,4-naphthoquinone **(483-55-6)**
Hydroxymethylnitrile **(107-16-4)**
2-Hydroxy-5-methyl-2'-nitroazobenzene **(1435-71-8)**
2-(Hydroxymethyl)-2-nitro-1,3-propanediol **(126-11-4)**
2-Hydroxymethyl-2-nitropropane-1,3-diol **(126-11-4)**
2-Hydroxymethyl-N-nitrosothiazolidine **(92134-93-5)**
2-Hydroxymethyl-5-norbornene **(95-12-5)**
5-Hydroxymethyl-2-norbornene **(95-12-5)**
17-β-Hydroxy-18-methyl-19-nor-17-α-pregn-4-en-20-yn-3-one **(797-63-7)**
N-(Hydroxymethyl)octadecanamide **(3370-35-2)**
7-Hydroxy-4-methyl-2-oxo-2H-1-benzopyran **(90-33-5)**
4-Hydroxy-4-methyl-pentan-2-on (German, Dutch) **(123-42-2)**
4-Hydroxy-4-methyl pentan-2-one **(123-42-2)**
4-Hydroxy-4-methylpentanone-2 **(123-42-2)**
4-Hydroxy-4-methyl-2-pentanone (OSHA) **(123-42-2)**
4-Hydroxy-4-methyl-2-pentanone peroxide **(54693-46-8)**
2-Hydroxymethylphenol **(90-01-7)**
o-(Hydroxymethyl)phenol **(90-01-7)**
p-Hydroxy-N-(1-methyl-2-phenoxyethyl)norephedrine **(395-28-8)**
N-(2-Hydroxy-5-methylphenyl)acetamide **(6375-17-3)**
N-(4-((2-Hydroxy-5-methylphenyl)azo)phenyl)acetamide **(2832-40-8)**
2-(2-Hydroxy-5-methylphenyl)benzotriazole **(2440-22-4)**
1-(2-Hydroxy-5-methylphenyl)ethanone **(1450-72-2)**
N-(1-(2-Hydroxy-1-methyl-2-phenylethyl)-3-methyl-4-piperidinyl)-N-phenylpropanamide **(78995-14-9)**
o-(Hydroxymethyl)phenyl β-D-glucopyranoside **(138-52-3)**
2-Hydroxy-2-methyl-1-phenyl-1-propanone **(7473-98-5)**
4-Hydroxy-α-(1-((1-methyl-3-phenylpropyl)amino)ethyl)benzemethanol **(447-41-6)**
p-Hydroxy-α-(1-((1-methyl-3-phenylpropyl)amino)ethyl)-benzyl alcohol **(447-41-6)**
5-(1-Hydroxy-2-(1-methyl-3-phenylpropylamino)ethyl)salicylamide **(36894-69-6)**
p-Hydroxy-N-(1-methyl-3-phenylpropyl)norephedrine **(447-41-6)**
3-Hydroxy-N-(2-methylphenyl)-4-((2,4,5-trichlorophenyl)-azo)-2-naphtha enecarboxamide **(6535-46-2)**
γ-(Hydroxymethylphosphinyl)-L-α-aminobutyryl-L-alanyl-L-alanine **(35597-43-4)**
(Hydroxymethyl)phosphonic acid **(2617-47-2)**
3-Hydroxy-2-methyl-5-((phosphonooxy)methyl)-4-pyridinecarboxaldehyde monohydrate **(41468-25-1)**
Hydroxymethylphthalimide **(118-29-6)**
N-(Hydroxymethyl)phthalimide **(118-29-6)**
17-Hydroxy-6-methylpregna-4,6-diene-3,20-dione acetate **(595-33-5)**
17-Hydroxy-6-α-methylpregn-4-ene-3,20-dione acetate **(71-58-9)**
17-Hydroxy-6-methylpregn-4-ene-3,20-dione acetate **(1172-82-3)**
17-α-Hydroxy-6-α-methylpregn-4-ene-3,20-dione acetate **(71-58-9)**
17-α-Hydroxy-6-α-methylprogesterone acetate **(71-58-9)**
1-Hydroxymethylpropane **(78-83-1)**
2-Hydroxy-2-methylpropanoic acid **(594-61-6)**
N-(Hydroxymethyl)-2-propenamide **(924-42-5)**
2-Hydroxy-2-methylpropionitrile **(75-86-5)**
2-Hydroxy-2-methylpropiophenone **(7473-98-5)**
(6-α,17-β)-17-Hydroxy-6-methyl-17-(1-propynyl)-androst-4-en-3-one **(79-64-1)**
17-β-Hydroxy-6-α-methyl-17-(1-propynyl)androst-4-en-3-one **(79-64-1)**
3-Hydroxy-2-methyl-4H-pyran-4-one **(118-71-8)**
2-(Hydroxymethyl)pyridine **(586-98-1)**
3-(Hydroxymethyl)pyridine **(100-55-0)**
3-Hydroxy-6-methylpyridine **(1121-78-4)**
4-(Hydroxymethyl)pyridine **(586-95-8)**
5-Hydroxy-6-methyl-3,4-pyridinedicarbinol hydrochloride **(58-56-0)**
5-Hydroxy-6-methyl-3,4-pyridinedimethanol **(65-23-6)**
5-Hydroxy-6-methyl-3,4-pyridinedimethanol hydrochloride **(58-56-0)**
3-Hydroxy-1-methylpyridinium bromide dimethylcarbamate (ester) **(101-26-8)**
4-Hydroxy-2-methyl-N-(2-pyridyl)-2H-1,2-benzothiazin-3-caboxyamid-1,1-dioxid (German) **(36322-90-4)**
4-Hydroxy-2-methyl-N-(2-pyridyl)-2H-1,2-benzothiazine-3-carboxamide-1,1-dioxide **(36322-90-4)**
3-Hydroxy-2-methyl-4-pyrone **(118-71-8)**
3-Hydroxy-2-methyl-γ-pyrone **(118-71-8)**
1-(Hydroxymethyl)-2-pyrrolidinone **(15438-71-8)**
1-Hydroxymethyl-2-pyrrolidinone **(15438-71-8)**
N-(Hydroxymethyl)-2-pyrrolidinone **(15438-71-8)**
1-(Hydroxymethyl)-2-pyrrolidone **(15438-71-8)**
N-(Hydroxymethyl)-2-pyrrolidone **(15438-71-8)**
8-Hydroxy-2-methylquinoline **(826-81-3)**
3-Hydroxy-2-methyl-4-quinolinecarboxylic acid **(117-57-7)**
3-Hydroxy-4-((4-methyl-2-sulfophenyl)azo)-2-naphthalene-carboxylic acid, calcium salt **(5281-04-9)**
3-Hydroxy-4-((3-methyl-1-(3-sulfophenyl)-5-hydroxy-4-pyrazolyl)-azo)-1-naphthalenesulfonic acid, sodium salt, chromium complex **(52677-44-8)**
2-Hydroxymethyltetrahydropyran **(100-72-1)**
2-Hydroxy-4-(methylthio)butanenitrile **(17773-41-0)**
2-Hydroxy-4-(methylthio)butanoic acid **(583-91-5)**
5-Hydroxymethyl-3-(m-tolyl)-2-oxazolidinone **(29218-27-7)**
3-α-Hydroxy-8-methyl-1-α-h,5-α-h-tropanium bromide 2-propyl-valerate **(80-50-2)**
3-α-Hydroxy-8-methyl-1-α,5-α-H-tropanium bromide (+-)tropate **(2870-71-5)**
(Hydroxymethyl)urea **(1000-82-4)**
1-(Hydroxymethyl)urea **(1000-82-4)**
N-(Hydroxymethyl)urea **(1000-82-4)**
N-Hydroxymocovina (Czech) **(127-07-1)**
Hydroxymuconic semialdehyde **(3270-98-2)**
Hydroxymycin **(7542-37-2)**
3-Hydroxymyristic acid **(1961-72-4)**
α-Hydroxymyristic acid **(2507-55-3)**
β-Hydroxymyristic acid **(1961-72-4)**
5-Hydroxy-1,4-naftochinon (Czech) **(481-39-0)**
1-Hydroxynaphthalene **(90-15-3)**
2-Hydroxynaphthalene **(135-19-3)**
α-Hydroxynaphthalene **(90-15-3)**
β-Hydroxynaphthalene **(135-19-3)**
2-Hydroxy-3-naphthalenecarboxanilide **(92-77-3)**
3-Hydroxy-2-naphthalenecarboxanilide **(92-77-3)**
1-Hydroxy-2-naphthalenecarboxylic acid **(86-48-6)**
6-Hydroxy-2-naphthalenecarboxylic acid **(16712-64-4)**
2-Hydroxy-6,8-naphthalenedisulfonic acid **(118-32-1)**
2-Hydroxynaphthalene-6,8-disulfonic acid **(118-32-1)**
3-Hydroxy-2,7-naphthalenedisulfonic acid **(148-75-4)**
7-Hydroxy-1,3-naphthalenedisulfonic acid **(118-32-1)**
1-Hydroxy-2-naphthalenesulfonic acid **(567-18-0)**
1-Hydroxy-4-naphthalenesulfonic acid **(84-87-7)**
1-Hydroxynaphthalene-4-sulfonic acid **(84-87-7)**
2-Hydroxy-6-naphthalenesulfonic acid **(93-01-6)**
2-Hydroxynaphthalene-7-sulfonic acid **(92-40-0)**
4-Hydroxy-1-naphthalenesulfonic acid **(84-87-7)**
6-Hydroxy-2-naphthalenesulfonic acid **(93-01-6)**
7-Hydroxy-2-naphthalenesulfonic acid **(92-40-0)**
4-((4-Hydroxy-1-naphthalenyl)azo)benzenesulfonic acid, monosodium salt **(523-44-4)**
4-((4-Hydroxy-1-naphthalenyl)azo)benzenesulphonic acid, monosodium salt **(523-44-4)**
4-((2-Hydroxy-1-naphthalenyl)azo)1-naphthalenesulfonic acid sodium salt **(1658-56-6)**
2-(3-Hydroxy-2-naphthamido)anisole **(135-62-6)**
2-Hydroxy-3-naphthanilide **(92-77-3)**
3-Hydroxy-1H,3H-naphtho(1,8-cd)pyran-1-one **(5656-90-6)**
1-Hydroxy-2-naphthoic acid **(86-48-6)**

3-Hydroxy-2-naphthoic acid **(92-70-6)**
2-Hydroxy-3-naphthoic acid anilide **(92-77-3)**
3-Hydroxy-2-naphthoic acid o-aniside **(135-62-6)**
2-Hydroxy-3-naphthoic anilide **(92-77-3)**
β-Hydroxynaphthoic anilide **(92-77-3)**
3-Hydroxy-2-naphthoic o-anisidide **(135-62-6)**
3-Hydroxy-2-naphthoic-α-naphthalide **(132-68-3)**
6-Hydroxy-1-naphthol **(575-44-0)**
2-Hydroxy-1,4-naphthoquinone **(83-72-7)**
2-Hydroxynaphthoquinone **(83-72-7)**
5-Hydroxy-1,4-naphthoquinone **(481-39-0)**
1-(2',3'-Hydroxynaphthoylamino)-2-methoxybenzene **(135-62-6)**
1-(2',3'-Hydroxynaphthoylamino)-2-methylbenzene **(135-61-5)**
1-(2',3'-Hydroxynaphthoylamino)naphthalene **(132-68-3)**
3-Hydroxy-2-naphthoylanilide **(92-77-3)**
1-Hydroxy-2-naphthylamine **(41772-23-0)**
N-Hydroxy-1-naphthylamine **(607-30-7)**
2-Hydroxy-1-naphthylamine hydrochloride **(1198-27-2)**
p-((4-Hydroxy-1-naphthyl)azo)benzenesulfonic acid, monosodium salt **(523-44-4)**
p-((2-Hydroxy-1-naphthyl)azo)benzenesulfonic acid sodium salt **(633-96-5)**
p-((4-Hydroxy-1-naphthyl)azo)benzenesulfonic acid, sodium salt **(523-44-4)**
p-((4-Hydroxy-1-naphthyl)azo)benzenesulphonic acid, monosodium salt **(523-44-4)**
p-((4-Hydroxy-1-naphthyl)azo)benzenesulphonic acid, sodium salt **(523-44-4)**
2-(2-Hydroxy-1-naphthylazo)benzoic acid **(29128-56-1)**
Hydroxyneopentyl hydroxypivalate **(1115-20-4)**
2-Hydroxy-5-nitroaniline **(99-57-0)**
4-Hydroxy-3-nitroaniline **(119-34-6)**
4-Hydroxy-4'-nitroazobenzene **(1435-60-5)**
4-Hydroxy-3-nitrobenzaldehyde **(3011-34-5)**
2-Hydroxynitrobenzene **(88-75-5)**
3-Hydroxynitrobenzene **(554-84-7)**
4-Hydroxynitrobenzene **(100-02-7)**
Hydroxynitrobenzene **(25154-55-6)**
m-Hydroxynitrobenzene **(554-84-7)**
4-Hydroxy-3-nitrobenzenearsonic acid **(121-19-7)**
2-Hydroxy-5-nitrobenzoic acid **(96-97-9)**
4-Hydroxy-3-nitrobenzoic acid **(616-82-0)**
4-Hydroxy-3-nitrobenzoic acid methyl ester **(99-42-3)**
2-Hydroxy-5-nitrometanilic acid **(96-67-3)**
3-Hydroxy-4'-nitro-2-naphthanilide chloroacetate **(135-61-5)**
4-Hydroxy-3-nitrophenylarsonic acid **(121-19-7)**
4-Hydroxy-3-(1-(4-nitrophenyl)-3-oxobutyl)-2H-1-benzopyran-2-one **(152-72-7)**
N-Hydroxy-N-nitroso-benzenamine, ammonium salt **(135-20-6)**
trans-4-Hydroxy-1-nitroso-L-proline **(30310-80-6)**
3-Hydroxy-4-((5-nitro-o-tolyl)azo)-2-naphthanilide **(6448-95-9)**
2-Hydroxy-nonachlorodiphenyl ether **(35245-80-8)**
3-Hydroxy-nonachlorodiphenyl ether **(42255-14-1)**
4-Hydroxynonanoic acid, γ-lactone **(104-61-0)**
7-Hydroxynorbornadiene **(822-80-0)**
N-Hydroxy-5-norbornene-2,3-dicarboximide **(21715-90-2)**
Hydroxynorephedrine **(54-49-9)**
m-Hydroxy norephedrine **(54-49-9)**
(17-α)-17-Hydroxy-19-norpregn-4-en-20-yn-3-one **(68-22-4)**
(17-α)-17-Hydroxy-19-norpregn-5(10)-en-20-yn-3-one **(68-23-5)**
17-Hydroxy(17-α)-19-norpregn-5(10)-en-20-yn-3-one **(68-23-5)**
17-Hydroxy-17-α-19-norpregn-4-en-20-yn-3-one **(68-22-4)**
17-Hydroxy-19-nor-17-α-pregn-4-en-20-yn-3-one **(68-22-4)**
17-Hydroxy-19-nor-17-α-pregn-5(10)-en-20-yn-3-one **(68-23-5)**
17-β-Hydroxy-19-norpregn-4-en-20-yn-3-one **(68-22-4)**
17-Hydroxy-19-nor-17-α-pregn-4-en-20-yn-3-one acetate **(51-98-9)**
17-β-Hydroxy-19-nor-17-α-pregn-4-en-20-yn-3-one acetate **(51-98-9)**
12-Hydroxyoctadecanoic acid **(106-14-9)**
Hydroxyoctadecanoic acid **(1330-70-7)**
12-Hydroxy-cis-9-octadecenoic acid **(141-22-0)**
12-Hydroxy-9-octadecenoic acid, methyl ester **(141-24-2)**
12-Hydroxy-9-octadecenoic acid, monopotassium salt **(7492-30-0)**
1-Hydroxy-6-octadecyloxy-2-naphthoic acid **(38134-94-0)**
1-Hydroxyoctane **(111-87-5)**
5-Hydroxyoctanoic acid lactone **(104-50-7)**
2-Hydroxy-4-(octyloxy)benzophenone **(1843-05-6)**
2-(2'-Hydroxy-5'-t-octylphenyl)benzotriazole **(3147-75-9)**
16-α-Hydroxyoestradiol **(50-27-1)**
3-Hydroxy-1,3,5(10)-oestratrien-17-one **(53-16-7)**
3-Hydroxy-oestra-1,3,5(10)-trien-17-one **(53-16-7)**
2-Hydroxy-4-oktyloxybenzofenon (Czech) **(1843-05-6)**
γ-Hydroxy-β-oxobutane **(513-86-0)**
4-Hydroxy-3-(3-oxo-1-fenyl-butyl) cumarine (Dutch) **(81-81-2)**
2-Hydroxy-6-oxohexa-2,4-dienoate semialdehyde **(3270-98-2)**
2-Hydroxy-6-oxo-2,4-hexadienoic acid **(3270-98-2)**
para-Hydroxy-2-oxo-phenylacethydroxymic acid chloride **(34911-46-1)**
4-Hydroxy-3-(3-oxo-1-phenylbutyl)-2H-1-benzopyran-2-one **(81-81-2)**
4-Hydroxy-3-(3-oxo-1-phenyl-butyl)-cumarin (German) **(81-81-2)**
3-Hydroxy-4-oxo-4H-pyran-2,6-dicarboxylic acid **(497-59-6)**
13-Hydroxy-3-oxo-13,17-secoandrosta-1,4-dien-17-oic acid δ-lactone **(968-93-4)**
6-(10-Hydroxy-6-oxo-trans-1-undecenyl)-β-resorcylic acid-N-lactone **(17924-92-4)**
16-Hydroxypalmitic acid **(506-13-8)**
ω-Hydroxypalmitic acid **(506-13-8)**
15-Hydroxypentadecanoic acid, lactone **(106-02-5)**
15-Hydroxypentadecanoic acid-ε-lactone **(106-02-5)**
5-Hydroxypentanal **(4221-03-8)**
3-Hydroxy-2-pentanedioic acid, dimethyl ester, dimethyl phosphate **(122-10-1)**
4-Hydroxypentanoic acid lactone **(108-29-2)**
5-Hydroxy-2-pentanone acetate **(5185-97-7)**
Hydroxypethidine **(468-56-4)**
Hydroxypethidinum (Latin) **(468-56-4)**
9-Hydroxyphenanthrene **(484-17-3)**
p-Hydroxyphenethyl alcohol **(501-94-0)**
4-Hydroxyphenethylamine **(51-67-2)**
β-Hydroxyphenethylamine **(7568-93-6)**
p-Hydroxy-β-phenethylamine **(51-67-2)**
p-Hydroxyphenethylamine **(51-67-2)**
2-(6-(β-Hydroxy-phenethyl)-1-methyl-2-piperidyl)acetophenone **(90-69-7)**
2-(6-(β-Hydroxyphenethyl)-1-methyl-2-piperidyl)acetophenone hydrochloride **(63990-84-1)**
p-Hydroxyphenetole **(622-62-8)**
4-Hydroxyphenobarbital **(379-34-0)**
2-Hydroxyphenol **(120-80-9)**
3-Hydroxyphenol **(108-46-3)**
m-Hydroxyphenol **(108-46-3)**
o-Hydroxyphenol **(120-80-9)**
p-Hydroxyphenol **(123-31-9)**
7-Hydroxy-3H-phenoxazin-3-one 10-oxide **(550-82-3)**
(4-Hydroxyphenoxy)acetic acid **(1878-84-8)**
4-Hydroxyphenoxyacetic acid **(1878-84-8)**
1-Hydroxy-2-phenoxyethane **(122-99-6)**
α-Hydroxy-3-phenoxytoluene, acetate **(50789-44-1)**
2-Hydroxy-N-phenylacetamide **(4746-61-6)**
N-(4-Hydroxyphenyl)acetamide **(103-90-2)**
m-Hydroxyphenyl acetate **(102-29-4)**
(4-Hydroxyphenyl)acetic acid **(156-38-7)**
(p-Hydroxyphenyl)acetic acid **(156-38-7)**
2-Hydroxyphenylacetic acid **(614-75-5)**
3-Hydroxyphenylacetic acid **(621-37-4)**
α-Hydroxyphenylacetic acid **(90-64-2)**
meta-Hydroxyphenylacetic acid **(621-37-4)**
(4-Hydroxyphenyl)acetonitrile **(14191-95-8)**
p-Hydroxyphenylacetonitrile **(14191-95-8)**
2-Hydroxy-2-phenylacetophenone **(119-53-9)**
α-Hydroxy-α-phenylacetophenone **(119-53-9)**
β-(4-Hydroxyphenyl)acrylic acid **(7400-08-0)**

p-Hydroxyphenylacrylic acid (7400-08-0)
α-(4-Hydroxyphenyl)-β-aminoethane (51-67-2)
1-(p-Hydroxyphenyl)-2-aminoethanol (104-14-3)
7-Hydroxy-8-(phenylazo)-1,3-naphthalenedisulfonic acid, disodium salt (1936-15-8)
7-Hydroxy-8-(phenylazo)-1,3-naphthalenedisulphonic acid, disodium salt (1936-15-8)
α-Hydroxy-α-phenylbenzeneacetic acid 2-(diethylamino)ethyl ester (302-40-9)
α-Hydroxy-α-phenylbenzeneacetic acid, 1-methyl-3-piperidinyl ester (3321-80-0)
3-Hydroxyphenyl benzoate (136-36-7)
2-(2-Hydroxyphenyl)benzothiazole (3411-95-8)
2-(o-Hydroxyphenyl)-benzoxazole (835-64-3)
p-Hydroxyphenyl benzyl ether (103-16-2)
1-(4-Hydroxyphenyl)-2-bromoethanone (2491-38-5)
p-Hydroxyphenylbutazone (129-20-4)
p-Hydroxyphenyl-n-butylamine (103-62-8)
(3-Hydroxyphenyl)dimethylamine (99-07-0)
(4-Hydroxyphenyl)dimethylsulfonium chloride (1005-35-2)
(p-Hydroxyphenyl)dimethylsulfonium chloride (1005-35-2)
4-Hydroxyphenylethanol (501-94-0)
p-Hydroxyphenylethanolamine (104-14-3)
1-(3-Hydroxyphenyl)ethan-1-one (121-71-1)
1-(3-Hydroxyphenyl)ethanone (121-71-1)
1-(4-Hydroxyphenyl)ethanone (99-93-4)
2-(p-Hydroxyphenyl)ethylamine (51-67-2)
3-Hydroxyphenylethylamine (588-05-6)
4-Hydroxyphenylethylamine (51-67-2)
β-Hydroxy-β-phenylethylamine (7568-93-6)
β-Hydroxyphenylethylamine (51-67-2)
p-Hydroxyphenylethylamine (51-67-2)
2-(6-(2-Hydroxy-2-phenylethyl)-1-methyl-2-piperidinyl)-1-phenylethanone (90-69-7)
N-(1-(2-Hydroxy-2-phenyl)ethyl-4-piperidyl)-N-phenyl-propanamide, Its optical isomers, salts, and salts of isomers (78995-10-5)
β-(meta-Hydroxyphenyl)hydracrylic acid (3247-75-4)
3-(3-Hydroxyphenyl)-3-hydroxypropanoic acid (3247-75-4)
Hydroxyphenylmercury (100-57-2)
l-1-(m-Hydroxyphenyl)-2-methylaminoethanol (59-42-7)
l-1-(m-Hydroxyphenyl)-2-methyl-aminoethanol hydrochloride (61-76-7)
l-(3-Hydroxyphenyl)-N-methylethanolamine (59-42-7)
o-Hydroxyphenyl methyl ketone (118-93-4)
p-Hydroxyphenyl methyl ketone (99-93-4)
1-(4-Hydroxyphenyl)-2-(1-methyl-2-phenoxyethylamino)propanol (395-28-8)
1-(p-Hydroxyphenyl)-2-(1'-methyl-2'-phenoxyethylamino)-propanol-2-hydrochloride (395-28-8)
1-(p-Hydroxyphenyl)-2-(1'-methyl-3'-phenylpropylamino)-1-propanol (447-41-6)
1-(4-(3-Hydroxyphenyl)-1-methyl-4-piperidinyl)-1-propanone (469-79-4)
3-Hydroxy-N-phenyl-2-naphthalenecarboxamide (92-77-3)
N-p-Hydroxyphenyl-2-naphthylamine (93-45-8)
N-p-Hydroxyphenyl-β-naphthylamine (93-45-8)
p-Hydroxyphenyl-2-naphthylamine (93-45-8)
p-Hydroxyphenyl-β-naphthylamine (93-45-8)
N-(4-Hydroxyphenyl)octadecanamide (103-99-1)
1-(p-Hydroxyphenyl)octane (1806-26-4)
1-(p-Hydroxyphenyl)-2-phenyl-4-butyl-3,5-pyrazolidinedione (129-20-4)
1-p-Hydroxyphenyl-2-phenyl-3,5-dioxo-4-N-butylpyrazolidine (129-20-4)
(2-Hydroxyphenyl)phenylmethanone (117-99-7)
(4-Hydroxyphenyl)phenylmethanone (1137-42-4)
β-(p-Hydroxyphenyl)phloropropiophenone (60-82-2)
3-(4-Hydroxyphenyl)propanoic acid (501-97-3)
1-(4-Hydroxyphenyl)-1-propanone (70-70-2)
p-Hydroxyphenyl-1-propanone (70-70-2)
3-(4-Hydroxyphenyl)-2-propenoic acid (7400-08-0)
o-Hydroxyphenylpropionamide (22367-76-6)
2-Hydroxy-2-phenylpropionic acid (515-30-0)
3-(4'-Hydroxyphenyl)propionic acid (501-97-3)
3-(4-Hydroxyphenyl)propionic acid (501-97-3)
3-(p-Hydroxyphenyl)propionic acid (501-97-3)
3-Hydroxy-3-phenylpropionic acid (3480-87-3)
4-Hydroxyphenylpropionic acid (501-97-3)
α-Hydroxy-α-phenylpropionic acid (515-30-0)
β-(p-Hydroxyphenyl)propionic acid (501-97-3)
β-Hydroxyphenylpropionic acid (3480-87-3)
p-Hydroxyphenylpropionic acid (501-97-3)
p-Hydroxyphenylpyruvic acid (156-39-8)
N-(4-Hydroxyphenyl)retinamide (65646-68-6)
4-Hydroxyphenylsulfonic acid (98-67-9)
2-((N-(m-Hydroxyphenyl)-p-toluidino)methyl)-2-imidazoline (50-60-2)
β-(p-Hydroxyphenyl)-2,4,6-trihydroxypropiophenone (60-82-2)
(m-Hydroxyphenyl)trimethylammonium bromide dimethylcarbamate (114-80-7)
3-Hydroxyphenyltrimethylammonium bromide dimethylcarbamic ester (114-80-7)
(m-Hydroxyphenyl)trimethylammonium dimethylcarbamate (ester) (59-99-4)
(3-Hydroxyphenyl)urea (701-82-6)
N,N'-((Hydroxyphosphinylidene)bis(oxy-2,1-ethanediyl))bis-(N-ethylheptadecafluoro-1-octanesulfonamide), ammonium salt (30381-98-7)
N-Hydroxyphthalimide (524-38-9)
3-Hydroxypicolinic acid (874-24-8)
3-Hydroxypivalaldehyde (597-31-9)
Hydroxypivalaldehyde (597-31-9)
Hydroxypivaldehyde (597-31-9)
Hydroxypivalic acid (4835-90-9)
Hydroxypivalic acid neopentyl glycol ester (1115-20-4)
3-(Hydroxypivaloyloxy)-2,2-dimethylpropanol (1115-20-4)
Hydroxypivalyl hydroxypivalate (1115-20-4)
Hydroxypolyethoxydodecane (9002-92-0)
17-Hydroxypregn-4-ene-3,20-dione (68-96-2)
17-Hydroxypregn-4-ene-3,20-dione hexanoate (630-56-8)
17-Hydroxyprogesterone (68-96-2)
17-α-Hydroxyprogesterone (68-96-2)
17-α-Hydroxy progesterone n-caproate (630-56-8)
17-α-Hydroxyprogesterone caproate (630-56-8)
Hydroxyprogesterone caproate (630-56-8)
17-α-Hydroxyprogesterone hexanoate (630-56-8)
4-Hydroxy-L-proline (51-35-4)
4-L-Hydroxyproline (51-35-4)
Hydroxy-L-proline (51-35-4)
Hydroxyproline (51-35-4)
L-4-Hydroxyproline (51-35-4)
L-Hydroxyproline (51-35-4)
δ-Hydroxyproline (51-35-4)
trans-4-Hydroxy-L-proline (51-35-4)
trans-4-Hydroxyproline (51-35-4)
trans-Hydroxyproline (51-35-4)
trans-L-Hydroxyproline (51-35-4)
Hydroxyproline, (L) (51-35-4)
m-Hydroxypropadrine (54-49-9)
1-Hydroxypropane (71-23-8)
2-Hydroxypropane (67-63-0)
3-Hydroxypropanenitrile (109-78-4)
3-Hydroxy-1-propanesulfonic acid γ-sultone (1120-71-4)
3-Hydroxy-1-propanesulphonic acid sulfone (1120-71-4)
3-Hydroxy-1-propanesulphonic acid sultone (1120-71-4)
2-Hydroxy-1,2,3-propanetricarboxylic acid (77-92-9)
2-Hydroxy-1,2,3-propanetricarboxylic acid calcium salt (2:3) (813-94-5)
2-Hydroxy-1,2,3-propanetricarboxylic acid, copper salt (1:2) (866-82-0) (866-82-0)
2-Hydroxy-1,2,3-propanetricarboxylic acid, diammonium salt (3012-65-5)
2-Hydroxy-1,2,3-propanetricarboxylic acid, tributyl ester (77-94-1)
2-Hydroxy-1,2,3-propanetricarboxylic acid, trioctadecyl ester (7775-50-0)
2-Hydroxypropannitril (Czech) (78-97-7)
(+-)-2-Hydroxypropanoic acid (598-82-3)

2-Hydroxypropanoic acid **(50-21-5)**
2-Hydroxypropanoic acid **(598-82-3)**
2-Hydroxypropanoic acid calcium salt **(814-80-2)**
2-Hydroxypropanoic acid methyl ester **(547-64-8)**
2-Hydroxy-propanoic acid 1-methylethyl ester **(617-51-6)**
2-Hydroxypropanoic acid strontium salt **(29870-99-3)**
Hydroxypropazine **(7374-53-0)**
3-Hydroxy-2-propenal sodium salt **(24382-04-5)**
3-Hydroxypropene **(107-18-6)**
4-(3-Hydroxy-1-propenyl)-2-methoxyphenol **(458-35-5)**
2-Hydroxypropionic acid **(50-21-5)**
α-Hydroxypropionic acid **(50-21-5)**
α-Hydroxypropionic acid **(598-82-3)**
3-Hydroxypropionic acid lactone **(57-57-8)**
2-Hydroxypropionitrile **(78-97-7)**
3-Hydroxypropionitrile **(109-78-4)**
β-Hydroxypropionitrile **(109-78-4)**
4-Hydroxypropiophenone **(70-70-2)**
Hydroxypropiophenone **(70-70-2)**
p-Hydroxypropiophenone **(70-70-2)**
2-(2-(2-Hydroxypropoxy)propoxy)-1-propanol **(24800-44-0)**
Hydroxypropyl acrylate **(25584-83-2)**
β-Hydroxypropyl acrylate **(999-61-1)**
2-Hydroxypropyl acrylate (ACGIH,OSHA) **(999-61-1)**
2-Hydroxy-1-propylamine **(78-96-6)**
2-Hydroxypropylamine **(78-96-6)**
3-Hydroxypropylamine **(156-87-6)**
Hydroxypropylated diphenylolpropane **(116-37-0)**
(3-Hydroxypropyl)benzene **(122-97-4)**
α-Hydroxypropylbenzene **(93-54-9)**
p-Hydroxypropyl benzoate **(94-13-3)**
2-Hydroxypropyl bromide **(19686-73-8)**
3-Hydroxypropyl bromide **(627-18-9)**
Hydroxypropyl-cellulose **(9004-64-2)**
N-(2-Hydroxypropyl)dodecanamide **(142-54-1)**
Hydroxypropyl ether of cellulose **(9004-64-2)**
N-(2-Hydroxypropyl)methacrylamide polymer **(40704-75-4)**
2-Hydroxypropyl methacrylate **(923-26-2)**
Hydroxypropyl methacrylate **(27813-02-1)**
β-Hydroxypropyl methacrylate **(923-26-2)**
3-Hydroxypropyl methyl sulfide **(505-10-2)**
(Z)-N-(2-Hydroxypropyl)-9-octadecenamide **(111-05-7)**
N-(2-Hydroxypropyl)-9-octadecenamide **(111-05-7)**
S-(2-Hydroxypropyl)thiomethanesulfonate **(29803-57-4)**
(2-Hydroxypropyl)trimethylammonium chloride acetate **(62-51-1)**
3-Hydroxypseudocumene **(2416-94-6)**
6-Hydroxypurine **(68-94-0)**
4-Hydroxy-1H-pyrazolo(3,4-d)pyrimidine **(315-30-0)**
4-Hydroxypyrazolo(3,4-d)pyrimidine **(315-30-0)**
4'-Hydroxypyrazolol(3,4-d)pyrimidine **(315-30-0)**
4-Hydroxy-3,4-pyrazolopyrimidine **(315-30-0)**
4-Hydroxypyrazolopyrimidine **(315-30-0)**
4-Hydroxypyrazolyl(3,4-d)pyrimidine **(315-30-0)**
1-Hydroxypyrene **(5315-79-7)**
8-Hydroxypyrene-1,3,6-trisulfonic acid sodium salt **(6358-69-6)**
8-Hydroxy-1,3,6-pyrenetrisulfonic acid trisodium salt **(6358-69-6)**
6-Hydroxy-3(2H)-pyridazinone **(123-33-1)**
6-Hydroxy-3-(2H)-pyridazinone diethanolamine **(5716-15-4)**
3-Hydroxy-2-pyridinamine **(16867-03-1)**
2-Hydroxypyridine **(142-08-5)**
3-Hydroxypyridine **(109-00-2)**
4-Hydroxypyridine **(626-64-2)**
β-Hydroxypyridine **(109-00-2)**
γ-Hydroxypyridine **(626-64-2)**
3-Hydroxy-2-pyridinecarboxylic acid **(874-24-8)**
3-Hydroxypyridine-2-thiol **(23003-22-7)**
1-Hydroxy-2(1H)-pyridinethionato sodium **(15922-78-8)**
1-Hydroxy-2(1H)-pyridinethione, sodium salt **(15922-78-8)**
(1-Hydroxy-2-pyridinethione), sodium salt, tech. **(3811-73-2)**
2-Hydroxy-1-(2-pyridylazo)naphthalene **(85-85-8)**
4-Hydroxypyrimidine **(4562-27-0)**
6-Hydroxypyrimidine **(4562-27-0)**
4-Hydroxy-2(1H)-pyrimidinethione **(141-90-2)**
4-Hydroxy-2-pyrrolidinecarboxylic acid **(51-35-4)**
8-Hydroxyqinaldine **(826-81-3)**
4-Hydroxyquinaldic acid **(492-27-3)**
Hydroxyquinaldine **(826-81-3)**
4-Hydroxyquinaldinic acid **(492-27-3)**
Hydroxyquinol **(533-73-3)**
Hydroxyquinoleine (French) **(1321-40-0)**
2-Hydroxyquinoline **(59-31-4)**
8-Hydroxyquinoline **(148-24-3)**
α-Hydroxyquinoline **(59-31-4)**
4-Hydroxyquinoline-2-carboxylic acid **(492-27-3)**
8-Hydroxyquinoline citrate **(134-30-5)**
8-Hydroxyquinoline copper complex **(10380-28-6)**
8-Hydroxyquinoline sulfate **(134-31-6)**
2-(3-Hydroxy-2-quinolyl)-1,3-indanedione **(7576-65-0)**
5-Hydroxyresorcinol **(108-73-6)**
1'-Hydroxysafrole **(5208-87-7)**
1'-Hydroxysafrole-2',3'-oxide **(59901-91-6)**
4-Hydroxysalicyladehyde **(95-01-2)**
3-Hydroxysalicylic acid **(303-38-8)**
4-Hydroxysalicylic acid **(89-86-1)**
5-Hydroxysalicylic acid **(490-79-9)**
6-Hydroxysalicylic acid **(303-07-1)**
p-Hydroxysalicylic acid **(89-86-1)**
12-Hydroxysenecionan-11,16-dione **(130-01-8)**
Hydroxysenkirkine **(26782-43-4)**
12-Hydroxystearic acid **(106-14-9)**
12-Hydroxystearic acid, methyl ester **(141-23-1)**
13-Hydroxystevane **(5749-44-0)**
4-Hydroxystyrene **(2628-17-3)**
p-Hydroxystyrene **(2628-17-3)**
p-Hydroxystyrene polymer **(24979-70-2)**
Hydroxysuccinic acid **(6915-15-7)**
α-Hydroxysuccinic acid **(6915-15-7)**
7-Hydroxy-8-((4-sulfo-1-naphthalenyl)azo)-, disodium salt **(5858-93-5)**
3-Hydroxy-4-((1-sulfo-2-naphthalenyl)azo)-2-naphthalene-carboxylic acid, calcium salt(1:1) **(6417-83-0)**
4-Hydroxy-3-((4-sulfo-1-naphthalenyl)azo)-1-naphthalenesulfonic acid, disodium salt **(3567-69-9)**
3-Hydroxy-4-((4-sulfo-1-naphthalenyl)azo)-2,7-naphthlenedi-sulfonic acid, trisodium salt **(915-67-3)**
2-Hydroxy-4-sulfo-1-naphthylamine **(116-63-2)**
3-Hydroxy-4-((4-sulfo-1-naphthyl)azo)-2,7-naphthalenedi-sulfonic acid, trisodium salt **(915-67-3)**
2-Hydroxy-5-sulfoaniline **(98-37-3)**
2-((2-Hydroxy-5-sulfonamidophenyl)azo)-N-phenyl-3-oxobutan-amide **(21811-92-7)**
2-Hydroxy-5-((5-sulfonaphth-2-yl)azo)benzoic acid, disodium salt **(10114-96-2)**
2-Hydroxy-5-((8-sulfonaphth-2-yl)azo)benzoic acid, disodium salt **(10114-97-3)**
6-Hydroxy-5-((3-sulfophenyl)azo)-2-naphthalenesulfonic acid, disodium salt **(2347-72-0)**
6-Hydroxy-5-((4-sulfophenyl)azo)-2-naphthalenesulfonic acid, disodium salt **(2783-94-0)**
6-Hydroxy-5-((p-sulfophenyl)azo)-2-naphthalenesulfonic acid, disodium salt **(2783-94-0)**
3-Hydroxy-4-((2-sulfo-p-tolyl)azo)-2-naphthalenecarboxylic acid, calcium salt (1:1) **(5281-04-9)**
4-Hydroxy-3-((5-sulfo-2,4-xylyl)azo)-1-naphthalenesulfonic acid, disodium salt **(4548-53-2)**

3-Hydroxy-4-((4-sulpho-1-naphthalenyl)azo)-2,7-naphthalene-disulphonic acid, trisodium salt **(915-67-3)**
4-Hydroxy-3-((4-sulpho-1-naphthalenyl)azo)-1-naphthalene-sulphonic acid, disodium salt **(3567-69-9)**
3-Hydroxy-4-((4-sulpho-1-naphthyl)azo)-2,7-naphthalenedi-sulphonic acid, trisodium salt **(915-67-3)**
6-Hydroxy-5-((4-sulphophenyl)azo)-2-naphthalenesulphonic acid, disodium salt **(2783-94-0)**
6-Hydroxy-5-((p-sulphophenyl)azo)-2-naphthalenesulphonic acid, disodium salt **(2783-94-0)**
4-Hydroxy-3-((5-sulpho-2,4-xylyl)azo)-1-naphthalenesulphonic acid, disodium salt **(4548-53-2)**
5-Hydroxytetracycline **(79-57-2)**
5-Hydroxytetracycline hydrochloride **(2058-46-0)**
2-Hydroxytetradecanoic acid **(2507-55-3)**
3-Hydroxytetradecanoic acid **(1961-72-4)**
β-Hydroxytetradecanoic acid **(1961-72-4)**
4-Hydroxy-3-(1,2,3,4-tetrahydro-1-naftyl)-cumarine (Dutch) **(5836-29-3)**
4-Hydroxy-3-(1,2,3,4-tetrahydro-1-naphthyl)-cumarin **(5836-29-3)**
14-Hydroxy-6-β-thebainol 4-methyl ether **(3176-03-2)**
2-Hydroxytoluene **(95-48-7)**
3-Hydroxytoluene **(108-39-4)**
4-Hydroxytoluene **(106-44-5)**
Hydroxytoluene **(100-51-6)**
α-Hydroxytoluene **(100-51-6)**
m-Hydroxytoluene **(108-39-4)**
o-Hydroxytoluene **(95-48-7)**
p-Hydroxytoluene **(106-44-5)**
α-Hydroxy-α-toluic acid **(90-64-2)**
Hydroxytoluole (German) **(1319-77-3)**
2-(m-Hydroxy-N-p-tolylanilinomethyl)-2-imidazoline **(50-60-2)**
4'-((6-Hydroxy-m-tolyl)azo)acetanilide **(2832-40-8)**
Hydroxytributylstannane 4,4-dimethyloctanoate **(28801-69-6)**
β-Hydroxytricarballylic acid **(77-92-9)**
4-Hydroxy-2,5,6-trichloro-1,3-benzenedicarbonitrile **(28343-61-5)**
1-Hydroxy-2,2,2-trichloro-ethyle phosphonate de dimethyle (French) **(52-68-6)**
1-Hydroxy-2,2,2-trichloroethylphosphonic acid dimethyl ester **(52-68-6)**
4-Hydroxy-2,5,6-trichloroisophthalonitrile **(28343-61-5)**
2-Hydroxytriethylamine **(100-37-8)**
1-Hydroxy-2,4,5-trimethylbenzene **(496-78-6)**
1-Hydroxy-3,4,5-trimethylbenzene **(527-54-8)**
6-Hydroxy-β,2,7-trimethyl-5-benzofuranacrylic acid γ-lactone **(3902-71-4)**
2-Hydroxy-2,6,6-trimethylcyclohexylideneacetic acid γ-lactone **(15356-74-8)**
5,5'-((2-Hydroxytrimethylene)dioxy)bis(4-oxo-4H-1-benzo-pyran-2-carboxylic acid) disodium salt **(15826-37-6)**
5,5'-(2-Hydroxytrimethylenedioxy)bis(4-oxochromene-2-carboxylic acid) **(16110-51-3)**
3-Hydroxy-4((2,4,5-trimethylphenyl)azo)-2,7-naphthalenedi-sulfonic acid, disodium salt **(3564-09-8)**
3-Hydroxy-4-((2,4,5-trimethylphenyl)azo)-2,7-naphthalenedi-sulphonic acid, disodium salt **(3564-09-8)**
Hydroxytrimethylstannane **(56-24-6)**
Hydroxytrimethyltin **(56-24-6)**
3-Hydroxy-2,2,6-trimethyl-6-vinyltetrahydropyran **(14049-11-7)**
2-Hydroxy-1,3,5-trinitrobenzene **(88-89-1)**
3-Hydroxy-2,4,6-trinitrophenol **(82-71-3)**
(T-4)-Hydroxytriphenylborate(1-) sodium **(12113-07-4)**
Hydroxytriphenylsilane **(791-31-1)**
Hydroxytriphenylstannane **(76-87-9)**
Hydroxytriphenyltin **(76-87-9)**
3 β-Hydroxy-2 β-tropanecarboxylic acid **(481-37-8)**
5-Hydroxytryptamine **(50-67-9)**
5-Hydroxytryptophan **(56-69-9)**
Hydroxytryptophan **(56-69-9)**
l-5-Hydroxytryptophan **(4350-09-8)**
5-Hydroxytryptophane **(56-69-9)**
5-Hydroxytrytophan **(56-69-9)**
3-Hydroxytyramine **(51-61-6)**
l-o-Hydroxytyrosine **(59-92-7)**
4-Hydroxyundecanoic acid **(7779-60-4)**
4-Hydroxyundecanoic acid lactone **(104-67-6)**
4-Hydroxyundecanoic acid, γ-lactone **(104-67-6)**
6-Hydroxyuracil **(67-52-7)**
Hydroxyurea **(127-07-1)**
N-Hydroxyurea **(127-07-1)**
Hydroxyurea (D4) **(127-07-1)**
4-Hydroxyvaleric acid lactone **(108-29-2)**
3-Hydroxy-4-(2,4-xylylazo)-3,7-naphthalenedisulfonic acid, disodium salt **(3761-53-3)**
3-Hydroxy-4-(2,4-xylylazo)-3,7-naphthalenedisulphonic acid, disodium salt **(3761-53-3)**
17-Hydroxyyohimban-16-carboxylic acid methyl ester **(146-48-5)**
o-Hydroxyzimtsaure-lacton (German) **(91-64-5)**
Hydroxyzine **(68-88-2)**
Hydrozobenzene **(530-50-7)**
Hydrure de lithium (French) **(7580-67-8)**
Hydura **(127-07-1)**
Hydurea **(127-07-1)**
Hyflavin **(83-88-5)**
Hygromull **(9011-05-6)**
Hygroton **(77-36-1)**
Hygroton-reserpine **(50-55-5)**
Hylemox **(563-12-2)**
Hylene M50 **(101-68-8)**
Hylene T **(584-84-9)**
Hylene-T **(26471-62-5)**
Hylene TCPA **(584-84-9)**
Hylene TCPA **(91-08-7)**
Hylene TIC **(91-08-7)**
Hylene TLC **(584-84-9)**
Hylene TM **(584-84-9)**
Hylene TM **(91-08-7)**
Hylene TM-65 **(584-84-9)**
Hylene TM-65 **(91-08-7)**
Hylene TRF **(584-84-9)**
Hylene TRF **(91-08-7)**
Hylenta **(113-98-4)**
Hylite LF **(1854-26-8)**
Hylutin **(630-56-8)**
Hymecromon **(90-33-5)**
Hymecromone **(90-33-5)**
Hymecromone O,O-diethyl phosphorothioate **(299-45-6)**
Hymenovin **(57377-32-9)**
Hymenoxon **(57377-32-9)**
Hymenoxone **(57377-32-9)**
Hymexazol **(10004-44-1)**
Hyminal **(72-44-6)**
Hymol PM **(9004-74-4)**
Hymorphan **(466-99-9)**
Hyocine F hydrobromide **(114-49-8)**
Hyonic PE 260 **(9036-19-5)**
Hyonic PE-250 **(9002-93-1)**
Hyosan **(97-23-4)**
Hyosceine **(51-34-3)**
(-)-Hyoscine **(51-34-3)**
Hyoscine **(51-34-3)**
Hyoscine bromide **(114-49-8)**
(-)-Hyoscine hydrobromide **(114-49-8)**
Hyoscine hydrobromide **(114-49-8)**
l-Hyoscine hydrobromide **(114-49-8)**
Hyoscine methyl bromide **(155-41-9)**
(+)-Hyoscyamine **(13269-35-7)**
(-)-Hyoscyamine **(101-31-5)**

1-Hyoscyamine	(101-31-5)
Hyoscyamine	(101-31-5)
Hyoscyamine, (+)	(13269-35-7)
Hyoscyamine, (-)	(101-31-5)
dl-Hyoscyamine	(51-55-8)
Hyoscyamine methylbromide	(2870-71-5)
Hyoscyamus Oil	(8001-41-0)
Hyoscyamus Oil-Expressed	(8001-41-0)
Hyoscyine hydrobromide	(114-49-8)
Hyosol	(51-34-3)
Hyozid	(54-85-3)
Hypalox II	(1344-28-1)
Hypcol	(72-44-6)
Hyperabsolute Benzoin, Siam	(9000-05-9)
Hyperabsolute Geranium	(8000-46-2)
Hyperazin	(304-20-1)
Hyperazine	(304-20-1)
Hyperbutal	(77-28-1)
Hypercal B	(50-55-5)
Hypercillin	(6130-64-9)
Hypernephrin	(51-43-4)
Hyperol	(124-43-6)
Hyperpax	(41372-08-1)
Hyperpax	(555-30-6)
Hypertane Forte	(50-55-5)
Hypertenain	(15825-70-4)
Hypertensan	(50-55-5)
Hyphylline	(479-18-5)
Hypnaletten	(50-06-6)
Hypnogen	(50-06-6)
Hypnogene	(57-44-3)
Hypnolone	(50-06-6)
Hypnoltol	(50-06-6)
Hypnon	(2011-67-8)
Hypnon	(98-86-2)
Hypnone	(98-86-2)
Hypnorex	(554-13-2)
Hypo	(10102-17-7)
Hypo	(7772-98-7)
Hypochlorite de lithium (French)	(13840-33-0)
Hypochlorous acid, calcium salt	(7778-54-3)
Hypochlorous acid, calcium salt, (Dry mixture)	(7778-54-3)
Hypochlorous acid, calcium salt, Dry mixture with 10% to 39% available chlorine	(7778-54-3)
Hypochlorous acid, 2,3-dichloro-6-methoxyphenyl ester (9CI)	(108544-90-7)
Hypochlorous acid, lithium salt (8CI,9CI)	(13840-33-0)
Hypochlorous acid, lithium salt, Mixture with sodium chloride [UN 1471]	(13840-33-0)
Hypochlorous acid, sodium salt	(7681-52-9)
Hypochlorous acid, sodium salt, pentahydrate	(10022-70-5)
Hypochylin	(138-15-8)
Hypodermacid	(52-68-6)
Hypoglycin	(156-56-9)
Hypoglycin A	(156-56-9)
Hypoglycine	(156-56-9)
Hypoglycine A	(156-56-9)
Hypoiodous acid, 2,2'-dimethyl-5,5'-bis(1-methylethyl)(1,1'-biphenyl)-4,4'-diyl ester (9CI)	(552-22-7)
Hyponitrite	(14448-38-5)
Hyponitrous acid	(14448-38-5)
Hyponitrous acid anhydride	(10024-97-2)
Hyponitrous acid, disodium salt	(14448-38-5)
α-Hypophamine	(50-56-6)
β-Hypophamine	(11000-17-2)
Hypophenon	(70-70-2)
Hypophosphoric acid, disodium salt (9CI)	(7782-95-8)
Hypophosphorous acid	(6303-21-5)
Hypophthalin	(304-20-1)
Hypophthalin	(86-54-4)
Hypophyseal Growth Hormone	(9002-72-6)
Hyporenin	(51-43-4)
Hypos	(304-20-1)
Hypothiazid	(58-93-5)
Hypothiazide	(58-93-5)
Hypotrol	(76-73-3)
Hypoxanthine	(68-94-0)
Hypoxanthine enol	(68-94-0)
Hypoxanthine nucleoside	(58-63-9)
Hypoxanthine ribonucleoside	(58-63-9)
Hypoxanthine D-riboside	(58-63-9)
Hypoxanthine riboside	(58-63-9)
Hypoxanthine, thio-	(50-44-2)
Hypoxanthosine	(58-63-9)
Hypro	(51-35-4)
Hyproval-PA	(630-56-8)
Hypno-tablinetten	(50-06-6)
Hyptor	(72-44-6)
Hyptor base	(72-44-6)
Hypyrin	(23413-80-1)
Hyre	(83-88-5)
Hyroxon	(630-56-8)
Hysco	(114-49-8)
Hyscylene P	(101-48-4)
Hysteps	(50-06-6)
Hystl	(9003-17-2)
Hystl B 300	(9003-17-2)
Hystl B 1000	(9003-17-2)
Hystl B 2000	(9003-17-2)
Hystl B 3000	(9003-17-2)
Hystrene 80	(57-11-4)
Hystrene 4516	(57-11-4)
Hystrene 5016	(57-11-4)
Hystrene 7018	(57-11-4)
Hystrene 8016	(57-10-3)
Hystrene 9014	(544-63-8)
Hystrene 9016	(57-10-3)
Hystrene 9512	(143-07-7)
Hystrene 9718	(57-11-4)
Hytakerol	(67-96-9)
Hytech	(557-05-1)
Hyton	(2152-34-3)
Hytone	(50-23-7)
Hytone Lotion	(50-23-7)
Hytox	(2631-40-5)
Hytrol O	(108-94-1)
Hyvar	(314-42-1)
Hyvar X	(314-40-9)
Hyvar X Bromacil	(314-40-9)
Hyvar X-L	(314-40-9)
Hyvar X-WS	(314-40-9)
Hyvar X Weed Killer	(314-40-9)
Hyvarex	(314-40-9)
Hyvermectin	(70288-86-7)
Hyvis 200	(9003-27-4)
Hyvis 2000	(9003-27-4)
Hyvis 30	(9003-27-4)
Hyvisc	(9004-61-9)
Hyxobamine	(13422-51-0)
Hyzyd	(54-85-3)
I 337A	(56-75-7)
I-1431	(68-41-7)
IA	(118-48-9)
IA	(64-69-7)
3-IAA	(87-51-4)

Name	CAS No.
IAA	(87-51-4)
IAB	(1397-89-3)
IA-But	(50-33-9)
I.A.I.	(54-85-3)
I Acid	(87-02-5)
I Acid Urea	(134-47-4)
IBA	(133-32-4)
IBDU	(6104-30-9)
IBN	(542-56-3)
IBP	(13286-32-3)
IBP	(26087-47-8)
IBZ	(366-70-1)
I-Butyl bromide	(78-77-3)
ICG	(3599-32-4)
ICI	(80-08-0)
ICI 543	(9003-07-0)
ICI 28257	(637-07-0)
ICI 29661	(5221-49-8)
ICI 35868	(2078-54-8)
ICI 38174	(51-02-5)
ICI 45520	(525-66-6)
ICI 50123	(5534-95-2)
ICI 50172	(6673-35-4)
ICI 66082	(29122-68-7)
ICI 146814	(68085-85-8)
ICI-32865	(1954-28-5)
ICI-58834	(46817-91-8)
ICI-CF 2	(71-55-6)
ICIG 1109	(13010-47-4)
ICIG 1110	(13909-09-6)
I.C.I. Hydrochloride	(51-02-5)
ICI-PP 557	(52645-53-1)
ICI-PP 563	(68085-85-8)
ICN-1229	(36791-04-5)
ICR 10	(4213-45-0)
ICR 170	(146-59-8)
ICR 502	(16238-56-5)
ICRF-159	(21416-87-5)
ID 480	(1172-18-5)
ID 540	(3900-31-0)
IDA	(142-73-4)
ID 480 Dihydrochloride	(1172-18-5)
IDPN	(111-94-4)
IEM	(30674-80-7)
IEM-1-15	(1967-16-4)
IF	(6012-97-1)
IFC	(122-42-9)
IF (Fumigant)	(6012-97-1)
IFK	(122-42-9)
IG 11	(7782-42-5)
IGE (OSHA)	(4016-14-2)
IH 773B	(69806-50-4)
IHSA I-125	(9048-46-8)
IHSA I-131	(9048-46-8)
II-C-2	(359-83-1)
IK 2	(661-19-8)
IKhS 1	(9011-06-7)
IL 6001	(739-71-9)
IM	(50-49-7)
IMC 3950	(28249-77-6)
IMD 760	(1249-84-9)
IMI 115	(7440-32-6)
IMOL S 140	(1330-78-5)
IMPA	(1832-54-8)
o-IMPC	(114-26-1)
IMPF	(107-44-8)
IMS	(926-06-7)
IN 1179	(16752-77-5)
IN-73	(54-85-3)
INAH	(54-85-3)
INF 3355	(61-68-7)
INH	(54-85-3)
INPC	(122-42-9)
IOP	(51-42-3)
IP	(193-39-5)
IP 50	(34123-59-6)
IP-82	(15687-27-1)
IPA	(121-91-5)
IPC	(122-42-9)
IPD	(3458-22-8)
IPDI	(4098-71-9)
IPN	(110-46-3)
IPN	(626-17-5)
IPO 63	(13104-21-7)
IPO 8	(961-11-5)
IPPC	(122-42-9)
IPSP	(5827-05-4)
IPT	(50512-35-1)
IPT (Pesticide)	(50512-35-1)
IPX	(2634-33-5)
IQ	(76180-96-6)
2341 I.S.	(111-94-4)
I-Sedrin	(299-42-3)
IT 40	(9003-53-6)
IT 931	(3347-22-6)
IT 3456	(2536-31-4)
IT Extra	(14807-96-6)
IVE	(109-53-5)
IXPER 25M	(14452-57-4)
IZ 914	(113-53-1)
Iambolen	(25038-59-9)
Iaractan	(113-59-7)
Ibenzmethyzine	(671-16-9)
Ibenzmethyzine hydrochloride	(366-70-1)
Ibenzmethyzin hydrochloride	(366-70-1)
Ibiamox	(26787-78-0)
Ibifur	(139-91-3)
Ibiodral	(58-73-1)
Ibiofural	(59-87-0)
Ibiosuc	(139-05-9)
Ibioton	(113-92-8)
Ibiozedrine	(60-13-9)
(-)-Ibogaine	(83-74-9)
Ibogaine (8CI)	(83-74-9)
Ibogamine, 12-methoxy- (9CI)	(83-74-9)
Ibufen	(15687-27-1)
Ibuprocin	(15687-27-1)
Ibuprofen	(15687-27-1)
Iceland Spar	(13397-26-7)
Ichtho-bellol	(55-48-1)
Icipen	(132-98-9)
Icoral B	(54-49-9)
Ictalis simple	(57-41-0)
Idalene	(4268-36-4)
Idantoil	(57-41-0)
Idantoil	(630-93-3)
Idantoin	(57-41-0)
Idantoinal	(630-93-3)
Idarac	(23779-99-9)
Ido-C	(50-81-7)
Ido-Tebin	(54-85-3)
Idocyl novum	(54-21-7)
Idomethine	(53-86-1)
Idoserp	(50-55-5)

Idragin **(50-78-2)**
Idralazina (Italian) **(86-54-4)**
Idrazide dell'acido isonicotinico (Italian) **(54-85-3)**
Idrazil **(54-85-3)**
Idrazina idrata (Italian) **(7803-57-8)**
Idrazina solfato (Italian) **(10034-93-2)**
Idrianol **(61-76-7)**
Idrobil **(519-95-9)**
Idrobutazina **(129-20-4)**
Idrochinone (Italian) **(123-31-9)**
Idroepar **(519-95-9)**
Idroestril **(56-53-1)**
Idrogeno solforato (Italian) **(7783-06-4)**
Idrogestene **(630-56-8)**
Idrogrisevit **(13422-51-0)**
Idrokin **(73-49-4)**
Idromorfinolo **(2183-56-4)**
Idroperossido di cumene (Italian) **(80-15-9)**
Idroperossido di cumolo (Italian) **(80-15-9)**
Idroperossido di tetralina (Italian) **(771-29-9)**
Idrossido di stagno trifenile (Italian) **(76-87-9)**
1',1-(2-Idrossietil)-4,3-(2-cloro-10-fenotiazil)propilpiperazina (Italian) **(58-39-9)**
4-Idrossi-4-metil-pentan-2-one (Italian) **(123-42-2)**
4-Idrossi-3-(3-oxo-1-fenil-butil)-cumarine (Italian) **(81-81-2)**
Idrossipetidina **(468-56-4)**
4-Idrossi-3-(1,2,3,4-tetraidro-1-naftil)-cumarina (Italian) **(5836-29-3)**
Idrossizina **(68-88-2)**
Idrossocobalamina **(13422-51-0)**
Idrotiazide **(58-93-5)**
Idryl **(206-44-0)**
Idsoserp **(50-55-5)**
Iergigan **(60-87-7)**
Ieroin **(561-27-3)**
Ifenec **(24169-02-6)**
Ifibrium **(58-25-3)**
Ifosfamid **(3778-73-2)**
Ifosfamide **(3778-73-2)**
Igepal CA **(9036-19-5)**
Igepal CA 520 **(9036-19-5)**
Igepal CA-630 **(9002-93-1)**
Igepal CO-630 **(9016-45-9)**
Igepon T **(137-20-2)**
Igepon T 33 **(137-20-2)**
Igepon T 51 **(137-20-2)**
Igepon T 77 **(137-20-2)**
Igepon T-43 **(137-20-2)**
Igepon T-71 **(137-20-2)**
Igepon T-73 **(137-20-2)**
Igepon TE **(137-20-2)**
Igran **(886-50-0)**
Igran 50 **(886-50-0)**
Igran Special **(37341-11-0)**
Igroton **(77-36-1)**
Ikaclomin **(50-41-9)**
Ikada Rhodamine B **(81-88-9)**
Iketon Yellow Extra **(60-11-7)**
Ikterosan **(132-60-5)**
Ikurin **(7773-06-0)**
Ilcocillin P **(6130-64-9)**
Iletin **(9004-10-8)**
Ilitia **(59-02-9)**
Ilixathin **(153-18-4)**
Illoxan **(51338-27-3)**
Illoxol **(60-57-1)**
Ilopan **(81-13-0)**
Ilotycin **(114-07-8)**
Iloxan **(51338-27-3)**
Ilvin **(86-22-6)**
Imagon **(54-05-7)**
Imagotan **(14759-06-9)**
Imazalil **(35554-44-0)**
Imazapyr **(81334-34-1)**
Imazapyr-isopropylammonium **(81510-83-0)**
Imazaquin **(81335-37-7)**
Imazethapyr **(81335-77-5)**
Imbrilon **(53-86-1)**
Imecromone **(90-33-5)**
Imesonal **(76-73-3)**
Imferon **(9004-66-4)**
Imida-Lab **(50-35-1)**
Imidan **(732-11-6)**
Imidan (Peyta) **(50-35-1)**
4H-Imidazo(1,5-a)(1,4)benzodiazepine, 8-chloro-6-(2-fluorophenyl)-1-methyl- **(59467-70-8)**
5H-Imidazo(2,1-a)isoindol-5-ol, 5-(4-chlorophenyl)-2,3-dihydro **(22232-71-9)**
Imidazo(2,1-b)thiazole, 2,3,5,6-tetrahydro-6-phenyl-, (S) **(14769-73-4)**
3H-Imidazo(2,1-c)-1,2,4-dithiazole-3-thione, 5,6-dihydro **(33813-20-6)**
1H-Imidazo(4,5-c)pyridine **(272-97-9)**
Imidazo(4,5-d)imidazole-2,5(1H,3H)-dione, 1,3,4,6-tetrachlorotetrahydro- (9CI) **(776-19-2)**
Imidazo(4,5-d)imidazole-2,5(1H,3H)-dione, tetrahydro- (9CI) **(496-46-8)**
7H-Imidazo(4,5-d)pyrimidine **(120-73-0)**
3H-Imidazo(4,5-f)quinoline, 2-amino-3,4-dimethyl **(77094-11-2)**
3H-Imidazo(4,5-f)quinoline, 2-amino-3-methyl **(76180-96-6)**
3H-Imidazo(4,5-f)quinoxaline, 2-amino-3,8-dimethyl **(77500-04-0)**
Imidazo(5,1-f)(1,2,4)triazine-2,7-diamine, 4,5-dimethyl- (9CI) **(50473-86-4)**
Imidazol **(288-32-4)**
1H-Imidazol-5-amine, 1,2-dimethyl-4-nitro- (9CI) **(21677-57-6)**
1H-Imidazol-5-amine, 1-methyl-4-nitro- (9CI) **(4531-54-8)**
Imidazole **(288-32-4)**
Imidazole Mustard **(5034-77-5)**
Imidazole, 1-allyl- (8CI) **(31410-01-2)**
Imidazole, 4-(2-amino-1,1-dimethylethyl)- (8CI) **(21150-01-6)**
Imidazole, 5-amino-1,2-dimethyl-4-nitro- (8CI) **(21677-57-6)**
Imidazole, 4-(2-aminoethyl)- **(51-45-6)**
Imidazole, 5-amino-1-methyl-4-nitro- (7CI,8CI) **(4531-54-8)**
1H-Imidazole, 4-bromo-1,2-dimethyl-5-nitro- (9CI) **(21431-58-3)**
1H-Imidazole, 5-bromo-1,2-dimethyl-4-nitro- (9CI) **(21117-52-2)**
Imidazole, 4-bromo-1,2-dimethyl-5-nitro- (8CI) **(21431-58-3)**
Imidazole, 5-bromo-1,2-dimethyl-4-nitro- (8CI) **(21117-52-2)**
1H-Imidazole, 2-bromo-1-methyl-4-nitro- (9CI) **(16681-63-3)**
1H-Imidazole, 4-bromo-1-methyl-5-nitro- (9CI) **(59177-47-8)**
1H-Imidazole, 4-bromo-2-methyl-5-nitro- (9CI) **(18874-52-7)**
Imidazole, 2-bromo-1-methyl-4-nitro- (8CI) **(16681-63-3)**
Imidazole, 4-bromo-2-methyl-5-nitro- (8CI) **(18874-52-7)**
1H-Imidazole, 2-bromo-4-nitro- (9CI) **(65902-59-2)**
1H-Imidazole, 4-bromo-5-nitro- (9CI) **(6963-65-1)**
Imidazole, 4-bromo-5-nitro- (8CI) **(6963-65-1)**
Imidazole, 1-butyl **(4316-42-1)**
1H-Imidazole, 1-butyl- (9CI) **(4316-42-1)**
Imidazole-5-butyric acid, α-ethyl-β-(hydroxymethyl)-1-methyl-, γ-lactone **(92-13-7)**
2-Imidazolecarbaldehyde **(10111-08-7)**
1H-Imidazole-2-carboxaldehyde (9CI) **(10111-08-7)**
Imidazole-2-carboxaldehyde (7CI,8CI) **(10111-08-7)**
Imidazole-4-carboxamide, 5-(3,3-bis(2-chloroethyl)-1-triazeno) **(5034-77-5)**
1H-Imidazole-4-carboxamide, 5-(3,3-bis(2-chloroethyl)-1-triazenyl)- (9CI) **(5034-77-5)**
Imidazole-4-carboxamide, 5-(3,3-dimethyl-1-triazeno) **(4342-03-4)**
1H-Imidazole-4-carboxamide, 5-(3,3-dimethyl-1-triazenyl)- (9CI) **(4342-03-4)**
1H-Imidazole-1-carboxylic acid, 2,3-dihydro-3-methyl-2-thioxo, ethyl ester **(22232-54-8)**
1H-Imidazole, 4-chloro-1,2-dimethyl-5-nitro- (9CI) **(91027-94-0)**

1H-Imidazole, 5-chloro-1,2-dimethyl-4-nitro- (9CI) **(91027-93-9)**
Imidazole, 1-(o-chloro-α,α-diphenylbenzyl) **(23593-75-1)**
1H-Imidazole, 2-chloro-1-methyl-4-nitro **(63634-21-9)**
Imidazole, 4-chloro-1-methyl-5-nitro **(4897-31-8)**
Imidazole, 5-chloro-1-methyl-4-nitro **(4897-25-0)**
1H-Imidazole, 2-chloro-1-methyl-5-nitro- (9CI) **(86072-07-3)**
1H-Imidazole, 4-chloro-2-methyl-5-nitro- (9CI) **(63662-67-9)**
Imidazole, 4(or 5)-chloro-5(or 4)-nitro- **(57531-38-1)**
1H-Imidazole, 2-chloro-4-nitro- (9CI) **(57531-37-0)**
1H-Imidazole, 4-chloro-5-nitro- (9CI) **(57531-38-1)**
1H-Imidazole, 2-(2-chlorophenyl)-4,5-diphenyl- (9CI) **(1707-67-1)**
1H-Imidazole, 1-((2-chlorophenyl)diphenylmethyl)- **(23593-75-1)**
1H-Imidazole, 1-(2-((4-chlorophenyl)methoxy)-2-(2,4-dichlorophenyl)ethyl)-, mononitrate (9CI) **(24169-02-6)**
Imidazole, 1-(2,4-dichloro-β-((p-chlorobenzyl)oxy)-phenethyl)-, mononitrate (8CI) **(24169-02-6)**
1H-Imidazole, 1-(2-(2,4-dichlorophenyl)-2-(2-propenyloxy)ethyl) **(35554-44-0)**
1H-Imidazole, 4,5-dihydro-2-methyl- **(534-26-9)**
1H-Imidazole, 4,5-dihydro-2-(1-naphthalenylmethyl)-, monohydrochloride **(550-99-2)**
1H-Imidazole, 4,5-dihydro-, 2-nortall-oil alkyl derivs. **(61791-36-4)**
1H-Imidazole, 4,5-dihydro-2-phenyl- **(936-49-2)**
1H-Imidazole, 1,2-dimethyl- (9CI) **(1739-84-0)**
Imidazole, 1,2-dimethyl- (8CI) **(1739-84-0)**
1H-Imidazole, 1,2-dimethyl-4,5-dinitro- (9CI) **(19183-17-6)**
Imidazole, 1,2-dimethyl-4,5-dinitro- (8CI) **(19183-17-6)**
Imidazole, 1,2-dimethyl-4-nitro **(13230-04-1)**
Imidazole, 1,2-dimethyl-5-nitro **(551-92-8)**
Imidazole, 1,5-dimethyl-4-nitro **(7464-68-8)**
1H-Imidazole, 1,2-dimethyl-5-nitro- (9CI) **(551-92-8)**
1H-Imidazole, 1,2-dimethyl-5-nitro-, monohydrochloride (9CI) **(25332-20-1)**
Imidazole, 2,4-dinitro **(5213-49-0)**
Imidazole, 2,5-dinitro- **(5213-49-0)**
1H-Imidazole, 4,5-dinitro- (9CI) **(19183-14-3)**
Imidazole, 4,5-dinitro- (8CI) **(19183-14-3)**
1H-Imidazole-4-ethanamine **(51-45-6)**
1H-Imidazole-1-ethanamine, 4,5-dihydro-, 2-nortall-oil alkyl derivs. **(68442-97-7)**
1H-Imidazole-4-ethanamine, β,β-dimethyl- (9CI) **(21150-01-6)**
1H-Imidazole-1-ethanamine, 2-(8-heptadecenyl)-4,5-dihydro- (9CI) **(3528-63-0)**
1H-Imidazole-1-ethanamine, 2-(heptadecenyl)-2,3-dihydro- (9CI) **(27476-93-3)**
1H-Imidazole-1-ethanol, α-(chloromethyl)-2-methyl-5-nitro- **(16773-42-5)**
Imidazole-1-ethanol, α-(chloromethyl)-2-methyl-5-nitro **(16773-42-5)**
1H-Imidazole-1-ethanol, 4,5-dihydro-, 2-nortall oil **(61791-39-7)**
1H-Imidazole-1-ethanol, 4,5-dihydro-, 2-nortall-oil alkyl derivs. **(61791-39-7)**
1H-Imidazole-1-ethanol, 4,5-dihydro-2-undecyl- (9CI) **(136-99-2)**
1H-Imidazole-1-ethanol, 4,5-dihydro-2-undecyl, carboxymethylated, sodium salt **(68647-44-9)**
1H-Imidazole-1-ethanol, 2-dodecyl-4,5-dihydro- (9CI) **(16058-17-6)**
1H-Imidazole-1-ethanol, 2-dodecyl-4,5-dihydro-, monohydrochloride (9CI) **(71242-00-7)**
1H-Imidazole-1-ethanol, 2-(heptadecenyl)-4,5-dihydro- **(27136-73-8)**
1H-Imidazole-1-ethanol, 2-(8-heptadecenyl)-4,5-dihydro-, monoacetate (Salt) (9CI) **(3388-72-5)**
1H-Imidazole-1-ethanol, 2-heptadecyl-4,5-dihydro-, monoacetate (Salt) (9CI) **(28832-11-3)**
1H-Imidazole-1-ethanol, 2-heptyl-4,5-dihydro- (9CI) **(36060-61-4)**
Imidazole-1-ethanol, α-(methoxymethyl)-2-nitro **(13551-87-6)**
1H-Imidazole-1-ethanol, α-(methoxymethyl)-2-nitro- (9CI) **(13551-87-6)**
1H-Imidazole-1-ethanol, α-methyl- (9CI) **(37788-55-9)**
Imidazole-1-ethanol, 2-methyl-5-nitro **(443-48-1)**
1H-Imidazole, 1-ethenyl- (9CI) **(1072-63-5)**
4-Imidazoleethylamine **(51-45-6)**
5-Imidazoleethylamine **(51-45-6)**
Imidazole-4-ethylamine **(51-45-6)**
1H-Imidazole, 2-ethyl-4,5-dihydro-4-methyl- (9CI) **(931-35-1)**
1H-Imidazole, 2-ethyl-4-methyl- (9CI) **(931-36-2)**
Imidazole, 2-ethyl-4-methyl- (8CI) **(931-36-2)**
1H-Imidazole, 5-methoxy-1,2-dimethyl-4-nitro- (9CI) **(35687-44-6)**
1H-Imidazole, 4-methoxy-2-methyl-5-nitro- (9CI) **(35687-42-4)**
1H-Imidazole, 5-methoxy-1-methyl-3-nitro- (9CI) **(35687-41-3)**
1H-Imidazole, 4-methoxy-5-nitro- (9CI) **(68019-78-3)**
Imidazole, 1-methyl **(616-47-7)**
Imidazole, 2-methyl **(693-98-1)**
Imidazole, 1-methyl-2,4-dinitro- **(5213-50-3)**
1H-Imidazole, 1-methyl-2,4-dinitro- (9CI) **(5213-50-3)**
1H-Imidazole, 1-methyl-4,5-dinitro- (9CI) **(19183-15-4)**
1H-Imidazole, 2-methyl-4,5-dinitro- (9CI) **(19183-16-5)**
Imidazole, 1-methyl-4,5-dinitro- (8CI) **(19183-15-4)**
Imidazole, 2-methyl-4,5-dinitro- (8CI) **(19183-16-5)**
Imidazole, 1-methyl-2-mercapto- **(60-56-0)**
Imidazole, 1-methyl-2-nitro **(1671-82-5)**
Imidazole, 1-methyl-4-nitro **(3034-41-1)**
Imidazole, 1-methyl-5-nitro **(3034-42-2)**
Imidazole, 2-methyl-4-nitro- **(696-23-1)**
Imidazole, 2-methyl-5-nitro **(696-23-1)**
Imidazole, 4-methyl-5-nitro **(14003-66-8)**
Imidazole, 5-methyl-4-nitro- **(14003-66-8)**
Imidazole, 2-nitro **(527-73-1)**
Imidazole, 4-nitro **(3034-38-6)**
1H-Imidazole, nitro- (9CI) **(36877-68-6)**
1H-Imidazole-1-(2-propanol) **(37788-55-9)**
1H-Imidazole, 1-(2-propenyl)- (9CI) **(31410-01-2)**
Imidazole-2-thiol, 1-methyl **(60-56-0)**
Imidazole-2-thio, 1-methyl- **(60-56-0)**
Imidazole-2(3H)-thione, 4,5-dihydro- **(96-45-7)**
Imidazole, 2,4,5-tribromo **(2034-22-2)**
1H-Imidazole, 2,4,5-tribromo-1-(4-chlorobenzoyl)- (9CI) **(15287-32-8)**
Imidazole, 2,4,5-tribromo-1-(p-chlorobenzoyl)- (8CI) **(15287-32-8)**
1H-Imidazole, 2,4,5-tribromo-1-propyl- (9CI) **(31250-78-9)**
Imidazole, 2,4,5-tribromo-1-propyl- (8CI) **(31250-78-9)**
1H-Imidazole, 2,4,5-trichloro-1-methyl- (9CI) **(873-25-6)**
Imidazole, 2,4,5-trichloro-1-methyl- (8CI) **(873-25-6)**
1-Imidazolidinecarboxamide, 3-(3,5-dichlorophenyl)-N-(1-methylethyl)-2,4-dioxo **(36734-19-7)**
1-Imidazolidinecarboxamide, N-isobutyl-2-oxo- (8CI) **(30979-48-7)**
1-Imidazolidinecarboxamide, N-(2-methylpropyl)-2-oxo- (9CI) **(30979-48-7)**
2,4-Imidazolidinedione (9CI) **(461-72-3)**
2,4-Imidazolidinedione, 1,3-bis(2-hydroxyethyl)-5,5-dimethyl- (9CI) **(26850-24-8)**
2,4-Imidazolidinedione, 1,3-bis(hydroxymethyl)-5,5-dimethyl- (9CI) **(6440-58-0)**
2,4-Imidazolidinedione, 3-bromo-1-chloro-5,5-dimethyl- **(126-06-7)**
2,4-Imidazolidinedione, 1,3-dibromo-5,5-dimethyl- (9CI) **(77-48-5)**
2,4-Imidazolidinedione, 1,3-dichloro-5,5-dimethyl- (9CI) **(118-52-5)**
2,4-Imidazolidinedione, 5,5-diethyl- (9CI) **(5455-34-5)**
2,4-Imidazolidinedione, 5,5-dimethyl- **(77-71-4)**
2,4-Imidazolidinedione, 5,5-dimethyl-3-(2-(oxiranylmethoxy)-propyl)-1-(oxiranylmethyl)- (9CI) **(32568-89-1)**
2,4-Imidazolidinedione, 5,5-diphenyl-, monosodium salt **(630-93-3)**
2,4-Imidazolidinedione, 5-ethyl-5-methyl-1,3-bis(oxiranylmethyl)- **(15336-82-0)**
2,4-Imidazolidinedione, 5-ethyl-5-methyl-1,3-bis(oxiranylmethyl)-, Homopolymer (9CI) **(68012-07-7)**
2,4-Imidazolidinedione, 5-ethyl-5-(2-methylbutyl)-1,3-bis-(oxiranylmethyl)- (9CI) **(68444-05-3)**
2,4-Imidazolidinedione, 5-ethyl-3-methyl-5-phenyl- **(50-12-4)**
2,4-Imidazolidinedione, 3-(2-hydroxyethyl)-5,5-dimethyl- (9CI) **(29071-93-0)**
2,4-Imidazolidinedione, 1-(hydroxymethyl)-5,5-dimethyl- (9CI) **(116-25-6)**
2,4-Imidazolidinedione, 5-(2-(methylthio)ethyl)- (9CI) **(13253-44-6)**
2,4-Imidazolidinedione, 1-nitro- **(2825-15-2)**
2,4-Imidazolidinedione, 1-(((5-nitro-2-furanyl)methylene)amino)- **(67-20-9)**
2,4-Imidazolidinedione, 3,3'-(2-(oxiranylmethoxy)-1,3-propanediyl)bis(5,5-dimethyl-1-(oxiranylmethyl)- (9CI) **(38304-52-8)**
2-Imidazolidinethione **(96-45-7)**

2-Imidazolidinethione, 4-methyl **(2122-19-2)**
Imidazolidin-2-on-1-carbonsaeure-isobutylamid **(67292-92-6)**
Imidazolidin-2-on-1-carbonsaeure-isobutylamid (German) **(30979-48-7)**
2-Imidazolidinone **(120-93-4)**
2-Imidazolidinone, 1-(2-aminoethyl)- (9CI) **(6281-42-1)**
2-Imidazolidinone, 1,3-bis(hydroxymethyl)- (9CI) **(136-84-5)**
2-Imidazolidinone, 1,3-bis(hydroxymethyl)-4,5-dihydroxy **(1854-26-8)**
2-Imidazolidinone, 3-(5-tert-butyl-1,3,4-thiadiazol-2-yl)-4-hydroxy-1-methyl **(55511-98-3)**
2-Imidazolidinone, 4,5-dihydroxy- (9CI) **(3720-97-6)**
2-Imidazolidinone, 4,5-dihydroxy-1,3-bis(hydroxymethyl)-, methylated **(68411-81-4)**
2-Imidazolidinone, 4,5-dihydroxy-1,3-bis(methoxymethyl)- (9CI) **(3001-61-4)**
2-Imidazolidinone, 4,5-dimethoxy-1,3-bis(methoxymethyl)- (9CI) **(4356-60-9)**
2-Imidazolidinone, 4,5-dimethoxy-1,3-dimethyl- **(13464-10-3)**
2-Imidazolidinone, 3-(5-(1,1-dimethylethyl)-1,3,4-thiadiazol-2-yl)-4-hydroxy-1-methyl- (9CI) **(55511-98-3)**
2-Imidazolidinone, 1-(2-hydroxyethyl)- (9CI) **(3699-54-5)**
2-Imidazolidinone, 1-(((5-nitro-2-furanyl)methylene)amino)- **(555-84-0)**
2-Imidazolidinone, 1-((5-nitrofurfurylidene)amino) **(555-84-0)**
2-Imidazolidinone, 1-(5-nitro-2-thiazolyl) **(61-57-4)**
2-Imidazolidone **(120-93-4)**
4-Imidazoline-1-carboxylic acid, 3-methyl-2-thioxo-, ethyl ester **(22232-54-8)**
2-Imidazoline, 2-(2,6-dichloroanilino) **(4205-90-7)**
2-Imidazoline-1-ethanol, 2-(8-heptadecenyl) **(95-38-5)**
2-Imidazoline-1-ethanol, 2-heptadecyl **(95-19-2)**
2-Imidazoline-1-ethanol, 2-undecyl- **(136-99-2)**
2-Imidazoline, 2-ethyl-4-methyl- (7CI,8CI) **(931-35-1)**
2-Imidazoline, 2-heptadecyl-, acetate **(556-22-9)**
2-Imidazoline, 2-heptadecyl-1-hydroxyethyl- **(95-19-2)**
2-Imidazoline, 2-heptadecyl-, monoacetate **(556-22-9)**
2-Imidazoline, 1-(2-hydroxyethyl)-2-(tall oil alkyl)- **(61791-39-7)**
2-Imidazoline, 2-((N-(m-hydroxyphenyl)-p-toluidino)methyl)- **(50-60-2)**
Imidazoline, 2-mercapto- **(96-45-7)**
2-Imidazoline, 2-(1-naphthylmethyl) **(835-31-4)**
2-Imidazoline, 2-(1-naphthylmethyl)-, monohydrochloride **(550-99-2)**
2-Imidazoline, 2-phenyl **(936-49-2)**
4-Imidazolin-2-one **(5918-93-4)**
Imidazolin-2-one, 1-isobutylcarbamoyl **(30979-48-7)**
Imidazolin-2-one, 1-isobutylcarbamoyl- **(67292-92-6)**
1H-Imidazolium, 1-(carboxymethyl)-4,5-dihydro-1(or 3)-(2-hydroxyethyl)-2-undecyl-, hydroxide, monosodium salt (9CI) **(68647-44-9)**
Imidazolium compounds, 2-C13-17-alkyl-1-(2-C14-18-amidoethyl)-4,5-dihydro-3-methyl, Me sulfates **(72623-82-6)**
Imidazolium compounds, 1-benzyl-4,5-dihydro-1-(hydroxyethyl)-2-norcoco alkyl, chlorides **(61791-52-4)**
Imidazolium compounds, 1-benzyl-4,5-dihydro-1-(hydroxyethyl)-2-nortall-oil alkyl, chlorides **(68309-34-2)**
Imidazolium compounds, 4,5-dihydro-1-methyl-2-nortallow alkyl-1-(2-tallow amidoethyl), Me sulfates **(68122-86-1)**
Imidazolium compounds, 1(or 3)-(carboxymethyl)-4,5-dihydro-1-(hydroxethyl)-2-norcoco alkyl, hydroxides, monosodium salts **(68647-53-0)**
1H-Imidazolium, 1,1-didodecyldihydro-, chloride (9CI) **(71729-96-9)**
1H-Imidazolium, 1-((dodecylthio)methyl)-3-methyl-, chloride (9CI) **(68279-02-7)**
1H-Imidazolium, 1-ethyl-2-(heptadecenyl)-4,5-dihydro-1-(2-hydroxyethyl)-, ethyl sulfate (Salt) (9CI) **(26266-76-2)**
1H-Imidazolium, 1-methyl-3-((octylthio)methyl)-, chloride (9CI) **(68279-00-5)**
4H-Imidazol-4-one, 2-amino-1,5-dihydro-1-methyl- (9CI) **(60-27-5)**
2H-Imidazol-2-one, 1,3-dihydro- (9CI) **(5918-93-4)**
2-(4-Imidazolyl)ethylamine **(51-45-6)**
β-Imidazolyl-4-ethylamine **(51-45-6)**
Imidene **(50-35-1)**
Imidin **(835-31-4)**
Imino-β,β'-dipropionitrile **(111-94-4)**
Imidobenzyle **(50-49-7)**
Imidocarbonic acid, (diethoxyphosphinyl)dithio-, cyclic ethylene ester **(947-02-4)**
Imidocarbonic acid, phosphonodithio-, cyclic ethylene p,p-diethyl ester **(947-02-4)**
Imidocarbonic acid, phosphonodithio-, cyclic methylene p,p-diethyl ester **(21548-32-3)**
Imidocarbonic acid, phosphonodithio-, p,p-diethyl cyclic ethylene ester **(947-02-4)**
4,4'-(Imidocarbonyl)bis(N,N-dimethylamine), monohydrochloride **(2465-27-2)**
4,4'-(Imidocarbonyl)bis(N,N-dimethylaniline) **(492-80-8)**
Imidocarbonyl chloride, (6-chloro-s-triazine-2,4-diyl)bis- (8CI) **(30863-30-0)**
Imidocarbonyl chloride, (4,6-dichloro-s-triazin-2-yl)- (7CI,8CI) **(877-83-8)**
Imidocarbonyl chloride, phenyl **(622-44-6)**
Imidodicarbonic diamide (9CI) **(108-19-0)**
Imidodicarbonic diamide, N,N',2-tris(6-isocyanatohexyl)- **(4035-89-6)**
Imidodicarbonimidic diamide **(56-03-1)**
Imidodicarbonitrile **(504-66-5)**
3,3'-Imidodi-1-propanol, dimethanesulfonate (ester), hydrochloride **(3458-22-8)**
Imidodisulfuric acid, methoxy-, disodium salt (9CI) **(63450-73-7)**
Imidodisulfuric acid, methoxy-, monosodium salt (9CI) **(67874-55-9)**
Imidole **(109-97-7)**
Iminazole **(288-32-4)**
4,5'-Iminobis(4-benzamidoanthraquinone) **(128-89-2)**
Iminobis(acetic acid) **(142-73-4)**
N,N'-Iminobis(dipropylenedistearamide) **(13998-73-7)**
2,2'-Iminobisethanol **(111-42-2)**
2,2'-Iminobisethanol acetate **(23251-72-1)**
2,2'-Iminobisethanol acetate (Salt) **(23251-72-1)**
Iminobis(octamethylene)diguanidine **(13516-27-3)**
1,1'-(Iminobis(octamethylene))diguanidine triacetate **(57520-17-9)**
1,1'-Iminobis-2-propanol **(110-97-4)**
3,3'-Iminobis(propylamine) **(56-18-8)**
Iminobis(propylamine) **(56-18-8)**
Iminobispropylamine **(56-18-8)**
3,3'-Iminobispropylamine [UN 2269] **(56-18-8)**
4,4'-((4-Imino-2,5-cyclohexadien-1-ylidene)methylene)dianiline monohydrochloride **(569-61-9)**
2,2'-Iminodi-N-nitrosoethanol **(1116-54-7)**
2,2'-Iminodiacetic acid **(142-73-4)**
Iminodiacetic acid **(142-73-4)**
4,4'-Iminodianiline **(537-65-5)**
Iminodibenzyl **(494-19-9)**
Iminodiethanoic acid **(142-73-4)**
2,2'-Iminodiethanol **(111-42-2)**
Iminodioctan disodny (Czech) **(928-72-3)**
Iminodioctan sodny (Czech) **(928-72-3)**
1,1'-Iminodi-2-propanol **(110-97-4)**
3,3'-Iminodipropionitrile **(111-94-4)**
β,β'-Iminodipropionitrile **(111-94-4)**
β,β-Iminodipropionitrile **(111-94-4)**
2H-10, 4a-Iminoethanophenanthren-6-ol, 1,3,4,9, 10,10a-hexahydro-11-methyl-, dl- **(297-90-5)**
2H-10,4a-Iminoethanophenanthren-6-ol, 1,3,4,9,10,10a-hexahydro-11-methyl-, l- **(77-07-6)**
9H-9,9c-Iminoethanophenanthro(4,5-bcd)furan-3,5-diol, 4a,5,7a,8-tetrahydro-12-methyl- **(57-27-2)**
4,5-Iminophenanthrene **(203-65-6)**
2(3H)-Imino-9-β-D-ribofuranosyl-9H-purin-6(1H)-one **(118-00-3)**
2,2'-Iminostilbene **(256-96-2)**
Iminostilbene **(256-96-2)**
2-Iminothiazolidine **(1779-81-3)**
Iminourea **(113-00-8)**
2-Imino-5-phenyl-4-oxazolidinone **(2152-34-3)**
Imipramina (Italian) **(50-49-7)**
Imipramine **(50-49-7)**
Imipramine, demethyl- **(50-47-5)**
Imiprin **(50-49-7)**

Imizin (50-49-7)
Imizine (50-49-7)
Imizinum (50-49-7)
Imizolamid (30979-48-7)
Immedial Black AT (1326-82-5)
Immedial Black GN (1326-82-5)
Immedial Black MF (1326-82-5)
Immedial Carbon B (1326-82-5)
Immedial Carbon BO (1326-82-5)
Immedial Carbon BR (1326-82-5)
Immedial Carbon CBO (1326-82-5)
Immedial Carbon CM (1326-82-5)
Immedial Carbon CMR (1326-82-5)
Immedial Carbon LP (1326-82-5)
Immedial Carbon MLB (1326-82-5)
Immedial Carbon NGD (1326-82-5)
Immedial Carbon RRC (1326-82-5)
Immedial Indone RR (1327-57-7)
Immedial Indone Violet B (1327-57-7)
Immedial MFS Grains (1326-82-5)
Immedial Pinting Black B Paste (1326-82-5)
Immenoctal (76-73-3)
Immenox (76-73-3)
Imnudorm (77-75-8)
Impact Sopra (87676-93-5)
Impedex (137-40-6)
Imperator (52315-07-8)
Imperatorin (482-44-0)
Imperial Green (12002-03-8)
Imperon Fixer T (545-55-1)
Impervotar (8007-45-2)
Impingement Black (1333-86-4)
Imposil (9004-66-4)
Impramine (50-49-7)
Impruvol (128-37-0)
Imsil (14808-60-7)
Imsol A (67-63-0)
Imugan (20856-57-9)
Imuran (446-86-6)
Imurek (446-86-6)
Imurel (446-86-6)
Imutex (288-32-4)
Imvite I.G.B.A (1302-78-9)
Imwitor 191 (31566-31-1)
Imwitor 900K (31566-31-1)
In-Sone (53-03-2)
Inacid (53-86-1)
Inactive Limonene (138-86-3)
Inakor (1912-24-9)
Inakor (8073-77-6)
Inakor T (8073-77-6)
Inamil (526-08-9)
Inappin (548-73-2)
Inapsin (548-73-2)
Inapsine (548-73-2)
Inbuton (339-43-5)
Incense LS (8001-21-6)
Incidol (94-36-0)
Incortin (50-04-4)
Incortin (53-06-5)
Incortin-H (50-23-7)
Incracide 10A (7758-98-7)
Incracide E 51 (7758-98-7)
Increcel (999-81-5)
Incromide LCL (142-78-9)
Incromide LI (142-54-1)
Incromide LMI (142-54-1)
Incromide LMM (142-78-9)
Incromide OPD (93-83-4)
Incromide SM (111-57-9)
Incromine Oxide OD 50 (14351-50-9)
Incromine Oxide S (2571-88-2)
Incromine SB (7651-02-7)
Incronam 30 (61789-40-0)
Incronol SLS (151-21-3)
Incropol L-7 (3055-97-8)
Incropol L2 (3055-93-4)
Incropol L7-90 (3055-97-8)
Incroquat CR CON (67762-27-0)
Incroquat CR Concentrate (67762-27-0)
Incroquat O-50 (37139-99-4)
Indo-Rectolmin (53-86-1)
Indo-Tablinen (53-86-1)
s-Indacene, 1,2,3,5,6,7-hexahydro-1,1,4,5,5,8-hexamethyl- (9CI) (17465-59-7)
s-Indacene, 1,2,3,5,6,7-hexahydro-1,1,4,7,7,8-hexamethyl- (9CI) (17465-58-6)
Indacin (53-86-1)
Indalca AG (9000-30-0)
Indalca AG-BV (9000-30-0)
Indalca AG-HV (9000-30-0)
Indalone (532-34-3)
Indamine (537-65-5)
Indan (496-11-7)
Indan, 1,2-dimethyl- (17057-82-8)
Indan, 1,6-dimethyl- (17059-48-2)
Indan, 4,6-dimethyl (1685-82-1)
Indan, dimethyl- (53563-67-0)
Indan, 1,3-dimethyl- (8CI) (4175-53-5)
Indan, 4,7-dimethyl- (8CI) (6682-71-9)
Indan, 4,6-dinitro-1,1,3,3,5-pentamethyl (116-66-5)
1,2-Indandiol, cis- (8CI) (4647-42-1)
1,2-Indandiol, trans- (8CI) (4647-43-2)
1,3-Indandione (606-23-5)
1,3-Indandione, 2-((p-chlorophenyl)phenylacetyl)- (3691-35-8)
1,3-Indandione, 2,2-dihydroxy (485-47-2)
1,3-Indandione, 2-diphenylacetyl (82-66-6)
1,3-Indandione, 2-isovaleryl (83-28-3)
1,3-Indandione, 2-isovaleryl-, ion(1-), calcium (8CI) (23710-76-1)
1,3-Indandione, 2-(p-methoxyphenyl) (117-37-3)
1,3-Indandione, 2-phenyl (83-12-5)
1,3-Indandione, 2-pivaloyl (83-26-1)
1,3-Indandione, 2-pivalyl- (83-26-1)
Indane (496-11-7)
Indane (Alkane) (9CI) (56573-11-6)
Indan, 2-ethyl- (56147-63-8)
Indan, 1-ethyl- (8CI) (4830-99-3)
Indan, hexahydro- (8CI) (496-10-6)
Indan, 3a,4,5,6,7,7a.β.-hexahydro-2.α.-isopropyl-2,3a.β.,4.β.- tri methyl-, (+)- (8CI) (31230-13-4)
Indan, 1-methyl- (8CI) (767-58-8)
Indan, 2-methyl- (8CI) (824-63-5)
Indan, 4-methyl- (8CI) (824-22-6)
Indan, 5-methyl- (8CI) (874-35-1)
Indan, methyl- (8CI) (27133-93-3)
Indan, 1-methyl-3-phenyl- (8CI) (6416-39-3)
5-Indanol (1470-94-6)
Indan-5-ol (1470-94-6)
1-Indanol (8CI) (6351-10-6)
4-Indanol, 7-chloro (145-94-8)
1-Indanol, 3,3-dimethyl- (7CI) (38393-92-9)
2-Indanone (615-13-4)
Indan-1-one (83-33-0)
α-Indanone (83-33-0)
1-Indanone (8CI) (83-33-0)
Indanone (VAN) (30286-23-8)

Name	CAS No.
Indanone (VAN)	**(83-33-0)**
1-Indanone, 5,6-dimethyl- (6CI,8CI)	**(16440-97-4)**
2-Indanone, 1,1,3,3-tetramethyl- (6CI,7CI,8CI)	**(5689-12-3)**
Indan, 1,1,2,3,3-pentamethyl-	**(1203-17-4)**
Indanthren Blue	**(81-77-6)**
Indanthren Blue BC	**(130-20-1)**
Indanthren Blue GP	**(81-77-6)**
Indanthren Blue GPT	**(81-77-6)**
Indanthren Blue RPT	**(81-77-6)**
Indanthren Blue RS	**(81-77-6)**
Indanthren Blue RSN	**(81-77-6)**
Indanthren Blue RSP	**(81-77-6)**
Indanthren Brilliant Blue R	**(81-77-6)**
Indanthren Brilliant Green B	**(128-58-5)**
Indanthren Brilliant Green FFB	**(128-58-5)**
Indanthren Brilliant Green FFB Extra pure	**(128-58-5)**
Indanthren Brilliant Orange GR	**(4424-06-0)**
Indanthren Brilliant Pink R	**(2379-74-0)**
Indanthren Brilliant Pink RB	**(2379-74-0)**
Indanthren Brilliant Pink RP	**(2379-74-0)**
Indanthren Brilliant Pink RS	**(2379-74-0)**
Indanthren Brilliant Violet 4R	**(1324-55-6)**
Indanthren Brilliant Violet RR	**(1324-55-6)**
Indanthren Bronze BR	**(2475-33-4)**
Indanthren Brown BR	**(2475-33-4)**
Indanthren Brown GR	**(2475-33-4)**
Indanthren Gold Orange G	**(128-70-1)**
Indanthren Golden Orange G	**(128-70-1)**
Indanthren Golden Orange GLP	**(128-70-1)**
Indanthren Golden Yellow	**(128-66-5)**
Indanthren Navy Blue G	**(6424-76-6)**
Indanthren Olive R	**(2379-81-9)**
Indanthren Printing Blue FRS	**(81-77-6)**
Indanthren Printing Blue KRS	**(81-77-6)**
Indanthren Printing Violet F 4R	**(1324-55-6)**
Indanthren Printing Yellow	**(128-66-5)**
Indanthren Rubine R	**(4203-77-4)**
Indanthren Rubine RS	**(4203-77-4)**
Indanthrene	**(81-77-6)**
Indanthrene Blue	**(81-77-6)**
Indanthrene Blue BC	**(130-20-1)**
Indanthrene Blue BCF	**(130-20-1)**
Indanthrene Blue GP	**(81-77-6)**
Indanthrene Blue RP	**(81-77-6)**
Indanthrene Blue RS	**(81-77-6)**
Indanthrene Blue RSA	**(81-77-6)**
Indanthrene Blue RSN	**(81-77-6)**
Indanthrene Brilliant Green B	**(128-58-5)**
Indanthrene Brilliant Green BN	**(128-58-5)**
Indanthrene Brilliant Green FFB	**(128-58-5)**
Indanthrene Brilliant Orange GR	**(4424-06-0)**
Indanthrene Brilliant Orange GRP	**(4424-06-0)**
Indanthrene Brilliant Violet 4R	**(1324-55-6)**
Indanthrene Brilliant Violet RR	**(1324-55-6)**
Indanthrene Brown BR	**(2475-33-4)**
Indanthrene Gold Orange G	**(128-70-1)**
Indanthrene Golden Orange G	**(128-70-1)**
Indanthrene Golden Orange GA	**(128-70-1)**
Indanthrene Golden Yellow	**(128-66-5)**
Indanthrene Navy Blue G	**(6424-76-6)**
Indanthrene Olive R	**(2379-81-9)**
Indanthrene Rubine R	**(4203-77-4)**
Indanthrone	**(81-77-6)**
Indan, 1,1,3-trimethyl- (8CI)	**(2613-76-5)**
Indan, 4,5,7-trimethyl- (8CI)	**(6682-06-0)**
Indan, 1,1,2-trimethyl-3-phenyl- (8CI)	**(33508-02-0)**
Indan, 1,1,3-trimethyl-3-phenyl- (8CI)	**(3910-35-8)**
Indan, 1,2,3-trimethyl-1-phenyl-, stereoisomer (8CI)	**(33603-39-3)**
Indan, 1,2,3-trimethyl-1-phenyl-, stereoisomer (8CI)	**(33611-16-4)**
1,2,3-Indantrione, 2-hydrate	**(485-47-2)**
1,2,3-Indantrione monohydrate	**(485-47-2)**
Indar	**(16227-10-4)**
1H-Indazol-3-amine (9CI)	**(874-05-5)**
1H-Indazole	**(271-44-3)**
Indazole	**(271-44-3)**
1H-Indazole, 5-nitro	**(5401-94-5)**
Indema	**(83-12-5)**
Inden	**(95-13-6)**
Indene (ACGIH,OSHA) (8CI)	**(95-13-6)**
1H-Indene (9CI)	**(95-13-6)**
1H-Indene-3-acetic acid, 5-fluoro-2-methyl-1-((4-(methylsulfinyl)-phenyl)methylene)-, (Z)	**(38194-50-2)**
1H-Indene-4-carboxylic acid, 2,3-dihydro-1,1-dimethyl-, ethyl ester (9CI)	**(55591-12-3)**
1H-Indene-1,7-dicarboxaldehyde, 2,3,5,6-tetrahydro-1,3,3,6-tetra-methyl-, (1S-cis)- (9CI)	**(97165-23-6)**
Indene, 2,3-dihydro-	**(496-11-7)**
1H-Indene, 2,3-dihydro-1,2-dimethyl- (9CI)	**(17057-82-8)**
1H-Indene, 2,3-dihydro-1,3-dimethyl- (9CI)	**(4175-53-5)**
1H-Indene, 2,3-dihydro-1,6-dimethyl- (9CI)	**(17059-48-2)**
1H-Indene, 2,3-dihydro-4,7-dimethyl- (9CI)	**(6682-71-9)**
1H-Indene, 2,3-dihydrodimethyl- (9CI)	**(53563-67-0)**
1H-Indene, 2,3-dihydro-1-methyl- (9CI)	**(767-58-8)**
1H-Indene, 2,3-dihydro-2-methyl- (9CI)	**(824-63-5)**
1H-Indene, 2,3-dihydro-4-methyl- (9CI)	**(824-22-6)**
1H-Indene, 2,3-dihydro-5-methyl- (9CI)	**(874-35-1)**
1H-Indene, 2,3-dihydromethyl- (9CI)	**(27133-93-3)**
1H-Indene, 2,3-dihydro-1-methyl-3-phenyl- (9CI)	**(6416-39-3)**
1H-Indene, 2,3-dihydro-1,1,2,3,3-pentamethyl- (9CI)	**(1203-17-4)**
1H-Indene, 2,3-dihydro-1,1,3,3,5-pentamethyl- (9CI)	**(81-03-8)**
1H-Indene, 2,3-dihydro-1,1,3-trimethyl- (9CI)	**(2613-76-5)**
1H-Indene, 2,3-dihydro-1,4,7-trimethyl- (9CI)	**(54340-87-3)**
1H-Indene, 2,3-dihydro-1,5,7-trimethyl- (9CI)	**(54340-88-4)**
1H-Indene, 2,3-dihydro-4,5,7-trimethyl- (9CI)	**(6682-06-0)**
1H-Indene, 2,3-dihydrotrimethyl- (9CI)	**(36541-18-1)**
1H-Indene, 2,3-dihydro-1,1,2-trimethyl-3-phenyl- (9CI)	**(33508-02-0)**
1H-Indene, 2,3-dihydro-1,1,3-trimethyl-3-phenyl- (9CI)	**(3910-35-8)**
1H-Indene, 1,1-dimethyl- (9CI)	**(18636-55-0)**
Indene, 1,1-dimethyl- (7CI,8CI)	**(18636-55-0)**
1H-Indene, 1,2-dimethyl-, (S)- (9CI)	**(53204-57-2)**
1H-Indene, 5-(1,1-dimethylethyl)-2,3-dihydro-1,1-dimethyl- (9CI)	**(38393-97-4)**
1H-Indene-1,2-diol, 2,3-dihydro-, cis- (9CI)	**(4647-42-1)**
1H-Indene-1,2-diol, 2,3-dihydro-, trans- (9CI)	**(4647-43-2)**
1H-Indene-1,2-diol, 2,3-dihydro-2-methyl-, cis- (9CI)	**(56588-40-0)**
1H-Indene-1,3(2H)-dione	**(606-23-5)**
1H-Indene-1,3(2H)-dione, 2-benzo(f)quinolin-3-yl-, disulfo deriv. (9CI)	**(1324-04-5)**
1H-Indene-1,3(2H)-dione, 2-(4-bromo-3-hydroxy-2-quinolinyl)- (9CI)	**(10319-14-9)**
1H-Indene-1,3(2H)-dione, 2-(diphenylacetyl)-, ion(1-), sodium (9CI)	**(42721-99-3)**
1H-Indene-1,3(2H)-dione, 2-(3-hydroxy-2-quinolinyl)- (9CI)	**(7576-65-0)**
1H-Indene-1,3(2H)-dione, 2-phenyl- (9CI)	**(83-12-5)**
1H-Indene-1,3(2H)-dione, 2-(2-quinolinyl)- (9CI)	**(83-08-9)**
1H-Indene-5-ethanol, 2,3-dihydro-β,1,1,2,3,3-hexamethyl- (9CI)	**(1217-08-9)**
1H-Indene, 1-ethyl-2,3-dihydro- (9CI)	**(4830-99-3)**
1H-Indene, 2-ethyl-2,3-dihydro- (9CI)	**(56147-63-8)**
1H-Indene, 1-ethyl-2,3-dihydro-1-methyl- (9CI)	**(56298-75-0)**
1H-Indene, 1-ethylidene- (9CI)	**(2471-83-2)**
Indene, 1-ethylidene- (8CI)	**(2471-83-2)**
1H-Indene, 1-ethylideneoctahydro-, trans- (9CI)	**(56324-70-0)**
1H-Indene, 2,3,4,5,6,7-hexahydro-1,1,2,3,3-pentamethyl- (9CI)	**(33704-59-5)**
Indene, methyl-	**(29036-25-7)**
1H-Indene, 1-methyl- (9CI)	**(767-59-9)**

1H-Indene, 2-methyl- (9CI) **(2177-47-1)**
1H-Indene, 3-methyl- (9CI) **(767-60-2)**
1H-Indene, methyl- (9CI) **(29036-25-7)**
Indene, 1-methyl- (8CI) **(767-59-9)**
Indene, 2-methyl- (8CI) **(2177-47-1)**
Indene, 3-methyl- (8CI) **(767-60-2)**
1H-Indene, methyldinitro- (9CI) **(116211-91-7)**
1H-Indene, octahydro- (9CI) **(496-10-6)**
1H-Indene, octahydro-2,2,4,4,7,7-hexamethyl-, trans- (9CI) **(54832-83-6)**
1H-Indene, octahydro-2,3a,4-trimethyl-2-(1-methylethyl)-, (2.α.,3a.β.,4.β.,7a.β.)-(+)- (9CI) **(31230-13-4)**
1H-Indene, 3a,4,7,7a-tetrahydro- **(3048-65-5)**
Indene, 3a,4,7,7a-tetrahydro **(3048-65-5)**
13H-Indeno(2,1-a)anthracen-13-one (9CI) **(86854-04-8)**
8H-Indeno(1,2-a)anthracen-8-one (9CI) **(86853-94-3)**
7H-Indeno(1,2-a)phenanthren-7-one (9CI) **(86853-95-4)**
11H-Indeno(2,1-a)phenanthren-11-one (8CI,9CI) **(4599-92-2)**
7H-Indeno(1,2-a)pyren-7-one (9CI) **(86854-20-8)**
11H-Indeno(2,1-a)pyren-11-one (8CI,9CI) **(7267-90-5)**
13H-Indeno(1,2-b)anthracen-13-one (9CI) **(86853-99-8)**
12H-Indeno(1,2-b)phenanthren-12-one (9CI) **(86854-00-4)**
8H-Indeno(2,1-b)phenanthren-8-one (9CI) **(86853-98-7)**
5H-Indeno(1,2-b)pyridine (8CI,9CI) **(244-99-5)**
3H-Indeno(2,1,7-cde)pyren-3-one (9CI) **(86862-68-2)**
Indeno(1,2,3-cd)pyrene **(193-39-5)**
13H-Indeno(1,2-c)phenanthren-13-one (9CI) **(86854-03-7)**
9H-Indeno(2,1-c)phenanthren-9-one (9CI) **(86853-93-2)**
Indeno(1,2,3-de)isoquinoline (9CI) **(7148-92-7)**
Indeno(1,2,3-de)quinoline (8CI,9CI) **(206-55-3)**
13H-Indeno(1,2-e)pyren-13-one (9CI) **(86854-21-9)**
4H-Indeno(7,1,2-ghi)chrysen-4-one (9CI) **(86854-14-0)**
Indeno(1,2,3-ij)isoquinoline (8CI,9CI) **(206-56-4)**
1H-Inden-1-ol, 2,3-dihydro- (9CI) **(6351-10-6)**
1H-Inden-4-ol, 2,3-dihydro-7-chloro- (9CI) **(145-94-8)**
1H-Inden-1-ol, 2,3-dihydro-3,3-dimethyl- (9CI) **(38393-92-9)**
13H-Indeno(1,2-l)phenanthren-13-one (9CI) **(86853-96-5)**
1H-Inden-1-one, 2-chloro-2,3-dihydro-, (.+-.)- (9CI) **(73908-22-2)**
2H-Inden-2-one, 1,3-dihydro- **(615-13-4)**
1H-Inden-1-one, 2,3-dihydro- (9CI) **(83-33-0)**
Indenone, dihydro- (9CI) **(30286-23-8)**
1H-Inden-1-one, 2,3-dihydro-5,6-dimethyl- (9CI) **(16440-97-4)**
Indenone, 2,3-dihydromethyl- (9CI) **(87259-53-8)**
2H-Inden-2-one, 1,3-dihydro-1,1,3,3-tetramethyl- (9CI) **(5689-12-3)**
4H-Inden-4-one, 1,2,3,5,6,7-hexahydro-1,1,2,3,3-pentamethyl- (9CI) **(33704-61-9)**
Indenopyrene (9CI) **(72254-06-9)**
13H-Indeno(2,1,7-qra)naphthacen-13-one (9CI) **(86854-19-5)**
Independence Red **(2425-85-6)**
Inderal **(525-66-6)**
Indi **(41708-76-3)**
Indian Berry **(124-87-8)**
Indian Cannabis **(8063-14-7)**
Indian Gum **(9000-01-5)**
Indian Hemp **(8063-14-7)**
Indian Poke **(65072-04-0)**
Indian Red **(1309-37-1)**
Indian Red **(18454-12-1)**
Indian Saffron **(458-37-7)**
Indian Turmeric **(458-37-7)**
Indicine **(480-82-0)**
Indicine N-oxide **(41708-76-3)**
Indigo-Karmin (German) **(860-22-0)**
Indigenous Peanut Oil **(8002-03-7)**
Indigo **(482-89-3)**
Indigo Blue **(482-89-3)**
Indigo Blue 2B **(2602-46-2)**
Indigo Carmine **(860-22-0)**
Indigo Carmine A **(860-22-0)**
Indigo Carmine AC **(860-22-0)**
Indigo Carmine BP **(860-22-0)**
Indigo Carmine (Biological stain) **(860-22-0)**
Indigo Carmine Conc. FQ **(860-22-0)**
Indigo Carmine Disodium salt **(860-22-0)**
Indigo Carmine Powder **(860-22-0)**
Indigo Carmine X **(860-22-0)**
Indigo Ciba **(482-89-3)**
Indigo Ciba SL **(482-89-3)**
Indigo Disulfonate (Biological stain) **(860-22-0)**
Indigo Extract **(860-22-0)**
Indigo J **(482-89-3)**
Indigo N **(482-89-3)**
Indigo Nac **(482-89-3)**
Indigo Nacco **(482-89-3)**
Indigo P **(482-89-3)**
Indigo PLN **(482-89-3)**
Indigo Powder W **(482-89-3)**
Indigo Pure BASF **(482-89-3)**
Indigo Pure BASF Powder K **(482-89-3)**
Indigo Synthetic **(482-89-3)**
Indigo VS **(482-89-3)**
Indigo Yellow **(520-18-3)**
Indigotin-5,5'-disulfonic acid disodium salt **(860-22-0)**
Indigotin **(482-89-3)**
Indigotin I **(860-22-0)**
5,5'-Indigotindisulfonic acid **(483-20-5)**
5,5'-Indigotindisulfonic acid **(860-22-0)**
Indigotine **(860-22-0)**
Indigotine B **(860-22-0)**
Indigotine Blue LZ **(860-22-0)**
Indigotine Conc. Powder **(860-22-0)**
Indigotine Disodium salt **(860-22-0)**
Indigotine Extra Pure A **(860-22-0)**
Indigotine I **(860-22-0)**
Indigotine Lake **(860-22-0)**
Indigotine N **(860-22-0)**
Indion **(83-12-5)**
Indisulfat (German) **(13464-82-9)**
Indium (ACGIH,OSHA) **(7440-74-6)**
Indium bromide (9CI) **(13465-09-3)**
Indium-chloride **(10025-82-8)**
Indium-citrate **(4194-69-8)**
Indium-nitrate **(13770-61-1)**
Indium phosphide (9CI) **(22398-80-7)**
Indium-sulfate **(13464-82-9)**
Indium trichloride **(10025-82-8)**
Indo Blue B-I **(130-20-1)**
Indo Maroon Lake RV 6666 **(4203-77-4)**
Indoblack GR **(6428-31-5)**
Indobufen **(63610-08-2)**
Indocarmine F **(860-22-0)**
Indocid **(53-86-1)**
Indocin **(53-86-1)**
Indocyanine Green **(3599-32-4)**
Indocybin **(520-52-5)**
Indofast Orange OV 5983 **(4424-06-0)**
Indofast Violet Lake **(1324-55-6)**
Indol (German) **(120-72-9)**
Indol-4(5H)-one, 3-ethyl-6,7-dihydro-2-methyl-5-(morpholinomethyl) **(7416-34-4)**
Indole **(120-72-9)**
1H-Indole-1-acetaldehyde, 2,3-dihydro-3,3-dimethyl-2-(2-oxoethylidene)- (9CI) **(63455-65-2)**
1H-Indole-3-acetic acid **(87-51-4)**
3-Indoleacetic acid **(87-51-4)**

β-Indole-3-acetic acid **(87-51-4)**
β-Indoleacetic acid **(87-51-4)**
Indole-3-acetic acid, 1-(p-chlorobenzoyl)-5-methoxy-2-methyl **(53-86-1)**
Indole-3-alanine **(73-22-3)**
Indole, 3-(2-aminoethyl) **(61-54-1)**
1H-Indole, 5-bromo- (9CI) **(10075-50-0)**
1H-Indole-3-butanoic acid **(133-32-4)**
Indole butyric **(133-32-4)**
3-Indolebutyric acid **(133-32-4)**
Indole butyric acid **(133-32-4)**
β-Indolebutyric acid **(133-32-4)**
γ-(Indole-3)-butyric acid **(133-32-4)**
1H-Indole-7-carboxylic acid, 3-(1-(7-((hexadecylsulfonyl)amino)-1H-indol3-yl)3-oxo-1H,3H-naphtho(1,8-cd)pyran-1-yl)- (9CI) **(37921-74-7)**
1H-Indole, 1-(4-chlorobenzoyl)-5-methoxy-2-methyl- **(6260-97-5)**
Indole, 1-(p-chlorobenzoyl)-5-methoxy-2-methyl- **(6260-97-5)**
Indole, 3-(2-(diethylamino)ethyl)- **(61-51-8)**
1H-Indole, 2,3-dihydro- **(496-15-1)**
1H-Indole, 1,2-dimethyl- (9CI) **(875-79-6)**
Indole, 1,2-dimethyl- (8CI) **(875-79-6)**
Indole, 3-(2-(dimethylamino)ethyl) **(61-50-7)**
Indole-2,3-dione **(91-56-5)**
1H-Indole-3-ethanamine **(61-54-1)**
Indole, 5-methoxy **(1006-94-6)**
1H-Indole, 1-methoxy- (9CI) **(54698-11-2)**
Indole, 3-methyl **(83-34-1)**
1H-Indole, 1-methyl- (9CI) **(603-76-9)**
1H-Indole, 2-methyl- (9CI) **(95-20-5)**
1H-Indole, 5-methyl- (9CI) **(614-96-0)**
1H-Indole, methyl- (9CI) **(27323-28-0)**
Indole, 1-methyl- (8CI) **(603-76-9)**
Indole, 2-methyl- (8CI) **(95-20-5)**
Indole, 5-methyl- (8CI) **(614-96-0)**
Indole, methyl- (8CI) **(27323-28-0)**
1H-Indole, 1-methyl-2-phenyl- (9CI) **(3558-24-5)**
1H-Indole, methylphenyl- (9CI) **(64844-52-6)**
Indole, 1-methyl-2-phenyl- (8CI) **(3558-24-5)**
1H-Indole, 1-methyl-2-phenyl-3-(2-thiazolylazo)- (9CI) **(34367-95-8)**
1H-Indole, nitro- (9CI) **(60544-75-4)**
1H-Indole, 2-phenyl- (9CI) **(948-65-2)**
Indole, 2-phenyl- (8CI) **(948-65-2)**
1H-Indole-3-propionic acid **(830-96-6)**
Indolepropionic acid **(830-96-6)**
β-Indolepropionic acid **(830-96-6)**
Indole-3-propionic acid, α-amino- **(73-22-3)**
1H-Indole-5-sulfonic acid, 2,3-dihydro-2,3-dioxo-, monosodium salt (9CI) **(80789-74-8)**
1H-Indole-5-sulfonic acid, 2-(1,3-dihydro-3-oxo-5-sulfo-2H-indol-2-ylidene)-2,3-dihydro-3-oxo- (9CI) **(483-20-5)**
Indol-3-ethylamine **(61-54-1)**
Indole, trimethyl- **(30642-36-5)**
1H-Indole, trimethyl- (9CI) **(30642-36-5)**
3H-Indole, 2,3,3-trimethyl- (9CI) **(1640-39-7)**
Indoline **(496-15-1)**
δ2,α-Indolineacetaldehyde, 1,3,3-trimethyl- (8CI) **(84-83-3)**
2,3-Indolinedione **(91-56-5)**
Indoline, 2-methyl **(6872-06-6)**
Indoline, 2-methylene-1,3,3-trimethyl **(118-12-7)**
5-Indolinesulfonic acid, 2,3-dioxo-, sodium salt (7CI) **(80789-74-8)**
2-Indolinone, 1,3,3-trimethyl- (8CI) **(20200-86-6)**
3H-Indolium, 2-(2-(4-(diethylamino)phenyl)ethenyl)-1,3,3-trimethyl-, chloride (9CI) **(6359-45-1)**
3H-Indolium, 2-(((4-methoxyphenyl)methylhydrazono)methyl)-1,3,3-trimethyl-, methyl sulfate (9CI) **(54060-92-3)**
3H-Indolium, 2-(2-(2-methylindol-3-yl)vinyl)-1,3,3-trimethyl-, chloride **(3056-93-7)**
3H-Indolium, 1,3,3-trimethyl-2-(2-(1-methyl-2-phenyl-1H-indol-3-yl)ethenyl)-, chloride (9CI) **(4657-00-5)**
3H-Indolium, 1,3,3-trimethyl-2-(2-(1-methyl-2-phenylindol-3-yl)-vinyl)-, chloride (8CI) **(4657-00-5)**
Indolizidine **(13618-93-4)**
Indolizine **(274-40-8)**
Indolizine, octahydro- (8CI,9CI) **(13618-93-4)**
1H-Indol-4-ol (9CI) **(2380-94-1)**
1H-Indolol (9CI) **(69594-78-1)**
Indol-4-ol (8CI) **(2380-94-1)**
Indol-5-ol, 3-(2-aminoethyl) **(50-67-9)**
1H-Indol-5-ol, 3-(2-(dimethylamino)ethyl)- (9CI) **(487-93-4)**
Indol-5-ol, 3-(2-(dimethylamino)ethyl)- (8CI) **(487-93-4)**
Indol-4-ol, 3-(2-(dimethylamino)ethyl)-, dihydrogen phosphate **(520-52-5)**
1H-Indolol, methyl- (9CI) **(116211-85-9)**
3H-Indol-3-one, 5-bromo-2-(1,3-dihydro-3-oxo-2H-indol-2-ylidene)-1,2-dihydro- (9CI) **(6492-73-5)**
3H-Indol-3-one, 5,7-dibromo-2-(5,7-dibromo-1,3-dihydro-3-oxo-2H-indol-2-ylidene)-1,2-dihydro- (9CI) **(2475-31-2)**
2H-Indol-2-one, 1,3-dihydro-1-acetyl-3,3-bis(4-(acetyloxy)phenyl) **(18869-73-3)**
3H-Indol-3-one, 2(1,3-dihydro-3-oxo-2H-indol-2-ylidene)-1,2-dihydro- (9CI) **(482-89-3)**
2H-Indol-2-one, 1,3-dihydro-1,3,3-trimethyl- (9CI) **(20200-86-6)**
3-Indolylacetic acid **(87-51-4)**
Indol-3-ylacetic acid **(87-51-4)**
Indolyl-3-acetic acid **(87-51-4)**
α-Indol-3-yl-acetic acid **(87-51-4)**
β-Indolylacetic acid **(87-51-4)**
1-β-3-Indolylalanine **(73-22-3)**
3-Indolyl-γ-butyric acid **(133-32-4)**
4-(3-Indolyl)butyric acid **(133-32-4)**
4-(Indol-3-yl)butyric acid **(133-32-4)**
Indolyl-3-butyric acid **(133-32-4)**
γ-(3-Indolyl)butyric acid **(133-32-4)**
γ-(Indol-3-yl)butyric acid **(133-32-4)**
2-(3-Indolyl)ethylamine **(61-54-1)**
3-(3-Indolyl)propanoic acid **(830-96-6)**
Indomecol **(53-86-1)**
Indomed **(53-86-1)**
Indomee **(53-86-1)**
Indometacin **(53-86-1)**
Indometacine **(53-86-1)**
Indometacyna (Polish) **(53-86-1)**
Indomethacin **(53-86-1)**
Indomethacine **(53-86-1)**
Indomethazine **(53-86-1)**
Indometicina (Spanish) **(53-86-1)**
Indon **(83-12-5)**
Indonaphthene **(95-13-6)**
1-Indone **(83-33-0)**
Indopol **(9003-29-6)**
Indopol H 50 **(9003-29-6)**
Indopol H 100 **(9003-29-6)**
Indopol H 300 **(9003-29-6)**
Indopol H 1900 **(9003-27-4)**
Indopol L 14 **(9003-29-6)**
Indopol L 10 **(9003-29-6)**
Indopol L 100 **(9003-29-6)**
Indoptic **(53-86-1)**
Indothrin **(52645-53-1)**
Indoxine KL **(2429-73-4)**
Induline R **(60-09-3)**
Industrene 105 **(112-80-1)**
Industrene 205 **(112-80-1)**
Industrene 206 **(112-80-1)**
Industrene 4516 **(57-10-3)**
Industrene 5016 **(57-11-4)**
Industrene 8718 **(57-11-4)**

Industrene 9018 (57-11-4)
Industrol N 3 (9003-11-6)
Inerteen (1336-36-3)
Inestra (57-63-6)
Inetol (51-02-5)
Inexit (58-89-9)
Inezin (21722-85-0)
Infamil (129-20-4)
Infammil (129-20-4)
Inferno (78-53-5)
Infiltrina (67-68-5)
Inflatine (90-69-7)
Inflazon (53-86-1)
Infrocin (53-86-1)
Infron (50-14-6)
Ingalan (76-38-0)
Ingalan (Russian) (76-38-0)
Ingalipt (8000-48-4)
Inhalan (76-38-0)
Inhibine (7722-84-1)
Inhibisol (71-55-6)
Inhiston (132-20-7)
Inhiston (86-21-5)
Inicardio (59-26-7)
Initiating Explosive Lead Azide, Dextrinated type only (13424-46-9)
Initiating Explosive Lead Mononitroresorcinate (51317-24-9)
Initiating Explosive Nitro Mannite (15825-70-4)
Initiating Explosive-guanyl nitrosamine guanyl tetrazene (DOT) (109-27-3)
Initiating Explosive-tetrazene (DOT) (109-27-3)
Initiating explosive fulminate of mercury (DOT) (628-86-4)
Inizid (54-85-3)
Ink Blue 6B (2610-05-1)
Ink Orange JSN (1936-15-8)
Ink Red JSN (3734-67-6)
Innovar-Vet (548-73-2)
Innovan (548-73-2)
Innovar (548-73-2)
Innoxalon (389-08-2)
Ino (58-63-9)
Inochrome Red 3J (6408-31-7)
Inofal (14759-06-9)
Inophylline (317-34-0)
Inopsin (548-73-2)
Inorganic bromides (24959-67-9)
Inosie (58-63-9)
Inosine (58-63-9)
β-Inosine (58-63-9)
Inosine, 2-amino- (118-00-3)
Inosine, 6-(methylthio)- (342-69-8)
myo-Inosistol hexakisphosphate (83-86-3)
Inosital (87-89-8)
Inositene (87-89-8)
Inosithexaphosphorsaure (German) (83-86-3)
Inositina (87-89-8)
i-Inositol (87-89-8)
iso-Inositol (87-89-8)
meso-Inositol (87-89-8)
myo-Inositol (9CI) (87-89-8)
Inositol (VAN) (87-89-8)
myo-Inositol, 1-(dihydrogen phosphate) (9CI) (573-35-3)
Inositol, 1-(dihydrogen phosphate), myo- (8CI) (573-35-3)
myo-Inositol, hexakis(dihydrogen phosphate), calcium salt (1:6) (7776-28-5)
Inositol, hexakis(dihydrogen phosphate) calcium salt(1:6), myo- (7776-28-5)
Inositol, hexakis(dihydrogen phosphate), myo (83-86-3)
Inositol hexaphosphate (83-86-3)
myo-Inositol hexaphosphate (83-86-3)
Inositol hexaphosphate calcium salt (7776-28-5)
Inositol, meso (87-89-8)
Inositol monophosphate (573-35-3)
Inositol, myo- (8CI) (87-89-8)
Inostral (15826-37-6)
Inoval (548-73-2)
Inovitan PP (98-92-0)
Insariotoxin (21259-20-1)
Insect Powder (8003-34-7)
6-12-Insect Repellent (94-96-2)
Insect Repellent 448 (1444-64-0)
Insecticide 1,179 (16752-77-5)
Insecticide ACC 4124 (2463-84-5)
Insecticide-Nematicide 1410 (23135-22-0)
Insecticide No. 497 (60-57-1)
Insecticide No. 4049 (121-75-5)
Insecticide 3960-X14 (8001-50-1)
Insectigas D (62-73-7)
Insectol (7696-12-0)
Insectophene (115-29-7)
Insegar (72490-01-8)
Insitolum (87-89-8)
Insolat (143-81-7)
Insoluble saccharin (81-07-2)
Insomnol (77-75-8)
Inspir (616-91-1)
Insubeta (8013-17-0)
Insulamina (59-92-7)
Insular (9004-10-8)
Insulin (9004-10-8)
Insulin Injection (9004-10-8)
Insulton (50-12-4)
Insulyl (9004-10-8)
Insumin (1172-18-5)
Insyst-D (298-04-4)
Intal (15826-37-6)
Intalbut (50-33-9)
Intalpram (50-49-7)
Intasedol (143-81-7)
Inteban SP (53-86-1)
Intense Blue (860-22-0)
Interchem Acetate Blue B (2475-46-9)
Interchem Acetate Blue B (86722-66-9)
Interchem Acetate Blue NBN (2475-46-9)
Interchem Acetate Blue NBN (86722-66-9)
Interchem Acetate Blue RBN (2475-46-9)
Interchem Acetate Blue RBN (86722-66-9)
Interchem Acetate Blue WNBN (2475-46-9)
Interchem Acetate Blue WNBN (86722-66-9)
Interchem Acetate Developed Black (539-17-3)
Interchem Acetate Pink 3B (2872-48-2)
Interchem Acetate Pink BLF (116-85-8)
Interchem Acetate Red Violet RRLF (128-95-0)
Interchem Acetate Violet 6B (1220-94-6)
Interchem Acetate Violet R (128-95-0)
Interchem Acetate Yellow G (2832-40-8)
Interchem Direct Black Z (1937-37-7)
Interchem Disperse Yellow GH (2832-40-8)
Interchem Hisperse Pink BH (116-85-8)
Interchem Hisperse Violet 2RH (128-95-0)
Interflo (9002-88-4)
Intermediate Catalytically Cracked Distillate (64741-60-2)
Intermediate Light Oil (Coal) (65996-78-3)
Intermediate Naphthalene (65996-82-9)
International Orange 2221 (18454-12-1)
Interox (7722-84-1)
Interpina (50-55-5)
Intexan CTC 29 (112-02-7)

Intexan LB-50	**(8001-54-5)**
Intexan SB-85	**(122-19-0)**
Intexsan CPC	**(123-03-5)**
Intexsan CTC 29	**(112-02-7)**
Intexsan CTC 50	**(112-02-7)**
Intexsan SB-85	**(122-19-0)**
Intolex	**(73-48-3)**
Intrabutazone	**(50-33-9)**
Intracid Fast Orange G	**(1936-15-8)**
Intracid Green F	**(4680-78-8)**
Intracid Pure Blue L	**(3844-45-9)**
Intracid Pure Blue V	**(129-17-9)**
Intracid Violet 4BNS	**(1694-09-3)**
Intracron Scarlet 4G-P	**(12238-07-2)**
Intramycetin	**(56-75-7)**
Intranefrin	**(51-43-4)**
Intraperse Yellow GBA	**(2832-40-8)**
Intrasperse Yellow GBA Extra	**(2832-40-8)**
Intrasporin	**(50-59-9)**
Intrathion	**(640-15-3)**
Intration	**(640-15-3)**
Intraval	**(76-75-5)**
Intravat Blue GF	**(130-20-1)**
Intromene	**(133-67-5)**
Invenol	**(339-43-5)**
Inverdex	**(8013-17-0)**
Inversal	**(539-21-9)**
Invertogen	**(8013-17-0)**
Inverton 245	**(93-76-5)**
Invertose	**(8013-17-0)**
Invert sugar	**(8013-17-0)**
Invisi-Gard	**(114-26-1)**
Iod-Ethamine	**(5700-49-2)**
Iocetamic acid	**(16034-77-8)**
Iodamide	**(440-58-4)**
Iode (French)	**(7553-56-2)**
Iodenterol	**(130-26-7)**
Iodeosin	**(15905-32-5)**
Iodic acid, calcium salt (9CI)	**(7789-80-2)**
Iodic acid, lead(2+) salt (8CI,9CI)	**(25659-31-8)**
Iodic acid, potassium salt	**(7758-05-6)**
Iodic acid, sodium salt	**(7681-55-2)**
Iodide (8CI,9CI)	**(20461-54-5)**
Iodinated glycerol	**(5634-39-9)**
Iodine-131	**(24267-56-9)**
Iodine (ACGIH,OSHA)	**(7553-56-2)**
Iodine Crystals	**(7553-56-2)**
Iodine(I) azide	**(14696-82-3)**
Iodine, Isotope of mass 129 (8CI,9CI)	**(15046-84-1)**
Iodine, Isotope of mass 132 (8CI,9CI)	**(14683-16-0)**
Iodine, Isotope of mass 133 (8CI,9CI)	**(14834-67-4)**
Iodine Sublimed	**(7553-56-2)**
Iodine azide	**(14696-82-3)**
Iodine azide, Dry	**(14696-82-3)**
Iodine-chloride	**(7790-99-0)**
Iodine-cyanide	**(506-78-5)**
Iodine-fluoride	**(7783-66-6)**
Iodine, ion (I(1-))	**(20461-54-5)**
Iodine, Isotope of mass 131 (8CI,9CI)	**(10043-66-0)**
Iodine monochloride [UN 1792]	**(7790-99-0)**
Iodine pentafluoride [UN 2495]	**(7783-66-6)**
Iodio (Italian)	**(7553-56-2)**
Iodistol	**(552-22-7)**
3,5-Iodo-L-tyrosine	**(300-39-0)**
Iodoacetamide	**(1397-89-3)**
Iodoacetate	**(64-69-7)**
Iodoacetic acid	**(64-69-7)**
3-Iodoaniline	**(626-01-7)**
4-Iodoaniline	**(540-37-4)**
m-Iodoaniline	**(626-01-7)**
o-Iodoaniline	**(615-43-0)**
p-Iodoaniline	**(540-37-4)**
Iodoazide	**(14696-82-3)**
2-Iodobenzenamine	**(615-43-0)**
Iodobenzene	**(591-50-4)**
p-Iodobenzenesulfonamide	**(825-86-5)**
3-Iodobenzenesulfonyl chloride	**(50702-38-0)**
4-Iodobenzenesulfonyl chloride	**(98-61-3)**
Iodobenzene-p-sulfonyl chloride	**(98-61-3)**
p-Iodobenzenesulfonyl chloride	**(98-61-3)**
4-Iodobenzenesulphonyl chloride	**(98-61-3)**
3-Iodobenzoic acid	**(618-51-9)**
o-Iodobenzoic acid	**(88-67-5)**
p-Iodobenzoic acid	**(619-58-9)**
2-Iodobenzoic acid anilide	**(15310-01-7)**
3-Iodobenzyl alcohol	**(57455-06-8)**
m-Iodobenzyl alcohol	**(57455-06-8)**
4-Iodo-1,1'-biphenyl	**(1591-31-7)**
4-Iodobiphenyl	**(1591-31-7)**
1-Iodobutane	**(542-69-8)**
2-Iodobutane [UN 2390]	**(513-48-4)**
Iodochlorhydroxyquin	**(130-26-7)**
Iodochlorhydroxyquinol	**(130-26-7)**
Iodochlorhydroxyquinoline	**(130-26-7)**
7-Iodo-5-chloro-8-hydroxyquinoline	**(130-26-7)**
7-Iodo-5-chloroxine	**(130-26-7)**
Iodochloroxyquinoline	**(130-26-7)**
Iodocyclohexane	**(626-62-0)**
1-Iododecane	**(2050-77-3)**
Iododocosanoic acid calcium salt	**(1319-91-1)**
1-Iodododecane	**(4292-19-7)**
Iodoenterol	**(130-26-7)**
Iodoethane	**(75-03-6)**
2-Iodoethanol	**(624-76-0)**
Iodoethanol	**(624-76-0)**
2-(1-Iodoethyl)-1,3-dioxolane-4-methanol	**(5634-39-9)**
Iodofenophos	**(18181-70-9)**
Iodofenphos	**(18181-70-9)**
4-Iodofluorobenzene	**(352-34-1)**
p-Iodofluorobenzene	**(352-34-1)**
Iodoform (ACGIH,OSHA)	**(75-47-8)**
Iodogorgonic acid	**(66-02-4)**
Iodoheptafluoropropane	**(27636-85-7)**
1-Iodoheptane	**(4282-40-0)**
1-Iodohexadecane	**(544-77-4)**
1-Iodohexane	**(638-45-9)**
Iodohydromol	**(552-22-7)**
4-(3-Iodo-4-hydroxyphenoxy)-3,5-diiodophenylalanine	**(6893-02-3)**
Iodometano (Italian)	**(74-88-4)**
Iodomethane	**(74-88-4)**
Iodomethanesulfonic acid sodium salt	**(126-31-8)**
(Iodomethyl)benzene	**(620-05-3)**
1-Iodo-2-methylbenzene	**(615-37-2)**
1-Iodo-3-methylbenzene	**(625-95-6)**
1-Iodo-4-methylbenzene	**(624-31-7)**
1-Iodo-2-methylpropane	**(513-38-2)**
1-Iodonaphthalene	**(90-14-2)**
α-Iodonaphthalene	**(90-14-2)**
1-Iodo-3-nitrobenzene	**(645-00-1)**
1-Iodo-4-nitrobenzene	**(636-98-6)**
Iodonium, (2-carboxyphenyl)phenyl-, hydroxide, inner salt (9CI)	**(1488-42-2)**
Iodonium, diphenyl-, hexafluoroarsenate(1-) (9CI)	**(62613-15-4)**
Iodonium, diphenyl-, hexafluorophosphate(1-) (9CI)	**(58109-40-3)**
1-Iodononane	**(4282-42-2)**

1-Iodooctadecane	**(629-93-6)**
1-Iodooctane	**(629-27-6)**
2-Iodooctane	**(557-36-8)**
Iodopanic acid	**(96-83-3)**
Iodopanoic acid	**(96-83-3)**
1-Iodopentane	**(628-17-1)**
2-Iodopentane	**(637-97-8)**
2-Iodophenol	**(533-58-4)**
3-Iodophenol	**(626-02-8)**
4-Iodophenol	**(540-38-5)**
m-Iodophenol	**(626-02-8)**
o-Iodophenol	**(533-58-4)**
p-Iodophenol	**(540-38-5)**
2-Iodo-N-phenylbenzamide	**(15310-01-7)**
Iodophenylmethane	**(620-05-3)**
Iodophos	**(18181-70-9)**
1-Iodopropane	**(107-08-4)**
2-Iodopropane	**(75-30-9)**
Iodopropane	**(26914-02-3)**
Iodopropanes [UN 2392]	**(26914-02-3)**
3-Iodopropene	**(556-56-9)**
3-Iodopropionic acid	**(141-76-4)**
3-Iodopropylene	**(556-56-9)**
Iodopropylidene glycerol	**(5634-39-9)**
3-Iodo-2-propynyl butylcarbamate	**(55406-53-6)**
3-Iodo-2-propynyl 2,4,5-trichlorophenyl ether	**(777-11-7)**
Iodoquinol	**(83-73-8)**
Iodosol	**(552-22-7)**
1-Iodotetradecane	**(19218-94-1)**
2-Iodothiophene	**(3437-95-4)**
Iodothymol (VAN)	**(552-22-7)**
α-Iodotoluene	**(620-05-3)**
m-Iodotoluene	**(625-95-6)**
Iodotributylstannane	**(7342-47-4)**
1-Iodoundecane	**(4282-44-4)**
5-Iodouracil	**(696-07-1)**
Iodure de dithiazanine (French)	**(514-73-8)**
Iodure de mercure (French)	**(15385-57-6)**
Iodure de methyle (French)	**(74-88-4)**
Ioduril	**(7681-82-5)**
Ioduro de ditiazanina (Spanish)	**(514-73-8)**
Iofendylate	**(99-79-6)**
Iomesan	**(50-65-7)**
Iomezan	**(50-65-7)**
Iomide	**(5700-49-2)**
Ion, Oxygen	**(11062-77-4)**
Ionet MO-400	**(9004-96-0)**
Ionet S 60	**(1338-41-6)**
Ionet S-80	**(1338-43-8)**
Ionet S 85	**(26266-58-0)**
Ionol	**(128-37-0)**
Ionol 1	**(128-37-0)**
Ionol 6	**(1817-68-1)**
Ionol (Antioxidant)	**(128-37-0)**
Ionol CP	**(128-37-0)**
Ionole	**(128-37-0)**
(e)-β-Ionone	**(79-77-6)**
α-Ionone	**(127-41-3)**
β-Ionone	**(14901-07-6)**
β-Ionone	**(79-77-6)**
trans-β-Ionone	**(79-77-6)**
Ionone, allyl α-	**(79-78-7)**
α-Ionone, isomethyl-	**(127-51-5)**
β-Ionone, methyl-	**(127-43-5)**
Ionone, methyl- (9CI)	**(1335-46-2)**
Ionox 100	**(88-26-6)**
Ionox 220	**(118-82-1)**
Ionox 100 Antioxidant	**(88-26-6)**
Ionox 220 Antioxidant	**(118-82-1)**
Ionox WSP	**(77-62-3)**
Iopanoic acid	**(96-83-3)**
Iopezite	**(7778-50-9)**
Iophendylate	**(99-79-6)**
Iopodic acid	**(5587-89-3)**
Ioquin suspension	**(83-73-8)**
Iosol	**(552-22-7)**
Iothymol	**(552-22-7)**
Iotox	**(1689-83-4)**
Ioxynil	**(1689-83-4)**
Ioxynil octanoate	**(3861-47-0)**
Ipabutona	**(129-20-4)**
Ipaner	**(94-75-7)**
Ipazine	**(1912-25-0)**
Ipe-Tobacco Wood	**(84-79-7)**
Iphosphamid	**(3778-73-2)**
Iphosphamide	**(3778-73-2)**
Ipitox	**(52645-53-1)**
Ipnofil	**(72-44-6)**
Ipodate	**(5587-89-3)**
Ipoglicone	**(64-77-7)**
Ipognox 89	**(123-28-4)**
Ipolina	**(304-20-1)**
4-Ipomeanol	**(32954-58-8)**
Ipomeanol	**(32954-58-8)**
Ipotensivo	**(64-55-1)**
Ipral sodium	**(57-33-0)**
Iprindole	**(5560-72-5)**
Iprobenfos	**(26087-47-8)**
Iprodione	**(36734-19-7)**
Ipropethidine	**(561-76-2)**
Iproveratril	**(52-53-9)**
Ipsilon	**(60-32-2)**
Ipsoflame	**(50-33-9)**
Ipsotian	**(57-53-4)**
Iro-Jex	**(9004-66-4)**
Iradicav	**(7681-49-4)**
Iragen Red L-U	**(81-88-9)**
β-Iraldeine	**(127-43-5)**
Iramil	**(50-49-7)**
Irax	**(9002-88-4)**
Iredale Yellow RD	**(1325-37-7)**
Irenal	**(7778-74-7)**
Irenat	**(7778-74-7)**
Iretin	**(147-94-4)**
Iretin	**(69-74-9)**
Irgachrome Orange OS	**(18454-12-1)**
Irgalite 1104	**(7440-44-0)**
Irgalite Blue BGL	**(147-14-8)**
Irgalite Blue BLP	**(147-14-8)**
Irgalite Blue CPV 2	**(147-14-8)**
Irgalite Blue CPV 3	**(147-14-8)**
Irgalite Blue GLSM	**(147-14-8)**
Irgalite Blue LGLD	**(147-14-8)**
Irgalite Brilliant Blue MRS	**(1325-87-7)**
Irgalite Bronze Red CL	**(17372-87-1)**
Irgalite Bronze Red CL	**(548-26-5)**
Irgalite Fast Brilliant Blue BL	**(147-14-8)**
Irgalite Fast Brilliant Green 3GL	**(1328-53-6)**
Irgalite Fast Brilliant Green GL	**(1328-53-6)**
Irgalite Fast Red 2GL	**(3468-63-1)**
Irgalite Fast Red P4R	**(2425-85-6)**
Irgalite Fast Scarlet RND	**(2425-85-6)**
Irgalite Fast Yellow PG	**(2512-29-0)**
Irgalite Fast Yellow PG Transparent	**(2512-29-0)**

Irgalite Green GLN (1328-53-6)
Irgalite Orange F2G (15793-73-4)
Irgalite Orange P (3520-72-7)
Irgalite Orange PG (3520-72-7)
Irgalite Orange PX (3520-72-7)
Irgalite Red 4B (5858-81-1)
Irgalite Red BRL (1103-38-4)
Irgalite Red CBN (5160-02-1)
Irgalite Red CBR (5160-02-1)
Irgalite Red CBT (5160-02-1)
Irgalite Red 2G (3468-63-1)
Irgalite Red 2GW (3468-63-1)
Irgalite Red MBC (5160-02-1)
Irgalite Red PRR (2814-77-9)
Irgalite Red PV2 (2425-85-6)
Irgalite Red PV8 (3468-63-1)
Irgalite Red RL (1248-18-6)
Irgalite Red RNPX (2425-85-6)
Irgalite Rubine PB (5858-81-1)
Irgalite Scarlet RB (2425-85-6)
Irgalite Victoria Blue TRCN (1325-87-7)
Irgalite Violet TCR (1325-82-2)
Irgalite Yellow BO (6358-85-6)
Irgalite Yellow BST (6358-85-6)
Irgalite Yellow BTR (6358-85-6)
Irgalite Yellow G (2512-29-0)
Irgalite Yellow GNS (2512-29-0)
Irgalite Yellow GTN (2512-29-0)
Irgalite Yellow PV 2 (2512-29-0)
Irgalon (64-02-8)
Irganaphthol RK (135-62-6)
Irganaphthol RM (135-65-9)
Irganox 1010 (6683-19-8)
Irganox 1076 (2082-79-3)
Irganox 1906 (2082-79-3)
Irganox PS 800 (123-28-4)
Irgaplast Blue RBP (147-14-8)
Irgaplast CH 300 (1254-78-0)
Irgaplast Orange G (3520-72-7)
Irgasan (3380-34-5)
Irgasan DP300 (3380-34-5)
Irgastab T 4 (78-04-6)
Irgastab T 150 (78-04-6)
Irgastab T 290 (78-04-6)
Irgazin Blue 3GT (574-93-6)
Iridil (129-20-4)
Iridium(IV) chloride (10025-97-5)
Iridium tetrachloride (10025-97-5)
Iridocin (536-33-4)
Iridozin (536-33-4)
Irifan (52-31-3)
Irish Gum (9000-07-1)
Irish Moss Extract (9000-07-1)
Irish Moss Gelose (9000-07-1)
Irisol Base (81-48-1)
Irisone acetate (50-04-4)
Irium (151-21-3)
Irmin (50-49-7)
Iron-Dextran Complex (9004-66-4)
Iron(III) arsenate (1:1) (10102-49-5)
Iron(III) o-arsenite pentahydrate (63989-69-5)
Iron(III) cacodylate (5968-84-3)
Iron(III) chloride (7705-08-0)
Iron(III) dimethyldithiocarbamate (14484-64-1)
Iron(III) ferrocyanide (14038-43-8)
Iron(III) oxide (1309-37-1)
Iron(III) oxide (1317-61-9)
Iron(III) sulfate (10028-22-5)
Iron(II) arsenate (3:2) (10102-50-8)
Iron(II) chloride (1:2) (7758-94-3)
Iron(II) oxalate (516-03-0)
Iron(II) oxide (1345-25-1)
Iron(II) sulfate (7720-78-7)
Iron(II) sulfate (1:1) (7720-78-7)
Iron(II) sulfate (1:1), heptahydrate (7782-63-0)
Iron(3+) NTA (16448-54-7)
Irocaine (51-05-8)
Iron(3+) chloride, hexahydrate (10025-77-1)
Iron(2+) chloride, tetrahydrate (13478-10-9)
Iron-dextrin Complex (9004-51-7)
Iron(3+) ferrocyanide (14038-43-8)
Iron-fluoride (7783-50-8)
Iroini (561-27-3)
Iron(2+) lactate (5905-52-2)
Iromin (299-29-6)
Iron (7439-89-6)
Iron Blue (14038-43-8)
Iron Carbohydrate Complex (9004-51-7)
Iron Dextran Injection (9004-66-4)
Iron Dextrin Injection (9004-51-7)
Iron, Furnace (65996-67-0)
Iron II carbonate (563-71-3)
Iron, Isotope of mass 55 (8CI,9CI) (14681-59-5)
Iron, Isotope of mass 59 (8CI,9CI) (14596-12-4)
Iron Mass, Spent (DOT) (1332-37-2)
Iron (2+) NTA (68391-67-3)
Iron Ore (1317-60-8)
Iron Oxide Pigments (1309-37-1)
Iron Oxide Red (1309-37-1)
Iron Oxide Red 130B (1332-37-2)
Iron Red (1309-37-1)
Iron Sesquioxide (1309-37-1)
Iron Sponge, Spent [UN 1376] (1332-37-2)
Iron Sugar (8047-67-4)
Iron Vitriol (7720-78-7)
Iron alloy, Base, Fe,P (Ferrophosphorus) (8049-19-2)
Iron alloy, Fe,Ni (9CI) (11148-32-6)
Iron alloy, nonbase, Ni,Fe,Mo (9CI) (37304-37-3)
Ironate (7782-63-0)
Iron, (N-(2-(bis(carboxymethyl)amino)ethyl)-N-(2-hydroxyethyl)glycinato(3-)) (9CI) (17084-02-5)
Iron, (N,N-bis(carboxymethyl)glycinato(3-)-N,O,O',O'')-, (T-4)- (9CI) (16448-54-7)
Iron bis(cyclopentadiene) (102-54-5)
Iron, bis(dimethylcarbamodithioato-S,S')- (9CI) (15339-38-5)
Iron, bis(dimethyldithiocarbamato)- (8CI) (VAN) (15339-38-5)
Iron bromide (FeBr) (6CI,7CI,8CI,9CI) (12514-32-8)
Iron cacodylate (5968-84-3)
Iron (2+) carbonate (563-71-3)
Iron carbonate (563-71-3)
Iron carbonyl (DOT) (13463-40-6)
Iron chloride (7705-08-0)
Iron chloride, Solid (DOT) (7705-08-0)
Iron (III), chloride, hexahydrate (10025-77-1)
Iron (II) chloride, tetrahydrate (13478-10-9)
Iron chloride tetrahydrate (13478-10-9)
Iron chromite (1308-31-2)
Iron cyanide (14038-43-8)
Iron dextran (9004-66-4)
Iron dichloride (7758-94-3)
Iron dichloride tetrahydrate (13478-10-9)
Iron (III) dichromate (10294-53-8)
Iron dicyclopentadienyl (102-54-5)
Iron dimethylarsonate (5968-84-3)

Iron dimethyldithiocarbamate **(14484-64-1)**
α-Irone **(79-69-6)**
Iron 2-ethylhexanoate **(19583-54-1)**
Iron ferrocyanide **(14038-43-8)**
Iron gluconate **(299-29-6)**
Iron hexacyanoferrate **(14038-43-8)**
Iron hydrogenated dextran **(9004-66-4)**
Iron hydroxide (9CI) **(18624-44-7)**
Iron, ion (Fe(3+)) (8CI,9CI) **(20074-52-6)**
Iron monooxide **(1345-25-1)**
Iron monosulfate **(7720-78-7)**
Iron monosulfide **(1317-37-9)**
Iron monoxide **(1345-25-1)**
Iron nickel sulfide (Fe9Ni9S16) (9CI) **(53809-87-3)**
Iron nitrate **(10421-48-4)**
Iron (III) nitrate, Anhydrous **(10421-48-4)**
Iron nitrilotriacetate **(16448-54-7)**
Iron-nitrilotriacetate chelate **(16448-54-7)**
Iron, (nitrilotriacetato) **(16448-54-7)**
Iron(3+) octadecanoate **(555-36-2)**
Ironorm Injection **(9004-66-4)**
Iron oxalate **(516-03-0)**
Iron(2+) oxalate **(516-03-0)**
Iron-oxide **(1332-37-2)**
Iron oxide (ACGIH,OSHA) **(1309-37-1)**
Iron oxide (9CI) **(1317-61-9)**
Iron oxide (9CI) **(1345-25-1)**
Iron oxide, Spent [UN 1376] **(1332-37-2)**
Iron oxide, saccharated **(8047-67-4)**
Iron pentacarbonyl (ACGIH,OSHA) [UN 1994] **(13463-40-6)**
Iron persulfate **(10028-22-5)**
Iron phosphate **(10045-86-0)**
Iron phosphide (9CI) **(1310-43-6)**
Iron protochloride **(7758-94-3)**
Iron protosulfate **(7720-78-7)**
Iron protosulfide **(1317-37-9)**
Iron protoxalate **(516-03-0)**
Iron pyrophosphate **(10058-44-3)**
Iron saccharate **(8047-67-4)**
Iron selenide (9CI) **(1310-32-3)**
Iron sesquichloride, Solid (DOT) **(7705-08-0)**
Iron sesquisulfate **(10028-22-5)**
Iron sodium pyrophosphate **(10045-87-1)**
Iron sorbitex **(1338-16-5)**
Iron-sorbitol-citrate **(1338-16-5)**
Iron-sorbitol-citric acid **(1338-16-5)**
Iron sulfate (1:1) **(7720-78-7)**
Iron sulfate (2:3) **(10028-22-5)**
Iron(2+) sulfate **(7720-78-7)**
Iron(2+) sulfate (1:1) **(7720-78-7)**
Iron(3+) sulfate **(10028-22-5)**
Iron sulfide (9CI) **(1317-37-9)**
Iron sulfuret **(1317-37-9)**
Iron tersulfate **(10028-22-5)**
Iron trichloride **(7705-08-0)**
Iron trichloride hexahydrate **(10025-77-1)**
Iron trifluoride **(7783-50-8)**
Iron trinitrate **(10421-48-4)**
Iron, tris(dimethylcarbamodithioato-S,S')-, (OC-6-11)- **(14484-64-1)**
Iron tris(dimethyldithiocarbamate) **(14484-64-1)**
Iron, tris(dimethyldithiocarbamato) **(14484-64-1)**
Iron, tris(dimethyldithiocarbamato)- **(14484-64-1)**
Iron vitrol **(7782-63-0)**
Iroquine **(54-05-7)**
Irospan **(7720-78-7)**
Irosul **(7720-78-7)**
Irosul **(7782-63-0)**
Irox (Gador) **(299-29-6)**
Irradiated ergosta-5,7,22-trien-3-β-ol **(50-14-6)**
Irrathene R **(9002-88-4)**
Irtran 2 **(1314-98-3)**
Irtran 3 **(7789-75-5)**
Isadrine **(51-30-9)**
Isadrine-hydrochloride **(51-30-9)**
Isalizina **(51-12-7)**
Isamin **(60-13-9)**
Isamin **(91-84-9)**
Isano oil **(8001-86-3)**
Isariotoxin **(21259-20-1)**
Isarol **(526-08-9)**
Isatex **(18869-73-3)**
Isatic acid lactam **(91-56-5)**
Isatidine **(15503-86-3)**
Isatin **(91-56-5)**
Isatinic acid anhydride **(91-56-5)**
Isatoic acid anhydride **(118-48-9)**
Isatoic anhydride **(118-48-9)**
Isazofos **(42509-80-8)**
Isazophos **(42509-80-8)**
Isceon 22 **(75-45-6)**
Isceon 113 **(76-13-1)**
Isceon 122 **(75-71-8)**
Isceon 131 **(75-69-4)**
Iscobrome **(74-83-9)**
Iscobrome D **(106-93-4)**
Iscothan **(39300-45-3)**
Iscothane **(39300-45-3)**
Iscotin **(54-85-3)**
Iscovesco **(56-53-1)**
Isethion **(70-18-8)**
Isethionic acid sodium salt **(1562-00-1)**
Isicaina **(137-58-6)**
Isicaine **(137-58-6)**
Isicetin **(56-75-7)**
Isidrina **(54-85-3)**
Iskia-C **(134-03-2)**
Islanditoxin **(12663-46-6)**
Ismafast Blue 4GL **(2503-73-3)**
Ismazide **(54-85-3)**
Ismelin **(55-65-2)**
Ismicetina **(56-75-7)**
Ismipur **(50-44-2)**
Isogamma Acid **(87-02-5)**
Iso.PPC. **(122-42-9)**
Isoacetophorone **(78-59-1)**
Isoaldehyde C-10 **(1321-89-7)**
Isoalloxazine, 7,8-dimethyl-10-d-ribityl- **(83-88-5)**
Isoalloxazine, 7,8-dimethyl-10-(d-ribo-2,3,4,5-tetrahydroxypentyl)- **(83-88-5)**
Isoamidone I **(467-85-6)**
Isoamidone II **(466-40-0)**
Isoamin **(60-13-9)**
Isoamycin **(300-62-9)**
Isoamyl acetate (ACGIH,OSHA) **(123-92-2)**
Isoamyl alcohol (ACGIH) **(584-02-1)**
Isoamyl alcohol (ACGIH) [UN 1105] **(123-51-3)**
Isoamyl alcohol, primary (OSHA) **(123-51-3)**
Isoamyl alcohol, secondary (OSHA) **(6032-29-7)**
Isoamyl aldehyde **(590-86-3)**
Isoamyl alkohol (Czech) **(123-51-3)**
Iso-amylalkohol (German) **(123-51-3)**
Isoamylamine **(107-85-7)**
γ-Isoamylamine **(107-85-7)**
Isoamylamine hydrochloride **(541-23-1)**
Isoamyl aminoformate **(543-86-2)**

Isoamyl benzyl ether **(122-73-6)**
Isoamyl bromide **(107-82-4)**
Isoamyl butanoate **(106-27-4)**
Isoamyl butylate **(106-27-4)**
Isoamyl-n-butyrate **(106-27-4)**
Isoamyl butyrate (DOT) **(106-27-4)**
Isoamyl caproate **(2198-61-0)**
Isoamyl caprylate **(2035-99-6)**
Isoamyl carbamate **(543-86-2)**
Isoamyl chloride **(107-84-6)**
Isoamyl cyanide **(542-54-1)**
Isoamylene **(26760-64-5)**
β-Iso-amylene **(513-35-9)**
Isoamylene alcohol **(625-69-4)**
β-Isoamylene oxide **(5076-19-7)**
8-Isoamylenoxypsoralen **(482-44-0)**
Isoamylester kyseliny octove (Czech) **(123-92-2)**
Isoamylester kyseliny salicylove (Czech) **(87-20-7)**
Isoamyl ethanoate **(123-92-2)**
Isoamyl ether **(544-01-4)**
5-Isoamyl-5-ethylbarbituric acid **(57-43-2)**
Isoamylethylbarbituric acid **(57-43-2)**
Isoamyl formate [UN 1109] **(110-45-2)**
Isoamyl hexanoate **(2198-61-0)**
Isoamylhydride **(78-78-4)**
Isoamyl o-hydroxybenzoate **(87-20-7)**
Isoamyl isovalerate **(659-70-1)**
Isoamylkyanid (Czech) **(542-54-1)**
Isoamyl methanoate **(110-45-2)**
Isoamyl methyl ketone **(110-12-3)**
Isoamyl nitrite **(110-46-3)**
Isoamyl octanoate **(2035-99-6)**
Isoamylol **(123-51-3)**
Isoamyl oxide **(544-01-4)**
Isoamyl propionate **(105-68-0)**
Isoamyl salicylate **(87-20-7)**
Isoamyne **(60-13-9)**
Isoanethole **(140-67-0)**
trans-Isoasarone **(2883-98-9)**
D-Isoascorbic acid **(89-65-6)**
Isoascorbic acid **(89-65-6)**
Isobac **(70-30-4)**
Isobac 20 **(5736-15-2)**
Isobac 20 **(70-30-4)**
Isobamate **(78-44-4)**
Isobarb **(57-33-0)**
Isobenzan **(297-78-9)**
5-Isobenzofurancarboxylic acid, 1,3-dihydro-3-oxo-, nonyl ester (9CI) **(54699-44-4)**
Isobenzofuran, 1,3-dihydro-1,3-dioxo- **(85-44-9)**
1,3-Isobenzofurandione **(85-44-9)**
1,3-Isobenzofurandione, Polymer with 2,5-furandione and 1,2-propanediol (9CI) **(25037-66-5)**
1,3-Isobenzofurandione, 5,5'-carbonylbis- (9CI) **(2421-28-5)**
1,3-Isobenzofurandione, 4,7-dimethyl- (9CI) **(5463-50-3)**
1,3-Isobenzofurandione, hexahydro- (9CI) **(85-42-7)**
1,3-Isobenzofurandione, hexahydro-, Polymer with (chloromethyl)oxirane (9CI) **(31095-87-1)**
1,3-Isobenzofurandione, hexahydromethyl- (9CI) **(25550-51-0)**
1,3-Isobenzofurandione, hydroxy- (9CI) **(116211-86-0)**
1,3-Isobenzofurandione, hydroxymethyl- (9CI) **(116211-88-2)**
1,3-Isobenzofurandione, 5-methyl- (9CI) **(19438-61-0)**
1,3-Isobenzofurandione, methyl- (9CI) **(30140-42-2)**
1,3-Isobenzofurandione, 5,5'-((1-methylethylidene)bis-(4,1-phenyleneoxy))bis- (9CI) **(38103-06-9)**
1,3-Isobenzofurandione, 5-nitro- (9CI) **(5466-84-2)**
1,3-Isobenzofurandione, 5,5'-sulfonylbis- (9CI) **(2540-99-0)**
1,3-Isobenzofurandione, 4,5,6,7-tetrabromo- (9CI) **(632-79-1)**
1,3-Isobenzofurandione, 4,5,6,7-tetrachloro- (9CI) **(117-08-8)**
1,3-Isobenzofurandione, 3a,4,7,7a-tetrahydro- **(85-43-8)**
1,3-Isobenzofurandione, 3a,4,7,7a-tetrahydro-4-methyl- (9CI) **(5333-84-6)**
1,3-Isobenzofurandione, tetrahydromethyl- (9CI) **(11070-44-3)**
4-Isobenzofuranol, octahydro-3a,7a-dimethyl-, (3a.α.,4.β.,7a.α.)-(.+-.)- (9CI) **(54382-58-0)**
1(3H)-Isobenzofuranone (9CI) **(87-41-2)**
1(3H)-Isobenzofuranone, 3,3-bis(4-hydroxyphenyl)- **(77-09-8)**
1(3H)-Isobenzofuranone, 6-(dimethylamino)-3,3-bis(4-(dimethylamino)phenyl)- (9CI) **(1552-42-7)**
1(3H)-Isobenzofuranone, 3-ethoxy- (9CI) **(16824-02-5)**
1(3H)-Isobenzofuranone, 3-methyl- (9CI) **(3453-64-3)**
1(3H)-Isobenzofuranone, 5-methyl- (9CI) **(54120-64-8)**
1(3H)-Isobenzofuranone, 3-propylidene **(17369-59-4)**
1(3H)-Isobenzofuranone, 3-(2-quinolinylmethylene)- (9CI) **(6365-50-0)**
1(3H)-Isobenzofuranone, 3a,4,5,7a-tetrahydro-4-hydroxy-3a,7a-dimethyl-, (3a.α.,4.β.,7a.α.)-(.+-.)- (9CI) **(54346-06-4)**
Isobicina **(54-85-3)**
Iso-bid **(87-33-2)**
Isobide **(652-67-5)**
Isobiuret (VAN) **(108-19-0)**
DL-Isoborneol **(124-76-5)**
Isoborneol **(124-76-5)**
Isoborneol, DL **(124-76-5)**
Isoborneol, acetate (8CI) **(125-12-2)**
Isoborneol, propionate (8CI) **(2756-56-1)**
Isoborneol, thiocyanatoacetate **(115-31-1)**
Isobornyl acetate **(125-12-2)**
Isobornyl acrylate **(5888-33-5)**
Isobornyl alcohol **(124-76-5)**
Isobornylester kyseliny thiokyanatooctove (Czech) **(115-31-1)**
Isobornyl methacrylate **(7534-94-3)**
Isobornyl thiocyanatoacetate **(115-31-1)**
Isobornyl thiocyanoacetate **(115-31-1)**
Isobutaldehyde **(78-84-2)**
Isobutanal **(78-84-2)**
Isobutanal diethyl acetal **(1741-41-9)**
Isobutandiol-2-amine **(115-69-5)**
Isobutane [UN 1969] **(75-28-5)**
Isobutanethiol **(513-44-0)**
Isobutanol [UN 1212] **(78-83-1)**
Isobutanol-2-amine **(124-68-5)**
Isobutanolamine **(124-68-5)**
Isobutazina **(129-20-4)**
Isobutenal **(78-85-3)**
Isobutene **(115-11-7)**
Isobutene Homopolymer **(9003-27-4)**
Isobutene Polymer **(9003-27-4)**
Isobutene, octafluoro- **(382-21-8)**
Isobutene trimer **(7756-94-7)**
Isobutenylcarbinol **(763-32-6)**
Isobutenyl chloride **(563-47-3)**
Isobutenyl methyl ketone **(141-79-7)**
Isobutil **(129-20-4)**
Isobutinyl-N-(3-chlorphenyl)-carbamat (German) **(1967-16-4)**
2-Isobutoxyethanol **(4439-24-1)**
N-Isobutoxymethylacrylamide **(16669-59-3)**
1-Isobutoxy-2-propanol **(23436-19-3)**
Isobutyl 2,4-D **(1713-15-1)**
Isobutyl 2,4,5-T **(4938-72-1)**
Isobutyl acetate (ACGIH,OSHA) [UN 1213] **(110-19-0)**
Isobutyl acrylate, Inhibited (DOT) **(106-63-8)**
Isobutyl acrylate [UN 2527] **(106-63-8)**
Isobutyl adipate **(141-04-8)**
Isobutyl-alcohol **(78-83-1)**
Isobutyl alcohol (ACGIH,DOT,OSHA) **(78-83-1)**

Isobutylaldehyde **(78-84-2)**
Isobutyl aldehyde (DOT) **(78-84-2)**
Isobutylalkohol (Czech) **(78-83-1)**
Iso-butylallylbarbituric acid **(77-26-9)**
Isobutylallylbarturic acid **(77-26-9)**
Isobutylamide **(563-83-7)**
Isobutylamine **(78-81-9)**
Isobutylamine [UN 1214] **(78-81-9)**
Isobutylamine, hydrochloride (8CI) **(5041-09-8)**
Isobutylbenzene **(538-93-2)**
Isobutyl bromide **(78-77-3)**
Isobutyl butanoate **(539-90-2)**
Isobutyl butyrate **(539-90-2)**
Isobutyl-n-butyrate **(539-90-2)**
Isobutyl caproate **(105-79-3)**
Isobutyl carbamate **(543-28-2)**
1-Isobutylcarbamoyl-imidazolin-2-one **(30979-48-7)**
1-Isobutylcarbamoyl-imidazolin-2-one **(67292-92-6)**
Isobutylcarbinol **(123-51-3)**
Isobutyl cellosolve **(4439-24-1)**
Isobutyl chloride **(513-36-0)**
Isobutyl chlorocarbonate **(543-27-1)**
Isobutyl chloroformate **(543-27-1)**
3-Isobutyl-1-cyclohexene **(4104-56-7)**
3-Isobutylcyclohexene **(4104-56-7)**
1-Isobutyl-2,5-dimethylbenzene **(55669-88-0)**
Isobutyl N,N-dimethylolcarbamate **(52304-17-3)**
Isobutyl dimethylolcarbamate **(52304-17-3)**
Isobutyldiurea **(6104-30-9)**
Isobutylene Homopolymer **(9003-27-4)**
Isobutylene Polymer **(9003-27-4)**
Isobutylene Resin **(9003-27-4)**
Isobutylene [UN 1055] **(115-11-7)**
Isobutylene dichloride **(594-37-6)**
Isobutylenediurea **(6104-30-9)**
Isobutylene/isoprene copolymer **(9010-85-9)**
Isobutylene, isoprene polymer, brominated **(68441-14-5)**
Isobutyleneoxide **(558-30-5)**
Isobutylester kyseliny akrylove (Czech) **(106-63-8)**
Isobutylester kyseliny isomaselne (Czech) **(97-85-8)**
Isobutylester kyseliny methakrylove (Czech) **(97-86-9)**
Isobutylester kyseliny mravenci (Czech) **(542-55-2)**
Isobutylester kyseliny octove (Czech) **(110-19-0)**
Iso-butyl formate **(542-55-2)**
Isobutyl formate [UN 2393] **(542-55-2)**
Isobutylglycerol, nitro- **(126-11-4)**
Isobutyl heptyl ketone **(123-18-2)**
Isobutyl hexanoate **(105-79-3)**
4-Isobutylhydratropic acid **(15687-27-1)**
p-Isobutylhydratropic acid **(15687-27-1)**
Isobutyl o-hydroxybenzoate **(87-19-4)**
Isobutyl N-(hydroxymethyl) carbamate **(67953-32-6)**
2,2'-Isobutylidenebis(4,6-dimethylphenol) **(33145-10-7)**
1,1'-Isobutylidenebisurea **(6104-30-9)**
Isobutylidenediurea **(6104-30-9)**
Isobutyl iodide **(513-38-2)**
Isobutyl isobutyrate [UN 2528] **(97-85-8)**
Isobutyl isopentanoate **(589-59-3)**
Isobutyl isovalerate **(589-59-3)**
Isobutyljodid (Czech) **(513-38-2)**
Isobutyl ketone **(108-83-8)**
Isobutyl mercaptan **(513-44-0)**
Isobutyl α-methacrylate **(97-86-9)**
Isobutyl methacrylate, Inhibited (DOT) **(97-86-9)**
Isobutyl methacrylate [UN 2283] **(97-86-9)**
2-Isobutyl-3-methoxypyrazine **(24683-00-9)**
3-Isobutyl-2-methoxypyrazine **(24683-00-9)**
Isobutylmethylcarbinol **(108-11-2)**
Isobutyl 2-methyl-4-chlorophenoxyacetate **(1713-11-7)**
Isobutyl-methylketon (Czech) **(108-10-1)**
Isobutyl methyl ketone **(108-10-1)**
Isobutyl methyl ketone peroxide **(37206-20-5)**
Isobutyl methyl ketone peroxide, No more than 62% in solution **(37206-20-5)**
Isobutylmethylmethanol **(108-11-2)**
2-Isobutyl-3-methylpyrazine **(13925-06-9)**
.α.-Isobutylnaphthalene **(16727-91-6)**
1-Isobutylnaphthalene **(16727-91-6)**
Isobutyl nitrate **(543-29-3)**
Isobutyl nitrite **(542-56-3)**
Isobutyl oleate, sulfate, sodium salt **(67859-39-6)**
Isobutyl palmitate **(110-34-9)**
Isobutyl phenylacetate **(102-13-6)**
Isobutyl phenylethanoate **(102-13-6)**
2-(4-Isobutylphenyl)propanoic acid **(15687-27-1)**
2-(p-Isobutylphenyl)propionic acid **(15687-27-1)**
α-(4-Isobutylphenyl)propionic acid **(15687-27-1)**
α-p-Isobutylphenylpropionic acid **(15687-27-1)**
Isobutyl phosphate **(126-71-6)**
Isobutylphosphinic acid **(54423-73-3)**
Isobutylphosphonous acid **(54423-73-3)**
Isobutyl 2-propenoate **(106-63-8)**
Isobutyl propenoate **(106-63-8)**
Isobutyl propionate [UN 2394] **(540-42-1)**
Isobutyl salicylate **(87-19-4)**
Isobutyl stearate **(646-13-9)**
Isobutyl α-toluate **(102-13-6)**
Isobutyltrichlorosilane **(18169-57-8)**
Isobutyltrimethoxysilane **(18395-30-7)**
Isobutyltrimethylmethane **(540-84-1)**
Isobutyl vinyl ether **(109-53-5)**
Isobutyral **(78-84-2)**
Isobutyraldehyd (Czech) **(78-84-2)**
Isobutyraldehyde **(78-84-2)**
Isobutyraldehyde (DOT) **(78-84-2)**
Isobutyraldehyde, O-(methylcarbamoyl)oxime (8CI) **(10520-38-4)**
Isobutyraldehyde dimethyl acetal (6CI) **(41632-89-7)**
Isobutyraldehyde, oxime **(151-00-8)**
Isobutyramide (8CI) **(563-83-7)**
Isobutyric-acid **(79-31-2)**
Isobutyric acid [UN 2529] **(79-31-2)**
Isobutyric acid, α-(p-chlorophenoxy)-, ethyl ester **(637-07-0)**
Isobutyric acid, ethyl ester **(97-62-1)**
Isobutyric acid, hexyl ester **(2349-07-7)**
Isobutyric acid, 3-hydroxy-2,2,4-trimethylpentyl ester benzoate (8CI) **(22527-63-5)**
Isobutyric acid, isobutyl ester **(97-85-8)**
Isobutyric acid, 1-isopropyl-2,2-dimethyltrimethylene ester **(6846-50-0)**
Isobutyric acid, isopropyl ester **(617-50-5)**
Isobutyric acid, methyl ester **(547-63-7)**
Isobutyric acid, 2-phenoxyethyl ester (8CI) **(103-60-6)**
Isobutyric acid, propyl ester (8CI) **(644-49-5)**
Isobutyric aldehyde **(78-84-2)**
Isobutyric-anhydride **(97-72-3)**
Isobutyric anhydride [UN 2530] **(97-72-3)**
Isobutyrimidic acid **(563-83-7)**
Isobutyrone **(565-80-0)**
Isobutyronitrile [UN 2284] **(78-82-0)**
Isobutyrophenone (8CI) **(611-70-1)**
Isobutyroyl peroxide **(3437-84-1)**
Isobutyryl aldehyde **(78-84-2)**
N-Isobutyrylglycine **(58695-42-4)**
Isobutyryl peroxide **(3437-84-1)**
Isobutyryl peroxide, Maximum concentration 52% in solution **(3437-84-1)**
Isocaine-Asid **(51-05-8)**

Name	CAS No.
Isocaine-Heisler	**(51-05-8)**
Isocamphol	**(124-76-5)**
Isocaproaldehyde	**(1119-16-0)**
γ-Isocaprolactone	**(3123-97-5)**
Isocapronitrile	**(542-54-1)**
Isocarb	**(114-26-1)**
Isocarbamid	**(30979-48-7)**
Isocarbamid	**(67292-92-6)**
Isocarbamide	**(30979-48-7)**
Isocarbostyril (VAN) (8CI)	**(491-30-5)**
Isocarnox	**(7085-19-0)**
Isocarvomenthol, 1-hydroxy- (VAN)	**(33669-76-0)**
Isocetyl alcohol	**(36311-34-9)**
Isocetyl stearate	**(25339-09-7)**
Isochinolin (Czech)	**(119-65-3)**
Isochloorthion (Dutch)	**(2463-84-5)**
Isochlorthion	**(2463-84-5)**
Isochron	**(87-33-2)**
Isochrysene	**(217-59-4)**
Isocid	**(54-85-3)**
Isocidene	**(54-85-3)**
Isocil	**(314-42-1)**
Isocillin	**(132-98-9)**
Isocinchomeronic acid	**(100-26-5)**
Isocinchomeronic acid, dipropyl ester	**(136-45-8)**
Isocinchomeronyl dipropylester	**(136-45-8)**
Isocineole	**(470-67-7)**
Isocodeine, dihydro-	**(795-38-0)**
Isocotin	**(54-85-3)**
Isocoumarin	**(496-14-0)**
Isocoumarin, 3,4-dihydro-6,8-dihydroxy-3-(6-methyl-tetrahydro-2H-pyran-2-yl)-	**(35818-31-6)**
Isocrotonic acid	**(503-64-0)**
Isocrotonolactone	**(497-23-4)**
Isocrotyl chloride	**(513-37-1)**
Isocumene	**(103-65-1)**
Isocyanate (9CI)	**(71000-82-3)**
Isocyanate de methyle (French)	**(624-83-9)**
2-Isocyanato-1,3-bis(1-methylethyl)benzene	**(28178-42-9)**
Isocyanatocyclohexane	**(3173-53-3)**
Isocyanatoethane	**(109-90-0)**
2-Isocyanatoethyl methacrylate	**(30674-80-7)**
β-Isocyanatoethyl methacrylate	**(30674-80-7)**
Iso-cyanatomethane	**(624-83-9)**
1-Isocyanato-2-methylbenzene	**(614-68-6)**
3-Isocyanatomethyl-3,5,5-trimethylcyclohexylisocyanate	**(4098-71-9)**
1-Isocyanatopropane	**(110-78-1)**
Isocyanic acid	**(75-13-8)**
Isocyanic acid, butyl ester	**(111-36-4)**
Isocyanic acid, m-chlorophenyl ester	**(2909-38-8)**
Isocyanic acid, p-chlorophenyl ester	**(104-12-1)**
Isocyanic acid, cyclohexyl ester	**(3173-53-3)**
Isocyanic acid, 3,4-dichlorophenyl ester	**(102-36-3)**
Isocyanic acid, diester with 1,6-hexanediol	**(822-06-0)**
Isocyanic acid, 3,3'-dimethoxy-4,4'-biphenylene ester	**(91-93-0)**
Isocyanic acid, 3,3'-dimethyl-4,4'-biphenylene ester	**(91-97-4)**
Isocyanic acid, ethyl ester	**(109-90-0)**
Isocyanic acid, hexamethylene ester	**(822-06-0)**
Isocyanic acid meta-isopropenyl-α,α-dimethylbenzyl ester	**(2094-99-7)**
Isocyanic acid, p-isopropenyl-.α.,.α.-dimethylbenzyl ester (7CI,8CI)	**(2889-58-9)**
Isocyanic acid, methylenedi-4,1-cyclohexylene ester	**(5124-30-1)**
Isocyanic acid, methylenedicyclohexylene ester	**(28605-81-4)**
Isocyanic acid, methylenedi-p-phenylene ester	**(101-68-8)**
Isocyanic acid, methylene(3,5,5-trimethyl-3,1-cyclohexylene) ester	**(4098-71-9)**
Isocyanic acid, methyl ester	**(624-83-9)**
Isocyanic acid, 2-methyl-meta-phenylene ester	**(91-08-7)**
Isocyanic acid, 4-methyl-m-phenylene ester	**(584-84-9)**
Isocyanic acid, methyl-m-phenylene ester	**(26471-62-5)**
Isocyanic acid, methylphenylene ester	**(26471-62-5)**
Isocyanic acid, methylphenylene ester	**(584-84-9)**
Isocyanic acid, methylphenylene ester (8CI)	**(1321-38-6)**
Isocyanic acid, 1,5-naphthylene ester	**(3173-72-6)**
Isocyanic acid, octadecyl ester	**(112-96-9)**
Isocyanic acid, p-phenylenediisopropylidene ester	**(2778-41-8)**
Isocyanic acid, phenyl ester	**(103-71-9)**
Isocyanic acid, propyl ester	**(110-78-1)**
Isocyanic acid, α,α,α',α'-tetramethyl-p-xylylene ester (8CI)	**(2778-41-8)**
Isocyanic acid, triester with 1,3,5-tris(6-hydroxyhexyl)biuret	**(4035-89-6)**
Isocyanic acid, (m-trifluoromethylphenyl) ester	**(329-01-1)**
Isocyanide	**(57-12-5)**
Isocyanoethane	**(624-79-3)**
1-Isocyano-p-toluene	**(7175-47-5)**
2-Isocyano-2,4,4-trimethylpentane	**(14542-93-9)**
Isocyanuric acid	**(108-80-5)**
Isocyanuric acid, dichloro-	**(2782-57-2)**
Isocyanuric acid, dichloro-, potassium salt	**(2244-21-5)**
Isocyanuric acid, dichloro-, sodium salt	**(2893-78-9)**
Isocyanuric acid tris(2-hydroxyethyl) ester	**(839-90-7)**
Isocyanuric chloride	**(87-90-1)**
Isocyanuric dichloride	**(2782-57-2)**
Isodecanal (9CI)	**(1321-89-7)**
Isodecanamine, N,N-diisodecyl- (9CI)	**(35723-89-8)**
Isodecane (8CI,9CI)	**(34464-38-5)**
Isodecanediamine (9CI)	**(67874-35-5)**
Isodecanoic acid (9CI)	**(26403-17-8)**
Isodecanol	**(25339-17-7)**
Isodecanol, hydrogen phosphorodithioate, zinc salt (9CI)	**(25103-54-2)**
3-Isodecoxypropylamine, acetate	**(28701-67-9)**
Isodecyl ZDDP	**(25103-54-2)**
Isodecyl acrylate	**(1330-61-6)**
Isodecyl-alcohol	**(25339-17-7)**
Isodecyl alcohol, acrylate	**(1330-61-6)**
Isodecylaldehyde	**(1321-89-7)**
Isodecyldiamine	**(67874-35-5)**
Isodecyl diphenyl phosphate	**(29761-21-5)**
Isodecyl isooctyl hexanedioate	**(31474-57-4)**
Isodecyl methacrylate	**(29964-84-9)**
Isodecyl octadecanoate	**(31565-38-5)**
Isodecyl octyl phthalate	**(1330-96-7)**
3-(Isodecyloxy)-1-propanamine	**(30113-45-2)**
3-(Isodecyloxy)-1-propanamine acetate	**(28701-67-9)**
3-(Isodecyloxy)propylamine	**(30113-45-2)**
3-Isodecyloxypropylamine	**(30113-45-2)**
Isodecyloxypropylamine	**(30113-45-2)**
Isodecyl propenoate	**(1330-61-6)**
Isodecyl stearate	**(31565-38-5)**
Isodemeton-Sulfone	**(2496-91-5)**
Isodemeton	**(126-75-0)**
Isodextropimaric acid	**(471-74-9)**
Isodiazine	**(1912-25-0)**
Isodienestrol	**(84-17-3)**
Isodin	**(604-75-1)**
Isodine	**(25655-41-8)**
Isodiphenylbenzene	**(92-06-8)**
Isodiprene	**(13466-78-9)**
Isododecane (9CI)	**(31807-55-3)**
Isodrin	**(465-73-6)**
Isodur	**(6104-30-9)**
Isodurene	**(527-53-7)**
β-Isodurylic acid	**(480-63-7)**
Isoenanthic acid	**(628-46-6)**
Isoendoxan	**(3778-73-2)**
d-Isoephedrine	**(90-82-4)**

Isoestragole (104-46-1)
Isoethadione (115-67-3)
Isoeugenol (97-54-1)
cis-Isoeugenol (5912-86-7)
trans-Isoeugenol (5932-68-3)
1,3,4-Isoeugenol methyl ether (93-16-3)
Isoeugenyl methyl ether (93-16-3)
Isofedrol (134-72-5)
Isofedrol (299-42-3)
Isofenphos (25311-71-1)
Isoflav (553-30-0)
Isoflav Base (92-62-6)
Isofluorophate (55-91-4)
Isoflurane (26675-46-7)
Isoflurophate (55-91-4)
Isoforon (78-59-1)
Isoforone (Italian) (78-59-1)
Isofosfamide (3778-73-2)
Isoftalodinitril (Czech) -(626-17-5)
Isoftalonitril (Czech) (626-17-5)
Isoheptadecanol (9CI) (57289-07-3)
Isoheptane (591-76-4)
Isoheptane [UN 1206] (31394-54-4)
Isoheptanoic acid (8CI,9CI) (1330-19-4)
Isohexadecanoic acid (VAN)(9CI) (32844-67-0)
Isohexadecanol (9CI) (36311-34-9)
Isohexanal (1119-16-0)
Isohexane (107-83-5)
Isohexane [UN 1208] (107-83-5)
Isohexanoic acid (Mixed isomers) (646-07-1)
Iso-hexanol (626-89-1)
Isohexanol (626-89-1)
Isohexene (27236-46-0)
Isohexeno (Spanish) (27236-46-0)
Isohexenyl cyclohexenyl carboxaldehyde (37677-14-8)
Isohexyl alcohol (105-30-6)
Isohexyl-alcohol (626-89-1)
sek.Isohexylester kyseliny octove (Czech) (108-84-9)
Isohol (67-63-0)
Isohomogenol (93-16-3)
Isoindazole (271-44-3)
2H-Isoindole-2-carboxylic acid, 1,3-dihydro-1,3-dioxo-, ethyl ester (9CI) (22509-74-6)
1H-Isoindole-5-carboxylic acid, 2,3-dihydro-1,3-dioxo-2-phenyl- (9CI) (4649-27-8)
1,3-Isoindoledione (85-41-6)
Isoindole-1,3-dione (85-41-6)
1H-Isoindole-1,3(2H)-dione, 2-(2-bromoethyl)- (9CI) (574-98-1)
1H-Isoindole-1,3(2H)-dione, 2-(3-bromopropyl)- (9CI) (5460-29-7)
1H-Isoindole-1,3(2H)-dione, 2-butyl- (9CI) (1515-72-6)
1H-Isoindole-1,3(2H)-dione, 2-chloro- (9CI) (3481-09-2)
1H-Isoindole-1,3(2H)-dione, 2-(chloromethyl)- (9CI) (17564-64-6)
1H-Isoindole-1,3(2H)-dione, 2-(cyclohexylthio)- (17796-82-6)
1H-Isoindole-1,3(2H)-dione, 2-(2,6-dioxo-3-piperidinyl)- (9CI) (50-35-1)
1H-Isoindole-1,3(2H)-dione, 2,2'-(1,2-ethanediyl)bis(4,5,6,7-tetrabromo- (9CI) (32588-76-4)
1H-Isoindole-1,3(2H)-dione, 2-((1-ethyl-2-methylpropyl)thio)- (9CI) (88683-38-9)
1H-Isoindole-1,3(2H)-dione, 2-(hydroxymethyl)- (118-29-6)
1H-Isoindole-1,3(2H)-dione, 2-methyl- (9CI) (550-44-7)
1H-Isoindole-1,3(2H)-dione, 2,2'-(methylenedi-4,1-phenylene)bis-(4,5,6,7-tetrabromo- (9CI) (32588-74-2)
1H-Isoindole-1,3(2H)-dione, 5,5'-((1-methylethylidene)bis-(4,1-phenyleneoxy))bis(2-methyl- (9CI) (54395-52-7)
1H-Isoindole-1,3(2H)-dione, 2-methyl-5-nitro- (41663-84-7)
1H-Isoindole-1,3(2H)-dione, 5-nitro- (89-40-7)
1H-Isoindole-1,3(2H)-dione, 4-nitro- (9CI) (603-62-3)
1H-Isoindole-1,3(2H)-dione, 2-(oxiranylmethyl)- (9CI) (5455-98-1)
1H-Isoindole-1,3(2H)-dione, potassium salt (9CI) (1074-82-4)
1H-Isoindole-1,3(2H)-dione, 4,5,6,7-tetrachloro-2-(2-(4,5,6,7-tetrachloro-2,3-dihydro-1,3-dioxo-1H-inden-2-yl)-8-quinolinyl)- (9CI) (30125-47-4)
1H-Isoindole-1,3(2H)-dione, 3-α,4,7,7-α-tetrahydro- (9CI) (85-40-5)
1H-Isoindole-1,3(2H)-dione, 3a,4,7,7a-tetrahydro-2-propyl- (9CI) (2021-20-7)
1H-Isoindole-1,3(2H)-dione, 3a,4,7,7a-tetrahydro-2-((1,1,2,2-tetrachloroethyl)thio)-, cis- (2939-80-2)
1H-Isoindole-1,3(2H)-dione, 3a,4,7,7a-tetrahydro-2-((trichloromethyl)thio)- (133-06-2)
1H-Isoindole-1,3(2H)-dione, 2-((trichloromethyl)thio)- (133-07-3)
1H-Isoindole-1,3(2h)-dione, potassium salt (1074-82-4)
2-Isoindolinecarboxylic acid, 1,3-dioxo-, ethyl ester (22509-74-6)
5-Isoindolinecarboxylic acid, 1,3-dioxo-2-phenyl- (8CI) (4649-27-8)
Isoindoline, 1,3-diimino (3468-11-9)
1,3-Isoindolinedione (85-41-6)
1H-Isoindol-1-one, 2,3-dihydro- (9CI) (480-91-1)
1H-Isoindol-1-one, 2,3-dihydro-3-(2-hydroxy-3-((1-methylethyl)-amino)propoxy)-2-phenyl-, (R*,S*)-(+-)-, (E)-2-butene-dioate (1:1) (Salt) (9CI) (70096-14-9)
1H-Isoindol-1-one, 2,3-dihydro-3-thioxo- (9CI) (18138-18-6)
1H-Isoindol-1-one, 3,3'-((2-methyl-1,3-phenylene)diimino)bis-(4,5,6,7-tetrachloro- (9CI) (5045-40-9)
1H-Isoindol-1-one, 3,3'-(1,4-phenylenediimino)bis(4,5,6,7-tetrachloro- (9CI) (5590-18-1)
1H-Isoindol-1-one, 4,5,6,7-tetrachloro-3-((3-methyl-4-((4-((4,5,6,7-tetrachloro-1-oxo-1H-isoindol-3-yl)amino)-phenyl)azo)phenyl)amino)- (9CI) (40716-47-0)
Isoket (87-33-2)
Isokhusenic acid (16202-79-2)
Isol (107-41-5)
Isol Aryl Yellow G (2512-29-0)
Isol Aryl Yellow GX (2512-29-0)
Isol Aryl Yellow G Transp (2512-29-0)
Isol Benzidine Yellow G (6358-85-6)
Isol Benzidine Yellow G 2537 (6358-85-6)
Isol Benzidine Yellow G Special (6358-85-6)
Isol Benzidine Yellow Gapropyl (6358-85-6)
Isol Benzidine Yellow Gbpropyl (6358-85-6)
Isol Bona Rubine BK (5858-81-1)
Isol Bona Rubine BKS (5858-81-1)
Isol Bona Rubine KBK (5858-81-1)
Isol Fast Blue B (147-14-8)
Isol Fast Red 2G (3468-63-1)
Isol Fast Red HB (2425-85-6)
Isol Fast Red R (2814-77-9)
Isol Fast Red RG (2814-77-9)
Isol Fast Red RN2B (2425-85-6)
Isol Fast Red RNB (2425-85-6)
Isol Fast Red RN2G (2425-85-6)
Isol Fast Red RNG (2425-85-6)
Isol Fast Yellow G (2512-29-0)
Isol Fast Yellow G Transp (2512-29-0)
Isol Lake Red LCS 12527 (5160-02-1)
Isol Phthalo Blue E 7543 (147-14-8)
Isol Red 3BK (1103-38-4)
Isol Red LCR 2517 (5160-02-1)
Isol Red Toner GB (1103-38-4)
Isol Red Toner RB (1103-38-4)
Isol Red Toner RC (1103-39-5)
Isol Ruby BK (5858-81-1)
Isol Ruby BKS (5858-81-1)
Isol Tobias Red 3BK (1103-38-4)
Isol Tobias Red GB (1103-38-4)
Isol Tobias Red RB (1103-38-4)
Isol Tobias Red RC (1103-39-5)

Isol Toluidine Red HB	(2425-85-6)
Isol Toluidine Red RN2B	(2425-85-6)
Isol Toluidine Red RNB	(2425-85-6)
Isol Toluidine Red RN2G	(2425-85-6)
Isol Toluidine Red RNG	(2425-85-6)
Isolamid	(30979-48-7)
Isolan	(119-38-0)
Isolane (French)	(119-38-0)
D-Isoleucine	(319-78-8)
Isoleucine	(73-32-5)
L-Isoleucine	(73-32-5)
Isoleucine, D	(319-78-8)
Isoleucine, L	(73-32-5)
Isolyn	(54-85-3)
Isomalathion	(3344-12-5)
Isomelamine	(108-78-1)
Isomenthone	(491-07-6)
Isomeprobamate	(78-44-4)
B-Isomer	(319-85-7)
Isomeric chlorthion	(2463-84-5)
Isomerine	(25523-97-1)
Isomerization butane isomer hydrocarbon stream	(68477-99-6)
Isomerization naphtha	(64741-70-4)
Isometadona (Spanish)	(466-40-0)
Isometadone	(466-40-0)
Isometasystox	(919-86-8)
Isometasystox sulfone	(17040-19-6)
Isomethadone	(466-40-0)
Isomethadone I	(467-85-6)
Isomethadone II	(466-40-0)
Isomethadonum (Latin)	(466-40-0)
Iso-α-methyl ionone	(127-51-5)
Isomethyl-α-ionone	(127-51-5)
α-Isomethylionone	(127-51-5)
Isomethylsystox	(919-86-8)
Isomethylsystox sulfone	(17040-19-6)
Isomethylsystox sulfoxide	(301-12-2)
Isomin	(50-35-1)
Isomyl	(57-43-2)
Isomyn	(300-62-9)
Isomyn	(60-13-9)
Isomyst	(110-27-0)
Isomytal	(57-43-2)
Isona	(6381-77-7)
Isonal	(115-38-8)
Isonal	(77-02-1)
Isonal (Roussel)	(115-38-8)
1,3-αβ-Isonaphthofurandione	(5343-99-7)
Isonaphthoic acid	(93-09-4)
Isonaphthol	(135-19-3)
Isonate	(101-68-8)
Isonate 125M	(101-68-8)
Isonate 125 MF	(101-68-8)
Isonerit	(54-85-3)
Isonex	(54-85-3)
Isoniacid	(54-85-3)
Isoniazid	(54-85-3)
Isoniazide	(54-85-3)
Isonicazide	(54-85-3)
Isonicid	(54-85-3)
Isonico	(54-85-3)
Isonicotan	(54-85-3)
Isonicoteine	(581-50-0)
Isonicotil	(54-85-3)
Isonicotinaldehyde	(872-85-5)
Isonicotinaldehyde, oxime	(696-54-8)
Isonicotinamide	(1453-82-3)
Isonicotinamide, 2-ethylthio	(536-33-4)
Isonicotinhydrazid	(54-85-3)
Isonicotinic-acid	(55-22-1)
Isonicotinic acid amide	(1453-82-3)
Isonicotinic acid, 2-(2-(benzylcarbamoyl)ethyl)hydrazide	(51-12-7)
Isonicotinic acid, 2-(2-(benzylcarbamoyl)ethyl)hydrazide (8CI)	(51-12-7)
Isonicotinic acid, 2,6-dihydroxy	(99-11-6)
Isonicotinic acid, ethyl ester	(1570-45-2)
Isonicotinic acid hydrazide	(54-85-3)
Isonicotinic acid, methyl ester (8CI)	(2459-09-8)
Isonicotinic acid nitrile	(100-48-1)
Isonicotinic-acid-hydrazide	(54-85-3)
Isonicotinic aldehyde	(872-85-5)
Isonicotinonitrile (8CI)	(100-48-1)
N-Isonicotinoyl-N'(β-N-benzylcarboxamidoethyl)hydrazine	(51-12-7)
Isonicotinoyl hydrazide	(54-85-3)
Isonicotinoylhydrazine	(54-85-3)
Isonicotinsaeurehydrazid (German)	(54-85-3)
Isonicotinyl hydrazide	(54-85-3)
Isonide	(54-85-3)
Isonidrin	(54-85-3)
Isonikazid	(54-85-3)
Isonilex	(54-85-3)
Isonin	(54-85-3)
Isonindon	(54-85-3)
Isonipecaine	(57-42-1)
Isonipecaine hydrochloride	(50-13-5)
Isonipecotic acid, 1-(p-aminophenethyl)-4-phenyl-, ethyl ester	(144-14-9)
Isonipecotic acid, 1-(2-(benzyloxy)ethyl)-4-phenyl-, ethyl ester	(3691-78-9)
Isonipecotic acid, 1-(3-cyano-3,3-diphenylpropyl)-4-phenyl-	(28782-42-5)
Isonipecotic acid, 1-(3-cyano-3,3-diphenylpropyl)-4-phenyl-, ethyl ester	(915-30-0)
Isonipecotic acid, 1-(2-(2-hydroxyethoxy)ethyl)-4-phenyl-, ethyl ester (8CI)	(469-82-9)
Isonipecotic acid, 4-(m-hydroxyphenyl)-1-methyl-, ethyl ester	(468-56-4)
Isonipecotic acid, 1-methyl-4-phenyl-, ethyl ester	(57-42-1)
Isonipecotic acid, 1-methyl-4-phenyl-, ethyl ester, hydrochloride	(50-13-5)
Isonipecotic acid, 1-methyl-4-phenyl-, isopropyl ester (8CI)	(561-76-2)
Isonipecotic acid, 1-(2-morpholinoethyl)-4-phenyl-, ethyl ester (6CI, 8CI)	(469-81-8)
Isonipecotic acid, 4-phenyl-1-(2-((tetrahydrofurfuryl)oxy)ethyl)-, ethyl ester	(2385-81-1)
Isonirit	(54-85-3)
Isoniton	(54-85-3)
Isonitox	(2587-90-8)
Isonitropropane	(79-46-9)
Isonitrosoacetone	(306-44-5)
β-Isonitrosopropane	(127-06-0)
Isonizide	(54-85-3)
Isononane [UN 1920]	(34464-40-9)
Isononanoic acid (VAN)(9CI)	(26896-18-4)
Isononanoic acid, lead salt (9CI)	(27253-41-4)
Isononanol (9CI)	(27458-94-2)
Isononyl alcohol	(2430-22-0)
Isononyl alcohol (8CI)	(27458-94-2)
Isooctadecanoic acid (VAN)(9CI)	(30399-84-9)
Isooctanamine, N,N-diisooctyl- (9CI)	(25549-16-0)
Isooctanamine, N,N-diisooctyl- (Mixed isomers)	(25549-16-0)
Isooctane [UN 1261]	(26635-64-3)
Isooctane [UN 1262]	(540-84-1)
Isooctanoic acid (9CI)	(25103-52-0)
Isooctanol	(26952-21-6)
Isooctanol, dihydrogen phosphate	(26403-12-3)
Isooctene	(11071-47-9)
Isoocteno (Spanish)	(11071-47-9)
Isooctopherone	(78-59-1)
Isooctyl ZDDP	(28629-66-5)

Isooctyl acid phosphate **(12645-53-3)**
Isooctyl adipate **(1330-86-5)**
Isooctyl-alcohol **(26952-21-6)**
Isooctyl alcohol (ACGIH,OSHA) **(26952-21-6)**
Isooctyl alcohol, (2,4-dichlorophenoxy)acetate **(25168-26-7)**
Isooctyl alcohol, dihydrogen phosphate (8CI) **(26403-12-3)**
Isooctyl alcohol, maleate (2:1) **(1330-76-3)**
Isooctyl 2,4-dichlorophenoxyacetate **(25168-26-7)**
Isooctyl 4-(2,4-dichlorophenoxy)butyrate **(1320-15-6)**
Isooctyl 2-(2,4-dichlorophenoxy)propionate **(28631-35-8)**
Isooctyldinitrophenol **(37224-61-6)**
Isooctyldinitrophenol, isomer unspecified **(37224-61-6)**
Isooctyl isodecyl adipate **(31474-57-4)**
Isooctyl linoleate **(67874-38-8)**
Isooctyl mercaptoacetate **(25103-09-7)**
Isooctyl 2-(2-methyl-4-chlorophenoxy)propionate **(28473-03-2)**
Isooctyl 9-octadecenoate **(26761-50-2)**
Isooctylphenoxypolyethoxyethanol **(30776-59-1)**
Isooctyl phosphate **(12645-53-3)**
Isooctyl phthalate **(27554-26-3)**
Isooctyl picloram **(26952-20-5)**
Isooctyl 2-propenoate **(29590-42-9)**
Isooctyl 2-propenoate homopolymer **(9036-63-9)**
Isooctyl thioglycolate **(25103-09-7)**
Isooktylester kyseliny 2,4-dichlorfenoxyoctove (Czech) **(25168-26-7)**
Isopal **(142-91-6)**
Isoparaffinic hydrocarbons **(64771-72-8)**
Isoparathion **(597-88-6)**
Isopentaldehyde **(590-86-3)**
Isopentane [UN 1265] **(78-78-4)**
Isopentanoic acid (DOT) **(503-74-2)**
Isopentanol **(123-51-3)**
2-Isopentanoyl-1,3-indanedione **(83-28-3)**
Isopentene [UN 2371] **(26760-64-5)**
Isopentene [UN 2371] **(513-35-9)**
Isopentene [UN 2371] **(563-45-1)**
Isopentene [UN 2371] **(563-46-2)**
8-Isopentenyloxypsoralene **(482-44-0)**
Isopentyl acetate **(123-92-2)**
Isopentyl alcohol **(123-51-3)**
Isopentyl alcohol, acetate **(123-92-2)**
Isopentyl alcohol, formate **(110-45-2)**
Isopentyl alcohol, nitrite **(110-46-3)**
Isopentyl alcohol, propionate **(105-68-0)**
Isopentylamine (8CI) **(107-85-7)**
Isopentylamine, hydrochloride **(541-23-1)**
Isopentyl bromide **(107-82-4)**
Isopentyl butanoate **(106-27-4)**
Isopentyl butyrate **(106-27-4)**
Isopentyl chloride **(107-84-6)**
Isopentyl cyanide **(542-54-1)**
Isopentyl ether (8CI) **(544-01-4)**
Isopentyl formate **(110-45-2)**
Isopentyl hexanoate **(2198-61-0)**
Isopentyl-n-hexanoate **(2198-61-0)**
Isopentyl-2-hydroxyphenyl methanoate **(87-20-7)**
Isopentyl isovalerate **(659-70-1)**
Isopentyl methyl ketone **(110-12-3)**
Isopentyl nitrite **(110-46-3)**
Isopentyl octanoate **(2035-99-6)**
Isopentyl propionate **(105-68-0)**
Isopentyl salicylate **(87-20-7)**
Isopestox **(371-86-8)**
Isophan **(300-42-5)**
Isophen **(300-42-5)**
Isophen **(973-21-7)**
Isophen (Pesticide) **(973-21-7)**
Isophenergan **(60-87-7)**
Isophenicol **(56-75-7)**
Isophenphos **(25311-71-1)**
Isophorol, dihydro- **(116-02-9)**
Isophoron **(78-59-1)**
α-Isophoron **(78-59-1)**
α-Isophorone **(78-59-1)**
β-Isophorone **(471-01-2)**
Isophorone (ACGIH,OSHA) **(78-59-1)**
Isophorone (PEL:REL) **(78-59-1)**
Isophoronediamine [UN 2289] **(2855-13-2)**
Isophorone diamine diisocyanate **(4098-71-9)**
Isophoronediisocyanate (ACGIH,OSHA) [UN 2290] **(4098-71-9)**
Isophorone epoxide **(10276-21-8)**
Isophorone oxide **(10276-21-8)**
Isophosphamide **(3778-73-2)**
Isophrin **(59-42-7)**
Isophrine **(61-76-7)**
Isophrin hydrochloride **(61-76-7)**
Isophthalamide, 2,4,5,6-tetrachloro- **(1786-86-3)**
Isophthalic-acid **(121-91-5)**
Isophthalic acid, bis(2,3-epoxypropyl)ester **(7195-43-9)**
Isophthalic acid, bis(2-ethylhexyl) ester **(137-89-3)**
Isophthalic acid chloride **(99-63-8)**
Isophthalic acid, diallyl ester (8CI) **(1087-21-4)**
Isophthalic acid dichloride **(99-63-8)**
Isophthalic acid, diethyl ester **(636-53-3)**
Isophthalic acid, di-(2-ethylhexyl)ester **(137-89-3)**
Isophthalic acid, dihydrazide (8CI) **(2760-98-7)**
Isophthalic acid, dimethyl ester **(1459-93-4)**
Isophthalic acid, diphenyl ester (8CI) **(744-45-6)**
Isophthalic acid, 4-hydroxy **(636-46-4)**
Isophthalic acid, 4-hydroxy-5-methoxy- (8CI) **(2134-91-0)**
Isophthalic acid, 5-nitro- (8CI) **(618-88-2)**
Isophthalic acid, 5-nitro-, dimethyl ester (8CI) **(13290-96-5)**
Isophthalic acid-resorcinol copolymer SRU **(26637-46-7)**
Isophthalic acid-resorcinol polymer, SRU **(26637-46-7)**
Isophthalodinitrile **(626-17-5)**
Isophthalonitrile **(626-17-5)**
Isophthalonitrile, 2,4,5,6-tetrachloro- **(1897-45-6)**
Isophthalonitrile, tetrachloro **(1897-45-6)**
Isophthaloyl-chloride **(99-63-8)**
Isophthaloyl chloride-resorcinol polymer, SRU **(26637-46-7)**
Isophthaloyl dichloride **(99-63-8)**
Isophthalyl chloride **(99-63-8)**
Isophthalyl dichloride **(99-63-8)**
Isophytol **(505-32-8)**
Isophytol, dehydro- **(29171-23-1)**
Isopimaric acid **(5835-26-7)**
δ8(14)-Isopimaric acid **(471-74-9)**
Isopimaric acid (D) **(471-74-9)**
Isopredioxin **(42255-14-1)**
Isopredon **(53-34-9)**
Isopregnenone **(152-62-5)**
Isoprenaline chloride **(51-30-9)**
Isoprenaline hydrochloride **(51-30-9)**
Isoprene D **(9003-31-0)**
Isoprene (DOT) **(78-79-5)**
Isoprene, Inhibited [UN 1218] **(78-79-5)**
Isoprene, Polymers **(9003-31-0)**
Isoprene oligomer **(9003-31-0)**
Isoprene polymer **(9003-31-0)**
Isoprenylaluminum **(24683-32-7)**
Isoprocarb **(2631-40-5)**
Isoprocarbe **(2631-40-5)**
Isoprocil (French) **(314-42-1)**
Isopromedol **(64-39-1)**

Name	CAS
Isopromethazine	**(60-87-7)**
Isopropalin	**(33820-53-0)**
Isopropaline	**(33820-53-0)**
Isopropanethiol	**(75-33-2)**
Isopropanol [UN 1219]	**(67-63-0)**
Isopropanolamine	**(78-96-6)**
mono-Iso-propanolamine	**(78-96-6)**
Isopropanolamine 2,4-dichlorophenoxyacetate	**(6365-72-6)**
Isopropanolamine dodecylbenzene sulfonate	**(42504-46-1)**
Isopropene cyanide	**(126-98-7)**
Isopropenil-benzolo (Italian)	**(98-83-9)**
Isopropenyl acetate [UN 2403]	**(108-22-5)**
Isopropenyl acetylene	**(78-80-8)**
Isopropenylacetylene	**(78-80-8)**
Isopropenyl-benzeen (Dutch)	**(98-83-9)**
Isopropenylbenzene [UN 2303]	**(98-83-9)**
Isopropenyl-benzol (German)	**(98-83-9)**
Isopropenyl carbinol	**(513-42-8)**
4-Isopropenyl-1-cyclohexene-1-carboxaldehyde	**(2111-75-3)**
4-Isopropenyl-cyclohex-1-ene-1-methanol	**(536-59-4)**
3-Isopropenyl-α,α-dimethylbenzyl isocyanate	**(2094-99-7)**
4-Isopropenyl-.α.,.α.-dimethylbenzyl isocyanate	**(2889-58-9)**
p-Isopropenyl-.α.,.α.-dimethylbenzyl isocyanate	**(2889-58-9)**
p-Isopropenyldimethylbenzylisocyanate	**(2889-58-9)**
Isopropenylester kyseliny octove (Czech)	**(108-22-5)**
(+)-4-Isopropenyl-1-methylcyclohexene	**(5989-27-5)**
Isopropenyl methyl ketone	**(814-78-8)**
Isopropenyl methyl ketone-styrene copolymer	**(25191-48-4)**
Isopropenylnitrile	**(126-98-7)**
Isopropilamina (Italian)	**(75-31-0)**
2-Isopropilamino-4-metilamino-6-metiltio-1,3,5-triazina (Italian)	**(1014-69-3)**
Isopropilbenzene (Italian)	**(98-82-8)**
Isopropile (acetato di) (Italian)	**(108-21-4)**
Isopropil-N-fenil-carbammato (Italian)	**(122-42-9)**
N-Isopropilftalimmide (Italian)	**(304-17-6)**
(1-Isopropil-3-metil-1H-pirazol-5-il)-N,N-dimetil-carbammato (Italian)	**(119-38-0)**
(E)-O-2-Isopropoxy-carbonyl-1-methylvinyl O-methyl ethyl-phosphoramidothioate	**(31218-83-4)**
O-(1-Isopropoxycarbonyl-1-propen-2-yl)-O-methyl N-ethyl-phosphoramidothionate	**(31218-83-4)**
4-Isopropoxydiphenylamine	**(101-73-5)**
p-Isopropoxydiphenylamine	**(101-73-5)**
2-Isopropoxyethanol (ACGIH,OSHA)	**(109-59-1)**
3'-Isopropoxy-2-methylbenzanilide	**(55814-41-0)**
(Isopropoxymethyl)oxirane	**(4016-14-2)**
Isopropoxymethylphosphoryl fluoride	**(107-44-8)**
Isopropoxy pentachlorobutadiene	**(68334-67-8)**
2-Isopropoxyphenol	**(4812-20-8)**
o-Isopropoxyphenol	**(4812-20-8)**
N-(4-Isopropoxyphenyl)aniline	**(101-73-5)**
2-Isopropoxyphenyl-N-methylcarbamat (German)	**(114-26-1)**
2-Isopropoxyphenyl N-methylcarbamate	**(114-26-1)**
2-Isopropoxyphenyl methylcarbamate	**(114-26-1)**
o-Isopropoxyphenyl N-methylcarbamate	**(114-26-1)**
o-Isopropoxyphenyl methylcarbamate	**(114-26-1)**
o-Isopropoxyphenyl methyl((2-methylphenyl)thio)carbamate	**(50539-85-0)**
2-Isopropoxypropane	**(108-20-3)**
Isopropyl 2,4-D ester	**(94-11-1)**
Isopropylacetaat (Dutch)	**(108-21-4)**
Isopropylacetat (German)	**(108-21-4)**
Isopropyl acetate (ACGIH,OSHA) [UN 1220]	**(108-21-4)**
Isopropyl (acetate d') (French)	**(108-21-4)**
Isopropylacetic acid	**(503-74-2)**
Isopropylacetone	**(108-10-1)**
2'-Isopropylacetophenone	**(2142-65-6)**
Isopropyl acid phosphate, Solid (DOT)	**(1623-24-1)**
Isopropyl acid phosphate [UN 1793]	**(1623-24-1)**
Isopropyl acrylamide	**(2210-25-5)**
N-Isopropylacrylamide	**(2210-25-5)**
Isopropyl acrylate	**(689-12-3)**
Isopropyl adipate	**(6938-94-9)**
Isopropyl aerofloat	**(107-56-2)**
Isopropyl alcohol (ACGIH,OSHA) [UN 1219]	**(67-63-0)**
Isopropyl alcohol, titanium(4+) salt	**(546-68-9)**
Isopropyl aldehyde	**(78-84-2)**
Iso-propylalkohol (German)	**(67-63-0)**
Isopropylallylbarbituric acid	**(77-02-1)**
Isopropylamid kyseliny akrylove (Czech)	**(2210-25-5)**
Isopropylamine (ACGIH,DOT,OSHA) [UN 1221]	**(75-31-0)**
Isopropylamine 2,4-dichlorophenoxyacetate	**(5742-17-6)**
Isopropylamine, 2-(2,5-dimethoxyphenyl)-, hydrochloride	**(2801-68-5)**
Isopropylamine, β-(2,5-dimethoxyphenyl)-, hydrochloride	**(24973-25-9)**
Isopropylamine dodecylbenzenesulfonate	**(26264-05-1)**
1-Isopropylamine-3-(1-naphthyloxy)-2-propanol	**(525-66-6)**
Isopropylamino-4-azido-6-methylthio-1,3,5-triazin (German)	**(4658-28-0)**
4-Isopropylaminodiphenylamine	**(101-72-4)**
2-Isopropylaminoethanol	**(109-56-8)**
Isopropylaminoethanol	**(109-56-8)**
N-Isopropylaminoethanol	**(109-56-8)**
Isopropylamino-O-ethyl-(4-methylmercapto-3-methylphenyl)-phosphate	**(22224-92-6)**
1-(Isopropylamino)-2-hydroxy-3-(o-(allyloxy)phenoxy)propane	**(6452-71-7)**
2-Isopropylamino-4-methoxy-6-methylamino-s-triazine	**(3035-45-8)**
2-Isopropylamino-4-(3-methoxypropylamino)-6-methylthio-1,3,5-triazin (German)	**(841-06-5)**
2-Isopropylamino-4-(3-methoxypropylamino)-6-methylthio-s-triazine	**(841-06-5)**
4-Isopropylamino-6-(3'-methoxypropylamino)-2-methythio-1,3,5-triazin (German)	**(841-06-5)**
2-Isopropylamino-4-methylamino-6-methylmercapto-s-triazine	**(1014-69-3)**
2-Isopropylamino-4-methylamino-6-methylthio-1,3,5-triazine	**(1014-69-3)**
α-(Isopropylaminomethyl)-3,4-dihydroxybenzyl alcohol hydrochloride	**(51-30-9)**
α-((Isopropylamino)methyl)-2-naphthalenemethanol hydrochloride	**(51-02-5)**
2-Isopropylamino-1-(2-naphthyl)ethanol hydrochloride	**(51-02-5)**
1-Isopropylamino-3-(1-naphthyloxy)-2-propanol	**(525-66-6)**
2-Isopropyl aniline	**(643-28-7)**
4-Isopropylaniline	**(99-88-7)**
o-Isopropylaniline	**(643-28-7)**
p-Isopropylaniline	**(99-88-7)**
N-Isopropylaniline (ACGIH)	**(643-28-7)**
n-Isopropylaniline (OSHA)	**(768-52-5)**
Isopropylarterenol hydrochloride	**(51-30-9)**
4-Isopropylbenzaldehyde	**(122-03-2)**
p-Isopropylbenzaldehyde	**(122-03-2)**
Isopropylbenzeen (Dutch)	**(98-82-8)**
Isopropyl benzene	**(98-82-8)**
Isopropylbenzene [UN 1918]	**(98-82-8)**
p-Isopropylbenzenecarboxaldehyde	**(122-03-2)**
Isopropylbenzene hydroperoxide	**(80-15-9)**
Isopropylbenzene peroxide	**(80-43-3)**
2-Isopropylbenzimidazole	**(5851-43-4)**
4-Isopropylbenzoate	**(536-66-3)**
Isopropyl benzoate	**(939-48-0)**
4-Isopropylbenzoic acid	**(536-66-3)**
p-Isopropylbenzoic acid	**(536-66-3)**
Isopropylbenzol	**(98-82-8)**
Isopropyl-benzol (German)	**(98-82-8)**
3-Isopropyl-1H-2,1,3-benzothiadiazin-4(3H)-one-2,2-dioxide	**(25057-89-0)**
3-Isopropyl-1H-2,1,3-benzothiadiazin-4(3H)-one-2,2-dioxide, sodium salt	**(50723-80-3)**
3-Isopropyl-2,1,3-benzothiadiazinon-(4)-2,2-dioxid (German)	**(25057-89-0)**
Isopropyl N-benzoyl-N-(3-chloro-4-fluorophenyl)-2-amino-	

Name	CAS No.
2-Isopropyl-2-methyl-1,3-dioxolane	**(4405-16-7)**
3-Isopropyl-6-methylene-1-cyclohexene	**(555-10-2)**
4-Isopropyl-1-methylene-2-cyclohexene	**(555-10-2)**
N-Isopropyl-4,4'-methylenedianiline	**(10029-31-9)**
N-Isopropylmethylenedianiline	**(10029-31-9)**
Isopropyl methylfluorophosphate	**(107-44-8)**
N-Isopropyl-p-(2-methylhydrazinomethyl)benzamide hydrochloride	**(366-70-1)**
n-Isopropyl-α-(2-methylhydrazino)-p-toluamide	**(671-16-9)**
N-Isopropyl-α-(2-methylhydrazino)-p-toluamide hydrochloride	**(366-70-1)**
p-Isopropyl-α-methylhydrocinnamaldehyde	**(103-95-7)**
p-Isopropyl-α-methylhydrocinnamic aldehyde	**(103-95-7)**
2-Isopropyl-4-methyl-6-hydroxypyrimidine	**(2814-20-2)**
Isopropyl methyl ketone	**(563-80-4)**
Isopropyl methyl ketone ethylene ketal	**(4405-16-7)**
2-(4-Isopropyl-4-methyl-5-oxo-2-imidazolin-2-yl)nicotinic acid	**(81334-34-1)**
7-Isopropyl-1-methylphenanthrene	**(483-65-8)**
2-Isopropyl-5-methylphenol	**(89-83-8)**
2-Isopropyl-6-methylphenol	**(3228-04-4)**
3-Isopropyl-6-methylphenol	**(499-75-2)**
p-Isopropyl-α-methylphenylpropyl aldehyde	**(103-95-7)**
Isopropyl methylphosphonic acid	**(1832-54-8)**
Isopropyl methylphosphonofluoridate	**(107-44-8)**
o-Isopropyl methylphosphonofluoridate	**(107-44-8)**
Isopropyl-methyl-phosphoryl fluoride	**(107-44-8)**
Isopropyl 2-methylpropanoate	**(617-50-5)**
Isopropyl-2-(1-methyl-n-propyl)-4,6-dinitrophenyl carbonate	**(973-21-7)**
N-Isopropyl-2-methyl-2-propyl-1,3-propanediol dicarbamate	**(78-44-4)**
(1-Isopropyl-3-methyl-1H-pyrazol-5-yl)-N,N-dimethyl-carbamaat (Dutch)	**(119-38-0)**
1-Isopropyl-3-methyl-5-pyrazolyl dimethylcarbamate	**(119-38-0)**
1-Isopropyl-3-methylpyrazolyl-(5)-dimethylcarbamate	**(119-38-0)**
Isopropylmethylpyrazolyl dimethylcarbamate	**(119-38-0)**
1-Isopropyl-3-methyl-5-pyrazolylester kyseliny dimethyl-karbaminove (Czech)	**(119-38-0)**
O-2-Isopropyl-4-methylpyrimidyl-O,O-diethyl phosphorothioate	**(333-41-5)**
Isopropylmethylpyrimidyl diethyl thiophosphate	**(333-41-5)**
Isopropyl myristate	**(110-27-0)**
1-Isopropylnaphthalene	**(6158-45-8)**
2-Isopropylnaphthalene	**(2027-17-0)**
Isopropyl nitrate [UN 1222]	**(1712-64-7)**
Isopropyl nitrite	**(541-42-4)**
Isopropylnorepinephrine-hydrochloride	**(51-30-9)**
Isopropyloctadecylamine	**(13329-71-0)**
Isopropyl oleate	**(112-11-8)**
2-Isopropyl-3-oxazolidineethanol	**(28770-01-6)**
13-Isopropyl-7-oxopodocarpa-8,11,13-trien-15-oic acid	**(18684-55-4)**
3-Isopropyloxypropylene oxide	**(4016-14-2)**
Isopropyl palmitate	**(142-91-6)**
Isopropyl percarbonate	**(105-64-6)**
Isopropyl percarbonate, Stabilized (DOT)	**(105-64-6)**
Isopropyl percarbonate, Unstabilized (DOT)	**(105-64-6)**
Isopropyl peroxydicarbonate	**(105-64-6)**
Isopropyl peroxydicarbonate, Not more than 52% in solution (DOT)	**(105-64-6)**
Isopropyl peroxydicarbonate, Technically pure (DOT)	**(105-64-6)**
2-Isopropylphenol	**(88-69-7)**
3-Isopropylphenol	**(618-45-1)**
4-Isopropylphenol	**(99-89-8)**
Isopropylphenol	**(25168-06-3)**
m-Isopropylphenol	**(618-45-1)**
o-Isopropylphenol	**(88-69-7)**
p-Isopropylphenol	**(99-89-8)**
Isopropylphenol methylcarbamate	**(2631-40-5)**
m-Isopropylphenol N-methylcarbamate	**(64-00-6)**
Isopropyl-N-phenyl-carbamat (German)	**(122-42-9)**
Isopropyl phenylcarbamate	**(122-42-9)**
Isopropyl-N-phenylcarbamate	**(122-42-9)**
o-Isopropyl N-phenyl carbamate	**(122-42-9)**
N-(4-Isopropylphenyl)-N',N'-dimethylharnstoff (German)	**(34123-59-6)**
N-(4-Isopropylphenyl)-N',N'-dimethylurea	**(34123-59-6)**
N-4-Isopropylphenyl-N,N-dimethylurea	**(34123-59-6)**
Isopropylphenyl diphenyl phosphate	**(28108-99-8)**
Isopropylphenylenediamine	**(31626-02-5)**
2-Isopropyl-phenyl-N-methylcarbamate	**(2631-40-5)**
3-Isopropylphenyl methylcarbamate	**(64-00-6)**
m-Isopropylphenyl N-methylcarbamate	**(64-00-6)**
m-Isopropylphenyl methylcarbamate	**(64-00-6)**
o-Isopropylphenyl N-methylcarbamate	**(2631-40-5)**
o-Isopropylphenyl methyl ketone	**(2142-65-6)**
N-Isopropyl-N'-phenyl-p-phenylenediamine	**(101-72-4)**
Isopropyl phenyl phosphate	**(46355-07-1)**
Isopropyl-N-phenylurethan (German)	**(122-42-9)**
Isopropyl phosphite, tri-	**(116-17-6)**
Isopropyl phosphonate	**(1809-20-7)**
Isopropylphosphoramidothioic acid, O-2,4-dichlorophenyl O-methyl ester	**(299-85-4)**
Isopropyl phosphoric acid, Solid (DOT)	**(1623-24-1)**
Isopropyl phosphorofluoridate	**(55-91-4)**
13-Isopropylpodocarpa-7,13-dien-15-oic acid	**(514-10-3)**
13-Isopropylpodocarpa-8,11,13-trien-15-oic acid	**(1740-19-8)**
Isopropyl potassium xanthate	**(140-92-1)**
Iso-propyl propanoate	**(637-78-5)**
Isopropyl 2-propenoate	**(689-12-3)**
Iso-propyl propionate	**(637-78-5)**
Isopropyl propionate [UN 2409]	**(637-78-5)**
3-Isopropylpyrocatechol	**(2138-48-9)**
Isopropyl salicylate O-ester with O-ethylisopropylphosphoramido-thioate	**(25311-71-1)**
Isopropyl 9-(sodiumsulfooxy)octadecanoate	**(14350-72-2)**
Isopropyl stearate	**(112-10-7)**
Isopropyl sulfate	**(2973-10-6)**
Isopropyl sulfide (8CI)	**(625-80-9)**
S-2-Isopropylthioethyl O,O-dimethyl phosphorodithioate	**(36614-38-7)**
Isopropylthiol	**(75-33-2)**
N-Isopropyl-2-thionaphthoic amide	**(64142-01-4)**
2-Isopropylthioxanthone	**(5495-84-1)**
Isopropyl orthotitanate	**(546-68-9)**
Isopropyl titanate(IV)	**(546-68-9)**
p-Isopropyltoluene	**(99-87-6)**
Isopropyl triethanolamine titanate	**(36673-16-2)**
1-Isopropylurea	**(691-60-1)**
Isopropylurea	**(691-60-1)**
n-Isopropylurea	**(691-60-1)**
O-Isopropyl xanthate	**(108-25-8)**
Isopropylxanthic acid	**(108-25-8)**
Isopropylxanthic acid, sodium salt	**(140-93-2)**
Isopropylxanthogenan sodny (Czech)	**(140-93-2)**
4-Isopropyl-m-xylene	**(4706-89-2)**
4,4'-Isopropylylidenebis(2,6-dibromophenol)	**(79-94-7)**
Isoproterenol hydrochloride	**(51-30-9)**
Isoproterenol monohydrochloride	**(51-30-9)**
Isoprothiolane	**(50512-35-1)**
Isoproturon	**(34123-59-6)**
Isopsoralin	**(523-50-2)**
Isoptin	**(52-53-9)**
Isopto-Atropine	**(51-55-8)**
Isopto Carbachol	**(51-83-2)**
Isopto-Carpine	**(54-71-7)**
Isopto Cetamide	**(144-80-9)**
Isopto Fenicol	**(56-75-7)**
Isopto Hyoscine	**(51-34-3)**
Isopulegol	**(7786-67-6)**
Isopurine	**(120-73-0)**
8-Isoquinolinamine, 1,2,3,4-tetrahydro-2-methyl-4-phenyl- (9CI)	**(24526-64-5)**
Isoquinoline	**(119-65-3)**

Isoquinoline, 1-benzyl- (8CI) **(6907-59-1)**
Isoquinoline, 1-((3,4-dimethoxyphenyl)methyl)-6,7-dimethoxy **(58-74-2)**
Isoquinoline, 1-((3,4-dimethoxyphenyl)methyl)-6,7-dimethoxy-, hydrochloride **(61-25-6)**
Isoquinoline, 6,7-dimethoxy-1-veratryl- **(58-74-2)**
Isoquinoline, 6,7-dimethoxy-1-veratryl-, hydrochloride **(61-25-6)**
Isoquinoline, 3-methyl **(1125-80-0)**
Isoquinoline, 1-methyl- (9CI) **(1721-93-3)**
Isoquinoline, 1-(phenylmethyl)- (9CI) **(6907-59-1)**
Isoquinoline, 1,2,3,4-tetrahydro-8-amino-2-methyl-4-phenyl **(24526-64-5)**
Isoquinolinium, 2-dodecyl-, chloride (9CI) **(71732-96-2)**
Isoquinolinium, 2-(phenylmethyl)-, chloride **(35674-56-7)**
1-Isoquinolinol **(491-30-5)**
Isoquinolin-1-one **(491-30-5)**
1(2H)-Isoquinolinone (9CI) **(491-30-5)**
1(2H)-Isoquinolone **(491-30-5)**
Isorbid **(87-33-2)**
Isordil **(87-33-2)**
Isordil Tembids **(87-33-2)**
Isoren **(77-36-1)**
α-D-Isosaccharinic acid **(1518-54-3)**
β-D-Isosaccharinic acid **(1518-56-5)**
Isosafrole **(120-58-1)**
Isosafrole n-octylsulfoxide **(120-62-7)**
Isosafrole, octyl sulfoxide **(120-62-7)**
Isoscopil **(114-49-8)**
Isoserine (8CI) **(565-71-9)**
(+)-D-Isosorbide **(652-67-5)**
Isosorbide **(652-67-5)**
Isosorbide dinitrate **(87-33-2)**
Isosteareth-2 **(52292-17-8)**
Isosteareth-3 **(52292-17-8)**
Isosteareth-10 **(52292-17-8)**
Isosteareth-12 **(52292-17-8)**
Isosteareth-20 **(52292-17-8)**
Isosteareth-22 **(52292-17-8)**
Isosteareth-50 **(52292-17-8)**
Isostearic acid **(30399-84-9)**
Isostearic acid **(5638-12-0)**
Isostilbene **(645-49-8)**
Isosumithion **(3344-14-7)**
Isosystox **(126-75-0)**
Isotactic Polypropylene **(9003-07-0)**
Isotebe **(54-85-3)**
Isotebezid **(54-85-3)**
5-Isothiazolamine, 3-methyl- (9CI) **(24340-76-9)**
Isothiazole, 5-amino-3-methyl- (8CI) **(24340-76-9)**
Isothiazole, 3-methyl- (8CI,9CI) **(693-92-5)**
4-Isothiazolidinecarboxylic acid, 2-nitroso- (9CI) **(94751-62-9)**
Isothiazolinone chloride **(55965-84-9)**
4-Isothiazolin-3-one, 5-chloro-2-methyl- **(26172-55-4)**
4-Isothiazolin-3-one, 2-methyl- (8CI) **(2682-20-4)**
4-Isothiazolin-3-one, 2-octyl **(26530-20-1)**
3(2H)-Isothiazolone, 5-chloro-2-methyl- **(26172-55-4)**
3(2H)-Isothiazolone, 5-chloro-2-methyl-, Mixt. with 2-Methyl-3(2H)-isothiazolone **(55965-84-9)**
3(2H)-Isothiazolone, 2-methyl- (9CI) **(2682-20-4)**
3(2H)-Isothiazolone, 2-octyl- (9CI) **(26530-20-1)**
Isothin **(536-33-4)**
Isothioate **(36614-38-7)**
Isothiocyanate d'allyle (French) **(57-06-7)**
Isothiocyanate de methyle (French) **(556-61-6)**
1-Isothiocyanate-naphthalene **(551-06-4)**
Isothiocyanatoethane **(542-85-8)**
Isothiocyanatomethane **(556-61-6)**
2-Isothiocyanato-2-methylpropane **(590-42-1)**
1-Isothiocyanatonaphthalene **(551-06-4)**
3-Isothiocyanato-1-propene **(57-06-7)**
Isothiocyanic acid **(3129-90-6)**
Isothiocyanic acid, allyl ester **(57-06-7)**
Isothiocyanic acid, benzyl ester **(622-78-6)**
Isothiocyanic acid, 4,6-bis(ethylamino)-s-triazin-2-yl ester (8CI) **(30360-07-7)**
Isothiocyanic acid, 4,6-bis(isopropylamino)-s-triazin-2-yl ester (8CI) **(30360-08-8)**
Isothiocyanic acid, 4,6-bis(trichloromethyl)-s-triazin-2-yl ester (8CI) **(30863-25-3)**
Isothiocyanic acid, 6-(o-chloroanilino)-s-triazine-2,4-diyl ester (8CI) **(30362-22-2)**
Isothiocyanic acid, 6-(p-chloroanilino)-s-triazine-2,4-diyl ester (8CI) **(30362-23-3)**
Isothiocyanic acid, p-chlorophenyl ester **(2131-55-7)**
Isothiocyanic acid, cyclohexyl ester **(1122-82-3)**
Isothiocyanic acid, 4,6-dichloro-s-triazin-2-yl ester (6CI,7CI,8CI) **(4267-15-6)**
Isothiocyanic acid, ethyl ester **(542-85-8)**
Isothiocyanic acid, lead(2+) salt **(592-87-0)**
Isothiocyanic acid, methyl ester **(556-61-6)**
Isothiocyanic acid, 1-naphthyl ester **(551-06-4)**
Isothiocyanic acid, 1,4-phenylenedi- **(4044-65-9)**
Isothiocyanic acid p-phenylene ester **(4044-65-9)**
Isothiocyanic acid, phenyl ester **(103-72-0)**
Isothionate **(36614-38-7)**
Isothiourea **(62-56-6)**
Isothiouronium chloride, benzyl- **(538-28-3)**
(-)-3-Isothujone **(546-80-5)**
Isothymol **(499-75-2)**
Isotiamida **(536-33-4)**
Isotinyl **(54-85-3)**
Isotiocianato di metile (Italian) **(556-61-6)**
Isotox **(58-89-9)**
Isotrate **(87-33-2)**
Isotretinoin **(4759-48-2)**
Isotridecanoic acid (9CI) **(25448-24-2)**
Isotridecanol **(27458-92-0)**
Isotridecyl-alcohol **(27458-92-0)**
Isotridecyl octadecanoate **(31565-37-4)**
Isotridecyl stearate **(31565-37-4)**
Isotron 11 **(75-69-4)**
Isotron 12 **(75-71-8)**
Isotron 22 **(75-45-6)**
Isoundecane (8CI,9CI) **(34464-43-2)**
Isourea **(57-13-6)**
Isourethane **(9009-54-5)**
Isouron **(55861-78-4)**
Isoval **(83-28-3)**
Isovaleral **(590-86-3)**
Isovaleraldehyde **(590-86-3)**
Isovaleraldehyde, diethyl acetal **(3842-03-3)**
Isovaleramide **(541-46-8)**
8-Isovalerate **(21259-20-1)**
Isovalerianic **(503-74-2)**
Isovalerianic acid **(503-74-2)**
Isovaleric-acid **(503-74-2)**
Isovaleric acid, allyl ester **(2835-39-4)**
Isovaleric acid, butyl ester (8CI) **(109-19-3)**
Isovaleric acid, ethyl ester **(108-64-5)**
Isovaleric acid, isobutyl ester **(589-59-3)**
Isovaleric acid, isopentyl ester **(659-70-1)**
Isovaleric acid, methyl ester **(556-24-1)**
Isovaleric acid, propyl ester **(557-00-6)**
Isovaleric acid, sodium salt (8CI) **(539-66-2)**
Isovaleric aldehyde **(590-86-3)**
Isovaleric amide **(541-46-8)**
Isovalerone **(108-83-8)**

2-Isovalerylindan-1,3-dione (83-28-3)
2-Isovaleryl-1,3-indandione (83-28-3)
Isovaleryl indandione (83-28-3)
2-Isovaleryl-1,3-indandione calcium salt (23710-76-1)
2-Isovaleryl-1,3-indanedione (83-28-3)
Isovanillic acid (645-08-9)
Isovanillin (621-59-0)
Isovanilline (621-59-0)
Isovitamin C (89-65-6)
Isoxaben (82558-50-7)
Isoxamin (127-69-5)
Isoxanthine (69-89-6)
Isoxanthopterin (529-69-1)
Isoxathion (18854-01-8)
p-Isoxazine, tetrahydro- (110-91-8)
Isoxazole (9CI) (288-14-2)
4,5-Isoxazoledione, 3-methyl-, 4-((o-chlorophenyl)hydrazone) (5707-69-7)
4,5-Isoxazoledione, 3-methyl-, 4-((3-chlorophenyl)hydrazone) (9CI) (5707-73-3)
4,5-Isoxazoledione, 3-methyl-, 4-((m-chlorophenyl)hydrazone) (8CI) (5707-73-3)
Isoxazole, 3-hydroxy-5-methyl (10004-44-1)
Isoxazole, 5,5'-(1,3-propanediyl)bis- (9CI) (37704-51-1)
3-Isoxazolidine, 4-amino-, (R)- (9CI) (68-41-7)
3-Isoxazolidinone, 4-amino-, D (68-41-7)
3-Isoxazolidinone, 2-((2-chlorophenyl)methyl)-4,4-dimethyl (81777-89-1)
3-Isoxazolol, 5-(aminomethyl) (2763-96-4)
3-Isoxazolol, 5-methyl- (10004-44-1)
Isoxsuprine (395-28-8)
Isoxyl (55861-78-4)
Isozide (54-85-3)
Isozizanoic acid (16202-79-2)
Isozyd (54-85-3)
Ispenoral (132-98-9)
Isravin (8048-52-0)
Istin (117-10-2)
Istizin (117-10-2)
Isuprel (51-30-9)
Isuprel hydrochloride (51-30-9)
Iszilin (9004-10-8)
Itachigarden (10004-44-1)
Itaconic acid (97-65-4)
Italchin (69-05-6)
Itamid (105-60-2)
Itamid (25038-54-4)
Itamid 250 (105-60-2)
Itamid 250 (25038-54-4)
Itamide 25 (25038-54-4)
Itamide 25 (105-60-2)
Itamide 35 (105-60-2)
Itamide 35 (25038-54-4)
Itamide 250 (25038-54-4)
Itamide 250 (105-60-2)
Itamide 250G (105-60-2)
Itamide 250G (25038-54-4)
Itamide 350 (105-60-2)
Itamide 350 (25038-54-4)
Itamide S (105-60-2)
Itamide S (25038-54-4)
Itamidone (58-15-1)
Itinerol (569-65-3)
Itiocide (536-33-4)
Itir (66-02-4)
Itobarbital (77-26-9)
Itopaz (563-12-2)
Ituran (67-20-9)
Iuglon (481-39-0)
Ivalon (50-00-0)
Ivalon (9002-89-5)
Ivarlan 3000 (8006-54-0)
Ivarlan 3001 (8006-54-0)
Ivarlan 3100 (8006-54-0)
Ivarlan 3406 (61790-81-6)
Ivarlan 3407 (61790-81-6)
Ivaugan (58-93-5)
Ivermectin (9CI) (70288-86-7)
Iversal (539-21-9)
Ivertol (539-21-9)
Iviron (8047-67-4)
Ivoran (50-29-3)
Ivorit (111-60-4)
Ivosit (2813-95-8)
Ivosit (88-85-7)
Ivy LS (8001-21-6)
Ixodex (50-29-3)
Izadrin (51-30-9)
Izoacridina (90-45-9)
Izoforon (Polish) (78-59-1)
Izopropylowy eter (Polish) (108-20-3)
Izosystox (Czech) (126-75-0)
J 2FP (9000-30-0)
J 164 (9004-62-0)
J 400 (9003-07-0)
J 700 (9003-07-0)
J-38 (2540-82-1)
J Acid (87-02-5)
JB 318 (3567-12-2)
JB 336 (3321-80-0)
177 J.D. (60-32-2)
JF 5705F (52315-07-8)
JGD 1800 (9003-07-0)
JISC 3108 (7429-90-5)
JISC 3110 (7429-90-5)
JMD 4500 (9003-07-0)
JP 9 (82863-50-1)
JP-10 (2825-82-3)
J Soft C 4 (122-19-0)
J-Sul (127-69-5)
JZF (74-31-7)
Ja-Fa IPM (110-27-0)
Ja-Fa IPP (142-91-6)
Jack Bean Urease (9002-13-5)
Jack Wilson Chloro 51 (Oil) (101-21-3)
Jacobine (6870-67-3)
Jacodine (480-81-9)
Jacutin (58-89-9)
Jacutin (608-73-1)
Jade Green Base (128-58-5)
Jado (7440-44-0)
Jaguar (9000-30-0)
Jaguar 6000 (9000-30-0)
Jaguar A 20 B (9000-30-0)
Jaguar A 20D (9000-30-0)
Jaguar A 40F (9000-30-0)
Jaguar Gum A-20-D (9000-30-0)
Jaguar No.124 (9000-30-0)
Jaguar Plus (9000-30-0)
Jaikin (1303-96-4)
Jalan (2212-67-1)
Janupap (103-90-2)
Japan Agar (9002-18-0)
Japan Blue No. 404 (147-14-8)
Japan Brown No. 201 (1320-07-6)
Japan Camphor (76-22-2)

Name	CAS No.
Japan Isinglass	**(9002-18-0)**
Japan Red 201	**(5858-81-1)**
Japan Red No. 202	**(5281-04-9)**
Japan Red No. 220	**(6417-83-0)**
Japan Red No. 505	**(2610-11-9)**
Japan Yellow 201	**(2321-07-5)**
Japan Yellow 203	**(8004-92-0)**
Japan Yellow No. 201	**(2321-07-5)**
Japanese Camphor	**(464-49-3)**
Japanese Camphor Oil	**(8008-51-3)**
Japanese, Oil of Camphor	**(8008-51-3)**
Japanese Wood Oil	**(8001-20-5)**
Japanol Black BHK	**(2429-73-4)**
Japanol Brilliant Blue 6BKX	**(2610-05-1)**
Japanol Brown M	**(2429-82-5)**
Japanol Violet J	**(2586-60-9)**
Jasad	**(7440-66-6)**
Jasminaldehyde	**(122-40-7)**
Jasmolin I or II	**(8003-34-7)**
(Z)-Jasmone	**(488-10-8)**
Jasmone	**(488-10-8)**
cis-Jasmone	**(488-10-8)**
Jatroneural	**(117-89-5)**
Jatropur	**(396-01-0)**
Jaune AB	**(85-84-7)**
Jaune OB	**(131-79-3)**
Jaune Orange S	**(2783-94-0)**
Jaune Soleil	**(2783-94-0)**
Jaune de Beurre (French)	**(60-11-7)**
Jaune de Quinoleine (French)	**(8004-92-0)**
Java Amaranth	**(915-67-3)**
Java Chrome Violet B	**(2092-55-9)**
Java Metanil Yellow G	**(587-98-4)**
Java Naphtol Red G	**(3734-67-6)**
Java Orange 2G	**(1936-15-8)**
Java Orange I	**(523-44-4)**
Java Orange II	**(633-96-5)**
Java Ponceau 2R	**(3761-53-3)**
Java Rubine N	**(3567-69-9)**
Java Scarlet 3R	**(2611-82-7)**
Javanicin	**(476-45-9)**
Jayflex DTDP	**(119-06-2)**
Jaysol	**(64-17-5)**
Jaysol S	**(64-17-5)**
Jeffamine AP-20	**(101-77-9)**
Jeffamine AP22	**(25214-70-4)**
Jeffamine AP27	**(25214-70-4)**
Jeffersol DB	**(112-34-5)**
Jeffersol EB	**(111-76-2)**
Jeffersol EE	**(110-80-5)**
Jeffersol EM	**(109-86-4)**
Jeffox	**(25322-68-3)**
Jeffox	**(25322-69-4)**
Jeffox FF 200	**(9003-11-6)**
Jeffox Ol 2700	**(37286-64-9)**
Jeffox PPG 2000	**(9003-11-6)**
Jellin	**(67-73-2)**
Jen-Diril	**(58-93-5)**
Jenacain	**(51-05-8)**
Jenacain	**(59-46-1)**
Jenacaine	**(59-46-1)**
Jenacillin O	**(54-35-3)**
Jerva acid	**(99-32-1)**
Jervaic acid	**(99-32-1)**
Jervasic acid	**(99-32-1)**
Jervine	**(469-59-0)**
Jestryl	**(51-83-2)**
Jetrium	**(357-56-2)**
Jetrium R	**(357-56-2)**
Jeweler's Rouge	**(1309-37-1)**
Jiffy Grow	**(133-32-4)**
Jod (German, Polish)	**(7553-56-2)**
Jodamid (German)	**(440-58-4)**
4-Jodbenzoesaeure (German)	**(619-58-9)**
1-Jodbutan (Czech)	**(542-69-8)**
2-Jodbutan (Czech)	**(513-48-4)**
Jodcyan	**(506-78-5)**
Jodethamine	**(5700-49-2)**
Jodethan (Czech)	**(75-03-6)**
2-Jodfenol (Czech)	**(533-58-4)**
o-Jodfenol (Czech)	**(533-58-4)**
Jodfenphos	**(18181-70-9)**
Jodgorgon	**(66-02-4)**
Jodgorgosaeure	**(66-02-4)**
Jodid sodny (Czech)	**(7681-82-5)**
Jodoform (Czech)	**(75-47-8)**
1-Jodoktan (Czech)	**(629-27-6)**
Jodomiron	**(440-58-4)**
1-Jodpentan (Czech)	**(628-17-1)**
3-Jodphenol (German)	**(626-02-8)**
1-Jodpropan (Czech)	**(107-08-4)**
2-Jodpropan (Czech)	**(75-30-9)**
Johnkolor	**(88-82-4)**
Joker	**(66441-11-0)**
Jolt	**(13194-48-4)**
Jod-methan (German)	**(74-88-4)**
1-Jod-2-methylpropan (Czech)	**(513-38-2)**
Jon-Trol	**(144-21-8)**
Jonit	**(4044-65-9)**
Jonnix	**(3337-71-1)**
Jood (Dutch)	**(7553-56-2)**
Joodmethaan (Dutch)	**(74-88-4)**
Jopagnost	**(96-83-3)**
Jorchem 400 ML	**(25322-68-3)**
Jordamide 201	**(93-83-4)**
Jordamide DAPLM	**(3179-80-4)**
Jordamide DAPSA	**(7651-02-7)**
Jordamide LAA	**(3179-80-4)**
Jordamox CDA	**(7128-91-8)**
Jordamox CDA-40	**(7128-91-8)**
Jordamox ODA	**(14351-50-9)**
Jordamox SDA	**(2571-88-2)**
Jordanol SL-300	**(151-21-3)**
Jordaphos 236	**(9004-80-2)**
Jordaphos JS-61	**(9046-01-9)**
Jordaquat JO-50	**(37139-99-4)**
Jordaquat ODBAC	**(37139-99-4)**
Jortaine	**(107-43-7)**
Jortaine CAB-35	**(61789-40-0)**
Jortaine CFA-35	**(61789-40-0)**
Jortaine LMAB	**(4292-10-8)**
Jortaine OB	**(871-37-4)**
Joy Powder	**(561-27-3)**
Jubenon R	**(1344-28-1)**
Judean Pitch	**(8052-42-4)**
Juglane	**(481-39-0)**
Juglon	**(481-39-0)**
Juglone	**(481-39-0)**
Julin's Carbon Chloride	**(118-74-1)**
Jully	**(814-71-1)**
Julodin	**(29975-16-4)**
Juniper-Berry-Oil	**(8002-68-4)**
Juniper-Tar	**(8013-10-3)**
Junipen	**(475-20-7)**

Name	CAS No.
Junipene	(475-20-7)
Juniper Berry Oil	(8002-68-4)
Juniper Berry Oil, Terpenes	(8002-68-4)
Juniper Oil	(8002-68-4)
Juniperberry Oil	(8002-68-4)
Juniperic acid	(506-13-8)
Junlon 110	(9003-01-4)
Jurimer AC 10H	(9003-01-4)
Jurimer AC 10P	(9003-01-4)
Jusonin	(144-55-8)
Juston-Wirkstoff	(2152-34-3)
Juva-K	(58-27-5)
Juvamycetin	(56-75-7)
Juvason	(53-03-2)
Juvocaine	(51-05-8)
06K	(2797-51-5)
Iso-K	(22071-15-4)
K 0	(9011-05-6)
K-9	(3811-49-2)
K15	(9003-39-8)
K-17	(50-35-1)
K 17	(9011-05-6)
K 17(Polymer)	(9011-05-6)
K 21 (Silicone)	(63148-62-9)
K25	(9003-39-8)
K30	(9003-39-8)
K 52	(112-80-1)
K 55E	(9003-55-8)
K60	(9003-39-8)
K90	(9003-39-8)
K115	(9003-39-8)
K 257	(7440-44-0)
K 300	(9003-07-0)
K-300	(68-41-7)
K-373	(2955-38-6)
K385	(9011-05-6)
K 331	(63148-62-9)
K 411-02	(9011-05-6)
K 525	(9003-53-6)
K6-30	(12001-29-5)
K 1875	(555-89-5)
K3917	(846-50-4)
K-3920	(63610-08-2)
K 4710	(469-79-4)
K 6451	(80-33-1)
K 8870	(9011-05-6)
K 22023	(299-85-4)
K62-105	(21609-90-5)
KS-4	(7439-92-1)
KB-53	(154-23-4)
KBE 903	(13822-56-5)
KB (Polymer)	(9003-53-6)
K-Brite	(7775-14-6)
KC 200	(37353-62-1)
KC-300	(37353-63-2)
KC-400	(12737-87-0)
KC-500	(37317-41-2)
KCA Acetate Fast Yellow G	(2832-40-8)
KCA Acid Milling Yellow M	(6375-55-9)
KCA Foodcol Amaranth A	(915-67-3)
KCA Foodcol Sunset Yellow FCF	(2783-94-0)
KCA Foodcol Tartrazine PF	(1934-21-0)
KCA Light Fast Brown BR	(16071-86-6)
KCA Methyl Orange	(547-58-0)
KCA Silk Red G	(3567-65-5)
KCA Tartrazine PF	(1934-21-0)
KCH 749	(9006-03-5)
KCH 770	(9006-03-5)
KCH 771	(9006-03-5)
KCH 1116	(9006-03-5)
KCO-3001	(62850-32-2)
KD 83	(107-64-2)
KDNBF	(29267-75-2)
KDP	(7778-77-0)
KE	(53-06-5)
KE 77	(63148-62-9)
KER 710	(1302-74-5)
KF 96	(63148-62-9)
KF 96-100	(63148-62-9)
KF 96-500	(63148-62-9)
KF-32	(27355-22-2)
KF-1820	(359-83-1)
KFS	(9011-05-6)
23K (Fiber)	(63428-84-2)
K-GA	(125-67-7)
K-Gran	(584-08-7)
KH 360	(13463-67-7)
KHE 0145	(2631-40-5)
K-IAO	(299-27-4)
K III	(534-52-1)
K IV	(534-52-1)
KK 60	(20115-34-8)
KKM 43	(55965-84-9)
KL-001	(1812-30-2)
KLT 40	(25038-59-9)
K-Lor	(7447-40-7)
K-Lyte	(298-14-6)
K-Lyte/Cl	(7447-40-7)
4K-2M	(94-74-6)
KM	(59-01-8)
KM	(9003-53-6)
KM 2	(9011-05-6)
KM 200	(4719-04-4)
KM 102 (9CI)	(54392-02-8)
KMH	(123-33-1)
KS-M 0.3P	(9011-05-6)
KM (Polymer)	(9003-53-6)
KM 2 (Polymer)	(9011-05-6)
KMTs	(9000-11-7)
KMTs 212	(9004-32-4)
KMTs 300	(9004-32-4)
KMTs 500	(9004-32-4)
KMTs 600	(9004-32-4)
KMVA	(1460-34-0)
KM 9k	(63148-62-9)
KM (the antibiotic)	(59-01-8)
K 120 N	(9011-14-7)
K1-N	(7681-11-0)
KO 08	(63148-62-9)
KO 7	(1302-74-5)
KO 811	(63148-62-9)
KOW	(106-14-9)
K-Obiol	(52918-63-5)
KO (fatty acid)	(67254-79-9)
K-Othrin	(52918-63-5)
K-Othrine	(22431-62-5)
KP 2	(82-68-8)
KP 5	(9003-27-4)
KP 140	(78-51-3)
KP 201	(84-61-7)
KPMK	(7235-40-7)
K Phenethicillin	(132-93-4)
K-Pin	(1918-02-1)
K115 (Polyamide)	(9003-39-8)

Name	CAS No.
K25 (Polymer)	**(9003-39-8)**
K30 (Polymer)	**(9003-39-8)**
K60 (Polymer)	**(9003-39-8)**
K-Predne-Dome	**(7447-40-7)**
K Preparation	**(502-55-6)**
06K-Quinone	**(2797-51-5)**
KR 220	**(63148-62-9)**
KR 2537	**(9003-53-6)**
KRMD 58	**(10326-21-3)**
KRO 1	**(9003-55-8)**
KRO 2	**(9003-55-8)**
KS 4B	**(78-04-6)**
KS 11	**(9011-05-6)**
KS 30P	**(105-60-2)**
KS 30P	**(25038-54-4)**
KS 35	**(9011-05-6)**
KS 68M	**(9011-05-6)**
KS 607A	**(63148-62-9)**
KS 700	**(63148-62-9)**
KS 705F	**(63148-62-9)**
KS 770	**(63148-62-9)**
KS 773	**(63148-62-9)**
KS 774	**(63148-62-9)**
KT 35	**(1332-40-7)**
KTIOL 15	**(67254-79-9)**
KTIOL 77	**(67254-79-9)**
K-Thrombyl	**(58-27-5)**
KTsA	**(20227-92-3)**
KU 10	**(1302-74-5)**
KU 5-3	**(1302-74-5)**
KU 13-O32-C	**(1085-98-9)**
KUE 13032c	**(1085-98-9)**
KUE 13183b	**(731-27-1)**
K-Vitan	**(58-27-5)**
06K-50W	**(2797-51-5)**
KW-066	**(1069-66-5)**
KW-125	**(23214-92-8)**
KWD 2019	**(23031-25-6)**
KWG 0599	**(55179-31-2)**
KW 600S	**(1343-88-0)**
KX 52	**(39362-66-8)**
KZ 3M	**(409-21-2)**
KZ 5M	**(409-21-2)**
KZ 7M	**(409-21-2)**
K-Zinc	**(1314-13-2)**
KA 101	**(1344-28-1)**
Kabat	**(40596-69-8)**
Kabipenin	**(6130-64-9)**
Kabivitrum	**(439-14-5)**
Kacha Haldi	**(458-37-7)**
Kadethrin	**(58769-20-3)**
cis-Kadethrin	**(58769-20-3)**
Kadmium (German)	**(7440-43-9)**
Kadmiumchlorid (German)	**(10108-64-2)**
Kadmiumstearat (German)	**(2223-93-0)**
Kadmu tlenek (Polish)	**(1306-19-0)**
Kadol	**(50-33-9)**
Kadox 15	**(1314-13-2)**
Kadox 72	**(1314-13-2)**
Kadox-25	**(1314-13-2)**
Kaempferol	**(520-18-3)**
Kaempferol-3-O-gal-rham-7-O-rham	**(301-19-9)**
Kaempherol	**(520-18-3)**
Kaergona	**(58-27-5)**
Kafa	**(62-44-2)**
Kafar Copper	**(7440-50-8)**
Kafil	**(52645-53-1)**
Kafil super	**(52315-07-8)**
Kafocin	**(3577-01-3)**
Kaiser Chemicals 11	**(76-13-1)**
Kaiser Chemicals 12	**(75-71-8)**
Kajos	**(866-84-2)**
Kaken	**(66-81-9)**
Kakerbin	**(9002-64-6)**
Kako Blue B Salt	**(119-90-4)**
Kako Red B Base	**(97-52-9)**
Kako Red RL Base	**(99-52-5)**
Kako Red TR Base	**(95-69-2)**
Kako Tartrazine	**(1934-21-0)**
Kakodylan dodny (Czech)	**(124-65-2)**
Kalex	**(64-02-8)**
Kalgan	**(156-51-4)**
Kalignost	**(143-66-8)**
Kaliksir	**(866-84-2)**
Kalitabs	**(7447-40-7)**
Kalium-β	**(299-27-4)**
Kalium-Cyanid (German)	**(151-50-8)**
Kaliumarsenit (German)	**(10124-50-2)**
Kaliumcarbonat (German)	**(584-08-7)**
Kaliumchloraat (Dutch)	**(3811-04-9)**
Kaliumchlorat (German)	**(3811-04-9)**
Kaliumcyanat (German)	**(590-28-3)**
Kaliumdichromat (German)	**(7778-50-9)**
Kaliumhydroxid (German)	**(1310-58-3)**
Kaliumhydroxyde (Dutch)	**(1310-58-3)**
Kaliumnitrat (German)	**(7757-79-1)**
Kaliumpermanganaat (Dutch)	**(7722-64-7)**
Kaliumpermanganat (German)	**(10118-76-0)**
Kaliumpermanganat (German)	**(7722-64-7)**
Kallocryl K	**(9011-14-7)**
Kallodent 222	**(9011-14-7)**
Kallodent Clear	**(9011-14-7)**
Kalmettumsomniferum	**(15879-93-3)**
Kalmin	**(62-44-2)**
Kalmocaps	**(58-25-3)**
Kalo	**(7778-44-1)**
Kalpur TE	**(4719-04-4)**
Kalziumarseniat (German)	**(7778-44-1)**
Kalziumzyklamate (German)	**(139-06-0)**
Kam 1000	**(57-11-4)**
Kam 2000	**(57-11-4)**
Kam 3000	**(57-11-4)**
Kamaver	**(56-75-7)**
Kambamine Red TR	**(95-69-2)**
Kambamine Scarlet GG Base	**(95-82-9)**
Kambothol AS	**(92-77-3)**
Kambothol ASBS	**(135-65-9)**
Kambothol ASG	**(91-96-3)**
Kambothol ASOL	**(135-62-6)**
Kambothol ASPH	**(92-74-0)**
2-Kamfanon (Czech)	**(76-22-2)**
Kamfochlor	**(8001-35-2)**
Kampfer (German)	**(76-22-2)**
Kampherol	**(520-18-3)**
Kamposan	**(16672-87-0)**
Kampstoff "Lost"	**(505-60-2)**
Kanamicina (Italian)	**(59-01-8)**
Kanamycin	**(59-01-8)**
Kanamycin A	**(59-01-8)**
Kanamytrex	**(59-01-8)**
Kandiset	**(81-07-2)**
Kaneace PA 20	**(9011-14-7)**
Kanechlor	**(1336-36-3)**
Kanechlor 300	**(1336-36-3)**

Kanechlor 300 (37353-63-2)
Kanechlor 400 (12737-87-0)
Kanechlor 400 (1336-36-3)
Kanechlor 500 (37317-41-2)
Kanechlor 500 (61788-33-8)
Kanechlor C (61788-33-8)
Kanechlor 200 (9CI) (37353-62-1)
Kanekrol 500 (25429-29-2)
Kanepar (85-34-7)
Kanone (58-27-5)
Kansai Direct Fast Yellow A (1325-37-7)
Kantharidin (German) (56-25-7)
Kantrex (59-01-8)
Kanzo (Japanese) (68916-91-6)
Kanzou (Chinese) (68916-91-6)
Kaon-Cl (7447-40-7)
Kaon-Cl 10 (7447-40-7)
Kaon-Cl Tabs (7447-40-7)
Kaochlor (7447-40-7)
Kaolin (OSHA) (1332-58-7)
Kaolin, calcined (66402-68-4)
Kaolinite (H4Al2Si2O9), Mixt. contg. (9CI) (8047-42-5)
Kaon Elixir (299-27-4)
Kaopaous (1332-58-7)
Kaophills-2 (1332-58-7)
Kaopolite (1302-76-7)
Kappaxan (58-27-5)
Kappaxin (58-27-5)
Kapreomycin (11003-38-6)
Kaprilon (63428-84-2)
e-Kaprolaktam (Czech) (105-60-2)
ε-Kaprolakton (Czech) (502-44-3)
Kaprolit (105-60-2)
Kaprolit (25038-54-4)
Kaprolit B (105-60-2)
Kaprolit B (25038-54-4)
Kaprolon (105-60-2)
Kaprolon (25038-54-4)
Kaprolon B (105-60-2)
Kaprolon B (25038-54-4)
Kapromin (25038-54-4)
Kapromine (105-60-2)
Kapron (105-60-2)
Kapron (25038-54-4)
Kapron (63428-83-1)
Kapron A (105-60-2)
Kapron A (25038-54-4)
Kapron B (105-60-2)
Kapron B (25038-54-4)
Kapronaldehyd (Czech) (66-25-1)
Kaptan (133-06-2)
Kaptax (Czech) (149-30-4)
Kari-Rinse (7681-49-4)
Karakhol (33878-50-1)
Karamate (8018-01-7)
Karathane (39300-45-3)
Karathane WD (39300-45-3)
Karathene (39300-45-3)
Karbachol (Czech) (51-83-2)
Karbadox (Czech) (6804-07-5)
Karbam Black (14484-64-1)
Karbam White (137-30-4)
Karbamol (9011-05-6)
Karbamol B/M (9011-05-6)
Karbamol TsEM (136-84-5)
Karbamoylcholin chlorid (Czech) (51-83-2)
Karbanil (Czech) (103-71-9)
Karbanilid (Czech) (102-07-8)
Karbaryl (Polish) (63-25-2)
Karbaspray (63-25-2)
Karbation (137-42-8)
Karbation (6734-80-1)
Karbatox (63-25-2)
Karbatox 75 (63-25-2)
Karbazid (Czech) (497-18-7)
N-Karbetoksi-ftalimid (Yugoslavian) (22509-74-6)
Karbitol (Czech) (111-90-0)
Karbitolacetat (Czech) (112-15-2)
Karbofenothion (Czech) (786-19-6)
Karbofos (121-75-5)
Karbofuranu (Polish) (1563-66-2)
Karbosep (63-25-2)
Karbromal (77-65-6)
Karbutilate (4849-32-5)
Karcon (58-27-5)
Kardiamid (59-26-7)
Kardiazol (54-95-5)
Kardonyl (59-26-7)
3-Karen (Czech) (13466-78-9)
Kareon (58-27-5)
Karidium (7681-49-4)
Karigel (7681-49-4)
Karion (50-70-4)
Karlan (299-84-3)
Karmesin (3567-69-9)
Karmex (330-54-1)
Karmex DW (330-54-1)
Karmex Diuron Herbicide (330-54-1)
Karmex Monuron Herbicide (150-68-5)
Karmex W. Monuron Herbicide (150-68-5)
Karnitin (541-15-1)
Karotin (Czech,Sweden) (7235-40-7)
Karphos (18854-01-8)
Karsan (50-00-0)
Karsil (2533-89-3)
Kartryl (77-65-6)
Kasebon (973-21-7)
Kasil (1312-76-1)
Kasil 6 (1312-76-1)
Kassia Oel (German) (8007-80-5)
Kastone (7722-84-1)
Katagrippe (938-73-8)
Katamin AB (8001-54-5)
Katamine AB (122-19-0)
Katamine AB (8001-54-5)
Katchung Oil (8002-03-7)
Katechol (Czech) (120-80-9)
Kathon 886 (55965-84-9)
Kathon Biocide (55965-84-9)
Kathon CG (55965-84-9)
Kathon LP Preservative (26530-20-1)
Kathon LX (55965-84-9)
Kathon 886MW (55965-84-9)
Kathon RH 886 (55965-84-9)
Kathon SP 70 (26530-20-1)
Kathon 886 W (55965-84-9)
Kathon WT (55965-84-9)
Kathro (57-88-5)
Kativ-G (58-27-5)
Katigen Blue BC3R Conc CF (1327-57-7)
Katigen Blue BC8R Conc CF (1327-57-7)
Katigen Blue BCR Conc CF (1327-57-7)
Katigen Deep Black NND-CF (1326-82-5)
Katigen Deep Black RND-CF (1326-82-5)

Katine **(492-39-7)**
Kativ N **(84-80-0)**
Katlex **(54-31-9)**
Katonil **(62-37-3)**
Katorin **(299-27-4)**
Kauran-13-ol (8CI,9CI) **(5749-44-0)**
Kauresin K244 **(9011-05-6)**
Kaurit 420 **(9011-05-6)**
Kaurit 285 FL **(9011-05-6)**
Kaurit S **(140-95-4)**
Kaurit W **(141-07-1)**
Kauritil **(1332-40-7)**
Kautschin **(138-86-3)**
Kavadel **(78-34-2)**
Kay Ciel **(7447-40-7)**
Kayabutyl C **(3457-61-2)**
Kayacarbon BIC **(2372-21-6)**
Kayacion Blue P 3R **(12236-92-9)**
Kayacyl Blue BR **(2666-17-3)**
Kayacyl Sky Blue R **(4368-56-3)**
Kayafume **(74-83-9)**
Kayaku Acid Brilliant Scarlet 3R **(2611-82-7)**
Kayaku Alizarine Sky Blue R **(4368-56-3)**
Kayaku Amaranth **(915-67-3)**
Kayaku Benzopurpurine 4B **(992-59-6)**
Kayaku Blue B Base **(119-90-4)**
Kayaku Blue B Salt **(119-90-4)**
Kayaku Congo Red **(573-58-0)**
Kayaku Direct **(2602-46-2)**
Kayaku Direct Black BH **(2429-73-4)**
Kayaku Direct Brown M **(2429-82-5)**
Kayaku Direct Dark Green B **(3626-28-6)**
Kayaku Direct Deep Black EX **(1937-37-7)**
Kayaku Direct Deep Black GX **(1937-37-7)**
Kayaku Direct Deep Black S **(1937-37-7)**
Kayaku Direct Leather Black EX **(1937-37-7)**
Kayaku Direct Scarlet 3B **(6358-29-8)**
Kayaku Direct Sky Blue 5B **(2429-74-5)**
Kayaku Direct Sky Blue 6B **(2610-05-1)**
Kayaku Direct Special Black AAX **(1937-37-7)**
Kayaku Direct Violet LN **(6426-67-1)**
Kayaku Fast Red 3GL Base **(89-63-4)**
Kayaku Food Colour Red No. 102 **(2611-82-7)**
Kayaku Food Colour Red No. 2 **(915-67-3)**
Kayaku Food Colour Yellow No. 4 **(1934-21-0)**
Kayaku Indone Violet BC **(1327-57-7)**
Kayaku Red B Base **(97-52-9)**
Kayaku Red RL Base **(99-52-5)**
Kayaku Red Salt 3GL **(89-63-4)**
Kayaku Roccelline **(1658-56-6)**
Kayaku Scarlet G Base **(99-55-8)**
Kayaku Scarlet GG Base **(95-82-9)**
Kayaku Sulphur Black BBR 200 **(1326-82-5)**
Kayaku Sulphur Black BBX **(1326-82-5)**
Kayaku Sulphur Black BDX **(1326-82-5)**
Kayaku Sulphur Black BNX **(1326-82-5)**
Kayaku Sulphur Black 3BX **(1326-82-5)**
Kayaku Sulphur Black BX **(1326-82-5)**
Kayaku Sulphur Black BX 200 **(1326-82-5)**
Kayaku Sulphur Black BX 200 Flakes **(1326-82-5)**
Kayaku Sulphur Black G **(1326-82-5)**
Kayaku Sulphur Black TB **(1326-82-5)**
Kayaku Sulphur Blue BK **(1327-57-7)**
Kayaku Sulphur Blue BN **(1327-57-7)**
Kayaku Sulphur Blue BP **(1327-57-7)**
Kayaku Sulphur Blue FBB **(1327-57-7)**
Kayaku Sulphur Blue 4R **(1327-57-7)**
Kayaku Sulphur Blue RC **(1327-57-7)**
Kayaku Sulphur Blue RN **(1327-57-7)**
Kayaku Sulphur Blue RS **(1327-57-7)**
Kayaku Sulphur Blue TFB **(1327-57-7)**
Kayaku Sulphur Blue TFR **(1327-57-7)**
Kayaku Sulphur Indone Violet **(1327-57-7)**
Kayaku Tartrazine **(1934-21-0)**
Kayalon Fast Blue BR **(2475-45-8)**
Kayalon Fast Blue FN **(2475-46-9)**
Kayalon Fast Blue FN **(86722-66-9)**
Kayalon Fast Violet BB **(1220-94-6)**
Kayalon Fast Violet BR **(82-33-7)**
Kayalon Fast Yellow G **(2832-40-8)**
Kayalon Fast Yellow 4R **(6300-37-4)**
Kayalon Fast Yellow RR **(119-15-3)**
Kayalon Polyester Blue EBL-E **(12217-79-7)**
Kayalon Polyester Yellow 4R-E **(6300-37-4)**
Kayalon Polyester Yellow RF **(6300-37-4)**
Kayanol Milling Red PG **(3567-65-5)**
Kayanol Milling Red RS **(6459-94-5)**
Kayanol Red PG **(3567-65-5)**
Kayaphor LSK **(24019-80-5)**
Kayaphos **(7292-16-2)**
Kayarus Black G **(6428-31-5)**
Kayarus Black G Conc. **(6428-31-5)**
Kayarus Supra Brown BRS **(16071-86-6)**
Kayaset Yellow G **(2832-40-8)**
Kayasol Black B **(1326-82-5)**
Kayazinon **(333-41-5)**
Kayazol **(333-41-5)**
Kaydol **(8012-95-1)**
Kayklot **(58-27-5)**
Kaykot **(58-27-5)**
Kayphosnac **(7292-16-2)**
Kayquinone **(58-27-5)**
Kaytrate **(78-11-5)**
Kazoe **(26628-22-8)**
Kedavon **(50-35-1)**
Keestar **(9005-25-8)**
Kefenid **(22071-15-4)**
Kefglycin **(3577-01-3)**
Keflex **(15686-71-2)**
Keflodin **(50-59-9)**
Keflordin **(50-59-9)**
Keforal **(15686-71-2)**
Kefspor **(50-59-9)**
Keimstop **(101-99-5)**
Kelacid **(9005-32-7)**
Kelco Gel LV **(9005-38-3)**
Kelcoloid **(9005-37-2)**
Kelcosol **(9005-38-3)**
Kelecin **(8002-43-5)**
Kelene **(75-00-3)**
Kelevan **(4234-79-1)**
Kelgin **(9005-38-3)**
Kelgin F **(9005-38-3)**
Kelgin HV **(9005-38-3)**
Kelgin LV **(9005-38-3)**
Kelgin XL **(9005-38-3)**
Kelgum **(9005-38-3)**
Kelmar **(9005-36-1)**
Kelmar Improved **(9005-36-1)**
Keloform **(94-09-7)**
Keloform P **(94-12-2)**
Kelset **(9005-38-3)**
Kelsize **(9005-38-3)**
Keltane **(115-32-2)**

Name	CAS No.
Keltex	**(9005-38-3)**
Kelthane	**(115-32-2)**
para,para'-Kelthane	**(115-32-2)**
Kelthane A	**(115-32-2)**
Kelthane Dust Base	**(115-32-2)**
Kelthanethanol	**(115-32-2)**
Keltone	**(9005-38-3)**
Kemamide S	**(124-26-5)**
Kemamide W 40	**(110-30-5)**
Kemamine P 989	**(112-90-3)**
Kemate	**(101-05-3)**
Kemdazin	**(10605-21-7)**
Kemester 105	**(112-62-9)**
Kemester 115	**(112-62-9)**
Kemester 205	**(112-62-9)**
Kemester 213	**(112-62-9)**
Kemester 5652	**(103-23-1)**
Kemester 9018	**(112-61-8)**
Kemicetina	**(56-75-7)**
Kemicetine	**(56-75-7)**
Kemikal	**(1305-62-0)**
Kemitracin-50	**(142-30-3)**
Kemodrin	**(300-42-5)**
Kemolate	**(732-11-6)**
Kempferol	**(520-18-3)**
Kemplex 100	**(64-02-8)**
Kempore	**(123-77-3)**
Kempore 125	**(123-77-3)**
Kempore R 125	**(123-77-3)**
Kempore 50XPT	**(18039-42-4)**
Kenachrome Blue 2R	**(3567-69-9)**
Kenacort-A	**(76-25-5)**
Kenacort	**(124-94-7)**
Kenalog	**(76-25-5)**
Kenamide B	**(3061-75-4)**
Kenamide E	**(112-84-5)**
Kenamide E-180	**(10094-45-8)**
Kenamide P-181	**(16260-09-6)**
Kenapon	**(75-99-0)**
Kendall's Compound B	**(50-22-6)**
Kendall's Compound F	**(50-23-7)**
Kendall's compound E	**(53-06-5)**
Kephton	**(84-80-0)**
Kepinol	**(8064-90-2)**
Kepone	**(143-50-0)**
Kepone-2-one, decachlorooctahydro-	**(143-50-0)**
Kerabit	**(128-37-0)**
Keralyt	**(69-72-7)**
Kerasalicyl	**(54-21-7)**
Keratin	**(68238-35-7)**
Kerb	**(23950-58-5)**
Kerb 50W	**(23950-58-5)**
Kermac 600W (Mineral Seal Oil)	**(64742-46-7)**
Kerocaine	**(51-05-8)**
Keromet MD	**(94-91-7)**
Keropur	**(3813-05-6)**
Kerosal	**(54-21-7)**
Kerosene (Petroleum), Hydrotreated	**(64742-47-8)**
Kerosene [UN 1223]	**(8008-20-6)**
Kerosine	**(8008-20-6)**
Kerosine (Petroleum)	**(8008-20-6)**
Kerosine, Petroleum, hydrodesulfurized	**(64742-81-0)**
Keselan	**(52-86-8)**
Kessco 40	**(31566-31-1)**
Kessco BSC	**(123-95-5)**
Kessco ICS	**(25339-09-7)**
Kesscocide	**(551-06-4)**
Kesscoflex	**(117-83-9)**
Kesscoflex BS	**(123-95-5)**
Kesscoflex MCP	**(117-82-8)**
Kesscoflex TRA	**(102-76-1)**
Kessco isopropyl myristate	**(110-27-0)**
Kessco isopropyl palmitate	**(142-91-6)**
Kesscomir	**(110-27-0)**
Kessobamate	**(57-53-4)**
Kessodanten	**(57-41-0)**
Kessodrate	**(302-17-0)**
Kestrel (Pesticide)	**(52645-53-1)**
Kestrone	**(53-16-7)**
Ketaject	**(1867-66-9)**
Ketalar	**(1867-66-9)**
Ketalgin	**(76-99-3)**
Ketaman	**(50-34-0)**
Ketamine	**(1867-66-9)**
(+-)-Ketamine hydrochloride	**(1867-66-9)**
Ketamine hydrochloride	**(1867-66-9)**
Ketanest	**(1867-66-9)**
Ketanrift	**(315-30-0)**
Ketaset	**(1867-66-9)**
Ketavet	**(1867-66-9)**
Ketazolam	**(27223-35-4)**
Ketazolamum (Latin)	**(27223-35-4)**
Ketene (ACGIH,OSHA)	**(463-51-4)**
Ketene, dimer	**(674-82-8)**
Kethamed	**(2152-34-3)**
Ketjen B	**(1344-28-1)**
Ketjenblack EC	**(1333-86-4)**
δ-Keto 153	**(53494-70-5)**
Keto-Ethylene	**(463-51-4)**
3-Keto-L-gulofuranolactone	**(50-81-7)**
3-(3-Keto-7-α-acetylthio-17-β-hydroxy-4-androsten-17-α-yl)-propionic acid lactone	**(52-01-7)**
β-Ketoadipate	**(689-31-6)**
3-Ketoadipic	**(689-31-6)**
β-Ketoadipic acid	**(689-31-6)**
Ketobemidone	**(469-79-4)**
Ketobemidonum	**(469-79-4)**
4-Ketobenzotriazine	**(90-16-4)**
Ketobun-A	**(315-30-0)**
β-Ketobutyranilide	**(102-01-2)**
3-Ketobutyrate	**(541-50-4)**
3-Ketocarbofuran	**(16709-30-1)**
Ketocycloheptane	**(502-42-1)**
Ketocyclopentane	**(120-92-3)**
4-Ketocyclophosphamide	**(27046-19-1)**
7-Ketodehydroabietic acid	**(18684-55-4)**
Ketodestrin	**(53-16-7)**
3-Keto-1,5-dimethyl-4-dimethylamino-2-phenyl-2,3-dihydropyrazole	**(58-15-1)**
δ-Ketoendrin	**(53494-70-5)**
α-Ketoglutaric acid	**(328-50-7)**
Ketoheptamethylene	**(502-42-1)**
Ketohexamethylene	**(108-94-1)**
2-Ketohexamethyleneimine	**(105-60-2)**
2-Ketohexamethylenimine	**(105-60-2)**
Ketohydroxy-estratriene	**(53-16-7)**
Ketohydroxyestrin	**(53-16-7)**
Ketohydroxyestrin benzoate	**(2393-53-5)**
Ketohydroxyoestrin	**(53-16-7)**
Ketohydroxyoestrin benzoate	**(2393-53-5)**
2,3-Ketoindoline	**(91-56-5)**
α-Ketoisocaproic acid	**(816-66-0)**
Ketolar	**(1867-66-9)**
Ketole	**(120-72-9)**
Ketolin-H	**(120-32-1)**

α-Keto-γ-methiolbutyrate (583-92-6)
α-Ketomethionine (583-92-6)
γ-Keto-β-methoxy-δ-methylene-δ^{α}-hexenoic acid (90-65-3)
2-Keto-4-methylthiobutyric acid (583-92-6)
2-Keto-3-methylvaleric acid (1460-34-0)
α-Keto-β-methyl-n-valeric acid (1460-34-0)
α-Keto-β-methylvaleric acid (1460-34-0)
Ketone, 4-biphenylyl methyl (92-91-1)
Ketone, butyl methyl (591-78-6)
Ketone, t-butyl methyl (75-97-8)
Ketone, 1-cyclohepten-1-yl methyl (6CI,7CI,8CI) (14377-11-8)
Ketone, 1-cyclohexen-1-yl methyl (8CI) (932-66-1)
Ketone, cyclohexyl methyl (8CI) (823-76-7)
Ketone, cyclopentyl methyl (8CI) (6004-60-0)
Ketone, cyclopropyl methyl (8CI) (765-43-5)
Ketone, 3,5-diiodo-4-hydroxyphenyl 2-ethyl-3-benzofuranyl (68-90-6)
Ketone, dimethyl (67-64-1)
Ketone, diphenyl (119-61-9)
Ketone, ethyl 4-(m-hydroxyphenyl)-1-methylpiperidyl (469-79-4)
Ketone, ethyl methyl (78-93-3)
Ketone, ethyl phenyl (93-55-0)
Ketone, ethyl vinyl (1629-58-9)
Ketone, ferrocenyl methyl (1271-55-2)
Ketone, 2-furyl α-hydroxyfurfuryl (552-86-3)
Ketone, 2-furyl methyl (1192-62-7)
Ketone, heptyl methyl (821-55-6)
Ketone, α-hydroxybenzyl phenyl (119-53-9)
Ketone, 4-hydroxy-3-thienyl methyl (7CI,8CI) (5556-16-1)
Ketone, isobutyl methyl (108-10-1)
Ketone, methyl isoamyl (110-12-3)
Ketone, methyl isopropenyl (814-78-8)
Ketone, methyl 5-methylpyrazinyl (8CI) (22047-27-4)
Ketone, methyl 2-naphthyl (93-08-3)
Ketone, methyl pentyl (110-43-0)
Ketone, methyl phenyl (98-86-2)
Ketone, methyl pyrazinyl (22047-25-2)
Ketone, methyl 3-pyridyl (350-03-8)
Ketone, methyl 4-pyridyl (1122-54-9)
Ketone, methyl 2-pyridyl (8CI) (1122-62-9)
Ketone, methyl pyrrol-2-yl (8CI) (1072-83-9)
Ketone, methyl 2-thienyl (88-15-3)
Ketone, methyl 3-thienyl (8CI) (1468-83-3)
Ketone, methyl 2,2,3-trimethylcyclopentyl (8CI) (17983-22-1)
Ketone, methyl vinyl (78-94-4)
Ketone, phenyl 4-pyridyl (14548-46-0)
Ketone propane (67-64-1)
Ketone, 3-pyridyl 3-(N-methyl-N-nitrosamino)propyl (64091-91-4)
Ketones, C14 (68458-86-6)
Ketonox (1338-23-4)
7-Ketooctanoic acid (14112-98-2)
Ketopentamethylene (120-92-3)
4-Ketopentanoic acid butyl ester (2052-15-5)
4-Ketopentenal (5729-47-5)
Ketoprofen (22071-15-4)
Ketopron (22071-15-4)
β-Ketopropane (67-64-1)
1-Ketopropionaldehyde (78-98-8)
2-Ketopropionaldehyde (78-98-8)
α-Ketopropionaldehyde (78-98-8)
α-Ketopropionic acid (127-17-3)
2-Ketosuccinic acid (328-42-7)
Ketosuccinic acid (328-42-7)
2-Keto-4-thiomethylbutyrate (583-92-6)
2-Ketothiomethylbutyric acid (583-92-6)
L-3-Ketothreohexuronic acid lactone (50-81-7)
2-Keto-1,7,7-trimethylnorcamphane (76-22-2)
4-Ketovaleric acid (123-76-2)
γ-Ketovaleric acid (123-76-2)
4-Ketovaleryl acetate (5185-97-7)
Ketrax (14769-73-4)
Kevadon (50-35-1)
Key-Serpine (50-55-5)
Key-Tusscapine (128-62-1)
Kh15N55M16 (11114-92-4)
Kh15N55M16V (11114-92-4)
KhS 596 (9011-06-7)
KhTsA (20736-64-5)
Khaladon 22 (75-45-6)
Khellinol (478-42-2)
Khinalizarin (81-61-8)
Khladon 113 (76-13-1)
Khladon 114B2 (124-73-2)
Khloridin (58-14-0)
Khlortrianizen (569-57-3)
KHP 2 (1344-28-1)
Khromolan (15242-96-3)
Kiatrium (439-14-5)
Kidoline (51-43-4)
Kieselguhr (61790-53-2)
Kieselguhr, Soda ash flux-calcined (68855-54-9)
Kiezelfluorwaterstofzuur (Dutch) (16961-83-4)
Kikuthrin (27223-49-0)
Kilcop 53 (1332-03-2)
Kilcop 53 (7758-98-7)
Kildip (120-36-5)
Kilex (1332-40-7)
Kilex (1332-65-6)
Kilex 3 (93-79-8)
Kilex Lindane (57-74-9)
Kill-All (7784-46-5)
Kill-Ko Rat (117-52-2)
Kill-Ko Rat Killer (82-66-6)
Killax (107-49-3)
Killeen (9000-07-1)
Kill kantz (86-88-4)
Killmaster (2921-88-2)
Kilmag (7778-44-1)
Kilmite 40 (107-49-3)
Kiloseb (88-85-7)
Kilprop (7085-19-0)
Kilprop (93-65-2)
Kilrat (1314-84-7)
Kilsem (94-74-6)
Kilval (2275-23-2)
Kinadion (84-80-0)
Kineks (80-35-3)
Kinetin (525-79-1)
Kinetin (Plant hormone) (525-79-1)
Kinex (80-35-3)
Kingcot (9004-34-6)
King's Gold (1303-33-9)
King's Green (12002-03-8)
King's Yellow (1303-33-9)
King's Yellow (7758-97-6)
Kinic acid (77-95-2)
Kinidin duretter (747-45-5)
Kinidin durules (747-45-5)
Kinilentin (747-45-5)
2-Kinoprene (42588-37-4)
Kinoprene (42588-37-4)
Kinurenic acid (492-27-3)
Kipca (58-27-5)
Kipca, Oil soluble (58-27-5)
Kiresuto B (139-33-3)

Kiresuto NTB	**(15467-20-6)**
Kiresuto P	**(140-01-2)**
Kirkstigmine bromide	**(114-80-7)**
Kirticopper	**(1344-67-8)**
Kitazin	**(13286-32-3)**
Kitazin L	**(26087-47-8)**
Kitazin P	**(26087-47-8)**
Kitine	**(50-55-5)**
Kiton Blue AR	**(2650-18-2)**
Kiton Brilliant Orange G	**(1934-20-9)**
Kiton Crimson 2R	**(3567-69-9)**
Kiton Fast Orange G	**(1936-15-8)**
Kiton Fast Yellow A	**(6373-74-6)**
Kiton Green F	**(4680-78-8)**
Kiton Green FC	**(4680-78-8)**
Kiton Magenta A	**(3244-88-0)**
Kiton Orange II	**(633-96-5)**
Kiton Orange MNO	**(587-98-4)**
Kiton Ponceau 4G	**(1934-20-9)**
Kiton Ponceau 2R	**(3761-53-3)**
Kiton Ponceau R	**(3761-53-3)**
Kiton Pure Blue L	**(2650-18-2)**
Kiton Pure Blue V	**(129-17-9)**
Kiton Pure Blue V.FQ	**(129-17-9)**
Kiton Red A	**(5858-93-5)**
Kiton Red 2G	**(3734-67-6)**
Kiton Red G	**(3734-67-6)**
Kiton Rubine R	**(3567-69-9)**
Kiton Rubine S	**(915-67-3)**
Kiton Scarlet 4R	**(2611-82-7)**
Kiton Scarlet 2RC	**(3761-53-3)**
Kiton Violet 4BNS	**(1694-09-3)**
Kiton Yellow MS	**(587-98-4)**
Kiton Yellow T	**(1934-21-0)**
Kivatin	**(50-34-0)**
Kiwa Grounder G	**(91-96-3)**
Kiwi Lustr 277	**(90-43-7)**
Klavi Kordal	**(55-63-0)**
Kleer-Lot	**(61-82-5)**
Kleesalz (German)	**(127-95-7)**
Klerat	**(56073-10-0)**
Klimanosid	**(50-55-5)**
Klimoral	**(50-27-1)**
Klingtite	**(86-87-3)**
Klinit	**(87-99-0)**
Klion	**(443-48-1)**
Kloben	**(555-37-3)**
Kloben Neburon	**(555-37-3)**
Klofiran	**(637-07-0)**
Klondike Yellow X-2261	**(1328-53-6)**
Klop	**(76-06-2)**
Kloramin	**(127-65-1)**
Kloramin	**(55-86-7)**
Kloramine-T	**(127-65-1)**
Klorex	**(7775-09-9)**
Klorinol	**(2122-77-2)**
Klorita	**(56-75-7)**
Klorocid S	**(56-75-7)**
Klorokin	**(54-05-7)**
Klorproman	**(69-09-0)**
Klorpromex	**(69-09-0)**
Klort	**(57-53-4)**
Klotrix	**(7447-40-7)**
Klottone	**(58-27-5)**
Klucel	**(9004-64-2)**
Knittex ASL	**(140-95-4)**
Knittex LE	**(1854-26-8)**
Knittex TC	**(9011-05-6)**
Knittex TS	**(9011-05-6)**
Knockmate	**(14484-64-1)**
Knollide	**(7681-11-0)**
Knoxweed	**(759-94-4)**
Ko 18	**(102-30-7)**
Koaxin	**(58-27-5)**
Kobalt (German, Polish)	**(7440-48-4)**
Kobalt chlorid (German)	**(7646-79-9)**
Kobaltocen (Czech)	**(1277-43-6)**
Koban	**(2593-15-9)**
Kobasic	**(1332-03-2)**
Kobasic	**(7758-98-7)**
Kobu	**(82-68-8)**
Kobutol	**(82-68-8)**
Kochineal Red A for Food	**(2611-82-7)**
Kocide	**(20427-59-2)**
Kocide	**(7704-34-9)**
Kocide 101	**(20427-59-2)**
Kocide 220	**(20427-59-2)**
Kocide 404	**(20427-59-2)**
Kocide SD	**(20427-59-2)**
Kodaflex	**(60-01-5)**
Kodaflex DBS	**(109-43-3)**
Kodaflex DOA	**(103-23-1)**
Kodaflex DOP	**(117-81-7)**
Kodaflex DOTP	**(6422-86-2)**
Kodaflex TOTM	**(3319-31-1)**
Kodaflex TXIB	**(6846-50-0)**
Kodaflex Triacetin	**(102-76-1)**
Kodak CD-3	**(25646-71-3)**
Kodak LR 115	**(9004-70-0)**
Kofein (Czech)	**(58-08-2)**
Koffein (German)	**(58-08-2)**
Kohlendioxyd (German)	**(124-38-9)**
Kohlendisulfid (schwefelkohlenstoff) (German)	**(75-15-0)**
Kohlenmonoxid (German)	**(630-08-0)**
Kohlenoxyd (German)	**(630-08-0)**
Kohlensaure (German)	**(124-38-9)**
Kokain	**(50-36-2)**
Kokan	**(50-36-2)**
Kokayeen	**(50-36-2)**
Kokotine	**(58-89-9)**
Kolamin (Czech)	**(141-43-5)**
Kolchamin	**(477-30-5)**
Kolchicin (Czech)	**(477-30-5)**
Kolkamin	**(477-30-5)**
Kolklot	**(58-27-5)**
2,4,6-Kollidin (Czech)	**(108-75-8)**
Kollidon	**(9003-39-8)**
Kollidon 17	**(9003-39-8)**
Kollidon 25	**(9003-39-8)**
Kollidon 30	**(9003-39-8)**
Kolofog	**(7704-34-9)**
Kolospray	**(7704-34-9)**
Kolphos	**(56-38-2)**
Kolpon	**(53-16-7)**
Komplexon	**(64-02-8)**
Komplexon I	**(139-13-9)**
Komplexon II	**(60-00-4)**
Komplexon III	**(139-33-3)**
Komplexon IV	**(482-54-2)**
Konakion	**(84-80-0)**
Kondremul	**(8012-95-1)**
Konesta	**(76-03-9)**
Konlax	**(577-11-7)**
Kontrast-U	**(126-31-8)**

Name	CAS No.
Koolmonoxyde (Dutch)	**(630-08-0)**
Koolstofdisulfide (zwavelkoolstof) (Dutch)	**(75-15-0)**
Koolstofoxychloride (Dutch)	**(75-44-5)**
Kop Karb	**(12069-69-1)**
Kop-Mite	**(510-15-6)**
Kop-Thiodan	**(115-29-7)**
Kop-Thion	**(121-75-5)**
Kopfume	**(106-93-4)**
Koplen 2	**(9003-53-6)**
Kopolymer butadien styrenovy (Czech)	**(9003-55-8)**
Koprez 87-110	**(9011-05-6)**
Koprol	**(77-09-8)**
Koprosterin (German)	**(360-68-9)**
Kopsol	**(50-29-3)**
Koro-Sulf	**(127-69-5)**
Korad	**(9011-14-7)**
Korax	**(2425-66-3)**
Korax	**(600-25-9)**
Korazol	**(54-95-5)**
Korazole	**(54-95-5)**
Kordiamin	**(59-26-7)**
Korglykon	**(508-75-8)**
Korium	**(97-23-4)**
Korlan	**(299-84-3)**
Korlane	**(299-84-3)**
Korobon	**(7782-42-5)**
Korodil	**(87-33-2)**
Korostan Red G	**(10169-02-5)**
Korotrin	**(9002-61-3)**
Korum	**(103-90-2)**
Korund	**(1302-74-5)**
Kosate	**(577-11-7)**
Kosmink	**(1333-86-4)**
Kosmobil	**(1333-86-4)**
Kosmolak	**(1333-86-4)**
Kosmos	**(1333-86-4)**
Kosmotherm	**(1333-86-4)**
Kosmovar	**(1333-86-4)**
Kosol	**(9037-22-3)**
Kostat P 650/5	**(61791-14-8)**
Koster Keunen Candelilla Wax	**(8006-44-8)**
Koster Keunen Carnauba Wax	**(8015-86-9)**
Kostil	**(9003-54-7)**
Kostil 235	**(9003-54-7)**
Kostil AN (ATX) 2010	**(9003-54-7)**
Kotion	**(122-14-5)**
Koyoside	**(15096-52-3)**
Kralac 1155	**(9003-54-7)**
Krasten 052	**(9003-53-6)**
Krasten 1.4	**(9003-53-6)**
Krasten SB	**(9003-53-6)**
Kratedyn	**(299-42-3)**
Kreatin	**(57-00-1)**
Krebiozon	**(57-00-1)**
Krecalvin	**(62-73-7)**
Kregasan	**(137-26-8)**
Kremart	**(36335-67-8)**
Krenite	**(25954-13-6)**
Krenite Brush Control Agent	**(25954-13-6)**
Krenite (Obs.)	**(2312-76-7)**
Krenite (Obs.)	**(534-52-1)**
Kresamone	**(534-52-1)**
p-Kresidin (Czech)	**(120-71-8)**
m-Kresol	**(108-39-4)**
p-Kresol	**(106-44-5)**
o-Kresol (German)	**(95-48-7)**
Kresole (German)	**(1319-77-3)**
Kresolen (Dutch)	**(1319-77-3)**
o-Kresol-glycidaether (German)	**(2210-79-9)**
Kresonite-E	**(534-52-1)**
Krezidine	**(120-71-8)**
Krezol (Polish)	**(1319-77-3)**
Krezone	**(94-74-6)**
Krezonite	**(2312-76-7)**
Krezotol	**(2312-76-7)**
Krezotol 50	**(534-52-1)**
Krezotol DNOC	**(2312-76-7)**
Kriplex	**(15307-79-6)**
Kriptin	**(91-84-9)**
Krisolamine	**(128-95-0)**
Kristall-Violett (German)	**(548-62-9)**
Kristallose	**(128-44-9)**
Krokydolith (German)	**(12001-28-4)**
Krolor Yellow KY 788D	**(1344-37-2)**
Kromal Blue OB	**(1325-87-7)**
Kromal Blue RBS	**(1325-87-7)**
Kromal Violet R	**(1325-82-2)**
Kromfax Solvent	**(111-48-8)**
Kromon Green B	**(3165-93-3)**
Kromon Helio Fast Red	**(2425-85-6)**
Kromon Helio Fast Red YS	**(2425-85-6)**
Kromon Lake Orange Toner	**(633-96-5)**
Kromon Orange G	**(3520-72-7)**
Kromon Permanent Red 4B	**(5858-81-1)**
Kromon Red R	**(2814-77-9)**
Kromon Sodium Lithol	**(1248-18-6)**
Kromon Yellow G	**(2512-29-0)**
Kromon Yellow GXT Conc	**(6358-85-6)**
Kromon Yellow MTB	**(6358-85-6)**
Kronisol	**(117-83-9)**
Kronitex	**(1330-78-5)**
Kronitex 200	**(96300-97-9)**
Kronitex CDP	**(26444-49-5)**
Kronitex 50 (9CI)	**(67426-57-7)**
Kronitex 100 (9CI)	**(66797-44-2)**
Kronitex 300 (9CI)	**(67426-58-8)**
Kronitex KP-140	**(78-51-3)**
Kronitex TOF	**(78-42-2)**
Kronitex TXP (9CI)	**(76775-00-3)**
Kronos	**(13463-67-7)**
Kronos Titanium Dioxide	**(13463-67-7)**
Krotenal	**(97-77-8)**
Krotilin	**(2971-38-2)**
Krotiline	**(2971-38-2)**
Krotonaldehyd (Czech)	**(4170-30-3)**
Krotylalkohol (Czech)	**(6117-91-5)**
Krotylchlorid (Czech)	**(591-97-9)**
Krovar II	**(314-40-9)**
Krumkil	**(117-52-2)**
Kryocide	**(15096-52-3)**
Kryogenin	**(103-03-7)**
Kryolith (German)	**(15096-52-3)**
Krypton	**(7439-90-9)**
Krypton, Compressed [UN 1056]	**(7439-90-9)**
Krypton, Refrigerated liquid [UN 1970]	**(7439-90-9)**
Krysid	**(86-88-4)**
Krysid pi	**(86-88-4)**
Krystallin	**(62-53-3)**
Krzewotoks	**(93-79-8)**
Krzewotox	**(93-79-8)**
Ksylen (Polish)	**(1330-20-7)**
Kubacron	**(133-67-5)**
Kue 2079A	**(33439-45-1)**
Kuemmel Oil (German)	**(8000-42-8)**

Kumachlor (Czech) (81-82-3)
Kumader (81-81-2)
Kumadu (81-81-2)
Kumarin (Czech) (91-64-5)
Kumatox (117-52-2)
m-Kumenylester kyseliny methylkarbaminove (Czech) (64-00-6)
Kumenylhydroperoxid (Czech) (80-15-9)
Kumiai (1129-41-5)
Kumoran (66-76-2)
Kumulus (7704-34-9)
Kumulus S (7704-34-9)
Kupfercarbonat (German) (12069-69-1)
Kupferoxychlorid (German) (1332-40-7)
Kupferoxydul (German) (1317-39-1)
Kupferron (Czech) (135-20-6)
Kupfersulfat (German) (7758-98-7)
Kupfersulfat (German) (7758-99-8)
Kupfersulfat-pentahydrat (German) (7758-99-8)
Kupfervitriol (German) (7758-99-8)
Kuprablau (20427-59-2)
Kupratsin (12122-67-7)
Kupricol (1332-40-7)
Kuprikol (1332-40-7)
Kuprite (1317-39-1)
Kuralon VP (9002-89-5)
Kuran (93-72-1)
Kurare OM 100 (9003-20-7)
Kurare PVA 205 (9002-89-5)
Kurare Poval 1700 (9002-89-5)
Kurate Poval 120 (9002-89-5)
Kurbamol TsEM (136-84-5)
Kurehalon A0 (9011-06-7)
Kurkumin (Czech) (458-37-7)
Kuromatsuen (475-20-7)
Kuromatsuene (475-20-7)
Kuron (93-72-1)
Kurosal (93-72-1)
Kurosal G (93-72-1)
Kurosal SL (2818-16-8)
Kurosal SL (93-72-1)
Kusa-Tohru (7775-09-9)
Kusagard (55635-13-7)
Kusakira (14491-59-9)
Kusatol (7775-09-9)
Kusnarin (389-08-2)
Kvercetin (Czech) (117-39-5)
Kwas benzydynodwukaroksylowy (Polish) (2130-56-5)
Kwas 1-(p-chlorobenzoilo)-2-metylo-5-metoksy-3-indolilo-octowy (Polish) (53-86-1)
Kwas 4-chloro-2-metylofenoksyoctowy (Polish) (94-74-6)
Kwas 4-chloro-2-metylofenoksypropionowy (Polish) (93-65-2)
Kwas 2,4-dwuchlorofenoksyoctowy (Polish) (94-75-7)
Kwas 2,4-dwuchlorofenoksypropionowy (Polish) (120-36-5)
Kwas dwumetylo-dwutiofosforowy (Polish) (756-80-9)
Kwas metaniowy (Polish) (64-18-6)
Kwas 2,4,5-trojchlorofenoksyoctowy (Polish) (93-76-5)
Kwas 2,4,5-trojchlorofenoksypropionowy (Polish) (93-72-1)
Kwasu 2,4-dwuchlorofenoksyoctowego (Polish) (94-75-7)
Kwell (58-89-9)
Kwells (114-49-8)
Kwik (Dutch) (7439-97-6)
Kwik-Kil (57-24-9)
Kwiksan (62-38-4)
Kwit (563-12-2)
Kyanacetamid (Czech) (107-91-5)
Kyanacethydrazid (Czech) (140-87-4)
Kyanid sodny (Czech) (143-33-9)
Kyanid stribrny (Czech) (506-64-9)
Kyanite (1302-76-7)
Kyanmethylester kyseliny octove (Czech) (1001-55-4)
2-Kyan-4-nitroanilin (Czech) (17420-30-3)
Kyanol (62-53-3)
Kyanostribrnan draselny (Czech) (506-61-6)
Kyanurchlorid (Czech) (108-77-0)
Kyaphenine (493-77-6)
Kylar (1596-84-5)
Kynex (80-35-3)
Kynurenic acid (492-27-3)
Kynurenine (343-65-7)
Kynuronic acid (492-27-3)
Kyocristine (2068-78-2)
Kyonate (333-20-0)
Kypchlor (57-74-9)
Kypfarin (81-81-2)
Kypfos (121-75-5)
Kypman 80 (12427-38-2)
Kypthion (56-38-2)
Kypzin (12122-67-7)
Kyselina p-terc.Butylbenzoova (Czech) (98-73-7)
Kyselina C (Czech) (131-27-1)
Kyselina H (Czech) (90-20-0)
Kyselina I (Czech) (87-02-5)
Kyselina abietova (Czech) (514-10-3)
Kyselina 2-acetoxybenzoova (Czech) (50-78-2)
Kyselina acetylsalicylova (Czech) (50-78-2)
Kyselina adipova (Czech) (124-04-9)
Kyselina akrylova (Czech) (79-10-7)
Kyselina amidosulfonova (Czech) (5329-14-6)
Kyselina amino-I (118-33-2)
Kyselina 4-aminoanisol-3-sulfonova (Czech) (13244-33-2)
Kyselina 4-aminoazobenzen-3,4'-disulfonova (Czech) (101-50-8)
Kyselina o-aminobenzoova (Czech) (118-92-3)
Kyselina p-aminobenzoova (Czech) (150-13-0)
Kyselina 3-amino-2,5-dichlorbenzoova (Czech) (133-90-4)
Kyselina ω-aminokapronova (Czech) (60-32-2)
Kyselina 4-aminolistova (Czech) (54-62-6)
Kyselina 4-amino-N^{10}-methylpteroylglutamova (Czech) (59-05-2)
Kyselina 1-amino-8-naftol-3,6-disulfonova (Czech) (90-20-0)
Kyselina 8-amino-1-naftol-3,6-disulfonova (Czech) (90-20-0)
Kyselina 2-amino-5-naftol-7-sulfonova (Czech) (87-02-5)
Kyselina 6-amino-1-naftol-3-sulfonova (Czech) (87-02-5)
Kyselina 2-amino-4-nitrofenol-6-sulfonova (Czech) (96-67-3)
Kyselina 2-amino-6-nitrofenol-4-sulfonova (Czech) (96-93-5)
Kyselina 4-amino-4'-nitrostilben-2,2'-disulfonova (Czech) (119-72-2)
Kyselina 4-aminopteroylglutamova (Czech) (54-62-6)
Kyselina p-aminosalicylova (Czech) (65-49-6)
Kyselina anilin-3-sulfonova (Czech) (121-47-1)
Kyselina anthranilova (Czech) (118-92-3)
Kyselina arsanilova (Czech) (98-50-0)
Kyselina askorbova (Czech) (50-81-7)
Kyselina 4,4'-azo-bis-(4-kyanvalerova) (Czech) (2638-94-0)
Kyselina benzenarsonova (Czech) (98-05-5)
Kyselina benzensulfonova (Czech) (98-11-3)
Kyselina benzidin-2,2'-disulfonova (Czech) (117-61-3)
Kyselina benzoova (Czech) (65-85-0)
Kyselina 4-(N,N-bis-(2-chlorethyl)-p-aminofenyl)maselna (Czech) (305-03-3)
Kyselina bromoctova (Czech) (79-08-3)
Kyselina 4-(2-(1-butenyl)karbonyl)-2,3-dichlorfenoxyoctova (Czech) (58-54-8)
Kyselina o-chlorbenzoova (Czech) (118-91-2)
Kyselina 4-chlorfenoxyoctova (Czech) (122-88-3)
Kyselina 4-(4-chlor-2-methylfenoxy)maselna (Czech) (94-81-5)
Kyselina 4-chlor-2-methylfenoxyoctova (Czech) (94-74-6)
Kyselina 2-(4-chlor-2-methylfenoxy)propionova (Czech) (93-65-2)
Kyselina chloroctova (Czech) (79-11-8)

Kyselina 2-chloro-4-nitrobenzoova (Czech) **(99-60-5)**
Kyselina 2-chlor-4-toluidin-5-sulfonova (Czech) **(88-51-7)**
Kyselina citrakonova (Czech) **(498-23-7)**
Kyselina citrazinova (Czech) **(99-11-6)**
Kyselina citronova (Czech) **(77-92-9)**
Kyselina cleve (Czech) **(119-79-9)**
Kyselina 1,2-cyklohexylendiamintetraoctova (Czech) **(482-54-2)**
Kyselina 9-decen-1-karboxylova (Czech) **(112-38-9)**
Kyselina dehydroacetova (Czech) **(520-45-6)**
Kyselina dehydroacetova sodna sul (Czech) **(4418-26-2)**
Kyselina 4-desoxy-4-amino-N^{10}-methyllistova (Czech) **(59-05-2)**
Kyselina 2,4-diaminobenzensulfonova (Czech) **(88-63-1)**
Kyselina N-(p-((2,4-diamino-6-pteridinylmethyl)amino)-benzoyl)-l(+)-glutamova (Czech) **(54-62-6)**
Kyselina N-(p-((2,4-diamino-6-pteridinylmethyl)methylamino)-benzoyl)-l-glutamova (Czech) **(59-05-2)**
Kyselina 4-(2,4-dichlorfenoxy)maselna (Czech) **(94-82-6)**
Kyselina 2,4-dichlorfenoxyoctova (Czech) **(94-75-7)**
Kyselina 2-(2,4-dichlorfenoxy)propionova (Czech) **(120-36-5)**
Kyselina dichlorisokyanurova (Czech) **(2782-57-2)**
Kyselina 3,6-dichlor-2-methoxybenzoova (Czech) **(1918-00-9)**
Kyselina 2,5-dichlor-4-(3'-methyl-5'-pyrazolon-1'-yl)benzen-sulfonova (Czech) **(84-57-1)**
Kyselina 2,5-dichlor-3-nitrobenzoova (Czech) **(88-86-8)**
Kyselina dichloroctova (Czech) **(79-43-6)**
Kyselina 3,6-dichlorpikolinova (Czech) **(1702-17-6)**
Kyselina 2,2-dichlorpropionova (Czech) **(75-99-0)**
Kyselina p-N,N-dichlorsulfamoylbenzoova (Czech) **(80-13-7)**
Kyselina 5,5-diethylbarbiturova (Czech) **(57-44-3)**
Kyselina O,O-diethyldithiofosforecna (Czech) **(298-06-6)**
Kyselina di-(2-ethylhexyl)fosforecna (Czech) **(298-07-7)**
Kyselina diethyloctova (Czech) **(88-09-5)**
Kyselina 2,5-dihydroxybenzoova (Czech) **(490-79-9)**
Kyselina 2,3-dihydroxybutandiova (Czech) **(87-69-4)**
Kyselina 3,5-dikarboxybenzensulfonova sodny (Czech) **(6362-79-4)**
Kyselina O,O-dimethyldithiofosforcna (Czech) **(756-80-9)**
Kyselina 2,2-dimethylpropionova (Czech) **(75-98-9)**
Kyselina 4,4'-dinitrostilben-2,2'-disulfonova (Czech) **(128-42-7)**
Kyselina 3-dodecylguanidinooctova (Czech) **(2439-10-3)**
Kyselina dusicne (Czech) **(7697-37-2)**
Kyselina dusite (Czech) **(7782-77-6)**
Kyselina 3,6-endomethylen-3,4,5,6,7,7-hexachlor-δ^4-tetra-hydroftalova (Czech) **(115-28-6)**
Kyselina ethakrynova (Czech) **(58-54-8)**
Kyselina 3-(2-ethylbutoxy)propionova (Czech) **(10213-74-8)**
Kyselina ethylendiamintetraoctova (Czech) **(60-00-4)**
Kyselina 2-ethylkapronova (Czech) **(149-57-5)**
Kyselina 1,3-fenylendiamin-4-sulfonova (Czech) **(88-63-1)**
Kyselina 2-fenyl-2-hydroxyethanova (Czech) **(90-64-2)**
Kyselina fenyloctova (Czech) **(103-82-2)**
Kyselina ftalova (Czech) **(88-99-3)**
Kyselina fumarova (Czech) **(110-17-8)**
Kyselina 2-furoova (Czech) **(88-14-2)**
Kyselina gallova (Czech) **(149-91-7)**
Kyselina gentisinova (Czech) **(490-79-9)**
Kyselina glykolova (Czech) **(79-14-1)**
Kyselina glyoxylova (Czech) **(298-12-4)**
Kyselina heptafluormaselna (Czech) **(375-22-4)**
Kyselina heptan-3-karboxylova (Czech) **(149-57-5)**
Kyselina 3-hepten-3-karboxylova (Czech) **(5309-52-4)**
Kyselina 1,2,3,4,7,7-hexachlorbicyklo(2,2,1)hept-2-en-5,6-di-karboxylova (Czech) **(115-28-6)**
Kyselina 2-hydroxybenzoova (Czech) **(69-72-7)**
Kyselina 3-hydroxybenzoova (Czech) **(99-06-9)**
Kyselina 4-hydroxybenzoova (Czech) **(99-96-7)**
Kyselina hydroxybutandiova (Czech) **(6915-15-7)**
Kyselina 3-hydroxy-2-naftoova (Czech) **(92-70-6)**
Kyselina 4-hydroxy-3-nitrofenylarsonova (Czech) **(121-19-7)**
Kyselina hydroxyoctova (Czech) **(79-14-1)**
Kyselina 12-hydroxy-9-oktadecenova (Czech) **(141-22-0)**
Kyselina 2-hydroxypropanova (Czech) **(50-21-5)**
Kyselina 2-hydroxy-1,2,3-propantrikarbonova (Czech) **(77-92-9)**
Kyselina 4-indol-3-ylmaselina (Czech) **(133-32-4)**
Kyselina 3-indolyloctova (Czech) **(87-51-4)**
Kyselina isoftalova (Czech) **(121-91-5)**
Kyselina isomaselna (Czech) **(79-31-2)**
Kyselina isovalerova (Czech) **(503-74-2)**
Kyselina jablecna (Czech) **(6915-15-7)**
Kyselina jantarova (Czech) **(110-15-6)**
Kyselina o-jodbenzoova (Czech) **(88-67-5)**
Kyselina jodoctova (Czech) **(64-69-7)**
Kyselina kakodylova (Czech) **(75-60-5)**
Kyselina kapronova (Czech) **(142-62-1)**
Kyselina kaprylova (Czech) **(124-07-2)**
Kyselina N-karbamylarsanilova (Czech) **(121-59-5)**
Kyselina krotonova (Czech) **(3724-65-0)**
Kyselina kyanoctova (Czech) **(372-09-8)**
Kyselina kyanurova (Czech) **(108-80-5)**
Kyselina listova (Czech) **(59-30-3)**
Kyselina maleinova (Czech) **(110-16-7)**
Kyselina malonova (Czech) **(141-82-2)**
Kyselina mandlova (Czech) **(90-64-2)**
Kyselina maselna (Czech) **(107-92-6)**
Kyselina merkaptooctova (Czech) **(68-11-1)**
Kyselina metanilova (Czech) **(121-47-1)**
Kyselina methakrylova (Czech) **(79-41-4)**
Kyselina methansulfonova (Czech) **(75-75-2)**
Kyselina 2-methoxybenzoova (Czech) **(579-75-9)**
Kyselina 4-methoxybenzoova (Czech) **(100-09-4)**
Kyselina methylarsonova (Czech) **(124-58-3)**
Kyselina 3-methyl-2-butenova (Czech) **(541-47-9)**
Kyselina 2-methylvalerova (Czech) **(97-61-0)**
Kyselina mlecna (Czech) **(50-21-5)**
Kyselina mravenci (Czech) **(64-18-6)**
Kyselina 2-naftalensulfonova (Czech) **(120-18-3)**
Kyselina nafthionova (Czech) **(84-86-6)**
Kyselina 2-naftol-6-sulfonova (Czech) **(93-01-6)**
Kyselina 2-naftylamin-1,5-disulfonova (Czech) **(117-62-4)**
Kyselina 2-naftylamin-4,8-disulfonova (Czech) **(131-27-1)**
Kyselina 1-naftylamin-4-sulfonova (Czech) **(84-86-6)**
Kyselina 1-naftylamin-6-sulfonova (Czech) **(119-79-9)**
Kyselina 2-naftylamin-1-sulfonova (Czech) **(81-16-3)**
Kyselina N-1-naftylftalamova (Czech) **(132-66-1)**
Kyselina 1-naftyloctova (Czech) **(86-87-3)**
Kyselina nikotinova (Czech) **(59-67-6)**
Kyselina nitrilotrioctova (Czech) **(139-13-9)**
Kyselina 4-nitro-2-aminofenol-6-sulfonova (Czech) **(96-67-3)**
Kyselina 6-nitro-2-aminofenol-4-sulfonova (Czech) **(96-93-5)**
Kyselina 4-nitro-4'-aminostilben-2,2'-disulfonova (Czech) **(119-72-2)**
Kyselina 3-nitrobenzensulfonova (Czech) **(98-47-5)**
Kyselina nitrobenzen-m-sulfonova (Czech) **(98-47-5)**
Kyselina p-nitrobenzoova (Czech) **(62-23-7)**
Kyselina 4-nitrotoluen-2-sulfonova (Czech) **(121-03-9)**
Kyselina octova (Czech) **(64-19-7)**
Kyselina pantothenova (Czech) **(79-83-4)**
Kyselina paraskorbova (Czech) **(10048-32-5)**
Kyselina penicilova (Czech) **(90-65-3)**
Kyselina 1,3-pentadien-1-karboxylova (Czech) **(110-44-1)**
Kyselina peroxyoctova (Czech) **(79-21-0)**
Kyselina pikrova (Czech) **(88-89-1)**
Kyselina pivalova (Czech) **(75-98-9)**
Kyselina propionova (Czech) **(79-09-4)**
Kyselina 2-propylvalerova (Czech) **(99-66-1)**
Kyselina pyroslizova (Czech) **(88-14-2)**

Kyselina ricinolova (Czech)	**(141-22-0)**
Kyselina salicylova (Czech)	**(69-72-7)**
Kyselina schaferova (Czech)	**(93-01-6)**
Kyselina skoricove (Czech)	**(621-82-9)**
Kyselina sorbova (Czech)	**(110-44-1)**
Kyselina stavelova (Czech)	**(144-62-7)**
Kyselina sulfaminova (Czech)	**(5329-14-6)**
Kyselina sulfanilova (Czech)	**(121-57-3)**
Kyselina sulfooctova (Czech)	**(123-43-3)**
Kyselina sulfo-tobiasova (Czech)	**(117-62-4)**
Kyselina tereftalova (Czech)	**(100-21-0)**
Kyselina 1,2,5,6-tetrahydrobenzoova (Czech)	**(4771-80-6)**
Kyselina 3,3-thiodipropionova (Czech)	**(111-17-1)**
Kyselina β,β'-thiodipropionova (Czech)	**(111-17-1)**
Kyselina thioglykolova (Czech)	**(68-11-1)**
Kyselina thiooctova (Czech)	**(507-09-5)**
Kyselina tobiasova (Czech)	**(81-16-3)**
Kyselina p-toluensulfonova (Czech)	**(104-15-4)**
Kyselina 4-toluidin-3-sulfonova (Czech)	**(88-44-8)**
Kyselina N-m-tolylftalamova (Czech)	**(85-72-3)**
Kyselina trichloisokyanurova (Czech)	**(87-90-1)**
Kyselina 2,3,6-trichlorbenzoova (Czech)	**(50-31-7)**
Kyselina 2,3,6-trichlorbenzoova dimethylamonna sul (Czech)	**(3426-62-8)**
Kyselina 2-(2,4,5-trichlorfenoxy)propionova (Czech)	**(93-72-1)**
Kyselina 2,3,6-trichlorfenyloctova (Czech)	**(85-34-7)**
Kyselina 3,5,6-trichlor-2-methoxybenzoova (Czech)	**(2307-49-5)**
Kyselina trichloroctova (Czech)	**(76-03-9)**
Kyselina 2,2,3-trichlorpropionova (Czech)	**(3278-46-4)**
Kyselina trifluoroctova (Czech)	**(76-05-1)**
Kyselina 3,4,5-trihydroxybenzoova (Czech)	**(149-91-7)**
Kyselina 2,3,5-trijodbenzoova (Czech)	**(88-82-4)**
Kyselina undecylenova (Czech)	**(112-38-9)**
Kyselina valerova (Czech)	**(109-52-4)**
Kyselina vinna (Czech)	**(87-69-4)**
Kyslicnik di-n-butylcinicity (Czech)	**(818-08-6)**
Kyslicnik tri-n-butylcinicity (Czech)	**(56-35-9)**
L 11/6	**(298-02-2)**
L 14	**(9003-29-6)**
L 14 (Polymer]	**(9003-29-6)**
L 16/184	**(2497-07-6)**
L 43	**(63148-62-9)**
L 45 (Silicone)	**(63148-62-9)**
L 67	**(721-50-6)**
L 100	**(9003-29-6)**
L 195	**(9011-05-6)**
L 343	**(2275-18-5)**
L 546	**(63148-62-9)**
L 1811	**(702-54-5)**
L 1945	**(54-85-3)**
L 2329	**(68-90-6)**
L 33379	**(1524-88-5)**
L 54521	**(1861-40-1)**
L-01748	**(514-73-8)**
L-3	**(7440-22-4)**
L-310	**(8001-26-1)**
L-395	**(60-51-5)**
L-561	**(2597-03-7)**
L-3003 (Russian)	**(25791-96-2)**
L-5103	**(13292-46-1)**
L-34314	**(957-51-7)**
L-36352	**(1582-09-8)**
L16	**(7429-90-5)**
LA	**(478-94-4)**
LA	**(9002-92-0)**
LA 01	**(9004-34-6)**
LA 1	**(146-22-5)**
LA 6	**(1344-28-1)**
LA 7	**(9002-92-0)**
LA 96A	**(603-50-9)**
LA 956	**(2152-34-3)**
LA (Alcohol)	**(9002-92-0)**
2329 LABAZ	**(68-90-6)**
LA-III	**(439-14-5)**
LA-III	**(478-94-4)**
LAS-Na	**(68411-30-3)**
LAS	**(42615-29-2)**
LAS 99	**(27176-87-0)**
LAS, sodium salt	**(68411-30-3)**
LA XVII	**(1812-30-2)**
LB 502	**(54-31-9)**
L-Blau 3	**(129-17-9)**
L-Blau 1 (German)	**(81-77-6)**
L-Blau 2 (German)	**(860-22-0)**
LB-Rot 1	**(16423-68-0)**
LBX 5	**(147-14-8)**
LCB 150	**(9003-17-2)**
LCR	**(57-22-7)**
"L," Carpserp	**(50-55-5)**
LD 400	**(9002-88-4)**
LD 600	**(9002-88-4)**
LDA	**(120-40-1)**
LDE	**(120-40-1)**
LD Norgestrel (French)	**(6533-00-2)**
LDPE 4	**(9002-88-4)**
29060 LE	**(143-67-9)**
29060-LE	**(865-21-4)**
LE 79-519	**(52645-53-1)**
LEA	**(3055-94-5)**
LEO 299	**(4891-15-0)**
LFA 2043	**(36734-19-7)**
LG 56	**(25791-96-2)**
LG 61	**(2511-10-6)**
LG 56 (Polymer)	**(25791-96-2)**
LH 3012	**(12071-83-9)**
L-5-HTP	**(4350-09-8)**
LH 30/Z	**(12071-83-9)**
LIPAL 15S	**(9004-99-3)**
LIV 1176	**(16091-18-2)**
L.J. 206	**(638-23-3)**
LK 36	**(3089-11-0)**
LM 91	**(3691-35-8)**
LM 94	**(90-33-5)**
LM 637	**(28772-56-7)**
LM-637	**(28772-56-7)**
LM-2717	**(22316-47-8)**
L 3MB1	**(118-82-1)**
LMR 100	**(14807-96-6)**
LM Seed Protectant	**(86-85-1)**
L. Orange Z2010	**(2783-94-0)**
LPC	**(104-74-5)**
L.P.G.	**(68476-85-7)**
LPG	**(1538-09-6)**
L.P.G. (ACGIH,OSHA)	**(68476-85-7)**
LPG Ethyl mercaptan 1010	**(75-08-1)**
LPT	**(9011-14-7)**
LPT 1	**(9011-14-7)**
LR 115	**(9004-70-0)**
L-Red 3	**(915-67-3)**
L-Red 5	**(5850-44-2)**
L Red Z 3000	**(3257-28-1)**
L. Red Z 3040	**(3567-69-9)**
L.S. 3394	**(56-35-9)**
LS 061A	**(9003-53-6)**
LS 299	**(4891-15-0)**

LS 4442 **(639-58-7)**
LS 74783 **(39148-24-8)**
LS 80.1213 **(62476-59-9)**
D-LSD **(50-37-3)**
LSD **(50-37-3)**
LSD-25 **(50-37-3)**
LS 1028E **(9003-53-6)**
LSO-M **(9011-14-7)**
LSO-M 4B **(9011-14-7)**
L-T3 **(6893-02-3)**
L-T4 **(51-48-9)**
L-TRP **(73-22-3)**
LTX (9CI) **(75602-99-2)**
LV 50 **(9003-29-6)**
LW 3170 **(1977-10-2)**
LX 1065 **(64742-16-1)**
LX 14-0 **(2691-41-0)**
LX 3 **(9004-99-3)**
LYP 97 **(105-74-8)**
LYP 97F **(105-74-8)**
LZ-MB 1 **(118-82-1)**
Lab **(9001-98-3)**
Lab Ferment (German) **(9001-98-3)**
Labazene **(1069-66-5)**
Labilite **(23564-05-8)**
Labopal **(57-41-0)**
Labrol EC **(4301-50-2)**
Lacceric acid **(3625-52-3)**
L'acide oleique (French) **(112-80-1)**
Lacolin **(72-17-3)**
Lacqren 506 **(9003-53-6)**
Lacqren 550 **(9003-53-6)**
Lacqten 1020 **(9002-88-4)**
Lacquer Orange V **(2646-17-5)**
Lacquer Orange V 3G **(2051-85-6)**
Lacquer Orange VG **(842-07-9)**
Lacquer Orange VR **(3118-97-6)**
Lacquer Pink S **(509-34-2)**
Lacquer Red V **(85-83-6)**
Lacquer Red V3B **(6368-72-5)**
Lacquer Red VS **(85-83-6)**
Lactamide, N-ethyl-, carbanilate **(16118-45-9)**
Lactamide, N-ethyl-, carbanilate (ester) (8CI) **(16118-45-9)**
Lactams, polyamides (VAN) **(63428-84-2)**
DL-Lactate **(598-82-3)**
Lactate d'antimoine (French) **(58164-88-8)**
Lactate d'ethyle (French) **(97-64-3)**
Lactato de antimonio (Spanish) **(58164-88-8)**
DL-Lactic acid **(50-21-5)**
DL-Lactic acid **(598-82-3)**
Lactic acid **(598-82-3)**
Lactic-acid **(50-21-5)**
Lactic acid, antimony salt **(58164-88-8)**
Lactic acid, butyl ester **(138-22-7)**
Lactic acid, calcium salt (2:1) **(814-80-2)**
Lactic acid, copper(2+) salt (2:1) (8CI) **(814-81-3)**
Lactic acid, ethyl ester **(97-64-3)**
Lactic acid, iron(2+) salt (2:1) **(5905-52-2)**
Lactic acid, isopropyl ester **(617-51-6)**
Lactic acid, 2-methyl- (8CI) **(594-61-6)**
Lactic acid, methyl ester (8CI) **(547-64-8)**
Lactic acid, monosodium salt **(72-17-3)**
Lactic acid sodium salt **(72-17-3)**
Lactic acid, sodium zirconium salt (4:4:1) **(10377-98-7)**
Lactic acid, strontium salt (2:1) **(29870-99-3)**
Lactic acid, tris(2-hydroxyethyl)(phenylmercuri)ammonium deriv. **(23319-66-6)**
Lactic acid, ion(1-), tris(2-hydroxyethyl)(phenylmercurio)-ammonium **(23319-66-6)**
Lactin **(63-42-3)**
Lactine **(9004-99-3)**
Lactobacillus lactis dorner factor **(68-19-9)**
Lactobaryt **(7727-43-7)**
Lactobiose **(63-42-3)**
Lactocaine **(51-05-8)**
Lactofen **(77501-63-4)**
Lactoflavin **(83-88-5)**
Lactoflavine **(83-88-5)**
Lactogen **(9002-62-4)**
Lactogenic Hormone **(9002-62-4)**
Lactonitrile **(78-97-7)**
Lactonitrile, 2-methyl **(75-86-5)**
D-Lactose **(63-42-3)**
Lactose **(63-42-3)**
Lactosomatotropic Hormone **(9002-62-4)**
Ladakamycin **(320-67-2)**
Laevoral **(7660-25-5)**
Laevosan **(7660-25-5)**
Laevulic acid **(123-76-2)**
Laevulinic acid **(123-76-2)**
Lageracetal **(5921-80-2)**
Lagistase **(476-66-4)**
Laidlaw U-San-O Moth Proofing Spray **(16919-19-0)**
Lake Basic Violet **(1325-82-2)**
Lake Black Extra **(11099-03-9)**
Lake Blue B Base **(119-90-4)**
Lake Developer A **(92-77-3)**
Lake Fast Blue BS **(81-77-6)**
Lake Fast Blue GGS **(81-77-6)**
Lake Orange A **(633-96-5)**
Lake Orange II YS **(633-96-5)**
Lake Ponceau **(3761-53-3)**
Lake Red 1520 **(5160-02-1)**
Lake Red BK Base **(95-79-4)**
Lake Red C **(5160-02-1)**
Lake Red C 18287 **(5160-02-1)**
Lake Red C 18958 **(5160-02-1)**
Lake Red C 21245 **(5160-02-1)**
Lake Red C 27200 **(5160-02-1)**
Lake Red C 27217 **(5160-02-1)**
Lake Red C 27218 **(5160-02-1)**
Lake Red C Amine **(88-53-9)**
Lake Red C Barium Toner **(5160-02-1)**
Lake Red CC **(5160-02-1)**
Lake Red CCT **(5160-02-1)**
Lake Red CR **(5160-02-1)**
Lake Red CRLC-232 (Barium) **(5160-02-1)**
Lake Red C Toner 8195 **(5160-02-1)**
Lake Red C Toner 8366 **(5160-02-1)**
Lake Red GB Barium salt **(5160-02-1)**
Lake Red G Base **(89-62-3)**
Lake Red 2GL **(3468-63-1)**
Lake Red 4R **(2425-85-6)**
Lake Red R **(1248-18-6)**
Lake Red 4RII **(2425-85-6)**
Lake Red RL **(1248-18-6)**
Lake Red RRG **(5160-02-1)**
Lake Red Toner C **(5160-02-1)**
Lake Red Toner LCLL **(5160-02-1)**
Lake Scarlet G Base **(99-55-8)**
Lake Scarlet GG Base **(95-82-9)**
Lake Scarlet R **(3761-53-3)**
Lake Scarlet 2RBN **(3761-53-3)**
Lake Yellow **(1934-21-0)**

Laktonitril (Czech) **(78-97-7)**
Labetalol **(36894-69-6)**
Lam **(616-45-5)**
Lamacit CA **(9004-99-3)**
Lambast **(23184-66-9)**
Lambast **(845-52-3)**
Lambeth **(9003-07-0)**
Lambraten **(60-23-1)**
Lambrol **(4301-50-2)**
Lamdiol **(50-28-2)**
Lamidon **(15687-27-1)**
Lamitex **(9005-38-3)**
Lamoryl **(126-07-8)**
Lamprecid **(88-30-2)**
Lanadin **(79-01-6)**
Lanain **(8006-54-0)**
Lanalin **(8006-54-0)**
Lanaperl Blue B **(2666-17-3)**
Lanaperl Fast Red 3G **(3567-65-5)**
Lanaperl Red G **(3567-65-5)**
Lanastil **(63428-84-2)**
Lanatoxin **(71-63-6)**
Lanazine **(300-42-5)**
Lance **(51487-69-5)**
Lanco Wax PP 1362D **(9003-07-0)**
Lancol **(143-28-2)**
Landalgine **(9005-32-7)**
Landisan **(151-38-2)**
Landplaster **(13397-24-5)**
2,3,5-Landrin **(2655-15-4)**
Landrin **(12407-86-2)**
Landrin **(2686-99-9)**
Landrin B **(2655-15-4)**
Lanesin **(8006-54-0)**
Lanesta **(145-94-8)**
Lanette Wax KS **(112-72-1)**
Lanette Wax-S **(151-21-3)**
Lanettes **(9002-92-0)**
Lanex **(2164-17-2)**
Langford **(1332-58-7)**
Laniazid **(54-85-3)**
Lanichol **(8006-54-0)**
Lanicor **(20830-75-5)**
Laniol **(8006-54-0)**
Laniozid **(54-85-3)**
Lankroflex ED 3 **(106-84-3)**
Lannagol LF **(9004-96-0)**
Lannate L **(16752-77-5)**
Lannate (OSHA) **(16752-77-5)**
Lanodoxin **(83-73-8)**
Lanol **(57-88-5)**
Lanolin **(8006-54-0)**
Lanolin anhydrous USP **(8006-54-0)**
Lanolin, ethoxylated **(61790-81-6)**
Lanolin fatty acids, isopropyl esters **(63393-93-1)**
Lanolin oil **(70321-63-0)**
Lanolin oil **(8006-54-0)**
Lanophyllin **(58-55-9)**
Lanoxin **(20830-75-5)**
Lanray **(34622-58-7)**
Lanstan **(2425-66-3)**
Lanstan **(600-25-9)**
Lantadene A **(467-81-2)**
Lantadene B **(467-82-3)**
Lanthanum-chloride **(10099-58-8)**
Lanthanum (9CI) **(7439-91-0)**
Lanthanum, Isotope of mass 140 (8CI,9CI) **(13981-28-7)**
Lanthanum fluoride (9CI) **(13709-38-1)**
Lanthanum(III) nitrate, hexahydrate (1:3:6) **(10277-43-7)**
Lanthanum oxide (9CI) **(1312-81-8)**
Lanthanum, tris(ethanedioato(2-))di- (9CI) **(537-03-1)**
Lanthanum-nitrate **(10099-59-9)**
Lantrol **(8006-54-0)**
Lanum **(8006-54-0)**
Lanvis **(154-42-7)**
Lapachic acid **(84-79-7)**
Lapachol **(84-79-7)**
Lapachol Wood **(84-79-7)**
Lapaquin **(54-05-7)**
Laplen **(9011-06-7)**
Laprol 263 **(25791-96-2)**
Laprol 503 **(25791-96-2)**
Laprol 702 **(25322-69-4)**
Laprol 1003 **(25791-96-2)**
Laprol 1502 **(9003-11-6)**
Laprol 1601 **(9003-11-6)**
Laprol 3003 **(25791-96-2)**
Laprol 5003 **(25791-96-2)**
Laprol 1502-2-70 **(9003-11-6)**
Laprol 5003-2B10 **(9003-11-6)**
Laptran **(5131-24-8)**
Lapyrium chloride **(6272-74-8)**
Laradopa **(59-92-7)**
Lard Factor **(68-26-8)**
Lard Oil **(8016-28-2)**
Larex **(9011-05-6)**
Largactil **(50-53-3)**
Largactil monohydrochloride **(69-09-0)**
Largactilothiazine **(50-53-3)**
Largactyl **(50-53-3)**
Largaktyl **(69-09-0)**
Larixic acid **(118-71-8)**
Larixin **(15686-71-2)**
Larixinic acid **(118-71-8)**
Larodopa **(59-92-7)**
Laroks (Polish) **(330-55-2)**
Laroscorbine **(50-81-7)**
Laroxil **(50-48-6)**
Laroxyl **(50-48-6)**
Larten **(57-53-4)**
Larvacide **(76-06-2)**
Larvatrol **(68038-71-1)**
Larvin **(59669-26-0)**
Lasan **(480-22-8)**
Lasex **(54-31-9)**
Lasilix **(54-31-9)**
Lasiocarpine **(303-34-4)**
Lasix **(54-31-9)**
Lasochron **(58-89-9)**
Lasodex **(317-34-0)**
Lasso **(15972-60-8)**
Lasso Micro-Tech **(15972-60-8)**
Lastan **(2425-66-3)**
Lastar A **(2403-88-5)**
Latex **(9016-00-6)**
Latex SVKh **(9011-06-7)**
Latexol Fast Blue SD **(81-77-6)**
Latexol Fast Orange J **(3520-72-7)**
Latexol Red J **(2814-77-9)**
Latexol Scarlet R **(5160-02-1)**
Latexol Scarlet R Solupowder **(5160-02-1)**
Lathanol **(1847-58-1)**
Lathanol LAL **(1847-58-1)**
Lathanol-LAL 70 **(1847-58-1)**

Name	CAS No.
Latka 42 (Czech)	**(81-81-2)**
Latka 104 (Czech)	**(76-44-8)**
Latka 118 (Czech)	**(309-00-2)**
Latka 269 (Czech)	**(72-20-8)**
Latka 333 (Czech)	**(83-26-1)**
Latka 497 (Czech)	**(60-57-1)**
Latka 604 (Czech)	**(117-80-6)**
Latka 612 (Czech)	**(94-96-2)**
Latka 666 (Czech)	**(608-73-1)**
Latka 711 (Czech)	**(465-73-6)**
Latka 923 (Czech)	**(97-16-5)**
Latka 1068 (Czech)	**(57-74-9)**
Latka 1080 (Czech)	**(62-74-8)**
Latka 1836 (Czech)	**(311-47-7)**
Latka 3956 (Czech)	**(8001-35-2)**
Latka 4049 (Czech)	**(121-75-5)**
Latka 6249 (Czech)	**(3278-46-4)**
Latka 7744 (Czech)	**(63-25-2)**
Latka 3960x14 (Czech)	**(8001-50-1)**
Latusate	**(115-44-6)**
Latyl Blue BCN	**(12217-79-7)**
Latyl Cerise N	**(17418-58-5)**
Laudicon	**(466-99-9)**
Laudran di-n-butylcinicity (Czech)	**(77-58-7)**
Laughing Gas	**(10024-97-2)**
Lauramide	**(1120-16-7)**
Lauramide DEA	**(120-40-1)**
Lauramide MEA	**(142-78-9)**
Lauramide MIPA	**(142-54-1)**
Lauramidopropyl dimethylamine	**(3179-80-4)**
Lauramidopropyldimethylamine	**(3179-80-4)**
Lauramidopropyl betaine	**(4292-10-8)**
Lauraminopropionic acid	**(1462-54-0)**
Lauran sodny (Czech)	**(629-25-4)**
Laurel Camphor	**(76-22-2)**
Laureth-1	**(4536-30-5)**
Laureth-2	**(3055-93-4)**
Laureth-4	**(5274-68-0)**
Laureth-5	**(3055-95-6)**
Laureth-7	**(3055-97-8)**
Laureth	**(9002-92-0)**
Laureth 9	**(9002-92-0)**
Laureth 12	**(9002-92-0)**
Laureth 23	**(9002-92-0)**
Lauric-acid	**(143-07-7)**
Lauric acid, Compd with 2,2'-iminodiethanol (1:1)	**(7487-79-8)**
Lauric acid amide	**(1120-16-7)**
Lauric acid, ammonium salt	**(2437-23-2)**
Lauric acid, barium cadmium salt	**(15337-60-7)**
Lauric acid, cadmium salt (2:1)	**(2605-44-9)**
Lauric acid chloride	**(112-16-3)**
Lauric acid, dibutylstannylene deriv.	**(77-58-7)**
Lauric acid, dibutylstannylene salt	**(77-58-7)**
Lauric acid, dibutyltin deriv.	**(77-58-7)**
Lauric acid diethanolamide	**(120-40-1)**
Lauric acid diethanolamine condensate	**(120-40-1)**
Lauric acid diethanolamine salt	**(7487-79-8)**
Lauric acid, 2,3-epoxypropyl ester	**(1984-77-6)**
Lauric acid, ester with dimethyl (2,2,2-trichloro-1-hydroxyethyl)-phosphonate (8CI)	**(4414-15-7)**
Lauric acid ester with 2-hydroxyethyl thiocyanate	**(301-11-1)**
Lauric acid ethanolamide	**(142-78-9)**
Lauric acid, 2-ethoxyethyl ester (8CI)	**(106-13-8)**
Lauric acid, ethyl ester (8CI)	**(106-33-2)**
Lauric acid, 2-(2-hydroxyethoxy)ethyl ester (8CI)	**(141-20-8)**
Lauric acid, methyl ester (8CI)	**(111-82-0)**
Lauric acid monoethanolamide	**(142-78-9)**
Lauric acid monoethanolamine	**(142-78-9)**
Lauric acid, sodium salt	**(629-25-4)**
Lauric acid, 2-thiocyanatoethyl ester	**(301-11-1)**
Lauric acid triglyceride	**(538-24-9)**
Lauric acid triglycerin ester	**(538-24-9)**
Lauric alcohol	**(112-53-8)**
Lauric diethanolamide	**(120-40-1)**
Lauric diethanolamine	**(7487-79-8)**
Lauric 3-dimethylaminopropylamide	**(3179-80-4)**
Lauric ethylolamide	**(142-78-9)**
Lauric N-(2-hydroxyethyl)amide	**(142-78-9)**
Lauric monoethanolamide	**(142-78-9)**
Lauridit LM	**(142-78-9)**
Laurine	**(107-75-5)**
Laurinic alcohol	**(112-53-8)**
Laurin, tri- (8CI)	**(538-24-9)**
Lauristyl diglycol ether sulfate sodium salt	**(3088-31-1)**
Lauro-Sebum	**(141-20-8)**
Lauromacrogol	**(9002-92-0)**
Lauromacrogol 400	**(9002-92-0)**
Lauronitrile	**(2437-25-4)**
Lauropal (9CI)	**(58968-32-4)**
Laurostearic acid	**(143-07-7)**
Laurox	**(105-74-8)**
3-Lauroylamidopropyl betaine	**(4292-10-8)**
Lauroylaminopropyldimethylaminoacetate	**(4292-10-8)**
Lauroyl chloride	**(112-16-3)**
N-(Lauroylcolamenoformylmethyl)pyridinium chloride	**(6272-74-8)**
Lauroyl diethanolamide	**(120-40-1)**
Lauroyl(dimethyl amino propyl)amine	**(3179-80-4)**
N-Lauroyl-3-(dimethylamino)propylamine	**(3179-80-4)**
Lauroyl isopropanolamide	**(142-54-1)**
Lauroyl monoethanolamide	**(142-78-9)**
Lauroyl-peroxide	**(105-74-8)**
Lauroyl peroxide (DOT)	**(105-74-8)**
Lauroyl peroxide, Not more than 42% (DOT)	**(105-74-8)**
Lauroyl peroxide, Technically pure (DOT)	**(105-74-8)**
N-Lauroylsarcosine, sodium	**(137-16-6)**
Lauroylsarcosine sodium salt	**(137-16-6)**
N-Lauroylsarcosine, sodium salt	**(137-16-6)**
Laurtrimonium chloride	**(112-00-5)**
Laurydol	**(105-74-8)**
Lauryl 24	**(112-53-8)**
Lauryl acetate	**(112-66-3)**
Lauryl acrylate	**(2156-97-0)**
n-Lauryl acrylate	**(2156-97-0)**
Lauryl alcohol	**(112-53-8)**
Lauryl alcohol	**(27342-88-7)**
Lauryl alcohol, ethoxylated	**(9002-92-0)**
Lauryl alcohol hexa(oxyethylene) ethanol	**(3055-97-8)**
Lauryl alcohol mono(oxyethylene) ethanol	**(3055-93-4)**
n-Lauryl alcohol, primary	**(112-53-8)**
Lauryl alcohol sulfate, diethanolamine salt	**(143-00-0)**
Lauryl alcohol tetra(oxyethylene) ethanol	**(3055-95-6)**
Lauryl alcohol triglycol ether	**(3055-94-5)**
Lauryl alcohol tri(oxyethylene) ethanol	**(5274-68-0)**
Lauryl aldehyde	**(112-54-9)**
Laurylamidoethanol	**(142-78-9)**
N-Laurylamidopropyl-N,N-dimethylbetaine	**(4292-10-8)**
N-(3-Laurylamidopropyl)-N,N-dimethyl-N-(2-hydroxy-3-sulfopropyl)ammonium, inner salt	**(19223-55-3)**
Laurylamine	**(124-22-1)**
Laurylamine hydrochloride	**(929-73-7)**
Laurylamine, N-methyl-N-nitroso-	**(55090-44-3)**
(3-Laurylaminopropyl)dimethylaminoacetic acid, hydroxide, inner salt	**(4292-10-8)**
Laurylammonium hydrochloride	**(929-73-7)**

Lauryl ammonium sulfate	**(2235-54-3)**	Lawsone	**(481-39-0)**
Laurylbenzenesulfonic acid	**(27176-87-0)**	Lawsone	**(83-72-7)**
Laurylbetain	**(683-10-3)**	Lawsonite	**(25038-59-9)**
Lauryl betaine	**(683-10-3)**	Laxagen	**(18869-73-3)**
Lauryl bromide	**(143-15-7)**	Laxagetten	**(18869-73-3)**
Lauryl chloride	**(112-52-7)**	Laxanorm	**(117-10-2)**
Lauryl diethanolamide	**(120-40-1)**	Laxans	**(603-50-9)**
Lauryldiethanolamine	**(1541-67-9)**	Laxanthreen	**(117-10-2)**
Lauryl diethylenediaminoglycine	**(6843-97-6)**	Laxinate	**(577-11-7)**
Lauryl diethylene glycol ether sulfonate sodium	**(3088-31-1)**	Laxinate 100	**(577-11-7)**
Lauryl dihydrogen phosphate	**(2627-35-2)**	Laxipur	**(117-10-2)**
Lauryldimethylamine	**(112-18-5)**	Laxipurin	**(117-10-2)**
N-Lauryldimethylamine	**(112-18-5)**	Laxogen	**(77-09-8)**
Lauryldimethylamine oxide	**(1643-20-5)**	Layor Carang	**(9002-18-0)**
Lauryldimethylbetaine	**(683-10-3)**	Lazo	**(15972-60-8)**
Lauryldimethyldichlorobenzylammonium chloride	**(102-30-7)**	Ld Rubber Red 16913	**(5160-02-1)**
Lauryl dimethyl glycine	**(683-10-3)**	Le-100	**(77-81-6)**
Lauryl disodium sulfosuccinate	**(26838-05-1)**	Lea-Cov	**(7681-49-4)**
Laurylester kyseliny methakrylove (Czech)	**(142-90-5)**	Lead-210	**(14255-04-0)**
Laurylethanolamide	**(142-78-9)**	Lead-214	**(15067-28-4)**
Lauryl ethoxylate	**(4536-30-5)**	Lead-carbonate	**(598-63-0)**
Lauryl glycidyl ether	**(2461-18-9)**	Lead (ACGIH)	**(7439-92-1)**
Laurylguanidine acetate	**(2439-10-3)**	Lead Bottoms	**(7446-14-2)**
Lauryl hexaethoxylate	**(3055-96-7)**	Lead Brown	**(1309-60-0)**
Lauryl hydroxyethyl imidazoline	**(136-99-2)**	Lead Chromate (ACGIH)	**(7758-97-6)**
N-Lauryl-β-iminodipropionate sodium salt	**(3655-00-3)**	Lead Chromate(VI)	**(7758-97-6)**
N-Lauryl-β-iminodipropionic acid, sodium salt	**(14960-06-6)**	Lead Dross	**(69029-52-3)**
Lauryl mercaptan	**(112-55-0)**	Lead Dross (Containing 3% or more free acid)	**(69029-52-3)**
m-Lauryl mercaptan	**(112-55-0)**	Lead Dross (DOT)	**(7446-14-2)**
Lauryl methacrylate	**(142-90-5)**	Lead Flake	**(7439-92-1)**
Lauryl monoethoxylate	**(4536-30-5)**	Lead(IV) arsenate	**(10102-48-4)**
N-Laurylmorpholine	**(1541-81-7)**	Lead Kettle Dross (Secondary nonferrous plant)	**(69029-52-3)**
N-Lauryl, myristyl β-aminopropionic acid	**(1462-54-0)**	Lead (2+) NTA	**(53113-59-0)**
Lauryl pentachlorophenate	**(3772-94-9)**	Lead Oxide Red	**(1314-41-6)**
Lauryl phosphate	**(12751-23-4)**	Lead Oxide Yellow	**(1317-36-8)**
Lauryl polyethylene glycol ether	**(9002-92-0)**	Lead Refinery Copper Dross	**(69029-52-3)**
Lauryl poly(oxyethylene) ether	**(9002-92-0)**	Lead Refinery Pyrite Dross	**(69029-52-3)**
1-Laurylpyridinium chloride	**(104-74-5)**	Lead Refining Caustic Dross	**(69029-52-3)**
Laurylpyridinium chloride	**(104-74-5)**	Lead S2	**(7439-92-1)**
Laurylpyridinium iodide	**(3026-66-2)**	Lead Scrap	**(69029-52-3)**
Laurylsiran sodny (Czech)	**(151-21-3)**	Lead Skimmings	**(69029-52-3)**
Lauryl sodium sulfate	**(151-21-3)**	Lead Smelter Copper Dross	**(69029-52-3)**
Lauryl sulfate	**(151-41-7)**	Lead Smelter Dross	**(69029-52-3)**
Lauryl sulfate ammonium salt	**(2235-54-3)**	Lead Smelter Lead Dross	**(69029-52-3)**
Lauryl sulfate diethanolamine salt	**(143-00-0)**	Lead Wolframate	**(7759-01-5)**
Lauryl sulfate, sodium salt	**(151-21-3)**	Lead acetate	**(15347-57-6)**
Lauryl sulfobetaine	**(14933-08-5)**	Lead(2+) acetate	**(301-04-2)**
Lauryl sulfuric acid	**(151-41-7)**	Lead(II) acetate	**(301-04-2)**
Lauryl sulphate	**(151-41-7)**	Lead acetate, Basic	**(1335-32-6)**
Lauryl sultaine	**(14933-08-5)**	Lead acetate [UN 1616]	**(301-04-2)**
Lauryl 3,3'-thiodipropionate	**(123-28-4)**	Lead acetate (II), trihydrate	**(6080-56-4)**
Lauryl thioglycolate	**(3746-39-2)**	Lead acetate trihydrate	**(6080-56-4)**
Lauryl triethoxylate	**(3055-94-5)**	Lead acid arsenate	**(7784-40-9)**
Lauryltriglycol ether	**(3055-94-5)**	Lead antimonate	**(13510-89-9)**
Lauryl trimethyl ammonium chloride	**(112-00-5)**	Lead antimonide	**(12266-38-5)**
Lauryltrimethylammonium chloride	**(10182-91-9)**	Lead arsenate	**(10102-48-4)**
Lauryltrimethylammonium chloride	**(112-00-5)**	Lead arsenate (ACGIH) [UN 1617]	**(3687-31-8)**
Lausit	**(53-86-1)**	Lead arsenate, Solid (DOT)	**(7645-25-2)**
Lautarite	**(7789-80-2)**	Lead arsenate (Standard)	**(7784-40-9)**
Lauxtol	**(87-86-5)**	Lead arsenate [UN 1617]	**(7645-25-2)**
Lauxtol A	**(87-86-5)**	Lead arsenate [UN 1617]	**(7784-40-9)**
Lav	**(8007-45-2)**	Lead arsenite, Solid	**(70910-35-9)**
Lavamenthe	**(90-42-6)**	Lead azide (8CI,9CI)	**(13424-46-9)**
Lavatar	**(8007-45-2)**	Lead azide, Containing, by weight, at least 20% water or mixture of alcohol and water [UN 0129]	**(13424-46-9)**
Lavon	**(2717-15-9)**	Lead azide (Dry)	**(13424-46-9)**
Lavsan	**(25038-59-9)**	Lead, bis(acetato)tetrahydroxytri	**(1335-32-6)**
Lawn-Keep	**(94-75-7)**		

Lead, bis(acetato-O)tetrahydroxytri- **(1335-32-6)**
Lead, bis(carbonato)dihydroxytri- **(1319-46-6)**
Lead, bis(carbonato(2-))dihydroxytri- (9CI) **(1319-46-6)**
Lead, bis(dimethyldithiocarbamato) **(19010-66-3)**
Lead, bis(dipentylcarbamodithioato-S,S')-, (T-4)- (9CI) **(36501-84-5)**
Lead, bis(octadecanoato)dioxodi- (9CI) **(56189-09-4)**
Lead borate **(14720-53-7)**
Lead borosilicate glass enamel flux **(65997-17-3)**
Lead bromide (9CI) **(10031-22-8)**
Lead bromide (PbBr) (6CI,8CI,9CI) **(15576-47-3)**
Lead bromide chloride (8CI,9CI) **(13778-36-4)**
Lead, butyltrimethyl- **(54964-75-9)**
Lead caprylate **(15696-43-2)**
Lead(2+) carbonate **(598-63-0)**
Lead carbonate hydroxide **(1319-46-6)**
Lead (II) chloride **(7758-95-4)**
Lead(2+) chloride **(7758-95-4)**
Lead chloride (DOT) **(7758-95-4)**
Lead chloride silicate (9CI) **(39390-00-6)**
Lead chromate, Basic **(18454-12-1)**
Lead chromate, Red **(18454-12-1)**
Lead chromate(VI) oxide **(18454-12-1)**
Lead chromate oxide **(1344-38-3)**
Lead chromate oxide **(18454-12-1)**
Lead chromate silicate **(11113-70-5)**
Lead citrate **(512-26-5)**
Lead citrate (VAN) **(512-26-5)**
Lead cyanamide **(20837-86-9)**
Lead cyanide **(592-05-2)**
Lead(II)cyanide **(592-05-2)**
Lead cyclohexanecarboxylate **(50825-29-1)**
Lead diacetate **(301-04-2)**
Lead diacetate trihydrate **(6080-56-4)**
Lead diamyldithiocarbamate **(36501-84-5)**
Lead dibasic acetate **(301-04-2)**
Lead dichloride **(7758-95-4)**
Lead, dichlorodiethyl- **(13231-90-8)**
Lead, diethyldimethyl- **(1762-27-2)**
Lead difluoride **(7783-46-2)**
Lead, dihydroxy(2,4,6-trinitro-1,3-benzenediolato(2-))di- (9CI) **(12403-82-6)**
Lead diiodide **(10101-63-0)**
Lead(2+), dimethyl- (9CI) **(21774-13-0)**
Lead dimethyldithiocarbamate **(19010-66-3)**
Lead(2+), dimethyl-, ion (8CI) **(21774-13-0)**
Lead dinitrate **(10099-74-8)**
Lead dioxide [UN 1872] **(1309-60-0)**
Lead diperchlorate **(13637-76-8)**
Lead dithiocyanate **(592-87-0)**
Lead 2-ethylhexanoate **(16996-40-0)**
Lead 2-ethylhexanoate **(301-08-6)**
Lead(2+) 2-ethylhexanoate **(301-08-6)**
Lead 2-ethylhexoate **(16996-40-0)**
Lead, ethyltrimethyl- **(1762-26-1)**
Lead fluoborate (DOT) **(13814-96-5)**
Lead(II) fluoride **(7783-46-2)**
Lead fluoride (DOT) **(7783-46-2)**
Lead(II) fluorosilicate **(25808-74-6)**
Lead formate **(811-54-1)**
Lead(2+) formate **(811-54-1)**
Lead hydrogen phosphate **(15845-52-0)**
Lead hydroxide (8CI,9CI) **(19783-14-3)**
Lead hydroxide carbonate **(1319-46-6)**
Lead (II) hydroxide salicylate **(87903-39-7)**
Lead, hydroxy(2-hydroxybenzoato-O1,O2)- (9CI) **(87903-39-7)**
Lead hydroxysalicylate **(87903-39-7)**
Lead hyposulfite **(26265-65-6)**
Lead inorganic (OSHA) **(7439-92-1)**
Lead iodate **(25659-31-8)**
Lead(II) iodide **(10101-63-0)**
Lead iodide (9CI) **(10101-63-0)**
Lead, ion (Pb(2+)) (8CI,9CI) **(14280-50-3)**
Lead isononanoate **(27253-41-4)**
Lead isothiocyanate **(592-87-0)**
Lead linoleate **(16996-51-3)**
Lead methacrylate **(1068-61-7)**
Lead(2+) 2-methyl-2-propenoate **(1068-61-7)**
Lead (II) methylthiolate **(35029-96-0)**
Lead molybdate **(10190-55-3)**
Lead molybdenum oxide (9CI) **(10190-55-3)**
Lead monobasic salicylate **(87903-39-7)**
Lead monobromide **(15576-47-3)**
Lead mononitroresorcinate **(51317-24-9)**
Lead mononitroresorcinate (Dry) **(51317-24-9)**
Lead monooxide **(1317-36-8)**
Lead monosubacetate **(1335-32-6)**
Lead myristate **(20403-41-2)**
Lead naphthenate **(50825-29-1)**
Lead naphthenate **(61790-14-5)**
Lead neobate **(12034-88-7)**
Lead neodecanoate **(27253-28-7)**
Lead niobate **(12034-88-7)**
Lead niobium oxide (8CI,9CI) **(12034-88-7)**
Lead nitrate **(10099-74-8)**
Lead nitrate **(13826-65-8)**
Lead(2+) nitrate **(10099-74-8)**
Lead(II) nitrate **(10099-74-8)**
Lead(II) nitrate (1:2) **(10099-74-8)**
Lead nitrate [UN 1469] **(10099-74-8)**
Lead nitroresorcinate **(51317-24-9)**
Lead(2+) octadecanoate **(1072-35-1)**
Lead octanoate **(15696-43-2)**
Lead octoate **(15696-43-2)**
Lead oleate **(1120-46-3)**
Lead(II) oleate (1:2) **(1120-46-3)**
Lead oxalate **(814-93-7)**
Lead oxide **(1314-41-6)**
Lead oxide **(1317-36-8)**
Lead(II) oxide **(1317-36-8)**
Lead oxide Brown **(1309-60-0)**
Lead oxide (8CI,9CI) **(1314-27-8)**
Lead oxide (VAN)(9CI) **(1335-25-7)**
Lead oxide sulfate (8CI,9CI) **(12036-76-9)**
Lead(2+) perchlorate **(13637-76-8)**
Lead perchlorate [UN 1470] **(13637-76-8)**
Lead peroxide (DOT) **(1309-60-0)**
Lead orthophosphate **(7446-27-7)**
Lead phosphate **(7446-27-7)**
Lead phosphate (3:2) **(7446-27-7)**
Lead(2+) phosphate **(7446-27-7)**
Lead(II) phosphate (3:2) **(7446-27-7)**
Lead phthalate **(6838-85-3)**
Lead picrate **(25721-38-4)**
Lead picrate (Dry) **(25721-38-4)**
Lead orthoplumbate **(1314-41-6)**
Lead protoxide **(1317-36-8)**
Lead pyrophosphate **(13453-66-2)**
Lead(2+) pyrophosphate **(13453-66-2)**
Lead β-resorcylate **(41453-50-3)**
Lead sebacate **(29473-77-6)**
Lead selenate **(7446-15-3)**
Lead selenide (8CI,9CI) **(12069-00-0)**
Lead selenite **(7488-51-9)**
Lead sesquioxide **(1314-27-8)**
Lead silicate **(11120-22-2)**

Lead silicate sulfate (9CI)	**(67711-86-8)**
Lead silicon fluoride	**(25808-74-6)**
Lead stearate	**(1072-35-1)**
Lead stearate	**(56189-09-4)**
Lead stearate	**(7428-48-0)**
Lead styphnate	**(15245-44-0)**
Lead styphnate (Dry)	**(15245-44-0)**
Lead subacetate	**(1335-32-6)**
Lead subacetate soln.	**(8006-24-4)**
Lead subcarbonate	**(1319-46-6)**
Lead sulfate	**(12036-76-9)**
Lead sulfate	**(15739-80-7)**
Lead(II) sulfate (1:1)	**(7446-14-2)**
Lead sulfate, Solid, Containing more than 3% free acid [UN 1794]	**(7446-14-2)**
Lead sulfate (With more than 3% free acid) [UN 1794]	**(12036-76-9)**
Lead sulfocyanate	**(592-87-0)**
Lead sulphate, With more than 3% free acid (DOT)	**(7446-14-2)**
Lead superoxide	**(1309-60-0)**
Lead tantalate	**(12065-68-8)**
Lead tantalum oxide ($PbTa_2O_6$) (8CI,9CI)	**(12065-68-8)**
Lead tartrate	**(815-84-9)**
Lead(II) tartrate (1:1)	**(815-84-9)**
Lead tellurate	**(13845-35-7)**
Lead telluride (8CI,9CI)	**(1314-91-6)**
Lead tellurite	**(13845-35-7)**
Lead tetraacetate	**(546-67-8)**
Lead tetradecanoate	**(20403-41-2)**
Lead, tetraethyl-	**(78-00-2)**
Lead, tetramethyl-	**(75-74-1)**
Lead tetraoxide	**(1314-41-6)**
Lead, tetrapropyl-	**(3440-75-3)**
Lead thiocyanate	**(592-87-0)**
Lead(2+) thiocyanate	**(592-87-0)**
Lead(II) thiocyanate	**(592-87-0)**
Lead thiosulfate	**(13478-50-7)**
Lead thiosulfate	**(26265-65-6)**
Lead titanate	**(12060-00-3)**
Lead titanium oxide ($PbTiO_3$) (8CI,9CI)	**(12060-00-3)**
Lead titanium trioxide	**(12060-00-3)**
Lead, tributyl-, chloride	**(13302-14-2)**
Lead, triethyl-, chloride	**(1067-14-7)**
Lead, triethylhydro- (7CI)	**(5224-23-7)**
Lead, triethylmethyl-	**(1762-28-3)**
Lead, trimethyl-	**(7442-13-9)**
Lead, trimethyl-, chloride	**(1520-78-1)**
Lead trinitroresorcinate	**(15245-44-0)**
Lead trioxide	**(1314-27-8)**
Lead tripropyl	**(6618-03-7)**
Lead tungstate	**(7759-01-5)**
Lead tungstate (IV)	**(7759-01-5)**
Lead tungsten oxide (9CI)	**(7759-01-5)**
Lead tungsten tetraoxide	**(7759-01-5)**
Lead vanadate	**(10099-79-3)**
Lead vanadium oxide (9CI)	**(10099-79-3)**
Lead zirconate	**(12060-01-4)**
Lead zirconium oxide (8CI,9CI)	**(12060-01-4)**
Leaf Alcohol	**(928-96-1)**
Leaf Aldehyde	**(505-57-7)**
Leaf Green	**(1308-38-9)**
Lealgin compositum	**(52-86-8)**
Lead-monoxide	**(1317-36-8)**
Leandin	**(140-87-4)**
Lead-sulfide	**(1314-87-0)**
Lead-tetroxide	**(1314-41-6)**
Leather Blue G	**(129-17-9)**
Leather Fast Red B	**(6459-94-5)**
Leather Green B	**(4680-78-8)**
Leather Green SF	**(5141-20-8)**
Leather Orange HR	**(532-82-1)**
Leather Orange Extra	**(633-96-5)**
Leather Pure Blue HB	**(61-73-4)**
Leather Red G	**(3734-67-6)**
Leather Red HT	**(477-73-6)**
Lebaycid	**(55-38-9)**
Le captane (French)	**(133-06-2)**
Lecithin	**(8002-43-5)**
Lecithin, Soybean	**(8002-43-5)**
Lecithins	**(8002-43-5)**
Lecithol	**(8002-43-5)**
Lectopam	**(1812-30-2)**
Ledercillin	**(6130-64-9)**
Ledercillin VK	**(132-98-9)**
Lederkyn	**(80-35-3)**
Lederle 2246	**(119-47-1)**
Ledermycin	**(127-33-3)**
Ledertrexate	**(59-05-2)**
Le dinitrocresol-4,6 (French)	**(534-52-1)**
Ledon 11	**(75-69-4)**
Ledon 12	**(75-71-8)**
Ledon 113	**(76-13-1)**
Ledon 114	**(76-14-2)**
Ledopa	**(59-92-7)**
Ledosten	**(702-54-5)**
Lefebar	**(50-06-6)**
Leftose	**(9001-63-2)**
Legumex	**(94-81-5)**
Legumex D	**(94-82-6)**
Legumex DB	**(94-74-6)**
Legumex Extra	**(3813-05-6)**
Legurame	**(16118-45-9)**
Legurame	**(16118-49-3)**
Lehydan	**(57-41-0)**
Leinoleic acid	**(60-33-3)**
Leiormone	**(11000-17-2)**
Leipzig Yellow	**(7758-97-6)**
Leivasom	**(52-68-6)**
Lekamin	**(817-09-4)**
Lekutherm 2159	**(5493-45-8)**
Lekutherm Hardener H	**(85-42-7)**
Lekutherm X 100	**(5493-45-8)**
Lemac	**(9003-20-7)**
Lemac 1000	**(9003-20-7)**
Lemascorb	**(50-81-7)**
Lembrol	**(439-14-5)**
Lemiserp	**(50-55-5)**
Lemoflur	**(7681-49-4)**
Lemol	**(9002-89-5)**
Lemol 5-88	**(9002-89-5)**
Lemol 5-98	**(9002-89-5)**
Lemol 12-88	**(9002-89-5)**
Lemol 16-98	**(9002-89-5)**
Lemol 24-98	**(9002-89-5)**
Lemol 30-98	**(9002-89-5)**
Lemol 51-98	**(9002-89-5)**
Lemol 60-98	**(9002-89-5)**
Lemol 75-98	**(9002-89-5)**
Lemol GF-60	**(9002-89-5)**
Lemon-Oil	**(8008-56-8)**
Lemon-Petitgrain-Oil	**(8008-56-8)**
Lemon Chrome	**(10294-40-3)**
Lemon Chrome A 3G	**(1344-37-2)**
Lemon Chrome C 4G	**(1344-37-2)**
Lemon Yellow	**(10294-40-3)**
Lemon Yellow	**(7758-97-6)**

Name	CAS No.
Lemon Yellow A	(1934-21-0)
Lemon Yellow A Geigy	(1934-21-0)
Lemon Yellow ZN 3	(8004-92-0)
Lemonene	(92-52-4)
Lemongrass Oil	(8007-02-1)
Lemongrass Oil West Indian	(8007-02-1)
Lemonol	(106-24-1)
Lenacil	(2164-08-1)
Lendine	(58-89-9)
Lenigallol	(525-52-0)
Lenisarin	(502-55-6)
Lenitral	(55-63-0)
Lenotan	(8064-77-5)
Lentac	(80-35-3)
Lentagran	(55512-33-9)
Lentanet	(51-63-8)
Lenticillin	(6130-64-9)
Lentin	(51-83-2)
Lentine (French)	(51-83-2)
Lentocillin	(1538-09-6)
Lentopen	(6130-64-9)
Lentopenil	(1538-09-6)
Lentox	(58-89-9)
Leodrin	(60-13-9)
Leodrine	(135-09-1)
Leomypen	(1538-09-6)
Leonal	(50-06-6)
Leostesin	(137-58-6)
Leostigmine bromide	(114-80-7)
Lepargylic acid	(123-99-9)
Lepazol	(54-95-5)
Lepenil	(57-53-4)
Lepetown	(57-53-4)
Lephebar	(50-06-6)
Lepicron	(59669-26-0)
Lepidin	(491-35-0)
4-Lepidine	(491-35-0)
Lepidine	(491-35-0)
Lepidine, hydrochloride	(3007-43-0)
Lepidine, 2-phenyl- (8CI)	(4789-76-8)
Lepidine, 1,2,3,4-tetrahydro-	(19343-78-3)
Lepidin hydrochlorid (German)	(3007-43-0)
Lepimidin	(125-33-7)
Lepinal	(50-06-6)
Lepinaletten	(50-06-6)
Lepitoin	(57-41-0)
Lepitoin	(630-93-3)
Lepitoin sodium	(630-93-3)
Lepsin	(57-41-0)
Lepsiral	(125-33-7)
Leptamin	(59-26-7)
Leptanal	(548-73-2)
Leptazol	(54-95-5)
Leptazole	(54-95-5)
Leptofen	(548-73-2)
Lepton	(702-54-5)
Leptophos	(21609-90-5)
Leptophos phenol	(1940-42-7)
Lerbek	(2971-90-6)
Lercigan	(60-87-7)
Lerenox	(495-73-8)
Lergigan	(58-33-3)
Lergigan	(60-87-7)
Lergitin	(961-71-7)
Lergotrile	(36945-03-6)
Lergotrilo (Spanish)	(36945-03-6)
Lergotrilum (Latin)	(36945-03-6)
Leritin	(144-14-9)
Leritine	(144-14-9)
Leropropoxyphene	(2338-37-6)
Lertus	(22071-15-4)
Lesan	(140-56-7)
Lescopine bromide	(155-41-9)
Lestemp	(103-90-2)
Lethalaire G-52	(107-49-3)
Lethalaire G-54	(56-38-2)
Lethalaire G-57	(3689-24-5)
Lethalaire G-58	(80-33-1)
Lethalaire G-59	(152-16-9)
Lethane	(112-56-1)
Lethane 60	(301-11-1)
Lethane 384	(112-56-1)
Lethane 384 Regular	(112-56-1)
Lethelmin	(92-84-2)
Lethidrome	(62-67-9)
Lethidrone	(62-67-9)
Lethox	(786-19-6)
Lethurin	(79-01-6)
L'ethylene thiouree (French)	(96-45-7)
Letidron	(62-67-9)
Letyl	(57-53-4)
Leuco-4	(73-24-5)
Leucaenine	(500-44-7)
Leucaenol	(500-44-7)
Leucamine	(107-85-7)
Leucarsone	(121-59-5)
Leucena glauca α-amino acid	(500-44-7)
Leucenine	(500-44-7)
Leucenol	(500-44-7)
Leucethane	(51-79-6)
Leucin (German)	(61-90-5)
(+-)-Leucine	(328-39-2)
D-Leucine	(328-38-1)
L-Leucine	(61-90-5)
Leucine	(61-90-5)
ε-Leucine	(60-32-2)
DL-Leucine (9CI)	(328-39-2)
Leucine, DL- (8CI)	(328-39-2)
Leucine, D	(328-38-1)
Leucine, L	(61-90-5)
L-Leucine-p-nitroanilide	(4178-93-2)
Leucogentian violet	(603-48-5)
Leucoindigo	(6537-68-4)
Leucol	(91-22-5)
Leucoline	(119-65-3)
Leucoline	(91-22-5)
Leucopin	(149-29-1)
Leucosol Golden Yellow	(128-66-5)
Leucosulfan	(55-98-1)
Leucothane	(51-79-6)
Leucovyl PA 1302	(9003-22-9)
Leukaemomycin C	(20830-81-3)
Leukeran	(305-03-3)
Leukeran	(50-44-2)
Leukerin	(50-44-2)
Leukersan	(305-03-3)
Leukol	(91-22-5)
Leukomyan	(56-75-7)
Leukomycin	(56-75-7)
Leukoran	(305-03-3)
Leukorrosin C	(3129-91-7)
Leuna M	(94-74-6)
Leupurin	(50-44-2)
Leurocristine	(57-22-7)

Leurocristine sulfate (1:1) (2068-78-2)
Leurocristine sulfate (1:1) (Salt) (2068-78-2)
Leurosine (23360-92-1)
Leutosan (2279-64-3)
(-)-Levallorphan (152-02-3)
Levallorphan (152-02-3)
Levamisol (14769-73-4)
Levamisole (14769-73-4)
Levanol Red GG (6459-94-5)
Levanox Green GA (1308-38-9)
Levanox Green GA (Hydrated chromic oxide) (1308-38-9)
Levanox Red 130A (1309-37-1)
Levanox White RKB (13463-67-7)
Levanxene (846-50-4)
Levanxol (846-50-4)
Levargin (1119-34-2)
Levarterenol (51-41-2)
Levatrom (637-07-0)
Levegal PT (23287-26-5)
Levenol PW (9005-00-9)
Levenol WZ (9014-90-8)
Levetamin (300-42-5)
Levium (439-14-5)
Levoarterenol (51-41-2)
Levodopa (59-92-7)
Levofalan (148-82-3)
Levofuraltadona (Spanish) (3795-88-8)
Levofuraltadone (3795-88-8)
Levofuraltadonum (Latin) (3795-88-8)
Levoglutamid (56-85-9)
Levoglutamide (56-85-9)
Levomethorphan (125-70-2)
Levomethorphane (French) (125-70-2)
Levomethorphanum (Latin) (125-70-2)
Levometorfano (Spanish) (125-70-2)
Levomicetina (56-75-7)
Levomoramida (Spanish) (5666-11-5)
Levomoramide (5666-11-5)
Levomoramidum (Latin) (5666-11-5)
Levomycetin (56-75-7)
Levomysol (14769-73-4)
Levonor (60-13-9)
Levonoradrenaline (51-41-2)
Levonorepinephrine (51-41-2)
Levopa (59-92-7)
Levophed (51-41-2)
Levophenacylmorphan (10061-32-2)
Levopimaric acid (79-54-9)
Levopropossifene (2338-37-6)
Levopropoxifeno (Spanish) (2338-37-6)
Levopropoxiphenum (2338-37-6)
Levopropoxyphenum (Latin) (2338-37-6)
Levoramide (5666-11-5)
Levorenin (51-43-4)
Levorenine (51-43-4)
Levorphan (77-07-6)
Levorphanol (77-07-6)
Levothyroxine (51-48-9)
Levsin (101-31-5)
Levugen (7660-25-5)
Levulic acid (123-76-2)
Levulinic-acid (123-76-2)
Levulinic acid, butyl ester (2052-15-5)
Levulinic acid, ethyl ester (539-88-8)
Levulose (57-48-7)
Levulose (7660-25-5)
Lewis-Red Devil Lye (1310-73-2)
Lewisite (541-25-3)
Lewisite (Arsenic compound) (541-25-3)
Lexamine 22 (16889-14-8)
Lexatol (61-76-7)
Lexgard bronopol (52-51-7)
Lexibiotico (15686-71-2)
Lexomil (1812-30-2)
Lexone (21087-64-9)
Lexotan (1812-30-2)
Lexotanil (1812-30-2)
Ley-Cornox (3813-05-6)
Leymin (3813-05-6)
Leymin (65280-19-5)
Leyspray (94-74-6)
Leytosan (2279-64-3)
Leytosan (62-38-4)
Liatris (8024-14-4)
Liatrix Oleoresin (8024-14-4)
Libavius fuming spirit (7646-78-8)
Liberetas (439-14-5)
Libiolan (57-53-4)
Librax (58-25-3)
Librinin (58-25-3)
Libritabs (58-25-3)
Librium (58-25-3)
Licareol acetate (115-95-7)
Lichenic acid (110-17-8)
Lichtgruen (German) (5141-20-8)
Licorice (68916-91-6)
Licorice Extract (68916-91-6)
Licorice Root (68916-91-6)
Licorice Root Extract (68916-91-6)
Licorice, ext. (68916-91-6)
Lida-Mantle (137-58-6)
Lidamycin creme (1405-10-3)
Lidanar (5588-33-0)
Lidanil (5588-33-0)
Lidenal (58-89-9)
Lidocaine (137-58-6)
Lidol (50-13-5)
Lidol (57-42-1)
Lifeampil (69-53-4)
Lifeampil (7177-48-2)
Lifene (86-34-0)
Light Camphor Oil (8008-51-3)
Light Coker Naphtha (Petroleum) (64741-74-8)
Light Crude Oil Distillate (68410-05-9)
Light Fast Yellow ES (6373-74-6)
Light Gas Oil (64741-58-8)
Light Green CF (5141-20-8)
Light Green FCF Yellowish (5141-20-8)
Light Green FS (5141-20-8)
Light Green G (5141-20-8)
Light Green 2GN (5141-20-8)
Light Green Lake (5141-20-8)
Light Green N (569-64-2)
Light Green N (Biological stain) (569-64-2)
Light Green S (5141-20-8)
Light Green SF (5141-20-8)
Light Green SFA (5141-20-8)
Light Green SFD (5141-20-8)
Light Green SF Yellowish (5141-20-8)
Light Green Yellowish (5141-20-8)
Light Oil of Camphor (8008-51-3)
Light Olefin Feed (68477-83-8)
Light Orange Chrome (18454-12-1)
Light Orange R (3468-63-1)

Light Red	(1309-37-1)
Light Red RB	(1103-38-4)
Light Red RCA	(1103-39-5)
Light Red RCN	(1103-38-4)
Light Red RS	(1248-18-6)
Light SF Yellowish (Biological stain)	(5141-20-8)
Light Yellow	(2512-29-0)
Light Yellow JB	(6358-85-6)
Light Yellow JBO	(6358-85-6)
Light Yellow JBT	(6358-85-6)
Light catalytically cracked distillate	(64741-59-9)
Light catalytically cracked naphtha	(64741-55-5)
Light clarified oil solvent extract (Petroleum)	(68783-03-9)
Light steam-cracked naphtha	(64742-83-2)
Light steam cracked aromatic naphtha (C6) concentrate (Petroleum)	(64742-83-2)
Light steam cracked naphtha piperylene concentrate (Petroleum)	(64742-83-2)
Light thermal cracked C4-C5 naphtha and gas oil distillate	(64741-74-8)
Light thermally cracked distillate	(64741-82-8)
Light thermally cracked naphtha	(64741-74-8)
Light vacuum distillate (Light vacuum gas oil)	(64741-58-8)
Light vacuum gas oil (Petroleum)	(64741-58-8)
Lightfast Yellow	(2512-29-0)
Lighthouse Chrome Blue 2R	(3567-69-9)
Light hydrocracked naphtha	(64741-69-1)
Light naphthenic distillate	(64741-52-2)
Light naphthenic distillate, Solvent extract	(64742-03-6)
Light naphthenic distillates (Petroleum)	(64741-52-2)
Light normal paraffin concentrate (Petroleum)	(64771-72-8)
Light normal paraffins (Petroleum)	(64771-72-8)
Light oil, Coal, Coke-oven	(65996-78-3)
Light paraffinic distillate	(64741-50-0)
Light paraffinic distillate, Solvent extract	(64742-05-8)
Light reformed naphtha	(64741-63-5)
Light straight-run naphtha	(64741-46-4)
Lignasan	(2235-25-8)
Lignin (9CI)	(9005-53-2)
Lignin Liquor	(8002-26-4)
Lignin, alkali (9CI)	(8068-05-1)
Lignin, chlorinated	(8068-02-8)
Lignite coal tar	(65996-90-9)
Lignocaine	(137-58-6)
Lignocellulose (9CI)	(11132-73-3)
Lignoceric acid	(557-59-5)
Lignoceric alcohol	(506-51-4)
Lignocerol	(506-51-4)
Lignoceryl alcohol	(506-51-4)
Lignosan BLP	(52316-55-9)
Lignosite 458	(8061-51-6)
Lignosite 854	(8061-51-6)
Lignosol D 10	(8061-51-6)
Lignosol XD	(8061-51-6)
Lignosulfonates	(8062-15-5)
Lignosulfonic acid (9CI)	(8062-15-5)
Lignosulfonic acid, calcium salt (9CI)	(8061-52-7)
Lignosulfonic acid, chromium iron salt (9CI)	(8075-74-9)
Lignosulfonic acid, chromium salt (9CI)	(9066-50-6)
Lignosulfonic acid, ferro chromium salt	(8075-74-9)
Lignosulfonic acid, iron chromium salt	(8075-74-9)
Lignosulfonic acid, sodium salt	(8061-51-6)
Lignosulfonic acid, sodium salt (9CI)	(8061-51-6)
Lignyl acetate	(128-51-8)
Ligroin	(8032-32-4)
Ligroine	(8032-32-4)
Lihocin	(999-81-5)
Likuden	(126-07-8)
Lilial	(80-54-6)
Lilion	(63428-84-2)
Lilly 01516	(60-87-7)
Lilly 1516	(60-87-7)
Lilly 22451	(309-36-4)
Lilly 34,314	(957-51-7)
Lilly 36,352	(1582-09-8)
Lilly 37231	(2068-78-2)
Lilly 39435	(3577-01-3)
Lilly 40602	(50-59-9)
Lilly 99638	(53994-73-3)
Lilly 109514	(51022-71-0)
Lilly/Miller Microcop Fungicide	(1332-03-2)
Lilo	(77-09-8)
Lilyal	(80-54-6)
Lilyl aldehyde	(107-75-5)
Limas	(554-13-2)
Limax	(9002-91-9)
Limbial	(604-75-1)
Lime	(1305-78-8)
Lime, Burned	(1305-78-8)
Lime Fractionated, Spent Pulping Liquor, Precipitate	(8061-52-7)
Lime, Unslaked (DOT)	(1305-78-8)
Lime Water	(1305-62-0)
Lime acetate	(62-54-4)
Lime chloride	(7778-54-3)
Limed Rosin	(9007-13-0)
Lime nitrogen (DOT)	(156-62-7)
Lime pyrolignite	(62-54-4)
Limestone (OSHA)	(1317-65-3)
Lime sulfur	(1344-81-6)
Lime sulphur	(1344-81-6)
(+)-R-Limonene	(5989-27-5)
(-)-Limonene	(5989-54-8)
D-(+)-Limonene	(5989-27-5)
Limonene	(138-86-3)
d-Limonene	(5989-27-5)
dl-Limonene	(138-86-3)
l-Limonene	(5989-54-8)
Limonene dimercaptan	(4802-20-4)
Limonene dioxide	(96-08-2)
Limonene oxide	(470-82-6)
Limonene polymer	(9003-73-0)
d-Limoneno (Spanish)	(5989-27-5)
Limovet	(9002-91-9)
Linalol	(78-70-6)
Linalol acetate	(115-95-7)
Linalool	(78-70-6)
Linalool acetate	(115-95-7)
Linalool oxide	(5989-33-3)
Linalool oxide C	(39028-58-5)
Linalool oxide D	(14009-71-3)
Linalool oxide (VAN)(9CI)	(1365-19-1)
Linalool oxide (α-ethenyl-α,3,3-trimethyloxiranepropanol + α-methyl-α-(4-methyl-3-pentenyl)oxiranemethanol)	(1365-19-1)
Linalool tetrahydride	(78-69-3)
Linalyl acetate	(115-95-7)
Linalyl alcohol	(78-70-6)
Linalyl o-aminobenzoate	(7149-26-0)
Linalyl anthranilate	(7149-26-0)
Linamarin	(554-35-8)
Linampheta	(317-34-0)
Linampheta	(60-13-9)
Linaris	(8064-90-2)
Linarodin	(100-06-1)
Linasen	(50-06-6)
Linco 4 (9CI)	(88997-61-9)

Lincocin (154-21-2)
Lincoln Green Toner B 15-2900 (569-64-2)
Lincoln Red 1002 (2814-77-9)
Lincolnensin (154-21-2)
Lincomycin (154-21-2)
Lincomycine (French) (154-21-2)
Lindafor (58-89-9)
Lindagam (58-89-9)
Lindagrain (58-89-9)
Lindagranox (58-89-9)
Lindan (62-73-7)
α-Lindane (319-84-6)
β-Lindane (319-85-7)
δ-Lindane (319-86-8)
γ-Lindane (58-89-9)
Lindane (ACGIH,DOT,OSHA) (58-89-9)
Lindapoudre (58-89-9)
Lindatox (58-89-9)
Lindatox (938-73-8)
Lindex (58-89-9)
Lindol (1330-78-5)
Lindosep (58-89-9)
Line LS (8001-21-6)
Line Rider (93-76-5)
Linear Alkylbenzenesulfonate, sodium salt (68411-30-3)
Linear(C12-C14)alkanol, ethoxylated, sulfated, sodium salt (68891-38-3)
Linear (C8-C22) alkyl alcohol, Ethoxylated (69013-19-0)
Linear (C12-C15) alkyl alcohols, ethoxylated (68131-39-5)
Linear (C12-C14) alkyl alcohols, ethoxylated, propoxylated (68439-51-0)
Linear (C12 and C14) alkyl alcohols, ethoxylated (68439-50-9)
Linear alchohol ethoxylated C12Eq (68131-40-8)
Linear alkylbenzene sulfonate (42615-29-2)
Linear alkylbenzene sulphonate (42615-29-2)
Linear octadecene (27070-58-2)
Linear primary(C12-C15)alcohol, ethoxylate (68131-39-5)
Linear primary alcohol (C12-C15) ethoxylate (68131-39-5)
Linear trans quinacridone (1047-16-1)
Linear random secondary alcohol (C11-C15) ethoxylate (68131-40-8)
Linear secondary(C11-C15)alcohol, ethoxylate (68131-40-8)
Linear secondary alcohol (C11-C15) ethoxylate (68131-40-8)
Linestrenol (52-76-6)
Linevol 79 (66587-56-2)
Linevol 911 (66455-17-2)
Linex 4L (330-55-2)
Linfadol (357-56-2)
Linfolizin (305-03-3)
Linfolysin (305-03-3)
Lingel (50-33-9)
Lingraine (379-79-3)
Lingran (379-79-3)
Lingusorbs (57-83-0)
Linoleamide DEA (27883-12-1)
9,12-Linoleic acid (60-33-3)
Linoleic acid (60-33-3)
Linoleic acid, methyl ester (8CI) (112-63-0)
Linolelaidic acid (506-21-8)
α-Linolenic acid (463-40-1)
Linolenic acid (8CI) (463-40-1)
Linoral (57-63-6)
Linormone (94-74-6)
Linorox (330-55-2)
Linseed-Oil (8001-26-1)
Linseed oil, Polymer with glycerol, pentaerythritol, phthalic anhydride and styrene (66071-63-4)
Linseed oil, Polymer with glycerol, phthalic anhydride and tung oil (66071-18-9)
Linseed oil, tung oil, glycerin, phthalic anhydride polymer (66071-18-9)
Linseed oil, tung oil, glycerin, phthalic anhydride resin (66071-18-9)
Linseed oil, tung oil, phthalic anhydride, glycerol resin (66071-18-9)
Linseed, tung oil, phthalic anhydride, glycerol, alkyd resin (66071-18-9)
Linton (52-86-8)
Lintox (58-89-9)
Linurex (330-55-2)
Linuron (330-55-2)
Linuron (Herbicide) (330-55-2)
Linvur (58-89-9)
Lio-Atropin (55-48-1)
Lionogen Blue R (81-77-6)
Lionol Blue E (147-14-8)
Lionol Blue ER (147-14-8)
Lionol Blue ES (147-14-8)
Lionol Blue ESP (147-14-8)
Lionol Blue GLA (147-14-8)
Lionol Blue KL (147-14-8)
Lionol Blue KW (574-93-6)
Lionol Blue NCB Toner (147-14-8)
Lionol Blue SM (147-14-8)
Lionol Blue SN (147-14-8)
Liothyronin (6893-02-3)
L-Liothyronine (6893-02-3)
Liothyronine (6893-02-3)
Lioxin (121-33-5)
Lipal 4LA (9002-92-0)
Lipal 15-DS (9005-08-7)
Lipal 30W (9004-96-0)
Lipal 400-OL (9004-96-0)
Lipal OA (9004-98-2)
Lipamid (637-07-0)
Lipan (534-52-1)
Liparite (7789-75-5)
Lipavil (637-07-0)
Lipavlon (637-07-0)
Liphadione (3691-35-8)
Lipid Crimson (85-83-6)
Lipide 500 (637-07-0)
Lipidsenker (637-07-0)
Lipo-Diazine (68-35-9)
Lipo EGMS (111-60-4)
Lipo GMS 410 (31566-31-1)
Lipo GMS 450 (31566-31-1)
Lipo GMS 600 (31566-31-1)
Lipo-Hepin (9005-49-6)
Lipo-Levazine (68-35-9)
Lipo-Lutin (57-83-0)
Lipocol L-1 (4536-30-5)
Lipocol L-4 (9002-92-0)
Lipocol L 12 (9002-92-0)
Lipocol L-23 (9002-92-0)
Lipofacton (637-07-0)
Lipoic acid (62-46-4)
α-Lipoic acid (62-46-4)
Lipomid (637-07-0)
α-Liponic acid (62-46-4)
Liponorm (637-07-0)
α-Liponsaeure (German) (62-46-4)
Lipopill (122-09-8)
Liporeduct (637-07-0)
Liporil (637-07-0)
Liposan (62-46-4)
Liposid (637-07-0)
Liposorb L-20 (9005-64-5)
Liposorb O (1338-43-8)
Liposorb O-20 (1338-43-8)
Liposorb O-20 (9005-65-6)

Liposorb P **(26266-57-9)**
Liposorb S **(1338-41-6)**
Liposorb S-20 **(1338-41-6)**
Liposorb S-20 **(9005-67-8)**
Liposorb SQ0 **(8007-43-0)**
Liposorb TO **(26266-58-0)**
Lipothion **(62-46-4)**
Lipotril **(67-48-1)**
Liprin **(637-07-0)**
Liprinal **(637-07-0)**
Liptan **(15687-27-1)**
Liqua-Tox **(81-81-2)**
Liquacillin **(61-33-6)**
Liquadiazine **(68-35-9)**
Liquaemin **(9005-49-6)**
Liquagesic **(103-90-2)**
Liquamar **(435-97-2)**
Liquamycin **(60-54-8)**
Liquamycin injectable **(2058-46-0)**
Liquefied Petroleum Gas (DOT,OSHA) **(68476-85-7)**
Liquefied Petroleum Gas [UN 1075] **(106-97-8)**
Liquefied Petroleum Gas [UN 1075] **(75-28-5)**
Liquefied Petroleum Gas [UN 1075] **(115-11-7)**
Liquemin **(9005-49-6)**
Liqui-Cee **(50-81-7)**
Liqui-San **(86-85-1)**
Liqui-Stik **(86-87-3)**
Liquibarine **(7727-43-7)**
Liquid Bright Platinum **(7440-06-4)**
Liquid Camphor **(8008-51-3)**
Liquid Derris **(83-79-4)**
Liquid Lanolin Oil **(8006-54-0)**
Liquid Pitch Oil **(8001-58-9)**
Liquid Rosin **(8002-26-4)**
Liquidambar Styraciflua **(1401-55-4)**
Liquid ethyene **(74-85-1)**
Liquidow **(10043-52-4)**
Liquid paraffin **(8012-95-1)**
Liquigel **(21645-51-2)**
Liquimeth **(63-68-3)**
Liquiphene **(62-38-4)**
Liquiprin **(65-45-2)**
Liquital **(50-06-6)**
Liquophylline **(58-55-9)**
Liquorice **(1405-86-3)**
Liranol **(58-40-2)**
Liranox **(93-65-2)**
Liro CIPC **(101-21-3)**
Lirohex **(107-49-3)**
Liromat **(1634-78-2)**
Liromate **(14484-64-1)**
Liromatin **(900-95-8)**
Lironion **(14214-32-5)**
Lironox **(94-80-4)**
Liropon **(75-99-0)**
Liroprem **(87-86-5)**
Lirobetarex **(150-68-5)**
Lirostanol **(900-95-8)**
Lirotan **(12122-67-7)**
Lirothion **(56-38-2)**
Lisacort **(53-03-2)**
Liskonum **(554-13-2)**
Lissamine Amaranth AC **(915-67-3)**
Lissamine Blue AR **(2666-17-3)**
Lissamine Fast Yellow AE **(6373-74-6)**
Lissamine Green G **(4680-78-8)**
Lissamine Green SF **(5141-20-8)**
Lissamine Lake Green SF **(5141-20-8)**
Lissamine Red 2G **(3734-67-6)**
Lissamine Turquoise VN **(129-17-9)**
Lissamine Ultra Blue AR **(2666-17-3)**
Lissapol NX **(9016-45-9)**
Lissolamin V **(57-09-0)**
Lissolamine **(57-09-0)**
Lissolamine A **(57-09-0)**
Lissolamine V **(57-09-0)**
Lisulfen **(80-35-3)**
Litac **(9003-54-7)**
Litac C 100P **(9003-54-7)**
Litaler **(127-07-1)**
Litalir **(127-07-1)**
Litex CA **(9003-55-8)**
Lithalure **(4485-12-5)**
Lithamide **(7782-89-0)**
Lithane **(554-13-2)**
Litharge **(1317-36-8)**
Litharge Pure **(1317-36-8)**
Litharge Yellow L-28 **(1317-36-8)**
Lithate 2,4-D **(3766-27-6)**
Lithic acid **(69-93-2)**
Lithicarb **(554-13-2)**
Lithidrone **(62-67-9)**
Lithinate **(554-13-2)**
Lithium 2,4-D **(3766-27-6)**
Lithium [UN 1415] **(7439-93-2)**
Lithium acetate **(546-89-4)**
Lithium alanate **(16853-85-3)**
Lithium aluminohydride **(16853-85-3)**
Lithium aluminum hydride [UN 1410] **(16853-85-3)**
Lithium aluminum hydride, ethereal [UN 1411] **(16853-85-3)**
Lithium aluminum tetrahydride **(16853-85-3)**
Lithium amide (DOT) **(7782-89-0)**
Lithium amide, Powdered (DOT) **(7782-89-0)**
Lithium borate **(12007-60-2)**
Lithium borohydride **(16949-15-8)**
Lithium bromacil **(53404-19-6)**
Lithium-bromide **(7550-35-8)**
Lithium, butyl- (9CI) **(109-72-8)**
Lithium carbonate **(554-13-2)**
Lithium carbonate (2:1) **(554-13-2)**
Lithium-chloride **(7447-41-8)**
Lithium chromate **(14307-35-8)**
Lithium chromate(VI) **(14307-35-8)**
Lithium 2,4-dichlorophenoxyacetate **(3766-27-6)**
Lithium dichromate **(13843-81-7)**
Lithium, (1,1-dimethylethyl)- (9CI) **(594-19-4)**
Lithium docosanoate **(4499-91-6)**
Lithium ethylene glycoxide **(23248-23-9)**
Lithium ferrosilicon [UN 2830] **(64082-35-5)**
Lithium-fluoride **(7789-24-4)**
Lithium fluorure (French) **(7789-24-4)**
Lithium formate **(556-63-8)**
Lithium-hydride **(7580-67-8)**
Lithium hydride (ACGIH,OSHA) [UN 1414] **(7580-67-8)**
Lithium hydride, In fused solid form [UN 2805] **(7580-67-8)**
Lithium hydroxide (Li(OH)) (9CI) **(1310-65-2)**
Lithium hydroxide, Solution [UN 2679] **(1310-65-2)**
Lithium 2-hydroxyethoxide **(23248-23-9)**
Lithium, 12-hydroxyoctadecanoate sebacate complexes **(68815-49-6)**
Lithium hypochlorite **(13840-33-0)**
Lithium hypochlorite, Dry [UN 1471] **(13840-33-0)**
Lithium hypochlorite compound, Dry, Containing more than 39% available chlorine [UN 1471] **(13840-33-0)**
Lithium iodide (9CI) **(10377-51-2)**

Lithium-iron-silicon (64082-35-5)
Lithium metal (DOT) (7439-93-2)
Lithium metal, In cartridges (DOT) (7439-93-2)
Lithium, (1-methylpropyl)- (9CI) (598-30-1)
Lithium nitrate [UN 2722] (7790-69-4)
Lithium nitride (8CI,9CI) (26134-62-3)
Lithium octadecanoate (4485-12-5)
Lithium oxide (9CI) (12057-24-8)
Lithium peroxide [UN 1472] (12031-80-0)
Lithium silicate (10102-24-6)
Lithium silicon (68848-64-6)
Lithium stearate (ACGIH) (4485-12-5)
Lithium sulfate (2:1) (10377-48-7)
Lithium sulphate (10377-48-7)
Lithium tetraborate (12007-60-2)
Lithium tetrahydroaluminate (16853-85-3)
Lithium tetrahydroaluminate(1-) (16853-85-3)
Lithium tetrahydroborate (16949-15-8)
Lithobid (554-13-2)
Lithocholic acid (434-13-9)
Lithofor Brown A (6416-57-5)
Lithograpic Stone (1317-65-3)
Lithol Fast Scarlet RN (2425-85-6)
Lithol Red (1248-18-6)
Lithol Red 3580 (1248-18-6)
Lithol Red 17676 (1248-18-6)
Lithol Red 18959 (1103-38-4)
Lithol Red 19592 (1103-39-5)
Lithol Red 22060 (1103-38-4)
Lithol Red 27965 (1103-38-4)
Lithol Red B (1248-18-6)
Lithol Red Barium Toner (1103-38-4)
Lithol Red CA (1103-39-5)
Lithol Red Calcium Toner (1103-39-5)
Lithol Red GGS (5850-90-8)
Lithol Red 3GS (1248-18-6)
Lithol Red Lake (1248-18-6)
Lithol Red R (1248-18-6)
Lithol Red RB Extra (1248-18-6)
Lithol Red RC Extra (1103-39-5)
Lithol Red RL 151 (1248-18-6)
Lithol Red RS (1248-18-6)
Lithol Red Sodium salt (1248-18-6)
Lithol Red Toner (1248-18-6)
Lithol Red Toner 3BX (1103-39-5)
Lithol Rubin B (5858-81-1)
Lithol Rubin B Ca (5281-04-9)
Lithol Rubine (5858-81-1)
Lithol Rubine BNA (5858-81-1)
Lithol Toner (1248-18-6)
Lithol Toner Extra Light 5000 (1248-18-6)
Lithol Toner Sodium salt RT 314 (1248-18-6)
Lithol Toner YA 8003 (1248-18-6)
Litholite (4485-12-5)
Lithonate (554-13-2)
Lithosol Deep Blue V (482-89-3)
Lithosol Fast Pink SVP (2379-74-0)
Lithosol Orange R Base (99-55-8)
Lithosol Scarlet Base M (89-62-3)
Lithosol Scarlet Base MB (89-62-3)
Lithosol Scarlet Base MBW (89-62-3)
Lithosol Scarlet Base MW (89-62-3)
Lithostat (546-88-3)
Lithotabs (554-13-2)
Liticon (359-83-1)
Liviatin (564-25-0)
Livonal (93-54-9)
Lixa-Beta (67-03-8)
Lixophen (50-06-6)
Lld Factor (68-19-9)
Lloncefal (50-59-9)
Lna 21-1000 (9003-54-7)
Lo-Bax (7778-54-3)
Lobak (80-77-3)
Lobamine (59-51-8)
Lobelin (90-69-7)
(-)-Lobeline (90-69-7)
Lobeline (90-69-7)
α-Lobeline (90-69-7)
Lobeline hydrochloride (63990-84-1)
Lobetrin (637-07-0)
Lobnico (90-69-7)
Lobron (63990-84-1)
Localyn (67-73-2)
Locron Extra (12042-91-0)
Locron Flakes (12042-91-0)
Locron Powder (12042-91-0)
Locron Solution (12042-91-0)
Locust-Bean-Gum (9000-40-2)
Locuturine (486-84-0)
Lodestone Yellow YB-57 (6358-85-6)
Loflazepate d'ethyle (French) (29177-84-2)
Loflazepato de etilo (Spanish) (29177-84-2)
Loha (7439-89-6)
Loisol (52-68-6)
Lombristop (148-79-8)
Lomidin (140-64-70)
Lomidine (140-64-70)
Lomidine isoethionate (140-64-70)
Lomudal (15826-37-6)
Lomudas (15826-37-6)
Lomustine (13010-47-4)
Lomustine, methyl- (13909-09-6)
Lonacol (12122-67-7)
Lonamin (122-09-8)
Lonarid (103-90-2)
London Purple, Solid (DOT) (8012-74-6)
London Purple [UN 1621] (8012-74-6)
Longacilina (1538-09-6)
Longanoct (77-28-1)
Longatin (128-62-1)
Longicil (1538-09-6)
Longifene (569-65-3)
Longifolen (475-20-7)
(+)-Longifolene (475-20-7)
Longifolene (475-20-7)
d-Longifolene (475-20-7)
β-Longilobine (480-54-6)
Longin (80-35-3)
Lonocol (12122-67-7)
Lonocol M (12427-38-2)
Lontrel (1702-17-6)
Lontrel 3 (1702-17-6)
Looplure Inhibitor (20056-92-2)
Loperamide (53179-11-6)
Lopezite (8CI) (27020-65-1)
Lopezite ($K_2(Cr_2O_7)$) (9CI) (27020-65-1)
Lopirin (62571-86-2)
Loprazolam (61197-73-7)
Loprazolamum (Latin) (61197-73-7)
Lopres (304-20-1)
Lopress (304-20-1)
Lopurin (315-30-0)
Loramine AMB 13 (107-43-7)

Loramine MY 228	**(142-58-5)**
Loramine S 280	**(111-57-9)**
Lorax	**(846-49-1)**
Lorazepam	**(846-49-1)**
Lorex	**(330-55-2)**
Lorexane	**(58-89-9)**
Lorfan	**(152-02-3)**
Loridine	**(50-59-9)**
Lorinal	**(302-17-0)**
Lorkaril	**(9003-54-7)**
Lormetazepam	**(848-75-9)**
Lormetazepamum (Latin)	**(848-75-9)**
Lormin	**(302-22-7)**
Lorol	**(112-53-8)**
Lorol 5	**(112-53-8)**
Lorol 7	**(112-53-8)**
Lorol 11	**(112-53-8)**
Lorol 20	**(111-87-5)**
Lorol 22	**(112-30-1)**
Lorol 24	**(36653-82-4)**
Lorol 28	**(112-92-5)**
Loromisan	**(56-75-7)**
Loromisin	**(56-75-7)**
Lorothidol	**(97-18-7)**
Lorothiodol	**(97-18-7)**
Lorox	**(314-42-1)**
Lorox	**(330-55-2)**
Lorox Linuron Weed Killer	**(330-55-2)**
Loroxide	**(94-36-0)**
Lorphen	**(113-92-8)**
Lorsban	**(2921-88-2)**
Lorsilan	**(846-49-1)**
Lorvec	**(7159-34-4)**
Lorvek	**(7159-34-4)**
Losantin	**(7778-54-3)**
S-Lost	**(505-60-2)**
N-Lost (German)	**(55-86-7)**
Losungsmittel APV	**(111-90-0)**
Lothymol	**(552-22-7)**
Lotrimin	**(23593-75-1)**
Loturine	**(486-84-0)**
Lotusate	**(115-44-6)**
Loubarb	**(143-81-7)**
Lovosa	**(9004-32-4)**
Lovosa TN	**(9004-32-4)**
Lovosa 20alk.	**(9004-32-4)**
Lovozal	**(14255-88-0)**
Low-boiling Butene-2	**(624-64-6)**
Lowedex	**(51-63-8)**
Loweserp	**(50-55-5)**
Lowetrate	**(78-11-5)**
Lowpstron	**(54-31-9)**
Loxanol 95	**(143-28-2)**
Loxanol K	**(36653-82-4)**
Loxanol K Extra	**(36653-82-4)**
Loxanol M	**(143-28-2)**
Loxanol V	**(112-72-1)**
Loxapine	**(1977-10-2)**
Loxiol G 10	**(25496-72-4)**
Loxiol G 21	**(106-14-9)**
Loxuran	**(1642-54-2)**
Loxynil (German)	**(1689-83-4)**
Ltan	**(117-10-2)**
Lu 274	**(126-81-8)**
Lubergal	**(50-06-6)**
Lubrokal	**(50-06-6)**
Lubrol 12A9	**(9002-92-0)**
Lubrol EA	**(110-30-5)**
Lubrol PX	**(9002-92-0)**
Lubrol TSC 5110	**(7775-50-0)**
Lucalox	**(1344-28-1)**
Lucamide	**(938-73-8)**
Lucanthone Metabolite	**(3105-97-3)**
Lucel	**(9004-32-4)**
Lucel ADA	**(123-77-3)**
Lucel (Polysaccharide)	**(9004-32-4)**
Lucidol	**(94-36-0)**
Lucidol Deltax	**(1338-23-4)**
Lucite	**(9011-14-7)**
Lucite 30	**(9011-14-7)**
Lucite 47	**(9011-14-7)**
Lucite 120	**(9011-14-7)**
Lucite 129	**(9011-14-7)**
Lucite 130	**(9011-14-7)**
Lucite 140	**(9011-14-7)**
Lucite 147	**(9011-14-7)**
Lucite 180	**(9011-14-7)**
Lucite 2013	**(25608-33-7)**
Lucorteum Sol	**(57-83-0)**
Lucosil	**(144-82-1)**
Lucovyl GA 8502	**(9003-22-9)**
Lucovyl MA 6028	**(9003-22-9)**
Lucovyl PA 1208	**(9003-22-9)**
Ludigol F,60	**(127-68-4)**
Ludiomil	**(10262-69-8)**
Ludox	**(7631-86-9)**
Ludox CL	**(1344-28-1)**
Lufyllin	**(479-18-5)**
Lukooil M 100	**(63148-62-9)**
Lukooil M 200	**(63148-62-9)**
Lukosan A 311	**(63148-62-9)**
Lukosan M 02	**(63148-62-9)**
Lukosan M 07	**(63148-62-9)**
Lulamin	**(50-35-1)**
Lulamin	**(91-80-5)**
Lullamin	**(135-23-9)**
Lullamin	**(91-80-5)**
Lumatex Blue B	**(147-14-8)**
Lumbrical	**(110-85-0)**
Lumen	**(50-06-6)**
Lumesettes	**(50-06-6)**
Lumesyn	**(50-06-6)**
Lumicrease Blue 4GL	**(2610-05-1)**
Lumicrease Sky Blue 6GUL	**(2610-05-1)**
Lumilar 100	**(25038-59-9)**
Luminal	**(50-06-6)**
Luminal sodium	**(57-30-7)**
Luminol	**(521-31-3)**
Lumirelax	**(532-03-6)**
Lumirror	**(25038-59-9)**
Lumirror 38S	**(25038-59-9)**
Lumofridetten	**(50-06-6)**
Luna Yellow	**(6358-31-2)**
Lunar Caustic	**(7761-88-8)**
Lunipax	**(1172-18-5)**
Lupareen	**(9003-07-0)**
Luperco	**(80-43-3)**
Luperco	**(94-36-0)**
Luperco CST	**(133-14-2)**
Luperco 231G	**(6731-36-8)**
Luperco 231XL	**(6731-36-8)**
Luperco 231XLP	**(6731-36-8)**
Luperox	**(80-43-3)**
Luperox 118	**(2618-77-1)**

Luperox 231	**(6731-36-8)**
Luperox FL	**(94-36-0)**
Luperox 500R	**(80-43-3)**
Luperox 500T	**(80-43-3)**
Lupersol	**(1338-23-4)**
Lupersol 8	**(109-13-7)**
Lupersol 10	**(26748-41-4)**
Lupersol 11	**(927-07-1)**
Lupersol 70	**(107-71-1)**
Lupersol 118	**(2618-77-1)**
Lupersol 231	**(6731-36-8)**
Lupersol DDA 30	**(1338-23-4)**
Lupersol DDM	**(1338-23-4)**
Lupersol DNF	**(1338-23-4)**
Lupersol DSW	**(1338-23-4)**
Lupersol 1OM75	**(26748-41-4)**
Lupersol TBIC	**(2372-21-6)**
Lupersol TBIC-M75	**(2372-21-6)**
Lupersol Delta-X	**(1338-23-4)**
Lupersol 228Z	**(3179-56-4)**
Lupetazine	**(106-58-1)**
Luphenil	**(50-06-6)**
Lupine LS	**(8001-21-6)**
Lupinidine	**(90-39-1)**
Lupolen 1010H	**(9002-88-4)**
Lupolen 1800H	**(9002-88-4)**
Lupolen 1800S	**(9002-88-4)**
Lupolen 1810H	**(9002-88-4)**
Lupolen 4261A	**(9002-88-4)**
Lupolen 6011H	**(9002-88-4)**
Lupolen 6011l	**(9002-88-4)**
Lupolen 6042D	**(9002-88-4)**
Lupolen KR 1032	**(9002-88-4)**
Lupolen KR 1051	**(9002-88-4)**
Lupolen KR 1257	**(9002-88-4)**
Lupolen L 6041D	**(9002-88-4)**
Lupolen N	**(9002-88-4)**
Luprosil	**(79-09-4)**
Lurafix Blue FFR	**(2475-46-9)**
Luramin	**(50-06-6)**
Luran	**(9003-54-7)**
Luran 378P	**(9003-54-7)**
Luran 368R	**(9003-54-7)**
Lurat	**(117-52-2)**
Lurazol Black BA	**(1937-37-7)**
Lurazol Deep Blue EB	**(8005-03-6)**
Lurazol Orange E	**(633-96-5)**
Lurazol Red E	**(1658-56-6)**
Lurgo	**(60-51-5)**
Luride	**(7681-49-4)**
Luride Lozi-Tabs	**(7681-49-4)**
Luride-SF	**(7681-49-4)**
Luridine	**(123-41-1)**
Luronit	**(9004-61-9)**
Lusil	**(63-74-1)**
Lusmit	**(123-28-4)**
Lustran	**(9003-54-7)**
Lustran 28	**(9003-54-7)**
Lustran A	**(9003-54-7)**
Lustran A 21	**(9003-54-7)**
Lustran A 2121	**(9003-54-7)**
Lustran LNA 21	**(9003-54-7)**
Lustran SAN	**(9003-54-7)**
Lustrex	**(9003-53-6)**
Lustrex H 77	**(9003-53-6)**
Lustrex HH 101	**(9003-53-6)**
Lustrex HP 77	**(9003-53-6)**
Lustrex HT 88	**(9003-53-6)**
Lutate	**(630-56-8)**
Luteal Hormone	**(57-83-0)**
Luteine	**(57-83-0)**
Luteinique	**(57-83-0)**
Luteinizing hormone-releasing factor (Pig), 6-(3-(2-naphthalenyl)-D-alanine)-, acetate (Salt), hydrate	**(86220-42-0)**
Luteoantine	**(9002-68-0)**
Luteocrin	**(630-56-8)**
Luteocrin Depot	**(630-56-8)**
Luteodyn	**(57-83-0)**
Luteogan	**(57-83-0)**
Luteohormone	**(57-83-0)**
Luteol	**(57-83-0)**
Luteosan	**(57-83-0)**
(-)-Luteoskyrin	**(21884-44-6)**
Luteoskyrin	**(21884-44-6)**
Luteostab	**(57-83-0)**
Luteotrophin	**(9002-62-4)**
Luteotropic Hormone	**(9002-62-4)**
Luteotropic Hormone LTH	**(9002-62-4)**
Luteotropin	**(9002-62-4)**
Luteovis	**(57-83-0)**
Lutetia Fast Blue RS	**(81-77-6)**
Lutetia Fast Emerald J	**(1328-53-6)**
Lutetia Fast Orange 3R	**(2814-77-9)**
Lutetia Fast Orange R	**(3468-63-1)**
Lutetia Fast Red 3R	**(2425-85-6)**
Lutetia Fast Scarlet RF	**(2425-85-6)**
Lutetia Fast Scarlet RJN	**(2425-85-6)**
Lutetia Orange J	**(3520-72-7)**
Lutetia Orange 2JR	**(1934-20-9)**
Lutetia Orange 3JR	**(633-96-5)**
Lutetia Percyanine BRS	**(147-14-8)**
Lutetia Red CLN	**(5160-02-1)**
Lutetia Red CLN-ST	**(5160-02-1)**
Lutetia Red R	**(1248-18-6)**
Lutetium (9CI)	**(7439-94-3)**
Lutetium-chloride	**(10099-66-8)**
Lutex	**(57-83-0)**
2,4-Lutidine	**(108-47-4)**
2,5-Lutidine	**(589-93-5)**
2,6-Lutidine	**(108-48-5)**
3,4-Lutidine	**(583-58-4)**
α,α'-Lutidine	**(108-48-5)**
2,3-Lutidine (8CI)	**(583-61-9)**
Lutidine (8CI)	**(27175-64-0)**
2,6-Lutidine, 4-ethyl- (6CI,7CI)	**(36917-36-9)**
2,6-Lutidine N-oxide	**(1073-23-0)**
2,6-Lutidine, 1-oxide (8CI)	**(1073-23-0)**
3,5-Lutidine, 1-oxide (8CI)	**(3718-65-8)**
Lutidon	**(57-83-0)**
Lutinyl	**(302-22-7)**
Luto-Metrodiol	**(297-76-7)**
Lutocyclin	**(57-83-0)**
Lutocyclin M	**(57-83-0)**
Lutocylin	**(57-83-0)**
Lutocylol	**(57-83-0)**
Lutoform	**(57-83-0)**
Lutogan	**(79-64-1)**
Lutogyl	**(57-83-0)**
Lutopron	**(630-56-8)**
Lutosan	**(79-64-1)**
Lutosol	**(67-63-0)**
Lutren	**(57-83-0)**
Lutrol	**(17109-49-8)**
Lutrol	**(25322-68-3)**

Name	CAS No.
Lutrol 9	**(25322-68-3)**
Lutrol-9	**(107-21-1)**
Lutromone	**(57-83-0)**
Luviskol	**(9003-39-8)**
Luviskol K30	**(9003-39-8)**
Luviskol k90	**(9003-39-8)**
Luxan Black R	**(101-54-2)**
Luxistelm	**(640-15-3)**
Luxol-Fast-Blue	**(1328-51-4)**
Luxol Fast Blue MBS	**(1328-51-4)**
Lyamine	**(657-27-2)**
Lycedan	**(61-19-8)**
Lycine	**(107-43-7)**
Lycoid DR	**(9000-30-0)**
Lycopersicin	**(17406-45-0)**
Lydol	**(50-13-5)**
Lye	**(1310-58-3)**
Lye	**(1310-73-2)**
Lye (DOT)	**(1310-73-2)**
Lygomme CDS	**(9000-07-1)**
Lym 42	**(9003-07-0)**
Lymphchin	**(553-27-5)**
Lymphocin	**(553-27-5)**
Lymphoquin	**(553-27-5)**
Lymphoscan	**(1345-04-6)**
Lynenol	**(52-76-6)**
Lynestrenol	**(52-76-6)**
Lynestrol	**(68-23-5)**
Lynoestrenol	**(52-76-6)**
Lynoral	**(57-63-6)**
Lyobex	**(128-62-1)**
Lyopect	**(3688-66-2)**
Lyophrin	**(51-42-3)**
Lyophrin	**(51-43-4)**
Lyovac cosmegen	**(50-76-0)**
Lyovit-H	**(13422-51-0)**
Lysalgo	**(61-68-7)**
Lysergamid	**(50-37-3)**
Lysergamide	**(478-94-4)**
Lysergamide, N,N-diethyl-	**(50-37-3)**
Lysergaure diethylamid	**(50-37-3)**
Lysergic acid	**(82-58-6)**
Lysergic acid amide	**(478-94-4)**
D-Lysergic acid diethylamide	**(50-37-3)**
Lysergic acid diethylamide-25	**(50-37-3)**
D-Lysergic acid 1-hydroxymethylethylamide	**(60-79-7)**
Lysergic acid propanolamide	**(60-79-7)**
D-Lysergic acid-l,2-propanolamide	**(60-79-7)**
Lysergide	**(50-37-3)**
Lysergsauerediaethylamid	**(50-37-3)**
Lysine	**(56-87-1)**
l-(+)-Lysine	**(56-87-1)**
DL-Lysine (9CI)	**(70-54-2)**
Lysine, DL- (8CI)	**(70-54-2)**
l-Lysine (9CI)	**(56-87-1)**
Lysine, L	**(56-87-1)**
Lysine acid	**(56-87-1)**
DL-Lysine, dihydrochloride (9CI)	**(617-68-5)**
Lysine, dihydrochloride, DL- (8CI)	**(617-68-5)**
D-Lysine hydrochloride	**(7274-88-6)**
L-Lysine hydrochloride	**(657-27-2)**
Lysine hydrochloride	**(657-27-2)**
Lysine, hydrochloride, D	**(7274-88-6)**
L-Lysine, hydrochloride (VAN)(9CI)	**(10098-89-2)**
L-Lysine, monohydrochloride	**(657-27-2)**
Lysine monohydrochloride	**(657-27-2)**
Lysine, monohydrochloride, L	**(657-27-2)**
Lysococcine	**(63-74-1)**
Lysoform	**(50-00-0)**
Lysozyme	**(9001-63-2)**
Lysozyme G	**(9001-63-2)**
Lysuron	**(315-30-0)**
Lyteca	**(103-90-2)**
Lyteca Syrup	**(103-90-2)**
Lythidathion	**(2669-32-1)**
Lytispasm	**(80-50-2)**
Lytron 5202	**(9003-55-8)**
L-Lyxoascorbic acid	**(50-81-7)**
01M	**(142-22-3)**
4M20	**(897-55-2)**
M 1	**(7440-50-8)**
M 2	**(9011-05-6)**
M 2 (Polymer)	**(9011-05-6)**
M 3	**(7440-50-8)**
M 3P	**(63428-84-2)**
M 3/158	**(17040-19-6)**
M 4	**(7440-50-8)**
M 13/20	**(9002-89-5)**
M 40	**(94-74-6)**
M 60	**(9011-05-6)**
M 60 (Formaldehyde polymer)	**(9011-05-6)**
M 70	**(9011-05-6)**
M 70 (Polymer)	**(9011-05-6)**
M 73	**(1420-04-8)**
M 81	**(640-15-3)**
M 90/20	**(9003-11-6)**
M 99 Reckitt	**(13764-49-3)**
M 140	**(57-74-9)**
M 176	**(67-68-5)**
M 410	**(57-74-9)**
M 1028	**(777-11-7)**
M 2060	**(4301-50-2)**
M 3180	**(13121-70-5)**
M 3196	**(5598-13-0)**
M 3724	**(57213-69-1)**
M 4021	**(64700-56-7)**
M 5055	**(8001-35-2)**
M 8164	**(72391-46-9)**
M 9834	**(71626-11-4)**
M 10797	**(77732-09-3)**
M-1	**(541-25-3)**
M-6	**(56961-25-2)**
M-74	**(298-04-4)**
M-1261	**(299-86-5)**
M-2452	**(5131-24-8)**
M-3432	**(36756-79-3)**
M-9500	**(51-18-3)**
M. 99	**(14521-96-1)**
M. 5050	**(14357-78-9)**
M40 & 80	**(298-00-0)**
MA	**(72-63-9)**
MA 1623	**(18039-42-4)**
MA-1214	**(112-53-8)**
MA300A	**(78-04-6)**
MAA	**(124-58-3)**
1-MA-4-AA (Russian)	**(1220-94-6)**
MAAC	**(108-84-9)**
MAA sodium salt	**(144-21-8)**
MABA	**(99-05-8)**
MA 40 (9CI)	**(12710-10-0)**
MA 100 (Carbon)	**(7440-44-0)**
MAG	**(7558-63-6)**
MAH	**(123-33-1)**
MAM	**(590-96-5)**

MAM AC (592-62-1)
MAMN (56856-83-8)
MAM acetate (592-62-1)
1-MA-4OEAA (Russian) (2475-46-9)
MAOH (108-11-2)
MAP (71-58-9)
MAPO (57-39-6)
MAPP (OSHA) (59355-75-8)
MAS (2533-82-6)
2MB (9002-98-6)
M&B 800 (140-64-7)
M&B 3046 (6062-26-6)
M&B 8873 (1689-83-4)
M&B 10,064 (1689-84-5)
M&B 10731 (1689-84-5)
M&B 10731 (1689-99-2)
M&B 11,461 (3861-47-0)
M+B 693 (144-83-2)
M+B 760 (72-14-0)
MB (74-83-9)
MB 2878 (10433-59-7)
MB 3046 (94-81-5)
MB 8882 (3337-70-0)
MB 9057 (3337-71-1)
MB 10064 (1689-84-5)
7-MBA (2541-69-7)
MBA (51-75-2)
MBAO (126-85-2)
MBAO Hydrochloride (302-70-5)
MBA hydrochloride (55-86-7)
MB 1 (Antioxidant) (VAN) (118-82-1)
MBBA (21739-91-3)
MBBA (26227-73-6)
MBBH 766 (16090-02-1)
MBC (10605-21-7)
MBC (17804-35-2)
MBC 33 (8004-09-9)
M-B-C Fumigant (8004-09-9)
MBC-P (52316-55-9)
MBCP (21609-90-5)
MBDZ (31431-39-7)
MBH (366-70-1)
MBK (55-48-1)
MBK (591-78-6)
MBNA (7068-83-9)
MBOCA (101-14-4)
MBOCA (OSHA) (101-14-4)
MBOT (838-88-0)
MBP (131-70-4)
MBR 12325 (53780-34-0)
MBR 6033 (47000-92-0)
MBR 6168 (26419-73-8)
MBR 8251 (37924-13-3)
MBT (149-30-4)
MBTS (120-78-5)
MBTS Rubber Accelerator (120-78-5)
MBX (74-83-9)
MBY (115-19-5)
MS-Benzanthrone (82-05-3)
20-MC (56-49-5)
2M-4C (94-74-6)
3-MC (56-49-5)
6-MC (92-48-8)
MC (56-49-5)
MC (7487-94-7)
MC 338 (1836-77-7)
MC 474 (2595-54-2)
MC 1053 (973-21-7)
MC 1415 (115-76-4)
MC 1478 (1836-77-7)
MC 2188 (24934-91-6)
MC 2420 (29173-31-7)
MC 6063 (51282-69-0)
MC 7181 (51937-92-9)
MC 10978 (62476-59-9)
MC-4379 (42576-02-3)
3-MCA (56-49-5)
MCA (56-49-5)
MCA (79-11-8)
MCA-600 (1079-33-0)
4-(MCB) (94-81-5)
MCB (108-90-7)
MCC (1918-18-9)
MC 3761 (9CI) (12680-10-3)
MC 5127 (9CI) (12680-12-5)
MC Defoliant (10326-21-3)
MCE (77-81-6)
MCF (79-22-1)
2M-4CH (94-74-6)
MCH 52 (9011-05-6)
MCN 1025 (991-42-4)
MCN 2559 (26171-23-3)
MCN 2783-21-98 (64092-48-4)
MCN-JR-4749 (548-73-2)
MCO 8000 (9004-67-5)
2M-4CP (93-65-2)
MCP (94-74-6)
MCP 875 (357-56-2)
2,4-MCPA (94-74-6)
MCPA (94-74-6)
MCPA Butyl (1713-12-8)
MCPA-Isooctyl (26544-20-7)
MCPA Na salt (3653-48-3)
MCPA-Thioethyl (25319-90-8)
MCPA diethanolamine salt (20405-19-0)
MCPA sodium salt (3653-48-3)
2,4-MCPB (94-81-5)
4MCPB (94-81-5)
MCPB (6062-26-6)
MCPB (94-81-5)
γ-MCPB (94-81-5)
MCPB-ethyl (10443-70-6)
MCP Butyl ester (1713-12-8)
MCP-Butyric (94-81-5)
4-(MCPD) (6062-26-6)
MCPE (36220-29-8)
2-MCPP (93-65-2)
MCPP (7085-19-0)
MCPP (93-65-2)
MCPP 2,4-D (93-65-2)
MCPP-D-4 (7085-19-0)
MCPP-D-4 (93-65-2)
MCPP-K-4 (7085-19-0)
MCPP-K-4 (93-65-2)
MCPP potassium salt (1929-86-8)
MC 20000S (9004-67-5)
MCS 1043 (9CI) (58543-15-0)
MCT (12079-65-1)
MCT (538-23-8)
MC 4000 cP (9004-67-5)
MN-Cellulose (9004-34-6)
M1 (Copper) (7440-50-8)
M2 (Copper) (7440-50-8)
M3 (Copper) (7440-50-8)

Name	CAS
M4 (Copper)	(7440-50-8)
69276 MD	(29218-27-7)
L-(α-MD)	(555-30-6)
MDA	(101-77-9)
MDA	(27496-82-8)
MDA	(4764-17-4)
MDA 150	(25214-70-4)
MDA 220	(25214-70-4)
3'-MDAB	(55-80-1)
MDAB	(55-80-1)
MDBA	(1918-00-9)
MDBCP	(10474-14-3)
MDEA	(105-59-9)
MDI	(101-68-8)
MDI 193	(2259-96-3)
MDI (OSHA)	(101-68-8)
MDMA	(54946-52-0)
MDMH	(116-25-6)
MDM hydantoin	(116-25-6)
M-Diphar	(12427-38-2)
2-ME	(60-24-2)
50ME	(9003-22-9)
ME 277	(113-42-8)
ME-1700	(72-54-8)
MEA	(141-43-5)
MEA	(60-23-1)
MEA Hydrochloride	(2002-24-6)
MEB	(12427-38-2)
MEB 6447	(43121-43-3)
ME4 Brominal	(1689-84-5)
MECB	(111-77-3)
MECS	(109-86-4)
MED 6	(63428-84-2)
MED-T	(63428-84-2)
MEE	(50-50-0)
M.E.G.	(107-21-1)
MEGA	(64-55-1)
MEHP	(4376-20-9)
MEK (OSHA)	(78-93-3)
MEK-Oxime	(96-29-7)
MEKP (OSHA)	(1338-23-4)
MEK peroxide	(1338-23-4)
MEMA	(502-39-6)
MEMC	(123-88-6)
ME-MDA	(838-88-0)
MEP	(104-90-5)
MEP	(122-14-5)
MEP (Pesticide)	(122-14-5)
MER-41	(50-41-9)
META	(108-62-3)
META	(9002-91-9)
MF	(9011-05-6)
MF 1	(9011-05-6)
MF 17	(9011-05-6)
MF 27	(9011-05-6)
MF-344	(2593-15-9)
MFA	(144-49-0)
MFH	(758-17-8)
MFI	(107-44-8)
MFM	(15625-89-5)
MFPS 1	(9011-05-6)
MFR 4	(9003-07-0)
MF Resin	(9011-05-6)
MG 1	(7782-42-5)
MG 18370	(122-09-8)
MG 18570	(122-09-8)
MGK 11	(126-15-8)
MGK 326	(136-45-8)
MGK-264	(113-48-4)
MGK Diethyltoluamide	(134-62-3)
MGK Dog and Cat Repellent	(112-12-9)
MGK R-326	(136-45-8)
MGK Repellent-326	(136-45-8)
MGK Repellent 1,207	(3569-57-1)
MGK Repellent 11	(126-15-8)
MGK Repellent 874	(3547-33-9)
MGK Repellent R-874	(3547-33-9)
MGK Repellent II	(126-15-8)
M7-Giftkoerner	(7446-18-6)
MH	(123-33-1)
MH 101-2	(9011-14-7)
MH 30	(123-33-1)
MH 4	(9003-07-0)
MH-40	(123-33-1)
MHA-FA	(583-91-5)
MHA acid	(583-91-5)
MH 36 Bayer	(123-33-1)
MHT (VAN)	(137-00-8)
217 MI	(513-10-0)
3-MI	(83-34-1)
MI	(87-89-8)
N-6-MI	(932-83-2)
MIA	(64-69-7)
MIAK	(110-12-3)
M.I.B.C.	(105-30-6)
MIBC	(108-11-2)
MIBK	(108-10-1)
MIBK	(141-79-7)
3-MIC	(108-11-2)
MIC	(108-11-2)
MIC	(556-61-6)
MIC	(624-83-9)
MI-Gee	(75-11-6)
MIH	(366-70-1)
MIH	(671-16-9)
MIH Hydrochloride	(366-70-1)
MIK	(108-10-1)
MIL-B-4394-B	(74-97-5)
MIO 40GN	(1332-37-2)
MIPA-Dodecylbenzenesulfonate	(42504-46-1)
MIPK	(563-80-4)
MIT	(556-61-6)
MITC	(556-61-6)
M50-50 Injection	(14357-78-9)
M99 Injection	(14521-96-1)
MJ 5022	(1229-35-2)
MJ 5022	(1982-37-2)
MJF 9325	(3778-73-2)
MK 56	(98-96-4)
MK 125	(50-02-2)
MK 141	(129-03-3)
MK 184	(113-59-7)
MK 231	(38194-50-2)
MK 240	(438-60-8)
MK 351	(555-30-6)
MK 360	(148-79-8)
MK 933	(70288-86-7)
MK-188	(26538-44-3)
MK-351	(41372-08-1)
MK-595	(58-54-8)
MK7	(2124-57-4)
MK. B51	(555-30-6)
MK.B51	(41372-08-1)
2M-4KH	(94-74-6)

MKH 52	**(9011-05-6)**
2M 4KHP	**(93-65-2)**
2M-4KH Sodium salt	**(3653-48-3)**
MKP	**(7778-77-0)**
2M 4KhM	**(94-81-5)**
ML 33F	**(1338-43-8)**
ML 55F	**(1338-43-8)**
ML 97	**(13171-21-6)**
ML 97	**(297-99-4)**
MLT	**(121-75-5)**
MM2A	**(9003-07-0)**
MMA	**(85-91-6)**
3M MBR 6168	**(26419-73-8)**
MMC	**(115-09-3)**
MMC	**(50-07-7)**
MMD	**(502-39-6)**
MMDA	**(13674-05-0)**
MME	**(150-76-5)**
MME	**(80-62-6)**
MMH	**(60-34-4)**
4-MMPD	**(615-05-4)**
4-MMPD Sulphate	**(39156-41-7)**
MMS	**(66-27-3)**
MMT	**(12108-13-3)**
MMTs-BTR	**(9004-67-5)**
MMTP	**(3120-74-9)**
MNA	**(99-09-2)**
MNA	**(110-41-8)**
MNA	**(614-00-6)**
MNBK	**(591-78-6)**
MNF 166	**(30979-48-7)**
MNF O 166	**(67292-92-6)**
1-MNG	**(624-43-1)**
MNG	**(70-25-7)**
MNNG	**(70-25-7)**
MNPA	**(22204-53-1)**
MNPN	**(60153-49-3)**
MNPT	**(89-62-3)**
MNQ	**(58-27-5)**
MNT	**(99-08-1)**
MNU	**(615-53-2)**
MNU	**(684-93-5)**
MNUN	**(615-53-2)**
MO	**(1836-77-7)**
MO 338	**(1836-77-7)**
MO-500	**(13738-63-1)**
MOB	**(131-57-7)**
MOCA	**(101-14-4)**
MO 55F	**(9005-65-6)**
MOF	**(76-38-0)**
MON 0459	**(34494-03-6)**
MON 0573	**(1071-83-6)**
MON 820	**(5994-61-6)**
5-MOP	**(484-20-8)**
8-MOP	**(298-81-7)**
MOPA	**(104-01-8)**
MOS-708	**(1079-33-0)**
6 MP	**(50-44-2)**
8-MP	**(298-81-7)**
MP	**(50-44-2)**
MP 1	**(1302-74-5)**
MP 12-50	**(14807-96-6)**
MP 25-38	**(14807-96-6)**
MP 40-27	**(14807-96-6)**
MP 45-26	**(14807-96-6)**
MP 620	**(16034-77-8)**
MP79	**(52645-53-1)**
3-MPA	**(5332-73-0)**
3MPA	**(107-96-0)**
MPA	**(53-36-1)**
MPA	**(71-58-9)**
M.P. Chlorcaps T.D.	**(113-92-8)**
MPEG	**(9004-74-4)**
MPEG 5000	**(9004-74-4)**
MPF 2	**(9011-05-6)**
MPG 025	**(9004-74-4)**
MPG 081	**(9004-74-4)**
MPG 6	**(7782-42-5)**
M-PHDM	**(3006-93-7)**
MPK	**(107-87-9)**
MPK 90	**(9003-39-8)**
MPMC	**(2425-10-7)**
MPMT	**(845-52-3)**
MPP	**(55-38-9)**
MPPP	**(13147-09-6)**
MP 1 (Refractory)	**(1302-74-5)**
MPT	**(99-75-2)**
M-Parathion	**(298-00-0)**
M-Pentynol	**(77-75-8)**
M-Pyrol	**(872-50-4)**
M3R	**(7440-50-8)**
MRC 910	**(36734-19-7)**
MRD 108	**(467-60-7)**
MRL 41	**(50-41-9)**
3-MS	**(83-40-9)**
6-MS	**(567-61-3)**
M3S	**(7440-50-8)**
MS 1	**(2997-92-4)**
MS 4A	**(1318-02-1)**
MS 5A	**(1318-02-1)**
MS 33	**(1338-41-6)**
MS 53	**(723-46-6)**
MS 1053	**(2595-54-2)**
MS 1143	**(2595-54-2)**
6-MSA	**(567-61-3)**
MS 1 (Catalyst)	**(2997-92-4)**
MSD 803	**(75330-75-5)**
MS 33F	**(1338-41-6)**
MSF	**(558-25-8)**
MSG	**(142-47-2)**
MSMA	**(2163-80-6)**
MST	**(14807-96-6)**
MT 14-411	**(24166-13-0)**
MT 101	**(52570-16-8)**
MTB 51	**(53-46-3)**
MTBHQ	**(1948-33-0)**
MTC	**(63-99-0)**
M&T Chemicals 1222-45	**(1066-45-1)**
MTD	**(10265-92-6)**
MTD	**(95-80-7)**
MTMC	**(1129-41-5)**
M.T.Mucorettes	**(58-18-4)**
MTN	**(620-22-4)**
MTQ	**(72-44-6)**
MTR 1-80	**(933-87-9)**
MTU	**(56-04-2)**
MTX	**(59-05-2)**
MTX sodium	**(15475-56-6)**
MTs	**(9004-67-5)**
MV 119A	**(3347-22-6)**
δ-MVE	**(72-33-3)**
MVNA	**(4549-40-0)**
MW 217	**(7613-16-3)**
2M-4X	**(3653-48-3)**

MX (77439-76-0)
MX 4500 (9003-53-6)
MX 5514 (9003-53-6)
MX 5516 (9003-53-6)
MX 5517-02 (9003-53-6)
MXDA (1477-55-0)
MY 301 (93-14-1)
MY/68 (5141-20-8)
MYTAB (1119-97-7)
Mabertin (846-50-4)
Mablin (55-98-1)
Macarol (77-75-8)
Macasirool (54-31-9)
Macbal (2655-14-3)
Macbar (122-14-5)
Mace (1341-24-8)
Mace (Lacrimator) (532-27-4)
Mace Oil (8007-12-3)
Mach-Nic (54-11-5)
Machete (23184-66-9)
Machete (Herbicide) (23184-66-9)
Machette (23184-66-9)
Mackreazid (140-87-4)
Macleyine (130-86-9)
Maclicine (3116-76-5)
Macquer's Salt (7784-41-0)
Macrabin (68-19-9)
Macrodantin (67-20-9)
Macrodiol (50-28-2)
Macrogol 1000 (25322-68-3)
Macrogol 4000 (25322-68-3)
Macrogol 400 BPC (107-21-1)
Macrogol Ester 2000 (9004-99-3)
Macrogol Ester 400 (9004-99-3)
Macrogol Oleate 600 (9004-96-0)
Macrogol Stearate 400 (9004-99-3)
Macrogol Stearate 2000 (9004-99-3)
Macrol (50-28-2)
Macrolex Red G (82-38-2)
Macro-lex Red G (82-38-2)
Macrondray (94-75-7)
Macropaque (7727-43-7)
Macrospherical 95 (12042-91-0)
Maculotoxin (4368-28-9)
Madar (1088-11-5)
Madhurin (128-44-9)
Madiol (57-53-4)
Madlexin (15686-71-2)
Madrine (300-42-5)
Mafu (62-73-7)
Mafu Strip (62-73-7)
Magacrom Violet N (2092-55-9)
Magadi soda (533-96-0)
Magbond (1302-78-9)
Magcal (1309-48-4)
Magchem 100 (1309-48-4)
Magcyl (9003-11-6)
Magecol (1333-86-4)
Magenta (632-99-5)
Magenta Base (3248-93-9)
Magenta DP (632-99-5)
Magenta E (632-99-5)
Magenta G (632-99-5)
Magenta I (632-99-5)
Magenta PN (632-99-5)
Magenta Powder N (632-99-5)
Magenta S (632-99-5)
Magenta Supertine (632-99-5)
Magic Methyl (421-20-5)
Magistery of Bismuth (1304-85-4)
Maglite (1309-48-4)
Magmasil (1343-88-0)
Magmaster (546-93-0)
Magnacat (7439-96-5)
Magnacide (107-02-8)
Magnacide H (107-02-8)
Magnesate(2-), ((N,N'-1,2-ethanediylbis(N-(carboxymethyl) glycinato))(4-)-N,N',O,O',ON,ON')-, disodium, (OC-6-21)-(9CI) (14402-88-1)
Magnesia (1309-42-8)
Magnesia (1309-48-4)
Magnesia Magma (1309-42-8)
Magnesia USTA (1309-48-4)
Magnesio, Escorias de, humedas o Calientes (Spanish) (69011-63-8)
Magnesio (Italian) (7439-95-4)
Magnesite (7760-50-1)
Magnesite (OSHA) (546-93-0)
Magnesium-chloride (7786-30-3)
Magnesium (7439-95-4)
Magnesium Borings [UN 1869] (7439-95-4)
Magnesium Clippings [UN 1869] (7439-95-4)
Magnesium Dross (69011-63-8)
Magnesium Dross, Hot (69011-63-8)
Magnesium Dross, Wet (69011-63-8)
Magnesium Dross (Wet or hot) (69011-63-8)
Magnesium Gold Purple (7440-57-5)
Magnesium Hydroxide Gel (1309-42-8)
Magnesium Metal [UN 1869] (7439-95-4)
Magnesium Pellets [UN 1869] (7439-95-4)
Magnesium Powdered [UN 1418] (7439-95-4)
Magnesium Ribbons [UN 1869] (7439-95-4)
Magnesium Turnings [UN 1869] (7439-95-4)
Magnesium acetate (142-72-3)
Magnesium aluminum hydroxide carbonate (11097-59-9)
Magnesium arsenate, Solid (DOT) (10103-50-1)
Magnesium arsenate [UN 1622] (10103-50-1)
Magnesium arsenate phosphor (10103-50-1)
Magnesium basic carbonate (7760-50-1)
Magnesium benzoate (553-70-8)
Magnesium, bis(2-hydroxybenzoato-O1,O2)-, (T-4)- (9CI) (18917-89-0)
Magnesium, bromo(hexahydro-2H-azepin-2-onato-N)- (9CI) (17091-31-5)
Magnesium(II) carbonate (1:1) (546-93-0)
Magnesium carbonate ($MgCO_3$) dihydrate (68973-26-2)
Magnesium carbonate hydroxide (7760-50-1)
Magnesium, (carbonato(2-))hexadecahydroxybis(aluminum)hexa-(9CI) (11097-59-9)
Magnesium chlorate [UN 2723] (10326-21-3)
Magnesium chloride, hexahydrate (7791-18-6)
Magnesium, chlorocyclohexyl- (9CI) (931-51-1)
Magnesium, chloroethyl- (9CI) (2386-64-3)
Magnesium, chloromethyl- (9CI) (676-58-4)
Magnesium, chloro(1-methylpropyl)- (9CI) (15366-08-2)
Magnesium, chlorophenyl- (9CI) (100-59-4)
Magnesium chromate (13423-61-5)
Magnesium dalapon (29110-22-3)
Magnesium diacetate (142-72-3)
Magnesium dichlorate (10326-21-3)
Magnesium dichloride hexahydrate (7791-18-6)
Magnesium 2,2-dichloropropanoate (29110-22-3)
Magnesium dichromate (14104-85-9)
Magnesium dihydroxide (1309-42-8)
Magnesium, dimethyl (2999-74-8)
Magnesium ethoxide (2414-98-4)
Magnesium ethoxyethoxide (14064-03-0)

Magnesium ethylate (2414-98-4)
Magnesium fluorosilicate (16949-65-8)
Magnesium fluosilicate (16949-65-8)
Magnesiumfosfide (Dutch) (12057-74-8)
Magnesium hydrate (1309-42-8)
Magnesium hydroxide (9CI) (1309-42-8)
Magnesium lauryl sulfate (3097-08-3)
Magnesium mesotrisilicate (14987-04-3)
Magnesium, methyl-, bromide (ethyl ether solution) (75-16-1)
Magnesium monododecyl sulfate (3097-08-3)
Magnesium(II) nitrate (1:2) (10377-60-3)
Magnesium nitrate [UN 1474] (10377-60-3)
Magnesium oxide (1309-42-8)
Magnesium oxide (ACGIH,OSHA) (1309-48-4)
Magnesium perchlorate [UN 1475] (10034-81-8)
Magnesium peroxide (8CI,9CI) (14452-57-4)
Magnesium peroxide, Solid (14452-57-4)
Magnesium orthophosphate (7757-87-1)
Magnesium phosphate (7757-87-1)
Magnesium phosphate, neutral (7757-87-1)
Magnesium phosphate, tribasic (7757-87-1)
Magnesium phosphide (8CI,9CI) (12057-74-8)
Magnesium 2-propenoate (5698-98-6)
Magnesium silicate (1343-88-0)
Magnesium silicate (14987-04-3)
Magnesium silicon oxide (9CI) (14987-04-3)
Magnesium sodium ethylenediaminetetraacetate (14402-88-1)
Magnesium stearate (ACGIH) (557-04-0)
Magnesium styphnate (13255-27-1)
Magnesium sulfate (1:1) (7487-88-9)
Magnesium sulfate adduct of 2,2-dithio-bis-pyridine 1-oxide (43143-11-9)
Magnesium sulphate (7487-88-9)
Magnesium, tetrakis(carbonato(2-))dihydroxypenta- (9CI) (7760-50-1)
Magnesium, (3,7,11,15-tetramethyl-2-hexadecenyl 9-ethenyl-14-ethyl-21-(methoxycarbonyl)-4,8,13,18-tetramethyl-20-oxo-3-phorbinepropanoato(2-)-N23,N24,N25,N26)-,(SP-4-2-(3S-(3α(2E,7S*,11S*),4β,21β)))- (9CI) (479-61-8)
Magnesium trisilicate (14987-04-3)
Magnesium trisilicate USP (14987-04-3)
Magnesium-oxide (1309-48-4)
Magnesol (1343-88-0)
Magneson (74-39-5)
Magneson I (74-39-5)
Magnetic 70, 90, and 95 (7704-34-9)
Magnezon I (74-39-5)
Magnezu Tlenek (Polish) (1309-48-4)
Magnosil (14987-04-3)
Magnus 101 (9CI) (88997-62-0)
Magox (1309-48-4)
Magox 85 (1309-48-4)
Magox 90 (1309-48-4)
Magox 95 (1309-48-4)
Magox 98 (1309-48-4)
Magox op (1309-48-4)
Magron (10326-21-3)
Magsalyl (54-21-7)
Magsorbent (1343-88-0)
Mahogany EMBL (2429-82-5)
Maintain 3 (123-33-1)
Maintain A (2536-31-4)
Maintain CF125 (2536-31-4)
Maipedopa (59-92-7)
Maise Oil (8001-30-7)
Maizena (9005-25-8)
Majol PLX (9004-32-4)
Majsolin (125-33-7)
Majudin (484-20-8)
Makarol (56-53-1)
Maki (28772-56-7)
Malachit-Grun (German) (569-64-2)
Malachite (12069-69-1)
Malachite Green (569-64-2)
Malachite Green A (569-64-2)
Malachite Green AN (569-64-2)
Malachite Green B (569-64-2)
Malachite Green CP (569-64-2)
Malachite Green Chloride (569-64-2)
Malachite Green Crystals (569-64-2)
Malachite Green Crystals BPC (569-64-2)
Malachite Green G (633-03-4)
Malachite Green Hydrochloride (569-64-2)
Malachite Green (Indicator) (569-64-2)
Malachite Green J3E (569-64-2)
Malachite Green Oxalate (2437-29-8)
Malachite Green Powder (569-64-2)
Malachite Green WS (569-64-2)
Malachite Lake Green A (569-64-2)
Malacid (58-14-0)
Malacide (121-75-5)
Malafor (121-75-5)
Malagran (121-75-5)
Malakill (121-75-5)
Malamar (121-75-5)
Malamar 50 (121-75-5)
Malamine, hexakis(methoxymethyl) (3089-11-0)
Malaoxon (1634-78-2)
Malaoxone (1634-78-2)
Malaphele (121-75-5)
Malaphos (121-75-5)
Malaquin (54-05-7)
Malaren (54-05-7)
Malarex (54-05-7)
Malaricida (69-05-6)
Malasol (121-75-5)
Malaspray (121-75-5)
Malathion (ACGIH,OSHA) (121-75-5)
Malathion E50 (121-75-5)
Malathion LV Concentrate (121-75-5)
Malathion-O-Analog (1634-78-2)
Malathion ULV Concentrate (121-75-5)
Malathiozoo (121-75-5)
Malathon (121-75-5)
Malathyl LV Concentrate & ULV Concentrate (121-75-5)
Malation (Polish) (121-75-5)
Malatol (121-75-5)
Malatox (121-75-5)
Malayan Camphor (507-70-0)
Malazide (123-33-1)
Maldison (121-75-5)
Maleamate (557-24-4)
Maleamic acid (8CI) (557-24-4)
Maleic-acid (110-16-7)
Maleic acid [NA 2215] (110-16-7)
Maleic acid anhydride (108-31-6)
Maleic acid, bis(1,3-dimethylbutyl) ester (105-52-2)
Maleic acid, bis(2-ethylhexyl)ester (142-16-5)
Maleic acid, diallyl ester (999-21-3)
Maleic acid, dibutyl ester (105-76-0)
Maleic acid, di(1,3-dimethylbutyl) ester (105-52-2)
Maleic acid, didodecyl ester (8CI) (2915-52-8)
Maleic acid, diethyl ester (141-05-9)
Maleic acid, dihexyl ester (105-52-2)
Maleic acid, diisodecyl ester (1330-76-3)
Maleic acid, diisooctyl ester (1330-76-3)

Maleic acid, dimethyl- (8CI) **(488-21-1)**
Maleic acid, dimethyl ester **(624-48-6)**
Maleic acid, dioctyl ester **(2915-53-9)**
Maleic acid, dipentyl ester **(10099-71-5)**
Maleic acid, ethylmethyl- **(41654-09-5)**
Maleic acid hydrazide **(123-33-1)**
Maleic acid, methyl- **(498-23-7)**
Maleic acid, monoethyl ester (8CI) **(3990-03-2)**
Maleic acid, mono(2-(N-(2-hydroxyethyl)anilino)ethyl) ester (8CI) **(15772-26-6)**
Maleic-anhydride **(108-31-6)**
Maleic anhydride (ACGIH,DOT,OSHA) **(108-31-6)**
Maleic anhydride (8CI) **(3675-13-6)**
Maleic anhydride, Solid or molten (DOT) **(108-31-6)**
Maleic anhydride adduct of butadiene **(85-43-8)**
Maleic anhydride and 1,3-pentadiene adduct **(5333-84-6)**
Maleic anhydride, dimethyl **(766-39-2)**
Maleic anhydride, methyl- **(616-02-4)**
Maleic anhydride, phthalic anhydride, propylene glycol terpolymer **(25037-66-5)**
Maleic anhydride, propylene glycol, 1,2-benzenedicarboxylic anhydride polymer **(25037-66-5)**
Maleic anhydride, propylene glycol, phthalic anhydride polymer **(25037-66-5)**
Maleic hydrazide **(123-33-1)**
Maleic hydrazide 30% **(123-33-1)**
Maleic hydrazide diethanolamine salt **(5716-15-4)**
Maleic hydrazide potassium salt **(28382-15-2)**
Maleic hydrazine **(123-33-1)**
Maleic monoamide **(557-24-4)**
Maleic monoperoxy acid, 1-tert-butyl ester **(1931-62-0)**
Maleic monoperoxy acid, 1-tert-butyl ester, Not more than 55% in solution **(1931-62-0)**
Maleimide, 2,3-dimethyl- (8CI) **(17825-86-4)**
Maleimide, 2-ethyl-3-methyl- (6CI,7CI,8CI) **(20189-42-8)**
Maleimide, N,N'-(methylenedi-p-phenylene)di- **(13676-54-5)**
Maleimide, N-phenyl **(941-69-5)**
Maleimide, N,N'-(m-phenylene)di **(3006-93-7)**
Malein 30 **(123-33-1)**
Maleinanhydrid (Czech) **(108-31-6)**
Maleinic acid **(110-16-7)**
Maleinsaeurehydrazid (German) **(123-33-1)**
Malenic acid **(110-16-7)**
Maleonitrile, diamino- (8CI) **(1187-42-4)**
Maleoylacetic acid **(24740-88-3)**
N,N-Maleoylhydrazine **(123-33-1)**
Malestrone **(58-18-4)**
Malestrone (AMPS) **(58-22-0)**
Malex **(62-44-2)**
Maleylacetate **(24740-88-3)**
Malgesic **(50-33-9)**
Malic-acid **(6915-15-7)**
Malic acid, DL- (8CI) **(617-48-1)**
Malic acid, 3-hydroxy- **(87-69-4)**
Malil **(52-43-7)**
Malilum **(52-43-7)**
Malipur **(133-06-2)**
Malivan **(1165-48-6)**
Malix **(115-29-7)**
Mallofeen **(136-40-3)**
Mallophene **(136-40-3)**
Mallophene **(94-78-0)**
Mallorepine **(767-98-6)**
Mallorol **(50-52-2)**
Malmed **(121-75-5)**
Malocid **(58-14-0)**
Malocide **(58-14-0)**
Malogen **(58-18-4)**
Malogen L.A.200 **(315-37-7)**
Malomalic ether **(7408-18-6)**
Malonal **(57-44-3)**
Malonaldehyde **(542-78-9)**
Malonaldehyde, bis(dimethyl acetal) (8CI) **(102-52-3)**
Malonaldehyde, ion(1-), sodium **(24382-04-5)**
Malonaldehyde sodium salt **(24382-04-5)**
Malonaldehyde tetramethyl acetal **(102-52-3)**
Malonaldehydic acid, phenyl-, methyl ester **(5894-79-1)**
Malonamic acid, N-(2-carboxy-3,3-dimethyl-7-oxo-4-thia-1-azabicyclo-(3.2.0)hept-6-yl)-2- phenyl **(4697-36-3)**
Malonamide, 2-ethyl-2-phenyl- (8CI) **(7206-76-0)**
Malonamide nitrile **(107-91-5)**
Malonamonitrile **(107-91-5)**
Malondialdehyde **(542-78-9)**
Malonic-acid **(141-82-2)**
Malonic acid, acetamido-, diethyl ester **(1068-90-2)**
Malonic acid, butyl- (8CI) **(534-59-8)**
Malonic acid, butyl(2-methoxyethyl)-, diethyl ester (8CI) **(20591-91-7)**
Malonic acid, chloro- (8CI) **(600-33-9)**
Malonic acid, decyl- (8CI) **(4372-29-6)**
Malonic acid, diethyl ester **(105-53-3)**
Malonic acid, dimethyl- (8CI) **(595-46-0)**
Malonic acid, dimethyl-, diethyl ester (8CI) **(1619-62-1)**
Malonic acid, dimethyl ester **(108-59-8)**
Malonic acid dinitrile **(109-77-3)**
Malonic acid, ethyl- (8CI) **(601-75-2)**
Malonic acid, ethyl-, diethyl ester (8CI) **(133-13-1)**
Malonic acid ethyl ester nitrile **(105-56-6)**
Malonic acid, ethyl(2-methoxyethyl)-, diethyl ester (8CI) **(20591-89-3)**
Malonic acid, ethylmethyl-, diethyl ester (8CI) **(2049-70-9)**
Malonic acid, isobutyl-, diethyl ester **(10203-58-4)**
Malonic acid, isopropyl-, diethyl ester (8CI) **(759-36-4)**
Malonic acid, isopropyl(2-methoxyethyl)-, diethyl ester (7CI,8CI) **(20721-77-1)**
Malonic acid, (2-methoxyethyl)-, diethyl ester (7CI,8CI) **(6335-02-0)**
Malonic acid, (2-methoxyethyl)methyl-, diethyl ester (8CI) **(20721-76-0)**
Malonic acid, (2-methoxyethyl)propyl-, diethyl ester (7CI,8CI) **(20591-90-6)**
Malonic acid, methyl **(516-05-2)**
Malonic acid, methyl-, diethyl ester (8CI) **(609-08-5)**
Malonic acid, methylpropyl-, diethyl ester (6CI) **(55898-43-6)**
Malonic acid, octyl- (8CI) **(760-55-4)**
Malonic acid, pentyl- (8CI) **(616-88-6)**
Malonic acid, propyl- (8CI) **(616-62-6)**
Malonic acid, propyl-, diethyl ester (8CI) **(2163-48-6)**
Malonic acid, sodium salt **(141-95-7)**
Malonic acid, thallium salt (1:2) **(2757-18-8)**
Malonic aldehyde **(542-78-9)**
Malonic dialdehyde **(542-78-9)**
Malonic dinitrile **(109-77-3)**
Malonic ester **(105-53-3)**
Malonic mononitrile **(372-09-8)**
Malonitrile hydrazide **(140-87-4)**
Malonodialdehyde **(542-78-9)**
Malononitrile (8CI) **(4341-85-9)**
Malononitrile [UN 2647] **(109-77-3)**
Malononitrile, o-chlorobenzylidene **(2698-41-1)**
Malononitrile, dimethyl- (6CI,7CI,8CI) **(7321-55-3)**
Malononitrile hydrazide **(140-87-4)**
Malonyldialdehyde **(542-78-9)**
Malonylurea **(67-52-7)**
Maloprim **(58-14-0)**
Maloprim **(80-08-0)**
Maloran **(13360-45-7)**
Malphos **(121-75-5)**
Malt, Ext **(8002-48-0)**
Malt Extract **(8002-48-0)**
Malt Extract, Powder **(8002-48-0)**

Malt Sugar **(69-79-4)**
α-Malt Sugar **(69-79-4)**
Malta Red X 2284 **(6471-49-4)**
Malted Barley Extract **(8002-48-0)**
Maltine **(8002-48-0)**
Maltobiose **(69-79-4)**
Maltodextrin (9CI) **(9050-36-6)**
Maltol **(118-71-8)**
Maltonic acid **(526-95-4)**
D-Maltose **(69-79-4)**
Maltose **(69-79-4)**
Maltox **(121-75-5)**
Maltox MLT **(121-75-5)**
Malvalic acid **(503-05-9)**
Malvic acid **(503-05-9)**
Malzid **(123-33-1)**
Mamallet-A **(58-15-1)**
Mammex **(59-87-0)**
Mammotropin **(9002-62-4)**
Man-Gro **(7785-87-7)**
Manadrin **(299-42-3)**
Manam **(12427-38-2)**
Manchester Yellow **(605-69-6)**
Mancofol **(8018-01-7)**
Mancokar **(8064-42-4)**
Mancozeb **(8018-01-7)**
Mandarin G **(633-96-5)**
Mandarin Petitgrain Oil **(8014-17-3)**
Mandelaldehyde, p-chloro- (8CI) **(34025-32-6)**
Mandelic-acid **(90-64-2)**
Mandelic acid, p-chloro- (8CI) **(492-86-4)**
Mandelic acid, ethyl ester **(774-40-3)**
Mandelic acid, α-methyl- (8CI) **(515-30-0)**
Mandelic acid, .α.-methyl-, acetate, (+)- (8CI) **(10487-92-0)**
Mandelic acid nitrile **(532-28-5)**
Mandelic acid, 3,3,5-trimethylcyclohexyl ester **(456-59-7)**
Mandelic acid, 3d-tropanyl ester **(87-00-3)**
Mandelonitrile **(532-28-5)**
D(-)-Mandelonitrile-β-D-gentiobioside **(29883-15-6)**
Mandelonitrile-β-gentiobioside **(29883-15-6)**
D-Mandelonitrile-β-D-glucosido-6-β-D-glucoside **(29883-15-6)**
Mandelsaeureaethylester (German) **(774-40-3)**
Mandelyltropeine **(87-00-3)**
Mandelytropeine **(87-00-3)**
Manderol **(72391-46-9)**
Mandokef **(34444-01-4)**
Mandrin **(299-42-3)**
Maneb **(12427-38-2)**
Maneb 80 **(12427-38-2)**
Maneb, Stabilized against self heating [UN 2968] **(12427-38-2)**
Maneb, With not less than 60% Maneb [UN 2210] **(12427-38-2)**
Maneb ZL4 **(12427-38-2)**
Maneb-Zinc **(8018-01-7)**
Maneb-Zineb-Komplex (German) **(8018-01-7)**
Maneb-Zineb-Mischkomplex (German) **(8018-01-7)**
Maneba **(12427-38-2)**
Manebe 80 **(12427-38-2)**
Manebe (French) **(12427-38-2)**
Manebgan **(12427-38-2)**
Manesan **(12427-38-2)**
Manex **(12427-38-2)**
Manexin **(15825-70-4)**
Mangaanbioxyde (Dutch) **(1313-13-9)**
Mangaandioxyde (Dutch) **(1313-13-9)**
Mangaan (II)-(N,N'-ethyleen-bis(dithiocarbamaat)) (Dutch) **(12427-38-2)**
Mangan (Polish) **(7439-96-5)**
Mangan(II)-(N,N-aethylen-bis(dithiocarbamat)) (German) **(12427-38-2)**
Mangan (II)-(N,N'-aethylen-bis(dithiocarbamate)) (German) **(12427-38-2)**
Manganate(2-), ((N,N'-1,2-ethanediylbis(N-(carboxymethyl)-glycinato))(4-)-N,N',O,O',ON,ON')-, disodium, (OC-6-21)-(9CI) **(15375-84-5)**
Manganate(2-), ((ethylenedinitrilo)tetraacetato)-, disodium (8CI) **(15375-84-5)**
Mangandioxid (German) **(1313-13-9)**
Manganese **(7439-96-5)**
Manganese(2+) **(16397-91-4)**
Manganese(II) **(16397-91-4)**
Manganese Alloy, Base, Mn 74-82, Fe 8-19, C 6.9-C7.5, Si O-1.2, P O-O.4 (ASTM A99), Exothermic **(12604-53-4)**
Manganese Black **(1313-13-9)**
Manganese Green **(1344-43-0)**
Manganese (IV) oxide **(1313-13-9)**
Manganese (Mn2+) **(16397-91-4)**
Manganese acetate **(638-38-0)**
Manganese(2+) acetate **(638-38-0)**
Manganese(II) acetate **(638-38-0)**
Manganese acetate (Mn(OAc)2) **(638-38-0)**
Manganese acetate tetrahydrate **(6156-78-1)**
Manganese(II) acetate tetrahydrate **(6156-78-1)**
Manganese binoxide **(1313-13-9)**
Manganese (biossido di) (Italian) **(1313-13-9)**
Manganese (bioxyd de) (French) **(1313-13-9)**
Manganese carbonate **(598-62-9)**
Manganese carbonate (1:1) **(598-62-9)**
Manganese(2+) carbonate **(598-62-9)**
Manganese(2+) carbonate (1:1) **(598-62-9)**
Manganese(II) carbonate **(598-62-9)**
Manganese, (carbonato(2-)-O)-, monohydrogen (9CI) **(68013-64-9)**
Manganese cation **(16397-91-4)**
Manganese chloride **(7773-01-5)**
Manganese(II) chloride (1:2) **(7773-01-5)**
Manganese chloride (MnCl) (6CI,7CI,9CI) **(50646-06-5)**
Manganese(II) chloride, tetrahydrate **(13446-34-9)**
Manganese, (4-((5-chloro-4-methyl-2-sulfophenyl)azo)-3-hydroxy-2-naphthalenecarboxylato(2-))- (9CI) **(12688-94-7)**
Manganese, (4-((5-chloro-2-sulfo-p-tolyl)azo)-3-hydroxy-2-naphthoato(2-))- **(12688-94-7)**
Manganese citrate **(5968-88-7)**
Manganese, cyclopentadienyltricarbonyl **(12079-65-1)**
Manganese cyclopentadienyl tricarbonyl (ACGIH,OSHA) **(12079-65-1)**
Manganese diacetate **(638-38-0)**
Manganese diacetate, tetrahydrate **(6156-78-1)**
Manganese dichloride **(7773-01-5)**
Manganese dichloride tetrahydrate **(13446-34-9)**
Manganese (diossido di) (Italian) **(1313-13-9)**
Manganese-dioxide **(1313-13-9)**
Manganese (dioxyde de) (French) **(1313-13-9)**
Manganese disodium ethylene diamine tetraacetate **(15375-84-5)**
Manganese, ((1,2-ethanediylbis(carbamodithioato))(2-))-, Mixt. with ((1,2-ethanediylbis(carbamodithioato))(2-))zinc and 2(or 4)-isooctyl-4,6(or 2,6)-dinitrophenyl 2-butenoate (9CI) **(8064-42-4)**
Manganese ethylene-1,2-bisdithiocarbamate **(12427-38-2)**
Manganese ethylene bis-dithiocarbamate (DOT) **(12427-38-2)**
Manganese, (ethylenebis(dithiocarbamato)) **(12427-38-2)**
Manganese (II) ethylene di(dithiocarbamate) **(12427-38-2)**
Manganese 2-ethylhexanoate **(15956-58-8)**
Manganese(II) fluoride **(7782-64-1)**
Manganese fluorure (French) **(7782-64-1)**
Manganese gluconate **(6485-39-8)**
Manganese ion(2+) **(16397-91-4)**
Manganese(2+) ion **(16397-91-4)**
Manganese(II) ion **(16397-91-4)**
Manganese, ion (Mn2+) **(16397-91-4)**
Manganese manganate **(1317-34-6)**
Manganese, (methylcyclopentadienyl)tricarbonyl- **(12108-13-3)**

Manganese monochloride (50646-06-5)
Manganese monooxide (1344-43-0)
Manganese, monosulfate, monohydrate (10034-96-5)
Manganese monoxide (1344-43-0)
Manganese neodecanoate (27253-32-3)
Manganese oxide (1313-13-9)
Manganese(3+) oxide (1317-34-6)
Manganese(II) oxide (1344-43-0)
Manganese(III) oxide (1317-34-6)
Manganese-oxide (1317-35-7)
Manganese peroxide (1313-13-9)
Manganese phosphate (10124-54-6)
Manganese sisquioxide (1317-34-6)
Manganese sulfate (1:1) (7785-87-7)
Manganese(2+) sulfate (1:1) (7785-87-7)
Manganese(II) sulfate (1:1) (7785-87-7)
Manganese sulfate monohydrate (10034-96-5)
Manganese(2+) sulfate monohydrate (10034-96-5)
Manganese sulphate (7785-87-7)
Manganese superoxide (1313-13-9)
Manganese tetroxide (ACGIH,OSHA) (1317-35-7)
Manganese, tricarbonyl methylcyclopentadienyl (12108-13-3)
Manganese, tricarbonyl-pi-cyclopentadienyl- (12079-65-1)
Manganese trioxide (1317-34-6)
Manganese, tris(2,4-pentanedionato)- (8CI) (14284-89-0)
Manganese, tris(2,4-pentanedionato-O,O')-, (OC-6-11)- (9CI) (14284-89-0)
Manganic oxide (1317-34-6)
Mangan nitridovany (Czech) (7439-96-5)
Manganomanganic oxide (1317-35-7)
Manganous acetate (638-38-0)
Manganous acetate tetrahydrate (6156-78-1)
Manganous carbonate (598-62-9)
Manganous chloride (7773-01-5)
Manganous chloride tetrahydrate (13446-34-9)
Manganous ethylenebis(dithiocarbamate) (12427-38-2)
Manganous ion (16397-91-4)
Manganous oxide (1344-43-0)
Manganous sulfate (7785-87-7)
Manganous sulfate monohydrate (10034-96-5)
Mangan-zink-aethylendiamin-bis-dithio-carbamat (German) (8018-01-7)
Manhexin (15825-70-4)
Manicole (15825-70-4)
Manil Fast Yellow AN (1325-37-7)
Manite (15825-70-4)
Manna Sugar (69-65-8)
Mannex (15825-70-4)
Mannit-Lost (German) (576-68-1)
Mannit-Mustard (German) (576-68-1)
Mannite (69-65-8)
D-Mannitol (69-65-8)
Mannitol (87-78-5)
Mannitol, D (69-65-8)
Mannitol Mustard (551-74-6)
Mannitol Mustard (576-68-1)
Mannitol Mustard Dihydrochloride (551-74-6)
Mannitol Nitrogen Mustard (551-74-6)
Mannitol Nitrogen Mustard (576-68-1)
Mannitol, 1,6-bis((2-chloroethyl)amino)-1,6-dideoxy-, D-, (576-68-1)
Mannitol, 1,6-bis((2-chloroethyl)amino)-1,6-dideoxy-, dihydrochloride, D (551-74-6)
Mannitol, 1,6-dibromo-1,6-dideoxy-, D (488-41-5)
Mannitol, 1,6-dideoxy-1,6-dithiocyanato-, (D) (73928-09-3)
Mannitol hexanitrate (15825-70-4)
D-Mannitol, hexanitrate (9CI) (15825-70-4)
Mannitol hexanitrate, Containing, by weight, at least 40% water [UN 0133] (15825-70-4)
Mannitoli hexanitras (Latin) (15825-70-4)
Mannitolo esanitrato (15825-70-4)
Mannitrin (15825-70-4)
Mannityli nitras (15825-70-4)
Mannitylium hexanitricum (15825-70-4)
Mannityl nitrate (15825-70-4)
Mannogranol (551-74-6)
Mannomustine (551-74-6)
Mannomustine (576-68-1)
Mannomustine dihydrochloride (551-74-6)
Mannopyranoside, quercetin-3 6-deoxy-, α-l- (522-12-3)
Mannopyranoside, strophanthidin-3 6-deoxy-, α-l- (508-75-8)
D-Mannose (9CI) (3458-28-4)
Mannose, D- (8CI) (3458-28-4)
D-Mannose, 3-amino-3,6-dideoxy-, Mixt. with 1-(4-(methylamino)phenyl)ethanone (8065-41-6)
L-Mannose, 6-deoxy- (9CI) (3615-41-6)
Manolene 6050 (9002-88-4)
Manoseb (8018-01-7)
Manoxal OT (577-11-7)
Manoxol DT (9CI) (57608-28-3)
Manoxol OT (577-11-7)
Manro PTSA 65 E (104-15-4)
Manro PTSA 65 H (104-15-4)
Manro PTSA 65 LS (104-15-4)
Mansonone C (5574-34-5)
Manta (40596-69-8)
Mantheline (53-46-3)
Manucol (9005-38-3)
Manucol DM (9005-38-3)
Manucol KMF (9005-38-3)
Manucol SS/LD2 (9005-38-3)
Manufactured Iron Oxides (1309-37-1)
Manugel F 331 (9005-38-3)
Manutex (9005-38-3)
Manutex F (9005-38-3)
Manutex RS (9005-38-3)
Manutex RS 1 (9005-38-3)
Manutex RS-5 (9005-38-3)
Manutex SA/KP (9005-38-3)
Manutex SH/LH (9005-38-3)
Manzate (12427-38-2)
Manzate 200 (12427-38-2)
Manzate 200 (8018-01-7)
Manzate D (12427-38-2)
Manzate Maneb Fungicide (12427-38-2)
Manzeb (12427-38-2)
Manzeb (8018-01-7)
Manzin (12427-38-2)
Manzin 80 (8018-01-7)
Mao-Rem (156-51-4)
Maoa (72-44-6)
Maolate (886-74-8)
Mapine (487-93-4)
Maple Amaranth (915-67-3)
Maple Brilliant Blue FCF (2650-18-2)
Maple Erythrosine (16423-68-0)
Maple Indigo Carmine (860-22-0)
Maple Lactone (80-71-7)
Maple Ponceau 3R (3564-09-8)
Maple Ponceau SX (4548-53-2)
Maple Sunset Yellow FCF (2783-94-0)
Maple Tartrazol Yellow (1934-21-0)
Mapolose M25 (9004-67-5)
Mapolose 60SH50 (9004-67-5)
Maposol (137-42-8)
Maposol (6734-80-1)
Mappine (487-93-4)

Name	CAS No.
Maprofix 563	**(151-21-3)**
Maprofix ES	**(9004-82-4)**
Maprofix LK	**(151-21-3)**
Maprofix NH	**(2235-54-3)**
Maprofix Neu	**(151-21-3)**
Maprofix 60S	**(9004-82-4)**
Maprofix TLS	**(139-96-8)**
Maprofix TLS 65	**(139-96-8)**
Maprofix TLS 500	**(139-96-8)**
Maprofix WAC	**(151-21-3)**
Maprofix WAC-LA	**(151-21-3)**
Maprosyl 30	**(137-16-6)**
Maprosyl O	**(110-25-8)**
Maprotiline	**(10262-69-8)**
Maprotylina (Polish)	**(10262-69-8)**
Maqbarl	**(2655-14-3)**
Mar Bate	**(57-53-4)**
Mar-Frin	**(81-81-2)**
Maracell C	**(8061-51-6)**
Maracell E	**(8061-51-6)**
Maralate	**(72-43-5)**
Maranhist	**(91-84-9)**
Maranta	**(9005-25-8)**
Maranyl F 114	**(105-60-2)**
Maranyl F 114	**(25038-54-4)**
Maranyl F 124	**(105-60-2)**
Maranyl F 124	**(25038-54-4)**
Maranyl F 500	**(105-60-2)**
Maranyl F 500	**(25038-54-4)**
Marasperse B	**(8061-51-6)**
Marasperse CBS	**(8061-51-6)**
Marasperse N	**(8061-51-6)**
Marasperse N 22	**(8061-51-6)**
Marasperse N 22 Dispersant	**(8061-51-6)**
Marazine	**(82-92-8)**
Marble (OSHA)	**(1317-65-3)**
Marbon 8000a	**(9003-55-8)**
Marbon 9200	**(9003-55-8)**
Marcoumar	**(435-97-2)**
Marcumar	**(435-97-2)**
Marevan	**(129-06-6)**
Marevan (sodium salt)	**(129-06-6)**
Marex	**(569-65-3)**
Marezine	**(82-92-8)**
Marfotoks	**(2595-54-2)**
Margaric acid	**(506-12-7)**
Margonil	**(57-53-4)**
Margonovine	**(60-79-7)**
Maricaine	**(137-58-6)**
Marihuana	**(8063-14-7)**
Marijuana	**(8063-14-7)**
Marimet 45	**(1344-95-2)**
Marinco H	**(1309-42-8)**
Marinco H 1241	**(1309-42-8)**
Marine Blue A 8021	**(1325-87-7)**
Marinol	**(8001-54-5)**
Marisilan	**(69-53-4)**
Maritime Pine LS	**(8001-21-6)**
Maritus Yellow	**(605-69-6)**
Marks 4-CPA	**(122-88-3)**
Markure Ul2	**(78-04-6)**
Marlamid M 18	**(111-57-9)**
Marlate	**(72-43-5)**
Marlex 9	**(9002-88-4)**
Marlex 50	**(9002-88-4)**
Marlex 60	**(9002-88-4)**
Marlex 960	**(9002-88-4)**
Marlex 6003	**(9002-88-4)**
Marlex 6009	**(9002-88-4)**
Marlex 6015	**(9002-88-4)**
Marlex 6050	**(9002-88-4)**
Marlex 6060	**(9002-88-4)**
Marlex 9400	**(9003-07-0)**
Marlex EHM 6001	**(9002-88-4)**
Marlex HGH 050-01	**(9003-07-0)**
Marlex M 309	**(9002-88-4)**
Marlex TR 704	**(9002-88-4)**
Marlex TR 880	**(9002-88-4)**
Marlex TR 885	**(9002-88-4)**
Marlex TR 906	**(9002-88-4)**
Marlipal 1217	**(9002-92-0)**
Marlipal 1850	**(9005-00-9)**
Marlon AS 3	**(27176-87-0)**
Marlophen 820	**(9002-93-1)**
Marlophen 830 (9CI)	**(75882-11-0)**
Marlophen P	**(9004-78-8)**
Marlophen P 7	**(9004-78-8)**
Marmag	**(1309-48-4)**
Marmelosin	**(482-44-0)**
Marmer	**(330-54-1)**
Marnitension simple	**(50-55-5)**
Maroxol-50	**(51-28-5)**
Mars Brown	**(1309-37-1)**
Mars Red	**(1309-37-1)**
Marsh Gas	**(74-82-8)**
Marshal	**(55285-14-8)**
Martinal	**(21645-51-2)**
Martin's Mar-Frin	**(81-81-2)**
Martonite	**(598-31-2)**
Martoxin	**(1344-28-1)**
Martrate-45	**(78-11-5)**
Marukarez R 100A	**(64742-16-1)**
Marukarez R 100B	**(64742-16-1)**
Maruzen M	**(24979-70-2)**
Marvex	**(62-73-7)**
Marvinol VP 56	**(9003-22-9)**
Marzin	**(8018-01-7)**
Marzine	**(82-92-8)**
Mascagnite	**(7783-20-2)**
Maschitt	**(58-93-5)**
Masenate	**(57-85-2)**
Masenone	**(58-18-4)**
Maseptol	**(99-76-3)**
Masoten	**(52-68-6)**
Masposol	**(137-42-8)**
Massicot	**(1317-36-8)**
Massicotite	**(1317-36-8)**
Mastestona	**(58-18-4)**
Mastiphen	**(56-75-7)**
Mastofuran	**(59-87-0)**
Matacil	**(2032-59-9)**
Mataven	**(43222-48-6)**
Mataven	**(52756-25-9)**
Mate LS	**(8001-21-6)**
Mathe	**(557-05-1)**
Mathylamine, hydrochloride (9CI)	**(593-51-1)**
Matox	**(67485-29-4)**
Matricaria Camphor	**(76-22-2)**
Matrigon	**(1702-17-6)**
Matsutake alcohol (Japanese)	**(3391-86-4)**
Matting Acid (DOT)	**(7664-93-9)**
Matulane	**(366-70-1)**
Matulane	**(671-16-9)**
Maurylene	**(9003-07-0)**

Name	CAS
Maveran	(81-81-2)
Maviserpin	(50-55-5)
Mavrik	(69409-94-5)
Mavrik HR	(69409-94-5)
Maxatase	(9014-01-1)
Maxforce	(67485-29-4)
Maxibalin	(965-90-2)
Maxibolin	(965-90-2)
Maxidex	(50-02-2)
Maxilon Blue 5G	(55840-82-9)
Maxilon Blue GRL	(12270-13-2)
Maxinutril	(115-77-5)
Maxipen	(132-93-4)
Maxitate	(15825-70-4)
Maxiton	(51-63-8)
Maxiton sulfate	(51-63-8)
Maxolon	(364-62-5)
Maxvis 2000	(9003-27-4)
May & Baker S-4084	(2636-26-2)
Maycor	(87-33-2)
Maydol	(8001-30-7)
Mayserpine	(50-55-5)
2-Maythic acid	(93-09-4)
Mayvat Brown BR	(2475-33-4)
Mayvat Golden Yellow	(128-66-5)
Mayvat Jade Green	(128-58-5)
Mayvat Olive AR	(2379-81-9)
Mazide 30	(5716-15-4)
Mazindol	(22232-71-9)
Mazola Oil	(8001-30-7)
Mazoten	(52-68-6)
McN-JR-15,403-11	(28782-42-5)
McN-JR-1625	(52-86-8)
Mcnamee	(1332-58-7)
MeBr	(74-83-9)
Me-CCNU	(13909-09-6)
MeCsAc	(110-49-6)
3'-Me-DAB	(55-80-1)
6-Me-GLU-P-2	(67730-11-4)
Me2NMor	(1456-28-6)
Me-Parathion	(298-00-0)
2,5-Me2-THF	(1003-38-9)
Meadow Green	(12002-03-8)
Mearlmaid	(73-40-5)
Measurin	(50-78-2)
Meat Sugar	(87-89-8)
Mebadin	(4914-30-1)
Meballymal	(76-73-3)
Mebaral	(115-38-8)
Mebendazole	(31431-39-7)
Mebenil	(7055-03-0)
Meberal	(115-38-8)
Mebichloramine	(55-86-7)
Mebubarbital	(57-33-0)
Mebubarbital	(76-74-4)
Mebubarbital sodium	(57-33-0)
Mebumal natrium	(57-33-0)
Mebumal sodium	(57-33-0)
Mebutal	(143-81-7)
Mebutamat	(64-55-1)
Mebutamate	(64-55-1)
Mebutamato (Spanish)	(64-55-1)
Mebutamatum (Latin)	(64-55-1)
Mebutina	(64-55-1)
Mecadox	(6804-07-5)
Mecarbam	(2595-54-2)
Mecarbame	(2595-54-2)
Mecarol	(77-75-8)
Mecarphon	(29173-31-7)
Mecarphos	(29173-31-7)
Mechlorethamine	(51-75-2)
Mechlorethamine hydrochloride	(55-86-7)
Mechlorethamine oxide	(126-85-2)
Mechlorethamine oxide hydrochloride	(302-70-5)
Mechlorprop	(7085-19-0)
Mecholyl	(62-51-1)
Mecholyl chloride	(62-51-1)
Meclizine	(569-65-3)
Meclocualona (Spanish)	(340-57-8)
Mecloqualon	(340-57-8)
Mecloqualone	(340-57-8)
Mecloqualonum (Latin)	(340-57-8)
Mecloretamina (Italian)	(51-75-2)
Meclozine	(569-65-3)
Mecodin	(76-99-3)
Mecodrin	(300-62-9)
Mecodrin	(60-13-9)
Mecomec	(7085-19-0)
Mecomec	(93-65-2)
Meconic acid	(497-59-6)
Mecopeop	(93-65-2)
Mecoper	(93-65-2)
Mecopex	(1929-86-8)
Mecopex	(93-65-2)
Mecoprop	(7085-19-0)
Mecoprop	(93-65-2)
Mecoprop diethanolamine salt	(1432-14-0)
Mecoprop potassium salt	(1929-86-8)
Mecoturf	(93-65-2)
Mecprop	(93-65-2)
Mecramine	(60-23-1)
Mecrilat	(137-05-3)
Mecrothene F	(9002-88-4)
Mecryl	(69-05-6)
Mecrylate	(137-05-3)
Medamycin	(64-75-5)
Medapan	(509-86-4)
Medaron	(67-45-8)
Medarsed	(125-40-6)
Medazepam	(2898-12-6)
Medazepamum (Latin)	(2898-12-6)
Medazepol	(2898-12-6)
Medemanol	(15825-70-4)
Medex	(51-63-8)
Medfalan	(13045-94-8)
Medi-Calgon	(10124-56-8)
Medialan LL-99	(137-16-6)
Medialanic acid (VAN)	(110-25-8)
Mediamid	(59-26-7)
Mediamycetine	(56-75-7)
Mediben	(1918-00-9)
360 Medical Fluid	(9006-65-9)
Medicel	(80-35-3)
Medidryl	(58-73-1)
Mediflavin	(8048-52-0)
Mediflor FC 43	(311-89-7)
Medifuran	(139-91-3)
Medihaler-EPI	(51-42-3)
Medihaler-EPI	(51-43-4)
Medilla	(90-33-5)
Medinoterb	(3996-59-6)
Mediquil	(78-44-4)
Medithionat	(919-76-6)
Medium Blue	(81-77-6)

Medium Blue EMBL	**(3861-73-2)**
Medomet	**(41372-08-1)**
Medomet	**(555-30-6)**
Medomin	**(509-86-4)**
Medomine	**(509-86-4)**
Medon	**(126-81-8)**
α-Medopa	**(41372-08-1)**
Medopren	**(41372-08-1)**
Medopren	**(555-30-6)**
Medphalan	**(13045-94-8)**
Medrol	**(83-43-2)**
Medrol ADT Pak	**(83-43-2)**
Medrol Dosepak	**(83-43-2)**
Medrol acetate	**(53-36-1)**
Medrone	**(83-43-2)**
Medrosteron	**(58-19-5)**
Medroxyacetate progesterone	**(71-58-9)**
Medroxyprogesteron	**(520-85-4)**
Medroxyprogesterone	**(520-85-4)**
Medroxyprogesterone 17-acetate	**(1172-82-3)**
Medroxyprogesterone acetate	**(1172-82-3)**
Medroxyprogesterone acetate	**(71-58-9)**
Meerschaum	**(14987-04-3)**
Meerschaum	**(18307-23-8)**
Meetco	**(78-93-3)**
Mefedina	**(50-13-5)**
Mefenamic acid	**(61-68-7)**
Mefenaminsaeure (German)	**(61-68-7)**
Mefenorex	**(17243-57-1)**
Mefenorexum (Latin)	**(17243-57-1)**
Mefentanyl	**(42045-86-3)**
Mefenterdrin	**(100-92-5)**
Mefentermin	**(100-92-5)**
Mefluidide	**(53780-34-0)**
Mefluidide, potassium salt	**(83601-83-6)**
Megabion	**(68-19-9)**
Megace	**(595-33-5)**
Megacillin Oral	**(132-98-9)**
Megacillin Suspension	**(1538-09-6)**
Megacillin Tablets	**(113-98-4)**
Megadiuril	**(58-93-5)**
Megalectil	**(653-03-2)**
Megalovel	**(68-19-9)**
Megapen	**(6130-64-9)**
Megaphen	**(50-53-3)**
Megaphen	**(53-60-1)**
Megaphen	**(69-09-0)**
Megasedan	**(2898-12-6)**
Megatox	**(640-19-7)**
Megestrol	**(3562-63-8)**
Megestrol acetate	**(595-33-5)**
Megestrolo	**(3562-63-8)**
Megestrolum (Latin)	**(3562-63-8)**
Megestryl acetate	**(595-33-5)**
Meglumina (Spanish)	**(6284-40-8)**
Meglumine	**(6284-40-8)**
Megluminum (Latin)	**(6284-40-8)**
Mehendi	**(83-72-7)**
Mehltaumittel	**(31717-87-0)**
Meikatex 5000NG60	**(9003-20-7)**
Meisei Acemyl Diazo Black B	**(539-17-3)**
Meisei Teryl Diazo Black CR	**(539-17-3)**
Meisei Teryl Diazo Blue HR	**(119-90-4)**
Meisi Fast Red RL Base	**(99-52-5)**
Melabon	**(62-44-2)**
Meladinin	**(298-81-7)**
Meladinine	**(298-81-7)**
Melaforte	**(62-44-2)**
Melamine	**(108-78-1)**
Melamine, Polymer with formaldehyde	**(9003-08-1)**
Melamine, N2,N4-bis(p-chlorophenyl)- (7CI,8CI)	**(30360-18-0)**
Melamine, N2,N4-bis(p-chlorophenyl)-N6-ethyl- (8CI)	**(30360-22-6)**
Melamine, (p-chlorophenyl)- (8CI)	**(30360-12-4)**
Melamine, (o-chlorophenyl)- (6CI,8CI)	**(30360-11-3)**
Melamine, N2-(p-chlorophenyl)-N4,N6-diethyl- (8CI)	**(30360-20-4)**
Melamine, N2-(p-chlorophenyl)-N4-isopropyl- (8CI)	**(30360-17-9)**
Melamine, N2,N4-diallyl- (6CI,7CI,8CI)	**(30360-15-7)**
Melamine, N(2),N(4)-diethyl-N(6)-isopropyl- (8CI)	**(30360-19-1)**
Melamine, N(2),N(4)-diisopropyl- (8CI)	**(16274-44-5)**
Melamine, N^2,N^2-dimethyl	**(1985-46-2)**
Melamine, N2,N4-dimethyl-N2,N4-diphenyl- (8CI)	**(30377-20-9)**
Melamine, N2,N4-diphenyl- (6CI,7CI,8CI)	**(5606-18-8)**
Melamine, ethyl- (8CI)	**(5606-23-5)**
Melamine, hexaethyl- (8CI)	**(2827-49-8)**
Melamine, hexamethyl	**(645-05-6)**
Melamine, isopropyl- (8CI)	**(16274-81-0)**
Melamine, N2-methyl-N2-phenyl- (6CI,7CI,8CI)	**(30360-14-6)**
Melamine, (p-nitrophenyl)- (8CI)	**(30360-13-5)**
Melamine, pentamethyl	**(16268-62-5)**
Melamine, phenyl- (8CI)	**(5606-27-9)**
Melamine, N2,N2,N4,N4-tetraethyl- (6CI,7CI,8CI)	**(5606-20-2)**
Melamine, N2,N4,N6-triallyl- (7CI,8CI)	**(30360-21-5)**
Melamine, N2,N4,N6-tribenzoyl- (7CI)	**(5637-84-3)**
Melamine, N(2),N(4),N(6)-triethyl- (8CI)	**(16268-92-1)**
Melamine, triethylene-	**(51-18-3)**
Melamine, N(2),N(4),N(6)-triisopropyl- (8CI)	**(5465-03-2)**
Melamine, N^2,N^4,N^6-trimethyl	**(2827-46-5)**
Melamine, N2,N4,N6-tri-1-naphthyl- (8CI)	**(30360-24-8)**
Melamine, N(2),N(4),N(6)-tris(o-chlorophenyl)- (8CI)	**(2272-28-8)**
Melamine, N2,N4,N6-tris(p-chlorophenyl)- (7CI,8CI)	**(2748-40-5)**
Melamine, N^2,N^4,N^6-tris(hydroxymethyl)-	**(1017-56-7)**
Melan 11	**(9011-05-6)**
Melan Black	**(2519-30-4)**
Melanex	**(565-33-3)**
Melaniline	**(102-06-7)**
Melanol CL	**(151-21-3)**
Melanol CL 30	**(151-21-3)**
Melanol LP20T	**(139-96-8)**
Melantherine BH	**(2429-73-4)**
Melantherine BHX	**(2429-73-4)**
Melatonin	**(73-31-4)**
Melatonine	**(73-31-4)**
Meldane	**(56-72-4)**
Meldian	**(94-20-2)**
Meldone	**(56-72-4)**
Meleril	**(50-52-2)**
Meletin	**(117-39-5)**
Melfalan	**(148-82-3)**
Meliform	**(25038-59-9)**
Melilot LS	**(8001-21-6)**
Melilotal	**(122-00-9)**
Melilotin	**(119-84-6)**
Melilotine	**(119-84-6)**
Melilotol	**(119-84-6)**
Melin	**(153-18-4)**
Melinex	**(25038-59-9)**
Melinex O	**(25038-59-9)**
Melinite	**(88-89-1)**
Melipan	**(3771-19-5)**
Melipax	**(8001-35-2)**
Melipramin	**(50-49-7)**
Melipramine	**(50-49-7)**
Melissic acid	**(506-50-3)**
Melitase	**(94-20-2)**

Melitoxin (66-76-2)
Mellaril (50-52-2)
Mellarit (50-52-2)
Mellerette (50-52-2)
Melleretten (50-52-2)
Melleril (50-52-2)
Mellic acid (517-60-2)
Mellite 825 (16091-18-2)
Mellitic acid (517-60-2)
Mellophanic acid (476-73-3)
Mellose (9004-67-5)
Melogel (9005-25-8)
Melonex (565-33-3)
Meloxine (298-81-7)
Melphalan (148-82-3)
Melphalen (148-82-3)
Melpintol (77-75-8)
Melprex (2439-10-3)
Melsedin Base (72-44-6)
Melsomin (72-44-6)
Meltox (31717-87-0)
Meltrol (834-28-6)
Meluna (9005-25-8)
Mema (151-38-2)
Membrane Mobility Agent A(2)C (54050-62-3)
Memine (509-67-1)
Menachinonum natrium bisulfurosum (130-37-0)
Menadion (58-27-5)
Menadione (58-27-5)
Menadione sodio bisolfito (130-37-0)
Menadioni natrii bisulfis (Latin) (130-37-0)
Menadioni natrii hydrogensulfis (130-37-0)
Menagen (53-16-7)
Menaphtam (63-25-2)
Menaphthon (58-27-5)
Menaphthone (58-27-5)
Menaphtone (58-27-5)
Menaquinone (58-27-5)
Menaquinone 0 (58-27-5)
Menaquinone 7 (2124-57-4)
Menaquinone K7 (2124-57-4)
Menazon (78-57-9)
Mendel (57-53-4)
Mendi (83-72-7)
Mendiaxon (90-33-5)
Mendrin (72-20-8)
Menformon (53-16-7)
Menhydrinate (523-87-5)
Menidazole (696-23-1)
Menidrabol (434-22-0)
Meniphos (7786-34-7)
Menite (7786-34-7)
Menolyn (57-63-6)
Menonasal (136-47-0)
Menostilbeen (56-53-1)
Menotrophin (9002-68-0)
Menotropins (9002-68-0)
Menta-Bal (115-38-8)
para-Mentha-1,8-dien-7-al (2111-75-3)
1,8(9)-p-Menthadiene (138-86-3)
2-p-Menthadiene (555-10-2)
d-p-Mentha-1,8-diene (5989-27-5)
p-Mentha-1,3-diene (99-86-5)
p-Mentha-1,4-diene (99-85-4)
p-Mentha-1,5-diene (99-83-2)
p-Mentha-1,8-diene (138-86-3)
p-Mentha-1,8-diene (5989-27-5)
p-Mentha-1,8-diene, DL- (138-86-3)
m-Mentha-1,8-diene, (+-)- (8CI) (499-03-6)
p-Mentha-1(7),2-diene (8CI) (555-10-2)
p-Mentha-1,8-diene, (+-)- (8CI) (7705-14-8)
p-Mentha-1,8-diene, Polymers (9003-73-0)
m-Mentha-1(7),8-diene, (R)-(-)- (8CI) (13837-95-1)
p-Mentha-1,8-diene, (S)-(-) (5989-54-8)
p-Mentha-1,8-dien-7-ol (536-59-4)
p-Mentha-6,8-dien-2-ol, acetate (97-42-7)
p-Mentha-6,8-dien-2-ol, l (99-48-9)
6,8(9)-p-Menthadien-2-one (99-49-0)
d-p-Mentha-6,8,(9)-dien-2-one (2244-16-8)
l-6,8(9)-p-Menthadien-2-one (6485-40-1)
p-Mentha-6,8-dien-2-one (99-49-0)
p-Mentha-6,8-dien-2-one, (-)- (6485-40-1)
p-Mentha-6,8-dien-2-one, (R)-(-) (6485-40-1)
p-Mentha-6,8-dien-2-one, (S)-(+) (2244-16-8)
p-Mentha-6,8-dien-2-one, oxime (31198-76-2)
1-p-Mentha-6(8,9)-dien-2-yl acetate (97-42-7)
1,8-p-Menthadienyl-6-oxime (31198-76-2)
(-)-trans-2-p-Mentha-1,8-dien-3-yl-5-pentylresorcinol (13956-29-1)
cis-p-Menthane (6069-98-3)
p-Menthane (138-86-3)
p-Menthane (99-82-1)
m-Menthane (8CI) (16580-24-8)
p-Menthane, cis- (8CI) (6069-98-3)
p-Menthane, trans- (8CI) (1678-82-6)
Menthane diamine (80-52-4)
p-Menthane-1,8-diamine (80-52-4)
Menthane, 1,2:8,9-diepoxy- (96-08-2)
p-Menthane, 1,2:8,9-diepoxy (96-08-2)
p-Menthane-1,2-diol (33669-76-0)
p-Menthane-1,8-diol (8CI) (80-53-5)
p-Menthane-1,8-diol monohydrate (2451-01-6)
p-Menthane-2,9-dithiol (8CI) (4802-20-4)
p-Menthane, 1,4-epoxy (470-67-7)
p-Menthane, 1,8-epoxy (470-82-6)
p-Menthane hydroperoxide (80-47-7)
p-Menthane-8-hydroperoxide (80-47-7)
dl-3-p-Menthanol (15356-70-4)
p-Menthan-3-ol (89-78-1)
p-Menthan-8-ol (498-81-7)
p-Menthan-3-one, trans (89-80-5)
p-Menthanone (10458-14-7)
p-Menthan-3-one (8CI) (10458-14-7)
p-Menthan-3-one, (Z) (491-07-6)
4(8)-p-Menthene (1124-27-2)
p-Menth-4(8)-ene (7CI,8CI) (1124-27-2)
m-Menth-1(7)-ene, (R)-(-)- (8CI) (13837-71-3)
p-Menth-2-ene, 1,4-epidioxy (512-85-6)
1-para-Menthen-4-ol (562-74-3)
8(9)-p-Menthen-3-ol (7786-67-6)
p-Menth-1-en-3-ol (491-04-3)
p-Menth-8-en-1-ol (138-87-4)
p-Menth-8-en-3-ol (7786-67-6)
para-Menth-1-en-4-ol (562-74-3)
t-Menth-1-en-8-ol (138-87-4)
p-Menth-1-en-8-ol (8CI) (98-55-5)
p-Menth-3-en-1-ol (8CI) (586-82-3)
p-Menth-1-en-8-ol, (S)-(-)- (10482-56-1)
p-Menth-1-en-8-ol, acetate (80-26-2)
p-Menth-1-en-3-one (89-81-6)
p-Menth-8-en-2-one (7764-50-3)
(+-)-Menthol (15356-70-4)
(-)-Menthol (2216-51-5)
(1R,3R,4S)-(-)-Menthol (2216-51-5)
(L)-Menthol (2216-51-5)

Name	CAS No.
(R)-(-)-Menthol	**(2216-51-5)**
1-Menthol	**(2216-51-5)**
3-p-Menthol	**(15356-70-4)**
L-Menthol	**(2216-51-5)**
Menthol	**(1490-04-6)**
Menthol	**(89-78-1)**
Menthol, cis-1,3-trans-1,4-(+-)-	**(15356-70-4)**
dl-Menthol	**(15356-70-4)**
l-Menthol	**(89-78-1)**
Menthol, (1R,3R,4S)-(-)- (8CI)	**(2216-51-5)**
Menthol, acetate	**(16409-45-3)**
Menthol racemic	**(15356-70-4)**
Menthol racemique (French)	**(15356-70-4)**
Menthone	**(10458-14-7)**
Menthone	**(89-80-5)**
p-Menthone	**(89-80-5)**
trans-Menthone	**(89-80-5)**
Menthyl acetate	**(16409-45-3)**
dl-Menthyl acetate	**(16409-45-3)**
Menthyl acetate racemic	**(16409-45-3)**
(-)-Menthyl alcohol	**(2216-51-5)**
p-Menthyl hydroperoxide	**(3031-73-0)**
Mentor 28	**(63148-53-8)**
Meobal	**(2425-10-7)**
Meonal	**(77-28-1)**
Meonine	**(59-51-8)**
Meothrin	**(39515-41-8)**
Mepacrine	**(83-89-6)**
Mepacrine dihydrochloride	**(69-05-6)**
Mepacrine hydrochloride	**(69-05-6)**
Mepadin	**(50-13-5)**
Mepamtin	**(57-53-4)**
Mepantin	**(57-53-4)**
Meparfynol	**(77-75-8)**
Mepatar	**(2058-46-0)**
Mepaton	**(298-00-0)**
Mepavlon	**(57-53-4)**
Mepentamato	**(77-75-8)**
Mepentil	**(77-75-8)**
Meperidine	**(57-42-1)**
Meperidine hydrochloride	**(50-13-5)**
Mephabutazone	**(50-33-9)**
Mephacyclin	**(64-75-5)**
Mephadryl	**(58-73-1)**
Mephalan	**(148-82-3)**
Mephanac	**(94-74-6)**
Mephaserpin	**(50-55-5)**
Mephedine	**(50-13-5)**
Mephenamic acid	**(61-68-7)**
Mephenaminic acid	**(61-68-7)**
Mephenmetrazine	**(634-03-7)**
Mephenterdrine	**(100-92-5)**
Mephenterdrinum	**(100-92-5)**
Mephentermine	**(100-92-5)**
Mephentoin	**(50-12-4)**
Mephenytoin	**(50-12-4)**
Mephetedrine	**(100-92-5)**
Mephine	**(100-92-5)**
Mepho-D	**(300-42-5)**
Mephobarbital	**(115-38-8)**
Mephobarbitone	**(115-38-8)**
Mephosfolan	**(950-10-7)**
Mephospholan	**(950-10-7)**
Mephytal	**(115-38-8)**
Mephyton	**(84-80-0)**
Mepidon	**(467-85-6)**
Mepiosine	**(57-53-4)**
Mepiquat chloride	**(24307-26-4)**
Mepiramine	**(91-84-9)**
Meposed	**(57-53-4)**
Mepranil	**(57-53-4)**
Meprin	**(57-53-4)**
Meprindon	**(57-53-4)**
Mepro	**(7085-19-0)**
Mepro	**(93-65-2)**
Meprobam	**(57-53-4)**
Meprobamat (German)	**(57-53-4)**
Meprobamate	**(57-53-4)**
Meprobamato (Italian)	**(57-53-4)**
Meproban	**(57-53-4)**
Meprocompren	**(57-53-4)**
Meprocon CMC	**(57-53-4)**
Meprodil	**(57-53-4)**
Meprodiol	**(57-53-4)**
Meprofen	**(22071-15-4)**
Meprol	**(57-53-4)**
Meproleaf	**(57-53-4)**
Mepron	**(57-53-4)**
Mepronil	**(55814-41-0)**
Mepronil (Pesticide)	**(55814-41-0)**
Meprosa	**(57-53-4)**
Meprosan	**(57-53-4)**
Meprosin	**(57-53-4)**
Meprospan	**(57-53-4)**
Meprotabs	**(57-53-4)**
Meprotan	**(57-53-4)**
Meprovan	**(57-53-4)**
Meprozine	**(57-53-4)**
Meptox	**(298-00-0)**
Meptran	**(57-53-4)**
Mepyramin (German)	**(91-84-9)**
Mepyramine	**(91-84-9)**
Mepyramine maleate	**(59-33-6)**
Mepyrapone	**(54-36-4)**
Mepyren	**(91-84-9)**
Mequin	**(72-44-6)**
Mequinol	**(150-76-5)**
Meractinomycin	**(50-76-0)**
Meraklon	**(9003-07-0)**
Merantine Blue EG	**(3844-45-9)**
Merantine Blue VF	**(129-17-9)**
Merantine Green G	**(4680-78-8)**
Merantine Green SF	**(5141-20-8)**
Meratran	**(467-60-7)**
Merbentul	**(569-57-3)**
Mercaleukin	**(50-44-2)**
Mercamine	**(60-23-1)**
Mercaptamine	**(60-23-1)**
Mercaptan amylique (French)	**(110-66-7)**
Mercaptan methylique (French)	**(74-93-1)**
Mercaptan methylique perchlore (French)	**(594-42-3)**
Mercaptazole	**(60-56-0)**
Mercaptoacetate	**(68-11-1)**
2-Mercaptoacetic acid	**(68-11-1)**
Mercaptoacetic acid	**(68-11-1)**
α-Mercaptoacetic acid	**(68-11-1)**
Mercaptoacetic acid calcium derivative	**(814-71-1)**
Mercaptoacetic acid ethyl ester	**(623-51-8)**
Mercaptoacetic acid 2-ethylhexyl ester	**(7659-86-1)**
Mercaptoacetic acid sodium salt	**(367-51-1)**
4-(Mercaptoacetyl)morpholine O,O-dimethyl phosphorodithioate	**(144-41-2)**
β-Mercaptoalanine	**(52-90-4)**
6-Mercapto-2-aminopurine	**(154-42-7)**
2-Mercapto-5-amino-1,3,4-thiadiazole	**(2349-67-9)**

4-Mercaptoaniline **(1193-02-8)**
o-Mercaptoaniline **(137-07-5)**
p-Mercaptoaniline **(1193-02-8)**
2-Mercaptobarbituric acid **(504-17-6)**
2-Mercaptobenzimidazole **(583-39-1)**
o-Mercaptobenzoesaeure (German) **(147-93-3)**
o-Mercaptobenzoic acid **(147-93-3)**
2-Mercaptobenzoimidazole **(583-39-1)**
Mercaptobenzoimidazole **(583-39-1)**
2-Mercaptobenzothiazole **(149-30-4)**
Mercaptobenzothiazole **(149-30-4)**
2-Mercaptobenzothiazole disulfide **(120-78-5)**
2-Mercaptobenzothiazole potassium salt **(7778-70-3)**
2-Mercapto-benzothiazole, sodium **(2492-26-4)**
2-Mercaptobenzothiazole sodium deriv **(2492-26-4)**
2-Mercaptobenzothiazole, sodium salt **(2492-26-4)**
Mercaptobenzothiazole sodium salt **(2492-26-4)**
2-Mercaptobenzothiazole zinc salt **(155-04-4)**
Mercaptobenzothiazol, sodium salt solution **(2492-26-4)**
2-Mercaptobenzothiazyl disulfide **(120-78-5)**
2-Mercaptobutane **(513-53-1)**
Mercaptocyclopentane **(1679-07-8)**
1-Mercaptodecane **(143-10-2)**
Mercaptodiacetic acid **(123-93-3)**
Mercaptodimethur **(2032-65-7)**
1-Mercaptododecane **(112-55-0)**
2-Mercaptoethanol **(60-24-2)**
Mercaptoethanol **(60-24-2)**
β-Mercaptoethanol **(60-24-2)**
2-Mercaptoethanol adipate **(10194-00-0)**
2-Mercaptoethanol monosodium salt **(37482-11-4)**
(2-Mercaptoethyl)amine **(60-23-1)**
β-Mercaptoethylamine **(60-23-1)**
N-(2-Mercaptoethyl)benzenesulfonamide S-(O,O-diisopropyl phosphorodithioate) **(741-58-2)**
2-Mercaptoethyl oleate **(59118-78-4)**
2-Mercaptoethyl tetradecanoate **(29946-28-9)**
(2-Mercaptoethyl)trimethylammonium iodide S-ester with O,O-diethyl phosphorothioate **(513-10-0)**
Mercaptofos (Russian) **(298-03-3)**
1-Mercaptoglycerol **(96-27-5)**
6-Mercaptoguanine **(154-42-7)**
2-Mercapto-4-hydroxy-6-methylpyrimidine **(56-04-2)**
2-Mercapto-4-hydroxy-6-n-propylpyrimidine **(51-52-5)**
2-Mercapto-3-hydroxypyridine **(23003-22-7)**
2-Mercapto-4-hydroxypyrimidine **(141-90-2)**
2-Mercaptoimidazoline **(96-45-7)**
Mercaptomerin **(20223-84-1)**
(Mercaptomethyl)benzene **(100-53-8)**
3-(Mercaptomethyl)-1,2,3-benzotriazin-4(3H)-one O,O-dimethyl phosphorodithioate S-ester **(86-50-0)**
2-Mercapto-1-methylimidazole **(60-56-0)**
1-(3-Mercapto-2-methyl-1-oxopropyl)-l-proline **(62571-86-2)**
1-(D-3-Mercapto-2-methyl-1-oxopropyl)-l-proline (S,S) **(62571-86-2)**
N-(Mercaptomethyl)phthalimide S-(O,O-dimethyl phosphorodithioate) **(732-11-6)**
D-3-Mercapto-2-methylpropanoyl-l-proline **(62571-86-2)**
(2S)-1-(3-Mercapto-2-methylpropionyl)-l-proline **(62571-86-2)**
1-((2S)-3-Mercapto-2-methylpropionyl)-l-proline **(62571-86-2)**
2-Mercapto-6-methyl-4-pyrimidone **(56-04-2)**
2-Mercapto-6-methylpyrimid-4-one **(56-04-2)**
2-Mercaptonaphthalene **(91-60-1)**
β-Mercaptonaphthalene **(91-60-1)**
1-Mercaptooctane **(111-88-6)**
5-Mercapto-1-phenyltetrazole **(86-93-1)**
Mercaptophenyltetrazole **(86-93-1)**
Mercaptophos **(55-38-9)**
Mercaptophos **(8065-48-3)**
2-Mercaptopropane **(75-33-2)**
1-Mercapto-2,3-propanediol **(96-27-5)**
3-Mercapto-1,2-propanediol **(96-27-5)**
3-Mercaptopropanoic acid, 2-ethyl-2-(hydroxymethyl)-1,3-propanediol triester **(33007-83-9)**
3-Mercaptopropanol **(107-03-9)**
3-Mercaptopropionic acid **(107-96-0)**
Mercaptopropionic acid, dibutyltin salt **(78-06-8)**
3-Mercaptopropionic acid methyl ester **(2935-90-2)**
β-Mercaptopropylamine **(598-36-7)**
2-Mercapto-6-propyl-4-pyrimidone **(51-52-5)**
2-Mercapto-6-propylpyrimid-4-one **(51-52-5)**
3-Mercaptopropyltrimethoxysilane **(4420-74-0)**
γ-Mercaptopropyltrimethoxysilane **(4420-74-0)**
6-Mercaptopurin **(50-44-2)**
Mercaptopurin (German) **(50-44-2)**
6-Mercaptopurine **(50-44-2)**
Mercaptopurine **(50-44-2)**
6-Mercaptopurine monohydrate **(6112-76-1)**
2-Mercaptopyridine monoxide **(1121-31-9)**
2-Mercaptopyridine-N-oxide sodium salt **(3811-73-2)**
2-Mercapto-4-pyrimidinol **(141-90-2)**
2-Mercapto-4(1H)-pyrimidinone **(141-90-2)**
2-Mercapto-4-pyrimidone **(141-90-2)**
2-Mercaptopyrimid-4-one **(141-90-2)**
Mercaptosuccinic acid diethyl ester **(121-75-5)**
Mercaptosuccinic acid diethyl ester, S-ester with O,S-dimethylphosphorodithioate **(3344-12-5)**
7-Mercapto-1,3,4,6-tetrazaindene **(50-44-2)**
Mercaptothion **(121-75-5)**
Mercaptotion (Spanish) **(121-75-5)**
α-Mercaptotoluene **(100-53-8)**
m-Mercaptotoluene **(108-40-7)**
o-Mercaptotoluene **(137-06-4)**
p-Mercaptotoluene **(106-45-6)**
D-3-Mercaptovaline **(52-67-5)**
D-Mercaptovaline **(52-67-5)**
Mercapturic acid **(616-91-1)**
Mercapturic acid, (R)- **(616-91-1)**
Mercapurin **(50-44-2)**
Mercasolyl **(60-56-0)**
Mercate 5 **(89-65-6)**
Mercate 20 **(6381-77-7)**
Mercazole **(60-56-0)**
Mercazolyl **(60-56-0)**
Merchlorate **(123-88-6)**
Merchlorethamine **(55-86-7)**
Merckogel OR **(9003-20-7)**
Merckogen 6000 **(9003-20-7)**
Mercloran **(62-37-3)**
Mercol 25 **(25155-30-0)**
Mercol 30 **(25155-30-0)**
Mercoral **(62-37-3)**
Mercuram **(137-26-8)**
Mercuran **(151-38-2)**
Mercurate(1-), (N,N-bis(carboxymethyl)glycinato(3-)-N,O,O',O'')-, hydrogen, (T-4)- (9CI) **(53113-61-4)**
Mercurate(2-), (orthoborato(3-)-o)phenyl-, dihydrogen **(102-98-7)**
Mercurate (2-), tetraiodo-, dipotassium **(7783-33-7)**
Mercure (French) **(7439-97-6)**
Mercuriacetate **(1600-27-7)**
Mercurialin **(74-89-5)**
Mercuric acetate (DOT) **(1600-27-7)**
Mercuric ammonium chloride, Solid (DOT) **(10124-48-8)**
Mercuric basic sulfate **(1312-03-4)**
Mercuric benzoate **(583-15-3)**

Mercuric benzoate, Solid (DOT) **(583-15-3)**
Mercuric bichloride **(7487-94-7)**
Mercuric bromide, Solid (DOT) **(7789-47-1)**
Mercuric chloride, Solid (DOT) **(7487-94-7)**
Mercuric chloride [UN 1624] **(7487-94-7)**
Mercuric chloride, ammoniated **(10124-48-8)**
Mercuric cyanide (DOT) **(592-04-1)**
Mercuric cyanide, Solid (DOT) **(592-04-1)**
Mercuric diacetate **(1600-27-7)**
Mercuric iodide **(7774-29-0)**
Mercuric iodide, Red **(7774-29-0)**
Mercuric iodide, Solid (DOT) **(7774-29-0)**
Mercuric iodide, Solution (DOT) **(7774-29-0)**
Mercuric nitrate [UN 1625] **(10045-94-0)**
Mercuric oleate **(1191-80-6)**
Mercuric oleate, Solid (DOT) **(1191-80-6)**
Mercuric oxide **(21908-53-2)**
Mercuric oxide, Red **(21908-53-2)**
Mercuric oxide, Solid (DOT) **(21908-53-2)**
Mercuric oxide, Yellow **(21908-53-2)**
Mercuric oxycyanide **(1335-31-5)**
Mercuric oxycyanide, Solid (Desensitized) [UN 1642] **(1335-31-5)**
Mercuric potassium cyanide **(591-89-9)**
Mercuric potassium cyanide, Solid (DOT) **(591-89-9)**
Mercuric potassium cyanide [UN 1626] **(591-89-9)**
Mercuric potassium iodide **(7783-33-7)**
Mercuric potassium iodide, Solid (DOT) **(7783-33-7)**
Mercuric salicylate **(5970-32-1)**
Mercuric salicylate, Solid **(5970-32-1)**
Mercuric subsulfate, Solid [NA 2025] **(1312-03-4)**
Mercuric sulfate **(7783-35-9)**
Mercuric sulfate, Solid (DOT) **(7783-35-9)**
Mercuric sulfocyanate **(592-85-8)**
Mercuric sulfo cyanate, Solid (DOT) **(592-85-8)**
Mercuric sulfocyanide **(592-85-8)**
Mercuric sulphate (DOT) **(7783-35-9)**
Mercuric thiocyanate **(592-85-8)**
Mercuric thiocyanate, Solid (DOT) **(592-85-8)**
Mercurio (Italian) **(7439-97-6)**
Mercuriphenyl acetate **(62-38-4)**
Mercuriphenyl chloride **(100-56-1)**
Mercuriphenyl nitrate **(55-68-5)**
Mercurisalicylic acid **(5970-32-1)**
Mercurol (DOT) **(12002-19-6)**
Mercurothiolate **(54-64-8)**
Mercurous acetate (DOT) **(631-60-7)**
Mercurous acetate, Solid (DOT) **(631-60-7)**
Mercurous azide (DOT) **(38232-63-2)**
Mercurous bromide (DOT) **(10031-18-2)**
Mercurous bromide, Solid (DOT) **(10031-18-2)**
Mercurous chloride **(10112-91-1)**
Mercurous gluconate **(63937-14-4)**
Mercurous gluconate, Solid **(63937-14-4)**
Mercurous iodide **(15385-57-6)**
Mercurous iodide, Solid **(15385-57-6)**
Mercurous nitrate **(7782-86-7)**
Mercurous nitrate, Solid (DOT) **(10415-75-5)**
Mercurous nitrate [UN 1627] **(10415-75-5)**
Mercurous oxide, Black, Solid (DOT) **(15829-53-5)**
Mercurous sulfate **(7783-36-0)**
Mercurous sulfate, Solid (DOT) **(7783-36-0)**
Mercurous sulphate (DOT) **(7783-36-0)**
Mercury (ACGIH,OSHA) **(7439-97-6)**
Mercury Cadmium Reds **(1345-09-1)**
Mercury biniodide **(7774-29-0)**
Mercury(I) bromide (1:1) **(10031-18-2)**
Mercury(I) chromate **(13465-34-4)**
Mercury(I) gluconate **(63937-14-4)**
Mercury(I) nitrate (1:1) **(10415-75-5)**
Mercury(I) oxide **(15829-53-5)**
Mercury(I) sulfate **(7783-36-0)**
Mercury, Metallic (DOT) **(7439-97-6)**
Mercury (2+) NTA **(53113-61-4)**
Mercury(2+) acetate **(1600-27-7)**
Mercury(II) acetate **(1600-27-7)**
Mercury acetate [UN 1629] **(1600-27-7)**
Mercury acetate [UN 1629] **(631-60-7)**
Mercury (II) acetate, phenyl- **(62-38-4)**
Mercury, (acetato)ethyl **(109-62-6)**
Mercury, (acetato)(2-methoxyethyl) **(151-38-2)**
Mercury, (acetato)phenyl **(62-38-4)**
Mercury, acetoxy(2-methoxyethyl)- **(151-38-2)**
Mercury, acetoxyphenyl- **(62-38-4)**
Mercury acetylide **(68833-55-6)**
Mercury-amide-chloride **(10124-48-8)**
Mercury amine chloride **(10124-48-8)**
Mercury ammoniated **(10124-48-8)**
Mercury ammonium chloride [UN 1630] **(10124-48-8)**
Mercury-azide **(38232-63-2)**
Mercury(II) benzoate **(583-15-3)**
Mercury benzoate [UN 1631] **(583-15-3)**
Mercury bichloride **(7487-94-7)**
Mercury, bis(acetato)(mu-(3',6'-dihydroxy-2',7'-fluorandiyl))di **(3570-80-7)**
Mercury, bis(4-methylphenyl)- (9CI) **(537-64-4)**
Mercury, bis(thiocyanato)- **(592-85-8)**
Mercury bisulfate **(7783-35-9)**
Mercury bisulphate (DOT) **(7783-35-9)**
Mercury(II) bromide (1:2) **(7789-47-1)**
Mercury bromide [UN 1634] **(7789-47-1)**
Mercury bromides [UN 1634] **(10031-18-2)**
Mercury, (p-carboxyphenyl)chloro **(59-85-8)**
Mercury, ((o-carboxyphenyl)thio)ethyl-, sodium salt **(54-64-8)**
Mercury, (3-(3-carboxy-2,2,3-trimethylcyclopentanecarboxamido)-2-methoxypropyl) (hydrogen mercaptoacetato) **(20223-84-1)**
Mercury(II) chloride **(7487-94-7)**
Mercury-chloride **(10112-91-1)**
Mercury, chloroethyl **(107-27-7)**
Mercury, chloro(2-methoxyethyl) **(123-88-6)**
Mercury, chloro(2-methoxy-3-ureidopropyl) **(62-37-3)**
Mercury, chloromethyl **(115-09-3)**
Mercury, chlorophenyl **(100-56-1)**
Mercury, chloropropyl **(2440-40-6)**
Mercury, chloro(4-sulfophenyl)- **(554-77-8)**
Mercury (II) chromate **(13444-75-2)**
Mercury(II) cyanide **(592-04-1)**
Mercury cyanide [UN 1636] **(592-04-1)**
Mercury cyanide oxide (9CI) **(1335-31-5)**
Mercury cyanide, Solid **(37020-93-2)**
Mercury, (3-cyanoguanidino)methyl **(502-39-6)**
Mercury cyanomethyl **(2597-97-9)**
Mercury diacetate **(1600-27-7)**
Mercury, diethyl **(627-44-1)**
Mercury, (dihydrogen borato)phenyl- **(102-98-7)**
Mercury diiodide **(15385-57-6)**
Mercury, dimethyl **(593-74-8)**
Mercury, diphenyl **(587-85-9)**
Mercury, diphenyl(mu-((tetrapropenyl)butanedioato(2-)-O:O'))di- (9CI) **(27236-65-3)**
Mercury dithiocyanate **(592-85-8)**
Mercury, di-p-tolyl- (8CI) **(537-64-4)**
Mercury, (ethoxyethyl)hydroxy- (8CI,9CI) **(26983-51-7)**
Mercury, ethyl(1,4,5,6,7,7-hexachloro-5-norbornene-2,3-dicarboximidato) **(2597-93-5)**
Mercury, ethyl(4-mercaptobenzenesulfonato-S^{4})-, sodium salt **(5964-24-9)**

Mercury, ethyl(2-mercaptobenzoate-s)-, sodium salt **(54-64-8)**
Mercury, ethyl(4-methyl-N-phenylbenzenesulfonamidato-n)- (9CI) **(517-16-8)**
Mercury, ethyl(N-phenyl-p-toluenesulfonamidato)- **(517-16-8)**
Mercury, ethyl(N-phenyl-p-toluenesulfonamido)- **(517-16-8)**
Mercury, ethyl(phosphato(1-)) **(2235-25-8)**
Mercury, ethyl((p-sulfophenyl)thio)-, sodium salt **(5964-24-9)**
Mercury, ethyl(p-toluenesulfonanilidato) **(517-16-8)**
Mercury fulminate, Containing, by weight, at least 20% water [UN 0135] **(628-86-4)**
Mercury fulminate (DOT) **(628-86-4)**
Mercury fulminate (Dry) **(628-86-4)**
Mercury fulminate (Wet) **(628-86-4)**
Mercury gluconate **(63937-14-4)**
Mercury, hydroxymethyl **(1184-57-2)**
Mercury, hydroxyphenyl **(100-57-2)**
Mercury, (2-hydroxypropanoato)phenyl- (9CI) **(122-64-5)**
Mercury iodide [UN 1638] **(15385-57-6)**
Mercury(II) iodide [UN 1638] **(7774-29-0)**
Mercury, ion (Hg(2+)) (8CI,9CI) **(14302-87-5)**
Mercuryl acetate **(1600-27-7)**
Mercury, (lactato)phenyl- (8CI) **(122-64-5)**
Mercury, (lactoyloxy)phenyl- **(122-64-5)**
Mercury(1+), methyl- (9CI) **(22967-92-6)**
Mercury, methyl- (8CI,9CI) **(16056-34-1)**
Mercurymethylchloride **(115-09-3)**
Mercury(1+), methyl-, ion **(22967-92-6)**
Mercury, methyl(8-quinolinolato) **(86-85-1)**
Mercury, methyl(8-quinolyloxy)- **(86-85-1)**
Mercury monoacetate **(631-60-7)**
Mercury nitrate **(10045-94-0)**
Mercury(II) nitrate (1:2) **(10045-94-0)**
Mercury, nitratophenyl **(55-68-5)**
Mercury nitride **(12136-15-1)**
Mercury(1+), (2,2',2''-nitrilotriethanol)phenyl-, lactate (Salt) (8CI) **(23319-66-6)**
Mercury nucleate, Solid (DOT) **(12002-19-6)**
Mercury nucleate [UN 1639] **(12002-19-6)**
Mercury, (9-octadecenoato-O)phenyl-, (Z)- (9CI) **(104-60-9)**
Mercury-oleate **(1191-80-6)**
Mercury oleate [UN 1640] **(1191-80-6)**
Mercury, (oleato)phenyl- **(104-60-9)**
Mercury(II) oxide **(21908-53-2)**
Mercury oxide [UN 1641] **(15829-53-5)**
Mercury oxide [UN 1641] **(21908-53-2)**
Mercury(II) oxide cyanide **(1335-31-5)**
Mercury-oxide-sulfate **(1312-03-4)**
Mercury oxycyanide **(1335-31-5)**
Mercury perchloride **(7487-94-7)**
Mercury pernitrate **(10045-94-0)**
Mercury persulfate **(7783-35-9)**
Mercury, phenylureido **(2279-64-3)**
Mercury(II) potassium iodide **(7783-33-7)**
Mercury potassium iodide [UN 1643] **(7783-33-7)**
Mercury protoiodide **(15385-57-6)**
Mercury, (8-quinolinolato)methyl- **(86-85-1)**
Mercury salicylate **(5970-32-1)**
Mercury, (salicylato(2-))- **(5970-32-1)**
Mercury subchloride **(10112-91-1)**
Mercury subsalicylate **(5970-32-1)**
Mercury(II) sulfate (1:1) **(7783-35-9)**
Mercury sulfates [UN 1645] **(7783-35-9)**
Mercury(II)thiocyanate **(592-85-8)**
Mercury thiocyanate [UN 1646] **(592-85-8)**
Mereprine **(562-10-7)**
Merex **(143-50-0)**
Merfalan **(531-76-0)**
Merfamin **(54-64-8)**
Merfazin **(100-56-1)**
Merfen **(102-98-7)**
Mergal AF **(79-07-2)**
Merge **(2163-80-6)**
Merge 823 **(2163-80-6)**
Mergital LM 11 **(9002-92-0)**
Mergon **(517-16-8)**
Mergon D **(517-16-8)**
Meri-C **(50-81-7)**
Merian **(526-08-9)**
Meridil **(113-45-1)**
Merilid **(62-37-3)**
Merita Earth **(458-37-7)**
Merizone **(50-33-9)**
2-Merkaptobenzimidazol (Czech) **(583-39-1)**
Merkaptobenzimidazol (Czech) **(583-39-1)**
2-Merkaptobenzotiazol (Polish) **(149-30-4)**
2-Merkaptobenzthiazol (Czech) **(149-30-4)**
2-Merkaptoimidazolin (Czech) **(96-45-7)**
6-Merkaptopurin (Czech) **(50-44-2)**
6-Merkaptopurin, monohydrat (Czech) **(6112-76-1)**
Merkaptopuryna (Polish) **(50-44-2)**
Merkazin **(7287-19-6)**
Merkazolil **(60-56-0)**
Mermeth **(57-68-1)**
Mern **(50-44-2)**
Meron (VAN) **(63428-84-2)**
Meronidal **(443-48-1)**
Meropenin **(87-08-1)**
Meroxyl **(548-62-9)**
Meroxyl-Wander **(548-62-9)**
Meroxylan **(548-62-9)**
Meroxylan-Wander **(548-62-9)**
Merpafol **(2939-80-2)**
Merpan **(133-06-2)**
Merpelan AZ **(30979-48-7)**
Merpelan AZ **(67292-92-6)**
Merphalan **(531-76-0)**
o-Merphalan **(531-76-0)**
Merphen **(102-98-7)**
Merphenyl nitrate **(55-68-5)**
Merphos **(150-50-5)**
Merpol **(75-21-8)**
Merpol HC **(9004-98-2)**
Merpoxen OLF 80 **(9004-98-2)**
Merrillite **(7440-66-6)**
Mersolite **(62-38-4)**
Mersolite 1 **(100-57-2)**
Mersolite 2 **(100-56-1)**
Mersolite 7 **(55-68-5)**
Mersolite 8 **(62-38-4)**
Mertec **(148-79-8)**
Mertect **(148-79-8)**
Mertect 160 **(148-79-8)**
Mertestate **(58-22-0)**
Merthiolate **(54-64-8)**
Merthiolate salt **(54-64-8)**
Merthiolate sodium **(54-64-8)**
Mertionin **(59-51-8)**
Mertorgan **(54-64-8)**
Mervamine **(577-11-7)**
Merzonin **(54-64-8)**
Merzonin sodium **(54-64-8)**
Merzonin, sodium salt **(54-64-8)**
Mesamate **(2163-80-6)**
Mesamate-400 **(2163-80-6)**
Mesamate-600 **(2163-80-6)**

Mesamate Concentrate (2163-80-6)
Mesamate H.C. (2163-80-6)
Mesantoin (50-12-4)
Mesaton (59-42-7)
Mesatone (59-42-7)
Mescalin (German) (54-04-6)
Mescaline (54-04-6)
Mescline (54-04-6)
Mescomine (6106-46-3)
Mescopil (155-41-9)
Mesentol (77-67-8)
Meserein (34807-41-5)
Mesidin (Czech) (88-05-1)
Mesidine (88-05-1)
Mesidine hydrochloride (6334-11-8)
Mesidin hydrochloride (6334-11-8)
Mesitaldehyde (487-68-3)
Mesitoic acid (480-63-7)
Mesitol (527-60-6)
Mesityl alcohol (527-60-6)
Mesitylaldehyde (487-68-3)
Mesitylamine (88-05-1)
Mesitylamine hydrochloride (6334-11-8)
Mesitylbenzene (3976-35-0)
Mesitylcarbinol (4170-90-5)
Mesitylene (108-67-8)
Mesitylene, 2-amino- (88-05-1)
Mesitylene, 2,2'-azodi- (5692-66-0)
2-Mesitylenecarboxaldehyde (487-68-3)
Mesitylenecarboxaldehyde (487-68-3)
Mesitylene, 2-chloro- (8CI) (1667-04-5)
Mesitylene, 2,4-dinitro- (8CI) (608-50-4)
2-Mesitylenesulfonic acid, 4,4'-(1,4-anthraquinonylenediimino)di-, disodium salt (8CI) (4474-24-2)
2-Mesitylenesulfonyl chloride (773-64-8)
Mesitylenic acid (499-06-9)
Mesityloxid (German) (141-79-7)
Mesityl oxide (ACGIH,OSHA) [UN 1229] (141-79-7)
Mesityloxyde (Dutch) (141-79-7)
Mesmar (57-53-4)
Mesocain (137-58-6)
Mesoinosit (87-89-8)
Mesoinosite (87-89-8)
Mesoinositol (87-89-8)
Mesol (87-89-8)
Mesomile (16752-77-5)
Mesontoin (50-12-4)
Mesoranil (4658-28-0)
Mesoridazine (5588-33-0)
Mesothorium 1 (15262-20-1)
Mesovit (87-89-8)
Mesoxalylcarbamide (50-71-5)
Mesoxalylcarbamide monohydrate (2244-11-3)
Mesoxalylurea (50-71-5)
Mesoxalylurea monohydrate (2244-11-3)
Mesterone (58-18-4)
Mestinon (101-26-8)
Mestinone bromide (101-26-8)
Mestranol (72-33-3)
Mestrenol (72-33-3)
Mesulfenos (3761-41-9)
Mesural (58-25-3)
Mesurol (2032-65-7)
Mesuximide (77-41-8)
3-Mesylbutanone O-methylcarbamoyloxime (34681-23-7)
Mesyl chloride (124-63-0)
Mesylith (54-05-7)
2-Mesyl-2-methylpropionaldehyde O-methylcarbamoyloxime (1646-88-4)
Met-Spar (7789-75-5)
Meta Black (1937-37-7)
Metabarbital (50-11-3)
Metabolin (67-03-8)
Metabolite C (80-08-0)
Metabolite I (129-20-4)
Metabrom (2104-96-3)
Metabromsalan (2577-72-2)
Metabromsalanum (Latin) (2577-72-2)
Metace (569-57-3)
Metacen (53-86-1)
Metacetaldehyde (108-62-3)
Metacetamol (621-42-1)
Metacetin (51-66-1)
Metacetone (96-22-0)
Metacetonic acid (79-09-4)
Metachalon (Czech) (72-44-6)
Metachlor (15972-60-8)
Metachlore (51218-45-2)
Metacholine chloride (62-51-1)
Metacid (298-00-0)
Metacid 50 (298-00-0)
Metacide (298-00-0)
Metacil (56-04-2)
Metaclopramide (364-62-5)
Metaclopromide (364-62-5)
Metacortandracin (53-03-2)
Metacortandralone (50-24-8)
Metacrate (1129-41-5)
Metacrefos (36335-67-8)
Metadee (50-14-6)
Metadelphene (134-62-3)
Metadiazine (289-95-2)
Metadifluorobenzene (372-18-9)
Metaforming (86290-81-5)
Metafos (10361-03-2)
Metafos (298-00-0)
Metafume (74-83-9)
Metahexamide (565-33-3)
Metahomomenthyl salicylate (118-56-9)
Metahydrin (133-67-5)
Metaisoseptox (919-86-8)
Metaisosystox (919-86-8)
Metaisosystox-solfon 20 315 (17040-19-6)
Metaisosystoxsulfoxide (301-12-2)
Metakril 40BM (25608-33-7)
Metakril 80BM (25608-33-7)
Metakrylan metylu (Polish) (80-62-6)
Metalaxil (57837-19-1)
Metalaxyl (57837-19-1)
Metalaxyl-Mancozeb Mixt. (75701-74-5)
Metalcaptase (52-67-5)
Metaldehyd (German) (108-62-3)
Metaldehyde (108-62-3)
Metaldehyde (9002-91-9)
Metaldehyde (VAN) (108-62-3)
Metaldeide (Italian) (108-62-3)
Metalid (621-42-1)
Metalkamate (8065-36-9)
Metallac (557-05-1)
Metallic arsenic (7440-38-2)
Metallic mercury (7439-97-6)
Metallic osmium (7440-04-2)
Metallon E 5010 (28825-96-9)
Metallothionein (Agaricus campestris bisporus copper-binding peptide moiety reduced) (98526-74-0)

Metallothionein (Neurospora crassa copper-binding peptide moiety reduced), 12-L-threonine-14-L-alanine-16a-endo-L-glutamine-18-L-threonine-21-g lycine-24-de-L-serine- (9CI) **(98526-74-0)**
Metam **(6734-80-1)**
Metam-Fluid BASF **(137-42-8)**
Metamfetamina **(300-42-5)**
Metamid **(105-60-2)**
Metamid **(25038-54-4)**
Metamidofos estrella **(10265-92-6)**
Metamina **(300-42-5)**
Metamine **(300-42-5)**
Metamiton **(41394-05-2)**
Metamitron (German) **(41394-05-2)**
Metamphetamin **(300-42-5)**
Metam-sodium (Dutch, French, German, Italian) **(137-42-8)**
Metamsustac **(300-42-5)**
Metana **(7429-90-5)**
Metana Aluminum Paste **(7429-90-5)**
Metanabol **(72-63-9)**
Metandienon **(72-63-9)**
Metandienone **(72-63-9)**
Metandienonum **(72-63-9)**
Metandren **(58-18-4)**
Metandrostenolon **(72-63-9)**
Metandrostenolone **(72-63-9)**
Metanephrin **(51-43-4)**
Metanex **(1333-86-4)**
Metanfetamina **(300-42-5)**
Metanil Yellow **(4005-68-9)**
Metanil Yellow **(587-98-4)**
Metanil Yellow 1955 **(587-98-4)**
Metanil Yellow C **(587-98-4)**
Metanil Yellow E **(587-98-4)**
Metanil Yellow Extra **(587-98-4)**
Metanil Yellow F **(587-98-4)**
Metanil Yellow G **(587-98-4)**
Metanil Yellow Griesbach **(587-98-4)**
Metanil Yellow K **(587-98-4)**
Metanil Yellow KRSU **(587-98-4)**
Metanil Yellow M3X **(587-98-4)**
Metanil Yellow O **(587-98-4)**
Metanil Yellow PL **(587-98-4)**
Metanil Yellow S **(587-98-4)**
Metanil Yellow Supra P **(587-98-4)**
Metanil Yellow VS **(587-98-4)**
Metanil Yellow WS **(587-98-4)**
Metanil Yellow Y **(587-98-4)**
Metanil Yellow YK **(587-98-4)**
Metanilamide **(98-18-0)**
Metanilamide, 4-hydroxy- (8CI) **(98-32-8)**
Metanile Yellow O **(587-98-4)**
Metanilic-acid **(121-47-1)**
Metanilic acid, 4-chloro- (8CI) **(98-36-2)**
Metanilic acid, 6-chloro- (8CI) **(88-43-7)**
Metanilic acid, 5-chloro-2-hydroxy- (8CI) **(88-23-3)**
Metanilic acid, 4-hydroxy- (8CI) **(98-37-3)**
Metanilic acid, 2-hydroxy-5-nitro **(96-67-3)**
Metanolo (Italian) **(67-56-1)**
Metantyl **(53-46-3)**
Metaoksedrin **(61-76-7)**
Metaoxedrin **(59-42-7)**
Metaoxedrin **(61-76-7)**
Metaoxedrine **(59-42-7)**
Metaoxedrinum **(61-76-7)**
Metaoxon **(2255-17-6)**
Metaphenylenediamine **(108-45-2)**
Metaphor **(298-00-0)**
Metaphos **(298-00-0)**
Metaphosphoric acid, calcium sodium salt **(23209-59-8)**
Metaphosphoric acid, hexasodium salt **(10124-56-8)**
Metaphosphoric acid, sodium salt **(10361-03-2)**
Metaphosphoric acid, tetrasodium salt **(13396-41-3)**
Metaphosphoric acid, trisodium salt **(7785-84-4)**
Metaphyllin **(317-34-0)**
Metaphyllin **(51-63-8)**
Metaphylline **(317-34-0)**
Metaplex NO **(9011-14-7)**
Metaplex 4002t **(9011-14-7)**
Metapoxide **(57-39-6)**
Metaqualon **(72-44-6)**
Metaquest A **(60-00-4)**
Metaquest B **(139-33-3)**
Metaquest C **(64-02-8)**
Metaradrine **(54-49-9)**
(-)-Metaraminol **(54-49-9)**
1-Metaraminol **(54-49-9)**
Metaraminol **(54-49-9)**
Metartril **(53-86-1)**
Metasap 576 **(557-05-1)**
Metasap XX **(637-12-7)**
Metasol **(86-85-1)**
Metasol 30 **(62-38-4)**
Metasol BT **(102-98-7)**
Metasol MMH **(86-85-1)**
Metasol TK-100 **(148-79-8)**
Metason **(108-62-3)**
Metastenol **(72-63-9)**
Metasympatol **(59-42-7)**
Metasynephrine **(59-42-7)**
Metasystemox **(301-12-2)**
Metasystemox R **(301-12-2)**
Metasystox **(8022-00-2)**
Metasystox Forte **(919-86-8)**
Metasystox (I) **(919-86-8)**
Metasystox J **(919-86-8)**
Metasystox-R **(301-12-2)**
Metatensin **(50-55-5)**
Metathio E-50 **(122-14-5)**
Metathion **(122-14-5)**
Metathion E 50 **(122-14-5)**
Metathione **(122-14-5)**
Metathionine **(122-14-5)**
Metathionine E50 **(122-14-5)**
Metathion, S-methyl isomer **(3344-14-7)**
Metation **(122-14-5)**
Metation E50 **(122-14-5)**
Metatolylcarbamide **(63-99-0)**
Metatolylenediamine dihydrochloride **(636-23-7)**
N-Metatolyl phthalamic acid **(85-72-3)**
Metatrexan **(59-05-2)**
Metatyl **(55-55-0)**
Metaupon **(112-80-1)**
Metaupon Paste **(137-20-2)**
Metawanadanem sodowym (Polish) **(13718-26-8)**
Metaxalone **(1665-48-1)**
Metaxan **(53-46-3)**
Metaxite **(12001-29-5)**
Metaxon **(94-74-6)**
Metaxone **(3653-48-3)**
Metazachlor **(67129-08-2)**
Metazachlore **(67129-08-2)**
Metazalone **(1665-48-1)**
Metazin **(3089-11-0)**
Metazin **(57-68-1)**

Name	CAS
Metazin	(67704-68-1)
Metazine	(3089-11-0)
Metazine	(67704-68-1)
Metazine (Pesticide)	(67704-68-1)
Metazocina	(3734-52-9)
Metazocine	(3734-52-9)
Metazocinum (Latin)	(3734-52-9)
Metazol	(86-85-1)
Metazolo	(60-56-0)
Metazolone	(1665-48-1)
Metazoxolone	(5707-73-3)
Metebanyl	(3176-03-2)
Metelilachlor	(51218-45-2)
Metenix	(17560-51-9)
Metepa	(57-39-6)
Meterazine	(58-38-8)
Metfenossidiolo	(93-14-1)
Metflorylthiazidine	(135-09-1)
Methaanthiol (Dutch)	(74-93-1)
Methabenzthiazuron	(18691-97-9)
Methabol	(434-07-1)
Methacetin	(51-66-1)
Methacetone	(96-22-0)
Methachlor	(15972-60-8)
Methachlorphenprop	(14437-17-3)
Methacholine chloride	(62-51-1)
Methacholinium chloride	(62-51-1)
Methacide	(108-88-3)
Methacil	(56-04-2)
Methacon	(135-23-9)
Methacraldehyde [UN 2396]	(78-85-3)
Methacrolein	(78-85-3)
Methacrolein dimer	(1920-21-4)
Methacrylaldehyde dimer	(1920-21-4)
Methacrylamide (8CI)	(79-39-0)
Methacrylate de butyle (French)	(97-88-1)
Methacrylate de methyle (French)	(80-62-6)
Methacryl chloride	(920-46-7)
Methacrylic-acid	(79-41-4)
Methacrylic acid (ACGIH,OSHA)	(79-41-4)
Methacrylic acid, Inhibited [UN 2531]	(79-41-4)
Methacrylic acid, allyl ester	(96-05-9)
Methacrylic acid amide	(79-39-0)
Methacrylic acid anhydride	(760-93-0)
Methacrylic acid, 2-(tert-butylamino)ethyl ester	(3775-90-4)
Methacrylic acid, butyl ester	(97-88-1)
Methacrylic acid, tert-butyl ester	(585-07-9)
Methacrylic acid, butyl ester, Polymer with methyl methacrylate (8CI)	(25608-33-7)
Methacrylic acid chloride	(920-46-7)
Methacrylic acid, cyclohexyl ester (8CI)	(101-43-9)
Methacrylic acid, decyl ester (8CI)	(3179-47-3)
Methacrylic acid, 2,3-dibromopropyl ester	(3066-70-4)
Methacrylic acid 3,4-dichloroanilide	(2164-09-2)
Methacrylic acid, diester with tetraethylene glycol	(109-17-1)
Methacrylic acid, diester with triethylene glycol	(109-16-0)
Methacrylic acid, 2-(diethylamino)ethyl ester	(105-16-8)
Methacrylic acid, 2-(diisopropylamino)ethyl ester	(16715-83-6)
Methacrylic acid, 2-(dimethylamino)ethyl ester	(2867-47-2)
Methacrylic acid, dodecyl ester	(142-90-5)
Methacrylic acid, 2,3-epoxypropyl ester	(106-91-2)
Methacrylic acid, ethylene ester	(97-90-5)
Methacrylic acid, ethyl ester	(97-63-2)
Methacrylic acid, 2-ethylhexyl ester	(688-84-6)
Methacrylic acid, hexadecyl ester	(2495-27-4)
Methacrylic acid, hexyl ester (8CI)	(142-09-6)
Methacrylic acid, 2-hydroxyethyl ester	(868-77-9)
Methacrylic acid, 2-hydroxypropyl ester	(923-26-2)
Methacrylic acid, ion(1-) (8CI)	(18358-13-9)
Methacrylic acid, isobornyl ester	(7534-94-3)
Methacrylic acid, isobutyl ester	(97-86-9)
Methacrylic acid, isodecyl ester	(29964-84-9)
Methacrylic acid, isopropyl ester	(4655-34-9)
Methacrylic acid, lauryl ester	(142-90-5)
Methacrylic acid, methyl ester	(80-62-6)
Methacrylic acid methyl ester polymers	(9011-14-7)
Methacrylic acid, monoester with 1,2-propanediol	(27813-02-1)
Methacrylic acid, 3,3,4,4,5,5,6,6,6-nonafluorohexyl ester	(1799-84-4)
Methacrylic acid, oxydiethylene ester	(2358-84-1)
Methacrylic acid, propyl ester	(2210-28-8)
Methacrylic acid, sodium salt	(5536-61-8)
Methacrylic acid, stearyl ester	(32360-05-7)
Methacrylic acid, tetradecyl ester	(2549-53-3)
Methacrylic acid, tetrahydrofurfuryl ester (8CI)	(2455-24-5)
Methacrylic acid, tetramethylene ester	(2082-81-7)
Methacrylic acid, tridecyl ester	(2495-25-2)
Methacrylic acid, 2,2,2-trifluoroethyl ester	(352-87-4)
Methacrylic acid, 3-(trimethoxysilyl)propyl ester	(2530-85-0)
Methacrylic aldehyde	(78-85-3)
Methacrylic amide	(79-39-0)
Methacrylic anhydride	(760-93-0)
Methacrylic chloride	(920-46-7)
Methacrylonitrile	(126-98-7)
γ-Methacryloxypropyltrimethoxysilane	(2530-85-0)
Methacryloyl anhydride	(760-93-0)
Methacryloyl-chloride	(920-46-7)
Methacryloyloxyethyl isocyanate	(30674-80-7)
Methacrylsaeurebutylester (German)	(97-88-1)
Methacrylsaeuremethyl ester (German)	(80-62-6)
Methacrylyl chloride	(920-46-7)
Methacycline	(914-00-1)
Methacycline amphoteric	(914-00-1)
Methacycline base	(914-00-1)
Methadol	(545-90-4)
αMethadol	(17199-54-1)
β-Methadol	(17199-55-2)
Methadon	(76-99-3)
Methadone	(76-99-3)
Methadyl acetate	(509-74-0)
Methaform	(57-15-8)
Methafurylene Fumarate	(5429-41-4)
Methahexamide	(565-33-3)
Methakrylaldehyd (Czech)	(78-85-3)
Methaldehyde	(50-00-0)
Methallenestril	(517-18-0)
Methallenestrol	(517-18-0)
Methallyl alcohol [UN 2614]	(513-42-8)
Methallyl chloride	(563-47-3)
α-Methallyl chloride	(563-52-0)
β-Methallyl chloride	(563-47-3)
γ-Methallyl chloride	(591-97-9)
Methallyl cyanide	(4786-19-0)
Methallylkyanid (Czech)	(4786-19-0)
2-Methallyl-6-nitrophenol	(13414-58-9)
Methallyl 2-nitrophenyl ether	(13414-54-5)
Metham (German)	(137-42-8)
Methamazole	(60-56-0)
Metham dihydrate	(6734-80-1)
Methamidophos	(10265-92-6)
Methamin	(100-97-0)
Methaminodiazepoxide	(58-25-3)
Methampex	(300-42-5)
Methamphetamine hydrochloride	(300-42-5)
Methamphin	(300-42-5)

Name	CAS
Metham sodium	**(137-42-8)**
Methan	**(144-54-7)**
Methanal	**(50-00-0)**
Methanamide	**(75-12-7)**
Methanamine (9CI)	**(74-89-5)**
Methanamine, Compd. with trinitromethane (1:1)	**(14147-71-8)**
Methanamine, N,N-dimethyl	**(75-50-3)**
Methanamine, N,N-dimethyl-, Compd. with Borane (1:1)	**(75-22-9)**
Methanamine, N,N-dimethyl-, nitrate (9CI)	**(25238-43-1)**
Methanamine, N,N-dimethyl-, N-oxide (9CI)	**(1184-78-7)**
Methanamine, N-ethyl-N-methyl-	**(598-56-1)**
Methanamine, N-hydroxy-	**(593-77-1)**
Methanamine, N-hydroxy-N-methyl-, hydrochloride (9CI)	**(16645-06-0)**
Methanamine, N-methoxy- (9CI)	**(1117-97-1)**
Methanamine, N-methyl- (9CI)	**(124-40-3)**
Methanamine, N-methyl-, hydrochloride (9CI)	**(506-59-2)**
Methanamine, N-methyl-, sulfate (1:1) (9CI)	**(23307-05-3)**
Methanamine, nitrate (9CI)	**(22113-87-7)**
Methanaminium, 1-carboxy-N,N,N-trimethyl-, chloride	**(590-46-5)**
Methanaminium, 1-carboxy-N,N,N-trimethyl-, hydroxide, inner salt, Compd. with 2,2,2-trichloro-1,1-ethanediol (1:1); (2) chloral hydrate betaine (1:1) compound	**(2218-68-0)**
Methanaminium, N-(4-((2-chlorophenyl)(4-(dimethylamino)phenyl)-methylene)-2,5-cyclohexadien-1-ylidene)-N-methyl-, chloride (9CI)	**(3521-06-0)**
Methanaminium, N-(4-((4-(dimethylamino)phenyl)(4-(ethylamino)-1-naphthalenyl)methylene)- 2,5-cyclohexadien-1-ylidene)-N-methyl-, chloride	**(2185-86-6)**
Methanaminium, N-(4-((4-(dimethylamino)phenyl)(4-(phenylamino)-1-naphthalenyl)methylene)-2,5-cyclohexadien-1-ylidene)-N-methyl-, chloride (9CI)	**(2580-56-5)**
Methanaminium, N,N,N-trimethyl-, bromide (9CI)	**(64-20-0)**
Methanaminium, N,N,N-trimethyl-, chloride (9CI)	**(75-57-0)**
Methanaminium, N,N,N-trimethyl-, hydroxide	**(75-59-2)**
Methandienone	**(72-63-9)**
Methandrolone	**(72-63-9)**
Methandrostenolone	**(72-63-9)**
Methane Base	**(101-61-1)**
Methane, Compressed [UN 1971]	**(74-82-8)**
Methane, Refrigerated liquid [UN 1972]	**(74-82-8)**
Methane [UN 1971]	**(74-82-8)**
Methanearsonic-acid	**(124-58-3)**
Methanearsonic acid, calcium salt (2:1)	**(5902-95-4)**
Methanearsonic acid, disodium salt	**(144-21-8)**
Methanearsonic acid, monosodium salt	**(2163-80-6)**
Methanearsonous acid, Compd. with octylamine (1:1)	**(6379-37-9)**
Methanearsonous acid, dithio-, bis(anhydrosulfide) with dimethyl-dithiocarbamic acid	**(2445-07-0)**
Methane, azoxy	**(25843-45-2)**
Methane, bis(4-amino-3-methylphenyl)-	**(838-88-0)**
Methane, bis(2-(2-butoxyethoxy)ethoxy)	**(143-29-3)**
Methane, 2,2'-bis(6-t-butyl-p-cresyl)	**(119-47-1)**
Methane, bis(2-chloroethoxy)	**(111-91-1)**
Methane, bis(p-chlorophenoxy)	**(555-89-5)**
Methane, bis(4-chlorophenyl)	**(101-76-8)**
Methane, bis(chlorophenyl)- (6CI,7CI,8CI)	**(25249-39-2)**
Methane, bis(p-(dimethylamino)phenyl)-	**(101-61-1)**
Methane, bis(p-(2,3-epoxypropoxy)phenyl)- (8CI)	**(2095-03-6)**
Methane, bis(methylthio)- (9CI)	**(1618-26-4)**
Methane, bis(p-nitrophenyl)- (8CI)	**(1817-74-9)**
Methane, bis(2,3,5-trichloro-6-hydroxyphenyl)	**(70-30-4)**
Methane, bromo	**(74-83-9)**
Methane, bromochloro	**(74-97-5)**
Methane, bromochlorodifluoro	**(353-59-3)**
Methane, bromochloroiodo- (9CI)	**(34970-00-8)**
Methane, bromodichloro	**(75-27-4)**
Methane, bromodichlorofluoro- (8CI,9CI)	**(353-58-2)**
Methane, bromodiphenyl	**(776-74-9)**
Methane, bromotrichloro	**(75-62-7)**
Methane, bromotrifluoro	**(75-63-8)**
Methanecarbonitrile	**(75-05-8)**
Methanecarbothiolic acid	**(507-09-5)**
Methanecarboxamide	**(60-35-5)**
Methanecarboxylic acid	**(64-19-7)**
Methane, chloro	**(74-87-3)**
Methane, chloro(p-chlorophenyl)phenyl- (8CI)	**(134-83-8)**
Methane, chlorodibromo	**(124-48-1)**
Methane, chlorodifluoro	**(75-45-6)**
Methane, chlorodiphenyl- (8CI)	**(90-99-3)**
Methane, chlorofluoro	**(593-70-4)**
Methane, chloroiodo- (8CI,9CI)	**(593-71-5)**
Methane, (4-chlorophenyl)phenyl-	**(831-81-2)**
Methane, (p-chlorophenyl)phenyl- (8CI)	**(831-81-2)**
Methane, chlorotrifluoro	**(75-72-9)**
Methane, chlorotrinitro	**(1943-16-4)**
Methane, chlorotriphenyl	**(76-83-5)**
Methane, cyano-	**(75-05-8)**
Methane, cyclohexylcyclopentyl- (8CI)	**(4431-89-4)**
Methane, cyclohexyl(4-isopropylcyclohexyl)- (7CI)	**(54965-61-6)**
Methane-d (8CI,9CI)	**(676-49-3)**
Methanediamine, N,N,N',N'-tetramethyl- (8CI,9CI)	**(51-80-9)**
Methane, diazo	**(334-88-3)**
Methane, diazodiphenyl-	**(883-40-9)**
Methane, dibromo	**(74-95-3)**
Methane, dibromodichloro- (9CI)	**(594-18-3)**
Methane, dibromodifluoro	**(75-61-6)**
Methane, dibromoiodo- (9CI)	**(593-94-2)**
Methane, dibutoxy- (8CI)	**(2568-90-3)**
Methane, di-sec-butoxy- (7CI,8CI)	**(2568-92-5)**
Methanedicarboxylic acid	**(141-82-2)**
Methanedicarboxylic acid, diethyl ester	**(105-53-3)**
Methane dichloride	**(75-09-2)**
Methane, dichloro	**(75-09-2)**
Methane, dichlorodifluoro	**(75-71-8)**
Methane, dichlorodiphenyl- (8CI)	**(2051-90-3)**
Methane, dichlorofluoro	**(75-43-4)**
Methane, dichloroiodo- (8CI,9CI)	**(594-04-7)**
Methane, dicyano-	**(109-77-3)**
Methane, diethoxy	**(462-95-3)**
Methane, difluoro- (9CI)	**(75-10-5)**
Methane, diiodo	**(75-11-6)**
Methane, dimethoxy	**(109-87-5)**
Methane, dinitro-	**(625-76-3)**
Methanediol, dinitrate	**(38483-28-2)**
Methane, diphenyl	**(101-81-5)**
Methanedisulfonic acid (9CI)	**(503-40-2)**
Methane, ethoxy	**(540-67-0)**
Methane, fluoro	**(593-53-3)**
Methane, fluorotrichloro-	**(75-69-4)**
Methane, iodo	**(74-88-4)**
Methane, isothiocyanato	**(556-61-6)**
Methane, (methyldithio)(methylthio)-	**(42474-44-2)**
Methane, nitro	**(75-52-5)**
Methane, nitrotribromo	**(464-10-8)**
Methane, oxybis(chloro-	**(542-88-1)**
Methane, oxybis(methoxy- (9CI)	**(628-90-0)**
Methane, phenyl-	**(108-88-3)**
Methane, phenyl-3,4-xylyl- (8CI)	**(13540-56-2)**
Methanephosphonic acid dimethyl ester	**(756-79-6)**
Methane quinone	**(82-38-2)**
Methanesulfanenamide, 1,1-dichloro-N-((dimethylamino)sulfonyl)-1-fluoro-N-(4-methylphenyl)-	**(731-27-1)**
Methanesulfenamide, 1,1,1-trichloro-N-(4-chloro-6-(ethylamino)-s-triazin-2-yl)-N-ethyl- (7CI,8CI)	**(3028-00-0)**

Methanesulfenamide, 1,1,1-trichloro-N-ethyl-N-(4-(ethylamino)-6-(trichloromethyl)-s-triazin-2-yl)- (8CI) **(30377-15-2)**
Methanesulfenamide, 1,1,1-trichloro-N-(4-methyl-6-(trichloromethyl)-s-triazin-2-yl)- (8CI) **(30357-74-5)**
Methanesulfenyl chloride, trichloro **(594-42-3)**
Methanesulfinic acid, aminoimino- (9CI) **(1758-73-2)**
Methanesulfinic acid, hydroxy-, monosodium salt **(149-44-0)**
Methane, sulfinylbis- **(67-68-5)**
Methane-D3, sulfinylbis- (9CI) **(2206-27-1)**
Methanesulfon-m-anisidide, 4'-(9-acridinylamino) **(51264-14-3)**
Methanesulfonamide, N-(4-(9-acridinylamino)-3-methoxyphenyl)- **(51264-14-3)**
Methanesulfonamide, N-(2-(4-amino-N-ethyl-m-toluidino)-ethyl)-, sulfate (2:3) **(25646-71-3)**
Methanesulfonamide, N-(2-((4-amino-3-methylphenyl)ethylamino)-ethyl)-, sulfate (2:3) (9CI) **(25646-71-3)**
Methanesulfonamide, N-(2-(ethyl(3-methyl-4-nitrosophenyl)amino)-ethyl)- (9CI) **(56046-62-9)**
Methanesulfonamide, N-(2-(ethyl(3-methylphenyl)amino)-ethyl), sodium salt (9CI) **(27159-90-6)**
Methanesulfonamide, N-(4-(1-hydroxy-2-((1-methylethyl)amino)-ethyl)phenyl)- (9CI) **(3930-20-9)**
Methanesulfonamide, N-(4-phenylsulfonyl-o-tolyl)-1,1,1-trifluoro **(37924-13-3)**
Methanesulfonanilide, 4'-(9-acridinylamino)-3'-methoxy- **(51264-14-3)**
Methanesulfonanilide, 4'-(1-hydroxy-2-(isopropylamino)ethyl) **(3930-20-9)**
Methanesulfonic-acid **(75-75-2)**
Methanesulfonic acid, anilino- (8CI) **(103-06-0)**
Methanesulfonic acid, anilino-, monosodium salt **(26021-90-9)**
Methanesulfonic acid, o-anisidino- (8CI) **(93-13-0)**
Methanesulfonic acid, (bis(2-hydroxyethyl)amino)-, monosodium salt (8CI) **(25857-20-9)**
Methanesulfonic acid chloride **(124-63-0)**
Methanesulfonic acid, ethyl ester (8CI,9CI) **(62-50-0)**
Methanesulfonic acid, hydroxy-, sodium salt **(870-72-4)**
Methanesulfonic acid, iodo-, sodium salt **(126-31-8)**
Methanesulfonic acid, isopropyl ester **(926-06-7)**
Methanesulfonic acid, ((2-methoxyphenyl)amino)- (9CI) **(93-13-0)**
Methanesulfonic acid, methyl ester **(66-27-3)**
Methanesulfonic acid, 1-methylethyl ester **(926-06-7)**
Methanesulfonic acid, ((2-methylphenyl)amino)- (9CI) **(94-57-5)**
Methanesulfonic acid, phenyl- **(100-87-8)**
Methanesulfonic acid, (phenylamino)- (9CI) **(103-06-0)**
Methanesulfonic acid, (phenylamino)-, monosodium salt (9CI) **(26021-90-9)**
Methanesulfonic acid, tetramethylene ester **(55-98-1)**
Methanesulfonic acid, o-toluidino- (8CI) **(94-57-5)**
Methanesulfonic acid, trifluoro- **(1493-13-6)**
Methanesulfonic acid, trifluoro-, methyl ester (9CI) **(333-27-7)**
Methanesulfonothioic acid, sodium salt (9CI) **(1950-85-2)**
Methane, sulfonylbis- (9CI) **(67-71-0)**
Methane, sulfonylbis(trichloro- (9CI) **(3064-70-8)**
Methanesulfonyl chloride (9CI) **(124-63-0)**
Methanesulfonyl-fluoride **(558-25-8)**
Methanesulfonyl fluoride, trifluoro- (9CI) **(335-05-7)**
Methanesulfuryl chloride **(124-63-0)**
Methanesulphonic acid ethyl ester **(62-50-0)**
Methanesulphonic acid methyl ester **(66-27-3)**
Methanesulphonyl chloride **(124-63-0)**
Methanesulphonyl fluoride **(558-25-8)**
Methane, tetrabromide **(558-13-4)**
Methane, tetrabromo- **(558-13-4)**
Methane tetrachloride **(56-23-5)**
Methane, tetrachloro- **(56-23-5)**
Methane, tetrafluoro- **(75-73-0)**
Methane, tetraiodo- **(507-25-5)**
Methane tetramethylol **(115-77-5)**
Methane, tetranitro **(509-14-8)**
N,N'-Methanetetraylbiscyclohexanamine **(538-75-0)**
N,N'-Methanetetraylbis(2-methylbenzenamine) **(1215-57-2)**
Methane, (thioarsenoso)- **(2533-82-6)**
Methane, thiocyanato- **(556-64-9)**
Methanethiol **(74-93-1)**
Methanethiol (OSHA) **(74-93-1)**
Methanethiol, ((p-chlorophenyl)thio)-, s-ester with O,O-di-methyl phosphorodithioate **(953-17-3)**
Methanethiol, ((2,5-dichlorophenyl)thio)-, S-ester with O,O-di-ethyl phosphorodithioate **(2275-14-1)**
Methanethiol, ((2,5-dichlorophenyl)thio)-, S-ester with O,O-di-methyl phosphorodithioate **(3735-23-7)**
Methanethiol, (ethylsulfinyl)-, S-ester with O,O-diisopropyl-phosphorodithioate **(5827-05-4)**
Methanethiol, (ethylsulfonyl)-, S-ester with O,O-diethyl phosphorodithioate **(2588-04-7)**
Methanethiol, (ethylthio)-, S-ester with O,O-diethyl phosphoro-dithioate **(298-02-2)**
Methanethiol, (isopropylthio)-, S-ester with O,O-diethyl phosphoro-dithioate **(78-52-4)**
Methanethiol, lead(2+) salt (9CI) **(35029-96-0)**
Methanethiol, (methylthio)- (8CI,9CI) **(29414-47-9)**
Methanethiol, phenyl- **(100-53-8)**
Methanethiol, sodium salt (9CI) **(5188-07-8)**
Methanethiol, trichloro- **(75-70-7)**
Methane, tribromo **(75-25-2)**
Methane, tribromochloro- (8CI,9CI) **(594-15-0)**
Methane trichloride **(67-66-3)**
Methane, trichloro- **(67-66-3)**
Methane, trichlorofluoro **(75-69-4)**
Methane, trichloronitro-, (Flammable mixture) [UN 1583] **(76-06-2)**
Methane, trichloronitro-, Mixt. with Bromomethane **(8004-09-9)**
Methane, triethoxy- **(122-51-0)**
Methane, trifluoro **(75-46-7)**
Methane, trifluoroiodo- (9CI) **(2314-97-8)**
Methane, trifluoronitro- (8CI,9CI) **(335-02-4)**
Methane, triiodo **(75-47-8)**
Methane, trimethoxy- **(149-73-5)**
Methane, trimethylolnitro- **(126-11-4)**
Methane, trinitro **(517-25-9)**
Methane, triphenyl- (8CI) **(519-73-3)**
Methanide **(53-46-3)**
Methanimidamide, N'-(4-chloro-2-methylphenyl)-N,N-dimethyl- **(6164-98-3)**
Methanimidamide, N'-(4-chloro-2-methylphenyl)-N-methyl-N-((methylthio)methyl)-, monohydrochloride (9CI) **(34863-74-6)**
Methanimidamide, N'-(4-chloro-2-methylphenyl)-, monohydro-chloride **(19750-95-9)**
Methanimidamide, N,N-dimethyl-, N'-(3-(((methylamino)carbon-yl)oxy)phenyl)- (9CI) **(22259-30-9)**
Methanimidamide, N,N-dimethyl-N'-(3-(((methylamino)carbonyl)-oxy)phenyl)-, monohydrochloride **(23422-53-9)**
Methanimidamide, N,N-dimethyl-N'-(2-methyl-4-(((methylamino)-carbonyl)oxy)phenyl)- **(17702-57-7)**
Methanimidamide, N,N-dimethyl-N'-(5-(2-(5-nitro-2-furanyl)-ethenyl)-1,3,4-oxadiazol-2-yl)- **(25962-77-0)**
Methanimidamide, N'-(2,4-dimethylphenyl)-N-(((2,4-dimethyl-phenyl)imino)methyl)-N-methyl- **(33089-61-1)**
1H-3aα,6-Methanoazulene-3-carboxylic acid, 2,3β,4,5,6β,7-hexa-hydro-7,7,8-trimethyl- (8CI) **(16202-79-2)**
1H-3a,6-Methanoazulene-3-carboxylic acid, 2,3,4,5,6,7-hexa-hydro-7,7,8-trimethyl, (3S-(3α,3aα,6α))- (9CI) **(16202-79-2)**
1,4-Methanoazulene, decahydro-4,8,8-trimethyl-9-methylene- **(475-20-7)**
1,4-Methanoazulene, decahydro-4,8,8-trimethyl-9-methylene-, (1S,3aR,4S,8aS)-(+)- (8CI) **(475-20-7)**
1,4-Methanoazulene, decahydro-4,8,8-trimethyl-9-methylene-, (1S-(1α,3aβ,4α,8aβ))- (9CI) **(475-20-7)**
1H-3a,7-Methanoazulene, 2,3,4,7,8,8a-hexahydro-3,6,8,8-tetra-methyl-, (3R-(3-α,3a-β, 7-β,8a-α)) **(469-61-4)**
1H-3a,7-Methanoazulene, octahydro-1,4,9,9-tetramethyl-

(8CI,9CI) (VAN) **(19078-35-4)**
1H-3a,7-Methanoazulen-6-ol, octahydro-3,6,8,8-tetramethyl-, acetate **(77-54-3)**
1H-3a,7-Methanoazulen-6-ol, octahydro-3,6,8,8-tetramethyl-, (3R-(3α,3aβ,6α,7β,8aα))- (9CI) **(77-53-2)**
2,6-Methano-3-benzazocin-8-ol, 1,2,3,4,5,6-hexahydro-6,11-dimethyl-3-(3-methyl-2-butenyl) **(359-83-1)**
2,6-Methano-3-benzazocin-8-ol, 1,2,3,4,5,6-hexahydro-6,11-dimethyl-3-phenethyl- **(127-35-5)**
1,4-Methanobenzocyclooctene, 7,8-dibromo-1,2,3,4,11,11-hexachloro-1,4,4a,5,6,7,8,9,10,10a-decahydro- (9CI) **(51936-55-1)**
2,4-Methano-2H-bisoxireno(a,f)indene, octahydro- **(81-21-0)**
7,9a-Methano-9aH-cyclopenta(b)heptalene-2,4,8,11,11a,12(1H)-hexol, dodecahydro-1,1,4,8- tetramethyl-, 12-acetate (2S,3aS,4R,4aR,7R,8R,9aS,11R,11aR,12R) **(4720-09-6)**
3a,7-Methano-3aH-cyclopentacyclooctene, 1,4,5,6,7,8,9,9a-octahydro-1,1,7-trimethyl- **(469-92-1)**
3a,7-Methano-3aH-cyclopentacyclooctene, 1,4,5,6,7,8,9,9a-octahydro-1,1,7β-trimethyl- (8CI) **(469-92-1)**
3a,7-Methano-3aH-cyclopentacyclooctene, 1,4,5,6,7,8,9,9a-octahydro-1,1,7-trimethyl-, (3aR-(3aα,7α,9aβ))- (9CI) **(469-92-1)**
6,12-Methano-12H-dibenzo(d,g)(1,3)dioxocin-3-ol, 8-methoxy-6-(p-methoxyphenyl)-13-methyl-, acetate (8CI) **(2652-25-7)**
7,14-Methano-2H,6H-dipyrido(1,2-a:1',2'-e)(1,5)diazocine, dodecahydro- **(90-39-1)**
Methanoic acid **(64-18-6)**
4,7-Methanoindan, 1,2:5,6-diepoxy-3a,4,5,6,7,7a-hexahydro- **(81-21-0)**
4,7-Methanoindan, 1,2:5,6-diepoxyhexahydro- **(81-21-0)**
4,7-Methanoindan, 1,4,5,6,7,8,8-heptachloro-2,3-epoxy-3a,4,7,7a-tetrahydro **(1024-57-3)**
4,7-Methanoindan, 3a,4,5,6,7,7a-hexahydro-1,2:5,6-diepoxy **(81-21-0)**
4,7-Methanoindan, hexahydro-, exo **(2825-82-3)**
4,7-Methanoindan, 1α,2α,3α,4β,5,6,7β,8,8-nonachloro-3aα,4,7,7aα-tetrahydro- (8CI) **(5103-73-1)**
4,7-Methanoindan, 1,2,4,5,6,7,8,8-octachloro-2,3-epoxy-3a,4,7,7a-tetrahydro-, exo,endo **(27304-13-8)**
4,7-Methanoindan, 1,2,4,5,6,7,8,8-octachloro-3a,4,7,7a-tetrahydro **(57-74-9)**
4,7-Methanoindan, 1-α,2-α,4-β,5,6,7-β,8,8-octachloro-3a-α,4,7,7a-α- tetrahydro **(5103-71-9)**
4,7-Methanoindan, 2,2,4,5,6,7,8,8-octachloro-3a,4,7,7a-tetrahydro **(5566-34-7)**
4,7-Methanoindan, 3a,4,7,7a-tetrahydro-2,3-epoxy-1,2,4,5,6,7,8,8-octachloro-, exo,endo- **(27304-13-8)**
4,7-Methanoindan, 3a-α,4,7,7a-α-tetrahydro-2-α,4-β,5,6,7-β,8,8-heptachloro **(14168-01-5)**
4,7-Methanoindan, 3a-β,4,7,7a-β-tetrahydro-1-β,2-α,4-α,5,6,7-α,8,8- octachloro **(5103-74-2)**
4,7-Methanoindene-6-carboxylic acid, 3a,4,5,6,7,7a-hexahydro-, ethyl ester **(17511-60-3)**
4,7-Methano-1H-indenediol, 4,5,6,7,8,8-hexachloro-3a,4,7,7a-tetrahydro- (9CI) **(39660-14-5)**
4,7-Methano-1H-indene, 1,1,2,3,3a,4,5,6,7,7a,8,8-dodecachloro-3a,4,7,7a-tetrahydro- **(14979-34-1)**
4,7-Methanoindene, 1,1,2,3,3a,4,5,6,7,7a,8,8-dodecachloro-3a,4,7,7a-tetrahydro- **(14979-34-1)**
4,7-Methanoindene, 1,4,5,6,7,8,8-heptachloro-3a,4,7,7a-tetrahydro **(76-44-8)**
4,7-Methanoindene, 4,5,6,7,8,8-hexachloro-3a,4,7,7a-tetrahydro- **(3734-48-3)**
4,7-Methanoindene, 4,5,6,7,8,8-hexachloro-$\delta^{1,5}$-tetrahydro **(3734-48-3)**
4,7-Methano-1H-indene, 1,2,3,4,5,6,7,8,8-nonachlor-2,3,3a,4,7,7a-hexahydro-, (1-α,2-β, 3-α,3a-α,4-β,7-β,7a-α) **(39765-80-5)**
4,7-Methano-1H-indene, 1,2,3,4,5,6,7,8,8-nonachloro-2,3,3a,4,7,7a-hexahydro- **(3734-49-4)**
4,7-Methano-1H-indene, 1,2,2,4,5,6,7,8,8-nonachloro-2,3,3a,4,7,7a-hexahydro-, (1α,3aα,4β,7β,7aα)- (9CI) **(98318-97-9)**
4,7-Methano-1H-indene, 1,2,3,4,5,6,7,8,8-nonachloro-2,3,3a,4,7,7a-hexahydro-, (1α,2α,3α,3aα,4β,7β,7aα)- (9CI) **(5103-73-1)**
4,7-Methano-1H-indene, 1,2,3,4,5,6,7,8,8-nonachlorooctahydro- (9CI) **(115384-94-6)**
4,7-Methano-1H-indene, 1,2,3,4,5,6,7,8,8-nonachloro-3a,4,7,7a-tetrahydro- (9CI) **(21641-70-3)**
4,7-Methanoindene, 1,2,3,4,5,6,7,8,8-nonachloro-3a,4,7,7a-tetrahydro- (8CI) (VAN) **(21641-70-3)**
4,7-Methano-1H-indene, 1,2,4,5,6,7,8,8-octachloro-2,3,3a,4,7,7a-hexahydro- **(57-74-9)**
4,7-Methano-1H-indene, 2,2,4,5,6,7,8,8-octachloro-2,3,3a,4,7,7a-hexahydro- (9CI) **(5566-34-7)**
4,7-Methano-1H-indene, octahydro- (9CI) **(6004-38-2)**
4,7-Methano-1H-indene, octahydro-, (3a-α,4-β,7-β,7a-α)- (9CI) **(2825-82-3)**
4,7-Methano-1H-indene, octahydrodimethyl- (9CI) **(30496-78-7)**
4,7-Methanoindene,3a,4,7,7a-tetrahydro **(77-73-6)**
4,7-Methanoindene, 3a,4,7,7a-tetrahydrodimethyl **(26472-00-4)**
4,7-Methano-1H-indene, 3a,4,7,7a-tetrahydrodimethyl- (9CI) **(26472-00-4)**
2,4-Methano-2H-indeno(1,2-b:5,6-b')bisoxirene, octahydro- **(81-21-0)**
4,7-Methanoinden, 1,2,4,5,6,7,8,9-octachloro-3a,4,7,7a-tetrahydro- **(12789-03-6)**
4,7-Methano-1H-indenol, 4,5,6,7,8,8-hexachloro-3a,4,7,7a-tetrahydro- (9CI) **(12408-14-9)**
4,7-Methanoindenol, 4,5,6,7,8,8-hexachloro-3a,4,7,7a-tetrahydro- (8CI) **(12408-14-9)**
4,7-Methanoinden-1-ol, 4,5,6,7,8,8-hexachloro-3a,4,7,7a-tetrahydro-, endo,exo **(24009-05-0)**
4,7-Methanoinden-6-ol, 3a,4,5,6,7,7a-hexahydro-, acetate **(5413-60-5)**
4,7-Methano-1H-inden-5-ol, 3a,4,5,6,7,7a-hexahydro-, acetate (9CI) **(2500-83-6)**
4,7-Methano-1H-inden-6-ol, 3a,4,5,6,7,7a-hexahydro-, acetate (9CI) **(5413-60-5)**
4,6-Methano-6H,14H-indolo(3,2,1-ij)oxepino(2,3,4-de)pyrrolo (2,3-h)qu inoline, strychnidin-10-one deriv. (9CI) **(52748-69-3)**
4,7-Methanoisobenzofuran-1,3-dione, 4,5,6,7,8,8-hexachloro-3a,4,7,7a-tetrahydro- (9CI) **(115-27-5)**
4,7-Methanoisobenzofuran, 1,3,4,5,6,7,8,8-octachloro-1,3,3a,4,7,7a-hexahydro **(297-78-9)**
4,7-Methano-1H-isoindole-1,3(2H)-dione, 5,6-dibromohexahydro-2-phenyl- (9CI) **(40703-79-5)**
4,7-Methano-1H-isoindole-1,3(2H)-dione, 2,2'-(1,2-ethanediyl)bis-(5,6-dibromohexahydro- (9CI) **(52907-07-0)**
4,7-Methano-1H-isoindole-1,3(2H)-dione, 2,2'-(1,2-ethanediyl)-bis(3a,4,7,7a-tetrahydro- (9CI) **(25502-52-7)**
4,7-Methano-1H-isoindole-1,3(2H)-dione, 3a,4,7,7a-tetrahydro-2-hydroxy- (9CI) **(21715-90-2)**
Methanol **(67-56-1)**
p-Methan-7-ol, (cis)- **(13828-37-0)**
Methanol (DOT) **(67-56-1)**
Methanolacetonitrile **(109-78-4)**
N-Methanolacrylamide **(924-42-5)**
Methanol, benzyl- **(60-12-8)**
Methanol, butoxy- (9CI) **(3085-35-6)**
Methanol, (butylnitrosoamino)-, acetate (ester) **(56986-36-8)**
Methanol, chloro-, benzoate **(5335-05-7)**
Methanol, cyclohexyl- **(100-49-2)**
Methanol, dibutoxy- (9CI) **(54518-04-6)**
Methanol, (dimethylamino)- (9CI) **(14002-21-2)**
Methanol, (4-(1,1-dimethylethyl)phenoxy)-, acetate (9CI) **(54889-98-4)**
Methanol, ethynyl- **(107-19-7)**
Methanol, (2-furyl)- **(98-00-0)**
Methanol, methoxy- (9CI) **(4461-52-3)**
Methanol, (methyl-ONN-azoxy) **(590-96-5)**
Methanol, (methyl-ONN-azoxy)-, acetate (ester) **(592-62-1)**
Methanol, (methylnitrosoamino)-, acetate (ester) **(56856-83-8)**
Methanol, methylphenyl- **(98-85-1)**
Methanol, ((6-(2-(5-nitro-2-furanyl)ethenyl)-1,2,4-triazin-3-yl)-imino)bis- **(794-93-4)**
Methanol, ((6-(2-(5-nitro-2-furyl)vinyl)-as-triazin-3-yl)imino)di **(794-93-4)**
Methanol, (nitrosopropylamino)-, acetate (ester) **(66017-91-2)**
Methanol, oxiranyl- **(556-52-5)**

Methanol, phenyl- **(100-51-6)**
Methanol, (phenylmethoxy)- (9CI) **(14548-60-8)**
Methanol, sodium salt **(124-41-4)**
Methanol, sulfonylbis- (9CI) **(87954-49-2)**
2-Methanol tetrahydropyran **(100-72-1)**
Methanol, (s-triazine-2,4,6-triyltriimino)tri **(1017-56-7)**
Methanol, trichloro-, chloroformate **(503-38-8)**
Methanol, trimethyl- **(75-65-0)**
Methanol, triphenyl- (8CI) **(76-84-6)**
Methanol, tris(4-aminophenyl)- **(467-62-9)**
4,7-Methano-2,3,8-methenocyclopent(a)indene, dodecahydro-, stereoisomer **(66289-74-5)**
4,7-Methano-2,3,8-methenocyclopent(a)indene, dodecahydro-, stereoisomer, Mixt. with methylcyclohexane and (3aα,4β,7β,7aα)-octahydro-4,7-methano-1H-indene (9CI) **(82863-50-1)**
1,4-Methanonaphthalene, 1,2,3,4,9,9-hexachloro-5,6-epoxy-1,4,4a,5,6,7,8,8a-octahydro-, stereoisomer (8CI) **(21858-40-2)**
4,7-Methanonaphth(1,2-b)oxirene, 4,5,6,7,8,8-hexachloro-1a,2,3,3a,4,7,7a,7b-octahydro-, (1aα,3aα,4α,7α,7aα,7bα)-(9CI) **(21858-40-2)**
Methanone, (2-amino-5-chlorophenyl)phenyl **(719-59-5)**
Methanone, bis((1,1'-biphenyl)yl)- (9CI) **(72776-75-1)**
Methanone, bis(4-(diethylamino)phenyl)- (9CI) **(90-93-7)**
Methanone, bis(2,4-dihydroxyphenyl)- (9CI) **(131-55-5)**
Methanone, bis(2-hydroxy-4-methoxyphenyl)- (9CI) **(131-54-4)**
Methanone, bis(4-methoxyphenyl)- (9CI) **(90-96-0)**
Methanone, bis(3-methylphenyl)- (9CI) **(2852-68-8)**
Methanone, (5-chloro-2-hydroxyphenyl)phenyl- (9CI) **(85-19-8)**
Methanone, (5-chloro-2-methylphenyl)phenyl- (9CI) **(33184-55-3)**
Methanone, (2-chlorophenyl)(4-chlorophenyl)- (9CI) **(85-29-0)**
Methanone, (4-chlorophenyl)(2-chloro-5-(trifluoromethyl)-phenyl)- (9CI) **(95998-69-9)**
Methanone, (2-chlorophenyl)(4-fluorophenyl)- (9CI) **(1806-23-1)**
Methanone, (4-chlorophenyl)phenyl- (9CI) **(134-85-0)**
Methanone, (2,4-dichlorophenyl)(1,3-dimethyl-5-(((4-methyl-phenyl)sulfonyl)oxy)-1H-pyrazol-4-yl) **(58011-68-0)**
Methanone, (4-(dimethylamino)phenyl)phenyl- (9CI) **(530-44-9)**
Methanone, diphenyl-, monochloro monohydroxy deriv. (9CI) **(55299-12-2)**
Methanone, (4-(dodecyloxy)-2-hydroxyphenyl)phenyl- (9CI) **(2985-59-3)**
Methanone, (4-ethylphenyl)phenyl- (9CI) **(18220-90-1)**
Methanone, (1-hydroxycyclohexyl)phenyl- (9CI) **(947-19-3)**
Methanone, (2-hydroxy-4-(2-hydroxyethoxy)phenyl)phenyl-(9CI) **(16909-78-7)**
Methanone, (2-hydroxy-4-methoxyphenyl)(2-hydroxyphenyl)-(9CI) **(131-53-3)**
Methanone, (2-hydroxy-4-methoxyphenyl)phenyl- **(131-57-7)**
Methanone, (4-hydroxy-3-methylphenyl)(3-methylphenyl)- (9CI) **(62064-85-1)**
Methanone, (2-hydroxy-5-nonylphenyl)phenyl-, oxime (9CI) **(37339-32-5)**
Methanone, (2-hydroxyphenyl)phenyl- (9CI) **(117-99-7)**
Methanone, (4-hydroxyphenyl)phenyl- (9CI) **(1137-42-4)**
Methanone, (4-nitrophenyl)phenyl- (9CI) **(1144-74-7)**
Methanone, phenyl(2,3,4-trihydroxyphenyl)- (9CI) **(1143-72-2)**
2,5-Methano-2H-oxireno(a)indene, 2,3,4,5,6,7,7-heptachloro-1a,1b,5,5a,6,6a-hexahydro- **(1024-57-3)**
1,5-Methano-8H-pyrido(1,2-a)(1,5)diazocin-8-one, 1,2,3,4,5,6-hexahydro- **(485-35-8)**
Methantheline bromide **(53-46-3)**
Methanthelinium bromide **(53-46-3)**
Methanthine bromide **(53-46-3)**
Methanthiol (German) **(74-93-1)**
Methaphoxide **(57-39-6)**
Methapyrapone **(54-36-4)**
Methapyrilene **(91-79-2)**
Methapyrilene **(91-80-5)**
Methapyrilene hydrochloride **(135-23-9)**
Methaqualone **(72-44-6)**
Methaqualoneinone **(72-44-6)**
Methar **(144-21-8)**
Methar 30 **(144-21-8)**
Metharbital **(50-11-3)**
Metharbitone **(50-11-3)**
Metharbutal **(50-11-3)**
Metharcylic acid, 2-isocyanatoethyl ester **(30674-80-7)**
Metharsan **(144-21-8)**
Metharsinat **(144-21-8)**
Methasan **(137-30-4)**
Methasol Fast Blue **(1328-51-4)**
Methaxalonum **(1665-48-1)**
Methazate **(137-30-4)**
Methazine **(53-86-1)**
Methazine **(67704-68-1)**
Methazolamide **(554-57-4)**
Methazole **(20354-26-1)**
Methazonic acid **(5653-21-4)**
Methdilazine **(1982-37-2)**
Methdilazine hydrochloride **(1229-35-2)**
Methedrinal **(300-42-5)**
Methedrine **(300-42-5)**
Methedrine hydrochloride **(300-42-5)**
Methelina **(53-46-3)**
Methenamic acid **(61-68-7)**
Methenamide **(554-57-4)**
Methenamine **(100-97-0)**
1,3,4-Metheno-1H-cyclobuta(cd)pentalene, 1,1a,2,2,3,3a,4,5,5,5a,5b,6- dodecachlorooctahydro **(2385-85-5)**
1,3,4-Metheno-1H-cyclobuta(cd)pentalene, dodecachlorooctahydro- **(2385-85-5)**
1,2,4-Metheno-1H-cyclobuta(cd)pentalene, 1,3,4,5,5,5a,6-hepta-chlorooctahydro- (8CI,9CI) **(21161-58-0)**
1,3,4-Metheno-1H-cyclobuta(c,d)-pentalene-2-levulinic acid, 1,1a,3,3a,4,5,5a,5b,6- decachlorooctahydro-2-hydroxy-, ethyl ester **(4234-79-1)**
1,3,4-Metheno-1H-cyclobuta(cd)pentalene, 1,1a,2,2,3,3a,4,5,5,5a,5-undecachlorooctahydro **(39801-14-4)**
1,3,4-Metheno-1H-cyclobuta(cd)pentalen-2-ol, 1,1a,3,3a,4,5,5,5a,5b,6-decachlorooctahydro- **(1034-41-9)**
1,3,4-Metheno-2H-cyclobuta(cd)pentalen-2-one, 1,1a,3,3a,4,5,5,5a,5b,6-decachloroctahydro **(143-50-0)**
2,4,7-Metheno-1H-cyclopenta(a)pentalene, 1,1,2,3,3a,7a-hexachloro-2,3,3a,3b,4,6a,7,7a-octahydro-, (2α,3α,3aα,3bα,4β,6aα,7β,7aα)-(9CI) **(13350-71-5)**
2,4,7-Metheno-1H-cyclopenta(a)pentalene, 1,1,2,3,3a,7a-hexa-chloro-2,3,3a,3b,4,6a,7,7a-octahydro-, stereoisomer (8CI) **(13350-71-5)**
2,4,7-Metheno-1H-cyclopenta(a)pentalene, 1,1,2,3,3a,7a-hexa-chloro-5,6-epoxydecahydro-, stereoisomer **(13366-73-9)**
2,5,7-Metheno-3H-cyclopenta(a)pentalen-3-one, 3b,4,5,6,6,6a-hexa-chlorodecahydro-, (2-α,3a-β,3b-β,4-β,5-β,6a-β,7-α,7a-β,8R*) **(53494-70-5)**
1,2,4-Methenecyclopenta(c,d)pentalene-r-carboxaldehyde, 2,2a,3,3,4,7-hexachlorodecahydro **(7421-93-4)**
1,2,4-Methenocyclopenta(cd)pentalene-5-carboxaldehyde, 2,2a,3,3,4,7-hexachlorodecahydro-, (1α,2β,2aβ,4β,4aβ,5β,6aβ,6bβ,7R*)- **(7421-93-4)**
N,N'-Methenyl-o-phenylenediamine **(51-17-2)**
Methenyl tribromide **(75-25-2)**
Methenyl trichloride **(67-66-3)**
Methergine **(113-42-8)**
Methexamide **(565-33-3)**
Methexenyl **(56-29-1)**
Methforylthiazidine **(135-09-1)**
Methiacil **(56-04-2)**
Methiamazole **(60-56-0)**
Methiamitron (French) **(41394-05-2)**
Methibenzuron **(18691-97-9)**
Methicil **(56-04-2)**
Methicillin **(61-32-5)**

Name	CAS
Methidathion	(950-37-8)
Methidathion 50S	(950-37-8)
Methilanin	(59-51-8)
Methimazol	(60-56-0)
Methimazole	(60-56-0)
Methiocarb	(2032-65-7)
Methiocarbe	(2032-65-7)
Methiochlor	(19679-38-0)
Methiocil	(56-04-2)
Methiodal sodium	(126-31-8)
Methional	(3268-49-3)
(+-)-Methionine	(59-51-8)
D-Methionine	(348-67-4)
Methionine	(63-68-3)
l-(-)-Methionine	(63-68-3)
l-Methionine	(63-68-3)
DL-Methionine (9CI)	(59-51-8)
Methionine, D	(348-67-4)
Methionine, DL	(59-51-8)
Methionine, L	(63-68-3)
Methionine hydroxy analog	(583-91-5)
Methionine, seleno	(1464-42-2)
Methionine sulfoximine	(1982-67-8)
Methionine-DL-sulfoximine, DL-	(1982-67-8)
Methionol	(505-10-2)
Methocarbamol	(532-03-6)
Methocel 10	(9004-67-5)
Methocel 15	(9004-67-5)
Methocel 181	(9004-67-5)
Methocel 400	(9004-67-5)
Methocel 4000	(9004-67-5)
Methocel A	(9004-67-5)
Methocel CHG	(9004-67-5)
Methocel 400CPS	(9004-67-5)
Methocel 4000CPS	(9004-67-5)
Methocel MC	(9004-67-5)
Methocel MC 25	(9004-67-5)
Methocel MC 8000	(9004-67-5)
Methocel MC4000	(9004-67-5)
Methocel SM 100	(9004-67-5)
Methochlopramide	(364-62-5)
Methocillin S	(61-72-3)
Methoclopramide	(364-62-5)
Methoflurane	(76-38-0)
Methogas	(74-83-9)
Methohexital	(151-83-7)
Methohexital	(309-36-4)
Methohexital sodium	(309-36-4)
Methohexitalum (Latin)	(151-83-7)
Methohexitone sodium	(309-36-4)
Methoidal sodium	(126-31-8)
Methoin	(50-12-4)
Metholcarb	(1129-41-5)
Metholene 2095	(110-42-9)
Metholene 2216	(112-39-0)
Metholene 2218	(112-61-8)
Metholene 2296	(111-82-0)
Metholeneat 2495	(124-10-7)
Metholone	(58-19-5)
Methometon	(1771-07-9)
Methomyl (ACGIH,OSHA)	(16752-77-5)
Methomyl oxime	(13749-94-5)
Methon	(126-81-8)
Methone	(126-81-8)
Methophylline	(317-34-0)
Methopirapone	(54-36-4)
Methoplain	(41372-08-1)
Methoplain	(555-30-6)
Methoprene	(40596-69-8)
Methoproptryne	(841-06-5)
Methoprotryn	(841-06-5)
Methopterin	(59-05-2)
Methopyrapone	(54-36-4)
Methopyrinine	(54-36-4)
Methopyrone	(54-36-4)
Methoquine	(69-05-6)
Methorfinan (Czech)	(297-90-5)
Methorphan	(125-70-2)
d-Methorphan	(125-71-3)
δ-Methorphan	(125-71-3)
l-Methorphan	(125-70-2)
Methorphinan	(297-90-5)
Methosarb	(17021-26-0)
Methoscopylamine bromide	(155-41-9)
Methotextrate	(59-05-2)
Methotrexat	(59-05-2)
Methotrexate	(59-05-2)
Methotrexate, dichloro-	(528-74-5)
Methotrexate sodium	(15475-56-6)
Methoxa-Dome	(298-81-7)
Methoxalen	(298-81-7)
Methoxamine hydrochloride	(61-16-5)
Methoxane	(76-38-0)
Methoxcide	(72-43-5)
Methoxo	(72-43-5)
Methoxolone	(1665-48-1)
Methoxon	(3653-48-3)
Methoxone	(3653-48-3)
Methoxone	(7085-19-0)
Methoxone	(93-65-2)
Methoxone	(94-74-6)
Methoxone M	(1929-86-8)
Methoxsalen	(298-81-7)
Methoxy-DDT	(72-43-5)
Methoxy Simazine	(673-04-1)
4'-Methoxyacetanilide	(51-66-1)
4-Methoxyacetanilide	(51-66-1)
p-Methoxyacetanilide	(51-66-1)
2-Methoxyacetic acid	(625-45-6)
Methoxyacetic acid	(625-45-6)
2'-Methoxyacetoacetanilide	(92-15-9)
2-Methoxyacetoacetanilide	(92-15-9)
o-Methoxyacetoacetanilide	(92-15-9)
4-Methoxyacetofenon (Czech)	(100-06-1)
1-Methoxyacetone	(5878-19-3)
Methoxyacetone	(5878-19-3)
4'-Methoxyacetophenone	(100-06-1)
4-Methoxyacetophenone	(100-06-1)
p-Methoxyacetophenone	(100-06-1)
5-Methoxy-N-acetyltryptamine	(73-31-4)
2-Methoxy-aethanol (German)	(109-86-4)
2-Methoxyaethylacetat (German)	(110-49-6)
Methoxyaethylquecksilberchlorid (German)	(123-88-6)
p-Methoxyallylbenzene	(140-67-0)
2-Methoxy-4-allylphenol	(97-53-0)
Methoxyamine	(67-62-9)
Methoxyamine, hydrochloride	(593-56-6)
Methoxyamine, N-methyl-	(1117-97-1)
2-Methoxy-1-aminobenzene	(90-04-0)
4-Methoxy-1-aminobenzene	(104-94-9)
2-Methoxy-1-aminobenzene hydrochloride	(134-29-2)
4-Methoxy-1-aminobenzene hydrochloride	(20265-97-8)
4-Methoxy-2-aminobenzothiazole	(5464-79-9)
6-Methoxy-2-aminobenzothiazole	(1747-60-0)

(+-)-p-Methoxyamphetamine **(23239-32-9)**
4-Methoxyamphetamine **(23239-32-9)**
p-Methoxyamphetamine **(23239-32-9)**
2-Methoxyanilid kyseliny acetoctove (Czech) **(92-15-9)**
2-Methoxyaniline **(90-04-0)**
3-Methoxyaniline **(536-90-3)**
4-Methoxyaniline **(104-94-9)**
Methoxyaniline **(29191-52-4)**
o-Methoxyaniline **(90-04-0)**
p-Methoxyaniline **(104-94-9)**
2-Methoxyaniline hydrochloride **(134-29-2)**
4-Methoxyaniline hydrochloride **(20265-97-8)**
o-Methoxyaniline hydrochloride **(134-29-2)**
p-Methoxyaniline hydrochloride **(20265-97-8)**
2-Methoxy-p-anisidine hydrochloride **(54150-69-5)**
4-Methoxy-o-anisidine hydrochloride **(54150-69-5)**
3-Methoxyanisole **(151-10-0)**
p-Methoxyanisole **(150-78-7)**
1-Methoxyanthraquinone **(82-39-3)**
2-Methoxyazobenzene **(6319-21-7)**
o-Methoxyazobenzene **(6319-21-7)**
2-Methoxybenzaldehyde **(135-02-4)**
3-Methoxybenzaldehyde **(591-31-1)**
4-Methoxybenzaldehyde **(123-11-5)**
6-Methoxybenzaldehyde **(135-02-4)**
m-Methoxybenzaldehyde **(591-31-1)**
o-Methoxybenzaldehyde **(135-02-4)**
p-Methoxybenzaldehyde **(123-11-5)**
2-Methoxybenzenamine **(90-04-0)**
3-Methoxybenzenamine **(536-90-3)**
4-Methoxybenzenamine **(104-94-9)**
Methoxybenzene **(100-66-3)**
2-Methoxybenzeneacetic acid **(93-25-4)**
4-Methoxybenzeneacetic acid **(104-01-8)**
p-Methoxybenzeneacetonitrile **(104-47-2)**
4-Methoxybenzeneamine **(104-94-9)**
2-Methoxybenzeneamine hydrochloride **(134-29-2)**
4-Methoxybenzeneamine hydrochloride **(20265-97-8)**
2-Methoxybenzenecarboxaldehyde **(135-02-4)**
4-Methoxy-1,3-benzenediamine **(615-05-4)**
4-Methoxy-1,3-benzenediamine sulfate **(39156-41-7)**
4-Methoxy-1,3-benzenediamine sulfate (1:1) **(39156-41-7)**
4-Methoxy-1,3-benzenediamine sulphate **(39156-41-7)**
4-Methoxybenzenemethanol **(105-13-5)**
ar-Methoxybenzenemethanol **(1331-81-3)**
4-Methoxybenzenemethanol acetate **(104-21-2)**
3-Methoxy-1,2-benzisothiazole **(40991-38-6)**
2-Methoxy-4H-1,2,3-benzodioxaphosphorine-2-sulfide **(3811-49-2)**
7-Methoxy-1,3-benzodioxole-5-carboxaldehyde **(5780-07-4)**
2-Methoxybenzoic acid **(579-75-9)**
3-Methoxybenzoic acid **(586-38-9)**
4-Methoxybenzoic acid **(100-09-4)**
m-Methoxybenzoic acid **(586-38-9)**
o-Methoxybenzoic acid **(119-36-8)**
o-Methoxybenzoic acid **(579-75-9)**
p-Methoxybenzoic acid **(100-09-4)**
Methoxybenzoyl chloride **(100-07-2)**
4-Methoxybenzyl acetate **(104-21-2)**
p-Methoxybenzyl acetate **(104-21-2)**
p-Methoxybenzyl alcohol **(105-13-5)**
p-Methoxybenzyl alcohol acetate **(104-21-2)**
p-Methoxybenzyl cyanide **(104-47-2)**
N-(p-Methoxybenzyl)-N',N'-dimethyl-N-2-pyridylethylenediamine **(91-84-9)**
N-p-Methoxybenzyl-N',N'-dimethyl-N-α-pyridylethylenediamine **(91-84-9)**
N-p-Methoxybenzyl-N'-N'-dimethyl-N-α-pyridylethylenediamine maleate **(59-33-6)**
N-p-Methoxybenzyl-N',N'-dimethyl-N-2-pyrimidinylethylene diamine hydrochloride **(63-56-9)**
p-Methoxybenzyl formate **(104-01-8)**
4-Methoxybenzylidene-4'-n-butylaniline **(26227-73-6)**
p-Methoxybenzyl-α-pyridyl-dimethyl-aethylendiamin (German) **(91-84-9)**
2-(p-Methoxybenzyl)-3,4-pyrrolidinediol 3-acetate **(22862-76-6)**
2-Methoxy biphenyl **(86-26-0)**
4-Methoxybiphenyl **(613-37-6)**
p-Methoxybiphenyl **(613-37-6)**
2-Methoxy-4,6-bis(ethylamino)-s-triazine **(673-04-1)**
2-Methoxy-4,6-bis(isopropylamino)-1,3,5-triazine **(1610-18-0)**
2-Methoxy-4,6-bis(isopropylamino)-s-triazine **(1610-18-0)**
2-Methoxy-4,6-bis((3-methoxypropyl)amino)-s-triazine **(1771-07-9)**
2-Methoxybromobenzene **(578-57-4)**
4-Methoxybromobenzene **(104-92-7)**
o-Methoxybromobenzene **(578-57-4)**
p-Methoxybromobenzene **(104-92-7)**
1-Methoxybutane **(628-28-4)**
α-Methoxybutane **(628-28-4)**
1-Methoxy-2-butanol **(53778-73-7)**
3-Methoxy-1-butanol **(2517-43-3)**
4-Methoxy-1-butanol **(111-32-0)**
3-Methoxybutyl acetate **(4435-53-4)**
2-Methoxy-4-sec-butylamino-6-aethylamino-s-triazin (German) **(26259-45-0)**
2-Methoxy-4-tert-butylamino-6-aethylamino-s-triazin (German) **(33693-04-8)**
3-Methoxybutylester kyseliny octove (Czech) **(4435-53-4)**
4-Methoxybutyl methyl ketone **(29006-00-6)**
4-Methoxy-2-tert-butylphenol **(121-00-6)**
2-Methoxy-3-sec-butylpyrazine **(24168-70-5)**
2-(Methoxy-carbonylamino)-benzimidazol (German) **(10605-21-7)**
2-(Methoxycarbonylamino)-benzimidazole **(10605-21-7)**
4-((4-((4-((Methoxycarbonyl)amino)phenyl)azo)-3-methyl-phenyl)azo)phenol **(6465-02-7)**
3-Methoxycarbonylaminophenyl N-3'-methylphenylcarbamate **(13684-63-4)**
2-(Methoxycarbonyl)aniline **(134-20-3)**
2-Methoxycarbonylbenzaldehyde **(1571-08-0)**
4-(Methoxycarbonyl)benzaldehyde **(1571-08-0)**
4-Methoxycarbonylbenzoic acid **(1679-64-7)**
Methoxycarbonyl chloride **(79-22-1)**
Methoxycarbonylethylene **(96-33-3)**
S-(N-Methoxycarbonyl-N-methylcarbamoylmethyl)dimethyl phosphonothiolothionate **(29173-31-7)**
3-Methoxycarbonyl-N-(3'-methylphenyl)-carbamat (German) **(13684-63-4)**
(2-Methoxycarbonyl-1-methyl-vinyl)-dimethyl-fosfaat (Dutch) **(7786-34-7)**
(2-Methoxycarbonyl-1-methyl-vinyl)-dimethyl-phosphat (German) **(7786-34-7)**
2-Methoxycarbonyl-1-methylvinyl dimethyl phosphate **(7786-34-7)**
1-Methoxycarbonyl-1-propen-2-yl dimethyl phosphate **(7786-34-7)**
Methoxychlor **(72-43-5)**
o,p'-Methoxychlor **(30667-99-3)**
o,p-Methoxychlor **(30667-99-3)**
p,p'-Methoxychlor **(72-43-5)**
Methoxychlor 2 EC **(72-43-5)**
Methoxychlore **(72-43-5)**
2-Methoxy-6-chloro-9-(4-bis(2-chloroethyl)amino-1-methyl-butylamino)acridine dihydrochloride **(4213-45-0)**
2-Methoxy-6-chloro-9-(4-diethylamino-1-methylbutylamino)-acridine dihydrochloride **(69-05-6)**
2-Methoxy-6-chloro-9-diethylaminopentylaminoacridine **(83-89-6)**
2-Methoxy-6-chloro-9-(3-(ethyl-2-chloroethyl)aminopropyl-amino)acridine dihydrochloride **(146-59-8)**
2-Methoxy-6-chloro-9-(3-(ethyl-2-chloroethyl)aminopropyl-amino)acridine dihydrochloride **(4213-45-0)**
2-Methoxy-5-chloroprocainamide **(364-62-5)**
6'-Methoxycinchonan-9-ol **(56-54-2)**
(9S)-6'-Methoxycinchonan-9-ol sulfate (1:1) (Salt) **(747-45-5)**
(9S)-6'-Methoxycinchonan-9-ol sulfate (2:1) (Salt) **(50-54-4)**
8α,9R-6'-Methoxycinchonan-9-ol, sulfate (1:1) salt **(549-56-4)**
6-Methoxycinchonine **(130-95-0)**

2-Methoxycinnamaldehyde **(1504-74-1)**
o-Methoxy cinnamaldehyde **(1504-74-1)**
o-Methoxycinnamic aldehyde **(1504-74-1)**
Methoxycitronellal methyl ether **(3613-30-7)**
2-Methoxy-p-cresol **(93-51-6)**
1-Methoxycyclohexane **(931-56-6)**
Methoxycyclohexane **(931-56-6)**
2-Methoxy-3,6-dichlorobenzoic acid **(1918-00-9)**
2-Methoxy-3,6-dichlorobenzoic acid diethanolamine salt **(25059-78-3)**
2-Methoxy-3,6-dichlorobenzoic acid sodium salt **(1982-69-0)**
Methoxydiglycol **(111-77-3)**
2-Methoxy-3,4-dihydro-2H-pyran **(4454-05-1)**
5-Methoxy-2-(dimethoxyphosphinylthiomethyl)pyrone-4 **(2778-04-3)**
4-Methoxy-3,3'-dimethylbenzophenone (8CI) **(41295-28-7)**
1-(8-Methoxy-4,8-dimethylnonyl)-4-(1-methylethyl)benzene **(53905-38-7)**
1-Methoxy-2,4-dinitrobenzene **(119-27-7)**
Methoxydiphenylarsine **(24582-54-5)**
Methoxy dipropylene glycol **(13429-07-7)**
Methoxydiuron **(330-55-2)**
2-Methoxyethanol (ACGIH,OSHA) **(109-86-4)**
2-Methoxyethanol, acetate **(110-49-6)**
2-Methoxyethanol, acrylate **(3121-61-7)**
Methoxy ether of propylene glycol **(107-98-2)**
3-Methoxy-17-α-ethinylestradiol **(72-33-3)**
3-Methoxy-17-α-ethinyloestradiol **(72-33-3)**
2-(2-Methoxyethoxy)ethanol **(111-77-3)**
2-(2-Methoxyethoxy)ethanol acetate **(629-38-9)**
2-(2-(2-Methoxyethoxy)ethoxy)ethanol **(112-35-6)**
2-(2-Methoxyethoxy)ethyl acetate **(629-38-9)**
2-(2-Methoxyethoxy)ethylester kyseliny octove (Czech) **(629-38-9)**
2-(2-Methoxy)ethoxyethyl-8-(2-n-octylcyclopropyl)octanoate **(54050-62-3)**
3-(2-Methoxyethoxy)propanenitrile **(35633-50-2)**
Methoxyethoxypropionitrile **(35633-50-2)**
2-Methoxy-ethyl acetaat (Dutch) **(110-49-6)**
N-(2-Methoxyethyl)acetamide **(5417-42-5)**
2-Methoxyethyl acetate (ACGIH,OSHA) **(110-49-6)**
2-Methoxyethyl 12-acetoxy-9-octadecenoate **(140-05-6)**
2-Methoxyethyl acetyl ricinoleate **(140-05-6)**
2-Methoxyethyl acrylate **(3121-61-7)**
Methoxyethyl acrylate **(3121-61-7)**
2-Methoxyethylamine **(109-85-3)**
2-Methoxy-4-ethylamino-6-isopropylamino-s-triazine **(1610-17-9)**
S-(2-((2-Methoxyethyl)amino-2-oxoethyl) O,O-dimethyl) phosphorodithioate **(919-76-6)**
(1-Methoxyethyl)benzene **(4013-34-7)**
2-Methoxyethyl carbamate **(1616-88-2)**
S-(N-2-Methoxyethylcarbamoylmethyl)dimethyl phophoro-thiolothionate **(919-76-6)**
2-Methoxyethyl chloride **(627-42-9)**
2-Methoxyethyle, acetate de (French) **(110-49-6)**
2-Methoxyethylester kyseliny acetylricinolove (Czech) **(140-05-6)**
2-Methoxyethylester kyseliny octove (Czech) **(110-49-6)**
2-Methoxyethyl formate **(628-82-0)**
2-Methoxyethyl hexadecanoate **(111-07-9)**
Methoxyethyl mercuric acetate **(151-38-2)**
(β-Methoxyethyl)mercuric chloride **(123-88-6)**
2-Methoxyethylmercuric chloride **(123-88-6)**
Methoxyethyl mercuric chloride **(123-88-6)**
Methoxyethylmercury acetate **(151-38-2)**
2-Methoxyethylmercury chloride **(123-88-6)**
Methoxyethylmercury chloride **(123-88-6)**
β-Methoxyethylmercury chloride **(123-88-6)**
2-Methoxyethylmerkuriacetat (Czech) **(151-38-2)**
2-Methoxyethylmerkurichlorid (Czech) **(123-88-6)**
4-(2-Methoxyethyl)morpholine **(10220-23-2)**
2-Methoxyethyl phthalate **(117-82-8)**
2-Methoxy-3-ethylpyrazine **(25680-58-4)**
2-Methoxyethyl vinyl ether **(1663-35-0)**
3-Methoxy-17-α-ethynylestradiol **(72-33-3)**
3-Methoxyethynylestradiol **(72-33-3)**
3-Methoxy-17-α-ethynyl-1,3,5(10)-estratrien-17-β-ol **(72-33-3)**
3-Methoxy-17-α-ethynyloestradiol **(72-33-3)**
3-Methoxy-17-ethynyloestradiol-17-β **(72-33-3)**
3-Methoxyethynyloestradiol **(72-33-3)**
3-Methoxy-17-α-ethynyl-1,3,5(10)-oestratrien-17-β-ol **(72-33-3)**
Methoxyfenac **(3004-74-8)**
Methoxyfluoran **(76-38-0)**
Methoxyfluorane **(76-38-0)**
2-Methoxyfluorene **(2523-46-8)**
p-Methoxyfluorobenzene **(459-60-9)**
Methoxyflurane **(76-38-0)**
5-Methoxy-6,7-furanocoumarin **(484-20-8)**
8-Methoxy-(furano-3'.2':6.7-coumarin) **(298-81-7)**
8-Methoxy-2',3',6,7-furocoumarin **(298-81-7)**
8-Methoxy-4',5',6,7-furocoumarin **(298-81-7)**
4-Methoxy-7H-furo(3,2-g)(1)benzopyran-7-one **(484-20-8)**
9-Methoxy-7H-furo(3,2-g)benzopyran-7-one **(298-81-7)**
6-Methoxyhexan-2-one **(29006-00-6)**
Methoxyhydrastine **(128-62-1)**
2-Methoxy-1-hydroxy-4-allylbenzene **(97-53-0)**
3-Methoxy-4-hydroxybenzaldehyde **(121-33-5)**
3-Methoxy-4-hydroxybenzoic acid **(121-34-6)**
4-Methoxy-2-hydroxybenzophenone **(131-57-7)**
3-Methoxy-4-hydroxy-benzylacetone **(122-48-5)**
β-Methoxy-β'-hydroxydiethyl ether **(111-77-3)**
Methoxyhydroxyethane **(109-86-4)**
2'-Methoxy-2-hydroxy-3-naphthanilide **(135-62-6)**
3-Methoxy-4-hydroxyphenylpyruvic acid **(1081-71-6)**
3-Methoxy-4-hydroxystyrene polymer **(31853-85-7)**
3-Methoxy-4-hydroxytoluene **(93-51-6)**
5-Methoxyindole **(1006-94-6)**
Methoxy-5 indole (French) **(1006-94-6)**
2-Methoxy-3-isobutylpyrazine **(24683-00-9)**
1-(2-Methoxyisopropoxy)-2-propanol **(34590-94-8)**
2-Methoxy-4-isopropylamino-6-diethylamino-s-triazine **(3004-70-4)**
2-Methoxy-4-isopropylamino-6-ethylamino-s-triazine **(1610-17-9)**
2-Methoxy-3-isopropylpyrazine **(25773-40-4)**
Methoxylamine hydrochloride **(593-56-6)**
Methoxylene **(135-23-9)**
N-Methoxymethanamine **(1117-97-1)**
Methoxymethanol **(4461-52-3)**
Methoxy-methiochlor **(34197-16-5)**
2-Methoxy-N-(2-methoxyethyl)ethanamine **(111-95-5)**
4-((5-Methoxy-4-((4-methoxyphenyl)azo)-2-methylphenyl)-azo)benzenesulfonic acid, sodium salt **(68555-86-2)**
4-Methoxy-N-(4-methoxyphenyl)-α-(trichloromethyl)benzene-methanamine **(38766-64-2)**
Methoxymethylamine **(1117-97-1)**
N-Methoxy-N-methylamine **(1117-97-1)**
N-Methoxymethylamine **(1117-97-1)**
2-Methoxy-4-methylamino-6-isopropylamino-s-triazine **(3035-45-8)**
2-Methoxy-5-methylaniline **(120-71-8)**
4-Methoxy-2-methylaniline **(102-50-1)**
2-Methoxy-2'-methylazobenzene **(29268-78-8)**
2-Methoxy-5-methylbenzenamine **(120-71-8)**
4-Methoxy-2-methylbenzenamine **(102-50-1)**
1-Methoxy-3-methylbenzene **(100-84-5)**
(+-)-4-Methoxy-α-methyl-benzeneethanamine **(23239-32-9)**
3-Methoxy-3-methyl-1-butanol **(56539-66-3)**
1-Methoxy-1-methyl-3-(3,4-dichlorophenyl)urea **(330-55-2)**
5-Methoxy-3,4-methylenedioxyamphetamine **(13674-05-0)**
1(or 2)-(2-Methoxymethylethoxy)propanol **(34590-94-8)**
1-(2-Methoxy-1-methylethoxy)-2-propanol **(20324-32-7)**
2-Methoxy-3-(1-methylethyl)pyrazine **(25773-40-4)**

3-Methoxy-α-methyl-4,5-methylenedioxyphenethylamine **(13674-05-0)**
β-Methoxy-α-methyl-4,5-(methylenedioxy)phenethyl amine **(13674-05-0)**
Methoxymethyl methyl ketone **(5878-19-3)**
(+)-6-Methoxy-α-methyl-2-naphthaleneacetic acid **(22204-53-1)**
(S)-6-Methoxy-α-methyl-2-naphthalene acetic acid **(22204-53-1)**
4-Methoxy-N-methylnaphthalimide **(3271-05-4)**
1-Methoxy-4-methyl-2-nitrobenzene **(119-10-8)**
5-Methoxy-1-methyl-4-nitroimidazole **(35687-41-3)**
α-(Methoxymethyl)-2-nitro-1H-imidazole-1-ethanol **(13551-87-6)**
α-(Methoxymethyl)-2-nitroimidazole-1-ethanol **(13551-87-6)**
(Methoxymethyl)oxirane **(930-37-0)**
3-Methoxy-5-methyl-4-oxo-2,5-hexadienoic acid **(90-65-3)**
N-(7-Methoxy-3-methyl-4-oxo-2-phenyl-4H-chromen-8-yl)-methyl-N,N-dimethylamine **(1165-48-6)**
4-Methoxy-4-methyl-2-pentanone **(107-70-0)**
4-Methoxy-4-methylpentan-2-one [UN 2293] **(107-70-0)**
(+-)-p-Methoxy-α-methylphenethylamine **(23239-32-9)**
4-Methoxy-α-methylphenethylamine **(23239-32-9)**
2-Methoxy-4-methylphenol **(93-51-6)**
N-(2-Methoxy-5-methylphenyl)acetamide **(6962-44-3)**
(4-Methoxy-3-methylphenyl)(3-methylphenyl)methanone (9CI) **(41295-28-7)**
2-Methoxy-3-(1-methylpropyl)pyrazine **(24168-70-5)**
2-Methoxy-3-(2-methylpropyl)pyrazine **(24683-00-9)**
2-Methoxy-6-methylpyrazine **(2882-21-5)**
p-Methoxy-β-methylstyrene **(104-46-1)**
N-(4-(Methoxymethyl)-1-(2-(2-thienyl)ethyl)-4-piperidinyl)-N-phenylpropanamide **(56030-54-7)**
N-(4-(Methoxymethyl)-1-(2-(2-thienyl)ethyl)-4-piperidyl)-propionanilide **(56030-54-7)**
(Methoxy(methylthio)phosphinothioyl)butanedioic acid diethyl ester **(3344-12-5)**
N-(Methoxy(methylthio)phosphinoyl)acetamide **(30560-19-1)**
Methoxyn **(300-42-5)**
1-Methoxynaphthalene **(2216-69-5)**
2-Methoxynaphthalene **(93-04-9)**
β-Methoxynaphthalene **(93-04-9)**
3-(6-Methoxy-2-naphthyl)-2,2-dimethylpentanoic acid **(517-18-0)**
(+)-2-(Methoxy-2-naphthyl)-propionic acid **(22204-53-1)**
d-2-(6-Methoxy-2-naphthyl)propionic acid **(22204-53-1)**
(+)-2-(Methoxy-2-naphthyl)-propionsaeure (German) **(22204-53-1)**
D-2-(6'-Methoxy-2'-naphthyl)-propionsaeure (German) **(22204-53-1)**
4-Methoxy-2-nitroanilin (Czech) **(96-96-8)**
2-Methoxy-4-nitroaniline **(97-52-9)**
2-Methoxy-5-nitroaniline **(99-59-2)**
4-Methoxy-2-nitroaniline **(96-96-8)**
4'-Methoxy-4-nitroazobenzene **(29418-59-5)**
2-Methoxy-5-nitrobenzenamine **(99-59-2)**
1-Methoxy-2-nitrobenzene **(91-23-6)**
2-Methoxynitrobenzene **(91-23-6)**
3-Methoxynitrobenzene **(555-03-3)**
4-Methoxynitrobenzene **(100-17-4)**
m-Methoxynitrobenzene **(555-03-3)**
p-Methoxynitrobenzene **(100-17-4)**
2-Methoxy-5-nitrobenzenediazonium **(27165-17-9)**
5-Methoxy-4-nitroimidazole **(68019-78-3)**
2-((2-Methoxy-4-nitrophenyl)azo)-o-acetoacetanisidide **(6358-31-2)**
2-((4-Methoxy-2-nitrophenyl)azo)-o-acetoacetanisidide **(6528-34-3)**
2-Methoxy-4-((4-nitrophenyl)azo)benzenamine **(101-52-0)**
4-((2-Methoxy-4-((4-nitrophenyl)azo)phenyl)azo)phenol **(19800-42-1)**
4-Methoxy-3-nitro-N-phenylbenzamide **(97-32-5)**
4-Methoxy-2-nitrophenylthiocyanate **(59607-71-5)**
(17-α)-3-Methoxy-19-norpregna-1,3,5(10)-trien-20-yn-17-ol **(72-33-3)**
3-Methoxy-17-α-19-norpregna-1,3,5(10)-trien-20-yn-17-ol **(72-33-3)**
3-Methoxy-19-nor-17-α-pregna-1,3,5(10)-trien-20-yn-17-ol **(72-33-3)**
(E,Z)-(2S,3S)-4-Methoxy-3-(1-octenyl-ONN-azoxy)-2-butanol **(23315-05-1)**
4-Methoxy-3-(1-octenylazoxy)-2-butanol **(499-48-9)**
S-5-Methoxy-4-oxopyran-2-ylmethyl dimethyl phosphorothioate **(2778-04-3)**
S-((5-Methoxy-2-oxo-1,3,4-thiadiazol-3(2H)-yl)methyl) O,O-dimethyl phosphorodithioate **(950-37-8)**
Methoxypectin **(9000-69-5)**
3-Methoxypentane **(36839-67-5)**
5-Methoxy-1-pentanol **(4799-62-6)**
2-Methoxyphenol **(90-05-1)**
3-Methoxyphenol **(150-19-6)**
m-Methoxyphenol **(150-19-6)**
o-Methoxyphenol **(90-05-1)**
p-Methoxyphenol **(150-76-5)**
4-Methoxyphenol (ACGIH,OSHA) **(150-76-5)**
Methoxyphenone **(41295-28-7)**
(o-Methoxyphenoxy)acetic acid **(1878-85-9)**
3-(2-Methoxyphenoxy)-1-glyceryl carbamate **(532-03-6)**
3-(o-Methoxyphenoxy)-2-hydroxypropyl carbamate **(532-03-6)**
2-(p-Methoxyphenoxymethyl)oxirane **(2211-94-1)**
N-(4-(4-Methoxyphenoxy)phenyl)-N,N-dimethylurea **(14214-32-5)**
3-(2-Methoxyphenoxy)-1,2-propanediol **(93-14-1)**
3-(o-Methoxyphenoxy)-1,2-propanediol **(93-14-1)**
3-o-Methoxyphenoxypropane 1:2-diol **(93-14-1)**
3-(o-Methoxyphenoxy)-1,2-propanediol 1-carbamate **(532-03-6)**
(o-Methoxyphenyl)acetic acid **(93-25-4)**
4-Methoxyphenylacetic acid **(104-01-8)**
p-Methoxyphenylacetic acid **(104-01-8)**
4-Methoxyphenylacetonitrile **(104-47-2)**
p-Methoxyphenylacetonitrile **(104-47-2)**
β-(o-Methoxyphenyl)acrolein **(1504-74-1)**
o-Methoxyphenylamine **(90-04-0)**
p-Methoxyphenylamine **(104-94-9)**
o-Methoxyphenylamine hydrochloride **(134-29-2)**
p-Methoxyphenylamine hydrochloride **(20265-97-8)**
3-(o-Methoxyphenylaminocarbonyl)-2-naphthol **(135-62-6)**
((2-Methoxyphenyl)amino)methanesulfonic acid **(93-13-0)**
2-Methoxyphenyl bromide **(578-57-4)**
4-Methoxyphenyl bromide **(104-92-7)**
o-Methoxyphenyl bromide **(578-57-4)**
p-Methoxyphenyl bromide **(104-92-7)**
3-(2-Methoxyphenylcarbamoyl)-2-naphthol **(135-62-6)**
1-(2-Methoxy-5-phenylcarbamoylphenylazo)-2-hydroxy-3-(3-nitrophenylcarbamoyl)naphthalene **(6448-96-0)**
p-Methoxyphenylcarbanilate **(19219-48-8)**
2-Methoxy-p-phenylenediamine **(5307-02-8)**
4-Methoxy-m-phenylenediamine **(615-05-4)**
p-Methoxy-m-phenylenediamine **(615-05-4)**
4-Methoxy-m-phenylenediamine sulfate **(39156-41-7)**
4-Methoxy-m-phenylenediamine sulphate **(39156-41-7)**
p-Methoxy-m-phenylenediamine sulphate **(39156-41-7)**
9-(p-Methoxyphenyl)fluorene **(21846-08-2)**
o-Methoxyphenyl glyceryl ether **(93-14-1)**
Methoxyphenyl glycidyl ether **(2211-94-1)**
p-Methoxyphenyl glycidyl ether **(2211-94-1)**
2-(p-Methoxyphenyl)-1,3-indandione **(117-37-3)**
2-(p-Methoxyphenyl)indane-1,3-dione **(117-37-3)**
2-(4-Methoxyphenyl)-1H-indene-1,3(2H)-dione **(117-37-3)**
2-p-Methoxyphenylmethyl-3-acetoxy-4-hydroxypyrrolidine **(22862-76-6)**
4-Methoxyphenyl methyl ketone **(100-06-1)**
p-Methoxyphenyl methyl ketone **(100-06-1)**
2-(p-Methoxyphenyl)-2-(p-(methylthio)phenyl)-1,1,1-trichloroethane **(34197-16-5)**
4-Methoxyphenyl N-phenylcarbamate **(19219-48-8)**
4-Methoxyphenyl phenylcarbamate **(19219-48-8)**
(E)-3-(4-Methoxyphenyl)-2-propenoic acid **(943-89-5)**
3-(4-Methoxyphenyl)-2-propenoic acid **(830-09-1)**
3-(4-Methoxyphenyl)-2-propenoic acid 2-ethoxyethyl ester **(104-28-9)**
3-(4-Methoxyphenyl)-2-propenoic acid, 2-ethylhexyl ester **(5466-77-3)**
Methoxy polyethylene glycol 350 **(9004-74-4)**
Methoxy polyethylene glycol 550 **(9004-74-4)**

Name	CAS No.
Methoxy polyethylene glycol 750 (9004-74-4)	**(9004-74-4)**
Methoxypoly(ethylene glycol)	**(9004-74-4)**
O-Methoxypolyethylene glycol	**(9004-74-4)**
α,ω-Methoxypoly(ethylene oxide)	**(24991-55-7)**
2-Methoxy-4-prop-2-enylphenol	**(97-53-0)**
1-Methoxypropane	**(557-17-5)**
α-Methoxy propane	**(557-17-5)**
Methoxypropanediol	**(93-14-1)**
3-Methoxypropannitril (Czech)	**(110-67-8)**
1-Methoxy-2-propanol	**(107-98-2)**
2-Methoxy-1-propanol	**(1589-47-5)**
Methoxy-1-propanol	**(28677-93-2)**
1-Methoxy-2-propanone	**(5878-19-3)**
Methoxy-2-propanone	**(5878-19-3)**
Methoxypropazine	**(1610-18-0)**
2-Methoxypropene	**(116-11-0)**
1-Methoxy-4-(1-propenyl)benzene	**(104-46-1)**
1-Methoxy-4-(2-propenyl)benzene	**(140-67-0)**
1-Methoxy-4-propenylbenzene	**(104-46-1)**
4-Methoxypropenylbenzene	**(104-46-1)**
2-Methoxy-4-(2-propenyl)phenol	**(97-53-0)**
2-Methoxy-4-propenylphenol	**(97-54-1)**
cis-2-Methoxy-4-propenylphenol	**(5912-86-7)**
trans-2-Methoxy-4-propenylphenol	**(5932-68-3)**
3-Methoxypropionic acid methyl ester	**(3852-09-3)**
β-Methoxypropionic acid, methyl ester	**(3852-09-3)**
3-Methoxypropionitrile	**(110-67-8)**
1-(2-Methoxypropoxy)-2-propanol	**(13429-07-7)**
2-Methoxy-1-propyl acetate	**(70657-70-4)**
2-Methoxypropylacetate-1	**(70657-70-4)**
3-Methoxypropylamine	**(5332-73-0)**
1-Methoxy-4-propylbenzene	**(104-45-0)**
3-Methoxypropylene oxide	**(930-37-0)**
2-Methoxy-4-propylphenol	**(2785-87-7)**
5-Methoxypsoralen	**(484-20-8)**
8-Methoxypsoralen	**(298-81-7)**
9-Methoxypsoralen	**(298-81-7)**
8-Methoxypsoralene	**(298-81-7)**
Methoxypyrazine	**(3149-28-8)**
N^1-(6-Methoxy-3-pyridazinyl)sulfanilamide	**(80-35-3)**
S-((5-Methoxy-4H-pyron-2-yl)-methyl)-O,O-dimethyl-monothio-fosfaat (Dutch)	**(2778-04-3)**
S-((5-Methoxy-4H-pyron-2-yl)-methyl)-O,O-dimethyl-monothio-phosphat (German)	**(2778-04-3)**
S-(5-Methoxy-4-pyron-2-ylmethyl) dimethyl phosphorothiolate	**(2778-04-3)**
6-Methoxyquinoline	**(5263-87-6)**
α-(6-Methoxy-4-quinolyl)-5-vinyl-2-quinuclidinemethanol	**(130-95-0)**
α-(6-Methoxy-4-quinolyl)-5-vinyl-2-quinuclidinemethanol	**(56-54-2)**
3-Methoxysalicylaldehyde	**(148-53-8)**
3-Methoxysalicylic acid	**(877-22-5)**
6-Methoxy-3-sulfanilamidopyridazine	**(80-35-3)**
3-Methoxy-6-sulfanylamidopyridazine	**(80-35-3)**
4-Methoxy-2-sulfoaniline	**(13244-33-2)**
2-Methoxy-3,4,5,6-tetrachlorophenol	**(2539-17-5)**
2-Methoxytetrachlorophenol	**(2539-17-5)**
Methoxythiocarbonyl chloride	**(2812-72-8)**
3-Methoxytoluene	**(100-84-5)**
4-Methoxytoluene	**(104-93-8)**
α-Methoxytoluene	**(538-86-3)**
m-Methoxytoluene	**(100-84-5)**
p-Methoxytoluene	**(104-93-8)**
4-Methoxy-m-toluidine	**(120-71-8)**
2-Methoxy-1,3,5-triazine	**(17635-40-4)**
2-Methoxy-3,5,6-trichlorobenzoic acid	**(2307-49-5)**
2-Methoxy-trichlorophenol	**(57057-83-7)**
Methoxytriglycol	**(112-35-6)**
Methoxytrimethylsilane	**(1825-61-2)**
N-Methoxyurethane	**(1616-88-2)**
6-Methoxy-α-(5-vinyl-2-quinuclidinyl)-4-quinolinemethanol	**(56-54-2)**
Methozin	**(60-80-0)**
Methphenoxydiol	**(93-14-1)**
Methriol	**(77-85-0)**
Methscopolamine bromide	**(155-41-9)**
Methsuximide	**(77-41-8)**
Methulose	**(9004-67-5)**
Methural	**(140-95-4)**
Methurin (Russian)	**(140-95-4)**
Methvtiolo (Italian)	**(74-93-1)**
Methychlothiazide	**(135-07-9)**
Methyclothiazide	**(135-07-9)**
Methycyclothiazide	**(135-07-9)**
Methyl-CCNU	**(13909-09-6)**
trans-Methyl-CCNU	**(13909-09-6)**
Methyl Cellulose-A	**(9004-67-5)**
3'-Methyl-DAB	**(55-80-1)**
Methyl-DBCP	**(10474-14-3)**
α-Methyl-L-DOPA	**(555-30-6)**
Methyl-Demeton-O	**(867-27-6)**
Methyl 2,4-D ester	**(1928-38-7)**
Methyl Dopa Sesquihydrate	**(41372-08-1)**
Methyl-E 605	**(298-00-0)**
Methyl-E-600	**(950-35-6)**
Methyl Ester of Wood Rosin	**(127-25-3)**
Methyl Green	**(82-94-0)**
Methyl Green Chloride	**(82-94-0)**
6-Methyl MP-Riboside	**(342-69-8)**
Methyl Mustard Oil	**(556-61-6)**
Methyl Orange	**(547-58-0)**
Methyl Orange B	**(547-58-0)**
Methyl PCT	**(2524-03-0)**
Methyl Red	**(493-52-7)**
Methyl Systox	**(8022-00-2)**
Methyl Topsin	**(23564-05-8)**
Methyl Tuads	**(137-26-8)**
Methyl Violet	**(8004-87-3)**
Methyl Violet 10B	**(548-62-9)**
Methyl Violet 2B	**(8004-87-3)**
Methyl Violet BB	**(8004-87-3)**
Methyl Violet 10BD	**(548-62-9)**
Methyl Violet 10BK	**(548-62-9)**
Methyl Violet 10BN	**(548-62-9)**
Methyl Violet 5BNO	**(548-62-9)**
Methyl Violet 10BNS	**(548-62-9)**
Methyl Violet 5BO	**(548-62-9)**
Methyl Violet 10BO	**(548-62-9)**
Methyl Violet FN	**(8004-87-3)**
Methyl Violet Lake	**(1325-82-2)**
Methyl Violet N	**(8004-87-3)**
Methyl Violet PMA Lake VA-4150	**(1325-82-2)**
Methyl-Violett (German)	**(8004-87-3)**
Methyl Viologen (2+)	**(4685-14-7)**
Methyl Yellow	**(60-11-7)**
Methyl Zimate	**(137-30-4)**
Methyl Zineb	**(137-30-4)**
Methyl Ziram	**(137-30-4)**
Methyl abietate	**(127-25-3)**
Methylacetaat (Dutch)	**(79-20-9)**
Methylacetaldehyde	**(123-38-6)**
Methylacetamide	**(79-16-3)**
N-Methylacetamide	**(79-16-3)**
2'-Methylacetanilide	**(120-66-1)**
2-Methylacetanilide	**(120-66-1)**
3'-Methylacetanilide	**(537-92-8)**
3-Methylacetanilide	**(537-92-8)**

4'-Methylacetanilide **(103-89-9)**
4-Methylacetanilide **(103-89-9)**
m-Methylacetanilide **(537-92-8)**
o-Methylacetanilide **(120-66-1)**
p-Methylacetanilide **(103-89-9)**
5'-Methyl-o-acetanisidide **(6962-44-3)**
Methylacetaphos **(2088-72-4)**
Methylacetat (German) **(79-20-9)**
Methyl acetate (ACGIH,OSHA) [UN 1231] **(79-20-9)**
Methyl acetic acid **(79-09-4)**
Methylacetic anhydride **(123-62-6)**
2'-Methylacetoacetanilide **(93-68-5)**
Methylacetoacetate **(105-45-3)**
3-Methylacetoin **(115-22-0)**
Methylacetoin **(115-22-0)**
Methyl acetone **(8013-65-8)**
Methyl acetone (DOT) **(78-93-3)**
Methylacetone (French) **(8013-65-8)**
2'-Methylacetophenone **(577-16-2)**
2-Methylacetophenone **(577-16-2)**
4'-Methylacetophenone **(122-00-9)**
o-Methylacetophenone **(577-16-2)**
p-Methylacetophenone **(122-00-9)**
Methyl acetophos **(2088-72-4)**
Methylacetopyronone **(520-45-6)**
Methyl acetoxon **(2088-72-4)**
Methyl(acetoxymethyl)nitrosamine **(56856-83-8)**
Methyl 12-acetoxy-9-octadecenoate **(140-03-4)**
Methyl 12-acetoxyoleate **(140-03-4)**
6-Methyl-17-α-acetoxypregna-4,6-diene-3,20-dione **(595-33-5)**
6-α-Methyl-17-α-acetoxypregn-4-ene-3,20-dione **(71-58-9)**
6-α-Methyl-17-acetoxy progesterone **(71-58-9)**
6-α-Methyl-17-α-acetoxyprogesterone **(71-58-9)**
Methyl acetylacetate **(105-45-3)**
Methyl acetylacetonate **(105-45-3)**
Methyl 2-(acetylamino)benzoate **(2719-08-6)**
Methyl N-acetylanthranilate **(2719-08-6)**
1-Methyl-4-acetylbenzene **(122-00-9)**
Methylacetyl choline **(62-51-1)**
β-Methylacetylcholine chloride **(62-51-1)**
Methyl acetylene (ACGIH,OSHA) **(74-99-7)**
Methyl acetylene and propadiene mixture, Stabilized [UN 1060] **(59355-75-8)**
Methyl acetylene and propadiene mixtures **(59355-75-8)**
Methylacetylene et propadiene en melange (French) **(59355-75-8)**
Methyl acetylene propadiene, Stabilized **(59355-75-8)**
Methylacetylene-propadiene, Stabilized **(59355-75-8)**
Methyl acetylene propadiene mixture **(59355-75-8)**
Methyl acetylene-propadiene mixture (ACGIH,OSHA) **(59355-75-8)**
2'-Methylacetylphenone **(577-16-2)**
2-Methyl-3-acetylpyrazine **(23787-80-6)**
Methyl acetyl ricinoleate **(140-03-4)**
2-Methylacrolein **(78-85-3)**
α-Methylacrolein **(78-85-3)**
β-Methyl acrolein **(123-73-9)**
β-Methylacrolein **(4170-30-3)**
β-Methyl acrolein (DOT) **(4170-30-3)**
Methylacrylaat (Dutch) **(96-33-3)**
Methylacrylaldehyde **(78-85-3)**
2-Methylacrylamide **(79-39-0)**
N-Methylacrylamide **(1187-59-3)**
Methyl-acrylat (German) **(96-33-3)**
Methyl acrylate (ACGIH,OSHA) **(96-33-3)**
Methyl acrylate, Inhibited [UN 1919] **(96-33-3)**
3-Methylacrylic acid **(3724-65-0)**
α-Methylacrylic acid **(79-41-4)**
β-Methylacrylic acid **(3724-65-0)**
α-Methylacrylic acid, 3,4-dichloroanilide **(2164-09-2)**
α-Methyl acrylic amide **(79-39-0)**
α-Methylacrylonitrile **(126-98-7)**
β-Methylacrylonitrile **(4786-20-3)**
Methylacrylonitrile (ACGIH,OSHA) **(126-98-7)**
Methylacryloyl chloride **(920-46-7)**
α-Methylacryloyl chloride **(920-46-7)**
1-Methyladenosine **(15763-06-1)**
N1-Methyladenosine **(15763-06-1)**
Methyl adipate **(627-93-0)**
4-Methylaesculetin **(529-84-0)**
Methylaethylnitrosamin (German) **(10595-95-6)**
Methylal (ACGIH,OSHA) [UN 1234] **(109-87-5)**
Methyl alcohol (ACGIH,OSHA) [UN 1230] **(67-56-1)**
Methyl aldehyde **(50-00-0)**
Methylalkohol (German) **(67-56-1)**
1-Methylallene **(590-19-2)**
Methylallene **(590-19-2)**
Methylallyl acetone **(25659-22-7)**
2-Methyl-allylchlorid (German) **(563-47-3)**
1-Methylallyl chloride **(563-52-0)**
2-Methylallyl chloride **(563-47-3)**
α-Methylallyl chloride **(563-52-0)**
β-Methylallyl chloride **(563-47-3)**
γ-Methylallyl chloride **(591-97-9)**
Methyl allyl chloride [UN 2554] **(563-47-3)**
1-Methyl-5-allyl-5-(1-methyl-2-pentynyl)barbituric acid sodium salt **(309-36-4)**
2-(2-Methylallyl)-6-nitrophenol **(13414-58-9)**
Methylallylnitrosamin (German) **(4549-43-3)**
Methylallylnitrosamine **(4549-43-3)**
Methyl aluminium sesquibromide (DOT) **(12263-85-3)**
Methyl aluminium sesquichloride **(12542-85-7)**
Methyl aluminum sesquibromide (DOT) **(12263-85-3)**
Methyl aluminum sesquichloride **(12542-85-7)**
Methylaluminum sesquichloride **(12542-85-7)**
Methylamine (ACGIH,OSHA) **(74-89-5)**
Methylamine, Anhydrous [UN 1061] **(74-89-5)**
Methylamine, Aqueous solution UN [1235] **(74-89-5)**
Methylamine, Compd. with trinitromethane **(14147-71-8)**
Methylamine, 1-acetoxy-n-butyl-N-nitroso- **(56986-36-8)**
Methylamine, N-acetoxymethyl-N-nitroso- **(56856-83-8)**
Methylamine, N,N-bis(3-aminopropyl)- **(105-83-9)**
Methylamine, 1-(2-furyl)- **(617-89-0)**
Methylamine, N-methoxy- **(1117-97-1)**
Methylaminen (Dutch) **(74-89-5)**
Methylamine nitrate **(22113-87-7)**
Methylamine nitroform **(14147-71-8)**
Methylamine, m-phenylenebis **(1477-55-0)**
Methylamine, 1,1,1-tris(hydroxymethyl)- **(77-86-1)**
2-(Methylamino)acetaldehyde dimethyl acetal **(122-07-6)**
Methylaminoacetaldehyde dimethyl acetal **(122-07-6)**
N-Methylaminoacetaldehyde dimethyl acetal **(122-07-6)**
(Methylamino)acetic acid **(107-97-1)**
N-Methylaminoacetic acid **(107-97-1)**
4-Methyl-2-aminoanisole **(120-71-8)**
1-(Methylamino)-9,10-anthracenedione **(82-38-2)**
1-(Methylamino)-9,10-anthraquinone **(82-38-2)**
1-(Methylamino)anthraquinone **(82-38-2)**
1-(N-Methylamino)-9,10-anthraquinone **(82-38-2)**
1-(N-Methylamino)anthraquinone **(82-38-2)**
α-Methylaminoanthraquinone **(82-38-2)**
(Methylamino)benzene **(100-61-8)**
1-Methyl-2-aminobenzene **(95-53-4)**
2-Methyl-1-aminobenzene **(95-53-4)**
N-Methylaminobenzene **(100-61-8)**
1-Methyl-2-aminobenzene hydrochloride **(636-21-5)**
2-Methyl-1-aminobenzene hydrochloride **(636-21-5)**

Methyl N-(4-aminobenzenesulfonyl)carbamate **(3337-71-1)**
Methyl 4-aminobenzenesulphonyl carbamate **(3337-71-1)**
Methyl 2-aminobenzoate **(134-20-3)**
Methyl o-aminobenzoate **(134-20-3)**
Methyl p-aminobenzoate **(619-45-4)**
2-Methylaminobenzothiazole **(16954-69-1)**
4-Methyl-2-aminobenzothiazole **(1477-42-5)**
N-(((Methylamino)carbonyl)oxy)ethanimidothioic acid methyl ester **(16752-77-5)**
N-Methylaminodiglycol **(105-59-9)**
(Methylamino)ethanoic acid **(107-97-1)**
1-Methyl-2-aminoethanol **(78-96-6)**
2-Methylaminoethanol **(109-83-1)**
N-Methylaminoethanol **(109-83-1)**
β-(Methylamino)ethanol **(109-83-1)**
1-(Methylamino)-4-ethanolaminoanthraquinone **(86722-66-9)**
1-Methylamino-4-ethanolaminoanthraquinone **(2475-46-9)**
l-Methylaminoethanolcathechol hydrochloride **(55-31-2)**
m-Methylaminoethanolphenol **(59-42-7)**
m-Methylaminoethanolphenol hydrochloride **(61-76-7)**
(-)-α-(1-Methylaminoethyl)benzyl alcohol **(299-42-3)**
α-(1-(Methylamino)ethyl)benzyl alcohol **(90-82-4)**
l-α-(1-Methylaminoethyl)benzyl alcohol **(299-42-3)**
1-α-(1-(Methylamino)ethyl)benzyl alcohol sulfate **(134-72-5)**
1-(Methylamino)-4-(2-hydroxyethylamino)anthraquinone **(86722-66-9)**
1-(Methylamino)-4-(β-hydroxyethylamino)anthraquinone **(86722-66-9)**
1-Methylamino-4-(β-hydroxyethylamino)anthraquinone **(2475-46-9)**
2-Methylamino-4-isopropylamino-6-chloro-s-triazine **(3004-71-5)**
2-Methyl-4-amino-6-methoxy-s-triazine **(1668-54-8)**
2-Methylamino methyl benzoate **(85-91-6)**
1-Methylamino-4-(4-methylphenylamino)anthraquinone **(128-85-8)**
2-Methylamino-2-methyl-1-phenylpropane **(100-92-5)**
2-Methylamino-4-methylthio-6-isopropylamino-1,3,5-triazine **(1014-69-3)**
p-(Methylamino)nitrobenzene **(100-15-2)**
N-Methyl-amino-2-nitro-4-N',N'-bis-(2-hydroxyethyl)-amino-benzene **(2784-94-3)**
2,2'-((4-(Methylamino)-3-nitrophenyl)imino)diethanol **(2784-94-3)**
1-Methylamino-4-oxyethylaminoanthraquinone (Russian) **(2475-46-9)**
4-(Methylamino)phenol **(150-75-4)**
p-(Methylamino)phenol **(150-75-4)**
Methyl-p-aminophenol sulfate **(55-55-0)**
p-Methylaminophenolsulfate **(55-55-0)**
d-psi-2-Methylamino-1-phenyl-1-propanol **(90-82-4)**
l-2-Methylamino-1-phenylpropanol **(299-42-3)**
Methyl ((4-aminophenyl)sulfonyl)carbamate **(3337-71-1)**
Methyl 4-aminophenylsulphonyl carbamate **(3337-71-1)**
3-Methyl-4-amino-6-phenyl-1,2,4-triazin(4H)-on (German) **(41394-05-2)**
5-(3-Methylaminopropyl)-5H-dibenzo(a,d)cycloheptene **(438-60-8)**
5-(3-(Methylamino)propylidene)dibenzo(a,e)cyclohepta(1,5)diene **(72-69-5)**
5-(3-Methylaminopropylidene)-10,11-dihydro-5H-dibenzo-(a,d)cycloheptene **(72-69-5)**
Methylaminopropyliminodibenzyl **(50-47-5)**
Methylaminopterin **(59-05-2)**
Methylaminopterinum **(59-05-2)**
1-Methyl-3-amino-5H-pyrido(4,3-b)indole **(62450-07-1)**
1-(Methylamino)-4-p-toluidinoanthraquinone **(128-85-8)**
Methylamphetamine **(300-42-5)**
Methylamphetamine hydrochloride **(300-42-5)**
Methyl amyl acetate **(108-84-9)**
Methylamyl acetate [UN 1233] **(108-84-9)**
Methyl amyl alcohol **(108-11-2)**
Methyl amyl alcohol **(54972-97-3)**
Methylamyl alcohol **(105-30-6)**
Methyl amyl alcohol (OSHA) **(108-11-2)**
Methyl amyl carbinol **(543-49-7)**
Methyl-amyl-cetone (French) **(110-43-0)**
Methyl n-amyl ketone (ACGIH,OSHA) **(110-43-0)**
Methyl amyl ketone [UN 1110] **(110-43-0)**
Methylamylnitrosamin (German) **(13256-07-0)**
Methyl-N-amylnitrosamine **(13256-07-0)**
Methylamylnitrosamine **(13256-07-0)**
17-Methyl-2'H-5α-androst-2-eno(3,2-c)pyrazol-17β-ol **(10418-03-8)**
17-Methyl-2H-5α-androst-2-eno(3,2-c)pyrazol-17β-ol **(10418-03-8)**
4'-Methylangelicin **(78982-40-8)**
5-Methylangelicin **(73459-03-7)**
9-Methylangelicin **(78982-40-8)**
2-Methylaniline **(95-53-4)**
3-Methylaniline **(108-44-1)**
4-Methylaniline **(106-49-0)**
Methylaniline **(100-61-8)**
m-Methylaniline **(108-44-1)**
o-Methylaniline **(95-53-4)**
p-Methylaniline **(106-49-0)**
N-Methylaniline (ACGIH) [UN 2294] **(100-61-8)**
2-(N-Methylaniline)ethanol **(93-90-3)**
2-Methylaniline hydrochloride **(636-21-5)**
4-Methylaniline hydrochloride **(540-23-8)**
o-Methylaniline hydrochloride **(636-21-5)**
2-(N-Methylanilino)ethanol **(93-90-3)**
3-(N-Methylanilino)propionitrile **(94-34-8)**
β-(N-Methylanilino)propionitrile **(94-34-8)**
β-N-Methylanilinopropionitrile **(94-34-8)**
Methyl p-anisate **(121-98-2)**
2-Methyl-p-anisidine **(102-50-1)**
5-Methyl-o-anisidine **(120-71-8)**
3-Methylanisole **(100-84-5)**
m-Methylanisole **(100-84-5)**
p-Methylanisole **(104-93-8)**
1-Methylanthracene **(610-48-0)**
2-Methylanthracene **(613-12-7)**
9-Methylanthracene **(779-02-2)**
2-Methyl-9,10-anthracenedione **(84-54-8)**
3-Methyl-1,8,9-anthracenetriol **(491-59-8)**
3-Methylanthralin **(491-59-8)**
Methyl anthranilate **(134-20-3)**
N-Methylanthranilic acid, methyl ester **(85-91-6)**
2-Methylanthraquinone **(84-54-8)**
β-Methylanthraquinone **(84-54-8)**
2-Methyl-1-anthraquinonylamine **(82-28-0)**
N-Methyl-1-anthraquinonylamine **(82-38-2)**
Methyl aphoxide **(57-39-6)**
Methylarsenic acid **(124-58-3)**
Methylarsenic acid, sodium salt **(2163-80-6)**
Methylarsenic dimethyl dithiocarbamate **(2445-07-0)**
Methylarsenic sulfide **(2533-82-6)**
Methyl arsine-bis(dimethyldithiocarbamate) **(2445-07-0)**
Methylarsine dichloride **(593-89-5)**
Methylarsine sulfide **(2533-82-6)**
Methylarsinic acid **(124-58-3)**
Methylarsinic sulfide **(2533-82-6)**
Methylarsinic sulphide **(2533-82-6)**
Methylarsonat disodny (Czech) **(144-21-8)**
Methylarsonat monosodny (Czech) **(2163-80-6)**
Methylarsonic acid **(124-58-3)**
Methyl-arsonous dichloride **(593-89-5)**
Methylarsonous dichloride **(593-89-5)**
Methylarterenol **(51-43-4)**
Methyl aspartylphenylalanate **(22839-47-0)**
Methylatropine **(2870-71-5)**
Methylatropine bromide **(2870-71-5)**
N-Methylatropine bromide **(2870-71-5)**
8-Methylatropinium bromide **(2870-71-5)**
Methylatropinium bromide **(2870-71-5)**
endo-8-Methyl-8-azabicyclo(3.2.1)octan-3-ol **(120-29-6)**

exo-8-Methyl-8-azabicyclo(3.2.1)-octan-3-ol benzoate **(537-26-8)**
2-Methylazacyclopropane **(75-55-8)**
Methylazinphos **(86-50-0)**
2-Methylaziridine **(75-55-8)**
2-Methylazobenzene **(6676-90-0)**
Methylazoxymethanol **(590-96-5)**
Methylazoxymethanol acetate **(592-62-1)**
(Methyl-ONN-azoxy)methanol, acetate (ester) **(592-62-1)**
Methylazoxymethanol glucoside **(14901-08-7)**
Methylazoxymethanol-β-D-glucoside **(14901-08-7)**
Methylazoxymethyl acetate **(592-62-1)**
Methylazoxymethylester kyseliny octove (Czech) **(592-62-1)**
(Methyl-ONN-azoxy)-methyl-β-D-glucopyranoside **(14901-08-7)**
1-Methylbarbital **(50-11-3)**
Methylbarbital **(50-11-3)**
N-Methylbarbital **(50-11-3)**
Methyl behenate **(929-77-1)**
Methylben **(99-76-3)**
1-Methylbenz(a)anthracene **(2498-77-3)**
2-Methylbenz(a)anthracene **(2498-76-2)**
3-Methylbenz(a)anthracene **(2498-75-1)**
4-Methylbenz(a)anthracene **(316-49-4)**
7-Methylbenz(a)anthracene **(2541-69-7)**
8-Methylbenz(a)anthracene **(2381-31-9)**
9-Methylbenz(a)anthracene **(2381-16-0)**
10-Methylbenz(a)anthracene **(2381-15-9)**
11-Methylbenz(a)anthracene **(6111-78-0)**
12-Methylbenz(a)anthracene **(2422-79-9)**
2-Methylbenzaldehyde **(529-20-4)**
3-Methylbenzaldehyde **(620-23-5)**
4-Methylbenzaldehyde **(104-87-0)**
o-Methylbenzaldehyde **(529-20-4)**
p-Methylbenzaldehyde **(104-87-0)**
para-Methylbenzaldehyde **(104-87-0)**
Methyl benzaldehyde-4-carboxylate **(1571-08-0)**
2-Methylbenzamide **(527-85-5)**
4-Methylbenzamide **(619-55-6)**
N-Methylbenzamide **(613-93-4)**
o-Methylbenzamide **(527-85-5)**
1-(2-Methyl-5-benzamide)azo-2-hydroxy-3-naphthanalide **(16403-84-2)**
2-Methylbenzanilide **(7055-03-0)**
o-Methylbenzanilide **(7055-03-0)**
10-Methyl-1,2-benzanthracen (German) **(2541-69-7)**
1'-Methyl-1,2-benzanthracene **(2498-77-3)**
2'-Methyl-1,2-benzanthracene **(2498-76-2)**
4'-Methyl-1:2-benzanthracene **(316-49-4)**
5-Methyl-1,2-benzanthracene **(2381-31-9)**
6-Methyl-1,2-benzanthracene **(2381-16-0)**
7-Methyl-1,2-benzanthracene **(2381-15-9)**
8-Methyl-1:2-benzanthracene **(6111-78-0)**
9-Methyl-1,2-benzanthracene **(2422-79-9)**
10-Methyl-1,2-benzanthracene **(2541-69-7)**
N-Methylbenzazimide, dimethyldithiophosphoric acid ester **(86-50-0)**
2-Methyl-7H-benz(de)anthracen-7-one **(82-03-1)**
Methylbenzedrin **(300-42-5)**
Methyl 1H-benzemedazol-2-ylcarbamate **(10605-21-7)**
N-Methylbenzenamide **(613-93-4)**
2-Methylbenzenamine **(95-53-4)**
3-Methylbenzenamine **(108-44-1)**
4-Methylbenzenamine **(106-49-0)**
N-Methylbenzenamine **(100-61-8)**
ar-Methylbenzenamine **(26915-12-8)**
m-Methylbenzenamine **(108-44-1)**
o-Methylbenzenamine **(95-53-4)**
p-Methylbenzenamine **(106-49-0)**
2-Methylbenzenamine hydrochloride **(636-21-5)**
o-Methylbenzenamine hydrochloride **(636-21-5)**
Methylbenzene **(108-88-3)**
4-Methylbenzeneacetaldehyde **(104-09-6)**
N-Methylbenzeneacetamide **(6830-82-6)**
Methyl benzeneacetate **(101-41-7)**
4-Methylbenzeneacetic acid **(622-47-9)**
α-Methylbenzeneacetic acid **(492-37-5)**
α-Methylbenzeneacetonitrile **(1823-91-2)**
2-Methylbenzenecarbonitrile **(529-19-1)**
Methyl benzenecarboxylate **(93-58-3)**
2-Methyl-1,4-benzenediamine **(95-70-5)**
4-Methyl-1,3-benzenediamine **(95-80-7)**
5-Methyl-1,3-benzenediamine **(108-71-4)**
4-Methyl-1,3-benzenediamine monohydrochloride **(5459-85-8)**
2-Methyl-1,4-benzenediamine sulfate **(615-50-9)**
3-Methyl-1,2-benzenedicarboxylic acid **(37102-74-2)**
4-Methyl-1,2-benzenedicarboxylic acid **(4316-23-8)**
2-Methyl-1,3-benzenediol **(608-25-3)**
2-Methyl-1,4-benzenediol **(95-71-6)**
4-Methyl-1,3-benzenediol **(496-73-1)**
5-Methyl-1,3-benzenediol **(504-15-4)**
α-Methylbenzeneethanamine sulfate **(51-63-8)**
α-Methylbenzeneethaneamine **(300-62-9)**
(+-)-α-Methylbenzenemethanamine **(618-36-0)**
(R)-α-Methylbenzenemethanamine **(3886-69-9)**
(S)-α-Methylbenzenemethanamine **(2627-86-3)**
N-Methylbenzenemethanamine **(103-67-3)**
α-Methylbenzenemethanamine **(98-84-0)**
4-Methyl-benzenemethanol (9CI) **(589-18-4)**
α-Methylbenzenemethanol acetate **(93-92-5)**
Methyl benzenepropanoate **(103-25-3)**
4-Methylbenzenesulfinic acid **(536-57-2)**
2-Methylbenzenesulfonamide **(88-19-7)**
4-Methylbenzenesulfonamide **(70-55-3)**
ar-Methylbenzenesulfonamide **(1333-07-9)**
o-Methylbenzenesulfonamide **(88-19-7)**
p-Methylbenzenesulfonamide **(70-55-3)**
Methyl benzenesulfonate **(80-18-2)**
2-Methylbenzenesulfonic acid **(88-20-0)**
4-Methylbenzenesulfonic acid **(104-15-4)**
p-Methylbenzenesulfonic acid **(104-15-4)**
Methylbenzenesulfonic acid, sodium salt **(12068-03-0)**
2-Methylbenzenesulfonyl chloride **(133-59-5)**
4-Methylbenzenesulfonyl chloride **(98-59-9)**
p-Methylbenzenesulfonyl chloride **(98-59-9)**
4-Methylbenzenesulfonyl isocyanate **(4083-64-1)**
2-Methylbenzenethiol **(137-06-4)**
3-Methylbenzenethiol **(108-40-7)**
4-Methylbenzenethiol **(106-45-6)**
m-Methylbenzenethiol **(108-40-7)**
o-Methylbenzenethiol **(137-06-4)**
p-Methylbenzenethiol **(106-45-6)**
6-Methyl-1,2,4-benzenetriol **(767-81-7)**
Methylbenzethonii chloridum (Latin) **(25155-18-4)**
Methylbenzethonium **(25155-18-4)**
Methylbenzethonium chloride **(25155-18-4)**
2-Methylbenzimidazole **(615-15-6)**
Methyl-2-benzimidazole **(615-15-6)**
Methyl 2-benzimidazolecarbamate **(10605-21-7)**
Methyl (2-benzimidazole)carbamate phosphate **(52316-55-9)**
Methyl-2-benzimidazolecarbamate phosphate **(52316-55-9)**
Methyl benzimidazole-2-yl carbamate **(10605-21-7)**
Methyl benzimidazol-2-yl carbamate **(10605-21-7)**
2-Methyl-1,2-benzisothiazole-3(2H)-one-1,1-dioxide **(15448-99-4)**
2-Methylbenzoanilide **(7055-03-0)**
2-Methylbenzo(a)pyrene **(16757-82-7)**
3-Methylbenzo(a)pyrene **(16757-81-6)**
4-Methylbenzo(a)pyrene **(16757-83-8)**

6-Methylbenzo(a)pyrene **(2381-39-7)**
11-Methylbenzo(a)pyrene **(16757-80-5)**
12-Methylbenzo(a)pyrene **(4514-19-6)**
Methylbenzoate **(93-58-3)**
Methyl benzoate [UN 2938] **(93-58-3)**
6-Methylbenzo(b)furan-3(2H)-one **(20895-41-4)**
7-Methylbenzo(b)furan-3(2H)-one **(669-04-5)**
2-Methyl-1,4-benzochinon (Czech) **(553-97-9)**
3-Methylbenzo(f)quinoline **(85-06-3)**
2-Methylbenzoic acid **(118-90-1)**
3-Methylbenzoic acid **(99-04-7)**
Methylbenzoic acid **(25567-10-6)**
m-Methylbenzoic acid **(99-04-7)**
o-Methylbenzoic acid **(118-90-1)**
2-Methylbenzoic acid anilide **(7055-03-0)**
4-Methylbenzoic acid chloride **(874-60-2)**
4-Methylbenzoic acid methyl ester **(140-39-6)**
Methylbenzol **(108-88-3)**
N-4-Methylbenzolsulfonyl-N-butylurea **(64-77-7)**
2-Methylbenzonitrile **(529-19-1)**
4-Methylbenzonitrile **(104-85-8)**
o-Methylbenzonitrile **(529-19-1)**
p-Methylbenzonitrile **(104-85-8)**
Methyl-1,12-benzoperylene **(41699-09-6)**
4-Methyl benzophenone **(134-84-9)**
5-Methyl-3,4-benzopyrene **(2381-39-7)**
6-Methyl-1,2-benzopyrone **(92-48-8)**
6-Methylbenzopyrone **(92-48-8)**
2-Methyl-1,4-benzoquinone **(553-97-9)**
2-Methyl-p-benzoquinone **(553-97-9)**
2-Methylbenzoquinone-1,4 **(553-97-9)**
Methyl-1,4-benzoquinone **(553-97-9)**
Methyl-p-benzoquinone **(553-97-9)**
2-Methylbenzothiazole **(120-75-2)**
p-(6-Methylbenzothiazol-2-yl)aniline **(92-36-4)**
4-(6-Methyl-2-benzothiazolyl)benzenamine **(92-36-4)**
1-Methylbenzotriazole **(13351-73-0)**
5-Methyl-1,2,3-benzotriazole **(136-85-6)**
5-Methylbenzotriazole **(136-85-6)**
Methyl-1H-benzotriazole **(29385-43-1)**
4(or 5)-Methyl-1H-benzotriazole sodium salt **(64665-57-2)**
Methyl 5-benzoyl benzimidazole-2-carbamate **(31431-39-7)**
Methyl 5-benzoyl-2-benzimidazolecarbamate **(31431-39-7)**
Methyl 2-benzoylbenzoate **(606-28-0)**
4-Methylbenzoyl chloride **(874-60-2)**
p-Methylbenzoyl chloride **(874-60-2)**
Methyl N-benzoyl-N-(3-chloro-4-fluorophenyl)-2-amino-propionate **(52756-25-9)**
5-Methyl-3,4-benzpyrene **(2381-39-7)**
6-Methyl-3,4-benzpyrene **(16757-80-5)**
8-Methyl-3,4-benzpyrene **(16757-81-6)**
9-Methyl-3,4-benzpyrene **(16757-82-7)**
1-Methyl-3-(2-benzthiazolyl)urea **(1929-88-0)**
N-Methyl-N'-benzyhydrylpiperazine **(82-92-8)**
α-Methylbenzyl acetate **(93-92-5)**
α-Methylbenzyl acetoacetate **(40552-84-9)**
2-Methylbenzyl alcohol **(89-95-2)**
3-Methylbenzyl alcohol **(587-03-1)**
4-Methylbenzyl alcohol **(589-18-4)**
o-Methylbenzyl alcohol **(89-95-2)**
p-Methylbenzylalcohol **(589-18-4)**
α-Methylbenzyl alcohol [UN 2937] **(98-85-1)**
p-Methylbenzylalkohol (German) **(589-18-4)**
Methylbenzylamine **(103-67-3)**
N-Methyl-N-benzylamine **(103-67-3)**
N-Methylbenzylamine **(103-67-3)**
α-Methylbenzylamine **(98-84-0)**
Methylbenzylbis(hydrogenated tallow)ammonium bentonite **(68153-30-0)**
2-Methylbenzyl bromide **(89-92-9)**
3-Methylbenzyl bromide **(620-13-3)**
4-Methylbenzyl bromide **(104-81-4)**
α-Methylbenzyl bromide **(585-71-7)**
m-Methylbenzyl bromide **(620-13-3)**
o-Methylbenzyl bromide **(89-92-9)**
p-Methylbenzyl bromide **(104-81-4)**
2-Methylbenzyl chloride **(552-45-4)**
4-Methylbenzyl chloride **(104-82-5)**
o-Methylbenzyl chloride **(552-45-4)**
p-Methylbenzyl chloride **(104-82-5)**
α-Methylbenzyl 2-chloroacetoacetate **(68683-30-7)**
1-Methylbenzyl-3-(dimethoxyphosphinyloxo)isocrotonate **(7700-17-6)**
α-Methyl benzyl-3-(dimethoxy-phosphinyloxy)-cis-crotonate **(7700-17-6)**
α-Methylbenzyl dimethyl amine **(2449-49-2)**
Methyl benzyl ether **(538-86-3)**
α-Methyl benzyl ether **(93-96-9)**
α-Methylbenzyl ether **(538-86-3)**
α-Methylbenzyl ether **(93-96-9)**
1-Methyl-2-benzyl-hydrazine **(10309-79-2)**
α-Methylbenzyl 3-hydroxycrotonate dimethyl phosphate **(7700-17-6)**
Methyl benzyl ketone **(103-79-7)**
Methylbenzylnitrosamine **(937-40-6)**
N-Methyl-N-benzylnitrosamine **(937-40-6)**
Methyl-benzyl-nitrosoamin (German) **(937-40-6)**
Methylbicyclo(2.2.1)heptene-2,3-dicarboxylic anhydride isomers **(25134-21-8)**
2-Methylbiphenyl **(643-58-3)**
3-Methyl-1,1'-biphenyl **(643-93-6)**
4-Methyl-1,1'-biphenyl **(644-08-6)**
Methyl-1,1'-biphenyl **(28652-72-4)**
Methylbiphenyl **(28652-72-4)**
p-Methylbiphenyl **(644-08-6)**
Methylbis(3-aminopropyl)amine **(105-83-9)**
N-Methyl-bis-chloraethylamin (German) **(51-75-2)**
Methyl-bis-(β-chloraethyl)-amin-N-oxyd-hydrochlorid (German) **(302-70-5)**
N-Methyl-bis-β-chlorethylamine hydrochloride **(55-86-7)**
Methylbis(2-chloroethyl)amine **(51-75-2)**
Methylbis(β-chloroethyl)amine **(51-75-2)**
N-Methyl-bis(2-chloroethyl)amine **(51-75-2)**
N-Methyl-bis(β-chloroethyl)amine **(51-75-2)**
Methylbis(2-chloroethyl)amine hydrochloride **(55-86-7)**
Methylbis(β-chloroethyl)amine hydrochloride **(55-86-7)**
N-Methylbis(2-chloroethyl)amine hydrochloride **(55-86-7)**
Methyl-bis(β-chloroethyl)amine oxide **(126-85-2)**
Methylbis(β-chloroethyl)amine N-oxide **(126-85-2)**
Methylbis(β-chloroethyl)amine N-oxide hydrochloride **(302-70-5)**
N-Methylbis(2-chloroethyl)amine N-oxide hydrochloride **(302-70-5)**
Methylbis(dimethylthiocarbamoylthio)arsine **(2445-07-0)**
Methylbis(2-hydroxyethyl)amine **(105-59-9)**
Methylbis(2-hydroxyethyl)(tallowalkyl)ammonium chloride **(67784-77-4)**
2-Methyl-4,5-bis(hydroxymethyl)-3-hydroxypyridine **(65-23-6)**
2-Methyl-N,N-bis(2-methylpropyl)-1-propanamine **(1116-40-1)**
O-Methyl-O,O-bis-(p-nitrofenyl)thiofosfat (Czech) **(39004-94-9)**
Methylbis(phenylmethyl)benzene **(26898-17-9)**
2-Methyl-4,6-bis(trichloromethyl)-1,3,5-triazine **(949-42-8)**
N-Methyl-bis(2,4-xylyliminomethyl)amine **(33089-61-1)**
β-Methylbivinyl **(78-79-5)**
Methyl borate **(121-43-7)**
Methylbromfenvinphos **(13104-21-7)**
Methylbromid (German) **(74-83-9)**
Methyl bromide (ACGIH,OSHA) [UN 1062] **(74-83-9)**
Methyl bromide and more than 2% chloropicrin mixture, Liquid (DOT) **(8004-09-9)**
Methyl bromide-chloropicrin mixt. **(8004-09-9)**
Methyl α-bromoacetate **(96-32-2)**
Methyl bromoacetate **(96-32-2)**

Methyl bromoacetate [UN 2643] **(96-32-2)**
2-Methylbromobenzene **(95-46-5)**
3-Methylbromobenzene **(591-17-3)**
m-Methylbromobenzene **(591-17-3)**
Methyl 2-bromobenzoate **(610-94-6)**
Methyl 3-bromobenzoate **(618-89-3)**
Methyl 4-bromobenzoate **(619-42-1)**
O-Methyl-O-(4-bromo-2,5-dichlorophenyl)phenyl thiophosphonate **(21609-90-5)**
1-Methyl-5-bromo-4-nitroimidazole **(933-87-9)**
2-Methyl-5-bromo-4-nitroimidazole **(18874-52-7)**
Methyl p-bromophenyl ketone **(99-90-1)**
1-Methyl-2,3-butadiene **(591-95-7)**
1-Methylbutadiene **(504-60-9)**
2-Methylbutadiene **(78-79-5)**
2-Methyl-1,3-butadiene (DOT) **(78-79-5)**
2-Methyl-1,3-butadiene, Homopolymer **(9003-31-0)**
2-Methyl-1,3-butadiene polymer with 2-methyl-1-propene **(9010-85-9)**
2-Methyl-1-butanal **(96-17-3)**
2-Methylbutanal **(96-17-3)**
2-Methylbutanal-4 **(590-86-3)**
3-Methylbutanal **(590-86-3)**
α-Methylbutanal **(96-17-3)**
3-Methylbutanal, diethyl acetal **(3842-03-3)**
3-Methylbutanamide **(541-46-8)**
2-Methyl-1-butanamine **(96-15-1)**
3-Methyl-1-butanamine **(107-85-7)**
3-Methylbutanamine **(107-85-7)**
2-Methylbutane **(78-78-4)**
Methylbutanedioic acid **(498-21-5)**
2-Methyl-1-butanethiol **(1878-18-8)**
2-Methyl-2-butanethiol **(1679-09-0)**
3-Methyl-1-butanethiol **(541-31-1)**
3-Methyl-2-butanethiol **(2084-18-6)**
Methyl n-butanoate **(623-42-7)**
(+-)-2-Methylbutanoic acid **(600-07-7)**
2-Methylbutanoic acid **(116-53-0)**
3-Methylbutanoic acid **(503-74-2)**
2-Methylbutanoic acid, n-hexyl ester **(10032-15-2)**
2-Methylbutanoic acid, 2-methylbutyl ester **(2445-78-5)**
3-Methylbutanoic acid, 2-propenyl ester **(2835-39-4)**
2-Methyl butanol-1 **(137-32-6)**
2-Methyl butanol-2 **(75-85-4)**
2-Methyl-1-butanol **(137-32-6)**
2-Methyl-2-butanol **(75-85-4)**
2-Methyl-4-butanol **(123-51-3)**
2-Methylbutanol **(137-32-6)**
3-Methyl butanol **(123-51-3)**
3-Methyl-1-butanol **(123-51-3)**
3-Methyl-2-butanol **(598-75-4)**
3-Methylbutan-1-ol **(123-51-3)**
3-Methylbutan-3-ol **(75-85-4)**
3-Methyl-butanol-(3) (German) **(75-85-4)**
2-Methyl-1-butanol acetate **(624-41-9)**
3-Methyl-1-butanol hydrogen phosphorodithioate **(32650-55-8)**
3-Methylbutanol nitrite **(110-46-3)**
3-Methyl-2-butanone **(563-80-4)**
3-Methylbutan-2-one [UN 2397] **(563-80-4)**
(E)-2-Methyl-2-butenal **(497-03-0)**
E-2-Methyl-2-butenal **(497-03-0)**
trans-2-Methyl-2-butenal **(497-03-0)**
2-Methyl-2-butene **(513-35-9)**
2-Methylbutene **(26760-64-5)**
Methylbutene **(26760-64-5)**
Methyl butene (DOT) **(513-35-9)**
2-Methyl-1-butene [UN 2459] **(563-46-2)**
2-Methyl-2-butene [UN 2460] **(26760-64-5)**
2-Methyl-2-butene [UN 2460] **(513-35-9)**
3-Methyl-1-butene [UN 2561] **(563-45-1)**
2-Methyl-2-butenedioic acid **(498-23-7)**
cis-Methylbutenedioic acid **(498-23-7)**
(E)-2-Methyl-2-butenenitrile **(30574-97-1)**
(Z)-2-Methyl-2-butenenitrile **(20068-02-4)**
2-Methyl-3-butenenitrile **(16529-56-9)**
2-Methyl-cis-2-butenenitrile **(20068-02-4)**
cis-2-Methyl-2-butenenitrile **(20068-02-4)**
3-Methyl-3-butennitril (Czech) **(4786-19-0)**
cis-2-Methyl-2-butennitrile **(20068-02-4)**
Methyl 2-butenoate **(18707-60-3)**
3-Methyl-2-butenoic acid **(541-47-9)**
cis-2-Methyl-2-butenoic acid **(565-63-9)**
trans-2-Methyl-2-butenoic acid **(80-59-1)**
3-Methyl-2-butenoic acid 2-sec-butyl-4,6-dinitrophenyl ester **(485-31-4)**
3-Methyl-2-butenoic acid 2-(1-methylpropyl)-4,6-dinitrophenyl ester **(485-31-4)**
2-Methyl-2-buten-1-ol **(4675-87-0)**
2-Methyl-3-buten-2-ol **(115-18-4)**
3-Methyl-1-buten-3-ol **(115-18-4)**
3-Methyl-2-buten-1-ol **(556-82-1)**
3-Methyl-3-buten-1-ol **(763-32-6)**
Methylbutenol **(115-18-4)**
3-Methyl-buten-(1)-ol-(3) (German) **(115-18-4)**
3-Methyl-3-buten-2-on (German) **(814-78-8)**
2-Methyl-1-buten-3-one **(814-78-8)**
3-Methyl-3-buten-2-one **(814-78-8)**
3-Methyl-3-buten-2-one-methyl methacrylate copolymer **(51555-36-3)**
3-Methyl-3-buten-2-one-styrene copolymer **(25191-48-4)**
3-Methyl-3-buten-2-one-styrene polymer **(25191-48-4)**
3-Methyl-3-butenonitrile **(4786-19-0)**
cis-2-Methyl-2-butenonitrile **(20068-02-4)**
3-Methyl-2-butenyl chloride **(503-60-6)**
3-(3-Methyl-2-butenyl)-1,2,3,4,5,6-hexahydro-6,11-dimethyl-2,6-methano-3-benzazocin-8-ol **(359-83-1)**
9-((3-Methyl-2-butenyl)oxy)-7H-furo(3,2-g)(1)benzopyran-7-one **(482-44-0)**
2-Methyl-1-buten-3-yne **(78-80-8)**
3-Methyl-butin-(1)-ol-(3) (German) **(115-19-5)**
3'-Methylbuttergelb (German) **(55-80-1)**
1-Methylbutyl acetate **(626-38-0)**
3-Methyl-1-butyl acetate **(123-92-2)**
3-Methylbutyl acetate **(123-92-2)**
2-Methyl-butylacrylaat (Dutch) **(97-88-1)**
2-Methyl-butylacrylat (German) **(97-88-1)**
2-Methyl-butylacrylate **(97-88-1)**
Methylbutylamine **(110-68-9)**
N-(Methyl) butyl amine **(110-68-9)**
N-Methyl-n-butylamine **(110-68-9)**
N-Methylbutylamine [UN 2945] **(110-68-9)**
1-Methyl-4-tert-butylbenzene **(98-51-1)**
p-Methyl-tert-butylbenzene **(98-51-1)**
Methyl 5-butyl-2-benzimidazolecarbamate **(14255-87-9)**
3-Methylbutyl bromide **(107-82-4)**
3-Methylbutyl butyrate **(106-27-4)**
Methyl 1-(butylcarbamoyl)-2-benzimidazolylcarbamate **(17804-35-2)**
3-Methylbutyl chloride **(107-84-6)**
2-(1-Methylbutyl)-4,6-dinitrofenol (Dutch) **(4097-36-3)**
2-(1-Methyl-n-butyl)-4,6-dinitrophenol **(4097-36-3)**
Methyl-1,3-butylene glycol acetate **(4435-53-4)**
3-Methylbutyl ethanoate **(123-92-2)**
Methyl butyl ether **(628-28-4)**
Methyl n-butyl ether **(628-28-4)**
Methyl-tert-butylether [UN 2398] **(1634-04-4)**
3-Methylbutyl formate **(110-45-2)**
α-Methyl-p-(tert-butyl)hydrocinnamaldehyde **(80-54-6)**
3-Methylbutyl 2-hydroxybenzoate **(87-20-7)**

Methyl sec-butyl ketone **(565-61-7)**
Methyl t-butyl ketone **(75-97-8)**
Methyltert-butyl ketone **(75-97-8)**
Methyl n-butyl ketone (ACGIH,OSHA) **(591-78-6)**
2-Methylbutyl 2-methylbutanoate **(2445-78-5)**
2-Methylbutyl 2-methylbutyrate **(2445-78-5)**
3-Methylbutyl nitrite **(110-46-3)**
Methyl-butyl-nitrosamin (German) **(7068-83-9)**
Methyl-n-butylnitrosamine **(7068-83-9)**
Methylbutylnitrosamine **(7068-83-9)**
2-Methyl-2-butylpentanoic acid **(31080-37-2)**
3-Methyl-6-tert-butylphenol **(88-60-8)**
m-(1-Methylbutyl)phenyl methylcarbamate **(2282-34-0)**
3-(1-Methylbutyl)phenyl methylcarbamate mixed with 3-(1-ethylpropyl)phenyl methylcarbamate **(8065-36-9)**
α-Methyl, β-(p-tert-butylphenyl)propionaldehyde **(80-54-6)**
2-Methyl-2-sec-butyl-1,3-propanediol dicarbamate **(64-55-1)**
trans-2-Methyl-3-butyltetrahydrofuran **(36712-20-6)**
2-Methyl-3-butyn-2-amine **(2978-58-7)**
2-Methyl-3-butyn-2-ol **(115-19-5)**
2-Methylbutyn-3-ol-2 **(115-19-5)**
3-Methyl-1-butyn-3-ol **(115-19-5)**
2-Methylbutyraldehyde **(96-17-3)**
3-Methylbutyraldehyde **(590-86-3)**
α-Methylbutyraldehyde **(96-17-3)**
β-Methylbutyramide **(541-46-8)**
2-Methylbutyrate **(116-53-0)**
3-Methylbutyrate **(503-74-2)**
Methyl-n-butyrate **(623-42-7)**
Methyl butyrate [UN 1237] **(623-42-7)**
2-Methylbutyric acid **(116-53-0)**
3-Methylbutyric acid **(503-74-2)**
DL-2-Methylbutyric acid **(600-07-7)**
α-Methylbutyric acid **(116-53-0)**
β-Methylbutyric acid **(503-74-2)**
3-Methylbutyric acid, allyl ester **(2835-39-4)**
α-Methylbutyric aldehyde **(96-17-3)**
4-Methyl-γ-butyrolactone **(108-29-2)**
γ-Methyl-γ-butyrolactone **(108-29-2)**
2-Methylbutyronitrile **(18936-17-9)**
8-(3-Methylbutyryloxy)-diacetoxyscirpenol **(21259-20-1)**
Methyl-calminal **(115-38-8)**
Methyl caprate **(110-42-9)**
Methyl-n-caprate **(110-42-9)**
Methyl caprinate **(110-42-9)**
Methyl caproate **(106-70-7)**
1-Methylcaprolactam **(2549-67-9)**
4-Methylcaprolactam **(3623-05-0)**
N-Methylcaprolactam **(2549-67-9)**
β-Methylcaprolactam **(3623-05-0)**
Methyl capronate **(106-70-7)**
Methyl caprylate **(111-11-5)**
Methyl carbamate **(598-55-0)**
N-Methylcarbamate de 4-dimethylamino 3-methyl phenyle (French) **(2032-59-9)**
N-Methylcarbamate de 1-naphtyle (French) **(63-25-2)**
Methylcarbamate 3,4-dichlorobenzyl ester **(1966-58-1)**
Methylcarbamate 1-naphthalenol **(63-25-2)**
Methylcarbamate 1-naphthol **(63-25-2)**
Methylcarbamic acid o-sec-butylphenyl ester **(3766-81-2)**
Methylcarbamic acid m-cumenyl ester **(64-00-6)**
Methylcarbamic acid m-cym-5-yl ester **(2631-37-0)**
Methyl carbamic acid 2,3-dihydro-2,2-dimethyl-7-benzofuranyl ester **(1563-66-2)**
Methylcarbamic acid, 4-(dimethylamino)-3,5-xylyl ester **(315-18-4)**
Methylcarbamic acid ester with N'-(m-hydroxyphenyl)-N,N-dimethylformamidine **(22259-30-9)**
Methylcarbamic acid, ethyl ester **(105-40-8)**
Methylcarbamic acid 2,3-(isopropylidenedioxy)phenyl ester **(22781-23-3)**
Methyl carbamic acid 4-(methylthio)-3,5-xylyl ester **(2032-65-7)**
Methylcarbamic acid, 1-naphthyl ester **(63-25-2)**
Methylcarbamic acid phenyl ester **(1943-79-9)**
Methylcarbamic acid 2,6-pyridinediyldimethylene ester **(1882-26-4)**
Methylcarbamic acid m-tolyl ester **(1129-41-5)**
Methylcarbamic acid 3,4,5-trimethylphenyl ester **(2686-99-9)**
Methylcarbamic acid 3,4-xylyl ester **(2425-10-7)**
Methylcarbamic acid 3,5-xylyl ester **(2655-14-3)**
Methylcarbamodithioic acid **(144-54-7)**
S-Methylcarbamoylmethyl O,O-dimethyl phosphorodithioate **(60-51-5)**
N-((Methylcarbamoyl)oxy)thioacetimidic acid methyl ester **(16752-77-5)**
N-(4-Methylcarbamoyloxy-O-tolyl)-N,N-dimethylformamidine **(17702-57-7)**
1-(2-Methyl-5-carbamylphenylazo)-2-hydroxy-3-phenyl-carbamoylnaphthalene **(16403-84-2)**
Methyl carbazate **(6294-89-9)**
9-Methylcarbazole **(1484-12-4)**
N-Methylcarbazole **(1484-12-4)**
1-Methyl-4-carbethoxy-4-phenylhexamethyleneimine **(77-15-6)**
1-Methyl-4-carbethoxy-4-phenylhexamethylenimine **(77-15-6)**
1-Methyl-4-carbethoxy-4-phenylpiperidine hydrochloride **(50-13-5)**
Methylcarbinol **(64-17-5)**
Methyl carbitol **(111-77-3)**
Methyl carbitol acetate **(629-38-9)**
2-Methyl-β-carboline **(486-84-0)**
3-Methyl-4-carboline **(486-84-0)**
Methyl 4-carbomethoxybenzoate **(120-61-6)**
Methyl carbonate **(616-38-6)**
Methylcarbophenothion **(953-17-3)**
4-((2-Methyl-5-carboxamidophenyl)azo)-3-hydroxy-N-phenyl-2-naphthalenecarboxamide **(16403-84-2)**
17-β-(1-Methyl-3-carboxypropyl)ethiocholan-3-α-ol **(434-13-9)**
17-β-(1-Methyl-3-carboxypropyl)-etiocholane-3-α,12-α-diol **(83-44-3)**
3-Methylcatechol **(488-17-5)**
4-Methylcatechol **(452-86-8)**
Methylcatechol **(90-05-1)**
Methyl cellosolve (DOT,OSHA) **(109-86-4)**
Methyl cellosolve acetate (DOT,OSHA) **(110-49-6)**
Methyl cellosolve acetylricinoleate **(140-05-6)**
Methyl cellosolve acrylate **(3121-61-7)**
Methyl cellosolye acetaat (Dutch) **(110-49-6)**
Methylcellulose **(9004-67-5)**
Methylcellulose (1/2%) **(9004-67-5)**
Methyl cellulose ether **(9004-67-5)**
Methylcelosolv (Czech) **(109-86-4)**
Methylcelosolvacetat (Czech) **(110-49-6)**
Methyl centralite **(611-92-7)**
Methyl chavicol **(140-67-0)**
Methyl chemosept **(99-76-3)**
2-Methylchinolin (Czech) **(91-63-4)**
4-Methylchinolin hydrochlorid (German) **(3007-43-0)**
6-Methyl-chinoxalin-2,3-dithiol-cyclo-carbonat (German) **(2439-01-2)**
Methylchloorformiaat (Dutch) **(79-22-1)**
Methylchlorid (German) **(74-87-3)**
Methyl chloride (ACGIH,OSHA) [UN 1063] **(74-87-3)**
Methyl, chloro- (8CI,9CI) **(6806-86-6)**
Methyl chloroacetate [UN 2295] **(96-34-4)**
N-Methyl-2-chloroacetoacetamide **(4116-10-3)**
Methyl 2-chloroacetoacetate **(4755-81-1)**
Methylchloroacetylene **(7747-84-4)**
Methyl-2-chloroacrylate **(80-63-7)**
Methyl-α-chloroacrylate **(80-63-7)**
2-Methyl-4-chloroaniline **(95-69-2)**
4-Methyl-2-chloroaniline **(615-65-6)**
2-Methyl-4-chloroaniline hydrochloride **(3165-93-3)**
4-Methyl, 3-chloroaniline hydrochloride **(7745-89-3)**

1-Methyl-2-chlorobenzene **(95-49-8)**
2-Methylchlorobenzene **(95-49-8)**
Methyl 2-chlorobenzoate **(610-96-8)**
Methyl 3-chlorobenzoate **(2905-65-9)**
Methyl 4-chlorobenzoate **(1126-46-1)**
Methyl m-chlorobenzoate **(2905-65-9)**
O-Methyl O-2-chloro-4-tert-butylphenyl N-methylamidophosphate **(299-86-5)**
Methyl chlorocarbonate [UN 1238] **(79-22-1)**
Methyl 2-chloro-3-(4-chlorophenyl)propionate **(14437-17-3)**
Methyl 6-chloro-3,4-dimethoxybenzoate **(30714-88-6)**
4-Methyl-5-(β-chloroethyl)thiazole **(533-45-9)**
Methylchloroform **(71-55-6)**
Methyl chloroform (ACGIH,DOT,OSHA) **(71-55-6)**
Methyl chloroformate [UN 1238] **(79-22-1)**
Methyl 2-chloro-9-hydroxyfluorene-9-carboxylate **(2464-37-1)**
Methyl-2-chloro-9-hydroxyfluorene-9-carboxylate **(2536-31-4)**
Methylchloroisothiazolinone **(26172-55-4)**
1-Methyl-4-(chloromethyl)benzene **(104-82-5)**
Methyl chloromethyl ether, Anhydrous (DOT) **(107-30-2)**
Methylchloromethyl ether [UN 1239] **(107-30-2)**
Methyl 5-chloro-2-nitrobenzoate **(51282-49-6)**
1-Methyl-2-chloro-4-nitroimidazole **(63634-21-9)**
Methyl 2-chloro-3-oxobutanoate **(4755-81-1)**
3-Methyl-4-chlorophenol **(59-50-7)**
2-Methyl-4-chlorophenoxyacetic acid **(94-74-6)**
2-Methyl-4-chlorophenoxyacetic acid, 2-butoxyethyl ester **(19480-43-4)**
2-Methyl-4-chlorophenoxyacetic acid, butoxyethyl ester **(19480-43-4)**
2-Methyl-4-chlorophenoxyacetic acid n-butyl ester **(1713-12-8)**
(2-Methyl-4-chlorophenoxy)acetic acid, diethanolamine salt **(20405-19-0)**
2-Methyl-4-chlorophenoxyacetic acid diethanolamine salt **(20405-19-0)**
2-Methyl-4-chlorophenoxyacetic acid, dimethylamine salt **(2039-46-5)**
(2-Methyl-4-chlorophenoxy)acetic acid, sodium salt **(3653-48-3)**
2-Methyl-4-chlorophenoxybutyric acid **(94-81-5)**
4-(2-Methyl-4-chlorophenoxy)butyric acid **(94-81-5)**
γ-2-Methyl-4-chlorophenoxybutyric acid **(94-81-5)**
4-(2-Methyl-4-chlorophenoxy)butyric acid, sodium salt **(6062-26-6)**
2-(2-Methyl-4-chlorophenoxy)propionic acid **(93-65-2)**
2-Methyl-4-chlorophenoxy-α-propionic acid **(93-65-2)**
α-(2-Methyl-4-chlorophenoxy)propionic acid **(93-65-2)**
2-(2-Methyl-4-chlorophenoxy)propionic acid, diethanolamine salt **(1432-14-0)**
2-(2-Methyl-4-chlorophenoxy)propionic acid, dimethylamine salt **(32351-70-5)**
2-Methyl-4-chlorophenoxythiol acetic acid S-ethyl ester **(25319-90-8)**
N'-(2-Methyl-4-chlorophenyl)-N,N-dimethylformamidine **(6164-98-3)**
2-Methyl-3-(2-chlorophenyl)chinazolon-4 **(340-57-8)**
3-Methyl-4-((o-chlorophenyl)hydrazone)-4,5-isoxazoledione **(5707-69-7)**
3-Methyl-4-(o-chlorophenylhydrazono)-5-isoxazolone **(5707-69-7)**
2-Methyl-3-(o-chlorophenyl)-4-quinazolinone **(340-57-8)**
Methyl 4-chlorophenyl sulfide **(123-09-1)**
Methyl p-chlorophenyl sulfide **(123-09-1)**
Methyl 4-chlorophenyl sulfone **(98-57-7)**
Methyl 4-chlorophenyl sulfoxide **(934-73-6)**
Methyl chlorophos **(52-68-6)**
Methylchloropindol **(2971-90-6)**
Methyl 2-chloropropanoate **(17639-93-9)**
Methyl-2-chloropropionate [UN 2933] **(17639-93-9)**
Methyl 2-(chlorosulfonyl)benzoate **(26638-43-7)**
Methylchlorothiazide **(135-07-9)**
O-Methyl chlorothioformate **(2812-72-8)**
Methylchlorothion **(500-28-7)**
Methyl 6-chloroveratrate **(30714-88-6)**
4-(2-Methyl-4-chlorphenoxy)-buttersaeure (German) **(94-81-5)**
2-Methyl-4-chlorphenoxyessigsaeure (German) **(94-74-6)**
2-(2-Methyl-4-chlorphenoxy)-propionsaeure (German) **(93-65-2)**
N'-(2-Methyl-4-chlorphenyl)-formamidin-hydrochlorid (German) **(6164-98-3)**
Methylchlorpindol **(2971-90-6)**
Methyl chlorpyrifos **(5598-13-0)**
Methyl chlorpyriphos **(5598-13-0)**
Methylchlortetracycline **(127-33-3)**
20-Methylcholanthrene **(56-49-5)**
3-Methylcholanthrene **(56-49-5)**
Methylcholanthrene **(56-49-5)**
3-Methylcholanthren-1-ol **(3342-98-1)**
24 β-Methylcholesta-5,22-dien-3 β-ol **(474-67-9)**
6.β.-Methyl-5.α.-cholestan-3.β.-ol **(43217-65-8)**
6.β.-Methylcholestanol **(43217-65-8)**
24α-Methylcholesterol **(474-62-4)**
3-Methylchrysazin **(481-74-3)**
1-Methylchrysene **(3351-28-8)**
2-Methylchrysene **(3351-32-4)**
3-Methylchrysene **(3351-31-3)**
4-Methyl chrysene **(3351-30-2)**
5-Methylchrysene **(3697-24-3)**
6-Methylchrysene **(1705-85-7)**
α-Methylcinnamaldehyde **(101-39-3)**
Methyl cinnamate **(103-26-4)**
4-Methylcinnamic acid **(1866-39-3)**
p-Methylcinnamic acid **(1866-39-3)**
Methyl cinnamic aldehyde **(101-39-3)**
α-Methylcinnamic aldehyde **(101-39-3)**
Methyl cinnamylate **(103-26-4)**
α-Methylcinnimal **(101-39-3)**
Methylclothiazide **(135-07-9)**
Methylcolchicine **(477-30-5)**
6-Methylcoumarin **(92-48-8)**
6-Methylcoumarinic anhydride **(92-48-8)**
6-Methyl-m-cresol **(95-87-4)**
Methyl m-cresyl ether **(100-84-5)**
Methyl crotonate **(18707-60-3)**
(E)-2-Methylcrotonic acid **(80-59-1)**
3-Methylcrotonic acid **(541-47-9)**
β-Methylcrotonic acid **(541-47-9)**
trans-2-Methylcrotonic acid **(80-59-1)**
3-Methylcrotonic acid 2-sec-butyl-4,6-dinitrophenyl ester **(485-31-4)**
3-Methylcrotyl chloride **(503-60-6)**
Methyl cyanide [UN 1648] **(75-05-8)**
Methyl 2-cyanoacetate **(105-34-0)**
Methyl cyanoacetate **(105-34-0)**
Methyl α-cyanoacrylate **(137-05-3)**
Methyl cyanoacrylate **(137-05-3)**
Methyl 2-cyanoacrylate (ACGIH,OSHA) **(137-05-3)**
4-Methylcyanobenzene **(104-85-8)**
Methyl cyanocarbamate **(21729-98-6)**
Methyl cyanoethanoate **(105-34-0)**
N-Methyl-N-(2-cyanoethyl)aniline **(94-34-8)**
3-Methylcyanoguanidine **(1609-07-0)**
2-Methylcyclobutanone **(1517-15-3)**
2-Methylcyclohexanamine **(7003-32-9)**
N-Methylcyclohexanamine **(100-60-7)**
Methylcyclohexane (ACGIH,OSHA) **(108-87-2)**
Methyl cyclohexane [UN 2296] **(108-87-2)**
α-Methylcyclohexanemethanol **(1193-81-3)**
4-Methylcyclohexanol **(589-91-3)**
m-Methylcyclohexanol **(591-23-1)**
o-Methylcyclohexanol **(583-59-5)**
Methylcyclohexanol (ACGIH,DOT,OSHA) **(25639-42-3)**
Methyl cyclohexanols, Flash point not more than 60.5 degrees C [UN 2617] **(25639-42-3)**
2-Methyl-cyclohexanon (German, Dutch) **(583-60-8)**
2-Methylcyclohexanone **(583-60-8)**
3-Methylcyclohexanone **(591-24-2)**
4-Methylcyclohexanone **(589-92-4)**
Methylcyclohexanone **(1331-22-2)**
o-Methylcyclohexanone (ACGIH,OSHA) **(583-60-8)**
Methyl-3 cyclohexanone-1 (French) **(591-24-2)**

Methyl-4 cyclohexanone-1 (French) **(589-92-4)**
Methyl cyclohexanone [UN 2297] **(1331-22-2)**
1-Methylcyclohexene **(591-49-1)**
3-Methyl-1-cyclohexene **(591-48-0)**
3-Methylcyclohexene **(591-48-0)**
4-Methylcyclohexene **(591-47-9)**
1-Methyl-3-cyclohexenol **(33061-16-4)**
3-Methyl-2-cyclohexen-1-one **(1193-18-6)**
N-Methyl-5-cyclohexenyl-5-methylbarbituric acid **(56-29-1)**
2-Methylcyclohexylamine **(7003-32-9)**
Methylcyclohexylamine **(100-60-7)**
N-Methylcyclohexylamine **(100-60-7)**
p-Methylcyclohexyl bromide **(6294-40-2)**
Methyl N-cyclohexyl-2,5-dimethylfuran-3-carbohydroxamate **(60568-05-0)**
Methyl cyclohexyl ether **(931-56-6)**
1-(2-Methylcyclohexyl)-3-phenylurea **(1982-49-6)**
Methylcyclooctane **(1502-38-1)**
3-Methyl-1-cyclopentadecanone **(541-91-3)**
3-Methylcyclopentadecanone **(541-91-3)**
Methyl-1,3-cyclopentadiene **(26519-91-5)**
Methylcyclopentadiene **(26519-91-5)**
Methylcyclopentadiene dimer **(26472-00-4)**
Methylcyclopentadiene, dimer, hydrogenated **(30496-78-7)**
2-Methylcyclopentadienyl manganesetricarbonyl **(12108-13-3)**
2-Methylcyclopentadienyl manganese tricarbonyl (ACGIH) **(12108-13-3)**
Methylcyclopentadienyl manganese tricarbonyl (OSHA) **(12108-13-3)**
Methylcyclopentane (DOT) **(96-37-7)**
3-Methylcyclopentane-1,2-dione **(80-71-7)**
cis-2-Methylcyclopentanol **(25144-05-2)**
2-Methylcyclopentanone **(1120-72-5)**
Methylcyclopentenolone **(80-71-7)**
1-Methyl-1-cyclopenten-3-one **(2758-18-1)**
3-Methyl-2-cyclopenten-1-one **(2758-18-1)**
Methylcyclopropane **(594-11-6)**
Methylcyclothiazide **(135-07-9)**
Methylcyklopentadientrikarbonylmanganium (Czech) **(12108-13-3)**
Methyl-2-cysteamine **(598-36-7)**
S-Methyl-L-cysteine **(7728-98-5)**
S-Methylcysteine **(7728-98-5)**
4-Methyldaphnetin **(2107-77-9)**
N-Methyl-N-deacetylcolchicine **(477-30-5)**
Methyl decanoate **(110-42-9)**
Methyl decyl ketone **(6175-49-1)**
6-Methyl-6-dehydro-17-α-acetoxyprogesterone **(595-33-5)**
6-Methyl-δ^6-dehydro-17-α-acetoxyprogesterone **(595-33-5)**
6-Methyl-6-dehydro-17-α-acetylprogesterone **(595-33-5)**
N-Methyldemecolcine **(477-30-5)**
O-Methyldemeton **(867-27-6)**
Methyl demeton (ACGIH,OSHA) **(8022-00-2)**
Methyl demeton methyl **(2587-90-8)**
Methyl demeton-methyl **(8065-62-1)**
Methyl demeton thioester **(919-86-8)**
Methyl demeton-O-sulfoxide **(301-12-2)**
5-Methyldeoxyurindine **(50-89-5)**
N-Methyl-N-desacetylcolchicine **(477-30-5)**
Methyldesorphine **(16008-36-9)**
Methyldiaethylcarbinol (German) **(77-74-7)**
2-Methyl-4,6-diaminophenol **(15872-73-8)**
2-Methyl-4,6-diamino-1,3,5-triazine **(542-02-9)**
Methyl diazepinone **(439-14-5)**
10-Methyl-1,2:7,8-dibenzacridine **(59652-21-0)**
10-Methyl-3,4,5,6-dibenzacridine **(59652-20-9)**
9-Methyl-3,4,5,6-dibenzacridine **(59652-21-0)**
14-Methyl dibenz(a,j)acridine **(59652-20-9)**
7-Methyldibenz(c,h)acridine **(59652-21-0)**
2-Methyl-1,2-dibromo-3-chloropropane **(10474-14-3)**
Methyldibromoglutaronitrile **(35691-65-7)**
4-Methyl-2,6-di-terc. butylfenol (Czech) **(128-37-0)**
4-Methyl-2,6-di-tert-butylphenol **(128-37-0)**
Methyldi-tert-butylphenol **(128-37-0)**
Methyldichlorarsine **(593-89-5)**
O-Methyl-O-(2,4-dichlorfenyl)ester kyseliny isopropyl-amidothiofosforecne (Czech) **(299-85-4)**
Methyl dichloroacetate [UN 2299] **(116-54-1)**
Methyldichloroarsine **(593-89-5)**
Methyl 2,5-dichlorobenzoate **(2905-69-3)**
O-Methyl O-2,5-dichloro-4-bromophenyl phenylthiophosphonate **(21609-90-5)**
Methyl 3,4-dichlorocarbanilate **(1918-18-9)**
N-Methyl-2,2'-dichlorodiethylamine **(51-75-2)**
N-Methyl-2,2'-dichlorodiethylamine hydrochloride **(55-86-7)**
N-Methyl-2,2'-dichlorodiethylamine N-oxide hydrochloride **(302-70-5)**
Methyldi(2-chloroethyl)amine **(51-75-2)**
Methyldi(2-chloroethyl)amine hydrochloride **(55-86-7)**
Methyldi(β-chloroethyl)amine hydrochloride **(55-86-7)**
N-Methyl-di-2-chloroethylamine hydrochloride **(55-86-7)**
N-Methyl-di-2-chloroethylamine-N-oxide **(126-85-2)**
Methyldi(2-chloroethyl)amine N-oxide hydrochloride **(302-70-5)**
Methyl 5-(2',4'-dichloro-6'-fluorophenoxy)-2-nitrobenzoate **(51937-92-9)**
Methyl 5-(2,4-dichloro-6-fluorophenoxy)-2-nitrobenzoate **(51937-92-9)**
Methyl 3,5-dichloro-4-methoxybenzoate **(24295-27-0)**
Methyl 2,5-dichloro-3-nitrobenzoate **(34408-25-8)**
Methyl 2,5-dichloro-6-nitrobenzoate **(40188-83-8)**
Methyl 3,6-dichloro-2-nitrobenzoate **(40188-83-8)**
Methyl 5-(2,4-dichlorophenoxy)-2-nitrobenzoate **(42576-02-3)**
Methyl (RS)-2-(4-(2,4-dichlorophenoxy)phenoxy)propionate **(40843-25-2)**
Methyl 2-(4-(2,4-dichlorophenoxy)phenoxy)propionate **(51338-27-3)**
Methyl-N-(3,4-dichlorophenyl) carbamate **(1918-18-9)**
Methyldichlorophenylisoxazolylpenicillin **(3116-76-5)**
Methyl α,4-dichlorophenylpropanoate **(14437-17-3)**
Methyl 2-(4-(2,4-dichlorophenyoxy)phenoxy)propanoate **(40843-25-2)**
Methyldichlorosilane **(75-54-7)**
Methyl dichlorosilane [UN 1242] **(75-54-7)**
Methyl 3-(2,2-dichlorovinyl)-2,2-dimethylcyclopropanecarboxy-late **(61898-95-1)**
Methyl-dichlorsilan (Czech) **(75-54-7)**
N-Methyldicyclohexylamine **(7560-83-0)**
Methyldiethanolamine **(105-59-9)**
N-Methyldiethanolamine **(105-59-9)**
N-Methyldiethanolimine **(105-59-9)**
m-Methyl-N,N-diethylaniline **(91-67-8)**
3-Methyl-N,N-diethylbenzamide **(134-62-3)**
1-Methyl-4-diethylcarbamoylpiperazine citrate **(1642-54-2)**
Methyldiethylcarbinol **(77-74-7)**
Methyldiethylvinylsilane **(18292-29-0)**
Methyl dihydroabietate **(67893-02-1)**
5-Methyl-6,7-dihydro-5H-cyclopentapyrazine **(23747-48-0)**
5H-5-Methyl-6,7-dihydrocyclopentapyrazine **(23747-48-0)**
Methyl dihydrojasmonate **(24851-98-7)**
Methyldihydromorphine **(7732-92-5)**
6-Methyldihydromorphinone **(143-52-2)**
Methyldihydromorphinone **(143-52-2)**
6-Methyl-3,4-dihydro-1,2,3-oxathiazin-4-one 2,2-dioxide **(33665-90-6)**
2-Methyl-.δ.2-dihydropyran **(16015-11-5)**
6-Methyl-2,3-dihydro-4H-pyran **(16015-11-5)**
2-Methyl-5,6-dihydro-4,4-pyran-3-carbonsaeureanilid (German) **(24691-76-7)**
2-Methyl-5,6-dihydro-4-H-pyrane-3-carboxylic acid anilide **(24691-76-7)**
2-α-Methyldihydrotestosterone **(58-19-5)**
Methyl 2,4-dihydroxybenzoate **(2150-47-2)**
3-Methyl-2,5-dihydroxybenzoic acid **(5981-39-5)**
4-Methyl-5,7-dihydroxycoumarin **(2107-76-8)**
L-(-)-α-Methyl-β-(3,4-dihydroxyphenyl)alanine **(555-30-6)**
L-α-Methyl-3,4-dihydroxyphenylalanine **(555-30-6)**
α-Methyl-β-(3,4-dihydroxyphenyl)-l-alanine **(555-30-6)**
4-Methyl-2,5-dimethoxyamphetamine **(15588-95-1)**

4-Methyl-1,2-dimethoxybenzene (494-99-5)
Methyl 3,4-dimethoxybenzoate (2150-38-1)
4-Methyl-2,5-dimethoxy-α-methylphenethyl-amine (15588-95-1)
Methyl 3-(dimethoxyphosphinyloxy)crotonate (7786-34-7)
3'-Methyl-4-dimethylaminoazobenzen (Czech) (55-80-1)
2'-Methyl-4-dimethylaminoazobenzene (3731-39-3)
2-Methyl-N,N-dimethyl-4-aminoazobenzene (3731-39-3)
3'-Methyl-4-(N,N-dimethylamino)azobenzene (55-80-1)
3'-Methyl-4-dimethylaminoazobenzene (55-80-1)
3'-Methyl-N,N-dimethyl-4-aminoazobenzene (55-80-1)
m'-Methyl-p-dimethylaminoazobenzene (55-80-1)
o'-Methyl-p-dimethylaminoazobenzene (3731-39-3)
3'-Methyldimethylaminoazobenzol (German) (55-80-1)
Methyl 2-(dimethylamino)-N-(((methylamino)carbonyl)oxy)-2-oxoethanimidothioate (23135-22-0)
Methyl-4-dimethylamino-3,5-xylyl carbamate (315-18-4)
Methyl-4-dimethylamino-3,5-xylyl ester of carbamic acid (315-18-4)
o-Methyldimethylaniline (609-72-3)
Methyl 1-(dimethylcarbamoyl)-N-(methylcarbamoyloxy)thioformimidate (23135-22-0)
S-Methyl 1-(dimethylcarbamoyl)-N-((methylcarbamoyl)oxy)thioformimidate (23135-22-0)
Methyl 3-((((1,1-dimethylethyl)amino)carbonyl)amino)-2-butenoate (64346-47-0)
Methyl 1,1-dimethylethyl ether (1634-04-4)
Methyl N',N'-dimethyl-N-((methylcarbamoyl)oxy)-1-thiooxamimidate (23135-22-0)
Methyl 3,3-dimethyl-4-pentenoate (63721-05-1)
Methyl 2,2-dimethylpropanoate (598-98-1)
Methyl 2-(((((4,6-dimethyl-2-pyrimidinyl)amino)carbonyl)amino)-sulfonyl)benzoate (74222-97-2)
Methyl 3,6-dimethylresorcylate (4707-47-5)
N-Methyl O,O-dimethylthiolophosphoryl-5-thia-3-methyl-2-valeramide (2275-23-2)
4-Methyl-2,6-dinitroaniline (6393-42-6)
2-Methyl-3,5-dinitrobenzamide (148-01-6)
2-Methyl-3,5-dinitrobenzenamine (35572-78-2)
4-Methyl-3,5-dinitrobenzenamine (19406-51-0)
1-Methyl-2,4-dinitrobenzene (121-14-2)
1-Methyl-3,5-dinitro-benzene (618-85-9)
Methyldinitrobenzene (25321-14-6)
2-Methyl-3,5-dinitrobenzoic acid (28169-46-2)
1-Methyl-4,5-dinitroimidazole (19183-15-4)
2-Methyl-4,6-dinitrophenol (534-52-1)
3-Methyl-4,6-dinitrophenol (616-73-9)
2-Methyl-4,6-dinitro-phenol sodium salt (2312-76-7)
2-Methyl-4,6-dinitrophenol sodium salt (2312-76-7)
Methyl 3,5-dinitrosalicylate (22633-33-6)
N-Methyl-N,4-dinitrosoaniline (99-80-9)
N-Methyl-N,p-dinitrosoaniline (99-80-9)
N-Methyl-N,4-dinitrosobenzenamine (99-80-9)
N-Methyl-2,4-dinitro-N-(2,4,6-tribromophenyl)-6-(trifluoromethyl)benzenamine (63333-35-7)
N-Methyl-2,4-dinitro-N-2,4,6-tribromophenyl(-6)trifluoromethylbenzeneamin-E (63333-35-7)
4-Methyldioxalone-2 (108-32-7)
4-Methyl-1,3-dioxane (1120-97-4)
4-Methyl-m-dioxane (1120-97-4)
2-Methyl-1,3-dioxolane (497-26-7)
Methyl dioxolane (1331-09-5)
Methyldioxolane (497-26-7)
4-Methyl-1,3-dioxolan-2-one (108-32-7)
6-Methyl-1,11-dioxy-2-naphthacenecarboxamide (60-54-8)
N-Methyldiphenethylamine (552-82-9)
2-Methyldiphenhydramine (83-98-7)
o-Methyldiphenhydramine (83-98-7)
4-Methyldiphenyl (644-08-6)
p-Methyldiphenyl (644-08-6)
3-Methyldiphenylamine (1205-64-7)
Methyldiphenylamine (552-82-9)
N-Methyldiphenylamine (552-82-9)
Methyl diphenylarsinite (24582-54-5)
Methyldiphenylchlorosilane (144-79-6)
3-Methyldiphenyl ether (3586-14-9)
Methyl diphenyl ether (86-26-0)
Methyl diphenyl phosphate (115-89-9)
Methyldiphenylsilyl chloride (144-79-6)
2-Methyl-4,6-diphenyl-1,3,5-triazine (3599-62-0)
N'-Methyl-N,N-diphenylurea (13114-72-2)
S-Methyl O,O-dipropyl phosphorothioate (5301-73-5)
Methyldithiocarbamate (144-54-7)
N-Methyldithiocarbamate de sodium (French) (137-42-8)
Methyldithiocarbamic acid, sodium salt (137-42-8)
Methyldithiocarbanic acid (144-54-7)
Methyldithiokarbaman sodny (Czech) (137-42-8)
Methyldithiokarbaman sodny dihydrat (Czech) (6734-80-1)
6-Methyl-1,3-dithiolo(4,5-b)quinoxalin-2-one (2439-01-2)
2-Methyl-1,3-di(2,4-xylylimino)-2-azapropane (33089-61-1)
Methyl docosanoate (929-77-1)
2-Methyldodecane (1560-97-0)
Methyl dodecanoate (111-82-0)
Methyl n-dodecanoate (111-82-0)
(-)-2-Methyl-1-dodecanol (57289-26-6)
Methyl dodecylate (111-82-0)
Methyl dodecyl benzyl ammonium chloride (1399-80-0)
3-Methyl-n-dodecylthiomethylpyridinium chloride (70700-60-6)
(-)-Methyldopa (555-30-6)
L-Methyldopa (555-30-6)
L-α-Methyldopa (555-30-6)
Methyldopa (555-30-6)
α-Methyl dopa (555-30-6)
α-Methyldopa, L- (555-30-6)
Methyl dursban (5598-13-0)
Methyle (acetate de) (French) (79-20-9)
O,O-Methyleen-bis(4-chloorfenol) (Dutch) (97-23-4)
Methyleen-S,S'-bis(O,O-diethyl-dithiofosfaat) (Dutch) (563-12-2)
3,3'-Methyleen-bis(4-hydroxy-cumarine) (Dutch) (66-76-2)
Methyle (formiate de) (French) (107-31-3)
Methyl eicosanoate (1120-28-1)
S,S'-Methylen-bis(O,O-diaethyl-dithiophosphat) (German) (563-12-2)
3,3'-Methylen-bis(4-hydroxy-cumarin) (German) (66-76-2)
Methylenblau (German) (61-73-4)
1-(2,3-Methylendioxyfenyl)-2-(oktylsufinyl)propan (Czech) (120-62-7)
3,4-Methylendioxy-6-propylbenzyl-n-butyl-diaethylenglykolaether (German) (51-03-6)
Methylendirhodanid (Czech) (6317-18-6)
Methylendithiokyanat (Czech) (6317-18-6)
Methylene Base (101-61-1)
Methylene Blue (61-73-4)
Methylene Blue A (61-73-4)
Methylene Blue 2B (61-73-4)
Methylene Blue B (61-73-4)
Methylene Blue BB (61-73-4)
Methylene Blue BBA (61-73-4)
Methylene Blue BB (Zinc free) (61-73-4)
Methylene Blue BD (61-73-4)
Methylene Blue 2BF (61-73-4)
Methylene Blue 2BN (61-73-4)
Methylene Blue 2BP (61-73-4)
Methylene Blue BP (61-73-4)
Methylene Blue BPC (61-73-4)
Methylene Blue BX (61-73-4)
Methylene Blue BZ (61-73-4)
Methylene Blue Chloride (61-73-4)

Methylene Blue Chloride (Biological stain) **(61-73-4)**
Methylene Blue D **(61-73-4)**
Methylene Blue FZ **(61-73-4)**
Methylene Blue G **(61-73-4)**
Methylene Blue GZ **(61-73-4)**
Methylene Blue HGG **(61-73-4)**
Methylene Blue IAD **(61-73-4)**
Methylene Blue I (Medicinal) **(61-73-4)**
Methylene Blue JFA **(61-73-4)**
Methylene Blue (Medicinal) **(61-73-4)**
Methylene Blue N **(61-73-4)**
Methylene Blue NF (medicinal) **(61-73-4)**
Methylene Blue NZ **(61-73-4)**
Methylene Blue Polychrome **(61-73-4)**
Methylene Blue SG **(61-73-4)**
Methylene Blue SP **(61-73-4)**
Methylene Blue USP (Medicinal) **(61-73-4)**
Methylene Blue USP XII (Medicinal) **(61-73-4)**
Methylene Blue ZF **(61-73-4)**
Methylene Blue ZX **(61-73-4)**
Methylene Blue Zinc Free **(61-73-4)**
Methylene acetone **(78-94-4)**
Methylene bichloride **(75-09-2)**
2,2'-Methylenebiphenyl **(86-73-7)**
Methylenebisacrylamide **(110-26-9)**
N,N'-Methylenebis(acrylamide) **(110-26-9)**
Methylenebis(4-aminocyclohexane) **(1761-71-3)**
2,4'-Methylenebis(aniline) **(1208-52-2)**
4,4'-Methylenebisaniline **(101-77-9)**
Methylenebis(aniline) **(101-77-9)**
2,2'-Methylenebisbenzenamine **(6582-52-1)**
4,4'-Methylenebis(benzenamine), (chloromethyl)oxirane polymer **(28390-91-2)**
2,2'-Methylenebis(benzeneamine) **(6582-52-1)**
4,4'-Methylenebis(benzeneamine) **(101-77-9)**
2,2'-Methylenebis(6-tert-butyl-4-ethylphenol) **(88-24-4)**
2,2'-Methylene-bis(6-tert-butyl-4-methylphenol) **(119-47-1)**
4,4'-Methylene(bis)-chloroaniline **(101-14-4)**
4,4'-Methylenebis(2-chloroaniline) **(101-14-4)**
4,4'-Methylenebis(o-chloroaniline) **(101-14-4)**
Methylene 4,4'-bis(o-chloroaniline) **(101-14-4)**
Methylene-bis-orthochloroaniline **(101-14-4)**
p,p'-Methylenebis(α-chloroaniline) **(101-14-4)**
p,p'-Methylenebis(o-chloroaniline) **(101-14-4)**
4,4'-Methylene bis(2-chloroaniline) (ACGIH,OSHA) **(101-14-4)**
4,4'-Methylenebis-2-chlorobenzenamine **(101-14-4)**
2,2'-Methylenebis(4-chlorophenol) **(97-23-4)**
2,2'-Methylenebis(4-chlorophenol), disodium salt **(22232-25-3)**
(cis(cis))-4,4'-Methylenebiscyclohexanamine **(6693-31-8)**
4,4'-Methylenebis(cyclohexanamine) **(1761-71-3)**
4,4'-Methylenebis(cyclohexylamine) **(1761-71-3)**
Methylene bis-(4-cyclohexylisocyanate) **(5124-30-1)**
Methylene bis(4-cyclohexylisocyanate) (ACGIH,OSHA) **(5124-30-1)**
Methylene-S,S'-bis(O,O-diaethyl-dithiophosphat) (German) **(563-12-2)**
4,4'-Methylenebis(2,6-di-tert-butylphenol) **(118-82-1)**
Methylenebis(di-n-butylthiocarbamate) **(10254-57-6)**
4,4'-Methylenebis(N,N-diethylbenzenamine) **(135-91-1)**
4,4'-Methylenebis(diglycidyl aniline) **(28768-32-3)**
Methylenebis(dimethylamine) **(51-80-9)**
4,4'-Methylenebis(N,N-dimethylaniline) **(101-61-1)**
4,4'-Methylenebis(N,N-dimethyl)benzenamine **(101-61-1)**
2,2'-Methylenebis(4-ethyl-6-tert-butylphenol) **(88-24-4)**
3,3'-Methylenebis(4-hydroxy-2H-1-benzopyran-2-one) **(66-76-2)**
3,3'-Methylenebis(4-hydroxy-1,2-benzopyrone) **(66-76-2)**
3,3'-Methylenebis(4-hydroxycoumarin) **(66-76-2)**
3,3'-Methylene-bis(4-hydroxycoumarine) (French) **(66-76-2)**
1,1'-Methylenebis(isocyanatobenzene) **(26447-40-5)**
1,1-Methylenebis(4-isocyanatobenzene) **(101-68-8)**
Methylenebis(4-isocyanatobenzene) **(101-68-8)**
4,4'-Methylenebis(2-methylaniline) **(838-88-0)**
4,4'-Methylenebis(N-methylaniline) **(1807-55-2)**
4,4'-Methylenebis(2-methylbenzenamine) **(838-88-0)**
2,2''-Methylenebis(4-methyl-6-tert-butylphenol) **(119-47-1)**
2,2'-Methylenebis(6-(1-methylcyclohexyl)-p-cresol) **(77-62-3)**
2,2'-Methylenebis(4-methyl-6-(1-methylcyclohexyl)phenol) **(77-62-3)**
2,2'-Methylene-bis(4-methyl-6-nonylphenol) **(7786-17-6)**
1,1'-Methylenebis(2-naphthol) **(1096-84-0)**
1,1'-Methylenebis(4-nitrobenzene) **(1817-74-9)**
2,2'-Methylenebis(p-tert-octylphenol), calcium salt **(68527-62-8)**
1,1'-(Methylenebis(oxy))bisbutane **(2568-90-3)**
1,1'-(Methylenebis(oxy))bis(4-chloro)benzene **(555-89-5)**
4,4'-Methylenebisphenol **(620-92-8)**
Methylenebis(o-phenol), 3-propylene oxide ether **(54208-63-8)**
Methylenebis(4-phenylene isocyanate) **(101-68-8)**
Methylenebis(p-phenylene isocyanate) **(101-68-8)**
4,4'-Methylenebis(N-phenylenemaleimide) **(13676-54-5)**
4,4'-Methylenebis(phenyl isocyanate) **(101-68-8)**
Methylenebis(4-phenyl isocyanate) **(101-68-8)**
Methylenebis(p-phenyl isocyanate) **(101-68-8)**
p,p'-Methylenebis(phenyl isocyanate) **(101-68-8)**
Methylene bisphenyl isocyanate (ACGIH,OSHA) **(101-68-8)**
4,4'-Methylenebis(N-phenylmaleimide) **(13676-54-5)**
4,4'-Methylenebis(phenylmaleimide) **(13676-54-5)**
p,p'-Methylenebis(N-phenylmaleimide) **(13676-54-5)**
1,1'-(Methylenebis(sulfonyl))bisethene **(3278-22-6)**
Methylenebis(4-(3',4',5',6'-tetrabromophthalimido)benzene) **(32588-74-2)**
4,4'-Methylenebis(o-toluidine) **(838-88-0)**
2,2'-Methylenebis(3,4,6-trichlorophenol) **(70-30-4)**
2,2'-Methylenebis(3,4,6-trichlorophenol) sodium salt **(5736-15-2)**
Methylene bromide **(74-95-3)**
Methylenebutanedioic acid **(97-65-4)**
(4-(2-Methylenebutyryl)-2,3-dichlorophenoxy)acetic acid **(58-54-8)**
Methylenebutyryl phenoxyacetic acid **(58-54-8)**
Methylene chloride (ACGIH,DOT,OSHA) [UN 1593] **(75-09-2)**
Methylene chlorobromide **(74-97-5)**
Methylene cyanide **(109-77-3)**
Methylenecyclobutane **(1120-56-5)**
Methylenecyclohexane **(1192-37-6)**
4-Methylenecyclohexanemethanol **(1004-24-6)**
2-Methylenecyclopropanealanine **(156-56-9)**
2-Methylenecyclopropanylalanine **(156-56-9)**
β-(Methylenecyclopropyl)alanine **(156-56-9)**
N,N'-Methylenediacrylamide **(110-26-9)**
Methylenediamine, N,N,N',N'-tetramethyl- **(51-80-9)**
2,4'-Methylenedianiline **(1208-52-2)**
4,4'-Methylenedianiline **(101-77-9)**
Methylenedianiline **(101-77-9)**
p,p'-Methylenedianiline **(101-77-9)**
4,4-Methylenedianiline (ACGIH) **(101-77-9)**
4,4'-Methylenedianiline bismaleimide **(13676-54-5)**
4,4'-Methylenedianiline dihydrochloride **(13552-44-8)**
p,p'-Methylenedianiline dihydrochloride **(13552-44-8)**
Methylene dibromide **(74-95-3)**
Methylene dibutylcarbamodithioate **(10254-57-6)**
Methylene dichloride **(75-09-2)**
4,4'-Methylenedicyclohexanamine **(1761-71-3)**
4,4'-Methylenedicyclohexaneamine **(1761-71-3)**
4,4'-Methylenedicyclohexylamine **(1761-71-3)**
3,4-Methylene-dihydroxybenzaldehyde **(120-57-0)**
Methylene diiodide **(75-11-6)**
Methylene dimethyl ether **(109-87-5)**
Methylene dinitrate **(38483-28-2)**
1,5-Methylene-3,7-dinitroso-1,3,5,7-tetraazacyclooctane **(101-25-7)**
3,4-Methylenedioxy-allybenzene **(94-59-7)**
1,2-Methylenedioxy-4-allylbenzene **(94-59-7)**

3,4-Methylenedioxy-amphetamine **(4764-17-4)**
Methylenedioxyamphetamine **(4764-17-4)**
3,4-Methylenedioxybenzaldehyde **(120-57-0)**
1,2-Methylenedioxybenzene **(274-09-9)**
Methylenedioxybenzene **(274-09-9)**
3,4-Methylenedioxybenzoic acid **(94-53-1)**
3,4-Methylenedioxybenzyl acetate **(326-61-4)**
3,4-Methylenedioxybenzyl acetone **(3160-37-0)**
1,2-Methylenedioxy-4-(1-hydroxyallyl)benzene **(5208-87-7)**
3,4-Methylenedioxymethamphetamine **(54946-52-0)**
1,2-(Methylenedioxy)-4-(2-(octylsulfinyl)propyl)benzene **(120-62-7)**
2-(3,4-Methylenedioxyphenoxy)-3,6,9-trioxoundecane **(51-14-9)**
4-(3,4-Methylenedioxyphenyl)-2-butanone **(3160-37-0)**
1,2-Methylenedioxy-4-propenylbenzene **(120-58-1)**
3,4-Methylenedioxy-1-propenyl benzene **(120-58-1)**
1,2-(Methylenedioxy)-4-propylbenzene **(94-58-6)**
(3,4-Methylenedioxy-6-propylbenzyl) (butyl) diethylene glicol ether **(51-03-6)**
3,4-Methylenedioxy-6-propylbenzyl n-butyl diethyleneglycol ether **(51-03-6)**
4,4'-Methylene diphenol **(620-92-8)**
Methylenediphenyl-4,4'-diamidine **(63690-09-5)**
4,4'-Methylenediphenyl diisocyanate **(101-68-8)**
Methylenedi-p-phenylene-N,N'-bismaleimide **(13676-54-5)**
N,N'-(Methylenedi-4,1-phenylene)bismaleimide **(13676-54-5)**
Methylenedi-p-phenylene diisocyanate **(101-68-8)**
N,N'-(Methylenedi-p-phenylene)dimaleimide **(13676-54-5)**
4,4'-Methylenediphenylene isocyanate **(101-68-8)**
Methylenedi-p-phenylene isocyanate **(101-68-8)**
Methylene di(phenylene isocyanate) (DOT) **(101-68-8)**
4,4'-Methylenediphenyl isocyanate **(101-68-8)**
α-Methylene-diphenylmethane **(530-48-3)**
Methylenedirhodanid (German) **(6317-18-6)**
Methylenedisalicylic acid **(27496-82-8)**
Methylene dithiocyanate **(6317-18-6)**
4,4'-Methylene di-o-toluidine **(838-88-0)**
Methylene glycol **(50-00-0)**
Methylene glycol dinitrate **(38483-28-2)**
6-Methylene-5-hydroxytetracycline **(914-00-1)**
Methylene iodide **(75-11-6)**
3-Methylene-6-(1-methylethyl)cyclohexene **(555-10-2)**
3-Methylene-7-methyl-1,6-octadiene **(123-35-3)**
3-Methylene-7-methyl-1-octen-7-ol **(543-39-5)**
3-Methylene-7-methyl-1-octen-7-yl acetate **(1118-39-4)**
4-Methylene-2-oxetanone **(674-82-8)**
Methylene oxide **(50-00-0)**
6-Methylene-5-oxytetracycline **(914-00-1)**
3-Methylenepentane **(760-21-4)**
Methylenesuccinic acid **(97-65-4)**
S,S'-Methylene O,O,O',O'-tetraethyl phosphorodithioate **(563-12-2)**
2,5-endo-Methylene-δ^3-tetrahydrobenzyl acrylate **(95-39-6)**
8-Methylene-4,11,11-(trimethyl)bicyclo(7.2.0)undec-4-ene **(87-44-5)**
2-Methylene-1,3,3-trimethylindoline **(118-12-7)**
Methylenium ceruleum **(61-73-4)**
Methylenjodid (Czech) **(75-11-6)**
6-Methyl-3,4-epoxycyclohexylmethyl 6-methyl-3,4-epoxycyclohexane carboxylate **(141-37-7)**
4-Methyl-3,4-epoxypentan-2-one **(4478-63-1)**
2-Methyl-2,3-epoxypropyl acrylate **(19900-46-0)**
Methylergobasine **(113-42-8)**
Methylergobrevin **(113-42-8)**
Methylergometrin **(113-42-8)**
Methylergometrine **(113-42-8)**
Methylergonovin **(113-42-8)**
Methylergonovine **(113-42-8)**
4-Methylesculetin **(529-84-0)**
Methylesculetin **(529-84-0)**
4-Methylesculetol **(529-84-0)**
Methylester kiseliny octove (Czech) **(79-20-9)**
Methylester kyseliny acetoctove (Czech) **(105-45-3)**
Methylester kyseliny acetylricinolejove (Czech) **(140-03-4)**
Methylester kyseliny akrylove (Czech) **(96-33-3)**
Methylester kyseliny anthranilove (Czech) **(134-20-3)**
Methylester kyseliny benzoove (Czech) **(93-58-3)**
Methylester kyseliny bromoctove (Czech) **(96-32-2)**
Methylester kyseliny 2-chlorakrylove (Czech) **(80-63-7)**
Methylester kyseliny 2-chlor-3-p-chlorfenylpropionove (Czech) **(14437-17-3)**
Methylester kyseliny chlormravenci (Czech) **(79-22-1)**
Methylester kyseliny chloroctove (Czech) **(96-34-4)**
Methylester kyseliny chloruhlicite (Czech) **(79-22-1)**
Methylester kyseliny 3,4-dichlorkarbanilove (Czech) **(1918-18-9)**
Methylester kyseliny dusicne (Czech) **(598-58-3)**
Methylester kyseliny dusite (Czech) **(624-91-9)**
Methylester kyseliny fosforite (Czech) **(13590-71-1)**
Methylester kyseliny p-hydroxybenzoove (Czech) **(99-76-3)**
Methylester kyseliny isomaselne (Czech) **(547-63-7)**
Methylester kyseliny karbaminove (Czech) **(598-55-0)**
Methylester kyseliny kyanoctove (Czech) **(105-34-0)**
Methylester kyseliny methakrylove (Czech) **(80-62-6)**
Methylester kyseliny methansulfonove (Czech) **(66-27-3)**
Methylester kyseliny 3-methoxypropionove (Czech) **(3852-09-3)**
Methylester kyseliny mravenci (Czech) **(107-31-3)**
Methylester kyseliny orthomravenci (Czech) **(149-73-5)**
Methylester kyseliny propionove (Czech) **(554-12-1)**
Methylester kyseliny salicylove (Czech) **(119-36-8)**
Methylester kyseliny p-toluensulfonove (Czech) **(80-48-8)**
Methyl ester of p-hydroxybenzoic acid **(99-76-3)**
Methyl ester of methanesulfonic acid **(66-27-3)**
Methyl ester of methanesulphonic acid **(66-27-3)**
Methyl esters of fatty acids (C8-C12) **(67762-39-4)**
Methyle (sulfate de) (French) **(77-78-1)**
N-Methylethanamine **(624-78-2)**
N-Methylethanamine hydrochloride **(624-60-2)**
(((1-Methyl-1,2-ethanediyl)bis(carbamodithioato))(2-))zinc homopolymer **(12071-83-9)**
1-Methylethanethiol **(75-33-2)**
Methyl ethanoate **(79-20-9)**
Methylethanolamine **(109-83-1)**
N-Methylethanolamine **(109-83-1)**
Methylethene **(115-07-1)**
(1-Methylethenyl)benzene dimer **(6144-04-3)**
2-(1-Methylethenyl)-4,5-dihydrooxazole **(10471-78-0)**
Methyl-ether **(115-10-6)**
Methyl ether of propylene glycol **(1320-67-8)**
Methyl ethoxol **(109-86-4)**
Methylethoxychlor **(34197-05-2)**
2-(1-Methylethoxy)phenol **(4812-20-8)**
2-(1-Methylethoxy)phenol methylcarbamate **(114-26-1)**
n-2-(1-Methylethoxy)phenyl methyl-carbamate **(114-26-1)**
N-(3-(1-Methylethoxy)phenyl)-2-(trifluoromethyl)benzamide **(66332-96-5)**
1-(1-Methylethoxy)propane **(627-08-7)**
Methylethylacetaldehyde **(96-17-3)**
Methylethylacetic acid **(116-53-0)**
Methylethylacetylenylcarbinol **(77-75-8)**
trans-α-Methyl-β-ethylacrylonitrile **(31551-28-7)**
1-Methylethylamine **(75-31-0)**
α-(((1-Methylethyl)amino)methyl)-2-naphthalenemethanol, hydrochloride **(51-02-5)**
1-((1-Methylethyl)amino)-3-(2-(2-propenyloxy)phenoxy)-2-propanol **(6452-71-7)**
2-Methyl-6-ethyl aniline **(24549-06-2)**
4-(1-Methylethyl)benzaldehyde **(122-03-2)**
(1-Methylethyl)benzene **(98-82-8)**
4-Methylethylbenzene **(622-96-8)**
o-Methylethylbenzene **(611-14-3)**
p-Methylethylbenzene **(622-96-8)**

ar-(1-Methylethyl)benzenediamine **(31626-02-5)**
4-(1-Methylethyl)benzenesulfonic acid **(16066-35-6)**
(1-Methylethyl)benzenesulfonic acid, ammonium salt **(37475-88-0)**
4-(1-Methylethyl)benzoic acid **(536-66-3)**
3-(1-Methylethyl)-1H-2,1,3-benzothiazain-4(3H)-one, 2,2-dioxide **(25057-89-0)**
1-Methylethyl 4-bromo-α-(4-bromophenyl)-α-hydroxybenzene-acetate **(18181-80-1)**
Methylethylbromomethane **(78-76-2)**
(1-Methylethyl)carbamic acid 2-(((aminocarbonyl)oxy)methyl)-2-methylpentyl ester **(78-44-4)**
Methylethylcarbinol **(78-92-2)**
Methyl ethyl cellulose **(9004-59-5)**
Methylethylcellulose **(9004-59-5)**
1-Methylethyl 4-chloro-α-(4-chlorophenyl)-α-hydroxybenzene-acetate **(5836-10-2)**
1-Methylethyl(3-chlorophenyl)carbamate **(101-21-3)**
(1-Methylethyl)cyclohexane **(696-29-7)**
cis-4-(1-Methylethyl)cyclohexanemethanol **(13828-37-0)**
4-(1-Methylethyl)-2,6-dinitro-N,N-dipropylbenzenamine **(33820-53-0)**
2-Methyl-2-ethyl-1,3-dioxolane **(126-39-6)**
2-Methyl-2-ethyldioxolane **(126-39-6)**
1-Methylethyl dodecanoate **(10233-13-3)**
Methylethylene **(115-07-1)**
1-Methylethylene carbonate **(108-32-7)**
Methylethylene glycol **(57-55-6)**
Methyl ethylene oxide **(75-56-9)**
4-Methylethylenethiourea **(2122-19-2)**
2-Methylethylenimine **(75-55-8)**
Methylethylenimine **(75-55-8)**
Methyl ethyl ether (DOT) **(540-67-0)**
1-Methylethyl-2-((ethoxy((1-methylethyl)amino)phosphinothioyl)-oxy) benzoate **(25311-71-1)**
(E)-1-Methylethyl 3-(((ethylamino)methoxyphosphinothioyl)oxy)-2-butenoate **(31218-83-4)**
1-Methylethyl (E)-3-(((ethylamino)methoxyphosphinothioyl)oxy)-2-butenoate **(31218-83-4)**
Methylethylethylene **(109-68-2)**
sym-Methylethylethylene **(109-68-2)**
1-(Methylethyl)-ethyl 3-methyl-4-(methylthio)phenyl phosphor-amidate **(22224-92-6)**
1-Methylethyl 2-(1-ethylpropyl)-4,6-dinitrophenyl carbonate **(973-21-7)**
Methylethylethynylcarbinol **(77-75-8)**
2-Methyl-4-ethylhexane **(3074-75-7)**
3-Methyl-4-ethylhexane **(3074-77-9)**
1-Methylethyl 2-hydroxybenzoate **(607-85-2)**
4,4'-(1-Methylethylidene)biscyclohexanol **(80-04-6)**
4,4'-(1-Methylethylidene)biscyclohexanol, Polymer with (chloromethyl)oxirane **(30583-72-3)**
1,1'-(1-Methylethylidene)bis(3,5-dibromo-4-(2,3-dibromo-propoxy))benzene **(21850-44-2)**
4,4'-(1-Methylethylidene)bis(2,6-dibromophenol) **(79-94-7)**
4,4'-(1-Methylethylidene)bis(2,6-dibromophenol), (chloromethyl)-oxirane polymer **(40039-93-8)**
4,4'-(1-Methylethylidene)bis(2,6-dibromophenol), epichloro-hydrin polymer **(40039-93-8)**
4,4'-(1-Methylethylidene)bis(2,6-dichlorophenol) **(79-95-8)**
4,4'-(1-Methylethylidene)bisphenol, (chloromethyl)oxirane, 4,4'-(1-methylethylidene)bis(2,6-dibromophenol polymer **(26265-08-7)**
2,2'-((1-Methylethylidene)bis(4,1-phenyleneoxymethylene))-bisoxirane **(1675-54-3)**
2,2'-((1-Methylethyl)imino)bisethanol **(121-93-7)**
Methyl ethyl ketone (ACGIH,OSHA) [UN 1193] **(78-93-3)**
Methyl ethyl ketone hydroperoxide **(1338-23-4)**
Methyl ethyl ketone peroxide (ACGIH,OSHA) **(1338-23-4)**
Methyl ethyl ketone peroxide, In solution with not more than 9% by wt. active oxygen (DOT) **(1338-23-4)**
Methyl ethyl ketone peroxide, Maximum concentration 60% **(1338-23-4)**
Methylethylketonhydroperoxide **(1338-23-4)**
Methyl ethyl ketoxime **(96-29-7)**
Methylethylmaleimide **(20189-42-8)**
Methylethylmethane **(106-97-8)**
1-(1-Methylethyl)naphthalene **(6158-45-8)**
N-(1-Methylethyl)-4-nitrobenzenamine **(25186-43-0)**
1-(1-Methylethyl)-4-nitrobenzene **(1817-47-6)**
Methylethylnitrosamine **(10595-95-6)**
N,N-Methylethylnitrosamine **(10595-95-6)**
1-Methylethyl octadecanoate **(112-10-7)**
(Z)-N-(1-Methylethyl)-9-octadecenamide **(10574-01-3)**
1-Methylethyl-9-octadecenoate **(112-11-8)**
Methylethylolamine **(109-83-1)**
2-(1-Methylethyl)-3-oxazolidineethanol **(28770-01-6)**
2-Methyl-3-ethylpentane **(609-26-7)**
(1-Methylethyl)phenol **(25168-06-3)**
4-(1-Methylethyl)phenol **(99-89-8)**
Methyl ethylphenylarsinite **(24582-56-7)**
1-Methyl-5-ethyl-5-phenylbarbituric acid **(115-38-8)**
N-(1-Methylethyl)-N-phenyl-1,4-benzenediamine **(3085-82-3)**
1-(4-(1-Methylethyl)phenyl)ethanone **(645-13-6)**
3-Methyl-5,5-ethylphenylhydantoin **(50-12-4)**
3-Methyl-5-ethyl-5-phenylhydantoin **(50-12-4)**
2-(1-Methylethyl)phenyl methylcarbamate **(2631-40-5)**
(1-Methylethyl)phosphoramidothioic acid O-(2,4-dichloro-phenyl) O-methyl ester **(299-85-4)**
N-(1-Methylethyl)-2-propanamine **(108-18-9)**
1-Methylethyl 2-propenoate **(689-12-3)**
(1-Methylethyl)pyrazine **(29460-90-0)**
2-Methyl-5-ethylpyridine **(104-90-5)**
6-Methyl-3-ethylpyridine **(104-90-5)**
Methyl ethyl pyridine (DOT) **(104-90-5)**
2-Methyl-5-ethylpyridine [UN 2300] **(104-90-5)**
3-Methyl-3-ethylpyrrolidine-2,5-dione **(77-67-8)**
γ-Methyl-γ-ethyl-succinimide **(77-67-8)**
(1-Methylethyl)sulfamoyl chloride **(26118-67-2)**
2-(1-Methylethyl)-9H-thioxanthen-9-one **(5495-84-1)**
O-Methyl O-ethyl O-2,4,5-trichlorophenyl thiophosphate **(2633-54-7)**
3-Methylethynylestradiol **(72-33-3)**
18-Methyl-17-α-ethynyl-19-nortestosterone **(797-63-7)**
3-Methylethynyloestradiol **(72-33-3)**
Methyl eugenol **(93-15-2)**
S-Methyl fenitrooxon **(3344-14-7)**
S-Methyl fenitrothion **(3344-14-7)**
3-Methylfentanyl **(42045-86-3)**
α Methyl fentanyl **(79704-88-4)**
α-Methylfentanyl **(79704-88-4)**
2-Methylfluoranthene **(33543-31-6)**
3-Methylfluoranthene **(1706-01-0)**
1-Methylfluorene **(1730-37-6)**
9-Methylfluorene **(2523-37-7)**
Methyl fluoride [UN 2454] **(593-53-3)**
16-α-Methyl-9-α-fluoro-1-dehydrocortisol **(50-02-2)**
Methylfluoroform **(420-46-2)**
16-α-Methyl-9-α-fluoro-δ^1-hydrocortisone **(50-02-2)**
17-α-Methyl-9-α-fluoro-11-β-hydroxytesterone **(76-43-7)**
16-α-Methyl-9-α-fluoroprednisolone **(50-02-2)**
16α-Methyl-6α-fluoroprednisolone **(53-33-8)**
16-α-Methyl-9-α-fluoro-1,4-pregnadiene-11-β,17-α,21-triol-3,20-dione **(50-02-2)**
Methyl fluorosulfate **(421-20-5)**
Methyl fluorosulfonate **(421-20-5)**
16-α-Methyl-9-α-fluoro-11-β,17-α,21-trihydroxypregna-1,4-diene-3,20-dione **(50-02-2)**
Methylfluorphosphorsaeureisopropylester (German) **(107-44-8)**
Methylfluorphosphorsaeurepinakolylester (German) **(96-64-0)**
Methyl fluosulfonate **(421-20-5)**

Methylflurether (13838-16-9)
Methylformamide (123-39-7)
N-Methylformamide (123-39-7)
o-Methylformanilide (94-69-9)
Methyl orthoformate (149-73-5)
Methyl formate (ACGIH,OSHA) [UN 1243] (107-31-3)
Methylformiaat (Dutch) (107-31-3)
Methylformiat (German) (107-31-3)
Methyl formyl (534-15-6)
o-Methyl-N-formylaniline (94-69-9)
Methyl α-formylbenzeneacetate (5894-79-1)
Methyl 4-formylbenzoate (1571-08-0)
Methyl p-formylbenzoate (1571-08-0)
1-Methyl-1-formylhydrazide (758-17-8)
N-Methyl-N-formylhydrazine (758-17-8)
N-Methyl-N-formyl hydrazone of acetaldehyde (16568-02-8)
Methyl fosferno (298-00-0)
Methylfosfit (Czech) (13590-71-1)
Methylfosfonat (Czech) (13590-71-1)
5-Methyl-2-furaldehyde (620-02-0)
3-Methylfuran (930-27-8)
Methylfuran (DOT) (534-22-5)
2-Methylfuran [UN 2301] (534-22-5)
5-Methyl-2-furancarboxaldehyde (620-02-0)
Methyl furancarboxylate (1334-76-5)
5-Methyl-2(?H)-furanone (1333-38-6)
5-Methyl furfural (620-02-0)
5-Methyl-2-furfural (620-02-0)
5-Methylfurfuraldehyde (620-02-0)
5-Methyl-2H-furo(2,3-H)-1-benzopyran-2-one (73459-03-7)
Methyl 2-furyl ketone (1192-62-7)
3-O-Methylgallate (3934-84-7)
Methyl gallate (99-24-1)
3-O-Methylgallic acid (3934-84-7)
3-Methylgentisic acid (5981-39-5)
Methyl α-D-glucopyranoside (97-30-3)
α-Methyl D-glucose ether (97-30-3)
α-Methylglucoside (97-30-3)
α-Methylglucoside tetranitrate (13225-10-0)
Methyl glutarate (1119-40-0)
3-Methylglutaric acid (626-51-7)
Methyl glutaronitrile (4553-62-2)
α-Methylglycerol trinitrate (84002-64-2)
β.-Methylglycidyl acrylate (19900-46-0)
2-Methylglycidyl acrylate (19900-46-0)
Methyl glycidyl ether (930-37-0)
Methylglycine (107-97-1)
N-Methylglycine (107-97-1)
1-Methylglycocyamidine (60-27-5)
Methyl glycol (109-86-4)
Methyl glycol (57-55-6)
Methyl glycol acetate (110-49-6)
Methyl glycol monoacetate (110-49-6)
Methylglykol (German) (109-86-4)
Methylglykolacetat (German) (110-49-6)
Methylglyoxal (78-98-8)
Methyl glyoxylate (922-68-9)
4-Methylguaiacol (93-51-6)
p-Methylguaiacol (93-51-6)
Methylguanidin (German) (471-29-4)
Methylguanidine (471-29-4)
(α-Methylguanido)acetic acid (57-00-1)
N-Methyl-N-guanylglycine (57-00-1)
Methyl guthion (86-50-0)
Methyl heptadecanoate (1731-92-6)
2-Methylheptadecanoic acid (5638-12-0)
3-Methylheptane (589-81-1)
Methyl heptanoate (106-73-0)
2-Methylheptanoic acid (1188-02-9)
(.+-.)-3-Methyl-3-heptanol (598-06-1)
3-Methyl-2-heptanol (31367-46-1)
4-Methyl-3-heptanol (14979-39-6)
4-Methyl-4-heptanol (598-01-6)
5-Methyl-2-heptanol (54630-50-1)
6-Methyl-2-heptanol acetate (67952-57-2)
2-Methyl-3-heptanone (13019-20-0)
3-Methyl-5-heptanone (541-85-5)
4-Methyl-3-heptanone (6137-11-7)
5-Methyl-3-heptanone (541-85-5)
6-Methyl-2-heptanone (928-68-7)
6-Methyl-3-heptanone (624-42-0)
5-Methyl-3-heptanone (OSHA) (106-68-3)
2-Methyl-1-heptene (15870-10-7)
6-Methyl-1-heptene (5026-76-6)
6-Methyl-5-heptene-2-one (110-93-0)
6-Methyl-6-hepten-2-ol (1335-09-7)
Methylheptenol (1335-09-7)
4-Methyl-3-hepten-2-one (22319-25-1)
6-Methyl-5-hepten-2-one (110-93-0)
6-Methyl-5-hepten-2-one (409-02-9)
Methyl heptenone (110-93-0)
Methyl heptenone (409-02-9)
Methylheptenone (110-93-0)
1-Methylheptylamine (693-16-3)
(6-(1-Methyl-heptyl)-2,4-dinitro-fenyl)-crotonaat (Dutch) (39300-45-3)
2-(1-Methylheptyl)-4,6-dinitrofenylester kyseliny krotonove (Czech) (39300-45-3)
(E)-2-(1-Methylheptyl)-4,6-dinitrophenyl 2-butenoate (9CI) (131-72-6)
(6-(1-Methyl-heptyl)-2,3-dinitro-phenyl)-crotonat (German) (39300-45-3)
2-(1-Methylheptyl)-4,6-dinitrophenyl crotonate (131-72-6)
2-(1-Methylheptyl)-4,6-dinitrophenyl crotonate (39300-45-3)
Methyl heptyl ketone (821-55-6)
(((6-Methylheptyl)oxy)methyl)oxirane (68134-07-6)
(1-Methylheptyl)phenol (27985-70-2)
N-(1-Methylheptyl)-N'-phenyl-1,4-benzenediamine (15233-47-3)
N-(1-Methylheptyl)-N'-phenyl-p-phenylenediamine (15233-47-3)
Methylhexabarbital (56-29-1)
Methylhexabital (56-29-1)
Methyl hexadecanoate (112-39-0)
Methyl n-hexadecanoate (112-39-0)
14-Methylhexadecanoic acid (5918-29-6)
(E,E)-Methyl 2,4-hexadienoate (689-89-4)
Methylhexahydrophthalic anhydride (25550-51-0)
2-Methylhexane (591-76-4)
3-Methylhexane (589-34-4)
5-Methylhexanenitrile (19424-34-1)
Methyl hexanoate (106-70-7)
Methyl n-hexanoate (106-70-7)
2-Methylhexanoic acid (4536-23-6)
4-Methylhexanoic acid (1561-11-1)
3-Methyl-3-hexanol (597-96-6)
3-Methyl-hexanol-(3) (German) (597-96-6)
2-Methyl-3-hexanone (7379-12-6)
2-Methyl-5-hexanone (110-12-3)
4-Methyl-2-hexanone (105-42-0)
5-Methyl-2-hexanone (110-12-3)
5-Methylhexan-2-one [UN 2302] (110-12-3)
α-Methylhexanone isoxime (3623-05-0)
α-Methylhexanonisoxim (German) (3623-05-0)
(Z)-3-Methyl-2-hexene (10574-36-4)
3-Methyl-1-hexene (3404-61-3)
3-Methyl-2-hexene (17618-77-8)
3-Methyl-3-hexene (3404-65-7)
3-Methyl-cis-2-hexene (10574-36-4)

5-Methyl-2-hexene **(3404-62-4)**
cis-3-Methyl-2-hexene **(10574-36-4)**
5-Methyl-3-hexen-2-one **(5166-53-0)**
Methyl hexoate **(106-70-7)**
1-Methylhexylamine **(123-82-0)**
Methyl hexylate **(106-70-7)**
Methyl hexyl ketone **(111-13-7)**
α-((N-Methyl-N-homoveratryl)-γ-aminopropyl)-3,4-dimethoxy-phenylacetonitrile **(52-53-9)**
Methyl hydantoin **(50-12-4)**
1-Methylhydantoin-2-imide **(60-27-5)**
Methylhydrargyriumhydroxid (Czech) **(1184-57-2)**
Methyl hydrate **(67-56-1)**
1-Methylhydrazine **(60-34-4)**
Methylhydrazine (ACGIH,OSHA) [UN 1244] **(60-34-4)**
N-Methylhydrazinecarbothioamide **(6610-29-3)**
Methyl hydrazinecarboxylate **(6294-89-9)**
Methylhydrazine hydrochloride **(7339-53-9)**
4-((2-Methylhydrazino)methyl)-N-isopropylbenzamide **(671-16-9)**
p-(N'-Methylhydrazinomethyl)-N-isopropylbenzamide hydro-chloride **(366-70-1)**
Methyl hydride **(74-82-8)**
Methyl hydrocinnamate **(103-25-3)**
ortho-Methylhydrocinnamic acid **(22084-89-5)**
δ^1-6-α-Methylhydrocortisone **(83-43-2)**
Methyl hydrogen adipate **(627-91-8)**
Methyl, hydrogenated tallow diethylene triamine condensate, polyethoxylated, methyl sulfate **(68389-89-9)**
Methyl hydrogen glutarate **(1501-27-5)**
Methyl hydrogen succinate **(3878-55-5)**
Methyl hydrogen terephthalate **(1679-64-7)**
2-Methylhydroquinone **(95-71-6)**
Methyl-p-hydroquinone **(95-71-6)**
Methylhydroquinone **(95-71-6)**
Methylhydroxamic acid **(546-88-3)**
Methyl hydroxide **(67-56-1)**
2-Methyl-3-hydroxy-4-aminomethyl-5-hydroxymethylpyridene dihydrochloride **(524-36-7)**
17-α-Methyl-17-β-hydroxy-1,4-androstadien-3-one **(72-63-9)**
2-α-Methyl-17-β-hydroxy-5-α-androstan-3-one **(58-19-5)**
1-Methyl-4-hydroxybenzene **(106-44-5)**
p-Methylhydroxybenzene **(106-44-5)**
Methyl 3-hydroxybenzeneacetate **(42058-59-3)**
Methyl 4-hydroxybenzeneacetate **(14199-15-6)**
Methyl o-hydroxybenzoate **(119-36-8)**
Methyl p-hydroxybenzoate **(99-76-3)**
2-Methyl-3-hydroxy-4,5-bis(hydroxymethyl)pyridine **(65-23-6)**
2-Methyl-3-hydroxy-4,5-bis(hydroxymethyl)pyridine hydrochloride **(58-56-0)**
2-Methyl-2-hydroxybutan-3-one **(115-22-0)**
Methyl m-hydroxycarbanilate, m-methylcarbanilate **(13684-63-4)**
4-Methyl-7-hydroxycoumarin **(90-33-5)**
4-Methyl-7-hydroxycoumarin diethoxythiophosphate **(299-45-6)**
2-Methyl-3-hydroxy-4,5-dihydroxymethyl-pyridin (German) **(65-23-6)**
2-Methyl-3-hydroxy-4,5-di(hydroxymethyl)pyridine **(65-23-6)**
Methyl N-hydroxyethanimidothioate **(13749-94-5)**
Methyl(β-hydroxyethyl)amine **(109-83-1)**
4-(Methyl(β-hydroxyethyl)amino)-2',6'-dichloro-4'-nitroazo-benzene **(6232-56-0)**
N-Methyl-N-(2-hydroxyethyl)aniline **(93-90-3)**
N-Methyl-N-(hydroxyethyl)aniline **(93-90-3)**
N-Methyl-N-β-hydroxyethylaniline **(93-90-3)**
N-Methyl-N-hydroxyethylaniline **(93-90-3)**
2-Methyl-1-(2-hydroxyethyl)-5-nitroimidazole **(443-48-1)**
2-Methyl-3-(2-hydroxyethyl)-4-nitroimidazole **(443-48-1)**
4-Methyl-5-(2-hydroxyethyl)thiazole **(137-00-8)**
4-Methyl-5-(β-hydroxyethyl)thiazole **(137-00-8)**
2-Methyl-3-hydroxy-4-formyl-5-hydroxymethylpyridine hydrochloride **(65-22-5)**
1-Methyl-3-hydroxy-4-isopropylbenzene **(89-83-8)**
Methylhydroxylamine **(593-77-1)**
o-Methylhydroxylamine **(593-56-6)**
o-Methylhydroxylamine **(67-62-9)**
o-Methylhydroxylamine hydrochloride **(593-56-6)**
Methyl 2-hydroxy-3-methylbenzoate **(23287-26-5)**
17-α-Methyl-2-hydroxymethylene-17-hydroxy-5-α-androstan-3-one **(434-07-1)**
2-Methyl-2-hydroxymethyl-1,3-propanediol trinitrate **(3032-55-1)**
N-Methyl-3-hydroxymorphinan **(297-90-5)**
2-Methyl-3-hydroxy-1,4-naphthoquinone **(483-55-6)**
Methyl 4-hydroxy-3-nitrobenzoate **(99-42-3)**
Methyl 12-hydroxy-9-octadecenoate **(141-24-2)**
Methyl m-hydroxyphenylacetate **(42058-59-3)**
2-Methyl-2-p-hydroxyphenylbutane **(80-46-6)**
N-(3-Methyl-1-(2-hydroxy-2-phenyl)ethyl-4-piperidyl)-N-phenylpropanamide **(78995-14-9)**
Methyl-p-hydroxyphenyl ketone **(99-93-4)**
6-α-Methyl-17-α-hydroxyprogesterone **(520-85-4)**
6-Methyl-17-α-hydroxy-δ^6-progesterone acetate **(595-33-5)**
6-α-Methyl-17-α-hydroxyprogesterone acetate **(71-58-9)**
Methyl 2-hydroxypropanoate **(547-64-8)**
Methyl 2-hydroxypropionate **(547-64-8)**
5-Methyl-3-(2-hydroxypropyl)-2-oxazolidinone **(3375-84-6)**
2-Methyl-3-hydroxy-4-pyrone **(118-71-8)**
2-Methyl-8-hydroxyquinoline **(826-81-3)**
Methyl 12-hydroxystearate **(141-23-1)**
Methyl 12-hydroxy stearate, lithium salt **(53422-16-5)**
Methyl 12-hydroxystearate, lithium salt **(53422-16-5)**
5-Methyl-7-hydroxy-1,3,4-triazaindolizine **(2503-56-2)**
Methyl 3-β-hydroxy-1-α-H,5-α-H-tropan-2-β-carboxylate, benzoate (ester) HCl **(53-21-4)**
Methyl 3-β-hydroxy-1-α-h,5-α-h-tropane-2-β-carboxylate benzoate (ester) **(50-36-2)**
N-Methylhyoscine bromide **(155-41-9)**
N,N'-Methylidenebisacrylamide **(110-26-9)**
1,1',1''-Methylidynetrisbenzene **(519-73-3)**
1,1',1''-Methylidynetris(4-isocyanatobenzene) **(2422-91-5)**
1,1',1'-(Methylidynetris(oxy))tris(ethane) **(122-51-0)**
3-Methyl-3H-imidazo(4,5-f)quinolin-2-amine **(76180-96-6)**
1-Methylimidazole **(616-47-7)**
2-Methylimidazole **(693-98-1)**
α-Methyl-1H-imidazole-1-ethanol **(37788-55-9)**
1-Methylimidazole-2-thiol **(60-56-0)**
3-Methylimidazolidine-2-thione **(1908-87-8)**
4-Methyl-2-imidazolidinethione **(2122-19-2)**
2-Methyliminodiacetic acid **(4408-64-4)**
2,2'-(Methylimino)diethanol **(105-59-9)**
Methyliminodiethanol **(105-59-9)**
N-Methyl-2,2'-iminodiethanol **(105-59-9)**
N-Methyliminodiethanol **(105-59-9)**
N,N'-((Methylimino)dimethylidyne)di-2,4-xylidine **(33089-61-1)**
β-Methylimipramine **(739-71-9)**
1-Methyl-1H-indole **(603-76-9)**
1-Methylindole **(603-76-9)**
2-Methyl-1H-indole **(95-20-5)**
2-Methylindole **(95-20-5)**
3-Methyl-1H-indole **(83-34-1)**
3-Methylindole **(83-34-1)**
β-Methylindole **(83-34-1)**
2-Methylindoline **(6872-06-6)**
Methyl iodide (ACGIH,OSHA) [UN 2644] **(74-88-4)**
Methyl-β-ionone **(127-43-5)**
Methylionone **(1335-46-2)**
β-Methylionone **(127-43-5)**
Methylisamin **(300-42-5)**

Methylisoamyl acetate **(108-84-9)**
Methyl isoamyl ketone (ACGIH,OSHA) **(110-12-3)**
5-Methyl-1,3-isobenzofurandione **(19438-61-0)**
2-Methylisoborneol **(2371-42-8)**
Methyl isobutenyl ketone **(141-79-7)**
N-Methylisobutylamine **(625-43-4)**
Methyl isobutyl carbinol **(105-30-6)**
Methylisobutyl carbinol **(108-11-2)**
Methyl isobutyl carbinol (ACGIH,OSHA) [UN 2053] **(108-11-2)**
Methylisobutylcarbinol acetate **(108-84-9)**
Methylisobutylcarbinyl acetate **(108-84-9)**
Methyl-isobutyl-cetone (French) **(108-10-1)**
Methyl-isobutylkarbinol (Czech) **(108-11-2)**
Methylisobutylketon (Dutch, German) **(108-10-1)**
Methyl isobutyl ketone (ACGIH,OSHA) [UN 1245] **(108-10-1)**
Methyl isobutyl ketone peroxide **(37206-20-5)**
Methyl isobutyl ketone peroxide (In solution with more than 9% by weight active oxygen) **(37206-20-5)**
Methyl isobutyl ketone peroxide (In solution with not more than 9% by weight active oxygen) **(37206-20-5)**
Methyl isobutyl ketone peroxide, Solution with more than 9% by weight active oxygen **(37206-20-5)**
Methyl isobutyl ketone peroxide, Solution with not more than 9% by weight active oxygen **(37206-20-5)**
Methyl isobutyl ketone reaction product with hydrogen peroxide **(37206-20-5)**
Methyl isobutyrate **(547-63-7)**
Methylisocyanaat (Dutch) **(624-83-9)**
Methyl isocyanat (German) **(624-83-9)**
Methyl isocyanate (ACGIH,OSHA) [UN 2480] **(624-83-9)**
Methyl isocyanate, Solutions (DOT) **(624-83-9)**
Methyl isocyanide **(593-75-9)**
Methyl isoeugenol **(93-16-3)**
2-Methyl-1H-isoindole-1,3(2H)-dione **(550-44-7)**
Methyl-α-isoionone **(127-51-5)**
Methylisokyanat (Czech) **(624-83-9)**
Methylisomin **(300-42-5)**
Methylisomyn **(300-42-5)**
Methylisonitrile **(593-75-9)**
Methyl isopentanoate **(556-24-1)**
1-Methyl-4-isopropenylcyclohexan-3-ol **(7786-67-6)**
1-Methyl-4-isopropenyl-1-cyclohexene **(138-86-3)**
1-Methyl-4-isopropenyl-6-cyclohexen-2-ol **(99-48-9)**
d-1-Methyl-4-isopropenyl-6-cyclohexen-2-one **(2244-16-8)**
δ-1-Methyl-4-isopropenyl-6-cyclohexen-2-one **(99-49-0)**
l-1-Methyl-4-isopropenyl-6-cyclohexen-2-one **(6485-40-1)**
Methyl isopropenyl ketone **(814-78-8)**
Methyl isopropenyl ketone, Inhibited [UN 1246] **(814-78-8)**
Methyl isopropenyl ketone-methyl methacrylate copolymer **(51555-36-3)**
Methyl isopropenyl ketone-methyl methacrylate polymer **(51555-36-3)**
Methyl isopropenyl ketone-styrene copolymer **(25191-48-4)**
Methyl isopropenyl ketone-styrene polymer **(25191-48-4)**
N-Methyl-2-isopropoxyphenylcarbamate **(114-26-1)**
1-Methyl-4-isopropylbenzene **(99-87-6)**
p-Methylisopropyl benzene **(99-87-6)**
1-Methyl-2-(p-(isopropylcarbamoyl)benzyl)hydrazine **(671-16-9)**
1-Methyl-2-(p-isopropylcarbamoylbenzyl)hydrazine hydrochloride **(366-70-1)**
1-Methyl-2-p-(isopropylcarbamyol)benzohydrazine hydrochloride **(366-70-1)**
1-Methyl-4-isopropylcyclohexadiene-1,3 **(99-86-5)**
1-Methyl-4-isopropylcyclohexadiene-1,4 **(99-85-4)**
2-Methyl-5-isopropyl-1,3-cyclohexadiene **(99-83-2)**
1-Methyl-4-isopropylcyclohexane-8-ol **(498-81-7)**
5-Methyl-2-(isopropylcyclo)hexanone **(10458-14-7)**
1-Methyl-4-isopropyl-1-cyclohexen-3-ol **(491-04-3)**
1-Methyl-4-isopropyl-1-cyclohexen-3-one **(89-81-6)**
α-Methyl-p-isopropylhydrocinnamaldehyde **(103-95-7)**
4-Methyl-1-isopropylidenecyclohexane **(1124-27-2)**
Methyl isopropyl ketone (ACGIH,OSHA) **(563-80-4)**
3-Methyl-5-isopropyl N-methylcarbamate **(2631-37-0)**
1-Methyl-7-isopropylphenanthrene **(483-65-8)**
2-Methyl-5-isopropylphenol **(499-75-2)**
3-Methyl-6-isopropylphenol **(89-83-8)**
5-Methyl-2-isopropyl-1-phenol **(89-83-8)**
N-Methyl 3-isopropylphenyl carbamate **(64-00-6)**
N-Methyl m-isopropylphenyl carbamate **(64-00-6)**
(3-Methyl-5-isopropylphenyl)-N-methylcarbamat (German) **(2631-37-0)**
3-Methyl-5-isopropylphenyl-N-methylcarbamate **(2631-37-0)**
2-Methyl-3-(p-isopropylphenyl)propionaldehyde **(103-95-7)**
5-Methyl-2-isopropyl-3-pyrazolyl dimethylcarbamate **(119-38-0)**
3-Methylisoquinoline **(1125-80-0)**
Methyl isosystox **(919-86-8)**
2-Methyl-4-isothiazolin-3-one **(2682-20-4)**
Methylisothiazolinone **(2682-20-4)**
2-Methyl-4-isothiazolin-3-one calcium chloride **(57373-20-3)**
2-Methyl-3(2H)-isothiazolone **(2682-20-4)**
Methylisothiocyanaat (Dutch) **(556-61-6)**
Methyl-isothiocyanat (German) **(556-61-6)**
Methyl isothiocyanate [UN 2477] **(556-61-6)**
Methylisothiokyanat (Czech) **(556-61-6)**
S-Methylisothiourea hemisulfate **(867-44-7)**
S-Methylisothiourea sulfate (2:1) **(867-44-7)**
Methylisovalerate **(556-24-1)**
Methyl isovalerate [UN 2400] **(556-24-1)**
3-Methyl-4,5-isoxazoledione 4-((2-chlorophenyl)hydrazone) **(5707-69-7)**
N'-(5-Methyl-3-isoxazole)sulfanilamide **(723-46-6)**
5-Methyl-3(2H)-isoxazolone **(10004-44-1)**
N'-(5-Methyl-3-isoxazolyl)sulfanilamide **(723-46-6)**
N'-(5-Methylisoxazol-3-yl)sulphanilamide **(723-46-6)**
N^{1}-(5-Methyl-3-isoxazolyl)sulphanilamide **(723-46-6)**
Methylium, dimethoxy-, hexafluorophosphate(1-) (9CI) **(50318-32-6)**
Methyljodid (German) **(74-88-4)**
Methyljodide (Dutch) **(74-88-4)**
Methylkarbamat (Czech) **(598-55-0)**
Methyl karbitol (Czech) **(111-77-3)**
Methylkarbitolacetat (Czech) **(629-38-9)**
S-Methyl-α-ketobutyric acid **(583-92-6)**
Methyl ketone **(67-64-1)**
Methylkyanid (Czech) **(75-05-8)**
Methyl lactate **(547-64-8)**
2-Methyllactonitrile **(75-86-5)**
1-Methyl N-l-α-aspartyl-l-phenylalanine **(22839-47-0)**
Methyl laurate **(111-82-0)**
Methyl laurate α-sulfonic acid, sodium salt **(4016-21-1)**
Methyl laurinate **(111-82-0)**
α-Methyl-l-3,4-dihydroxyphenylalanine **(41372-08-1)**
α-Methyl-l-3,4-dihydroxyphenylalanine **(555-30-6)**
Methyl ledate **(19010-66-3)**
Methyl linoleate **(112-63-0)**
Methyllorazepam **(848-75-9)**
N-Methyllorazepam **(848-75-9)**
N-Methyl-lost (German) **(51-75-2)**
1-Methyllysergic acid butanolamide **(361-37-5)**
Methyllysergic acid butanolamide **(361-37-5)**
Methyl magnesium bromide, In ethyl ether [UN 1928] **(75-16-1)**
Methyl magnesium bromide in ethyl ether, Not over 40% concentration [UN 1928] **(75-16-1)**
Methylmagnesium chloride **(676-58-4)**
S-Methylmalathion **(3344-12-5)**
Methyl maleate **(624-48-6)**
Methylmaleic acid **(498-23-7)**
2-Methylmaleic anhydride **(616-02-4)**
3-Methylmaleic anhydride **(616-02-4)**
Methylmaleic anhydride **(616-02-4)**
α-Methylmaleic anhydride **(616-02-4)**
Methyl malonate **(108-59-8)**

Methylmalonic acid **(516-05-2)**
α-Methylmandelic acid **(515-30-0)**
Methylmercaptaan (Dutch) **(74-93-1)**
Methyl mercaptan (ACGIH,OSHA) [UN 1064] **(74-93-1)**
Methyl mercaptan sodium salt **(5188-07-8)**
Methyl 2-mercaptoacetate **(2365-48-2)**
Methylmercaptoacetate **(2365-48-2)**
2-Methylmercapto-4,6-bis(ethylamino)-s-triazine **(1014-70-6)**
2-Methylmercapto-4,6-bis(isopropylamino)-s-triazine **(7287-19-6)**
2-Methylmercapto-4,6-bis(3-methoxypropylamino)-s-triazine **(845-52-3)**
4-Methylmercapto-3,5-dimethylphenyl N-methylcarbamate **(2032-65-7)**
2-Methylmercapto-4-ethylamino-6-isopropylamino-s-triazine **(834-12-8)**
Methyl-mercaptofos teolery **(919-86-8)**
1-Methyl-2-mercaptoimidazole **(60-56-0)**
Methylmercaptoimidazole **(60-56-0)**
2-Methylmercapto-4-isopropylamino-6-ethylamino-s-triazine **(834-12-8)**
Methylmercapto-4-isopropylamino-6-methylamino-s-triazine **(1014-69-3)**
4-Methylmercapto-3-methylphenyl dimethyl thiophosphate **(55-38-9)**
2-Methylmercapto-10-(2-(n-methyl-2-piperidyl)ethyl)phenothiazine **(50-52-2)**
4-Methylmercapto-2-oxobutyrate **(583-92-6)**
4-Methylmercaptophenol **(1073-72-9)**
Methyl-mercaptophos **(8022-00-2)**
Methylmercaptophos **(867-27-6)**
Methyl 3-mercaptopropanoate **(2935-90-2)**
3-Methylmercapto-1-propanol **(505-10-2)**
D-2-Methyl-3-mercaptopropanoyl-l-proline **(62571-86-2)**
3-(Methylmercapto)propionaldehyde **(3268-49-3)**
β-(Methylmercapto)propionaldehyde **(3268-49-3)**
Methyl 3-mercaptopropionate **(2935-90-2)**
Methyl β-mercaptopropionate **(2935-90-2)**
Methylmercaptopropionic aldehyde **(3268-49-3)**
γ-Methylmercaptopropyl alcohol **(505-10-2)**
6-Methylmercaptopurine ribonucleoside **(342-69-8)**
6-Methylmercaptopurine riboside **(342-69-8)**
1-Methyl-5-mercapto-1,2,3,4-tetrazole **(13183-79-4)**
1-Methyl-5-mercaptotetrazole **(13183-79-4)**
4-Methylmercapto-3,5-xylyl methylcarbamate **(2032-65-7)**
Methylmercuric chloride **(115-09-3)**
Methylmercuric cyanide **(2597-97-9)**
Methylmercuric cyanoguanidine **(502-39-6)**
Methylmercuric dicyandiamide **(502-39-6)**
Methylmercuric hydroxide **(1184-57-2)**
8-(Methylmercurioxy)quinoline **(86-85-1)**
Methylmercury **(22967-92-6)**
Methylmercury(1+) **(22967-92-6)**
Methylmercury(II) cation **(22967-92-6)**
Methylmercury chloride **(115-09-3)**
Methylmercury cyanide **(2597-97-9)**
Methylmercury dicyandiamide **(502-39-6)**
Methylmercury hydroxide **(1184-57-2)**
Methylmercury β-hydroxyquinolate **(86-85-1)**
Methylmercury 8-hydroxyquinolinate **(86-85-1)**
Methylmercury ion **(22967-92-6)**
Methylmercury ion(1+) **(22967-92-6)**
Methylmercury nitrile **(2597-97-9)**
Methylmercury oxinate **(86-85-1)**
Methylmercury oxyquinolinate **(86-85-1)**
Methylmercury quinolinolate **(86-85-1)**
Methylmerkaptofos (Czech) **(8022-00-2)**
Methylmerkuri-8-chinolinolat (Czech) **(86-85-1)**
Methylmerkurichlorid (Czech) **(115-09-3)**
Methylmerkuridikyandiamid (Czech) **(502-39-6)**
Methylmerkurihydroxid (Czech) **(1184-57-2)**
α-Methylmescaline **(1082-88-8)**
Methyl mesylate **(66-27-3)**
Methylmethacrylaat (Dutch) **(80-62-6)**
Methyl-methacrylat (German) **(80-62-6)**
Methyl methacrylate (ACGIH,OSHA) **(80-62-6)**
Methyl methacrylate, Polymer with ethyl acrylate **(9010-88-2)**
Methyl methacrylate, ethyl acrylate polymer **(9010-88-2)**
Methyl methacrylate homopolymer **(9011-14-7)**
Methyl methacrylate-methyl 1-methylvinyl ketone copolymer **(51555-36-3)**
Methyl methacrylate monomer **(80-62-6)**
Methyl methacrylate monomer, Inhibited [UN 1247] **(80-62-6)**
Methyl methacrylate monomer, Uninhibited (DOT) **(80-62-6)**
Methyl methacrylate polymer **(9011-14-7)**
Methyl methacrylate resin **(9011-14-7)**
N-Methylmethanamine **(124-40-3)**
N-Methylmethanamine sulfate (1:1) **(23307-05-3)**
Methylmethane **(74-84-0)**
Methyl methanesulfonate **(66-27-3)**
Methyl methanesulphonate **(66-27-3)**
Methyl methanoate **(107-31-3)**
Methylmethansulfonat (German) **(66-27-3)**
Methyl methansulfonate **(66-27-3)**
Methyl methansulphonate **(66-27-3)**
Methylmethoxyamine **(1117-97-1)**
2-Methyl-4-methoxyaniline **(102-50-1)**
1-Methyl-3-methoxybenzene **(100-84-5)**
3-Methylmethoxybenzene **(100-84-5)**
4-Methyl-1-methoxybenzene **(104-93-8)**
Methyl p-methoxybenzoate **(121-98-2)**
3-Methyl-3-methoxybutanol **(56539-66-3)**
Methyl (((methoxymethylphosphinothioyl)thio)acetyl)methyl-carbamate **(29173-31-7)**
Methyl-2-(((((4-methoxy-6-methyl-1,3,5-triazin-2-yl)amino)-carbonyl)amino)sulfonyl)benzoate **(74223-64-6)**
N-Methyl-4-methoxynaphthalimide **(3271-05-4)**
Methyl 2-methoxy-2-phenoxyacetate **(24607-12-3)**
α-Methyl-ω-methoxypolydimethylsiloxane **(63148-62-9)**
Methyl α-methylacrylate **(80-62-6)**
Methyl methylacrylate **(80-62-6)**
Methyl methylaminobenzoate **(85-91-6)**
Methyl N-(((methylamino)carbonyl)oxy)ethanimidothioate **(16752-77-5)**
4-Methyl-3-methylaminophenol **(6265-13-0)**
2-Methyl-2-methylamino-1-phenylpropane **(100-92-5)**
Methyl N-methyl anthranilate **(85-91-6)**
Methyl 4-methylbenzenesulfonate **(80-48-8)**
Methyl p-methylbenzenesulfonate **(80-48-8)**
Methyl (4-methylbenzenesulfonyl)acetate **(50397-64-3)**
Methyl 3-methylbenzoate **(99-36-5)**
Methyl 4-methylbenzoate **(99-75-2)**
Methyl m-methylbenzoate **(99-36-5)**
Methyl methylbenzoate **(25567-11-7)**
Methyl p-methylbenzoate **(99-75-2)**
N-Methyl-N'-methyl-N'-(2-benzothiazolyl)urea **(18691-97-9)**
1-Methyl-5-(4-methylbenzoyl)-pyrrole-2-acetic acid **(26171-23-3)**
Methyl 2-methylbutanoate **(868-57-5)**
Methyl 3-methylbutanoate **(556-24-1)**
Methyl 3-methyl-2-butenoate **(924-50-5)**
3-Methyl-N-(3-methylbutyl)-1-butanamine **(544-00-3)**
Methyl 2-methylbutyl ketone **(105-42-0)**
Methyl 3-methylbutyrate **(556-24-1)**
Methyl-N-((methylcarbamoyl)oxy)thioacetimidate **(16752-77-5)**
S-Methyl N-(methylcarbamoyloxy)thioacetimidate **(16752-77-5)**
Methyl O-(methylcarbamoyl)thiolacethohydroxamate **(16752-77-5)**
cis-1-Methyl-2-methyl carbamoyl vinyl phosphate **(6923-22-4)**
Methyl (E)-2-methylcrotonate **(6622-76-0)**
Methyl α-methylcrotonate **(6622-76-0)**
Methyl trans-2-methylcrotonate **(6622-76-0)**
Methyl-(1-methyl-2-cyclohexylethyl)amine **(101-40-6)**
α-Methyl-3,4-(methylenedioxy)phenethylamine **(4764-17-4)**
1-Methyl-2-(3,4-methylenedioxyphenyl)ethyl octyl sulfoxide **(120-62-7)**
3-Methyl-4-methylenehexane **(3404-67-9)**

7-Methyl-3-methylene-1,6-octadiene **(123-35-3)**
2-Methyl-6-methylene-7-octen-2-yl acetate **(1118-39-4)**
2-Methyl-5-(1-methylethenyl)cyclohexanone **(7764-50-3)**
(+-)-1-Methyl-4-(1-methylethenyl)cyclohexene **(7705-14-8)**
2-Methyl-N-(3-(1-methylethoxy)phenyl)benzamide **(55814-41-0)**
Methyl(1-methylethyl)benzene **(25155-15-1)**
α-Methyl-2-(1-methylethyl)benzenepropanal **(6502-20-1)**
1-Methyl-4-(1-methylethyl)cyclohexane **(99-82-1)**
cis-1-Methyl-4-(1-methylethyl)cyclohexane **(6069-98-3)**
trans-1-Methyl-4-(1-methylethyl)cyclohexane **(1678-82-6)**
(1R-(1-α,2-β,5-α))-5-Methyl-2-(1-methylethyl)cyclohexanol **(2216-51-5)**
5-Methyl-2-(1-methylethyl)cyclohexanol **(1490-04-6)**
5-Methyl-2-(1-methylethyl)cyclohexanol **(89-78-1)**
5-Methyl-2-(1-methylethyl)cyclohexanone **(10458-14-7)**
5-Methyl-2-(1-methylethyl)cyclohexanone **(491-07-6)**
1-Methyl-4-(1-methylethyl)cyclohexene **(5502-88-5)**
3-Methyl-6-(1-methylethyl)-2-cyclohexen-1-ol **(491-04-3)**
1-Methyl-4-(1-methylethyl)-2,3-dioxabicyclo(2.2.2)Oct-5-ene **(512-85-6)**
2-Methyl (1-methylethylidene)bis(4,1-phenyleneoxy-2,1-ethane-diyl) ester of 2-propenoic acid **(24448-20-2)**
(+-)-exo-1-Methyl-4-(1-methylethyl)-2-((2-methylphenyl)-methoxy)-7-oxabicyclo(2.2.1)heptane **(87818-31-3)**
1-Methyl-7-(1-methylethyl)phenanthrene **(483-65-8)**
2-Methyl-5-(1-methylethyl)phenol **(499-75-2)**
5-Methyl-2-(1-methylethyl)phenol **(89-83-8)**
3-Methyl-5-(1-methylethyl)phenolmethylcarbamate **(2631-37-0)**
Methyl 3-methylhexyl ketone **(58654-67-4)**
N-Methyl-O-methylhydroxylamine **(1117-97-1)**
O-Methyl-N-methylhydroxylamine **(1117-97-1)**
Methyl 2-methyloctanoate **(2177-86-8)**
6-Methyl-2-(p-((3-methyl-5-oxo-2-pyrazolin-4-yl)azo)phenyl)-7-benzothiazolesulfonic acid, sodium salt **(21493-04-9)**
Methyl 4-methylpentanoate **(2412-80-8)**
2-Methyl-4-((2-methylphenyl)azo)benzenamine **(97-56-3)**
1-((2-Methyl-4-((2-methylphenyl)azo)phenyl)azo)-2-naphthalenol **(85-83-6)**
Methyl (3-methylphenyl)carbamothioic acid o-2-naphthalenyl ester **(2398-96-1)**
Methyl 3-methylphenyl ether **(100-84-5)**
3-Methyl-2-((1-methyl-2-phenyl-1H-indol-3-yl)azo)thiazolium chloride **(42373-04-6)**
2-Methyl-3-(2-methylphenyl)-4(3H)-quinazolinone **(72-44-6)**
2-Methyl-3-(2-methylphenyl)-4-quinazolinone **(72-44-6)**
N-Methyl-α,α-methylphenylsuccinimide **(77-41-8)**
N-Methyl-α-methyl-α-phenylsuccinimide **(77-41-8)**
Methyl 2-methyl-2-propenoate **(80-62-6)**
Methyl 1-methylpropyl ketone **(565-61-7)**
2-Methyl-2-(1-methylpropyl)-1,3-propanediol dicarbamate **(64-55-1)**
2-Methyl-3-(2-methylpropyl)pyrazine **(13925-06-9)**
2-Methyl-2-(methylsulfinyl)propanal O-((methylamino)-carbonyl)oxime **(1646-87-3)**
2-Methyl-2-(methylsulfinyl)propionaldehyde O-(methylcarbamoyl)-oxime **(1646-87-3)**
2-Methyl-2-(methylsulfonyl)propanal O-((methylamino)carbonyl)-oxime **(1646-88-4)**
2-Methyl-2-(methylsulfonyl)propionaldehyde O-(methylcarbamoyl)-oxime **(1646-88-4)**
3-Methyl-4-methylthiophenol **(3120-74-9)**
3-Methyl-4-(methylthio)phenyl methylcarbamate **(3566-00-5)**
2-Methyl-2-(methylthio)propanal, O-((methylamino)carbonyl)oxime **(116-06-3)**
2-Methyl-2-(methylthio)propanal oxime **(1646-75-9)**
2-Methyl-2-(methylthio)propionaldehyde O-(methylcarbamoyl)-oxime **(116-06-3)**
2-Methyl-2-methylthio-propionaldehyd-O-(N-methyl-carbamoyl)-oxim (German) **(116-06-3)**
Methyl 4-methylvalerate **(2412-80-8)**
Methyl 1-methylvinyl ketone-styrene copolymer **(25191-48-4)**
Methylmocovina (Czech) **(598-50-5)**
Methyl monobromoacetate **(96-32-2)**
Methyl monochloracetate **(96-34-4)**
Methyl monochloroacetate **(96-34-4)**
4-Methylmorfolin (Czech) **(109-02-4)**
Methylmorphine **(76-57-3)**
4-Methylmorpholine **(109-02-4)**
N-Methylmorpholine **(109-02-4)**
Methylmorpholine [UN 2535] **(109-02-4)**
N-Methyl-N-(morpholinothio)carbamic acid 2,3-dihydro-2,2-dimethyl-7-benzofuranyl ester **(55285-05-7)**
Methyl myristate **(124-10-7)**
Methyl myristate α-sulfonic acid **(29454-23-7)**
Methylnaftalen (Czech) **(1321-94-4)**
2-Methyl-1,4-naftochinon (Czech) **(58-27-5)**
N-Methyl-1-naftyl-carbamaat (Dutch) **(63-25-2)**
Methyl namate **(128-04-1)**
2-Methyl-1,4-naphthalendione **(58-27-5)**
1-Methylnaphthalene **(90-12-0)**
2-Methylnaphthalene **(91-57-6)**
Methylnaphthalene **(1321-94-4)**
α-Methylnaphthalene **(90-12-0)**
β-Methylnaphthalene **(91-57-6)**
N-Methyl- α.-naphthamide **(3400-33-7)**
N-Methyl-1-naphthamide **(3400-33-7)**
2-Methyl-1,4-naphthochinon (German) **(58-27-5)**
2-Methyl-1,4-naphthoquinone **(58-27-5)**
3-Methyl-1,4-naphthoquinone **(58-27-5)**
N-Methyl-1-naphthyl-carbamat (German) **(63-25-2)**
N-Methyl-1-naphthyl carbamate **(63-25-2)**
N-Methyl-α-naphthylcarbamate **(63-25-2)**
Methyl 2-naphthyl ether **(93-04-9)**
Methyl β-naphthyl ether **(93-04-9)**
Methyl 1-naphthyl ketone **(941-98-0)**
Methyl 2-naphthyl ketone **(93-08-3)**
Methyl α-naphthyl ketone **(941-98-0)**
Methyl β-naphthyl ketone **(93-08-3)**
α-Methyl naphthyl ketone **(941-98-0)**
β-Methyl naphthyl ketone **(93-08-3)**
N-(2-Methyl-1-naphthyl)maleimide **(70017-56-0)**
N-Methyl-α-naphthylurethan **(63-25-2)**
Methyl nicotinate **(93-60-7)**
N-Methylnicotinate **(535-83-1)**
N'-Methylnicotinic acid **(535-83-1)**
N-Methylnicotinic acid **(535-83-1)**
Methyl niran **(298-00-0)**
N-Methyl-p-nitraniline **(100-15-2)**
Methyl nitrate, Compd. with scopolamine (1:1) **(6106-46-3)**
Methyl nitrate (DOT) **(598-58-3)**
1-Methylnitrazepam **(2011-67-8)**
Methyl nitrite (DOT) **(624-91-9)**
2-Methyl-4-nitroaniline **(99-52-5)**
4-Methyl-2-nitroaniline **(89-62-3)**
4-Methyl-3-nitroaniline **(119-32-4)**
6-Methyl-3-nitroaniline **(99-55-8)**
N-Methyl-4-nitroaniline **(100-15-2)**
N-Methyl-p-nitroaniline **(100-15-2)**
2-Methyl-1-nitro-9,10-anthracenedione **(129-15-7)**
2-Methyl-1-nitroanthraquinone **(129-15-7)**
4-Methyl-3-nitrobenzamide **(19013-11-7)**
4-Methyl-2-nitrobenzenamine **(89-62-3)**
N-Methyl-4-nitrobenzenamine **(100-15-2)**
2-Methylnitrobenzene **(88-72-2)**
3-Methylnitrobenzene **(99-08-1)**
4-Methylnitrobenzene **(99-99-0)**
Methylnitrobenzene **(1321-12-6)**
m-Methylnitrobenzene **(99-08-1)**
o-Methylnitrobenzene **(88-72-2)**
p-Methylnitrobenzene **(99-99-0)**

2-Methyl-5-nitro-benzeneamine (99-55-8)
Methyl-4-nitrobenzenesulfonate (6214-20-6)
2-Methyl-5-nitrobenzenesulfonyl chloride (121-02-8)
Methyl N-(4-nitrobenzenesulphonyl)carbamate (3337-70-0)
Methyl-p-nitrobenzoate (619-50-1)
2-Methyl-3-nitrobenzoic acid (1975-50-4)
3-Methyl-2-nitrobenzoic acid (5437-38-7)
3-Methyl-4-nitrobenzoic acid (3113-71-1)
3-Methyl-6-nitrobenzoic acid (3113-72-2)
4-Methyl-3-nitrobenzoic acid (96-98-0)
5-Methyl-2-nitrobenzoic acid (3113-72-2)
Methyl 3-nitro-2,5-dichlorobenzoate (34408-25-8)
3'-Methyl-4-nitrodiphenylamine (15979-82-5)
1-Methyl-7-nitro-5-(2-fluorophenyl)-3H-1,4-benzodiazepin-2(1H)-one (1622-62-4)
1-Methyl-2-nitroimidazole (1671-82-5)
2-Methyl-4-nitroimidazole (696-23-1)
1-Methyl-4-nitro-1H-imidazole (9CI) (3034-41-1)
1-Methyl-5-nitro-1H-imidazole (9CI) (3034-42-2)
2-Methyl-5-nitro-1H-imidazole (9CI) (696-23-1)
4-Methyl-5-nitro-1H-imidazole (9CI) (14003-66-8)
2-Methyl-5-nitroimidazole-1-ethanol (443-48-1)
6-(1'-Methyl-4'-nitro-5'-imidazolyl)-mercaptopurine (446-86-6)
Methylnitroimidazolylmercaptopurine (446-86-6)
6-(1-Methyl-4-nitroimidazol-5-ylthio)purin (Czech) (446-86-6)
6-((1-Methyl-4-nitro-1H-imidazol-5-yl)thio)-1H-purine (446-86-6)
6-((1-Methyl-4-nitroimidazol-5-yl)thio)purine (446-86-6)
6-(1-Methyl-4-nitroimidazol-5-ylthio)purine (446-86-6)
6-(1-Methyl-p-nitro-5-imidazolyl)-thiopurine (446-86-6)
6-(Methyl-p-nitro-5-imidazolyl)-thiopurine (446-86-6)
2-Methyl-5-nitro-1H-isoindole-1,3(2H)-dione (41663-84-7)
1-Methyl-2-nitronaphthalene (63017-87-8)
2-Methyl-1-nitronaphthalene (881-03-8)
2-Methyl-3-nitronaphthalene (1204-72-4)
3-Methyl-2-nitronaphthalene (1204-72-4)
1-Methyl-3-nitro-1-nitrosoguanidine (70-25-7)
Methylnitronitrosoguanidine (70-25-7)
N-Methyl-N'-nitro-N-nitrosoguanidine (70-25-7)
N-Methyl-N'-nitro-N-nitrosoguanidine, Not exceeding 25 grams in one outside packaging [NA 1325] (70-25-7)
2-Methyl-4-nitrophenol (99-53-6)
3-Methyl-4-nitrophenol (2581-34-2)
3-(N-Methyl-p-((p-nitrophenyl)azo)anilino)propionitrile (31464-38-7)
2-((4-Methyl-2-nitrophenyl)azo)-3-oxo-N-phenylbutanamide (2512-29-0)
4-Methyl-2-((2-nitrophenyl)azo)phenol (1435-71-8)
1-Methyl-7-nitro-5-phenyl-1,3-dihydro-2H-1,4-benzodiazepin-2-one (2011-67-8)
3-Methyl-4-nitrophenyl dimethyl phosphate (2255-17-6)
Methyl m-nitrophenyl ether (555-03-3)
Methyl 3-nitrophenyl ketone (121-89-1)
Methyl-p-nitrophenyl ketone (100-19-6)
Methylnitrophos (122-14-5)
N-Methyl-4-nitrophthalimide (41663-84-7)
2-Methyl-2-nitropropane (594-70-7)
2-Methyl-2-nitropropane-1,3-diol (77-49-6)
2-Methyl-2-nitro-propanol nitrate (24884-69-3)
N-(2-Methyl-2-nitropropyl)-p-nitrosoaniline (24458-48-8)
N-(2-Methyl-2-nitropropyl)-4-nitrosobenzamine (24458-48-8)
3-Methyl-4-nitroquinoline 1-oxide (14073-00-8)
3-Methylnitrosaminopropionitrile (60153-49-3)
γ-(Methylnitrosamino)-3-pyridinebutyraldehyde (64091-90-3)
4-(Methylnitrosamino)-4-(3-pyridyl)butanal (64091-90-3)
4-(N-Methyl-N-nitrosamino)-4-(3-pyridyl)butanal (64091-90-3)
4-(N-Methyl-N-nitrosamino)-1-(3-pyridyl)-1-butanone (64091-91-4)
Methylnitrosoacetamid (German) (7417-67-6)
Methylnitrosoacetamide (7417-67-6)
N-Methyl-N-nitrosoacetamide (7417-67-6)
N-Methyl-N-nitrosoallylamine (4549-43-3)
4-(N-Methyl-N-nitrosoamino)-4-(3-pyridyl)-1-butanone (64091-91-4)
Methylnitrosoaniline (614-00-6)
N-Methyl-N-nitrosoaniline (614-00-6)
N-Methyl-N-nitrosobenzenamine (614-00-6)
N-Methyl-N-nitrosobenzylamine (937-40-6)
N-Methyl-N-nitrosobutylamine (7068-83-9)
N-Methyl-N-nitrosocarbamic acid, ethyl ester (615-53-2)
1-Methyl-N-nitrosodiethylamine (16339-04-1)
N-Methyl-N-nitroso-ethamine (10595-95-6)
N-Methyl-N-nitroso-ethenylamine (4549-40-0)
N-Methyl-N-nitrosoethylamine (10595-95-6)
N-Methyl-N-nitrosoethylcarbamate (615-53-2)
N-Methyl-N-nitrosoglycine (13256-22-9)
Methylnitroso-harnstoff (German) (684-93-5)
N-Methyl-N-nitroso-harnstoff (German) (684-93-5)
N-Methyl-N-nitrosolaurylamine (55090-44-3)
N-Methyl-N-nitrosomethanamine (62-75-9)
1-Methyl-1-nitrosomocovina (Czech) (684-93-5)
N-Methyl-N-nitrosonitroguanidin (German) (70-25-7)
1-Methyl-1-nitroso-3-nitroguanidine (70-25-7)
N-Methyl-N-nitroso-N'-nitroguanidine (70-25-7)
N-Methyl-N-nitroso-4-oxo-4-(3-pyridyl)butyl amine (64091-91-4)
N-Methyl-N-nitrosopentylamine (13256-07-0)
Methyl-(4-nitrosophenyl)nitrosamine (99-80-9)
3-Methylnitrosopiperidine (13603-07-1)
Methylnitroso-p-toluenesulfonamide (80-11-5)
N-Methyl-N-nitroso-p-toluenesulfonamide (80-11-5)
1-Methyl-1-nitrosourea (684-93-5)
Methylnitrosourea (684-93-5)
N-Methyl-N-nitrosourea (684-93-5)
Methylnitrosouree (French) (684-93-5)
Methylnitrosourethan (German) (615-53-2)
Methylnitrosourethane (615-53-2)
N-Methyl-N-nitroso-urethane (615-53-2)
N-Methyl-N-nitrosovinylamine (4549-40-0)
5-Methylnonane (15869-85-9)
Methyl nonanoate (1731-84-6)
Methyl-n-nonylacetaldehyde (110-41-8)
Methylnonylacetaldehyde (110-41-8)
Methylnonylacetic aldehyde (110-41-8)
Methyl n-nonyl ketone (112-12-9)
Methyl nonyl ketone (112-12-9)
8-Methylnonyl nonanoate (109-32-0)
Methyl 8-nonynoate (7003-48-7)
N-Methylnorapomorphine hydrochloride (314-19-2)
1-Methylnorharman (486-84-0)
Methyl 9-cis,12-cis-octadecadienoate (112-63-0)
Methyl cis,cis-9,12-octadecadienoate (112-63-0)
Methyl octadecadienoate (112-63-0)
Methyl trans,trans-9,11-octadecadienoate (13038-47-6)
Methyl trans-9,trans-11-octadecadienoate (13038-47-6)
Methyl octadecanoate (112-61-8)
(E)-Methyl 9-octadecenoate (1937-62-8)
Methyl (Z)-9-octadecenoate (112-62-9)
Methyl 6-octadecenoate (52355-31-4)
Methyl 9-octadecenoate (112-62-9)
Methyl cis-9-octadecenoate (112-62-9)
Methyl trans-10-octadecenoate (13038-45-4)
N-Methyl-N-octadecyl-1-octadecanamine (4088-22-6)
Methyl 10-octadecynoate (26543-36-2)
5-Methyl-2-octanamine (67953-04-2)
2-Methyloctane (3221-61-2)
3-Methyloctane (2216-33-3)
4-Methyloctane (2216-34-4)
Methyl n-octanoate (111-11-5)
Methyl octanoate (111-11-5)

2-Methyloctanoic acid methyl ester **(2177-86-8)**
3-Methyl-3-octanol **(5340-36-3)**
3-Methyloctan-3-ol **(5340-36-3)**
7-Methyl-1-octanol **(2430-22-0)**
3-Methyl-2-octanone **(6137-08-2)**
5-Methyl-2-octanone **(58654-67-4)**
Methyl octylate **(111-11-5)**
N-Methyl-N-octyl-1-decanamine **(22020-14-0)**
Methyl n-octyl ketone **(693-54-9)**
Methyl octyl ketone **(693-54-9)**
N-Methyl-N-octyl-1-octanamine **(4455-26-9)**
Methylol **(67-56-1)**
N-Methylolacrylamide **(924-42-5)**
N-Methylol-chloracetamide **(2832-19-1)**
N-Methylol dimethylphosphonopropionamide **(20120-33-6)**
Methyl oleate **(112-62-9)**
Methyl oleate, sulfated, sodium salt **(139-99-1)**
N-Methyl-N-oleoyltaurine **(97-80-3)**
N-Methyl-N-oleoyltaurine sodium salt **(137-20-2)**
Methylol formaldehyde **(141-46-8)**
N-Methylolformamide **(13052-19-2)**
Methylolmethacrylamide **(923-02-4)**
3-Methylolpentane **(97-95-0)**
2-Methylolphenol **(90-01-7)**
4-Methylol phenol **(623-05-2)**
o-Methylolphenol **(90-01-7)**
N-Methylolphthalimide **(118-29-6)**
Methylolpropane **(71-36-3)**
N-Methylol-2-pyrrolidinone **(15438-71-8)**
N-Methylolpyrrolidinone **(15438-71-8)**
N-Methylol-2-pyrrolidone **(15438-71-8)**
N-Methylolpyrrolidone **(15438-71-8)**
2-Methyloltetrahydro-1,4-pyran **(100-72-1)**
Methylolurea **(1000-82-4)**
N-Methylolurea **(1000-82-4)**
Methylolurea resin **(9011-05-6)**
Methylolureas **(1000-82-4)**
Methyloranz (Czech) **(547-58-0)**
4-Methylorcinol **(527-55-9)**
Methyl β-orcinolcarboxylate **(4707-47-5)**
Methyl oxalate **(553-90-2)**
6-Methyl-1,2,3-oxathiazin-4(3H)-one 2,2-dioxide **(33665-90-6)**
Methyloxazepam **(846-50-4)**
N-Methyloxazepam **(846-50-4)**
2-Methyloxetan **(2167-39-7)**
2-Methyloxetane **(2167-39-7)**
4-Methyl-2-oxetanone **(3068-88-0)**
2-Methyloxine **(826-81-3)**
Methyl oxirane **(75-56-9)**
Methyloxirane, Polymer with oxirane, monobutyl ether **(9038-95-3)**
Methyloxirane-oxirane copolymer **(9003-11-6)**
Methyloxirane-oxirane polymer **(9003-11-6)**
Methyl oxitol **(109-86-4)**
Methyl oxoacetate **(922-68-9)**
N-Methyl-3-oxobutanamide **(20306-75-6)**
22-((3-Methyl-1-oxo-2-butenyl)oxy)-3-oxo-olean-12-en-28-oic acid, (22,β)- **(467-82-3)**
22-((2-Methyl-1-oxo-2-butenyl)oxy)-3-oxo-olean-12-en-28-oic acid, 22-β(Z)- **(467-81-2)**
Methyl 3-oxobutyrate **(105-45-3)**
Methyl 9-oxodecanoate **(2575-07-7)**
4-Methyl-2-oxo-1,3-dioxolane **(108-32-7)**
6-Methyl-2-oxo-1,3-dithiolo(4,5-b)quinoxaline **(2439-01-2)**
N-Methyl-N-(1-oxododecyl)glycine sodium salt **(137-16-6)**
5-Methyl-4-oxohexanoic acid **(41654-04-0)**
2-Methyloxolane **(96-47-9)**
(Z)-N-Methyl-N-(1-oxo-9-octadecenyl)glycine **(110-25-8)**
N-Methyl-N-(1-oxo-9-octadecenyl)glycine **(110-25-8)**
3-Methyl-2-oxopentanoic acid **(1460-34-0)**
4-Methyl-2-oxopentanoic acid **(816-66-0)**
N-(2-(2-Methyl-4-oxopentyl))acrylamide **(2873-97-4)**
2-Methyl-2-((1-oxo-2-propenyl)amino)-1-propanesulfonic acid, sodium salt **(5165-97-9)**
Methyl 4-(N-(1-oxopropyl)-n-phenylamino)-1-(2-phenylethyl)-4-piperidinecarboxylate **(59708-52-0)**
Methyloxyammonium chloride **(593-56-6)**
Methyl p-oxybenzoate **(99-76-3)**
2-Methyl-3-oxy-γ-pyrone **(118-71-8)**
Methyl palmitate **(112-39-0)**
Methyl palmitate α-sulfonic acid **(58849-75-5)**
14-Methylpalmitic acid **(5918-29-6)**
Methylparaben **(99-76-3)**
Methylparafynol **(77-75-8)**
Methyl parahydroxybenzoate **(99-76-3)**
Methyl paraoxon **(950-35-6)**
Methyl parasept **(99-76-3)**
Methyl parathion (ACGIH,OSHA) **(298-00-0)**
Methyl parathion, Liquid [NA 3018] **(298-00-0)**
Methyl parathion mixture, Dry [NA 2783] **(298-00-0)**
Methyl pectin **(9000-69-5)**
Methyl pectinate **(9000-69-5)**
Methyl pelargonate **(1731-84-6)**
Methyl pentachlorophenate **(1825-21-4)**
Methyl pentachlorophenyl ester **(1825-21-4)**
Methyl pentachlorophenyl sulfide **(1825-19-0)**
4-Methylpentadecane **(2801-87-8)**
5-Methylpentadecane **(25117-33-3)**
Methyl pentadecanoate **(7132-64-1)**
2-Methyl-1,3-pentadiene **(1118-58-7)**
2-Methyl-1,4-pentadiene **(763-30-4)**
3-Methyl-1,3-pentadiene **(4549-74-0)**
3-Methyl-1,4-pentadiene **(1115-08-8)**
3-Methylpentadiene **(4549-74-0)**
4-Methyl-1,3-pentadiene **(926-56-7)**
Methylpentadiene **(1115-08-8)**
Methylpentadiene **(1118-58-7)**
Methylpentadiene **(926-56-7)**
2-Methylpentaldehyde **(123-15-9)**
Methylpentaldehyde **(73513-30-1)**
4-Methylpentanal **(1119-16-0)**
2-Methylpentane (DOT) **(107-83-5)**
3-Methylpentane (DOT) **(96-14-0)**
2-Methyl-1,5-pentanediamine **(15520-10-2)**
2-Methyl pentane-2,4-diol **(107-41-5)**
2-Methyl-1,3-pentanediol **(149-31-5)**
2-Methyl-2,4-pentanediol **(107-41-5)**
3-Methyl-1,5-pentanediol **(4457-71-0)**
4-Methylpentanenitrile **(542-54-1)**
3-Methyl-1,3,5-pentanetriol **(7564-64-9)**
Methyl pentanoate **(624-24-8)**
2-Methylpentanoic acid **(97-61-0)**
3-Methylpentanoic acid **(105-43-1)**
2-Methyl-2-pentanol **(590-36-3)**
2-Methyl-3-pentanol **(565-67-3)**
2-Methyl-4-pentanol **(108-11-2)**
2-Methylpentanol-1 **(105-30-6)**
3-Methyl-1-pentanol **(589-35-5)**
3-Methyl-2-pentanol **(565-60-6)**
3-Methyl-3-pentanol **(77-74-7)**
4-Methyl-2-pentanol **(108-11-2)**
4-Methylpentanol **(1320-98-5)**
4-Methylpentanol-2 **(108-11-2)**
Methyl-1-pentanol **(54972-97-3)**
3-Methyl-pentanol-(3) (German) **(77-74-7)**

Name	CAS No.
2-Methylpentan-2-ol [UN 2560]	**(590-36-3)**
4-Methyl-2-pentanol, acetate	**(108-84-9)**
2-Methyl-2-pentanol-4-one	**(123-42-2)**
4-Methyl-2-pentanon (Czech)	**(108-10-1)**
4-Methyl-pentan-2-on (Dutch, German)	**(108-10-1)**
2-Methyl-4-pentanone	**(108-10-1)**
3-Methyl-2-pentanone	**(565-61-7)**
4-Methyl-2-pentanone	**(108-10-1)**
4-Methyl-2-pentanone peroxide	**(37206-20-5)**
2-Methyl-2-penten-1-al	**(623-36-9)**
2-Methyl-2-pentenal	**(623-36-9)**
3-Methyl-4-pentenal	**(1777-33-9)**
4-Methyl-2-pentenal	**(5362-56-1)**
(Z)-3-Methyl-2-pentene	**(922-62-3)**
2-Methyl-1-pentene	**(27236-46-0)**
2-Methyl-2-pentene	**(625-27-4)**
2-Methyl-pentene	**(27236-46-0)**
2-Methyl-pentene-1	**(763-29-1)**
2-Methyl-pentene-2	**(625-27-4)**
2-Methylpentene	**(27236-46-0)**
2-Methylpentene	**(763-29-1)**
4-Methyl-1-pentene	**(27236-46-0)**
4-Methyl-1-pentene	**(691-37-2)**
4-Methyl-2-pentene	**(4461-48-7)**
4-Methylpentene	**(27236-46-0)**
cis-3-Methyl-2-pentene	**(922-62-3)**
2-Methyl-2-pentene-1-al	**(623-36-9)**
4-Methyl-2-pentene-1-al	**(5362-56-1)**
(E)-2-Methyl-2-pentenenitrile	**(31551-28-7)**
cis-2-Methyl-2-pentenenitrile	**(31551-28-7)**
4-Methyl-3-pentene-2-one	**(141-79-7)**
(E)-3-Methyl-3-pentenoic acid	**(41653-93-4)**
2-Methyl-2-pentenoic acid	**(3142-72-1)**
trans-.δ.3-3-Methylpentenoic acid	**(41653-93-4)**
2-Methyl-1-penten-3-ol	**(2088-07-5)**
4-Methyl-3-penten-2-on (Dutch, German)	**(141-79-7)**
2-Methyl-2-penten-4-one	**(141-79-7)**
2-Methyl-2-pentenone-4	**(141-79-7)**
3-Methyl-4-penten-2-one	**(758-87-2)**
4-Methyl-3-penten-2-one	**(141-79-7)**
4-(4-Methyl-3-penten-1-yl)-3-cyclohexen-1-carboxaldehyde	**(37677-14-8)**
1-(4-Methyl-3-pentenyl)-1-cyclohexene-4-carboxaldehyde	**(37677-14-8)**
4-(4-Methyl-3-penten-1-yl)-3-cyclohexene-1-carboxaldehyde	**(37677-14-8)**
4-(4-Methyl-3-pentenyl)-3-cyclohexene-1-carboxaldehyde	**(37677-14-8)**
4-(4-Methyl-3-pentenyl)-3-cyclohexenecarboxaldehyde	**(37677-14-8)**
4-(4-Methylpent-3-enyl)cyclohex-3-ene-1-carboxaldehyde	**(37677-14-8)**
3-Methyl-2-(cis-2-penten-1-yl)-2-cyclopenten-1-one	**(488-10-8)**
1-(4-Methyl-3-pentenyl)-4-formyl-1-cyclohexene	**(37677-14-8)**
3-Methylpentin-3-ol	**(77-75-8)**
Methylpentinol	**(77-75-8)**
3-Methyl-pentin-(1)-ol-(3) (German)	**(77-75-8)**
4-Methyl-2-pentyl acetate	**(108-84-9)**
1-Methylpentylamine	**(5329-79-3)**
2-Methyl-2,3-pentylene oxide	**(1192-22-9)**
Methyl pentyl ketone	**(110-43-0)**
Methyl-N-pentylnitrosamine	**(13256-07-0)**
Methyl pentyl sulfide	**(1741-83-9)**
3-Methyl-1-pentyn-3-ol	**(77-75-8)**
3-Methylpent-1-yn-3-ol	**(77-75-8)**
Methylpentynol	**(77-75-8)**
Methylpentynolum	**(77-75-8)**
1-Methylperylene	**(10350-33-1)**
1-Methylphenanthrene	**(832-69-9)**
2-Methylphenanthrene	**(2531-84-2)**
9-Methylphenanthrene	**(883-20-5)**
Methylphenazonium methosulfate	**(58-34-4)**
Methyl phencapton	**(3735-23-7)**
β-Methylphenethyl alcohol	**(1123-85-9)**
(+-)-α-Methylphenethylamine	**(300-62-9)**
α-Methylphenethylamine	**(60-15-1)**
dl-α-Methylphenethylamine	**(300-62-9)**
α-Methylphenethylamine, d-Form	**(51-64-9)**
(+)-α-Methylphenethylamine sulfate (2:1)	**(51-63-8)**
α-Methylphenethylamine sulfate, (+-)-	**(60-13-9)**
d-α-Methylphenethylamine sulfate	**(51-63-8)**
dextro-α-Methylphenethylamine sulfate	**(51-63-8)**
α-Methylphenethylamine sulfate, d-form	**(51-63-8)**
7-(2-((α-Methylphenethyl)amino)ethyl)theophylline	**(3736-08-1)**
N-(2-(Methylphenethylamino)propyl)propionanilide	**(552-25-0)**
Methyl(β-phenethyl)dichlorosilane	**(772-65-6)**
Methyl phenethyl ketone	**(2550-26-7)**
Methylphenidan	**(113-45-1)**
Methyl phenidate	**(113-45-1)**
Methylphenidate hydrochloride	**(298-59-9)**
Methyl phenidyl acetate	**(113-45-1)**
Methylphenidylacetate hydrochloride	**(298-59-9)**
Methyl phenkapton	**(3735-23-7)**
1-Methylphenobarbital	**(115-38-8)**
Methylphenobarbital	**(115-38-8)**
N-Methylphenobarbital	**(115-38-8)**
Methylphenobarbitone	**(115-38-8)**
2-Methylphenol	**(95-48-7)**
3-Methylphenol	**(108-39-4)**
4-Methylphenol	**(106-44-5)**
m-Methylphenol	**(108-39-4)**
o-Methylphenol	**(95-48-7)**
p-Methylphenol	**(106-44-5)**
N-Methylphenolbarbitol	**(115-38-8)**
4-Methylphenol methyl ether	**(104-93-8)**
Methylphenol potassium salt	**(12002-51-6)**
4-Methylphenol sodium salt	**(1121-70-6)**
Methylphenol sodium salt	**(34689-46-8)**
Methylphenol titanium(4+) salt	**(28503-70-0)**
Methyl phenoxyacetate	**(2065-23-8)**
1-Methyl-3-phenoxybenzene	**(3586-14-9)**
((Methylphenoxy)methyl)oxirane	**(26447-14-3)**
2-((2-Methylphenoxy)methyl)oxirane	**(2210-79-9)**
3-(2-Methylphenoxy)pyridazine	**(14491-59-9)**
α-Methylphensuximide	**(77-41-8)**
α-Methyl phenylacetaldehyde	**(93-53-8)**
2-Methylphenyl acetate	**(533-18-6)**
4-Methylphenyl acetate	**(140-39-6)**
Methyl phenylacetate	**(101-41-7)**
o-Methylphenyl acetate	**(533-18-6)**
p-Methylphenyl acetate	**(140-39-6)**
(4-Methylphenyl)acetic acid	**(622-47-9)**
α-Methylphenylacetic acid	**(492-37-5)**
p-Methylphenylacetic acid	**(622-47-9)**
Methylphenylamine	**(100-61-8)**
N-Methylphenylamine	**(100-61-8)**
2-(((3-Methylphenyl)amino)carbonyl)benzoic acid	**(85-72-3)**
2-(Methylphenylamino)ethanol	**(93-90-3)**
2-(N-Methyl-N-phenylamino)ethanol	**(93-90-3)**
((2-Methylphenyl)amino)methanesulfonic acid	**(94-57-5)**
3-(Methylphenylamino)propanenitrile	**(94-34-8)**
N-Methyl-N-phenylaniline	**(552-82-9)**
4-Methyl-6-(phenylazo)-1,3-benzenediamine	**(5042-54-6)**
1-((2-Methylphenyl)azo)-2-naphthalenamine	**(131-79-3)**
1-(2-Methylphenyl)azo-2-naphthalenamine	**(131-79-3)**
1-((2-Methylphenyl)azo)-2-naphthalenol	**(2646-17-5)**
1-(4'-Methylphenylazo)-2-naphthol-6-sulfonic acid	**(5859-07-4)**
1-(2-Methylphenyl)azo-2-naphthylamine	**(131-79-3)**
4-Methyl-2-(phenylazo)phenol	**(952-47-6)**
Methylphenylbarbituric acid	**(115-38-8)**

2-Methyl-N-phenylbenzamide **(7055-03-0)**
3-Methyl-N-phenylbenzenamine **(1205-64-7)**
4-Methyl-N-phenylbenzenamine **(620-84-8)**
N-Methyl-N-phenylbenzenamine **(552-82-9)**
o-Methylphenyl bromide **(95-46-5)**
2-Methyl-4-phenyl-2-butanol **(103-05-9)**
N-Methyl-ω-phenyl-t-butylamine **(100-92-5)**
3-Methyl-4-phenyl-butyric acid **(7315-68-6)**
3-(Methylphenyl)carbamic acid 3-((methoxycarbonyl)amino)-phenyl ester **(13684-63-4)**
N-Methyl-4-phenyl-4-carbethoxypiperidine **(57-42-1)**
N-Methyl-4-phenyl-4-carbethoxypiperidine hydrochloride **(50-13-5)**
Methylphenylcarbinol **(98-85-1)**
Methylphenylcarbinol acetate **(93-92-5)**
Methylphenylcarbinyl acetate **(93-92-5)**
1-Methyl-4-phenyl-4-carboethoxypiperidine hydrochloride **(50-13-5)**
1-Methyl-5-phenyl-7-chloro-1,3-dihydro-2H-1,4-benzo-diazepin-2-one **(439-14-5)**
Methylphenyldichlorosilane [UN 2437] **(149-74-6)**
Methylphenyl diphenyl phosphate **(26444-49-5)**
2-Methyl-p-phenylenediamine **(95-70-5)**
4-Methyl-m-phenylenediamine **(95-80-7)**
Methylphenylenediamine **(25376-45-8)**
1-Methyl-2,6-phenylenediaminebis(4,5,6,7-tetrachloroisoindolin-1-one-3-ylidene) **(5045-40-9)**
2-Methyl-p-phenylenediamine sulphate **(615-50-9)**
2-Methyl-meta-phenylene diisocyanate **(91-08-7)**
4-Methyl-phenylene diisocyanate **(584-84-9)**
Methyl-meta-phenylene diisocyanate **(26471-62-5)**
2-Methyl-meta-phenylene isocyanate **(91-08-7)**
4-Methyl-phenylene isocyanate **(584-84-9)**
Methylphenylene isocyanate **(26471-62-5)**
1-(4-Methylphenyl)ethanol **(536-50-5)**
1-(p-Methylphenyl)ethanol **(536-50-5)**
1-(Methylphenyl)ethanone **(26444-19-9)**
Methyl phenyl ether **(100-66-3)**
α-Methyl phenylethyl alcohol **(1123-85-9)**
N-Methyl-5-phenyl-5-ethylbarbital **(115-38-8)**
1-Methyl-5-phenyl-5-ethylbarbituric acid **(115-38-8)**
as-Methylphenylethylene **(98-83-9)**
3-Methyl-5,5-phenylethylhydantoin **(50-12-4)**
Methyl 2-phenylethyl ketone **(2550-26-7)**
Methyl phenylethyl ketone **(2550-26-7)**
(1-Methyl-1-phenylethyl)phenol **(27576-86-9)**
(1-Methyl-1-phenylethyl)phenol phosphate **(63302-98-7)**
N-(1-Methyl-2-phenylethyl)-γ-phenylbenzenepropanamine **(390-64-7)**
Methyl 1-phenylethyl-4-(N-phenylpropionamido)isonipecotate **(59708-52-0)**
N-(3-Methyl-1-(2-phenylethyl)-4-piperidinyl)-N-phenylpropan-amide **(42045-86-3)**
N-(3-Methyl-1-(2-phenylethyl)-4-piperidyl)-N-phenylpropanamide **(42045-86-3)**
N-(1-(α-Methyl-β-phenyl)ethyl-4-piperidyl) propionanilide **(79704-88-4)**
1-(1-Methyl-2-phenyl-ethyl)-4-(N-propanilido)piperidine **(79704-88-4)**
2-Methyl-1-(2-phenylethyl)pyridinium bromide **(10551-21-0)**
N-(2-Methylphenyl)formamide **(94-69-9)**
.β.-Methyl-.β.-phenylglycidic acid **(5669-15-8)**
3-Methyl-3-phenylglycidic acid **(5669-15-8)**
3-Methyl-3-phenylglycidic acid ethyl ester **(77-83-8)**
2-Methylphenyl glycidyl ether **(2210-79-9)**
Methylphenyl glycidyl ether **(2186-24-5)**
Methylphenylglyoxal **(579-07-7)**
1,1'-((4-Methylphenyl)imino)bis-2-propanol **(38668-48-3)**
1-Methyl-2-phenyl-1H-indole **(3558-24-5)**
2-(2-(1-Methyl-2-phenyl-1H-indol-3-yl)ethenyl)-1,3,3-trimethyl-3H-indolium chloride **(4657-00-5)**
4-Methylphenyl isocyanide **(7175-47-5)**
p-Methylphenyl isocyanide **(7175-47-5)**
1-Methyl-4-phenylisonipecotic acid, ethyl ester **(57-42-1)**
1-Methyl-4-phenylisonipecotic acid ethyl ester hydrochloride **(50-13-5)**
N-Methyl-β-phenylisopropylaminhydrochlorid (German) **(300-42-5)**
Methyl phenyl ketone **(98-86-2)**
4-Methylphenylmercaptan **(106-45-6)**
p-Methylphenylmercaptan **(106-45-6)**
3-Methylphenyl N-methylcarbamate **(1129-41-5)**
(4-Methylphenyl)methyl chloride **(104-82-5)**
Methyl-1-(phenylmethyl)pyridinium chloride **(26747-91-1)**
α-Methyl-α-phenyl N-methyl succinimide **(77-41-8)**
3-Methyl-2-phenylmorpholine **(134-49-6)**
1-Methyl-5-phenyl-7-nitro-1,3-dihydro-2H-1,4-benzodiazepin-2-one **(2011-67-8)**
2-Methylphenyl 4-nitrophenyl ether **(2444-29-3)**
Methylphenylnitrosamine **(614-00-6)**
o-Methylphenylol **(95-48-7)**
2-Methyl-2-phenyloxirane **(2085-88-3)**
N-(4-Methylphenyl)-3-oxobutanamide **(2415-85-2)**
2-Methyl-2-phenylpentane **(1985-57-5)**
3-Methylphenyl phenyl ether **(3586-14-9)**
m-Methylphenyl phenyl ether **(3586-14-9)**
Methyl phenyl phosphate **(115-89-9)**
Methylphenylphosphinic chloride **(5761-97-7)**
Methylphenylphosphinyl chloride **(5761-97-7)**
Methyl phenylphosphonate **(10088-45-6)**
1-Methyl-4-phenyl-piperidin-4-carbon-saeure-aethylester (German) **(57-42-1)**
Methyl α-phenyl-2-piperidineacetate hydrochloride **(298-59-9)**
Methyl phenylpiperidine carbonic acid ethyl ester **(57-42-1)**
1-Methyl-4-phenylpiperidine-4-carboxylic acid ethyl ester **(57-42-1)**
Methyl α-phenyl-α-(2-piperidyl)acetate **(113-45-1)**
Methyl phenyl polysiloxane **(2116-84-9)**
2-Methyl-1-phenylpropane **(538-93-2)**
2-Methyl-2-phenylpropane **(98-06-6)**
4-Methylphenyl propanoate **(7495-84-3)**
Methyl 3-phenylpropanoate **(103-25-3)**
3-(2-Methylphenyl)propanoic acid **(22084-89-5)**
2-Methyl-1-phenyl-1-propanone **(611-70-1)**
2-Methyl-3-phenyl-2-propenal **(101-39-3)**
Methyl 3-phenylpropenoate **(103-26-4)**
Methyl β-phenylpropionate **(103-25-3)**
2-Methyl-.β.-phenylpropionic acid **(22084-89-5)**
1-Methyl-4-phenyl-4-propionoxypiperidine **(13147-09-6)**
2-Methyl-2-phenylpropyl chloride **(515-40-2)**
4-Methyl-1-phenyl-3-pyrazolidinone **(2654-57-1)**
3-Methyl-1-phenyl-2-pyrazolin-5-one **(89-25-8)**
3-Methyl-1-phenyl-5-pyrazolone **(89-25-8)**
3-Methyl-1-phenyl-5-pyrazolyl dimethyl carbamate **(87-47-8)**
3-Methyl-1-phenylpyrazol-5-yl dimethyl carbamate **(87-47-8)**
1-Methyl-3-phenylpyrrolidin-2,5-dione **(86-34-0)**
Methylphenylsilane difluoride **(328-57-4)**
4-Methyl-3'-phenylspiro(3H-1,4-benzodiazepine-3,2'-oxirane)-2,5(1H,4H)-dione **(20007-87-8)**
Methylphenylsuccinimide **(86-34-0)**
N-Methyl-2-phenyl-succinimide **(86-34-0)**
N-Methyl-α-phenylsuccinimide **(86-34-0)**
Methyl phenylsulfide **(100-68-5)**
Methyl phenyl sulfone **(3112-85-4)**
p-Methylphenylsulfonic acid **(104-15-4)**
4-Methyl-N-(phenylsulfonyl)benzenesulfonamide **(14706-41-3)**
1-Methyl-2-phenyl-3-(2-thiazolylazo)-1H-indole **(34367-95-8)**
(2-Methylphenyl)thiourea **(614-78-8)**
N-Methyl-N'-phenyl thiourea **(2724-69-8)**
1-Methyl-3-phenyl-5-(3-(trifluoromethyl)phenyl)-4(1H)-pyri-dinone **(59756-60-4)**
1-Methyl-3-phenyl-5-(α,α,α-trifluoro-m-tolyl)-4-pyridone **(59756-60-4)**
1-Methyl-3-phenylurea **(1007-36-9)**
3-Methylphenylurea **(63-99-0)**

4-Methylphenylurea **(622-51-5)**
N-Methyl-N'-phenylurea **(1007-36-9)**
Methyl phosphate **(512-56-1)**
Methyl phosphate **(812-00-0)**
O-Methylphosphate **(812-00-0)**
Methyl phosphite **(121-45-9)**
Methyl phosphite **(13590-71-1)**
Methyl phosphonate **(13590-71-1)**
Methylphosphonate **(993-13-5)**
Methyl phosphonic acid, dimethyl ester **(756-79-6)**
Methyl phosphonic dichloride [NA 9206] **(676-97-1)**
Methylphosphonofluoridic acid, 3,3-dimethyl-2-butyl ester **(96-64-0)**
Methylphosphonofluoridic acid isopropyl ester **(107-44-8)**
Methylphosphonofluoridic acid 1-methylethyl ester **(107-44-8)**
Methylphosphonofluoridic acid 1,2,2-trimethylpropyl ester **(96-64-0)**
Methylphosphonothioic acid S-(2-(bis(methylethyl)amino)ethyl) O-ethyl ester **(50782-69-9)**
Methylphosphonothioic acid O,S-diethyl ester **(2511-10-6)**
Methyl phosphonothioic dichloride, Anhydrous **(676-98-2)**
Methylphosphoramidic acid, 4-t-butyl-2-chlorophenyl methyl ester **(299-86-5)**
Methylphosphorothioate((meo)2(mes)po) **(152-20-5)**
3'-Methylphthalanilic acid **(85-72-3)**
3-Methylphthalate **(37102-74-2)**
Methyl phthalate **(131-11-3)**
N-Methylphthalimide **(550-44-7)**
Methyl phthalyl ethyl glycolate **(85-71-2)**
2-Methyl-3-phytyl-1,4-naphthochinon (German) **(84-80-0)**
Methyl picrate **(606-35-9)**
Methyl pinacolyloxy phosphorylfluoride **(96-64-0)**
Methyl pinacolyl phosphonofluoridate **(96-64-0)**
1-Methylpiperazine **(109-01-3)**
N-Methylpiperazine **(109-01-3)**
10-(γ-(N'-Methylpiperazino)propyl)-2-trifluoromethylphenothiozine **(117-89-5)**
3-(4-Methylpiperazinyliminomethyl)-rifamycin SV **(13292-46-1)**
8-(((4-Methyl-1-piperazinyl)imino)methyl)rifamycin SV **(13292-46-1)**
8-(4-Methylpiperazinyliminomethyl) rifamycin SV **(13292-46-1)**
N-(γ-(4'-Methylpiperazinyl-1')propyl)-3-chlorophenothiazine **(58-38-8)**
1-(10-(3-(4-Methyl-1-piperazinyl)propyl)phenothiazin-2-yl)-1-butanone **(653-03-2)**
10-(3-(4-Methyl-1-piperazinyl)propyl)-2-(trifluoromethyl)phenothiazine **(117-89-5)**
2-Methylpiperidine **(109-05-7)**
n-Methylpiperidine **(626-67-5)**
1-Methylpiperidine [UN 2399] **(626-67-5)**
1-Methyl-2-piperidineethanol **(533-15-3)**
Methylpiperidine hydrochloride **(24307-26-4)**
N-(1-Methyl-2-piperidinoethyl)propionanilide **(129-83-9)**
(+-)-N-(1-Methyl-2-piperidinoethyl)-N-2-pyridylpropionamide **(15686-91-6)**
1-Methyl-2-piperidinone **(931-20-4)**
3-(2-Methylpiperidino)propyl 3,4-dichlorobenzoate **(3478-94-2)**
γ-(2-Methylpiperidino)propyl 3,4-dichlorobenzoate **(3478-94-2)**
N-(1-Methyl-2-(1-piperidinyl)ethyl)-N-2-pyridinylpropanamide **(15686-91-6)**
1-Methyl-2-piperidone **(931-20-4)**
N-Methyl-3-piperidyl benzilate **(3321-80-0)**
10-(2-(1-Methyl-2-piperidyl)ethyl)-2-methylsulfinyl phenothiazine **(5588-33-0)**
10-(2-(1-Methyl-2-piperidyl)ethyl)-2-methylsulfonylphenothiazine **(14759-06-9)**
10-(2-(1-Methyl-2-piperidyl)ethyl)-2-(methylthio)phenothiazine **(50-52-2)**
Methylpirimiphos **(29232-93-7)**
Methyl pivalate **(598-98-1)**
Methyl polyglycol **(9004-74-4)**
O-Methyl potassium phenylphosphonothioate **(67446-04-2)**
6-α-Methylprednisolone **(83-43-2)**
Methylprednisolone **(83-43-2)**
6-Methylprednisolone acetate **(53-36-1)**
6-α-Methylprednisolone acetate **(53-36-1)**
Methylprednisolone 21-acetate **(53-36-1)**
Methylprednisolone acetate **(53-36-1)**
16-β-Methyl-1,4-pregnadiene-9-α-fluoro-11-β,17-α,21-triol-3,20-dione **(378-44-9)**
6-Methyl-$\delta^{4,6}$-pregnadien-17-α-ol-3,20-dione acetate **(595-33-5)**
6-α-Methyl-4-pregnene-3,20-dion-17-α-ol acetate **(71-58-9)**
Methylpromazine **(84-96-8)**
Methylpropamine **(300-42-5)**
2-Methyl-1-propanal **(78-84-2)**
2-Methylpropanal **(78-84-2)**
Methyl propanal **(78-84-2)**
2-Methylpropanal dimethyl acetal **(41632-89-7)**
2-Methyl-1-propanal oxime **(151-00-8)**
2-Methylpropanamide **(563-83-7)**
N-Methylpropanamide **(1187-58-2)**
N-Methyl-1,3-propanediamine **(6291-84-5)**
1,1'-(1-Methyl-1,3-propanediyl)bisbenzene **(1520-44-1)**
1-Methyl-1-propanethiol **(513-53-1)**
2-Methyl-2-propanethiol **(75-66-1)**
Methyl propanoate **(554-12-1)**
2-Methylpropanoic acid **(79-31-2)**
1-Methyl propanol **(78-92-2)**
1-Methyl-1-propanol **(78-92-2)**
2-Methyl propanol **(78-83-1)**
2-Methyl-1-propanol **(78-83-1)**
2-Methyl-2-propanol **(75-65-0)**
2-Methylpropan-1-ol **(78-83-1)**
2-Methyl-2-propanol potassium salt **(865-47-4)**
4,4',4''-(1-Methyl-1-propanyl-3-ylidene)tris(2-(1,1-dimethylethyl)-5-methylphenol **(1843-03-4)**
2-Methylpropenal **(78-85-3)**
2-Methylpropenamide **(79-39-0)**
Methyl propenate **(96-33-3)**
2-Methyl-2-propene-1,1-diol diacetate **(10476-95-6)**
2-Methyl-1-propene homopolymer **(9003-27-4)**
2-Methylpropenenitrile **(126-98-7)**
2-Methylpropene polymer **(9003-27-4)**
2-Methyl-1-propene tetramer **(15220-85-6)**
Methyl propenoate **(96-33-3)**
Methyl-2-propenoate **(96-33-3)**
2-Methylpropenoic acid **(79-41-4)**
2-Methyl-2-propenoic acid anhydride **(760-93-0)**
2-Methylpropenoic acid chloride **(920-46-7)**
2-Methyl-2-propenoic acid homopolymer **(25087-26-7)**
2-Methyl-2-propenoic acid methyl ester **(80-62-6)**
2-Methyl-2-propenoic acid methyl ester homopolymer (9CI) **(9011-14-7)**
2-Methyl-2-propenoic acid, sodium salt **(5536-61-8)**
2-Methylpropenoyl chloride **(920-46-7)**
(2-Methyl-1-propenyl)benzene **(768-49-0)**
(2-Methylpropenyl)benzene **(768-49-0)**
Methyl propenyl ketone **(625-33-2)**
2-(2-Methyl-1-propenyl)-6-nitrophenol **(64061-59-2)**
2-(2-Methyl-2-propenyl)-6-nitrophenol **(13414-58-9)**
1-((2-Methyl-2-propenyl)oxy)-2-nitrobenzene **(13414-54-5)**
Methyl propiolate **(922-67-8)**
2-Methylpropionaldehyde **(78-84-2)**
α-Methylpropionaldehyde **(78-84-2)**
2-Methylpropionamide **(563-83-7)**
N-Methylpropionamide **(1187-58-2)**
Methyl propionate [UN 1248] **(554-12-1)**
2-Methylpropionic acid **(79-31-2)**
α-Methylpropionic acid **(79-31-2)**
N-Methylpropionic acid amide **(1187-58-2)**
2-Methylpropionitrile **(78-82-0)**
N-Methylpropionsaureamid (German) **(1187-58-2)**
Methyl 4-(N-propionyl-n-phenylamino)-1-(2-phenylethyl)-4-piperidine-carboxylate **(59708-52-0)**
2-(2-Methylpropoxy)-1,2-diphenylethanone **(22499-12-3)**

(1-Methyl-1-propoxyethyl)benzene **(24142-77-6)**
2-Methyl-1-propyl acetate **(110-19-0)**
2-Methylpropyl acetate **(110-19-0)**
Methylpropylacetic acid **(97-61-0)**
Methyl propyl acetylene **(764-35-2)**
z-Methylpropyl acrylate **(106-63-8)**
1-Methylpropyl alcohol **(78-92-2)**
2-Methylpropyl alcohol **(78-83-1)**
2-Methylpropyl 2-aminobenzoate **(7779-77-3)**
o-Methyl-α-propylaminopropionanilide **(721-50-6)**
Methyl propylate **(554-12-1)**
ar-(1-Methylpropyl)benzenamine **(68400-78-2)**
1-Methyl-2-propylbenzene **(1074-17-5)**
1-Methyl-3-propylbenzene **(1074-43-7)**
1-Methyl-4-propylbenzene **(1074-55-1)**
2-Methylpropyl benzeneacetate **(102-13-6)**
N-(1-Methylpropyl)-2-butanamine **(626-23-3)**
N-(2-Methylpropyl)-1-butanamine **(20810-06-4)**
2-Methylpropyl butyrate **(539-90-2)**
2-Methylpropyl carbamate **(543-28-2)**
Methyl propyl carbinol **(6032-29-7)**
1-Methylpropyl carbonochloridate **(17462-58-7)**
2-Methylpropyl carbonochloridate **(543-27-1)**
Methyl-propyl-cetone (French) **(107-87-9)**
1-Methylpropyl chloroformate **(17462-58-7)**
2-Methylpropyl chloroformate **(543-27-1)**
6-(1-Methyl-propyl)-2,4-dinitrofenol (Dutch) **(88-85-7)**
(6-(1-Methyl-propyl)-2,4-dinitro-fenyl)-3,3-dimethyl-acrylaat (Dutch) **(485-31-4)**
2-(1-Methylpropyl)-4,6-dinitrophenol **(88-85-7)**
2-(1-Methyl-n-propyl) 4,6-dinitrophenol, ammonium salt **(6365-83-9)**
2-(1-Methyl-n-propyl)-4,6-dinitrophenol triethanolamine salt **(6420-47-9)**
2-(1-Methylpropyl)-4,6-dinitrophenyl acetate **(2813-95-8)**
2-(1-Methylpropyl)-4,6-dinitrophenyl β,β-dimethacrylate **(485-31-4)**
(6-(1-Methyl-propyl)-2,4-dinitro-phenyl)-3,3-dimethyl-acrylat (German) **(485-31-4)**
2-(1-Methyl-2-propyl)-4,6-dinitrophenyl isopropylcarbonate **(973-21-7)**
Methyl propyl disulfide **(2179-60-4)**
β-Methylpropyl ethanoate **(110-19-0)**
2-Methyl-2-propylethanol **(105-30-6)**
Methyl n-propyl ether **(557-17-5)**
Methyl propyl ether [UN 2612] **(557-17-5)**
N-(1-Methylpropyl)glycine **(58695-42-4)**
2-Methylpropyl hexadecanoate **(110-34-9)**
2-Methylpropyl hexanoate **(105-79-3)**
2-Methylpropyl isobutyrate **(97-85-8)**
2-Methylpropyl isovalerate **(589-59-3)**
Methyl-n-propyl ketone **(107-87-9)**
Methylpropyl ketone **(107-87-9)**
Methyl propyl ketone (ACGIH,OSHA) [UN 1249] **(107-87-9)**
(1-Methylpropyl)lithium **(598-30-1)**
2-Methylpropyl methacrylate **(97-86-9)**
2-Methylpropyl 3-methylbutyrate **(589-59-3)**
8-Methyl-3-(2-propylpentanoyloxy)tropinium bromide **(80-50-2)**
N-(1-Methylpropyl)-N'-phenyl-1,4-benzenediamine **(788-17-0)**
2-(1-Methylpropyl)phenyl methylcarbamate **(3766-81-2)**
Methyl propyl phosphorothioate ((MeS)(PrO)2PO) (7CI) **(5301-73-5)**
2-Methyl-2-propyl-1,3-propanediol butylcarbamate carbamate **(4268-36-4)**
2-Methyl-2-propyl-1,3-propanediol carbamate isopropylcarbamate **(78-44-4)**
2-Methyl-2-N-propyl-1,3-propanediol dicarbamate **(57-53-4)**
5-(1-Methylpropyl)-5-(2-propenyl)-2,4,6(1H,3H,5H)-pyrimidinetrione **(115-44-6)**
2-Methylpropyl propionate **(540-42-1)**
2-Methyl-4-propylthiazole **(41981-63-9)**
2-Methyl-2-propyltrimethylene butylcarbamate carbamate **(4268-36-4)**
2-Methyl-2-propyltrimethylene carbamate **(57-53-4)**
1-Methyl-2-propynyl m-chlorocarbanilate **(1967-16-4)**
1-Methyl-2-propynyl m-chlorophenylcarbamate **(1967-16-4)**
1-Methylpropynyl 3-chlorophenylcarbamate **(1967-16-4)**
1-Methylpropynyl ester of 3-chlorophenylcarbamic acid **(1967-16-4)**
2-Methyl-5-(2-propynyl)-3-furylmethyl-cis-trans-chrysanthemate **(27223-49-0)**
6-α-Methyl-17-(1-propynyl)testosterone **(79-64-1)**
6-α-Methyl-17-α-propynyltestosterone **(79-64-1)**
6-Methylpurine **(2004-03-7)**
2-Methylpyrazine **(109-08-0)**
5-Methyl-2-pyrazinecarboxylic acid 4-oxide **(51037-30-0)**
1-(3-Methylpyrazinyl)ethanone **(23787-80-6)**
3-Methyl-2-pyrazolin-5-one **(108-26-9)**
3-Methyl-pyrazolon-(5) (German) **(108-26-9)**
3-Methylpyrazolyl-5-diethylphosphate **(108-34-9)**
Methylpyrazolyl diethylphosphate **(108-34-9)**
Methylpyrazolyl diethylthiophosphate **(108-35-0)**
5-Methyl-1H-pyrazol-3-yl dimethylcarbamate **(644-64-4)**
1-Methylpyrene **(2381-21-7)**
2-Methylpyrene **(3442-78-2)**
3-Methylpyrene **(2381-21-7)**
2-Methylpyridine **(109-06-8)**
3-Methylpyridine **(108-99-6)**
4-Methylpyridine **(108-89-4)**
Methylpyridine **(1333-41-1)**
α-Methylpyridine **(109-06-8)**
Methyl 4-pyridinecarboxylate **(2459-09-8)**
2-Methylpyridine 1-oxide **(931-19-1)**
3-Methylpyridine-1-oxide **(1003-73-2)**
4-Methylpyridine 1-oxide **(1003-67-4)**
1-Methylpyridinium chloride **(7680-73-1)**
Methyl pyridinium chloride **(7680-73-1)**
n-Methylpyridinium chloride **(7680-73-1)**
6-Methyl-3-pyridinol **(1121-78-4)**
3-Methyl-1H-pyrido(2,3-b)indol-2-amine **(68006-83-7)**
1-Methyl-9H-pyrido(3,4-b)indole **(486-84-0)**
7-Methylpyrido(3,4-c)psoralen **(85878-63-3)**
Methyl 3-pyridyl ketone **(350-03-8)**
Methyl 4-pyridyl ketone **(1122-54-9)**
Methyl β-pyridyl ketone **(350-03-8)**
Methyl pyridyl ketone **(350-03-8)**
1-Methyl-2-(3-pyridyl)pyrrole **(487-19-4)**
1-Methyl-2-(3-pyridyl)pyrrolidine **(54-11-5)**
l-1-Methyl-2-(3-pyridyl)-pyrrolidine sulfate **(65-30-5)**
4-Methylpyrimidine **(3438-46-8)**
N^1-(4-Methyl-2-pyrimidinyl)sulfanilamide sodium salt **(127-58-2)**
3-Methylpyrocatechol **(488-17-5)**
4-Methylpyrocatechol **(452-86-8)**
p-Methylpyrocatechol **(452-86-8)**
2-Methyl pyromeconic acid **(118-71-8)**
1-Methyl-1H-pyrrole **(96-54-8)**
1-Methylpyrrole **(96-54-8)**
Methylpyrrole **(96-54-8)**
N-Methylpyrrole **(96-54-8)**
1-Methyl-1H-pyrrole-2-carboxaldehyde **(1192-58-1)**
1-Methylpyrrolidine **(120-94-5)**
2-Methylpyrrolidine **(765-38-8)**
N-Methylpyrrolidine **(120-94-5)**
1-Methyl-2-pyrrolidinone **(872-50-4)**
1-Methyl-5-pyrrolidinone **(872-50-4)**
N-Methyl-2-pyrrolidinone **(872-50-4)**
N-Methylpyrrolidinone **(872-50-4)**
3-(N-Methylpyrrolidino)pyridine **(54-11-5)**
10-((1-Methyl-3-pyrrolidinyl)methyl)-phenothiazine **(1982-37-2)**
10-((1-Methyl-3-pyrrolidinyl)methyl)phenothiazine, hydrochloride **(1229-35-2)**
(S)-3-(1-Methyl-2-pyrrolidinyl)pyridine sulfate (2:1) **(65-30-5)**
1-Methyl-2-pyrrolidone **(872-50-4)**
Methylpyrrolidone **(872-50-4)**
N-Methyl-2-pyrrolidone **(872-50-4)**

N-Methyl-α-pyrrolidone **(872-50-4)**
N-Methylpyrrolidone **(872-50-4)**
(-)-3-(1-Methyl-2-pyrrolidyl)pyridine **(54-11-5)**
l-3-(1-Methyl-2-pyrrolidyl)pyridine **(54-11-5)**
l-3-(1-Methyl-2-pyrrolidyl)pyridine sulfate **(65-30-5)**
3-(1-Methyl-2-pyrrolyl)pyridine **(487-19-4)**
2-Methylquinoline **(91-63-4)**
3-Methylquinoline **(612-58-8)**
4-Methylquinoline **(491-35-0)**
6-Methylquinoline **(91-62-3)**
7-Methylquinoline **(612-60-2)**
8-Methylquinoline **(611-32-5)**
γ-Methylquinoline **(491-35-0)**
p-Methylquinoline **(91-62-3)**
2-Methylquinoline hydrochloride **(62763-89-7)**
4-Methylquinoline hydrochloride **(3007-43-0)**
2-Methyl-8-quinolinol **(826-81-3)**
1-Methyl-2(1H)-quinolinone **(606-43-9)**
1-Methyl-2-quinolone **(606-43-9)**
2-Methyl-1,4-quinone **(553-97-9)**
2-Methylquinoxaline **(7251-61-8)**
6-Methyl-2,3-quinoxaline dithiocarbonate **(2439-01-2)**
6-Methyl-2,3-quinoxalinedithiol cyclic S,S-dithiocarbonate **(2439-01-2)**
6-Methyl-2,3-quinoxalinedithiol cyclic carbonate **(2439-01-2)**
6-Methyl-2,3-quinoxalinedithiol cyclic dithiocarbonate **(2439-01-2)**
6-Methyl-quinoxaline-2,3-dithiolcyclocarbonate **(2439-01-2)**
Methylreserpate 3,4,5-trimethoxybenzoic acid **(50-55-5)**
Methyl reserpate 3,4,5-trimethoxybenzoic acid ester **(50-55-5)**
Methyl reserpate 3,4,5-trimethoxycinnamic acid ester **(24815-24-5)**
2-Methylresorcinol **(608-25-3)**
4-Methylresorcinol **(496-73-1)**
5-Methylresorcinol **(504-15-4)**
Methylrhodanid (German) **(556-64-9)**
6-Methyl-9-ribofuranosylpurine-6-thiol **(342-69-8)**
Methyl ricinoleate **(141-24-2)**
Methylrosanilinchlorid (German) **(548-62-9)**
Methylrosaniline chloride **(548-62-9)**
Methylrosanilinum chloratum **(548-62-9)**
N-Methylsaccharin **(15448-99-4)**
Methyl salicylate **(119-36-8)**
4-Methylsalicylic acid **(50-85-1)**
5-Methylsalicylic acid **(89-56-5)**
6-Methylsalicylic acid **(567-61-3)**
o-Methylsalicylic acid **(579-75-9)**
3-Methylsalycilic acid **(83-40-9)**
Methylscopolamine bromide **(155-41-9)**
Methylscopolamine hydrobromide **(155-41-9)**
Methyl scopolamine nitrate **(6106-46-3)**
N-Methylscopolammonium bromide **(155-41-9)**
Methyl sebacate **(106-79-6)**
Methyl selenac **(144-34-3)**
3-Methyl-1,2,5-selenadiazole **(17505-11-2)**
Methyl selenide (8CI) **(593-79-3)**
Methyl selenium **(593-79-3)**
Methylsenfoel (German) **(556-61-6)**
Methylsilanetriol sodium salt **(16589-43-8)**
Methyl silicate (ACGIH,OSHA) **(681-84-5)**
Methyl orthosilicate [UN 2606] **(681-84-5)**
Methyl silicone **(9016-00-6)**
o-Methylsinapic acid **(90-50-6)**
Methyl sorbate **(689-89-4)**
Methyl stearate **(112-61-8)**
Methyl stearate α-sulfonic acid **(3076-26-4)**
(E)-.α.-Methylstilbene **(833-81-8)**
trans-.α.-Methylstilbene **(833-81-8)**
trans-o-Methylstilbene **(14064-48-3)**
α-Methylstyreen (Dutch) **(98-83-9)**
2-Methylstyrene **(611-15-4)**
3-Methylstyrene **(100-80-1)**
Methylstyrene **(25013-15-4)**
α-Methylstyrene **(98-83-9)**
β-Methylstyrene **(637-50-3)**
m-Methylstyrene **(100-80-1)**
o-Methylstyrene **(611-15-4)**
ω-Methylstyrene **(637-50-3)**
p-Methylstyrene **(622-97-9)**
α-Methyl styrene (ACGIH,OSHA) **(98-83-9)**
α-Methylstyrene dimer **(6144-04-3)**
α-Methylstyrene epoxide **(2085-88-3)**
α-Methylstyrene oxide **(2085-88-3)**
α-Methyl-styrol (German) **(98-83-9)**
Methyl succinate **(106-65-0)**
2-Methylsuccinic acid **(498-21-5)**
Methylsuccinic acid **(498-21-5)**
Methylsuccinic anhydride **(4100-80-5)**
N-(2-(Methylsulfamido)ethyl)-N-ethyl-3-methyl-4-nitrosoaniline **(56046-62-9)**
5-Methyl-3-sulfanilamidoisoxazole **(723-46-6)**
5-Methyl-2-sulfanilamido-1,3,4-thiadiazole **(144-82-1)**
Methyl sulfanilyl carbamate **(3337-71-1)**
Methyl sulfanilylcarbamate, sodium salt **(2302-17-2)**
5-Methyl-3-sulfanylamidoisoxazole **(723-46-6)**
Methyl sulfate **(75-93-4)**
Methyl sulfate **(77-78-1)**
Methyl sulfate [UN 1595] **(77-78-1)**
Methyl sulfide, Compd. with borane (1:1) **(13292-87-0)**
Methyl sulfide [UN 1164] **(75-18-3)**
Methylsulfinylmethane **(67-68-5)**
Methyl sulfochloride **(124-63-0)**
Methyl sulfocyanate **(556-64-9)**
1-Methyl 2-sulfohexadecanoate **(58849-75-5)**
Methyl α-sulfomyristate, sodium salt **(4016-22-2)**
Methylsulfonal **(76-20-0)**
Methyl sulfone (8CI) **(67-71-0)**
Methylsulfonic acid, ethyl ester **(62-50-0)**
6-(Methylsulfonyl)-2-benzothiazolamine **(17557-67-4)**
3-(Methylsulfonyl)-2-butanone O-((methylamino)carbonyl)oxime (9CI) **(34681-23-7)**
3-Methylsulfonylbutanone O-methylcarbamoyloxime **(34681-23-7)**
3-(Methylsulfonyl)-2-butanone O-(methylcarbamoyl)oxime (8CI) **(34681-23-7)**
Methyl sulfonyl chloride **(124-63-0)**
Methylsulfonyl chloride **(124-63-0)**
4-(Methylsulfonyl)-2,6-dinitro-N,N-dipropylaniline **(4726-14-1)**
4-Methylsulfonyl-2,6-dinitro-N,N-dipropylaniline **(4726-14-1)**
4-(Methylsulfonyl)-2,6-dinitro-N,N-dipropylbenzenamine **(4726-14-1)**
4-(Methylsulfonyl)-2,6-dinitro-N,N-dipropylbenzeneamine **(4726-14-1)**
Methylsulfonyl methane **(67-71-0)**
2-Methylsulfonyl-O-(N-methyl-carbamoyl)-butanon-(3)-oxim (German) **(34681-23-7)**
1-Methyl 2-sulfooctadecanoate **(3076-26-4)**
1-Methyl 2-sulfotetradecanoate **(29454-23-7)**
Methyl-sulfoxide **(67-68-5)**
(Methyl sulfoxide)-D6 (8CI) **(2206-27-1)**
5-Methyl-3-sulphanil-amidoisoxazole **(723-46-6)**
Methyl sulphide [UN 1164] **(75-18-3)**
Methylsulphonal **(76-20-0)**
3-Methylsulphonylbutanone O-methylcarbamoyloxime **(34681-23-7)**
4-Methylsulphonyl-2,6-dinitro-N,N-dipropylaniline **(4726-14-1)**
Methylsystox **(867-27-6)**
1-Methyl-1-tallowalkylamidoethyl-2-tallowalkylimidazoline methosulfate **(68122-86-1)**
1-Methyl-1-(tallow alkyl amido)-2-nor(tallow alkyl)-2-imidazolinium, methylsulfate **(68122-86-1)**
1-Methyl-1-(2-tallowamidoethyl)-2-tallowimidazolinium methylsulfate **(68122-86-1)**

Methyl tallow diethylenetriamine condensate, polyethoxylated, methyl sulfate (68410-69-5)
N-Methyltaurine (107-68-6)
Methyl terephthalaldehydate (1571-08-0)
Methyl terephthalate (1679-64-7)
Methyl terephthaldehydate (1571-08-0)
17-Methyltestosteron (58-18-4)
17-Methyltestosterone (58-18-4)
17-α-Methyltestosterone (58-18-4)
Methyltestosterone (58-18-4)
δ'-17-Methyltestosterone (72-63-9)
δ^1-17-α-Methyltestosterone (72-63-9)
2-Methyl-3,4,5,6-tetrabromophenol (576-55-6)
Methyl 4,6,6,6-tetrachloro-3,3-dimethylhexanoate (64667-33-0)
Methyl n-tetradecanoate (124-10-7)
Methyl tetradecanoate (124-10-7)
2-Methyltetrahydrofuran (96-47-9)
Methyltetrahydrofuran (96-47-9)
2-Methyltetrahydrofuran-3-one (3188-00-9)
1-Methyl-5,6,7,8-tetrahydronaphthalene (2809-64-5)
2-Methyl-5,6,7,8-tetrahydronaphthalene (1680-51-9)
5-Methyl-1,2,3,4-tetrahydronaphthalene (2809-64-5)
6-Methyl-1,2,3,4-tetrahydronaphthalene (1680-51-9)
2-Methyl-2-(4-(1,2,3,4-tetrahydro-1-naphthalenyl)phenoxy)-propanoic acid (3771-19-5)
2-Methyl-2-(4-(1,2,3,4-tetrahydro-1-naphthyl)phenoxy)propanoic acid (3771-19-5)
2-Methyl-2-(p-(1,2,3,4-tetrahydro-1-naphthyl)phenoxy)propionic acid (3771-19-5)
α-Methyl-α-(p-1,2,3,4-tetrahydronaphth-1-ylphenoxy)propionic acid (3771-19-5)
Methyl tetrahydrophthalic anhydride (11070-44-3)
N-Methyltetrahydropyrrole (120-94-5)
4-Methyl-1,2,3,4-tetrahydroquinoline (19343-78-3)
6-Methyl-1,2,3,4-tetrahydroquinoline (91-61-2)
N-Methyl-tetrahydrothiamidinthione acetic acid (3655-88-7)
5-Methyltetralin (2809-64-5)
6-Methyltetralin (1680-51-9)
Methyltetralin (Czech) (31291-71-1)
2-Methyl-3-(3,7,11,15-tetramethyl-2-hexadecenyl)-1,4-naphthalenedione (84-80-0)
N-Methyl-n,2,4,6-tetranitroaniline (479-45-8)
1-Methyl-1H-tetrazole-5-thiol (13183-79-4)
N-Methyltetrazolethiol (13183-79-4)
Methyl 5-(2-thenoyl)-2-benzimidazolecarbamate (31430-18-9)
Methyltheobromide (58-08-2)
5-Methyl-1,2,3-thiadiazole (50406-54-7)
N^1-(5-Methyl-1,3,4-thiadiazol-2-yl)-sulfanilamide (144-82-1)
N-Methyl-3-thia-2-methyl-valeramid der O,O-dimethylthiol-phosphorsaeure (German) (2275-23-2)
4-Methylthiazole (693-95-8)
4-Methyl-5-thiazoleethanol (137-00-8)
4-Methyl-5-thiazolethanol (137-00-8)
3-Methyl-2-thiazolidinethione (1908-87-8)
Methyl (5-(2-thienylcarbonyl)-1H-benzimidazole-2-yl)carbamate (31430-18-9)
(Methylthio)acetaldehyde oxime (10533-67-2)
2-Methylthio-acetaldehyd-O-(methylcarbamoyl)-oxim (German) (16752-77-5)
(Methylthio)acetaldoxime (10533-67-2)
Methyl thioacetate (1534-08-3)
l-γ-Methylthio-α-aminobutyric acid (63-68-3)
(Methylthio)benzene (100-68-5)
2-(Methylthio)benzothiazole (615-22-5)
2-Methylthio-4,6-bis(ethylamino)-s-triazine (1014-70-6)
2-Methylthio-4,6-bis(isopropylamino)-s-triazine (7287-19-6)
2-Methylthio-4,6-bis(monoethylamino)-2-triazine (1014-70-6)
1-(Methylthio)butane (628-29-5)
3-(Methylthio)-2-butanone O-((methylamino)carbonyl)oxime (34681-10-2)
3-(Methylthio)-2-butanone O-(methylcarbamoyl)oxime (34681-10-2)
3-(Methylthio)butanone O-methylcarbamoyloxime (34681-10-2)
O-Methyl thiochloroformate (2812-72-8)
1-Methylthio-5-chlorosulfonylnaphthalene (53135-95-8)
4-(Methylthio)-m-cresol (3120-74-9)
Methyl thiocyanate (556-64-9)
4-Methylthio-3,5-dimethylphenyl methylcarbamate (2032-65-7)
(Methylthio)ethane (624-89-5)
2-Methylthio-4-ethylamino-6-tert-butylamino-s-triazine (886-50-0)
2-Methylthio-4-ethylamino-6-isopropylamino-s-triazine (834-12-8)
2-Methylthioethyl chloride (542-81-4)
2-Methylthioethyl O,O-dimethyl phosphorothioate (2587-90-8)
5-(2-(Methylthio)ethyl)-2,4-imidazolidinedione (13253-44-6)
Methylthiofanate (23564-05-8)
Methylthioglycolate (2365-48-2)
6-Methylthioinosine (342-69-8)
Methylthioinosine (342-69-8)
2-Methylthio-4-isopropylamino-6-methylamino-s-triazine (1014-69-3)
4-Methylthio-2-ketobutyric acid (583-92-6)
Methylthiokyanat (Czech) (556-64-9)
Methylthiomethane (75-18-3)
3-(Methylthio)-O-((methylamino)carbonyl)oxime-2-butanone (34681-10-2)
2-(Methylthio)-4-(methylamino)-6-(isopropylamino)-s-triazine (1014-69-3)
((Methylthio)methyl)benzene (766-92-7)
2-Methylthio-O-(N-methylcarbamoyl)-butanonoxim-3 (German) (34681-10-2)
2-((Methylthio)methyl)furan (1438-91-1)
Methylthionine (61-73-4)
Methylthionine chloride (61-73-4)
Methylthionium chloride (61-73-4)
4-Methylthio-2-oxobutanoate (583-92-6)
4-(Methylthio)-2-oxobutanoic acid (583-92-6)
Methylthiopentachlorobenzene (1825-19-0)
Methyl thioperoxydiphosphate (7CI) (5930-71-2)
Methyl thiophanate (23564-05-8)
2-Methylthiophene (554-14-3)
3-Methylthiophene (616-44-4)
2-Methylthiophenol (137-06-4)
3-Methylthiophenol (108-40-7)
4-(Methylthio)phenol (1073-72-9)
4-Methylthiophenol (106-45-6)
m-Methylthiophenol (108-40-7)
o-Methylthiophenol (137-06-4)
p-(Methylthio)phenol (1073-72-9)
p-Methylthiophenol (106-45-6)
4-Methylthiophenyldimethyl phosphate (3254-63-5)
4-Methylthiophenyl dipropyl phosphate (7292-16-2)
Methylthiophos (298-00-0)
3-(Methylthio)propanal (3268-49-3)
3-(Methylthio)-1-propanol (505-10-2)
3-(Methylthio)-1-propene (10152-76-8)
3-(Methylthio)propionaldehyde (3268-49-3)
β-(Methylthio)propionaldehyde (3268-49-3)
2-Methylthio-propionaldehyd-O-(methylcarbamoyl)-oxim (German) (16752-77-5)
6-(Methylthio)purine ribonucleoside (342-69-8)
6-Methylthiopurine riboside (342-69-8)
6-Methyl-2-thio-2,4-(1H3H)-pyrimidinedione (56-04-2)
4-Methyl-3-thiosemicarbazide (6610-29-3)
4-Methylthiosemicarbazide (6610-29-3)
Methylthiosemicarbazide (6610-29-3)
Methylthio-s-triazine (26292-91-1)
4-Methyl-2-thiouracil (56-04-2)
6-Methyl-2-thiouracil (56-04-2)
6-Methylthiouracil (56-04-2)
Methylthiouracil (56-04-2)
6-Methyl-2-thiouracyl (Czech) (56-04-2)
1-Methylthiourea (598-52-7)

Methyl thiourea (598-52-7)
S-Methylthiouronium sulfate (2:1) (867-44-7)
2-Methyl-9H-thioxanthen-9-one (15774-82-0)
3-Methyl-2-thioxanthine (28139-02-8)
2-Methylthioxanthone (15774-82-0)
3-Methyl-2-thioxo-4-imidazoline-1-carboxylic acid ethyl ester (22232-54-8)
5-Methyl-6-thioxotetrahydro-3-thiadiazineacetic acid (3655-88-7)
4-Methylthio-3,5-xylenol (7379-51-3)
4-(Methylthio)-3,5-xylyl methylcarbamate (2032-65-7)
Methyl thiram (137-26-8)
Methyl thiuramdisulfide (137-26-8)
Methyl tiglate (6622-76-0)
Methyltin (16408-15-4)
Methyltin trichloride (993-16-8)
Methyltin S,S',S''-tris(isooctyl mercaptoacetate) (54849-38-6)
Methyltin tris(isooctyl mercaptoacetate) (54849-38-6)
Methyltin tris(isooctyl thioglycolate) (54849-38-6)
Methyltin tris(2-mercaptoethyl oleate) (59118-79-5)
Methyl 3-toluate (99-36-5)
Methyl 4-toluate (99-75-2)
Methyl α-toluate (101-41-7)
Methyl m-toluate (99-36-5)
Methyl p-toluate (99-75-2)
Methyl toluene (1330-20-7)
o-Methyltoluene (95-47-6)
p-Methyltoluene (106-42-3)
N-Methyl-p-toluenesulfonamide (640-61-9)
Methyl toluene-4-sulfonate (80-48-8)
Methyl-p-toluenesulfonate (80-48-8)
α-Methyl-α-toluic aldehyde (93-53-8)
2-Methyl-p-toluidine (95-68-1)
4-Methyl-o-toluidine (95-68-1)
5-Methyl-o-toluidine (95-78-3)
6-Methyl-m-toluidine (95-78-3)
2-Methyl-p-toluidine hydrochloride (21436-96-4)
4-Methyl-o-toluidine hydrochloride (21436-96-4)
5-Methyl-o-toluidine hydrochloride (51786-53-9)
6-Methyl-m-toluidine hydrochloride (51786-53-9)
1-Methyl-5-p-toluoyl-pyrrole-2-acetic acid (26171-23-3)
Methyl 3-(m-tolylcarbamoyloxy)phenylcarbamate (13684-63-4)
Methyl-p-tolylcarbinol (536-50-5)
2-Methyl-3-o-tolyl-4(3H)-chinazolinon (German) (72-44-6)
2-Methyl-3-o-tolyl-4(3H)-chinazolone (72-44-6)
2-Methyl-3-(o-tolyl)-3,4-dihydro-4-quinazolinone (72-44-6)
Methyl m-tolyl ether (100-84-5)
Methyl p-tolyl ether (104-93-8)
Methyl p-tolyl ketone (122-00-9)
2-Methyl-3-tolyl-4-oxybensdiazine (72-44-6)
3-Methyl-1-p-tolyl-pyrazolin-5-one (86-92-0)
2-Methyl-3-o-tolyl-4(3H)-quinazolinone (72-44-6)
2-Methyl-3-(2-tolyl)quinazol-4-one (72-44-6)
2-Methyl-3-o-tolyl-4-quinazolone (72-44-6)
2-Methyl-3-o-tolyl-6-sulfamyl-7-chloro-1,2,3,4-tetrahydro-4-quinazolinone (17560-51-9)
Methyl (p-tolylsulfonyl)acetate (50397-64-3)
Methyl p-tosylate (80-48-8)
Methyl tosylate (80-48-8)
Methylcistox (867-27-6)
Methyltriacetoxysilane (4253-34-3)
2-Methyl-1,3,5-triazine (3599-87-9)
Methyl-s-triazine (3599-87-9)
6-Methyl-1,3,5-triazine-2,4-diamine (542-02-9)
5-Methyl-1,2,4-triazole(3,4-b)benzothiazole (41814-78-2)
5-Methyl-(1,2,4)triazolo(1,5-a)pyrimidin-7-ol (2503-56-2)
1-Methyl-N,N',N''-tributylsilanetriamine (16411-33-9)
Methyltricaprylylammonium chloride (5137-55-3)
Methyl trichloride (67-66-3)
Methyl trichloroacetate (598-99-2)
Methyl trichloroacetate [UN 2533] (598-99-2)
Methyltrichlorobenzene (30583-33-6)
Methyltrichloromethane (71-55-6)
Methyl 2,4,6-trichlorophenyl ether (87-40-1)
Methyltrichlorosilane [UN 1250] (75-79-6)
Methyltrichlorostannane (993-16-8)
Methyltrichlorotin (993-16-8)
Methyl-trichlorsilan (Czech) (75-79-6)
6-Methyltridecane (13287-21-3)
7-Methyltridecane (26730-14-3)
Methyl tridecanoate (1731-88-0)
7-Methyl-6-tridecene (24949-42-6)
Methyl triethanol ammonium silicate (12687-85-3)
Methyltriethanolammonium silicate (12687-85-3)
Methyltriethoxysilane (2031-67-6)
Methyltriethyllead (1762-28-3)
Methyltriethylplumbane (1762-28-3)
Methyl triflate (333-27-7)
Methyl trifluoride (75-46-7)
Methyl trifluoroacetate (431-47-0)
1-Methyl-2,2,2-trifluoroethanol (374-01-6)
Methyl trifluoromethane sulfonate (333-27-7)
Methyl trifluoromethanesulfonate (333-27-7)
Methyl trifluoromethyl ketone (421-50-1)
(N-4-Methyl-(((1,1,1-trifluoromethyl)sulfonyl)amino)phenyl)-acetamide (47000-92-0)
6-Methyl-1,3,8-trihydroxyanthraquinone (518-82-1)
Methyl 3,4,5-trihydroxybenzoate (99-24-1)
Methyl trimellitate (2459-10-1)
Methyltrimethanolmethane (77-85-0)
Methyl 18-O-(3,4,5-trimethoxycinnamoyl)reserpate (24815-24-5)
Methyltrimethoxysilane (1185-55-3)
2,2'-(1-Methyltrimethylenedioxy)bis(4-methyl-1,3,2-dioxaborinane) (14697-50-8)
2,2'-(1-Methyltrimethylenedioxy)bis(4-methyl-1,3,2-dioxaborinane) (2665-13-6)
Methyltrimethylene glycol (107-88-0)
1-Methyltrimethylene oxide (2167-39-7)
Methyltrimethylolmethane (77-85-0)
Methyltrioctylammonium chloride (5137-55-3)
Methyltris(cyclohexylamino)silane (15901-40-3)
Methyltris(2-ethylhexyloxycarbonylmethylthio)stannane (57583-34-3)
Methyltris(trimethylsiloxy)silane (17928-28-8)
Methyl trisulfide (8CI) (3658-80-8)
Methyl trithion (953-17-3)
8-Methyltropinium bromide 2-propylvalerate (80-50-2)
4-Methylumbelliferon (Czech) (90-33-5)
4-Methylumbelliferone (90-33-5)
β-Methylumbelliferone (90-33-5)
4-Methylumbelliferone-O,O-diethyl thiophosphate (299-45-6)
2-Methyl-1-undecanal (110-41-8)
2-Methylundecanal (110-41-8)
Methyl undecanoate (1731-86-8)
Methyl 10-undecenate (111-81-9)
Methyl undecenate (111-81-9)
Methyl 10-undecenoate (111-81-9)
Methyl undecenoate (111-81-9)
Methyl undecyl ketone (593-08-8)
4-Methyluracil (56-04-2)
4-Methyluracil (626-48-2)
5-Methyluracil (65-71-4)
6-Methyluracil (626-48-2)
1-Methylurea (598-50-5)
Methylurea (598-50-5)
n-Methylurea (598-50-5)
Methylurethan (598-55-0)

N-Methyl urethan (105-40-8)
Methylurethane (598-55-0)
4-Methyl valeraldehyde (1119-16-0)
α-Methylvaleraldehyde [UN 2367] (123-15-9)
Methyl n-valerate (624-24-8)
Methyl valerate (624-24-8)
Methyl valerianate (624-24-8)
2-Methylvaleric acid (97-61-0)
3-Methylvaleric acid (105-43-1)
α-Methylvaleric acid (97-61-0)
α-Methylvaleric acid, 3,4-dichloroanilide (2533-89-3)
4-Methylvaleronitrile (542-54-1)
4-o-Methylvanillin (120-14-9)
Methylvanillin (120-14-9)
trans-8-Methyl-N-vanillyl-6-nonenamide (404-86-4)
4-Methylveratrol (494-99-5)
4-Methylveratrole (494-99-5)
Methylvinyl acetate (108-22-5)
α-Methylvinyl bromide (557-93-7)
Methyl vinyl carbinol (598-32-3)
Methyl-vinyl-cetone (French) (78-94-4)
Methyl vinyl ether (107-25-5)
Methylvinylether ethylenglykolu (Czech) (1663-35-0)
Methylvinylketon (German) (78-94-4)
Methylvinyl ketone (78-94-4)
Methyl vinyl ketone, Inhibited (DOT) (78-94-4)
Methyl vinyl ketone [UN 1251] (78-94-4)
Methylvinylnitrosamin (German) (4549-40-0)
Methylvinylnitrosamine (4549-40-0)
2-Methyl-5-vinylpyrazine (13925-08-1)
2-Methyl-5-vinylpyridine (140-76-1)
4-Methyl-5-vinyl thiazole (1759-28-0)
Methylviolett (German) (548-62-9)
Methylviologen (1910-42-5)
Methyl viologen (Reduced) (1910-42-5)
Methyl viologen dichloride (1910-42-5)
7-Methylxanthin (552-62-5)
1-Methylxanthine (6136-37-4)
3-Methylxanthine (1076-22-8)
N-Methyl-N'-2,4-xylyl-N-(N-2,4-xylylformimidoyl)formamidine (33089-61-1)
Methylzineb (12071-83-9)
Methyprolon (125-64-4)
Methyprylon (125-64-4)
Methysergid (361-37-5)
Methysergide (361-37-5)
Meti-Derm (50-24-8)
Meticlorpindol (2971-90-6)
Meticortelone (50-24-8)
Meticorten (53-03-2)
Metifonate (52-68-6)
Metilacetona (Spanish) (8013-65-8)
Metilacrilato (Italian) (96-33-3)
Metilamil alcohol (Italian) (108-11-2)
Metilamine (Italian) (74-89-5)
Metilar (53-33-8)
Metilbenzetonio cloruro (25155-18-4)
3-Metil-butanolo (Italian) (123-51-3)
2-(1-Metil-butil)-4,6-dinitro-fenolo (Italian) (4097-36-3)
Metil cellosolve (Italian) (109-86-4)
2-Metilcicloesanone (Italian) (583-60-8)
Metilcloroformiato (Italian) (79-22-1)
2'-Metil-3'-dimetilamino-propil-5-iminodibenzile (Italian) (739-71-9)
N-Metil-ditiocarbammato di sodio (Italian) (137-42-8)
Metile (acetato di) (Italian) (79-20-9)
O,O-Metilen-bis(4-cloro-fenolo) (Italian) (97-23-4)
3,3'-Metilen-bis(4-idrossi-cumarina) (Italian) (66-76-2)
4,4-Metilene-bis-o-cloroanilina (Italian) (101-14-4)
(6-(1-Metil-epitl)-2,4-dinitro-fenil)-crotonato (Italian) (39300-45-3)
Metiletilchetone (Italian) (78-93-3)
(E)-1-Metiletil-3-(((etilamino)metoxifosfinotiol)oxi)-2-butenoato (Spanish) (31218-83-4)
Metil (formiato di) (Italian) (107-31-3)
Metilisobutilchetone (Italian) (108-10-1)
Metil isocianato (Italian) (624-83-9)
Metilmercaptano (Italian) (74-93-1)
Metilmercaptofosoksid (301-12-2)
Metil metacrilato (Italian) (80-62-6)
N-Metil-1-naftil-carbammato (Italian) (63-25-2)
Metilparafinolo (77-75-8)
Metilparation (Hungarian) (298-00-0)
Metilpentadieno (Spanish) (926-56-7)
4-Metilpentan-2-olo (Italian) (108-11-2)
4-Metilpentan-2-one (Italian) (108-10-1)
4-Metil-3-penten-2-one (Italian) (141-79-7)
3-Metil-pentin-3-ol (Italian) (77-75-8)
Metilpentinolo (77-75-8)
(6-(1-Metil-propil)-2,4-dinitro-fenil)-3,3-dimetil-acrilato (Italian) (485-31-4)
6-(1-Metil-propil)-2,4-dinitro-fenolo (Italian) (88-85-7)
α-Metil-stirolo (Italian) (98-83-9)
2-Metil-2-tiometil-propionaldeid-O-(N-metil-carbamoil)-ossima (Italian) (116-06-3)
6-Metil-tiouracile (Italian) (56-04-2)
Metiltriazotion (86-50-0)
Metindol (53-86-1)
Metione (59-51-8)
D-Metionien (Australian) (348-67-4)
Metipregnone (71-58-9)
Metiprilone (125-64-4)
Metiram (9CI) (9006-42-2)
Metirame zinc (French) (9006-42-2)
Metizol (60-56-0)
Metizolin (3813-05-6)
Metmercapturon (2032-65-7)
Metobromuron (3060-89-7)
Metocarbamol (532-03-6)
Metocarbamolo (532-03-6)
Metochin (69-05-6)
Metochlopramide (364-62-5)
Metoclol (364-62-5)
Metoclopramide (364-62-5)
Metoesital (151-83-7)
Metofane (76-38-0)
Metofenina (532-03-6)
Metohexital (Spanish) (151-83-7)
2-Metoksy-4-allilofenol (Polish) (97-53-0)
Metoksychlor (Polish) (72-43-5)
Metoksyetylowy alkohol (Polish) (109-86-4)
3-Metoksy-4-hydroksyacetofenon (Polish) (498-02-2)
S-2-Metoksy-1,3,4-tiadiazolo-5-on-N-metylo-O,O-dwumetylowy (Polish) (950-37-8)
Metol (55-55-0)
Metolachlor (51218-45-2)
Metolazone (17560-51-9)
Metolcarb (1129-41-5)
Metolose MC 8000 (9004-67-5)
Metolose 60SH (9004-67-5)
Metolose 60SH400 (9004-67-5)
Metolose SM 15 (9004-67-5)
Metolose SM 100 (9004-67-5)
Metolose SM 4000 (9004-67-5)
Metolquizolone (72-44-6)
Metomil (Italian) (16752-77-5)
Metopiron (54-36-4)

Metopirone (54-36-4)
Metopon (143-52-2)
Metopone (143-52-2)
Metoponum (Latin) (143-52-2)
(+-)-Metoprolol (37350-58-6)
Metoprolol (37350-58-6)
Metoprotryn (841-06-5)
Metoprotryne (841-06-5)
Metopryl (557-17-5)
Metopyrone (54-36-4)
Metoquin (69-05-6)
Metoquine (69-05-6)
(2-Metossicarbonil-1-metil-vinil)-dimetil-fosfato (Italian) (7786-34-7)
2-Metossietanolo (Italian) (109-86-4)
2-Metossietilacetato (Italian) (110-49-6)
Metossipropandiolo (93-14-1)
S-((5-Metossi-4H-piron-2-il)-metil)-O,O-dimetil-monotio-fosfato (Italian) (2778-04-3)
Metothyrine (60-56-0)
Metox (103-17-3)
Metox (72-43-5)
Metoxal (723-46-6)
Metoxfluran (76-38-0)
Metoxifluran (76-38-0)
Metoxon (101-21-3)
Metoxuron (19937-59-8)
Metoxyde (99-76-3)
Metractyl (57-53-4)
Metramac (78-53-5)
Metramak (78-53-5)
Metranil (78-11-5)
Metrazol (54-95-5)
Metrazole (54-95-5)
Metriben (2307-49-5)
Metribuzin (ACGIH,OSHA) (21087-64-9)
Metrifonate (52-68-6)
Metriol (77-85-0)
Metriol trinitrate (3032-55-1)
Metriphonate (52-68-6)
Metrisone (83-43-2)
Metrodiol (297-76-7)
Metrodiol diacetate (297-76-7)
Metrogen Red former KB Soln (95-79-4)
Metron (298-00-0)
Metron (86-21-5)
Metrone (58-18-4)
Metronidaz (443-48-1)
Metronidazol (443-48-1)
Metronidazole (443-48-1)
Metronidazolo (443-48-1)
Metroxedrine (61-76-7)
Metso 20 (6834-92-0)
Metso Beads 2048 (6834-92-0)
Metso Beads, Drymet (6834-92-0)
Metso Pentabead 20 (6834-92-0)
Metsuccimide (77-41-8)
Metsulfuron methyl (74223-64-6)
Metsulfuron methyl ester (74223-64-6)
Metural (140-95-4)
Metylal (Polish) (109-87-5)
Metyleno-bis-fenyloizocyjanian (Polish) (822-06-0)
Metylenu chlorek (Polish) (75-09-2)
Metylfenemal (115-38-8)
Metyloamina (Polish) (74-89-5)
Metylocykloheksan (Polish) (108-87-2)
Metylocykloheksanol (Polish) (25639-42-3)
Metylocykloheksanon (Polish) (1331-22-2)
Metyloetyloketon (Polish) (78-93-3)
Metylohydrazyna (Polish) (60-34-4)
Metyloizobutyloketon (Polish) (108-10-1)
1-Metylo 2 merkaptoimidazolem (Polish) (60-56-0)
N-Metylo-N'-nitro-N-nitrozoguanidyny (Polish) (70-25-7)
Metyloparation (Polish) (298-00-0)
Metylopropyloketon (Polish) (107-87-9)
Metylowy alkohol (Polish) (67-56-1)
Metylparation (Czech) (298-00-0)
Metylu bromek (Polish) (74-83-9)
Metylu chlorek (Polish) (74-87-3)
Metylu jodek (Polish) (74-88-4)
Metyna (115-38-8)
Metyrapon (54-36-4)
Metyrapone (54-36-4)
Mevastatin (73573-88-3)
Mevastatina (Spanish) (73573-88-3)
Mevastatine (French) (73573-88-3)
Mevastatinum (Latin) (73573-88-3)
Mevinfos (Dutch) (7786-34-7)
Mevinolin (75330-75-5)
Mevinox (7786-34-7)
(E)-Mevinphos (298-01-1)
cis-Mevinphos (338-45-4)
trans-Mevinphos (298-01-1)
Mevinphos (ACGIH,DOT,OSHA) (7786-34-7)
Mevinphos Mixture, Dry (DOT) (7786-34-7)
Mevinphos Mixture, Wet (DOT) (7786-34-7)
Mexacarbate (DOT) (315-18-4)
Mexene (137-30-4)
Mexide (83-79-4)
Mexidex (50-02-2)
Mexocine (127-33-3)
Mexoryl SD (15087-24-8)
Mexpectin (9000-69-5)
Meypralgin R/LV (9005-38-3)
Mezaronil (4658-28-0)
Mezaton (59-42-7)
Mezaton (61-76-7)
R(-)-Mezaton (59-42-7)
Mezcaline (54-04-6)
Mezclas de acido clorhidrico y acido nitrico (Spanish) (8007-56-5)
Mezclas de dioxido de carbono y oxido nitroso (Spanish) (53569-62-3)
Mezclas de dioxido de carbono y oxigeno (Spanish) (8063-77-2)
Mezclas estabilizadas de metilacetileno y propadieno (Spanish) (59355-75-8)
Mezcline (54-04-6)
Mezene (137-30-4)
Mezepan (2898-12-6)
Mezerein (34807-41-5)
Mezidine (88-05-1)
Mezineb (12071-83-9)
Mezlocillin (51481-65-3)
Mezolin (53-86-1)
Mezopur (20354-26-1)
Mezotox (1836-75-5)
Mezox K (72-43-5)
Mezuron (4658-28-0)
Mg 164 (1522-00-5)
Mglawik F (122-14-5)
Mglawik L (58-89-9)
Mhoromer (868-77-9)
Mhoromer BM801 (79-39-0)
Mit-C (50-07-7)
Mi-Pilo Ophth Sol (54-71-7)
Mio-Sed (80-77-3)
Miazine (289-95-2)
Miazole (288-32-4)

Name	CAS No.	Name	CAS No.
Mica	**(12003-38-2)**	Microsulfon	**(68-35-9)**
Mica (ACGIH,OSHA)	**(12001-26-2)**	Microtalco IT Extra	**(14807-96-6)**
Michler Ketone	**(90-94-8)**	Microtan Pirazolo	**(526-08-9)**
Michler's Base	**(101-61-1)**	Microtex Lake Red CR	**(5160-02-1)**
Michler's Hydride	**(101-61-1)**	Microthene	**(9002-88-4)**
Michler's Hydrol	**(119-58-4)**	Microthene 510	**(9002-88-4)**
p,p'-Michler's Hydrol	**(119-58-4)**	Microthene 704	**(9002-88-4)**
p,p'-Michler's Ketone	**(90-94-8)**	Microthene 710	**(9002-88-4)**
Michler's Methane	**(101-61-1)**	Microthene FN 500	**(9002-88-4)**
Michler's ethyl ketone	**(90-93-7)**	Microthene FN 510	**(9002-88-4)**
Michler's ketone	**(90-94-8)**	Microthene MN 754-18	**(9002-88-4)**
Micide	**(12122-67-7)**	Microtomic 280	**(110-30-5)**
Micloretin	**(56-75-7)**	Microtrim	**(8064-90-2)**
Micochlorine	**(56-75-7)**	Microzul	**(3691-35-8)**
Micoclorina	**(56-75-7)**	Midazolam	**(59467-70-8)**
Micofume	**(533-74-4)**	Midazolamum (Latin)	**(59467-70-8)**
Micol	**(57-09-0)**	Middle Chrome	**(1344-37-2)**
Micrest	**(56-53-1)**	Middle Chrome BHG	**(1344-37-2)**
Micro Ace K1	**(14807-96-6)**	Middle Coal Tar Distillate	**(65996-92-1)**
Micro Ace L1	**(14807-96-6)**	Middle Tar Distillate (Coal)	**(65996-92-1)**
Micro-Cel	**(1344-95-2)**	Mideton Fast Red Violet R	**(128-95-0)**
Micro-Cel A	**(1344-95-2)**	Midicel	**(80-35-3)**
Micro-Cel B	**(1344-95-2)**	Midikel	**(80-35-3)**
Micro-Cel C	**(1344-95-2)**	Midlon Fast Yellow E	**(12217-38-8)**
Micro-Cel E	**(1344-95-2)**	Midlon Red PG	**(3567-65-5)**
Micro-Cel T	**(1344-95-2)**	Midlon Red PRS	**(6459-94-5)**
Micro-Cel T26	**(1344-95-2)**	Midlon Yellow Propyl	**(6375-55-9)**
Micro-Cel T38	**(1344-95-2)**	Midone	**(125-33-7)**
Micro-Cel T41	**(1344-95-2)**	Miedzian	**(1332-40-7)**
Micro-Check 12	**(133-06-2)**	Miedzian 50	**(1332-40-7)**
Micro-Chek 11	**(26530-20-1)**	Mielevcin	**(55-98-1)**
Micro-Chek 11D	**(26530-20-1)**	Mielosan	**(55-98-1)**
Micro-Chek Skane	**(26530-20-1)**	Mielucin	**(55-98-1)**
Micro DDT 75	**(50-29-3)**	Mieobromol	**(488-41-5)**
Micro Dry	**(12042-91-0)**	Mierenzuur (Dutch)	**(64-18-6)**
Micro-Lex Green 5B	**(128-80-3)**	Mighty 150	**(91-20-3)**
Micro-Pen	**(54-35-3)**	Mighty RD1	**(91-20-3)**
Microcal 160	**(1344-95-2)**	Miike 20	**(1333-86-4)**
Microcal ET	**(1344-95-2)**	Mikacion Brilliant Red 2BS	**(17752-85-1)**
Microcetina	**(56-75-7)**	Mikacion Brilliant Red 5BS	**(17804-49-8)**
Microcide	**(79-07-2)**	Mikal	**(39148-24-8)**
Microcop	**(1332-40-7)**	Mikametan	**(53-86-1)**
Microdiol	**(50-28-2)**	Mikardol	**(78-11-5)**
Microest	**(56-53-1)**	Mikephor TB	**(16090-02-1)**
Microflotox	**(7704-34-9)**	Miketazol Developer NDF	**(135-61-5)**
Microgrit WCA	**(1344-28-1)**	Miketazol Developer NLF	**(135-62-6)**
Microlith Green G-FP	**(1328-53-6)**	Miketazol Developer ONS	**(92-70-6)**
Microlysin	**(76-06-2)**	Mikethren Brilliant Orange GR	**(4424-06-0)**
Micron White 5000A	**(14807-96-6)**	Mikethren Gold Orange G	**(128-70-1)**
Micron White 5000P	**(14807-96-6)**	Mikethren Navy Blue FRA	**(1324-54-5)**
Micron White 5000S	**(14807-96-6)**	Mikethrene Blue BC	**(130-20-1)**
Micronex	**(1333-86-4)**	Mikethrene Blue RSN	**(81-77-6)**
Micronor	**(68-22-4)**	Mikethrene Brilliant Blue R	**(81-77-6)**
Micropor	**(101-25-7)**	Mikethrene Brilliant Green B	**(128-58-5)**
Microsetile Blue EB	**(2475-45-8)**	Mikethrene Brilliant Green FFB	**(128-58-5)**
Microsetile Blue FF	**(2475-46-9)**	Mikethrene Brilliant Pink R	**(2379-74-0)**
Microsetile Blue FF	**(86722-66-9)**	Mikethrene Brown BR	**(2475-33-4)**
Microsetile Blue FFR	**(2475-46-9)**	Mikethrene Brown GR	**(2475-33-4)**
Microsetile Blue FFR	**(86722-66-9)**	Mikethrene Gold Orange G	**(128-70-1)**
Microsetile Diazo Black G	**(539-17-3)**	Mikethrene Gold Yellow	**(128-66-5)**
Microsetile Orange RA	**(82-28-0)**	Mikethrene Marine Blue G	**(6424-76-6)**
Microsetile Pink BN	**(116-85-8)**	Mikethrene Olive R	**(2379-81-9)**
Microsetile Violet B	**(1220-94-6)**	Mikethrene Orange GR	**(4424-06-0)**
Microsetile Violet 3R	**(128-95-0)**	Mikethrene Orange R	**(3263-31-8)**
Microsetile Yellow GR	**(2832-40-8)**	Miketon Brilliant Blue B	**(2475-46-9)**
Microsetile Yellow 2R	**(119-15-3)**	Miketon Brilliant Blue B	**(86722-66-9)**
Microsul	**(144-82-1)**	Miketon Fast Blue	**(2475-45-8)**

Miketon Fast Blue B (2475-45-8)
Miketon Fast Pink FF 3B (2872-48-2)
Miketon Fast Red Violet R (128-95-0)
Miketon Fast Violet B (82-33-7)
Miketon Fast Yellow G (2832-40-8)
Miketon Polyester Blue FBL (12217-79-7)
Miketon Polyester Red FB (17418-58-5)
Miketon Polyester Yellow 5R (6300-37-4)
Mikrofor N (101-25-7)
Mikrolour (9002-88-4)
Mikrolour (9003-07-0)
Mil-H-19457C (28777-70-0)
Mil-Col (5707-69-7)
Mil-Du-Rid (132-27-4)
Milbam (137-30-4)
Milban (137-30-4)
Milban (31717-87-0)
Milban F (123-28-4)
Milbedoce (68-19-9)
Milbol (115-32-2)
Milbol 49 (58-89-9)
Milchsaure (German) (50-21-5)
Milchsaure (German) (598-82-3)
Milcurb (23947-60-6)
Milcurb Super (23947-60-6)
Mildex (39300-45-3)
Mildmen (58-25-3)
Mildothane (23564-05-8)
Milecitan (55-98-1)
Milepsin (125-33-7)
Mileran (55-98-1)
Milestrol (56-53-1)
Milfaron (20856-57-9)
Milgo (23947-60-6)
Milgo E (23947-60-6)
Milk Acid (50-21-5)
Milk Of Magnesia (1309-42-8)
Milk Protein (9000-71-9)
Milk Sugar (63-42-3)
Milk White (7446-14-2)
Miller Drine (300-42-5)
Miller Nu Set (93-72-1)
Miller P.C. Weedkiller (590-28-3)
Millerite (16812-54-7)
Miller's Fumigrain (107-13-1)
Millicorten (50-02-2)
Millinese (94-20-2)
Milling Brilliant Scarlet GN (3567-65-5)
Milling Fast Red B (6459-94-5)
Milling Fast Red G (3567-65-5)
Milling Fast Red GL (3567-65-5)
Milling Fast Red PG (3567-65-5)
Milling Red A (10169-02-5)
Milling Red B (6459-94-5)
Milling Red BB (6459-94-5)
Milling Red J (3567-65-5)
Milling Red SWB (6459-94-5)
Milling Red SWG (3567-65-5)
Milling Scarlet DH (10169-02-5)
Milling Scarlet 2G (10169-02-5)
Milling Scarlet G (3567-65-5)
Milling Scarlet R (10169-02-5)
Milling Yellow 3G (6375-55-9)
Milling Yellow 3J (6375-55-9)
Milling Yellow RX (6375-55-9)
Mill scale, Ferrous metal (65996-74-9)
Milmer (10380-28-6)
Milo-Pro (139-40-2)
Milogard (139-40-2)
Milontin (86-34-0)
Milorganite (8CI,9CI) (8049-99-8)
Milori Blue (14038-43-8)
Milprem (57-53-4)
Milprex (2439-10-3)
Milstem (23947-60-6)
Milstem Seed dressing (23947-60-6)
Miltamato (57-53-4)
Miltann (57-53-4)
Miltaun (57-53-4)
Milton (7681-52-9)
Miltown (57-53-4)
Miltox (12122-67-7)
Miltox Special (12122-67-7)
Miltuan (57-53-4)
Miltwon (57-53-4)
Mimetina (60-13-9)
Mimosa-Tannin (1401-55-4)
Mimosin (500-44-7)
(L)-Mimosine (500-44-7)
L-Mimosine (500-44-7)
Mimosine (500-44-7)
Min-U-Gel 200 (12174-11-7)
Min-U-Gel 400 (12174-11-7)
Min-U-Gel FG (12174-11-7)
Min-U-Sil (14808-60-7)
Minacide (2631-37-0)
Minaphil (317-34-0)
Mineral Carbon (7782-42-5)
Mineral Fat (8009-03-8)
Mineral Fire Red 5DDS (12656-85-8)
Mineral Fire Red 5GS (12656-85-8)
Mineral Grease (Petrolatum) (8009-03-8)
Mineral Green (12002-03-8)
Mineral Jelly (8009-03-8)
Mineral Jelly No. 14 (8009-03-8)
Mineral Jelly No. 17 (8009-03-8)
Mineral-Oil (8012-95-1)
Mineral Oil, Petroleum condensates, Vacuum tower (64741-49-7)
Mineral Oil, Slab Oil (8042-47-5)
Mineral Orange (1314-41-6)
Mineral Pitch (8052-42-4)
Mineral Red (1314-41-6)
Mineral Wax (8009-03-8)
Mineral naphtha (71-43-2)
Mineral oil, Petroleum distillates, Acid-treated heavy naphthenic (64742-18-3)
Mineral oil, Petroleum distillates, Acid-treated heavy paraffinic (64742-20-7)
Mineral oil, Petroleum distillates, Acid-treated light naphthenic (64742-19-4)
Mineral oil, Petroleum distillates, Acid-treated light paraffinic (64742-21-8)
Mineral oil, Petroleum distillates, Heavy naphthenic (64741-53-3)
Mineral oil, Petroleum distillates, Heavy paraffinic (64741-51-1)
Mineral oil, Petroleum distillates, Hydrotreated heavy naphthenic (64742-52-5)
Mineral oil, Petroleum distillates, Hydrotreated heavy paraffinic (64742-54-7)
Mineral oil, Petroleum distillates, Hydrotreated light naphthenic (64742-53-6)
Mineral oil, Petroleum distillates, Hydrotreated light paraffinic (64742-55-8)
Mineral oil, Petroleum distillates, Light naphthenic (64741-52-2)
Mineral oil, Petroleum distillates, Light paraffinic (64741-50-0)
Mineral oil, Petroleum distillates, Solvent-dewaxed heavy naphthenic (64742-63-8)
Mineral oil, Petroleum distillates, Solvent-dewaxed heavy paraffinic (64742-65-0)
Mineral oil, Petroleum distillates, Solvent-dewaxed light naphthenic (64742-64-9)
Mineral oil, Petroleum distillates, Solvent-dewaxed light paraffinic (64742-56-9)

Mineral oil, Petroleum distillates, Solvent-refined heavy naphthenic **(64741-96-4)**
Mineral oil, Petroleum distillates, Solvent-refined heavy paraffinic **(64741-88-4)**
Mineral oil, Petroleum distillates, Solvent-refined light naphthenic **(64741-97-5)**
Mineral oil, Petroleum distillates, Solvent-refined light paraffinic **(64741-89-5)**
Mineral oil, Petroleum extracts, Heavy naphthenic distillate solvent **(64742-11-6)**
Mineral oil, Petroleum extracts, Heavy paraffinic distillate solvent **(64742-04-7)**
Mineral oil, Petroleum extracts, Light naphthenic distillate solvent **(64742-03-6)**
Mineral oil, Petroleum extracts, Light paraffinic distillate solvent **(64742-05-8)**
Mineral oil, Petroleum extracts, Residual oil solvent **(64742-10-5)**
Mineral oil, Petroleum naphthenic oils, Catalytic dewaxed heavy **(64742-68-3)**
Mineral oil, Petroleum naphthenic oils, Catalytic dewaxed light **(64742-69-4)**
Mineral oil, Petroleum paraffin oils, Catalytic dewaxed heavy **(64742-70-7)**
Mineral oil, Petroleum paraffin oils, Catalytic dewaxed light **(64742-71-8)**
Mineral oil, Petroleum residual oils, Acid-treated **(64742-17-2)**
Mineral oil sulfonic acids, sodium salts **(68608-26-4)**
Mineral-spirits **(64475-85-0)**
Minetoin **(57-41-0)**
Minetoin **(630-93-3)**
Minex **(40596-69-8)**
Mingit **(58-54-8)**
Minihist **(59-33-6)**
Minihist **(91-84-9)**
Miniplanor **(315-30-0)**
Minium **(1314-41-6)**
Minium Non Setting RL-95 **(1314-41-6)**
Minocyclin **(10118-90-8)**
Minocycline **(10118-90-8)**
Minophagen A **(1119-34-2)**
Mint-O-Mag **(1309-42-8)**
Mintaco **(311-45-5)**
Mintacol **(311-45-5)**
Mintal **(57-33-0)**
Mintesol **(148-79-8)**
Mintezol **(148-79-8)**
Mintussin **(2870-71-5)**
Minus **(9005-38-3)**
Minzil **(58-94-6)**
Minzolum **(148-79-8)**
Mioartrina **(64-55-1)**
Mioartrina **(78-44-4)**
Mioblock **(15500-66-0)**
Miocurin **(93-14-1)**
Miofilin **(317-34-0)**
Miolaxene **(532-03-6)**
Miolisodal **(78-44-4)**
Miolisodol **(78-44-4)**
Mio-pressin **(50-55-5)**
Mioratrina **(78-44-4)**
Miorelax **(93-14-1)**
Mioril **(78-44-4)**
Miorilas **(532-03-6)**
Miorilax **(80-77-3)**
Mioriodol **(78-44-4)**
Miostat **(51-83-2)**
Miotisal **(311-45-5)**
Miotisal A **(311-45-5)**
Miowas **(532-03-6)**
Mipafox [UN 2783] **(371-86-8)**
Mipax **(131-11-3)**
Mipc **(2631-40-5)**
Mipcin **(2631-40-5)**
Mipsin **(2631-40-5)**
Miracle **(94-75-7)**
Miradol **(15676-16-1)**
Miradon **(117-37-3)**
Miral **(42509-80-8)**
Miral 10 G **(42509-80-8)**
Miramel **(77-75-8)**
Miramid H 2 **(105-60-2)**
Miramid H 2 **(25038-54-4)**
Miramid WM 55 **(105-60-2)**
Miramid WM 55 **(25038-54-4)**
Miramine TOC **(61791-39-7)**
Mirapront **(122-09-8)**
Mirason 9 **(9002-88-4)**
Mirason 16 **(9002-88-4)**
Mirason M 15 **(9002-88-4)**
Mirason M 50 **(9002-88-4)**
Mirason M 68 **(9002-88-4)**
Mirason Neo 23H **(9002-88-4)**
Mirathen **(9002-88-4)**
Mirathen 1313 **(9002-88-4)**
Mirathen 1350 **(9002-88-4)**
Mirbane Oil **(98-95-3)**
Mirbane SU 118K **(9011-05-6)**
Mirbanil **(15676-16-1)**
Mireton Brilliant Blue B **(2475-46-9)**
Mireton Brilliant Blue B **(86722-66-9)**
Mirex **(2385-85-5)**
Mirlon **(63428-83-1)**
Miroserina **(68-41-7)**
Mirotin **(86-34-0)**
Miscleron **(637-07-0)**
Misodine **(125-33-7)**
Misolyne **(125-33-7)**
Misonidazole **(13551-87-6)**
Misoprostol **(62015-39-8)**
Missile **(13457-18-6)**
Mistron 139 **(14807-96-6)**
Mistron Frost P **(14807-96-6)**
Mistron RCS **(14807-96-6)**
Mistron 2SC **(14807-96-6)**
Mistron Star **(14807-96-6)**
Mistron Super Frost **(14807-96-6)**
Mistron Vapor **(14807-96-6)**
Mistura C **(51-83-2)**
Misulban **(55-98-1)**
Misulvan **(15676-16-1)**
Mito-C **(50-07-7)**
Mitaban **(33089-61-1)**
Mitac **(33089-61-1)**
Mitacil **(2032-59-9)**
Mitenon **(58-27-5)**
Mitenone **(58-27-5)**
Mithracin **(18378-89-7)**
Mithramycin **(18378-89-7)**
Mithramycin A **(18378-89-7)**
Miticide K-101 **(80-33-1)**
Mitigan **(115-32-2)**
Mitin **(3567-25-7)**
Mitin FF **(3567-25-7)**
Mition **(116-29-0)**
Mitis Green **(12002-03-8)**
Mitobronitol **(488-41-5)**
Mitocin-C **(50-07-7)**
Mitolac **(10318-26-0)**

Mitolactol **(10318-26-0)**
Mitomen **(126-85-2)**
Mitomen **(302-70-5)**
Mitomin **(126-85-2)**
Mitomycin **(50-07-7)**
Mitomycin A **(4055-39-4)**
Mitomycin B **(4055-40-7)**
Mitomycin-C **(50-07-7)**
Mitomycinum **(50-07-7)**
Mitomycyna C (Polish) **(50-07-7)**
Mitosan **(55-98-1)**
Mitostan **(55-98-1)**
Mitotane **(53-19-0)**
Mitox **(103-17-3)**
Mitoxan **(50-18-0)**
Mitoxan **(6055-19-2)**
Mitoxana **(3778-73-2)**
Mitoxine **(55-86-7)**
Mitramycin **(18378-89-7)**
Mitsui Alizarine B **(72-48-0)**
Mitsui Alizarine Red S **(130-22-3)**
Mitsui Auramine O **(2465-27-2)**
Mitsui Benzopurpurine 4BX **(992-59-6)**
Mitsui Blue B Base **(119-90-4)**
Mitsui Blue B Salt **(119-90-4)**
Mitsui Brilliant Green GX **(633-03-4)**
Mitsui Chrome Violet bc **(2092-55-9)**
Mitsui Congo Red **(573-58-0)**
Mitsui Crystal Violet **(548-62-9)**
Mitsui Direct Black BH **(2429-73-4)**
Mitsui Direct Black EX **(1937-37-7)**
Mitsui Direct Black GX **(1937-37-7)**
Mitsui Direct Blue 2BN **(2602-46-2)**
Mitsui Direct Brilliant Blue 6B **(2610-05-1)**
Mitsui Direct Brilliant Scarlet 8B **(6548-29-4)**
Mitsui Direct Brown M **(2429-82-5)**
Mitsui Direct Dark Green BX **(3626-28-6)**
Mitsui Direct Scarlet 3BX **(6358-29-8)**
Mitsui Direct Sky Blue 5B **(2429-74-5)**
Mitsui Direct Violet LN **(6426-67-1)**
Mitsui Indigo Carmine **(860-22-0)**
Mitsui Indigo Paste **(482-89-3)**
Mitsui Indigo Pure **(482-89-3)**
Mitsui Malachite Green **(569-64-2)**
Mitsui Metanil Yellow **(587-98-4)**
Mitsui Methylene Blue **(61-73-4)**
Mitsui Milling Scarlet G **(3567-65-5)**
Mitsui Naphthozol AS **(92-77-3)**
Mitsui Naphthozol BO **(132-68-3)**
Mitsui Naphthozol BS **(135-65-9)**
Mitsui Naphthozol D **(135-61-5)**
Mitsui Naphthozol G **(91-96-3)**
Mitsui Naphthozol OL **(135-62-6)**
Mitsui Polypro B 220 **(9003-07-0)**
Mitsui Red B Base **(97-52-9)**
Mitsui Red 3GL Base **(89-63-4)**
Mitsui Red GL Base **(89-62-3)**
Mitsui Red 3GL Salt **(89-63-4)**
Mitsui Red RL Base **(99-52-5)**
Mitsui Red TR Base **(95-69-2)**
Mitsui Rhodamine BX **(81-88-9)**
Mitsui Rhodamine 6GCP **(989-38-8)**
Mitsui Safranine **(477-73-6)**
Mitsui Scarlet G Base **(99-55-8)**
Mitsui Scarlet GG Base **(95-82-9)**
Mitsui Sulphur Black B **(1326-82-5)**
Mitsui Sulphur Black BC **(1326-82-5)**
Mitsui Sulphur Black BF **(1326-82-5)**
Mitsui Sulphur Black BO **(1326-82-5)**
Mitsui Sulphur Black BS **(1326-82-5)**
Mitsui Sulphur Black G **(1326-82-5)**
Mitsui Sulphur Black GF **(1326-82-5)**
Mitsui Sulphur Blue BC **(1327-57-7)**
Mitsui Sulphur Blue 3BN **(1327-57-7)**
Mitsui Sulphur Blue FB **(1327-57-7)**
Mitsui Sulphur Blue LC **(1327-57-7)**
Mitsui Sulphur Blue LCR **(1327-57-7)**
Mitsui Sulphur Blue 4R **(1327-57-7)**
Mitsui Sulphur Blue R **(1327-57-7)**
Mitsui Sulphur Blue RC **(1327-57-7)**
Mitsui Sulphur Blue RCP **(1327-57-7)**
Mitsui Sulphur Blue TFA **(1327-57-7)**
Mitsui Sulphur Blue TFB **(1327-57-7)**
Mitsui Tartrazine **(1934-21-0)**
Mitsui Tsuya Indigo RN **(6492-73-5)**
Mixed Vegetable Oil, Sulfated **(61790-19-0)**
Mixed fatty alcohols (C10-C16) **(67762-41-8)**
Mixed fatty alcohols (C6-C12) **(68603-15-6)**
Mixed hexyl, octyl, decyl phthalates **(68648-93-1)**
Mixture of Bulan and Prolan (2:1) **(8027-00-7)**
Mixture of p-Methenols **(8006-39-1)**
Mizodin **(125-33-7)**
Mizolin **(125-33-7)**
Mn2+ **(16397-91-4)**
MnEBD **(12427-38-2)**
Mor-Cran **(132-66-1)**
Mobam **(1079-33-0)**
Mobam phenol **(1079-33-0)**
Mobenol **(64-77-7)**
Mobil DBHP **(1809-19-4)**
Mobil MC-A-600 **(1079-33-0)**
Mobil V-C 9-104 **(13194-48-4)**
Mobilan **(53-86-1)**
Mobilawn **(97-17-6)**
Mocap **(13194-48-4)**
Mocovina (Czech) **(57-13-6)**
Modane **(117-10-2)**
Modane Soft **(577-11-7)**
Modenol **(50-55-5)**
Moderil **(24815-24-5)**
Moditen **(69-23-8)**
Modocoll 1200 **(9004-32-4)**
Modown **(42576-02-3)**
Modr Brilantni FCF (Czech) **(3844-45-9)**
Modr Brilantni Ostazinova S-R (Czech) **(4499-01-8)**
Modr Disperzni 3 (Czech) **(2475-46-9)**
Modr Disperzni 56 (Czech) **(12217-79-7)**
Modr Disperzni 72 (Czech) **(81-48-1)**
Modr Disperzni 73 (Czech) **(12222-78-5)**
Modr Evansova (Czech) **(314-13-6)**
Modr Ftalostanova 3G (Czech) **(3468-11-9)**
Modr Kypova 1 (Czech) **(482-89-3)**
Modr Kypova 4 (Czech) **(81-77-6)**
Modr Kypova 16 (Czech) **(6424-76-6)**
Modr Kypova 18 (Czech) **(1324-54-5)**
Modr Kysela 1 (Czech) **(129-17-9)**
Modr Kysela 9 (Czech) **(3844-45-9)**
Modr Kysela 41 (Czech) **(2666-17-3)**
Modr Kysela 74 (Czech) **(860-22-0)**
Modr Kysela 92 (Czech) **(3861-73-2)**
Modr Methylenova (Czech) **(61-73-4)**
Modr Namornicka Ostanthrenova G (Czech) **(6424-76-6)**
Modr Ostacetova LR (Czech) **(27312-17-0)**
Modr Ostacetova P3R (Czech) **(2475-46-9)**

Modr Ostacetova SE-LB (Czech)	**(4702-64-1)**
Modr Pigment 60 (Czech)	**(81-77-6)**
Modr Pigment 63 (Czech)	**(860-22-0)**
Modr Potravinarska 1 (Czech)	**(860-22-0)**
Modr Potravinarska 2 (Czech)	**(3844-45-9)**
Modr Potravinarska 3 (Czech)	**(129-17-9)**
Modr Potravinarska 4 (Czech)	**(81-77-6)**
Modr Prima 1 (Czech)	**(2610-05-1)**
Modr Prima 6 (Czech)	**(2602-46-2)**
Modr Prima 14 (Czech)	**(72-57-1)**
Modr Prima 15 (Czech)	**(2429-74-5)**
Modr Prima 53 (Czech)	**(314-13-6)**
Modr Rozpoustedlova 8 (Czech)	**(61-73-4)**
Modr Trypanova (Czech)	**(72-57-1)**
Modr Zasadita 9 (Czech)	**(61-73-4)**
Modrenal	**(13647-35-3)**
Modulex	**(1333-86-4)**
Modumate	**(4320-30-3)**
Moebiquin	**(83-73-8)**
Mogadan	**(146-22-5)**
Mogadon	**(146-22-5)**
Mogadone	**(146-22-5)**
Mogeton G	**(2797-51-5)**
Mogeton Granule	**(2797-51-5)**
Mogul	**(1333-86-4)**
Mogul L	**(1333-86-4)**
Mohican Red A-8008	**(5160-02-1)**
Mohr's Salt	**(10045-89-3)**
Moisturizing Extract SV	**(9048-46-8)**
Mokotex D 2602	**(9003-20-7)**
Mol-Iron	**(7782-63-0)**
Molo-Jel	**(8009-03-8)**
Molacco	**(1333-86-4)**
Molasses	**(68476-78-8)**
Molasses Alcohol	**(64-17-5)**
Molasses, Beet	**(68476-78-8)**
Molatoc	**(577-11-7)**
Molcer	**(577-11-7)**
Moldamin	**(1538-09-6)**
Moldex	**(99-76-3)**
Mole Death	**(57-24-9)**
Molecular Chlorine	**(7782-50-5)**
Molinal	**(50-06-6)**
Molinate	**(2212-67-1)**
Molindone	**(7416-34-4)**
Mollan O	**(117-81-7)**
Mollan S	**(103-23-1)**
Mollinox	**(72-44-6)**
Molluscicide Bayer 73	**(1420-04-8)**
Molmate	**(2212-67-1)**
Molofac	**(577-11-7)**
Molol	**(8012-95-1)**
Molten adipic acid	**(124-04-9)**
Molurame	**(137-30-4)**
Molybdate	**(7439-98-7)**
Molybdate-Orange	**(12656-85-8)**
Molybdate Red	**(12656-85-8)**
Molybdate, diammonium (9CI)	**(27546-07-2)**
Molybdate, hexaammonium (9CI)	**(12027-67-7)**
Molybden Red	**(12656-85-8)**
Molybdenum (ACGIH)	**(7439-98-7)**
Molybdenum, Isotope of mass 99 (8CI,9CI)	**(14119-15-4)**
Molybdenum Red	**(12656-85-8)**
Molybdenum alloy, nonbase, Ni,Fe,Mo (9CI)	**(37304-37-3)**
Molybdenum disulfide	**(1317-33-5)**
Molybdenum pentachloride [UN 2508]	**(10241-05-1)**
Molybdenum (IV) sulfide	**(1317-33-5)**
Molybdenum zinc oxide (9CI)	**(61583-60-6)**
Molybdenum-sulfide	**(1317-33-5)**
Molybdenum-trioxide	**(1313-27-5)**
Molybdic acid, diammonium salt	**(13106-76-8)**
Molybdic acid, disodium salt	**(7631-95-0)**
Molybdic acid, hexaammonium salt	**(12027-67-7)**
Molybdic anhydride	**(1313-27-5)**
Molybdic trioxide	**(1313-27-5)**
Molykote	**(1317-33-5)**
Momentol	**(8064-90-2)**
Momentum	**(103-90-2)**
Mon 139	**(38641-94-0)**
Mon 39	**(38641-94-0)**
Mon 4620	**(40164-67-8)**
Monacetylferrocene	**(1271-55-2)**
Monaco Red	**(5850-90-8)**
Monacrin	**(134-50-9)**
Monacrin	**(90-45-9)**
Monacrin hydrochloride	**(134-50-9)**
Monagyl	**(443-48-1)**
Monalide	**(7287-36-7)**
Monalube 29-78 (9CI)	**(82347-33-9)**
Monam	**(6734-80-1)**
Monamid 150-LW	**(120-40-1)**
Monamine	**(569-65-3)**
Monaquest	**(67-43-6)**
Monaquest ICA-120	**(139-89-9)**
Monarch	**(1333-86-4)**
Monarch	**(315-30-0)**
Monarch Green WD	**(1328-53-6)**
Monasirup	**(104-46-1)**
Monastral Blue	**(147-14-8)**
Monastral Blue B	**(147-14-8)**
Monastral Fast Blue	**(147-14-8)**
Monastral Fast Green BGNA	**(1328-53-6)**
Monastral Fast Green G	**(1328-53-6)**
Monastral Fast Green GD	**(1328-53-6)**
Monastral Fast Green GF	**(1328-53-6)**
Monastral Fast Green GFNP	**(1328-53-6)**
Monastral Fast Green GN	**(1328-53-6)**
Monastral Fast Green GNA	**(1328-53-6)**
Monastral Fast Green GTP	**(1328-53-6)**
Monastral Fast Green GV	**(1328-53-6)**
Monastral Fast Green 2GWD	**(1328-53-6)**
Monastral Fast Green GWD	**(1328-53-6)**
Monastral Fast Green GX	**(1328-53-6)**
Monastral Fast Green GXB	**(1328-53-6)**
Monastral Fast Green GYH	**(1328-53-6)**
Monastral Fast Green LGNA	**(1328-53-6)**
Monastral Fast Green 3Y	**(14302-13-7)**
Monastral Fast Green 6Y	**(14302-13-7)**
Monastral Fast Green 3YA	**(14302-13-7)**
Monastral Fast Green 6YA	**(14302-13-7)**
Monastral Green B	**(1328-53-6)**
Monastral Green B Pigment	**(1328-53-6)**
Monastral Green G	**(1328-53-6)**
Monastral Green GFN	**(1328-53-6)**
Monastral Green GH	**(1328-53-6)**
Monastral Green GN	**(1328-53-6)**
Monastral Green Y-GT 805D	**(14302-13-7)**
Monastral Red	**(1047-16-1)**
Monastral Red B	**(1047-16-1)**
Monastral Red Y	**(1047-16-1)**
Monastral Violet R	**(1047-16-1)**
Monate	**(2163-80-6)**
Monawet MD 70E	**(577-11-7)**
Monawet MO-70	**(577-11-7)**

Monochloromonofluoromethane	(593-70-4)
Monochloropentafluoroethane [UN 1020]	(76-15-3)
p-Monochlorophenyl phenyl sulfone	(80-00-2)
Monochloropropionic acid	(28554-00-9)
β-Monochloropropionic acid	(107-94-8)
Monochlorosulfuric acid	(7790-94-5)
Monochlorotetrafluoroethane (DOT)	(63938-10-3)
Monochloro-s-triazinetrione	(13057-78-8)
Monochlorotributyltin	(1461-22-9)
Monochlorotrifluoroethylene	(79-38-9)
Monochlorotrifluoromethane (DOT)	(75-72-9)
Monochromium oxide	(1333-82-0)
Monochromium trioxide	(1333-82-0)
Monocil 40	(6923-22-4)
"Monocite" methacrylate monomer	(80-62-6)
Monoclorobenzene (Italian)	(108-90-7)
Monocobalt oxide	(1307-96-6)
Monocopper monosulfide	(1317-40-4)
Monocortin	(53-33-8)
Monocresyl diphenyl phosphate	(26444-49-5)
Monocron	(6923-22-4)
Monocrotalin	(315-22-0)
Monocrotaline	(315-22-0)
E-Monocrotophos	(6923-22-4)
Monocrotophos	(919-44-8)
Monocrotophos (ACGIH,OSHA)	(6923-22-4)
Monocyanoacetic acid	(372-09-8)
Monocyclohexyltin acid	(22771-18-2)
Monodecyl phosphate	(3921-30-0)
Monodehydrosorbitol monooleate	(1338-43-8)
Monodemethylimipramine	(50-47-5)
Mono(2,3-dibromopropyl)phosphate	(5324-12-9)
Monodietilamide dell'acido succinico (Italian)	(1522-00-5)
Mono-digitoxid (German)	(71-63-6)
Monodigitoxoside	(71-63-6)
Monodion	(84-80-0)
Monododecyl hydrogen sulfate	(151-41-7)
Monododecyl phosphate	(2627-35-2)
Monododecyl sodium sulfate	(151-21-3)
Monodorm	(77-28-1)
Monodrin	(6923-22-4)
Monoethanolamine (DOT)	(141-43-5)
Monoethanolamine dicamba	(53404-28-7)
Monoethanolamine dodecylbenzenesulfonate	(26836-07-7)
Monoethanolamine hydrochloride	(2002-24-6)
Monoethanolamine lauric acid amide	(142-78-9)
Monoethanolamine myristic acid condensate	(142-58-5)
Monoethanolamine phosphate	(29868-05-1)
Monoethanolamine stearic acid amide	(111-57-9)
Monoethanolammonium dodecylbenzenesulfonate	(26836-07-7)
Monoethanolethylenediamine	(111-41-1)
N-Monoethylamide of O,O-dimethyldithiophosphorylacetic acid	(116-01-8)
Monoethylamine	(75-04-7)
2-N-Monoethylaminoethanol	(110-73-6)
Monoethylenediurea	(1852-14-8)
Monoethylene glycol	(107-21-1)
Monoethylene glycol dimethyl ether	(110-71-4)
Monoethyl ether of diethylene glycol	(111-90-0)
Mono(2-ethylhexyl)maleate	(7423-42-9)
Mono(2-ethylhexyl) phosphate	(1070-03-7)
Mono(2-ethylhexyl)phosphate	(1070-03-7)
Mono(2-ethylhexyl)phthalate	(4376-20-9)
Monoethylhexyl phthalate	(4376-20-9)
Mono(2-ethylhexyl)sulfate sodium salt	(126-92-1)
Monoethyl phosphate	(1623-14-9)
Monofen	(156-51-4)
Monoferrous acid citrate	(23383-11-1)
Monofluorazijnzuur (Dutch)	(144-49-0)
Monofluoressigsaure (German)	(144-49-0)
Monofluoressigsaures natrium (German)	(62-74-8)
Monofluoroacetamide	(640-19-7)
Monofluoroacetate	(144-49-0)
Monofluoroacetic acid	(144-49-0)
Monofluoroethane	(353-36-6)
Monofluoroethylene	(75-02-5)
Monofluorophosphoric acid	(13537-32-1)
Monofluorophosphoric acid, Anhydrous (DOT)	(13537-32-1)
Monofluorotrichloromethane	(75-69-4)
Monofuracin	(59-87-0)
Monofurfurylideneacetone	(623-15-4)
Monogen Y 100	(151-21-3)
Monogermane	(7782-65-2)
Monoglycerol p-aminobenzoate	(136-44-7)
Monoglyceryl oleate	(25496-72-4)
Mono-glycocard	(71-63-6)
Monoglyme	(110-71-4)
Mono-n-hexylamine	(111-26-2)
Monohexyl dihydrogen phosphate	(3900-04-7)
Monohexyl phosphate	(3900-04-7)
8-Monohydro mirex	(39801-14-4)
Monohydrotriethyletain (French)	(997-50-2)
Monohydroxybenzene	(108-95-2)
Monohydroxyethylpiperazine	(25154-38-5)
Monohydroxymethane	(67-56-1)
Mono(hydroxymethyl)urea	(1000-82-4)
Monoiodoacetate	(64-69-7)
Monoiodoacetic acid	(64-69-7)
Monoiodomethanesulfonic acid, sodium salt	(126-31-8)
Monoioduro di metile (Italian)	(74-88-4)
Monoisoamylamine	(107-85-7)
Monoisobutylamine	(78-81-9)
Monoisooctyl phosphate	(26403-12-3)
Monoisopropanolamine dodecylbenzenesulfonate	(42504-46-1)
Monoisopropanolamine lauric acid amide	(142-54-1)
Monoisopropanolamine oleic acid amide	(111-05-7)
N-Monoisopropylamide of O,O-diethyldithiophosphorylacetic acid	(2275-18-5)
Monoisopropylamine	(75-31-0)
Monoisopropylaminoethanol	(109-56-8)
Mono-isopropylammoniova sul (Czech)	(38641-94-0)
Monoisopropylbiphenyl	(25640-78-2)
Monoisopropyl ether of ethylene glycol	(109-59-1)
Monoisopropyl maleate	(924-83-4)
Monokrotofosz (Hungarian)	(6923-22-4)
Monolan 12000E80	(9003-11-6)
Monolan 8000E80	(9003-11-6)
Monolan PB	(9003-11-6)
Monolauryl dimethylamine	(112-18-5)
Monolinuron	(1746-81-2)
Monolite Fast Blue GS	(574-93-6)
Monolite Fast Blue 3R	(81-77-6)
Monolite Fast Blue 3RD	(81-77-6)
Monolite Fast Blue 2RV	(130-20-1)
Monolite Fast Blue RV	(81-77-6)
Monolite Fast Blue SRS	(81-77-6)
Monolite Fast Gold Orange GV	(128-70-1)
Monolite Fast Green GVSA	(1328-53-6)
Monolite Fast Navy Blue BV	(482-89-3)
Monolite Fast Orange G	(3520-72-7)
Monolite Fast Orange GA	(3520-72-7)
Monolite Fast Orange 2R	(3468-63-1)
Monolite Fast Paper Orange 2R	(3468-63-1)
Monolite Fast Red 2G	(3468-63-1)
Monolite Fast Red G	(2814-77-9)
Monolite Fast Red GA	(2814-77-9)

(2.XI.)-Monosodium D-gluco-heptonate **(31138-65-5)**
Monosodium gluconate **(527-07-1)**
Monosodium glutamate **(142-47-2)**
Monosodium l-glutamate **(142-47-2)**
α-Monosodium glutamate **(142-47-2)**
Monosodium D-glycero-D-gulo-heptonate **(13007-85-7)**
Monosodium glycolate **(2836-32-0)**
Monosodium methanearsonate **(2163-80-6)**
Monosodium methanearsonic acid **(2163-80-6)**
Monosodium methoxyimidodisulfurate **(67874-55-9)**
Monosodium methylarsonate **(2163-80-6)**
Monosodium 2,2'-methylenebis(3,4,6-trichlorophenate) **(5736-15-2)**
Monosodium O-methylhydroxylamine-N,N-disulfonate **(67874-55-9)**
Monosodium phosphate **(7558-80-7)**
Monosodium salt of 2,2'-methylene bis(3,4,6-trichlorophenol) **(5736-15-2)**
Monosodium 2-sulfanilamidothiazole **(144-74-1)**
Monosodium 4-sulfophthalate **(33562-89-9)**
Monosodium taurocholic acid **(145-42-6)**
Monosorb XP-4 **(7558-80-7)**
1-Monostearin **(123-94-4)**
Monostearin **(123-94-4)**
Monostearin **(31566-31-1)**
α-Monostearin **(123-94-4)**
Monostearin (L) **(123-94-4)**
1-Monostearoylglycerol **(123-94-4)**
Monostearyl acid phosphate **(2958-09-0)**
Monosteol **(1323-39-3)**
Monosteol TG **(1323-39-3)**
Monosulfonic acid F **(92-40-0)**
Monosulfur dichloride **(10545-99-0)**
Monosulfure de tetramethylthiurame (French) **(97-74-5)**
Monoten **(156-51-4)**
Monothioethyleneglycol **(60-24-2)**
Monothioglycerin **(96-27-5)**
Monothioglycerol **(96-27-5)**
α-Monothioglycerol **(96-27-5)**
Monothiophosphoric acid **(13598-51-1)**
Monothiophthalimide **(18138-18-6)**
Monothiuram **(97-74-5)**
Monotrichlor-aethyliden-α-glucose (German) **(15879-93-3)**
Mono-(trichloro)tetra(mono-potassium dichloro)-penta-s-triazine-trione **(34651-95-1)**
Monotridecyl phosphate **(5116-94-9)**
Monotropitin **(490-67-5)**
Monotropitoside **(490-67-5)**
Monovar **(6533-00-2)**
Monoxido sodico (Spanish) **(12401-86-4)**
Monoxol OT **(577-11-7)**
Monoxone **(3926-62-3)**
Monoxyde de sodium (French) **(12401-86-4)**
β-Monoxynaphthalene **(135-19-3)**
Monsanto 31675 **(3785-20-4)**
Monsanto CP 47114 **(122-14-5)**
Monsanto CP 51969 **(2104-96-3)**
Monsanto CP-16226 **(127-90-2)**
Monsanto CP-19699 **(13067-93-1)**
Monsanto CP-40294 **(2665-30-7)**
Monsanto CP-49674 **(13265-60-6)**
Monsur **(63-25-2)**
Montan 80 **(1338-43-8)**
Montane 40 **(26266-57-9)**
Montane 60 **(1338-41-6)**
Montane 83 **(8007-43-0)**
Montanox 80 **(9005-65-6)**
Montanyl alcohol **(557-61-9)**
Montar **(1336-36-3)**
Montecatini L-561 **(2597-03-7)**
Monthybase **(111-60-4)**
Monthyle **(111-60-4)**
Montmorillonite **(1302-78-9)**
Montmorillonite ((Al1.33-1.67Mg0.33-0.67)(Ca0-1Na0-1) 0.33Si4(OH)2O10.xH2O), calcined **(70892-59-0)**
Montmorillonite clay, calcined **(70892-59-0)**
Montmorillonite, cuprian (9CI) **(69402-28-4)**
Montopol La 20 **(2235-54-3)**
Montopol La Paste **(151-21-3)**
Montrek 6 **(9002-98-6)**
Montrel **(299-86-5)**
Montrose Propanil **(709-98-8)**
Monurex **(150-68-5)**
Monuron-TCA **(140-41-0)**
Monuron **(150-68-5)**
Monurox **(150-68-5)**
Monuruon **(150-68-5)**
Monuuron **(150-68-5)**
Monydrin **(154-41-6)**
Monzet **(2445-07-0)**
Moon **(56-81-5)**
Mopari **(62-73-7)**
Moplen **(9003-07-0)**
Moplen AD 50N **(9003-07-0)**
Moplen AS 50 **(9003-07-0)**
Moplen Q 51C **(9003-07-0)**
Moplen RO-QG 6015 **(9002-88-4)**
Moplen T 30G **(9003-07-0)**
Mopol M **(1317-33-5)**
Mopol S **(1317-33-5)**
Moramide **(357-56-2)**
Morbam **(57-53-4)**
Morbicid **(50-00-0)**
Morbusan **(115-38-8)**
Morepen **(7177-48-2)**
Morestan **(2439-01-2)**
Morestane **(2439-01-2)**
Morestin **(438-67-5)**
Moretane **(1176-44-9)**
Morfax **(95-32-9)**
Morfina (Italian) **(57-27-2)**
Morflex 510 **(3319-31-1)**
Morfothion (Dutch) **(144-41-2)**
Moringine **(100-46-9)**
Moriperan **(364-62-5)**
Morkit **(84-65-1)**
Morocide **(485-31-4)**
Moronal **(1400-61-9)**
Morosan **(439-14-5)**
Morpan CBP **(140-72-7)**
Morpan CHA **(112-02-7)**
Morpan T **(1119-97-7)**
Morphacetin **(561-27-3)**
Morphactin **(2536-31-4)**
Morpheridin **(469-81-8)**
Morpheridine **(469-81-8)**
Morphia **(57-27-2)**
Morphin (German) **(57-27-2)**
Morphina **(57-27-2)**
Morphinan, 17-allyl-3-hydroxy- **(152-02-3)**
Morphinan, 6-β,14-dihydroxy-3,4-dimethoxy-N-methyl- **(3176-03-2)**
Morphinan-3,6-α-diol, 17-allyl-7,8-didehydro-4,5-α-epoxy **(62-67-9)**
Morphinan-3,14-diol, 17-(cyclobutylmethyl) **(42408-82-2)**
Morphinan-3,6-α-diol, 7,8-didehydro-4,5-α-epoxy- **(466-97-7)**
Morphinan-3,6α-diol, 7,8-didehydro-4,5α-epoxy- (8CI) **(466-97-7)**
Morphinan-3,6-diol, 7,8-didehydro-4,5-epoxy-, (5α,6α)- (9CI) **(466-97-7)**
Morphinan-3,6-α-diol, 7,8-didehydro-4,5-α-epoxy-17-methyl **(57-27-2)**

Morphinan-3,6-diol, 7,8-didehydro-4,5-epoxy-17-methyl- (5α,6α)-, acetate (Salt) (9CI) **(596-15-6)**
Morphinan-3,6-α-diol, 7,8-didehydro-4,5-α-epoxy-17-methyl-, diacetate (ester) **(561-27-3)**
Morphinan-3,6-α-diol, 7,8-didehydro-4,5-α-epoxy-17-methyl-, diacetate (ester), hydrochloride **(1502-95-0)**
Morphinan-3,6-diol, 7,8-didehydro-4,5-epoxy-17-methyl- (5α,6α)-, (R-(R*,R*))-2,3-dihydroxybutanedioate (2:1) (Salt) (9CI) **(302-31-8)**
Morphinan-3,6-diol, 7,8-didehydro-4,5-epoxy-17-methyl-(5α,6α)-, (Z)-9-octadecenoate (Salt) (9CI) **(6033-05-2)**
Morphinan-3,6α-diol, 7,8-didehydro-4,5α-epoxy-17-methyl-, oleate (Salt) **(6033-05-2)**
Morphinan-3,6-α-diol, 7,8-didehydro-4,5-α-epoxy-17-methyl-, 17-oxide **(639-46-3)**
Morphinan-3,6-diol, 7,8-didehydro-4,5-epoxy-17-methyl-, (5-α,6-α)-, 17-oxide (9CI) **(639-46-3)**
Morphinan-3,6-α-diol, 7,8-didehydro-4, 5-α-epoxy-17-methyl-, sulfate **(64-31-3)**
Morphinan-3,6α-diol, 7,8-didehydro-4,5α-epoxy-17-methyl-, tartrate (2:1) (Salt) (8CI) **(302-31-8)**
Morphinan-6-β,14-diol, 3,4-dimethoxy-17-methyl- **(3176-03-2)**
Morphinan-3,6-α-diol, 4,5-α-epoxy-17-methyl- **(509-60-4)**
Morphinan-3,6α-diol, 4,5α-epoxy-17-methyl- **(509-60-4)**
Morphinan-3,6-diol, 4,5-epoxy-17-methyl-, (5α,6α)- (9CI) **(509-60-4)**
Morphinan, 3-hydroxy-17-methyl- **(77-07-6)**
Morphinan, 3-hydroxy-N-methyl-, (+-)- **(297-90-5)**
Morphinanium, 7,8-didehydro-3,6-α-dihydroxy-17,17-dimethyl-4,5-α-epoxy-, bromide **(125-23-5)**
Morphinanium, 7,8-didehydro-17,17-dimethyl-4,5-α-epoxy-6-α-hydroxy-3-methoxy-, bromide **(125-27-9)**
9-α,13-α,14-α-Morphinan, 3-methoxy-17-methyl **(125-71-3)**
Morphinan, 3-methoxy-17-methyl-, (9-α,13-α,14-α)- (9CI) **(125-71-3)**
Morphinan, 3-methoxy-17-methyl-, (+-)- (8CI,9CI) **(510-53-2)**
Morphinan, 3-methoxy-17-methyl-, l- **(125-70-2)**
Morphinan-3-ol, 17-allyl-, (-) **(152-02-3)**
Morphinan-6-α-ol, 3-(benzyloxy)-7,8-didehydro-4,5-α-epoxy-17-methyl- **(14297-87-1)**
Morphinan-6-α-ol, 7,8-didehydro-4,5-α-epoxy-3-ethoxy-17-methyl **(76-58-4)**
Morphinan-6-ol, 7,8-didehydro-4,5-epoxy-3-ethoxy-17-methyl-, (5-α,6-α)- (9CI) **(76-58-4)**
Morphinan-6-α-ol, 7,8-didehydro-4,5-α-epoxy-3-ethoxy-17-methyl-, hydrochloride **(125-30-4)**
Morphinan-6-α-ol, 7,8-didehydro-4,5-α-epoxy-3-ethoxy-17-methyl-, hydrochloride, dihydrate **(6746-59-4)**
Morphinan-6-α-ol, 7,8-didehydro-4,5-α-epoxy-3-methoxy **(467-15-2)**
Morphinan-6-ol, 7,8-didehydro-4,5-epoxy-3-methoxy-, (5-α,6-α)- (9CI) **(467-15-2)**
Morphinan-6-α-ol, 7,8-didehydro-4,5-α-epoxy-3-methoxy-17-methyl **(76-57-3)**
Morphinan-6-ol, 6,7-didehydro-4,5-α-epoxy-3-methoxy-17-methyl-, acetate **(466-90-0)**
Morphinan-6-ol, 6,7-didehydro-4,5-epoxy-3-methoxy-17-methyl-, acetate (ester), (5α)- (9CI) **(466-90-0)**
Morphinan-6-α-ol, 7,8-didehydro-4,5-α-epoxy-3-methoxy-17-methyl-, hydrochloride **(1422-07-7)**
Morphinan-6-α-ol, 7,8-didehydro-4,5-α-epoxy-3-methoxy-17-methyl-, nicotinate (Ester) **(3688-66-2)**
Morphinan-6-α-ol, 7,8-didehydro-4,5-α-epoxy-3-methoxy-17-methyl-, 17-oxide **(3688-65-1)**
Morphinan-6-ol, 7,8-didehydro-4,5-epoxy-3-methoxy-17-methyl-, 17-oxide, (5-α,6-α)- **(3688-65-1)**
Morphinan-6-α-ol, 7,8-didehydro-4,5-α-epoxy-3-methoxy-17-methyl-, phosphate (1:1) **(52-28-8)**
Morphinan-6-α-ol, 7,8-didehydro-4,5-α-epoxy-3-methoxy-17-methyl-, sulfate (2:1) (Salt) **(1420-53-7)**
Morphinan-6-α-ol, 7,8-didehydro-4,5-α-epoxy-17-methyl-3-(2-morpholinoethoxy)- **(509-67-1)**
Morphinan-6-ol, 6,7-didehydro-4,5α-epoxy-3-methoxy-17-methyl-, acetate (ester) (8CI) **(466-90-0)**
Morphinan-3-ol, 4,5-α-epoxy-6-α-methoxy-17-methyl- **(7732-92-5)**
Morphinan-6-α-ol, 4,5-α-epoxy-3-methoxy-17-methyl **(125-28-0)**
Morphinan-6-β-ol, 4,5-α-epoxy-3-methoxy-17-methyl **(795-38-0)**
Morphinan-6-α-ol, 4,5-α-epoxy-3-methoxy-17-methyl-, acetate **(3861-72-1)**
Morphinan-3-ol, 4,5-α-epoxy-17-methyl- **(427-00-9)**
Morphinan-3-ol, 17-methyl-, (-) **(77-07-6)**
Morphinan-3-ol, 17-phenethyl-, (-)- **(468-07-5)**
Morphinan-6-one, 3,14-dihydroxy-4,5-α-epoxy-17-methyl- **(76-41-5)**
Morphinan-6-one, 4,5α-epoxy-3,14-dihydroxy-17-methyl- (8CI) **(76-41-5)**
Morphinan-6-one, 4,5-epoxy-3,14-dihydroxy-17-methyl-, (5α)- (9CI) **(76-41-5)**
Morphinan-6-one, 4,5-α-epoxy-3,14-dihydroxy-17-(2-propenyl) **(465-65-6)**
Morphinan-6-one, 4,5-α-epoxy-3-hydroxy-5-β-17-dimethyl- **(143-52-2)**
Morphinan-6-one, 4,5-α-epoxy-14-hydroxy-3-methoxy-17-methyl **(76-42-6)**
Morphinan-6-one, 4,5-α-epoxy-3-hydroxy-17-methyl **(466-99-9)**
Morphinan-6-one, 4,5-α-epoxy-3-methoxy-17-methyl **(125-29-1)**
Morphinan, 6,7,8,14-tetradehydro-4,5-α-epoxy-3,6-dimethoxy-17-methyl **(115-37-7)**
Morphinan, 6,7,8,14-tetradehydro-4,5-epoxy-3,6-dimethoxy-17-methyl-, (5.α.)-, Compd. with 2,4,6-trinitrophenol (1:1) (9CI) **(5967-77-1)**
Morphinan-3,6-α,14-triol, 4,5-α-epoxy-17-methyl- **(2183-56-4)**
(-)-Morphine **(57-27-2)**
Morphine **(57-27-2)**
Morphine acetate **(596-15-6)**
Morphine, benzyl- **(14297-87-1)**
Morphine bromomethylate **(125-23-5)**
Morphine, 6-deoxy-7,8-dihydro- **(427-00-9)**
Morphine diacetate **(561-27-3)**
Morphine, dihydro- **(509-60-4)**
Morphine, 7,8-dihydro-14-hydroxy- **(2183-56-4)**
Morphine, ethyl- **(76-58-4)**
Morphine methylbromide **(125-23-5)**
Morphine-3-methyl ether **(76-57-3)**
Morphine monomethyl ether **(76-57-3)**
Morphine, 3-O-(2-morpholinoethyl)- **(509-67-1)**
Morphine, O^{3}-(2-morpholinoethyl)- **(509-67-1)**
Morphine oleate **(6033-05-2)**
Morphine oleate, 20% **(6033-05-2)**
Morphine oxide **(639-46-3)**
Morphine-N-oxide **(639-46-3)**
Morphine sulfate **(64-31-3)**
Morphine sulphate **(64-31-3)**
Morphine tartrate **(302-31-8)**
Morphinism **(57-27-2)**
Morphinone, dihydro- **(466-99-9)**
Morphinone, dihydro-14-hydroxy- **(76-41-5)**
Morphinone, dihydro-6-methyl- **(143-52-2)**
Morphinone, methyldihydro- **(143-52-2)**
Morphinum **(57-27-2)**
Morphium **(57-27-2)**
Morpholine (ACGIH,OSHA) [UN 2054] **(110-91-8)**
Morpholine, Aqueous, Mixture [NA 1760] **(110-91-8)**
4-Morpholineacetonitrile **(5807-02-3)**
Morpholine, 4-acetyl **(1696-20-4)**
Morpholine, 4-(2-aminoethyl)- **(2038-03-1)**
Morpholine, 4-aminopropyl **(123-00-2)**
Morpholine, N-aminopropyl- **(123-00-2)**
Morpholine, 4-(bis(1-aziridinyl)phosphinothioyl)- (9CI) **(2168-68-5)**
Morpholine, 4-(3-(bis(2-chloroethyl)amino)-4-methylbenzene-sulfonyl)- **(63905-03-3)**
Morpholine, 4-butyl- (9CI) **(1005-67-0)**
4-Morpholinecarboxaldehyde **(4394-85-8)**
Morpholine, 4-chloro- (9CI) **(23328-69-0)**
Morpholine, n-cyclododecyl-2,6-dimethyl-, acetate **(31717-87-0)**

Morpholine, 4-cyclododecyl-2,6-dimethyl-, acetate (8CI,9CI) **(31717-87-0)**
Morpholine 2,4-dichlorophenoxyacetate **(6365-73-7)**
Morpholine, 2,6-dimethyl **(141-91-3)**
Morpholine, 2,6-dimethyl-N-nitroso **(1456-28-6)**
Morpholine, 3,4-dimethyl-2-phenyl-, (+) **(634-03-7)**
Morpholine, 3,4-dimethyl-2-phenyl-, (2-R-trans)- (9CI) **(634-03-7)**
Morpholine, 2,6-dimethyl-N-tridecyl **(24602-86-6)**
Morpholine disulfide **(103-34-4)**
Morpholine, 4,4'-dithiodi **(103-34-4)**
Morpholine, 4-dodecyl- **(1541-81-7)**
4-Morpholineethanamine **(2038-03-1)**
4-Morpholineethanamine, N,N-dimethyl- (9CI) **(4385-05-1)**
4-Morpholineethanol **(622-40-2)**
Morpholine ethanol **(622-40-2)**
Morpholine, 2-((2-ethoxyphenoxy)methyl) **(46817-91-8)**
Morpholine, 4-ethyl **(100-74-3)**
Morpholine, 4,4'-(2-ethyl-2-nitro-1,3-propanediyl)bis-, Mixt. with 4-(2-nitrobutyl)morpholine (9CI) **(37304-88-4)**
Morpholine hydrochloride **(10024-89-2)**
Morpholine, hydrochloride **(10024-89-2)**
Morpholine, 4-(mercaptoacetyl)-, S-ester with O,O-dimethyl phosphorodithioate **(144-41-2)**
Morpholine, 4-(2-methoxyethyl)- (9CI) **(10220-23-2)**
Morpholine, 4-methyl **(109-02-4)**
Morpholine, N-methyl- **(109-02-4)**
Morpholine, 3-methyl-2-phenyl **(134-49-6)**
Morpholine, 4-((morpholinothiocarbonyl)thio) **(13752-51-7)**
Morpholine, 4-((4-morpholinylthio)thioxomethyl)- (9CI) **(13752-51-7)**
Morpholine, 4-nitro **(4164-32-3)**
Morpholine, 4-(2-nitrobutyl)- **(2224-44-4)**
Morpholine, 4-nitroso- **(59-89-2)**
Morpholine, N-nitroso **(59-89-2)**
Morpholine 9-octadecenoate **(1095-66-5)**
Morpholine oleate **(1095-66-5)**
Morpholine, 4,4'-(oxydi-2,1-ethanediyl)bis- (9CI) **(6425-39-4)**
Morpholine, 4-phenyl- (8CI,9CI) **(92-53-5)**
4-Morpholinepropanenitrile (9CI) **(4542-47-6)**
Morpholine, 4-trityl- **(1420-06-0)**
Morpholine, N-trityl **(1420-06-0)**
Morpholinium compounds, N-ethyl-N-soya alkyl, ethyl sulfates **(61791-34-2)**
Morpholinium, 4-ethyl-4-hexadecyl-, ethyl sulfate (9CI) **(78-21-7)**
Morpholino 2-benzothiazolyl disulfide **(95-32-9)**
Morpholinodisulfide **(103-34-4)**
2-(4-Morpholinodithio)benzothiazole **(95-32-9)**
2-(Morpholinodithio)benzothiazole **(95-32-9)**
3-(2-Morpholinoethyl)morphine **(509-67-1)**
O^3-(2-Morpholinoethyl)morphine **(509-67-1)**
β-Morpholinoethylmorphine **(509-67-1)**
levo-5-(Morpholinomethyl)-3-((5-nitrofurfurylidene)amino)-2-oxazolidinone hydrochloride **(3031-51-4)**
(-)-5-(Morpholinomethyl)-3-((5-nitrofurfurylidene)amino)-2-oxazolidinone **(3795-88-8)**
5-(Morpholinomethyl)-3-((5-nitrofurfurylidene)amino)-2-oxazolidinone **(3031-51-4)**
5-(Morpholinomethyl)-3-((5-nitrofurfurylidene)amino)-2-oxazolidinone **(3795-88-8)**
5-Morpholinomethyl-3-(5-nitro-2-furfurylidine-amino)-2-oxazolidinone **(139-91-3)**
Morpholinophosphonic acid dimethyl ester **(597-25-1)**
N-(3-Morpholinopropyl)-3-hydroxy-2-naphthoamide **(10155-47-2)**
N-(Morpholinosulfenyl)carbofuran **(55285-05-7)**
2-(Morpholinothio)benzothiazole **(102-77-2)**
2-(Morpholinothio)benzothiazole **(95-32-9)**
4-((Morpholinothiocarbonyl)thio)morpholine **(13752-51-7)**
N-Morpholinyl-2-benzothiazolyl disulfide **(95-32-9)**
4-Morpholinyl 2-benzothiazyl disulfide **(95-32-9)**
6-(4-Morpholinyl)-4,4-diphenyl-3-heptanone **(467-84-5)**
2-(4-Morpholinyldithio)benzothiazole **(95-32-9)**
3-(2-(4-Morpholinyl)ethyl)morphine **(509-67-1)**
Morpholinylethylmorphine **(509-67-1)**
Morpholinylmercaptobenzothiazole **(102-77-2)**
2-(4-Morpholinylthio)benzothiazole **(102-77-2)**
3-Morpholylaethylmorphin (German) **(509-67-1)**
Morphosan **(125-23-5)**
Morphothion **(144-41-2)**
Morphotox **(144-41-2)**
Morrocid **(485-31-4)**
Morsodren **(502-39-6)**
Morton EP 332 **(23422-53-9)**
Morton EP 333 **(19750-95-9)**
Morton EP-227 **(502-39-6)**
Morton EP-316 **(2631-37-0)**
Morton EP-161E **(556-61-6)**
Morton Soil-Drench-C **(502-39-6)**
Morton Soil Drench **(502-39-6)**
Mortopal **(107-49-3)**
Moryl **(51-83-2)**
Morzid **(2168-68-5)**
Mosanon **(9005-38-3)**
Mosatil **(62-33-9)**
Moscarda **(121-75-5)**
Moschus Korner (German) **(8015-62-1)**
Moschus ketone **(541-91-3)**
Moskene **(116-66-5)**
Moss Green **(12002-03-8)**
Mosten **(9003-07-0)**
Moth Balls **(91-20-3)**
Moth Flakes **(91-20-3)**
Motilyn **(81-13-0)**
Motiorange R **(842-07-9)**
Motirot G **(3118-97-6)**
Motirot 2R **(85-86-9)**
Motolon **(72-44-6)**
Motomco Tracking Powder **(83-28-3)**
Motor Benzol **(71-43-2)**
Motor Fuel (DOT) **(8006-61-9)**
Motor Fuels (VAN) **(86290-81-5)**
Motor Spirit (DOT) **(8006-61-9)**
Motox **(8001-35-2)**
Motrin **(15687-27-1)**
Mottenhexe **(67-72-1)**
Mouldrite A256 **(9011-05-6)**
Mountain Green **(12002-03-8)**
Mous-Con **(1314-84-7)**
Mouse Blues **(117-52-2)**
Mouse-Nots **(57-24-9)**
Mouse Pak **(81-81-2)**
Mouse-Rid **(57-24-9)**
Mouse-Tox **(57-24-9)**
Movinyl **(9003-20-7)**
Movinyl 50M **(9003-20-7)**
Movinyl 114 **(9003-20-7)**
Movinyl 801 **(9003-20-7)**
Mowilith 30 **(9003-20-7)**
Mowilith 50 **(9003-20-7)**
Mowilith 70 **(9003-20-7)**
Mowilith 90 **(9003-20-7)**
Mowilith D **(9003-20-7)**
Mowilith DV **(9003-20-7)**
Mowilith M70 **(9003-20-7)**
Mowiol **(9002-89-5)**
Mowiol N 30-88 **(9002-89-5)**
Mowiol N 50-98 **(9002-89-5)**
Mowiol N 70-98 **(9002-89-5)**

Moxie	(72-43-5)
Moxone	(94-75-7)
Mozambin	(72-44-6)
Mravencan amonny (Czech)	(540-69-2)
Mravencan draselny (Czech)	(590-29-4)
Mravencan methylnaty (Czech)	(107-31-3)
Mravencan sodny (Czech)	(141-53-7)
Mravencan vapenaty (Czech)	(544-17-2)
Mrowczan etylu (Polish)	(109-94-4)
Mszycol	(58-89-9)
Mucaine	(126-27-2)
Mucic acid	(526-99-8)
Mucidrina	(51-43-4)
Mucodyne	(638-23-3)
Mucolyticum	(616-91-1)
Mucolyticum lappe	(616-91-1)
Mucolytikum lappe	(616-91-1)
Mucomyst	(616-91-1)
Muconic acid	(505-70-4)
Mucopeptide glucohydrolase	(9001-63-2)
Mucorama	(154-41-6)
Mucosolvin	(616-91-1)
Mucostop	(93-14-1)
Mucoxin	(126-27-2)
Mugan	(63-25-2)
Muldamine	(36069-45-1)
Mulhouse White	(7446-14-2)
Mullein LS	(8001-21-6)
Mullite	(12068-56-3)
Mullite	(1302-76-7)
Mullite	(66402-68-4)
Mulsiferol	(50-14-6)
Mulsopaque	(99-79-6)
Multacodin	(125-29-1)
Multamat	(22781-23-3)
Multaun	(57-53-4)
Multergan	(58-34-4)
Multergan methyl sulfate	(58-34-4)
Multezin	(58-34-4)
Multichlor	(127-65-1)
Multicide	(7696-12-0)
Multimet	(22781-23-3)
Multin	(103-90-2)
Multiprop	(2536-31-4)
Multitok	(8075-80-7)
Mundial Black MO	(1326-82-5)
Murabba	(860-22-0)
Muracil	(56-04-2)
Muramidase	(9001-63-2)
Muratox	(2595-54-2)
Murex	(15879-93-3)
Murfos	(56-38-2)
Murfotox	(2595-54-2)
Murfulvin	(126-07-8)
Muriamic	(138-15-8)
Muriate of platinum	(10025-65-7)
Muriatic acid (DOT)	(7647-01-0)
Muriatic ether	(75-00-3)
Murine Ear Drops	(124-43-6)
Muriol	(3691-35-8)
Murotox	(2595-54-2)
Murphotox	(2595-54-2)
Murutox	(2595-54-2)
Murvesco	(80-38-6)
Murvin	(63-25-2)
Musaril	(10379-14-3)
Muscalure	(27519-02-4)
Muscamone	(27519-02-4)
(+)-Muscarine chloride	(2303-35-7)
L-(+)-Muscarine chloride	(2303-35-7)
Muscarine-chloride	(2303-35-7)
Muscatox	(56-72-4)
Muscimol	(2763-96-4)
Muscle adenylic acid	(61-19-8)
Muscone	(541-91-3)
Musculamine	(71-44-3)
Musk Ambrette	(83-66-9)
Musk Ketone	(81-14-1)
Musk Tibetene	(145-39-1)
Musk Xyldl	(81-15-2)
Musk Xylene	(81-15-2)
Muskalactone	(106-02-5)
Muskel	(80-77-3)
Muskel-Trancopal	(80-77-3)
Muskone	(541-91-3)
Mussel Poison Dihydrochloride	(35554-08-6)
Mussolinite	(14807-96-6)
Mustard Gas	(505-60-2)
Mustard Gas Sulfone	(471-03-4)
N-Mustard (German)	(55-86-7)
Mustard HD	(505-60-2)
Mustard Sulfone	(471-03-4)
Mustard, Sulfur	(505-60-2)
Mustard Vapor	(505-60-2)
Mustard chlorohydrin	(693-30-1)
Mustard oil	(57-06-7)
Mustargen	(51-75-2)
Mustargen	(55-86-7)
Mustargen hydrochloride	(55-86-7)
Mustine	(51-75-2)
Mustine hydrochlor	(55-86-7)
Mustine hydrochloride	(55-86-7)
Mustron	(302-70-5)
Musuet synthetic	(107-75-5)
Musuettine principle	(107-75-5)
Mutagen	(51-75-2)
Mutamycin(mitomycin for injection)	(50-07-7)
Mutamycin	(50-07-7)
Mutesa	(126-27-2)
Muthesa	(126-27-2)
Muthmann's liquid	(79-27-6)
Mutoxin	(50-29-3)
Mutton Tallow	(61789-97-7)
My-B-Den	(61-19-8)
Myacide SP	(1777-82-8)
Myacyne	(1404-04-2)
Myamin	(2152-34-3)
Myasul	(80-35-3)
Mybasan	(54-85-3)
Mycaifradin sulfate	(1405-10-3)
Mycanden	(777-11-7)
Mycardol	(78-11-5)
Mycelax	(23593-75-1)
Mycelex	(23593-75-1)
Mycelex G	(23593-75-1)
Mychel	(56-75-7)
Mycifradin	(1404-04-2)
Mycifradin	(1405-10-3)
Mycifradin-N	(1405-10-3)
Mycigient	(1405-10-3)
Myciguent	(1405-10-3)
Mycilan	(777-11-7)
E-Mycin	(114-07-8)
Mycinol	(56-75-7)

Mycoban (137-40-6)
Mycodifol (2939-80-2)
Mycofarm (69-57-8)
Mycoin (149-29-1)
Mycoin C (149-29-1)
Mycoin C3 (149-29-1)
Mycoine C3 (149-29-1)
Mycose (99-20-7)
Mycosin (149-29-1)
Mycosporin (23593-75-1)
Mycostatin (1400-61-9)
Mycostatin 20 (1400-61-9)
Mycotoxin F2 (17924-92-4)
Mycotoxin T-2 (21259-20-1)
Mycozol (148-79-8)
Mycronil (137-30-4)
Mydfrin (61-76-7)
Mydrial (60-15-1)
Mydriasin (2870-71-5)
Mydriatine (154-41-6)
Myebrol (488-41-5)
Myeleukon (55-98-1)
Myelobromol (488-41-5)
Myeloleukon (55-98-1)
Myelosan (55-98-1)
Myelotrast (126-31-8)
Mygal (51-12-7)
Mykoin BF 510 (1406-05-9)
Mykostin (50-14-6)
Mylar (25038-59-9)
Mylar A (25038-59-9)
Mylar C (25038-59-9)
Mylar C-25 (25038-59-9)
Mylar HS (25038-59-9)
Mylar T (25038-59-9)
Mylecytan (55-98-1)
Mylepsin (125-33-7)
Mylepsinum (125-33-7)
Myleran (55-98-1)
Mylipen (6130-64-9)
Mylodorm (57-43-2)
Mylofanol (132-60-5)
Mylon (Czech) (533-74-4)
Mylone (1689-83-4)
Mylone (533-74-4)
Mylone 85 (533-74-4)
Mylosar (320-67-2)
Mylosul (80-35-3)
Mynosedin (15687-27-1)
Myocaine (93-14-1)
Myocol (58-61-7)
Myocon (55-63-0)
Myodigin (71-63-6)
Myodil (99-79-6)
Myodyl (99-79-6)
Myofer 100 (9004-66-4)
Myoglycerin (55-63-0)
Myoinosite (87-89-8)
Myoinositol (87-89-8)
Myolastan (10379-14-3)
Myolaxene (532-03-6)
Myorelax (93-14-1)
Myoscaine (93-14-1)
Myosmine (532-12-7)
Myosthenine (51-43-4)
Myoston (61-19-8)
Myotrate "10" (78-11-5)
Myotriphos (56-65-5)
Myrac aldehyde (37677-14-8)
Myrcene (123-35-3)
Myrcenol (543-39-5)
Myrcenyl acetate (1118-39-4)
Myristaldehyde (124-25-4)
Myristamide MEA (142-58-5)
Myristamine oxide (3332-27-2)
Myristica Oil (8008-45-5)
Myristic-acid (544-63-8)
Myristic acid, butyl ester (8CI) (110-36-1)
Myristic acid, 9-ester with 1,1a-α,1b-β,4,4a,7a-α,7b,8,9,9a-decahydro-4a-β, 7b-α,9-β,9a-α-tetrahydroxy-3-(hydroxymethyl)-1,1,6,8-α-tetramethyl-5H- cyclopropa(3,4)benz(1,2-e)azulen-5-one, 9a-acetate (16561-29-8)
Myristic acid, 9-ester with 1,1aα,1bβ,4,4a,7aα,7b,8,9,9a-decahydro-4aβ,7bα,9β,9a-α-tetrahydroxy-3-(hydroxymethyl)-1,1,6,8α-tetramethyl-5H-cyclopropa(3,4)benz(1,2-e)azulen-5-one 9a-acetate, (+)- (8CI) (16561-29-8)
Myristic acid, ethyl ester (8CI) (124-06-1)
Myristic acid, isopropyl ester (110-27-0)
Myristic acid, methyl ester (8CI) (124-10-7)
Myristic acid, triethanolamine salt (41669-40-3)
Myristic acid triglyceride (555-45-3)
Myristic alcohol (112-72-1)
Myristic aldehyde (124-25-4)
Myristicin (607-91-0)
Myristic monoethanolamide (142-58-5)
Myristin (555-45-3)
Myristin, tri- (8CI) (555-45-3)
Myristo chromic chloride (15659-56-0)
Myristone (542-50-7)
Myristonitrile (8CI) (629-63-0)
Myristoyl monoethanolamide (142-58-5)
Myristyl acetate (638-59-5)
Myristyl alcohol (112-72-1)
Myristyl alcohol (Mixed isomers) (27196-00-5)
Myristyl bromide (112-71-0)
Myristyl dimethyl amine (112-75-4)
Myristyldimethylamine (112-75-4)
Myristyl dimethyl amine oxide (3332-27-2)
Myristyldimethylamine oxide (3332-27-2)
Myristyl methacrylate (2549-53-3)
Myristyl monoethanolamide (142-58-5)
Myristyl myristate (3234-85-3)
(Myristyloxymethyl)oxirane (38954-75-5)
Myristyl propionate (6221-95-0)
Myristylpyridinium chloride (2785-54-8)
Myristyl stearate (17661-50-6)
Myristyl sulfate, sodium salt (1191-50-0)
Myristyltrimethylammonium bromide (1119-97-7)
Myrj (9004-99-3)
Myrj 45 (9004-99-3)
Myrj 49 (9004-99-3)
Myrj 51 (9004-99-3)
Myrj 52 (9004-99-3)
Myrj 53 (9004-99-3)
Myrophine (467-18-5)
Myrticalorin (153-18-4)
Myrticolorin (153-18-4)
Mysedon (125-33-7)
Mysoline (125-33-7)
Mysorite (12172-73-5)
Mysteclin-F (1397-89-3)
Mystox WFA (132-27-4)
Myticolorin (153-18-4)
Mytomycin (50-07-7)

Mytrate **(51-43-4)**
Mytrol **(4301-50-2)**
Myvak **(127-47-9)**
Myvax **(127-47-9)**
Myvpack **(68-26-8)**
168N15 **(9003-53-6)**
50N **(9011-14-7)**
N 5 **(9004-57-3)**
N-5' **(53902-12-8)**
N-9 **(26027-38-3)**
N 20 (Polyamide) **(63428-84-2)**
N 40 (VAN) **(63428-84-2)**
N 50 **(9011-05-6)**
N-244 **(6012-92-6)**
N 252 **(55290-64-7)**
N 480 **(9003-11-6)**
N 521 **(533-74-4)**
N 714 **(113-59-7)**
N 714C **(113-59-7)**
N-869 **(137-42-8)**
N 1386 **(3064-70-8)**
N-1544A **(156-51-4)**
N-2596 **(2984-64-7)**
N 2788 **(333-43-7)**
N 2790 **(944-22-9)**
N 4000v **(9003-53-6)**
N 4,556 **(2164-09-2)**
4010 NA **(101-72-4)**
NA **(389-08-2)**
NA 22 **(96-45-7)**
NA 97 **(15500-66-0)**
NA 101 **(2782-91-4)**
NA 0150 [Pentaerythrite tetranitrate wetted with not less than 25 per cent water, by mass] **(78-11-5)**
NA 0473 [Nitrosoguanidine] **(674-81-7)**
NA 1051 (DOT) **(74-90-8)**
NA 1067 (DOT) **(10544-72-6)**
NA 1067 (DOT) **(10102-44-0)**
NA 1120 (DOT) **(71-36-3)**
NA 1204 (DOT) **(55-63-0)**
NA 1247 (DOT) **(80-62-6)**
NA 1270 (DOT) **(8002-05-9)**
NA 1324 (DOT) **(9004-70-0)**
NA 1325 **(12627-52-0)**
NA 1325 **(1345-04-6)**
NA 1325 [Smokeless powder for small arms (100 pounds or less)] **(70-25-7)**
NA 1325 (DOT) **(9004-70-0)**
NA 1344 [Picric acid, wet, with not less than 10% water] **(88-89-1)**
NA 1361 [Charcoal briquettes, shell, screenings, wood, etc.] **(16291-96-6)**
NA 1401 (DOT) **(7440-70-2)**
NA 1463 [Chromic acid, Solid] **(13530-68-2)**
NA 1463 [Chromic acid, Solid] **(1333-82-0)**
NA 1477 **(12436-94-1)**
NA 1479 (DOT) **(7778-50-9)**
NA 1479 (DOT) **(10588-01-9)**
NA 1479 **(3251-23-8)**
NA 1511 [Urea hydrogen peroxide] **(124-43-6)**
NA 1549 [Antimony trifluoride, Solid] **(7783-56-4)**
NA 1549 [Antimony tribromide, Solution] **(7789-61-9)**
NA 1556 (DOT) **(696-28-6)**
NA 1556 [Methyldichloroarsine] **(593-89-5)**
NA 1557 [Arsenic compounds, Solid, N.O.S.] **(7784-45-4)**
NA 1557 [Arsenic sulfide] **(1303-33-9)**
NA 1566 (DOT) **(7787-49-7)**
NA 1566 (DOT) **(7787-47-5)**
NA 1574 [Calcium arsenite, Solid] **(52740-16-6)**
NA 1574 [Calcium arsenite, Solid] **(27152-57-4)**
NA 1581 (DOT) **(8004-09-9)**
NA 1583 (DOT) **(76-06-2)**
NA 1648 (DOT) **(75-05-8)**
NA 1649 [Tetraethyl lead, Liquid] **(78-00-2)**
NA 1665 (DOT) **(25168-04-1)**
NA 1707 [Thallium sulfate, Solid] **(10031-59-1)**
NA 1709 (DOT) **(25376-45-8)**
NA 1759 (DOT) **(7772-99-8)**
NA 1759 [Ferrous chloride, Solid] **(7758-94-3)**
NA 1760 **(10043-01-3)**
NA 1760 **(503-74-2)**
NA 1760 **(676-98-2)**
NA 1760 **(7758-94-3)**
NA 1760 [Corrosive liquids, N.O.S.] **(109-52-4)**
NA 1760 (DOT) **(4985-85-7)**
NA 1760 **(13693-11-3)**
NA 1760 (DOT) **(1498-51-7)**
NA 1760 (DOT) **(929-06-6)**
NA 1760 [Corrosive liquid, N.O.S.] **(110-91-8)**
NA 1760 (DOT) **(993-43-1)**
NA 1778 (DOT) **(16961-83-4)**
NA 1791 (DOT) **(7681-52-9)**
NA 1794 **(69029-52-3)**
NA 1807 (DOT) **(1314-56-3)**
NA 1831 (DOT) **(8014-95-7)**
NA 1869 (DOT) **(7439-95-4)**
NA 1902 [Diisooctyl acid phosphate] **(298-07-7)**
NA 1942 [Ammonium nitrate fertilizers with not more than 0.2 percent carbon; which meet the definition in the Fertilizer Institute publication (Definition and Test Procedures for Ammonium Nitrate Fertilizer), dated May 8, 1971] **(6484-52-2)**
NA 1956 [Hexafluoropropylene oxide] **(428-59-1)**
NA 1961 [Ethane-Propane mixture, Refrigerated liquid (Cryogenic liquid)] **(68475-58-1)**
NA 1967 [Parathion and compressed gas mixture] **(56-38-2)**
NA 1980 **(58933-55-4)**
NA 1986 **(107-19-7)**
NA 1989 (DOT) **(100-52-7)**
NA 1993 [Diesel fuel] **(68334-30-5)**
NA 1999 [Asphalt, at or above its flashpoint] **(8052-42-4)**
NA 2020 (DOT) **(25167-82-2)**
NA 2020 (DOT) **(87-86-5)**
NA 2025 [Mercury compounds, Solid, N.O.S.] **(1312-03-4)**
NA 2054 (DOT) **(110-91-8)**
NA 2059 (DOT) **(9004-70-0)**
NA 2085 (DOT) **(94-36-0)**
NA 2091 **(30026-92-7)**
NA 2129 (DOT) **(762-16-3)**
NA 2131 (DOT) **(79-21-0)**
NA 2133 (DOT) **(105-64-6)**
NA 2134 (DOT) **(105-64-6)**
NA 2215 [Maleic acid] **(110-16-7)**
NA 2267 (DOT) **(2524-03-0)**
NA 2267 **(993-12-4)**
NA 2291 (DOT) **(13814-96-5)**
NA 2291 (DOT) **(7758-95-4)**
NA 2449 [Oxalates, water soluble] **(1113-38-8)**
NA 2465 (DOT) **(2244-21-5)**
NA 2468 [(Mono-(trichloro) tetra- (monopotassium dichloro)-penta-s-triazinetrione, Dry (containing over 39% available chlorine)] **(87-90-1)**
NA 2555 (DOT) **(9004-70-0)**
NA 2556 (DOT) **(9004-70-0)**
NA 2584 [Dodecylbenzenesulfonic acid] **(27176-87-0)**
NA 2626 (DOT) **(7790-93-4)**
NA 2672 (DOT) **(1336-21-6)**
NA 2683 (DOT) **(12124-99-1)**

Name	CAS No.
NA 2693 (DOT)	**(7681-57-4)**
NA 2693 (DOT)	**(7631-90-5)**
NA 2693 (DOT)	**(16731-55-8)**
NA 2693 (DOT)	**(10192-30-0)**
NA 2693 (DOT)	**(13780-03-5)**
NA 2757 (DOT)	**(63-25-2)**
NA 2757 (DOT)	**(1563-66-2)**
NA 2757 (DOT)	**(315-18-4)**
NA 2761 [Aldrin, Solid or dieldrin]	**(60-57-1)**
NA 2761	**(115-29-7)**
NA 2761 [Aldrin, Solid]	**(309-00-2)**
NA 2761	**(72-54-8)**
NA 2761	**(72-20-8)**
NA 2761	**(58-89-9)**
NA 2761 (DOT)	**(8001-35-2)**
NA 2762 [Aldrin, Liquid]	**(309-00-2)**
NA 2762	**(57-74-9)**
NA 2765 (DOT)	**(29990-39-4)**
NA 2765 (DOT)	**(93-76-5)**
NA 2765 (DOT)	**(94-75-7)**
NA 2771 (DOT)	**(137-26-8)**
NA 2783 (DOT)	**(86-50-0)**
NA 2783 (DOT)	**(7786-34-7)**
NA 2783	**(107-49-3)**
NA 2783 (DOT)	**(757-58-4)**
NA 2783 (DOT)	**(2921-88-2)**
NA 2783	**(121-75-5)**
NA 2783 [Methyl parathion solid]	**(298-00-0)**
NA 2783 [Organophosphorus pesticides, Solid, toxic, N.O.S.]	**(333-41-5)**
NA 2783 [Organophosphorus pesticides, Solid, toxic, N.O.S.]	**(563-12-2)**
NA 2783 [Parathion]	**(56-38-2)**
NA 2783 [Organophosphorus pesticides, Solid, toxic, N.O.S.]	**(2275-14-1)**
NA 2783	**(62-73-7)**
NA 2783	**(298-04-4)**
NA 2809 (DOT)	**(7439-97-6)**
NA 2811 [Selenium oxide]	**(14832-90-7)**
NA 2811	**(51317-24-9)**
NA 2811 (DOT)	**(7783-46-2)**
NA 2821 (DOT)	**(108-95-2)**
NA 2902 (DOT)	**(584-79-2)**
NA 2922 [Sodium hydrosulfide, solution]	**(16721-80-5)**
NA 2923 (DOT)	**(16721-80-5)**
NA 2924	**(11069-19-5)**
NA 2929 (DOT)	**(76-06-2)**
NA 3018 [Methyl parathion liquid]	**(298-00-0)**
NA 9011 (DOT)	**(79-92-5)**
NA 9018	**(27156-03-2)**
NA 9026 (DOT)	**(131-89-5)**
NA 9037 (DOT)	**(67-72-1)**
NA 9069	**(51-80-9)**
NA 9083 (DOT)	**(1111-78-0)**
NA 9084 (DOT)	**(506-87-6)**
NA 9088 (DOT)	**(13826-83-0)**
NA 9152 (DOT)	**(27774-13-6)**
NA 9163 [Zirconium sulfate]	**(14644-61-2)**
NA 9180 (DOT)	**(541-09-3)**
NA 9190	**(13446-10-1)**
NA 9191 [Chlorine dioxide, hydrate, frozen]	**(10049-04-4)**
NA 9202 [Carbon monoxide, refrigerated liquid (cryogenic liquid)]	**(630-08-0)**
NA 9206 [Methyl phosphonic dichloride]	**(676-97-1)**
α-NA	**(86-87-3)**
1-NAA	**(86-87-3)**
NAA	**(86-87-3)**
NAA 800	**(86-87-3)**
α-NAA	**(86-87-3)**
NAAM	**(86-86-2)**

Name	CAS No.
NAB	**(1133-64-8)**
NAB-930	**(56961-25-2)**
NAC	**(616-91-1)**
NAC	**(63-25-2)**
NACAP	**(2492-26-4)**
NAC-TB	**(616-91-1)**
NA-22-D	**(96-45-7)**
NAD	**(86-86-2)**
N4-Acetylsulfanilyl chloride	**(121-60-8)**
NAE	**(63148-62-9)**
NAH	**(59-67-6)**
5-NAN	**(602-87-9)**
NB2B	**(2602-46-2)**
NB Coat CR 503	**(9006-03-5)**
NBHA	**(3817-11-6)**
N.B. Mecoprop	**(93-65-2)**
NBN	**(544-16-1)**
NBS 706	**(9003-53-6)**
NBT	**(1594-56-5)**
NC 100	**(7439-89-6)**
NC 123	**(5588-33-0)**
NC 150	**(136-40-3)**
NC 150	**(94-78-0)**
NC 1667	**(1912-26-1)**
NC 2983	**(2338-25-2)**
NC 302	**(76578-14-8)**
NC 3363	**(3615-21-2)**
NC 5016	**(14255-88-0)**
NC 8438	**(26225-79-6)**
NC-262	**(60-51-5)**
NC-2962	**(2669-32-1)**
NC6897	**(22781-23-3)**
NCA	**(2533-89-3)**
NCI-96683	**(76578-14-8)**
NCI-C00044	**(309-00-2)**
NCI-C00055	**(133-90-4)**
NCI-C00066	**(86-50-0)**
NCI-C00077	**(133-06-2)**
NCI-C00099	**(57-74-9)**
NCI-C00102	**(1897-45-6)**
NCI-C00113	**(62-73-7)**
NCI-C00124	**(60-57-1)**
NCI-C00135	**(60-51-5)**
NCI-C00157	**(72-20-8)**
NCI-C00168	**(961-11-5)**
NCI-C00180	**(76-44-8)**
NCI-C00191	**(143-50-0)**
NCI-C00204	**(58-89-9)**
NCI-C00215	**(121-75-5)**
NCI-C00226	**(56-38-2)**
NCI-C00237	**(1918-02-1)**
NCI-C00259	**(8001-35-2)**
NCI-C00260	**(76-87-9)**
NCI-C00395	**(78-34-2)**
NCI-C00408	**(510-15-6)**
NCI-C00419	**(82-68-8)**
NCI-C00420	**(1836-75-5)**
NCI-C00431	**(1420-04-8)**
NCI-C00442	**(1582-09-8)**
NCI-C00453	**(95-06-7)**
NCI-C00464	**(50-29-3)**
NCI-C00475	**(72-54-8)**
NCI-C00486	**(115-32-2)**
NCI-C00497	**(72-43-5)**
NCI-C00500	**(96-12-8)**
NCI-C00511	**(107-06-2)**
NCI-C00522	**(106-93-4)**

NCI-C00533	(76-06-2)
NCI-C00544	(315-18-4)
NCI-C00555	(72-55-9)
NCI-C00566	(115-29-7)
NCI-C00588	(13171-21-6)
NCI-C00599	(13366-73-9)
NCI-C00920	(107-21-1)
NCI-C01445	(18662-53-8)
NCI-C01478	(303-34-4)
NCI-C01514	(23214-92-8)
NCI-C01536	(7008-42-6)
NCI-C01547	(3458-22-8)
NCI-C01558	(3546-10-9)
NCI-C01569	(320-67-2)
NCI-C01570	(22966-79-6)
NCI-C01581	(789-61-7)
NCI-C01592	(55-98-1)
NCI-C01616	(5034-77-5)
NCI-C01627	(21416-87-5)
NCI-C01638	(3778-73-2)
NCI-C01649	(52-24-4)
NCI-C01661	(63-92-3)
NCI-C01672	(136-40-3)
NCI-C01683	(58-14-0)
NCI-C01694	(536-33-4)
NCI-C01707	(139-65-1)
NCI-C01718	(80-08-0)
NCI-C01729	(73-22-3)
NCI-C01730	(118-92-3)
NCI-C01741	(114-86-3)
NCI-C01752	(94-20-2)
NCI-C01763	(64-77-7)
NCI-C01785	(98-96-4)
NCI-C01810	(366-70-1)
NCI-C01821	(138-89-6)
NCI-C01832	(6369-59-1)
NCI-C01843	(99-55-8)
NCI-C01854	(122-66-7)
NCI-C01854	(530-50-7)
NCI-C01865	(121-14-2)
NCI-C01876	(117-79-3)
NCI-C01887	(17026-81-2)
NCI-C01901	(82-28-0)
NCI-C01912	(94-52-0)
NCI-C01923	(129-15-7)
NCI-C01934	(99-59-2)
NCI-C01945	(619-17-0)
NCI-C01956	(86-57-7)
NCI-C01967	(602-87-9)
NCI-C01978	(1777-84-0)
NCI-C01989	(39156-41-7)
NCI-C01990	(101-61-1)
NCI-C02006	(90-94-8)
NCI-C02017	(103-85-5)
NCI-C02028	(1067-33-0)
NCI-C02039	(106-47-8)
NCI-C02040	(95-74-9)
NCI-C02051	(95-79-4)
NCI-C02073	(150-69-6)
NCI-C02084	(7632-00-0)
NCI-C02095	(628-94-4)
NCI-C02108	(60-35-5)
NCI-C02119	(57-13-6)
NCI-C02131	(592-31-4)
NCI-C02142	(628-02-4)
NCI-C02153	(622-51-5)
NCI-C02175	(91-93-0)
NCI-C02186	(2489-77-2)
NCI-C02200	(100-42-5)
NCI-C02211	(102-96-5)
NCI-C02222	(5307-14-2)
NCI-C02233	(101-54-2)
NCI-C02244	(156-10-5)
NCI-C02255	(54150-69-5)
NCI-C02299	(137-17-7)
NCI-C02302	(95-80-7)
NCI-C02335	(636-21-5)
NCI-C02368	(3165-93-3)
NCI-C02551	(1306-19-0)
NCI-C02653	(70-30-4)
NCI-C02664	(11097-69-1)
NCI-C02686	(67-66-3)
NCI-C02697	(8003-03-0)
NCI-C02711	(1306-23-6)
NCI-C02722	(7772-99-8)
NCI-C02733	(58-08-2)
NCI-C02766	(139-13-9)
NCI-C02799	(50-00-0)
NCI-C02813	(51-03-6)
NCI-C02824	(120-62-7)
NCI-C02835	(148-18-5)
NCI-C02846	(150-68-5)
NCI-C02857	(20941-65-5)
NCI-C02868	(72-56-0)
NCI-C02880	(86-30-6)
NCI-C02891	(19010-66-3)
NCI-C02904	(88-06-2)
NCI-C02915	(135-88-6)
NCI-C02926	(103-33-3)
NCI-C02937	(156-62-7)
NCI-C02959	(97-77-8)
NCI-C02960	(999-81-5)
NCI-C02971	(298-00-0)
NCI-C02982	(120-71-8)
NCI-C02993	(102-50-1)
NCI-C03009	(142-46-1)
NCI-C03010	(140-56-7)
NCI-C03021	(2243-62-1)
NCI-C03032	(2438-88-2)
NCI-C03043	(6109-97-3)
NCI-C03054	(118-52-5)
NCI-C03065	(121-66-4)
NCI-C03076	(504-88-1)
NCI-C03134	(64-17-5)
NCI-C03167	(18883-66-4)
NCI-C03247	(968-81-0)
NCI-C03258	(135-20-6)
NCI-C03269	(6358-85-6)
NCI-C03270	(126-72-7)
NCI-C03281	(1465-25-4)
NCI-C03292	(95-83-0)
NCI-C03305	(5131-60-2)
NCI-C03316	(6219-71-2)
NCI-C03327	(1156-19-0)
NCI-C03361	(92-87-5)
NCI-C03372	(96-45-7)
NCI-C03474	(446-86-6)
NCI-C03485	(305-03-3)
NCI-C03510	(87-29-6)
NCI-C03521	(95-14-7)
NCI-C03554	(79-34-5)
NCI-C03565	(128-66-5)
NCI-C03598	(128-37-0)
NCI-C03601	(85-44-9)

NCI No.	CAS No.
NCI-C03612	**(88-96-0)**
NCI-C03656	**(262-12-4)**
NCI-C03667	**(33857-26-0)**
NCI-C03678	**(3268-87-9)**
NCI-C03689	**(123-91-1)**
NCI-C03714	**(1746-01-6)**
NCI-C03736	**(142-04-1)**
NCI-C03736	**(62-53-3)**
NCI-C03747	**(134-29-2)**
NCI-C03758	**(20265-97-8)**
NCI-C03770	**(140-49-8)**
NCI-C03781	**(512-56-1)**
NCI-C03792	**(139-94-6)**
NCI-C03805	**(77-65-6)**
NCI-C03816	**(105-55-5)**
NCI-C03827	**(1596-84-5)**
NCI-C03838	**(6959-48-4)**
NCI-C03849	**(126-31-8)**
NCI-C03850	**(105-11-3)**
NCI-C03861	**(434-13-9)**
NCI-C03907	**(6959-47-3)**
NCI-C03918	**(57-97-6)**
NCI-C03930	**(624-18-0)**
NCI-C03941	**(99-56-9)**
NCI-C03952	**(89-25-8)**
NCI-C03963	**(119-34-6)**
NCI-C03974	**(150-38-9)**
NCI-C03985	**(542-75-6)**
NCI-C04035	**(93-71-0)**
NCI-C04126	**(1955-45-9)**
NCI-C04159	**(2784-94-3)**
NCI-C04240	**(13463-67-7)**
NCI-C04251	**(7440-32-6)**
NCI-C04273	**(18540-29-9)**
NCI-C04502	**(1271-19-8)**
NCI-C04524	**(1212-29-9)**
NCI-C04535	**(75-34-3)**
NCI-C04546	**(79-01-6)**
NCI-C04557	**(77-79-2)**
NCI-C04568	**(75-47-8)**
NCI-C04579	**(79-00-5)**
NCI-C04580	**(127-18-4)**
NCI-C04591	**(75-15-0)**
NCI-C04604	**(67-72-1)**
NCI-C04615	**(107-05-1)**
NCI-C04626	**(71-55-6)**
NCI-C04637	**(75-69-4)**
NCI-C04671	**(59-05-2)**
NCI-C04682	**(50-76-0)**
NCI-C04693	**(20830-81-3)**
NCI-C04706	**(50-07-7)**
NCI-C04717	**(4342-03-4)**
NCI-C04728	**(147-94-4)**
NCI-C04739	**(17433-31-7)**
NCI-C04740	**(13010-47-4)**
NCI-C04762	**(488-41-5)**
NCI-C04773	**(154-93-8)**
NCI-C04784	**(342-69-8)**
NCI-C04795	**(10318-26-0)**
NCI-C04819	**(1455-77-2)**
NCI-C04820	**(66-75-1)**
NCI-C04831	**(127-07-1)**
NCI-C04842	**(865-21-4)**
NCI-C04853	**(148-82-3)**
NCI-C04864	**(57-22-7)**
NCI-C04875	**(528-74-5)**
NCI-C04886	**(50-44-2)**
NCI-C04897	**(53-03-2)**
NCI-C04900	**(50-18-0)**
NCI-C04922	**(4465-94-5)**
NCI-C04933	**(53-19-0)**
NCI-C04944	**(531-76-0)**
NCI-C04955	**(13909-09-6)**
NCI-C05210	**(69-09-0)**
NCI-C05970	**(108-46-3)**
NCI-C06111	**(100-51-6)**
NCI-C06155	**(109-69-3)**
NCI-C06224	**(75-00-3)**
NCI-C06360	**(100-44-7)**
NCI-C06428	**(2385-85-5)**
NCI-C06462	**(26628-22-8)**
NCI-C06508	**(140-11-4)**
NCI-C07103	**(91-64-5)**
NCI-C07272	**(108-88-3)**
NCI-C08628	**(1634-78-2)**
NCI-C08640	**(116-06-3)**
NCI-C08651	**(55-38-9)**
NCI-C08662	**(56-72-4)**
NCI-C08673	**(333-41-5)**
NCI-C08684	**(101-05-3)**
NCI-C08695	**(2164-17-2)**
NCI-C08899	**(2757-90-6)**
NCI-C08991	**(1332-21-4)**
NCI-C08991	**(77536-68-6)**
NCI-C09007	**(12001-28-4)**
NCI-C50000	**(15356-70-4)**
NCI-C50011	**(119-53-9)**
NCI-C50022	**(127-69-5)**
NCI-C50033	**(7446-34-6)**
NCI-C50044	**(108-60-1)**
NCI-C50055	**(120-61-6)**
NCI-C50077	**(115-07-1)**
NCI-C50088	**(75-21-8)**
NCI-C50099	**(75-56-9)**
NCI-C50102	**(75-09-2)**
NCI-C50124	**(108-95-2)**
NCI-C50135	**(107-07-3)**
NCI-C50146	**(101-80-4)**
NCI-C50157	**(50-55-5)**
NCI-C50168	**(8001-29-4)**
NCI-C50191	**(151-21-3)**
NCI-C50204	**(126-92-1)**
NCI-C50226	**(17924-92-4)**
NCI-C50259	**(645-05-6)**
NCI-C50260	**(609-20-1)**
NCI-C50282	**(2185-92-4)**
NCI-C50317	**(15481-70-6)**
NCI-C50351	**(90-43-7)**
NCI-C50362	**(69-65-8)**
NCI-C50373	**(593-60-2)**
NCI-C50384	**(140-88-5)**
NCI-C50395	**(9000-30-0)**
NCI-C50419	**(9000-40-2)**
NCI-C50442	**(137-30-4)**
NCI-C50453	**(97-53-0)**
NCI-C50464	**(57-06-7)**
NCI-C50475	**(9002-18-0)**
NCI-C50533	**(584-84-9)**
NCI-C50544	**(68916-39-2)**
NCI-C50602	**(106-99-0)**
NCI-C50613	**(2432-99-7)**
NCI-C50635	**(80-05-7)**
NCI-C50646	**(105-60-2)**
NCI-C50657	**(131-17-9)**

NCI-C50668	**(101-68-8)**	NCI-C55163	**(8001-79-4)**
NCI-C50680	**(80-62-6)**	NCI-C55174	**(111-42-2)**
NCI-C50715	**(108-78-1)**	NCI-C55185	**(12002-43-6)**
NCI-C50737	**(21739-91-3)**	NCI-C55196	**(67-20-9)**
NCI-C50748	**(9000-01-5)**	NCI-C55209	**(144-62-7)**
NCI-C52459	**(630-20-6)**	NCI-C55210	**(83-79-4)**
NCI-C52733	**(117-81-7)**	NCI-C55221	**(7681-49-4)**
NCI-C52904	**(91-20-3)**	NCI-C55232	**(1330-20-7)**
NCI-C53587	**(63449-39-8)**	NCI-C55243	**(75-27-4)**
NCI-C53634	**(36355-01-8)**	NCI-C55254	**(124-48-1)**
NCI-C53781	**(2832-40-8)**	NCI-C55265	**(113-92-8)**
NCI-C53792	**(5160-02-1)**	NCI-C55276	**(71-43-2)**
NCI-C53838	**(1936-15-8)**	NCI-C55287	**(1163-19-5)**
NCI-C53849	**(3567-69-9)**	NCI-C55298	**(148-24-3)**
NCI-C53894	**(76-01-7)**	NCI-C55301	**(110-86-1)**
NCI-C53907	**(2783-94-0)**	NCI-C55312	**(68603-42-9)**
NCI-C53929	**(842-07-9)**	NCI-C55323	**(120-40-1)**
NCI-C54262	**(75-35-4)**	NCI-C55334	**(13961-86-9)**
NCI-C54364	**(39300-88-4)**	NCI-C55345	**(120-83-2)**
NCI-C54375	**(85-68-7)**	NCI-C55367	**(75-65-0)**
NCI-C54386	**(103-23-1)**	NCI-C55378	**(87-86-5)**
NCI-C54557	**(1937-37-7)**	NCI-C55403	**(1300-72-7)**
NCI-C54568	**(16071-86-6)**	NCI-C55425	**(111-30-8)**
NCI-C54579	**(2602-46-2)**	NCI-C55436	**(96-13-9)**
NCI-C54604	**(13552-44-8)**	NCI-C55447	**(1338-23-4)**
NCI-C54626	**(12656-85-8)**	NCI-C55458	**(81-49-2)**
NCI-C54637	**(1328-53-6)**	NCI-C55469	**(5634-39-9)**
NCI-C54659	**(1325-82-2)**	NCI-C55470	**(13983-17-0)**
NCI-C54660	**(5716-15-4)**	NCI-C55481	**(74-96-4)**
NCI-C54706	**(518-47-8)**	NCI-C55492	**(108-86-1)**
NCI-C54717	**(2835-39-4)**	NCI-C55505	**(1344-00-9)**
NCI-C54728	**(105-87-3)**	NCI-C55516	**(3296-90-0)**
NCI-C54739	**(569-61-9)**	NCI-C55527	**(106-88-7)**
NCI-C54740	**(597-25-1)**	NCI-C55538	**(7320-37-8)**
NCI-C54751	**(78-42-2)**	NCI-C55549	**(556-52-5)**
NCI-C54762	**(756-79-6)**	NCI-C55550	**(91-80-5)**
NCI-C54773	**(868-85-9)**	NCI-C55561	**(64-75-5)**
NCI-C54808	**(50-81-7)**	NCI-C55572	**(5989-27-5)**
NCI-C54819	**(513-37-1)**	NCI-C55583	**(1116-54-7)**
NCI-C54820	**(563-47-3)**	NCI-C55594	**(598-55-0)**
NCI-C54831	**(52-68-6)**	NCI-C55607	**(77-47-4)**
NCI-C54842	**(542-78-9)**	NCI-C55618	**(78-59-1)**
NCI-C54853	**(110-80-5)**	NCI-C55630	**(51-30-9)**
NCI-C54875	**(29718-44-3)**	NCI-C55641	**(61-76-7)**
NCI-C54875	**(9002-92-0)**	NCI-C55652	**(134-72-5)**
NCI-C54886	**(108-90-7)**	NCI-C55663	**(55-31-2)**
NCI-C54897	**(33229-34-4)**	NCI-C55674	**(643-22-1)**
NCI-C54900	**(2475-45-8)**	NCI-C55685	**(98-85-1)**
NCI-C54911	**(6373-74-6)**	NCI-C55696	**(108-30-5)**
NCI-C54922	**(2871-01-4)**	NCI-C55709	**(56-75-7)**
NCI-C54933	**(87-86-5)**	NCI-C55710	**(60-13-9)**
NCI-C54944	**(95-50-1)**	NCI-C55721	**(555-30-6)**
NCI-C54955	**(106-46-7)**	NCI-C55743	**(78-11-5)**
NCI-C54966	**(101-90-6)**	NCI-C55754	**(11084-85-8)**
NCI-C54977	**(96-09-3)**	NCI-C55765	**(57-41-0)**
NCI-C54988	**(78-00-2)**	NCI-C55776	**(15663-27-1)**
NCI-C54999	**(100-40-3)**	NCI-C55787	**(136-77-6)**
NCI-C55005	**(108-94-1)**	NCI-C55798	**(77-09-8)**
NCI-C55050	**(55566-30-8)**	NCI-C55801	**(103-90-2)**
NCI-C55061	**(124-64-1)**	NCI-C55812	**(92-48-8)**
NCI-C55072	**(115-28-6)**	NCI-C55823	**(77-06-5)**
NCI-C55107	**(532-27-4)**	NCI-C55834	**(123-31-9)**
NCI-C55118	**(2698-41-1)**	NCI-C55845	**(106-51-4)**
NCI-C55129	**(1643-20-5)**	NCI-C55856	**(120-80-9)**
NCI-C55130	**(75-25-2)**	NCI-C55867	**(99-49-0)**
NCI-C55141	**(78-87-5)**	NCI-C55878	**(96-48-0)**
NCI-C55152	**(1309-64-4)**	NCI-C55889	**(517-28-2)**

NCI-C55890 (119-84-6)
NCI-C55903 (298-81-7)
NCI-C55925 (58-93-5)
NCI-C55936 (54-31-9)
NCI-C55947 (509-14-8)
NCI-C55958 (99-57-0)
NCI-C55969 (548-62-9)
NCI-C55970 (121-88-0)
NCI-C55981 (123-77-3)
NCI-C55992 (100-02-7)
NCI-C56008 (523-47-7)
NCI-C56019 (52551-67-4)
NCI-C56031 (540-59-0)
NCI-C56042 (396-01-0)
NCI-C56064 (59-87-0)
NCI-C56075 (147-24-0)
NCI-C56086 (7177-48-2)
NCI-C56097 (57-66-9)
NCI-C56100 (1538-09-6)
NCI-C56111 (104-55-2)
NCI-C56122 (989-38-8)
NCI-C56133 (100-52-7)
NCI-C56144 (53-86-1)
NCI-C56155 (118-96-7)
NCI-C56166 (271-89-6)
NCI-C56177 (98-01-1)
NCI-C56188 (87-62-7)
NCI-C56199 (389-08-2)
NCI-C56202 (110-00-9)
NCI-C56213 (828-00-2)
NCI-C56224 (98-00-0)
NCI-C56235 (78-44-4)
NCI-C56246 (56-54-2)
NCI-C56257 (50-34-0)
NCI-C56279 (123-73-9)
NCI-C56280 (113-45-1)
NCI-C56291 (123-72-8)
NCI-C56304 (101-72-4)
NCI-C56315 (793-24-8)
NCI-C56326 (75-07-0)
NCI-C56337 (3081-14-9)
NCI-C56348 (5392-40-5)
NCI-C56359 (61-25-6)
NCI-C56360 (125-33-7)
NCI-C56382 (55-86-7)
NCI-C56393 (100-41-4)
NCI-C56406 (25013-15-4)
NCI-C56417 (10043-35-3)
NCI-C56428 (121-69-7)
NCI-C56439 (4016-14-2)
NCI-C56440 (684-16-2)
NCI-C56451 (9009-54-5)
NCI-C56462 (315-22-0)
NCI-C56473 (79-57-2)
NCI-C56484 (8007-12-3)
NCI-C56508 (121-19-7)
NCI-C56519 (149-30-4)
NCI-C56520 (1825-21-4)
NCI-C56531 (50-33-9)
NCI-C56542 (2624-17-1)
NCI-C56553 (544-16-1)
NCI-C56564 (404-86-4)
NCI-C56575 (470-82-6)
NCI-C56586 (303-47-9)
NCI-C56597 (57-50-1)
NCI-C56600 (57-68-1)
NCI-C56611 (51207-31-9)
NCI-C56633 (552-30-7)
NCI-C56644 (56-69-9)
NCI-C56655 (87-86-5)
NCI-C56666 (106-92-3)
NCI-C60015 (8024-37-1)
NCI-C60026 (137-09-7)
NCI-C60048 (84-66-2)
NCI-C60060 (67-62-9)
NCI-C60066 (593-77-1)
NCI-C60071 (2163-80-6)
NCI-C60082 (98-95-3)
NCI-C60093 (100-65-2)
NCI-C60102 (522-12-3)
NCI-C60106 (117-39-5)
NCI-C60117 (13494-80-9)
NCI-C60128 (115-96-8)
NCI-C60139 (106-87-6)
NCI-C60162 (81-11-8)
NCI-C60173 (7487-94-7)
NCI-C60184 (7647-10-1)
NCI-C60195 (673-06-3)
NCI-C60208 (75-38-7)
NCI-C60219 (75-44-5)
NCI-C60220 (96-18-4)
NCI-C60231 (79-11-8)
NCI-C60242 (50-31-7)
NCI-C60253A (12172-73-5)
NCI-C60286 (9005-65-6)
NCI-C60311 (7440-48-4)
NCI-C60322 (1854-26-8)
NCI-C60333 (924-42-5)
NCI-C60344 (7786-81-4)
NCI-C60355 (89-63-4)
NCI-C60366 (2425-85-6)
NCI-C60377 (6471-49-4)
NCI-C60388 (91-23-6)
NCI-C60399 (7439-97-6)
NCI-C60402 (107-15-3)
NCI-C60413 (510-15-6)
NCI-C60537 (99-99-0)
NCI-C60559 (148-65-2)
NCI-C60560 (109-99-9)
NCI-C60571 (110-54-3)
NCI-C60582 (9003-39-8)
NCI-C60606 (107-35-7)
NCI-C60628 (97-18-7)
NCI-C60639 (523-87-5)
NCI-C60640 (91-79-2)
NCI-C60651 (91-84-9)
NCI-C60662 (91-81-6)
NCI-C60673 (60-87-7)
NCI-C60684 (469-21-6)
NCI-C60695 (86-21-5)
NCI-C60708 (91-85-0)
NCI-C60719 (961-71-7)
NCI-C60720 (1982-37-2)
NCI-C60742 (81-55-0)
NCI-C60753 (606-22-4)
NCI-C60753 (97-02-9)
NCI-C60764 (20702-77-6)
NCI-C60786 (100-01-6)
NCI-C60800 (15242-96-3)
NCI-C60811 (7699-43-6)
NCI-C60822 (75-05-8)
NCI-C60844 (1817-73-8)
NCI-C60866 (109-79-5)
NCI-C60877 (28407-37-6)

NM 11	**(9002-89-5)**
NM 14	**(9002-89-5)**
NMA	**(25134-21-8)**
NMA	**(614-00-6)**
1-N-2-MA (Russian)	**(129-15-7)**
NMBA	**(7068-83-9)**
NMC 50	**(63-25-2)**
NMDDA	**(55090-44-3)**
NMEA	**(10595-95-6)**
NMH	**(684-93-5)**
NMO	**(126-85-2)**
NMOR	**(59-89-2)**
NMP	**(872-50-4)**
NMPD	**(77-49-6)**
NMS 03	**(63148-62-9)**
N-M-T	**(85-72-3)**
NMU	**(684-93-5)**
NMUT	**(615-53-2)**
NMVA	**(4549-40-0)**
NNA	**(64091-90-3)**
NNF-109	**(50512-35-1)**
NNF-136	**(66332-96-5)**
NNI 750	**(69327-76-0)**
NNK	**(64091-91-4)**
N-N-PIP	**(100-75-4)**
N-N-PYR	**(930-55-2)**
4-NO	**(99-56-9)**
NOHFAA	**(53-95-2)**
NONEX 28	**(9004-99-3)**
NONEX 29	**(9004-99-3)**
NONEX 53	**(9004-99-3)**
NONEX 54	**(9004-99-3)**
NONEX 63	**(9004-99-3)**
4-NOP	**(99-56-9)**
4-NOPD	**(99-56-9)**
1-NP	**(108-03-2)**
2-NP	**(5307-14-2)**
2-NP	**(79-46-9)**
NP 2	**(7440-02-0)**
NP-9	**(26027-38-3)**
NP 9 (Surfactant)	**(98113-10-1)**
NP-48	**(55635-13-7)**
NP 55	**(74051-80-2)**
NP 212	**(68-88-2)**
NPH-1091	**(2310-17-0)**
NPA	**(132-66-1)**
NPA	**(132-67-2)**
NPA-3	**(132-66-1)**
NPA-3	**(132-67-2)**
NPA, sodium salt	**(132-67-2)**
NPD	**(3244-90-4)**
2-N-p-PDA	**(5307-14-2)**
4-N-o-PDA	**(99-56-9)**
NPG	**(126-30-7)**
NPH 83	**(2275-23-2)**
NPIP	**(100-75-4)**
NPL 1	**(2152-34-3)**
NP-48 Na	**(55635-13-7)**
NPOH	**(25182-84-7)**
2-NPPD	**(5307-14-2)**
NPRO	**(7519-36-0)**
NPYR	**(930-55-2)**
50N (Polymer)	**(9011-14-7)**
4-NQO	**(56-57-5)**
NR.C 2294	**(555-30-6)**
NRC 910	**(36734-19-7)**
NRDC 104	**(10453-86-8)**
NRDC 107	**(28434-01-7)**
NRDC 143	**(52645-53-1)**
NRDC 146	**(52341-32-9)**
NRDC 148	**(52341-33-0)**
NRDC 149	**(52315-07-8)**
NRDC 156	**(52820-00-5)**
NRDC 158	**(52820-00-5)**
NRDC 159	**(71697-59-1)**
NRDC 160	**(52315-07-8)**
NRDC 161	**(52918-63-5)**
NRDC 166	**(52315-07-8)**
NS 11	**(1854-26-8)**
NS 2842	**(9003-20-7)**
NSAR	**(13256-22-9)**
NSB6B	**(2610-05-1)**
NSC-D-254157	**(54749-90-5)**
NSC-19	**(638-46-0)**
NSC-49	**(3333-15-1)**
NSC-91	**(89-95-2)**
NSC-104	**(3299-05-6)**
NSC-130	**(5325-20-2)**
NSC-168	**(130-13-2)**
NSC-176	**(499-83-2)**
NSC-177	**(100-26-5)**
NSC-178	**(490-11-9)**
NSC-183	**(96-97-9)**
NSC-185	**(66-81-9)**
NSC-212	**(121-63-1)**
NSC-240	**(5325-97-3)**
NSC-251	**(504-08-5)**
NSC-257	**(542-02-9)**
NSC-286	**(570-24-1)**
NSC-304	**(2749-59-9)**
NSC-339	**(140-79-4)**
NSC-402	**(405-50-5)**
NSC-423	**(94-75-7)**
NSC-439	**(92-85-3)**
NSC-504	**(129-35-1)**
NSC-521	**(5343-99-7)**
NSC-607	**(84-54-8)**
NSC-623	**(117-99-7)**
NSC-736	**(538-51-2)**
NSC-739	**(54-62-6)**
NSC-740	**(59-05-2)**
NSC-742	**(115-02-6)**
NSC-746	**(51-79-6)**
NSC-749	**(134-58-7)**
NSC-750	**(55-98-1)**
NSC-750 Sulphabutin	**(55-98-1)**
NSC-751	**(67-21-0)**
NSC-752	**(154-42-7)**
NSC-753	**(120-73-0)**
NSC-755	**(50-44-2)**
NSC-756	**(1468-26-4)**
NSC-757	**(64-86-8)**
NSC-759	**(51-17-2)**
NSC-762 hydrochloride	**(55-86-7)**
NSC-762	**(55-86-7)**
NSC-762	**(51-75-2)**
NSC-763	**(67-68-5)**
NSC-777	**(121-43-7)**
NSC-791	**(534-59-8)**
NSC-805	**(497-59-6)**
NSC-821	**(5337-72-4)**
NSC-822	**(536-75-4)**
NSC-829	**(126-39-6)**
NSC-845	**(1123-09-7)**

NSC-850	(143-10-2)	NSC-2438	(574-42-5)
NSC-851	(19484-26-5)	NSC-2440	(121-71-1)
NSC-852	(5337-93-9)	NSC-2475	(2164-13-8)
NSC-858	(100-12-9)	NSC-2525	(118-41-2)
NSC-900	(597-05-7)	NSC-2581	(83-33-0)
NSC-944	(590-50-1)	NSC-2600	(85-97-2)
NSC-945	(926-41-0)	NSC-2628	(92-53-5)
NSC-964	(100-71-0)	NSC-2632	(553-12-8)
NSC-978	(107-54-0)	NSC-2637	(2051-31-2)
NSC-1007	(759-36-4)	NSC-2651	(98-31-7)
NSC-1026	(52-52-8)	NSC-2667	(940-71-6)
NSC-1052	(5336-24-3)	NSC-2671	(78-38-6)
NSC-1073	(622-86-6)	NSC-2688	(574-98-1)
NSC-1075	(628-46-6)	NSC-2752	(110-17-8)
NSC-1110	(137-43-9)	NSC-2759	(89-62-3)
NSC-1115	(36413-60-2)	NSC-2765	(496-46-8)
NSC-1117	(5344-44-5)	NSC-2843	(132-65-0)
NSC-1128	(88-50-6)	NSC-2853	(121-53-9)
NSC-1254	(141-24-2)	NSC-2859	(505-10-2)
NSC-1267	(92-05-7)	NSC-2860	(615-53-2)
NSC-1273	(111-81-9)	NSC-2865	(76-08-4)
NSC-1313	(552-32-9)	NSC-2872	(134-85-0)
NSC-1314	(122-28-1)	NSC-2884	(78-33-1)
NSC-1390	(315-30-0)	NSC-2956	(127-27-5)
NSC-1463	(1133-80-8)	NSC-2998	(14167-59-0)
NSC-1469	(87-41-2)	NSC-3051	(123-39-7)
NSC-1487	(110-69-0)	NSC-3053	(50-76-0)
NSC-1491	(98-37-3)	NSC-3058	(95-14-7)
NSC-1532	(51-28-5)	NSC-3058	(95-14-7)
NSC-1533	(119-91-5)	NSC-3060	(13464-35-2)
NSC-1699	(134-47-4)	NSC-3060	(10124-50-2)
NSC-1704	(92-40-0)	NSC-3061	(58-14-0)
NSC-1735	(86-76-0)	NSC-3069	(56-75-7)
NSC-1746	(82-76-8)	NSC-3070	(56-53-1)
NSC-1773	(504-57-4)	NSC-3072	(26628-22-8)
NSC-1809	(94-71-3)	NSC-3073	(59-30-3)
NSC-1862	(104-38-1)	NSC-3088	(305-03-3)
NSC-1875	(92-91-1)	NSC-3094	(7227-91-0)
NSC-1895	(1455-77-2)	NSC-3096	(477-30-5)
NSC-1897	(110-21-4)	NSC-3131	(93-98-1)
NSC-1903	(142-45-0)	NSC-3138	(127-19-5)
NSC-1907	(536-66-3)	NSC-3148	(3370-35-2)
NSC-1966	(482-05-3)	NSC-3187	(70-69-9)
NSC-2019	(517-23-7)	NSC-3215	(80-44-4)
NSC-2042	(463-40-1)	NSC-3216	(80-42-2)
NSC-2066	(58-55-9)	NSC-3217	(515-46-8)
NSC-2073	(118-32-1)	NSC-3221	(85-29-0)
NSC-2083	(65-49-6)	NSC-3357	(97-65-4)
NSC-2093	(99-20-7)	NSC-3371	(126-00-1)
NSC-2098	(92-86-4)	NSC-3377	(111-57-9)
NSC-2100	(59-87-0)	NSC-3637	(374-01-6)
NSC-2101	(121-19-7)	NSC-3689	(480-91-1)
NSC-2107	(67-20-9)	NSC-3710	(111-11-5)
NSC-2139	(1142-15-0)	NSC-3713	(110-42-9)
NSC-2144	(103-74-2)	NSC-3717	(86-48-6)
NSC-2162	(98-03-3)	NSC-3721	(82-38-2)
NSC-2169	(527-85-5)	NSC-3790	(552-82-9)
NSC-2179	(497-03-0)	NSC-3806	(300-85-6)
NSC-2181	(78-32-0)	NSC-3868	(141-20-8)
NSC-2222	(552-22-7)	NSC-3875	(123-94-4)
NSC-2224	(104-87-0)	NSC-3894	(77-90-7)
NSC-2292	(5331-48-6)	NSC-3896	(140-24-9)
NSC-2293	(95-59-0)	NSC-3914	(74-39-5)
NSC-2319	(140-04-5)	NSC-3983	(138-60-3)
NSC-2397	(93-92-5)	NSC-3998	(554-95-0)
NSC-2413	(834-14-0)	NSC-4002	(556-08-1)
NSC-2428	(58-90-2)	NSC-4047	(103-43-5)

NSC-4049	(519-73-3)	NSC-5594	(3730-60-7)
NSC-4050	(76-84-6)	NSC-5595	(110-03-2)
NSC-4055	(563-04-2)	NSC-5596	(78-39-7)
NSC-4057	(523-31-9)	NSC-5604	(115-80-0)
NSC-4058	(506-52-5)	NSC-5613	(151-19-9)
NSC-4061	(538-24-9)	NSC-5619	(101-55-3)
NSC-4062	(555-45-3)	NSC-5630	(126-86-3)
NSC-4143	(300-39-0)	NSC-5651	(556-48-9)
NSC-4170	(58-27-5)	NSC-5710	(59-88-1)
NSC-4171	(93-04-9)	NSC-5751	(138-52-3)
NSC-4173	(398-23-2)	NSC-5937	(530-47-2)
NSC-4191	(90-96-0)	NSC-5953	(2373-80-0)
NSC-4193	(112-82-3)	NSC-6001	(18266-55-2)
NSC-4197	(112-39-0)	NSC-6077	(553-82-2)
NSC-4205	(506-46-7)	NSC-6078	(1647-26-3)
NSC-4397	(5396-91-8)	NSC-6081	(479-27-6)
NSC-4406	(1853-88-9)	NSC-6091D	(80-08-0)
NSC-4445	(105-97-5)	NSC-6091	(80-08-0)
NSC-4506	(98-58-8)	NSC-6093	(119-42-6)
NSC-4549	(107-47-1)	NSC-6108	(87-88-7)
NSC-4552	(623-93-8)	NSC-6123	(1129-50-6)
NSC-4647	(76-30-2)	NSC-6181	(533-98-2)
NSC-4652	(14024-48-7)	NSC-6187	(109-19-3)
NSC-4672	(459-60-9)	NSC-6197	(108-63-4)
NSC-4683	(102-92-1)	NSC-6199	(75-81-0)
NSC-4701	(2049-92-5)	NSC-6209	(625-92-3)
NSC-4706	(119-70-0)	NSC-6212	(88-87-9)
NSC-4732	(563-41-7)	NSC-6232	(1072-16-8)
NSC-4811	(103-49-1)	NSC-6247	(542-28-9)
NSC-4814	(110-36-1)	NSC-6255	(100-84-5)
NSC-4815	(111-06-8)	NSC-6261	(544-00-3)
NSC-4827	(79-54-9)	NSC-6266	(120-46-7)
NSC-4860	(489-98-5)	NSC-6288	(110-56-5)
NSC-4866	(576-55-6)	NSC-6295	(99-54-7)
NSC-4874	(632-79-1)	NSC-6315	(93-02-7)
NSC-4883	(129-96-4)	NSC-6316	(91-52-1)
NSC-4958	(16941-12-1)	NSC-6317	(93-03-8)
NSC-4976	(98-32-8)	NSC-6380	(106-10-5)
NSC-4983	(119-28-8)	NSC-6381	(27554-26-3)
NSC-5027	(111-82-0)	NSC-6385	(42343-35-1)
NSC-5029	(124-10-7)	NSC-6396	(52-24-4)
NSC-5064	(521-31-3)	NSC-6435	(471-74-9)
NSC-5185	(501-65-5)	NSC-6470	(555-84-0)
NSC-5215	(14861-06-4)	NSC-6495	(565-70-8)
NSC-5240	(93-37-8)	NSC-6497	(499-81-0)
NSC-5267	(128-70-1)	NSC-6528	(107-84-6)
NSC-5276	(498-21-5)	NSC-6602	(102-13-6)
NSC-5285	(36673-16-2)	NSC-6649	(122-40-7)
NSC-5312	(94-04-2)	NSC-6738	(62-73-7)
NSC-5313	(91-65-6)	NSC-6786	(143-15-7)
NSC-5332	(85-85-8)	NSC-6794	(104-66-5)
NSC-5344	(88-97-1)	NSC-6814	(112-86-7)
NSC-5354	(95-54-5)	NSC-6931	(112-00-5)
NSC-5356	(68-12-2)	NSC-6973	(109-49-9)
NSC-5366	(128-62-1)	NSC-6976	(88-65-3)
NSC-5366	(6035-40-1)	NSC-7111	(82-47-3)
NSC-5375	(96-73-1)	NSC-7128	(2049-92-5)
NSC-5376	(138-42-1)	NSC-7193	(83-38-5)
NSC-5386	(89-59-8)	NSC-7197	(99-64-9)
NSC-5390	(100-15-2)	NSC-7198	(99-88-7)
NSC-5402	(83-41-0)	NSC-7201	(575-44-0)
NSC-5403	(99-12-7)	NSC-7295	(506-93-4)
NSC-5406	(111-85-3)	NSC-7298	(594-89-8)
NSC-5446	(143-09-9)	NSC-7304	(116-53-0)
NSC-5534	(91-29-2)	NSC-7345	(15375-84-5)
NSC-5542	(112-89-0)	NSC-7378	(494-99-5)
NSC-5576	(115-22-0)	NSC-7415	(532-02-5)

NSC-10321	(456-22-4)	NSC-16057	(5463-50-3)
NSC-10342	(402-31-3)	NSC-16201	(123-79-5)
NSC-10348	(351-28-0)	NSC-16221	(81-78-7)
NSC-10366	(321-60-8)	NSC-16507	(94-97-3)
NSC-10433	(85-48-3)	NSC-16627	(505-70-4)
NSC-10441	(547-57-9)	NSC-16644	(586-89-0)
NSC-10451	(119-40-4)	NSC-16682	(306-08-1)
NSC-10456	(554-73-4)	NSC-16684	(5438-40-4)
NSC-10479	(107-68-6)	NSC-16797	(2613-76-5)
NSC-10483	(50-23-7)	NSC-16842	(81-33-4)
NSC-10770	(557-61-9)	NSC-16895	(554-13-2)
NSC-10815	(57-43-2)	NSC-16939	(104-52-9)
NSC-10969	(5396-91-8)	NSC-17012	(6641-64-1)
NSC-10973	(57-63-6)	NSC-17028	(42087-80-9)
NSC-11011	(81-83-4)	NSC-17041	(611-05-2)
NSC-11041	(151-32-6)	NSC-17071	(53306-54-0)
NSC-11135	(3333-15-1)	NSC-17262	(800-24-8)
NSC-11138	(18039-42-4)	NSC-17391	(328-50-7)
NSC-11207	(108-85-0)	NSC-17503	(2050-20-6)
NSC-11208	(81-08-3)	NSC-17524	(2132-80-1)
NSC-11212	(94-57-5)	NSC-17541	(99-90-1)
NSC-11271	(104-63-2)	NSC-17591	(315-37-7)
NSC-11440	(6164-47-2)	NSC-17592	(630-56-8)
NSC-11685	(1119-85-3)	NSC-17695	(13552-21-1)
NSC-11809	(6295-15-4)	NSC-17764	(141-94-6)
NSC-11905	(84-79-7)	NSC-17817	(86-20-4)
NSC-12016	(112-41-4)	NSC-18188	(80-97-7)
NSC-12165	(76-43-7)	NSC-18429	(553-27-5)
NSC-12169	(50-27-1)	NSC-18522	(552-86-3)
NSC-12173	(968-93-4)	NSC-18609	(300-57-2)
NSC-12216	(932-66-1)	NSC-18698	(56-05-3)
NSC-12352	(4440-33-9)	NSC-18728	(6966-10-5)
NSC-13064	(61-78-9)	NSC-18745	(18707-60-3)
NSC-13065	(98-74-8)	NSC-18950	(83-08-9)
NSC-13123	(60-27-5)	NSC-18964	(1460-16-8)
NSC-13127	(89-00-9)	NSC-19045	(76-41-5)
NSC-13875	(645-05-6)	NSC-19177	(15022-08-9)
NSC-13876	(6295-15-4)	NSC-19178	(2105-40-0)
NSC-14083	(57-92-1)	NSC-19185	(121-93-7)
NSC-14210	(531-76-0)	NSC-19598	(467-85-6)
NSC-14451	(489-01-0)	NSC-19675	(100-36-7)
NSC-14460	(17540-75-9)	NSC-19893	(51-21-8)
NSC-14662	(565-80-0)	NSC-19930	(480-96-6)
NSC-14856	(80-30-8)	NSC-19987	(83-43-2)
NSC-14984	(126-81-8)	NSC-20004	(99-36-5)
NSC-15039	(124-63-0)	NSC-20264	(61-19-8)
NSC-15117	(5434-82-2)	NSC-20526	(87-10-5)
NSC-15164	(505-54-4)	NSC-20527	(87-12-7)
NSC-15184	(542-50-7)	NSC-20743	(629-83-4)
NSC-15193	(154-17-6)	NSC-20891	(111-63-7)
NSC-15257	(97-91-6)	NSC-20953	(1123-84-8)
NSC-15258	(107-56-2)	NSC-20968	(101-43-9)
NSC-15282	(121-06-2)	NSC-20975	(3179-47-3)
NSC-15294	(110-93-0)	NSC-20990	(529-64-6)
NSC-15333	(89-74-7)	NSC-21130	(5441-52-1)
NSC-15336	(25870-62-6)	NSC-21302	(430-51-3)
NSC-15339	(81-14-1)	NSC-21402	(2349-67-9)
NSC-15346	(88-43-7)	NSC-21491	(4097-49-8)
NSC-15355	(89-29-2)	NSC-21630	(344-04-7)
NSC-15432	(68-23-5)	NSC-21635	(551-62-2)
NSC-15780	(29883-15-6)	NSC-21705	(140-07-8)
NSC-15912	(492-22-8)	NSC-21978	(13019-20-0)
NSC-15920	(56-09-7)	NSC-22050	(54340-87-3)
NSC-15962	(530-44-9)	NSC-22229	(115-27-5)
NSC-15976	(147-14-8)	NSC-22364	(136-45-8)
NSC-16044	(2396-60-3)	NSC-23056	(1731-81-3)
NSC-16045	(595-37-9)	NSC-23262	(137-00-8)

NSC	CAS
NSC-23516	(94-12-2)
NSC-23696	(565-59-3)
NSC-23737	(5458-59-3)
NSC-23759	(968-93-4)
NSC-23909	(684-93-5)
NSC-23966	(139-90-2)
NSC-23989	(20679-58-7)
NSC-24068	(525-52-0)
NSC-24169	(142-09-6)
NSC-24170	(103-44-6)
NSC-24343	(96-26-4)
NSC-24559	(18378-89-7)
NSC-24818	(518-28-5)
NSC-24846	(565-75-3)
NSC-24852	(124-43-6)
NSC-24860	(135-70-6)
NSC-24890	(614-95-9)
NSC-25084	(352-11-4)
NSC-25154	(54-91-1)
NSC-25505	(81-24-3)
NSC-25530	(623-55-2)
NSC-25872	(116-44-9)
NSC-25952	(505-48-6)
NSC-26136	(1805-32-9)
NSC-26154	(60-32-2)
NSC-26198	(434-07-1)
NSC-26265	(495-76-1)
NSC-26271	(50-18-0)
NSC-26271	(6055-19-2)
NSC-26317	(483-65-8)
NSC-26341	(83-31-8)
NSC-26345	(106-57-0)
NSC-26386	(71-58-9)
NSC-26429	(89-36-1)
NSC-26769	(6542-67-2)
NSC-26770	(5465-03-2)
NSC-26805	(62-50-0)
NSC-26812	(52-46-0)
NSC-26958	(15945-07-0)
NSC-26980	(50-07-7)
NSC-26987	(301-02-0)
NSC-27273	(491-30-5)
NSC-27274	(2518-72-1)
NSC-27435	(303-38-8)
NSC-27531	(15118-60-2)
NSC-27640	(50-91-9)
NSC-27788	(535-15-9)
NSC-27794	(102-52-3)
NSC-27929	(789-24-2)
NSC-28044	(597-76-2)
NSC-28342	(773-99-9)
NSC-28462	(1619-62-1)
NSC-28524	(5532-90-1)
NSC-28593	(19077-97-5)
NSC-28667	(99-42-3)
NSC-28693	(315-22-0)
NSC-28731	(94-45-1)
NSC-28952	(826-55-1)
NSC-29095	(826-55-1)
NSC-29215	(68-76-8)
NSC-29469	(473-72-3)
NSC-29558	(17564-64-6)
NSC-29630	(528-74-5)
NSC-29705	(1538-74-5)
NSC-29737	(5473-16-5)
NSC-30022	(538-75-0)
NSC-30032	(123-81-9)
NSC-30092	(6324-11-4)
NSC-30211	(817-09-4)
NSC-30235	(504-01-8)
NSC-30512	(16365-27-8)
NSC-30551	(118-82-1)
NSC-30635	(403-42-9)
NSC-30678	(38721-71-0)
NSC-31005	(949-87-1)
NSC-31007	(584-90-7)
NSC-31008	(501-60-0)
NSC-31011	(501-58-6)
NSC-31187	(543-28-2)
NSC-31400	(492-86-4)
NSC-31508	(90-51-7)
NSC-31511	(93-00-5)
NSC-31590	(2314-36-5)
NSC-31630	(1208-86-2)
NSC-31666	(7493-58-5)
NSC-31811	(9004-99-3)
NSC-32065	(127-07-1)
NSC-32074	(54-25-1)
NSC-32305	(115-58-2)
NSC-32364	(112-85-6)
NSC-32410	(101-40-6)
NSC-32427	(7781-98-8)
NSC-32606	(30031-64-2)
NSC-32617	(352-87-4)
NSC-32863	(823-22-3)
NSC-32865	(51-83-2)
NSC-32965	(14167-18-1)
NSC-33407	(85-19-8)
NSC-33417	(537-65-5)
NSC-33531	(84-60-6)
NSC-33535	(537-64-4)
NSC-33659	(6272-74-8)
NSC-33669	(316-42-7)
NSC-33669	(483-18-1)
NSC-33832	(50-81-7)
NSC-33906	(1523-06-4)
NSC-33926	(103-63-9)
NSC-34462	(66-75-1)
NSC-34533	(126-07-8)
NSC-34652	(50-12-4)
NSC-35011	(624-47-5)
NSC-35051	(13045-94-8)
NSC-35134	(102-87-4)
NSC-35142	(87-40-1)
NSC-35717	(10551-21-0)
NSC-35762	(6274-12-0)
NSC-35770	(50-41-9)
NSC-36365	(90-93-7)
NSC-36629	(3279-07-0)
NSC-36763	(106-02-5)
NSC-36935	(106-46-7)
NSC-37006	(24313-88-0)
NSC-37116	(93-97-0)
NSC-37168	(135-65-9)
NSC-37184	(120-00-3)
NSC-37187	(135-63-7)
NSC-37188	(135-61-5)
NSC-37202	(132-68-3)
NSC-37212	(91-96-3)
NSC-37224	(91-92-9)
NSC-37398	(769-11-9)
NSC-37409	(1878-87-1)
NSC-37448	(59-96-1)
NSC-37584	(128-91-6)

NSC-37725	**(52-76-6)**
NSC-37756	**(2827-49-8)**
NSC-38302	**(1127-75-9)**
NSC-38367	**(3321-64-0)**
NSC-38629	**(478-43-3)**
NSC-38693	**(2314-78-5)**
NSC-38721	**(53-19-0)**
NSC-38776	**(496-78-6)**
NSC-38857	**(767-59-9)**
NSC-38861	**(4830-99-3)**
NSC-38865	**(1636-39-1)**
NSC-39069	**(299-75-2)**
NSC-39084	**(446-86-6)**
NSC-39113	**(10139-47-6)**
NSC-39120	**(15448-99-4)**
NSC-39452	**(537-24-6)**
NSC-39470	**(378-44-9)**
NSC-39906	**(128-85-8)**
NSC-39966	**(825-41-2)**
NSC-40562	**(533-17-5)**
NSC-40774	**(342-69-8)**
NSC-40902	**(956-90-1)**
NSC-41053	**(100-93-6)**
NSC-41205	**(2042-14-0)**
NSC-41404	**(13183-09-0)**
NSC-41799	**(522-66-7)**
NSC-41820	**(505-47-5)**
NSC-41903	**(111-88-6)**
NSC-41917	**(26403-12-3)**
NSC-41924	**(21302-09-0)**
NSC-41927	**(5659-41-6)**
NSC-42111	**(114-38-5)**
NSC-42192	**(551-45-1)**
NSC-42647	**(6311-92-8)**
NSC-42722	**(72-63-9)**
NSC-42739	**(372-31-6)**
NSC-42752	**(434-45-7)**
NSC-42810	**(4789-76-8)**
NSC-42869	**(1694-31-1)**
NSC-42872	**(492-37-5)**
NSC-43025	**(454-92-2)**
NSC-43063	**(135-69-3)**
NSC-43193	**(10418-03-8)**
NSC-43206	**(6311-44-0)**
NSC-43243	**(332-77-4)**
NSC-43245	**(589-63-9)**
NSC-43316	**(85-06-3)**
NSC-43644	**(5606-27-9)**
NSC-43870	**(120-29-6)**
NSC-43944	**(509-34-2)**
NSC-44042	**(115-87-7)**
NSC-44059	**(550-44-7)**
NSC-44233	**(4403-61-6)**
NSC-44517	**(28693-00-7)**
NSC-44611	**(88-95-9)**
NSC-44690	**(2150-48-3)**
NSC-44754	**(13676-54-5)**
NSC-44869	**(6299-66-7)**
NSC-44899	**(31121-12-7)**
NSC-45173	**(92-77-3)**
NSC-45353	**(16645-06-0)**
NSC-45388	**(4342-03-4)**
NSC-45403	**(759-73-9)**
NSC-45530	**(4313-13-7)**
NSC-45624	**(10102-17-7)**
NSC-45762	**(13548-68-0)**
NSC-45929	**(19463-48-0)**
NSC-45930	**(62936-23-6)**
NSC-46100	**(122-63-4)**
NSC-46102	**(104-21-2)**
NSC-46106	**(106-23-0)**
NSC-46119	**(540-07-8)**
NSC-46127	**(110-41-8)**
NSC-46131	**(112-17-4)**
NSC-46150	**(101-86-0)**
NSC-46152	**(523-47-7)**
NSC-46246	**(2704-78-1)**
NSC-46248	**(473-72-3)**
NSC-46419	**(39589-98-5)**
NSC-46520	**(3140-73-6)**
NSC-46521	**(493-77-6)**
NSC-46590	**(104-82-5)**
NSC-46704	**(51-35-4)**
NSC-46707	**(556-33-2)**
NSC-46808	**(1647-26-3)**
NSC-46823	**(619-14-7)**
NSC-47001	**(471-47-6)**
NSC-47035	**(2480-86-6)**
NSC-47196	**(95-15-8)**
NSC-47439	**(53-34-9)**
NSC-47760	**(3476-90-2)**
NSC-47842	**(865-21-4)**
NSC-48467	**(88-60-8)**
NSC-48754	**(215-62-3)**
NSC-48909	**(122-64-5)**
NSC-49126	**(134-83-8)**
NSC-49150	**(89-33-8)**
NSC-49171	**(552-94-3)**
NSC-49172	**(303-07-1)**
NSC-49346	**(556-50-3)**
NSC-49360	**(15862-72-3)**
NSC-49417	**(462-60-2)**
NSC-49513	**(12122-67-7)**
NSC-49580	**(25619-56-1)**
NSC-49842	**(143-67-9)**
NSC-50256	**(66-27-3)**
NSC-50330	**(3470-97-1)**
NSC-50364	**(443-48-1)**
NSC-50574	**(3638-04-8)**
NSC-50649	**(97-32-5)**
NSC-50655	**(93-94-7)**
NSC-50659	**(96-98-0)**
NSC-50661	**(89-87-2)**
NSC-50680	**(135-62-6)**
NSC-50681	**(92-74-0)**
NSC-50685	**(137-52-0)**
NSC-50932	**(111-59-1)**
NSC-50952	**(112-11-8)**
NSC-51012	**(769-25-5)**
NSC-51100	**(931-86-2)**
NSC-51871	**(1700-02-3)**
NSC-52209	**(106-65-0)**
NSC-52221	**(502-41-0)**
NSC-52232	**(4629-58-7)**
NSC-52602	**(126-83-0)**
NSC-52695	**(4465-94-5)**
NSC-52966	**(575-43-9)**
NSC-52971	**(7144-65-2)**
NSC-53044	**(555-10-2)**
NSC-53659	**(2163-48-6)**
NSC-53816	**(4887-30-3)**
NSC-53832	**(506-50-3)**
NSC-53866	**(6659-45-6)**
NSC-53964	**(6938-45-0)**

NSC-54006 **(931-86-2)**
NSC-54015 **(94-65-5)**
NSC-54157 **(5809-41-6)**
NSC-54159 **(515-40-2)**
NSC-54255 **(6963-65-1)**
NSC-54357 **(279-49-2)**
NSC-54458 **(765-69-5)**
NSC-54656 **(4122-04-7)**
NSC-54990 **(536-40-3)**
NSC-54993 **(27578-60-5)**
NSC-55519 **(100-27-6)**
NSC-56443 **(287-27-4)**
NSC-56686 **(324-74-3)**
NSC-56763 **(111-19-3)**
NSC-56774 **(103-96-8)**
NSC-56901 **(71-33-0)**
NSC-56930 **(101-18-8)**
NSC-57030 **(26021-90-9)**
NSC-57107 **(112-52-7)**
NSC-57477 **(504-02-9)**
NSC-57546 **(136-84-5)**
NSC-57582 **(2272-28-8)**
NSC-57595 **(28804-67-3)**
NSC-57645 **(530-48-3)**
NSC-57755 **(89-20-3)**
NSC-57860 **(110-87-2)**
NSC-57905 **(4824-72-0)**
NSC-58039 **(14233-37-5)**
NSC-58374 **(253-66-7)**
NSC-58376 **(142-94-9)**
NSC-58378 **(18263-25-7)**
NSC-58775 **(146-22-5)**
NSC-58961 **(533-18-6)**
NSC-58973 **(492-27-3)**
NSC-59261 **(530-59-6)**
NSC-59388 **(571-61-9)**
NSC-59452 **(33058-12-7)**
NSC-59697 **(100-70-9)**
NSC-59702 **(98-36-2)**
NSC-59714 **(3400-45-1)**
NSC-59716 **(286-28-2)**
NSC-59775 **(79-94-7)**
NSC-59788 **(243-28-7)**
NSC-59832 **(573-98-8)**
NSC-59855 **(92-51-3)**
NSC-59876 **(501-94-0)**
NSC-59990 **(119-56-2)**
NSC-60003 **(6630-01-9)**
NSC-60023 **(101-71-3)**
NSC-60055 **(16824-02-5)**
NSC-60108 **(873-62-1)**
NSC-60145 **(89-92-9)**
NSC-60155 **(111-78-4)**
NSC-60284 **(91-43-0)**
NSC-60292 **(122-94-1)**
NSC-60297 **(101-32-6)**
NSC-60380 **(5598-13-0)**
NSC-60388 **(97-97-2)**
NSC-60551 **(623-47-2)**
NSC-60574 **(123-17-1)**
NSC-60585 **(86-28-2)**
NSC-60651 **(402-67-5)**
NSC-60681 **(100-48-1)**
NSC-60689 **(638-02-8)**
NSC-60702 **(7319-23-5)**
NSC-60720 **(456-42-8)**
NSC-60721 **(89-60-1)**
NSC-60722 **(118-69-4)**
NSC-60773 **(575-37-1)**
NSC-61373 **(10108-91-5)**
NSC-61406 **(5659-41-6)**
NSC-61779 **(571-58-4)**
NSC-61873 **(24654-08-8)**
NSC-61938 **(5659-41-6)**
NSC-61979 **(681-57-2)**
NSC-61989 **(108-08-7)**
NSC-61993 **(575-89-3)**
NSC-62035 **(3760-20-1)**
NSC-62066 **(4750-28-1)**
NSC-62078 **(3393-64-4)**
NSC-62085 **(77-76-9)**
NSC-62133 **(1463-17-8)**
NSC-62142 **(2049-94-7)**
NSC-62209 **(494-03-1)**
NSC-62222 **(126-71-6)**
NSC-62484 **(253-52-1)**
NSC-62486 **(125-12-2)**
NSC-62674 **(5673-07-4)**
NSC-62701 **(2050-23-9)**
NSC-62748 **(6961-73-5)**
NSC-62788 **(2216-51-5)**
NSC-62789 **(112-95-8)**
NSC-62794 **(7235-40-7)**
NSC-62840 **(26872-84-4)**
NSC-63044 **(13602-12-5)**
NSC-63112 **(1453-06-1)**
NSC-63121 **(751-38-2)**
NSC-63345 **(67-71-0)**
NSC-63701 **(606-58-6)**
NSC-63878 **(69-74-9)**
NSC-63878 **(147-94-4)**
NSC-64375 **(63-12-7)**
NSC-64753 **(86722-66-9)**
NSC-65329 **(13195-76-1)**
NSC-65426 **(504-63-2)**
NSC-65430 **(600-22-6)**
NSC-65440 **(96-54-8)**
NSC-65448 **(142-68-7)**
NSC-65581 **(77-85-0)**
NSC-65606 **(91-61-2)**
NSC-65612 **(527-35-5)**
NSC-65648 **(527-54-8)**
NSC-65881 **(126-58-9)**
NSC-65884 **(84-23-1)**
NSC-66186 **(123-18-2)**
NSC-66270 **(122-07-6)**
NSC-66307 **(14426-42-7)**
NSC-66318 **(6964-19-8)**
NSC-66406 **(115-17-3)**
NSC-66412 **(421-50-1)**
NSC-66432 **(127-51-5)**
NSC-66460 **(112-88-9)**
NSC-66473 **(52078-56-5)**
NSC-66475 **(19074-59-0)**
NSC-66492 **(565-61-7)**
NSC-66515 **(25653-16-1)**
NSC-66540 **(4049-81-4)**
NSC-66544 **(326-91-0)**
NSC-66547 **(33562-89-9)**
NSC-66555 **(99-51-4)**
NSC-66572 **(111-67-1)**
NSC-66847 **(50-35-1)**
NSC-66908 **(2518-72-1)**
NSC-66993 **(3877-19-8)**

NSC	CAS
NSC-67488	**(90-66-4)**
NSC-67574	**(57-22-7)**
NSC-67574	**(2068-78-2)**
NSC-68048	**(84-83-3)**
NSC-68072	**(35029-96-0)**
NSC-68382	**(355-68-0)**
NSC-68522	**(10203-58-4)**
NSC-68626	**(17433-31-7)**
NSC-68803	**(80-39-7)**
NSC-68982	**(5051-62-7)**
NSC-69110	**(13110-37-7)**
NSC-69188	**(500-44-7)**
NSC-69894	**(84-78-6)**
NSC-69963	**(13361-34-7)**
NSC-69994	**(84-64-0)**
NSC-70731	**(154-21-2)**
NSC-70762	**(1156-19-0)**
NSC-70955	**(500-99-2)**
NSC-71047	**(3902-71-4)**
NSC-71130	**(78-23-9)**
NSC-71207	**(541-88-8)**
NSC-71423	**(595-33-5)**
NSC-72005	**(101-20-2)**
NSC-72031	**(141-12-8)**
NSC-72329	**(7149-75-9)**
NSC-72372	**(253-82-7)**
NSC-72374	**(22047-25-2)**
NSC-72426	**(292-64-8)**
NSC-72739	**(89-96-3)**
NSC-72788	**(13402-02-3)**
NSC-72944	**(2973-77-5)**
NSC-72983	**(7254-11-7)**
NSC-73599	**(101-25-7)**
NSC-73701	**(22483-09-6)**
NSC-73707	**(32749-94-3)**
NSC-73712	**(3786-91-2)**
NSC-73718	**(7058-01-7)**
NSC-73906	**(563-78-0)**
NSC-73907	**(563-79-1)**
NSC-73914	**(674-76-0)**
NSC-73924	**(6094-02-6)**
NSC-73926	**(3769-23-1)**
NSC-73928	**(2738-19-4)**
NSC-73929	**(10574-36-4)**
NSC-73938	**(464-06-2)**
NSC-73940	**(14850-23-8)**
NSC-73943	**(565-77-5)**
NSC-73949	**(15870-10-7)**
NSC-73952	**(3726-47-4)**
NSC-73954	**(564-02-3)**
NSC-73956	**(1067-08-9)**
NSC-73961	**(124-11-8)**
NSC-73972	**(1587-04-8)**
NSC-73982	**(538-68-1)**
NSC-74119	**(558-37-2)**
NSC-74125	**(13269-52-8)**
NSC-74131	**(14686-13-6)**
NSC-74133	**(14686-14-7)**
NSC-74134	**(3404-72-6)**
NSC-74136	**(3404-73-7)**
NSC-74139	**(625-65-0)**
NSC-74142	**(762-63-0)**
NSC-74147	**(822-50-4)**
NSC-74150	**(562-49-2)**
NSC-74161	**(2207-03-6)**
NSC-74163	**(7145-20-2)**
NSC-74164	**(3404-78-2)**
NSC-74174	**(563-16-6)**
NSC-74179	**(2040-95-1)**
NSC-74187	**(1678-98-4)**
NSC-74191	**(104-72-3)**
NSC-74499	**(20017-67-8)**
NSC-74550	**(4920-92-7)**
NSC-74687	**(619-56-7)**
NSC-74700	**(130-20-1)**
NSC-74770	**(21184-58-7)**
NSC-74793	**(2164-13-8)**
NSC-74803	**(2388-12-7)**
NSC-74872	**(612-35-1)**
NSC-75061	**(141-07-1)**
NSC-75446	**(127-52-6)**
NSC-75520	**(70-00-8)**
NSC-75616	**(533-15-3)**
NSC-75857	**(301-10-0)**
NSC-75858	**(4730-22-7)**
NSC-76041	**(13679-75-9)**
NSC-76078	**(3481-09-2)**
NSC-76090	**(621-36-3)**
NSC-76239	**(2465-59-0)**
NSC-76482	**(1973-09-7)**
NSC-76559	**(19524-06-2)**
NSC-76583	**(118-97-8)**
NSC-76584	**(90-99-3)**
NSC-76674	**(473-55-2)**
NSC-77079	**(98-61-3)**
NSC-77125	**(13360-61-7)**
NSC-77135	**(18435-45-5)**
NSC-77213	**(671-16-9)**
NSC-77213	**(366-70-1)**
NSC-77381	**(526-95-4)**
NSC-77447	**(560-21-4)**
NSC-77452	**(493-01-6)**
NSC-77453	**(493-02-7)**
NSC-77518	**(439-14-5)**
NSC-77688	**(328-42-7)**
NSC-78319	**(112-75-4)**
NSC-78416	**(638-32-4)**
NSC-78417	**(513-53-1)**
NSC-78429	**(111-47-7)**
NSC-78431	**(111-95-5)**
NSC-78559	**(1172-18-5)**
NSC-78728	**(32582-63-1)**
NSC-78926	**(2877-14-7)**
NSC-78938	**(5379-19-1)**
NSC-79019	**(78-98-8)**
NSC-79037	**(13010-47-4)**
NSC-79060	**(638-26-6)**
NSC-79268	**(101-67-7)**
NSC-79389	**(637-07-0)**
NSC-79477	**(63-91-2)**
NSC-79865	**(1809-21-8)**
NSC-79866	**(81-19-6)**
NSC-79895	**(128-69-8)**
NSC-80588	**(886-59-9)**
NSC-80657	**(6575-09-3)**
NSC-81212	**(4533-96-4)**
NSC-81226	**(88-56-2)**
NSC-81263	**(81-10-7)**
NSC-81349	**(1192-79-6)**
NSC-81389	**(6682-71-9)**
NSC-82151	**(20830-81-3)**
NSC-82174	**(389-08-2)**
NSC-82196	**(5034-77-5)**
NSC-82261	**(1405-41-0)**

NSC-82289 (3225-97-6)
NSC-82319 (14284-89-0)
NSC-82356 (143-13-5)
NSC-82358 (18979-50-5)
NSC-82391 (70-47-3)
NSC-83467 (106-33-2)
NSC-83468 (112-71-0)
NSC-83547 (79-03-8)
NSC-83612 (504-53-0)
NSC-83613 (110-30-5)
NSC-83845 (13822-56-5)
NSC-83941 (13132-25-7)
NSC-84199 (10375-96-9)
NSC-84227 (15438-71-8)
NSC-84233 (577-16-2)
NSC-84241 (81-30-1)
NSC-85598 (18883-66-4)
NSC-85847 (10411-52-6)
NSC-85998 (18883-66-4)
NSC-86117 (19398-61-9)
NSC-86142 (10203-28-8)
NSC-86935 (230-17-1)
NSC-86978 (508-32-7)
NSC-86982 (3385-66-8)
NSC-87078 (372-20-3)
NSC-87419 (3173-53-3)
NSC-87522 (192-51-8)
NSC-88277 (456-49-5)
NSC-88293 (363-72-4)
NSC-88295 (345-35-7)
NSC-88301 (455-36-7)
NSC-88308 (1435-53-6)
NSC-88327 (455-24-3)
NSC-88347 (700-17-4)
NSC-88536 (17021-26-0)
NSC-88940 (451-13-8)
NSC-88985 (99-29-6)
NSC-89259 (239-35-0)
NSC-89289 (506-38-7)
NSC-89593 (487-93-4)
NSC-89696 (13511-38-1)
NSC-89698 (1918-79-2)
NSC-89716 (28469-92-3)
NSC-89737 (52196-74-4)
NSC-89756 (4064-06-6)
NSC-89936 (6870-67-3)
NSC-89945 (2318-18-5)
NSC-90717 (641-96-3)
NSC-90784 (135-48-8)
NSC-91460 (2131-42-2)
NSC-91463 (3905-64-4)
NSC-91500 (615-29-2)
NSC-91501 (617-29-8)
NSC-91523 (525-66-6)
NSC-91616 (94-34-8)
NSC-91724 (542-81-4)
NSC-92165 (15763-06-1)
NSC-92231 (143-23-7)
NSC-92338 (302-22-7)
NSC-92617 (1124-19-2)
NSC-92741 (565-60-6)
NSC-92742 (3970-35-2)
NSC-92762 (31367-46-1)
NSC-93744 (505-32-8)
NSC-93768 (506-51-4)
NSC-93786 (13290-96-5)
NSC-93794 (148-69-6)
NSC-93798 (38775-38-1)
NSC-93799 (4798-58-7)
NSC-93801 (2388-14-9)
NSC-93810 (2313-65-7)
NSC-93914 (13432-25-2)
NSC-93961 (144-79-6)
NSC-93976 (126-80-7)
NSC-93981 (112-63-0)
NSC-93984 (19781-72-7)
NSC-94002 (13450-90-3)
NSC-94100 (488-41-5)
NSC-94304 (2237-36-7)
NSC-94782 (14970-87-7)
NSC-95412 (4798-45-2)
NSC-95413 (100-73-2)
NSC-95441 (13909-09-6)
NSC-95796 (18268-76-3)
NSC-95810 (885-82-5)
NSC-96359 (3988-77-0)
NSC-96364 (16898-52-5)
NSC-96629 (91-67-8)
NSC-96635 (115-89-9)
NSC-96748 (473-72-3)
NSC-96755 (33669-76-0)
NSC-96885 (565-63-9)
NSC-96907 (878-13-7)
NSC-96995 (110-25-8)
NSC-97195 (87-79-6)
NSC-97240 (31198-76-2)
NSC-97299 (112-77-6)
NSC-97346 (17084-02-5)
NSC-97503 (821-09-0)
NSC-97522 (13389-42-9)
NSC-97575 (17219-94-2)
NSC-97579 (78-24-0)
NSC-99286 (4342-60-3)
NSC-99806 (99-54-7)
NSC-99856 (10409-78-6)
NSC-100281 (30358-19-1)
NSC-100284 (16370-63-1)
NSC-100615 (89-08-7)
NSC-100738 (156-39-8)
NSC-100740 (21715-90-2)
NSC-100902 (7443-70-1)
NSC-101580 (16947-63-0)
NSC-101862 (259-79-0)
NSC-102101 (97-30-3)
NSC-102627 (3458-22-8)
NSC-102764 (15877-57-3)
NSC-102776 (627-97-4)
NSC-102816 (320-67-2)
NSC-103152 (529-20-4)
NSC-104469 (3546-10-9)
NSC-104800 (10318-26-0)
NSC-104801 (21739-91-3)
NSC-105613 (119-75-5)
NSC-105776 (935-31-9)
NSC-106273 (60913-86-2)
NSC-106568 (738-70-5)
NSC-107430 (359-83-1)
NSC-107566 (3096-47-7)
NSC-107654 (1018-71-9)
NSC-108264 (33533-53-8)
NSC-109422 (205-39-0)
NSC-109494 (7225-67-4)
NSC-109555 (15427-93-7)
NSC-109724 (3778-73-2)

NSC-110364 **(14698-29-4)**
NSC-110431 **(135-07-9)**
NSC-110432 **(76-38-0)**
NSC-110708 **(93-25-4)**
NSC-111180 **(616-91-1)**
NSC-112228 **(512-26-5)**
NSC-112231 **(544-18-3)**
NSC-112232 **(544-19-4)**
NSC-112259 **(22966-79-6)**
NSC-112518 **(24410-19-3)**
NSC-112907 **(18771-50-1)**
NSC-113134 **(10458-14-7)**
NSC-113288 **(33284-52-5)**
NSC-113482 **(27241-31-2)**
NSC-113926 **(13292-46-1)**
NSC-113975 **(87-82-1)**
NSC-114133 **(369-77-7)**
NSC-114470 **(24979-70-2)**
NSC-114538 **(20007-87-8)**
NSC-114981 **(31603-77-7)**
NSC-115260 **(634-91-3)**
NSC-115447 **(128-81-4)**
NSC-115894 **(555-35-1)**
NSC-115944 **(13838-16-9)**
NSC-116342 **(5004-48-8)**
NSC-117261 **(142-30-3)**
NSC-117442 **(125-42-8)**
NSC-117860 **(466-90-0)**
NSC-117863 **(469-79-4)**
NSC-117865 **(509-60-4)**
NSC-117874 **(137-16-6)**
NSC-118131 **(570-74-1)**
NSC-118365 **(56-45-1)**
NSC-118417 **(107-71-1)**
NSC-119749 **(27032-78-6)**
NSC-119875 **(15663-27-1)**
NSC-120281 **(16088-73-6)**
NSC-121172 **(29366-72-1)**
NSC-121779 **(90-97-1)**
NSC-122023 **(2001-95-8)**
NSC-122237 **(33058-12-7)**
NSC-122456 **(17085-91-5)**
NSC-122699 **(2395-96-2)**
NSC-122758 **(302-79-4)**
NSC-123014 **(30667-99-3)**
NSC-123127 **(23214-92-8)**
NSC-123458 **(256-96-2)**
NSC-123956 **(38185-06-7)**
NSC-124034 **(9002-98-6)**
NSC-124514 **(51-12-7)**
NSC-125066 **(11056-06-7)**
NSC-125427 **(17811-28-8)**
NSC-125973 **(33069-62-4)**
NSC-126195 **(16889-14-8)**
NSC-126766 **(2987-87-3)**
NSC-127744 **(19730-04-2)**
NSC-127858 **(16024-56-9)**
NSC-127860 **(121-60-8)**
NSC-128078 **(3123-97-5)**
NSC-128153 **(13462-88-9)**
NSC-128218 **(105-42-0)**
NSC-129943 **(21416-87-5)**
NSC-131419 **(110-31-6)**
NSC-131956 **(13140-89-1)**
NSC-132303 **(16824-02-5)**
NSC-132,319 **(41708-76-3)**
NSC-132541 **(5737-13-3)**
NSC-133447 **(1007-32-5)**
NSC-133893 **(93-11-8)**
NSC-134422 **(9002-98-6)**
NSC-134434 **(3105-97-3)**
NSC-134774 **(823-22-3)**
NSC-134776 **(20628-36-8)**
NSC-134914 **(567-72-6)**
NSC-134990 **(14072-86-7)**
NSC-135002 **(1942-46-7)**
NSC-135004 **(24903-95-5)**
NSC-135500 **(116-75-6)**
NSC-136052 **(480-82-0)**
NSC-136288 **(13114-87-9)**
NSC-136548 **(6130-82-1)**
NSC-136557 **(10431-98-8)**
NSC-136559 **(18281-04-4)**
NSC-136806 **(16867-03-1)**
NSC-137774 **(288-14-2)**
NSC-137831 **(14010-23-2)**
NSC-137833 **(41114-00-5)**
NSC-138780 **(21259-20-1)**
NSC-138831 **(28345-91-7)**
NSC-139128 **(1667-04-5)**
NSC-139442 **(822-83-3)**
NSC-139652 **(16214-98-5)**
NSC-139877 **(31543-75-6)**
NSC-140128 **(349-88-2)**
NSC-140729 **(29927-08-0)**
NSC-141021 **(52-51-7)**
NSC-141555 **(21544-02-5)**
NSC-141688 **(14542-93-9)**
NSC-142005 **(340-57-8)**
NSC-143025 **(15932-66-8)**
NSC-143038 **(24448-89-3)**
NSC-143932 **(18317-90-3)**
NSC-143933 **(22039-38-9)**
NSC-144478 **(880-93-3)**
NSC-145234 **(938-09-0)**
NSC-146405 **(319-87-9)**
NSC-147799 **(14920-92-4)**
NSC-148309 **(80-70-6)**
NSC-148314 **(133-37-9)**
NSC-148338 **(139-89-9)**
NSC-148361 **(28623-46-3)**
NSC-148862 **(16712-64-4)**
NSC-150014 **(10034-93-2)**
NSC-150161 **(548-39-0)**
NSC-150808 **(475-20-7)**
NSC-150953 **(4505-54-8)**
NSC-151043 **(19248-13-6)**
NSC-151735 **(19184-65-7)**
NSC-152080 **(31556-45-3)**
NSC-152396 **(272-12-8)**
NSC-153111 **(35082-49-6)**
NSC-153180 **(2280-93-5)**
NSC-154850 **(622-32-2)**
NSC-155332 **(19393-92-1)**
NSC-155516 **(17619-97-5)**
NSC-155648 **(2198-20-1)**
NSC-157494 **(16220-58-9)**
NSC-157589 **(4984-01-4)**
NSC-158150 **(15763-57-2)**
NSC-158434 **(23708-56-7)**
NSC-158442 **(4377-41-7)**
NSC-158520 **(10203-33-5)**
NSC-158522 **(1534-27-6)**
NSC-158676 **(10105-38-1)**

NSC-159025	(15009-91-3)
NSC-159266	(379-34-0)
NSC-159292	(506-13-8)
NSC-159352	(6149-34-4)
NSC-159729	(13344-99-5)
NSC-163039	(36791-04-5)
NSC-163314	(5145-99-3)
NSC-163319	(4110-50-3)
NSC-163321	(78-73-9)
NSC-163322	(140-82-9)
NSC-163587	(4443-55-4)
NSC-163901	(13014-24-9)
NSC-163921	(64-55-1)
NSC-163961	(4423-94-3)
NSC-163995	(127-43-5)
NSC-164350	(22175-22-0)
NSC-164918	(118-56-9)
NSC-164940	(10381-75-6)
NSC-165652	(627-35-0)
NSC-165800	(20116-65-8)
NSC-166169	(51-80-9)
NSC-166334	(10556-98-6)
NSC-166354	(103-99-1)
NSC-166462	(17977-09-2)
NSC-166467	(24157-81-1)
NSC-166503	(120-56-9)
NSC-166667	(10191-18-1)
NSC-167086	(2235-83-8)
NSC-168527	(28169-46-2)
NSC-168933	(146-84-9)
NSC-169496	(14362-31-3)
NSC-170228	(512-69-6)
NSC-171732	(18508-00-4)
NSC-173214	(14055-02-8)
NSC-173943	(22268-16-2)
NSC-174063	(3741-00-2)
NSC-174082	(5363-64-4)
NSC-174207	(27134-26-5)
NSC-175283	(54934-71-3)
NSC-175822	(98-59-9)
NSC-176003	(1075-49-6)
NSC-176118	(553-72-0)
NSC-176136	(546-88-3)
NSC-176805	(13623-06-8)
NSC-178248	(54749-90-5)
NSC-179032	(1886-75-5)
NSC-180808	(4649-27-8)
NSC-180823	(10563-26-5)
NSC-186892	(97-69-8)
NSC-187676	(4672-26-8)
NSC-190361	(403-24-7)
NSC-190376	(15619-48-4)
NSC-190452	(1928-39-8)
NSC-190561	(587-64-4)
NSC-190562	(3547-07-7)
NSC-190935	(6164-98-3)
NSC-190939	(7696-12-0)
NSC-190939	(104-30-3)
NSC-190940	(97-77-8)
NSC-190945	(13265-60-6)
NSC-190978	(2597-03-7)
NSC-190981	(6988-21-2)
NSC-190986	(13593-03-8)
NSC-190987	(10265-92-6)
NSC-190998	(18181-70-9)
NSC-191020	(4419-57-2)
NSC-191025	(14255-88-0)
NSC-192745	(25168-73-4)
NSC-192746	(26446-38-8)
NSC-193373	(464-17-5)
NSC-194838	(26028-46-6)
NSC-195022	(10453-86-8)
NSC-195058	(5598-52-7)
NSC-195087	(18181-80-1)
NSC-195102	(19750-95-9)
NSC-195106	(22224-92-6)
NSC-195164	(21923-23-9)
NSC-196235	(39589-98-5)
NSC-196335	(9002-98-6)
NSC-202854	(25371-75-9)
NSC-202959	(513-74-6)
NSC-203106	(52670-79-8)
NSC-203306	(22633-33-6)
NSC-203323	(143-66-8)
NSC-204323	(656-31-5)
NSC-206315	(56-41-7)
NSC-207409	(938-09-0)
NSC-208959	(66-02-4)
NSC-209799	(940-71-6)
NSC-210804	(14171-89-2)
NSC-210913	(556-53-6)
NSC-211456	(13142-64-8)
NSC-211975	(4514-53-8)
NSC-212132	(2486-71-7)
NSC-212255	(141-85-5)
NSC-212544	(538-74-9)
NSC-215231	(78-21-7)
NSC-219884	(2400-02-4)
NSC-220215	(383-63-1)
NSC-220312	(102-04-5)
NSC-221122	(3123-97-5)
NSC-221154	(514-73-8)
NSC-222656	(4721-24-8)
NSC-223080	(35576-91-1)
NSC-224330	(474-62-4)
NSC-224419	(481-21-0)
NSC-226232	(21117-52-2)
NSC-226561	(555-48-6)
NSC-226830	(3149-65-3)
NSC-226920	(42975-18-8)
NSC-227210	(103-60-6)
NSC-227854	(73727-39-6)
NSC-227897	(2765-04-0)
NSC-227915	(618-41-7)
NSC-227945	(2486-70-6)
NSC-229358	(517-60-2)
NSC-229428	(111-62-6)
NSC-231371	(128-10-9)
NSC-231508	(1205-71-6)
NSC-231527	(537-47-3)
NSC-231629	(137-47-3)
NSC-233046	(10418-03-8)
NSC-234415	(30995-65-4)
NSC-236821	(938-09-0)
NSC-238159	(31430-18-9)
NSC-239116	(88-98-2)
NSC-239709	(563-84-8)
NSC-240567	(17752-85-1)
NSC-243115	(24169-02-6)
NSC-243680	(519-44-8)
NSC-243682	(527-55-9)
NSC-243743	(562-73-2)
NSC-243747	(609-46-1)
NSC-244460	(7642-04-8)

NSC	CAS
NSC-244854	(16106-59-5)
NSC-244871	(55976-13-1)
NSC-244887	(7443-52-9)
NSC-244909	(33467-76-4)
NSC-244920	(51149-70-3)
NSC-244937	(820-29-1)
NSC-245854	(7443-52-9)
NSC-246414	(6746-27-6)
NSC-249268	(15598-34-2)
NSC-249764	(83-74-9)
NSC-249835	(14995-38-1)
NSC-249992	(51264-14-3)
NSC-250665	(5145-01-7)
NSC-250971	(3675-13-6)
NSC-251008	(6742-54-7)
NSC-253011	(537-01-9)
NSC-261036	(13551-92-3)
NSC-261037	(13551-87-6)
NSC-261427	(17688-68-5)
NSC-263483	(623-69-8)
NSC-263518	(15774-82-0)
NSC-263780	(4161-60-8)
NSC-263827	(14548-01-7)
NSC-263840	(18017-73-7)
NSC-270042	(466-97-7)
NSC-270679	(333-27-7)
NSC-272271	(3718-65-8)
NSC-272281	(3095-95-2)
NSC-277452	(373-49-9)
NSC-283470	(23003-22-7)
NSC-288740	(1917-44-8)
NSC-290818	(29975-16-4)
NSC-293057	(546-56-5)
NSC-297936	(54350-48-0)
NSC-298102	(16274-81-0)
NSC-298103	(5606-23-5)
NSC-298536	(533-60-8)
NSC-309702	(36735-22-5)
NSC-309957	(24719-19-5)
NSC-309965	(136-81-2)
NSC-310005	(16919-19-0)
NSC-319997	(10291-28-8)
NSC-322921	(23491-45-4)
NSC-323990	(13355-96-9)
NSC-326243	(10190-68-8)
NSC-328430	(79-28-7)
NSC-329117	(59177-47-8)
NSC-334055	(68558-73-6)
NSC-338158	(27223-35-4)
NSC-338250	(53202-98-5)
NSC-342705	(5213-50-3)
NSC-344238	(3703-10-4)
NSC-345670	(23159-07-1)
NSC-345692	(23783-42-8)
NSC-347484	(16681-63-3)
NSC-348403	(13052-19-2)
NSC-349941	(488-93-7)
NSC-351138	(6659-45-6)
NSC-353895	(2396-63-6)
NSC-356717	(17784-12-2)
NSC-357087	(17092-92-1)
NSC-363752	(671-36-3)
NSC-370498	(22306-37-2)
NSC-372149	(1145-44-4)
NSC-375994	(17526-94-2)
NSC-376770	(85878-63-3)
NSC-381839	(64925-80-0)
NSC-401092	(569-51-7)
NSC-401113	(1199-77-5)
NSC-401609	(574-00-5)
NSC-401681	(7500-53-0)
NSC-401846	(515-30-0)
NSC-402438	(557-08-4)
NSC-402555	(541-46-8)
NSC-402999	(1583-67-1)
NSC-403110	(95-12-5)
NSC-403169	(7008-42-6)
NSC-403248	(499-80-9)
NSC-403292	(98-68-0)
NSC-403657	(78-59-1)
NSC-403804	(4505-54-8)
NSC-403839	(27583-37-5)
NSC-403840	(13014-18-1)
NSC-403856	(80-53-5)
NSC-403881	(5814-85-7)
NSC-403883	(77-53-2)
NSC-403888	(93-96-9)
NSC-404034	(103-62-8)
NSC-404086	(330-13-2)
NSC-404118	(87-89-8)
NSC-404177	(112-69-6)
NSC-404457	(122-70-3)
NSC-405015	(10264-17-2)
NSC-405072	(19525-59-8)
NSC-405124	(87-90-1)
NSC-405639	(625-77-4)
NSC-406128	(94-69-9)
NSC-406248	(547-64-8)
NSC-406279	(106-13-8)
NSC-406285	(111-03-5)
NSC-406536	(64919-15-9)
NSC-406547	(51479-36-8)
NSC-406584	(102-25-0)
NSC-406603	(115-84-4)
NSC-406702	(644-49-5)
NSC-406705	(493-09-4)
NSC-406799	(101-86-0)
NSC-406847	(94-22-4)
NSC-406892	(94-99-5)
NSC-406893	(102-47-6)
NSC-406894	(619-56-7)
NSC-406963	(142-50-7)
NSC-407158	(76-49-3)
NSC-407292	(60-82-2)
NSC-407311	(506-48-9)
NSC-407752	(34593-75-4)
NSC-407764	(614-27-7)
NSC-407822	(4824-72-0)
NSC-407829	(1805-32-9)
NSC-408419	(597-43-3)
NSC-408471	(13954-62-6)
NSC-408494	(116-37-0)
NSC-408852	(1067-71-6)
NSC-409411	(575-90-6)
NSC-409425	(3844-60-8)
NSC-409492	(1679-02-3)
NSC-409767	(94-80-4)
NSC-409777	(19434-42-5)
NSC-409780	(104-10-9)
NSC-409786	(102-86-3)
NSC-409886	(4395-79-3)
NSC-409962	(154-93-8)
NSC-511991	(5981-06-6)
NSC-512314	(97-09-6)

NSC-512726 (7169-34-8)
NSC-512922 (77-08-7)
NSC-513490 (18619-18-6)
NSC-518113 (24255-23-0)
NSC-519695 (95-32-9)
NSC-520345 (89-64-5)
NSC-521077 (286-99-7)
NSC-521917 (616-88-6)
NSC-522480 (25660-70-2)
NSC-522884 (14309-41-2)
NSC-523741 (27944-79-2)
NSC-524443 (20020-02-4)
NSC-525334 (3570-75-0)
NSC-526280 (2577-72-2)
NSC-526936 (56-95-1)
NSC-527017 (1397-89-3)
NSC-527179 (50-35-1)
NSC-527604 (70-51-9)
NSC-527913 (281-23-2)
NSC-527986 (3795-88-8)
NSC-528004 (23360-92-1)
NSC-528986 (69-53-4)
NT 907 (85-42-7)
NTN-8629 (34643-46-4)
NTA (139-13-9)
NTA (5064-31-3)
NTA, Compound with iron chloride (14695-88-6)
NTA, K3 (2399-85-1)
NTA, aluminium(3+) complex (19010-73-2)
NTA, antimony(3+) complex (46242-44-8)
NTA, barium salt (1:1) (2399-83-9)
NTA, beryllium potassium salt (1:1) (18983-72-7)
NTA, beryllium salt (1:1) (2399-81-7)
NTA, cadmium(2+) complex (18432-54-7)
NTA, calcium potassium salt (1:1:1) (2455-08-5)
NTA, calcium potassium salt (2:1:4) (23555-98-8)
NTA, calcium salt (14981-08-9)
NTA, calcium salt (1:1) (2399-94-2)
NTA, calcium salt (2:3) (62979-89-6) (62979-89-6)
NTA, calcium sodium salt (1:1:1) (60034-45-9)
NTA, cerium salt (29027-90-5)
NTA, cobalt(3+) complex (23319-51-9)
NTA, copper(2+) complex (15844-52-7)
NTA, copper(2+) complex ammonium salt (71484-80-5)
NTA, copper(2+) complex sodium salt (53108-47-7)
NTA, copper(2+) hydrogen complex (34831-02-2)
NTA, copper(2+) salt (1:1) (1188-47-2)
NTA, diammonium salt (71264-32-9)
NTA, dilithium salt (72629-49-3)
NTA, dipotassium salt (2399-86-2)
NTA, disodium ammonium salt (86892-89-9)
NTA, disodium salt, Compd. with oxo(dihydrogen nitrilo-acetato)bismuth (5798-43-6)
NTA, erbium(3+) salt (3:1) (10413-71-5)
NTA, holmium salt (28927-38-0)
NTA, indium(3+) complex (19456-58-7)
NTA, iron(2+) complex sodium salt (1:1:1) (61017-62-7)
NTA, lead(2+) potassium salt (1:1:1) (79915-08-5)
NTA, lead(2+) salt (1:1) (79849-02-8)
NTA, lead(2+) salt (2:3) (79915-09-6)
NTA, magnesium salt (73772-91-5)
NTA, magnesium salt (1:1) (1188-48-3)
NTA, manganese salt (36711-58-7)
NTA, mercury(2+) salt (2:3) (18105-03-8)
NTA, monoammonium salt (15934-02-8)
NTA, monopotassium salt (28444-53-3)
NTA, neodymium(3+) salt (1) (18946-94-6)
NTA, nickel(3+) complex (22965-60-2)
NTA, nickel(2+) hydrogen complex (34831-03-3)
NTA, potassium magnesium salt (1:1:1) (2399-88-4)
NTA, potassium salt (25817-24-7)
NTA, potassium strontium salt (1:1:1) (2399-89-5)
NTA, potassium strontium salt (2:4:1) (23555-96-6)
NTA, scandium(3+) salt (1:1) (3130-95-8)
NTA, sodium hydrate (18662-53-8)
NTA, strontium sodium salt (92988-11-9)
NTA, tin(2+) salt (53818-84-1)
NTA, triammonium salt (32685-17-9)
NTA, tricadmium(2+) complex (50648-02-7)
NTA, tripotassium salt (2399-85-1)
NTA, trisilver salt (92474-39-0)
NTA, yttrium(3+) salt (1:1) (15414-25-2)
NTA, zinc(3+) complex sodium salt (29507-58-2)
NTD 2 (15263-52-2)
NTG (55-63-0)
NTL (56391-56-1)
NTM (131-11-3)
NTN 811 (76608-88-3)
NTN 5006 (33857-23-7)
NTN 9306 (35400-43-2)
NTOI (61-57-4)
NTS 1 (56776-27-3)
NTS 62 (9004-70-0)
NTS 218 (9004-70-0)
NTS 222 (9004-70-0)
NTS 539 (9004-70-0)
NTS 542 (9004-70-0)
NU 445 (127-69-5)
NU 2206 (297-90-5)
NU-1196 (77-20-3)
NU-2121 (100-55-0)
NUZ (7446-20-0)
NVC 9025 (9002-88-4)
NW Acid (84-87-7)
NYAD (13983-17-0)
NYAD 10 (13983-17-0)
NYAD 325 (13983-17-0)
NYAD G (13983-17-0)
NYCOR (13983-17-0)
NYCOR 200 (13983-17-0)
NYCOR 300 (13983-17-0)
Na-C10LAS. (2627-06-7)
Na Frinse (7681-49-4)
NaH 80 (7646-69-7)
NaMBT (2492-26-4)
Na MCPA (3653-48-3)
Na-0101 T 1/8" (1333-83-1)
NaTA (650-51-1)
NaTCA (650-51-1)
Nabac (70-30-4)
Nabac 25 EC (70-30-4)
Nabam (111-54-6)
Nabam (142-59-6)
Nabame (French) (142-59-6)
Nabasan (142-59-6)
Nabilone (51022-71-0)
Nablen S 50 (9003-07-0)
Nabolin (58-18-4)
Nabor Orange G (3056-93-7)
Nabu (74051-80-2)
Nacm-Cellulose Salt (9004-32-4)
Nacarat (3567-69-9)
Nacarat A Export (3567-69-9)
Nacarat Extra Pure A (3567-69-9)

Naccanol NR **(25155-30-0)**
Naccanol SW **(25155-30-0)**
Nacconate 300 **(101-68-8)**
Nacconate-100 **(26471-62-5)**
Nacconate H 12 **(5124-30-1)**
Nacconate IOO **(584-84-9)**
Nacconol 40F **(25155-30-0)**
Nacconol 90F **(25155-30-0)**
Nacconol LAL **(1847-58-1)**
Nacconol 98SA **(27176-87-0)**
Nacconol 35SL **(25155-30-0)**
Nacelan Blue G **(2475-45-8)**
Nacelan Blue KLT **(2475-46-9)**
Nacelan Blue KLT **(86722-66-9)**
Nacelan Fast Yellow CG **(2832-40-8)**
Nacelan Pink 3B **(2872-48-2)**
Nacelan Pink B **(116-85-8)**
Nacelan Violet 4B **(1220-94-6)**
Nacelan Violet 4R **(128-95-0)**
Naclex **(135-09-1)**
Naclex **(91-33-8)**
Nacyclyl **(113-38-2)**
Nadazone **(50-33-9)**
Nadeine **(125-28-0)**
Nadic methyl anhydride **(25134-21-8)**
Nadisal **(54-21-7)**
Nadisan **(339-43-5)**
Nadizan **(339-43-5)**
Nadolol **(42200-33-9)**
Nadone **(108-94-1)**
Nadozone **(50-33-9)**
Nafarelin acetate **(86220-42-0)**
Nafcillin **(147-52-4)**
Nafeen **(7681-49-4)**
Nafenoic acid **(3771-19-5)**
Nafenopin **(3771-19-5)**
Nafpak **(7681-49-4)**
Naftalam (Czech) **(132-66-1)**
α-Naftalamin (Czech) **(134-32-7)**
β-Naftalamin (Czech) **(91-59-8)**
Naftalen (Polish) **(91-20-3)**
Naftalin-butil-solfonato (Italian) **(25417-20-3)**
β-Naftilamina (Italian) **(91-59-8)**
1-Naftilamina (Spanish) **(134-32-7)**
1-Naftil-tiourea (Italian) **(86-88-4)**
1,2-Naftochinon (Czech) **(524-42-5)**
1,4-Naftochinon (Czech) **(130-15-4)**
Naftoelan A **(92-77-3)**
2-Naftol (Dutch) **(135-19-3)**
β-Naftol (Dutch) **(135-19-3)**
2-Naftolo (Italian) **(135-19-3)**
β-Naftolo (Italian) **(135-19-3)**
Naftolo MBO **(132-68-3)**
Naftolo MBS **(135-65-9)**
Naftolo MD **(135-61-5)**
Naftolo MM **(92-77-3)**
Naftolo MOL **(135-62-6)**
2-Naftylamin-5,7-disulfonan sodny (Czech) **(118-33-2)**
1-Naftylamin (Czech) **(134-32-7)**
2-Naftylamin (Czech) **(91-59-8)**
α-Naftylamin (Czech) **(134-32-7)**
β-Naftylamin (Czech) **(91-59-8)**
1-Naftylamine (Dutch) **(134-32-7)**
2-Naftylamine (Dutch) **(91-59-8)**
4-(2-Naftylamino)fenol (Czech) **(93-45-8)**
1-Naftylester kyseliny methylkarbaminove (Czech) **(63-25-2)**
1-Naftylisothiokyanat (Czech) **(551-06-4)**
α-Naftyl-N-methylkarbamat (Czech) **(63-25-2)**
alfa-Naftyloamina (Polish) **(134-32-7)**
β-Naftyloamina (Polish) **(91-59-8)**
α-Naftylthiomocovina (Czech) **(86-88-4)**
1-Naftylthioureum (Dutch) **(86-88-4)**
Nafusaku **(86-87-3)**
Nagarse **(9014-01-1)**
Nagravon **(68-19-9)**
Nairit **(9010-98-4)**
Naixan **(22204-53-1)**
Nako Brown R **(123-30-8)**
Nako H **(106-50-3)**
Nako TEG **(591-27-5)**
Nako TGG **(108-46-3)**
Nako TMT **(95-80-7)**
Nako TRB **(90-15-3)**
Nako TSA **(39156-41-7)**
Nako Yellow 3GA **(95-55-6)**
Nakva **(133-67-5)**
Nalan RF **(3370-35-2)**
Nalcamine G-11 **(136-99-2)**
Nalcamine G-13 **(95-38-5)**
Nalco L-699 **(139-90-2)**
Nalco SPF-WTB 33 **(9003-11-6)**
Nalcoag **(7631-86-9)**
Nalco dispersant SPF-WTB 33 **(9003-11-6)**
Nalcon 243 **(533-74-4)**
Nalcrom **(15826-37-6)**
Naled (ACGIH) **(300-76-5)**
Naledu (Polish) **(300-76-5)**
Nalfloc 636 **(9003-01-4)**
Nalidic acid **(389-08-2)**
Nalidicron **(389-08-2)**
Nalidixic acid **(389-08-2)**
Nalidixin **(389-08-2)**
Nalidixinic acid **(389-08-2)**
Nalitucsan **(389-08-2)**
Nalix **(389-08-2)**
Nalkil **(314-40-9)**
Nalline **(62-67-9)**
Nalorfina **(62-67-9)**
Nalorphine **(62-67-9)**
Nalorphinium **(62-67-9)**
Nalox **(443-48-1)**
Naloxiphan **(152-02-3)**
Naloxone **(465-65-6)**
l-Naloxone **(465-65-6)**
Nalquat P **(26006-22-4)**
Nalurin **(389-08-2)**
Nalutron **(57-83-0)**
Nam **(98-92-0)**
Namate **(144-21-8)**
Named reagents and solutions, Goulard's ext. **(8006-24-4)**
Named solutions, Goulard's extract **(8006-24-4)**
Namekil **(108-62-3)**
Namekil **(9002-91-9)**
Namphen **(61-68-7)**
Namuron **(52-31-3)**
Nanchor **(299-84-3)**
Nandervit-N **(98-92-0)**
Nandrolon **(434-22-0)**
Nandrolone **(434-22-0)**
Nankai Acid Orange I **(523-44-4)**
Nankai Direct Fast Yellow A **(1325-37-7)**
Nanker **(299-84-3)**
Nankor **(299-84-3)**
Nanm **(62-67-9)**

Nansa 1042P **(27176-87-0)**
Nansa SSA **(27176-87-0)**
Naotin **(59-67-6)**
Napa **(103-90-2)**
Napacetin **(15687-27-1)**
Napafen **(103-90-2)**
Napap **(103-90-2)**
Napclor-G **(131-52-2)**
Napental **(57-33-0)**
Naphazoline **(835-31-4)**
Naphazoline hydrochloride **(550-99-2)**
Naphcon **(550-99-2)**
Naphcon Forte **(550-99-2)**
Naphid **(1338-24-5)**
2-Naphtalenol, 1-((2-chloro-4-nitrophenyl)azo)- **(2814-77-9)**
Naphtamine Blue 2B **(2602-46-2)**
Naphtamine Blue 2B **(72-57-1)**
Naphtamine Blue 3BX **(72-57-1)**
Naphtamine Blue 10G **(2429-74-5)**
Naphtamine Brown DC **(2429-82-5)**
Naphtamine Sky Blue DD **(2610-05-1)**
Naphtamine Violet N **(2586-60-9)**
Naphtanilide BO **(132-68-3)**
Naphtanilide BO Supra **(132-68-3)**
Naphtanilide BR **(91-92-9)**
Naphtanilide BS **(135-65-9)**
Naphtanilide D **(135-61-5)**
Naphtanilide D Supra **(135-61-5)**
Naphtanilide EL **(137-52-0)**
Naphtanilide G **(91-96-3)**
Naphtanilide KB **(135-63-7)**
Naphtanilide OL **(135-62-6)**
Naphtanilide OL Supra **(92-77-3)**
Naphtanilide Phenyl **(92-74-0)**
Naphtanilide Phenyl Supra **(92-74-0)**
Naphtanilide RC **(92-77-3)**
Naphtanilide RC Supra **(92-77-3)**
2-Naphth-o-anisidide, 5'-chloro-3-hydroxy- (8CI) **(137-52-0)**
2-Naphth-o-anisidide, 3-hydroxy- (8CI) **(135-62-6)**
Naphtazol A **(92-77-3)**
Naphtazol 3B **(132-68-3)**
Naphtazol B **(135-65-9)**
Naphtazol C **(135-63-7)**
Naphtazol D **(135-61-5)**
Naphtazol EL **(137-52-0)**
Naphtazol F **(135-62-6)**
Naphtazol J **(91-96-3)**
Naphtazol OP **(92-74-0)**
Naphtenate de cobalt (French) **(61789-51-3)**
Naphtha **(8030-30-6)**
Naphtha, Aromatic, High flash **(64742-95-6)**
Naphtha, Coal, Solvent-refining **(68476-79-9)**
Naphtha Coal Tar (OSHA) **(8030-30-6)**
Naphtha Distillate (DOT) **(8030-30-6)**
Naphtha (OSHA) [UN 2553] **(8030-30-6)**
Naphtha, Petroleum, Full-range alkylate **(64741-64-6)**
Naphtha, Petroleum, Full-range reformed **(68919-37-9)**
Naphtha, Petroleum, Heavy catalytic reformed **(64741-68-0)**
Naphtha, Petroleum, Heavy thermal cracked **(64741-83-9)**
Naphtha, Petroleum, Hydrodesulfurized Heavy **(64742-82-1)**
Naphtha, Petroleum, Isomerization **(64741-70-4)**
Naphtha, Petroleum, Light Hydrocracked **(64741-69-1)**
Naphtha, Petroleum, Light Steam-cracked **(64742-83-2)**
Naphtha, Petroleum, Light Thermal Cracked **(64741-74-8)**
Naphtha, Petroleum, Light alkylate **(64741-66-8)**
Naphtha (Petroleum), Light catalytic cracked **(64741-55-5)**
Naphtha, Petroleum, Light catalytic reformed **(64741-63-5)**
Naphtha (Petroleum), Light straight-run **(64741-46-4)**
Naphtha, Petroleum, Polymn. **(64741-72-6)**
Naphtha, Petroleum, Sweetened **(64741-87-3)**
Naphtha Petroleum [UN 1255] **(8030-30-6)**
Naphtha, Solvent [UN 1256] **(8030-30-6)**
Naphth(2,1-a)anthracene **(214-17-5)**
Naphthacene **(92-24-0)**
2-Naphthacenecarboxamide, 4,7-bis(dimethylamino)-1,4,4a,5,5a, 6,11,12a-octahydro-3,10,12,12a- tetrahydroxy-1,11-dioxo **(10118-90-8)**
2-Naphthacenecarboxamide, 7-chloro-4-(dimethylamino)-1,4,4a, 5,5a,6,11,12a-octahydro- 3,6,10,12,12a-pentahydroxy- 1,11-dioxo **(127-33-3)**
2-Naphthacenecarboxamide, 7-chloro-4-(dimethylamino)-1,4,4a, 5,5a,6,11,12a-octahydro- 3,6,10,12,12a-pentahydroxy- 6-methyl-1,11-dioxo **(57-62-5)**
2-Naphthacenecarboxamide, 4-(dimethylamino)-1,4,4a,5,5a, 6,11,12a-octahydro-3,5,6,10,12,12a- hexahydroxy-6-methyl- 1,11-dioxo **(79-57-2)**
2-Naphthacenecarboxamide, 4-(dimethylamino)-1,4,4a,5,5a, 6,11,12a-octahydro-3,5,6,10,12,12a- hexahydroxy-6-methyl- 1,11-dioxo-, monohydrochloride **(2058-46-0)**
2-Naphthacenecarboxamide, 4-(dimethylamino)-1,4,4a,5,5a, 6,11,12a-octahydro- 3,5,10,12,12a-pentahydroxy-6-methylene- 1,11-dioxo **(914-00-1)**
2-Naphthacenecarboxamide, 4-(dimethylamino)-1,4,4a,5,5a, 6,11,12a-octahydro-3,6,10,12,12a- pentahydroxy-6-methyl- 1,11-dioxo **(60-54-8)**
2-Naphthacenecarboxamide, 4-(dimethylamino)-1,4,4a,5,5a, 6,11,12a-octahydro-3,6,10,12,12a- pentahydroxy-6-methyl- 1,11-dioxo-, monohydrochloride **(64-75-5)**
2-Naphthacenecarboxamide, 4-α-S-(dimethylamino)-1,4,4a-α- 5,5a-α,6,11,12a- octahydro-3,5-α,10,12,12a-α-pentahydroxy- 6-α-methyl-1,11-dioxo **(564-25-0)**
5,12-Naphthacenedione (9CI) **(1090-13-7)**
5,12-Naphthacenedione, 10-((3-amino-2,3,6-trideoxy-α-l-lyxo-hexo- pyranosyl)oxy)- 7,8,9,10-tetrahydro-6,8,11-trihydroxy- 8-(hydroxyacetyl)-1-methoxy-, hydrochloride, (8s-cis) **(25316-40-9)**
Naphthaceno(2,1,12-qra)naphthacene (8CI,9CI) **(189-45-7)**
α-Naphthacridine **(225-51-4)**
α-Naphthal **(66-77-3)**
2-Naphthalamine **(91-59-8)**
Naphthalane **(91-17-8)**
1-Naphthaldehyde **(66-77-3)**
2-Naphthaldehyde **(66-99-9)**
α-Naphthaldehyde **(66-77-3)**
β-Naphthaldehyde **(66-99-9)**
2-Naphthalenamine **(91-59-8)**
Naphthalenamine (9CI) **(25168-10-9)**
1-Naphthalenamine, 4-((4-aminophenyl)azo)- (9CI) **(6054-48-4)**
1-Naphthalenamine, N-ethyl- (9CI) **(118-44-5)**
2-Naphthalenamine, N-ethyl-1-((4-(phenylazo)phenyl)azo)- (9CI) **(6368-72-5)**
1-Naphthalenamine hydrochloride **(552-46-5)**
2-Naphthalenamine, hydrochloride **(612-52-2)**
2-Naphthalenamine, 1-((4-nitrophenyl)azo)- (9CI) **(3025-77-2)**
1-Naphthalenamine, 4-(phenylazo) **(131-22-6)**
1-Naphthalenamine, N-((1,1,3,3-tetramethylbutyl)phenyl)- (9CI) **(51772-35-1)**
1-Naphthalenazo-2',4'-diaminobenzene **(6416-57-5)**
Naphthalene **(91-20-3)**
Naphthalene (ACGIH,OSHA) [UN 1334] **(91-20-3)**
Naphthalene, Crude or refined [UN 1334] **(91-20-3)**
Naphthalene Fast Orange 2G **(1936-15-8)**
Naphthalene Fast Orange 2GS **(1936-15-8)**
Naphthalene Green G **(4680-78-8)**
Naphthalene Ink Scarlet 4R **(2611-82-7)**
Naphthalene Lake Green G **(4680-78-8)**
Naphthalene Lake Orange G **(633-96-5)**
Naphthalene Lake Scarlet R **(3761-53-3)**

Naphthalene Leather Green G **(4680-78-8)**
Naphthalene Leather Scarlet G **(10169-02-5)**
Naphthalene Leather Yellow 2G **(6375-55-9)**
Naphthalene Leather Yellow GL **(1325-37-7)**
Naphthalene, Molten [UN 2304] **(91-20-3)**
Naphthalene Oil **(8001-58-9)**
Naphthalene Orange G **(633-96-5)**
Naphthalene Orange I **(523-44-4)**
Naphthalene Orange Solide GG **(1936-15-8)**
Naphthalene Red B **(3567-66-6)**
Naphthalene Red EA **(2302-96-7)**
Naphthalene Red J (6CI) **(1658-56-6)**
Naphthalene Red JS **(1658-56-6)**
Naphthalene Scarlet 4R **(2611-82-7)**
Naphthalene Scarlet R **(3761-53-3)**
Naphthalene Scarlet 4RS **(2611-82-7)**
1-Naphthaleneacetamide **(86-86-2)**
Naphthalene acetamide **(86-86-2)**
α-Naphthaleneacetamide **(86-86-2)**
β-Naphthaleneacetate **(581-96-4)**
1-Naphthaleneacetic acid **(86-87-3)**
2-Naphthaleneacetic acid **(581-96-4)**
Naphthalene-1-acetic acid **(86-87-3)**
Naphthaleneacetic acid **(86-87-3)**
α-Naphthaleneacetic acid **(86-87-3)**
β-Naphthaleneacetic acid **(581-96-4)**
Naphthaleneacetic acid (8CI,9CI) (VAN) **(26445-01-2)**
1-Naphthaleneacetic acid, ethyl ester (8CI,9CI) **(2122-70-5)**
2-Naphthaleneacetic acid, 6-methoxy-α-methyl-, (+)- (8CI) **(22204-53-1)**
2-Naphthaleneacetic acid, 6-methoxy-α-methyl-, (S)- (9CI) **(22204-53-1)**
1-Naphthaleneacetic acid, sodium salt (8CI,9CI) **(61-31-4)**
Naphthalene, 1-allyl- (8CI) **(2489-86-3)**
Naphthalene, 2-benzyl- (8CI) **(613-59-2)**
Naphthalene, 2,6-bis(1,1-dimethylethyl)- (9CI) **(3905-64-4)**
Naphthalene, 2,7-bis(1,1-dimethylethyl)- (9CI) **(10275-58-8)**
Naphthalene, 2,6-bis(1,1-dimethylethyl)-1,2,3,4-tetrahydro- (9CI) **(42981-76-0)**
Naphthalene, 2,6-bis(1-methylethyl)- (9CI) **(24157-81-1)**
Naphthalene, bis(1-methylethyl)- (9CI) **(38640-62-9)**
Naphthalene, 1,5-bis(methylsulfonyl)- (9CI) **(53135-94-7)**
Naphthalene, 1,5-bis(methylthio)- (8CI,9CI) **(10075-74-8)**
Naphthalene, 1-bromo **(90-11-9)**
Naphthalene, 2-bromo- (9CI) **(580-13-2)**
1-Naphthalenebutanoic acid (9CI) **(781-74-8)**
Naphthalene, 1-tert-butyl- (8CI) **(17085-91-5)**
Naphthalene, 2-butyl- (8CI,9CI) **(1134-62-9)**
Naphthalene, 2-tert-butyl- (6CI,7CI,8CI) **(2876-35-9)**
Naphthalene, 1-butyl- (6CI,7CI,8CI,9CI) **(1634-09-9)**
1-Naphthalenebutyric acid (6CI,7CI,8CI) **(781-74-8)**
1-Naphthalenecarbonitrile **(86-53-3)**
2-Naphthalenecarbonitrile (9CI) **(613-46-7)**
Naphthalenecarbonitrile, methyl- (9CI) **(77417-07-3)**
2-Naphthalenecarbothioamide, N-(1-methylethyl)- (9CI) **(64142-01-4)**
1-Naphthalenecarboxaldehyde **(66-77-3)**
2-Naphthalenecarboxaldehyde **(66-99-9)**
Naphthalenecarboxaldehyde (9CI) **(30678-61-6)**
Naphthalenecarboxaldehyde, methyl- (9CI) **(77468-37-2)**
2-Naphthalenecarboxamide, N-(o-(acetylamino)phenethyl)-1-hydroxy **(5254-41-1)**
2-Naphthalenecarboxamide, N-(4-(acetylamino)phenyl)-4-((5-(aminocarbonyl)-2-chlorophenyl)azo)-3-hydroxy- (9CI) **(12236-64-5)**
2-Naphthalenecarboxamide, 4-((5-aminocarbonyl-2-methylphenyl)-azo)-3-hydroxy-N-phenyl- **(16403-84-2)**
2-Naphthalenecarboxamide, 4-((4-(aminocarbonyl)phenyl)azo)-N-(2-ethoxyphenyl)-3-hydroxy- (9CI) **(2786-76-7)**
2-Naphthalenecarboxamide, 4-((4-(aminocarbonyl)phenyl)azo)-3-hydroxy-N-(2-methoxyphenyl)- (9CI) **(36968-27-1)**
2-Naphthalenecarboxamide, N-(4-(2,4-bis(1,1-dimethylpropyl)-phenoxy)butyl)-1-hydroxy- (9CI) **(32180-75-9)**
2-Naphthalenecarboxamide, N-(4-chloro-2,5-dimethoxyphenyl)-3-hydroxy-4-((2-methoxy-5-((phenylamino)carbonyl)phenyl)-azo)- (9CI) **(5280-68-2)**
2-Naphthalenecarboxamide, N-(5-chloro-2-methoxyphenyl)-3-hydroxy- (9CI) **(137-52-0)**
2-Naphthalenecarboxamide, N-(4-chloro-2-methylphenyl)-4-((4-chloro-2-methylphenyl)azo)-3-hydroxy- (9CI) **(6471-51-8)**
2-Naphthalenecarboxamide, N-(5-chloro-2-methylphenyl)-3-hydroxy- (9CI) **(135-63-7)**
2-Naphthalenecarboxamide, N,N'-(2-chloro-1,4-phenylene)bis-(4-((4-chloro-2-nitrophenyl)azo)-3-hydroxy- (9CI) **(35869-64-8)**
2-Naphthalenecarboxamide, N,N'-(2-chloro-1,4-phenylene)bis-(4-((2,5-dichlorophenyl)azo)-3-hydroxy- (9CI) **(5280-78-4)**
2-Naphthalenecarboxamide, N-(4-chlorophenyl)-3-hydroxy-4-((2-methyl-5-nitrophenyl)azo)- (9CI) **(6410-30-6)**
2-Naphthalenecarboxamide, 4,4'-((3,3'-dichloro(1,1'-biphenyl)-4,4'-diyl)bis(azo))bis(3-hydroxy-N-phenyl- (9CI) **(41709-76-6)**
2-Naphthalenecarboxamide, N,N'-(3,3'-dichloro(1,1'-biphenyl)-4,4'-diyl)bis(4-((2-chlorophenyl) azo)-3-hydroxy- (9CI) **(5280-74-0)**
2-Naphthalenecarboxamide, 4-((2,5-dichlorophenyl)azo)-3-hydroxy-N-(2-methoxyphenyl)- (9CI) **(6410-38-4)**
2-Naphthalenecarboxamide, 4-((2,5-dichlorophenyl)azo)-3-hydroxy-N-phenyl- (9CI) **(6041-94-7)**
2-Naphthalenecarboxamide, N-(2,3-dihydro-2-oxo-1H-benzimidazol-5-yl)-4-((2,5-dimethoxy-4-((methylamino)sulfonyl)phenyl)azo)-3-hydroxy- (9CI) **(12225-08-0)**
2-Naphthalenecarboxamide, N-(2,3-dihydro-2-oxo-1H-benzimidazol-5-yl)-3-hydroxy-4-((2-methoxy-5-((phenylamino)carbonyl)-phenyl)azo)- (9CI) **(12225-06-8)**
2-Naphthalenecarboxamide, N,N'-(3,3'-dimethoxy(1,1'-biphenyl)-4,4'-diyl)bis(3-hydroxy- (9CI) **(91-92-9)**
2-Naphthalenecarboxamide, N-(2-ethoxyphenyl)-3-hydroxy- (9CI) **(92-74-0)**
2-Naphthalenecarboxamide, 3-hydroxy-4-((2-methoxy-5-nitro-phenyl)azo)-N-(3-nitrophenyl) **(6471-49-4)**
2-Naphthalenecarboxamide, 3-hydroxy-N-(2-methoxyphenyl)- (9CI) **(135-62-6)**
2-Naphthalenecarboxamide, 3-hydroxy-4-((2-methoxy-5-((phenyl-amino)carbonyl)phenyl)azo)-N-(3-nitrophenyl)- (9CI) **(6448-96-0)**
2-Naphthalenecarboxamide, 3-hydroxy-4-((2-methyl-5-amino-carbonylphenyl)azo)-N-phenyl- **(16403-84-2)**
2-Naphthalenecarboxamide, 3-hydroxy-4-((2-methyl-5-nitro-phenyl)azo)-N-(2-methylphenyl)- (9CI) **(6655-84-1)**
2-Naphthalenecarboxamide, 3-hydroxy-4-((2-methyl-5-nitro-phenyl)azo)-N-phenyl- (9CI) **(6448-95-9)**
2-Naphthalenecarboxamide, 3-hydroxy-N-(2-methylphenyl)- (9CI) **(135-61-5)**
2-Naphthalenecarboxamide, 3-hydroxy-4-((2-methyl-5-phenyl-carboxyamide)azo)-N-phenyl- **(16403-84-2)**
2-Naphthalenecarboxamide, 3-hydroxy-N-(2-methylphenyl)-4-((2,4,5-trichlorophenyl)azo)- (9CI) **(6535-46-2)**
2-Naphthalenecarboxamide, 3-hydroxy-N-(3-(4-morpholinyl)-propyl)- (9CI) **(10155-47-2)**
2-Naphthalenecarboxamide, 3-hydroxy-N-1-naphthalenyl- (9CI) **(132-68-3)**
2-Naphthalenecarboxamide, 3-hydroxy-N-(3-nitrophenyl)- (9CI) **(135-65-9)**
2-Naphthalenecarboxamide, 3-hydroxy-N-phenyl- (9CI) **(92-77-3)**
2-Naphthalenecarboxamide, 3-hydroxy-4-((4-phenylcarboxy-amide)azo)-N-(2-methoxyphenyl)- **(36968-27-1)**
1-Naphthalenecarboxamide, N-methyl- (9CI) **(3400-33-7)**
2-Naphthalenecarboxamide, N,N'-1,4-phenylenebis(4-((2,5-di-chlorophenyl)azo)-3-hydroxy- (9CI) **(3905-19-9)**
2-Naphthalenecarboxylic acid **(93-09-4)**
Naphthalene-α-carboxylic acid **(86-55-5)**
Naphthalene-β-carboxylic acid **(93-09-4)**
1-Naphthalenecarboxylic acid (9CI) **(86-55-5)**
2-Naphthalenecarboxylic acid, 5-((4-carboxyphenyl)azo)-6-hydroxy- (9CI) **(69579-72-2)**
2-Naphthalenecarboxylic acid, 4-((4-chloro-5-methyl-2-sulfo phenyl)azo)-3-hydroxy-, calcium salt (1:1) (9CI) **(17852-99-2)**

Name	CAS RN
2-Naphthalenecarboxylic acid, 4-((5-chloro-4-methyl-2-sulfophenyl)azo)-3-hydroxy-, barium salt (1:1) (9CI)	**(7585-41-3)**
2-Naphthalenecarboxylic acid, 4-((5-chloro-4-methyl-2-sulfophenyl)azo)-3-hydroxy-, calcium salt (1:1) (9CI)	**(7023-61-2)**
2-Naphthalenecarboxylic acid, 4-((5-chloro-4-methyl-2-sulfophenyl)azo)-3-hydroxy-, manganese salt (1:1)	**(12688-94-7)**
2-Naphthalenecarboxylic acid, 4-((5-chloro-4-methyl-2-sulfophenyl)azo)-3-hydroxy-, manganese(2+) salt (1:1)	**(5280-66-0)**
2-Naphthalenecarboxylic acid, 4-((5-chloro-4-methyl-2-sulfophenyl)azo)-3-hydroxy-, strontium salt (1:1) (9CI)	**(15782-05-5)**
1-Naphthalenecarboxylic acid, 2,3-dihydroxy-	**(16715-77-8)**
1-Naphthalenecarboxylic acid, 2-ethoxy- (9CI)	**(2224-00-2)**
2-Naphthalenecarboxylic acid, 1-hydroxy- (9CI)	**(86-48-6)**
2-Naphthalenecarboxylic acid, 3-hydroxy- (9CI)	**(92-70-6)**
2-Naphthalenecarboxylic acid, 6-hydroxy- (9CI)	**(16712-64-4)**
2-Naphthalenecarboxylic acid, 1-hydroxy-4-(1-(4-hydroxy-3-(methoxycarbonyl)-1-naphthalenyl)-3-oxo-1H,3H-naphtho-(1,8-cd)pyran-1-yl)-6-(octadecyloxy)- (9CI)	**(38122-80-4)**
2-Naphthalenecarboxylic acid, 1-hydroxy-, methyl ester (9CI)	**(948-03-8)**
2-Naphthalenecarboxylic acid, 3-hydroxy-, methyl ester (9CI)	**(883-99-8)**
2-Naphthalenecarboxylic acid, 3-hydroxy-4-((4-methyl-2-sulfo phenyl)azo)- (9CI)	**(16014-23-6)**
2-Naphthalenecarboxylic acid, 3-hydroxy-4-((4-methyl-2-sulfophenyl)azo)-, barium salt (1:1) (9CI)	**(17852-98-1)**
2-Naphthalenecarboxylic acid, 3-hydroxy-4-((4-methyl-2-sulfophenyl)azo)-, calcium salt	**(5281-04-9)**
2-Naphthalenecarboxylic acid, 3-hydroxy-4-((4-methyl-2-sulfophenyl)azo)-, calcium salt (1:1) (9CI)	**(5281-04-9)**
2-Naphthalenecarboxylic acid, 3-hydroxy-4-((4-methyl-2-sulfophenyl)azo)-, disodium salt	**(5858-81-1)**
2-Naphthalenecarboxylic acid, 3-hydroxy-, monosodium salt (9CI)	**(14206-62-3)**
2-Naphthalenecarboxylic acid, 1-hydroxy-6-(octadecyloxy)- (9CI)	**(38134-94-0)**
2-Naphthalenecarboxylic acid, 6-hydroxy-5-(phenylazo)- (9CI)	**(69644-64-0)**
2-Naphthalenecarboxylic acid, 3-hydroxy-4-((1-sulfo-2-naphthalenyl)azo)-, calcium salt (1:1) (9CI)	**(6417-83-0)**
2-Naphthalenecarboxylic acid, 6-hydroxy-5-((4-sulfophenyl)-azo)- (9CI)	**(69644-65-1)**
4a(2H)-Naphthalenecarboxylic acid, octahydro-, cis- (8CI,9CI)	**(3021-73-6)**
Naphthalene, 1-chloro	**(90-13-1)**
Naphthalene, 2-chloro	**(91-58-7)**
Naphthalene, chloro- (8CI,9CI)	**(25586-43-0)**
Naphthalene, 1-chloro-2,4-dinitro	**(2401-85-6)**
Naphthalene, 1-chloromethyl	**(86-52-2)**
Naphthalene, α-chloromethyl-	**(86-52-2)**
Naphthalene, decahydro	**(91-17-8)**
Naphthalene, decahydro-, cis- (9CI)	**(493-01-6)**
Naphthalene, decahydro-, trans- (9CI)	**(493-02-7)**
Naphthalene, decahydro-2,3-dimethyl- (8CI,9CI)	**(1008-80-6)**
Naphthalene, decahydro-2,6-dimethyl- (8CI,9CI)	**(1618-22-0)**
Naphthalene, decahydro-1,8a-dimethyl-7-(1-methylethyl)-, (1R-(1α,4aβ,7β,8aα))- (9CI)	**(15404-63-4)**
Naphthalene, decahydro-1,4a-dimethyl-7-(1-methylethyl)-, (1S-(1α,4aα,7α,8aβ))- (9CI)	**(30824-81-8)**
Naphthalene, decahydro-1,4-dimethyl-5-octyl- (7CI,9CI)	**(54964-83-9)**
Naphthalene, decahydro-4a-methyl-1-methylene-7-(1-methylethenyl)-, (4aR-(4aα,7α,8aβ))- (9CI)	**(17066-67-0)**
1,5-Naphthalenediamine	**(2243-62-1)**
Naphthalene-1,8-diamine	**(479-27-6)**
1,8-Naphthalenediamine (9CI)	**(479-27-6)**
Naphthalene, 2,6-di-tert-butyl- (8CI)	**(3905-64-4)**
Naphthalene, 2,7-di-tert-butyl- (6CI,7CI,8CI)	**(10275-58-8)**
Naphthalenedicarboxaldehyde (9CI)	**(70848-82-7)**
1,8-Naphthalenedicarboximide	**(81-83-4)**
Naphthalene-1,2-dicarboxyic acid, 1,2,3,4-tetrahydro-3-methyl-6,7-methylenedioxy-, dipropyl ester	**(83-59-0)**
2,6-Naphthalenedicarboxylic acid (9CI)	**(1141-38-4)**
Naphthalenedicarboxylic acid, cyclic anhydride (9CI)	**(34314-32-4)**
2,6-Naphthalene dicarboxylic acid, dimethyl ester	**(840-65-3)**
2,6-Naphthalenedicarboxylic acid, dimethyl ester (9CI)	**(840-65-3)**
1,2-Naphthalenedicarboxylic anhydride (8CI)	**(5343-99-7)**
Naphthalenedicarboxylic anhydride (8CI)	**(34314-32-4)**
Naphthalene, 1,2-dichloro- (9CI)	**(2050-69-3)**
Naphthalene, 1,3-dichloro- (9CI)	**(2198-75-6)**
Naphthalene, 1,4-dichloro- (9CI)	**(1825-31-6)**
Naphthalene, 1,5-dichloro- (9CI)	**(1825-30-5)**
Naphthalene, 1,6-dichloro- (9CI)	**(2050-72-8)**
Naphthalene, 1,7-dichloro- (9CI)	**(2050-73-9)**
Naphthalene, 1,8-dichloro- (9CI)	**(2050-74-0)**
Naphthalene, 2,3-dichloro- (9CI)	**(2050-75-1)**
Naphthalene, 2,6-dichloro- (9CI)	**(2065-70-5)**
Naphthalene, 2,7-dichloro- (9CI)	**(2198-77-8)**
Naphthalene, dichloro- (9CI)	**(28699-88-9)**
Naphthalene, 1,2-dihydro	**(447-53-0)**
Naphthalene, dihydro- (8CI,9CI)	**(29828-28-2)**
Naphthalene, dihydromethyl- (9CI)	**(39292-53-0)**
Naphthalene, 1,2-dihydro-4-phenyl- (7CI,8CI,9CI)	**(7469-40-1)**
Naphthalene, 1,2-dihydro-1,4,6-trimethyl- (9CI)	**(55682-80-9)**
Naphthalene, 1,2-dihydro-2,5,8-trimethyl- (8CI)	**(30316-23-5)**
Naphthalene, 1,4-dihydro-2,5,8-trimethyl- (8CI)	**(30316-19-9)**
Naphthalene, 1,2-dihydro-1,5,8-trimethyl- (7CI,8CI,9CI)	**(4506-36-9)**
Naphthalene, 1,6-dihydroxy-	**(575-44-0)**
Naphthalene, 1-(2,3-dihydroxypropoxy)-	**(36112-95-5)**
1,5-Naphthalene diisocyanate	**(3173-72-6)**
Naphthalene diisocyanate	**(25551-28-4)**
Naphthalene, 1,5-diisocyanato- (9CI)	**(3173-72-6)**
Naphthalene, diisononyl- (9CI)	**(63512-64-1)**
Naphthalene, 2,6-diisopropyl-	**(24157-81-1)**
Naphthalene, diisopropyl-	**(38640-62-9)**
Naphthalene, 1,3-dimethyl	**(575-41-7)**
Naphthalene, dimethyl-	**(28804-88-8)**
Naphthalene, 1,2-dimethyl- (9CI)	**(573-98-8)**
Naphthalene, 1,4-dimethyl- (9CI)	**(571-58-4)**
Naphthalene, 1,5-dimethyl- (9CI)	**(571-61-9)**
Naphthalene, 1,6-dimethyl- (9CI)	**(575-43-9)**
Naphthalene, 1,7-dimethyl- (9CI)	**(575-37-1)**
Naphthalene, 2,3-dimethyl- (9CI)	**(581-40-8)**
Naphthalene, 2,6-dimethyl- (9CI)	**(581-42-0)**
Naphthalene, 2,7-dimethyl- (9CI)	**(582-16-1)**
Naphthalene, 1,8-dimethyl- (8CI,9CI)	**(569-41-5)**
Naphthalene, 1-(1,1-dimethylethyl)- (9CI)	**(17085-91-5)**
Naphthalene, 2-(1,1-dimethylethyl)- (9CI)	**(2876-35-9)**
Naphthalene, 1-(1,1-dimethylethyl)decahydro- (9CI)	**(56292-64-9)**
Naphthalene, 1,6-dimethyl-4-(1-methylethyl)- (9CI)	**(483-78-3)**
Naphthalene, 1,4-dimethyl-5-octyl- (7CI,9CI)	**(55000-53-8)**
Naphthalene, 1,3-dinitro	**(606-37-1)**
Naphthalene, 1,5-dinitro	**(605-71-0)**
Naphthalene, 1,8-dinitro	**(602-38-0)**
Naphthalene, dinitro	**(27478-34-8)**
1,2-Naphthalenediol	**(574-00-5)**
1,3-Naphthalenediol	**(132-86-5)**
1,4-Naphthalenediol	**(571-60-8)**
1,5-Naphthalenediol	**(83-56-7)**
2,3-Naphthalenediol	**(92-44-4)**
2,5-Naphthalenediol	**(575-44-0)**
2,7-Naphthalenediol	**(582-17-2)**
Naphthalene-2,7-diol	**(582-17-2)**
1,6-Naphthalenediol (9CI)	**(575-44-0)**
Naphthalenediol-2,7 (French)	**(582-17-2)**
1,4-Naphthalenediol, diacetate (9CI)	**(5697-00-7)**
2,3-Naphthalenediol, 5-(3-((1,1-dimethylethyl)amino)-2-hydroxypropoxy)-1,2,3,4-tetrahydro	**(42200-33-9)**
1,2-Naphthalenedione	**(524-42-5)**

1,4-Naphthalenedione

1,4-Naphthalenedione (130-15-4)
Naphthalenedione (9CI) (12679-43-5)
1,4-Naphthalenedione, 2-amino-3-chloro- (2797-51-5)
1,4-Naphthalenedione, 5,8-dihydroxy-6-methoxy-2-methyl-3-(2-oxopropyl)- (476-45-9)
1,2-Naphthalenedione, 3,8-dimethyl-5-(1-methylethyl)- (9CI) (5574-34-5)
1,4-Naphthalenedione, 2-(3,7,11,15,19,23,27-heptamethyl-2,6,10,14,18,22,26-octacosahept-aenyl)-3-methyl-, (All-E)- (2124-57-4)
1,4-Naphthalenedione, 5-hydroxy (481-39-0)
1,4-Naphthalenedione, 2-hydroxy- (9CI) (83-72-7)
1,4-Naphthalenedione, 2-methyl- (9CI) (58-27-5)
1,4-Naphthalenedione, 2-methyl-3-(3,7,11,15-tetramethyl-2-hexadecenyl) (84-80-0)
Naphthalenedione, nitro- (9CI) (80267-67-0)
2,6-Naphthalenedione, octahydro-1,1,8a-trimethyl-, cis- (9CI) (57289-16-4)
Naphthalene, 1,7-diphenyl- (8CI,9CI) (970-06-9)
1,5-Naphthalenedisulfochloride (1928-01-4)
1,5-Naphthalenedisulfonic acid (81-04-9)
Naphthalene-1,6-disulfonic acid (525-37-1)
Naphthalene-2,6-disulfonic acid (581-75-9)
2,6-Naphthalenedisulfonic acid (9CI) (581-75-9)
1,7-Naphthalenedisulfonic acid, 4-acetamido-5-hydroxy-6-((7-sulfo-4-((p-sulfophenyl)azo)-1-naphthyl)azo)-, tetrasodium salt (2519-30-4)
2,7-Naphthalenedisulfonic acid, 5-(acetylamino)-4-hydroxy-3-((2-methoxyphenyl)azo)-, disodium salt (9CI) (6625-46-3)
2,7-Naphthalenedisulfonic acid, 5-(acetylamino)-4-hydroxy-3-(phenylazo)-, disodium salt (3734-67-6)
1,3-Naphthalenedisulfonic acid, 7-amino (86-65-7)
1,5-Naphthalenedisulfonic acid, 2-amino (117-62-4)
1,5-Naphthalenedisulfonic acid, 3-amino (131-27-1)
Naphthalene-1,3-disulfonic acid, 6-amino (118-33-2)
2,7-Naphthalenedisulfonic acid, 5-amino-3-((4'-((7-amino-1-hydroxy-3-sulfo-2-naphthalenyl) azo)(1,1'-biphenyl)-4-yl)-azo)-4-hydroxy-, trisodium salt (2429-73-4)
2,7-Naphthalenedisulfonic acid, 4-amino-3-((4'-((2,4-diamino-5-methylphenyl)azo)(1,1'-biphenyl)-4-yl)azo)-5-hydroxy-6-(phenylazo)-, disodium salt (9CI) (2429-83-6)
2,7-Naphthalenedisulfonic acid, 4-amino-3-((4'-((2,4-diamino-phenyl)azo)(1,1'-biphenyl)-4-yl) azo)-5-hydroxy-6-(phenyl-azo)-, disodium salt (1937-37-7)
2,7-Naphthalenedisulfonic acid, 4-amino-3-((4-((2,4-di-aminophenyl)azo)phenyl)azo)-5-hydroxy-6-(phenylazo)-, disodium salt (9CI) (68877-33-8)
2,7-Naphthalenedisulfonic acid, 4-amino-3-((4'-((2,4-dihydroxy-phenyl)azo)-3,3'-dimethyl(1,1'-biphenyl)-4-yl)azo)-5-hydroxy-6-((4-sulfophenyl)azo)-, trisodium salt (9CI) (68318-35-4)
1,3-Naphthalenedisulfonic acid, 8-((4-(2-aminoethyl)-phenyl)azo)-7-hydroxy- (9CI) (78335-11-2)
2,7-Naphthalenedisulfonic acid, 4-amino-5-hydroxy (90-20-0)
1,3-Naphthalenedisulfonic acid, 4-amino-5-hydroxy- (9CI) (82-47-3)
1,3-Naphthalenedisulfonic acid, 4-amino-5-hydroxy-6-((4'-((2-hydroxy-1-naphthalenyl)azo)-3,3'-dimethoxy-(1,1'-biphenyl)-4-yl)azo)-, disodium salt (9CI) (2586-57-4)
2,7-Naphthalenedisulfonic acid, 5-amino-4-hydroxy-3-((2-hydroxy-5-nitrophenyl)azo)- (9CI) (13301-33-2)
2,6-Naphthalenedisulfonic acid, 4-amino-5-hydroxy-3-((4'-((4-hydroxyphenyl)azo)(1,1'- biphenyl)-4-yl)azo)-6-(phenylazo)-, disodium salt (3626-28-6)
2,7-Naphthalenedisulfonic acid, 4-amino-5-hydroxy-6-((4'-((4-hydroxyphenyl)azo)(1,1'-biphenyl)-4-yl)azo)-3-((4-nitrophenyl)azo)-, disodium salt (9CI) (4335-09-5)
2,7-Naphthalenedisulfonic acid, 4-amino-5-hydroxy-, mono-sodium salt (9CI) (5460-09-3)
2,7-Naphthalenedisulfonic acid, 4-amino-5-hydroxy-3-((4-nitro-phenyl)azo)-6-(phenylazo)-, disodium salt (1064-48-8)
2,7-Naphthalenedisulfonic acid, 4-amino-5-hydroxy-6-phenylazo-, disodium salt (3567-66-6)
1,5-Naphthalenedisulfonic acid, 3-((4-((4-((6-amino-1-hydroxy-3-sulfo-2-naphthalenyl)azo)-6-sulfo-1-naphthalenyl)azo)-1-naphthalenyl)azo)-, tetrasodium salt (9CI) (4399-55-7)
2,7-Naphthalenedisulfonic acid, 5-amino-4-hydroxy-3-((1-sulfo-2-naphthalenyl)azo)-, trisodium salt (9CI) (5045-23-8)
1,3-Naphthalenedisulfonic acid, 7-((4-amino-2-methylphenyl)azo)-(9CI) (2494-93-1)
1,3-Naphthalenedisulfonic acid, 7-amino-, monopotassium salt (9CI) (842-15-9)
1,5-Naphthalenedisulfonic acid, 2-amino-, monosodium salt (8CI) (19532-03-7)
1,3-Naphthalenedisulfonic acid, 7-((4-amino-o-tolyl)azo)- (8CI) (2494-93-1)
2,7-Naphthalenedisulfonic acid, 4-((4-anilino-5-sulfo-1-naphthyl)-azo)-5-hydroxy-, trisodium salt (3861-73-2)
2,7-Naphthalenedisulfonic acid, 3,3'-((4,4'-biphenylylene)bis-(azo))bis(5-amino-4-hydroxy-, tetrasodium salt (2602-46-2)
2,7-Naphthalenedisulfonic acid, 3,6-(bis(4-((2-hydroxyethyl)-sulfonyl)phenyl)bis(azo))- 5-amino-4-hydroxy-, di(hydrogen sulfate) ester, tetrasodium salt (17095-24-8)
1,5-Naphthalenedisulfonic acid, 3,3'-(carbonylbis(imino-(5-methoxy-2-methyl-4,1-phenylene)azo))bis-, tetrasodium salt (9CI) (6420-33-3)
1,5-Naphthalenedisulfonic acid, 3,3'-(carbonylbis(imino-(3-methoxy-4,1-phenylene)azo))bis-, tetrasodium salt (9CI) (28706-22-1)
1,5-Naphthalenedisulfonic acid, 3,3'-(carbonylbis(imino-(2-methyl-4,1-phenylene)azo))bis-, tetrasodium salt (9CI) (3214-47-9)
2,7-Naphthalenedisulfonic acid, 3-((5-chloro-2-hydroxyphenyl)-azo)-4,5-dihydroxy-, disodium salt (9CI) (1058-92-0)
2,7-Naphthalenedisulfonic acid, 5-((4-chloro-6-(methylphenyl amino)-1,3,5-triazin-2-yl)amino)-4-hydroxy-3-((4-methyl-2-sulfo-phenyl)azo)-, trisodium salt (9CI) (70210-46-7)
2,7-Naphthalenedisulfonic acid, 5-(((2,3-dichloro-6-quinoxalinyl) carbonyl)amino)-4-hydroxy-3-((2-sulfophenyl)azo)-, trisodium salt (9CI) (2407-13-8)
2,7-Naphthalenedisulfonic acid, 5-((4,6-dichloro-1,3,5-triazin-2-yl) amino)-4-hydroxy-3-((2-sulfophenyl)azo)-, trisodium salt (9CI) (17752-85-1)
2,7-Naphthalenedisulfonic acid, 5-((4,6-dichloro-s-triazin-2-yl) amino)-4-hydroxy-3- (phenylazo)-, disodium salt (17804-49-8)
1,5-Naphthalenedisulfonic acid, 2-((6-((4,6-dichloro-1,3,5-triazin-2-yl)methylamino)-1-hydroxy-3-sulfo-2-naphthalenyl)azo)-, trisodium salt (9CI) (70616-90-9)
1,3-Naphthalenedisulfonic acid, 6-(4,5-dihydro-3-methyl-5-oxo-1H-pyrazol-1-yl)- (9CI) (7277-87-4)
2,7-Naphthalenedisulfonic acid, 3,3'((3,3'-dihydroxy-(1,1'-biphenyl)-4,4'-diyl)bis(azo)bis(5-amino-4-hydroxy-, sodium salt, copper complex (28407-37-6)
2,7-Naphthalenedisulfonic acid, 4,5-dihydroxy-, disodium salt (9CI) (129-96-4)
1,3-Naphthalenedisulfonic acid, 7-((1,8-dihydroxy-3,6-disulfo-2-naphthyl)azo)-, tetrasodium salt (8CI) (29637-28-3)
2,7-Naphthalenedisulfonic acid, 4-((2,4-dihydroxy-5-((2-hydroxy-3,5-dinitrophenyl)azo)-3-((4-nitrophenyl)azo)phenyl)azo)-5-hydroxy-, disodium salt (9CI) (6637-87-2)
2,7-Naphthalenedisulfonic acid, 4,5-dihydroxy-3-(phenylazo)-, disodium salt (4197-07-3)
2,7-Naphthalenedisulfonic acid, 3,3'-((3,3'-dimethoxy(1,1'-bi phenyl)-4,4'-diyl)bis(azo))bis(5-amino-4-hydroxy-, sodium salt (9CI) (68966-50-7)
2,7-Naphthalenedisulfonic acid, 3,3'-((3,3'-dimethoxy-(1,1'-biphenyl)-4,4'-diyl)bis(azo))bis(4,5-dihydroxy-, tetrasodium salt (9CI) (4198-19-0)
6,8-Naphthalenedisulfonic acid, 3,3'-((3,3'-dimethoxy-4,4'-bi-phenylene)bis(azo))bis(5-amino-4-hydroxy-, tetrasodium salt (2610-05-1)
2,7-Naphthalenedisulfonic acid, 3,3'-((3,3'-dimethoxy-4,4'-bi-phenylylene)bis(azo))bis(5-amino-4-hydroxy-, tetrasodium salt (2429-74-5)
2,7-Naphthalenedisulfonic acid, 3,3'-((3,3'-dimethyl(1,1'-bi-phenyl)-4,4'-diyl)bis(azo))bis(4,5-dihydroxy-, tetrasodium

salt (9CI) **(2150-54-1)**
1,3-Naphthalenedisulfonic acid, 6,6'-((3,3'-dimethyl-4,4'-biphenylylene)bis(azo))bis (4-amino-5-hydroxy-, tetrasodium salt **(314-13-6)**
2,7-Naphthalenedisulfonic acid, 3,3'-((3,3'-dimethyl-4,4'-biphenylylene)bis(azo))bis(5- amino-4-hydroxy-, tetrasodium salt **(72-57-1)**
1,3-Naphthalenedisulfonic acid, 8-((3,3'-dimethyl-4'-((4-(((4-methylphenyl)sulfonyl)oxy) phenyl)azo)(1,1'-biphenyl)-4-yl)azo)-7-hydroxy-, disodium salt **(6459-94-5)**
2,7-Naphthalenedisulfonic acid, 3-((2,2'-dimethyl-4'-((4-(((4-methylphenyl)sulfonyl)oxy)phenyl)azo)(1,1'-biphenyl)-4-yl)azo)-4-hydroxy-, disodium salt (9CI) **(6358-57-2)**
2,7-Naphthalenedisulfonic acid, 3-((3,3'-dimethyl-4'-((4-((phenylsulfonyl)oxy)phenyl)azo)(1,1'-biphenyl)-4-yl)azo)-4-hydroxy-, disodium salt (9CI) **(71701-30-9)**
2,7-Naphthalenedisulfonic acid, disodium salt (9CI) **(1655-35-2)**
1,3-Naphthalenedisulfonic acid, 8-((2,5-disulfophenyl)-azo)-7-hydroxy- (9CI) **(78335-10-1)**
1,3-Naphthalenedisulfonic acid, 8-((4'-((4-ethoxyphenyl)azo)-(1,1'-biphenyl)-4-yl)azo)-7-hydroxy-, disodium salt (9CI) **(3530-19-6)**
1,3-Naphthalenedisulfonic acid, 8-((4'-((4-ethoxyphenyl)azo)-3,3'-dimethyl(1,1'-biphenyl)-4- yl)azo)-7-hydroxy-, disodium salt **(6358-29-8)**
2,7-Naphthalenedisulfonic acid, 4-((4-(ethylamino)-1-naphthalenyl)azo)-5-hydroxy- (9CI) **(78335-12-3)**
1,3-Naphthalenedisulfonic acid, 7-hydroxy- (9CI) **(118-32-1)**
2,7-Naphthalenedisulfonic acid, 3-hydroxy- (9CI) **(148-75-4)**
1,3-Naphthalenedisulfonic acid, 7-hydroxy-, dipotassium salt (9CI) **(842-18-2)**
2,7-Naphthalenedisulfonic acid, 3-hydroxy-, disodium salt **(135-51-3)**
1,3-Naphthalenedisulfonic acid, 7-hydroxy-8-((4'-((4-(((4-methyl phenyl)sulfonyl)oxy)phenyl) azo)(1,1'-biphenyl)-4-yl)azo)-, disodium salt **(3567-65-5)**
2,7-Naphthalenedisulfonic acid, 4-hydroxy-, monosodium salt (9CI) **(61931-87-1)**
1,3-Naphthalenedisulfonic acid, 7-hydroxy-8-(phenylazo)-, disodium salt **(1936-15-8)**
2,7-Naphthalenedisulfonic acid, 3-hydroxy-4-(phenylazo)-, disodium salt (9CI) **(5859-00-7)**
1,3-Naphthalenedisulfonic acid, 7-hydroxy-8-((4-(phenylazo)-phenyl)azo)-, disodium salt (9CI) **(5413-75-2)**
1,3-Naphthalenedisulfonic acid, 7-hydroxy-8-((4-sulfo-1-naphthyl) azo)-, trisodium salt **(2611-82-7)**
2,7-Naphthalenedisulfonic acid, 3-hydroxy-4-((4-sulfo-1-naphthyl) azo)-, trisodium salt **(915-67-3)**
1,3-Naphthalenedisulfonic acid, 7-hydroxy-8-((4-sulfophenyl)azo)-(9CI) **(2657-89-8)**
1,3-Naphthalenedisulfonic acid, 7-hydroxy-8-((p-sulfophenyl)azo)-(7CI,8CI) **(2657-89-8)**
2,7-Naphthalenedisulfonic acid, 3-hydroxy-4-((2,4,5-trimethylphenyl)azo)-, disodium salt **(3564-09-8)**
2,7-Naphthalenedisulfonic acid, 3-hydroxy-4-(2,4-xylylazo)-, disodium salt **(3761-53-3)**
1,5-Naphthalenedisulfonic acid, 3-((4-((((2-methoxy-4-((3-sulfophenyl)azo)phenyl)amino)carbonyl)amino)-2-methylphenyl)-azo)-, trisodium salt (9CI) **(28706-19-6)**
2,7-Naphthalenedisulfonic acid, 3-((4-nitrophenyl)azo)-6-(phenylazo)- (9CI) **(78335-13-4)**
1,3-Naphthalenedisulfonic acid, 7,7'-(ureylenebis((2-methyl-p-phenylene)azo))di-, tetrasodium salt (8CI) **(28706-21-0)**
1,5-Naphthalenedisulfonyl chloride (6CI,7CI,8CI) **(1928-01-4)**
1,5-Naphthalenedisulfonyl dichloride (9CI) **(1928-01-4)**
1,5-Naphthalenedithiol (6CI,7CI,9CI) **(5325-88-2)**
1,2-(1,8-Naphthalenediyl)benzene **(206-44-0)**
Naphthalene, dodecyl- (9CI) **(38641-16-6)**
Naphthalene-d8, 1,2,3,4-tetrahydro-1,2,3,4-d4- (9CI) **(75840-23-2)**
Naphthalene, 2,3-epoxydecahydro- (8CI) **(21399-51-9)**
1-Naphthaleneethanol (8CI,9CI) **(773-99-9)**
Naphthalene, 2-ethenyl- (9CI) **(827-54-3)**
Naphthalene, 2-ethoxy **(93-18-5)**
Naphthalene, 1-ethoxy- (9CI) **(5328-01-8)**
Naphthalene, 4a-ethoxydecahydro-, cis- (9CI) **(51953-10-7)**
Naphthalene, 1-ethyl **(1127-76-0)**
Naphthalene, 2-ethyl **(939-27-5)**
Naphthalene, ethyl- (8CI,9CI) **(27138-19-8)**
Naphthalene, ethyldecahydro- (8CI,9CI) **(25551-49-9)**
Naphthalene, ethyl-1,2,3,4-tetrahydro- (9CI) **(81598-29-0)**
Naphthalene, 2-ethyl-1,2,3,4-tetrahydro- (6CI,8CI,9CI) **(32367-54-7)**
Naphthalene, 6-ethyl-1,2,3,4-tetrahydro- (6CI,8CI,9CI) **(22531-20-0)**
Naphthalene, 1-fluoro **(321-38-0)**
Naphthalene, heptabromo- (9CI) **(55688-01-2)**
Naphthalene, heptachloro- (9CI) **(32241-08-0)**
Naphthalene, heptyl- (9CI) **(38622-51-4)**
Naphthalene, 1,2,3,4,6,7-hexabromo **(75625-24-0)**
Naphthalene, hexabromo **(56480-06-9)**
Naphthalene, hexachloro **(1335-87-1)**
Naphthalene, hexadecyl- (9CI) **(56388-47-7)**
Naphthalene, 1,2,3,5,6,8a-hexahydro-4,7-dimethyl-1-(1-methylethyl)-, (1S-cis)- (9CI) **(483-76-1)**
Naphthalene, 1,2,4a,5,8,8a-hexahydro-4,7-dimethyl-1-(1-methylethyl)-, (1S-(1α,4aβ,8aα))- (9CI) **(523-47-7)**
Naphthalene, hexahydrodimethylpropyl- (9CI) **(86825-83-4)**
Naphthalene, 1-iodo- (9CI) **(90-14-2)**
Naphthalene, 1-isobutyl- (6CI,8CI) **(16727-91-6)**
Naphthalene, 1-isocyano- (9CI) **(1984-04-9)**
Naphthalene, 2-isopropyl **(2027-17-0)**
Naphthalene, 1-isopropyl- (8CI) **(6158-45-8)**
Naphthalene, 4-isopropyl-1,6-dimethyl- (8CI) **(483-78-3)**
Naphthalene, 7-isopropyl-1-methyl- (7CI,8CI) **(490-65-3)**
Naphthalene, isothiocyanato- **(551-06-4)**
1-Naphthalenemethanamine (9CI) **(118-31-0)**
1-Naphthalenemethanaminium, N-dodecyl-N,N-dimethyl-, chloride (9CI) **(1733-96-6)**
1-Naphthalenemethanol **(4780-79-4)**
1-Naphthalenemethanol, α,α-bis(4-(dimethylamino)phenyl)-4-(phenylamino)- (9CI) **(6786-83-0)**
Naphthalenemethanol, α-((isopropylamino)methyl)-, hydrochloride **(51-02-5)**
Naphthalene, 1-methoxy- (9CI) **(2216-69-5)**
Naphthalene, 2-methoxy- (9CI) **(93-04-9)**
Naphthalene, 1-methyl **(90-12-0)**
Naphthalene, 2-methyl **(91-57-6)**
Naphthalene, methyl **(1321-94-4)**
Naphthalene, (1-methylethyl)- **(29253-36-9)**
Naphthalene, 1-(1-methylethyl)- (9CI) **(6158-45-8)**
Naphthalene, 1-methyl-4-(1-methyl-2-butenyl)- (9CI) **(38171-97-0)**
Naphthalene, 1-methyl-7-(1-methylethyl)- (9CI) **(490-65-3)**
Naphthalene, 1-methyl-2-nitro **(63017-87-8)**
Naphthalene, 2-methyl-3-nitro **(1204-72-4)**
Naphthalene, 1-methyl-3-nitro- (9CI) **(41037-13-2)**
Naphthalene, 1-methyl-5-nitro- (9CI) **(91137-27-8)**
Naphthalene, 1-methyl-7-nitro- (9CI) **(116530-07-5)**
Naphthalene, 2-methyl-6-nitro- (9CI) **(54357-08-3)**
Naphthalene, 2-methyl-7-nitro- (9CI) **(91137-28-9)**
Naphthalene, 6-methyl-1-nitro- (9CI) **(54755-20-3)**
Naphthalene, 7-methyl-1-nitro- (9CI) **(54755-21-4)**
Naphthalene, 1-methyl-4-nitro- (8CI,9CI) **(880-93-3)**
Naphthalene, 2-methyl-1-nitro (8CI,9CI) **(881-03-8)**
Naphthalene, 3-methyl-1-nitro- (8CI,9CI) **(13615-38-8)**
Naphthalene, methylphenyl- (9CI) **(97232-29-6)**
Naphthalene, 1-(2-methylpropyl)- (9CI) **(16727-91-6)**
Naphthalene, 2-methyl-1-propyl- (9CI) **(54774-89-9)**
Naphthalene, 1-(methylsulfonyl)-5-(methylthio)- (9CI) **(54616-10-3)**
Naphthalene, mononitro **(27254-36-0)**
1-Naphthalenenitrile **(86-53-3)**
2-Naphthalenenitrile **(613-46-7)**
Naphthalene, 1-nitro **(86-57-7)**

Naphthalene, 2-nitro **(581-89-5)**
Naphthalene, nitro- **(27254-36-0)**
Naphthalene, octachloro **(2234-13-1)**
Naphthalene, 1,2,3,4,4a,5,6,8a-octahydro-4a,8-dimethyl-2-(1-methylethenyl)-, (2R-(2α,4aα,8aβ))- (9CI) **(473-13-2)**
Naphthalene, 1,2,3,4,4a,5,6,8a-octahydro-7-methyl-4-methylene-1-(1-methylethyl)-, (1R-(1α,4aβ,8aα))- (9CI) **(1460-97-5)**
Naphthalene, octyl- (9CI) **(41638-56-6)**
Naphthalene, pentachloro **(1321-64-8)**
Naphthalene, 1-phenyl- (9CI) **(605-02-7)**
Naphthalene, 2-phenyl- (9CI) **(612-94-2)**
Naphthalene, phenyl- (9CI) **(35465-71-5)**
Naphthalene, (phenylmethyl)- (9CI) **(38620-92-7)**
Naphthalene, 2-(phenylmethyl)- (9CI) **(613-59-2)**
Naphthalene plant heavy gasoline **(64741-68-0)**
Naphthalene plant light gasoline **(64741-63-5)**
2-Naphthalenepropanoic acid, β-ethyl-6-methoxy-α,α-dimethyl- (9CI) **(517-18-0)**
Naphthalene, 1-(2-propenyl)- (9CI) **(2489-86-3)**
2-Naphthalenepropionic acid, β-ethyl-6-methoxy-α,α-dimethyl **(517-18-0)**
β-Naphthalenesulfochloride **(93-11-8)**
2-Naphthalenesulfonamide, 6-hydroxy- (8CI,9CI) **(6528-46-7)**
1-Naphthalenesulfonic acid, **(5858-93-5)**
2-Naphthalenesulfonic acid **(120-18-3)**
Naphthalene-2-sulfonic acid **(120-18-3)**
β-Naphthalenesulfonic acid **(120-18-3)**
1-Naphthalenesulfonic acid (9CI) **(85-47-2)**
2-Naphthalenesulfonic acid, 7-acetamido-4-hydroxy- (8CI) **(6334-97-0)**
2-Naphthalenesulfonic acid, 7-(acetylamino)-3-((4-(acetylamino)-phenyl)azo)-4-hydroxy-, monosodium salt (9CI) **(6222-63-5)**
2-Naphthalenesulfonic acid, 7-(acetylamino)-4-hydroxy- (9CI) **(6334-97-0)**
2-Naphthalenesulfonic acid, 6-(acetylamino)-4-hydroxy-3-((4-((2-(sulfooxy)ethyl)sulfonyl)phenyl)azo)-, disodium salt (9CI) **(20262-58-2)**
2-Naphthalenesulfonic acid, 3-((4-(acetylamino)phenyl) azo)-4-hydroxy-7-((((5-hydroxy-6-(phenylazo)-7-sulfo-2-naphthalenyl)amino)carbonyl)amino)-, disodium salt (9CI) **(3441-14-3)**
1-Naphthalenesulfonic acid, 2-amino **(81-16-3)**
1-Naphthalenesulfonic acid, 4-amino **(84-86-6)**
1-Naphthalenesulfonic acid, 5-amino **(84-89-9)**
2-Naphthalenesulfonic acid, 5-amino **(119-79-9)**
1-Naphthalenesulfonic acid, 8-amino- (9CI) **(82-75-7)**
2-Naphthalenesulfonic acid, 1-amino- (9CI) **(81-06-1)**
2-Naphthalenesulfonic acid, 6-amino- (9CI) **(93-00-5)**
2-Naphthalenesulfonic acid, 8-amino- (9CI) **(119-28-8)**
2-Naphthalenesulfonic acid, 6-amino-5-((4-chloro-2-(trifluoromethyl)phenyl)azo)-4-hydroxy-, monosodium salt (9CI) **(57741-47-6)**
2-Naphthalenesulfonic acid, 6-amino-5-((2-((cyclohexylmethylamino)sulfonyl)phenyl)azo)-4-hydroxy-, monosodium salt (9CI) **(32846-21-2)**
1-Naphthalenesulfonic acid, 4-amino-3-hydroxy **(116-63-2)**
2-Naphthalenesulfonic acid, 7-amino-4-hydroxy **(87-02-5)**
2-Naphthalenesulfonic acid, 6-amino-4-hydroxy- (9CI) **(90-51-7)**
2-Naphthalenesulfonic acid, 7-amino-4-hydroxy-3-((5-hydroxy-6-(phenylazo)-7-sulfo-2-naphthalenyl)azo)-, disodium salt (9CI) **(6227-02-7)**
1-Naphthalenesulfonic acid, 3-((4'-((6-amino-1-hydroxy-3-sulfo-2-naphthalenyl)azo)-3,3'-dimethoxy(1,1'-biphenyl)-4-yl)azo)-4-hydroxy-, disodium salt (9CI) **(6449-35-0)**
1-Naphthalenesulfonic acid, 4-amino-3-((4'-((1-hydroxy-4-sulfo-2-naphthalenyl)azo)-3,3'-dimethoxy(1,1'-biphenyl)-4-yl)azo)-, disodium salt (9CI) **(6428-94-0)**
2-Naphthalenesulfonic acid, 6-((7-amino-1-hydroxy-3-sulfo-2-naphthalenyl)azo)-3-((4-((4-amino-6(or 7)-sulfo-1-naphthalenyl)-azo)phenyl)azo)-4-hydroxy-, trisodium salt (9CI) **(8003-69-8)**
2-Naphthalenesulfonic acid, 7-amino-4-hydroxy-3-((4-((4-sulfophenyl)azo)phenyl)azo)-, disodium salt (9CI) **(6300-50-1)**
1-Naphthalenesulfonic acid, 2-amino-, monoammonium salt (9CI) **(68540-41-0)**
1-Naphthalenesulfonic acid, 4-amino-, monosodium salt (9CI) **(130-13-2)**
1-Naphthalenesulfonic acid, 8-anilino- (8CI) **(82-76-8)**
2-Naphthalenesulfonic acid, 7-anilino-4-hydroxy- (8CI) **(119-40-4)**
1-Naphthalenesulfonic acid, 8-anilino-, monosodium salt **(1445-19-8)**
2-Naphthalenesulfonic acid, 7-(benzoylamino)-4-hydroxy- (9CI) **(132-87-6)**
2-Naphthalenesulfonic acid, 7-(benzoylamino)-4-hydroxy-3-((4-((4-sulfophenyl)azo)phenyl)azo)-, disodium salt **(2610-11-9)**
1-Naphthalenesulfonic acid, 3,3'-(4,4'-biphenylenebis-(azo))bis(4-amino-, disodium salt **(573-58-0)**
2-Naphthalenesulfonic acid, 5,5'-(4,4'-biphenylylenebis-(azo))bis(6-amino-4-hydroxy-, disodium salt **(2586-60-9)**
Naphthalenesulfonic acid, bis(1-methylethyl)- (9CI) **(28757-00-8)**
Naphthalenesulfonic acid, bis(1-methylethyl)-, Compd. with cyclohexanamine (1:1) (9CI) **(68425-61-6)**
Naphthalenesulfonic acid, bis(1-methylethyl)-, sodium salt (9CI) **(1322-93-6)**
Naphthalenesulfonic acid, bis(2-methylpropyl)-, sodium salt (9CI) **(27213-90-7)**
Naphthalenesulfonic acid, butyl- (9CI) **(26761-78-4)**
Naphthalenesulfonic acid, butyl-, ammonium salt (9CI) **(27478-24-6)**
Naphthalenesulfonic acid, butyl-, sodium salt (9CI) **(25638-17-9)**
2-Naphthalenesulfonic acid, 7,7'-(carbonyldiimino)bis(4-hydroxy- (9CI) **(134-47-4)**
2-Naphthalenesulfonic acid, 7,7'-(carbonyldiimino)bis(4-hydroxy-, disodium salt (9CI) **(20324-87-2)**
2-Naphthalenesulfonic acid, 7,7'-(carbonyldiimino)bis(4-hydroxy-3-((2-sulfo-4-((4-sulfophenyl)azo)phenyl)azo)-, hexasodium salt (9CI) **(2610-10-8)**
2-Naphthalenesulfonic acid, 7,7'-(carbonyldiimino)bis(4-hydroxy-3-((6-sulfo-2-naphthalenyl)azo)-, Compd. with 2,2',2''-nitrilotris-(ethanol) (1:4) (9CI) **(36596-36-8)**
2-Naphthalenesulfonic acid, 7,7'-(carbonyldiimino)bis(4-hydroxy-3-((6-sulfo-2-naphthalenyl)azo)-, tetrasodium salt (9CI) **(28706-25-4)**
Naphthalene-2-sulfonic acid chloride **(93-11-8)**
1-Naphthalenesulfonic acid, 6-((5-chloro-2-hydroxy-3-sulfophenyl)azo)-5-hydroxy-, disodium salt (9CI) **(3624-68-8)**
2-Naphthalenesulfonic acid, 5-((4-chlorophenyl)azo)-6-hydroxy- (9CI) **(4531-71-9)**
2-Naphthalenesulfonic acid, 5-((p-chlorophenyl)azo)-6-hydroxy- (7CI,8CI) **(4531-71-9)**
2-Naphthalenesulfonic acid, 5-((4-cyanophenyl)azo)-6-hydroxy- (9CI) **(78335-09-8)**
2-Naphthalenesulfonic acid, 6-((2,4-diaminophenyl)azo)-3-((4-((4-((7-((2,4-diaminophenyl)azo)-1-hydroxy-3-sulfo-2-naphthalenyl)azo)phenyl)amino)-3-sulfophenyl)azo)-4-hydroxy-, trisodium salt (9CI) **(6473-13-8)**
1-Naphthalenesulfonic acid, 6-diazo-5,6-dihydro-5-oxo-, 4-benzoyl-1,2,3-benzenetriyl ester (9CI) **(5610-94-6)**
1-Naphthalenesulfonic acid, 6-diazo-5,6-dihydro-5-oxo-, sodium salt **(2657-00-3)**
Naphthalenesulfonic acid, dibutyl-, sodium salt **(25417-20-3)**
1-Naphthalenesulfonic acid, 3,3'-((3,3'-dichloro(1,1'-biphenyl)-4,4'-diyl)bis(azo))bis(4-amino-, disodium salt (9CI) **(6470-31-1)**
2-Naphthalenesulfonic acid, 7-((4,6-dichloro-1,3,5-triazin-2-yl)-amino)-4-hydroxy-3-((2-sulfophenyl)azo)-, disodium salt (9CI) **(18886-16-3)**
2-Naphthalenesulfonic acid, 1,4-dihydro-1,4-dioxo-, sodium salt (9CI) **(7045-83-2)**
2-Naphthalenesulfonic acid, 6,7-dihydroxy- (9CI) **(92-27-3)**
1-Naphthalenesulfonic acid, 4,4'-((2,4-dihydroxy-1,3-phenylene)-bis(azo))bis-, disodium salt (9CI) **(5850-16-8)**
1-Naphthalenesulfonic acid, 3-((7-((4'-((1,8-dihydroxy-4-sulfo-2-naphthalenyl)azo)-3,3'-dimethyl (1,1'-biphenyl)-4-yl)azo)-8-hydroxy-6-sulfo-2-naphthalenyl)azo)-4,5- dihydroxy-, trisodium salt (9CI) **(7082-31-7)**
Naphthalenesulfonic acid, diisopropyl-, sodium salt (8CI) **(1322-93-6)**
1-Naphthalenesulfonic acid, 3,3'-((3,3'-dimethoxy(1,1'-biphenyl)-4,4'-diyl)bis(azo))bis(4-amino-, disodium salt (9CI) **(2868-75-9)**
1-Naphthalenesulfonic acid, 3,3'-((3,3'-dimethoxy(1,1'-biphenyl)-

4,4'-diyl)bis(azo))bis(4-hydroxy-, disodium salt (9CI)	**(2429-71-2)**
1-Naphthalenesulfonic acid, 3,3'-((3,3'-dimethyl(1,1'-biphenyl)-4,4'-diyl)bis(azo))bis (4-amino-, disodium salt	**(992-59-6)**
1-Naphthalenesulfonic acid, 6-((2,4-dimethyl-6-sulfo-phenyl)azo)-5-hydroxy-, disodium salt	**(3257-28-1)**
Naphthalenesulfonic acid, dinonyl- (9CI)	**(25322-17-2)**
Naphthalenesulfonic acid, dinonyl-, barium salt (9CI)	**(25619-56-1)**
Naphthalenesulfonic acid, dinonyl-, calcium salt (9CI)	**(57855-77-3)**
Naphthalenesulfonic acid, dinonyl-, lithium salt (9CI)	**(28214-91-7)**
Naphthalenesulfonic acid, dinonyl-, sodium salt (9CI)	**(26834-28-6)**
Naphthalenesulfonic acid, dinonyl-, zinc salt (9CI)	**(28016-00-4)**
2-Naphthalenesulfonic acid, 6-hydroxy	**(93-01-6)**
1-Naphthalenesulfonic acid, 4-hydroxy- (9CI)	**(84-87-7)**
2-Naphthalenesulfonic acid, 1-hydroxy- (9CI)	**(567-18-0)**
2-Naphthalenesulfonic acid, 7-hydroxy- (9CI)	**(92-40-0)**
1-Naphthalenesulfonic acid, 4-hydroxy-3,4'-azodi-, disodium salt	**(3567-69-9)**
2-Naphthalenesulfonic acid, 4-hydroxy-7-((((5-hydroxy-6-((2-methoxyphenyl)azo)-7-sulfo-2-naphthalenyl)amino)-carbonyl)amino)-3-((2-methyl-4-sulfophenyl)azo)-, trisodium salt (9CI)	**(6420-44-6)**
2-Naphthalenesulfonic acid, 4-hydroxy-7-((((5-hydroxy-6-((2-methoxyphenyl)azo)-7-sulfo-2-naphthalenyl)amino)-carbonyl)amino)-3-((6-sulfo-2-naphthalenyl)azo)-, trisodium salt (9CI)	**(10114-24-6)**
2-Naphthalenesulfonic acid, 4-hydroxy-7-((((5-hydroxy-6-((2-methylphenyl)azo)-7-sulfo-2-naphthalenyl)amino)carbonyl) amino)-3-((6-sulfo-2-naphthalenyl)azo)-, trisodium salt (9CI)	**(6460-01-1)**
1-Naphthalenesulfonic acid, 3-hydroxy-4-((1-hydroxy-2-naphthalenyl)azo)-7-nitro-, monosodium salt (9CI)	**(1787-61-7)**
1-Naphthalenesulfonic acid, 3-hydroxy-4-((2-hydroxy-1-naphthalenyl)azo)-7-nitro-, monosodium salt (9CI)	**(3618-58-4)**
1-Naphthalenesulfonic acid, 3-hydroxy-4-(2-hydroxy-1-naphthylazo)-, sodium salt	**(2538-85-4)**
2-Naphthalenesulfonic acid, 4-hydroxy-3-((2-methoxy-5-methyl-4-((4-sulfophenyl)azo)phenyl)azo)-7-(phenylamino)-, disodium salt (9CI)	**(6227-14-1)**
2-Naphthalenesulfonic acid, 6-hydroxy-5-((4-methoxyphenyl)azo)-(9CI)	**(27959-50-8)**
2-Naphthalenesulfonic acid, 6-hydroxy-5-((p-methoxyphenyl)azo)-(8CI)	**(27959-50-8)**
1-Naphthalenesulfonic acid, 4-hydroxy-3-((2-methoxyphenyl)-azo)-, monosodium salt (9CI)	**(5858-39-9)**
2-Naphthalenesulfonic acid, 6-hydroxy-5-((6-methoxy-4-sulfo-m-tolyl)azo)-, disodium salt	**(25956-17-6)**
2-Naphthalenesulfonic acid, 6-hydroxy-5-((4-methylphenyl)azo)-(9CI)	**(5859-07-4)**
2-Naphthalenesulfonic acid, 6-hydroxy-, monopotassium salt (9CI)	**(833-66-9)**
2-Naphthalenesulfonic acid, 6-hydroxy-, monosodium salt (9CI)	**(135-76-2)**
1-Naphthalenesulfonic acid, 2-((2-hydroxy-1-naphthalenyl)azo)-, barium salt (2:1) (9CI)	**(1103-38-4)**
1-Naphthalenesulfonic acid, 2-((2-hydroxy-1-naphthalenyl)azo)-, calcium salt (2:1) (9CI)	**(1103-39-5)**
1-Naphthalenesulfonic acid, 2-((2-hydroxy-1-naphthalenyl)azo)-, monosodium salt	**(1248-18-6)**
1-Naphthalenesulfonic acid, 4-((2-hydroxy-1-naphthalenyl)azo)-, monosodium salt	**(1658-56-6)**
2-Naphthalenesulfonic acid, 6-hydroxy-5-((4-nitrophenyl)azo)-(9CI)	**(5859-04-1)**
2-Naphthalenesulfonic acid, 6-hydroxy-5-((p-nitrophenyl)azo)-(7CI,8CI)	**(5859-04-1)**
2-Naphthalenesulfonic acid, 4-hydroxy-7-(phenylamino)- (9CI)	**(119-40-4)**
1-Naphthalenesulfonic acid, 5-((4-hydroxyphenyl)amino)-8-(phenylamino)- (9CI)	**(82-31-5)**
2-Naphthalenesulfonic acid, 5-((1-hydroxy-6-(phenylamino)-3-sulfo-2-naphthalenyl)azo)-8-((6-sulfo-4-(- (3-sulfophenyl)azo)-1-naphthalenyl)azo)-, tetrasodium salt (9CI)	**(6428-60-0)**
1-Naphthalenesulfonic acid, 4-hydroxy-3-((4-((4-(phenylamino)-3-sulfophenyl)azo)-1-naphthalenyl) azo)-, disodium salt (9CI)	**(6406-45-7)**
2-Naphthalenesulfonic acid, 6-hydroxy-5-(phenylazo)- (9CI)	**(23481-33-6)**
2-Naphthalenesulfonic acid, 6-hydroxy-5-(phenylazo)-, mono-sodium salt	**(1934-20-9)**
1-Naphthalenesulfonic acid, 4-((2-hydroxy-6-sulfo-1-naphthalenyl) azo)-, disodium salt	**(2302-96-7)**
2-Naphthalenesulfonic acid, 6-hydroxy-5-((4-sulfophenyl)azo)-(9CI)	**(5859-11-0)**
2-Naphthalenesulfonic acid, 6-hydroxy-2-((4-sulfophenyl)azo)-, aluminum lake	**(15790-07-5)**
2-Naphthalenesulfonic acid, 6-hydroxy-5-((p-sulfophenyl)azo)-, disodium salt	**(2783-94-0)**
1-Naphthalenesulfonic acid, 4-hydroxy-3-((6-sulfo-2,4-xylyl)azo)-, disodium salt	**(4548-53-2)**
2-Naphthalenesulfonic acid, 6-hydroxy-5-(p-tolylazo)- (8CI)	**(5859-07-4)**
Naphthalenesulfonic acid, methylenebis-, disodium salt (9CI)	**(26545-58-4)**
Naphthalenesulfonic acid, methylenedi-, disodium salt	**(26545-58-4)**
1-Naphthalenesulfonic acid, 8-(phenylamino)- (9CI)	**(82-76-8)**
1-Naphthalenesulfonic acid, 8-(phenylamino)-, monoammonium salt	**(28836-03-5)**
1-Naphthalenesulfonic acid, 8-(phenylamino)-, monosodium salt (9CI)	**(1445-19-8)**
1-Naphthalenesulfonic acid, 8-(phenylamino)-5-((4-((5-sulfo-1-naphthalenyl)azo)-1-naphthalenyl)azo)-, disodium salt (9CI)	**(3071-73-6)**
1-Naphthalenesulfonic acid, 8-(phenylamino)-5-((4-((3-sulfo-phenyl)azo)-1-naphthalenyl)azo)-, disodium salt (9CI)	**(3351-05-1)**
2-Naphthalenesulfonic acid, sodium salt (9CI)	**(532-02-5)**
Naphthalenesulfonic acid, sodium salt (8CI,9CI)	**(1321-69-3)**
2-Naphthalenesulfonic acid, sodium salt, Polymer with formaldehyde (9CI)	**(29321-75-3)**
Naphthalene sulfonic acid, sodium salt solution	**(1321-69-3)**
2-Naphthalenesulfonic acid, 1,2,3,4-tetrahydro-2-methyl-1,4-dioxo-, sodium salt (9CI)	**(130-37-0)**
2-Naphthalenesulfonic acid, 7,7'-ureylenebis(4-hydroxy- (8CI)	**(134-47-4)**
β-Naphthalenesulfonic sodium salt	**(532-02-5)**
Naphthalene-2-sulfonyl chloride	**(93-11-8)**
β-Naphthalenesulfonyl chloride	**(93-11-8)**
2-Naphthalenesulfonyl chloride (9CI)	**(93-11-8)**
1-Naphthalenesulfonyl chloride, 6-diazo-5,6-dihydro-5-oxo- (9CI)	**(3770-97-6)**
1-Naphthalenesulfonyl chloride, 5-(methylthio)- (9CI)	**(53135-95-8)**
1,4,5,8-Naphthalenetetracarboxylic acid, 1,8:4,5-dianhydride (8CI)	**(81-30-1)**
Naphthalenetetracarboxylic dianhydride	**(81-30-1)**
Naphthalene, 1,2,3,4-tetrachloro-	**(20020-02-4)**
Naphthalene, tetrachloro	**(1335-88-2)**
Naphthalene 1,2,3,4-tetrahydride	**(119-64-2)**
Naphthalene, 1,2,3,4-tetrahydro	**(119-64-2)**
Naphthalene, 1,2,3,4-tetrahydro-1,8-dimethyl- (8CI,9CI)	**(25419-33-4)**
Naphthalene, 1,2,3,4-tetrahydro-2,3-dimethyl- (8CI,9CI)	**(21564-92-1)**
Naphthalene, 1,2,3,4-tetrahydro-2,7-dimethyl- (8CI,9CI)	**(13065-07-1)**
Naphthalene, 1,2,3,4-tetrahydro-6,7-dimethyl- (8CI,9CI)	**(1076-61-5)**
Naphthalene, 1,2,3,4-tetrahydro-1,1-dimethyl- (6CI,7CI,8CI,9CI)	**(1985-59-7)**
Naphthalene, 1,2,3,4-tetrahydro-1,6-dimethyl-4-(1-methyl-ethyl)- (9CI)	**(6617-49-8)**
Naphthalene, 1,2,3,4-tetrahydro-1,6-dimethyl-4-(1-methyl-ethyl)-, (1R-cis)- (9CI)	**(22339-23-7)**
Naphthalene, 1,2,3,4-tetrahydro-1,6-dimethyl-4-(1-methyl-ethyl)-, (1S-cis)- (9CI)	**(483-77-2)**
Naphthalene, 1,2,3,4-tetrahydro-5,8-dimethyl-1-octyl- (7CI,9CI)	**(55255-58-8)**
Naphthalene, 1,2,3,4-tetrahydro-4-isopropyl-1,6-dimethyl- (8CI)	**(6617-49-8)**
Naphthalene, 1,2,3,4-tetrahydromethyl	**(31291-71-1)**
Naphthalene, 1,2,3,4-tetrahydro-1-methyl- (8CI,9CI)	**(1559-81-5)**
Naphthalene, 1,2,3,4-tetrahydro-2-methyl- (8CI,9CI)	**(3877-19-8)**
Naphthalene, 1,2,3,4-tetrahydro-5-methyl- (6CI,7CI,8CI,9CI)	**(2809-64-5)**
Naphthalene, 1,2,3,4-tetrahydro-6-methyl- (6CI,7CI,8CI,9CI)	**(1680-51-9)**
Naphthalene, 1,2,3,4-tetrahydro-1-phenyl- (6CI,7CI,8CI,9CI)	**(3018-20-0)**
Naphthalene, 1,2,3,4-tetrahydro-2,2,5,7-tetramethyl- (8CI,9CI)	**(23342-25-8)**
Naphthalene, 1,2,3,4-tetrahydro-1,4,6-trimethyl- (7CI,8CI,9CI)	**(22824-32-4)**

Naphthalene, 1,2,3,4-tetrahydro-1,5,8-trimethyl- (6CI,8CI,9CI) **(21693-51-6)**
Naphthalene, 1,3,6,8-tetranitro **(28995-89-3)**
Naphthalene, 1,2,3,4-tetraphenyl- (8CI,9CI) **(751-38-2)**
Naphthalenetetrone (8CI,9CI) **(11063-25-5)**
Naphthalenetetrone, methyl- (9CI) **(116211-95-1)**
2-Naphthalenethiol **(91-60-1)**
Naphthalene-2-thiol **(91-60-1)**
β-Naphthalenethiol **(91-60-1)**
Naphthalene, trichloro **(1321-65-9)**
Naphthalene, 1,6,7-trimethyl- (9CI) **(2245-38-7)**
Naphthalene, 2,3,6-trimethyl- (9CI) **(829-26-5)**
Naphthalene, 1,2,8-trimethyl- (8CI,9CI) **(3876-97-9)**
Naphthalene, 1,3,6-trimethyl- (8CI,9CI) **(3031-08-1)**
Naphthalene, 1,3,7-trimethyl- (8CI,9CI) **(2131-38-6)**
Naphthalene, 1,3,8-trimethyl- (8CI,9CI) **(17057-91-9)**
Naphthalene, 1,4,5-trimethyl- (8CI,9CI) **(2131-41-1)**
Naphthalene, 1,4,6-trimethyl- (8CI,9CI) **(2131-42-2)**
Naphthalene, trimethyl- (8CI,9CI) **(28652-77-9)**
Naphthalene, 1,2,3-trimethyl-4-propenyl-, (E)- (8CI) **(26137-53-1)**
Naphthalene, 1,2,3-trimethyl-4-(1-propenyl)-, (Z)- (9CI) **(26130-84-7)**
Naphthalene, 1,2,3-trimethyl-4-propenyl-, (Z)- (8CI) **(26130-84-7)**
1,3,6-Naphthalenetrisulfonic acid, 8-hydroxy-7-((4'-((2-hydroxy-1-naphthalenyl)azo)(1,1'-biphenyl)-4-yl)azo)-, trisodium salt (9CI) **(6426-67-1)**
1,3,6-Naphthalenetrisulfonic acid, 7-hydroxy-8-((4-sulfo-1-naphthalenyl)azo)-, tetrasodium salt **(5850-44-2)**
1,3,6-Naphthalenetrisulfonic acid, sodium salt (9CI) **(19437-42-4)**
Naphthalene, 2-vinyl- (8CI) **(827-54-3)**
1-Naphthalenol **(90-15-3)**
2-Naphthalenol **(135-19-3)**
Naphthalenol **(9CI)**
1-Naphthalenol, acetate (9CI) **(830-81-9)**
1-Naphthalenol, 4-amino- (9CI) **(2834-90-4)**
2-Naphthalenol, 1-amino- (9CI) **(2834-92-6)**
2-Naphthalenol, amino- (9CI) **(95609-86-2)**
Naphthalenol, amino- (9CI) **(42884-33-3)**
1-Naphthalenol, 2-chloro- (9CI) **(606-40-6)**
1-Naphthalenol, 4-chloro- (9CI) **(604-44-4)**
2-Naphthalenol, decahydro **(825-51-4)**
Naphthalen-2-ol, decahydro- **(825-51-4)**
1-Naphthalenol, 2,4-dichloro- (9CI) **(2050-76-2)**
2-Naphthalenol, 1-(6-(2,2-dimethyl-6-methylenecyclohexyl)-4-methyl-3-hexenyl)decahydro-2,5,5,8a-tetramethyl-, (1R-(1α(3E,6(S*)),2β,4aβ,8aα))- (9CI) **(473-03-0)**
2-Naphthalenol, 6,6'-dithiobis- (9CI) **(6088-51-3)**
2-Naphthalenol, heptyl- (9CI) **(31215-04-0)**
2-Naphthalenol, 1-((2-hydroxy-3,5-dinitrophenyl)azo)- (9CI) **(4998-82-7)**
2-Naphthalenol, 1-((2-methoxy-5-methylphenyl)azo)- (9CI) **(6410-20-4)**
1-Naphthalenol, 2-methyl- (9CI) **(7469-77-4)**
2-Naphthalenol, 1-((4-methyl-2-nitrophenyl)azo) **(2425-85-6)**
Naphthalenol, nitro- (9CI) **(82322-43-8)**
2-Naphthalenol, 1-((4-nitrophenyl)azo)- **(6410-10-2)**
4a(2H)-Naphthalenol, octahydro-4,8a-dimethyl-, (4S-(4-α,4a-α,8a-β))- **(19700-21-1)**
2-Naphthalenol, 1-((4-(phenylazo)phenyl)azo) **(85-86-9)**
1-Naphthalenol, propanoate **(3121-71-9)**
2-Naphthalenol, 1-(2-pyridinylazo)- (9CI) **(85-85-8)**
1-Naphthalenol, 2,5,8-trimethyl- (9CI) **(33583-02-7)**
1(2H)-Naphthalenone, 3,4-dihydro **(529-34-0)**
1(2H)-Naphthalenone, 3,4-dihydro-6-methoxy- (9CI) **(1078-19-9)**
2(1H)-Naphthalenone, 4a,5,6,7,8,8a-hexahydro-4a,8a-dimethyl-, cis- (8CI,9CI) **(13485-66-0)**
2(3H)-Naphthalenone, 4,4a,5,6,7,8-hexahydro-4,4a-dimethyl-6-(1-methylethenyl)-, (4R-(4α,4aα,6β))- (9CI) **(4674-50-4)**
1(2H)-Naphthalenone, octahydro-4a,5-dimethyl-3-(1-methylethyl)-, (3.α.,4a.α.,5.α.,8a.α.)- (9CI) **(55332-02-0)**
2(1H)-Naphthalenone, octahydro-4a,5-dimethyl-3-(1-methylethyl)-, (3.α.,4a.α.,5.α.,8a.α.)- (9CI) **(55332-03-1)**
1(2H)-Naphthalenone, octahydro-4-hydroxy- (8CI,9CI) **(21766-50-7)**
2(1H)-Naphthalenone, octahydro-4a-methyl-7-(1-methylethyl)-, (4a.α.,7.β.,8a.β.)- (9CI) **(54594-42-2)**
2-Naphthalenoxyacetic acid **(120-23-0)**
2-((1-Naphthalenylamino)carbonyl)benzoic acid **(132-66-1)**
4-(1-Naphthalenylazo)-1,3-phenylenediamine **(6416-57-5)**
N-1-Naphthalenyl-1,2-ethanediamine dihydrochloride **(1465-25-4)**
1-(1-Naphthalenyl)ethanone **(941-98-0)**
1-(2-Naphthalenyl)ethanone **(93-08-3)**
1-(Naphthalenyl)ethanone **(1333-52-4)**
2-Naphthalenyl 2-hydroxypropanoate **(93-43-6)**
1-Naphthalenyl methylcarbamate **(63-25-2)**
α-Naphthalenylmethylcarbamate **(63-25-2)**
(β-Naphthalenyloxy)acetic acid **(120-23-0)**
2-(2-Naphthalenyloxy)-N-phenylpropanamide **(52570-16-8)**
1-Naphthalenylthiourea **(86-88-4)**
1,8-Naphthalic anhydride **(81-84-5)**
Naphthalidam **(134-32-7)**
Naphthalidine **(134-32-7)**
1,8-Naphthalimide **(81-83-4)**
Naphthalimide (8CI) **(81-83-4)**
Naphthalimide, 4-methoxy-N-methyl **(3271-05-4)**
Naphthalin **(91-20-3)**
Naphthaline **(91-20-3)**
α-Naphthalthioharnstoff (German) **(86-88-4)**
2-Naphthamide, 3-hydroxy-N-1-naphthyl- (8CI) **(132-68-3)**
1-Naphthamide, N-methyl- (7CI,8CI) **(3400-33-7)**
Naphthaminblau 3BX **(72-57-1)**
Naphthamine Blue 3BX **(72-57-1)**
Naphthamine Dark Green B **(3626-28-6)**
Naphthane **(91-17-8)**
Naphthanil AS-D **(135-61-5)**
Naphthanil AS **(92-77-3)**
Naphthanil BS **(135-65-9)**
Naphthanil Blue B Base **(119-90-4)**
Naphthanil G **(91-96-3)**
Naphthanil OL **(135-62-6)**
Naphthanil OP **(92-74-0)**
Naphthanil Red B Base **(97-52-9)**
Naphthanil Red 3G Base **(89-63-4)**
Naphthanil Red G Base **(89-62-3)**
Naphthanil Scarlet 2G Base **(95-82-9)**
Naphthanil Scarlet G Base **(99-55-8)**
Naphthanilid EL **(137-52-0)**
Naphthanilid KB **(135-63-7)**
2-Naphthanilide, 4-((5-carbamoyl-o-tolyl)azo)-3-hydroxy- (8CI) **(16403-84-2)**
2-Naphthanilide, 3-hydroxy- (8CI) **(92-77-3)**
2-Naphthanilide, 3-hydroxy-3'-nitro- (8CI) **(135-65-9)**
Naphthanthracene **(56-55-3)**
Naphthanthrone **(3074-00-8)**
Naphthanthrone **(82-05-3)**
$\delta^{5,7,9}$-Naphthantriene **(119-64-2)**
Naphtha, petroleum, catalytic reformed **(68955-35-1)**
Naphtha, petroleum, heavy catalytic cracked **(64741-54-4)**
Naphtha, petroleum, heavy straight-run **(64741-41-9)**
1,2-Naphthaquinone **(524-42-5)**
Naphthathenic soap **(61790-13-4)**
Naphthazine Rose 2G **(3734-67-6)**
Naphthazine Scarlet 2r **(3761-53-3)**
Naphthazine Yellow RP **(547-57-9)**
Naphth(2,3-b)oxirene, decahydro- (9CI) **(21399-51-9)**
Naphth(1,8-cd)-1,2-oxathiole, 2,2-dioxide (9CI) **(83-31-8)**
5H-Naphth(3,2,1-de)anthracen-5-one (9CI) **(62716-20-5)**
Naphth(1,2-d)(1,2,3)oxadiazole-5-sulfonic acid (9CI) **(84-23-1)**
Naphth(1,2-d)(1,2,3)oxadiazole-5-sulfonic acid, 7-nitro- (9CI) **(84-91-3)**
Naphth(1,2-d)(1,2,3)oxadiazole-5-sulfonic acid, 8-nitro- (9CI) **(130-59-6)**

Naphth(2,3-e)acephenanthrylene (9CI) **(205-97-0)**
Naphthemol **(130-13-2)**
Naphthene **(91-20-3)**
Naphthenic-acid **(1338-24-5)**
Naphthenic acid, calcium salt **(61789-36-4)**
Naphthenic acid, cobalt salt **(61789-51-3)**
Naphthenic acid, copper salt **(1338-02-9)**
Naphthenic acid, lead salt **(61790-14-5)**
Naphthenic acid, sodium salt solution **(61790-13-4)**
Naphthenic acids, sodium salts **(61790-13-4)**
Naphthenic acid, zinc salt **(12001-85-3)**
Naphthenic oils **(67254-74-4)**
Naphthenic oils (Petroleum), Catalytic dewaxed heavy(9CI) **(64742-68-3)**
Naphthenic oils (Petroleum), Catalytic dewaxed light (9CI) **(64742-69-4)**
Naphthenic oils, Petroleum, Complex dewaxed heavy **(64742-75-2)**
Naphthenic oils, Petroleum, Complex dewaxed light **(64742-76-3)**
1H-Naphth(2,3-f)isoindole-1,3,5,10(2H)-tetrone, 4,11-diamino- (9CI) **(128-81-4)**
1H-Naphth(2,3-f)isoindole-1,3,5,10(2H)-tetrone, 4,11-di- amino-2-butyl- (9CI) **(3176-88-3)**
1H-Naphth(2,3-f)isoindole-1,3,5,10(2H)-tetrone, 4,11-di- amino-2-(3-methoxypropyl)- (9CI) **(12217-80-0)**
1H-Naphth(2,3-f)isoindole-1,5,10-trione, 4,11-diamino-2,3-di- hydro-3-imino-2-(3-methoxypropyl)- (9CI) **(13418-49-0)**
Naphth(2,1-f)isoquinoline **(218-02-0)**
peri-Naphthindene **(203-80-5)**
Naphth(2',3':6,7)indolo(2,3-c)dinaphtho(2,3-a:2',3'-i)- carbazole-5,10,15,17,22,24-hexone **(2475-33-4)**
Naphthiomate T **(2398-96-1)**
Naphthionic Red A **(2302-96-7)**
1,4-Naphthionic acid **(84-86-6)**
Naphthionic acid **(84-86-6)**
1,4-Naphthionic monosodium salt **(130-13-2)**
Naphthionine **(130-13-2)**
Naphthizine **(835-31-4)**
lin-Naphthoanthracene **(135-48-8)**
Naphtho(2,3-a)pyrene **(196-42-9)**
Naphtho(2,3-b)-p-dithiin-2,3-dicarbonitrile, 5,10-dihydro- 5,10-dioxo **(3347-22-6)**
Naphtho(1,2-b)puran-2,8(3H,4H)-dione, 3a,5,5a,9b-tetra- hydro-3,5a,9-trimethyl- **(481-06-1)**
1H-Naphtho(2,1-b)pyran-1-one, 3-phenyl **(6051-87-2)**
Naphtho(2,3-b)pyrene **(196-42-9)**
Naphtho(1,2-b)thianaphthene **(239-35-0)**
1H,3H-Naphtho(1,8-cd)pyran-1,3-dione, methyl- (9CI) **(79075-22-2)**
1H,3H-Naphtho(1,8-cd)pyran-1-one, 3-hydroxy- (9CI) **(5656-90-6)**
Naphtho(1,2-c)furan-1,3-dione (9CI) **(5343-99-7)**
Naphtho(2,3-c)furan-1(3H)-one, 4-(3,4-dihydroxyphenyl)- 3a,4,9,9a-tetrahydro-6,7-dihydroxy- (8CI) **(2316-10-1)**
Naphtho(2,3-c)furan-1(3H)-one, 4-(3,4-dihydroxyphenyl)- 3a,4,9,9a-tetrahydro-6,7-dihydroxy-, (3aR-(3a.α.,4.α.,9a.β.))- (9CI) **(2316-10-1)**
Naphtho(1,2,3,4-def)chrysene **(192-65-4)**
8H-Naphtho(3,2,1,8-defg)chrysen-8-one (9CI) **(86854-23-1)**
6H-Naphtho(2,1,8,7-defg)naphthacen-6-one (9CI) **(86854-22-0)**
Naphtho(2,1,8-def)quinoline **(313-80-4)**
2H-Naphtho(1,2-d)triazole-6,8-disulfonic acid, 2-(4-nitrophenyl)- (9CI) **(130-34-7)**
2H-Naphtho(1,2-d)triazole, 2-(4-(2-phenylethenyl)-3-sulfophenyl)-, sodium salt **(6416-68-8)**
2H-Naphtho(1,2-d)triazole-5-sulfonic acid, 2,2'-(azobis((2-sulfo- 4,1-phenylene)-2,1-ethenediyl(3-sulfo-4,1-phenylene)))bis-, hexasodium salt (9CI) **(12222-60-5)**
Naphthoelan Navy Blue **(101-54-2)**
Naphthoelan Red B Base **(97-52-9)**
Naphthoelan Red RL Base **(99-52-5)**
Naphtho(2,3-e)pyrene **(193-09-9)**
β-Naphthoflavone **(6051-87-2)**
α-Naphthofluorene **(238-84-6)**
Naphtho(2,1-f)quinoline **(218-08-6)**
Naphtho(2,3-h)quinoline **(84-56-0)**
1-Naphthoic acid **(86-55-5)**
2-Naphthoic acid **(93-09-4)**
α-Naphthoic acid **(86-55-5)**
β-Naphthoic acid **(93-09-4)**
2-Naphthoic acid, 3-amino **(5959-52-4)**
1-Naphthoic acid, 2,3-dihydroxy **(16715-77-8)**
1-Naphthoic acid, 2-hydroxy **(2283-08-1)**
2-Naphthoic acid, 3-hydroxy **(92-70-6)**
2-Naphthoic acid, 1-hydroxy- (8CI) **(86-48-6)**
2-Naphthoic acid, 3-hydroxy-, methyl ester (8CI) **(883-99-8)**
Naphthoide AD **(135-61-5)**
Naphthoide AS **(92-77-3)**
Naphthoide BO **(132-68-3)**
Naphthoide BS **(135-65-9)**
Naphthoide G **(91-96-3)**
Naphthoide OL **(135-62-6)**
Naphtho(1,2-k)fluoranthene (8CI,9CI) **(238-04-0)**
Naphtho(2,3-k)fluoranthene (8CI,9CI) **(207-18-1)**
1-Naphthol **(90-15-3)**
2-Naphthol **(135-19-3)**
Naphthol **(1321-67-1)**
α-Naphthol **(90-15-3)**
β-Naphthol **(135-19-3)**
Naphthol AS-A **(92-77-3)**
Naphthol AS-BO **(132-68-3)**
Naphthol AS-BR **(91-92-9)**
Naphthol AS-BS **(135-65-9)**
Naphthol AS-BS Dispersible **(135-65-9)**
Naphthol AS-BS Supra **(135-65-9)**
Naphthol AS-CA **(137-52-0)**
Naphthol AS-CL **(137-52-0)**
Naphthol ACNA C **(92-77-3)**
Naphthol ACNA F **(132-68-3)**
Naphthol AS-D Dispersible **(135-61-5)**
Naphthol AS-D Supra **(135-61-5)**
Naphthol AS-EL **(137-52-0)**
Naphthol AS-G **(91-96-3)**
Naphthol AS-G Dispersible **(91-96-3)**
Naphthol AS-G Supra **(91-96-3)**
Naphthol AS-KB **(135-63-7)**
Naphthol AS-OL **(135-62-6)**
Naphthol AS-OP **(92-74-0)**
Naphthol AS-PH **(92-74-0)**
Naphthol AS-RC Supra **(137-52-0)**
Naphthol AS-RO **(92-74-0)**
Naphthol AS **(92-77-3)**
Naphthol AS D **(135-61-5)**
Naphthol AS Supra **(92-77-3)**
Naphthol AS-phenyl **(92-74-0)**
Naphthol AS-phenyl supra **(92-74-0)**
Naphthol B **(135-19-3)**
Naphthol B.O.N. **(92-70-6)**
Naphthol NEL **(137-52-0)**
Naphthol Orange **(523-44-4)**
Naphthol Orange **(633-96-5)**
α-Naphthol Orange **(523-44-4)**
β-Naphthol Orange **(633-96-5)**
2-Naphthol Orange II **(633-96-5)**
Naphthol Red B **(6471-49-4)**
Naphthol Red B **(915-67-3)**
Naphthol Red B 20-7575 **(6471-49-4)**
Naphthol Red C **(915-67-3)**
Naphthol Red D Toner 35-6001 **(6471-49-4)**

Name	CAS No.
Naphthol Red Deep 10459	**(6471-49-4)**
2-Naphthol Red J	**(1658-56-6)**
Naphthol Red LZS	**(915-67-3)**
Naphthol Red O	**(915-67-3)**
Naphthol Red S	**(915-67-3)**
Naphthol Red S Conc. Specially Pure	**(915-67-3)**
Naphthol Red SI	**(915-67-3)**
Naphthol Red S Specially Pure	**(915-67-3)**
Naphthol Yellow	**(605-69-6)**
1-Naphthol, acetate (8CI)	**(830-81-9)**
1-Naphthol, 2-amino	**(42884-33-3)**
1-Naphthol, 4-amino- (8CI)	**(2834-90-4)**
1-Naphthol, 2-amino-, hydrochloride	**(41772-23-0)**
2-Naphthol, 1-amino-, hydrochloride	**(1198-27-2)**
Naphtholate AS	**(92-77-3)**
Naphtholate AS Soln	**(92-77-3)**
1-Naphthol-2-carboxylic acid	**(86-48-6)**
1-Naphthol, 2-chloro- (8CI)	**(606-40-6)**
1-Naphthol, 4-chloro- (8CI)	**(604-44-4)**
2-Naphthol, 1-((2-chloro-4-nitrophenyl)azo)	**(2814-77-9)**
1-Naphthol, 2,4-dichloro- (8CI)	**(2050-76-2)**
2-Naphthol, 1-((2,5-dimethoxyphenyl)azo)	**(6358-53-8)**
1-Naphthol, 2,4-dinitro	**(605-69-6)**
2-Naphthol, 1-((2,4-dinitrophenyl)azo)	**(3468-63-1)**
2-Naphthol-3,6-disulfonic acid	**(148-75-4)**
2-Naphthol-6,8-disulfonic acid	**(118-32-1)**
β-Naphthol-γ-disulfonic acid	**(118-32-1)**
β-Naphtholdisulfonic acid G	**(118-32-1)**
2-Naphthol-3,6-disulfonic acid, sodium salt	**(135-51-3)**
2-Naphthol, 6,6'-dithiodi- (8CI)	**(6088-51-3)**
2-Naphthol ethyl ether	**(93-18-5)**
β-Naphthol ethyl ether	**(93-18-5)**
2-Naphthol, 1-mercapto-, 2-benzoate (8CI)	**(14814-89-2)**
2-Naphthol, 1-((6-methoxy-m-tolyl)azo)- (8CI)	**(6410-20-4)**
1-Naphthol, 2-methyl- (8CI)	**(7469-77-4)**
1-Naphthol N-methylcarbamate	**(63-25-2)**
2-Naphthol, 1,1'-methylenedi	**(1096-84-0)**
2-Naphthol methyl ether	**(93-04-9)**
β-Naphthol methyl ether	**(93-04-9)**
2-Naphthol, 1-((4-nitrophenyl)azo)	**(6410-10-2)**
1-Naphthol, 2-nitroso	**(132-53-6)**
2-Naphthol, 1-nitroso	**(131-91-9)**
2-Naphthol, 1-(phenylazo)	**(842-07-9)**
2-Naphthol, 1-(2-pyridylazo)- (8CI)	**(85-85-8)**
2-Naphthol, 7-sulfo-	**(92-40-0)**
2-Naphthol-6-sulfonamide	**(6528-46-7)**
1-Naphthol-4-sulfonic acid	**(84-87-7)**
2-Naphthol-6-sulfonic acid	**(93-01-6)**
2-Naphthol-7-sulfonic acid	**(92-40-0)**
α-Naphthol-4-sulfonic acid	**(84-87-7)**
β-Naphthol-6-sulfonic acid	**(93-01-6)**
β-Naphthol-7-sulfonic acid	**(92-40-0)**
β-Naphtholsulfonic acid S	**(93-01-6)**
2-Naphthol-6-sulfonic acid, 1-(p-nitrophenylazo)- (6CI)	**(5859-04-1)**
2-Naphthol, 1,2,3,4-tetrahydro	**(530-91-6)**
1-Naphthol, 1,2,3,4-tetrahydro-4,5,7-trimethyl- (6CI,8CI)	**(32281-79-1)**
2-Naphthol, 1-(o-tolylazo)	**(2646-17-5)**
2-Naphthol, 1-((4-(o-tolylazo)-o-tolyl)azo)	**(85-83-6)**
1-Naphthol, 2,5,8-trimethyl- (8CI)	**(33583-02-7)**
2-Naphthol, 1-(2,4-xylylazo)	**(3118-97-6)**
1-Naphthonitrile	**(86-53-3)**
2-Naphthonitrile	**(613-46-7)**
α-Naphthonitrile	**(86-53-3)**
β-Naphthonitrile	**(613-46-7)**
Naphtho(3',4':3,4)pyrene	**(189-96-8)**
Naphtho(2,1,8-qra)naphthacene (8CI,9CI)	**(196-42-9)**
β-Naphthoquinaldine	**(85-02-9)**
β-Naphthoquinaldine	**(85-06-3)**
α-Naphthoquinoline	**(230-27-3)**
Naphthoquinoline (9CI)	**(78361-94-1)**
1,2-Naphthoquinone	**(524-42-5)**
1,4-Naphthoquinone	**(130-15-4)**
α-Naphthoquinone	**(130-15-4)**
β-Naphthoquinone	**(524-42-5)**
1,4-Naphthoquinone, 3-acetonyl-5,8-dihydroxy-6-methoxy-2-methyl	**(476-45-9)**
1,4-Naphthoquinone, 2-amino-3-chloro	**(2797-51-5)**
1,4-Naphthoquinone, 2,3-dichloro	**(117-80-6)**
1,4-Naphthoquinone, 2-(3,7,11,15,19,23,27-hepta-methyl-2,6,10,14,18,22,26-octacosahept-aenyl)-3-methyl-, (All-E)-	**(2124-57-4)**
1,4-Naphthoquinone, 2-hydroxy	**(83-72-7)**
1,4-Naphthoquinone, 5-hydroxy-	**(481-39-0)**
1,4-Naphthoquinone, 8-hydroxy-	**(481-39-0)**
1,4-Naphthoquinone, 2-hydroxy-3-methyl	**(483-55-6)**
1,4-Naphthoquinone, 2-hydroxy-3-(3-methyl-2-butenyl)	**(84-79-7)**
1,2-Naphthoquinone, 5-isopropyl-3,8-dimethyl- (8CI)	**(5574-34-5)**
1,4-Naphthoquinone, 2-methyl	**(58-27-5)**
1,4-Naphthoquinone, 2-methyl-3-phytyl-	**(84-80-0)**
Naphthoresorcinol	**(132-86-5)**
Naphtho(1,2,3,4-rst)pentaphene (7CI,8CI,9CI)	**(191-20-8)**
Naphthosol Fast Red KB Base	**(95-79-4)**
Naphthosultone	**(83-31-8)**
α-Naphthothiourea	**(86-88-4)**
Naphthothiourea (OSHA) [UN 1651]	**(86-88-4)**
Naphtho(2,1,8-uva)pentacene (6CI,7CI,8CI,9CI)	**(196-45-2)**
2-Naphthoxyacetic acid	**(120-23-0)**
β-Naphthoxyacetic acid	**(120-23-0)**
2-(α-Naphthoxy)-N,N-diaethyl-propionsaeureamid (German)	**(15299-99-7)**
2-(α-Naphthoxy)-N,N-diethylpropionamide	**(15299-99-7)**
3-(α-Naphthoxy)-1,2-propanediol	**(36112-95-5)**
α-(2-Naphthoxy)propionanilide	**(52570-16-8)**
α-(β-Naphthoxy)propionanilide	**(52570-16-8)**
Naphtho(2,1,8-yxa)hexacene (7CI)	**(196-46-3)**
Naphtho(2,1,8-yza)hexacene (6CI,8CI,9CI)	**(196-46-3)**
2-Naphtho-o-phenetidide, 4-((p-carbamoylphenyl)azo)-3-hydroxy-	**(2786-76-7)**
2-Naphtho-o-phenetidide, 3-hydroxy- (8CI)	**(92-74-0)**
1-Naphtho-4-sulfonic acid	**(84-87-7)**
1-Naphththaleneacetic acid, ethyl ester	**(2122-70-5)**
2-Naphtho-o-toluidide, 5'-chloro-3-hydroxy- (8CI)	**(135-63-7)**
2-Naphtho-o-toluidide, 3-hydroxy- (8CI)	**(135-61-5)**
β-Naphthyl Orange	**(633-96-5)**
α-Naphthyl Red	**(131-22-6)**
1-Naphthylacetamide	**(86-86-2)**
α-Naphthylacetamide	**(86-86-2)**
a-Naphthyl acetate	**(830-81-9)**
α-Naphthyl acetate	**(830-81-9)**
α-Naphthylacetic	**(86-87-3)**
1-Naphthylacetic acid	**(86-87-3)**
2-(1-Naphthyl)acetic acid	**(86-87-3)**
2-(2-Naphthyl)acetic acid	**(581-96-4)**
Naphthylacetic acid	**(86-87-3)**
α-Naphthylacetic acid	**(86-87-3)**
β-Naphthylacetic acid	**(581-96-4)**
2-(1-Naphthyl)acetic acid ethyl ester	**(2122-70-5)**
α-(1-Naphthyl)acetonitrile	**(132-75-2)**
α-Naphthyl acetonitrile	**(132-75-2)**
β-Naphthyl alcohol	**(135-19-3)**
1-Naphthylaldehyde	**(66-77-3)**
α-Naphthylaldehyde	**(66-77-3)**
β-Naphthylaldehyde	**(66-99-9)**
1-Naphthylamin (German)	**(134-32-7)**
2-Naphthylamin (German)	**(91-59-8)**
β-Naphthylamin (German)	**(91-59-8)**

1-Naphthylamine **(134-32-7)**
2-Naphthylamine **(91-59-8)**
6-Naphthylamine **(91-59-8)**
β-Naphthylamine (ACGIH,OSHA) [UN 1650] **(91-59-8)**
Naphthylamine Blue **(72-57-1)**
Naphthylamine (8CI) **(25168-10-9)**
α-Naphthylamine (OSHA) [UN 2077] **(134-32-7)**
1-Naphthylamine, N-acetyl- **(86-86-2)**
2-Naphthylamine, N,N-bis(2-chloroethyl) **(494-03-1)**
1-Naphthylamine, N,N-dimethyl **(86-56-6)**
2-Naphthylamine-4,8-disulfonic acid **(131-27-1)**
2-Naphthylamine-6,8-disulfonic acid **(86-65-7)**
β-Naphthylamine-4,8-disulfonic acid **(131-27-1)**
β-Naphthylaminedisulfonic acid **(131-27-1)**
1-Naphthylamine, N-ethyl- (8CI) **(118-44-5)**
2-Naphthylamine, N-ethyl-1-((p-(phenylazo)phenyl)azo) **(6368-72-5)**
1-Naphthylamine hydrochloride **(552-46-5)**
2-Naphthylamine, hydrochloride **(612-52-2)**
α-Naphthylamine hydrochloride **(552-46-5)**
β-Naphthylamine hydrochloride **(612-52-2)**
Naphthylaminemonosulfonic acid S **(82-75-7)**
2-Naphthylamine mustard **(91-59-8)**
Naphthylamine mustard **(494-03-1)**
1-Naphthylamine, N-phenyl **(90-30-2)**
2-Naphthylamine, N-phenyl **(135-88-6)**
2-Naphthylamine, 1-(phenylazo) **(85-84-7)**
1-Naphthylamine-4-sulfonic acid **(84-86-6)**
1-Naphthylamine-6-sulfonic acid **(119-79-9)**
1-Naphthylamine-7-sulfonic acid **(119-28-8)**
1-Naphthylamine-8-sulfonic acid **(82-75-7)**
2-Naphthylamine-1-sulfonic acid **(81-16-3)**
2-Naphthylamine-6-sulfonic acid **(93-00-5)**
5-Naphthylamine-2-sulfonic acid **(119-79-9)**
6-Naphthylamine-2-sulfonic acid **(93-00-5)**
8-Naphthylamine-2-sulfonic acid **(119-28-8)**
α-Naphthylamine-8-sulfonic acid **(82-75-7)**
α-Naphthylamine-p-sulfonic acid **(84-86-6)**
2-Naphthylamine-1-sulfonic acid, ammonium salt **(68540-41-0)**
6-Naphthylamine-2-sulphonic acid **(93-00-5)**
2-Naphthylamine, 1-(o-tolylazo) **(131-79-3)**
4-(2-Naphthylamino)phenol **(93-45-8)**
p-(2-Naphthylamino)phenol **(93-45-8)**
N-(1-Naphthyl)aniline **(90-30-2)**
N-(2-Naphthyl)aniline **(135-88-6)**
4-(1-Naphthylazo)-m-phenylenediamine **(6416-57-5)**
2-Naphthylbis(2-chloroethyl)amine **(494-03-1)**
β-Naphthyl-bis-(β-chloroethyl)amine **(494-03-1)**
1-Naphthyl-1-butane **(1634-09-9)**
γ.-(1-Naphthyl)butyric acid **(781-74-8)**
4-(1-Naphthyl)butyric acid **(781-74-8)**
α-Naphthylcarboxaldehyde **(66-77-3)**
β-Naphthylcarboxaldehyde **(66-99-9)**
α-Naphthylcarboxylic acid **(86-55-5)**
β-Naphthyl-di-(2-chloroethyl)amine **(494-03-1)**
Naphthylene Yellow **(605-69-6)**
α-Naphthyleneacetic acid **(86-87-3)**
1,2-(1,8-Naphthylene)benzene **(206-44-0)**
1,4-Naphthylenediamine **(2243-61-0)**
1,5-Naphthylenediamine **(2243-62-1)**
1,8-Naphthylenediamine **(479-27-6)**
1,5-Naphthylene disulfonic acid **(81-04-9)**
Naphthyleneethylene **(83-32-9)**
α-Naphthylessigsaeure (German) **(86-87-3)**
N-(1-Naphthyl)ethylenediamine dihydrochloride **(1465-25-4)**
o-(2-Naphthyl)glycolic acid **(120-23-0)**
β-Naphthyl hydroxide **(135-19-3)**
1-Naphthylhydroxylamine **(607-30-7)**
N-1-Naphthylhydroxylamine **(607-30-7)**
1-Naphthyl iodide **(90-14-2)**
1-Naphthyl isocyanide (7CI,8CI) **(1984-04-9)**
Naphthylisoproterenol hydrochloride **(51-02-5)**
1-Naphthyl isothiocyanate **(551-06-4)**
α-Naphthyl isothiocyanate **(551-06-4)**
o-2-Naphthyl m,n-dimethylthiocarbanilate **(2398-96-1)**
2-Naphthyl mercaptan **(91-60-1)**
β-Naphthyl mercaptan **(91-60-1)**
1-Naphthyl N-methylcarbamate **(63-25-2)**
1-Naphthyl methylcarbamate **(63-25-2)**
α-Naphthyl N-methylcarbamate **(63-25-2)**
α-Naphthyl methylcarbamate **(63-25-2)**
2-Naphthyl methyl ether **(93-04-9)**
β-Naphthyl methyl ether **(93-04-9)**
2-(Naphthyl-(1')-methyl)imidazolin (German) **(835-31-4)**
2-(1-Naphthylmethyl)-2-imidazoline **(835-31-4)**
2-(α-Naphthylmethyl)-imidazoline **(835-31-4)**
α-Naphthylmethyl imidazoline **(835-31-4)**
2-(1-Naphthylmethyl)-2-imidazoline hydrochloride **(550-99-2)**
2-(1-Naphthylmethyl)imidazoline hydrochloride **(550-99-2)**
1-Naphthyl-N-methyl-karbamat (German) **(63-25-2)**
1-Naphthyl methyl ketone **(941-98-0)**
2-Naphthyl methyl ketone **(93-08-3)**
α-Naphthyl methyl ketone **(941-98-0)**
β-Naphthyl methyl ketone **(93-08-3)**
2-Naphthyl N-methyl-N-(3-tolyl)thionocarbamate **(2398-96-1)**
1-Naphthylnitrile **(86-53-3)**
α-Naphthylnitrile **(86-53-3)**
(2-Naphthyloxy)acetic acid **(120-23-0)**
2-(2-Naphthyloxy)propionanilide **(52570-16-8)**
2-Naphthylphenylamine **(135-88-6)**
β-Naphthylphenylamine **(135-88-6)**
2-Naphthyl-p-phenylenediamine **(93-46-9)**
N-1-Naphthylphthalamate **(132-66-1)**
N-1-Naphthylphthalamic acid **(132-66-1)**
α-Naphthylphthalamic acid **(132-66-1)**
N-1-Naphthylphthalamic acid sodium salt **(132-67-2)**
α-Naphthylphthalamic acid sodium salt **(132-67-2)**
N-1-Naphthyl-phthalamidsaeure (German) **(132-66-1)**
α-Naphthyl propionate **(3121-71-9)**
β-Naphthylsulfonic acid **(120-18-3)**
2-Naphthylsulfonyl chloride **(93-11-8)**
α-Naphthylthiocarbamide **(86-88-4)**
1-Naphthyl-thioharnstoff (German) **(86-88-4)**
2-Naphthyl thiol **(91-60-1)**
1-(1-Naphthyl)-2-thiourea **(86-88-4)**
1-Naphthyl thiourea **(86-88-4)**
N-(1-Naphthyl)-2-thiourea **(86-88-4)**
α-Naphthylthiourea (DOT,OSHA) **(86-88-4)**
1-Naphthyl-thiouree (French) **(86-88-4)**
1,7-Naphthyridine (6CI,7CI,8CI,9CI) **(253-69-0)**
1,8-Naphthyridine-3-carboxylic acid, 1-ethyl-
1,4-dihydro-7-methyl-4-oxo **(389-08-2)**
Naphtocard Fast Red C **(1658-56-6)**
Naphtocard Orange II **(633-96-5)**
Naphtocard Red 2G **(3734-67-6)**
Naphtocard Yellow O **(1934-21-0)**
Naphtoelan A **(92-77-3)**
Naphtoelan BO **(132-68-3)**
Naphtoelan BS **(135-65-9)**
Naphtoelan Blue BB Base **(120-00-3)**
Naphtoelan D **(135-61-5)**
Naphtoelan Fast Orange GC Base **(141-85-5)**
Naphtoelan Fast Orange GC Salt **(141-85-5)**
Naphtoelan Fast Red 3GL Base **(89-63-4)**
Naphtoelan Fast Red GL Base **(89-62-3)**

Naphtoelan Fast Red 3GL Salt (89-63-4)
Naphtoelan Fast Scarlet G Base (99-55-8)
Naphtoelan Fast Scarlet GG Base (95-82-9)
Naphtoelan Fast Scarlet G Salt (99-55-8)
Naphtoelan G (91-96-3)
Naphtoelan OL (135-62-6)
Naphtoelan Orange R Base (99-09-2)
Naphtoelan Phenyl (92-74-0)
Naphtoelan Red GG Base (100-01-6)
Naphtol AS-BO (132-68-3)
Naphtol AS-BOLL (132-68-3)
Naphtol AS-BR (91-92-9)
Naphtol AS-BS (135-65-9)
Naphtol AS-BS Supra (135-65-9)
Naphtol AS-CA (137-52-0)
Naphtol AS-CALL (137-52-0)
Naphtol AS-D (135-61-5)
Naphtol AS-D Supra (135-61-5)
Naphtol AS-G (91-96-3)
Naphtol AS-G Supra (91-96-3)
Naphtol AS-KB (135-63-7)
Naphtol AS-KG (106-49-0)
Naphtol AS-KGLL (106-49-0)
Naphtol AS-OL (135-62-6)
Naphtol AS-RC (137-52-0)
Naphtol AS (92-77-3)
Naphtol AS-phenyl (92-74-0)
2-Naphtol (French) (135-19-3)
β-Naphtol (German) (135-19-3)
Naphtolate AS Soln (92-77-3)
2-Naphtol-6-sulfosaure (German) (93-01-6)
Naphtox (86-88-4)
Naphyl-1-essigsaeure (German) (86-87-3)
Napolone (9004-67-5)
Napoton (58-25-3)
Napqi (50700-49-7)
Naprinol (103-90-2)
Naproanilide (52570-16-8)
Napropamide (15299-99-7)
Napropion (137-40-6)
Naprosine (22204-53-1)
Naprosyn (22204-53-1)
Naproxen (22204-53-1)
Naprux (22204-53-1)
Napst (9003-53-6)
Naptalam (132-66-1)
Naptalame (132-66-1)
Naptalam sodium (132-67-2)
Naptanilide BS Supra (135-65-9)
Naptazane (554-57-4)
Napthalene, 1,2,3,4-tetrachloro- (20020-02-4)
Naptholrot S (German) (915-67-3)
Napton (58-25-3)
Naptro (132-66-1)
Napvis 30 (9003-27-4)
Naqua (133-67-5)
Naquival (50-55-5)
Naramycin (66-81-9)
Naramycin A (66-81-9)
Narcein (131-28-2)
Narceine (131-28-2)
Narceol (140-39-6)
Narcogen (79-01-6)
Narcolo (357-56-2)
Narcompren (128-62-1)
Narcosan (56-29-1)
Narcosine (128-62-1)
Narcotane (151-67-7)
Narcotann Ne-Spofa (Russian) (151-67-7)
Narcotile (75-00-3)
1-α-Narcotine (128-62-1)
Narcotine (128-62-1)
Narcotine (6035-40-1)
Narcotussin (128-62-1)
Narcozep (1622-62-4)
Narcylen (74-86-2)
Nardelzine (156-51-4)
Nardil (156-51-4)
Nardil (51-71-8)
Narea (18530-56-8)
Narigix (389-08-2)
Naringeninic acid (501-98-4)
Naringenin-7-β-neohesperidoside (10236-47-2)
Naringin (10236-47-2)
Naringoside (10236-47-2)
Naritheracin (606-58-6)
Narkogen (79-01-6)
Narkosoid (79-01-6)
Narkotil (75-09-2)
Narsis (2898-12-6)
Nasdol (57-85-2)
Nasivine (1491-59-4)
Nasmil (15826-37-6)
Nasol (299-42-3)
Nastenon (434-07-1)
Natasol Blue B Salt (119-90-4)
Natasol Fast Orange GC Salt (141-85-5)
Natasol Fast Orange GR Salt (88-74-4)
Natasol Fast Red TR Salt (3165-93-3)
Nateretin (73-48-3)
Nathulane (366-70-1)
National 120-1207 (9003-20-7)
National Starch 1014 (9003-20-7)
Natrascorb (134-03-2)
Natrascorb (50-81-7)
Natrascorb injectable (50-81-7)
Natreen (139-05-9)
Natreen (81-07-2)
Natri-C (134-03-2)
Natrin (3570-61-4)
Natrin Herbicide (3570-61-4)
Natrionex (59-66-5)
Natriphene (132-27-4)
Natrium (7440-23-5)
Natriumacetat (German) (127-09-3)
Natriumaluminiumfluorid (German) (15096-52-3)
Natriumazid (26628-22-8)
Natriumazid (German) (26628-22-8)
Natriumbichromaat (Dutch) (10588-01-9)
Natriumchloraat (Dutch) (7775-09-9)
Natriumchlorat (German) (7775-09-9)
Natriumchlorid (German) (7647-14-5)
Natrium citricum (German) (144-33-2)
Natrium-2,4-dichlorphenoxyathylsulfat (German) (136-78-7)
Natriumdichromaat (Dutch) (10588-01-9)
Natriumdichromat (German) (10588-01-9)
Natrium-2,3:4,6-di-O-isopropyliden-2-keto-l-gulonat (German) (52508-35-7)
Natrium-N-methyl-dithiocarbamaat (Dutch) (137-42-8)
Natrium-N-methyl-dithiocarbamat (German) (137-42-8)
Natriumfluoracetaat (Dutch) (62-74-8)
Natriumfluoracetat (German) (62-74-8)
Natrium fluoride (7681-49-4)
Natriumglutaminat (German) (142-47-2)
Natriumhexafluoroaluminate (German) (15096-52-3)

Nefrecil	(136-40-3)	Nemazine	(92-84-2)
Nefrix	(58-93-5)	Nemazon	(96-12-8)
Nefrotest	(61-78-9)	Nembu-serpin	(50-55-5)
Neftin	(67-45-8)	Nembutal	(57-33-0)
Nefurthiazole	(3570-75-0)	Nembutal	(76-74-4)
Nefusan	(533-74-4)	Nembutal sodium	(57-33-0)
Negalip	(637-07-0)	Nemerol	(57-42-1)
Neganox DLTP	(123-28-4)	Nemispor	(8018-01-7)
Negashunt	(56-72-4)	Nendrin	(72-20-8)
Neggram	(389-08-2)	Nenesin	(77-65-6)
Negram	(389-08-2)	Neo	(50-35-1)
Neguvon	(52-68-6)	Neo-Antitensol	(50-55-5)
Neguvon A	(52-68-6)	Neo-Antitersol	(50-55-5)
Neirine	(463-88-7)	Neo-Asozin	(6585-53-1)
Nekal	(25417-20-3)	Neo-Atomid	(637-07-0)
Nekal WT-27	(577-11-7)	Neo-Atromid	(637-07-0)
Nekanil LN	(9036-19-5)	Neo-Avagal	(155-41-9)
Neklacid Azorubine W	(3567-69-9)	Neo-Calma	(68-88-2)
Neklacid Fast Light Orange GG	(1936-15-8)	Neo-Cebicure	(89-65-6)
Neklacid Fast Orange 2G	(1936-15-8)	Neo-Codema	(58-93-5)
Neklacid Fast Red A	(1658-56-6)	Neo-Corovas	(78-11-5)
Neklacid Fast Red E	(2302-96-7)	Neo-Cultol	(8012-95-1)
Neklacid Orange 1	(523-44-4)	Neo-Cytamen	(13422-51-0)
Neklacid Orange II	(633-96-5)	Neo-Dema	(58-94-6)
Neklacid Red A	(915-67-3)	Neo-Devomit	(82-92-8)
Neklacid Red E	(2302-96-7)	Neo-Ergotin	(379-79-3)
Neklacid Red 3R	(2611-82-7)	Neo-Farmadol	(129-20-4)
Neklacid Red 4R	(2611-82-7)	Neo-Fat 10	(334-48-5)
Neklacid Red RR	(3761-53-3)	Neo-Fat 12	(143-07-7)
Neklacid Red 6r	(5850-44-2)	Neo-Fat 12-43	(143-07-7)
Neklacid Rubine W	(3567-69-9)	Neo-Fat 14	(544-63-8)
Neklacid Yellow G	(547-57-9)	Neo-Fat 18	(57-11-4)
Neklacid Yellow T	(1934-21-0)	Neo-Fat 18-53	(57-11-4)
Neklamin Black BH	(2429-73-4)	Neo-Fat 18-54	(57-11-4)
Nektrohan	(315-30-0)	Neo-Fat 18-55	(57-11-4)
Nelbon	(146-22-5)	Neo-Fat 18-59	(57-11-4)
Nelipramin	(50-49-7)	Neo-Fat 18-61	(57-11-4)
Nellite	(1754-58-1)	Neo-Fat 90-04	(112-80-1)
Nem-A-Tak	(21548-32-3)	Neo-Fat 92-04	(112-80-1)
Nema	(127-18-4)	Neo-Fat 18-S	(57-11-4)
Nemabrom	(96-12-8)	Neo-Ferrum	(8047-67-4)
Nemacide	(97-17-6)	Neo-Fulcin	(126-07-8)
Nemacide VC-13	(97-17-6)	Neo Germ-I-Tol	(8001-54-5)
Nemacur	(22224-92-6)	Neo-Hibernex	(58-40-2)
Nemacur P	(22224-92-6)	Neos-Hidantoina	(57-41-0)
Nemafax	(23564-06-9)	Neo-Hombreol	(57-85-2)
Nemafene	(8003-19-8)	Neo-Hombreol-M	(58-18-4)
Nemafos	(297-97-2)	Neo-Istafene	(569-65-3)
Nemafume	(96-12-8)	Neo-Macrabin	(13422-51-0)
Nemagon	(96-12-8)	Neo-Mantle Creme	(1405-10-3)
Nemagon 20	(96-12-8)	Neo-Naclex	(73-48-3)
Nemagon 90	(96-12-8)	Neo-Navigan	(523-87-5)
Nemagon 206	(96-12-8)	Neo-Oestranol II	(130-80-3)
Nemagon 20G	(96-12-8)	Neo-Oestronol II	(130-80-3)
Nemagon Soil Fumigant	(96-12-8)	Neo-Ormonal	(76-43-7)
Nemagone	(96-12-8)	Neo-Pynamin	(7696-12-0)
Nemalite	(1309-42-8)	Neo-Rojamin	(13422-51-0)
Nemamort	(108-60-1)	Neo-Rontyl	(73-48-3)
Nemanax	(96-12-8)	Neo-Salicyl	(54-21-7)
Nemapan	(148-79-8)	Neo-Scabicidol	(58-89-9)
Nemapaz	(96-12-8)	Neo-Serp	(50-55-5)
Nemaphos	(297-97-2)	Neo So Sin Gin	(6585-53-1)
Nemaset	(96-12-8)	Neo So Sin Gin-S	(6585-53-1)
Nematocide	(297-97-2)	Neo-Spectra	(1333-86-4)
Nematocide	(96-12-8)	Neo Spectra II	(1333-86-4)
Nematox	(96-12-8)	Neo-Suprimal	(569-65-3)
Nemazene	(92-84-2)	Neo-Suprimel	(569-65-3)

Neo-Testis **(58-22-0)**
Neo-Tizide **(54-85-3)**
Neo-Tric **(443-48-1)**
Neo-Urofort **(500-42-5)**
Neo-Valdrin **(50-81-7)**
Neo-Vasophylline **(479-18-5)**
Neo-Zoline **(50-33-9)**
Neoabietic acid **(471-77-2)**
Neoamyl alcohol **(75-84-3)**
Neoantergan **(91-84-9)**
Neoantergan maleate **(59-33-6)**
Neoasycodile **(144-21-8)**
Neoban **(101-27-9)**
Neobar **(7727-43-7)**
Neobiotic **(1405-10-3)**
Neobor **(1303-96-4)**
Neobridal **(91-84-9)**
Neocaine **(51-05-8)**
Neocaine **(59-46-1)**
Neocardol **(54-95-5)**
A'-Neogammacer-17(21)-ene (8CI,9CI) **(546-99-6)**
A'-Neogammacerane **(471-62-5)**
A'-Neogammacerane, (21β)- (9CI) **(1176-44-9)**
A'-Neogammacerane, (17α)- (9CI) **(13849-96-2)**
Neochin **(54-05-7)**
Neocid **(50-29-3)**
Neocidol **(333-41-5)**
Neococcyl **(63-74-1)**
Neocodin **(509-67-1)**
Neocompensan **(9003-39-8)**
Neocryl A-1038 **(9003-01-4)**
Neocycline **(60-54-8)**
Neocycloheximide **(66-81-9)**
Neodecaneperoxoic acid, 1,1-dimethylpropyl ester (9CI) **(68299-16-1)**
Neodecaneperoxoic acid, 1-methyl-1-phenylethyl ester (9CI) **(26748-47-0)**
Neodecanoic-acid **(26896-20-8)**
Neodecanoic acid, calcium salt (9CI) **(27253-33-4)**
Neodecanoic acid, cobalt salt (9CI) **(27253-31-2)**
Neodecanoic acid, copper salt **(32276-75-8)**
Neodecanoic acid, 2,3-epoxypropyl ester **(26761-45-5)**
Neodecanoic acid, ethenyl ester (9CI) **(51000-52-3)**
Neodecanoic acid, lead salt (9CI) **(27253-28-7)**
Neodecanoic acid, manganese salt (9CI) **(27253-32-3)**
Neodecanoic acid, oxiranylmethyl ester (9CI) **(26761-45-5)**
Neodecanoic acid, vinyl ester **(51000-52-3)**
Neodecanoyl chloride (9CI) **(40292-82-8)**
A'-Neo-28,30-dinorgammacerane, (17α)- (9CI) **(65636-26-2)**
Neodol 91 **(66455-17-2)**
Neodol-12 **(68131-39-5)**
Neodorm **(76-74-4)**
Neodorm (new) **(76-74-4)**
Neodrine **(300-42-5)**
Neodurabolin **(965-90-2)**
Neodymium oxide (9CI) **(1313-97-9)**
Neoeserine bromide **(114-80-7)**
Neo-estrone **(57-63-6)**
Neofemergen **(60-79-7)**
Neofen **(129-20-4)**
Neoflumen **(58-93-5)**
Neofollin **(979-32-8)**
A'-Neo-17α-gammacerane (8CI) **(13849-96-2)**
A'-Neo-21αH-gammacerane (8CI) **(1176-44-9)**
Neoglaucit **(55-91-4)**
Neohesperidin dihydrochalcone **(20702-77-6)**
Neohesporidin dihydrochalcone **(20702-77-6)**
Neohetramine **(91-85-0)**
Neohetramine hydrochloride **(63-56-9)**
Neohexane [UN 1208] **(75-83-2)**
Neohexene **(558-37-2)**
Neohydrazid **(140-87-4)**
Neohydrin **(62-37-3)**
Neol **(126-30-7)**
Neolan Red GRE **(6408-31-7)**
Neolexina **(15686-71-2)**
Neo-Betalin 12 **(13422-51-0)**
Neolin **(1538-09-6)**
Neoloid **(8001-79-4)**
Neolutin **(630-56-8)**
Neomagnol **(127-52-6)**
Neomcin **(1404-04-2)**
Neomercazole **(22232-54-8)**
Neometantyl **(50-34-0)**
Neomix **(1405-10-3)**
Neomycin **(1404-04-2)**
Neomycin E **(7542-37-2)**
Neomycine sulfate **(1405-10-3)**
Neomycins sulfate **(1405-10-3)**
Neomycin sulfate **(1405-10-3)**
Neomycin, sulfate (Salt) (9CI) **(1405-10-3)**
Neon, Compressed [UN 1065] **(7440-01-9)**
Neon (DOT) **(7440-01-9)**
Neon, Refrigerated liquid [UN 1913] **(7440-01-9)**
Neonal **(77-28-1)**
Neonicotine **(494-52-0)**
Neonikotin **(494-52-0)**
A'-Neo-30-norgammacerane (9CI) **(36728-72-0)**
A'-Neo-30-norgammacerane, (21β)- (9CI) **(3258-87-5)**
A'-Neo-30-norgammacerane, 22-butyl-, (17α,22R)- (9CI) **(67069-26-5)**
A'-Neo-30-norgammacerane, 22-butyl-, (17α,22S)- (9CI) **(67069-16-3)**
A'-Neo-30-norgammacerane, 22-ethyl- (9CI) **(53584-62-6)**
A'-Neo-30-norgammacerane, 22-ethyl-, (17α,22R)- (9CI) **(60305-22-8)**
A'-Neo-30-norgammacerane, 22-ethyl-, (17α,22S)- (9CI) **(60305-23-9)**
A'-Neo-30-norgammacerane, (17α)- (9CI) **(53584-60-4)**
A'-Neo-30-norgammacerane, 22-propyl-, (17α,22R)- (9CI) **(67069-25-4)**
A'-Neo-30-norgammacerane, 22-propyl-, (17α,22S)- (9CI) **(67069-15-2)**
Neonyl Scarlet R **(3567-65-5)**
Neo-oestranol 1 **(56-53-1)**
Neooxedrine **(61-76-7)**
Neopantanoyl chloride **(3282-30-2)**
Neopellis **(97-18-7)**
Neopentane **(463-82-1)**
Neopentanetetrayl nitrate **(78-11-5)**
1,1',1'',1'''-(Neopentanetetrayltetraoxy)tetrakis(2,2,2-trichloroethanol) **(78-12-6)**
Neopentanoic acid **(75-98-9)**
Neopentanol **(75-84-3)**
Neopentyl alcohol **(75-84-3)**
Neopentyl bromide **(630-17-1)**
Neopentyl chloride **(753-89-9)**
Neopentyldiamine **(7328-91-8)**
Neopentylene glycol **(126-30-7)**
Neopentyl glycol **(126-30-7)**
Neopentyl glycol diacrylate **(2223-82-7)**
Neopentyl glycol dibenzoate **(4196-89-8)**
Neopentyl glycol diglycidyl ether **(17557-23-2)**
Neopentyl glycol dipelargonate **(15834-05-6)**
Neopentyl glycol monohydroxypivalate **(1115-20-4)**
Neopepulsan **(50-34-0)**
Neopharmedrine **(300-42-5)**
Neophedan **(63-98-9)**
Neophenal **(63-98-9)**
Neophryn **(61-76-7)**
Neophyiline **(317-34-0)**
Neophyl chloride **(515-40-2)**

Neophyllin (479-18-5)
Neophyllin M (479-18-5)
Neophylline (479-18-5)
Neophytadiene (504-96-1)
Neoplatin (15663-27-1)
Neopolen (9002-88-4)
Neopolen 30N (9002-88-4)
Neopolymer (64742-16-1)
Neopolymer 120 (64742-16-1)
Neopolymer 140 (64742-16-1)
Neopolymer 150 (64742-16-1)
Neopolymer 160 (64742-16-1)
Neopolymer 180 (64742-16-1)
Neopolymer L (64742-16-1)
Neopolymer L 120 (64742-16-1)
Neopolymer NP 150 (64742-16-1)
Neopolymer 170S (64742-16-1)
Neopon Lam (2235-54-3)
Neoprene (126-99-8)
Neoprene (9010-98-4)
Neoproc (6130-64-9)
Neopynamin forte (7696-12-0)
Neoram (1332-65-6)
Neoram Blu (1332-40-7)
Neoram Blu (1332-65-6)
Neorestamin (113-92-8)
Neoron (18181-80-1)
Neosabenyl (120-32-1)
Neosar (50-18-0)
Neosedyn (50-35-1)
Neosept V (70-30-4)
Neoserfin (50-55-5)
Neoserine bromide (114-80-7)
Neosetile Blue EB (2475-45-8)
Neosetile Pink BN (116-85-8)
Neosidantoina (57-41-0)
Neosiluol (7783-96-2)
Neosilvol (7783-96-2)
Neo-sinefrina (61-76-7)
Neoslowten (50-55-5)
Neosorb (50-70-4)
Neosorb 70/70 (50-70-4)
Neostenovasan (479-18-5)
Neostigmine (59-99-4)
Neostigmine bromide (114-80-7)
Neostigmine methyl bromide (114-80-7)
Neostrepsan (72-14-0)
Neosydyn (50-35-1)
Neosympatol (61-76-7)
Neosynephrine (59-42-7)
Neosynephrine (61-76-7)
Neosynephrine hydrochloride (61-76-7)
Neosynesine (61-76-7)
Neoteben (54-85-3)
Neotex (1333-86-4)
Neothyl (557-17-5)
Neothylline (479-18-5)
Neotilina (479-18-5)
Neotizol (1405-10-3)
Neotopsin (23564-05-8)
Neo-tran (57-53-4)
Neotran (555-89-5)
Neotridecanoic acid (8CI,9CI) (26403-14-5)
A'-Neo-22,29,30-trinorgammacerane, (17α)- (9CI) (53584-59-1)
Neovet (1405-10-3)
Neovitamin A acid (4759-48-2)
Neoxazol (127-69-5)
Neoxin (54-85-3)
Neozepam (146-22-5)
Neozex 45150 (9002-88-4)
Neozex 4010B (9002-88-4)
Neozon A (90-30-2)
Neozon D (135-88-6)
Neozone (135-88-6)
Neozone A (90-30-2)
Neozone D (135-88-6)
Nephentine (57-53-4)
Nephis (106-93-4)
Nephocarp (786-19-6)
Nephramid (59-66-5)
Nephramide (59-66-5)
Nephridine (51-43-4)
Nephrite (12172-67-7)
Neptazane (554-57-4)
Neptazaneat (554-57-4)
Neptune Blue BRA (2650-18-2)
Neptune Blue BRA Concentration (2650-18-2)
Neracid (133-06-2)
Neran Brilliant Green G (4680-78-8)
Neravan (143-81-7)
Nereb (12427-38-2)
Nereistoxin dicarbamate hydrochloride (15263-52-2)
Neriodin (15307-79-6)
Neriol (465-16-7)
Neriolin (465-16-7)
Neriostene (465-16-7)
Nerkol (62-73-7)
Nerobol (72-63-9)
Nerobolettes (72-63-9)
Nerol (106-25-2)
Nerol 2B (6406-45-7)
Nerol acetate (141-12-8)
Neroli Oil, Artifical (134-20-3)
(+)-Nerolidol (142-50-7)
Nerolidol (7212-44-4)
Nerolidol, cis-(+)- (142-50-7)
d-Nerolidol (142-50-7)
Nerolidol (VAN) (142-50-7)
Nerolin (93-18-5)
Nerolin II (93-18-5)
Nerolin New (93-18-5)
Neroline (93-18-5)
Nerosedyn (50-35-1)
Nervanaid B (64-02-8)
Nervanaid B acid (60-00-4)
Nervanaid B liquid (64-02-8)
Nervanid B (64-02-8)
Nervocidine (468-76-8)
Nervonus (57-53-4)
Neryl acetate (141-12-8)
Nesdonal (76-75-5)
Nesol (138-86-3)
Nesontil (604-75-1)
Nespor (12427-38-2)
Nessler Reagent (7783-33-7)
Neston (59-51-8)
Net (68-22-4)
Netagrone (94-75-7)
Netagrone 600 (94-75-7)
Netal (2104-96-3)
Netazol (94-74-6)
Nethalide hydrochloride (51-02-5)
Nethaqualone (72-44-6)
Netilmicin (56391-56-1)

Netocyd	(514-73-8)
Netsusarin	(58-15-1)
Neuchlonic	(146-22-5)
Neucoccin (German)	(2611-82-7)
Neufatin	(50-35-1)
Neufil	(479-18-5)
Neumandin	(54-85-3)
Neuperm GFN	(1854-26-8)
Neuperm ON	(136-84-5)
Neuraxin	(532-03-6)
Neurazine	(69-09-0)
Neurazol	(54-95-5)
Neuridine	(71-44-3)
Neurin	(463-88-7)
Neurine	(463-88-7)
Neurobarb	(50-06-6)
Neurocaine	(50-36-2)
Neurodyn	(50-35-1)
Neurofort	(500-42-5)
Neuronika	(50-78-2)
Neurosedin	(50-35-1)
Neurosedym	(50-35-1)
Neurotone	(93-14-1)
Neurotrast	(99-79-6)
Neurozina	(68-88-2)
Neut	(144-55-8)
Neuthion	(70-18-8)
Neutrafil	(479-18-5)
Neutrafillina	(479-18-5)
Neutral Acriflavine	(8048-52-0)
Neutral Acriflavine	(86-40-8)
Neutral Ammonium Chromate	(7788-98-9)
Neutral Ammonium Fluoride	(12125-01-8)
Neutral Berberine Sulfate	(316-41-6)
"Neutral" Magnesium Phosphate	(7757-87-1)
Neutral Potassium Chromate	(7789-00-6)
Neutral Proflavine Sulphate	(553-30-0)
Neutral Red PG	(3567-65-5)
Neutral Red R	(1658-56-6)
Neutral Soap Stock	(68952-95-4)
Neutral Soapstock	(68952-95-4)
Neutral Sodium Chromate	(7775-11-3)
Neutral Verdigris	(142-71-2)
Neutral Zinc Phosphate	(7779-90-0)
Neutralizing agents, Petroleum, Spent sodium hydroxide	(64742-40-1)
Neutraphyllin	(479-18-5)
Neutraphylline	(479-18-5)
Neutrazyme	(151-21-3)
Neutroflavin	(8048-52-0)
Neutroflavine	(8048-52-0)
Neutroflavine	(86-40-8)
Neutronyx 600	(9016-45-9)
Neutronyx 605	(9002-93-1)
Neutronyx 622	(9036-19-5)
Neutronyx 675	(9036-19-5)
Neutrosel Navy BN	(119-90-4)
Neutrosel Red TRVA	(3165-93-3)
Neutroxantina	(479-18-5)
Neuwied Green	(12002-03-8)
Nevaflor	(63428-84-2)
Nevanaid-B Powder	(150-38-9)
Nevax	(577-11-7)
Nevifos	(83733-82-8)
Nevifos 50	(83733-82-8)
Nevigramon	(389-08-2)
Neviki 79168	(83733-82-8)
Nevile and Winther's acid	(84-87-7)
Neville-winther acid	(84-87-7)
Nevin	(54-85-3)
Nevrodyn	(50-35-1)
New Coccin	(2611-82-7)
New Coccine	(2611-82-7)
New Coccine Extra Conc. A Export	(2611-82-7)
New Coccine Extra Pure A	(2611-82-7)
New Fuchsin	(3248-91-7)
New Green	(12002-03-8)
New Milstem	(23947-60-6)
New-Oestranol II	(130-80-3)
New Pink Bluish Geigy	(16423-68-0)
New Ponceau 4R	(3761-53-3)
New Red WO	(1658-56-6)
New Victoria Green Extra I	(569-64-2)
New Victoria Green Extra II	(569-64-2)
New Victoria Green Extra O	(569-64-2)
New York Red	(1248-18-6)
Newcol 1203	(9002-92-0)
Newcol 60	(1338-41-6)
Newcol 808	(9036-19-5)
Newcol 865	(9036-19-5)
Newcol 560SF	(9051-57-4)
Newcol 560SN	(9014-90-8)
Newphrine	(61-76-7)
Newpol 75H90000	(9038-95-3)
Newpol 50HB5100	(9038-95-3)
Newpol LB3000	(9003-13-8)
Nexagan	(2104-96-3)
Nexagan	(4824-78-6)
Nexen FB	(58-89-9)
Nexit-Stark	(58-89-9)
Nexion	(2104-96-3)
Nexion 40	(2104-96-3)
Nexion LC40	(2104-96-3)
Nexion 5g	(2104-96-3)
Nexit	(58-89-9)
Nexol-E	(58-89-9)
Nexoval	(101-21-3)
Ni 270	(7440-02-0)
Ni 0901-S	(7440-02-0)
Ni 4303T	(7440-02-0)
Nia 1240	(563-12-2)
Nia 5462	(115-29-7)
Nia Proof 08	(126-92-1)
Niacevit	(98-92-0)
Niacide	(14484-64-1)
Niacin	(59-67-6)
Niacinamide	(98-92-0)
Niadrin	(54-85-3)
Niagara 1240	(563-12-2)
Niagara 4512	(2307-68-8)
Niagara 4556	(2164-09-2)
Niagara 4562	(2533-89-3)
Niagara 5006	(1194-65-6)
Niagara 5462	(115-29-7)
Niagara 5767	(2778-04-3)
Niagara 5943	(1031-47-6)
Niagara 5961	(2425-66-3)
Niagara 5996	(1194-65-6)
Niagara 9044	(485-31-4)
Niagara 9241	(2310-17-0)
Niagara 10242	(1563-66-2)
Niagara Blue	(72-57-1)
Niagara Blue 2B	(2602-46-2)
Niagara Blue 3B	(72-57-1)
Niagara Blue 4B	(2429-74-5)

Name	CAS
Niagara Chrome Blue Black B	**(7082-31-7)**
Niagara NIA-9260	**(7696-12-0)**
Niagara NIA-10242	**(1563-66-2)**
Niagara NIA-24110	**(22431-62-5)**
Niagara P.A. Dust	**(54-11-5)**
Niagara Sky Blue	**(2429-74-5)**
Niagara Sky Blue 6B	**(2610-05-1)**
Niagara-Stik	**(86-87-3)**
Niagaramite	**(140-57-8)**
Niagaratran	**(80-33-1)**
Niagaril	**(73-48-3)**
Niagathal	**(117-08-8)**
Niagrathal	**(129-67-9)**
Niagrathal	**(145-73-3)**
Nialamida (Spanish)	**(51-12-7)**
Nialamide	**(51-12-7)**
Nialamidum (Latin)	**(51-12-7)**
Nialate	**(563-12-2)**
Nialk	**(79-01-6)**
Niamid	**(51-12-7)**
Niamidal	**(51-12-7)**
Niamide	**(51-12-7)**
Niamide	**(98-92-0)**
Niamine	**(59-26-7)**
Niaproof 4	**(1191-50-0)**
Niaquitil	**(51-12-7)**
Niax Catalyst Al	**(3033-62-3)**
Niax ESN	**(62765-93-9)**
Niax Flame Retardant 3 CF	**(115-96-8)**
Niax Isocyanate TDI	**(26471-62-5)**
Niax LG 240	**(25791-96-2)**
Niax LG 56	**(25791-96-2)**
Niax Polyol L-56	**(25791-96-2)**
Niax Polyol LG-168	**(25791-96-2)**
Niax TDI	**(584-84-9)**
Niax TDI	**(91-08-7)**
Niax TDI-P	**(584-84-9)**
Niax TDI-P	**(91-08-7)**
Niazol	**(550-99-2)**
Nibren Wax	**(1321-65-9)**
Nibrol	**(50-35-1)**
Nicacid	**(59-67-6)**
Nicamide	**(59-26-7)**
Nicamide	**(98-92-0)**
Nicamin	**(59-67-6)**
Nicamina	**(98-92-0)**
Nicamindon	**(98-92-0)**
Nicangin	**(59-67-6)**
Nicasir	**(98-92-0)**
Nicazide	**(54-85-3)**
Nicel	**(9004-67-5)**
Nicelate	**(389-08-2)**
Nicetal	**(54-85-3)**
Nicetamide	**(59-26-7)**
Nicethamide	**(59-26-7)**
Nichel (Italian)	**(7440-02-0)**
Nichel tetracarbonile (Italian)	**(13463-39-3)**
Nicizina	**(54-85-3)**
Nickel 270	**(7440-02-0)**
Nickel (ACGIH)	**(7440-02-0)**
Nickel Antimony Titanium Yellow Rutile	**(8007-18-9)**
Nickel Black	**(12125-56-3)**
Nickel, Compd with pi-cyclopentadienyl (1:2)	**(1271-28-9)**
Nickel (Dust)	**(7440-02-0)**
Nickel, Metal (OSHA)	**(7440-02-0)**
Nickel (2+) NTA	**(34831-03-3)**
Nickel Oxide Sinter 75	**(1313-99-1)**
Nickel Particles	**(7440-02-0)**
Nickel Rutile Yellow	**(8007-18-9)**
Nickel Sponge	**(7440-02-0)**
Nickel(II) acetate (1:2)	**(373-02-4)**
Nickel(II) acetate tetrahydrate	**(6018-89-9)**
Nickel-acetate-tetrahydrate	**(6018-89-9)**
Nickel alloy, Ni,Be	**(37227-61-5)**
Nickel alloy, base, Ni,Cu (9CI)	**(11102-90-2)**
Nickel alloy, base, Ni,Fe,Mo (9CI)	**(37304-37-3)**
Nickel alloy, nonbase, Al 60-66,Si 25-30,Ni 5-7,Al2O3 3-4 (SAS 1) (9CI)	**(12743-20-3)**
Nickel ammonium sulfate	**(15699-18-0)**
Nickelate(1-), (N,N-bis(carboxymethyl)glycinato(3-)-N,O,O',O'')-, hydrogen, (T-4)- (9CI)	**(34831-03-3)**
Nickel base, copper	**(11102-90-2)**
Nickel-beryllium Alloy	**(37227-61-5)**
Nickel, bis(2,3-bis(hydroxyimino)-N-phenylbutanamidato-N2,N3)- (9CI)	**(29204-84-0)**
Nickel, bis(1,5-cyclooctadiene)- (8CI)	**(1295-35-8)**
Nickel biscyclopentadiene	**(1271-28-9)**
Nickel, bis(dibutyldithiocarbamato)	**(13927-77-0)**
Nickel, bis(dimethyldithiocarbamato)	**(15521-65-0)**
Nickel, bis((1,2,5,6-eta)-1,5-cyclooctadiene)- (9CI)	**(1295-35-8)**
Nickel bis(p-octylphenol)sulfide	**(27574-34-1)**
Nickel bromide (9CI)	**(13462-88-9)**
Nickel, (1-butanamine)((2,2'-thiobis(4-(1,1,3,3-tetramethyl-butyl)phenolato))(2-)-O,O',S)- (9CI)	**(14516-71-3)**
Nickel(II) carbonate (1:1)	**(3333-67-3)**
Nickel carbonate hydroxide	**(12607-70-4)**
Nickel carbonate hydroxide	**(39430-27-8)**
Nickel, (carbonato(2-))tetrahydroxytri- (9CI)	**(12607-70-4)**
Nickel, (carbonato(2-))tetrahydroxytri-, tetrahydrate	**(39430-27-8)**
Nickel carbonyl	**(12612-55-4)**
Nickel carbonyl (ACGIH,OSHA) [UN 1259]	**(13463-39-3)**
Nickel carbonyle (French)	**(13463-39-3)**
Nickel chloride	**(7718-54-9)**
Nickel(2+) chloride	**(7718-54-9)**
Nickel(II) chloride (1:2)	**(7718-54-9)**
Nickel chloride (9CI)	**(7718-54-9)**
Nickel(II) chloride, hexahydrate (1:2:6)	**(7791-20-0)**
Nickel cyanide, Solid (DOT)	**(557-19-7)**
Nickel cyanide, [UN 1653]	**(557-19-7)**
Nickel diacetate tetrahydrate	**(6018-89-9)**
Nickel dibromide	**(13462-88-9)**
Nickel dibutyldithiocarbamate	**(13927-77-0)**
Nickel dichloride	**(7718-54-9)**
Nickel difluoride	**(10028-18-9)**
Nickel diformate dihydrate	**(15694-70-9)**
Nickel diiodide	**(13462-90-3)**
Nickel diperchlorate	**(13637-71-3)**
Nickel 2-ethylhexanoate	**(7580-31-6)**
Nickel(2+) 2-ethylhexanoate	**(4454-16-4)**
Nickel(II) fluoride (1:2)	**(10028-18-9)**
Nickel formate	**(15843-02-4)**
Nickel formate (Ni(HCO2)2) dihydrate	**(15694-70-9)**
Nickel(2+) formate dihydrate	**(15694-70-9)**
Nickel formate, dihydrate (7CI)	**(15694-70-9)**
Nickel hydroxide	**(11113-74-9)**
Nickel(II) hydroxide	**(12054-48-7)**
Nickel(III) hydroxide	**(12125-56-3)**
Nickelic hydroxide	**(12125-56-3)**
Nickelic oxide	**(1314-06-3)**
Nickel(2+) iodide	**(13462-90-3)**
Nickel iodide (9CI)	**(13462-90-3)**
Nickel monosulfate hexahydrate	**(10101-97-0)**
Nickel monosulfide	**(11113-75-0)**
Nickel monoxide	**(1313-99-1)**

Nickel nitrate (14216-75-2)
Nickel(II) nitrate (1:2) (13138-45-9)
Nickel nitrate [UN 2725] (13138-45-9)
Nickel(2+) nitrate, hexahydrate (13478-00-7)
Nickel(II) nitrate, hexahydrate (1:2:6) (13478-00-7)
Nickel nitrilotriacetic acid (34831-03-3)
Nickelocene (1271-28-9)
Nickelous acetate (373-02-4)
Nickelous acetate tetrahydrate (6018-89-9)
Nickelous bromide (13462-88-9)
Nickelous carbonate (3333-67-3)
Nickelous chloride (7718-54-9)
Nickelous fluoride (10028-18-9)
Nickelous hydroxide (12054-48-7)
Nickelous iodide (13462-90-3)
Nickelous oxide (1313-99-1)
Nickelous sulfate (7786-81-4)
Nickelous sulfide (11113-75-0)
Nickel oxide (1313-99-1)
Nickel oxide (1314-06-3)
Nickel(II) oxide (1:1) (1313-99-1)
Nickel(III) oxide (1314-06-3)
Nickel oxide (9CI) (12035-36-8)
Nickel oxide (9CI) (VAN) (11099-02-8)
Nickel oxide coated ceramic bonded zircon (66402-68-4)
Nickel oxide peroxide (1314-06-3)
Nickel perchlorate (13637-71-3)
Nickel(2+) perchlorate (13637-71-3)
Nickel(2+) perchlorate, hexahydrate (13520-61-1)
Nickel peroxide (1314-06-3)
Nickel, (phthalocyaninato(2-))- (8CI) (14055-02-8)
Nickel, (29H,31H-phthalocyaninato(2-)-N29,N30,N31,N32)-, (SP-4-1)- (9CI) (14055-02-8)
Nickel phthalocyanine (14055-02-8)
Nickel(II) phthalocyanine (14055-02-8)
Nickel phthalocyanine blue (14055-02-8)
Nickel protoxide (1313-99-1)
Nickel sisquioxide (1314-06-3)
Nickel subsulfide (12035-72-2)
Nickel subsulphide (12035-72-2)
Nickel (II) sulfamate (13770-89-3)
Nickel sulfate (7786-81-4)
Nickel sulfate(1:1) (7786-81-4)
Nickel(2+)sulfate(1:1) (7786-81-4)
Nickel(II) sulfate (7786-81-4)
Nickel(II) sulfate (1:1) (7786-81-4)
Nickel sulfate hexahydrate (10101-97-0)
Nickel(2+) sulfate hexahydrate (10101-97-0)
Nickel(II) sulfate hexahydrate (10101-97-0)
Nickel(II) sulfate hexahydrate (1:1:6) (10101-97-0)
Nickel sulfide (3:2) (12035-72-2)
Nickel(II) sulfide (11113-75-0)
Nickel-sulfide (11113-75-0)
Nickel sulfide (9CI) (16812-54-7)
α-Nickel sulfide (1:1) crystalline (11113-75-0)
α-Nickel sulfide (3:2) crystalline (12035-72-2)
Nickel sulphate hexahydrate (10101-97-0)
Nickel sulphide (12035-72-2)
Nickel tetracarbonyl (13463-39-3)
Nickel tetracarbonyle (French) (13463-39-3)
Nickel, tetrakis(tris(methylphenyl) phosphite-P)- (9CI) (35884-66-3)
Nickel, tetrakis(tritolyl phosphite)- (35884-66-3)
Nickel, ((2,2'-thiobis(4-(1,1,3,3-tetramethylbutyl)phenolato))-(2-)-O,O',S)- (9CI) (27574-34-1)
Nickel titanium oxide (9CI) (12653-76-8)
Nickel trioxide (1314-06-3)
Nickel tritadisulphide (12035-72-2)
Niclofen (1836-75-5)
Niclosamide (1420-04-8)
Niclosamide (50-65-7)
Nico (59-67-6)
Nico-400 (59-67-6)
Nico-Dust (54-11-5)
Nico-Fume (54-11-5)
Nico-Span (59-67-6)
Nicobid (59-67-6)
Nicobion (98-92-0)
Nicocap (59-67-6)
Nicochloran (58-89-9)
Nicocide (54-11-5)
Nicocidin (59-67-6)
Nicocodeine (3688-66-2)
Nicocodina (Spanish) (3688-66-2)
Nicocodine (3688-66-2)
Nicocodinum (Latin) (3688-66-2)
Nicocrisina (59-67-6)
Nicodan (59-67-6)
Nicodelmine (59-67-6)
Nicofort (98-92-0)
Nicogen (98-92-0)
Nicolane (128-62-1)
Nicolar (59-67-6)
Nicolen (67-45-8)
Nicollembal (9004-35-7)
Nicometh (93-60-7)
Nicomidol (98-92-0)
Nicomorfina (639-48-5)
Nicomorphine (639-48-5)
Nicomorphinum (Latin) (639-48-5)
Niconacid (59-67-6)
Niconat (59-67-6)
Niconazid (59-67-6)
Niconyl (54-85-3)
Nicor (59-26-7)
Nicordamin (59-26-7)
Nicorine (59-26-7)
Nicorol (54-31-9)
Nicorol (59-67-6)
Nicoryl (59-26-7)
Nicosan 2 (98-92-0)
Nicoside (59-67-6)
Nicosyl (59-67-6)
Nicota (98-92-0)
Nicotamide (98-92-0)
Nicotamin (59-67-6)
Nicotelline (494-04-2)
Nicotene (59-67-6)
Nicotibina (54-85-3)
Nicotibine (54-85-3)
Nicotil (59-67-6)
Nicotilamide (98-92-0)
Nicotililamido (98-92-0)
Nicotina (Italian) (54-11-5)
Nicotinaldehyde (500-22-1)
Nicotinamide (98-92-0)
Nicotinamide adenine dinucleotide phosphate (53-57-6)
Nicotinamide, N,N-diethyl (59-26-7)
(-)-Nicotine (54-11-5)
(R,S)-Nicotine (22083-74-5)
(S)-Nicotine (54-11-5)
DL-Nicotine (22083-74-5)
Nicotine (54-11-5)
Nicotine, (+-) (22083-74-5)
dl-β-Nicotine (22083-74-5)

l-Nicotine	(54-11-5)
Nicotine (ACGIH,OSHA) [UN 1654]	(54-11-5)
Nicotine, Liquid [UN 3144]	(54-11-5)
Nicotine, Solid [UN 1655]	(54-11-5)
Nicotine acid	(59-67-6)
Nicotine acid amide	(98-92-0)
Nicotine acid tartrate	(65-31-6)
Nicotinealdehyde	(500-22-1)
Nicotine alkaloid	(54-11-5)
Nicotine bitartrate	(65-31-6)
Nicotine, 1'-demethyl-1'-nitroso	(16543-55-8)
Nicotine, 1'-demethyl-1'-nitroso- (8CI)	(16543-55-8)
Nicotine, hydrochloride, (-)	(21361-93-3)
l-Nicotine hydrochloride	(21361-93-3)
Nicotine hydrochloride, Solution [UN 1656]	(2820-51-1)
Nicotine hydrochloride [UN 1656]	(2820-51-1)
Nicotine hydrochloride (d,l)	(2820-51-1)
(-)-Nicotine hydrogen tartrate	(65-31-6)
Nicotine hydrogen tartrate	(65-31-6)
Nicotine, monosalicylate	(29790-52-1)
Nicotine, 1'-nitroso-1'-demethyl-	(16543-55-8)
Nicotine salicylate	(29790-52-1)
Nicotine sulfate	(65-30-5)
Nicotine, sulfate (2:1)	(65-30-5)
Nicotine sulfate, Solid [UN 1658]	(65-30-5)
Nicotine sulfate, Solution [UN 1658]	(65-30-5)
Nicotine sulphate	(65-30-5)
Nicotine, tartrate (1:2)	(65-31-6)
Nicotine tartrate [UN 1659]	(65-31-6)
Nicotinic acid-benzyl chloride quat	(16214-98-5)
Nicotinic-acid	(59-67-6)
Nicotinic acid amide	(98-92-0)
Nicotinic acid, 6-amino- (8CI)	(3167-49-5)
Nicotinic acid, 7,8-didehydro-4,5-α-epoxy-3-methoxy-17-methyl-morphinan-6-α-yl ester	(3688-66-2)
Nicotinic acid diethylamide	(59-26-7)
Nicotinic acid, ester with codeine	(3688-66-2)
Nicotinic acid, methyl ester	(93-60-7)
Nicotinic acid N-methylbetaine	(535-83-1)
Nicotinic acid nitrile	(100-54-9)
Nicotinic acid, 1-oxide (8CI)	(2398-81-4)
Nicotinic acid, sodium salt	(54-86-4)
Nicotinic acid, 1,2,5,6-tetrahydro-1-nitroso-, methyl ester	(55557-02-3)
Nicotinic alcohol	(100-55-0)
Nicotinic aldehyde	(500-22-1)
Nicotinic amide	(98-92-0)
Nicotinipca	(59-67-6)
Nicotinonitrile	(100-54-9)
Nicotinonitrile, 1,2-dihydro-4-methyl-2-oxo	(524-40-3)
Nicotinoylhydrazine	(59-67-6)
Nicotinsaure (German)	(59-67-6)
Nicotinsaureamid (German)	(98-92-0)
Nicotinuric acid	(583-08-4)
Nicotinyl alcohol	(100-55-0)
Nicotion	(536-33-4)
Nicotol	(98-92-0)
Nicotylamide	(98-92-0)
3,2'-Nicotyrine	(487-19-4)
Nicotyrine	(487-19-4)
β-Nicotyrine	(487-19-4)
Nicouline	(83-79-4)
Nicoumalone	(152-72-7)
Nicovasan	(59-67-6)
Nicovasen	(59-67-6)
Nicovel	(59-67-6)
Nicovel	(98-92-0)
Nicovit	(98-92-0)
Nicovitol	(98-92-0)
Nicozide	(54-85-3)
Nicozymin	(98-92-0)
Nicyl	(59-67-6)
Nida	(443-48-1)
Nidantin	(14698-29-4)
Nidaton	(54-85-3)
Nidrazid	(54-85-3)
Nieraline	(51-43-4)
Niesymetryczna dwu metylohydrazyna (Polish)	(57-14-7)
Nifedin	(21829-25-4)
Nifedipine	(21829-25-4)
Nifelat	(21829-25-4)
Niflex	(91-53-2)
Nifos	(107-49-3)
Nifos T	(107-49-3)
Nifost	(107-49-3)
Nifucin	(59-87-0)
Nifulidone	(67-45-8)
Nifuradene	(555-84-0)
Nifuradine	(555-84-0)
Nifuran	(67-45-8)
Nifurantin	(67-20-9)
Nifurid	(59-87-0)
Nifuroquina (Spanish)	(57474-29-0)
Nifuroquine	(57474-29-0)
Nifuroquinum (Latin)	(57474-29-0)
Nifurthiazol	(3570-75-0)
Nifurthiazole	(3570-75-0)
Nifuzon	(59-87-0)
Niglin	(55-63-0)
Niglycon	(55-63-0)
Nigrosin	(8005-03-6)
Nigrosine	(8005-03-6)
Nigrosine B	(8005-03-6)
Nigrosine SB	(11099-03-9)
Nigrosine SSB	(11099-03-9)
Nigrosine SSBZ 14	(11099-03-9)
Nigrosine SSBZ 30	(11099-03-9)
Nigrosine SSJ	(11099-03-9)
Nigrosine SSJJ	(11099-03-9)
Nigrosine Spirit Soluble	(11099-03-9)
Nigrosine WL Water Soluble	(8005-03-6)
Nigrosine WN	(11099-03-9)
Nigrosine WSB	(8005-03-6)
Nihonthrene Blue RSN	(81-77-6)
Nihonthrene Brilliant Blue RCL	(130-20-1)
Nihonthrene Brilliant Blue RP	(81-77-6)
Nihonthrene Brilliant Green B	(128-58-5)
Nihonthrene Brilliant Green FFB	(128-58-5)
Nihonthrene Brilliant Pink R	(2379-74-0)
Nihonthrene Brilliant Violet 4R	(1324-55-6)
Nihonthrene Brilliant Violet RR	(1324-55-6)
Nihonthrene Brown BR	(2475-33-4)
Nihonthrene Brown GR	(2475-33-4)
Nihonthrene Fast Orange R	(3263-31-8)
Nihonthrene Golden Orange G	(128-70-1)
Nihonthrene Golden Yellow	(128-66-5)
Nihonthrene Navy Blue G	(6424-76-6)
Nihonthrene Olive R	(2379-81-9)
Nihonthrene Red BB	(4203-77-4)
Nikardin	(59-26-7)
Nikelocen (Czech)	(1271-28-9)
Niketamid	(59-26-7)
Nikethamide	(59-26-7)
Niketharol	(59-26-7)
Nikethyl	(59-26-7)

Niketilamid **(59-26-7)**
Nikion **(73-48-3)**
Nikkeltetracarbonyl (Dutch) **(13463-39-3)**
Nikkol BL **(9002-92-0)**
Nikkol BL 25 **(9002-92-0)**
Nikkol BL 42 **(9002-92-0)**
Nikkol BL 9EX **(9002-92-0)**
Nikkol BO **(9004-98-2)**
Nikkol MYO 2 **(9004-96-0)**
Nikkol MYO 10 **(9004-96-0)**
Nikkol MYS **(9004-99-3)**
Nikkol OP **(9036-19-5)**
Nikkol OTP 70 **(577-11-7)**
Nikkol S.C.S **(1120-01-0)**
Nikkol SLS **(151-21-3)**
Nikkol SNP **(9051-57-4)**
Nikkol SO 10 **(1338-43-8)**
Nikkol SO 15 **(8007-43-0)**
Nikkol SO-15 **(1338-43-8)**
Nikkol SO-30 **(1338-43-8)**
Nikkol SP10 **(26266-57-9)**
Nikkol SS 30 **(1338-41-6)**
Nikkol TO **(9005-65-6)**
Nikkol TO 10 **(9005-65-6)**
Niko-Tamin **(98-92-0)**
Nikorin **(59-26-7)**
Nikotin **(59-26-7)**
Nikotin (German) **(54-11-5)**
Nikotinbitartrat (German) **(65-31-6)**
Nikotinsaeureamid (German) **(98-92-0)**
Nikotinsulfat (German) **(65-30-5)**
Nikotyna (Polish) **(54-11-5)**
Nikozid **(54-85-3)**
Nilidrine **(447-41-6)**
Nilox **(125-40-6)**
Nilox **(51-63-8)**
Nilox PBNA **(135-88-6)**
Nilstat **(1400-61-9)**
Nilurid **(2016-88-8)**
Nimbecetin **(520-18-3)**
Nimco Cholesterol Base H **(57-88-5)**
Nimco Cholesterol Base No. 712 **(57-88-5)**
Nimetazepam **(2011-67-8)**
Nimetazepamum (Latin) **(2011-67-8)**
Nimitex **(3383-96-8)**
Nimitox **(3383-96-8)**
Nincaluicolflastine **(865-21-4)**
Ninhydrin **(485-47-2)**
Ninhydrin hydrate **(485-47-2)**
Ninol 4821 **(120-40-1)**
Ninol AA62 **(120-40-1)**
Ninol AA-62 Extra **(120-40-1)**
Ninol AA62 Extra **(143-07-7)**
Ninol 2012E **(68603-42-9)**
Ninol P-621 **(120-40-1)**
Niobe Oil **(93-58-3)**
Niobium-chloride **(10026-12-7)**
Niobium (9CI) **(7440-03-1)**
Niobium pentachloride **(10026-12-7)**
Niocinamide **(98-92-0)**
Nioform **(130-26-7)**
Niomil **(22781-23-3)**
Nionate **(299-29-6)**
Niong **(55-63-0)**
Niozymin **(98-92-0)**
Nip-A-Thin **(132-66-1)**
Nipa 49 **(121-79-9)**
Nipa No. 48 **(831-61-8)**
Nipabenzyl **(94-18-8)**
Nipabutyl **(94-26-8)**
Nipacide MX **(88-04-0)**
Nipagallin A **(831-61-8)**
Nipagallin P **(121-79-9)**
Nipagin **(99-76-3)**
Nipagin A **(120-47-8)**
Nipagin M **(99-76-3)**
Nipagin P **(94-13-3)**
Nipagina A **(120-47-8)**
Nipam **(2210-25-5)**
Nipantiox 1-F **(25013-16-5)**
Nipar S-20 **(79-46-9)**
Nipar S-20 solvent **(79-46-9)**
Nipar S-30 solvent **(79-46-9)**
Nipasol **(94-13-3)**
Nipasol M **(94-13-3)**
Nipasol P **(94-13-3)**
Nipaxon **(128-62-1)**
Nipazin A **(120-47-8)**
Nipazol **(94-13-3)**
Nipecotan **(144-14-9)**
Nipellen **(59-67-6)**
Niperyt **(78-11-5)**
Niperyth **(78-11-5)**
Niplen **(54-85-3)**
Nipodal **(58-38-8)**
Nipol 407 **(9003-55-8)**
Nippon Blue BB **(2602-46-2)**
Nippon Dark Green B **(3626-28-6)**
Nippon Deep Black **(1937-37-7)**
Nippon Deep Black GX **(1937-37-7)**
Nippon Deep Black 3RL **(2429-83-6)**
Nippon Deep Black RL **(2429-83-6)**
Nippon Deep Black RL Extra **(2429-83-6)**
Nippon Direct Sky Blue **(2429-74-5)**
Nippon Fast Yellow A **(1325-37-7)**
Nippon Kagaku Chrysoidine **(532-82-1)**
Nippon Kagaku Safranine GK **(477-73-6)**
Nippon Kagaku Safranine T **(477-73-6)**
Nippon Orange X-881 **(3468-63-1)**
Nippon Purpurine 8B **(6548-29-4)**
Nippon Soda **(55635-13-7)**
Nippon Violet LN **(6426-67-1)**
Nipsan **(333-41-5)**
Niquetamida **(59-26-7)**
Niran **(56-38-2)**
Niran **(57-74-9)**
Niran E-4 **(56-38-2)**
Niridazole **(61-57-4)**
Nirit **(1594-56-5)**
Nirvonal **(50-06-6)**
Nisentil **(77-20-3)**
Nisetamide **(59-26-7)**
Nisotin **(536-33-4)**
Nissan Cation S2-100 **(122-19-0)**
Nissan Cation AB **(112-03-8)**
Nissan Cation BB **(112-00-5)**
Nissan Cation M2-100 **(139-08-2)**
Nissan Cation PB 40 **(112-02-7)**
Nissan Nonion HS 206 **(9036-19-5)**
Nissan Nonion HS 208 **(9036-19-5)**
Nissan Nonion HS 210 **(9036-19-5)**
Nissan Nonion K 220 **(9002-92-0)**
Nissan Nonion OP 83 **(8007-43-0)**
Nissan Nonion OP 85 **(26266-58-0)**

Name	CAS No.
Nissan Nonion OP 85R	(26266-58-0)
Nissan Nonion OP 83RAT	(8007-43-0)
Nissan Nonion PP40	(26266-57-9)
Nissan Nonion PP 40R	(26266-57-9)
Nissan Nonion S 15	(9004-99-3)
Nissan Nonion SP 60	(1338-41-6)
Nissan Nymeen F 215	(61791-14-8)
Nissan Trax N300	(9014-90-8)
Nissan Unilube 25DE	(9003-11-6)
Nissan Unilube 75DE3800	(9003-11-6)
Nissan Unilube 750DE2620	(9003-11-6)
Nissan Unilube DE 60	(9003-11-6)
Nissan Unilube 70DP950B	(9003-11-6)
Nissan Unilube 50MB26X	(9003-11-6)
Nissan Unilube 50MB168X	(9003-11-6)
Nissan Uniox M 2000	(9004-74-4)
Nissan Uniox M 400	(9004-74-4)
Nissan Uniox M 550	(9004-74-4)
Nissan diapion S	(137-20-2)
Nissan diapon T	(137-20-2)
Nissen Black BGL	(1326-82-5)
Nissen Black BK	(1326-82-5)
Nissen Black BX	(1326-82-5)
Nisso BN 1000	(9003-17-2)
Nisso PB 100	(9003-17-2)
Nisso PB 3000	(9003-17-2)
Nisso PB 4000	(9003-17-2)
Nisso PB-B 4000	(9003-17-2)
Nisso PB-GQ 3000	(9003-17-2)
Nisso PR 2000	(9003-17-2)
Nisso TG 4400	(25791-96-2)
Nissocaine	(59-46-1)
Nitadon	(54-85-3)
Nitan	(2152-34-3)
Niteban	(54-85-3)
Niter	(7757-79-1)
Nithiazid	(139-94-6)
Nithiazide	(139-94-6)
Niticid	(1918-16-7)
Nitofen	(1836-75-5)
Nitol	(55-86-7)
Nitol "Takeda"	(55-86-7)
Niton	(10043-92-2)
Nitora	(55-63-0)
Nitosorg	(38641-94-0)
Nitrador	(534-52-1)
Nitrados	(146-22-5)
Nitrafen	(1836-75-5)
Nitraldone	(139-91-3)
Nitralin	(4726-14-1)
Nitraline	(4726-14-1)
Nitramide (8CI,9CI)	(7782-94-7)
Nitramin	(121-66-4)
Nitramin Ido	(121-66-4)
Nitramine	(121-66-4)
Nitramine	(479-45-8)
Nitramine	(7782-94-7)
Nitramyl	(110-46-3)
Nitramyl	(463-04-7)
Nitran	(1582-09-8)
Nitranilin	(99-09-2)
4-Nitraniline	(100-01-6)
m-Nitraniline	(99-09-2)
o-Nitraniline	(88-74-4)
p-Nitraniline	(100-01-6)
Nitranitol	(15825-70-4)
Nitraphen	(1836-75-5)
Nitrapyrin	(1929-82-4)
Nitrate (8CI,9CI)	(14797-55-8)
Nitrate d'amyle (French)	(1002-16-0)
Nitrate d'argent (French)	(7761-88-8)
Nitrate d'azidoethyle (French)	(53422-49-4)
Nitrate de baryum (French)	(10022-31-8)
Nitrate de guanidine (French)	(506-93-4)
Nitrate mercureux (French)	(10415-75-5)
Nitrate mercurique (French)	(10045-94-0)
Nitrate de plomb (French)	(10099-74-8)
Nitrate de propyle normal (French)	(627-13-4)
Nitrate salt of ethyl auramine	(43130-12-7)
Nitrate de sodium (French)	(7631-99-4)
Nitrate de strontium (French)	(10042-76-9)
Nitrate de zinc (French)	(7779-88-6)
Nitratine	(7631-99-4)
Nitration benzene	(71-43-2)
Nitrato de aluminio (Spanish)	(13473-90-0)
Nitrato de amilo (Spanish)	(1002-16-0)
Nitrato de azidoetilo (Spanish)	(53422-49-4)
Nitrato de guanidina (Spanish)	(506-93-4)
Nitrazepam	(146-22-5)
Nitrazol CF Extra	(100-01-6)
Nitre	(7757-79-1)
Nitre Cake	(7681-38-1)
Nitrenpax	(146-22-5)
Nitric-acid	(7697-37-2)
Nitric acid (ACGIH,DOT,OSHA)	(7697-37-2)
Nitric acid, Fuming (DOT)	(7697-37-2)
Nitric acid, Over 40% (DOT)	(7697-37-2)
Nitric acid, Red fuming [UN 2032]	(7697-37-2)
Nitric acid, aluminum salt	(13473-90-0)
Nitric acid, aluminum(3+) salt	(13473-90-0)
Nitric acid, ammonium calcium salt (9CI)	(15245-12-2)
Nitric acid, ammonium salt	(6484-52-2)
Nitric acid, anhydride with peroxyacetic acid	(2278-22-0)
Nitric acid, barium salt	(10022-31-8)
Nitric acid, beryllium salt	(13597-99-4)
Nitric acid, cadmium salt	(10325-94-7)
Nitric acid, cerium(3+) salt (8CI,9CI)	(10108-73-3)
Nitric acid, chromium(3+) salt, nonahydrate	(7789-02-8)
Nitric acid, cobalt(2+) salt	(10141-05-6)
Nitric acid, copper(2+)salt	(3251-23-8)
Nitric acid, ethyl ester	(625-58-1)
Nitric acid, 2-ethylhexyl ester (9CI)	(27247-96-7)
Nitric acid, heptyl ester (8CI,9CI)	(20633-12-9)
Nitric acid, hexyl ester (9CI)	(20633-11-8)
Nitric acid, ion(1-) (VAN)	(14797-55-8)
Nitric acid, iron(3+) salt	(10421-48-4)
Nitric acid, isobutyl ester (8CI)	(543-29-3)
Nitric acid, isopropyl ester	(1712-64-7)
Nitric acid, lanthanum(3+) salt, hexahydrate	(10277-43-7)
Nitric acid, lead(2+) salt	(10099-74-8)
Nitric acid, lithium salt	(7790-69-4)
Nitric acid, magnesium salt (2:1)	(10377-60-3)
Nitric acid, mercury(I) salt	(10415-75-5)
Nitric acid, mercury(II) salt	(10045-94-0)
Nitric acid, methyl ester	(598-58-3)
Nitric acid, 2-methylpropyl ester (9CI)	(543-29-3)
Nitric acid, nickel(II) salt	(13138-45-9)
Nitric acid, nickel(2+) salt, hexahydrate	(13478-00-7)
Nitric acid, octyl ester (8CI,9CI)	(629-39-0)
Nitric acid, pentyl ester	(1002-16-0)
Nitric acid, phenylmercury salt	(55-68-5)
Nitric acid, potassium salt	(7757-79-1)
Nitric acid, propyl ester	(627-13-4)
Nitric acid, silver(1+) salt	(7761-88-8)

Nitric acid, sodium salt **(7631-99-4)**
Nitric acid, strontium salt **(10042-76-9)**
Nitric acid, thallium(1+) salt **(10102-45-1)**
Nitric acid, thorium(4+) salt (8CI,9CI) **(13823-29-5)**
Nitric acid triester of glycerol **(55-63-0)**
Nitric acid, zinc salt **(7779-88-6)**
Nitric acid, zinc salt, hexahydrate **(10196-18-6)**
Nitric ether (DOT) **(625-58-1)**
Nitric oxide (ACGIH,OSHA) [UN 1660] **(10102-43-9)**
Nitric oxide and nitrogen tetroxide, Mixtures [UN 1975] **(63907-41-5)**
Nitridazole **(61-57-4)**
Nitrile acrilico (Italian) **(107-13-1)**
Nitrile acrylique (French) **(107-13-1)**
Nitrile adipico (Italian) **(111-69-3)**
Nitriles, tallow **(61790-28-1)**
Nitrile trichloracetique (French) **(545-06-2)**
Nitril kyseliny o-chlorbenzoove (Czech) **(873-32-5)**
Nitril kyseliny p-chlorbenzoove (Czech) **(623-03-0)**
Nitril kyseliny isoftalove (Czech) **(626-17-5)**
Nitril kyseliny malonove (Czech) **(109-77-3)**
Nitril kyseliny mandlove (Czech) **(532-28-5)**
Nitril kyseliny stearove (Czech) **(638-65-3)**
Nitril kyseliny β,β'-thiodipropionove (Czech) **(111-97-7)**
Nitril kyseliny m-toluylove (Czech) **(620-22-4)**
Nitril kyseliny p-toluylove (Czech) **(104-85-8)**
Nitriloacetic acid bismuth complex sodium salt **(5798-43-6)**
Nitriloacetic acid trisodium salt monohydrate **(18662-53-8)**
Nitriloacetonitrile **(460-19-5)**
Nitrilomalonamide **(107-91-5)**
Nitrilo(methylenephosphonic acid), pentasodium salt **(2235-43-0)**
3-Nitrilo-propionamide **(107-91-5)**
Nitrilotriacetic acid **(139-13-9)**
Nitrilotriacetic acid, Compound with iron chloride **(14695-88-6)**
Nitrilotriacetic acid, aluminium(3+) complex **(19010-73-2)**
Nitrilotriacetic acid and its salts **(10413-71-5)**
Nitrilotriacetic acid and its salts **(1188-47-2)**
Nitrilotriacetic acid and its salts **(1188-48-3)**
Nitrilotriacetic acid and its salts **(15467-20-6)**
Nitrilotriacetic acid and its salts **(18994-66-6)**
Nitrilotriacetic acid and its salts **(2399-81-7)**
Nitrilotriacetic acid and its salts **(2399-83-9)**
Nitrilotriacetic acid and its salts **(2399-89-5)**
Nitrilotriacetic acid and its salts **(2399-94-2)**
Nitrilotriacetic acid and its salts **(2455-08-5)**
Nitrilotriacetic acid and its salts **(3130-95-8)**
Nitrilotriacetic acid and its salts **(34831-02-2)**
Nitrilotriacetic acid and its salts **(34831-03-3)**
Nitrilotriacetic acid and its salts **(5798-43-6)**
Nitrilotriacetic acid and its salts **(60034-45-9)**
Nitrilotriacetic acid and its salts **(79849-02-8)**
Nitrilotriacetic acid, antimony(3+) complex **(46242-44-8)**
Nitrilotriacetic acid, beryllium potassium salt (1:1) **(18983-72-7)**
Nitrilotriacetic acid, cadmium(2+) complex **(18432-54-7)**
Nitrilotriacetic acid, calcium potassium salt (2:1:4) **(23555-98-8)**
Nitrilotriacetic acid, calcium salt **(14981-08-9)**
Nitrilotriacetic acid, calcium salt (2:3) **(62979-89-6)**
Nitrilotriacetic acid, cerium salt **(29027-90-5)**
Nitrilotriacetic acid, cobalt(3+) complex **(23319-51-9)**
Nitrilotriacetic acid, copper(2+) complex **(15844-52-7)**
Nitrilotriacetic acid, copper(2+) complex ammonium salt **(71484-80-5)**
Nitrilotriacetic acid, copper(2+) complex sodium salt **(53108-47-7)**
Nitrilotriacetic acid, diammonium salt **(71264-32-9)**
Nitrilotriacetic acid, dilithium salt **(72629-49-3)**
Nitrilotriacetic acid, dipotasium salt **(2399-86-2)**
Nitrilotriacetic acid, disodium ammonium salt **(86892-89-9)**
Nitrilotriacetic acid, disodium salt **(15467-20-6)**
Nitrilotriacetic acid, disodium salt, monohydrate **(23255-03-0)**
Nitrilotriacetic acid, holmium salt **(28927-38-0)**
Nitrilotriacetic acid, indium(3+) complex **(19456-58-7)**
Nitrilotriacetic acid, iron(2+) complex sodium salt (1:1:1) **(61017-62-7)**
Nitrilotriacetic acid, lead(2+) potassium salt (1:1:1) **(79915-08-5)**
Nitrilotriacetic acid, lead(2+) salt (2:3) **(79915-09-6)**
Nitrilotriacetic acid, magnesium salt **(73772-91-5)**
Nitrilotriacetic acid, manganese salt **(36711-58-7)**
Nitrilotriacetic acid, mercury(2+) salt (2:3) **(18105-03-8)**
Nitrilotriacetic acid, monoammonium salt **(15934-02-8)**
Nitrilotriacetic acid, monopotassium salt **(28444-53-3)**
Nitrilotriacetic acid, neodymium(3+) salt (1) **(18946-94-6)**
Nitrilotriacetic acid, nickel(3+) complex **(22965-60-2)**
Nitrilotriacetic acid, potassium magnesium salt (1:1:1) **(2399-88-4)**
Nitrilotriacetic acid, potassium salt **(25817-24-7)**
Nitrilotriacetic acid, potassium strontium salt (2:4:1) **(23555-96-6)**
Nitrilotriacetic acid, sodium salt **(10042-84-9)**
Nitrilotriacetic acid, strontium sodium salt **(92988-11-9)**
Nitrilotriacetic acid, tin(2+) salt **(53818-84-1)**
Nitrilotriacetic acid, triammonium salt **(32685-17-9)**
Nitrilotriacetic acid, tricadmium(2+) complex **(50648-02-7)**
Nitrilotriacetic acid, trisilver salt **(92474-39-0)**
Nitrilotriacetic acid, trisodium salt **(5064-31-3)**
Nitrilotriacetic acid, trisodium salt monohydrate **(18662-53-8)**
Nitrilotriacetic acid, yttrium(3+) salt (1:1) **(15414-25-2)**
Nitrilotriacetic acid, zinc(3+) complex sodium salt **(29507-58-2)**
2,2',2''-Nitrilotriethanol **(102-71-6)**
Nitrilo-2,2',2''-triethanol **(102-71-6)**
Nitrilotrimethanephosphonic acid **(6419-19-8)**
Nitrilotrimethylenephosphonic acid **(6419-19-8)**
Nitrilotri(methylenephosphonic acid), pentasodium salt **(2235-43-0)**
Nitrilotrimethylphosphonic acid **(6419-19-8)**
1,1',1''-Nitrilotri-2-propanol **(122-20-3)**
2,2',2''-Nitrilotrisacetonitrile **(7327-60-8)**
2,2',2''-Nitrilotrisethanol acetate **(14806-72-5)**
2,2',2''-Nitrilotrisethanol acetate (Salt) **(14806-72-5)**
2,2',2''-Nitrilo trisethanol 4-tert-butyl benzoate **(59993-86-1)**
2,2',2''-Nitrilotrisethanol formate **(24794-58-9)**
2,2',2''-Nitrilotrisethanol phosphate (Salt) **(10017-56-8)**
2,2',2''-Nitrilotris(ethanol) sulfate (Salt) **(7376-31-0)**
2,2',2''-Nitrilotrisethanol sulfate (Salt) **(7376-31-0)**
Nitrilotris(ethyl phosphate), sodium salt **(68171-29-9)**
Nitrilotris(methylenephosphonic acid) pentasodium salt **(2235-43-0)**
Nitrilotris(methylene phosphonic acid), sodium salt **(20592-85-2)**
(Nitrilotris(methylene))trisphosphonic acid **(6419-19-8)**
Nitrilotris(methylphosphonic acid) **(6419-19-8)**
Nitrin **(55-63-0)**
Nitrine **(55-63-0)**
Nitrine-TDC **(55-63-0)**
Nitrinol **(78-11-5)**
Nitrite **(1594-56-5)**
Nitrite de sodium (French) **(7632-00-0)**
Nitrito **(10102-44-0)**
3-Nitro-10 **(121-19-7)**
3-Nitro-20 **(121-19-7)**
3-Nitro-50 **(121-19-7)**
3-Nitro-80 **(121-19-7)**
Nitro Acid 100 Per Cent **(121-19-7)**
N-Nitro-DMA **(4164-28-7)**
Nitro-Dur **(55-63-0)**
Nitro Fast Blue 3GB **(128-85-8)**
Nitro Fast Green GB **(128-80-3)**
Nitro Fast Yellow SL **(8003-22-3)**
Nitro Kleenup **(51-28-5)**
Nitro-Sil **(7664-41-7)**
5-Nitroacenaphthene **(602-87-9)**
5-Nitroacenapthene **(602-87-9)**
2-Nitro-4-acetaminofenetol (Czech) **(1777-84-0)**

2'-Nitroacetanilide	**(552-32-9)**	7-Nitrobenz(a)anthracene	**(20268-51-3)**
3'-Nitroacetanilide	**(122-28-1)**	o-Nitrobenzacetonitrile	**(610-66-2)**
4'-Nitroacetanilide	**(104-04-1)**	2-Nitrobenzaldehyde	**(552-89-6)**
4-Nitroacetanilide	**(104-04-1)**	3-Nitrobenzaldehyde	**(99-61-6)**
m-Nitroacetanilide	**(122-28-1)**	4-Nitrobenzaldehyde	**(555-16-8)**
o-Nitroacetanilide	**(552-32-9)**	m-Nitrobenzaldehyde	**(99-61-6)**
p-Nitroacetanilide	**(104-04-1)**	o-Nitrobenzaldehyde	**(552-89-6)**
Nitroacetate	**(61201-44-3)**	p-Nitrobenzaldehyde	**(555-16-8)**
3-Nitroacetofenon (Czech)	**(121-89-1)**	2-Nitrobenzamide	**(610-15-1)**
3'-Nitro-p-acetophenetide	**(1777-84-0)**	3-Nitrobenzamide	**(645-09-0)**
3-Nitro-p-acetophenetide	**(1777-84-0)**	o-Nitrobenzamide	**(610-15-1)**
5-Nitro-p-acetophenetidide	**(1777-84-0)**	7-Nitrobenzanthracene	**(20268-51-3)**
3'-Nitro-p-acetophenetidin	**(1777-84-0)**	3-Nitrobenz(a)pyrene	**(70021-98-6)**
2'-Nitroacetophenone	**(577-59-3)**	6-Nitrobenz(a)pyrene	**(63041-90-7)**
3'-Nitroacetophenone	**(121-89-1)**	Nitrobenzeen (Dutch)	**(98-95-3)**
4'-Nitroacetophenone	**(100-19-6)**	Nitrobenzen (Polish)	**(98-95-3)**
m-Nitroacetophenone	**(121-89-1)**	3-Nitrobenzenamine	**(99-09-2)**
o-Nitroacetophenone	**(577-59-3)**	4-Nitrobenzenamine	**(100-01-6)**
p-Nitroacetophenone	**(100-19-6)**	4-Nitrobenzenamine monohydrochloride	**(15873-51-5)**
3-Nitro-N-acetylaniline	**(122-28-1)**	Nitrobenzene (ACGIH,OSHA) [UN 1662]	**(98-95-3)**
β-Nitroalcohol	**(625-48-9)**	Nitrobenzene, Liquid (DOT)	**(98-95-3)**
Nitroalmidon (Spanish)	**(9056-38-6)**	2-Nitrobenzeneacetonitrile	**(610-66-2)**
Nitroamidon (French)	**(9056-38-6)**	p-Nitrobenzeneazoresorcinol	**(74-39-5)**
.ω.-Nitro-o-aminoacetophenone	**(63892-06-8)**	m-Nitrobenzenecarboxylic acid	**(121-92-6)**
m-Nitroaminobenzene	**(99-09-2)**	p-Nitrobenzenecarboxylic acid	**(62-23-7)**
4-Nitro-2-aminofenol (Czech)	**(99-57-0)**	2-Nitro-1,4-benzenediamine	**(5307-14-2)**
p-Nitroaminofenol (Polish)	**(99-57-0)**	4-Nitro-1,2-benzenediamine	**(99-56-9)**
2-Nitro-4-aminophenol	**(119-34-6)**	5-Nitro-1,3-benzenediamine	**(5042-55-7)**
4-Nitro-2-aminophenol	**(99-57-0)**	2-Nitro-1,4-benzenediamine dihydrochloride	**(18266-52-9)**
o-Nitro-p-aminophenol	**(119-34-6)**	4-Nitro-1,2-benzenediamine dihydrochloride	**(6219-77-8)**
p-Nitro-o-aminophenol	**(99-57-0)**	2-Nitro-1,4-benzenediamine sulfate	**(68239-83-8)**
Nitroaminostilbene Disa	**(6634-82-8)**	2-Nitro-1,4-benzenediamine sulfate (1:1)	**(68239-83-8)**
5-Nitro-2-aminothiazole	**(121-66-4)**	4-Nitro-1,2-benzenediamine sulfate	**(68239-82-7)**
2-Nitro-4-aminotoluene	**(119-32-4)**	4-Nitro-1,2-benzenediamine sulfate (1:1)	**(68239-82-7)**
3-Nitro-4-aminotoluene	**(89-62-3)**	5-Nitro-1,3-benzenedicarboxylic acid	**(618-88-2)**
4-Nitro-2-aminotoluene	**(99-55-8)**	2-Nitro-1,3-benzenediol	**(601-89-8)**
p-Nitroanilina (Polish)	**(100-01-6)**	2-Nitrobenzeneethanol	**(15121-84-3)**
N-Nitroanilina (Spanish)	**(645-55-6)**	4-Nitrobenzeneethanol	**(100-27-6)**
2-Nitroaniline	**(88-74-4)**	2-Nitrobenzeneethanol acetate (ester)	**(833-43-2)**
3-Nitroaniline	**(99-09-2)**	4-Nitrobenzeneethanol acetate (ester)	**(104-30-3)**
4-Nitroaniline	**(100-01-6)**	2-Nitrobenzenesulfenyl chloride	**(7669-54-7)**
N-Nitroaniline	**(645-55-6)**	3-Nitrobenzenesulfonic acid	**(98-47-5)**
Nitroaniline	**(29757-24-2)**	4-Nitrobenzenesulfonic acid	**(138-42-1)**
p-Nitroaniline (ACGIH,OSHA) [UN 1661]	**(100-01-6)**	Nitrobenzenesulfonic acid	**(31212-28-9)**
N-Nitroaniline (French)	**(645-55-6)**	m-Nitrobenzenesulfonic acid	**(98-47-5)**
m-Nitroaniline [UN 1661]	**(99-09-2)**	p-Nitrobenzenesulfonic acid	**(138-42-1)**
o-Nitroaniline [UN 1661]	**(88-74-4)**	4-Nitrobenzenesulfonic acid chloride	**(98-74-8)**
4-Nitroaniline, 2,6-dichloro-	**(99-30-9)**	m-Nitrobenzenesulfonic acid sodium salt	**(127-68-4)**
4-Nitroaniline-2-sulfonic acid	**(96-75-3)**	2-Nitrobenzenesulfonyl chloride	**(1694-92-4)**
3-Nitro-p-anisanilide	**(97-32-5)**	3-Nitrobenzenesulfonyl chloride	**(121-51-7)**
4-Nitro-o-anisidine	**(97-52-9)**	4-Nitrobenzenesulfonyl chloride	**(98-74-8)**
5-Nitro-o-anisidine	**(99-59-2)**	m-Nitrobenzenesulfonyl chloride	**(121-51-7)**
p-Nitroanisol	**(100-17-4)**	p-Nitrobenzenesulfonyl chloride	**(98-74-8)**
2-Nitroanisole	**(91-23-6)**	Nitrobenzenesulphonic acid	**(31212-28-9)**
3-Nitroanisole	**(555-03-3)**	4-Nitrobenzenethiol	**(1849-36-1)**
4-Nitroanisole	**(100-17-4)**	p-Nitrobenzenethiol	**(1849-36-1)**
m-Nitroanisole [UN 2730]	**(555-03-3)**	Nitrobenzen-m-sulfonan sodny (Czech)	**(127-68-4)**
o-Nitroanisole [UN 2730]	**(91-23-6)**	5-Nitro-1H-benzimidazole	**(94-52-0)**
p-Nitroanisole [UN 2730]	**(100-17-4)**	6-Nitro-benzimidazole	**(94-52-0)**
5-Nitroanthracene	**(602-60-8)**	6-Nitrobenzisoxazole-3-carboxylate	**(42540-91-0)**
9-Nitroanthracene	**(602-60-8)**	1-Nitrobenzo(a)pyrene	**(70021-99-7)**
1-Nitroanthrachinon (Czech)	**(82-34-8)**	6-Nitrobenzo(a)pyrene	**(63041-90-7)**
4-Nitroanthranilic acid	**(619-17-0)**	2-Nitrobenzoic acid	**(552-16-9)**
1-Nitroanthraquinone	**(82-34-8)**	3-Nitrobenzoic acid	**(121-92-6)**
α-Nitroanthraquinone	**(82-34-8)**	4-Nitrobenzoic acid	**(62-23-7)**
4-Nitroazobenzene	**(2491-52-3)**	m-Nitrobenzoic acid	**(121-92-6)**
p-Nitroazobenzene	**(2491-52-3)**	o-Nitrobenzoic acid	**(552-16-9)**

p-Nitrobenzoic acid (62-23-7)
4-Nitrobenzoic acid chloride (122-04-3)
p-Nitrobenzoic acid chloride (122-04-3)
p-Nitrobenzoic acid, ethyl ester (99-77-4)
Nitrobenzol (98-95-3)
Nitrobenzol (DOT) (98-95-3)
Nitrobenzol, Liquid (DOT) (98-95-3)
3-Nitrobenzonitrile (619-24-9)
4-Nitrobenzonitrile (619-72-7)
m-Nitrobenzonitrile (619-24-9)
p-Nitrobenzonitrile (619-72-7)
3-Nitrobenzotrifluoride (98-46-4)
m-Nitrobenzotrifluoride [UN 2306] (98-46-4)
o-Nitrobenzotrifluoride [UN 2306] (384-22-5)
p-Nitrobenzotrifluoride [UN 2306] (402-54-0)
2-Nitrobenzoyl chloride (610-14-0)
3-Nitrobenzoyl chloride (121-90-4)
4-Nitrobenzoyl chloride (122-04-3)
m-Nitrobenzoyl chloride (121-90-4)
o-Nitrobenzoyl chloride (610-14-0)
p-Nitrobenzoyl chloride (122-04-3)
N-p-Nitrobenzylacetamide (56222-10-7)
2-Nitrobenzyl alcohol (612-25-9)
2-Nitrobenzyl chloride (612-23-7)
3-Nitrobenzyl chloride (619-23-8)
m-Nitrobenzyl chloride (619-23-8)
o-Nitrobenzyl chloride (612-23-7)
p-Nitrobenzyl chloride (100-14-1)
2-Nitrobenzyl nitrile (610-66-2)
2-Nitrobiphenyl (86-00-0)
3-Nitrobiphenyl (2113-58-8)
4-Nitrobiphenyl (92-93-3)
Nitrobiphenyl (28984-85-2)
m-Nitrobiphenyl (2113-58-8)
o-Nitrobiphenyl (86-00-0)
p-Nitrobiphenyl (92-93-3)
4-Nitrobiphenyl ether (620-88-2)
1-(4'-Nitro(1,1'-biphenyl)-4-yl)ethanone (135-69-3)
2-Nitro-1,1-bis(p-chlorophenyl)butane (117-26-0)
2-Nitro-1,1-bis(p-chlorophenyl)propane (117-27-1)
2-Nitro-4-bromoaniline (875-51-4)
2-Nitrobromobenzene (577-19-5)
3-Nitrobromobenzene (585-79-5)
4-Nitrobromobenzene (586-78-7)
o-Nitrobromobenzene (577-19-5)
o-Nitrobromobenzene, Liquid [UN 2732] (577-19-5)
p-Nitrobromobenzene, Liquid [UN 2732] (586-78-7)
m-Nitrobromobenzene, Solid [UN 2732] (585-79-5)
2-Nitro-2-bromo-1,3-propanediol (52-51-7)
1-Nitrobutane (627-05-4)
2-Nitrobutane (600-24-8)
2-Nitro-1-butanol (609-31-4)
1,1'-(2-Nitrobutylidene)bis(4-chlorobenzene) (117-26-0)
4-(2-Nitrobutyl)morpholine (2224-44-4)
N-(2-Nitrobutyl)morpholine (2224-44-4)
Nitrocarbol (75-52-5)
4-Nitrocatechol (3316-09-4)
R.S.Nitrocellulose (9004-70-0)
Nitrocellulose, Block, Wet with not less than 25% alcohol [UN 0342] (9004-70-0)
Nitrocellulose, Colloided, Granular or flake, Wet with not less than 20% alcohol or solvent (DOT) (9004-70-0)
Nitrocellulose, Colloided, Granular or flake, Wet with not less than 20% water (DOT) (9004-70-0)
Nitrocellulose, Containing at least 25% alcohol and not exceeding 12.6% nitrogen (DOT) (9004-70-0)
Nitrocellulose, Containing at least 25%, by weight, water (DOT) (9004-70-0)
Nitrocellulose (DOT) (9004-70-0)
Nitrocellulose, Dry (DOT) (9004-70-0)
Nitrocellulose, Dry or wetted with less than 25% alcohol [UN 0340] (9004-70-0)
Nitrocellulose, Dry or wetted with less than 25% water [UN 0340] (9004-70-0)
Nitrocellulose E950 (9004-70-0)
Nitrocellulose, In solution in flammable liquids (DOT) (9004-70-0)
Nitrocellulose, Wetted with more than 40% flammable liquids (DOT) (9004-70-0)
Nitrocellulose, Wet with less than 25% alcohol or 25% water [UN 0340] (9004-70-0)
Nitrocellulose, Wet with not less than 30% alcohol or solvent (DOT) (9004-70-0)
Nitrocellulose, Wet with not less than 20% water (9004-70-0)
4-Nitrochinolin N-oxid (swedish) (56-57-5)
p-Nitrochloorbenzeen (Dutch) (100-00-5)
Nitrochlor (1836-75-5)
2-Nitro-4-chloroaniline (89-63-4)
4-Nitro-2-chloroaniline (121-87-9)
1-Nitro-5-chloroanthraquinone (129-40-8)
Nitrochlorobenzene (25167-93-5)
m-Nitrochlorobenzene (121-73-3)
o-Nitrochlorobenzene (88-73-3)
p-Nitrochlorobenzene (ACGIH,OSHA) (100-00-5)
Nitrochlorobenzene, ortho, Liquid [UN 1578] (88-73-3)
Nitrochlorobenzene, meta, Solid (DOT) (121-73-3)
Nitrochlorobenzene, para, Solid [UN 1578] (100-00-5)
p-Nitrochlorobenzol (German) (100-00-5)
3-Nitro-4-chlorobenzotrifluoride [UN 2307] (121-17-5)
Nitrochloroform (76-06-2)
2-Nitro-4-chlorophenol (89-64-5)
p-Nitro-m-chlorophenyl dimethyl thionophosphate (500-28-7)
p-Nitro-o-chlorophenyl dimethyl thionophosphate (2463-84-5)
2-Nitro-4-chlorotoluene (89-59-8)
3-Nitro-4-chlorotoluene (89-60-1)
3-Nitro-4-chloro-α,α,α-trifluorotoluene (121-17-5)
6-Nitrochrysene (7496-02-8)
p-Nitroclorobenzene (Italian) (100-00-5)
Nitrocotton (9004-70-0)
2-Nitro-4-cresol (119-33-5)
2-Nitro-p-cresol (119-33-5)
4-Nitro-3-cresol (2581-34-2)
4-Nitro-m-cresol (2581-34-2)
4-Nitro-o-cresol (99-53-6)
Nitrocresol (12167-20-3)
Nitrocresolamine (82-33-7)
Nitrocresols [UN 2446] (12167-20-3)
Nitrocyclohexane (1122-60-7)
3-Nitro-6-((4-(N,N-diacetoxyethylamino)phenyl)azo)benzonitrile) (30124-94-8)
2-Nitro-1,4-diaminobenzene (5307-14-2)
4-Nitro-1,2-diaminobenzene (99-56-9)
1-Nitro-3,4-dichlorobenzene (99-54-7)
Nitro-p-dichlorobenzene (89-61-2)
3-Nitro-2,5-dichlorobenzoic acid (88-86-8)
4'-Nitro-2,4-dichlorodiphenyl ether (1836-75-5)
N-(2-((4-Nitro-2,6-dicyanophenyl)azo)-5-(diethylamino)phenyl)-acetamide (41642-51-7)
4-Nitrodifenylamin (Czech) (836-30-6)
4-Nitrodifenylether (Czech) (620-88-2)
N-Nitrodimethylamine (4164-28-7)
2'-Nitro-4-dimethylaminoazobenzene (3010-38-6)
3-Nitro-9-(3'-dimethylaminopropylamino)acridine (6237-24-7)
4-Nitro-N,N-dimethylaniline (100-23-2)
4-Nitrodimethylaniline (100-23-2)
p-Nitro-N,N-dimethylaniline (100-23-2)
p-Nitrodimethylaniline (100-23-2)

1-Nitro-2,4-dimethylbenzene **(89-87-2)**
4-Nitro-1,2-dimethylbenzene **(99-51-4)**
4-Nitro-1,3-dimethylbenzene **(89-87-2)**
Nitrodimethylbenzene **(25168-04-1)**
2-Nitrodiphenyl **(86-00-0)**
o-Nitrodiphenyl **(86-00-0)**
p-Nitrodiphenyl **(92-93-3)**
4-Nitrodiphenyl (ACGIH,OSHA) **(92-93-3)**
2-Nitrodiphenylamine **(119-75-5)**
4-Nitrodiphenylamine **(836-30-6)**
Nitrodiphenylamine **(119-75-5)**
o-Nitrodiphenylamine **(119-75-5)**
p-Nitrodiphenylamine **(836-30-6)**
4-Nitrodiphenyl ether **(620-88-2)**
p-Nitrodiphenyl ether **(620-88-2)**
4-Nitrodracylic acid **(62-23-7)**
Nitroetan (Polish) **(79-24-3)**
Nitroethane (ACGIH,OSHA) [UN 2842] **(79-24-3)**
2-Nitroethanol **(625-48-9)**
p-Nitroethylbenzene **(100-12-9)**
3-Nitro-9-ethylcarbazole **(86-20-4)**
3-Nitro-N-ethylcarbazole **(86-20-4)**
Nitroethyl nitrate **(4528-34-1)**
2-Nitro-2-ethyl-1,3-propanediol **(597-09-1)**
Nitrofan **(534-52-1)**
Nitrofen **(1836-75-5)**
Nitrofene (French) **(1836-75-5)**
m-Nitrofenol (Czech) **(554-84-7)**
o-Nitrofenol (Czech) **(88-75-5)**
p-Nitrofenol (Czech) **(100-02-7)**
4-Nitrofenol (Dutch) **(100-02-7)**
4-Nitro-1,2-fenylendiamin (Czech) **(99-56-9)**
4-Nitro-1,3-fenylendiamin (Czech) **(5131-58-8)**
1-Nitrofluoranthene **(13177-28-1)**
2-Nitrofluoranthene **(13177-29-2)**
3-Nitrofluoranthene **(892-21-7)**
4-Nitrofluoranthene **(892-21-7)**
2-Nitrofluorene **(607-57-8)**
Nitrofluorene **(55345-04-5)**
3-Nitro-9-fluorenone **(42135-22-8)**
3-Nitrofluorenone **(42135-22-8)**
4-Nitrofluorobenzene **(350-46-9)**
m-Nitrofluorobenzene **(402-67-5)**
p-Nitrofluorobenzene **(350-46-9)**
Nitroform **(517-25-9)**
Nitrofural **(59-87-0)**
5-Nitro-2-furaldehyde **(698-63-5)**
5-Nitrofuraldehyde semicarbazide **(59-87-0)**
6-Nitrofuraldehyde semicarbazide **(59-87-0)**
5-Nitro-2-furaldehyde semicarbazone **(59-87-0)**
Nitrofuraldehyde semicarbazone **(59-87-0)**
2-Nitrofuran **(609-39-2)**
5-Nitrofuran **(609-39-2)**
Nitrofuran **(609-39-2)**
5-Nitrofuran-2-aldehyde semicarbazone **(59-87-0)**
5-Nitro-2-furancarboxaldehyde semicarbazone **(59-87-0)**
5-Nitrofurancarboxylic acid **(645-12-5)**
Nitrofurantoin **(67-20-9)**
3-(((5-Nitro-2-furanyl)methylene)amino)-2-oxazolidinone **(67-45-8)**
2((5-Nitro-2-furanyl)methylene)hydrazinecarboxamide **(59-87-0)**
N-(4-(5-Nitro-2-furanyl)-2-thiazolyl)acetamide **(531-82-8)**
Nitrofurate **(645-12-5)**
Nitrofurazan **(59-87-0)**
Nitrofurazolidone **(67-45-8)**
Nitrofurazolidonum **(67-45-8)**
Nitrofurazone **(59-87-0)**
5-Nitrofurfural **(698-63-5)**
Nitrofurfural **(698-63-5)**
3-(5'-Nitrofurfuralamino)-2-oxazolidone **(67-45-8)**
5-Nitrofurfuraldehyde **(698-63-5)**
5-Nitro-2-furfuraldehyde semicarbazone **(59-87-0)**
5-Nitro-2-furfural semicarbazone **(59-87-0)**
5-Nitrofurfural semicarbazone **(59-87-0)**
1-((5-Nitrofurfurylidene)amino)hydantoin **(67-20-9)**
N-(5-Nitro-2-furfurylidene)-1-aminohydantoin **(67-20-9)**
N-(5-Nitrofurfurylidene)-1-aminohydantoin **(67-20-9)**
N-(5-Nitro-2-furfurylideneamino)-2-imidazoaidinone **(555-84-0)**
1-((5-Nitrofurfurylidene)amino)-2-imidazolidinone **(555-84-0)**
N-(5-Nitro-2-furfurylidene)-1-amino-2-imidazolidone **(555-84-0)**
N-(5-Nitro-2-furfurylidene)-3-aminooxazolidine-2-one **(67-45-8)**
3-((5-Nitrofurfurylidene)amino)-2-oxazolidone **(67-45-8)**
N-(5-Nitro-2-furfurylidene)-3-amino-2-oxazolidone **(67-45-8)**
(5-Nitro-2-furfurylideneamino)urea **(59-87-0)**
N-(6-(5-Nitrofurfurylidenemethyl)-1,2,4-triazin-3-yl)iminodi-methanol **(794-93-4)**
N-(5-Nitro-2-furfurylideno)-1-aminohydantoina (Polish) **(67-20-9)**
Nitrofurmethone **(139-91-3)**
Nitrofurmeton **(139-91-3)**
5-Nitrofuroic acid **(645-12-5)**
Nitrofuroxon **(67-45-8)**
2-(5-Nitro-2-furyl)-5-amino-1,3,4-thiadiazole **(712-68-5)**
5-(5-Nitro-2-furyl)-2-amino-1,3,4-thiadiazole **(712-68-5)**
3-((5-Nitrofurylidene)amino)-2-oxazolidone **(67-45-8)**
((3-(5-Nitro-2-furyl)-1-(2-(5-nitro-2-furyl)vinyl)allylidene)amino)-guanidine **(804-36-4)**
N-(4-(5-Nitro-2-furyl)-2-thiazolyl)acetamide **(531-82-8)**
N-(4-(5-Nitro-2-furyl)thiazol-2-yl)acetamide **(531-82-8)**
N-(4-(5-Nitro-2-furyl)-2-thiazolyl)formamid (German) **(24554-26-5)**
N-(4-(5-Nitro-2-furyl)-2-thiazolyl)formamide **(24554-26-5)**
((6-(2-(5-Nitro-2-furyl)vinyl)-as-triazin-3-yl)imino)dimethanol **(794-93-4)**
6-(5-Nitro-2-furylvinyl)-3-(dihydroxydimethylamino)-1,2,4-triazene **(794-93-4)**
N-(6-(2-(5-Nitro-2-furyl)vinyl)-1,2,4-triazin-3-yl)iminodimethanol **(794-93-4)**
Nitrogen, Compressed [UN 1066] **(7727-37-9)**
Nitrogen (DOT) **(7727-37-9)**
Nitrogen Gas **(7727-37-9)**
Nitrogen Mustard Amine Oxide **(126-85-2)**
Nitrogen Mustard N-Oxide **(126-85-2)**
Nitrogen Mustard Oxide **(126-85-2)**
Nitrogen, Refrigerated liquid [UN 1977] **(7727-37-9)**
Nitrogen bromide **(13973-87-0)**
Nitrogen chloride **(10025-85-1)**
Nitrogen chloride **(13973-88-1)**
Nitrogen-dioxide **(10102-44-0)**
Nitrogen dioxide (ACGIH,OSHA) **(10102-44-0)**
Nitrogen dioxide, Liquid (DOT) **(10102-44-0)**
Nitrogen dioxide, di- **(10544-72-6)**
Nitrogen fluoride **(7783-54-2)**
Nitrogen iodide **(13444-85-4)**
Nitrogen iodide **(14696-82-3)**
Nitrogen lime **(156-62-7)**
Nitrogen-monoxide **(10102-43-9)**
Nitrogen monoxide, Mixed with nitrogen tetroxide **(63907-41-5)**
Nitrogen mustard **(51-75-2)**
Nitrogen mustard hydrochloride **(55-86-7)**
Nitrogen mustard N-oxide hydrochloride **(302-70-5)**
Nitrogenol **(140-72-7)**
Nitrogen-oxide **(10024-97-2)**
Nitrogen oxide (9CI) (VAN) **(11104-93-1)**
Nitrogen oxychloride **(2696-92-6)**
Nitrogen peroxide **(10102-44-0)**
Nitrogen peroxide, Liquid [UN 1067] **(10102-44-0)**
Nitrogen selenide **(12033-59-9)**
Nitrogen sesquioxide **(10544-73-7)**
Nitrogen tetroxide (DOT) **(10544-72-6)**

Nitrogen tetroxide, Liquid [UN 1067] **(10544-72-6)**
Nitrogen trichloride **(10025-85-1)**
Nitrogen trifluoride (ACGIH,OSHA) [UN 2451] **(7783-54-2)**
Nitrogen triiodide **(13444-85-4)**
Nitrogen trioxide [UN 2421] **(10544-73-7)**
Nitroglicerina (Italian) **(55-63-0)**
Nitrogliceryna (Polish) **(55-63-0)**
Nitroglycerin (ACGIH,OSHA) **(55-63-0)**
Nitroglycerin, Liquid, Not desensitized [UN 0143] **(55-63-0)**
Nitroglycerin, Liquid, Desensitized [UN 0143] **(55-63-0)**
Nitroglycerin, Solution with 5% but not more than 10% nitroglycerin [UN 144] **(55-63-0)**
Nitroglycerine **(55-63-0)**
Nitroglycerine, Spirit of (1% to 5%) [UN 3064] **(55-63-0)**
Nitroglycerol **(55-63-0)**
Nitroglycol **(628-96-6)**
Nitroglykol (Czech) **(628-96-6)**
Nitroglyn **(55-63-0)**
Nitrogranulogen **(55-86-7)**
Nitrogranulogen hydrochloride **(55-86-7)**
2-Nitroguanidine **(556-88-7)**
Nitroguanidine **(556-88-7)**
α-Nitroguanidine **(556-88-7)**
Nitroguanidine, Containing less than 20% water [UN 0282] **(556-88-7)**
Nitroguanidine, Dry [UN 0282] **(556-88-7)**
Nitroguanidine, Wetted with not less than 20 per cent water, by mass [UN 1336] **(556-88-7)**
2-Nitro-2-heptene **(6065-14-1)**
3-Nitro-2-heptene **(6065-13-0)**
3-Nitro-3-heptene **(6187-24-2)**
2-Nitro-2-hexene **(6065-17-4)**
4-Nitrohippuric acid **(2645-07-0)**
para-Nitrohippuric acid **(2645-07-0)**
1-Nitro hydantoin **(2825-15-2)**
1-Nitrohydantoin **(2825-15-2)**
Nitrohydrochloric acid **(8007-56-5)**
Nitrohydrochloric acid, Diluted **(8007-56-5)**
2-Nitro-1-hydroxybenzene-4-arsonic acid **(121-19-7)**
3-Nitro-4-hydroxybenzenearsonic acid **(121-19-7)**
5-Nitro-2-hydroxybenzoic acid **(96-97-9)**
2-Nitro-2'-hydroxy-3',5'-bis(α,α-dimethylbenzyl)azobenzene **(70693-50-4)**
2-Nitro-2-(hydroxymethyl)-1,3-propanediol **(126-11-4)**
3-Nitro-4-hydroxyphenylarsonic acid **(121-19-7)**
2-Nitro-1H-imidazole **(527-73-1)**
2-Nitroimidazole **(527-73-1)**
4-Nitro-1H-imidazole (9CI) **(3034-38-6)**
1-(2-Nitro-1-imidazolyl)-3-hydroxy-2-propanol **(13551-92-3)**
1-(2-Nitro-1-imidazolyl)-3-methoxy-2-propanol **(13551-87-6)**
1-(2-Nitroimidazol-1-yl)-3-methoxypropan-2-ol **(13551-87-6)**
3-(2-Nitroimidazol-1-yl)-1,2-propanediol **(13551-92-3)**
5-Nitro-1H-indazole **(5401-94-5)**
5-Nitroindazole **(5401-94-5)**
Nitro isobutane triol trinitrate **(20820-44-4)**
Nitroisobutane triol trinitrate **(20820-44-4)**
Nitroisobutanetriol trinitrate **(20820-44-4)**
Nitroisobutyl glyceryl trinitrate **(20820-44-4)**
Nitroisobutyl glycol trinitrate **(20820-44-4)**
4-Nitro-1H-isoindole-1,3(2H)-dione **(603-62-3)**
5-Nitroisophthalic acid **(618-88-2)**
5-Nitroisophthalic acid, dimethyl ester **(13290-96-5)**
Nitroisopropane **(79-46-9)**
p-Nitroisopropylbenzene **(1817-47-6)**
Nitrol **(24458-48-8)**
Nitrol **(55-63-0)**
Nitrol **(86-57-7)**
Nitrol (Promoter) **(24458-48-8)**
Nitrolan **(55-63-0)**
Nitro-lent **(55-63-0)**
Nitroletten **(55-63-0)**
Nitrolim **(156-62-7)**
Nitrolime **(156-62-7)**
Nitrolingual **(55-63-0)**
Nitrolowe **(55-63-0)**
Nitrol (pharmaceutical) **(55-63-0)**
Nitro mannite **(15825-70-4)**
Nitromannite **(15825-70-4)**
Nitromannite (Dry) **(15825-70-4)**
Nitromannitol **(15825-70-4)**
Nitromel **(55-63-0)**
Nitrometan (Polish) **(75-52-5)**
Nitromethane (ACGIH,OSHA) [UN 1261] **(75-52-5)**
3-Nitro-6-methoxyaniline **(99-59-2)**
5-Nitro-2-methoxyaniline **(99-59-2)**
2-Nitro-4-methylaniline **(89-62-3)**
3-Nitro-4-methylaniline **(119-32-4)**
4-Nitro-N-methylaniline **(100-15-2)**
1-Nitro-2-methylanthraquinone **(129-15-7)**
3-Nitro-4-methylbenzoic acid **(96-98-0)**
4-Nitro-N-methylphthalimide **(41663-84-7)**
2-Nitro-2-methyl-1,3-propanediol **(77-49-6)**
2-Nitro-2-methyl-1-propanol **(76-39-1)**
2-Nitro-2-methylpropanol nitrate **(24884-69-3)**
Nitromin **(126-85-2)**
Nitromin Ido **(121-66-4)**
Nitromin hydrochloride **(302-70-5)**
4-Nitromorpholine **(4164-32-3)**
N-Nitromorpholine **(4164-32-3)**
Nitromuriatic acid **(8007-56-5)**
Nitron **(9004-70-0)**
Nitron Lavsan **(25038-59-9)**
Nitron (Nitrocellulose) **(9004-70-0)**
Nitron (Polyester) **(25038-59-9)**
1-Nitronaftalen (Czech) **(86-57-7)**
1-Nitronaphthalene **(86-57-7)**
2-Nitronaphthalene **(581-89-5)**
α-Nitronaphthalene **(86-57-7)**
β-Nitronaphthalene **(581-89-5)**
Nitronaphthalene [UN 2538] **(27254-36-0)**
5-Nitronaphthalene ethylene **(602-87-9)**
7-Nitronaphth(1,2-d)(1,2,3)oxadiazole-5-sulfonic acid **(84-91-3)**
Nitronet **(55-63-0)**
Nitrong **(55-63-0)**
2-Nitro-1-(4-nitrophenoxy)-4-(trifluoromethyl)benzene **(15457-05-3)**
4-Nitro-N-(4-nitrophenyl)benzamide **(6333-15-9)**
N'-Nitro-N-nitroso-N-methylguanidine **(70-25-7)**
3-Nitro-3-nonene **(6065-04-9)**
5-Nitro-4-nonene **(6065-01-6)**
2-Nitro-2-octene **(6065-11-8)**
3-Nitro-2-octene **(6065-10-7)**
3-Nitro-3-octene **(6065-09-4)**
5-Nitro-N-(2-oxo-3-oxazolidinyl)-2-furanmethanimine **(67-45-8)**
N-Nitropendimethalin **(73215-09-5)**
Nitropenta **(78-11-5)**
Nitropentaerythrite **(78-11-5)**
Nitropentaerythritol **(78-11-5)**
1-Nitropentane **(463-04-7)**
2-Nitro-2-pentene **(6065-19-6)**
3-Nitro-2-pentene **(6065-18-5)**
3-Nitro-3-pentene **(6065-18-5)**
3-Nitroperylene **(20589-63-3)**
Nitrophen **(1836-75-5)**
2-Nitrophenanthrene **(17024-18-9)**
Nitrophene **(1836-75-5)**
4-Nitrophenethyl alcohol **(100-27-6)**

p-Nitrophenethyl alcohol (100-27-6)
o-Nitrophenethyl alcohol, acetate (833-43-2)
p-Nitrophenethyl alcohol, acetate (104-30-3)
p-Nitrophenetol (German) (100-29-8)
p-Nitrophenetole (100-29-8)
2-Nitrophenol (88-75-5)
3-Nitrophenol (554-84-7)
4-Nitrophenol (100-02-7)
Nitrophenol (25154-55-6)
Nitrophenol, Mixed (25154-55-6)
m-Nitrophenol [UN 1663] (554-84-7)
o-Nitrophenol [UN 1663] (88-75-5)
p-Nitrophenol [UN 1663] (100-02-7)
p-Nitrophenol acetate (830-03-5)
3-(α-(p-Nitrophenol)-β-acetylethyl)-4-hydroxycoumarin (152-72-7)
Nitrophenolarsonic acid (121-19-7)
Nitrophenols (25154-55-6)
4-Nitrophenol sodium salt (824-78-2)
p-Nitrophenol sodium salt (824-78-2)
2-(4-Nitrophenoxy)ethanol (16365-27-8)
2-(p-Nitrophenoxy)ethanol (16365-27-8)
p-Nitrophenoxyethanol (16365-27-8)
4-Nitrophenyl (2395-99-5)
para-Nitrophenyl (2395-99-5)
N-(2-Nitrophenyl)acetamide (552-32-9)
N-(3-Nitrophenyl)acetamide (122-28-1)
N-(4-Nitrophenyl)acetamide (104-04-1)
2-Nitrophenyl acetate (610-69-5)
4-Nitrophenyl acetate (830-03-5)
p-Nitrophenyl acetate (830-03-5)
(2-Nitrophenyl)acetonitrile (610-66-2)
(o-Nitrophenyl)acetonitrile (610-66-2)
4'-(p-Nitrophenyl)acetophenone (135-69-3)
3-(α-(4'-Nitrophenyl)-β-acetylethyl)-4-hydroxycoumarin (152-72-7)
3-(α-p-Nitrophenyl-β-acetylethyl)-4-hydroxycoumarin (152-72-7)
Nitrophenylacetylethyl-4-hydroxycoumarine (152-72-7)
m-Nitrophenylamine (99-09-2)
p-Nitrophenylamine (100-01-6)
o-Nitro-N-phenylaniline (119-75-5)
4'-Nitrophenylazo-4-(1-cyanoethyl, (N-ethyl)phenylamine) (31482-56-1)
4-((4-Nitrophenyl)azo)benzenamine (730-40-5)
4-((4-Nitrophenyl)azo)-1,3-benzenediol (74-39-5)
2-(4-Nitrophenylazo)-5-(N,N-bis(acetoxyethyl)amino)benzanilide (29765-00-2)
N-(2-((4-Nitrophenyl)azo)-5-(N,N-bis(2-acetoxyethyl)amino)-phenyl)benzamide (29765-00-2)
4-((p-Nitrophenyl)azo)diphenylamine (2581-69-3)
4-((4-((p-Nitrophenyl)azo)-2-methoxyphenyl)azo)phenol (19800-42-1)
1-((4-Nitrophenyl)azo)-2-naphthalenamine (3025-77-2)
1-((4-Nitrophenyl)azo)-2-naphthol (6410-10-2)
4-((4-Nitrophenyl)azo)phenol (1435-60-5)
4-(p-Nitrophenylazo)resorcinol (74-39-5)
p-Nitrophenylazoresorcinol (74-39-5)
2-Nitro-N-phenylbenzenamine (119-75-5)
3-Nitrophenyl carbanilate (35289-89-5)
p-Nitrophenyl cyclohexanecarboxylate (13551-17-2)
p-Nitrophenyl cyclohexylcarboxylate (13551-17-2)
D-(-)-threo-1-p-Nitrophenyl-2-dichloracetamido-1,3-propanediol (56-75-7)
D-threo-1-(p-Nitrophenyl)-2-(dichloroacetylamino)-1,3-propanediol (56-75-7)
p-Nitrophenyl diethylphosphate (311-45-5)
7-Nitro-5-phenyl-2,3-dihydro-1H-1,4-benzodiazepin-2-one (146-22-5)
4-Nitrophenyl dihydrogen phosphate (330-13-2)
p-Nitrophenyl dihydrogen phosphate (330-13-2)
4-(2'-Nitrophenyl)-2,6-dimethyl-3,5-dicarbomethoxy-1,4-dihydropyridine (21829-25-4)
4-(2'-Nitrophenyl)-2,6-dimethyl-1,4-dihydropyridin-3,5-dicarbonsaeuredimethylester (German) (21829-25-4)
p-Nitrophenyldimethylthionophosphate (298-00-0)
2-Nitro-1,4-phenylenediamine (5307-14-2)
2-Nitro-p-phenylenediamine (5307-14-2)
4-Nitro-1,2-phenylenediamine (99-56-9)
4-Nitro-1,3-phenylenediamine (5131-58-8)
4-Nitro-o-phenylene-diamine (99-56-9)
Nitro-p-phenylenediamine (5307-14-2)
o-Nitro-p-phenylenediamine (5307-14-2)
p-Nitro-o-phenylenediamine (99-56-9)
4-Nitro-o-phenylenediamine HCl (6219-77-8)
2-Nitro-p-phenylenediamine, sulfate (68239-83-8)
4-Nitro-o-phenylenediamine, sulfate (68239-82-7)
1-Nitro-1-phenylethane (7214-61-1)
p-Nitrophenylethane (100-12-9)
2-(4-Nitrophenyl)ethanol (100-27-6)
2-(p-Nitrophenyl)ethanol (100-27-6)
1-(2-Nitrophenyl)ethanone (577-59-3)
5-Nitro-2-(2-phenylethenyl)benzenesulfonic acid, sodium salt (10359-69-0)
p-Nitrophenyl ether (101-63-3)
O-(4-Nitrophenyl) O-ethyl phenyl thiophosphonate (2104-64-5)
Nitrophenyl glycidyl ether (5255-75-4)
p-Nitrophenyl glycidyl ether (5255-75-4)
p-Nitrophenyl mercaptan (1849-36-1)
Nitrophenylmethane (1321-12-6)
o-Nitrophenyl methyl ether (91-23-6)
(3-Nitrophenyl) methyl ketone (121-89-1)
p-Nitrophenyl methyl ketone (100-19-6)
4-Nitrophenyl N-methyl-N-phenylcarbamate (49839-35-2)
p-Nitrophenyl 2-nitro-4-(trifluoromethyl) phenyl ether (15457-05-3)
5-(4-Nitrophenyl)-2,4-pentadienal (2608-48-2)
3-Nitrophenyl phenylcarbamate (35289-89-5)
m-Nitrophenyl phenylcarbamate (35289-89-5)
4-Nitrophenylphenyl ether (620-88-2)
p-Nitrophenylphenyl ether (620-88-2)
(4-Nitrophenyl)phenylmethanone (1144-74-7)
O-(4-Nitrophenyl) O-phenylmethylphosphonothioate (2665-30-7)
Nitrophenylphosphate (330-13-2)
1-Nitro-1-phenylpropane (5279-14-1)
N-(4-Nitrophenyl)-N'-(3-pyridinylmethyl)urea (53558-25-1)
p-Nitrophenylsulfonic acid (138-42-1)
3-Nitrophenylsulfonyl chloride (121-51-7)
4-Nitrophenylsulfonyl chloride (98-74-8)
m-Nitrophenylsulfonyl chloride (121-51-7)
p-Nitrophenylsulfonyl chloride (98-74-8)
2-Nitrophenylsulphenyl chloride (7669-54-7)
p-Nitrophenyl o-tolyl ether (2444-29-3)
p-Nitrophenyl 2,4,6-trichlorophenyl ether (1836-77-7)
4-Nitrophenyl α,α,α-trifluoro-2-nitro-p-tolyl ether (15457-05-3)
p-Nitrophenyl α,α,α-trifluoro-2-nitro-p-tolyl ether (15457-05-3)
Nitrophos (122-14-5)
6-Nitrophthalhydrazide (3682-19-7)
4-Nitrophthalic acid anhydride (5466-84-2)
4-Nitrophthalic anhydride (5466-84-2)
3-Nitrophthalimide (603-62-3)
4-Nitrophthalimide (89-40-7)
Nitropone C (88-85-7)
Nitropore (123-77-3)
Nitropore OBSH (80-17-1)
Nitropore OBSH (80-51-3)
Nitropropane (25322-01-4)
β-Nitropropane (79-46-9)
1-Nitropropane (ACGIH,OSHA) (108-03-2)
2-Nitropropane (ACGIH,OSHA) (79-46-9)
3-Nitropropanol (25182-84-7)
3-Nitropropionic acid (504-88-1)
β-Nitropropionic acid (504-88-1)
1,1'-(2-Nitropropylidene)bis(4-chlorobenzene) (117-27-1)
1-Nitropyrene (5522-43-0)

2-Nitropyrene (789-07-1)
3-Nitropyrene (5522-43-0)
4-Nitropyrene (57835-92-4)
Nitropyrene (63021-86-3)
2-Nitropyridine (15009-91-3)
4-Nitropyridine (1122-61-8)
4-Nitropyridine 1-oxide (1124-33-0)
4-Nitropyridine-N-oxide (1124-33-0)
5-Nitropyromucate (645-12-5)
5-Nitroquinaldic acid (525-47-3)
5-Nitroquinaldinic acid (525-47-3)
5-Nitroquinoline (607-34-1)
6-Nitroquinoline (613-50-3)
8-Nitroquinoline (607-35-2)
Nitroquinoline (12408-11-6)
5-Nitro-2-quinolinecarboxylic acid (525-47-3)
4-Nitroquinoline-1-oxide (56-57-5)
4-Nitroquinoline-N-oxide (56-57-5)
Nitrorectal (55-63-0)
Nitroresorcinato de plomo (Spanish) (51317-24-9)
3-Nitrosalicylic acid (85-38-1)
5-Nitrosalicylic acid (96-97-9)
Nitrosamide (9CI) (35576-91-1)
Nitrosamine (35576-91-1)
N-Nitroso-N-(1-acetoxymethyl)butylamine (56986-36-8)
N-Nitroso-N-(acetoxy)methyl-N-methylamine (56856-83-8)
N-Nitroso-N-(1-acetoxymethyl)propyl amine (66017-91-2)
N-Nitroso-N-(acetoxy)methyl-N-n-propylamine (66017-91-2)
N-Nitrosoaethylaethanolamin (German) (13147-25-6)
Nitrosoaethyldimethylharnstoff (50285-71-7)
N-Nitrosoallylmethylamine (4549-43-3)
N-Nitrosoaminodiethanol (1116-54-7)
4-(Nitrosoamino-N-methyl)-1-(3-pyridyl)-1-butanone (64091-91-4)
(+-)-1-Nitrosoanabasine (84237-39-8)
(+-)-N-Nitrosoanabasine (84237-39-8)
1-Nitrosoanabasine (1133-64-8)
N'-Nitrosoanabasine (1133-64-8)
N'-Nitrosoanabasine (84237-39-8)
N-Nitrosoanabasine (1133-64-8)
N'-Nitrosoanatabine (71267-22-6)
N-(p-Nitrosoanilinomethyl)-2-nitropropane (24458-48-8)
N-Nitrosoazacycloheptane (932-83-2)
Nitrosobenzene (586-96-9)
N-Nitrosobenzylmethylamine (937-40-6)
N-Nitrosobis(2-hydroxyethyl)amine (1116-54-7)
N-Nitrosobis(2-hydroxypropyl)amine (53609-64-6)
N-Nitrosobutylamine (56375-33-8)
N-Nitroso-n-butylethylamine (4549-44-4)
N-Nitroso-tert-butylethylamine (3398-69-4)
N-Nitroso-n-butyl-(4-hydroxybutyl)amine (3817-11-6)
N-Nitroso-n-butylmethylamine (7068-83-9)
1-Nitroso-1-butylurea (869-01-2)
n-Nitrosobutylurea (869-01-2)
1'-Nitroso-1'-demethylnicotine (16543-55-8)
N-Nitrosodiaethanolamin (German) (1116-54-7)
N-Nitroso-diaethylamine (German) (55-18-5)
N-Nitrosodiallyl amine (16338-97-9)
N-Nitroso-di-n-butylamine (924-16-3)
N-Nitrosodibutylamine (924-16-3)
Nitrosodibutylamine (924-16-3)
N-Nitrosodicyclohexylamine (947-92-2)
N-Nitrosodiethanolamine (1116-54-7)
N-Nitroso-N,N-diethylamine (55-18-5)
N-Nitrosodiethylamine (55-18-5)
Nitrosodiethylamine (55-18-5)
Nitroso-1,1-diethyl-3-methylurea (50285-72-8)
N-Nitrosodifenylamin (Czech) (86-30-6)
p-Nitrosodifenylamin (Czech) (156-10-5)
N-Nitroso-N,N-di(2-hydroxypropyl)amine (53609-64-6)
N-Nitrosodiisopropylamine (601-77-4)
N-Nitroso-N,N-dimethylamine (62-75-9)
Nitrosodimethylamine (62-75-9)
N-Nitrosodimethylamine (ACGIH,OSHA) (62-75-9)
4-Nitrosodimethylaniline (138-89-6)
p-Nitroso-N,N-dimethylaniline (138-89-6)
p-Nitrosodimethylaniline [UN 1369] (138-89-6)
Nitroso-1,1-dimethyl-3-ethylurea (50285-71-7)
N-Nitroso-2,6-dimethylmorpholine (1456-28-6)
Nitroso-2,6-dimethylmorpholine (1456-28-6)
N-Nitrosodi-n-pentylamine (13256-06-9)
4-Nitrosodiphenylamine (156-10-5)
N-Nitrosodiphenylamine (86-30-6)
Nitrosodiphenylamine (86-30-6)
p-Nitrosodiphenylamine (156-10-5)
N-Nitrosodi-n-propylamine (621-64-7)
N-Nitrosodipropylamine (621-64-7)
para-Nitro sodium phenolate (824-78-2)
N-Nitroso-N-ethyl aniline (612-64-6)
Nitrosoethylaniline (612-64-6)
N-Nitroso-N-ethylbiuret (32976-88-8)
N-Nitrosoethyl-n-butylamine (4549-44-4)
N-Nitrosoethyl-tert-butylamine (3398-69-4)
1-Nitroso-1-ethyl-3,3-dimethylurea (50285-71-7)
Nitrosoethyldimethylurea (50285-71-7)
N-Nitrosoethylethanolamine (13147-25-6)
N-Nitroso-N-ethyl-N-(2-hydroxyethyl)amine (13147-25-6)
N-Nitrosoethyl-2-hydroxyethylamine (13147-25-6)
N-Nitrosoethylisopropylamine (16339-04-1)
N-Nitrosoethylmethylamine (10595-95-6)
4-Nitroso-N-ethyl-N-(β-methylsulfonamidoethyl)-m-toluidine (56046-62-9)
Nitrosoethylurea (759-73-9)
N-Nitroso-N-ethylurethan (614-95-9)
Nitrosoethylurethan (614-95-9)
N-Nitroso-N-ethylvinylamine (13256-13-8)
N-Nitrosoethylvinylamine (13256-13-8)
4-Nitrosofenol (Czech) (104-91-6)
N-Nitrosofenylhydroxylamin amonny (Czech) (135-20-6)
2-Nitrosofluorene (2508-20-5)
Nitrosofluorene (2508-20-5)
Nitrosofolic acid (29291-35-8)
Nitrosoguanidin (German) (674-81-7)
N-Nitrosoguanidine (674-81-7)
Nitrosoguanidine, Initiating explosive (DOT) (674-81-7)
Nitrosoguanidine [NA 0473] (674-81-7)
N-Nitrosoguvacine (55557-01-2)
N-Nitrosoguvacoline (55557-02-3)
Nitrosoguvacoline (55557-02-3)
N-Nitrosohexahydroazepine (932-83-2)
N-Nitrosohexamethyleneimine (932-83-2)
Nitrosohexamethylenimine (932-83-2)
N-Nitrosohydroxylamine (14448-38-5)
N-Nitrosohydroxyproline (30310-80-6)
2,2'-(Nitrosoimino)bisethanol (1116-54-7)
Nitrosoimino diethanol (1116-54-7)
1-Nitroso-l-proline (7519-36-0)
N-Nitroso-l-proline (7519-36-0)
N-Nitroso-N-methylacetamide (7417-67-6)
N-Nitroso-N-methyl-N-acetoxymethylamine (56856-83-8)
N-Nitrosomethylallylamine (4549-43-3)
Nitrosomethylallylamine (4549-43-3)
4-(N-Nitroso-N-methylamino)-4-(3-pyridyl)butanal (64091-90-3)
4-(N-Nitrosomethylamino)-4-(3-pyridyl)-1-butanal (64091-90-3)
4-(N-Nitroso-N-methylamino)-1-(3-pyridyl)-1-butanone (64091-91-4)
N-Nitroso-N-methyl-n-amylamine (13256-07-0)

Nitrosomethyl-n-amylamine	(13256-07-0)
N-Nitroso-N-methylaniline	(614-00-6)
Nitrosomethylaniline	(614-00-6)
N-Nitrosomethylbenzylamine	(937-40-6)
N-Nitroso-N-methyl-N-n-butylamine	(7068-83-9)
N-Nitrosomethyl-n-butylamine	(7068-83-9)
Nitrosomethyl-n-butylamine	(7068-83-9)
N-Nitroso-N-methylcarbamide	(684-93-5)
Nitrosomethyldiaethylharnstoff	(50285-72-8)
1-Nitroso-1-methyl-3,3-diethylurea	(50285-72-8)
Nitrosomethyldiethylurea	(50285-72-8)
N-Nitroso-N-methyl-N-dodecylamin (German)	(55090-44-3)
Nitrosomethyl-n-dodecylamine	(55090-44-3)
N-Nitrosomethylethylamine	(10595-95-6)
N-Nitrosomethylglycine	(13256-22-9)
N-Nitroso-N-methyl-harnstoff (German)	(684-93-5)
N-Nitroso-N-methylnitroguanidine	(70-25-7)
N-Nitroso-N-methyl-4-nitroso-aniline	(99-80-9)
Nitrosomethyl-n-pentylamine	(13256-07-0)
N-Nitrosomethylphenylamine	(614-00-6)
N-Nitroso-N-methyl-4-tolylsulfonamide	(80-11-5)
1-Nitroso-1-methylurea	(684-93-5)
N-Nitroso-N-methylurea	(684-93-5)
Nitrosomethylurea	(684-93-5)
Nitrosomethylurethan (German)	(615-53-2)
N-Nitroso-N-methylurethane	(615-53-2)
Nitrosomethylurethane	(615-53-2)
N-Nitrosomethylvinylamine	(4549-40-0)
N-Nitrosomorfolin (Czech)	(59-89-2)
N-Nitrosomorpholin (German)	(59-89-2)
4-Nitrosomorpholine	(59-89-2)
N-Nitrosomorpholine	(59-89-2)
Nitrosomorpholine	(59-89-2)
1-Nitroso-2-naftol (Czech)	(131-91-9)
α-Nitroso-β-naftol (Czech)	(131-91-9)
1-Nitroso-2-naphthol	(131-91-9)
2-Nitroso-1-naphthol	(132-53-6)
Nitroso-β-naphthol	(131-91-9)
α-Nitroso-β-naphthol	(131-91-9)
1'-Nitrosonornicotine	(16543-55-8)
N'-Nitrosonornicotine	(16543-55-8)
N'-Nitrosonornicotine	(80508-23-2)
N-Nitrosonornicotine	(16543-55-8)
Nitrosonornicotine	(16543-55-8)
Nitrosonornicotine	(80508-23-2)
N-Nitrosoperhydroazepine	(932-83-2)
4-Nitrosophenol	(104-91-6)
Nitrosophenol	(104-91-6)
p-Nitrosophenol	(104-91-6)
4-Nitrosophenol sodium salt	(823-87-0)
4-Nitroso-N-phenylaniline	(156-10-5)
N-Nitroso-N-phenylaniline	(86-30-6)
p-Nitroso-N-phenylaniline	(156-10-5)
4-Nitroso-N-phenylbenzenamine	(156-10-5)
N-Nitrosophenylhydroxylamin ammonium salz (German)	(135-20-6)
N-Nitrosophenylhydroxylamine ammonium salt	(135-20-6)
1-Nitroso-3-pipecoline	(13603-07-1)
N-Nitroso-piperidin (German)	(100-75-4)
Nitrosopiperidin (German)	(100-75-4)
1-Nitrosopiperidine	(100-75-4)
N-Nitrosopiperidine	(100-75-4)
(+-)-3-(1-Nitroso-2-piperidinyl)pyridine	(84237-39-8)
(Nitrosopropylamino)methyl acetate	(66017-91-2)
N-Nitroso-N-propyl-1-propanamine	(621-64-7)
N-Nitroso-2-(3'-pyridyl)piperidine	(1133-64-8)
1-Nitroso-2-(3-pyridyl)pyrrolidine	(16543-55-8)
N-Nitrosopyrrolidin (German)	(930-55-2)
1-Nitrosopyrrolidine	(930-55-2)
N-Nitrosopyrrolidine	(930-55-2)
3-(1-Nitroso-2-pyrrolidinyl)pyridine	(16543-55-8)
Nitrosorbid	(87-33-2)
Nitrosorbide	(87-33-2)
N-Nitrososarcosine	(13256-22-9)
Nitroso sarkosin (German)	(13256-22-9)
1-Nitroso-1,2,5,6-tetrahydronicotinic acid methyl ester	(55557-02-3)
3-Nitrosothiazolidine-2-methanol	(92134-93-5)
Nitrosotriaethylharnstoff (German)	(50285-70-6)
N-Nitrosotriethylurea	(50285-70-6)
Nitrosotriethylurea	(50285-70-6)
N-Nitroso-trimethylharnstoff (German)	(3475-63-6)
N-Nitrosotrimethylurea	(3475-63-6)
Nitrosotrimethylurea	(3475-63-6)
1-Nitrosourea, 1-(2-chloroethyl)-3-cyclohexyl-	(13010-47-4)
Nitro-span	(55-63-0)
Nitrostabilin	(55-63-0)
Nitrostarch, Containing less than 20% water [UN 0146]	(9056-38-6)
Nitrostarch, Dry [UN 0146]	(9056-38-6)
Nitrostarch, Wet with not less than 30% alcohol or solvent	(9056-38-6)
Nitrostarch, Wet with not less than 20% water [UN 1337]	(9056-38-6)
Nitrostarch	(9056-38-6)
Nitrostat	(55-63-0)
Nitrostigmin (German)	(56-38-2)
Nitrostigmine	(56-38-2)
4-Nitrostilben (German)	(4003-94-5)
4-Nitrostilbene	(4003-94-5)
Nitrostygmine	(56-38-2)
β-Nitrostyrene	(102-96-5)
γ-Nitrostyrene	(102-96-5)
Nitrosyl chloride [UN 1069]	(2696-92-6)
Nitrosyl ethoxide	(109-95-5)
Nitrosyl hydroxide	(7782-77-6)
4-Nitro-2,3,5,6-tetrachloranisole	(2438-88-2)
Nitrothiamidazol	(61-57-4)
Nitrothiamidazole	(61-57-4)
Nitrothiazole	(61-57-4)
5-Nitro-2-thiazolylamine	(121-66-4)
1-(5-Nitro-2-thiazolyl)-2-imidazolidinone	(61-57-4)
1-(5-Nitro-2-thiazolyl)imidazolidin-2-one	(61-57-4)
1-(5-Nitro-2-thiazolyl)-2-imidazolinone	(61-57-4)
1-(5-Nitro-2-thiazolyl)-2-oxotetrahydroimidazol	(61-57-4)
1-(5-Nitro-2-thiazolyl)-2-oxotetrahydroimidazole	(61-57-4)
2-Nitrothiophene	(609-40-5)
4-Nitrothiophenol	(1849-36-1)
p-Nitrothiophenol	(1849-36-1)
2-Nitrotoluene	(88-72-2)
3-Nitrotoluene	(99-08-1)
4-Nitrotoluene	(99-99-0)
Nitrotoluene	(1321-12-6)
m-Nitrotoluene (ACGIH,OSHA) [UN 1664]	(99-08-1)
o-Nitrotoluene (ACGIH,OSHA) [UN 1664]	(88-72-2)
p-Nitrotoluene (ACGIH,OSHA) [UN 1664]	(99-99-0)
Nitrotoluenes	(1321-12-6)
5-Nitro-o-toluenesulfonyl chloride	(121-02-8)
4-Nitrotoluen-2-sulfochlorid (Czech)	(121-02-8)
4-Nitrotoluen-2-sulfonylchlorid (Czech)	(121-02-8)
3-Nitro-o-toluic acid	(1975-50-4)
3-Nitro-para-toluic acid	(96-98-0)
m-Nitro-p-toluic acid	(96-98-0)
3-Nitro-4-toluidin (Czech)	(119-32-4)
2-Nitro-p-toluidine	(89-62-3)
3-Nitro-4-toluidine	(89-62-3)
3-Nitro-o-toluidine	(603-83-8)
3-Nitro-p-toluidine	(119-32-4)
5-Nitro-4-toluidine	(119-32-4)

5-Nitro-o-toluidine	**(99-55-8)**
6-Nitro-m-toluidine	**(578-46-1)**
m-Nitro-p-toluidine	**(119-32-4)**
3-Nitrotoluol	**(99-08-1)**
4-Nitrotoluol	**(99-99-0)**
Nitrotoluol	**(1321-12-6)**
Nitrotribromomethane	**(464-10-8)**
4-Nitro-1,2,3-trichlorobenzene	**(17700-09-3)**
Nitrotrichloromethane	**(76-06-2)**
Nitrotrichloromethane (OSHA)	**(76-06-2)**
2-Nitro-3,4,6-trichlorophenol	**(82-62-2)**
2-Nitro-α,α,α-trifluoro-p-cresol	**(400-99-7)**
p-Nitro(trifluoromethyl)benzene	**(402-54-0)**
4-Nitro-2-(trifluoromethyl)chlorobenzene	**(777-37-7)**
4-Nitro-3-trifluoromethylphenol	**(88-30-2)**
m-Nitrotrifluorotoluene	**(98-46-4)**
m-Nitrotrifluortoluol (German)	**(98-46-4)**
3-Nitro-N^1,N^1,N^4-tris(2-hydroxyethyl)-	**(33229-34-4)**
Nitrous-acid	**(7782-77-6)**
Nitrous Fumes	**(7697-37-2)**
Nitrous acid, ammonium salt	**(13446-48-5)**
Nitrous acid, butyl ester	**(544-16-1)**
Nitrous acid, n-butyl ester	**(544-16-1)**
Nitrous acid, sec-butyl ester	**(924-43-6)**
Nitrous acid, calcium salt (9CI)	**(13780-06-8)**
Nitrous acid, ethyl ester	**(109-95-5)**
Nitrous acid, isobutyl ester	**(542-56-3)**
Nitrous acid, isopropyl ester	**(541-42-4)**
Nitrous acid, lead(2+) salt (8CI,9CI)	**(13826-65-8)**
Nitrous acid, 3-methylbutyl ester	**(110-46-3)**
Nitrous acid, methyl ester	**(624-91-9)**
Nitrous acid, 1-methylethyl ester (9CI)	**(541-42-4)**
Nitrous acid, 1-methyl propyl ester	**(924-43-6)**
Nitrous acid, 2-methylpropyl ester (9CI)	**(542-56-3)**
Nitrous acid, pentyl ester	**(463-04-7)**
Nitrous acid, 3-phenylpropyl ester (9CI)	**(28537-55-5)**
Nitrous acid, potassium salt	**(7758-09-0)**
Nitrous acid, n-propyl ester	**(543-67-9)**
Nitrous acid, propyl ester	**(543-67-9)**
Nitrous acid, sodium salt	**(7632-00-0)**
Nitrous anhydride	**(10544-73-7)**
Nitrous diphenylamide	**(86-30-6)**
Nitrous ether (DOT)	**(109-95-5)**
Nitrous ethyl ether	**(109-95-5)**
Nitrous oxide, Compressed [UN 1070]	**(10024-97-2)**
Nitrous oxide (DOT)	**(10024-97-2)**
Nitrous oxide, Refrigerated liquid [UN 2201]	**(10024-97-2)**
Nitrovarfarian	**(152-72-7)**
Nitrovin	**(804-36-4)**
Nitrowarfarin	**(152-72-7)**
Nitrox	**(298-00-0)**
Nitrox 80	**(298-00-0)**
Nitroxanthic acid	**(88-89-1)**
2-Nitro-m-xylene	**(81-20-9)**
3-Nitro-o-xylene	**(83-41-0)**
4-Nitro-1,3-xylene	**(89-87-2)**
4-Nitro-m-xylene	**(89-87-2)**
4-Nitro-o-xylene	**(99-51-4)**
5-Nitro-m-xylene	**(99-12-7)**
Nitro-p-xylene	**(89-58-7)**
Nitroxylene	**(25168-04-1)**
p-Nitro-o-xylene	**(99-51-4)**
para-Nitro-ortho-xylene	**(99-51-4)**
4-Nitro-2,6-xylenol	**(2423-71-4)**
Nitroxylol (DOT)	**(25168-04-1)**
Nitrozan K	**(99-80-9)**
Nitrozell retard	**(55-63-0)**
Nitrozone	**(59-87-0)**
Nitrumon	**(154-93-8)**
Nitrure de lithium (French)	**(26134-62-3)**
Nitrure de mercure (French)	**(12136-15-1)**
Nitrure de selenium (French)	**(12033-59-9)**
Nitruro de litio (Spanish)	**(26134-62-3)**
Nitruro de mercurio (Spanish)	**(12136-15-1)**
Nitruro de selenio (Spanish)	**(12033-59-9)**
Nitryl kwasu nikotynowego (Polish)	**(100-54-9)**
Nitto Acid Red PG	**(3567-65-5)**
Nitto Direct Sky Blue 5B	**(2429-74-5)**
Nitto Ester T-1570	**(26446-38-8)**
Nitto Roccelline	**(1658-56-6)**
Niuif-100	**(56-38-2)**
Nivachine	**(54-05-7)**
Nivalenol	**(23282-20-4)**
Nivalenol-4-O-acetate	**(23255-69-8)**
Nivaquine	**(54-05-7)**
Nivaquine B	**(54-05-7)**
Nivemycin	**(1404-04-2)**
Nivitin	**(50-70-4)**
Nix	**(140-93-2)**
Nix-Scald	**(91-53-2)**
Nixolen	**(9003-11-6)**
Nixolen NS 4	**(9003-11-6)**
Nixolen SL 8	**(9003-11-6)**
Nixolen SL 19	**(9003-11-6)**
Nixolen VS 13	**(9003-11-6)**
Nixolen VS 2600	**(9003-11-6)**
Nixon C/A	**(9004-35-7)**
Nixon E/C	**(9004-57-3)**
Nixon N/C	**(9004-70-0)**
Nizotin	**(536-33-4)**
No. Bunt	**(118-74-1)**
No. Bunt 40	**(118-74-1)**
No. Bunt 80	**(118-74-1)**
No. Bunt Liquid	**(118-74-1)**
No. 49 Conc. Benzidine Yellow	**(6358-85-6)**
No. 3 Conc. Bronze Scarlet	**(5160-02-1)**
No. 66 Conc Lithol Toner	**(1103-39-5)**
No. 56 Conc. Permanent Orange G	**(3520-72-7)**
No. 3 Conc. Scarlet	**(5160-02-1)**
No. 2 Diesel Fuel	**(68476-34-6)**
No. 34 Forthbrite Fast Violet	**(1325-82-2)**
No. 48 Forthbrite Fast Violet	**(1325-82-2)**
No. 59 Forthfast Benzidine Yellow	**(3520-72-7)**
No. 1 Forthfast Red R	**(2814-77-9)**
No. 2 Forthfast Scarlet	**(2425-85-6)**
No. 4 Fuel oil	**(68476-31-3)**
No. 156 Orange Chrome	**(18454-12-1)**
No. 177 Orange Lake	**(633-96-5)**
No-Pest	**(62-73-7)**
No-Pest Strip	**(62-73-7)**
No-Press	**(64-55-1)**
No Scald	**(122-39-4)**
No Scald DPA 283	**(122-39-4)**
Noan	**(439-14-5)**
Nobacid	**(552-94-3)**
Nobecutan	**(137-26-8)**
Nobedon	**(103-90-2)**
Nobedorm	**(72-44-6)**
Nobelium	**(10028-14-5)**
Nobfelon	**(15687-27-1)**
Nobfen	**(15687-27-1)**
Nobgen	**(15687-27-1)**
Nobilen	**(141-90-2)**
Nobitocin S	**(50-56-6)**

Noblen	(9003-07-0)
Noblen BC 8	(9003-07-0)
Noblen D 101	(9003-07-0)
Noblen D 501	(9003-07-0)
Noblen EBG	(9003-07-0)
Noblen FA 3	(9003-07-0)
Noblen FL	(9003-07-0)
Noblen FL 4	(9003-07-0)
Noblen FP	(9003-07-0)
Noblen FS 101	(9003-07-0)
Noblen FS 2011	(9003-07-0)
Noblen H	(9003-07-0)
Noblen H 101	(9003-07-0)
Noblen H 501	(9003-07-0)
Noblen HS	(9003-07-0)
Noblen JHHG	(9003-07-0)
Noblen JK-M	(9003-07-0)
Noblen MA 4	(9003-07-0)
Noblen MH 6	(9003-07-0)
Noblen MM 2A	(9003-07-0)
Noblen S 101	(9003-07-0)
Noblen SHG	(9003-07-0)
Noblen 2VH501	(9003-07-0)
Noblen W 101	(9003-07-0)
Noblen W 501	(9003-07-0)
Noblen W 502	(9003-07-0)
Noblen WF 464	(9003-07-0)
Nobormide	(991-42-4)
Nobrium	(2898-12-6)
Noca	(1323-39-3)
Nocbin	(97-77-8)
Nocceler NS	(95-31-8)
Noclon	(60-13-9)
Nocodazole	(31430-18-9)
Nocrac 200	(128-37-0)
Nocrac 224	(26780-96-1)
Nocrac NS 5	(88-24-4)
Nocrac NS 6	(119-47-1)
Noctan	(125-64-4)
Noctazepam	(604-75-1)
Noctec	(302-17-0)
Noctilene	(72-44-6)
Noctinal	(143-81-7)
Noctivane	(56-29-1)
Noctosediv	(50-35-1)
Noctyn	(509-86-4)
Noflamol	(1336-36-3)
No. 6 fuel oil	(68553-00-4)
Nogest	(71-58-9)
Nogos	(62-73-7)
Nogos 50	(62-73-7)
Nogos 50 EC	(62-73-7)
Nogos G	(62-73-7)
Nogram	(389-08-2)
Noigen 160	(9002-92-0)
Noigen 170	(9002-92-0)
Noigen EA 102	(9036-19-5)
Noigen EA 110	(9036-19-5)
Noigen EA 120	(9036-19-5)
Noigen EA 140	(9036-19-5)
Noigen EA 142	(9036-19-5)
Noigen EA 160	(9036-19-5)
Noigen EA 170	(9036-19-5)
Noigen ES 160	(9004-96-0)
Noigen ET 60	(9004-98-2)
Noigen ET 77	(9004-98-2)
Noigen ET 80	(9004-98-2)
Noigen ET 83	(9002-92-0)
Noigen ET 102	(9002-92-0)
Noigen ET 120	(9004-98-2)
Noigen ET 143	(9002-92-0)
Noigen ET 160	(9002-92-0)
Noigen ET 170	(9002-92-0)
Noigen ET 180	(9004-98-2)
Noigen ET 190	(9002-92-0)
Noigen P	(9002-92-0)
Noigen YX 400	(9002-92-0)
Noigen YX 500	(9002-92-0)
Noir Brillant BN (French)	(2519-30-4)
Noltran	(5598-13-0)
Noludar	(125-64-4)
Nolvasan	(55-56-1)
Nolvasan	(56-95-1)
Nomate PBW	(50933-33-0)
Nomate Shootgard	(16974-11-1)
Nomersan	(137-26-8)
Nometic	(972-02-1)
Nomifensin	(24526-64-5)
Nomifensine	(24526-64-5)
Non-Flocculating Green G 25	(1328-53-6)
Nonabromobiphenyl	(27753-52-2)
Nonabromodiphenyl ether	(63936-56-1)
Nonabromodiphenyl oxide	(63936-56-1)
Nonabromophenoxybenzene	(63936-56-1)
Nonachlor	(3734-49-4)
Nonachlor, cis-	(5103-73-1)
trans-Nonachlor	(39765-80-5)
2,2',3,3',4,4',5,5',6-Nonachlorobiphenyl	(40186-72-9)
2,2',3,3',4,5,5',6,6'-Nonachlorobiphenyl	(52663-77-1)
Nonachlorobiphenyl	(53742-07-7)
Nonachloropredioxin	(35245-80-8)
Nonacosane	(630-03-5)
Nonacosane, 3-methyl- (8CI,9CI)	(14167-67-0)
Nonacosanoic acid (8CI,9CI)	(4250-38-8)
Nonacosanoic acid, methyl ester (8CI,9CI)	(4082-55-7)
Nonactin, 5,14,23,32-tetrademethyl-5,14,23,32-tetraethyl	(33956-61-5)
Nonadecane (9CI)	(629-92-5)
Nonadecane, 1-bromo- (9CI)	(4434-66-6)
Nonadecanenitrile (9CI)	(28623-46-3)
1,2-Nonadecane oxide	(67860-04-2)
Nonadecane, 2,6,10,14-tetramethyl- (9CI)	(55124-80-6)
Nonadecanoic acid (9CI)	(646-30-0)
Nonadecanoic acid, ethyl ester (8CI,9CI)	(18281-04-4)
Nonadecanoic acid, methyl ester	(1731-94-8)
Nonadecanol (9CI)	(52783-43-4)
10-Nonadecanone (9CI)	(504-57-4)
Nonadecanonitrile	(28623-46-3)
1-Nonadecene (9CI)	(18435-45-5)
Nonadecenoic acid, (Z)- (8CI,9CI)	(31627-33-5)
2,4-Nonadienal	(6750-03-4)
(E)-1,3-Nonadiene	(56700-77-7)
trans-1,3-Nonadiene	(56700-77-7)
1,3-Nonadiene, (E)- (9CI)	(56700-77-7)
trans,cis-3,6-Nonadien-1-ol	(56805-23-3)
3,6-Nonadien-1-ol, (E,Z)- (9CI)	(56805-23-3)
3,8-Nonadien-2-one, (E)- (9CI)	(55282-90-1)
2,4-Nonadien-6-yn-1-ol, (E,E)- (9CI)	(43212-86-8)
3,5-Nonadien-7-yn-2-ol, (E,E)- (9CI)	(43142-43-4)
1,8-Nonadiyne (9CI)	(2396-65-8)
Nonaethylene-glycol	(3386-18-3)
1,1,2,2,3,3,4,4,4-Nonafluorobutane-1-sulfonic acid, potassium salt	(29420-49-3)
1,1,2,2,3,3,4,4,4-Nonafluoro-1-butanesulfonyl fluoride	(375-72-4)
3,3,4,4,5,5,6,6,6-Nonafluoro-1-hexanol	(2043-47-2)

3,3,4,4,5,5,6,6,6-Nonafluoro-1-hexene **(19430-93-4)**
Nonafluoro-N-(2-hydroxyethyl)-N-methyl-1-butanesulfonamide **(34454-97-2)**
1,1,1,2,2,3,3,4,4-Nonafluoro-6-iodohexane **(2043-55-2)**
Nonahexacontanoic acid (9CI) **(40710-32-5)**
γ-Nonalactone **(104-61-0)**
1-Nonaldehyde **(124-19-6)**
Nonalol **(143-08-8)**
1,4-Nonalolide **(104-61-0)**
Nonamethylene dimethanesulfonate **(4248-77-5)**
Nonamethylenimine **(4396-27-4)**
1-Nonanal **(124-19-6)**
Nonanal **(124-19-6)**
Nonanamide, N-(2-((2-((2-((2-aminoethyl)amino)ethyl)amino)-ethyl)amino)ethyl)- (9CI) **(26392-63-2)**
Nonanamide, N,N-bis(2-hydroxyethyl)- (9CI) **(3077-37-0)**
1-Nonanamine **(112-20-9)**
n-Nonane **(111-84-2)**
Nonane (ACGIH,OSHA) [UN 1920] **(111-84-2)**
Nonane, 1-bromo- (9CI) **(693-58-3)**
Nonane, 2-bromo-5-ethyl- (7CI,9CI) **(55162-38-4)**
1-Nonanecarboxylic acid **(334-48-5)**
4-Nonanecarboxylic acid **(31080-39-4)**
Nonane, 1-chloro- (9CI) **(2473-01-0)**
Nonane, chloro- (6CI,8CI,9CI) **(34314-35-7)**
Nonane, 1,1-dibromo- (9CI) **(62168-27-8)**
Nonane, 1,9-dibromo- (9CI) **(4549-33-1)**
Nonane, 1,9-dichloro- (9CI) **(821-99-8)**
Nonane, 1,1-dichloro- (7CI,8CI,9CI) **(821-88-5)**
Nonane, 1,1-difluoro- (9CI) **(62127-42-8)**
Nonane-1,9-dimethanesulfonate **(4248-77-5)**
Nonanedioic acid **(123-99-9)**
Nonanedioic acid, bis(2-methylpropyl) ester **(105-80-6)**
Nonanedioic acid, dibutyl ester (9CI) **(2917-73-9)**
Nonanedioic acid, diethyl ester (9CI) **(624-17-9)**
Nonanedioic acid, dihexyl ester **(109-31-9)**
Nonanedioic acid, diisodecyl ester (9CI) **(28472-97-1)**
Nonanedioic acid, diisooctyl ester (9CI) **(26544-17-2)**
Nonanedioic acid, dimethyl ester (9CI) **(1732-10-1)**
Nonanedioic acid, ditridecyl ester (9CI) **(26719-40-4)**
1,9-Nonanediol (9CI) **(3937-56-2)**
1,3-Nonanediol acetate **(1322-17-4)**
1,9-Nonanediol, dimethanesulfonate **(4248-77-5)**
Nonane-1,3-diol monoacetate **(1322-17-4)**
1,3-Nonanediol, monoacetate (9CI) **(1322-17-4)**
4,6-Nonanedione (9CI) **(14090-88-1)**
Nonane, 1-fluoro **(463-18-3)**
Nonane, 2,2,4,4,6,8,8-heptamethyl- (9CI) **(4390-04-9)**
Nonane, 1-iodo- (9CI) **(4282-42-2)**
Nonane, 2-methyl- (9CI) **(871-83-0)**
Nonane, 5-methyl- (9CI) **(15869-85-9)**
Nonane, 3-methyl- (8CI,9CI) **(5911-04-6)**
Nonane, 5-(1-methylpropyl)- (9CI) **(62185-54-0)**
Nonanenitrile **(2243-27-8)**
n-Nonanenitrile **(2243-27-8)**
Nonane, 1,1'-oxybis- (9CI) **(2456-27-1)**
Nonane, phenyl- (9CI) **(79554-39-5)**
Nonane, tetrachloro-1-(dichloropropoxy)- (9CI) **(99308-26-6)**
2,4,6,8-Nonanetetraenoic acid, 9-(4-methoxy-2,3,6-trimethyl-phenyl)-3,7-dimethyl-, ethyl ester, All-trans- **(54350-48-0)**
1-Nonanethiol (9CI) **(1455-21-6)**
tert-Nonanethiol (9CI) **(25360-10-5)**
Nonane, 2,2,3-trimethyl- (9CI) **(55499-04-2)**
Nonanoic-acid **(112-05-0)**
Nonanoic acid, 2,2-bis(((1-oxononyl)oxy)methyl)-1,3-propanediyl ester **(14450-05-6)**
Nonanoic acid, butyl ester (9CI) **(50623-57-9)**
Nonanoic acid, calcium salt (9CI) **(29813-38-5)**
Nonanoic acid, 2,2-dimethyl-1,3-propanediyl ester (9CI) **(15834-05-6)**
Nonanoic acid, 1,2-ethanediylbis(oxy-2,1-ethanediyl) ester **(106-06-9)**
Nonanoic acid, ethyl ester **(123-29-5)**
Nonanoic acid, 2-ethyl-2-(((1-oxononyl)oxy)methyl)-1,3-propanediyl ester (9CI) **(126-57-8)**
Nonanoic acid, 3-hydroxy-, D- (8CI) **(33796-87-1)**
Nonanoic acid, methyl ester **(1731-84-6)**
Nonanoic acid, 8-methylnonyl ester (9CI) **(109-32-0)**
Nonanoic acid, 2-methyl-2-(((1-oxononyl)oxy)methyl)-1,3-propanediyl ester (9CI) **(10535-50-9)**
Nonanoic acid, 4-oxo- (8CI,9CI) **(6064-52-4)**
Nonanoic acid, 9-oxo-, ethyl ester (9CI) **(3433-16-7)**
Nonanoic acid, 2-sulfo-, 1-methyl ester, sodium salt (9CI) **(65520-63-0)**
1-Nonanol **(143-08-8)**
3-Nonanol **(624-51-1)**
4-Nonanol **(5932-79-6)**
Nonan-1-ol **(143-08-8)**
2-Nonanol (9CI) **(628-99-9)**
Nonanol (9CI) **(28473-21-4)**
5-Nonanol (8CI,9CI) **(623-93-8)**
Nonanol (Mixed isomers) **(28473-21-4)**
Nonanol acetate **(143-13-5)**
2-Nonanol, 5-ethyl **(103-08-2)**
1-Nonanol, 5-methyl- (8CI,9CI) **(2768-16-3)**
1-Nonanol, 1-phenyl- **(57716-72-0)**
4-Nonanol, 2,6,8-trimethyl- (8CI,9CI) **(123-17-1)**
2-Nonanone **(821-55-6)**
3-Nonanone **(925-78-0)**
5-Nonanone **(502-56-7)**
Nonan-2-one **(821-55-6)**
Nonan-5-one **(502-56-7)**
1-Nonanone, 1-phenyl- (9CI) **(6008-36-2)**
Nonanone, trimethyl **(1331-50-6)**
4-Nonanone, 2,6,8-trimethyl- (9CI) **(123-18-2)**
Nonanonitrile **(2243-27-8)**
n-Nonanonitrile **(2243-27-8)**
Nonanophenone (8CI) **(6008-36-2)**
Nonanoyl chloride (9CI) **(764-85-2)**
Nonanoyl peroxide **(762-13-0)**
n-Nonanyl acetate **(143-13-5)**
3,6,9,12,15,18,21,24,27-Nonaoxanonacosan-1-ol, 29-(iso-nonylphenoxy)- (9CI) **(65455-72-3)**
3,6,9,12,15,18,21,24,27-Nonaoxanonacosan-1-ol, 29-(4-(1,1,3,3-tetramethylbutyl)phenoxy)- (9CI) **(2315-66-4)**
3,6,9,12,15,18,21,24,27-Nonaoxanonacosan-1-ol, 29-(p-(1,1,3,3-tetramethylbutyl)phenoxy)- (8CI) **(2315-66-4)**
Nonasiloxane, eicosamethyl- (9CI) **(2652-13-3)**
2,4,6,8-Nonatetraenoic acid, 9-(4-methoxy-2,3,6-tri-methylphenyl)-3,7-dimethyl-, ethyl ester, (All-E)- (9CI) **(54350-48-0)**
2,4,6,8-Nonatetraenoic acid, 9-(4-methoxy-2,3,6-trimethylphenyl)-, ethyl ester, (All-E-) **(54350-48-0)**
2,4,6,8-Nonatetraen-1-ol, 3,7-dimethyl-9-(2,6,6-trimethyl-1-cyclohexen-1-yl)-, (all-E)- **(68-26-8)**
2,4,6,8-Nonatetranoic acid, 3,7-dimethyl-9-(2,6,6-trimethyl-1-cyclohexen-1-yl)- **(302-79-4)**
2-Nonen-1-al **(2463-53-8)**
2-Nonenal **(2463-53-8)**
4-Nonenal, (E)- (8CI,9CI) **(2277-16-9)**
6-Nonenamide, N-((4-hydroxy-3-methoxyphenyl)methyl)-8-methyl-, (E) **(404-86-4)**
6-Nonenamide, 8-methyl-N-vanillyl-, (E)- (8CI) **(404-86-4)**
α-Nonene **(124-11-8)**
n-Non-1-ene **(124-11-8)**
1-Nonene (9CI) **(124-11-8)**
Nonene (9CI) **(27215-95-8)**
Nonene (Mixed isomers) **(27215-95-8)**
Nonene (Non-linear) **(27215-95-8)**

Name	CAS
3-Nonene, 3-nitro	**(6065-04-9)**
4-Nonene, 5-nitro	**(6065-01-6)**
1-Nonene, 4,6,8-trimethyl- (6CI,9CI)	**(54410-98-9)**
6-Nonenoic acid, (E)- (8CI,9CI)	**(31502-23-5)**
8-Nonenoic acid (6CI,8CI,9CI)	**(31642-67-8)**
8-Nonenoic acid, ethyl ester (7CI,9CI)	**(35194-39-9)**
3-Nonen-2-one (9CI)	**(14309-57-0)**
8-Nonen-2-one (8CI,9CI)	**(5009-32-5)**
7-Nonen-2-one, 4,8-dimethyl- (7CI,8CI,9CI)	**(3664-64-0)**
2-Nonen-4-one, 2-methyl- (8CI,9CI)	**(2903-23-3)**
α-Nonenyl aldehyde	**(2463-53-8)**
Nonex 25	**(9004-96-0)**
Nonex 30	**(9004-96-0)**
Nonex 32	**(1323-39-3)**
Nonex 52	**(9004-96-0)**
Nonex 64	**(9004-96-0)**
Nonex 80	**(9005-08-7)**
Nonex 411	**(106-11-6)**
Nonex 413	**(141-20-8)**
Nonflex RD	**(26780-96-1)**
Nonidet A 50 (9CI)	**(11106-35-7)**
Nonidet P 40	**(9036-19-5)**
Nonidet P40	**(9036-19-5)**
Noniolite AL 20	**(9002-92-0)**
Noniolite AO 5	**(9004-98-2)**
Noniolite AO 20	**(9004-98-2)**
Noniolite S 100	**(9004-99-3)**
Nonion 06	**(9004-96-0)**
Nonion HS 206	**(9036-19-5)**
Nonion HS 208	**(9036-19-5)**
Nonion O2	**(9004-96-0)**
Nonion O4	**(9004-96-0)**
Nonion OP80R	**(1338-43-8)**
Nonion PP40	**(26266-57-9)**
Nonion S 15	**(9004-99-3)**
Nonion S 2	**(9004-99-3)**
Nonion S 220	**(9005-00-9)**
Nonion S 4	**(9004-99-3)**
Nonion SP 60	**(1338-41-6)**
Nonion SP 60R	**(1338-41-6)**
Nonisol 200	**(9004-96-0)**
n-Nonoic acid	**(112-05-0)**
Nonox CL	**(93-46-9)**
Nonox D	**(135-88-6)**
Nonox DPPD	**(74-31-7)**
Nonox TBC	**(128-37-0)**
Nonox WSP	**(77-62-3)**
Nonox ZA	**(101-72-4)**
Nonoxynol	**(9016-45-9)**
Nonoxynol-4	**(7311-27-5)**
Nonoxynol-9	**(26027-38-3)**
Nonoxynol-9	**(98113-10-1)**
Nonplesin	**(511-45-5)**
Nonyl Nonoxynol-5	**(9014-93-1)**
Nonyl Nonoxynol-10	**(9014-93-1)**
Nonyl Nonoxynol-49	**(9014-93-1)**
Nonyl Nonoxynol-100	**(9014-93-1)**
Nonyl Nonoxynol-150	**(9014-93-1)**
Nonyl acetate	**(143-13-5)**
n-Nonyl acetate	**(143-13-5)**
Nonyl-alcohol	**(143-08-8)**
n-Nonyl alcohol	**(143-08-8)**
sec-Nonyl alcohol	**(108-82-7)**
1-Nonyl aldehyde	**(124-19-6)**
1-Nonylamine	**(112-20-9)**
Nonylbenzene	**(1081-77-2)**
.α.-Nonylbenzenemethanol	**(21078-95-5)**
α.-Nonylbenzyl alcohol	**(21078-95-5)**
4-Nonyl-2,6-bis(1-phenylethyl)phenol	**(15860-96-5)**
n-Nonyl bromide	**(693-58-3)**
Nonylcarbinol	**(112-30-1)**
Nonylcyclohexane	**(2883-02-5)**
Nonylene	**(124-11-8)**
n-Nonyl ethanoate	**(143-13-5)**
Nonyl ether (8CI)	**(2456-27-1)**
α-Nonylethylene	**(821-95-4)**
5-Nonyl-2-hydroxybenzophenoxime	**(37339-32-5)**
n-Nonylic acid	**(112-05-0)**
n-Nonyl iodide	**(4282-42-2)**
tert-Nonyl mercaptan	**(25360-10-5)**
Nonyl methacrylate	**(2696-43-7)**
Nonyl methyl ketone	**(112-12-9)**
ar-Nonyl-N-(nonylphenyl)benzenamine	**(36878-20-3)**
4-Nonylphenol	**(104-40-5)**
para Nonyl phenol	**(104-40-5)**
Nonyl phenol (Mixed isomers)	**(25154-52-3)**
Nonylphenol, ethoxylated and phosphated	**(51811-79-1)**
Nonyl phenol, ethoxylated, phosphated, sodium salt	**(37340-60-6)**
Nonylphenol ethoxylated, phosphated, sodium salt	**(37340-60-6)**
p-Nonylphenol, ethoxylate, sulfate, ammonium salt	**(31691-97-1)**
Nonyl phenol, glycidyl polyether	**(68072-38-8)**
Nonylphenol mono(oxyethylene) ethanol	**(27176-93-8)**
Nonylphenol phosphate (3:1)	**(26569-53-9)**
Nonylphenol phosphite (3:1)	**(26523-78-4)**
Nonylphenol, polyoxyethylene ether	**(9016-45-9)**
Nonylphenol polyoxyethylene sulfuric acid	**(9081-17-8)**
Nonylphenol sulfide	**(28503-85-7)**
Nonyl phenol sulfide solution	**(34992-00-2)**
(4-Nonylphenoxy)acetic acid	**(3115-49-9)**
2-(Nonylphenoxy)ethanol	**(27986-36-3)**
2-(2-(4-Nonylphenoxy)ethoxy)ethanol	**(20427-84-3)**
2-(2-(Nonylphenoxy)ethoxy)ethanol	**(27176-93-8)**
((4-Nonylphenoxy)methyl)oxirane	**(6178-32-1)**
Nonylphenoxypolyethoxyethanol - iodine complex	**(35860-86-7)**
Nonylphenoxypoly(ethyleneoxy)ethanol	**(26027-38-3)**
Nonylphenoxypolyethyleneoxy ethanol-iodine complex	**(11096-42-7)**
Nonylphenoxypoly(ethyleneoxy)ethyl ester of phosphoric acid, sodium salt	**(37340-60-6)**
Nonylphenyl diphenyl phosphate	**(38638-05-0)**
p-Nonylphenyl glycidyl ether	**(6178-32-1)**
Nonyl phenyl polyethylene glycol	**(9016-45-9)**
Nonyl phenyl polyethylene glycol ether	**(9016-45-9)**
Nonylphenyl sulfide	**(28503-85-7)**
tert-Nonylphenyl undecaethylene glycol ether	**(37281-58-6)**
Nonyl trichlorosilane	**(5283-67-0)**
Nonyltrichlorosilane [UN 1799]	**(5283-67-0)**
1-Nonyne (9CI)	**(3452-09-3)**
8-Nonynoic acid (8CI,9CI)	**(30964-01-3)**
8-Nonynoic acid, methyl ester (7CI,8CI,9CI)	**(7003-48-7)**
(+)-Nootkatane	**(15404-63-4)**
Nootkatane	**(15404-63-4)**
Nootkatane, (+)-	**(15404-63-4)**
Nootkatone	**(4674-50-4)**
Nopalcol 1-0	**(9004-96-0)**
Nopalcol 6-0	**(9004-96-0)**
Nopalcol 6-L	**(9004-81-3)**
Nopalcol 4-O	**(9004-96-0)**
Nopcaine	**(54-35-3)**
Nopcocide	**(1897-45-6)**
Nopcocide N-96	**(1897-45-6)**
Nopcocide N40D & N96	**(1897-45-6)**
Nopcogen 14-l	**(3352-87-2)**
Nopcote C 104	**(1592-23-0)**
Nopcowax 22-DS	**(110-30-5)**

Nopil **(8064-90-2)**
Nopinen **(127-91-3)**
Nopinene **(127-91-3)**
Nopinon **(24903-95-5)**
Nopinone **(24903-95-5)**
Noplen FL 6314 **(9003-07-0)**
Nopol (Polymer) **(9002-88-4)**
Nopol acetate **(128-51-8)**
Nopropiophenone **(90-84-6)**
Noptil **(50-06-6)**
Nopyl acetate **(128-51-8)**
Nor-Am EP 332 **(23422-53-9)**
Nor-Am EP 333 **(19750-95-9)**
Nor-Press 25 **(304-20-1)**
Nor-Q.D. **(68-22-4)**
Noracimetadol (Spanish) **(1477-39-0)**
Noracimetadolo **(1477-39-0)**
Noracymethadol **(1477-39-0)**
Noracymethadolum (Latin) **(1477-39-0)**
(-)-Noradrec **(51-41-2)**
Noradrenalin **(51-41-2)**
Noradrenalina (Italian) **(51-41-2)**
(-)-Noradrenaline **(51-41-2)**
D-(-)-Noradrenaline **(51-41-2)**
Noradrenaline **(51-41-2)**
l-Noradrenaline **(51-41-2)**
(+-)-Noradrenaline hydrochloride **(55-27-6)**
dl-Noradrenaline hydrochloride **(55-27-6)**
Noradrenline **(51-41-2)**
Noradrin **(300-42-5)**
Noral Aluminum **(7429-90-5)**
Noral Extra Fine Lining Grade **(7429-90-5)**
Noral Ink Grade Aluminum **(7429-90-5)**
Noral Non-Leafing Grade **(7429-90-5)**
Noram O **(112-90-3)**
Noramidone **(467-85-6)**
Noramitriptyline **(72-69-5)**
Noramox C **(61791-14-8)**
Norandrostenolon **(434-22-0)**
19-Norandrostenolone **(434-22-0)**
Norandrostenolone **(434-22-0)**
6a-β-Noraporphine-10,11-diol, 6-methyl-, hydrochloride **(314-19-2)**
Norartrinal **(51-41-2)**
Norazine **(3004-71-5)**
Norbilan **(536-50-5)**
Norboral **(339-43-5)**
Norbormide **(991-42-4)**
2,5-Norbornadiene **(121-46-0)**
Norbornadiene **(121-46-0)**
2,5-Norbornadiene, 1,2,3,4,7,7-hexachloro- (8CI) **(3389-71-7)**
2-Norbornanamine, N-ethyl-3-phenyl- **(1209-98-9)**
Norbornane, 2,2-dimethyl-3-methylene-, octachloro deriv. **(1319-80-8)**
Norbornane, 2,3-epoxy-, exo **(3146-39-2)**
Norbornane oxide, exo-2,3- **(3146-39-2)**
2-Norbornanol, 1,2,7,7-tetramethyl- **(2371-42-8)**
2-Norbornanol, 1,2,7,7-tetramethyl-, exo- (8CI) **(2371-42-8)**
2-Norbornanol, 1,3,3-trimethyl-, (-)-endo **(512-13-0)**
2-Norbornanone **(497-38-1)**
2-Norbornanone, endo-3-chloro-exo-6-cyano-, O-(methylcarbamoyl)oxime **(15271-41-7)**
2-Norbornanone, 1,3,3-trimethyl **(1195-79-5)**
2-Norbornanone, 1,3,3-trimethyl- (8CI) **(1195-79-5)**
2-Norbornanone, 1,3,3-trimethyl-, (1R,4S)-(+) **(4695-62-9)**
2-Norbornene **(498-66-8)**
5-Norbornene-2,3-dicarboximide, N-(2-ethylhexyl) **(113-48-4)**
5-Norbornene-2,3-dicarboximide, 1,4,5,6,7,7-hexachloro-N-(ethylmercuri)- **(2597-93-5)**
5-Norbornene-2,3-dicarboximide, N-hydroxy- **(21715-90-2)**
5-Norbornene-2,3-dicarboximide, 5-(α-hydroxy-α-2-pyridylbenzyl)-7-(α-2- pyridylbenzylidene) **(991-42-4)**
5-Norbornene-2,3-dicarboxylic acid, dimethyl ester, cis-endo-(8CI) **(39589-98-5)**
5-Norbornene-2,3-dicarboxylic acid, 1,4,5,6,7,7-hexachloro **(115-28-6)**
5-Norbornene-2,3-dicarboxylic anhydride, 1,4,5,6,7,7-hexachloro-(8CI) **(115-27-5)**
5-Norbornene-2,3-dicarboxylic anhydride, methyl **(25134-21-8)**
5-Norbornene-2,3-dimethanol, 1,4,5,6,7,7-hexachloro-, cyclic sulfate **(1031-07-8)**
5-Norbornene-2,3-dimethanol, 1,4,5,6,7,7-hexachloro-, cyclic sulfite **(115-29-7)**
5-Norbornene-2,3-dimethanol, 1,4,5,6,7,7-hexachloro-, cyclic sulfite, endo **(959-98-8)**
5-Norbornene-2,3-dimethanol, 1,4,5,6,7,7-hexachloro-, cyclic sulfite, exo **(33213-65-9)**
2-Norbornene, 2,3-dimethyl- (8CI) **(529-16-8)**
2-Norbornene, 5-ethylidene **(16219-75-3)**
5-Norbornene-2-methanol (8CI) **(95-12-5)**
5-Norbornene-2-methanol, acrylate **(95-39-6)**
5-Norbornene-2-methylolacrylate **(95-39-6)**
exo-Norbornene oxide **(3146-39-2)**
2-Norbornene, 1,2,3,4,7-pentachloro- **(5825-64-9)**
2-Norbornene, 1,2,3,4,7-pentachloro-, syn- (8CI) **(18317-90-3)**
2-Norbornene, 5-vinyl **(3048-64-4)**
2-Norbornen-7-ol (6CI,7CI) **(53783-87-2)**
Norbornylene **(498-66-8)**
Norcain **(94-09-7)**
Norcamphene **(498-66-8)**
Norcamphor **(497-38-1)**
Norcamphor, 1,7,7-trimethyl- **(76-22-2)**
Norcarane (6CI,8CI) **(286-08-8)**
Norcarane, 7,7-dimethyl-3-methylene- **(554-60-9)**
.δ.3-Norcarene **(16554-83-9)**
3-Norcarene (6CI,7CI,8CI) **(16554-83-9)**
3-Norcarene, 3,7,7-trimethyl- **(13466-78-9)**
Norcassamidine **(36150-73-9)**
Norcocaine **(18717-72-1)**
N-Norcodeine **(467-15-2)**
Norcodeine **(467-15-2)**
Norcodeine, N-methyl- **(76-57-3)**
.α.-Norconidendrin **(2316-10-1)**
Norcozine **(69-09-0)**
20-Norcrotalanan-11,15-dione, 14,19-dihydro-12,13-dihydroxy-, (13-α,14-α)- (9CI) **(315-22-0)**
Norden **(104-14-3)**
Nordhausen Acid (DOT) **(7664-93-9)**
Nordiazepam **(1088-11-5)**
Nordicol **(50-28-2)**
Nordihydroguaiaretic acid **(500-38-9)**
Nordihydroguairaretic acid **(500-38-9)**
Nordimaprit **(16111-27-6)**
Nordopan **(66-75-1)**
Nordox **(1317-39-1)**
Norea **(18530-56-8)**
Norea **(2163-79-3)**
Norephedrane **(300-62-9)**
Norephedrane **(60-13-9)**
dl-Norephedrine **(14838-15-4)**
psi-Norephedrine **(492-39-7)**
Norephedrine, deoxy- **(300-62-9)**
dl-Norephedrine hydrochloride **(154-41-6)**
Norephedrine, N-methyl- **(299-42-3)**
(-)-Norepinephrine **(51-41-2)**
Norepinephrine **(51-41-2)**
l-Norepinephrine **(51-41-2)**

Name	CAS
(+-)-Norepinephrine hydrochloride	**(55-27-6)**
dl-Norepinephrine hydrochloride	**(55-27-6)**
Norepirenamine	**(51-41-2)**
Nores	**(18530-56-8)**
19-Nor-ethindrone	**(68-22-4)**
Norethindrone	**(68-22-4)**
Norethindrone 17-acetate	**(51-98-9)**
Norethindrone acetate	**(51-98-9)**
Norethinodrel	**(68-23-5)**
19-Nor-17-ethinyltestosterone	**(68-22-4)**
19-Nor-ethinyl-4,5-testosterone	**(68-22-4)**
19-Nor-ethinyl-5,10-testosterone	**(68-23-5)**
19-Norethinyltestosterone	**(68-22-4)**
Norethinynodrel	**(68-23-5)**
Norethisteron	**(68-22-4)**
Norethisteron acetate	**(51-98-9)**
19-Norethisterone	**(68-22-4)**
Norethisterone	**(68-22-4)**
19-Norethisterone acetate	**(51-98-9)**
Norethisterone acetate	**(51-98-9)**
Norethyndron	**(68-22-4)**
Norethynodral	**(68-23-5)**
19-Norethynodrel	**(68-23-5)**
Norethynodrel	**(68-23-5)**
Norethynodrone	**(68-22-4)**
19-Nor-17-α-ethynylandrosten-17-β-ol-3-one	**(68-22-4)**
19-Nor-17-α-ethynyl-17-β-hydroxy-4-androsten-3-one	**(68-22-4)**
19-Nor-17-α-ethynyltestosterone	**(68-22-4)**
19-Norethynyltestosterone acetate	**(51-98-9)**
Norethynyltestosterone acetate	**(51-98-9)**
Norethysterone acetate	**(51-98-9)**
Noretone	**(3035-45-8)**
Norex	**(1982-47-4)**
Norfarnesane	**(6864-53-5)**
Norflurane	**(811-97-2)**
Norflurano (Spanish)	**(811-97-2)**
Norfluranum (Latin)	**(811-97-2)**
Norflurazon	**(27314-13-2)**
Norflurazone	**(27314-13-2)**
Norforms	**(62-38-4)**
Norgesic	**(62-44-2)**
(+-)-Norgestrel	**(6533-00-2)**
D-Norgestrel	**(797-63-7)**
Norgestrel	**(6533-00-2)**
α-Norgestrel	**(6533-00-2)**
d(-)-Norgestrel	**(797-63-7)**
d-Norgestrel	**(797-63-7)**
dl-Norgestrel	**(6533-00-2)**
Norgine	**(9005-32-7)**
Norguaiaretic acid, dihydro-	**(500-38-9)**
Norharman	**(244-63-3)**
Noridil	**(396-01-0)**
Norilgan-S	**(127-69-5)**
Norimipramine	**(50-47-5)**
Norisodrine hydrochloride	**(51-30-9)**
Norit	**(7440-44-0)**
5-Norkhellin	**(478-42-2)**
Norkool	**(107-21-1)**
15-Nor-5.β.,8.β.H,8.β.H,10.α.-labdan-14-al, 8,13-epoxy-19-hydroxy-, (13R)-(-)- (8CI)	**(17904-23-3)**
19-Nor-9-β,10-α-lanosta-5,23-diene-11,22-dione, 3-β,16-α,20,25-tetrahydroxy- 9-(hydroxymethyl)-, 25-acetate	**(5988-76-1)**
19-Nor-9-β,10-α-lanosta-5,23-diene-3,11,22-trione, 2-β,16-α,20,25- tetrahydroxy-9-methyl-, 25-acetate	**(6199-67-3)**
Norleucamine	**(110-58-7)**
L-(+)-Norleucine	**(327-57-1)**
Norleucine	**(327-57-1)**
ε-Norleucine	**(60-32-2)**
DL-Norleucine (9CI)	**(616-06-8)**
L-Norleucine (9CI)	**(327-57-1)**
Norleucine, DL- (8CI)	**(616-06-8)**
Norleucine, l	**(327-57-1)**
Norlevorfanol (Spanish)	**(1531-12-0)**
Norlevorphanol	**(1531-12-0)**
Norlevorphanolum (Latin)	**(1531-12-0)**
Norlutate	**(51-98-9)**
Norlutin	**(68-22-4)**
Norlutine acetate	**(51-98-9)**
Normi-Nox	**(72-44-6)**
Normadrine	**(300-42-5)**
Normal Lead Orthophosphate	**(7446-27-7)**
Normalip	**(637-07-0)**
Normal lead acetate	**(301-04-2)**
Normal paraffins	**(64771-72-8)**
Normat	**(637-07-0)**
Normedon	**(467-85-6)**
Normeperidine	**(77-17-8)**
Normersan	**(137-26-8)**
Normet	**(637-07-0)**
Normetadona (Spanish)	**(467-85-6)**
Normetadone	**(467-85-6)**
Normethadone	**(467-85-6)**
Normethadonum (Latin)	**(467-85-6)**
Normimycin V	**(56-75-7)**
Normison	**(846-50-4)**
Normiten	**(29122-68-7)**
Normocytin	**(68-19-9)**
Normodyne	**(36894-69-6)**
Normolipol	**(637-07-0)**
Normonson	**(113-18-8)**
Normorescina	**(24815-24-5)**
Normorfina (Spanish)	**(466-97-7)**
(-)-Normorphine	**(466-97-7)**
N-Normorphine	**(466-97-7)**
Normorphine	**(466-97-7)**
Normorphine, N-allyl-	**(62-67-9)**
Normorphine 3-methyl ether	**(467-15-2)**
Normorphinone, N-allyl-7,8-dihydro-14-hydroxy-, (-)-	**(465-65-6)**
Normorphinum (Latin)	**(466-97-7)**
Normosan	**(113-18-8)**
Normoson	**(113-18-8)**
Normuscone	**(502-72-7)**
Nornicotine, N-nitroso-	**(16543-55-8)**
Norocaine	**(59-46-1)**
Norodin	**(300-42-5)**
Norodrin	**(300-42-5)**
Norox BZP-250	**(94-36-0)**
Norox BZP-C-35	**(94-36-0)**
Norpethidine	**(77-17-8)**
Norphedrane	**(60-13-9)**
Norphen	**(104-14-3)**
Norphenazone	**(89-25-8)**
2-Norpinanone, 6,6-dimethyl- (8CI)	**(24903-95-5)**
2-Norpinene-2-ethanol, 6,6-dimethyl-, acetate	**(128-51-8)**
Norpipanona (Spanish)	**(561-48-8)**
Norpipanone	**(561-48-8)**
Norpipanonum (Latin)	**(561-48-8)**
Norprazepam	**(1088-11-5)**
19-Nor-17-α-pregna-1,3,5(10)-triene-20-yne-3,17-diol	**(57-63-6)**
(17-α)-19-Norpregna-1,3,5(10)-trien-20-yne-3,17,diol	**(57-63-6)**
19-Nor-17-α-pregna-1,3,5(10)-trien-20-yne-3,17-diol	**(57-63-6)**
17-α-19-Norpregna-1,3,5(10)-trien-20-yn-17-ol, 3-methoxy	**(72-33-3)**
Norpregneninlone	**(68-22-4)**
19-Nor-17-α-pregn-4-en-17-ol	**(965-90-2)**

Name	CAS
Nu Rexform	**(3687-31-8)**
Nu-Tone	**(86-87-3)**
Nuarsol	**(127-85-5)**
Nubarene	**(340-57-8)**
Nubian Yellow TB	**(131-22-6)**
Nubilon Orange R	**(633-96-5)**
Nuchar	**(7440-44-0)**
Nucidol	**(333-41-5)**
Nucin	**(481-39-0)**
Nucite	**(87-89-8)**
Nucleocardyl	**(58-61-7)**
Nuctalon	**(29975-16-4)**
Nudrin	**(16752-77-5)**
Nufluor	**(7681-49-4)**
Nujol	**(8012-95-1)**
Nullapon	**(64-02-8)**
Nullapon B	**(64-02-8)**
Nullapon BF-12	**(64-02-8)**
Nullapon BF-78	**(64-02-8)**
Nullapon BFC	**(64-02-8)**
Nullapon BFC Conc	**(64-02-8)**
Nullapon BFC Conc Beads	**(64-02-8)**
Nullapon BFC Liquid	**(64-02-8)**
Nullapon BF acid	**(60-00-4)**
Nullapon B acid	**(60-00-4)**
Nulomoline	**(8013-17-0)**
Numal	**(77-02-1)**
Number 2 Burner fuel	**(68476-30-2)**
Number 2 Fuel Oil	**(68476-30-2)**
Nunol	**(50-06-6)**
Nuocure 28	**(301-10-0)**
Nuodex V 1525	**(78-04-6)**
Nuoplaz	**(119-06-2)**
Nuoplaz DOP	**(117-81-7)**
Nuosept 95	**(6542-37-6)**
Nupercainal	**(85-79-0)**
Nupercaine	**(85-79-0)**
Nupol 1629	**(1565-94-2)**
Nupol 46-4005	**(1565-94-2)**
Nuredal	**(51-12-7)**
Nurelle	**(7159-34-4)**
Nurelle	**(95-95-4)**
Nurofen	**(15687-27-1)**
Nusyn-Noxfish	**(51-03-6)**
Nutek 7C	**(9003-11-6)**
Nutmeg Butter	**(8007-12-3)**
Nutmeg Oil	**(8008-45-5)**
Nutrasweet	**(22839-47-0)**
Nutri-Sperse Copper 50	**(1332-03-2)**
Nutrifos STP	**(7601-54-9)**
Nutrop	**(155-41-9)**
Nuva	**(62-73-7)**
Nuvacron	**(6923-22-4)**
Nuvacron 20	**(6923-22-4)**
Nuvacthen Depot	**(16960-16-0)**
Nuvan	**(62-73-7)**
Nuvan 7	**(62-73-7)**
Nuvan 100EC	**(62-73-7)**
Nuvanol	**(122-14-5)**
Nuvanol N	**(18181-70-9)**
Nuvapen	**(69-53-4)**
Nyacol	**(7631-86-9)**
Nyacol 830	**(7631-86-9)**
Nyacol 1430	**(7631-86-9)**
Nyacol A 1530	**(1309-64-4)**
Nyanthrene Blue BFP	**(130-20-1)**
Nyanthrene Brilliant Green B	**(128-58-5)**
Nyanthrene Brilliant Pink R	**(2379-74-0)**
Nyanthrene Brilliant Violet 4R	**(1324-55-6)**
Nyanthrene Brown RB	**(2475-33-4)**
Nyanthrene Golden Orange G	**(128-70-1)**
Nyanthrene Golden Yellow	**(128-66-5)**
Nyanthrene Olive R	**(2379-81-9)**
Nyanthrene Red G 2B	**(4203-77-4)**
Nyanza Sky Blue 6B	**(2610-05-1)**
Nyasol Red GGS	**(6408-31-7)**
Nyazin	**(51-12-7)**
Nyco Liquid Blue BF	**(1325-87-7)**
Nyco Liquid Red GF	**(989-38-8)**
Nyco Liquid Violet RF	**(1325-82-2)**
Nyco Super Blue B	**(1325-87-7)**
Nyco Super Violet 4R	**(1325-82-2)**
Nycoton	**(302-17-0)**
Nyctal	**(77-65-6)**
Nycton	**(302-17-0)**
Nydrazid	**(54-85-3)**
S-Nyl-P 42	**(9003-20-7)**
Nylidrin	**(447-41-6)**
Nylidrinum	**(447-41-6)**
Nylmerate	**(62-38-4)**
Nylon-6	**(25038-54-4)**
Nylofil Blue BLL	**(147-14-8)**
Nylomine Acid Blue B-B	**(2666-17-3)**
Nylomine Acid Red P4B	**(3567-69-9)**
Nylomine Acid Scarlet C-R	**(3567-65-5)**
Nylomine Acid Scarlet P-R	**(3567-65-5)**
Nylomine Acid Yellow B-RD	**(6373-74-6)**
Nylon	**(32131-17-2)**
Nylon	**(63428-83-1)**
Nylon A1035SF	**(105-60-2)**
Nylon A1035SF	**(25038-54-4)**
Nylon BCF 800	**(63428-84-2)**
Nylon CM 1031	**(105-60-2)**
Nylon CM 1031	**(25038-54-4)**
Nylon SI-N	**(32131-17-2)**
Nylon 6-12 Salt	**(13188-60-8)**
Nylon (VAN)	**(63428-84-2)**
Nylon X 1051	**(105-60-2)**
Nylon X 1051	**(25038-54-4)**
Nylon 2-nylon 6 copolymer	**(54590-59-9)**
Nylon 610 salt	**(6422-99-7)**
Nylon 66 salt	**(3323-53-3)**
Nyloquinone Blue 2J	**(2475-45-8)**
Nyloquinone Light Yellow 4JL	**(2832-40-8)**
Nyloquinone Orange JR	**(82-28-0)**
Nyloquinone Pure Blue	**(2475-46-9)**
Nyloquinone Pure Blue	**(86722-66-9)**
Nyloquinone Pure Blue R	**(2475-46-9)**
Nyloquinone Pure Blue R	**(86722-66-9)**
Nyloquinone Violet R	**(128-95-0)**
Nyloquinone Yellow 4J	**(2832-40-8)**
Nyloquinone Yellow 2R	**(119-15-3)**
Nylsuisse	**(63428-84-2)**
Nymcel S	**(9004-32-4)**
Nymcel SLC-T	**(9004-32-4)**
Nymcel ZSB 10	**(9004-32-4)**
Nymcel ZSB 16	**(9004-32-4)**
Nyscaps	**(91-84-9)**
Nysconitrine	**(55-63-0)**
Nyscozid	**(54-85-3)**
Nystan	**(1400-61-9)**
Nystatin	**(1400-61-9)**
Nystatine	**(1400-61-9)**
Nystatyna (Polish)	**(1400-61-9)**

Nystavescent **(1400-61-9)**
Nytal 200 **(14807-96-6)**
Nytal 400 **(14807-96-6)**
O 250 **(1338-43-8)**
O-2857 **(95-16-9)**
OA 20 **(9004-98-2)**
OA 8 **(9004-98-2)**
OAAT **(97-56-3)**
OAP **(999-97-3)**
1A-4OA (Russian) **(116-85-8)**
OBB **(27858-07-7)**
OH-BBN **(3817-11-6)**
OBPA **(58-36-6)**
OBSH **(80-51-3)**
OCDD **(3268-87-9)**
OCH **(4024-81-1)**
OCI 56 **(2893-78-9)**
OCPA **(5345-54-0)**
OCPNA **(121-87-9)**
OCS-21,944 **(3765-57-9)**
OCTA **(482-54-2)**
OCTD **(76379-67-4)**
ODA **(128-44-9)**
ODB **(50-50-0)**
ODB **(95-50-1)**
ODCB **(95-50-1)**
Z,Z-ODDA **(53120-27-7)**
ODP **(1330-96-7)**
OE3 **(50-27-1)**
OF 7 **(9036-19-5)**
OFHC Cu **(7440-50-8)**
OFNA-Perl Salt RRA **(3165-93-3)**
OG 1 **(9005-38-3)**
3-OHAA **(548-93-6)**
OHB **(65-45-2)**
OHB12 **(13422-51-0)**
OH-Duphar **(13422-51-0)**
OJI Malachite Green **(569-64-2)**
OK 622 **(1910-42-5)**
OK 7 **(9004-96-0)**
OKO **(62-73-7)**
OK Pre-Gel **(9005-25-8)**
OKTOL **(9003-27-4)**
OL 27-400 **(59865-13-3)**
OL 55F2 **(9004-98-2)**
OL 55F10 **(9004-98-2)**
OM 100 **(9003-20-7)**
OM 2424 **(2593-15-9)**
OM-1563 **(13463-41-7)**
OMCA **(69-23-8)**
OM Hidantoina simple **(57-41-0)**
OM-Hydantoine **(57-41-0)**
OM-Hydantoine sodium **(630-93-3)**
OMPA **(152-16-9)**
OMS 1 **(121-75-5)**
OMS 2 **(55-38-9)**
OMS 14 **(62-73-7)**
OMS 15 **(64-00-6)**
OMS-15 **(64-00-6)**
OMS-29 **(63-25-2)**
OMS-32 **(2631-40-5)**
OMS-33 **(114-26-1)**
OMS-47 **(315-18-4)**
OMS 16 **(50-29-3)**
OMS 19 **(56-38-2)**
OMS 20 **(87-47-8)**
OMS 37 **(115-90-2)**
OMS 43 **(122-14-5)**
OMS 62 **(119-38-0)**
OMS 75 **(300-76-5)**
OMS-93 **(2032-65-7)**
OMS 94 **(60-51-5)**
OMS 111 **(60-51-5)**
OMS 115 **(299-85-4)**
OMS 123 **(299-84-3)**
OMS 162 **(64-00-6)**
OMS-174 **(671-04-5)**
OMS 197 **(72-20-8)**
OMS 206 **(297-78-9)**
OMS-214 **(2463-84-5)**
OMS 217 **(500-28-7)**
OMS 219 **(2104-64-5)**
OMS 226 **(2636-26-2)**
OMS 227 **(2282-34-0)**
OMS 239 **(7700-17-6)**
OMS 244 **(786-19-6)**
OMS 252 **(116-01-8)**
OMS 253 **(141-66-2)**
OMS 410 **(944-22-9)**
OMS 412 **(327-98-0)**
OMS 466 **(72-43-5)**
OMS 468 **(584-79-2)**
OMS 469 **(333-41-5)**
OMS 479 **(644-64-4)**
OMS 570 **(115-29-7)**
OMS 578 **(327-98-0)**
OMS-597 **(2686-99-9)**
OMS-658 **(2104-96-3)**
OMS-659 **(4824-78-6)**
OMS-708 **(1079-33-0)**
OMS-712 **(2274-67-1)**
OMS-771 **(116-06-3)**
OMS-774 **(1224-63-1)**
OMS 639 **(315-18-4)**
OMS 646 **(947-02-4)**
OMS 716 **(2631-37-0)**
OMS 834 **(6923-22-4)**
OMS 844 **(950-37-8)**
OMS 869 **(2636-26-2)**
OMS-864 **(1563-66-2)**
OMS-968 **(2540-82-1)**
OMS-0971 **(2921-88-2)**
OMS-1155 **(5598-13-0)**
OMS-1206 **(10453-86-8)**
OMS-1211 **(18181-70-9)**
OMS-1344 **(1757-18-2)**
OMS-1394 **(22781-23-3)**
OMS 1017 **(76-87-9)**
OMS 1020 **(900-95-8)**
OMS 1056 **(973-21-7)**
OMS 1075 **(2597-03-7)**
OMS 1078 **(72-54-8)**
OMS 1168 **(5598-52-7)**
OMS 1170 **(14816-18-3)**
OMS 1243 **(14255-88-0)**
OMS 1325 **(13171-21-6)**
OMS 1328 **(470-90-6)**
OMS 1342 **(21923-23-9)**
OMS 1342 **(60238-56-4)**
OMS 1356 **(26258-70-8)**
OMS 1424 **(29232-93-7)**
OMS 1437 **(57-74-9)**
OMS 1438 **(21609-90-5)**
OMS 1502 **(31218-83-4)**

OMS 1696 **(41096-46-2)**
OMS 1697 **(40596-69-8)**
OMS 1804 **(35367-38-5)**
OMS 1806 **(38260-54-7)**
OMS 1988 **(52918-63-5)**
OMS 3023 **(66230-04-4)**
OMT **(137-20-2)**
OMU **(2163-69-1)**
ONCB **(88-73-3)**
ONT **(88-72-2)**
OP 13.2 **(9036-19-5)**
OP 17.7 **(9036-19-5)**
OP 30 **(9036-19-5)**
OP 115 **(9036-19-5)**
OP 1062 **(9036-19-5)**
OP (Chinese surfactant) **(9036-19-5)**
OPE 30 **(9002-93-1)**
OPE-3 **(9036-19-5)**
OPP **(90-43-7)**
OPP-Na **(132-27-4)**
OPP-Sodium **(132-27-4)**
OP 85R **(26266-58-0)**
OPSB **(9003-13-8)**
OPSPA **(2168-68-5)**
8-OQ **(148-24-3)**
OR 1191 **(13171-21-6)**
OR 1191 **(297-99-4)**
OR 1500 **(9003-20-7)**
ORG 485-50 **(52-76-6)**
ORG-483 **(965-90-2)**
ORG NA 97 **(15500-66-0)**
OS 20A **(9005-00-9)**
OS 1836 **(311-47-7)**
OS 1897 **(96-12-8)**
OS 2046 **(298-01-1)**
OS 2046 **(7786-34-7)**
OSBAC **(3766-81-2)**
OSDMP **(2511-10-6)**
OTBE (French) **(56-35-9)**
OTC **(79-57-2)**
OTOS **(13752-51-7)**
OTS **(88-19-7)**
OTS 11 **(15571-58-1)**
OU-B **(7440-44-0)**
O-V Statin **(1400-61-9)**
OX **(604-75-1)**
Oasil **(57-53-4)**
Oat LS **(8001-21-6)**
Obeline picrate **(131-74-8)**
Obesedrin **(51-63-8)**
Obesin **(101-40-6)**
Obesine **(101-40-6)**
Obesonil **(51-63-8)**
Obestat **(154-41-6)**
Oblevil **(77-75-8)**
Obliterol **(139-88-8)**
N-Oblivon **(77-75-8)**
Oblivon **(77-75-8)**
Oblivon C **(77-75-8)**
Obramycin **(32986-56-4)**
Obston **(577-11-7)**
Ocenol **(143-28-2)**
Oceol **(143-28-2)**
Ochratoxin A **(303-47-9)**
Ochratoxin A ethyl ester **(4865-85-4)**
Ochratoxin B **(4825-86-9)**
Ochratoxin C **(4865-85-4)**
Ochre **(1309-37-1)**
Ocimene **(29714-87-2)**
allo-Ocimene **(673-84-7)**
cis-β-Ocimene **(3338-55-4)**
Octa-Klor **(57-74-9)**
Octaacetylsucrose **(126-14-7)**
Octabromobiphenyl **(27858-07-7)**
ar,ar,ar,ar,ar',ar',ar',ar'-Octabromo-1,1'-biphenyl **(27858-07-7)**
Octabromodibenzo-p-dioxin **(2170-45-8)**
Octabromodibenzofuran **(103582-29-2)**
Octabromodiphenyl **(27858-07-7)**
Octacarbonyldicobalt **(10210-68-1)**
1,2,4,5,6,7,8,8-Octachloor-3a,4,7,7a-tetrahydro-4,7-endo-methano-indaan (Dutch) **(57-74-9)**
Octachlor **(57-74-9)**
Octachlor epoxide **(27304-13-8)**
1,2,4,5,6,7,8,8-Octachlor-2,3,3a,4,7,7a-hexahydro-4,7-methano-indane **(57-74-9)**
2,2',3,3',4,4',5,5'-Octachlorobiphenyl **(35694-08-7)**
2,2',3,3',4,4',5,6'-Octachlorobiphenyl **(42740-50-1)**
2,2',3,3',4,4',5,6-Octachlorobiphenyl **(52663-78-2)**
2,2',3,3',4,4',6,6'-Octachlorobiphenyl **(33091-17-7)**
2,2',3,3',4,5,5',6'-Octachlorobiphenyl **(52663-75-9)**
2,2',3,3',4,5,5',6-Octachlorobiphenyl **(68194-17-2)**
2,2',3,3',4,5,6,6'-Octachlorobiphenyl **(52663-73-7)**
2,2',3,3',5,5',6,6'-Octachlorobiphenyl **(2136-99-4)**
2,2',3,4,4',5,5',6-Octachlorobiphenyl **(52663-76-0)**
Octachlorobiphenyl **(55722-26-4)**
ar,ar,ar,ar,ar',ar',ar',ar'-Octachlorobiphenyl **(31472-83-0)**
1,1,1,2,3,4,4,4-Octachlorobutane **(18791-19-0)**
Octachlorocamphene **(1319-80-8)**
Octachlorocamphene **(8001-35-2)**
Octachlorocyclopentene **(706-78-5)**
Octachlorocyclotetraphosphazene **(2950-45-0)**
Octachlorodibenzo(b,e)(1,4)dioxin **(3268-87-9)**
1,2,3,4,6,7,8,9-Octachlorodibenzodioxin **(3268-87-9)**
Octachlorodibenzo-p-dioxin **(3268-87-9)**
Octachlorodibenzodioxin **(3268-87-9)**
Octachlorodibenzofuran **(39001-02-0)**
Octachlorodihydrodicyclopentadiene **(57-74-9)**
Octachlorodipropylether **(127-90-2)**
1,2,4,5,6,7,8,8-Octachloro-2,3,3a,4,7,7a-hexahydro-4,7-methano-1H-indene **(57-74-9)**
1,2,4,5,6,7,8,8-Octachloro-2,3,3a,4,7,7a-hexahydro-4,7-methanoindene **(57-74-9)**
1,3,4,5,6,8,8-Octachloro-1,3,3a,4,7,7a-hexahydro-4,7-methano-isobenzofuran **(297-78-9)**
Octachloro-hexahydro-methanoisobenzofuran **(297-78-9)**
1,2,4,5,6,7,8,8-Octachloro-3a,4,7,7a-hexahydro-4,7-methylene indane **(57-74-9)**
Octachloro-4,7-methanohydroindane **(57-74-9)**
1,2,4,5,6,7,8,8-Octachloro-4,7-methano-3a,4,7,7a-tetrahydroindane **(57-74-9)**
Octachloro-4,7-methanotetrahydroindane **(57-74-9)**
1,3,4,5,6,7,10,10-Octachloro-4,7-endo-methylene-4,7,8,9-tetra-hydrophthalan **(297-78-9)**
Octachloronaphthalene (ACGIH,OSHA) **(2234-13-1)**
1,3,4,5,6,7,8,8-Octachloro-2-oxa-3a,4,7,7a-tetrahydro-4,7-methanoindene **(297-78-9)**
Octachloropropane **(594-90-1)**
Octachlorostyrene **(29082-74-4)**
1,2,4,5,6,7,8,8-Octachloro-3a,4,7,7a-tetrahydro-4,7-methanoindan **(57-74-9)**
2,2,4,5,6,7,8,8-Octachloro-3a,4,7,7a-tetrahydro-4,7-methanoindan **(5566-34-7)**
1,2,4,5,6,7,8,8-Octachloro-3a,4,7,7a-tetrahydro-4,7-methanoindane **(57-74-9)**
1,2,4,5,6,7,10,10-Octachloro-4,7,8,9-tetrahydro-4,7-methyleneindane **(57-74-9)**
1,2,4,5,6,7,8,8-Octachlor-3a,4,7,7a-tetrahydro-4,7-endo-methano-indan (German) **(57-74-9)**
Octacide 264 **(113-48-4)**

n-Octacosane (630-02-4)
Octacosane (9CI) (630-02-4)
Octacosanoic acid, methyl ester (9CI) (55682-92-3)
Octacosanol (557-61-9)
Octacosanol-1 (557-61-9)
n-Octacosanol (557-61-9)
1-Octacosanol (9CI)2 (557-61-9)
1-Octacosanol, aluminum salt (9CI) (67905-27-5)
Octacosyl alcohol (557-61-9)
9,17-Octadecadienal, (Z)- (9CI) (56554-35-9)
6,9-Octadecadienamide, N,N-bis(2-hydroxyethyl)-, (Z,Z)- (9CI) (27883-12-1)
9,12-Octadecadien-1-amine, N,N-dimethyl-, (Z,Z)-, (2,4,5-trichlorophenoxy)acetate (9CI) (53404-88-9)
9,12-Octadecadien-1-amine, N,N-dimethyl-, (Z,Z)-, (2,4,5-trichlorophenoxy)acetate, Mixt. contg. (9CI) (55256-33-2)
9,12-Octadecadienoic acid (60-33-3)
9,12-Octadecadienoic acid, (E,E)- (506-21-8)
cis,cis-9,12-Octadecadienoic acid (60-33-3)
cis-9,cis-12-Octadecadienoic acid (60-33-3)
8,11-Octadecadienoic acid (8CI,9CI) (2197-52-6)
9,12-Octadecadienoic acid, (Z) (60-33-3)
9,12-Octadecadienoic acid (Z,Z)-, calcium salt (9CI) (19704-83-7)
9,12-Octadecadienoic acid (Z,Z)-, 2-chloroethyl ester (25525-76-2)
(Z,Z)-9,12-Octadecadienoic acid, copper salt (7721-15-5)
9,12-Octadecadienoic acid (Z,Z)-, copper salt (9CI) (7721-15-5)
9,12-Octadecadienoic acid (Z,Z)-, (dimethylstannylene)bis-(thio-2,1-ethanediyl) ester (9CI) (67859-64-7)
9,12-Octadecadienoic acid (Z,Z)-, isooctyl ester (9CI) (67874-38-8)
9,12-Octadecadienoic acid (Z,Z)-, lead salt (9CI) (16996-51-3)
9,12-Octadecadienoic acid, methyl ester (112-63-0)
9,12-Octadecadienoic acid, methyl ester, (Z,Z)- (112-63-0)
9,12-Octadecadienoic acid (Z,Z)-, methyl ester (9CI) (112-63-0)
9,11-Octadecadienoic acid, methyl ester, (E,E)- (8CI,9CI) (13038-47-6)
9,12-Octadecadienoic acid, 12-sulfo-, (?,Z)- (9CI) (68201-84-3)
(E,Z)-3,13-Octadecadien-1-ol acetate (53120-26-6)
(Z,Z) 3,13-Octadecadien-1-ol acetate (53120-27-7)
(Z,Z)-3,13-Octadecadien-1-ol acetate (53120-27-7)
3,13-Octadecadien-1-ol, acetate (53120-27-7)
3,13-Octadecadien-1-ol, acetate, (E,Z)- (53120-26-6)
3,13-Octadecadien-1-ol, acetate, (Z,Z)- (53120-27-7)
(E,Z)-3,13-Octadecadienyl acetate (53120-26-6)
2,5-Octadecadiynoic acid, methyl ester (9CI) (57156-91-9)
10,13-Octadecadiynoic acid, methyl ester (8CI,9CI) (18202-24-9)
Octadecafluorooctane (307-34-6)
Octadecamethyloctasiloxane (556-69-4)
Octadecanal (9CI) (638-66-4)
Octadecanamide (124-26-5)
Octadecanamide, N-acetyl- (9CI) (65882-23-7)
Octadecanamide, N-(2-aminoethyl)-N-(2-hydroxyethyl)- (9CI) (120-41-2)
Octadecanamide, N,N-bis(2-hydroxyethyl)- (9CI) (93-82-3)
Octadecanamide, N-(2-(diethylamino)ethyl)- (9CI) (16889-14-8)
Octadecanamide, N-(3-(dimethylamino)propyl)- (9CI) (7651-02-7)
Octadecanamide, N-(3-(dimethylamino)propyl)-, N-oxide (9CI) (25066-20-0)
Octadecanamide, N,N'-1,2-ethanediylbis- (9CI) (110-30-5)
Octadecanamide, N,N'-ethylenebis- (8CI) (110-30-5)
Octadecanamide, N-(2-hydroxyethyl)- (9CI) (111-57-9)
Octadecanamide, N-(2-((2-hydroxyethyl)amino)ethyl)- (9CI) (141-21-9)
Octadecanamide, N-(2-hydroxyethyl)-N-methyl-, hydrogen sulfate (ester), sodium salt (8CI) (26535-50-2)
Octadecanamide, N-(hydroxymethyl)- (8CI,9CI) (3370-35-2)
Octadecanamide, N-(4-hydroxyphenyl)- (9CI) (103-99-1)
Octadecanamide, N-(2-hydroxypropyl)-, hydrogen sulfate (ester), monosodium salt (8CI) (26577-87-7)
Octadecanamide, N,N'-(iminodi-3,1-propanediyl)bis- (9CI) (13998-73-7)
Octadecanamide, N-methyl-N-(2-(sulfooxy)ethyl)-, sodium salt (9CI) (26535-50-2)
Octadecanamide, N-(2-(sulfooxy)propyl)-, monosodium salt (9CI) (26577-87-7)
Octadecanamine acetate (2190-04-7)
1-Octadecanamine, N,N-dimethyl-, acetate (9CI) (19855-61-9)
1-Octadecanamine, N,N-dimethyl-, N-oxide (9CI) (2571-88-2)
1-Octadecanamine, N-methyl-N-octadecyl- (9CI) (4088-22-6)
1-Octadecanaminium, N,N-dimethyl-N-octadecyl-, chloride (9CI) (107-64-2)
1-Octadecanaminium, N,N-dimethyl-N-(3-(trimethoxysilyl)-propyl)-, chloride (9CI) (27668-52-6)
Octadecananilide, 4'-hydroxy- (8CI) (103-99-1)
n-Octadecane (593-45-3)
Octadecane (9CI) (593-45-3)
Octadecane, 1-bromo- (9CI) (112-89-0)
Octadecane, 1-chloro- (9CI) (3386-33-2)
Octadecane, 1-(ethenyloxy)- (9CI) (930-02-9)
Octadecane, 2,2,4,15,17,17-hexamethyl-7,12-bis(3,5,5-trimethyl-hexyl)- (6CI,9CI) (55470-97-8)
Octadecane, 1-iodo- (9CI) (629-93-6)
Octadecane, 2-methyl- (1560-88-9)
Octadecanenitrile (638-65-3)
Octadecane, 1-phenyl- (8CI) (4445-07-2)
1-Octadecanethiol (9CI) (2885-00-9)
Octadecane, 1-(2,4,6-trimethylcyclohexyl)- (6CI) (55282-34-3)
Octadecane, 1,1,1,2,2,3,3,4,4,5,5,6,6,7,7,8,8,9,9,10,10,11,11,12,12,13,13,14,14,15,15,16,16-tritriacontafluoro-18-iodo- (9CI) (65150-94-9)
(Octadecanoato-O)oxoaluminum (13419-15-3)
Octadecanoic acid (57-11-4)
n-Octadecanoic acid (57-11-4)
Octadecanoic acid, Compd. with 2,2',2''-nitrilotris(ethanol) (1:1) (9CI) (4568-28-9)
Octadecanoic acid, Monoester with 1,2,3-propanetriol (123-94-4)
Octadecanoic acid, aluminum salt (637-12-7)
Octadecanoic acid, ammonium salt (1002-89-7)
Octadecanoic acid, barium cadmium salt (4:1:1) (9CI) (1191-79-3)
Octadecanoic acid, barium salt (9CI) (6865-35-6)
Octadecanoic acid, 2-(bis(2-hydroxyethyl)amino)ethyl ester (10248-74-5)
Octadecanoic acid, 2,2-bis(hydroxymethyl)-1,3-propanediyl ester (9CI) (13081-97-5)
Octadecanoic acid, 2,2-bis(((1-oxooctadecyl)oxy)methyl)-1,3-propanediyl ester (9CI) (115-83-3)
Octadecanoic acid, 2-bromo- (9CI) (142-94-9)
Octadecanoic acid, 2-butoxyethyl ester (109-38-6)
Octadecanoic acid, butyl ester (9CI) (123-95-5)
Octadecanoic acid, cadmium salt (2223-93-0)
Octadecanoic acid, calcium salt (1592-23-0)
Octadecanoic acid, cerium salt (9CI) (10119-53-6)
Octadecanoic acid, chloride (112-76-5)
Octadecanoic acid, cobalt salt (9CI) (13586-84-0)
Octadecanoic acid, copper(2+) salt (660-60-6)
Octadecanoic acid, dichloro- (8CI,9CI) (31135-63-4)
Octadecanoic acid, 2,3-dihydroxypropyl ester (9CI) (123-94-4)
Octadecanoic acid, eicosyl ester (9CI) (22413-02-1)
Octadecanoic acid, 9,10-epoxy (2443-39-2)
Octadecanoic acid, 9,10-epoxy-, cis- (8CI) (24560-98-3)
Octadecanoic acid, 9,10-epoxy-, butyl ester (106-83-2)
Octadecanoic acid, 9,10-epoxy-, 2,3-epoxy-2-ethylhexyl ester (63907-12-0)
Octadecanoic acid, 9,10-epoxy-, 2-ethylhexyl ester (141-38-8)
Octadecanoic acid, 9,10-epoxy-12-hydroxy-, triester with glycerol (6CI,8CI) (106-81-0)
Octadecanoic acid, 9,10-epoxy-, octyl ester (106-84-3)
Octadecanoic acid, 1,2-ethanediyl ester (9CI) (627-83-8)
Octadecanoic acid, ethenyl ester (9CI) (111-63-7)
Octadecanoic acid, 2-ethylhexyl ester (9CI) (22047-49-0)
Octadecanoic acid, hydroxy- (9CI) (1330-70-7)
Octadecanoic acid, 10-hydroxy- (8CI,9CI) (638-26-6)
Octadecanoic acid, 9-hydroxy- (8CI,9CI) (3384-24-5)
Octadecanoic acid, 3-hydroxy-2,2-bis(hydroxymethyl)propyl ester (9CI) (78-23-9)

Octadecanoic acid, 12-hydroxy-, calcium salt (2:1) (9CI) **(3159-62-4)**
Octadecanoic acid, 9-hydroxy-, isopropyl ester, hydrogen sulfate, sodium salt (8CI) **(14350-72-2)**
Octadecanoic acid, 12-hydroxy-, methyl ester, lithium salt (9CI) **(53422-16-5)**
Octadecanoic acid, 12-hydroxy-, monolithium salt **(7620-77-1)**
Octadecanoic acid, iron(3+) salt (9CI) **(555-36-2)**
Octadecanoic acid, isodecyl ester (9CI) **(31565-38-5)**
Octadecanoic acid, isohexadecyl ester (9CI) **(25339-09-7)**
Octadecanoic acid, isotridecyl ester (9CI) **(31565-37-4)**
Octadecanoic acid, lead(2+) salt (9CI) **(1072-35-1)**
Octadecanoic acid, lithium salt **(4485-12-5)**
Octadecanoic acid, magnesium salt **(557-04-0)**
Octadecanoic acid, methyl ester **(112-61-8)**
Octadecanoic acid, 1-methylethyl ester (9CI) **(112-10-7)**
Octadecanoic acid, monoester with 1,2-propanediol **(1323-39-3)**
Octadecanoic acid, monoester with 1,2,3-propanetriol **(31566-31-1)**
Octadecanoic acid, nitrilotri-2,1-ethanediyl ester (9CI) **(3002-22-0)**
Octadecanoic acid, octadecyl ester (9CI) **(2778-96-3)**
Octadecanoic acid, oxydi-2,1-ethanediyl ester (9CI) **(109-30-8)**
Octadecanoic acid, pentachloro-, methyl ester (9CI) **(26638-28-8)**
Octadecanoic acid, potassium salt **(593-29-3)**
Octadecanoic acid, 1,2,3-propanetriyl ester **(555-43-1)**
Octadecanoic acid, silver(1+) salt (9CI) **(3507-99-1)**
Octadecanoic acid, sodium salt (9CI) **(822-16-2)**
Octadecanoic acid, strontium salt (9CI) **(10196-69-7)**
Octadecanoic acid, 2-sulfo-, 1-methyl ester (9CI) **(3076-26-4)**
Octadecanoic acid, 2-sulfo-, 1-methyl ester, sodium salt (9CI) **(4062-78-6)**
Octadecanoic acid, 9-(sulfooxy)-, 1-methyl ester, sodium salt (9CI) **(139-99-1)**
Octadecanoic acid, 10-(sulfooxy)-, 1-(2-methylpropyl) ester, sodium salt (9CI) **(67859-39-6)**
Octadecanoic acid, tetradecyl ester (9CI) **(17661-50-6)**
Octadecanoic acid, tin salt (9CI) **(7637-13-0)**
Octadecanoic acid, tridecyl ester (9CI) **(31556-45-3)**
Octadecanoic acid, triethanolamine salt **(4568-28-9)**
Octadecanoic acid, zinc salt **(557-05-1)**
1-Octadecanol **(112-92-5)**
Octadecanol **(112-92-5)**
n-Octadecanol **(112-92-5)**
Octadecanol (8CI,9CI) **(26762-44-7)**
1-Octadecanol, aluminum salt (8CI) **(3985-81-7)**
1-Octadecanol, monoether with polyethylene glycol **(9005-00-9)**
Octadecanonitrile **(638-65-3)**
Octadecanoylamidopropyldimethylamine **(7651-02-7)**
n-Octadecanoyl chloride **(112-76-5)**
Octadecanoyl chloride (9CI) **(112-76-5)**
5,7,11,13-Octadecatrayne-1,18-diol **(76379-67-4)**
9,12,15-Octadecatrienal (8CI,9CI) **(26537-71-3)**
(Z,Z,Z)-9,12,15-Octadecatrienoic acid **(463-40-1)**
all-cis-9,12,15-Octadecatrienoic acid **(463-40-1)**
cis,cis,cis-9,12,15-Octadecatrienoic acid **(463-40-1)**
9,12,15-Octadecatrienoic acid, (Z,Z,Z)- (9CI) **(463-40-1)**
8,11,14-Octadecatrienoic acid, (Z,Z,Z)- (8CI,9CI) **(4906-91-6)**
9,12,15-Octadecatrienoic acid, 2,3-bis(acetyloxy)propyl ester, (Z,Z,Z)- (9CI) **(55320-02-0)**
9,12,15-Octadecatrienoic acid, methyl ester, (Z,Z,Z)- **(301-00-8)**
17-Octadecenal (9CI) **(56554-86-0)**
2-Octadecenal (9CI) **(56554-96-2)**
3-Octadecenal (9CI) **(56554-99-5)**
4-Octadecenal (9CI) **(56554-98-4)**
9-Octadecenal (8CI,9CI) **(5090-41-5)**
9-Octadecenal, (Z)- (9CI) **(2423-10-1)**
(Z)-9-Octadecenamide **(301-02-0)**
9-Octadecenamide **(301-02-0)**
9-Octadecenamide, (Z)- (9CI) **(301-02-0)**
9-Octadecenamide, N-(2-((2-aminoethyl)amino)ethyl)-, (Z)- (9CI) **(15566-80-0)**
9-Octadecenamide, N,N-bis(2-hydroxyethyl)- **(93-83-4)**
9-Octadecenamide, N,N-bis(2-hydroxyethyl)-, (Z)- (9CI) **(93-83-4)**
9-Octadecenamide, N,N-dimethyl-, (Z)- (9CI) **(2664-42-8)**
9-Octadecenamide, N,N-dimethyl-, (Z)-, (80%) and related amides **(2664-42-8)**
9-Octadecenamide, N,N'-1,2-ethanediylbis-, (Z,Z)- (9CI) **(110-31-6)**
9-Octadecenamide, N-(2-hydroxypropyl)- **(111-05-7)**
9-Octadecenamide, N-(2-hydroxypropyl)-, (Z)- (9CI) **(111-05-7)**
9-Octadecenamide, N-(1-methylethyl)-, (Z)- (9CI) **(10574-01-3)**
9-Octadecen-1-amine (9CI) **(1838-19-3)**
9-Octadecen-1-amine, (Z)- (9CI) **(112-90-3)**
(Z)-9-Octadecen-1-amine acetate **(10460-00-1)**
9-Octadecen-1-amine, (Z)-, acetate (9CI) **(10460-00-1)**
9-Octadecen-1-amine, acetate (9CI) **(3811-68-5)**
9-Octadecen-1-amine, N,N-dimethyl-, N-oxide **(14351-50-9)**
9-Octadecen-1-amine, N,N-dimethyl-, N-oxide, (Z)- (9CI) **(14351-50-9)**
9-Octadecen-1-amine, N,N-dimethyl-, (Z)-, (2,4,5-trichlorophenoxy)acetate (9CI) **(53404-89-0)**
9-Octadecen-1-amine, N,N-dimethyl-, (Z)-, (2,4,5-trichlorophenoxy)acetate, Mixt. contg. (9CI) **(55256-33-2)**
9-Octadecen-1-amine, (Z)-, 2-(2,4,5-trichlorophenoxy)propanoate (9CI) **(53404-73-2)**
9-Octadecen-1-aminium, N,N-bis(2-hydroxyethyl)-N-methyl-, chloride, (Z)- (9CI) **(18448-65-2)**
9-Octadecen-1-aminium, N-(carboxymethyl)-N,N-dimethyl-, hydroxide, inner salt, (Z)- (9CI) **(871-37-4)**
α-Octadecene **(112-88-9)**
1-Octadecene (9CI) **(112-88-9)**
Octadecene (9CI) **(27070-58-2)**
(Z)-9-Octadecenenitrile **(112-91-4)**
9-Octadecenenitrile, (Z)- (9CI) **(112-91-4)**
9-Octadecene-1-sulfonic acid, sodium salt, (Z)- (9CI) **(15075-85-1)**
(Z)-(9-Octadecenoato-O)phenylmercury **(104-60-9)**
(Z)-(9-Octadecenoato-O)phenyl-mercury **(104-60-9)**
11-Octadecenoic acid, (E)- **(693-72-1)**
9,10-Octadecenoic acid **(112-80-1)**
9-Octadecenoic acid, (E)- **(112-79-8)**
9-Octadecenoic acid, cis- **(112-80-1)**
cis-9-Octadecenoic acid **(112-80-1)**
cis-Octadec-9-enoic acid **(112-80-1)**
cis-δ^9-Octadecenoic acid **(112-80-1)**
trans-9-Octadecenoic acid **(112-79-8)**
trans-Octadec-9-enoic acid **(112-79-8)**
trans-δ^9-Octadecenoic acid **(112-79-8)**
9-Octadecenoic acid (Z)-, Compd. with 2,2'-iminobis(ethanol) (1:1) **(13961-86-9)**
9-Octadecenoic acid (Z)-, Compd. with morpholine (1:1) (9CI) **(1095-66-5)**
9-Octadecenoic acid, Compd. with 2,2',2''-nitrilotris(ethanol) (1:1) **(2717-15-9)**
9-Octadecenoic acid (VAN)(9CI) **(2027-47-6)**
Octadecenoic acid (VAN)(9CI) **(26764-26-1)**
9-Octadecenoic acid, (Z) **(112-80-1)**
6-Octadecenoic acid, (Z)- (8CI,9CI) **(593-39-5)**
9-Octadecenoic acid (Z)-, Compd. with 1-butanamine (1:1) (9CI) **(26094-13-3)**
9-Octadecenoic acid (Z)-, Compd. with 2,2',2''-nitrilotris-(ethanol) (1:1) (9CI) **(2717-15-9)**
9-Octadecenoic acid (Z)-, Ester with 1,2,3-propanetriol (9CI) **(37220-82-9)**
9-Octadecenoic acid, 12-(acetyloxy)-, butyl ester **(140-04-5)**
9-Octadecenoic acid, 12-(acetyloxy)-, butyl ester, (R-(Z))- (9CI) **(140-04-5)**
9-Octadecenoic acid (Z)-, aluminum salt **(688-37-9)**
9-Octadecenoic acid, (Z)-, aluminum salt (3:1) **(688-37-9)**
9-Octadecenoic acid, amide(cis) **(301-02-0)**
9-Octadecenoic acid, ammonium salt **(544-60-5)**
9-Octadecenoic acid (Z)-, ammonium salt (9CI) **(544-60-5)**
9-Octadecenoic acid (Z)-, 2-(bis(2-hydroxyethyl)amino)ethyl ester (9CI) **(10277-04-0)**
9-Octadecenoic acid (Z)-, 2-butoxyethyl ester (9CI) **(109-39-7)**
9-Octadecenoic acid, butyl ester (Z) **(142-77-8)**

Name	CAS
9-Octadecenoic acid (Z)-, (butylstannylidyne)tris(thio-2,1-ethanediyl) ester (9CI)	(67361-76-6)
9-Octadecenoic acid (Z)-, cadmium salt (9CI)	(10468-30-1)
9-Octadecenoic acid, (Z)-, calcium salt (2:1)	(142-17-6)
9-Octadecenoic acid (Z)-, calcium salt (9CI)	(142-17-6)
9-Octadecenoic acid (Z)-, chromium salt (9CI)	(13308-40-2)
9-Octadecenoic acid (Z)-, cobalt salt (9CI)	(14666-94-5)
9-Octadecenoic acid (Z)-, Compd. with guanidine (1:1) (9CI)	(53048-47-8)
(Z)-9-Octadecenoic acid, copper salt	(10402-16-1)
9-Octadecenoic acid (Z)-, copper salt (9CI)	(10402-16-1)
9-Octadecenoic acid (Z)-, decyl ester (9CI)	(3687-46-5)
9-Octadecenoic acid, decyl ester	(3687-46-5)
9-Octadecenoic acid (Z)-, (dibutylstannylene)bis(thio-2,1-ethanediyl) ester (9CI)	(67361-77-7)
9-Octadecenoic acid (Z)-, diester with 1,2,3-propanetriol (9CI)	(25637-84-7)
9-Octadecenoic acid, diester with 1,2,3-propanetriol	(25637-84-7)
9-Octadecenoic acid (Z)-, 2,3-dihydroxypropyl ester (9CI)	(111-03-5)
9-Octadecenoic acid, 2,3-dihydroxypropyl ester	(111-03-5)
9-Octadecenoic acid (Z)-, 2,2-dimethyl-1-(1-methylethyl)-1,3-propanediyl ester (9CI)	(68201-79-6)
9-Octadecenoic acid (Z)-, (dimethylstannylene)bis(thio-2,1-ethanediyl) ester (9CI)	(67859-63-6)
9-Octadecenoic acid (Z)-, ethyl ester (9CI)	(111-62-6)
9-Octadecenoic acid (Z)-, 2-ethyl-2-(hydroxymethyl)-1,3-propanediyl ester (9CI)	(25111-05-1)
9-Octadecenoic acid (Z)-, 2-(2-hydroxyethoxy)ethyl ester (9CI)	(106-12-7)
9-Octadecenoic acid (Z)-, 3-hydroxy-2,2-bis(hydroxymethyl)-propyl ester (9CI)	(10332-32-8)
9-Octadecenoic acid, 12-hydroxy-, butyl ester, (R-(Z))-	(151-13-3)
9-Octadecenoic acid, 2-(2-hydroxyethoxy)ethyl ester	(106-12-7)
9-Octadecenoic acid, 12-hydroxy-, methyl ester	(141-24-2)
9-Octadecenoic acid, 12-hydroxy-, methyl ester, (R-(Z))- (9CI)	(141-24-2)
9-Octadecenoic acid, 12-hydroxy-, monoester with 1,2,3-propanetriol, (R-(Z))-	(1323-38-2)
9-Octadecenoic acid, 12-hydroxy-, (R-(Z))-, monoester with 1,2,3-propanetriol	(1323-38-2)
9-Octadecenoic acid, 12-hydroxy-, monopotassium salt	(7492-30-0)
9-Octadecenoic acid, 12-hydroxy-, monopotassium salt, (R-(Z))- (9CI)	(7492-30-0)
9-Octadecenoic acid, 12-hydroxy-, (Z)-	(141-22-0)
9-Octadecenoic acid (Z)-, isooctyl ester (9CI)	(26761-50-2)
9-Octadecenoic acid (Z)-, lead(2+) salt (9CI)	(1120-46-3)
9-Octadecenoic acid, (Z)-, magnesium salt (2:1)	(1555-53-9)
9-Octadecenoic acid (Z)-, magnesium salt (9CI)	(1555-53-9)
9-Octadecenoic acid (Z)-, 2-mercaptoethyl ester (9CI)	(59118-78-4)
9-Octadecenoic acid (Z)-, 2-mercaptoethyl ester, Reaction products with dichlorodimethylstannane, sodium sulfide-(Na_2S) and trichloromethylstannane	(68442-12-6)
(Z)-9-Octadecenoic acid, methyl ester	(112-62-9)
9-Octadecenoic acid, methyl ester, (E)-	(1937-62-8)
11-Octadecenoic acid, methyl ester (9CI)	(52380-33-3)
10-Octadecenoic acid, methyl ester, (E)- (8CI,9CI)	(13038-45-4)
6-Octadecenoic acid, methyl ester (7CI,9CI)	(52355-31-4)
9-Octadecenoic acid (Z)-, 1-methyl-1,2-ethanediyl ester (9CI)	(105-62-4)
9-Octadecenoic acid (Z)-, 1-methylethyl ester (9CI)	(112-11-8)
9-Octadecenoic acid, 1-methylethyl ester	(112-11-8)
9-Octadecenoic acid (Z)-, 2-methylpropyl ester (9CI)	(10024-47-2)
9-Octadecenoic acid (Z)-, (methylstannylidyne)tris(thio-2,1-ethanediyl) ester (9CI)	(59118-79-5)
9-Octadecenoic acid (Z)-, monoester with 1,2-propanediol (9CI)	(1330-80-9)
9-Octadecenoic acid, monoester with 1,2-propanediol	(1330-80-9)
9-Octadecenoic acid (Z)-, monoester with 1,2,3-propanetriol (9CI)	(25496-72-4)
9-Octadecenoic acid, monoester with 1,2,3-propanetriol	(111-03-5)
9-Octadecenoic acid(cis), nitrile(cis)	(112-91-4)
9-Octadecenoic acid, 9-octadecenyl ester	(3687-45-4)
9-Octadecenoic acid (Z)-, 9-octadecenyl ester, (Z)- (9CI)	(3687-45-4)
9-Octadecenoic acid, 12-(oxiranylmethoxy)-, 1,2,3-propanetriyl ester, Homopolymer (9CI)	(74398-71-3)
9-Octadecenoic acid (Z)-, potassium salt	(143-18-0)
9-Octadecenoic acid (Z)-, 1,2,3-propanetriyl ester (9CI)	(122-32-7)
9-Octadecenoic acid, 1,2,3-propanetriyl ester	(122-32-7)
9-Octadecenoic acid (Z)-, propyl ester (9CI)	(111-59-1)
9-Octadecenoic acid (Z)-, 2-sulfoethyl ester, sodium salt (9CI)	(142-15-4)
9-Octadecenoic acid, 2-sulfoethyl ester, sodium salt	(142-15-4)
9-Octadecenoic acid (Z)-, sulfonated, sodium salts	(68443-05-0)
9-Octadecenoic acid, (sulfooxy)-, 1-butyl ester (9CI)	(38621-44-2)
9-Octadecen-1-ol, cis-	(143-28-2)
cis-9-Octadecen-1-ol	(143-28-2)
9-Octadecen-1-ol, (Z)	(143-28-2)
9-Octadecen-1-ol, dihydrogen phosphate (9CI)	(24613-61-4)
9-Octadecen-1-ol, dihydrogen phosphate, (Z)- (9CI)	(7722-71-6)
9-Octadecen-1-ol, hydrogen phosphate, (Z,Z)- (9CI)	(14450-07-8)
9-Octadecen-1-ol, hydrogen sulfate, sodium salt, (Z)- (9CI)	(1847-55-8)
(Z)-9-Octadecenoyl chloride	(112-77-6)
9-Octadecenoyl chloride, (Z)- (9CI)	(112-77-6)
cis-9-Octadecenylamine	(112-90-3)
9-Octadecenylamine (8CI)	(1838-19-3)
Octadecenylamine, N,N-dimethyl- (8CI)	(28061-69-0)
9-Octadecenylamine, (Z)	(112-90-3)
cis-9-Octadecenylamine, acetate	(10460-00-1)
(Z)-3-(9-Octadecenylamino)propanenitrile	(26351-32-6)
N-(9-Octadecenyl)-3-aminopropionitrile	(26351-32-6)
(Z)-N-9-Octadecenylhexadecanamide	(16260-09-6)
N-9-Octadecenyl hexadecanamide	(16260-09-6)
2,2'-(9-Octadecenylimino)bisethanol	(25307-17-9)
α,α'-((9-Octadecenylimino)di-2,1-ethanediyl)bis(ω-hydroxypoly-(oxy-1,2-ethanediyl)-, (Z)-	(26635-93-8)
(Z)-((9-Octadecenyloxy)methyl)oxirane	(60501-41-9)
(Z)-N-9-Octadecenyl-1,3-propanediamine	(7173-62-8)
N-cis-9-Octadecenyl-1,3-propanediamine monoglucomate	(83542-86-3)
Octadecenylsuccinic anhydride	(28777-98-2)
17-Octadecen-14-yn-1-ol (8CI)	(18202-28-3)
Octadecyl acrylate	(4813-57-4)
n-Octadecyl acrylate	(4813-57-4)
Octa decyl alcohol	(112-92-5)
n-Octadecyl alcohol	(112-92-5)
Octadecylamine	(124-30-1)
n-Octadecylamine	(124-30-1)
Octadecylamine, acetate	(2190-04-7)
Octadecylamine, N,N-dimethyl	(124-28-7)
Octadecylamine-hydrochloride	(1838-08-0)
Octadecylamine, N-isopropyl	(13329-71-0)
4-(Octadecylamino)-4-oxo-2-sulfobutanedioic acid, disodium salt	(14481-60-8)
Octadecylbenzene	(4445-07-2)
n-Octadecyl-N-benzyl-N,N-dimethylammonium chloride	(122-19-0)
Octadecyl bromide	(112-89-0)
n-Octadecyl bromide	(112-89-0)
Octadecyl chloride	(3386-33-2)
Octadecyl cyanide	(28623-46-3)
Octadecylcyclohexane	(4445-06-1)
Octadecyl 3,5-di-tert-butyl-4-hydroxyhydrocinnamate	(2082-79-3)
Octadecyl 3-(3,5-di-tert-butyl-4-hydroxyphenyl)propionate	(2082-79-3)
Octadecyldimethylamine oxide	(2571-88-2)
Octadecyldimethylbenzylammonium chloride	(122-19-0)
Octadecyldimethyl(3-(trimethoxysilyl)propyl)ammonium chloride	(27668-52-6)
(Z)-N-Octadecyl-13-docosenamide	(10094-45-8)
N-Octadecyl-13-docosenamide	(10094-45-8)
Octadecylene α-	(112-88-9)
Octadecyl glycidyl ether	(16245-97-9)
2,2'-(Octadecylimino)bisethanol	(10213-78-2)
1,1'-(Octadecylimino)bis-2-propanol	(28137-64-6)
Octadecyl iodide	(629-93-6)
Octadecyl isocyanate	(112-96-9)

n-Octadecyl mercaptan **(2885-00-9)**
Octadecyl 2-methyl-2-propenoate **(32360-05-7)**
((Octadecyloxy)methyl)oxirane **(16245-97-9)**
Octadecylphosphonic acid, dimethyl ester **(25371-54-4)**
Octadecyl polyoxyethylene ether **(9005-00-9)**
Octadecyl 2-propenoate **(4813-57-4)**
Octadecyl sodium sulfate **(1120-04-3)**
Octadecyl stearate **(2778-96-3)**
N-Octadecyl-N-(sulfosuccinyl)aspartic acid, tetrasodium salt **(3401-73-8)**
N-Octadecylterephthalamate, monosodium salt **(5994-45-6)**
N-Octadecylterephthalamic acid, monosodium salt **(5994-45-6)**
Octadecyltrichlorosilane [UN 1800] **(112-04-9)**
Octadecyltrimethylammonium chloride **(112-03-8)**
10-Octadecynoic acid, methyl ester (8CI,9CI) **(26543-36-2)**
1-Octadedecylamine acetate **(2190-04-7)**
2,6-Octadienal, 3,7-dimethyl **(5392-40-5)**
2,6-Octadienal, 3,7-dimethyl-, (E) **(141-27-5)**
2,6-Octadienal, 3,7-dimethyl-, (Z)- **(106-26-3)**
2,6-Octadienal, 3,7-dimethyl-, diethyl acetal **(7492-66-2)**
2,6-Octadienal, 3,7-dimethyl-, oxime **(13372-77-5)**
1,6-n-Octadiene **(3710-41-6)**
1,7-Octadiene **(3710-30-3)**
2,7-Octadiene **(3710-41-6)**
1,6-Octadiene (6CI,7CI,8CI,9CI) **(3710-41-6)**
1,5-Octadiene, 1,2-dichloro- (9CI) **(83682-61-5)**
2,6-Octadiene, 1,1-diethoxy-3,7-dimethyl- (9CI) **(7492-66-2)**
3,5-Octadiene, 4,5-diethyl-, (E,Z)- (8CI,9CI) **(21293-02-7)**
2,6-Octadiene, 1,1-dimethoxy-3,7-dimethyl-, (cis and trans) **(7549-37-3)**
1,6-Octadiene, 3,7-dimethyl **(2436-90-0)**
1,6-Octadiene, 7-methyl-3-methylene **(123-35-3)**
1,6-Octadiene, 7-methyl-2-methylene-, Reaction products with hydrochloric acid **(68921-34-6)**
1,6-Octadiene, 7-methyl-3-methylene-, acetylated **(68412-04-4)**
2,6-Octadienenitrile, 3,7-dimethyl-, (E) **(5585-39-7)**
2,6-Octadienenitrile, 3,7-dimethyl- (9CI) **(5146-66-7)**
2,6-Octadienenitrile, 3,7-dimethyl-, (Z)- (9CI) **(31983-27-4)**
2,6-Octadienoic acid, 3,7-dimethyl **(459-80-3)**
2,7-Octadien-1-ol, acetate **(3491-27-8)**
1,6-Octadien-3-ol, 3,7-dimethyl **(78-70-6)**
2,6-Octadien-1-ol, 3,7-dimethyl-, trans- **(106-24-1)**
2,6-Octadien-1-ol, 3,7-dimethyl- (9CI) **(624-15-7)**
4,6-Octadien-3-ol, 3,7-dimethyl- (9CI) **(18479-54-4)**
2,6-Octadien-1-ol, 3,7-dimethyl-, (E) **(106-24-1)**
2,6-Octadien-1-ol, 3,7-dimethyl-, (Z) **(106-25-2)**
1,6-Octadien-3-ol, 3,7-dimethyl-, acetate **(115-95-7)**
2,6-Octadien-1-ol, 3,7-dimethyl-, acetate, trans- **(105-87-3)**
2,6-Octadien-1-ol, 3,7-dimethyl-, acetate, (E) **(105-87-3)**
2,6-Octadien-1-ol, 3,7-dimethyl-, acetate, (Z)- (9CI) **(141-12-8)**
1,6-Octadien-3-ol, 3,7-dimethyl-, o-aminobenzoate **(7149-26-0)**
1,6-Octadien-3-ol, 3,7-dimethyl-, 2-aminobenzoate (9CI) **(7149-26-0)**
2,6-Octadien-1-ol, 3,7-dimethyl-, formate, (E) **(105-86-2)**
3,7-Octadien-2-one, (E)- (8CI,9CI) **(25172-06-9)**
3,5-Octadien-2-one, (E,E)- (8CI,9CI) **(30086-02-3)**
4,5-Octadien-3-one, 2,2,7,7-tetramethyl- (8CI,9CI) **(19377-97-0)**
2,7-Octadienyl acetate **(3491-27-8)**
Octafluorobut-2-ene [UN 2422] **(360-89-4)**
Octafluoro-sec-butene **(382-21-8)**
Octafluorobutene-2 **(360-89-4)**
Octafluorocyclobutane [UN 1976] **(115-25-3)**
Octafluoroisobutylene **(382-21-8)**
Octafluoropropane [UN 2424] **(76-19-7)**
Octahydrodimethyl-4,7-methano-1H-indene **(30496-78-7)**
1,2,3,4,5,6,7,8-Octahydro-1,4-dimethyl-7-(1-methylethylidene)-azulene, (1S,cis) **(88-84-6)**
1,2,4,5,6,7,8,8-Octahydro-2,3,3a,4,7,7a-hexahydro-4,7-methan-oindan **(57-74-9)**
Octahydro-4,7-methano-1H-indene **(6004-38-2)**
δ-Octalactone **(698-76-0)**
γ-Octalactone **(104-50-7)**
Octalene **(309-00-2)**
Octalox **(60-57-1)**
Octamethyl **(152-16-9)**
1,1,1,3,5,7,7,7-Octamethyl-3,5-bis(6,7-epoxy-4-oxaheptyl)tetra-siloxane **(69155-42-6)**
Octamethylcyclotetrasiloxane **(556-67-2)**
Octamethyl-difosforzuur-tetramide (Dutch) **(152-16-9)**
1,1,1,3,5,7,7,7-Octamethyl-3,5-diphenyltetrasiloxane **(13270-97-8)**
Octamethyldiphosphoramide **(152-16-9)**
Octamethyl-diphosphorsaeure-tetramid (German) **(152-16-9)**
Octamethylene dicyanide **(1871-96-1)**
Octamethylpyrophosphoramide **(152-16-9)**
Octamethyl pyrophosphortetramide **(152-16-9)**
Octamethyl tetramido pyrophosphate **(152-16-9)**
Octamethyltrisiloxane **(107-51-7)**
Octamidophos **(152-16-9)**
1-Octanal **(124-13-0)**
Octanal **(124-13-0)**
Octanaldehyde **(124-13-0)**
Octanal, 3,7-dimethyl- (VAN)(9CI) **(5988-91-0)**
1-Octanal, 3,7-dimethyl-7-hydroxy **(107-75-5)**
1-Octanal, 3,7-dimethyl-7-methoxy **(3613-30-7)**
Octanal, 7-hydroxy-3,7-dimethyl- **(107-75-5)**
Octanal, 2-(phenylmethylene)- (9CI) **(101-86-0)**
Octanamide (9CI) **(629-01-6)**
Octanamide, N,N-bis(2-hydroxyethyl)- (9CI) **(3077-30-3)**
Octanamide, N,N-dimethyl **(1118-92-9)**
Octanamide, 2,2,3,3,4,4,5,5,6,6,7,7,8,8,8-pentadecafluoro-N,N-bis(2-hydroxyethyl)- (9CI) **(42268-97-3)**
1-Octanamine **(111-86-4)**
2-Octanamine **(693-16-3)**
1-Octanamine, N,N-dimethyl- (9CI) **(7378-99-6)**
1-Octanamine, N,N-dioctyl **(1116-76-3)**
2-Octanamine, 5-methyl- (9CI) **(67953-04-2)**
1-Octanamine, N-methyl-N-octyl- (9CI) **(4455-26-9)**
1-Octanamine, N-octyl- (9CI) **(1120-48-5)**
1-Octanaminium, N,N-dimethyl-N-octyl-, chloride (9CI) **(5538-94-3)**
1-Octanaminium, N-methyl-N,N-dioctyl-, chloride (9CI) **(5137-55-3)**
1-Octanaminium, N,N,N-trimethyl-, chloride (9CI) **(10108-86-8)**
Octan amylu (Polish) **(628-63-7)**
Octan barnaty (Czech) **(543-80-6)**
Octan n-butylu (Polish) **(123-86-4)**
Octan cyklohexylaminu (Czech) **(58695-41-3)**
Octan draselny (Czech) **(127-08-2)**
Octane (ACGIH,OSHA) [UN 1262] **(111-65-9)**
Octane, 1-bromo **(111-83-1)**
1-Octanecarboxylic acid **(112-05-0)**
Octane, 1-chloro- (9CI) **(111-85-3)**
Octane, 2-chloro- (9CI) **(628-61-5)**
Octane, chloro- (8CI,9CI) **(26655-49-2)**
Octane, 1-cyclohexyl- (8CI) **(1795-15-9)**
Octane, 2-cyclohexyl- (7CI,8CI) **(2883-05-8)**
1,8-Octanediamine, dihydrochloride **(7613-16-3)**
Octane-1,8-diamine dihydrochloride **(7613-16-3)**
Octane, 1,1-dibromo- (9CI) **(62168-26-7)**
1,8-Octanedicarboxylic acid **(111-20-6)**
Octane, 1,1-dichloro- (8CI,9CI) **(20395-24-8)**
Octane, 1,2:7,8-diepoxy **(2426-07-5)**
Octane, 1,1-difluoro- (9CI) **(61350-03-6)**
Octane, 2,3-dimethyl- (9CI) **(7146-60-3)**
Octane, 2,6-dimethyl- (9CI) **(2051-30-1)**
Octane, 2,7-dimethyl- (8CI,9CI) **(1072-16-8)**
Octanedinitrile (9CI) **(629-40-3)**
1,8-Octanedioic acid **(505-48-6)**
Octanedioic acid (9CI) **(505-48-6)**

Octanedioic acid, diethyl ester (9CI) **(2050-23-9)**
Octanedioic acid, dimethyl ester (9CI) **(1732-09-8)**
Octanedioic acid, 2-ethyl- (8CI,9CI) **(3971-33-3)**
1,2-Octanediol (9CI) **(1117-86-8)**
1,2-Octanediol, 3,7-dimethyl **(107-74-4)**
2,3-Octanedione (8CI,9CI) **(585-25-1)**
3,5-Octanedione, 6,6,7,7,8,8,8-heptafluoro-2,2-dimethyl- (9CI) **(17587-22-3)**
2,7-Octanedione, 4,4,5,5-tetramethyl- (8CI,9CI) **(17663-27-3)**
Octane, 1,1'-dioxybis- **(19102-74-0)**
Octane, 1,2-epoxy **(2984-50-1)**
Octane, 1,4-epoxy- **(1004-29-1)**
Octane, 1-(ethenylsulfonyl)- (9CI) **(28345-91-7)**
Octane, 1-fluoro **(463-11-6)**
Octane, 1,1,1,2,3,3,4,4,5,5,6,6,7,7,8,8,8-hexadecafluoro-2,7-bis(trifluoromethyl)- (9CI) **(3021-63-4)**
Octane, hexadecafluoro-2,7-bis(trifluoromethyl)- (8CI) **(3021-63-4)**
Octane, 1-iodo **(629-27-6)**
Octane, 2-iodo- (9CI) **(557-36-8)**
Octane, 2-methyl- (9CI) **(3221-61-2)**
Octane, 3-methyl- (9CI) **(2216-33-3)**
Octane, 4-methyl- (9CI) **(2216-34-4)**
Octanenitrile **(124-12-9)**
Octane, octadecafluoro **(307-34-6)**
Octane, 1,1'-oxybis- (9CI) **(629-82-3)**
Octane, 1-phenyl- (8CI) **(2189-60-8)**
1-Octanesulfonamide, N-butyl-1,1,2,2,3,3,4,4,5,5,6,6,7,7,8,8,8-heptadecafluoro-N-(2-hydroxyethyl)- (9CI) **(2263-09-4)**
1-Octanesulfonamide, N-(3-(dimethylamino)propyl)-1,1,2,2,3,3,4,4,5,5,6,6,7,7,8,8,8-heptadecafluoro- (9CI) **(13417-01-1)**
1-Octanesulfonamide, N-ethyl-1,1,2,2,3,3,4,4,5,5,6,6,7,7,8,8,8-heptadecafluoro-N-(2-hydroxyethyl)- (9CI) **(1691-99-2)**
1-Octanesulfonamide, 1,1,2,2,3,3,4,4,5,5,6,6,7,7,8,8,8-heptadecafluoro-N-(2-hydroxyethyl)-N-methyl- (9CI) **(24448-09-7)**
1-Octanesulfonamide, 1,1,2,2,3,3,4,4,5,5,6,6,7,7,8,8,8-heptadecafluoro-N-methyl- (9CI) **(31506-32-8)**
1-Octanesulfonamide, N,N'-(phosphinicobis(oxy-2,1-ethanediyl))-bis(N-ethyl-1,1,2,2,3,3,4,4,5,5,6,6,7,7,8,8,8-heptadecafluoro-, ammonium salt (9CI) **(30381-98-7)**
1-Octanesulfonic acid, 1,1,2,2,3,3,4,4,5,5,6,6,7,7,8,8,8-heptadecafluoro-, potassium salt (9CI) **(2795-39-3)**
1-Octanesulfonic acid, sodium salt (9CI) **(5324-84-5)**
1-Octanesulfonyl chloride (9CI) **(7795-95-1)**
1-Octanesulfonyl fluoride (9CI) **(40630-63-5)**
1-Octanesulfonyl fluoride, 1,1,2,2,3,3,4,4,5,5,6,6,7,7,8,8,8-heptadecafluoro- (9CI) **(307-35-7)**
1-Octanethiol **(94805-33-1)**
Octane-1-thiol **(111-88-6)**
tert-Octanethiol **(141-59-3)**
1-Octanethiol (9CI) **(111-88-6)**
2-Octanethiol (9CI) **(3001-66-9)**
Octan etoksyetylu (Polish) **(111-15-9)**
Octane, 1,2,3-trichloro- (9CI) **(85269-46-1)**
Octane, 1,1,1,2,2,3,3,4,4,5,5,6,6-tridecafluoro-8-iodo- (9CI) **(2043-57-4)**
Octane, 2,3,7-trimethyl- (9CI) **(62016-34-6)**
Octan etylu (Polish) **(141-78-6)**
Octan fenylrtutnaty (Czech) **(62-38-4)**
Octan kobaltnaty (Czech) **(6147-53-1)**
Octan manganaty (Czech) **(638-38-0)**
Octan mednaty (Czech) **(142-71-2)**
Octan metylu (Polish) **(79-20-9)**
Octanoic-acid **(124-07-2)**
Octanoic acid, aluminum salt (9CI) **(6028-57-5)**
Octanoic acid, 2,2-bis(((1-oxooctyl)oxy)methyl)-1,3-propanediyl ester (9CI) **(3008-50-2)**
Octanoic acid, 8-bromo- (8CI,9CI) **(17696-11-6)**
Octanoic acid, butyl ester (9CI) **(589-75-3)**
Octanoic acid, cadmium salt (2:1) **(2191-10-8)**
Octanoic acid, (4-cyano-2,6-diiodo)phenyl ester **(3861-47-0)**
Octanoic acid, decyl ester (9CI) **(2306-89-0)**
Octanoic acid, diester with triethylene glycol **(106-10-5)**
Octanoic acid, 3,6-dimethyl- (8CI,9CI) **(4812-29-7)**
Octanoic acid, 2,2-dimethyl-, copper(2+) salt (9CI) **(32276-75-8)**
Octanoic acid, 4,6-dimethyl-, methyl ester (7CI) **(2553-96-0)**
Octanoic acid, 4,6-dimethyl-, methyl ester, (4S,6S)-(+)- (8CI) **(2553-96-0)**
Octanoic acid, 2,2-dimethyl-, phenylmethyl ester **(81325-79-3)**
Octanoic acid, 4,4-dimethyl-, tributylstannyl ester **(28801-69-6)**
Octanoic acid, (1,2-dithioxo-1,2-ethanediyl)bis(imino-2,1-ethanediyl) ester (9CI) **(24928-72-1)**
Octanoic acid, 1,2-ethanediylbis(oxy-2,1-ethanediyl) ester (9CI) **(106-10-5)**
Octanoic acid, ethylenebis(oxyethylene) ester (8CI) **(106-10-5)**
Octanoic acid, ethyl ester **(106-32-1)**
Octanoic acid, 3-hexenyl ester, (Z)- (9CI) **(61444-41-5)**
Octanoic acid, 3-hydroxy- (8CI,9CI) **(14292-27-4)**
Octanoic acid, isopentyl ester **(2035-99-6)**
Octanoic acid, isopropyl ester (8CI) **(5458-59-3)**
Octanoic acid, lead salt (8CI,9CI) **(15696-43-2)**
Octanoic acid, 6-methyl- (8CI,9CI) **(504-99-4)**
Octanoic acid, methyl ester (9CI) **(111-11-5)**
Octanoic acid, 1-methylethyl ester (9CI) **(5458-59-3)**
Octanoic acid, 2-methyl-, methyl ester (6CI,7CI,8CI,9CI) **(2177-86-8)**
Octanoic acid, 1-methyltridecyl ester (9CI) **(55193-79-8)**
Octanoic acid, 1-naphthalenyl ester (9CI) **(4483-62-9)**
Octanoic acid, 1-naphthyl ester (8CI) **(4483-62-9)**
Octanoic acid, octyl ester (9CI) **(2306-88-9)**
Octanoic acid, pentadecafluoro **(335-67-1)**
Octanoic acid, pentadecafluoro-, ammonium salt **(3825-26-1)**
Octanoic acid, potassium salt **(764-71-6)**
Octanoic acid, 1,2,3-propanetriyl ester **(538-23-8)**
Octanoic acid, sodium salt **(1984-06-1)**
Octanoic acid triglyceride **(538-23-8)**
Octanoin, tri- (8CI) **(538-23-8)**
1-Octanol **(111-87-5)**
2-Octanol **(123-96-6)**
3-Octanol **(589-98-0)**
Octanol **(111-87-5)**
Octanol-3 **(589-98-0)**
n-Octanol **(111-87-5)**
4-Octanol (9CI) **(589-62-8)**
2-Octanol, (+-)- (8CI,9CI) **(4128-31-8)**
Octanol (9CI) (VAN) **(29063-28-3)**
1-Octanol acetate **(112-14-1)**
1-Octanol, aluminum salt (9CI) **(14624-13-6)**
1-Octanol, 2-butyl **(3913-02-8)**
1-Octanol, 3,7-dimethyl **(106-21-8)**
2-Octanol, 2,6-dimethyl- **(18479-57-7)**
3-Octanol, 3,7-dimethyl **(78-69-3)**
3-Octanol, 3,6-dimethyl- (9CI) **(151-19-9)**
1-Octanol, 2,7-dimethyl- (8CI,9CI) **(15250-22-3)**
1-Octanol, 3,7-dimethyl-7-hydroxy- **(107-74-4)**
Octanolide-1,4 **(104-50-7)**
1-Octanol, 7-methyl **(2430-22-0)**
3-Octanol, 3-methyl **(5340-36-3)**
1-Octanol, 3,3,4,4,5,5,6,6,7,7,8,8,8-tridecafluoro- (9CI) **(647-42-7)**
2-Octanone **(111-13-7)**
3-Octanone **(106-68-3)**
Octanone (9CI) **(27457-18-7)**
4-Octanone (8CI,9CI) **(589-63-9)**
1-Octanone, 1-(2,5-dihydroxyphenyl)- (9CI) **(4693-19-0)**
2-Octanone, 3-methyl **(6137-08-2)**
2-Octanone, 5-methyl **(58654-67-4)**
4-Octanone, 7-methyl- (8CI,9CI) **(20809-46-5)**
2-Octanone, 1-nitro- (8CI,9CI) **(16067-01-9)**
Octanonitrile **(124-12-9)**
Octanoyl chloride (9CI) **(111-64-8)**

Name	CAS
Octanoyl fluoride, pentadecafluoro- (9CI)	**(335-66-0)**
n-Octanoyl peroxide (DOT)	**(762-16-3)**
n-Octanoyl peroxide, Technically pure (DOT)	**(762-16-3)**
Octan propylu (Polish)	**(109-60-4)**
Octan sodny (Czech)	**(127-09-3)**
Octan winylu (Polish)	**(108-05-4)**
n-Octanyl acetate	**(112-14-1)**
Octan zinecnaty (Czech)	**(5970-45-6)**
3,6,9,12,15,18,21,24-Octaoxadotetracontan-1-ol (8CI,9CI)	**(13149-87-6)**
3,6,9,12,15,18,21,24-Octaoxahexacosane-1,26-diol	**(3386-18-3)**
1,1,3,3,5,5,7,7-Octaphenylcyclotetrasiloxane	**(546-56-5)**
Octaphenylcyclotetrasiloxane	**(546-56-5)**
Octaphenyltetracyclosiloxane	**(546-56-5)**
Octapol 100	**(9036-19-5)**
Octasiloxane, octadecamethyl- (9CI)	**(556-69-4)**
Octasiloxane, 1,1,1,3,3,5,7,7,9,11,11,13,15,15,15-pentadeca-methyl-5,9,13-triphenyl- (9CI)	**(60617-40-5)**
Octatensin	**(55-65-2)**
Octatenzine	**(55-65-2)**
1,3,5,7-Octatetrayne (9CI)	**(6165-96-4)**
Octatetrayne (8CI)	**(6165-96-4)**
Octatriacontane	**(7194-85-6)**
1,3,6-Octatriene, (E,E)- (8CI,9CI)	**(22038-69-3)**
Octatriene, dimethyl	**(29714-87-2)**
1,3,6-Octatriene, 3,7-dimethyl- (9CI)	**(13877-91-3)**
1,3,6-Octatriene, 3,7-dimethyl-, (E)- (8CI,9CI)	**(3779-61-1)**
2,4,6-Octatriene, 2,6-dimethyl- (VAN) (9CI)	**(673-84-7)**
1,3,6-Octatriene, 3,7-dimethyl-, (Z)- (9CI)	**(3338-55-4)**
1,5,7-Octatrien-3-ol, 3,7-dimethyl-, (R-(E))- (9CI)	**(20053-88-7)**
1,5,7-Octatrien-3-ol, 3,7-dimethyl-, (E)-(R)-(-)- (8CI)	**(20053-88-7)**
Octatropine methylbromide	**(80-50-2)**
2-Octenal	**(2363-89-5)**
Octenal (8CI,9CI)	**(25447-69-2)**
2-Octenal, 2-butyl- (8CI,9CI)	**(13019-16-4)**
6-Octenal, 3,7-dimethyl- (9CI)	**(106-23-0)**
6-Octenal, 3,7-dimethyl-, (S)- (9CI)	**(5949-05-3)**
3-Octen-2-amine, N,N-dimethyl-, (E)- (9CI)	**(55956-31-5)**
(E)-4-Octene	**(14850-23-8)**
1-n-Octene	**(111-66-0)**
Octene-1	**(111-66-0)**
Octene-2	**(111-67-1)**
α-Octene	**(111-66-0)**
n-1-Octene	**(111-66-0)**
n-trans-4-Octene	**(14850-23-8)**
trans-4-Octene	**(14850-23-8)**
trans-n-4-Octene	**(14850-23-8)**
1-Octene (9CI)	**(111-66-0)**
2-Octene (9CI)	**(111-67-1)**
4-Octene, (E)- (9CI)	**(14850-23-8)**
Octene (9CI)	**(25377-83-7)**
2-Octene, (E)- (8CI,9CI)	**(13389-42-9)**
4-Octene (8CI,9CI)	**(592-99-4)**
2-Octene, Mixed cis & trans	**(111-67-1)**
2-Octene (Mixed cis, trans isomers)	**(111-67-1)**
2-Octene, (Z)- (8CI,9CI)	**(7642-04-8)**
1-Octene, 3,7-dimethyl- (8CI,9CI)	**(4984-01-4)**
1-Octene, 7,8-epoxy- (8CI)	**(19600-63-6)**
2-Octene, 2-nitro	**(6065-11-8)**
2-Octene, 3-nitro	**(6065-10-7)**
3-Octene, 3-nitro	**(6065-09-4)**
(Z)-3-Octenoic acid	**(5169-51-7)**
cis-Oct-3-enoic acid	**(5169-51-7)**
3-Octenoic acid (9CI)	**(1577-19-1)**
4-Octenoic acid (6CI,7CI,8CI,9CI)	**(18294-89-8)**
3-Octenoic acid, (Z)- (9CI)	**(5169-51-7)**
2-Octenoic acid, ethyl ester, (E)- (9CI)	**(7367-82-0)**
3-Octenoic acid, methyl ester, (Z)- (9CI)	**(69668-85-5)**
(E)-2-Octen-1-ol	**(18409-17-1)**
(Z)-5-Octen-1-ol	**(64275-73-6)**
1-Octen-3-ol	**(3391-86-4)**
2-Octen-1-ol, (E)-	**(18409-17-1)**
3-Octen-1-ol	**(18185-81-4)**
cis-5-Octen-1-ol	**(64275-73-6)**
trans-3-Octen-2-ol	**(57648-55-2)**
3-Octen-2-ol, (E)- (9CI)	**(57648-55-2)**
7-Octen-4-ol (9CI)	**(53907-72-5)**
5-Octen-1-ol, (Z)- (9CI)	**(64275-73-6)**
3-Octen-1-ol, acetate, (Z)- (9CI)	**(69668-83-3)**
6-Octen-1-ol, 3,7-dimethyl	**(106-22-9)**
6-Octen-3-ol, 3,7-dimethyl-	**(18479-51-1)**
7-Octen-2-ol, 2,6-dimethyl	**(18479-58-8)**
6-Octen-1-ol, 3,7-dimethyl-, (+-)- (9CI)	**(26489-01-0)**
6-Octen-2-ol, 2,6-dimethyl- (9CI)	**(30385-25-2)**
2-Octen-8-ol, 2,6-dimethyl-, acetate	**(150-84-5)**
6-Octen-1-ol, 3,7-dimethyl-, acetate	**(150-84-5)**
7-Octen-2-ol, 2,6-dimethyl-, formate (9CI)	**(25279-09-8)**
6-Octen-1-ol, 3,7-dimethyl-, propanoate	**(141-14-0)**
3-Octen-2-ol, 2-methyl-, (Z)- (8CI,9CI)	**(18521-07-8)**
7-Octen-2-ol, 2-methyl-6-methylene	**(543-39-5)**
7-Octen-2-ol, 2-methyl-6-methylene-, acetate	**(1118-39-4)**
3-Octen-2-one (9CI)	**(1669-44-9)**
7-Octen-2-one (8CI,9CI)	**(3664-60-6)**
2-Octen-4-one, 2-methoxy- (8CI)	**(24985-48-6)**
3-Octen-2-one, 7-methyl- (9CI)	**(33046-81-0)**
2-Octen-4-one, 2-methyl- (8CI,9CI)	**(19860-71-0)**
1-Octenyl succinic anhydride	**(7757-96-2)**
1-Octenylsuccinic anhydride	**(7757-96-2)**
6-Octen-1-yn-3-ol, 3,7-dimethyl-	**(29171-20-8)**
Octhilinone	**(26530-20-1)**
Octic acid	**(124-07-2)**
Octilin	**(111-87-5)**
Octocrilene	**(6197-30-4)**
Octocrileno (Spanish)	**(6197-30-4)**
Octocrilenum (Latin)	**(6197-30-4)**
Octocrylene	**(6197-30-4)**
Octofen	**(134-31-6)**
Octogen	**(2691-41-0)**
n-Octoic acid	**(124-07-2)**
Octoil	**(117-81-7)**
Octoil S	**(122-62-3)**
Octone	**(4024-81-1)**
Octopamine	**(104-14-3)**
Octowy aldehyd (Polish)	**(75-07-0)**
Octowy bezwodnik (Polish)	**(108-24-7)**
Octowy kwas (Polish)	**(64-19-7)**
Octoxinol	**(9002-93-1)**
Octoxynol	**(9002-93-1)**
Octoxynol 3	**(9002-93-1)**
Octoxynol 9	**(9002-93-1)**
3-Octoxypropane-1-amine	**(15930-66-2)**
3-Octoxypropanenitrile	**(16728-49-7)**
Octrizol (Spanish)	**(3147-75-9)**
Octrizole	**(3147-75-9)**
Octrizolum (Latin)	**(3147-75-9)**
4-Octyn-3,6-diol, 3,6-dimethyl	**(78-66-0)**
Octyl Phenol EO (16)	**(9036-19-5)**
1-Octyl acetate	**(112-14-1)**
Octyl acetate	**(103-09-3)**
Octyl acetate	**(112-14-1)**
n-Octyl acetate	**(112-14-1)**
β-Octyl acrolein	**(4826-62-4)**
N-tert-Octylacrylamide	**(4223-03-4)**
Octyl acrylate	**(103-11-7)**
Octyl acrylate	**(2499-59-4)**

Name	CAS No.
Octyl adipate	**(103-23-1)**
Octyl adipate (VAN)	**(123-79-5)**
Octyl-alcohol	**(111-87-5)**
n-Octyl alcohol	**(111-87-5)**
Octyl alcohol acetate	**(112-14-1)**
Octyl alcohol, normal-primary	**(111-87-5)**
n-Octyl aldehyde	**(124-13-0)**
2-Octylamine	**(693-16-3)**
Octylamine	**(111-86-4)**
n-Octylamine	**(111-86-4)**
Octylamine 2,4-dichlorophenoxyacetate	**(2212-53-5)**
Octylamine, N,N-dimethyl- (8CI)	**(7378-99-6)**
Octylamine, methanearsonate (1:1)	**(6379-37-9)**
Octyl 4-aminobenzoate	**(14309-41-2)**
2-(Octylamino)ethanol	**(32582-63-1)**
Octylammonium methanearsonate	**(6379-37-9)**
Octylammonium methylarsonate	**(6379-37-9)**
Octylbenzene	**(2189-60-8)**
n-Octylbenzene	**(2189-60-8)**
Octylbenzenesulfonic acid	**(25321-43-1)**
n-Octyl bicycloheptene dicarboximide	**(113-48-4)**
n-Octylbicyclo-(2.2.1)-5-heptene-2,3-dicarboximide	**(113-48-4)**
n-Octyl bromide	**(111-83-1)**
Octyl butyl phthalate	**(84-78-6)**
γ-n-Octyl-γ-n-butyrolactone	**(2305-05-7)**
Octyl carbinol	**(143-08-8)**
1-Octyl chloride	**(111-85-3)**
Octyl chloride	**(111-85-3)**
n-Octyl chloride	**(111-85-3)**
sec-Octyl chloride	**(26655-49-2)**
Octyl cresoxyethoxyethyl dimethyl benzyl ammonium chloride	**(25155-18-4)**
1-Octyl cyanide	**(2243-27-8)**
Octyl cyanide	**(2243-27-8)**
n-Octyl cyanide	**(2243-27-8)**
Octylcyclohexane	**(1795-15-9)**
2-Octyl-1-cyclopropene-1-heptanoic acid	**(503-05-9)**
Octyl decanoate	**(2306-92-5)**
Octyl decyl adipate	**(110-29-2)**
n-Octyl decyl adipate	**(110-29-2)**
Octyl decyl dimethyl ammonium chloride	**(32426-11-2)**
Octyldecylmethylamine	**(22020-14-0)**
Octyl decyl phthalate	**(119-07-3)**
n-Octyl n-decyl phthalate	**(119-07-3)**
2-Octyl 2,4-dichlorophenoxyacetate	**(1917-97-1)**
Octyl 2,4-dichlorophenoxyacetate	**(1928-44-5)**
Octyl dihydrogen phosphate	**(3991-73-9)**
Octyl dimethyl PABA	**(21245-02-3)**
Octyl dimethyl p-aminobenzoate	**(21245-02-3)**
Octyl-dimethyl-benzylammonium chloride	**(959-55-7)**
4-Octyldiphenylamine	**(4175-37-5)**
Octyl diphenyl phosphate	**(115-88-8)**
2-Octyldodecanol	**(5333-42-6)**
Octyldodecanol	**(5333-42-6)**
2-Octyldodecyl alcohol	**(5333-42-6)**
Octyl dodecyl dimethyl ammonium chloride	**(10361-16-7)**
Octyldodecyldimethylammonium chloride	**(10361-16-7)**
1-Octylene	**(111-66-0)**
Octylene	**(111-66-0)**
Octylene	**(25377-83-7)**
α-Octylene	**(111-66-0)**
Octylene epoxide	**(2984-50-1)**
Octylene glycol	**(94-96-2)**
Octyl 9,10-epoxystearate	**(106-84-3)**
Octyl epoxytallate	**(61788-72-5)**
Octyl ester of 2,4-D	**(1928-44-5)**
N-Octylethanolamine	**(32582-63-1)**
Octyl-ether	**(629-82-3)**
n-Octylic acid	**(124-07-2)**
1,1'-(Octylimino)bis-2-propanol	**(28482-15-7)**
1-Octyl iodide	**(629-27-6)**
1-n-Octyl iodide	**(629-27-6)**
Octyl iodide	**(629-27-6)**
n-Octyl iodide	**(629-27-6)**
Octyl isodecyl phthalate	**(1330-96-7)**
n-Octylisosafrole sulfoxide	**(120-62-7)**
2-Octyl-4-isothiazolin-3-one	**(26530-20-1)**
Octyl mercaptan	**(111-88-6)**
n-Octyl mercaptan	**(111-88-6)**
t-Octyl mercaptan	**(141-59-3)**
Octyl methacrylate	**(2157-01-9)**
Octyl methoxycinnamate	**(5466-77-3)**
Octyl methyl ketone	**(693-54-9)**
Octyl 2-methyl-2-propenoate	**(2157-01-9)**
Octyl nitrate	**(629-39-0)**
N-Octyl-1-octanamine	**(1120-48-5)**
Octyl octanoate	**(2306-88-9)**
Octyl 3-octyloxiraneoctanoate	**(106-84-3)**
4-Octyl-N-(4-octylphenyl)benzenamine	**(101-67-7)**
Octyl oleate epoxide	**(106-84-3)**
Octyloxirane	**(2404-44-6)**
cis-3-Octyl-oxiraneoctanoic acid	**(2443-39-2)**
4-(Octyloxy)benzoic acid	**(2493-84-7)**
3-(Octyloxy)-1-propanamine	**(15930-66-2)**
3-Octyloxypropanamine	**(15930-66-2)**
3-(Octyloxy)propanenitrile	**(16728-49-7)**
3-Octyloxypropanenitrile	**(16728-49-7)**
Octyl palmitate	**(29806-73-3)**
Octyl peroxide	**(19102-74-0)**
2-Octylphenol	**(949-13-3)**
4-Octylphenol	**(1806-26-4)**
Octylphenol	**(27193-28-8)**
o-Octylphenol	**(949-13-3)**
o-tert-Octylphenol	**(3884-95-5)**
p-Octylphenol	**(1806-26-4)**
p-tert-Octylphenol	**(140-66-9)**
Octylphenol EO (3)	**(9036-19-5)**
Octylphenol EO (10)	**(9036-19-5)**
Octyl phenol EO (20)	**(9036-19-5)**
Octyl phenol condensed with 1 mole ethylene oxide	**(1322-97-0)**
Octyl phenol condensed with 3 moles ethylene oxide	**(9036-19-5)**
Octyl phenol condensed with 8-10 moles ethylene oxide	**(9036-19-5)**
Octyl phenol condensed with 12-13 moles ethylene oxide	**(9002-93-1)**
Octyl phenol condensed with 16 moles ethylene oxide	**(9036-19-5)**
Octyl phenol condensed with 20 moles ethylene oxide	**(9036-19-5)**
Octyl phenoxy ethanol	**(1322-97-0)**
Octylphenoxyethanol	**(1322-97-0)**
p-tert-Octylphenoxyethoxyethyldimethylbenzylammonium chloride	**(121-54-0)**
Octylphenoxypoly(ethoxyethanol)	**(9036-19-5)**
Octylphenoxypolyethoxyethanol	**(9036-19-5)**
p-tert-Octylphenoxypolyethoxyethanol	**(9002-93-1)**
tert-Octylphenoxypoly(ethoxyethanol)	**(9036-19-5)**
p-Octylphenoxypolyethoxyethanol - iodine complex	**(53404-04-9)**
Octylphenoxypoly(ethyleneoxy)ethanol	**(9036-19-5)**
tert-Octylphenoxy poly(oxyethylene)ethanol	**(9036-19-5)**
4-Octyl-N-phenylbenzenamine	**(4175-37-5)**
tert-Octylphenyl-α-naphthylamine	**(51772-35-1)**
Octyl phosphate	**(39407-03-9)**
Octyl phthalate	**(117-81-7)**
Octyl phthalate	**(117-84-0)**
n-Octyl phthalate	**(117-84-0)**
Octyl potassium phosphate	**(19045-79-5)**
Octyl 2-propenoate	**(2499-59-4)**
Octyl sebacate	**(122-62-3)**
Octyl sodium sulfate	**(142-31-4)**

Octyl stearate	(22047-49-0)
5-(2-(Octylsulfinyl)propyl)-1,3-benzodioxole	(120-62-7)
n-Octylsulfoxide of isosafrole	(120-62-7)
2-(Octylthio)ethanol	(3547-33-9)
1-Octyl thiol	(111-88-6)
1-Octylthiol	(111-88-6)
Octylthiol	(111-88-6)
Octyl trichlorosilane	(5283-66-9)
Octyltrichlorosilane [UN 1801]	(5283-66-9)
Octyltrichlorostannane	(3091-25-6)
Octyl(triethoxy)silane	(2943-75-1)
1-Octyne	(629-05-0)
2-Octyne (8CI,9CI)	(2809-67-8)
4-Octyne (8CI,9CI)	(1942-45-6)
Octynediol, dimethyl- (6CI,8CI,9CI)	(1321-87-5)
3-Octyne, 2,2,7-trimethyl- (9CI)	(55402-13-6)
1-Octyn-3-ol, 4-ethyl- (9CI)	(5877-42-9)
Ocuseptine	(137-40-6)
Ocusol	(61-76-7)
Odiston	(117-96-4)
Odoripon Al 95	(151-21-3)
Oekolp	(56-53-1)
Oenanthal	(111-71-7)
Oenanthaldehyde	(111-71-7)
Oenanthic acid	(111-14-8)
Oenanthic aldehyde	(111-71-7)
Oenanthic ether	(106-30-9)
Oenanthol	(111-71-7)
Oenanthylic acid	(111-14-8)
Oesipos	(8006-54-0)
Oestergon	(50-28-2)
17-β-OH-Oestradiol	(50-28-2)
3,17-β-Oestradiol	(50-28-2)
Oestradiol	(50-28-2)
Oestradiol-17-α	(57-91-0)
Oestradiol-17-β	(50-28-2)
α-Oestradiol	(50-28-2)
β-Oestradiol	(50-28-2)
cis-Oestradiol	(50-28-2)
d-Oestradiol	(50-28-2)
Oestradiol Mustard	(22966-79-6)
Oestradiol R	(50-28-2)
17-β-Oestradiol 3-benzoate	(50-50-0)
Oestradiol 3-benzoate	(50-50-0)
Oestradiol benzoate	(50-50-0)
β-Oestradiol 3-benzoate	(50-50-0)
β-Oestradiol benzoate	(50-50-0)
17-β-Oestradiol dipropionate	(113-38-2)
3,17-β-Oestradiol dipropionate	(113-38-2)
Oestradiol dipropionate	(113-38-2)
Oestradiol-3,17-dipropionate	(113-38-2)
β-Oestradiol dipropionate	(113-38-2)
Oestradiol monobenzoate	(50-50-0)
Oestradiol phosphate Polymer	(28014-46-2)
Oestradiol polyester with phosphoric acid	(28014-46-2)
Oestradiol valerate	(979-32-8)
Oestraform (BDH)	(50-50-0)
Oestrasid	(84-17-3)
17-β-Oestra-1,3,5(10)-triene-3,17-diol	(50-28-2)
Oestra-1,3,5(10)-triene-3,17-β-diol	(50-28-2)
1,3,5(10)-Oestratriene-3,17-β-diol 3-benzoate	(50-50-0)
(16-α,17-β)-Oestra-1,3,5(10)-triene-3,16,17-triol	(50-27-1)
1,3,5-Oestratriene-3-β,16-α,17-β-triol	(50-27-1)
Oestra-1,3,5(10)-triene-3,16-α,17-β-triol	(50-27-1)
1,3,5(10)-Oestratrien-3-ol-17-one	(53-16-7)
1,3,5-Oestratrien-3-ol-17-one	(53-16-7)
δ-1,3,5-Oestratrien-3-β-ol-17-one	(53-16-7)
Oestratriol	(50-27-1)
Oestrenolon	(434-22-0)
Oestrin	(53-16-7)
16-α,17-β-Oestriol	(50-27-1)
3,16-α,17-β-Oestriol	(50-27-1)
Oestriol	(50-27-1)
Oestrodiene	(84-17-3)
Oestrodienol	(84-17-3)
Oestroform	(53-16-7)
Oestrogenine	(56-53-1)
Oestroglandol	(50-28-2)
Oestrogynaedron	(130-80-3)
Oestrogynal	(50-28-2)
Oestrol Vetag	(56-53-1)
Oestromenin	(56-53-1)
Oestromensil	(56-53-1)
Oestromensyl	(56-53-1)
Oestromienin	(56-53-1)
Oestromon	(56-53-1)
Oestronbenzoat (German)	(2393-53-5)
Oestrone	(53-16-7)
Oestrone-3-sulphate sodium salt	(438-67-5)
Oestroperos	(53-16-7)
Oestroral	(84-17-3)
Oestrovis	(84-17-3)
Oets	(9004-62-0)
Off	(134-62-3)
Offitril	(129-20-4)
Ofnack	(119-12-0)
Ofnak	(119-12-0)
Oftalent	(56-75-7)
Oftalfrine	(61-76-7)
Oftanol	(25311-71-1)
Ofunack	(119-12-0)
Ogeen 515	(31566-31-1)
Ogeen GRB	(31566-31-1)
Ogeen M	(31566-31-1)
Ogeen MAV	(31566-31-1)
Ogen	(7280-37-7)
Ohio 347	(13838-16-9)
Ohio Red	(1248-18-6)
Ohmefentanyl	(78995-14-9)
Ohton	(524-84-5)
Oil Blue	(1317-40-4)
Oil Blue A	(14233-37-5)
Oil Camphor Sassafrassy	(8008-51-3)
Oil Cedar	(8000-27-9)
Oil Citrus Reticulata	(8014-17-3)
Oil Dispersant BP 1100X	(39403-84-4)
Oil-Dri	(1302-76-7)
Oil, Edible	(68553-81-1)
Oil, Edible	(68915-86-6)
Oil, Edible	(68956-68-3)
Oil, Edible	(8002-11-7)
Oil, Edible	(8002-13-9)
Oil, Edible	(8023-79-8)
Oil Eucalyptus	(8000-48-4)
Oil Eucalyptus Globulus or Macarthuri	(8000-48-4)
Oil, Fuel, No. 6	(68553-00-4)
Oil-Furnace Black	(1333-86-4)
Oil Garlic	(592-88-1)
Oil Green	(1308-38-9)
Oil Mandarin	(8014-17-3)
Oil, Misc.	(68132-21-8)
Oil, Misc.	(70321-63-0)
Oil, Misc.	(8002-16-2)
Oil, Misc.	(8002-24-2)

Oil, Misc. **(8002-26-4)**
Oil, Misc. **(8016-35-1)**
Oil Mist (ACGIH) **(8012-95-1)**
Oil Mist, Mineral (OSHA) **(8012-95-1)**
Oil Orange **(842-07-9)**
Oil Orange 2311 **(842-07-9)**
Oil Orange 31 **(842-07-9)**
Oil Orange R-14 **(842-07-9)**
Oil Orange 2B **(842-07-9)**
Oil Orange E **(842-07-9)**
Oil Orange EP **(842-07-9)**
Oil Orange 4G **(2051-85-6)**
Oil Orange G **(2051-85-6)**
Oil Orange G **(842-07-9)**
Oil Orange KB **(3118-97-6)**
Oil Orange MO **(2051-85-6)**
Oil Orange MON **(2051-85-6)**
Oil Orange MON Extra **(2051-85-6)**
Oil Orange N Extra **(3118-97-6)**
Oil Orange O'pel **(2646-17-5)**
Oil Orange Opel **(2646-17-5)**
Oil Orange PEL **(842-07-9)**
Oil Orange 2R **(3118-97-6)**
Oil Orange R **(3118-97-6)**
Oil Orange R **(842-07-9)**
Oil Orange SS **(2646-17-5)**
Oil Orange TX **(2646-17-5)**
Oil Orange 7078-V **(842-07-9)**
Oil Orange X **(3118-97-6)**
Oil Orange XO **(2646-17-5)**
Oil Orange XO **(3118-97-6)**
Oil Orange Z-7078 **(842-07-9)**
Oil Pink **(1229-55-6)**
Oil Red **(85-86-9)**
Oil Red 3 **(85-83-6)**
Oil Red 7 **(85-83-6)**
Oil Red 47 **(85-83-6)**
Oil Red 282 **(85-83-6)**
Oil Red 6566 **(85-86-9)**
Oil Red A **(85-83-6)**
Oil Red APT **(85-83-6)**
Oil Red AS **(85-86-9)**
Oil Red 2B **(85-83-6)**
Oil Red 3B **(85-83-6)**
Oil Red 3B **(85-86-9)**
Oil Red 4B **(85-83-6)**
Oil Red B **(85-86-9)**
Oil Red BB **(85-83-6)**
Oil Red BS **(481-39-0)**
Oil Red BS **(85-83-6)**
Oil Red D **(85-83-6)**
Oil Red ED **(85-83-6)**
Oil Red F **(85-83-6)**
Oil Red 3G **(85-86-9)**
Oil Red G **(85-86-9)**
Oil Red GO **(85-83-6)**
Oil Red GRO **(3118-97-6)**
Oil Red IV **(85-83-6)**
Oil Red O **(3118-97-6)**
Oil Red O **(85-86-9)**
Oil Red PEL **(85-83-6)**
Oil Red RC **(85-83-6)**
Oil Red RO **(3118-97-6)**
Oil Red RR **(85-83-6)**
Oil Red S **(85-83-6)**
Oil Red TAX **(85-83-6)**
Oil Red XO **(3118-97-6)**
Oil Red ZD **(85-83-6)**
Oil Red ZMQ **(82-38-2)**
Oil Rose Geranium Algerian **(8000-46-2)**
Oil Scarlet **(3118-97-6)**
Oil Scarlet **(85-83-6)**
Oil Scarlet **(85-86-9)**
Oil Scarlet 48 **(85-83-6)**
Oil Scarlet 371 **(3118-97-6)**
Oil Scarlet APYO **(3118-97-6)**
Oil Scarlet AS **(85-86-9)**
Oil Scarlet BL **(3118-97-6)**
Oil Scarlet 6G **(3118-97-6)**
Oil Scarlet G **(85-86-9)**
Oil Scarlet L **(3118-97-6)**
Oil Scarlet Y **(3118-97-6)**
Oil Scarlet YS **(3118-97-6)**
Oil-Sol. Aniline Yellow **(60-09-3)**
Oil-Sol. Yellow ZH **(2051-85-6)**
Oil Soluble Aniline Yellow **(60-09-3)**
Oil Violet **(6368-72-5)**
Oil Violet IRS **(81-48-1)**
Oil Violet R **(128-95-0)**
Oil Violet ZIRS **(81-48-1)**
Oil Yellow **(60-11-7)**
Oil Yellow **(97-56-3)**
Oil Yellow 20 **(60-11-7)**
Oil Yellow 21 **(97-56-3)**
Oil Yellow 2625 **(60-11-7)**
Oil Yellow 2635 **(2481-94-9)**
Oil Yellow 2681 **(97-56-3)**
Oil Yellow 7463 **(60-11-7)**
Oil Yellow A **(85-84-7)**
Oil Yellow A **(97-56-3)**
Oil Yellow AAB **(60-09-3)**
Oil Yellow AB **(60-09-3)**
Oil Yellow AB Pure **(85-84-7)**
Oil Yellow AN **(60-09-3)**
Oil Yellow APC **(952-47-6)**
Oil Yellow AT **(97-56-3)**
Oil Yellow B **(60-09-3)**
Oil Yellow BB **(60-11-7)**
Oil Yellow C **(97-56-3)**
Oil Yellow D **(60-11-7)**
Oil Yellow DE **(2481-94-9)**
Oil Yellow DEA **(2481-94-9)**
Oil Yellow DN **(60-11-7)**
Oil Yellow E190 **(2481-94-9)**
Oil Yellow ENC **(2481-94-9)**
Oil Yellow FF **(60-11-7)**
Oil Yellow FN **(60-11-7)**
Oil Yellow 2G **(60-09-3)**
Oil Yellow 2G **(60-11-7)**
Oil Yellow G **(60-11-7)**
Oil Yellow G-2 **(60-11-7)**
Oil Yellow GA **(2481-94-9)**
Oil Yellow G Extra **(2051-85-6)**
Oil Yellow GG **(2051-85-6)**
Oil Yellow GG **(60-11-7)**
Oil Yellow GR **(60-11-7)**
Oil Yellow I **(97-56-3)**
Oil Yellow II **(60-11-7)**
Oil Yellow N **(60-11-7)**
Oil Yellow NB **(2481-94-9)**
Oil Yellow OB **(131-79-3)**
Oil Yellow OB **(85-84-7)**
Oil Yellow OB Pure **(131-79-3)**
Oil Yellow Pel **(60-11-7)**

Oil Yellow 2R	**(97-56-3)**
Oil Yellow R	**(60-09-3)**
Oil Yellow S	**(60-11-7)**
Oil Yellow SIS	**(8003-22-3)**
Oil Yellow T	**(97-56-3)**
Oil of Angelica Seed	**(8015-64-3)**
Oil of Anise	**(8007-70-3)**
Oil of Aniseed	**(104-46-1)**
Oil of Bergamot, Rectified	**(8007-75-8)**
Oil of Camphor Rectified	**(8008-51-3)**
Oil of Camphor White	**(8008-51-3)**
Oil of Canada Fleabane	**(8007-27-0)**
Oil of Caraway	**(8000-42-8)**
Oil of Cassia	**(8007-80-5)**
Oil of Cedar Wood	**(8000-27-9)**
Oil of Chinese Cinnamon	**(8007-80-5)**
Oil of Cinnamon	**(8007-80-5)**
Oil of Cinnamon, Ceylon	**(8007-80-5)**
Oil of Citronella	**(8000-29-1)**
Oil of Erigeron	**(8007-27-0)**
Oil of Eucalyptus	**(8000-48-4)**
Oil of Fennel	**(8006-84-6)**
Oil of Fleabane	**(8007-27-0)**
Oil of Garlic	**(8000-78-0)**
Oil of Geranium	**(8000-46-2)**
Oil-of-Grapefruit	**(8016-20-4)**
Oil of Hartshorn	**(8001-85-2)**
Oil of Juniper	**(8002-68-4)**
Oil of Juniper Berry	**(8002-68-4)**
Oil of Lemon	**(8008-56-8)**
Oil of Lemon Grass	**(8007-02-1)**
Oil of Lemongrass	**(8007-02-1)**
Oil of Lemongrass, West Indian	**(8007-02-1)**
Oil of Mace	**(8007-12-3)**
Oil of Mirbane (DOT)	**(98-95-3)**
Oil of Mustard, Artificial	**(57-06-7)**
Oil of Mustard BPC 1949	**(57-06-7)**
Oil of Myrbane	**(98-95-3)**
Oil of Myristica	**(8008-45-5)**
Oil of Niobe	**(93-58-3)**
Oil of Nutmeg	**(8008-45-5)**
Oil of Nutmeg, Expressed	**(8007-12-3)**
Oil-of-Orange	**(8008-57-9)**
Oil of Palma Christi	**(8001-79-4)**
Oil of Pelargonium	**(8000-46-2)**
Oil of Pennyroyal	**(8007-44-1)**
Oil of Pine	**(8002-09-3)**
Oil of Rapeseed	**(8002-13-9)**
Oil of Rose Geranium	**(8000-46-2)**
Oil of Santal	**(8006-87-9)**
Oil of Spearmint	**(8008-79-5)**
Oil of Sweet Orange	**(8008-57-9)**
Oil of Turmeric	**(8024-37-1)**
Oil of Turpentine	**(8006-64-2)**
Oil of Turpentine, Rectified	**(8006-64-2)**
Oil of Vitriol (DOT)	**(7664-93-9)**
Oil of Wintergreen	**(119-36-8)**
Oils, Ambrette	**(8015-62-1)**
Oils, Amyris	**(8015-65-4)**
Oils, Angelica	**(8015-64-3)**
Oils, Anise	**(8007-70-3)**
Oils, Cedarwood	**(8000-27-9)**
Oils, Cinnamon	**(8007-80-5)**
Oils, Citronella	**(8000-29-1)**
Oils, Crocus Sativus	**(8022-19-3)**
Oils, Croton	**(8001-28-3)**
Oils, Cumin	**(8014-13-9)**
Oils, Curcuma	**(8024-37-1)**
Oils, Eucalyptus	**(8000-48-4)**
Oils, Fir	**(8021-28-1)**
Oils, Fish	**(8016-13-5)**
Oils, Fleabane	**(8007-27-0)**
Oils, Fleabane, Erigeron Canadensis	**(8007-27-0)**
Oils, Garlic	**(8000-78-0)**
Oils, Geranium	**(8000-46-2)**
Oils, Glyceridic, Oiticica	**(8016-35-1)**
Oils, Glyceridic, Palm kernel	**(8023-79-8)**
Oils, Juniper	**(8002-68-4)**
Oils, Juniperus Communis	**(8002-68-4)**
Oils, Lanolin	**(70321-63-0)**
Oils, Lanolin	**(8006-54-0)**
Oils, Lard	**(8016-28-2)**
Oils, Lemongrass	**(8007-02-1)**
Oils, Mace	**(8007-12-3)**
Oils, Neat's-foot	**(8002-64-0)**
Oils, Nutmeg	**(8008-45-5)**
Oils, Palm	**(8002-75-3)**
Oils, Palm Kernel	**(8023-79-8)**
Oils, Pennyroyal, Hedeoma Pulegioides	**(8007-44-1)**
Oils, Perilla	**(68132-21-8)**
Oils, Petitgrain	**(8014-17-3)**
Oils, Pine	**(8002-09-3)**
Oils, Raisin	**(68915-86-6)**
Oils, Rice Bran	**(68553-81-1)**
Oils, Saffron	**(8022-19-3)**
Oils, Sandalwood	**(8006-87-9)**
Oils, Sesame	**(8008-74-0)**
Oils, Sperm	**(8002-24-2)**
Oils, Spikenard	**(8022-22-8)**
Oils, Sunflower Seed	**(8001-21-6)**
Oils, Vegetable	**(68956-68-3)**
Oils, Vegetable, Sulfated	**(61790-19-0)**
Oils, Walnut	**(8024-09-7)**
Oils, camphor	**(8008-51-3)**
Oil soluble petroleum sulfonate, sodium salt	**(68608-26-4)**
Oil soluble petroleum sulfonates, sodium salts	**(68608-26-4)**
Oil soluble polyolefin phenol	**(68610-06-0)**
Oils, sassafras, hydrogenated	**(61790-23-6)**
Oiticica Oil	**(8016-35-1)**
Okasa-Mascul	**(57-85-2)**
Okenite	**(13983-17-0)**
Okiten G 23	**(9002-88-4)**
Okodon	**(2152-34-3)**
Oktadecylamin (Czech)	**(124-30-1)**
Oktadekannitril (Czech)	**(638-65-3)**
Oktamethyl (Czech)	**(152-16-9)**
Oktamethylcyklotetrasiloxan (Czech)	**(556-67-2)**
Oktamethylendiamin hydrochlorid (Czech)	**(7613-16-3)**
Oktamethylendikyanid (Czech)	**(1871-96-1)**
Oktamidofos (Czech)	**(152-16-9)**
Oktan (Polish)	**(111-65-9)**
Oktanen (Dutch)	**(111-65-9)**
terc. Oktanthiol (Czech)	**(141-59-3)**
Oktaterr	**(57-74-9)**
Oktedrin	**(60-13-9)**
1-Okten-3-ol (Czech)	**(3391-86-4)**
Oktogen	**(2691-41-0)**
Oktol 600	**(9003-29-6)**
Oktylenoxid (Czech)	**(2984-50-1)**
Oktylester 2,4-dichlorfenoxyoctove (Czech)	**(1928-44-5)**
p-terc.Oktylfenol (Czech)	**(140-66-9)**
2-(Oktylthio)ethanol (Czech)	**(3547-33-9)**
Okultin M	**(94-74-6)**
Okultin MP	**(7085-19-0)**

Olamine (141-43-5)
Olate Flakes (143-19-1)
Olcadil (24166-13-0)
Old 01 (9003-01-4)
Oleo-Coll LP (8002-43-5)
Oleal Orange R (842-07-9)
Oleal Orange SS (2646-17-5)
Oleal Red BB (85-83-6)
Oleal Yellow 2G (60-11-7)
Olealdehyde (8CI) (2423-10-1)
Olealkonium chloride (37139-99-4)
Oleamide (301-02-0)
Oleamide DEA (93-83-4)
Oleamide MIPA (111-05-7)
Oleamide, N,N-dimethyl- (8CI) (2664-42-8)
Oleamide, N,N'-ethylenebis- (8CI) (110-31-6)
Oleamine (112-90-3)
Oleamine Oxide (14351-50-9)
Oleandrin (465-16-7)
Oleandrina (Spanish) (465-16-7)
Oleandrine (465-16-7)
Olean-12-en-30-oic acid, 3-β-hydroxy-11-oxo-, hydrogen succinate (5697-56-3)
Olean-12-en-28-oic acid, 22-β-hydroxy-3-oxo-, 3-methylcrotonate (467-82-3)
Olean-12-en-28-oic acid, 22-β-hydroxy-3-oxo-, 2-methylcrotonate, (Z) (467-81-2)
Oleate of mercury (1191-80-6)
Olefiant Gas (74-85-1)
α Olefins (Petroleum), (C11-C12) cut (64743-02-8)
α Olefins (Petroleum), (C11-C14) cut (64743-02-8)
α Olefins (Petroleum), (C15-C20) cut (64743-02-8)
α Olefins (Petroleum), (C18-C20) cut (64743-02-8)
α Olefins (Petroleum), C10 cut (64743-02-8)
α-Olefin sulfonate (72674-05-6)
α-Olefin sulphonate (72674-05-6)
Oleic acid (112-80-1)
Oleic acid, Compd. with butylamine (1:1) (26094-13-3)
Oleic acid, Compd. with 2,2'-iminodiethanol (1:1) (13961-86-9)
Oleic acid, Compd with 2,2',2''-nitrilotriethanol (1:1) (2717-15-9)
Oleic acid aluminum salt (688-37-9)
Oleic acid amide (301-02-0)
Oleic acid, butyl ester (142-77-8)
Oleic acid chloride (112-77-6)
Oleic acid, chromium salt (8CI) (13308-40-2)
Oleic acid diethanolamide (93-83-4)
Oleic acid diethanolamine (1:1) (13961-86-9)
Oleic acid, diethylenetriamine amide (15566-80-0)
Oleic acid, 2,3-epoxypropyl ester (5431-33-4)
Oleic acid-ethylenediamine condensate (110-31-6)
Oleic acid, ethyl ester (8CI) (111-62-6)
Oleic acid glycerol monoester (25496-72-4)
Oleic acid glycidyl ester (5431-33-4)
Oleic acid, 12-hydroxy- (141-22-0)
Oleic acid, isopropylamide (10574-01-3)
Oleic acid, isopropyl ester (8CI) (112-11-8)
Oleic acid, isopropyl ester, sulfated, sodium salt (14350-72-2)
Oleic acid lead salt (1120-46-3)
Oleic acid, lead(2+) salt (2:1) (1120-46-3)
Oleic acid, methyl ester, cis (112-62-9)
Oleic acid monoglyceride (25496-72-4)
Oleic acid, morpholine salt (1095-66-5)
Oleic acid, morpholine soap (1095-66-5)
Oleic acid nitrile (112-91-4)
Oleic acid polyglyceride (9007-48-1)
Oleic acid poly(oxyethylene) ester (9004-96-0)
Oleic acid, potassium salt (143-18-0)
Oleic acid, propyl ester (8CI) (111-59-1)
Oleic acid, sodium salt (143-19-1)
Oleic acid, triethanolamine salt (2717-15-9)
Oleic acid, triethanolamine soap (2717-15-9)
Oleic acid triglyceride (122-32-7)
Oleic acid, zinc salt (557-07-3)
Oleic diethanolamide (93-83-4)
Oleic monoisopropanolamide (111-05-7)
Oleic sarcosine (110-25-8)
Oleic triglyceride (122-32-7)
Olein (122-32-7)
Oleinamine (112-90-3)
Oleinic acid (112-80-1)
Olein, mono (25496-72-4)
Olein, 1-mono- (8CI) (111-03-5)
Oleinol 7 (9004-98-2)
Olein, tri- (122-32-7)
Olej Napedowy III (Polish) (68334-30-5)
Oleoakarithion (786-19-6)
Oleocuivre (1317-39-1)
Oleofac (2275-18-5)
Oleofos 20 (56-38-2)
Oleogesaprim (1912-24-9)
Oleol (143-28-2)
Oleol 18 (9004-98-2)
Oleomycetin (56-75-7)
Oleonitrile (112-91-4)
Oleo nordox (1317-39-1)
Oleoparaphene (56-38-2)
Oleoparathion (56-38-2)
Oleophosphothion (121-75-5)
Oleosumifene (122-14-5)
Oleovitamin A (68-26-8)
Oleovitamin D (50-14-6)
Oleovitamin D3 (67-97-0)
Oleovofotox (298-00-0)
Oleox 5 (9004-96-0)
Oleoyl Sarcosine (110-25-8)
Oleoyl chloride (8CI) (112-77-6)
1-Oleoylglycerol (111-03-5)
Oleoylglycerol (25496-72-4)
Oleoyl N-methylaminoacetic acid (110-25-8)
N-Oleoyl-N-methyltaurine (97-80-3)
Oleoylmethyltaurine sodium salt (137-20-2)
Oleoylnitrile (112-91-4)
N-Oleoylsarcosine (110-25-8)
Oleoyl sarcosine (110-25-8)
Oleoylsarcosine (110-25-8)
Olepal I (9004-96-0)
Olepal III (9004-96-0)
Oletac 100 (9003-07-0)
Oletetrin (60-54-8)
Oleth (9004-98-2)
Oleth-2 (9004-98-2)
Oleth-3 (9004-98-2)
Oleth-4 (9004-98-2)
Oleth-5 (9004-98-2)
Oleth-6 (9004-98-2)
Oleth-7 (9004-98-2)
Oleth-8 (9004-98-2)
Oleth-9 (9004-98-2)
Oleth-10 (9004-98-2)
Oleth-12 (9004-98-2)
Oleth-15 (9004-98-2)
Oleth-16 (9004-98-2)
Oleth-20 (9004-98-2)
Oleth-23 (9004-98-2)
Oleth-25 (9004-98-2)
Oleth-44 (9004-98-2)

Oleth-50	(9004-98-2)
Oleum Abietis	(8002-09-3)
Oleum (DOT)	(8014-95-7)
Oleum Sinapis Volatile	(57-06-7)
Oleum Tiglii	(8001-28-3)
Oleum Vitis Viniferae	(106-30-9)
Oleyl Alcohol EO (2)	(9004-98-2)
Oleyl Alcohol EO (10)	(9004-98-2)
Oleyl alcohol	(143-28-2)
Oleyl alcohol condensed with 2 moles ethylene oxide	(9004-98-2)
Oleyl alcohol condensed with 20 moles ethylene oxide	(9004-98-2)
Oleyl amide	(301-02-0)
Oleylamide	(301-02-0)
Oleylamin (German)	(112-90-3)
Oleyl amine	(112-90-3)
Oleylamine	(1838-19-3)
Oleyl amine, acetate	(10460-00-1)
Oleylamine acetate	(10460-00-1)
Oleylamine, acetic acid salt	(10460-00-1)
Oleylamine, ethoxylated	(26635-93-8)
Oleyl dimethyl amine oxide	(14351-50-9)
N-Oleyl-N,N-dimethylbenzylammonium chloride	(37139-99-4)
Oleyl dimethyl benzyl ammonium chloride	(37139-99-4)
Oleyl dimethyl glycine	(871-37-4)
Oleyldimethylbetaine	(871-37-4)
1-Oleylglycerol	(111-03-5)
Oleyl betaine	(871-37-4)
Oleyl methylaminoethanoic acid	(110-25-8)
Oleyl N-methylglycine	(110-25-8)
Oleylmethyltaurine	(97-80-3)
Oleylmonoglyceride	(25496-72-4)
Oleyl oleate	(3687-45-4)
Oleylonitrile	(112-91-4)
(Oleyloxymethyl)oxirane	(60501-41-9)
Oleyl palmitamide	(16260-09-6)
Oleylpolyoxethylene-glycol-ether	(9004-98-2)
N-Oleyl-1,3-propylenediamine 2,4-dichlorophenoxyacetate	(2212-59-1)
Oleyl sarcosine	(110-25-8)
Oleyl triglyceride	(122-32-7)
Olicine	(25496-72-4)
Oligo Z	(9080-79-9)
Oligoether L 1502-2-30	(9003-11-6)
Oligotetramethylene glycol dimethacrylate	(2082-81-7)
Olin	(3696-28-4)
Olin MO. 2174	(52-46-0)
Olin Mathieson 2,424	(2593-15-9)
Olio di Croton (Italian)	(8001-28-3)
Olitref	(1582-09-8)
Oliv Ostanthrenovy R (Czech)	(2379-81-9)
Olive-Oil	(8001-25-0)
Olive Tree LS	(8001-21-6)
Olive alcohol	(143-28-2)
Olivetol	(500-66-3)
Olmagran	(135-09-1)
Ololiuqui	(478-94-4)
Olosot	(77-75-8)
Olothorb	(9005-65-6)
Olow (Polish)	(7439-92-1)
Olpisan	(82-68-8)
Oltitox	(63-25-2)
Olvadon	(77-75-8)
Omacide 24	(15922-78-8)
Omadine	(1121-31-9)
Omadine MDS	(43143-11-9)
Omadine sodium	(15922-78-8)
Omadine-sodium	(15922-78-8)
Omadine zinc	(13463-41-7)
Omaflora	(109-84-2)
Omaha	(7439-92-1)
Omaha & Grant	(7439-92-1)
Omain	(477-30-5)
Omaine	(477-30-5)
Omait	(2312-35-8)
Omal	(88-06-2)
Omchlor	(118-52-5)
Omega 127	(90-33-5)
Omega Chrome Blue FB	(3567-69-9)
Omega Chrome Dark Violet D	(2092-55-9)
Omegamycin	(60-54-8)
Omethoat	(1113-02-6)
Omethoate	(1113-02-6)
Omifin	(50-41-9)
Omite	(2312-35-8)
Omni-Passin	(514-73-8)
Omnipen	(69-53-4)
Omnitox	(58-89-9)
Omnizole	(148-79-8)
Omnyl	(72-44-6)
Ompacide	(152-16-9)
Ompatox	(152-16-9)
Ompax	(152-16-9)
Omperan	(15676-16-1)
Omsat	(8064-90-2)
Omtan	(297-78-9)
Omyalen	(13397-26-7)
Onco-Carbide	(127-07-1)
Onco-Carbide	(88-19-7)
Oncodazole	(31430-18-9)
Oncoredox	(68-76-8)
Oncostatin K	(50-76-0)
Oncotepa	(52-24-4)
Oncothio-tepa	(52-24-4)
Oncotiotepa	(52-24-4)
Oncovedex	(68-76-8)
Oncovin	(2068-78-2)
Oncovin	(57-22-7)
Onex	(80-33-1)
Onium compounds, morpholinium, 4-ethyl-4-soya alkyl, Et sulfates	(61791-34-2)
Onkovin	(2068-78-2)
Onozuka P 500	(9004-34-6)
Ontracic 800	(1610-18-0)
Ontrack	(1610-18-0)
Ontrack 8E	(51218-45-2)
Ontrack-WE-2	(1610-18-0)
Onyx	(14808-60-7)
Onyx BTC (Onyx Oil & Chem Co)	(8001-54-5)
Onyx Wax EL	(111-57-9)
Onyxide 200	(4719-04-4)
Onyxide 500	(52-51-7)
Onyxol 345	(120-40-1)
Ooporphyrin	(553-12-8)
Op-Isophrin	(61-76-7)
Op-Sulfa 30	(144-80-9)
Op-Thal-Zin	(7733-02-0)
Opal Blue SS	(2152-64-9)
Opaline Green G 1	(1328-53-6)
Opalon 400	(9003-22-9)
Opclor	(56-75-7)
Opelor	(56-75-7)
Operidine	(50-13-5)
Opex	(101-25-7)
Ophthalamin	(68-26-8)
Ophthel-S	(144-80-9)

Ophthochlor	**(56-75-7)**
Ophtochlor	**(56-75-7)**
Opian	**(128-62-1)**
Opian	**(6035-40-1)**
Opianine	**(128-62-1)**
Opinsul	**(80-35-3)**
Opium	**(8008-60-4)**
Opium, Crude	**(8008-60-4)**
Oplossingen (Dutch)	**(50-00-0)**
Opogard	**(8066-11-3)**
Oppanol B	**(9003-27-4)**
Oppanol B 3	**(9003-27-4)**
Oppanol B 15	**(9003-27-4)**
Oppanol B 100	**(9003-27-4)**
Oppanol B 150	**(9003-27-4)**
Oppanol B 200	**(9003-27-4)**
Opren	**(51234-28-7)**
Optal	**(71-23-8)**
Optalidon	**(77-26-9)**
Optamine PC 5	**(61791-14-8)**
Optanol Fast Scarlet GN	**(3567-65-5)**
Optanol Scarlet GS	**(10169-02-5)**
Optanol Yellow R	**(6375-55-9)**
Optarket	**(57-53-4)**
Optazol	**(67-45-8)**
Optef	**(50-23-7)**
Opticor	**(54-95-5)**
Optinoxan	**(72-44-6)**
Option	**(66441-23-4)**
Optiphyllin	**(58-55-9)**
Optochinidin	**(747-45-5)**
Optunal	**(23422-53-9)**
Ora-Testryl	**(76-43-7)**
Orabet	**(64-77-7)**
Orabolin	**(965-90-2)**
Oracef	**(15686-71-2)**
Oracet Red 3B	**(116-85-8)**
Oracet Sapphire Blue G	**(2475-45-8)**
Oracet Violet B	**(1220-94-6)**
Oracet Violet BN	**(1220-94-6)**
Oracet Violet 2R	**(128-95-0)**
Oracil-VK	**(132-98-9)**
Oracillin	**(87-08-1)**
Oradexon	**(50-02-2)**
Oradian	**(94-20-2)**
Oradil	**(77-36-1)**
Oradiol	**(57-63-6)**
Oraflex	**(51234-28-7)**
Orafuran	**(67-20-9)**
Oragest	**(71-58-9)**
Oraldrina	**(60-13-9)**
Oralin	**(64-77-7)**
Oralith Brilliant Pink R	**(2379-74-0)**
Oralith Orange PG	**(3520-72-7)**
Oralith Red	**(2814-77-9)**
Oralith Red 2GL	**(3468-63-1)**
Oralith Red P4R	**(2425-85-6)**
Oralith Red SR Water Soluble	**(1248-18-6)**
Oralopen	**(132-93-4)**
Oralsterone	**(76-43-7)**
Oramid	**(65-45-2)**
1008 Orange	**(1934-20-9)**
1333 Orange	**(523-44-4)**
1370 Orange	**(1936-15-8)**
11048 Orange	**(3468-63-1)**
11550 Orange	**(633-96-5)**
L-Orange 1	**(2347-72-0)**
L-Orange 2	**(2783-94-0)**
Orange #10	**(1936-15-8)**
Orange 3	**(547-58-0)**
Orange AB	**(2657-89-8)**
Orange A L'huile	**(842-07-9)**
Orange Acid G	**(547-57-9)**
Orange BPC	**(1936-15-8)**
Orange Base Ciba II	**(88-74-4)**
Orange Base Ciba IV	**(141-85-5)**
Orange Base Irga I	**(99-09-2)**
Orange Base Irga II	**(88-74-4)**
Orange Base Irga IV	**(141-85-5)**
Orange Base NGC	**(141-85-5)**
Orange Chrome	**(18454-12-1)**
Orange Extra N	**(633-96-5)**
Orange Extra P	**(633-96-5)**
Orange 2G	**(1936-15-8)**
Orange G	**(1934-20-9)**
Orange G	**(1936-15-8)**
Orange G	**(3520-72-7)**
Orange G BPC	**(1936-15-8)**
Orange G (Biological stain)	**(1936-15-8)**
Orange GC Base	**(108-42-9)**
Orange GCS Salt	**(141-85-5)**
Orange GC Salt	**(141-85-5)**
Orange G Dye	**(1936-15-8)**
Orange GG	**(1936-15-8)**
Orange GGN	**(2347-72-0)**
Orange G (Indicator)	**(1936-15-8)**
Orange GMP	**(1936-15-8)**
Orange GRS Salt	**(88-74-4)**
Orange GS	**(554-73-4)**
Orange I	**(523-44-4)**
Orange I Extra conc. A export	**(523-44-4)**
Orange II	**(633-96-5)**
Orange IIC	**(633-96-5)**
Orange III	**(547-58-0)**
Orange IIP	**(633-96-5)**
Orange II R	**(2783-94-0)**
Orange IIS	**(633-96-5)**
Orange IISM	**(633-96-5)**
Orange II Special for Lacquer	**(633-96-5)**
Orange II for Lakes	**(633-96-5)**
Orange IM	**(523-44-4)**
Orange I, Sodium salt	**(523-44-4)**
Orange IV	**(554-73-4)**
Orange 2 Insoluble	**(842-07-9)**
Orange Insoluble OLG	**(3118-97-6)**
Orange Insoluble OLG	**(842-07-9)**
Orange Insoluble RR	**(3118-97-6)**
Orange LZS	**(1934-20-9)**
Orange 4 Lake	**(554-73-4)**
Orange Leaf Oil, Bitter	**(8014-17-3)**
Orange Leaf Water, Absolute	**(8014-17-3)**
Orange N	**(554-73-4)**
Orange Nitrate Chrome	**(18454-12-1)**
Orange No. 1	**(2347-72-0)**
Orange No. 203	**(3468-63-1)**
Orange No. 205	**(633-96-5)**
Orange OT	**(2646-17-5)**
Orange OT*	**(2646-17-5)**
Orange Oil	**(8008-57-9)**
Orange Oil KB	**(3118-97-6)**
Orange PEL	**(842-07-9)**
Orange Pal	**(2783-94-0)**
Orange Pigment X	**(3468-63-1)**
Orange 3RA Soluble in Grease	**(842-07-9)**

Orange R Fat Soluble (842-07-9)
Orange RGL Conc. Specially Pure (2783-94-0)
Orange RN (1934-20-9)
Orange 3R Soluble in Grease (2646-17-5)
Orange Resenole No. 3 (842-07-9)
Orange Root (458-37-7)
Orange S (523-44-4)
Orange SS (2646-17-5)
Orange Salt Ciba II (88-74-4)
Orange Salt Ciba IV (141-85-5)
Orange Salt Irga II (88-74-4)
Orange Salt Irga IV (141-85-5)
Orange Salt NGC (141-85-5)
Orange 2 Sodium salt (633-96-5)
Orange Soluble A L'huile (842-07-9)
Orange Toner GRT (633-96-5)
Orange Y (633-96-5)
Orange YA (633-96-5)
Orange YZ (633-96-5)
Orange Yellow S (2783-94-0)
Orange Yellow S.AF (2783-94-0)
Orange Yellow S.FQ (2783-94-0)
Orange ZH (3056-93-7)
Orange lead (1314-41-6)
Orange oil (60-12-8)
Oranger Crystals (93-08-3)
Oranges (51-63-8)
Oranil (339-43-5)
Oranyl (339-43-5)
Oranz Akridinova (Czech) (494-38-2)
Oranz Disperzni 11 (Czech) (82-28-0)
Oranz GG (Czech) (1936-15-8)
Oranz G (Polish) (1936-15-8)
Oranz I (Czech) (523-44-4)
Oranz III (Czech) (547-58-0)
Oranz Kypova 5 (Czech) (3263-31-8)
Oranz Kysela 7 (Czech) (633-96-5)
Oranz Kysela 10 (Czech) (1936-15-8)
Oranz Kysela 20 (Czech) (523-44-4)
Oranz Kysela 52 (Czech) (547-58-0)
Oranz Methylova (Czech) (547-58-0)
Oranz Potravinarska 3 (Czech) (2051-85-6)
Oranz Potravinarska 4 (Czech) (1936-15-8)
Oranz Rozpoustedlova 1 (Czech) (2051-85-6)
Oranz Rozpoustedlova 2 (Czech) (2646-17-5)
Oranz Rozpoustedlova 3 (Czech) (495-54-5)
Oranz Rozpoustedlova 7 (Czech) (3118-97-6)
Oranz Rozpoustedlova 15 (Czech) (494-38-2)
Oranz SS (Czech) (2646-17-5)
Oranz Viktoria (Czech) (534-52-1)
Oranz Zasadita 2 (Czech) (495-54-5)
Orapen (132-98-9)
Orasone (53-03-2)
Oraspor (51762-05-1)
Orasthin (50-56-6)
Orasulin (339-43-5)
Oratestin (76-43-7)
Oratrast (7727-43-7)
Oratren (87-08-1)
Oratrol (120-97-8)
Oraviron (58-18-4)
Orbencarb (34622-58-7)
Orbinamon (5591-45-7)
Orbisan (4418-66-0)
Orbon (31566-31-1)
Orcanon (56-04-2)
Orced (2782-57-2)
Orchard Brand Ziram (137-30-4)
Orchiol (57-85-2)
Orchistin (57-85-2)
Orcin (504-15-4)
Orcinol (504-15-4)
Orcinol, 5-methylresorcinol (504-15-4)
Ordimel (968-81-0)
Ordinary Lactic Acid (598-82-3)
Ordinary azoxybenzene (495-48-7)
Ordinary lactic acid (50-21-5)
Ordram (2212-67-1)
Oremet (7440-32-6)
Oresol (93-14-1)
Oreson (93-14-1)
Orestol (130-80-3)
Orestralyn (57-63-6)
Orestrayln (57-63-6)
Oretic (58-93-5)
Oreton (57-85-2)
Oreton (58-22-0)
Oreton-F (58-22-0)
Oreton-M (58-18-4)
Oreton methyl (58-18-4)
Oreton propionate (57-85-2)
Orezan (64-77-7)
Orga-414 (61-82-5)
Orgabolin (965-90-2)
Orgaboral (965-90-2)
Orgadine (5700-49-2)
Orgametil (52-76-6)
Orgametril (52-76-6)
Orgametrol (52-76-6)
Orgamid RMNOCD (105-60-2)
Orgamid RMNOCD (25038-54-4)
Orgamide (105-60-2)
Orgamide (25038-54-4)
Organex (58-08-2)
Organic Glass E 2 (9011-14-7)
Organidin (5634-39-9)
Organol Blue J (128-85-8)
Organol Bordeaux B (6368-72-5)
Organol Brilliant Blue J (128-85-8)
Organol Brown 2R (6416-57-5)
Organol Fast Green J (128-80-3)
Organol Green J (128-80-3)
Organol Orange (842-07-9)
Organol Orange 2J (2051-85-6)
Organol Orange 2R (2646-17-5)
Organol 2R (6416-57-5)
Organol Red B (85-83-6)
Organol Red BS (85-86-9)
Organol Scarlet (85-86-9)
Organol Yellow (60-09-3)
Organol Yellow 2A (60-09-3)
Organol Yellow ADM (60-11-7)
Organol Yellow AP (1689-82-3)
Organol Yellow 2T (97-56-3)
Orgaseptine (63-74-1)
Orgasol 1002D Natural Cos (32131-17-2)
Orgasol 1002D White 5 Cos (32131-17-2)
Orgasol 20030 White 5 Cos (32131-17-2)
Orgastyptin (50-27-1)
Oricur (62-37-3)
Orient Basic Magenta (632-99-5)
Orient Oil Orange PS (842-07-9)
Orient Oil Red RR (85-83-6)
Orient Oil Yellow GG (60-11-7)

Orient Oil Yellow GGS	**(2481-94-9)**
Orient Spirit Black AB	**(11099-03-9)**
Orient Spirit Black SB	**(11099-03-9)**
Oriental Berry	**(124-87-8)**
Orientomycin	**(68-41-7)**
Orimon	**(92-84-2)**
Orinase	**(64-77-7)**
Orinaz	**(64-77-7)**
Orion Blue 3B	**(72-57-1)**
Orisul	**(526-08-9)**
Orisulf	**(526-08-9)**
Orizon	**(9002-88-4)**
Orizon 805	**(9002-88-4)**
Orlandin	**(69975-77-5)**
Orlevol	**(57-53-4)**
Orlon (Trademark)	**(37243-36-0)**
Orlutate	**(51-98-9)**
Orlycycline	**(60-54-8)**
Ornalin	**(50471-44-8)**
Ornamental Weeder	**(133-90-4)**
Ornamental Weeder 4G	**(133-90-4)**
Ornidazole	**(16773-42-5)**
(S)-Ornithine	**(70-26-8)**
Ornithine	**(70-26-8)**
l-(-)-Ornithine	**(70-26-8)**
Ornithine, L	**(70-26-8)**
Ornitrol	**(1249-84-9)**
Orodin	**(65-86-1)**
Orolevol	**(57-53-4)**
Oronite	**(28629-66-5)**
Oronite 6	**(9003-29-6)**
Oronol	**(12192-57-3)**
Orotic-acid	**(65-86-1)**
Orotonin	**(65-86-1)**
Orotsaure (German)	**(65-86-1)**
Oroturic	**(65-86-1)**
Orotyl	**(65-86-1)**
Oroxin	**(15686-71-2)**
Orphenadine	**(83-98-7)**
Orphenadrin	**(83-98-7)**
Orphenadrine	**(83-98-7)**
Orphenol	**(132-27-4)**
Orpiment	**(1303-33-9)**
Orpizin	**(500-42-5)**
Orquisteron	**(58-22-0)**
Orsile	**(73-48-3)**
Orsin	**(106-50-3)**
Ortazol	**(67-45-8)**
Ortedrine	**(300-62-9)**
Ortenal	**(60-13-9)**
Orthamine	**(95-54-5)**
Orthanilic acid	**(88-21-1)**
Orthedrin	**(60-13-9)**
Orthene	**(30560-19-1)**
Orthene-755	**(30560-19-1)**
Orthesin	**(94-09-7)**
Ortho 4355	**(300-76-5)**
Ortho 5,353	**(2282-34-0)**
Ortho 5353	**(8065-36-9)**
Ortho 5865	**(2425-06-1)**
Ortho-5865	**(2939-80-2)**
Ortho 8890	**(24201-58-9)**
Ortho 9006	**(10265-92-6)**
Ortho 124120	**(30560-19-1)**
Ortho C-1 Defoliant & Weed Killer	**(7775-09-9)**
Ortho N-4 Dust	**(54-11-5)**
Ortho N-5 Dust	**(54-11-5)**
Ortho Earwig Bait	**(16893-85-9)**
Ortho Grass Killer	**(122-42-9)**
Ortho-Klor	**(57-74-9)**
Ortho-LM Apple Spray	**(86-85-1)**
Ortho LM Concentrate	**(86-85-1)**
Ortho LM Seed Protectant	**(86-85-1)**
Ortho L10 Dust	**(3687-31-8)**
Ortho L10 Dust	**(7784-40-9)**
Ortho L40 Dust	**(7784-40-9)**
Ortho MC	**(10326-21-3)**
Ortho Malathion	**(121-75-5)**
Ortho-Mite	**(140-57-8)**
Ortho N-4 and N-5 Dusts	**(54-11-5)**
Ortho P-G Bait	**(12002-03-8)**
Ortho Paraquat Cl	**(1910-42-5)**
Ortho RE-5353	**(2282-34-0)**
Ortho Weevil Bait	**(16893-85-9)**
Orthoacetic acid, triethyl ester (8CI)	**(78-39-7)**
Ortho-p-anisic acid, trimethyl ester (6CI,7CI,8CI)	**(4316-33-0)**
Orthoarsenic acid	**(7778-39-4)**
Orthobencarb	**(34622-58-7)**
Orthobenzyl-p-chlorophenol	**(120-32-1)**
Orthobenzylparachlorophenol	**(120-32-1)**
Orthoborate	**(14213-97-9)**
Orthoboric acid	**(10043-35-3)**
Orthochloroparanisidine	**(5345-54-0)**
Orthocide	**(133-06-2)**
Orthocide 7.5	**(133-06-2)**
Orthocide 50	**(133-06-2)**
Orthocide 406	**(133-06-2)**
Orthocresol	**(95-48-7)**
Orthodiazine	**(289-80-5)**
Orthodibrom	**(300-76-5)**
Orthodibromo	**(300-76-5)**
Orthodichlorobenzene	**(95-50-1)**
Orthodichlorobenzol	**(95-50-1)**
Orthodifluorobenzene	**(367-11-3)**
Orthoformic acid, cyclic 1,3,5-cyclohexanetriyl ester (7CI)	**(281-32-3)**
Orthoformic acid, cyclic ester with 1,3,5-cyclohexanetriol	**(281-32-3)**
Orthoformic acid, cyclic ethylene methyl ester (8CI)	**(19693-75-5)**
Orthoformic acid, ethyl ester	**(122-51-0)**
Orthoformic acid, triethyl ester	**(122-51-0)**
Orthoformic acid, trimethyl ester	**(149-73-5)**
Orthohydroxybenzoic acid	**(69-72-7)**
Orthohydroxydiphenyl	**(90-43-7)**
Ortholeum 162	**(7057-92-3)**
Orthomravencan ethylnaty (Czech)	**(122-51-0)**
Orthomravencan methylnaty (Czech)	**(149-73-5)**
Orthonal	**(72-44-6)**
Orthonitroaniline (DOT)	**(88-74-4)**
Orthophaltan	**(133-07-3)**
Orthophenanthroline	**(66-71-7)**
Orthophenylphenol	**(90-43-7)**
Orthophos	**(56-38-2)**
Ortho phosphate defoliant	**(78-48-8)**
Orthophosphoric acid	**(7664-38-2)**
Orthophosphorus acid	**(13598-36-2)**
Orthopropionic acid ethyl ester	**(115-80-0)**
Orthopropionic acid, triethyl ester (8CI)	**(115-80-0)**
Orthorix	**(1344-81-6)**
Orthosan MB	**(122-19-0)**
Orthoserpina	**(50-55-5)**
Orthosil	**(6834-92-0)**
Orthosilicate	**(15191-85-2)**
Orthotoluic acid	**(118-90-1)**
Orthotran	**(80-33-1)**
Orthovanilline	**(148-53-8)**

Orthoxenol **(90-43-7)**
Ortisporina **(15686-71-2)**
Ortizon **(124-43-6)**
Ortofen **(15307-79-6)**
Ortonal **(72-44-6)**
Ortran **(30560-19-1)**
Ortril **(30560-19-1)**
Orudis **(22071-15-4)**
Oruvail **(22071-15-4)**
Orvagil **(443-48-1)**
Orvinylcarbinol **(107-18-6)**
Orvus Wa Paste **(151-21-3)**
Oryzalin **(19044-88-3)**
Oryzanin **(59-43-8)**
Oryzanine **(59-43-8)**
Orzan S **(8061-51-6)**
Osacyl **(65-49-6)**
Osbon AC **(79-21-0)**
Oscine **(51-34-3)**
Oscophen **(8003-03-0)**
Osdaran **(13356-08-6)**
Osiren **(52-01-7)**
Osmic acid **(20816-12-0)**
Osmitrol **(69-65-8)**
Osmium **(7440-04-2)**
Osmium tetroxide (ACGIH,OSHA) [UN 2471] **(20816-12-0)**
Osmosol extra **(71-23-8)**
Osocide **(133-06-2)**
Ospen **(87-08-1)**
Ospeneff **(132-98-9)**
Ossalin **(7681-49-4)**
Ossiamina **(302-70-5)**
Ossian **(14698-29-4)**
Ossichlorin **(302-70-5)**
Ossido di mesitile (Italian) **(141-79-7)**
Ossimorfone **(76-41-5)**
Ossin **(7681-49-4)**
Ossipurinolo **(2465-59-0)**
Ostacet Brilliant Red E-LB **(17418-58-5)**
Ostacet Orange E-R **(12217-83-3)**
Ostacet Yellow P2G **(2832-40-8)**
Ostanthren Blue BCL **(130-20-1)**
Ostanthren Blue RS **(81-77-6)**
Ostanthren Blue RSN **(81-77-6)**
Ostanthren Blue RSZ **(81-77-6)**
Ostanthren Brilliant Green FFB **(128-58-5)**
Ostanthren Brown BR **(2475-33-4)**
Ostanthren Green FFB **(128-58-5)**
Ostanthren Olive R **(2379-81-9)**
Ostanthren Orange GR **(4424-06-0)**
Ostanthrene Blue RS **(81-77-6)**
Ostanthrene Brown BR **(2475-33-4)**
Ostanthrene Orange GR **(4424-06-0)**
Ostazin Black H-N **(12225-26-2)**
Ostazin Brilliant Red H-B **(12238-00-5)**
Ostazin Brilliant Red S 5b **(17804-49-8)**
Ostelin **(50-14-6)**
Osteobond **(9011-14-7)**
Osteobond Surgical Bone Cement **(9011-14-7)**
Osvan **(8001-54-5)**
Oswego Orange X 2065 **(3520-72-7)**
Osyritin **(153-18-4)**
Osyritrin **(153-18-4)**
Osyrol **(52-01-7)**
Otachron **(56-75-7)**
Otacril **(58-54-8)**
Otan **(96-26-4)**
Oterben **(64-77-7)**
Otetryn **(2058-46-0)**
Othrine **(52918-63-5)**
Otifuril **(139-91-3)**
Otobiotic **(1405-10-3)**
Otofuran **(59-87-0)**
Otophen **(56-75-7)**
Otosone-F **(50-23-7)**
Otracid **(80-33-1)**
Ottacide **(133-53-9)**
Ottafact **(59-50-7)**
Ottani (Italian) **(111-65-9)**
Ottasept **(88-04-0)**
Ottasept extra **(88-04-0)**
1,2,4,5,6,7,8,8-Ottochloro-3a,4,7,7a-tetraidro-4,7-endo-metano-indano (Italian) **(57-74-9)**
Ottometil-pirofosforammide (Italian) **(152-16-9)**
Ouabagenin-l-rhamnosid (German) **(630-60-4)**
Ouabagenin l-rhamnoside **(630-60-4)**
Ouabain **(630-60-4)**
Ouabaine **(630-60-4)**
Oubain **(630-60-4)**
Outflank **(52645-53-1)**
Outflank-Stockade **(52645-53-1)**
Outfox **(22936-86-3)**
Outmine **(1314-13-2)**
Ovaban **(595-33-5)**
Ovadofos **(122-14-5)**
Ovadziak **(58-89-9)**
Ovahormon **(50-28-2)**
Ovahormon benzoate **(50-50-0)**
Ovasterol **(50-28-2)**
Ovasterol-B **(50-50-0)**
Ovastevol **(50-28-2)**
Ovatoxion **(6164-98-3)**
Ovatran **(80-33-1)**
Ovatron **(80-33-1)**
Ovesterin **(50-27-1)**
Ovestin **(50-27-1)**
Ovestinon **(50-27-1)**
Ovestrion **(50-27-1)**
Ovex **(50-50-0)**
Ovex **(53-16-7)**
Ovex **(80-33-1)**
Ovidip **(31218-83-4)**
Ovifollin **(53-16-7)**
Ovitelmin **(31431-39-7)**
Ovochlor **(80-33-1)**
Ovociclina **(50-28-2)**
Ovocyclin **(50-28-2)**
Ovocyclin M **(50-50-0)**
Ovocyclin-MB **(50-50-0)**
Ovocyclin-P **(113-38-2)**
Ovocyclin benzoate **(50-50-0)**
Ovocyclin dipropionate **(113-38-2)**
Ovocycline **(50-28-2)**
Ovocylin **(50-28-2)**
Ovotox **(80-33-1)**
Ovotran **(80-33-1)**
Ovovitellin **(8002-43-5)**
Ovulen 50 **(297-76-7)**
Owadofos **(122-14-5)**
Owadziak **(58-89-9)**
Owispol GF **(9003-53-6)**
Oxaalzuur (Dutch) **(144-62-7)**
D-homo-17a-Oxaandrosta-1,4-diene-3,17-dione (9CI) **(968-93-4)**
2-Oxa-5-α-androstan-3-one, 17-β-hydroxy-17-methyl **(53-39-4)**

2-Oxaandrostan-3-one, 17-hydroxy-17-methyl-, (5-α,17-β)- (9CI) **(53-39-4)**
10H-9-Oxaanthracene **(92-83-1)**
1-Oxa-4-azacyclohexane **(110-91-8)**
1-Oxa-2-azacyclopentadiene **(288-14-2)**
1-Oxa-3-azaindene **(273-53-0)**
3-Oxa-9-azatricyclo(3.3.1.O^{2,4})nonan-7-ol, 9-methyl-, tropate (ester) **(51-34-3)**
2-Oxabicyclo(2.2.2)octane, 1,3,3-trimethyl- **(470-82-6)**
7-Oxabicyclo(2.2.1)hept-5-ene-2,3-dicarboxylic acid, dipotassium salt, (endo,endo)- (9CI) **(59985-42-1)**
7-Oxabicyclo(4.1.0)heptane **(286-20-4)**
7-Oxabicyclo(2.2.1)heptane (8CI,9CI) **(279-49-2)**
7-Oxabicyclo(4.1.0)heptane-3-carboxylic acid, 4-methyl-, (4-methyl-7-oxabicyclo(4.1.0) Hept-3-yl)methyl ester **(141-37-7)**
7-Oxabicyclo(4.1.0)heptane-3-carboxylic acid, 7-oxabicyclo-(4.1.0)hept-3-ylmethyl ester **(2386-87-0)**
7-Oxabicyclo(4.1.0)heptane-3-carboxylic acid, 7-oxabicyclo-(4.1.0)hept-3-ylmethyl ester, Homopolymer (9CI) **(25085-98-7)**
7-Oxabicyclo(2.2.1)heptane-2,3-dicarboxylic acid **(145-73-3)**
7-Oxabicyclo(2.2.1)heptane-2,3-dicarboxylic acid, Compd. with N,N-dimethyl-1-tridecanamine N-oxide (1:1) (9CI) **(35493-90-4)**
7-Oxabicyclo(2.2.1)heptane-2,3-dicarboxylic acid, dipotassium salt (8CI,9CI) **(2164-07-0)**
7-Oxabicyclo(2.2.1)heptane-2,3-dicarboxylic acid, disodium salt **(129-67-9)**
7-Oxabicyclo(2.2.1)heptane-2,3-dicarboxylic anhydride, 2,3-dimethyl **(56-25-7)**
7-Oxabicyclo(4.1.0)heptane, 3-(epoxyethyl) **(106-87-6)**
7-Oxabicyclo(4.1.0)heptane, 4-(1,2-epoxy-1-methylethyl)-1-methyl- **(96-08-2)**
7-Oxabicyclo(2.2.1)heptane, 1-isopropyl-4-methyl- (6CI) **(470-67-7)**
7-Oxabicyclo(2.2.1)heptane, 1-methyl-4-(1-methylethyl)- (9CI) **(470-67-7)**
7-Oxabicyclo(2.2.1)heptane, 1-methyl-4-(1-methylethyl)-2-((2-methylphenyl)methoxy)-, exo-(+-) **(87818-31-3)**
7-Oxabicyclo(4.1.0)heptane, 3-oxiranyl-, Homopolymer **(25086-25-3)**
2-Oxabicyclo(2.2.1)heptane, 1,3,3,7-tetramethyl-, (1R,4S,7S)-(+)-(8CI) **(15404-57-6)**
7-Oxabicyclo(4.1.0)heptane, 3-vinyl **(106-86-5)**
7-Oxabicyclo(4.1.0)heptan-2-one, 4,4,6-trimethyl-(6CI,7CI,8CI,9CI) **(10276-21-8)**
6-Oxabicyclo(3.1.0)hexane **(285-67-6)**
6-Oxabicyclo(3.1.0)hexane, 2,2'-oxybis- (8CI,9CI) **(2386-90-5)**
13-Oxabicyclo(10.1.0)trideca-4,8-diene, 2,6,10-trimethyl- (8CI) **(14840-89-2)**
13-Oxabicyclo(10.1.0)tridecane (9CI) **(286-99-7)**
Oxacyclobutane **(503-30-0)**
Oxacycloheptane **(592-90-5)**
Oxacyclohexadecan-2-one (9CI) **(106-02-5)**
Oxacyclohexane **(142-68-7)**
Oxacyclopentadiene **(110-00-9)**
Oxacyclopentane **(109-99-9)**
Oxacyclopropane **(75-21-8)**
Oxacyclotetradecan-2-one (9CI) **(1725-04-8)**
2H-Oxacyclotetradec(2,3-d)isoindole-2,18(5H)-dione, 16-benzyl-6,7,8,9,10,12a,13,14,15,15a, 16,17-dodecahydro-5,13-dihydroxy-9,15-dimethyl-14-methylene-, (E)-(5S,9R,12aS,13S,15S,15aS,16aS,18aS) **(14930-96-2)**
2-Oxa-7,10-diaza-3-silatridecan-13-oic acid, 3,3-dimethoxy-, methyl ester (9CI) **(1067-66-9)**
4H-1,3,5-Oxadiazin-4-one, tetrahydro-3,5-bis(hydroxymethyl)-(9CI) **(7327-69-7)**
4H-1,3,5-Oxadiazin-4-one, tetrahydro-3,5-bis(methoxymethyl)-(9CI) **(7388-44-5)**
1,3,4-Oxadiazole, 2-(((dimethylamino)methylene)amino)-5-(2-(5-nitro-2-furyl)vinyl)- **(25962-77-0)**
1,3,4-Oxadiazole, 2-((dimethylamino)methylimino)-5-(2-(5-nitro-2-furyl)vinyl)-, (E)- **(55738-54-0)**
1,2,4-Oxadiazolidine-3,5-dione, 2-(3,4-dichlorophenyl)-4-methyl **(20354-26-1)**
δ^2-1,3,4-Oxadiazolin-5-one, 2-tert-butyl-4-(2,4-dichloro-5-isopropyloxyphenyl) **(19666-30-9)**
Oxadiazon **(19666-30-9)**
Oxadieldrin **(61217-08-1)**
Oxadihydroaldrin **(61167-23-5)**
8-Oxa-3,5-dithia-4-stannatetradecanoic acid, 10-ethyl-4,4-dimethyl-7-oxo-, 2-ethylhexyl ester (9CI) **(57583-35-4)**
8-Oxa-3,5-dithia-4-stannatetradecanoic acid, 10-ethyl-4,4-dioctyl-7-oxo-, 2-ethylhexyl ester **(15571-58-1)**
Oxadixyl **(77732-09-3)**
Oxaf **(155-04-4)**
Oxafuradene **(555-84-0)**
3-Oxa-1-heptanol **(111-76-2)**
Oxaidin **(71-33-0)**
Oxaine **(126-27-2)**
Oxal **(107-22-2)**
Oxalacetic acid **(328-42-7)**
Oxalaldehyde **(107-22-2)**
Oxalamide **(471-46-5)**
Oxalate d'argent (French) **(533-51-7)**
Oxalato de plata (Spanish) **(533-51-7)**
Oxaldihydrazide **(996-98-5)**
Oxaldiimidic acid, dithio- **(79-40-3)**
Oxalgon **(9003-11-6)**
Oxalhydrazide **(996-98-5)**
Oxalic-acid **(144-62-7)**
Oxalic acid (ACGIH,OSHA) **(144-62-7)**
Oxalic acid bishydrazide **(996-98-5)**
Oxalic acid, chromium(2+) salt (1:1) **(814-90-4)**
Oxalic acid, copper(2+) salt (1:1) (8CI) **(814-91-5)**
Oxalic acid diamide **(471-46-5)**
Oxalic acid, diammonium salt **(1113-38-8)**
Oxalic acid, diammonium salt, monohydrate (8CI) **(6009-70-7)**
Oxalic acid, dibutyl ester (8CI) **(2050-60-4)**
Oxalic acid, diethyl ester **(95-92-1)**
Oxalic acid, dihydrazide **(996-98-5)**
Oxalic acid, dimethyl ester (8CI) **(553-90-2)**
Oxalic acid dinitrile **(460-19-5)**
Oxalic acid disilver salt **(533-51-7)**
Oxalic acid, disilver(1+) salt **(533-51-7)**
Oxalic acid, disodium salt **(62-76-0)**
Oxalic acid hydrazide **(996-98-5)**
Oxalic acid, iron(2+) salt (1:1) **(516-03-0)**
Oxalic acid monoamide **(471-47-6)**
Oxalic acid, monopotassium salt **(127-95-7)**
Oxalic acid, polyester with 1,4-butanediol (8CI) **(34090-00-1)**
Oxalic acid silver salt (1:2) **(533-51-7)**
Oxalic acid, strontium salt (1:1) (8CI) **(814-95-9)**
Oxalic acid, tin(2+) salt (1:1) (8CI) **(814-94-8)**
Oxalic dihydrazide **(996-98-5)**
Oxalic hydrazide **(996-98-5)**
Oxalic nitrile **(460-19-5)**
Oxalid **(129-20-4)**
Oxalonitrile **(460-19-5)**
Oxaloyl dihydrazide **(996-98-5)**
Oxaloylhydrazide **(996-98-5)**
Oxalsaeure (German) **(144-62-7)**
Oxalyl chloride, ethyl ester **(4755-77-5)**
Oxalyl cyanide **(460-19-5)**
Oxalyl dihydrazide **(996-98-5)**
Oxalyl hydrazide **(996-98-5)**
Oxalylhydrazine **(996-98-5)**
Oxamate, (aminocarbonyl)- **(471-47-6)**
Oxamic acid **(471-47-6)**
Oxamicina (Italian) **(68-41-7)**
Oxamid (Czech) **(471-46-5)**
Oxamide **(471-46-5)**
Oxamide, N,N'-dipicryl- **(29135-62-4)**
Oxamide, dithio **(79-40-3)**

Oxamimidic acid **(471-46-5)**
Oxamimidic acid, N',N'-dimethyl-N-((methylcarbamoyl)oxy)-1-methylthio **(23135-22-0)**
Oxammonium **(7803-49-8)**
Oxammonium hydrochloride **(5470-11-1)**
Oxammonium sulfate **(10039-54-0)**
Oxamycin **(68-41-7)**
Oxamyl **(23135-22-0)**
Oxan 600 **(637-07-0)**
Oxanal Fast Red SW **(130-22-3)**
Oxanal Yellow T **(1934-21-0)**
Oxandrolone **(53-39-4)**
Oxane **(75-21-8)**
Oxane (VAN) **(142-68-7)**
Oxanol O 18 **(9004-98-2)**
Oxanthrene **(262-12-4)**
Oxantin **(96-26-4)**
N-(3-Oxapentamethylene)-N',N''-diethylenethiophosphoramide **(2168-68-5)**
3-Oxapentanedioic acid **(110-99-6)**
3-Oxa-1,5-pentanediol **(111-46-6)**
3-Oxapentane-1,5-diol **(111-46-6)**
Oxaprim **(8064-90-2)**
1-Oxa-2-stanna-3-thiacyclohexan-6-one, 2,2-dibutyl- **(78-06-8)**
2-Oxa-4-thia-7-aza-3-phosphaoctan-8-oic acid, 3,7-dimethyl-6-oxo-, methyl ester, 3-sulfide **(29173-31-7)**
1,4-Oxathiane **(15980-15-1)**
Oxathiane **(15980-15-1)**
1,2-Oxathiane, 2,2-dioxide **(1633-83-6)**
6H-1,3,2-Oxathiastannin-6-one, 2,2-dibutyldihydro **(78-06-8)**
6H-1,3,2-Oxathiastannin-6-one, dihydro-2,2-dioctyl **(3033-29-2)**
1,2,3-Oxathiazin-4(3H)-one, 6-methyl-, 2,2-dioxide **(33665-90-6)**
1,2-Oxathietane, 3,3,4,4-tetrafluoro-, 2,2-dioxide (9CI) **(697-18-7)**
1,4-Oxathiin-3-carboxamide, 5,6-dihydro-2-methyl-N-phenyl **(5234-68-4)**
1,4-Oxathiin-3-carboxamide, 5,6-dihydro-2-methyl-N-phenyl-, 4-oxide (9CI) **(17757-70-9)**
1,4-Oxathiin-3-carboxanilide, 5,6-dihydro-2-methyl- **(5234-68-4)**
1,4-Oxathiin-3-carboxanilide, 5,6-dihydro-2-methyl-, 4,4-dioxide **(5259-88-1)**
1,4-Oxathiin-3-carboxanilide, 5,6-dihydro-2-methyl-, 4-oxide (8CI) **(17757-70-9)**
1,4-Oxathiin, 2,3-dihydro-5-carboxanilido-6-methyl- **(5234-68-4)**
1,4-Oxathiin, 2,3-dihydro-5-carboxanilido-6-methyl-, 4,4-dioxide **(5259-88-1)**
1,2-Oxathiolane 2,2-dioxide **(1120-71-4)**
1,2-Oxathrolane 2,2-dioxide **(1120-71-4)**
Oxatone **(96-26-4)**
5-Oxatricyclo(8.2.0.0^{4,6})dodecane, 4,12,12-trimethyl-9-methylene-, (1R,4R,6R,10S) **(1139-30-6)**
3-Oxatricyclo(3.2.1.0^{2,4})octane, (1-α,2-β,4-β,5-α)- (9CI) **(3146-39-2)**
3-Oxatricyclo(4.1.1.02,4)octane, 2,7,7-trimethyl- (9CI) **(1686-14-2)**
2H-1,3,2-Oxazaphosphorin-2-amine, N,3-bis(2-chloroethyl)-tetrahydro-, 2-oxide (9CI) **(3778-73-2)**
2H-1,3,2-Oxazaphosphorin-2-amine, N,N-bis(2-chloroethyl)-tetrahydro-, 2-oxide (9CI) **(50-18-0)**
2-H-1,3,2-Oxazaphosphorinane **(50-18-0)**
2H-1,3,2-Oxazaphosphorine, 2-(bis(2-chloroethyl)amino)-tetrahydro-, 2-oxide **(50-18-0)**
1,3,2-Oxazaphosphorine, 3-(2-chloroethyl)-2-((2-chloroethyl)-amino)tetrahydro-, 2-oxide **(3778-73-2)**
2H-1,3,2-Oxazaphosphorine, tetrahydro-2-(bis(2-chloroethyl)-amino)-, 2-oxide, monohydrate **(6055-19-2)**
Oxazepam **(604-75-1)**
1,2-Oxazetidine, 3,3,4,4-tetrafluoro-2-(pentafluoroethyl)- (8CI) **(360-46-3)**
Oxazimedrine **(134-49-6)**
2H-1,3-Oxazine-2,4(3H)-dione, 5,5-diethyldihydro **(702-54-5)**
2H-1,4-Oxazine, tetrahydro- **(110-91-8)**
4H-1,4-Oxazine, tetrahydro- **(110-91-8)**
4H-(1,3)Oxazino(3,2-d)(1,4)benzodiazepine-4,7(6H)-dione, 11-chloro-8,12b-dihydro-2,8-dimethyl-12b-phenyl- (8CI,9CI) **(27223-35-4)**
4H-(1,3)Oxazino(3,2-d)(1,4)benzodiazepine-4,7(6h)-dione, 8,12b-dihydro-11-chloro-2,8-dimethyl-12b-phenyl- **(27223-35-4)**
Oxazolam **(24143-17-7)**
Oxazolamum (Latin) **(24143-17-7)**
Oxazolazepam **(24143-17-7)**
1,3-Oxazole **(288-42-6)**
Oxazole (8CI,9CI) **(288-42-6)**
Oxazole, 2,5-dihydro-2,4-dimethyl- (9CI) **(77311-02-5)**
Oxazole, 4,5-dihydro-2-(1-methylethenyl)- (9CI) **(10471-78-0)**
4,4(5H)-Oxazoledimethanol, 2-(heptadecenyl)- **(28984-69-2)**
Oxazole, 4,5-dimethyl- (8CI,9CI) **(20662-83-3)**
Oxazole, 4,5-dimethyl-2-(1-methylethyl)- (9CI) **(19519-45-0)**
Oxazole, 2-ethyl-4,5-dihydro- (9CI) **(10431-98-8)**
Oxazole, 2-isopropyl-4,5-dimethyl- (8CI) **(19519-45-0)**
Oxazole, 2,2'-(1,4-phenylene)bis(5-phenyl- (9CI) **(1806-34-4)**
Oxazole, 2,2'-p-phenylenebis(5-phenyl- (8CI) **(1806-34-4)**
Oxazole, 2,4,5-trimethyl- **(20662-84-4)**
Oxazole, trimethyl- (8CI,9CI) **(20662-84-4)**
Oxazolidin **(129-20-4)**
Oxazolidin-Geigy **(129-20-4)**
Oxazolidine A **(51200-87-4)**
Oxazolidine-5-carboxylic acid, 3-(3,5-dichlorophenyl)-2,4-dioxo-5-methyl-, ethyl ester **(72391-46-9)**
Oxazolidine, 4,4-dimethyl **(51200-87-4)**
2,4-Oxazolidinedione, 3-(3,5-dichlorophenyl)-5,5-dimethyl **(24201-58-9)**
2,4-Oxazolidinedione, 3-(3,5-dichlorophenyl)-5-methyl-5-vinyl **(50471-44-8)**
2,4-Oxazolidinedione, 5-ethyl-3,5-dimethyl **(115-67-3)**
3-Oxazolidineethanol (9CI) **(20073-50-1)**
3-Oxazolidineethanol, 2-(1-methylethyl)- (9CI) **(28770-01-6)**
2-Oxazolidinone, 5-hydroxymethyl-3-(m-tolyl) **(29218-27-7)**
2-Oxazolidinone, 3-(2-hydroxypropyl)-5-methyl-(6CI,7CI,8CI,9CI) **(3375-84-6)**
4-Oxazolidinone, 2-imino-5-phenyl **(2152-34-3)**
2-Oxazolidinone, 5-(morpholinomethyl)-3-((5-nitrofurfurylidene)-amino) **(139-91-3)**
2-Oxazolidinone, 5-(morpholinomethyl)-3-((5-nitrofurfurylidene)-amino), hydrochloride, L- (8CI) **(13146-28-6)**
2-Oxazolidinone, 5-(morpholinomethyl)-3-((5-nitrofurfurylidene)-amino)-, (-)-, (8CI) **(3795-88-8)**
2-Oxazolidinone, 5-(morpholinomethyl)-3-((5-nitrofurfurylidene)-amino)-, L-, monohydrochloride **(3031-51-4)**
2-Oxazolidinone, 5-(4-morpholinylmethyl)-3-(((5-nitro-2-furanyl)-methylene) amino)-, (S)-, (9CI) **(3795-88-8)**
2-Oxazolidinone, 5-(4-morpholinylmethyl)-3-(((5-nitro-2-furanyl)-methylene)amino), hydrochloride, (S)- (9CI) **(13146-28-6)**
2-Oxazolidinone, 5-(4-morpholinylmethyl)-3-(((5-nitro-2-furanyl)-methylene)amino)-, (-)- **(3795-88-8)**
2-Oxazolidinone, 3-(((5-nitro-2-furanyl)methylene)amino)- **(67-45-8)**
2-Oxazolidinone, 3-(5-nitrofurfurylidine-amino) **(67-45-8)**
2-Oxazolidinone, 5-((3,5-xylyloxy)methyl) **(1665-48-1)**
2-Oxazoline, 2-ethyl- **(10431-98-8)**
3-Oxazoline, 2,4,5-trimethyl **(22694-96-8)**
1H,3H,5H-Oxazolo(3,4-c)oxazole-7a(7H)-methanol (9CI) **(6542-37-6)**
1H,3H,5H-Oxazolo(3,4-c)oxazole, 7a-ethyldihydro- (9CI) **(7747-35-5)**
Oxazolo(3,2-d)(1,4)benzodiazepin-6(5H)-one, 2,3,7,11b-tetra hydro-10-bromo-11b-(2-fluoro phenyl)- **(59128-97-1)**
Oxazolo(3,2-d)(1,4)benzodiazepin-6(5h)-one, 10-chloro-11b-(o-chlorophenyl)-2,3,7,11b-tetrahydro- **(24166-13-0)**
1,3,4-Oxazol-2(3H)-one, 3-(2,4-dichloro-5-(1-methylethoxy) phenyl)-5-(1,1-dimethylethyl)- **(19666-30-9)**
4H-1,3,2-Oxazophosphorin-4-one, 2-(bis(2-chloroethyl)amino)-tetrahydro-, 2-oxide **(27046-19-1)**
Oxcord **(21829-25-4)**
(-)-m-Oxedrine **(59-42-7)**
m-Oxedrine **(59-42-7)**
m-Oxedrine **(61-76-7)**
Oxepane (9CI) **(592-90-5)**

2-Oxepanone (8CI,9CI) **(502-44-3)**
2-Oxepanone, Homopolymer (9CI) **(24980-41-4)**
Oxetacaine **(126-27-2)**
Oxetan **(503-30-0)**
Oxetane **(503-30-0)**
Oxetane, 3,3-bis(chloromethyl) **(78-71-7)**
2-Oxetanecarboxylic acid, 3-amino-, (2R-cis)- **(94818-85-6)**
Oxetane, 3,3-dimethyl- (9CI) **(6921-35-3)**
Oxetane, 2-ethyl-3-methyl- (9CI) **(53778-62-4)**
Oxetane, 2-methyl- (9CI) **(2167-39-7)**
Oxetane, 2,3,4-trimethyl- (9CI) **(53778-61-3)**
2-Oxetanone **(57-57-8)**
2-Oxetanone, Homopolymer (9CI) **(25037-58-5)**
2-Oxetanone, 3,3-dimethyl **(1955-45-9)**
2-Oxetanone, 4-methyl **(3068-88-0)**
2-Oxetanone, 4-methyl-, (+-)- (9CI) **(36536-46-6)**
2-Oxetanone, 4-methylene **(674-82-8)**
2-Oxetanone, 4-methyl-, Homopolymer (9CI) **(36486-76-7)**
2-Oxetanone, polyesters (8CI) **(25037-58-5)**
Oxethacaina (Italian) **(126-27-2)**
Oxethacaine **(126-27-2)**
Oxethazaine **(126-27-2)**
Oxethazine **(126-27-2)**
Oxetin **(94818-85-6)**
Oxi-Fenibutol **(129-20-4)**
Oxiamin **(58-63-9)**
Oxibutinina (Spanish) **(5633-20-5)**
Oxibutol **(129-20-4)**
Oxicarboxin **(5259-88-1)**
Oxicob **(1332-40-7)**
Oxidate LE **(93-58-3)**
Oxidation Base 10 **(106-50-3)**
Oxidation Base 10A **(624-18-0)**
Oxidation Base 12A **(39156-41-7)**
Oxidation Base 22 **(5307-14-2)**
Oxidation Base 25 **(119-34-6)**
Oxidation base **(83-56-7)**
Oxide of chromium **(1308-38-9)**
Oxidimethiin **(55290-64-7)**
10,10'-Oxidiphenoxarsine **(58-36-6)**
Oxidized l-cysteine **(56-89-3)**
Oxidoethane **(75-21-8)**
α,β-Oxidoethane **(75-21-8)**
1,8-Oxido-p-menthane **(470-82-6)**
exo-2,3-Oxidonorbornane **(3146-39-2)**
Oxifenylbutazon **(129-20-4)**
Oxifuradene **(555-84-0)**
Oxilapine **(1977-10-2)**
Oxilube 50/150 **(9003-11-6)**
Oxilube 50000 **(9003-11-6)**
Oxime copper **(10380-28-6)**
Oximetholonum **(434-07-1)**
Oximetolona **(434-07-1)**
2-Oximino-3-butanone **(57-71-6)**
Oximinophenylacetonitrile **(825-52-5)**
Oximorfona (Spanish) **(76-41-5)**
Oximorphonum **(76-41-5)**
Oxin **(148-24-3)**
1-Oxindene **(271-89-6)**
Oxine **(148-24-3)**
Oxine copper **(10380-28-6)**
Oxine cuivre **(10380-28-6)**
Oxine sulfate **(134-31-6)**
Oxiniacic acid **(2398-81-4)**
p-Oxinozon **(93-45-8)**
Oxipethidine **(468-56-4)**
Oxipethidinum **(468-56-4)**
Oxiphenbutazone **(129-20-4)**
Oxipurinol **(2465-59-0)**
Oxipurinolum (Latin) **(2465-59-0)**
Oxiraan (Dutch) **(75-21-8)**
Oxiran **(75-21-8)**
Oxirane **(75-21-8)**
Oxirane, Polymer with methyloxirane **(9003-11-6)**
Oxirane, (((1,1'-biphenyl)-2-yloxy)methyl)- (9CI) **(7144-65-2)**
Oxirane, 2,2'-(1,4-butanediyl)bis- (9CI) **(2426-07-5)**
Oxirane, 2,2'-(1,4-butanediylbis(oxymethylene))bis- (9CI) **(2425-79-8)**
Oxirane, 2,2'-(1,4-butanediylbis(oxymethylene))bis-, Homopolymer (9CI) **(29611-97-0)**
Oxirane, ((2-butoxyethoxy)methyl)- (9CI) **(13483-47-1)**
Oxirane, 2-butyl-3-methyl- (9CI) **(14925-96-3)**
Oxirane-carboxaldehyde **(765-34-4)**
Oxiranecarboxylic acid, (2,4-bis(1-methylethyl)phenyl)-, ethyl ester (9CI) **(1334-99-2)**
Oxiranecarboxylic acid, 2-methyl-, decahydro-8-hydroxy-3,6-bis-(methylene)-2-oxospiro(azuleno(4,5-b)furan-9(2H),2'-oxiran)-4-yl ester, (3aR-(3aα,4α(S*),6aα,8β,9α,9aα,9bβ))- (9CI) **(11024-67-2)**
Oxiranecarboxylic acid, 3-methyl-3-phenyl- (9CI) **(5669-15-8)**
Oxiranecarboxylic acid, 3-methyl-3-phenyl-, ethyl ester, trans- (9CI) **(19464-92-7)**
Oxiranecarboxylic acid, 3-phenyl-, ethyl ester (9CI) **(121-39-1)**
Oxirane, chloro- (9CI) **(7763-77-1)**
Oxirane, (chloromethyl)- **(106-89-8)**
Oxirane, 2-(chloromethyl) **(106-89-8)**
Oxirane, (chloromethyl)-, (+-)- (9CI) **(13403-37-7)**
Oxirane, 2,2'-(1,4-cyclohexanediylbis(methyleneoxy-methylene))bis- (9CI) **(14228-73-0)**
Oxirane, 2-decyl-3-(5-methylhexyl)-, cis- **(29804-22-6)**
Oxirane, ((2,4-dibromo-6-methylphenoxy)methyl)- (9CI) **(75150-13-9)**
Oxirane, ((2,6-dibromo-4-methylphenoxy)methyl)- (9CI) **(22421-59-6)**
Oxirane, ((2,4-dibromophenoxy)methyl)- (9CI) **(20217-01-0)**
Oxirane, ((1,2-dibromopropoxy)methyl)- (9CI) **(35243-89-1)**
Oxirane, 2,3-dimethyl- (9CI) **(3266-23-7)**
Oxirane, 2,3-dimethyl-, cis- (9CI) **(1758-33-4)**
Oxirane, 2,3-dimethyl-, trans- (9CI) **(21490-63-1)**
Oxirane, ((1,3-dimethylbutoxy)methyl)- (9CI) **(68134-06-5)**
Oxirane, ((1,3-dimethylbutyloxy)methyl)- **(68134-06-5)**
Oxirane, ((1,1-dimethylethoxy)methyl) **(7665-72-7)**
Oxirane, 3-(1,1-dimethylethyl)-2,2-dimethyl- (9CI) **(96-06-0)**
Oxirane, (((1,1-dimethylethyl)phenoxy)methyl)- (9CI) **(26447-45-0)**
Oxirane, ((4-(1,1-dimethylethyl)phenoxy)methyl)- (9CI) **(3101-60-8)**
Oxirane, ((4-(1,1-dimethylethyl)phenoxy)methyl)-, Homopolymer (9CI) **(29298-03-1)**
Oxirane, 2,2-dimethyl-3-(3-methyl-2,4-pentadienyl)- (9CI) **(69103-20-4)**
Oxirane, 2,2'-((2,2-dimethyl-1,3-propanediyl)bis(oxy-methylene))bis- (9CI) **(17557-23-2)**
Oxirane, 2,2-dimethyl-3-propyl- (9CI) **(17612-35-0)**
Oxirane, 2,3-diphenyl- (9CI) **(17619-97-5)**
Oxirane, dodecyl- **(3234-28-4)**
Oxirane, ((dodecyloxy)methyl)- (9CI) **(2461-18-9)**
Oxirane, 2,2'-(1,2-ethanediylbis(oxymethylene))bis- (9CI) **(2224-15-9)**
Oxirane, 2,2',2'',2'''-(1,2-ethanediylidenetetrakis(4,1-phenylene-oxymethylene))tetrakis- (9CI) **(7328-97-4)**
Oxirane, 2,2',2'',2'''-(1,2-ethanediylidenetetrakis(phenylene-oxymethylene))tetrakis- (9CI) **(27043-37-4)**
Oxirane, (ethoxymethyl)- (9CI) **(4016-11-9)**
Oxirane, (((2-ethylhexyl)oxy)methyl)- (9CI) **(2461-15-6)**
Oxirane, 2-ethyl-3-propyl- (9CI) **(53897-32-8)**
Oxirane, 2-ethyl-3-propyl-, trans- (9CI) **(56052-95-0)**
Oxirane, heptadecyl- (9CI) **(67860-04-2)**
Oxirane, hexadecyl- (9CI) **(7390-81-0)**
Oxirane, ((hexadecyloxy)methyl)- (9CI) **(15965-99-8)**
Oxirane, 2,2',2''-(1,2,6-hexanetriyltris(oxymethylene))tris- (9CI) **(68959-23-9)**
Oxirane, 5-hexenyl- (9CI) **(19600-63-6)**

Oxirane, ((hexyloxy)methyl)- (9CI) **(5926-90-9)**
Oxiranemethanamine, N,N'-(methylenedi-4,1-phenylene)bis-(N-(oxiranylmethyl)- (9CI) **(28768-32-3)**
Oxiranemethanamine, N-(4-(oxiranylmethoxy)phenyl)-N-(oxiranylmethyl)- (9CI) **(5026-74-4)**
Oxiranemethanamine, N-(oxiranylmethyl)-N-phenyl- (9CI) **(2095-06-9)**
Oxiranemethanol, Polymer with nonylphenol (9CI) **(68072-38-8)**
Oxiranemethanol, acetate (9CI) **(6387-89-9)**
Oxirane, (methoxymethyl)- (9CI) **(930-37-0)**
Oxirane, ((4-methoxyphenoxy)methyl)- (9CI) **(2211-94-1)**
Oxirane, methyl- **(75-56-9)**
Oxirane, methyl-, Polymer with oxirane (9CI) **(9003-11-6)**
Oxirane, methyl-, Polymer with oxirane, ether with (1,2-ethanediyldinitrilo)tetrakis(propanol) (4:1) (9CI) **(11111-34-5)**
Oxirane, methyl-, Polymer with oxirane, ether with 1,2,3-propanetriol (3:1) (9CI) **(9082-00-2)**
Oxirane, methyl-, Polymer with oxirane, monobutyl ester (9CI) **(9038-95-3)**
Oxirane, methyl-, Polymer with oxirane, monobutyl ether, Compd. with iodine (9CI) **(68610-00-4)**
Oxirane, (1-methylbutyl)- (9CI) **(53229-39-3)**
Oxirane, 2,2'-(methylenebis(2,1-phenyleneoxymethylene))bis-(9CI) **(54208-63-8)**
Oxirane, 2,2'-(methylenebis(4,1-phenyleneoxymethylene))bis-(9CI) **(2095-03-6)**
Oxirane, 2,2'-(methylenebis(phenyleneoxymethylene))bis- (9CI) **(39817-09-9)**
Oxirane, 2,2'-(methylenebis(2,1-phenyleneoxymethylene))bis-, Homopolymer (9CI) **(58145-38-3)**
Oxirane, ((1-methylethoxy)methyl)- (9CI) **(4016-14-2)**
Oxirane, 2,2'-((1-methylethylidene)bis(4,1-phenyleneoxy-(1-(butoxymethyl)-2,1-ethanediyl)oxymethylene))bis- (9CI) **(71033-08-4)**
Oxirane, 2,2'-((1-methylethylidene)bis(4,1-phenyleneoxy-methylene))bis-, Homopolymer (9CI) **(25085-99-8)**
Oxirane, 2,2'-((1-methylethylidene)bis(4,1-phenyleneoxy-3,1-propanediyloxy-4,1-phenylene(1-methylethylidene)-4,1-phenyleneoxymethylene))bis- (9CI) **(72319-24-5)**
Oxirane, ((6-methylheptyloxy)methyl)- **(68134-07-6)**
Oxirane, (((6-methylheptyl)oxy)methyl)- (9CI) **(68134-07-6)**
Oxirane, 2-methyl-2-(1-methylpropyl)- (9CI) **(42328-43-8)**
Oxirane, 2-methyl-2-(2-methylpropyl)- (9CI) **(53897-31-7)**
Oxirane-methyloxirane copolymer **(9003-11-6)**
Oxirane-methyloxirane polymer **(9003-11-6)**
Oxirane ((methylphenoxy)methyl)- (9CI) **(26447-14-3)**
Oxirane, ((2-methylphenoxy)methyl)- (9CI) **(2210-79-9)**
Oxirane, ((4-methylphenoxy)methyl)- (9CI) **(2186-24-5)**
Oxirane, 2-methyl-2-phenyl **(2085-88-3)**
Oxirane, ((4-(1-methyl-1-phenylethyl)phenoxy)methyl)- (9CI) **(61578-04-9)**
Oxirane, 2-methyl-3-propyl-, cis- (9CI) **(6124-90-9)**
Oxirane, mono((C12-14-alkyloxy)methyl) derivs. **(68609-97-2)**
Oxirane, mono((C6-12-alkyloxy)methyl) derivs. **(68987-80-4)**
Oxirane, mono((C8-10-alkyloxy)methyl) derivs. **(68609-96-1)**
Oxirane, ((4-nitrophenoxy)methyl)- (9CI) **(5255-75-4)**
Oxirane, ((4-nonylphenoxy)methyl)- (9CI) **(6178-32-1)**
Oxirane, ((9-octadecenyloxy)methyl)-, (Z)- (9CI) **(60501-41-9)**
Oxirane, ((octadecyloxy)methyl)- (9CI) **(16245-97-9)**
Oxiraneoctanoic acid, 3-(2-hydroxyoctyl)-, 1,2,3-propanetriyl ester (9CI) **(106-81-0)**
Oxiraneoctanoic acid, 3-octyl-, cis- (9CI) **(24560-98-3)**
Oxiraneoctanoic acid, 3-octyl-, octyl ester (9CI) **(106-84-3)**
Oxirane, octyl- (9CI) **(2404-44-6)**
Oxirane, ((octyloxy)methyl)- (9CI) **(3385-66-8)**
Oxirane, 2,2'-(((2-(oxiranylmethoxy)phenyl)methylene)bis-(4,1-phenyleneoxymethylene))bis- (9CI) **(67786-03-2)**
Oxirane, 2,2'-(oxybis(2,1-ethanediyloxymethylene))bis- (9CI) **(4206-61-5)**
Oxirane, 2,2'-oxybis(methylene)bis-2-methyl- **(7487-28-7)**
Oxirane, 2,2'-(oxybis((methyl-2,1-ethanediyl)oxymethylene))-bis- (9CI) **(41638-13-5)**
Oxirane, pentadecyl- (9CI) **(22092-38-2)**
Oxirane, 2,2'-(2,5,8,11,14-pentaoxapentadecane-1,15-diyl)bis-(9CI) **(17626-93-6)**
Oxirane, (phenoxymethyl)- **(122-60-1)**
Oxirane, phenyl- **(96-09-3)**
Oxirane, 2,2''-(1,4-phenylenebis(oxymethylene))bis- **(2425-01-6)**
Oxirane, 2,2'-(1,3-phenylenebis(oxymethylene))bis- **(101-90-6)**
Oxirane, 2,2'-(1,4-phenylenebis(oxymethylene))bis- (9CI) **(2425-01-6)**
Oxirane, 2,2',2''-(1,2,3-propanetriyltris(oxymethylene))tris-(9CI) **(13236-02-7)**
Oxirane, 2,2',2''-(1,2,3-propanetriyltris(oxymethylene))tris-, Homopolymer (9CI) **(31305-91-6)**
Oxirane, 2,2',2''-(1-propanyl-3-ylidenetris(4,1-phenyleneoxy-methylene))tris- (9CI) **(6130-72-9)**
Oxirane, ((2-propenyloxy)methyl) **(106-92-3)**
Oxirane, 2,2',2''-(propylidynetris(4,1-phenyleneoxy-methylene))tris- (9CI) **(68517-02-2)**
Oxirane, tetrachloro- (9CI) **(16650-10-5)**
Oxirane, ((tetradecyloxy)methyl)- (9CI) **(38954-75-5)**
Oxirane, tetramethyl- (9CI) **(5076-20-0)**
Oxirane, 2,2'-(3,7,7,11-tetramethyl-2,5,9,12-tetraoxatri-decane-1,13-diyl)bis- (9CI) **(87257-05-4)**
Oxirane, 2,2'-(2,5,8,11-tetraoxadodecane-1,12-diyl)bis- (9CI) **(1954-28-5)**
Oxirane, trichloro- (9CI) **(16967-79-6)**
Oxirane, (trichloromethyl)- (9CI) **(3083-23-6)**
Oxirane, tridecyl- (9CI) **(18633-25-5)**
Oxirane, trifluoro(trifluoromethyl)- **(428-59-1)**
Oxirane, trimethyl- (9CI) **(5076-19-7)**
Oxiranylmethyl ester of octadecanoic acid **(7460-84-6)**
Oxiranylmethyl ester of 9-octadecenoic acid **(5431-33-4)**
3-Oxiranyl-7-oxabicyclo(4.1.0)heptene **(106-87-6)**
Oxirene, dihydro- **(75-21-8)**
Oxirene, 2,2,2-trichloroethyl- **(3083-25-8)**
Oxitetracyclin **(79-57-2)**
Oxitol **(110-80-5)**
Oxitosona-50 **(434-07-1)**
Oxiuran **(548-62-9)**
Oxivor **(1332-40-7)**
Oxlopar **(2058-46-0)**
2-Oxo-3-acetyltetrahydrofuran **(517-23-7)**
3'-(3-Oxo-7-α-acetylthio-17-β-hydroxyandrost-4-en-17-β-yl)-propionic acid lactone **(52-01-7)**
3-Oxoadipic acid **(689-31-6)**
β-Oxoadipic acid **(689-31-6)**
Oxo aluminium stearate **(13419-15-3)**
Oxobemin **(13422-51-0)**
7-Oxobenz(de)anthracene **(82-05-3)**
α-Oxobenzeneacetonitrile **(613-90-1)**
γ-Oxobenzenebutanoic acid **(2051-95-8)**
α-Oxobenzenepropanoic acid **(156-06-9)**
2-Oxo-1,2-benzopyran **(91-64-5)**
Oxoboi **(14698-29-4)**
Oxobutanedioic acid **(328-42-7)**
3-Oxobutanoic acid ethyl ester **(141-97-9)**
3-Oxobutanoic acid methyl ester **(105-45-3)**
γ-Oxo-α-butylene **(78-94-4)**
Oxobutyrate **(541-50-4)**
3-Oxobutyric acid **(541-50-4)**
3-Oxo-N-(2-chlorophenylbutanamide) **(93-70-9)**
4,5-Oxochrysene **(86853-91-0)**
N-(4-Oxo-2,5-cyclohexadien-1-ylidene)acetamide **(50700-49-7)**
N-(4-Oxo-2,5-cyclohexadienylidene)acetamide **(50700-49-7)**
((4-Oxo-2,5-cyclohexadien-1-ylidene)amino)guanidine thiosemicarbazone **(539-21-9)**
5-(3 or 6-Oxo-1-cyclohexen-1-yl)-5-ethylbarbituric acid **(25104-37-4)**
4-Oxocyclophosphamide **(27046-19-1)**
7-Oxodehydroabietic acid **(18684-55-4)**
Oxodiacetic acid **(110-99-6)**

2-Oxo-2,5-dihydrofuran (497-23-4)
3-Oxo-2,3-dihydro-1H-indene-1-acetic acid (38194-50-2)
α-Oxodiphenylmethane (119-61-9)
α-Oxoditane (119-61-9)
Oxodolin (77-36-1)
1-Oxo-2-(p-((α-ethyl)carboxymethyl)phenyl)isoindoline (63610-08-2)
γ-Oxo-8-fluoranthenebutanoic acid (519-95-9)
γ-Oxo-8-fluoranthenebutyric acid (519-95-9)
7-Oxo-7H-furo(3,2-g)(1)benzopyran-6-carboxylic acid ethyl ester (20073-24-9)
α-Oxoglutaric acid (328-50-7)
2-Oxoglutaric acid dimethyl ester (13192-04-6)
3-Oxo-L-gulofuranolactone (50-81-7)
3-Oxo-L-gulofuranolactone (enol form) (50-81-7)
α-(1-Oxohexadecyl)-ω-hydroxypoly(oxy-1,2-ethanediyl) (9004-94-8)
2-Oxohexamethyleneimine (105-60-2)
2-Oxohexamethylenimine (105-60-2)
5-Oxohexanoic acid (3128-06-1)
5-Oxohexyl acetate (4305-26-4)
17-((1-Oxohexyl)oxy)pregn-4-ene-3,20-dione (630-56-8)
17α-Oxo-d-homo-1,4-androstadiene-3,17-dione (968-93-4)
Oxolamine (arcum) (13422-51-0)
Oxolane (109-99-9)
2-Oxolanone (96-48-0)
Oxole (110-00-9)
Oxolinic acid (14698-29-4)
Oxomethane (50-00-0)
α-Oxomethionine (583-92-6)
6-Oxo-3-methoxy-N-methyl-4,5-epoxymorphinan (125-29-1)
4-Oxo-5-methylhexanoic acid (41654-04-0)
3-Oxo-N-(2,4-methylphenyl)butanamide (97-36-9)
α-Oxo-γ-methylthiobutyric acid (583-92-6)
2-Oxo-3-methylvaleric acid (1460-34-0)
Oxonium, trimethyl-, (OC-6-11)-hexachloroantimonate(1-) (9CI) (54075-76-2)
5-Oxononane (502-56-7)
α-(1-Oxo-9-octadecenyl)-ω-methoxypoly(oxy-1,2-ethanediyl), (Z) (34397-99-4)
7-Oxooctanoic acid (14112-98-2)
8-Oxopentadecane (818-23-5)
2-Oxo-1,5-pentanedioic acid (328-50-7)
2-Oxopentanedioic acid (328-50-7)
4-Oxopentanoic acid (123-76-2)
4-Oxopentyl acetate (5185-97-7)
5-Oxo-L-proline (98-79-3)
5-Oxo-L-prolyl-L-histidyl-L-tryptophyl-L-seryl-L-tyrosyl-3-(2-naphthyl)-D-alanyl-L-leucyl-L-arginyl-L-prolylglycinamide acetate (Salt) hydrate (86220-42-0)
2-Oxopropanal (78-98-8)
2-Oxopropanoic acid (127-17-3)
2-Oxopropionic acid (127-17-3)
4-(2-Oxopropyl)benzoic acid (15482-54-9)
4-((1-Oxopropyl)phenylamino)-1-(2-phenylethyl)-4-piperidinecarboxylic acid methyl ester (59708-52-0)
6-Oxopurine (68-94-0)
4-Oxo-1,4-pyran-2,6-dicarboxylic acid (99-32-1)
2-Oxopyridine (142-08-5)
2-Oxopyrrolidine (616-45-5)
Oxosuccinic acid (328-42-7)
Oxosumithion (2255-17-6)
4-Oxo-2,2,6,6-tetramethylpiperidine (826-36-8)
2-Oxo-4-thiomethylbutyric acid (583-92-6)
4-Oxo-2-thionothiazolidine (141-84-4)
4-Oxo-2-thiothiazolidin (Czech) (141-84-4)
6-Oxoundecane (927-49-1)
4-Oxovaleric acid (123-76-2)
22-Oxovincaleukoblastine (57-22-7)
Oxprenolol (6452-71-7)
Oxsoralen (298-81-7)
Oxy-5 (94-36-0)
Oxy-10 (94-36-0)
1,4-Oxy Acid (84-87-7)
Oxy Acid Black Base (101-54-2)
Oxy COC (1332-65-6)
Oxy Chek 114 (119-47-1)
Oxy DBCP (96-12-8)
Oxy-NH2 (126-85-2)
Oxy Wash (94-36-0)
p-Oxyacetophenone (99-93-4)
β-Oxyaethyl-morpholin (German) (622-40-2)
Oxyamine (302-70-5)
3-Oxyanthranilic acid (548-93-6)
p-Oxybenzaldehyde (123-08-0)
Oxybenzene (108-95-2)
4,4'-Oxybenzenesulfonylchloride (121-63-1)
p-Oxybenzoesaeureaethylester (German) (120-47-8)
p-Oxybenzoesaure (German) (99-96-7)
p-Oxybenzoesauremethylester (German) (99-76-3)
p-Oxybenzoesaurepropylester (German) (94-13-3)
Oxybenzone (131-57-7)
Oxybenzopyridine (148-24-3)
Oxybisacetic acid (110-99-6)
Oxybis(4-aminobenzene) (101-80-4)
4,4'-Oxybisaniline (101-80-4)
p,p'-Oxybis(aniline) (101-80-4)
4,4'-Oxybisbenzenamine (101-80-4)
p,p'-Oxybisbenzene disulfonylhydrazide (80-51-3)
1,1'-Oxybisbenzene heptabromo deriv. (68928-80-3)
1,1'-Oxybisbenzene hexabromo deriv. (36483-60-0)
1,1'-Oxybisbenzene hexachloro deriv. (31242-93-0)
1,1'-Oxybisbenzene octabromo deriv. (32536-52-0)
1,1'-Oxybisbenzene pentabromo deriv. (32534-81-9)
4,4'-Oxybis(benzenesulfonyl chloride) (121-63-1)
4,4'-Oxybisbenzenesulfonyl chloride (121-63-1)
Oxybis(4-benzenesulfonyl chloride) (121-63-1)
Oxybis(benzenesulfonyl chloride) (121-63-1)
p,p'-Oxybis(benzenesulfonyl chloride) (121-63-1)
Oxybis(benzenesulfonylhydrazide) (80-51-3)
p,p'-Oxybis(benzenesulfonyl hydrazide) (80-51-3)
1,1'-Oxybisbenzene tetrabromo deriv. (40088-47-9)
1,1'-Oxybisbenzene tribromo deriv. (49690-94-0)
1,1'-Oxybis(2-bromoethane) (5414-19-7)
1,1'-Oxybis(butane) (142-96-1)
2,2'-Oxybisbutanedioic acid (7408-18-6)
4,4'-Oxybis(2-chloroaniline) (28434-86-8)
4,4'-Oxybis(2-chloro-benzenamine) (28434-86-8)
1,1'-Oxybis(1-chloroethane) (6986-48-7)
1,1'-Oxybis(2-chloro)ethane (111-44-4)
Oxybis(chloromethane) (542-88-1)
1,1'-Oxybis(3-chloropropane) (629-36-7)
2,2'-Oxybis(1-chloropropane) (108-60-1)
2,2'-Oxybis(2-chloropropane) (39638-32-9)
1,1'-Oxybisdecane (2456-28-2)
1,1'-Oxybisdodecane (4542-57-8)
1,1'-Oxybis(dodecylbenzene) (69834-19-1)
Oxybis(dodecylbenzene) (69834-19-1)
Oxybis(dodecylbenzenesulfonic acid) (30260-73-2)
2,2'-Oxybisethanol (111-46-6)
1,1'-Oxybisethene (109-93-3)
1,1'-Oxybis(methylbenzene) (28299-41-4)
1,1'-Oxybis(3-methylbutane) (544-01-4)
2,2'-(Oxybis(methylenesulfonyl))bisethanol (36724-43-3)
1,1'-(Oxybis(methylenesulfonyl))bisethene (26750-50-5)
2,2'-(Oxybis(methylenethio))bisethanol (36727-72-7)
1,1'-Oxybis(2-methylpropane) (628-55-7)
Oxybis(4-nitrobenzene) (101-63-3)
2,2'-Oxybis-6-oxabicyclo(3.1.0)hexane (2386-90-5)

10,10'-Oxybis-10H-phenoxarsine (58-36-6)
10,10'-Oxybisphenoxarsine (58-36-6)
10-10' Oxybisphenoxyarsine (58-36-6)
Oxybispropanol dibenzoate (27138-31-4)
3,3'-Oxybis(1-propene) (557-40-4)
Oxybis(tributyltin) (56-35-9)
2,2'-Oxybis(4,4,6-trimethyl-1,3,2-dioxaborinane) (14697-50-8)
Oxybis(trimethylsilane) (107-46-0)
Oxybutanal (107-89-1)
β-Oxybutene (3266-23-7)
Oxybutynin (5633-20-5)
Oxybutynine (French) (5633-20-5)
Oxybutyninum (Latin) (5633-20-5)
γ-Oxybutyric acid (692-29-5)
Oxybutyric aldehyde (107-89-1)
Oxycarbon sulfide (463-58-1)
Oxycarbophos (1634-78-2)
Oxycarboxin (5259-88-1)
Oxycarboxine (5259-88-1)
Oxychinolin (148-24-3)
o-Oxychinolin (German) (148-24-3)
Oxychlordan (27304-13-8)
Oxychlordane (27304-13-8)
Oxychlorid fosforecny (Czech) (10025-87-3)
Oxychlorue de cuivre (French) (1332-40-7)
Oxychlorure chromique (French) (14977-61-8)
Oxycil (7775-09-9)
Oxyclor (1332-40-7)
Oxycodeinone (76-42-6)
Oxycodon (76-42-6)
Oxycodone (76-42-6)
Oxycolor (548-62-9)
7-Oxycoumarin (93-35-6)
Oxycur (1332-40-7)
N-Oxyd-Lost (German) (126-85-2)
N-Oxyd-Mustard (German) (126-85-2)
Oxyde d'allyle et de glycidyle (French) (106-92-3)
Oxyde cuivreux (French) (1317-39-1)
Oxyde de baryum (French) (1304-28-5)
Oxyde de calcium (French) (1305-78-8)
Oxyde de carbone (French) (630-08-0)
Oxyde de chlorethyle (French) (111-44-4)
Oxyde de mercure (French) (21908-53-2)
Oxyde de mesityle (French) (141-79-7)
Oxyde de propylene (French) (75-56-9)
Oxyde de tributyletain (56-35-9)
Oxyde d'ethyle (French) (60-29-7)
Oxydemeton-metile (Italian) (301-12-2)
Oxydemetonmethyl (301-12-2)
Oxyde nitrique (French) (10102-43-9)
Oxydess (300-42-5)
2,2'-Oxydiacetic acid (110-99-6)
Oxydiacetic acid (110-99-6)
Oxydiamine Brown 3GN (2586-58-5)
4,4-Oxydianiline (101-80-4)
Oxydianiline (101-80-4)
p,p'-Oxydianiline (101-80-4)
Oxydiazepam (846-50-4)
Oxydiazol (20354-26-1)
4,4'-Oxydibenzenesulfonyl chloride (121-63-1)
1,1'-Oxydi-4-chlorobutane (6334-96-9)
4,4'-(Oxydi-2,1-ethanediyl)bismorpholine (6425-39-4)
Oxydi-2,1-ethanediyl octadecanoate (109-30-8)
2,2'-Oxydiethankarbonitril (Czech) (1656-48-0)
2,2'-Oxydiethanol (111-46-6)
2,2'-Oxydiethanol dicarbamate (5952-26-1)
Oxydiethanolic acid (110-99-6)
N-Oxydiethyl-2-benzthiazolsulfenamid (Czech) (102-77-2)
N-Oxydiethyl-2-benzthiazolsulfenamid (Czech) (95-32-9)
Oxydiethylene acrylate (4074-88-8)
N-(Oxydiethylene)benzothiazole-2-sulfenamide (102-77-2)
Oxydiethylenebis(alkyl*-dimethyl ammonium chloride) *(Derived from coconut oil fatty acids) (68607-28-3)
Oxydiethylene bis(chloroformate) (106-75-2)
Oxydiethylene chloroformate (106-75-2)
Oxydiethylene diacrylate (4074-88-8)
Oxydiethylenedicarbonic acid diallyl ester (142-22-3)
(Oxydiethyleneglycol)bis(coco alkyl)dimethyl ammonium chloride (68607-28-3)
Oxydiethylene methacrylate (2358-84-1)
N-Oxydiethylene thiocarbamyl-N-oxydiethylene sulfenamide (13752-51-7)
Oxydimethylquinazine (60-80-0)
Oxydiphenyl (101-84-8)
4,4'-Oxydiphenylamine (101-80-4)
p-Oxydiphenylamine (122-37-2)
Oxydi-p-phenylenediamine (101-80-4)
1,1'-Oxydi-2-propanol (110-98-5)
Oxydipropanol phosphite (3:1) (36788-39-3)
3,3'-Oxydipropionitrile (1656-48-0)
β,β'-Oxydipropionitrile (1656-48-0)
Oxydisulfoton (2497-07-6)
Oxydol (7722-84-1)
Oxydrene (300-42-5)
Oxydrin (300-42-5)
Oxyethylenated dodecyl alcohol (9002-92-0)
Oxyethylene-oxypropylene polymer (9003-11-6)
Oxyethylidenediphosphonic acid (2809-21-4)
1-(β-Oxyethyl)-2-methyl-5-nitroimidazole (443-48-1)
Oxyfed (300-42-5)
Oxyfluorfen (42874-03-3)
Oxyfluorfene (42874-03-3)
Oxyfume (75-21-8)
Oxyfume 12 (75-21-8)
Oxyfuradene (555-84-0)
Oxygen, Compressed [UN 1072] (7782-44-7)
Oxygen (DOT) (7782-44-7)
Oxygen Ion (11062-77-4)
Oxygen, Refrigerated liquid [UN 1073] (7782-44-7)
Oxygen-carbon dioxide, Mixture (8063-77-2)
Oxygen-difluoride (7783-41-7)
Oxygen difluoride (ACGIH,OSHA) [UN 2190] (7783-41-7)
Oxygen fluoride (7783-41-7)
Oxyhydrochinon (German) (533-73-3)
Oxyhydroquinone (533-73-3)
Oxyject 100 (2058-46-0)
Oxylan (57-41-0)
Oxylite (94-36-0)
Oxymag (1309-48-4)
Oxymetazoline (1491-59-4)
Oxymetebanol (3176-03-2)
Oxymethalone (434-07-1)
Oxymethansulfinsaeuren natrium (German) (149-44-0)
Oxymethazoline (1491-59-4)
Oxymethebanol (3176-03-2)
Oxymethenolone (434-07-1)
Oxymetholone (434-07-1)
Oxy-2 methoxy-3 benzaldehyde (French) (148-53-8)
Oxy-3 methoxy-4 benzaldehyde (French) (621-59-0)
Oxymethurea (140-95-4)
Oxymethylene (50-00-0)
5-Oxymethylfurfurole (67-47-0)
Oxymethylphthalimide (118-29-6)
Oxymetozoline (1491-59-4)
Oxymorphine (76-41-5)

Oxymorphone (76-41-5)
Oxymorphonum (Latin) (76-41-5)
Oxymuriate of Potash (3811-04-9)
Oxymycin (68-41-7)
Oxymykoin (79-57-2)
β-Oxynaphtoic acid (16715-77-8)
Oxyneurine (107-43-7)
Oxyozyl (548-62-9)
Oxyparathion (311-45-5)
Oxyphenbutazone (129-20-4)
Oxyphenic acid (120-80-9)
Oxyphenobutazone (129-20-4)
Oxyphenylbutazone (129-20-4)
p-Oxypropiophenone (70-70-2)
Oxypropyldiphenylolpropane (116-37-0)
Oxypsoralen (298-81-7)
Oxypurinol (2465-59-0)
4-Oxypyrimidine (4562-27-0)
8-Oxyquinoline (148-24-3)
Oxyquinoline (148-24-3)
Oxyquinoline sulfate (134-31-6)
Oxyquinolinoleate de cuivre (French) (10380-28-6)
5-Oxyresorcinol (108-73-6)
Oxyritin (153-18-4)
Oxystin (50-56-6)
Oxyterracin (79-57-2)
Oxyterracine (79-57-2)
Oxyterracyne (79-57-2)
Oxytetracycline (79-57-2)
Oxytetracycline amphoteric (79-57-2)
Oxytetracycline hydrochloride (2058-46-0)
Oxytetracycline, 6-methylene- (914-00-1)
Oxythane (555-89-5)
Oxythioquinox (2439-01-2)
Oxytocin (50-56-6)
Oxytol acetate (111-15-9)
m-Oxytoluene (108-39-4)
o-Oxytoluene (95-48-7)
p-Oxytoluene (106-44-5)
Oxytril (1689-83-4)
Oxytril M (1689-84-5)
Oxyurea (127-07-1)
Oxyurea (88-19-7)
Ozhilon (63428-84-2)
Ozide (1314-13-2)
Ozlo (1314-13-2)
Ozolamid (30979-48-7)
Ozon (Polish) (10028-15-6)
Ozone (ACGIH,OSHA) (10028-15-6)
23P (1344-00-9)
6020P (9002-88-4)
P 1 (3806-34-6)
P 3 (14807-96-6)
P 3 (Mineral) (14807-96-6)
P 3 Vetralat (27176-87-0)
P 10 (2152-34-3)
P 11H (9003-01-4)
P 20 (9003-27-4)
P 21 (128-37-0)
P 30BF (21645-51-2)
P-33 (1333-86-4)
P-40 (13410-01-0)
P-50 (69-53-4)
P68 (1333-86-4)
P 85 (9003-27-4)
P 118 (9003-27-4)
P-165 (115-02-6)
P-170 (9003-20-7)
P 200 (9003-27-4)
P 252 (141-94-6)
P 474 (2595-54-2)
P 607 (94-20-2)
P-974 (2310-17-0)
P 1133 (51-12-7)
P1250 (1333-86-4)
P 1393 (91-33-8)
P 1487 (37304-88-4)
P1496 (26538-44-3)
P 1531 (156-51-4)
P 1570 (26446-38-8)
P 2020T (9002-88-4)
P 2050T (9002-88-4)
P 2070P (9002-88-4)
P-2292 (17433-31-7)
P 2525 (346-18-9)
P-2647 (63-12-7)
P 4007EU (9002-88-4)
P 4007T (9002-88-4)
P-5048 (79-64-1)
P 6500 (9003-07-0)
P1 3419 (55512-33-9)
PA (132-66-1)
PA (90-65-3)
PA (91-40-7)
PA 130 (9002-88-4)
PA 144 (18378-89-7)
PA 190 (9002-88-4)
PA 520 (9002-88-4)
PA 560 (9002-88-4)
PA 6 (105-60-2)
PA 6 (25038-54-4)
PA 94 (68-41-7)
PAA (122-78-1)
PAA-25 (9003-01-4)
PAAB (556-08-1)
PAB (555-06-6)
PABA (150-13-0)
PABAVJT (555-06-6)
PABH-T (63428-84-2)
PACM 20 (1761-71-3)
PAD 522 (9002-88-4)
PADAN (15263-52-2)
PADAN 4 G (15263-52-2)
PAG (9003-11-6)
PAG 1 (9003-11-6)
PAG 2 (9003-11-6)
PAG 1 (Polyglycol) (9003-11-6)
PAHA (61-78-9)
PAH (Amino acid) (61-78-9)
PAH (VAN) (61-78-9)
PAL (63-91-2)
L-PAM (148-82-3)
PA 11M (9003-01-4)
PAMN (66017-91-2)
PAN (2278-22-0)
PAN (90-30-2)
PANA (90-30-2)
PAN (Indicator) (85-85-8)
PAN (Polymer) (25014-41-9)
PAN (VAN) (25014-41-9)
PAN (VAN) (85-85-8)
PAP (123-30-8)
PAP (2597-03-7)
PAP-1 (7429-90-5)

Name	CAS No.
PAPP	**(70-69-9)**
PA 6 (Polymer)	**(105-60-2)**
PA 6 (Polymer)	**(25038-54-4)**
PATAP	**(15263-52-2)**
PAZ 33	**(9002-98-6)**
P 2010B	**(9002-88-4)**
P-4657-B	**(5591-45-7)**
PB	**(51-03-6)**
PB 150	**(9003-27-4)**
PB 2110	**(9003-29-6)**
PBA	**(2312-73-4)**
PBA (Growth stimulant)	**(2312-73-4)**
PBA, dimethylamine salt	**(1338-32-5)**
2,2',3,4,4',5,5'-PBB	**(67733-52-2)**
2,2',4,5,5'-PBB	**(67888-96-4)**
PBB	**(59536-65-1)**
PBB	**(67774-32-7)**
PBC200	**(9003-17-2)**
PBDZ	**(14255-87-9)**
PBNA	**(135-88-6)**
PBS	**(57-30-7)**
PBX(AF) 108	**(121-82-4)**
PBXW 108(E)	**(121-82-4)**
PBZ	**(50-33-9)**
PBZ	**(91-81-6)**
3P Bamate	**(57-53-4)**
1PC	**(64-00-6)**
PC	**(64-10-8)**
PC 621	**(927-60-6)**
PCA	**(1698-60-8)**
PCA	**(52-67-5)**
PCA	**(527-20-8)**
PCA	**(98-79-3)**
2',3,4-PCB	**(38444-86-9)**
2,2',3,3',4,4',5,5',6-PCB	**(40186-72-9)**
2,2',3,3',4,4',5,5'-PCB	**(35694-08-7)**
2,2',3,3',4,4',5,6'-PCB	**(42740-50-1)**
2,2',3,3',4,4',5,6-PCB	**(52663-78-2)**
2,2',3,3',4,4',5-PCB	**(35065-30-6)**
2,2',3,3',4,4',6-PCB	**(52663-71-5)**
2,2',3,3',4,5',6'-PCB	**(52663-70-4)**
2,2',3,3',4,5',6,6'-PCB	**(40186-71-8)**
2,2',3,3',4,5',6-PCB	**(40186-70-7)**
2,2',3,3',4,5'-PCB	**(52663-66-8)**
2,2',3,3',4,5,5',6'-PCB	**(52663-75-9)**
2,2',3,3',4,5,5',6,6'-PCB	**(52663-77-1)**
2,2',3,3',4,5,5',6-PCB	**(68194-17-2)**
2,2',3,3',4,5,5'-PCB	**(52663-74-8)**
2,2',3,3',4,5,6'-PCB	**(38411-25-5)**
2,2',3,3',4,5,6,6'-PCB	**(52663-73-7)**
2,2',3,3',4,5,6-PCB	**(68194-16-1)**
2,2',3,3',4,5-PCB	**(55215-18-4)**
2,2',3,3',4,6'-PCB	**(38380-05-1)**
2,2',3,3',4-PCB	**(52663-62-4)**
2,2',3,3',5,5',6,6'-PCB	**(2136-99-4)**
2,2',3,3',5,5',6-PCB	**(52663-67-9)**
2,2',3,3',5,6'-PCB	**(52744-13-5)**
2,2',3,3',5,6,6'-PCB	**(52663-64-6)**
2,2',3,3',5,6-PCB	**(52704-70-8)**
2,2',3,3'-PCB	**(38444-93-8)**
2,2',3,4',5',6-PCB	**(38380-04-0)**
2,2',3,4',5'-PCB	**(41464-51-1)**
2,2',3,4',5,5',6-PCB	**(52663-68-0)**
2,2',3,4',5,5'-PCB	**(51908-16-8)**
2,2',3,4',5,6-PCB	**(68194-13-8)**
2,2',3,4',5-PCB	**(68194-07-0)**
2,2',3,4',6'-PCB	**(60233-25-2)**
2,2',3,4',6,6'-PCB	**(68194-08-1)**
2,2',3,4',6-PCB	**(68194-05-8)**
2,2',3,4,4',5',6-PCB	**(52663-69-1)**
2,2',3,4,4',5,5',6-PCB	**(52663-76-0)**
2,2',3,4,4',5,5'-PCB	**(35065-29-3)**
2,2',3,4,4',5,6'-PCB	**(60145-23-5)**
2,2',3,4,4',5,6-PCB	**(74472-47-2)**
2,2',3,4,4',5-PCB	**(35694-06-5)**
2,2',3,4,4',6'-PCB	**(59291-64-4)**
2,2',3,4,4',6,6'-PCB	**(74472-48-3)**
2,2',3,4,4'-PCB	**(65510-45-4)**
2,2',3,4,5',6-PCB	**(68194-14-9)**
2,2',3,4,5'-PCB	**(38380-02-8)**
2,2',3,4,5,5',6-PCB	**(52712-05-7)**
2,2',3,4,5,5'-PCB	**(52712-04-6)**
2,2',3,4,5-PCB	**(55312-69-1)**
2,2',3,4,6'-PCB	**(73575-57-2)**
2,2',3,4,6-PCB	**(55215-17-3)**
2,2',3,4-PCB	**(52663-59-9)**
2,2',3,5'-PCB	**(41464-39-5)**
2,2',3,5,5',6-PCB	**(52663-63-5)**
2,2',3,5,5'-PCB	**(52663-61-3)**
2,2',3,5-PCB	**(70362-46-8)**
2,2',3,6'-PCB	**(41464-47-5)**
2,2',3-PCB	**(38444-78-9)**
2,2',4,4',5-PCB	**(38380-01-7)**
2,2',4,4',6-PCB	**(39485-83-1)**
2,2',4,5-PCB	**(70362-47-9)**
2,2',4,6-PCB	**(62796-65-0)**
2,2',4-PCB	**(37680-66-3)**
2,2',5,6'-PCB	**(41464-41-9)**
2,2',5-PCB	**(37680-65-2)**
2,2',6-PCB	**(38444-73-4)**
2,3',4',5'-PCB	**(70362-48-0)**
2,3',4',5-PCB	**(32598-11-1)**
2,3',4',6-PCB	**(41464-46-4)**
2,3',4,4',5'-PCB	**(65510-44-3)**
2,3',4,4',5-PCB	**(31508-00-6)**
2,3',4,5'-PCB	**(73575-52-7)**
2,3',4,5-PCB	**(73575-53-8)**
2,3',4-PCB	**(55712-37-3)**
2,3',5'-PCB	**(37680-68-5)**
2,3',5,5'-PCB	**(41464-42-9)**
2,3',5-PCB	**(38444-81-4)**
2,3',6-PCB	**(38444-76-7)**
2,3,3',4',5,5',6-PCB	**(69782-91-8)**
2,3,3',4',5-PCB	**(70424-68-9)**
2,3,3',4'-PCB	**(41464-43-9)**
2,3,3',4,4',5',6-PCB	**(74472-50-7)**
2,3,3',4,4',5'-PCB	**(69782-90-7)**
2,3,3',4,4',5,6-PCB	**(41411-64-7)**
2,3,3',4,4',6-PCB	**(74472-42-7)**
2,3,3',4,5-PCB	**(70424-69-0)**
2,3,3',5,5',6-PCB	**(74472-46-1)**
2,3,3',5,6-PCB	**(74472-36-9)**
2,3,3',6-PCB	**(74472-33-6)**
2,3,3'-PCB	**(38444-84-7)**
2,3,4',5-PCB	**(74472-34-7)**
2,3,4'-PCB	**(38444-85-8)**
2,3,4,4',5,6-PCB	**(41411-63-6)**
2,3,4,4'-PCB	**(33025-41-1)**
2,3,4,5,6-PCB	**(18259-05-7)**
2,3,4,5-PCB	**(33284-53-6)**
2,3,4-PCB	**(55702-46-0)**
2,3,5,6-PCB	**(33284-54-7)**
2,3,5-PCB	**(55720-44-0)**
2,3,6-PCB	**(55702-45-9)**

2,4',5-PCB	**(16606-02-3)**
2,4',6-PCB	**(38444-77-8)**
2,4,4',5-PCB	**(32690-93-0)**
2,4,5-PCB	**(15862-07-4)**
2,4,6-PCB	**(35656-92-6)**
3,3',4,5-PCB	**(70362-49-1)**
3,3',4-PCB	**(37680-69-6)**
3,3',5-PCB	**(38444-87-0)**
3,4',5-PCB	**(38444-88-1)**
3,4'-PCB	**(2974-90-5)**
3,4,4',5-PCB	**(70362-50-4)**
3,4,4'-PCB	**(38444-90-5)**
PCB	**(11097-69-1)**
PCB	**(1336-36-3)**
PCB	**(671-16-9)**
PCBA	**(16022-69-8)**
PCB Hydrochloride	**(366-70-1)**
PCBS	**(80-38-6)**
PCBs	**(1336-36-3)**
PCBy	**(2201-39-0)**
PCC	**(3867-15-0)**
PCC	**(8001-35-2)**
γ-PCCH	**(319-94-8)**
P.C. 80 Crabgrass Killer	**(590-28-3)**
PCE	**(2201-15-2)**
PCEO	**(16650-10-5)**
PCHO	**(123-63-7)**
PCI	**(80-38-6)**
PCL	**(77-47-4)**
PCL 700	**(24980-41-4)**
PCL-700	**(24980-41-4)**
PCM	**(594-42-3)**
PCMC	**(59-50-7)**
PCMX	**(88-04-0)**
PCNB	**(82-68-8)**
PCP	**(608-93-5)**
PCP	**(77-10-1)**
PCP	**(87-86-5)**
PCPA	**(122-88-3)**
PCP (Anesthetic)	**(77-10-1)**
PCPB	**(80-38-6)**
PCPBS	**(80-38-6)**
PCPCBS	**(80-33-1)**
PCP hydrochloride	**(956-90-1)**
PCPI	**(104-12-1)**
PCPY	**(2201-39-0)**
PCS (9CI)	**(62601-62-1)**
PCT	**(17760-93-9)**
PCT	**(61788-33-8)**
PCTAS	**(1825-19-0)**
PCTP	**(133-49-3)**
P-D	**(88-96-0)**
PD 5	**(7786-34-7)**
P.D.A.B.	**(60-11-7)**
p-PDA HCl	**(624-18-0)**
PDB	**(106-46-7)**
1,2-PDC	**(108-32-7)**
PDC	**(563-54-2)**
PDCB	**(106-46-7)**
PDD 6040I	**(35367-38-5)**
p-PD HCl	**(624-18-0)**
PDMT	**(7227-91-0)**
m-PDN	**(626-17-5)**
o-PDN	**(91-15-6)**
PDP	**(136-40-3)**
PDP	**(511-45-5)**
PDQ	**(94-81-5)**
PDT	**(7227-91-0)**
PDU	**(101-42-8)**
PE	**(115-77-5)**
PE 512	**(9002-88-4)**
PE 617	**(9002-88-4)**
PE-1	**(1322-97-0)**
PE-11	**(6872-06-6)**
PEB1	**(50-29-3)**
PEBC	**(1114-71-2)**
PEDG	**(114-86-3)**
P.E.G. 400	**(25322-68-3)**
P.E.G. 1000	**(25322-68-3)**
P.E.G. 1500	**(25322-68-3)**
P.E.G. 4000	**(25322-68-3)**
P.E.G. 6000	**(25322-68-3)**
PEG 42	**(9004-99-3)**
PEG-1 Cetyl Ether	**(9004-95-9)**
PEG-2 Cetyl Ether	**(9004-95-9)**
PEG-4 Cetyl Ether	**(9004-95-9)**
PEG-5 Cetyl Ether	**(9004-95-9)**
PEG-6 Cetyl Ether	**(9004-95-9)**
PEG-10 Cetyl Ether	**(9004-95-9)**
PEG-12 Cetyl Ether	**(9004-95-9)**
PEG-15 Cetyl Ether	**(9004-95-9)**
PEG-16 Cetyl Ether	**(9004-95-9)**
PEG-20 Cetyl Ether	**(9004-95-9)**
PEG-24 Cetyl Ether	**(9004-95-9)**
PEG-25 Cetyl Ether	**(9004-95-9)**
PEG-30 Cetyl Ether	**(9004-95-9)**
PEG-45 Cetyl Ether	**(9004-95-9)**
PEG-2 Cocamine	**(61791-14-8)**
PEG-3 Cocamine	**(61791-14-8)**
PEG-5 Cocamine	**(61791-14-8)**
PEG-10 Cocamine	**(61791-14-8)**
PEG-15 Cocamine	**(61791-14-8)**
PEG 1540DS	**(9005-08-7)**
PEG 6000DS	**(9005-08-7)**
PEG-4 Decyl Ether Phosphate	**(9004-80-2)**
PEG-5 Dinonyl Phenyl Ether	**(9014-93-1)**
PEG-10 Dinonyl Phenyl Ether	**(9014-93-1)**
PEG-49 Dinonyl Phenyl Ether	**(9014-93-1)**
PEG-100 Dinonyl Phenyl Ether	**(9014-93-1)**
PEG-150 Dinonyl Phenyl Ether	**(9014-93-1)**
PEG-4 Dioleate	**(9005-07-6)**
PEG-6 Dioleate	**(9005-07-6)**
PEG-6-32 Dioleate	**(9005-07-6)**
PEG-8 Dioleate	**(9005-07-6)**
PEG-10 Dioleate	**(9005-07-6)**
PEG-12 Dioleate	**(9005-07-6)**
PEG-20 Dioleate	**(9005-07-6)**
PEG-32 Dioleate	**(9005-07-6)**
PEG-75 Dioleate	**(9005-07-6)**
PEG-150 Dioleate	**(9005-07-6)**
PEG-2 Distearate	**(109-30-8)**
PEG-3 Distearate	**(9005-08-7)**
PEG-4 Distearate	**(9005-08-7)**
PEG-6 Distearate	**(9005-08-7)**
PEG-8 Distearate	**(9005-08-7)**
PEG-9 Distearate	**(9005-08-7)**
PEG-12 Distearate	**(9005-08-7)**
PEG-20 Distearate	**(9005-08-7)**
PEG-32 Distearate	**(9005-08-7)**
PEG-75 Distearate	**(9005-08-7)**
PEG-150 Distearate	**(9005-08-7)**
PEG-175 Distearate	**(9005-08-7)**
PEG n-Dodecyl Ether	**(9002-92-0)**
PEG-12 Glyceryl Ether	**(31694-55-0)**

Name	CAS No.
PEG-26 Glyceryl Ether	**(31694-55-0)**
PEG-2 Isostearyl Ether	**(52292-17-8)**
PEG-3 Isostearyl Ether	**(52292-17-8)**
PEG-10 Isostearyl Ether	**(52292-17-8)**
PEG-12 Isostearyl Ether	**(52292-17-8)**
PEG-20 Isostearyl Ether	**(52292-17-8)**
PEG-22 Isostearyl Ether	**(52292-17-8)**
PEG-50 Isostearyl Ether	**(52292-17-8)**
PEG-5 Lanolin	**(61790-81-6)**
PEG-20 Lanolin	**(61790-81-6)**
PEG-24 Lanolin	**(61790-81-6)**
PEG-30 Lanolin	**(61790-81-6)**
PEG-50 Lanolin	**(61790-81-6)**
PEG-60 Lanolin	**(61790-81-6)**
PEG-85 Lanolin	**(61790-81-6)**
PEG-100 Lanolin	**(61790-81-6)**
PEG-2 Laurate	**(141-20-8)**
PEG-2 Lauryl Ether	**(3055-93-4)**
PEG-4 Lauryl Ether	**(5274-68-0)**
PEG-5 Lauryl Ether	**(3055-95-6)**
PEG-7 Lauryl Ether	**(3055-97-8)**
PEG 200MO	**(9004-96-0)**
PEG 600MO	**(9004-96-0)**
PEG 1000MO	**(9004-96-0)**
PEG 1000MS	**(9004-99-3)**
PEG 600MS	**(9004-99-3)**
PEG-4 Nonyl Phenyl Ether	**(7311-27-5)**
PEG-9 Nonyl Phenyl Ether	**(9016-45-9)**
PEG-9 Octyl Phenyl Ether	**(9002-93-1)**
PEG-2 Oleate	**(106-12-7)**
PEG-6 Oleate	**(9004-96-0)**
PEG-20 Oleate	**(9004-96-0)**
PEG-32 Oleate	**(9004-96-0)**
PEG-2 Oleyl Ether	**(9004-98-2)**
PEG-3 Oleyl Ether	**(9004-98-2)**
PEG-4 Oleyl Ether	**(9004-98-2)**
PEG-5 Oleyl Ether	**(9004-98-2)**
PEG-6 Oleyl Ether	**(9004-98-2)**
PEG-7 Oleyl Ether	**(9004-98-2)**
PEG-8 Oleyl Ether	**(9004-98-2)**
PEG-9 Oleyl Ether	**(9004-98-2)**
PEG-10 Oleyl Ether	**(9004-98-2)**
PEG-12 Oleyl Ether	**(9004-98-2)**
PEG-15 Oleyl Ether	**(9004-98-2)**
PEG-16 Oleyl Ether	**(9004-98-2)**
PEG-20 Oleyl Ether	**(9004-98-2)**
PEG-23 Oleyl Ether	**(9004-98-2)**
PEG-25 Oleyl Ether	**(9004-98-2)**
PEG-44 Oleyl Ether	**(9004-98-2)**
PEG-50 Oleyl Ether	**(9004-98-2)**
PEG/PPG-24/24 Glycerine	**(9082-00-2)**
PEG-6 Palmitate	**(9004-94-8)**
PEG-18 Palmitate	**(9004-94-8)**
PEG-20 Palmitate	**(9004-94-8)**
PEG-5 Stearate	**(9004-99-3)**
PEG-6 Stearate	**(9004-99-3)**
PEG-6-32 Stearate	**(9004-99-3)**
PEG-7 Stearate	**(9004-99-3)**
PEG-8 Stearate	**(9004-99-3)**
PEG-9 Stearate	**(9004-99-3)**
PEG-10 Stearate	**(9004-99-3)**
PEG-12 Stearate	**(9004-99-3)**
PEG-14 Stearate	**(9004-99-3)**
PEG-18 Stearate	**(9004-99-3)**
PEG-20 Stearate	**(9004-99-3)**
PEG-25 Stearate	**(9004-99-3)**
PEG-30 Stearate	**(9004-99-3)**
PEG-32 Stearate	**(9004-99-3)**
PEG-35 Stearate	**(9004-99-3)**
PEG-36 Stearate	**(9004-99-3)**
PEG-40 Stearate	**(9004-99-3)**
PEG-45 Stearate	**(9004-99-3)**
PEG-50 Stearate	**(9004-99-3)**
PEG-75 Stearate	**(9004-99-3)**
PEG-90 Stearate	**(9004-99-3)**
PEG-100 Stearate	**(9004-99-3)**
PEG-120 Stearate	**(9004-99-3)**
PEG-150 Stearate	**(9004-99-3)**
PEG-2 Stearyl Ether	**(9005-00-9)**
PEG-4 Stearyl Ether	**(9005-00-9)**
PEG-7 Stearyl Ether	**(9005-00-9)**
PEG-10 Stearyl Ether	**(9005-00-9)**
PEG-11 Stearyl Ether	**(9005-00-9)**
PEG-13 Stearyl Ether	**(9005-00-9)**
PEG-15 Stearyl Ether	**(9005-00-9)**
PEG-16 Stearyl Ether	**(9005-00-9)**
PEG-20 Stearyl Ether	**(9005-00-9)**
PEG-25 Stearyl Ether	**(9005-00-9)**
PEG-27 Stearyl Ether	**(9005-00-9)**
PEG-30 Stearyl Ether	**(9005-00-9)**
PEG-40 Stearyl Ether	**(9005-00-9)**
PEG-50 Stearyl Ether	**(9005-00-9)**
PEG-100 Stearyl Ether	**(9005-00-9)**
PEG-4 Tallate	**(61791-00-2)**
PEG-8 Tallate	**(61791-00-2)**
PEG-10 Tallate	**(61791-00-2)**
PEG-12 Tallate	**(61791-00-2)**
PEG-16 Tallate	**(61791-00-2)**
PEG-20 Tallate	**(61791-00-2)**
PEG-6 Tridecyl Ether Phosphate	**(9046-01-9)**
PEHA	**(4067-16-7)**
PEI	**(9002-98-6)**
PEI-7	**(9002-98-6)**
PEI-15	**(9002-98-6)**
PEI-30	**(9002-98-6)**
PEI-45	**(9002-98-6)**
PEI 75	**(60-51-5)**
PEI-1000	**(9002-98-6)**
PEI-1500	**(9002-98-6)**
PEI-2500	**(9002-98-6)**
PEMA	**(7206-76-0)**
PEMA (Amide)	**(7206-76-0)**
PEN 100	**(9002-88-4)**
PEN 200	**(132-93-4)**
PEP	**(28014-46-2)**
PEP 211	**(9002-88-4)**
PEPAP	**(64-52-8)**
PER	**(127-18-4)**
PERC	**(127-18-4)**
PES 100	**(9002-88-4)**
PES 200	**(9002-88-4)**
PET	**(78-11-5)**
PETA	**(3524-68-3)**
PETN	**(78-11-5)**
PF 38	**(13674-87-8)**
PF 80	**(9003-11-6)**
PF-3	**(55-91-4)**
PFIB	**(382-21-8)**
PFOA	**(335-67-1)**
PFPA	**(60-17-3)**
PG	**(504-63-2)**
PG 12	**(57-55-6)**
PG 50	**(7782-42-5)**
PGA	**(21645-51-2)**

PGBE 2,4,5-T **(62922-39-8)**
PGDN **(6423-43-4)**
PGE (OSHA) **(122-60-1)**
PGF2-α **(551-11-1)**
l-PGF2-α **(551-11-1)**
PH 40-21 **(1982-55-4)**
PH 60-40 **(35367-38-5)**
PHC **(114-26-1)**
PHP **(2201-39-0)**
PHP **(70-70-2)**
PIB 100 **(9003-27-4)**
PIN **(2104-64-5)**
PITC **(103-72-0)**
PK 4 **(105-60-2)**
PK 4 **(25038-54-4)**
PKA **(105-60-2)**
PKA **(25038-54-4)**
PK-C **(14807-96-6)**
PK-Merz **(768-94-5)**
PK-N **(14807-96-6)**
PKhNB **(82-68-8)**
PLASKON 8201HS **(25038-54-4)**
PLK **(58-89-9)**
PLP **(54-47-7)**
PLTU **(2122-19-2)**
PM 241 **(86-21-5)**
PM245 **(961-71-7)**
PM 255 **(58-73-1)**
PM 334 **(86-34-0)**
PM 396 **(77-41-8)**
PM 671 **(77-67-8)**
PMA **(16561-29-8)**
PMA **(23239-32-9)**
PMA **(62-38-4)**
PMAC **(62-38-4)**
PMAL **(62-38-4)**
PMAS **(62-38-4)**
PMA (Tumor promoter) **(16561-29-8)**
PMA Violet 3 **(1325-82-2)**
PMC **(100-56-1)**
PMDT **(3030-47-5)**
PMF 600 **(63148-62-9)**
PMFP **(96-64-0)**
PMMA **(9011-14-7)**
PMMA-A **(9011-14-7)**
PM_1MM^3 **(10448-09-6)**
PMO 10 **(104-60-9)**
PMP **(732-11-6)**
PMP **(83-28-3)**
PMPA **(616-52-4)**
PMP Sodium gluconate **(527-07-1)**
PMS **(58-34-4)**
PMS 1.5 **(63148-62-9)**
PMS 154A **(63148-62-9)**
PMS 200A **(63148-62-9)**
PMS 1000A **(63148-62-9)**
PMS No. 1 **(9004-99-3)**
PMS No. 2 **(9004-99-3)**
PMS (Siloxane) **(63148-62-9)**
PMT **(54-95-5)**
PN **(1018-71-9)**
PN/135 **(2152-34-3)**
PN6 **(52-46-0)**
PNA **(100-01-6)**
PNAP **(100-19-6)**
PNB **(92-93-3)**
PNCB **(100-00-5)**
PNOA **(97-52-9)**
PNOT **(99-55-8)**
PNS 25 **(63148-62-9)**
PNT **(99-99-0)**
PO-Dimethoate **(1113-02-6)**
POE 20 Sorbitan monolaurate **(9005-64-5)**
POOA **(9004-96-0)**
PO 64P **(74082-93-2)**
POP **(70-70-2)**
POPN **(5796-89-4)**
POPOP **(1806-34-4)**
P=O-Rogor **(1113-02-6)**
PO Systox Sulfone **(2496-91-5)**
PP 005 **(79241-46-6)**
PP 009 **(69806-50-4)**
PP 062 **(23103-98-2)**
PP 1 **(9003-07-0)**
PP 1151 **(9003-07-0)**
PP 2 **(9003-07-0)**
PP 4 **(9003-07-0)**
PP 450 **(87676-93-5)**
PP 557 **(52645-53-1)**
PP 563 **(68085-85-8)**
PP 581 **(56073-10-0)**
PP 612 **(9004-35-7)**
PP 613 **(9004-35-7)**
PP 628 **(9004-35-7)**
PP 910 **(2074-50-2)**
PP-65-25 **(3134-12-1)**
PP148 **(1910-42-5)**
PP149 **(23947-60-6)**
PP175 **(78-57-9)**
PP211 **(23505-41-1)**
PP296 **(75736-33-3)**
PP383 **(52315-07-8)**
PP511 **(29232-93-7)**
PP618 **(69327-76-0)**
PP781 **(5707-69-7)**
PP993 **(79538-32-2)**
PPAL 6 **(27941-08-8)**
PPC 3 **(2631-40-5)**
PPD **(106-50-3)**
PPE 2 **(9002-88-4)**
PP Factor **(59-67-6)**
P.P. Factor-pellagra preventive factor **(59-67-6)**
PP-Faktor **(98-92-0)**
P.P.G. 1000 **(25322-69-4)**
P.P.G. 1025 **(25322-69-4)**
P.P.G. 1200 **(25322-69-4)**
P.P.G. 150 **(25322-69-4)**
P.P.G. 1800 **(25322-69-4)**
P.P.G. 2025 **(25322-69-4)**
P.P.G. 3025 **(25322-69-4)**
P.P.G. 400 **(25322-69-4)**
P.P.G. 4025 **(25322-69-4)**
P.P.G. 425 **(25322-69-4)**
P.P.G. 750 **(25322-69-4)**
PPG-12-Buteth-16 **(9038-95-3)**
PPG-15-Buteth-20 **(9038-95-3)**
PPG-2-Buteth-3 **(9038-95-3)**
PPG-20-Buteth-30 **(9038-95-3)**
PPG-24-Buteth-27 **(9038-95-3)**
PPG-28-Buteth-35 **(9038-95-3)**
PPG-3-Buteth-5 **(9038-95-3)**
PPG-33-Buteth-45 **(9038-95-3)**
PPG-5-Buteth-7 **(9038-95-3)**
PPG-7-Buteth-10 **(9038-95-3)**

Name	CAS No.
PPG-9-Buteth-12	(9038-95-3)
PPG-14 Butyl Ether	(9003-13-8)
PPG-15 Butyl Ether	(9003-13-8)
PPG-16 Butyl Ether	(9003-13-8)
PPG-18 Butyl Ether	(9003-13-8)
PPG-22 Butyl Ether	(9003-13-8)
PPG-24 Butyl Ether	(9003-13-8)
PPG-30 Butyl Ether	(9003-13-8)
PPG-33 Butyl Ether	(9003-13-8)
PPG-4 Butyl Ether	(9003-13-8)
PPG-40 Butyl Ether	(9003-13-8)
PPG-5 Butyl Ether	(9003-13-8)
PPG-53 Butyl Ether	(9003-13-8)
PPG-9 Butyl Ether	(9003-13-8)
PPG Diol 3000EO	(9003-11-6)
PPG-17 Dioleate	(26571-49-3)
PPG-20-Glycereth-30	(9082-00-2)
PPG-24-Glycereth-24	(9082-00-2)
PPG-66-Glycereth-12	(9082-00-2)
PPG-2 Methyl Ether	(34590-94-8)
PPG-26 Oleate	(31394-71-5)
PPG-36 Oleate	(31394-71-5)
PPMP	(13147-09-6)
PPPS	(3761-60-2)
PP 1 (Polymer)	(9003-07-0)
PPSD 30	(9003-07-0)
PPTC	(1929-77-7)
PQD	(105-11-3)
PR 144	(9003-07-0)
PR 703-78	(9011-05-6)
PRB-8	(21342-85-8)
PRD 49	(63428-84-2)
PRD 49-1	(63428-84-2)
PRD 49III	(63428-84-2)
PRL-3191	(30812-87-4)
PRO 21	(9003-11-6)
PRX 1195	(9003-53-6)
3PS	(105-99-7)
PS	(76-06-2)
PS 072	(9003-11-6)
PS 1	(1344-28-1)
PS 1	(9003-53-6)
PS 2	(9003-53-6)
PS 7	(1312-76-1)
PS 197	(63148-62-9)
PS 200	(9003-53-6)
PS 209	(9003-53-6)
PS 802	(693-36-7)
PS 2011	(9003-07-0)
PS 1 (Alumina)	(1344-28-1)
PS-B	(9003-53-6)
PSB	(9003-27-4)
PSB (Aliphatic polymer)	(9003-27-4)
PSB-C	(9003-53-6)
PSB-S-E	(9003-53-6)
PSB-S	(9003-53-6)
PSB-S 40	(9003-53-6)
PSC Co-Op Weevil Bait	(16893-85-9)
PS 3H	(9003-20-7)
PS 454H	(9003-53-6)
PSL	(21609-90-5)
PSML	(9005-64-5)
PSP	(143-74-8)
PSP	(5827-05-4)
PSP 204	(5827-05-4)
PSP (Indicator)	(143-74-8)
PS 2 (Polymer)	(9003-53-6)
PS 5 (Polymer)	(9003-53-6)
PS-SU2	(9003-55-8)
PSV-L	(9003-53-6)
PSV-L 1	(9003-53-6)
PSV-L 2	(9003-53-6)
PSV-L 1S	(9003-53-6)
PT 360	(2152-34-3)
cisPT(II)	(15663-27-1)
PTAB	(23319-66-6)
PTAP	(80-46-6)
PTBP	(98-54-4)
PTC	(103-85-5)
PTH	(9002-64-6)
PTMS	(88-44-8)
PTMSA	(88-44-8)
PTS 2	(9002-88-4)
PTSA	(6528-53-6)
PTU	(103-85-5)
PTU	(51-52-5)
PTU (Thyreostatic)	(51-52-5)
PTZ	(54-95-5)
PUD (Herbicide)	(101-42-8)
PVA	(9002-89-5)
PVA 008	(9002-89-5)
PVAE	(9003-20-7)
PVBR	(25951-54-6)
PVC Cordo	(9003-22-9)
P.V. Carbachol	(51-83-2)
PV Fast Blue A 2R	(147-14-8)
PV Fast Blue B	(147-14-8)
PV Fast Blue B 2G	(147-14-8)
PV-Fast Green G	(1328-53-6)
PV Fast Orange GRL	(4424-06-0)
PV Fast Red E 3B	(1047-16-1)
PV Fast Red E 5B	(1047-16-1)
PVK	(132-98-9)
PV-Orange G	(3520-72-7)
PV Orange HL	(12236-62-3)
PVP	(9003-39-8)
PVP 1	(9003-39-8)
PVP 2	(9003-39-8)
PVP 3	(9003-39-8)
PVP 4	(9003-39-8)
PVP 5	(9003-39-8)
PVP 6	(9003-39-8)
PVP 7	(9003-39-8)
PVP 40	(9003-39-8)
PVP-K 15	(9003-39-8)
PVP-K 30	(9003-39-8)
PVP-K 60	(9003-39-8)
PVP-K 90	(9003-39-8)
PVPP	(9003-39-8)
PVP 8T	(9002-88-4)
PVP-iodine	(25655-41-8)
PVS 4	(9002-89-5)
PX 104	(84-74-2)
PX 114	(89-19-0)
PX 118	(1330-96-7)
PX 208	(1330-86-5)
PX 338	(89-04-3)
PX 404	(109-43-3)
PX 438	(122-62-3)
PX 914	(84-78-6)
PX-138	(117-84-0)
PX-202	(110-29-2)
PX-238	(103-23-1)
PX-800	(8013-07-8)

PX-806	**(61788-72-5)**
PXC 3391	**(9003-07-0)**
PXC 8639	**(9003-07-0)**
PXO	**(58-36-6)**
PY 100	**(9002-88-4)**
PY2763	**(9003-53-6)**
PY 61H	**(1072-71-5)**
PZh2M	**(7439-89-6)**
PZhO	**(7439-89-6)**
Pas-C	**(65-49-6)**
Pano-Drench 4	**(502-39-6)**
Paarlan	**(33820-53-0)**
Pabanol	**(150-13-0)**
Pabestrol	**(130-80-3)**
Pabestrol	**(56-53-1)**
Pabestrol D	**(130-80-3)**
Pabs	**(63-74-1)**
PAC	**(56-38-2)**
Pacemo	**(103-90-2)**
6-β,7-α,9-α,11-α-Pachycarpine	**(90-39-1)**
Pacienx	**(604-75-1)**
Pacifan	**(57-33-0)**
Pacinol	**(69-23-8)**
Pacitran	**(439-14-5)**
Pacitron	**(73-22-3)**
Pacol	**(56-38-2)**
Padaryl	**(98-52-2)**
Padding Brown J	**(10190-66-6)**
Padding Brown N	**(10190-66-6)**
Padimate O	**(21245-02-3)**
Padisal	**(58-34-4)**
Padophene	**(92-84-2)**
Paidazolo	**(526-08-9)**
Paint White	**(1304-85-4)**
Painters Naphtha	**(8032-32-4)**
Paisley 750	**(9003-07-0)**
Paisley Polymer	**(9003-07-0)**
Pakhtaran	**(2164-17-2)**
Pal-P	**(54-47-7)**
Palacet Yellow GN	**(2832-40-8)**
Palacos	**(9011-14-7)**
Palacos R	**(9011-14-7)**
Palacrin	**(69-05-6)**
Palanil Blue GL	**(12222-78-5)**
Palanil Blue R	**(12217-79-7)**
Palanil Blue RT	**(12222-79-6)**
Palanil Red BF	**(17418-58-5)**
Palanil Scarlet BRE	**(58051-96-0)**
Palanil Violet 3B	**(82-33-7)**
Palanil Violet 6R	**(2872-48-2)**
Palanil Yellow G	**(2832-40-8)**
Palanil Yellow 5R	**(6300-37-4)**
Palanil Yellow 5RX	**(6300-37-4)**
Palanthrene Blue BCA	**(130-20-1)**
Palanthrene Blue GPT	**(81-77-6)**
Palanthrene Blue GPZ	**(81-77-6)**
Palanthrene Blue RPT	**(81-77-6)**
Palanthrene Blue RPZ	**(81-77-6)**
Palanthrene Blue RSN	**(81-77-6)**
Palanthrene Brilliant Blue R	**(81-77-6)**
Palanthrene Brilliant Orange GR	**(4424-06-0)**
Palanthrene Brilliant Pink R	**(2379-74-0)**
Palanthrene Brown BR	**(2475-33-4)**
Palanthrene Gold Orange G	**(128-70-1)**
Palanthrene Golden Yellow	**(128-66-5)**
Palanthrene Jade Green	**(128-58-5)**
Palanthrene Jade Green Supra	**(128-58-5)**
Palanthrene Navy Blue G	**(6424-76-6)**
Palanthrene Olive R	**(2379-81-9)**
Palanthrene Orange R	**(3263-31-8)**
Palanthrene Printing Blue KRS	**(81-77-6)**
Palanthrene Red G 2B	**(4203-77-4)**
Palapent	**(57-33-0)**
Palatin Fast Red GREN	**(6408-31-7)**
Palatine Fast Red GREN	**(6408-31-7)**
Palatinol A	**(84-66-2)**
Palatinol AH	**(117-81-7)**
Palatinol C	**(84-74-2)**
Palatinol DN	**(28553-12-0)**
Palatinol IC	**(84-69-5)**
Palatinol M	**(131-11-3)**
Palatinol N	**(28553-12-0)**
Palatinol Z	**(26761-40-0)**
Palatinol bb	**(85-68-7)**
Palatone	**(118-71-8)**
Palavale	**(24169-02-6)**
Pale Gentian	**(72968-42-4)**
Pale Orange Chrome	**(18454-12-1)**
Palestrol	**(56-53-1)**
Palfadonna	**(357-56-2)**
Palinum	**(52-31-3)**
Paliogen Red BG	**(1047-16-1)**
Paliuroside	**(153-18-4)**
Palladium (9CI)	**(7440-05-3)**
Palladium chloride	**(7647-10-1)**
Palladium(2+) chloride	**(7647-10-1)**
Palladous chloride	**(7647-10-1)**
Pallethrine	**(584-79-2)**
Palm Butter	**(8002-75-3)**
Palm Kernel Oil	**(8023-79-8)**
Palm Nut Oil	**(8023-79-8)**
Palm Oil	**(8002-75-3)**
Palm Oil	**(8023-79-8)**
Palm Oil (From seed)	**(8023-79-8)**
Palmitamine oxide	**(7128-91-8)**
Palmitic-acid	**(57-10-3)**
Palmitic acid, aluminum salt (8CI)	**(555-35-1)**
Palmitic acid, butyl ester (8CI)	**(111-06-8)**
Palmitic acid chloride	**(112-67-4)**
Palmitic acid, ethyl ester (8CI)	**(628-97-7)**
Palmitic acid, 2-ethylhexyl ester	**(29806-73-3)**
Palmitic acid, hexadecyl ester	**(540-10-3)**
Palmitic acid, 16-hydroxy-	**(506-13-8)**
Palmitic acid, isobutyl ester	**(110-34-9)**
Palmitic acid, isopropyl ester	**(142-91-6)**
Palmitic acid, methyl ester (8CI)	**(112-39-0)**
Palmitic acid, sodium salt	**(408-35-5)**
Palmitic acid sucrose monoester	**(26446-38-8)**
Palmitic sucrose ester	**(26446-38-8)**
Palmitoleic acid	**(2091-29-4)**
Palmitoleic acid	**(373-49-9)**
Palmitolinoleic acid	**(373-49-9)**
Palmitonitrile (8CI)	**(629-79-8)**
Palmitoyl chloride (8CI)	**(112-67-4)**
12-o-Palmitoyl-16-hydroxyphorbol-13-acetate	**(53202-98-5)**
Palmityl acetate	**(629-70-9)**
Palmityl acrylate	**(13402-02-3)**
Palmityl alcohol	**(36653-82-4)**
Palmitylamine	**(143-27-1)**
Palmityl chloride	**(112-67-4)**
Palmityl dimethyl amine	**(112-69-6)**
Palmityl dimethylamine oxide	**(7128-91-8)**
Palmityldimethylbetaine	**(693-33-4)**
Palmityl palmitate	**(540-10-3)**

Palmityltrimethylammonium chloride	(112-02-7)	Panoram 75	(137-26-8)
Palmotoxin-Bo	(39450-10-7)	Panoram D-31	(60-57-1)
Palmotoxin G0 (9CI)	(39450-11-8)	Panorama	(112-30-1)
Palmotoxin G(O)	(39450-11-8)	Panosine	(58-27-5)
Palmotoxin Go	(39450-11-8)	Panospray 30	(502-39-6)
Palonyl	(57-63-6)	Panoxyl	(94-36-0)
Paltet	(64-75-5)	Pansoil	(2593-15-9)
Palusan	(69-05-6)	Panstreptin	(5490-27-7)
Palygorskit (German)	(12174-11-7)	Pantalgine	(50-13-5)
Palygorskite	(12174-11-7)	Pantas	(50-34-0)
Palygorskite (8CI,9CI) (VAN)	(12174-11-7)	Pantelmin	(31431-39-7)
Pamacel Yellow G-3	(2832-40-8)	Pantenol (Spanish)	(16485-10-2)
Pamacyl	(65-49-6)	Pantenolo	(16485-10-2)
Pamazone	(68-88-2)	Pantestin	(57-85-2)
Pamine	(155-41-9)	Pantheline	(50-34-0)
Pamine bromide	(155-41-9)	D(+)-Panthenol	(81-13-0)
Pamisan	(62-38-4)	D-Panthenol	(81-13-0)
Pamisyl	(65-49-6)	Panthenol	(16485-10-2)
Pamolyn	(112-80-1)	Panthenol	(81-13-0)
Pamosol 2 Forte	(12122-67-7)	Panthenolum (Latin)	(16485-10-2)
Pamprin	(62-44-2)	Panther Creek Bentonite	(1302-78-9)
Pan-Tranquil	(57-53-4)	Pantherin	(2763-96-4)
Panacelan	(551-11-1)	Panthion	(56-38-2)
Panacide	(97-23-4)	Panthoderm	(81-13-0)
Panacur	(43210-67-9)	Panthoject	(137-08-6)
Panadol	(103-90-2)	Panthoject	(867-81-2)
Panadon	(81-13-0)	Pantholic-l	(58-63-9)
Panam	(63-25-2)	Pantholin	(137-08-6)
Panaplate	(62-73-7)	Pantocaine hydrochloride	(136-47-0)
Panaron	(70-18-8)	Pantocid	(80-13-7)
Panazon	(804-36-4)	Pantol	(81-13-0)
Panazone	(804-36-4)	Pantomicina	(114-07-8)
Pancal	(137-08-6)	Pantonsiletten	(8048-52-0)
Pancalma	(57-53-4)	Pantopaque	(99-79-6)
Pancid	(127-69-5)	Pantoprim	(8064-90-2)
Pancil	(26530-20-1)	Pantosediv	(50-35-1)
Pancil-T	(26530-20-1)	Pantothenate calcium	(137-08-6)
Pancreatin	(8049-47-6)	(+)-Pantothenic acid	(79-83-4)
Pancridine	(553-30-0)	D-Pantothenic acid	(79-83-4)
Pancuronium bromide	(15500-66-0)	Pantothenic acid	(79-83-4)
P and G Emulsifier 104	(151-21-3)	Pantothenic acid, D	(79-83-4)
Pandrinox	(502-39-6)	(+)-Pantothenic acid calcium salt	(137-08-6)
Panediol	(57-53-4)	Pantothenic acid, calcium salt	(137-08-6)
Panets	(103-90-2)	Pantothenic acid, calcium salt (2:1), (+)	(137-08-6)
Panex	(103-90-2)	Pantothenic acid, calcium salt, (+)-	(137-08-6)
Panflavin	(8048-52-0)	Pantothenic acid, monosodium salt, D-	(867-81-2)
Panflavin	(86-40-8)	Pantothenic acid, sodium salt	(867-81-2)
Panfuran-S	(794-93-4)	Pantothenol	(81-13-0)
Pangul	(50-35-1)	d-Pantothenol	(81-13-0)
Panithal	(65-45-2)	(+-)-Pantothenyl alcohol	(16485-10-2)
Panivarfin	(129-06-6)	D(+)-Pantothenyl alcohol	(81-13-0)
Pankalma	(57-53-4)	Pantothenyl alcohol	(81-13-0)
Panmycin	(60-54-8)	d-Pantothenyl alcohol	(81-13-0)
Panmycin hydrochloride	(64-75-5)	Pantovernil	(56-75-7)
Panocon	(62850-32-2)	Pantozol 1	(7700-17-6)
Panoctine triacetate	(57520-17-9)	Pantozol 2	(2633-54-7)
Panodrin A-13	(502-39-6)	Panurin	(58-93-5)
Panofen	(103-90-2)	Panwarfin	(129-06-6)
Panogen	(151-38-2)	Pap	(136-40-3)
Panogen	(502-39-6)	Pap H	(61-25-6)
Panogen 15	(502-39-6)	Papanerin-HCl (German)	(61-25-6)
Panogen 43	(502-39-6)	Papanerine	(58-74-2)
Panogen M	(151-38-2)	Papavarine chlorhydrate	(61-25-6)
Panogen PX	(502-39-6)	Papaverina (Italian)	(58-74-2)
Panogen Turf Fungicide	(502-39-6)	Papaverine	(58-74-2)
Panogen Turf Spray	(502-39-6)	Papaverine chlorohydrate	(61-25-6)
Panoral	(53994-73-3)	Papaverine hydrochloride	(61-25-6)

Papaverine monohydrochloride **(61-25-6)**
Paper Black BA **(1937-37-7)**
Paper Black RW **(2429-83-6)**
Paper Black T **(1937-37-7)**
Paper Blue 6B **(2610-05-1)**
Paper Blue R **(548-62-9)**
Paper Deep Black C **(1937-37-7)**
Paper Deep Black R **(2429-83-6)**
Paper Red HRR **(3761-53-3)**
Paper Red 4b **(992-59-6)**
Paper Scarlet 3BX **(6358-29-8)**
Paper Yellow RF **(1325-37-7)**
Papthion **(2597-03-7)**
Papyex **(7782-42-5)**
Para **(106-50-3)**
Para Crystals **(106-46-7)**
Para M **(116-85-8)**
Para-Magenta **(569-61-9)**
Para Orange **(2783-94-0)**
Para Red **(6410-10-2)**
Paraacetaldehyde **(123-63-7)**
Paraaminodiphenyl **(92-67-1)**
Parabar 441 **(128-37-0)**
Paraben **(94-13-3)**
Paraben **(99-76-3)**
Parabis **(97-23-4)**
Parabromdylamine **(86-22-6)**
Parabromodylamine **(86-22-6)**
Parabromotoluene **(106-38-7)**
Paracab II **(9004-36-8)**
Paracain **(51-05-8)**
Paracetaldehyde **(123-63-7)**
Paracetamol **(103-90-2)**
Paracetamole **(103-90-2)**
Paracetamolo (Italian) **(103-90-2)**
Paracetanol **(103-90-2)**
Paracetophenetidin **(62-44-2)**
Parachloramine **(569-65-3)**
Parachlorocidum **(50-29-3)**
Parachlorophenol **(106-48-9)**
Parachlorophenoxyacetic acid **(122-88-3)**
3-(Parachlorophenyl)-1,1-dimethylurea trichloroacetate **(140-41-0)**
Parachlorophenyl-parachlorobenzene-sulfonate **(80-33-1)**
Parachlorostyrene **(1073-67-2)**
Paracide **(106-46-7)**
Paracodin **(125-28-0)**
Paracodine **(125-28-0)**
Paracort **(53-03-2)**
Paracortol **(50-24-8)**
Paracotol **(50-24-8)**
Paracresyl acetate **(140-39-6)**
Paracymene **(99-87-6)**
Paracymol **(99-87-6)**
Paraderil **(83-79-4)**
Paradi **(106-46-7)**
Paradiazine **(290-37-9)**
Paradichlorbenzol (German) **(106-46-7)**
Paradichlorobenzene **(106-46-7)**
Paradichlorobenzol **(106-46-7)**
Paradione **(115-67-3)**
(0)-Paradol **(122-48-5)**
Paradone Blue RC **(130-20-1)**
Paradone Blue RS **(81-77-6)**
Paradone Brilliant Blue R **(81-77-6)**
Paradone Brilliant Orange GR **(4424-06-0)**
Paradone Brilliant Orange GR New **(4424-06-0)**
Paradone Brilliant PNK R **(2379-74-0)**
Paradone Dark Blue RFW **(1324-54-5)**
Paradone Golden Orange G **(128-70-1)**
Paradone Golden Yellow **(128-66-5)**
Paradone Jade Green B **(128-58-5)**
Paradone Jade Green B New, BX New XS New **(128-58-5)**
Paradone Jade Green BX **(128-58-5)**
Paradone Jade Green XS **(128-58-5)**
Paradone Navy Blue G **(6424-76-6)**
Paradone Olive Green B **(95-55-6)**
Paradone Olive R **(2379-81-9)**
Paradone Printing Blue FRS **(81-77-6)**
Paradone Red Brown 2RD **(2475-33-4)**
Paradormalene **(91-80-5)**
Paradow **(106-46-7)**
Paradust **(56-38-2)**
Paraffin **(8002-74-2)**
Paraffin Jelly **(8009-03-8)**
Paraffin oil **(8012-95-1)**
Paraffin oils and hydrocarbon oils, chloro **(85422-92-0)**
Paraffin oils (Petroleum), Catalytic dewaxed heavy (9CI) **(64742-70-7)**
Paraffin oils (Petroleum), Catalytic dewaxed light (9CI) **(64742-71-8)**
Paraffin sulfonate **(68608-15-1)**
Paraffin, Sulfonated, Sodium salt **(68608-15-1)**
Paraffin wax fume (ACGIH,OSHA) **(8002-74-2)**
Paraffin wax sulfonic acids, Sodium salts **(68608-15-1)**
Paraffin waxes and hydrocarbon waxes, Chlorinated (C12, 60% chlorine) **(63449-39-8)**
Paraffin waxes and hydrocarbon waxes, Chlorinated (C23, 43% chlorine) **(63449-39-8)**
Paraffinic hydrocarbons (C14-C30) **(74664-93-0)**
Paraffins, Petroleum, Normal C5-20 **(64771-72-8)**
Paraffins, Petroleum, Normal C>10 **(64771-71-7)**
Paraffins (Petroleum), Normal C> 10, chloro **(97553-43-0)**
Paraform **(30525-89-4)**
Paraform **(50-00-0)**
Paraformaldehyde, Formaldehyde, Phenol polymer **(9003-35-4)**
Paraformaldehyde, Phenol polymer **(9003-35-4)**
Paraformaldehyde [UN 2213] **(30525-89-4)**
Paraformaldehyde, rosin, phenol polymer **(67700-45-2)**
Paraformaldehyde-urea polymer **(9011-05-6)**
Paraformaldehyde-urea resin **(9011-05-6)**
Parafuchsin (German) **(569-61-9)**
Parafuchsine **(569-61-9)**
Parafuchsine, nitric acid salt **(61467-64-9)**
Paraglas **(9011-14-7)**
Parahexyl **(117-51-1)**
Parahydroxybenzaldehyde **(123-08-0)**
Paral **(123-63-7)**
Paralctin **(9002-62-4)**
Paraldehyd (German) **(123-63-7)**
Paraldehyde Black RW **(2429-83-6)**
Paraldehyde [UN 1264] **(123-63-7)**
Paraldehyde and ammonia reaction product **(68391-11-7)**
Paraldeide (Italian) **(123-63-7)**
Paralkan **(8001-54-5)**
Paralytic Shellfish Poison Dihydrochloride **(35554-08-6)**
Paramal **(59-33-6)**
Paramandelic acid **(90-64-2)**
Paramar **(56-38-2)**
Paramar 50 **(56-38-2)**
Parametadione **(115-67-3)**
Parametasona (Spanish) **(53-33-8)**
Parametasone **(53-33-8)**
Paramethadione **(115-67-3)**
Paramethasone **(53-33-8)**
Paramethasonum (Latin) **(53-33-8)**
Paramethoxy-amphetamine **(23239-32-9)**

Paramethyl phenol (106-44-5)
Paramette (62-44-2)
Paramezone (53-33-8)
Paramid (80-35-3)
Paramid Supra (80-35-3)
Paramine Black B (1937-37-7)
Paramine Black BH (2429-73-4)
Paramine Black E (1937-37-7)
Paramine Blue 2B (2602-46-2)
Paramine Blue 3B (72-57-1)
Paramine Fast Brown M (2429-82-5)
Paramine Fast Scarlet 3B (6358-29-8)
Paramine Fast Violet N (2586-60-9)
Paramine Fast Yellow GR (1325-37-7)
Paramine Sky Blue FF (2610-05-1)
Paramine Yellow 2R (1325-37-7)
Paraminol (150-13-0)
Paraminopropiophenone (70-69-9)
Paraminyl (91-84-9)
Paraminyl maleate (59-33-6)
Paramorfan (509-60-4)
Paramorphan (466-99-9)
Paramorphine (115-37-7)
Paramoth (106-46-7)
Paramycin (65-49-6)
Paranaphthalene (120-12-7)
Paranate (150-13-0)
Paranephrin (51-43-4)
Paranitroaniline, Solid (DOT) (100-01-6)
Paranitrofenol (Dutch) (100-02-7)
Paranitrofenolo (Italian) (100-02-7)
Paranitrophenol (French,German) (100-02-7)
Paranitrosodimethylanilide (138-89-6)
Paranol (112-30-1)
Paranol (123-30-8)
Paranol Fast Brown BRL (16071-86-6)
Paranol Fast Red 8BL (2610-11-9)
Paranox 441 (128-37-0)
Paranten (439-14-5)
Paranuggets (106-46-7)
Paraoxon-methyl (950-35-6)
Paraoxon (311-45-5)
Paraoxone (311-45-5)
Paraoxyphenyl aminoethanol (104-14-3)
Parapan (103-90-2)
Parapest M-50 (298-00-0)
Parapet 60N (9011-14-7)
Paraphenolazo aniline (60-09-3)
Paraphenylen-diamine (106-50-3)
Paraphos (56-38-2)
Paraplex (9CI) (52232-24-3)
Paraplex G 53 (27941-08-8)
Paraplex P 543 (9011-14-7)
Paraplex P 681 (9011-14-7)
Paraquat (ACGIH) (1910-42-5)
Paraquat (ACGIH) (2074-50-2)
Paraquat (ACGIH,OSHA) (4685-14-7)
Paraquat Cl (1910-42-5)
Paraquat I (2074-50-2)
Paraquat bis(methyl sulfate) (2074-50-2)
Paraquat chloride (1910-42-5)
Paraquat dication (4685-14-7)
Paraquat, dichloride (1910-42-5)
Paraquat dimethyl sulfate (2074-50-2)
Paraquat dimethyl sulphate (2074-50-2)
Paraquat ion (4685-14-7)
Pararosaniline (479-73-2)
Pararosaniline (569-61-9)
Pararosaniline base (467-62-9)
Pararosaniline base, N,N',N''-triphenyl- (23681-60-9)
Pararosaniline chloride (569-61-9)
Pararosaniline, N,N,N',N',N'',N''-hexamethyl-, chloride (548-62-9)
Pararosaniline hydrochloride (569-61-9)
Parasal (65-49-6)
Parasalicil (65-49-6)
Parasalindon (65-49-6)
Parascorbic acid (10048-32-5)
Parasepiolite (14987-04-3)
Parasept (94-13-3)
Parasept (94-26-8)
Parasept (99-76-3)
Parasol (20427-59-2)
Parasorbic acid (10048-32-5)
(+)-Parasorbinsaeure (German) (10048-32-5)
Paraspan (155-41-9)
Paraspen (103-90-2)
Parastarin (111-60-4)
Parasulfondichloramido benzoic acid (80-13-7)
Paratac (9003-27-4)
Parataf (298-00-0)
Paratartaric acid (133-37-9)
Parathene (56-38-2)
Parathesin (94-09-7)
Parathion (ACGIH,OSHA) [NA 2783] (56-38-2)
Parathion, Liquid (DOT) (56-38-2)
Parathion S (3270-86-8)
Parathion-aethyl (German) (56-38-2)
Parathion and compressed gas mixture [NA 1967] (56-38-2)
Parathion-ethyl (56-38-2)
Parathion methyl (298-00-0)
Parathion methyl Homolog (298-00-0)
Parathion-metile (Italian) (298-00-0)
Parathion mixture, Dry (56-38-2)
Parathion mixture, Liquid (DOT) (56-38-2)
Parathormone (9002-64-6)
Parathyrin (9002-64-6)
Parathyroid Hormone (9002-64-6)
Paratodol (62-44-2)
Paratox (298-00-0)
Parawet (56-38-2)
Paraxanthine (611-59-6)
Paraxenol (92-69-3)
Paraxin (56-75-7)
Parazene (106-46-7)
Parbenate (10287-53-3)
Parbendazole (14255-87-9)
Parbocyl-rev (54-21-7)
Parclay (1332-58-7)
Parda (59-92-7)
Pardopa (59-92-7)
Paregoric (8029-99-0)
Paregoric Elixir (8029-99-0)
Parel (Fiber) (63428-84-2)
Parencillin (6130-64-9)
Parenteral (68-88-2)
Parentracin (1405-87-4)
Parepectolin (8029-99-0)
Parest (72-44-6)
Pareth 15 (68131-40-8)
Pareth-15-3 (68131-40-8)
Pareth-15-5 (68131-40-8)
Pareth-15-7 (68131-40-8)
Pareth-15-9 (68131-40-8)
Pareth-15-20 (68131-40-8)

Pareth-15-30 **(68131-40-8)**
Pareth-15-40 **(68131-40-8)**
Pareth-25-2 **(68131-39-5)**
Pareth-25-3 **(68131-39-5)**
Pareth-25-4 **(68131-39-5)**
Pareth-25-5 **(68131-39-5)**
Pareth-25-7 **(68131-39-5)**
Pareth-25-9 **(68131-39-5)**
Pareth-25-12 **(68131-39-5)**
Parfuran **(67-20-9)**
Pargonyl **(7542-37-2)**
Paridine Red LCL **(5160-02-1)**
Paridol **(99-76-3)**
Parinol **(17781-31-6)**
Paris Blue **(14038-43-8)**
Paris Green **(12002-03-8)**
Paris Green, Solid (DOT) **(12002-03-8)**
Paris Red **(1314-41-6)**
Paris Violet R **(8004-87-3)**
Paris Yellow **(7758-97-6)**
Parkemed **(61-68-7)**
Parkibleu **(72-57-1)**
Parkipan **(72-57-1)**
Parkophyllin **(58-55-9)**
Parkosed **(77-65-6)**
Parkotal **(50-06-6)**
Parks **(511-45-5)**
Parks 12 **(511-45-5)**
Parks 12 Hommel **(511-45-5)**
Parlodion **(9004-70-0)**
Parlon **(9006-03-5)**
Parlon 300CP **(9006-03-5)**
Parlon S 5 **(9006-03-5)**
Parlon S 10 **(9006-03-5)**
Parlon S 20 **(9006-03-5)**
Parlon S 125 **(9006-03-5)**
Parlon S 300 **(9006-03-5)**
Parmal **(91-84-9)**
Parmetol **(59-50-7)**
Parmidin **(1882-26-4)**
Parmidine **(1882-26-4)**
Parmidine R **(1882-26-4)**
Parminal **(72-44-6)**
Parmol **(103-90-2)**
Parmone **(86-87-3)**
Parnithrene Brilliant Green FFB **(128-58-5)**
Parnon **(17781-31-6)**
Parodyne **(60-80-0)**
Paroil Chlorez **(63449-39-8)**
Parol **(59-50-7)**
Parol **(8012-95-1)**
Paroleine **(8012-95-1)**
Paromomycin **(7542-37-2)**
Paromomycin I **(7542-37-2)**
Paromomycine **(7542-37-2)**
Parosept **(94-18-8)**
Paroxan **(311-45-5)**
Paroxon **(70-70-2)**
Paroxypropione **(70-70-2)**
Parrot Green **(12002-03-8)**
Parrycop **(1332-40-7)**
Parsol 1789 **(70356-09-1)**
Partel **(514-73-8)**
Partergin **(113-42-8)**
Parterol **(67-96-9)**
Parton-M **(298-00-0)**
Partocon **(50-56-6)**
Partox **(3691-35-8)**
Partrex **(64-75-5)**
Partron M **(298-00-0)**
Parvolex **(616-91-1)**
Parzate **(12122-67-7)**
Parzate **(142-59-6)**
Parzate C **(12122-67-7)**
Parzate Liquid **(142-59-6)**
Parzate Zineb **(12122-67-7)**
Parzate zineb **(12122-67-7)**
Parzone **(125-28-0)**
Para-Pas **(65-49-6)**
Pas **(65-49-6)**
Pasa **(65-49-6)**
Pasalon **(65-49-6)**
Pasara **(65-49-6)**
Pasco **(1314-13-2)**
Pasco **(7440-66-6)**
Pascorbic **(65-49-6)**
Pasem **(65-49-6)**
Paseptol **(94-13-3)**
Pasexon 100T **(527-07-1)**
Pask **(65-49-6)**
Pasmed **(65-49-6)**
Pasnodia **(65-49-6)**
Pasolac **(65-49-6)**
Passiflora Extract **(8057-62-3)**
Passiflora Incarnata, Extract **(8057-62-3)**
Passiflorae Incarnatae Extractum **(8057-62-3)**
Passiflorin **(486-84-0)**
Passionflower-Extract **(8057-62-3)**
Pasta Caffaro **(1332-65-6)**
Patent Blue **(129-17-9)**
Patent Blue AE **(2650-18-2)**
Patent Blue V **(129-17-9)**
Patent Blue VF **(129-17-9)**
Patent Blue VF-CF **(129-17-9)**
Patent Blue VF Special **(129-17-9)**
Patent Blue VS **(129-17-9)**
Patent Blue 2Y **(2650-18-2)**
Patent Green **(12002-03-8)**
Patentblau V (German) **(129-17-9)**
Pathclear **(1910-42-5)**
Pathocidin **(134-58-7)**
Pathocidine **(134-58-7)**
Patoran **(3060-89-7)**
Patrovine **(64-95-9)**
Pattonex **(3060-89-7)**
Patulin **(149-29-1)**
Patuline **(149-29-1)**
Paucimycin **(7542-37-2)**
Pavabid **(61-25-6)**
Pavisoid **(434-07-1)**
Pavulon **(15500-66-0)**
Paxate **(439-14-5)**
Paxel **(439-14-5)**
Paxilon **(20354-26-1)**
Paxin **(57-53-4)**
Paxipam **(23092-17-3)**
Paxistil **(68-88-2)**
Paxisyn **(146-22-5)**
Paxital **(143-81-7)**
Paxon 3204 **(9003-27-4)**
Paxyl **(113-59-7)**
Pay-Off **(40487-42-1)**
Pay-Off **(70124-77-5)**
Paycite **(149-32-6)**

Payze (21725-46-2)
Payzone (804-36-4)
Pazital (2898-12-6)
Pcon (89-63-4)
Pcona (89-63-4)
Pea (60-12-8)
β-Pea (60-12-8)
Peach-Thin (132-66-1)
Peace Pill (956-90-1)
Peach aldehyde (104-67-6)
Peaches (60-13-9)
Peach lactone (104-67-6)
Peacock Blue X-1756 (2650-18-2)
Peacoline Blue (147-14-8)
Peanut-Oil (8002-03-7)
Pear Oil (123-92-2)
Pear Oil (628-63-7)
Pearl Ash (3811-04-9)
Pearl Ash (584-08-7)
Pearlpuss (9000-07-1)
Pearl stearic (57-11-4)
Pearly Gates (50-37-3)
Pearsall (7446-70-0)
Pebulate (1114-71-2)
Pecan Shell Powder (8002-03-7)
Pecilocerin A (1986-70-5)
Pecta-diazine, Suspension (68-35-9)
Pectalgine (9005-38-3)
Pectic acid (9CI) (9046-40-6)
Pectin (9000-69-5)
Pectinate (9000-69-5)
Pectinic acid (9000-69-5)
Pectin, Mixt. contg. (9CI) (8047-42-5)
Pectins (9000-69-5)
Pectin, sodium salt (9005-59-8)
Pectolin (509-67-1)
Pediaflor (7681-49-4)
Pediamycin (1264-62-6)
Pedident (7681-49-4)
Pedinex (French) (131-89-5)
Pedipen (132-98-9)
Pedituss (621-42-1)
Pedraczak (58-89-9)
Pedric (103-90-2)
Peeracid Orange II (633-96-5)
Peeramine Black E (1937-37-7)
Peeramine Black GXOO (1937-37-7)
Peeramine Congo Red (573-58-0)
Peeramine Fast Brown BRL (16071-86-6)
Peeramine Stilbene Yellow GA (1325-37-7)
Peerless (1332-58-7)
Peerless (1333-86-4)
Pegnol L 12 (9002-92-0)
Pegnol L 20 (9002-92-0)
Pegosperse 100L (141-20-8)
Pegosperse 100 LN (141-20-8)
Pegosperse 400MO (9004-96-0)
Pegosperse S 9 (9004-99-3)
Pegoterate (25038-59-9)
Pehanorm (77-86-1)
Pehnaminosulf (140-56-7)
Peladow (10043-52-4)
Pelagol BA (39156-41-7)
Pelagol CD (624-18-0)
Pelagol D (106-50-3)
Pelagol DA (615-05-4)
Pelagol DR (106-50-3)
Pelagol EG (591-27-5)
Pelagol 3GA (95-55-6)
Pelagol Grey (39156-41-7)
Pelagol Grey C (120-80-9)
Pelagol Grey CD (624-18-0)
Pelagol Grey D (106-50-3)
Pelagol Grey GG (95-55-6)
Pelagol Grey J (95-80-7)
Pelagol Grey L (615-05-4)
Pelagol Grey P Base (123-30-8)
Pelagol Grey RS (108-46-3)
Pelagol Grey SLA (39156-41-7)
Pelagol J (95-80-7)
Pelagol L (615-05-4)
Pelagol P Base (123-30-8)
Pelagol RS (108-46-3)
Pelagol SLA (39156-41-7)
Pelargic acid (112-05-0)
Pelargidenolon (520-18-3)
Pelargidenolon 1497 (520-18-3)
Pelargol (106-21-8)
Pelargon (Russian) (112-05-0)
Pelargon (VAN) (63428-84-2)
Pelargonic acid (112-05-0)
Pelargonic acid, diethanolamide (3077-37-0)
Pelargonic acid, diethanolamine amide (3077-37-0)
Pelargonic acid, diethanolamine soap (3077-37-0)
Pelargonic alcohol (143-08-8)
Pelargonic aldehyde (124-19-6)
Pelargonic diethanolamide (3077-37-0)
Pelargonitrile (2243-27-8)
Pelargonium Oil (8000-46-2)
Pelargononitrile (2243-27-8)
Pelargonoyl peroxide (762-13-0)
Pelargonyl acetate (143-13-5)
Pelargonyl peroxide (762-13-0)
Pelargonyl peroxide (Technically pure) (762-13-0)
Pelargonyl peroxide, Technically pure (762-13-0)
Pelaspan 333 (9003-53-6)
Pelaspan ESP 109s (9003-53-6)
Pelazid (54-85-3)
Pelidorm (77-65-6)
Pelikan C 11/1431a (7440-44-0)
Pellagramin (59-67-6)
Pellagra preventive factor (59-67-6)
Pellagrin (59-67-6)
Pellcafs (51-63-8)
Pellcap (51-63-8)
Pellcaps (51-63-8)
Pelletex (1333-86-4)
Pellidol (83-63-6)
Pellidole (83-63-6)
Pellon 2505 (9003-07-0)
Pellon 2506 (9003-07-0)
Pellon FT 2140 (9003-07-0)
Pellon P 6 (63428-84-2)
Pellugel (9000-07-1)
Pellugel ID (9000-07-1)
Pellugel ID (9062-07-1)
Pelmin (98-92-0)
Pelmine (98-92-0)
Pelonin (59-67-6)
Pelonin amide (98-92-0)
Pelor (7783-28-0)
Pelson (146-22-5)
Pelt (23564-06-9)
Pelt 14 (23564-05-8)

Pelt-44 **(23564-05-8)**
Pelt Sol **(23564-06-9)**
Peltol BR **(101-54-2)**
Peltol BR II **(101-54-2)**
Peltol D **(106-50-3)**
Peluces **(52-86-8)**
Pemal **(77-67-8)**
Pemalin **(77-67-8)**
Pemetesan **(54-95-5)**
Pemolina (Italian) **(2152-34-3)**
Pemoline **(2152-34-3)**
Pen A **(7177-48-2)**
PenCB **(32598-14-4)**
Pen-Di-Ben **(1538-09-6)**
Pen-Fifty **(6130-64-9)**
Pen-Oral **(87-08-1)**
Pen V **(87-08-1)**
Pen-V-K Powder **(132-98-9)**
Pen-Vee **(87-08-1)**
Pen-Vee-K **(132-98-9)**
Pen-Vee-K Powder **(132-98-9)**
Pen-a-brasive **(69-57-8)**
Penadur **(1538-09-6)**
Penadur L-A **(1538-09-6)**
Penagen **(132-98-9)**
Penalev **(113-98-4)**
D-Penamine **(52-67-5)**
Penaquacaine G **(6130-64-9)**
Penatin **(149-29-1)**
Penbar **(57-33-0)**
Penbristol **(69-53-4)**
Penbritin **(69-53-4)**
Penbritin paediatric **(69-53-4)**
Penbritin syrup **(69-53-4)**
Penbrock **(69-53-4)**
Pencal **(7778-44-1)**
Pencard **(78-11-5)**
Penchlorol **(87-86-5)**
Pencil Green SF **(5141-20-8)**
Pencillic acid **(90-65-3)**
Pencillin G **(61-33-6)**
Pencogel **(9000-07-1)**
Pencompren **(132-98-9)**
Pendepon **(1538-09-6)**
Penderol **(115-76-4)**
Pendimethalin **(40487-42-1)**
Pendimethaline **(40487-42-1)**
Penditan **(1538-09-6)**
Penduran **(1538-09-6)**
Peneteck **(8012-95-1)**
Penetiazol **(54-95-5)**
Penetrasol **(54-95-5)**
Penetratsol **(54-95-5)**
Penford Gum 380 **(9005-25-8)**
Penicidin **(149-29-1)**
(S)-Penicillamin **(52-67-5)**
Penicillamin **(52-67-5)**
D-Penicillamine **(52-67-5)**
Penicillamine **(52-67-5)**
Penicillanic acid, 6-phenoxyacetamido **(87-08-1)**
Penicillic acid **(90-65-3)**
Penicillin (9CI) **(1406-05-9)**
Penicillin G **(61-33-6)**
Penicillin G, Benzathine **(1538-09-6)**
Penicillin G, Compd. with N,N'-dibenzylethylenediamine (2:1) **(1538-09-6)**
Penicillin G, Compd. with 2-(diethylamino)ethyl p-aminobenzoate **(6130-64-9)**
Penicillin G Potassium **(113-98-4)**
Penicillin G Potassium salt **(113-98-4)**
Penicillin G Procaine **(6130-64-9)**
Penicillin-G, monosodium salt **(69-57-8)**
Penicillin G procaine **(54-35-3)**
Penicillin G salt of N,N'-dibenzylethylenediamine **(1538-09-6)**
Penicillin G, sodium **(69-57-8)**
Penicillin G, sodium salt **(69-57-8)**
Penicillin-Strep **(5490-27-7)**
Penicillin V **(87-08-1)**
Penicillin V potassium **(132-98-9)**
Penicillin V potassium salt **(132-98-9)**
Penicillin, (aminophenylmethyl)- **(69-53-4)**
Penicillinic acid, benzyl- **(61-33-6)**
Penicillin phenoxymethyl **(87-08-1)**
Penicillin potassium phenoxymethyl **(132-98-9)**
Penicline **(69-53-4)**
Penidural **(1538-09-6)**
Penidure **(1538-09-6)**
Penilaryn **(69-57-8)**
Penilente **(1538-09-6)**
Penisem **(113-98-4)**
Penite **(7784-46-5)**
Penitracin **(1405-87-4)**
Penizillin (German) **(1406-05-9)**
Penlator **(6130-64-9)**
Penn Salt TD-183 **(6012-97-1)**
Pennac **(136-30-1)**
Pennac CBS **(95-33-0)**
Pennac CRA **(96-45-7)**
Pennac MBT Powder **(149-30-4)**
Pennac MS **(97-74-5)**
Pennac TBBS **(95-31-8)**
Pennac ZT **(155-04-4)**
Pennamine **(94-75-7)**
Pennamine D **(94-75-7)**
Penncap E **(56-38-2)**
Penncap-M **(298-00-0)**
Pennfloat M **(112-55-0)**
Pennfloat S **(112-55-0)**
Pennsalt TD-72 **(2595-54-2)**
Pennsoline Soft Yellow **(8009-03-8)**
Pennwalt C-4852 **(122-14-5)**
Pennwhite **(7681-49-4)**
Pennyroyal **(8007-44-1)**
Pennyroyal Oil **(8007-44-1)**
Pennyroyal Oil, European **(8007-44-1)**
Pennzone B **(109-46-6)**
Pennzone E **(105-55-5)**
Penoxalin **(40487-42-1)**
Penoxaline **(40487-42-1)**
Penoxyn **(40487-42-1)**
Penphene **(6012-97-1)**
Penphene **(8001-35-2)**
Penreco **(8012-95-1)**
Penreco White **(8009-03-8)**
Pensig **(132-93-4)**
Pensive **(57-53-4)**
Pensyn **(7177-48-2)**
Pent-Acetate **(628-63-7)**
Pent-Acetate 28 **(628-63-7)**
Penta **(87-86-5)**
Penta-Kil **(87-86-5)**
Penta Ready **(87-86-5)**
Penta WR **(87-86-5)**
Pentaaldol **(597-31-9)**
3,9,14,20,25-Pentaazatriacontane-2,10,13,21,24-pentone, 30-amino-3,14,25-trihydroxy- **(70-51-9)**

1,4,7,10,13-Pentaazatridecane **(112-57-2)**
Pentabarbital sodium **(57-33-0)**
Pentabarbitone **(76-74-4)**
Pentaborane(9) **(19624-22-7)**
Pentaborane (ACGIH,OSHA) [UN 1380] **(19624-22-7)**
Pentabromfenol (Czech) **(608-71-9)**
2,2',4,5,5'-Pentabromobiphenyl **(67888-96-4)**
1,2,3,4,5-Pentabromo-6-chlorocyclohexane **(87-84-3)**
Pentabromodibenzo(b,e)(1,4)dioxin **(103456-36-6)**
Pentabromodibenzo-p-dioxin **(103456-36-6)**
Pentabromodibenzofuran **(68795-14-2)**
Pentabromodiphenyl ether **(32534-81-9)**
Pentabromodiphenyl oxide **(32534-81-9)**
2,3,4,5,6-Pentabromoethylbenzene **(85-22-3)**
Pentabromoethylbenzene **(85-22-3)**
Pentabromomethylbenzene **(87-83-2)**
Pentabromophenol **(608-71-9)**
Pentabromophenoxybenzene **(32534-81-9)**
Pentabromophenyl ether **(1163-19-5)**
Pentabromo(tetrabromophenoxy)benzene **(63936-56-1)**
2,3,4,5,6-Pentabromotoluene **(87-83-2)**
Pentabromotoluene **(87-83-2)**
Pentabromprop (9CI) **(61288-32-2)**
Pentac **(2227-17-0)**
Pentac WP **(2227-17-0)**
Pentacalcium monohydroxyorthophosphate **(1306-06-5)**
Pentacarbonyliron **(13463-40-6)**
Pentacard **(54-95-5)**
Pentacene (9CI) **(135-48-8)**
Pentachloorethaan (Dutch) **(76-01-7)**
Pentachloorfenol (Dutch) **(87-86-5)**
Pentachloraethan (German) **(76-01-7)**
Pentachlore **(2307-68-8)**
Pentachlorethane (French) **(76-01-7)**
Pentachlorin **(50-29-3)**
Pentachlornitrobenzol (German) **(82-68-8)**
Pentachloroacetone **(1768-31-6)**
2,2,2',4',5'-Pentachloroacetophenone **(1203-86-7)**
Pentachloroacetophenone **(1203-86-7)**
α,α,2,4,5-Pentachloroacetophenone **(1203-86-7)**
Pentachloroaminobenzene **(527-20-8)**
2,3,4,5,6-Pentachloroaniline **(527-20-8)**
Pentachloroaniline **(527-20-8)**
2,3,4,5,6-Pentachloroanisole **(1825-21-4)**
Pentachloroanisole **(1825-21-4)**
Pentachloroantimony **(7647-18-9)**
2,3,4,5,6-Pentachlorobenzenamine **(527-20-8)**
Pentachlorobenzene **(608-93-5)**
Pentachloro-benzenethiol **(133-49-3)**
Pentachlorobenzonitrile **(20925-85-3)**
2,2',3,3',4-Pentachlorobiphenyl **(52663-62-4)**
2,2',3,3',6-Pentachlorobiphenyl **(52663-60-2)**
2,2',3,4',5'-Pentachlorobiphenyl **(41464-51-1)**
2,2',3,4',5-Pentachlorobiphenyl **(68194-07-0)**
2,2',3,4',6'-Pentachlorobiphenyl **(60233-25-2)**
2,2',3,4',6-Pentachlorobiphenyl **(68194-05-8)**
2,2',3,4,4'-Pentachlorobiphenyl **(65510-45-4)**
2,2',3,4,5'-Pentachlorobiphenyl **(38380-02-8)**
2,2',3,4,5-Pentachlorobiphenyl **(55312-69-1)**
2,2',3,4,6'-Pentachlorobiphenyl **(73575-57-2)**
2,2',3,4,6-Pentachlorobiphenyl **(55215-17-3)**
2,2',3,5',6-Pentachlorobiphenyl **(38379-99-6)**
2,2',3,5,5'-Pentachlorobiphenyl **(52663-61-3)**
2,2',4,4',5-Pentachlorobiphenyl **(38380-01-7)**
2,2',4,4',6-Pentachlorobiphenyl **(39485-83-1)**
2,3',4,4',5'-Pentachlorobiphenyl **(65510-44-3)**
2,3',4,4',5-Pentachlorobiphenyl **(31508-00-6)**
2,3',4,5,5'-Pentachlorobiphenyl **(68194-12-7)**
2,3,3',4',5-Pentachlorobiphenyl **(70424-68-9)**
2,3,3',4',6-Pentachlorobiphenyl **(38380-03-9)**
2,3,3',4,4'-Pentachlorobiphenyl **(32598-14-4)**
2,3,3',4,5-Pentachlorobiphenyl **(70424-69-0)**
2,3,3',5,6-Pentachlorobiphenyl **(74472-36-9)**
2,3,4,3',4'-Pentachlorobiphenyl **(32598-14-4)**
2,3,4,4'5-Pentachlorobiphenyl **(74472-37-0)**
2,3,4,5,6-Pentachlorobiphenyl **(18259-05-7)**
2,4,5,2',5'-Pentachlorobiphenyl **(37680-73-2)**
3,3',4,4',5-Pentachloro-1,1'-biphenyl **(57465-28-8)**
3,4,2',3',4'-Pentachlorobiphenyl **(32598-14-4)**
3,4,5,3',4'-Pentachlorobiphenyl **(57465-28-8)**
Pentachlorobiphenyl **(25429-29-2)**
1,2,2,3,3-Pentachlorobutane **(83293-82-7)**
2,2,3,4,4-Pentachloro-3-butenoic acid, n-butyl ester **(75147-20-5)**
γ-Pentachlorocyclohexene **(319-94-8)**
1,2,3,4,5-Pentachlorocyclopentadiene **(25329-35-5)**
Pentachlorocyclopentadiene **(25329-35-5)**
Pentachloro-2,4-cyclopentadien-1-yl **(25329-35-5)**
Pentachlorocyclopropane **(6262-51-7)**
1,2,3,4,7-Pentachlorodibenzo-para-dioxin **(39227-61-7)**
1,2,3,4,6-Pentachlorodibenzo(b,e)(1,4)dioxin **(67028-19-7)**
1,2,3,6,7-Pentachlorodibenzo(b,e)(1,4)dioxin **(71925-15-0)**
1,2,3,6,8-Pentachlorodibenzo(b,e)(1,4)dioxin **(71925-16-1)**
1,2,3,6,9-Pentachlorodibenzo(b,e)(1,4)dioxin **(82291-34-7)**
1,2,3,7,9-Pentachlorodibenzo(b,e)(1,4)dioxin **(71925-17-2)**
1,2,3,8,9-Pentachlorodibenzo(b,e)(1,4)dioxin **(71925-18-3)**
1,2,4,6,7-Pentachlorodibenzo(b,e)(1,4)dioxin **(82291-35-8)**
1,2,4,6,8-Pentachlorodibenzo(b,e)(1,4)dioxin **(71998-76-0)**
1,2,4,6,9-Pentachlorodibenzo(b,e)(1,4)dioxin **(82291-36-9)**
1,2,4,7,9-Pentachlorodibenzo(b,e)(1,4)dioxin **(82291-37-0)**
1,2,4,8,9-Pentachlorodibenzo(b,e)(1,4)dioxin **(82291-38-1)**
Pentachlorodibenzo(b,e)(1,4)dioxin **(36088-22-9)**
1,2,3,4,7-Pentachlorodibenzodioxin **(39227-61-7)**
1,2,3,4,6-Pentachlorodibenzo-p-dioxin **(67028-19-7)**
1,2,3,6,7-Pentachlorodibenzo-p-dioxin **(71925-15-0)**
1,2,3,6,8-Pentachlorodibenzo-p-dioxin **(71925-16-1)**
1,2,3,6,9-Pentachlorodibenzo-p-dioxin **(82291-34-7)**
1,2,3,7,8-Pentachlorodibenzo-p-dioxin **(40321-76-4)**
1,2,3,7,9-Pentachlorodibenzo-p-dioxin **(71925-17-2)**
1,2,3,8,9-Pentachlorodibenzo-p-dioxin **(71925-18-3)**
1,2,4,6,7-Pentachlorodibenzo-p-dioxin **(82291-35-8)**
1,2,4,6,8-Pentachlorodibenzo-p-dioxin **(71998-76-0)**
1,2,4,6,9-Pentachlorodibenzo-p-dioxin **(82291-36-9)**
1,2,4,7,8-Pentachlorodibenzo-p-dioxin **(58802-08-7)**
1,2,4,7,9-Pentachlorodibenzo-p-dioxin **(82291-37-0)**
1,2,4,8,9-Pentachlorodibenzo-p-dioxin **(82291-38-1)**
Pentachlorodibenzo-p-dioxin **(36088-22-9)**
1,2,3,4,6-Pentachlorodibenzofuran **(83704-47-6)**
1,2,3,4,7-Pentachlorodibenzofuran **(83704-48-7)**
1,2,3,4,8-Pentachlorodibenzofuran **(67517-48-0)**
1,2,3,4,9-Pentachlorodibenzofuran **(83704-49-8)**
1,2,3,6,8-Pentachlorodibenzofuran **(83704-51-2)**
1,2,3,6,9-Pentachlorodibenzofuran **(83704-52-3)**
1,2,3,7,8-Pentachlorodibenzofuran **(57117-41-6)**
1,2,3,7,9-Pentachlorodibenzofuran **(83704-53-4)**
1,2,3,8,9-Pentachlorodibenzofuran **(83704-54-5)**
1,2,4,6,7-Pentachlorodibenzofuran **(83704-50-1)**
1,2,4,6,8-Pentachlorodibenzofuran **(69698-57-3)**
1,2,4,6,9-Pentachlorodibenzofuran **(70648-24-7)**
1,2,4,7,8-Pentachlorodibenzofuran **(58802-15-6)**
1,2,4,7,9-Pentachlorodibenzofuran **(71998-74-8)**
1,2,4,8,9-Pentachlorodibenzofuran **(70648-23-6)**
1,2,6,7,9-Pentachlorodibenzofuran **(70872-82-1)**
1,3,4,6,7-Pentachlorodibenzofuran **(83704-36-3)**
1,3,4,6,8-Pentachlorodibenzofuran **(83704-55-6)**

1,3,4,6,9-Pentachlorodibenzofuran **(70648-15-6)**
1,3,4,7,8-Pentachlorodibenzofuran **(58802-16-7)**
1,3,4,7,9-Pentachlorodibenzofuran **(70648-20-3)**
1,3,6,7,8-Pentachlorodibenzofuran **(70648-21-4)**
1,4,6,7,8-Pentachlorodibenzofuran **(83704-35-2)**
2,3,4,6,8-Pentachlorodibenzofuran **(67481-22-5)**
2,3,4,7,8-Pentachlorodibenzofuran **(57117-31-4)**
Pentachlorodiphenyl **(25429-29-2)**
Pentachloro diphenyl ether **(42279-29-8)**
Pentachloro diphenyl oxide **(42279-29-8)**
Pentachloroethane [UN 1669] **(76-01-7)**
Pentachlorofenol **(87-86-5)**
Pentachloromethoxybenzene **(1825-21-4)**
1,1,2,3,4-Pentachloro-4-(1-methylethoxy)-1,3-butadiene **(68334-67-8)**
Pentachloro(methylthio)benzene **(1825-19-0)**
Pentachloronaphthalene (ACGIH,OSHA) **(1321-64-8)**
Pentachloronitrobenzene **(82-68-8)**
Pentachlorophenate **(87-86-5)**
Pentachlorophenate sodium **(131-52-2)**
2,3,4,5,6-Pentachlorophenol **(87-86-5)**
Pentachlorophenol (ACGIH,OSHA) **(87-86-5)**
Pentachlorophenol, DP-2 **(87-86-5)**
Pentachlorophenol, Dowicide EC-7 **(87-86-5)**
Pentachlorophenol, Technical **(87-86-5)**
Pentachlorophenol laurate **(3772-94-9)**
Pentachlorophenol potassium salt **(7778-73-6)**
Pentachlorophenol, sodium salt **(131-52-2)**
Pentachlorophenoxy sodium **(131-52-2)**
Pentachlorophenyl acetate **(1441-02-7)**
Pentachlorophenyl chloride **(118-74-1)**
Pentachlorophenyl dodecanoate **(3772-94-9)**
Pentachlorophenyl methyl ether **(1825-21-4)**
Pentachlorophenyl methyl sulfide **(1825-19-0)**
1,1,2,2,3-Pentachloropropane **(16714-68-4)**
1,1,1,3,3-Pentachloropropanone **(1768-31-6)**
1,1,2,3,3-Pentachloropropene **(1600-37-9)**
1,1,2,3,3-Pentachloropropylene **(1600-37-9)**
2,3,4,5,6-Pentachloropyridine **(2176-62-7)**
Pentachloropyridine **(2176-62-7)**
Pentachloro(2,2,3,3-tetrafluoropropoxy)cyclotriphosphazene **(59700-57-1)**
Pentachlorothioanisole **(1825-19-0)**
Pentachlorothiophenol **(133-49-3)**
2,3,4,5,6-Pentachlorotoluene **(877-11-2)**
2,4,α,α,α-Pentachlorotoluene **(13014-18-1)**
Pentachlorotoluene **(877-11-2)**
α,α,α,3,4-Pentachlorotoluene **(13014-24-9)**
Pentachlorozincate(3-) triammonium **(14639-98-6)**
Pentachlorphenol (German) **(87-86-5)**
Pentachlorthiofenol (Czech) **(133-49-3)**
Pentachlorure d'antimoine (French) **(7647-18-9)**
Pentacin **(12111-24-9)**
Pentacine **(12111-24-9)**
Pentacloroetano (Italian) **(76-01-7)**
Pentaclorofenolo (Italian) **(87-86-5)**
Pentacon **(87-86-5)**
Pentacor **(54-95-5)**
Pentacosane (9CI) **(629-99-2)**
Pentacosanoic acid **(506-38-7)**
Pentacosanoic acid, methyl ester (9CI) **(55373-89-2)**
3,5-Pentadecadien-2-one, 6,10,14-trimethyl- (8CI,9CI) **(1604-32-6)**
7,10-Pentadecadiynoic acid (8CI) **(22117-06-2)**
1H-Pentadecafluoroheptane **(375-83-7)**
1,1,2,2,3,3,4,4,5,5,6,6,7,7,7-Pentadecafluoro-1-heptanesulfonyl fluoride **(335-71-7)**
Pentadecafluoro-n-octanoic acid **(335-67-1)**
Pentadecafluorooctanoic acid **(335-67-1)**
Pentadecafluorooctanoyl fluoride **(335-66-0)**
Pentadecalactone **(106-02-5)**
2-Pentadecalone **(106-02-5)**
Pentadecanal **(2765-11-9)**
Pentadecanal (9CI) **(2765-11-9)**
1-Pentadecanamine (9CI) **(2570-26-5)**
1-Pentadecanaminium, 1-carboxy-N,N,N-trimethyl-, hydroxide, inner salt (9CI) **(16545-85-0)**
Pentadecane **(629-62-9)**
n-Pentadecane **(629-62-9)**
Pentadecane, 1-bromo- (9CI) **(629-72-1)**
1-Pentadecanecarboxylic acid **(57-10-3)**
Pentadecane, 1-chloro- (9CI) **(4862-03-7)**
Pentadecane, chloro- (8CI,9CI) **(34214-86-3)**
Pentadecane, 2-methyl- (9CI) **(1560-93-6)**
Pentadecane, 3-methyl- (8CI,9CI) **(2882-96-4)**
Pentadecane, 6-methyl- (8CI,9CI) **(10105-38-1)**
Pentadecane, 7-methyl- (8CI,9CI) **(6165-40-8)**
Pentadecane, 8-methyl- (8CI,9CI) **(22306-28-1)**
Pentadecane, 4-methyl- (6CI,7CI,8CI,9CI) **(2801-87-8)**
Pentadecane, 5-methyl- (6CI,7CI,8CI,9CI) **(25117-33-3)**
1,2-Pentadecane oxide **(18633-25-5)**
Pentadecane, 1-phenyl- (8CI) **(2131-18-2)**
Pentadecane, 2,6,10,14-tetramethyl **(1921-70-6)**
Pentadecane, 2,6,10-trimethyl- (9CI) **(3892-00-0)**
Pentadecanoic-acid **(1002-84-2)**
Pentadecanoic acid, ethyl ester (9CI) **(41114-00-5)**
Pentadecanoic acid, 15-hydroxy-, xi-lactone **(106-02-5)**
Pentadecanoic acid, methyl ester **(7132-64-1)**
Pentadecanoic acid, 14-methyl-, methyl ester (8CI,9CI) **(5129-60-2)**
Pentadecanol **(629-76-5)**
1-Pentadecanol (9CI) **(629-76-5)**
Pentadecanol (8CI,9CI) **(31389-11-4)**
1,15-Pentadecanolide **(106-02-5)**
Pentadecanolide **(106-02-5)**
8-Pentadecanone **(818-23-5)**
Pentadecan-8-one **(818-23-5)**
2-Pentadecanone, 6,10,14-trimethyl-, (6R,10R)- (8CI) **(16825-16-4)**
2-Pentadecanone, 6,10,14-trimethyl-, (R-(R*,R*))- (9CI) **(16825-16-4)**
2-Pentadecanone, 6,10,14-trimethyl- (VAN)(9CI) **(502-69-2)**
Pentadecene,1- **(13360-61-7)**
1-Pentadecene (9CI) **(13360-61-7)**
Pentadecyclic acid **(1002-84-2)**
1-Pentadecylamine **(2570-26-5)**
Pentadecylamine **(2570-26-5)**
n-Pentadecylamine **(2570-26-5)**
Pentadecylbenzene **(2131-18-2)**
n-Pentadecylbenzene **(2131-18-2)**
Pentadecylbenzenesulfonic acid **(61215-89-2)**
Pentadecylcyclohexane **(6006-95-7)**
Pentadecyl methacrylate **(6140-74-5)**
Pentadecyl 2-methyl-2-propenoate **(6140-74-5)**
Pentadecyloxirane **(22092-38-2)**
3-Pentadecylphenol **(501-24-6)**
3-n-Pentadecylphenol **(501-24-6)**
m-Pentadecylphenol **(501-24-6)**
2-(3-Pentadecylphenoxy)butanoic acid **(14230-52-5)**
(Z)-1,3-Pentadiene **(1574-41-0)**
1,3-Pentadiene **(504-60-9)**
1,3-Pentadiene, (E) **(2004-70-8)**
cis-1,3-Pentadiene **(1574-41-0)**
trans-1,3-Pentadiene **(2004-70-8)**
1,4-Pentadiene (9CI) **(591-93-5)**
1,2-Pentadiene (8CI,9CI) **(591-95-7)**
1,3-Pentadiene, (Z)- (9CI) **(1574-41-0)**
1,3-Pentadiene-1-carboxaldehyde **(142-83-6)**
1,3-Pentadiene-1-carboxylic acid **(110-44-1)**
Pentadiene, dichloro- (9CI) **(61626-71-9)**

1,3-Pentadiene, 1,1-dichloro-4-methyl- (9CI) **(55667-43-1)**
1,4-Pentadiene, 2-methyl **(763-30-4)**
1,3-Pentadiene, 2-methyl- (9CI) **(1118-58-7)**
1,3-Pentadiene, 4-methyl- (9CI) **(926-56-7)**
1,4-Pentadiene, 3-methyl- (9CI) **(1115-08-8)**
1,3-Pentadiene, 3-methyl- (7CI,8CI,9CI) **(4549-74-0)**
1,3-Pentadiene, 2-methyl- (85%), and 4-methyl-1,3-pentadiene (15%) **(1118-58-7)**
2,4-Pentadienoic acid, 5-hydroxy-, δ-lactone **(504-31-4)**
2,4-Pentadienoic acid, 2,3,4,5,5-pentachloro- (8CI,9CI) **(5659-41-6)**
1,4-Pentadien-3-one, 1,5-di-2-furanyl- (9CI) **(886-77-1)**
Pentan-2,4-dione [UN 2310] **(123-54-6)**
Pentadorm **(77-75-8)**
Pentaerythrite **(115-77-5)**
Pentaerythrite tetranitrate [NA 0150] **(78-11-5)**
Pentaerythrite tetranitrate, Desensitized, wet [NA 0150] **(78-11-5)**
Pentaerythrite tetranitrate, Desensitized with not less than 15% phlegmatizer [NA 0150] **(78-11-5)**
Pentaerythrite tetranitrate, Dry [NA 0150] **(78-11-5)**
Pentaerythrite tetranitrate, With not less than 7% wax [UN 0411] **(78-11-5)**
Pentaerythritol (ACGIH,OSHA) **(115-77-5)**
Pentaerythritol chlorol **(78-12-6)**
Pentaerythritol diacrylate **(53417-29-1)**
Pentaerythritol dibromide **(3296-90-0)**
Pentaerythritol dibromohydrin **(3296-90-0)**
Pentaerythritol dichlorohydrin **(115-69-5)**
Pentaerythritol monooleate **(10332-32-8)**
Pentaerythritol monostearate **(78-23-9)**
Pentaerythritol, phthalic anhydride, soya oil polymer **(66070-60-8)**
Pentaerythritol, phthalic anhydride, soybean oil polymer **(66070-60-8)**
Pentaerythritol, phthalic anhydride, soybean oil resin **(66070-60-8)**
Pentaerythritol tetraacrylate **(4986-89-4)**
Pentaerythritol, tetrabenzoate **(4196-86-5)**
Pentaerythritol tetracaprylate **(3008-50-2)**
Pentaerythritol, tetranitrate **(78-11-5)**
Pentaerythritol, tetranitrate, Containing at least 25% water/at least 15% phlegmatizer [NA 0150] **(78-11-5)**
Pentaerythritol tetranitrate, Diluted **(78-11-5)**
Pentaerythritol tetrastearate **(115-83-3)**
Pentaerythritol triacrylate **(3524-68-3)**
Pentaerythritol tribromide **(1522-92-5)**
Pentaerythritol tribromohydrin **(1522-92-5)**
Pentaerythrityl tetrastearate **(115-83-3)**
Pentaethylbenzene **(605-01-6)**
Pentaethylene-glycol **(4792-15-8)**
Pentaethylene glycol, monobutyl ether **(23601-39-0)**
Pentaethylenehexamine **(4067-16-7)**
Pentafin **(78-11-5)**
2,3,4,5,6-Pentafluoroaniline **(771-60-8)**
Pentafluoroaniline **(771-60-8)**
Pentafluoroantimony **(7783-70-2)**
1,2,3,4,5-Pentafluorobenzene **(363-72-4)**
Pentafluorobenzene **(363-72-4)**
Pentafluorobenzenethiol- **(771-62-0)**
Pentafluorobromobenzene **(344-04-7)**
Pentafluorochlorobenzene **(344-07-0)**
Pentafluoroethyl iodide **(354-64-3)**
Pentafluoroiodine **(7783-66-6)**
Pentafluoroiodoethane **(354-64-3)**
Pentafluoromonochloroacetone **(79-53-8)**
Pentafluorophenol **(771-61-9)**
Pentafluorophenylamine **(771-60-8)**
Pentafluorophenyl bromide **(344-04-7)**
Pentafluorophenyl chloride **(344-07-0)**
2,2,3,3,3-Pentafluoro-1-propanol **(422-05-9)**
2,2,3,3,3-Pentafluoropropanol **(422-05-9)**
Pentafluorothiophenol **(771-62-0)**
Pentagastrin **(5534-95-2)**
Pentagen **(82-68-8)**
Pentagin **(359-83-1)**
Pentaglycerine **(77-85-0)**
Pentaglycerol **(77-85-0)**
Pentaglycerol (9CI) **(51555-31-8)**
Pentahydroxycaproic acid **(526-95-4)**
3,3',4',5,7-Pentahydroxyflavanone **(480-18-2)**
3,5,7,3',4'-Pentahydroxyflavon **(117-39-5)**
3,5,7,3',4'-Pentahydroxyflavone **(117-39-5)**
3,3',4',5,7-Pentahydroxyflavone-3-l-rhamnoside **(522-12-3)**
3,3',4',5,7-Pentahydroxyflavone-3-rutinoside **(153-18-4)**
2,3,4,5,6-Pentahydroxyhexanoic acid **(526-95-4)**
L-1,3,4,5,6-Pentahydroxyhexan-2-one **(87-79-6)**
N,N,N',N'',N''-Penta(2-hydroxypropyl)diethylenetriamine **(17121-34-5)**
Pentahydroxy-tigliadienone-monoacetate(c)monomyristate(b) **(16561-29-8)**
Pentaiodobenzene **(608-96-8)**
Pental **(57-33-0)**
γ-Pentalactone **(108-29-2)**
Pentalarm **(110-66-7)**
Pentaldol **(597-31-9)**
1-Pentalenecarboxylic acid, 1.α.,2,3,3a.β.,4,5,6,6a.β.-octahydro-3.β.-methyl- (8CI) **(28645-03-6)**
Pentalene, octahydro-1-methyl- (8CI,9CI) **(32273-77-1)**
Pentaleno(1,2-b)oxirene, octahydro-, (1a.α.,1b.β.,4a.α.,5a.α.)-(9CI) **(55449-70-2)**
Pentalin **(76-01-7)**
Pentam 300 **(140-64-70)**
Pentamethazol **(54-95-5)**
Pentamethazolum **(54-95-5)**
Pentamethylbenzene **(700-12-9)**
1,1,4,7,7-Pentamethyldiethylenetriamine **(3030-47-5)**
N,N,N',N',N''-Pentamethyldiethylenetriamine **(3030-47-5)**
Pentamethyldiethylenetriamine **(3030-47-5)**
1,1,3,3,5-Pentamethyl-4,6-dinitroindane **(116-66-5)**
Pentamethylene **(287-92-3)**
Pentamethylene bromide **(111-24-0)**
1,5-Pentamethylenediamine **(462-94-2)**
Pentamethylenediamine **(462-94-2)**
Pentamethylene dibromide **(111-24-0)**
p,p'-(Pentamethylenedioxy)dibenzamidine bis(β-hydroxyethanesulfonate) **(140-64-70)**
Pentamethylene glycol **(111-29-5)**
Pentamethyleneimine **(110-89-4)**
Pentamethylene oxide **(142-68-7)**
Pentamethylene sulfide **(1613-51-0)**
Pentamethylenetetramine, dinitroso- **(101-25-7)**
Pentamethylenetetrazal **(54-95-5)**
Pentamethylenetetrazol **(54-95-5)**
1,5-Pentamethylenetetrazole **(54-95-5)**
Pentamethylene-1,5-tetrazole **(54-95-5)**
Pentamethylenetetrazole **(54-95-5)**
Pentamethylenimine **(110-89-4)**
2,2,4,6,6-Pentamethylheptane **(13475-82-6)**
2,2,4,6,6-Pentamethyl-3-heptene **(123-48-8)**
2,4,4,6,6-Pentamethyl-1-heptene **(14031-86-8)**
2,4,4,6,6-Pentamethylheptene-1 **(14031-86-8)**
1,1,2,3,3-Pentamethylindan **(1203-17-4)**
1,1,2,3,3-Pentamethylindane **(1203-17-4)**
Pentamethylmelamine **(16268-62-5)**
Pentamethylphenol **(2819-86-5)**
2,2,4,6,8-Pentamethyl-4,6,8-triphenylcyclotetrasiloxane **(10448-10-9)**
Pentametilentetrazolo (Italian) **(54-95-5)**
Pentamidine diisethionate **(140-64-70)**
Pentamidine isethionate **(140-64-70)**
Pentamull 6 **(78-23-9)**
Pentamycetin **(56-75-7)**

Name	CAS
Pentanenitrile, 2,2'-azobis(2,4-dimethyl-	**(4419-11-8)**
Pentane, 1,1'-oxybis- (9CI)	**(693-65-2)**
1,2,3,4,5-Pentanepentol	**(488-81-3)**
Pentane, 1-phenyl-	**(538-68-1)**
1-Pentanephosphonic acid	**(4672-26-8)**
1-Pentanesulfonyl fluoride, 1,1,2,2,3,3,4,4,5,5,5-undecafluoro- (8CI)	**(375-81-5)**
Pentane, 2,2,3,3-tetramethyl- (9CI)	**(7154-79-2)**
Pentane, 2,2,4,4-tetramethyl- (9CI)	**(1070-87-7)**
Pentane, 2,3,3,4-tetramethyl- (7CI,8CI,9CI)	**(16747-38-9)**
1-Pentanethiol	**(110-66-7)**
2-Pentanethiol, 2,4,4-trimethyl	**(141-59-3)**
Pentane, 2,2,4-trimethyl	**(540-84-1)**
Pentane, 2,2,3-trimethyl- (9CI)	**(564-02-3)**
Pentane, 2,3,3-trimethyl- (9CI)	**(560-21-4)**
Pentane, 2,3,4-trimethyl- (9CI)	**(565-75-3)**
Pentane, trimethyl- (8CI,9CI)	**(29222-48-8)**
1,3,5-Pentanetriol, 3-methyl- (9CI)	**(7564-64-9)**
Pentani (Italian)	**(109-66-0)**
Pentanitrine	**(78-11-5)**
Pentanitroanilina (seca) (Spanish)	**(21985-87-5)**
2,3,4,5,6-Pentanitroaniline	**(21985-87-5)**
Pentanitroaniline (Dry)	**(21985-87-5)**
Pentanitroaniline, Dry	**(21985-87-5)**
Pentanochlor	**(2307-68-8)**
Pentanoic acid	**(109-52-4)**
n-Pentanoic acid	**(109-52-4)**
tert-Pentanoic acid	**(75-98-9)**
Pentanoic acid, 5-amino-4-oxo- (9CI)	**(106-60-5)**
Pentanoic acid, 4,4'-azobis(4-cyano- (9CI)	**(2638-94-0)**
Pentanoic acid, 4,4-bis((1,1-dimethylethyl)dioxy)-, butyl ester	**(995-33-5)**
Pentanoic acid, 5-bromo- (9CI)	**(2067-33-6)**
Pentanoic acid, 2-chloro- (9CI)	**(6155-96-0)**
Pentanoic acid, 5-chloro- (9CI)	**(1119-46-6)**
Pentanoic acid, 2,5-diamino-, (S)-	**(70-26-8)**
Pentanoic acid, 4,4-dimethyl- (9CI)	**(1118-47-4)**
Pentanoic acid, 2-ethyl- (9CI)	**(20225-24-5)**
Pentanoic acid, ethyl ester (9CI)	**(539-82-2)**
Pentanoic acid, 5-hydroxy- (9CI)	**(13392-69-3)**
Pentanoic acid, 5-hydroxy-, δ-lactone	**(542-28-9)**
Pentanoic acid, 3-methyl- (9CI)	**(105-43-1)**
Pentanoic acid, methyl ester (9CI)	**(624-24-8)**
Pentanoic acid, 4-methyl-, methyl ester (9CI)	**(2412-80-8)**
Pentanoic acid, 4-methyl-4-nitro-, ethyl ester (9CI)	**(23102-02-5)**
Pentanoic acid, 3-methyl-2-oxo- (9CI)	**(1460-34-0)**
Pentanoic acid, 4-methyl-2-oxo- (9CI)	**(816-66-0)**
Pentanoic acid, 4-methyl-, pentyl ester (9CI)	**(25415-71-8)**
Pentanoic acid, (4-methylphenyl)- (9CI)	**(59094-71-2)**
Pentanoic acid, 4-oxo-, butyl ester (9CI)	**(2052-15-5)**
Pentanoic acid, 4-oxo-, ethyl ester (9CI)	**(539-88-8)**
Pentanoic acid, pentyl ester (9CI)	**(2173-56-0)**
Pentanoic acid, 2-propyl-, sodium salt	**(1069-66-5)**
Pentanoic acid, sodium salt (9CI)	**(6106-41-8)**
Pentanoic acid, 2,2,4,4-tetramethyl- (9CI)	**(3302-12-3)**
Pentanoic acid, 2,2,4,4-tetramethyl-, phenylmethyl ester	**(81325-80-6)**
1-Pentanol	**(71-41-0)**
2-Pentanol	**(6032-29-7)**
3-Pentanol	**(584-02-1)**
Pentan-1-ol	**(71-41-0)**
Pentan-3-ol	**(584-02-1)**
Pentanol	**(71-41-0)**
Pentanol-1	**(71-41-0)**
Pentanol-2	**(6032-29-7)**
Pentanol-3	**(584-02-1)**
n-Pentanol	**(71-41-0)**
tert-Pentanol	**(75-85-4)**
1-Pentanol acetate	**(628-63-7)**
2-Pentanol, acetate (8CI, 9CI)	**(626-38-0)**
1-Pentanol, 2,2-diethyl- (8CI,9CI)	**(14202-62-1)**
1-Pentanol, 2,3-dimethyl	**(10143-23-4)**
1-Pentanol, 2,2-dimethyl- (9CI)	**(2370-12-9)**
1-Pentanol, 2,4-dimethyl- (9CI)	**(6305-71-1)**
1-Pentanol, 4,4-dimethyl- (9CI)	**(3121-79-7)**
2-Pentanol, 2,3-dimethyl- (9CI)	**(4911-70-0)**
2-Pentanol, 2,4-dimethyl- (9CI)	**(625-06-9)**
3-Pentanol, 2,2-dimethyl- (9CI)	**(3970-62-5)**
3-Pentanol, 2,4-dimethyl- (9CI)	**(600-36-2)**
3-Pentanol, 2,3-dimethyl- (8CI,9CI)	**(595-41-5)**
3-Pentanol, 3-ethyl	**(597-49-9)**
1-Pentanol, 2-ethyl-4-methyl	**(106-67-2)**
3-Pentanol, 3-ethyl-2-methyl- (8CI,9CI)	**(597-05-7)**
4-Pentanolide	**(108-29-2)**
1-Pentanol, 5-methoxy- (6CI,7CI,8CI,9CI)	**(4799-62-6)**
2-Pentanol, 5-methoxy-2-methyl- (9CI)	**(55724-04-4)**
1-Pentanol, 2-methyl	**(105-30-6)**
2-Pentanol, 2-methyl	**(590-36-3)**
2-Pentanol, 4-methyl	**(108-11-2)**
3-Pentanol, 2-methyl	**(565-67-3)**
3-Pentanol, 3-methyl	**(77-74-7)**
4-Pentanol, 2-methyl-	**(108-11-2)**
Pentanol, 4-methyl	**(1320-98-5)**
1-Pentanol, 3-methyl- (9CI)	**(589-35-5)**
1-Pentanol, 4-methyl- (9CI)	**(626-89-1)**
1-Pentanol, methyl- (9CI)	**(54972-97-3)**
2-Pentanol, 3-methyl- (9CI)	**(565-60-6)**
2-Pentanol, 4-methyl-, acetate	**(108-84-9)**
2-Pentanol, 4-methyl-, hydrogen phosphorodithioate (9CI)	**(6028-47-3)**
1-Pentanol, 4-methyl-, hydrogen phosphorodithioate, zinc salt	**(15874-15-4)**
1-Pentanol, 2,2,4-trimethyl	**(123-44-4)**
3-Pentanol, 2,2,3-trimethyl- (8CI,9CI)	**(7294-05-5)**
3-Pentanol, 2,2,4-trimethyl- (8CI,9CI)	**(5162-48-1)**
3-Pentanol, 2,3,4-trimethyl- (6CI,7CI,8CI,9CI)	**(3054-92-0)**
2-Pentanone	**(107-87-9)**
3-Pentanone	**(96-22-0)**
Pentanone-3	**(96-22-0)**
2-Pentanone (OSHA)	**(107-87-9)**
2-Pentanone, 5-(acetyloxy)- (9CI)	**(5185-97-7)**
2-Pentanone, cyclic 1,2-ethanediyl acetal	**(4352-98-1)**
3-Pentanone, 1,5-di-2-furyl	**(6075-11-2)**
3-Pentanone, 2,4-dimethyl- (9CI)	**(565-80-0)**
2-Pentanone, 4,4-dimethyl- (8CI,9CI)	**(590-50-1)**
3-Pentanone, 1,5-diphenyl- (8CI,9CI)	**(5396-91-8)**
2-Pentanone, 3,4-epoxy-3,4-dimethyl- (8CI)	**(15120-99-7)**
2-Pentanone, 3,4-epoxy-4-methyl	**(4478-63-1)**
Pentanone, 1-(3-furyl)-4-hydroxy	**(32954-58-8)**
2-Pentanone, 4-hydroxy- (8CI,9CI)	**(4161-60-8)**
2-Pentanone, 5-hydroxy-, acetate	**(5185-97-7)**
2-Pentanone, 4-hydroxy-4-methyl	**(123-42-2)**
2-Pentanone, 4-hydroxy-4-methyl-, peroxide (9CI)	**(54693-46-8)**
2-Pentanone, 4-hydroxy-4-methyl-, peroxide, More than 57% in solution	**(54693-46-8)**
2-Pentanone, 4-hydroxy-4-methyl-, peroxide, Not more than 57% in solution	**(54693-46-8)**
2-Pentanone, 4-methoxy-4-methyl	**(107-70-0)**
2-Pentanone, 4-methyl	**(108-10-1)**
2-Pentanone, 3-methyl- (9CI)	**(565-61-7)**
3-Pentanone, 2-methyl- (8CI,9CI)	**(565-69-5)**
2-Pentanone, 4-methyl-, peroxide (9CI)	**(37206-20-5)**
2-Pentanone, 4-methyl-, peroxide, With more than 9% by weight active oxygen	**(37206-20-5)**
2-Pentanone, 4-methyl-, peroxide, With not more than 9% by weight active oxygen	**(37206-20-5)**
2-Pentanone, 5-phenyl- (8CI,9CI)	**(2235-83-8)**
3-Pentanone, 2,2,4,4-tetramethyl- (9CI)	**(815-24-7)**

Pentanoyl chloride (9CI) **(638-29-9)**
Pentantin **(50-13-5)**
Pentanyl **(437-38-7)**
5,8,11,14,17-Pentaoxaheneicosane **(112-98-1)**
3,6,9,12,15-Pentaoxaheptacosan-1-ol (9CI) **(3055-95-6)**
3,6,9,12,15-Pentaoxaheptadecane **(4353-28-0)**
3,6,9,12,15-Pentaoxaheptadecane-1,17-diol **(2615-15-8)**
3,6,9,12,15-Pentaoxaheptadecan-1-ol, 17-(4-(1,1,3,3-tetramethylbutyl)phenoxy)- (9CI) **(2497-58-7)**
3,6,9,12,15-Pentaoxaheptadecan-1-ol, 17-(p-(1,1,3,3-tetramethylbutyl)phenoxy)- (8CI) **(2497-58-7)**
4,7,10,13,16-Pentaoxanonadecane, 1,2:18,19-diepoxy- (8CI) **(17626-93-6)**
2,5,8,11,14-Pentaoxapentadecane **(143-24-8)**
Pentaphen **(80-46-6)**
Pentaphenate **(131-52-2)**
Pentaphene (8CI,9CI) **(222-93-5)**
Pentaric acid, calcium salt (9CI) **(68568-63-8)**
Pentarit **(78-11-5)**
Pentasiloxane, 1,1,1,3,5,5,7,9,9,9-decamethyl-3,7-diphenyl- (8CI,9CI) **(20252-66-8)**
Pentasiloxane, dodecamethyl **(141-63-9)**
Pentasiloxane, 1,1,1,3,3,5,7,7,9,9,9-undecamethyl-5-phenyl- (9CI) **(60587-10-2)**
Pentasodium DTPA **(140-01-2)**
Pentasodium aminotrimethylene phosphonate **(2235-43-0)**
Pentasodium aminotris(methylphosphonic acid) **(2235-43-0)**
Pentasodium diethylenetriaminepentaacetate **(140-01-2)**
Pentasodium diethylenetriaminepentaacetic acid **(140-01-2)**
Pentasodium diethylenetriaminepentacetate **(140-01-2)**
Pentasodium nitrilotris(methylenephosphonate) **(2235-43-0)**
Pentasodium(nitrilotris(methylene))triphosphonate **(2235-43-0)**
Pentasodium (nitrilotris(methylene))trisphosphonate **(2235-43-0)**
Pentasodium pentetate **(140-01-2)**
Pentasodium triphosphate **(7758-29-4)**
Pentasodium tripolyphosphate **(7758-29-4)**
Pentasol **(71-41-0)**
Pentasol **(87-86-5)**
Pentasulfure de phosphore (French) **(1314-80-3)**
Pentatriacontane (9CI) **(630-07-9)**
18-Pentatriacontanone (9CI) **(504-53-0)**
Penta-s-triazinetrione **(30622-37-8)**
1,3,5,7,9,2,4,6,8,10-Pentazapentaphosphecine, 2,2,4,4,6,6,8,8,10,10-decachloride **(13596-41-3)**
1,3,5,7,9,2,4,6,8,10-Pentazapentaphosphecine, 2,2,4,4,6,6,8,8,10,10-decachloro-2,2,4,4,6,6,8,8,10,10-decahydro- (8CI,9CI) **(13596-41-3)**
Pentazocine **(359-83-1)**
Pentazol **(54-95-5)**
Pentazolum **(54-95-5)**
Pentazone **(25057-89-0)**
Pentech **(50-29-3)**
Pentek **(115-77-5)**
Pentemesan **(54-95-5)**
2-Pentenal **(764-39-6)**
4-Pentenal **(2100-17-6)**
4-Pentenal, 2-ethyl- (6CI,7CI,8CI,9CI) **(5204-80-8)**
2-Pentenal, 2-methyl **(623-36-9)**
2-Pentenal, 4-methyl- (9CI) **(5362-56-1)**
3-Pentenal, 4-methyl- (8CI,9CI) **(5362-50-5)**
4-Pentenal, 3-methyl- (6CI,7CI,8CI,9CI) **(1777-33-9)**
2-Pentenal, 4-oxo **(5729-47-5)**
2-Pentenal, 2,4,4-trimethyl- (6CI,9CI) **(53907-61-2)**
(E)-2-Pentene **(646-04-8)**
(Z)-2-Pentene **(627-20-3)**
2-trans-Pentene **(646-04-8)**
3-Pentene **(109-68-2)**
Pentene **(25377-72-4)**
cis-2-Pentene **(627-20-3)**
cis-Pentene **(627-20-3)**
trans-2-Pentene **(646-04-8)**
1-Pentene (9CI) **(109-67-1)**
2-Pentene (9CI) **(109-68-2)**
2-Pentene, (E)- (9CI) **(646-04-8)**
2-Pentene, (Z)- (9CI) **(627-20-3)**
1-Pentene, 1-chloro **(21450-13-5)**
2-Pentene-1,1,1-d3, 2,4-dimethyl-, (E)- (9CI) **(69432-96-8)**
1-Pentene, 2,4-dimethyl- (9CI) **(2213-32-3)**
1-Pentene, 3,4-dimethyl- (9CI) **(7385-78-6)**
1-Pentene, 4,4-dimethyl- (9CI) **(762-62-9)**
2-Pentene, 3,4-dimethyl- (9CI) **(24910-63-2)**
2-Pentene, 4,4-dimethyl-, (E)- (9CI) **(690-08-4)**
Pentene, 2,2-dimethyl- (9CI) **(50819-06-2)**
1-Pentene, 2,3-dimethyl- (8CI,9CI) **(3404-72-6)**
1-Pentene, 3,3-dimethyl- (8CI,9CI) **(3404-73-7)**
2-Pentene, 2,3-dimethyl- (8CI,9CI) **(10574-37-5)**
2-Pentene, 2,4-dimethyl- (8CI,9CI) **(625-65-0)**
2-Pentene, 4,4-dimethyl-, (Z)- (8CI,9CI) **(762-63-0)**
2-Pentene-1,1,1-d3, 2-methyl-, (E)- (9CI) **(69432-95-7)**
2-Pentene, 1-ethoxy-4,4-dimethyl- (9CI) **(55702-60-8)**
1-Pentene, 3-ethyl- (9CI) **(4038-04-4)**
2-Pentene, 3-ethyl- (9CI) **(816-79-5)**
1-Pentene, 2-methyl **(763-29-1)**
2-Pentene, 2-methyl **(625-27-4)**
1-Pentene, 3-methyl- (9CI) **(760-20-3)**
1-Pentene, 4-methyl- (9CI) **(691-37-2)**
2-Pentene, 3-methyl-, (E)- (9CI) **(616-12-6)**
Pentene, 2-methyl- (9CI) **(27236-46-0)**
Pentene, methyl- (9CI) **(37275-41-5)**
2-Pentene, 3-methyl- (8CI,9CI) **(922-61-2)**
2-Pentene, 4-methyl-, (E)- (8CI,9CI) **(674-76-0)**
2-Pentene, 3-methyl-, (Z)- (9CI) **(922-62-3)**
2-Pentene, 4-methyl-, (Z)- (9CI) **(691-38-3)**
(E)-2-Pentenenitrile **(26294-98-4)**
(Z)-2-Pentenenitrile **(25899-50-7)**
cis-2-Pentenenitrile **(25899-50-7)**
2-Pentenenitrile, (E)- (9CI) **(26294-98-4)**
3-Pentenenitrile (9CI) **(4635-87-4)**
4-Pentenenitrile (9CI) **(592-51-8)**
2-Pentenenitrile (8CI,9CI) **(13284-42-9)**
2-Pentenenitrile, (Z)- (9CI) **(25899-50-7)**
2-Pentenenitrile, 2-methyl-, (E)- (9CI) **(31551-28-7)**
2-Pentene, 2-nitro **(6065-19-6)**
2-Pentene, 3-nitro **(6065-18-5)**
3-Pentene, 3-nitro- **(6065-18-5)**
2-Pentene, 5-(pentyloxy)- (9CI) **(34061-80-8)**
2-Pentene, 5-(pentyloxy)-, (E)- (9CI) **(56052-85-8)**
2-Pentene, 3-phenyl-, (E)- (8CI) **(4165-86-0)**
1-Pentene, 3-phenyl- (7CI,8CI) **(19947-22-9)**
2-Pentene, 3-phenyl-, (Z)- (8CI) **(4165-78-0)**
1-Pentene, 2,2,4-trimethyl- **(11071-47-9)**
1-Pentene, 2,4,4-trimethyl **(107-39-1)**
2-Pentene, 2,4,4-trimethyl- **(107-40-4)**
Pentene, 2,4,4-trimethyl **(25167-70-8)**
2-Pentene, 3,4,4-trimethyl- (9CI) **(598-96-9)**
Pentene, trimethyl- (9CI) **(61665-19-8)**
1-Pentene, 2,3,3-trimethyl- (8CI,9CI) **(560-23-6)**
2-Pentene, 2,3,4-trimethyl- (8CI,9CI) **(565-77-5)**
(Z)-3-Pentenoic acid **(33698-87-2)**
2-Pentenoic acid **(626-98-2)**
4-Pentenoic acid **(591-80-0)**
cis-2-Pentenoic acid **(16666-42-5)**
2-Pentenoic acid, (E)- (8CI,9CI) **(13991-37-2)**
3-Pentenoic acid (8CI,9CI) **(5204-64-8)**
3-Pentenoic acid, (E)- (8CI,9CI) **(1617-32-9)**

Name	CAS Registry Number
3-Pentenoic acid, (Z)- (9CI)	**(33698-87-2)**
2-Pentenoic acid, (Z)- (8CI,9CI)	**(16666-42-5)**
2-Pentenoic acid, 4,4-dimethyl- (6CI,7CI,9CI)	**(6945-35-3)**
Penten-4-oic acid, 3,3-dimethyl, methyl ester	**(63721-05-1)**
4-Pentenoic acid, 3,3-dimethyl-, methyl ester (9CI)	**(63721-05-1)**
2-Pentenoic acid, 4-hydroxy-4-methyl-, lactone	**(20019-64-1)**
2-Pentenoic acid, 4-hydroxy-4-methyl-, methyl ester, (E)- (9CI)	**(5739-83-3)**
2-Pentenoic acid, 2-methoxy-4-methyl-, methyl ester (9CI)	**(56009-36-0)**
2-Pentenoic acid, 2-methyl- (9CI)	**(3142-72-1)**
3-Pentenoic acid, 3-methyl-, (E)- (9CI)	**(41653-93-4)**
4-Pentenoic acid, nitrile	**(592-51-8)**
3-Pentenoic acid, 3,4,5,5,5-pentachloro-2-oxo- (9CI)	**(99165-95-4)**
3-Pentenoic acid, 3,4,5,5-tetrachloro-2-oxo- (9CI)	**(99165-94-3)**
3-Pentenoic acid, 5,5,5,?-tetrachloro-2-oxo- (9CI)	**(99165-91-0)**
3-Pentenoic acid, 5,5,5-trichloro-2-oxo- (9CI)	**(99165-97-6)**
3-Pentenoic acid, 5,5,?-trichloro-2-oxo- (9CI)	**(99165-90-9)**
(Z)-2-Penten-1-ol	**(1576-95-0)**
3-Penten-2-ol	**(1569-50-2)**
1-Penten-3-ol (9CI)	**(616-25-1)**
4-Penten-2-ol (9CI)	**(625-31-0)**
2-Penten-1-ol (8CI,9CI)	**(20273-24-9)**
4-Penten-1-ol (8CI,9CI)	**(821-09-0)**
2-Penten-1-ol, (Z)- (9CI)	**(1576-95-0)**
2-Penten-1-ol, 5-(2,3-dimethyltricyclo(2.2.1.0^{2,6})hept-3-yl)-2-methyl-, (R(Z))	**(115-71-9)**
1-Penten-3-ol, 2-methyl	**(2088-07-5)**
1-Penten-3-ol, 4-methyl- (8CI,9CI)	**(4798-45-2)**
(E)-3-Penten-2-one	**(3102-33-8)**
1-Penten-3-one	**(1629-58-9)**
3-Penten-2-one	**(625-33-2)**
trans-3-Penten-2-one	**(3102-33-8)**
3-Penten-2-one, (E)- (8CI,9CI)	**(3102-33-8)**
4-Penten-2-one, 3-cyclohexyl- (9CI)	**(55702-54-0)**
3-Penten-2-one, 3,4-dimethyl- (8CI,9CI)	**(684-94-6)**
3-Penten-2-one, 3-ethyl-4-methyl- (7CI,8CI,9CI)	**(22287-11-2)**
3-Penten-2-one, 4-methyl	**(141-79-7)**
4-Penten-2-one, 4-methyl-	**(3744-02-3)**
1-Penten-3-one, 2-methyl- (8CI,9CI)	**(25044-01-3)**
4-Penten-2-one, 3-methyl- (8CI,9CI)	**(758-87-2)**
1-Penten-3-one, 1-(2,6,6-trimethyl-1-cyclohexen-1-yl)- (9CI)	**(127-43-5)**
1-Penten-3-one, 1-(2,6,6-trimethyl-2-cyclohexen-1-yl)-, (R-(E))- (9CI)	**(127-42-4)**
4-Pentenonitrile	**(592-51-8)**
1-Penten-4-yn-3-ol, 1-chloro-3-ethyl	**(113-18-8)**
Pentestan-80	**(78-11-5)**
Pentetate trisodium calcium	**(12111-24-9)**
Pentetic acid	**(67-43-6)**
Pentetrate unicelles	**(78-11-5)**
Pentetrazol	**(54-95-5)**
Pentetrazole	**(54-95-5)**
Penthamil	**(12111-24-9)**
Penthamil	**(67-43-6)**
Penthazine	**(92-84-2)**
Penthiobarbital	**(76-75-5)**
Penthiophane	**(1613-51-0)**
Penthrane	**(76-38-0)**
Penthrit	**(78-11-5)**
Penthrite	**(78-11-5)**
Pentid	**(113-98-4)**
Pentids	**(113-98-4)**
Pentiformic acid	**(142-62-1)**
Pentilen	**(69-05-6)**
R-Pentine	**(542-92-7)**
Pentinimid	**(77-67-8)**
Pentinol	**(77-75-8)**
Pentitol	**(488-81-3)**
Pentitrate	**(78-11-5)**
Pentobarbital	**(115-58-2)**
Pentobarbital	**(76-74-4)**
Pentobarbital sodium	**(57-33-0)**
Pentobarbitone	**(76-74-4)**
Pentobarbitone sodium	**(57-33-0)**
Pentobarbiturate	**(76-74-4)**
Pentobarbituric acid	**(76-74-4)**
Pentofran	**(50-47-5)**
Pentole	**(542-92-7)**
Pentonal	**(57-33-0)**
Pentone	**(57-33-0)**
D-threo-Pentonic acid, 3-deoxy- (8CI,9CI)	**(21569-63-1)**
D-erythro-Pentonic acid, 3-deoxy-2-C-(hydroxymethyl)- (8CI,9CI)	**(1518-54-3)**
D-threo-Pentonic acid, 3-deoxy-2-C-(hydroxymethyl)- (8CI,9CI)	**(1518-56-5)**
D-erythro-Pentose, 2-deoxy-	**(533-67-5)**
Pentothal	**(76-75-5)**
Pentothiobarbital	**(76-75-5)**
Pentoxyverine	**(77-23-6)**
Pentran	**(76-38-0)**
Pentrane	**(76-38-0)**
Pentrate	**(78-11-5)**
Pentrazol	**(54-95-5)**
Pentrex	**(69-53-4)**
Pentrexl	**(69-53-4)**
Pentriol	**(78-11-5)**
Pentritol	**(78-11-5)**
Pentrolone	**(54-95-5)**
Pentrozol	**(54-95-5)**
Pentryate	**(78-11-5)**
Pentryate 80	**(78-11-5)**
Pentydorm	**(77-75-8)**
Pentydrom	**(77-75-8)**
Pentyl	**(57-33-0)**
1-Pentyl acetate	**(628-63-7)**
2-Pentyl acetate	**(626-38-0)**
Pentyl acetate	**(628-63-7)**
n-Pentyl acetate	**(628-63-7)**
Pentyl-alcohol	**(71-41-0)**
sec-Pentyl alcohol	**(6032-29-7)**
tert-Pentyl alcohol	**(75-85-4)**
Pentyl alcohol, nitrite	**(463-04-7)**
1-Pentylamine	**(110-58-7)**
Pentylamine	**(110-58-7)**
n-Pentylamine	**(110-58-7)**
Pentylamine, 1-methyl- (8CI)	**(5329-79-3)**
Pentylamine, N-methyl-N-nitroso	**(13256-07-0)**
Pentylamine, pentyl-	**(2050-92-2)**
2-Pentyl-9,10-anthracenedione	**(13936-21-5)**
4-Pentylbenzaldehyde	**(6853-57-2)**
Pentylbenzene	**(538-68-1)**
n-Pentylbenzene	**(538-68-1)**
sec-Pentylbenzene	**(29316-05-0)**
tert-Pentylbenzene	**(2049-95-8)**
Pentyl benzoate	**(2049-96-9)**
n-Pentyl benzoate	**(2049-96-9)**
1-Pentyl bromide	**(110-53-2)**
Pentyl bromide	**(110-53-2)**
n-Pentyl bromide	**(110-53-2)**
Pentyl butyrate	**(540-18-1)**
Pentyl caproate	**(540-07-8)**
3-Pentylcarbinol	**(97-95-0)**
Pentylcarbinol	**(111-27-3)**
sec-Pentylcarbinol	**(97-95-0)**
Pentyl chloride	**(543-59-9)**
Pentylcinnamaldehyde	**(122-40-7)**
α-Pentylcinnamaldehyde	**(122-40-7)**

Pentylcyclohexane (4292-92-6)
2-Pentylcyclopentanone (4819-67-4)
N-(3-Pentyl)-3,4-dimethyl-2,6-dinitroaniline (40487-42-1)
1-Pentylene (109-67-1)
Pentylene (25377-72-4)
1,5-Pentylene glycol (111-29-5)
Pentylenetetrazol (54-95-5)
Pentylenetetrazole (54-95-5)
Pentyl-ether (693-65-2)
Pentyl formate (638-49-3)
m-Pentyl formate (638-49-3)
Pentylformic acid (142-62-1)
2-Pentylfuran (3777-69-3)
2-n-Pentylfuran (3777-69-3)
Pentyl hexanoate (540-07-8)
tert-Pentyl hydroperoxide (3425-61-4)
1-Pentyl iodide (628-17-1)
Pentyl iodide (628-17-1)
n-Pentyl iodide (628-17-1)
Pentyl ketone (927-49-1)
Pentyl mercaptan (110-66-7)
Pentyl methyl sulfide (1741-83-9)
Pentyl nitrite (463-04-7)
Pentyl 4-nitrobenzoate (14309-42-3)
1-(Pentyloxy)hexane (32357-83-8)
4-(Pentyloxy)phenol (18979-53-8)
Pentyl pentanoate (2173-56-0)
2-Pentyl-phenol (136-81-2)
2-Pentylphenol (136-81-2)
o-Pentylphenol (136-81-2)
p-t-Pentylphenol (80-46-6)
Pentyl propanoate (624-54-4)
O-Pentyl S-2-propenyl carbonodithioate (2956-12-9)
n-Pentyl propionate (624-54-4)
6-Pentyl-2H-pyran-2-one (27593-23-3)
2-Pentylpyridine (2294-76-0)
5-Pentylresorcinol (500-66-3)
5-n-Pentylresorcinol (500-66-3)
Pentyl salicylate (2050-08-0)
2-Pentylthiophene (4861-58-9)
Pentyltrichlorosilane (107-72-2)
3-Pentyl-6,6,9-trimethyl-6a,7,8,10a-tetrahydro-6H-dibenzo-(b,d)pyran-1-ol (1972-08-3)
n-Pentyl valerate (2173-56-0)
Pentyl vinyl ether (5363-63-3)
Pentymal (57-43-2)
Pentymalum (57-43-2)
1-Pentyne (627-19-0)
2-Pentyne (8CI,9CI) (627-21-4)
1-Pentyn-3-ol, 3-methyl (77-75-8)
3-Pentyn-2-one, 5,5-diethoxy- (9CI) (55402-04-5)
Pentyrest (77-75-8)
Penvikal (132-98-9)
Penwar (87-86-5)
Penzal N 300 (54-35-3)
Peony (1658-56-6)
Peppermint-Oil (8006-90-4)
Peppermint Camphor (89-78-1)
Peppermint LS (8001-21-6)
Peprosan (1332-40-7)
Pepsdol (138-15-8)
Pepsidol (138-15-8)
Peptavlon (5534-95-2)
Peptazin BAFD (135-57-9)
Peptisant 1O (135-57-9)
Pepton 22 (135-57-9)
Peptones (73049-73-7)
Peracetic acid (79-21-0)
Peracetic acid, Solution not over 43% acid and not over 6% hydrogen peroxide (DOT) (79-21-0)
Peragal ST (9003-39-8)
Perandren (57-85-2)
Perandren (58-22-0)
Peratox (87-86-5)
Perawin (127-18-4)
Perazil (14362-31-3)
Perbenzoate de butyle tertiaire (French) (614-45-9)
Perboric acid, sodium salt (7632-04-4)
Perbulate (1929-77-7)
Perbunan C (9010-98-4)
Perbutyl H (75-91-2)
Perca Orange gr (633-96-5)
Percapyl (62-37-3)
Percarbamid (124-43-6)
Percarbamide (124-43-6)
Perchloorethyleen, per (Dutch) (127-18-4)
Perchlor (127-18-4)
Perchloraethylen, per (German) (127-18-4)
Perchlorate de baryum (French) (13465-95-7)
Perchlorate de magnesium (French) (10034-81-8)
Perchlorate de sodium (French) (7601-89-0)
Perchlorate d'ethyle (French) (22750-93-2)
Perchlorethylene (127-18-4)
Perchlorethylene, per (French) (127-18-4)
Perchloric-acid (7601-90-3)
Perchloric acid, More than 50% but not more than 72% strength [UN 1873] (7601-90-3)
Perchloric acid, More than 72% strength (DOT) (7601-90-3)
Perchloric acid, Not over 50% acid [UN 1802] (7601-90-3)
Perchloric acid, ammonium salt (7790-98-9)
Perchloric acid, barium salt (8CI,9CI) (13465-95-7)
Perchloric acid, ethyl ester (22750-93-2)
Perchloric acid, lead(2+) salt (13637-76-8)
Perchloric acid, magnesium salt (10034-81-8)
Perchloric acid, nickel(2+) salt (9CI) (13637-71-3)
Perchloric acid, nickel(2+) salt, hexahydrate (13520-61-1)
Perchloric acid, potassium salt (1:1) (7778-74-7)
Perchloric acid, silver(1+) salt (9CI) (7783-93-9)
Perchloric acid, sodium salt (7601-89-0)
Perchloric acid, strontium salt (13450-97-0)
Perchloric acid, trichloromethyl ester (67632-66-0)
Perchloride of mercury (7487-94-7)
Perchlorinemethylmercaptan (75-70-7)
Perchlormethylmerkaptan (Czech) (594-42-3)
Perchlorobenzene (118-74-1)
Perchlorobutadiene (87-68-3)
Perchlorocyclopentadiene (77-47-4)
Perchlorodihomocubane (2385-85-5)
Perchloroethane (67-72-1)
Perchloroethylene (ACGIH,DOT,OSHA) (127-18-4)
Perchloromethane (56-23-5)
Perchloromethanethiol (75-70-7)
Perchloromethyl mercaptan (75-70-7)
Perchloromethyl mercaptan (ACGIH,OSHA) (594-42-3)
Perchloromethylmercaptan [UN 1670] (594-42-3)
Perchloron (7778-54-3)
Perchloron (8CI) (8007-32-7)
Perchloropentacyclo(5.2.1.$0^{2,6}$.$0^{3,9}$.$0^{5,8}$)Decane (2385-85-5)
Perchloropentacyclodecane (2385-85-5)
Perchloropropane (594-90-1)
Perchloropyridine (2176-62-7)
Perchloroterephthaloyl chloride (719-32-4)
Perchlorothiophene (6012-97-1)
Perchlorure d'antimoine (French) (7647-18-9)

Perchlorure de fer (French) **(7705-08-0)**
Perchloryl-fluoride **(7616-94-6)**
Perchloryl fluoride (ACGIH,OSHA) **(7616-94-6)**
Percin **(54-85-3)**
Perclene **(127-18-4)**
Perclene D **(127-18-4)**
Perclorato barico (Spanish) **(13465-95-7)**
Perclorato de etilo (Spanish) **(22750-93-2)**
Perclorato de tetraetilamonio (Spanish) **(2567-83-1)**
Percloroetilene (Italian) **(127-18-4)**
Percobarb **(62-44-2)**
Percobarb **(76-42-6)**
Percodan **(76-42-6)**
Percol 352 (9CI) **(106946-88-7)**
Percol E 10 (9CI) **(106946-89-8)**
Percolate **(732-11-6)**
Percomon **(60-13-9)**
Percoral **(59-26-7)**
Percosolve **(127-18-4)**
Percutacrine **(57-83-0)**
Percutacrine androgenique **(58-22-0)**
Percutatrine oestrogenique iscovesco **(56-53-1)**
Percutina **(67-73-2)**
Perdeutero-tetracosane **(16416-32-3)**
Pere-Col **(1332-65-6)**
Perebral **(456-59-7)**
Perecot **(1317-39-1)**
Peregal O 20 **(9002-92-0)**
Peregal ST **(9003-39-8)**
Peregin **(52645-53-1)**
Peregin W **(52645-53-1)**
Peremesin **(569-65-3)**
Perenox **(1317-39-1)**
Perequietil **(57-53-4)**
Perequil **(57-53-4)**
Perfecta **(8009-03-8)**
Perfecta **(8012-95-1)**
Perfecthion **(60-51-5)**
Perfekthion **(60-51-5)**
Perfektion **(60-51-5)**
Perfenazina (Italian) **(58-39-9)**
Perflan **(34014-18-1)**
Perfluidone **(37924-13-3)**
Perfluoride **(16984-48-8)**
Perfluoroacetic acid **(76-05-1)**
Perfluoroacetic anhydride **(407-25-0)**
Perfluoroacetyl(2-thenoyl)methane **(326-91-0)**
Perfluoroammonium octanoate **(3825-26-1)**
Perfluorobenzene **(392-56-3)**
Perfluorobut-2-ene **(360-89-4)**
Perfluorobutane **(355-25-9)**
Perfluoro-2-butene (DOT) **(360-89-4)**
Perfluorocaprylic acid **(335-67-1)**
Perfluoroctanoic acid **(335-67-1)**
Perfluorocyclobutane **(115-25-3)**
Perfluorocyclohexane **(355-68-0)**
1H,1H,2H,2H-Perfluorodecanol **(678-39-7)**
Perfluoroethane **(76-16-4)**
Perfluoroethene **(116-14-3)**
Perfluoroethylene **(116-14-3)**
1H-Perfluoroheptane **(375-83-7)**
Perfluoro-n-heptane **(335-57-9)**
Perfluoroheptane **(335-57-9)**
Perfluoroheptanecarboxylic acid **(335-67-1)**
Perfluoro-n-hexane **(355-42-0)**
Perfluorohexane **(355-42-0)**
Perfluoroisobutylene **(382-21-8)**
Perfluoromethane **(75-73-0)**
Perfluoromethylcyclohexane **(355-02-2)**
Perfluoro(methyloxirane) **(428-59-1)**
Perfluorooctane **(307-34-6)**
n-Perfluorooctane **(307-34-6)**
Perfluorooctanesulfonic acid, potassium salt **(2795-39-3)**
N-Perfluorooctanesulfonyl fluoride **(307-35-7)**
Perfluorooctanesulfonyl fluoride **(307-35-7)**
Perfluorooctanoic acid **(335-67-1)**
1H,1H,2H,2H-Perfluorooctanol **(647-42-7)**
Perfluorooctylsulfonyl fluoride **(307-35-7)**
Perfluoropentane **(678-26-2)**
Perfluoropropane **(76-19-7)**
Perfluoropropene **(116-15-4)**
Perfluoropropylene **(116-15-4)**
Perfluoropropylene oxide **(428-59-1)**
Perfluoro(propyl vinyl ether) **(1623-05-8)**
Perfluorotoluene **(434-64-0)**
Perfluorotributylamine **(311-89-7)**
Perfmid **(34014-18-1)**
Pergacid Violet 2B **(1694-09-3)**
Pergacid Violet 3B **(1694-09-3)**
Pergantene **(7681-49-4)**
Pergillin **(74798-20-2)**
Pergitral **(78-11-5)**
Perglottal **(55-63-0)**
Pergut S 10 **(9006-03-5)**
Pergut S 20 **(9006-03-5)**
Pergut S 40 **(9006-03-5)**
Pergut S 90 **(9006-03-5)**
Perhexa 3M **(6731-36-8)**
Perhydrit **(124-43-6)**
Perhydroanthracene **(6596-35-6)**
Perhydroazepine **(111-49-9)**
2-Perhydroazepinone **(105-60-2)**
Perhydrogeraniol **(106-21-8)**
Perhydrol **(7722-84-1)**
Perhydrol-Urea **(124-43-6)**
Perhydronaphthalene **(91-17-8)**
cis-Perhydronaphthalene **(493-01-6)**
trans-Perhydronaphthalene **(493-02-7)**
Perhydroxyl radical **(3170-83-0)**
Peri acid **(82-75-7)**
Peri acid, phenyl- **(82-76-8)**
Periactin **(129-03-3)**
Periactine **(129-03-3)**
Periactinol **(129-03-3)**
Perichlor **(78-12-6)**
Perichloral **(78-12-6)**
Periclase **(1309-48-4)**
Pericler **(78-12-6)**
Periclor **(78-12-6)**
Pericyclocamphanone **(875-99-0)**
β-Pericyclocamphanone **(875-99-0)**
Peridamol **(58-32-2)**
Peridex-la **(78-11-5)**
Peri-dinaphthalene **(198-55-0)**
Periethylenenaphthalene **(83-32-9)**
Perilax **(532-03-6)**
Perilene **(198-55-0)**
Perilla Oil **(68132-21-8)**
Perilla Seed Oil **(68132-21-8)**
Perilla alcohol **(536-59-4)**
Perilla aldehyde **(2111-75-3)**
Perillal **(2111-75-3)**
Perillaldehyde **(2111-75-3)**
Perillol **(536-59-4)**

Perillyl alcohol	(536-59-4)
Perillyl aldehyde	(2111-75-3)
1H-Perimidine, 2,3-dihydro-2,2-dimethyl-6-((4-(phenylazo)-1-naphthyl)azo)	(4197-25-5)
7-Perinaphthenone	(548-39-0)
Perinaphthenone	(548-39-0)
Perinaphthindene	(203-80-5)
trans-Perinone	(4424-06-0)
Periodin	(7778-74-7)
Periograf	(1306-06-5)
Periphermin	(83-63-6)
Periplanone B	(61228-92-0)
Periston	(9003-39-8)
Periston-N	(9003-39-8)
Peritan NA	(8061-51-6)
Peritonan	(59-50-7)
Peritrate	(78-11-5)
Perityl	(78-11-5)
Perk	(127-18-4)
Perkadox 14	(2212-81-9)
Perkadox 14/40	(2212-81-9)
Perkadox 14/96	(2212-81-9)
Perkadox 14/40C	(2781-00-2)
Perkadox 14C	(2212-81-9)
Perkadox SE 10	(762-12-9)
Perkadox SE 8	(762-16-3)
Perkadox U 14/40	(2781-00-2)
Perkadox 24W40	(26322-14-5)
Perke	(51-63-8)
Perklone	(127-18-4)
Perko cleaner (9CI)	(57036-00-7)
Perlandrol L	(151-21-3)
Perlankrol RN	(9014-90-8)
Perlatan	(53-16-7)
Perlatum 310	(8009-03-8)
Perlatum 315	(8009-03-8)
Perlatum 320	(8009-03-8)
Perlatum 325	(8009-03-8)
Perlex Paste 500	(7446-27-7)
Perlex Paste 600A	(7446-27-7)
Perlite	(12427-27-9)
Perlite, Containing < 1% quartz	(12427-27-9)
Perliton Blue B	(2475-45-8)
Perliton Blue FFR	(2475-46-9)
Perliton Blue FFR	(86722-66-9)
Perliton Orange 3R	(82-28-0)
Perliton Pink 3B	(116-85-8)
Perliton Red Violet FFB	(2872-48-2)
Perliton Violet B	(82-33-7)
Perliton Violet 3R	(128-95-0)
Perliton Yellow G	(2832-40-8)
Perliton Yellow RR	(119-15-3)
Perlon	(63428-83-1)
Perlopal	(77-75-8)
Perlutex	(71-58-9)
Perm-A-Chlor	(79-01-6)
Perm-A-Clor	(79-01-6)
Perma Kleer	(67-43-6)
Perma Kleer 100	(64-02-8)
Perma Kleer 140	(140-01-2)
Perma Kleer 80	(139-89-9)
Perma Kleer 50 Crystals	(64-02-8)
Perma Kleer 80 Crystals	(139-89-9)
Perma Kleer 80, Crystals	(139-89-9)
Perma Kleer 50 Crystals Disodium Salt	(139-33-3)
Perma Kleer Di Crystals	(139-33-3)
Perma Kleer Tetra CP	(64-02-8)
Perma Kleer 50, Trisodium salt	(150-38-9)
Perma Kleer 50 acid	(60-00-4)
Permacide	(87-86-5)
Permafresh 183	(1854-26-8)
Permafresh 477	(140-95-4)
Permafresh 113B	(1854-26-8)
Permafresh LF	(1854-26-8)
Permafresh LH	(1854-26-8)
Permafresh LKS	(1854-26-8)
Permagard	(87-86-5)
Permagel	(12174-11-7)
Permanax 45	(26780-96-1)
Permanax 120	(793-24-8)
Permanax TQ	(26780-96-1)
Permanent Green Toner GT-376	(1328-53-6)
Permanent Orange	(3468-63-1)
Permanent Orange DN Toner	(3468-63-1)
Permanent Orange G	(3520-72-7)
Permanent Orange G Extra	(3520-72-7)
Permanent Orange GG	(3468-63-1)
Permanent Orange HD	(3468-63-1)
Permanent Orange HL	(12236-62-3)
Permanent Orange Toner RA-5650	(3468-63-1)
Permanent Pink	(2379-74-0)
Permanent Purple	(1325-82-2)
Permanent Purple Toner	(1325-82-2)
Permanent Red 4B	(5858-81-1)
Permanent Red BFR	(2814-77-9)
Permanent Red E3B	(1047-16-1)
Permanent Red E5B	(1047-16-1)
Permanent Red F	(2814-77-9)
Permanent Red F 6R	(5858-81-1)
Permanent Red F 3RK70	(2786-76-7)
Permanent Red F 5RK	(2786-76-7)
Permanent Red GG	(3468-63-1)
Permanent Red 4R	(2425-85-6)
Permanent Red R	(2814-77-9)
Permanent Red R Extra	(2814-77-9)
Permanent Red RG Extra	(2814-77-9)
Permanent Red Toner R	(2814-77-9)
Permanent Victoria Blue Toner	(1325-87-7)
Permanent White	(1314-13-2)
Permanent White	(7727-43-7)
Permanent Yellow	(10294-40-3)
Permanent Yellow DHG	(6358-85-6)
Permanent Yellow GHG	(6358-85-6)
Permanent Yellow 2K	(119-15-3)
Permanent Yellow, Lead Free	(6358-31-2)
Permanganate d'ammonium (French)	(13446-10-1)
Permanganate de potassium (French)	(7722-64-7)
Permanganate de sodium (French)	(10101-50-5)
Permanganate of potash (DOT)	(7722-64-7)
Permanganato amonico (Spanish)	(13446-10-1)
Permanganic acid, ammonium salt	(13446-10-1)
Permanganic acid, barium salt	(7787-36-2)
Permanganic acid, potassium salt	(7722-64-7)
Permanganic acid, sodium salt	(10101-50-5)
Permansa Orange	(3468-63-1)
Permansa Red	(2814-77-9)
Permapen	(1538-09-6)
Permasan	(87-86-5)
Permasect-25EC	(52645-53-1)
Permasect	(52645-53-1)
Permaton Orange XL 45-3015	(3468-63-1)
Permaton Red XL 20-7015	(2814-77-9)
Permatone Orange	(3468-63-1)
Permatox 100 (9CI)	(89286-97-5)

Permatox DP-2 **(87-86-5)**
Permatox Penta **(87-86-5)**
Permek N **(1338-23-4)**
(+-)-cis-Permethrin **(52341-33-0)**
1RS,cis-Permethrin **(52341-33-0)**
Permethrin **(52645-53-1)**
cis-Permethrin **(61949-76-6)**
Permetrin (Hungarian) **(52645-53-1)**
Permetrina (Portuguese) **(52645-53-1)**
Permicort **(50-23-7)**
Permital **(523-87-5)**
Permite **(87-86-5)**
Permitil **(69-23-8)**
Permitrene (Hungarian) **(52645-53-1)**
Permlastic **(9080-49-3)**
Permonid **(427-00-9)**
Pernaemon **(68-19-9)**
Pernaevit **(68-19-9)**
Pernipuron **(68-19-9)**
Pernithrene Blue BC **(130-20-1)**
Pernithrene Blue RS **(81-77-6)**
Pernithrene Brilliant Green GG **(128-58-5)**
Pernithrene Olive R **(2379-81-9)**
Pernox **(52-86-8)**
Perone **(7722-84-1)**
Perone 30 **(7722-84-1)**
Perone 35 **(7722-84-1)**
Perone 50 **(7722-84-1)**
Perosin **(12122-67-7)**
Perosin 75B **(12122-67-7)**
Perossido di benzoile (Italian) **(94-36-0)**
Perossido di butile terziario (Italian) **(110-05-4)**
Perossido di idrogeno (Italian) **(7722-84-1)**
Perovex **(57-63-6)**
Peroxan **(7722-84-1)**
Peroxi-2-etilhexanoato de terc-butilo (Spanish) **(62695-55-0)**
Peroxiacetato de terc-butilo (Spanish) **(107-71-1)**
Peroxicrotonato de terc-butilo (Spanish) **(23474-91-1)**
Peroxide **(7722-84-1)**
Peroxide, acetyl benzoyl **(644-31-5)**
Peroxide, acetyl benzoyl, Maximum concentration 45% in solution **(644-31-5)**
Peroxide, acetyl cyclohexylsulfonyl (8CI,9CI) **(3179-56-4)**
Peroxide, acetyl cyclohexylsulfonyl, More than 82% wetted with less than 12% water **(3179-56-4)**
Peroxide, acetyl cyclohexylsulfonyl, Not more than 32% in solution **(3179-56-4)**
Peroxide, acetyl cyclohexylsulfonyl, Not more than 82% wetted with not less than 12% water **(3179-56-4)**
Peroxide, bis(3-carboxypropionyl) **(123-23-9)**
Peroxide, bis(p-chlorobenzoyl) **(94-17-7)**
Peroxide, bis(p-chlorobenzoyl)-, Not more than 52% as a paste or in solution **(94-17-7)**
Peroxide, bis(2,4-dichlorobenzoyl)-, More than 75% with water **(133-14-2)**
Peroxide, bis(2,4-dichlorobenzoyl)-, Not more than 52% as a paste or in solution **(133-14-2)**
Peroxide, bis(2,4-dichlorobenzoyl)-, Not more than 75% with water **(133-14-2)**
Peroxide, bis(α,α-dimethylbenzyl) **(80-43-3)**
Peroxide, bis(2-methyl-1-oxopropyl) (9CI) **(3437-84-1)**
Peroxide, bis(1-oxodecyl) **(762-12-9)**
Peroxide, bis(1-oxododecyl)- **(105-74-8)**
Peroxide, bis(1-oxononyl) (9CI) **(762-13-0)**
Peroxide, bis(1-oxooctyl) (9CI) **(762-16-3)**
Peroxide, bis(1-oxopropyl) **(3248-28-0)**
Peroxide, tert-butyl α,α-dimethylbenzyl **(3457-61-2)**
Peroxide, sec-butylidenebis(tert-butyl (8CI) **(2167-23-9)**
Peroxide, cyclohexylidenebis((1,1-dimethylethyl) (9CI) **(3006-86-8)**
Peroxide, dibenzoyl **(94-36-0)**
Peroxide, 1,1-dimethylethyl 1-methyl-1-phenylethyl (9CI) **(3457-61-2)**
Peroxide, 1-hydroperoxycyclohexyl 1-hydroxycyclohexyl **(78-18-2)**
Peroxide, (1-methylpropylidene)bis((1,1-dimethylethyl) (9CI) **(2167-23-9)**
Peroxide, octanoyl **(762-16-3)**
Peroxide, (1,3-phenylenebis(1-methylethylidene))bis((1,1-dimethylethyl) (9CI) **(2212-81-9)**
Peroxide, (1,4-phenylenebis(1-methylethylidene))bis((1,1-dimethylethyl) (9CI) **(2781-00-2)**
Peroxide, (phenylenebis(1-methylethylidene))bis(1,1-dimethylethyl)- **(25155-25-3)**
Peroxide, (p-phenylenediisopropylidene)bis(tert-butyl **(2781-00-2)**
Peroxide, (phenylenediisopropylidene)bis(tert-butyl **(25155-25-3)**
Peroxide, (m-phenylenediisopropylidene)bis(tert-butyl (8CI) **(2212-81-9)**
Peroxide, (3,3,5-trimethylcyclohexylidene)bis(tert-butyl (8CI) **(6731-36-8)**
Peroxide, (3,3,5-trimethylcyclohexylidene)bis(tert-butyl-, Not more than 57% in solution **(6731-36-8)**
Peroxide, (3,3,5-trimethylcyclohexylidene)bis(tert-butyl-, Not more than 58% with inert solid **(6731-36-8)**
Peroxide, (3,3,5-trimethylcyclohexylidene)bis((1,1-dimethylethyl)-(9CI) **(6731-36-8)**
Peroxidicarbonato de dibencilo (Spanish) **(2144-45-8)**
Peroxidicarbonato de di-(4-terc-butilciclohexilo) (Spanish) **(15520-11-3)**
Peroxidicarbonato de di-n-butilo (Spanish) **(16215-49-9)**
Peroxidicarbonato de dicetilo (Spanish) **(26322-14-5)**
Peroxidicarbonato de diciclohexilo (Spanish) **(1561-49-5)**
Peroxidicarbonato de diestearilo (Spanish) **(52326-66-6)**
Peroxidicarbonato de dietilo (Spanish) **(14666-78-5)**
Peroxidicarbonato de dimiristilo (Spanish) **(53220-22-7)**
Peroxido de acetil ciclohexano sulfonilo (Spanish) **(3179-56-4)**
Peroxido de diacetonalcohol (Spanish) **(54693-46-8)**
Peroxido de di-(1-hidroxiciclohexilo) (Spanish) **(2407-94-5)**
Peroxido de diisobutirilo (Spanish) **(3437-84-1)**
Peroxido de dipropionilo (Spanish) **(3248-28-0)**
Peroxido de estroncio (Spanish) **(1314-18-7)**
Peroxido de metilisobutilcetona (Spanish) **(37206-20-5)**
1,4-Peroxido-p-menthene-2 **(512-85-6)**
Peroximon F 40 **(2781-00-2)**
Peroximon F 100 **(2781-00-2)**
Peroxoacetic acid **(79-21-0)**
Peroxyacetate de tert-butyle (French) **(107-71-1)**
Peroxyacetic-acid **(79-21-0)**
Peroxyacetic acid (8CI) **(79-21-0)**
Peroxyacetic acid, More than 43% with more than 6% hydrogen peroxide (DOT) **(79-21-0)**
Peroxyacetic acid, t-butyl ester **(107-71-1)**
Peroxyacetic acid, tert-butyl ester (8CI) **(107-71-1)**
Peroxyacetic acid, tert-butyl ester, More than 76% in solution **(107-71-1)**
Peroxyacetic acid, tert-butyl ester, More than 52% to a maximum concentration of 76% **(107-71-1)**
Peroxyacetic acid, tert-butyl ester, Not more than 52% in solution **(107-71-1)**
Peroxyacetyl-nitrate **(2278-22-0)**
Peroxybenzoic acid, tert-butyl ester **(614-45-9)**
Peroxybenzoic acid, tert-butyl ester, Not more than 75% in solution **(614-45-9)**
Peroxybenzoic acid, tert-butyl ester, Not more than 50% with inert inorganic solid **(614-45-9)**
Peroxybenzoic acid, m-chloro- (8CI) **(937-14-4)**
Peroxybenzoic acid, m-chloro-, Maximum concentration 86% **(937-14-4)**
Peroxybenzoic acid, 1,1,4,4-tetramethyltetramethylene ester **(2618-77-1)**
Peroxybenzoic acid, 1,1,4,4-tetramethyltetramethylene ester (8CI) **(2618-77-1)**
Peroxybenzoic acid, 1,1,4,4-tetramethyltetramethylene ester, Not more than 82% with water **(2618-77-1)**
Peroxybenzoyl nitrate **(32368-69-7)**
Peroxycarbonic acid, OO-tert-butyl O-isopropyl ester **(2372-21-6)**
Peroxycrotonate de tert-butyle (French) **(23474-91-1)**

Peroxycrotonic acid, tert-butyl ester, Not more than 76% in solution **(23474-91-1)**
Peroxyde d'acetyle et de cyclohexane sulfonyle (French) **(3179-56-4)**
Peroxyde de baryum (French) **(1304-29-6)**
Peroxyde de benzoyle (French) **(94-36-0)**
Peroxyde de bis (hydroxy-1 cyclohexyle) (French) **(2407-94-5)**
Peroxyde de butyle tertiaire (French) **(110-05-4)**
Peroxyde de diacetone-alcool (French) **(54693-46-8)**
Peroxyde de diisobutyryle (French) **(3437-84-1)**
Peroxyde de lauroyle (French) **(105-74-8)**
Peroxyde de methylisobutylcetone (French) **(37206-20-5)**
Peroxyde de plomb (French) **(1309-60-0)**
Peroxyde de strontium (French) **(1314-18-7)**
Peroxyde d'hydrogene (French) **(7722-84-1)**
Peroxydicarbonate de bis (tert-butyl-4 cyclohexyle) (French) **(15520-11-3)**
Peroxydicarbonate de cetyle (French) **(26322-14-5)**
Peroxydicarbonate de dibenzyle (French) **(2144-45-8)**
Peroxydicarbonate de dicyclohexyle (French) **(1561-49-5)**
Peroxydicarbonate de dimyristyle (French) **(53220-22-7)**
Peroxydicarbonate de di-n-butyle (French) **(16215-49-9)**
Peroxydicarbonate d'ethyle (French) **(14666-78-5)**
Peroxydicarbonate d'isopropyle (French) **(105-64-6)**
Peroxydicarbonate d'octadecyle (French) **(52326-66-6)**
Peroxydicarbonic acid, bis(4-(1,1-dimethylethyl)cyclohexyl) ester (9CI) **(15520-11-3)**
Peroxydicarbonic acid, bis(2-ethylhexyl) ester **(16111-62-9)**
Peroxydicarbonic acid, bis(1-methylethyl) ester **(105-64-6)**
Peroxydicarbonic acid, dibenzyl ester **(2144-45-8)**
Peroxydicarbonic acid, dibenzyl ester, More than 87% with water **(2144-45-8)**
Peroxydicarbonic acid, di-sec-butyl ester **(19910-65-7)**
Peroxydicarbonic acid, dibutyl ester, More than 52% in solution **(16215-49-9)**
Peroxydicarbonic acid, dibutyl ester, Not more than 27% in solution **(16215-49-9)**
Peroxydicarbonic acid, dibutyl ester, Not more than 52% in solution **(16215-49-9)**
Peroxydicarbonic acid, dicetyl ester **(26322-14-5)**
Peroxydicarbonic acid, dicyclohexyl ester (8CI,9CI) **(1561-49-5)**
Peroxydicarbonic acid, dicyclohexyl ester, Not more than 91% with water **(1561-49-5)**
Peroxydicarbonic acid, diethyl ester, More than 27% in solution **(14666-78-5)**
Peroxydicarbonic acid, diethyl ester, Not more than 27% in solution **(14666-78-5)**
Peroxydicarbonic acid, di(2-ethylhexyl) ester **(16111-62-9)**
Peroxydicarbonic acid, di(2-ethylhexyl) ester, Not more than 77% in solution **(16111-62-9)**
Peroxydicarbonic acid, dihexadecyl ester (8CI,9CI) **(26322-14-5)**
Peroxydicarbonic acid, dihexadecyl ester, Not more than 42% in water **(26322-14-5)**
Peroxydicarbonic acid, diisopropyl ester **(105-64-6)**
Peroxydicarbonic acid, diisopropyl ester, Not more than 52% in solution **(105-64-6)**
Peroxydicarbonic acid, diisotridecyl ester **(82065-80-3)**
Peroxydicarbonic acid, dimyristyl ester **(53220-22-7)**
Peroxydicarbonic acid, dioctadecyl ester, Not more than 85% with stearyl alcohol **(52326-66-6)**
Peroxydicarbonic acid, dipropyl ester **(16066-38-9)**
Peroxydicarbonic acid, ditetradecyl ester (9CI) **(53220-22-7)**
Peroxydicarbonic acid, ditetradecyl ester, Not more than 22% in water **(53220-22-7)**
1,1'-Peroxydicyclohexanol **(2407-94-5)**
Peroxydisulfuric acid, diammonium salt **(7727-54-0)**
Peroxydisulfuric acid, dipotassium salt **(7727-21-1)**
Peroxydisulfuric acid, disodium salt **(7775-27-1)**
Peroxyhexanoic acid, 2-ethyl-, tert-pentyl ester **(686-31-7)**
Peroxyhexanoic acid, 2-ethyl-, 1,1,4,4-tetramethyltetramethylene ester **(13052-09-0)**
Peroxyisobutyric acid, tert-butyl ester, More than 77% in solution **(109-13-7)**
Peroxyisobutyric acid, tert-butyl ester, Not more than 52% in solution **(109-13-7)**
Peroxyisobutyric acid, tert-butyl ester, More than 52% but not more than 77% in solution **(109-13-7)**
Peroxylauric acid (8CI) **(2388-12-7)**
Peroxyl radical **(3170-83-0)**
Peroxymonosulfuric acid, monopotassium salt (8CI,9CI) **(10058-23-8)**
Peroxyneodecanoic acid, tert-butyl ester **(26748-41-4)**
Peroxyneodecanoic acid, tert-butyl ester, Not more than 77% in solution **(26748-41-4)**
Peroxyphthalic acid, di-tert-butyl ester (8CI) **(2155-71-7)**
Peroxypivalic acid, tert-butyl ester **(927-07-1)**
Peroxypivalic acid, tert-butyl ester, Not more than 77% in solution **(927-07-1)**
Peroxypropionyl nitrate **(5796-89-4)**
Peroyxde de dipropionyle (French) **(3248-28-0)**
Perozin **(12122-67-7)**
Perozine **(12122-67-7)**
Perphenazin **(58-39-9)**
Perphenazine **(58-39-9)**
Perphenazine acetate **(84-06-0)**
Perphinol 45/100 **(9004-99-3)**
Perquietil **(57-53-4)**
Persadox **(94-36-0)**
Persantin **(58-32-2)**
Persantinat **(637-07-0)**
Persantine **(58-32-2)**
Persec **(127-18-4)**
Persia-Perazol **(106-46-7)**
Persian Berry Lake **(518-82-1)**
Persian Orange **(633-96-5)**
Persian Orange Lake **(633-96-5)**
Persian Orange X **(633-96-5)**
Persian Red **(18454-12-1)**
Persicol **(104-67-6)**
Persisten **(702-54-5)**
Persistol **(51-18-3)**
Persistol HO 1/193 **(51-18-3)**
Persistol HOE 1/193 **(51-18-3)**
Perskleran **(50-55-5)**
Perspex **(9011-14-7)**
Persulfate d'ammonium (French) **(7727-54-0)**
Persulfate de sodium (French) **(7775-27-1)**
Perthane **(72-56-0)**
Pertofran **(50-47-5)**
Pertonal **(62-44-2)**
Pertranquil **(57-53-4)**
Pertrofane **(50-47-5)**
Peruscabin **(120-51-4)**
Peruviol **(142-50-7)**
Pervagal **(50-34-0)**
Pervitin **(300-42-5)**
Perylene **(198-55-0)**
3,10-Perylenedione, 1,2,12a,12b-tetrahydro-1,4,9,12a-tetra-hydroxy-, (1α,12aβ,12bα)-(+)- **(88899-62-1)**
Perylene, methyl- (9CI) **(64031-91-0)**
Perylene, 3-methyl- (8CI,9CI) **(24471-47-4)**
Perylene, 1-methyl- (6CI,7CI,9CI) **(10350-33-1)**
Perylene, 3-nitro- **(20589-63-3)**
Perylenetetracarboxylic acid dianhydride **(128-69-8)**
3,4,9,10-Perylenetetracarboxylic acid diimide **(81-33-4)**
3,4:9,10-Perylenetetracarboxylic anhydride **(128-69-8)**
Perylenetetracarboxylic anhydride **(128-69-8)**
3,4,9,10-Perylenetetracarboxylic dianhydride **(128-69-8)**
Perylene-3,4,9,10-tetracarboxylic 3,4:9,10-dianhydride **(128-69-8)**
Perylene-3,4,9,10-tetracarboxylic dianhydride **(128-69-8)**
3,4,9,10-Perylenetetracarboxylic 3,4:9,10-dianhydride (8CI) **(128-69-8)**
3,4,9,10-Perylenetetracarboxylic 3,4:9,10-diimide (8CI) **(81-33-4)**

Perylimid **(81-33-4)**
Perylo(3,4-cd:9,10-c'd')dipyran-1,3,8,10-tetrone (9CI) **(128-69-8)**
Pestan **(2595-54-2)**
Z 3 (Pesticide) **(140-89-6)**
Pestmaster **(106-93-4)**
Pestmaster EDB-85 **(106-93-4)**
Pestmaster (Obs.) **(74-83-9)**
Peston XV **(371-86-8)**
Pestox **(152-16-9)**
Pestox 3 **(152-16-9)**
Pestox 14 **(115-26-4)**
Pestox 15 **(371-86-8)**
Pestox 66 **(152-16-9)**
Pestox 101 **(311-45-5)**
Pestox III **(152-16-9)**
Pestox IV **(115-26-4)**
Pestox Plus **(56-38-2)**
Pestox XIV **(115-26-4)**
Pestox XV **(371-86-8)**
Petantin hydrochloride **(50-13-5)**
Petasitenine **(60102-37-6)**
Petasitenine (Neutral) **(60102-37-6)**
Petavlon **(5534-95-2)**
Petazol **(54-95-5)**
Petcoal 140 **(64742-16-1)**
Petcoal LX **(64742-16-1)**
Peterphyllin **(317-34-0)**
Petezol **(54-95-5)**
Pethidine **(57-42-1)**
Pethidine chloride **(50-13-5)**
Pethidine, hydrochloride **(50-13-5)**
Pethidineter **(57-42-1)**
Pethion **(56-38-2)**
Petidin **(50-13-5)**
Petimin **(14987-04-3)**
Petinimid **(77-67-8)**
Petinutin **(77-41-8)**
Petitgrain Oil **(8014-17-3)**
Petitgrain Oil Saponified **(8014-17-3)**
Petnidan **(77-67-8)**
Petogasrin **(5534-95-2)**
Petrac ZN-41 **(557-05-1)**
Petrazole **(54-95-5)**
Petrichloral **(78-12-6)**
Petrichloralum (Latin) **(78-12-6)**
Petricloral (Spanish) **(78-12-6)**
Petrisul **(80-35-3)**
Petrofin 100 **(9003-29-6)**
Petrogalar **(8012-95-1)**
Petrohol **(67-63-0)**
Petrol **(8002-05-9)**
Petrol (DOT) **(8006-61-9)**
Petrol Orange Y **(842-07-9)**
Petrol (VAN) **(86290-81-5)**
Petrol Yellow C **(8003-22-3)**
Petrol Yellow WT **(60-11-7)**
Petrolatum **(8009-03-8)**
Petrolatum Amber **(8009-03-8)**
Petrolatum, Liquid **(8012-95-1)**
Petrolatum USP **(8009-03-8)**
Petrolatum White **(8009-03-8)**
Petrolatum acid sulfonate **(61789-85-3)**
Petrolatum sulfonic acids, sodium salts **(68918-07-0)**
Petroleum-Derived Naphtha **(8030-30-6)**
Petroleum **(8002-05-9)**
Petroleum Asphalt **(8052-42-4)**
Petroleum Benzin **(8030-30-6)**
Petroleum Bitumen **(8052-42-4)**
Petroleum Crude, Crude Oil [UN 1267] **(8002-05-9)**
Petroleum Distillates (Naphtha) **(8030-30-6)**
Petroleum Gas Liquefied **(68476-85-7)**
Petroleum Gases **(68476-85-7)**
Petroleum Gases, Liquefied **(68476-85-7)**
Petroleum Jelly **(8009-03-8)**
Petroleum Naphtha (DOT) **(8030-30-6)**
Petroleum Oil [UN 1270] **(8002-05-9)**
Petroleum Pitch **(68187-58-6)**
Petroleum Pitch **(8052-42-4)**
Petroleum Roofing Tar **(8052-42-4)**
Petroleum Spirit [UN 1271] **(8032-32-4)**
Petroleum coke, Uncalcined (DOT) **(64741-79-3)**
(Petroleum) decene **(64741-72-6)**
Petroleum distillates, Clay-treated heavy naphthenic **(64742-44-5)**
Petroleum distillates, Clay-treated light naphthenic **(64742-45-6)**
Petroleum distillates, Hydrotreated heavy naphthenic **(64742-52-5)**
Petroleum distillates, Solvent-dewaxed heavy paraffinic **(64742-65-0)**
(Petroleum) dodecene **(64741-72-6)**
Petroleum ether (DOT) **(8032-32-4)**
Petroleum gases, Liquefied [UN 1075] **(74-98-6)**
(Petroleum) heptene **(64741-72-6)**
(Petroleum) nonene **(64741-72-6)**
(Petroleum) octene **(64741-72-6)**
Petroleum products, C5-12, Reclaimed, Wastewater treatment **(68956-70-7)**
Petroleum products, Refinery gases **(68607-11-4)**
Petroleum spirits **(64475-85-0)**
Petroleum sulfonates **(61789-85-3)**
Petroleumsulfonate, sodium salt **(68608-26-4)**
Petroleum sulfonic acid, monosodium salt **(68608-26-4)**
Petroleum sulfonic acid, sodium salt **(68608-26-4)**
Petroselinic acid **(593-39-5)**
Petrosin 150 **(64742-16-1)**
Petrosin 80 **(64742-16-1)**
Petrosin K **(64742-16-1)**
Petrosin PR 120 **(64742-16-1)**
Petrosin (VAN) **(64742-16-1)**
Petrothene **(9002-88-4)**
Petrothene LB 861 **(9002-88-4)**
Petrothene LC 731 **(9002-88-4)**
Petrothene LC 941 **(9002-88-4)**
Petrothene NA 219 **(9002-88-4)**
Petrothene NA 227 **(9002-88-4)**
Petrothene XL 6301 **(9002-88-4)**
Pettitgrain Oil **(8014-17-3)**
Petydyna (Polish) **(57-42-1)**
Petzinol **(79-01-6)**
Pevaryl **(24169-02-6)**
Pevikon C 870 **(9003-22-9)**
Peviston **(9003-39-8)**
Peviton **(59-67-6)**
Peyote **(11006-96-5)**
Peyrone's Chloride **(15663-27-1)**
Pfefferminz Oel (German) **(8006-90-4)**
Pfiklor **(7447-40-7)**
Pfizer 1393 **(91-33-8)**
Pfizer-E **(643-22-1)**
Pfizerpen **(113-98-4)**
Pfizerpen A **(69-53-4)**
Pfizerpen-AS **(6130-64-9)**
Pfizerpen VK **(132-98-9)**
Pflanzol **(58-89-9)**
Ph/778 **(712-68-5)**
Phacetur **(63-98-9)**
Phaldrone **(302-17-0)**
Phalloidin **(17466-45-4)**

Phalloidine (17466-45-4)
Phalloin (28227-92-1)
Phaltan (133-07-3)
Phanantin (57-41-0)
Phanatine (57-41-0)
Phanodorm (52-31-3)
Phanodorn (52-31-3)
Phargan (60-87-7)
Pharlon (979-32-8)
Pharmacillin (61-33-6)
Pharmacine Yellow R (6375-55-9)
Pharmagel A (9000-70-8)
Pharmagel AdB (9000-70-8)
Pharmagel B (9000-70-8)
Pharmaglo Red G (10169-02-5)
Pharmamedrine (60-13-9)
Pharmanil Scarlet Y (3567-65-5)
Pharmanthrene Golden Yellow (128-66-5)
Pharmasorb-Colloidal (12174-11-7)
Pharmatex Yellow G (6375-55-9)
Pharmazoid Red KB (95-79-4)
Pharmedrine (60-13-9)
Pharmetten (50-06-6)
Pharoid (40596-69-8)
Pharos 100.1 (9003-55-8)
Phaseolunatin (554-35-8)
Phaseomannite (87-89-8)
Phaseomannitol (87-89-8)
Phasolon (2310-17-0)
Phe-Mer-Nite (55-68-5)
Phebuzin (50-33-9)
Phebuzine (50-33-9)
Phedoxe (300-42-5)
Phedrisox (300-42-5)
Phelazin (156-51-4)
Phellandrene, β (555-10-2)
α-Phellandrene (99-83-2)
β-Phellandrene (555-10-2)
Phemeride (121-54-0)
Phemerol (121-54-0)
Phemerol chloride (121-54-0)
Phemersol chloride (121-54-0)
Phemetone (115-38-8)
Phemithyn (121-54-0)
Phemiton (115-38-8)
Phemitone (115-38-8)
Phen-Bar (50-06-6)
Phen-Buta-Vet (50-33-9)
Phenacalum (63-98-9)
Phenacemide (63-98-9)
Phenacemidum (Latin) (63-98-9)
Phenacereum (63-98-9)
Phenacet (62-44-2)
Phenacetaldehyde dimethyl acetal (101-48-4)
Phenacetin (62-44-2)
para-Phenacetin (62-44-2)
Phenacetine (62-44-2)
Phenacetinum (62-44-2)
Phenacetur (63-98-9)
Phenacetylcarbamide (63-98-9)
Phenacetyl chloride (103-80-0)
Phenacetylurea (63-98-9)
Phenachlor (88-06-2)
Phenacide (8001-35-2)
Phenacite (13598-00-0)
Phenacite (15191-85-2)
Phenacitin (62-44-2)
Phenacon (62-44-2)
Phenactyl (50-53-3)
Phenacyl bromide [UN 2645] (70-11-1)
Phenacyl chloride (OSHA) (532-27-4)
Phenadone (76-99-3)
Phenador-X (92-52-4)
Phenadoxone (467-84-5)
Phenadoxonum (Latin) (467-84-5)
Phenaemal (50-06-6)
Phenakite (15191-85-2)
Phenakite (8CI,9CI) (13598-00-0)
Phenalco (55-68-5)
1H-Phenalene (9CI) (203-80-5)
Phenalene (8CI) (203-80-5)
Phenaleno(1,9-gh)quinoline (189-92-4)
Phenalenone (548-39-0)
1H-Phenalen-1-one (9CI) (548-39-0)
Phenalen-1-one (8CI) (548-39-0)
Phenalgene (103-84-4)
Phenalgin (103-84-4)
Phenalone (2152-34-3)
Phenalzine (156-51-4)
Phenalzine dihydrogen sulfate (156-51-4)
Phenalzine hydrogen sulphate (156-51-4)
Phenamine (60-13-9)
Phenamine Black BCN-CF (1937-37-7)
Phenamine Black CL (1937-37-7)
Phenamine Black E (1937-37-7)
Phenamine Black E 200 (1937-37-7)
Phenamine Black RW (2429-83-6)
Phenamine Blue BB (2602-46-2)
Phenamine Brilliant Blue 6B (2610-05-1)
Phenamine Brown D 3G (2586-58-5)
Phenamine Dark Green B (3626-28-6)
Phenamine Fast Brown T (2429-81-4)
Phenamine Fast Brown TWC (2429-81-4)
Phenamine Fast Scarlet 4BGP (93-00-5)
Phenamine Purpurine 4B (992-59-6)
Phenamine Scarlet 3B (6358-29-8)
Phenamine Sky Blue A (2429-74-5)
Phenamine Viscose Black RR (6428-31-5)
Phenamiphos (22224-92-6)
Phenamizol (490-55-1)
Phenamizole (490-55-1)
Phenampromid (129-83-9)
Phenampromide (129-83-9)
Phenampromidum (Latin) (129-83-9)
9,10-Phenanthraquinone (84-11-7)
Phenanthren (German) (85-01-8)
2-Phenanthrenamine, 7-nitro- (9CI) (62245-47-0)
Phenanthrene (85-01-8)
9-Phenanthrenecarbonitrile (9CI) (2510-55-6)
1-Phenanthrenecarboxaldehyde, 1,2,3,4,4a,9,10,10a-octahydro-1,4a-dimethyl-7-(1-methylethyl)-, (1S-(1α,4aα,10aβ))- (9CI) (24035-50-5)
1-Phenanthrenecarboxylic acid, 1,2,3,4,4a,4b,5,6,10,10a-decahydro-1,4a-dimethyl-7-(1-methylethyl)-, sodium salt, (1R-(1α,4aβ,4bα,10aα))- (9CI) (14351-66-7)
1-Phenanthrenecarboxylic acid, 1,2,3,4,4a,4b,5,9,10,10a-decahydro-1,4a-dimethyl-7-(1-methylethyl)-, (1R-(1α,4aβ,4bα,10aα))- (9CI) (79-54-9)
1-Phenanthrenecarboxylic acid, 1,2,3,4,4a,4b,5,6,7,8,10,10a-dodecahydro-1,4a-dimethyl-7-(1-methylethyl)-, methyl ester (9CI) (67893-02-1)
1-Phenanthrenecarboxylic acid, 1,2,3,4,4a,4b,5,6,7,9,10,10a-dodecahydro-1,4a-dimethyl-7-(1-methylethyl)-, 1,2,3-propanetriyl ester, (1R-(1α,4aβ,4bα,10aα))- (9CI) (125-93-9)
1-Phenanthrenecarboxylic acid, 1,2,3,4,4a,4b,5,6,7,9,10,10a-dodecahydro-1,4a-dimethyl-7-(1-methylethylidene)-, methyl ester, (1R-(1α,4aβ,4bα,10aα))- (9CI) (3310-97-2)

1-Phenanthrenecarboxylic acid, chloro-1,2,3,4,4a,9,10,10a-octahydro-1,4a-dimethyl-7-(1-methylethyl)-, (1R-(1α,4aβ,10aα))- (9CI) **(57055-38-6)**
1-Phenanthrenecarboxylic acid, 1,2,3,4,4a,5,6,9,10,10a-decahydro-1,4a-dimethyl-7-(1-methylethyl)-, (1R-(1α,4aβ,10aα))- (9CI) **(1945-53-5)**
1-Phenanthrenecarboxylic acid, 6,8-dichloro-1,2,3,4,4a,9,10,10a-octahydro-1,4a-dimethyl-7-(1-methylethyl)-, (1R-(1α,4aβ,10aα))- (9CI) **(65281-77-8)**
1-Phenanthrenecarboxylic acid, dichloro-1,2,3,4,4a,9,10,10a-octa hydro-1,4a-dimethyl-7-(1-methylethyl)-, (1R-(1α,4aβ,10aα))- (9CI) **(57055-39-7)**
1-Phenanthrenecarboxylic acid, 7-ethenyl-1,2,3,4,4a,4b,5,6,7,8,10,10a-dodecahydro-1,4a,7-trimethyl-, (1R-(1α,4aβ,4bα,7α,10aα))- (9CI) **(5835-26-7)**
1-Phenanthrenecarboxylic acid, 7-ethenyl-1,2,3,4,4a,4b,5,6,7,8,10,10a-dodecahydro-1,4a,7-trimethyl-, methyl ester, (1R-(1α,4aβ,4bα,7α,10aα))- (9CI) **(1686-62-0)**
1-Phenanthrenecarboxylic acid, 7-ethenyl-1,2,3,4,4a,4b,5,6,7,9,10,10a-dodecahydro-1,4a,7-trimethyl-, (1R-(1α,4a β,4b α,7α,10a α))- **(471-74-9)**
1-Phenanthrenecarboxylic acid, 7-ethenyl-1,2,3,4,4a,4b,5,6,7,9,10,10a-dodecahydro-1,4a,7-trimethyl-, (1R-(1α,4aβ,4bα,7β,10aα))- (9CI) **(127-27-5)**
1-Phenanthrenecarboxylic acid, 7-ethenyl-1,2,3,4,4a,4b,5,6,7,9,10,10a-dodecahydro-1,4a,7-trimethyl-, methyl ester, (1R-(1α,4aβ,4bα,7β,10aα))- (9CI) **(3730-56-1)**
1-Phenanthrenecarboxylic acid, 1,2,3,4,4a,9,10,10a-octahydro-1,4a-dimethyl-, (1S-(1.α.,4a.α.,10a.α.))- (9CI) **(57345-30-9)**
1-Phenanthrenecarboxylic acid, 1,2,3,4,4a,9,10,10a-octahydro-1,4a-dimethyl-, methyl ester, (1R-(1α,4aβ,10aα))- (9CI) **(3650-04-2)**
1-Phenanthrenecarboxylic acid, 1,2,3,4,4a,9,10,10a-octahydro-1,4a-dimethyl-7-(1-methylethyl)-, (1S-(1α,4aα,10aβ))- (9CI) **(5155-70-4)**
1-Phenanthrenecarboxylic acid, 1,2,3,4,4a,9,10,10a-octahydro-1,4a-dimethyl-7-(1-methylethyl)-,methyl ester, (1R-(1α,4aβ,10aα))- (9CI) **(1235-74-1)**
1-Phenanthrenecarboxylic acid, tetradecahydro-9-hydroxy-7-(2-(2-(methylamino)ethoxy)-2-oxoethylidene)-1,4a,8-trimethyl-, methyl ester **(36150-73-9)**
Phenanthrene, 9-chloro **(947-72-8)**
Phenanthrene, 2-chloro- (8CI,9CI) **(24423-11-8)**
Phenanthrene, dichloro- (9CI) **(59116-88-0)**
Phenanthrene, 9,10-dichloro- (8CI,9CI) **(17219-94-2)**
Phenanthrene, 9,10-dihydro- (9CI) **(776-35-2)**
Phenanthrene, 9,10-dihydro-9,10-dioxo- **(84-11-7)**
Phenanthrene, 1,4-dimethyl **(22349-59-3)**
Phenanthrene, 2,7-dimethyl- (9CI) **(1576-69-8)**
Phenanthrene, 3,10-dimethyl- (9CI) **(66291-33-6)**
Phenanthrene, 3,6-dimethyl- (9CI) **(1576-67-6)**
Phenanthrene, 3,9-dimethyl- (9CI) **(66291-32-5)**
Phenanthrene, dimethyl- (9CI) **(29062-98-4)**
Phenanthrene, 1,3-dimethyl- (8CI,9CI) **(16664-45-2)**
Phenanthrene, 1,6-dimethyl- (8CI,9CI) **(20291-74-1)**
Phenanthrene, 1,7-dimethyl- (8CI,9CI) **(483-87-4)**
Phenanthrene, 1,8-dimethyl- (8CI,9CI) **(7372-87-4)**
Phenanthrene, 2,10-dimethyl- (8CI,9CI) **(2497-54-3)**
Phenanthrene, 2,3-dimethyl- (8CI,9CI) **(3674-65-5)**
Phenanthrene, 2,6-dimethyl- (8CI,9CI) **(17980-16-4)**
Phenanthrene, 2,9-dimethyl- (8CI,9CI) **(17980-09-5)**
Phenanthrene, dimethylnitro- (9CI) **(80182-27-0)**
9,10-Phenanthrenedione **(84-11-7)**
Phenanthrene, 9-ethyl- (9CI) **(3674-75-7)**
Phenanthrene, ethyl- (9CI) **(30997-38-7)**
Phenanthrene, 2-ethyl- (8CI,9CI) **(3674-74-6)**
Phenanthrene, 4,5-imino- **(203-65-6)**
Phenanthrene, 7-isopropyl-1-methyl- (8CI) **(483-65-8)**
1-Phenanthrenemethanamine, 1,2,3,4,4a,9,10,10a-octahydro-1,4a-dimethyl-7-(1-methylethyl)-, (1R-(1α,4aβ,10aα))- (9CI) **(1446-61-3)**
1-Phenanthrenemethanamine, 1,2,3,4,4a,9,10,10a-octahydro-1,4a-dimethyl-7-(1-methylethyl)-, acetate, (1R-(1α,4aβ,10aα))- (9CI) **(2026-24-6)**
1-Phenanthrenemethanol, tetradecahydro-1,4a-dimethyl-7-(1-methylethyl)- **(13393-93-6)**
Phenanthrene, methoxy- (9CI) **(61128-87-8)**
Phenanthrene, 1-methyl **(832-69-9)**
Phenanthrene, 2-methyl **(2531-84-2)**
Phenanthrene, 9-methyl **(883-20-5)**
Phenanthrene, 3-methyl- (9CI) **(832-71-3)**
Phenanthrene, 4-methyl- (9CI) **(832-64-4)**
Phenanthrene, methyl- (8CI,9CI) **(31711-53-2)**
Phenanthrene, 1-methyl-7-(1-methylethyl)- (9CI) **(483-65-8)**
Phenanthrene, methylnitro- (9CI) **(80191-44-2)**
Phenanthrene, 2-nitro **(17024-18-9)**
Phenanthrene, nitro- (9CI) **(68455-92-5)**
Phenanthrene, 1,2,3,4,5,6,7,8-octahydro- (8CI,9CI) **(5325-97-3)**
Phenanthrene, octyl- (9CI) **(76501-51-4)**
Phenanthrene, 1-phenyl- (8CI,9CI) **(4325-76-2)**
9,10-Phenanthrenequinone **(84-11-7)**
Phenanthrenequinone **(84-11-7)**
Phenanthrene, tetradecahydro- (9CI) **(5743-97-5)**
Phenanthrene, 1,2,3,4-tetramethyl **(4466-77-7)**
Phenanthrene, 2,4,5,7-tetramethyl- (9CI) **(7396-38-5)**
Phenanthrene, tetramethyl- (9CI) **(71607-70-0)**
Phenanthrene, 3,4,5,6-tetramethyl- (8CI,9CI) **(7343-06-8)**
Phenanthrene, 2,3,5-trimethyl- (8CI,9CI) **(3674-73-5)**
Phenanthrene, trimethyl- (8CI,9CI) **(30232-26-9)**
Phenanthrene, 1,2,4-trimethyl- (6CI,7CI,8CI,9CI) **(23189-64-2)**
9-Phenanthrenol **(484-17-3)**
Phenanthrenol (9CI) **(30774-95-9)**
3-Phenanthrenol, 4b,5,6,7,8,8a,9,10-octahydro-4b,8,8-trimethyl-, (4bS-trans)- (9CI) **(15340-76-8)**
Phenanthrenol, dinitro- (9CI) **(116212-00-1)**
Phenanthrenol, nitro- (9CI) **(116211-98-4)**
6-Phenanthridine **(229-87-8)**
Phenanthridine **(229-87-8)**
Phenanthridine, 1,2,3,4,7,8,9,10-octahydro-6-pentyl- (9CI) **(10594-03-3)**
Phenanthridine N-oxide **(14548-01-7)**
Phenanthridine, 5-oxide (9CI) **(14548-01-7)**
Phenanthridinium, 3,8-diamino-5-ethyl-6-phenyl-, bromide **(1239-45-8)**
Phenanthro(4,5-bcd)thiophene (8CI,9CI) **(30796-92-0)**
9-Phenanthrol **(484-17-3)**
Phenanthrol **(30774-95-9)**
1,10-Phenanthroline **(66-71-7)**
1,10-o-Phenanthroline **(66-71-7)**
5,6-Phenanthroline **(230-17-1)**
β-Phenanthroline **(66-71-7)**
o-Phenanthroline **(66-71-7)**
1,10-Phenanthroline, 2,9-dichloro-, copper complex (9CI) **(69742-55-8)**
Phenantoin **(50-12-4)**
Phenantrin **(85-01-8)**
Phenaphen **(62-44-2)**
Phenaphen plus **(62-44-2)**
Phenarol **(80-77-3)**
Phenarone **(63-98-9)**
Phenarsazine chloride **(578-94-9)**
Phenarsazine, 10-chloro-5,10-dihydro **(578-94-9)**
Phenasal **(50-65-7)**
Phenathyl **(50-53-3)**
Phenatine **(57-41-0)**
Phenatoine **(57-41-0)**
Phenatol **(2275-14-1)**
Phenatox **(8001-35-2)**
Phenazarsine chloride **(578-94-9)**

Phenazetin **(62-44-2)**
Phenazetina **(62-44-2)**
Phenazine **(92-82-0)**
2,3-Phenazinediamine (9CI) **(655-86-7)**
Phenazine, 2,3-diamino- (8CI) **(655-86-7)**
Phenazinium, 3,7-diamino-2,8-dimethyl-5-phenyl-, chloride **(477-73-6)**
Phenazite **(15191-85-2)**
Phenazo **(136-40-3)**
Phenazo Black BH **(2429-73-4)**
Phenazocine **(127-35-5)**
Phenazocinum (Latin) **(127-35-5)**
Phenazodine **(136-40-3)**
Phenazodine **(94-78-0)**
Phenazon **(1698-60-8)**
Phenazon **(60-80-0)**
Phenazone **(60-80-0)**
Phenazone (VAN) **(230-17-1)**
Phenazone (Pharmaceutical) **(60-80-0)**
Phenazonum **(60-80-0)**
Phenazopyridine **(94-78-0)**
Phenazopyridine hydrochloride **(136-40-3)**
Phenazopyridinium chloride **(136-40-3)**
Phenbenzamine **(961-71-7)**
Phenbenzamine hydrochloride **(2045-52-5)**
Phenbutazol **(50-33-9)**
Phencapton [NA 2783] **(2275-14-1)**
Phencen **(58-33-3)**
Phencyclidine **(77-10-1)**
Phencyclidine hydrochloride **(956-90-1)**
Phendal **(2597-03-7)**
Phendextro **(25523-97-1)**
(+)-Phendimetrazine **(634-03-7)**
Phendimetrazine **(634-03-7)**
Phendipham **(13684-63-4)**
Phendon **(103-90-2)**
Phene **(71-43-2)**
Phenedina **(62-44-2)**
Phenedrine **(300-62-9)**
Phenedrine **(60-13-9)**
Pheneene Germicidal Solution and Tincture **(8001-54-5)**
Phenegic **(92-84-2)**
Phenelzin **(156-51-4)**
Phenelzine **(51-71-8)**
Phenelzine acid sulfate **(156-51-4)**
Phenelzine bisulphate **(156-51-4)**
Phenelzine sulfate **(156-51-4)**
Phenelzine sulphate **(156-51-4)**
Phenemal **(50-06-6)**
Phenemalum **(57-30-7)**
(v-Phenenyltris(oxyethylene))tris(triethylammonium)iodide **(65-29-2)**
Phenergan **(60-87-7)**
Phenergan hydrochloride **(58-33-3)**
Phenesterin **(3546-10-9)**
Phenesterine **(3546-10-9)**
Phenestrin **(3546-10-9)**
Phenethanol **(60-12-8)**
Phenethecillin potassium **(132-93-4)**
Phenethicillin K **(132-93-4)**
Phenethicillin K salt **(132-93-4)**
Phenethicillin potassium salt **(132-93-4)**
Phenethidine **(156-43-4)**
β-Phenethybiguanide **(114-86-3)**
2-Phenethyl acetate **(103-45-7)**
Phenethyl acetate **(103-45-7)**
β-Phenethyl acetate **(103-45-7)**
2-Phenethyl alcohol **(60-12-8)**
Phenethyl-alcohol **(60-12-8)**
α-Phenethyl alcohol **(98-85-1)**
β-Phenethyl alcohol **(60-12-8)**
Phenethyl alcohol, p-amino- (8CI) **(104-10-9)**
Phenethyl alcohol, α,α-dimethyl **(100-86-7)**
Phenethyl alcohol, ar,ar-dimethyl- (6CI,7CI) **(27577-96-4)**
Phenethyl alcohol, α,α-dimethyl-, acetate **(151-05-3)**
Phenethyl alcohol, β-methyl **(1123-85-9)**
Phenethyl alcohol, p-nitro- (8CI) **(100-27-6)**
Phenethyl alcohol, p-nitro-, acetate (8CI) **(104-30-3)**
Phenethyl alcohol, propionate **(122-70-3)**
Phenethyl alcohol, p-toluenesulfonate (6CI) **(4455-09-8)**
Phenethyl alcohol, .α.,.α.,.β.-trimethyl- (6CI,7CI,8CI) **(3280-08-8)**
Phenethylamine **(64-04-0)**
β-Phenethylamine **(64-04-0)**
Phenethylamine, N-benzyl-N,α-dimethyl-, (+) **(156-08-1)**
Phenethylamine, N-benzyl-N,α-dimethyl-, hydrochloride, (+)- (8CI) **(5411-22-3)**
Phenethylamine, 4-bromo-2,5-dimethoxy-α-methyl-, hydrobromide, DL- **(53581-53-6)**
Phenethylamine, o-chloro-α,α-dimethyl- **(10389-73-8)**
Phenethylamine, p-chloro-α,α-dimethyl **(461-78-9)**
Phenethylamine, N-(3-chloropropyl)-α-methyl- **(17243-57-1)**
Phenethylamine, 2,5-dimethoxy-4-ethyl-α-methyl- **(15588-95-1)**
Phenethylamine, 2,5-dimethoxy-α-methyl-, hydrochloride (6CI,8CI) **(24973-25-9)**
Phenethylamine, N,α-dimethyl-, (+) **(537-46-2)**
Phenethylamine, α,α-dimethyl **(122-09-8)**
Phenethylamine, N,α-dimethyl-, hydrochloride **(300-42-5)**
Phenethylamine, N-(3,3-diphenylpropyl)-α-methyl **(390-64-7)**
Phenethylamine, N-ethyl-α-methyl- **(457-87-4)**
Phenethylamine, N-ethyl-α-methyl-m-(trifluoromethyl) **(458-24-2)**
Phenethylamine, β-hydroxy- **(7568-93-6)**
Phenethylamine, p-hydroxy- **(51-67-2)**
Phenethylamine, p-methoxy-α-methyl-, (+-)- **(23239-32-9)**
Phenethylamine, 3-methoxy-α-methyl-4,5-(methylenedioxy)- **(13674-05-0)**
Phenethylamine, α-methyl **(60-15-1)**
Phenethylamine, α-methyl, (+-) **(300-62-9)**
Phenethylamine, α-methyl-, (+) **(51-64-9)**
Phenethylamine, α-methyl-, d- **(51-64-9)**
Phenethylamine, α-methyl-, D- **(51-64-9)**
Phenethylamine, α-methyl-3-methoxy-4,5-(methylenedioxy)- **(13674-05-0)**
Phenethylamine, α-methyl-3,4-(methylenedioxy) **(4764-17-4)**
Phenethylamine, α-methyl-, sulfate (2:1), (+) **(51-63-8)**
Phenethylamine, α-methyl-, sulfate (2:1), (+-) **(60-13-9)**
Phenethylamine,α-methyl-, sulfate, (+) **(51-63-8)**
Phenethylamine, α-methyl-3,4,5-trimethoxy- **(1082-88-8)**
Phenethylamine, 3,4,5-trimethoxy **(54-04-6)**
Phenethylamine, 3,4,5-trimethoxy-α-methyl- **(1082-88-8)**
Phenethylamine, N,α,α-trimethyl **(100-92-5)**
β-Phenethyl-o-aminobenzoate **(133-18-6)**
Phenethyl anthranilate **(133-18-6)**
1-Phenethylbiguanide **(114-86-3)**
β-Phenethylbiguanide **(114-86-3)**
1-Phenethylbiguanide hydrochloride **(834-28-6)**
N'-β-Phenethylbiguanide hydrochloride **(834-28-6)**
N[1]-β-Phenethylbiguanide hydrochloride **(834-28-6)**
Phenethylbiguanide hydrochloride **(834-28-6)**
1-Phenethyl bromide **(585-71-7)**
2-Phenethyl bromide **(103-63-9)**
Phenethyl bromide **(103-63-9)**
α-Phenethyl bromide **(585-71-7)**
β-Phenethyl bromide **(103-63-9)**
Phenethylcarbamid (German) **(150-69-6)**
Phenethyldiguanide **(114-86-3)**
Phenethylene **(100-42-5)**
Phenethylene oxide **(96-09-3)**
N'-β-Phenethylformamidinylliminourea **(114-86-3)**

Phenethylhydrazine	(51-71-8)
Phenethylhydrazine sulfate (1:1)	(156-51-4)
Phenethylhydrazine sulphate	(156-51-4)
Phenethyl methyl ketone	(2550-26-7)
Phenethyl phenylacetate	(102-20-5)
1-(2-Phenethyl)-4-phenyl-4-acetoxypiperidine	(64-52-8)
1-Phenethyl-4-phenyl-4-piperidinol acetate (ester)	(64-52-8)
1-Phenethyl-2-picolinium bromide	(10551-21-0)
N-Phenethyl-α-picolinium bromide	(10551-21-0)
2-Phenethyl propionate	(122-70-3)
Phenethyl propionate	(122-70-3)
1-Phenethyl-4-n-propionylanilinopiperidine	(437-38-7)
N-Phenethyl-4-(n-propionylanilino)piperidine	(437-38-7)
Phenethyl α-toluate	(102-20-5)
Phenethyl 4-toluenesulfonate	(4455-09-8)
Phenethyl p-toluenesulfonate	(4455-09-8)
Phenethyl tosylate	(4455-09-8)
Pheneticillin potassium	(132-93-4)
p-Phenetidin	(156-43-4)
m-Phenetidine	(621-33-0)
p-Phenetidine	(156-43-4)
p-Phenetidine [UN 2311]	(156-43-4)
p-Phenetidine, N-acetyl-	(62-44-2)
m-Phenetidine, N,N-diethyl- (8CI)	(1864-92-2)
p-Phenetolcarbamid (German)	(150-69-6)
Phenetolcarbamide	(150-69-6)
p-Phenetolcarbamide	(150-69-6)
Phenetole	(103-73-1)
p-Phenetolecarbamide	(150-69-6)
Phenetole, β-chloro- (8CI)	(622-86-6)
Phenetole, 2,4-dinitro	(610-54-8)
Phenetole, m-methyl- (8CI)	(621-32-9)
Phenetole, p-nitro-	(100-29-8)
p-Phenetylurea	(150-69-6)
Phenformin	(114-86-3)
Phenformin HCl No. 9113	(834-28-6)
Phenformine	(114-86-3)
Phenformin, hydrochloride	(834-28-6)
Phenformix	(114-86-3)
Phenhydan	(630-93-3)
Phenhydren	(83-12-5)
Phenic acid	(108-95-2)
Phenicarb	(63-98-9)
Phenicarbazide	(103-03-7)
Phenicol	(93-54-9)
Phenidin	(62-44-2)
Phenidone	(92-43-3)
Phenidylate	(113-45-1)
Phenilone	(2152-34-3)
Phenin	(62-44-2)
Phenindione	(83-12-5)
Pheniramine	(86-21-5)
Pheniramine maleate	(132-20-7)
Phenisatin	(18869-73-3)
Phenisobromolate	(18181-80-1)
(+-)-Phenisopropylamine sulfate	(60-13-9)
Phenisopropylamine sulfate	(60-13-9)
Phenistan	(61-76-7)
Phenitoin	(57-41-0)
Phenitol	(55-68-5)
Phenitrothion	(122-14-5)
Phenkapton	(2275-14-1)
Phenkaptone	(2275-14-1)
Phenline	(156-51-4)
Phenmad	(62-38-4)
Phenmedipham	(13684-63-4)
Phenmediphame	(13684-63-4)
Phenmerzyl nitrate	(55-68-5)
Phenmethyl-trimethylammonium iodide	(4525-46-6)
Phenmetrazin	(134-49-6)
Phenmetrazine	(134-49-6)
Phenmiazine	(253-82-7)
Pheno Black EP	(1937-37-7)
Pheno Black SGN	(1937-37-7)
Pheno Blue 2B	(2602-46-2)
Pheno Fast Scarlet 4B	(6358-29-8)
Pheno Fast Scarlet 9B	(2610-11-9)
Pheno Fast Yellow 95	(1325-37-7)
Pheno Navy Blue	(2429-73-4)
Pheno Sky Blue 6BX	(2610-05-1)
Pheno Violet N	(2586-60-9)
Phenobal	(50-06-6)
Phenobal sodium	(57-30-7)
Phenobarbital	(50-06-6)
Phenobarbital elixir	(57-30-7)
Phenobarbital na	(57-30-7)
Phenobarbital sodium	(57-30-7)
Phenobarbital sodium salt	(57-30-7)
Phenobarbitone	(50-06-6)
Phenobarbitone sodium	(57-30-7)
Phenobarbitone sodium salt	(57-30-7)
Phenobarbituric acid	(50-06-6)
Phenobarbyl	(50-06-6)
Phenobenzoron	(3134-12-1)
Phenobenzuron	(3134-12-1)
Phenochlor	(1336-36-3)
Phenoclor	(1336-36-3)
Phenoclor DP6	(11096-82-5)
Phenoctyl	(362-29-8)
α-Phenodiazine	(253-66-7)
β-Phenodiazine	(253-52-1)
Phenodioxin	(262-12-4)
Phenodoxone	(467-84-5)
Phenodyn	(156-51-4)
Phenodyne	(156-51-4)
Phenodyne	(62-44-2)
Phenoformine hydrochloride	(834-28-6)
Phenohep	(67-72-1)
Phenol (ACGIH,DOT,OSHA)	(108-95-2)
Phenol, 4-Chloro-2-(phenylmethyl)- (9CI)	(120-32-1)
Phenol, Liquid or solution (liquid tar acid containing over 50% phenol) [UN 2821]	(108-95-2)
Phenol, Molten [UN 2312]	(108-95-2)
Phenol, Polymer with formaldehyde	(9003-35-4)
Phenol, Polymer with formaldehyde, bisphenol A and epichlorohydrin	(40216-08-8)
Phenol, Polymer with paraformaldehyde	(9003-35-4)
Phenol Red	(143-74-8)
Phenol, Solid [UN 1671]	(108-95-2)
Phenol, 2-acetamido-	(614-80-2)
Phenol, p-acetamido-	(103-90-2)
Phenol acetate	(122-79-2)
Phenol alcohol	(108-95-2)
Phenol, o-allyl	(1745-81-9)
Phenol, 4-allyl-2-methoxy	(97-53-0)
Phenol, 4-allyl-2-methoxy-, acetate	(93-28-7)
Phenol, 4-allyl-2-methoxy-, formate (Ester)	(10031-96-6)
Phenol, aluminum salt (9CI)	(15086-27-8)
Phenol, m-amino	(591-27-5)
Phenol, o-amino	(95-55-6)
Phenol, p-amino	(123-30-8)
Phenol, 2-amino-4-chloro-5-nitro	(6358-07-2)
Phenol, 2-amino-6-chloro-4-nitro- (9CI)	(6358-09-4)
Phenol, 2-amino-4-chloro-6-nitro- (8CI,9CI)	(6358-08-3)

Name	CAS No.
Phenol, o-chloro	**(95-57-8)**
Phenol, p-chloro	**(106-48-9)**
Phenol, chloro- (8CI,9CI)	**(25167-80-0)**
Phenol, p-chloro-, acetate	**(876-27-7)**
Phenol, 2-(5-chloro-2H-benzotriazol-2-yl)-4,6-bis(1,1-dimethylethyl)- (9CI)	**(3864-99-1)**
Phenol, 2-(5-chloro-2H-benzotriazol-2-yl)-6-(1,1-dimethylethyl)-4-methyl- (9CI)	**(3896-11-5)**
Phenol, m-chloro-, carbanilate (6CI)	**(16400-09-2)**
Phenol, 4-chloro-2-cyclopentyl- (8CI,9CI)	**(13347-42-7)**
Phenol, 5-chloro-2-(2,4-dichlorophenoxy)-	**(3380-34-5)**
Phenol, 4-chloro-2,6-dimethoxy- (9CI)	**(108545-00-2)**
Phenol, 4-chloro-3,5-dimethyl-	**(88-04-0)**
Phenol, 2-chloro-4-(1,1-dimethylethyl)- (9CI)	**(98-28-2)**
Phenol, 2-chloro-4-(1,1-dimethylethyl)-6-nitro- (9CI)	**(14593-28-3)**
Phenol, 2-chloro-4,5-dimethyl-, methyl carbamate (9CI)	**(671-04-5)**
Phenol, 2-chloro-4,6-dinitro	**(946-31-6)**
Phenol, 4-chloro-2,6-dinitro- (9CI)	**(88-87-9)**
Phenol, 4-chloro-2-(2,4-dinitroanilino)- (7CI,8CI)	**(6358-18-5)**
Phenol, 4-chloro-2-((2,4-dinitrophenyl)amino)- (9CI)	**(6358-18-5)**
Phenol, 4-chloro-2-ethyl	**(18979-90-3)**
Phenol, 2-chloro-4-ethyl- (8CI,9CI)	**(18980-00-2)**
Phenol, 2-chloro-4-heptyl- (8CI,9CI)	**(18980-06-8)**
Phenol, 4-chloro-2-heptyl- (8CI,9CI)	**(18979-96-9)**
Phenol, 2-chloro-4-methoxy- (8CI,9CI)	**(18113-03-6)**
Phenol, 4-chloro-2-methoxy- (8CI,9CI)	**(16766-30-6)**
Phenol, 2-(2-chloro-1-methoxyethoxy)-, methylcarbamate	**(51487-69-5)**
Phenol, 2-chloro-5-methyl	**(615-74-7)**
Phenol, (chloromethyl)- (9CI)	**(30915-79-8)**
Phenol, 2-chloro-4-methyl- (9CI)	**(6640-27-3)**
Phenol, 2-chloromethyl- (9CI)	**(68137-05-3)**
Phenol, 4-chloro-3-methyl- (9CI)	**(59-50-7)**
Phenol, chloromethyl- (9CI)	**(1321-10-4)**
Phenol, 3-(chloromethyl)-6-(1,1-dimethylethyl)-2,4-dimethyl- (9CI)	**(23500-79-0)**
Phenol, 4-chloro-5-methyl-2-nitro- (9CI)	**(7147-89-9)**
Phenol, 4-chloro-2-methyl-, sodium salt (9CI)	**(52106-86-2)**
Phenol, 2-chloro-4-nitro	**(619-08-9)**
Phenol, 3-chloro-4-nitro	**(491-11-2)**
Phenol, 2-chloro-6-nitro- (9CI)	**(603-86-1)**
Phenol, 4-chloro-2-nitro- (9CI)	**(89-64-5)**
Phenol, 5-chloro-2-nitro- (9CI)	**(611-07-4)**
Phenol, 2-chloro-4-nitro-, O-ester with O,O-dimethyl phosphorothioate	**(2463-84-5)**
Phenol, 3-chloro-4-nitro-, O-ester with O,O-dimethyl phosphorothioate	**(500-28-7)**
Phenol, 2-chloro-6-nitro-4-(1,1,3,3-tetramethylbutyl)- (8CI,9CI)	**(17199-21-2)**
Phenol, 2-chloro-4-nonyl-	**(60044-33-9)**
Phenol, 4-chloro-3-pentadecyl- (8CI,9CI)	**(6964-19-8)**
Phenol, 2-chloro-4-phenyl-	**(92-04-6)**
Phenol, 2-chloro-4-(phenylazo)	**(6657-05-2)**
Phenol, 4-chloro-2-(phenylmethyl)-, potassium salt (9CI)	**(35471-49-9)**
Phenol, 4-chloro-2-(phenylmethyl)-, sodium salt (9CI)	**(3184-65-4)**
Phenol, 2-chloro-4-(1,1,3,3-tetramethylbutyl)- (9CI)	**(17199-24-5)**
Phenol, 2-cyclohexyl- (9CI)	**(119-42-6)**
Phenol, 3-cyclohexyl- (9CI)	**(1943-95-9)**
Phenol, 4-cyclohexyl- (9CI)	**(1131-60-8)**
Phenol, cyclohexyl- (9CI)	**(26570-85-4)**
Phenol, o-cyclohexyl- (8CI)	**(119-42-6)**
Phenol, p-cyclohexyl- (8CI)	**(1131-60-8)**
Phenol, 2-cyclohexyl-4,6-dinitro	**(131-89-5)**
Phenol, 6-cyclohexyl-2,4-dinitro-	**(131-89-5)**
Phenol, 4-cyclohexyl-2,6-dinitro- (8CI,9CI)	**(4097-58-9)**
Phenol, 2-cyclohexyl-4,6-dinitro-, sodium salt (8CI,9CI)	**(130-60-9)**
Phenol, 4-cyclopentyl-	**(1518-83-8)**
Phenol, decyl- (9CI)	**(27157-66-0)**
Phenol, 2,4-diamino	**(95-86-3)**
Phenol, 2,4-diamino-, dihydrochloride	**(137-09-7)**
Phenol, 2,4-diamino-, hydrochloride (8CI,9CI)	**(29849-01-2)**
Phenol, 2,4-diamino-6-methyl- (9CI)	**(15872-73-8)**
Phenol, 2,4-diamino-6-methyl-, hydrochloride (9CI)	**(65879-44-9)**
Phenol, diamyl-	**(28652-04-2)**
Phenol, 2,4-dibromo	**(615-58-7)**
Phenol, 2,3-dibromo- (9CI)	**(57383-80-9)**
Phenol, 2,6-dibromo- (9CI)	**(608-33-3)**
Phenol, 3,4-dibromo- (9CI)	**(615-56-5)**
Phenol, dibromo- (9CI)	**(28514-45-6)**
Phenol, 2,5-dibromo- (8CI,9CI)	**(28165-52-8)**
Phenol, 3,5-dibromo- (8CI,9CI)	**(626-41-5)**
Phenol, 2,6-dibromo-4-chloro- (9CI)	**(5324-13-0)**
Phenol, 2,4-dibromo-6-chloro- (8CI,9CI)	**(4526-56-1)**
Phenol, dibromochloromethyl- (9CI)	**(86006-44-2)**
Phenol, 2,4-dibromo-6-methyl- (9CI)	**(609-22-3)**
Phenol, dibromomethyl- (9CI)	**(86006-42-0)**
Phenol, 2,6-dibromo-4-nitro	**(99-28-5)**
Phenol, 2,4-di-sec-butyl	**(1849-18-9)**
Phenol, 2,4-di-tert-butyl	**(96-76-4)**
Phenol, 2,6-di-tert-butyl	**(128-39-2)**
Phenol, 2,5-di-tert-butyl- (8CI)	**(5875-45-6)**
Phenol, 3,5-di-tert-butyl- (8CI)	**(1138-52-9)**
Phenol, 2,6-di-tert-butyl-4-ethyl- (8CI)	**(4130-42-1)**
Phenol, (2,2'-di-tert-butyl-4,4'-isopropylene)di	**(79-96-9)**
Phenol, 2,6-di-tert-butyl-4-methoxy- (8CI)	**(489-01-0)**
Phenol, 2,3-dichloro	**(576-24-9)**
Phenol, 2,4-dichloro	**(120-83-2)**
Phenol, 2,5-dichloro	**(583-78-8)**
Phenol, 2,6-dichloro	**(87-65-0)**
Phenol, 3,4-dichloro	**(95-77-2)**
Phenol, 3,5-dichloro	**(591-35-5)**
Phenol, dichloro-	**(25167-81-1)**
Phenol, 2,4-dichloro-, benzenesulfonate	**(97-16-5)**
Phenol, 3,4-dichloro-2,6-dimethoxy- (9CI)	**(35869-50-2)**
Phenol, dichloro-2,5-dimethoxy- (9CI)	**(108548-71-6)**
Phenol, dichloro-2,6-dimethoxy- (9CI)	**(75248-88-3)**
Phenol, 2,4-dichloro-3,5-dimethyl- (9CI)	**(133-53-9)**
Phenol, 2,6-dichloro-4-(1,1-dimethylethyl)- (9CI)	**(34593-75-4)**
Phenol, 2,4-dichloro-, O-ester with O-methyl isopropylphosphoramidothioate	**(299-85-4)**
Phenol, 3,4-dichloro-, O-ester with O-methyl methylphosphoramidothioate	**(18181-70-9)**
Phenol, 2,4-dichloro-, O-ester with O,O-diethyl phosphorothioate	**(97-17-6)**
Phenol, 3,5-dichloro-4-ethyl-2,6-dimethoxy- (9CI)	**(108545-02-4)**
Phenol, 4,5-dichloro-2-methoxy	**(2460-49-3)**
Phenol, 2,3-dichloro-4-methoxy- (9CI)	**(39542-65-9)**
Phenol, 3,4-dichloro-2-methoxy- (9CI)	**(77102-94-4)**
Phenol, 3,5-dichloro-2-methoxy- (9CI)	**(56680-89-8)**
Phenol, 3,5-dichloro-4-methoxy- (9CI)	**(56680-68-3)**
Phenol, 3,6-dichloro-2-methoxy- (9CI)	**(77102-93-3)**
Phenol, dichloro-2-methoxy- (9CI)	**(65724-16-5)**
Phenol, 2,4-dichloro-6-methoxy- (8CI,9CI)	**(16766-31-7)**
Phenol, 2,6-dichloro-4-methoxy- (7CI,8CI,9CI)	**(2423-72-5)**
Phenol, 2,4-dichloro-6-methyl- (9CI)	**(1570-65-6)**
Phenol, 2,6-dichloro-4-methyl- (9CI)	**(2432-12-4)**
Phenol, 2,4-dichloro-6-nitro	**(609-89-2)**
Phenol, 2,6-dichloro-4-nitro- (9CI)	**(618-80-4)**
Phenol, 4,5-dichloro-2-nitro- (9CI)	**(39224-65-2)**
Phenol, 2,5-dichloro-4-nitro- (8CI,9CI)	**(5847-57-4)**
Phenol, 2,4-dichloro-6-nitro-, sodium salt	**(64047-88-7)**
Phenol, 2,6-dichloro-4-octyl	**(73986-52-4)**
Phenol, 2,5-dichloro-, potassium salt (9CI)	**(68938-81-8)**
Phenol, 2,4-dichloro-, sodium salt (9CI)	**(3757-76-4)**
Phenol, 2,5-dichloro-, sodium salt (9CI)	**(52166-72-0)**
Phenol, 2,4-dichloro-6-(2,3,4,6-tetrachlorophenoxy)- (9CI)	**(94888-10-5)**
Phenol, 2,6-dichloro-4-(2,3,4,6-tetrachlorophenoxy)- (9CI)	**(90986-11-1)**

Phenol, 4,5-dichloro-2-(2,4,5-trichlorophenoxy) **(61639-90-5)**
Phenol, 2,4-dichloro-6-(2,4,6-trichlorophenoxy)- (9CI) **(94888-09-2)**
Phenol, 2,6-dichloro-4-(2,4,6-trichlorophenoxy)- (9CI) **(90986-10-0)**
Phenol, didodecyl- (9CI) **(25482-47-7)**
Phenol, diethyl- (8CI,9CI) **(26967-65-7)**
Phenol, m-(diethylamino) **(91-68-9)**
Phenol, 3-(diethylamino)- (9CI) **(91-68-9)**
Phenol, 5-(diethylamino)-4-methyl-2-nitroso- (9CI) **(6265-09-4)**
Phenol, 5-(diethylamino)-2-nitroso- (8CI,9CI) **(6358-20-9)**
Phenol,5-(diethylamino)-2-nitroso-, monohydrochloride **(25953-06-4)**
Phenol 4,4'-(1,2-diethyl-1,2-ethenediyl)bis-, (E)- **(56-53-1)**
Phenol, 4,4'-(1,2-diethyl-1,2-ethenediyl)bis-, bis(dihydrogen phosphate), (E)- **(522-40-7)**
Phenol, 4,4'-(1,2-diethylethylene)di-, meso **(84-16-2)**
Phenol, 4,4'-(1,2-diethyl-1,2-ethylenediyl)bis-, (R*,S*)- (9CI) **(84-16-2)**
Phenol, 4,4'-(1,2-diethylidene-1,2-ethanediyl)bis- (9CI) **(84-17-3)**
Phenol, 4,4'-(diethylideneethylene)di **(84-17-3)**
Phenol, 3-(((4,5-dihydro-1H-imidazol-2-yl)methyl)(4-methylphenyl)amino) **(50-60-2)**
Phenol, 2,6-diiodo-4-nitro **(305-85-1)**
Phenol, diisobutyl- **(27515-66-8)**
Phenol, 2,4-diisopropyl **(2934-05-6)**
Phenol, 2,5-diisopropyl- **(35946-91-9)**
Phenol, 2,6-diisopropyl **(2078-54-8)**
Phenol, 3,5-diisopropyl- (8CI) **(26886-05-5)**
Phenol, 2,6-dimethoxy **(91-10-1)**
Phenol, 3,4-dimethoxy- (9CI) **(2033-89-8)**
Phenol, 3,5-dimethoxy- (9CI) **(500-99-2)**
Phenol, 2,6-dimethoxy-4-vinyl-, Polymers (8CI) **(31872-14-7)**
Phenol, 2,3-dimethyl- **(526-75-0)**
Phenol, dimethyl- **(1300-71-6)**
Phenol, m-(dimethylamino) **(99-07-0)**
Phenol, ((dimethylamino)methyl)- (9CI) **(25338-55-0)**
Phenol, 3-(dimethylamino)-4-methyl- (9CI) **(119-31-3)**
Phenol, 5-(dimethylamino)-4-methyl-2-nitroso- (9CI) **(6265-11-8)**
Phenol, p-(α,α-dimethylbenzyl) **(599-64-4)**
Phenol, (1,1-dimethylethyl)- (9CI) **(27178-34-3)**
Phenol, 3-(1,1-dimethylethyl)- (9CI) **(585-34-2)**
Phenol, 4-(1,1-dimethylethyl)- (9CI) **(98-54-4)**
Phenol, (1,1-dimethylethyl)-2,5-dimethyl- (9CI) **(31391-49-8)**
Phenol, (1,1-dimethylethyl)dimethyl- (9CI) **(36812-13-2)**
Phenol, 4-(1,1-dimethylethyl)-2,5-dimethyl- (9CI) **(17696-37-6)**
Phenol, 4-(1,1-dimethylethyl)-2,6-dimethyl- (9CI) **(879-97-0)**
Phenol, 4-(1,1-dimethylethyl)-2,6-dinitro- (9CI) **(4097-49-8)**
Phenol, 2-(1,1-dimethylethyl)-4-ethyl- (9CI) **(96-70-8)**
Phenol, 2-(1,1-dimethylethyl)-5-ethyl- (9CI) **(4237-25-6)**
Phenol, 4-(1,1-dimethylethyl)-2-ethyl- (9CI) **(63452-61-9)**
Phenol, (1,1-dimethylethyl)-4-methoxy **(25013-16-5)**
Phenol, (1,1-dimethylethyl)-2-methoxy- (9CI) **(53894-31-8)**
Phenol, (1,1-dimethylethyl)-3-methyl- (9CI) **(1333-13-7)**
Phenol, (1,1-dimethylethyl)-4-methyl- (9CI) **(25567-40-2)**
Phenol, 2-(1,1-dimethylethyl)-5-methyl- (9CI) **(88-60-8)**
Phenol, 4-(1,1-dimethylethyl)-3-methyl- (9CI) **(2219-72-9)**
Phenol, 6-(1,1-dimethylethyl)-3-methyl-2,4-dinitro- **(3996-59-6)**
Phenol, 4-(1,1-dimethylethyl)-2-nitro- (9CI) **(3279-07-0)**
Phenol, (1,1-dimethylethyl)-, phosphate (3:1) **(28777-70-0)**
Phenol, 4-(1,1-dimethylethyl)-, phosphate (3:1) (9CI) **(78-33-1)**
Phenol, 4-(1,1-dimethylethyl)-, sodium salt (9CI) **(5787-50-8)**
Phenol, 2,6-dimethyl-, Homopolymer (9CI) **(25134-01-4)**
Phenol, 3,4-dimethyl-, methylcarbamate (9CI) **(2425-10-7)**
Phenol, 2,4-dimethyl-6-(1-methylcyclohexyl)- (9CI) **(77-61-2)**
Phenol, 4,4'-dimethylmethylenedi- **(80-05-7)**
Phenol, 3,5-dimethyl-4-(methylthio)- (9CI) **(7379-51-3)**
Phenol, 3,4-dimethyl-, phosphate (3:1) **(3862-11-1)**
Phenol, 2,4-dimethyl-, phosphate (3:1) (9CI) **(3862-12-2)**
Phenol, 2,5-dimethyl-, phosphate (3:1) (9CI) **(19074-59-0)**
Phenol, 2,6-dimethyl-, phosphate (3:1) (9CI) **(121-06-2)**
Phenol, 3,5-dimethyl-, phosphate (3:1) (9CI) **(25653-16-1)**
Phenol, dimethyl-, phosphate (3:1) (9CI) **(25155-23-1)**
Phenol, 2-(1,1-dimethylpropyl)- (9CI) **(3279-27-4)**
Phenol, 4-(1,1-dimethylpropyl)-, potassium salt (9CI) **(53404-18-5)**
Phenol, 4-(1,1-dimethylpropyl)-, sodium salt (9CI) **(31366-95-7)**
Phenol, 2,3-dinitro **(66-56-8)**
Phenol, 2,4-dinitro **(51-28-5)**
Phenol, 2,5-dinitro **(329-71-5)**
Phenol, 2,6-dinitro **(573-56-8)**
Phenol, 3,4-dinitro **(577-71-9)**
Phenol, 3,5-dinitro **(586-11-8)**
Phenol, α-dinitro- **(51-28-5)**
Phenol, dinitro **(25550-58-7)**
Phenol, γ-dinitro- **(329-71-5)**
Phenol, dinitro-, Wetted with at least 15% water **(25550-58-7)**
Phenol, 2,4-dinitro-, acetate (ester) (9CI) **(4232-27-3)**
Phenol, o-(2,4-dinitroanilino) **(6358-23-2)**
Phenol, p-(2,4-dinitroanilino) **(119-15-3)**
Phenol, 2,6-dinitro-4-isopropyl- **(4097-47-6)**
Phenol, 2,6-dinitro-4-octyl- **(4097-33-0)**
Phenol, 2,4-dinitro-6-phenyl- **(731-92-0)**
Phenol, 2,4-dinonyl- (9CI) **(137-99-5)**
Phenol, dinonyl- (9CI) **(1323-65-5)**
Phenol, dinonyl- (Mixture of isomers) **(1323-65-5)**
Phenol, dinonyl-, phosphite (3:1) (9CI) **(1333-21-7)**
Phenol, dioctyl- (9CI) **(29988-16-7)**
Phenol, 2,4-di-tert-pentyl **(120-95-6)**
Phenol, dipentyl **(28652-04-2)**
Phenol, 2,4-dipentyl- (8CI,9CI) **(138-00-1)**
Phenol, 2-(4,6-diphenyl-1,3,5-triazin-2-yl)- (9CI) **(3202-86-6)**
Phenol, o-(4,6-diphenyl-s-triazin-2-yl)- (7CI,8CI) **(3202-86-6)**
Phenol, dodecyl-, Mixed isomers **(27193-86-8)**
Phenol, dodecyl-, hydrogen phosphorodithioate (9CI) **(30304-41-7)**
Phenol, dodecyl-, hydrogen phosphorodithioate, zinc salt (9CI) **(54261-67-5)**
Phenole (German) **(108-95-2)**
Phenol, 4-ethenyl- (9CI) **(2628-17-3)**
Phenol, 4-ethenyl-2,6-dimethoxy-, Homopolymer (9CI) **(31872-14-7)**
Phenol, 4-ethenyl-, Homopolymer (9CI) **(24979-70-2)**
Phenol, 4-ethenyl-2-methoxy-, Homopolymer (9CI) **(31853-85-7)**
Phenol, p-ethoxy **(622-62-8)**
Phenol, 2-ethoxy- (9CI) **(94-71-3)**
Phenol, 4-ethoxy- (9CI) **(622-62-8)**
Phenol, o-ethoxy- (8CI) **(94-71-3)**
Phenol, o-ethyl **(90-00-6)**
Phenol, p-ethyl **(123-07-9)**
Phenol, 3-ethyl- (9CI) **(620-17-7)**
Phenol, m-ethyl- (8CI) **(620-17-7)**
Phenol, ethyl- (8CI,9CI) **(25429-37-2)**
Phenol, m-(ethylamino) **(621-31-8)**
Phenol, 3-(ethylamino)-4-methyl **(120-37-6)**
Phenol-ethylene oxide adduct **(9004-78-8)**
Phenol, 4-ethyl-2-methoxy- (9CI) **(2785-89-9)**
Phenol, 3-ethyl-5-methyl- (9CI) **(698-71-5)**
Phenol, ethylmethyl- (9CI) **(30230-52-5)**
Phenol, 4-(1-ethyl-1-methylhexyl)- (9CI) **(52427-13-1)**
Phenol, 3-(1-ethylpropyl)-, methylcarbamate **(672-04-8)**
Phenol, m-(1-ethylpropyl)-, methylcarbamate (8CI) **(672-04-8)**
Phenol, 4-ethyl-, sodium salt (9CI) **(19277-91-9)**
Phenol, o-fluoro **(367-12-4)**
Phenol, p-fluoro **(371-41-5)**
Phenol, 3-fluoro- (9CI) **(372-20-3)**
Phenol, m-fluoro- (8CI) **(372-20-3)**
Phenol, 2-fluoro-4-nitro- (8CI,9CI) **(403-19-0)**
Phenol, formaldehyde, biphenol A, epichlorohydrin polymer **(40216-08-8)**
Phenol, formaldehyde, bisphenol A, epichlorohydrin polymer **(40216-08-8)**
Phenol-formaldehyde copolymer **(9003-35-4)**
Phenol, formaldehyde, gum rosin polymer **(67700-45-2)**

Phenol, formaldehyde polymer **(9003-35-4)**
Phenol-formaldehyde resin **(9003-35-4)**
Phenol, formaldehyde, rosin polymer **(67700-45-2)**
Phenol-glycerinaether (German) **(538-43-2)**
Phenol glycerol ether **(538-43-2)**
Phenol glyceryl ether **(538-43-2)**
Phenol-glycidaether (German) **(122-60-1)**
Phenol glycidyl ether **(122-60-1)**
Phenol, heptadecyl- (9CI) **(86812-27-3)**
Phenol, 4-heptyl- (9CI) **(1987-50-4)**
Phenol, heptyl- (8CI,9CI) **(26997-02-4)**
Phenol, 4-(heptyloxy)- (9CI) **(13037-86-0)**
Phenol, 2-hexadecyl- (9CI) **(25401-86-9)**
Phenol, 4-hexadecyl- (9CI) **(2589-78-8)**
Phenol, hexahydro- **(108-93-0)**
Phenol, 4-hexyl- (9CI) **(2446-69-7)**
Phenol, 2-hexyl-4,6-dinitro- (8CI,9CI) **(4099-65-4)**
Phenol, 4-(hexyloxy)- (9CI) **(18979-55-0)**
Phenol, m-hydroxy- **(108-46-3)**
Phenol, 2,2'-((((2-hydroxy-5-octylphenyl)methyl)imino)bis-(2,1-ethanediyliminomethylene))bis(4-octyl-, calcium salt (9CI) **(68568-82-1)**
Phenol, 4-(3-hydroxy-1-propenyl)-2-methoxy- **(458-35-5)**
Phenol, m-(N-(2-imidazolin-2-ylmethyl)-p-toluidino)- **(50-60-2)**
Phenol, 3-iodo- **(626-02-8)**
Phenol, m-iodo **(626-02-8)**
Phenol, o-iodo **(533-58-4)**
Phenol, p-iodo **(540-38-5)**
Phenol, isobutylenated **(68610-06-0)**
Phenol, isobutylenated, phosphate (3:1) **(68937-40-6)**
Phenol, 4-isocyanato-, phosphorothioate (3:1) (ester) (VAN)(9CI) **(4151-51-3)**
Phenol, 2-isononyl-4-methyl- (9CI) **(28983-26-8)**
Phenol, isooctyldinitro- (9CI) **(37224-61-6)**
Phenol, o-isopropoxy-, methylcarbamate **(114-26-1)**
Phenol, m-isopropyl **(618-45-1)**
Phenol, o-isopropyl **(88-69-7)**
Phenol, p-isopropyl **(99-89-8)**
Phenol, isopropylated, phosphate (3:1) **(68937-41-7)**
Phenol, 4-isopropyl-2,6-dinitro **(4097-47-6)**
Phenol, 4,4'-isopropylidenebis(2,6-dibromo- (8CI) **(79-94-7)**
Phenol, 4,4'-isopropylidenebis(dibromo- (VAN) **(79-94-7)**
Phenol, 4,4'-isopropylidenebis(2,6-dichloro **(79-95-8)**
Phenol, 4,4'-isopropylidenebis(2,3,5,6-tetrabromo-, diacetate (8CI) **(34372-18-4)**
Phenol, 4,4'-isopropylidenedi **(80-05-7)**
Phenol, 4,4'-isopropylidenedi-, Dimer with 1-chloro-2,3-epoxypropane **(25068-38-6)**
Phenol, 4,4'-isopropylidenedi-, Monomer with 1-chloro-2,3-epoxypropane **(25068-38-6)**
Phenol, 4,4'-isopropylidenedi-, Polymer with 1-chloro-2,3-epoxypropane **(25068-38-6)**
Phenol, 4,4'-isopropylidenedi-, disodium salt **(2444-90-8)**
Phenol, 4,4'-isopropylidenedi-, tetramer with 1-chloro-2,3-epoxypropane **(25068-38-6)**
Phenol, 2-isopropyl-5-methyl- **(89-83-8)**
Phenol, 2-isopropyl-6-methyl **(3228-04-4)**
Phenol, 3-isopropyl-6-methyl- **(499-75-2)**
Phenol, 5-isopropyl-2-methyl- **(499-75-2)**
Phenol, m-isopropyl-, methylcarbamate **(64-00-6)**
Phenol, o-isopropyl-, methylcarbamate **(2631-40-5)**
Phenol, p-isopropyl-, phosphate (3:1) (8CI) **(2502-15-0)**
Phenol, m-methoxy **(150-19-6)**
Phenol, o-methoxy **(90-05-1)**
Phenol, p-methoxy **(150-76-5)**
Phenol, 3-methoxy- (9CI) **(150-19-6)**
Phenol, p-methoxy-, carbanilate (6CI,8CI) **(19219-48-8)**
Phenol, 2-methoxy-4-methyl- **(93-51-6)**
Phenol, 4-methoxy-2-nitro- (8CI,9CI) **(1568-70-3)**
Phenol, 2-methoxy-4-nitro-, acetate (ester) (9CI) **(67851-29-0)**
Phenol, 2-methoxy-5-nitro-, acetate (ester) (9CI) **(53606-41-0)**
Phenol, 4-((2-methoxy-4-((4-nitrophenyl)azo)phenyl)azo)- (9CI) **(19800-42-1)**
Phenol, 4-methoxy-, phenylcarbamate (9CI) **(19219-48-8)**
Phenol, 2-methoxy-4-(2-propenyl)- **(97-53-0)**
Phenol, 2-methoxy-4-propenyl **(97-54-1)**
Phenol, 2-methoxy-4-propenyl-, (E) **(5932-68-3)**
Phenol, 2-methoxy-4-propenyl-, (Z) **(5912-86-7)**
Phenol, 2-methoxy-4-propyl **(2785-87-7)**
Phenol, 2-methoxy-3,4,5,6-tetrachloro **(2539-17-5)**
Phenol, 2-methoxy-trichloro **(57057-83-7)**
Phenol, 6-methoxy-2,3,4-trichloro **(2668-24-8)**
Phenol, 3-methoxy-2,4,6-trimethyl- (9CI) **(34883-05-1)**
Phenol, 2-methoxy-4-vinyl-, Polymers (8CI) **(31853-85-7)**
Phenol, 2-methyl- (9CI) **(95-48-7)**
Phenol, 3-methyl- (9CI) **(108-39-4)**
Phenol, 4-methyl- (9CI) **(106-44-5)**
Phenol, methyl- (9CI) **(1319-77-3)**
Phenol, o-(2-methylallyl)- (8CI) **(20944-88-1)**
Phenol, 2-(2-methylallyl)-6-nitro- (8CI) **(13414-58-9)**
Phenol, p-(methylamino) **(150-75-4)**
Phenol, p-(methylamino)-, sulfate (2:1) (Salt) **(55-55-0)**
Phenol, p-methylamino-, sulfate (Salt) **(55-55-0)**
Phenol, 4-methyl-2,6-bis(1-phenylethyl)- (9CI) **(1817-68-1)**
Phenol, 4-(1-methylbutyl)- (9CI) **(94-06-4)**
Phenol, p-(1-methylbutyl)- (8CI) **(94-06-4)**
Phenol, 2-(1-methylbutyl)-4,6-dinitro **(4097-36-3)**
Phenol, m-(1-methylbutyl)-, methylcarbamate **(2282-34-0)**
Phenol, 2-methyl-4,6-dinitro- (9CI) **(534-52-1)**
Phenol, 2-methyldinitro- (9CI) **(1335-85-9)**
Phenol, 3-methyl-2,4-dinitro- (9CI) **(1817-66-9)**
Phenol, 4-methyl-2,3-dinitro- (9CI) **(68191-07-1)**
Phenol, 2-methyl-4,6-dinitro-, sodium salt (9CI) **(2312-76-7)**
Phenol, 4,4'-methylenebis **(620-92-8)**
Phenol, 2,2'-methylenebis(6-tert-butyl-4-ethyl **(88-24-4)**
Phenol, 2,2'-methylenebis(4-chloro **(97-23-4)**
Phenol, 2,2'-methylenebis(4-chloro-, disodium salt (9CI) **(22232-25-3)**
Phenol, 4,4'-methylenebis(2,6-di-tert-butyl- (8CI) **(118-82-1)**
Phenol, 4,4'-methylenebis(2,6-bis(1,1-dimethylethyl)- (9CI) **(118-82-1)**
Phenol, 2,2'-methylenebis(4-methyl-6-(1-methylcyclohexyl)- (9CI) **(77-62-3)**
Phenol, 2,2'-methylenebis(4-methyl-6-nonyl- (9CI) **(7786-17-6)**
Phenol, 2,2'-methylenebis(4-(1,1,3,3-tetramethylbutyl)- (9CI) **(27725-17-3)**
Phenol, 2,2'-methylenebis(4-(1,1,3,3-tetramethylbutyl)-, calcium salt (9CI) **(68527-62-8)**
Phenol, 2,2'-methylenebis(3,4,6-trichloro **(70-30-4)**
Phenol, 2,2'-methylenebis(3,4,6-trichloro-, sodium salt **(5736-15-2)**
Phenol, 4,4'-methylenedi- **(620-92-8)**
Phenol, 2,2'-((1-methyl-1,2-ethanediyl)bis(nitrilomethylidyne))-bis- (9CI) **(94-91-7)**
Phenol, 2-(1-methylethoxy)- (9CI) **(4812-20-8)**
Phenol, 3-(1-methylethyl)- **(618-45-1)**
Phenol, 4-(1-methylethyl)- **(99-89-8)**
Phenol, (1-methylethyl)- (9CI) **(25168-06-3)**
Phenol, 2-(1-methylethyl)- (9CI) **(88-69-7)**
Phenol, 4,4'-(1-methylethylidene)bis-, Polymer with 2,2'-((1-methylethylidene)bis(4,1-phenyleneoxymethylene))-bis(oxirane) (9CI) **(25036-25-3)**
Phenol, 4,4'-(1-methylethylidene)bis(2,6-dibromo- (9CI) **(79-94-7)**
Phenol, 4,4'-(1-methylethylidene)bis(2,6-dibromo-, Polymer with (chloromethyl)oxirane and 4,4'-(1-methylethylidene)-bis(phenol) (9CI) **(26265-08-7)**
Phenol, 4,4'-(1-methylethylidene)bis(2,6-dibromo-, Polymer with (chloromethyl)oxirane (9CI) **(40039-93-8)**
Phenol, 4,4'-(1-methylethylidene)bis(2,6-dichloro- **(79-95-8)**
Phenol, 4,4'-(1-methylethylidene)bis-, disodium salt (9CI) **(2444-90-8)**
Phenol, 4,4'-(1-methylethylidene)bis-, tetrabromo deriv. **(79-94-7)**
Phenol, 4,4'-(1-methylethylidene)bis(2,3,5,6-tetrabromo-, diacetate

(9CI) **(34372-18-4)**
Phenol, 3-(1-methylethyl)-, methylcarbamate (9CI) **(64-00-6)**
Phenol, (1-methylethyl)-, phosphate (3:1) (9CI) **(26967-76-0)**
Phenol, 3-(1-methylethyl)-, phosphate (3:1) (9CI) **(72668-27-0)**
Phenol, 4-(1-methylethyl)-, phosphate (3:1) (9CI) **(2502-15-0)**
Phenol, (1-methylheptyl)- (9CI) **(27985-70-2)**
Phenol, 2-(1-methylheptyl)-4,6-dinitro- (8CI,9CI) **(3687-22-7)**
Phenol, 2-(1-methylheptyl)-4,6-dinitro-, crotonate (ester) **(39300-45-3)**
Phenol, 4-methyl-3-(methylamino)- (9CI) **(6265-13-0)**
Phenol, 2-methyl-5-(1-methylethyl)- (9CI) **(499-75-2)**
Phenol, 5-methyl-2-(1-methylethyl)- (9CI) **(89-83-8)**
Phenol, 3-methyl-4-(methylthio)- (9CI) **(3120-74-9)**
Phenol, 3-methyl-, monochloro deriv. (9CI) **(54548-50-4)**
Phenol, 3-methyl-2-nitro- (9CI) **(4920-77-8)**
Phenol, 4-methyl-3-nitro- (9CI) **(2042-14-0)**
Phenol, 5-methyl-2-nitro- (9CI) **(700-38-9)**
Phenol, methyl-4-nitro- (9CI) **(58882-68-1)**
Phenol, 4-methyl-2-((2-nitrophenyl)azo)- (9CI) **(1435-71-8)**
Phenol, 4-(1-methyl-1-phenethyl)- (9CI) **(599-64-4)**
Phenol, 4-methyl-2-(phenylazo)- (9CI) **(952-47-6)**
Phenol, 2-methyl-4-((4-(phenylazo)phenyl)azo) **(6300-37-4)**
Phenol, (1-methyl-1-phenylethyl)- (9CI) **(27576-86-9)**
Phenol, (1-methyl-1-phenylethyl)-, phosphate (9CI) **(63302-98-7)**
Phenol, p-(5-(5-(4-methyl-1-piperazinyl)-2-benzimidazolyl)-2-benzimidazolyl)-, trihydrochloride **(23491-45-4)**
Phenol, 4-(5-(4-methyl-1-piperazinyl)(2,5'-bi-1H-benzimidazol)-2'-yl)-, trihydrochloride **(23491-45-4)**
Phenol, methyl-, potassium salt (9CI) **(12002-51-6)**
Phenol, 4,4',4''-(1-methyl-1-propanyl-3-ylidene)tris(2-(1,1-dimethylethyl)-5-methyl- (9CI) **(1843-03-4)**
Phenol, 2-(2-methyl-2-propenyl)- (9CI) **(20944-88-1)**
Phenol, 2-(2-methyl-1-propenyl)-6-nitro- (9CI) **(64061-59-2)**
Phenol, 2-(1-methylpropyl)-4,6-dinitro-, acetate (ester) (9CI) **(2813-95-8)**
Phenol, 2,2'-(2-methylpropylidene)bis(4,6-dimethyl- (9CI) **(33145-10-7)**
Phenol, 2-(1-methylpropyl)-, methylcarbamate **(3766-81-2)**
Phenol, 4-(1-methylpropyl)-2-nitro- (9CI) **(3555-18-8)**
Phenol, 4-methyl-, sodium salt (9CI) **(1121-70-6)**
Phenol, methyl-, sodium salt (9CI) **(34689-46-8)**
Phenol, p-(methylsulfinyl)-, O-ester with O,O-diethyl phosphorothioate **(115-90-2)**
Phenol, p-(methylthio) **(1073-72-9)**
Phenol, p-(methylthio)-, dimethyl phosphate **(3254-63-5)**
Phenol, methyl-, titanium(4+) salt (9CI) **(28503-70-0)**
Phenol, methyl-, tribromo deriv. (9CI) **(65436-87-5)**
Phenol, 3-methyl-2,4,6-trinitro- (9CI) **(602-99-3)**
Phenol, p-(2-naphthylamino) **(93-45-8)**
Phenol, m-nitro **(554-84-7)**
Phenol, nitro- **(25154-55-6)**
Phenol, o-nitro **(88-75-5)**
Phenol, p-nitro **(100-02-7)**
Phenol, m-nitro-, acetate **(1523-06-4)**
Phenol, m-nitro-, carbanilate (6CI) **(35289-89-5)**
Phenol, p-nitro-, dihydrogen phosphate **(330-13-2)**
Phenol, p-nitro-, ester with diethyl phosphate **(311-45-5)**
Phenol, p-nitro-, O-ester with O,O-diethylphosphorothioate **(56-38-2)**
Phenol, p-nitro-, O-ester with O,O-dimethylphosphorothioate **(298-00-0)**
Phenol, p-nitro-, O-ester with O-ethyl phenyl phosphonothioate **(2104-64-5)**
Phenol, 2-nitro-4-phenyl- **(885-82-5)**
Phenol, 2-nitro-4-(phenylazo)- (9CI) **(55936-40-8)**
Phenol, 4-((4-nitrophenyl)azo)- (9CI) **(1435-60-5)**
Phenol, p-nitroso **(104-91-6)**
Phenol, 4-nitroso- (9CI) **(104-91-6)**
Phenol, 4-nitro-, sodium salt **(824-78-2)**
Phenol, p-nitro-, sodium salt **(824-78-2)**
Phenol, 4-nitroso-, sodium salt **(823-87-0)**
Phenol, 2-nitro-3,4,6-trichloro **(82-62-2)**
Phenol, 4-nitro-3-(trifluoromethyl)- **(88-30-2)**
Phenol, nonyl **(25154-52-3)**
Phenol, p-nonyl **(104-40-5)**
Phenol, 2-nonyl- (9CI) **(136-83-4)**
Phenol, o-nonyl- (8CI) **(136-83-4)**
Phenol, nonyl-, barium salt (9CI) **(28987-17-9)**
Phenol, 4-nonyl-2,6-bis(1-phenylethyl)- (9CI) **(15860-96-5)**
Phenol, nonyl-, calcium salt (9CI) **(30977-64-1)**
Phenol, p-nonyl-, monoether with polyethylene glycol **(26027-38-3)**
Phenol, nonyl-, phosphate (3:1) (9CI) **(26569-53-9)**
Phenol, nonyl-, phosphite (3:1) (9CI) **(26523-78-4)**
Phenol, o-octyl **(949-13-3)**
Phenol, p-(tert-octyl)- **(140-66-9)**
Phenol, p-octyl- **(1806-26-4)**
Phenol, 2-octyl- (9CI) **(949-13-3)**
Phenol, 4-octyl- (9CI) **(1806-26-4)**
Phenol, octyl (8CI) **(27193-28-8)**
Phenol, 2,4(or 2,6)-dibromo-, Homopolymer (9CI) **(69882-11-7)**
Phenol, 2(or 4)-methyl-3-nitro- (9CI) **(68137-09-7)**
Phenol, 3(or 4)-methyl-2-nitro- (9CI) **(68137-08-6)**
Phenol, pentabromo **(608-71-9)**
Phenol, pentachloro **(87-86-5)**
Phenol, pentachloro-, acetate **(1441-02-7)**
Phenol, pentachloro-, potassium salt (8CI,9CI) **(7778-73-6)**
Phenol, pentachloro-, sodium salt **(131-52-2)**
Phenol, 3-pentadecyl- (9CI) **(501-24-6)**
Phenol, m-pentadecyl- (8CI) **(501-24-6)**
Phenol, pentafluoro **(771-61-9)**
Phenol, pentamethyl- (9CI) **(2819-86-5)**
Phenol, p-(tert-pentyl) **(80-46-6)**
Phenol, p-pentyl **(14938-35-3)**
Phenol, 2-pentyl- (9CI) **(136-81-2)**
Phenol, o-pentyl- (8CI) **(136-81-2)**
Phenol, m-sec-pentyl-, methylcarbamate **(2282-34-0)**
Phenol, 4-(pentyloxy)- (9CI) **(18979-53-8)**
Phenol, m-phenoxy **(713-68-8)**
Phenol, 4-phenoxy- (9CI) **(831-82-3)**
Phenol, p-phenoxy- (8CI) **(831-82-3)**
Phenol, 3-(2-phenoxyethoxy)- (9CI) **(36429-48-8)**
Phenol, o-phenyl- **(90-43-7)**
Phenol, phenyl- **(1322-20-9)**
Phenol, 4-(phenylamino)- **(122-37-2)**
Phenol, 3-(phenylamino)- (9CI) **(101-18-8)**
Phenol, 4-((4-(phenylamino)phenyl)amino)- (9CI) **(101-74-6)**
Phenol, p-(phenylazo) **(1689-82-3)**
Phenol, 2-(phenylazo)- (9CI) **(2362-57-4)**
Phenol, o-(phenylazo)- (8CI) **(2362-57-4)**
Phenol, 4-((4-(phenylazo)-1-naphthalenyl)azo)- (9CI) **(6253-10-7)**
Phenol, 4-((4-(phenylazo)phenyl)azo)- (9CI) **(6250-23-3)**
Phenol, 2-(phenylmethyl)- **(28994-41-4)**
Phenol, 4-(phenylmethyl)- (9CI) **(101-53-1)**
Phenol, 4-(phenylmethyl)-, carbamate (9CI) **(101-71-3)**
Phenol, o-phenyl-, sodium deriv. **(132-27-4)**
Phenolphthalein **(77-09-8)**
Phenolphthalein, 4,5,6,7-tetrabromo-3',3''-disulfo-, disodium salt **(71-67-0)**
Phenol, potassium salt (9CI) **(100-67-4)**
Phenol, 4-propoxy- (9CI) **(18979-50-5)**
Phenol, p-propoxy- (8CI) **(18979-50-5)**
Phenol, o-propyl **(644-35-9)**
Phenol, p-propyl **(645-56-7)**
Phenol, 4,4'-(2-pyridinylmethylene)bis-, diacetate (ester) **(603-50-9)**
Phenol, 4,4'-(2-pyridylmethylene)di-, diacetate (ester) **(603-50-9)**
Phenol, rosin, formaldehyde polymer **(67700-45-2)**
Phenol, rosin, formaldehyde resin **(67700-45-2)**
Phenol sodium **(139-02-6)**
Phenol sodium salt **(139-02-6)**
Phenol, sodium salt, (Solid) **(139-02-6)**
Phenolsulfonephthalein **(143-74-8)**

Phenolsulfonic acid (1333-39-7)
1-Phenol-4-sulfonic acid zinc salt (127-82-2)
Phenolsulfonphthalein (143-74-8)
Phenol, 4,4'-sulfonyldi (80-09-1)
Phenolsulphonic acid, Liquid [UN 1803] (1333-39-7)
Phenolsulphonphthalein (143-74-8)
Phenol, 2,3,4,6-tetrabromo- (8CI,9CI) (14400-94-3)
Phenol, 2,3,4,5-tetrabromo-6-methoxy- (7CI,9CI) (35488-17-6)
Phenol, 2,3,4,5-tetrabromo-6-methyl- (9CI) (576-55-6)
Phenoltetrabromophthaleinsulfonate (71-67-0)
Phenol, 2,3,4,5-tetrachloro (4901-51-3)
Phenol, 2,3,4,6-tetrachloro (58-90-2)
Phenol, 2,3,5,6-tetrachloro (935-95-5)
Phenol, tetrachloro (25167-83-3)
Phenol, 2,3,4,6-tetrachloro- (8CI,9CI) (58-90-2)
Phenol, 2,3,5,6-tetrachloro-4-nitro- (8CI,9CI) (4824-72-0)
Phenol, 2,3,4,5-tetrachloro-6-(pentachlorophenoxy) (35245-80-8)
Phenol, 2,3,4,6-tetrachloro-5-(pentachlorophenoxy) (42255-14-1)
Phenol, tetrachloro-, potassium salt (9CI) (53535-27-6)
Phenol, tetrachloro-, sodium salt (8CI,9CI) (25567-55-9)
Phenol, tetramethyl- (527-35-5)
Phenol, 2,3,5,6-tetramethyl- (9CI) (527-35-5)
Phenol, p-(1,1,3,3-tetramethylbutyl) (140-66-9)
Phenol, (1,1,3,3-tetramethylbutyl)- (9CI) (27193-28-8)
Phenol, 2-(1,1,3,3-tetramethylbutyl)- (9CI) (3884-95-5)
Phenol, 4-(2,2,3,3-tetramethylbutyl)- (9CI) (54932-78-4)
Phenol, 2,3,4,6-tetranitro- (641-16-7)
Phenol, thio- (108-98-5)
Phenol, 2,2'-thiobis(4-chloro (97-24-5)
Phenol, 2,2'-thiobis(4-chloro-6-methyl)- (4418-66-0)
Phenol, 2,2'-thiobis(4-chloro-6-methyl- (9CI) (4418-66-0)
Phenol, 2,2'-thiobis(4,6-dichloro (97-18-7)
Phenol, 2,2'-thiobis(6-(1,1-dimethylethyl)-4-methyl- (9CI) (90-66-4)
Phenol, 4,4'-thiobis(2-(1,1-dimethylethyl)-5-methyl- (9CI) (96-69-5)
Phenol, 4,4'-thiobis(3-(1,1-dimethylethyl)-5-methyl- (9CI) (3818-54-0)
Phenol, thiobis(dodecyl-, calcium salt (1:1) (9CI) (26998-97-0)
Phenol, thiobis(nonyl- (28503-85-7)
Phenol, 2,2'-thiobis(4-(1,1,3,3-tetramethylbutyl)- (9CI) (3294-03-9)
Phenol, thiobis((tetrapropenyl)-, magnesium salt (9CI) (68974-78-7)
Phenol, thiobis(tetrapropylene- (9CI) (68815-67-8)
Phenol, 4,4'-thiodi (2664-63-3)
Phenol, 4,4'-thiodi-, O,O-diester with O,O-dimethyl phosphorothioate (3383-96-8)
Phenol, 2,4,6-tribromo (118-79-6)
Phenol, tribromo- (9CI) (25376-38-9)
Phenol, 2,4,6-tribromo-, carbonate (2:1) (9CI) (67990-32-3)
Phenol, tribromodimethyl- (9CI) (58170-32-4)
Phenol, 2,3,4-tribromo-6-methoxy- (9CI) (38926-85-1)
Phenol, 2,4,6-tri-tert-butyl- (8CI) (732-26-3)
Phenol, 2,3,6-trichloro (933-75-5)
Phenol, 2,4,5-trichloro (95-95-4)
Phenol, 2,4,6-trichloro (88-06-2)
Phenol, 3,4,5-trichloro (609-19-8)
Phenol, trichloro (25167-82-2)
Phenol, trichloro-, copper(2+) salt (25267-55-4)
Phenol, 2,4,5-trichloro-, O-ester with O,O-dimethyl phosphorothioate (299-84-3)
Phenol, 2,4,5-trichloro-, O-ester with O-ethyl ethylphosphonothioate (327-98-0)
Phenol, 3,4,6-trichloro-2-methoxy- (9CI) (60712-44-9)
Phenol, 2,3,5-trichloro-6-methoxy- (8CI,9CI) (938-23-8)
Phenol, 2,4,6-trichloro-3-methyl- (9CI) (551-76-8)
Phenol, 2,3,6-trichloro-4-nitro- (8CI,9CI) (20404-02-8)
Phenol, 2,4,5-trichloro-, sodium salt (136-32-3)
Phenol, 2,3,4-trichloro-6-(2,3,4,6-tetrachlorophenoxy)- (9CI) (94888-12-7)
Phenol, 2,3,6-trichloro-4-(2,3,4,6-tetrachlorophenoxy)- (9CI) (94888-13-8)
Phenol, 2,3,4-trichloro-6-(2,4,6-trichlorophenoxy)- (9CI) (94897-81-1)
Phenol, 2,3,6-trichloro-4-(2,4,6-trichlorophenoxy)- (9CI) (94888-11-6)
Phenol, 2,4,5-trichloro-, zinc salt (9CI) (136-24-3)
Phenol, m-trifluoromethyl- (88-30-2)
Phenol, 3-(trifluoromethyl)- (9CI) (98-17-9)
Phenol, 4,4'-(2,2,2-trifluoro-1-(trifluoromethyl)ethylidene)bis- (9CI) (1478-61-1)
Phenol, 2,4,6-triiodo (609-23-4)
Phenol, 2,3,5-trimethyl- (9CI) (697-82-5)
Phenol, 2,3,6-trimethyl- (9CI) (2416-94-6)
Phenol, 2,4,5-trimethyl- (9CI) (496-78-6)
Phenol, 2,4,6-trimethyl- (9CI) (527-60-6)
Phenol, 3,4,5-trimethyl- (9CI) (527-54-8)
Phenol, trimethyl- (9CI) (26998-80-1)
Phenol, 2,3,5-trimethyl-, methylcarbamate (2655-15-4)
Phenol, 2,4,6-trimethyl-, phosphate (3:1) (9CI) (56444-79-2)
Phenol, 4-(trimethylsilyl)- (9CI) (13132-25-7)
Phenol trinitrate (88-89-1)
Phenol, 2,4,6-trinitro- (88-89-1)
Phenol, 2,4,6-trinitro-, ammonium salt (9CI) (131-74-8)
Phenol, 2,4,6-trinitro-, lead salt (9CI) (25721-38-4)
Phenol, 2,4,6-trinitro-, silver(1+) salt (9CI) (146-84-9)
Phenol, 2,4,6-tris(dimethylaminomethyl) (90-72-2)
Phenol, 2,4,6-tris(1,1-dimethylethyl)- (9CI) (732-26-3)
Phenoluric (50-06-6)
Phenolurio (50-06-6)
Phenol, p-vinyl (2628-17-3)
Phenol, p-vinyl-, Polymers (24979-70-2)
Phenomercuric acetate (62-38-4)
Phenomet (50-06-6)
Phenomorphan (468-07-5)
Phenomorphane (French) (468-07-5)
Phenomorphanum (Latin) (468-07-5)
Phenonyl (50-06-6)
Pheno-m-penicillin (132-93-4)
Phenopenicillin (87-08-1)
Phenoperidine (562-26-5)
Phenoperidinum (Latin) (562-26-5)
Phenophan (132-60-5)
Phenopiazine (91-19-0)
Phenoplaste Organol Red B (85-83-6)
Phenopromin (51-63-8)
Phenopromin (60-13-9)
Phenopropamine (150-59-4)
Phenopyridine (148-24-3)
Phenopyrine (50-33-9)
Phenoquin (132-60-5)
Phenosan (92-84-2)
Phenosane (1698-60-8)
Phenostat A (900-95-8)
Phenostat-C (639-58-7)
Phenostat-H (76-87-9)
Phenosuccimide (86-34-0)
Phenotan (2813-95-8)
Phenotan (88-85-7)
Phenothiazine (92-84-2)
Phenothiazine (ACGIH,OSHA) (92-84-2)
Phenothiazine, 2-chloro-10-(3-(dimethylamino)propyl) (50-53-3)
Phenothiazine, 2-chloro-10-(3-(dimethylamino)propyl)-, monohydrochloride (69-09-0)
10H-Phenothiazine, 2-chloro-10-(3-(4-methyl-1-piperazinyl)propyl)- (58-38-8)
Phenothiazine, 2-chloro-10-(3-(4-methyl-1-piperazinyl)propyl) (58-38-8)
3H-Phenothiazine, 7-(dimethylamino)-3-(methylimino)-, 3-methochloride (61-73-4)
Phenothiazine, 10-(3-(dimethylamino)-2-methylpropyl) (84-96-8)
Phenothiazine, 10-(2-dimethylaminopropyl) (60-87-7)
Phenothiazine, 10-(3-(dimethylamino)propyl) (58-40-2)
Phenothiazine, 10-(3-(dimethylamino)propyl)-, hydrochloride (53-60-1)

Phenothiazine, 10-(2-(dimethylamino)propyl)-, monohydrochloride **(58-33-3)**
Phenothiazine, 10-(3-(dimethylamino)propyl)-, monohydrochloride **(53-60-1)**
Phenothiazine, 10-(3-(dimethylamino)propyl)-2-(trifluoromethyl) **(146-54-3)**
10H-Phenothiazine-10-ethanamine, N,N,α-trimethyl-, monohydrochloride **(58-33-3)**
10H-Phenothiazine-10-ethanaminium, N,N,N,α-tetramethyl-, methyl sulfate **(58-34-4)**
10H-Phenothiazine, 2-(ethylthio)-10-(3-(4-methyl-1-piperazinyl)-propyl)- **(1420-55-9)**
Phenothiazine, 2-(ethylthio)-10-(3-(4-methyl-1-piperazinyl)propyl) **(1420-55-9)**
Phenothiazine hydrochloride **(69-09-0)**
Phenothiazine, 10-(3-(4-(2-hydroxyethyl)-1-piperazinyl)propyl)-2-(trifluoromethyl)- **(69-23-8)**
Phenothiazine, 10-(3-(4-methyl-1-piperazinyl)propyl)-2-(trifluoromethyl) **(117-89-5)**
10H-Phenothiazine, 10-(2-(1-methyl-2-piperidinyl)ethyl)-2-(methylthio)- **(50-52-2)**
10H-Phenothiazine, 10-[2-(1-methyl-2-piperidinyl)ethyl]-2-(methylthio)-, 5-oxide (9CI) **(7776-05-8)**
Phenothiazine, 10-(2-(1-methyl-2-piperidyl)ethyl)-2-(methylsulfinyl) **(5588-33-0)**
Phenothiazine, 10-(2-(1-methyl-2-piperidyl)ethyl)-2-methylsulfonyl-, **(14759-06-9)**
Phenothiazine, 10-((1-methyl-2-piperidyl)ethyl)-2-(methylthio) **(50-52-2)**
Phenothiazine, 10-[2-(1-methyl-2-piperidyl)ethyl]-2-(methylthio)-, 5-oxide (7CI, 8CI) **(7776-05-8)**
Phenothiazine, 10-((1-methyl-3-pyrrolidinyl)methyl) **(1982-37-2)**
Phenothiazine, 10-((1-methyl-3-pyrrolidinyl)methyl)-, hydrochloride **(1229-35-2)**
10H-Phenothiazine-10-propanamine, 2-chloro-N,N-dimethyl- **(50-53-3)**
10H-Phenothiazine-10-propanamine, 2-chloro-N,N-dimethyl-, monohydrochloride **(69-09-0)**
10H-Phenothiazine-10-propanamine, N,N-dimethyl- **(58-40-2)**
10H-Phenothiazine-10-propanamine, N,N-dimethyl-, monohydrochloride **(53-60-1)**
10H-Phenothiazine-10-propanamine, N,N-dimethyl-2-(trifluoromethyl)- **(146-54-3)**
10H-Phenothiazine-10-propanamine, N,N,β-trimethyl- **(84-96-8)**
Phenothiazin-5-ium, 3,7-bis(dimethylamino)-, chloride **(61-73-4)**
3H-Phenothiazin-3-one (9CI) **(581-30-6)**
N-(β-(10-Phenothiazinyl)propyl)trimethylammonium methyl sulfate **(58-34-4)**
Phenothiocarb **(62850-32-2)**
Phenothiol **(25319-90-8)**
Phenothioxin **(262-20-4)**
Phenothrin **(26002-80-2)**
Phenotolamine **(50-60-2)**
Phenoturic **(50-06-6)**
Phenoverm **(92-84-2)**
Phenovis **(92-84-2)**
Phenox **(94-75-7)**
Phenoxaksine oxide **(58-36-6)**
Phenoxalin **(40487-42-1)**
Phenoxarsine oxide **(58-36-6)**
10H-Phenoxarsine, 10,10'-oxybis- (9CI) **(58-36-6)**
Phenoxarsine, 10,10'-oxydi- **(58-36-6)**
Phenoxathiin **(262-20-4)**
Phenoxathine **(262-20-4)**
Phenoxathrin **(262-20-4)**
Phenoxazin-5-ium, 3-(diethylamino)-7-(dimethylamino)-2-methyl-, chloride, Compd. with zinc chloride (8CI) **(7199-02-2)**
Phenoxazin-5-ium, 3-(diethylamino)-7-(dimethylamino)-2-methyl-, chloride, Compd. with zinc chloride ($ZnCl_2$) (9CI) **(7199-02-2)**
3H-Phenoxazin-3-one, 7-hydroxy-, 10-oxide **(550-82-3)**
Phenoxazole **(2152-34-3)**
Phenoxethol **(122-99-6)**
Phenoxetol **(122-99-6)**
Phenoxthin **(262-20-4)**
Phenoxur **(92-84-2)**
6-Phenoxyacetamidopenicillanic acid **(87-08-1)**
Phenoxyacetate sodium **(3598-16-1)**
Phenoxyacetic acid **(122-59-8)**
Phenoxyaethylpenicillin K-salz (German) **(132-93-4)**
4-Phenoxyaniline **(139-59-3)**
p-Phenoxyaniline **(139-59-3)**
Phenoxyarsine, 10,10'-oxybis- **(58-36-6)**
3-Phenoxybenzaldehyde **(39515-51-0)**
m-Phenoxybenzaldehyde **(39515-51-0)**
Phenoxybenzamide hydrochloride **(63-92-3)**
Phenoxybenzamine **(59-96-1)**
Phenoxybenzamine hydrochloride **(63-92-3)**
Phenoxybenzene **(101-84-8)**
Phenoxybenzene-4,4'-disulfonyl chloride **(121-63-1)**
3-Phenoxybenzenemethanol acetate **(50789-44-1)**
3-Phenoxybenzoic acid **(3739-38-6)**
m-Phenoxybenzoic acid **(3739-38-6)**
3-Phenoxybenzyl alcohol **(13826-35-2)**
m-Phenoxybenzyl alcohol **(13826-35-2)**
3-Phenoxybenzyl bromide **(51632-16-7)**
3-Phenoxybenzyl D-Z/E chrysanthemate **(26002-80-2)**
3-Phenoxybenzyl (+-)-3-(2,2-dichlorovinyl)-2,2-dimethylcyclopropanecarboxylate **(52645-53-1)**
3-Phenoxybenzyl dl-cis/trans-3-(2,2-dichlorovinyl)-2,2-dimethyl-1-cyclopropanecarboxylate **(52645-53-1)**
p-(Phenoxy)bromobenzene **(101-55-3)**
p-Phenoxybromobenzene **(101-55-3)**
S-4-Phenoxybutyl dimethylthiocarbamate **(62850-32-2)**
4-Phenoxybutyric acid **(6303-58-8)**
Phenoxydiamino-s-triazine **(1467-72-7)**
Phenoxydiaminotriazine **(1467-72-7)**
3-Phenoxy-1,2-epoxypropane **(122-60-1)**
Phenoxyessigsaure natrium salz (German) **(3598-16-1)**
Phenoxyethanoic acid **(122-59-8)**
2-Phenoxyethanol **(122-99-6)**
Phenoxyethanol **(122-99-6)**
2-(2-Phenoxyethoxy)ethanol **(104-68-7)**
Phenoxyethyl alcohol **(122-99-6)**
Phenoxyethyl isobutyrate **(103-60-6)**
2-Phenoxyethyl 2-methylpropanoate **(103-60-6)**
α-Phenoxyethylpenicillin **(132-93-4)**
α-Phenoxyethylpenicillin potassium **(132-93-4)**
α-Phenoxyethylpenicillin potassium salt **(132-93-4)**
N-Phenoxyisopropyl-N-benzyl-β-chloroethylamine **(59-96-1)**
N-2-Phenoxyisopropyl-N-benzyl-chloroethylamine hydrochloride **(63-92-3)**
N-Phenoxyisopropyl-N-benzyl-β-chloroethylamine hydrochloride **(63-92-3)**
Phenoxylene **(3653-48-3)**
Phenoxylene 50 **(94-74-6)**
Phenoxylene Plus **(94-74-6)**
Phenoxylene Super **(94-74-6)**
Phenoxymethylenepenicillinic acid **(87-08-1)**
Phenoxymethylpenicillin **(87-08-1)**
D-α-Phenoxymethylpenicillinate k salt **(132-98-9)**
Phenoxymethylpenicillin potassium **(132-98-9)**
4-Phenoxyphenol **(831-82-3)**
m-Phenoxyphenol **(713-68-8)**
N-(2-(p-Phenoxyphenoxy)ethyl)carbamic acid **(72490-01-8)**
(2-(4-Phenoxyphenoxy)ethyl)carbamic acid ethyl ester **(72490-01-8)**
(3-Phenoxyphenyl)methanol **(13826-35-2)**
(3-Phenoxyphenyl)methyl 3-(2,2-dichlorethenyl)-2,2-dimethylcyclopropanecarboxylate **(52645-53-1)**
1-Phenoxy-2,3-propanediol **(538-43-2)**
3-Phenoxy-1,2-propanediol **(538-43-2)**
1-Phenoxy-2-propanol **(770-35-4)**
2-Phenoxypropanol **(4169-04-4)**

Phenoxypropene oxide **(122-60-1)**
2-Phenoxypropyl alcohol **(4169-04-4)**
2-(Phenoxy-2-propylamino)-1-(p-hydroxyphenyl)-1-propanol hydrochloride **(395-28-8)**
Phenoxypropylene oxide **(122-60-1)**
Phenoxythrin **(26002-80-2)**
Phenoxytol **(122-99-6)**
3-Phenoxytoluene **(3586-14-9)**
m-Phenoxytoluene **(3586-14-9)**
Phenozin **(127-82-2)**
Phenozone **(60-80-0)**
Phenpiazine **(91-19-0)**
Phenprocoumarol **(435-97-2)**
Phenprocoumarole **(435-97-2)**
Phenprocoumon **(435-97-2)**
Phenpromin **(60-13-9)**
Phenpropamine **(150-59-4)**
Phensedyl **(60-87-7)**
Phensuximide **(86-34-0)**
Phentalamine **(50-60-2)**
Phentanyl **(437-38-7)**
Phentermine **(122-09-8)**
Phenthiazine **(92-84-2)**
Phenthoate **(2597-03-7)**
Phentin acetate **(900-95-8)**
Phentinoacetate **(900-95-8)**
Phentoin **(57-41-0)**
Phentolamine **(50-60-2)**
Phenudin **(2275-14-1)**
Phenuron **(63-98-9)**
Phenurone **(63-98-9)**
Phenutal **(63-98-9)**
Phenvalerate **(51630-58-1)**
Phenychol **(93-54-9)**
Phenyl-Idium **(136-40-3)**
Phenyl-Idium 200 **(136-40-3)**
Phenyl J Acid **(119-40-4)**
Phenyl Mustard Oil **(103-72-0)**
Phenylacetaldehyde **(122-78-1)**
Phenylacetaldehyde dimethyl acetal **(101-48-4)**
2-Phenylacetamide **(103-81-1)**
N-Phenylacetamide **(103-84-4)**
α-Phenylacetamide **(103-81-1)**
Phenylacetamidopenicillanic acid **(61-33-6)**
4'-Phenylacetanilide **(4075-79-0)**
p-Phenylacetanilide **(4075-79-0)**
Phenyl acetate **(122-79-2)**
Phenylacetate sodium salt **(114-70-5)**
N'-Phenylacethydrazide **(114-83-0)**
Phenylacetic acid **(103-82-2)**
ω-Phenylacetic acid **(103-82-2)**
Phenylacetic acid amide **(103-81-1)**
Phenylacetic acid chloride **(103-80-0)**
Phenylacetic acid, ethyl ester **(101-97-3)**
Phenylacetic acid, isobutyl ester **(102-13-6)**
Phenylacetic acid, methyl ester **(101-41-7)**
Phenylacetic acid, phenethyl ester **(102-20-5)**
Phenylacetic acid sodium salt **(114-70-5)**
Phenylacetic aldehyde **(122-78-1)**
N-Phenylacetoacetamide **(102-01-2)**
4-Phenylacetoacetic acid **(25832-09-1)**
Phenylacetone **(103-79-7)**
α-Phenylacetone **(103-79-7)**
2-Phenylacetonitrile **(140-29-4)**
Phenylacetonitrile **(140-29-4)**
Phenylacetonitrile, Liquid [UN 2470] **(140-29-4)**
2-Phenylacetophenone **(451-40-1)**
4'-Phenylacetophenone **(92-91-1)**
4-Phenylacetophenone **(92-91-1)**
p-Phenylacetophenone **(92-91-1)**
3-(α-Phenyl-β-acetylaethyl)-4-hydroxycumarin (German) **(81-81-2)**
Phenyl-β-acetylamine **(103-81-1)**
α-Phenylacetyl chloride **(103-80-0)**
Phenylacetyl chloride [UN 2577] **(103-80-0)**
Phenylacetylene **(536-74-3)**
3-(1'-Phenyl-2'-acetylethyl)-4-hydroxycoumarin **(81-81-2)**
3-(α-Phenyl-β-acetylethyl)-4-hydroxycoumarin **(81-81-2)**
(Phenyl-1 acetyl-2 ethyl) 3-hydroxy-4 coumarine (French) **(81-81-2)**
Phenyl acetyl nitrile **(140-29-4)**
(Phenylacetyl)urea **(63-98-9)**
α-Phenylacetylurea **(63-98-9)**
Phenylacetyluree (French) **(63-98-9)**
3-Phenylacrolein **(104-55-2)**
Phenylacrolein **(104-55-2)**
3-Phenylacrylic acid **(621-82-9)**
Phenylacrylic acid **(621-82-9)**
tert-β-Phenylacrylic acid **(621-82-9)**
trans-3-Phenylacrylic acid **(140-10-3)**
α.-Phenylacrylonitrile **(495-10-3)**
2-Phenylacrylonitrile **(495-10-3)**
3-Phenylacrylophenone **(94-41-7)**
β-Phenylacrylophenone **(94-41-7)**
β-Phenylacryloyl chloride **(102-92-1)**
β-Phenylaethylamin (German) **(64-04-0)**
1-Phenyl-2-aethylamino-propan (German) **(457-87-4)**
Phenyl-aethyl-barbitursaeure natrium (German) **(57-30-7)**
1-Phenylaethylbiguanid hydrochlorid (German) **(834-28-6)**
Phenylaethylcarbinol (German) **(93-54-9)**
Phenyl-aethyl-glutarsaeureimid (German) **(77-21-4)**
5-Phenyl-5-aethylhexahydropyrimidindion-(4,6) (German) **(125-33-7)**
Phenylaethyl-hydrazin **(156-51-4)**
(+-)-Phenylalanine **(150-30-1)**
(-)-β-Phenylalanine **(63-91-2)**
(L)-Phenylalanine **(63-91-2)**
(S)-Phenylalanine **(63-91-2)**
3-Phenyl-L-alanine **(63-91-2)**
3-Phenylalanine **(63-91-2)**
D-Phenylalanine **(673-06-3)**
D-β-Phenylalanine **(673-06-3)**
DL-3-Phenylalanine **(150-30-1)**
DL-β-Phenyl-α-alanine **(150-30-1)**
DL-β-Phenylalanine **(150-30-1)**
L-Phenylalanine **(63-91-2)**
L-β-Phenylalanine **(63-91-2)**
Phenyl-α-alanine **(63-91-2)**
β-Phenyl-L-alanine **(63-91-2)**
β-Phenyl-α-alanine **(63-91-2)**
β-Phenylalanine **(63-91-2)**
β-Phenylalanine, dl- **(150-30-1)**
DL-Phenylalanine (9CI) **(150-30-1)**
Phenylalanine (9CI) **(63-91-2)**
β-Phenyl-α-alanine, L- **(63-91-2)**
l-Phenylalanine, N-l-α-aspartyl-, 1-methyl ester (9CI) **(22839-47-0)**
DL-Phenylalanine, N-benzoyl-N-(3-chloro-4-fluorophenyl)-(9CI) **(72274-16-9)**
DL-Phenylalanine, 4-(bis(2-chloroethyl)amino)- **(531-76-0)**
L-Phenylalanine, 4-(bis(2-chloroethyl)amino)- **(148-82-3)**
D-Phenylalanine, 4-(bis(2-chloroethyl)amino)- (9CI) **(13045-94-8)**
L-Phenylalanine, N-((5-chloro-3,4-dihydro-8-hydroxy-3-methyl-1-oxo-1H-2-benzopyran-7-yl)carbonyl)-, ethyl ester, (R)- **(4865-85-4)**
DL-Phenylalanine, 4-fluoro- (9CI) **(51-65-0)**
Phenylalanine, 4-fluoro- (9CI) **(60-17-3)**
L-Phenylalanine, 3-hydroxy- (9CI) **(587-33-7)**
D-Phenylalanine mustard **(13045-94-8)**

DL-Phenylalanine mustard (531-76-0)
L-Phenylalanine mustard (148-82-3)
Phenylalanine mustard (148-82-3)
Phenylalanine nitrogen mustard (148-82-3)
DL-Phenylalanin-lost (German) (531-76-0)
Phenylalanin-lost (German) (531-76-0)
Phenylalaninum (Latin) (63-91-2)
3-Phenylallyl alcohol (104-54-1)
γ-Phenylallyl alcohol (104-54-1)
Phenylamine (62-53-3)
Phenylamine hydrochloride (142-04-1)
N-Phenyl-p-aminoaniline (101-54-2)
1-Phenyl-2-amino-athan (German) (64-04-0)
4-(Phenylamino)benzenediazonium sulfate (1:1) (4477-28-5)
2-(Phenylamino)benzoic acid (91-40-7)
4-(Phenylamino)butane (1126-78-9)
1-Phenyl-4-amino-5-chloropyridaz-6-one (1698-60-8)
1-Phenyl-4-amino-5-chloropyridazin-6-one (1698-60-8)
1-Phenyl-4-amino-5-chloro-6-pyridazone (1698-60-8)
1-Phenyl-4-amino-5-chloropyridazone-6 (1698-60-8)
1-Phenyl-4-amino-5-chlorpyridaz-6-one (1698-60-8)
1-Phenyl-4-amino-5-chlorpyridazon-(6) (German) (1698-60-8)
2,2'-(Phenylamino)diethanol (120-07-0)
4-Phenylaminodiphenylamine (74-31-7)
p-Phenylaminodiphenylamine (74-31-7)
1-Phenyl-2-aminoethane (64-04-0)
2-(Phenylamino)ethanol (122-98-5)
(Phenylamino)methanesulfonic acid (103-06-0)
Phenylaminomethanesulfonic acid, monosodium salt (26021-90-9)
(Phenylamino)methanesulfonic acid, sodium salt (26021-90-9)
2-Phenylaminonaphthalene (135-88-6)
1-(Phenylamino)-8-naphthalenesulfonic acid (82-76-8)
8-(Phenylamino)-1-naphthalenesulfonic acid (82-76-8)
3-(Phenylamino)phenol (101-18-8)
4-Phenylaminophenol (122-37-2)
N-Phenyl-p-aminophenol (122-37-2)
Phenyl-p-aminophenol (122-37-2)
4-((4-(Phenylamino)phenyl)amino)phenol (101-74-6)
1-Phenyl-2-amino-propan (German) (60-15-1)
d-1-Phenyl-2-aminopropan (German) (51-64-9)
1-Phenyl-2-aminopropane (60-15-1)
d-1-Phenyl-2-aminopropane (51-64-9)
dl-1-Phenyl-2-aminopropane (300-62-9)
DL-1-Phenyl-2-aminopropane sulfate (60-13-9)
d-1-Phenyl-2-aminopropane sulfate (51-63-8)
dextro-1-Phenyl-2-amino-propane sulfate (51-63-8)
dl-1-Phenyl-2-aminopropanol-1 (14838-15-4)
DL-1-Phenyl-2-amino-1-propanol monohydrochloride (154-41-6)
3-Phenyl-5-amino-1,2,4-triazolyl-(1)-(N,N'-tetramethyl) diamidophosphonate (1031-47-6)
1-Phenylazo-β-naphthol (842-07-9)
2-Phenylaniline (90-41-5)
N-Phenylaniline (122-39-4)
o-Phenylaniline (90-41-5)
p-Phenylaniline (92-67-1)
2-Phenylanisole (86-26-0)
o-Phenyl anisole (86-26-0)
N-Phenylanthranilic acid (91-40-7)
Phenylanthranilic acid (91-40-7)
Phenyl arsenic acid (98-05-5)
Phenylarsinedichloride (696-28-6)
Phenylarsonic acid (98-05-5)
3-Phenylazo-2,6-diaminopyridine hydrochloride (136-40-3)
β-Phenylazo-α,α'-diaminopyridine hydrochloride (136-40-3)
Phenylazo-α,α'-diaminopyridine monohydrochloride (136-40-3)
2-Phenylazo-4-methyl-1-phenol (952-47-6)
2-Phenylazo-4-methylphenol (952-47-6)
1-Phenylazo-2-naphthol (842-07-9)
1-Phenylazo-2-naphthol-6-carboxylic acid (69644-64-0)
1-Phenylazo-2-naphthol-6,8-disulfonic acid, disodium salt (1936-15-8)
1-Phenylazo-2-naphthol-6,8-disulphonic acid, disodium salt (1936-15-8)
1-Phenylazo-2-naphthylamine (85-84-7)
4-Phenylazo-1-naphthylamine (131-22-6)
Phenylazo (136-40-3)
Phenylazo Tablets (136-40-3)
4'-Phenylazoacetanilide (4128-71-6)
p-Phenylazoacetanilide (4128-71-6)
4-(Phenylazo)aniline (60-09-3)
p-(Phenylazo)aniline (60-09-3)
4-(Phenylazo)benzenamine (60-09-3)
4-(Phenylazo)-1,3-benzenediamine, monohydrochloride (532-82-1)
4-(Phenylazo)benzenesulfonic acid (2484-88-0)
4-(Phenylazo)benzoic acid (1562-93-2)
4-(Phenylazo)biphenyl (7466-42-4)
Phenylazodiaminopyridine hydrochloride (136-40-3)
4-(Phenylazo)-N,N-dimethylaniline (60-11-7)
4-Phenylazodiphenyl (7466-42-4)
1-(Phenylazo)-2-naphthalenamine (85-84-7)
1-(Phenylazo)-2-naphthalenol (842-07-9)
4-((4-(Phenylazo)-1-naphthalenyl)azo)phenol (6253-10-7)
Phenylazo α-naphthylamine (131-22-6)
4-Phenylazophenol (1689-82-3)
p-Phenylazophenol (1689-82-3)
p-Phenylazophenylamine (60-09-3)
1-((p-Phenylazo)phenyl)azo-2-naphthol (85-86-9)
1-((4-(Phenylazo)phenyl)azo)-2-naphthalenol (85-86-9)
4-((4-(Phenylazo)phenyl)azo)phenol (6250-23-3)
4-(Phenylazo)-m-phenylenediamine, monohydrochloride (532-82-1)
3-(Phenylazo)-2,6-pyridinediamine (94-78-0)
3-(Phenylazo)-2,6-pyridinediamine, hydrochloride (136-40-3)
Phenylazopyridine hydrochloride (136-40-3)
4-(Phenylazo)resorcinol (2051-85-6)
Phenylazo tablet (94-78-0)
(Phenylazo-4-phenylazo)-1-ethylamino-2-naphthalene (6368-72-5)
1-(4-Phenylazo-phenylazo)-2-ethylaminonaphthalene (6368-72-5)
4-Phenylazo-m-phenylenediamine hydrochloride (532-82-1)
N-Phenylbenzamide (93-98-1)
N-Phenylbenzenamine (122-39-4)
Phenylbenzene (92-52-4)
4-Phenyl-1,2-benzenediol (92-05-7)
9-Phenyl-3,4-benzfluorene (32377-10-9)
2-Phenyl-1H-benzimidazole-5-sulfonic acid (27503-81-7)
2-Phenylbenzimidazole-5-sulfonic acid (27503-81-7)
Phenyl benzoate (93-99-2)
7-Phenyl-7H-benzo(c)fluorene (32377-10-9)
1-Phenyl-2-benzoylethylene (94-41-7)
p-Phenylbenzyl alcohol (3597-91-9)
α-Phenylbenzylcyanide (86-29-3)
N-Phenyl-N-benzyl-N',N'-dimethylethylenediamine hydrochloride (2045-52-5)
Phenyl benzyl ketone (451-40-1)
4-Phenylbiphenyl (92-94-4)
Phenylbis(2-chloroethylamine) (553-27-5)
Phenyl bis(chloromethyl)phosphinate (14212-98-7)
Phenyl bromide (108-86-1)
Phenyl, 3-bromo- (9CI) (2973-44-6)
Phenyl, m-bromo- (8CI) (2973-44-6)
1-Phenyl-1-bromoethane (585-71-7)
2-Phenyl-1-bromoethane (103-63-9)
4-Phenylbut-2-enoic acid (2243-52-9)
N-Phenylbutanamide (1129-50-6)
1-Phenylbutane (104-51-8)
2-Phenylbutane (135-98-8)
1-Phenyl-1-butanone (495-40-9)
4-Phenyl-2-butanone (2550-26-7)

4-Phenylbutan-2-one **(2550-26-7)**
Phenylbutaz **(50-33-9)**
Phenylbutazon (German) **(50-33-9)**
Phenylbutazone **(50-33-9)**
Phenylbutazonum **(50-33-9)**
1-Phenyl-2-butene **(1560-06-1)**
1-Phenylbutene-2 **(1560-06-1)**
4-Phenyl-1-butene **(768-56-9)**
4-Phenylbutene-1 **(768-56-9)**
1-Phenyl-3-buten-2-one **(37442-55-0)**
4-Phenyl-3-buten-2-one **(122-57-6)**
Phenylbuttersaeure-lost (German) **(305-03-3)**
2-Phenyl-tert-butylamine **(122-09-8)**
Phenyl-sec-butyl norsuprifen **(447-41-6)**
2-Phenylbutyric acid **(90-27-7)**
3-Phenylbutyric acid **(4593-90-2)**
α-Phenyl butyric acid **(90-27-7)**
Phenylbutyric acid nitrogen mustard **(305-03-3)**
N-Phenylcarbamate d'isopropyle (French) **(122-42-9)**
Phenylcarbamic acid **(501-82-6)**
Phenylcarbamic acid carboxyethyl ester **(73622-98-7)**
N-Phenylcarbamic acid, isopropyl ester **(122-42-9)**
Phenylcarbamic acid, 1-methylethyl ester **(122-42-9)**
Phenylcarbamide **(64-10-8)**
3-(N-Phenylcarbamoyl)-2-naphthol **(92-77-3)**
2-Phenyl-carbamoyloxy-N-aethyl-propionamid (German) **(16118-49-3)**
(Phenylcarbamoyloxy)-2-N-ethylpropionamide **(16118-49-3)**
O-(N-Phenylcarbamoyl)-propanonoxim (German) **(2828-42-4)**
O-(N-Phenylcarbamoyl)propanonoxime **(2828-42-4)**
1-Phenyl-3-carbethoxypyrazolone **(89-33-8)**
Phenylcarbimide **(103-71-9)**
Phenylcarbinol **(100-51-6)**
Phenyl carbitol **(104-68-7)**
Phenyl carbonate **(102-09-0)**
Phenyl carbonimide **(103-71-9)**
N-Phenylcarbonimidic dichloride **(622-44-6)**
Phenylcarbonylaminoacetic acid **(495-69-2)**
Phenylcarboxyamide **(55-21-0)**
Phenyl carboxylic acid **(65-85-0)**
Phenyl carbylamine chloride **(622-44-6)**
Phenylcarbylamine chloride [UN 1672] **(622-44-6)**
4-Phenylcatechol **(92-05-7)**
Phenyl cellosolve **(122-99-6)**
Phenyl chloride **(108-90-7)**
N-Phenylchloroacetamide **(587-65-5)**
Phenyl chloroacetate **(620-73-5)**
Phenyl chlorocarbonate **(1885-14-9)**
Phenyl chloroform **(98-07-7)**
Phenyl chloroformate **(1885-14-9)**
Phenylchloroformate [UN 2746] **(1885-14-9)**
Phenyl chloromercury **(100-56-1)**
Phenylchloromethylketone **(532-27-4)**
1-Phenyl-2-chloro-2-methyl-1-propanone **(7473-99-6)**
2-Phenyl-6-chlorophenol **(85-97-2)**
4-Phenyl-2-chlorophenol **(92-04-6)**
2-(2-Phenyl-2-(4-chlorophenyl)acetyl)-1,3-indandione **(3691-35-8)**
Phenylcholon **(93-54-9)**
2-Phenylcinchonic acid **(132-60-5)**
2-Phenylcinchoninic acid **(132-60-5)**
α-Phenyl-p-cresol **(101-53-1)**
4-Phenylcrotonic acid **(2243-52-9)**
Phenyl cyanide **(100-47-0)**
3-Phenyl-1,4-cyclohexadiene **(4794-05-2)**
Phenylcyclohexane **(827-52-1)**
2-Phenyl cyclohexanol **(1444-64-0)**
Phenylcyclohexene **(31017-40-0)**
1-Phenylcyclohexylamine **(2201-24-3)**
1-(1-Phenylcyclohexyl)piperidine **(77-10-1)**
1-(1-Phenylcyclohexyl)piperidine hydrochloride **(956-90-1)**
1-(1-Phenylcyclohexyl)-pyrrolidine **(2201-39-0)**
1-(1-Phenylcyclohexyl)pyrrolidine **(2201-39-0)**
1-Phenyldecane **(104-72-3)**
1-Phenyl-1-decanol **(21078-95-5)**
1-Phenyl-1-decanone **(6048-82-4)**
2-Phenyl-4,6-diamino-1,3,5-triazine **(91-76-9)**
2-Phenyl-4,6-diamino-s-triazine **(91-76-9)**
2-Phenyldiazenecarboxamide **(103-03-7)**
Phenyldiazonium hexafluorophosphate **(369-58-4)**
Phenyl dichlorarsine **(696-28-6)**
Phenyldichloroarsine **(696-28-6)**
Phenyl dichloroarsine (DOT) **(696-28-6)**
Phenyl dichlorophosphate **(770-12-7)**
Phenyldichlorophosphine **(644-97-3)**
Phenyl-5,6-dichloro-2-trifluoromethyl-benzimidazole-1-carboxylate **(14255-88-0)**
Phenyl didecyl phosphite **(1254-78-0)**
N-Phenyldiethanolamine **(120-07-0)**
Phenyl diethanolamine **(120-07-0)**
N-Phenyldiethanolamine bis(2-methoxy-3,6-dichlorobenzoate) **(56141-00-5)**
1-Phenyl-2-diethylamino-1-propanone **(90-84-6)**
1-Phenyl-3,5-diethyl-2-propyl-1,2-dihydropyridine **(34562-31-7)**
1-Phenyl-3-(O,O-diethyl-thionophosphoryl)-1,2,4-triazole **(24017-47-8)**
1-Phenyl-3,4-dihydronaphthalene **(7469-40-1)**
2-Phenyl-1,3-diketohydrindene **(83-12-5)**
Phenyldimazone **(467-85-6)**
Phenyl dimethicone **(9005-12-3)**
N-Phenyl-N'-(1,3-dimethyl butyl)-para-phenylenediamine **(793-24-8)**
Phenyldimethylcarbinol **(617-94-7)**
1-Phenyl-2,3-dimethyl-4-dimethylaminopyrazol-5-one **(58-15-1)**
1-Phenyl-2,3-dimethyl-4-dimethylaminopyrazolone-5 **(58-15-1)**
Phenyldimethylfluorosilane **(454-57-9)**
1-Phenyl-3,5-dimethyl-4-nitroso-pyrazol (German) **(715-99-1)**
1-Phenyl-1-(3,4-dimethylphenyl)ethane **(6196-95-8)**
o-Phenyl-N,N'-dimethyl phosphorodiamidate **(1754-58-1)**
1-Phenyl-2,3-dimethylpyrazole-5-one **(60-80-0)**
1-Phenyl-2,3-dimethyl-5-pyrazolone **(60-80-0)**
Phenyldimethylsilane **(766-77-8)**
1-Phenyl-3,3-dimethyltriazene **(7227-91-0)**
1-Phenyl-3,3-dimethyl-triazine **(7227-91-0)**
Phenyl-dimethyl-triazine **(7227-91-0)**
1-Phenyl-3,3-dimethylurea **(101-42-8)**
3-Phenyl-1,1-dimethylurea **(101-42-8)**
N-Phenyl-N',N'-dimethylurea **(101-42-8)**
3-Phenyl-1,1-dimethylurea, trichloroacetate **(4482-55-7)**
4-Phenyldiphenyl **(92-94-4)**
1-Phenyl-2-(1',1'-diphenylpropyl-3'-amino)propane **(390-64-7)**
Phenyl-disulfide **(882-33-7)**
Phenyldodecan (German) **(123-01-3)**
1-Phenyldodecane **(123-01-3)**
Phenyl-drane **(61-76-7)**
1-Phenyleicosane **(2398-68-7)**
Phenylen **(83-12-5)**
N,N'-p-Phenylenebis(4-((2,5-dichlorophenyl)azo)-3-hydroxy-2-naphthalenecarboxamide) **(3905-19-9)**
1,1'-(1,4-Phenylene)bisethanone **(1009-61-6)**
p-Phenylenebis(β-hydroxyethyl) ether **(104-38-1)**
N,N'-(m-Phenylene)bismaleimide **(3006-93-7)**
m-Phenylenebis(methylamine) **(1477-55-0)**
2,2'-(1,3-Phenylenebis(oxymethylene))bisoxirane **(101-90-6)**
2,2'-(1,4-Phenylene)bis(5-phenyloxazole) **(1806-34-4)**
1,1'-(m-Phenylene)bis-1H-pyrrole-2,5-dione (9CI) **(3006-93-7)**
1,2-Phenylenediamine **(95-54-5)**
1,3-Phenylenediamine **(108-45-2)**
1,4-Phenylenediamine **(106-50-3)**

m-Phenylenediamine (108-45-2)
o-Phenylenediamine (95-54-5)
p-Phenylenediamine (106-50-3)
p-Phenylenediamine (ACGIH,OSHA) [UN 1673] (106-50-3)
Phenylenediamine (8CI) (25265-76-3)
m-Phenylenediamine [UN 1673] (108-45-2)
o-Phenylenediamine [UN 1673] (95-54-5)
p-Phenylenediamine, N-(p-aminophenyl)- (537-65-5)
m-Phenylenediamine, 4-((p-aminophenyl)azo)- (8CI) (6364-34-7)
p-Phenylenediamine, N,N'-bis(1,4-dimethylpentyl) (3081-14-9)
p-Phenylenediamine, N,N'-bis(1-ethyl-3-methylpentyl) (139-60-6)
p-Phenylenediamine, N,N-bis(2-hydroxyethyl)-N'-methyl-2-nitro- (2784-94-3)
p-Phenylenediamine, N,N'-bis(1-methylheptyl)- (8CI) (103-96-8)
1,4-Phenylenediaminebis(4,5,6,7-tetrachloroisoindolin-1-one-3-ylidene) (5590-18-1)
o-Phenylenediamine, 4-butyl- (3663-23-8)
m-Phenylenediamine, 4-chloro (5131-60-2)
o-Phenylenediamine, 4-chloro (95-83-0)
p-Phenylenediamine, 2-chloro (615-66-7)
p-Phenylenediamine, 2-chloro-, dihydrochloride (615-46-3)
m-Phenylenediamine,4-chloro-, sulfate (68239-80-5)
p-Phenylenediamine, 2-chloro-, sulfate (6219-71-2)
p-Phenylenediamine, N-cyclohexyl-N'-phenyl- (101-87-1)
p-Phenylenediamine, N,N'-di-sec-butyl (101-96-2)
p-Phenylenediamine, 2,5-dichloro (20103-09-7)
p-Phenylenediamine, 2,6-dichloro (609-20-1)
p-Phenylenediamine, N,N'-dicyclohexyl (4175-38-6)
p-Phenylenediamine, N,N-diethyl (93-05-0)
p-Phenylenediamine, N,N-diethyl-, hydrochloride (2198-58-5)
p-Phenylenediamine, N,N-diethyl-3-methyl-,hydrochloride (2051-79-8)
1,3-Phenylenediamine dihydrochloride (541-69-5)
1,4-Phenylenediamine dihydrochloride (624-18-0)
m-Phenylenediamine, dihydrochloride (541-69-5)
o-Phenylenediamine, dihydrochloride (615-28-1)
p-Phenylenediamine, dihydrochloride (624-18-0)
p-Phenylenediamine, N,N'-dimethyl (105-10-2)
p-Phenylenediamine, N,N-dimethyl (99-98-9)
p-Phenylenediamine, N-(1,3-dimethylbutyl)-N'-phenyl- (793-24-8)
p-Phenylenediamine, N-(1,4-dimethylpentyl)-N'-phenyl- (3081-01-4)
p-Phenylenediamine, N,N'-(di-2-naphthyl) (93-46-9)
p-Phenylenediamine, N-(2,4-dinitrophenyl)- (8CI) (6373-73-5)
p-Phenylenediamine, N,N'-di-sec-octyl- (8CI) (28633-36-5)
p-Phenylenediamine, N,N'-diphenyl (74-31-7)
o-Phenylenediamine, 4-ethoxy- (1197-37-1)
m-Phenylenediamine, 4-ethoxy-, dihydrochloride (67801-06-3)
m-Phenylenediamine hydrochloride (541-69-5)
p-Phenylenediamine hydrochloride (624-18-0)
p-Phenylenediamine, N-isopropyl-N'-phenyl (101-72-4)
Phenylenediamine, meta, Solid (DOT) (108-45-2)
p-Phenylenediamine, 2-methoxy (5307-02-8)
m-Phenylenediamine, 4-methoxy- (8CI) (615-05-4)
m-Phenylenediamine, 4-methoxy-, dihydrochloride (614-94-8)
m-Phenylenediamine, 4-methoxy-, sulfate (39156-41-7)
p-Phenylenediamine, monohydrochloride (540-24-9)
m-Phenylenediamine, 4-(1-naphthylazo) (6416-57-5)
m-Phenylenediamine, 4-nitro (5131-58-8)
m-Phenylenediamine, 5-nitro- (5042-55-7)
o-Phenylenediamine, 4-nitro (99-56-9)
p-Phenylenediamine, 2-nitro (5307-14-2)
o-Phenylenediamine, 4-nitro-, dihydrochloride (6219-77-8)
p-Phenylenediamine, 2-nitro-, dihydrochloride (18266-52-9)
p-Phenylenediamine, N-(p-nitrophenyl)- (8CI) (6149-34-4)
p-Phenylenediamine, 3-nitro-N^1,N^1,N^4-tris(2-hydroxyethyl)- (33229-34-4)
Phenylenediamine, para, Solid (DOT) (106-50-3)
p-Phenylenediamine, N-phenyl (101-54-2)
m-Phenylenediamine, 4-(phenylazo) (495-54-5)
m-Phenylenediamine, 4-(phenylazo)-, hydrochloride (532-82-1)
p-Phenylenediamine, N-phenyl-N'-cyclohexyl (101-87-1)
1,3-Phenylenediamine sulfate (541-70-8)
m-Phenylenediamine sulfate (541-70-8)
m-Phenylenediamine, sulfate (1:1) (541-70-8)
p-Phenylenediamine sulfate (16245-77-5)
1,3-Phenylenediamine-4-sulfonic acid (88-63-1)
m-Phenylenediamine-4-sulfonic acid (88-63-1)
m-Phenylenediaminesulfonic acid (88-63-1)
p-Phenylenediamine, N,N,N',N'-tetramethyl (100-22-1)
m-Phenylenediamine, 2,4,6-trinitro- (8CI) (1630-08-6)
p-Phenylene diazide (2294-47-5)
p-Phenylenedicarbonyl dichloride (100-20-9)
m-Phenylenedichloride (541-73-1)
(Phenylenediisopropylidene)bis(tert-butylperoxide) (25155-25-3)
Phenylene-1,4-diisothiocyanate (4044-65-9)
N,N'-(m-Phenylenedimaleimide) (3006-93-7)
o-Phenylenediol (120-80-9)
2,2'-(Phenylenedioxy)diethanol (104-38-1)
2,2'-(p-Phenylenedioxy)diethanol (104-38-1)
2,2'-(1,4-Phenylenebis(oxy))bisethanol (104-38-1)
1,10-(1,2-Phenylene)pyrene (193-39-5)
1,10-(ortho-Phenylene)pyrene (193-39-5)
2,3-Phenylenepyrene (193-39-5)
2,3-o-Phenylenepyrene (193-39-5)
o-Phenylenepyrene (193-39-5)
Phenylene thiocyanate (4044-65-9)
o-Phenylenethiourea (583-39-1)
(-)-Phenylephrine (59-42-7)
Phenylephrine (59-42-7)
R(-)-Phenylephrine (59-42-7)
D-(-)-Phenylephrine hydrochloride (61-76-7)
Phenylephrine hydrochloride (61-76-7)
1-Phenyl-1,2-epoxyethane (96-09-3)
Phenyl 2,3-epoxypropyl ether (122-60-1)
Phenylessigsaure natrium-salz (German) (114-70-5)
Phenylethanal (122-78-1)
Phenylethane (100-41-4)
Phenylethanediol (93-56-1)
Phenylethane-p-sulfonate (98-69-1)
1-Phenylethanol (98-85-1)
2-Phenylethanol (60-12-8)
β-Phenylethanol (60-12-8)
N-Phenylethanolamine (122-98-5)
Phenyl ethanolamine (122-98-5)
1-Phenylethanone (98-86-2)
Phenylethene (100-42-5)
2-(2-Phenylethenyl)-1,3-dioxolane (5660-60-6)
Phenyl ether (ACGIH,OSHA) (101-84-8)
Phenyl ether, Polymer with formaldehyde (8CI) (26007-63-6)
Phenyl ether-biphenyl Mixture (OSHA) (8004-13-5)
Phenyl ether, 4-bromo- (101-55-3)
Phenyl ether dichloro (28675-08-3)
Phenyl ether, hexachloro deriv. (8CI) (31242-93-0)
Phenyl ether, 2'-hydroxy-2,4,4'-trichloro- (3380-34-5)
Phenyl ether pentachloro (42279-29-8)
Phenyl ether tetrachloro (31242-94-1)
Phenyl ether, tetrachloro deriv. (8CI) (42279-29-8)
cis-2-(1-Phenylethoxy)carbonyl-1-methylvinyl dimethylphosphate (7700-17-6)
1-Phenylethyl acetate (93-92-5)
2-Phenylethyl acetate (103-45-7)
α-Phenylethyl acetate (93-92-5)
β-Phenylethyl acetate (103-45-7)
sec-Phenylethyl acetate (93-92-5)
2-Phenylethyl alcohol (60-12-8)
Phenylethyl alcohol (60-12-8)
β-Phenylethyl alcohol (60-12-8)
1-Phenylethylamine (98-84-0)

Name	CAS No.
2-Phenylethylamine	**(64-04-0)**
Phenylethylamine	**(64-04-0)**
α-Phenylethylamine	**(98-84-0)**
β-Phenylethylamine	**(64-04-0)**
ω-Phenylethylamine	**(64-04-0)**
β-Phenylethylamine sulfate	**(5471-08-9)**
2-Phenylethyl-o-aminobenzoate	**(133-18-6)**
2-Phenyl-3-ethylaminobicyclo(2.2.1)heptane	**(1209-98-9)**
1-Phenyl-2-ethylaminopropane	**(457-87-4)**
α-Phenyl-β-ethylaminopropane	**(457-87-4)**
2-Phenylethyl anthranilate	**(133-18-6)**
Phenylethyl anthranilate	**(133-18-6)**
Phenylethylbarbiturate	**(50-06-6)**
5-Phenyl-5-ethylbarbituric acid	**(50-06-6)**
Phenyl-ethyl-barbituric acid	**(50-06-6)**
Phenylethylbarbituric acid, sodium salt	**(57-30-7)**
(Phenylethyl)benzene	**(38888-98-1)**
Phenylethylbenzylamine	**(92-59-1)**
Phenylethylbiguanide	**(114-86-3)**
N-β-Phenylethyl biguanide hydrochloride	**(834-28-6)**
1-Phenylethyl bromide	**(585-71-7)**
2-Phenylethyl bromide	**(103-63-9)**
α-Phenylethyl bromide	**(585-71-7)**
β-Phenylethyl bromide	**(103-63-9)**
Phenylethyl bromide (VAN)	**(103-63-9)**
Phenylethyl carbamate	**(101-99-5)**
N-Phenyl-1-(ethylcarbamoyl-1)-ethylcarbamate, D isomer	**(16118-49-3)**
α-Phenylethyl chloride	**(672-65-1)**
3-Phenyl-3-ethyl-2,6-diketopiperidine	**(77-21-4)**
1-Phenylethyl 3-(dimethoxyphosphinoyloxy)isocrotonate	**(7700-17-6)**
Phenylethyl dimethyl carbinol	**(103-05-9)**
3-Phenyl-3-ethyl-2,6-dioxopiperidine	**(77-21-4)**
Phenylethylene	**(100-42-5)**
Phenylethylene (OSHA)	**(100-42-5)**
1-Phenylethylene glycol	**(93-56-1)**
Phenylethylene glycol	**(93-56-1)**
Phenylethylene oxide	**(96-09-3)**
N-Phenyl-N-ethylethanolamine	**(92-50-2)**
Phenylethylethanolamine	**(92-50-2)**
Phenyl ethyl ether	**(103-73-1)**
2-Phenyl-2-ethylglutaric acid imide	**(77-21-4)**
α-Phenyl-α-ethylglutaric acid imide	**(77-21-4)**
α-Phenyl-α-ethylglutarimide	**(77-21-4)**
5-Phenyl-5-ethyl-hexahydropyrimidine-4,6-dione	**(125-33-7)**
2-Phenylethylhydrazine	**(51-71-8)**
β-Phenylethylhydrazine	**(51-71-8)**
β-Phenylethylhydrazine dihydrogen sulfate	**(156-51-4)**
2-Phenylethylhydrazine dihydrogen sulphate	**(156-51-4)**
Phenylethylhydrazine dihydrogen sulphate	**(156-51-4)**
2-Phenylethylhydrazine hydrogen sulphate	**(156-51-4)**
β-Phenylethylhydrazine hydrogen sulphate	**(156-51-4)**
β-Phenylethylhydrazine sulfate	**(156-51-4)**
2-Phenylethylhydrazine sulphate	**(156-51-4)**
Phenylethylhydrazine sulphate	**(156-51-4)**
Phenyl ethyl ketone	**(93-55-0)**
Phenylethylmalonamide	**(7206-76-0)**
Phenylethylmalondiamide	**(7206-76-0)**
Phenylethylmalonylurea	**(50-06-6)**
5-Phenyl-5-ethyl-3-methylbarbituric acid	**(115-38-8)**
2-Phenylethyl p-methylbenzenesulfonate	**(4455-09-8)**
Phenylethylmethylhydantoin	**(50-12-4)**
β-Phenylethyl methyl ketone	**(2550-26-7)**
3-Phenyl-N-ethyl-2-norbornanamine	**(1209-98-9)**
1-Phenylethyl 3-oxobutanoate	**(40552-84-9)**
2-Phenylethyl phenylacetate	**(102-20-5)**
Phenylethyl phenylacetate	**(102-20-5)**
β-Phenylethyl phenylacetate	**(102-20-5)**
1-(2-Phenylethyl)-4-phenyl-4-acetoxypiperidine	**(64-52-8)**
N-(1-(2-Phenylethyl)-4-piperidyl)-N-(4-fluorophenyl)-propanamide its optical isomers, salts and salts of isomers	**(90736-23-5)**
2-Phenylethyl propanoate	**(122-70-3)**
2-Phenylethyl propionate	**(122-70-3)**
Phenylethyl propionate	**(122-70-3)**
2-Phenylethyl α-toluate	**(102-20-5)**
.β.-Phenylethyl tosylate	**(4455-09-8)**
2-Phenylethyl tosylate	**(4455-09-8)**
(2-Phenylethyl)trichlorosilane	**(940-41-0)**
Phenyletten	**(50-06-6)**
Phenyl fluoride	**(462-06-6)**
Phenylfluoroform	**(98-08-8)**
N-Phenylformamide	**(103-70-8)**
Phenyl formamide	**(103-70-8)**
Phenylformic acid	**(65-85-0)**
Phenyl-α-glycerol ether	**(538-43-2)**
Phenylglyceryl ether	**(538-43-2)**
Phenyl glycidyl ether (ACGIH,OSHA)	**(122-60-1)**
N-Phenylglycine monopotassium salt	**(19525-59-8)**
N-Phenylglycine monosodium salt	**(10265-69-7)**
N-Phenylglycine potassium salt	**(19525-59-8)**
Phenyl glycol	**(93-56-1)**
Phenylglycolic acid	**(90-64-2)**
o-Phenylglycolic acid	**(122-59-8)**
Phenylglycydyl ether	**(122-60-1)**
Phenylglyoxal diethyl acetal	**(6175-45-7)**
Phenylglyoxylonitrile oxime O,O-diethyl phosphorothioate	**(14816-18-3)**
1-Phenylheptadecane	**(14752-75-1)**
4-Phenylheptane	**(2132-86-7)**
1-Phenylhexadecane	**(1459-09-2)**
2-Phenylhexane	**(6031-02-3)**
3-Phenylhexane	**(4468-42-2)**
2-Phenylhydracrylic acid	**(529-64-6)**
2-Phenylhydracrylic acid 3-α-tropanyl ester	**(51-55-8)**
Phenyl hydrate	**(108-95-2)**
2-Phenylhydrazide, carbamic acid	**(103-03-7)**
Phenylhydrazin (German)	**(100-63-0)**
Phenylhydrazine (ACGIH,OSHA) [UN 2572]	**(100-63-0)**
Phenylhydrazine-HCl	**(59-88-1)**
1-Phenylhydrazine carboxamide	**(103-03-7)**
2-Phenylhydrazinecarboxamide	**(103-03-7)**
Phenylhydrazine hydrochloride	**(59-88-1)**
Phenylhydrazine hydrochloride (VAN)	**(59-88-1)**
Phenylhydrazine monohydrochloride	**(59-88-1)**
Phenylhydrazin hydrochlorid (German)	**(59-88-1)**
Phenylhydrazinium chloride	**(59-88-1)**
Phenyl hydride	**(71-43-2)**
Phenylhydroquinone	**(1079-21-6)**
Phenylhydroxamic acid	**(495-18-1)**
Phenyl hydroxide	**(108-95-2)**
Phenylhydroxyacetic acid	**(90-64-2)**
threo-1-Phenyl-1-hydroxy-2-aminopropane	**(492-39-7)**
N-Phenylhydroxylamine	**(100-65-2)**
β-Phenylhydroxylamine	**(100-65-2)**
Phenyl hydroxymercury	**(100-57-2)**
1-Phenyl-2-hydroxy-2-methylpropan-1-one	**(7473-98-5)**
N-Phenyl-N-hydroxy-N'-methylurea	**(6263-38-3)**
1-Phenyl-2-(p-hydroxyphenyl)-3,5-dioxo-4-butylpyrazolidine	**(129-20-4)**
Phenyl 2-hydroxyphenyl ketone	**(117-99-7)**
2-Phenyl-2-hydroxypropionic acid	**(515-30-0)**
Phenylic acid	**(108-95-2)**
Phenylic alcohol	**(108-95-2)**
2-Phenyl-2-imidazoline	**(936-49-2)**
Phenylimidocarbonyl chloride	**(622-44-6)**
N-Phenylimidophosgene	**(622-44-6)**
3,3'-(Phenylimino)bispropanenitrile	**(1555-66-4)**

N-Phenyliminocarbonyl dichloride (622-44-6)
Phenyliminocarbonyl dichloride (622-44-6)
2,2'-(Phenylimino)diethanol (120-07-0)
2,2'-Phenyliminodiethanol diacetate (19249-34-4)
5-Phenyl-2-imino-4-oxazolidinone (2152-34-3)
5-Phenyl-2-imino-4-oxooxazolidine (2152-34-3)
2-Phenyl-1,3-indandione (83-12-5)
2-Phenylindan-1,3-dione (83-12-5)
Phenylindione (83-12-5)
2-Phenyl-1H-indole (948-65-2)
α-Phenylindole (948-65-2)
Phenyline (83-12-5)
Phenyl iodide (591-50-4)
Phenyl isocyanate [UN 2487] (103-71-9)
Phenylisocyanide (931-54-4)
Phenylisohydantoin (2152-34-3)
Phenylisonitrile dichloride (622-44-6)
3-Phenylisopentanoic acid (1010-48-6)
2-Phenylisopropanol (617-94-7)
β-Phenylisopropylamin (German) (60-15-1)
(Phenylisopropyl)amine (60-15-1)
β-Phenylisopropylamine (60-15-1)
β-Phenyl isopropylamine sulfate (60-13-9)
d-β-Phenylisopropylamine sulfate (51-63-8)
dextro-β-Phenylisopropylamine sulfate (51-63-8)
N-Phenyl isopropyl carbamate (122-42-9)
N-Phenyl-N'-isopropyl-p-phenylenediamine (101-72-4)
Phenyl isothiocyanate (103-72-0)
β-Phenylisovaleric acid (1010-48-6)
Phenyl ketone (119-61-9)
α-Phenyllactic acid (515-30-0)
Phenyllin (83-12-5)
Phenylmagnesium chloride (100-59-4)
N-Phenylmaleimide (941-69-5)
Phenyl mercaptan (ACGIH,OSHA) [UN 2337] (108-98-5)
Phenyl mercaptan, p-chloro- (106-54-7)
1-Phenyl-5-mercapto-1,2,3,4-tetrazole (86-93-1)
1-Phenyl-5-mercaptotetrazole (86-93-1)
Phenylmercaptotetrazole (86-93-1)
Phenylmercuriacetate (62-38-4)
Phenyl mercuric acetate (62-38-4)
Phenylmercuric acetate [UN 1674] (62-38-4)
Phenylmercuric ammonium acetate (53404-67-4)
Phenylmercuric ammonium propionate (53404-68-5)
Phenyl mercuric chloride (100-56-1)
Phenylmercuric dimethyldithiocarbamate (32407-99-1)
Phenylmercuric hydroxide [UN 1894] (100-57-2)
Phenylmercuric lactate (122-64-5)
Phenylmercuric monoethanol ammonium acetate (5822-97-9)
Phenylmercuric nitrate [UN 1895] (55-68-5)
Phenylmercuric oleate (104-60-9)
Phenylmercuric triethanolammonium lactate (23319-66-6)
Phenylmercuric urea (2279-64-3)
Phenylmercuritriethanolammonium lactate (23319-66-6)
Phenylmercuriurea (2279-64-3)
Phenylmercury acetate (62-38-4)
Phenylmercury borate (102-98-7)
Phenylmercury chloride (100-56-1)
Phenylmercury hydroxide (100-57-2)
Phenylmercury lactate (122-64-5)
Phenylmercury nitrate (55-68-5)
Phenylmercury oleate (104-60-9)
Phenylmercury triethanolamine lactate (23319-66-6)
Phenyl mercury urea (2279-64-3)
2-Phenylmesitylene (3976-35-0)
Phenylmethanal (100-52-7)
Phenylmethane (108-88-3)
Phenylmethanesulfonic acid (100-87-8)
Phenylmethanethiol (100-53-8)
Phenylmethanol (100-51-6)
Phenyl, 3-methoxy- (9CI) (18815-11-7)
Phenyl, 4-methoxy- (9CI) (2396-03-4)
Phenyl, m-methoxy- (8CI) (18815-11-7)
Phenyl, p-methoxy- (8CI) (2396-03-4)
N-((Phenylmethoxy)carbonyl)-L-aspartic acid (1152-61-0)
(Phenylmethoxy)methanol (14548-60-8)
4-(Phenylmethoxy)phenol (103-16-2)
3-(Phenylmethoxy)propanenitrile (6328-48-9)
Phenylmethyl acetate (140-11-4)
Phenylmethyl alcohol (100-51-6)
(Phenylmethyl)amine (100-46-9)
N-Phenylmethylamine (100-61-8)
1-Phenyl-2-methylamine-propanol-1-sulfate (134-72-5)
2-((Phenylmethyl)amino)ethanol (104-63-2)
d-1-Phenyl-2-methylaminopropan (German) (537-46-2)
d-1-Phenyl-2-methylaminopropane (537-46-2)
1-Phenyl-2-methylaminopropanol (299-42-3)
1-Phenyl-3-methyl-5-aminopyrazole (1131-18-6)
N-(Phenylmethyl)benzenemethanamine (103-49-1)
Phenylmethylcarbinol (98-85-1)
Phenylmethylcarbinyl acetate (93-92-5)
1-Phenyl-5-methyl-8-chloro-1,2,4,5-tetrahydro-2,4-dioxo-3H-1,5-benzodiazepine (22316-47-8)
N-Phenyl-N'-(2-methylcyclohexyl)urea (1982-49-6)
Phenylmethyldichlorosilane (149-74-6)
Phenylmethyldiketone (579-07-7)
N-(Phenylmethyl)dimethylamine (103-83-3)
2-(Phenylmethylene)heptanal (122-40-7)
2-(Phenylmethylene)octanal (101-86-0)
Phenylmethyl ethanol amine (93-90-3)
Phenylmethylethanolamine (93-90-3)
Phenyl methyl ether (100-66-3)
Phenylmethyl 4-hydroxybenzoate (94-18-8)
2-(Phenylmethyl)isoquinolinium chloride (35674-56-7)
Phenyl methyl ketone (98-86-2)
Phenylmethyl mercaptan (100-53-8)
Phenylmethyl methyl ketone (103-79-7)
Phenylmethyl 2-methyl-2-propenoate (2495-37-6)
2-Phenyl-3-methylmorpholine (134-49-6)
Phenylmethylnitrosamine (614-00-6)
(Phenylmethyl)penicillin (61-33-6)
(Phenylmethyl)penicillinic acid (61-33-6)
4-(Phenylmethyl)phenol (101-53-1)
Phenyl N-methyl-N-phenylcarbamate (13599-69-4)
Phenyl methylphenylcarbamate (13599-69-4)
Phenylmethyl propanoate (122-63-4)
N-(Phenylmethyl)-1H-purin-6-amine (1214-39-7)
1-Phenyl-4-methyl-3-pyrazolidone (2654-57-1)
1-Phenyl-3-methyl-5-pyrazolone (89-25-8)
1-Phenyl-3-methylpyrazolone-5 (89-25-8)
1-Phenyl-3-methyl-5-pyrazolyl N,N-dimethyl carbamate (87-47-8)
1-(Phenylmethyl)pyridinium chloride (2876-13-3)
1-(Phenylmethyl)quinolinium chloride (15619-48-4)
Phenyl methyl sulfone (3112-85-4)
dl-2-Phenyl-3-methyltetrahydro-1,4-oxazine (134-49-6)
Phenyl-mobuzon (50-33-9)
Phenylmonoglycol ether (122-99-6)
Phenyl monomethylcarbamate (1943-79-9)
4-Phenylmorpholine (92-53-5)
N-Phenylmorpholine (92-53-5)
Phenyl morpholine (92-53-5)
Phenylmorpholine (92-53-5)
3-Phenyl-1H-naphtho(2,1-b)pyran-1-one (6051-87-2)
3-Phenyl-7-(1,2-2H-naphthotriazolyl)coumarin (3333-62-8)

N-Phenyl-1-naphthylamine **(90-30-2)**
N-Phenyl-2-naphthylamine **(135-88-6)**
N-Phenyl-α-naphthylamine **(90-30-2)**
Phenyl-2-naphthylamine **(135-88-6)**
Phenyl-α-naphthylamine **(90-30-2)**
Phenyl-β-naphthylamine **(135-88-6)**
Phenyl-β-naphthylamine **(28258-64-2)**
Phenylnaphthylamine **(90-30-2)**
α-Phenylnaphthylamine **(90-30-2)**
N-Phenyl-β-naphthylamine (ACGIH) **(135-88-6)**
N-Phenylnitramine **(645-55-6)**
Phenylnitramine **(645-55-6)**
N-Phenyl-o-nitroaniline **(119-75-5)**
4-Phenyl-nitrobenzene **(92-93-3)**
p-Phenyl-nitrobenzene **(92-93-3)**
1-Phenyl-1-nitroethane **(7214-61-1)**
N-Phenyl-p-nitrosoaniline **(156-10-5)**
1-Phenyloctadecane **(4445-07-2)**
Phenylon **(60-80-0)**
Phenylone **(60-80-0)**
N-(3'-Phenylo-2-propylo)-1,1-diphenylo-3-propyloamine (Polish) **(390-64-7)**
1-Phenyloxirane **(96-09-3)**
2-Phenyloxirane **(96-09-3)**
Phenyl oxirane **(96-09-3)**
4-Phenyl-3-oxobutyric acid **(25832-09-1)**
1-Phenyl-3-oxopyrazolidine **(92-43-3)**
1-Phenyl-5-oxo-2-pyrazoline-3-carboxylic acid, ethyl ester **(89-33-8)**
β-Phenyl-γ-oxypropionsaeure-tropyl-ester (German) **(51-55-8)**
S-Phenyl parathion **(3270-86-8)**
1-Phenylpentadecane **(2131-18-2)**
1-Phenyl-n-pentane **(538-68-1)**
1-Phenylpentane **(538-68-1)**
Phenylpentane **(538-68-1)**
1-Phenyl-2,4-pentanedione **(3318-61-4)**
4-Phenylpentanoic acid **(16433-43-5)**
5-Phenylpentanoic acid **(2270-20-4)**
Phenylpentanoic acid **(2270-20-4)**
(E)-3-Phenyl-2-pentene **(4165-86-0)**
(Z)-3-Phenyl-2-pentene **(4165-78-0)**
3-Phenyl-1-pentene **(19947-22-9)**
cis-3-Phenyl-2-pentene **(4165-78-0)**
trans-3-Phenyl-2-pentene **(4165-86-0)**
Phenyl perchloryl **(118-74-1)**
Phenylperi acid **(82-76-8)**
Phenyl peri acid, ammonium salt **(28836-03-5)**
Phenyl phenacyl ketone **(120-46-7)**
2-Phenylphenol **(90-43-7)**
4-Phenylphenol **(92-69-3)**
Phenylphenol **(90-43-7)**
o-Phenylphenol **(90-43-7)**
p-Phenylphenol **(92-69-3)**
o-Phenylphenol potassium salt **(13707-65-8)**
2-Phenylphenol sodium salt **(132-27-4)**
o-Phenylphenol, sodium salt **(132-27-4)**
α-Phenylphenylacetonitrile **(86-29-3)**
N-Phenyl-p-phenylenediamine **(101-54-2)**
Phenyl 2-phenylvinyl ketone **(94-41-7)**
Phenylphosphine (ACGIH,OSHA) **(638-21-1)**
Phenylphosphine dichloride **(644-97-3)**
Phenylphosphinic acid **(1779-48-2)**
Phenylphosphonic acid **(1571-33-1)**
Phenylphosphonic acid dioctyl ester **(1754-47-8)**
Phenylphosphonic acid, monomethyl ester **(10088-45-6)**
Phenylphosphonic dichloride **(824-72-6)**
Phenylphosphonochloridothioic acid O-ethyl ester **(5075-13-8)**
Phenylphosphonothioate, O-ethyl-O-p-nitrophenyl- **(2104-64-5)**
Phenylphosphonothioic acid O-(4-bromo-2,5-dichlorophenyl)
O-methyl ester **(21609-90-5)**
Phenylphosphonothioic acid O-ethyl ester O-ester with
p-hydroxybenzonitrile **(13067-93-1)**
Phenylphosphonothioic acid O-ethyl O-p-nitrophenyl ester **(2104-64-5)**
Phenylphosphonous acid dichloride **(644-97-3)**
Phenylphosphonous dichloride **(644-97-3)**
Phenyl phosphorodichloridate **(770-12-7)**
Phenyl phosphorus dichloride [UN 2798] **(644-97-3)**
Phenyl phosphorus dichloride [UN 2798] **(824-72-6)**
Phenyl phosphorus thiodichloride [UN 2799] **(3497-00-5)**
Phenyl phthalate **(84-62-8)**
1-Phenylpiperazine **(92-54-6)**
N-Phenylpiperazine **(92-54-6)**
1-Phenylpiperidine **(4096-20-2)**
α-Phenyl-2-piperidineacetic acid methyl ester **(113-45-1)**
2-Phenyl-1-propanal **(93-53-8)**
2-Phenylpropanal **(93-53-8)**
1-Phenylpropane **(103-65-1)**
2-Phenylpropane **(98-82-8)**
1-Phenyl-1,2-propanediol **(1855-09-0)**
1-Phenyl-1,2-propanedione **(579-07-7)**
3-Phenyl-2,3-propanedione **(579-07-7)**
Phenyl propanoate **(637-27-4)**
2-Phenylpropanoic acid **(492-37-5)**
3-Phenylpropanoic acid **(501-52-0)**
Phenylpropanoic acid **(501-52-0)**
1-Phenyl-1-propanol **(93-54-9)**
1-Phenylpropanol **(93-54-9)**
2-Phenylpropan-1-ol **(1123-85-9)**
3-Phenyl-1-propanol **(122-97-4)**
3-Phenylpropanol **(122-97-4)**
γ-Phenylpropanol **(122-97-4)**
Phenylpropanolamine hydrochloride **(154-41-6)**
1-Phenyl-1-propanone **(93-55-0)**
1-Phenyl-2-propanone **(103-79-7)**
(E)-3-Phenylpropenal **(14371-10-9)**
3-Phenyl-2-propenal **(104-55-2)**
3-Phenylpropenal **(104-55-2)**
3-Phenyl-2-propenal monopentyl deriv. **(122-40-7)**
2-Phenylpropene **(98-83-9)**
3-Phenyl-1-propene **(637-50-3)**
β-Phenylpropene **(98-83-9)**
2-Phenylpropene oxide **(2085-88-3)**
3-Phenyl-2-propenoic acid **(621-82-9)**
3-Phenylpropenoic acid **(621-82-9)**
3-Phenyl-2-propenoic acid phenylmethyl ester **(103-41-3)**
3-Phenyl-2-propen-1-ol **(104-54-1)**
3-Phenyl-2-propenoyl chloride **(102-92-1)**
3-Phenyl-2-propen-1-yl anthranilate **(87-29-6)**
3-Phenyl-2-propenylanthranilate **(87-29-6)**
2-Phenyl propionaldehyde **(93-53-8)**
α-Phenyl propionaldehyde **(93-53-8)**
2-Phenylpropionic acid **(492-37-5)**
3-Phenylpropionic acid **(501-52-0)**
Phenylpropionic acid **(501-52-0)**
α-Phenylpropionic acid **(492-37-5)**
β-Phenylpropionic acid **(501-52-0)**
β-Phenylpropionic acid methyl ester **(103-25-3)**
1-Phenylpropyl alcohol **(93-54-9)**
2-Phenylpropyl alcohol **(1123-85-9)**
3-Phenylpropyl alcohol **(122-97-4)**
Phenylpropyl alcohol **(122-97-4)**
β-Phenylpropyl alcohol **(1123-85-9)**
γ-Phenylpropyl alcohol **(122-97-4)**
2-Phenylpropylene **(98-83-9)**
β-Phenylpropylene **(98-83-9)**
Phenyl propyl ketone **(495-40-9)**

3-Phenylpropyl nitrite (28537-55-5)
3-(1'-Phenyl-propyl)-4-oxycoumarin (German) (435-97-2)
3-Phenylpropyl 3-phenyl-2-propenoate (122-68-9)
4-(3-Phenylpropyl)pyridine (2057-49-0)
Phenylpseudohydantoin (2152-34-3)
6-Phenyl-2,4,7-pteridinetriamine (396-01-0)
1-Phenyl-3-pyrazolidinone (92-43-3)
1-Phenyl-3-pyrazolidone (92-43-3)
N'-(1-Phenylpyrazol-5-yl)sulfanilamide (526-08-9)
2-Phenylpyridine (1008-89-5)
3-Phenylpyridine (1008-88-4)
4-Phenylpyridine (939-23-1)
o-Phenylpyridine (1008-89-5)
p-Phenylpyridine (939-23-1)
Phenyl(2-pyridyl)(β-N,N-dimethylaminomethyl) methane maleate (132-20-7)
1-Phenyl-1-(2-pyridyl)-3-dimethylaminopropane (86-21-5)
1-Phenyl-1-(2-pyridyl)-3-dimethylaminopropane maleate (132-20-7)
3-Phenyl-3-(2-pyridyl)-N,N-dimethylpropylanine (86-21-5)
Phenyl 4-pyridyl ketone (14548-46-0)
Phenyl-2-pyridylmethyl-β-N,N-dimethylaminoethyl ether (469-21-6)
Phenyl2-pyridylmethyl-β-N,N-dimethylaminoethyl ether succinate (562-10-7)
4-Phenylpyrocatechol (92-05-7)
Phenylpyruvic acid (156-06-9)
Phenylquecksilberacetat (German) (62-38-4)
Phenylquecksilberchlorid (German) (100-56-1)
2-Phenyl-4-quinolinecarboxylic acid (132-60-5)
2-Phenylquinoline-4-carboxylic acid (132-60-5)
Phenylrhodanid (German) (5285-87-0)
Phenyl salicylate (118-55-8)
1-Phenylsemicarbazide (103-03-7)
4-Phenylsemicarbazide (537-47-3)
Phenylsemicarbazide (103-03-7)
Phenylsenfoel (German) (103-72-0)
1-Phenylsilatrane (2097-19-0)
Phenylsilatrane (2097-19-0)
Phenylsilicon trichloride (98-13-5)
Phenyl simethicone (9005-12-3)
α-Phenylstyrene (530-48-3)
Phenyl styryl ketone (94-41-7)
1-Phenyl-5-sulfanilamidopyrazole (526-08-9)
Phenyl-sulfide (139-66-2)
Phenylsulfinic acid (618-41-7)
Phenylsulfohydrazide (80-17-1)
Phenyl-sulfone (127-63-9)
Phenylsulfonic acid (98-11-3)
1-(Phenylsulfonyl)aziridine (10302-15-5)
Phenylsulfonyl chloride (98-09-9)
Phenylsulfonyl hydrazide (80-17-1)
Phenylsulfonylhydrazine (80-17-1)
(Phenylsulfonyl)methane (3112-85-4)
4-(Phenylsulfoxyethyl)-1,2-diphenyl-3,5-pyrazolidinedione (57-96-5)
Phenyl sulphone (127-63-9)
Phenylsuximide (86-34-0)
5'-Phenyl-m-terphenyl (612-71-5)
1-Phenyltetradecane (1459-10-5)
1-Phenyl-1,2,3,4-tetrahydronaphthalene (3018-20-0)
1-Phenyltetralin (3018-20-0)
5-Phenyl tetrazole (18039-42-4)
5-Phenyl-1H-tetrazole (18039-42-4)
5-Phenyl-2H-tetrazole (18039-42-4)
5-Phenyltetrazole (VAN) (18039-42-4)
1-Phenyltetrazole-5-thiol (86-93-1)
3-Phenyl-1,2,4-thiadiazol-5-amine (17467-15-1)
N-Phenyl-N'-1,2,3-thiadiazol-5-yl-urea (51707-55-2)
N-Phenyl-N'-1,2,5-thiadiazol-3-ylurea (71769-74-9)
N-Phenyl-4-thiazolecarboxamidine hydrochloride (13631-64-6)
5-Phenyl-2,4-thiazolediamine (490-55-1)
Phenylthiocarbamide (103-85-5)
2-Phenylthiochroman (5961-99-9)
Phenyl thiocyanate (5285-87-0)
4-Phenylthiomorpholine-1,1-dioxide (17688-68-5)
Phenylthionophosphonic dichloride (3497-00-5)
Phenylthiophosphonate de O-ethyle et O-4-nitrophenyle (French) (2104-64-5)
4-Phenyl-3-thiosemicarbazide (5351-69-9)
1-Phenyl-2-thiourea (103-85-5)
1-Phenylthiourea (103-85-5)
N-Phenylthiourea (103-85-5)
Phenylthiourea (103-85-5)
α-Phenylthiourea (103-85-5)
Phenyltin trichloride (1124-19-2)
Phenyltoluene (28652-72-4)
(N-Phenyl-p-toluenesulfonamido)ethylmercury (517-16-8)
Phenyl m-tolyl ether (3586-14-9)
Phenyl p-tolyl ketone (134-84-9)
6-Phenyl-2,4,7-triaminopteridine (396-01-0)
2-Phenyl-1,3,5-triazine (1722-18-5)
2-Phenyl-s-triazine (1722-18-5)
Phenyl-1,3,5-triazine (1722-18-5)
1-Phenyl-1,2,4-triazolyl-3-(O,O-diethylthionophosphate) (24017-47-8)
Phenyltrichloromethane (98-07-7)
Phenyl(trichloromethyl)carbinol (2000-43-3)
N-Phenyl-N-(trichloromethylsulfenyl)benzene sulfonamide (2280-49-1)
Phenyl trichlorosilane (98-13-5)
Phenyltrichlorosilane [UN 1804] (98-13-5)
Phenyltriethoxysilane (780-69-8)
Phenyl trifluoromethyl ketone (434-45-7)
N-Phenyl-α,α,α-trifluorotoluidide (101-23-5)
Phenyl(2,3,4-trihydroxyphenyl)methanone (1143-72-2)
Phenyl trimethicone (2116-84-9)
Phenyltrimethoxysilane (2996-92-1)
Phenyltris(trimethylsiloxyl)silane (2116-84-9)
1-Phenylundecane (6742-54-7)
1-Phenylurea (64-10-8)
N-Phenylurea (64-10-8)
Phenylurea (64-10-8)
N-Phenylurethane (101-99-5)
Phenylurethan(e) (101-99-5)
Phenyl valerate (20115-23-5)
3-Phenylvaleric acid (5669-17-0)
4-Phenylvaleric acid (16433-43-5)
5-Phenylvaleric acid (2270-20-4)
Phenylvaleric acid (2270-20-4)
γ-Phenylvaleric acid (2270-20-4)
Phenyral (50-06-6)
Phenyrit (63-98-9)
Phenytoin (57-41-0)
Phenytoin sodium (630-93-3)
Phenytoinum sodium (630-93-3)
Phermernite (55-68-5)
Phetadex (51-63-8)
Phetidine (57-42-1)
Phetylureum (63-98-9)
Phiasol (104-28-9)
Philblack (1333-86-4)
Philblack N 550 (1333-86-4)
Philblack N 765 (1333-86-4)
Philblack O (1333-86-4)
Philex (79-01-6)
Philips-Duphar PH 60-40 (35367-38-5)
Philips-Duphar V-101 (2227-13-6)
R-874 Phillips (3547-33-9)
Phillips R-11 (126-15-8)
Phillips Repellent 11 (126-15-8)
Philodorm (52-31-3)

Philopon **(300-42-5)**
Philosopher's Wool **(1314-13-2)**
Philostigmin bromide **(114-80-7)**
Phisodan **(70-30-4)**
Phisohex **(70-30-4)**
Phix **(62-38-4)**
Phixia **(107-75-5)**
Phloretic acid **(501-97-3)**
Phloretin **(60-82-2)**
Phloretol **(60-82-2)**
Phlorhizin **(60-81-1)**
Phloridzin **(60-81-1)**
Phloridzine **(60-81-1)**
Phlorizin **(60-81-1)**
Phlorizine **(60-81-1)**
Phlorizoside **(60-81-1)**
Phloroglucin **(108-73-6)**
Phloroglucine **(108-73-6)**
Phloroglucinol **(108-73-6)**
Phloroglucinol dimethyl ether **(500-99-2)**
Phloroglucinol trimethyl ether **(621-23-8)**
Phlorol **(90-00-6)**
Phlorrhizin **(60-81-1)**
Phlox Red Toner X-1354 **(17372-87-1)**
Phloxin **(6441-77-6)**
Phloxine **(6441-77-6)**
Phloxine 2G **(3734-67-6)**
Phloxine G **(3734-67-6)**
Phloxine K **(6441-77-6)**
Phloxine Red 20-7600 **(17372-87-1)**
Phloxine Red 20-7600 **(548-26-5)**
Phloxine Toner B **(17372-87-1)**
Phob **(50-06-6)**
Pholate **(52-46-0)**
Pholcodin **(509-67-1)**
Pholcodine **(509-67-1)**
Pholcodinum (Latin) **(509-67-1)**
Phomenone **(55785-58-5)**
Phomin **(14930-96-2)**
Phomopsin **(64925-80-0)**
Phomopsin-A **(64925-80-0)**
Phonurit **(59-66-5)**
Phorat (German) **(298-02-2)**
Phorate (ACGIH,OSHA) **(298-02-2)**
Phorate-10G **(298-02-2)**
Phorate oxon sulfoxide **(2588-05-8)**
Phorate sulfone **(2588-04-7)**
Phorate sulfoxide **(2588-05-8)**
Phoratoxon sulfoxide **(2588-05-8)**
Phorbol **(17673-25-5)**
Phorbol acetate, myristate **(16561-29-8)**
Phorbol-13-acetate, 12-o-palmitoyl-16-hydroxy **(53202-98-5)**
Phorbol monoacetate monomyristate **(16561-29-8)**
Phorbol 12-myristate 13-acetate **(16561-29-8)**
Phorbol myristate acetate **(16561-29-8)**
Phorbol 12-tetradecanoate 13-acetate **(16561-29-8)**
Phorbyol **(8001-79-4)**
Phordene **(2008-39-1)**
Phoron (German) **(504-20-1)**
Phorone **(504-20-1)**
β-Phorone **(471-01-2)**
Phortox **(93-76-5)**
Phos-Flur **(7681-49-4)**
Phosacetim **(4104-14-7)**
Phosacetime **(4104-14-7)**
Phosaden **(61-19-8)**
Phosalon **(2310-17-0)**
Phosalone **(2310-17-0)**
Phosazetim **(4104-14-7)**
Phoschlor **(52-68-6)**
Phoschlor R50 **(52-68-6)**
cis-Phosdrin **(298-01-1)**
trans-Phosdrin **(338-45-4)**
Phosdrin (OSHA) **(7786-34-7)**
Phosethyl Aluminum **(39148-24-8)**
Phosfene **(7786-34-7)**
Phosfleur **(115-78-6)**
Phosflex 112 **(26444-49-5)**
Phosflex 179-C **(78-30-8)**
Phosflex 300 (9CI) **(64176-84-7)**
Phosflex 400 (9CI) **(64176-85-8)**
Phosflex 31P **(96300-97-9)**
Phosflex T-Bep **(78-51-3)**
Phosflex Z (9CI) **(100179-07-5)**
Phosfolan **(947-02-4)**
Phosfon **(115-78-6)**
Phosfon D **(115-78-6)**
Phosgen (German) **(75-44-5)**
Phosgene (ACGIH,OSHA) [UN 1076] **(75-44-5)**
Phosgene, thio- **(463-71-8)**
Phoshorothioic acid, S-(2-(ethylamino)-2-oxoethyl) O,O-dimethyl ester **(116-01-8)**
Phoskil **(56-38-2)**
Phosmet **(732-11-6)**
Phosmethylan **(83733-82-8)**
Phosnic 390 **(940-71-6)**
3-Phosphabicyclo(4.4.0)decane, p-chloro-2,4-dioxa-5-methyl-p-thiono **(2921-31-5)**
9-Phosphabicyclo(3.3.1)nonane (9CI) **(13887-02-0)**
9-Phosphabicyclo(4.2.1)nonane (9CI) **(13396-80-0)**
9-Phosphabicyclo(3.3.1)nonane, 9-eicosyl- (9CI) **(13887-00-8)**
9-Phosphabicyclo(4.2.1)nonane, 9-eicosyl- (9CI) **(13886-99-2)**
Phosphacol **(311-45-5)**
Phosphaden **(61-19-8)**
Phosphalugel **(7784-30-7)**
Phosphamid **(60-51-5)**
Phosphamide **(60-51-5)**
(E)-Phosphamidon **(297-99-4)**
(Z)-Phosphamidon **(23783-98-4)**
Phosphamidon **(13171-21-6)**
cis-Phosphamidon **(23783-98-4)**
trans-Phosphamidon **(297-99-4)**
Phosphate 100 **(2778-04-3)**
Phosphate (9CI) **(14265-44-2)**
Phosphate de O,O-diethyle et de O-2-chloro-1-(2,4-dichlorophenyl) vinyle (French) **(470-90-6)**
Phosphate de diethyle et de 3-methyl-5-pyrazolyle (French) **(108-34-9)**
Phosphate de O,O-dimethle et de O-(1,2-dibromo-2,2-dichlorethyle) (French) **(300-76-5)**
Phosphate de dimethyle et de (2-chloro-2-diethylcarbamoyl-1-methyl-vinyle) (French) **(13171-21-6)**
Phosphate de dimethyle et de 2,2-dichlorovinyle (French) **(62-73-7)**
Phosphate de dimethyle et de 2-dimethylcarbamoyl 1-methyl vinyle (French) **(141-66-2)**
Phosphate de dimethyle et de 2-methoxycarbonyl-1 methylvinyle (French) **(7786-34-7)**
Phosphate de dimethyle et de 2-methylcarbamoyl 1-methyl vinyle (French) **(6923-22-4)**
Phosphated, ethoxylated nonylphenol **(51811-79-1)**
Phosphated 2-ethyl hexanol **(12645-31-7)**
Phosphate de tricresyle (French) **(1330-78-5)**
Phosphated stearyl alcohol **(3037-89-6)**
Phosphate esters of coal tar or petroleum-derived cresylic acid **(68952-35-2)**
Phosphate esters of coal tar or petroleum derived cresylic acid **(68952-33-0)**

Name	CAS No.
Phosphate(1-) hexafluoro-, hydrogen	**(16940-81-1)**
Phosphate rock and phosphorite, calcined	**(65996-94-3)**
Phosphatidylcholine	**(8002-43-5)**
Phosphemol	**(56-38-2)**
Phosphene (French)	**(7786-34-7)**
Phosphenol	**(56-38-2)**
Phosphentaside	**(61-19-8)**
Phosphestrol	**(522-40-7)**
Phosphine (ACGIH,OSHA) [UN 2199]	**(7803-51-2)**
Phosphine, cyclohexyl- (8CI,9CI)	**(822-68-4)**
Phosphine, dicyclohexyl- (9CI)	**(829-84-5)**
Phosphine, diphenylpropyl- (8CI,9CI)	**(7650-84-2)**
Phosphine oxide, bis(dimethylamino)fluoro-	**(115-26-4)**
Phosphine oxide, bis(1,1-dimethylethyl)- (9CI)	**(684-19-5)**
Phosphine oxide, di-tert-butyl- (7CI,8CI)	**(684-19-5)**
Phosphine oxide, difluoro-	**(14939-34-5)**
Phosphine oxide, fluorobis(isopropylamino)-	**(371-86-8)**
Phosphine oxide, fluoroisopropoxymethyl-	**(107-44-8)**
Phosphine oxide, fluoromethyl(1,2,2-trimethylpropoxy)-	**(96-64-0)**
Phosphine oxide, tributyl	**(814-29-9)**
Phosphine oxide, trioctyl- (9CI)	**(78-50-2)**
Phosphine oxide, triphenyl- (9CI)	**(791-28-6)**
Phosphine oxide, tris(1-aziridinyl)	**(545-55-1)**
Phosphine oxide, tris(2-methyl-1-aziridinyl)	**(57-39-6)**
Phosphine, phenyl	**(638-21-1)**
Phosphine sulfide, bis(1-aziridinyl)morpholino	**(2168-68-5)**
Phosphine sulfide, tris(1-aziridinyl)	**(52-24-4)**
Phosphine, tributyl	**(998-40-3)**
Phosphine, triethyl- (9CI)	**(554-70-1)**
Phosphine, trioctyl- (9CI)	**(4731-53-7)**
Phosphine, triphenyl	**(603-35-0)**
Phosphine, tris(dipropylene glycol)-	**(36788-39-3)**
Phosphinic acid (9CI)	**(6303-21-5)**
Phosphinic acid, Compd. with 2-(4-thiazolyl)-1H-benzimidazole (1:1)	**(28558-32-9)**
Phosphinic acid, ammonium salt (8CI,9CI)	**(7803-65-8)**
Phosphinic acid, bis(chloromethyl)-, allyl ester (8CI)	**(14212-97-6)**
Phosphinic acid, bis(chloromethyl)-, sec-butyl ester (8CI)	**(24767-66-6)**
Phosphinic acid, bis(chloromethyl)-, butyl ester (8CI,9CI)	**(14590-60-4)**
Phosphinic acid, bis(chloromethyl)-, ethyl ester (7CI,8CI,9CI)	**(13274-84-5)**
Phosphinic acid, bis(chloromethyl)-, methyl ester (8CI,9CI)	**(14212-91-0)**
Phosphinic acid, bis(chloromethyl)-, phenyl ester (8CI,9CI)	**(14212-98-7)**
Phosphinic acid, bis(chloromethyl)-, 2-propenyl ester (9CI)	**(14212-97-6)**
Phosphinic acid, bis(iodomethyl)-, butyl ester (8CI)	**(17052-18-5)**
Phosphinic acid, bis(iodomethyl)-, ethyl ester (8CI)	**(17052-17-4)**
Phosphinic acid, bis(phenoxymethyl)-, ethyl ester (8CI,9CI)	**(21993-11-3)**
Phosphinic acid, (bromomethyl)(chloromethyl)-, ethyl ester (8CI)	**(24327-56-8)**
Phosphinic acid, calcium salt (9CI)	**(7789-79-9)**
Phosphinic acid, (chloromethyl)(2-cyanoethyl)-, ethyl ester (8CI)	**(21310-38-3)**
Phosphinic acid, (chloromethyl)ethyl-, ethyl ester (8CI,9CI)	**(24327-58-0)**
Phosphinic acid, (chloromethyl)(iodomethyl)-, ethyl ester (8CI)	**(17052-15-2)**
Phosphinic acid, diethyl-, ethyl ester (6CI,7CI,8CI,9CI)	**(4775-09-1)**
Phosphinic acid, ethylhexyl-, ethyl ester (8CI)	**(24327-59-1)**
Phosphinic acid, methyl-, ethyl ester (8CI,9CI)	**(16391-07-4)**
Phosphinic acid, (2-methylpropyl)- (9CI)	**(54423-73-3)**
Phosphinic acid, phenyl- (9CI)	**(1779-48-2)**
Phosphinic acid, phenyl-, sodium salt (9CI)	**(4297-95-4)**
Phosphinic acid, potassium salt (9CI)	**(7782-87-8)**
Phosphinic chloride, tert-butyl(p-tert-butylphenyl)- (8CI)	**(25097-44-3)**
Phosphinic chloride, tert-butylphenyl- (7CI)	**(4923-85-7)**
Phosphinic chloride, tert-butylphenyl- (8CI)	**(4923-85-7)**
Phosphinic chloride, (1,1-dimethylethyl)(4-(1,1-dimethylethyl)-phenyl)- (9CI)	**(25097-44-3)**
Phosphinic chloride, (1,1-dimethylethyl)phenyl- (9CI)	**(4923-85-7)**
Phosphinic chloride, methylphenyl- (7CI,8CI,9CI)	**(5761-97-7)**
Phosphinic fluoride, (1,1-dimethylethyl)phenyl- (9CI)	**(55236-56-1)**
Phosphinic fluoride, methylphenyl- (7CI,8CI,9CI)	**(657-37-4)**
Phosphinoselenoic acid, (1,1-dimethylethyl)phenyl-, (R)- (9CI)	**(51584-27-1)**
Phosphinoselenoic acid, (1,1-dimethylethyl)phenyl-, (S)- (9CI)	**(51584-28-2)**
Phosphinothioic chloride, dimethyl-	**(993-12-4)**
1,1',1''-Phosphinothioylidynetrisaziridine	**(52-24-4)**
Phosphinothricin	**(35597-44-5)**
Phosphinothricin	**(51276-47-2)**
Phosphinothricylalanylalanine	**(35597-43-4)**
Phosphinous chloride, diphenyl- (9CI)	**(1079-66-9)**
1,1',1''-Phosphinylidynetrisaziridine	**(545-55-1)**
1,1',1''-Phosphinylidynetris(2-methyl)azridine	**(57-39-6)**
O-Phosphoethanolamine	**(1071-23-4)**
Phosphoethanolamine	**(1071-23-4)**
α-Phosphoglycerol	**(57-03-4)**
Phosphogypsum	**(13397-24-5)**
Phospholan	**(947-02-4)**
Phospholeum	**(8017-16-1)**
Phospholine (The pharmaceutical)	**(513-10-0)**
Phospholine iodide	**(513-10-0)**
Phospholutein	**(8002-43-5)**
Phosphon	**(115-78-6)**
Phosphon D	**(115-78-6)**
Phosphonacetic acid	**(4408-78-0)**
Phosphone D	**(115-78-6)**
Phosphonic-acid	**(13598-36-2)**
Phosphonic acid, acetonyl-, diethyl ester (8CI)	**(1067-71-6)**
Phosphonic acid, (1-(acetyloxy)-2,2,2-trichloroethyl)-, diphenyl ester	**(74548-80-4)**
Phosphonic acid, (aminocarbonyl-14C)-, monoethyl ester, mono-ammonium salt (9CI)	**(69975-80-0)**
Phosphonic acid, (aminomethyl)- (9CI)	**(1066-51-9)**
Phosphonic acid, ((3,5-bis(1,1-dimethylethyl)-4-hydroxyphenyl) methyl)-, monoethyl ester, nickel(2+) salt (2:1) (9CI)	**(30947-30-9)**
Phosphonic acid, bis(2-ethylhexyl) ester	**(3658-48-8)**
Phosphonic acid, ((bis(2-hydroxyethyl)amino)methyl)-, diethyl ester (9CI)	**(2781-11-5)**
Phosphonic acid, bis(1-methylethyl) ester	**(1809-20-7)**
Phosphonic acid, butyl- (8CI,9CI)	**(3321-64-0)**
Phosphonic acid, 1-(butylamino)cyclohexyl-, dibutyl ester	**(51249-05-9)**
Phosphonic acid, butyl-, dibutyl ester	**(78-46-6)**
Phosphonic acid, (2-(((2-chloroethoxy)(2-chloroethyl)phosphinyl)-oxy)ethyl)-, bis(2-chloroethyl) ester (9CI)	**(58823-09-9)**
Phosphonic acid, (2-chloroethyl)	**(16672-87-0)**
Phosphonic acid, (2-chloroethyl)-, bis(2-chloroethyl) ester (9CI)	**(6294-34-4)**
Phosphonic acid, didodecyl ester (9CI)	**(21302-09-0)**
Phosphonic acid, diisopropyl ester	**(1809-20-7)**
Phosphonic acid, dimethyl ester	**(868-85-9)**
Phosphonic acid, di-9-octadecenyl ester (Z,Z)-	**(25088-57-7)**
Phosphonic acid, diphenyl ester (9CI)	**(4712-55-4)**
Phosphonic acid, dipropyl ester (8CI,9CI)	**(1809-21-8)**
Phosphonic acid, dodecyl	**(5137-70-2)**
Phosphonic acid, (1,2-ethanediylbis(nitrilobis(methylene)))-tetrakis- (9CI)	**(1429-50-1)**
Phosphonic acid, (1,2-ethanediylbis(nitrilobis(methylene)))-tetrakis-, hexasodium salt (9CI)	**(15142-96-8)**
Phosphonic acid, (1,2-ethanediylbis(nitrilobis(methylene)))-tetrakis-, tetrapotassium salt (9CI)	**(68188-96-5)**
Phosphonic acid, ethenyl- (9CI)	**(1746-03-8)**
Phosphonic acid, ethenyl-, bis(2-((butoxymethylphosphinyl)oxy)-ethyl) ester	**(53529-45-6)**
Phosphonic acid, ethyl- (8CI,9CI)	**(6779-09-5)**
Phosphonic acid, ethyl-, diethyl ester (9CI)	**(78-38-6)**
Phosphonic acid, ethyl-, ethyl ester	**(7305-61-5)**
Phosphonic acid, ethyl-, monoethyl ester	**(7305-61-5)**
Phosphonic acid, (1,6-hexanediylbis(nitrilobis(methylene)))-tetrakis- (9CI)	**(23605-74-5)**
Phosphonic acid, hexyl- (8CI,9CI)	**(4721-24-8)**
Phosphonic acid, (1-hydroxyethane-1,1-diyl)di-, disodium salt	**(7414-83-7)**

Phosphonic acid, 1-hydroxy-1,1-ethanediyl ester **(2809-21-4)**
Phosphonic acid, (1-hydroxyethylidene)bis- **(2809-21-4)**
Phosphonic acid, (1-hydroxyethylidene)bis-, dipotassium salt **(21089-06-5)**
Phosphonic acid, (1-hydroxyethylidene)bis-, dipotassium salt (9CI) **(21089-06-5)**
Phosphonic acid, (1-hydroxyethylidene) bis-, potassium salt **(67953-76-8)**
Phosphonic acid, (1-hydroxyethylidene)bis-, potassium salt (9CI) **(67953-76-8)**
Phosphonic acid, (1-hydroxyethylidene)bis-, sodium salt (9CI) **(29329-71-3)**
Phosphonic acid, (1-hydroxyethylidene)di **(2809-21-4)**
Phosphonic acid, (1-hydroxyethylidene)di-, disodium salt **(7414-83-7)**
Phosphonic acid, (((2-hydroxyethyl)imino)bis(methylene))bis- (9CI) **(5995-42-6)**
Phosphonic acid, (hydroxymethyl)- (9CI) **(2617-47-2)**
Phosphonic acid, (3-((hydroxymethyl)amino)-3-oxopropyl)-, dimethyl ester (9CI) **(20120-33-6)**
Phosphonic acid, (2-((hydroxymethyl)carbamoyl)ethyl)-, dimethyl ester **(20120-33-6)**
Phosphonic acid, (1-hydroxy-2,2,2-trichloroethyl)-, dimethyl eşter **(52-68-6)**
Phosphonic acid, methyl- (9CI) **(993-13-5)**
Phosphonic acid, methyl-, bimol. monoanhydride, diethyl ester (8CI) **(32288-17-8)**
Phosphonic acid, methyl-, bis(1-methylethyl) ester **(1445-75-6)**
Phosphonic acid, methyl-, bis(1,2,2-trimethylpropyl) ester (8CI,9CI) **(7040-58-6)**
Phosphonic acid, methyl-, dibutyl ester (8CI,9CI) **(2404-73-1)**
Phosphonic acid, methyl-, diethyl ester **(683-08-9)**
Phosphonic acid, methyl-, diisopropyl ester **(1445-75-6)**
Phosphonic acid, methyl-, dimethyl ester **(756-79-6)**
Phosphonic acid, methyl-, dioctyl ester (8CI,9CI) **(1832-68-4)**
Phosphonic acid, methyl-, dipentyl ester (8CI,9CI) **(1000-36-8)**
Phosphonic acid, methyl-, isopropyl phenyl ester (8CI) **(33684-08-1)**
Phosphonic acid, methyl-, 1-methylethyl phenyl ester (9CI) **(33684-08-1)**
Phosphonic acid, methyl-, monoethyl ester (8CI,9CI) **(1832-53-7)**
Phosphonic acid, (2-methyl-3-oxopentyl)-, diethyl ester (8CI,9CI) **(16965-90-5)**
Phosphonic acid, monoethyl ester, aluminum salt (3:1) **(39148-24-8)**
Phosphonic acid, monomethyl ester **(13590-71-1)**
Phosphonic acid, morpholino-, dimethyl ester **(597-25-1)**
Phosphonic acid, 4-morpholinyl-, dimethyl ester (9CI) **(597-25-1)**
Phosphonic acid, (nitrilotris(methylene))tri **(6419-19-8)**
Phosphonic acid, (nitrilotris(methylene))tri-, pentasodium salt **(2235-43-0)**
Phosphonic acid, (nitrilotris(methylene))tris-, pentasodium salt (9CI) **(2235-43-0)**
Phosphonic acid, (nitrilotris(methylene))tris-, potassium salt (9CI) **(27794-93-0)**
Phosphonic acid, (nitrilotris(methylene))tris-, sodium salt (9CI) **(20592-85-2)**
Phosphonic acid, octadecyl-, dimethyl ester (9CI) **(25371-54-4)**
Phosphonic acid, (3-oxobutyl)-, diethyl ester (7CI,8CI,9CI) **(1067-90-9)**
Phosphonic acid, (2-oxopropyl)-, diethyl ester (9CI) **(1067-71-6)**
Phosphonic acid, pentyl- (8CI,9CI) **(4672-26-8)**
Phosphonic acid, phenyl **(1571-33-1)**
Phosphonic acid, phenyl-, dioctyl ester **(1754-47-8)**
Phosphonic acid, (((phosphonomethyl)imino)bis(2,1-ethanediyl-nitrilobis(methylene)))tetrakis- (9CI) **(15827-60-8)**
Phosphonic acid, (((phosphonomethyl)imino)bis(2,1-ethanediyl-nitrilobis(methylene)))tetrakis-, sodium salt (9CI) **(22042-96-2)**
Phosphonic acid, (((phosphonomethyl)imino)bis(6,1-hexanediyl-nitrilobis(methylene)))tetrakis- (9CI) **(34690-00-1)**
Phosphonic acid, (((phosphonomethyl)imino)bis(6,1-hexanediyl-nitrilobis(methylene)))tetrakis-, sodium salt (9CI) **(35657-77-3)**
Phosphonic acid, propyl **(4672-38-2)**
Phosphonic acid, propyl-, monoethyl ester **(21921-96-0)**
Phosphonic acid, (2,2,2-trichloro-1-hydroxyethyl)-, dimethyl ester **(52-68-6)**
Phosphonic acid, (2,2,2-trichloro-1-hydroxyethyl)-, dimethyl ester, butyrate **(126-22-7)**
Phosphonic diamide, P-(5-amino-3-phenyl-1H-1,2,4-triazol-1-yl)-N,N,N',N'-tetramethyl- (8CI,9CI) **(1031-47-6)**
Phosphonic dichloride, methyl **(676-97-1)**
Phosphonic dichloride, phenyl- (9CI) **(824-72-6)**
Phosphonic difluoride **(14939-34-5)**
Phosphonitrile chloride, Homopolymer (9CI) **(25034-79-1)**
Phosphonitrile chloride, cyclic pentamer **(13596-41-3)**
Phosphonitrile chloride, cyclic trimer **(940-71-6)**
Phosphonitrilic chloride **(2950-45-0)**
Phosphonitrilic chlorides **(25034-79-1)**
Phosphonium, benzyltriphenyl-, chloride (8CI) **(1100-88-5)**
Phosphonium, 1,2-ethanediylbis(tris(2-cyanoethyl)-, dibromide **(10310-38-0)**
Phosphonium, ethylenebis(tris(2-cyanoethyl)-, dibromide **(10310-38-0)**
Phosphonium, ethyltriphenyl-, acetate (9CI) **(35835-94-0)**
Phosphonium, ethyltriphenyl-, iodide **(4736-60-1)**
Phosphonium, tetrabutyl-, bromide **(3115-68-2)**
Phosphonium, tetrabutyl-, chloride **(2304-30-5)**
Phosphonium, tetrakis(hydroxymethyl)-, acetate **(7580-37-2)**
Phosphonium, tetrakis(hydroxymethyl)-, acetate mixed with tetrakis(hydroxymethyl) phosphonium dihydrogen phosphate (76:24) **(55818-96-7)**
Phosphonium, tetrakis(hydroxymethyl)-, chloride **(124-64-1)**
Phosphonium, tetrakis(hydroxymethyl)-, ethanedioate (2:1) (Salt) (9CI) **(52221-67-7)**
Phosphonium, tetrakis(hydroxymethyl)-, hydroxide **(512-82-3)**
Phosphonium, tetrakis(hydroxymethyl)-, phosphate (3:1) (Salt) (9CI) **(22031-17-0)**
Phosphonium, tetrakis(hydroxymethyl)-, sulfate (2:1) **(55566-30-8)**
Phosphonium, tetrakis(hydroxymethyl)-, sulfate (2:1) (Salt), Polymer with urea (9CI) **(63502-25-0)**
Phosphonium, tributyl(2,4-dichlorobenzyl)-, chloride **(115-78-6)**
Phosphonium, triphenyl(phenylmethyl)-, chloride (9CI) **(1100-88-5)**
Phosphonoacetic acid **(4408-78-0)**
2-Phosphonobutane-1,2,4-tricarbonic acid **(37971-36-1)**
2-Phosphono-1,2,4-butanetricarboxylic acid **(37971-36-1)**
2-Phosphonobutane-1,2,4-tricarboxylic acid **(37971-36-1)**
Phosphonochloridothioic acid, ethyl-, O-ethyl ester (9CI) **(1497-68-3)**
Phosphonochloridothioic acid, phenyl-, O-ethyl ester **(5075-13-8)**
Phosphonodithioic acid, ethyl-, S-(p-chlorophenyl) O-ethyl ester **(2984-64-7)**
Phosphonodithioic acid, ethyl-, O-ethyl S-phenyl ester **(944-22-9)**
Phosphonodithioic acid, ethyl-, O-ethyl S-(p-tolyl) ester **(333-43-7)**
Phosphonodithioic acid, methyl-, S-((N-methoxy-carbonyl)-N-methylcarbamoyl) methyl O-methyl ester **(29173-31-7)**
Phosphonofluoridic acid, ethyl-, ethyl ester (6CI,7CI,8CI,9CI) **(650-20-4)**
Phosphonofluoridic acid, methyl-, isopropyl ester **(107-44-8)**
Phosphonofluoridic acid, methyl-, 1,2,2-trimethylpropyl ester **(96-64-0)**
N-(Phosphonomethyl)glycine **(1071-83-6)**
Phosphonothioic acid, S-benzyl O-ethyl ester **(21722-85-0)**
Phosphonothioic acid, chloro-, O,O-diethyl ester **(2524-04-1)**
Phosphonothioic acid, chloro-, O,O-dimethyl ester **(2524-03-0)**
Phosphonothioic acid, diethylparanitrophenyl ester **(311-45-5)**
Phosphonothioic acid, (1,2-dihydro-1,3-dioxo-2H-isoindol-2-yl)-, O,O-diethyl ester **(5131-24-8)**
Phosphonothioic acid, ethyl-, O-(2-chloro-1-(2,5-dichlorophenyl)-ethenyl) O-methyl ester **(41491-52-5)**
Phosphonothioic acid, ethyl-, O-(1-(2,4-dichlorophenyl)ethenyl) O-methyl ester (9CI) **(38338-57-7)**
Phosphonothioic acid, ethyl-, O-ethyl O-(2,4,5-trichlorophenyl) ester **(327-98-0)**
Phosphonothioic acid, methyl-, S-(2-(bis(1-methylethyl)amino)-ethyl) O-ethyl ether **(50782-69-9)**
Phosphonothioic acid, methyl-, S-(2-(bis(1-methylethyl)amino)-ethyl) ester (9CI) **(73207-98-4)**
Phosphonothioic acid, methyl-, O-(2-chloro-1-(2,5-dichloro-phenyl)ethenyl) O-methyl ester (9CI) **(56549-12-3)**
Phosphonothioic acid, methyl-, S-(2-(diethylamino)ethyl) O-ethyl ester **(21770-86-5)**
Phosphonothioic acid, methyl-, O,S-diethyl ester **(2511-10-6)**
Phosphonothioic acid, methyl-, O,O-diethyl ester (8CI,9CI) **(6996-81-2)**
Phosphonothioic acid, methyl-, S-(2-(diisopropylamino)ethyl)

O-ethyl ester **(50782-69-9)**
Phosphonothioic acid, methyl-, O,O-dipropyl ester (8CI,9CI) **(25371-75-9)**
Phosphonothioic acid, methyl-, O-ethyl ester (8CI,9CI) **(18005-40-8)**
Phosphonothioic acid, methyl-, O-ethyl O-(4-(methylthio)phenyl) ester **(2703-13-1)**
Phosphonothioic acid, methyl-, O-(4-nitrophenyl) O-phenyl ester (9CI) **(2665-30-7)**
Phosphonothioic acid, methyl-, O-(p-nitrophenyl) O-phenyl ester (8CI) **(2665-30-7)**
Phosphonothioic acid, phenyl-, O-(4-bromo-2,5-dichlorophenyl) O-methyl ester **(21609-90-5)**
Phosphonothioic acid, phenyl-, O-(4-cyanophenyl) ester (9CI) **(61073-10-7)**
Phosphonothioic acid, phenyl-, O-(4-cyanophenyl) O-ethyl ester **(13067-93-1)**
Phosphonothioic acid, phenyl-, O-ethyl ester, O-ester with p-hydroxybenzonitrile **(13067-93-1)**
Phosphonothioic acid, phenyl-, O-ethyl O-(p-nitrophenyl)ester **(2104-64-5)**
Phosphonothioic acid, phenyl-, O-methyl ester, potassium salt (9CI) **(67446-04-2)**
Phosphonothioic acid, phthalimido-, O,O-diethyl ester **(5131-24-8)**
Phosphonothioic dichloride, ethyl **(993-43-1)**
Phosphonothioic dichloride, methyl-, Anhydrous **(676-98-2)**
Phosphonothioic dichloride, phenyl **(3497-00-5)**
Phosphonous acid, (1,1'-biphenyl)-4,4'-diylbis-, tetrakis-(2,4-bis(1,1-dimethylethyl)phenyl) ester (9CI) **(38613-77-3)**
Phosphonous dichloride, phenyl **(644-97-3)**
Phosphopyridoxal **(54-47-7)**
Phosphopyridoxal monohydrate **(41468-25-1)**
Phosphopyron **(2778-04-3)**
Phosphopyrone **(2778-04-3)**
Phosphoramide, N,N',N''-triethylene- **(545-55-1)**
Phosphoramidic acid, 4-tert-butyl-2-chlorophenylphosphoramidate **(299-86-5)**
Phosphoramidic acid, 1,3-dithietan-2-ylidene-, diethyl ester **(21548-32-3)**
Phosphoramidic acid, 1,3-dithiolan-2-ylidene-, diethyl ester **(947-02-4)**
Phosphoramidic acid, isopropyl-, ethyl 4-(ethylsulfinyl)-m-tolyl ester (8CI) **(31972-43-7)**
Phosphoramidic acid, isopropyl-, ethyl 4-(methylsulfonyl)-m-tolyl ester (8CI) **(31972-44-8)**
Phosphoramidic acid, isopropyl-, 4-(methylthio)-m-tolyl ethyl ester **(22224-92-6)**
Phosphoramidic acid, methyl-, 4-tert-butyl-2-chlorophenyl methyl ester **(299-86-5)**
Phosphoramidic acid, methyl-, 2-chloro-4-(1,1-dimethylethyl)-phenyl methyl ester **(299-86-5)**
Phosphoramidic acid, (4-methyl-1,3-dithiolan-2-ylidene)-, diethyl ester **(950-10-7)**
Phosphoramidic acid, (1-methylethyl)-, ethyl 3-methyl-4-(methylsulfinyl)phenyl ester (9CI) **(31972-43-7)**
Phosphoramidic acid, (1-methylethyl)-, ethyl 3-methyl-4-(methylsulfonyl)phenyl ester (9CI) **(31972-44-8)**
Phosphoramidic acid, (1-methylethyl)-, ethyl (3-methyl-4-(methylthio)phenyl) ester **(22224-92-6)**
Phosphoramidic acid, methylphenyl-, dibutyl ester (9CI) **(52670-79-8)**
Phosphoramidocyanidic acid, dimethyl-, ethyl ester **(77-81-6)**
Phosphoramidothioic acid, acetimidoyl-, O,O-bis(p-chloro-phenyl) ester (8CI) **(4104-14-7)**
Phosphoramidothioic acid, N-acetyl-, O,S-dimethyl ester **(30560-19-1)**
Phosphoramidothioic acid, acetyl-, O,O-dimethyl ester (9CI) **(42072-27-5)**
Phosphoramidothioic acid, N-(sec-butyl)-, O-ethyl O-(6-nitro-m-tolyl) ester **(36335-67-8)**
Phosphoramidothioic acid, O,S-dimethyl ester **(10265-92-6)**
Phosphoramidothioic acid, (1-iminoethyl)-, O,O-bis(4-chloro-phenyl) ester **(4104-14-7)**
Phosphoramidothioic acid, isopropyl-, O-(2,4-dichlorophenyl) O-methyl ester **(299-85-4)**
Phosphoramidothioic acid, isopropyl-, O-ethyl O-(2-isopropoxy-carbonylphenyl) ester **(25311-71-1)**
Phosphoramidothioic acid, isopropyl-, O-ethyl O-(2-nitro-p-tolyl) ester **(33857-23-7)**
Phosphoramidothioic acid, isopropyl-, O-ethyl O-(6-nitro-m-tolyl) ester (8CI) **(14151-45-2)**
Phosphoramidothioic acid, (1-methylethyl)-, O-ethyl O-(4-methyl-2-nitrophenyl) ester (9CI) **(33857-23-7)**
Phosphoramidothioic acid, (1-methylethyl)-, O-ethyl O-(5-methyl-2-nitrophenyl) ester (9CI) **(14151-45-2)**
Phosphoramidothioic acid, (1-methylpropyl)-, O-ethyl O-(5-methyl-2-nitrophenyl) ester (9CI) **(36335-67-8)**
Phosphorane, bis(4-methoxyphenyl)phenylbis(2,2,2-trifluoro-ethoxy)- (9CI) **(102040-55-1)**
Phosphorane, (4-chlorophenyl)tetrakis(2,2,2-trifluoroethoxy)-(9CI) **(102040-62-0)**
Phosphorane, diethoxy(2-methylphenyl)bis(2,2,2-trifluoro-ethoxy)- (9CI) **(102040-46-0)**
Phosphorane, diethoxyphenylbis(2,2,2-trifluoroethoxy)- (9CI) **(102040-45-9)**
Phosphorane, diphenyltris(2,2,2-trifluoroethoxy)- (9CI) **(76943-21-0)**
Phosphorane, (2-methoxyphenyl)diphenylbis(2,2,2-trifluoro-ethoxy)- (9CI) **(102040-44-8)**
Phosphorane, (4-methoxyphenyl)diphenylbis(2,2,2-trifluoro-ethoxy)- (9CI) **(102040-51-7)**
Phosphorane, (4-methoxyphenyl)tetrakis(2,2,2-trifluoroethoxy)-(9CI) **(102040-59-5)**
Phosphorane, (4-methylphenyl)diphenylbis(2,2,2-trifluoroethoxy)-(9CI) **(102040-53-9)**
Phosphorane, (3-methylphenyl)tetrakis(2,2,2-trifluoroethoxy)-(9CI) **(102040-61-9)**
Phosphorane, (4-methylphenyl)tetrakis(2,2,2-trifluoroethoxy)-(9CI) **(102040-58-4)**
Phosphorane, pentachloro **(10026-13-8)**
Phosphorane, pentakis(2,2,2-trifluoroethoxy)- (9CI) **(71181-76-5)**
Phosphorane, phenyltetrakis(2,2,2-trifluoroethoxy)- (9CI) **(63325-06-4)**
Phosphorane, triphenylbis(2,2,2-trifluoroethoxy)-, (TB-5-11)-(9CI) **(67696-25-7)**
Phosphorane, tris(3-chlorophenyl)bis(2,2,2-trifluoroethoxy)-(9CI) **(102040-57-3)**
Phosphorane, tris(4-chlorophenyl)bis(2,2,2-trifluoroethoxy)-(9CI) **(102040-54-0)**
Phosphorane, tris(4-fluorophenyl)bis(2,2,2-trifluoroethoxy)-(9CI) **(102040-56-2)**
Phosphorane, tris(4-methoxyphenyl)bis(2,2,2-trifluoroethoxy)-(9CI) **(102040-49-3)**
Phosphorane, tris(1-methylethoxy)bis(2,2,2-trifluoroethoxy)-(9CI) **(102040-47-1)**
Phosphorane, tris(3-methylphenyl)bis(2,2,2-trifluoroethoxy)-(9CI) **(102040-52-8)**
Phosphorane, tris(4-methylphenyl)bis(2,2,2-trifluoroethoxy)-(9CI) **(102040-50-6)**
Phosphore (French) **(7723-14-0)**
Phosphore blanc (French) **(7723-14-0)**
Phosphore(pentachlorure de) (French) **(10026-13-8)**
Phosphore(trichlorure de) (French) **(7719-12-2)**
Phosphoric acid (ACGIH,OSHA) [UN 1805] **(7664-38-2)**
Phosphoric acid, Liquid (DOT) **(7664-38-2)**
Phosphoric acid, Mixed decyl and Et and octyl esters **(68412-60-2)**
Phosphoric acid, Solid (DOT) **(7664-38-2)**
Phosphoric acid, aluminum salt (1:1) **(7784-30-7)**
Phosphoric acid, aluminum sodium salt (9CI) **(7785-88-8)**
Phosphoric acid, 2-aminoethyl phenyl ester **(1071-23-4)**
Phosphoric acid, beryllium salt (1:1) **(13598-15-7)**
Phosphoric acid, 2-biphenylyl diphenyl ester **(132-29-6)**
Phosphoric acid, 2,2-bis(bromomethyl)-3-chloropropyl bis-(2-chloro-1-(chloromethyl)ethyl) ester **(66108-37-0)**
Phosphoric acid, bis(p-tert-butylphenyl) phenyl ester (8CI) **(115-87-7)**
Phosphoric acid, bis(2-chloro-1-(chloromethyl)ethyl) 2,3-dichloro-propyl ester (9CI) **(68460-03-7)**
Phosphoric acid, 2,2-bis(chloromethyl)-1,3-propanediyl tetrakis-(2-chloroethyl) ester (9CI) **(38051-10-4)**

Phosphoric acid, bis((1,1-dimethylethyl)phenyl) phenyl ester (9CI) **(65652-41-7)**
Phosphoric acid, bis(4-(1,1-dimethylethyl)phenyl) phenyl ester (9CI) **(115-87-7)**
Phosphoric acid, bis(2,5-dimethylphenyl) 2,6-dimethylphenyl ester (9CI) **(73179-49-4)**
Phosphoric acid, bis(2,5-dimethylphenyl) phenyl ester (9CI) **(72121-83-6)**
Phosphoric acid, bis(2,6-dimethylphenyl) phenyl ester (9CI) **(23666-93-5)**
Phosphoric acid, bis(2,5-dimethylphenyl) 2,4,6-trimethylphenyl ester (9CI) **(73179-47-2)**
Phosphoric acid, bis(2,6-dimethylphenyl) 2,4,6-trimethylphenyl ester (9CI) **(73179-37-0)**
Phosphoric acid, bis(2-ethylhexyl) ester **(298-07-7)**
Phosphoric acid, bis(2-ethylhexyl)phenyl ester **(16368-97-1)**
Phosphoric acid, bis(1-methylethyl)phenyl diphenyl ester (9CI) **(58570-87-9)**
Phosphoric acid, bis(1-methylethyl) phenyl ester (9CI) **(51496-03-8)**
Phosphoric acid, bis((1-methylethyl)phenyl) phenyl ester (9CI) **(28109-00-4)**
Phosphoric acid, bis(2-(1-methylethyl)phenyl) phenyl ester (9CI) **(69500-29-4)**
Phosphoric acid, bis((1-methyl-1-phenylethyl)phenyl) phenyl ester (9CI) **(63302-95-4)**
Phosphoric acid, bis(methylphenyl) phenyl ester (9CI) **(26446-73-1)**
Phosphoric acid, bis(4-nitrophenyl) ester (9CI) **(645-15-8)**
Phosphoric acid, bis(p-nitrophenyl) ester (8CI) **(645-15-8)**
Phosphoric acid, bis(nonylphenyl) phenyl ester (9CI) **(63302-94-3)**
Phosphoric acid, 2-bromo-1-(2,4-dichlorophenyl)ethenyl dimethyl ester (9CI) **(13104-21-7)**
Phosphoric acid, 2-bromo-1-(2,4-dichlorophenyl)vinyl dimethyl ester **(13104-21-7)**
Phosphoric acid, butyl diphenyl ester **(2752-95-6)**
Phosphoric acid, butyl ester **(12788-93-1)**
Phosphoric acid, (p-tert-butylphenyl) diphenyl ester **(981-40-8)**
Phosphoric acid, butylphenyl diphenyl ester (9CI) **(75675-48-8)**
Phosphoric acid, sec-butylphenyl diphenyl ester (8CI) **(28109-02-6)**
Phosphoric acid calcium(2+) salt (2:3) **(7758-87-4)**
Phosphoric acid, calcium salt (1:1) (9CI) **(7757-93-9)**
Phosphoric acid, calcium salt (2:3) (9CI) **(7758-87-4)**
Phosphoric acid, calcium salt, Hydrate (2:1:1) **(7758-23-8)**
Phosphoric acid, 2-chloro-1-(2,4-dichlorophenyl)ethenyl dimethyl ester **(2274-67-1)**
Phosphoric acid, 2-chloro-1-(2,4-dichlorophenyl)vinyl diethyl ester **(470-90-6)**
Phosphoric acid, 2-chloro-1-(2,4-dichlorophenyl)vinyl dimethyl ester **(2274-67-1)**
Phosphoric acid, 2-chloro-3-(diethylamino)-1-methyl-3-oxo-1-propenyl dimethyl ester, (E) **(297-99-4)**
Phosphoric acid, 2-chloro-3-(diethylamino)-1-methyl-3-oxo-1-propenyldimethyl ester, (Z)- (9CI) **(23783-98-4)**
Phosphoric acid, 2-chloro-1-(2,4,5-trichlorophenyl)ethenyl dimethyl ester **(961-11-5)**
Phosphoric acid, 2-chloro-1-(2,4,5-trichlorophenyl)ethenyl dimethyl ester, (Z)- **(22248-79-9)**
Phosphoric acid, 2-chloro-1-(2,4,5-trichlorophenyl)vinyl dimethyl ester **(961-11-5)**
Phosphoric acid, 2-chloro-1-(2,4,5-trichlorophenyl)vinyl dimethyl ester, (Z) **(22248-79-9)**
Phosphoric acid, 2-chlorovinyl diethyl ester **(311-47-7)**
Phosphoric acid chromium (III) salt **(7789-04-0)**
Phosphoric acid, chromium(3+) salt (1:1) **(7789-04-0)**
Phosphoric acid, compound with 2-aminoethanol **(29868-05-1)**
Phosphoric acid, copper(2+) salt **(7798-23-4)**
Phosphoric acid, copper salt (9CI) **(10103-48-7)**
Phosphoric acid, copper(2+) salt (2:3) (9CI) **(7798-23-4)**
Phosphoric acid, 4-cyanophenyl dimethyl ester **(61090-94-6)**
Phosphoric acid, diammonium salt **(7783-28-0)**
Phosphoric acid, 1,2-dibromo-2,2-dichloroethyl dimethyl ester **(300-76-5)**
Phosphoric acid, dibutyl ester **(107-66-4)**
Phosphoric acid, dibutyl ester, potassium salt (9CI) **(25238-98-6)**
Phosphoric acid, dibutyl phenyl ester **(2528-36-1)**
Phosphoric acid, 2,2-dichloroethenyl dimethyl ester **(62-73-7)**
Phosphoric acid, 2,2-dichloroethenyl dimethyl ester, Mixt-. with O,O-diethyl O-(3,5,6-trichloro-2-pyridinyl) phosphorothioate (9CI) **(70840-42-5)**
Phosphoric acid, 2,2-dichlorovinyl diethyl ester **(72-00-4)**
Phosphoric acid, 2,2-dichlorovinyl dimethyl ester **(62-73-7)**
Phosphoric acid, didodecyl ester (9CI) **(7057-92-3)**
Phosphoric acid, diethyl ester (9CI) **(598-02-7)**
Phosphoric acid, diethyl ester, mercury(1+) salt (8CI) **(21504-45-0)**
Phosphoric acid, diethyl-, 3-methylpyrazole-5-yl ester **(108-34-9)**
Phosphoric acid, diethyl-(3-methyl-5-pyrazolyl) ester **(108-34-9)**
Phosphoric acid diethyl 4-nitrophenyl ester **(311-45-5)**
Phosphoric acid, diethyl p-nitrophenyl ester **(311-45-5)**
Phosphoric acid, diethyl 5(or 3)-methylpyrazole-3(or 5)-yl ester **(108-34-9)**
Phosphoric acid, diethyl pentyl ester (8CI,9CI) **(20195-08-8)**
Phosphoric acid, diethyl phenyl ester (9CI) **(2510-86-3)**
Phosphoric acid, diisodecyl phenyl ester (9CI) **(51363-64-5)**
Phosphoric acid, diisooctyl ester **(27215-10-7)**
Phosphoric acid, diisotridecyl ester (9CI) **(27073-01-4)**
Phosphoric acid 3-(dimethylamino)-1-methyl-3-oxo-1-propenyl dimethyl ester **(141-66-2)**
Phosphoric acid, dimethyl ester **(813-78-5)**
Phosphoric acid, dimethyl ester, ester with 2-chloro-N,N-diethyl-3-hydroxycrotonamide **(13171-21-6)**
Phosphoric acid, dimethyl ester, ester with 2-chloro-N,N-diethyl-3-hydroxycrotonamide, (Z)- (8CI) **(23783-98-4)**
Phosphoric acid, dimethyl ester, ester with dimethyl 3-hydroxy-glutaconate **(122-10-1)**
Phosphoric acid, dimethyl ester, ester with (E)-3-hydroxy-N,N-dimethylcrotonamide **(141-66-2)**
Phosphoric acid, dimethyl ester, ester with cis-3-hydroxy-N,N-dimethylcrotonamide **(141-66-2)**
Phosphoric acid, dimethyl ester, ester with p-hydroxybenzonitrile **(61090-94-6)**
Phosphoric acid, dimethyl ester, ester with (E)-3-hydroxy-N-methylcrotonamide **(6923-22-4)**
Phosphoric acid, dimethyl ester, ester with cis-3-hydroxy-N-methylcrotonamide **(6923-22-4)**
Phosphoric acid, dimethyl ester, ester with methyl 3-hydroxy-crotonate **(7786-34-7)**
Phosphoric acid, (1,1-dimethylethyl)phenyl diphenyl ester (9CI) **(56803-37-3)**
Phosphoric acid, 2-(1,1-dimethylethyl)phenyl diphenyl ester (9CI) **(83242-23-3)**
Phosphoric acid, 4-(1,1-dimethylethyl)phenyl diphenyl ester, Mixt. with triphenyl phosphate (9CI) **(96300-96-8)**
Phosphoric acid, dimethyl 1-methyl-N,N-(dimethylamino)-3-oxo-1-propenyl ester, (E)- (9CI) **(141-66-2)**
Phosphoric acid, dimethyl 1-methyl-3-(methylamino)-2-oxo-1-propenyl ester, (Z) **(919-44-8)**
Phosphoric acid, dimethyl 1-methyl-3-(methylamino)-3-oxo-1-propenyl ester, (E)- **(6923-22-4)**
Phosphoric acid, dimethyl p-(methylthio)phenyl ester **(3254-63-5)**
Phosphoric acid, dimethyl p-nitrophenyl ester **(950-35-6)**
Phosphoric acid, dimethyl-4-nitrophenyl ester (9CI) **(950-35-6)**
Phosphoric acid, dimethyl 4-nitro-m-tolyl ester **(2255-17-6)**
Phosphoric acid, 2,5-dimethylphenyl bis(2,6-dimethylphenyl) ester (9CI) **(73179-48-3)**
Phosphoric acid, 2,5-dimethylphenyl bis(2,4,6-trimethylphenyl) ester (9CI) **(73179-38-1)**
Phosphoric acid, 2,6-dimethylphenyl bis(2,4,6-trimethylphenyl) ester (9CI) **(73195-13-8)**
Phosphoric acid, 2,5-dimethylphenyl 2,6-dimethylphenyl phenyl ester (9CI) **(73179-45-0)**
Phosphoric acid, 2,5-dimethylphenyl 2,6-dimethylphenyl 2,4,6-trimethylphenyl ester (9CI) **(73179-46-1)**
Phosphoric acid, 2,5-dimethylphenyl diphenyl ester (9CI) **(73179-40-5)**
Phosphoric acid, 2,6-dimethylphenyl diphenyl ester (9CI) **(23666-94-6)**
Phosphoric acid, 2,5-dimethylphenyl phenyl 2,4,6-trimethyl-

phenyl ester (9CI) **(73179-42-7)**
Phosphoric acid, 2,6-dimethylphenyl phenyl 2,4,6-trimethylphenyl ester (9CI) **(73179-41-6)**
Phosphoric acid, dimethyl 3,5,6-trichloro-2-pyridyl ester **(5598-52-7)**
Phosphoric acid, dioctadecyl ester (9CI) **(3037-89-6)**
Phosphoric acid, dioctyl ester (9CI) **(3115-39-7)**
Phosphoric acid, dipentyl ester **(3138-42-9)**
Phosphoric acid, diphenyl ester **(3138-42-9)**
Phosphoric acid, diphenyl ester (9CI) **(838-85-7)**
Phosphoric acid, diphenyl p-tolyl ester **(78-31-9)**
Phosphoric acid, diphenyl tolyl ester **(26444-49-5)**
Phosphoric acid, diphenyl 2,4,6-trimethylphenyl ester (9CI) **(73179-43-8)**
Phosphoric acid, dipotassium salt (9CI) **(7758-11-4)**
Phosphoric acid, dipropyl ester **(1804-93-9)**
Phosphoric acid, dipropyl 4-methylthiophenyl ester **(7292-16-2)**
Phosphoric acid, disodium salt **(7558-79-4)**
Phosphoric acid, disodium salt, dodecahydrate **(10039-32-4)**
Phosphoric acid, dodecyl ester (9CI) **(12751-23-4)**
Phosphoric acid, esters with 2-ethylhexanol **(12645-31-7)**
Phosphoric acid, 1,2-ethanediyl tetrakis(2-chloroethyl) ester (9CI) **(33125-86-9)**
Phosphoric acid, (ethoxylated tridecyl alcohol) esters **(9046-01-9)**
Phosphoric acid, ethyl bis(4-nitrophenyl) ester (9CI) **(905-14-6)**
Phosphoric acid, ethyl bis(p-nitrophenyl) ester (8CI) **(905-14-6)**
Phosphoric acid, 2-ethylhexyl diphenyl ester **(1241-94-7)**
Phosphoric acid, 2-ethylhexyl ester (9CI) **(12645-31-7)**
Phosphoric acid, 2-ethylhexyl esters **(12645-31-7)**
Phosphoric acid, 3-ethylphenyl diphenyl ester (9CI) **(52784-49-3)**
Phosphoric acid hexamethyltriamide **(680-31-9)**
Phosphoric acid, 4-hexylphenyl diphenyl ester (9CI) **(64532-96-3)**
Phosphoric acid, hexylphenyl diphenyl ester (9CI) **(69682-29-7)**
Phosphoric acid, iron(3+) salt (1:1) (9CI) **(10045-86-0)**
Phosphoric acid, isodecyl diphenyl ester (9CI) **(29761-21-5)**
Phosphoric acid, isooctyl ester (9CI) **(12645-53-3)**
Phosphoric acid, isopropyl ester **(1623-24-1)**
Phosphoric acid, lead(2+) salt (2:3) **(7446-27-7)**
Phosphoric acid, lead(2+) salt (1:1) (8CI,9CI) **(15845-52-0)**
Phosphoric acid, magnesium salt (2:3) (9CI) **(7757-87-1)**
Phosphoric acid, manganese salt (9CI) **(10124-54-6)**
Phosphoric acid, (1-methoxycarboxypropen-2-yl) dimethyl ester **(7786-34-7)**
Phosphoric acid, methyl bis(4-nitrophenyl) ester **(799-87-1)**
Phosphoric acid, methyl diphenyl ester (9CI) **(115-89-9)**
Phosphoric acid, ar'-(1-methylethyl)-(1,1'-biphenyl)yl diphenyl ester (9CI) **(50851-28-0)**
Phosphoric acid, 1-methylethyl diphenyl ester (9CI) **(60763-39-5)**
Phosphoric acid, (1-methylethyl)phenyl diphenyl ester (9CI) **(28108-99-8)**
Phosphoric acid, 2-(1-methylethyl)phenyl diphenyl ester (9CI) **(64532-94-1)**
Phosphoric acid, 4-(1-methylethyl)phenyl diphenyl ester (9CI) **(55864-04-5)**
Phosphoric acid, 2-(1-methylethyl)phenyl diphenyl ester, Mixt. with triphenyl phosphate (9CI) **(96300-97-9)**
Phosphoric acid, methylfluoro-, isopropyl ester **(107-44-8)**
Phosphoric acid, 4-methylphenyl diphenyl ester **(78-31-9)**
Phosphoric acid, methylphenyl diphenyl ester (9CI) **(26444-49-5)**
Phosphoric acid, (1-methyl-1-phenylethyl)phenyl diphenyl ester (9CI) **(34364-42-6)**
Phosphoric acid, (1-methyl-1-phenylethyl)phenyl nonylphenyl phenyl ester (9CI) **(63340-28-3)**
Phosphoric acid, p-(methylthio)phenyl dipropyl ester **(7292-16-2)**
Phosphoric acid, 4-(methylthio)phenyl dipropyl ester (9CI) **(7292-16-2)**
Phosphoric acid, monoammonium salt (9CI) **(7722-76-1)**
Phosphoric acid, monobutyl ester (9CI) **(1623-15-0)**
Phosphoric acid, monobutyl ester, dipotassium salt (9CI) **(26290-70-0)**
Phosphoric acid, monobutyl ester, monopotassium salt (8CI) **(25238-99-7)**
Phosphoric acid, mono(4-cyanophenyl) monomethyl ester (9CI) **(31328-16-2)**
Phosphoric acid, monodecyl ester **(3921-30-0)**
Phosphoric acid, mono(2,3-dibromopropyl) ester **(5324-12-9)**
Phosphoric acid, monododecyl ester (9CI) **(2627-35-2)**
Phosphoric acid, monoethyl ester **(1623-14-9)**
Phosphoric acid, mono(2-ethylhexyl) ester (9CI) **(1070-03-7)**
Phosphoric acid, mono(2-ethylhexyl) ester, disodium salt (9CI) **(15505-13-2)**
Phosphoric acid, mono(2-ethylhexyl) ester, sodium salt (9CI) **(31044-12-9)**
Phosphoric acid, monohexyl ester (9CI) **(3900-04-7)**
Phosphoric acid, monoisooctyl ester (9CI) **(26403-12-3)**
Phosphoric acid, monomethyl ester **(812-00-0)**
Phosphoric acid, monomethyl ester, monoester with p-hydroxybenzonitrile (8CI) **(31328-16-2)**
Phosphoric acid, mono(1-methylethyl) monophenyl ester (9CI) **(46355-07-1)**
Phosphoric acid, mono(4-nitrophenyl) ester (9CI) **(330-13-2)**
Phosphoric acid, mono(p-nitrophenyl) ester (8CI) **(330-13-2)**
Phosphoric acid, monooctadecyl ester (9CI) **(2958-09-0)**
Phosphoric acid, monooctyl ester **(3991-73-9)**
Phosphoric acid, monooctyl ester, dipotassium salt (9CI) **(19045-79-5)**
Phosphoric acid, monopentyl ester (9CI) **(2382-76-5)**
Phosphoric acid, monophenyl ester (9CI) **(701-64-4)**
Phosphoric acid, monopotassium salt (9CI) **(7778-77-0)**
Phosphoric acid, monopropyl ester (9CI) **(1623-06-9)**
Phosphoric acid, 4-nitrophenyl diphenyl ester (9CI) **(10359-36-1)**
Phosphoric acid, p-nitrophenyl diphenyl ester (8CI) **(10359-36-1)**
Phosphoric acid, 4-nitrophenyl ester (9CI) **(12778-12-0)**
Phosphoric acid, 4-nonylphenyl diphenyl ester (9CI) **(64532-97-4)**
Phosphoric acid, nonylphenyl diphenyl ester (9CI) **(38638-05-0)**
Phosphoric acid, octyl diphenyl ester **(115-88-8)**
Phosphoric acid, octyl ester (9CI) **(39407-03-9)**
Phosphoric acid, pentyl ester **(12789-46-7)**
Phosphoric acid, phenyl bis(2,4,6-trimethylphenyl) ester (9CI) **(73179-44-9)**
Phosphoric acid, sodium salt (9CI) **(7632-05-5)**
Phosphoric acid, titanium(4+) salt (4:3) (9CI) **(15578-51-5)**
Phosphoric acid, triallyl ester **(1623-19-4)**
Phosphoric acid, tributyl ester **(126-73-8)**
Phosphoric acid, 2,2,2-trichloroethyl ester **(306-52-5)**
Phosphoric acid, tri-o-cresyl ester **(78-30-8)**
Phosphoric acid, triethanolamine salt **(10017-56-8)**
Phosphoric acid, triethylene imide **(545-55-1)**
Phosphoric acid, triethyleneimine (DOT) **(545-55-1)**
Phosphoric acid, triethyl ester **(78-40-0)**
Phosphoric acid, triisobutyl ester (8CI) **(126-71-6)**
Phosphoric acid, trimethyl ester **(512-56-1)**
Phosphoric acid, trinonyl ester (8CI,9CI) **(13018-37-6)**
Phosphoric acid, trioctyl ester **(1806-54-8)**
Phosphoric acid, triphenethyl ester (8CI) **(5770-08-1)**
Phosphoric acid, triphenyl ester **(115-86-6)**
Phosphoric acid, tripotassium salt (9CI) **(7778-53-2)**
Phosphoric acid, tri-2-propenyl ester **(1623-19-4)**
Phosphoric acid, tripropyl ester **(513-08-6)**
Phosphoric acid, tris(2-butoxyethyl) ester **(78-51-3)**
Phosphoric acid, tris(tert-butylphenyl) ester **(28777-70-0)**
Phosphoric acid, tris(2-chloroethyl)ester **(115-96-8)**
Phosphoric acid, tris(2-chloro-1-methylethyl) ester **(13674-84-5)**
Phosphoric acid, tris(decyl) ester (9CI) **(4200-55-9)**
Phosphoric acid, tris(2,3-dibromopropyl) ester **(126-72-7)**
Phosphoric acid, tris(1,3-dichloro-2-propyl)ester **(13674-87-8)**
Phosphoric acid, tris(2,3-dichloropropyl) ester **(78-43-3)**
Phosphoric acid, tris(2,4-dimethylphenyl)ester **(3862-12-2)**
Phosphoric acid, tris(2,5-dimethylphenyl)ester **(19074-59-0)**
Phosphoric acid, tris(3,4-dimethylphenyl)ester **(3862-11-1)**
Phosphoric acid, tris(3,5-dimethylphenyl)ester **(25653-16-1)**
Phosphoric acid, tris(dimethylphenyl)ester **(121-06-2)**
Phosphoric acid, tris(2-ethylhexyl) ester **(78-42-2)**
Phosphoric acid, tris(2-methylphenyl) ester **(78-30-8)**
Phosphoric acid, tris(methylphenyl) ester **(1330-78-5)**
Phosphoric acid, tris(3-methylphenyl) ester (9CI) **(563-04-2)**
Phosphoric acid, tris(4-methylphenyl) ester (9CI) **(78-32-0)**
Phosphoric acid, tris(2-methylpropyl) ester (9CI) **(126-71-6)**
Phosphoric acid, tris(nitrophenyl) ester (9CI) **(60337-47-5)**

Phosphorothioic acid, O,S-diethyl-O-(p-nitrophenyl) ester **(597-88-6)**
Phosphorothioic acid, O,O-diethyl-S-(4-nitrophenyl) ester (9CI) **(3270-86-8)**
Phosphorothioic acid, O,S-diethyl O-(4-nitrophenyl) ester (9CI) **(597-88-6)**
Phosphorothioic acid, O,O-diethyl O-phenyl ester **(32345-29-2)**
Phosphorothioic acid, O,O-diethyl O-(5-phenyl-3-isoxazolyl) ester **(18854-01-8)**
Phosphorothioic acid, O,O-diethyl O-(1-phenyl-1,2,4-triazolyl) ester **(24017-47-8)**
Phosphorothioic acid, O,O-diethyl O-(2-propyl-4-methyl-6-pyrimidyl) ester **(5826-91-5)**
Phosphorothioic acid, O,O-diethyl O-2-pyrazinyl ester **(297-97-2)**
Phosphorothioic acid, O,O-diethyl O-pyrazinyl ester **(297-97-2)**
Phosphorothioic acid, O,O-diethyl O-(2-quinoxalinyl) ester **(13593-03-8)**
Phosphorothioic acid, O,O-diethyl O-(3,5,6-trichloro-2-pyridyl) ester **(2921-88-2)**
Phosphorothioic acid, O,O-diisopropyl S-methyl ester (8CI) **(22907-64-8)**
Phosphorothioic acid, O-(2-(dimethylamino)-6-methyl-4-pyrimidinyl) O,O-diethyl ester **(5221-49-8)**
Phosphorothioic acid, O,O-dimethyl ester **(1112-38-5)**
Phosphorothioic acid, dimethyl ester (9CI) **(59401-04-6)**
Phosphorothioic acid, O,O-dimethyl ester, O,O-diester with 4,4'-thiodiphenol **(3383-96-8)**
Phosphorothioic acid, O,O-dimethyl ester, S-ester with 1,2-bis(methoxycarbonyl)ethanethiol **(1634-78-2)**
Phosphorothioic acid, O,O-dimethyl ester, S-ester with ethyl mercaptoacetate **(2088-72-4)**
Phosphorothioic acid, O,O-dimethyl ester, o-ester with p-hydroxy-N,N-dimethylbenzene- sulfonamide **(52-85-7)**
Phosphorothioic acid, O,O-dimethyl ester, O-ester with p-hydroxybenzenesulfonamide **(115-93-5)**
Phosphorothioic acid, O,O-dimethyl ester, O-ester with p-hydroxybenzonitrile **(2636-26-2)**
Phosphorothioic acid, O,O-dimethyl ester, S-ester with 2-((2-mercaptoethyl)thio)-N- methylpropionamide **(2275-23-2)**
Phosphorothioic acid, O,O-dimethyl ester, S-ester with 2-(mercaptomethyl)-5-methoxy- 4H-pyran-4-one **(2778-04-3)**
Phosphorothioic acid, O,O-dimethyl ester, S-ester with 2-mercapto-N-methylacetamide **(1113-02-6)**
Phosphorothioic acid, O,O-dimethyl ester, potassium salt (8CI,9CI) **(28523-79-7)**
Phosphorothioic acid, O,O-dimethyl O-(6-ethoxy-2-ethyl-4-pyrimidinyl) ester **(38260-54-7)**
Phosphorothioic acid, O,O-dimethyl S-(2-(ethylsulfinyl)ethyl) ester **(301-12-2)**
Phosphorothioic acid, O,O-dimethyl S-(2-(ethylsulfonyl)ethyl) ester **(17040-19-6)**
Phosphorothioic acid, O,O-dimethyl S-(2-(ethylthio)ethyl) ester (919-86-8) **(919-86-8)**
Phosphorothioic acid, O,O-dimethyl O-(2-(ethylthio)ethyl) ester, Mixed with O,O- dimethyl s-(2-(ethylthio)ethyl) ester (7:3) **(8022-00-2)**
Phosphorothioic acid, O,O-dimethyl S-(2-(methylamino)-2-oxoethyl) ester **(1113-02-6)**
Phosphorothioic acid, O,O-dimethyl O-(3-methyl-4-(methylsulfinyl)phenyl) ester (9CI) **(3761-41-9)**
Phosphorothioic acid, O,O-dimethyl O-(3-methyl-4-(methylsulfonyl)phenyl) ester (9CI) **(3761-42-0)**
Phosphorothioic acid O,O-dimethyl O-(3-methyl-4-(methylthio)-phenyl) ester **(55-38-9)**
Phosphorothioic acid, O,O-dimethyl O-(3-methyl-4-nitrophenyl) ester **(122-14-5)**
Phosphorothioic acid, O,S-dimethyl O-(3-methyl-4-nitrophenyl) ester (9CI) **(3344-14-7)**
Phosphorothioic acid, O,O-dimethyl O-(4-(methylsulfinyl)-m-tolyl) ester **(3761-41-9)**
Phosphorothioic acid, O,O-dimethyl O-(4-(methylsulfonyl)-m-tolyl) ester **(3761-42-0)**
Phosphorothioic acid, O,O-dimethyl S-(2-(methylthio)ethyl) ester **(2587-90-8)**
Phosphorothioic acid, O,O-dimethyl O-(2-(methylthio)ethyl) ester, Mixt. with O,O-dimethyl S-(2-(methylthio)ethyl) phosphorothioate (9CI) **(8065-62-1)**
Phosphorothioic acid, O,O-dimethyl O-(4-(methylthio)-m-tolyl) ester **(55-38-9)**
Phosphorothioic acid, O,O-dimethyl O-(4-nitrophenyl) ester **(298-00-0)**
Phosphorothioic acid, O,O-dimethyl O-(p-nitrophenyl) ester **(298-00-0)**
Phosphorothioic acid, O,S-dimethyl O-(p-nitrophenyl) ester **(597-89-7)**
Phosphorothioic acid, O,O-dimethyl O-(4-nitro-m-tolyl) ester **(122-14-5)**
Phosphorothioic acid, O,S-dimethyl O-(4-nitro-m-tolyl) ester **(3344-14-7)**
Phosphorothioic acid, O,O-dimethyl O-(2,4,5-trichlorophenyl) ester **(299-84-3)**
Phosphorothioic acid, O,O-dimethyl O-(3,5,6-trichloro-2-pyridyl) ester **(5598-13-0)**
Phosphorothioic acid, S-ethyl O,O-bis(2-methylphenyl) ester (9CI) **(95114-67-3)**
Phosphorothioic acid, S-ethyl O,O-bis(3-methylphenyl) ester (9CI) **(95114-68-4)**
Phosphorothioic acid, S-ethyl O,O-bis(4-methylphenyl) ester (9CI) **(95114-69-5)**
Phosphorothioic acid, S-ethyl O,O-bis(4-nitrophenyl) ester (9CI) **(16604-76-5)**
Phosphorothioic acid, S-ethyl O,O-bis(p-nitrophenyl) ester (8CI) **(16604-76-5)**
Phosphorothioic acid, S-ethyl O,O-bis(pentachlorophenyl) ester (9CI) **(95114-66-2)**
Phosphorothioic acid, S-ethyl O,O-bis(2,4,5-trichlorophenyl) ester (9CI) **(95150-15-5)**
Phosphorothioic acid, S-ethyl O,O-bis(2,3,5-trimethylphenyl) ester (9CI) **(95114-78-6)**
Phosphorothioic acid, S-ethyl O,O-diphenyl ester (8CI,9CI) **(16611-66-8)**
Phosphorothioic acid, O-ethyl O-methyl O-(2,4,5-trichlorophenyl) ester **(2633-54-7)**
Phosphorothioic acid, S-(2-(ethylsulfinyl)ethyl) O,O-dimethyl ester **(301-12-2)**
Phosphorothioic acid, O-(2-(ethylthio)ethyl) O,O-dimethyl ester **(867-27-6)**
Phosphorothioic acid, S-(2-(ethylthio)ethyl) O,O-dimethyl ester **(919-86-8)**
Phosphorothioic acid, S-methyl O,O-bis(1-methylethyl) ester (9CI) **(22907-64-8)**
Phosphorothioic acid, S-methyl O,O-dipropyl ester (8CI,9CI) **(5301-73-5)**
Phosphorothioic acid, O-methyl ester, O-ester with p-hydroxy-benzonitrile (8CI) **(31328-15-1)**
Phosphorothioic acid, O,O'-(thiodi-4,1-phenylene) O,O,O',O'-tetra-methyl ester **(3383-96-8)**
Phosphorothioic acid triethylenetriamide **(52-24-4)**
Phosphorothioic acid, O,O,O-triethyl ester **(126-68-1)**
Phosphorothioic acid, O,O,S-triethyl ester **(1186-09-0)**
Phosphorothioic acid, O,O,O-trimethyl ester **(152-18-1)**
Phosphorothioic acid, O,O,S-trimethyl ester **(152-20-5)**
Phosphorothioic triamide, N,N',N''-tri-1,2-ethanediyl- **(52-24-4)**
Phosphorothioic triamide, N,N',N''-triethylene- **(52-24-4)**
Phosphorothioic trichloride **(3982-91-0)**
Phosphorothionic trichloride **(3982-91-0)**
Phosphorotrithioic acid, S,S,S-tributyl ester **(78-48-8)**
Phosphorotrithious acid, S,S,S-tributyl ester **(150-50-5)**
Phosphorotrithious acid, tributyl ester **(150-50-5)**
Phosphorotrithious acid, tridodecyl ester (9CI) **(1656-63-9)**
Phosphorous acid **(10294-56-1)**
Phosphorous acid **(13598-36-2)**
Phosphorous acid, ortho [UN 2834] **(13598-36-2)**
Phosphorous acid (VAN)(9CI) **(10294-56-1)**
Phosphorous acid, beryllium salt **(13598-15-7)**
Phosphorous acid, bis(2,4-dichlorophenoxyethyl) ester, Mixed with tris(2,4-dichloro phenoxyethyl)phosphite **(8005-49-0)**
Phosphorous acid, bis(2-ethylhexyl) phenyl ester (9CI) **(3164-60-1)**
Phosphorous acid, cyclic neopentanetetrayl bis(2,4-di-tert-butylphenyl)ester **(26741-53-7)**
Phosphorous acid, decyl diphenyl ester **(3287-06-7)**

Phosphorous acid, didecyl phenyl ester (9CI) **(1254-78-0)**
Phosphorous acid, diethyl ester **(762-04-9)**
Phosphorous acid, diisodecyl phenyl ester (9CI) **(25550-98-5)**
Phosphorous acid, 2-(1,1-dimethylethyl)-4-(1-(3-(1,1-dimethylethyl)-4-hydroxyphenyl)-1-methylethyl)phenyl bis(4-nonylphenyl) ester (9CI) **(20227-53-6)**
Phosphorous acid, dinonylphenyl bis(nonylphenyl) ester (9CI) **(54771-30-1)**
Phosphorous acid, 2-ethylhexyl diphenyl ester (9CI) **(15647-08-2)**
Phosphorous acid, isodecyl diphenyl ester (9CI) **(26544-23-0)**
Phosphorous acid, isooctyl diphenyl ester (9CI) **(26401-27-4)**
Phosphorous acid, tributyl ester **(102-85-2)**
Phosphorous acid, tri-p-cresyl ester **(620-42-8)**
Phosphorous acid, tridodecyl ester (9CI) **(3076-63-9)**
Phosphorous acid, triethyl ester **(122-52-1)**
Phosphorous acid, triisodecyl ester (9CI) **(25448-25-3)**
Phosphorous acid, triisooctyl ester **(25103-12-2)**
Phosphorous acid, triisopropyl ester **(116-17-6)**
Phosphorous acid, trimethyl ester **(121-45-9)**
Phosphorous acid, triphenyl ester **(101-02-0)**
Phosphorous acid, tris(2-chloroethyl) ester **(140-08-9)**
Phosphorous acid, tris(decyl) ester (9CI) **(2929-86-4)**
Phosphorous acid, tris(2-(2,4-dichlorophenoxy)ethyl ester **(94-84-8)**
Phosphorous acid, tris(dipropylene glycol) ester **(36788-39-3)**
Phosphorous acid, tris(2-ethylhexyl) ester **(301-13-3)**
Phosphorous acid, tris(1-methylethyl) ester **(116-17-6)**
Phosphorous acid, tris(methylphenyl) ester (9CI) **(25586-42-9)**
Phosphorous bromide (DOT) **(7789-60-8)**
Phosphorous chloride **(7719-12-2)**
Phosphorous oxybromide **(7789-59-5)**
Phosphorous pentachloride **(10026-13-8)**
Phosphorous sesquisulfide **(1314-85-8)**
Phosphorous sulfochloride **(3982-91-0)**
Phosphorous thiochloride **(3982-91-0)**
Phosphorous trichloride sulfide **(3982-91-0)**
Phosphorous (white) **(7723-14-0)**
Phosphorpentachlorid (German) **(10026-13-8)**
Phosphorsaeureloesungen (German) **(7664-38-2)**
Phosphortrichlorid (German) **(7719-12-2)**
Phosphorus-31 **(7723-14-0)**
Phosphorus-chloride **(7719-12-2)**
Phosphorus-oxide **(1314-56-3)**
Phosphorus **(7723-14-0)**
Phosphorus (8CI,9CI) **(7723-14-0)**
Phosphorus, Isotope of mass 32 (8CI,9CI) **(14596-37-3)**
Phosphorus(V) oxide **(1314-56-3)**
Phosphorus acid, cyclic ethylene 2-hydroxyethyl ester **(1073-75-2)**
Phosphorus bromide (DOT) **(7789-60-8)**
Phosphorus chloride (DOT) **(7719-12-2)**
Phosphorus-sesquisulfide **(1314-85-8)**
Phosphorus furnace slag **(65997-17-3)**
Phosphorus heptasulfide **(12037-82-0)**
Phosphorus heptasulphide, Free from yellow or white phosphorus [UN 1339] **(12037-82-0)**
Phosphorus oxybromide, Molten [UN 2576] **(7789-59-5)**
Phosphorus oxybromide, Solid (DOT) **(7789-59-5)**
Phosphorus oxybromide [UN 1939] **(7789-59-5)**
Phosphorus oxychloride (ACGIH,OSHA) [UN 1810] **(10025-87-3)**
Phosphorus oxytrichloride **(10025-87-3)**
Phosphorus pentachloride (ACGIH,OSHA) [UN 1806] **(10026-13-8)**
Phosphorus pentachloride, Solid (DOT) **(10026-13-8)**
Phosphorus pentafluoride [UN 2198] **(7647-19-0)**
Phosphorus pentaoxide **(1314-56-3)**
Phosphorus pentasulfide (ACGIH,DOT,OSHA) **(1314-80-3)**
Phosphorus pentasulphide, Free from yellow or white phosphorus [UN 1340] **(1314-80-3)**
Phosphorus pentoxide [UN 1807] **(1314-56-3)**
Phosphorus perchloride **(10026-13-8)**
Phosphorus persulfide **(1314-80-3)**
Phosphorus, red (DOT) **(7723-14-0)**
Phosphorus sesquisulfide (DOT) **(1314-85-8)**
Phosphorus sesquisulphide, Free from yellow or white phosphorus [UN 1341] **(1314-85-8)**
Phosphorus sulfide **(12037-82-0)**
Phosphorus (III) sulfide (IV) **(1314-85-8)**
Phosphorus tribromide [UN 1808] **(7789-60-8)**
Phosphorus trichloride (ACGIH,OSHA) [UN 1809] **(7719-12-2)**
Phosphorus trihydride **(7803-51-2)**
Phosphorus trihydroxide **(13598-36-2)**
Phosphorus trioxide [UN 2578] **(1314-24-5)**
Phosphorus trisulfide **(12165-69-4)**
Phosphorus trisulphide, Free from yellow or white phosphorus [UN 1343] **(12165-69-4)**
Phosphorus-sulfide **(1314-80-3)**
Phosphorus, white (DOT) **(7723-14-0)**
Phosphorwasserstoff (German) **(7803-51-2)**
Phosphoryethanolamine **(1071-23-4)**
Phosphoryl-bromide **(7789-59-5)**
Phosphoryl chloride (DOT) **(10025-87-3)**
Phosphorylethanolamine **(1071-23-4)**
Phosphoryl hexamethyltriamide **(680-31-9)**
O-Phosphoryl-4-hydroxy-N,N-dimethyltryptamine **(520-52-5)**
4-Phosphoryloxy-ω-N,N-dimethyltryptamine **(520-52-5)**
5-Phosphorylribose-1-pyrophosphate **(97-55-2)**
Phosphoryl tribromide **(7789-59-5)**
Phosphostigmine **(56-38-2)**
Phosphotex **(7722-88-5)**
Phosphothion **(121-75-5)**
Phosphotox E **(563-12-2)**
Phosphure de magnesium (French) **(12057-74-8)**
Phosphure de zinc (French) **(1314-84-7)**
Phosphures d'alumium (French) **(20859-73-8)**
Phostoxin **(20859-73-8)**
Phosvel **(21609-90-5)**
Phosvel phenol **(1940-42-7)**
Phosvin **(1314-84-7)**
Phosvit **(62-73-7)**
Photoaldrin **(13350-71-5)**
Photodieldrin **(13366-73-9)**
Photoheptachlor **(33442-83-0)**
Photol **(55-55-0)**
Photomirex **(39801-14-4)**
Photothidiazuron **(71769-74-9)**
Phoxim **(14816-18-3)**
Phoxime **(14816-18-3)**
Phoxin **(14816-18-3)**
Phozalon **(2310-17-0)**
Phph **(92-52-4)**
Phrenazol **(54-95-5)**
Phrenazone **(54-95-5)**
Phrilon **(63428-83-1)**
Phtalaldehydes (French) **(643-79-8)**
Phthalaldehyde **(643-79-8)**
p-Phthalaldehyde **(623-27-8)**
Phthalamate **(85-72-3)**
Phthalamic acid (8CI) **(88-97-1)**
Phthalamic acid, N-1-naphthyl **(132-66-1)**
Phthalamic acid, N-1-naphthyl-, monosodium salt **(132-67-2)**
Phthalamic acid, N-1-naphthyl-, sodium salt **(132-67-2)**
Phthalamide **(88-96-0)**
Phthalamide acid **(88-97-1)**
Phthalamidic acid **(88-97-1)**
Phthalamodine **(77-36-1)**
Phthalamudine **(77-36-1)**
5-Phthalanacarboxylic acid, 1,3-dioxo- **(552-30-7)**

1,3-Phthalandione	**(85-44-9)**
Phthalandione	**(85-44-9)**
Phthalanilic acid, 3'-methoxy- (8CI)	**(19336-97-1)**
Phthalanilic acid, 3'-methyl	**(85-72-3)**
1-Phthalanone	**(87-41-2)**
1-Phthalazinamine (9CI)	**(19064-69-8)**
Phthalazine (9CI)	**(253-52-1)**
Phthalazine, 1-amino- (8CI)	**(19064-69-8)**
1,4-Phthalazinedione, 5-amino-2,3-dihydro- (9CI)	**(521-31-3)**
1,4-Phthalazinedione, 2,3-dihydro-6-nitro	**(3682-19-7)**
Phthalazine, 1-hydrazino-, monohydrochloride	**(304-20-1)**
Phthalazine, 1-hydrazino	**(86-54-4)**
1(2H)Phthalazinone	**(119-39-1)**
Phthalazinone	**(119-39-1)**
1(2H)-Phthalazinone hydrazone	**(86-54-4)**
1(2H)-Phthalazinone, hydrazone hydrochloride	**(304-20-1)**
1(2H)-Phthalazinone, 4-methyl- (8CI,9CI)	**(5004-48-8)**
Phthalazone	**(119-39-1)**
Phthalic-acid	**(88-99-3)**
m-Phthalic acid	**(121-91-5)**
o-Phthalic acid	**(88-99-3)**
Phthalic acid, (C7-C11) alkyl esters	**(68648-91-9)**
Phthalic acid anhydride	**(85-44-9)**
Phthalic acid, benzyl butyl ester	**(85-68-7)**
Phthalic acid, bis(2-butoxyethyl) ester	**(117-83-9)**
Phthalic acid, bis(2,3-epoxypropyl) ester	**(7195-45-1)**
Phthalic acid, bis(2-ethylhexyl) ester	**(117-81-7)**
Phthalic acid, bis(2-methoxyethyl) ester	**(117-82-8)**
Phthalic acid, bis(1-methylheptyl) ester	**(131-15-7)**
Phthalic acid, bis(6-methylheptyl)ester	**(27554-26-3)**
Phthalic acid, bis(2-octyl) ester	**(131-15-7)**
Phthalic acid, butoxycarbonylmethyl butyl ester	**(85-70-1)**
Phthalic acid, 2-butoxyethyl butyl ester (8CI)	**(33374-28-6)**
Phthalic acid, butyl cyclohexyl ester (8CI)	**(84-64-0)**
Phthalic acid, butyl decyl ester	**(89-19-0)**
Phthalic acid, butyl ester, butyl glycolate	**(85-70-1)**
Phthalic acid, butyl ester, ester with butyl glycolate	**(85-70-1)**
Phthalic acid, butyl 2-ethylhexyl ester	**(85-69-8)**
Phthalic acid, butyl isobutyl ester (8CI)	**(17851-53-5)**
Phthalic acid, butyl isodecyl ester	**(42343-36-2)**
Phthalic acid, butyl octyl ester (8CI)	**(84-78-6)**
Phthalic acid, 4-chloro- (8CI)	**(89-20-3)**
Phthalic acid, cyclic oxydiethylene ester	**(13988-26-6)**
Phthalic acid, decyl hexyl ester	**(25724-58-7)**
Phthalic acid, decyl octyl ester	**(119-07-3)**
Phthalic acid, dialkyl(C7-C11) ester	**(68515-42-4)**
Phthalic acid, dialkyl(C7-C9) ester	**(68515-41-3)**
Phthalic acid, dialkyl(C7) ester	**(68515-44-6)**
Phthalic acid, dialkyl(C9) ester	**(68515-45-7)**
Phthalic acid, diallyl ester	**(131-17-9)**
o-Phthalic acid, diallyl ester	**(131-17-9)**
Phthalic acid, diallyl ester, Polymer with ethyl acrylate and methacrylic acid (8CI)	**(28411-49-6)**
o-Phthalic acid diamide	**(88-96-0)**
Phthalic acid, dibenzyl ester (8CI)	**(523-31-9)**
Phthalic acid, dibutyl ester	**(84-74-2)**
Phthalic acid, dicapryl ester	**(131-15-7)**
Phthalic acid dichloride	**(88-95-9)**
Phthalic acid, dicyclohexyl ester	**(84-61-7)**
Phthalic acid, didecyl ester	**(84-77-5)**
Phthalic acid, didodecyl ester	**(2432-90-8)**
Phthalic acid, diethyl ester	**(84-66-2)**
Phthalic acid, diglycidyl ester	**(7195-45-1)**
Phthalic acid, diheptyl ester	**(3648-21-3)**
Phthalic acid, dihexyl ester	**(84-75-3)**
Phthalic acid, diisobutyl ester	**(84-69-5)**
Phthalic acid, diisodecyl ester	**(26761-40-0)**
Phthalic acid, diisohexyl ester (8CI)	**(146-50-9)**
Phthalic acid, diisononyl ester	**(28553-12-0)**
Phthalic acid, diisooctyl ester	**(27554-26-3)**
Phthalic acid, diisopropyl ester	**(605-45-8)**
Phthalic acid, di(methoxyethyl) ester	**(117-82-8)**
Phthalic acid, dimethyl ester	**(131-11-3)**
Phthalic acid dinitrile	**(91-15-6)**
Phthalic acid, dinonyl ester	**(84-76-4)**
Phthalic acid, di-2-octyl ester	**(131-15-7)**
Phthalic acid, dioctyl ester	**(117-81-7)**
Phthalic acid, dioctyl ester	**(117-84-0)**
Phthalic acid, dipentyl ester	**(131-18-0)**
Phthalic acid, diphenyl ester	**(84-62-8)**
Phthalic acid, dipropyl ester	**(131-16-8)**
Phthalic acid, distearyl ester	**(14117-96-5)**
Phthalic acid, ditridecyl ester	**(119-06-2)**
Phthalic acid, diundecyl ester	**(3648-20-2)**
Phthalic acid, ethyl ester, ester with ethyl glycolate	**(84-72-0)**
Phthalic acid, ethyl methyl ester (8CI)	**(34006-77-4)**
Phthalic acid, 3-fluoro- (8CI)	**(1583-67-1)**
Phthalic acid, hexahydro-, bis(2,3-epoxypropyl) ester	**(5493-45-8)**
Phthalic acid, hexahydro-, diglycidyl ester	**(5493-45-8)**
Phthalic acid, hexahydro-3,6-endo-oxy-	**(145-73-3)**
Phthalic acid, isodecyl ester	**(31047-64-0)**
Phthalic acid, isodecyl isooctyl ester	**(42343-35-1)**
Phthalic acid, isodecyl octyl ester	**(1330-96-7)**
Phthalic acid, isooctyl ester	**(30849-48-0)**
Phthalic acid, methyl-	**(30497-87-1)**
Phthalic acid, 4-methyl- (8CI)	**(4316-23-8)**
Phthalic acid, 4-methyl-, dimethyl ester (8CI)	**(20116-65-8)**
Phthalic acid, methyl ester	**(131-11-3)**
Phthalic acid monoamide	**(88-97-1)**
Phthalic acid, monobutyl ester	**(131-70-4)**
Phthalic acid, mono(p-tert-butyl-.α.-methylbenzyl) ester (8CI)	**(33533-56-1)**
Phthalic acid, monododecyl ester (8CI)	**(21577-80-0)**
Phthalic acid, mono-(2-ethylhexyl) ester	**(4376-20-9)**
Phthalic acid, mono(p-methoxy-.α.-methylbenzyl) ester (8CI)	**(33533-57-2)**
Phthalic acid, mono(α-methylbenzyl) ester (8CI)	**(33533-53-8)**
Phthalic acid, monomethyl ester (8CI)	**(4376-18-5)**
Phthalic acid, monomethyl ester, ester with ethyl glycolate	**(85-71-2)**
Phthalic acid, mono(p,α-dimethylbenzyl) ester (8CI)	**(23005-56-3)**
Phthalic acid, 4-sulfo- (8CI)	**(89-08-7)**
Phthalic acid, 4-sulfo-, monosodium salt	**(33562-89-9)**
Phthalic acid, tetrachloro	**(632-58-6)**
Phthalic acid, tetrahydroabietyl alcohol diester	**(36388-36-0)**
Phthalic-anhydride	**(85-44-9)**
Phthalic anhydride (ACGIH,OSHA)	**(85-44-9)**
Phthalic anhydride, Polymer with maleic anhydride and propylene glycol	**(25037-66-5)**
Phthalic anhydride, Solid or molten (DOT)	**(85-44-9)**
Phthalic anhydride, 4,4'-carbonyldi- (8CI)	**(2421-28-5)**
Phthalic anhydride, coconut oil, glycerin polymer	**(66070-87-9)**
Phthalic anhydride, 3,6-dimethyl- (8CI)	**(5463-50-3)**
Phthalic anhydride, glycerin, coconut oil polymer	**(66070-87-9)**
Phthalic anhydride, glycerin, soybean oil polymer	**(66070-61-9)**
Phthalic anhydride, glycerin, soybean oil resin	**(66070-61-9)**
Phthalic anhydride, glycerin, tall oil acids resin	**(66070-71-1)**
Phthalic anhydride, glycerol, linseed oil, tung oil polymer	**(66071-18-9)**
Phthalic anhydride, hexyl, octyl, decyl esters	**(68648-93-1)**
Phthalic anhydride, maleic anhydride, and propylene glycol polymer	**(25037-66-5)**
Phthalic anhydride, 4-nitro	**(5466-84-2)**
Phthalic anhydride, pentaerythritol, soybean oil resin	**(66070-60-8)**
Phthalic anhydride, soya oil, glycerin polymer	**(66070-61-9)**
Phthalic anhydride, soya oil, glycerin resin	**(66070-61-9)**
Phthalic anhydride, soybean oil, glycerin polymer	**(66070-61-9)**
Phthalic anhydride, 4,4'-sulfonyldi- (8CI)	**(2540-99-0)**

Phthalic anhydride, tall oil fatty acids, glycerin polymer **(66070-71-1)**
Phthalic anhydride, tetrabromo **(632-79-1)**
Phthalic anhydride, tetrabromo- (8CI) **(632-79-1)**
Phthalic anhydride, tetrachloro **(117-08-8)**
Phthalic anhydride, 1,2,3,6-tetrahydro- **(85-43-8)**
Phthalic chloride **(88-95-9)**
Phthalic dichloride **(88-95-9)**
m-Phthalic dichloride **(99-63-8)**
o-Phthalic imide **(85-41-6)**
Phthalic monoamide **(88-97-1)**
Phthalide **(27355-22-2)**
Phthalide (8CI) **(87-41-2)**
Phthalide 3,3,-bis(p-hydroxyphenyl)- **(77-09-8)**
Phthalide, 6,7-dimethoxy-3-(5,6,7,8-tetrahydro-6-methyl-1,3-dioxolo(4,5-g)isoquinolin-5-yl)- **(118-08-1)**
Phthalide, 3-ethoxy- (8CI) **(16824-02-5)**
Phthalide, 3-methyl- (8CI) **(3453-64-3)**
Phthalide, 4,5,6,7-tetrachloro **(27355-22-2)**
Phthalimetten **(77-09-8)**
Phthalimid (German) **(85-41-6)**
Phthalimide **(85-41-6)**
Phthalimide, N-(2-bromoethyl)- (8CI) **(574-98-1)**
Phthalimide, N-(3-bromopropyl)- (8CI) **(5460-29-7)**
Phthalimide, N-butyl- (8CI) **(1515-72-6)**
Phthalimide, N-chloro- (8CI) **(3481-09-2)**
Phthalimide, N-(chloromethyl)- **(17564-64-6)**
Phthalimide, N-chloromethyl- **(17564-64-6)**
Phthalimide, N-(cyclohexylthio) **(17796-82-6)**
Phthalimide, N-(2,6-dioxo-3-piperidyl)- (8CI) **(50-35-1)**
Phthalimide, N-(2,3-epoxypropyl) **(5455-98-1)**
Phthalimide, N-hydroxy **(524-38-9)**
Phthalimide, N-(hydroxymethyl) **(118-29-6)**
Phthalimide, N-isopropyl **(304-17-6)**
Phthalimide, N-(mercaptomethyl)-, S-ester with O,O-dimethyl phosphorodithioate **(732-11-6)**
Phthalimide, N-methyl- (8CI) **(550-44-7)**
Phthalimide, 4-nitro **(89-40-7)**
Phthalimide, potassium salt (8CI) **(1074-82-4)**
Phthalimide, thio- (6CI,8CI) **(18138-18-6)**
Phthalimide, N-((trichloromethyl)thio) **(133-07-3)**
Phthalimidimide **(3468-11-9)**
Phthalimidine (8CI) **(480-91-1)**
Phthalimido O,O-dimethyl phosphorodithioate **(732-11-6)**
β-Phthalimidoethyl bromide **(574-98-1)**
3-Phthalimidoglutarimide **(50-35-1)**
α-(N-Phthalimido)glutarimide **(50-35-1)**
α-Phthalimidoglutarimide **(50-35-1)**
Phthalimidomethyl alcohol **(118-29-6)**
Phthalimidomethyl O,O-dimethyl phosphorodithioate **(732-11-6)**
(Phthalocyaninato(2-))nickel **(14055-02-8)**
2,1-Phthalocyaninato nickel **(14055-02-8)**
Phthalocyanine **(574-93-6)**
Phthalocyanine Blue **(147-14-8)**
Phthalocyanine Blue 01206 **(3468-11-9)**
Phthalocyanine Brilliant Green **(1328-53-6)**
29H,31H-Phthalocyanine (9CI) **(574-93-6)**
Phthalocyanine B 4ZU **(147-14-8)**
Phthalocyanine Green **(1328-53-6)**
Phthalocyanine Green 6G **(14302-13-7)**
Phthalocyanine Green LX **(1328-53-6)**
Phthalocyanine Green V **(1328-53-6)**
Phthalocyanine Green VFT 1080 **(1328-53-6)**
Phthalocyanine Green WDG 47 **(1328-53-6)**
Phthalocyanine VK **(147-14-8)**
Phthalocyanine 2ZU **(147-14-8)**
Phthalocyaninetrisulfonic acid, copper complex **(30638-09-6)**
Phthalocyaninetrisulfonyl chloride copper(II) complex **(27121-30-8)**
Phthalodinitrile **(91-15-6)**
o-Phthalodinitrile **(91-15-6)**
m-Phthalodinitrile (ACGIH,OSHA) **(626-17-5)**
Phthalogen **(3468-11-9)**
Phthalol **(84-66-2)**
Phthalonic acid **(528-46-1)**
Phthalonitrile **(91-15-6)**
12-Phthaloperinone **(6925-69-5)**
12H-Phthaloperin-12-one (9CI) **(6925-69-5)**
Phthalophos **(732-11-6)**
m-Phthaloyl chloride **(99-63-8)**
p-Phthaloyl chloride **(100-20-9)**
Phthaloyl chloride (8CI) **(88-95-9)**
Phthaloyl dichloride **(88-95-9)**
p-Phthaloyl dichloride **(100-20-9)**
N-Phthaloylglutamimide **(50-35-1)**
Phthalsaeureanhydrid (German) **(85-44-9)**
Phthalsaeurediaethylester (German) **(84-66-2)**
Phthalsaeuredimethylester (German) **(131-11-3)**
Phthaltan **(133-07-3)**
(+-)-cis/trans-Phthalthrin **(7696-12-0)**
Phthalthrin **(7696-12-0)**
d-Phthalthrin **(7696-12-0)**
Phthalyl chloride **(88-95-9)**
Phthalyl dichloride **(88-95-9)**
N-Phthalylglutamic acid imide **(50-35-1)**
N-Phthalyl-glutaminsaeure-imid (German) **(50-35-1)**
α-N-Phthalylglutaramide **(50-35-1)**
Phthiocol **(483-55-6)**
Phthisen **(54-85-3)**
1(2H)-Phthlazinone, hydrazone, monohydrochloride **(304-20-1)**
Phyban **(2163-80-6)**
Phyban H.C. **(2163-80-6)**
Phycitol **(149-32-6)**
Phygon **(117-80-6)**
Phygon Paste **(117-80-6)**
Phygon Seed Protectant **(117-80-6)**
Phygon XL **(117-80-6)**
Phylcardin **(317-34-0)**
Phyllemblin **(831-61-8)**
Phyllindon **(317-34-0)**
Phyllochinon (German) **(84-80-0)**
Phyllocontin **(317-34-0)**
Phyllopyrrole **(520-69-4)**
Phylloquinone **(84-80-0)**
α-Phylloquinone **(84-80-0)**
trans-Phylloquinone **(84-80-0)**
Phymone **(86-87-3)**
Phyol **(9002-72-6)**
Phyomone **(86-87-3)**
Phyone **(9002-72-6)**
Physeptone **(76-99-3)**
Physex **(9002-61-3)**
Physostigmine **(57-47-6)**
Physostigmine SO_4 **(64-47-1)**
Physostigmine, salicylate (1:1) **(57-64-7)**
Physostigmine sulfate **(64-47-1)**
Physostigmine sulphate **(64-47-1)**
Physostol **(57-47-6)**
Physostol salicylate **(57-64-7)**
Phytane **(638-36-8)**
Phytar **(75-60-5)**
Phytar 138 **(75-60-5)**
Phytar 560 **(124-65-2)**
Phytar 560 **(75-60-5)**
Phytic acid **(83-86-3)**
Phyto-Bordeaux **(1332-03-2)**

Phyto-Bordeaux	(7758-98-7)
Phytogermine	(59-02-9)
Phytol	(150-86-7)
trans-Phytol	(150-86-7)
Phytol ketone	(16825-16-4)
Phytomelin	(153-18-4)
Phytomenadione	(84-80-0)
Phytomycin	(3810-74-0)
Phytonadione	(84-80-0)
Phytone	(16825-16-4)
Phytosol	(327-98-0)
Phytox	(12122-67-7)
Pianadalin	(77-65-6)
Pianizol	(9011-05-6)
Piaponon	(57-83-0)
Piatherm	(9011-05-6)
Piatherm D	(9011-05-6)
Piazine	(290-37-9)
Pic-Clor	(76-06-2)
Piccodiene 2025	(64742-16-1)
Piccoflex	(9003-54-7)
Piccolastic	(9003-53-6)
Piccolastic A	(9003-53-6)
Piccolastic A 5	(9003-53-6)
Piccolastic A 25	(9003-53-6)
Piccolastic A 50	(9003-53-6)
Piccolastic A 75	(9003-53-6)
Piccolastic C 125	(9003-53-6)
Piccolastic D	(9003-53-6)
Piccolastic D-100	(9003-53-6)
Piccolastic D 125	(9003-53-6)
Piccolastic D 150	(9003-53-6)
Piccolastic E 75	(9003-53-6)
Piccolastic E 100	(9003-53-6)
Piccolastic E 200	(9003-53-6)
Piccopale	(64742-16-1)
Piccopale 100	(64742-16-1)
Piccopale 100BHT	(64742-16-1)
Piccopale 200HM	(64742-16-1)
Piccopale 100SF	(64742-16-1)
Picene	(213-46-7)
Picene, 2,9-dimethyl- (8CI,9CI)	(1679-02-3)
Picene, methyl- (8CI,9CI)	(30283-95-5)
Picene, 1,2,3,4-tetrahydro-1,2-dimethyl- (9CI)	(74229-81-5)
Picene, 1,2,3,4-tetrahydro-2,2,9-trimethyl- (8CI,9CI)	(1242-76-8)
Piceol	(99-93-4)
Picfume	(76-06-2)
Pichloram K	(2545-60-0)
Pichloram potassium salt	(2545-60-0)
Pichtosin	(125-12-2)
Pichtosine	(125-12-2)
Picket	(52645-53-1)
Picket G	(52645-53-1)
Picloram (ACGIH,OSHA)	(1918-02-1)
Picloram potassium salt	(2545-60-0)
Picolinaldehyde (8CI)	(1121-60-4)
Picolinaldehyde, oxime	(873-69-8)
Picolinaldoxime	(873-69-8)
Picolinamide (8CI)	(1452-77-3)
2-Picoline	(109-06-8)
3-Picoline	(108-99-6)
4-Picoline	(108-89-4)
Picoline	(109-06-8)
α-Picoline	(109-06-8)
γ-Picoline	(108-89-4)
Picoline [UN 2313]	(1333-41-1)
m-Picoline [UN 2313]	(108-99-6)
o-Picoline [UN 2313]	(109-06-8)
p-Picoline [UN 2313]	(108-89-4)
α-Picolinealdoxime	(873-69-8)
3-Picoline, 6-amino	(1603-41-4)
4-Picoline, 2-chloro-6-methoxy-α,α,α-trichloro	(7159-34-4)
2-Picoline-4,5-dimethanol, 3-hydroxy-	(65-23-6)
2-Picoline, 5-ethyl	(104-90-5)
Picoline, ethyl-	(27987-10-6)
2-Picoline, 6-ethyl- (8CI)	(1122-69-6)
3-Picoline, 5-ethyl- (8CI)	(3999-78-8)
2-Picoline, 1-oxide (8CI)	(931-19-1)
2-Picoline, 5-vinyl-	(140-76-1)
Picolinic-acid	(98-98-6)
α-Picolinic acid	(55-22-1)
Picolinic acid, 4-amino-3,5,6-trichloro	(1918-02-1)
Picolinic acid, 4-amino-3,5,6-trichloro-, isooctyl ester (8CI)	(26952-20-5)
Picolinic acid, 4-amino-3,5,6-trichloro-, monopotassium salt	(2545-60-0)
Picolinic acid, 5-butyl-	(536-69-6)
Picolinic acid, 6-chloro	(4684-94-0)
Picolinic acid, 3,6-dichloro	(1702-17-6)
Picolinic acid, ethyl ester	(2524-52-9)
Picolinic acid nitrile	(100-70-9)
2-Picolinium, 1-phenethyl-, bromide (8CI)	(10551-21-0)
α-Picolinium-β-phenylethyl bromide	(10551-21-0)
Picolinonitrile (8CI)	(100-70-9)
α-Picolyl alcohol	(586-98-1)
β-Picolyl alcohol	(100-55-0)
γ-Picolyl alcohol	(586-95-8)
2-Picolyl chloride hydrochloride	(6959-47-3)
Picragol	(146-84-9)
Picral	(88-89-1)
Picramic acid	(91-96-3)
Picramic acid	(96-91-3)
Picramic acid, sodium salt	(831-52-7)
Picramic acid, sodium salt, Dry	(831-52-7)
Picramic acid, zirconium salt, Dry	(63868-82-6)
Picramic acid, zirconium salt (Wet)	(63868-82-6)
Picramide	(489-98-5)
Picrate d'argent (French)	(146-84-9)
Picrate de plomb (French)	(25721-38-4)
Picrate of ammonia (DOT)	(131-74-8)
Picrato de plata (Spanish)	(146-84-9)
Picrato de plomo (Spanish)	(25721-38-4)
Picratol	(131-74-8)
Picric-acid	(88-89-1)
Picric acid (ACGIH,OSHA) (8CI)	(88-89-1)
Picric acid, ammonium salt	(131-74-8)
Picric acid, Dry (DOT)	(88-89-1)
Picric acid, lead salt (8CI)	(25721-38-4)
Picric acid, silver(1+) salt (8CI)	(146-84-9)
Picric acid, Wetted with at least 10% water [NA 1344]	(88-89-1)
Picric acid, Wetted with at least 30% water [UN 1344]	(88-89-1)
Picric acid, Wetted with 10% to 30% water	(88-89-1)
Picric acid, Wet with not less than 10% water, over 25 pounds [NA 1344]	(88-89-1)
Picride	(76-06-2)
Picrite (The explosive)	(556-88-7)
Picronitric acid	(88-89-1)
Picrosirius red	(2610-10-8)
Picrotin, Compd. with picrotoxinin (1:1)	(124-87-8)
Picrotol	(146-84-9)
Picrotoxin	(124-87-8)
Picrotoxine	(124-87-8)
Picrotoxinin, Compd. with picrotin (1:1)	(124-87-8)
Picryl chloride	(88-88-0)
Picrylmethylnitramine	(479-45-8)
Picrylnitromethylamine	(479-45-8)

Pictol (55-55-0)
Pid (82-66-6)
Pid (83-12-5)
Pidolic acid (98-79-3)
Pieciochlorek fosforu (Polish) (10026-13-8)
Pied Piper Mouse Seed (57-24-9)
Pielik (94-75-7)
Pielik E (2702-72-9)
Pielika (Polish) (2702-72-9)
Pietil (14698-29-4)
Pig-Wrack (9000-07-1)
Piglet Pro-Gen V (127-85-5)
Pigment Anthraquinone Deep Blue (81-77-6)
Pigment Black 7 (1333-86-4)
Pigment Blue 1 (1325-87-7)
Pigment Blue 15 (147-14-8)
Pigment Blue 27 (14038-43-8)
Pigment Blue 60 (81-77-6)
Pigment Blue 61 (1324-76-1)
Pigment Blue Anthraquinone (81-77-6)
Pigment Blue Anthraquinone V (81-77-6)
Pigment Blue Green Phthalocyanine U (574-93-6)
Pigment Deep Blue Anthraquinone (81-77-6)
Pigment Fast Blue B (147-14-8)
Pigment Fast Green G (1328-53-6)
Pigment Fast Green GN (1328-53-6)
Pigment Fast Orange (3468-63-1)
Pigment Fast Orange G (3520-72-7)
Pigment Green 7 (1328-53-6)
Pigment Green 15 (7758-97-6)
Pigment Green 38 (14302-13-7)
Pigment Green Phthalocyanine (1328-53-6)
Pigment Green Phthalocyanine V (1328-53-6)
Pigment Lake Red BFC (5160-02-1)
Pigment Lake Red CD (5160-02-1)
Pigment Lake Red LC (5160-02-1)
Pigment Orange 31 (5280-74-0)
Pigment Orange ERH (3520-72-7)
Pigment Orange G (3520-72-7)
Pigment Orange ZH (3520-72-7)
Pigment Ponceau R (3761-53-3)
Pigment Quinacridone Red (1047-16-1)
Pigment Red 3 (2425-85-6)
Pigment Red 4 (2814-77-9)
Pigment Red 23 (6471-49-4)
Pigment Red 49:1 (1103-38-4)
Pigment Red 49:2 (1103-39-5)
Pigment Red 57 (5858-81-1)
Pigment Red 57:1 (5281-04-9)
Pigment Red 63:1 (6417-83-0)
Pigment Red 69 (5850-90-8)
Pigment Red 112 (6535-46-2)
Pigment Red BH (6471-49-4)
Pigment Red CD (5160-02-1)
Pigment Red RL (2425-85-6)
Pigment Rubine B (5858-81-1)
Pigment Rubine BCL (5858-81-1)
Pigment Ruby (2425-85-6)
Pigment Ruby ZH (2814-77-9)
Pigment Scarlet (2425-85-6)
Pigment Scarlet B (2425-85-6)
Pigment Scarlet N (2425-85-6)
Pigment Scarlet R (2425-85-6)
Pigment Scarlet (Russian) (2425-85-6)
Pigment Scarlet ZH (2814-77-9)
Pigment Sky Blue (147-14-8)
Pigment Violet #19 (1047-16-1)
Pigment Violet 19 (1047-16-1)
Pigment Yellow 1 (2512-29-0)
Pigment Yellow 3 (6486-23-3)
Pigment Yellow 12 (6358-85-6)
Pigment Yellow 13 (5102-83-0)
Pigment Yellow 73 (13515-40-7)
Pigment Yellow 74 (6358-31-2)
Pigment Yellow GT (6358-85-6)
Pigmex (103-16-2)
Pikrinezuur (Dutch) (88-89-1)
Pikrinsaeure (German) (88-89-1)
Pikrynowy kwas (Polish) (88-89-1)
Pilate Fast Red GREN (6408-31-7)
Pileric acid (111-16-0)
Pillardrin (6923-22-4)
Pillaron (10265-92-6)
Pillarquat (1910-42-5)
Pillartan (2939-80-2)
Pillarxone (1910-42-5)
Pillarzo (15972-60-8)
Pilocar (54-71-7)
Pilocarpin (92-13-7)
Pilocarpine (92-13-7)
Pilocarpine, hydrochloride (54-71-7)
Pilocarpine, monohydrochloride (54-71-7)
Pilocarpine muriate (54-71-7)
Pilocarpol (92-13-7)
Pilocel (54-71-7)
Pilomiotin (54-71-7)
Pilot (76578-14-8)
Pilot 447 (90-33-5)
Pilot HD-90 (25155-30-0)
Pilot SF-40 (25155-30-0)
Pilot SF-60 (25155-30-0)
Pilot SF-96 (25155-30-0)
Pilot SF-40B (25155-30-0)
Pilot SF-40FG (25155-30-0)
Pilot SP-60 (25155-30-0)
Pilovisc (54-71-7)
Pilpophen (60-87-7)
Pimacol-Sol (86-87-3)
Pimal (57-53-4)
Pimanthrene (483-87-4)
Pimara-8(14),15-dien-19-oic acid (127-27-5)
Pimaric acid (127-27-5)
α-Pimaric acid (127-27-5)
β-Pimaric acid (79-54-9)
d-Pimaric acid (127-27-5)
δ8(14)-Pimaric acid (127-27-5)
l-Pimaric acid (79-54-9)
Pimaric acid, d (127-27-5)
Pimelic-acid (111-16-0)
Pimelic acid, diethyl ester (8CI) (2050-20-6)
Pimelic acid dinitrile (646-20-8)
Pimelic ketone (108-94-1)
Pimelin ketone (108-94-1)
Pimelonitrile (646-20-8)
Pimeton (673-04-1)
Piminodina (Spanish) (13495-09-5)
Piminodine (13495-09-5)
Piminodinum (Latin) (13495-09-5)
Pinacol (76-09-5)
Pinacolin (75-97-8)
Pinacoline (75-97-8)
Pinacolone (75-97-8)
Pinacolone cyclic carbonate (19424-29-4)
Pinacoloxymethylphosphoryl fluoride (96-64-0)

Pinacolyl alcohol (6CI) **(464-07-3)**
Pinacolyl methylfluorophosphonate **(96-64-0)**
Pinacolyl methylphosphonic acid **(616-52-4)**
Pinacolyl methylphosphonofluoridate **(96-64-0)**
Pinacolyl methylphosphonofluoride **(96-64-0)**
Pinacolyloxy methylphosphoryl fluoride **(96-64-0)**
Pinakolin **(75-97-8)**
Pinakolin (German) **(75-97-8)**
Pinakon **(107-41-5)**
cis-Pinane **(6876-13-7)**
Pinane (8CI) **(473-55-2)**
Pinane, 2,3-epoxy- (8CI) **(1686-14-2)**
Pinane hydroperoxide, Solution, Not over 45% peroxide (DOT) **(28324-52-9)**
Pinane hydroperoxide, Technically pure (DOT) **(28324-52-9)**
Pinane, stereoisomer (8CI) **(6876-13-7)**
2-Pinanol **(473-54-1)**
cis-2-Pinanol **(4948-28-1)**
Pinanyl hydroperoxide **(28324-52-9)**
Pinanyl hydroperoxide, Technically pure (DOT) **(28324-52-9)**
Pinazepam **(52463-83-9)**
Pinazepamum (Latin) **(52463-83-9)**
Pindione **(83-12-5)**
Pindon (Dutch) **(83-26-1)**
Pindone (ACGIH,OSHA) **(83-26-1)**
Pindone, Liquid (DOT) **(83-26-1)**
Pindone, Solid (DOT) **(83-26-1)**
Pine Oil [UN 1272] **(8002-09-3)**
Pine Tar **(8011-48-1)**
Pineapple ketone **(3658-77-3)**
2(10)-Pinene **(127-91-3)**
2-Pinene **(80-56-8)**
α-Pinene **(80-56-8)**
β-Pinene **(127-91-3)**
Pinene (DOT) **(1330-16-1)**
α-Pinene [UN 2368] **(1330-16-1)**
α-Pinene [UN 2368] **(80-56-8)**
α-Pinene epoxide **(1686-14-2)**
2-Pinene-10-methyl acetate **(128-51-8)**
2-Pinene oxide **(1686-14-2)**
α-Pinene oxide **(1686-14-2)**
2-Pinen-4-one, (1R,5R)-(+) **(18309-32-5)**
Pinhole AK 2 **(123-77-3)**
Pinocamphone **(547-60-4)**
β-Pinone **(24903-95-5)**
Pinonic acid **(473-72-3)**
Pinoran **(14214-32-5)**
Pio **(2152-34-3)**
Piodel **(54-47-7)**
Pioloform F **(9003-20-7)**
Piombo tetra-etile (Italian) **(78-00-2)**
Piombo tetra-metile (Italian) **(75-74-1)**
Pioxol **(2152-34-3)**
No-Pip **(100-75-4)**
Pip-Pip **(98-77-1)**
Pipecolate **(535-75-1)**
Pipecolic-acid **(535-75-1)**
Pipecolic acid, ethyl ester (8CI) **(15862-72-3)**
α-Pipecolin **(109-05-7)**
2-Pipecoline, 1-(3-aminopropyl)- (8CI) **(25560-00-3)**
3-Pipecoline, 1-nitroso **(13603-07-1)**
Pipecolinic acid **(535-75-1)**
α-Pipecolinic acid **(535-75-1)**
Piperalin **(3478-94-2)**
Piperazidine **(110-85-0)**
Piperazin (German) **(110-85-0)**
Piperazine, Anhydrous **(110-85-0)**
Piperazine [UN 2579] **(110-85-0)**
Piperazine, 1-(2-aminoethyl) **(140-31-8)**
Piperazine, 1-(2-aminoethyl)- (8CI) **(140-31-8)**
Piperazine, 1,4-bis(3-bromo-1-oxopropyl)- **(54-91-1)**
Piperazine, 1,4-bis(3-bromopropionyl) **(54-91-1)**
Piperazine, 1,4-bis(1-formamido-2,2,2-trichloroethyl) **(26644-46-2)**
Piperazine, N,N'-bis(2-hydroxyethyl)- **(122-96-3)**
1-Piperazinecarboxamide, N,N-diethyl-4-methyl-, citrate (1:1) **(1642-54-2)**
1-Piperazinecarboxamide, N,N-diethyl-4-methyl-, 2-hydroxy-1,2,3-propanetricarboxylate **(1642-54-2)**
Piperazine, 1-(p-chloro-α-phenylbenzyl)-4-methyl **(82-93-9)**
Piperazine, 1-(p-chloro-α-phenylbenzyl)-4-(m-methylbenzyl) **(569-65-3)**
Piperazine, 1-(p-chloro-α-phenylbenzyl)-4-methyl-, hydrochloride **(14362-31-3)**
Piperazine, 1-(p-chloro-α-phenylbenzyl)-4-methyl-, monohydrochloride (8CI) **(14362-31-3)**
Piperazine, 1-((4-chlorophenyl)phenylmethyl)-4-methyl-, monohydrochloride (9CI) **(14362-31-3)**
1,4-Piperazinediethanamine (9CI) **(6531-38-0)**
1,4-Piperazinediethanamine, N-(2-aminoethyl)- (9CI) **(31295-54-2)**
1,4-Piperazinediethanol **(122-96-3)**
Piperazine dihydrochloride (ACGIH,OSHA) **(142-64-3)**
1,4-Piperazinedimethanol (9CI) **(3312-58-1)**
Piperazine, 1,4-dimethyl **(106-58-1)**
Piperazine, 2,5-dimethyl **(106-55-8)**
Piperazine, 1,4-dimethyl-, (R-(R*,R*))-2,3-dihydroxybutanedioate (1:1) (9CI) **(133-35-7)**
Piperazine, 1,4-dimethyl-, tartrate **(133-35-7)**
Piperazine, 1,4-dimethyl-, tartrate (1:1) (8CI) **(133-35-7)**
Piperazine, 1,4-dinitroso **(140-79-4)**
2,5-Piperazinedione (9CI) **(106-57-0)**
2,6-Piperazinedione, 4,4'-(1-methyl-1,2-ethanediyl)bis-, (+-)- (9CI) **(21416-87-5)**
2,6-Piperazinedione, 4,4'-propylenedi-, (+-) **(21416-87-5)**
2,6-Piperazinedione-4,4'-propylene dioxopiperazine **(21416-87-5)**
Piperazine, 1-(diphenylmethyl)-4-methyl **(82-92-8)**
N,N'-(Piperazinediylbis(2,2,2-trichloroethylidene)) bis(formamide) **(26644-46-2)**
Piperazine estrone sulfate **(7280-37-7)**
1-Piperazineethanamine (9CI) **(140-31-8)**
Piperazine, 1,1'-(1,2-ethanediyl)bis- (9CI) **(19479-83-5)**
1-Piperazineethanol **(103-76-4)**
Piperazineethanol (9CI) **(25154-38-5)**
1-Piperazineethanol, 4-(3-(2-chlorophenothiazin-10-yl)propyl) **(58-39-9)**
1-Piperazineethanol, 4-(3-(2-chloro-10h-phenothiazin-10-yl)propyl)-, acetate (ester) (9CI) **(84-06-0)**
1-Piperazineethanol, 4-(3-(2-chlorophenothiazin-10-yl)propyl)-, acetate (ester) **(84-06-0)**
1-Piperazineethanol, 4-(3-(2-(trifluoromethyl)phenothiazin-10-yl)-propyl) **(69-23-8)**
1-Piperazineethanol, 4-(3-(2-(trifluoromethyl)-9H-thioxanthen-9-ylidene)propyl)-, (Z)- (9CI) **(53772-82-0)**
Piperazine hydrochloride **(142-64-3)**
Piperazine, 1-methyl **(109-01-3)**
Piperazine, 1-phenyl **(92-54-6)**
1-Piperazinepropanamine, γ-methyl- (9CI) **(90853-14-8)**
Piperidic acid **(56-12-2)**
Piperidin (German) **(110-89-4)**
δ^3-Piperidine **(694-05-3)**
Piperidine [UN 2401] **(110-89-4)**
2-Piperidineacetic acid, α-phenyl-, methyl ester **(113-45-1)**
2-Piperidineacetic acid, α-phenyl-, methyl ester, hydrochloride **(298-59-9)**
Piperidine, 1-acetyl **(618-42-8)**
Piperidine, 1-(2-aminoethyl)- (8CI) **(27578-60-5)**
Piperidine, 2-(2-aminoethyl)- (8CI) **(15932-66-8)**
Piperidine, 1-(3-aminopropyl)- (8CI) **(3529-08-6)**
Piperidine, 1-(5-(1,3-benzodioxol-5-yl)-1-oxo-2,4-pentadienyl)-, (E,E)- (9CI) **(94-62-2)**
1-Piperidinebutanamide, 4-(4-chlorophenyl)-4-hydroxy-N,N-di-

methyl-α,α-diphenyl (53179-11-6)
1-Piperidinebutanol, α,α-diphenyl (972-02-1)
1-Piperidinecarbodithioic acid, Compd. with piperidine (98-77-1)
1-Piperidinecarbothioic acid, 2,6-dimethyl-, S-ethyl ester (9CI) (107348-46-9)
1-Piperidinecarbothioic acid, S-ethyl ester (8CI,9CI) (6961-73-5)
1-Piperidinecarbothioic acid, 2-methyl-, S-ethyl ester (9CI) (52372-17-5)
1-Piperidinecarbothioic acid, 3-methyl-, S-ethyl ester (9CI) (107348-45-8)
1-Piperidinecarbothioic acid, 4-methyl-, S-ethyl ester (9CI) (67587-12-6)
2-Piperidinecarboxylic acid (9CI) (535-75-1)
4-Piperidinecarboxylic acid, 1-(2-(4-aminophenyl)ethyl)-4-phenyl-, ethyl ester (144-14-9)
4-Piperidinecarboxylic acid, 1-(3-cyano-3,3-diphenylpropyl)-4-phenyl- (28782-42-5)
4-Piperidinecarboxylic acid, 1-(3-cyano-3,3-diphenylpropyl)-4-phenyl-, ethyl ester (9CI) (915-30-0)
2-Piperidinecarboxylic acid, ethyl ester (9CI) (15862-72-3)
4-Piperidinecarboxylic acid, 4-(m-hydroxyphenyl)-1-methyl-, ethyl ester (468-56-4)
4-Piperidinecarboxylic acid, 1-methyl-4-phenyl-, ethyl ester, hydrochloride (50-13-5)
4-Piperidinecarboxylic acid, 1-methyl-4-phenyl-, 1-methylethyl ester (9CI) (561-76-2)
4-Piperidinecarboxylic acid, 1-[2-(4-morpholinyl)ethyl]-4-phenyl-, ethyl ester (9CI) (469-81-8)
4-Piperidinecarboxylic acid, 4((1-oxopropyl)phenylamino)-1-(2-phenylethyl)-, methyl ester (59708-52-0)
4-Piperidinecarboxylic acid, 4-((1-oxopropyl)phenylamino)-1-(2-phenylethyl)-, methyl ester, 2-hydroxy-1,2,3-propanetricarboxylate (1:1) (59708-52-0)
4-Piperidinecarboxylic acid, 4-phenyl-1-(2-(phenylmethoxy)-ethyl)-, ethyl ester (3691-78-9)
1-Piperidinecarboxylic acid, 6-(trimethylammonio)thymyl ester, chloride (2438-53-1)
Piperidine, 4-(5H-dibenzo(a,d)cyclohepten-5-ylidene)-1-methyl (129-03-3)
Piperidine, 1,2-dimethyl- (8CI,9CI) (671-36-3)
2,4-Piperidinedione, 3,3-diethyl-5-methyl (125-64-4)
1-Piperidineethanamine (9CI) (27578-60-5)
2-Piperidineethanamine (9CI) (15932-66-8)
2-Piperidineethanol (9CI) (1484-84-0)
2-Piperidineethanol, 1-methyl- (9CI) (533-15-3)
Piperidine, 1-ethyl (766-09-6)
Piperidine, 5-ethyl-2-methyl (104-89-2)
Piperidine, 1-formyl (2591-86-8)
Piperidine, 1,1'-(hexathiodicarbonothioyl)bis- (9CI) (971-15-3)
Piperidine, hydrochloride (6091-44-7)
4-Piperidinemethanamine (9CI) (7144-05-0)
2-Piperidinemethanol, α,α-diphenyl- (467-60-7)
Piperidine, 1-methyl (626-67-5)
Piperidine, 2-methyl (109-05-7)
Piperidine, 1-methyl-2-(3-pyridyl)- (8CI) (19730-04-2)
Piperidine, 1-nitro- (8CI,9CI) (7119-94-0)
Piperidine, 1-nitroso (100-75-4)
Piperidine, 1-phenyl- (9CI) (4096-20-2)
Piperidine, 1-(1-phenylcyclohexyl) (77-10-1)
Piperidine, 1-(1-phenylcyclohexyl)-, hydrochloride (956-90-1)
Piperidine, 1-piperoyl-, (E,E) (94-62-2)
1-Piperidinepropanamine (9CI) (3529-08-6)
1-Piperidinepropanamine, 2-methyl- (9CI) (25560-00-3)
Piperidine, 4,4'-(1,3-propanediyl)bis- (9CI) (16898-52-5)
2-Piperidinepropanol (8CI,9CI) (24448-89-3)
1-Piperidinepropanol, α-cyclohexyl-α-phenyl (144-11-6)
1-Piperidinepropanol, α,α-diphenyl (511-45-5)
1-Piperidinepropanol, 2-methyl-, 3,4-dichlorobenzoate (Ester) (3478-94-2)
Piperidine, 1-propyl (5470-02-0)
Piperidine, 2-propyl-, (S) (458-88-8)
Piperidine, 2-(3-pyridyl)- (494-52-0)
Piperidine, 1,1'-(tetrathiodicarbonothioyl)bis (120-54-7)
Piperidine, 1-(1-(2-thienyl)cyclohexyl)- (21500-98-1)
Piperidine, 4,4'-trimethylenedi- (8CI) (16898-52-5)
Piperidinic acid (56-12-2)
Piperidinium, 1,1'-(2-β,16-β-(3-α,17-β-dihydroxy-5-α-androstanylene)) bis(1-methyl-, dibromide, diacetate (15500-66-0)
Piperidinium, 1,1-dimethyl-, chloride (24307-26-4)
Piperidinium pentamethylenedithiocarbamate (98-77-1)
1-Piperidinocyclohexanecarbonitrile (3867-15-0)
Piperidinocyclohexanecarbonitrile (3867-15-0)
3-Piperidinol (9CI) (6859-99-0)
4-Piperidinol, 3-allyl-1-methyl-4-phenyl-, propionate (ester) (8CI) (469-81-8)
4-Piperidinol, 1,3-dimethyl-4-phenyl-, propanoate (ester), trans- (9CI) (468-59-7)
4-Piperidinol, 1,3-dimethyl-4-phenyl-, propionate (77-20-3)
4-Piperidinol, 1,3-dimethyl-4-phenyl-, propionate (ester), stereoisomer (468-59-7)
4-Piperidinol, 1-methyl-4-phenyl-, propanoate (Ester) (13147-09-6)
4-Piperidinol, 1-methyl-4-phenyl-3-(2-propenyl)-, propanoate (ester) (9CI) (25384-17-2)
4-Piperidinol, 1-phenethyl-4-phenyl-, acetate (ester) (64-52-8)
4-Piperidinol, 4-phenyl-1-(2-phenylethyl)-, acetate (ester) (64-52-8)
4-Piperidinol, 2,2,6,6-tetramethyl- (9CI) (2403-88-5)
4-Piperidinol, 2,2,6-trimethyl-, benzoate (ester) (500-34-5)
4-Piperidinol, 1,2,5-trimethyl-4-phenyl-, propionate (ester) (64-39-1)
Piperidinone (9CI) (27154-43-4)
2-Piperidinone, 1-methyl- (9CI) (931-20-4)
4-Piperidinone, 2,2,6,6-tetramethyl- (9CI) (826-36-8)
1-(1-Piperidinyl)-cyclohexanecarbonitrile (3867-15-0)
3-(2-Piperidinyl)pyridine (494-52-0)
Piperidon (German) (675-20-7)
2-Piperidone (675-20-7)
Piperidone (8CI) (27154-43-4)
Piperidone-2 (French) (675-20-7)
2-Piperidone, 1-methyl- (8CI) (931-20-4)
4-Piperidone, 2,2,6,6-tetramethyl (826-36-8)
α-(2-Piperidyl)benzhydrol (467-60-7)
l-3-(2'-Piperidyl)pyridine (494-52-0)
Piperilona (Spanish) (2531-04-6)
Piperin (94-62-2)
Piperine (94-62-2)
Piperitol (491-04-3)
Piperitol (monoterpene) (491-04-3)
Piperitone (89-81-6)
Piperolinic acid (535-75-1)
Piperonal (120-57-0)
Piperonal bis(2-(2-butoxyethoxy)ethyl)acetal (5281-13-0)
Piperonaldehyde (120-57-0)
Piperonol (495-76-1)
Piperonyl acetate (326-61-4)
Piperonyl acetone (3160-37-0)
Piperonyl alcohol (8CI) (495-76-1)
Piperonyl alcohol, α-vinyl- (5208-87-7)
Piperonyl aldehyde (120-57-0)
Piperonyl butoxide (51-03-6)
Piperonyl cyclonene (119-89-1)
Piperonylcyklonen (Czech) (119-89-1)
Piperonylic acid (8CI) (94-53-1)
Piperonyl sulfoxide (120-62-7)
Piperophos (24151-93-7)
1-Piperoylpiperidine (94-62-2)
Pipersal (57-42-1)
Piperylene (504-60-9)
trans-Piperylene (2004-70-8)
Piperylone (2531-04-6)
Piperylonum (Latin) (2531-04-6)
Pipobroman (54-91-1)

Pipolphen (60-87-7)
α-Pipradol (467-60-7)
Pipradol (Spanish) (467-60-7)
Pipradrol (467-60-7)
Pipradrolo (467-60-7)
Pipradrolum (Latin) (467-60-7)
Pipron (3478-94-2)
Piprotal (5281-13-0)
Pipsyl chloride (98-61-3)
Pirabutina (129-20-4)
Piracaps (64-75-5)
Piraflogin (129-20-4)
Piramidon (58-15-1)
Piramox (26787-78-0)
Pirarreumol "B" (50-33-9)
Pirazinon (5826-91-5)
Pirazoxon (Italian) (108-34-9)
Piretrina 1 (Portuguese) (121-21-1)
Piribenzil (91-81-6)
Pirid (136-40-3)
Pirid (94-78-0)
Piridacil (136-40-3)
Piridazol (144-83-2)
Piridina (Italian) (110-86-1)
Piridinol carbamato (Spanish) (1882-26-4)
Piridisir (68-35-9)
Piridol (58-15-1)
Piridolan (302-41-0)
Piridolo (80-35-3)
Piridosal (50-13-5)
Piridosal (57-42-1)
Piridrol (467-60-7)
Piriex (113-92-8)
Pirimecidan (58-14-0)
Pirimetamina (Spanish) (58-14-0)
Pirimicarb (23103-98-2)
Pirimifosethyl (23505-41-1)
Pirimifosmethyl (29232-93-7)
Pirimiphos-ethyl (23505-41-1)
Pirimiphos-methyl (29232-93-7)
Pirimor (23103-98-2)
Pirinitramide (302-41-0)
Piria's acid (84-86-6)
Piristin (154-69-8)
Piriton (113-92-8)
Piriton (132-22-9)
Piritramida (Spanish) (302-41-0)
Piritramide (302-41-0)
Piritramidum (Latin) (302-41-0)
Pirmazin (57-68-1)
Piroan (58-32-2)
Pirocard Green 491 (1689-82-3)
Pirod (66-22-8)
Pirofos (3689-24-5)
Pirolen 120A (64742-16-1)
Piromidina (58-15-1)
Pirosolvina (938-73-8)
Pirovalerona (Spanish) (3563-49-3)
Piroxicam (36322-90-4)
Pirprofen (31793-07-4)
Pirydyna (Polish) (110-86-1)
Pisichergina (300-42-5)
Pistac CC (9003-07-0)
Pistac L (9003-07-0)
Pitayine (56-54-2)
Pitch, Coal Tar (65996-93-2)
Pitch, Petroleum, Arom. (68187-58-6)
Pitch (61789-60-4)
Pitch (65996-93-2)
Pitocin (50-56-6)
Piton S (50-56-6)
Pitressin (11000-17-2)
Pitrex (2398-96-1)
Pittchlor (7778-54-3)
Pittcide (7778-54-3)
Pittclor (7778-54-3)
Pittsburgh PX-138 (117-81-7)
Pituitary-Growth-Hormone (9002-72-6)
Pituitary Lactogenic Hormone (9002-62-4)
Pivacin (83-26-1)
Pival (83-26-1)
Pivalaldehyde (8CI) (630-19-3)
Pivalaldehyde, O-(methylcarbamoyl)oxime (8CI) (6062-02-8)
Pivalamide (8CI) (754-10-9)
Pivaldehyde (630-19-3)
Pivaldion (Italian) (83-26-1)
Pivaldione (French) (83-26-1)
Pivalic-acid (75-98-9)
Pivalic acid chloride (3282-30-2)
Pivalic acid, ethyl ester (8CI) (3938-95-2)
Pivalic acid lactone (1955-45-9)
Pivalic acid, phenyl ester (8CI) (4920-92-7)
Pivalic anhydride (1538-75-6)
Pivalolactone (1955-45-9)
Pivalolyl chloride (3282-30-2)
Pivalonitrile (8CI) (630-18-2)
Pivaloyl chloride (DOT) (3282-30-2)
2-Pivaloyl-indaan-1,3-dion (Dutch) (83-26-1)
2-Pivaloyl-indan-1,3-dion (German) (83-26-1)
2-Pivaloyl-1,3-indandione (83-26-1)
2-Pivaloylindane-1,3-dione (83-26-1)
1-Pivaloyl-1,2,4-triazole (60718-52-7)
Pivalyl chloride (3282-30-2)
2-Pivalyl-1,3-indandione (83-26-1)
2-Pivalyl-1,3-indandione (OSHA) (83-26-1)
Pivalyl valone (83-26-1)
Pivalyn (83-26-1)
Pix (24307-26-4)
Pix Carbonis (8007-45-2)
Pix Lithanthracis (8007-45-2)
Pixalbol (8007-45-2)
P 40701 (9002-88-4)
Placidal (77-75-8)
Placidil (113-18-8)
Placidol (68-88-2)
Placidol E (84-66-2)
Placidon (57-53-4)
Placidyl (113-18-8)
Placitate (57-53-4)
Planadalin (77-65-6)
Planavin (4726-14-1)
Planavin 75 (4726-14-1)
Planavit C (50-81-7)
Planetol (55-55-0)
Planium (9002-88-4)
Planocaine (51-05-8)
Planofix (86-87-3)
Planofixe (86-87-3)
Planomide (72-14-0)
Planotox (1929-73-3)
Planotox (94-75-7)
Plant Dithio Aerosol (3689-24-5)
Plant Pin (34681-23-7)
Plant Protection PP511 (29232-93-7)

Plantain LS (8001-21-6)
Plantdrin (6923-22-4)
Plantfume 103 Smoke Generator (3689-24-5)
Plantgard (94-75-7)
Plantifog 160M (12427-38-2)
Plantomycin (3810-74-0)
Plantulin (139-40-2)
Plantvax (5259-88-1)
Plantvax 20 (5259-88-1)
Planuin (4726-14-1)
Planum (846-50-4)
Plasdone (9003-39-8)
Plasdone K 29-32 (9003-39-8)
Plasdone K-26/28 (9003-39-8)
Plasdone XL (9003-39-8)
Plasil (364-62-5)
Plaskin 8200 (105-60-2)
Plaskin 8200 (25038-54-4)
Plaskon 201 (105-60-2)
Plaskon 201 (25038-54-4)
Plaskon 8201 (105-60-2)
Plaskon 8201 (25038-54-4)
Plaskon 8205 (105-60-2)
Plaskon 8205 (25038-54-4)
Plaskon 8207 (105-60-2)
Plaskon 8207 (25038-54-4)
Plaskon 8252 (105-60-2)
Plaskon 8252 (25038-54-4)
Plaskon 8202C (105-60-2)
Plaskon 8201HS (105-60-2)
Plaskon PP 60-002 (9002-88-4)
Plaskon XP 607 (105-60-2)
Plaskon XP 607 (25038-54-4)
Plaskon 8202c (25038-54-4)
Plasmin (9CI) (9001-90-5)
Plasmin (Human) (9CI) (9004-09-5)
Plasmosan (9003-39-8)
Plastacele (9004-35-7)
Plastanox 2246 (119-47-1)
Plastanox LTDP (123-28-4)
Plastanox LTDP Antioxidant (123-28-4)
Plastanox STDP (693-36-7)
Plastanox STDP Antioxidant (693-36-7)
Plastanox 425 Antioxidant (88-24-4)
Plastazote X 1016 (9002-88-4)
Plastflow (110-30-5)
Plasthall 503 (142-77-8)
Plastibest 20 (12001-29-5)
Plasticizer BDP (89-19-0)
Plasticizer BOP (84-78-6)
Plasticizer 3GH (95-08-9)
Plasticizer 4GO (18268-70-7)
Plasticizer OBP (84-78-6)
Plastics, epoxy (61788-97-4)
Plastifix PC (9010-98-4)
Plastol Orange G (3520-72-7)
Plastol Rubine BC (5858-81-1)
Plastolein 9050 (109-31-9)
Plastolein 9051 (109-31-9)
Plastolein 9058 (103-24-2)
Plastolein 9214 (61788-72-5)
Plastolein 9765 (27941-08-8)
Plastolein 9051 DHNZ (109-31-9)
Plastolein 9050 DHZ (10332-40-8)
Plastolein 9058 DOZ (103-24-2)
Plastomoll DOA (103-23-1)
Plastomoll Na (151-32-6)
Plastopal BT (9011-05-6)
Plastoresin Orange F 3A (2051-85-6)
Plastoresin Orange F4A (842-07-9)
Plastoresin Red F (85-83-6)
Plastoresin Red RC (1658-56-6)
Plastoresin Red SR (1248-18-6)
Plastoresin Violet 5BO (548-62-9)
Plastronga (9002-88-4)
Plastylene MA 2003 (9002-88-4)
Plastylene MA 7007 (9002-88-4)
Platformate (64741-63-5)
Plath-Lyse (97-23-4)
Platiblastin (15663-27-1)
cis-Platin (15663-27-1)
Platin (German) (7440-06-4)
Platinate(2-), hexachloro-, diammonium, (OC-6-11)- (16919-58-7)
Platinate(2-), hexachloro-, dihydrogen (8CI) (16941-12-1)
Platinate(2-), hexachloro-, dihydrogen, (OC-6-11)- (9CI) (16941-12-1)
Platinex (15663-27-1)
Platinic ammonium chloride (16919-58-7)
Platinic chloride (VAN) (16941-12-1)
Platinol (15663-27-1)
Platinol AH (117-81-7)
Platinol DOP (117-81-7)
Platinous chloride (10025-65-7)
cis-Platinous diaminodichloride (15663-27-1)
cis-Platinous diammine dichloride (15663-27-1)
Platinum-chloride (10025-65-7)
cis-Platinum (15663-27-1)
cis-Platinum(II) (15663-27-1)
Platinum (ACGIH) (7440-06-4)
Platinum Black (7440-06-4)
Platinum (IV) chloride (13454-96-1)
Platinum(IV) tetrachloride (13454-96-1)
Platinum Sponge (7440-06-4)
Platinum chloride (16941-12-1)
Platinum chloride (H2PtCl6) (16941-12-1)
cis-Platinum(II) diamminedichloride (15663-27-1)
trans-Platinum(II)diamminedichloride (14913-33-8)
Platinum (II), diamminedichloro-, trans (14913-33-8)
Platinum(II), diamminedichloro-, cis (15663-27-1)
Platinum, diamminedichloro-, (SP-4-2)- (9CI) (15663-27-1)
Platinum metal (OSHA) (7440-06-4)
Platinum tetrachloride (13454-96-1)
Plaxidol (68-88-2)
Plecyamin (68-19-9)
Plegomazin (50-53-3)
Plegomazin (69-09-0)
Plenastril (434-07-1)
Plenur (554-13-2)
Plessy's Green (Hemiheptahydrate) (7789-04-0)
Plex 8572-F (9011-14-7)
Plexene D (140-01-2)
Plexiglas (9011-14-7)
Plexigum M 920 (9011-14-7)
Plexigum PM 381 (25608-33-7)
Plexisol PM 709 (25608-33-7)
Pliabrac 521 (9CI) (98913-83-8)
Plictran (13121-70-5)
Plidan (439-14-5)
Plimasine (113-45-1)
Plinol B (4099-07-4)
Plioflex (9003-55-8)
Pliogrip (9009-54-5)
Pliolite (25339-57-5)
Pliolite 151 (9003-55-8)
Pliolite 160 (9003-55-8)

Pliolite 491 **(9003-55-8)**
Pliolite 55B **(9003-55-8)**
Pliolite S 50 **(9003-55-8)**
Pliolite S5 **(9003-55-8)**
Pliolite S 5A **(9003-55-8)**
Pliolite S-5B **(9003-55-8)**
Pliolite S-5C **(9003-55-8)**
Pliolite S 5D **(9003-55-8)**
Pliolite S-5E **(9003-55-8)**
Pliovac AO **(9003-22-9)**
Pliovic AO **(9003-22-9)**
Plisulfan **(526-08-9)**
Pliva **(55-86-7)**
Plletia **(58-33-3)**
Plomb fluorure (French) **(7783-46-2)**
Plondrel **(5131-24-8)**
Plucker **(86-87-3)**
Plumbago **(7440-44-0)**
Plumbago **(7782-42-5)**
Plumbago (Graphite) **(7782-42-5)**
Plumbane, acetoxytriphenyl **(1162-06-7)**
Plumbane, (acetyloxy)triphenyl- (9CI) **(1162-06-7)**
Plumbane, butyldiethylmethyl- (9CI) **(65122-13-6)**
Plumbane, butylethyldimethyl- (9CI) **(65122-14-7)**
Plumbane, butyltriethyl- (9CI) **(64346-32-3)**
Plumbane, butyltrimethyl- (9CI) **(54964-75-9)**
Plumbane, chlorotributyl **(13302-14-2)**
Plumbane, chlorotriethyl **(1067-14-7)**
Plumbane, chlorotrimethyl **(1520-78-1)**
Plumbane, dichlorodiethyl **(13231-90-8)**
Plumbane, dichlorodimethyl- (8CI,9CI) **(1520-77-0)**
Plumbane, diethyldimethyl- (9CI) **(1762-27-2)**
Plumbane, ethyltrimethyl- (9CI) **(1762-26-1)**
Plumbane, tetrabutyl- (9CI) **(1920-90-7)**
Plumbane, tetraethyl **(78-00-2)**
Plumbane, tetramethyl **(75-74-1)**
Plumbane, tetrapropyl **(3440-75-3)**
Plumbane, tributylchloro- **(13302-14-2)**
Plumbane, tributylethyl- (9CI) **(65151-10-2)**
Plumbane, triethyl- (8CI,9CI) **(5224-23-7)**
Plumbane, triethylchloro- **(1067-14-7)**
Plumbane, triethylmethyl- (9CI) **(1762-28-3)**
Plumbane, trimethyl **(7442-13-9)**
Plumbane, tripropyl **(6618-03-7)**
Plumbate(1-), (N,N-bis(carboxymethyl)glycinato-(3-)-N,O,O',O''), hydrogen, (T-4)- (9CI) **(53113-59-0)**
Plumboplumbic oxide **(1314-41-6)**
Plumbous acetate **(301-04-2)**
Plumbous acetate **(6080-56-4)**
Plumbous chloride **(7758-95-4)**
Plumbous chromate **(7758-97-6)**
Plumbous fluoride **(7783-46-2)**
Plumbous iodide **(10101-63-0)**
Plumbous nitrate **(10099-74-8)**
Plumbous oxide **(1317-36-8)**
Plumbous phosphate **(7446-27-7)**
Plumbous sulfide **(1314-87-0)**
Plumbylium, ethyldimethyl- (9CI) **(103730-90-1)**
Plumbylium, triethyl- (9CI) **(14570-15-1)**
Plumbylium, trimethyl- (9CI) **(14570-16-2)**
Plumbylium, tripropyl- (9CI) **(44910-38-5)**
Pluracol 686 **(9003-11-6)**
Pluracol E-200 **(25322-68-3)**
Pluracol E-300 **(25322-68-3)**
Pluracol E-400 **(25322-68-3)**
Pluracol E-600 **(25322-68-3)**
Pluracol E-1500 **(25322-68-3)**
Pluracol E-4000 **(25322-68-3)**
Pluracol E-6000 **(25322-68-3)**
Pluracol GP 430 **(25791-96-2)**
Pluracol P-410 **(25322-68-3)**
Pluracol P-710 **(25322-68-3)**
Pluracol P-1010 **(25322-68-3)**
Pluracol P-2010 **(25322-68-3)**
Pluracol P-3010 **(25322-68-3)**
Pluracol P-4010 **(25322-68-3)**
Pluracol V **(9003-11-6)**
Pluracol W 3520N **(9038-95-3)**
Plurafac RA 43 **(9002-92-0)**
Plurafac RA 30 (9CI) **(39316-51-3)**
Plurafac RA 40 (9CI) **(56590-81-9)**
Plurazol **(144-83-2)**
Pluriol L 64 **(9003-11-6)**
Pluriol SC 9361 **(9003-11-6)**
Plurol oleique **(9007-48-1)**
Pluryl **(73-48-3)**
Pluryle **(73-48-3)**
Plusuril **(73-48-3)**
Plutonium **(7440-07-5)**
Plutonium, Isotope of mass 238 (8CI,9CI) **(13981-16-3)**
Plutonium, Isotope of mass 243 (8CI,9CI) **(15706-37-3)**
Plyamine HD 1129A **(9011-05-6)**
Plyamine P 364BL **(9011-05-6)**
Plyamul 40-155 **(9003-20-7)**
Plyamul 40-350 **(9003-20-7)**
Plyctran **(13121-70-5)**
Plymouth IPP **(142-91-6)**
Plymoutm IPM **(110-27-0)**
Pmacetate **(62-38-4)**
Po-Systox **(126-75-0)**
Poast **(74051-80-2)**
Podocarpa-7,13-dien-15-oic acid, 13-isopropyl **(514-10-3)**
Podocarpa-8(14),12-dien-15-oic acid, 13-isopropyl- (8CI) **(79-54-9)**
Podocarpa-8,13-dien-15-oic acid, 13-isopropyl- (8CI) **(1945-53-5)**
Podocarpa-7,13-dien-15-oic acid, 13-isopropyl-, methyl ester **(127-25-3)**
Podocarpane-δ(13,α)-acetic acid, 3β-hydroxy-14α-methyl-7-oxo-, 2-(dimethylamino)ethyl ester, (E)- **(468-76-8)**
8-β-Podocarpan-16-oic acid, 13-(carboxymethylene)-3-β-hydroxy-14-methyl-7-oxo-, 13- (2-(dimethylamino)ethyl) 16-methyl ester, hydrochloride **(23451-24-3)**
Podocarpa-8,11,13-trien-16-al, 13-isopropyl- (8CI) **(24035-50-5)**
Podocarpa-8,11,13-trien-15-amine, 13-isopropyl- (8CI) **(1446-61-3)**
Podocarpa-8,11,13-trien-15-amine, 13-isopropyl-, acetate (8CI) **(2026-24-6)**
5.β.-Podocarpa-8,11,13-trien-16-oic acid (7CI) **(57345-30-9)**
Podocarpa-8,11,13-trien-15-oic acid, 13-isopropyl **(1740-19-8)**
Podocarpa-8,11,13-trien-16-oic acid, 13-isopropyl- (8CI) **(5155-70-4)**
Podocarpa-8,11,13-trien-15-oic acid, 13-isopropyl-, methyl ester (8CI) **(1235-74-1)**
Podocarpa-8,11,13-trien-15-oic acid, 13-isopropyl-7-oxo **(18684-55-4)**
Podocarpa-8,11,13-trien-15-oic acid, methyl ester (8CI) **(3650-04-2)**
Podocarpa-8,11,13-trien-12-ol (8CI) **(15340-76-8)**
Podocarp-8(14)-en-15-oic acid, 13-isopropylidene **(471-77-2)**
Podocarp-8(14)-en-15-oic acid, 13-isopropylidene-, methyl ester (8CI) **(3310-97-2)**
Podocarp-8(14)-en-15-oic acid, 13α-methyl-13-vinyl- (8CI) **(127-27-5)**
Podocarp-8(14)-en-15-oic acid, 13α-methyl-13-vinyl-, methyl ester (8CI) **(3730-56-1)**
Podocarp-7-en-15-oic acid, 13β-methyl-13-vinyl-, (-)- (8CI) **(5835-26-7)**
Podocarp-8(14)-en-15-oic acid, 13β-methyl-13-vinyl- (8CI) **(471-74-9)**
Podocarp-7-en-15-oic acid, 13β-methyl-13-vinyl-, methyl ester (8CI) **(1686-62-0)**
Podophyllinic acid lactone **(518-28-5)**
Podophyllotoxin **(518-28-5)**
Point Two **(7681-49-4)**

Pokilocerin A (1986-70-5)
Polaax (112-92-5)
Polacaritox (116-29-0)
Polamidon (76-99-3)
Polamidone (76-99-3)
Polan Navy Blue E 2R (4368-56-3)
Polana (63428-84-2)
Polar Red G (3567-65-5)
Polar Red G Supra (3567-65-5)
Polar Red RS (6459-94-5)
Polaramine (25523-97-1)
Polaris (2439-99-8)
Polaronil (132-22-9)
Polaronil (German) (113-92-8)
Polaxal Violet 6B (1694-09-3)
Polcominal (50-06-6)
Poleon (389-08-2)
Polfos (13104-21-7)
Polfoschlor (52-68-6)
Policapran (25038-54-4)
Policar MZ (8018-01-7)
Policar S (8018-01-7)
Polidocanol (9002-92-0)
Poliflogil (129-20-4)
Poligeenan (53973-98-1)
Poligeenane (French) (53973-98-1)
Poligeenano (Spanish) (53973-98-1)
Poligeenanum (Latin) (53973-98-1)
Polignate Sodium (8061-51-6)
Poligostyrene (9003-53-6)
Polik (777-11-7)
Polikarbatsin (Russian) (9006-42-2)
Polinalin (58-15-1)
Poliseptil (72-14-0)
Polisin (7287-19-6)
Polisol S-3 (9003-20-7)
Politen (9002-88-4)
Politen I 020 (9002-88-4)
Poliuron (73-48-3)
Polival (148-79-8)
Pollacid (57-09-0)
Polnoks R (26780-96-1)
Polonium (7440-08-6)
Polonium 214 (15735-67-8)
Polonium-210 (13981-52-7)
Polonium-218 (15422-74-9)
Polopiryna (50-78-2)
Poloxal Red 2B (3567-69-9)
Poloxalcol (9003-11-6)
Poloxalene (9003-11-6)
Poloxalene 2930 (9003-11-6)
Poloxalene L 64 (9003-11-6)
Poloxalkol (9003-11-6)
Poloxamer-iodone (26617-87-8)
Poloxamine 304 (11111-34-5)
Poloxamine 504 (11111-34-5)
Poloxamine 701 (11111-34-5)
Poloxamine 702 (11111-34-5)
Poloxamine 704 (11111-34-5)
Poloxamine 707 (11111-34-5)
Poloxamine 901 (11111-34-5)
Poloxamine 904 (11111-34-5)
Poloxamine 908 (11111-34-5)
Poloxamine 1101 (11111-34-5)
Poloxamine 1102 (11111-34-5)
Poloxamine 1104 (11111-34-5)
Poloxamine 1301 (11111-34-5)
Poloxamine 1302 (11111-34-5)
Poloxamine 1304 (11111-34-5)
Poloxamine 1307 (11111-34-5)
Poloxamine 1501 (11111-34-5)
Poloxamine 1502 (11111-34-5)
Poloxamine 1504 (11111-34-5)
Poloxamine 1508 (11111-34-5)
Polphos (13104-21-7)
Polsil 350 (63148-62-9)
Polsil OM 1000 (63148-62-9)
Polvo arsenical (Spanish) (8028-73-7)
Poly (7758-29-4)
Poly-Em 12 (9002-88-4)
Poly-Em 40 (9002-88-4)
Poly-Em 41 (9002-88-4)
Poly G 400 (25322-68-3)
Poly-G Series (25322-68-3)
Poly-G WT 9150 (9003-11-6)
Poly-G WT 90000 (9003-11-6)
Poly-Giron (50-35-1)
Poly-Solv (111-90-0)
Poly-Solv DB (112-34-5)
Poly-Solv DM (111-77-3)
Poly-Solv EB (111-76-2)
Poly-Solv EE (110-80-5)
Poly-Solv EE Acetate (111-15-9)
Poly-Solv EM (109-86-4)
Poly-Solv TB (143-22-6)
Poly-Solv TE (112-50-5)
Poly-Solv TM (112-35-6)
Poly-Solv TPM (10213-77-1)
Poly-Solve MPM (107-98-2)
Poly-Zole AZDN (78-67-1)
Polyacetaldehyde (9002-91-9)
Polyacrylamide (9003-05-8)
Poly (acrylamide-dimethylaminoethyl methacrylate, dimethyl sulfate quat) (26006-22-4)
Polyacrylate (9003-01-4)
Poly(acrylic acid) (9003-01-4)
Polyacrylonitrile (25014-41-9)
Polyacrylonitrile (37243-36-0)
Poly(acrylonitrile), Fibers (25014-41-9)
Polyaethylen (German) (9002-88-4)
Polyaethylenglykole #200 (German) (25322-68-3)
Polyaethylenglykole #238 (German) (25322-68-3)
Polyaethylenglykole #282 (German) (25322-68-3)
Polyaethylenglykole #300 (German) (25322-68-3)
Polyaethylenglykole #400 (German) (25322-68-3)
Polyaethylenglykole #600 (German) (25322-68-3)
Polyaethylenglykole #810 (German) (25322-68-3)
Polyaethylenglykole #1000 (German) (25322-68-3)
Polyaethylenglykole #1250 (German) (25322-68-3)
Polyaethylenglykole #1500 (German) (25322-68-3)
Polyaethylenglykole #1540 (German) (25322-68-3)
Polyaethylenglykole #4000 (German) (25322-68-3)
Polyaethylenglykole #6000 (German) (25322-68-3)
Polyaethylenglykole #10000 (German) (25322-68-3)
Poly(allyl glycidyl ether) (25639-25-2)
Polyaluminium chloride (1327-41-9)
Polyamid (German) (63428-83-1)
Polyamide 6 (25038-54-4)
Polyamide PK 4 (105-60-2)
Polyamide PK 4 (25038-54-4)
Polyamide fibers (63428-84-2)
Polyamine T (25214-70-4)
Poly(ε-aminocaproic acid) (25038-54-4)
Polyarylate R 1 (26637-46-7)

Polyaziridine (9002-98-6)
Polybor-chlorate (9011-70-5)
Polybor (1303-96-4)
Polybor 3 (1333-73-9)
Polybrominated biphenyl (36355-01-8)
Polybrominated biphenyl (67774-32-7)
Polybrominated biphenyl (FF-1) (67774-32-7)
Polybrominated biphenyls (13654-09-6)
Polybrominated biphenyls (59536-65-1)
Polybrominated biphenyls (61288-13-9)
Polybrominated salicylanilide (87-10-5)
Polybromoethylene (25951-54-6)
Polybuden 300H (9003-29-6)
Poly-1,3-butadiene (9003-17-2)
Polybutadiene (9003-17-2)
Polybutadiene-polystyrene copolymer (9003-55-8)
Poly(1,4-butanediol terephthalate) (26062-94-2)
Polybutene (9003-27-4)
Polybutene (9003-29-6)
Polybutene SH 015 (9003-29-6)
Polybutylene (9003-27-4)
Polybutylene (9003-29-6)
Poly(1,4-butylene terephthalate) (26062-94-2)
Poly(butylene terephthalate) (26062-94-2)
Poly-.β.-butyrolactone (36486-76-7)
Poly(ε-caproamide) (25038-54-4)
Polycaproamide (25038-54-4)
Poly(ε-caprolactam) (25038-54-4)
Polycaprolactam (25038-54-4)
Poly(ε-caprolactone) (24980-41-4)
Polycaprolactone 700 (25248-42-4)
Polycarbacin (9006-42-2)
Polycarbacine (9006-42-2)
Polycarbazin (9006-42-2)
Polycarbazine (9006-42-2)
Polycat 8 (98-94-2)
Polychlorcamphene (8001-35-2)
Polychlorinated biphenyl (Aroclor 1016) (12674-11-2)
Polychlorinated biphenyl (Aroclor 1221) (11104-28-2)
Polychlorinated biphenyl (Aroclor 1232) (11141-16-5)
Polychlorinated biphenyl (Aroclor 1242) (53469-21-9)
Polychlorinated biphenyl (Aroclor 1248) (12672-29-6)
Polychlorinated biphenyl (Aroclor 1254) (11097-69-1)
Polychlorinated biphenyl (Aroclor 1260) (11096-82-5)
Polychlorinated biphenyl (Aroclor 1268) (11100-14-4)
Polychlorinated biphenyl (Aroclor 4465) (11120-29-9)
Polychlorinated biphenyl (Kanechlor 300) (37353-63-2)
Polychlorinated biphenyl (Kanechlor 400) (12737-87-0)
Polychlorinated biphenyl (Kanechlor 500) (37317-41-2)
Polychlorinated-biphenyls (1336-36-3)
Polychlorinated biphenyls [UN 2315] (1336-36-3)
Polychlorinated camphenes (8001-35-2)
Polychlorinated terphenyl (61788-33-8)
Polychlorinated-triphenyl (17760-93-9)
Polychlorinated triphenyl (Aroclor 5442) (12642-23-8)
Polychlorinated triphenyl (Aroclor 5460) (11126-42-4)
Polychlorobenzoic acid (12002-27-6)
Polychlorobenzoic acid, dimethylamine salt (1338-32-5)
Polychlorobiphenyl (1336-36-3)
Poly(2-chloro-1,3-butadiene) (9010-98-4)
Poly(2-chlorobutadiene) (9010-98-4)
Polychlorocamphene (8001-35-2)
Polychloro copper phthalocyanine (1328-53-6)
Polychloroprene (9010-98-4)
Polychlorotriphenyl (17760-93-9)
Polycillin (69-53-4)
Polycillin (7177-48-2)
Polycizer 162 (117-84-0)
Polycizer 332 (123-95-5)
Polycizer 532 (119-07-3)
Polycizer 562 (119-07-3)
Polycizer 632 (105-97-5)
Polycizer 962-BPA (119-06-2)
Polycizer DBP (84-74-2)
Polycizer DBS (109-43-3)
Polyclar AT (9003-39-8)
Polyclar H (9003-39-8)
Polyclar L (9003-39-8)
Polyclene (120-36-5)
Polyco (9003-04-7)
Polyco 953 (9003-20-7)
Polyco 2116 (9003-20-7)
Polyco 2134 (9003-20-7)
Polyco 2410 (9003-55-8)
Polyco 2415 (9003-55-8)
Polyco 2611 (9011-06-7)
Polyco 117FR (9003-20-7)
Polyco 220NS (9003-53-6)
Polycor Dark Green S (3626-28-6)
Polycor Red GS (10169-02-5)
Polycor Yellow R (1325-37-7)
Polycron (41198-08-7)
Polycycline (60-54-8)
Polycycline hydrochloride (64-75-5)
Polydesis (9002-89-5)
Poly(diallyl isophthalate) (25035-78-3)
Poly(difluoromethylene), α-fluoro-ω-(2-((2-methyl-1-oxo-2-propenyl)oxy)ethyl)- (9CI) (65530-66-7)
Poly(1,2-dihydro-2,2,4-trimethylquinoline) (26780-96-1)
Poly(3,5-dimethoxy-4-hydroxystyrene) (31872-14-7)
Polydimethylcyclosiloxane (69430-24-6)
Polydimethyl silicone oil (63148-62-9)
Poly(dimethylsiloxane) (63148-62-9)
Polydimethyl-siloxane (9016-00-6)
Polydimethylsiloxane, methyl end-blocked (63148-62-9)
Polydimethyl siloxy cyclics (69430-24-6)
Polydipentene (9003-73-0)
Poly(di-2-propenyl 1,3-benzenedicarboxylate) (25035-78-3)
Polyester TGM 3 (109-16-0)
Poly(estradiol phosphate) (28014-46-2)
Polyethers, epoxy resins (61788-97-4)
Polyethoxylated (C12-C15) linear primary saturated alcohols (68131-39-5)
Polyethoxylated dodecanol (9002-92-0)
Polyethoxylated isooctadecanol (52292-17-8)
Polyethoxypolypropoxyethanol - iodine complex (26617-87-8)
(Polyethyl)benzenes (64742-94-5)
Polyethylene (9002-88-4)
Polyethylene AS (9002-88-4)
Polyethylene 600 dibenzoate (9004-86-8)
Polyethylene, chlorosulfonated (68037-39-8)
Polyethylene glycol 20M 25322-68-3)
Polyethylene glycol #200 25322-68-3)
Polyethylene glycol #238 25322-68-3)
Polyethylene glycol #282 25322-68-3)
Polyethylene glycol #300 25322-68-3)
Polyethylene glycol #350 25322-68-3)
Polyethylene glycol #400 25322-68-3)
Polyethylene glycol 425 25322-68-3)
Polyethylene glycol #600 25322-68-3)
Polyethylene glycol #810 25322-68-3)
Polyethylene glycol #1000 25322-68-3)
Polyethylene glycol 1200 25322-68-3)
Polyethylene glycol #1250 25322-68-3)
Polyethylene glycol 1500 25322-68-3)

Name	CAS
Polyethylene glycol #1540	25322-68-3)
Polyethylene glycol 2000	25322-68-3)
Polyethylene glycol 4000	25322-68-3)
Polyethylene glycol 6000	25322-68-3)
Polyethylene glycol #10000	25322-68-3)
Polyethylene glycol 2000000	25322-68-3)
Polyethylene glycol 4000000	25322-68-3)
Polyethylene glycol 5000000	25322-68-3)
Polyethylene glycol, (C12-C15) alkyl ethers	(68131-39-5)
Polyethylene glycol E 600	(25322-68-3)
Polyethylene glycol butyl ether	(9004-77-7)
Polyethylene glycol (5) cetyl ether	(9004-95-9)
Polyethylene glycol (15) cetyl ether	(9004-95-9)
Polyethylene glycol (16) cetyl ether	(9004-95-9)
Polyethylene glycol (24) cetyl ether	(9004-95-9)
Polyethylene glycol (25) cetyl ether	(9004-95-9)
Polyethylene glycol (30) cetyl ether	(9004-95-9)
Polyethylene glycol (45) cetyl ether	(9004-95-9)
Polyethylene glycol 100 cetyl ether	(9004-95-9)
Polyethylene glycol 200 cetyl ether	(9004-95-9)
Polyethylene glycol 300 cetyl ether	(9004-95-9)
Polyethylene glycol 500 cetyl ether	(9004-95-9)
Polyethylene glycol 600 cetyl ether	(9004-95-9)
Polyethylene glycol 1000 cetyl ether	(9004-95-9)
Polyethylene glycol (3) coconut amine	(61791-14-8)
Polyethylene glycol (5) coconut amine	(61791-14-8)
Polyethylene glycol (15) coconut amine	(61791-14-8)
Polyethylene glycol 100 coconut amine	(61791-14-8)
Polyethylene glycol 500 coconut amine	(61791-14-8)
Polyethylene glycol 200 decyl ether phosphate	(9004-80-2)
Polyethylene glycol dibenzoate	(9004-86-8)
Polyethylene glycol 220 dibenzoate	(9004-86-8)
Polyethylene glycol 200 di(2-ethylhexoate)	(18268-70-7)
Poly(ethylene glycol dimethacrylate)	(25721-76-0)
Polyethylene glycol dimethacrylate	(25852-47-5)
Polyethylene glycol dimethyl ether	(24991-55-7)
Polyethylene glycol (5) dinonyl phenyl ether	(9014-93-1)
Polyethylene glycol (49) dinonyl phenyl ether	(9014-93-1)
Polyethylene glycol (100) dinonyl phenyl ether	(9014-93-1)
Polyethylene glycol (150) dinonyl phenyl ether	(9014-93-1)
Polyethylene glycol 500 dinonyl phenyl ether	(9014-93-1)
Polyethylene glycol dioleate	(9005-07-6)
Polyethylene glycol 200 dioleate	(9005-07-6)
Polyethylene glycol 300 dioleate	(9005-07-6)
Polyethylene glycol 400 dioleate	(9005-07-6)
Polyethylene glycol 500 dioleate	(9005-07-6)
Polyethylene glycol 600 dioleate	(9005-07-6)
Polyethylene glycol 1000 dioleate	(9005-07-6)
Polyethylene glycol 1500 dioleate	(9005-07-6)
Polyethylene glycol 1540 dioleate	(9005-07-6)
Polyethylene glycol 4000 dioleate	(9005-07-6)
Polyethylene glycol 6000 dioleate	(9005-07-6)
Polyethylene glycol distearate	(9005-08-7)
Polyethylene glycol 3 distearate	(9005-08-7)
Polyethylene glycol 100 distearate	(109-30-8)
Polyethylene glycol 175 distearate	(9005-08-7)
Polyethylene glycol 200 distearate	(9005-08-7)
Polyethylene glycol 300 distearate	(9005-08-7)
Polyethylene glycol 400 distearate	(9005-08-7)
Polyethylene glycol 450 distearate	(9005-08-7)
Polyethylene glycol 600 distearate	(9005-08-7)
Polyethylene glycol 1000 distearate	(9005-08-7)
Polyethylene glycol distearate #1000	(9005-08-7)
Polyethylene glycol 1540 distearate	(9005-08-7)
Polyethylene glycol 4000 distearate	(9005-08-7)
Polyethylene glycol 6000 distearate	(9005-08-7)
Polyethylene glycol dodecyl ether	(9002-92-0)
Polyethylene glycol, dodecyl, tetradecyl, hexadecyl ether	(68551-12-2)
Polyethylene glycol (26) glyceryl ether	(31694-55-0)
Polyethylene glycol 600 glyceryl ether	(31694-55-0)
Polyethylene glycol (3) isostearyl ether	(52292-17-8)
Polyethylene glycol (22) isostearyl ether	(52292-17-8)
Polyethylene glycol (50) isostearyl ether	(52292-17-8)
Polyethylene glycol 100 isostearyl ether	(52292-17-8)
Polyethylene glycol 500 isostearyl ether	(52292-17-8)
Polyethylene glycol 600 isostearyl ether	(52292-17-8)
Polyethylene glycol 1000 isostearyl ether	(52292-17-8)
Polyethylene glycol, isotridecyl ether	(9043-30-5)
Polyethylene glycol (5) lanolin	(61790-81-6)
Polyethylene glycol (24) lanolin	(61790-81-6)
Polyethylene glycol-27 lanolin	(61790-81-6)
Polyethylene glycol (30) lanolin	(61790-81-6)
Polyethylene glycol-40 lanolin	(61790-81-6)
Polyethylene glycol (50) lanolin	(61790-81-6)
Polyethylene glycol-75 lanolin	(61790-81-6)
Polyethylene glycol (85) lanolin	(61790-81-6)
Polyethylene glycol (60) lanolin	(61790-81-6)
Polyethylene glycol (100) lanolin	(61790-81-6)
Polyethylene glycol 1000 lanolin	(61790-81-6)
Polyethylene glycol lauryl ether	(9002-92-0)
Polyethylene glycol (5) lauryl ether	(3055-95-6)
Polyethylene glycol (7) lauryl ether	(3055-97-8)
Polyethylene glycol 100 lauryl ether	(3055-93-4)
Polyethylene glycol 200 lauryl ether	(5274-68-0)
Polyethylene glycol, linear (C12-C15)alkyl alcohols ether	(68131-39-5)
Polyethylene glycol methyl ether	(9004-74-4)
Polyethylene glycol, monobutyl ester	(9004-77-7)
Polyethylene glycol monocaprylate	(42131-42-0)
Polyethylene glycol monocetyl ether	(9004-95-9)
Polyethylene glycol monoester of tall oil	(61791-00-2)
Polyethylene glycol, monoester with tall oil acids	(61791-00-2)
Polyethylene glycol monoether with p-tert-octylphenyl	(9002-93-1)
Polyethylene glycol mono(2-ethylhexyl) ether	(26468-86-0)
Polyethylene glycol 100 monolaurate	(141-20-8)
Polyethylene glycol monoleyl ether	(9004-98-2)
Polyethylene glycol monomethyl ether	(9004-74-4)
Polyethylene glycol mono(4-tert-nonylphenyl)ether	(53496-16-5)
Polyethylene glycol mono(p-tert-nonylphenyl) ether	(53496-16-5)
Polyethylene glycol mono(4-octylphenyl) ether	(9002-93-1)
Polyethylene glycol mono(4-tert-octylphenyl) ether	(9002-93-1)
Polyethylene glycol mono(octylphenyl) ether	(9036-19-5)
Polyethylene glycol mono(p-tert-octylphenyl) ether	(9002-93-1)
Polyethylene glycol monooleate	(9004-96-0)
Polyethylene glycol 100 monooleate	(106-12-7)
Polyethylene glycol (18) monopalmitate	(9004-94-8)
Polyethylene glycol 300 monopalmitate	(9004-94-8)
Polyethylene glycol 1000 monopalmitate	(9004-94-8)
Polyethylene glycol (5) monostearate	(9004-99-3)
Polyethylene glycol (7) monostearate	(9004-99-3)
Polyethylene glycol (14) monostearate	(9004-99-3)
Polyethylene glycol (18) monostearate	(9004-99-3)
Polyethylene glycol (25) monostearate	(9004-99-3)
Polyethylene glycol (30) monostearate	(9004-99-3)
Polyethylene glycol (35) monostearate	(9004-99-3)
Polyethylene glycol (45) monostearate	(9004-99-3)
Polyethylene glycol (50) monostearate	(9004-99-3)
Polyethylene glycol (90) monostearate	(9004-99-3)
Polyethylene glycol (100) monostearate	(9004-99-3)
Polyethylene glycol (120) monostearate	(9004-99-3)
Polyethylene glycol 300 monostearate	(9004-99-3)
Polyethylene glycol 400 monostearate	(9004-99-3)
Polyethylene glycol 450 monostearate	(9004-99-3)
Polyethylene glycol 500 monostearate	(9004-99-3)
Polyethylene glycol 600 monostearate	(9004-99-3)

Polyethylene glycol 1000 monostearate (9004-99-3)
Polyethylene glycol 1500 monostearate (9004-99-3)
Polyethylene glycol 1540 monostearate (9004-99-3)
Polyethylene glycol 1800 monostearate (9004-99-3)
Polyethylene glycol 2000 monostearate (9004-99-3)
Polyethylene glycol 4000 monostearate (9004-99-3)
Polyethylene glycol 6000 monostearate (9004-99-3)
Polyethylene glycol monostearate #200 (9004-99-3)
Polyethylene glycol monostearate #400 (9004-99-3)
Polyethylene glycol monostearate #1000 (9004-99-3)
Polyethylene glycol monostearate #6000 (9004-99-3)
Polyethylene-glycol-monostearate (9004-99-3)
Polyethylene glycol monotallate (61791-00-2)
Polyethylene glycol (16) monotallate (61791-00-2)
Polyethylene glycol 200 monotallate (61791-00-2)
Polyethylene glycol 400 monotallate (61791-00-2)
Polyethylene glycol 500 monotallate (61791-00-2)
Polyethylene glycol 600 monotallate (61791-00-2)
Polyethylene glycol 1000 monotallate (61791-00-2)
Polyethylene glycol mono(p-(1,1,3,3-tetramethylbutyl)phenyl) ether (9002-93-1)
Polyethylene glycol 200 nonyl phenyl ether (7311-27-5)
Polyethylene glycol 450 nonyl phenyl ether (9016-45-9)
Polyethylene glycol octylphenol ether (9002-93-1)
Polyethylene glycol octylphenyl ether (9036-19-5)
Polyethylene glycol 450 octyl phenyl ether (9002-93-1)
Polyethylene glycol p-octylphenyl ether (9002-93-1)
Polyethylene glycol p-tert-octylphenyl ether (9002-93-1)
Polyethylene glycol oleate (9004-96-0)
Polyethylene glycol oleyl ether (9004-98-2)
Polyethylene glycol (3) oleyl ether (9004-98-2)
Polyethylene glycol (5) oleyl ether (9004-98-2)
Polyethylene glycol (7) oleyl ether (9004-98-2)
Polyethylene glycol (15) oleyl ether (9004-98-2)
Polyethylene glycol (16) oleyl ether (9004-98-2)
Polyethylene glycol (23) oleyl ether (9004-98-2)
Polyethylene glycol (25) oleyl ether (9004-98-2)
Polyethylene glycol (44) oleyl ether (9004-98-2)
Polyethylene glycol (50) oleyl ether (9004-98-2)
Polyethylene glycol 100 oleyl ether (9004-98-2)
Polyethylene glycol 200 oleyl ether (9004-98-2)
Polyethylene glycol (300) oleyl ether (9004-98-2)
Polyethylene glycol 400 oleyl ether (9004-98-2)
Polyethylene glycol 450 oleyl ether (9004-98-2)
Polyethylene glycol 500 oleyl ether (9004-98-2)
Polyethylene glycol 600 oleyl ether (9004-98-2)
Polyethylene glycol 1000 oleyl ether (9004-98-2)
Polyethyleneglycol palmitate (9004-94-8)
Polyethylene glycol phenyl ether (9004-78-8)
Polyethylene glycol stearate (9004-99-3)
Polyethylene glycol (7) stearyl ether (9005-00-9)
Polyethylene glycol (11) stearyl ether (9005-00-9)
Polyethylene glycol (13) stearyl ether (9005-00-9)
Polyethylene glycol (15) stearyl ether (9005-00-9)
Polyethylene glycol (16) stearyl ether (9005-00-9)
Polyethylene glycol (25) stearyl ether (9005-00-9)
Polyethylene glycol (27) stearyl ether (9005-00-9)
Polyethylene glycol (30) stearyl ether (9005-00-9)
Polyethylene glycol (50) stearyl ether (9005-00-9)
Polyethylene glycol 100 stearyl ether (9005-00-9)
Polyethylene glycol (100) stearyl ether (9005-00-9)
Polyethylene glycol 200 stearyl ether (9005-00-9)
Polyethylene glycol 500 stearyl ether (9005-00-9)
Polyethylene glycol 1000 stearyl ether (9005-00-9)
Polyethylene glycol 2000 stearyl ether (9005-00-9)
Polyethylene glycol sulfate monododecyl ether sodium salt (9004-82-4)
Polyethylene glycol tallate (61791-00-2)
Polyethylene glycol, tall oil ester (61791-00-2)
Polyethylene glycol, tall oil fatty acid polymer (61791-00-2)
Polyethylene glycol p-1,1,3,3-tetramethylbutylphenyl ether (9002-93-1)
Polyethylene glycol, tridecyl ether, phosphate, potassium salt (68186-36-7)
Polyethylene glycol 300 tridecyl ether phosphate (9046-01-9)
Polyethyleneglycols monostearate (9004-99-3)
Polyethyleneimine (9002-98-6)
Poly(ethylene oxide) (25322-68-3)
Polyethylene oxide, dehydroabietylamine polymer (51344-62-8)
Poly(ethylene oxide) dodecyl ether (9002-92-0)
Poly(ethylene oxide) ether with (C12-C15)linear primary alcohols (68131-39-5)
Polyethylene oxide monooleate (9004-96-0)
Poly(ethylene oxide)octylphenyl ether (9036-19-5)
Poly(ethylene oxide) oleate (9004-96-0)
Polyethylene oxide-polypropylene oxide (9003-11-6)
Polyethylene oxide-polypropylene oxide copolymer (9003-11-6)
N-Polyethylenepolyamine-N-oleylamine hydrochloride (67905-86-6)
Polyethylene-polypropylene glycol (9003-11-6)
Polyethylene Resins (9002-88-4)
Polyethylene terephthalate (25038-59-9)
Polyethylene terephthalate film (25038-59-9)
Polyethylenimine (9002-98-6)
Polyethylenimine 7 (9002-98-6)
Polyethylenimine 15 (9002-98-6)
Polyethylenimine 30 (9002-98-6)
Polyethylenimine 45 (9002-98-6)
Polyethylenimine 1000 (9002-98-6)
Polyethylenimine 1500 (9002-98-6)
Polyethylenimine 2500 (9002-98-6)
Polyethylenimine (10,000) (9002-98-6)
Polyethylenimine (20,000) (9002-98-6)
Polyethylenimine (35,000) (9002-98-6)
Polyethylenimine (40,000) (9002-98-6)
Polyfer (9004-66-4)
Polyfibron 120 (9004-32-4)
Polyflex (9003-53-6)
Polyfoam Plastic Sponge (9009-54-5)
Polyfoam Sponge (9009-54-5)
Polyfon (8061-51-6)
Polyfon F (8061-51-6)
Polyfon H (8061-51-6)
Polyfon HUN (8061-51-6)
Polyfon O (8061-51-6)
Polyfon T (8061-51-6)
Polyfox P 20 (9003-20-7)
Polyfox PO (9003-20-7)
Polygard (58968-53-9)
Polygeenan (53973-98-1)
Polyglycerin (25618-55-7)
Polyglycerol monooleate (9007-48-1)
Polyglycerol oleate (9007-48-1)
Polyglyceryl oleate (9007-48-1)
Polyglycol 1000 (25322-68-3)
Polyglycol 4000 (25322-68-3)
Polyglycol E-300 (25322-68-3)
Polyglycol E-1000 (25322-68-3)
Polyglycol E-4000 (25322-68-3)
Polyglycol E-4000 USP (25322-68-3)
Polyglycol P-425 (25322-68-3)
Polyglycol P-1200 (25322-68-3)
Polyglycol P-2000 (25322-69-4)
Polyglycolic acid (26124-68-5)
Polyglycol-laurate (9004-81-3)
Polyglycollic acid (26124-68-5)
Polyglycol monooleate (9004-96-0)
Polyglycol oleate (9004-96-0)
Polyglycol-oleate (9004-96-0)

Name	CAS No.
Poly(glycyl-.ε.-aminocaproic acid)	**(39610-34-9)**
Polygon	**(7758-29-4)**
Polygorskite	**(12174-11-7)**
Polygripan	**(50-35-1)**
Poly(3-hydroxybutyrate)	**(26063-00-3)**
Poly-β-hydroxybutyrate	**(26063-00-3)**
Poly-β-hydroxybutyric acid	**(26063-00-3)**
Poly(4-hydroxy-3,5-dimethoxystyrene)	**(31872-14-7)**
Poly(4-hydroxy-3-methoxystyrene)	**(31853-85-7)**
Poly(N-(2-hydroxypropyl)methacrylamide)	**(40704-75-4)**
Poly(4-hydroxystyrene)	**(24979-70-2)**
Poly(p-hydroxystyrene)	**(24979-70-2)**
Poly(iminocarbonyliminohexamethylene)	**(25035-67-0)**
Poly(iminocarbonylimino-1,6-hexanediyl) (9CI)	**(25035-67-0)**
Poly(iminocarbonylpentamethylene)	**(25038-54-4)**
Poly(imino(1,6-dioxo-1,6-hexanediyl)imino-1,6-hexanediyl) (9CI)	**(32131-17-2)**
Poly(imino(1-oxo-1,2-ethanediyl)imino(1-oxo-1,6-hexanediyl)) (9CI)	**(39610-34-9)**
Poly(imino(1-oxo-1,6-hexanediyl)) (9CI)	**(25038-54-4)**
Polyiodide	**(20461-54-5)**
Polyisobutene	**(9003-27-4)**
Polyisobutene	**(9003-29-6)**
Polyisobutenyl tetraethylenepentamine succinimide	**(67762-72-5)**
Polyisobutylene	**(9003-27-4)**
Polyisobutylene	**(9003-29-6)**
Polyisobutylene PSG	**(9003-27-4)**
Polyisoprene	**(9003-31-0)**
cis-1,4-Polyisoprene	**(9003-31-0)**
trans-1,4-Polyisoprene	**(9003-31-0)**
trans-Polyisoprene	**(9003-31-0)**
Polylimonene	**(9003-73-0)**
Polylin No. 515	**(60-33-3)**
Polylon 13-5	**(9003-11-6)**
Polymannuronic acid	**(9005-32-7)**
Polymarcin	**(9006-42-2)**
Polymarcine	**(9006-42-2)**
Polymarsin	**(9006-42-2)**
Polymarzin	**(9006-42-2)**
Polymarzine	**(9006-42-2)**
Polymat	**(9006-42-2)**
Polymerization Feed	**(68476-54-0)**
Polymerization naphtha	**(64741-72-6)**
Polymerized dipentene	**(9003-73-0)**
Polymer of propylene glycol, maleic anhydride, phthalic anhydride	**(25037-66-5)**
Polymethacrylic acid	**(25087-26-7)**
Poly(methibis(hydroxymethyl)ureylene)amer	**(9011-05-6)**
Poly(3-methoxy-4-hydroxystyrene)	**(31853-85-7)**
Poly(2-methyl-1,3-butadiene)	**(9003-31-0)**
Poly-1-methylbutenylene	**(9003-31-0)**
Polymethylenepolyphenyl-isocyanate	**(9016-87-9)**
Polymethylhydrogensiloxane, trimethylsiloxy end blocked	**(63148-57-2)**
Poly(methylhydrosiloxane)	**(63148-57-2)**
Polymethylmethacrylate	**(9011-14-7)**
Poly(2-methylpropene)	**(9003-27-4)**
Poly(.β.-methyl-.β.-propiolactone)	**(36486-76-7)**
Poly(.β.-methylpropiolactone)	**(36486-76-7)**
Polymin FL	**(9002-98-6)**
Polymin G 35	**(9002-98-6)**
Polymin HS	**(9002-98-6)**
Polymin Waterfree	**(9002-98-6)**
Polymine D	**(121-54-0)**
Polymist A12	**(9002-88-4)**
Polymo Green FBH	**(1328-53-6)**
Polymo Green FGH	**(1328-53-6)**
Polymo Orange GR	**(3520-72-7)**
Polymo Red FGN	**(2425-85-6)**
Polymon Blue G	**(574-93-6)**
Polymon Blue LBS	**(147-14-8)**
Polymon Blue 3R	**(81-77-6)**
Polymon Green 6G	**(1328-53-6)**
Polymon Green G	**(1328-53-6)**
Polymon Green GN	**(1328-53-6)**
Polymone	**(120-36-5)**
Polymul CS 81	**(9002-88-4)**
Polymyxin E	**(1066-17-7)**
Polynoxylin	**(9011-05-6)**
Polyoestradiol phosphate	**(28014-46-2)**
Polyoil 110	**(9003-17-2)**
Polyoil 130	**(9003-17-2)**
Poly(α-olefins)	**(68527-08-2)**
Polyox	**(25322-68-3)**
Polyoxiethylene (6) alkyl (13) ether	**(24938-91-8)**
Poly(1-(2-oxo-1-pyrrolidinyl)ethylene)	**(9003-39-8)**
Poly(1-(2-oxo-1-pyrrolidinyl)ethylene)iodine complex	**(25655-41-8)**
Poly(oxy-1,4-butanediyl), α-hydro-ω-hydroxy- (9CI)	**(25190-06-1)**
Poly(oxy(dibromophenylene)) (9CI)	**(74082-93-2)**
Poly(oxy(dimethylsilylene))	**(63148-62-9)**
Poly(oxy(dimethylsilylene))	**(9016-00-6)**
Polyoxy(dimethylsilylene), α-(trimethylsilyl)-ω-hydroxy	**(63148-62-9)**
Poly(oxy(dimethylsilylene)), α-(trimethylsilyl)-ω-methyl-	**(9006-65-9)**
Poly(oxy-1,2-ethanediyl), α-(C12-C18) alkyl-ω-hydroxy-	**(68213-23-0)**
Poly(oxy-1,2-ethanediyl), α-benzoyl-ω-(benzoyloxy)- (9CI)	**(9004-86-8)**
Poly(oxy-1,2-ethanediyl), α-(2-(bis(2-aminoethyl)methylammonio)-ethyl)-ω-hydroxy-, N,N'-bis(hydrogenated tallow acyl) derivs., Me sulfates (Salts)	**(68389-89-9)**
Poly(oxy-1,2-ethanediyl), α-(2-(bis(2-aminoethyl)methyl-ammonio)ethyl)-ω-hydroxy-, N,N'-ditallow acyl derivs., Me sulfates (Salts)	**(68410-69-5)**
Poly(oxy-1,2-ethanediyl), α-(bis(octyloxy)phosphinyl)-ω-hydroxy- (9CI)	**(57344-02-2)**
Poly(oxy-1,2-ethanediyl), α-butyl-ω-hydroxy- (9CI)	**(9004-77-7)**
Poly(oxy-1,2-ethanediyl), α,α'-2-butyne-1,4-diylbis(ω-hydroxy- (9CI)	**(32167-31-0)**
Poly(oxy-1,2-ethanediyl), .α.-(2-(decyl((4-(1,1-dimethylethyl)-phenyl)sulfonyl)amino)ethyl)-.ω.-hydroxy- (9CI)	**(89697-89-2)**
Poly(oxy-1,2-ethanediyl), α-decyl-ω-hydroxy- (9CI)	**(26183-52-8)**
Poly(oxy-1,2-ethanediyl), .α.-(2-(((4-(1,1-dimethyethyl)-phenyl)sulfonyl)hexylamino)ethyl)-.ω.-hydroxy- (9CI)	**(89697-80-3)**
Poly(oxy-1,2-ethanediyl), .α.-(2-(((4-(1,1-dimethylethyl)phenyl)-sulfonyl)octylamino)ethyl)-.ω.-hydroxy- (9CI)	**(89697-83-6)**
Poly(oxy-1,2-ethanediyl), α-(dinonylphenyl)-ω-hydroxy- (9CI)	**(9014-93-1)**
Poly(oxy-1,2-ethanediyl), α-(dinonylphenyl)-ω-hydroxy-, phosphate (9CI)	**(39464-64-7)**
Poly(oxy-1,2-ethanediyl), α-(2-(dodecylamino)ethyl)-ω-hydroxy- (9CI)	**(25190-01-6)**
Poly(oxy-1,2-ethanediyl), α-dodecyl-ω-hydroxy-	**(9002-92-0)**
Poly(oxy-1,2-ethanediyl), α-(4-dodecylphenyl)-ω-hydroxy- (9CI)	**(26401-47-8)**
Poly(oxy-1,2-ethanediyl), .α.-(4-tert-dodecylphenyl)-.ω.-hydroxy- (9CI)	**(67881-24-7)**
Poly(oxy-1,2-ethanediyl), α-(2-(tert-dodecylthio)ethyl)-ω-hydroxy- (9CI)	**(9004-83-5)**
Poly(oxy-1,2-ethanediyl), α-(2-ethylhexyl)-ω-hydroxy- (9CI)	**(26468-86-0)**
Poly(oxy-1,2-ethanediyl), α-hexadecyl-ω-hydroxy- (9CI)	**(9004-95-9)**
Poly(oxy-1,2-ethanediyl), α-hexyl-ω-hydroxy- (9CI)	**(31726-34-8)**
Poly(oxy-1,2-ethanediyl), α-hydro-ω-hydroxy-	**(25322-68-3)**
Poly(oxy-1,2-ethanediyl), α-hydro-ω-hydroxy- (9CI)	**(25322-68-3)**
Poly(oxy-1,2-ethanediyl), α-hydro-ω-(oxiranylmethoxy)-, ether with 2-ethyl-2-(hydroxymethyl)-1,3-propanediol (3:1) (9CI)	**(52495-71-3)**
Poly(oxy-1,2-ethanediyl), α-isooctadecyl-ω-hydroxy- (9CI)	**(52292-17-8)**
Poly(oxy-1,2-ethanediyl), α-isotridecyl-ω-hydroxy- (9CI)	**(9043-30-5)**
Poly(oxy-1,2-ethanediyl), α-methyl-ω-methoxy- (9CI)	**(24991-55-7)**

Poly(oxy-1,2-ethanediyl), .α.-methyl-.ω.-hydroxy- (9CI) **(9004-74-4)**
Poly(oxy-1,2-ethanediyl), α-(2-methyl-1-oxo-2-propenyl)-ω-((2-methyl-1-oxo-2-propenyl)oxy)- (9CI) **(25852-47-5)**
Poly(oxy-1,2-ethanediyl), α,α',α''-(nitrilotri-2,1-ethanediyl)tris-(ω-hydroxy- (9CI) **(36936-60-4)**
Poly(oxy-1,2-ethanediyl), α-(nonylphenyl)-ω-hydroxy-, phosphate (9CI) **(51811-79-1)**
Poly(oxy-1,2-ethanediyl), α-(nonylphenyl)-, ω-hydroxy, phosphate, sodium salt **(37340-60-6)**
Poly(oxy-1,2-ethanediyl), α-(nonylphenyl)-ω-hydroxy, phosphate, sodium salt **(37340-60-6)**
Poly(oxy-1,2-ethanediyl), α-(nonylphenyl)-ω-hydroxy-, phosphate, sodium salt **(37340-60-6)**
Poly(oxy-1,2-ethanediyl), α-(nonylphenyl)-ω-hydroxy-, phosphate, sodium salt (9CI) **(37340-60-6)**
Poly(oxy-1,2-ethanediyl)-α-(nonylphenyl)-ω-hydroxy-, phosphate, sodium salt **(37340-60-6)**
Poly(oxy-1,2-ethanediyl)-α-(nonylphenyl)-ω-hydroxy-phosphate, sodium salt **(37340-60-6)**
Poly(oxy-1,2-ethanediyl), .α.-(4-tert-nonylphenyl)-.ω.-hydroxy- (9CI) **(53496-16-5)**
Poly(oxy-1,2-ethanediyl), .α.-(tert-nonylphenyl)-.ω.-hydroxy- (9CI) **(37281-58-6)**
Polyoxy-1,2-ethanediyl, α-((Z)-9-octadecenyl-ω-hydroxy- **(9004-98-2)**
Poly(oxy-1,2-ethanediyl), α-9-octadecenyl-ω-hydroxy-, (Z)- **(9004-98-2)**
Poly(oxy-1,2-ethanediyl), α-9-octadecenyl-ω-hydroxy-, (Z)- (9CI) **(9004-98-2)**
Poly(oxy-1,2-ethanediyl), α,α'-((9-octadecenylimino)di-2,1-ethanediylbis(ω-hydroxy-, (Z)- (9CI) **(26635-93-8)**
Poly(oxy-1,2-ethanediyl), α-octadecyl-ω-hydroxy- (9CI) **(9005-00-9)**
Poly(oxy-1,2-ethanediyl), α,α'-((octadecylimino)di-2,1-ethanediyl)bis(ω-hydroxy- (9CI) **(26635-92-7)**
Poly(oxy-1,2-ethanediyl), α,α'-((((1,2,3,4,4a,9,10,10a-octahydro-1,4a-dimethyl-7-(1-methylethyl)-1-phenanthrenyl)-methyl)imino)di-2,1-ethanediyl)bis(ω-hydroxy-, (1R-(1α,4aβ,10aα))- (9CI) **(51344-62-8)**
Poly(oxy-1,2-ethanediyl), α,α'-((octyloxy)phosphinylidene)-bis(ω-hydroxy- (9CI) **(57344-01-1)**
Poly(oxy-1,2-ethanediyl), α,α'-(((1-oxododecyl)imino)di-2,1-ethanediyl)bis(ω-hydroxy- (9CI) **(31587-78-7)**
Poly(oxy-1,2-ethanediyl), α-(1-oxohexadecyl)-ω-hydroxy- (9CI) **(9004-94-8)**
Poly(oxy-1,2-ethanediyl), α-(1-oxononyl)-ω-hydroxy- (9CI) **(31621-91-7)**
Poly(oxy-1,2-ethanediyl), α-(2-((1-oxo-9-octadecenyl)amino)-ethyl)-ω-hydroxy-, (Z)- (9CI) **(26027-37-2)**
Poly(oxy-1,2-ethanediyl), α-(1-oxo-9-octadecenyl)-ω-hydroxy-, (Z)- (9CI) **(9004-96-0)**
Poly(oxy-1,2-ethanediyl), α-(1-oxo-9-octadecenyl)-ω-methoxy-, (Z)- (9CI) **(34397-99-4)**
Poly(oxy-1,2-ethanediyl), α-(1-oxo-9-octadecenyl)-ω-((1-oxo-9-octadecenyl)oxy)-, (Z,Z)- (9CI) **(9005-07-6)**
Poly(oxy-1,2-ethanediyl), α-(1-oxooctadecyl)-ω-hydroxy- (9CI) **(9004-99-3)**
Poly(oxy-1,2-ethanediyl), α-1-(oxooctadecyl)-ω-hydroxy- (9CI) **(9004-99-3)**
Poly(oxy-1,2-ethanediyl), α-(1-oxooctadecyl)-ω-((1-oxooctadecyl)oxy)- (9CI) **(9005-08-7)**
Poly(oxy-1,2-ethanediyl), α-(1-oxooctyl)-ω-hydroxy- (9CI) **(42131-42-0)**
Poly(oxy-1,2-ethanediyloxycarbonyl-1,4-phenylenecarbonyl) (9CI) **(25038-59-9)**
Poly(oxy-1,2-ethanediyl), α-phenyl-ω-hydroxy- (9CI) **(9004-78-8)**
Poly(oxy-1,2-ethanediyl), α-phosphono-ω-(decyloxy)- (9CI) **(9004-80-2)**
Poly(oxy-1,2-ethanediyl), α,α',α''-1,2,3-propanetriyltris-(ω-hydroxy- (9CI) **(31694-55-0)**
Poly(oxy-1,2-ethanediyl), α-sulfo-ω-(dodecyloxy)-, sodium salt (9CI) **(9004-82-4)**
Poly(oxy-1,2-ethanediyl), .α.-sulfo-.ω.-(2-(hexadecyldimethyl-ammonio)ethoxy)-, hydroxide, inner salt (9CI) **(73131-17-6)**
Poly(oxy-1,2-ethanediyl), α-sulfo-ω-hydroxy-, C10-12-alkyl ethers, ammonium salts **(68890-88-0)**
Poly(oxy-1,2-ethanediyl), α-sulfo-ω-hydroxy-, C10-16-alkyl ethers, ammonium salts **(67762-19-0)**
Poly(oxy-1,2-ethanediyl), α-sulfo-ω-hydroxy-, C12-18-alkyl ethers, ammonium salts **(68610-22-0)**
Poly(oxy-1,2-ethanediyl), α-sulfo-ω-hydroxy-, C10-16-alkyl ethers, magnesium salts **(67762-21-4)**
Poly(oxy-1,2-ethanediyl), α-sulfo-ω-hydroxy-, C10-16-alkyl ethers, sodium salts **(68585-34-2)**
Poly(oxy-1,2-ethanediyl), α-sulfo-ω-hydroxy-, C12-14-alkyl ethers, sodium salts **(68891-38-3)**
Poly(oxy-1,2-ethanediyl), α-sulfo-ω-(nonylphenoxy)- (9CI) **(9081-17-8)**
Poly(oxy-1,2-ethanediyl), α-sulfo-ω-(nonylphenoxy)-, ammonium salt **(9051-57-4)**
Poly(oxy-1,2-ethanediyl), α-sulfo-ω-(4-nonylphenoxy)-, ammonium salt (9CI) **(31691-97-1)**
Poly(oxy-1,2-ethanediyl), α-sulfo-ω-(nonylphenoxy)-, sodium salt **(9014-90-8)**
Poly(oxy-1,2-ethanediyl), α-sulfo-ω-(octylphenoxy)-, branched, sodium salt **(69011-84-3)**
Poly(oxy-1,2-ethanediyl), α-sulfo-ω-(tetradecyloxy)-, ammonium salt (9CI) **(27731-61-9)**
Poly(oxy-1,2-ethanediyl), α-((1,1,3,3-tetramethylbutyl)-phenyl)-ω-hydroxy- (9CI) **(9036-19-5)**
Poly(oxy-1,2-ethanediyl), α-(4-(1,1,3,3-tetramethylbutyl)-phenyl)-ω-hydroxy- (9CI) **(9002-93-1)**
Poly(oxy-1,2-ethanediyl), α-((1,1,3,3-tetramethylbutyl) phenyl)-ω-hydroxy-, phosphate (9CI) **(52623-95-7)**
Poly(oxy-1,2-ethanediyl), α-tridecyl-ω-hydroxy-, phosphate (9CI) **(9046-01-9)**
Poly(oxy-1,2-ethanediyl)-α-tridecyl-ω-hydroxy-, phosphated, potassium salt **(68186-36-7)**
Poly(oxy-1,2-ethanediyl), α-tridecyl-ω-hydroxy-, phosphate, potassium salt (9CI) **(68186-36-7)**
Polyoxyethylated cetyl alcohol **(9004-95-9)**
Polyoxyethylated stearyl alcohol **(9005-00-9)**
Polyoxyethylated-vegetable-oil **(9004-98-2)**
N-Polyoxyethylated-N-octadecylamine **(26635-92-7)**
N-Polyoxyethylated-N-oleylamine hydrochloride **(26635-93-8)**
Polyoxyethylenated poly(oxypropylene) **(9003-11-6)**
Polyoxyethylene (75) **(25322-68-3)**
Polyoxyethylene 1500 **(25322-68-3)**
Polyoxyethylene 2000000 **(25322-68-3)**
Polyoxyethylene 5000000 **(25322-68-3)**
Polyoxyethylene (2) cetyl ether **(9004-95-9)**
Polyoxyethylene (4) cetyl ether **(9004-95-9)**
Polyoxyethylene (5) cetyl ether **(9004-95-9)**
Polyoxyethylene (6) cetyl ether **(9004-95-9)**
Polyoxyethylene (20) cetyl ether **(9004-95-9)**
Polyoxyethylene (24) cetyl ether **(9004-95-9)**
Polyoxyethylene (25) cetyl ether **(9004-95-9)**
Polyoxyethylene (30) cetyl ether **(9004-95-9)**
Polyoxyethylene (45) cetyl ether **(9004-95-9)**
Polyoxyethylene (2) coconut amine **(61791-14-8)**
Polyoxyethylene (3) coconut amine **(61791-14-8)**
Polyoxyethylene (5) coconut amine **(61791-14-8)**
Polyoxyethylene (10) coconut amine **(61791-14-8)**
Polyoxyethylene (15) coconut amine **(61791-14-8)**
Polyoxyethylene (4) decyl ether phosphate **(9004-80-2)**
Polyoxyethylene dibenzoate **(9004-86-8)**
Polyoxyethylene dimethyl ether **(24991-55-7)**
Poly(oxyethylene(dimethyliminio)ethylene(dimethylimino)-ethylene dichloride) **(31512-74-0)**
Polyoxyethylene (5) dinonyl phenyl ether **(9014-93-1)**
Polyoxyethylene (10) dinonyl phenyl ether **(9014-93-1)**
Polyoxyethylene (49) dinonyl phenyl ether **(9014-93-1)**
Polyoxyethylene (100) dinonyl phenyl ether **(9014-93-1)**
Polyoxyethylene (150) dinonyl phenyl ether **(9014-93-1)**
Polyoxyethylene (4) dioleate **(9005-07-6)**
Polyoxyethylene (6) dioleate **(9005-07-6)**
Polyoxyethylene (8) dioleate **(9005-07-6)**

Polyoxyethylene (10) dioleate **(9005-07-6)**
Polyoxyethylene (12) dioleate **(9005-07-6)**
Polyoxyethylene (20) dioleate **(9005-07-6)**
Polyoxyethylene (32) dioleate **(9005-07-6)**
Polyoxyethylene (75) dioleate **(9005-07-6)**
Polyoxyethylene (150) dioleate **(9005-07-6)**
Polyoxyethylene 1500 dioleate **(9005-07-6)**
Polyoxyethylene (2) distearate **(109-30-8)**
Polyoxyethylene (3) distearate **(9005-08-7)**
Polyoxyethylene (4) distearate **(9005-08-7)**
Polyoxyethylene (6) distearate **(9005-08-7)**
Polyoxyethylene (8) distearate **(9005-08-7)**
Polyoxyethylene (9) distearate **(9005-08-7)**
Polyoxyethylene (12) distearate **(9005-08-7)**
Polyoxyethylene (20) distearate **(9005-08-7)**
Polyoxyethylene (32) distearate **(9005-08-7)**
Polyoxyethylene (75) distearate **(9005-08-7)**
Polyoxyethylene (150) distearate **(9005-08-7)**
Polyoxyethylene (175) distearate **(9005-08-7)**
Polyoxyethylene (12) glyceryl ether **(31694-55-0)**
Polyoxyethylene (26) glyceryl ether **(31694-55-0)**
Polyoxyethylene glyceryl ether **(31694-55-0)**
Polyoxyethylene (2) isostearyl ether **(52292-17-8)**
Polyoxyethylene (3) isostearyl ether **(52292-17-8)**
Polyoxyethylene (10) isostearyl ether **(52292-17-8)**
Polyoxyethylene (12) isostearyl ether **(52292-17-8)**
Polyoxyethylene (20) isostearyl ether **(52292-17-8)**
Polyoxyethylene (22) isostearyl ether **(52292-17-8)**
Polyoxyethylene (50) isostearyl ether **(52292-17-8)**
Polyoxyethylene (5) lanolin **(61790-81-6)**
Polyoxyethylene (20) lanolin **(61790-81-6)**
Polyoxyethylene (24) lanolin **(61790-81-6)**
Polyoxyethylene (30) lanolin **(61790-81-6)**
Polyoxyethylene (50) lanolin **(61790-81-6)**
Polyoxyethylene (60) lanolin **(61790-81-6)**
Polyoxyethylene (85) lanolin **(61790-81-6)**
Polyoxyethylene (100) lanolin **(61790-81-6)**
Polyoxyethylene lauramide **(31587-78-7)**
Polyoxyethylene lauric alcohol **(9002-92-0)**
Polyoxyethylene lauryl alcohol **(9002-92-0)**
Polyoxyethylene (2) lauryl ether **(3055-93-4)**
Polyoxyethylene (4) lauryl ether **(5274-68-0)**
Polyoxyethylene (5) lauryl ether **(3055-95-6)**
Polyoxyethylene (7) lauryl ether **(3055-97-8)**
Polyoxyethylene (2) monolaurate **(141-20-8)**
Poly(oxyethylene) monolauryl ether **(9002-92-0)**
Polyoxyethylene mono(octylphenyl) ether **(9002-93-1)**
Polyoxyethylene monooctylphenyl ether **(9036-19-5)**
Poly(oxyethylene) monooleate **(9004-96-0)**
Polyoxyethylene (2) monooleate **(106-12-7)**
Polyoxyethylene monooleyl ether **(9004-98-2)**
Polyoxyethylene (6) monopalmitate **(9004-94-8)**
Polyoxyethylene (18) monopalmitate **(9004-94-8)**
Polyoxyethylene (20) monopalmitate **(9004-94-8)**
Polyoxyethylene monostearate **(9004-99-3)**
Polyoxyethylene (5) monostearate **(9004-99-3)**
Polyoxyethylene (6) monostearate **(9004-99-3)**
Polyoxyethylene (7) monostearate **(9004-99-3)**
Polyoxyethylene (8) monostearate **(9004-99-3)**
Polyoxyethylene-8-monostearate **(9004-99-3)**
Polyoxyethylene (9) monostearate **(9004-99-3)**
Polyoxyethylene (10) monostearate **(9004-99-3)**
Polyoxyethylene (12) monostearate **(9004-99-3)**
Polyoxyethylene (14) monostearate **(9004-99-3)**
Polyoxyethylene (18) monostearate **(9004-99-3)**
Polyoxyethylene (20) monostearate **(9004-99-3)**
Polyoxyethylene (25) monostearate **(9004-99-3)**
Polyoxyethylene (30) monostearate **(9004-99-3)**
Polyoxyethylene (32) monostearate **(9004-99-3)**
Polyoxyethylene (35) monostearate **(9004-99-3)**
Polyoxyethylene (36) monostearate **(9004-99-3)**
Polyoxyethylene (40) monostearate **(9004-99-3)**
Polyoxyethylene (45) monostearate **(9004-99-3)**
Polyoxyethylene (50) monostearate **(9004-99-3)**
Polyoxyethylene (75) monostearate **(9004-99-3)**
Polyoxyethylene (90) monostearate **(9004-99-3)**
Polyoxyethylene (100) monostearate **(9004-99-3)**
Polyoxyethylene (120) monostearate **(9004-99-3)**
Polyoxyethylene (150) monostearate **(9004-99-3)**
Polyoxyethylene 1500 monostearate **(9004-99-3)**
Polyoxyethylene (4) monotallate **(61791-00-2)**
Polyoxyethylene (8) monotallate **(61791-00-2)**
Polyoxyethylene (10) monotallate **(61791-00-2)**
Polyoxyethylene (12) monotallate **(61791-00-2)**
Polyoxyethylene (16) monotallate **(61791-00-2)**
Polyoxyethylene (20) monotallate **(61791-00-2)**
Polyoxyethylene nonylphenol **(9016-45-9)**
Polyoxyethylene (4) nonyl phenyl ether **(7311-27-5)**
Polyoxyethylene (9) nonyl phenyl ether **(9016-45-9)**
Polyoxyethylene p-tert-nonylphenyl ether **(53496-16-5)**
Poly(oxyethylene)octylphenol ether **(9036-19-5)**
Poly(oxyethylene)octylphenyl ether **(9036-19-5)**
Poly(oxyethylene)p-tert-octylphenyl ether **(9002-93-1)**
Polyoxyethylene (9) octylphenyl ether **(9002-93-1)**
Polyoxyethylene (13) octylphenyl ether **(9002-93-1)**
Poly(oxyethylene) oleate **(9004-96-0)**
Poly(oxyethylene) oleic acid ester **(9004-96-0)**
Polyoxyethylene (2) oleyl ether **(9004-98-2)**
Polyoxyethylene (3) oleyl ether **(9004-98-2)**
Polyoxyethylene (4) oleyl ether **(9004-98-2)**
Polyoxyethylene (5) oleyl ether **(9004-98-2)**
Polyoxyethylene (6) oleyl ether **(9004-98-2)**
Polyoxyethylene (7) oleyl ether **(9004-98-2)**
Polyoxyethylene (8) oleyl ether **(9004-98-2)**
Polyoxyethylene (9) oleyl ether **(9004-98-2)**
Polyoxyethylene (10) oleyl ether **(9004-98-2)**
Polyoxyethylene (12) oleyl ether **(9004-98-2)**
Polyoxyethylene (15) oleyl ether **(9004-98-2)**
Polyoxyethylene (16) oleyl ether **(9004-98-2)**
Polyoxyethylene (20) oleyl ether **(9004-98-2)**
Polyoxyethylene (23) oleyl ether **(9004-98-2)**
Polyoxyethylene (25) oleyl ether **(9004-98-2)**
Polyoxyethylene (44) oleyl ether **(9004-98-2)**
Polyoxyethylene (50) oleyl ether **(9004-98-2)**
Polyoxyethylene oxypropylene **(9003-11-6)**
Poly(oxyethyleneoxyterephthaloyl) **(25038-59-9)**
Polyoxyethylene phenyl ether **(9004-78-8)**
Polyoxyethylene-polyoxypropylene **(9003-11-6)**
Polyoxyethylene-polyoxypropylene copolymer **(9003-11-6)**
Poly(oxyethylene) poly(oxypropylene) glycol **(9003-11-6)**
Polyoxyethylene (12) polyoxypropylene (66) glyceryl ether **(9082-00-2)**
Polyoxyethylene (24) polyoxypropylene (24) glyceryl ether **(9082-00-2)**
Polyoxyethylene (30) polyoxypropylene (20) glyceryl ether **(9082-00-2)**
Polyoxyethylene (3) polyoxypropylene (2) monobutyl ether **(9038-95-3)**
Polyoxyethylene (5) polyoxypropylene (3) monobutyl ether **(9038-95-3)**
Polyoxyethylene (7) polyoxypropylene (5) monobutyl ether **(9038-95-3)**
Polyoxyethylene (10) polyoxypropylene (7) monobutyl ether **(9038-95-3)**
Polyoxyethylene (12) polyoxypropylene (9) monobutyl ether **(9038-95-3)**
Polyoxyethylene (16) polyoxypropylene (12) monobutyl ether **(9038-95-3)**
Polyoxyethylene (20) polyoxypropylene (15) monobutyl ether **(9038-95-3)**
Polyoxyethylene (27) polyoxypropylene (24) monobutyl ether **(9038-95-3)**
Polyoxyethylene (30) polyoxypropylene (20) monobutyl ether **(9038-95-3)**
Polyoxyethylene (35) polyoxypropylene (28) monobutyl ether **(9038-95-3)**
Polyoxyethylene (45) polyoxypropylene (33) monobutyl ether **(9038-95-3)**

Poly(oxyethylene)-poly(oxypropylene) polymer **(9003-11-6)**
Polyoxyethylene-polyoxypropylene polymer **(9003-11-6)**
Polyoxyethylenepropylene glycol ether **(9003-11-6)**
Polyoxyethylene (20) sorbitan monolaurate **(9005-64-5)**
Polyoxyethylene sorbitan monolaurate **(9005-64-5)**
Polyoxyethylene sorbitan monooleate **(9005-65-6)**
Polyoxyethylene 20 sorbitan monopalmitate **(9005-66-7)**
Polyoxyethylene sorbitan monopalmitate **(9005-66-7)**
Polyoxyethylene 20 sorbitan monostearate **(9005-67-8)**
Polyoxyethylene sorbitan monostearate **(9005-67-8)**
Polyoxyethylene sorbitan oleate **(9005-65-6)**
Polyoxyethylene sorbitol oleate- laurate **(53466-71-0)**
Polyoxyethylene 50 stearate **(9004-99-3)**
Polyoxyethylene (2) stearyl ether **(9005-00-9)**
Polyoxyethylene (4) stearyl ether **(9005-00-9)**
Polyoxyethylene (6) tridecyl ether phosphate **(9046-01-9)**
Polyoxyethylene (7) stearyl ether **(9005-00-9)**
Polyoxyethylene (8) stearate **(9004-99-3)**
Polyoxyethylene (8)stearate **(9004-99-3)**
Polyoxyethylene (10) stearyl ether **(9005-00-9)**
Polyoxyethylene (11) stearyl ether **(9005-00-9)**
Polyoxyethylene (13) stearyl ether **(9005-00-9)**
Polyoxyethylene (15) stearyl ether **(9005-00-9)**
Polyoxyethylene (16) stearyl ether **(9005-00-9)**
Polyoxyethylene (20) stearyl ether **(9005-00-9)**
Polyoxyethylene (25) stearyl ether **(9005-00-9)**
Polyoxyethylene (27) stearyl ether **(9005-00-9)**
Polyoxyethylene (30) stearyl ether **(9005-00-9)**
Polyoxyethylene (40) stearyl ether **(9005-00-9)**
Polyoxyethylene (50) stearyl ether **(9005-00-9)**
Polyoxyethylene (100) stearyl ether **(9005-00-9)**
Polyoxyl 10 oleyl ether **(9004-98-2)**
Polyoxyl 8 stearate **(9004-99-3)**
Polyoxyl 40 stearate **(9004-99-3)**
Polyoxyl 50 stearate **(9004-99-3)**
Poly(oxymethylene) (9CI) **(9002-81-7)**
Poly(oxy(methyl-1,2-ethanediyl)), α-(2-aminomethylethyl)-ω-(2-aminomethylethoxy)- (9CI) **(9046-10-0)**
Poly(oxy(methyl-1,2-ethanediyl)), α-butyl-ω-hydroxy **(9003-13-8)**
Poly(oxy(methyl-1,2-ethanediyl)), α-butyl-ω-hydroxy- (9CI) **(9003-13-8)**
Poly(oxy(methyl-1,2-ethanediyl)), α-hydro-ω-hydroxy- (9CI) **(25322-69-4)**
Poly(oxy(methyl-1,2-ethanediyl)), α-hydro-ω-hydroxy-, ether with 2-ethyl-2-(hydroxymethyl)-1,3-propanediol (3:1) (9CI) **(25723-16-4)**
Poly(oxy(methyl-1,2-ethanediyl)), α-hydro-ω-((1-oxo-2-propenyl)-oxy)-, ether with 2-ethyl-2-(hydroxymethyl)-1,3-propanediol (3:1) (9CI) **(53879-54-2)**
Poly(oxy(methyl-1,2-ethanediyl)), α-methyl-ω-hydroxy **(37286-64-9)**
Poly(oxy(methyl-1,2-ethanediyl)), α-(1-oxo-9-octadecenyl)-ω-hydroxy-, (Z)- (9CI) **(31394-71-5)**
Poly(oxy(methyl-1,2-ethanediyl)), α-(1-oxo-9-octadecenyl)-ω-((1-oxo-9-octadecenyl)oxy)-, (Z,Z)- (9CI) **(26571-49-3)**
Poly(oxy(methyl-1,2-ethanediyl)oxy(1,6-dioxo-1,6-hexanediyl)) (9CI) **(27941-08-8)**
Poly(oxy(methyl-1,2-ethanediyl)), α,α',α''-1,2,3-propanetriyltris-(ω-hydroxy- **(25791-96-2)**
Poly(oxy(methyl-1,2-ethanediyl)), α,α',α''-1,2,3-propanetriyltris-(ω-(oxiranylmethoxy)- (9CI) **(37237-76-6)**
Poly(oxy(methylethylene)oxyadipoyl) (8CI) **(27941-08-8)**
Poly(oxy(methyl-1,3-propanediyl)oxy(1,6-dioxo-1,6-hexanediyl)) (9CI) **(27941-09-9)**
Poly(oxy(methyltrimethylene)oxyadipoyl) (8CI) **(27941-09-9)**
Poly(oxy-1,3-phenyleneoxycarbonyl-1,3-phenylenecarbonyl) (9CI) **(26637-46-7)**
Poly(oxy-m-phenyleneoxyisophthaloyl) (8CI) **(26637-46-7)**
Poly(oxypropylene) butyl ether **(9003-13-8)**
Polyoxypropylene (4) butyl ether **(9003-13-8)**
Polyoxypropylene (5) butyl ether **(9003-13-8)**
Polyoxypropylene (9) butyl ether **(9003-13-8)**
Polyoxypropylene (14) butyl ether **(9003-13-8)**
Polyoxypropylene (15) butyl ether **(9003-13-8)**
Polyoxypropylene (16) butyl ether **(9003-13-8)**
Polyoxypropylene (18) butyl ether **(9003-13-8)**
Polyoxypropylene (22) butyl ether **(9003-13-8)**
Polyoxypropylene (24) butyl ether **(9003-13-8)**
Polyoxypropylene (30) butyl ether **(9003-13-8)**
Polyoxypropylene (33) butyl ether **(9003-13-8)**
Polyoxypropylene (40) butyl ether **(9003-13-8)**
Polyoxypropylene (53) butyl ether **(9003-13-8)**
Polyoxypropylene (17) dioleate **(26571-49-3)**
Poly(oxypropylene)glyceryl ether **(25791-96-2)**
Polyoxypropylene glycol butyl monoether **(9003-13-8)**
Polyoxypropylene (2) methyl ether **(13429-07-7)**
Polyoxypropylene monobutyl ether **(9003-13-8)**
Polyoxypropylene (26) monooleate **(31394-71-5)**
Polyoxypropylene (36) monooleate **(31394-71-5)**
Polyoxypropylene oleate **(31394-71-5)**
Polyoxypropylene-polyoxyethylene copolymer **(9003-11-6)**
Polyoxypropylene (20) polyoxyethylene (30) glyceryl ether **(9082-00-2)**
Polyoxypropylene (24) polyoxyethylene (24) glyceryl ether **(9082-00-2)**
Polyoxypropylene (66) polyoxyethylene (12) glyceryl ether **(9082-00-2)**
Polyoxypropylene (2) polyoxyethylene (3) monobutyl ether **(9038-95-3)**
Polyoxypropylene (3) polyoxyethylene (5) monobutyl ether **(9038-95-3)**
Polyoxypropylene (5) polyoxyethylene (7) monobutyl ether **(9038-95-3)**
Polyoxypropylene (7) polyoxyethylene (10) monobutyl ether **(9038-95-3)**
Polyoxypropylene (9) polyoxyethylene (12) monobutyl ether **(9038-95-3)**
Polyoxypropylene (12) polyoxyethylene (16) monobutyl ether **(9038-95-3)**
Polyoxypropylene (15) polyoxyethylene (20) monobutyl ether **(9038-95-3)**
Polyoxypropylene (20) polyoxyethylene (30) monobutyl ether **(9038-95-3)**
Polyoxypropylene (24) polyoxyethylene (27) monobutyl ether **(9038-95-3)**
Polyoxypropylene (28) polyoxyethylene (35) monobutyl ether **(9038-95-3)**
Polyoxypropylene (33) polyoxyethylene (45) monobutyl ether **(9038-95-3)**
Poly(1,3-phenylene isophthalate) **(26637-46-7)**
Polyphenylmethyl siloxane **(2116-84-9)**
Polyphlogin **(132-60-5)**
Polyphos **(10124-56-8)**
Polyphosphoric acid **(8017-16-1)**
Polyphosphoric acids **(8017-16-1)**
Polyphosphoric acids, 2-ethoxyethyl esters **(68554-00-7)**
Polyphosphoric acids, sodium salts **(68915-31-1)**
Polyphosphoric acids, zinc salts **(68607-18-1)**
Polypro 1014 **(9003-07-0)**
Polypro B 220 **(9003-07-0)**
Polypro G 400P **(9003-07-0)**
Polypro J 600 **(9003-07-0)**
Polypro J 400P **(9003-07-0)**
Polypropene **(9003-07-0)**
Poly(2-propyl-m-dioxane-4,6-diylene) **(63148-65-2)**
Polypropylene **(9003-07-0)**
Polypropylene glycol 150 **(25322-69-4)**
Polypropylene glycol #400 **(25322-69-4)**
Polypropylene glycol #425 **(25322-69-4)**
Polypropylene glycol #750 **(25322-69-4)**
Polypropylene glycol #1000 **(25322-69-4)**
Polypropylene glycol 1025 **(25322-69-4)**
Polypropylene glycol #1200 **(25322-69-4)**
Polypropylene glycol #1800 **(25322-69-4)**
Polypropylene glycol 2000 **(25322-69-4)**
Polypropylene glycol 2025 **(25322-69-4)**
Polypropylene glycol 3025 **(25322-69-4)**
Polypropylene glycol 4025 **(25322-69-4)**
Polypropylene-glycol **(25322-69-4)**
Poly(propylene glycol adipate) **(25101-03-5)**
Polypropylene glycol butyl ether **(9003-13-8)**
Polypropylene glycol (4) butyl ether **(9003-13-8)**

Polypropylene glycol (5) butyl ether	(9003-13-8)
Polypropylene glycol (9) butyl ether	(9003-13-8)
Polypropylene glycol (14) butyl ether	(9003-13-8)
Polypropylene glycol (15) butyl ether	(9003-13-8)
Polypropylene glycol (16) butyl ether	(9003-13-8)
Polypropylene glycol (18) butyl ether	(9003-13-8)
Polypropylene glycol (22) butyl ether	(9003-13-8)
Polypropylene glycol (24) butyl ether	(9003-13-8)
Polypropylene glycol (30) butyl ether	(9003-13-8)
Polypropylene glycol (33) butyl ether	(9003-13-8)
Polypropylene glycol (40) butyl ether	(9003-13-8)
Polypropylene glycol (53) butyl ether	(9003-13-8)
Polypropylene glycol (17) dioleate	(26571-49-3)
Polypropylene glycol dioleate	(26571-49-3)
Polypropylene glycol-ethylene oxide copolymer	(9003-11-6)
Polypropylene glycol (2) methyl ether	(13429-07-7)
Polypropylene glycol methyl ether	(37286-64-9)
Polypropylene glycol monobutyl ether	(9003-13-8)
Polypropylene glycol monobutylether	(9003-13-8)
Polypropylene glycol monomethylether	(37286-64-9)
Polypropylene glycol (26) monooleate	(31394-71-5)
Polypropylene glycol (36) monooleate	(31394-71-5)
Poly(propylene oxide-ethylene oxide)	(9003-11-6)
Poly(propylene oxide), monobutyl ether	(9003-13-8)
Polypropylenglykol (Czech)	(25322-69-4)
Polyquaternium-11	(53633-54-8)
Polyquaternium-14	(27103-90-8)
Polyquaternium-5	(26006-22-4)
Polyram	(9006-42-2)
Polyram 80	(9006-42-2)
Polyram Combi	(9006-42-2)
Polyram M	(12427-38-2)
Polyram Ultra	(137-26-8)
Polyram 80WP	(9006-42-2)
Polyram Z	(12122-67-7)
Polysan	(9003-54-7)
Polysilane	(9006-65-9)
Polysilicic acid, ethyl ester	(11099-06-2)
Polysiloxane	(9011-19-2)
Polysion N 22	(9002-88-4)
Polysizer 173	(9002-89-5)
Polysizer W 2300	(27941-08-8)
Polysizer W 2600	(27941-08-8)
Polysol 1000	(9003-20-7)
Polysol 1200	(9003-20-7)
Polysol PS 10	(9003-20-7)
Polysol S 5	(9003-20-7)
Polysol S 6	(9003-20-7)
Polysol 1000AX	(9003-20-7)
Polysorban 80	(9005-65-6)
Polysorbate 20	(9005-64-5)
Polysorbate 40	(9005-66-7)
Polysorbate 60	(9005-67-8)
Polysorbate 80	(9005-65-6)
Polysorbate 81	(9005-65-6)
Polysorbate 80 B.P.C.	(9005-65-6)
Polysorbate 80, U.S.P.	(9005-65-6)
Polystate	(9004-99-3)
Polystate B	(9004-99-3)
Polystep A 13	(27176-87-0)
Polystrol D	(9003-53-6)
Polystyrene	(9003-53-6)
Polystyrene BW	(9003-53-6)
Polystyrene Beads, Expandable, evolving flammable vapor. [UN 2211]	(9003-53-6)
Polystyrene Latex	(9003-53-6)
Polystyrene-acrylonitrile	(9003-54-7)
Poly(styrenesulfonate) sodium salt	(9080-79-9)
Polystyrol	(9003-53-6)
Polysulfide	(9080-49-3)
Polysulfure de calcium (French)	(1344-81-6)
Polytac	(9003-07-0)
Polytal 4641	(14807-96-6)
Polytal 4725	(14807-96-6)
Polytar Bath	(8007-45-2)
Polytetrafluoroethylene Decomposition Products	(93763-70-3)
Poly((6-((1,1,3,3-tetramethylbutyl)amino)-1,3,5-triazine-2,4-diyl)-((2,2,6,6-tetramethyl-4- piperidinyl)imino)-1,6-hexanediyl-((2,2,6,6-tetramethyl-4-piperidinyl)imino))	(71878-19-8)
Polytetramethylene glycol	(25190-06-1)
Poly(tetramethylene terephthalate)	(26062-94-2)
Polytetramethylene terephthalate	(24968-12-5)
Polytex 973	(9003-01-4)
Polythene	(9002-88-4)
Polythiazide	(346-18-9)
Polytox	(120-36-5)
Polytrin	(52315-07-8)
Polyurax G 3000	(25791-96-2)
Polyurea 6	(25035-67-0)
Polyurethane Ester Foam	(9009-54-5)
Polyurethane Ether Foam	(9009-54-5)
Polyurethane-Foam	(9009-54-5)
Polyurethane Sponge	(9009-54-5)
Polyurethane polymer	(68400-67-9)
Poly(ureylenehexamethylene) (8CI)	(25035-67-0)
Polyvel G	(64742-16-1)
Polyvel GP 65	(64742-16-1)
Polyvel M	(64742-16-1)
Polyvidone	(9003-39-8)
Polyvinol	(9002-89-5)
Polyvinyl Butyral Resins	(63148-65-2)
Poly(vinylacetate)	(9003-20-7)
Polyvinyl acetate chloride	(34149-92-3)
Polyvinyl-alcohol	(9002-89-5)
Polyvinylbromide	(25951-54-6)
Polyvinylbutyral (Czech)	(63148-65-2)
Poly(n-vinylbutyrolactam)	(9003-39-8)
Poly-N-vinylcarbazole	(25067-59-8)
Polyvinylchloride acetate	(34149-92-3)
Polyvinyl chloride-polyvinyl acetate	(9003-22-9)
Poly(4-vinylguaiacol)	(31853-85-7)
Poly(vinylguaiacol)	(31853-85-7)
Polyvinylidene fluoride	(24937-79-9)
Polyvinyl octadecyl carbamate	(36671-85-9)
Poly(4-vinylphenol)	(24979-70-2)
Poly(p-vinylphenol)	(24979-70-2)
Poly(1-vinylpyrrolidinone)	(9003-39-8)
Poly(n-vinylpyrrolidinone)	(9003-39-8)
Poly(vinylpyrrolidinone)	(9003-39-8)
Poly(1-vinyl-2-pyrrolidinone) Hueper's polymer No.1	(9003-39-8)
Poly(1-vinyl-2-pyrrolidinone) Hueper's polymer No.2	(9003-39-8)
Poly(1-vinyl-2-pyrrolidinone) Hueper's polymer No.3	(9003-39-8)
Poly(1-vinyl-2-pyrrolidinone) Hueper's polymer No.4	(9003-39-8)
Poly(1-vinyl-2-pyrrolidinone) Hueper's polymer No.5	(9003-39-8)
Poly(1-vinyl-2-pyrrolidinone) Hueper's polymer No.6	(9003-39-8)
Poly(1-vinyl-2-pyrrolidinone) Hueper's polymer No.7	(9003-39-8)
Poly(1-vinyl-2-pyrrolidinone) homopolymer	(9003-39-8)
Polyvinylpyrrolidone	(9003-39-8)
Polyviol	(9002-89-5)
Polyviol M 13/140	(9002-89-5)
Polyviol MO 5/140	(9002-89-5)
Polyviol W 25/140	(9002-89-5)
Polyviol W 40/140	(9002-89-5)
Polyvis 2000CH	(9003-29-6)

Polyvis OO **(9003-29-6)**
Polyvis 06SH **(9003-27-4)**
Polyvis 015SH **(9003-29-6)**
Polyvis 30SH **(9003-27-4)**
Polyvis 150SH **(9003-27-4)**
Polyvis 200SH **(9003-27-4)**
Polywax 1000 **(9002-88-4)**
Pomadex **(51-63-8)**
Pomalus acid **(6915-15-7)**
Pomarsol **(137-26-8)**
Pomarsol Forte **(137-26-8)**
Pomarsol Z Forte **(137-30-4)**
Pomasol **(137-26-8)**
Pomex **(63-25-2)**
Pomoline **(2152-34-3)**
Ponalar **(61-68-7)**
Ponceau BNA **(3761-53-3)**
Ponceau De Xylidine **(3761-53-3)**
Ponceau FR **(3761-53-3)**
Ponceau 4G **(1934-20-9)**
Ponceau G **(3761-53-3)**
Ponceau GR **(3761-53-3)**
Ponceau Insoluble OLG **(3118-97-6)**
Ponceau Insoluble OLG **(85-86-9)**
Ponceau J **(3761-53-3)**
Ponceau MX **(3761-53-3)**
Ponceau NR **(3761-53-3)**
Ponceau PXM **(3761-53-3)**
Ponceau 2R **(3761-53-3)**
Ponceau 3R **(3564-09-8)**
Ponceau 4R **(2611-82-7)**
Ponceau 6R **(5850-44-2)**
Ponceau R **(3761-53-3)**
Ponceau 6RA **(5850-44-2)**
Ponceau 4R Aluminum Lake **(2611-82-7)**
Ponceau 2R (Biological stain) **(3761-53-3)**
Ponceau R (Biological stain) **(3761-53-3)**
Ponceau 2RE **(3761-53-3)**
Ponceau 4RE **(2611-82-7)**
Ponceau 4RE.FQ **(2611-82-7)**
Ponceau 2R Extra A Export **(3761-53-3)**
Ponceau 4RF **(2611-82-7)**
Ponceau RG **(3761-53-3)**
Ponceau 2RL **(3761-53-3)**
Ponceau 3R Lake **(3564-09-8)**
Ponceau 3RN **(3564-09-8)**
Ponceau RN **(3564-09-8)**
Ponceau 6RPA **(5850-44-2)**
Ponceau RR **(3761-53-3)**
Ponceau RR Type 8019 **(3761-53-3)**
Ponceau RS **(3761-53-3)**
Ponceau 3R Sodium salt **(3564-09-8)**
Ponceau 6R Specially Pure **(5850-44-2)**
Ponceau 4RT **(2611-82-7)**
Ponceau 2RX **(3761-53-3)**
Ponceau Red **(3761-53-3)**
Ponceau Red 6R **(5850-44-2)**
Ponceau Red R **(3761-53-3)**
Ponceau SX **(4548-53-2)**
Ponceau SX Lake **(4548-53-2)**
Ponceau Xylidine **(3761-53-3)**
Ponceau Xylidine (Biological stain) **(3761-53-3)**
Poncyl **(126-07-8)**
Pondex **(2152-34-3)**
Ponecil **(69-53-4)**
Ponolith Fast Violet 4RN **(1324-55-6)**
Ponolith Orange Y **(3520-72-7)**
Ponolith Yellow Y **(6358-31-2)**
Ponoxylan **(9011-05-6)**
Ponsol Blue BCS **(130-20-1)**
Ponsol Blue GZ **(81-77-6)**
Ponsol Blue RCL **(81-77-6)**
Ponsol Blue RPC **(81-77-6)**
Ponsol Brilliant Blue R **(81-77-6)**
Ponsol Brown RBT **(2475-33-4)**
Ponsol Golden Orange G **(128-70-1)**
Ponsol Golden Orange GD **(128-70-1)**
Ponsol Jade Green Supra D **(1324-54-5)**
Ponsol Navy Blue **(6424-76-6)**
Ponsol Navy Blue D **(6424-76-6)**
Ponsol Navy Blue RA **(1324-54-5)**
Ponsol Navy Blue RAD **(1324-54-5)**
Ponsol Olive AR **(2379-81-9)**
Ponsol Olive ARD **(2379-81-9)**
Ponsol RP **(81-77-6)**
Ponsol Red 2B **(4203-77-4)**
Ponsol Red 2BD **(4203-77-4)**
Ponstan **(61-68-7)**
Ponstan forte **(61-68-7)**
Ponstel **(61-68-7)**
Ponstil **(61-68-7)**
Ponstyl **(61-68-7)**
Pontachrome Violet SW **(2092-55-9)**
Pontacyl Brilliant Blue **(129-17-9)**
Pontacyl Brilliant Blue V **(129-17-9)**
Pontacyl Carmine 2G **(3734-67-6)**
Pontacyl Fast Blue R **(3861-73-2)**
Pontacyl Fast Red AS **(1658-56-6)**
Pontacyl Green BL **(4680-78-8)**
Pontacyl Rubine R **(3567-69-9)**
Pontacyl Scarlet RR **(2611-82-7)**
Pontacyl Sky Blue 4BX **(2429-74-5)**
Pontal **(61-68-7)**
Pontalite **(9011-14-7)**
Pontamine Black E **(1937-37-7)**
Pontamine Black EBN **(1937-37-7)**
Pontamine Black RRX **(2429-83-6)**
Pontamine Blue BB **(2602-46-2)**
Pontamine Blue 3BX **(72-57-1)**
Pontamine Bond Blue B **(28407-37-6)**
Pontamine Brilliant Violet RN **(6426-67-1)**
Pontamine Brown BCW **(2429-81-4)**
Pontamine Brown BT **(2429-81-4)**
Pontamine Brown D 3GN **(2586-58-5)**
Pontamine Brown NCR **(2586-58-5)**
Pontamine Deep Blue BH **(2429-73-4)**
Pontamine Developer TN **(95-80-7)**
Pontamine Diazo Black BHSW **(2429-73-4)**
Pontamine Fast Blue 7GLN **(28407-37-6)**
Pontamine Fast Brown BRL **(16071-86-6)**
Pontamine Fast Brown NP **(16071-86-6)**
Pontamine Fast Red 8BLX **(2610-11-9)**
Pontamine Fast Scarlet 4BA **(93-00-5)**
Pontamine Green S **(3626-28-6)**
Pontamine Navy Blue BFN **(7082-31-7)**
Pontamine Scarlet 3B **(6358-29-8)**
Pontamine Sky Blue **(2610-05-1)**
Pontamine Sky Blue 5BX **(2429-74-5)**
Pontamine Sky Blue 6BX **(2610-05-1)**
Pontamine Sky Blue 6BX Greenish **(2610-05-1)**
Pontamine Sky Blue 6x **(2610-05-1)**
Pontamine Violet N **(2586-60-9)**
Poppy Oil **(8002-11-7)**
Poppy Seed Oil **(8002-11-7)**

Name	CAS No.
Poprolin	**(9003-07-0)**
Populnetin	**(520-18-3)**
Poraminar	**(1344-28-1)**
Porekal	**(866-84-2)**
Porex P	**(2440-22-4)**
Porocel	**(1318-16-7)**
Porocel O	**(1318-16-7)**
Porofor-BSH-Pulver	**(80-17-1)**
Porofor 505	**(123-77-3)**
Porofor 57	**(78-67-1)**
Porofor ADC/R	**(123-77-3)**
Porofor BSH	**(80-17-1)**
Porofor ChKhZ 21	**(123-77-3)**
Porofor ChKhZ 9	**(80-17-1)**
Porofor ChKhZ-18	**(101-25-7)**
Porofor ChKhZ 21R	**(123-77-3)**
Porolen	**(9002-88-4)**
Porophor B	**(101-25-7)**
Porophor N	**(78-67-1)**
Porous	**(7784-28-3)**
21H,23H-Porphine-2,18-dipropanoic acid, 7,12-diethenyl-3,8,13,17-tetramethyl- (9CI)	**(553-12-8)**
2,18-Porphinedipropionic acid, 3,8,13,17-tetramethyl-7,12-divinyl- (8CI)	**(553-12-8)**
Portland Cement (ACGIH,OSHA)	**(65997-15-1)**
Portland Stone	**(1317-65-3)**
Posse	**(55285-14-8)**
Postafen	**(569-65-3)**
Postafene	**(569-65-3)**
Poster Red	**(1103-38-4)**
Posterior pituitary extract	**(50-56-6)**
Postinor	**(797-63-7)**
Potaba	**(138-84-1)**
Potablan	**(7287-36-7)**
Potalium	**(299-27-4)**
Potasan	**(299-45-6)**
Potasan-G-Liquid	**(299-45-6)**
Potash	**(584-08-7)**
Potash alum	**(10043-67-1)**
Potash chlorate (DOT)	**(3811-04-9)**
Potasiocarbonilo (Spanish)	**(12397-35-2)**
Potasoral	**(299-27-4)**
Potassa	**(1310-58-3)**
Potasse Caustique (French)	**(1310-58-3)**
Potassio (chlorato di) (Italian)	**(3811-04-9)**
Potassio (idrossido di) (Italian)	**(1310-58-3)**
Potassio (permanganato di) (Italian)	**(7722-64-7)**
n-Potassiophthalimide	**(1074-82-4)**
Potassium Chromium Alum	**(10141-00-1)**
Potassium D-gluconate	**(299-27-4)**
Potassium D-gluconate	**(35087-77-5)**
Potassium, (Liquid alloy)	**(7440-09-7)**
Potassium, Metal (DOT)	**(7440-09-7)**
Potassium, Metal alloys [UN 1420]	**(7440-09-7)**
Potassium, Metallic (DOT)	**(7440-09-7)**
Potassium, Metal liquid alloy (DOT)	**(7440-09-7)**
Potassium [UN 2257]	**(7440-09-7)**
Potassium Water Glass	**(1312-76-1)**
Potassium acetate	**(127-08-2)**
Potassium acid arsenate	**(7784-41-0)**
Potassium acid fluoride	**(7789-29-9)**
Potassium acid phosphate	**(7778-77-0)**
Potassium acid phthalate	**(877-24-7)**
Potassium acid sulfate	**(7646-93-7)**
Potassium acid sulfite	**(7773-03-7)**
Potassium acid tartrate	**(868-14-4)**
Potassium acrylate	**(10192-85-5)**
Potassium alginate	**(9005-36-1)**
Potassium alum	**(10043-67-1)**
Potassium aluminum sulfate (1:1:2)	**(10043-67-1)**
Potassium 4-tert-amylphenate	**(53404-18-5)**
Potassium p-tert-amylphenate	**(53404-18-5)**
Potassium p-tert-amylphenolate	**(53404-18-5)**
Potassium amylxanthate	**(2720-73-2)**
Potassium amylxanthogenate	**(2720-73-2)**
Potassium n-amylxanthogenate	**(2720-73-2)**
Potassium antimonyl D,L-tartrate	**(64070-12-8)**
Potassium antimonyl L-tartrate	**(11071-15-1)**
Potassium antimonyl d-tartrate	**(28300-74-5)**
Potassium antimonyl tartrate	**(28300-74-5)**
Potassium antimony tartrate	**(28300-74-5)**
Potassium arsenate, Solid (DOT)	**(7784-41-0)**
Potassium arsenate [UN 1677]	**(7784-41-0)**
Potassium arsenate, monobasic	**(7784-41-0)**
Potassium arsenite	**(13464-35-2)**
Potassium arsenite, Solid (DOT)	**(10124-50-2)**
Potassium arsenite [UN 1678]	**(10124-50-2)**
Potassium-azide	**(20762-60-1)**
Potassium benzeneacetate	**(13005-36-2)**
Potassium benzoate	**(582-25-2)**
Potassium 2-benzyl-4-chlorophenate	**(35471-49-9)**
Potassium o-benzyl-p-chlorophenate	**(35471-49-9)**
Potassium 2-benzyl-4-chlorophenolate	**(35471-49-9)**
Potassium o-benzyl-p-chlorophenolate	**(35471-49-9)**
Potassium benzylpenicillin	**(113-98-4)**
Potassium benzylpenicillin G	**(113-98-4)**
Potassium benzylpenicillinate	**(113-98-4)**
Potassium bicarbonate	**(298-14-6)**
Potassium bichromate	**(7778-50-9)**
Potassium bifluoride	**(7789-29-9)**
Potassium bifluoride, Solid	**(7789-29-9)**
Potassium bifluoride, Solution [UN 1811]	**(7789-29-9)**
Potassium biphosphate	**(7758-11-4)**
Potassium biphosphate	**(7778-77-0)**
Potassium biphthalate	**(877-24-7)**
Potassium bis(2-hydroxyethyl)dithiocarbamate	**(23746-34-1)**
Potassium-bisulfate	**(7646-93-7)**
Potassium bisulfite	**(7773-03-7)**
Potassium bisulphate	**(7646-93-7)**
Potassium bitartrate	**(868-14-4)**
Potassium borohydride [UN 1870]	**(13762-51-1)**
Potassium bromate [UN 1484]	**(7758-01-2)**
Potassium-bromide	**(7758-02-3)**
Potassium 4-tert-butylphenate	**(3130-29-8)**
Potassium caprylate	**(764-71-6)**
Potassium carbonate (2:1)	**(584-08-7)**
Potassium carbonyl	**(12397-35-2)**
Potassium carbonyle (French)	**(12397-35-2)**
Potassium chlorate [UN 1485]	**(3811-04-9)**
Potassium chlorate, aqueous Solution [UN 2427]	**(3811-04-9)**
Potassium (chlorate de) (French)	**(3811-04-9)**
Potassium-chloride	**(7447-40-7)**
Potassium 4-chloro-2-cyclopentylphenate	**(35471-38-6)**
Potassium 2-chloro-4-phenylphenate	**(18128-16-0)**
Potassium 4-chloro-2-phenylphenate	**(53404-21-0)**
Potassium 6-chloro-2-phenylphenate	**(18128-17-1)**
Potassium 3-(2-chloro-4-(trifluoromethyl)phenoxy)benzoate	**(72252-48-3)**
Potassium chromate	**(7789-00-6)**
Potassium chromate (VI)	**(7789-00-6)**
Potassium chromic sulfate	**(10141-00-1)**
Potassium chromic sulphate	**(10141-00-1)**
Potassium chromium Alum	**(7788-99-0)**
Potassium citrate	**(866-84-2)**
Potassium cresylate	**(12002-51-6)**

Potassium cyanate (590-28-3)
Potassium-cyanide (151-50-8)
Potassium cyanide (ACGIH) [UN 1680] (151-50-8)
Potassium cyanide, Solid [UN 1680] (151-50-8)
Potassium cyanide, Solution [UN 1680] (151-50-8)
Potassium cyclamate (7758-04-5)
Potassium N-cyclohexylsulfamate (7758-04-5)
Potassium cyclohexyl sulfamate (7758-04-5)
Potassium decanoate (13040-18-1)
Potassium dibasic phosphate (7758-11-4)
Potassium dichloroisocyanurate (2244-21-5)
Potassium dichloro isocyanurate [UN 2465] (2244-21-5)
Potassium 2,5-dichlorophenolate (68938-81-8)
Potassium 4,6-dichloro-2-phenylphenate (53404-30-1)
Potassium dichloro-s-triazinetrione (2244-21-5)
Potassium dichloro-s-triazinetrione, Dry, containing more than 39% available chlorine [UN 2465] (2244-21-5)
Potassium dichromate (DOT) (7778-50-9)
Potassium dichromate (VI) (7778-50-9)
Potassium O,O-dihexyl dithiophosphate (3287-87-4)
Potassium dihydrogen arsenate (7784-41-0)
Potassium dihydrogen orthophosphate (7778-77-0)
Potassium dihydrogen phosphate (7778-77-0)
Potassium 1,2-dihydro-3,6-pyridazinedione (28382-15-2)
Potassium dimethylbenzenesulfonate (30346-73-7)
Potassium dimethyl dithiocarbamate (128-03-0)
Potassium dinitrobenzofuroxan (29267-75-2)
Potassium dioxide (12030-88-5)
Potassium diphosphate (7778-77-0)
Potassium disulphatochromate (III) (10141-00-1)
Potassium dodecanoate (10124-65-9)
Potassium dodecylbenzenesulfonate (27177-77-1)
Potassium ethyl dithiocarbonate (140-89-6)
Potassium o-ethyl dithiocarbonate (140-89-6)
Potassium ethylenediaminetetraacetate (53404-51-6)
Potassium 2-ethylhexanoate (3164-85-0)
Potassium ethylxanthate (140-89-6)
Potassium ethyl xanthogenate (140-89-6)
Potassium fluoride (8CI,9CI) (7789-29-9)
Potassium fluoride, Solution (DOT) (7789-23-3)
Potassium fluoride [UN 1812] (7789-23-3)
Potassium fluorozirconate (16923-95-8)
Potassium fluorure (French) (7789-23-3)
Potassium fluotantalate (16924-00-8)
Potassium fluozirconate (16923-95-8)
Potassium formate (590-29-4)
Potassium gibberellate (125-67-7)
Potassium gluconate (299-27-4)
Potassium glycerophosphate (1335-34-8)
Potassium-heptafluorotantalate (16924-00-8)
Potassium heptanoate (16761-12-9)
Potassium hexafluorozirconate (16923-95-8)
Potassium hexafluorozirconate(IV) (16923-95-8)
Potassium hydrate (DOT) (1310-58-3)
Potassium hydrogen arsenate (7784-41-0)
Potassium hydrogen difluoride (7789-29-9)
Potassium hydrogen fluoride (7789-29-9)
Potassium hydrogen fluoride, Solution (7789-29-9)
Potassium hydrogen oxalate (127-95-7)
Potassium hydrogen phosphate (7758-11-4)
Potassium hydrogen phosphate (7778-77-0)
Potassium hydrogen sulfate, Solid (DOT) (7646-93-7)
Potassium hydrogen sulfite (7773-03-7)
Potassium hydrogen sulphate [UN 2509] (7646-93-7)
Potassium hydrogen tartrate (868-14-4)
Potassium-hydroxide (1310-58-3)
Potassium hydroxide (ACGIH,OSHA) (1310-58-3)
Potassium hydroxide, Dry, Solid, Flake, Bead, or Granular [UN 1813] (1310-58-3)
Potassium hydroxide, Liquid or solution [UN 1814] (1310-58-3)
Potassium hydroxyacetate (1932-50-9)
Potassium hydroxyacetate (25904-89-6)
Potassium (hydroxyde de) (French) (1310-58-3)
Potassium N-hydroxymethyl-N-methyldithiocarbamate (51026-28-9)
Potassium hyperchloride (7778-74-7)
Potassium hypophosphite (7782-87-8)
Potassium hypophosphite, monobasic (7782-87-8)
Potassium iodate (7758-05-6)
Potassium-iodide (7681-11-0)
Potassium isocyanate (590-28-3)
Potassium isopropylxanthate (140-92-1)
Potassium isopropyl xanthogenate (140-92-1)
Potassium isothiocyanate (333-20-0)
Potassium laurate (10124-65-9)
Potassium maleic hydrazide (28382-15-2)
Potassium mercuric iodide (7783-33-7)
Potassium metaarsenite (10124-50-2)
Potassium metabisulfite (DOT) (16731-55-8)
Potassium metasilicate (1312-76-1)
Potassium methacrylate (6900-35-2)
Potassium methylbenzenesulfonate (30526-22-8)
Potassium N-methyldithiocarbamate (137-41-7)
Potassium methylphenoxymethylpenicillin (132-93-4)
Potassium O-methyl phenylphosphonothioate (67446-04-2)
Potassium 2-methyl-2-propenoate (6900-35-2)
Potassium monochloride (7447-40-7)
Potassium monohydrogen difluoride (7789-29-9)
Potassium monohydrogen phosphate (7758-11-4)
Potassium monophosphate (7758-11-4)
Potassium monosulfide (1312-73-8)
Potassium myristate (13429-27-1)
Potassium 1-naphthaleneacetate (15165-79-4)
Potassium nitrate [UN 1486] (7757-79-1)
Potassium nitrite (1:1) (7758-09-0)
Potassium nitrite [UN 1488] (7758-09-0)
Potassium N-(α-(nitroethyl)benzyl)ethylenediamine (53404-62-9)
Potassium cis-9-octadecenoic acid (143-18-0)
Potassium octanoate (764-71-6)
Potassium octatitanate (12056-53-0)
Potassium oleate (143-18-0)
Potassium oxymuriate (3811-04-9)
Potassium penicillin G (113-98-4)
Potassium penicillin V (132-98-9)
Potassium penicillin V salt (132-98-9)
Potassium pentachlorophenate (7778-73-6)
Potassium pentylxanthate (2720-73-2)
Potassium pentyl xanthogenate (2720-73-2)
Potassium perchlorate [UN 1489] (7778-74-7)
Potassium permanganate [UN 1490] (7722-64-7)
Potassium (permanganate de) (French) (7722-64-7)
Potassium peroxide [UN 1491] (17014-71-0)
Potassium peroxydisulfate (7727-21-1)
Potassium peroxydisulphate (7727-21-1)
Potassium peroxymonosulfate (10058-23-8)
Potassium peroxymonosulfuric acid (10058-23-8)
Potassium persulfate (ACGIH) [UN 1492] (7727-21-1)
Potassium persulphate (DOT) (7727-21-1)
Potassium phenethicillin (132-93-4)
Potassium (1-phenoxyethyl)penicillin (132-93-4)
Potassium α-phenoxyethyl penicillin (132-93-4)
Potassium phenoxymethylpenicillin (132-98-9)
Potassium 6-(α-phenoxypropionamido)penicillanate (132-93-4)
Potassium 2-phenylphenate (13707-65-8)
Potassium orthophosphate (7778-53-2)

Potassium phosphate	(7778-53-2)
Potassium phosphate	(7778-77-0)
Potassium phosphate NF XII	(7758-11-4)
Potassium phosphate, dibasic	(7758-11-4)
Potassium orthophosphate, dihydrogen	(7778-77-0)
Potassium orthophosphate, mono-H	(7758-11-4)
Potassium phosphate, monobasic	(7778-77-0)
Potassium phosphate, tribasic	(7778-53-2)
Potassium phosphinate	(7782-87-8)
Potassium phthalimidate	(1074-82-4)
Potassium phthalimide	(1074-82-4)
Potassium picloram	(2545-60-0)
Potassium polymannuronate	(9005-36-1)
Potassium polysilicate	(1312-76-1)
Potassium polysulfide	(37199-66-9)
Potassium 2-propenoate	(10192-85-5)
Potassium pyrophosphate	(7320-34-5)
Potassium-pyrosulfite	(16731-55-8)
Potassium rhodanate	(333-20-0)
Potassium rhodanide	(333-20-0)
Potassium ricinoleate	(7492-30-0)
Potassium salt of benzylpenicillin	(113-98-4)
Potassium salt of sorrel	(127-95-7)
Potassium silicate	(1312-76-1)
Potassium silicate solution	(1312-76-1)
Potassium-silver-cyanide	(506-61-6)
Potassium sodium alloy	(11135-81-2)
Potassium-sodium, alloy	(11135-81-2)
Potassium sodium tartrate	(304-59-6)
Potassium sorbate	(24634-61-5)
Potassium sorbate	(590-00-1)
Potassium stearate (ACGIH)	(593-29-3)
Potassium sulfate	(7646-93-7)
Potassium sulfate (2:1)	(7778-80-5)
Potassium sulfide (2:1)	(1312-73-8)
Potassium sulfide (9CI)	(37199-66-9)
Potassium sulfide (DOT)	(1312-73-8)
Potassium sulfide (2:1), Hydrated, containing at least 30% water	(1312-73-8)
Potassium sulfite	(10117-38-1)
Potassium sulfite, hydrogen	(7773-03-7)
Potassium sulfocyanate	(333-20-0)
Potassium sulphide, Anhydrous or containing less than 30% water [UN 1382]	(1312-73-8)
Potassium sulphide, Hydrated, containing not less than 30% water [UN 1847]	(1312-73-8)
Potassium superoxide (K(O2)) (9CI)	(12030-88-5)
Potassium tantalum fluoride	(16924-00-8)
Potassium tartrate	(868-14-4)
Potassium tetrachlorophenate	(53535-27-6)
Potassium tetracyanomercurate (II)	(591-89-9)
Potassium tetradecanoate	(13429-27-1)
Potassium tetraiodomercurate (II)	(7783-33-7)
Potassium tetrathionate	(13932-13-3)
Potassium thiocyanate	(333-20-0)
Potassium thiocyanide	(333-20-0)
Potassium thiosulfate	(10233-00-8)
Potassium titanate	(12030-97-6)
Potassium-titanium-oxide	(12056-53-0)
Potassium toluene sulfonate	(30526-22-8)
Potassium troclosene	(2244-21-5)
Potassium xanthate	(140-89-6)
Potassium xanthogenate	(140-89-6)
Potassium xylene sulfonate	(30346-73-7)
Potassium xylenesulfonate	(30346-73-7)
Potassium zinc chromate	(11103-86-9)
Potassium zinc chromate hydroxide	(11103-86-9)
Potassium zinc chromate oxide	(12433-50-0)
Potassium zirconium fluoride	(16923-95-8)
Potassium zirconium hexafluoride	(16923-95-8)
Potassuril	(299-27-4)
Potato Alcohol	(64-17-5)
Potavescent	(7447-40-7)
Potcrate	(3811-04-9)
Potentiated acid glutaraldehyde	(111-30-8)
Potide	(7681-11-0)
Potomac Red	(5160-02-1)
Pounce	(52645-53-1)
Poussiere arsenicale (French)	(8028-73-7)
Poval 117	(9002-89-5)
Poval 120	(9002-89-5)
Poval 203	(9002-89-5)
Poval 205	(9002-89-5)
Poval 217	(9002-89-5)
Poval 1700	(9002-89-5)
Poval C 17	(9002-89-5)
Povidone	(9003-39-8)
Povidone (USP XIX)	(9003-39-8)
Povidone-iodine	(25655-41-8)
Powder Base 900	(1314-13-2)
Powder Green	(12002-03-8)
Powder and Root	(83-79-4)
Powdered opium	(8008-60-4)
Poyamin	(68-19-9)
Pr 168	(9003-11-6)
Pracarbamin	(51-79-6)
Pracarbamine	(51-79-6)
Practalol	(6673-35-4)
Practolol	(6673-35-4)
Pradupen	(61-33-6)
Praecirheumin	(50-33-9)
Praedyn	(9002-61-3)
Praktololu (Polish)	(6673-35-4)
Pralumin	(52-31-3)
Pramex	(52645-53-1)
Pramindole	(5560-72-5)
Pramitol	(1610-18-0)
Prandiol	(58-32-2)
Praparat 5968	(304-20-1)
Praseodymium (9CI)	(7440-10-0)
Praseodymium carbonate	(5895-45-4)
Praseodymium oxalate	(3269-10-1)
Praseodymium oxide (9CI)	(11113-81-8)
Praseodymium, tris(ethanedioato(2-))di- (9CI)	(3269-10-1)
Praxiten	(604-75-1)
Prazepam	(2955-38-6)
Prazepamum (Latin)	(2955-38-6)
Prazepine	(50-49-7)
Prazil	(50-53-3)
Prazin	(58-40-2)
Prazine	(58-40-2)
Praziquantel	(55268-74-1)
Prazosin	(19216-56-9)
Pre-San	(741-58-2)
Prean	(64-55-1)
Prebane	(886-50-0)
Preceptin	(9002-93-1)
Precipitated Barium Sulphate	(7727-43-7)
Precipitated Silica	(1343-98-2)
Precipitated Sulfur	(7704-34-9)
Preciptated silica	(112926-00-8)
Precision Cleaning Agent	(354-58-5)
Precor	(40596-69-8)
Precort	(53-03-2)
Precortancyl	(50-24-8)

Name	CAS No.
Precortisyl	(50-24-8)
Predent	(7681-49-4)
Predne-Dome	(50-24-8)
Prednelan	(50-24-8)
Predni-Sediv	(50-35-1)
Prednicen-M	(53-03-2)
Prednilonga	(53-03-2)
Prednis	(50-24-8)
Prednisolon F	(50-02-2)
Prednisolone	(50-24-8)
Prednisolone F	(50-02-2)
Prednisolone, 6α-fluoro-	(53-34-9)
Prednisolone, methyl-	(83-43-2)
Prednison	(53-03-2)
Prednisone	(53-03-2)
Prednizon	(53-03-2)
Predonin	(50-24-8)
Predonine	(50-24-8)
Preeglone	(85-00-7)
Prefar	(741-58-2)
Prefemin	(54-31-9)
Prefix	(1918-13-4)
Preflan	(34014-18-1)
Prefmid	(34014-18-1)
Preforan	(15457-05-3)
Prefox	(2941-55-1)
Prefrin	(61-76-7)
Pregard	(26399-36-0)
1,4-Pregnadiene-17-α,21-diol-3,11,20-trione	(53-03-2)
9-β,10-α-Pregna-4,6-diene-3,20-dione	(152-62-5)
Pregna-4,6-diene-3,20-dione, (9-β,10-α)- (9CI)	(152-62-5)
Pregna-4,6-diene-3,20-dione, 17-(acetoxy)-6-chloro-	(302-22-7)
Pregna-1,4-diene-3,20-dione, 21-(acetyloxy)-11,17-dihydroxy-6-methyl-, (6-α,11-β)-	(53-36-1)
Pregna-4,6-diene-3,20-dione, 6-chloro-17-hydroxy-, acetate	(302-22-7)
Pregna-1,4-diene-3,20-dione, 6,9-difluoro-11,12-dihydroxy-16,17-((1-methylethylidene) bis(oxy))-, (6-α,11-β,16-α)	(67-73-2)
Pregna-1,4-diene-3,20-dione, 9-fluoro-11,16,17,21-tetrahydroxy-, (11-β,16-α)-	(124-94-7)
Pregna-1,4-diene-3,20-dione, 9-fluoro-11-β,16-α,17,21-tetrahydroxy	(124-94-7)
Pregna-1,4-diene-3,20-dione, 9-fluoro-11-β,16-α,17,21-tetra-hydroxy-, cyclic 16,17-acetal with acetone	(76-25-5)
Pregna-1,4-diene-3,20-dione, 6α-fluoro-11β,17,21-trihydroxy- (8CI)	(53-34-9)
Pregna-1,4-diene-3,20-dione, 6-fluoro-11,17,21-trihydroxy-, (6α,11β)- (9CI)	(53-34-9)
Pregna-1,4-diene-3,20-dione, 6α-fluoro-11β,17,21-trihydroxy-16α-methyl-	(53-33-8)
Pregna-1,4-diene-3,20-dione, 9-fluoro-11-β,17,21-trihydroxy-16-α-methyl	(50-02-2)
Pregna-1,4-diene-3,20-dione, 9-fluoro-11-β,17,21-trihydroxy-16-β-methyl	(378-44-9)
Pregna-1,4-diene-3,20-dione, 6-fluoro-11,17,21-trihydroxy-16-methyl-, (6α,11β,16α)-	(53-33-8)
Pregna-4,6-diene-3,20-dione, 17-hydroxy-6-methyl-	(3562-63-8)
Pregna-4,6-diene-3,20-dione, 17-hydroxy-6-methyl-, acetate	(595-33-5)
Pregna-1,4-diene-3,20-dione, 6-α-methyl-11-β-17,21-trihydroxy	(83-43-2)
Pregna-1,4-diene-3,20-dione, 11-β,17,21-trihydroxy	(50-24-8)
Pregna-1,4-diene-3,20-dione, 11-β,17,21-trihydroxy-6-α-methyl-	(83-43-2)
Pregna-1,4-diene-3,20-dione, 11-β,17,21-trihydroxy-6-α-methyl-, 21-acetate	(53-36-1)
1,4-Pregnadiene-3,20-dione-11-β,17-α,21-triol	(50-24-8)
1,4-Pregnadiene-11-β,17-α,21-triol-3,20-dione	(50-24-8)
Pregna-1,4-diene-3,11,20-trione, 17,21-hydroxy	(53-03-2)
1,4-Pregnadien-11-β,17-α,21-triol-3,20-dione	(50-24-8)
17-α-2,4-Pregnadien-20-yno(2,3-d)isoxazol-17-ol	(17230-88-5)
17-α-Pregna-2,4-dien-20-yno(2,3-d)isoxazol-17-ol	(17230-88-5)
Pregna-2,4-dien-20-yno(2,3-d)isoxazol-17-ol, (17-α)- (9CI)	(17230-88-5)
5α-Pregnane (8CI)	(641-85-0)
Pregnane, (5α)- (9CI)	(641-85-0)
Pregna-1,4,6-triene-3,20-dione, 6-chloro-11,17,21-trihydroxy-, (11β)-	(5251-34-3)
Pregn-4-en-17α,21-diol-3,11,20-trione	(53-06-5)
3,20-Pregnene-4	(57-83-0)
17-α-Pregn-4-ene-21-carboxylic acid, 1-hydroxy-7-α-mercapto-3-oxo-α-lactone	(52-01-7)
17-α-Pregn-4-ene-21-carboxylic acid, 17-hydroxy-7-α-mercapto-3-oxo-, γ- lactone acetate	(52-01-7)
4-Pregnene-11-β,21-diol-3,20-dione	(50-22-6)
4-Pregnene-17α,21-diol-3,11,20-trione	(53-06-5)
δ^4-Pregnene-17α,21-diol-3,11,20-trione	(53-06-5)
4-Pregnene-17,α,21-diol-3,11,20-trione 21-acetate	(50-04-4)
4-Pregnene-3,20-dione	(57-83-0)
Pregn-4-ene-3,20-dione	(57-83-0)
Pregnene-3,20-dione	(57-83-0)
Pregnenedione	(57-83-0)
δ^4-Pregnene-3,20-dione	(57-83-0)
(6-α)-Pregn-4-ene-3,20-dione, 17-(acetyloxy)-6-methyl	(71-58-9)
Pregn-4-ene-3,20-dione, 11-β,21-dihydroxy-	(50-22-6)
Pregn-4-ene-3,20-dione, 11,21-dihydroxy-, (11-β)- (9CI)	(50-22-6)
Pregn-4-ene-3,20-dione, 6-α-fluoro-11-β,16-α,17,21-tetrahydroxy-, cyclic 16,17- acetal with acetone	(1524-88-5)
Pregn-4-ene-3,20-dione, 9-fluoro-11-β,17,21-trihydroxy	(127-31-1)
Pregn-4-ene-3,20-dione, 17-hydroxy	(68-96-2)
Pregn-4-ene-3,20-dione, 17-hydroxy-, hexanoate	(630-56-8)
Pregn-4-ene-3,20-dione, 17-hydroxy-6-α-methyl	(520-85-4)
Pregn-4-ene-3,20-dione, 17-hydroxy-6-methyl-, (6-α)- (9CI)	(520-85-4)
Pregn-4-ene-3,20-dione, 17-hydroxy-6-methyl-, acetate	(1172-82-3)
Pregn-4-ene-3,20-dione, 17-((1-oxohexyl)oxy)-	(630-56-8)
Pregn-4-ene-3,20-dione, 11,17,21-trihydroxy-, (11-β)-	(50-23-7)
4-Pregnene-11-β,17-α,21-triol 3,20-dione	(50-23-7)
Pregn-4-ene-3,11,20-trione, 21-(acetyloxy)-17-hydroxy- (9CI)	(50-04-4)
Pregn-4-ene-3,11,20-trione, 17,21-dihydroxy-	(53-06-5)
Pregn-4-ene-3,11,20-trione, 17,21-dihydroxy-, 21-acetate	(50-04-4)
17-α-Pregn-4-en-20-yno(2,3-d)isoxazol-17-ol	(17230-88-5)
17-α-Pregn-4-en-20-yn-3-one, 6-α,21-dimethyl-17-hydroxy-	(79-64-1)
Pregnyl	(9002-61-3)
Prehnitene	(488-23-3)
Prehnitic acid	(479-47-0)
Prehnitol	(488-23-3)
Prehnitylic acid	(1076-47-7)
Prelital	(143-81-7)
Preludin	(134-49-6)
Premalin	(1746-81-2)
Premalin	(330-55-2)
Premalox	(122-42-9)
Premazine	(122-34-9)
Premerge	(88-85-7)
Premerge 3	(88-85-7)
Premerge Plus	(132-66-1)
Premerge Plus	(25013-16-5)
Premgard	(10453-86-8)
Preminex	(64-55-1)
Premix	(98-50-0)
Premocillin	(6130-64-9)
Premodrin	(300-42-5)
Prenderol	(115-76-4)
Prendiol	(115-76-4)
Prenimon	(68-76-8)
Prenol	(556-82-1)
Prenolone	(50-24-8)
Prentox	(51-03-6)
Prentox	(55-38-9)
Prentox	(62-73-7)
Prentox	(83-79-4)

Name	CAS No.
Prentox Malathion 95% Spray	(121-75-5)
Prenyl alcohol	(556-82-1)
Prenylamine	(390-64-7)
Prenyl chloride	(503-60-6)
Prepalin	(68-26-8)
Preparation 125	(1836-75-5)
Preparation 6424	(54-85-3)
Preparation AF	(100-97-0)
Preparation K	(502-55-6)
Prepodyne	(26617-87-8)
Prequil	(57-53-4)
Preseed	(1113-14-0)
Preserv-O-Sote	(8001-58-9)
Preserval B	(94-18-8)
Preserval B	(94-26-8)
Preserval Butylique	(94-26-8)
Preserval M	(99-76-3)
Preserval P	(94-13-3)
Presfersul	(7782-63-0)
Presinol	(41372-08-1)
Presinol	(555-30-6)
Presolisin	(41372-08-1)
Presolisin	(555-30-6)
Presoxin	(50-56-6)
Prespersion, 75 urea	(57-13-6)
Pressimedin	(50-55-5)
Pressomin hydrochloride	(61-16-5)
Pressonex	(54-49-9)
Presulin	(2235-54-3)
Prevangor	(78-11-5)
Prevenol	(101-21-3)
Prevenol	(87-86-5)
Prevenol 56	(101-21-3)
Prevental	(97-23-4)
Preventol	(101-21-3)
Preventol	(97-23-4)
Preventol 1	(136-32-3)
Preventol 56	(101-21-3)
Preventol CMK	(59-50-7)
Preventol GD	(97-23-4)
Preventol GDC	(97-23-4)
Preventol I	(95-95-4)
Preventol O Extra	(90-43-7)
Preventol-ON	(132-27-4)
Preventol ON & ON Extra	(132-27-4)
Preventol WB (9CI)	(99752-90-6)
Previcur	(19622-19-6)
Prevocell EO	(9003-11-6)
Preweed	(101-21-3)
Preza	(561-27-3)
Prezervit	(533-74-4)
Priadel	(554-13-2)
Pridinol	(511-45-5)
Pridoxine	(65-23-6)
Pri-n-eicosyl alcohol	(629-96-9)
Prilepsin	(125-33-7)
Prilocaine	(721-50-6)
Priltox	(87-86-5)
Primabalt RP	(13422-51-0)
Primacione	(125-33-7)
Primaclone	(125-33-7)
Primacol	(86-87-3)
Primacone	(125-33-7)
Primagram	(51218-45-2)
Primakton	(125-33-7)
Primal	(539-21-9)
Primal ASE 60	(9003-01-4)
Primal (VAN)	(57657-42-8)
Primary ammonium phosphate	(7722-76-1)
Primary amyl acetate	(628-63-7)
Primary amyl alcohol	(71-41-0)
Primary coco amine ethylene oxide adduct	(61791-14-8)
Primary copper phosphate	(10103-48-7)
Primary decyl alcohol	(112-30-1)
Primary isobutyl iodide	(513-38-2)
Primary octyl alcohol	(111-87-5)
Primary sodium phosphate	(7558-80-7)
Primatene mist	(51-43-4)
Primatol	(1610-17-9)
Primatol	(1610-18-0)
Primatol	(1912-24-9)
Primatol A	(1912-24-9)
Primatol 25E	(1610-18-0)
Primatol M	(5915-41-3)
Primatol-M80	(5915-41-3)
Primatol P	(139-40-2)
Primatol Q	(7287-19-6)
Primatol S	(122-34-9)
Primaze	(1912-24-9)
Primazin	(57-68-1)
Primextra	(51218-45-2)
Primicarbe	(23103-98-2)
Primicid	(23505-41-1)
Primidon	(125-33-7)
Primidone	(125-33-7)
Primin	(119-38-0)
Primine	(58-33-3)
Primofol	(50-28-2)
Primogonyl	(9002-61-3)
Primogyn	(57-63-6)
Primogyn B	(50-50-0)
Primogyn Boleosum	(50-50-0)
Primogyn C	(57-63-6)
Primogyn I	(50-50-0)
Primogyn M	(57-63-6)
Primol 355	(8012-95-1)
Primol D	(8012-95-1)
Primolut	(57-83-0)
Primolut Depot	(630-56-8)
Primotec	(23505-41-1)
Primotest	(58-22-0)
Primoteston	(58-22-0)
Primperan	(364-62-5)
Primrose Chrome	(1344-37-2)
Primrose Yellow	(13530-65-9)
Prinadol	(127-35-5)
Princep	(122-34-9)
Princillin	(7177-48-2)
Principal Bile Pigment	(635-65-4)
Principen	(69-53-4)
Principen	(7177-48-2)
Prinicid	(23505-41-1)
Prinsyl	(133-53-9)
Printel's	(9003-53-6)
Printex	(1333-86-4)
Printex 60	(1333-86-4)
Printop	(122-34-9)
Prioderm	(121-75-5)
Priospen	(132-93-4)
Prisilidine	(77-20-3)
Prist	(109-86-4)
Pristacin	(123-03-5)
Pristane	(1921-70-6)
Privine	(835-31-4)

Privine hydrochloride	(550-99-2)
Prizole hydrochloride	(550-99-2)
No-Pro	(7519-36-0)
Pro	(67-63-0)
Pro-Ban M	(50-35-1)
Pro-Dexter	(51-63-8)
Pro-Dorm	(72-44-6)
Pro-Drone	(53905-38-7)
Pro-Gen	(92-62-6)
Pro-Gen	(98-50-0)
Pro-Gen 227	(98-50-0)
Pro-Gen Sodium	(127-85-5)
Pro-Gibb	(77-06-5)
Prop-Job	(709-98-8)
Pro-Nox Fish	(83-79-4)
Pro-Pen	(6130-64-9)
Pro-Sonil	(52-31-3)
Proazaimine	(60-87-7)
Proazamine	(60-87-7)
Probamato	(57-53-4)
Probamyl	(57-53-4)
Proban	(115-93-5)
Pro-banthine	(50-34-0)
Probe	(20354-26-1)
Probecid	(57-66-9)
Probedryl	(58-73-1)
Proben	(57-66-9)
Probenecid	(57-66-9)
Probenecid acid	(57-66-9)
Probenemid	(57-66-9)
Probese-P	(134-49-6)
Procain	(59-46-1)
Procainamide	(51-06-9)
Procaine	(59-46-1)
Procaine amide	(51-06-9)
Procaine, base	(59-46-1)
Procaine benzylpenicillinate	(54-35-3)
Procaine benzylpenicillinate	(6130-64-9)
Procaine hydrochloride	(51-05-8)
Procaine penicillin	(6130-64-9)
Procaine penicillin G	(54-35-3)
Procaine penicillin G	(6130-64-9)
Procalmadiol	(57-53-4)
Procalmadol	(57-53-4)
Procalmidol	(57-53-4)
Procamide	(51-06-9)
Procanodia	(6130-64-9)
Procarbamide	(57-53-4)
Procarbazin (German)	(366-70-1)
Procarbazin (German)	(671-16-9)
Procarbazine	(671-16-9)
Procarbazine hydrochloride	(366-70-1)
Procardia	(21829-25-4)
Procardine	(59-26-7)
Procasil	(51-52-5)
Procene UF 1.5	(9002-88-4)
Processed Lanolin	(8006-54-0)
Prochloroperazine	(58-38-8)
Prochlorpemazine	(58-38-8)
Prochlorperazine	(58-38-8)
Prochlorpromazine	(58-38-8)
Procion Black H-N	(12225-26-2)
Procion Blue (9CI)	(85568-72-5)
Procion Brilliant Blue H 3R	(12236-92-9)
Procion Brilliant Red 2BS	(17752-85-1)
Procion Brilliant Red 5BS	(17804-49-8)
Procion Brilliant Red M 2B	(17752-85-1)
Procion Brilliant Red M 5B	(17804-49-8)
Procion Brilliant Red M-2BS	(17752-85-1)
Procion Brilliant Red MX 5B	(17804-49-8)
Procion Golden Yellow H-R	(12225-84-2)
Procion Golden Yellow HRS	(12225-84-2)
Procion Red 2BS	(17752-85-1)
Procion Red MX 5B	(17804-49-8)
Procion Yellow H 3R	(12225-84-2)
Procit	(60-87-7)
Procol OA-25	(9004-98-2)
Proconazole	(60207-90-1)
Procorman	(59-26-7)
Proctin	(9005-38-3)
Proctocort	(50-23-7)
Procutene	(101-20-2)
Procyazine	(32889-48-8)
Procymidone	(32809-16-8)
Procytox	(50-18-0)
Procytox	(6055-19-2)
Prodalumnol	(7784-46-5)
Prodalumnol double	(7784-46-5)
Prodan	(16893-85-9)
Prodaram	(137-30-4)
Prodectine	(1882-26-4)
Prodel	(152-62-5)
Prodhybas N	(111-60-4)
Prodhybase ethyl	(111-60-4)
Prodhygine	(822-16-2)
Prodhyphore B	(9004-96-0)
α-Prodine	(77-20-3)
β-Prodine	(468-59-7)
β-Prodinol	(468-59-7)
Prodipate	(6938-94-9)
Prodix	(68-96-2)
Prodixamon	(50-34-0)
Prodox	(68-96-2)
Prodox 131	(88-69-7)
Prodox 133	(99-89-8)
Prodox 146	(96-76-4)
Prodox 156	(120-95-6)
Prodox 340	(1879-09-0)
Prodox 146A-85X	(96-76-4)
Prodoxol	(14698-29-4)
Prodromine	(509-67-1)
Product 308	(36653-82-4)
Product 5022	(1982-37-2)
Product No. 75	(151-21-3)
Product No. 161	(151-21-3)
Profam	(122-42-9)
Profamina	(300-62-9)
Profamina	(60-13-9)
Profarmil	(50-35-1)
Profax	(9003-07-0)
Profax 6301	(9003-07-0)
Profax 6401	(9003-07-0)
Profax 6423	(9003-07-0)
Profax 6501	(9003-07-0)
Profax 6523	(9003-07-0)
Profax 6601	(9003-07-0)
Profax 6723	(9003-07-0)
Profax 6823	(9003-07-0)
Profax A 60-008	(9002-88-4)
Profax 6523F	(9003-07-0)
Profax PCO 72	(9003-07-0)
Profecundin	(59-02-9)
Profemin	(54-31-9)
Profenid	(22071-15-4)

Profenil	(150-59-4)
Profenofos	(41198-08-7)
Profenone	(70-70-2)
Proferrin	(8047-67-4)
Profetamine	(60-13-9)
Proflavin	(92-62-6)
Proflavine	(92-62-6)
Proflavine dihydrochloride	(531-73-7)
Proflavine hemisulphate	(1811-28-5)
Proflavine hydrochloride	(952-23-8)
Proflavine monohydrochloride	(952-23-8)
Proflavine (sulfate)	(553-30-0)
Proflavine sulphate	(553-30-0)
Proflavin sulfate	(553-30-0)
Profluralin	(26399-36-0)
Profluraline	(26399-36-0)
Profoliol	(50-28-2)
Profoliol	(92-62-6)
Profoliol-B	(92-62-6)
Proformiphen	(92-62-6)
Profume A	(76-06-2)
Profume (Obs.)	(74-83-9)
Profundal	(77-26-9)
Profundol	(115-44-6)
Profundol	(92-62-6)
Profura	(92-62-6)
Progallin A	(831-61-8)
Progallin P	(121-79-9)
Progarmed	(92-62-6)
Pro-gastron	(50-34-0)
Progekan	(57-83-0)
Progen 90	(98-50-0)
Progesic	(92-62-6)
Progestasert	(57-83-0)
Progesterol	(57-83-0)
Progesterone	(57-83-0)
β-Progesterone	(57-83-0)
Progesterone Retard Pharlon	(630-56-8)
Progesterone caproate	(630-56-8)
Progesterone, 17-α-hydroxy-6-α-methyl-, acetate	(71-58-9)
Progesteronum	(57-83-0)
Progestin	(57-83-0)
Progestone	(57-83-0)
Progynon-DH	(50-28-2)
Progynon-DP	(113-38-2)
Progynon-Depot	(979-32-8)
Progynon	(50-28-2)
Progynon	(979-32-8)
Progynon B	(50-50-0)
Progynon Benzoate	(50-50-0)
Progynon C	(57-63-6)
Progynova	(979-32-8)
Proheptadiene	(50-48-6)
Proheptazine	(77-14-5)
Prokarbol	(534-52-1)
Prokayvit	(58-27-5)
Proksanol	(9003-11-6)
Prolactin	(9002-62-4)
Prolamine	(123-75-1)
Prolan	(117-27-1)
Prolan B	(9002-68-0)
Prolan (CSC)	(117-27-1)
Prolate	(732-11-6)
Prolidon	(57-83-0)
(L)-Proline	(147-85-3)
DL-Proline (9CI)	(609-36-9)
Proline, DL- (8CI)	(609-36-9)
Proline, L	(147-85-3)
L-Proline, 4-hydroxy-	(51-35-4)
L-Proline, 4-hydroxy-, trans- (9CI)	(51-35-4)
Proline, 4-hydroxy-, L- (8CI)	(51-35-4)
Proline, 4-hydroxy- (VAN)	(51-35-4)
Proline, 4-hydroxy-1-nitroso-, L	(30310-80-6)
l-Proline, 1-(3-mercapto-2-methyl-1-oxopropyl)-, (S)-	(62571-86-2)
Proline, N-nitroso-, L	(7519-36-0)
L-Proline, 5-oxo- (9CI)	(98-79-3)
Prolixin	(69-23-8)
Prolixine	(69-23-8)
Proloid	(9010-34-8)
Prolongal	(9004-66-4)
Prolongine	(57-66-9)
Proloprim	(738-70-5)
Proluton Depot	(630-56-8)
Prolutone	(57-83-0)
Promacid	(69-09-0)
Promactil	(50-53-3)
Promamide	(23950-58-5)
Promantine	(58-33-3)
Promapar	(69-09-0)
Promar	(82-66-6)
Promassol	(539-21-9)
Promate	(57-53-4)
Promato	(57-53-4)
Promaxon P60	(1344-95-2)
Promazil	(50-53-3)
Promazin	(58-40-2)
Promazina (Italian)	(58-40-2)
Promazinamide	(60-87-7)
Promazine	(58-40-2)
Promazine hydrochloride	(53-60-1)
Promecarb	(2631-37-0)
Promecarbe	(2631-37-0)
Promedol	(64-39-1)
Promeran	(62-37-3)
Promet	(65907-30-4)
Promet 660SCO	(65907-30-4)
Prometasin	(60-87-7)
Prometazin	(60-87-7)
Promethazine	(60-87-7)
Promethazine N-(2'-dimethylamino-2'-methylethyl)phenothiazine hydrochloride	(58-33-3)
Promethazine hydrochloride	(58-33-3)
Promethiazin (German)	(58-33-3)
Promethiazine	(60-87-7)
Promethryn	(7287-19-6)
Prometon	(1610-18-0)
Prometone	(1610-18-0)
Prometrex	(7287-19-6)
Prometrin	(7287-19-6)
Prometryn	(7287-19-6)
Prometryne	(7287-19-6)
Promezathine	(60-87-7)
Promiben	(50-49-7)
Promidione	(36734-19-7)
Promilan	(63428-84-2)
Prominal	(115-38-8)
Promptonal	(50-06-6)
Promul 5080	(106-11-6)
Promyr	(110-27-0)
Pronamide	(23950-58-5)
Prondol	(5560-72-5)
Pronestyl	(51-06-9)
Pronetalol hydrochloride	(51-02-5)
Pronethalol hydrochloride	(51-02-5)

oxy) derivs., chlorides (68187-63-3)
2-Propanaminium, N-methyl-N-(1-methylethyl)-N-(2-((9H-xanthen-9-ylcarbonyl)oxy)ethyl)- (298-50-0)
1-Propanaminium, N,N,N-trimethyl-3-((1-oxododecyl)amino)-, methyl sulfate (9CI) (10595-49-0)
1-Propanaminium, N,N,N-tripropyl-, iodide (9CI) (631-40-3)
Propane (OSHA) [UN 1978] (74-98-6)
Propane, 1-(allyloxy)-2,3-epoxy- (106-92-3)
Propane, 2-amino- (75-31-0)
Propane, 1-(2-biphenylyloxy)-2,3-epoxy- (8CI) (7144-65-2)
Propane, 1,3-bis(carbamoylthio)-2-(N,N-dimethylamino)-, hydrochloride (15263-52-2)
Propane, 1,1-bis(p-chlorophenyl)-2-nitro (117-27-1)
Propane, 1,1-bis(p-chlorophenyl)-2-nitro- mixed with 1,1-bis-(p-chlorophenyl)-2-nitro butane (1:2) (8027-00-7)
Propane, 1,3-bis(diazo)- (5239-06-5)
Propane, (+-)-1,2-bis(3,5-dioxopiperazin-1-yl)- (21416-87-5)
Propane, 1,3-bis(2,3-epoxypropoxy)-2,2-dimethyl (17557-23-2)
Propane, 2,2-bis(p-(2,3-epoxypropoxy)phenyl) (1675-54-3)
Propane, 1,1-bis(p-ethoxyphenyl)-2,2-dimethyl (27955-87-9)
Propane, 1,1-bis(p-ethoxyphenyl)-2-nitro (26258-70-8)
Propane, 2,2-bis(ethylsulfonyl) (115-24-2)
Propane, 2,2-bis(p-hydroxyphenyl)- (80-05-7)
Propane, 1,1-bis(p-methoxyphenyl)-2,2-dimethyl (4741-74-6)
Propane, 1,1-bis(p-methoxyphenyl)-2-nitro- (8CI) (34197-26-7)
Propane, 2,2-bis(methylthio)- (9CI) (6156-18-9)
Propane, 1-bromo (106-94-5)
Propane, 2-bromo (75-26-3)
Propane, bromo- (8CI,9CI) (26446-77-5)
Propane, 1-bromo-3-chloro (109-70-6)
Propane, 2-bromo-1-chloro- (9CI) (3017-95-6)
Propane, bromochloro- (9CI) (34652-54-5)
Propane, 1-bromo-2-chloro- (8CI,9CI) (3017-96-7)
Propane, 3-bromo-1,1-dichloro- (9CI) (36668-45-8)
Propane, 1-bromo-2,3-dichloro- (8CI,9CI) (33037-07-9)
Propane, 2-bromo-1,2-dichloro- (7CI,8CI,9CI) (17759-88-5)
Propane, 1-bromo-2,2-dimethyl- (9CI) (630-17-1)
Propane, 3-bromo-1,2-epoxy (3132-64-7)
Propane, 1-bromo-2-methyl (78-77-3)
Propane, 2-bromo-2-methyl (507-19-7)
Propane, 1-bromo-1,1,2,2-tetrafluoro- (9CI) (70192-84-6)
Propane, 1-butoxy-2,3-epoxy (2426-08-6)
Propane, 1-tert-butoxy-2,3-epoxy- (7665-72-7)
Propane, 1-(p-tert-butylphenoxy)-2,3-epoxy (3101-60-8)
Propane, 1-(tert-butylphenoxy)-2,3-epoxy- (7CI,8CI) (26447-45-0)
Propane, 1-(p-tert-butylphenoxy)-2,3-epoxy-, Polymers (8CI) (29298-03-1)
1-Propanecarboxylic acid (107-92-6)
Propane, 1-chloro (540-54-5)
Propane, 2-chloro (75-29-6)
Propane, 1-chloro-2,3-dibromo- (96-12-8)
Propane, 3-chloro-1,2-dibromo-2-methyl (10474-14-3)
Propane, 1-chloro-2,2-dimethyl- (9CI) (753-89-9)
Propane, 1-chloro-2,3-epoxy (106-89-8)
Propane, 1-chloro-2,3-epoxy-, (+-)- (8CI) (13403-37-7)
Propane, 1-chloro-1,1,2,2,3,3-hexafluoro (422-55-9)
Propane, 2-chloro-2-methyl (507-20-0)
Propane, 1-chloro-2-methyl- (9CI) (513-36-0)
Propane, 1-chloro-1-nitro (600-25-9)
Propane, 1-chloro-2-nitro (2425-66-3)
Propane, 2-chloro-2-nitro (594-71-8)
Propane, 3-chloro-1,1,1,2,2-pentafluoro- (8CI) (422-02-6)
Propane, 1-chloro-1,1,2,2-tetrafluoro- (8CI) (421-75-0)
Propane, 3-chloro-1,1,1-trifluoro (460-35-5)
Propane, 1-cyclopropyl- (6CI,7CI,8CI) (2415-72-7)
Propane-1,1,1,3,3,3-d6, 2-chloro-2-(methyl-d3)- (8CI,9CI) (918-20-7)
1,3-Propanedial (542-78-9)
Propanedial (542-78-9)
1,3-Propanedialdehyde (542-78-9)
Propanedial, ion(1-), sodium (9CI) (24382-04-5)
Propanediamide, 2-ethyl-2-phenyl (7206-76-0)
1,2-Propanediamine (78-90-0)
1,3-Propanediamine (109-76-2)
1,3-Propanediamine, N-(2-aminoethyl)- (9CI) (13531-52-7)
1,3-Propanediamine, N-(3-aminopropyl)- (56-18-8)
1,3-Propanediamine, N-(3-aminopropyl)-N-butyl- (9CI) (1555-68-6)
1,3-Propanediamine, N,N-bis(3-(dimethylamino)propyl)-N',N'-dimethyl- (9CI) (33329-35-0)
1,3-Propanediamine, N-(2-chloroethyl)-N'-(6-chloro-2-methoxy-9-acridinyl)-N-ethyl-, dihydrochloride, hydrate (146-59-8)
1,3-Propanediamine, N-cyclohexyl- (9CI) (3312-60-5)
1,3-Propanediamine, N'-cyclohexyl-N,N-dimethyl- (9CI) (71326-18-6)
1,3-Propanediamine, N-cyclohexyl-N'-methyl- (9CI) (90853-13-7)
1,3-Propanediamine, N,N-diethyl (104-78-9)
1,3-Propanediamine, N,N-dimethyl (109-55-7)
1,3-Propanediamine, 2,2-dimethyl- (9CI) (7328-91-8)
1,3-Propanediamine, N-(3-(dimethylamino)propyl)-N,N',N'-trimethyl- (9CI) (3855-32-1)
1,3-Propanediamine, N-(1,1-dimethylethyl)- (9CI) (52198-64-8)
1,3-Propanediamine, N,N-dimethyl-N'-(3-nitro-9-acridinyl)- (9CI) (6237-24-7)
1,3-Propanediamine, N,N'-diphenyl- (9CI) (104-69-8)
1,3-Propanediamine, N-docosyl- (9CI) (15268-40-3)
1,3-Propanediamine, N,N''-1,2-ethanediylbis- (9CI) (10563-26-5)
1,3-Propanediamine, N,N''-ethylenebis- (10563-26-5)
1,3-Propanediamine, N-methyl- (9CI) (6291-84-5)
1,3-Propanediamine, N-9-octadecenyl-, (Z)- (9CI) (7173-62-8)
1,3-Propanediamine, N-9-octadecenyl-, (Z)-, mono((2,4,5-trichlorophenoxy)acetate) (9CI) (53404-87-8)
1,3-Propanediamine, N-tallow-, diacetate (68911-78-4)
Propane, 1,2-dibromo (78-75-1)
Propane, 1,3-dibromo (109-64-8)
Propane, 2,2-dibromo- (9CI) (594-16-1)
Propane, 1,2-dibromo-3-chloro (96-12-8)
Propane, 1,1-dibromo-2-chloro- (9CI) (55162-35-1)
Propane, dibromochloro- (9CI) (67708-83-2)
Propane, 1,2-dibromo-1-chloro-2-methyl- (9CI) (69036-12-0)
Propane, 1,3-dibromo-2,2-dimethyl- (9CI) (5434-27-5)
Propane, 1,2-dibromo-2-methyl (594-34-3)
1,2-Propanedicarboxylic acid (498-21-5)
1,3-Propanedicarboxylic acid (110-94-1)
Propane, 1,1-dichloro (78-99-9)
Propane, 1,2-dichloro (78-87-5)
Propane, 1,3-dichloro (142-28-9)
Propane, dichloro (26638-19-7)
Propane, 2,2-dichloro- (9CI) (594-20-7)
Propane, 1,2-dichloro-, Mixt. with 1,3-Dichloropropene and isothiocyanatomethane (8066-01-1)
Propane, 1,2-dichloro-2-methyl (594-37-6)
Propane, dichloro- mixed with propene, dichloro (8003-19-8)
Propane, 2,2-diethoxy- (126-84-1)
Propane, 1,1-diethoxy- (9CI) (4744-08-5)
Propane, 1,1-diethoxy-2-methyl- (1741-41-9)
Propane-diethyl sulfone (115-24-2)
Propane, 3-(difluoromethoxy)-1,1,1,2,2-pentafluoro- (9CI) (56860-81-2)
1,3-Propanedimercaptan (109-80-8)
Propane, 1,1-dimethoxy- (9CI) (4744-10-9)
Propane, 2,2-dimethoxy- (9CI) (77-76-9)
Propane, 1,2-dimethoxy- (8CI,9CI) (7778-85-0)
Propane, 1,1-dimethoxy-2-methyl- (9CI) (41632-89-7)
Propane, 2,2-dimethyl (463-82-1)
Propane, 1-(N,N-dimethylamino)-2-chloro-, hydrochloride (4584-49-0)
Propane, 2-((1,1-dimethylethyl)sulfonyl)-2-methyl- (9CI) (1886-75-5)
Propanedinitrile, ((2-chlorophenyl)methylene) (2698-41-1)
Propanedinitrile, ((chlorophenyl)methylene)- (9CI) (35254-70-7)
Propanedinitrile, dimethyl- (9CI) (7321-55-3)

Propanedinitrile, (1-ethoxyethylidene)- (9CI) **(5417-82-3)**
Propanedinitrite **(109-77-3)**
Propanedioic acid **(141-82-2)**
Propanedioic acid, ((3,5-bis(1,1-dimethylethyl)-4-hydroxyphenyl)-methyl)butyl-, bis(1,2,2,6,6- pentamethyl-4-piperidinyl) ester **(63843-89-0)**
Propanedioic acid, butyl- (9CI) **(534-59-8)**
Propanedioic acid, butyl(2-methoxyethyl)-, diethyl ester (9CI) **(20591-91-7)**
Propanedioic acid, butylmethyl-, diethyl ester (9CI) **(55114-29-9)**
Propanedioic acid, (carboxymethoxy) **(55203-12-8)**
Propanedioic acid, (carboxymethoxy)-, trisodium salt (9CI) **(41999-58-0)**
Propanedioic acid, chloro- (9CI) **(600-33-9)**
Propanedioic acid, decyl- (9CI) **(4372-29-6)**
Propanedioic acid, diethyl ester **(105-53-3)**
Propanedioic acid, dimethyl- (9CI) **(595-46-0)**
Propanedioic acid, dimethyl-, diethyl ester (9CI) **(1619-62-1)**
Propanedioic acid, dimethyl ester (9CI) **(108-59-8)**
Propanedioic acid, (1,1-dimethylethyl)methyl-, diethyl ester (9CI) **(53268-44-3)**
Propanedioic acid, dithallium salt (9CI) **(2757-18-8)**
Propanedioic acid, 1,3-dithiolan-2-ylidene-, bis(1-methylethyl) ester **(50512-35-1)**
Propanedioic acid, ethyl- (9CI) **(601-75-2)**
Propanedioic acid, ethyl-, diethyl ester (9CI) **(133-13-1)**
Propanedioic acid, ethyl(2-methoxyethyl)-, diethyl ester (9CI) **(20591-89-3)**
Propanedioic acid, ethylmethyl-, diethyl ester (9CI) **(2049-70-9)**
Propanedioic acid, (2-methoxyethyl)-, diethyl ester (9CI) **(6335-02-0)**
Propanedioic acid, (2-methoxyethyl)methyl-, diethyl ester (9CI) **(20721-76-0)**
Propanedioic acid, (2-methoxyethyl)(1-methylethyl)-, diethyl ester (9CI) **(20721-77-1)**
Propanedioic acid, (2-methoxyethyl)propyl-, diethyl ester (9CI) **(20591-90-6)**
Propanedioic acid, methyl-, diethyl ester (9CI) **(609-08-5)**
Propanedioic acid, (1-methylethyl)-, diethyl ester (9CI) **(759-36-4)**
Propanedioic acid, methyl(1-methylethyl)-, diethyl ester (9CI) **(58447-69-1)**
Propanedioic acid, methyl(2-methylpropyl)-, diethyl ester (9CI) **(58447-70-4)**
Propanedioic acid, (2-methylpropyl)-, diethyl ester (9CI) **(10203-58-4)**
Propanedioic acid, methylpropyl-, diethyl ester (9CI) **(55898-43-6)**
Propanedioic acid, octyl- (9CI) **(760-55-4)**
Propanedioic acid, pentyl- (9CI) **(616-88-6)**
Propanedioic acid, propyl- (9CI) **(616-62-6)**
Propanedioic acid, propyl-, diethyl ester (9CI) **(2163-48-6)**
1,2-Propanediol **(57-55-6)**
1,3-Propanediol **(504-63-2)**
Propane-1,2-diol **(57-55-6)**
Propane-1,3-diol **(504-63-2)**
1,2(or 3)-Propanediol, 1-acrylate **(25584-83-2)**
1,2-Propanediol, 1-acrylate **(999-61-1)**
1,2-Propanediol, 3-allyloxy **(123-34-2)**
Propanediol, (allyloxy) **(25136-53-2)**
1,2-Propanediol, 3-amino **(616-30-8)**
1,3-Propanediol, 2-amino-2-ethyl- (9CI) **(115-70-8)**
1,3-Propanediol, 2-amino-2-(hydroxymethyl) **(77-86-1)**
1,3-Propanediol, 2-amino-2-methyl **(115-69-5)**
1,3-Propanediol, 2-((benzoyloxy)methyl)-2-methyl-, dibenzoate (9CI) **(4196-87-6)**
1,3-Propanediol, 2,2-bis((benzoyloxy)methyl)-, dibenzoate **(4196-86-5)**
1,3-Propanediol, 2,2-bis(bromomethyl) **(3296-90-0)**
1,3-Propanediol, 2,2-bis((3-hydroxy-2,2-bis(hydroxymethyl)-propoxy)methyl)- (9CI) **(78-24-0)**
1,3-Propanediol, 2,2-bis(hydroxymethyl)- **(115-77-5)**
1,3-Propanediol, 2,2-bis((nitrooxy)methyl)-, dinitrate (ester) **(78-11-5)**
1,2-Propanediol, 3-bromo **(4704-77-2)**
1,3-Propanediol, 2-(bromomethyl)-2-(hydroxymethyl)- (9CI) **(19184-65-7)**
1,3-Propanediol, 2-bromo-2-nitro- (8CI,9CI) **(52-51-7)**
1,3-Propanediol, 2-butyl-2-ethyl- (8CI,9CI) **(115-84-4)**
1,3-Propanediol, 2-sec-butyl-2-methyl-, dicarbamate (8CI) **(64-55-1)**
1,2-Propanediol carbonate **(108-32-7)**
1,2-Propanediol, 3-chloro **(96-24-2)**
1,3-Propanediol, 2-chloro- (8CI,9CI) **(497-04-1)**
1,2-Propanediol, 3-(p-chlorophenoxy)-, 1-carbamate **(886-74-8)**
1,2-Propanediol cyclic carbonate **(108-32-7)**
1,2-Propanediol, diacetate **(623-84-7)**
1,2-Propanediol, dibenzoate (9CI) **(19224-26-1)**
1,3-Propanediol, 2-(2,4-dichlorophenoxy)-1-(3,4-dimethoxy-phenyl)- (9CI) **(75217-44-6)**
1,3-Propanediol, 2,2-diethyl **(115-76-4)**
1,3-Propanediol, 1-(3,4-dimethoxyphenyl)-2-(3,5-dimethyl-4-(methylthio)phenoxy)- (9CI) **(78749-45-8)**
1,3-Propanediol, 1-(3,4-dimethoxyphenyl)-2-(4-methoxy-phenoxy)- (9CI) **(22676-00-2)**
1,3-Propanediol, 1-(3,4-dimethoxyphenyl)-2-(4-(methyl-thio)phenoxy)- (9CI) **(75383-83-4)**
1,3-Propanediol, 1-(3,4-dimethoxyphenyl)-2-(4-nitrophenoxy)-(9CI) **(81826-15-5)**
1,3-Propanediol, 1-(3,4-dimethoxyphenyl)-2-phenoxy- (9CI) **(75217-43-5)**
1,3-Propanediol, 2,2-dimethyl **(126-30-7)**
1,3-Propanediol, 2,2-dimethyl-, diacrylate **(2223-82-7)**
1,3-Propanediol, 2,2-dimethyl-, dibenzoate (9CI) **(4196-89-8)**
1,2-Propanediol, dinitrate **(6423-43-4)**
1,2-Propanediol, 3-(dodecylamino)- (9CI) **(821-91-0)**
Propanediol, (2,3-epoxypropoxy)- (8CI) **(32555-29-6)**
1,3-Propanediol, 2-ethyl-2-(hydroxymethyl) **(77-99-6)**
1,3-Propanediol, 2-ethyl-2-(hydroxymethyl)-, Polymer with (chloromethyl)oxirane (9CI) **(30499-70-8)**
1,3-Propanediol, 2-ethyl-2-(hydroxymethyl)-, cyclic phosphate **(1005-93-2)**
1,3-Propanediol, 2-ethyl-2-(hydroxymethyl)-, cyclic phosphite (1:1) (8CI) **(824-11-3)**
1,3-Propanediol, 2-ethyl-2-(hydroxymethyl)-, diallyl ether **(682-09-7)**
1,3-Propanediol, 2-ethyl-2-(hydroxymethyl)-, triacrylate **(15625-89-5)**
1,3-Propanediol, 2-ethyl-2-hydroxymethyl-, trimethacrylate **(3290-92-4)**
1,2-Propanediol, 3-(ethyl(3-methylphenyl)amino)- (9CI) **(92-11-5)**
1,3-Propanediol, 2-ethyl-2-nitro- (9CI) **(597-09-1)**
1,3-Propanediol formal **(505-22-6)**
1,3-Propanediol, 1-(4-hydroxy-3-methoxyphenyl)-2-(2-methoxy-phenoxy)- **(7382-59-4)**
1,3-Propanediol, 2-(hydroxymethyl)-2-methyl- (9CI) **(77-85-0)**
1,3-Propanediol, 2-(hydroxymethyl)-2-methyl-, tribenzoate **(4196-87-6)**
1,3-Propanediol, 2-(hydroxymethyl)-2-nitro **(126-11-4)**
1,3-Propanediol, 2-(hydroxymethyl)-2-nitro-, trinitrate (ester) **(20820-44-4)**
1,2-Propanediol, maleic anhydride, phthalic anhydride polymer **(25037-66-5)**
1,2-Propanediol, 3-mercapto **(96-27-5)**
1,2-Propanediol, 3-(2-methoxyphenoxy)- **(93-14-1)**
1,2-Propanediol, 3-(o-methoxyphenoxy) **(93-14-1)**
1,2-Propanediol, 3-(o-methoxyphenoxy)-, 1-carbamate **(532-03-6)**
1,3-Propanediol, 2-methyl-2-(1-methylpropyl)-, dicarbamate (9CI) **(64-55-1)**
1,2-Propanediol, 2-methyl, monomethacrylate **(27813-02-1)**
1,3-Propanediol, 2-methyl-2-nitro **(77-49-6)**
1,3-Propanediol, 2-methyl-2-((nitrooxy)methyl)-, dinitrate (ester) (9CI) **(3032-55-1)**
1,3-Propanediol, 2-methyl-2-propyl-, butylcarbamate carbamate **(4268-36-4)**
1,3-Propanediol, 2-methyl-2-propyl-, carbamate isopropyl-carbamate (ester) **(78-44-4)**
1,3-Propanediol, 2-methyl-2-propyl-, dicarbamate **(57-53-4)**
1,2-Propanediol, monobutyl ether **(29387-86-8)**
1,2-Propanediol, monomethyl ether **(1320-67-8)**
1,2-Propanediol, monostearate **(1323-39-3)**
1,2-Propanediol, 3-(1-naphthyloxy) **(36112-95-5)**
1,2-Propanediol, 3-(2-nitroimidazol-1-yl) **(13551-92-3)**
1,2-Propanediol, 3-(2-nitro-1H-imidazol-1-yl)-, (9CI) **(13551-92-3)**
1,3-Propanediol, 2-nitro-2-((nitrooxy)methyl)-, dinitrate (ester) (9CI) **(20820-44-4)**
Propanediol, (oxiranylmethoxy)- (9CI) **(32555-29-6)**
Propanediol, oxybis- (9CI) (VAN) **(51266-87-6)**
Propanediol, oxybis- (9CI) (VAN) **(59113-36-9)**
1,3-Propanediol, 2,2'-(oxybis(methylene))bis(2-(hydroxymethyl)-(9CI) **(126-58-9)**

Name	CAS No.
1,2-Propanediol, 3,3'-oxydi-, tetranitrate	**(20600-96-8)**
1,2-Propanediol, 3-phenoxy	**(538-43-2)**
1,2-Propanediol, 1-phenyl- (9CI)	**(1855-09-0)**
Propanediol, (2-propenyloxy)-	**(25136-53-2)**
1,3-Propanedione	**(542-78-9)**
Propanedione	**(78-98-8)**
1,3-Propanedione, 1-(4-(1,1-dimethylethyl)phenyl)-3-(4-methoxy-phenyl)- (9CI)	**(70356-09-1)**
1,3-Propanedione, 1,3-diphenyl- (9CI)	**(120-46-7)**
1,2-Propanedione, 1-oxime	**(306-44-5)**
1,2-Propanedione, 1-phenyl- (9CI)	**(579-07-7)**
Propane, 1,2-diphenyl-	**(5814-85-7)**
Propane, 1,3-diphenyl- (8CI)	**(1081-75-0)**
Propane, 2,2-diphenyl- (8CI)	**(778-22-3)**
1,3-Propanedithiol	**(109-80-8)**
1,3-Propanedithiol, 2-(dimethylamino)-, dicarbamate (ester), hydrochloride	**(15263-52-2)**
4,4'-(1,3-Propanediyl)bispiperidine	**(16898-52-5)**
1,2-Propanediyl carbonate	**(108-32-7)**
Propane, 1-(dodecyloxy)-2,3-epoxy-	**(2461-18-9)**
Propane, 1,3-epithio-	**(287-27-4)**
Propane, 1,2-epoxy	**(75-56-9)**
Propane, epoxy-	**(75-56-9)**
Propane, 1,3-epoxy-2,2-dimethyl	**(6921-35-3)**
Propane, 1,2-epoxy-3-ethoxy	**(4016-11-9)**
Propane, 1,2-epoxy-3-((2-ethylhexyl)oxy)	**(2461-15-6)**
Propane, 1,2-epoxy-3-fluoro	**(503-09-3)**
Propane, 1,2-epoxy-1,1,2,3,3,3-hexafluoro	**(428-59-1)**
Propane, 1,2-epoxy-3-(hexyloxy)	**(5926-90-9)**
Propane, 1,2-epoxy-3-isopropoxy	**(4016-14-2)**
Propane, 1,2-epoxy-3-methoxy	**(930-37-0)**
Propane, 1,2-epoxy-2-methyl	**(558-30-5)**
Propane, 1,2-epoxy-3-(p-nitrophenoxy)	**(5255-75-4)**
Propane, 1,2-epoxy-3-(octyloxy)- (8CI)	**(3385-66-8)**
Propane, 1,2-epoxy-3-phenoxy	**(122-60-1)**
Propane, 1,2-epoxy-3-propoxy	**(3126-95-2)**
Propane, 1,2-epoxy-3-(p-tolyloxy)	**(2186-24-5)**
Propane, 1,2-epoxy-3-(tolyloxy)	**(26447-14-3)**
Propane, 1,2-epoxy-3-(o-tolyoxy)	**(2210-79-9)**
Propane, 1,2-epoxy-3,3,3-trichloro	**(3083-23-6)**
Propane, 1,1'-(1,2-ethanediylbis(sulfonyl))bis- (9CI)	**(3563-34-6)**
Propane, 1,1'-(1,2-ethenediylbis(sulfonyl))bis-, (E)- (9CI)	**(1113-14-0)**
Propane, 2-(ethenyloxy)- (9CI)	**(926-65-8)**
Propane, 1-ethoxy- (9CI)	**(628-32-0)**
Propane, 2-ethoxy- (9CI)	**(625-54-7)**
Propane, 1-(1-ethoxyethoxy)-	**(20680-10-8)**
Propane, 2-(2-ethoxyethoxy)-2-methyl- (9CI)	**(51422-54-9)**
Propane, 2-ethoxy-2-methyl- (9CI)	**(637-92-3)**
Propane, 1-(ethylthio)- (9CI)	**(4110-50-3)**
Propane, 2-(ethylthio)- (9CI)	**(5145-99-3)**
Propane, 2-fluoro- (8CI,9CI)	**(420-26-8)**
Propane, 2-fluoro-2-methyl- (9CI)	**(353-61-7)**
Propane, 1,1,1,2,2,3,3-heptachloro- (8CI,9CI)	**(594-89-8)**
Propane, heptafluoroiodo	**(27636-85-7)**
Propane, 1,1,1,2,2,3,3-heptafluoro-3-((trifluoroethenyl)oxy)- (9CI)	**(1623-05-8)**
Propane, 1,1,1,3,3,3-hexachloro- (7CI,8CI,9CI)	**(3607-78-1)**
Propane, 1-iodo	**(107-08-4)**
Propane, 2-iodo	**(75-30-9)**
Propane, iodo	**(26914-02-3)**
Propane, 1-iodo-2-methyl	**(513-38-2)**
Propane, 1-isocyanato-	**(110-78-1)**
Propane, 2-(2-isopropylidene-3-methylcyclopropyl)-, cis- (8CI)	**(24524-52-5)**
Propane, 2-isothiocyanato-2-methyl- (9CI)	**(590-42-1)**
Propane, 1-methoxy- (9CI)	**(557-17-5)**
Propane, 2-methoxy- (9CI)	**(598-53-8)**
Propane, 1-methoxy-2-methyl- (9CI)	**(625-44-5)**
Propane, 2-methoxy-2-methyl- (9CI)	**(1634-04-4)**
Propane, 2-methyl	**(75-28-5)**
Propane, 1-(1-methylethoxy)- (9CI)	**(627-08-7)**
Propane, 2-methyl-2-nitro- (9CI)	**(594-70-7)**
Propane, 2-methyl-2-nitroso-, dimer (8CI,9CI)	**(6841-96-9)**
Propane, 1-(methylthio)- (9CI)	**(3877-15-4)**
Propane, 2-(methylthio)- (9CI)	**(1551-21-9)**
Propane, 1-((2-(methylthio)ethyl)thio)- (9CI)	**(76229-76-0)**
Propanenitrile	**(107-12-0)**
Propanenitrile, 3-((2-(acetyloxy)ethyl)(4-((2,6-dichloro-4-nitrophenyl)azo)phenyl)amino)- (9CI)	**(5261-31-4)**
Propanenitrile, 3-((2-(acetyloxy)ethyl)phenylamino)- (9CI)	**(22031-33-0)**
Propanenitrile, 2-amino-2-methyl- (9CI)	**(19355-69-2)**
Propanenitrile, 3-(butyl(4-((4-nitrophenyl)azo)phenyl)amino)- (9CI)	**(69472-19-1)**
Propanenitrile, 2-((4-chloro-6-(ethylamino)-1,3,5-triazin-2-yl)-amino)-2-methyl-	**(21725-46-2)**
Propanenitrile, 3-((4-((2-chloro-4-nitrophenyl)azo)-3-methyl-phenyl)ethylamino)- (9CI)	**(16586-43-9)**
Propanenitrile, 3-((4-((2-chloro-4-nitrophenyl)azo)phenyl)-(2-hydroxyethyl)amino)- (9CI)	**(6657-33-6)**
Propanenitrile, 3,3'-((4-((2-chloro-4-nitrophenyl)azo)-phenyl)imino)bis- (9CI)	**(4058-30-4)**
Propanenitrile, 3-(decyloxy)- (9CI)	**(16728-51-1)**
Propanenitrile, 3-((4-((2,6-dichloro-4-nitrophenyl)azo)-phenyl)ethylamino)- (9CI)	**(13301-61-6)**
Propanenitrile, 2,2-dimethyl- (9CI)	**(630-18-2)**
Propanenitrile, 3-(dimethylamino)-, Mixt. with 2,2'-oxybis-(N,N-dimethylethanamine) (9CI)	**(62765-93-9)**
Propanenitrile, 3,3'-(1,2-ethanediyldiimino)bis- (9CI)	**(3217-00-3)**
Propanenitrile, 3-ethoxy- (9CI)	**(2141-62-0)**
Propanenitrile, 3-(ethyl(4-formyl-3-methylphenyl)amino)- (9CI)	**(119-97-1)**
Propanenitrile, 3-((2-ethylhexyl)oxy)-	**(10213-75-9)**
Propanenitrile, 3-(ethyl(3-methyl-4-((6-(methylsulfonyl)-2-benzothiazolyl)azo)phenyl)amino)- (9CI)	**(16588-67-3)**
Propanenitrile, 3-(ethyl(3-methyl-4-((6-nitro-2-benzo-thiazolyl)azo)phenyl)amino)- (9CI)	**(16586-42-8)**
Propanenitrile, 3-(ethyl(3-methylphenyl)amino)- (9CI)	**(148-69-6)**
Propanenitrile, 3-(ethyl(4-((6-nitro-2-benzo-thiazolyl)azo)phenyl)amino)- (9CI)	**(25510-81-0)**
Propanenitrile, 3-(ethyl(4-((4-nitrophenyl)azo)phenyl)amino)- (9CI)	**(31482-56-1)**
Propanenitrile, 2-(β-D-glucopyranosyloxy)-2-methyl	**(554-35-8)**
Propanenitrile, 3-hydroxy-	**(109-78-4)**
Propanenitrile, 3-((2-hydroxyethyl)(3-methyl-4-((4-nitrophenyl)-azo)phenyl)amino)- (9CI)	**(6054-58-6)**
Propanenitrile, 2-hydroxy-2-methyl-	**(75-86-5)**
Propanenitrile, 3-hydroxy-2-methyl- (9CI)	**(2567-01-3)**
Propanenitrile, 3,3'-iminobis-	**(111-94-4)**
Propanenitrile, 3-(2-methoxyethoxy)- (9CI)	**(35633-50-2)**
Propanenitrile, 2-methyl	**(78-82-0)**
Propanenitrile, 3-(methyl(4-((4-nitrophenyl)azo)phenyl)amino)- (9CI)	**(31464-38-7)**
Propanenitrile, 3-(methylnitrosoamino)-	**(60153-49-3)**
Propanenitrile, 3-(methylphenylamino)- (9CI)	**(94-34-8)**
Propanenitrile, 3-((1-methyl-2-phenylethyl)amino)-, (+-)- (9CI)	**(15686-61-0)**
Propanenitrile, 3-(4-((5-nitro-2-thiazolyl)azo)(2-phenylethyl)-amino)- (9CI)	**(19745-44-9)**
Propanenitrile, 3-(9-octadecenylamino)-, (Z)- (9CI)	**(26351-32-6)**
Propanenitrile, 3-(octyloxy)- (9CI)	**(16728-49-7)**
Propanenitrile, 3,3'-oxybis-	**(1656-48-0)**
Propanenitrile, 3,3'-(phenylimino)bis- (9CI)	**(1555-66-4)**
Propanenitrile, 3-(phenylmethoxy)- (9CI)	**(6328-48-9)**
Propane, 1-nitro	**(108-03-2)**
Propane, 2-nitro	**(79-46-9)**
Propane, nitro- (8CI,9CI)	**(25322-01-4)**
Propane, octachloro- (6CI,7CI,8CI,9CI)	**(594-90-1)**
Propane, octafluoro	**(76-19-7)**

α,γ-Propane oxide **(503-30-0)**
Propane, 2,2'-oxybis **(108-20-3)**
Propane, 1,1'-oxybis- (9CI) **(111-43-3)**
Propane, 2,2'-oxybis(1-chloro- **(108-60-1)**
Propane, 1,1'-oxybis(3-chloro- (9CI) **(629-36-7)**
Propane, 2,2'-oxybis(2-chloro- (9CI) **(39638-32-9)**
Propane, 1,1'-oxybis(dichloro- (9CI) **(99342-08-2)**
Propane, 1,1'-oxybis(2-methyl- (9CI) **(628-55-7)**
Propane, 2,2'-oxybis(2-methyl- (9CI) **(6163-66-2)**
Propane, 1,1,2,2,3-pentachloro **(16714-68-4)**
Propane, 1,1,2,3,3-pentachloro- (8CI,9CI) **(15104-61-7)**
Propane, 1,1,1,2,2-pentafluoro- (8CI,9CI) **(1814-88-6)**
Propaneperoxoic acid, 2,2-dimethyl-, 1,1-dimethylpropyl ester (9CI) **(29240-17-3)**
Propaneperoxoic acid, 2-methyl-, 1,1-dimethylethyl ester **(109-13-7)**
Propane, 2-phenyl **(98-82-8)**
1-Propanesulfonic acid, 3-chloro-2-hydroxy-, monosodium salt (9CI) **(126-83-0)**
1-Propanesulfonic acid, 2,3-dimercapto **(74-61-3)**
1-Propanesulfonic acid, 3-(((dimethylamino)thioxomethyl)thio)-, sodium salt (9CI) **(18880-36-9)**
1-Propanesulfonic acid, 3,3'-dithiobis-, disodium salt (9CI) **(27206-35-5)**
1-Propanesulfonic acid, 2-methyl-2-((1-oxo-2-propenyl)amino)- (9CI) **(15214-89-8)**
1-Propanesulfonic acid, 2-methyl-2-((1-oxo-2-propenyl)amino)-, Homopolymer (9CI) **(27119-07-9)**
1-Propanesulfonic acid, 2-methyl-2-((1-oxo-2-propenyl)amino)-, monosodium salt (9CI) **(5165-97-9)**
1-Propanesulfonic acid-3-hydroxy-γ-sultone **(1120-71-4)**
1,3-Propane sultone **(1120-71-4)**
Propanesultone **(1120-71-4)**
Propane sultone (ACGIH) **(1120-71-4)**
Propane, 1,2,2,3-tetrabromo- (6CI,7CI,9CI) **(54268-02-9)**
Propane, 1,1,1,3-tetrachloro **(1070-78-6)**
Propane, 1,1,1,2-tetrachloro- (9CI) **(812-03-3)**
Propane, 1,1,2,3-tetrachloro- (9CI) **(18495-30-2)**
Propane, 1,2,2,3-tetrachloro- (9CI) **(13116-53-5)**
Propane, 1,1,2,2-tetrachloro- (8CI,9CI) **(13116-60-4)**
Propane, 1,1,2,2-tetrafluoro- (9CI) **(40723-63-5)**
Propane, 1,1,3,3-tetramethoxy- (9CI) **(102-52-3)**
Propane, 1,1'-thiobis- (9CI) **(111-47-7)**
Propane, 2,2'-thiobis- (9CI) **(625-80-9)**
Propane, 2,2'-thiobis(2-methyl- (9CI) **(107-47-1)**
Propane-1-thiol **(107-03-9)**
Propanethiol **(107-03-9)**
1-Propanethiol [UN 2402] **(107-03-9)**
2-Propanethiol [UN 2402] **(75-33-2)**
2-Propanethiol, 1-amino- (9CI) **(598-36-7)**
1-Propanethiol, 2-methyl **(513-44-0)**
2-Propanethiol, 2-methyl **(75-66-1)**
Propanethiol, trimethoxysilyl **(4420-74-0)**
Propane, 1,2,3-tribromo **(96-11-7)**
1,2,3-Propanetricarboxylic acid, 2-(acetyloxy)-, tributyl ester (9CI) **(77-90-7)**
1,2,3-Propanetricarboxylic acid, 2-(acetyloxy)-, triethyl ester (9CI) **(77-89-4)**
1,2,3-Propanetricarboxylic acid, 2-hydroxy- **(77-92-9)**
1,2,3-Propanetricarboxylic acid, 2-hydroxy-, ammonium salt (9CI) **(7632-50-0)**
1,2,3-Propanetricarboxylic acid, 2-hydroxy-, calcium salt (2:3) (9CI) **(813-94-5)**
1,2,3-Propanetricarboxylic acid, 2-hydroxy-, calcium salt (9CI) **(7693-13-2)**
1,2,3-Propanetricarboxylic acid, 2-hydroxy-, cobalt(2+) salt (2:3) (9CI) **(866-81-9)**
1,2,3-Propanetricarboxylic acid, 2-hydroxy-, copper(2+) salt (1:2) **(9CI)**
1,2,3-Propanetricarboxylic acid, 2-hydroxy-, diammonium salt **(3012-65-5)**
1,2,3-Propanetricarboxylic acid, 2-hydroxy-, disodium salt (9CI) **(144-33-2)**
1,2,3-Propanetricarboxylic acid, 2-hydroxy-, ion(3-) (9CI) **(126-44-3)**
1,2,3-Propanetricarboxylic acid, 2-hydroxy-, iron(2+) salt **(23383-11-1)**
1,2,3-Propanetricarboxylic acid, 2-hydroxy-, lead(2+) salt (2:3) (9CI) **(512-26-5)**
1,2,3-Propanetricarboxylic acid, 2-hydroxy-, manganese(3+) salt (1:1) (9CI) **(5968-88-7)**
1,2,3-Propanetricarboxylic acid, 2-hydroxy-, monosodium salt (9CI) **(18996-35-5)**
1,2,3-Propanetricarboxylic acid, 2-hydroxy-, sodium salt (9CI) **(994-36-5)**
1,2,3-Propanetricarboxylic acid, 2-hydroxy-, tributyl ester (9CI) **(77-94-1)**
1,2,3-Propanetricarboxylic acid, 2-hydroxy-, triethyl ester (9CI) **(77-93-0)**
1,2,3-Propanetricarboxylic acid, 2-hydroxy-, trioctadecyl ester (9CI) **(7775-50-0)**
1,2,3-Propanetricarboxylic acid, 2-hydroxy-, tripotassium salt (9CI) **(866-84-2)**
1,2,3-Propanetricarboxylic acid, 2-hydroxy-, trisodium salt (9CI) **(68-04-2)**
1,2,3-Propanetricarboxylic acid, 2-hydroxy-, zinc salt (2:3) (9CI) **(546-46-3)**
Propane, 1,1,1-trichloro **(7789-89-1)**
Propane, 1,1,2-trichloro **(598-77-6)**
Propane, 1,2,3-trichloro **(96-18-4)**
Propane, 1,2,2-trichloro (8CI,9CI) **(3175-23-3)**
Propane, trichloro- (6CI,7CI,8CI,9CI) **(25735-29-9)**
Propane, 1,1,1-trichloro-2,3-epoxy- **(3083-23-6)**
Propane, 1,1,1-triethoxy- (9CI) **(115-80-0)**
Propane, 1,1,1-trimethoxy-2,2-dimethyl- (9CI) **(97419-16-4)**
Propane, 1,1,1-trimethoxy-2-methyl- (9CI) **(52698-46-1)**
1,2,3-Propanetriol **(56-81-5)**
1,2,3-Propanetriol, Homopolymer (9CI) **(25618-55-7)**
1,2,3-Propanetriol, Homopolymer, (Z)-9-octadecenoate **(9007-48-1)**
1,2,3-Propanetriol, Polymer with (chloromethyl)oxirane (9CI) **(25038-04-4)**
1,2,3-Propanetriol, Polymer with methyloxirane and oxirane **(9082-00-2)**
1,2,3-Propanetriol, Polymer with oxirane **(31694-55-0)**
1,2,3-Propanetriol, 1-(4-aminobenzoate) **(136-44-7)**
1,2,3-Propanetriol, p-aminobenzoate **(136-44-7)**
1,2,3-Propanetriol, diacetate (9CI) **(25395-31-7)**
1,2,3-Propanetriol, 1-(dihydrogen phosphate) (9CI) **(57-03-4)**
1,2,3-Propanetriol, 1,3-dinitrate (9CI) **(623-87-0)**
1,2,3-Propanetriol, ethoxylated **(31694-55-0)**
1,2,3-Propanetriol, 1-(4-hydroxy-3-methoxyphenyl)- (8CI,9CI) **(1208-42-0)**
1,2,3-Propanetriol, 1,3-isobenzofurandione, coconut oil polymer **(66070-87-9)**
1,2,3-Propanetriol, 1,3-isobenzofurandione, soybean oil polymer **(66070-61-9)**
1,2,3-Propanetriol, mono(dihydrogen phosphate), disodium salt (9CI) **(1334-74-3)**
1,2,3-Propanetriol, mono(dihydrogen phosphate), potassium salt **(1335-34-8)**
1,2,3-Propanetriol, 1-nitrate (9CI) **(624-43-1)**
1,2,3-Propanetriol, 1-phenyl-, (S-(R*,S*))- (9CI) **(16354-95-3)**
1,2,3-Propanetriol, 1-phenyl-, D-erythro- (8CI) **(16354-95-3)**
1,2,3-Propanetriol, 1-propanoate (9CI) **(624-47-5)**
1,2,3-Propanetriol triacetate **(102-76-1)**
1,2,3-Propanetriol, tribenzoate (9CI) **(614-33-5)**
1,2,3-Propanetriol tridodecanoate **(538-24-9)**
1,2,3-Propanetriol tri(12-hydroxystearate) **(112-85-6)**
1,2,3-Propanetriol, trinitrate **(55-63-0)**
Propanetriol trinitrate **(55-63-0)**
1,2,3-Propanetriol trioctadecanoate **(555-43-1)**
Propane, 1,2,3-tris(chloromethoxy) **(38571-73-2)**
Propane, 1,2,3-tris(2,3-epoxypropoxy)- **(13236-02-7)**
Propane, 1,2,3-tris(2,3-epoxypropoxy)-, Polymers (8CI) **(31305-91-6)**
Propane, 1,1,3-tris(p-(2,3-epoxypropoxy)phenyl)- (7CI,8CI) **(6130-72-9)**
1,2,3-Propanetriyl dodecanoate **(538-24-9)**
1,2,3-Propanetriyl nitrate **(55-63-0)**
1,2,3-Propanetriyl octadecanoate **(555-43-1)**
1,2,3-Propanetriyl tetradecanoate **(555-45-3)**
α,α',α''-1,2,3-Propanetriyltris(ω-(2,3-epoxypropoxy)poly-(oxypropylene)) **(37237-76-6)**
2,2',2''-(1,2,3-Propanetriyltris(oxymethylene))trisoxirane **(13236-02-7)**
Propanex **(709-98-8)**
Propanid **(709-98-8)**
Propanide **(709-98-8)**
Propanil **(709-98-8)**
Propanimidamide, 2,2'-azobis(2-methyl-, dihydrochloride (9CI) **(2997-92-4)**

2-Propanimine, N-chloro- (9CI) **(34508-68-4)**
Propannitril (Czech) **(107-12-0)**
Propanoic acid **(75-98-9)**
Propanoic acid, 2-amino- **(56-41-7)**
Propanoic acid, 3-amino- **(107-95-9)**
Propanoic acid, 2-amino-, (S)- **(56-41-7)**
Propanoic acid, 3-((amino(bis(2-chloroethyl)amino)phosphinyl)-oxy)- **(22788-18-7)**
Propanoic acid, 3-amino-2-hydroxy- (9CI) **(565-71-9)**
Propanoic acid, 2-amino-3-hydroxy-, (S)- **(56-45-1)**
Propanoic acid, 3,3'-(bis((2-(isooctyloxy)-2-oxoethyl)-thio)stannylene)bis-, dibutyl ester (9CI) **(63397-60-4)**
Propanoic acid, 2-bromo-2-methyl-, ethyl ester (9CI) **(600-00-0)**
Propanoic acid, butyl ester (9CI) **(590-01-2)**
Propanoic acid, calcium salt (9CI) **(4075-81-4)**
Propanoic acid, chloro- (9CI) **(28554-00-9)**
Propanoic acid, 2-(4-((6-chloro-2-benzothiazolyl)oxy)phenoxy)-, ethyl ester **(66441-11-0)**
Propanoic acid, 2-(4-((6-chloro-2-benzothiazolyl)oxy)phenoxy)-, ethyl ester, (+-)- (9CI) **(93921-16-5)**
Propanoic acid, 2-(4-((6-chloro-2-benzoxazolyl)oxy)phenoxy)-, ethyl ester (9CI) **(66441-23-4)**
Propanoic acid, 2-(4-((6-chloro-2-benzoxazolyl)oxy)phenoxy)-, ethyl ester, (+-) **(66441-23-4)**
Propanoic acid, 3-chloro-2-(chloromethyl)-2-methyl- (9CI) **(67329-11-7)**
Propanoic acid, 3-chloro-2,2-dimethyl- (9CI) **(13511-38-1)**
Propanoic acid, 2-chloro-, ethyl ester (9CI) **(535-13-7)**
Propanoic acid, 3-chloro-2-hydroxy-2-methyl-, decahydro-8-hydroxy-3,6-bis(methylene)-2-oxospiro(azuleno(4,5-b)furan-9(2H), 2'-oxiran)-4-yl ester, (3ar-(3a-α,4-α(S*),6a-α,8-β, 9-α,9a-α,9b-β)) **(41787-75-1)**
Propanoic acid, 2-chloro-, 1-methyl-2-(2-methylpropoxy)ethyl ester (9CI) **(67969-81-7)**
Propanoic acid, 2-(4-chloro-2-methylphenoxy)- **(93-65-2)**
Propanoic acid, 2-(4-chloro-2-methylphenoxy)-, (+-)- **(7085-19-0)**
Propanoic acid, 2-(4-chloro-2-methylphenoxy)-, butyl ester (9CI) **(1713-14-0)**
Propanoic acid, 2-(4-chloro-2-methylphenoxy)-, potassium salt (9CI) **(1929-86-8)**
Propanoic acid, 2-(3-chlorophenoxy)- (9CI) **(101-10-0)**
Propanoic acid, 2-(4-chlorophenoxy)- (9CI) **(3307-39-9)**
Propanoic acid, 2-(4-chlorophenoxy)-2-methyl-, ethyl ester **(637-07-0)**
Propanoic acid, 2-(4-chlorophenoxy)-2-methyl-, methyl ester (9CI) **(55162-41-9)**
Propanoic acid, 2-(4-((6-chloro-2-quinoxalinyl)oxy)phenoxy)-, ethyl ester **(76578-14-8)**
Propanoic acid, 2-(4-((3-chloro-5-(trifluoromethyl)-2-pyridinyl)-oxy)phenoxy)- (9CI) **(69806-34-4)**
Propanoic acid, 2-(4-((3-chloro-5-(trifluoromethyl)-2-pyridinyl)-oxy)phenoxy)-, 2-ethoxyethyl ester **(87237-48-7)**
Propanoic acid, 2-(4-((3-chloro-5-(trifluoromethyl)-2-pyridinyl)-oxy)phenoxy)-, methyl ester **(69806-40-2)**
Propanoic acid, 2,3-dibromo- (9CI) **(600-05-5)**
Propanoic acid, 3,3-dichloro-2,2-dimethyl- (9CI) **(64855-18-1)**
Propanoic acid, 2,2-dichloro-, magnesium salt (9CI) **(29110-22-3)**
Propanoic acid, 2,3-dichloro-2-methyl- (9CI) **(10411-52-6)**
Propanoic acid, 2-(3,4-dichlorophenoxy)- (9CI) **(3307-41-3)**
Propanoic acid, 2-(2,4-dichlorophenoxy)-, 2-butoxyethyl ester (9CI) **(53404-31-2)**
Propanoic acid, 2-(2,4-dichlorophenoxy)-, Mixt. with N-2-benzo-thiazolyl-N,N'-dimethylurea (9CI) **(39283-72-2)**
Propanoic acid, 2-(4-(2,4-dichlorophenoxy)phenoxy)- **(40843-25-2)**
Propanoic acid, 2,2-dimethyl-, anhydride (9CI) **(1538-75-6)**
Propanoic acid, 2,2-dimethyl-, ethyl ester (9CI) **(3938-95-2)**
Propanoic acid, 2,2-dimethyl-, methyl ester (9CI) **(598-98-1)**
Propanoic acid, 2,2-dimethyl-, phenyl ester (9CI) **(4920-92-7)**
Propanoic acid, 2,2-dimethyl-, 2,4,6-trimethylphenyl ester (9CI) **(54644-40-5)**
Propanoic acid, 2,3-diphenyl- **(3333-15-1)**
Propanoic acid, 3,3'-dithiobis-, dimethyl ester (9CI) **(15441-06-2)**
Propanoic acid, 3-((3-(dodecyloxy)-3-oxopropyl)thio)-, octadecyl ester (9CI) **(13103-52-1)**
Propanoic acid, 3-(dodecylthio)-, 2,2-bis((3-(dodecylthio)-1-oxopropoxy)methyl)-1,3-propanediyl ester (9CI) **(29598-76-3)**
Propanoic acid, ethenyl ester **(105-38-4)**
Propanoic acid, 3-ethoxy-3-imino-, ethyl ester (9CI) **(27317-59-5)**
Propanoic acid, 3-ethoxy-3-imino-, ethyl ester, hydrochloride (9CI) **(2318-25-4)**
Propanoic acid, heptyl ester (9CI) **(2216-81-1)**
Propanoic acid, hexyl ester **(2445-76-3)**
Propanoic acid, 2-hydroxy- **(50-21-5)**
Propanoic acid, 2-hydroxy-, (+-)- (9CI) **(598-82-3)**
Propanoic acid, 2-hydroxy-, antimony(3+) salt (3:1) **(58164-88-8)**
Propanoic acid, 2-hydroxy-, butyl ester (9CI) **(138-22-7)**
Propanoic acid, 2-hydroxy-, calcium salt **(814-80-2)**
Propanoic acid, 3-hydroxy-2,2-dimethyl- (9CI) **(4835-90-9)**
Propanoic acid, 3-hydroxy-2,2-dimethyl-, 3-hydroxy-2,2-dimethyl propyl ester, diacrylate **(30145-51-8)**
Propanoic acid, 3-hydroxy-2,2-dimethyl-, 3-hydroxy-2,2-dimethyl-propyl ester (9CI) **(1115-20-4)**
Propanoic acid, 3-hydroxy-2,2-dimethyl-, monosodium salt (9CI) **(56974-57-3)**
Propanoic acid, 3-hydroxy-2-(hydroxymethyl)-2-methyl- (9CI) **(4767-03-7)**
Propanoic acid, 3-hydroxy-2-(hydroxymethyl)-2-methyl-, 2-phenylhydrazide (9CI) **(17872-56-9)**
Propanoic acid, 2-hydroxy-2-methyl- (9CI) **(594-61-6)**
Propanoic acid, 2-hydroxy-, methyl ester (9CI) **(547-64-8)**
Propanoic acid, 2-hydroxy-, 1-methylethyl ester **(617-51-6)**
Propanoic acid, 2-hydroxy-, monosodium salt **(72-17-3)**
Propanoic acid, 2-hydroxy-, 2-naphthalenyl ester (9CI) **(93-43-6)**
Propanoic acid, 2-hydroxy-, trianhydride with antimonic acid **(58164-88-8)**
Propanoic acid, 3,3'-iminobis- **(505-47-5)**
Propanoic acid, 3-mercapto-, 2,2-bis((3-mercapto-1-oxopropoxy)-methyl)-1,3-propanediyl ester (9CI) **(7575-23-7)**
Propanoic acid, 3-mercapto-, ethyl ester (9CI) **(5466-06-8)**
Propanoic acid, 3-mercapto-, 2-ethyl-2-((3-mercapto-1-oxopro-poxy)methyl)-1,3-propanediyl ester (9CI) **(33007-83-9)**
Propanoic acid, 3-mercapto-, methyl ester (9CI) **(2935-90-2)**
Propanoic acid, 2-methyl-, ammonium salt (9CI) **(22228-82-6)**
Propanoic acid, 2-methyl-, butyl ester **(97-87-0)**
Propanoic acid, methyl ester **(554-12-1)**
Propanoic acid, 1-methylethyl ester (9CI) **(637-78-5)**
Propanoic acid, 2-methyl-, hexyl ester (9CI) **(2349-07-7)**
Propanoic acid, 2-methyl-, 3-hydroxy-2,2,4-trimethylpentyl ester (9CI) **(77-68-9)**
Propanoic acid, 2-methyl-, 1-methylethyl ester (9CI) **(617-50-5)**
Propanoic acid, 2-methyl-, 2-methylpropyl ester (9CI) **(97-85-8)**
Propanoic acid, methyl-, monoester with 2,2,4-trimethyl-1,3-pentanediol **(25265-77-4)**
Propanoic acid, 2-methyl-, 2-phenoxyethyl ester (9CI) **(103-60-6)**
Propanoic acid, 4-methylphenyl ester **(7495-84-3)**
Propanoic acid, 2-methyl-, propyl ester (9CI) **(644-49-5)**
Propanoic acid, 2-methylpropyl ester (9CI) **(540-42-1)**
Propanoic acid, 3-nitro- (9CI) **(504-88-1)**
Propanoic acid, 2-oxo- (9CI) **(127-17-3)**
Propanoic acid, 2-oxo-, ethyl ester (9CI) **(617-35-6)**
Propanoic acid, 2-oxo-, methyl ester (9CI) **(600-22-6)**
Propanoic acid, 2-oxo-, sodium salt (9CI) **(113-24-6)**
Propanoic acid, pentyl ester (9CI) **(624-54-4)**
Propanoic acid, phenyl ester (9CI) **(637-27-4)**
Propanoic acid, 2-phenylethyl ester (9CI) **(122-70-3)**
Propanoic acid, phenylmethyl ester (9CI) **(122-63-4)**
Propanoic acid, sodium salt **(137-40-6)**
Propanoic acid, 2,2,3,3-tetrafluoro-3-methoxy-, methyl ester (9CI) **(755-73-7)**
Propanoic acid, 3,3'-thiobis-, didodecyl ester **(123-28-4)**
Propanoic acid, 3,3'-thiobis-, dimethyl ester (9CI) **(4131-74-2)**
Propanoic acid, 3,3'-thiobis-, dioctadecyl ester (9CI) **(693-36-7)**

Propanoic acid, 3,3'-thiobis-, ditetradecyl ester (9CI) **(16545-54-3)**
Propanoic acid, 3,3'-thiobis-, ditridecyl ester (9CI) **(10595-72-9)**
Propanoic acid, 2-(2,4,5-trichlorophenoxy)-, Compd. with 2,2'-iminobis(ethanol) (1:1) (9CI) **(51170-59-3)**
Propanoic acid, 2-(2,4,5-trichlorophenoxy)-, aluminum complex (9CI) **(69622-82-8)**
Propanoic acid, 2-(2,4,5-trichlorophenoxy)-, Compd. with N,N-diethylethanamine (1:1) (9CI) **(53404-74-3)**
Propanoic acid, 2-(2,4,5-trichlorophenoxy)-, Compd. with (Z)-9-octadecen-1-amine (1:1) (9CI) **(53404-73-2)**
Propanoic acid, 2-(2,4,5-trichlorophenoxy)-, Compd. with 1-amino-2-propanol (1:1) (9CI) **(53404-13-0)**
Propanoic acid, 2-(2,4,5-trichlorophenoxy)-, Compd. with 1,1'-iminobis(2-propanol) (1:1) (9CI) **(53404-09-4)**
Propanoic acid, 2-(2,4,5-trichlorophenoxy)-, Compd. with N-methylmethanamine (1:1) (9CI) **(55617-85-1)**
Propanoic acid, 2-(2,4,5-trichlorophenoxy)-, Compd. with 1,1',1''-nitrilotris(2-propanol) (1:1) (9CI) **(53404-75-4)**
Propanoic acid, 2-(2,4,5-trichlorophenoxy)-, ester with butoxypropanol (9CI) **(28903-26-6)**
Propanoic acid, 2-(2,4,5-trichlorophenoxy)-, ester with 1(or 2)-(2-methylpropoxy)propanol (9CI) **(53466-84-5)**
Propanoic acid, 2-(2,4,5-trichlorophenoxy)-, 2-ethylhexyl ester (9CI) **(53404-76-5)**
Propanoic acid, 2-(2,4,5-trichlorophenoxy)-, 2-ethyl-4-methylpentyl ester (9CI) **(53404-10-7)**
Propanoic acid, 2-(2,4,5-trichlorophenoxy)-, isooctyl ester (9CI) **(32534-95-5)**
Propanoic acid, 2-(2,4,5-trichlorophenoxy)-, 1-methylheptyl ester (9CI) **(53404-14-1)**
Propanoic acid, 2-(2,4,5-trichlorophenoxy)-, methyl-2-(methyl-2-(methyl-2-(2-methylpropoxy)ethoxy)ethoxy)ethyl ester (9CI) **(53535-30-1)**
Propanoic acid, 2-(2,4,5-trichlorophenoxy)-, methyl-2-(methyl-2-(2-methylpropoxy)ethoxy)ethyl ester (9CI) **(53535-26-5)**
Propanoic acid, 2-(2,4,5-trichlorophenoxy)-, potassium salt (9CI) **(2818-16-8)**
Propanoic acid, 2-(2,4,5-trichlorophenoxy)-, sodium salt (9CI) **(37913-89-6)**
Propanoic acid, 2-(4-((5-(trifluoromethyl)-2-pyridinyl)oxy)-phenoxy)-, butyl ester **(69806-50-4)**
Propanoic acid, 2-(4-((5-(trifluoromethyl)-2-pyridinyl)oxy)-phenoxy)-, butyl ester, (R) **(79241-46-6)**
Propanoic acid, zinc salt (9CI) **(557-28-8)**
1-Propanol **(71-23-8)**
2-Propanol **(67-63-0)**
Propan-2-ol **(67-63-0)**
Propanol **(71-23-8)**
Propanol-1 **(71-23-8)**
n-Propan-2-ol **(67-63-0)**
n-Propanol **(71-23-8)**
i-Propanol (German) **(67-63-0)**
Propanol [UN 1274] **(71-23-8)**
Propanol, allyloxy **(1331-17-5)**
2-Propanol, 1-(2-allyloxyphenoxy)-3-(isopropylamino) **(6452-71-7)**
2-Propanol, 1-(o-allylphenoxy)-3-(isopropylamino) **(13655-52-2)**
1,3-Propanolamine **(156-87-6)**
3-Propanolamine **(156-87-6)**
Propanolamine **(156-87-6)**
1-Propanol, 3-amino **(156-87-6)**
2-Propanol, 1-amino **(78-96-6)**
2-Propanol, 1-amino-, Compd. with (2,4,5-trichlorophenoxy)acetic acid (1:1) **(1319-72-8)**
1-Propanol, 2-amino-2-methyl **(124-68-5)**
1-Propanol, 2-amino-2-methyl-, hydrochloride (9CI) **(3207-12-3)**
2-Propanol, 1-((3-aminophenyl)amino)-3-phenoxy- (9CI) **(38353-82-1)**
2-Propanol, 1-amino-, (2,4,5-trichlorophenoxy)acetate (Salt) (9CI) **(1319-72-8)**
2-Propanol, 1-amino-, (2,4,5-trichlorophenoxy)propanoate (Salt) (9CI) **(53404-13-0)**
Propanol, bis(2,3-epoxypropoxy)- (8CI) **(27043-36-3)**
2-Propanol, 1,3-bis(oxiranylmethoxy)- (9CI) **(3568-29-4)**
Propanol, bis(oxiranylmethoxy)- (9CI) **(27043-36-3)**
1-Propanol, 3-((4,6-bis(trichloromethyl)-s-triazin-2-yl)amino)- (8CI) **(24803-12-1)**
1-Propanol, 3-bromo **(627-18-9)**
2-Propanol, 1-bromo **(19686-73-8)**
1-Propanol, 2-bromo- (8CI,9CI) **(598-18-5)**
1-Propanol, 3-bromo-2,2-bis(bromoethyl) **(1522-92-5)**
1-Propanol, 2-bromo-3-chloro- (9CI) **(73727-39-6)**
2-Propanol, 1-bromo-3-chloro- (9CI) **(4540-44-7)**
2-Propanol, 1-butoxy **(5131-66-8)**
2-Propanol, 1-(2-butoxyethoxy) **(124-16-3)**
2-Propanol, 1-(2-butoxy-1-methoxy) **(29911-28-2)**
2-Propanol, 1-(3-butoxypropoxy)- (9CI) **(35075-24-2)**
2-Propanol, 1-(p-t-butylphenoxy)-, 2-chloroethyl sulfite **(140-57-8)**
2-Propanol, 1-(2-(p-tert-butylphenoxy)-1-methylethoxy)-, 2-chloroethylsulfite **(3761-60-2)**
1-Propanol, 2-chloro **(78-89-7)**
1-Propanol, 3-chloro **(627-30-5)**
2-Propanol, 1-chloro **(127-00-4)**
1-Propanol, 2-chloro-, phosphate (3:1) (9CI) **(6145-73-9)**
1-Propanol, 3-chloro-, phosphate (3:1) (8CI,9CI) **(1067-98-7)**
1-Propanol, 2-chloro-, phosphate (3:1), Mixed with 1-chloro-2-propanol phosphate (3:1) **(26248-87-3)**
1-Propanol, 3,3'-((6-chloro-s-triazine-2,4-diyl)diimino)di- (8CI) **(30084-25-4)**
2-Propanol, 1-(cyclohexylamino) **(103-00-4)**
1-Propanol, 3-(cyclohexylamino)- (8CI,9CI) **(31121-12-7)**
1-Propanol, 3-cyclooctyl- **(16782-30-2)**
2-Propanol, 1,3-diamino- (8CI,9CI) **(616-29-5)**
1-Propanol, 2,3-dibromo **(96-13-9)**
2-Propanol, 1,3-dibromo **(96-21-9)**
1-Propanol, 2,3-dibromo-, dihydrogen phosphate **(5324-12-9)**
1-Propanol, 2,3-dibromo-, phosphate (1:1) **(5324-12-9)**
1-Propanol, 2,3-dibromo-, phosphate (3:1) **(126-72-7)**
2-Propanol, 1,3-dibromo-, phosphate (3:1) **(18713-51-4)**
2-Propanol, 1-dibutylamino- **(2109-64-0)**
2-Propanol, 1-(dibutylamino)- (8CI,9CI) **(2109-64-0)**
1-Propanol, 2,3-dichloro **(616-23-9)**
2-Propanol, 1,3-dichloro **(96-23-1)**
Propanol, dichloro- (9CI) **(26545-73-3)**
1-Propanol, 2,2-dichloro- (7CI,9CI) **(63151-11-1)**
1-Propanol, 3,3-dichloro- (7CI,9CI) **(83682-72-8)**
2-Propanol, 1,3-dichloro-, phosphate (3:1) **(13674-87-8)**
Propanol, 2,3-dichloro-, phosphate (3:1) **(78-43-3)**
1-Propanol, 1,3-dichloro-, phosphate (3:1) (9CI) **(40120-74-9)**
1-Propanol, dichloro-, phosphate (3:1) (8CI,9CI) **(26604-51-3)**
2-Propanol, 1,3-difluoro-, benzenesulfonate (7CI,8CI) **(882-71-3)**
1-Propanol, 2,3-dimercapto **(59-52-9)**
2-Propanol, 1,3-dimethoxy- (8CI,9CI) **(623-69-8)**
1-Propanol, 2,2-dimethyl- **(75-84-3)**
1-Propanol, 2,2-dimethyl-, acetate (8CI,9CI) **(926-41-0)**
2-Propanol, 1-(dimethylamino) **(108-16-7)**
1-Propanol, 2-dimethylamino-2-methyl **(7005-47-2)**
2-Propanol, 1-(1,3-dimethylbutoxy)- (9CI) **(54340-89-5)**
2-Propanol, 1-((1,1-dimethylethyl)amino)-3-((4-(4-morpholinyl)-1,2,5-thiadiazol-3-yl)oxy)-, (S)- (9CI) **(26839-75-8)**
1-Propanol, 2,2-dimethyl-, nitrate (8CI,9CI) **(926-42-1)**
1-Propanol, 2,2-dimethyl-, tribromo deriv. (9CI) **(36483-57-5)**
1-Propanol, 3,3-diphenyl- (8CI) **(20017-67-8)**
2-Propanol, 1,1'-(dodecylimino)bis- (9CI) **(1541-66-8)**
2-Propanol, 1-(tert-dodecylthio)- (9CI) **(67124-09-8)**
Propanole (German) **(71-23-8)**
Propanolen (Dutch) **(71-23-8)**
1-Propanol, 2,3-epoxy **(556-52-5)**
1-Propanol, 2,3-epoxy-, acetate (8CI) **(6387-89-9)**
1-Propanol, 2,3-epoxy-, acrylate **(106-90-1)**
1-Propanol, 2,3-epoxy-, methacrylate **(106-91-2)**
2-Propanol, 1',1',1'',1'''-(1,2-ethanediyldinitrilo)tetrakis- **(102-60-3)**

1-Propanol, 2-ethoxy **(19089-47-5)**
1-Propanol, 3-ethoxy **(111-35-3)**
2-Propanol, 1-ethoxy- (8CI,9CI) **(1569-02-4)**
2-Propanol, 1-(2-(2-ethoxy-1-methylethoxy)-1-methylethoxy)- (8CI,9CI) **(20178-34-1)**
2-Propanol, 1,1',1'',1'''-(ethylenedinitrilo)tetra **(102-60-3)**
Propanol, 1-ethyl-2-methylene- **(2088-07-5)**
2-Propanol, 1,1'-(hexadecylimino)bis- (9CI) **(2269-21-8)**
2-Propanol, 1,1,1,3,3,3-hexafluoro **(920-66-1)**
1-Propanol, hexafluoro- (9CI) **(53520-89-1)**
2-Propanol, 1,1'-((2-((2-hydroxyethyl)(2-hydroxypropyl)-amino)ethyl)imino)bis- (9CI) **(139-90-2)**
2-Propanol, 1,1'-((2-((2-hydroxyethyl)(2-hydroxypropyl)-amino)ethyl)imino)di- (8CI) **(139-90-2)**
1-Propanol, 2-((hydroxymethyl)amino)-2-methyl- (9CI) **(52299-20-4)**
1-Propanol, 2-(2-hydroxypropoxy)- (8CI,9CI) **(106-62-7)**
2-Propanol, 1,1',1'',1'''-(((2-hydroxypropyl)imino)bis(2,1-ethanediylnitrilo))tetrakis- (9CI) **(17121-34-5)**
Propanoli (Italian) **(71-23-8)**
3-Propanolide **(57-57-8)**
Propanolide **(57-57-8)**
2-Propanol, 1,1'-iminobis- (9CI) **(110-97-4)**
2-Propanol, 1,1'-iminobis-, N-coco alkyl derivs. **(68516-06-3)**
2-Propanol, 1,1'-iminobis-, 2-(2,4,5-trichlorophenoxy)propanoate (Salt) (9CI) **(53404-09-4)**
2-Propanol, 1,1'-iminodi **(110-97-4)**
1-Propanol, 3,3'-iminodi-, dimethanesulfonate (ester), hydrochloride **(3458-22-8)**
2-Propanol, 1-isobutoxy **(23436-19-3)**
Propanol, 1-isopropyl- **(565-67-3)**
2-Propanol, 1-(isopropylamino)-3-(1-naphthyloxy) **(525-66-6)**
2-Propanol, 1,1'-(isopropylidenebis(p-phenyleneoxy))di- (8CI) **(116-37-0)**
2-Propanol, 1-methoxy **(107-98-2)**
1-Propanol, methoxy- (9CI) **(28677-93-2)**
1-Propanol, 2-methoxy- (8CI,9CI) **(1589-47-5)**
Propanol, methoxy-, acetate (9CI) **(84540-57-8)**
2-Propanol, 1-(4-(2-methoxyethyl)phenoxy)-3-((1-methylethyl)-amino)-, (+-) **(37350-58-6)**
Propanol, (2-methoxymethylethoxy)- **(34590-94-8)**
1-Propanol, 2-(2-methoxy-1-methylethoxy)- (9CI) **(55956-21-3)**
2-Propanol, 1-(2-methoxy-1-methylethoxy)- (9CI) **(20324-32-7)**
2-Propanol, 2-(2-methoxy-1-methylethoxy)- (7CI,9CI) **(55956-22-4)**
2-Propanol, 1-(2-(2-methoxy-1-methylethoxy)-1-methylethoxy) **(20324-33-8)**
Propanol, (2-(2-methoxymethylethoxy)methylethoxy)- (9CI) **(25498-49-1)**
1-Propanol, 2-(2-methoxypropoxy)- (8CI,9CI) **(13588-28-8)**
2-Propanol, 1-(2-methoxypropoxy)- (8CI,9CI) **(13429-07-7)**
1-Propanol, 2-(2-(2-methoxypropoxy)propoxy) **(10213-77-1)**
Propanol, 3-(3-(3-methoxypropoxy)propoxy) **(25498-49-1)**
1-Propanol, 3-(3-(3-methoxypropoxy)propoxy)- (9CI) **(13133-29-4)**
1-Propanol, 2-methyl- **(78-83-1)**
2-Propanol, 2-methyl- **(75-65-0)**
Propanol, ((1-methyl-1,2-ethanediyl)bis(oxy))bis- (9CI) **(24800-44-0)**
2-Propanol, 1-((1-methylethyl)amino)-3-(2-(2-propenyloxy)-phenoxy)- **(6452-71-7)**
2-Propanol, 1-((1-methylethyl)amino)-3-(2-(2-propenyl)phenoxy)- **(13655-52-2)**
2-Propanol, 1,1'-((1-methylethylidene)bis(4,1-phenyleneoxy))bis- (9CI) **(116-37-0)**
1-Propanol, 2-methyl-2-nitro **(76-39-1)**
1-Propanol, 2-methyl-2-nitro-, nitrate **(24884-69-3)**
2-Propanol, 1,1'-((4-methylphenyl)imino)bis- (9CI) **(38668-48-3)**
2-Propanol, 2-methyl-, potassium salt (9CI) **(865-47-4)**
2-Propanol, 1-(1-methyl-2-(2-propenyloxy)ethoxy)- (9CI) **(55956-25-7)**
2-Propanol, 1-(1-methylpropoxy)- (9CI) **(53907-95-2)**
2-Propanol, 1-(2-methylpropoxy)- (9CI) **(23436-19-3)**
1-Propanol, 3-(methylthio)- (9CI) **(505-10-2)**
2-Propanol, 1,1'-((6-methyl-s-triazine-2,4-diyl)diimino)di- (8CI) **(26322-44-1)**
1-Propanol, 3,3'-((6-methyl-s-triazine-2,4-diyl)diimino)di- (7CI,8CI) **(5943-83-9)**
2-Propanol, 2-methyl-1,1,1-trichloro- **(57-15-8)**
1-Propanol, 3-((4-methyl-6-(trichloromethyl)-s-triazin-2-yl)-amino)- (8CI) **(24803-62-1)**
2-Propanol, 1,1',1''-nitrilotri **(122-20-3)**
2-Propanol, 1,1',1''-nitrilotris-, 2-(2,4,5-trichlorophenoxy)propanoate (Salt) (9CI) **(53404-75-4)**
2-Propanol nitrite **(541-42-4)**
Propanol nitrite **(543-67-9)**
1-Propanol, 3-nitro **(25182-84-7)**
2-Propanol, 1,1'-nitrosoiminodi- **(53609-64-6)**
2-Propanol, N-nitroso-1,1'-iminodi- **(53609-64-6)**
2-Propanol, 1,1'-(octadecylimino)bis- (9CI) **(28137-64-6)**
2-Propanol, 1,1'-(octylimino)bis- (9CI) **(28482-15-7)**
2-Propanol, 1,1'-(octylimino)di- **(28482-15-7)**
Propanolol **(525-66-6)**
Propanolol glycol **(36112-95-5)**
Propanolone **(78-98-8)**
Propanol, 1(or 2)-(2-methoxymethylethoxy)- (9CI) **(34590-94-8)**
Propanol, oxybis **(25265-71-8)**
Propanol, oxybis-, dibenzoate (9CI) **(27138-31-4)**
2-Propanol, 1,1'-oxydi **(110-98-5)**
Propanol, oxydi-, phosphite (3:1) **(36788-39-3)**
1-Propanol, 2,2,3,3,3-pentafluoro **(422-05-9)**
1-Propanol, 2-phenoxy **(4169-04-4)**
2-Propanol, 1-phenoxy- (9CI) **(770-35-4)**
1-Propanol, 3-phenyl **(122-97-4)**
1-Propanol, 3-phenyl-, nitrite (7CI,8CI) **(28537-55-5)**
1-Propanol, 3-(2-propenyloxy)-2,2-bis((2-propenyloxy)methyl)- **(1471-17-6)**
2-Propanol, 1-propoxy **(1569-01-3)**
Propanol, n-propoxy- **(30136-13-1)**
1-Propanol, 2-propoxy- (8CI,9CI) **(10215-30-2)**
3-Propanolpyridine **(2859-67-8)**
1-Propanol, 2,2,3,3-tetrafluoro **(76-37-9)**
2-Propanol, 1,1,1-tribromo-2-methyl- (9CI) **(76-08-4)**
2-Propanol, 1-((2,3,6-trichlorobenzyl)oxy)- (7CI,8CI) **(1861-44-5)**
2-Propanol, 1,1,1-trichloro-2-methyl **(57-15-8)**
1-Propanol, 3,3'-((6-(trichloromethyl)-s-triazine-2,4-diyl)-diimino)di- (8CI) **(26235-07-4)**
2-Propanol, 1,1'-((6-(trichloromethyl)-s-triazine-2,4-diyl)-diimino)di- (8CI) **(26235-08-5)**
2-Propanol, 1-((2,3,6-trichlorophenyl)methoxy)- (9CI) **(1861-44-5)**
2-Propanol, 1,1,1-trifluoro- (9CI) **(374-01-6)**
1-Propanol, 3-(trimethoxysilyl)-, methacrylate **(2530-85-0)**
1-Propanol, zirconium(4+) salt (9CI) **(23519-77-9)**
2-Propanone **(67-64-1)**
Propanone **(67-64-1)**
1-Propanone, 1-(2-(β-D-glucopyranosyloxy)-4,6-dihydroxyphenyl)-3-(4-hydroxyphenyl) **(60-81-1)**
2-Propanone, 1-(acetyloxy)- (9CI) **(592-20-1)**
1-Propanone, 1-(4-aminophenyl)- (9CI) **(70-69-9)**
2-Propanone, bromo **(598-31-2)**
2-Propanone, 1-bromo- (9CI) **(598-31-2)**
2-Propanone, 1-chloro **(78-95-5)**
1-Propanone, 2-chloro-1-(4-ethylphenyl)-2-methyl- (9CI) **(55012-69-6)**
1-Propanone, 2-chloro-2-methyl-1-phenyl- (9CI) **(7473-99-6)**
2-Propanone, 1-chloro-1,1,3,3,3-pentafluoro **(79-53-8)**
2-Propanone, 1-(1-cyclohexen-1-yl)- (9CI) **(768-50-3)**
2-Propanone, 1-cyclohexylidene- **(874-68-0)**
2-Propanone-1,1,1,3,3,3-d6 (9CI) **(666-52-4)**
2-Propanone, 1-(4,6-diamino-s-triazin-2-yl)- (8CI) **(30354-98-4)**
2-Propanone, 1,1-dichloro **(513-88-2)**
2-Propanone, 1,3-dichloro **(534-07-6)**
2-Propanone, 1-(2,6-dichloro-4-hydroxy-3,5-dimethoxyphenyl)- (9CI) **(75315-46-7)**
2-Propanone, 1-(2,6-dichloro-3,4,5-trimethoxyphenyl)- (9CI) **(75315-56-9)**
2-Propanone, 1,3-dihydroxy **(96-26-4)**

2-Propanone, 1,1-dimethoxy- (9CI) **(6342-56-9)**
1-Propanone, 1-(3,4-dimethoxyphenyl) **(1835-04-7)**
1-Propanone, 1-(10-(2-(dimethylamino)propyl)-10H-phenothiazin-2-yl)- **(362-29-8)**
1-Propanone, 1-(10-(2-(dimethylamino)propyl)phenothiazin-2-yl)- **(362-29-8)**
2-Propanone, 1-(5-((2,4-dinitrophenyl)thio)-2,4-dinitro-2,4-cyclohexadien-1-yl)-, ion(1-), potassium (9CI) **(69742-90-1)**
2-Propanone, 1,3-diphenyl- (9CI) **(102-04-5)**
1-Propanone, 1,2-di-3-pyridyl-2-methyl- **(54-36-4)**
2-Propanone, 1-fluoro- (8CI,9CI) **(430-51-3)**
2-Propanone, 1,1,1,3,3,3-hexachloro **(116-16-5)**
2-Propanone, hexachloro- **(116-16-5)**
2-Propanone, 1,1,1,3,3,3-hexafluoro **(684-16-2)**
2-Propanone, hexafluoro-, sesquihydrate **(13098-39-0)**
2-Propanone, hexafluoro-, trihydrate **(34202-69-2)**
2-Propanone, 1-hydroxy **(116-09-6)**
2-Propanone, 1-hydroxy-, acetate (8CI) **(592-20-1)**
1-Propanone, 1-(4-hydroxy-3-methoxyphenyl)- (9CI) **(1835-14-9)**
2-Propanone, 1-(4-hydroxy-3-methoxyphenyl)- (9CI) **(2503-46-0)**
1-Propanone, 2-hydroxy-2-methyl-1-phenyl- (9CI) **(7473-98-5)**
1-Propanone, 1-(4-(3-hydroxyphenyl)-1-methyl-4-piperidinyl)- (9CI) **(469-79-4)**
1-Propanone, 1-(4-(m-hydroxyphenyl)-1-methyl-4-piperidyl)- (8CI) **(469-79-4)**
1-Propanone, 3-(4-hydroxyphenyl)-1-(2,4,6-trihydroxyphenyl)- (9CI) **(60-82-2)**
2-Propanone, 1-methoxy **(5878-19-3)**
1-Propanone, 2-methyl-1,2-di-3-pyridinyl- (9CI) **(54-36-4)**
1-Propanone, 2-methyl-1,2-di-3-pyridyl **(54-36-4)**
1-Propanone, 2-methyl-1-(4-(methylthio)phenyl)-2-(4-morpholinyl)- (9CI) **(71868-10-5)**
1-Propanone, 1-(4-methylphenyl)- (9CI) **(5337-93-9)**
1-Propanone, 2-methyl-1-phenyl- (9CI) **(611-70-1)**
2-Propanone, 1-monochloro-1,1,3,3,3-pentafluoro- **(79-53-8)**
2-Propanone oxime **(127-06-0)**
2-Propanone, 1,1,1,3,3-pentachloro **(1768-31-6)**
1-Propanone, 1-phenyl- **(93-55-0)**
2-Propanone, 1-phenyl **(103-79-7)**
2-Propanone, o-((phenylamino)carbonyl)oxime (9CI) **(2828-42-4)**
2-Propanone, 1,1,1,3-tetrachloro **(16995-35-0)**
2-Propanone, 1,1,3,3-tetrachloro **(632-21-3)**
1-Propanone, 1-(2-thienyl)- (9CI) **(13679-75-9)**
2-Propanone, 1,1,1-trichloro **(918-00-3)**
2-Propanone, 1,1,3-trichloro **(921-03-9)**
2-Propanone, 1,1,1-trifluoro- (9CI) **(421-50-1)**
Propanonitrile, 3-(ethyl(4-((2,6-dichloro-4-nitrophenyl)-azo)phenyl)amino)- **(13301-61-6)**
Propanosedyl **(93-14-1)**
Propanoyl bromide, 2-bromo- (9CI) **(563-76-8)**
Propanoyl chloride (9CI) **(79-03-8)**
Propanoyl chloride, 2-chloro- (9CI) **(7623-09-8)**
Propanoyl chloride, 3-chloro-2,2-dimethyl- (9CI) **(4300-97-4)**
Propanoyl chloride, 2,2-dimethyl- (9CI) **(3282-30-2)**
Propanoyl fluoride, 2,3,3,3-tetrafluoro-2-(heptafluoropropoxy)- (9CI) **(2062-98-8)**
Propanoyl fluoride, 2,3,3,3-tetrafluoro-2-(1,1,2,3,3,3-hexafluoro-2-(1,1,2,2-tetrafluoro-2-(fluorosulfonyl)ethoxy)propoxy)- (9CI) **(4089-58-1)**
Propanoyl fluoride, 2,3,3,3-tetrafluoro-2-(trifluoromethoxy)- (9CI) **(2927-83-5)**
Propantel **(50-34-0)**
Propantheline **(298-50-0)**
Propantheline bromide **(50-34-0)**
Propanthelinium **(298-50-0)**
Propanthelinum **(298-50-0)**
Propaphen **(69-09-0)**
Propaphenin **(50-53-3)**
Propaphenin hydrochloride **(69-09-0)**
Propaphos **(7292-16-2)**
Propargite **(2312-35-8)**
Propargyl alcohol (ACGIH,DOT,OSHA) **(107-19-7)**
Propargylaldehyde **(624-67-9)**
Propargyl bromide **(106-96-7)**
5-Propargylfurfuryl chrysanthemate **(23031-38-1)**
5-Propargyl-2-furylmethyl dl-cis,trans-chrysanthemate **(23031-38-1)**
Propargylic acid **(471-25-0)**
Propargylidenecyclohexane **(2806-45-3)**
3-Propargyloxyphenyl-N-methyl-carbamate **(3692-90-8)**
Proparthrin **(27223-49-0)**
Propasa **(65-49-6)**
Propasin **(139-40-2)**
Propasol B **(29387-86-8)**
Propasol Solvent B **(5131-66-8)**
Propasol Solvent M **(107-98-2)**
Propasol Solvent P **(29387-86-8)**
Propasol solvent P **(1569-01-3)**
Propaste 6708 **(126-92-1)**
Propaste D **(143-00-0)**
Propaste T **(139-96-8)**
Propathene **(9003-07-0)**
Propathene 101/24 **(9003-07-0)**
Propathene 22/44 **(9003-07-0)**
Propathene GSE 108 **(9003-07-0)**
Propathene GSE 180 **(9003-07-0)**
Propathene GWE 21 **(9003-07-0)**
Propathene GW 522 M **(9003-07-0)**
Propathene GW 601M **(9003-07-0)**
Propathene GY 702M **(9003-07-0)**
Propathene 112/00/Grey 9897 **(9003-07-0)**
Propathene HF 20 **(9003-07-0)**
Propathene HW 70GR **(9003-07-0)**
Propathene HWM 25 **(9003-07-0)**
Propathene LWF 31 **(9003-07-0)**
Propathene LY 542M **(9003-07-0)**
Propathene O **(9003-07-0)**
Propathene PXC 3830 **(9003-07-0)**
Propathene PXC 4515 **(9003-07-0)**
Propathene PXC 8639 **(9003-07-0)**
Propathene PXC 9617 **(9003-07-0)**
Propax **(604-75-1)**
Propazin **(139-40-2)**
Propazine **(139-40-2)**
Propazine (Herbicide) **(139-40-2)**
Propazyl **(94-12-2)**
Propellant 12 **(75-71-8)**
Propellant 22 **(75-45-6)**
Propellant 114 **(76-14-2)**
Propellant C318 **(115-25-3)**
2-Propenal **(107-02-8)**
Prop-2-en-1-al **(107-02-8)**
Propenal **(107-02-8)**
2-Propenal, 2-chloro- (9CI) **(683-51-2)**
2-Propenal dimer **(100-73-2)**
2-Propenal, 3-(4-(1,1-dimethylethyl)phenyl)-2-methyl- (9CI) **(13586-68-0)**
2-Propenal, 3-(2-furanyl)- (9CI) **(623-30-3)**
2-Propenal, 3-(4-hydroxy-3-methoxyphenyl)- (9CI) **(458-36-6)**
2-Propenal, 3-(2-methoxyphenyl)- **(1504-74-1)**
2-Propenal, 2-methyl-3-phenyl- (9CI) **(101-39-3)**
2-Propenal, 3-phenyl- (9CI) **(104-55-2)**
2-Propenal, 3-phenyl-, (E)- (9CI) **(14371-10-9)**
2-Propenal, 3-phenyl-, monopentyl deriv. **(122-40-7)**
Propenamide **(79-06-1)**
2-Propenamide (9CI) **(79-06-1)**
2-Propenamide, Homopolymer **(9003-05-8)**
2-Propenamide, N-(butoxymethyl)- (9CI) **(1852-16-0)**
2-Propenamide, N-(3,4-dichlorophenyl)-2-methyl- (9CI) **(2164-09-2)**

Name	CAS No.
2-Propenamide, N,N-dimethyl- (9CI)	**(2680-03-7)**
2-Propenamide, N-(3-(dimethylamino)propyl)-2-methyl- (9CI)	**(5205-93-6)**
2-Propenamide, N-(1,1-dimethylethyl)- (9CI)	**(107-58-4)**
2-Propenamide, N-(1,1-dimethyl-3-oxobutyl)-	**(2873-97-4)**
2-Propenamide, N-(hydroxymethyl)-	**(924-42-5)**
2-Propenamide, N-(hydroxymethyl)-2-methyl- (9CI)	**(923-02-4)**
2-Propenamide, 2-methyl	**(79-39-0)**
2-Propenamide, N-methyl- (9CI)	**(1187-59-3)**
2-Propenamide, N,N'-methylenebis-	**(110-26-9)**
2-Propenamide, N-(1-methylethyl)- (9CI)	**(2210-25-5)**
2-Propenamide, N-((2-methylpropoxy)methyl)- (9CI)	**(16669-59-3)**
2-Propenamide, N-(1,1,3,3-tetramethylbutyl)- (9CI)	**(4223-03-4)**
2-Propenamine	**(107-11-9)**
2-Propen-1-amine (9CI)	**(107-11-9)**
2-Propen-1-amine, 3-(4-bromophenyl)-N,N-dimethyl-3-(3-pyridinyl)-, (Z)	**(56775-88-3)**
2-Propen-1-amine, N-N-di-2-propenyl- (9CI)	**(102-70-5)**
2-Propen-1-amine, N-ethyl-2-methyl- (9CI)	**(18328-90-0)**
2-Propen-1-amine, N-methyl-N-nitroso- (9CI)	**(4549-43-3)**
2-Propen-1-amine, N-2-propenyl	**(124-02-7)**
2-Propen-1-aminium, N,N-dimethyl-N-2-propenyl-, chloride (9CI)	**(7398-69-8)**
Propene	**(115-07-1)**
1-Propene (9CI)	**(115-07-1)**
Propene, Polymers, tetramer	**(25378-22-7)**
Propene acid	**(79-10-7)**
1-Propene, 3,3'-((2,2-bis((2-propenyloxy)methyl)-1,3-propanediyl)-bis(oxy))bis-	**(1471-18-7)**
Propene, 2-bromo	**(557-93-7)**
Propene, 3-bromo	**(106-95-6)**
1-Propene, 2-bromo- (9CI)	**(557-93-7)**
1-Propene, 3-bromo-1-chloro- (9CI)	**(3737-00-6)**
Propene, 3-bromo-1-chloro- (8CI)	**(3737-00-6)**
1-Propene, 3-chloro-	**(107-05-1)**
Propene, 1-chloro	**(590-21-6)**
Propene, 2-chloro	**(557-98-2)**
Propene, 3-chloro	**(107-05-1)**
Propene, 3-chloro-2-(chloromethyl)	**(1871-57-4)**
1-Propene, 3-(2-chloroethoxy)-1,3-dichloro-	**(84987-77-9)**
1-Propene, 1-chloro-2-methyl-	**(513-37-1)**
Propene, 1-chloro-2-methyl	**(513-37-1)**
Propene, 3-chloro-2-methyl	**(563-47-3)**
1-Propene, 2,3-dibromo	**(513-31-5)**
2-Propene-1,2-dicarboxylic acid	**(97-65-4)**
1-Propene, dichloro	**(26952-23-8)**
Propene, 1,1-dichloro	**(563-58-6)**
Propene, 1,2-dichloro	**(563-54-2)**
Propene, 1,3-dichloro	**(542-75-6)**
Propene, 1,3-dichloro-, (E)	**(10061-02-6)**
Propene, 2,3-dichloro	**(78-88-6)**
1-Propene, 1,1-dichloro- (9CI)	**(563-58-6)**
1-Propene, 1,3-dichloro-, (E)- (9CI)	**(10061-02-6)**
1-Propene, 3,3-dichloro- (9CI)	**(563-57-5)**
Propene, 3,3-dichloro- (8CI)	**(563-57-5)**
1-Propene, 1,3-dichloro-, Mixt. with 1,2-Dichloropropane and isothiocyanatomethane	**(8066-01-1)**
Propene, 1,3-dichloro-, (Z)	**(10061-01-5)**
1-Propene, 1,3-dichloro-, (Z)- (9CI)	**(10061-01-5)**
1-Propene, 1,3-dichloro-2-(chloromethyl)- (9CI)	**(13245-65-3)**
1-Propene, 3,3-dichloro-2-(chloromethyl)- (9CI)	**(60845-51-4)**
Propene, 1,3-dichloro-2-(chloromethyl)- (7CI,8CI)	**(13245-65-3)**
Propene, 1,3-dichloro-2-methyl	**(3375-22-2)**
1-Propene, 1,3-dichloro-2-methyl- (9CI)	**(3375-22-2)**
1-Propene, 1-(3,4-dimethoxyphenyl)-	**(93-16-3)**
2-Propene-1,1-diol diacetate	**(869-29-4)**
2-Propene-1,1-diol, diacetate (8CI,9CI)	**(869-29-4)**
2-Propene-1,1-diol, 2-methyl-, diacetate	**(10476-95-6)**
1-Propene, 3-(ethenyloxy)- (9CI)	**(3917-15-5)**
1-Propene, 3-ethoxy- (9CI)	**(557-31-3)**
1-Propene, 2-fluoro- (9CI)	**(1184-60-7)**
1-Propene, 3-fluoro- (9CI)	**(818-92-8)**
Propene, 2-fluoro- (8CI)	**(1184-60-7)**
Propene, 3-fluoro- (8CI)	**(818-92-8)**
Propene, hexachloro	**(1888-71-7)**
Propene, hexafluoro	**(116-15-4)**
Propene, hexafluoro-, cyclic dimer	**(13429-24-8)**
1-Propene, 1,1,2,3,3,3-hexafluoro-, dimer (9CI)	**(13429-24-8)**
1-Propene homopolymer (9CI)	**(9003-07-0)**
Propene, 3-iodo	**(556-56-9)**
1-Propene, 3-iodo- (9CI)	**(556-56-9)**
Propene, 3-isothiocyanato-	**(57-06-7)**
1-Propene, 2-methoxy	**(116-11-0)**
Propene, 1-(p-methoxyphenyl)-	**(104-46-1)**
Propene, 2-methyl	**(115-11-7)**
1-Propene, 2-methyl-, Homopolymer (9CI)	**(9003-27-4)**
Propene, 2-methyl-, Polymers	**(9003-27-4)**
1-Propene, 2-methyl-, dimer (9CI)	**(18923-87-0)**
Propene, 2-methyl-, dimer (8CI)	**(18923-87-0)**
1-Propene, 2-methyl-3-phenyl-	**(3290-53-7)**
1-Propene, 2-methyl-, tetramer (9CI)	**(15220-85-6)**
1-Propene, 1-(methylthio)- (9CI)	**(10152-77-9)**
1-Propene, 3-(methylthio)- (9CI)	**(10152-76-8)**
Propene, 2-methyl-, trimer	**(7756-94-7)**
1-Propene, 2-methyl-, trimer (9CI)	**(7756-94-7)**
2-Propenenitrile	**(107-13-1)**
Propenenitrile	**(107-13-1)**
2-Propenenitrile, Homopolymer (9CI)	**(25014-41-9)**
2-Propenenitrile, Polymer with 1,3-butadiene and ethenyl-benzene (9CI)	**(9003-56-9)**
2-Propenenitrile, Polymer with ethenylbenzene (9CI)	**(9003-54-7)**
2-Propenenitrile, 2-methyl	**(126-98-7)**
2-Propenenitrile, 2,3,3-trichloro- (9CI)	**(16212-28-5)**
2-Propene-1-ol	**(107-18-6)**
Propene oxide	**(75-56-9)**
1-Propene, 1,1,2,3,3-pentachloro-	**(1600-37-9)**
Propene, 1,1,2,3,3-pentachloro	**(1600-37-9)**
1-Propene, 1,1,3,3,3-pentafluoro-2-trifluoromethyl	**(382-21-8)**
1-Propene, 3-phenyl-	**(300-57-2)**
1-Propene, 2-(phenylmethyl)-	**(3290-53-7)**
Propene-polymers	**(9003-07-0)**
1-Propene, 3-propoxy- (9CI)	**(1471-03-0)**
2-Propene-1-sulfonic acid, sodium salt (9CI)	**(2495-39-8)**
Propene, 1,1,2,3-tetrachloro	**(10436-39-2)**
1-Propene, 1,3,3,3-tetrachloro-2-(trichloromethyl)- (9CI)	**(83682-39-7)**
1-Propene, tetramer	**(25378-22-7)**
Propene, tetramer	**(25378-22-7)**
Propene, tetramer	**(6842-15-5)**
1-Propene, tetramer (9CI)	**(6842-15-5)**
1-Propene, 3,3'-thiobis- (9CI)	**(592-88-1)**
2-Propene-1-thiol (9CI)	**(870-23-5)**
2-Propene-1-thiol, 2-chloro-, diethyldithiocarbamate	**(95-06-7)**
2-Propene-1-thiol, 2,3-dichloro-, diisopropylcarbamate	**(2303-16-4)**
2-Propene-1-thiol, 2,3,3-trichloro-, diisopropylcarbamate	**(2303-17-5)**
1-Propene, 1,1,3-tribromo- (9CI)	**(36417-14-8)**
Propene, 1,1,3-tribromo- (6CI)	**(36417-14-8)**
(Z)-1-Propene-1,2,3-tricarboxylic acid	**(585-84-2)**
1-Propene-1,2,3-tricarboxylic acid	**(499-12-7)**
1-Propene-1,2,3-tricarboxylic acid, (Z)- (9CI)	**(585-84-2)**
Propene, 1,2,3-trichloro	**(96-19-5)**
Propene, 2,3,3-trichloro-	**(37077-84-2)**
1-Propene, 1,1,2-trichloro- (9CI)	**(21400-25-9)**
1-Propene, 1,1,3-trichloro- (9CI)	**(2567-14-8)**
1-Propene, 2,3,3-trichloro- (9CI)	**(37077-84-2)**
1-Propene, 3,3,3-trichloro- (9CI)	**(2233-00-3)**
Propene, 1,1,2-trichloro- (8CI)	**(21400-25-9)**

Propene, 1,1,3-trichloro- (8CI) **(2567-14-8)**
Propene, 1,3,3-trichloro-2-(dichloromethyl)- (8CI) **(14129-82-9)**
1-Propene, 1,1,3-trichloro-2-methyl- (9CI) **(31702-33-7)**
1-Propene, 3,3,3-trichloro-2-methyl- (9CI) **(4749-27-3)**
Propene, 1,1,3-trichloro-2-methyl- (8CI) **(31702-33-7)**
Propene, 3,3,3-trichloro-2-methyl- (8CI) **(4749-27-3)**
1-Propene, 3,3,3-trichloro-2-(trichloromethyl)- (9CI) **(83682-38-6)**
Propene, 3,3,3-trifluoro **(677-21-4)**
1-Propene, 3,3,3-trifluoro- (9CI) **(677-21-4)**
Propene, 3,3,3-trifluoro-2-(trifluoromethyl) **(382-10-5)**
Propene, trimer **(13987-01-4)**
Propenoic acid **(79-10-7)**
2-Propenoic acid (9CI) **(79-10-7)**
2-Propenoic acid, Homopolymer (9CI) **(9003-01-4)**
2-Propenoic acid, Polymer with 2-methyl-2-((1-oxo-2-propenyl)-amino)-1-propanesulfonic acid (9CI) **(40623-75-4)**
2-Propenoic acid, ammonium salt (9CI) **(10604-69-0)**
2-Propenoic acid, 3-(1,3-benzodioxol-5-yl)- (9CI) **(2373-80-0)**
2-Propenoic acid, 2-(4-benzoyl-3-hydroxyphenoxy)ethyl ester (9CI) **(16432-81-8)**
2-Propenoic acid, 2-(4-benzoyl-3-hydroxyphenoxy)ethyl ester, Homopolymer (9CI) **(29963-76-6)**
2-Propenoic acid, bicyclo(2,2,1)hept-5-en-2-ylmethyl ester (9CI) **(95-39-6)**
2-Propenoic acid, 2,2-bis(hydroxymethyl)-1,3-propanediyl ester (9CI) **(53417-29-1)**
2-Propenoic acid, 2,2-bis(((1-oxo-2-propenyl)oxy)methyl)-1,3-propanediyl ester (9CI) **(4986-89-4)**
2-Propenoic acid, 2-bromoethyl ester (9CI) **(4823-47-6)**
2-Propenoic acid, 3-(3-bromophenyl)- (9CI) **(32862-97-8)**
2-Propenoic acid, butyl ester **(141-32-2)**
2-Propenoic acid, butyl ester, Polymer with ethenyl acetate and N-(hydroxymethyl)-2-propenamide (9CI) **(26428-41-1)**
2-Propenoic acid, 2-(butyl((heptadecafluorooctyl)sulfonyl)amino)-ethyl ester (9CI) **(383-07-3)**
2-Propenoic acid, 2-carboxyethyl ester (9CI) **(24615-84-7)**
2-Propenoic acid, 2-chloro- **(598-79-8)**
2-Propenoic acid, 3-chloro-, (E)- (9CI) **(2345-61-1)**
2-Propenoic acid, chloro- (9CI) **(26952-44-3)**
2-Propenoic acid, 3-chloro-, (Z)- (9CI) **(1609-93-4)**
2-Propenoic acid, 2-chloroethyl ester (9CI) **(2206-89-5)**
2-Propenoic acid, 3-(3-chloro-4-hydroxy-5-methoxyphenyl)- (9CI) **(5438-40-4)**
2-Propenoic acid, 2-chloro-, methyl ester (9CI) **(80-63-7)**
2-Propenoic acid, 3-(4-chlorophenyl)- (9CI) **(1615-02-7)**
2-Propenoic acid, 2-cyano-3,3-diphenyl-, 2-ethylhexyl ester (9CI) **(6197-30-4)**
2-Propenoic acid, 2-cyano-, ethyl ester (9CI) **(7085-85-0)**
2-Propenoic acid, 2-cyanoethyl ester (9CI) **(106-71-8)**
2-Propenoic acid, decyl ester (9CI) **(2156-96-9)**
2-Propenoic acid, 2,3-dibromopropyl ester (9CI) **(19660-16-3)**
2-Propenoic acid, 3,3-dichloro- (9CI) **(1561-20-2)**
2-Propenoic acid, dichloro- (9CI) **(99165-89-6)**
2-Propenoic acid, 3-(2,4-dichlorophenyl)- (9CI) **(1201-99-6)**
2-Propenoic acid, 2-(diethylamino)ethyl ester (9CI) **(2426-54-2)**
2-Propenoic acid, 3-(3,4-dihydroxyphenyl)- (9CI) **(331-39-5)**
2-Propenoic acid, 3-(3,4-dimethoxyphenyl)- (9CI) **(2316-26-9)**
2-Propenoic acid, 2-(dimethylamino)ethyl ester (9CI) **(2439-35-2)**
2-Propenoic acid, 1,1-dimethylethyl ester (9CI) **(1663-39-4)**
2-Propenoic acid, 3-(2,2-dimethyl-1-oxo-3-((1-oxo-2-propenyl)-oxy)propoxy)-2,2-dimethylpropyl ester (9CI) **(30145-51-8)**
2-Propenoic acid, 2,2-dimethyl-1,3-propanediyl ester (9CI) **(2223-82-7)**
2-Propenoic acid, 2,2-dinitropropyl ester (9CI) **(17977-09-2)**
2-Propenoic acid, dodecyl ester (9CI) **(2156-97-0)**
2-Propenoic acid, 1,2-ethanediylbis(oxy-2,1-ethanediyl) ester (9CI) **(1680-21-3)**
2-Propenoic acid, 1,2-ethanediyl ester, Homopolymer (9CI) **(28158-16-9)**
2-Propenoic acid, 2-(2-ethoxyethoxy)ethyl ester (9CI) **(7328-17-8)**
2-Propenoic acid, 2-ethoxyethyl ester **(106-74-1)**
2-Propenoic acid, 2-ethylbutyl ester (9CI) **(3953-10-4)**
2-Propenoic acid, ethyl ester **(140-88-5)**
2-Propenoic acid, 2-ethylhexyl ester (9CI) **(103-11-7)**
2-Propenoic acid, 2-ethyl-2-(((1-oxo-2-propenyl)oxy)methyl)-1,3-propanediyl ester (9CI) **(15625-89-5)**
2-Propenoic acid, 2(((heptadecafluorooctyl)sulfonyl)butylamino)-ethyl ester **(383-07-3)**
2-Propenoic acid, 2-(((heptadecafluorooctyl)sulfonyl)methylamino)-ethyl ester (9CI) **(25268-77-3)**
2-Propenoic acid, hexadecyl ester (9CI) **(13402-02-3)**
2-Propenoic acid, 3a,4,5,6,7,7a-hexahydro-4,7-methano-1H-indenyl ester (9CI) **(33791-58-1)**
2-Propenoic acid, 2-((3a,4,5,6,7,7a-hexahydro-4,7-methano-1H-inden-6-yl)oxy)ethyl ester (9CI) **(65983-31-5)**
2-Propenoic acid, 1,6-hexanediyl ester **(13048-33-4)**
2-Propenoic acid, hexyl ester (9CI) **(2499-95-8)**
2-Propenoic acid, 2-hydroxyethyl ester (9CI) **(818-61-1)**
2-Propenoic acid, 2-hydroxy-, homopolymer, sodium salt (9CI) **(37956-57-3)**
2-Propenoic acid, 3-(4-hydroxy-3-methoxyphenyl)- (9CI) **(1135-24-6)**
2-Propenoic acid, 3-(4-hydroxy-3-methoxyphenyl)-, (E)- (9CI) **(537-98-4)**
2-Propenoic acid, 3-(4-hydroxy-3-methoxyphenyl)-, (Z)- (9CI) **(1014-83-1)**
2-Propenoic acid, 2-(hydroxymethyl)-2-(((1-oxo-2-propenyl)-oxy)methyl)-1,3-propanediyl ester **(3524-68-3)**
2-Propenoic acid, 3-(2-hydroxyphenyl)-, (E)- (9CI) **(614-60-8)**
2-Propenoic acid, 3-(3-hydroxyphenyl)- (9CI) **(588-30-7)**
2-Propenoic acid, 3-(4-hydroxyphenyl)- (9CI) **(7400-08-0)**
2-Propenoic acid, 3-(4-hydroxyphenyl)-, (E)- (9CI) **(501-98-4)**
2-Propenoic acid, 3-(hydroxyphenyl)- (9CI) **(25429-38-3)**
2-Propenoic acid, 2-hydroxypropyl ester (9CI) **(999-61-1)**
2-Propenoic acid, ion(1-) (9CI) **(10344-93-1)**
2-Propenoic acid, isodecyl ester (9CI) **(1330-61-6)**
2-Propenoic acid, isooctyl ester **(29590-42-9)**
2-Propenoic acid, isooctyl ester, homopolymer (9CI) **(9036-63-9)**
2-Propenoic acid, magnesium salt (9CI) **(5698-98-6)**
2-Propenoic acid, 2-methoxyethyl ester **(3121-61-7)**
trans-2-Propenoic acid, 3-(4-methoxyphenyl)- **(943-89-5)**
2-Propenoic acid, 3-(4-methoxyphenyl)- (9CI) **(830-09-1)**
2-Propenoic acid, 3-(4-methoxyphenyl)-, (E)- (9CI) **(943-89-5)**
Propenoic acid, 3-(4-methoxyphenyl)-, 2-ethoxyethyl ester **(104-28-9)**
2-Propenoic acid, 3-(4-methoxyphenyl)-, 2-ethoxyethyl ester (9CI) **(104-28-9)**
2-Propenoic acid, 3-(4-methoxyphenyl)-, 2-ethylhexyl ester (9CI) **(5466-77-3)**
2-Propenoic acid, 3-(4-(((4-methoxyphenyl)methylene)-amino)phenyl)-, 2-methylbutyl ester, (S-(E,E))- (9CI) **(24140-30-5)**
2-Propenoic acid, 2-methyl-, **(376-14-7)**
2-Propenoic acid, 2-methyl-, anhydride (9CI) **(760-93-0)**
2-Propenoic acid, 2-methyl-, Homopolymer (9CI) **(25087-26-7)**
2-Propenoic acid, 2-methyl-, 2-(bis(1-methylethyl)amino)-ethyl ester (9CI) **(16715-83-6)**
2-Propenoic acid, 2-methyl-, 1,4-butanediyl ester (9CI) **(2082-81-7)**
2-Propenoic acid, 2-methyl-, butyl ester **(97-88-1)**
2-Propenoic acid, 2-methyl-, butyl ester, Polymer with ethenyl-benzene (9CI) **(25213-39-2)**
2-Propenoic acid, 2-methyl-, butyl ester, Polymer with methyl 2-methyl-2-propenoate (9CI) **(25608-33-7)**
2-Propenoic acid, 2-methyl-, cyclohexyl ester (9CI) **(101-43-9)**
2-Propenoic acid, 2-methyl-, decyl ester (9CI) **(3179-47-3)**
2-Propenoic acid, 2-methyl-, 2,3-dibromopropyl ester **(3066-70-4)**
2-Propenoic acid, 2-methyl-, 2-(diethylamino)ethyl ester (9CI) **(105-16-8)**
2-Propenoic acid, 2-methyl-, 2-(dimethylamino)ethyl ester, Homopolymer (9CI) **(25154-86-3)**
2-Propenoic acid, 2-methyl-, 2-(dimethylamino)ethyl ester, Polymer with 1-ethenyl-2-pyrrolidinone, Compd. with diethyl sulfate (9CI) **(53633-54-8)**
2-Propenoic acid, 2-methyl-, 2-(dimethylamino)ethyl ester, acetate (9CI) **(10018-87-8)**
2-Propenoic acid, 2-methyl-, 1,1-dimethylethyl ester (9CI) **(585-07-9)**
2-Propenoic acid, 2-methyl-, 2,2-dimethyl-1,3-propanediyl ester (9CI) **(1985-51-9)**
2-Propenoic acid, 2-methyl-, eicosyl ester (9CI) **(45294-18-6)**

Name	CAS No.
2-Propenoic acid, methyl ester	**(96-33-3)**
Propenoic acid methyl ester	**(96-33-3)**
2-Propenoic acid, 2-methyl-, 1,2-ethanediylbis(oxy-2,1-ethanediyl) ester (9CI)	**(109-16-0)**
2-Propenoic acid, (1-methyl-1,2-ethanediyl)bis(oxy(methyl-2,1-ethanediyl)) ester	**(42978-66-5)**
2-Propenoic acid, 2-methyl-, 1,2-ethanediyl ester (9CI)	**(97-90-5)**
2-Propenoic acid, 2-methyl-, 1,2-ethanediyl ester, Homopolymer (9CI)	**(25721-76-0)**
2-Propenoic acid, 2-methyl-, 2-ethoxyethyl ester (9CI)	**(2370-63-0)**
2-Propenoic acid, 2-methyl-, ethyl ester	**(97-63-2)**
2-Propenoic acid, 1-methylethyl ester (9CI)	**(689-12-3)**
2-Propenoic acid, 2-methyl-, 3,3,4,4,5,5,6,6,7,7,8,8,9,9,10,10,11,11,12,12,12-heneicosafluorododecyl ester (9CI)	**(2144-54-9)**
2-Propenoic acid, 2-methyl-, 3,3,4,4,5,5,6,6,7,7,8,8,9,9,10,10,10-heptadecafluorodecyl ester (9CI)	**(1996-88-9)**
2-Propenoic acid, 2-methyl-, 2-(((heptadecafluorooctyl)sulfonyl)-methylamino)ethyl ester (9CI)	**(14650-24-9)**
2-Propenoic acid, 2-methyl-, hexadecyl ester (9CI)	**(2495-27-4)**
2-Propenoic acid, 2-methyl-, 1,6-hexanediyl ester (9CI)	**(6606-59-3)**
2-Propenoic acid, 2-methyl-, hexyl ester (9CI)	**(142-09-6)**
2-Propenoic acid, 2-methyl-, 2-hydroxymethylethyl ester	**(27813-02-1)**
2-Propenoic acid, 2-methyl-, 2-hydroxypropyl ester (9CI)	**(923-26-2)**
2-Propenoic acid, 2-methyl-, ion(1-) (9CI)	**(18358-13-9)**
2-Propenoic acid, 2-methyl-, 2-isocyanatoethyl ester	**(30674-80-7)**
2-Propenoic acid, 2-methyl-, lead(2+) salt (9CI)	**(1068-61-7)**
2-Propenoic acid, 2-methyl-, methyl ester	**(80-62-6)**
2-Propenoic acid, 2-methyl-, methyl ester, Polymer with ethyl 2-propenoate (9CI)	**(9010-88-2)**
2-Propenoic acid, 2-methyl-, methyl ester, Polymer with ethyl 2-propenoate, N-(hydroxymethyl)-2-propenamide and 2-propenamide (9CI)	**(30394-81-1)**
2-Propenoic acid, 2-methyl-, methyl ester, Polymer with 3-methyl-3-buten-2-one (9CI)	**(51555-36-3)**
2-Propenoic acid, 2-methyl-, (1-methyl-1,2-ethanediyl)bis(oxy-(methyl-2,1-ethanediyl)) ester (9CI)	**(51247-87-1)**
2-Propenoic acid, 2-methyl-, 1-methylethyl ester (9CI)	**(4655-34-9)**
2-Propenoic acid, 2-methyl-, (1-methylethylidene)bis(4,1-phenyleneoxy(2-hydroxy-3,1- propanediyl)) ester	**(1565-94-2)**
2-Propenoic acid, 2-methyl-, (1-methylethylidene)bis(4,1-phenyleneoxy-2,1-ethanediyl) ester (9CI)	**(24448-20-2)**
2-Propenoic acid, 2-methyl-, 1-methyl-1,3-propanediyl ester (9CI)	**(1189-08-8)**
2-Propenoic acid, 2-methyl-, 2-methylpropyl ester	**(97-86-9)**
2-Propenoic acid, 2-methyl-, 2-methylpropyl ester, Polymer with ethenylbenzene (9CI)	**(25667-93-0)**
2-Propenoic acid, 2-methyl-, monoester with 1,2-propanediol (9CI)	**(27813-02-1)**
2-Propenoic acid, 2-methyl-, 3,3,4,4,5,5,6,6,6-nonafluorohexyl ester (9CI)	**(1799-84-4)**
2-Propenoic acid, 2-methyl-, octadecyl ester	**(32360-05-7)**
2-Propenoic acid, 2-methyl-, octyl ester	**(2157-01-9)**
2-Propenoic acid, 2-methyl-, 2-((2,3,3a,4,7,7a(or 3a,4,5,6,7,7a)-hexahydro-4,7-methano-1H-indenyl)oxy)ethyl ester (9CI)	**(68586-19-6)**
2-Propenoic acid, (2-methyloxiranyl)methyl ester (9CI)	**(19900-46-0)**
2-Propenoic acid, 2-methyl-, oxybis(2,1-ethanediyloxy-2,1-ethanediyl) ester	**(109-17-1)**
2-Propenoic acid, 2-methyl-, oxydi-2,1-ethanediyl ester (9CI)	**(2358-84-1)**
2-Propenoic acid, 2-methyl-, 3,3,4,4,5,5,6,6,7,7,8,8,9,9,10,10,11,11,12,12,13,13,14,14,14-pentacosafluorotetradecyl ester (9CI)	**(6014-75-1)**
2-Propenoic acid, 2-methyl-, pentadecyl ester (9CI)	**(6140-74-5)**
2-Propenoic acid, 2-methyl-3-phenyl- (9CI)	**(1199-77-5)**
2-Propenoic acid, 3-(4-methylphenyl)- (9CI)	**(1866-39-3)**
2-Propenoic acid, 2-methyl-, phenylmethyl ester (9CI)	**(2495-37-6)**
2-Propenoic acid, 2-methyl-, potassium salt (9CI)	**(6900-35-2)**
2-Propenoic acid, 1-methyl-1,3-propanediyl ester (9CI)	**(19485-03-1)**
2-Propenoic acid, 2-methyl-, 2-propenyl ester	**(96-05-9)**
2-Propenoic acid, 2-methylpropyl ester (9CI)	**(106-63-8)**
2-Propenoic acid, 2-methyl-, sodium salt	**(5536-61-8)**
2-Propenoic acid, 2-methyl-, 2-sulfoethyl ester (9CI)	**(10595-80-9)**
2-Propenoic acid, 2-methyl-, 2-sulfoethyl ester, sodium salt (9CI)	**(1804-87-1)**
2-Propenoic acid, 2-methyl-, tetradecyl ester (9CI)	**(2549-53-3)**
2-Propenoic acid, 2-methyl-, (tetrahydro-2-furanyl)methyl ester (9CI)	**(2455-24-5)**
2-Propenoic acid, 2-methyl-, 3-(trichlorosilyl)propyl ester (9CI)	**(7351-61-3)**
2-Propenoic acid, 2-methyl-, 3,3,4,4,5,5,6,6,7,7,8,8,8-tridecafluorooctyl ester (9CI)	**(2144-53-8)**
2-Propenoic acid, 2-methyl-, tridecyl ester (9CI)	**(2495-25-2)**
2-Propenoic acid, 2-methyl-, 2,2,2-trifluoroethyl ester (9CI)	**(352-87-4)**
2-Propenoic acid, 2-methyl-, 3-(trimethoxysilyl)propyl ester	**(2530-85-0)**
2-Propenoic acid, 2-methyl-, 1,7,7-trimethylbicyclo-(2.2.1)hept-2-yl ester, exo- (9CI)	**(7534-94-3)**
2-Propenoic acid, monoester with 1,2-propanediol (9CI)	**(25584-83-2)**
2-Propenoic acid, octadecyl ester	**(4813-57-4)**
2-Propenoic acid, octyl ester (9CI)	**(2499-59-4)**
2-Propenoic acid, oxiranylmethyl ester (9CI)	**(106-90-1)**
2-Propenoic acid, oxybis(2,1-ethanediyloxy-2,1-ethanediyl)ester	**(17831-71-9)**
2-Propenoic acid, oxybis(methyl-2,1-ethanediyl) ester	**(57472-68-1)**
2-Propenoic acid, oxydi-2,1-ethanediyl ester (9CI)	**(4074-88-8)**
2-Propenoic acid, 3-phenyl-	**(621-82-9)**
2-Propenoic acid, 3-phenyl-, (E)- (9CI)	**(140-10-3)**
2-Propenoic acid, 3-phenyl-, butyl ester	**(538-65-8)**
2-Propenoic acid, 3,3'-(1,4-phenylene)bis-, diethyl ester (9CI)	**(17088-28-7)**
2-Propenoic acid, 3-phenyl-, ethyl ester (9CI)	**(103-36-6)**
2-Propenoic acid, 3-phenyl-, methyl ester (9CI)	**(103-26-4)**
2-Propenoic acid, 3-phenyl-, phenylmethyl ester (9CI)	**(103-41-3)**
2-Propenoic acid, 3-phenyl-, 3-phenylpropyl ester (9CI)	**(122-68-9)**
2-Propenoic acid, 3-phenyl-, propyl ester (9CI)	**(7778-83-8)**
2-Propenoic acid, potassium salt (9CI)	**(10192-85-5)**
2-Propenoic acid, 2-propenyl ester (9CI)	**(999-55-3)**
2-Propenoic acid, propyl ester (9CI)	**(925-60-0)**
2-Propenoic acid, sodium salt (9CI)	**(7446-81-3)**
2-Propenoic acid, (tetrahydro-2-furanyl)methyl ester (9CI)	**(2399-48-6)**
2-Propenoic acid, tetrahydrofurfuryl ester	**(2399-48-6)**
2-Propenoic acid, 3a,4,7,7a-tetrahydro-4,7-methano-1H-indenyl ester	**(50976-02-8)**
2-Propenoic acid, 2,3,3-trichloro- (9CI)	**(2257-35-4)**
2-Propenoic acid, tridecyl ester (9CI)	**(3076-04-8)**
2-Propenoic acid, 3-(3,4,5-trimethoxyphenyl)- (9CI)	**(90-50-6)**
2-Propenoic acid, 1,7,7-trimethylbicyclo(2.2.1)hept-2-yl ester, exo- (9CI)	**(5888-33-5)**
2-Propenoic acid, zinc salt (9CI)	**(14643-87-9)**
1-Propen-3-ol	**(107-18-6)**
1-Propenol-3	**(107-18-6)**
2-Propen-1-ol	**(107-18-6)**
Propen-1-ol-3	**(107-18-6)**
Propenol	**(107-18-6)**
1-Propen-2-ol, acetate	**(108-22-5)**
2-Propen-1-ol, 2-chloro	**(5976-47-6)**
2-Propen-1-ol, 3-chloro	**(29560-84-7)**
2-Propen-1-ol, 2-methyl	**(513-42-8)**
Propenol, 1-methyl	**(598-32-3)**
2-Propen-1-ol, 3-phenyl-	**(104-54-1)**
2-Propen-1-one	**(107-02-8)**
2-Propen-1-one, 1,3-diphenyl-	**(94-41-7)**
2-Propen-1-one, 3-phenyl-1-(2-thienyl)- (8CI,9CI)	**(3988-77-0)**
Propenoyl chloride	**(814-68-6)**
2-Propenoyl chloride (9CI)	**(814-68-6)**
2-Propenoyl chloride, 2-methyl- (9CI)	**(920-46-7)**
2-Propenoyl chloride, 3-phenyl- (9CI)	**(102-92-1)**
Propenyl	**(60-13-9)**
1-Propen-2-yl acetate	**(108-22-5)**
2-Propenylacrylic acid	**(110-44-1)**
2-Propenyl alcohol	**(107-18-6)**
Propenyl alcohol	**(107-18-6)**

4-Propenylanisole (104-46-1)
p-1-Propenylanisole (104-46-1)
p-Propenyl anisole (104-46-1)
trans-p-Propenylanisole (4180-23-8)
2-Propenylbenzene (300-57-2)
Propenyl benzene (637-50-3)
5-(1-Propenyl)-1,3-benzodioxole (120-58-1)
5-(2-Propenyl)-1,3-benzodioxole (94-59-7)
4-Propenylcatechol methylene ether (120-58-1)
2-Propenyl chloride (107-05-1)
Propenyl chloride (590-21-6)
1-Propenyl cyanide (4786-20-3)
2-Propenyl disulphide (2179-57-9)
Propenyl ether (557-40-4)
4-Propenylguaiacol (97-54-1)
2-Propenyl n-hexanoate (123-68-2)
2-Propenyl (hydroxymethyl)carbamate (24935-97-5)
2-Propenyl isothiocyanate (57-06-7)
2-Propenyl isovalerate (2835-39-4)
p-Propenylmethoxybenzene (104-46-1)
4-(2-Propenyl)-2-methoxyphenyl formate (10031-96-6)
2-Propenyl 3-methylbutanoate (2835-39-4)
4-Propenyl-1,2-methylenedioxybenzene (120-58-1)
(2-Propenyloxy)benzenemethanol (28655-62-1)
p-Propenylphenyl methyl ether (104-46-1)
2-Propenyl 2-propenoate (999-55-3)
2-Propenyl sulphide (592-88-1)
(2-Propenyl)thiourea (109-57-9)
N-2-Propenylurea (557-11-9)
4-Propenyl veratrole (93-16-3)
Properidina (Spanish) (561-76-2)
Properidine (561-76-2)
Properidinum (atin) (561-76-2)
Properidol (548-73-2)
Propesin (6CI) (94-12-2)
Propesine (94-12-2)
Propetamphos (31218-83-4)
Propham (122-42-9)
Prophame (122-42-9)
Prophenatin (15307-79-6)
Prophenpyridamine (86-21-5)
Prophenpyridamine maleate (132-20-7)
Prophos (13194-48-4)
Propiconazole (60207-90-1)
Propilentiourea (Italian) (2122-19-2)
Propilizon (83-59-0)
Propilthiouracil (51-52-5)
6-Propil-tiouracile (Italian) (51-52-5)
Propine (74-99-7)
Propineb (12071-83-9)
Propinebe (12071-83-9)
Propioaldehyde (624-67-9)
γ-Propiobutyrolactone (105-21-5)
Propiocine (114-07-8)
Propioguanamine, .β.-cyano- (4784-14-9)
Propiokan (57-85-2)
1,3-Propiolactone (57-57-8)
3-Propiolactone (57-57-8)
Propiolactone (57-57-8)
β-Propiolactone (57-57-8)
β-Propiolakton (Czech) (57-57-8)
Propiolaldehye (624-67-9)
Propiolic-acid (471-25-0)
Propiolic acid, (m-chlorophenyl)- (8CI) (7396-28-3)
Propiolic acid, (o-chlorophenyl)- (8CI) (24654-08-8)
Propiolic acid, (p-chlorophenyl)- (8CI) (3240-10-6)
Propiolic acid, ethyl ester (8CI) (623-47-2)
Propiolic acid, methyl ester (922-67-8)
Propiomazina (Spanish) (362-29-8)
Propiomazine (362-29-8)
Propiomazinum (Latin) (362-29-8)
Propionaldehyde (123-38-6)
Propionaldehyde [UN 1275] (123-38-6)
Propionaldehyde, β-(4-tert-butylphenyl)-α-methyl- (80-54-6)
Propionaldehyde, 3-chloro- (8CI) (19434-65-2)
Propionaldehyde, 2,3-dichloro-2-methyl (10141-22-7)
Propionaldehyde, diethyl acetal (8CI) (4744-08-5)
Propionaldehyde, dimethyl acetal (8CI) (4744-10-9)
Propionaldehyde, 2,3-epoxy- (765-34-4)
Propionaldehyde, 3-ethoxy (2806-85-1)
Propionaldehyde, ethylhydrazone (7CI,8CI) (7422-92-6)
Propionaldehyde, 3-hydroxy-2,2-dimethyl- (597-31-9)
Propionaldehyde, 2-keto (78-98-8)
Propionaldehyde, 2-methyl- (78-84-2)
Propionaldehyde, 2-methyl-2-(methylsulfinyl)-, O-(methylcarbamoyl)oxime (1646-87-3)
Propionaldehyde, 2-methyl-2-(methylsulfonyl)-, O-(methylcarbamoyl)oxime (1646-88-4)
Propionaldehyde, 2-methyl-2-(methylthio)-, O-(methylcarbamoyl)-oxime (116-06-3)
Propionaldehyde, 2-methyl-2-(methylthio)-, oxime (1646-75-9)
Propionaldehyde, 3-(methylthio) (3268-49-3)
Propionaldehyde, 2-oxo- (78-98-8)
Propionaldehyde, 2-phenyl- (93-53-8)
Propionaldehyde, 2,2,3-trichloro (7789-90-4)
Propionaldehyde, 3,3,3-trifluoro (460-40-2)
Propionamide (79-05-0)
Propionamide, N-(4-amino-6-(trichloromethyl)-s-triazin-2-yl)-(8CI) (30355-96-5)
Propionamide, N,N-diethyl-2-(1-naphthyloxy) (15299-99-7)
Propionamide, 3,3'-dithiobis(N-octyl- (8CI) (33312-01-5)
Propionamide, 3-((2-hydroxyethyl)amino)- (8CI) (27076-30-8)
Propionamide, N-methyl (1187-58-2)
Propionamide, N-(1-methyl-2-piperidinoethyl)-N-2-pyridyl-, (+-)- (15686-91-6)
Propionamidine, 2,2'-azobis(2-methyl-, dihydrochloride (2997-92-4)
Propionamidine, 2,2-dichloro-, Compd. with 4,6-bis(1,1-dichloroethyl)-s-triazin-2-ol (1:1) (8CI) (30886-04-5)
Propionamidine, 2,2-dichloro-, Compd. with 4,6-bis(1,1-dichloroethyl)-s-triazin-2-ol (6CI) (30886-04-5)
Propionamidine, 2,2,3-trichloro-, Compd. with 4,6-bis-(1,1,2-trichloroethyl)-s-triazin-2-ol (1:1) (8CI) (30886-05-6)
Propionanilide (620-71-3)
Propionanilide, 3',4'-dichloro (709-98-8)
Propionanilide, N-(2-(methylphenethylamino)propyl)- (552-25-0)
Propionanilide, N-(1-methyl-2-piperidinoethyl)- (129-83-9)
Propionanilide, N-(1-phenethyl-4-piperidyl) (437-38-7)
Propionan sodny (Czech) (137-40-6)
Propionate de methyle (French) (554-12-1)
Propionate d'ethyle (French) (105-37-3)
Propione (96-22-0)
Propionic Acid Grain Preserver (79-09-4)
Propionic-acid (79-09-4)
Propionic acid (ACGIH,OSHA) [UN 1848] (79-09-4)
Propionic acid, Solution (DOT) (79-09-4)
Propionic acid, Solution containing not less than 80% acid (DOT) (79-09-4)
Propionic acid, 3-acetyl- (123-76-2)
Propionic acid, 3-(acetyl-(3-amino-2,4,6-triiodophenyl)amino)-2-methyl (16034-77-8)
Propionic acid, 3-allyl-1-methyl-4-phenyl-4-piperidyl ester (6CI) (469-81-8)
Propionic acid amide (79-05-0)
Propionic acid, α-amino-β-methylamino- (17463-44-4)
Propionic acid, α-amino-β-methylamino-, DL (17463-44-4)
Propionic acid, 2-amino-3-indol-3-yl- (73-22-3)
Propionic acid anhydride (123-62-6)

Propionic acid, 3-benzoyl- (8CI) **(2051-95-8)**
Propionic acid, 2-(N-benzoyl-N-(3,4-dichlorophenyl))amino-, ethyl ester **(22212-55-1)**
Propionic acid, 2-(3-benzoylphenyl) **(22071-15-4)**
Propionic acid, benzyl ester (8CI) **(122-63-4)**
Propionic acid, 2,2-bis(chloromethyl)- **(67329-11-7)**
Propionic acid, 2-bromo **(598-72-1)**
Propionic acid, 3-bromo **(590-92-1)**
Propionic acid, butyl ester **(590-01-2)**
Propionic acid, calcium salt **(4075-81-4)**
Propionic acid chloride **(79-03-8)**
Propionic acid, 2-chloro **(598-78-7)**
Propionic acid, 3-chloro **(107-94-8)**
Propionic acid, α-chloro- **(598-78-7)**
Propionic acid, chloro- **(28554-00-9)**
Propionic acid, 2-chloro-3-(4-chlorophenyl)-, methyl ester **(14437-17-3)**
Propionic acid, 3-chloro-2,2-dimethyl- (8CI) **(13511-38-1)**
Propionic acid, 2-chloro-, ethyl ester **(535-13-7)**
Propionic acid, 2-(N-(3-chloro-4-fluorophenyl)benzamido)-, methyl ester **(52756-25-9)**
Propionic acid, 2-chloro-, methyl ester **(17639-93-9)**
Propionic acid, 2-(4-chloro-2-methylphenoxy) **(93-65-2)**
Propionic acid, 2-(p-chlorophenoxy)- (8CI) **(3307-39-9)**
Propionic acid, 2-(m-chlorophenoxy)-2-methyl **(17413-73-9)**
Propionic acid, 2-(p-chlorophenoxy)-2-methyl **(882-09-7)**
Propionic acid, 2-(4-chlorophenoxy)-2-methyl-, ethyl ester **(637-07-0)**
Propionic acid, 2-(p-chlorophenoxy)-2-methyl-, ethyl ester **(637-07-0)**
Propionic acid, 2-(p-chlorophenoxy)-2-methyl-, methyl ester **(55162-41-9)**
Propionic acid, 2-((4-chloro-o-tolyl)oxy) **(93-65-2)**
Propionic acid, 2-((4-chloro-o-tolyl)oxy)-, (+-) **(7085-19-0)**
Propionic acid, 2-((4-chloro-o-tolyl)oxy)-, Compd. with 2,2'-iminodiethanol (1:1) **(1432-14-0)**
Propionic acid, 2-((4-chloro-o-tolyl)oxy)-, butyl ester (8CI) **(1713-14-0)**
Propionic acid, 2-((4-chloro-o-tolyl)oxy)-, potassium salt **(1929-86-8)**
Propionic acid, 3-cyclopentyl- **(140-77-2)**
Propionic acid, 2,3-dibromo **(600-05-5)**
Propionic acid, 2,2-dichloro **(75-99-0)**
Propionic acid, 2,3-dichloro **(565-64-0)**
Propionic acid 3,4-dichloroanilide **(709-98-8)**
Propionic acid, 2-(2,4-dichlorophenoxy) **(120-36-5)**
Propionic acid, 2-(3,4-dichlorophenoxy)- (8CI) **(3307-41-3)**
Propionic acid, 2-(4-(2,4-dichlorophenoxy)phenoxy)-, methyl ester **(51338-27-3)**
Propionic acid, 2,2-dichloro-, sodium salt **(127-20-8)**
Propionic acid, 2,2-dichloro-, 2-(2,4,5-trichlorophenoxy)ethyl ester **(136-25-4)**
Propionic acid, 2,2-dimethyl- **(75-98-9)**
Propionic acid, α-1,3-dimethyl-4-phenyl-4-piperidyl ester **(77-20-3)**
Propionic acid, 3,3'-(dioxydicarbonyl)di-, Maximum concentration 72% **(123-23-9)**
Propionic acid, 2,3-diphenyl- (8CI) **(3333-15-1)**
Propionic acid, 3,3'-(dodecylimino)di-, disodium salt **(3655-00-3)**
Propionic acid, ethoxy- **(1331-11-9)**
Propionic acid, 3-ethoxy-, ethyl ester **(763-69-9)**
Propionic acid, 3-(2-ethylbutoxy)- **(10213-74-8)**
Propionic acid, ethyl ester **(105-37-3)**
Propionic acid, 3-fluoro **(461-56-3)**
Propionic acid, heptyl ester (8CI) **(2216-81-1)**
Propionic acid, 2-hydroxy- **(50-21-5)**
Propionic acid, 3-hydroxy-, β-lactone **(57-57-8)**
Propionic acid, 3,3'-iminodi- (8CI) **(505-47-5)**
Propionic acid, 3-iodo **(141-76-4)**
Propionic acid, isobutyl ester **(540-42-1)**
Propionic acid, isopentyl ester **(105-68-0)**
Propionic acid, isopropyl ester **(637-78-5)**
Propionic acid, 3-mercapto **(107-96-0)**
Propionic acid, 3-mercapto-, ethyl ester (8CI) **(5466-06-8)**
Propionic acid, 3-mercapto-, methyl ester (8CI) **(2935-90-2)**
Propionic acid, 3-methoxy-, methyl ester **(3852-09-3)**
Propionic acid, 2-(6-methoxy-2-naphthyl)-, (+) **(22204-53-1)**
Propionic acid, 2-methyl- **(79-31-2)**
Propionic acid, 2-(2-methyl-4-chlorophenoxy)- **(93-65-2)**
Propionic acid, 2-methylene- **(79-41-4)**
Propionic acid, methyl ester **(554-12-1)**
Propionic acid, 2-methyl-, ethyl ester **(97-62-1)**
Propionic acid, 2-methyl-, monoester with 2,2,4-trimethyl-1,3-pentanediol **(25265-77-4)**
Propionic acid, 2-methyl-2-(p-(1,2,3,4-tetrahydro-1-naphthyl)-phenoxy) **(3771-19-5)**
Propionic acid, 3-nitro **(504-88-1)**
Propionic acid, pentyl ester (8CI) **(624-54-4)**
Propionic acid, phenethyl ester (8CI) **(122-70-3)**
Propionic acid, 2-phenylcarbamoyloxy- **(73622-98-7)**
Propionic acid, 2-phenylethyl ester **(122-70-3)**
Propionic acid, propyl ester **(106-36-5)**
Propionic acid, sodium salt **(137-40-6)**
Propionic acid, 3,3'-thiodi **(111-17-1)**
Propionic acid, 3,3'-thiodi-, didodecyl ester **(123-28-4)**
Propionic acid, 3,3'-thiodi-, dioctadecyl ester (8CI) **(693-36-7)**
Propionic acid, 2,2,3-trichloro **(3278-46-4)**
Propionic acid, (2,4,5-trichlorophenoxy) **(29990-39-4)**
Propionic acid, 2-(2,4,5-trichlorophenoxy) **(93-72-1)**
Propionic acid, 2-(2,4,5-trichlorophenoxy)-, butoxypropanol ester (8CI) **(28903-26-6)**
Propionic acid, 2-(2,4,5-trichlorophenoxy)-, 3-butoxypropyl ester (8CI) **(25537-26-2)**
Propionic acid, 2-(2,4,5-trichlorophenoxy)-, Compd. with 2-aminoethanol (1:1) (8CI) **(7374-47-2)**
Propionic acid, 2-(2,4,5-trichlorophenoxy)-, ester with butoxypropanol **(28903-26-6)**
Propionic acid, 2-(2,4,5-trichlorophenoxy)-, 2-ethylhexyl ester **(53404-76-5)**
Propionic acid, 2-(2,4,5-trichlorophenoxy)-, isooctyl ester **(32534-95-5)**
Propionic acid, 2-(2,4,5-trichlorophenoxy)-, potassium salt (8CI) **(2818-16-8)**
Propionic acid, 3-(2,4,5-triethoxybenzoyl) **(41826-92-0)**
Propionic acid, 2-(p-((5-(trifluoromethyl)-2-pyridyl)oxy)-phenoxy)-, butyl ester **(69806-50-4)**
Propionic acid, vinyl ester **(105-38-4)**
Propionic acid, zinc salt **(557-28-8)**
Propionic aldehyde **(123-38-6)**
Propionic amide **(79-05-0)**
Propionic-anhydride **(123-62-6)**
Propionic anhydride [UN 2496] **(123-62-6)**
Propionic chloride **(79-03-8)**
Propionic ether **(105-37-3)**
Propionic nitrile **(107-12-0)**
Propionin, 1-mono- (8CI) **(624-47-5)**
Propionin, tri- (8CI) **(139-45-7)**
Propionitrile [UN 2404] **(107-12-0)**
Propionitrile, 3-amino **(151-18-8)**
Propionitrile, 2,2'-azobis(2-methyl **(78-67-1)**
Propionitrile, 3-((4,6-bis(trichloromethyl)-s-triazin-2-yl)amino)-(8CI) **(24848-40-6)**
Propionitrile, 3-((4,6-bis(trichloromethyl)-s-triazin-2-yl)-sec-butylamino)- (8CI) **(24863-52-3)**
Propionitrile, 3-((4,6-bis(trichloromethyl)-s-triazin-2-yl)butyl-amino)- (8CI) **(24863-51-2)**
Propionitrile, 3-((4,6-bis(trichloromethyl)-s-triazin-2-yl)ethyl-amino)- (8CI) **(24848-42-8)**
Propionitrile, 3-((4,6-bis(trichloromethyl)-s-triazin-2-yl)methyl-amino)- (8CI) **(24848-41-7)**
Propionitrile, 3-((4,6-bis(trichloromethyl)-s-triazin-2-yl)propyl-amino)- (8CI) **(24848-43-9)**
Propionitrile, 3-(butyl(4-methyl-6-(trichloromethyl)-s-triazin-2-yl)amino)- (8CI) **(24848-47-3)**
Propionitrile, 3-(sec-butyl(4-methyl-6-(trichloromethyl)-s-triazin-

2-yl)amino)- (8CI) (24848-48-4)
Propionitrile, 3-(butyl(4-pentyl-6-(trichloromethyl)-s-triazin-2-yl)-amino)- (8CI) (24848-61-1)
Propionitrile, 3-(sec-butyl(4-pentyl-6-(trichloromethyl)-s-triazin-2-yl)amino)- (8CI) (24848-62-2)
Propionitrile, 3-(butyl(4-propyl-6-(trichloromethyl)-s-triazin-2-yl)-amino)- (8CI) (24848-54-2)
Propionitrile, 3-(sec-butyl(4-propyl-6-(trichloromethyl)-s-triazin-2-yl)amino)- (8CI) (24863-54-5)
Propionitrile, 3-chloro (542-76-7)
Propionitrile, 2-chloro-3-(3-chloro-o-tolyl) (21342-85-8)
Propionitrile, 2-(4-chloro-6-(cyclopropylamino)-s-triazin-2-yl-amino)-2-methyl (32889-48-8)
Propionitrile, 2-((4-chloro-6-(ethylamino)-s-triazin-2-yl)amino)-2-methyl (21725-46-2)
Propionitrile, 3-(dimethylamino) (1738-25-6)
Propionitrile, 3-ethoxy- (8CI) (2141-62-0)
Propionitrile, 3,3'-(ethylenediimino)di- (8CI) (3217-00-3)
Propionitrile, 3-(2-ethylhexyloxy) (10213-75-9)
Propionitrile, 3-(ethyl(4-methyl-6-(trichloromethyl)-s-triazin-2-yl)amino)- (8CI) (24848-46-2)
Propionitrile, 3-(ethyl(4-pentyl-6-(trichloromethyl)-s-triazin-2-yl)amino)- (8CI) (24863-55-6)
Propionitrile, 3-(ethyl(4-propyl-6-(trichloromethyl)-s-triazin-2-yl)amino)- (8CI) (24848-52-0)
Propionitrile, 3-(N-ethyl-m-toluidino)- (8CI) (148-69-6)
Propionitrile, 2-hydroxy- (78-97-7)
Propionitrile, 3-hydroxy- (109-78-4)
Propionitrile, 3-(N-(2-hydroxyethyl)anilino) (92-64-8)
Propionitrile, 3-(N-(2-hydroxyethyl)-4-((p-nitrophenyl)azo)-m-toluidino)- (8CI) (6054-58-6)
Propionitrile, 3-((2-hydroxyethyl)phenylamino)- (92-64-8)
Propionitrile, 3,3'-iminodi (111-94-4)
Propionitrile, 3-methoxy (110-67-8)
Propionitrile, 3-(N-methylanilino)- (8CI) (94-34-8)
Propionitrile, 3-(methyl(4-methyl-6-(trichloromethyl)-s-triazin-2-yl)amino)- (8CI) (24848-45-1)
Propionitrile, 3-(methylnitrosamino) (60153-49-3)
Propionitrile, 3-(methyl(4-pentyl-6-(trichloromethyl)-s-triazin-2-yl)amino)- (8CI) (24848-59-7)
Propionitrile, 3-((α-methylphenethyl)amino)-, (+-)- (15686-61-0)
Propionitrile, 3-(methyl(4-propyl-6-(trichloromethyl)-s-triazin-2-yl)amino)- (8CI) (24848-51-9)
Propionitrile, 3,3'-((6-methyl-s-triazine-2,4-diyl)diimino)di- (8CI) (30368-97-9)
Propionitrile, 3,3',3'',3'''-((6-methyl-s-triazine-2,4-diyl)dinitrilo)tetra- (8CI) (30355-52-3)
Propionitrile, 3-((4-methyl-6-(trichloromethyl)-s-triazin-2-yl)-amino)- (8CI) (24848-44-0)
Propionitrile, 3-((4-methyl-6-(trichloromethyl)-s-triazin-2-yl)-propylamino)- (8CI) (24863-53-4)
Propionitrile, 3,3'-oxydi (1656-48-0)
Propionitrile, 3-((4-pentyl-6-(trichloromethyl)-s-triazin-2-yl)-propylamino)- (8CI) (24848-60-0)
Propionitrile, 3,3'-(phenylimino)di- (8CI) (1555-66-4)
Propionitrile, 3,3',3'',3'''-((6-phenyl-s-triazine-2,4-diyl)-dinitrilo)tetra- (7CI,8CI) (3786-23-0)
Propionitrile, 3-(propyl(4-propyl-6-(trichloromethyl)-s-triazin-2-yl)-amino)- (8CI) (24848-53-1)
Propionitrile, 3,3'-thiodi (111-97-7)
Propionitrile, 3,3'-((6-(trichloromethyl)-s-triazine-2,4-diyl)-diimino)di- (8CI) (26235-11-0)
Propionitrile, 3-(trichlorosilyl) (1071-22-3)
Propionitrile, 3-(triethoxysilyl)- (919-31-3)
Propionohydroxamic acid, N-(5-(3-((5-aminopentyl)hydroxy-carbamoyl)propionamido)pentyl)- 3-((5-(N-hydroxyacetamido)-pentyl)carbamoyl) (70-51-9)
β-Propionolactone (57-57-8)
Propiononitrile (107-12-0)
o-Propionotoluidide, 2-(propylamino) (721-50-6)
2',6'-Propionoxylidide, 2-amino (41708-72-9)
Propionylbenzene (93-55-0)
Propionyl chloride (8CI) (79-03-8)
Propionyl chloride, 3-chloro-2,2-dimethyl- (8CI) (4300-97-4)
3-Propionyl-10-dimethylaminoisopropylphenothiazine (362-29-8)
2-Propionyl-10-(2-(dimethylamino)propyl)phenothiazine (362-29-8)
Propionyl eryhthromycin (134-36-1)
Propionyl fluoride, tetrafluoro-2-(heptafluoropropoxy)- (2062-98-8)
Propionyl fluoride, tetrafluoro-2-(hexafluoro-2-(tetrafluoro-2-(fluorosulfonyl)ethoxy)propoxy)- (4089-58-1)
Propionyl oxide (123-62-6)
4-Propionyloxy-4-phenyl-N-methylpiperidine (13147-09-6)
Propionyl peroxide (8CI) (3248-28-0)
Propionyl peroxide (More than 28% in solution) (3248-28-0)
Propionyl peroxide, More than 28% in solution (3248-28-0)
Propionyl peroxide (Not more than 28% in solution) (3248-28-0)
Propionyl peroxide, Not more than 28% in solution (3248-28-0)
p-Propionylphenol (70-70-2)
Propionylpromethazine (362-29-8)
2-Propionylthiophene (13679-75-9)
Propiophenone (93-55-0)
Propiophenone, 4-amino- (70-69-9)
Propiophenone, 4'-amino- (8CI) (70-69-9)
Propiophenone, 2-diethylamino (90-84-6)
Propiophenone, 3',4'-dimethoxy- (1835-04-7)
Propiophenone, 4'-hydroxy (70-70-2)
Propiophenone, 4'-methyl- (8CI) (5337-93-9)
Propiophenone, 2',4',6'-trihydroxy-3-(p-hydroxyphenyl)- (8CI) (60-82-2)
Propiothetin, dimethyl- (7314-30-9)
Propioveratrone (1835-04-7)
Propiram (15686-91-6)
Propiramo (Spanish) (15686-91-6)
Propiramum (Latin) (15686-91-6)
Propisamine (300-62-9)
Propisamine (60-13-9)
Propitocaine (721-50-6)
Propofol (2078-54-8)
Propogon (114-26-1)
Propoksuru (Polish) (114-26-1)
Propol (67-63-0)
Propolin (9003-07-0)
Propon (93-72-1)
Proponex-Plus (7085-19-0)
Proponex-Plus (93-65-2)
Propophane (9003-07-0)
Propotox M (114-26-1)
Propoxur (ACGIH,OSHA) (114-26-1)
Propoxure (114-26-1)
4-(Propoxycarbonyl)aniline (94-12-2)
2-Propoxyethanol (2807-30-9)
2-(2-Propoxyethoxy)ethanol (6881-94-3)
(2-(1-Propoxyethoxy)ethyl)benzene (7493-57-4)
(Propoxy methyl)oxirane (3126-95-2)
(L)-Propoxyphene (2338-37-6)
Propoxyphene, (+)- (469-62-5)
d-Propoxyphene (469-62-5)
4-Propoxyphenol (18979-50-5)
p-Propoxyphenol (18979-50-5)
Propoxyphenyl (622-85-5)
1-Propoxy-2-propanol (1569-01-3)
n-Propoxypropanol (30136-13-1)
n-Propoxypropanol (Mixed isomers) (30136-13-1)
3-Propoxy-1-propene (1471-03-0)
Propranalol (525-66-6)
Propranolol (525-66-6)

β-Propriolactone (ACGIH,OSHA) **(57-57-8)**
β-Proprolactone **(57-57-8)**
Proprop **(75-99-0)**
2-Propyn-1-amine **(2450-71-7)**
Propycil **(51-52-5)**
Propyl Butex **(94-13-3)**
1-Propyl acetate **(109-60-4)**
2-Propyl acetate **(108-21-4)**
n-Propyl acetate (ACGIH,OSHA) [UN 1276] **(109-60-4)**
Propyl acetate (DOT) **(109-60-4)**
Propylacetic acid **(109-52-4)**
N-Propyl-N-(acetoxymethyl)nitrosamine **(66017-91-2)**
Propyl acetoxymethylnitrosamine **(66017-91-2)**
β-Propyl acrolein **(6728-26-3)**
Propyl acrylate **(925-60-0)**
n-Propyl acrylate **(925-60-0)**
S-Propyl-N-aethyl-N-butyl-thiocarbamat (German) **(1114-71-2)**
1-Propyl alcohol **(71-23-8)**
2-Propyl alcohol **(67-63-0)**
n-Propyl alcohol **(71-23-8)**
sec-Propyl alcohol **(67-63-0)**
n-Propyl alcohol (ACGIH,OSHA) [UN 1274] **(71-23-8)**
Propyl alcohol [UN 1274] **(71-23-8)**
Propyl alcohol, zirconium(4+) salt **(23519-77-9)**
Propyl aldehyde **(123-38-6)**
i-Propylalkohol (German) **(67-63-0)**
n-Propyl alkohol (German) **(71-23-8)**
1-Propylamine **(107-10-8)**
2-Propylamine **(75-31-0)**
n-Propylamine **(107-10-8)**
sec-Propylamine **(75-31-0)**
Propylamine [UN 1277] **(107-10-8)**
Propylamine, N-(1-acetoxymethyl)-N-nitroso- **(66017-91-2)**
Propylamine, N,N-bis(2-chloroethyl) **(621-68-1)**
Propylamine, 3-(diethoxymethylsilyl) **(3179-76-8)**
Propylamine, 3-(N,N-dimethylamino)- **(109-55-7)**
Propylamine, N,N-dimethyl-3-(dibenzo(b,e)thiepin-δ-sup(11(6H),γ)) **(113-53-1)**
Propylamine, 1-ethyl- (8CI) **(616-24-0)**
Propylamine, 3-((2-ethylhexyl)oxy) **(5397-31-9)**
Propylamine hydrochloride **(556-53-6)**
n-Propylamine hydrochloride **(556-53-6)**
Propylamine, hydrochloride (8CI) **(556-53-6)**
Propylamine, 3,3'-iminobis- **(56-18-8)**
Propylamine, 3-isodecoxy- **(30113-45-2)**
Propylamine, 3-methoxy **(5332-73-0)**
Propylamine, 1-methyl **(13952-84-6)**
Propylamine, N-methyl- (8CI) **(627-35-0)**
Propylamine, N-nitroso-N-di- **(621-64-7)**
Propylamine, 3-(triethoxysilyl)- **(919-30-2)**
Propylamine, 3-(trimethoxysilyl)- **(13822-56-5)**
Propyl 4-aminobenzoate **(94-12-2)**
Propyl p-aminobenzoate **(94-12-2)**
n-Propyl p-aminobenzoate **(94-12-2)**
2-(Propylamino)ethanol **(16369-21-4)**
α-n-Propyl-amino-2-methylpropionanilide **(721-50-6)**
1-Propylammonium chloride **(556-53-6)**
Propylammonium chloride **(556-53-6)**
Propylan 3 **(25791-96-2)**
4-Propylaniline **(2696-84-6)**
4-n-Propylaniline **(2696-84-6)**
p-Propylaniline **(2696-84-6)**
p-n-Propylaniline **(2696-84-6)**
4-Propylanisole **(104-45-0)**
4-n-Propylanisole **(104-45-0)**
p-n-Propyl anisole **(104-45-0)**
Propyl aseptoform **(94-13-3)**
N-Propylbenzenamine **(622-80-0)**
n-Propylbenzene **(103-65-1)**
n-Propyl benzene [UN 2364] **(103-65-1)**
Propyl benzenesulfonate **(80-42-2)**
Propyl benzoate **(2315-68-6)**
5-Propyl-1,3-benzodioxole **(94-58-6)**
Propylbis(β-chloroethyl)amine **(621-68-1)**
Propyl bromide **(106-94-5)**
Propyl butanoate **(105-66-8)**
S-Propyl butylethylthiocarbamate **(1114-71-2)**
Propyl butyrate **(105-66-8)**
Propylcain **(94-12-2)**
Propyl caproate **(626-77-7)**
Propyl carbamate **(627-12-3)**
n-Propyl carbamate **(627-12-3)**
Propylcarbinol **(71-36-3)**
n-Propylcarbinyl chloride **(109-69-3)**
Propyl cellosolve **(2807-30-9)**
Propyl chemosept **(94-13-3)**
Propyl chloride [UN 1278] **(540-54-5)**
1-Propyl-3-(p-chlorobenzenesulfonyl)urea **(94-20-2)**
N-Propyl-N'-(p-chlorobenzenesulfonyl)urea **(94-20-2)**
Propyl chlorocarbonate **(109-61-5)**
N-Propyl-N-(2-chloroethyl)-2,6-dinitro-4-trifluoromethylaniline **(33245-39-5)**
N-Propyl-N-(2-chloroethyl)-α,α,α-trifluoro-2,6-dinitro-p-toluidine **(33245-39-5)**
Propyl chloroformate **(109-61-5)**
n-Propyl chloroformate [UN 2740] **(109-61-5)**
S-Propyl chlorothioformate **(13889-92-4)**
N-Propyl-N'-p-chlorphenylsulfonylcarbamide **(94-20-2)**
n-Propyl cinnamate **(7778-83-8)**
Propyl cyanide **(109-74-0)**
2-Propylcyclohexanone **(94-65-5)**
Propylcyclopentane **(2040-96-2)**
Propylcyclopropane **(2415-72-7)**
N-Propylcyclopropanemethanamine **(26389-60-6)**
Propyl p,p'-dichlorobenzilate **(5836-10-2)**
Propyl (3-(dimethylamino)propyl)carbamate **(24579-73-5)**
Propyl 3-(dimethylamino)propylcarbamate **(24579-73-5)**
2-Propyl-3,6-dimethylpyrazine **(18433-97-1)**
n-Propyl-di-n-propylthiolcarbamate **(1929-77-7)**
S-Propyl dipropylthiocarbamate **(1929-77-7)**
Propyl N,N-dipropylthiolcarbamate **(1929-77-7)**
n-Propyl disulfide **(629-19-6)**
Propyl disulfide (8CI) **(629-19-6)**
1-Propylene **(115-07-1)**
Propylene [UN 1077] **(115-07-1)**
3-Propyleneacrolein **(142-83-6)**
Propylene aldehyde **(107-02-8)**
Propylene aldehyde **(123-73-9)**
Propylenebis(dithiocarbamato)zinc **(12071-83-9)**
1,2-Propylene carbonate **(108-32-7)**
Propylene carbonate **(108-32-7)**
Propylene chloride **(78-87-5)**
Propylenechlorohydrin **(78-89-7)**
α-Propylene chlorohydrin **(127-00-4)**
sec-Propylene chlorohydrin **(127-00-4)**
Propylene chlorohydrin [UN 2611] **(78-89-7)**
1,3-Propylenediamine **(109-76-2)**
Propylene diamine **(78-90-0)**
Propylenediamine **(78-90-0)**
1,2-Propylenediamine [UN 2258] **(78-90-0)**
Propylene dibromide **(78-75-1)**
Propylenedicarboxylic acid **(97-65-4)**
Propylene dichloride **(563-54-2)**
α,β-Propylene dichloride **(78-87-5)**
Propylene dichloride (ACGIH,OSHA) **(78-87-5)**
Propylene dichloride [UN 1279] **(26638-19-7)**
Propylene dimer **(16813-72-2)**

Propylene dinitrate (6423-43-4)
Propylene epoxide (75-56-9)
1,2-Propylene glycol (57-55-6)
1,3-Propylene glycol (504-63-2)
Propylene glycol (57-55-6)
α-Propyleneglycol (57-55-6)
β-Propylene glycol (504-63-2)
Propylene glycol, Adipic acid resin (25101-03-5)
Propylene glycol acrylate (25584-83-2)
Propylene-glycol-alginate (9005-37-2)
Propylene glycol, allyl ether (1331-17-5)
Propylene glycol butoxy ether (29387-86-8)
Propylene glycol n-butyl ether (5131-66-8)
Propylene glycol butyl ether 2,4-dichlorophenoxyacetate (1320-18-9)
Propylene glycol cyclic carbonate (108-32-7)
Propylene glycol diacetate (623-84-7)
α-Propylene glycol diacetate (623-84-7)
1,2-Propylene glycol dinitrate (6423-43-4)
Propylene glycol 1,2-dinitrate (6423-43-4)
Propylene glycol dinitrate (ACGIH,OSHA) (6423-43-4)
Propylene glycol ethyl ether (1569-02-4)
Propylene glycol isobutyl ether (23436-19-3)
Propylene glycol, maleic anhydride, phthalic anhydride polymer (25037-66-5)
Propylene glycol methyl ether (107-98-2)
Propylene glycol monoacrylate (25584-83-2)
Propylene glycol monoacrylate (999-61-1)
Propylene glycol β-monoethyl ether (111-35-3)
Propylene glycol monoethyl ether, α (19089-47-5)
Propylene glycol monoethyl ether, β (111-35-3)
α-Propylene glycol monomethyl ether (107-98-2)
Propylene glycol monomethyl ether (ACGIH,OSHA) (107-98-2)
Propyleneglycol monomethyl ether acetate (108-65-6)
Propylene glycol monostearate (1323-39-3)
Propylene glycol oleate (1330-80-9)
Propylene glycol phenyl ether (4169-04-4)
Propylene glycol, phthalic anhydride, maleic anhydride polymer (25037-66-5)
Propylene glycol, phthalic anhydride, maleic anhydride resin (25037-66-5)
Propylene glycol n-propyl ether (1569-01-3)
Propylene glycol USP (57-55-6)
Propylene, hexafluoro- (116-15-4)
1,2-Propyleneimine (75-55-8)
Propylene imine (ACGIH,OSHA) (75-55-8)
Propylene imine, Inhibited [UN 1921] (75-55-8)
α-Propylene mono-n-butyl ether (29387-86-8)
1,2-Propylene oxide (75-56-9)
1,3-Propylene oxide (503-30-0)
Propylene oxide (ACGIH,OSHA) [UN 1280] (75-56-9)
Propylene oxide-ethylene oxide copolymer (9003-11-6)
Propylene oxide, ethylene oxide, glycerol adduct (9082-00-2)
Propylene oxide-ethylene oxide polymer (9003-11-6)
Propylene oxide-glycerol polymer (25791-96-2)
Propylene oxide hexafluoride (428-59-1)
Propylene oxide-methanol Adduct (37286-64-9)
Propylene phenoxetol (770-35-4)
Propylene polymer (9003-07-0)
Propylenester kyseliny uhlicite (Czech) (108-32-7)
Propylene tetramer (25378-22-7)
Propylene tetramer [UN 2850] (6842-15-5)
Propylene thiourea (2122-19-2)
Propylene trimer (124-11-8)
Propylene trimer (13987-01-4)
Propylene trimer (27215-95-8)
1,2-Propylenglykol (German) (57-55-6)
Propylenglykol-monomethylaether (German) (107-98-2)
1,2-Propylenimine (75-55-8)
Propylenimine (75-55-8)
Propylenthioharnstoff (German) (2122-19-2)
Propylester kyseliny dusicne (Czech) (627-13-4)
Propylester kyseliny gallove (Czech) (121-79-9)
Propylester kyseliny p-hydroxybenzoove (Czech) (94-13-3)
Propylester kyseliny karbaminove (Czech) (627-12-3)
Propylester kyseliny maselne (Czech) (105-66-8)
Propylester kyseliny mravenci (Czech) (110-74-7)
Propylester kyseliny octove (Czech) (109-60-4)
Propylester kyseliny propionove (Czech) (106-36-5)
Propylester kyseliny skoricove (Czech) (7778-83-8)
n-Propyl ester of 3,4,5-trihydroxybenzoic acid (121-79-9)
Propyl-ether (111-43-3)
β-Propyl-α-ethylacrolein (123-05-7)
Propyl N-ethyl-n-butylthiocarbamate (1114-71-2)
Propyl-ethylbutylthiocarbamate (1114-71-2)
Propylethyl-n-butylthiocarbamate (1114-71-2)
S-(n-Propyl)-N-ethyl-N-n-butylthiocarbamate (1114-71-2)
n-Propyl-N-ethyl-N-(n-butyl)thiocarbamate (1114-71-2)
N-Propyl-N-ethyl-N-(n-butyl)thiolcarbamate (1114-71-2)
Propyl ethylbutylthiolcarbamate (1114-71-2)
Propylethylene (109-67-1)
Propyl ethyl ether (628-32-0)
n-Propyl formate (110-74-7)
Propyl formate [UN 1281] (110-74-7)
Propylformic acid (107-92-6)
Propyl gallate (121-79-9)
n-Propyl gallate (121-79-9)
Propyl glycidyl ether (3126-95-2)
4-Propylguaiacol (2785-87-7)
p-Propylguaiacol (2785-87-7)
p-n-Propylguaiacol (2785-87-7)
2-Propylheptanic acid (31080-39-4)
2-Propylheptanol (10042-59-8)
2-Propylheptansaeure (German) (31080-39-4)
Propylhexadrine (101-40-6)
Propyl hexanoate (626-77-7)
Propylhexedrin (101-40-6)
Propylhexedrine (101-40-6)
Propyl hexyl ketone (624-16-8)
Propyl hydride (74-98-6)
Propyl 4-hydroxybenzoate (94-13-3)
Propyl p-hydroxybenzoate (94-13-3)
n-Propyl p-hydroxybenzoate (94-13-3)
Propylic alcohol (71-23-8)
Propylic aldehyde (123-38-6)
Propylidene chloride (78-99-9)
3-Propylidenephthalide (17369-59-4)
Propylidene phthalide (17369-59-4)
i-Propyl iodide (75-30-9)
n-Propyl iodide (107-08-4)
1-Propyl isocyanate (110-78-1)
Propyl isocyanate (110-78-1)
n-Propyl isocyanate [UN 2482] (110-78-1)
Propyl isome (83-59-0)
n-Propylisome (83-59-0)
Propyl isomer (83-59-0)
n-Propyl isomer (83-59-0)
Propyl isovalerate (557-00-6)
Propyl ketone (123-19-3)
Propylkyanid (Czech) (109-74-0)
2-Propylmercaptan (75-33-2)
Propyl mercaptan (107-03-9)
n-Propyl mercaptan (107-03-9)
Propyl mercaptan [UN 2402] (107-03-9)
Propylmercuric chloride (2440-40-6)
Propylmercury chloride (2440-40-6)
Propyl methacrylate (2210-28-8)
n-Propyl methacrylate (2210-28-8)

2-Propyl methanesulphonate	**(926-06-7)**
Propyl methanoate	**(110-74-7)**
Propylmethanol	**(71-36-3)**
p-Propylmethoxybenzene	**(104-45-0)**
1-Propyl-3-methoxy-4-hydroxybenzene	**(2785-87-7)**
Propyl 3-methylbutyrate	**(557-00-6)**
Propylmethylcarbinylethyl barbituric acid sodium salt	**(57-33-0)**
4-Propyl-1,2-methylenedioxybenzene	**(94-58-6)**
Propyl monosulfide	**(111-47-7)**
Propyl nitrate	**(627-13-4)**
n-Propyl nitrate (ACGIH,OSHA) [UN 1865]	**(627-13-4)**
Propyl nitrite	**(543-67-9)**
n-Propyl-nitrite	**(543-67-9)**
Propylnitrosaminomethyl acetate	**(66017-91-2)**
Propyl 9-octadecenoate	**(111-59-1)**
Propyl oleate	**(111-59-1)**
19-Propylorvinol	**(14521-96-1)**
Propylorvinol	**(14521-96-1)**
Propylorvinol hydrochloride	**(13764-49-3)**
Propylowy alkohol (Polish)	**(71-23-8)**
Propyloxirane	**(1003-14-1)**
Propylparaben	**(94-13-3)**
Propylparasept	**(94-13-3)**
2-Propylpentanoic acid	**(99-66-1)**
2-Propylpentanoic acid sodium salt	**(1069-66-5)**
2-Propylpentanoyltropinium methylbromide	**(80-50-2)**
n-Propyl percarbonate	**(16066-38-9)**
o-Propylphenol	**(644-35-9)**
p-Propylphenol	**(645-56-7)**
Propyl phenyl ether	**(622-85-5)**
Propyl phenyl ketone	**(495-40-9)**
N-2-Propyl-N'-phenyl-p-phenylenediamine	**(101-72-4)**
1-Propylphosphonic acid	**(4672-38-2)**
Propyl phosphorodichloridodithioate	**(5390-61-4)**
2-Propylpiperidine	**(458-88-8)**
Propylpiperidine	**(5470-02-0)**
β-Propylpiperidine	**(458-88-8)**
6-(Propylpiperonyl)-butyl carbityl ether	**(51-03-6)**
6-Propylpiperonyl butyl diethylene glycol ether	**(51-03-6)**
N-Propyl-1-propanamine	**(142-84-7)**
Propyl propanoate	**(106-36-5)**
Propyl 2-propenoate	**(925-60-0)**
Propyl propionate	**(106-36-5)**
n-Propyl propionate	**(106-36-5)**
4-Propylpyridine	**(1122-81-2)**
Propyl sulfide (8CI)	**(111-47-7)**
5-Propylthiazole	**(52414-82-1)**
Propyl thioarsenite (7CI)	**(5582-57-0)**
6-Propyl-2-thio-2,4(1H,3H)pyrimidinedione	**(51-52-5)**
Propyl thiopyrophosphate	**(3244-90-4)**
Propyl-thiorist	**(51-52-5)**
Propyl-thiorit	**(51-52-5)**
4-Propyl-2-thiouracil	**(51-52-5)**
6-Propyl-2-thiouracil	**(51-52-5)**
6-Propylthiouracil	**(51-52-5)**
6-n-Propyl-2-thiouracil	**(51-52-5)**
6-n-Propylthiouracil	**(51-52-5)**
Propylthiouracil	**(51-52-5)**
Propyl-thyracil	**(51-52-5)**
n-Propyltrichlorosilane	**(141-57-1)**
Propyltrichlorosilane [UN 1816]	**(141-57-1)**
Propyl 3,4,5-trihydroxybenzoate	**(121-79-9)**
n-Propyl 3,4,5-trihydroxybenzoate	**(121-79-9)**
5-Propyl-4-(2,5,8-trioxa-dodecyl)-1,3-benzodioxol (German)	**(51-03-6)**
Propyl urethane	**(627-12-3)**
2-Propylvaleric acid	**(99-66-1)**
2-Propylvaleric acid sodium salt	**(1069-66-5)**
Propynal	**(624-67-9)**
2-Propynal (9CI)	**(624-67-9)**
Propyne	**(74-99-7)**
Propyne, Mixed with propadiene	**(59355-75-8)**
Propyne (OSHA)	**(74-99-7)**
Propyne, 3-bromo	**(106-96-7)**
1-Propyne, 3-bromo-, Mixt. with bromomethane and trichloronitromethane (9CI)	**(8000-21-3)**
1-Propyne, 1-chloro- (9CI)	**(7747-84-4)**
1-Propyne, 3-chloro- (9CI)	**(624-65-7)**
Propyne, 3-chloro- (8CI)	**(624-65-7)**
Propyne, 1-chloro- (6CI,7CI,8CI)	**(7747-84-4)**
1-Propyne, 3-cyclohexylidene-	**(2806-45-3)**
1-Propyne, 3,3-dichloro- (9CI)	**(25523-14-2)**
Propyne, 3,3-dichloro- (8CI)	**(25523-14-2)**
1-Propyne-3-ol	**(107-19-7)**
2-Propynoic acid	**(471-25-0)**
Propynoic acid	**(471-25-0)**
Propynoic acid, (2-chlorophenyl)-	**(24654-08-8)**
Propynoic acid, (3-chlorophenyl)-	**(7396-28-3)**
Propynoic acid, (4-chlorophenyl)-	**(3240-10-6)**
2-Propynoic acid, 2-(4-chlorophenyl)- (9CI)	**(3240-10-6)**
2-Propynoic acid, 3-(2-chlorophenyl)- (9CI)	**(24654-08-8)**
2-Propynoic acid, 3-(3-chlorophenyl)- (9CI)	**(7396-28-3)**
2-Propynoic acid, ethyl ester (9CI)	**(623-47-2)**
Propynoic acid, methyl ester	**(922-67-8)**
2-Propynyl alcohol	**(107-19-7)**
2-Propynylamine	**(2450-71-7)**
Propynylidenecyclohexane	**(2806-45-3)**
2-(2-Propynyloxy)ethanol	**(3973-18-0)**
3-(2-Propynyloxy)phenyl-N-methylcarbamate	**(3692-90-8)**
(E,E)-2-Propynyl 3,7,11-trimethyl-2,4-dodecadienoate	**(42588-37-4)**
2-Propynyl (E,E)-3,7,11-trimethyl-2,4-dodecadienoate	**(42588-37-4)**
2-Propyn-1-ol	**(107-19-7)**
Propyon	**(114-26-1)**
Propyphyllin	**(479-18-5)**
Propythiouracil	**(51-52-5)**
Propyzamide	**(23950-58-5)**
Proquanil	**(57-53-4)**
Proralone-MOP	**(298-81-7)**
Prorex	**(58-33-3)**
Prorex	**(60-87-7)**
Proscomide	**(155-41-9)**
Proscorbin	**(50-81-7)**
Proseptal	**(63-74-1)**
Proseptine	**(63-74-1)**
Proseptol	**(63-74-1)**
Proserine	**(114-80-7)**
Proserine bromide	**(114-80-7)**
Prosevor 85	**(63-25-2)**
Prosta-5,13-dien-1-oic acid, (5Z,9-α,11-α,13E,15S)-9,11,15-trihydroxy	**(551-11-1)**
Prosta-5,13-dien-1-oic acid, 9,11,15-trihydroxy-, (5Z,9-α,11-α,13E,15S)- (9CI)	**(551-11-1)**
Prostaglandin F2-α	**(551-11-1)**
Prostaglandin F2a	**(551-11-1)**
l-Prostaglandin F2-α	**(551-11-1)**
Prostalmon F	**(551-11-1)**
Prostarmon F	**(551-11-1)**
Prostearin	**(1323-39-3)**
Prostigmin	**(59-99-4)**
Prostigmin bromide	**(114-80-7)**
Prostigmine	**(59-99-4)**
Prostigmine bromide	**(114-80-7)**
Prostin F2-α	**(551-11-1)**
Prostrumyl	**(56-04-2)**
Protaben P	**(94-13-3)**

Protacell 8	(9005-38-3)
Protachem 630	(9016-45-9)
Protachem GMS	(31566-31-1)
Protachem SMP	(26266-57-9)
Protachem SOC	(8007-43-0)
Protachem STO	(26266-58-0)
Protactyl	(58-40-2)
Protagent	(9003-39-8)
Protamine	(9012-00-4)
Protamines	(9012-00-4)
Protanabol	(434-07-1)
Protanal	(9005-38-3)
Protasorb L-20	(9005-64-5)
Protasorb O-20	(9005-65-6)
Protatek	(9005-38-3)
Protazine	(58-33-3)
Protazine	(60-87-7)
Protect	(81-84-5)
Protectona	(56-53-1)
Proteins, Milk	(9000-71-9)
Proteins, Thaumatins	(53850-34-3)
Protesine DMU	(140-95-4)
Protex	(25013-16-5)
Protex	(9003-20-7)
Protex (Polymer)	(9003-20-7)
Prothazin	(60-87-7)
Prothazin methosulfate	(58-34-4)
Protheophylline	(479-18-5)
Prothiaden	(113-53-1)
Prothiaden Spofa	(113-53-1)
Prothiocarb	(19622-19-6)
Prothiophos	(34643-46-4)
Prothiucil	(51-52-5)
Prothiurone	(51-52-5)
Prothoate	(2275-18-5)
Prothrin	(23031-38-1)
Prothromadin	(129-06-6)
Prothromadin	(81-81-2)
Prothycil	(51-52-5)
Prothyran	(51-52-5)
Protioamphetamine	(60-15-1)
Protiural	(51-52-5)
Protivar	(53-39-4)
Protoat (Hungarian)	(2275-18-5)
Protobolin	(72-63-9)
Protocatechualdehyde	(139-85-5)
Protocatechualdehyde dimethyl ether	(120-14-9)
Protocatechualdehyde, methyl-	(121-33-5)
Protocatechuecaldehyde dimethyl ether	(120-14-9)
Protocatechuic-acid	(99-50-3)
Protocatechuic acid methylene ether	(94-53-1)
Protocatechuic acid, 3-methyl ester	(121-34-6)
Protocatechuic aldehyde	(139-85-5)
Protocatechuic aldehyde dimethyl ether	(120-14-9)
Protocatechuic aldehyde ethyl ether	(121-32-4)
Protocatechuic aldehyde methylene ether	(120-57-0)
Protochlorure d'iode (French)	(7790-99-0)
Protocol C	(1854-26-8)
Protomon	(9005-34-9)
Protopet	(8012-95-1)
Protopet, Alba	(8009-03-8)
Protopet, White 3C	(8009-03-8)
Protopet, White 2L	(8009-03-8)
Protopet, White 1S	(8009-03-8)
Protopet, Yellow 2A	(8009-03-8)
Protopet, Yellow 1E	(8009-03-8)
Protophenicol	(982-57-0)
Protopine	(130-86-9)
Protopine, hydrochloride	(6164-47-2)
Protoporphyrin	(553-12-8)
Protoporphyrin IX	(553-12-8)
Protoporphyrin IX (VAN)	(553-12-8)
Protoporphyrin IX, zinc chelate	(15442-64-5)
Protopyrin	(938-73-8)
Protox Type 166	(1314-13-2)
Protox Type 167	(1314-13-2)
Protox Type 168	(1314-13-2)
Protox Type 169	(1314-13-2)
Protox Type 267	(1314-13-2)
Protox Type 268	(1314-13-2)
Protoxyl	(127-85-5)
Protran	(57-53-4)
Protriptyline	(438-60-8)
Protryptyline	(438-60-8)
Provera	(520-85-4)
Provera	(71-58-9)
Provera dosepak	(71-58-9)
Provigan	(60-87-7)
Provitamin D	(57-87-4)
Provitamin D	(57-88-5)
Provitamin D_2	(57-87-4)
Provitamin D_3	(434-16-2)
Provitar	(53-39-4)
Prowl	(40487-42-1)
Prox DW	(1854-26-8)
Proxagesic	(469-62-5)
Proxan	(2634-33-5)
Proxanol	(9003-11-6)
Proxanol 158	(9003-11-6)
Proxanol 168	(9003-11-6)
Proxanol 186	(9003-11-6)
Proxanol 224	(9003-11-6)
Proxanol 228	(9003-11-6)
Proxanol P 168	(9003-11-6)
Proxanol P 268	(9003-11-6)
Proxanol TsL 3	(9003-11-6)
Proxan sodium	(140-93-2)
Proxel	(2634-33-5)
Proxel CRL (9CI)	(54392-15-3)
Proxel PL	(2634-33-5)
Proxel Press Paste	(2634-33-5)
Proxen	(22204-53-1)
Proxigel	(124-43-6)
Proximpham (German)	(2828-42-4)
Proxitane 4002	(79-21-0)
Proxol	(52-68-6)
Proxypham	(2828-42-4)
Prozil	(50-53-3)
Prozin	(50-53-3)
Prozin	(8073-77-6)
Prozin 50	(8073-77-6)
Prozinex	(139-40-2)
Prozoin	(79-09-4)
Prulet	(77-09-8)
Prunolide	(104-61-0)
Prussian Blue	(14038-43-8)
Prussian Brown	(1309-37-1)
Prussic acid	(74-90-8)
Prussic acid (DOT)	(74-90-8)
Prussic acid, Unstabilized	(74-90-8)
Prussite	(460-19-5)
Prym E	(136-84-5)
Prynachlor	(21267-72-1)
Prynachlore	(21267-72-1)

Prysoline (125-33-7)
Przedziorkofos (Polish) (2275-14-1)
Pseudoacetic acid (79-09-4)
Pseudobutylbenzene (98-06-6)
Pseudobutylene (107-01-7)
Pseudocarene (554-60-9)
Pseudochelerythrine (2447-54-3)
Pseudocumene (95-63-6)
Pseudocumenol (496-78-6)
Pseudocumidine (137-17-7)
Pseudocumidine hydrochloride (21436-97-5)
Pseudocumohydroquinone (700-13-0)
Pseudocumol (95-63-6)
Pseudocyanuric acid (108-80-5)
(+)-Pseudoephedrine (90-82-4)
Pseudoephedrine (90-82-4)
l-(+)-Pseudoephedrine (90-82-4)
Pseudoephedrine, l-(+) (90-82-4)
Pseudohexyl alcohol (97-95-0)
3H-Pseudoindolium, 2-(2-(2,4-dimethoxyanilino)vinyl)-1,3,3-trimethyl- chloride (4208-80-4)
Pseudoisatin (91-56-5)
Pseudokumidin (Czech) (137-17-7)
Pseudomethylionone (26651-96-7)
Pseudonorephedrine (492-39-7)
Pseudopinen (127-91-3)
Pseudopinene (127-91-3)
Pseudotheophylline (58-55-9)
Pseudothiourea (62-56-6)
Pseudothymine (626-48-2)
Pseudotropanol benzilate (537-26-8)
Pseudotropine benzoate (537-26-8)
Pseudotropine, benzoate (ester) (537-26-8)
Pseudourea (57-13-6)
Pseudourea, 2-benzyl-2-thio-, hydrochloride (538-28-3)
Pseudourea, 2-benzyl-2-thio-, monohydrochloride (538-28-3)
Pseudourea, 2-(2-(dimethylamino)ethyl)-2-thio-, dihydrochloride (16111-27-6)
Pseudourea, 2-methyl-2-thio-, sulfate (2:1) (867-44-7)
Pseudourea, 2-thio-2-benzyl-, hydrochloride (538-28-3)
Pseudourea, 3-ureido- (110-21-4)
Pseudoxanthine (69-89-6)
Psichergina (300-42-5)
Psicodisten (51-12-7)
Psicopan (300-42-5)
Psicopax (604-75-1)
Psicopax (846-49-1)
Psicosan (58-25-3)
Psilocibin (520-52-5)
Psilocibina (Spanish) (520-52-5)
Psilocin phosphate ester (520-52-5)
Psilocybin (520-52-5)
Psilocybine (520-52-5)
Psilocybinum (Latin) (520-52-5)
Psilocyn (520-52-5)
Psilotsibin (520-52-5)
Psiquergina (300-42-5)
Psiquium (2898-12-6)
Psoradrate (480-22-8)
Psoralen (66-97-7)
Psoralene (66-97-7)
Psoriacid-Stift (1143-38-0)
Psoriacid-Stift (480-22-8)
Psoriacide (480-22-8)
Psychamine A 66 (134-49-6)
Psychedrine (300-62-9)
Psychedrine (60-13-9)
Psychedrinum (60-13-9)
Psychedryna (60-13-9)
Psychergine (300-42-5)
Psychodrine (51-63-8)
Psycholiquid (50-35-1)
Psychoson (14759-06-9)
Psychotablets (50-35-1)
Psychoton (60-13-9)
Psychoton (60-15-1)
Psychozine (69-09-0)
Psykoton (300-42-5)
Pt-01 (15663-27-1)
Ptaquiloside (87625-62-5)
Pteglu (59-30-3)
Pteramina (Czech) (54-62-6)
4,7(3H,8H)-Pteridinedione, 2-amino (529-69-1)
4,7(1H,8H)-Pteridinedione, 2-amino- (9CI) (529-69-1)
2,4,7-Pteridinetriamine, 6-phenyl- (9CI) (396-01-0)
Pteridine, 2,4,7-triamino-6-phenyl (396-01-0)
4-Pteridinol, 2-amino-6-((p-((1,3-dicarboxypropyl)carbamoyl)anilino)methyl)- (59-30-3)
Pterofen (396-01-0)
Pterophene (396-01-0)
Pteroyl-l-glutamic acid (59-30-3)
Pteroylglutamic acid (59-30-3)
Pteroyl-l-monoglutamic acid (59-30-3)
Pteroylmonoglutamic acid (59-30-3)
Pulon (63428-84-2)
Pulsan (77732-09-3)
Pulvis conservans (94-13-3)
Puradin (67-45-8)
Puragel (9000-70-8)
Puralin (137-26-8)
Purapuridine (126-17-0)
Purasan-SC-10 (62-38-4)
Puratized (23319-66-6)
Puratized Agricultural Spray (23319-66-6)
Puratized N5E (23319-66-6)
Puratizedat Agricultural Spray (23319-66-6)
Puratronic chromium chloride (10025-73-7)
Puratronic chromium trioxide (1333-82-0)
Puraturf (23319-66-6)
Puraturf 10 (62-38-4)
Pure Chromium Oxide Green 59 (1308-38-9)
Pure Chrysoidine YBH (532-82-1)
Pure Chrysoidine YD (532-82-1)
Pure Eosine YY (17372-87-1)
Pure Lemon Chrome 24882 (1344-37-2)
Pure Lemon Chrome 3GN (1344-37-2)
Pure Lemon Chrome HL 3G (1344-37-2)
Pure Lemon Chrome L 3G (1344-37-2)
Pure Lemon Chrome L 3GS (1344-37-2)
Pure Lemon Chrome l3GS (7758-97-6)
Pure Middle Chrome 24883 (1344-37-2)
Pure Middle Chrome LG (1344-37-2)
Pure Orange Chrome M (18454-12-1)
Pure Orange Chrome Y (18454-12-1)
Pure Orange II S (633-96-5)
Pure Primrose Chrome 24880 (1344-37-2)
Pure Primrose Chrome 24881 (1344-37-2)
Pure Primrose Chrome L 6G (1344-37-2)
Pure Primrose Chrome L 10G (1344-37-2)
Pure Sky Blue 6B (2610-05-1)
Pure Zinc Chrome (13530-65-9)
Pure Zinc Chrome A (13530-65-9)
Pure Zinc Yellow (13530-65-9)
Purex (7647-14-5)
Purga (77-09-8)

Purgen **(77-09-8)**
Purgophen **(77-09-8)**
Purifilin **(479-18-5)**
Purimethol **(50-44-2)**
1H-Purin-2-amine **(452-06-2)**
1H-Purin-6-amine **(73-24-5)**
1H-Purin-2-amine, 6-chloro- (9CI) **(10310-21-1)**
1H-Purin-6-amine, N-(phenylmethyl)- (9CI) **(1214-39-7)**
1H-Purin-6-amine, phosphate **(52175-10-7)**
7H-Purine **(120-73-0)**
9H-Purine **(120-73-0)**
Purine **(120-73-0)**
β-Purine **(120-73-0)**
1H-Purine, 6-amino **(73-24-5)**
Purine, 2-amino **(452-06-2)**
Purine, 6-amino- **(73-24-5)**
Purine, 2-amino-6-chloro **(10310-21-1)**
9H-Purine, 6-benzylamino-9-tetrahydropyran-2-yl **(2312-73-4)**
9H-Purine, 1,6-dihydro-6-imino- **(73-24-5)**
9H-Purine-2,6-diol **(69-89-6)**
Purine-2,6-diol **(69-89-6)**
2,6(1,3)-Purinedion **(69-89-6)**
9H-Purine-2,6-(1H,3H)-dione **(69-89-6)**
Purine-2,6-(1H,3H)-dione **(69-89-6)**
1H-Purine-2,6-dione, 8-bromo-3,7-dihydro-1,3-dimethyl- (9CI) **(10381-75-6)**
1H-Purine-2,6-dione, 8-chloro-3,7-dihydro- (9CI) **(13548-68-0)**
1H-Purine-2,6-dione, 3,7-dihydro-1,3-dimethyl- (9CI) **(58-55-9)**
1H-Purine-2,6-dione, 3,7-dihydro-1,7-dimethyl- (9CI) **(611-59-6)**
1H-Purine-2,6-dione, 3,7-dihydro-1,3-dimethyl-, Compd. with 1,2-ethanediamine (2:1) (9CI) **(317-34-0)**
1H-Purine-2,8(3H,6H)-dione, 7,9-dihydro-6-thioxo- (9CI) **(2002-60-0)**
1H-Purine-2,6-dione, 3,7-dihydro-1,3,7-trimethyl- **(58-08-2)**
1H-Purine-2,6-dione, 7-(2,3-dihydroxypropyl)-3,7-dihydro-1,3-dimethyl **(479-18-5)**
Purine, 6-mercapto- **(50-44-2)**
Purine, 6-methyl **(2004-03-7)**
1H-Purine, 6-methyl- (9CI) **(2004-03-7)**
9H-Purine, 9-methyl-6-(methylthio)- (8CI,9CI) **(1127-75-9)**
1H-Purine, 6-((1-methyl-4-nitro-1H-imidazol-5-yl)thio)- **(446-86-6)**
Purine, 6-((1-methyl-4-nitroimidazol-5-yl)thio) **(446-86-6)**
9H-Purine, 6-(methylthio)-9-β-d-ribofuranosyl **(342-69-8)**
3H-Purine-6-thiol **(50-44-2)**
6-Purinethiol **(50-44-2)**
Purine-6-thiol **(50-44-2)**
Purinethiol **(50-44-2)**
Purine-6-thiol, 2-amino- **(154-42-7)**
Purine-6-thiol, 6-methyl-9-ribofuranosyl- **(342-69-8)**
Purine-6-thiol, monohydrate **(6112-76-1)**
Purine-6(1H)-thione, 2-amino **(154-42-7)**
9H-Purine-6(1H)-thione, 2-amino-9-(2-deoxy-β-D-erythro-pentofuranosyl) **(789-61-7)**
6H-Purine-6-thione, 2-amino-1,7-dihydro- (9CI) **(154-42-7)**
6H-Purine-6-thione, 1,7-dihydro- (9CI) **(50-44-2)**
6H-Purine-6-thione, 1,7-dihydro-, monohydrate (9CI) **(6112-76-1)**
Purinethol **(50-44-2)**
1H-Purine, 2,6,8-trichloro- (9CI) **(2562-52-9)**
1H-Purine-2,6,8(3H)-trione, 7,9-dihydro- (9CI) **(69-93-2)**
1H-Purine-2,6,8(3H)-trione, 4,9-dihydro-3-methyl- (9CI) **(605-99-2)**
9H-Purin-6-ol **(68-94-0)**
Purin-6-ol **(68-94-0)**
6(1H)-Purinone **(68-94-0)**
Purin-6(3H)-one **(68-94-0)**
6H-Purin-6-one, 1,7-dihydro **(68-94-0)**
6H-Purin-6-one, 1,7-dihydro-2-(methylamino)- (9CI) **(10030-78-1)**
6H-Purin-6-one, 1,2,3,7-tetrahydro-3-methyl-2-thioxo- (9CI) **(28139-02-8)**
2H-Purin-2-one, 1,3,6,7-tetrahydro-6-thioxo- **(2002-59-7)**
6H-Purin-6-one, 1,2,3,7-tetrahydro-2-thioxo **(2487-40-3)**
Purin-6-thiol, monohydrat (Czech) **(6112-76-1)**
6H-Purin-6-thion, monohydrat (Czech) **(6112-76-1)**
Purivel **(19937-59-8)**
Purocyclina **(60-54-8)**
Purodigin **(71-63-6)**
Purostrophan **(630-60-4)**
Purple Lake **(1325-82-2)**
Purple 4R **(4548-53-2)**
Purple Red **(2611-82-7)**
Purple SX **(2611-82-7)**
Purpurid **(71-63-6)**
Purpurin **(81-54-9)**
Purpurin 4B **(992-59-6)**
Purpurine **(81-54-9)**
Purpurine 4B **(992-59-6)**
Purpurocatechol **(533-75-5)**
Purserpine **(50-55-5)**
Putrescin **(110-60-1)**
Putrescine **(110-60-1)**
Py 11 **(56-05-3)**
Pybuthrin **(51-03-6)**
Pycazide **(54-85-3)**
Pydoxal **(54-47-7)**
Pydraul 50E (9CI) **(66594-31-8)**
Pydraul 115E (9CI) **(66594-32-9)**
Pydrin **(51630-58-1)**
Pyknolepsinum **(77-67-8)**
Pymafed **(59-33-6)**
Pymafed **(91-84-9)**
Pynacolyl methylfluorophosphonate **(96-64-0)**
Pynamin **(584-79-2)**
Pynamin-Forte **(584-79-2)**
Pynosect **(10453-86-8)**
Pyoktanin **(548-62-9)**
Pyoverm **(548-62-9)**
Pyquiton **(55268-74-1)**
No-Pyr **(930-55-2)**
Pyra **(91-84-9)**
Pyrabenzamine **(154-69-8)**
Pyrabutol **(50-33-9)**
Pyracarbolid **(24691-76-7)**
Pyracarbolide **(24691-76-7)**
Pyracetol **(525-52-0)**
Pyracort D **(61-76-7)**
Pyracryl Orange Y **(532-82-1)**
Pyradex **(2303-16-4)**
Pyradone **(58-15-1)**
Pyralan **(87-47-8)**
Pyralene **(1336-36-3)**
Pyralene 1476 **(25429-29-2)**
Pyralene 1498 **(26914-33-0)**
Pyralin **(9004-70-0)**
Pyramal **(91-84-9)**
Pyra maleate **(59-33-6)**
Pyramid 120 **(1312-76-1)**
Pyramid Royal Blue Toner **(1325-87-7)**
Pyramid Violet Toner NX **(1325-82-2)**
Pyramidon **(58-15-1)**
Pyramidone **(58-15-1)**
Pyramin **(1698-60-8)**
Pyramin RB **(1698-60-8)**
Pyramine **(1698-60-8)**
2H-Pyran-δ(sup 3(6H),α)-acetic acid, 2,4-dihydroxy-, 3,4-lactone **(149-29-1)**
Pyran aldehyde **(100-73-2)**
2H-Pyran-2-carboxaldehyde, 3,4-dihydro- (8CI,9CI) **(100-73-2)**
2H-Pyran-2-carboxaldehyde, 3,4-dihydro-2,5-dimethyl- (8CI) **(1920-21-4)**
4H-Pyran-3-carboxylic acid, 5-((decahydro-6-hydroxy-5-

(3-hydroxy-3-methylbutyl)-5,8a-dimethyl-2-methylene-1-naphthalenyl)methyl)-6-methoxy-2-methyl-4-oxo-, methyl ester, (1α,4aα,5α,6α,8aβ)- (9CI) **(61235-00-5)**
2H-Pyran-2-carboxylic acid, 3,6-dihydro-4,5-dimethyl-, methyl ester (7CI,8CI,9CI) **(24588-61-2)**
2H-Pyran-6-carboxylic acid, 3,4-dihydro-2,2-dimethyl-4-oxo-, butyl ester **(532-34-3)**
4H-Pyran-2,6-dicarboxylic acid, 3-hydroxy-4-oxo- (9CI) **(497-59-6)**
4H-Pyran-2,6-dicarboxylic acid, 4-oxo (8CI) **(99-32-1)**
2H-Pyran, 3,4-dihydro- (9CI) **(110-87-2)**
1,4-Pyran, 2,3-dihydro-2,5-dimethyl-2-formyl **(1920-21-4)**
2H-Pyran, 3,4-dihydro-2-methoxy- (9CI) **(4454-05-1)**
2H-Pyran, 3,4-dihydro-6-methyl- (6CI,7CI,8CI,9CI) **(16015-11-5)**
4H-Pyran, 5,6-dihydro-2-methyl-3-(phenylcarbamoyl) **(24691-76-7)**
2H-Pyran-2,4(3H)-dione, 3-acetyl-6-methyl **(520-45-6)**
2H-Pyran-2,4(3H)-dione, 3-acetyl-6-methyl-, sodium salt **(4418-26-2)**
2H-Pyran-2,6(3H)-dione, dihydro- **(108-55-4)**
2H-Pyran, 2-ethoxy-3,4-dihydro **(103-75-3)**
Pyranilamine maleate **(59-33-6)**
Pyranine **(6358-69-6)**
Pyranine Concentrated **(6358-69-6)**
Pyraninyl **(59-33-6)**
Pyranisamine **(91-84-9)**
Pyranisamine maleate **(59-33-6)**
2H-Pyran-2-methanol, tetrahydro **(100-72-1)**
Pyran-2-methanol, tetrahydro- **(100-72-1)**
Pyran, 2-(2-methyl-1-propenyl)-4-methyltetrahydro- **(16409-43-1)**
4H,5H-Pyrano(4,3-b)pyran-4,5-dione, 2,3-dihydro-3-α-hydroxy-2-β-methyl-7-propenyl **(10088-95-6)**
7H-Pyrano(2,3-c)acridin-7-one, 3,12-dihydro-6-methoxy-3,3,12-trimethyl **(7008-42-6)**
Pyranol **(1336-36-3)**
Pyranol 1499 **(25323-68-6)**
2H-Pyran-3-ol, 6-ethenyl-tetrahydro-2,2,6-trimethyl- **(14049-11-7)**
2H-Pyran-3-ol, 6-ethenyltetrahydro-2,2,6-trimethyl- (9CI) **(14049-11-7)**
2H-Pyran-3-ol, 6-ethenyltetrahydro-2,2,6-trimethyl-, cis- (9CI) **(14009-71-3)**
2H-Pyran-3-ol, 6-ethenyltetrahydro-2,2,6-trimethyl-, trans- (9CI) **(39028-58-5)**
2H-Pyran-4-ol, tetrahydro-3-pentyl-, acetate (9CI) **(18871-14-2)**
2H-Pyran-3-ol, tetrahydro-2,2,6-trimethyl-6-vinyl- **(14049-11-7)**
2H-Pyran-3-ol, tetrahydro-2,2,6-trimethyl-6-vinyl-, cis- (8CI) **(14009-71-3)**
2H-Pyran-2-one (8CI,9CI) **(504-31-4)**
4H-Pyran-4-one (8CI,9CI) **(108-97-4)**
2H-Pyran-2-one, 3-acetyl-4-hydroxy-6-methyl- **(520-45-6)**
2H-Pyran-2-one, 6-butyltetrahydro- **(3301-94-8)**
2H-Pyran-2-one, 5,6-dihydro-6-methyl **(10048-32-5)**
2H-Pyran-2-one, 6-heptyltetrahydro **(713-95-1)**
4H-Pyran-4-one, 3-hydroxy-2-methyl **(118-71-8)**
2H-Pyran-2-one, 6-pentyl- (9CI) **(27593-23-3)**
2H-Pyran-2-one, tetrahydro- (9CI) **(542-28-9)**
2H-Pyran-2-one, tetrahydro-3,6-dimethyl- (8CI,9CI) **(3720-22-7)**
2H-Pyran-2-one, tetrahydro-6-methyl- (8CI,9CI) **(823-22-3)**
2H-Pyran-2-one, tetrahydro-6-(2-pentenyl)-, (Z)- (9CI) (VAN) **(100428-67-9)**
2H-Pyran-2-one, tetrahydro-6-pentyl **(705-86-2)**
2H-Pyran-2-one, tetrahydro-6-propyl- (9CI) **(698-76-0)**
2H-Pyran-2-one, tetrahydro-4,6,6-trimethyl- (8CI,9CI) **(20628-36-8)**
4H-Pyran-4-one, 5-((trimethylsilyl)oxy)-2-(((trimethylsilyl)oxy)methyl)- (9CI) **(55557-21-6)**
2H-Pyran, 2-oxo- **(504-31-4)**
4H-Pyran, 4-oxo- **(108-97-4)**
Pyrantel **(15686-83-6)**
2H-Pyran, tetrahydro- (9CI) **(142-68-7)**
Pyran, tetrahydro-2-(2-methyl-1-propenyl)-4-methyl **(16409-43-1)**
2H-Pyran, tetrahydro-2-(12-pentadecynyloxy)- (9CI) **(56666-38-7)**
8,16-Pyranthrenedione (9CI) **(128-70-1)**
8,16-Pyranthrenedione, dibromo- (9CI) **(1324-35-2)**
Pyranthron **(128-70-1)**
Pyranthrone **(128-70-1)**
Pyranton **(123-42-2)**
Pyrapap **(621-42-1)**
Pyraphen **(62-44-2)**
Pyrathyn **(135-23-9)**
Pyrathyn **(91-80-5)**
Pyrazalone Orange NP 215 **(3520-72-7)**
Pyrazinamide **(98-96-4)**
Pyrazine **(290-37-9)**
Pyrazineamide **(98-96-4)**
Pyrazinecarboxamide **(98-96-4)**
Pyrazinecarboxamide, N-amidino-3,5-diamino-6-chloro-, monohydrochloride (8CI) **(2016-88-8)**
Pyrazinecarboxamide, 3,5-diamino-N-(aminoiminomethyl)-6-chloro-, monohydrochloride (9CI) **(2016-88-8)**
Pyrazine carboxylamide **(98-96-4)**
2-Pyrazinecarboxylic acid, 5-methyl-, 4-oxide **(51037-30-0)**
Pyrazine, 2,3-diethyl- (9CI) **(15707-24-1)**
Pyrazine, 2,5-diethyl- (8CI,9CI) **(13238-84-1)**
Pyrazine, 2,3-diethyl-5-methyl- (9CI) **(18138-04-0)**
Pyrazine, 2,5-diethyl-3-methyl- (8CI,9CI) **(32736-91-7)**
Pyrazine, 3,5-diethyl-2-methyl- (8CI,9CI) **(18138-05-1)**
Pyrazine, 2,3-dimethyl **(5910-89-4)**
Pyrazine, 2,5-dimethyl **(123-32-0)**
Pyrazine, 2,6-dimethyl **(108-50-9)**
Pyrazine, 2,6-dimethyl-3-ethyl- **(13925-07-0)**
Pyrazine, 3,5-dimethyl-2-ethyl **(13925-07-0)**
Pyrazine, 2,5-dimethyl-3-propyl- (9CI) **(18433-97-1)**
Pyrazine, 2-ethenyl-6-ethyl- (9CI) **(32736-90-6)**
Pyrazine, 2-ethenyl-5-methyl- (9CI) **(13925-08-1)**
Pyrazine, ethyl **(13925-00-3)**
Pyrazine, 3-ethyl-2,5-dimethyl- (9CI) **(13360-65-1)**
Pyrazine, 5-ethyl-2,3-dimethyl- (8CI,9CI) **(15707-34-3)**
Pyrazine, 2-ethyl-3-methoxy- (9CI) **(25680-58-4)**
Pyrazine, 2-ethyl-3-methyl **(15707-23-0)**
Pyrazine, 2-ethyl-5-methyl **(13360-64-0)**
Pyrazine, 2-ethyl-6-methyl- (8CI,9CI) **(13925-03-6)**
Pyrazine, 2-ethyl-6-vinyl- (8CI) **(32736-90-6)**
Pyrazine hexahydride **(110-85-0)**
Pyrazine, hexahydro- **(110-85-0)**
Pyrazine, methoxy- (9CI) **(3149-28-8)**
Pyrazine, 2-methoxy-6-methyl- (9CI) **(2882-21-5)**
Pyrazine, 2-methoxy-3-(1-methylethyl)- (9CI) **(25773-40-4)**
Pyrazine, 2-methoxy-3-(1-methylpropyl)- (9CI) **(24168-70-5)**
Pyrazine, 2-methoxy-3-(2-methylpropyl)- (9CI) **(24683-00-9)**
Pyrazine, 2-methyl **(109-08-0)**
Pyrazine, (1-methylethyl)- (9CI) **(29460-90-0)**
Pyrazine, 2-methyl-3-(2-methylpropyl)- (9CI) **(13925-06-9)**
Pyrazine, sulfanilamido- **(116-44-9)**
Pyrazine, tetramethyl **(1124-11-4)**
Pyrazine, trimethyl **(14667-55-1)**
4H-Pyrazino(2,1-a)isoquinolin-4-one, 2-(cyclohexylcarbonyl)-1,2,3,6,7,11b-hexahydro **(55268-74-1)**
Pyrazinoic acid amide **(98-96-4)**
Pyrazinol, O-ester with O,O-diethyl phosphorothioate **(297-97-2)**
Pyrazinon **(5826-91-5)**
1-Pyrazinylethanone **(22047-25-2)**
N1-2-Pyrazinylsulfanilamide **(116-44-9)**
Pyrazodine **(136-40-3)**
Pyrazofen **(136-40-3)**
Pyrazofen **(94-78-0)**
Pyrazol Blue 3B **(72-57-1)**
Pyrazol Fast Black GS **(6428-31-5)**
Pyrazol Fast Brilliant Blue VP **(2610-05-1)**
Pyrazol Fast Brown BRL **(16071-86-6)**
Pyrazol Fast Flavine 5G **(10190-68-8)**
Pyrazol Fast Red 5BL **(2610-11-9)**
Pyrazol Fast Red 8BL **(2610-11-9)**

Pyrazolate (58011-68-0)
Pyrazole (288-13-1)
1H-Pyrazole-5-amine, 3-methyl-1-phenyl (1131-18-6)
Pyrazole, 5-amino-3-methyl-1-phenyl- (1131-18-6)
Pyrazole-1-carboxamide, 3-hydroxy-N,N,5-trimethyl-, dimethylcarbamate (ester) (644-64-4)
1H-Pyrazole-3-carboxylic acid, 4,4'-((3,3'-dichloro(1,1'-biphenyl)-4,4'-diyl)bis(azo))bis(4,5-dihydro-5-oxo-1-phenyl-, diethyl ester (9CI) (6358-87-8)
1H-Pyrazole-3-carboxylic acid, 4,5-dihydro-5-oxo-1-phenyl-, ethyl ester (9CI) (89-33-8)
1H-Pyrazole-3-carboxylic acid, 5-hydroxy-1-(4-sulfophenyl)-(9CI) (21951-33-7)
Pyrazole-3-carboxylic acid, 5-hydroxy-1-(p-sulfophenyl)-4-(p-sulfophenyl)azo-, trisodium salt (1934-21-0)
Pyrazole, 3,5-dimethyl-4-nitroso-1-phenyl (715-99-1)
1H-Pyrazole, 5-oxo-4,5-dihydro-3-methyl-4-((3-((phenylamino)-sulfonyl)-4-methylphenyl)azo)-1-(4-sulfophenyl)-, sodium salt (6359-85-9)
1H-Pyrazole, 1,3,5-trimethyl- (9CI) (1072-91-9)
Pyrazole, 1,3,5-trimethyl- (8CI) (1072-91-9)
Pyrazolidin (50-33-9)
3,5-Pyrazolidinedione, 4-butyl-1,2-diphenyl (50-33-9)
3,5-Pyrazolidinedione, 4-butyl-1-(p-hydroxyphenyl)-2-phenyl (129-20-4)
3,5-Pyrazolidinedione, 1,2-diphenyl-4-(2-(phenylsulfinyl)ethyl) (57-96-5)
3-Pyrazolidinone, 4,4-dimethyl-1-phenyl- (9CI) (2654-58-2)
3-Pyrazolidinone, 4-(hydroxymethyl)-4-methyl-1-phenyl- (9CI) (13047-13-7)
3-Pyrazolidinone, 4-methyl-1-phenyl- (9CI) (2654-57-1)
3-Pyrazolidinone, 1-phenyl (92-43-3)
Pyrazoline Brown BRL (16071-86-6)
Pyrazoline G (89-36-1)
Pyrazoline Red 8BL (2610-11-9)
2-Pyrazoline-3-carboxylic acid, 5-oxo-1-phenyl-, ethyl ester (8CI) (89-33-8)
2-Pyrazolin-5-one, 3-amino-1-(2,4,6-trichlorophenyl)- (27241-31-2)
3-Pyrazolin-5-one, 1-(p-bromophenyl)-2,3-dimethyl- (6CI,8CI) (603-65-6)
2-Pyrazolin-5-one, 1,3-dimethyl- (8CI) (2749-59-9)
3-Pyrazolin-5-one, 4-(dimethylamino)-2,3-dimethyl-1-phenyl- (58-15-1)
3-Pyrazolin-5-one, 4-ethyl-1-(1-methyl-4-piperidyl)-3-phenyl-(8CI) (2531-04-6)
2-Pyrazolin-5-one, 3-methyl (108-26-9)
2-Pyrazolin-5-one, 3-methyl-1-phenyl (89-25-8)
2-Pyrazolin-5-one, 3-methyl-1-p-tolyl (86-92-0)
1H-Pyrazolium, 1,2-dimethyl-3,5-diphenyl- (9CI) (49866-87-7)
1H-Pyrazolium, 1,2-dimethyl-3,5-diphenyl-, methyl sulfate (43222-48-6)
1H-Pyrazolo(3,4-d)pyrimidin-4,6(5H,7H)-dione (2465-59-0)
1H-Pyrazolo(3,4-d)pyrimidine-4,6-diol (2465-59-0)
1H-Pyrazolo(3,4-d)pyrimidine-4,6(5H,7H)-dione (9CI) (2465-59-0)
H-Pyrazolo(3,4-d)pyrimidin-4-ol (315-30-0)
4H-Pyrazolo(3,4-d)pyrimidin-4-one (315-30-0)
1H-Pyrazol-4-ol, 4-(2,4-dichlorobenzoyl)-1,3-dimethyl-, p-toluenesulfonate (58011-68-0)
Pyrazol-5-ol, 1-isopropyl-3-methyl-, dimethylcarbamate (119-38-0)
Pyrazol-3(or 5)-ol, 5(or 3)-methyl-, diethyl phosphate (108-34-9)
Pyrazol-5-ol, 3-methyl-1-phenyl-, dimethyl carbamate (ester) (87-47-8)
Pyrazolone Orange YB 3 (3520-72-7)
3H-Pyrazol-3-one, 5-amino-2,4-dihydro-2-(2,4,6-trichlorophenyl)- (9CI) (27241-31-2)
3H-Pyrazol-3-one, 2-(4-bromophenyl)-1,2-dihydro-1,5-dimethyl- (9CI) (603-65-6)
3H-Pyrazol-3-one, 5-((2-chloro-5-nitrophenyl)amino)-2,4-dihydro-2-(2,4,6-trichlorophenyl)- (9CI) (30707-68-7)
3H-Pyrazol-3-one, 4,4'-((3,3'-dichloro(1,1'-biphenyl)-4,4'-diyl) bis(azo))bis(2,4-dihydro-5-methyl-2-(4-methylphenyl)- (9CI) (15793-73-4)
3H-Pyrazol-3-one, 2,4-dihydro-2,5-dimethyl- (9CI) (2749-59-9)
3H-Pyrazol-3-one, 1,2-dihydro-1,5-dimethyl-2-phenyl- (9CI) (60-80-0)
3H-Pyrazol-3-one, 4-(dimethylamino)-1,2-dihydro-1,5-dimethyl-2-phenyl- (9CI) (58-15-1)
3H-Pyrazol-3-one, 4-ethyl-1,2-dihydro-2-(1-methyl-4-piperidinyl)-5-phenyl- (9CI) (2531-04-6)
5-Pyrazolone, 3-methyl-1-phenyl- (89-25-8)
Pyrazolo-(1,5a)pyrimidine, 2-(O,O-diethyl-thionophosphoryl)-5-methyl-6-ethoxycarbonyl- (13457-18-6)
Pyrazon (1698-60-8)
Pyrazone (1698-60-8)
Pyrazonl (1698-60-8)
Pyrazophos (13457-18-6)
Pyrazophyl (60-80-0)
Pyrazothion (108-35-0)
Pyrazoxon (108-34-9)
Pyrazoxone (108-34-9)
Pyreazid (54-85-3)
Pyredal (136-40-3)
Pyremethamine (58-14-0)
Pyren (German) (129-00-0)
1-Pyrenamine (1606-67-3)
Pyrene (129-00-0)
β-Pyrene (129-00-0)
Pyrenecarbonitrile (9CI) (83536-57-6)
Pyrenecarboxaldehyde (9CI) (68967-09-9)
Pyrene, 1-chloro (34244-14-9)
Pyrene, dichloro- (9CI) (68402-20-0)
Pyrene, dihydro- (9CI) (28779-32-0)
Pyrene, 2,7-dimethyl- (15679-24-0)
Pyrene, 1,3-dinitro (75321-20-9)
Pyrene, 1,6-dinitro (42397-64-8)
Pyrene, 1,8-dinitro (42397-65-9)
Pyrene, dinitro (78432-19-6)
1,6-Pyrenedione (1785-51-9)
1,8-Pyrenedione (2304-85-0)
3,10-Pyrenedione (2304-85-0)
3,8-Pyrenedione (1785-51-9)
Pyrene, 1-methyl (2381-21-7)
Pyrene, 2-methyl- (3442-78-2)
Pyrene, 4-methyl- (8CI,9CI) (3353-12-6)
Pyrene, methyl- (8CI,9CI) (27577-90-8)
Pyrene, 1-nitro (5522-43-0)
Pyrene, 2-nitro (789-07-1)
Pyrene, 4-nitro- (57835-92-4)
Pyrene, nitro (63021-86-3)
1,6-Pyrenequinone (1785-51-9)
1,8-Pyrenequinone (2304-85-0)
3,10-Pyrenequinone (2304-85-0)
3,8-Pyrenequinone (1785-51-9)
Pyrene, 1,3,6,8-tetrabromo- (9CI) (128-63-2)
Pyrene, 1,3,6,8-tetranitro (28767-61-5)
1,3,6,8-Pyrene tetrasulfonate (6528-53-6)
1,3,6,8-Pyrenetetrasulfonic acid (9CI) (6528-53-6)
Pyrene, 1,3,6-trinitro (75321-19-6)
1,3,6-Pyrenetrisulfonic acid, 8-hydroxy-, trisodium salt (6358-69-6)
Pyrenidine (194-03-6)
Pyreno(4,5-e)pyrene (385-13-7)
1-Pyrenol (9CI) (5315-79-7)
Pyrenoline (189-92-4)
Pyrenon (51-03-6)
Pyrenone 606 (51-03-6)
Pyreno(1',2':1,2)pyrene (385-13-7)
Pyrequan (15686-83-6)
Pyresin (584-79-2)
Pyresyn (584-79-2)
Pyretherm (10453-86-8)
Pyrethia (60-87-7)
Pyrethiazine (60-87-7)
Pyrethrin (121-29-9)
Pyrethrin I (121-21-1)
Pyrethrin II (121-29-9)

Pyrethrin I or II

Pyrethrin I or II **(8003-34-7)**
Pyrethrins **(8003-34-7)**
Pyrethrolone, chrysanthemum dicarboxlic acid methyl ester ester **(121-29-9)**
Pyrethrolone, chrysanthemum monocarboxylic acid ester **(121-21-1)**
Pyrethrolone ester of chrsanthemumdicarboxylic acid monomethyl ester **(121-29-9)**
(+)-Pyrethronyl (+)-trans-chrysanthemate **(121-21-1)**
(+)-Pyrethronyl (+)-pyrethrate **(121-29-9)**
Pyrethrum (ACGIH,OSHA) **(8003-34-7)**
Pyrethrum (Insecticide) **(8003-34-7)**
Pyretrin II **(121-29-9)**
Pyribenzamine **(91-81-6)**
Pyribenzamine hydrochloride **(154-69-8)**
Pyribenzamine monohydrochloride **(154-69-8)**
Pyricardyl **(59-26-7)**
Pyricidin **(54-85-3)**
Pyriclor **(1970-40-7)**
Pyridacil **(136-40-3)**
Pyridacil **(94-78-0)**
Pyridafenthion **(119-12-0)**
Pyridamal-100 **(113-92-8)**
Pyridaphenthion **(119-12-0)**
Pyridate **(55512-33-9)**
Pyridazine **(289-80-5)**
3,6-Pyridazinedione, 1,2-dihydro **(123-33-1)**
3,6-Pyridazinedione, 1,2-dihydro-, Compd. with 2,2'-Iminodiethanol (1:1) **(5716-15-4)**
3,6-Pyridazinedione, 1,2-dihydro-, monopotassium salt **(28382-15-2)**
Pyridazine, 3-(2-methylphenoxy) **(14491-59-9)**
3(2H)-Pyridazinone, 5-amino-4-chloro-2-phenyl **(1698-60-8)**
3(2H)-Pyridazinone, 4-chloro-5-(methylamino)-2-(α,α,α-trifluoro-m-tolyl) **(27314-13-2)**
3(2H)-Pyridazinone, 6-hydroxy-2-phenyl-, O-ester with O,O-diethyl phosphorothioate **(119-12-0)**
Pyridenal **(136-40-3)**
Pyridene **(136-40-3)**
Pyridiate **(136-40-3)**
Pyridicin **(54-85-3)**
Pyridin-2,5-dicarbonsaeure-di-n-propylester (German) **(136-45-8)**
Pyridimine phosphate **(29232-93-7)**
Pyridin (German) **(110-86-1)**
3-Pyridinaldehyde **(500-22-1)**
2-Pyridinaldoxime **(873-69-8)**
3-Pyridinamine **(462-08-8)**
4-Pyridinamine **(504-24-5)**
α-Pyridinamine **(504-29-0)**
4-Pyridinamine, 3-bromo- (9CI) **(13534-98-0)**
2-Pyridinamine, 5-chloro- **(1072-98-6)**
2-Pyridinamine, 5-methyl- **(1603-41-4)**
2-Pyridinamine, N-2-pyridinyl- (9CI) **(1202-34-2)**
Pyridine (ACGIH,OSHA) [UN 1282] **(110-86-1)**
4-Pyridineacetic acid (9CI) **(28356-58-3)**
Pyridine, 3-acetyl- **(350-03-8)**
Pyridine, 4-acetyl- **(1122-54-9)**
Pyridine, 1-acetyl-1,2,3,4-tetrahydro- (8CI,9CI) **(19615-27-1)**
3-Pyridinealdehyde **(500-22-1)**
4-Pyridinealdehyde **(872-85-5)**
p-Pyridinealdehyde **(872-85-5)**
Pyridine-2-aldoximate **(873-69-8)**
2-Pyridinealdoxime **(873-69-8)**
4-Pyridinealdoxime **(696-54-8)**
Pyridine-2-aldoxime **(873-69-8)**
Pyridine-4-aldoxime **(696-54-8)**
Pyridine, alkyl derivs. **(68391-11-7)**
Pyridine, 2-amino **(504-29-0)**
Pyridine, 3-amino **(462-08-8)**
Pyridine, 4-amino **(504-24-5)**
Pyridine, 4-amino-3-bromo- (6CI,7CI,8CI) **(13534-98-0)**
Pyridine, 2-amino-4,6-dimethyl **(5407-87-4)**
Pyridine, 2-amino-6-methyl **(1824-81-3)**
Pyridine bases **(68391-11-7)**
Pyridine, 4-benzoyl- **(14548-46-0)**
Pyridine, 2-benzyl **(101-82-6)**
Pyridine, 4-benzyl **(2116-65-6)**
Pyridine, 3-benzyl- (8CI) **(620-95-1)**
Pyridine, 2-(benzyl(2-(dimethylamino)ethyl)amino) **(91-81-6)**
Pyridine, 2-(benzyl(2-(dimethylamino)ethyl)amino)-, hydrochloride **(154-69-8)**
Pyridine, 2-(benzyl(2-(dimethylamino)ethyl)amino)-, monohydrochloride **(154-69-8)**
Pyridine, 2-bromo **(109-04-6)**
Pyridine, 3-bromo- (9CI) **(626-55-1)**
Pyridine, 4-bromo- (9CI) **(1120-87-2)**
Pyridine, 2-(p-bromo-α-(2-(dimethylamino)ethyl)benzyl)- **(86-22-6)**
Pyridine, 4-bromo-, hydrochloride (8CI,9CI) **(19524-06-2)**
3-Pyridinebutanal, γ-(methylnitrosoamino)- **(64091-90-3)**
Pyridine, 3-butyl **(539-32-2)**
Pyridine, 4-tert-butyl- (8CI) **(3978-81-2)**
Pyridine-3-carbaldehyde **(500-22-1)**
Pyridine-4-carbaldehyde **(872-85-5)**
Pyridine-2-carbinol **(586-98-1)**
Pyridine-3-carbinol **(100-55-0)**
β-Pyridinecarbonaldehyde **(500-22-1)**
Pyridine-3-carbonic acid **(59-67-6)**
2-Pyridinecarbonitrile (9CI) **(100-70-9)**
3-Pyridinecarbonitrile (9CI) **(100-54-9)**
4-Pyridinecarbonitrile (9CI) **(100-48-1)**
2-Pyridinecarbonitrile, 4-amino-3,5,6-trichloro- (9CI) **(14143-60-3)**
3-Pyridinecarbonitrile, 5-((4-chloro-2-nitrophenyl)azo)-1-ethyl-1,2-dihydro-6-hydroxy-4-methyl-2-oxo- (9CI) **(70528-90-4)**
3-Pyridinecarbonitrile, 1,4-dihydro-1-methyl-4-oxo- **(767-98-6)**
3-Pyridinecarbonitrile, 1-ethyl-1,2-dihydro-6-hydroxy-4-methyl-2-oxo- (9CI) **(28141-13-1)**
3-Pyridinecarbonitrile, 1-(2-ethylhexyl)-1,2-dihydro-6-hydroxy-4-methyl-5-((2-nitrophenyl)azo)-2-oxo- (9CI) **(51249-07-1)**
2-Pyridinecarbonitrile, 3,4,5,6-tetrachloro- (9CI) **(17824-83-8)**
2-Pyridinecarboxaldehyde (9CI) **(1121-60-4)**
3-Pyridinecarboxaldehyde (9CI) **(500-22-1)**
4-Pyridinecarboxaldehyde (9CI) **(872-85-5)**
4-Pyridinecarboxaldehyde, 3-hydroxy-5-(hydroxymethyl)-2-methyl- **(66-72-8)**
4-Pyridinecarboxaldehyde, 3-hydroxy-2-methyl-5-((phosphonooxy) methyl)- **(54-47-7)**
4-Pyridinecarboxaldehyde, 3-hydroxy-2-methyl-5-((phosphonooxy)-methyl)-, monohydrate **(41468-25-1)**
2-Pyridinecarboxaldehyde, oxime **(873-69-8)**
3-Pyridinecarboxaldehyde, oxime, (E)- (9CI) **(51892-16-1)**
γ-Pyridinecarboxamide **(1453-82-3)**
2-Pyridinecarboxamide (9CI) **(1452-77-3)**
4-Pyridinecarboxamide (9CI) **(1453-82-3)**
Pyridine-3-carboxydiethylamide **(59-26-7)**
3-Pyridinecarboxylic acid **(59-67-6)**
4-Pyridinecarboxylic acid **(55-22-1)**
Pyridine-3-carboxylic acid **(59-67-6)**
Pyridine-β-carboxylic acid **(59-67-6)**
α-Pyridinecarboxylic acid **(98-98-6)**
o-Pyridinecarboxylic acid **(98-98-6)**
2-Pyridinecarboxylic acid (9CI) **(98-98-6)**
Pyridinecarboxylic acid (8CI,9CI) **(32075-31-3)**
2-Pyridinecarboxylic acid, 2-acetylhydrazide (9CI) **(17433-31-7)**
3-Pyridinecarboxylic acid amide **(98-92-0)**
Pyridine-3-carboxylic acid amide **(98-92-0)**
3-Pyridinecarboxylic acid, 6-amino- (9CI) **(3167-49-5)**
2-Pyridinecarboxylic acid, 4-aminodichloro- (9CI) **(66280-95-3)**
2-Pyridinecarboxylic acid, 4-amino-3,5-dichloro-6-hydroxy- (9CI) **(38116-59-5)**

2-Pyridinecarboxylic acid, 4-amino-3,5,6-trichloro-, Compd. with 1,1',1''-nitrilotris(2-propanol) (1:1), Mixt. with 1,1',1''-nitrilotris(2-propanol) (2,4-dichlorophenoxy)acetate (Salt) (9CI) **(8067-55-8)**
2-Pyridinecarboxylic acid, 4-amino-3,5,6-trichloro-, isooctyl ester (9CI) **(26952-20-5)**
2-Pyridinecarboxylic acid, 4-amino-3,5,6-trichloro-, mono-potassium salt (9CI) **(2545-60-0)**
2-Pyridinecarboxylic acid, 5-butyl **(536-69-6)**
2-Pyridinecarboxylic acid, 6-chloro- **(4684-94-0)**
2-Pyridinecarboxylic acid, 3,6-dichloro- (9CI) **(1702-17-6)**
Pyridine-3-carboxylic acid diethylamide **(59-26-7)**
4-Pyridinecarboxylic acid, ethyl ester **(1570-45-2)**
2-Pyridinecarboxylic acid, ethyl ester (9CI) **(2524-52-9)**
4-Pyridinecarboxylic acid, hydrazide **(54-85-3)**
4-Pyridinecarboxylic acid, methyl ester (9CI) **(2459-09-8)**
2-Pyridinecarboxylic acid, nitrile **(100-70-9)**
4-Pyridinecarboxylic acid 1-oxide **(13602-12-5)**
3-Pyridinecarboxylic acid, 1-oxide (9CI) **(2398-81-4)**
4-Pyridinecarboxylic acid, 1-oxide (9CI) **(13602-12-5)**
4-Pyridinecarboxylic acid 2-(3-oxo-3-((phenylmethyl)-amino)propyl)hydrazide (9CI) **(51-12-7)**
2-Pyridinecarboxylic acid, 3,4,5,6-tetrachloro- (9CI) **(10469-09-7)**
3-Pyridinecarboxylic acid, 1,2,5,6-tetrahydro-1-nitroso- **(55557-01-2)**
3-Pyridinecarboxylic acid, 1,2,5,6-tetrahydro-1-nitroso-, methyl ester **(55557-02-3)**
Pyridine-carboxylique-2 (French) **(98-98-6)**
Pyridine-carboxylique-3 (French) **(59-67-6)**
Pyridine, 2-chloro **(109-09-1)**
Pyridine, 3-chloro **(626-60-8)**
Pyridine, 4-chloro **(626-61-9)**
Pyridine, chloro- (8CI,9CI) **(29154-12-9)**
Pyridine, 2-(p-chloro-α-(2-(dimethylamino)ethyl)benzyl) **(132-22-9)**
Pyridine, 2-(p-chloro-α-(2-(dimethylamino)ethyl)benzyl)- **(25523-97-1)**
Pyridine, 2-(p-chloro-α-(2-(dimethylamino)ethyl)benzyl)-, maleate (1:1) **(113-92-8)**
Pyridine, 4-chloro-, hydrochloride (9CI) **(7379-35-3)**
Pyridine, 2-chloromethyl-, hydrochloride **(6959-47-3)**
Pyridine, 3-chloromethyl-, hydrochloride **(6959-48-4)**
Pyridine, 2-((5-chloro-2-thenyl)(2-(dimethylamino)ethyl)amino) **(148-65-2)**
Pyridine, 2-chloro-6-(trichloromethyl) **(1929-82-4)**
Pyridine, 2-chloro-5-(trichloromethyl)- (9CI) **(69045-78-9)**
2,6-Pyridinediamine, 3-(phenylazo)- **(94-78-0)**
2,6-Pyridinediamine, 3-(phenylazo)-, hydrochloride (9CI) **(10393-51-8)**
2,6-Pyridinediamine, 3-(phenylazo)-, monohydrochloride **(136-40-3)**
Pyridine, 2,3-diamino **(452-58-4)**
Pyridine, 2,5-diamino **(4318-76-7)**
Pyridine, 2,6-diamino **(141-86-6)**
Pyridine, 3,4-diamino **(54-96-6)**
Pyridine, 2,6-diamino-3-(phenylazo) **(94-78-0)**
Pyridine, 2,6-diamino-3-(phenylazo)-, hydrochloride (6CI,7CI, 8CI) **(10393-51-8)**
Pyridine, 2,6-diamino-3-(phenylazo)-, monohydrochloride **(136-40-3)**
Pyridine, 3,5-dibromo- (8CI.9CI) **(625-92-3)**
2,3-Pyridinedicarbonitrile (8CI,9CI) **(17132-78-4)**
Pyridine-2,3-dicarboxylic acid **(89-00-9)**
Pyridine-2,5-dicarboxylic acid **(100-26-5)**
Pyridine-3,5-dicarboxylic acid **(499-81-0)**
2,3-Pyridinedicarboxylic acid (9CI) **(89-00-9)**
2,4-Pyridinedicarboxylic acid (9CI) **(499-80-9)**
2,5-Pyridinedicarboxylic acid (9CI) **(100-26-5)**
2,6-Pyridinedicarboxylic acid (9CI) **(499-83-2)**
3,4-Pyridinedicarboxylic acid (9CI) **(490-11-9)**
3,5-Pyridinedicarboxylic acid (9CI) **(499-81-0)**
3,5-Pyridinedicarboxylic acid, 1,4-dihydro-2,6-dimethyl-4-(2-nitrophenyl)-, dimethyl ester **(21829-25-4)**
2,6-Pyridinedicarboxylic acid, 1,4-dihydro-4-oxo- (9CI) **(138-60-3)**
2,5-Pyridinedicarboxylic acid, dipropyl ester (8CI,9CI) **(136-45-8)**
Pyridine, 2,3-dichloro **(2402-77-9)**
Pyridine, 2,5-dichloro **(16110-09-1)**
Pyridine, 2,6-dichloro **(2402-78-0)**
Pyridine, 3,5-dichloro **(2457-47-8)**
Pyridine, 2,4-dichloro- (8CI,9CI) **(26452-80-2)**
Pyridine, 2,3-dichloro-5-(trichloromethyl)- (9CI) **(69045-83-6)**
Pyridine, 3,5-dichloro-2-(trichloromethyl)- (9CI) **(1128-16-1)**
Pyridine, 3,6-dichloro-2-(trichloromethyl)- (9CI) **(1817-13-6)**
Pyridine, 3,5-diethyl-1,2-dihydro-1-phenyl-2-propyl- (9CI) **(34562-31-7)**
Pyridine, diethylnitroso- (9CI) **(69481-32-9)**
2,6-Pyridinedimethanol, bis(methylcarbamate) (ester) (9CI) **(1882-26-4)**
3,4-Pyridinedimethanol, 5-hydroxy-6-methyl- **(65-23-6)**
3,4-Pyridinedimethanol, 5-hydroxy-6-methyl-, α^3-(dihydrogen phosphate) (9CI) **(447-05-2)**
3,4-Pyridinedimethanol, 5-hydroxy-6-methyl-, hydrochloride **(58-56-0)**
Pyridine, 2,3-dimethyl- (9CI) **(583-61-9)**
Pyridine, 2,4-dimethyl- (9CI) **(108-47-4)**
Pyridine, 2,5-dimethyl- (9CI) **(589-93-5)**
Pyridine, 2,6-dimethyl- (9CI) **(108-48-5)**
Pyridine, 3,4-dimethyl- (9CI) **(583-58-4)**
Pyridine, 3,5-dimethyl- (9CI) **(591-22-0)**
Pyridine, dimethyl- (9CI) **(27175-64-0)**
Pyridine, 4-(dimethylamino) **(1122-58-3)**
Pyridine, 2-(α-(2-(dimethylamino)ethoxy)-α-methylbenzyl) **(469-21-6)**
Pyridine, 2-(α-(2-(dimethylamino)ethoxy)-α-methylbenzyl)-, succinate (1:1) **(562-10-7)**
Pyridine, 2-(α-(2-(dimethylamino)ethyl)benzyl) **(86-21-5)**
Pyridine, 2-(α-(2-(dimethylamino)ethyl)benzyl)-, maleate (1:1) **(132-20-7)**
Pyridine, 2-((2-(dimethylamino)ethyl)furfurylamino)-, fumarate (1:1) (8CI) **(5429-41-4)**
Pyridine, 2-((2-(dimethylamino)ethyl)(p-methoxybenzyl)amino) **(91-84-9)**
Pyridine, 2-((2-(dimethylamino)ethyl)(p-methoxybenzyl)amino)-, maleate (1:1) **(59-33-6)**
Pyridine, 2-((2-(dimethylamino)ethyl)-2-thenylamino) **(91-80-5)**
Pyridine, 2-((2-(dimethylamino)ethyl)-3-thenylamino) **(91-79-2)**
Pyridine, 2-((2-(dimethylamino)ethyl)-2-thenylamino)-, monohydro-chloride **(135-23-9)**
Pyridine, 2-((2-(dimethylamino)ethyl)-3-thenylamino)-, monohydro-chloride **(958-93-0)**
Pyridine, 4-(1,1-dimethylethyl)- (9CI) **(3978-81-2)**
Pyridine, 2,6-dimethyl-, 1-oxide (9CI) **(1073-23-0)**
Pyridine, 3,5-dimethyl-, 1-oxide (9CI) **(3718-65-8)**
2,3-Pyridinediol **(16867-04-2)**
2,4-Pyridinediol, 3-butyl-6-methyl- (6CI,8CI) **(6967-70-0)**
Pyridine, diphenyl- (9CI) **(56842-43-4)**
Pyridine, 2,2'-dithiodi-, 1,1'-dioxide **(3696-28-4)**
4-Pyridineethanol **(5344-27-4)**
Pyridine-2-ethanol **(103-74-2)**
2-Pyridineethanol (9CI) **(103-74-2)**
Pyridine, 3-ethenyl- **(1121-55-7)**
Pyridine, 2-ethenyl-5-ethyl- (9CI) **(5408-74-2)**
Pyridine, 5-ethenyl-2-methyl **(140-76-1)**
Pyridine, 2-ethyl- (9CI) **(100-71-0)**
Pyridine, 3-ethyl- (9CI) **(536-78-7)**
Pyridine, 4-ethyl- (9CI) **(536-75-4)**
Pyridine, 4-ethyl-2,6-dimethyl- (9CI) **(36917-36-9)**
Pyridine, 5-ethyl-2-methyl- **(104-90-5)**
Pyridine, 2-ethyl-6-methyl- (9CI) **(1122-69-6)**
Pyridine, 3-ethyl-5-methyl- (9CI) **(3999-78-8)**
Pyridine, ethylmethyl- (9CI) **(27987-10-6)**
Pyridine, 5-ethyl-2-vinyl- (8CI) **(5408-74-2)**
Pyridine, 4-formyl-, oxime **(696-54-8)**
Pyridine, hexahydro- **(110-89-4)**
Pyridine, hexahydro-N-nitroso- **(100-75-4)**
Pyridine, 3-hexyl- (9CI) **(6311-92-8)**
Pyridine, hydrochloride **(628-13-7)**

Pyridine, 2-(2-hydroxyethyl)- **(103-74-2)**
Pyridine, 2,2'-iminodi- (8CI) **(1202-34-2)**
2-Pyridinemethanol **(586-98-1)**
3-Pyridinemethanol **(100-55-0)**
4-Pyridinemethanol **(586-95-8)**
3-Pyridinemethanol, 4-(aminomethyl)-5-hydroxy-6-methyl-, dihydrochloride **(524-36-7)**
3-Pyridinemethanol, 4-(aminomethyl)-5-hydroxy-6-methyl-, .α.-(dihydrogen phosphate), dihydrate (9CI) **(84878-64-8)**
3-Pyridinemethanol, α,α-bis(p-chlorophenyl) **(17781-31-6)**
Pyridine methochloride **(7680-73-1)**
Pyridine, 3-methoxy- (8CI,9CI) **(7295-76-3)**
Pyridine, 2-methyl- **(109-06-8)**
Pyridine, 4-methyl **(108-89-4)**
Pyridine, 3-methyl- (9CI) **(108-99-6)**
Pyridine, 4-methylbenzenesulfonate (9CI) **(24057-28-1)**
Pyridine, methylnitro- (9CI) **(116211-92-8)**
Pyridine, 3-methyl-, 1-oxide **(1003-73-2)**
Pyridine, 4-methyl-, 1-oxide **(1003-67-4)**
Pyridine, 2-methyl-, 1-oxide (9CI) **(931-19-1)**
Pyridine, methylphenyl- (9CI) **(64828-54-2)**
Pyridine, 2-(1-(4-methylphenyl)-3-(1-pyrrolidinyl)-1-propenyl)-, (E)- (9CI) **(486-12-4)**
Pyridine, 2-(1-(4-methylphenyl)-3-(1-pyrrolidinyl)-1-propenyl)-, monohydrochloride, monohydrate, (E)- **(6138-79-0)**
Pyridine, 3-(1-methyl-2-pyrrolidinyl)- **(54-11-5)**
Pyridine, 3-(1-methyl-2-pyrrolidinyl)-, (+-)- (9CI) **(22083-74-5)**
Pyridine, 3-(1-methyl-2-pyrrolidinyl)-, (S)- (9CI) **(54-11-5)**
Pyridine, 3-(1-methyl-2-pyrrolidinyl)-, (S)-, (R-(R*,R*))-2,3-dihydroxybutanedioate (1:2) **(65-31-6)**
Pyridine, 3-(1-methyl-2-pyrrolidinyl)-, (S)-, sulfate (2:1) **(65-30-5)**
Pyridine, 3-(1-methyl-2-pyrrolyl) **(487-19-4)**
Pyridine, 3-(1-methyl-1H-pyrrol-2-yl)- (9CI) **(487-19-4)**
Pyridine, 2-methyl-5-vinyl- **(140-76-1)**
3-Pyridinenitrile **(100-54-9)**
4-Pyridinenitrile **(100-48-1)**
Pyridine, 2-nitro- (9CI) **(15009-91-3)**
Pyridine, 4-nitro- (9CI) **(1122-61-8)**
Pyridine, 4-nitro-, 1-oxide (8CI,9CI) **(1124-33-0)**
Pyridine, 3-(1-nitroso-2-piperidinyl)-, (+-)- **(84237-39-8)**
Pyridine, 3-(1-nitroso-2-piperidinyl)-, (S)- (9CI) **(1133-64-8)**
Pyridine, 3-(1-nitroso-2-pyrrolidinyl)-, (S)- **(16543-55-8)**
Pyridine, 3-(1-nitroso-2-pyrrolidinyl)-, (S)- **(80508-23-2)**
Pyridine, 3-(1-nitroso-2-pyrrolidinyl)-, (S)- (9CI) **(16543-55-8)**
Pyridine N-oxide **(694-59-7)**
Pyridine-1-oxide **(694-59-7)**
Pyridine, 2,3,4,5,6-pentachloro **(2176-62-7)**
Pyridine, 2-pentyl- (9CI) **(2294-76-0)**
Pyridine perchlorate **(15598-34-2)**
Pyridine, perchlorate (8CI,9CI) **(15598-34-2)**
Pyridine, 2-phenyl **(1008-89-5)**
Pyridine, 4-phenyl **(939-23-1)**
Pyridine, 3-phenyl- (9CI) **(1008-88-4)**
Pyridine, phenyl- (9CI) **(52642-16-7)**
Pyridine, 2-(phenylmethyl)- (9CI) **(101-82-6)**
Pyridine, 3-(phenylmethyl)- (9CI) **(620-95-1)**
Pyridine, 4-(3-phenylpropyl)- (9CI) **(2057-49-0)**
Pyridine, 3-(2-piperidyl)- **(494-52-0)**
2-Pyridinepropanamine, γ-(4-bromophenyl)-N,N-dimethyl- (9CI) **(86-22-6)**
2-Pyridinepropanamine, γ-(4-chlorophenyl)-N,N-dimethyl- (9CI) **(132-22-9)**
2-Pyridinepropanamine, γ-(4-chlorophenyl)-N,N-dimethyl-, (S)- **(25523-97-1)**
2-Pyridinepropanamine, γ-(4-chlorophenyl)-N,N-dimethyl-, (Z)-2-butenedioate (1:1) (9CI) **(113-92-8)**
1(4H)-Pyridinepropanoic acid, α-amino-3-hydroxy-4-oxo-, (S)- (9CI) **(500-44-7)**
3-Pyridinepropanol **(2859-67-8)**
Pyridine, 4-propyl- (9CI) **(1122-81-2)**
Pyridine, 2-(3-(1-pyrrolidinyl)-1-p-tolylpropenyl)-, (E) **(486-12-4)**
Pyridine, 3-(1-pyrrolin-2-yl) **(532-12-7)**
Pyridines, polyalkylated, higher boiling fraction **(68391-11-7)**
Pyridines, polyalkylated, lower boiling fraction **(68391-11-7)**
Pyridines, polyalkylated: polyalkylated pyridines **(68391-11-7)**
Pyridine, 2,3,4,5-tetrachloro **(2808-86-8)**
Pyridine, 2,3,5,6-tetrachloro **(2402-79-1)**
Pyridine, tetrachloro- (9CI) **(33752-16-8)**
Pyridine, 2,3,4,5-tetrachloro-6-(trichloromethyl)- (9CI) **(1134-04-9)**
Pyridine, 1,2,3,6-tetrahydro **(694-05-3)**
Pyridine, 3-(tetrahydro-1-methylpyrrol-2-yl) **(54-11-5)**
Pyridine, 2,3,5,6-tetramethyl- (8CI,9CI) **(3748-84-3)**
2-Pyridinethiol, 1-oxide **(1121-31-9)**
2-Pyridinethiol, 1-oxide, sodium salt **(3811-73-2)**
2-Pyridinethiol, N-oxide, sodium salt **(3811-73-2)**
2-Pyridinethiol-1-oxide, zinc salt **(13463-41-7)**
2(1H)-Pyridinethione, 3-hydroxy- (9CI) **(23003-22-7)**
Pyridine, 2,3,5-trichloro **(16063-70-0)**
Pyridine, 2,3,6-trichloro **(6515-09-9)**
Pyridine, 2,4,6-trichloro **(16063-69-7)**
Pyridine, 2,3,5-trichloro-4-(n-propylsulfonyl)- **(38827-35-9)**
Pyridine, 3,4,5-trichloro-2-(trichloromethyl)- (9CI) **(1201-30-5)**
Pyridine, 2,4,6-trimethyl **(108-75-8)**
Pyridine, trimethyl **(29611-84-5)**
Pyridine, 2,3,6-trimethyl- (9CI) **(1462-84-6)**
Pyridine, 2,3,4-trimethyl- (8CI,9CI) **(2233-29-6)**
Pyridine, 2,3,5-trimethyl- (8CI,9CI) **(695-98-7)**
Pyridine, 2,4,5-trimethyl- (8CI,9CI) **(1122-39-0)**
Pyridine, 2-vinyl **(100-69-6)**
Pyridine, 4-vinyl **(100-43-6)**
Pyridinium, 1-benzyl-3-carboxy-, chloride **(16214-98-5)**
Pyridinium, 3-carboxy-1-methyl-, hydroxide, inner salt (8CI) **(535-83-1)**
Pyridinium, 1-((2-carboxy-8-oxo-7-(2-(2-thienyl)acetamido)-5-thia-1-azabicyclo(4.2.0)Oct- 2-en-3-yl)methyl)-, hydroxide, inner salt **(50-59-9)**
Pyridinium, 3-carboxy-1-(phenylmethyl)-, chloride (9CI) **(16214-98-5)**
Pyridinium, 3-carboxy-1-(phenylmethyl)-, hydroxide, inner salt (9CI) **(15990-43-9)**
Pyridinium chloride **(628-13-7)**
Pyridinium, 1-(2-((4-((2-chloro-4-nitrophenyl)azo)phenyl)ethylamino)ethyl)-, chloride (9CI) **(36986-04-6)**
Pyridinium, 1-decyl-, chloride **(1609-21-8)**
Pyridinium, 3-(((dimethylamino)carbonyl)oxy)-1-methyl-, bromide **(101-26-8)**
Pyridinium, 3,5-dimethyl-1-((octylthio)methyl)-, chloride (9CI) **(70700-62-8)**
Pyridinium, 1-dodecyl-, chloride **(104-74-5)**
Pyridinium, 1-dodecyl-, iodide **(3026-66-2)**
Pyridinium, 1-((dodecylthio)methyl)-, chloride (9CI) **(68315-17-3)**
Pyridinium, 1-((dodecylthio)methyl)-3,5-dimethyl-, chloride (9CI) **(70700-63-9)**
Pyridinium, 1-((dodecylthio)methyl)-3-methyl-, chloride **(70700-60-6)**
Pyridinium, 1-ethyl-, bromide **(1906-79-2)**
Pyridinium, 1-heptyl-3-((hydroxyimino)methyl)-, iodide (9CI) **(66290-87-7)**
Pyridinium, 1-hexadecyl-, bromide **(140-72-7)**
Pyridinium, 1-hexadecyl-, chloride **(123-03-5)**
Pyridinium, 1-(2-hydroxyethylcarbamoylmethyl)-, chloride, dodecanoate **(6272-74-8)**
Pyridinium, 3-hydroxy-1-methyl-, bromide, dimethylcarbamate (ester) **(101-26-8)**
Pyridinium, 1-methyl-, chloride **(7680-73-1)**
Pyridinium, 3-methyl-1-((octylthio)methyl)-, chloride (9CI) **(70700-59-3)**
Pyridinium, 2-methyl-1-(2-phenylethyl)-, bromide (9CI) **(10551-21-0)**
Pyridinium, methyl-1-(phenylmethyl)-, chloride (9CI) **(26747-91-1)**
Pyridinium monochloride **(628-13-7)**
Pyridinium, 1-((octylthio)methyl)-, chloride (9CI) **(68278-98-8)**
Pyridinium, 1-(2-oxo-2-((2-((1-oxododecyl)oxy)ethyl)-amino)-ethyl)-, chloride **(6272-74-8)**
Pyridinium perchlorate **(15598-34-2)**
Pyridinium, 1-(phenylmethyl)-, chloride (9CI) **(2876-13-3)**

Pyridinium, 1-(3-sulfopropyl)-, hydroxide, inner salt (9CI) **(15471-17-7)**
Pyridinium, 1-tetradecyl-, chloride (9CI) **(2785-54-8)**
2-Pyridinol **(142-08-5)**
3-Pyridinol **(109-00-2)**
4-Pyridinol **(626-64-2)**
3-Pyridinol, 2-amino- (9CI) **(16867-03-1)**
Pyridinol carbamate **(1882-26-4)**
4-Pyridinol, 3,5-dichloro-2,6-dimethyl **(2971-90-6)**
3-Pyridinol, 6-methyl **(1121-78-4)**
4-Pyridinol, 2,3,5-trichloro **(1970-40-7)**
2-Pyridinol, 3,5,6-trichloro-, dihydrogen phosphorothioate (ester) (9CI) **(66651-97-6)**
2-Pyridinol, 3,5,6-trichloro-, O-ester with O,O-diethyl phosphorothioate **(2921-88-2)**
2-Pyridinone **(142-08-5)**
2(1H)-Pyridinone (9CI) **(142-08-5)**
2(1H)-Pyridinone, 3-butyl-4-hydroxy-6-methyl- (9CI) **(6967-70-0)**
2(1H)-Pyridinone, 3-hydroxy- (9CI) **(16867-04-2)**
4(1H)-Pyridinone, 1-methyl-3-phenyl-5-(3-(trifluoromethyl)-phenyl)- **(59756-60-4)**
2(1H)-Pyridinone, 3,5,6-trichloro-, sodium salt (9CI) **(37439-34-2)**
1-(2-Pyridinylazo)-2-naphthalenol **(85-85-8)**
3-Pyridinylcarbonimidodithioic acid butyl (4-(1,1-dimethyl-ethyl)phenyl)methyl ester **(51308-54-4)**
1-(2-Pyridinyl)ethanone **(1122-62-9)**
2-Pyridinylmethanol **(586-98-1)**
N-2-Pyridinyl-2-pyridinamine **(1202-34-2)**
Pyridipca **(58-56-0)**
Pyridium **(136-40-3)**
Pyridium **(94-78-0)**
Pyridivite **(136-40-3)**
2-Pyridiylamine **(109-12-6)**
10H-Pyrido(3,2-b)(1,4)benzothiazine, 10-(3-(diethylamino)-propyl) **(67465-67-2)**
1H-Pyrido(2,3-b)indol-2-amine **(26148-68-5)**
5H-Pyrido(4,3-b)indol-3-amine, 1,4-dimethyl- **(62450-06-0)**
5H-Pyrido(4,3-b)indol-3-amine, 1,4-dimethyl-, acetate **(68808-54-8)**
5H-Pyrido(4,3-b)indol-3-amine, 1,4-dimethyl-, acetate (9CI) **(75104-43-7)**
5H-Pyrido(4,3-b)indol-3-amine, 1,4-dimethyl-, monoacetate **(68808-54-8)**
5H-Pyrido(4,3-b)indol-3-amine, 1-methyl- **(62450-07-1)**
5H-Pyrido(4,3-b)indol-3-amine, 1-methyl-, monoacetate **(72254-58-1)**
9H-Pyrido(3,4-b)indole **(244-63-3)**
9H-Pyrido(2,3-b)indole, 2-amino **(26148-68-5)**
5H-Pyrido(4,3-b)indole, 3-amino-1,4-dimethyl **(62450-06-0)**
5H-Pyrido(4,3-b)indole, 3-amino-1,4-dimethyl-, acetate **(68808-54-8)**
5H-Pyrido(4,3-b)indole, 3-amino-1-methyl **(62450-07-1)**
9H-Pyrido(2,3-b)indole, 2-amino-3-methyl **(68006-83-7)**
5H-Pyrido(4,3-b)indole, 3-amino-1-methyl-, acetate **(72254-58-1)**
9H-Pyrido(3,4-b)indole, 1-methyl **(486-84-0)**
Pyrido(3,4-c)-7-methylpsoralen **(85878-63-3)**
Pyrido(3,4-c)psoralen **(85878-62-2)**
Pyrido(2,3-d)pyrimidin-4-ol **(24410-19-3)**
Pyrido(2,3-d)pyrimidin-4(1H)-one (8CI,9CI) **(24410-19-3)**
3-Pyridol **(109-00-2)**
2(1H)-Pyridone **(142-08-5)**
2-Pyridone **(142-08-5)**
α-Pyridone **(142-08-5)**
Pyridone-2 (French) **(142-08-5)**
2(1H)-Pyridone, 3-hydroxy **(16867-04-2)**
Pyrido(2',3':4)pyrene **(189-92-4)**
1,7-Pyridopyridine **(253-69-0)**
Pyridostigmine bromide **(101-26-8)**
Pyridoxal **(66-72-8)**
Pyridoxal P **(54-47-7)**
Pyridoxaldehyde **(66-72-8)**
Pyridoxaldehyde phosphate **(54-47-7)**
Pyridoxal, 5-(dihydrogen phosphate) **(54-47-7)**
Pyridoxal, 5-(dihydrogen phosphate), monohydrate **(41468-25-1)**
Pyridoxal, hydrochloride **(65-22-5)**
Pyridoxal monophosphate **(54-47-7)**
Pyridoxal-5-monophosphoric acid ester monohydrate **(41468-25-1)**
Pyridoxal 5-phosphate **(54-47-7)**
Pyridoxal phosphate **(54-47-7)**
Pyridoxal-5'-phosphate **(54-47-7)**
Pyridoxal-5'-phosphate, monohydrate **(41468-25-1)**
Pyridoxal-5-phosphate, monohydrate **(41468-25-1)**
Pyridoxamine **(85-87-0)**
Pyridoxamine-P **(529-96-4)**
Pyridoxamine-dihydrochloride **(524-36-7)**
Pyridoxamine phosphate **(529-96-4)**
Pyridoxin **(65-23-6)**
Pyridoxine **(65-23-6)**
Pyridoxine, hydrochloride **(58-56-0)**
Pyridoxine 5'-phosphate **(447-05-2)**
Pyridoxine 5-phosphate **(447-05-2)**
Pyridoxine phosphate **(447-05-2)**
Pyridoxinium chloride **(58-56-0)**
Pyridoxinum hydrochloricum (Hungarian) **(58-56-0)**
Pyridoxol **(65-23-6)**
Pyridoxol, 5-(dihydrogen phosphate) **(447-05-2)**
Pyridoxol-hydrochloride **(58-56-0)**
Pyridoxol 5'-phosphate **(447-05-2)**
Pyridoxol 5-phosphate **(447-05-2)**
Pyridoxylamine dihydrochloride **(524-36-7)**
Pyridoxyl phosphate **(54-47-7)**
Pyridoxyl phosphate monohydrate **(41468-25-1)**
Pyridrol **(467-60-7)**
Pyridrole **(467-60-7)**
2-Pyridylaldehyde **(1121-60-4)**
3-Pyridylaldehyde **(500-22-1)**
2-Pyridylaldoxime **(873-69-8)**
4-(Pyridyl-2-amidosulfonyl)-3'-carboxy-4'-hydroxyazobenzene **(599-79-1)**
2-Pyridylamine **(504-29-0)**
3-Pyridylamine **(462-08-8)**
4-Pyridylamine **(504-24-5)**
α-Pyridylamine **(504-29-0)**
1-(2-Pyridylazo)-2-hydroxynaphthalene **(85-85-8)**
1-(2-Pyridylazo)-2-naphthol **(85-85-8)**
N[1]-α-Pyridyl-N[1]-benzyl-N,N-dimethyl ethylenediamine monohydro-chloride **(154-69-8)**
1-(3-Pyridyl)butane **(539-32-2)**
2-Pyridylcarbinol **(586-98-1)**
3-Pyridylcarbinol **(100-55-0)**
4-Pyridylcarbinol **(586-95-8)**
3-Pyridylcarbonitrile **(100-54-9)**
3-Pyridylcarboxaldehyde **(500-22-1)**
2-(2-Pyridyl)ethanol **(103-74-2)**
2-Pyridylmethanol **(586-98-1)**
3-Pyridylmethanol **(100-55-0)**
4-Pyridylmethanol **(586-95-8)**
2-Pyridylmethylchloride hydrochloride **(6959-47-3)**
4,4'-(2-Pyridylmethylene)diphenol diacetate **(603-50-9)**
1-(3-Pyridylmethyl)-3-(4-nitrophenyl)urea **(53558-25-1)**
N-3-Pyridylmethyl-N'-p-nitrophenylurea **(53558-25-1)**
β-Pyridyl-α-N-methylpyrrolidine **(54-11-5)**
2-Pyridyl nitrile **(100-70-9)**
(-)-2-(3'-Pyridyl)piperidine **(494-52-0)**
2-(3'-Pyridyl) piperidine **(494-52-0)**
1-(1-Pyridyl)propane-3-sulfonate **(15471-17-7)**
4-(4-Pyridyl)pyridine **(553-26-4)**
5-(4-(2-Pyridylsulfamoyl)phenylazo)-2-hydroxybenzoic acid **(599-79-1)**
5-(p-(2-Pyridylsulfamoyl)phenylazo)salicylic acid **(599-79-1)**
N[1]-2-Pyridylsulfanilamide **(144-83-2)**
N-2-Pyridylsulfanilamide **(144-83-2)**

N-(α-Pyridyl)-N-(α-thenyl)-N',N'-dimethylethylenediamine **(91-80-5)**
N-(α-Pyridyl)-N-(β-thenyl)-N',N'-dimethylethylenediamine **(91-79-2)**
N-(2-Pyridyl)-N-(2-thienyl)-N,N'-dimethyl-ethylenediamine hydrochloride **(135-23-9)**
Pyrilamine **(91-84-9)**
Pyrilamine maleate **(59-33-6)**
Pyrilax **(603-50-9)**
Pyrimal **(68-35-9)**
Pyrimethamine **(58-14-0)**
2-Pyrimidinamine (9CI) **(109-12-6)**
5-Pyrimidinamine, 1,3-bis(2-ethylhexyl)hexahydro-5-methyl- (9CI) **(141-94-6)**
2-Pyrimidinamine, 4,6-dichloro- (9CI) **(56-05-3)**
4-Pyrimidinamine, 2,6-dimethoxy- (9CI) **(3289-50-7)**
2-Pyrimidinamine, 4,6-dimethyl- (9CI) **(767-15-7)**
Pyrimidine **(289-95-2)**
5-Pyrimidineacetic acid, 4-chloro-2-methyl-, ethyl ester **(14273-76-8)**
Pyrimidine, 2-amino **(109-12-6)**
Pyrimidine, 5-amino-1,3-bis(2-ethylhexyl)hexahydro-5-methyl- (8CI) **(141-94-6)**
Pyrimidine, 2-amino-4-chloro-6-methyl **(5600-21-5)**
Pyrimidine, 2-amino-4,6-dichloro- (8CI) **(56-05-3)**
Pyrimidine, 2-amino-4,6-dimethyl- (8CI) **(767-15-7)**
Pyrimidine, 2-amino-4-methyl **(108-52-1)**
Pyrimidine, 5-bromo- (9CI) **(4595-59-9)**
4-Pyrimidinecarboxylic acid, 1,2,3,6-tetrahydro-2,6-dioxo- (9CI) **(65-86-1)**
4-Pyrimidinecarboxylic acid, 1,2,3,6-tetrahydro-2,6-dioxo-, calcium salt (2:1) (9CI) **(22454-86-0)**
Pyrimidine, 2-chloro-4-(dimethylamino)-6-methyl **(535-89-7)**
2,4-Pyrimidinediamine, 6-chloro-5-(methylsulfonyl)- (9CI) **(78744-33-9)**
2,4-Pyrimidinediamine, 6-chloro-5-(methylthio)- (9CI) **(68925-41-7)**
4,6-Pyrimidinediamine, 2-chloro-5-(methylthio)- (9CI) **(70958-50-8)**
2,4-Pyrimidinediamine, 5-(p-chlorophenyl)-6-ethyl **(58-14-0)**
Pyrimidinediamine, 2(or 6)-chloro-5-(methylthio)- (9CI) **(83623-05-6)**
2,4-Pyrimidinediamine, 5-((3,4,5-trimethoxyphenyl)-methyl)- **(738-70-5)**
Pyrimidine, 2,4-diamino-5-(3,4,5-trimethoxybenzyl) **(738-70-5)**
Pyrimidine, 2,4-diamino-5-(3,4,5-trimethoxybenzyl)- and N'-(5-methyl-3-isoxazolyl) sulfanilamide **(8064-90-2)**
Pyrimidine, 4,5-dichloro-6-methyl-2-(methylsulfonyl)- (8CI,9CI) **(17901-16-5)**
Pyrimidine, 2,4-diethoxy- (8CI,9CI) **(20461-60-3)**
Pyrimidine, 4,6-dimethyl- (9CI) **(1558-17-4)**
Pyrimidine, 2-((2-(dimethylamino)ethyl)(p-methoxybenzyl)amino) **(91-85-0)**
Pyrimidine, 2-((2-(dimethylamino)ethyl)(p-methoxybenzyl)-amino)-, hydrochloride **(63-56-9)**
2,4-Pyrimidinediol **(66-22-8)**
2,4-Pyrimidinedione **(66-22-8)**
2,4(1H,3H)-Pyrimidinedione (9CI) **(66-22-8)**
2,4(1H,3H)-Pyrimidinedione, 5-(bis(2-chloroethyl)amino)- **(66-75-1)**
2,4(1H,3H)-Pyrimidinedione, 5-bromo- (9CI) **(51-20-7)**
2,4(1H,3H)-Pyrimidinedione, 5-bromo-6-methyl-3-(1-methylethyl)- (9CI) **(314-42-1)**
2,4(1H,3H)-Pyrimidinedione, 5-bromo-6-methyl-3-(1-methylpropyl)-, lithium salt (9CI) **(53404-19-6)**
2,4(1H,3H)-Pyrimidinedione, 5-bromo-6-methyl-3-(1-methylpropyl)-, sodium salt (9CI) **(69484-12-4)**
2,4(1H,3H)-Pyrimidinedione, 5-chloro- **(1820-81-1)**
2,4(1H,3H)-Pyrimidinedione, 5-chloro-3-(1,1-dimethylethyl)-6-methyl- **(5902-51-2)**
2,4(1H,3H)-Pyrimidinedione, 1-(2-deoxy-β-D-ribofuranosyl)-5-(trifluoromethyl)- **(70-00-8)**
4,6(1H,5H)-Pyrimidinedione, dihydro-2-thioxo- (9CI) **(504-17-6)**
2,4(1H,3H)-Pyrimidinedione, 3-(1,1-dimethylethyl)-6-methyl-, sodium salt (9CI) **(65086-97-7)**
4,6,(1H,5H)-Pyrimidinedione, 5-ethyldihydro-5-(1-methylbutyl)-2-thioxo- (9CI) **(76-75-5)**
4,6(1H,5H)-Pyrimidinedione, 5-ethyldihydro-5-phenyl **(125-33-7)**
2,4(1H,3H)-Pyrimidinedione, 5-fluoro- **(51-21-8)**
2,4(1H,3H)-Pyrimidinedione, 6-methyl- (9CI) **(626-48-2)**
2,4(1H,3H)-Pyrimidinedione, 6-methyl-3-(1-methylpropyl)-, sodium salt (9CI) **(65208-42-6)**
5-Pyrimidinemethanol, α-(2-chlorophenyl)-α-(4-chlorophenyl) **(60168-88-9)**
5-Pyrimidinemethanol, α-cyclopropyl-α-(p-methoxyphenyl) **(12771-68-5)**
5-Pyrimidinemethanol, α-(2,4-dichlorophenyl)-α-phenyl **(26766-27-8)**
Pyrimidine, 4-methyl- (9CI) **(3438-46-8)**
Pyrimidine, 5-(4-(pentyloxy)phenyl)-2-(4-pentylphenyl)- (9CI) **(34913-07-0)**
Pyrimidine, 2-sulfanilamido- **(68-35-9)**
Pyrimidine, 1,4,5,6-tetrahydro-1-methyl-2-(2-(2-thienyl)vinyl)-, (E) **(15686-83-6)**
Pyrimidine, 1,4,5,6-tetrahydro-1,2,4-trimethyl- (9CI) **(53517-92-3)**
2,4,5,6(1H,3H)-Pyrimidinetetrone **(50-71-5)**
2,4,5,6(1H,3H)-Pyrimidinetetrone hydrate **(2244-11-3)**
2,4,5,6(1H,3H)-Pyrimidinetetrone, monohydrate **(2244-11-3)**
2,4,6-Pyrimidinetriamine (9CI) **(1004-38-2)**
Pyrimidine, 2,4,6-triamino- (8CI) **(1004-38-2)**
2,4,6-Pyrimidinetriol **(67-52-7)**
Pyrimidinetriol **(67-52-7)**
2,4,6(1H,3H,5H)-Pyrimidinetrione **(67-52-7)**
2,4,6-Pyrimidinetrione **(67-52-7)**
2,4,6(1H,3H,5H)-Pyrimidinetrione, 5-butyl-5-ethyl- (9CI) **(77-28-1)**
2,4,6(1H,3H,5H)-Pyrimidinetrione, 5-(1-cyclohepten-1-yl)-5-ethyl- (9CI) **(509-86-4)**
2,4,6(1H,3H,5H)-Pyrimidinetrione, 5-(1-cyclohexen-1-yl)-1,5-dimethyl- **(56-29-1)**
2,4,6(1H,3H,5H)-Pyrimidinetrione, 5,5-diethyl- **(57-44-3)**
2,4,6(1H,3H,5H)-Pyrimidinetrione, 5,5-diethyl-1-methyl- (9CI) **(50-11-3)**
2,4,6(1H,3H,5H)-Pyrimidinetrione, 5,5-dihydroxy- (9CI) **(3237-50-1)**
2,4,6(1H,3H,5H)-Pyrimidinetrione, 5,5-di-2-propenyl- (9CI) **(52-43-7)**
2,4,6(1H,3H,5H)-Pyrimidinetrione, 5-ethyl- (9CI) **(2518-72-1)**
2,4,6(1H,3H,5H)-Pyrimidinetrione, 5-ethyl-5-(1-methyl-1-butenyl)-(9CI) **(125-42-8)**
2,4,6(1H,3H,5H)-Pyrimidinetrione, 5-ethyl-5-(1-methylbutyl)- (9CI) **(76-74-4)**
2,4,6(1H,3H,5H)-Pyrimidinetrione, 5-ethyl-5-(3-methylbutyl)- (9CI) **(57-43-2)**
2,4,6(1H,3H,5H)-Pyrimidinetrione, 5-ethyl-5-(1-methylbutyl)-, monosodium salt **(57-33-0)**
2,4,6(1H,3H,5H)-Pyrimidinetrione, 5-ethyl-1-methyl-5-phenyl-(9CI) **(115-38-8)**
2,4,6(1H,3H,5H)-Pyrimidinetrione, 5-ethyl-5-(1-methylpropyl)-(9CI) **(125-40-6)**
2,4,6(1H,3H,5H)-Pyrimidinetrione, 5-ethyl-5-(1-methylpropyl)-, monosodium salt (9CI) **(143-81-7)**
2,4,6(1H,3H,5H)-Pyrimidinetrione, 5-ethyl-5-pentyl- (9CI) **(115-58-2)**
2,4,6(1H,3H,5H)-Pyrimidinetrione, 5-ethyl-5-phenyl- **(50-06-6)**
2,4,6(1H,3H,5H)-Pyrimidinetrione, 5-ethyl-5-phenyl-, monosodium salt (9CI) **(57-30-7)**
2,4,6(1H,3H,5H)-Pyrimidinetrione, 5,5'-(1H-isoindole-1,3(2H)-diylidene)bis- (9CI) **(36888-99-0)**
2,4,6(1H,3H,5H)-Pyrimidinetrione, 5-(1-methylethyl)-5-(2-propenyl)- **(77-02-1)**
2,4,6(1H,3H,5H)-Pyrimidinetrione, 1-methyl-5-(1-methyl-2-pentynyl)-5-(2-propenyl)-, (+-)- **(151-83-7)**
2,4,6(1H,3H,5H)-Pyrimidinetrione, 5-(2-methylpropyl)-5-(2-propenyl)- **(77-26-9)**
2,4,6(1H,3H,5H)-Pyrimidinetrione, 5-(1-methylpropyl)-5-(2-propenyl)- (9CI) **(115-44-6)**
2,4,6(1H,3H,5H)-Pyrimidinitrione, 5-(1-methylbutyl)-5-(2-propenyl)- (9CI) **(76-73-3)**
4-Pyrimidinol **(4562-27-0)**
6-Pyrimidinol, 2-isopropyl-4-methyl **(2814-20-2)**
4-Pyrimidinol, 2-isopropyl-6-methyl-, O-ester with O,O-diethyl phosphorothioate **(333-41-5)**
4-Pyrimidinol, 2-mercapto- **(141-90-2)**
4(1H)-Pyrimidinone **(4562-27-0)**
2(1H)-Pyrimidinone, 4-amino **(71-30-7)**
2(1H)-Pyrimidinone, 4-amino-1-β-D-arabinofuranosyl- (9CI) **(147-94-4)**
2(1H)-Pyrimidinone, 4-amino-1-β-D-arabinofuranosyl-,

monohydrochloride (69-74-9)
4(1H)-Pyrimidinone, 2-amino-6-hydroxy- (9CI) (56-09-7)
2(1H)-Pyrimidinone, 4-amino-1-β-D-ribofuranosyl (65-46-3)
4(1H)-Pyrimidinone, 5-butyl-2-(ethylamino)-6-methyl (23947-60-6)
4(1H)-Pyrimidinone, 2,3-dihydro-6-methyl-2-thioxo- (56-04-2)
4(1H)-Pyrimidinone, 2,3-dihydro-5-methyl-2-thioxo- (9CI) (636-26-0)
4(1H)-Pyrimidinone, 2,3-dihydro-6-propyl-2-thioxo- (51-52-5)
4(3H)-Pyrimidinone, 2-(2-hydroxy-1-methylethyl)-6-methyl- (8CI) (28175-98-6)
4(3H)-Pyrimidinone, 2-isopropyl-6-methyl (2814-20-2)
2(1H)-Pyrimidinone, tetrahydro-5,5-dimethyl-, (3-(4-(trifluoromethyl)phenyl)-1-(2-(4-(trifluoromethyl)phenyl)ethenyl)-2-propenylidene)hydrazone (67485-29-4)
2(1H)-Pyrimidinone, tetrahydro-4-hydroxy-1-β-D-ribofuranosyl- (9CI) (18771-50-1)
2,4,5,6-Pyrimidintetron (Czech) (50-71-5)
N^1-2-Pyrimidinyl-sulfanilamide (68-35-9)
Pyrimido(1,2-a)azepine, 2,3,4,6,7,8,9,10-octahydro- (9CI) (6674-22-2)
Pyrimido(1,2-a)indol-10-ol, 10-(3-chlorophenyl)-2,3,4,10-tetrahydro- (37751-39-6)
Pyrimido(5,4-d)pyrimidine, 2,6-bis(bis(2-hydroxyethyl)-amino)-4,8-dipiperidino- (58-32-2)
Pyrimidone Medi-Pets (125-33-7)
Pyriminil (53558-25-1)
Pyriminyl (53558-25-1)
Pyrimiphos methyl (29232-93-7)
Pyrimital (5221-49-8)
Pyrimithate (5221-49-8)
Pyrimor (23103-98-2)
Pyrinamine (154-69-8)
Pyrinamine Base (91-81-6)
Pyrinazine (103-90-2)
Pyrine-2-aldoximate (873-69-8)
Pyrinex (2921-88-2)
Pyrinistab (91-80-5)
Pyrinistol (91-80-5)
Pyrinuron (53558-25-1)
Pyripyridium (136-40-3)
Pyripyridium (94-78-0)
Pyrisept (123-03-5)
Pyristan (61-76-7)
Pyrite (9CI) (1309-36-0)
Pyrithen (148-65-2)
Pyrithione zinc (13463-41-7)
Pyrizidin (54-85-3)
Pyrizin (136-40-3)
Pyrkappl (15676-16-1)
Pyro-Carb 406 (7782-42-5)
Pyroace (1018-71-9)
Pyroacetic acid (67-64-1)
Pyroacetic ether (67-64-1)
Pyrobenzol (71-43-2)
Pyrobenzole (71-43-2)
Pyrocatechin (120-80-9)
Pyrocatechine (120-80-9)
Pyrocatechinic acid (120-80-9)
Pyrocatechitol (931-17-9)
Pyrocatechol (OSHA) (120-80-9)
Pyrocatechol, 4-(2-aminoethyl) (51-61-6)
Pyrocatechol, 3-bromo- (8CI) (14381-51-2)
Pyrocatechol, 4-tert-butyl (98-29-3)
Pyrocatechol, 4-chloro (2138-22-9)
Pyrocatechol, 4,5-dichloro (3428-24-8)
Pyrocatechol, 3,5-dichloro- (8CI) (13673-92-2)
Pyrocatechol, 3,6-dichloro- (8CI) (3938-16-7)
Pyrocatechol, dichloro- (8CI) (25167-85-5)
Pyrocatechol, 3,5-dimethyl- (6CI,7CI,8CI) (2785-75-3)
Pyrocatechol dimethyl ether (91-16-7)
Pyrocatechol, 4,4'-(2,3-dimethyltetramethylene)di (500-38-9)
Pyrocatechol ethylene ether (493-09-4)
Pyrocatechol, 3-isopropyl (2138-48-9)
Pyrocatechol, 4-isopropyl (2138-43-4)
Pyrocatechol, 3-methoxy- (8CI) (934-00-9)
Pyrocatechol, 3-methyl (488-17-5)
Pyrocatechol, 4-methyl (452-86-8)
Pyrocatechol, 3-phenyl- (1133-63-7)
Pyrocatechol, tetrabromo (488-47-1)
Pyrocatechol, tetrachloro (1198-55-6)
Pyrocatechol, 3,4,5-tribromo- (7CI) (2747-17-3)
Pyrocatechol, 3,4,6-trichloro (32139-72-3)
2-Pyrocatechuic acid (303-38-8)
Pyrocatechuic acid (120-80-9)
Pyrocatechuic acid (303-38-8)
o-Pyrocatechuic acid (8CI) (303-38-8)
o-Pyrocatechuic acid, 4-isopropyl- (8CI) (19420-61-2)
o-Pyrocatechuic acid, 4-methyl- (8CI) (3929-89-3)
Pyrocellulose (9004-34-6)
Pyrochol (83-44-3)
Pyrocinchonic acid (488-21-1)
Pyroclor 5 (25429-29-2)
Pyrod (66-22-8)
Pyrodin (114-83-0)
Pyrodine (114-83-0)
Pyrodone (113-48-4)
Pyrogallic acid (87-66-1)
Pyrogallol (87-66-1)
Pyrogallol 1,3-dimethyl ether (91-10-1)
Pyrogallol dimethylether (91-10-1)
Pyrogallol triacetate (525-52-0)
Pyrogallol, triacetate (8CI) (525-52-0)
Pyrogallol, trichloro- (56961-21-8)
L-Pyroglutamic acid (98-79-3)
Pyroguaiac acid (90-05-1)
Pyrokatechin (Czech) (120-80-9)
Pyrokatechol (Czech) (120-80-9)
2-Pyrol (616-45-5)
Pyrolan (87-47-8)
Pyroligneous acids, Reaction products with et alc., distillates (8030-89-5)
Pyrollnitrin (1018-71-9)
Pyrolusite Brown (1313-13-9)
Pyrolysis fuel oil (69013-21-4)
Pyromellitic acid (89-05-4)
Pyromellitic acid anhydride (89-32-7)
Pyromellitic acid dianhydride (89-32-7)
Pyromellitic anhydride (89-32-7)
Pyromellitic dianhydride (89-32-7)
Pyromijin (54-47-7)
Pyromucic acid (88-14-2)
Pyromucic aldehyde (98-01-1)
Pyron (55512-33-9)
Pyronalorange (842-07-9)
Pyronalrot B (85-86-9)
Pyronalrot R (3118-97-6)
2-Pyrone (504-31-4)
4-Pyrone (108-97-4)
α-Pyrone (504-31-4)
γ-Pyrone-2,6-dicarboxylic acid (99-32-1)
.α.-Pyronene (6CI) (514-94-3)
Pyronin B (2150-48-3)
Pyronine B (2150-48-3)
Pyronine B (BY) (2150-48-3)
Pyropentylene (542-92-7)
Pyrophosphate (7722-88-5)
Pyrophosphate de tetraethyle (French) (107-49-3)

Pyrophosphoramide, octamethyl (152-16-9)
Pyrophosphoric acid, bis(2-ethylhexyl) ester (26836-28-2)
Pyrophosphoric acid, calcium salt (1:2) (7790-76-3)
Pyrophosphoric acid, copper salt (10102-90-6)
Pyrophosphoric acid, disodium salt (7758-16-9)
Pyrophosphoric acid, iron(3+) salt (3:4) (10058-44-3)
Pyrophosphoric acid, iron(3+) sodium salt (1:1:1) (10045-87-1)
Pyrophosphoric acid octamethyltetraamide (152-16-9)
Pyrophosphoric acid, tetraethyl ester (107-49-3)
Pyrophosphoric acid, tetrasodium salt (7722-88-5)
Pyrophosphorodithioic acid, O,O,O,O-tetraethyl ester (3689-24-5)
Pyrophosphorodithioic acid, tetraethyl ester (3689-24-5)
Pyrophosphoryltetrakisdimethylamide (152-16-9)
Pyrophyllite (1327-36-2)
Pyroplast 2 (64742-16-1)
Pyroplast (VAN) (64742-16-1)
Pyroracemic acid (127-17-3)
Pyroracemic aldehyde (78-98-8)
Pyroset Flame Retardant TKP (55818-96-7)
Pyroset TKO (55566-30-8)
Pyroset TKP (55818-96-7)
Pyrosulfurous acid, dipotassium salt (16731-55-8)
Pyrosulfurous acid, disodium salt (7681-57-4)
Pyro sulfuryl chloride (7791-27-7)
Pyrosulfuryl chloride [UN 1817] (7791-27-7)
Pyrosulphuric acid (8014-95-7)
Pyrotartaric acid (498-21-5)
Pyrotartaric acid nitrile (544-13-8)
Pyrotone Red Toner RA-5520 (2814-77-9)
Pyrotropblau (72-57-1)
Pyrovalerone (3563-49-3)
Pyrovaleronum (Latin) (3563-49-3)
Pyrovatex 3805 (20120-33-6)
Pyrovatex CP (20120-33-6)
Pyroxychlor (7159-34-4)
Pyroxychlore (7159-34-4)
Pyroxylic spirit (67-56-1)
Pyroxylin (9004-70-0)
Pyroxylin Plastic (DOT) (9004-70-0)
Pyroxylin Rods (DOT) (9004-70-0)
Pyroxylin Rolls (DOT) (9004-70-0)
Pyroxylin Scrap (DOT) (9004-70-0)
Pyroxylin Sheets (DOT) (9004-70-0)
Pyroxylin Tubes (DOT) (9004-70-0)
Pyrro(b)monazole (288-32-4)
Pyrrol (109-97-7)
Pyrrolamidol (357-56-2)
Pyrrolamidolum (357-56-2)
Pyrrole (109-97-7)
1H-Pyrrole-2-acetic acid, 5-(4-chlorobenzoyl)-1,4-dimethyl-, sodium salt, dihydrate (64092-48-4)
1H-Pyrrole-2-acetic acid, 1-methyl-5-(4-methylbenzoyl)- (9CI) (26171-23-3)
Pyrrole-2-acetic acid, 1-methyl-5-p-toluoyl (26171-23-3)
1H-Pyrrole, 1-acetyl-2-bromo- (9CI) (84455-06-1)
1H-Pyrrole, 1-acetyl-2-chloro- (9CI) (84455-05-0)
1H-Pyrrole-2-carboxaldehyde (9CI) (1003-29-8)
Pyrrole-2-carboxaldehyde (8CI) (1003-29-8)
1H-Pyrrole-2-carboxaldehyde, 1-methyl (1192-58-1)
1H-Pyrrole-2-carboxaldehyde, 5-methyl- (9CI) (1192-79-6)
Pyrrole-2-carboxaldehyde, 5-methyl- (8CI) (1192-79-6)
2-Pyrrolecarboxylic acid (634-97-9)
1H-Pyrrole-2-carboxylic acid (9CI) (634-97-9)
Pyrrole-2-carboxylic acid (8CI) (634-97-9)
Pyrrole, 3-chloro-4-(3-chloro-2-nitrophenyl) (1018-71-9)
1H-Pyrrole, 2,5-dimethyl- (9CI) (625-84-3)
1H-Pyrrole, dimethyl- (9CI) (49813-61-8)
Pyrrole, 2,5-dimethyl- (8CI) (625-84-3)
1H-Pyrrole-2,5-dione, 3,4-dimethyl- (9CI) (17825-86-4)
1H-Pyrrole-2,5-dione, 3-ethyl-4-methyl- (9CI) (20189-42-8)
1H-Pyrrole-2,5-dione, 1,1'-(methylenedi-4,1-phenylene)bis- (9CI) (13676-54-5)
1H-Pyrrole-2,5-dione, 1,1'-(phenylene)bis- (3006-93-7)
1H-Pyrrole, diphenyl- (9CI) (103837-23-6)
1H-Pyrrole, 3-ethyl-2,4,5-trimethyl- (9CI) (520-69-4)
Pyrrole, 3-ethyl-2,4,5-trimethyl- (8CI) (520-69-4)
Pyrrole, 1-furfuryl (1438-94-4)
1H-Pyrrole, 1-methyl- (9CI) (96-54-8)
1H-Pyrrole, 2-methyl- (9CI) (636-41-9)
1H-Pyrrole, 3-methyl- (9CI) (616-43-3)
Pyrrole, 1-methyl- (8CI) (96-54-8)
Pyrrole, 2-methyl- (8CI) (636-41-9)
Pyrrole, 3-methyl- (8CI) (616-43-3)
1H-Pyrrole, 1-phenyl- (9CI) (635-90-5)
Pyrrole, 1-phenyl- (8CI) (635-90-5)
Pyrrole, tetrahydro- (123-75-1)
Pyrrole, tetrahydro-N-nitroso- (930-55-2)
1H-Pyrrole, 2,3,5-trimethyl- (9CI) (2199-41-9)
Pyrrole, 2,3,5-trimethyl- (6CI,7CI,8CI) (2199-41-9)
Pyrrolidine [UN 1922] (123-75-1)
Pyrrolidine, 1-(3-aminopropyl)- (23159-07-1)
Pyrrolidine analog of phencyclidine (2201-39-0)
1-Pyrrolidinecarboxamide, 2,5-dimethyl-N-phenyl-, cis- (9CI) (34484-77-0)
1-Pyrrolidinecarboxylic acid, 1-(D-3-mercapto-2-methyl-1-propionyl)-, l-(S,S) (62571-86-2)
3,4-Pyrrolidinediol, 2-(p-methoxybenzyl)-, 3-acetate, (2S,3R,4R)- (22862-76-6)
2,5-Pyrrolidinedione (123-56-8)
2,5-Pyrrolidinedione, 1-(2-((2-((2-((2-aminoethyl)amino)ethyl)-amino)ethyl)amino)ethyl)-, monopolyisobutenyl derivs. (67762-72-5)
2,5-Pyrrolidinedione, 1-(2-((2-((2-((2-aminoethyl)amino)ethyl)-amino)ethyl)amino)ethyl)-, monopolyisobutenyl derivs., reaction products with molybdenum oxide (MoO_3), sulfurized (72269-41-1)
2,5-Pyrrolidinedione, 1-chloro (128-09-6)
2,5-Pyrrolidinedione, 1-ethyl- (9CI) (2314-78-5)
2,5-Pyrrolidinedione, 3-ethyl-1,3-dimethyl- (9CI) (13861-99-9)
2,5-Pyrrolidinedione, hydroxy- (9CI) (116211-83-7)
2,5-Pyrrolidinedione, hydroxymethyl- (9CI) (116211-84-8)
2,5-Pyrrolidinedione, 1-methyl-3-phenyl- (9CI) (86-34-0)
2,5-Pyrrolidinedione, 1-propyl- (9CI) (3470-97-1)
Pyrrolidine, 1-(2,2-diphenyl-3-methyl-4-morpholinobutyryl)-, (+) (357-56-2)
Pyrrolidine, 1-(2,2-diphenyl-3-methyl-4-morpholinobutyryl)-, (-)- (5666-11-5)
Pyrrolidine, 1-(2-(4-(1-(4-methoxyphenyl)-2-nitro-2-phenylethenyl)-phenoxy)ethyl)-, (Z)- (9CI) (52235-18-4)
Pyrrolidine, 1-methyl (120-94-5)
Pyrrolidine, 2-methyl- (9CI) (765-38-8)
Pyrrolidine, 1-(3-methyl-4-(4-morpholinyl)-1-oxo-2,2-diphenyl-butyl)-, (+-)- (9CI) (545-59-5)
Pyrrolidine, 1-(3-methyl-4-(4-morpholinyl)-1-oxo-2,2-diphenyl-butyl)-, (R)- (9CI) (5666-11-5)
Pyrrolidine, 1-methyl-2-(3-pyridal)- (54-11-5)
Pyrrolidine, 1-methyl-2-(3-pyridyl)-, sulfate (65-30-5)
Pyrrolidine, 1-nitroso (930-55-2)
Pyrrolidine, 1-(1-oxopentyl)- (9CI) (4419-57-2)
Pyrrolidine, 1-(1-phenylcyclohexyl)- (8CI,9CI) (2201-39-0)
1-Pyrrolidinepropanamine (9CI) (23159-07-1)
Pyrrolidine, 1-valeryl- (8CI) (4419-57-2)
2-Pyrrolidinone (616-45-5)
α-Pyrrolidinone (616-45-5)
2-Pyrrolidinone, 3-chloro-4-(chloromethyl)-1-(3-(trifluoromethyl)phenyl) (61213-25-0)
2-Pyrrolidinone, 1-(3-chloro-4-methylphenyl)-3-methyl- (9CI) (2884-69-7)
2-Pyrrolidinone, 1-(3-chloro-p-tolyl)-3-methyl- (8CI) (2884-69-7)
2-Pyrrolidinone, 1-(3,4-dichlorophenyl)-3-methyl- (8CI,9CI) (2860-64-2)
2-Pyrrolidinone, 1-ethenyl-, Homopolymer, compd. with iodine (25655-41-8)
2-Pyrrolidinone, 1-ethenyl, Homopolymer (9003-39-8)
2-Pyrrolidinone, 1-ethyl (2687-91-4)

2-Pyrrolidinone, 1-ethyl-4-(2-morpholinoethyl)-3,3-diphenyl (309-29-5)
2-Pyrrolidinone, 1-ethyl-4-(2-(4-morpholinyl)ethyl)-3,3-diphenyl- (9CI) (309-29-5)
2-Pyrrolidinone, 1-(2-hydroxyethyl) (3445-11-2)
2-Pyrrolidinone, 1-(hydroxymethyl)- (9CI) (15438-71-8)
2-Pyrrolidinone, 1-methyl (872-50-4)
2-Pyrrolidinone, 1-methyl-5-(3-pyridinyl)-, (S)- (9CI) (486-56-6)
2-Pyrrolidinone, 1-vinyl (88-12-0)
2-Pyrrolidinone, 1-vinyl-, Polymers (9003-39-8)
3-Pyrrolidinopropylamine (23159-07-1)
3-(1-Pyrrolidinyl)propylamine (23159-07-1)
(E)-2-(3-(1-Pyrrolidinyl)-1-p-tolylpropenyl)pyridine monohydrochloride monohydrate (6138-79-0)
Pyrrolidon (German) (616-45-5)
2-Pyrrolidone (616-45-5)
Pyrrolidone (616-45-5)
α-Pyrrolidone (616-45-5)
2-Pyrrolidone-5-carboxylic acid (98-79-3)
Pyrrolidonecarboxylic acid (98-79-3)
Pyrrolizidine (643-20-9)
1H-Pyrrolizine, hexahydro- (9CI) (643-20-9)
1H-Pyrrolizine-7-methanol, 2,3-dihydro-1-hydroxy-, (S) (26400-24-8)
1H-Pyrrolizine-7-methanol, 2,3,5,7a-tetrahydro-1-hydroxy-, (1S-cis) (520-63-8)
1H-Pyrrolizine-7-methanol, 2,3,5,7a-tetrahydro-1-hydroxy-, (1R-trans)- (9CI) (480-85-3)
Pyrrolnitrin (1018-71-9)
Pyrrolo(1,2-a)pyridine (274-40-8)
Pyrrolo(2,3-b)indole, 1,2,3,3a,8,8a-hexahydro-5-hydroxy-1,3a,8-trimethyl-, methylcarbamate (ester), (3as-cis)-, sulfate (2:1) (64-47-1)
1H-Pyrrolo(3,2-b)pyridine (272-49-1)
1H-Pyrrolo(2,3-b)pyridine, 2,3-dihydro- (6CI,8CI,9CI) (10592-27-5)
1H,10H-Pyrrolo(1,2-c)purine-10,10-diol, 3a,4,8,9-tetrahydro-2,6-diamino-4-(((aminocarbonyl) oxy)methyl)-, dihydrochloride, (3as-(3a-α,4-α,10ar*)) (35554-08-6)
1H-Pyrrolo(2,3-c)pyridine (271-29-4)
1H-Pyrrolo(3,2-c)pyridine (271-34-1)
7H-Pyrrolo(2,3-d)pyrimidine-5-carbonitrile, 4-amino-7-β-D-ribofuranosyl (606-58-6)
Pyrrolylene (106-99-0)
1-(1H-Pyrrol-2-yl)ethanone (1072-83-9)
Pyrroxate (62-44-2)
Pyruvaldehyde (78-98-8)
Pyruvaldehyde, 1-oxime (306-44-5)
Pyruvaldehydoxim (Czech) (306-44-5)
Pyruvic-acid (127-17-3)
Pyruvic acid, ethyl ester (8CI) (617-35-6)
Pyruvic acid, p-hydroxyphenyl (156-39-8)
Pyruvic acid, (p-hydroxyphenyl)- (8CI) (156-39-8)
Pyruvic acid, methyl ester (8CI) (600-22-6)
Pyruvic acid, sodium salt (113-24-6)
Pyruvic aldehyde (78-98-8)
Pyruvophenone (579-07-7)
Pyrvinium (7187-62-4)
Pyrvinum (7187-62-4)
Pysococcine (63-74-1)
pPzeidan (50-29-3)
6Q8 (132-66-1)
Q-137 (72-56-0)
Q9-5700 (27668-52-6)
QBH (495-73-8)
QCB (608-93-5)
QCD 84924 (1972-08-3)
QDO (105-11-3)
Q-D 86P (107-64-2)
QDX (13980-00-2)
Q-Loid A 30 (1344-28-1)
QM 867 (9CI) (96420-93-8)
QO THFA (97-99-4)
QX 2168 (9011-06-7)
QZ 2 (72-44-6)
Qamlin (52645-53-1)
Qiana (63428-84-2)
Qidamp (69-53-4)
Qidmycin (643-22-1)
Qidpen G (113-98-4)
Qidpen VK (132-98-9)
Qidtet (64-75-5)
Qikron (80-06-8)
Qingxintong (1197-09-7)
Quaalude (72-44-6)
Quadracycline (64-75-5)
Quadriphenyl (135-70-6)
Quadrol (102-60-3)
Quadronal (62-44-2)
Quamonium (57-09-0)
Quaname (57-53-4)
Quanane (57-53-4)
Quanil (57-53-4)
Quantalan (11041-12-6)
Quantril (63-12-7)
Quantrovanil (121-32-4)
Quantryl (63-12-7)
Quarton 14 BCL (139-08-2)
α-Quartz (14808-60-7)
Quartz Glass (60676-86-0)
Quartz Sand (60676-86-0)
Quarzsand (German) (60676-86-0)
Quaternario CPC (123-03-5)
Quaternario LPC (104-74-5)
Quaternary ammonium compounds, alkylbenzyldimethyl, chlorides (8001-54-5)
Quaternary ammonium compounds, C14-18-alkyltrimethyl, bromides (68424-92-0)
Quaternary ammonium compounds, C12-18-alkyltrimethyl, chlorides (68391-03-7)
Quaternary ammonium compounds, benzyl-C12-18-alkyldimethyl, chlorides (68391-01-5)
Quaternary ammonium compounds, benzylbis(hydrogenated tallow alkyl)methyl, chlorides, Compds. with bentonite (68153-30-0)
Quaternary ammonium compounds, benzylcoco alkylbis(hydroxyethyl), chlorides (61789-68-2)
Quaternary ammonium compounds, benzyl(hydrogenated tallow alkyl)dimethyl, chlorides, Compds. with hectorite (71011-26-2)
Quaternary ammonium compounds, bis(hydrogenated tallow alkyl)dimethyl, chlorides, Reaction products with bentonite (68953-58-2)
Quaternary ammonium compounds, bis(hydrogenated tallow alkyl)dimethyl, chlorides, Compds. with hectorite (71011-27-3)
Quaternary ammonium compounds, bis(hydrogenated tallow alkyl)dimethyl, salts with bentonite (68953-58-2)
Quaternary ammonium compounds, bis(hydroxyethyl)methyltallow alkyl, chlorides (67784-77-4)
Quaternary ammonium compounds, (carboxymethyl)(3-cocoamidopropyl)dimethyl, hydroxides, inner salts (61789-40-0)
Quaternary ammonium compounds, coco alkylbis(hydroxyethyl)methyl, chlorides (70750-47-9)
Quaternary ammonium compounds, di-C12-20-alkyldimethyl, chlorides (68514-95-4)
Quaternary ammonium compounds, dicoco alkyl dimethyl, chlorides (61789-77-3)
Quaternary ammonium compounds, dicoco alkyldimethyl, chlorides (61789-77-3)
Quaternary ammonium compounds, dimethyl ditallow alkyl,

chlorides **(68783-78-8)**
Quaternary ammonium compounds, dimethylditallow alkyl, chlorides **(68783-78-8)**
Quaternary ammonium compounds, ethyldimethylsoya alkyl, Et sulfates **(68308-67-8)**
Quaternary ammonium compounds, ethyldimethylsoya alkyl, bromides **(61788-99-6)**
Quaternary ammonium compounds, (oxydi-2,1-ethanediyl)bis-(coco alkyldimethyl, dichlorides **(68607-28-3)**
Quaternary ammonium compounds, trimethylsoya alkyl, chlorides **(61790-41-8)**
Quaternary ammonium compounds, trimethyltallow alkyl-, chloride **(8030-78-2)**
Quaternium-1 **(8001-54-5)**
Quaternium-2 **(61791-34-2)**
Quaternium 5 **(107-64-2)**
Quaternium-9 **(61790-41-8)**
Quaternium-10 **(112-03-8)**
Quaternium 13 **(1119-97-7)**
Quaternium 15 **(4080-31-3)**
Quaternium-12 **(7173-51-5)**
Quaternium-14 **(27479-28-3)**
Quaternium-17 **(16919-58-7)**
Quaternium-18 bentonite **(68953-58-2)**
Quaternium-23 **(53633-54-8)**
Quaternium-24 **(32426-11-2)**
Quaternium-25 **(78-21-7)**
Quaternium-28 **(1330-85-4)**
Quaternium-34 **(61789-77-3)**
Quaternium-39 **(26006-22-4)**
Quaternium-47 **(3401-74-9)**
Quaternium-48 **(68783-78-8)**
Quaternol 1 **(122-19-0)**
1,1':2',1'':2'',1'''-Quaterphenyl (9CI) **(641-96-3)**
1,1':3',1'':3'',1'''-Quaterphenyl (9CI) **(1166-18-3)**
1,1':4',1'':4'',1'''-Quaterphenyl (9CI) **(135-70-6)**
Quaterphenyl (9CI) **(29036-02-0)**
m-Quaterphenyl (8CI) **(1166-18-3)**
o-Quaterphenyl (8CI) **(641-96-3)**
p-Quaterphenyl (8CI) **(135-70-6)**
Quaterphenyl, decachloro- (9CI) **(89590-79-4)**
Quaterphenyl, hexachloro- (9CI) **(89590-81-8)**
Quaterphenyl, octachloro- (9CI) **(89590-80-7)**
Quatrachlor **(121-54-0)**
Quatrex **(64-75-5)**
Quazepam **(36735-22-5)**
Quazepamum (Latin) **(36735-22-5)**
Quazo Puro (Italian) **(14808-60-7)**
Quebracho-Tannin **(1401-55-4)**
Quebrachin **(146-48-5)**
Quebrachine **(146-48-5)**
Quebrachol **(83-46-5)**
Quebracho wood extract **(1401-55-4)**
Quecksilber (German) **(7439-97-6)**
Quecksilber chlorid (German) **(7487-94-7)**
Quecksilberoxid (German) **(15829-53-5)**
Quecksilberoxid (German) **(21908-53-2)**
Quecodur AE **(136-84-5)**
Quel **(12771-68-5)**
Queletox **(55-38-9)**
Quelicin **(306-40-1)**
Quellada **(58-89-9)**
Quemicetina **(56-75-7)**
Quen **(604-75-1)**
Quercetin **(117-39-5)**
Quercetin, 3-(6-deoxy-α-l-mannopyranoside) **(522-12-3)**
Quercetin, 3-(6-O-(6-deoxy-α-l-mannopyranosyl)-β-D-gluco-pyranoside) **(153-18-4)**
Quercetin, dihydro- **(480-18-2)**
Quercetine **(117-39-5)**
Quercetin 3-rhamnoglucoside **(153-18-4)**
Quercetin rhamnoglucosine **(153-18-4)**
Quercetin, 3-(6-O-α-l-rhamnopyranosyl-β-D-glucopyranoside) **(153-18-4)**
Quercetin-3-l-rhamnoside **(522-12-3)**
Quercetin 3-rutinoside **(153-18-4)**
Quercetol **(117-39-5)**
Quercetrin **(522-12-3)**
Quercetrin-3-O-rham **(522-12-3)**
Quercimelin **(522-12-3)**
Quercitin **(117-39-5)**
Quercitrin **(522-12-3)**
Quercitroside **(522-12-3)**
Queroplex **(51-21-8)**
Quertine **(117-39-5)**
Questex **(64-02-8)**
Questex 4 **(64-02-8)**
Questex 4H **(60-00-4)**
Questiomycin B **(95-55-6)**
Questran **(11041-12-6)**
Quetimid **(50-35-1)**
Quetinil **(439-14-5)**
Quiatril **(439-14-5)**
Quick **(3691-35-8)**
Quick Silver **(7439-97-6)**
Quicklime (DOT) **(1305-78-8)**
Quickphos **(20859-73-8)**
Quicksan **(62-38-4)**
Quicksan 20 **(62-38-4)**
Quickset Extra **(1338-23-4)**
Quickset Super **(1338-23-4)**
Quiebar **(143-81-7)**
Quiescin **(50-55-5)**
Quietidon **(57-53-4)**
Quietoplex **(50-35-1)**
Quievita **(439-14-5)**
Quilan **(1861-40-1)**
Quilibrex **(604-75-1)**
Quillaja saponin **(1393-03-9)**
Quillajasaponin **(1393-03-9)**
Quilon S **(15242-96-3)**
Quilone **(546-89-4)**
Quilonum retard **(554-13-2)**
Quimar **(9004-07-3)**
Quimotrase **(9004-07-3)**
Quinacetophenone **(490-78-8)**
Quinachlor **(54-05-7)**
Quinacridone **(1047-16-1)**
Quinacridone Red **(1047-16-1)**
Quinacridone Red MC **(1047-16-1)**
Quinacridone Violet **(1047-16-1)**
Quinacridone Violet MC **(1047-16-1)**
Quinacridone, 4,11-dichloro- **(3089-16-5)**
Quinacridone, 4,11-dichloro-6,13-dihydro- **(15715-19-2)**
Quinacridone, 6,13-dihydro-2,9-dimethyl- **(13796-22-0)**
Quinacrine **(83-89-6)**
Quinacrine Mustard **(4213-45-0)**
Quinacrine Mustard Dihydrochloride **(4213-45-0)**
Quinacrine dihydrochloride **(69-05-6)**
Quinacrine hydrochloride **(69-05-6)**
Quinactine **(83-89-6)**
Quinadome **(83-73-8)**
Quinagamin **(54-05-7)**
Quinagamine **(54-05-7)**
Quinalbarbital **(76-73-3)**

Name	CAS
Quinalbarbitone	**(76-73-3)**
Quinaldic acid, 4-hydroxy- (8CI)	**(492-27-3)**
Quinaldic acid, 5-nitro	**(525-47-3)**
Quinaldine	**(91-63-4)**
Quinaldofur	**(57474-29-0)**
Quinalizarin	**(81-61-8)**
Quinalizarine	**(81-61-8)**
Quinalphos	**(13593-03-8)**
Quinambicide	**(130-26-7)**
Quinate	**(77-95-2)**
Quinazarin Green	**(128-80-3)**
Quinazine	**(91-19-0)**
Quinazoline (9CI)	**(253-82-7)**
Quinazoline, 4-amino-6,7-dimethoxy-2-(4-(2-furoyl)piperazin-1-yl)	**(19216-56-9)**
6-Quinazolinesulfonamide, 7-chloro-2-ethyl-1,2,3,4-tetrahydro-4-oxo-	**(73-49-4)**
6-Quinazolinesulfonamide, 7-chloro-1,2,3,4-tetrahydro-2-methyl-4-oxo-3-o-tolyl-	**(17560-51-9)**
6-Quinazolinesulfonamide, 1,2,3,4-tetrahydro-7-chloro-2-methyl-4-oxo-3-o-tolyl	**(17560-51-9)**
4(3H)-Quinazolinone, 3-(2-chlorophenyl)-2-methyl- (9CI)	**(340-57-8)**
4(3H)-Quinazolinone, 3-(o-chlorophenyl)-2-methyl- (8CI)	**(340-57-8)**
4(3H)-Quinazolinone, 2-methyl-3-(2-methylphenyl)-	**(72-44-6)**
4(3H)-Quinazolinone, 2-methyl-3-o-tolyl	**(72-44-6)**
Quinercyl	**(54-05-7)**
Quinetazona (Spanish)	**(73-49-4)**
Quinethazon	**(73-49-4)**
Quinethazone	**(73-49-4)**
Quinethazonum (Latin)	**(73-49-4)**
Quinhydrone	**(106-34-3)**
(-)-Quinic acid	**(77-95-2)**
Quinic acid	**(77-95-2)**
Quinic acid (VAN)	**(36413-60-2)**
Quinicardine	**(50-54-4)**
Quinicardine	**(56-54-2)**
Quinidate	**(50-54-4)**
Quinidex	**(50-54-4)**
Quinidex	**(56-54-2)**
(+)-Quinidine	**(56-54-2)**
Quinidine	**(56-54-2)**
α-Quinidine	**(485-71-2)**
Quinidine Yellow KT	**(8004-92-0)**
Quinidine bisulfate	**(747-45-5)**
Quinidine hydrochloride	**(1668-99-1)**
Quinidine, monohydrochloride	**(1668-99-1)**
Quinidine monosulfate	**(50-54-4)**
Quinidine-N-oxide	**(70116-00-6)**
Quinidine sulfate	**(50-54-4)**
Quinidine, sulfate (1:1) (Salt)	**(747-45-5)**
Quinidine sulfate (2:1) (Salt)	**(50-54-4)**
Quinidine sulphate	**(50-54-4)**
Quinilon	**(54-05-7)**
(-)-Quinine	**(130-95-0)**
Quinine	**(130-95-0)**
β-Quinine	**(56-54-2)**
Quinine bisulfate	**(549-56-4)**
Quinine bisulfate	**(804-63-7)**
Quinine chloride	**(130-89-2)**
Quinine, 10,11-dihydro-	**(522-66-7)**
Quinine hydrobromide	**(549-49-5)**
Quinine hydrochloride	**(130-89-2)**
Quinine hydrogen sulfate	**(804-63-7)**
Quinine lactate	**(749-49-5)**
Quinine, monohydrobromide	**(549-49-5)**
Quinine, monohydrochloride	**(130-89-2)**
Quinine, monolactate (Salt) (8CI)	**(749-49-5)**
Quinine, monophosphinate (Salt) (8CI)	**(6119-53-5)**
Quinine muriate	**(130-89-2)**
Quinine, sulfate	**(804-63-7)**
Quininone (8CI)	**(84-31-1)**
Quinitex	**(50-54-4)**
Quinitol	**(556-48-9)**
Quinizarin	**(81-64-1)**
Quinizarin Green SS	**(128-80-3)**
Quinizarine	**(81-64-1)**
Quinizarine Green Base	**(128-80-3)**
Quino(2,3-b)acridine-6,7,13,14(5H,12H)-tetrone (9CI)	**(1503-48-6)**
Quino(2,3-b)acridine-7,14-dione, 2,9-dichloro-5,12-dihydro-(9CI)	**(3089-17-6)**
Quino(2,3-b)acridine-7,14-dione, 4,11-dichloro-5,12-dihydro-(9CI)	**(3089-16-5)**
Quino(2,3-b)acridine-7,14-dione, 4,11-dichloro-5,6,12,13-tetrahydro- (9CI)	**(15715-19-2)**
Quino(2,3-b)acridine-7,14-dione, 5,12-dihydro- (9CI)	**(1047-16-1)**
Quino(2,3-b)acridine-7,14-dione, 5,12-dihydro-2,9-dimethyl- (9CI)	**(980-26-7)**
Quino(2,3-b)acridine-7,14-dione, 5,6,12,13-tetrahydro- (9CI)	**(5862-38-4)**
Quino(2,3-b)acridine-7,14-dione, 5,6,12,13-tetrahydro-2,9-dimethyl- (9CI)	**(13796-22-0)**
Quinofen	**(132-60-5)**
Quinofop-Ethyl	**(76578-14-8)**
8-Quinol	**(148-24-3)**
Quinol	**(123-31-9)**
β-Quinol	**(123-31-9)**
Quinol (VAN)	**(4323-21-1)**
Quinol dimethyl ether	**(150-78-7)**
5-Quinolinamine	**(611-34-7)**
2-Quinolinamine (9CI)	**(580-22-3)**
6-Quinolinamine (9CI)	**(580-15-4)**
Quinoline	**(91-22-5)**
Quinoline [UN 2656]	**(91-22-5)**
Quinoline Yellow	**(8004-92-0)**
Quinoline Yellow A Spirit Soluble	**(8003-22-3)**
Quinoline Yellow Base	**(8003-22-3)**
Quinoline Yellow 2SF	**(83-08-9)**
Quinoline Yellow SS	**(8003-22-3)**
Quinoline Yellow Spirit Soluble	**(8003-22-3)**
3-Quinolineamine	**(580-17-6)**
Quinoline, 2-amino	**(580-22-3)**
Quinoline, 3-amino	**(580-17-6)**
Quinoline, 5-amino	**(611-34-7)**
Quinoline, 8-amino	**(578-66-5)**
Quinoline, 6-amino- (8CI)	**(580-15-4)**
Quinoline-N-benzyl chloride quaternary	**(15619-48-4)**
Quinoline, 3-bromo- (9CI)	**(5332-24-1)**
4-Quinolinecarboxamide, 2-butoxy-N-(2-(diethylamino)ethyl)-(9CI)	**(85-79-0)**
3-Quinolinecarboxylic acid, 2-(4,5-dihydro-4-methyl-4-(1-methylethyl)-5-oxo-1H-imidazol-2-yl)	**(81335-37-7)**
2-Quinolinecarboxylic acid, 4-hydroxy- (9CI)	**(492-27-3)**
4-Quinolinecarboxylic acid, 3-hydroxy-2-methyl- (9CI)	**(117-57-7)**
4-Quinolinecarboxylic acid, 2-phenyl- (9CI)	**(132-60-5)**
Quinoline, 7-chloro-4-((4-(diethylamino)-1-methylbutyl)amino)	**(54-05-7)**
Quinoline, 2-(chloromethyl)- (8CI,9CI)	**(4377-41-7)**
Quinoline, 4,7-dichloro	**(86-98-6)**
Quinoline, diethyl- (9CI)	**(68228-10-4)**
Quinoline, 1,2-dihydro-2,2,4-trimethyl	**(147-47-7)**
Quinoline, 1,2-dihydro-2,2,4-trimethyl-, homopolymer	**(26780-96-1)**
Quinoline, 2,4-dimethyl- (9CI)	**(1198-37-4)**
Quinoline, 2,6-dimethyl- (9CI)	**(877-43-0)**
Quinoline, 2,7-dimethyl- (9CI)	**(93-37-8)**
Quinoline, 2,3-dimethyl- (8CI,9CI)	**(1721-89-7)**
Quinoline, 2,8-dimethyl- (8CI,9CI)	**(1463-17-8)**
Quinoline, 5,8-dimethyl- (8CI,9CI)	**(2623-50-9)**

Quinoline, dimethyl- (8CI,9CI)

Name	CAS No.
Quinoline, dimethyl- (8CI,9CI)	**(28351-04-4)**
Quinoline, 3,4-dimethyl- (6CI,7CI,8CI,9CI)	**(2436-92-2)**
Quinoline, 4-(p-(dimethylamino)styryl)	**(897-55-2)**
Quinoline, 6-dodecyl-1,2-dihydro-2,2,4-trimethyl	**(89-28-1)**
Quinoline, 6-ethoxy-1,2-dihydro-2,2,4-trimethyl	**(91-53-2)**
Quinoline, ethyl- (9CI)	**(53123-73-2)**
Quinoline, 7-ethyl- (6CI,8CI,9CI)	**(7661-47-4)**
Quinoline, ethylmethyl- (9CI)	**(76602-24-9)**
Quinoline, 3-ethyl-2-propyl- (9CI)	**(3290-24-2)**
Quinoline, 4-(hydroxyamino)-, 1-oxide	**(4637-56-3)**
Quinoline, 6-methoxy- (9CI)	**(5263-87-6)**
Quinoline, 2-methyl-	**(91-63-4)**
Quinoline, 3-methyl	**(612-58-8)**
Quinoline, 4-methyl-	**(491-35-0)**
Quinoline, 6-methyl	**(91-62-3)**
Quinoline, 7-methyl	**(612-60-2)**
Quinoline, 8-methyl	**(611-32-5)**
Quinoline, methyl- (8CI,9CI)	**(27601-00-9)**
Quinoline, 2-methyl-, hydrochloride (9CI)	**(62763-89-7)**
Quinoline, 4-methyl-, hydrochloride (9CI)	**(3007-43-0)**
Quinoline, 8-((methylmercuri)oxy)-	**(86-85-1)**
Quinoline, 3-methyl-4-nitro-, 1-oxide	**(14073-00-8)**
Quinoline, 4-methyl-2-phenyl- (9CI)	**(4789-76-8)**
Quinoline, 6-methyl-2-phenyl- (8CI,9CI)	**(27356-46-3)**
Quinoline, 5-nitro	**(607-34-1)**
Quinoline, 6-nitro	**(613-50-3)**
Quinoline, 8-nitro	**(607-35-2)**
Quinoline, nitro	**(12408-11-6)**
Quinoline, 4-nitro-, 1-oxide	**(56-57-5)**
Quinoline, phenyl- (9CI)	**(72776-77-3)**
Quinoline, sulfate (1:1) (8CI,9CI)	**(530-66-5)**
Quinoline-8-sulfonic acid	**(85-48-3)**
8-Quinolinesulfonic acid (9CI)	**(85-48-3)**
Quinoline, 1,2,3,4-tetrahydro- (9CI)	**(635-46-1)**
Quinoline, 1,2,3,4-tetrahydro-4-methyl-	**(19343-78-3)**
Quinoline, 1,2,3,4-tetrahydro-6-methyl- (9CI)	**(91-61-2)**
Quinoline, trimethyl- (9CI)	**(51366-52-0)**
Quinoline, 2,3,4-trimethyl- (7CI,8CI,9CI)	**(2437-72-1)**
Quinolinic acid	**(89-00-9)**
Quinolinium, 1-benzyl-, chloride (8CI)	**(15619-48-4)**
Quinolinium, 1-((1,3-dihydro-1,3,3-trimethyl-2H-indol-2-ylidene)ethylidene)-1,2,3,4-tetrahydro-6-methoxy-, chloride (9CI)	**(27326-17-6)**
Quinolinium, 6-(dimethylamino)-2-(2-(2,5-dimethyl-1-phenyl-1H-pyrrol-3-yl)ethenyl)-1-methyl-	**(7187-62-4)**
Quinolinium, 6-(dimethylamino)-2-(2-(2,5-dimethyl-1-phenyl-pyrrol-3-yl)vinyl)-1-methyl-	**(7187-62-4)**
Quinolinium, 1-(phenylmethyl)-, chloride (9CI)	**(15619-48-4)**
2-Quinolinol	**(59-31-4)**
8-Quinolinol	**(148-24-3)**
Quinolinol	**(1321-40-0)**
8-(Quinolinolato)methyl mercury	**(86-85-1)**
8-Quinolinol benzoate	**(7091-57-8)**
8-Quinolinol, 5-chloro	**(130-16-5)**
8-Quinolinol, 5-chloro-7-iodo	**(130-26-7)**
8-Quinolinol citrate	**(134-30-5)**
8-Quinolinol, copper (II) chelate	**(10380-28-6)**
8-Quinolinol, 5,7-dichloro-, Mixt. with 5-chloro-8-quinolinol and 7-chloro-8-quinolinol	**(8067-69-4)**
8-Quinolinol, 5,7-diiodo	**(83-73-8)**
8-Quinolinol, hydrogen sulfate (2:1)	**(134-31-6)**
8-Quinolinol, hydrogen sulfate (ester) (9CI)	**(2149-36-2)**
8-Quinolinol, 2-hydroxy-1,2,3-propanetricarboxylate (1:1) (Salt) (9CI)	**(134-30-5)**
8-Quinolinol, mercury complex	**(86-85-1)**
8-Quinolinol, 2-methyl	**(826-81-3)**
8-Quinolinol sulfate	**(134-31-6)**
8-Quinolinol, sulfate (2:1) (Salt)	**(134-31-6)**
8-Quinolinol, sulfate (Salt) (8CI,9CI) (VAN)	**(3819-18-9)**
2-Quinolinone	**(59-31-4)**
2(1H)-Quinolinone (9CI)	**(59-31-4)**
2(1H)-Quinolinone, 1-methyl- (9CI)	**(606-43-9)**
2(1H)-Quinolone	**(59-31-4)**
2-Quinolone	**(59-31-4)**
α-Quinolone	**(59-31-4)**
Quinolor	**(8067-69-4)**
Quinolor Compound	**(94-36-0)**
2-(2-Quinolyl)-1,3-indandione	**(83-08-9)**
2-(2-Quinolyl)-1,3-indandione disulfonic acid disodium salt	**(8004-92-0)**
2-(2-Quinolyl)-1,3-indanedione	**(83-08-9)**
Quinomethionate	**(2439-01-2)**
Quinondo	**(10380-28-6)**
o-Quinone	**(583-63-1)**
p-Quinone	**(106-51-4)**
Quinone (ACGIH,OSHA)	**(106-51-4)**
Quinone chlorimide	**(637-61-6)**
Quinone dioxime	**(105-11-3)**
p-Quinone dioxime	**(105-11-3)**
Quinone monooxime	**(637-62-7)**
p-Quinone monooxime	**(637-62-7)**
Quinone monoxime	**(104-91-6)**
p-Quinone monoxime	**(637-62-7)**
Quinone 4-oxime	**(637-62-7)**
Quinone oxime	**(104-91-6)**
para-Quinone oxime	**(105-11-3)**
Quinone oxime benzoylhydrazone	**(495-73-8)**
Quinophan	**(132-60-5)**
Quinophen	**(132-60-5)**
Quinophenol	**(148-24-3)**
Quinophthalone (VAN) (8CI)	**(83-08-9)**
Quinora	**(50-54-4)**
Quinoscan	**(54-05-7)**
Quinoseptyl	**(80-35-3)**
Quinothionate	**(93-75-4)**
Quinoxaline	**(91-19-0)**
Quinoxaline, 2,3-dichloro	**(2213-63-0)**
2,3-Quinoxalinedimethanol, diacetate, 1,4-dioxide	**(10103-89-6)**
2,3-Quinoxalinedimethanol, 1,4-dioxide	**(17311-31-8)**
2,3-Quinoxalinedithiol, cyclic trithiocarbonate	**(93-75-4)**
2,3-Quinoxalinedithiol, 6-methyl-, cyclic carbonate	**(2439-01-2)**
2,3-Quinoxalinedithiol, 6-methyl-, cyclic dithiocarbonate (ester)	**(2439-01-2)**
Quinoxaline-2,3-diyl trithiocarbonate	**(93-75-4)**
Quinoxaline, 2-methyl	**(7251-61-8)**
Quinoxaline, 5,6,7,8-tetrahydro- (9CI)	**(34413-35-9)**
N-(2-Quinoxalinyl)sulfanilamide	**(59-40-5)**
N'-2-Quinoxalylsulfanilamide	**(59-40-5)**
Quinoxidine	**(10103-89-6)**
Quinsorb 010	**(131-56-6)**
Quintar	**(117-80-6)**
Quintar 540F	**(117-80-6)**
Quintess-N	**(1405-10-3)**
Quintocene	**(82-68-8)**
Quintol	**(64742-16-1)**
Quintol B 1000	**(9003-17-2)**
Quintomycin C	**(7542-37-2)**
Quintone A 100	**(64742-16-1)**
Quintone C 200S	**(64742-16-1)**
Quintone D 100	**(64742-16-1)**
Quintone M 100	**(64742-16-1)**
Quintone N 180	**(64742-16-1)**
Quintone RX 05	**(64742-16-1)**
Quintone U 185	**(64742-16-1)**
Quintox	**(60-57-1)**
Quintox	**(82-68-8)**

Quintozen	(82-68-8)
Quintozene	(82-68-8)
Quintrate	(78-11-5)
2-Quinuclidinemethanol, α-(6-methoxy-4-quinolyl)-5-vinyl-	(130-95-0)
2-Quinuclidinemethanol, α-(6-methoxy-4-quinolyl)-5-vinyl-	(56-54-2)
2-Quinuclidinemethanol, α-4-quinolyl-5-vinyl-	(118-10-5)
2-Quinuclidinemethanol, α-4-quinolyl-5-vinyl-	(485-71-2)
2-Quinuclidinemethanol, α-(5-vinyl-2-quinolyl)-	(118-10-5)
3-Quinuclidinone (8CI)	(3731-38-2)
Quinurenic acid	(492-27-3)
Quivet	(57-53-4)
Quizalofop-Ethyl	(76578-14-8)
Quolac Ex-Ub	(151-21-3)
5R04	(12001-29-5)
88-R	(140-57-8)
R 1	(26637-46-7)
R 2	(1696-17-9)
R 3	(9003-53-6)
R 8 (Fungicide)	(502-39-6)
R 8	(502-39-6)
R 10	(56-23-5)
R 12 [UN 1028]	(75-71-8)
R 14	(75-73-0)
R 20 (Refrigerant)	(67-66-3)
R 20	(67-66-3)
R 22 (DOT)	(75-45-6)
R 23	(75-46-7)
R 30	(75-09-2)
R 31	(593-70-4)
R 31 (Refrigerant)	(593-70-4)
R 40	(74-87-3)
R 40B1	(74-83-9)
R 54	(488-41-5)
R 113 (Halocarbon)	(76-13-1)
R 113	(76-13-1)
R 114	(76-14-2)
R 114B2	(124-73-2)
R 124	(2837-89-0)
R 125	(354-33-6)
R 133a	(75-88-7)
R 134	(359-35-3)
R 142	(25497-29-4)
R 143a	(420-46-2)
R 161	(353-36-6)
R 400	(7542-37-2)
R 610	(545-59-5)
R 717	(7664-41-7)
R 875	(357-56-2)
R 898	(5666-11-5)
R 1007	(3101-60-8)
R 1406	(562-26-5)
R 1504	(732-11-6)
R 1513	(2642-71-9)
R 1582	(86-50-0)
R 1625	(52-86-8)
R 2063	(1134-23-2)
R 2170	(301-12-2)
R 3365	(302-41-0)
R 3612	(9003-53-6)
R 4263	(437-38-7)
R 4318	(23779-99-9)
R 4749	(548-73-2)
R 4845	(15301-48-1)
R 6700	(297-78-9)
R 9985	(59-05-2)
R 10688	(9003-20-7)
R 14827	(24169-02-6)
R 15,175	(78-57-9)
R 15,403	(28782-42-5)
R 17635	(31431-39-7)
R 17934	(31430-18-9)
R 23979	(35554-44-0)
R 30,730	(56030-54-7)
R 40244	(61213-25-0)
R 42211	(23505-41-1)
R-11	(126-15-8)
R-13 [UN 1022]	(75-72-9)
R-47	(817-09-4)
R-104	(9003-55-8)
R-152a	(25497-28-3)
R-242	(80-00-2)
R-246	(51-18-3)
R-326	(136-45-8)
R-874	(3547-33-9)
R-1303	(786-19-6)
R-1492	(953-17-3)
R-1607	(1929-77-7)
R-1608	(759-94-4)
R-1910	(2008-41-5)
R-2061	(1114-71-2)
R-4461	(741-58-2)
R-4572	(2212-67-1)
R-4845	(15301-48-1)
R-5,158	(78-53-5)
R-7465	(15299-99-7)
R-7475	(15299-99-7)
R-13423	(3116-76-5)
R-25788	(37764-25-3)
R-28,644	(60207-31-0)
R-33799	(59708-52-0)
R48	(494-03-1)
R50	(50-29-3)
R968	(9003-01-4)
R Salt	(135-51-3)
13-RA	(4759-48-2)
13-cis-RA	(4759-48-2)
RA 8	(58-32-2)
β-RA	(302-79-4)
R/AMP	(13292-46-1)
RAV 7	(142-22-3)
R-242-B	(80-00-2)
RB	(56-38-2)
RB 1509	(13010-47-4)
RBA 777	(88-04-0)
R-C 318	(115-25-3)
RC 102	(9003-11-6)
RC 146	(3688-66-2)
RC 5629	(112-18-5)
RC-173	(1317-25-5)
RC Comonomer DBM	(105-76-0)
RC Comonomer DOF	(141-02-6)
RC Comonomer Diom	(1330-76-3)
RC Comonomer Dom	(142-16-5)
RC 172DBM	(1344-28-1)
RCH 1000	(9002-88-4)
RC Plasticizer B-17	(123-95-5)
RC Plasticizer DOP	(117-81-7)
RCRA waste number P001	(81-81-2)
RCRA waste number P002	(591-08-2)
RCRA waste number P003	(107-02-8)
RCRA waste number P004	(309-00-2)
RCRA waste number P005	(107-18-6)
RCRA waste number P006	(20859-73-8)
RCRA waste number P007	(2763-96-4)

RCRA waste number P008 (121-21-1)
RCRA waste number P008 (504-24-5)
RCRA waste number P009 (131-74-8)
RCRA waste number P010 (7778-39-4)
RCRA waste number P011 (1303-28-2)
RCRA waste number P012 (1327-53-3)
RCRA waste number P013 (542-62-1)
RCRA waste number P014 (108-98-5)
RCRA waste number P015 (7440-41-7)
RCRA waste number P016 (542-88-1)
RCRA waste number P017 (598-31-2)
RCRA waste number P018 (357-57-3)
RCRA waste number P020 (88-85-7)
RCRA waste number P021 (592-01-8)
RCRA waste number P022 (75-15-0)
RCRA waste number P023 (107-20-0)
RCRA waste number P024 (106-47-8)
RCRA waste number P026 (5344-82-1)
RCRA waste number P027 (542-76-7)
RCRA waste number P028 (100-44-7)
RCRA waste number P029 (544-92-3)
RCRA waste number P030 (57-12-5)
RCRA waste number P031 (460-19-5)
RCRA waste number P033 (506-77-4)
RCRA waste number P034 (131-89-5)
RCRA waste number P036 (696-28-6)
RCRA waste number P037 (60-57-1)
RCRA waste number P038 (692-42-2)
RCRA waste number P039 (298-04-4)
RCRA waste number P040 (297-97-2)
RCRA waste number P040 (5826-91-5)
RCRA waste number P041 (311-45-5)
RCRA waste number P042 (51-43-4)
RCRA waste number P043 (55-91-4)
RCRA waste number P044 (60-51-5)
RCRA waste number P045 (39196-18-4)
RCRA waste number P046 (122-09-8)
RCRA waste number P047 (534-52-1)
RCRA waste number P048 (51-28-5)
RCRA waste number P049 (541-53-7)
RCRA waste number P050 (115-29-7)
RCRA waste number P051 (72-20-8)
RCRA waste number P054 (151-56-4)
RCRA waste number P056 (7782-41-4)
RCRA waste number P057 (640-19-7)
RCRA waste number P058 (62-74-8)
RCRA waste number P059 (76-44-8)
RCRA waste number P060 (465-73-6)
RCRA waste number P062 (757-58-4)
RCRA waste number P063 (74-90-8)
RCRA waste number P064 (624-83-9)
RCRA waste number P065 (628-86-4)
RCRA waste number P066 (16752-77-5)
RCRA waste number P067 (75-55-8)
RCRA waste number P068 (60-34-4)
RCRA waste number P069 (75-86-5)
RCRA waste number P070 (116-06-3)
RCRA waste number P071 (298-00-0)
RCRA waste number P072 (86-88-4)
RCRA waste number P073 (13463-39-3)
RCRA waste number P074 (557-19-7)
RCRA waste number P075 (54-11-5)
RCRA waste number P076 (10102-43-9)
RCRA waste number P077 (100-01-6)
RCRA waste number P078 (10102-44-0)
RCRA waste number P081 (55-63-0)
RCRA waste number P082 (62-75-9)
RCRA waste number P084 (4549-40-0)
RCRA waste number P085 (152-16-9)
RCRA waste number P087 (20816-12-0)
RCRA waste number P088 (129-67-9)
RCRA waste number P088 (145-73-3)
RCRA waste number P089 (56-38-2)
RCRA waste number P092 (62-38-4)
RCRA waste number P093 (103-85-5)
RCRA waste number P094 (298-02-2)
RCRA waste number P095 (75-44-5)
RCRA waste number P096 (7803-51-2)
RCRA waste number P097 (52-85-7)
RCRA waste number P098 (151-50-8)
RCRA waste number P099 (506-61-6)
RCRA waste number P101 (107-12-0)
RCRA waste number P102 (107-19-7)
RCRA waste number P103 (630-10-4)
RCRA waste number P104 (506-64-9)
RCRA waste number P105 (26628-22-8)
RCRA waste number P106 (143-33-9)
RCRA waste number P107 (1314-96-1)
RCRA waste number P108 (57-24-9)
RCRA waste number P109 (3689-24-5)
RCRA waste number P110 (78-00-2)
RCRA waste number P111 (107-49-3)
RCRA waste number P112 (509-14-8)
RCRA waste number P113 (1314-32-5)
RCRA waste number P114 (12039-52-0)
RCRA waste number P115 (10031-59-1)
RCRA waste number P115 (7446-18-6)
RCRA waste number P116 (79-19-6)
RCRA waste number P118 (594-42-3)
RCRA waste number P119 (7803-55-6)
RCRA waste number P120 (1314-62-1)
RCRA waste number P121 (557-21-1)
RCRA waste number P122 (1314-84-7)
RCRA waste number P123 (8001-35-2)
RCRA waste number U001 (75-07-0)
RCRA waste number U002 (67-64-1)
RCRA waste number U003 (75-05-8)
RCRA waste number U004 (98-86-2)
RCRA waste number U005 (53-96-3)
RCRA waste number U006 (75-36-5)
RCRA waste number U007 (79-06-1)
RCRA waste number U008 (79-10-7)
RCRA waste number U009 (107-13-1)
RCRA waste number U010 (50-07-7)
RCRA waste number U011 (61-82-5)
RCRA waste number U012 (62-53-3)
RCRA waste number U014 (492-80-8)
RCRA waste number U015 (115-02-6)
RCRA waste number U016 (225-51-4)
RCRA waste number U017 (98-87-3)
RCRA waste number U018 (56-55-3)
RCRA waste number U019 (71-43-2)
RCRA waste number U020 (98-09-9)
RCRA waste number U021 (92-87-5)
RCRA waste number U022 (50-32-8)
RCRA waste number U023 (98-07-7)
RCRA waste number U024 (111-91-1)
RCRA waste number U025 (111-44-4)
RCRA waste number U026 (494-03-1)
RCRA waste number U027 (108-60-1)
RCRA waste number U028 (117-81-7)
RCRA waste number U029 (74-83-9)
RCRA waste number U030 (101-55-3)
RCRA waste number U031 (71-36-3)

RCRA waste number U032 **(13765-19-0)**
RCRA waste number U033 **(353-50-4)**
RCRA waste number U034 **(75-87-6)**
RCRA waste number U035 **(305-03-3)**
RCRA waste number U036 **(12789-03-6)**
RCRA waste number U036 **(57-74-9)**
RCRA waste number U038 **(510-15-6)**
RCRA waste number U039 **(59-50-7)**
RCRA waste number U041 **(106-89-8)**
RCRA waste number U042 **(110-75-8)**
RCRA waste number U043 **(75-01-4)**
RCRA waste number U044 **(67-66-3)**
RCRA waste number U045 **(74-87-3)**
RCRA waste number U046 **(107-30-2)**
RCRA waste number U047 **(91-58-7)**
RCRA waste number U048 **(95-57-8)**
RCRA waste number U049 **(3165-93-3)**
RCRA waste number U050 **(218-01-9)**
RCRA waste number U051 **(8001-58-9)**
RCRA waste number U051 **(8021-39-4)**
RCRA waste number U052 **(106-44-5)**
RCRA waste number U052 **(108-39-4)**
RCRA waste number U052 **(1319-77-3)**
RCRA waste number U052 **(95-48-7)**
RCRA waste number U053 **(114-80-7)**
RCRA waste number U053 **(123-73-9)**
RCRA waste number U053 **(4170-30-3)**
RCRA waste number U055 **(98-82-8)**
RCRA waste number U056 **(110-82-7)**
RCRA waste number U057 **(108-94-1)**
RCRA waste number U058 **(50-18-0)**
RCRA waste number U059 **(20830-81-3)**
RCRA waste number U060 **(72-54-8)**
RCRA waste number U061 **(50-29-3)**
RCRA waste number U062 **(2303-16-4)**
RCRA waste number U063 **(53-70-3)**
RCRA waste number U064 **(189-55-9)**
RCRA waste number U066 **(96-12-8)**
RCRA waste number U067 **(106-93-4)**
RCRA waste number U068 **(74-95-3)**
RCRA waste number U069 **(84-74-2)**
RCRA waste number U070 **(106-46-7)**
RCRA waste number U071 **(106-46-7)**
RCRA waste number U071 **(541-73-1)**
RCRA waste number U072 **(106-46-7)**
RCRA waste number U073 **(91-94-1)**
RCRA waste number U074 **(764-41-0)**
RCRA waste number U075 **(75-71-8)**
RCRA waste number U076 **(75-34-3)**
RCRA waste number U077 **(107-06-2)**
RCRA waste number U078 **(75-35-4)**
RCRA waste number U079 **(156-60-5)**
RCRA waste number U080 **(75-09-2)**
RCRA waste number U081 **(120-83-2)**
RCRA waste number U082 **(87-65-0)**
RCRA waste number U083 **(563-54-2)**
RCRA waste number U083 **(78-87-5)**
RCRA waste number U084 **(542-75-6)**
RCRA waste number U085 **(1464-53-5)**
RCRA waste number U086 **(1615-80-1)**
RCRA waste number U087 **(3288-58-2)**
RCRA waste number U088 **(84-66-2)**
RCRA waste number U089 **(56-53-1)**
RCRA waste number U090 **(94-58-6)**
RCRA waste number U091 **(119-90-4)**
RCRA waste number U092 **(124-40-3)**
RCRA waste number U093 **(60-11-7)**
RCRA waste number U094 **(57-97-6)**
RCRA waste number U095 **(119-93-7)**
RCRA waste number U096 **(80-15-9)**
RCRA waste number U097 **(79-44-7)**
RCRA waste number U098 **(57-14-7)**
RCRA waste number U099 **(540-73-8)**
RCRA waste number U101 **(105-67-9)**
RCRA waste number U102 **(131-11-3)**
RCRA waste number U103 **(77-78-1)**
RCRA waste number U105 **(121-14-2)**
RCRA waste number U106 **(606-20-2)**
RCRA waste number U107 **(117-84-0)**
RCRA waste number U108 **(123-91-1)**
RCRA waste number U109 **(122-66-7)**
RCRA waste number U110 **(142-84-7)**
RCRA waste number U111 **(621-64-7)**
RCRA waste number U112 **(141-78-6)**
RCRA waste number U113 **(140-88-5)**
RCRA waste number U114 **(111-54-6)**
RCRA waste number U115 **(75-21-8)**
RCRA waste number U116 **(96-45-7)**
RCRA waste number U117 **(60-29-7)**
RCRA waste number U118 **(97-63-2)**
RCRA waste number U119 **(62-50-0)**
RCRA waste number U120 **(206-44-0)**
RCRA waste number U122 **(50-00-0)**
RCRA waste number U123 **(64-18-6)**
RCRA waste number U124 **(110-00-9)**
RCRA waste number U125 **(98-01-1)**
RCRA waste number U126 **(765-34-4)**
RCRA waste number U127 **(118-74-1)**
RCRA waste number U128 **(87-68-3)**
RCRA waste number U129 **(58-89-9)**
RCRA waste number U130 **(77-47-4)**
RCRA waste number U131 **(67-72-1)**
RCRA waste number U132 **(70-30-4)**
RCRA waste number U133 **(302-01-2)**
RCRA waste number U134 **(7664-39-3)**
RCRA waste number U135 **(7783-06-4)**
RCRA waste number U136 **(75-60-5)**
RCRA waste number U137 **(193-39-5)**
RCRA waste number U138 **(74-88-4)**
RCRA waste number U139 **(9004-66-4)**
RCRA waste number U140 **(78-83-1)**
RCRA waste number U141 **(120-58-1)**
RCRA waste number U142 **(143-50-0)**
RCRA waste number U143 **(303-34-4)**
RCRA waste number U144 **(301-04-2)**
RCRA waste number U145 **(7446-27-7)**
RCRA waste number U146 **(1335-32-6)**
RCRA waste number U147 **(108-31-6)**
RCRA waste number U148 **(123-33-1)**
RCRA waste number U149 **(109-77-3)**
RCRA waste number U150 **(148-82-3)**
RCRA waste number U151 **(7439-97-6)**
RCRA waste number U152 **(126-98-7)**
RCRA waste number U153 **(74-93-1)**
RCRA waste number U154 **(67-56-1)**
RCRA waste number U155 **(91-80-5)**
RCRA waste number U156 **(79-22-1)**
RCRA waste number U157 **(56-49-5)**
RCRA waste number U158 **(101-14-4)**
RCRA waste number U159 **(78-93-3)**
RCRA waste number U160 **(1338-23-4)**
RCRA waste number U161 **(108-10-1)**
RCRA waste number U162 **(80-62-6)**
RCRA waste number U163 **(70-25-7)**

RCRA waste number U164 (56-04-2)
RCRA waste number U165 (91-20-3)
RCRA waste number U166 (130-15-4)
RCRA waste number U167 (134-32-7)
RCRA waste number U168 (91-59-8)
RCRA waste number U169 (98-95-3)
RCRA waste number U170 (100-02-7)
RCRA waste number U171 (79-46-9)
RCRA waste number U172 (924-16-3)
RCRA waste number U173 (1116-54-7)
RCRA waste number U174 (55-18-5)
RCRA waste number U176 (759-73-9)
RCRA waste number U177 (684-93-5)
RCRA waste number U178 (615-53-2)
RCRA waste number U179 (100-75-4)
RCRA waste number U180 (930-55-2)
RCRA waste number U181 (99-55-8)
RCRA waste number U182 (123-63-7)
RCRA waste number U183 (608-93-5)
RCRA waste number U184 (76-01-7)
RCRA waste number U185 (82-68-8)
RCRA waste number U186 (504-60-9)
RCRA waste number U187 (62-44-2)
RCRA waste number U188 (108-95-2)
RCRA waste number U189 (1314-80-3)
RCRA waste number U190 (85-44-9)
RCRA waste number U191 (109-06-8)
RCRA waste number U192 (23950-58-5)
RCRA waste number U193 (1120-71-4)
RCRA waste number U194 (107-10-8)
RCRA waste number U196 (110-86-1)
RCRA waste number U197 (106-51-4)
RCRA waste number U200 (50-55-5)
RCRA waste number U201 (108-46-3)
RCRA waste number U202 (81-07-2)
RCRA waste number U203 (94-59-7)
RCRA waste number U204 (7446-08-4)
RCRA waste number U204 (7783-00-8)
RCRA waste number U205 (7446-34-6)
RCRA waste number U205 (7488-56-4)
RCRA waste number U206 (18883-66-4)
RCRA waste number U207 (95-94-3)
RCRA waste number U208 (630-20-6)
RCRA waste number U209 (79-34-5)
RCRA waste number U210 (127-18-4)
RCRA waste number U211 (56-23-5)
RCRA waste number U212 (58-90-2)
RCRA waste number U213 (109-99-9)
RCRA waste number U214 (563-68-8)
RCRA waste number U215 (6533-73-9)
RCRA waste number U216 (7791-12-0)
RCRA waste number U217 (10102-45-1)
RCRA waste number U218 (62-55-5)
RCRA waste number U219 (62-56-6)
RCRA waste number U220 (108-88-3)
RCRA waste number U221 (25376-45-8)
RCRA waste number U221 (95-80-7)
RCRA waste number U222 (636-21-5)
RCRA waste number U223 (26471-62-5)
RCRA waste number U223 (584-84-9)
RCRA waste number U225 (75-25-2)
RCRA waste number U226 (71-55-6)
RCRA waste number U227 (110-80-5)
RCRA waste number U227 (79-00-5)
RCRA waste number U228 (79-01-6)
RCRA waste number U230 (95-95-4)
RCRA waste number U231 (88-06-2)
RCRA waste number U232 (93-76-5)
RCRA waste number U233 (93-72-1)
RCRA waste number U234 (99-35-4)
RCRA waste number U235 (126-72-7)
RCRA waste number U236 (72-57-1)
RCRA waste number U237 (66-75-1)
RCRA waste number U238 (51-79-6)
RCRA waste number U239 (1330-20-7)
RCRA waste number U240 (94-75-7)
RCRA waste number U242 (87-86-5)
RCRA waste number U243 (1888-71-7)
RCRA waste number U244 (137-26-8)
RCRA waste number U246 (506-68-3)
RCRA waste number U247 (72-43-5)
RCRA waste number U328 (95-53-4)
RCRA waste number U353 (106-49-0)
RCRA waste number U359 (110-80-5)
RCRA waste number U359 (79-00-5)
R.D. 1403 (15676-16-1)
R.D. 2786 (91-84-9)
R.D. 13621 (15687-27-1)
R.D. 27419 (33089-61-1)
RD 174 (26544-38-7)
RD 406 (120-36-5)
RD 1572 (1239-45-8)
RD 2195 (103-17-3)
RD 2454 (405-30-1)
RD 4593 (7085-19-0)
RD 4593 (93-65-2)
RD 7693 (3813-05-6)
RD 14639 (2655-19-8)
RD-6584 (99-30-9)
RDGE (101-90-6)
RDX (121-82-4)
RDX 58456 (9CI) (87397-71-5)
RE 9659 (2282-34-0)
RE 12420 (30560-19-1)
RE-4355 (300-76-5)
RE-5353 (2282-34-0)
RE-45550 (99422-01-2)
REC 7/0267 (1165-48-6)
11411 Red (81-88-9)
REP (1696-17-9)
R-E-S (50-55-5)
RF 10 (9004-70-0)
RFNA (7697-37-2)
RG 600 (12001-29-5)
RH 315 (23950-58-5)
RH 886 (50815-77-5)
RH 893 (26530-20-1)
RH 0994 (59010-86-5)
RH 6201 (62476-59-9)
RH-124 (16227-10-4)
RH-787 (53558-25-1)
RH-2915 (42874-03-3)
RIC 272 (72-44-6)
RIP-15830 (3861-47-0)
RISA-125 (9048-46-8)
RISA-131 (9048-46-8)
RJ 5 (66289-74-5)
RL-50 (142-47-2)
RMC (7439-95-4)
RMD 4500 (9003-54-7)
ROE 101 (125-33-7)
ROP 500 F (36734-19-7)
3.697 R.P. (65-29-2)
13.057 R.P. (20830-81-3)

143 RP (9003-39-8)
866 R.P. (69-05-6)
2339 RP (961-71-7)
2512 R.P. (140-64-70)
2750 R.P. (91-81-6)
3277 R.P. (58-33-3)
3277 RP (60-87-7)
3354 R.P. (58-34-4)
3389 R.P. (60-87-7)
4182 R.P. (60-87-7)
4560 R.P. (50-53-3)
4753 R.P. (58-14-0)
5015 RP (54-85-3)
6140 RP (58-38-8)
7162 RP (739-71-9)
7175 RP (2778-04-3)
8595 R.P. (551-92-8)
10257 R.P. (68-76-8)
11,561 RP (16118-49-3)
19583 RP (22071-15-4)
32545 RP (39148-24-8)
RP 2090 (72-14-0)
RP 2145 (144-82-1)
RP 2275 (57-67-0)
R.P. 2512 (140-64-70)
RP 2616 (68-35-9)
RP 2786 (91-84-9)
RP 3276 (58-40-2)
RP 3377 (54-05-7)
RP 3554 (58-34-4)
RP 3697 (65-29-2)
RP 4560 (53-60-1)
RP 7522 (80-35-3)
RP 7623 (117-89-5)
RP 8167 (563-12-2)
RP 8532 (696-23-1)
RP 8823 (443-48-1)
RP-9895 (2275-23-2)
RP 10192 (127-33-3)
R.P. 10,465 (2275-23-2)
RP 11,974 (2310-17-0)
RP 13057 (20830-81-3)
RP-16272 (1689-99-2)
RP 17623 (19666-30-9)
R-25788 Plus EPTC (51990-04-6)
RP 26019 (36734-19-7)
RPA 2 (91-60-1)
RPA No. 2 (91-60-1)
RPH (148-79-8)
2339 R.P. Hydrochloride (2045-52-5)
4560 RP Hydrochloride (69-09-0)
2786 R.P. Maleate (59-33-6)
4R Purple (4548-53-2)
1R,cis-RU 15525 (58769-20-3)
RS (9004-70-0)
RS 141 (6164-98-3)
RS 1280 (302-22-7)
RS-1401 AT (67-73-2)
RS-3540 (22204-53-1)
RS-4691 (5251-34-3)
RS-94991-298 (86220-42-0)
RSU 3071 (59177-47-8)
RTC (36791-04-5)
RTCA (36791-04-5)
RTEC (Polish) (7439-97-6)
RU-11484 (28434-01-7)
RU 11679 (22431-62-5)
RU 15525 (58769-20-3)
RU 15750 (23779-99-9)
RU 16121 (28434-00-6)
RU 22950 (52820-00-5)
RU 22974 (52918-63-5)
RU 25472 (66841-25-6)
RU 25474 (66841-25-6)
RU 27998 (52315-07-8)
RU 43501 (80845-12-1)
RU-4723 (22316-47-8)
RU-EF-TB (54-85-3)
RV-12165 (2894-67-9)
RV225-5B (9003-20-7)
RVK (79-40-3)
RVM-FG (12174-11-7)
RX 6029M (52485-79-7)
R-3-Zon (50-33-9)
Rabcide (27355-22-2)
Rabcon (1441-02-7)
Rabon (22248-79-9)
Rabond (22248-79-9)
Race-acetylmethadol (509-74-0)
Racemethadol (545-90-4)
Racemethorphan (510-53-2)
Racemethorphane (French) (510-53-2)
Racemethorphanum (297-90-5)
Racemethorphanum (Latin) (510-53-2)
Racemetorfano (Spanish) (510-53-2)
Racemic acid (133-37-9)
Racemic clomiphene citrate (50-41-9)
Racemic dromoran (297-90-5)
Racemic lactic acid (50-21-5)
Racemic lactic acid (598-82-3)
Racemic mandelic acid (90-64-2)
Racemic tartaric acid (133-37-9)
Racemoramida (Spanish) (545-59-5)
Racemoramide (545-59-5)
Racemoramidum (Latin) (545-59-5)
Racemorfano (Spanish) (297-90-5)
Racemorphan (297-90-5)
Racemorphane (French) (297-90-5)
Racemorphanum (Latin) (297-90-5)
Racephen (60-13-9)
Racer (61213-25-0)
Racryl (9003-01-4)
Racumin (5836-29-3)
Racusan (60-51-5)
Rad-E-Cate (124-65-2)
Rad-E-Cate 16 (124-65-2)
Rad-E-Cate 25 (124-65-2)
Rad-E-Cate 25 (75-60-5)
Rad-E-Cate 35 (124-65-2)
Radapon (127-20-8)
Radapon (75-99-0)
Radar (60207-90-1)
Radazin (1912-24-9)
Raddle (1309-37-1)
Radedorm (146-22-5)
Radepur (58-25-3)
Radia 7267 (627-83-8)
Radiasurf 7125 (1338-39-2)
Radiasurf 7145 (9005-66-7)
Radiasurf 7155 (1338-43-8)
Radiatox (63428-84-2)
Radical, Superoxide (11062-77-4)
Radicalisin (23537-16-8)
Radicinin (10088-95-6)

Radiocin (67-73-2)
Radiographol (126-31-8)
Radiol Germicidal Solution (124-03-8)
Radiostol (50-14-6)
Radium A (15422-74-9)
Radium B (15067-28-4)
Radium C (14733-03-0)
Radium C (15735-67-8)
Radium (9CI) (7440-14-4)
Radium D (14255-04-0)
Radium E (14331-79-4)
Radium Emanation (10043-92-2)
Radium F (13981-52-7)
Radium, Isotope of mass 226 (8CI,9CI) (13982-63-3)
Radium, Isotope of mass 228 (8CI,9CI) (15262-20-1)
Radizine (1912-24-9)
Radocon (122-34-9)
Radokor (122-34-9)
Radon (10043-92-2)
Radon and Compounds (10043-92-2)
Radonil (723-46-6)
Radonna (54-31-9)
Radosan (151-38-2)
Radox (93-71-0)
Radoxone TL (61-82-5)
Radsterin (50-14-6)
Rafex (534-52-1)
Rafex 35 (534-52-1)
D-Raffinose (512-69-6)
d-(+)-Raffinose (512-69-6)
Raffinose (8CI) (512-69-6)
Rafluor (7681-49-4)
Ragadan (34783-40-9)
Raisin Extract (68915-86-6)
Raisin Seed Oil (68915-86-6)
Rakuto Amaranth (915-67-3)
Rakuto Brilliant Scarlet 3R (2611-82-7)
Ralabol (26538-44-3)
Ralgro (26538-44-3)
Ralone (26538-44-3)
Ram-320 (2183-56-4)
Ram-326 (3176-03-2)
Ramapo (1328-53-6)
Ramik (82-66-6)
Ramizol (61-82-5)
Ramor (7440-28-0)
Rampart (298-02-2)
Ramrod (1918-16-7)
Ramrod 65 (1918-16-7)
Ramucide (3691-35-8)
Ranac (3691-35-8)
Ranachrome 4 (529-69-1)
Rancinamycin IV (139-85-5)
Randolectil (653-03-2)
Randonos (9002-61-3)
Randox (93-71-0)
Randox-T (8005-43-4)
Raney Alloy (7440-02-0)
Raney Copper (7440-50-8)
Raney Nickel (7440-02-0)
Rangasil (31793-07-4)
Rangoon Yellow (6358-85-6)
Ranidil (66357-35-5)
Ranitidine hydrochloride (66357-35-5)
Rankotex (7085-19-0)
Rankotex (93-65-2)
Rantox T (93-71-0)
Raoline (122-32-7)
Rapacodin (125-28-0)
Rape Oil (8002-13-9)
Rapeseed Oil (8002-13-9)
Raphatox (534-52-1)
Raphetamine (300-62-9)
Raphone (94-74-6)
Rapid (23103-98-2)
Rapisol (577-11-7)
Rare earth oxides (68188-83-0)
Rasayanchlor (23184-66-9)
Rasayansulfan (115-29-7)
Raschit (59-50-7)
Raschit K (59-50-7)
Rasen-Anicon (59-50-7)
Rasikal (7775-09-9)
Rasorite 65 (1330-43-4)
Raspberin (65-45-2)
Raspberry Red for Jellies (915-67-3)
Rastinon (64-77-7)
Rat & Mice Bait (81-81-2)
Rat-A-Way (117-52-2)
Rat-A-Way (81-81-2)
Rat-B-Gon (81-81-2)
Rat-Gard (81-81-2)
Rat-Kill (81-81-2)
Rat-Mix (81-81-2)
Rat-Nip (7723-14-0)
Rat-O-Cide #2 (81-81-2)
Rat-Ola (81-81-2)
Rat-Trol (81-81-2)
Rat-Tu (86-88-4)
Ratafin (117-52-2)
Ratak (56073-10-0)
Ratak Plus (56073-10-0)
Ratbane 1080 (62-74-8)
Raticate (991-42-4)
Ratilan (81-82-3)
Ratimus (28772-56-7)
Ratindan 1 (82-66-6)
Rato (538-23-8)
Ratomet (3691-35-8)
Ratorex (81-81-2)
Ratox (10031-59-1)
Ratox (81-81-2)
Ratoxin (81-81-2)
Ratron (81-81-2)
Ratron G (81-81-2)
Rats-No-More (81-81-2)
Ratsul Soluble (129-06-6)
Rattengiftkonserve (7446-18-6)
Rattex (91-64-5)
Rattrack (86-88-4)
Rattunal (81-81-2)
Raucap (50-55-5)
Raucumin 57 (5836-29-3)
Raudiford (50-55-5)
Raudixin (50-55-5)
Raudixoid (50-55-5)
Raugal (50-55-5)
Raulen (50-55-5)
Rauloycin (50-55-5)
Rauloydin (50-55-5)
Raumanon (54-85-3)
Raumorine (50-55-5)
Raunervil (50-55-5)
Raunorine (50-55-5)

Raunormin (Orzan)	(50-55-5)
Raunova	(50-55-5)
Raupasil	(50-55-5)
Raupoid	(50-55-5)
Raupyrol	(24815-24-5)
Raurescine	(24815-24-5)
Raurine	(50-55-5)
Rausan	(50-55-5)
Rau-sed	(50-55-5)
Rausedan	(50-55-5)
Rausedil	(50-55-5)
Rausedyl	(50-55-5)
Rauserpen-alk	(50-55-5)
Rauserpin	(50-55-5)
Rauserpin-alk	(50-55-5)
Rauserpine	(50-55-5)
Rauserpol	(50-55-5)
Rausingle	(50-55-5)
Rautrin	(50-55-5)
Rauvilid	(50-55-5)
Rauvlid	(50-55-5)
Rauwasedin	(50-55-5)
Rauwilid	(50-55-5)
Rauwiloid	(50-55-5)
Rauwiloid+	(50-55-5)
Rauwipur	(50-55-5)
Rauwoleaf	(50-55-5)
25Rauwopur'byk'	(50-55-5)
Ravage	(55511-98-3)
Ravap	(62-73-7)
Raven	(1333-86-4)
Raven 30	(1333-86-4)
Raven 420	(1333-86-4)
Raven 500	(1333-86-4)
Raven 8000	(1333-86-4)
Raviac	(3691-35-8)
Raviflex 43	(9003-20-7)
Ravyon	(63-25-2)
Raw Shale Oil	(68308-34-9)
Raw, Straight run gasoline	(68606-11-1)
Rawilid	(50-55-5)
Rax	(81-81-2)
Ray-Gluciron	(299-29-6)
Raybar	(7727-43-7)
Raylig 260LR	(8061-51-6)
Rayon Black G	(6428-31-5)
Rayon Black GSN	(6428-31-5)
Rayon Black M	(6428-31-5)
Rayon Fast Black B	(6428-31-5)
Rayona	(63428-84-2)
Rayophane	(9004-34-6)
Rayox	(13463-67-7)
Raythesin	(94-12-2)
Rayweb Q	(9004-34-6)
Razide	(54-85-3)
Raziosulfa	(526-08-9)
Razol	(3691-35-8)
Razol Dock Killer	(94-74-6)
Razoxin	(21416-87-5)
Reacid	(140-87-4)
Reaction product of bis(p-1,1,3,3-tetramethylbutyl-phenol)-2,2'-sulfide with nickel salts	(27574-34-1)
Reactivan	(1209-98-9)
Reactive Black 8	(12225-26-2)
Reactive Blue 19	(2580-78-1)
Reactive Brilliant Red 5SKH	(17804-49-8)
Reactive Scarlet 2SKh	(12238-07-2)
Readex	(93-75-4)
Readpret KPN	(1854-26-8)
Reagens Ehrlichovo (Czech)	(100-10-7)
Reanimil	(1165-48-6)
Reax 45A	(8061-51-6)
Reax 80C	(8061-51-6)
Reax 85A	(8061-51-6)
Reazid	(140-87-4)
Reazide	(140-87-4)
Rebelate	(60-51-5)
Rebemid	(1696-17-9)
Rebonex	(1333-86-4)
Rebramin	(68-19-9)
Rebugen	(15687-27-1)
Recinnamine	(24815-24-5)
Recipin	(50-55-5)
Recitensina	(24815-24-5)
Recoil	(77732-09-3)
Recolip	(637-07-0)
Recolite Fast Red RBL	(2425-85-6)
Recolite Fast Red RL	(2425-85-6)
Recolite Fast Red RYL	(2425-85-6)
Recolite Orange G	(3520-72-7)
Recolite Red LYS	(1248-18-6)
Recolite Red Lake C	(5160-02-1)
Recolite Red Lake CR	(5160-02-1)
Recolite Royal Blue BDS	(1325-87-7)
Recolite Royal Blue BTS	(1325-87-7)
Recolite Violet RDS	(1325-82-2)
Recolite Violet RTS	(1325-82-2)
Recolite Yellow BG	(6358-85-6)
Recolite Yellow BGT	(6358-85-6)
Reconox	(92-84-2)
Recop	(1332-40-7)
Recop	(1332-65-6)
Recordati	(51-63-8)
Rectalad-Aminophylline	(317-34-0)
Recthormone oestradiol	(50-50-0)
Recthormone testosterone	(57-85-2)
Rectodelt	(53-03-2)
Rectofasa	(50-33-9)
Rectoid	(50-23-7)
Rectules	(302-17-0)
Recycle catalytic cracked slurry oil	(64741-62-4)
1302 Red	(915-67-3)
1306 Red	(4548-53-2)
1379 Red	(3734-67-6)
1424 Red	(3567-66-6)
1427 Red	(16423-68-0)
1508 Red	(915-67-3)
1578 Red	(2611-82-7)
1671 Red	(16423-68-0)
1695 Red	(3761-53-3)
1793 Red	(1103-39-5)
1860 Red	(5160-02-1)
1869 Red	(2302-96-7)
1883 Red	(1103-38-4)
11070 Red	(5858-81-1)
11391 Red	(1658-56-6)
11427 Red	(3567-66-6)
11445 Red	(17372-87-1)
11484 Red	(2379-74-0)
11554 Red	(1309-37-1)
11731 Red	(17372-87-1)
11935 Red	(1248-18-6)
11959 Red	(3567-69-9)
12094 Red	(2814-77-9)

12101 Red	**(4548-53-2)**
12418 Red	**(632-99-5)**
111440 Red	**(85-86-9)**
Red #14	**(3567-69-9)**
Red 1860	**(5160-02-1)**
Red 11938	**(5160-02-1)**
Red 10B	**(3567-66-6)**
Red B	**(3118-97-6)**
Red 3B Acid	**(94-76-8)**
Red 4B Acid	**(88-44-8)**
Red B Base	**(97-52-9)**
Red Base Ciba IX	**(3165-93-3)**
Red Base Ciba IX	**(95-69-2)**
Red Base Ciba V	**(97-52-9)**
Red Base Ciba VI	**(89-63-4)**
Red Base Ciba VII	**(89-62-3)**
Red Base Ciba X	**(99-52-5)**
Red Base 3 GL	**(89-63-4)**
Red Base IRGA IX	**(3165-93-3)**
Red Base Irga IX	**(95-69-2)**
Red Base Irga V	**(97-52-9)**
Red Base Irga VI	**(89-63-4)**
Red Base Irga VII	**(89-62-3)**
Red Base Irga X	**(99-52-5)**
Red Base NB	**(97-52-9)**
Red Base NGL	**(89-62-3)**
Red Base NRL	**(99-52-5)**
Red Base NTR	**(95-69-2)**
Red Cedarwood Oil	**(8000-27-9)**
Red Copper Oxide	**(1317-39-1)**
Red Dye No. 2	**(915-67-3)**
Red E for Food	**(2302-96-7)**
Red Extract R	**(2814-77-9)**
Red Fuming Nitric Acid	**(7697-37-2)**
Red 2G	**(3734-67-6)**
Red 2G Base	**(100-01-6)**
Red 3G Base	**(89-63-4)**
Red G Base	**(89-62-3)**
Red 3GS Salt	**(89-63-4)**
Red 3G Salt	**(89-63-4)**
Red G Salt	**(89-62-3)**
Red GTL	**(14097-03-1)**
Red Gum	**(8000-48-4)**
Red 16913H	**(5160-02-1)**
Red Iron Ore	**(1317-60-8)**
Red Iron Oxide	**(1309-37-1)**
Red J	**(1658-56-6)**
Red KB Base	**(95-79-4)**
Red Lake R-91	**(5160-02-1)**
Red Lake CR-1	**(5160-02-1)**
Red Lake CM 20-5650	**(5160-02-1)**
Red Lake C Toner	**(5160-02-1)**
Red Lake C Toner 20-5650	**(5160-02-1)**
Red Lake C Toner RA-5190	**(5160-02-1)**
Red Lake Camine	**(88-53-9)**
Red Lake 89865N	**(4548-53-2)**
Red Lead	**(1314-41-6)**
Red Lead Chromate	**(18454-12-1)**
Red Lead Oxide	**(1314-41-6)**
Red Mercuric Iodide	**(7774-29-0)**
Red No. 1	**(4548-53-2)**
Red No. 2	**(915-67-3)**
Red No. 4	**(4548-53-2)**
Red No. 5	**(3118-97-6)**
Red No. 40	**(25956-17-6)**
Red No. 202 (Japan)	**(5281-04-9)**
Red No. 205	**(1248-18-6)**
Red No 206	**(1103-39-5)**
Red No 207	**(1103-38-4)**
Red No. 213	**(81-88-9)**
Red No. 220 (Japan)	**(6417-83-0)**
Red No. 227	**(3567-66-6)**
Red No. 228	**(2814-77-9)**
Red No. 506	**(1658-56-6)**
Red Ochre	**(1309-37-1)**
Red Oil	**(112-80-1)**
Red Oxide	**(1309-37-1)**
Red Oxide D3452	**(1309-37-1)**
Red Oxide D6984	**(1309-37-1)**
Red Oxide of Iron	**(1309-37-1)**
Red Oxide of Mercury	**(21908-53-2)**
Red Precipitate	**(21908-53-2)**
Red R	**(3761-53-3)**
Red RL Base	**(99-52-5)**
Red 3R soluble in grease	**(85-83-6)**
Red Salt Ciba IX	**(3165-93-3)**
Red Salt Ciba VI	**(89-63-4)**
Red Salt Ciba VII	**(89-62-3)**
Red Salt IRGA IX	**(3165-93-3)**
Red Salt Irga VI	**(89-63-4)**
Red Salt Irga VII	**(89-62-3)**
Red Salt NBGL	**(89-63-4)**
Red Scarlet	**(5160-02-1)**
Red-Seal-9	**(1314-13-2)**
Red Squill	**(507-60-8)**
Red TR Base	**(95-69-2)**
Red TRS Salt	**(3165-93-3)**
Red Toner EBA	**(1103-39-5)**
Red Toner YTA	**(1103-38-4)**
Red Toner Z	**(5160-02-1)**
Red ZH	**(85-86-9)**
Redamina	**(68-19-9)**
Redax	**(86-30-6)**
Reddon	**(93-76-5)**
Reddox	**(93-76-5)**
Red for Lake C	**(5160-02-1)**
Red for Lake C Toner RA-5190	**(5160-02-1)**
Red for Lake Toner RA-5190	**(5160-02-1)**
Red for Lakes J	**(3761-53-3)**
Redi-Flow	**(7727-43-7)**
Redisol	**(68-19-9)**
αRedisol	**(13422-51-0)**
Redoxon	**(50-81-7)**
Red phosphorus	**(7723-14-0)**
Redskin	**(57-06-7)**
Reduced D-penicillamine	**(52-67-5)**
Reduced Michler's Ketone	**(101-61-1)**
Reduced glutathione	**(70-18-8)**
Reduced penicillamine	**(52-67-5)**
Reductone	**(7775-14-6)**
Reducymol	**(12771-68-5)**
Reduton	**(93-14-1)**
Reed Amine 400	**(2008-39-1)**
Reed LV 2,4-D	**(25168-26-7)**
Reed LV 400 2,4-D	**(25168-26-7)**
Reed LV 600 2,4-D	**(25168-26-7)**
Refchole	**(5989-27-5)**
Refined Solvent Naphtha	**(8032-32-4)**
Refinery gas	**(68607-11-4)**
Reflex Blue AGL	**(1324-76-1)**
Reflex Blue AGM	**(1324-76-1)**
Reflex Blue R	**(1324-76-1)**
Reflexyn	**(532-03-6)**
Refobacin	**(1405-41-0)**

Name	CAS
Reformin	(59-26-7)
Reformin	(62-44-2)
Refrigerant 112	(76-12-0)
Refrigerant 112a	(76-11-9)
Refrigerant 113	(76-13-1)
Refusal	(97-77-8)
Regal	(1333-86-4)
Regal 99	(1333-86-4)
Regal 300	(1333-86-4)
Regal 330	(1333-86-4)
Regal 400R	(1333-86-4)
Regal 600	(1333-86-4)
Regal SRF	(1333-86-4)
Regardin	(637-07-0)
Regelan	(637-07-0)
Regelan N	(637-07-0)
Regenon	(90-84-6)
Regent	(1333-86-4)
Regianin	(481-39-0)
Regim 8	(88-82-4)
Regin 8	(88-82-4)
Reginon	(90-84-6)
Region	(144-80-9)
Regitin	(50-60-2)
Regitine	(50-60-2)
Regitipe	(50-60-2)
Reglon	(85-00-7)
Reglone	(85-00-7)
Reglox	(85-00-7)
Regonal	(101-26-8)
Regonol	(9000-30-0)
Regroton	(50-55-5)
Regulaid	(9003-11-6)
Regulox	(123-33-1)
Regulox 50 W	(123-33-1)
Regulox W	(123-33-1)
Regutol	(577-11-7)
Rehmannic acid	(467-81-2)
Rehormin	(59-26-7)
Reichstein's F	(477-30-5)
Reichstein's Substance FA	(53-06-5)
Reichstein's Substance H	(50-22-6)
Reichstein's Substance M	(50-23-7)
Rein Guarin	(9000-30-0)
Reise-Engletten	(523-87-5)
Rekawan	(7447-40-7)
Rela	(78-44-4)
Relact	(146-22-5)
Relaminal	(439-14-5)
Relan β	(73-48-3)
Relanium	(439-14-5)
Relasom	(78-44-4)
Relax	(439-14-5)
Relax	(532-03-6)
Relax	(78-44-4)
Relaxan	(65-29-2)
Relaxin	(9002-69-1)
Relaxyl-G	(93-14-1)
Relbapiridina	(144-83-2)
Reldan	(5598-13-0)
Releasil 8	(63148-62-9)
Releasin	(9002-69-1)
Relestrid	(532-03-6)
Reliton Yellow C	(2832-40-8)
Reliton Yellow R	(119-15-3)
Reliveran	(364-62-5)
Relon	(63428-84-2)
Relon P	(105-60-2)
Relon P	(25038-54-4)
Relutin	(630-56-8)
Remaderm Yellow HPR	(587-98-4)
Remalan Brilliant Blue R	(2580-78-1)
Remasan Chloroble M	(12427-38-2)
Remazin	(129-20-4)
Remazol Black B	(17095-24-8)
Remazol Brilliant Blue R	(2580-78-1)
Remazol Brilliant Red 5B	(12226-12-9)
Remazol Brilliant Yellow GL	(12237-16-0)
Remazol Golden Yellow G	(20317-19-5)
Remazol Golden Yellow GGL	(20317-19-5)
Remazol Printing Rhodamine BB	(72979-85-2)
Remazol Turquoise Blue B	(12236-86-1)
Remazol Turquoise Blue G	(12236-86-1)
Remazol Yellow FG	(12226-63-0)
Remazol Yellow GNL	(12226-50-5)
Remeflin	(1165-48-6)
Remestan	(846-50-4)
Remicyclin	(64-75-5)
Remid	(315-30-0)
Remko	(7439-89-6)
Remol TRF	(90-43-7)
Remonol	(102-29-4)
Remsed	(58-33-3)
Remtal	(1912-26-1)
Remyline Ac	(9005-25-8)
Remyolan	(65-29-2)
Ren O-Sal	(121-19-7)
Renacit 1	(91-60-1)
Renacit II	(3773-14-6)
Renafur	(555-84-0)
Renagladin	(51-43-4)
Renaglandin	(51-43-4)
Renaglandulin	(51-43-4)
Renal AC	(123-30-8)
Renal EG	(591-27-5)
Renal MD	(95-80-7)
Renal PF	(106-50-3)
Renal SLA	(39156-41-7)
Renal SO	(6219-71-2)
Renaleptine	(51-43-4)
Renalina	(51-43-4)
Renardin	(2318-18-5)
Renardine	(2318-18-5)
Renazide	(2259-96-3)
Renborin	(439-14-5)
Renegade	(67375-30-8)
Renese	(346-18-9)
Renese R	(50-55-5)
Renex 714 (9CI)	(69913-55-9)
Rengasil	(31793-07-4)
Rennase	(9001-98-3)
Rennin	(9001-98-3)
Renoform	(51-43-4)
Renol Blue B 2G-H	(147-14-8)
Renol Chrome Yellow Y 2G	(1344-37-2)
Renol Chrome Yellow Y 2RS	(1344-37-2)
Renolblau 3B	(72-57-1)
Renon	(77-36-1)
Renoquid	(17784-12-2)
Renostypricin	(51-43-4)
Renostypticin	(51-43-4)
Renostyptin	(51-43-4)
Renosulfan	(127-69-5)
Renselin	(112-38-9)

Name	CAS
Renstamin	(59-33-6)
Rentovet	(113-59-7)
Renyl MV	(105-60-2)
Renyl MV	(25038-54-4)
Reofos 95	(25155-23-1)
Reofos 50 (9CI)	(63848-94-2)
Reomax	(58-54-8)
Reomol DOA	(103-23-1)
Reomol DOP	(117-81-7)
Reomol D 79P	(117-81-7)
Reoplex 400	(27941-08-8)
Reorganin	(93-14-1)
Reostral	(57-53-4)
Repairsin	(9011-14-7)
Repel	(134-62-3)
Repellent 612	(94-96-2)
Repeltin	(84-96-8)
Rephoxitin	(35607-66-0)
Repicin	(73-48-3)
Repin (8CI)	(11024-67-2)
Repoc	(9002-88-4)
Repoise	(653-03-2)
Reposo-TMD	(315-37-7)
Repper 333	(136-45-8)
Repper-Det	(134-62-3)
Repromix	(71-58-9)
Repudin-Special	(134-62-3)
Requtol	(577-11-7)
Reranil	(118-75-2)
76 Res	(9003-01-4)
76 Res	(9003-20-7)
Resacetophenone	(89-84-9)
β-Resacetophenone	(89-84-9)
Resaltex	(50-55-5)
Resamin 155F	(9011-05-6)
Resamin HW 505	(9011-05-6)
Resamine Fast Orange G	(3520-72-7)
Resamine Red RB	(1248-18-6)
Resamine Red RC	(1248-18-6)
Resamine Rubine BC	(5858-81-1)
Rescisan	(24815-24-5)
Resarit 4000	(9011-14-7)
Resazoin	(550-82-3)
Resazurin	(550-82-3)
Resazurine	(550-82-3)
Resbuthrin	(28434-01-7)
Rescaloid	(24815-24-5)
Rescamin	(24815-24-5)
Rescidan	(24815-24-5)
Rescin	(24815-24-5)
Rescinnamine	(24815-24-5)
Rescinpal	(24815-24-5)
Rescitens	(24815-24-5)
Rescue Squad	(7681-49-4)
Rese-LAR	(50-55-5)
Resedin	(50-55-5)
Resedrex	(50-55-5)
Resedril	(50-55-5)
Reser-AR	(50-55-5)
Reserbal	(50-55-5)
Resercaps	(50-55-5)
Resercen	(50-55-5)
Resercrine	(50-55-5)
Reserfia	(50-55-5)
Reserjen	(50-55-5)
Reserlor	(50-55-5)
Reserp	(50-55-5)
Reserp (Wander)	(50-55-5)
Reserpal	(50-55-5)
Reserpamed	(50-55-5)
Reserpanca	(50-55-5)
Reserpene	(50-55-5)
Reserpex	(50-55-5)
Reserpidefe	(50-55-5)
Reserpil	(50-55-5)
Reserpin	(50-55-5)
Reserpina	(50-55-5)
Reserpine	(50-55-5)
Reserpinene	(24815-24-5)
Reserpinin	(24815-24-5)
Reserpinine	(24815-24-5)
Reserpinum	(50-55-5)
Reserpka	(50-55-5)
Reserpoid	(50-55-5)
Reserpur	(50-55-5)
Resersana	(50-55-5)
Reserutin	(50-55-5)
Resiatric	(50-55-5)
Residine	(50-55-5)
Residual Fuel Oils, Heavy	(68476-31-3)
Residual Fuel Oils, Heavy	(68533-00-4)
Residual(Heavy) Fuel Oil	(68476-33-5)
Residual oils (Petroleum), Acid-treated (9CI)	(64742-17-2)
Residual oils, Petroleum, Clay-treated	(64742-41-2)
Residual oils, Petroleum, Hydrotreated	(64742-57-0)
Residual oils, Petroleum, Solvent-dewaxed	(64742-62-7)
Residual oils, Petroleum, Solvent-refined	(64742-01-4)
Residual oils, Petroleum, Solvent deasphalted	(64741-95-3)
Residual oil solvent extract	(64742-10-5)
Residues, Coal, Solvent-refining (SRC) filtration	(68410-72-0)
Residues, Petroleum, Atm. Tower	(64741-45-3)
Residues, Petroleum, Steam-cracked	(64742-90-1)
Residues, Petroleum, Thermal Cracked	(64741-80-6)
Residues, Petroleum, Vacuum	(64741-56-6)
Resil	(93-14-1)
Resimene X 970	(9011-05-6)
Resimene X 975	(9011-05-6)
Resimene X 980	(9011-05-6)
Resimine 975	(9011-05-6)
Resin 4301	(9003-22-9)
S-Resin AER 20	(9011-05-6)
Resin Benjamin	(9000-05-9)
Resin Benzoin	(9000-05-9)
Resin Scarlet 2R	(3118-97-6)
Resina X	(9011-05-6)
Resin acids and Rosin acids, Cobalt salts	(68956-82-1)
Resin acids and Rosin acids, Zinc salts	(9010-69-9)
Resin acids and rosin acids, copper salts	(9007-39-0)
Resin acid, zinc salt	(9010-69-9)
Resinate de cobalt (French)	(68956-82-1)
Resinato de cobalto (Spanish)	(68956-82-1)
Resine	(50-55-5)
Resinol Brown RRN	(6416-57-5)
Resinol Orange G	(2051-85-6)
Resinol Orange R	(842-07-9)
Resinol RRN	(6416-57-5)
Resinol Red 2B	(85-83-6)
Resinol Yellow GR	(60-11-7)
Resipal	(24815-24-5)
Resiren Blue TB	(2475-44-7)
Resiren Red TB	(17418-58-5)
Resiren Violet TR	(128-95-0)
Resiren Yellow TG	(2832-40-8)
Resisan	(99-30-9)

Resistab (63-56-9)
Resistamine (154-69-8)
Resistamine (91-81-6)
Resistoflex (9002-89-5)
Resistox (56-72-4)
Resitox (56-72-4)
Reskinnamin (24815-24-5)
(+)-trans-Resmethrin (28434-01-7)
Resmethrin (10453-86-8)
d-trans-Resmethrin (28434-01-7)
Resmetrina (Portuguese) (10453-86-8)
Resmin (147-24-0)
Resmit (2898-12-6)
Resoacetophenone (89-84-9)
Resobantin (53-46-3)
Resocalm (50-55-5)
Resochen (54-05-7)
Resochin (54-05-7)
Resoform Orange G (842-07-9)
Resoform Orange R (3118-97-6)
Resoform Red G (85-83-6)
Resoform Yellow GGA (60-11-7)
Resoidan (87-33-2)
Resolin Blue BSL (12222-78-5)
Resolin Blue FBL (12217-79-7)
Resolin Blue GRL (12222-79-6)
Resolin Navy Blue GLS (50922-60-6)
Resolin Red FB (17418-58-5)
Resolin Red FBE (17418-58-5)
Resolin Yellow 5R (6300-37-4)
Resolvable tartaric acid (133-37-9)
Resomine (50-55-5)
Resoquina (54-05-7)
Resoquine (54-05-7)
Resorcin (108-46-3)
Resorcin Brown (1320-07-6)
Resorcin Yellow (547-57-9)
Resorcin acetate (102-29-4)
β-Resorcinaldehyde (95-01-2)
Resorcine (108-46-3)
Resorcine Blue (9CI) (VAN) (87495-30-5)
Resorcine Brown J (1300-73-8)
Resorcine Brown R (1300-73-8)
Resorcine Yellow (547-57-9)
Resorcine Yellow O Extra (547-57-9)
Resorcin monoacetate (102-29-4)
Resorcinol (ACGIH,OSHA) [UN 2876] (108-46-3)
Resorcinol Yellow A (547-57-9)
Resorcinol, 4-acetyl- (89-84-9)
Resorcinol bis(2,3-epoxypropyl)ether (101-90-6)
Resorcinol, 4-chloro (95-88-5)
Resorcinol, diglycidyl (101-90-6)
Resorcinol diglycidyl ether (101-90-6)
Resorcinol, dihydro- (504-02-9)
Resorcinol, 4,5-dimethyl- (8CI) (527-55-9)
Resorcinol dimethyl ether (151-10-0)
Resorcinol, 2,4-dinitro- (8CI) (519-44-8)
Resorcinol, 4-ethyl- (8CI) (2896-60-8)
Resorcinol, 4-hexyl (136-77-6)
β-Resorcinolic acid (89-86-1)
Resorcinol, 2-p-mentha-1,8-dien-3-yl-5-pentyl-, (-)-(E) (13956-29-1)
Resorcinol, 5-methyl (504-15-4)
Resorcinol, 2-methyl- (8CI) (608-25-3)
Resorcinol methyl ether (150-19-6)
Resorcinol, monoacetate (102-29-4)
Resorcinol, monobenzoate (136-36-7)
Resorcinol monomethyl ether (150-19-6)
Resorcinol, 4-((p-nitrophenyl)azo)- (8CI) (74-39-5)
Resorcinol, 5-pentyl (500-66-3)
Resorcinolphthalein (2321-07-5)
Resorcinol phthalein sodium (518-47-8)
Resorcinol, tetrachloro (28520-00-5)
Resorcinol, 2,4,6-trinitro (82-71-3)
Resorcinol yellow (547-57-9)
Resorcinyl diglycidyl ether (101-90-6)
Resorcitate (102-29-4)
Resorcitol (504-01-8)
β-Resorcylaldehyde (95-01-2)
2,6-Resorcylic acid (303-07-1)
α-Resorcylic acid (99-10-5)
β-Resorcylic acid (89-86-1)
γ-Resorcylic acid (8CI) (303-07-1)
β-Resorcylic acid, 3,6-dimethyl-, methyl ester (4707-47-5)
Resorcylic acid, 6-(10-hydroxy-6-oxo-1-undecenyl)-, mu-lactone, trans- (17924-92-4)
β-Resorcylic acid, methyl ester (2150-47-2)
β.-Resorcylic acid, 6-propyl- (7CI,8CI) (4707-50-0)
β-Resorcylic aldehyde (95-01-2)
Resorin Red FBE (17418-58-5)
Resotropin (100-97-0)
Resoxol (127-69-5)
Respaire (616-91-1)
Respenyl (93-14-1)
Resperin (50-55-5)
Resperine (50-55-5)
Respil (93-14-1)
Respital (50-55-5)
Responsar (68359-37-5)
Respramin (60-32-2)
Rest-On (91-80-5)
Restamin (147-24-0)
Restamin (58-73-1)
Restamine (58-73-1)
Restenil (57-53-4)
Restenyl (57-53-4)
Restinal (57-53-4)
Restinil (57-53-4)
Restoril (846-50-4)
Restran (50-55-5)
Restrol (84-17-3)
Restropin (155-41-9)
Restryl (91-80-5)
Resulfon (57-67-0)
Resyl (93-14-1)
Resyn 25-1014 (9003-20-7)
Resyn 25-1025 (9003-20-7)
Retacel (999-81-5)
Retalon (84-17-3)
Retamid (80-35-3)
Retard (123-33-1)
Retarder AK (85-44-9)
Retarder BA (65-85-0)
Retarder Esen (85-44-9)
Retarder J (86-30-6)
Retarder PD (85-44-9)
Retarder W (69-72-7)
Retardex (65-85-0)
Retardillin (54-35-3)
Retarpen (1538-09-6)
Retasulfin (80-35-3)
Reten (483-65-8)
Retene (483-65-8)
Retensin (65-29-2)
Retil (9CI) (79485-04-4)

Retin-A **(302-79-4)**
Retinamide, N-(4-hydroxyphenyl) **(65646-68-6)**
13-cis-Retinoic acid **(4759-48-2)**
Retinoic acid **(302-79-4)**
Retinoic acid, 13-cis **(4759-48-2)**
Retinoic acid, all-trans **(302-79-4)**
all-trans-Retinoic acid **(302-79-4)**
β-Retinoic acid **(302-79-4)**
β-all-trans-Retinoic acid **(302-79-4)**
trans-Retinoic acid **(302-79-4)**
Retinoic acid p-hydroxyphenylamide **(65646-68-6)**
11 cis Retinol **(11103-57-4)**
Retinol **(11103-57-4)**
Retinol **(68-26-8)**
all-trans Retinol **(68-26-8)**
Retinol, acetate **(127-47-9)**
Retinol, all trans **(68-26-8)**
Retinyl acetate **(127-47-9)**
all-trans-Retinyl acetate **(127-47-9)**
Retozide **(54-85-3)**
Retrangor **(68-90-6)**
Retro-6-dehydroprogesterone **(152-62-5)**
Retrone **(152-62-5)**
cis-Retronecic acid ester of retronecine **(480-54-6)**
cis-Retronecic acid ester of retronecine-N-oxide **(15503-86-3)**
Retronecin **(480-85-3)**
(+)-Retronecine **(480-85-3)**
Retronecine **(480-85-3)**
δ^6-Retroprogesterone **(152-62-5)**
Retrorsine **(480-54-6)**
Retrorsine oxide **(15503-86-3)**
Retrorsine, N-oxide **(15503-86-3)**
Retrovir **(30516-87-1)**
Retrovitamin A **(68-26-8)**
Retzolate 1075 **(9014-90-8)**
Retzolate 60 **(9004-82-4)**
Reudo **(50-33-9)**
Reudox **(50-33-9)**
Reumachlor **(54-05-7)**
Reumacide **(53-86-1)**
Reumaquin **(54-05-7)**
Reumasyl **(50-33-9)**
Reumazin **(50-33-9)**
Reumazol **(50-33-9)**
Reumox **(129-20-4)**
Reumune **(50-33-9)**
Reumuzol **(50-33-9)**
Reupolar **(50-33-9)**
Revac **(577-11-7)**
Revacryl A 191 **(9003-01-4)**
Reveal NM **(8061-51-6)**
Reveal SM **(8061-51-6)**
Reveal SM 5 **(8061-51-6)**
Reveal WM **(8061-51-6)**
Revenge **(75-99-0)**
Revidex **(51-63-8)**
Revonal **(72-44-6)**
Rewomid DLMS **(120-40-1)**
Rewomid DL 203/S **(120-40-1)**
Rewomid L 203 **(142-78-9)**
Rewopol HV-9 **(9016-45-9)**
Rewopol NL-2 **(9004-82-4)**
Rewopol NLS 30 **(151-21-3)**
Rewopol TLS 40 **(139-96-8)**
Rewoquat B18 **(112-00-5)**
Rexall 413S **(9003-07-0)**
Rexan **(80-77-3)**
Rexcel **(9004-34-6)**
Rexene **(15708-41-5)**
Rexene **(9003-07-0)**
Rexene 106 **(9003-54-7)**
Rexolite 1422 **(9003-53-6)**
Reychler's Acid **(3144-16-9)**
Rezerpin **(50-55-5)**
Rezifilm **(137-26-8)**
Rezipas **(65-49-6)**
Rh **(7440-16-6)**
Rhamnol **(83-46-5)**
Rhamnolutein **(520-18-3)**
Rhamnolutin **(520-18-3)**
Rhamnopyranose, L- (8CI) **(3615-41-6)**
Rhamnoside, quercetin-3 **(522-12-3)**
Rhamnoside, strophanthidin-3, α-l- **(508-75-8)**
Rheic acid **(478-43-3)**
Rhein **(478-43-3)**
Rhematan **(132-60-5)**
Rhenium (9CI) **(7440-15-5)**
Rhenocure CA **(102-08-9)**
Rheodol SP 030 **(26266-58-0)**
Rheodol SP-P 10 **(26266-57-9)**
Rheonine B **(81-88-9)**
Rheum Emodin **(518-82-1)**
Rheumin **(132-60-5)**
Rheumin tabletten **(50-78-2)**
Rhinalator **(60-13-9)**
Rhinall **(61-76-7)**
Rhinantin **(550-99-2)**
Rhinathiol **(638-23-3)**
Rhinazine **(835-31-4)**
Rhinoperd **(550-99-2)**
Rhizobitoxine **(37658-95-0)**
Rhizoctol **(2533-82-6)**
Rhizoctol combi **(8066-69-1)**
Rhizoctol slurry **(8066-69-1)**
Rhizopin **(87-51-4)**
Rhizopon A **(87-51-4)**
Rhocya **(333-20-0)**
Rhodacryst **(68-19-9)**
Rhodallin **(109-57-9)**
Rhodalline **(109-57-9)**
Rhodamine **(81-88-9)**
Rhodamine B **(81-88-9)**
Rhodamine B 20-7470 **(81-88-9)**
Rhodamine B500 **(81-88-9)**
Rhodamine BA **(81-88-9)**
Rhodamine BA Export **(81-88-9)**
Rhodamine B Base **(509-34-2)**
Rhodamine B Base Extra **(509-34-2)**
Rhodamine B Chloride **(81-88-9)**
Rhodamine B Extra **(81-88-9)**
Rhodamine B Extra Base **(509-34-2)**
Rhodamine B Extra M 310 **(81-88-9)**
Rhodamine B Extra S **(81-88-9)**
Rhodamine BF **(81-88-9)**
Rhodamine BL **(81-88-9)**
Rhodamine B Lactone **(509-34-2)**
Rhodamine BN **(81-88-9)**
Rhodamine BS **(81-88-9)**
Rhodamine BX **(81-88-9)**
Rhodamine BXL **(81-88-9)**
Rhodamine BXP **(81-88-9)**
Rhodamine Base B Extra **(509-34-2)**
Rhodamine B500 hydrochloride **(81-88-9)**
Rhodamine, Blue Shade **(81-88-9)**

Name	CAS
Rhodamine 69DN Extra	**(989-38-8)**
Rhodamine FB	**(81-88-9)**
Rhodamine FB CI	**(81-88-9)**
Rhodamine F4G	**(989-38-8)**
Rhodamine F5G	**(989-38-8)**
Rhodamine F5G Chloride	**(989-38-8)**
Rhodamine GDN	**(989-38-8)**
Rhodamine 4GH	**(989-38-8)**
Rhodamine 4GD	**(989-38-8)**
Rhodamine 5GL	**(989-38-8)**
Rhodamine 5GDN	**(989-38-8)**
Rhodamine 6G Extra	**(989-38-8)**
Rhodamine 6G Extra base	**(989-38-8)**
Rhodamine 6GEX Ethyl ester	**(989-38-8)**
Rhodamine 6GX	**(989-38-8)**
Rhodamine 6G Lake	**(989-38-8)**
Rhodamine 6GH	**(989-38-8)**
Rhodamine 6G	**(989-38-8)**
Rhodamine 6GBN	**(989-38-8)**
Rhodamine 6GB	**(989-38-8)**
Rhodamine 6G (Biological stain)	**(989-38-8)**
Rhodamine 6GCP	**(989-38-8)**
Rhodamine 6 GDN	**(989-38-8)**
Rhodamine 6GD	**(989-38-8)**
Rhodamine 6 GDN Extra	**(989-38-8)**
Rhodamine J	**(989-38-8)**
Rhodamine 6JH	**(989-38-8)**
Rhodamine 7JH	**(989-38-8)**
Rhodamine Lake Red B	**(81-88-9)**
Rhodamine Lake Red 6G	**(989-38-8)**
Rhodamine O	**(81-88-9)**
Rhodamine S Lactone	**(509-34-2)**
Rhodamine S (Russian)	**(81-88-9)**
Rhodamine WT	**(37299-86-8)**
Rhodamine Y 20-7425	**(989-38-8)**
Rhodamine 6ZH	**(989-38-8)**
Rhodamine ZH	**(989-38-8)**
Rhodamine, tetraethyl-	**(81-88-9)**
Rhodandinitrobenzol	**(1594-56-5)**
Rhodanic acid	**(141-84-4)**
Rhodanid	**(1762-95-4)**
Rhodanide	**(1762-95-4)**
Rhodanide	**(333-20-0)**
Rhodanin (Czech)	**(141-84-4)**
Rhodanine	**(141-84-4)**
Rhodanine, 3-(p-chlorophenyl)-5-methyl	**(6012-92-6)**
Rhodaninic acid	**(141-84-4)**
Rhodia	**(94-75-7)**
Rhodia-6200	**(78-53-5)**
Rhodia RP 11974	**(2310-17-0)**
Rhodiachlor	**(76-44-8)**
Rhodiacid	**(137-30-4)**
Rhodiacide	**(563-12-2)**
Rhodiacuivre	**(1332-40-7)**
Rhodiacuivre	**(1332-65-6)**
Rhodianebe	**(12427-38-2)**
Rhodiasol	**(56-38-2)**
Rhodiatox	**(56-38-2)**
Rhodiatrox	**(56-38-2)**
D-Rhodinal	**(106-23-0)**
Rhodinal (VAN)	**(106-23-0)**
Rhodine	**(50-78-2)**
Rhodinol	**(106-22-9)**
Rhodium (ACGIH)	**(7440-16-6)**
Rhodium chloride	**(10049-07-7)**
Rhodium(III) chloride (1:3)	**(10049-07-7)**
Rhodium metal (OSHA)	**(7440-16-6)**
Rhodium phosphide (RhP3) (6CI,7CI,9CI)	**(12202-48-1)**
Rhodium trichloride	**(10049-07-7)**
Rhodium triphosphide	**(12202-48-1)**
Rhodocide	**(563-12-2)**
Rhodofix	**(86-87-3)**
Rhodol	**(55-55-0)**
Rhodolne	**(9003-53-6)**
Rhodopas	**(9003-20-7)**
Rhodopas 010	**(9003-20-7)**
Rhodopas 5425	**(9003-20-7)**
Rhodopas 6000	**(9003-22-9)**
Rhodopas A 10	**(9003-20-7)**
Rhodopas AM 041	**(9003-20-7)**
Rhodopas AX	**(9003-22-9)**
Rhodopas AX 30/10	**(9003-22-9)**
Rhodopas AX 85/15	**(9003-22-9)**
Rhodopas B	**(9003-20-7)**
Rhodopas BB	**(9003-20-7)**
Rhodopas HV 2	**(9003-20-7)**
Rhodopas M	**(9003-20-7)**
Rhodopas 5000SMR	**(9003-20-7)**
Rhodorsil CAF 3B	**(63148-62-9)**
Rhodorsil Oils 70045	**(69430-24-6)**
Rhodorsil Oils 70047	**(9006-65-9)**
Rhodorsil Oils 70641 V 200	**(2116-84-9)**
Rhodotoxin	**(4720-09-6)**
Rhodoviol	**(9002-89-5)**
Rhodoviol 16/200	**(9002-89-5)**
Rhodoviol 4/125	**(9002-89-5)**
Rhodoviol 4-125P	**(9002-89-5)**
Rhodoviol R 16/20	**(9002-89-5)**
Rhoduline Orange	**(494-38-2)**
Rhoduline Orange N	**(494-38-2)**
Rhoduline Orange NO	**(494-38-2)**
Rhoduline Orange NO	**(65-61-2)**
Rhomellose	**(9004-67-5)**
Rhomenc	**(94-74-6)**
Rhomene	**(94-74-6)**
Rhonite R 1	**(136-84-5)**
Rhonox	**(94-74-6)**
Rhoplex AC-33 (Rohm and Haas)	**(97-63-2)**
Rhoplex B 85	**(9011-14-7)**
Rhoplex (9CI)	**(57657-42-8)**
Rhotex GS	**(9003-04-7)**
Rhothane	**(72-54-8)**
Rhothane D-3	**(72-54-8)**
Rhubarb Yellow	**(478-43-3)**
Rhyuno Oil	**(94-59-7)**
Rianil	**(569-57-3)**
Riball	**(315-30-0)**
Ribamidyl	**(36791-04-5)**
Ribavirin	**(36791-04-5)**
Ribavirina (Spanish)	**(36791-04-5)**
Ribavirine (French)	**(36791-04-5)**
Ribavirinum (Latin)	**(36791-04-5)**
Ribena	**(50-81-7)**
D-Ribo-hexonic acid, 3-deoxy- (9CI)	**(498-43-1)**
Ribipca	**(83-88-5)**
Ribitol	**(488-81-3)**
Ribo-azauracil	**(54-25-1)**
Ribo-azuracil	**(54-25-1)**
Riboderm	**(83-88-5)**
Riboflavin	**(83-88-5)**
Riboflavine	**(83-88-5)**
Riboflavinequinone	**(83-88-5)**
Ribofuranose, 5-(dihydrogen phosphate) 1-(trihydrogen diphosphate)	**(97-55-2)**
Ribofuranose, 5-(dihydrogen phosphate) 1-(trihydrogen	

pyrophosphate) (97-55-2)
Ribofuranose, 5-phosphate 1-pyrophosphate (97-55-2)
Ribofuranoside, guanine-9, β-D- (118-00-3)
9-β-D-Ribofuranosidoadenine (58-61-7)
2-β-D-Ribofuranosyl-as-triazine-3,5(2H,4H)-dione (54-25-1)
1-β-Ribofuranosylcytosine (65-46-3)
9-β-D-Ribofuranosylguanine (118-00-3)
9-β-D-Ribofuranosyl-9H-purin-6-amine (58-61-7)
2-β-D-Ribofuranosyl-1,2,4-triazin-3,5(2H,4H)-dion (Czech) (54-25-1)
2-β-D-Ribofuranosyl-1,2,4-triazine-3,5(2H,4H)-dione (54-25-1)
1-β-D-Ribofuranosyl-1,2,4-triazole-3-carboxamide (36791-04-5)
1-β-D-Ribofuranosyl-1H-1,2,4-triazole-3-carboxamide (36791-04-5)
1-β-D-Ribofuranosyluracil (58-96-8)
Ribonosine (58-63-9)
D-Ribose (50-69-1)
Ribose, 5-(dihydrogen phosphate) 1-(trihydrogen diphosphate) (97-55-2)
Ribose, 5-(dihydrogen phosphate) 1-(trihydrogen pyrophosphate) (97-55-2)
β-D-Ribosyl-6-methylthiopurine (342-69-8)
Rice Bran (68553-81-1)
Rice Bran Oil (68553-81-1)
Rice Starch (9005-25-8)
Rice Syn Wax (8001-78-3)
Richamide 6310 (120-40-1)
Richonate 45B (25155-30-0)
Richonate 60B (25155-30-0)
Richonate 1850 (25155-30-0)
Richonic acid B (27176-87-0)
Richonol A (151-21-3)
Richonol AF (151-21-3)
Richonol AM (2235-54-3)
Richonol C (151-21-3)
Richonol T (139-96-8)
Ricid (13286-32-3)
Ricid II (26087-47-8)
Ricid P (26087-47-8)
Ricidine (524-40-3)
Ricifon (52-68-6)
Ricin (9009-86-3)
Ricin, Reconstituted (9009-86-3)
Ricine (9009-86-3)
Ricinic acid (141-22-0)
Ricinine (524-40-3)
Ricinoleic-acid (141-22-0)
Ricinoleic acid, butyl ester, acetate (8CI) (140-04-5)
Ricinoleic acid, 2-methoxyethyl ester, acetate (140-05-6)
Ricinoleic acid, methyl ester (8CI) (141-24-2)
Ricinoleic acid, methyl ester, acetate (140-03-4)
Ricinoleic acid, monopotassium salt (7492-30-0)
Ricinoleic acid, sodium salt (5323-95-5)
Ricinolic acid (141-22-0)
Ricinus Oil (8001-79-4)
Ricirus Oil (8001-79-4)
Ricketon (67-97-0)
Ricon 100 (9003-55-8)
Ricon 109 (9003-55-8)
Ricon 181 (9003-55-8)
Ricon 182 (9003-55-8)
Ricon 183 (9003-55-8)
Ricon A-84 (9003-55-8)
Ricortex (50-04-4)
Ricycline (64-75-5)
Riddelliin (23246-96-0)
Riddelliine (23246-96-0)
Riddelline (23246-96-0)
Ridomil (57837-19-1)
Ridomil 2E (57837-19-1)
Ridomil Fitorex (75701-74-5)
Ridomil MZ (75701-74-5)
Rifa (13292-46-1)
Rifa acid Orange II (633-96-5)
Rifadin (13292-46-1)
Rifadine (13292-46-1)
Rifagen (13292-46-1)
Rifaldazin (13292-46-1)
Rifaldazine (13292-46-1)
Rifaldin (13292-46-1)
Rifamate (13292-46-1)
Rifamate (54-85-3)
Rifampicin (13292-46-1)
Rifampicin SV (13292-46-1)
Rifampicine (French) (13292-46-1)
Rifampicinum (13292-46-1)
Rifampin (13292-46-1)
Rifamycin AMP (13292-46-1)
Rifamycin, 3-(((4-methyl-1-piperazinyl)imino)methyl)- (13292-46-1)
Rifaprodin (13292-46-1)
Rifinah (13292-46-1)
Rifobac (13292-46-1)
Rifoldin (13292-46-1)
Rifoldine (13292-46-1)
Rifomycin SV, 8-(N-(4-methyl-1-piperazinyl)formidoyl) (13292-46-1)
Riforal (13292-46-1)
Rigedal (87-33-2)
Rigenicid (536-33-4)
Rigetamin (379-79-3)
Rigidex (9002-88-4)
Rigidex 35 (9002-88-4)
Rigidex 50 (9002-88-4)
Rigidex Type 2 (9002-88-4)
Rigidil (58-73-1)
Rigidyl (58-73-1)
Rikelate calcium (62-33-9)
Rikemal O 71D (25496-72-4)
Rikemal OL 100 (25496-72-4)
Rikemal S 250 (1338-41-6)
Riker 545 (500-42-5)
Riker 595 (653-03-2)
Riker 601 (68-76-8)
Rilansyl (80-77-3)
Rilaquil (80-77-3)
Rilassol (80-77-3)
Rilax (80-77-3)
Rilentol (51-83-2)
Rillasol (80-77-3)
Rilof (24151-93-7)
Rilon (63428-84-2)
Rilsan BHV Nat Cos (32131-17-2)
Rimactan (13292-46-1)
Rimactane (13292-46-1)
Rimactazid (13292-46-1)
Rimactizid (13292-46-1)
Rimicid (54-85-3)
Rimidin (60168-88-9)
Rimifon (54-85-3)
Rimiterol (32953-89-2)
Rimiterolum (Latin) (32953-89-2)
Rimitsid (54-85-3)
Rimso-50 (67-68-5)
Rinatiol (638-23-3)
Rinderine (6029-84-1)
Rineptil (123-82-0)
Riomitsin (79-57-2)
Ripcord (52315-07-8)
Ripenthol (129-67-9)

Name	CAS No.
Riposon	(77-75-8)
Ripost	(77732-09-3)
Rise	(33671-46-4)
Riselect	(709-98-8)
Riseptin	(102-30-7)
Riserpa	(50-55-5)
Risocaine	(94-12-2)
Ristat	(121-19-7)
Riston	(9011-14-7)
Ritalin	(113-45-1)
Ritalin	(298-59-9)
Ritaline	(113-45-1)
Ritalin hydrochloride	(298-59-9)
Ritcher Works	(113-45-1)
Ritmenal	(57-41-0)
Ritosept	(70-30-4)
Ritsifon	(52-68-6)
Ritussin	(93-14-1)
Rivadorm	(76-74-4)
Rivadorn	(57-33-0)
Rivaite	(13983-17-0)
Rivased	(50-55-5)
Rivasin	(50-55-5)
Rivomycin	(56-75-7)
Rivotril	(1622-61-3)
Rizaben	(53902-12-8)
Ro 1-4849	(435-97-2)
Ro 1-5130	(101-26-8)
Ro-1-5155	(100-55-0)
Ro 1-5431	(297-90-5)
Ro 1-6463	(125-64-4)
Ro-1-7700	(152-02-3)
Ro-1-9213	(68-41-7)
Ro 1-9334	(4914-30-1)
Ro 2-7113	(25384-17-2)
Ro 2-9757	(51-21-8)
Ro 2-9915	(2022-85-7)
Ro 4-0403	(113-59-7)
Ro 4-2130	(723-46-6)
Ro-4-3780	(4759-48-2)
Ro 4-5360	(146-22-5)
Ro 4-6316	(59-92-7)
Ro 4-6467	(671-16-9)
Ro 4-6467	(366-70-1)
Ro 4-8180	(1622-61-3)
Ro 4-9253	(1812-30-2)
Ro 5-0360	(50-91-9)
Ro 5-2180	(1088-11-5)
Ro 5-2807	(439-14-5)
Ro 5-3059	(146-22-5)
Ro 5-3350	(1812-30-2)
Ro 5-3438	(3900-31-0)
Ro 5-4023	(1622-61-3)
Ro 5-4200	(1622-62-4)
Ro 5-5345	(846-50-4)
Ro 5-5516	(848-75-9)
Ro 5-6789	(604-75-1)
Ro 5-6901	(1172-18-5)
Ro-5-6901/3	(17617-23-1)
Ro 05-9129	(527-73-1)
Ro 5-9963	(13551-92-3)
Ro 7-0207	(16773-42-5)
Ro 7-0582	(13551-87-6)
Ro 7-6145	(52508-35-7)
Ro 10-3108	(57342-02-6)
Ro 10-9359	(54350-48-0)
Ro 12-1989	(1783-84-2)
Ro 13-5223	(72490-01-8)
Ro 215535	(32222-06-3)
Ro-Ampen	(69-53-4)
Ro-Ampen	(7177-48-2)
Ro-Cillin	(132-93-4)
Ro-Cycline	(64-75-5)
Ro-Deth	(81-81-2)
Ro-Dex	(57-24-9)
Ro-Hydrazide	(58-93-5)
Ro-Ko	(83-79-4)
Ro-Neet	(1134-23-2)
Ro-Sulfiram	(97-77-8)
Roaccutane	(4759-48-2)
Roach Salt	(7681-49-4)
Road Asphalt, Liquid, Tars or oil (DOT)	(8052-42-4)
Road Asphalt [UN 1999]	(8052-42-4)
Road Tar, Liquid [UN 1999]	(8052-42-4)
Road Tar [UN 1999]	(8052-42-4)
Robamate	(57-53-4)
Robarb	(57-43-2)
Robaxan	(532-03-6)
Robaxin	(532-03-6)
Robaxine	(532-03-6)
Robaxisal-PH	(62-44-2)
Robaxon	(532-03-6)
Robicillin VK	(132-98-9)
Robigenin	(520-18-3)
Robigram	(637-07-0)
Robimycin	(114-07-8)
Robinax	(532-03-6)
Robinin	(301-19-9)
Robiselin	(54-85-3)
Robisellin	(54-85-3)
Robitet	(60-54-8)
Robitussin	(93-14-1)
Robizon-V	(50-33-9)
Roboral	(434-07-1)
Rocaltrol	(32222-06-3)
Roccal	(63449-41-2)
Rocceline	(1658-56-6)
Roccelline	(1658-56-6)
Roccelline A	(1658-56-6)
Roccelline G	(1658-56-6)
Roccelline K	(1658-56-6)
Roccelline KG	(1658-56-6)
Roccelline L	(1658-56-6)
Roccelline NS	(1658-56-6)
Roccelline S	(1658-56-6)
Rochelle Salt	(304-59-6)
Rocillin-VK	(132-98-9)
Rock Candy	(57-50-1)
Rock Oil	(8002-05-9)
Rock Salt	(7647-14-5)
Rocol X 7119	(7782-42-5)
Rodafarin	(81-81-2)
Rodanca	(333-20-0)
Rodanin	(141-84-4)
Rodanin S-62 (Czech)	(96-45-7)
Rodatox 60	(1594-56-5)
Rodentin	(5836-29-3)
Rodex	(640-19-7)
Rodex	(81-81-2)
Rodex Blox	(81-81-2)
Rodinal	(123-30-8)
Rodinol	(106-22-9)
Rodinolone	(124-94-7)
Rodiuran	(135-09-1)

Rodocid (563-12-2)
Rodol D (106-50-3)
Rody (39515-41-8)
Roeridorm (113-18-8)
Rofen 240 (8075-80-7)
Rofob 3 (637-12-7)
Rogitine (50-60-2)
Rogodial (2597-03-7)
Rogodial (60-51-5)
Rogor (60-51-5)
Rogor 40 (60-51-5)
Rogor 20L (60-51-5)
Rogor L (60-51-5)
Rogor P (60-51-5)
Rogue (709-98-8)
Rohagit SD 15 (9003-01-4)
Rohydra (147-24-0)
Rohypnol (1622-62-4)
Roidenin (15687-27-1)
Rokacet (9004-96-0)
Rokacet O 7 (9004-96-0)
Rokacet S 10 (9004-99-3)
Rokafenol O (9036-19-5)
Rokanol L (9002-92-0)
Rokanol L 2 (9002-92-0)
Rokanol L 4 (9002-92-0)
Rokanol L 6 (9002-92-0)
Rokanol L 10 (9002-92-0)
Rokanol L 10/80 (9002-92-0)
Rokanol O (9004-98-2)
Rokon (149-30-4)
Rokopol 30P9 (9003-11-6)
Rol (22248-79-9)
Rolamid CD (120-40-1)
Rolamid CM (142-78-9)
Rolazine (304-20-1)
Roliciclidina (Spanish) (2201-39-0)
Rolicyclidine (2201-39-0)
Rolicyclidinum (Latin) (2201-39-0)
Roll-Fruct (16672-87-0)
Rolserp (50-55-5)
Romacryl (9011-14-7)
Roman Chamomile LS (8001-21-6)
Roman Vitriol (7758-98-7)
Roman Vitriol (7758-99-8)
Romanthrene Blue FRS (81-77-6)
Romantrene Blue FBC (130-20-1)
Romantrene Blue FRS (81-77-6)
Romantrene Blue GGSL (81-77-6)
Romantrene Blue RSZ (81-77-6)
Romantrene Brilliant Blue FR (81-77-6)
Romantrene Brilliant Blue R (81-77-6)
Romantrene Brilliant Green FB (128-58-5)
Romantrene Brilliant Green FFB (128-58-5)
Romantrene Brilliant Pink FR (2379-74-0)
Romantrene Brown FBR (2475-33-4)
Romantrene Brown FGR (2475-33-4)
Romantrene Golden Orange FG (128-70-1)
Romantrene Golden Yellow (128-66-5)
Romantrene Navy Blue FG (6424-76-6)
Romantrene Navy Blue FRA (1324-54-5)
Romantrene Navy Blue G (6424-76-6)
Romergan (60-87-7)
Rometin (130-26-7)
Romilar (125-71-3)
Romopal LN (9002-92-0)
Romosol (12192-57-3)
Romotal (321-64-2)
Romphenil (56-75-7)
Romulgin O (9005-65-6)
Rondar (604-75-1)
Rondomycin (914-00-1)
Ronfalin S (9003-54-7)
Rongalit (149-44-0)
Rongalit C (149-44-0)
Rongalite (149-44-0)
Rongalite C (149-44-0)
Roniacol (100-55-0)
Ronilan (50471-44-8)
Ronin (144-83-2)
Ronit (1134-23-2)
Ronnel (ACGIH,OSHA) (299-84-3)
Ronone (83-79-4)
Ronstar (19666-30-9)
Ronton (77-67-8)
Rontyl (135-09-1)
Rontyl (148-56-1)
Ronyl (2152-34-3)
Rootone (133-32-4)
Rootone (86-86-2)
Rootone (86-87-3)
Rootone F (133-32-4)
Ropol (9002-88-4)
Ropol 24 (9004-99-3)
Ropothene OB.03-110 (9002-88-4)
Roptazol (67-45-8)
Roquine (54-05-7)
Rorasul (446-86-6)
Rorasul (599-79-1)
Rorer 148 (72-44-6)
Rosacetol (90-17-5)
Rosanil (709-98-8)
Rosaniline (632-99-5)
para-Rosaniline (3248-93-9)
Rosaniline Base (3248-93-9)
p-Rosaniline HCl (569-61-9)
Rosaniline chloride (632-99-5)
Rosaniline hydrochloride (632-99-5)
p-Rosaniline nitrate (61467-64-9)
Rosanilinium chloride (632-99-5)
Roscopenin (132-98-9)
Roscorbic (50-81-7)
Rose Crystals (90-17-5)
Rose Ether (122-99-6)
Rose Geranium Oil Algerian (8000-46-2)
Rose Oil (60-12-8)
Rose Oxide Levo (16409-43-1)
Rose Quartz (14808-60-7)
Rosemide (54-31-9)
Rosensthiel (1344-43-0)
Roses (60-13-9)
Rosetone (86-86-2)
Rosex (81-81-2)
Rosin Oil (8002-16-2)
Rosin Oil (85026-55-7)
Rosin, Polymer with formaldehyde and phenol (67700-45-2)
Rosin Soap (Disproportionated) Solution (85026-55-7)
Rosin, copper hydroxide reaction product (9007-39-0)
Rosin, formaldehyde, phenol polymer (67700-45-2)
Rosin modified phenol, formaldehyde polymer (67700-45-2)
Rosin modified phenolic resin of phenol and formaldehyde (67700-45-2)
Rosin, phenol, formaldehyde polymer (67700-45-2)
Rosin, phenol, formaldehyde reaction product (67700-45-2)
β-Rosorcaldehyde (95-01-2)

Name	CAS No.
Rospan	(5836-10-2)
Rospin	(5836-10-2)
Rot B	(3118-97-6)
Rot C	(85-86-9)
Rot G	(85-86-9)
Rot GG Fettloeslich	(3118-97-6)
L-Rot 3 (German)	(915-67-3)
L-Rot 4 (German)	(2611-82-7)
Rotary kiln produced expanded shale lightweight aggregate	(68476-95-9)
Rotate	(22781-23-3)
Rotax	(149-30-4)
Rotefive	(83-79-4)
Rotefour	(83-79-4)
Rotenon	(83-79-4)
Rotenona (Spanish)	(83-79-4)
(-)-Rotenone	(83-79-4)
Rotenone (ACGIH,OSHA)	(83-79-4)
Rotenone, dihydro-	(6659-45-6)
Rotersept	(55-56-1)
Rotessenol	(83-79-4)
Rothane	(72-54-8)
Rotocide	(83-79-4)
Rotox	(74-83-9)
Rouge (OSHA)	(1309-37-1)
Rouge cerasine	(85-86-9)
Rouge de Cochenille A (French)	(2611-82-7)
Rough & Ready Mouse Mix	(81-81-2)
Rougoxin	(20830-75-5)
Roulone	(72-44-6)
Roundup	(38641-94-0)
Rouqualone	(72-44-6)
Routrax	(148-56-1)
Rovlinka	(6988-21-2)
Rovokil	(13194-48-4)
Rovral	(36734-19-7)
Rowalind	(500-22-1)
Rowmate	(1966-58-1)
Rowmate	(62046-37-1)
Rowtate	(34484-77-0)
Roxarsone	(121-19-7)
Roxel	(50-55-5)
Roxifen	(54-85-3)
Roxinoid	(50-55-5)
Roxion	(60-51-5)
Roxion U.A.	(60-51-5)
Roxosul	(127-69-5)
Roxosul Tablets	(127-69-5)
Roxoxol	(127-69-5)
Roxynoid	(50-55-5)
Royal Blue	(50-37-3)
Royal MBTS	(120-78-5)
Royal MH	(28382-15-2)
Royal MH-30	(123-33-1)
Royal Slo-Gro	(123-33-1)
Royal Spectra	(1333-86-4)
Royal TMTD	(137-26-8)
Royal Victoria Blue CP 637	(1325-87-7)
Royaltac	(112-30-1)
Royaltac-85	(112-30-1)
Royaltac M-2	(112-30-1)
Rozevin	(865-21-4)
Roztozol	(116-29-0)
R 1 (Polyester)	(26637-46-7)
R 10 (Refrigerant)	(56-23-5)
R 14 (Refrigerant)	(75-73-0)
Rubatone	(50-33-9)
"522" Rubber Accelerator	(98-77-1)
Rubber Red 16913R	(5160-02-1)
Rubber Red R Extra	(2814-77-9)
Rubber, chlorinated	(9006-03-5)
Rubber, natural, chlorinated	(9006-03-5)
Rubeane	(79-40-3)
Rubeanic acid	(79-40-3)
Rubene	(92-24-0)
Rubens Brown	(1317-34-6)
Ruberon	(2235-25-8)
Rubescence Red MT-21	(6471-49-4)
Rubesol	(68-19-9)
Rubiazol A	(63-74-1)
Rubidium [UN 1423]	(7440-17-7)
Rubidium chromate	(13446-72-5)
Rubidium dichromate	(13446-73-6)
Rubidium metal (DOT)	(7440-17-7)
Rubidium metal, In cartridges (DOT)	(7440-17-7)
Rubidomycin	(20830-81-3)
Rubidomycine	(20830-81-3)
Rubigan	(60168-88-9)
Rubigine	(7664-39-3)
Rubigo	(1309-37-1)
Rubinate 44	(101-68-8)
Rubinate TDI	(26471-62-5)
Rubinate TDI 80/20	(26471-62-5)
Rubinate TDI 80/20	(584-84-9)
Rubine Red RR 1253	(5858-81-1)
Rubine S (6CI)	(3244-88-0)
Rubitox	(2310-17-0)
Rubomycin C	(20830-81-3)
Rubomycin C 1	(20830-81-3)
Rubout	(77182-82-2)
Rubramin	(68-19-9)
Rubramin PC	(68-19-9)
Rubratoxin A	(22467-31-8)
Rubratoxin B	(21794-01-4)
Rubrine C	(107-43-7)
Rubripca	(68-19-9)
Rubrocitol	(68-19-9)
Rubrum Scarlatinum	(85-83-6)
Rucaina	(137-58-6)
Rucoflex Plasticizer DOA	(103-23-1)
Rudotel	(2898-12-6)
Ruelene	(299-86-5)
Ruelene Drench	(299-86-5)
Ruelene 25E	(299-86-5)
Rufol	(144-82-1)
(+)-Rugulosin	(23537-16-8)
Rugulosin	(23537-16-8)
Rukseam	(50-29-3)
Rulene	(299-86-5)
Rum Ether	(8030-89-5)
Rumapax	(129-20-4)
Rumensin	(22373-78-0)
Rumestrol 1	(56-53-1)
Rumestrol 2	(56-53-1)
Rumetan	(1314-84-7)
Runa RH20	(13463-67-7)
Runcatex	(93-65-2)
Ruocid	(57-67-0)
Ruphos	(78-34-2)
Rutabion	(153-18-4)
Rutgers 612	(94-96-2)
Ruthenium (9CI)	(7440-18-8)
Ruthenium, Isotope of mass 106 (8CI,9CI)	(13967-48-1)
Rutile	(13463-67-7)
Rutile (TiO_2) (9CI)	(1317-80-2)

Rutile titanium dioxide **(1317-80-2)**
Rutin **(153-18-4)**
Rutinic acid **(153-18-4)**
Rutinoside, 2-(3,4-dihydroxyphenyl)-5,7-dihydroxy-4-oxo-4H-1-benzopyran-3-yl **(153-18-4)**
Rutinoside, quercetin-3, β- **(153-18-4)**
Rutiox CR **(13463-67-7)**
Rutoside **(153-18-4)**
Rutozyd **(153-18-4)**
Rutralin **(33629-47-9)**
Ryanexel **(15662-33-6)**
Ryania **(15662-33-6)**
Ryania Powder **(15662-33-6)**
Ryania Speciosa **(15662-33-6)**
Ryania Speciosa, Powdered stems of **(15662-33-6)**
Ryanicide **(15662-33-6)**
Ryanodine **(15662-33-6)**
Rycelan **(19044-88-3)**
Rycelon **(19044-88-3)**
Ryfen **(102-98-7)**
Rylam **(63-25-2)**
Rynacrom **(15826-37-6)**
Ryomycin **(79-57-2)**
Ryoto Sugar Ester S-1170 **(25168-73-4)**
Ryoto Sugar Ester S-1570 **(25168-73-4)**
Ryoto Sugar Ester S-1670 **(25168-73-4)**
Ryser **(50-55-5)**
Rystal **(621-42-1)**
Rythritol **(78-11-5)**
Ryzelan **(19044-88-3)**
40S **(9004-99-3)**
60S **(9004-99-3)**
62S **(9005-08-7)**
413S **(9003-07-0)**
S 1 **(7782-42-5)**
S 1 **(76-06-2)**
S 1 (Graphite) **(7782-42-5)**
S 9 **(623-48-3)**
S 28 (Pesticide) **(36335-67-8)**
S 28 **(36335-67-8)**
S 75M **(9004-32-4)**
S 112A **(122-14-5)**
S 140 **(50-13-5)**
S 151 **(111-60-4)**
S 173 **(9003-53-6)**
S 276 **(298-04-4)**
S 309 **(2497-07-6)**
S 400 **(7758-29-4)**
S 421 **(127-90-2)**
S 541 **(9004-99-3)**
S 767 **(115-90-2)**
S 816 **(11114-92-4)**
S 820 **(9003-17-2)**
S 1004 **(9004-99-3)**
S 1006 **(9004-96-0)**
S 1009 **(9005-08-7)**
S 1012 **(9004-99-3)**
S 1013 **(9005-08-7)**
S 1016 **(9004-99-3)**
S 1042 **(9004-99-3)**
S 1054 **(9004-99-3)**
S 1096R **(25496-72-4)**
S 1096 **(25496-72-4)**
S 1097 **(25496-72-4)**
S 1116 **(9004-99-3)**
S 1132 **(9004-96-0)**
S 1530 **(2011-67-8)**
S 1544 **(156-51-4)**
S 1600 **(565-33-3)**
S 1752 **(55-38-9)**
S 1844 **(66230-04-4)**
S 1942 **(2104-96-3)**
S 2225 **(4824-78-6)**
S 2539 **(26002-80-2)**
S 2571 **(14151-45-2)**
S 2803 **(9003-22-9)**
S 2846 **(36335-67-8)**
S 2940 **(2597-03-7)**
S 2957 **(21923-23-9)**
S 2957 **(60238-56-4)**
S 3206 **(39515-41-8)**
S 4084 **(2636-26-2)**
S 4087 **(13067-93-1)**
S 4400 **(327-98-0)**
S 4706 **(333-43-7)**
S 5602 **(51630-58-1)**
S 5602 A α **(66230-04-4)**
S 5660 **(122-14-5)**
S 6437 **(15686-71-2)**
S 6900 **(2540-82-1)**
S 7131 **(32809-16-8)**
S 7173 **(5251-93-4)**
S 10165 **(709-98-8)**
S 7481F1 **(59865-13-3)**
S 22012 **(1929-88-0)**
S-14 **(115-26-4)**
S-82 **(67-99-2)**
S-578 **(50370-12-2)**
S-805 **(1977-10-2)**
S-847 **(101-27-9)**
S-1046 **(2425-10-7)**
S-1102A **(122-14-5)**
S-1358 **(51308-54-4)**
S-3151 **(52645-53-1)**
S-6000 **(2759-71-9)**
S-6115 **(22936-86-3)**
S-6538 **(13593-03-8)**
S-6,999 **(991-42-4)**
S-9115 **(22936-86-3)**
S-15076 **(2941-55-1)**
S51 **(58-73-1)**
S115 **(59-67-6)**
S6176 **(2941-55-1)**
S6876 **(1113-02-6)**
SA **(69-72-7)**
SA 1500 **(637-12-7)**
SA 42-548 **(22232-71-9)**
SA 546 **(58-36-6)**
SA III **(57-68-1)**
SAA **(552-94-3)**
SA-50 Brand Neutral Copper Fungicide **(1332-03-2)**
SADH **(1596-84-5)**
SAG 100 **(63148-62-9)**
SAH 22 **(13770-96-2)**
SAH 283 **(63148-62-9)**
SAH 288 **(63148-62-9)**
SAIB **(126-13-6)**
SAN 6706 **(23576-23-0)**
SAN 6706-3197 **(23576-23-0)**
SAN-C **(9003-54-7)**
S.A. R.L. **(439-14-5)**
S.A.S.-500 **(599-79-1)**
SAS **(7784-28-3)**
SAS 1 **(12743-20-3)**

SASP **(599-79-1)**
SAX **(69-72-7)**
SB 5833 **(36104-80-0)**
S.B.A. **(78-92-2)**
SBA 0108E **(10361-37-2)**
SB 475K **(9003-53-6)**
SBO **(577-11-7)**
SBP-1382 **(10453-86-8)**
SBP-1390 **(28434-01-7)**
SBP-1513 **(52645-53-1)**
SBP-1513TEC **(52645-53-1)**
S.B. Penick 1382 **(10453-86-8)**
SBS **(9003-55-8)**
SBS (Block polymer) **(9003-55-8)**
SBS Copolymer **(9003-55-8)**
S-3466-C **(33956-61-5)**
SC10 **(54-96-6)**
SC-110 **(62-38-4)**
SC 1674 **(519-95-9)**
SC 2910 **(53-46-3)**
SC-3171 **(50-34-0)**
SC 3900 **(13822-56-5)**
SC-4642 **(68-23-5)**
SC 4725 **(72-33-3)**
SC 9420 **(52-01-7)**
SC 10295 **(443-48-1)**
SC10363 **(595-33-5)**
SC 11585 **(53-39-4)**
SC 11800 **(297-76-7)**
SC 12937 **(1249-84-9)**
SC 15090 **(83-67-0)**
SC 15983 **(52-01-7)**
SCH 412 **(50-11-3)**
SCH 7307 **(78-44-4)**
SCH 9724 **(1405-41-0)**
SCH 10159 **(306-52-5)**
SCH 11527 **(50-59-9)**
SCH 20569 **(56391-56-1)**
SCTZ **(533-45-9)**
40 SD **(25311-71-1)**
SD 188 **(9003-53-6)**
SD-345 **(869-29-4)**
SD 354 **(9003-55-8)**
SD-1750 **(62-73-7)**
SD 1836 **(311-47-7)**
SD 1897 **(96-12-8)**
SD 2614 **(53494-70-5)**
SD 3562 **(141-66-2)**
SD 4072 **(470-90-6)**
SD 4294 **(7700-17-6)**
SD-4314 **(115-29-7)**
SD 4402 **(297-78-9)**
SD 4741 **(152-18-1)**
SD 4901 **(1214-39-7)**
SD 5220 **(9003-07-0)**
SD 5532 **(57-74-9)**
SD 7442 **(7421-93-4)**
SD 7859 **(470-90-6)**
SD 7961 **(1918-13-4)**
SD 8280 **(2274-67-1)**
SD 8339 **(2312-73-4)**
SD 8447 **(22248-79-9)**
SD 8530 **(2686-99-9)**
SD 8530 **(12407-86-2)**
SD 8591 **(5918-93-4)**
SD 8786 **(2655-15-4)**
SD-8803 **(1224-63-1)**
SD 8988 **(13104-21-7)**
SD 9098 **(1757-18-2)**
SD 9129 **(6923-22-4)**
SD 11831 **(4726-14-1)**
SD 14114 **(13356-08-6)**
SD 14999 **(16752-77-5)**
SD 15418 **(21725-46-2)**
SD 18303 **(13350-71-5)**
SD 30,053 **(22212-55-1)**
SD 35651 **(58842-20-9)**
SD 41706 **(39515-41-8)**
SD 43775 **(51630-58-1)**
SDA **(68-35-9)**
SDA 04-005-00 **(67701-06-8)**
SDA 13-060-00 **(68603-15-6)**
SDA 15-060-00 **(67762-41-8)**
SDA 15-067-01 **(67762-19-0)**
SDA 15-067-04 **(68585-34-2)**
SDA 15-080-00 **(68584-22-5)**
SDA 16-070-00 **(69227-21-0)**
SDA 17-043-00 **(67700-99-6)**
SDA 19-065-00 **(68439-49-6)**
SDA 22-101-04 **(69011-84-3)**
SD Alcohol 23-hydrogen **(64-17-5)**
S DC 200 **(63148-62-9)**
SDD **(149-45-1)**
SDDC **(128-04-1)**
SDEH **(1615-80-1)**
SDIC **(2893-78-9)**
SDMH **(540-73-8)**
SDM No.5 **(15825-70-4)**
SDM No. 23 **(78-11-5)**
SDM No. 35 **(78-11-5)**
SDP 640 **(9002-88-4)**
SDPH **(630-93-3)**
SD 345 (Polymer) **(9003-55-8)**
SD 8988 (Shell) **(13104-21-7)**
SER (IUPAC Abbrev) **(56-45-1)**
SET **(54-64-8)**
SETD **(94-19-9)**
SEX (Explosive) **(13980-00-2)**
SF 60 **(121-75-5)**
SF 96 **(63148-62-9)**
SF 96-100 **(63148-62-9)**
SF 97(50) **(63148-62-9)**
SF-337 **(134-58-7)**
SF 1293 **(35597-43-4)**
SF-6505 **(10004-44-1)**
S6F Histyrene Resin **(9003-55-8)**
SFK 70 **(9011-05-6)**
S/G 84 **(1309-42-8)**
SG-67 **(60676-86-0)**
SG-67 **(7631-86-9)**
SGM 36 **(63148-62-9)**
SH 015 **(9003-29-6)**
SH 200 **(63148-62-9)**
SH 261 **(93-54-9)**
SH 850 **(6533-00-2)**
SH 926 **(440-58-4)**
SH 6188 **(63148-62-9)**
SH 8708 **(63148-62-9)**
SH 70850 **(6533-00-2)**
SHMP **(10124-56-8)**
SHS **(1120-01-0)**
SI **(127-69-5)**
SI **(7439-92-1)**
SI-6711 **(18854-01-8)**

SIM	**(723-46-6)**	SMD 3500	**(9003-53-6)**
SK 65	**(469-62-5)**	SMDC	**(137-42-8)**
SK 75	**(9011-05-6)**	SMDC	**(6734-80-1)**
SK 1150	**(134-58-7)**	SMIF	**(14797-65-0)**
SK 6882	**(52-24-4)**	SMOP	**(80-35-3)**
SK 20501	**(50-18-0)**	SMP	**(80-35-3)**
SK 22591	**(127-07-1)**	S Mustard	**(505-60-2)**
SK 27702	**(154-93-8)**	S. N. 390	**(69-05-6)**
SK-100	**(817-09-4)**	S.N. 112	**(68-35-9)**
SK-598	**(302-70-5)**	S.N. 12870	**(67-99-2)**
SK-3818	**(545-55-1)**	SN 20	**(9003-54-7)**
SK-15673	**(148-82-3)**	SN 20A	**(9003-54-7)**
SK-19849	**(66-75-1)**	SN 25	**(9003-54-7)**
SK1133	**(51-18-3)**	SN 46	**(131-89-5)**
SK&F 29044	**(14255-87-9)**	SN 390	**(69-05-6)**
SK&F No. 478-A	**(972-02-1)**	SN 513	**(57520-17-9)**
SK-Ampicillin	**(69-53-4)**	SN 4075	**(13684-63-4)**
SK-Apap	**(103-90-2)**	SN 6718	**(54-05-7)**
SK-Bisacodyl	**(603-50-9)**	SN 7618	**(54-05-7)**
SK-Chlorothiazide	**(58-94-6)**	SN 34615	**(2631-37-0)**
SK-Dexamethasone	**(50-02-2)**	SN 35830	**(7287-36-7)**
SK-Digoxin	**(20830-75-5)**	SN 36056	**(23422-53-9)**
SK-Erythromycin	**(643-22-1)**	SN 36268	**(6164-98-3)**
SKF 385	**(155-09-9)**	SN 38107	**(13684-56-5)**
SKF 478	**(972-02-1)**	SN 49537	**(51707-55-2)**
SKF 688A	**(63-92-3)**	SN 81742	**(74051-80-2)**
SKF 1498	**(60-87-7)**	SN-38584	**(13684-63-4)**
SKF 3195	**(500-42-5)**	SN 38210 (9CI)	**(70781-06-5)**
SKF 5137	**(357-56-2)**	SN 38212 (9CI)	**(70781-07-6)**
SKF 6574	**(127-35-5)**	SN-41703	**(19622-19-6)**
SKF 8542	**(396-01-0)**	S.N.G	**(55-63-0)**
SKF 10056	**(12192-57-3)**	SN 20P	**(9003-54-7)**
SKF 14463	**(83-46-5)**	SNP	**(56-38-2)**
SKF 18667	**(9003-11-6)**	SNP 2	**(9003-54-7)**
SKF 29044	**(14255-87-9)**	SO	**(7439-92-1)**
SKF 91487	**(16111-27-6)**	SO 15	**(8007-43-0)**
SKF 92334	**(51481-61-9)**	SO 95	**(9011-14-7)**
SKF-2601	**(50-53-3)**	SO 120	**(9011-14-7)**
SKG	**(7440-44-0)**	SO 140	**(9011-14-7)**
SKI 24464	**(684-93-5)**	SOK	**(671-04-5)**
SKLN 1	**(7782-42-5)**	SOPP	**(132-27-4)**
SK-106N	**(55-63-0)**	SOS	**(142-31-4)**
SK-Niacin	**(59-67-6)**	75 SP	**(30560-19-1)**
SK-Penicillin VK	**(132-98-9)**	SP	**(526-08-9)**
SK 1000 (Petroleum resin)	**(64742-16-1)**	SP 60	**(9003-20-7)**
SK-Phenobarbital	**(50-06-6)**	SP 60 (Ester)	**(9003-20-7)**
SK-Prednisone	**(53-03-2)**	SP 104	**(1972-08-3)**
SK-Reserpine	**(50-55-5)**	SP 489	**(9011-06-7)**
SKS 85	**(9003-55-8)**	SP-500 Nylon Powder	**(32131-17-2)**
SK-Soxazole	**(127-69-5)**	SP-1103	**(7696-12-0)**
SKT	**(7440-44-0)**	SP 2100	**(63148-62-9)**
SKT (Adsorbent)	**(7440-44-0)**	SPA	**(14148-99-3)**
SK-Tetracycline	**(60-54-8)**	SPE 2792	**(117-37-3)**
SK-Tetracycline	**(64-75-5)**	SPF	**(94-26-8)**
SK-Tolbutamide	**(64-77-7)**	SPP	**(526-08-9)**
SK-Triamcinolone	**(124-94-7)**	SPP (Coating)	**(64742-16-1)**
SK 75V	**(9011-05-6)**	SPS 600	**(9003-53-6)**
1000SL	**(2809-21-4)**	SPT 50 CPS	**(9004-57-3)**
SL-90	**(93-14-1)**	SPX 5338	**(9004-30-2)**
SLS	**(151-21-3)**	SQ 1089	**(127-07-1)**
SM 2013	**(63148-62-9)**	SQ 1489	**(137-26-8)**
SM 2061	**(63148-62-9)**	SQ 3277	**(15922-78-8)**
SM 2138	**(63148-62-9)**	SQ 4609	**(1214-39-7)**
SM 5512	**(63148-62-9)**	SQ 8388	**(52-46-0)**
SM 8708	**(63148-62-9)**	SQ 9453	**(67-68-5)**
SMA	**(3926-62-3)**	SQ 9538	**(968-93-4)**
SMCA	**(3926-62-3)**	SQ 11725	**(42200-33-9)**

Name	CAS
SQ 14,225	**(62571-86-2)**
SQ 15500	**(95-54-5)**
SQ 16,401	**(8067-69-4)**
SQ 21977	**(342-69-8)**
SQ 22451	**(452-06-2)**
SR 73	**(1420-04-8)**
SR 203	**(2455-24-5)**
SR 206	**(97-90-5)**
SR 209	**(109-17-1)**
SR 247	**(2223-82-7)**
SR 351	**(15625-89-5)**
SR 1354	**(13551-87-6)**
SR 1530	**(13551-92-3)**
SR 720-22	**(17560-51-9)**
SR406	**(133-06-2)**
SRA 3886	**(22224-92-6)**
SRA 5172	**(10265-92-6)**
SRA 7312	**(13593-03-8)**
SRA 7847	**(17109-49-8)**
SRA 12869	**(25311-71-1)**
SRC mineral residue	**(68410-72-0)**
SRC naphtha	**(68476-79-9)**
SRC wash solvent	**(68410-09-3)**
SRI 859	**(684-93-5)**
SRI 1354	**(13551-87-6)**
SRI 1666	**(3844-60-8)**
SRI 1720	**(154-93-8)**
SRI 2200	**(13010-47-)**
SRI 2489	**(5034-77-5)**
SRM 705	**9003-53-6)**
SRM 706	**9003-53-6)**
SRM 1475	**9002-88-4)**
SRM 1476	**9002-88-4)**
SS	**(9004-70-0)**
SS 578	**(83-73-8)**
SS 1451	**(93-75-4)**
SS 2074	**(2439-01-2)**
SS 11946	**(50512-35-1)**
SS Acid	**(82-47-3)**
SSC-85213	**(8063-14-7)**
SSH 43	**(55861-78-4)**
SST-101	**(87-33-2)**
S.T. 37	**(136-77-6)**
ST 1	**(9011-14-7)**
ST 155	**(302-22-7)**
ST 90	**(9003-53-6)**
STAC	**(112-03-8)**
ST52-ASTA	**(522-40-7)**
STCA	**(650-51-1)**
STH	**(9002-72-6)**
STP	**(15588-95-1)**
STPP	**(7758-29-4)**
ST 1 (Polymer)	**(9011-14-7)**
STR	**(18883-66-4)**
STRZ	**(18883-66-4)**
STS	**(1191-50-0)**
STS	**(139-88-8)**
ST 30UL	**(9003-53-6)**
STX Dihydrochloride	**(35554-08-6)**
STZ	**(18883-66-4)**
SU 2000	**(7440-44-0)**
SU-4885	**(54-36-4)**
SU 5864	**(55-65-2)**
SU 5879	**(58-93-5)**
SU 21524	**(31793-07-4)**
SU-13437	**(3771-19-5)**
SUM 3170	**(1977-10-2)**
SV 102	**(577-11-7)**
SV 7000	**(9003-29-6)**
SVKh 1	**(9011-06-7)**
SVKh 40	**(9011-06-7)**
SVO 9	**(9005-65-6)**
SW 400	**(1344-95-2)**
SW-751	**(58011-68-0)**
SW-6701	**(14491-59-9)**
SW-6721	**(14491-59-9)**
SWP	**(85-60-9)**
SWP (Antioxidant)	**(85-60-9)**
SWS 03314	**(63148-62-9)**
SX Purple	**(2611-82-7)**
SY: Atonin O	**(50-56-6)**
Saatbeizfungizid (German)	**(118-74-1)**
Sabacide	**(8051-02-3)**
Sabadilla	**(8051-02-3)**
Sabadilla alkaloids	**(8051-02-3)**
Sabane Dust	**(8051-02-3)**
Sabari	**(569-65-3)**
Sabinene	**(3387-41-5)**
Sacarina	**(81-07-2)**
Saccharated ferric oxide	**(8047-67-4)**
Saccharated iron	**(8047-67-4)**
Saccharimide	**(81-07-2)**
Saccharin	**(81-07-2)**
Saccharin Insoluble	**(81-07-2)**
Saccharina	**(81-07-2)**
Saccharin acid	**(81-07-2)**
Saccharin ammonium	**(6381-61-9)**
Saccharinate ammonium (French)	**(6381-61-9)**
Saccharin calcium	**(6485-34-3)**
Saccharine	**(81-07-2)**
Saccharine soluble	**(128-44-9)**
Saccharinnatrium	**(128-44-9)**
Saccharinol	**(81-07-2)**
Saccharinose	**(81-07-2)**
Saccharin, sodium	**(128-44-9)**
Saccharin, sodium salt	**(128-44-9)**
Saccharin soluble	**(128-44-9)**
Saccharoidum natricum	**(128-44-9)**
Saccharol	**(81-07-2)**
Saccharolactic acid	**(526-99-8)**
Saccharose	**(57-50-1)**
Saccharose acetate isobutyrate	**(126-13-6)**
Saccharose monostearate	**(25168-73-4)**
Saccharosemonostearate	**(25168-73-4)**
Saccharose stearate	**(25168-73-4)**
Saccharosonic acid	**(89-65-6)**
Saccharum	**(57-50-1)**
Saccharum Lactin	**(63-42-3)**
Saceril	**(57-41-0)**
Saceril	**(630-93-3)**
Sacerno	**(50-12-4)**
Sacharin (Czech)	**(81-07-2)**
Sachsischblau	**(860-22-0)**
Sachtolith	**(1314-98-3)**
Sadanyl	**(57-53-4)**
Sadofos	**(121-75-5)**
Sadofos 30	**(121-75-5)**
Sadophos	**(121-75-5)**
Sadoplon	**(137-26-8)**
Sadoreum	**(53-86-1)**
Saeure fluoride (German)	**(7782-41-4)**
Safaritone Yellow G	**(2832-40-8)**
Safe-N-Dri	**(1302-76-7)**
Safflower-Oil	**(8001-23-8)**

Saffron Oil **(8022-19-3)**
Saffron Yellow **(605-69-6)**
Safranin **(477-73-6)**
Safranin T **(477-73-6)**
Safranine **(477-73-6)**
Safranine A **(477-73-6)**
Safranine B **(477-73-6)**
Safranine G **(477-73-6)**
Safranine GF **(477-73-6)**
Safranine J **(477-73-6)**
Safranine O **(477-73-6)**
Safranine OK **(477-73-6)**
Safranine Superfine G **(477-73-6)**
Safranine T **(477-73-6)**
Safranine TH **(477-73-6)**
Safranine TN **(477-73-6)**
Safranine Y **(477-73-6)**
Safranine YN **(477-73-6)**
Safranine ZH **(477-73-6)**
Safrol **(94-59-7)**
Safrole **(94-59-7)**
Safrole MF **(94-59-7)**
Safrole, dihydro- **(94-58-6)**
Safrotin **(31218-83-4)**
Safrotin **(77491-30-6)**
Safsan **(16893-85-9)**
Sagatal **(57-33-0)**
Sah **(90-02-8)**
Sahrekkusu **(9003-54-7)**
Saiclate **(456-59-7)**
Saiodin **(1319-91-1)**
Saiphos **(78-57-9)**
Saisan **(5707-69-7)**
Saitomycin **(2005-98-3)**
Sajodin **(1319-91-1)**
Sakunol SN 08 **(9003-20-7)**
Sakurai No. 864 **(3458-22-8)**
Sal **(90-01-7)**
Sal Ammonia **(12125-02-9)**
Sal Ammoniac **(12125-02-9)**
Sal Enixum **(7646-93-7)**
Sal Tartar **(868-18-8)**
Salacetin **(50-78-2)**
Salachlor **(141-53-7)**
Salad Oil **(68956-68-3)**
Salamid **(65-45-2)**
Salamide **(65-45-2)**
Salammonite **(12125-02-9)**
Salazopyrin **(599-79-1)**
Salazosulfapyridine **(599-79-1)**
Salcetogen **(50-78-2)**
Salcomin **(14167-18-1)**
Salcomine **(14167-18-1)**
Salcomine Powder **(14167-18-1)**
Sal de Merck **(53-21-4)**
Sal ethyl **(118-61-6)**
Saletin **(50-78-2)**
Salgydal **(62-44-2)**
Saliamin **(65-45-2)**
Salical **(552-94-3)**
Salicilamide (Italian) **(65-45-2)**
Salicilato de mercurio (Spanish) **(5970-32-1)**
Salicilato de nicotina (Spanish) **(29790-52-1)**
Salicil, 5,5'-dibromo **(523-88-6)**
Salicim **(65-45-2)**
Salicin (8CI) **(138-52-3)**
Salicine **(138-52-3)**
Salicoside **(138-52-3)**
Salicyladehyde **(90-02-8)**
Salicylal **(90-02-8)**
Salicyl alcohol **(90-01-7)**
Salicylaldehyde **(90-02-8)**
Salicylaldehyde, dichloro- **(91930-03-9)**
Salicylaldehyde ethylenediimine cobalt **(14167-18-1)**
Salicylaldehyde methyl ether **(135-02-4)**
Salicylamid **(65-45-2)**
Salicylamide **(65-45-2)**
Salicylanilide **(87-17-2)**
Salicylanilide, 3,5-dibromo- (8CI) **(2577-72-2)**
Salicylanilide, 4',5-dibromo- (8CI) **(87-12-7)**
Salicylanilide, 2',5-dichloro-4'-nitro **(50-65-7)**
Salicylanilide, 2',5-dichloro-4'-nitro-, Compd. with 2-aminoethanol (1:1) **(1420-04-8)**
Salicylanilide, 2',5-dichloro-4-nitro-, ethanolamine salt **(1420-04-8)**
Salicylanilide, tribromo **(1322-38-9)**
Salicylanilide, 3,4',5-tribromo- (8CI) **(87-10-5)**
Salicylate **(63-36-5)**
Salicylate anion **(63-36-5)**
Salicylate de mercure (French) **(5970-32-1)**
Salicylate de nicotine (French) **(29790-52-1)**
Salicylate ion **(63-36-5)**
Salicylazosulfapyridine **(599-79-1)**
Salicylic-acid **(69-72-7)**
m-Salicylic acid **(99-06-9)**
p-Salicylic acid **(99-96-7)**
Salicylic acid, acetate **(50-78-2)**
Salicylic acid, 4-amino **(65-49-6)**
Salicylic acid, 3-amino-5-sulfo- (8CI) **(6201-86-1)**
Salicylic acid, 5-amino-3-sulfo- (8CI) **(6201-87-2)**
Salicylic acid, benzyl ester **(118-58-1)**
Salicylic acid, bimolecular ester **(552-94-3)**
Salicylic acid, bismuth basic salt **(14882-18-9)**
Salicylic acid, p-tert-butylphenyl ester **(87-18-3)**
Salicylic acid, 5-chloro **(321-14-2)**
Salicylic acid, Compd. with physostigmine (1:1) **(57-64-7)**
Salicylic acid, 3,5-dichloro **(320-72-9)**
Salicylic acid, 3,6-dichloro **(3401-80-7)**
Salicylic acid, 3,5-diiodo **(133-91-5)**
Salicylic acid, diiodo- (8CI) **(1321-04-6)**
Salicylic acid, 3,5-dinitro **(609-99-4)**
Salicylic acid, 3,5-dinitro-, methyl ester **(22633-33-6)**
Salicylic acid, ethyl ester **(118-61-6)**
Salicylic acid, m-homomenthyl ester **(118-56-9)**
Salicylic acid, 5-hydroxy- **(490-79-9)**
Salicylic acid, ion(1-) **(63-36-5)**
Salicylic acid, isobutyl ester **(87-19-4)**
Salicylic acid, isopentyl ester **(87-20-7)**
Salicylic acid, isopropyl ester, O-ester with O-ethyl isopropylphosphoramidothioate **(25311-71-1)**
Salicylic acid, isopropyl ester, ethyl isopropylphosphoramidate (8CI) **(31120-85-1)**
Salicylic acid, methylenedi **(27496-82-8)**
Salicylic acid, methyl ester **(119-36-8)**
Salicylic acid methyl ether **(579-75-9)**
Salicylic acid, 5-((4'-((3-methyl-5-oxo-1-(p-sulfophenyl)-2-pyrazolin-4-yl)azo)-4-biphenylyl)azo)-, disodium salt (8CI) **(13164-93-7)**
Salicylic acid, monosodium salt **(54-21-7)**
Salicylic acid, 3-nitro **(85-38-1)**
Salicylic acid, 5-nitro- (8CI) **(96-97-9)**
Salicylic acid, pentyl ester **(2050-08-0)**
Salicylic acid, phenyl ester **(118-55-8)**
Salicylic acid, 5-((p-(2-pyridylsulfamoyl)phenyl)azo) **(599-79-1)**
Salicylic acid, sodium salt **(54-21-7)**
Salicylic acid, 5-sulfo **(97-05-2)**

Name	CAS No.
Salicylic acid, 3,3,5-trimethylcyclohexyl ester (8CI)	(118-56-9)
Salicylic aldehyde	(90-02-8)
Salicylic ether	(118-61-6)
Salicylic ethyl ester	(118-61-6)
Salicylohydroxamic acid, 5-bromo- (8CI)	(5798-94-7)
Salicylonitrile (8CI)	(611-20-1)
m-Salicylotoluidide, 3,5-dibromo-α,α,α-trifluoro- (8CI)	(4776-06-1)
Salicyloylsalicylic acid	(552-94-3)
Salicylsalicylic acid	(552-94-3)
o-Salicylsalicylic acid	(552-94-3)
Salicylsulfonic acid	(97-05-2)
Saligenin	(90-01-7)
Saligenin-β-D-glucopyranoside	(138-52-3)
Saligenol	(90-01-7)
Salina	(552-94-3)
Saline	(7647-14-5)
Salipur	(65-45-2)
Salisan	(58-94-6)
Salisil	(1343-88-0)
Salisod	(54-21-7)
Salithion	(3811-49-2)
Salithion-Sumitomo	(3811-49-2)
Salix	(54-31-9)
Salizell	(65-45-2)
Salmiac	(12125-02-9)
Salmon Red Geigy	(3257-28-1)
Salol	(118-55-8)
Saloxium	(552-94-3)
Salpetersaure (German)	(7697-37-2)
Salpeterzuuroplossingen (Dutch)	(7697-37-2)
Salrin	(65-45-2)
Salsalate	(552-94-3)
Salsonin	(54-21-7)
Salt	(7647-14-5)
Salt Cake	(7757-82-6)
Salt of Saturn	(301-04-2)
Salt of Tarter	(3811-04-9)
Saltpeter	(7757-79-1)
Salufer	(16893-85-9)
Salunil	(58-94-6)
Salupres	(50-55-5)
Salural	(73-48-3)
Salures	(73-48-3)
Saluretil	(58-94-6)
Saluretin	(77-36-1)
Saluric	(58-94-6)
Salurin	(133-67-5)
Saluron	(135-09-1)
Salutensin	(50-55-5)
Salvacard	(59-26-7)
Salvacorin	(59-26-7)
Salvo	(75-60-5)
Salvo	(94-75-7)
Salvo (Czech)	(533-74-4)
Salvo Liquid	(65-85-0)
Salvo Powder	(65-85-0)
Salymid	(65-45-2)
Salysal	(552-94-3)
Salzburg Vitriol	(7758-99-8)
Sam	(65-45-2)
Samarium (9CI)	(7440-19-9)
Samarium oxide (9CI)	(12060-58-1)
Samaron Blue FBL	(12217-79-7)
Samaron Brilliant Violet B	(82-33-7)
Samaron Navy Blue GR	(50922-60-6)
Samaron Pink FBL	(17418-58-5)
Samaron Red Violet F3B	(2872-48-2)
Samaron Yellow PA3	(2832-40-8)
Samaron Yellow 5RL	(6300-37-4)
Samid	(65-45-2)
Samuron	(1014-69-3)
San 52 139 I	(31218-83-4)
San 155	(31895-22-4)
San 155 I	(31895-22-4)
San 197 I	(38260-54-7)
San 230	(640-15-3)
San 244 I	(2540-82-1)
San 322I	(31218-83-4)
San 371	(77732-09-3)
San 371F	(77732-09-3)
San 1551	(31895-22-4)
San 6538 I	(13593-03-8)
San 6626 I	(13593-03-8)
San 6913 I	(2540-82-1)
San 6915H	(19937-59-8)
San 7102H	(19937-59-8)
San 7107 I	(2540-82-1)
San 9789	(27314-13-2)
San 9789 H	(27314-13-2)
San 52139	(31218-83-4)
San-Cyan	(917-61-3)
San-EI Amaranth	(915-67-3)
San-EI Indigo Carmine	(860-22-0)
San-Ei Brilliant Scarlet 3R	(2611-82-7)
San-Ei Tartrazine	(1934-21-0)
Sanalgine	(62-44-2)
Sanamid	(63-74-1)
Sanaseed	(57-24-9)
Sanatol AS-G	(91-96-3)
Sanatol BO	(132-68-3)
Sanatol BR	(91-92-9)
Sanatol G	(91-96-3)
Sanatrichom	(443-48-1)
Sancap	(4147-51-7)
Sanclomycine	(60-54-8)
Sancyclan	(456-59-7)
Sand	(14808-60-7)
Sand Acid (DOT)	(16961-83-4)
Sandal	(115-71-9)
Sandalwood Oil	(8006-87-9)
(-)-Sandaracopimaric acid	(471-74-9)
Sandaracopimaric acid	(471-74-9)
Sandesin	(58-61-7)
Sandimmun	(59865-13-3)
Sandimmune	(59865-13-3)
Sandin EU	(123-94-4)
Sandix	(1314-41-6)
Sandocryl Blue BRL	(61-73-4)
Sandocryl Blue B-RLE	(12270-13-2)
Sandocryl Orange B-G	(3056-93-7)
Sandofan	(77732-09-3)
Sandolin	(534-52-1)
Sandolin A	(534-52-1)
Sandopel Black EX	(1937-37-7)
Sandopel Dark Green B	(3626-28-6)
Sandoptal	(77-26-9)
Sandorin Blue 2GLS	(147-14-8)
Sandorin Green 8GLS	(14302-13-7)
Sandormin	(50-35-1)
Sandothrene Blue NGR	(130-20-1)
Sandothrene Blue NRSC	(81-77-6)
Sandothrene Blue NRSN	(81-77-6)
Sandothrene Brilliant Green NBF	(128-58-5)
Sandothrene Brilliant Pink R	(2379-74-0)

Sandothrene Brown NBR (2475-33-4)
Sandothrene Dark Blue NR (1324-54-5)
Sandothrene Golden Orange NG (128-70-1)
Sandothrene Golden Yellow (128-66-5)
Sandothrene Olive N2R (2379-81-9)
Sandothrene Orange R (3263-31-8)
Sandothrene Printing Yellow (128-66-5)
Sandothrene Red N 6B (4203-77-4)
Sandothrene Violet N 4R (1324-55-6)
Sandothrene Violet N 2RB (1324-55-6)
Sandothrene Violet 4R (1324-55-6)
Sandoz 6538 (13593-03-8)
Sandoz 6706 (23576-23-0)
Sandoz 52139 (31218-83-4)
Sandoz S-6900 (2540-82-1)
Sandril (50-55-5)
Sandron (50-55-5)
Sanedrine (299-42-3)
Sanekis P 213 (8061-51-6)
Sanekis P 552 (8061-51-6)
Sanepil (57-41-0)
Sanfuran (59-87-0)
Sang γ (58-89-9)
Sanguinarin (2447-54-3)
Sanguinarine (2447-54-3)
Sanguiritrin (2447-54-3)
Saniclor 30 (82-68-8)
Sanipriol-4 (65-49-6)
Sanitized SPG (62-38-4)
Sanlose SN 20A (9004-32-4)
Sanmarton (51630-58-1)
Sanmorin OT 70 (577-11-7)
Sannix PL 910 (9003-11-6)
Sanocid (118-74-1)
Sanocide (118-74-1)
Sanodiazine (68-35-9)
Sanoflavin (553-30-0)
Sanohidrazina (54-85-3)
Sanoma (78-44-4)
Sanoquin (54-05-7)
Sanorex (22232-71-9)
Sanorin (550-99-2)
Sanorin (835-31-4)
Sanorin-Spofa (550-99-2)
Sanquinon (117-80-6)
Sanrex (9003-54-7)
Sanrex C (9003-54-7)
Sanrex SAN-H (9003-54-7)
Sanrex Sanc (9003-54-7)
Sansel Orange G (842-07-9)
Sansocizer DINP (28553-12-0)
Sanspor (2425-06-1)
Sanspor (2939-80-2)
Santal Oil (8006-87-9)
β-Santalene (511-59-1)
(+)-α-Santalol (115-71-9)
cis-α-Santalol (115-71-9)
d-α-Santalol (115-71-9)
Santalol A (115-71-9)
Santar (21908-53-2)
Santavy's Substance F (477-30-5)
Santene (529-16-8)
Santheose (83-67-0)
Santicizer 3 (80-39-7)
Santicizer B-16 (85-70-1)
Santicizer E-15 (84-72-0)
Santicizer 140 (9CI) (12765-56-9)
Santicizer 140 (26444-49-5)
Santicizer 141 (1241-94-7)
Santicizer 143 (9CI) (70323-51-2)
Santicizer 148 (96300-95-7)
Santicizer 154 (96300-96-8)
Santicizer 160 (85-68-7)
Santicizer 275 (9CI) (78690-83-2)
Santicizer 409 (9CI) (39409-52-4)
Santicizer 711 (3648-20-2)
Santicizer 711 (9CI) (39393-37-8)
Santicizer 1H (80-30-8)
Santicizer M-17 (85-71-2)
Santicizer 141 (Monsanto) (1241-94-7)
Santobane (50-29-3)
Santobrite (131-52-2)
Santobrite (87-86-5)
Santocel (7631-86-9)
Santochlor (106-46-7)
Santocure (95-33-0)
Santocure MOR (102-77-2)
Santocure NS (95-31-8)
Santoflex 13 (793-24-8)
Santoflex 17 (139-60-6)
Santoflex 36 (101-72-4)
Santoflex 77 (3081-14-9)
Santoflex 217 (103-96-8)
Santoflex A (91-53-2)
Santoflex AW (91-53-2)
Santoflex DD (89-28-1)
Santoflex IC (106-50-3)
Santogard PVI (17796-82-6)
Santomerse 3 (25155-30-0)
Santomerse D (1322-98-1)
Santomerse No. 1 (25155-30-0)
Santomerse No. 85 (25155-30-0)
Santonin (481-06-1)
α-Santonin (481-06-1)
l-α-Santonin (481-06-1)
Santoninic anhydride (481-06-1)
Santonox (96-69-5)
Santonox BM (96-69-5)
Santonox R (96-69-5)
Santophen (120-32-1)
Santophen 1 (120-32-1)
Santophen 20 (87-86-5)
Santophen 1 Flake (120-32-1)
Santophen I (120-32-1)
Santophen I Germicide (120-32-1)
Santophen 1 Solution (120-32-1)
Santoquin (91-53-2)
Santoquine (91-53-2)
Santoquin emulsion (1709-70-2)
Santoquin mixture 6 (1709-70-2)
Santotherm (1336-36-3)
Santotherm FR (1336-36-3)
Santouar A (79-74-3)
Santovar A (79-74-3)
Santowax M (92-06-8)
Santowax P (92-94-4)
Santowhite (85-60-9)
Santowhite Crystals (96-69-5)
Santowhite powder (85-60-9)
Santox (2104-64-5)
Santox (96-69-5)
Sanvex (15263-52-2)
Sanwax 161P (9002-88-4)
Sanyo Benzidine Orange (3520-72-7)

Sanyo Benzidine Yellow-B (6358-85-6)
Sanyo Carmine l2B (72-48-0)
Sanyo Cyanine Green (1328-53-6)
Sanyo Fast Blue BB Base (120-00-3)
Sanyo Fast Blue Salt B (119-90-4)
Sanyo Fast Orange GC Base (141-85-5)
Sanyo Fast Red 10B (6471-49-4)
Sanyo Fast Red B Base (97-52-9)
Sanyo Fast Red GL Base (89-62-3)
Sanyo Fast Red NN (1103-38-4)
Sanyo Fast Red RL Base (99-52-5)
Sanyo Fast Red Salt 3GL (89-63-4)
Sanyo Fast Red Salt RL (99-52-5)
Sanyo Fast Red Salt TR (3165-93-3)
Sanyo Fast Red TR Base (95-69-2)
Sanyo Fast Scarlet GG Base (95-82-9)
Sanyo Gum Orange A (633-96-5)
Sanyo Lacquer Red RN (1103-38-4)
Sanyo Lake Red C (5160-02-1)
Sanyo Lithol Red R (1103-38-4)
Sanyo Permanent Orange D 213 (4424-06-0)
Sanyo Permanent Orange D 616 (4424-06-0)
Sanyo Phthalocyanine Green FB Pure (1328-53-6)
Sanyo Phthalocyanine Green F6G (1328-53-6)
Sanyo Scarlet Pure (2425-85-6)
Sanyo Scarlet Pure No. 1000 (2425-85-6)
Sanyo Threne Blue IRN (81-77-6)
Sanyothrene Brillant Pink IR (2379-74-0)
Saolan (119-38-0)
Sapamine COB-ST (VAN) (16889-14-8)
Sapecron (470-90-6)
Saphicol (78-57-9)
Saphizon-DP (78-57-9)
Saphizon (78-57-9)
Saphos (78-57-9)
l-Sapietic acid (79-54-9)
Sapilent (739-71-9)
Sapogen T (97-80-3)
Sapona Red Lake RL-6280 (6471-49-4)
Saponite (1319-41-1)
Saponite ((Mg0.5-1Fe0-0.5)3(Si3.67Al0.33)(Na0-0.33Ca0-0.17)-(OH)2O10.4H2O) (9CI) (1319-41-1)
Sappilan (80-33-1)
Sappiran (80-33-1)
Saprecon C (470-90-6)
Saprol (26644-46-2)
Saran 683 (9011-06-7)
Saran 746 (9011-06-7)
Saran Resin 683 (9011-06-7)
Sarcell TEL (9004-32-4)
Sarcine (68-94-0)
Sarclex (330-55-2)
Sarcoclorin (148-82-3)
Sarcoclorin (531-76-0)
DL-Sarcolysin (531-76-0)
L-Sarcolysin (148-82-3)
p-L-Sarcolysin (148-82-3)
D-Sarcolysine (13045-94-8)
DL-Sarcolysine (531-76-0)
L-Sarcolysine (148-82-3)
Sarcolysine (148-82-3)
Sarcoset GM (1854-26-8)
Sarcosin (107-97-1)
Sarcosine (107-97-1)
Sarcosine, N-lauroyl-, sodium salt (8CI) (137-16-6)
Sarcosine, N-nitroso (13256-22-9)
Sarcosine, N-oleoyl- (8CI) (110-25-8)
Sarcosinic acid (107-97-1)
Sarcosyl NL (137-16-6)
Sarcosyl NL 30 (137-16-6)
Saridon (62-44-2)
Sarin (107-44-8)
Sarin II (107-44-8)
Sarkin (68-94-0)
Sarkine (68-94-0)
L-Sarkolysin (148-82-3)
Sarkolysin (148-82-3)
Sarkosyl NL (137-16-6)
Sarkosyl NL 30 (137-16-6)
Sarkosyl NL 35 (137-16-6)
Sarkosyl NL 97 (137-16-6)
Sarkosyl NL 100 (137-16-6)
Sarkosyl O (110-25-8)
Sarlach GN (Czech) (3257-28-1)
Sarlach R (Czech) (85-83-6)
Sarodormin (77-21-4)
Sarolex (333-41-5)
Saromet (439-14-5)
Sarpagan (50-55-5)
Sarpagen (50-55-5)
Sarpifan HP 1 (9003-22-9)
Sartomer SR 203 (2455-24-5)
Sartomer SR 206 (97-90-5)
Sartomer SR 351 (15625-89-5)
Sartosona (15686-71-2)
Sasapirin (552-94-3)
Sasapyrin (552-94-3)
Sasapyrine (552-94-3)
Sasapyrinum (552-94-3)
Sasil (1344-00-9)
Satecid (1918-16-7)
Satiagel GS 350 (11114-20-8)
Satiagel GS350 (9000-07-1)
Satiagum 3 (9000-07-1)
Satiagum Standard (9000-07-1)
Satin White Pigment (12004-14-7)
Satin white (11070-82-9)
Satisfar (38260-54-7)
Satol (143-28-2)
Satox 20WSC (52-68-6)
Saturn (28249-77-6)
Saturn Brown LBR (16071-86-6)
Saturn Red (1314-41-6)
Saturn Red B (2610-11-9)
Saturn Red F 3B (2610-10-8)
Saturno (28249-77-6)
Saure des phytins (German) (83-86-3)
Sauteralgyl (50-13-5)
Sauterazid (54-85-3)
Sauterzid (54-85-3)
Savacotyl (98-92-0)
Saventrine (51-30-9)
Saviac (3691-35-8)
Savory Oil (Summer variety) (8016-68-0)
Saxin (128-44-9)
Saxin (81-07-2)
Saxitoxin Dihydrochloride (35554-08-6)
Saxitoxin Hydrochloride (35554-08-6)
Saxol (8012-95-1)
Saxoline (8009-03-8)
Saxosozine (127-69-5)
Sayfor (78-57-9)
Sayfos (78-57-9)
Sayphos (78-57-9)

Saytex 102 (1163-19-5)
Saytex BCL 462 (3322-93-8)
Saytex 115 (9CI) (117148-85-3)
Saytex 102E (1163-19-5)
Sazol (16977-58-5)
Sazzio (9005-32-7)
ScH 15719W (36894-69-6)
Scabanca (120-51-4)
Scaldip (122-39-4)
Scanbutazone (50-33-9)
Scandium (9CI) (7440-20-2)
Scandium, Isotope of mass 46 (8CI,9CI) (13967-63-0)
Scarclex (330-55-2)
Scarlet B Fat Soluble (85-86-9)
Scarlet Base Ciba I (95-82-9)
Scarlet Base Ciba II (99-55-8)
Scarlet Base GG (95-82-9)
Scarlet Base Irga II (99-55-8)
Scarlet Base Irga I (Free Base) (95-82-9)
Scarlet Base NGG (95-82-9)
Scarlet Base NSP (99-55-8)
Scarlet 2G Base (95-82-9)
Scarlet G Base (99-55-8)
Scarlet GN (3257-28-1)
Scarlet GN Specially Pure (3257-28-1)
Scarlet Pigment RN (2425-85-6)
Scarlet 2R (3761-53-3)
Scarlet 6R (5850-44-2)
Scarlet R (3761-53-3)
Scarlet 2RB (3761-53-3)
Scarlet 2RL (3761-53-3)
Scarlet 2RL Bluish (3761-53-3)
Scarlet R (Michaelis) (85-83-6)
Scarlet RRA (3761-53-3)
Scarlet Red (85-83-6)
Scarlet Red, Biebrich (85-83-6)
Scarlet TR Base (87-60-5)
Scarlet Toner Y (5160-02-1)
Scatole (83-34-1)
Scepter (81335-37-7)
Sch 12041 (23092-17-3)
Sch 16134 (36735-22-5)
Sch 4831 (378-44-9)
Schaeffer's Acid (93-01-6)
Schaeffer's β-Acid (93-01-6)
Schaeffer's β-Naphtholsulfonic acid (93-01-6)
Schaeffer's acid, monopotassium salt (833-66-9)
Schalstein (13983-17-0)
Scharlach B (842-07-9)
Scharlachrot (85-83-6)
Scheeles Green (10290-12-7)
Scheele's Mineral (10290-12-7)
Scheelite (7759-01-5)
Schemergin (50-33-9)
Schercemol DIA (6938-94-9)
Schercomid MME (142-58-5)
Schering 12041 (23092-17-3)
Schering 34615 (2631-37-0)
Schering 36056 (23422-53-9)
Schering 36103 (17702-57-7)
Schering 36268 (19750-95-9)
Schering 36268 (6164-98-3)
Schering 38107 (13684-56-5)
Schering-35830 (7287-36-7)
Schering-38584 (13684-63-4)
Scherisolon (50-24-8)
Scheroson (50-04-4)
Scheroson (53-06-5)
Scheroson F (50-23-7)
Schinopsis Lorentzii Tannin (1401-55-4)
Schiwanox (57-43-2)
Schleimsaure (526-99-8)
Schneckokorn (9002-91-9)
Schnex-Schneckentod (9002-91-9)
Schollkopf's acid (VAN) (82-75-7)
Schradan (152-16-9)
Schradane (French) (152-16-9)
Schultz-Tab. No. 779 (German) (569-61-9)
Schultenite (7784-40-9)
Schultenite (8CI) (14758-11-3)
Schultenite (Pb(HAsO4)) (9CI) (14758-11-3)
Schultz No. 31 (85-86-9)
Schultz No. 39 (1936-15-8)
Schultz No. 95 (3761-53-3)
Schultz No. 185 (523-44-4)
Schultz No. 541 (85-83-6)
Schultz No. 737 (1934-21-0)
Schultz No. 770 (2650-18-2)
Schultz No. 770 (3844-45-9)
Schultz No. 887 (16423-68-0)
Schultz No. 918 (8004-92-0)
Schultz No. 1038 (61-73-4)
Schultz No. 1228 (81-77-6)
Schultz Nr. 208 (German) (3567-69-9)
Schultz Nr. 212 (German) (915-67-3)
Schultz Nr. 213 (German) (2611-82-7)
Schultz Nr. 826 (German) (129-17-9)
Schultz Nr. 1309 (German) (860-22-0)
Schungite (7782-42-5)
Schuttgelb (518-82-1)
L-Schwarz 1 (2519-30-4)
Schwefel-Lost (505-60-2)
Schwefeldioxyd (German) (7446-09-5)
Schwefelkohlenstoff (German) (75-15-0)
Schwefelsaeureloesungen (German) (7664-93-9)
Schwefelwasserstoff (German) (7783-06-4)
Schweflige saure (German) (7782-99-2)
Schweinfurt Green (12002-03-8)
Schweinfurtergrun (12002-03-8)
Schweinfurth Green (12002-03-8)
Scifluorfen (62476-59-9)
Scillirosid (507-60-8)
Scilliroside (507-60-8)
Scilliroside, (3-β,6-β)- (507-60-8)
Scilliroside, (3β,6β)- (507-60-8)
Scillirosidin + glucose (German) (507-60-8)
Scinnamina (24815-24-5)
Scintillar (106-42-3)
Scirpenol, 8-(3-methylbutyryloxy)-diacetoxy- (21259-20-1)
Sclair 11K (9002-88-4)
Sclair 19A (9002-88-4)
Sclair 59C (9002-88-4)
Sclair 59 (9002-88-4)
Sclair 79D (9002-88-4)
Sclair 96A (9002-88-4)
Sclair 19X6 (9002-88-4)
Sclair 2911 (9002-88-4)
Sclaventerol (67-45-8)
Sclex (24201-58-9)
Scolazil (57-53-4)
Sconatex (75-35-4)
Sconatex (9003-22-9)
Scopamin (114-49-8)
Scoparon (120-08-1)

Scoparone (120-08-1)
Scopine tropate (51-34-3)
(-)-Scopolamine (51-34-3)
Scopolamine (51-34-3)
Scopolamine, Compd. with methyl nitrate (1:1) (6106-46-3)
Scopolamine aminoxide hydrobromide (6106-81-6)
(-)-Scopolamine bromide (114-49-8)
Scopolamine bromide (114-49-8)
(-)-Scopolamine hydrobromide (114-49-8)
Scopolamine hydrobromide (114-49-8)
Scopolamine methobromide (114-49-8)
(-)-Scopolamine methyl bromide (155-41-9)
Scopolamine methylbromide (155-41-9)
Scopolamine methyl nitrate (155-41-9)
Scopolamine, N-oxide, hydrobromide (6106-46-3)
Scopolaminium bromide (6106-81-6)
Scopolammonium bromide (114-49-8)
Scopos (114-49-8)
Scorbacid (114-49-8)
Scorbu-C (50-81-7)
Scorch (50-81-7)
Scories de magnesium (Humides ou chaudes) (French) (7778-39-4)
Scot (69011-63-8)
Scotch PAR (561-27-3)
Scotcil (25038-59-9)
Scrobin (113-98-4)
Scurenaline (637-07-0)
Scurocaine (51-43-4)
Scurocaine (51-05-8)
Scutl (59-46-1)
Scyan (62-38-4)
Scyllite (540-72-7)
Se-Methylselenomethionine (87-89-8)
Sea-Legs (7728-97-4)
Sea Salt (569-65-3)
Seacyl Violet R (7647-14-5)
Seakem 3 & LCM (128-95-0)
Seakem Carrageenin (9000-07-1)
Searlequin (9000-07-1)
Seaspen PF (83-73-8)
Seatrem (9000-07-1)
Seawater Magnesia (9000-07-1)
Sebacic-acid (1309-48-4)
Sebacic acid, Compd. with 1,6-Hexanediamine (1:1) (111-20-6)
Sebacic acid, bis(2-ethylhexyl)ester (6422-99-7)
Sebacic acid, dibenzyl ester (8CI) (122-62-3)
Sebacic acid, dibutyl ester (140-24-9)
Sebacic acid dichloride (109-43-3)
Sebacic acid, diethyl ester (111-19-3)
Sebacic acid dimethyl ester (110-40-7)
Sebacic acid, dimethyl ester (8CI) (106-79-6)
Sebacic acid, 1,6-hexanediol polymer (106-79-6)
Sebacil (26745-88-0)
Sebaconitrile (14816-18-3)
Sebacoyl chloride (8CI) (1871-96-1)
Sebacoyl dichloride (111-19-3)
Sebacyl chloride (111-19-3)
Sebakan hexamethylendiaminu (Czech) (111-19-3)
Sebaquin (6422-99-7)
Sebical (83-73-8)
Sebizon (97-59-6)
Sebuthylazine (144-80-9)
Secacornin (7286-69-3)
Secagyn (60-79-7)
Secbubarbital (379-79-3)
Secbubarbital sodium (125-40-6)
Secbumeton (143-81-7)
(26259-45-0)
Secbutabarbital (125-40-6)
Secbutobarbitone (125-40-6)
Secbutobarbitone sodium (143-81-7)
Sechvitan (54-47-7)
13,17-Secoandrosta-1,4-dien-17-oic acid, 13-hydroxy-3-oxo-, δ-lactone (8CI) (968-93-4)
Secobarbital (76-73-3)
Secobarbitone (76-73-3)
7,13a-Secoberbin-13a-one, 7-methyl-2,3:9,10-bis(methylenedioxy) (130-86-9)
7,13a-Secoberbin-13a-one, 7-methyl-2,3:9,10-bis(methylenedioxy)-, hydrochloride (8CI) (6164-47-2)
(5Z,7E)-9,10-Secochesta-5,7,10(19)-triene-1-α,3-β,25-triol (32222-06-3)
9,10-Secocholesta-5,7,10(19)-triene-1,3,25-triol, (1-α,3-β,5Z,7E)- (9CI) (32222-06-3)
9,10-Secocholesta-5,7,10(19)-trien-3-β-ol (67-97-0)
9,10,Secoergosta-5,7,10(19),22-tetraen 3-β-ol (50-14-6)
9,10-Secoergosta-5,7,22-trien-3-ol, (3-β,5e,7e,10-α,22e)- (9CI) (67-96-9)
Secometrin (60-79-7)
Seconal (76-73-3)
Secondary ammonium arsenate (7784-44-3)
Secondary ammonium phosphate (7783-28-0)
Secondary calcium phosphate (7757-93-9)
Secondary sludge (68188-15-8)
Secopal OP 20 (9036-19-5)
Secorbate (50-81-7)
4,8-Secosenecionan-8,11,16-trione, 12-hydroxy-4-methyl- (2318-18-5)
Secrosteron (79-64-1)
Sectacyl Violet Propyl (128-95-0)
Secumbeton (26259-45-0)
Secupan (379-79-3)
Security (7778-44-1)
Security (7784-40-9)
Seda-Bute (143-81-7)
Seda-Tablinen (50-06-6)
Sedabamate (57-53-4)
Sedabar (50-06-6)
Sedafamen (634-03-7)
Sedaform (57-15-8)
Sedalis Sedi-Lab (50-35-1)
Sedantoinal (50-12-4)
Sedanyl (57-53-4)
Sedapercut (77-75-8)
Sedaraupin (50-55-5)
Sedaraupina (50-55-5)
Seda-recipin (50-55-5)
Seda-salurepin (50-55-5)
Sedatin (60-80-0)
Sedatine (60-80-0)
Sedazil (57-53-4)
Sederaupin (50-55-5)
Sedestran (56-53-1)
Sedetine (31566-31-1)
Sedetol (111-60-4)
Sedeval (57-44-3)
Sedicat (50-06-6)
Sedimide (50-35-1)
Sedin (50-35-1)
Sedipam (439-14-5)
Sedisperil (50-35-1)
Sedizorin (50-06-6)
Sedlyn (50-06-6)
Sednotic (57-43-2)
Sedofen (50-06-6)
Sedolin (60-13-9)
Sedometil (41372-08-1)
Sedometil (555-30-6)
Sedonal (50-06-6)

Sedonettes (50-06-6)
Sedoneural (7647-15-6)
Sedophen (50-06-6)
Sedoquil (57-53-4)
Sedoselecta (57-53-4)
Sedoval (50-35-1)
Sedresan (123-88-6)
Sedserp (50-55-5)
Seduksen (439-14-5)
Sedural (136-40-3)
Sedural (94-78-0)
Seduxen (439-14-5)
Seed Dressing Universal (517-16-8)
Seedox (22781-23-3)
Seedrin (309-00-2)
Seedtox (62-38-4)
Seekay Wax (1321-65-9)
Sefacin (50-59-9)
Seffein (63-25-2)
Sefril (38821-53-3)
Segnale Light Green G (1328-53-6)
Segnale Light Orange G (3520-72-7)
Segnale Light Orange GR (1934-20-9)
Segnale Light Orange PG (3520-72-7)
Segnale Light Orange RN (3468-63-1)
Segnale Light Orange RNG (3468-63-1)
Segnale Light Red 2B (2425-85-6)
Segnale Light Red B (2425-85-6)
Segnale Light Red BR (2425-85-6)
Segnale Light Red C4R (2425-85-6)
Segnale Light Red PRG (2814-77-9)
Segnale Light Red RL (2425-85-6)
Segnale Light Rubine RG (6471-49-4)
Segnale Light Turquoise NCG (147-14-8)
Segnale Light Turquoise NFG (147-14-8)
Segnale Light Turquoise NFR (147-14-8)
Segnale Light Turquoise PAG (147-14-8)
Segnale Light Turquoise SR (147-14-8)
Segnale Light Yellow 2GR (6358-85-6)
Segnale Light Yellow 2GRT (6358-85-6)
Segnale Red LC (5160-02-1)
Segnale Red LCG (5160-02-1)
Segnale Red LCL (5160-02-1)
Segnale Red 3R (5858-81-1)
Segnale Red R (1248-18-6)
Segontin (390-64-7)
Seguril (54-31-9)
Seignette Salt (304-59-6)
Sel-Tox SSO2 and SS-20 (13410-01-0)
Selecron (41198-08-7)
Selective (6365-83-9)
Seleno-DL-cysteine (18312-66-8)
Selektin (7287-19-6)
Selekton B 2 (139-33-3)
Selen (Polish) (7782-49-2)
1,2,5-Selenadiazole, 3-methyl- (8CI,9CI) (17505-11-2)
Selenate (7782-49-2)
Selenate (9CI,8CI) (14124-68-6)
Selenazolidine, 4-methyl- (7CI,8CI,9CI) (6474-16-4)
Selene (57-53-4)
Selenic acid, Compd. with hydrazine (73506-32-8)
Selenic acid, Liquid (DOT) (7783-08-6)
Selenic acid [UN 1905] (7783-08-6)
Selenic acid, disodium salt (13410-01-0)
Selenic acid, lead(2+) salt (1:1) (9CI) (7446-15-3)
Selenide, diethyl- (627-53-2)
Selenide, dimethyl (593-79-3)
Seleninyl-chloride (7791-23-3)
Selenious-acid (7783-00-8)
Selenious acid, disodium salt (10102-18-8)
Selenious acid, lead(2+) salt (1:1) (9CI) (7488-51-9)
Selenious acid, monosodium salt (7782-82-3)
Selenious anhydride (7446-08-4)
Selenite (9CI,8CI) (14124-67-5)
Selenium-cystine (18312-66-8)
Selenium-fluoride (7783-79-1)
Selenium (ACGIH) (7782-49-2)
Selenium Base (7782-49-2)
Selenium (Colloidal) (7782-49-2)
Selenium Dust (7782-49-2)
Selenium Elemental (7782-49-2)
Selenium Homopolymer (7782-49-2)
Selenium(IV) dioxide (1:2) (7446-08-4)
Selenium(IV) disulfide (1:2) (7488-56-4)
Selenium, Isotope of mass 75 (8CI,9CI) (14265-71-5)
Selenium alloy (7782-49-2)
Selenium chloride oxide (7791-23-3)
Selenium diethyldithiocarbamate (136-92-5)
Selenium diethyldithiocarbamate (5456-28-0)
Selenium dimethyldithiocarbamate (144-34-3)
Selenium dioxide (7446-08-4)
Selenium dioxide (7783-00-8)
Selenium disulphide [UN 2657] (7488-56-4)
Selenium hexafluoride (ACGIH,OSHA) [UN 2194] (7783-79-1)
Selenium hydride (7783-07-5)
Selenium metal powder, Non-pyrophoric (DOT) (7782-49-2)
Selenium metal powder [UN 2658] (7782-49-2)
Selenium mononitride (12033-59-9)
Selenium monosulfide (7446-34-6)
Selenium nitride (12033-59-9)
Selenium oxide (14832-90-7)
Selenium oxide (7446-08-4)
Selenium oxychloride [UN 2879] (7791-23-3)
Selenium sulfide (7488-56-4)
Selenium-sulfide (7446-34-6)
Selenium sulphide (7446-34-6)
Selenium, tetrakis(diethyldithiocarbamato) (5456-28-0)
Selenium, tetrakis(dimethyldithiocarbamato) (144-34-3)
1,1'-Selenobisdodecane (5819-01-2)
d,l-Selenocysteine (18312-66-8)
Selenomethionine (1464-42-2)
Selenonium, (3-amino-3-carboxypropyl)dimethyl- (9CI) (7728-97-4)
Selenourea (630-10-4)
Selensulfid (German) (7446-34-6)
Selephos (56-38-2)
Selexol (24991-55-7)
Self Rock Moss (9000-07-1)
Selfer (58-63-9)
Selinane (30824-81-8)
(-)-α-Selinene (473-13-2)
α-Selinene (473-13-2)
α-Selinene, (-)- (473-13-2)
β-Selinene (17066-67-0)
Selinon (534-52-1)
Sella Fast Red RS (6459-94-5)
Selsun (7488-56-4)
Selsun Blue (7488-56-4)
Seltz-K (866-84-2)
Sembrina (41372-08-1)
Sembrina (555-30-6)
Semdoxan (50-18-0)
Semdoxan (6055-19-2)
Semeron (1014-69-3)
Semevin (59669-26-0)

Semicarbazide (57-56-7)
Semicarbazide, 4-amino- (497-18-7)
Semicarbazide, 1-carbamoyl- (110-21-4)
Semicarbazide chloride (563-41-7)
Semicarbazide hydrochloride (563-41-7)
Semicarbazide, 1-(1-hydroxyformimidoyl)- (110-21-4)
Semicarbazide, 4-methyl-3-thio- (8CI) (6610-29-3)
Semicarbazide, monohydrochloride (563-41-7)
Semicarbazide, monohydrochloride (8CI) (563-41-7)
Semicarbazide, 4-phenyl- (8CI) (537-47-3)
Semicarbazide, 4-phenyl-3-thio (5351-69-9)
Semicarbazide, 3-thio- (79-19-6)
Semicarbazide, thio (79-19-6)
Semicillin (69-53-4)
p-Semidine (101-54-2)
Semikarbazid (Czech) (57-56-7)
Semikarbazon 5-nitrofurfuralu (Polish) (59-87-0)
Semikon (135-23-9)
Semikon (91-80-5)
Semikon Hydrochloride (135-23-9)
Semopen (132-93-4)
Semoxydrine (300-42-5)
Semustine (13909-09-6)
Senarmontite (12412-52-1)
Senarmontite (1309-64-4)
Sencephalin (15686-71-2)
Sencor (21087-64-9)
Sencoral (21087-64-9)
Sencorer (21087-64-9)
Sencorex (21087-64-9)
Sendoxan (50-18-0)
Sendoxan (6055-19-2)
Sendran (114-26-1)
Senduxan (50-18-0)
Senduxan (6055-19-2)
Seneca Oil (8002-05-9)
Senecioic acid (541-47-9)
Senecioic acid 2-sec butyl-4,6-dinitrophenyl ester (485-31-4)
Senecionan-11,16-dione, 13,19-didehydro-12,18-dihydroxy- (9CI) (23246-96-0)
Senecionan-11,16-dione, 13,19-didehydro-12-hydroxy- (9CI) (480-81-9)
Senecionan-11,16-dione, 12,18-dihydroxy- (480-54-6)
Senecionan-11,16-dione, 12,18-dihydroxy-, 4-oxide (9CI) (15503-86-3)
Senecionan-11,16-dione, 15,20-epoxy-15,20-dihydro-12-hydroxy-, (15-α,20s)- (9CI) (6870-67-3)
Senecionan-11,16-dione, 12-hydroxy (130-01-8)
Senecionanium, 8,12-dihydroxy-4-methyl-11,16-dioxo (2318-18-5)
Senecionanium, 8,12,18-trihydroxy-4-methyl-11,16-dioxo (26782-43-4)
Senecionin (130-01-8)
Senecionine (130-01-8)
Seneciphyllin (480-81-9)
Seneciphylline (480-81-9)
Senegal Gum (9000-01-5)
Senf Oel (German) (57-06-7)
Senkirkin (2318-18-5)
Senkirkine (2318-18-5)
Senkirkine (Neutral) (2318-18-5)
Senkirkine, hydroxy- (26782-43-4)
Sentonil (437-38-7)
Sentry (7778-54-3)
Sentry Cyclomethicone (69430-24-6)
Sentry Dimethicone (9006-65-9)
Sentry Grain Preserver (79-09-4)
Seominal (50-55-5)
Separan MG 700 (9CI) (54578-21-1)
Separol 29 (9003-11-6)
Separol WF 34 (9003-11-6)
Separol WF 41 (9003-11-6)
Sepazon (24166-13-0)
Sephadex (9014-76-0)
Sephadex G-25 (9014-76-0)
Sephadex G-75 (9014-76-0)
Sephadex G-100 (9014-76-0)
Sephadex G-200 (9014-76-0)
Sephadex Gels (9014-76-0)
Sephadex LH-20 (9014-76-0)
Sepian N 050 (9003-54-7)
Sepian N 6000HL (9003-54-7)
Sepiolite (14987-04-3)
Sepiolite (18307-23-8)
Seppic MMD (94-74-6)
Seprisan (140-72-7)
Sepsinol (139-91-3)
Septamide Album (63-74-1)
Septanilam (63-74-1)
Septeal (18472-51-0)
Septene (63-25-2)
Septicol (56-75-7)
Septinal (63-74-1)
Septiphene (120-32-1)
Septipulmon (144-83-2)
Septisol (70-30-4)
Septochol (83-44-3)
Septofen (70-30-4)
Septolix (63-74-1)
Septoplex (63-74-1)
Septoplix (63-74-1)
Septos (99-76-3)
Septra (723-46-6)
Septra (8064-90-2)
Septran (723-46-6)
Septran (8064-90-2)
Septrim (8064-90-2)
Septrin (723-46-6)
Septrin (8064-90-2)
Sepyron (456-59-7)
Seq 100 (60-00-4)
Sequestrene (64-02-8)
Sequestrene 30A (64-02-8)
Sequestrene AA (60-00-4)
Sequestrene Na 3 (150-38-9)
Sequestrene Na 4 (64-02-8)
Sequestrene NaFe Iron Chelate (15708-41-5)
Sequestrene ST (64-02-8)
Sequestrene sodium 2 (139-33-3)
Sequestrene trisodium (150-38-9)
Sequestrene trisodium salt (150-38-9)
Sequestric acid (60-00-4)
Sequestrol (60-00-4)
Seradix (133-32-4)
Seral (77-75-8)
Seranex (62-44-2)
Seraphos (31218-83-4)
Serax (604-75-1)
Sereen (114-49-8)
Serenace (52-86-8)
Serenack (439-14-5)
Serenal (24143-17-7)
Serenal (604-75-1)
Serenamin (439-14-5)
Serenase (52-86-8)
Serenelfi (52-86-8)
Serenid (604-75-1)
Serenid-D (604-75-1)
Serenil (113-18-8)

Serenium (2898-12-6)
Serensil (113-18-8)
Serentil (5588-33-0)
Serenzin (439-14-5)
Serepax (604-75-1)
Seresta (604-75-1)
Serfin (50-55-5)
Serfolia (50-55-5)
Sergosin (126-31-8)
Seribak (5736-15-2)
Seril (57-53-4)
Serilene Brilliant Red 2BL (17418-58-5)
Serilene Fast Pink BT (82-38-2)
Serilene Red 2BL (17418-58-5)
Serina (Spanish (56-45-1)
Serinal (72391-46-9)
L-(-)-Serine (56-45-1)
Serine (56-45-1)
L-Serine (9CI) (56-45-1)
Serine, L- (8CI) (56-45-1)
l-Serine, diazoacetate (115-02-6)
Serine, diazoacetate (ester) (115-02-6)
l-Serine diazoacetate (ester) (115-02-6)
Serine, diazoacetate (ester), l- (115-02-6)
Serinum (Latin) (56-45-1)
Serinyl Blue 2G (2475-45-8)
Serinyl Blue 3G (2475-45-8)
Serinyl Blue 3GN (2475-45-8)
Serinyl Hosiery Blue (2475-46-9)
Serinyl Hosiery Blue (86722-66-9)
Serinyl Hosiery Blue BG (2475-46-9)
Serinyl Hosiery Blue BG (86722-66-9)
Serinyl Hosiery Yellow GD (2832-40-8)
Seriplas Red X3B (2872-48-2)
Seriplas Yellow GD (2832-40-8)
Serisol Blilliant Red X 3B (2872-48-2)
Serisol Brilliant Blue BG (2475-46-9)
Serisol Brilliant Blue BG (86722-66-9)
Serisol Brilliant Blue BP (2475-46-9)
Serisol Brilliant Blue BP (86722-66-9)
Serisol Brilliant Blue FF (2475-46-9)
Serisol Brilliant Blue FF (86722-66-9)
Serisol Brilliant Blue G (2475-44-7)
Serisol Brilliant Violet 2R (128-95-0)
Serisol Fast Blue BGLW (15791-78-3)
Serisol Fast Red 2B (116-85-8)
Serisol Fast Violet 6B (1220-94-6)
Serisol Fast Violet B (82-33-7)
Serisol Fast Yellow A (119-15-3)
Serisol Fast Yellow GD (2832-40-8)
Serisol Fast Yellow N 5RD (6300-37-4)
Serisol Fast Yellow 2RGL (12223-97-1)
Serisol Orange YL (82-28-0)
Seristan Black B (1937-37-7)
Seritox 50 (120-36-5)
Serlabo (7235-40-7)
Sermaka (1524-88-5)
Sernas (52-86-8)
Sernel (52-86-8)
Sernevin (15676-16-1)
Sernyl (956-90-1)
Sernylan (956-90-1)
Sernyl hydrochloride (956-90-1)
Seroalbumina humana iodada (125 I) (Spanish) (9048-46-8)
Seroalbumina humana iodada (131 I) (Spanish) (9048-46-8)
Seroalbuminum humanum iodinatum (125 I) (Latin) (9048-46-8)
Seroalbuminum humanum iodinatum (131 I) (Latin) (9048-46-8)
Seroalbuminum humanum jodinatum (125 I) (9048-46-8)
Serofinex (637-07-0)
Serolfia (50-55-5)
Seromycin (68-41-7)
Serotinex (637-07-0)
Serotonin (50-67-9)
Serotonine (50-67-9)
Serp (50-55-5)
Serp-AFD (50-55-5)
Serpalan (50-55-5)
Serpaloid (50-55-5)
Serpaneurona (50-55-5)
Serpanray (50-55-5)
Serpasil (50-55-5)
Serpasil Apresoline (50-55-5)
Serpasil Apresoline No. 2 (304-20-1)
Serpasil-Esidrex (50-55-5)
Serpasil-Esidrex K (50-55-5)
Serpasil-Esidrex No. 1 (50-55-5)
Serpasil-Esidrex No. 2 (50-55-5)
Serpasil Premix (50-55-5)
Serpasol (50-55-5)
Serpate (50-55-5)
Serpatone (50-55-5)
Serpax (50-55-5)
Serpax (604-75-1)
Serpazil (50-55-5)
Serpazol (50-55-5)
Serpedin (50-55-5)
Serpen (50-55-5)
Serpena (50-55-5)
Serpentil (50-55-5)
Serpentin (50-55-5)
Serpentina (50-55-5)
Serpentine (12001-29-5)
Serpentine (1332-21-4)
Serpentine chrysotile (12001-29-5)
Serpentine (Pharbil) (50-55-5)
Serpicon (50-55-5)
Serpil (50-55-5)
Serpiloid (50-55-5)
Serpilum (50-55-5)
Serpine (50-55-5)
Serpine (Pharmaceutical) (50-55-5)
Serpipur (50-55-5)
Serpivite (50-55-5)
Serplex K (50-55-5)
Serpogen (50-55-5)
Serpoid (50-55-5)
Serpone (50-55-5)
Serpresan (50-55-5)
Serpyrit (50-55-5)
Serral (56-53-1)
Sertabs (50-55-5)
Sertan (125-33-7)
Sertens (50-55-5)
Sertensin (50-55-5)
Sertina (50-55-5)
Sertinon (536-33-4)
Sertofran (50-47-5)
Serum Tryptase (9001-90-5)
Serum-albumine humaine iodee (125 I) (French) (9048-46-8)
Serum-albumine humaine iodee (131 I) (French) (9048-46-8)
Serum albumin, iodinated (125I) human (9048-46-8)
Serum albumin, iodinated (131I) human (9048-46-8)
Serum albuminum radio-iodatum (131 I) (9048-46-8)
Serum proteins (9048-46-8)

Serva Blue G (6104-58-1)
N-Serve (1929-82-4)
N-Serve Nitrogen Stabilizer (1929-82-4)
Servisone (53-03-2)
Ses (136-78-7)
Sesagard (7287-19-6)
Sesame Oil (8008-74-0)
Sesamex (51-14-9)
Sesone (ACGIH,OSHA) (136-78-7)
Sesoxane (51-14-9)
Sesquiethylaluminum chloride (12075-68-2)
Sesquimustard (3563-36-8)
Sesquimustard Q (3563-36-8)
Sesquisulfure de phosphore (French) (1314-85-8)
Sesquiterpene mixture from guaiacwood oil (68647-67-6)
Sesquiterpenes and sesquiterpenoids, Guaiac wood-oil (68647-67-6)
Sestron (150-59-4)
Sestron Base (150-59-4)
Setacyl Blue BN (2475-46-9)
Setacyl Blue BN (86722-66-9)
Setacyl Blue BS (2475-44-7)
Setacyl Blue FG (2475-46-9)
Setacyl Blue FG (86722-66-9)
Setacyl Blue 2GS (2475-45-8)
Setacyl Blue 2GS II (2475-45-8)
Setacyl Blue RF (2475-46-9)
Setacyl Blue RF (86722-66-9)
Setacyl Brilliant Blue (2475-46-9)
Setacyl Brilliant Blue (86722-66-9)
Setacyl Brilliant Blue BG (2475-46-9)
Setacyl Brilliant Blue BG (86722-66-9)
Setacyl Diazo Navy R (119-90-4)
Setacyl Pink 3B (116-85-8)
Setacyl Red P-3B (2872-48-2)
Setacyl Violet P-R (128-95-0)
Setacyl Violet R (128-95-0)
Setacyl Yellow G (2832-40-8)
Setacyl Yellow 2GN (2832-40-8)
Setacyl Yellow P-BS (119-15-3)
Setacyl Yellow P-2GL (2832-40-8)
Sethadil (94-19-9)
Sethoxydim (74051-80-2)
Sethoxydime (French) (74051-80-2)
Setile Violet 3R (128-95-0)
Setoglaucine (3521-06-0)
Setonil (439-14-5)
Setran (57-53-4)
Seudenone (1193-18-6)
Sevacarb (1333-86-4)
Seval (1333-86-4)
Sevenal (50-06-6)
Sevicaine (51-05-8)
Sevimol (63-25-2)
Sevin (63-25-2)
Sevin 4 (63-25-2)
Sevin (OSHA) (63-25-2)
Sevinol (69-23-8)
Sevinon (112-38-9)
Sevron Blue 2G (12217-41-3)
Sevron Blue 5G (55840-82-9)
Sevron Blue 5GNF (55840-82-9)
Sevron Orange G (3056-93-7)
Sevron Red GL (14097-03-1)
Sewin (63-25-2)
Sexadien (84-17-3)
Sexocretin (56-53-1)
Sextone (108-94-1)
Sextone B (108-87-2)
Sextra (8008-74-0)
Shale, Expanded (68476-95-9)
Shale, Expanded, Aggregates (68334-37-2)
Shale Oil [UN 1288] (68308-34-9)
Shale Oils, Crude (68308-34-9)
Shalvaton (57-53-4)
Shamrox (94-74-6)
Sharcillin (6130-64-9)
Sharstop 204 (128-04-1)
Shawinigan Acetylene Black (1333-86-4)
Shed-A-Leaf (7775-09-9)
Shed-A-Leaf "L" (7775-09-9)
Shell 40 (94-80-4)
Shell 300 (9003-53-6)
Shell 345 (869-29-4)
Shell 4072 (470-90-6)
Shell 4402 (297-78-9)
Shell 5520 (9003-07-0)
Shell Atrazine Herbicide (1912-24-9)
Shell Carbon (1333-86-4)
Shell Gold (7440-57-5)
Shell MIBK (108-10-1)
Shell OS 1836 (311-47-7)
Shell SD 345 (869-29-4)
Shell SD 4294 (7700-17-6)
Shell SD 9129 (6923-22-4)
Shell SD-3562 (141-66-2)
Shell SD-5532 (57-74-9)
Shell SD-8280 (2274-67-1)
Shell SD-8530 (2686-99-9)
Shell SD-8591 (5918-93-4)
Shell SD-8803 (1224-63-1)
Shell SD-8988 (13104-21-7)
Shell SD-9098 (1757-18-2)
Shell SD-14114 (13356-08-6)
Shell Silver (7440-22-4)
Shell Unkrautted A (107-18-6)
Shell WL 1650 (297-78-9)
Shelloyne H (66289-74-5)
Shellsol 140 (111-84-2)
Shellsol (9CI) (12676-97-0)
Shelon (63428-84-2)
Sherpa (52315-07-8)
Sherstat TBS (87-10-5)
Sherwood Green A 4436 (1328-53-6)
Shigatox (57-67-0)
Shigrodin (50-33-9)
Shikimate (138-59-0)
Shikimic acid (138-59-0)
Shikimole (94-59-7)
Shikiso Acid Anthracene Red G (10169-02-5)
Shikiso Acid Brilliant Blue 6B (2390-59-2)
Shikiso Acid Fast Yellow MR (6375-55-9)
Shikiso Acid Red PG (3567-65-5)
Shikiso Amaranth (915-67-3)
Shikiso Direct Dark Green B (3626-28-6)
Shikiso Direct Scarlet 3B (6358-29-8)
Shikiso Direct Sky Blue 5B (2429-74-5)
Shikiso Direct Sky Blue 6B (2610-05-1)
Shikiso Metanil Yellow (587-98-4)
Shikiso Roccelline (1658-56-6)
Shikomol (94-59-7)
Shimmerex (62-38-4)
Shin-Naito S (50-35-1)
Shinkolite (9011-14-7)
Shinnibrol (50-35-1)

Shinnippon Fast Red B Base (97-52-9)
Shinnippon Fast Red GG Base (100-01-6)
Shinnippon Fast Red 3GL Base (89-63-4)
Shinnippon Fast Red GL Base (89-62-3)
Shirahagen S (76280-91-6)
Shirlan Extra (87-17-2)
Shiso Oil (68132-21-8)
Shoallomer (9003-07-0)
Shoallomer FA 120 (9003-07-0)
Shoallomer FA 530 (9003-07-0)
Shoallomer MA 210 (9003-07-0)
Shock-Ferol (50-14-6)
Shodine 500 (37099-12-0)
Shodine 508 (37099-12-0)
Sholex 5003 (9002-88-4)
Sholex 5100 (9002-88-4)
Sholex 6000 (9002-88-4)
Sholex 6002 (9002-88-4)
Sholex F 171 (9002-88-4)
Sholex F 6050C (9002-88-4)
Sholex F 6080C (9002-88-4)
Sholex 4250HM (9002-88-4)
Sholex L 131 (9002-88-4)
Sholex S 6008 (9002-88-4)
Sholex Super (9002-88-4)
Sholex XMO 314 (9002-88-4)
Shorbic acid (1135-23-5)
Short-Stop E (886-50-0)
Shortstop (886-50-0)
Showa Fast Red B Base (97-52-9)
Shoxin (991-42-4)
Shungite (7782-42-5)
Siacarb (28249-77-6)
Siam Benzoin (9000-05-9)
Siarczan sparteiny (Polish) (299-39-8)
Siarki chlorek (Polish) (10025-67-9)
Siarki dwutlenek (Polish) (7446-09-5)
Siarkowodor (Polish) (7783-06-4)
Sibazon (439-14-5)
Sibephyllin (479-18-5)
Sibephylline (479-18-5)
Sibol (56-53-1)
Sibutol (55179-31-2)
Sicarol (24691-76-7)
Sicilian Blue A 7021 (1325-87-7)
Sicilian Cerise Toner A-7127 (81-88-9)
Sico Fat Yellow P (2481-94-9)
Sico Lake Red 2L (5160-02-1)
Sicol 150 (117-81-7)
Sicol 160 (85-68-7)
Sicol 184 (26761-40-0)
Sicol 250 (103-23-1)
Siderite, magnesian (9CI) (102785-99-9)
Siduron (1982-49-6)
Siegle Blue Extract D 448 (1325-87-7)
Siegle Bluish Violet Extract DH47 (1325-82-2)
Siegle Fast Blue LBGO (147-14-8)
Siegle Fast Green G (1328-53-6)
Siegle Orange 2S (3468-63-1)
Siegle Orange S (3520-72-7)
Siegle Red 1 (2425-85-6)
Siegle Red B (2425-85-6)
Siegle Red BB (2425-85-6)
Sienna (1309-37-1)
Siferrit (1332-37-2)
Sificetina (56-75-7)
Sigacalm (604-75-1)
Sigaprin (8064-90-2)
Sigma (23564-05-8)
Sigmacell (9004-34-6)
Sigmadyn (2152-34-3)
Sigmafon (64-55-1)
Sigmamycin (60-54-8)
Signal Orange Orange Y-17 (3468-63-1)
Signal Red (1248-18-6)
Signopam (846-50-4)
Siguran (569-65-3)
Sil-Co-Sil (14808-60-7)
Silak M 10 (63148-62-9)
Silanamine, 1,1,1-trimethyl-N-(trimethylsilyl)- (9CI) (999-97-3)
Silane SC 3900 (13822-56-5)
Silane [UN 2203] (7803-62-5)
Silane Y-4086 (3388-04-3)
Silane-Y-4087 (2530-83-8)
Silane, allyltrichloro (107-37-9)
Silane, (4-aminobutyl)diethoxymethyl- (3037-72-7)
Silane, (3-(2-aminoethyl)aminopropyl)trimethoxy- (1760-24-3)
Silane, (3-aminopropyl)diethoxymethyl- (3179-76-8)
Silane, (3-aminopropyl)triethoxy- (919-30-2)
Silane, γ-aminopropyltriethoxy- (919-30-2)
Silane, bromo- (13465-73-1)
Silane, butyltrichloro (7521-80-4)
Silane, 1-butynyltrimethyl- (9CI) (62108-37-6)
Silane, chlorodiethoxy (6485-91-2)
Silane, chlorodiethyl- (8CI,9CI) (1609-19-4)
Silane, chlorodimethyl- (9CI) (1066-35-9)
Silane, chlorodimethylphenyl- (9CI) (768-33-2)
Silane, chlorodimethylvinyl- (1719-58-0)
Silane, chloroethenyldimethyl- (9CI) (1719-58-0)
Silane, chloromethyldiphenyl- (9CI) (144-79-6)
Silane, chloromethyl(trichloro)- (1558-25-4)
Silane, chlorophenyltrichloro- (26571-79-9)
Silane, (3-chloropropyl)dimethoxy(3-(oxiranylmethoxy)-propyl)- (9CI) (71808-64-5)
Silane, (3-chloropropyl)trimethoxy- (9CI) (2530-87-2)
Silane, chlorotriethyl- (9CI) (994-30-9)
Silane, chlorotrimethyl (75-77-4)
Silane, (2-cyanoethyl)trichloro- (1071-22-3)
Silane, (2-cyanoethyl)triethoxy (919-31-3)
Silane, (3-cyclohexenyl)trichloro- (10137-69-6)
Silane, cyclohexyltrichloro- (98-12-4)
Silane, dibenzyldichloro- (18414-36-3)
Silane, dichloro (4109-96-0)
Silane, dichlorobis(phenylmethyl)- (18414-36-3)
Silane, dichloro(chloromethyl)methyl- (9CI) (1558-33-4)
Silane, dichlorodiethyl (1719-53-5)
Silane, dichlorodimethyl (75-78-5)
Silane, dichlorodiphenyl (80-10-4)
Silane, dichlorodi-2-propenyl- (9CI) (3651-23-8)
Silane, dichloroethyl (1789-58-8)
Silane, dichloroethylphenyl (1125-27-5)
Silane, dichloroethylvinyl (10138-21-3)
Silane, dichloromethyl (75-54-7)
Silane, dichloromethylphenyl (149-74-6)
Silane, dichloromethyl(2-phenylethyl)- (9CI) (772-65-6)
Silane, dichloromethyl(3,3,3-trifluoropropyl) (675-62-7)
Silane, dichloromethylvinyl (124-70-9)
Silane, (dichlorophenyl)trichloro (27137-85-5)
Silane, diethoxymethyl(3-(oxiranylmethoxy)propyl)- (9CI) (2897-60-1)
Silane, diethylmethylvinyl- (6CI,7CI,8CI) (18292-29-0)
Silane, difluorodimethyl- (9CI) (353-66-2)
Silane, difluoromethylphenyl- (6CI,7CI,8CI,9CI) (328-57-4)
Silane, dimethoxydiphenyl- (9CI) (6843-66-9)
Silane, dimethylbis(octadecyloxy)- (9CI) (29043-70-7)

Name	CAS
Silane, dimethyldiphenoxy- (8CI,9CI)	**(3440-02-6)**
Silane, dimethylphenyl- (9CI)	**(766-77-8)**
Silanediol, dimethyl- (9CI)	**(1066-42-8)**
Silanediol, methylphenyl- (8CI,9CI)	**(3959-13-5)**
Silane, (dodecyloxy)trimethyl- (8CI,9CI)	**(6221-88-1)**
Silane, dodecyltrichloro	**(4484-72-4)**
Silane, β-(3,4-epoxycyclohexyl)ethyltrimethoxy	**(3388-04-3)**
Silane, 3-(2,3-epoxypropoxy)propyltrimethoxy	**(2530-83-8)**
Silane, ethenyldiethylmethyl- (9CI)	**(18292-29-0)**
Silane, ethoxydimethyl(3-(oxiranylmethoxy)propyl)- (9CI)	**(17963-04-1)**
Silane, ethoxytrimethyl	**(1825-62-3)**
Silane, ethyltrichloro	**(115-21-9)**
Silane, ethyltriethoxy	**(78-07-9)**
Silane, fluorodimethylphenyl- (7CI,8CI,9CI)	**(454-57-9)**
Silane, fluorotrimethyl- (9CI)	**(420-56-4)**
Silane, (hexadecyloxy)trimethyl- (7CI,8CI,9CI)	**(6221-90-5)**
Silane, hexadecyltrichloro	**(5894-60-0)**
Silane, hexyltrichloro	**(928-65-4)**
Silane, (3-hydroxypropyl)trimethoxy-, methacrylate	**(2530-85-0)**
Silane, hydroxytriphenyl	**(791-31-1)**
Silane, 3-mercaptopropyltrimethoxy-	**(4420-74-0)**
Silane, methoxytrimethyl- (9CI)	**(1825-61-2)**
Silane, methyltriacetoxy	**(4253-34-3)**
Silane, methyltrichloro	**(75-79-6)**
Silane, methyltriethoxy	**(2031-67-6)**
Silane, methyltrimethoxy	**(1185-55-3)**
Silane, nonyltrichloro	**(5283-67-0)**
Silane, octadecyltrichloro	**(112-04-9)**
Silane, octyltrichloro	**(5283-66-9)**
Silane, oxybis(trimethyl-	**(107-46-0)**
Silane, pentyltrichloro	**(107-72-2)**
Silane, phenyltrichloro-	**(98-13-5)**
Silane, phenyltriethoxy	**(780-69-8)**
Silane, phenyltrimethoxy	**(2996-92-1)**
Silane, propyltrichloro	**(141-57-1)**
Silane, tetrachloro-	**(10026-04-7)**
Silane, tetraethoxy-	**(78-10-4)**
Silane, tetraethyl- (9CI)	**(631-36-7)**
Silane, tetrafluoro-	**(7783-61-1)**
Silane, tetramethyl	**(75-76-3)**
Silanetriamine, N,N,N',N',N'',N''-hexamethyl	**(15112-89-7)**
Silanetriamine, N,N',N''-tributyl-1-methyl- (9CI)	**(16411-33-9)**
Silanetriamine, N,N',N''-tricyclohexyl-1-methyl- (9CI)	**(15901-40-3)**
Silane, trichloro	**(10025-78-2)**
Silane, trichloroallyl-	**(107-37-9)**
Silane, trichloro(chloromethyl)- (8CI,9CI)	**(1558-25-4)**
Silane, trichloro(3-chloropropyl)- (9CI)	**(2550-06-3)**
Silane, trichloro-3-cyclohexen-1-yl- (9CI)	**(10137-69-6)**
Silane, trichlorocyclohexyl- (8CI,9CI)	**(98-12-4)**
Silane, trichlorododecyl-	**(4484-72-4)**
Silane, trichloroethenyl	**(75-94-5)**
Silane, trichloroethyl-	**(115-21-9)**
Silane, trichlorohexadecyl-	**(5894-60-0)**
Silane, trichlorohexyl-	**(928-65-4)**
Silane, trichloromethyl-	**(75-79-6)**
Silane, trichloro(2-methylpropyl)- (9CI)	**(18169-57-8)**
Silane, trichlorononyl-	**(5283-67-0)**
Silane, trichlorooctadecyl-	**(112-04-9)**
Silane, trichlorooctyl-	**(5283-66-9)**
Silane, trichloropentyl-	**(107-72-2)**
Silane, trichlorophenethyl	**(940-41-0)**
Silane, trichlorophenyl	**(98-13-5)**
Silane, trichloropropyl-	**(141-57-1)**
Silane, trichloro(tetrachlorophenyl)- (9CI)	**(33434-63-8)**
Silane, trichlorovinyl-	**(75-94-5)**
Silane, triethoxy	**(998-30-1)**
Silane, triethoxyethyl-	**(78-07-9)**
Silane, triethoxymethyl-	**(2031-67-6)**
Silane, triethoxyoctyl- (9CI)	**(2943-75-1)**
Silane, triethoxyphenyl-	**(780-69-8)**
Silane, triethoxyvinyl	**(78-08-0)**
Silane, triethyl- (9CI)	**(617-86-7)**
Silane, triethylfluoro- (8CI,9CI)	**(358-43-0)**
Silane, trimethoxy	**(2487-90-3)**
Silane, trimethoxymethyl-	**(1185-55-3)**
Silane, trimethoxy(2-methylpropyl)- (9CI)	**(18395-30-7)**
Silane, trimethoxyvinyl	**(2768-02-7)**
Silane, trimethylchloro-	**(75-77-4)**
Silane, trimethylethoxy-	**(1825-62-3)**
Silane, trimethyl((1-methyl-2-propynyl)oxy)- (8CI,9CI)	**(17869-76-0)**
Silanetriol, ethenyl-, triacetate (9CI)	**(4130-08-9)**
Silanetriol, ethyl-, triacetate (9CI)	**(17689-77-9)**
Silanetriol, methyl-, sodium salt (9CI)	**(16589-43-8)**
Silane, tris(2-methoxyethoxy)vinyl	**(1067-53-4)**
Silane, vinyl trichloro A-150	**(75-94-5)**
Silanol, diphenylmethyl	**(778-25-6)**
Silantin	**(57-41-0)**
Silar 10:100	**(63148-62-9)**
Silatrane, phenyl-	**(2097-19-0)**
Silbephylline	**(479-18-5)**
Silber (German)	**(7440-22-4)**
Silbernitrat	**(7761-88-8)**
Silbesan	**(54-05-7)**
Silbione	**(69430-24-6)**
Silbione Oils 70047	**(9006-65-9)**
Silene EF	**(1344-95-2)**
Silesian EM	**(136-84-5)**
Silica, Amorphous-diatomaceous earth (ACGIH)	**(61790-53-2)**
Silica, Amorphous-diatomaceous earth uncalcined; Containing < 1% quartz	**(61790-53-2)**
Silica, Amorphous	**(7631-86-9)**
Silica, Amorphous, Diatomaceous earth (Containing less than 1% crystalline silica)	**(68855-54-9)**
Silica, Amorphous, diatomaceous earth (OSHA)	**(61790-53-2)**
Silica, Amorphous fumed	**(7631-86-9)**
Silica, Amporphous-precipitated silica	**(112926-00-8)**
Silica, Amporphous-silica gel	**(112926-00-8)**
Silica, Crystalline - Fused	**(60676-86-0)**
Silica, Crystalline-cristobalite (ACGIH,OSHA)	**(14464-46-1)**
Silica, Crystalline quartz (OSHA)	**(14808-60-7)**
Silica, Crystalline-tridymite (ACGIH,OSHA)	**(15468-32-3)**
Silica, Crystalline-tripoli (ACGIH,OSHA)	**(1317-95-9)**
Silica Flour (Powdered crystalline silica)	**(14808-60-7)**
Silica, Fused (ACGIH,OSHA)	**(60676-86-0)**
Silica, Quartz (ACGIH)	**(14808-60-7)**
Silica gel	**(1343-98-2)**
Silica gel	**(63231-67-4)**
Silica hydrate	**(1343-98-2)**
Silicane	**(7803-62-5)**
Silicane, chlorotrimethyl-	**(75-77-4)**
Silicane, trichloroethyl-	**(115-21-9)**
Silicate (9CI)	**(12627-13-3)**
Silicate, Mica	**(12001-26-2)**
Silicate, Portland Cement	**(65997-15-1)**
Silicate d'ethyle (French)	**(78-10-4)**
Silicate(2-), hexafluoro-, barium	**(17125-80-3)**
Silicate(2-), hexafluoro-, barium (1:1) (9CI)	**(17125-80-3)**
Silicate, hexafluoro-, diammonium	**(16919-19-0)**
Silicate(2-), hexafluoro-, diammonium (9CI)	**(16919-19-0)**
Silicate(2-), hexafluoro-, diammonium + silica	**(16919-19-0)**
Silicate(2-), hexafluoro-, dihydrogen	**(16961-83-4)**
Silicate(2-), hexafluoro-, disodium	**(16893-85-9)**
Silicate(2-), hexafluoro-, lead(2+) (1:1) (9CI)	**(25808-74-6)**
Silicate(2-), hexafluoro-, lead(II) salt, dihydrate	**(25808-74-6)**

Name	CAS No.
Silicate(2-), hexafluoro-, magnesium (1:1) (9CI)	**(16949-65-8)**
Silicate(2-), hexafluoro-, zinc	**(16871-71-9)**
Silicate tetraethylique (French)	**(78-10-4)**
Silici-chloroforme (French)	**(10025-78-2)**
Silicic acid (H2SiO3), disodium salt	**(6834-92-0)**
Silicic acid (H4SiO4), tetrapropyl ester (9CI)	**(682-01-9)**
Silicic acid (VAN) (9CI)	**(1343-98-2)**
Silicic acid, aluminum calcium sodium salt (9CI)	**(1344-01-0)**
Silicic acid, aluminum salt	**(1302-76-7)**
Silicic acid, aluminum salt (9CI)	**(1335-30-4)**
Silicic acid, aluminum sodium salt (9CI)	**(1344-00-9)**
Silicic acid, beryllium salt	**(15191-85-2)**
Silicic acid, beryllium salt (9CI)	**(58500-38-2)**
Silicic acid, beryllium zinc salt	**(39413-47-3)**
Silicic acid, calcium salt	**(10101-39-0)**
Silicic acid, calcium salt	**(1344-95-2)**
Silicic acid, calcium salt (1:3)	**(12168-85-3)**
Silicic acid, calcium salt (1:1) (8CI,9CI)	**(10101-39-0)**
Silicic acid, chromium lead salt (9CI)	**(11113-70-5)**
Silicic acid, dilithium salt (9CI)	**(10102-24-6)**
Silicic acid, ethyl ester (9CI)	**(11099-06-2)**
Silicic acid, hexaethyl ester (9CI)	**(2157-42-8)**
Silicic acid, lead salt (9CI)	**(11120-22-2)**
Silicic acid, magnesium salt (1:1)	**(1343-88-0)**
Silicic acid, magnesium salt (1:2)	**(14987-04-3)**
Silicic acid, magnesium salt (9CI)	**(1343-88-0)**
Silicic acid, methyl ester of ortho-	**(681-84-5)**
Silicic acid, potassium salt (9CI)	**(1312-76-1)**
Silicic acid, sodium salt (9CI)	**(1344-09-8)**
Silicic acid, tetra(2-ethylbutyl) ester	**(78-13-7)**
Silicic acid, tetraethyl ester	**(78-10-4)**
Silicic acid, tetraethyl ester, Homopolymer	**(11099-06-2)**
Silicic acid, tetraethylester polymer	**(11099-06-2)**
Silicic acid, tetrakis(1-methylpropyl) ester (9CI)	**(5089-76-9)**
Silicic acid, tetramethyl ester	**(681-84-5)**
Silicic acid, tetrasodium salt (9CI)	**(13472-30-5)**
Silicic acid, tricalcium salt	**(12168-85-3)**
Silicic anhydride	**(14808-60-7)**
Silicic anhydride	**(7631-86-9)**
Silicio(tetracloruro di)	**(10026-04-7)**
Siliciumchloroform (German)	**(10025-78-2)**
Siliciumtetrachlorid (German)	**(10026-04-7)**
Siliciumtetrachloride (Dutch)	**(10026-04-7)**
Silicium(tetrachlorure de) (French)	**(10026-04-7)**
Silicochloroform	**(10025-78-2)**
Silicofluoric acid (DOT)	**(16961-83-4)**
Silicon (ACGIH,DOT,OSHA)	**(7440-21-3)**
Silicon alloy, nonbase, Al 60-66,Si 25-30,Ni 5-7,Al2O3 3-4 (SAS 1) (9CI)	**(12743-20-3)**
Silicon-carbide	**(409-21-2)**
Silicon carbide (ACGIH,OSHA)	**(409-21-2)**
Silicon chloride (DOT)	**(10026-04-7)**
Silicon dioxide	**(60676-86-0)**
Silicone 360	**(63148-62-9)**
Silicone A-172	**(1067-53-4)**
Silicone A-174	**(2530-85-0)**
Silicone A-186	**(3388-04-3)**
Silicone A-187	**(2530-83-8)**
Silicone A-189	**(4420-74-0)**
Silicone A-1100	**(919-30-2)**
Silicone A-1120	**(1760-24-3)**
Silicone COM 10000	**(69430-24-6)**
Silicone COM 16520	**(69430-24-6)**
Silicone COM 20000	**(69430-24-6)**
Silicone COM 27510	**(69430-24-6)**
Silicone COM 29010	**(69430-24-6)**
Silicone DC 200	**(63148-62-9)**
Silicone DC 360	**(63148-62-9)**
Silicone DC 360 Fluid	**(63148-62-9)**
Silicone Fluid PD 5	**(2116-84-9)**
Silicone Fluid PK 20	**(2116-84-9)**
Silicone Fluids M	**(9006-65-9)**
Silicone Y-6607	**(67762-92-9)**
Silicone dioxide	**(60676-86-0)**
Silicone oils	**(63148-62-9)**
Silicone Release L 45	**(63148-62-9)**
Silicon-fluoride	**(7783-61-1)**
Silicon fluoride barium salt	**(17125-80-3)**
Silicon hexafluoride dihydride	**(16961-83-4)**
Silicon monocarbide	**(409-21-2)**
Silicon oxide, di- (Sand)	**(14808-60-7)**
Silicon phenyl trichloride	**(98-13-5)**
Silicon powder, Amorphous [UN 1346]	**(7440-21-3)**
Silicon sodium fluoride	**(16893-85-9)**
Silicon tetrachloride [UN 1818]	**(10026-04-7)**
Silicon tetrafluoride [UN 1859]	**(7783-61-1)**
Silicon tetrahydride (ACGIH,OSHA)	**(7803-62-5)**
Silicon zinc fluoride	**(16871-71-9)**
Siligaz	**(63148-62-9)**
Silikill	**(7631-86-9)**
Silikon Antifoam FD 62	**(63148-62-9)**
Silmos T	**(1344-95-2)**
Silmurin	**(507-60-8)**
Silochrome	**(63231-67-4)**
Silogomma Orange G	**(3520-72-7)**
Silogomma Red RLL	**(2425-85-6)**
Silol 350	**(63148-62-9)**
Silol 5	**(63148-62-9)**
Silon	**(63428-83-1)**
Silon (Polyamide)	**(63428-84-2)**
Silopol Orange R	**(3468-63-1)**
Silopol Red G	**(2814-77-9)**
Siloprene C 1	**(63148-62-9)**
Siloprene C 2	**(63148-62-9)**
Siloprene C 18	**(63148-62-9)**
Silosan	**(29232-93-7)**
Silosol Orange RN	**(3468-63-1)**
Silosol Red GN	**(2814-77-9)**
Silosol Red RBN	**(2425-85-6)**
Silosol Red RN	**(2425-85-6)**
Silosuper Pink B	**(989-38-8)**
Silosuper Violet R	**(1325-82-2)**
Silotermo Carmine G	**(5858-81-1)**
Silotermo Orange G	**(3520-72-7)**
Siloton Orange GR	**(1934-20-9)**
Siloton Orange GT	**(3520-72-7)**
Siloton Orange RL	**(3468-63-1)**
Siloton Red BRLL	**(2425-85-6)**
Siloton Red R	**(1248-18-6)**
Siloton Red RLL	**(2425-85-6)**
Siloton Red 2g	**(2814-77-9)**
Siloton Rubine 2B	**(5858-81-1)**
Siloton Rubine B	**(5858-81-1)**
Siloton Yellow GTX	**(6358-85-6)**
Siloton Yellow 3GX	**(6358-85-6)**
Silotras Brown TRN	**(6416-57-5)**
Silotras Orange TR	**(842-07-9)**
Silotras Red T3B	**(85-83-6)**
Silotras Scarlet TB	**(85-86-9)**
Silotras Yellow T2G	**(60-11-7)**
Silotras Yellow TSG	**(2832-40-8)**
Siloxane and silicones, dimethyl	**(63148-62-9)**
Siloxanes and silicones, di-Me	**(63148-62-9)**
Siloxanes and silicones, Me hydrogen	**(63148-57-2)**

Siloxanes and silicones, Me 3,3,3-trifluoropropyl (63148-56-1)
Siloxanes and silicones, di-Me, Me Ph (63148-52-7)
Siloxanes and silicones, dimethyl (63148-62-9)
Siloxanes and silicones, dimethyl, (dimethylamino)- terminated (67762-92-9)
Siloxanes and silicones, methyl hydrogen (63148-57-2)
Silpian 2 (63148-62-9)
Silpian E 2 (63148-62-9)
Siltex (60676-86-0)
Silundum (409-21-2)
Silvadene (68-35-9)
Silvan (Czech) (534-22-5)
Silvanol (58-89-9)
Silver (ACGIH,OSHA) (7440-22-4)
Silver Atom (7440-22-4)
Silver Birch LS (8001-21-6)
Silver Bond B (14808-60-7)
Silver Graphite (7782-42-5)
Silver(I) chloride (7783-90-6)
Silver(I) nitrate (1:1) (7761-88-8)
Silver, Isotope of mass 110 (8CI,9CI) (14391-76-5)
Silver Matt Powder (7440-31-5)
Silver White (1319-46-6)
Silver acetylide (13092-75-6)
Silver acetylide, Dry (13092-75-6)
Silver-azide (13863-88-2)
Silver azide, Dry (DOT) (13863-88-2)
Silver bromide (9CI) (7785-23-1)
Silver chloride (9CI) (7783-90-6)
Silver chlorite (7783-91-7)
Silver chlorite (Dry) (7783-91-7)
Silver chlorite, Dry (7783-91-7)
Silver chromate (7784-01-2)
Silver-cyanide (506-64-9)
Silver cyanide [UN 1684] (506-64-9)
Silver difluoride (7783-95-1)
Silver(1+) docosanoate (2489-05-6)
Silver eicosanoate (24687-57-8)
Silver(II) fluoride (7783-95-1)
Silver fluoride (8CI,9CI) (7775-41-9)
Silver fulminate (5610-59-3)
Silver fulminate, Dry (5610-59-3)
Silver(1+) iodide (7783-96-2)
Silver iodide (9CI) (7783-96-2)
Silver monochloride (7783-90-6)
Silver monoiodide (7783-96-2)
Silver(1+) nitrate (7761-88-8)
Silver nitrate [UN 1493] (7761-88-8)
Silver(1+) octadecanoate (3507-99-1)
Silver oxalate (533-51-7)
Silver oxalate (Dry) (533-51-7)
Silver oxalate, Dry (533-51-7)
Silver(1+) perchlorate (7783-93-9)
Silver 3-phenylpropanoate (75112-79-7)
Silver picrate (146-84-9)
Silver picrate, Dry (146-84-9)
Silver picrate, Wetted with at least 30% water [UN 1347] (146-84-9)
Silver, (picryloxy)- (146-84-9)
Silver potassium cyanide (506-61-6)
Silver salt (131-08-8)
Silver sodium zeolite (1318-02-1)
Silver sulfide (9CI) (21548-73-2)
Silver thiuronium acrylate copolymer (53404-00-5)
Silver trifluoroacetate (2966-50-9)
Silver(1+) trifluoroacetate (2966-50-9)
Silvex (93-72-1)
Silvex, diethanolamine salt (51170-59-3)
Silvex isooctyl ester (32534-95-5)
Silvex, potassium salt (2818-16-8)
Silvex, sodium salt (37913-89-6)
Silvi-RHAP (93-72-1)
Silvisar (124-65-2)
Silvisar 510 (75-60-5)
Silvisar 550 (2163-80-6)
Silyl bromide (13465-73-1)
Silylene (9CI) (13825-90-6)
Simadex (122-34-9)
Siman (2898-12-6)
Simanex (122-34-9)
Simatin(e) (77-67-8)
Simazin (122-34-9)
Simazine (122-34-9)
Simazine 80W (122-34-9)
Simazol (61-82-5)
Simethicone (8050-81-5)
Simeton (673-04-1)
Simetone (673-04-1)
Simetryn (1014-70-6)
Simetryne (1014-70-6)
Simpamina (60-13-9)
Simpamina-D (51-63-8)
Simpamine (60-13-9)
Simpatedrin (300-62-9)
Simpatedrin (60-13-9)
Simpla (2893-78-9)
Simulol 330 M (9002-92-0)
Simulsol 72 (9005-00-9)
Simulsol 76 (9005-00-9)
Simulsol 92 (9004-98-2)
Simulsol 96 (9004-98-2)
Simulsol 98 (9004-98-2)
Simulsol M (9004-99-3)
Simulsol P 4 (9002-92-0)
Simulsol P 23 (9002-92-0)
Sinafid M-48 (298-00-0)
Sinalar (67-73-2)
Sinalost (817-09-4)
Sinapic acid (530-59-6)
Sinbar (5902-51-2)
Sincaline (123-41-1)
Sinciclan (130-80-3)
Sincoumar (152-72-7)
Sincurarine (65-29-2)
Sindesvel (72-44-6)
Sindrenina (51-43-4)
Sinedal (62-44-2)
α-Sinensal (17909-77-2)
Sinepyrin 222 (113-48-4)
Sinesalin (73-48-3)
Sinesalin composition (50-55-5)
Sinestrol (84-16-2)
Sinituho (87-86-5)
Sinkalin (123-41-1)
Sinkaline (123-41-1)
Sinkumar (152-72-7)
Sinnoester OGC (25496-72-4)
Sinnopal OP 8 (9036-19-5)
Sinnopon LS 100 (151-21-3)
Sinnopon LS 95 (151-21-3)
Sinnoquat BL 80 (7281-04-1)
Sinnoquat BL 95 (7281-04-1)
Sinnozon NCX 70 (26264-06-2)
Sinomin (723-46-6)
Sinopon (2235-54-3)
Sinoratox (60-51-5)

Sinox	**(2312-76-7)**
Sinox	**(534-52-1)**
Sinox General	**(88-85-7)**
Sinox W	**(6365-83-9)**
Sintapon L	**(151-21-3)**
Sintestrol	**(56-53-1)**
Sinthrom	**(152-72-7)**
Sinthrome	**(152-72-7)**
Sintomicetina	**(56-75-7)**
Sintomicetine R	**(56-75-7)**
Sintosian	**(548-73-2)**
Sintrom	**(152-72-7)**
Sintroma	**(152-72-7)**
Sintyal	**(77-75-8)**
Sinubid	**(62-44-2)**
Sinuron	**(330-55-2)**
Sinutab	**(62-44-2)**
Sinutab II	**(62-44-2)**
Sionit	**(50-70-4)**
Sionite	**(50-70-4)**
Sionon	**(50-70-4)**
Siosan	**(50-70-4)**
Sipas 60	**(105-60-2)**
Sipas 60	**(25038-54-4)**
Sipcaplant	**(23564-05-8)**
Sipcasan	**(23564-05-8)**
Sipcavit	**(23564-05-8)**
Siperin	**(52315-07-8)**
Sipex Bos	**(126-92-1)**
Sipex OLS	**(142-31-4)**
Sipex OP	**(151-21-3)**
Sipex SB	**(151-21-3)**
Sipex SD	**(151-21-3)**
Sipex SP	**(151-21-3)**
Sipex UB	**(151-21-3)**
Sipol O	**(143-28-2)**
Sipol S	**(112-92-5)**
Sipol l8	**(111-87-5)**
Sipol l10	**(112-30-1)**
Sipol l12	**(112-53-8)**
Sipomer DAM	**(999-21-3)**
Sipomer DMM	**(624-48-6)**
Sipon ES	**(9004-82-4)**
Sipon LA 30	**(2235-54-3)**
Sipon LD	**(143-00-0)**
Sipon LES 25	**(9004-82-4)**
Sipon LS	**(151-21-3)**
Sipon LS 100	**(151-21-3)**
Sipon LSB	**(151-21-3)**
Sipon LT	**(139-96-8)**
Sipon LT-40	**(139-96-8)**
Sipon LT-6	**(139-96-8)**
Sipon PD	**(151-21-3)**
Sipon WD	**(151-21-3)**
Siponic F 300	**(9036-19-5)**
Siponic F 400	**(9036-19-5)**
Siponic L	**(9002-92-0)**
Siponic L 3	**(9002-92-0)**
Siponic L-4	**(9002-92-0)**
Siponic L 7-90	**(9002-92-0)**
Siponic L 10	**(9002-92-0)**
Siponic L 15	**(9002-92-0)**
Siponic L 150	**(9002-92-0)**
Siponic Y 500	**(9004-98-2)**
Siponol S	**(112-92-5)**
Siprol l22	**(2235-54-3)**
Siptox I	**(121-75-5)**
Siragan	**(54-05-7)**
Siran hydrazinu (Czech)	**(10034-93-2)**
Sirenin	**(33156-92-2)**
Sirenin (Etocarpus)	**(33156-92-2)**
Sirius Red 4B	**(2610-11-9)**
Sirius Red 4BA	**(2610-11-9)**
Sirius Red F 3B	**(2610-10-8)**
Sirius Red F 3BA	**(2610-10-8)**
Sirius Supra Blue FGL-CF stain	**(6527-70-4)**
Sirius Supra Blue 4G	**(2610-05-1)**
Sirius Supra Brown BRL	**(16071-86-6)**
Sirius Supra Brown BRS	**(16071-86-6)**
Sirius Supra Brown G	**(6854-81-5)**
Sirius Supra Orange 7GL	**(50814-31-8)**
Sirius Supra Yellow 5G	**(10190-68-8)**
Sirlene	**(57-55-6)**
Sirmate	**(1966-58-1)**
Sirmate	**(62046-37-1)**
Sirnik amonny (Czech)	**(12124-99-1)**
Sirnik fosforecny (Czech)	**(1314-80-3)**
Sirokal	**(299-27-4)**
Siromycin	**(606-58-6)**
Sirotol	**(93-14-1)**
Sirup	**(50-99-7)**
Sistan	**(137-42-8)**
Sistral	**(2152-34-3)**
Sisuril	**(135-09-1)**
β-Sitosterin	**(83-46-5)**
β-Sitosterol	**(83-46-5)**
γ-Sitosterol	**(83-47-6)**
Sixty-three Special E.C. Insecticide	**(298-00-0)**
Sixty-three Special E.C. Insecticide	**(56-38-2)**
Sk-Bamate	**(57-53-4)**
Sk-Chloral hydrate	**(302-17-0)**
Sk-Diphenhydramine	**(147-24-0)**
Sk-Penicillin G	**(113-98-4)**
Skabilan	**(502-55-6)**
Skane M8	**(26530-20-1)**
Skatol	**(83-34-1)**
Skatole	**(83-34-1)**
ω-Skatole carboxylic acid	**(87-51-4)**
Skedule	**(302-22-7)**
Skekhg	**(106-89-8)**
Skelaxin	**(1665-48-1)**
Skellysolve F	**(8032-32-4)**
Skellysolve G	**(8032-32-4)**
Skerolip	**(637-07-0)**
Skimmetin	**(93-35-6)**
Skimmetine	**(93-35-6)**
Skino #1	**(110-69-0)**
Skino #2	**(96-29-7)**
Skiodan	**(126-31-8)**
Sklero-Tablinen	**(637-07-0)**
Sklero-Tabuls	**(637-07-0)**
Skleromex	**(637-07-0)**
Skleromexe	**(637-07-0)**
Skopolate	**(6106-46-3)**
Skopyl	**(6106-46-3)**
Skorexol (9CI)	**(83589-99-5)**
Sky Blue 4B	**(2429-74-5)**
Sky Blue 5B	**(2429-74-5)**
Sky Blue 6B	**(2610-05-1)**
Skydrol LD (9CI)	**(55962-27-1)**
Slab Oil (9CI)	**(8042-47-5)**
Slack wax, Petroleum	**(64742-61-6)**
Slag, Skims and fines from brass furnace (Secondary nonferrous plant)	**(69012-26-6)**

Slags, Brass-manufg. (69012-26-6)
Slags, Coal (68476-96-0)
Slags, Ferrous metal, Blast furnace (65996-69-2)
Slags, Steelmaking (65996-71-6)
Slags, Type metal smelting (69029-87-4)
Slaked Lime (1305-62-0)
Sleepan (50-35-1)
Sleepwell (91-80-5)
Slimacide V 10 (20679-58-7)
Slime-Trol (7166-19-0)
Slimes and sludges, Papermaking, Secondary (68188-15-8)
Slimicide (107-02-8)
Slimicide E (3064-70-8)
Slip-eze (301-02-0)
Slipro (50-35-1)
Slo-Gro (123-33-1)
Slo-Gro (5716-15-4)
Slo-Phyllin (58-55-9)
Slosul (80-35-3)
Slovanik (9003-11-6)
Slovanik 610 (9003-11-6)
Slovanik 630 (9003-11-6)
Slovanik 660 (9003-11-6)
Slovanik M (9003-11-6)
Slovanik PV 670 (9003-11-6)
Slovanik T 310 (9003-11-6)
Slovanik T 320 (9003-11-6)
Slovanik T 630 (9003-11-6)
Slovasol A (9003-11-6)
Slovasol MKS 16 (9004-96-0)
Slovasol S (9004-99-3)
Slovasol SF (9002-92-0)
Slovaton U (9002-92-0)
Slow-Fe (9004-98-2)
Slow-K (7720-78-7)
Slowten (7447-40-7)
Slug-Tox (67-03-8)
Slugit (108-62-3)
Smeesana (9002-91-9)
Smelter Dross (Lead) (86-88-4)
Smidan (69029-52-3)
Smite (732-11-6)
Smite (26628-22-8)
Smithion (9CI) (3761-60-2)
Smoke Brown G (95686-15-0)
Smoke Orange R (81-54-9)
Smooth Muscle Inhibitory Factor (81-64-1)
Smut-Go (14797-65-0)
Snapping Hazel (118-74-1)
Sneezing Gas (68916-39-2)
Snieciotox (712-48-1)
Snip (118-74-1)
Snip Fly (644-64-4)
Snip Fly Bands (644-64-4)
Snomelt (644-64-4)
Snow Algin H (10043-52-4)
Snow Algin L (9005-38-3)
Snow Algin M (9005-38-3)
Snow Tex (9005-38-3)
Snow Tex (1302-76-7)
Snow White (1332-58-7)
Snow White (1314-13-2)
Snowcal 5SW (8009-03-8)
Snowflake crystals (1304-85-4)
Snowit (533-96-0)
So-Flo (14808-60-7)
Soamin (7681-49-4)
(127-85-5)

Soap (68952-95-4)
Soap Yellow F (2321-07-5)
Soaps, Stocks, Vegetable-oil, Acidulated (68952-95-4)
Soapstock, Acidulated (68952-95-4)
Sobenate (532-32-1)
Sobiodopa (59-92-7)
Sobital (577-11-7)
Sobril (604-75-1)
Socarex (9002-88-4)
Soda Mint (144-55-8)
Soda Niter (7631-99-4)
Soda alum (10102-71-3)
Soda alum (7784-28-3)
Soda ash (497-19-8)
Soda chlorate (DOT) (7775-09-9)
Soda lime (8006-28-8)
Soda lime, Solid (8006-28-8)
Soda lye (1310-73-2)
Sodamide (7782-92-5)
Sodanit (7784-46-5)
Sodantoin (57-41-0)
Sodanton (57-41-0)
Sodanton (630-93-3)
Soda phosphate (7558-79-4)
Sodar (144-21-8)
Sodascorbate (134-03-2)
Sodasorb (8006-28-8)
Sodio (clorato di) (Italian) (7775-09-9)
Sodio (dicromato di) (Italian) (10588-01-9)
Sodio, fluoracetato di (Italian) (62-74-8)
Sodio(idrossido di) (Italian) (1310-73-2)
Sodio (perclorato di) (Italian) (7601-89-0)
Sodio(tricloroacetato di) (Italian) (650-51-1)
Sodital (57-33-0)
Sodium [UN 1428] (7440-23-5)
Sodium Amazolene (3861-73-2)
Sodium Anazolene (3861-73-2)
Sodium Base Spent Sulfite Liquor (8061-51-6)
Sodium Blue VRS (129-17-9)
Sodium CMC (9004-32-4)
Sodium CM-Cellulose (9004-32-4)
Sodium C14-17 alcohol sulfonate (68037-49-0)
Sodium, Compd. with thallium (1:1) (8CI,9CI) (12136-93-5)
Sodium C14-17 sec alcohol sulfonate (68037-49-0)
Sodium 2,4-D (2702-72-9)
Sodium DBDT (136-30-1)
Sodium EDTA (64-02-8)
Sodium Edetate (64-02-8)
Sodium Eosinate (17372-87-1)
Sodium Eosine (17372-87-1)
Sodium Equilin Sulfate (16680-47-0)
Sodium erythorbate (6381-77-7)
Sodium Feredetate (15708-41-5)
Sodium ferric EDTA (15708-41-5)
Sodium iron EDTA (15708-41-5)
Sodium Laureth-12 Sulfate (9004-82-4)
Sodium Laureth-5 Sulfate (9004-82-4)
Sodium Laureth-7 Sulfate (9004-82-4)
Sodium Lauriminodipropionate (14960-06-6)
Sodium Lithol (1248-18-6)
Sodium Lithol Red (1248-18-6)
Sodium Lithol Red 20-4018 (1248-18-6)
Sodium L-(+)-tartrate (868-18-8)
Sodium MBT (2492-26-4)
Sodium MCPA (3653-48-3)
Sodium NPA (132-67-2)
Sodium PCP (131-52-2)

Name	CAS
Sodium Patent Blue V	(129-17-9)
Sodium Prussiate Yellow	(13601-19-9)
Sodium Saltpeter	(7631-99-4)
Sodium Sesquicarbonate	(533-96-0)
Sodium Silicate Glass	(1344-09-8)
Sodium Silicate Solution	(1344-09-8)
Sodium TCA	(650-51-1)
Sodium TCA Inhibited	(650-51-1)
Sodium acetate	(127-09-3)
Sodium acetate, acid	(126-96-5)
Sodium, ((3-(α-acetonylbenzyl)-2-oxo-2H-1-benzopyran-4-yl)oxy)-	(129-06-6)
Sodium acid acetate	(126-96-5)
Sodium acid arsenate	(7778-43-0)
Sodium acid arsenate, heptahydrate	(10048-95-0)
Sodium acid carbonate	(144-55-8)
Sodium acid hypophosphate	(7782-95-8)
Sodium acid methanearsonate	(2163-80-6)
Sodium acid phosphate	(7558-80-7)
Sodium acid pyrophosphate	(7758-16-9)
Sodium acid sulfate	(7681-38-1)
Sodium acid sulfate, Solid (DOT)	(7681-38-1)
Sodium acid sulfate, Solution (DOT)	(7681-38-1)
Sodium acid sulfite	(7631-90-5)
Sodium acrylate	(7446-81-3)
Sodium adipate	(7486-38-6)
Sodium alginate	(9005-38-3)
Sodium alizarinesulfonate	(130-22-3)
Sodium alizarin-3-sulfonate	(130-22-3)
Sodium-dl-5-allyl-1-methyl-5-(1-methyl-2-pentynyl)barbiturate	(309-36-4)
Sodium allylsulfonate	(2495-39-8)
Sodium alum	(10102-71-3)
Sodium alum	(7784-28-3)
Sodium aluminate	(1302-42-7)
Sodium aluminate, Solid [UN 2812]	(12042-47-6)
Sodium aluminate, Solution [UN 1819]	(12042-47-6)
Sodium aluminate solution	(1302-42-7)
Sodium aluminofluoride	(15096-52-3)
Sodium aluminosilicate	(1318-02-1)
Sodium aluminosilicate	(1344-00-9)
Sodium aluminum dioxide	(1302-42-7)
Sodium aluminum fluoride	(15096-52-3)
Sodium aluminum hydride [UN 2835]	(13770-96-2)
Sodium aluminum oxide	(1302-42-7)
Sodium aluminum phosphate	(7785-88-8)
Sodium aluminum silicate	(1344-00-9)
Sodium aluminum sulfate	(10102-71-3)
Sodium aluminum sulfate	(7784-28-3)
Sodium aluminum sulfate dodecahydrate	(7784-28-3)
Sodium aluminum tetrahydride	(13770-96-2)
Sodium a-dl-1-methyl-5-allyl-5-(1-methyl-2-pentynyl)barbiturate	(309-36-4)
Sodium amide (DOT)	(7782-92-5)
Sodium aminarsonate	(127-85-5)
Sodium p-aminobenzenearsonate	(127-85-5)
Sodium 4-aminobenzoate	(555-06-6)
Sodium p-aminobenzoate	(555-06-6)
Sodium 1-aminonaphthalene-4-sulfonate	(130-13-2)
Sodium aminophenol arsonate	(127-85-5)
Sodium p-aminophenylarsonate	(127-85-5)
Sodium aminotriacetate	(10042-84-9)
Sodium 4-tert-amylphenate	(31366-95-7)
Sodium p-tert-amylphenate	(31366-95-7)
Sodium p-tert-amylphenolate	(31366-95-7)
Sodium-analine arsonate	(127-85-5)
Sodium anilarsonate	(127-85-5)
Sodium anilinomethanesulfonate	(26021-90-9)
Sodium-2-anthrachinonesulphonate	(131-08-8)
Sodium anthraquinone-1,5-disulfonate	(853-35-0)
Sodium 2-anthraquinonesulfonate	(131-08-8)
Sodium 9,10-anthraquinone-2-sulfonate	(131-08-8)
Sodium anthraquinone-1-sulfonate	(128-56-3)
Sodium anthraquinone-α-sulfonate	(128-56-3)
Sodium β-anthraquinonesulfonate	(131-08-8)
Sodium p-arsanilate	(127-85-5)
Sodium arsanilate [UN 2473]	(127-85-5)
Sodium arsenate	(13464-38-5)
Sodium arsenate	(7778-43-0)
Sodium orthoarsenate	(13464-38-5)
Sodium orthoarsenate	(7631-89-2)
Sodium arsenate [UN 1685]	(7631-89-2)
Sodium arsenate, dibasic	(7778-43-0)
Sodium arsenate dibasic, Anhydrous	(7778-43-0)
Sodium arsenate, dibasic, heptahydrate	(10048-95-0)
Sodium arsenate heptahydrate	(10048-95-0)
Sodium arseniate	(10048-95-0)
Sodium arsenite	(7784-46-5)
Sodium arsenite, Aqueous solutions [UN 1686]	(7784-46-5)
Sodium arsenite, Liquid (Solutions) [UN 1686]	(13464-37-4)
Sodium arsenite, Liquid (solution) (DOT)	(7784-46-5)
Sodium arsenite (Na_3AsO_3) (6CI,7CI)	(13464-37-4)
Sodium arsenite, Solid [UN 2027]	(7784-46-5)
Sodium arsonilate	(127-85-5)
Sodium L-ascorbate	(134-03-2)
Sodium ascorbate	(134-03-2)
Sodium asulam	(2302-17-2)
Sodium azide (ACGIH,DOT,OSHA)	(26628-22-8)
Sodium azide (ACGIH,OSHA) [UN 1687] (9CI)	(26628-22-8)
Sodium azo-α-naphtholsulfanilate	(523-44-4)
Sodium azo-α-naphtholsulphanilate	(523-44-4)
Sodium azoresorcinolsulfanilate	(547-57-9)
Sodium, azoture de (French)	(26628-22-8)
Sodium, azoturo di (Italian)	(26628-22-8)
Sodium bentazon	(50723-80-3)
Sodium benzeneacetate	(114-70-5)
Sodium benzenesulfinate	(873-55-2)
Sodium benzene sulfochloramide	(127-52-6)
Sodium benzenesulfochloramine	(127-52-6)
Sodium benzenesulfonate	(515-42-4)
Sodium benzenesulfonchloramide	(127-52-6)
Sodium benzenesulfonchloramine	(127-52-6)
Sodium benzenethiolate	(930-69-8)
Sodium 1,2 benzisothiazolin-3-one-1,1-dioxide	(128-44-9)
Sodium benzoate	(532-32-1)
Sodium benzoic acid	(532-32-1)
Sodium o-benzosulfimide	(128-44-9)
Sodium benzosulfochloramide	(127-52-6)
Sodium benzosulfonate	(515-42-4)
Sodium 2-benzosulphimide	(128-44-9)
Sodium benzosulphimide	(128-44-9)
Sodium o-benzosulphimide	(128-44-9)
Sodium 2-benzothiazolesulfonate	(21465-51-0)
Sodium 2-benzothiazolethioate	(2492-26-4)
Sodium 2-benzothiazolethiolate	(2492-26-4)
Sodium benzothiazolethiolate	(2492-26-4)
Sodium 2(3H)-benzothiazolethionate	(2492-26-4)
Sodium, (2-benzothiazolylthio)-	(2492-26-4)
Sodium 2-benzyl-4-chlorophenate	(3184-65-4)
Sodium o-benzyl-p-chlorophenate	(3184-65-4)
Sodium o-benzyl-p-chlorophenolate	(3184-65-4)
Sodium benzylpenicillin	(69-57-8)
Sodium benzylpenicillin G	(69-57-8)
Sodium benzylpenicillinate	(69-57-8)
Sodium biborate	(1303-96-4)
Sodium biborate	(1330-43-4)
Sodium biborate decahydrate	(1303-96-4)

Sodium bicarbonate (1:1) (144-55-8)
Sodium bichromate (10588-01-9)
Sodium bifluoride, Solid (DOT) (1333-83-1)
Sodium bifluoride, Solution (DOT) (1333-83-1)
Sodium (1,1'-biphenyl)-2-olate (132-27-4)
Sodium 2-biphenylolate (132-27-4)
Sodium, (2-biphenylyloxy)- (132-27-4)
Sodium biphosphate (7558-80-7)
Sodium biphosphate anhydrous (7558-80-7)
Sodium bis(2-ethylhexyl) sulfosuccinate (577-11-7)
Sodium N,N-bis-2-hydroxyethyl glycinate (139-41-3)
Sodium 2,3:4,6-bis-O-(1-methylethylidine)-α-L-xylo-2-hexulo-furanosonate (52508-35-7)
Sodium bispropylacetate (1069-66-5)
Sodium-bisulfate (7681-38-1)
Sodium bisulfate, Fused (7681-38-1)
Sodium bisulfate, Solid (DOT) (7681-38-1)
Sodium bisulfate, Solution (DOT) (7681-38-1)
Sodium bisulfide (16721-80-5)
Sodium bisulfite (1:1) (7631-90-5)
Sodium bisulfite (ACGIH,OSHA) (7631-90-5)
Sodium bisulfite, Solid (7631-90-5)
Sodium bisulfite, Solution [UN 2693] (7631-90-5)
Sodium bisulphite (7631-90-5)
Sodium borate (1333-73-9)
Sodium borate (7632-04-4)
Sodium borate, decahydrate (1303-96-4)
Sodium borate, pentahydrate (11130-12-4)
Sodium borohydride [UN 1426] (16940-66-2)
Sodium bromacil (69484-12-4)
Sodium bromate [UN 1494] (7789-38-0)
Sodium-bromide (7647-15-6)
Sodium bromosulfalein (71-67-0)
Sodium bromosulfophthalein (71-67-0)
Sodium bromsulphalein (71-67-0)
Sodium bromsulphthalein (71-67-0)
Sodium butabarbital (143-81-7)
Sodium butanedioate (14047-56-4)
Sodium butanoate (156-54-7)
Sodium butobarbitone (143-81-7)
Sodium 1-butoxide (2372-45-4)
Sodium butoxide (2372-45-4)
Sodium 2-butoxyethoxide (52663-57-7)
Sodium butylate (2372-45-4)
Sodium n-butylate (2372-45-4)
Sodium 5-sec-butyl-5-ethylbarbiturate (143-81-7)
Sodium 4-tert-butylphenate (5787-50-8)
Sodium butyrate (156-54-7)
Sodium n-butyrate (156-54-7)
Sodium cacodylate [UN 1688] (124-65-2)
Sodium calcium aluminosilicate (1344-01-0)
Sodium calcium aluminosilicate, hydrated (1344-01-0)
Sodium calcium magnesium polyphosphate (65997-17-3)
Sodium calcium magnesium silica polyphosphate (65997-17-3)
Sodium calcium polyphosphate (65997-17-3)
Sodium calcium silicoaluminate (1344-01-0)
Sodium calcium zinc silica polyphosphate (65997-17-3)
Sodium caprate (1002-62-6)
Sodium caprinate (1002-62-6)
Sodium caprylate (1984-06-1)
Sodium capryl sulfate (142-31-4)
Sodium carbolate (139-02-6)
Sodium carbonate (2:1) (497-19-8)
Sodium carbonate stabilized sulfonated polystyrene sodium salt (9080-79-9)
Sodium carboxmethylcellulose (9004-32-4)
Sodium carboxyl methyl cellulose (9004-32-4)
Sodium carboxymethyl cellulose (9004-32-4)
Sodium carboxymethyl hydroxyethylcellulose (9004-30-2)
Sodium cefacetril (23239-41-0)
Sodium cellulose glycolate (9004-32-4)
Sodium cephacetrile (23239-41-0)
Sodium cetyl sulfate (1120-01-0)
Sodium chloramben (1954-81-0)
Sodium chloramine T (127-65-1)
Sodium chloramphenicol succinate (982-57-0)
Sodium chlorate, Aqueous solution [UN 2428] (7775-09-9)
Sodium chlorate [UN 1495] (7775-09-9)
Sodium(chlorate de) (French) (7775-09-9)
Sodium-chloride (7647-14-5)
Sodium chlorite, Solution containing more than 5% available chlorine [UN 1908] (7758-19-2)
Sodium chlorite, Solution not exceeding 42% sodium chlorite (DOT) (7758-19-2)
Sodium chlorite [UN 1496] (7758-19-2)
Sodium chloroacetate [UN 2659] (3926-62-3)
Sodium N-chlorobenzenesulfonamide (127-52-6)
Sodium, (N-chlorobenzenesulfonamido)- (127-52-6)
Sodium 5-(4-chlorobenzoyl)-1,4-dimethyl-1H-pyrrole-2-acetate dihydrate (64092-48-4)
Sodium 5-chloro-2-(4-chloro-2-(3-(3,4-dichlorophenyl)-ureido)-phenoxy)benzenesulfonate (3567-25-7)
Sodium 4-chloro-2-cyclopentylphenate (53404-20-9)
Sodium 3-chloro-2-hydroxy-1-propanesulfonate (126-83-0)
Sodium 3-chloro-2-hydroxypropanesulfonate (126-83-0)
Sodium-3-chloro-2-hydroxypropane sulfonate (126-83-0)
Sodium 3-chloro-2-hydroxypropylsulfonate (126-83-0)
Sodium (4-chloro-2-methylphenoxy)acetate (3653-48-3)
Sodiumchlorooxypropanesulfonate (126-83-0)
Sodium 2-chloro-4-phenylphenate (31366-97-9)
Sodium 4-chloro-2-phenylphenate (10605-10-4)
Sodium 6-chloro-2-phenylphenate (10605-11-5)
Sodium-2-chloro-4-phenyl phenate (10605-10-4)
Sodium, (chloro(phenylsulfonyl)amino)- (127-52-6)
Sodium 5-(2-chloro-4-(trifluoromethyl)phenoxy)-2-nitrobenzoate (62476-59-9)
Sodium chromate (10588-01-9)
Sodium chromate (7775-11-3)
Sodium chromate(VI) (7775-11-3)
Sodium chromate decahydrate (13517-17-4)
Sodium chromotropate (129-96-4)
Sodium citrate (144-33-2)
Sodium citrate (18996-35-5)
Sodium citrate (68-04-2)
Sodium citrate anhydrous (68-04-2)
Sodium coumadin (129-06-6)
Sodium cresolate (34689-46-8)
Sodium p-cresolate (1121-70-6)
Sodium p-cresoxide (1121-70-6)
Sodium cresylate (34689-46-8)
Sodium cromoglycate (15826-37-6)
Sodium cromolyn (15826-37-6)
Sodium cumeneazo-β-naphthol disulfonate (3564-09-8)
Sodium cumeneazo-β-naphthol disulphonate (3564-09-8)
Sodium cuprocyanide, Solid [UN 2316] (14264-31-4)
Sodium cuprocyanide, Solution [UN 2317] (14264-31-4)
Sodium cyanate (917-61-3)
Sodium-cyanide (143-33-9)
Sodium cyanide (ACGIH) (143-33-9)
Sodium cyanide, Solid [UN 1689] (143-33-9)
Sodium cyanide, Solution (70152-47-5)
Sodium cyanide, Solution [UN 1689] (143-33-9)
Sodium 7-(2-cyanoacetamido)cephalosporanic acid (23239-41-0)
Sodium cyanurate (2624-17-1)
Sodium cyclamate (139-05-9)
Sodium cyclohexanesulfamate (139-05-9)

Name	CAS No.
Sodium cyclohexanesulphamate	(139-05-9)
Sodium cyclohexyl amidosulphate	(139-05-9)
Sodium N-cyclohexyl-N-palmitoyltaurate - iodine complex	(53404-81-2)
Sodium cyclohexyl sulfamate	(139-05-9)
Sodium cyclohexyl sulphamate	(139-05-9)
Sodium cyclohexylsulphamidate	(139-05-9)
Sodium dalapon	(127-20-8)
Sodium decanoate	(1002-62-6)
Sodium-n-decanoate	(1002-62-6)
Sodium decanoic acid	(1002-62-6)
Sodium decavanadate	(12200-88-3)
Sodium decylbenzenesulfonamide	(1322-98-1)
Sodium 4-decylbenzenesulfonate	(2627-06-7)
Sodium decylbenzenesulfonate	(1322-98-1)
Sodium p-decylbenzenesulfonate	(2627-06-7)
Sodium decyl sulfate	(142-87-0)
Sodium dedt	(148-18-5)
Sodium dehydroacetate	(4418-26-2)
Sodium dehydroacetic acid	(4418-26-2)
Sodium diacetate	(126-96-5)
Sodium dibutyldithiocarbamate	(136-30-1)
Sodium dibutylnaphthalene sulfate	(25417-20-3)
Sodium dibutylnaphthylsulfonate	(25417-20-3)
Sodium dibutylnapthalenesulfonate	(25417-20-3)
Sodium di-sec-butyl phosphorodithioate	(33619-92-0)
Sodium dicamba	(1982-69-0)
Sodium dichlorisocyanurate	(2893-78-9)
Sodium (o-(2,6-dichloroanilino)phenyl)acetate	(15307-79-6)
Sodium 2,2-dichlorobutyrate	(2517-16-0)
Sodium dichlorocyanurate	(2893-78-9)
Sodium dichloroisocyanurate (DOT)	(2893-78-9)
Sodium dichloroisocyanurate dihydrate	(51580-86-0)
Sodium 3,6-dichloro-2-methoxybenzoate	(1982-69-0)
Sodium 2,4-dichlorophenoxyacetate	(2702-72-9)
Sodium 4-(2,4-dichlorophenoxy)butyrate	(10433-59-7)
Sodium 2-(2,4-dichlorophenoxy)ethyl sulfate	(136-78-7)
Sodium 2,4-dichlorophenoxyethyl sulphate	(136-78-7)
Sodium (o-((2,6-dichlorophenyl)amino)phenyl)acetate	(15307-79-6)
Sodium 2,4-dichlorophenyl cellosolve sulfate	(136-78-7)
Sodium 2,2-dichloropropionate	(127-20-8)
Sodium α,α-dichloropropionate	(127-20-8)
1-Sodium-3,5-dichloro-1,3,5-triazine-2,4,6-trione	(2893-78-9)
1-Sodium-3,5-dichloro-s-triazine-2,4,6-trione	(2893-78-9)
Sodium dichloro-s-triazinetrione (Dry, Containing more than 39% available chlorine)	(54579-28-1)
Sodium dichloro-s-triazinetrione, Dry, Containing more than 39% available chlorine [UN 2465]	(2893-78-9)
Sodium dichloro-s-triazinetrione dihydrate	(51580-86-0)
Sodium dichromate (DOT)	(10588-01-9)
Sodium dichromate(VI)	(10588-01-9)
Sodium(dichromate de) (French)	(10588-01-9)
Sodium N,N-diethyldithiocarbamate	(148-18-5)
Sodium diethyldithiocarbamate	(148-18-5)
Sodium diethyldithiocarbamate trihydrate	(20624-25-3)
Sodium diethylene glycol dodecyl ether sulfate	(3088-31-1)
Sodium diethylenetriaminepentaacetate	(140-01-2)
Sodium di-(2-ethylhexyl) sulfosuccinate	(577-11-7)
Sodium difluoroacetate	(2218-52-2)
Sodium dihexyl phosphorodithioate	(26537-89-3)
Sodium 1,4-dihexyl sulfobutanedioate	(3006-15-3)
Sodium dihydrogen citrate	(18996-35-5)
Sodium dihydrogen hypophosphate	(7782-95-8)
Sodium dihydrogen phosphate (1:2:1)	(7558-80-7)
Sodium dihydroxyethylglycinate	(139-41-3)
Sodium O,O-diisopropyl dithiophosphate	(27205-99-8)
Sodium diisopropyldithiophosphate	(27205-99-8)
Sodium diisopropyl phosphorodithioate	(27205-99-8)
Sodium 4-(dimethylamino)benzenediazosulfonate	(140-56-7)
Sodium p-(dimethylamino)benzenediazosulfonate	(140-56-7)
Sodium 4-(dimethylamino)benzenediazosulphonate	(140-56-7)
Sodium p-(dimethylamino)benzenediazosulphonate	(140-56-7)
Sodium (4-(dimethylamino)phenyl)diazenesulfonate	(140-56-7)
Sodium dimethylarsinate	(124-65-2)
Sodium dimethylarsonate	(124-65-2)
Sodium dimethylbenzenesulfonate	(1300-72-7)
Sodium N,N-dimethyldithiocarbamate	(128-04-1)
Sodium N,O-dimethylhydroxylamine-N-sulfonate	(5390-94-3)
Sodium 4-(2,4-dinitroanilino)diphenylamine-2-sulfonate	(6373-74-6)
Sodium dinitro-o-cresolate	(2312-76-7)
Sodium 4,6-dinitro-o-cresoxide	(2312-76-7)
Sodium 4,6-dinitro-o-cresylate	(2312-76-7)
Sodium dioctyl sulfosuccinate	(577-11-7)
Sodium dioctyl sulphosuccinate	(577-11-7)
Sodium dioxide	(1313-60-6)
Sodium dioxyethylenedodecyl ether sulfate	(3088-31-1)
Sodium 1,8-dioxynaphthalene-3,6-disulfonate	(129-96-4)
Sodium diphacinone	(42721-99-3)
Sodium p-diphenylamino-azobenzenesulfonate	(554-73-4)
Sodium diphenyl-4,4'-bis-azo-2''-8''-amino-1''-naphthol-3'',6''-disulphonate	(2602-46-2)
Sodium diphenyldiazo-bis(α-naphthylaminesulfonate)	(573-58-0)
Sodium diphenylhydantoin	(630-93-3)
Sodium 5,5-diphenylhydantoinate	(630-93-3)
Sodium diphenyl hydantoinate	(630-93-3)
Sodium 5,5-diphenyl-2,4-imidazolidinedione	(630-93-3)
Sodium α,α-dipropylacetate	(1069-66-5)
Sodium dipropylacetate	(1069-66-5)
Sodium n-dipropylacetate	(1069-66-5)
Sodium dithiocarbamate	(4384-81-0)
Sodium dithionate	(14970-71-9)
Sodium dithionite [UN 1384]	(7775-14-6)
Sodium ditolyldisazobis-8-amino-1-naphthol-3,6-disulfonate	(72-57-1)
Sodium ditolyldisazobis-8-amino-1-naphthol-3,6-disulphonate	(72-57-1)
Sodium 1-dodecanesulfonate	(2386-53-0)
Sodium dodecanoate	(629-25-4)
Sodium 4-dodecylbenzenesulfonate	(2211-98-5)
Sodium 4-sec-dodecylbenzenesulfonate	(68628-60-4)
Sodium dodecylbenzenesulfonate	(25155-30-0)
Sodium dodecylbenzenesulfonate, Dry	(25155-30-0)
Sodium dodecylbenzenesulfonate - iodine complex	(53467-01-9)
Sodium dodecyl diphenyl oxide sulfonate	(53467-00-8)
Sodium N-dodecyliminodipropionate	(3655-00-3)
Sodium dodecylpoly(oxyethylene) sulfate	(9004-82-4)
Sodium dodecyl sulfate	(151-21-3)
Sodium n-dodecyl sulfate	(151-21-3)
Sodium dodecyl sulphate	(151-21-3)
Sodium edetate	(17421-79-3)
Sodium epichlorohydrinsulfonate	(126-83-0)
Sodium estrone sulfate	(438-67-5)
Sodium estrone-3-sulfate	(438-67-5)
Sodium etasulfate	(126-92-1)
Sodium ethaminal	(57-33-0)
Sodium ethanolate	(141-52-6)
Sodium ethasulfate	(126-92-1)
Sodium ethenesulfonate	(3039-83-6)
Sodium 4-ethenylbenzenesulfonate	(2695-37-6)
Sodium ethenylbenzenesulfonate	(27457-28-9)
Sodium ethidronate	(7414-83-7)
Sodium ethoxide	(141-52-6)
Sodium ethydronate	(7414-83-7)
Sodium ethylate	(141-52-6)
Sodium 4-ethylbenzenesulfonate	(14995-38-1)
Sodium ethylbenzenesulfonate	(30995-65-4)
Sodium p-ethylbenzenesulfonate	(14995-38-1)

Sodium 5-ethyl-5-sec-butylbarbiturate **(143-81-7)**
Sodium ethylenediaminetetraacetate **(17421-79-3)**
Sodium ethylenediaminetetraacetate **(64-02-8)**
Sodium ethylenediaminetetraacetic acid **(64-02-8)**
Sodium 4-ethyl-1-(3-ethylpentyl)-1-octyl sulfate **(3282-85-7)**
Sodium 2-ethylhexanoate **(19766-89-3)**
Sodium(2-ethylhexyl)alcohol sulfate **(126-92-1)**
Sodium 2-ethylhexyl sulfate **(126-92-1)**
Sodium 2-ethylhexylsulfosuccinate **(577-11-7)**
Sodium ethylmercuric thiosalicylate **(54-64-8)**
Sodium p-((ethylmercuri)thio)benzenesulfonate **(5964-24-9)**
Sodium o-(ethylmercurithio)benzoate **(54-64-8)**
Sodium ethylmercurithiosalicylate **(54-64-8)**
Sodium 5-ethyl-5-(1-methylbutyl)barbiturate **(57-33-0)**
Sodium 5-ethyl-5-(1-methylpropyl)barbiturate **(143-81-7)**
Sodium 7-ethyl-2-methyl-4-undecanol sulfate **(139-88-8)**
Sodium 7-ethyl-2-methylundecyl-4-sulfate **(139-88-8)**
Sodium p-ethylphenoxide **(19277-91-9)**
Sodium 5-ethyl-5-phenylbarbiturate **(57-30-7)**
Sodium etidronate **(7414-83-7)**
Sodium ferrocyanide **(13601-19-9)**
Sodium fluoacetate **(62-74-8)**
Sodium fluoacetic acid **(62-74-8)**
Sodium-fluoaluminate **(15096-52-3)**
Sodium fluoracetate **(62-74-8)**
Sodium fluorescein **(518-47-8)**
Sodium fluoresceinate **(518-47-8)**
Sodium-fluoride **(1333-83-1)**
Sodium fluoride, Solid (DOT) **(7681-49-4)**
Sodium fluoride, Solution (DOT) **(7681-49-4)**
Sodium fluoride [UN 1690] **(7681-49-4)**
Sodium fluoride cyclic dimer **(7681-49-4)**
Sodium fluoroacetate (ACGIH,OSHA) [UN 2629] **(62-74-8)**
Sodium fluoroacetate de (French) **(62-74-8)**
Sodium fluorophosphate (Na_2PO_3F) **(10163-15-2)**
Sodium fluorosilicate [UN 2674] **(16893-85-9)**
Sodium fluorure (French) **(7681-49-4)**
Sodium fluosilicate **(16893-85-9)**
Sodium formaldehyde bisulfite **(870-72-4)**
Sodium formaldehyde sulfoxylate **(149-44-0)**
Sodium formate **(141-53-7)**
Sodium 2-formylbenzenesulfonate **(1008-72-6)**
Sodium fumarin **(34490-93-2)**
Sodium D-gluconate **(14906-97-9)**
Sodium D-gluconate **(527-07-1)**
Sodium gluconate **(527-07-1)**
L(+) Sodium glutamate **(142-47-2)**
Sodium L-glutamate **(142-47-2)**
Sodium glutamate **(142-47-2)**
Sodium glycolate cellulose **(9004-32-4)**
Sodium glyphosate **(34494-03-6)**
Sodium heptylbenzenesulfonate **(33660-91-2)**
Sodium hexacyanoferate(II) **(13601-19-9)**
Sodium hexacyanoferrate (II) **(13601-19-9)**
Sodium hexadecanoate **(408-35-5)**
Sodium hexadecyl sulfate **(1120-01-0)**
Sodium hexafluoroaluminate **(15096-52-3)**
Sodium hexafluorosilicate **(16893-85-9)**
Sodium hexafluosilicate **(16893-85-9)**
Sodium hexametaphosphate **(10124-56-8)**
Sodium hexanedioate **(23311-84-4)**
Sodium hexanoate **(10051-44-2)**
Sodium hyaluronate **(9004-61-9)**
Sodium hydrate (DOT) **(1310-73-2)**
Sodium hydride [UN 1427] **(7646-69-7)**
Sodium hydrofluoride **(7681-49-4)**
Sodium hydrogen carbonate **(144-55-8)**
Sodium hydrogen diacetate **(126-96-5)**
Sodium hydrogen difluoride **(1333-83-1)**
Sodium hydrogen fluoride [UN 2439] **(1333-83-1)**
Sodium hydrogen phosphate **(7558-79-4)**
Sodium hydrogen sulfate, Solid [UN 1821] **(7681-38-1)**
Sodium hydrogen sulfate, Solution [UN 2837] **(7681-38-1)**
Sodium hydrogen sulfide **(16721-80-5)**
Sodium hydrogen sulfite **(7631-90-5)**
Sodium hydrogen sulfite, Solid (DOT) **(7631-90-5)**
Sodium hydrogen sulfite, Solution (DOT) **(7631-90-5)**
Sodium hydrogen trioxoselenite **(7782-82-3)**
Sodium hydroselenite **(7782-82-3)**
Sodium hydrosulfide **(16721-80-5)**
Sodium hydrosulfide, Solution [NA 2922] **(16721-80-5)**
Sodium hydrosulfite [UN 1384] **(7775-14-6)**
Sodium hydrosulphide, Solid (DOT) **(16721-80-5)**
Sodium hydrosulphide, Solid with not < 25% water of crystallization [UN 2949] **(16721-80-5)**
Sodium hydrosulphide, With less than 25% water of crystallization [UN 2318] **(16721-80-5)**
Sodium hydrosulphite **(7775-14-6)**
Sodium-hydroxide **(1310-73-2)**
Sodium hydroxide (ACGIH, DOT,OSHA) **(1310-73-2)**
Sodium hydroxide, Liquid [UN 1824] **(1310-73-2)**
Sodium hydroxide, Solid [UN 1823] **(1310-73-2)**
Sodium hydroxide, Solution [UN 1824] **(1310-73-2)**
Sodium α-hydroxyacetate **(2836-32-0)**
Sodium hydroxyacetate **(2836-32-0)**
Sodium o-hydroxybenzoate **(54-21-7)**
Sodium 2-hydroxy-3-chloropropanesulfonate **(126-83-0)**
Sodium(hydroxyde de) (French) **(1310-73-2)**
Sodium 2-hydroxydiphenyl **(132-27-4)**
Sodium 2-hydroxy-1-ethanesulfonate **(1562-00-1)**
Sodium 2-hydroxyethanesulfonate **(1562-00-1)**
Sodium β-hydroxyethanesulfonate **(1562-00-1)**
Sodium 2-hydroxyethanesulfonic acid **(1562-00-1)**
Sodium m-(2-hydroxyethyl)ethylenediaminetriacetate **(53404-54-9)**
Sodium β-hydroxyethyl sulfide **(37482-11-4)**
Sodium 2-hydroxyethylsulfonate **(1562-00-1)**
Sodium hydroxyethylsulfonate **(1562-00-1)**
Sodium hydroxymethanesulfinate **(149-44-0)**
Sodium hydroxymethanesulfonate **(75-92-3)**
Sodium 1-hydroxypyridine-2-thione **(15922-78-8)**
Sodium hydroxytriphenylborate **(12113-07-4)**
Sodium hypochlorite **(10022-70-5)**
Sodium hypochlorite (DOT) **(7681-52-9)**
Sodium hypochlorite phosphate (Na13(ClO)(PO4)4) (9CI) **(11084-85-8)**
Sodium-hypophosphite **(7681-53-0)**
Sodium hyposulfite **(10102-17-7)**
Sodium hyposulfite **(7772-98-7)**
Sodium 5,5'-indigotidisulfonate **(860-22-0)**
Sodium indigotindisulfonate **(860-22-0)**
Sodium iodate **(7681-55-2)**
Sodium-iodide **(7681-82-5)**
Sodium iodine **(7681-82-5)**
Sodium iodomethanesulfonate **(126-31-8)**
Sodium isethionate **(1562-00-1)**
Sodium isoascorbate **(6381-77-7)**
Sodium isocyanate **(917-61-3)**
Sodium isocyanurate **(2624-17-1)**
Sodium O-isopropyl dithiocarbonate **(140-93-2)**
Sodium O-isopropyl xanthate **(140-93-2)**
Sodium isopropylxanthate **(140-93-2)**
Sodium isopropylxanthogenate **(140-93-2)**
Sodium isothiocyanate **(540-72-7)**
Sodium lactate **(72-17-3)**
Sodium laurate **(629-25-4)**

Sodium N-lauroylsarcosinate **(137-16-6)**
Sodium lauroyl sarcosinate **(137-16-6)**
Sodium lauroylsarcosinate **(137-16-6)**
Sodium N-lauroylsarcosine **(137-16-6)**
Sodium lauroylsarcosine **(137-16-6)**
Sodium lauryl alcohol diglycol ether sulfate **(3088-31-1)**
Sodium laurylbenzenesulfonate **(25155-30-0)**
Sodium lauryl di(oxyethyl) sulfate **(3088-31-1)**
Sodium-lauryl-ether-sulfate **(9004-82-4)**
Sodium lauryl ethoxysulphate **(15826-16-1)**
Sodiumlaurylglycolether sulfate **(3088-31-1)**
Sodium N-lauryl-β-iminodipropionate **(14960-06-6)**
Sodium N-lauryl-β-iminodipropionate **(3655-00-3)**
Sodium lauryloxyethoxyethyl sulfate **(3088-31-1)**
Sodium (lauryloxypolyethoxy)ethyl sulfate **(9004-82-4)**
Sodium laurylpoly(oxyethylene) sulfate **(9004-82-4)**
Sodium lauryl sulfate **(151-21-3)**
Sodium lauryl sulfoacetate **(1847-58-1)**
Sodium lauryl sulphate **(151-21-3)**
Sodium lauryl trioxyethylene sulfate **(13150-00-0)**
Sodium ligninsulfonate **(8061-51-6)**
Sodium lignosulfite **(8061-51-6)**
Sodium lignosulfonate **(8061-51-6)**
Sodium lignosulfonic acid **(8061-51-6)**
Sodium luminal **(57-30-7)**
Sodium malonate **(141-95-7)**
Sodium malondialdehyde **(24382-04-5)**
Sodium mercaptan **(16721-80-5)**
Sodium mercaptide **(16721-80-5)**
Sodium mercaptoacetate **(367-51-1)**
Sodium 2-mercaptobenzothiazolate **(2492-26-4)**
Sodium mercaptobenzothiazolate **(2492-26-4)**
Sodium 2-mercaptobenzothiazole **(2492-26-4)**
Sodium mercaptobenzothiazole **(2492-26-4)**
Sodium 2-mercaptobenzothiazol solution **(2492-26-4)**
Sodium merthiolate **(54-64-8)**
Sodium metaaluminate **(1302-42-7)**
Sodium metaarsenate **(7631-89-2)**
Sodium metaarsenite **(7784-46-5)**
Sodium metabisulfite (ACGIH,DOT,OSHA) **(7681-57-4)**
Sodium metabisulphite **(7681-57-4)**
Sodium metaborate **(7775-19-1)**
Sodium metal (DOT) **(7440-23-5)**
Sodium metaphosphate **(10361-03-2)**
Sodium metasilicate **(6834-92-0)**
Sodium metasilicate, Anhydrous **(6834-92-0)**
Sodium metavanadate **(13718-26-8)**
Sodium methacrylate **(5536-61-8)**
Sodium methanalsulfoxylate **(149-44-0)**
Sodium methanearsonate **(144-21-8)**
Sodium methanearsonate **(2163-80-6)**
Sodium methanesulfonothioate **(1950-85-2)**
Sodium metharsonate **(144-21-8)**
Sodium methiodal **(126-31-8)**
Sodium methohexital **(309-36-4)**
Sodium methohexitone **(309-36-4)**
Sodium methoxide **(124-41-4)**
Sodium 2-methoxy-3,6-dichlorobenzoate **(1982-69-0)**
Sodium methoxymethylsulfamate **(5390-94-3)**
Sodium methylarsonate **(144-21-8)**
Sodium methylate [UN 1431] **(124-41-4)**
Sodium methylate, Dry (DOT) **(124-41-4)**
Sodium 4-methylbenzenesulfinate **(824-79-3)**
Sodium methylbenzenesulfonate **(12068-03-0)**
Sodium p-methylbenzenesulfonate **(657-84-1)**
Sodium (2-methyl-4-chlorophenoxy)acetate **(3653-48-3)**
Sodium 4-(2-methyl-4-chlorophenoxy)butyrate **(6062-26-6)**
Sodium N-methyldithiocarbamate **(137-42-8)**
Sodium methyldithiocarbamate **(137-42-8)**
Sodium N-methyldithiocarbamate dihydrate **(6734-80-1)**
Sodium 2,2'-methylenebis(4-chlorophenate) **(10254-48-5)**
Sodium 2-methyl-7-ethylundecanol-4-sulfate **(139-88-8)**
Sodium 2-methyl-7-ethylundecyl sulfate-4 **(139-88-8)**
Sodium 2-(N-methyloleamido)ethane-1-sulfonate **(137-20-2)**
Sodium N-methyl-N-oleoyltaurate **(137-20-2)**
Sodium N-methyl-N-oleoyltaurate **(7346-80-7)**
Sodium methyl oleoyl taurate **(137-20-2)**
Sodium 4-methylphenolate **(1121-70-6)**
Sodium p-methylphenolate **(1121-70-6)**
Sodium 4-methylphenoxide **(1121-70-6)**
Sodium p-methylphenoxide **(1121-70-6)**
Sodium 2-methyl-2-propenoate **(5536-61-8)**
Sodium methylsiliconate **(16589-43-8)**
Sodium (1-methylundecyl)benzenesulfonate **(27987-00-4)**
Sodium molybdate **(7631-95-0)**
Sodium molybdate(VI) **(7631-95-0)**
Sodium monensin **(22373-78-0)**
Sodium monochloracetate **(3926-62-3)**
Sodium monododecyl sulfate **(151-21-3)**
Sodium monofluoride **(7681-49-4)**
Sodium monofluoroacetate **(62-74-8)**
Sodium monofluorophosphate **(10163-15-2)**
Sodium monohexadecyl sulfate **(1120-01-0)**
Sodium monohydrogen phosphate (2:1:1) **(7558-79-4)**
Sodium monohydrogen phosphate dodecahydrate (2:1:1:12) **(10039-32-4)**
Sodium monoiodide **(7681-82-5)**
Sodium monoiodomethanesulfonate **(126-31-8)**
Sodium monolauryl sulfate **(151-21-3)**
Sodium monooctadecyl sulfate **(1120-04-3)**
Sodium monostearyl sulfate **(1120-04-3)**
Sodium monosulfide **(1313-82-2)**
Sodium monoxide **(12401-86-4)**
Sodium monoxide, Solid **(12401-86-4)**
Sodium myreth sulfate **(25446-80-4)**
Sodium myristyl ether sulfate **(25446-80-4)**
Sodium myristyl sulfate **(1191-50-0)**
Sodium 1-naphthaleneacetate **(61-31-4)**
Sodium 2-naphthalenesulfonate **(532-02-5)**
Sodium β-naphthalenesulfonate **(532-02-5)**
Sodium naphthalene-2-sulfonate **(532-02-5)**
Sodium naphthalene-6-sulfonate **(532-02-5)**
Sodium naphthalenesulfonate **(1321-69-3)**
Sodium naphthalene sulfonate solution **(1321-69-3)**
Sodium naphthalene-2-sulphonate **(532-02-5)**
Sodium naphthionate **(130-13-2)**
Sodium 1-naphthylamine-4-sulfonate **(130-13-2)**
Sodium α-naphthylamine-4-sulfonate **(130-13-2)**
Sodium N-1-naphthylphthalamate **(132-67-2)**
Sodium N-1-naphthylphthalamic acid **(132-67-2)**
Sodium nembutal **(57-33-0)**
Sodium-22 neoprene accelerator **(96-45-7)**
Sodium nicotinate **(54-86-4)**
Sodium(I) nitrate (1:1) **(7631-99-4)**
Sodium nitrate [UN 1498] **(7631-99-4)**
Sodium nitriloacetate **(10042-84-9)**
Sodium nitrilotriacetate **(10042-84-9)**
Sodium nitrilotriacetate **(5064-31-3)**
Sodium (nitrilotriacetato)calciate (7CI) **(60034-45-9)**
Sodium nitrite [UN 1500] **(7632-00-0)**
Sodium 3-nitrobenzoate **(827-95-2)**
Sodium nitrophenate **(824-78-2)**
Sodium p-nitrophenol **(824-78-2)**
Sodium 4-nitrophenolate **(824-78-2)**
Sodium p-nitrophenolate **(824-78-2)**

Sodium 4-nitrophenoxide (824-78-2)
Sodium p-nitrophenoxide (824-78-2)
Sodium nonylbenzenesulfonate (26856-61-1)
Sodium norsulfazole (144-74-1)
Sodium octadecanoate (822-16-2)
Sodium octadecyl sulfate (1120-04-3)
Sodium N-octadecylterephthalamate (5994-45-6)
Sodium 1-octanesulfonate (5324-84-5)
Sodium n-octanoate (1984-06-1)
Sodium octanoate (1984-06-1)
Sodium octoxynol-2 ethane sulfonate (2917-94-4)
Sodium octyl sulfate (142-31-4)
Sodium octyl sulphate (142-31-4)
Sodium oleate (143-19-1)
Sodium oleoyl isethionate (142-15-4)
Sodium N-oleoyl-N-methylataurine (137-20-2)
Sodium N-oleoyl-N-methyltaurate (137-20-2)
Sodium oleylmethyltauride (137-20-2)
Sodium omadine (15922-78-8)
Sodium oxalate (62-76-0)
Sodium oxide (12401-86-4)
Sodium oxide (Na2-O2) (1313-60-6)
Sodium oxide sulfide (7772-98-7)
Sodium 2-oxopropanoate (113-24-6)
Sodium oxymethanesulfinic acid (149-44-0)
Sodium palmitate (408-35-5)
Sodium palmityl sulfate (1120-01-0)
Sodium D-pantothenate (867-81-2)
Sodium pantothenate (867-81-2)
Sodium pantothenic acid (867-81-2)
Sodium paratoluene sulphonate (657-84-1)
Sodium pectinate (9005-59-8)
Sodium penicillin (69-57-8)
Sodium penicillin G (69-57-8)
Sodium penicillin II (69-57-8)
Sodium-pent (57-33-0)
Sodium pentabarbital (57-33-0)
Sodium pentabarbitone (57-33-0)
Sodium pentachlorophenate [UN 2567] (131-52-2)
Sodium pentachlorophenol (131-52-2)
Sodium pentachlorophenolate (131-52-2)
Sodium pentachlorophenoxide (131-52-2)
Sodium, (pentachlorophenoxy)- (131-52-2)
Sodium pentachlorphenate (131-52-2)
Sodium pentadecanecarboxylate (408-35-5)
Sodium pentanedioate (32224-61-6)
Sodium pentobarbital (57-33-0)
Sodium pentobarbitone (57-33-0)
Sodium pentobarbiturate (57-33-0)
Sodium perborate (7632-04-4)
Sodium perchlorate [UN 1502] (7601-89-0)
Sodium (perchlorate de) (French) (7601-89-0)
Sodium permanganate [UN 1503] (10101-50-5)
Sodium peroxide [UN 1504] (1313-60-6)
Sodium peroxoborate (7632-04-4)
Sodium peroxydisulfate (7775-27-1)
Sodium persulfate (ACGIH) [UN 1505] (7775-27-1)
Sodium persulphate (DOT) (7775-27-1)
Sodium petroleum sulfonate (68608-26-4)
Sodium phenate (139-02-6)
Sodium phenobarbital (57-30-7)
Sodium phenobarbitone (57-30-7)
Sodium phenolate, Solid [UN 2497] (139-02-6)
Sodium phenolsulfonate (1300-51-2)
Sodium phenol tetrabromophthalein (71-67-0)
Sodium phenoxide (139-02-6)
Sodium phenoxyacetate (3598-16-1)
Sodium phenylacetate (114-70-5)
Sodium phenylethylbarbiturate (57-30-7)
Sodium phenylethylmalonylurea (57-30-7)
Sodium phenylmercaptide (930-69-8)
Sodium 2-phenylphenate (132-27-4)
Sodium o-phenylphenate (132-27-4)
Sodium ortho phenylphenate (132-27-4)
Sodium o-phenylphenol (132-27-4)
Sodium o-phenylphenolate (132-27-4)
Sodium o-phenylphenoxide (132-27-4)
Sodium phenylphosphinate (4297-95-4)
Sodium phenylsulfide (930-69-8)
Sodium phenylsulfonate (515-42-4)
Sodium phenylthiolate (930-69-8)
Sodium phenytoin (630-93-3)
Sodium phosphate (7601-54-9)
Sodium phosphate (7632-05-5)
Sodium phosphate, Anhydrous (7601-54-9)
Sodium phosphate (Na5P3O10) (7758-29-4)
Sodium phosphate, dibasic (10140-65-5)
Sodium phosphate, dibasic (7558-79-4)
Sodium phosphate dodecahydrate (10101-89-0)
Sodium phosphate, monobasic (7558-80-7)
Sodium phosphate, tribasic (10361-89-4)
Sodium phosphate, tribasic (7601-54-9)
Sodium phosphate tribasic dodecahydrate (10101-89-0)
Sodium phosphide (87835-45-8)
Sodium phosphoaluminate (1344-06-5)
Sodium phosphorofluoridate (10163-15-2)
Sodium phosphorofluridate (10163-15-2)
Sodium picramate, Dry or containing less than 20% water [UN 0235] (831-52-7)
Sodium picramate, Wet (with at least 20% water) [UN 1349] (831-52-7)
Sodium polignate (8061-51-6)
Sodium-poly-acrylate (9003-04-7)
Sodium polyethylene glycol (5) lauryl ether sulfate (9004-82-4)
Sodium polyethylene glycol (7) lauryl ether sulfate (9004-82-4)
Sodium polyethylene glycol 600 lauryl ether sulfate (9004-82-4)
Sodium polymannuronate (9005-38-3)
Sodium poly(oxyethylene) lauryl ether sulfate (9004-82-4)
Sodium polyoxyethylene (5) lauryl ether sulfate (9004-82-4)
Sodium polyoxyethylene (7) lauryl ether sulfate (9004-82-4)
Sodium polyoxyethylene (12) lauryl ether sulfate (9004-82-4)
Sodium polyoxyethylene tridecyl sulfate (25446-78-0)
Sodium polyphosphate (68915-31-1)
Sodium polystyrenesulfonate (9080-79-9)
Sodium polysulfide (1344-08-7)
Sodium-potassium alloy (11135-81-2)
Sodium potassium alloy, Liquid (11135-81-2)
Sodium potassium alloy, Solid (11135-81-2)
Sodium potassium alloys (11135-81-2)
Sodium potassium (dl) tartrate (304-59-6)
Sodium potassium tartrate (304-59-6)
Sodium 2-propene-1-sulfonate (2495-39-8)
Sodium 2-propenoate (7446-81-3)
Sodium propionate (137-40-6)
Sodium 2-propylpentanoate (1069-66-5)
Sodium 2-propylvalerate (1069-66-5)
Sodium 2-pyridinethiol-1-oxide (15922-78-8)
Sodium pyridinethione (15922-78-8)
Sodium pyrithione (15922-78-8)
Sodium pyrithione (3811-73-2)
Sodium pyroarsenate (13464-42-1)
Sodium pyroborate (1303-96-4)
Sodium pyroborate decahydrate (1303-96-4)
Sodium pyrophosphate (7722-88-5)
Sodium pyrophosphate (7758-16-9)

Sodium pyrosulfate	(7681-38-1)
Sodium pyrosulfite	(7681-57-4)
Sodium rhodanate	(540-72-7)
Sodium rhodanide	(540-72-7)
Sodium ricinoleate	(5323-95-5)
Sodium saccharide	(128-44-9)
Sodium saccharin	(128-44-9)
Sodium saccharinate	(128-44-9)
Sodium saccharine	(128-44-9)
Sodium salicylate	(54-21-7)
Sodium salicylic acid	(54-21-7)
Sodium salt of acifluorfen	(62476-59-9)
Sodium salt of cacodylic acid	(124-65-2)
Sodium salt of carboxymethylcellulose	(9004-32-4)
Sodium salt of N-(3,4-dichlorophenyl)-N'-2(2-sulfo-4-chloro-phenoxy)-5-chlorophenyl urea	(3567-25-7)
Sodium salt of dichloro-s-triazinetrione	(2893-78-9)
Sodium salt of N,N-diethyldithiocarbamic acid	(148-18-5)
Sodium salt of 4,6-dinitro-o-cresol	(2312-76-7)
Sodium salt of ethylenediaminetetraacetic acid	(64-02-8)
Sodium salt of hydroxy-o-carboxy-phenyl-fluorone	(518-47-8)
Sodium salt of β-naphthalenesulfonic acid	(532-02-5)
Sodium salt of phosphoric acid	(7632-05-5)
Sodium salt of sulfonated naphthaleneformaldehyde condensate	(9084-06-4)
Sodium salt of 2,4,5-trichlorophenol	(136-32-3)
Sodium selenate	(13410-01-0)
Sodium selenite (7CI)	(7782-82-3)
Sodium selenite [UN 2630]	(10102-18-8)
Sodium β-silicate	(1344-09-8)
Sodium silicate	(1344-09-8)
Sodium silicate	(6834-92-0)
Sodium silicoaluminate	(1344-00-9)
Sodium silicofluoride (DOT)	(16893-85-9)
Sodium silicon fluoride	(16893-85-9)
Sodium sorbate	(7757-81-5)
Sodium sotradecol	(1191-50-0)
Sodium sotradecol	(139-88-8)
Sodium stannous tartrate	(72378-89-3)
Sodium stearate (ACGIH)	(822-16-2)
Sodium stearyl sulfate	(1120-04-3)
Sodium 2-(4-styryl-3-sulfophenyl)-2H-naphtho-(1,2-d)-triazole	(6416-68-8)
Sodium sucaryl	(139-05-9)
Sodium sulfamerazine	(127-58-2)
Sodium 2-sulfanilamidothiazole	(144-74-1)
Sodium sulfate (2:1)	(7757-82-6)
Sodium sulfate anhydrous	(7757-82-6)
Sodium sulfathiazole	(144-74-1)
Sodium sulfhydrate	(16721-80-5)
Sodium sulfide	(1344-08-7)
Sodium-sulfide	(16721-80-5)
Sodium sulfide (Na2(Sx)) (9CI)	(1344-08-7)
Sodium sulfide, Not less than 25% water of crystallization	(16721-80-5)
Sodium sulfide (Anhydrous) [UN 1385]	(1313-82-2)
Sodium sulfite	(7757-83-7)
Sodium sulfite (2:1)	(7757-83-7)
Sodium sulfite anhydrous	(7757-83-7)
Sodium sulfobromophthalein	(71-67-0)
Sodium sulfocyanate	(540-72-7)
Sodium sulfocyanide	(540-72-7)
Sodium sulfodi-(2-ethylhexyl)-sulfosuccinate	(577-11-7)
Sodium sulfonate oleic acid	(68443-05-0)
Sodium sulfoxylate	(7775-14-6)
Sodium sulfoxylate formaldehyde	(149-44-0)
Sodium sulhydrate	(7631-90-5)
Sodium sulphamerazine	(127-58-2)
Sodium sulphate	(7757-82-6)
Sodium sulphathiazole	(144-74-1)
Sodium sulphide	(1313-82-2)
Sodium sulphide, Anhydrous or containing less than 30% water [UN 1385]	(1313-82-2)
Sodium sulphite	(7757-83-7)
Sodium sulphobromophthalein	(71-67-0)
Sodium superoxide	(12034-12-7)
Sodium tartrate	(526-94-3)
Sodium tartrate	(868-18-8)
Sodium taurocholate	(145-42-6)
Sodium tauroglycocholate	(11006-55-6)
Sodium tellurate	(10102-20-2)
Sodium tellurate(IV)	(10102-20-2)
Sodium tellurite	(10102-20-2)
Sodium tetraborate	(1303-96-4)
Sodium tetraborate	(1330-43-4)
Sodium tetraborate (Na2B4O7)	(1330-43-4)
Sodium tetraborate decahydrate	(1303-96-4)
Sodium tetrachlorophenate	(25567-55-9)
Sodium, (tetrachlorophenoxy)-	(25567-55-9)
Sodium tetradecenesulfonate	(11066-21-0)
Sodium 4-tetradecylbenzenesulfonate	(1797-33-7)
Sodium tetradecylbenzenesulfonate	(28348-61-0)
Sodium tetradecyl sulfate	(1191-50-0)
Sodium tetrahydroaluminate(1-)	(13770-96-2)
Sodium tetrahydroborate(1-)	(16940-66-2)
Sodium tetrahydro-3,5-dimethyl-2H-1,3,5-thiadiazine-2-thione	(53404-60-7)
Sodium tetrametaphosphate	(13396-41-3)
Sodium tetraphenylborate	(143-66-8)
Sodium tetraphenylborate(1-)	(143-66-8)
Sodium tetraphenylboride	(143-66-8)
Sodium tetraphenylboride(1-)	(143-66-8)
Sodium tetraphenylboron	(143-66-8)
Sodium tetrapropylbenzene sulfonate	(11067-82-6)
Sodium, (N[1]-2-thiazolylsulfanilamido)-	(144-74-1)
Sodium thimerfonate	(5964-24-9)
Sodium thiocyanate	(540-72-7)
Sodium thiocyanide	(540-72-7)
Sodium thioglycolate	(367-51-1)
Sodium thioglycollate	(367-51-1)
Sodium thiomethylate	(5188-07-8)
Sodium thiophenate	(930-69-8)
Sodium thiophenolate	(930-69-8)
Sodium thiophenoxide	(930-69-8)
Sodium thiophenylate	(930-69-8)
Sodium thiosulfate	(7772-98-7)
Sodium thiosulfate anhydrous	(7772-98-7)
Sodium thiosulfate, pentahydrate	(10102-17-7)
Sodium thiosulphate	(7772-98-7)
Sodium 4-toluenesulfinate	(824-79-3)
Sodium p-toluenesulfinate	(824-79-3)
Sodium p-toluenesulfonate	(657-84-1)
Sodium toluenesulfonate	(12068-03-0)
Sodium p-toluenesulfonylchloramide	(127-65-1)
Sodium p-tolylsulfinate	(824-79-3)
Sodium p-tolyl sulfonate	(657-84-1)
Sodium tosylate	(657-84-1)
Sodium tosylchloramide	(127-65-1)
Sodium (trichloracetate de) (French)	(650-51-1)
Sodium trichloroacetate	(650-51-1)
Sodium 2,3,6-trichlorobenzeneacetate	(2439-00-1)
Sodium 2,4,5-trichlorophenate	(136-32-3)
Sodium, (2,4,5-trichlorophenoxy)-	(136-32-3)
Sodium 2,4,5-trichlorophenoxyethyl sulfate	(3570-61-4)
Sodium 2,3,6-trichlorophenylacetate	(2439-00-1)
Sodium trideceth sulfate	(25446-78-0)
Sodium n-tridecylbenzenesulfonate	(26248-24-8)
Sodium tridecylbenzene sulfonate	(26248-24-8)

Sodium tridecylbenzenesulfonate **(26248-24-8)**
Sodium tridecyl ether sulfate **(25446-78-0)**
Sodium tridecyl sulfate **(3026-63-9)**
Sodium tridecyl tri(oxyethyl) sulfate **(25446-78-0)**
Sodium trimetaphosphate **(7785-84-4)**
Sodium triphosphate **(7758-29-4)**
Sodium triphosphate (Na5P3O10) **(7758-29-4)**
Sodium tripolyphosphate **(7758-29-4)**
Sodium tungstate **(13472-45-2)**
Sodium undecylbenzenesulfonate **(27636-75-5)**
Sodium valproate **(1069-66-5)**
Sodium vanadate **(13718-26-8)**
Sodium versenate **(139-33-3)**
Sodium warfarin **(129-06-6)**
Sodium xylenesulfonate **(1300-72-7)**
Sodium zinc EDTA **(14025-21-9)**
Sodium zinc cyanide **(15333-24-1)**
Sodium zinc polyphosphate **(65997-17-3)**
Sodium zinc potassium polyphosphate **(65997-17-3)**
Sodium zirconium lactate **(10377-98-7)**
Sodium zomepirac **(64092-48-4)**
Sodiuretic **(73-48-3)**
Sodizole **(127-69-5)**
Sodna sul kyseliny cis-β-4-methoxybenzoyl-β-bromakrylove (Czech) **(21739-91-3)**
Sodothiol **(10102-17-7)**
Sodothiol **(7772-98-7)**
Sofril **(7704-34-9)**
Sofro **(2152-34-3)**
Softenil **(50-35-1)**
Softenon **(50-35-1)**
Softil **(577-11-7)**
Stercofuge **(9005-36-1)**
Sohnhofen Stone **(1317-65-3)**
Soil Fungicide 1823 **(2675-77-6)**
Soil Stabilizer 661 **(9003-55-8)**
Soilbrom **(106-93-4)**
Soilbrom-40 **(106-93-4)**
Soilbrom-85 **(106-93-4)**
Soilbrom-90 **(106-93-4)**
Soilbrom-90EC **(106-93-4)**
Soilbrom-100 **(106-93-4)**
Soilbrome-85 **(106-93-4)**
Soilfume **(106-93-4)**
Soilsin **(2235-25-8)**
Sok **(63-25-2)**
Sol **(9011-14-7)**
Sol 90 **(9011-14-7)**
Sol 95 **(9011-14-7)**
Sol Phenobarbital **(57-30-7)**
Sol Phenobarbitone **(57-30-7)**
Sol. Sulfur Blue 10 **(4368-56-3)**
Solacen **(4268-36-4)**
Solacin **(4268-36-4)**
Solactol **(97-64-3)**
Soladren **(51-43-4)**
Solaesthin **(75-09-2)**
Solakryl BMX **(25608-33-7)**
Solamin **(121-54-0)**
Solamine **(121-54-0)**
Solamine Fast Yellow 5G **(10190-68-8)**
Solan **(2307-68-8)**
Solancarpidine **(126-17-0)**
Solanesol **(13190-97-1)**
Solanex **(8066-11-3)**
Solanid-5-ene, 3-β-((O-6-deoxy-α-l-mannopyranosyl-(1-2)-O-(β-D-glucopyranosyl- (1-3))-β-d-galactopyranosyl)oxy) **(20562-02-1)**
Solanidine-S **(126-17-0)**
α-Solanin **(20562-02-1)**
Solanine **(20562-02-1)**
α-Solanine **(20562-02-1)**
Solanthrene Blue B **(130-20-1)**
Solanthrene Blue F-SBA **(130-20-1)**
Solanthrene Blue RS **(81-77-6)**
Solanthrene Blue RSN **(81-77-6)**
Solanthrene Brilliant Green B **(128-58-5)**
Solanthrene Brilliant Green BN **(128-58-5)**
Solanthrene Brilliant Green FF **(128-58-5)**
Solanthrene Brilliant Orange JR **(4424-06-0)**
Solanthrene Brilliant Pink R **(2379-74-0)**
Solanthrene Brilliant Pink RF **(2379-74-0)**
Solanthrene Brilliant Violet F 2R **(1324-55-6)**
Solanthrene Brilliant Yellow **(128-66-5)**
Solanthrene Brown BR **(2475-33-4)**
Solanthrene Brown JR **(2475-33-4)**
Solanthrene Olive R **(2379-81-9)**
Solanthrene Orange F-J **(128-70-1)**
Solanthrene Orange J **(128-70-1)**
Solanthrene R for sugar **(81-77-6)**
Solantin **(57-41-0)**
Solantine Brown BRL **(16071-86-6)**
Solantine Red 8BL **(2610-11-9)**
Solantine Yellow 8GL **(10190-68-8)**
Solantoin **(57-41-0)**
Solantoin **(630-93-3)**
Solantyl **(57-41-0)**
Solantyl **(630-93-3)**
Solapret **(9003-08-1)**
Solar 40 **(25155-30-0)**
Solar 90 **(25155-30-0)**
Solar Black G **(6428-31-5)**
Solar Blue 4GL **(2610-05-1)**
Solar Blue UMN 57-6692 **(1325-87-7)**
Solar Brown PL **(16071-86-6)**
Solar Coating Violet RMN47-3012 **(1325-82-2)**
Solar Fast Red 3g **(3734-67-6)**
Solar Flavine 5G **(10190-68-8)**
Solar Light Orange GX **(1936-15-8)**
Solar Orange **(633-96-5)**
Solar Orange IV **(554-73-4)**
Solar Red B **(2610-11-9)**
Solar Red O **(915-67-3)**
Solar Rubine **(3567-69-9)**
Solar Violet 5BN **(1694-09-3)**
Solar Violet RCL **(1325-82-2)**
Solar Violet RMN47-3612 **(1325-82-2)**
Solar Winter Ban **(57-55-6)**
Solarchem O **(21245-02-3)**
Solasan 500 **(137-42-8)**
Solasod-5-en-3-β-ol **(126-17-0)**
Solasodine **(126-17-0)**
Solatene **(7235-40-7)**
Solbar **(7727-43-7)**
Solbrol A **(120-47-8)**
Solbrol B **(94-26-8)**
Solbrol M **(99-76-3)**
Solbrol P **(94-13-3)**
Solbrol Z **(94-18-8)**
Solcain **(137-58-6)**
Soldep **(52-68-6)**
Soleal **(96-26-4)**
Solestro **(50-50-0)**
Solevione anastress **(57-53-4)**
Solex **(1344-95-2)**

Solex Brown R (16071-86-6)
Solex Canary Yellow 5G (10190-68-8)
Solfac (68359-37-5)
Solfacitina (17784-12-2)
Solfarin (81-81-2)
Solfast Green (1328-53-6)
Solfast Green 63102 (1328-53-6)
Solfast Victoria Blue CP 476 (1325-87-7)
Solfato di 8-ossichinolina (134-31-6)
Solfo Black B (51-28-5)
Solfo Black BB (51-28-5)
Solfo Black 2B supra (51-28-5)
Solfo Black G (51-28-5)
Solfo Black SB (51-28-5)
Solfo Serpine (50-55-5)
Solfoton (50-06-6)
Solfuro di carbonio (Italian) (75-15-0)
Solganal (12192-57-3)
Solganal B (12192-57-3)
Solgard (23505-41-1)
Solgol (42200-33-9)
Solicam (27314-13-2)
Solid Crotonic Acid (3724-65-0)
Solid Green (633-03-4)
Solid Green Crystals O (569-64-2)
Solid Green FCF (2353-45-9)
Solid Green O (569-64-2)
Solid Red A (1658-56-6)
Solindene Orange R (3263-31-8)
Solius Light Brown BRLL (16071-86-6)
Solius Light Brown BRS (16071-86-6)
Solius Light Yellow 5G (10190-68-8)
Solius Red 4B (2610-11-9)
Soliwax (577-11-7)
Solka-Fil (9004-34-6)
Solka-Floc (9004-34-6)
Solka-Floc BW (9004-34-6)
Solka-Floc BW 20 (9004-34-6)
Solka-Floc BW 100 (9004-34-6)
Solka-Floc BW 200 (9004-34-6)
Solka-Floc BW 2030 (9004-34-6)
Solketal (100-79-8)
Solliculin (53-16-7)
Solmethine (75-09-2)
Solo (132-66-1)
Solochrome Blue FB (3567-69-9)
Solochrome Violet R (2092-55-9)
Solochrome Violet RS (2092-55-9)
Solocrom Violet RS (2092-55-9)
Soloid (9003-20-7)
Solophenyl Brilliant Blue BL (6527-70-4)
Solophenyl Yellow 7GL (10190-68-8)
Solosin (58-55-9)
Solozone (1313-60-6)
Solprene 300 (9003-55-8)
Solprene 303 (9003-55-8)
Solprina (54-05-7)
Solpyron (50-78-2)
Sol sodowa kwasu laurylobenzenosulfonowego (Polish) (25155-30-0)
Solsol Needles (151-21-3)
Soluble Fluorescein (518-47-8)
Soluble Fluoresceine (518-47-8)
Soluble Gluside (128-44-9)
Soluble Gun Cotton (9004-70-0)
Soluble Indigo (860-22-0)
Soluble Pentobarbital (57-33-0)
Soluble Phenobarbital (57-30-7)
Soluble Phenobarbitone (57-30-7)
Soluble Phenytoin (630-93-3)
Soluble Potash Glass (1312-76-1)
Soluble Potash Water Glass (1312-76-1)
Soluble Saccharin (128-44-9)
Soluble Sulfamerazine (127-58-2)
Soluble Sulfathiazole (144-74-1)
Soluble Vandyke Brown (1317-34-6)
Solufilin (479-18-5)
Solufyllin (479-18-5)
Soluglacit (311-45-5)
Solumedine (127-58-2)
Solumin FP 85SD (9014-90-8)
Solunaptol A (92-77-3)
Solunaptol ANL (132-68-3)
Solunaptol FRL (135-62-6)
Solunaptol MNL (135-65-9)
Solunaptol OT (135-61-5)
Solunaptol YL (91-96-3)
Soluphyllin (479-18-5)
Solusol-75% (577-11-7)
Solusol-100% (577-11-7)
Solution concentree T271 (61-82-5)
Solution potassium iodohydragyrate (7783-33-7)
Solvanol (84-66-2)
Solvanom (131-11-3)
Solvant Yellow 33 (8003-22-3)
Solvar (9002-89-5)
Solvarone (131-11-3)
Solvat 14 (3012-37-1)
Solvent 111 (71-55-6)
Solvent Black 5 (11099-03-9)
Solvent Blue 11 (128-85-8)
Solvent Blue 36 (14233-37-5)
Solvent Extracted Sunflower Oil (8001-21-6)
Solvent Green 3 (128-80-3)
Solvent Green 7 (6358-69-6)
Solvent Orange 1 (2051-85-6)
Solvent Orange 15 (494-38-2)
Solvent Red 1 (1229-55-6)
Solvent Red 19 (6368-72-5)
Solvent Red 49 (509-34-2)
Solvent Sunflower Oil (8001-21-6)
Solvent Violet 26 (2872-48-2)
Solvent Yellow 1 (60-09-3)
Solvent Yellow 14 (842-07-9)
Solvent Yellow 44 (2478-20-8)
Solvent deasphalted residual oil (64741-95-3)
Solvent-dewaxed heavy naphthenic distillate (64742-63-8)
Solvent-dewaxed heavy paraffinic distillate (64742-65-0)
Solvent-dewaxed light naphthenic distillate (64742-64-9)
Solvent-dewaxed light paraffinic distillate (64742-56-9)
Solvent ether (60-29-7)
Solvent naphtha, Petroleum, Heavy arom. (64742-94-5)
Solvent naphtha, Petroleum, Light arom. (64742-95-6)
Solvent naphtha, Petroleum, Medium aliph. (64742-88-7)
Solvent-refined heavy naphthenic distillate (64741-96-4)
Solvent-refined heavy paraffinic distillate (64741-88-4)
Solvent-refined light naphthenic distillate (64741-97-5)
Solvent-refined light paraffinic distillate (64741-89-5)
Solvesso (9CI) (51109-97-8)
Solvic 523KC (9003-22-9)
Solvic PA 513 (9003-22-9)
Solvic 513PB (9003-22-9)
Solvinc PA 513 (9003-22-9)
Solvirex (298-04-4)
Solvolsol (111-90-0)

Solyacord (59-26-7)
Soma (62-44-2)
Soma (78-44-4)
Somacton (9002-72-6)
Somadril (78-44-4)
Somalgit (78-44-4)
Somalia Blue G (128-85-8)
Somalia Orange A2R (3118-97-6)
Somalia Orange I (842-07-9)
Somalia Orange 2R (3118-97-6)
Somalia Red III (85-86-9)
Somalia Red IV (85-83-6)
Somalia Yellow A (60-11-7)
Somalia Yellow 2G (60-09-3)
Somalia Yellow R (97-56-3)
Soman (96-64-0)
Somanil (78-44-4)
Somar (144-21-8)
Somatotrophic Hormone (9002-72-6)
Somatotrophin (9002-72-6)
Somatotropic Hormone (9002-72-6)
Somatotropin (9002-72-6)
Somberol (72-44-6)
Sombucaps (56-29-1)
Sombulex (56-29-1)
Sombutol (50-06-6)
Somelin (59128-97-1)
Sometam (137-42-8)
Somilan (55283-68-6)
Somio (15879-93-3)
Somipront (67-68-5)
Somlan (1172-18-5)
Somnafac (72-44-6)
Somnal (57-43-2)
Somnalert (56-29-1)
Somnased (146-22-5)
Somnesin (77-75-8)
Somnevrin (533-45-9)
Somni Sed (302-17-0)
Somnibel (146-22-5)
Somnicaps (135-23-9)
Somnite (146-22-5)
Somnolens (50-06-6)
Somnoletten (50-06-6)
Somnomed (72-44-6)
Somnopentyl (57-33-0)
Somnos (302-17-0)
Somnosan (50-06-6)
Somonal (50-06-6)
Somonil (950-37-8)
Somophyllin (317-34-0)
Somophyllin O (317-34-0)
Sonacide (111-30-8)
Sonacon (439-14-5)
Sonaform (52-31-3)
Sonal (72-44-6)
Sonalan (55283-68-6)
Sonalen (55283-68-6)
Sonapax (50-52-2)
Sonate (127-85-5)
Sonazine (69-09-0)
δ-Sone (53-03-2)
Sonebon (146-22-5)
Sonerile (77-28-1)
Soneryl (77-28-1)
Sonistan (57-33-0)
Sonnolin (146-22-5)
Sonnormon (77-75-8)
Sonojell #4 (8009-03-8)
Sonojell #9 (8009-03-8)
Sonolure (16974-11-1)
Sontec (302-17-0)
Sontobarbital nabitone (57-33-0)
Sopaquin (54-05-7)
Sopental (57-33-0)
Sophoretin (117-39-5)
Sophorin (153-18-4)
Sophorine (485-35-8)
Sopor (72-44-6)
Soprabel (3687-31-8)
Soprabel (7784-40-9)
Sopracol (5707-69-7)
Sopracol 781 (5707-69-7)
Sopranebe (12427-38-2)
Soprathion (56-38-2)
Soprathion (563-12-2)
Soprofor S 70 (26264-06-2)
Sorate-10 (87-33-2)
Sorate-5 (87-33-2)
Sorba-Spray Mn (7785-87-7)
Sorbaldehyde (142-83-6)
Sorbangil (87-33-2)
Sorban sodny (Czech) (7757-81-5)
Sorbax STS (26658-19-5)
Sorbester P 17 (1338-43-8)
Sorbex M (50-70-4)
Sorbex R (50-70-4)
Sorbex RP (50-70-4)
Sorbex S (50-70-4)
Sorbex X (50-70-4)
Sorbic Oil (10048-32-5)
Sorbic-acid (110-44-1)
Sorbic acid, calcium salt (7492-55-9)
Sorbic acid, ethyl ester (8CI) (2396-84-1)
Sorbic acid, glycidyl ester dimer (63915-78-6)
Sorbic acid, potassium salt (590-00-1)
Sorbic acid, potassium salt, (E,E) (24634-61-5)
Sorbic acid, sodium salt (7757-81-5)
Sorbic alcohol (111-28-4)
Sorbic aldehyde (142-83-6)
Sorbicolan (50-70-4)
Sorbide (87-33-2)
Sorbide nitrate (87-33-2)
Sorbidilat (87-33-2)
Sorbidinitrate (87-33-2)
Sorbilande (50-70-4)
Sorbimacrogol Oleate (9005-65-6)
Sorbimacrogol Oleate 300 (9005-65-6)
Sorbimacrogol laurate 300 (9005-64-5)
Sorbin (87-79-6)
Sorbinic alcohol (111-28-4)
L-Sorbinose (87-79-6)
Sorbinose (87-79-6)
Sorbistat (110-44-1)
Sorbistat-K (590-00-1)
Sorbistat potassium (24634-61-5)
Sorbistat-potassium (590-00-1)
Sorbit (50-70-4)
Sorbital 0 20 (9005-65-6)
Sorbitan C (1338-41-6)
Sorbitan (9CI) (12441-09-7)
Sorbitan O (1338-43-8)
Sorbitan Palmitate (26266-57-9)
Sorbitan, Sesquioleate (8007-43-0)

Sorbitan, dioctadecanoate	(36521-89-8)
Sorbitan distearate	(36521-89-8)
Sorbitan, esters, trioctadecanoate	(26658-19-5)
Sorbitani tristearas (Latin)	(26658-19-5)
Sorbitan, monododecanoate	(1338-39-2)
Sorbitan, monododecanote, poly(oxy-1,2-ethanediyl) derivatives	(9005-64-5)
Sorbitan, monohexadecanoate, poly(oxy-1,2-ethanediyl) derivs.	(9005-66-7)
Sorbitan, monolaurate	(1338-39-2)
Sorbitan, monolaurate polyoxyethylene deriv.	(9005-64-5)
Sorbitan, monooctadecanoate (9CI)	(1338-41-6)
Sorbitan, monooctadecanoate, poly(oxy-1,2-ethanediyl) derivs.	(9005-67-8)
Sorbitan, monooleate	(1338-43-8)
Sorbitan, monooleate polyoxyethylene deriv.	(9005-65-6)
Sorbitan monooleic acid ester	(1338-43-8)
Sorbitan, monopalmitate	(26266-57-9)
Sorbitan, monopalmitate polyoxyethylene deriv.	(9005-66-7)
Sorbitan, monostearate	(1338-41-6)
Sorbitan, monostearate polyoxyethylene deriv.	(9005-67-8)
Sorbitan, (Z)-9-octadecenoate (2:3)	(8007-43-0)
Sorbitan oleate	(1338-43-8)
Sorbitan stearate	(1338-41-6)
Sorbitan, trioctadecanoate (9CI)	(26658-19-5)
Sorbitan, tri-9-octadecenoate, (Z,Z,Z)- (9CI)	(26266-58-0)
Sorbitan trioleate	(26266-58-0)
Sorbitan, tris(9-octadecenoate), (Z)	(26266-58-0)
Sorbitan tristearate	(26658-19-5)
Sorbitan tristearate. (Compound usually contains also associated fatty acids.)	(26658-19-5)
Sorbite	(50-70-4)
D-(-)-Sorbitol	(50-70-4)
D-Sorbitol	(50-70-4)
Sorbitol	(50-70-4)
Sorbitol syrup C	(50-70-4)
Sorbitrate	(87-33-2)
Sorbo	(50-70-4)
Sorbo-Calcian	(62-54-4)
Sorbo-Calcion	(62-54-4)
Sorbol	(50-70-4)
Sorbon S 60	(1338-41-6)
Sorbon T 80	(9005-65-6)
Sorbonit	(87-33-2)
L-(-)-Sorbose	(87-79-6)
Sorbose	(87-79-6)
L-Sorbose (9CI)	(87-79-6)
Sorbose, L- (VAN) (8CI)	(87-79-6)
Sorbostyl	(50-70-4)
Sorbyl alcohol	(111-28-4)
Sorensen's Potassium Phosphate	(7778-77-0)
Sorethytan (20) Monooleate	(9005-65-6)
Sorgen 30	(8007-43-0)
Sorgen 40	(1338-43-8)
Sorgen 50	(1338-41-6)
Sorgen 70	(26266-57-9)
Sorghum Gum	(9005-25-8)
Sorgoprim	(5915-41-3)
Stercorin	(360-68-9)
Sorlate	(9005-65-6)
Sormetal	(62-33-9)
Sorquat	(87-33-2)
Sorrel salt	(127-95-7)
Sorvilande	(50-70-4)
Sosigon	(359-83-1)
Sosol	(127-69-5)
Sospitan	(1882-26-4)
Sotalol	(3930-20-9)
Sotipox	(52-68-6)
Sotradecol	(139-88-8)
Sotyl	(57-33-0)
Souchet	(458-37-7)
Soudan I	(842-07-9)
Soudan II	(3118-97-6)
Soudan III	(85-86-9)
Souframine	(92-84-2)
Soup	(55-63-0)
Southern Bentonite	(1302-78-9)
Sovcaine	(85-79-0)
Soverin	(72-44-6)
Soviet Technical Herbicide 2M-4C	(94-74-6)
Sovinexion	(2104-96-3)
Soviol	(9003-20-7)
Sovol	(1336-36-3)
Sovprene	(9010-98-4)
Sowbug & Cutworm Bait	(12002-03-8)
Sowbug Cutworm Control	(12002-03-8)
Sowell	(57-53-4)
Soxamide	(127-69-5)
Soxinal PZ	(137-30-4)
Soxinol PZ	(137-30-4)
Soxisol	(127-69-5)
Soxitabs	(127-69-5)
Soxo	(127-69-5)
Soxomide	(127-69-5)
Soxysympamine	(300-42-5)
Soya-Bean Oil	(8001-22-7)
1-((Soya alkyl)amino)-3-aminopropane	(61791-67-1)
(Soya alkyl) dimethyl ethyl ammonium bromide	(61788-99-6)
(Soya alkyl)ethyl dimethyl ammonium ethyl sulfates	(68308-67-8)
N-(Soya alkyl)-N-ethyl morpholinium ethyl sulfate	(61791-34-2)
(Soya alkyl) trimethyl ammonium chloride	(61790-41-8)
N-(Soya alkyl)-N,N,N-trimethyl ammonium chloride	(61790-41-8)
Soyadimethylethylammonium, ethylsulfate	(68308-67-8)
Soyaethyl Morpholinium Ethosulfate	(61791-34-2)
N-Soya-N-ethyl morpholinium ethosulfate	(61791-34-2)
Soya, glycerin, phthalic anhydride alkyd resin	(66070-61-9)
Soya oil, Polymer with pentaerythritol, glycerin and phthalic anhydride	(66070-93-7)
Soya oil, glycerin, phthalic anhydride polymer	(66070-61-9)
Soya oil, glycerol, phthalic anhydride polymer	(66070-61-9)
Soya oil, glycerol, phthalic anhydride, polymer	(66070-61-9)
Soya oil, glycerol, phthalic anhydride resin	(66070-61-9)
Soya oil, pentaerythritol, glycerol, phthalic anhydride polymer	(66070-93-7)
Soya oil, pentaerythritol, phthalic anhydride polymer	(66070-60-8)
Soya oil, pentaerythritol, phthalic anhydride resin	(66070-60-8)
Soya oil, phthalic anhydride, glycerol, pentaerythritol polymer	(66070-93-7)
Soya oil, phthalic anhydride, glycerol polymer	(66070-61-9)
Soya oil, phthalic anhydride, glycerol resin	(66070-61-9)
Soya oil, phthalic anhydride, pentaerythritol, glycerine resin	(66070-93-7)
Soya oil, phthalic anhydride, pentaerythritol, glycerin polymer	(66070-93-7)
Soya oil, phthalic anhydride, pentaerythritol polymer	(66070-60-8)
Soya oil, phthalic anhydride, pentaerythritol resin	(66070-60-8)
Soya trimethyl ammonium chloride	(61790-41-8)
Soybean Lecithin	(8002-43-5)
Soybean-Oil	(8001-22-7)
Soybean Oil Degummed	(8001-22-7)
Soybean oil, Polymer with glycerol and phthalic anhydride	(66070-61-9)
Soybean oil, Polymer with glycerol, pentaerythritol and phthalic anhydride	(66070-93-7)
Soybean oil, Polymer with pentaerythritol and phthalic anhydride	(66070-60-8)
Soybean oil acids, phthalic anhydride, pentaerythritol, glycerol resin	(66070-93-7)
Soybean oil, glycerine, pentaerythritol, phthalic anhydride polymer	(66070-93-7)
Soybean oil, glycerine, phthalic anhydrided polymer	(66070-61-9)
Soybean oil, glycerine, phthalic anhydride polymer	(66070-61-9)

Soybean oil, glycerin, pentaerythritol, phthalic anhydride polymer	(66070-93-7)
Soybean oil, glycerin, pentaerythritol, phthalic anhydride resin	(66070-93-7)
Soybean oil, glycerin, phthalic anhydride, pentaerythritol polymer	(66070-93-7)
Soybean oil, glycerin, phthalic anhydride polymer	(66070-61-9)
Soybean oil, glycerin, phthalic anhydride resin	(66070-61-9)
Soybean oil, glycerin, tetramethylolmethane, phthalic anhydride resin	(66070-93-7)
Soybean oil, glycerol, pentaerythritol phthalic anhydride resin	(66070-93-7)
Soybean oil, glycerol, phthalic anhydride alkyd resin	(66070-61-9)
Soybean oil, glycerol, phthalic anhydride polymer	(66070-61-9)
Soybean oil, glycerol, phthalic anhydride resin	(66070-61-9)
Soybean oil modified, glycerin, pentaerythritol, phthalic anhydride alkyd resin	(66070-93-7)
Soybean oil modified, glycerin, phthalic anhydride alkyd resin	(66070-61-9)
Soybean oil modified, pentaerythritol, phthalic anhydride alkyd resin	(66070-60-8)
Soybean oil, pentaerythritol, glycerine, phthalic anhydride polymer	(66070-93-7)
Soybean oil, pentaerythritol, glycerin, phthalic anhydride alkyd resin	(66070-93-7)
Soybean oil, pentaerythritol, glycerin, phthalic anhydride polymer	(66070-93-7)
Soybean oil, pentaerythritol, glycerol, phthalic anhydride resin	(66070-93-7)
Soybean oil, pentaerythritol, phthalic anhydride alkyd resin	(66070-60-8)
Soybean oil, pentaerythritol, phthalic anhydride, benzoic acid, trimethylolethane resin	(66070-93-7)
Soybean oil, pentaerythritol, phthalic anhydride, glycerin polymer	(66070-93-7)
Soybean oil, pentaerythritol, phthalic anhydride, glycerin resin	(66070-93-7)
Soybean oil, pentaerythritol, phthalic anhydride polymer	(66070-60-8)
Soybean oil, pentaerythritol, phthalic anhydride resin	(66070-60-8)
Soybean oil, phthalic anhydride, glycerine polymer	(66070-61-9)
Soybean oil, phthalic anhydride, glycerin, pentaerythritol polymer	(66070-93-7)
Soybean oil, phthalic anhydride glycerin polymer	(66070-61-9)
Soybean oil, phthalic anhydride, glycerin polymer	(66070-61-9)
Soybean oil, phthalic anhydride, glycerin resin	(66070-61-9)
Soybean oil, phthalic anhydride, glycerol, pentaerythritol polymer	(66070-93-7)
Soybean oil, phthalic anhydride, glycerol, pentaerythritol resin	(66070-93-7)
Soybean oil, phthalic anhydride, glycerol polymer	(66070-61-9)
Soybean oil, phthalic anhydride, glycerol resin	(66070-61-9)
Soybean oil, phthalic anhydride, pentaerythritol alkyd resin	(66070-60-8)
Soybean oil, phthalic anhydride, pentaerythritol, glycerin polymer	(66070-93-7)
Soybean oil, phthalic anhydride, pentaerythritol, glycerin resin	(66070-93-7)
Soybean oil, phthalic anhydride, pentaerythritol, glycerol polymer	(66070-93-7)
Soybean oil, phthalic anhydride, pentaerythritol, glycerol resin	(66070-93-7)
Soybean oil, phthalic anhydride, pentaerythritol polymer	(66070-60-8)
Soybean oil, phthalic anhydride, pentaerythritol resin	(66070-60-8)
Soybean oil, tetramethylolmethane, phthalic anhydride polymer	(66070-60-8)
Soyex	(15457-05-3)
Soytrimonium Chloride	(61790-41-8)
Span 20	(1338-39-2)
Span 40	(26266-57-9)
Span 55	(1338-41-6)
Span 60	(1338-41-6)
Span 65	(26658-19-5)
Span 80	(1338-43-8)
Span 83	(8007-43-0)
Span 85	(26266-58-0)
Spanbolet	(57-68-1)
Spanestrin	(57-63-6)
Spanish White	(1304-85-4)
Spanon	(6164-98-3)
Spanone	(6164-98-3)
Spantran	(57-53-4)
Sparic	(88-85-7)
Sparine	(58-40-2)
Sparine hydrochloride	(53-60-1)
(-)-Sparteine	(90-39-1)
Sparteine	(90-39-1)
Sparteine, (-)	(90-39-1)
l-Sparteine	(90-39-1)
Sparteine, sulfate	(299-39-8)
Spartocin	(299-39-8)
Spartose OM-22	(9004-34-6)
Spasepilin	(50-06-6)
Spasmaverine	(150-59-4)
Spasmedal	(50-13-5)
Spasmione	(456-59-7)
Spasmocyclon	(456-59-7)
Spasmocyclone	(456-59-7)
Spasmodolin	(50-13-5)
Spasmolytin	(64-95-9)
Spavit	(59-02-9)
Spearmint-Oil	(8008-79-5)
Special Black 1V & V	(1333-86-4)
Special Blue X 2137	(6368-72-5)
Special M	(300-92-5)
Special Orange GR	(633-96-5)
Special Orange H	(633-96-5)
Special Schwarz	(1333-86-4)
Special Termite Fluid	(95-50-1)
Specifen	(389-08-2)
Specilline G	(61-33-6)
Speckelon	(63428-84-2)
Spectra-Sorb UV 9	(131-57-7)
Spectra-Sorb UV 24	(131-53-3)
Spectra-Sorb UV 5411	(3147-75-9)
Spectracide	(333-41-5)
Spectrar	(67-63-0)
Spectrolene Blue B	(119-90-4)
Spectrolene Blue BB	(120-00-3)
Spectrolene Red KB	(95-79-4)
Spectrolene Red RL	(99-52-5)
Spectrolene Scarlet 2G	(95-82-9)
Specular Iron	(1309-37-1)
Speed	(300-42-5)
Spencer 401	(105-60-2)
Spencer 401	(25038-54-4)
Spencer 601	(105-60-2)
Spencer 601	(25038-54-4)
Spencer S-6538	(13593-03-8)
Spencer S-6900	(2540-82-1)
Spenkel	(9009-54-5)
Spenlite	(9009-54-5)
Spent caustic	(64742-40-1)
Spent caustic containing an average of 6.9 wt. % cresylate acid	(64742-40-1)
Spent pulping liquor, Furfural polymer	(2786-76-7)
Spent sulfuric acid (DOT)	(7664-93-9)
Spergon	(118-75-2)
Spergon I	(118-75-2)
Spergon Technical	(118-75-2)
Sperlox-S	(7704-34-9)
Sperlox-Z	(12122-67-7)
Sperm Oil	(8002-24-2)
Sperm Oil, Acidulated	(8002-24-2)
Spermaceti	(8002-23-1)
Spermaceti Wax	(8002-23-1)
Spermaceti Wax, Refined	(8002-23-1)
Spermicide 741	(9004-78-8)

Spermidine (124-20-9)
Spermine (71-44-3)
Spermine, puriss (71-44-3)
Spersul (7704-34-9)
Spersul Thiovit (7704-34-9)
S 35 (pesticide) (52372-17-5)
Spezialfett 118 (555-43-1)
Spheroidine (4368-28-9)
Spheron (1333-86-4)
Spheron 6 (1333-86-4)
Spherosil (63231-67-4)
Sphygmogenin (51-43-4)
Spidox (102-98-7)
Spidoxol (102-98-7)
Spike (34014-18-1)
Spikenard Oil (8022-22-8)
Spinacen (111-02-4)
Spinacene (111-02-4)
Spinocaine (59-46-1)
Spiresis (52-01-7)
Spiridon (52-01-7)
Spirit Black SB (11099-03-9)
Spirit Blue (2152-64-9)
Spirit Nigrosine SSB (11099-03-9)
Spirit Orange (842-07-9)
Spirit Yellow I (842-07-9)
Spirit of Hartshorn (7664-41-7)
Spirit of Turpentine (8006-64-2)
Spirits of Salt (DOT) (7647-01-0)
Spirits of Turpentine (8006-64-2)
Spirits of Wine (64-17-5)
Spiro(2-azabicyclo(2.2.2)octane-5,2'-(2H)pyran)-6'(3'H)-one, 2,4'-dimethyl-, (1R-(1α,4α,5α))- (9CI) (3329-91-7)
Spiro(9H-benzo(a)fluorene-9,2'(3'H)-furo(3,2-b)-pyridin)-11(1H)-one, 2,3,3'a,4,4',5',6,6',6a,6b,7,7',7'a,8,11a,11b-hexadecahydro-3-hydroxy- 3',6',10,11b-tetramethyl (469-59-0)
Spiro(3H-1,4-benzodiazepine-3,2'-oxirane)-2,5(1H,4H)-dione, 4-methyl-3'-phenyl-, cis-(-)- (9CI) (20007-87-8)
Spiro(benzofuran-2(3H),1'-(2)cyclohexene)-3,4'-dione, 7-chloro-2'4,6-trimethoxy-6'-β- methyl (126-07-8)
Spiro(benzothiazole-2(4H),1'-cyclohexane), 5,6,7,7a-tetrahydro- (9CI) (6675-28-1)
1,1'-Spirobi(indane-5,6-diol), 3,3,3',3'-tetramethyl- (77-08-7)
1,1'-Spirobi(indan)-5,5',6,6'-tetrol, 3,3,3',3'-tetramethyl- (8CI) (77-08-7)
1,1'-Spirobi(1H-indene)-5,5',6,6'-tetrol, 2,2',3,3'-tetrahydro-3,3,3',3'-tetramethyl- (9CI) (77-08-7)
1,1'-Spirobi(1H-inden)-5,5',6,6'-tetrol, 2,2',3,3'-tetrahydro-3,3,3',3'-tetramethyl- (77-08-7)
Spiroctanie (52-01-7)
Spiro(17H-cyclopenta(a)phenanthrene-17,2'-(3'H)-furan) (52-01-7)
Spiro(17H-cyclopenta(a)phenanthrene-17,2'(5'H)furan), pregn-4-ene-21-carboxylic acid deriv. (52-01-7)
Spiro(cyclopropane-1,5'-(5H)inden)-3'(2'H)-one, 1',3'-α,4',7'-α-tetrahydro-7'-α-(β-d-glucopyranosyloxy)-4'-hydroxy-2',4',6'-trimethyl (87625-62-5)
Spiro(cyclotetrasiloxane-2,6'-(6H)dibenz(c,e)(1,2)oxasilin), 4,4,6,6,8,8-hexamethyl- (9CI) (40169-27-5)
Spiro(cyclotetrasiloxane-2,1'-(1)silaacenaphthene), 4,4,6,6,8,8-hexamethyl- (8CI) (13093-12-4)
Spiro(cyclotetrasiloxane-2,1'(2'H)-(1)silaacenaphthylene), 4,4,6,6,8,8-hexamethyl- (9CI) (13093-12-4)
Spiro(1,3-dioxane-5,3'-(7)oxabicyclo(4.1.0)heptane), 2-(7-oxabicyclo(4.1.0)hept-3-yl)- (9CI) (3388-03-2)
Spiro(5.6)dodecane (6CI,7CI,8CI,9CI) (181-15-7)
Spirofulvin (126-07-8)
Spiro(isobenzofuran-1(3H),9'-(9H)xanthen)-3-one, 2'-amino-6'-(diethylamino)- (9CI) (24460-06-8)
Spiro(isobenzofuran-1(3H),9'-(9H)xanthen)-3-one, 3',6'-bis-(diethylamino)- (9CI) (509-34-2)
Spiro(isobenzofuran-1(3H),9'-(9H)xanthen)-3-one, 2'-(bis-(phenylmethyl)amino)-6'-(diethylamino)- (9CI) (34372-72-0)
Spiro(isobenzofuran-1(3H),9'-(9H)xanthen)-3-one, 6'-(diethyl-amino)-3'-methyl-2'-(phenylamino)- (9CI) (29512-49-0)
Spiro(isobenzofuran-1(3H),9'-(9H)xanthen)-3-one, 3',6'-di-hydroxy- (2321-07-5)
Spiro(isobenzofuran-1(3H),9'-(9H)xanthen)-3-one, 3',6'-di-hydroxy- (9CI) (2321-07-5)
Spiro(isobenzofuran-1(3H),9'-(9H)xanthen)-3-one, 3',6'-di-hydroxy-, disodium salt (518-47-8)
Spiro(isobenzofuran-1(3H),9'-(9H)xanthen)-3-one, 3',6'-di-hydroxy-2',4',5',7'-tetraiodo- (15905-32-5)
Spiro(isobenzofuran-1(3H),9'-(9H)xanthen)-3-one, 3',6'-di-hydroxy-2',4',5',7'-tetraiodo-, aluminum salt (9CI) (12227-78-0)
Spiro(isobenzofuran-1(3H),9'-(9H)xanthen)-3-one, 2',4',5',7'-tetrabromo-3',6'-dihydroxy-, 2Na (17372-87-1)
Spiro(isobenzofuran-1(3H),9'-(9H)xanthen)-3-one, 4,5,6,7-tetra-chloro-3',6'-dihydroxy- (9CI) (6262-21-1)
Spirolactone (52-01-7)
Spirolakton (52-01-7)
Spirolang (52-01-7)
Spirone (52-01-7)
Spironolactone (52-01-7)
Spironolactone A (52-01-7)
Spiro(2.2)pentane (157-40-4)
Spiropentane (6CI,7CI,8CI,9CI) (157-40-4)
Spiropentane, butyl- (7CI,8CI,9CI) (6191-90-8)
Spiro(5.5)undec-2-ene, 3,7,7-trimethyl-11-methylene-, (R)- (8CI,9CI) (18431-82-8)
Spirt (64-17-5)
Splotin (15676-16-1)
Spofadazine (80-35-3)
Spofadrizine (68-35-9)
Spolapret OS (20120-33-6)
Spongiofort (9000-70-8)
Spongocytidine (147-94-4)
Spongocytidine hydrochloride (69-74-9)
Spontox (93-76-5)
Spor-Kil (62-38-4)
Sporiline (2398-96-1)
Sporostatin (126-07-8)
Spotrete (137-26-8)
Spotrete-F (137-26-8)
Spotted Alder (68916-39-2)
Spotton (55-38-9)
Spracal (7778-44-1)
Spray-Dermis (59-87-0)
Spray-Foral (59-87-0)
Spray-Hormite (2702-72-9)
Spray-Tox (58769-20-3)
Spray-Trol Brand Roden-Trol (81-81-2)
Sprayset MEKP (1338-23-4)
Spring-Bak (142-59-6)
Spritz-Hormin/2,4-D (94-75-7)
Spritz-Hormit (2702-72-9)
Spritz-Hormit/2,4-D (94-75-7)
Spritz-Rapidin (58-89-9)
Spritzlindane (58-89-9)
Sprout Nip (101-21-3)
Sprout-Nip EC (101-21-3)
Sprout/Off (123-33-1)
Sprout-Stop (123-33-1)
Spruehpflanzol (58-89-9)
Spud-Nic (101-21-3)
Spud-Nie (101-21-3)

Spulmako-Lax **(77-09-8)**
Spur **(69409-94-5)**
Spurge **(88-85-7)**
Squalane **(111-01-3)**
Squalen **(111-02-4)**
(E,E,E,E)-Squalene **(111-02-4)**
Squalene **(111-02-4)**
Squalene **(7683-64-9)**
trans-Squalene **(111-02-4)**
Squibb **(630-56-8)**
Sra Golden Yellow VIII **(119-15-3)**
Srolex **(333-41-5)**
StE 360.7 **(39362-66-8)**
St. John's Bread **(9000-40-2)**
Sta-Fast **(93-72-1)**
Sta-Fresh 615 **(128-04-1)**
Sta-RX 1500 **(9005-25-8)**
Δ-Stab **(50-24-8)**
Stabamine Blue BB **(120-00-3)**
Stabamine Scarlet GG **(95-82-9)**
Stabicillin **(87-08-1)**
Stabilan **(999-81-5)**
Stabilene **(9003-13-8)**
Stabilene Fly Repellent **(9003-13-8)**
Stabilisator C **(102-08-9)**
Stabilisator VH **(64-10-8)**
Stabilizator AR **(135-88-6)**
Stabilized ethyl parathion **(56-38-2)**
Stabilizer D-22 **(77-58-7)**
Stabilizer DLT **(123-28-4)**
Stabilizer Mark 1178 **(26523-78-4)**
Stabilizer Mark 328 **(1843-03-4)**
Stabilizer VH **(64-10-8)**
Stabillin VK Syrup 125 **(132-98-9)**
Stabillin VK Syrup 62.5 **(132-98-9)**
Stabilor CMH **(142-78-9)**
Stabinex NW 7PS **(1344-95-2)**
Stabinol **(94-20-2)**
Stabisol **(14882-18-9)**
Stable Red KB Base **(95-79-4)**
Stabogel **(9005-08-7)**
Stadadorm **(57-43-2)**
Stafast **(86-87-3)**
Staflen E 650 **(9002-88-4)**
Staflex 500 **(119-07-3)**
Staflex BOP **(84-78-6)**
Staflex DBEA **(141-18-4)**
Staflex DBM **(105-76-0)**
Staflex DBP **(84-74-2)**
Staflex DBS **(109-43-3)**
Staflex DOA **(103-23-1)**
Staflex DOP **(117-81-7)**
Staflex DOS **(122-62-3)**
Staflex DOX **(103-24-2)**
Staflex DTDP **(119-06-2)**
Staflex NODA **(110-29-2)**
Staflex TOTM **(3319-31-1)**
Stagno (tetracloruro di) (Italian) **(7646-78-8)**
Stainless Steel **(12597-68-1)**
Stam **(709-98-8)**
Stam F 34 **(709-98-8)**
Stam LV 10 **(709-98-8)**
Stam M-4 **(709-98-8)**
Stam Supernox **(709-98-8)**
Stamine **(91-84-9)**
Stampede **(709-98-8)**
Stampede 3E **(709-98-8)**
Stamylan 900 **(9002-88-4)**
Stamylan 1000 **(9002-88-4)**
Stamylan 1700 **(9002-88-4)**
Stamylan 8200 **(9002-88-4)**
Stamylan 8400 **(9002-88-4)**
Stan-Guard 156 **(15546-11-9)**
Stan-Mag Magnesium carbonate **(546-93-0)**
Stananne, trichloromethyl- (8CI,9CI) **(993-16-8)**
Stanazol **(10418-03-8)**
Stanazolol **(10418-03-8)**
Stancleret 157 **(78-04-6)**
Standacol Carmoisine **(3567-69-9)**
Standacol Orange G **(1936-15-8)**
Standacol Sunset Yellow FCF **(2783-94-0)**
Standak **(1646-88-4)**
Standamidd LD **(120-40-1)**
Standamul 1616 **(540-10-3)**
Standamul 7061 **(25339-09-7)**
Standamul DIPA **(6938-94-9)**
Standamul LA 2 **(9002-92-0)**
Standapol 112 Conc **(151-21-3)**
Standapol ES 2 **(9004-82-4)**
Standapol ES 3 **(9004-82-4)**
Standapol TLS 40 **(139-96-8)**
Standapol WA-AC **(151-21-3)**
Standapol WAQ **(151-21-3)**
Standapol WAQ Special **(151-21-3)**
Standapol WAS 100 **(151-21-3)**
Standard Lead Arsenate **(7784-40-9)**
Stanephrin **(61-76-7)**
Stangen **(91-84-9)**
Stangen maleate **(59-33-6)**
Stann OMF **(16091-18-2)**
Stann RC 40F **(78-04-6)**
Stannane, acetoxytributyl **(56-36-0)**
Stannane, acetoxytriphenyl **(900-95-8)**
Stannane, acetoxytripropyl **(3267-78-5)**
Stannane, (acetyloxy)triphenyl- (9CI) **(900-95-8)**
Stannane, benzoyloxytributyl **(4342-36-3)**
Stannane, bis(butoxymaleoyloxy)dibutyl **(15546-16-4)**
Stannane, bis(dodecanoyloxy) di-n-butyl- **(77-58-7)**
Stannane, bis(dodecanoyloxy)dioctyl- **(3648-18-8)**
Stannane, bis(dodecylthio)dimethyl- (9CI) **(51287-84-4)**
Stannane, bis(2-ethylhexanoyloxy)dibutyl **(2781-10-4)**
Stannane, bis(isooctyloxycarbonylmethylthio)dibutyl **(25168-24-5)**
Stannane, bis(isooctyloxycarbonylmethylthio)dimethyl **(26636-01-1)**
Stannane, bis(isooctyloxycarbonylmethylthio)dioctyl **(26401-97-8)**
Stannane, bis(lauroyloxy)dibutyl- **(77-58-7)**
Stannane, bis(lauroyloxy)dioctyl- **(3648-18-8)**
Stannane, bis(methoxymaleoyloxy)dibutyl **(15546-11-9)**
Stannane, bis(monobutoxymaleoyloxy)dibutyl- **(15546-16-4)**
Stannane, butyl- (8CI,9CI) **(2406-65-7)**
Stannane, butylchlorodihydroxy- (9CI) **(13355-96-9)**
Stannane, butylhydroxyoxo **(2273-43-0)**
Stannane, butyloxo- (9CI) **(51590-67-1)**
Stannane, butyltrichloro **(1118-46-3)**
Stannane, butyltris((2-ethyl-1-oxohexyl)oxy)- (9CI) **(23850-94-4)**
Stannane, butyltris(isooctyloxycarbonylmethylthio) **(25852-70-4)**
Stannane, chlorotributyl **(1461-22-9)**
Stannane, chlorotriethyl **(994-31-0)**
Stannane, chlorotrimethyl **(1066-45-1)**
Stannane, chlorotriphenyl **(639-58-7)**
Stannane, chlorotripropyl **(2279-76-7)**
Stannane, cyclohexylhydroxyoxo **(22771-18-2)**
Stannane, diacetoxydibutyl **(1067-33-0)**
Stannane, dibutyl **(1002-53-5)**
Stannane, dibutylbis((3-carboxyacryloyl)oxy)-, dimethyl ester

Name	CAS No.
(Z,Z)- (8CI)	**(15546-11-9)**
Stannane, dibutylbis(dodecylthio)- (9CI)	**(1185-81-5)**
Stannane, dibutylbis((2-ethylhexanoyl)oxy)-	**(2781-10-4)**
Stannane, dibutylbis((2-ethyl-1-oxohexyl)oxy)- (9CI)	**(2781-10-4)**
Stannane, dibutylbis(lauroyloxy)	**(77-58-7)**
Stannane, dibutylbis((1-oxooctadecyl)oxy)- (9CI)	**(5847-55-2)**
Stannane, dibutylbis(stearoyloxy)	**(5847-55-2)**
Stannane, dibutyldichloro	**(683-18-1)**
Stannane, dibutyldifluoro	**(563-25-7)**
Stannane, dibutyldimethyl- (8CI,9CI)	**(1528-00-3)**
Stannane, dibutyloxo	**(818-08-6)**
Stannane, dibutylthioxo	**(4253-22-9)**
Stannane, dichlorodimethyl	**(753-73-1)**
Stannane, dichlorodioctyl	**(3542-36-7)**
Stannane, dichlorodiphenyl	**(1135-99-5)**
Stannane, dicyclohexyloxo	**(22771-17-1)**
Stannane, didodecanoyloxydioctyl-	**(3648-18-8)**
Stannane, difluorodimethyl	**(3582-17-0)**
Stannane, dimethylbis((1-oxoneodecyl)oxy)- (9CI)	**(68928-76-7)**
Stannane, (4,4-dimethyloctanoyloxy)tributyl-	**(28801-69-6)**
Stannane, dimethylthioxo- (9CI)	**(13269-74-4)**
Stannane, dioctylbis(lauroyloxy)-	**(3648-18-8)**
Stannane, dioctyldichloro-	**(3542-36-7)**
Stannane, dioctyldidodecanoyloxy-	**(3648-18-8)**
Stannane, dioctyldi(lauroyloxy)	**(3648-18-8)**
Stannane, dioctyloxo	**(870-08-6)**
Stannanediylium, chloromethyl- (9CI)	**(23066-18-4)**
Stannanediylium, dimethyl- (9CI)	**(16408-14-3)**
Stannane, fluorotributyl	**(1983-10-4)**
Stannane, fluorotriphenyl	**(379-52-2)**
Stannane, hydroxytrimethyl	**(56-24-6)**
Stannane, hydroxytriphenyl	**(76-87-9)**
Stannane, iodotributyl	**(7342-47-4)**
Stannane, (isopropylsuccinyloxy)tributyl-	**(53404-82-3)**
Stannane, (linoleoyloxy)tributyl-	**(24124-25-2)**
Stannane, methyltrichloro	**(993-16-8)**
Stannane methyltris((carboxymethyl)thio)tris isooctyl ester	**(54849-38-6)**
Stannane, methyltris(2-ethylhexyloxycarbonylmethylthio)	**(57583-34-3)**
Stannane, (neodecanoyloxy)tributyl-	**(28801-69-6)**
Stannane, octyltrichloro	**(3091-25-6)**
Stannane, oxodioctyl-	**(870-08-6)**
Stannane, tetrabutyl	**(1461-25-2)**
Stannane, tetraethyl	**(597-64-8)**
Stannane, tetramethyl	**(594-27-4)**
Stannane, tetraoctyl	**(3590-84-9)**
Stannane, tetraphenyl- (9CI)	**(595-90-4)**
Stannane, tributyl- (8CI,9CI)	**(688-73-3)**
Stannane, tributylchloro-	**(1461-22-9)**
Stannane, tri-n-butyl-, hydride	**(688-73-3)**
Stannane, tributyliodo-	**(7342-47-4)**
Stannane, tributyl(linoleoyloxy)-	**(24124-25-2)**
Stannane, tributyl(methacryloyloxy)-	**(2155-70-6)**
Stannane, tributylmethyl- (8CI,9CI)	**(1528-01-4)**
Stannane, tributyl((2-methyl-1-oxo-2-propenyl)oxy)- (9CI)	**(2155-70-6)**
Stannane, tributyl(neodecanoyloxy)-	**(28801-69-6)**
Stannane, tri-n-butyl-, oxide	**(56-35-9)**
Stannane, tributyl((1-oxo-9,12-octadecadienyl)oxy)-, (Z,Z)- (9CI)	**(24124-25-2)**
Stannane, trichlorooctyl-	**(3091-25-6)**
Stannane, trichlorophenyl- (8CI,9CI)	**(1124-19-2)**
Stannane, tricyclohexylhydroxy	**(13121-70-5)**
Stannane, triethyl	**(997-50-2)**
Stannane, triethylhydroxy- (8CI,9CI)	**(994-32-1)**
Stannane, trimethyl- (9CI)	**(1631-73-8)**
Stannane, triphenyl	**(892-20-6)**
Stannane, tris(((isooctylthio)acetyl)oxy)methyl-	**(54849-38-6)**
Stannic chloride	**(7646-78-8)**
Stannic chloride, Anhydrous [UN 1827]	**(7646-78-8)**
Stannic iodide	**(7790-47-8)**
Stannic phosphide (8CI,9CI)	**(25324-56-5)**
Stannochlor	**(7772-99-8)**
Stannous chloride	**(7772-99-8)**
Stannous chloride, Solid (DOT)	**(7772-99-8)**
Stannous 2-ethylhexanoate	**(301-10-0)**
Stannous 2-ethylhexoate	**(301-10-0)**
Stannous fluoride	**(7783-47-3)**
Stannous octoate	**(301-10-0)**
Stannous oxalate	**(814-94-8)**
Stannous potassium tartrate	**(73926-79-1)**
Stannyl, chlorodimethyl- (9CI)	**(41079-92-9)**
Stannylene, dimethyl- (8CI, 9CI)	**(23120-99-2)**
Stannylium, tributyl- (9CI)	**(36643-28-4)**
Stannylium, trimethyl- (9CI)	**(5089-96-3)**
Stannyl, tributyl- (8CI,9CI)	**(20763-88-6)**
Stanomycetin	**(56-75-7)**
Stanozide	**(54-85-3)**
Stanozol	**(10418-03-8)**
Stanozolo	**(10418-03-8)**
Stanozolol	**(10418-03-8)**
Stanozololum (Latin)	**(10418-03-8)**
Stansin	**(127-69-5)**
Stanzamine	**(154-69-8)**
Star	**(56-81-5)**
Star Anise Oil	**(8007-70-3)**
Staramic 747	**(9005-25-8)**
Starch, Corn	**(9005-25-8)**
Starch, carboxymethyl ether (9CI)	**(9057-06-1)**
α-Starch	**(9005-25-8)**
Starch (OSHA)	**(9005-25-8)**
Starch hydrogen 1-octenylsuccinate, dextrinized	**(68070-94-0)**
Starchlor	**(57-74-9)**
Starch, nitrate (9CI)	**(9056-38-6)**
Star dust	**(50-36-2)**
Starfol BS-100	**(123-95-5)**
Starfol GMS 450	**(31566-31-1)**
Starfol GMS 600	**(31566-31-1)**
Starfol GMS 900	**(31566-31-1)**
Starfol IPM	**(110-27-0)**
Starfol IPP	**(142-91-6)**
Starifen	**(50-06-6)**
Starilettae	**(50-06-6)**
Starlex L	**(1344-95-2)**
Starlicide	**(7745-89-3)**
Starsol No. 1	**(9000-01-5)**
Statac B	**(64742-16-1)**
Statex	**(1333-86-4)**
Statex N 550	**(1333-86-4)**
Stathion	**(56-38-2)**
Statomin	**(91-84-9)**
Statomin maleate	**(59-33-6)**
Stauffer 4644	**(70-11-1)**
Stauffer ASP-51	**(3244-90-4)**
Stauffer B-10094	**(13067-93-1)**
Stauffer Captan	**(133-06-2)**
Stauffer Ferbam	**(14484-64-1)**
Stauffer MV-119A	**(3347-22-6)**
Stauffer N 521	**(533-74-4)**
Stauffer N 2790	**(944-22-9)**
Stauffer N-2596	**(2984-64-7)**
Stauffer N-2788	**(333-43-7)**
Stauffer N-3049	**(327-98-0)**
Stauffer R 1504	**(732-11-6)**
Stauffer R 1608	**(759-94-4)**
Stauffer R-1,303	**(786-19-6)**
Stauffer R-1492	**(953-17-3)**

Stauffer R-1910 (2008-41-5)
Stauffer R-2061 (1114-71-2)
Stauffer R-4,572 (2212-67-1)
Stauffer R-25788 (37764-25-3)
Stavelan cinaty (Czech) (814-94-8)
Stavelan sodny (Czech) (62-76-0)
Stavinor (4485-12-5)
Stavinor 30 (1592-23-0)
Stavinor 40 (6865-35-6)
Stavinor 1300SN (78-04-6)
Stavinor SN 1300 (78-04-6)
Stavinor ZN-E (557-05-1)
Stavox (128-37-0)
Stay-Flo (7681-49-4)
Stazepin (298-46-4)
Steam Cracked Narrow Cut Naphtha (Petroleum) (64742-83-2)
Steam Cracked Residuum Pitch (Petroleum) (64742-90-1)
Steam-cracked Residue (64742-90-1)
Stear Yellow JB (60-11-7)
Stearaldehyde (8CI) (638-66-4)
Stearalkonium chloride (122-19-0)
Stearamide (124-26-5)
Stearamide DEA (93-82-3)
Stearamide MEA (111-57-9)
Stearamidoethyl diethylamine (16889-14-8)
Stearamidoethyl ethanolamine (141-21-9)
Stearamidopropyl dimethylamine (7651-02-7)
Stearamidopropyldimethylamine (7651-02-7)
Stearamine oxide (2571-88-2)
Stearamyl (111-57-9)
Stearate chromic chloride (15242-96-3)
Stearato chromic chloride (15242-96-3)
Stearato-chromic chloride complex (15242-96-3)
Stearatochromium chloride (15242-96-3)
Steareth (9005-00-9)
Steareth-2 (9005-00-9)
Steareth-4 (9005-00-9)
Steareth-7 (9005-00-9)
Steareth-10 (9005-00-9)
Steareth-11 (9005-00-9)
Steareth-13 (9005-00-9)
Steareth-15 (9005-00-9)
Steareth-16 (9005-00-9)
Steareth-20 (9005-00-9)
Steareth-25 (9005-00-9)
Steareth-27 (9005-00-9)
Steareth-30 (9005-00-9)
Steareth-40 (9005-00-9)
Steareth-50 (9005-00-9)
Steareth-100 (9005-00-9)
Stearethate 40 (9004-99-3)
Stearex beads (57-11-4)
Stearic-acid (57-11-4)
Stearic acid, aluminium salt (7047-84-9)
Stearic acid, aluminum salt (637-12-7)
Stearic acid, ammonium salt (1002-89-7)
Stearic acid, barium cadmium salt (4:1:1) (1191-79-3)
Stearic acid, barium salt (6865-35-6)
Stearic acid, butyl ester (123-95-5)
Stearic acid, cadmium salt (2223-93-0)
Stearic acid, calcium salt (1592-23-0)
Stearic acid chloride (112-76-5)
Stearic acid diethanolamide (93-82-3)
Stearic acid-N,N-diethylethylenediamine condensate (16889-14-8)
Stearic acid, 3-dimethylaminopropylamide (7651-02-7)
Stearic acid, 9,10-epoxy-, allyl ester (123-36-4)
Stearic acid, 2,3-epoxypropyl ester (7460-84-6)
Stearic acid, ethylenediamine diamide (110-30-5)
Stearic acid, ethylene ester (8CI) (627-83-8)
Stearic acid, ethyl ester (111-61-5)
Stearic acid, 2-ethylhexyl ester (22047-49-0)
Stearic acid, glycidyl ester (7460-84-6)
Stearic acid, 12-hydroxy (106-14-9)
Stearic acid-p-hydroxyanilide (103-99-1)
Stearic acid, 3-hydroxy-2,2-bis(hydroxymethyl)propyl ester (8CI) (78-23-9)
Stearic acid, 2-(2-hydroxyethoxy)ethyl ester (106-11-6)
Stearic acid, 2-hydroxyethyl ester (111-60-4)
Stearic acid, 12-hydroxy-, methyl ester (141-23-1)
Stearic acid, isobutyl ester (646-13-9)
Stearic acid, isohexadecyl ester (25339-09-7)
Stearic acid, isopropyl ester (112-10-7)
Stearic acid, lead salt (7428-48-0)
Stearic acid, lithium salt (4485-12-5)
Stearic acid, magnesium salt (557-04-0)
Stearic acid, methyl ester (112-61-8)
Stearic acid, monoester with ethylene glycol (111-60-4)
Stearic acid, monoester with glycerol (31566-31-1)
Stearic acid, monoester with pentaerythritol (78-23-9)
Stearic acid, monoester with 1,2-propanediol (1323-39-3)
Stearic acid monoethanolamide (111-57-9)
Stearic acid 1-monoglyceride (123-94-4)
Stearic acid α-monoglyceride (123-94-4)
Stearic acid, pentaerythritol ester (1:1) (78-23-9)
Stearic acid, polyethylene glycol diester (9005-08-7)
Stearic acid, potassium salt (593-29-3)
Stearic acid, sodium salt (822-16-2)
Stearic acid, stearyl ester (2778-96-3)
Stearic acid, tetradecyl ester (17661-50-6)
Stearic acid, tridecyl ester (31556-45-3)
Stearic acid, triester with 2,2',2''-nitrilotriethanol (3002-22-0)
Stearic acid, triethanolamine soap (4568-28-9)
Stearic acid triglyceride (555-43-1)
Stearic acid triglycerin ester (555-43-1)
Stearic acid, vinyl ester (8CI) (111-63-7)
Stearic acid, zinc salt (557-05-1)
Stearic N-(aminoethyl)ethanolamide (141-21-9)
Stearic chloride (112-76-5)
Stearic 3-dimethylaminopropylamide (7651-02-7)
Stearic ethanolamide (111-57-9)
Stearic ethylolamide (111-57-9)
Stearic monoethanolamine (111-57-9)
Stearic monoglyceride (31566-31-1)
Stearic triglyceride (555-43-1)
Stearin (555-43-1)
Stearin, 1-mono- (8CI) (123-94-4)
Stearin, tri- (555-43-1)
Stearix Brown 4R (60-09-3)
Stearix Orange (842-07-9)
Stearix Red 4B (85-83-6)
Stearix Red 4S (85-83-6)
Stearix Scarlet (85-86-9)
Stearoguanamine (2533-20-2)
Stearol (112-92-5)
Stearone (504-53-0)
Stearonitrile (638-65-3)
Stearophanic acid (57-11-4)
Stearox 6 (9004-99-3)
Stearox 920 (9004-99-3)
N-Stearoyl-4-aminophenol (103-99-1)
N-Stearoyl-p-aminophenol (103-99-1)
Stearoyl-p-aminophenol (103-99-1)
Stearoyl chloride (112-76-5)
Stearoyl diethanolamide (93-82-3)
N-Stearoylethanolamine (111-57-9)

Name	CAS
Stearoylethanolamine	(111-57-9)
N-Stearoyl-N'-(2-hydroxyethyl)ethylenediamine	(141-21-9)
N-Stearoyl-N'-(β-hydroxyethyl)ethylenediamine	(141-21-9)
Stearoyl monoethanolamide	(111-57-9)
Stearoylmonoethanolamide	(111-57-9)
3-Stearoyloxy-1,2-propanediol	(123-94-4)
Stearoyl triglyceride	(555-43-1)
Stearyl alcohol	(112-92-5)
Stearyl alcohol, ethoxylated	(9005-00-9)
Stearylamidopropyl-N,N-dimethylamine, oxide	(25066-20-0)
Stearylamine	(124-30-1)
Stearylamine, ethoxylated	(26635-92-7)
Stearyl bromide	(112-89-0)
Stearyl chloride	(112-76-5)
Stearyl dimethylamine oxide	(2571-88-2)
Stearyldimethylbenzylammonium chloride	(122-19-0)
Stearyl erucamide	(10094-45-8)
Stearyl stearate	(2778-96-3)
Stearyltrimethylammonium chloride	(112-03-8)
Steawhite	(14807-96-6)
Stebac	(122-19-0)
Stecker ASC-4	(87-10-5)
Steclin	(60-54-8)
Steclin	(64-75-5)
Steclin hydrochloride	(64-75-5)
Stedbac	(122-19-0)
Stee:PM	(25038-54-4)
Steel, (API X52) (9CI)	(39362-66-8)
Steel, Stainless	(12597-68-1)
Steel manufacture, Chemicals	(65997-19-5)
Steelon	(105-60-2)
Steinamid DL 203 S	(120-40-1)
Steinamid L 203	(142-78-9)
Steinapol NL 3	(9004-82-4)
Steinapol NLS 90	(151-21-3)
Steinapol NOS 25	(9014-90-8)
Steinapol SBF 12	(26838-05-1)
Steinapol TLS 40	(139-96-8)
Steinbuhl Yellow	(10294-40-3)
Steladone	(470-90-6)
Stelazine	(117-89-5)
Stellacardiol	(54-95-5)
Stellacyl	(62-44-2)
Stellamine	(59-26-7)
Stellazine	(117-89-5)
Stellite 8	(11114-92-4)
Stellite 8A	(11114-92-4)
Stellite 21	(11114-92-4)
Stellite 23	(11114-92-4)
Stellite 25	(11114-92-4)
Stellite 27	(11114-92-4)
Stellite 30	(11114-92-4)
Stellite 31	(11114-92-4)
Stellite 36	(11114-92-4)
Stellite C	(11114-92-4)
Stellon Pink	(9011-14-7)
Stemetil	(58-38-8)
Stemphylone	(10088-95-6)
Stenolon	(58-18-4)
Stenolon	(72-63-9)
Stenolone	(72-63-9)
Stenosine	(144-21-8)
Stenovasan	(317-34-0)
Stensolo	(57-53-4)
Stental	(50-06-6)
Stental extentabs	(50-06-6)
Stepa	(52-24-4)
Stepan d-70	(142-91-6)
Stepan D-50	(110-27-0)
Stepanate X	(1300-72-7)
Stepanol DEA	(143-00-0)
Stepanol ME	(151-21-3)
Stepanol ME Dry	(151-21-3)
Stepanol ME Dry AW	(151-21-3)
Stepanol Methyl	(151-21-3)
Stepanol Methyl Dry AW	(151-21-3)
Stepanol T 28	(151-21-3)
Stepanol WA	(151-21-3)
Stepanol WA-100	(151-21-3)
Stepanol WAC	(151-21-3)
Stepanol WA Paste	(151-21-3)
Stepanol WAQ	(151-21-3)
Stepanol WAT	(139-96-8)
Steraffine	(112-92-5)
Steral	(70-30-4)
Steramide	(144-80-9)
Sterandryl	(57-85-2)
Sterane	(50-24-8)
Steraskin	(70-30-4)
Sterazine	(68-35-9)
Sterculia-Gum	(9000-36-6)
Steri/Sol (VAN)	(141-94-6)
Stericol	(1300-71-6)
Sterido	(55-56-1)
Sterigmatocystin	(10048-13-2)
Sterilate	(141-94-6)
Sterile Tetracaine Hydrochloride	(136-47-0)
Sterilizing gas ethylene oxide 100%	(75-21-8)
Sterinol	(7281-04-1)
Sterinolu (Polish)	(7281-04-1)
Steriseal Liquid #40	(128-04-1)
Sterisil	(141-94-6)
Sterisol hand disinfectant	(67-63-0)
Sterling	(1333-86-4)
Sterling	(7647-14-5)
Sterling AM	(2235-54-3)
Sterling N 765	(1333-86-4)
Sterling NS	(1333-86-4)
Sterling SO 1	(1333-86-4)
Sterling WA Paste	(151-21-3)
Sterling WAT	(139-96-8)
Sterling WAQ-CH	(151-21-3)
Sterling WAQ-Cosmetic	(151-21-3)
Sternite 30	(9003-53-6)
Sternite ST 30VL	(9003-53-6)
Sterogenol	(140-72-7)
Sterogyl	(50-14-6)
Sterolamide	(102-71-6)
Sterolone	(50-24-8)
Steronyl	(58-18-4)
p-(Steroylamino)phenol	(103-99-1)
Stesolid	(439-14-5)
Stesolin	(439-14-5)
Stibic anhydride	(1314-60-9)
Stibilium	(56-53-1)
Stibine (ACGIH,OSHA) [UN 2676]	(7803-52-3)
Stibine, trichloro-	(10025-91-9)
Stibine, trifluoro- (9CI)	(7783-56-4)
Stibine, triiodo- (9CI)	(7790-44-5)
Stibine, triphenyl	(603-36-1)
Stibium	(7440-36-0)
Stibnite	(1317-86-8)
(Stibylidynetrithio)triacetic acid, triisooctyl ester	(27288-44-4)
Stickmonoxyd (German)	(10102-43-9)

Stickstoffdioxid (German) (10102-44-0)
Stickstofflost (55-86-7)
Stickstoffwasserstoffsaeure (German) (7782-79-8)
Stigmanol bromide (114-80-7)
Stigmasta-3,5-diene, (24xi)- (9CI) (79897-80-6)
(24S)-5,22-Stigmastadien-3β-ol (83-48-7)
Stigmasta-5,22-dien-3-ol, (3β)- (83-48-7)
Stigmasta-5,22-dien-3β-ol (8CI) (83-48-7)
Stigmasta-5,24(28)-dien-3-ol, (3β)- (9CI) (VAN) (18472-36-1)
Stigmasta-5,24(28)-dien-3β-ol (8CI) (VAN) (18472-36-1)
Stigmasta-5,22-dien-3-ol, (3β,22E)- (9CI) (83-48-7)
Stigmasta-7,24(28)-dien-3-ol, 4-methyl-, (3β,5α)- (9CI) (VAN) (120056-15-7)
Stigmasta-3,5,24(28)-triene (9CI) (86709-50-4)
Stigmast-5-en-3-β-ol (83-46-5)
Stigmast-5-en-3-ol, (3-β)- (9CI) (83-46-5)
Stigmast-5-en-3-ol, (3β,24S)- (9CI) (83-47-6)
Stigmast-5-en-3β-ol, (24S)- (8CI) (83-47-6)
Stigmasterin (83-48-7)
Stigmasterol (83-48-7)
β-Stigmasterol (83-48-7)
Stigmosan bromide (114-80-7)
Stik (86-87-3)
Stikstofdioxyde (Dutch) (10102-44-0)
Stil (56-53-1)
Stil-Rol (56-53-1)
Stilben (German) (588-59-0)
4-Stilbenamine, 4'-nitro- (8CI) (4629-58-7)
(E)-Stilbene (103-30-0)
(Z)-Stilbene (645-49-8)
Stilbene (588-59-0)
Stilbene 3 (27344-41-8)
Stilbene, (E) (103-30-0)
cis-Stilbene (645-49-8)
trans-Stilbene (103-30-0)
Stilbene Yellow TK (1325-37-7)
Stilbene Yellow TR (1325-37-7)
Stilbene, (Z)- (8CI) (645-49-8)
4,4'-Stilbenebis(diazonium), 2,2'-disulfo-, dichloride (13954-62-6)
Stilbene, .α.-chloro- (6CI,7CI,8CI) (1460-06-6)
Stilbene dichloride (5963-49-5)
Stilbene, .α.,.α.'-dichloro-, (E)- (8CI) (951-86-0)
Stilbene, 4,4'-dimethoxy- (8CI) (4705-34-4)
4,4'-Stilbenediol, α,α'-diethyl (56-53-1)
4,4'-Stilbenediol, α,α'-diethyl-, (E)- (56-53-1)
4,4'-Stilbenediol,2,2'-diethyl- (56-53-1)
4,4'-Stilbenediol, α,α'-diethyl-, bis(dihydrogen phosphate), (E) (522-40-7)
4,4'-Stilbenediol, α,α'-diethyl-, dipropionate, (E) (130-80-3)
4,4'-Stilbenediol, α,α'-diethyl-, dipropionate, trans- (130-80-3)
2,2'-Stilbenedisulfonic acid, 4-amino-4'-nitro (119-72-2)
2,2'-Stilbenedisulfonic acid, 4,4'-bis((4-anilino-6-bis(2-hydroxyethyl)amino-s-triazin-2-yl) amino)-, disodium salt (4193-55-9)
2,2'-Stilbenedisulfonic acid, 4,4'-bis((4-anilino-6-((2-hydroxyethyl)methylamino)-s-triazin- 2-yl)amino)-, disodium salt (13863-31-5)
2,2'-Stilbenedisulfonic acid, 4,4'-bis((4-anilino-6-((2-hydroxypropyl)amino)-s-triazin-2-yl)amino-, disodium salt (32694-95-4)
2,2'-Stilbenedisulfonic acid, 4,4'-bis((4-anilino-6-morpholino-s-triazin-2-yl)amino)-, disodium salt (16090-02-1)
2,2'-Stilbenedisulfonic acid, 4,4'-bis((4-bis(2-hydroxyethyl)-amino)-6-(p-sulfoanilino)-s-triazin-2-yl)amino)-, tetrasodium salt (16470-24-9)
2,2'-Stilbenedisulfonic acid, 4,4'-bis((4-((2-carbamoylethyl)-(2-hydroxyethyl)amino)-6-(p-sulfoanilino)-s-triazin-2-yl)-amino)-, tetrasodium salt (8CI) (29637-52-3)
2,2'-Stilbenedisulfonic acid, 4,4'-bis((4-morpholino-6-o-toluidino-s-triazin-2-yl)amino)-, 2Na (24019-80-5)
2,2'-Stilbenedisulfonic acid, 4,4'-diamino (81-11-8)
2,2'-Stilbenedisulfonic acid, 4,4'-diamino-, disodium salt (7336-20-1)
2,2'-Stilbenedisulfonic acid, 4,4'-dinitro (128-42-7)
2,2'-Stilbenedisulfonic acid, 4,4'-dinitro-, disodium salt (3709-43-1)
Stilbene, 2,2'-imino- (256-96-2)
Stilbene, .α.-methyl-, (E)- (8CI) (833-81-8)
Stilbene, 3-methyl-, (E)- (8CI) (14064-48-3)
Stilbene, α-methyl- (8CI) (779-51-1)
Stilbene, 4-nitro (4003-94-5)
Stilbene oxide (17619-97-5)
2-Stilbenesulfonic acid, 4-(2H-naphtho(1,2-d)triazol-2-yl)-, sodium salt (6416-68-8)
Stilbestrol (56-53-1)
Stilbestrol, diethyl- (56-53-1)
Stilbestrol, diethyl dipropionate (130-80-3)
Stilbestrol, dihydro- (84-16-2)
Stilbestrol diphosphate (522-40-7)
Stilbestrol dipropionate (130-80-3)
Stilbestrol propionate (130-80-3)
Stilbestronate (130-80-3)
Stilbestrone (56-53-1)
Stilbetin (56-53-1)
Stilboefral (56-53-1)
Stilboestroform (56-53-1)
Stilboestrol (130-80-3)
Stilboestrol (56-53-1)
Stilboestrol DP (130-80-3)
Stilboestrol dipropionate (130-80-3)
Stilbofax (130-80-3)
Stilbofolin (56-53-1)
Stilbofollin (56-53-1)
Stilbol (56-53-1)
Stilciclina (64-75-5)
Stilkap (56-53-1)
Stillcardiol (54-95-5)
Stilny (1088-11-5)
Stilofil (63428-84-2)
Stilon (105-60-2)
Stilon (25038-54-4)
Stilon Standard A (63428-84-2)
Stilon Standard C (63428-84-2)
Stilphostrol (522-40-7)
Stilronate (130-80-3)
Stimdex (300-42-5)
Stiminol (59-26-7)
Stimul (2152-34-3)
Stimulan (60-13-9)
Stimulex (300-42-5)
Stimulin (59-26-7)
Stimulina (56-85-9)
Stinerval (156-51-4)
Stinerval (51-71-8)
Stinging Nettle LS (8001-21-6)
Stink Damp (7783-06-4)
Stipend (2921-88-2)
Stipine (9005-38-3)
Stiptanon (50-27-1)
Stirofos (22248-79-9)
Stirolo (Italian) (100-42-5)
Stirophos (22248-79-9)
Stockade (52315-07-8)
Stockade (52645-53-1)
Stoddard-Solvent (8052-41-3)
Stoddard Solvent (ACGIH,OSHA) (8052-41-3)
Stomacain (126-27-2)
Stomoxin (52645-53-1)
Stomoxin P (52645-53-1)
Stomp (40487-42-1)
Stomp 330D (40487-42-1)
Stomp 330E (40487-42-1)

Stone Red	(1309-37-1)
Stop-Drop	(86-87-3)
Stop-Scald	(91-53-2)
Stopaethyl	(97-77-8)
Stopethyl	(97-77-8)
Stopetyl	(97-77-8)
Stopgerme-S	(101-21-3)
Stopmold B	(132-27-4)
Stopspot	(100-56-1)
Stopton Album	(63-74-1)
Storite	(148-79-8)
Stovaine	(644-26-8)
Stove Black	(7782-42-5)
Stovine	(644-26-8)
Straight-Run Gas Oil	(64741-43-1)
Straight-Run Kerosene	(8008-20-6)
Straight-chain alkyl benzene sulfonate	(68411-30-3)
Straight Orange G	(1936-15-8)
Straight Run Gas Oil (Petroleum)	(64741-43-1)
Stramid	(63-74-1)
Strathion	(56-38-2)
Strawberry Red A Geigy	(2611-82-7)
Strawberry aldehyde	(77-83-8)
Strazine	(1912-24-9)
Strel	(709-98-8)
Strep-Gran	(3810-74-0)
Strepamide	(63-74-1)
Strepcen	(57-92-1)
Strepcin	(3810-74-0)
Strepsan	(63-74-1)
Strepsulfat	(3810-74-0)
Streptagol	(63-74-1)
Streptamid	(63-74-1)
D-Streptamine, O-3-amino-3-deoxy-α-D-glucopyranosyl-(1-6)-O-(6-amino-6-deoxy-α- D-glucopyranosyl-(1-4))-2-deoxy	(59-01-8)
D-Streptamine, O-3-amino-3-deoxy-α-D-glucopyranosyl-(1-6)-O-(6-amino-6-deoxy-α-D-glucopyranosyl-(1,4))-N^1-(4-amino-2-hydroxy-1-oxobutyl)-2-deoxy-, (S)	(37517-28-5)
Streptamine, O-3-amino-3-deoxy-α-D-glucopyranosyl-(1-4)-O-(2,6-diamino-2,3,6-trideoxy-α-d-ribohexopyranosyl-(1-6))-2-deoxy-, D	(32986-56-4)
D-Streptamine, O-3-deoxy-4-c-methyl-3-(methylamino)-β-l-arabinopyranosyl(1-6)-O-(2,6-diamino-2,3,4,6-tetradeoxy-α-d-glycerohex-4-enopyranosyl(1-4))-2-deoxy-N^1-ethyl	(56391-56-1)
D-Streptamine, O-2-deoxy-2-(methylamino)-α-L-glucopyranosyl-(1-2)-O-5-deoxy-3-C-(hydroxymethyl)-α-L-lyxofuranosyl-(1-4)-N,N'-bis(aminoiminomethyl)-, sulfate (2:3) (Salt) (9CI)	(5490-27-7)
D-Streptamine, O-2-deoxy-2-(methylamino)-.α.-L-glucopyranosyl-(1.fwdarw.2)-O-5- deoxy-3-C-(hydroxymethyl)-.α.-L-lyxo-furanosyl-(1.fwdarw.4)-N,N'- bis(aminoiminomethyl)-, Mixt. with aluminum hydroxide (Al(OH)3), kaolinite ($H_4Al_2SiO_9$) and pectin (9CI	(8047-42-5)
Streptamine, O-2,6-diamino-2,6-dideoxy-β-l-idopyranosyl-(1-3)-O-β-D-ribofuranosyl- (1-5)-O-(2-amino-2-deoxy-α-D-gluco-pyranosyl-(1-4))-2-deoxy	(7542-37-2)
Streptamine, O-β-d-mannopyranosyl-(1-4)-2-deoxy-2-(methyl-amino)-α-l-glucopyranosyl- (1-2)-5-deoxy-O-3-c-(hydroxy-methyl)-α-l-lyxofuranosyl-(1-4)-n,N'-diamidino-, D	(128-46-1)
Streptasol	(63-74-1)
Streptocid Album	(63-74-1)
Streptocide White	(63-74-1)
Streptoclase	(63-74-1)
Streptocom	(63-74-1)
Streptol	(63-74-1)
Streptomicina (Italian)	(57-92-1)
Streptomyces Peucetius	(20830-81-3)
Streptomycin	(57-92-1)
Streptomycin A	(57-92-1)
Streptomycin, dihydro-	(128-46-1)
Streptomycin, dihydro-, sulfate (2:3) (Salt) (8CI)	(5490-27-7)
Streptomycine	(57-92-1)
Streptomycin sesquisulfate	(3810-74-0)
Streptomycin sulfate	(3810-74-0)
Streptomycin, sulfate (2:3) (Salt)	(3810-74-0)
Streptomycin sulphate B.P	(3810-74-0)
Streptomycinum	(57-92-1)
Streptomyzin (German)	(57-92-1)
Strepton	(63-74-1)
Streptorex	(3810-74-0)
Streptosil	(63-74-1)
Streptosilpyridine	(144-83-2)
Streptosilthiazole	(72-14-0)
Streptozocin	(18883-66-4)
Streptozol	(63-74-1)
Streptozone	(63-74-1)
Streptozoticin	(18883-66-4)
Streptozotocin	(18883-66-4)
Streptrocide	(63-74-1)
Strepvet	(3810-74-0)
Streunex	(58-89-9)
Striadyne	(56-65-5)
Stricnina (Italian)	(57-24-9)
Stricylon	(550-99-2)
Striped Alder	(68916-39-2)
Stripmix	(9004-35-7)
Strobane	(71-55-6)
Strobane	(8001-50-1)
Strobane T-90	(8001-35-2)
Strobane-T	(8001-35-2)
Stromba	(10418-03-8)
Strombaject	(10418-03-8)
Strombine	(4408-64-4)
Stroncylate	(526-26-1)
Strontium (9CI)	(7440-24-6)
Strontium Chromate X-2396	(7789-06-2)
Strontium Yellow	(7789-06-2)
Strontium acetate	(543-94-2)
Strontium arsenite	(10378-48-0)
Strontium arsenite	(91724-16-2)
Strontium arsenite, Solid	(91724-16-2)
Strontium arsenite tetrahydrate	(10378-48-0)
Strontium carbonate	(1633-05-2)
Strontium chlorate, Solid [UN 1506]	(7791-10-8)
Strontium chlorate, Wet [UN 1506]	(7791-10-8)
Strontium-chloride	(10476-85-4)
Strontium chromate	(7789-06-2)
Strontium chromate (1:1)	(7789-06-2)
Strontium chromate 12170	(7789-06-2)
Strontium chromate A	(7789-06-2)
Strontium chromate (VI)	(7789-06-2)
Strontium dioxide	(1314-18-7)
Strontium diperchlorate	(13450-97-0)
Strontium ferrite	(12023-91-5)
Strontium-fluoride	(7783-48-4)
Strontium hexaferrite	(12023-91-5)
Strontium hydroxide (9CI)	(18480-07-4)
Strontium iron oxide	(12023-91-5)
Strontium, Isotope of mass 89 (8CI,9CI)	(14158-27-1)
Strontium, Isotope of mass 90 (8CI,9CI)	(10098-97-2)
Strontium lactate	(29870-99-3)
Strontium monosulfide	(1314-96-1)
Strontium(II) nitrate (1:2)	(10042-76-9)
Strontium nitrate [UN 1507]	(10042-76-9)
Strontium octadecanoate	(10196-69-7)

Strontium perchlorate [UN 1508] **(13450-97-0)**
Strontium peroxide [UN 1509] (9CI) **(1314-18-7)**
Strontium salicylate **(526-26-1)**
Strontium sulfide (SrS) (9CI) **(1314-96-1)**
Strontium sulphide **(1314-96-1)**
Strontium titanate **(12060-59-2)**
Strontolac **(29870-99-3)**
Strophanthidin, 3-(6-deoxy-α-l-mannopyranoside) **(508-75-8)**
Strophanthidin α-l-rhamnoside **(508-75-8)**
g-Strophanthin **(630-60-4)**
Strophanthin G **(630-60-4)**
Strophoperm **(630-60-4)**
Strumacil **(56-04-2)**
Strumazole **(60-56-0)**
Strux **(9004-35-7)**
Strychinine sulfate **(60-41-3)**
Strychnidin-10-one **(57-24-9)**
Strychnidin-10-one, Compd. with arsenic acid (H_3AsO_4) (1:1) (9CI) **(10476-82-1)**
Strychnidin-10-one, 2,3-dimethoxy- (9CI) **(357-57-3)**
Strychnidin-10-one, monohydriodide (9CI) **(52748-69-3)**
Strychnidin-10-one, sulfate (2:1) **(60-41-3)**
Strychnin (German) **(57-24-9)**
Strychnine (ACGIH,OSHA) [UN 1692] **(57-24-9)**
Strychnine, Compd. with arsenic acid (H_3AsO_4) (1:1) (8CI) **(10476-82-1)**
Strychnine, Liquid [UN 1692] **(57-24-9)**
Strychnine, Solid [UN 1692] **(57-24-9)**
Strychnine, 2,3-dimethoxy- **(357-57-3)**
Strychnine hydriodide **(52748-69-3)**
Strychnine, mononitrate **(66-32-0)**
Strychnine nitrate **(66-32-0)**
Strychnine sulfate **(60-41-3)**
Strychnine, sulfate (2:1) **(60-41-3)**
Strychnos **(57-24-9)**
Stryon 686 **(9003-53-6)**
Stryptirenal **(51-43-4)**
Studafluor **(7681-49-4)**
Stuntman **(123-33-1)**
Stylon **(105-60-2)**
Stylon **(25038-54-4)**
Styphnate de plomb (French) **(15245-44-0)**
Styphnic acid **(82-71-3)**
Styptirenal **(51-43-4)**
Styrafoil **(9003-53-6)**
Styragel **(9003-53-6)**
Styrallyl alcohol **(98-85-1)**
Styralyl acetate **(93-92-5)**
Styralyl alcohol **(98-85-1)**
Styrax benzoin **(9000-05-9)**
Styreen (Dutch) **(100-42-5)**
Styren (Czech) **(100-42-5)**
Styren-acrylonitrile polymer **(9003-54-7)**
Styrene **(100-42-5)**
Styrene (OSHA) **(100-42-5)**
Styrene Polymers **(9003-53-6)**
Styrene, Polymer with 3-methyl-3-buten-2-one (8CI) **(25191-48-4)**
Styrene-acrylonitrile copolymer **(9003-54-7)**
Styrene, β-bromo **(103-64-0)**
Styrene, o-bromo- (8CI) **(2039-88-5)**
Styrene, p-bromo- (8CI) **(2039-82-9)**
Styrene-1,3-butadiene copolymer **(9003-55-8)**
Styrene-butadiene copolymer **(9003-55-8)**
Styrene-butadiene polymer **(9003-55-8)**
Styrene, butyl methacrylate polymer **(25213-39-2)**
Styrene, 3-chloro- **(2039-85-2)**
Styrene, 4-chloro- **(1073-67-2)**
Styrene, chloro **(1331-28-8)**
Styrene, o-chloro **(2039-87-4)**
Styrene, m-chloro- (8CI) **(2039-85-2)**
Styrene, p-chloro- (8CI) **(1073-67-2)**
Styrene, α,β-dichloro **(6607-45-0)**
Styrene, 2,4-dichloro- (8CI) **(2123-27-5)**
Styrene, 2,5-dichloro- (8CI) **(1123-84-8)**
Styrene, 2,6-dichloro- (8CI) **(28469-92-3)**
Styrene, 2,3-dichloro- (7CI,8CI) **(2123-28-6)**
Styrene, 2,4-dimethyl **(2234-20-0)**
Styrene, 2,5-dimethyl **(2039-89-6)**
Styrene, 3,5-dimethyl **(5379-20-4)**
Styrene, 2,6-dimethyl- (8CI) **(2039-90-9)**
Styrene, 3,4-dimethyl- (8CI) **(27831-13-6)**
Styrene, dimethyl- (8CI) **(27576-03-0)**
Styrene epoxide **(96-09-3)**
Styrene, m-ethyl **(7525-62-4)**
Styrene, o-ethyl- (8CI) **(7564-63-8)**
Styrene, p-ethyl- (8CI) **(3454-07-7)**
Styrene glycol **(93-56-1)**
Styrene, α,β,2,3,4,5,6-heptachloro-, (E)- (8CI) **(29086-38-2)**
Styrene, α,β,2,3,4,5,6-heptachloro-, (Z)- (8CI) **(29086-39-3)**
Styrene, p-isopropyl-α-methyl- (8CI) **(2388-14-9)**
Styrene, α-methyl **(98-83-9)**
Styrene, m-methyl **(100-80-1)**
Styrene, methyl **(25013-15-4)**
Styrene, o-methyl **(611-15-4)**
Styrene, p-methyl **(622-97-9)**
Styrene, methyl- (8CI) **(1319-73-9)**
Styrene, α-methyl-, dimer **(6144-04-3)**
Styrene-methyl isopropenyl ketone polymer **(25191-48-4)**
Styrene, monomer (ACGIH) **(100-42-5)**
Styrene monomer, Inhibited [UN 2055] **(100-42-5)**
Styrene, β-nitro **(102-96-5)**
Styrene oxide **(96-09-3)**
Styrene-7,8-oxide **(96-09-3)**
Styrene-polymer **(9003-53-6)**
Styrene polymer with 1,3-butadiene **(9003-55-8)**
Styrene, β,2,5-trichloro- **(54935-00-1)**
Styrene trimer **(28213-80-1)**
Styrene, 2,4,6-trimethyl- (8CI) **(769-25-5)**
Styrex C **(9003-53-6)**
Styrocell PM **(9003-53-6)**
Styrofan 2D **(9003-53-6)**
Styroflex **(9003-53-6)**
Styrofoam **(9003-53-6)**
Styrol (German) **(100-42-5)**
Styrole **(100-42-5)**
Styrolene **(100-42-5)**
Styrolux **(9003-53-6)**
Styrolyl alcohol **(93-56-1)**
Styron **(100-42-5)**
Styron **(9003-53-6)**
Styron 440A **(9003-53-6)**
Styron 470A **(9003-53-6)**
Styron 475 **(9003-53-6)**
Styron 492 **(9003-53-6)**
Styron 666 **(9003-53-6)**
Styron 678 **(9003-53-6)**
Styron 679 **(9003-53-6)**
Styron 683 **(9003-53-6)**
Styron 685 **(9003-53-6)**
Styron 690 **(9003-53-6)**
Styron 69021 **(9003-53-6)**
Styron 475D **(9003-53-6)**
Styron GP **(9003-53-6)**
Styron 666K27 **(9003-53-6)**
Styron PS 3 **(9003-53-6)**

Styron T 679 (9003-53-6)
Styron 666U (9003-53-6)
Styron 666V (9003-53-6)
Styrone (104-54-1)
Styropian (9003-53-6)
Styropian FH 105 (9003-53-6)
Styropol (100-42-5)
Styropol HT 500 (9003-53-6)
Styropol IBE (9003-53-6)
Styropol JQ 300 (9003-53-6)
Styropol KA (9003-53-6)
Styropor (100-42-5)
Styropor (9003-53-6)
Styryl carbinol (104-54-1)
2-Styryl-1,3-dioxolane (5660-60-6)
Styryl methyl ketone (122-57-6)
Styryl oxide (96-09-3)
Sualen (63428-84-2)
Subacetate lead (1335-32-6)
Subamycin (64-75-5)
Subari (569-65-3)
Subdue (57837-19-1)
Subdue 2E (57837-19-1)
Subdue 5SP (57837-19-1)
Suberic acid (505-48-6)
Suberic acid, diethyl ester (8CI) (2050-23-9)
Suberon (502-42-1)
Suberone (502-42-1)
Suberone oxime (1121-92-2)
Suberonisoxim (German) (1121-92-2)
Suberonitrile (629-40-3)
Subicard (78-11-5)
Subitex (88-85-7)
Sublimat (Czech) (7487-94-7)
Sublimed Sulfur (7704-34-9)
Sublingula (9005-49-6)
DS Substance (50-67-9)
Substance DS (50-67-9)
Substance F (477-30-5)
Substanz DS (50-67-9)
Subtilisin (ACGIH) (9CI) (9014-01-1)
Subtilisin BPN' (9014-01-1)
Subtilisin Carlsburg (9014-01-1)
Subtilisin Novo (9014-01-1)
Subtilopeptidase A (9014-01-1)
Subtilopeptidase B (9014-01-1)
Subtilopeptidase BPN' (9014-01-1)
Subtilopeptidase C (9014-01-1)
Subtosan (9003-39-8)
Sucaryl (100-88-9)
Sucaryl acid (100-88-9)
Sucaryl calcium (139-06-0)
Sucaryl sodium (139-05-9)
Succaril (128-44-9)
Succaril (139-05-9)
Succimal (77-67-8)
Succimitin (77-67-8)
Succinaldehydic-acid (692-29-5)
Succinamic acid (8CI) (638-32-4)
Succinamic acid, 3-amino-N-(α-carboxyphenethyl)-, N-methyl ester, stereoisomer (22839-47-0)
Succinamic acid, N,N-diethyl (1522-00-5)
Succinamide (8CI) (110-14-5)
Succinchlorimide (128-09-6)
Succindiamide (110-14-5)
Succinic-acid (110-15-6)
Succinic acid anhydride (108-30-5)
Succinic acid, cadmium salt (1:1) (141-00-4)
Succinic acid, chloro- (8CI) (16045-92-4)
Succinic acid, cobalt(2+) salt (1:1) (3267-76-3)
Succinic acid diamide (110-14-5)
Succinic acid, dibenzyl ester (8CI) (103-43-5)
Succinic acid di-n-butyl ester (141-03-7)
Succinic acid, dibutyl ester (141-03-7)
Succinic acid dichloride (543-20-4)
Succinic acid, 2,2-dichloro- (8CI) (15519-38-7)
Succinic acid, diester with choline (306-40-1)
Succinic acid, diethyl ester (123-25-1)
Succinic acid, 2,3-dihydroxy- (87-69-4)
Succinic acid, 2,2-dimethyl- (8CI) (597-43-3)
Succinic acid, dimethyl ester (8CI) (106-65-0)
Succinic acid 2,2-dimethylhydrazide (1596-84-5)
Succinic acid dinitrile (110-61-2)
Succinic acid, hydroxy- (6915-15-7)
Succinic acid, O-isopropyl-O'-tributylstannyl ester (53404-82-3)
Succinic acid, mercapto-, diethyl ester, S-ester with O,O-dimethylphosphorodithioate (121-75-5)
Succinic acid, mercapto-, diethyl ester, S-ester with O,O-dimethylphosphorothioate (1634-78-2)
Succinic acid, mercapto-, diethyl ester, S-ester with O,S-dimethyl phosphorodithioate (3344-12-5)
Succinic acid, mercapto-, monoethyl ester, S-ester with O,O-dimethyl phosphorodithioate (8CI) (35884-76-5)
Succinic acid, methyl- (8CI) (498-21-5)
Succinic acid, methylene- (8CI) (97-65-4)
Succinic acid, mono(2,2-dimethylhydrazide) (1596-84-5)
Succinic acid, monoethyl ester, monoester with erythromycin (1264-62-6)
Succinic acid, monomethyl ester (8CI) (3878-55-5)
Succinic acid, mono(2,5,7,8-tetramethyl-2-(4,8,12-trimethyltridecyl)-6-chromanyl) ester, (+)- (8CI) (4345-03-3)
Succinic acid peroxide (DOT) (123-23-9)
Succinic acid peroxide, Technically pure (DOT) (123-23-9)
Succinic acid semialdehyde (692-29-5)
Succinic acid, sulfo- (8CI) (5138-18-1)
Succinic acid, sulfo-, 1,4-bis(2-ethylhexyl) ester, sodium salt (577-11-7)
Succinic acid, sulfo-, 1,4-dicyclohexyl ester, sodium salt (23386-52-9)
Succinic acid, sulfo-, 1,4-dioctyl ester (2373-23-1)
Succinic acid, sulfo-, 1,4-dioctyl ester, sodium salt (1639-66-3)
Succinic acid, sulfo-, 1,4-ditridecyl ester, sodium salt (8CI) (2673-22-5)
Succinic acid, sulfo-, dodecyl ester, disodium salt (6CI) (26838-05-1)
Succinic acid, sulfo-, monododecyl ester, disodium salt (8CI) (26838-05-1)
Succinic acid, sulfo-, monododecyl ester, disodium salt (8CI) (36409-57-1)
Succinic acid, tetrahydroxy- (8CI) (76-30-2)
Succinic amide (110-14-5)
Succinic-anhydride (108-30-5)
Succinic anhydride, dodecenyl (25377-73-5)
Succinic anhydride, methyl- (8CI) (4100-80-5)
Succinic anhydride, (1-octenyl)- (8CI) (7757-96-2)
Succinic anhydride, (tetrapropenyl) (26544-38-7)
Succinic chloride (543-20-4)
Succinic 1,1-dimethyl hydrazide (1596-84-5)
Succinic dinitrile (110-61-2)
Succinic imide (123-56-8)
Succinic peroxide (123-23-9)
Succinic semialdehyde (692-29-5)
Succinimide (123-56-8)
Succinimide-Sauba (123-56-8)
Succinimide, N-chloro- (128-09-6)
Succinimide, N,2-dimethyl-2-phenyl (77-41-8)
Succinimide, N-ethyl- (8CI) (2314-78-5)
Succinimide, 2-ethyl-N,2-dimethyl- (8CI) (13861-99-9)
Succinimide, 2-ethyl-2-methyl (77-67-8)
Succinimide, N-methyl-2-phenyl (86-34-0)
Succinimide, N-propyl- (8CI) (3470-97-1)

Succinochlorimide (128-09-6)
Succinocholine (306-40-1)
Succinodinitrile (110-61-2)
Succinonitrile (110-61-2)
Succinonitrile, tetramethyl (3333-52-6)
Succinoyl chloride (543-20-4)
Succinoylcholine (306-40-1)
Succinylbischoline (306-40-1)
Succinyl-chloride (543-20-4)
Succinyl choline (306-40-1)
Succinyl dichloride (543-20-4)
Succinyldicholine (306-40-1)
Succinyl oxide (108-30-5)
Succinyl peroxide (123-23-9)
Succitimal (86-34-0)
Sucedaneo de trementina (Spanish) (9005-90-7)
Suchar 681 (7440-44-0)
Sucker-Stuff (123-33-1)
Sucol B (110-63-4)
Suconox-18 (103-99-1)
Suconox 18, 4'-hydroxy- (103-99-1)
Sucra (128-44-9)
Sucralfate (54182-58-0)
Sucralfato (Spanish) (54182-58-0)
Sucralfatum (Latin) (54182-58-0)
Sucrapan P (26446-38-8)
Sucraphen (61-76-7)
Sucre Edulcor (81-07-2)
Sucrets (136-77-6)
Sucrette (81-07-2)
Sucrodet (26446-38-8)
Sucrofer (8047-67-4)
Sucrol (150-69-6)
Sucrosa (139-05-9)
Sucrose (57-50-1)
Sucrose (OSHA) (57-50-1)
Sucrose acetate isobutyrate (126-13-6)
Sucrose acetoisobutyrate (126-13-6)
Sucrose benzoate (12738-64-6)
Sucrose, diacetate hexaisobutyrate (126-13-6)
Sucrose laurate (25339-99-5)
Sucrose monolaurate (25339-99-5)
Sucrose monopalmitate (26446-38-8)
Sucrose, monopalmitate (8CI) (26446-38-8)
Sucrose, monostearate (8CI) (25168-73-4)
Sucrose monostearic acid ester (25168-73-4)
Sucrose, octaacetate (126-14-7)
Sucrose octakis(hydrogen sulfate) aluminum complex (54182-58-0)
Sucrose palmitate (VAN) (26446-38-8)
Sucrose palmitic acid ester (26446-38-8)
Sucrose, propylene oxide, ethylene oxide polymer (26301-10-0)
Sucrose stearate (25168-73-4)
Sucrose stearic acid ester (25168-73-4)
Sucrun 7 (139-05-9)
Sudafed (90-82-4)
Sudan 1 (842-07-9)
Sudan 2 (3118-97-6)
Sudan AX (3118-97-6)
Sudan 7B (6368-72-5)
Sudan Black B (4197-25-5)
Sudan Blue (6994-46-3)
Sudan Blue GA (128-85-8)
Sudan Brown RR (6416-57-5)
Sudan Brown YR (6416-57-5)
Sudan G (2051-85-6)
Sudan G (85-86-9)
Sudan GG (60-11-7)
Sudan G III (85-86-9)
Sudan Green 4B (128-80-3)
Sudan I (842-07-9)
Sudan II (3118-97-6)
Sudan III (85-86-9)
Sudan III (G) (85-86-9)
Sudan IV (85-83-6)
Sudan J (842-07-9)
Sudan Orange (3118-97-6)
Sudan Orange G (2051-85-6)
Sudan Orange R (842-07-9)
Sudan Orange RA (842-07-9)
Sudan Orange RA New (842-07-9)
Sudan Orange RPA (3118-97-6)
Sudan Orange RRA (3118-97-6)
Sudan P (85-83-6)
Sudan P III (85-86-9)
Sudan Red (3118-97-6)
Sudan Red 7B (6368-72-5)
Sudan Red 4BA (85-83-6)
Sudan Red BB (85-83-6)
Sudan Red BBA (85-83-6)
Sudan Red III (85-86-9)
Sudan Red IV (85-83-6)
Sudan Scarlet 6G (3118-97-6)
Sudan X (3118-97-6)
Sudan Yellow (60-11-7)
Sudan Yellow 1339 (119-75-5)
Sudan Yellow AR (2051-85-6)
Sudan Yellow GG (60-11-7)
Sudan Yellow GGA (60-11-7)
Sudan Yellow GGN (2481-94-9)
Sudan Yellow R (60-09-3)
Sudan Yellow RA (60-09-3)
Sudan Yellow RRA (97-56-3)
Sudanrot 7B (6368-72-5)
Suessette (139-05-9)
Suesstoff (150-69-6)
Suestamin (139-05-9)
Sufentanil (56030-54-7)
Sufentanilum (Latin) (56030-54-7)
Suffix (22212-55-1)
Suffix 25 (22212-55-1)
Sugai Brilliant Scarlet 3R (2611-82-7)
Sugai Chrysoidine (532-82-1)
Sugai Congo Red (573-58-0)
Sugai Fast Scarlet G Base (99-55-8)
Sugai Tartrazine (1934-21-0)
Suganyl (57-67-0)
Sugar (57-50-1)
Sugar, Grape (50-99-7)
Sugar, Invert (9CI) (8013-17-0)
Sugarin (139-05-9)
Sugar of Lead (301-04-2)
Sugaron (139-05-9)
Sugee 2 (9CI) (53988-71-9)
Sugracillin (113-98-4)
Sukcinonitril (Czech) (110-61-2)
Sul-Spansion (94-19-9)
Sul-Spantab (94-19-9)
Suladyne (136-40-3)
Sulamyd (144-80-9)
Sul anilinova (Czech) (142-04-1)
Sulbio (127-69-5)
Sulcatol (4630-06-2)
Sulcofuron (3567-25-7)
Sulcolon (599-79-1)

Name	CAS No.
Sulema (Russian)	(7487-94-7)
Sulerex	(19937-59-8)
Sulestrex	(7280-37-7)
Sulf-10	(144-80-9)
Sulfa-Perlongit	(94-19-9)
Sulfabenzpyrazine	(59-40-5)
Sulfabid	(526-08-9)
Sulfabutin	(55-98-1)
Sulfacet	(144-80-9)
Sulfacetamide	(144-80-9)
Sulfacetimide	(144-80-9)
Sulfacid Brilliant Blue 6J	(129-17-9)
Sulfacid Brilliant Green 1B	(4680-78-8)
Sulfacid Light Orange J	(1936-15-8)
Sulfacitina (Spanish)	(17784-12-2)
Sulfacitine	(17784-12-2)
Sulfacitinum (Latin)	(17784-12-2)
Sulfacox	(59-40-5)
Sulfactin	(59-52-9)
Sulfacyl	(144-80-9)
Sulfacytine	(17784-12-2)
Sulfadene	(149-30-4)
Sulfadiazene	(68-35-9)
Sulfadiazine	(68-35-9)
Sulfadimerazine	(57-68-1)
Sulfadimesine	(57-68-1)
Sulfadimethyldiazine	(57-68-1)
Sulfadimethylisoxazole	(127-69-5)
Sulfadimethylpyrimidine	(57-68-1)
Sulfadimetine	(57-68-1)
Sulfadimezine	(57-68-1)
Sulfadimidine	(57-68-1)
Sulfadine	(57-68-1)
Sulfaethidiole	(94-19-9)
Sulfaethidol	(94-19-9)
Sulfaethidole	(94-19-9)
Sulfaethylthiadiazole	(94-19-9)
Sulfafenazolo (Italian)	(526-08-9)
Sulfafurazol	(127-69-5)
Sulfafurazole	(127-69-5)
Sulfagan	(127-69-5)
Sulfagen	(127-69-5)
Sulfaguanidine	(57-67-0)
Sulfaguine	(57-67-0)
Sulfa-isodimerazine	(57-68-1)
Sulfaisodimidine	(57-68-1)
Sulfaisoxazole	(127-69-5)
Sulfalar	(127-69-5)
Sulfalex	(80-35-3)
Sulfaline	(59-40-5)
Sulfallate	(95-06-7)
Sulfalone	(126-33-0)
Sulfamate	(7773-06-0)
Sulfamate dammonium	(7773-06-0)
Sulfamerazine sodium	(127-58-2)
Sulfametazyny (Polish)	(57-68-1)
Sulfamethalazole	(723-46-6)
Sulfamethazine	(57-68-1)
Sulfamethiazine	(57-68-1)
Sulfamethin	(57-68-1)
Sulfamethizol	(144-82-1)
Sulfamethizole	(144-82-1)
Sulfamethoxazol	(723-46-6)
Sulfamethoxazole	(723-46-6)
Sulfamethoxazol-trimethoprim	(8064-90-2)
3-Sulfa-6-methoxypyridazine	(80-35-3)
Sulfamethoxypyridazine	(80-35-3)
Sulfamethylisoxazole	(723-46-6)
Sulfamethylizole	(144-82-1)
Sulfamethylthiadiazole	(144-82-1)
Sulfametoxipiridazine	(80-35-3)
Sulfamezathine	(57-68-1)
Sulfamic-acid [UN 2967]	(5329-14-6)
Sulfamic acid, cobalt(2+) salt (2:1) (8CI,9CI)	(14017-41-5)
Sulfamic acid, cyclohexyl- (9CI)	(100-88-9)
Sulfamic acid, cyclohexyl-, calcium salt (2:1) (9CI)	(139-06-0)
Sulfamic acid, cyclohexyl-, monopotassium salt	(7758-04-5)
Sulfamic acid, methoxymethyl-, sodium salt (8CI)	(5390-94-3)
Sulfamic acid, monoammonium salt	(7773-06-0)
Sulfamide, N,N'-bis(4-nitrophenyl)-	(19757-13-2)
Sulfamide, N-((dichlorofluoromethyl)thio)-N',N'-dimethyl- N-phenyl	(1085-98-9)
Sulfamide, N-((dichlorofluoromethyl)thio)-N',N'-dimethyl- N-(p-tolyl)	(731-27-1)
Sulfamidic acid	(5329-14-6)
p-Sulfamidoaniline	(63-74-1)
Sulfamidyl	(63-74-1)
Sulfamine	(63-74-1)
Sulfaminsaure (German)	(7773-06-0)
Sulfamoyl chloride, (1-methylethyl)- (9CI)	(26118-67-2)
N-(5-Sulfamoyl-1,3,4-thiadiazol-2-yl)acetamide	(59-66-5)
Sulfamul	(72-14-0)
7-Sulfamyl-6-trifluoromethyl-3,4-dihydro-1,2,4-benzothia- diazine 1,1-dioxide	(135-09-1)
Sulfan	(7446-11-9)
Sulfan Blue	(129-17-9)
Sulfana	(63-74-1)
Sulfanalone	(63-74-1)
Sulfanil	(63-74-1)
Sulfanilacetamide	(144-80-9)
Sulfanilamide	(63-74-1)
Sulfanilamide, N^1-acetyl-	(144-80-9)
Sulfanilamide, N^1-amidino	(57-67-0)
Sulfanilamide, N^1-(3,4-dimethyl-5-isoxazolyl)	(127-69-5)
Sulfanilamide, N^1-(4,6-dimethyl-2-pyrimidinyl)	(57-68-1)
Sulfanilamide, 3,5-dinitro-N^4,N^4-dipropyl	(19044-88-3)
Sulfanilamide, N4-(p-ethoxyphenyl)-3-nitro- (8CI)	(22025-44-1)
Sulfanilamide, N(1)-(1-ethyl-1,2-dihydro-2-oxo-4-pyrimidinyl)-	(17784-12-2)
Sulfanilamide, N1-(1-ethyl-1,2-dihydro-2-oxo-4-pyrimidinyl)- (8CI)	(17784-12-2)
Sulfanilamide, N-(5-ethyl-1,3,4-thiadiazol-2-yl)	(94-19-9)
3-Sulfanilamide-6-methoxypyridazine	(80-35-3)
Sulfanilamide, N^1-(6-methoxy-3-pyridazinyl)	(80-35-3)
Sulfanilamide, N'-(5-methyl-3-isoxazolyl)	(723-46-6)
Sulfanilamide, N^1-(4-methyl-2-pyrimidinyl)-, monosodium salt	(127-58-2)
Sulfanilamide, N^1-(5-methyl-1,3,4-thiadiazol-2-yl)	(144-82-1)
Sulfanilamide, N^1-(1-phenylpyrazol-5-yl)- (8CI)	(526-08-9)
Sulfanilamide, N1-(pyrazinyl)- (8CI)	(116-44-9)
Sulfanilamide, N^1-2-pyridyl-	(144-83-2)
Sulfanilamide, N^1-2-pyrimidinyl	(68-35-9)
Sulfanilamide, N^1-2-quinoxalinyl	(59-40-5)
Sulfanilamide, N^1-2-thiazolyl	(72-14-0)
Sulfanilamide, N^1-2-thiazolyl-, monosodium salt	(144-74-1)
Sulfanilamide, N^1-2-thiazolyl-, N^1-sodium deriv.	(144-74-1)
2-Sulfanilamidoaethylthiodiazol (German)	(94-19-9)
5-Sulfanilamido-3,4-dimethylisoxazole	(127-69-5)
2-Sulfanilamido-4,6-dimethylpyrimidine	(57-68-1)
3-Sulfanilamido-6-methoxypyridazine	(80-35-3)
6-Sulfanilamido-3-methoxypyridazine	(80-35-3)
3-Sulfanilamido-5-methylisoxazole	(723-46-6)
2-Sulfanilamido-5-methyl-1,3,4-thiadiazole	(144-82-1)
5-Sulfanilamido-1-phenylpyrazole	(526-08-9)
2-Sulfanilamidopyrazine	(116-44-9)
2-Sulfanilamidopyridin (German)	(144-83-2)

Name	CAS
2-Sulfanilamidopyrimidin (German)	**(68-35-9)**
Sulfanilamidopyrimidine	**(68-35-9)**
2-Sulfanilamidoquinoxaline	**(59-40-5)**
2-Sulfanilamidothiazol (German)	**(72-14-0)**
2-Sulfanilamidothiazole	**(72-14-0)**
2-Sulfanilamidothiazole sodium salt	**(144-74-1)**
Sulfanilanilide, 3-nitro-N4-phenyl- (8CI)	**(5124-25-4)**
Sulfanilazetamid (German)	**(144-80-9)**
Sulfanilguanidine	**(57-67-0)**
Sulfanilic-acid	**(121-57-3)**
m-Sulfanilic acid	**(121-47-1)**
o-Sulfanilic acid	**(88-21-1)**
Sulfanilic acid, 2,5-dichloro- (8CI)	**(88-50-6)**
Sulfanilic acid, N-(4,6-dichloro-s-triazin-2-yl)-	**(16110-89-7)**
Sulfanilsaeure (German)	**(121-57-3)**
N-Sulfanilylacetamide	**(144-80-9)**
2-Sulfanilyl aminopyridine	**(144-83-2)**
2-Sulfanilylaminopyrimidine	**(68-35-9)**
2-(Sulfanilylamino)thiazole	**(72-14-0)**
N^1-Sulfanilyl-N^2-butylcarbamide	**(339-43-5)**
N^1-Sulfanilyl-N^2-butylurea	**(339-43-5)**
N-Sulfanilyl N'butyluree (French)	**(339-43-5)**
Sulfanilyl chloride, N-acetyl- (8CI)	**(121-60-8)**
Sulfanilylguanidine	**(57-67-0)**
N-Sulfanilyl-l-ethylcytosine	**(17784-12-2)**
Sulfanthrene Orange R	**(3263-31-8)**
Sulfanthrene Orange RS	**(3263-31-8)**
Sulfaphenazole	**(526-08-9)**
Sulfaphenazon	**(526-08-9)**
Sulfaphenylpipazol	**(526-08-9)**
Sulfaphenylpyrazole	**(526-08-9)**
Sulfapol	**(25155-30-0)**
Sulfapolar	**(127-69-5)**
Sulfapolu (Polish)	**(25155-30-0)**
Sulfapyrazine	**(116-44-9)**
Sulfapyridazine	**(80-35-3)**
2-Sulfapyridine	**(144-83-2)**
Sulfapyridine	**(144-83-2)**
Sulfapyrimidin (German)	**(68-35-9)**
2-Sulfapyrimidine	**(68-35-9)**
Sulfapyrimidine	**(68-35-9)**
Sulfaquinoxaline	**(59-40-5)**
Sulfasalazine	**(599-79-1)**
Sulfasan	**(103-34-4)**
Sulfasan	**(127-69-5)**
Sulfasan	**(502-55-6)**
Sulfasan R	**(103-34-4)**
Sulfasan R powder	**(103-34-4)**
Sulfasol	**(127-69-5)**
Sulfasoxazole	**(127-69-5)**
Sulfate (8CI,9CI)	**(14808-79-8)**
Sulfate d'atropine (French)	**(55-48-1)**
Sulfated Castor Oil	**(8002-33-3)**
Sulfate de cuivre (French)	**(7758-98-7)**
Sulfate de methyle (French)	**(77-78-1)**
Sulfate de nicotine (French)	**(65-30-5)**
Sulfate de plomb (French)	**(7446-14-2)**
Sulfate de zinc (French)	**(7733-02-0)**
Sulfate dimethylique (French)	**(77-78-1)**
Sulfated isopropyl oleate, sodium salt	**(14350-72-2)**
Sulfate mercurique (French)	**(7783-35-9)**
Sulfatep	**(3689-24-5)**
Sulfathiazol	**(72-14-0)**
Sulfathiazole	**(72-14-0)**
Sulfathiazole sodium	**(144-74-1)**
Sulfatom ammoniya (Russian)	**(7783-20-2)**
Sulfatropinol	**(55-48-1)**
Sulfazan R	**(103-34-4)**
Sulfazin	**(127-69-5)**
Sulfazine	**(68-35-9)**
Sulfdurazin	**(80-35-3)**
Sulfenamid DC	**(4979-32-2)**
Sulfenamide M	**(102-77-2)**
Sulfenamide TS	**(95-33-0)**
Sulfenax MOB (Czech)	**(102-77-2)**
Sulfenax MOB (Czech)	**(95-32-9)**
Sulfenazone	**(127-52-6)**
Sulfenone	**(80-00-2)**
Sulfentanil	**(56030-54-7)**
Sulferrous	**(7720-78-7)**
Sulferrous	**(7782-63-0)**
Sulfetal L 95	**(151-21-3)**
Sulfidal	**(7704-34-9)**
Sulfide (8CI,9CI)	**(18496-25-8)**
Sulfide, benzyl methyl (8CI)	**(766-92-7)**
Sulfide, benzyl phenyl (8CI)	**(831-91-4)**
Sulfide, bis(p-aminophenyl)	**(139-65-1)**
Sulfide, bis(2-carboxyethyl)	**(111-17-1)**
Sulfide, bis(2-chloroethyl)	**(505-60-2)**
Sulfide, bis(2-cyanoethyl)	**(111-97-7)**
Sulfide, bis(dimethylthiocarbamoyl)	**(97-74-5)**
Sulfide, bis(2-hydroxyethyl)	**(111-48-8)**
Sulfide, bis(4-hydroxyphenyl)	**(2664-63-3)**
Sulfide, butyl ethyl (8CI)	**(638-46-0)**
Sulfide, p-chlorobenzyl p-chlorophenyl	**(103-17-3)**
Sulfide, p-chlorobenzyl p-fluorophenyl	**(405-30-1)**
Sulfide, chloroethyl ethyl	**(693-07-2)**
Sulfide, 2-chloroethyl 2-hydroxyethyl	**(693-30-1)**
Sulfide, 2-chloroethyl methyl (8CI)	**(542-81-4)**
Sulfide, p-chlorophenyl methyl	**(123-09-1)**
Sulfide, p-chlorophenyl 2,4,5-trichlorophenyl	**(2227-13-6)**
Sulfide, dibenzyl	**(538-74-9)**
Sulfide, diphenyl	**(139-66-2)**
Sulfide, ethyl isopropyl (8CI)	**(5145-99-3)**
Sulfide, ethyl phenyl (8CI)	**(622-38-8)**
Sulfide, ethyl propyl (8CI)	**(4110-50-3)**
Sulfide, isobutyl p-tolyl (7CI)	**(54576-37-3)**
Sulfide, isopropyl methyl (8CI)	**(1551-21-9)**
Sulfide, methyl pentachlorophenyl	**(1825-19-0)**
Sulfide, methyl pentyl (6CI,7CI,8CI)	**(1741-83-9)**
Sulfide, methyl propenyl (8CI)	**(10152-77-9)**
Sulfide, methyl propyl (8CI)	**(3877-15-4)**
Sulfidine	**(144-83-2)**
Sulfidotrichlorid fosforecny (Czech)	**(3982-91-0)**
Sulfimel DOS	**(577-11-7)**
Sulfinpyrazine	**(57-96-5)**
Sulfinpyrazone	**(57-96-5)**
Sulfinylbis(methane)	**(67-68-5)**
Sulfinyl chloride	**(7719-09-7)**
Sulfisin	**(127-69-5)**
Sulfisomezole	**(723-46-6)**
Sulfisomidin	**(57-68-1)**
Sulfisomidine	**(57-68-1)**
Sulfisonazole	**(127-69-5)**
Sulfisoxazol	**(127-69-5)**
Sulfisoxazole	**(127-69-5)**
Sulfite (8CI,9CI)	**(14265-45-3)**
Sulfite Cellulose	**(9004-34-6)**
Sulfite Liquors and Cooking Liquors, Black, Vanillin	**(68514-06-7)**
Sulfix 6	**(25608-33-7)**
Sulfizin	**(127-69-5)**
Sulfizol	**(127-69-5)**
Sulfizole	**(127-69-5)**
Sulfo Green J	**(5141-20-8)**

Sulfo-Merthiolate **(5964-24-9)**
Sulfoacetic acid **(123-43-3)**
Sulfoacetic acid 1-dodecyl ester, sodium salt **(1847-58-1)**
Sulfoacetic acid dodecyl ester S-sodium salt **(1847-58-1)**
2-Sulfoanthraquinone sodium salt **(131-08-8)**
4-Sulfo-1,2-benzenedicarboxylic acid **(89-08-7)**
Sulfobenzide **(127-63-9)**
o-Sulfobenzimide **(81-07-2)**
2-Sulfobenzoic acid **(632-25-7)**
3-Sulfobenzoic acid **(121-53-9)**
4-Sulfobenzoic acid **(636-78-2)**
m-Sulfobenzoic acid **(121-53-9)**
o-Sulfobenzoic acid **(632-25-7)**
para-Sulfobenzoic acid **(636-78-2)**
o-Sulfobenzoic acid anhydride **(81-08-3)**
o-Sulfobenzoic acid, cyclic anhydride **(81-08-3)**
o-Sulfobenzoic acid imide **(81-07-2)**
Sulfobenzoic anhydride **(81-08-3)**
o-Sulfobenzoic anhydride **(81-08-3)**
Sulfobromophthalein **(71-67-0)**
Sulfobromophthalein sodium **(71-67-0)**
Sulfobromphthalein **(71-67-0)**
Sulfobutanedioic acid, 4-isodecyl ester, disodium salt **(37294-49-8)**
Sulfobutanedioic acid, monooctadecyl ester, disodium salt **(14481-60-8)**
Sulfocarb **(1646-88-4)**
Sulfocarbanilide **(102-08-9)**
Sulfocarbolic acid **(1333-39-7)**
Sulfocide **(120-62-7)**
Sulfocidin **(63-74-1)**
Sulfocidine **(63-74-1)**
(?,Z)-12-Sulfo-9,12-octadecadienoic acid **(68201-84-3)**
Sulfodimesin **(57-68-1)**
Sulfodimezine **(57-68-1)**
2-Sulfododecanoic acid **(3054-88-4)**
Sulfodor (Czech) **(352-93-2)**
Sulfoethanoic acid **(123-43-3)**
Sulfoethyl methacrylate **(10595-80-9)**
2-Sulfoethyl 2-methyl-2-propenoate **(10595-80-9)**
2-Sulfoethyl 9-octadecenoate, sodium salt **(142-15-4)**
Sulfoguanidine **(57-67-0)**
Sulfoguenil **(57-67-0)**
5-Sulfoisophthalic acid, bis(2-hydroxyethyl) ester, sodium salt **(24019-46-3)**
5-Sulfoisophthalic acid 5-sodium salt **(33562-89-9)**
Sulfolan **(126-33-0)**
Sulfolane **(126-33-0)**
.α.-Sulfolauric acid **(3054-88-4)**
3-Sulfolene **(77-79-2)**
Sulfol-3-ene **(77-79-2)**
β-Sulfolene **(77-79-2)**
12-Sulfolinoleic acid **(68201-84-3)**
N-(Sulfomethyl)aniline **(103-06-0)**
Sulfometuron **(74223-56-6)**
Sulfometuron methyl **(74222-97-2)**
Sulfona **(80-08-0)**
Sulfona-MAE **(80-08-0)**
Sulfonal **(115-24-2)**
Sulfonamide **(63-74-1)**
Sulfonamide P **(63-74-1)**
2-Sulfonamidothiazole **(72-14-0)**
2-Sulfo-1,4-naphthoquinone, sodium salt **(7045-83-2)**
1-(4-Sulfo-1-naphthylazo)-2-naphthol-3,6-disulfonic acid trisodium salt **(915-67-3)**
2-(4-Sulfo-1-naphthylazo)-1-naphthol-4-sulfonic acid, disodium salt **(3567-69-9)**
Sulfonated Castor Oil **(8002-33-3)**
Sulfonated Vegetable Oil **(61790-19-0)**
Sulfonated castor oil, sodium salt **(68187-76-8)**
Sulfonated lignin sodium salt **(8061-51-6)**
Sulfonated oleic acid, sodium salt **(68443-05-0)**
Sulfonated petroleum, sodium salt **(68608-26-4)**
o-Sulfonbenzoic acid imide sodium salt **(128-44-9)**
p-Sulfondichloramidobenzoic acid **(80-13-7)**
Sulfone UCB **(80-08-0)**
Sulfone, bis(2-chloroethyl) **(471-03-4)**
Sulfone, bis(p-chlorophenyl) **(80-07-9)**
Sulfone, bis(trichloromethyl) (8CI) **(3064-70-8)**
Sulfone, 2-chloroethyl phenyl (8CI) **(938-09-0)**
Sulfone, p-chlorophenyl methyl **(98-57-7)**
Sulfone, p-chlorophenyl phenyl **(80-00-2)**
Sulfone, p-chlorophenyl 2,4,5-trichlorophenyl **(116-29-0)**
Sulfone, demeton-S-methyl- **(17040-19-6)**
Sulfone, divinyl- **(77-77-0)**
Sulfone, methyl phenyl **(3112-85-4)**
Sulfone, methyl 2,3,5,6-tetrachlorophenyl **(98555-82-9)**
Sulfone, methyl p-tolyl (8CI) **(3185-99-7)**
Sulfone, methyl 2,4,5-trichlorophenyl (8CI) **(4163-80-8)**
Sulfone, octyl vinyl (8CI) **(28345-91-7)**
Sulfone, 2,4,4',5-tetrachlorodiphenyl **(116-29-0)**
Sulfonethylmethane **(76-20-0)**
Sulfonic acid, α-alkene **(72674-05-6)**
Sulfonic acid, monochloride **(7790-94-5)**
Sulfonic acids, C10-18-alkane, sodium salts **(68037-49-0)**
Sulfonic acids, alkane, sodium salts **(68608-15-1)**
Sulfonic acids, petrolatum, sodium salts **(68918-07-0)**
Sulfonic acids, petroleum **(61789-85-3)**
Sulfonic acids, petroleum, magnesium salts **(61789-87-5)**
Sulfonic acids, petroleum, sodium salts **(68608-26-4)**
Sulfonimide **(2425-06-1)**
Sulfonine Acid Blue R **(3861-73-2)**
Sulfonine Red G **(3567-65-5)**
Sulfonine Red GN **(3567-65-5)**
Sulfonine Red GS **(3567-65-5)**
Sulfonine Red SG **(3567-65-5)**
Sulfonine Yellow CSR **(6375-55-9)**
Sulfonium, bis(2-hydroxyethyl)-2-(2-chloroethylthio)ethyl-, chloride **(64036-91-5)**
Sulfonium, (2-carboxyethyl)dimethyl-, hydroxide, inner salt (8CI,9CI) **(7314-30-9)**
Sulfonium, (p-hydroxyphenyl)dimethyl-, chloride **(1005-35-2)**
Sulfonium, (4-hydroxyphenyl)dimethyl-, chloride (9CI) **(1005-35-2)**
Sulfonmethane **(115-24-2)**
α-Sulfo-ω-(nonylphenoxy)-poly(oxy-1,2-ethanediyl) **(9081-17-8)**
Sulfonphthal **(143-74-8)**
1,1'-Sulfonylbis(4-aminobenzene) **(80-08-0)**
4,4'-Sulfonylbisaniline **(80-08-0)**
4,4'-Sulfonylbisbenzamine **(80-08-0)**
p,p-Sulfonylbisbenzamine **(80-08-0)**
p,p-Sulfonylbisbenzenamine **(80-08-0)**
2,2'-Sulfonylbisethanol **(2580-77-0)**
Sulfonylbismethane **(67-71-0)**
Sulfonylbis(trichloromethane) **(3064-70-8)**
4,4'-Sulfonyldianiline **(80-08-0)**
p,p'-Sulfonyldianiline **(80-08-0)**
4,4'-Sulfonyldiphenol **(80-09-1)**
3-(Sulfonyl)-O-((methylamino)carbonyl)oxime-2-butanone **(34681-23-7)**
N-(Sulfonyl-p-methylbenzene)-N'-n-butylurea **(64-77-7)**
(Sulfooxy)-9-octadecenoic acid, 1-butyl ester **(38621-44-2)**
1-p-Sulfophenylazo-2-hydroxynaphthalene-6-sulfonate, disodium salt **(2783-94-0)**
4-p-Sulfophenylazo-1-naphthol monosodium salt **(523-44-4)**
1-p-Sulfophenylazo-2-naphthol-6-sulfonic acid, disodium salt **(2783-94-0)**
1-(2'-Sulfophenylazo)-2-naphthol **(6616-62-2)**
4-Sulfophenylmethallyl ether **(1208-67-9)**
p-Sulfophenyl methallyl ether, sodium salt **(1208-67-9)**

4-Sulfophthalic acid **(89-08-7)**
4-Sulfophthalic acid monosodium salt **(33562-89-9)**
Sulfopon WA 1 **(151-21-3)**
Sulfopon WA 2 **(151-21-3)**
Sulfopon WA 3 **(151-21-3)**
Sulfopon WA 1 Special **(151-21-3)**
3-Sulfopropyl N,N-dimethyldithiocarbamate, sodium salt **(18880-36-9)**
γ,γ'-Sulfopropyldisulfide, disodium salt **(27206-35-5)**
1-(3-Sulfopropyl)pyridinium hydroxide, inner salt **(15471-17-7)**
N-3-Sulfopropylpyridinium betaine **(15471-17-7)**
Sulfoquanidine **(57-67-0)**
Sulforidazine **(14759-06-9)**
Sulforon **(7704-34-9)**
5-Sulfosalicylic acid **(97-05-2)**
Sulfosalicylic acid **(97-05-2)**
Sulfosuccinic acid, bis(2-ethylhexyl)ester sodium salt **(577-11-7)**
Sulfosuccinic acid, 1,4-dicyclohexyl ester, sodium salt **(23386-52-9)**
Sulfosuccinic acid, dicyclohexyl ester, sodium salt **(23386-52-9)**
Sulfosuccinic acid 1,4-dioctyl ester sodium salt **(1639-66-3)**
Sulfosuccinic acid monododecyl ester disodium salt **(26838-05-1)**
Sulfosuccinic acid monolauryl ester disodium salt **(26838-05-1)**
Sulfotep (ACGIH,OSHA) **(3689-24-5)**
Sulfotepp **(3689-24-5)**
α-Sulfo-tetradecyloxypoly(oxy-1,2-ethanediyl), ammonium salt **(27731-61-9)**
Sulfotex WA **(151-21-3)**
Sulfotex Wala **(151-21-3)**
Sulfothiorine **(10102-17-7)**
Sulfotrim **(8064-90-2)**
Sulfotrimin **(8064-90-2)**
Sulfox-Cide **(120-62-7)**
Sulfoxide **(120-62-7)**
Sulfoxide, bis(2-chloroethyl) **(5819-08-9)**
Sulfoxide, 2-chloroethyl vinyl **(40709-82-8)**
Sulfoxide, p-chlorophenyl methyl **(934-73-6)**
Sulfoxide, 3-chloropropyl n-octyl- **(3569-57-1)**
Sulfoxide, 3-chloropropyl octyl **(3569-57-1)**
Sulfoxide, α-methyl-3,4-(methylenedioxy)phenethyl octyl **(120-62-7)**
Sulfoxide, methyl phenyl (8CI) **(1193-82-4)**
Sulfoximine, S-(3-amino-3-carboxypropyl)-S-methyl-, DL **(1982-67-8)**
Sulfoxol **(127-69-5)**
Sulfoxyl **(120-62-7)**
Sulfoxyl **(94-36-0)**
2-(6-Sulfo-2,4-xylylazo)-1-naphthol-4-sulfonic acid, disodium salt **(4548-53-2)**
Sulfoxyphenylpyrazolidine **(57-96-5)**
Sulfozona **(80-35-3)**
Sulframin 85 **(25155-30-0)**
Sulframin 1298 **(27176-87-0)**
Sulframin Acid 1298 **(27176-87-0)**
Sulframin 40 Flakes **(25155-30-0)**
Sulframin 90 Flakes **(25155-30-0)**
Sulframin 40 Granular **(25155-30-0)**
Sulframin 40RA **(25155-30-0)**
Sulframin 1238 Slurry **(25155-30-0)**
Sulframin 1250 Slurry **(25155-30-0)**
Sulfstat **(144-82-1)**
Sulftech **(7757-83-7)**
Sulfur Flower (DOT) **(7704-34-9)**
Sulfur Half-Mustard **(693-30-1)**
Sulfur Mustard **(505-60-2)**
Sulfur Mustard Gas **(505-60-2)**
Sulfur, Solid (DOT) **(7704-34-9)**
Sulfur [UN 1350] **(7704-34-9)**
Sulfur chloride **(10545-99-0)**
Sulfur chloride **(14989-32-3)**
Sulfur chloride (8CI,9CI) **(10545-99-0)**
Sulfur-chloride [UN 1828] **(10025-67-9)**
Sulfur chloride(di) (DOT) **(10025-67-9)**
Sulfur chloride(mono) [UN 1828] **(10545-99-0)**
Sulfur chloride oxide **(7719-09-7)**
Sulfur-decafluoride **(5714-22-7)**
Sulfur-dichloride **(10545-99-0)**
Sulfur difluoride monoxide **(7783-42-8)**
Sulfur difluoride oxide **(7783-42-8)**
Sulfur-dioxide **(7446-09-5)**
Sulfur dioxide (ACGIH,DOT,OSHA) **(7446-09-5)**
Sulfur dioxide Solution **(7782-99-2)**
Sulfure d'ammonium (French) **(9080-17-5)**
Sulfure de 4-chlorobenzyle et de 4-chlorophenyle (French) **(103-17-3)**
Sulfure de methyle (French) **(75-18-3)**
Sulfure de sodium (French) **(1344-08-7)**
Sulfureted hydrogen **(7783-06-4)**
Sulfur-fluoride **(2551-62-4)**
Sulfur hexafluoride (ACGIH,OSHA) [UN 1080] **(2551-62-4)**
Sulfur hydride **(7783-06-4)**
Sulfuric acid (ACGIH,OSHA) [UN 1830] **(7664-93-9)**
Sulfuric acid, Fuming [UN 1831] **(8014-95-7)**
Sulfuric acid, Spent [UN 1832] **(7664-93-9)**
Sulfuric acid, aluminum ammonium salt (2:1:1) (9CI) **(7784-25-0)**
Sulfuric acid, aluminum ammonium salt (2:1:1), dodecahydrate **(7784-25-0)**
Sulfuric acid, aluminum ammonium salt (2:1:1), dodecahydrate **(7784-26-1)**
Sulfuric acid, aluminum potassium salt (2:1:1) (9CI) **(10043-67-1)**
Sulfuric acid, aluminum salt (3:2) **(10043-01-3)**
Sulfuric acid, aluminum sodium salt (2:1:1) (9CI) **(10102-71-3)**
Sulfuric acid, ammonium cobalt(2+) salt (2:2:1) (9CI) **(13596-46-8)**
Sulfuric acid, ammonium iron(2+) salt (2:2:1) (9CI) **(10045-89-3)**
Sulfuric acid, ammonium nickel(2+) salt (2:2:1) **(15699-18-0)**
Sulfuric acid, ammonium zinc salt (2:2:1), hexahydrate **(7783-24-6)**
Sulfuric acid, barium salt (1:1) **(7727-43-7)**
Sulfuric acid, beryllium salt (1:1) **(13510-49-1)**
Sulfuric acid, beryllium salt (1:1), tetrahydrate **(7787-56-6)**
Sulfuric acid, cadmium(2+) salt **(10124-36-4)**
Sulfuric acid, calcium salt (1:1) **(7778-18-9)**
Sulfuric acid, chromium(3+) potassium salt (2:1:1) **(10141-00-1)**
Sulfuric acid, chromium(3+)potassium salt(2:1:1), dodecahydrate **(7788-99-0)**
Sulfuric acid, chromium(3+) salt (3:2) **(10101-53-8)**
Sulfuric acid, cobalt(2+) salt (1:1) **(10124-43-3)**
Sulfuric acid, cobalt(2+) salt (1:1), heptahydrate (8CI,9CI) **(10026-24-1)**
Sulfuric acid, copper complex (9CI) **(12400-75-8)**
Sulfuric acid, copper(2+) salt (1:1) **(7758-98-7)**
Sulfuric acid, copper salt, basic (9CI) **(1344-73-6)**
Sulfuric acid, copper(2+) salt, pentahydrate **(7758-99-8)**
Sulfuric acid, copper(2+) salt (1:1), pentahydrate (8CI,9CI) **(7758-99-8)**
Sulfuric acid, decyl ester, sodium salt **(142-87-0)**
Sulfuric acid, diammonium salt **(7783-20-2)**
Sulfuric acid, diethyl ester **(64-67-5)**
Sulfuric acid, diisopropyl ester **(2973-10-6)**
Sulfuric acid, dilithium salt **(10377-48-7)**
Sulfuric acid, dimethyl ester **(77-78-1)**
Sulfuric acid, dipotassium salt **(7778-80-5)**
Sulfuric acid, disodium salt **(7757-82-6)**
Sulfuric acid, dithallium(1+) salt (8ci,9CI) **(7446-18-6)**
Sulfuric acid, dodecyl ester, triethanolamine salt **(139-96-8)**
Sulfuric acid, 5,8,11-heptadecatrienyl methyl ester (9CI) **(56554-67-7)**
Sulfuric acid, indium salt **(13464-82-9)**
Sulfuric acid, ion(2-) **(14808-79-8)**
Sulfuric acid, iron salt (1:1) **(7720-78-7)**
Sulfuric acid, iron(2+) salt (1:1) **(7720-78-7)**
Sulfuric acid, iron(3+) salt (3:2) **(10028-22-5)**
Sulfuric acid, lauryl ester, ammonium salt **(2235-54-3)**
Sulfuric acid, lead(2+) salt (1:1) **(7446-14-2)**
Sulfuric acid, lead salt (9CI) **(15739-80-7)**
Sulfuric acid, lithium salt (1:2) **(10377-48-7)**
Sulfuric acid, magnesium salt (1:1), Compd. with 2,2'-dithiobis-(pyridine) 1,1'-oxide **(43143-11-9)**

Sulfuric acid, manganese (II) salt (1:1) (8CI) **(7785-87-7)**
Sulfuric acid, manganese(2+) salt (1:1), monohydrate **(10034-96-5)**
Sulfuric acid, mercury(2+) salt (1:1) **(7783-35-9)**
Sulfuric acid, monoammonium salt **(7803-63-6)**
Sulfuric acid, monodecyl ester, sodium salt **(142-87-0)**
Sulfuric acid, monododecyl ester **(151-41-7)**
Sulfuric acid, monododecyl ester, Compd. with 2,2'-iminobis-(ethanol) (1:1) (9CI) **(143-00-0)**
Sulfuric acid, monododecyl ester, Compd. with 2,2'-imino-diethanol(1:1) **(143-00-0)**
Sulfuric acid, monododecyl ester, ammonium salt **(2235-54-3)**
Sulfuric acid, monododecyl ester, Compd. with 2,2',2''-nitrilo-triethanol (1:1) **(139-96-8)**
Sulfuric acid, monododecyl ester, Compd. with 2,2',2''-nitrilotris-(ethanol) **(139-96-8)**
Sulfuric acid, monododecyl ester, magnesium salt (9CI) **(3097-08-3)**
Sulfuric acid, monododecyl ester, sodium salt **(151-21-3)**
Sulfuric acid, mono(2-(2-(2-(dodecyloxy)ethoxy)ethoxy)ethyl) ester, sodium salt **(13150-00-0)**
Sulfuric acid, mono(2-(2-(dodecyloxy)ethoxy)ethyl) ether sodium salt **(3088-31-1)**
Sulfuric acid, mono(2-ethylhexyl)ester, sodium salt (8CI) **(126-92-1)**
Sulfuric acid, monomethyl ester, sodium salt (9CI) **(512-42-5)**
Sulfuric acid, monooctadecyl ester (9CI) **(143-03-3)**
Sulfuric acid, monooctadecyl ester, ammonium salt (9CI) **(4696-46-2)**
Sulfuric acid, monooctadecyl ester, sodium salt **(1120-04-3)**
Sulfuric acid, monooctyl ester, sodium salt **(142-31-4)**
Sulfuric acid, monopotassium salt **(7646-93-7)**
Sulfuric acid, monosodium salt **(7681-38-1)**
Sulfuric acid, monotetradecyl ester, Compd. with 2,2',2''-nitrilo-tris(ethanol) (1:1) **(4492-78-8)**
Sulfuric acid, monotetradecyl ester, sodium salt **(1191-50-0)**
Sulfuric acid, myristyl ester, sodium salt **(1191-50-0)**
Sulfuric acid, nickel(2+) salt (1:1) **(7786-81-4)**
Sulfuric acid, nickel(2+)salt **(7786-81-4)**
Sulfuric acid, nickel(2+) salt, hexahydrate **(10101-97-0)**
Sulfuric acid, strontium salt (1:1) (9CI) **(7759-02-6)**
Sulfuric acid, thallium salt **(10031-59-1)**
Sulfuric acid, thallium(1+) salt (1:2) **(7446-18-6)**
Sulfuric acid, thallium(2+) salt **(63906-56-9)**
Sulfuric acid, titanium(4+) salt (2:1) (9CI) **(13693-11-3)**
Sulfuric acid, zinc salt (1:1) **(7446-19-7)**
Sulfuric acid, zinc salt (1:1) **(7733-02-0)**
Sulfuric acid, zinc salt, Basic (9CI) **(68813-94-5)**
Sulfuric acid, zinc salt (1:1), heptahydrate **(7446-20-0)**
Sulfuric acid, zirconium(4+) salt (2:1) **(14644-61-2)**
Sulfuric anhydride (DOT) **(7446-11-9)**
Sulfuric chlorohydrin **(7790-94-5)**
Sulfuric oxide **(7446-11-9)**
Sulfuric oxychloride **(7791-25-5)**
Sulfuric oxyfluoride **(2699-79-8)**
Sulfurine **(144-82-1)**
Sulfur, ion (S(2-)) **(18496-25-8)**
Sulfur monochloride **(12771-08-3)**
Sulfur monochloride (ACGIH,OSHA) **(10025-67-9)**
Sulfuro amonico (Spanish) **(9080-17-5)**
Sulfuro sodico (Spanish) **(1344-08-7)**
Sulfurous acid [UN 1833] **(7782-99-2)**
Sulfurous acid anhydride **(7446-09-5)**
Sulfurous acid, 2-(p-tert-butylphenoxy)cyclohexyl-2-propynyl ester **(2312-35-8)**
Sulfurous acid, 2-(2-(p-tert-butylphenoxy)-1-methylethoxy)-1-methylethyl-, 2-chloroethyl ester (8CI) **(3761-60-2)**
Sulfurous acid, 2-(p-t-butylphenoxy)-1-methylethyl-2-chloro-ethyl ester **(140-57-8)**
Sulfurous acid, 2-(p-tert-butylphenoxy)-1-methylethyl-2-chloro-ethyl ester **(140-57-8)**
Sulfurous acid, 2-chloroethyl-, 2-(2-(4-(1,1-dimethylethyl)-phenoxy)-1-methylethoxy)-1-methylethyl ester (9CI) **(3761-60-2)**
Sulfurous acid, 2-chloroethyl-, 2-(4-(1,1-dimethylethyl)-phenoxy)-1-methylethyl ester **(140-57-8)**
Sulfurous acid, cyclic ester with 1,4,5,6,7,7-hexachloro-5-norbornene-2,3-dimethanol **(115-29-7)**
Sulfurous acid, diammonium salt (9CI) **(10196-04-0)**
Sulfurous acid, diethyl ester (8CI,9CI) **(623-81-4)**
Sulfurous acid, 2-(4-(1,1-dimethylethyl)phenoxy)cyclohexyl 2-propynyl ester **(2312-35-8)**
Sulfurous acid, dipotassium salt (9CI) **(10117-38-1)**
Sulfurous acid, ion(2-) **(14265-45-3)**
Sulfurous acid, monoammonium salt **(10192-30-0)**
Sulfurous acid, monopotassium salt (9CI) **(7773-03-7)**
Sulfurous acid, monosodium salt **(7631-90-5)**
Sulfurous acid, potassium salt **(10117-38-1)**
Sulfurous acid, sodium salt (1:2) **(7757-83-7)**
Sulfurous acid, zinc salt (1:1) (9CI) **(13597-44-9)**
Sulfurous anhydride **(7446-09-5)**
Sulfurous dichloride **(7719-09-7)**
Sulfurous oxide **(7446-09-5)**
Sulfurous oxychloride **(7719-09-7)**
Sulfurous oxyfluoride **(7783-42-8)**
Sulfur oxide **(7446-09-5)**
Sulfur pentafluoride (ACGIH,OSHA) **(5714-22-7)**
Sulfur phosphide **(1314-80-3)**
Sulfur selenide **(7446-34-6)**
Sulfur subchloride **(10025-67-9)**
Sulfur tetrafluoride (ACGIH,OSHA) [UN 2418] **(7783-60-0)**
Sulfur trioxide (DOT) **(7446-11-9)**
Sulfur trioxide, Stabilized [UN 1829] **(7446-11-9)**
Sulfuryl chloride [UN 1834] **(7791-25-5)**
Sulfuryl-fluoride **(2699-79-8)**
Sulfuryl fluoride (ACGIH,OSHA) [UN 2191] **(2699-79-8)**
Sulgin **(57-67-0)**
Sterculic acid **(738-87-4)**
Sulindac **(38194-50-2)**
Sterculinic acid **(738-87-4)**
Sulisobenzone **(4065-45-6)**
Sulka **(1344-81-6)**
Sulkol **(7704-34-9)**
Sulmet **(57-68-1)**
Sulodyne **(136-40-3)**
Sulourea **(62-56-6)**
Suloxsol **(127-69-5)**
Sulphabutin **(55-98-1)**
Sulphacetamide **(144-80-9)**
Sulphadiazine **(68-35-9)**
Sulphadimethylisoxazole **(127-69-5)**
Sulphadimethylpyrimidine **(57-68-1)**
Sulphadimidine **(57-68-1)**
Sulphadione **(80-08-0)**
Sulphafuraz **(127-69-5)**
Sulphafurazol **(127-69-5)**
Sulphafurazole **(127-69-5)**
Sulphafurazolum **(127-69-5)**
Sulphaguanidine **(57-67-0)**
Sulphaisoxazole **(127-69-5)**
Sulphamethalazole **(723-46-6)**
Sulphamethazine **(57-68-1)**
Sulphamethizole **(144-82-1)**
Sulphamethoxazol **(723-46-6)**
Sulphamethoxazole **(723-46-6)**
Sulpha-methoxizole **(723-46-6)**
Sulphamethoxypyridazine **(80-35-3)**
Sulphamethylisoxazole **(723-46-6)**
Sulphamic acid **(5329-14-6)**

Name	CAS No.
Sulphan Blue	**(129-17-9)**
Sulphanilamide	**(63-74-1)**
5-Sulphanilamido-3,4-dimethyl-isoxazole	**(127-69-5)**
3-Sulphanilamido-5-methylisoxazole	**(723-46-6)**
Sulphanilic acid	**(121-57-3)**
Sulphapyridine	**(144-83-2)**
Sulphasalazine	**(599-79-1)**
Sulphasil	**(144-80-9)**
Sulphathiazole	**(72-14-0)**
Sulpheimide	**(2425-06-1)**
Sulphenazole	**(526-08-9)**
Sulphenone	**(80-00-2)**
Sulphental	**(143-74-8)**
Sulphinpyrazone	**(57-96-5)**
Sulphisomezole	**(723-46-6)**
Sulphisoxazol	**(127-69-5)**
Sulpho Green 2B	**(4680-78-8)**
para-Sulphobenzoic acid	**(636-78-2)**
2-Sulphobenzoic imide	**(81-07-2)**
Sulphobenzoic imide calcium salt	**(6485-34-3)**
Sulphobenzoic imide, sodium salt	**(128-44-9)**
Sulphobromophthalein	**(71-67-0)**
Sulphobromophthalein sodium	**(71-67-0)**
Sulphocarbonic anhydride	**(75-15-0)**
Sulphofurazole	**(127-69-5)**
Sulpholane	**(126-33-0)**
Sulphon Acid Blue R	**(3861-73-2)**
Sulphon Acid Blue RA	**(3861-73-2)**
Sulphon-Mere	**(80-08-0)**
Sulphon Yellow RS-CF	**(6375-55-9)**
1-(4-Sulpho-1-naphthylazo)-2-naphthol-3,6-disulphonic acid, trisodium salt	**(915-67-3)**
1-(4-Sulpho-1-naphthylazo)-2-naphthol-6,8-disulphonic acid, trisodium salt	**(2611-82-7)**
2-(4-Sulpho-1-naphthylazo)-1-naphthol-4-sulphonic acid, disodium salt	**(3567-69-9)**
Sulphonol Red PG	**(3567-65-5)**
Sulphonol Red R	**(6459-94-5)**
Sulphonthal	**(143-74-8)**
1,1'-Sulphonylbis(4-aminobenzene)	**(80-08-0)**
4,4'-Sulphonylbisbenzamine	**(80-08-0)**
p,p-Sulphonylbisbenzamine	**(80-08-0)**
4,4'-Sulphonylbisbenzenamine	**(80-08-0)**
p,p-Sulphonylbisbenzenamine	**(80-08-0)**
4,4'-Sulphonyldianiline	**(80-08-0)**
Sulphonyldianiline	**(80-08-0)**
p,p-Sulphonyldianiline	**(80-08-0)**
4-p-Sulphophenylazo-1-naphthol, monosodium salt	**(523-44-4)**
1-p-Sulphophenylazo-2-naphthol-6-sulphonic acid, disodium salt	**(2783-94-0)**
Sulphos	**(56-38-2)**
Sulphoxaline	**(126-33-0)**
Sulphoxide	**(120-62-7)**
2-(6-Sulpho-2,4-xylylazo)-1-naphthol-4-sulphonic acid, disodium salt	**(4548-53-2)**
Sulphur (DOT)	**(7704-34-9)**
Sulphur, Lump or powder (DOT)	**(7704-34-9)**
Sulphur, Molten [UN 2448]	**(7704-34-9)**
Sulphur Mustard	**(505-60-2)**
Sulphur Mustard Gas	**(505-60-2)**
Sulphur dioxide	**(7446-09-5)**
Sulphur dioxide, Liquefied [UN 1079]	**(7446-09-5)**
Sulphuric acid	**(7664-93-9)**
Sulphuric acid, Fuming (DOT)	**(8014-95-7)**
Sulphuric acid, cadmium salt (1:1)	**(10124-36-4)**
Sulphurous acid (DOT)	**(7782-99-2)**
Sulphuryl chloride (DOT)	**(7791-25-5)**
Sulphuryl fluoride (DOT)	**(2699-79-8)**

Name	CAS No.
Sulpirid	**(15676-16-1)**
Sulpiride	**(15676-16-1)**
Sulprim	**(8064-90-2)**
Sulprofos (ACGIH,OSHA)	**(35400-43-2)**
Sulpyrid	**(15676-16-1)**
Sulquin	**(59-40-5)**
Sulsol	**(7704-34-9)**
Sulsoxin	**(127-69-5)**
Sultirene	**(80-35-3)**
Sulzol	**(72-14-0)**
Sumapen VK	**(132-98-9)**
Sumatra Benzoin	**(9000-05-9)**
Sumatra Camphor	**(507-70-0)**
Sumetrolim	**(8064-90-2)**
Sumiacryl Blue 6G	**(55840-82-9)**
Sumiacryl Orange G	**(3056-93-7)**
Sumiacryl Red G	**(14097-03-1)**
Sumiacryl Red GT	**(14097-03-1)**
Sumi-alfa	**(66230-04-4)**
Sumi-α	**(66230-04-4)**
Sumicidin	**(51630-58-1)**
Sumicidin A α	**(66230-04-4)**
Sumicure M	**(101-77-9)**
Sumifly	**(51630-58-1)**
Sumikaron Blue S-BG	**(12222-78-5)**
Sumikaron Blue E-BL	**(12217-79-7)**
Sumikaron Blue E-FBL	**(12217-79-7)**
Sumikaron Blue E-GRL	**(12222-79-6)**
Sumikaron Blue R	**(12217-79-7)**
Sumikaron Red E-FBL	**(17418-58-5)**
Sumikaron Red S-GG	**(58051-96-0)**
Sumikathene	**(9002-88-4)**
Sumikathene F 101-1	**(9002-88-4)**
Sumikathene F 210-3	**(9002-88-4)**
Sumikathene F 702	**(9002-88-4)**
Sumikathene G 201	**(9002-88-4)**
Sumikathene G 202	**(9002-88-4)**
Sumikathene G 701	**(9002-88-4)**
Sumikathene G 801	**(9002-88-4)**
Sumikathene G 806	**(9002-88-4)**
Sumikathene Hard 2052	**(9002-88-4)**
Sumilex	**(32809-16-8)**
Sumilight Black G	**(6428-31-5)**
Sumilight Red 4B	**(2610-11-9)**
Sumilight Supra Brown BRS	**(16071-86-6)**
Sumilit BBM	**(85-60-9)**
Sumilit PCX	**(9003-22-9)**
Sumilizer BBM	**(85-60-9)**
Sumilizer BHT	**(128-37-0)**
Sumilizer WX	**(96-69-5)**
Sumilizer WX-R	**(96-69-5)**
Sumine 2015	**(103-83-3)**
Sumine 2079	**(98-84-0)**
Suminol Brilliant Scarlet DH	**(10169-02-5)**
Suminol Fast Yellow G	**(12217-38-8)**
Suminol Levelling Sky Blue R	**(4368-56-3)**
Suminol Milling Red RS	**(6459-94-5)**
Suminol Milling Yellow MR	**(6375-55-9)**
Suminol Red PG	**(3567-65-5)**
Sumioxon	**(2255-17-6)**
Sumioxone	**(2255-17-6)**
Sumipex B-LG	**(9011-14-7)**
Sumipex B-MH	**(9011-14-7)**
Sumipex-B MHD	**(9011-14-7)**
Sumipex LG	**(9011-14-7)**
Sumipex LO	**(9011-14-7)**
Sumipex MHO	**(9011-14-7)**

Name	CAS No.
Sumipower	(51630-58-1)
Sumirez 614	(9011-05-6)
Sumisclex	(32809-16-8)
Sumital	(57-43-2)
Sumitekkusu Rejin 810	(9011-05-6)
Sumitex 260	(9011-05-6)
Sumitex 810	(9011-05-6)
Sumitex FSK	(1854-26-8)
Sumitex H 10	(9002-89-5)
Sumitex NF 113	(9011-05-6)
Sumitex NS	(1854-26-8)
Sumitex NS 1SPE	(1854-26-8)
Sumitex Resin 810	(9011-05-6)
Sumithian	(122-14-5)
Sumithion	(122-14-5)
Sumithion S-isomer	(3344-14-7)
Sumithrin	(26002-80-2)
Sumitol	(26259-45-0)
Sumitol 80W	(26259-45-0)
Sumitomo S-1102A	(122-14-5)
Sumitomo Fast Scarlet G	(10169-02-5)
Sumitomo Light Green SF Yellowish	(5141-20-8)
Sumitomo Patent Pure Blue VX	(129-17-9)
Sumitomo S 4084	(2636-26-2)
Sumitomo SP-1103	(7696-12-0)
Sumitomo Wool Blue SBC	(860-22-0)
Sumitox	(121-75-5)
Sumitrin	(26002-80-2)
Summit	(55219-65-3)
Sumox	(26787-78-0)
Sumycin	(64-75-5)
Sun Block Gel	(13110-37-7)
Sun Orange A Geigy	(2783-94-0)
Sun Screen Gel	(13110-37-7)
Sun Yellow	(2783-94-0)
Sun Yellow A-CE	(2783-94-0)
Sun Yellow A-FDC	(2783-94-0)
Sun Yellow Extra Conc. A Export	(2783-94-0)
Sun Yellow Extra Pure A	(2783-94-0)
Sun Yellow FCF	(2783-94-0)
Sun Yellow G	(1325-37-7)
Sunaptic Acid B	(1338-24-5)
Sunaptic Acid C	(1338-24-5)
Sunbrella	(150-13-0)
Sunburst Red	(1248-18-6)
Sunchromine Violet B	(2092-55-9)
Suncide	(114-26-1)
Sunflower LS	(8001-21-6)
Sunflower Oil	(8001-21-6)
Sunflower Seed Oil	(8001-21-6)
Sunitomo S 4084	(2636-26-2)
Sunlife TN	(1401-55-4)
Sunmorl S 1500	(9005-00-9)
Sunoxol	(134-31-6)
Sunset Yellow	(2783-94-0)
Sunset Yellow BSS	(2783-94-0)
Sunset Yellow FCF	(2783-94-0)
Sunset Yellow FCF Supra	(2783-94-0)
Sunset Yellow FU	(2783-94-0)
Sunset Yellow FU Supra	(2783-94-0)
Sunset Yellow Lake	(2783-94-0)
Sunsoft O 30B	(25496-72-4)
Sunsolt WA	(9014-90-8)
Suntol	(56-72-4)
Sunwax 151	(9002-88-4)
Sup 'R Flo	(12427-38-2)
Sup'Operats	(28772-56-7)
Supacal	(41826-92-0)
Supeol	(25496-72-4)
Super Ad-IT	(27236-65-3)
Super Amide L-9A	(120-40-1)
Super Amide L-9C	(120-40-1)
Super Anahist	(62-44-2)
Super-Caid	(28772-56-7)
Super-Carbovar	(1333-86-4)
Super Cobalt	(7440-48-4)
Super Cosan	(7704-34-9)
Super Crab-E-Rad-Calar	(5902-95-4)
Super D	(8068-77-7)
Super D Weedone	(93-76-5)
Super D Weedone	(94-75-7)
Super Dal-E-Rad-Calar	(5902-95-4)
Super Dal-E-Rad	(5902-95-4)
Super-De-Sprout	(123-33-1)
Super-Dent	(7681-49-4)
Super Dylan	(9002-88-4)
Super Fast Red GRE	(6408-31-7)
Super Hartolan	(57-88-5)
Super Mosstox	(97-23-4)
Super Prodan	(16893-85-9)
Super-Rozol	(28772-56-7)
Super-Spectra	(1333-86-4)
Super Sprout Stop	(123-33-1)
Super Sucker-Stuff	(123-33-1)
Super Sucker-Stuff HC	(123-33-1)
Super VMP	(8030-30-6)
Superacryl AE	(9011-14-7)
Superacryl O	(9011-14-7)
Superan Blue AR	(2666-17-3)
Superba	(1333-86-4)
Supercel 3000	(57-13-6)
Superchlone CR 5	(9006-03-5)
Superchlone CR 10	(9006-03-5)
Superchlone CR 20	(9006-03-5)
Superchrome Violet B	(2092-55-9)
Supercol	(9000-40-2)
Supercol G.F.	(9000-30-0)
Supercol U Powder	(9000-30-0)
Supercortil	(53-03-2)
Superflake Anhydrous	(10043-52-4)
Superfloc	(25038-59-9)
Superfloc 127 (9CI)	(39393-20-9)
Supergan	(50-55-5)
Superian Yellow R	(6373-74-6)
Superior Dri-Die	(16919-19-0)
Superlan Astrol B Base	(128-85-8)
Superloid	(9005-34-9)
Superlysoform	(50-00-0)
Superman Maneb F	(12427-38-2)
Supernox	(709-98-8)
Superol	(134-31-6)
Superol	(56-81-5)
Superol Red C RT-265	(5160-02-1)
Superormone Concentre	(94-75-7)
Superox	(94-36-0)
Superoxide	(11062-77-4)
Superoxide Anion	(11062-77-4)
Superoxide Radical	(11062-77-4)
Superoxido potasico (Spanish)	(12030-88-5)
Superoxido sodico (Spanish)	(12034-12-7)
Superoxol	(7722-84-1)
Superoxyde de potassium (French)	(12030-88-5)
Superoxyde de sodium (French)	(12034-12-7)
Superpalite	(76-02-8)

Superphosphoric acid	(8017-16-1)
Superprednol	(50-02-2)
Super rodiatox	(56-38-2)
Superseptil	(57-68-1)
Supersorbon IV	(7440-44-0)
Supersorbon S 1	(7440-44-0)
Supertah	(8007-45-2)
Supertite 10	(1306-06-5)
Supona	(470-90-6)
Supone	(470-90-6)
Supotran	(80-77-3)
Suppoptanox	(125-42-8)
Sup'r Flo	(330-54-1)
Sup'r-Flo Ferbam Flowable	(14484-64-1)
Supra	(1309-37-1)
Supracapsulin	(51-43-4)
Supracet Brilliant Blue BG	(2475-46-9)
Supracet Brilliant Blue BG	(86722-66-9)
Supracet Brilliant Blue 2GN	(2475-45-8)
Supracet Brilliant Red 2B	(116-85-8)
Supracet Brilliant Violet 3R	(128-95-0)
Supracet Deep Blue R	(2475-45-8)
Supracet Diazo Black A	(539-17-3)
Supracet Fast Blue 2G	(2475-44-7)
Supracet Fast Pink 3B	(2872-48-2)
Supracet Fast Violet B	(82-33-7)
Supracet Fast Yellow G	(2832-40-8)
Supracet Fast Yellow 2R	(119-15-3)
Supracet Fast Yellow 4R	(6300-37-4)
Supracet Orange R	(82-28-0)
Supracet Pink R	(82-38-2)
Supracet Violet 2B	(1220-94-6)
Supracet Yellow RR	(119-15-3)
Supracid	(950-37-8)
Supracide	(950-37-8)
Supradin	(51-43-4)
Supralgin	(62-44-2)
Supramike	(7727-43-7)
Supramycin	(64-75-5)
Supranefran	(51-43-4)
Supranephrane	(51-43-4)
Supranephrine	(51-43-4)
Supranephrin solution	(55-31-2)
Supranol	(51-43-4)
Supranol Cyanine G	(6104-58-1)
Supranol Fast Red GG	(6459-94-5)
Supranol Fast Scarlet GN	(3567-65-5)
Supranol Red PBX-CF	(6459-94-5)
Supranol Red PG-CF	(3567-65-5)
Supranol Red R	(6459-94-5)
Supranol Scarlet BN	(3567-65-5)
Supranol Scarlet GS	(10169-02-5)
Supranol Yellow R	(6375-55-9)
Suprarenaline	(51-43-4)
Suprarenin	(51-42-3)
Suprarenin	(51-43-4)
Suprarenin hydrochloride	(55-31-2)
Suprasil	(60676-86-0)
Suprathen	(9002-88-4)
Suprathen C 100	(9002-88-4)
Suprazo Brown BRL	(16071-86-6)
Suprazo Red 4B	(2610-11-9)
Suprel	(51-43-4)
Supremal	(523-87-5)
Supreme	(14807-96-6)
Supreme Dense	(14807-96-6)
Suprexcel Brown BRL	(16071-86-6)
Suprexcel Red 8BL	(2610-11-9)
Suprifen-PSB	(447-41-6)
Suprimal	(569-65-3)
Suprin	(8064-90-2)
Supronic B 50	(9003-11-6)
Supronic B 75	(9003-11-6)
Supronic E 400	(9003-11-6)
Supronic E 800	(9003-11-6)
Suprotan	(80-77-3)
Surchlor	(7681-52-9)
Surco SXS	(1300-72-7)
Surcopur	(709-98-8)
Sure-Set	(122-88-3)
Surecide	(13067-93-1)
Surem	(146-22-5)
Surenine	(51-43-4)
Surfactant WK	(9002-92-0)
Surflan	(19044-88-3)
Surflo HS 1	(9003-11-6)
Surfonic JL 80X (9CI)	(120026-55-3)
Surfonic LF 17 (9CI)	(70431-21-9)
Surfynol 104	(126-86-3)
Surfynol 104A	(126-86-3)
Surfynol 61	(107-54-0)
Surfynol 104E	(126-86-3)
Surfynol (VAN)	(107-54-0)
Surgex	(51-12-7)
Surgex	(814-71-1)
Surgical Simplex	(9011-14-7)
Surgical Simplex P	(9011-14-7)
Surgi-cen	(70-30-4)
Surgi-cin	(70-30-4)
Surinam Greenheart Wood	(84-79-7)
Surirene	(80-35-3)
Surmontil	(739-71-9)
Surmontil 315695	(739-71-9)
Surofene	(70-30-4)
Surpass	(1929-77-7)
Surplix	(50-49-7)
Surpur	(709-98-8)
Surrenine	(51-43-4)
Sursumid	(15676-16-1)
Su seguro carpidor	(1582-09-8)
Suspen	(132-98-9)
Suspendol	(315-30-0)
Susphrine	(51-43-4)
Sustamycin	(64-75-5)
Sustane	(128-37-0)
Sustane	(1948-33-0)
Sustane	(25013-16-5)
Sustane BHT	(128-37-0)
Sustane 1-F	(25013-16-5)
Sustanone	(58-22-0)
Sustar	(47000-92-0)
Sustar 2S	(47000-92-0)
Susvin	(6923-22-4)
Sutan	(2008-41-5)
Suticide	(57-09-0)
Suturamid	(63428-84-2)
Suxamethonium	(306-40-1)
Suxemethonium	(306-40-1)
Suxil	(110-61-2)
Suxilep	(77-67-8)
Suximal	(77-67-8)
Suxin	(77-67-8)
Suxinutin	(77-67-8)
Suy-B 2	(7439-89-6)

Suzorite mica **(12001-26-2)**
Suzu **(900-95-8)**
Suzu H **(76-87-9)**
Svitpren **(9010-98-4)**
Swanox BHT **(128-37-0)**
Swartziol **(520-18-3)**
Swascol 3L **(151-21-3)**
Swascol 4L **(151-21-3)**
Swascol 1P **(151-21-3)**
Swasolve ETB **(7580-85-0)**
Swat **(122-10-1)**
Swebate **(3383-96-8)**
Sweden: Karotin **(7235-40-7)**
Swedish Black Lead **(7782-42-5)**
Swedish Green **(10290-12-7)**
Swedish Green **(12002-03-8)**
Sweet-Gum **(1401-55-4)**
Sweeney's Ant-Go **(7631-89-2)**
Sweep **(1897-45-6)**
Sweep **(1910-42-5)**
Sweet Almond LS **(8001-21-6)**
Sweet Birch Oil **(119-36-8)**
Sweet Dipeptide **(22839-47-0)**
Sweeta **(128-44-9)**
Sweetened hydrotreated light aromatic solvent naphtha **(64741-87-3)**
Swep **(1918-18-9)**
Swiss Blue **(61-73-4)**
Sybol 2 **(29232-93-7)**
Sydnone, 4-bromo-3-phenyl- (8CI,9CI) **(13183-09-0)**
Sykose **(128-44-9)**
Sykose **(81-07-2)**
Sylantoic **(57-41-0)**
Sylantoic **(630-93-3)**
Sylgard 1107-69 **(63148-62-9)**
Sylgard 182-62B **(63148-62-9)**
Sylgard 184-31B **(63148-62-9)**
Sylgard 184-36B **(63148-62-9)**
Syllit **(2439-10-3)**
Sylodex **(12001-29-5)**
Syloid **(63231-67-4)**
Syloid 63 **(63231-67-4)**
Syloid 266 **(63231-67-4)**
Syltherm 800 **(63148-62-9)**
Sylvan **(534-22-5)**
Sylvic acid **(514-10-3)**
Symazine **(122-34-9)**
Symclosen **(87-90-1)**
Symclosene **(87-90-1)**
Symetryczna dwumetylohydrazyna (Polish) **(540-73-8)**
Symetryne **(1014-70-6)**
Symmetrel **(768-94-5)**
Symmetric dimethylurea **(96-31-1)**
Sympametin **(60-13-9)**
Sympamin **(51-63-8)**
Sympamina-D **(51-63-8)**
Sympamine **(300-62-9)**
Sympamine **(60-13-9)**
Sympatedrine **(300-62-9)**
Sympatedrine **(60-13-9)**
Sympatektoman **(71-91-0)**
Sympathin E **(51-41-2)**
Sympathin I **(51-43-4)**
m-Sympathol **(59-42-7)**
m-Sympathol **(61-76-7)**
meta-Sympathol **(61-76-7)**
Sympatholytin **(51-50-3)**
m-Sympatol **(59-42-7)**
meta-Sympatol **(61-76-7)**
Symphytine **(22571-95-5)**
Symuler Eosin Toner **(17372-87-1)**
Symuler Fast Blue 6011 **(81-77-6)**
Symuler Fast Orange GRD **(4424-06-0)**
Symuler Fast Pyrazolone Orange G **(3520-72-7)**
Symuler Fast Scarlet 4R **(2425-85-6)**
Symuler Fast Violet R **(1324-55-6)**
Symuler Fast Yellow GF **(6358-85-6)**
Symuler Lake Red C **(5160-02-1)**
Symuler Orange Lake 43 **(633-96-5)**
Symuler Red 2R Ba Salt **(1103-38-4)**
Symulex Blue BF **(1325-87-7)**
Symulex Magenta F **(81-88-9)**
Symulex Pink F **(81-88-9)**
Symulex Rhodamine B Toner F **(81-88-9)**
Symulon Acid Brilliant Scarlet 3R **(2611-82-7)**
Symulon Acid Fast Yellow MR **(6375-55-9)**
Symulon Acid Orange II **(633-96-5)**
Symulon Acid Red PG **(3567-65-5)**
Symulon Chrome Violet B **(2092-55-9)**
Symulon Direct Black BH **(2429-73-4)**
Symulon Metanil Yellow **(587-98-4)**
Symulon Orange GC Base **(141-85-5)**
Symulon Red B Base **(97-52-9)**
Symulon Red 3GL Salt **(89-63-4)**
Symulon Red RL Base **(99-52-5)**
Symulon Scarlet 2G Base **(95-82-9)**
Symulon Scarlet G Base **(99-55-8)**
Synacril Blue 5G **(55840-82-9)**
Synacril Blue G **(12270-13-2)**
Synacril Blue R **(12217-41-3)**
Synacryl Fast Red 2G **(14097-03-1)**
Synacryl Red 2G **(14097-03-1)**
Synacthen **(16960-16-0)**
Synadrin **(390-64-7)**
Synalar **(67-73-2)**
Synalgos-DC **(62-44-2)**
Synalgos **(60-87-7)**
Synalogos **(62-44-2)**
Synamol **(67-73-2)**
Synandone **(67-73-2)**
Synandrets **(58-18-4)**
Synandrol **(57-85-2)**
Synandrol F **(58-22-0)**
Synandrone **(67-73-2)**
Synandrotabs **(58-18-4)**
Synapause **(50-27-1)**
Synapen **(132-93-4)**
Synaphorin **(9002-61-3)**
Synarel **(86220-42-0)**
Synasal **(61-76-7)**
Synasteron **(434-07-1)**
Syncaine **(51-05-8)**
Syncal **(81-07-2)**
Syncelose **(9004-67-5)**
Syncillin **(132-93-4)**
Syncl **(15686-71-2)**
Syncoumar **(152-72-7)**
Syncumar **(152-72-7)**
Syncurarine **(65-29-2)**
Syndiol **(50-28-2)**
Syndiotactic Polypropylene **(9003-07-0)**
Syndrox **(300-42-5)**
Synelaudine **(50-13-5)**
m-Synephrine **(59-42-7)**
meta-Synephrine **(61-76-7)**

meta-Synephrine hydrochloride	**(61-76-7)**
Synepirin 222	**(113-48-4)**
Synergid R	**(57-66-9)**
Synergist 264	**(113-48-4)**
Synerone	**(57-85-2)**
Synerpenin	**(132-93-4)**
Synestrin	**(130-80-3)**
Synestrin	**(56-53-1)**
Synestrol	**(84-16-2)**
Synestrol	**(84-17-3)**
Synethenate	**(61-76-7)**
Synfuels (VAN)	**(86290-81-5)**
Syngesterone	**(57-83-0)**
Syngestrets	**(57-83-0)**
Syngum D 46D	**(9000-30-0)**
Syngynon	**(630-56-8)**
Synhexyl	**(117-51-1)**
Synistamin	**(113-92-8)**
Synkay	**(58-27-5)**
Synklor	**(57-74-9)**
Synoestron	**(130-80-3)**
Synotol L-60	**(120-40-1)**
Synovex S	**(57-83-0)**
Synox 5LT	**(119-47-1)**
Synox TBC	**(98-29-3)**
Synpenin	**(69-53-4)**
Synperonic NX	**(9016-45-9)**
Synperonic OP	**(9036-19-5)**
Synperonic OP 10	**(9036-19-5)**
Synpitan	**(50-56-6)**
Synpol 1500	**(9003-55-8)**
Synpor	**(9004-70-0)**
Synpren-Fish	**(51-03-6)**
Synpro stearate	**(1592-23-0)**
Synpro stearate	**(557-05-1)**
Synsac	**(67-73-2)**
Synsatedrine	**(60-13-9)**
Synstigmin bromide	**(114-80-7)**
Synstigmine	**(51-44-5)**
Syntapon	**(151-21-3)**
Syntapon L	**(151-21-3)**
Syntapon L Pasta (Czech)	**(151-21-3)**
Syntar	**(8007-45-2)**
Syntarpen	**(61-72-3)**
Syntase 62	**(131-57-7)**
Syntase 100	**(131-56-6)**
Syntedril	**(58-73-1)**
Synten Yellow 2G	**(2832-40-8)**
Synten Yellow P 2R	**(119-15-3)**
Syntes 12A	**(64-02-8)**
Syntestrin	**(130-80-3)**
Syntestrine	**(130-80-3)**
Syntexan	**(67-68-5)**
Synthaline Green	**(1328-53-6)**
Synthecillin	**(132-93-4)**
Synthecilline	**(132-93-4)**
Synthemul 90-588	**(9003-01-4)**
Synthetic 3956	**(8001-35-2)**
Synthetic Eugenol	**(97-53-0)**
Synthetic Indigo	**(482-89-3)**
Synthetic Indigo TS	**(482-89-3)**
Synthetic Iron oxide	**(1309-37-1)**
Synthetic Mica	**(12003-38-2)**
Synthetic Mustard Oil	**(57-06-7)**
Synthetic Natural Gas	**(8006-14-2)**
Synthetic Pyrethrins	**(584-79-2)**
Synthetic Wintergreen Oil	**(119-36-8)**
Synthetic fibers, Polymeric, Polyamides	**(63428-84-2)**
Synthetic glycerin	**(56-81-5)**
Synthetic mordenite	**(66402-68-4)**
Synthetic oxytocin	**(50-56-6)**
Synthex P	**(84-80-0)**
Synthoestrin	**(56-53-1)**
Synthofolin	**(56-53-1)**
Synthomycetin	**(56-75-7)**
Synthomycetine	**(56-75-7)**
Synthomycine	**(56-75-7)**
Synthophylline	**(479-18-5)**
Synthos	**(7758-87-4)**
Synthostigmine bromide	**(114-80-7)**
Synthovo	**(84-16-2)**
Synthrin	**(10453-86-8)**
Syntocin	**(50-56-6)**
Syntocinon	**(50-56-6)**
Syntocinone	**(50-56-6)**
Syntodril	**(58-73-1)**
Syntofolin	**(56-53-1)**
Syntol HD (9CI)	**(83590-03-8)**
Syntolutan	**(57-83-0)**
Syntometrine	**(60-79-7)**
Synton Yellow 2G	**(2832-40-8)**
Syntopherol	**(59-02-9)**
Syntostigmin	**(51-44-5)**
Syntostigmin bromide	**(114-80-7)**
Syntostigmine bromide	**(114-80-7)**
Syntostigmin (injection)	**(51-44-5)**
Syntostigmin (tablet)	**(114-80-7)**
Syntrogene	**(84-16-2)**
Syntrom	**(152-72-7)**
Syntron B	**(64-02-8)**
Syntron C	**(140-01-2)**
Syphos	**(78-57-9)**
Syraprim	**(738-70-5)**
Syrfynol 104	**(126-86-3)**
Syrfynol 82	**(1321-87-5)**
Syringaldehyde	**(134-96-3)**
Syringealdehyde	**(134-96-3)**
Syringic acid	**(530-57-4)**
Syringic aldehyde	**(134-96-3)**
Syringol	**(91-10-1)**
Syringyl alcohol	**(530-56-3)**
Syringylaldehyde	**(134-96-3)**
Syrups, Corn	**(8029-43-4)**
Syrups, Hydrolyzed starch	**(8029-43-4)**
Sys 67ME	**(3653-48-3)**
Sys 67MPROP	**(1929-86-8)**
Systam	**(152-16-9)**
Systemox	**(8065-48-3)**
Systicide	**(144-41-2)**
Systodin	**(50-54-4)**
Systogene	**(51-67-2)**
Systophos	**(152-16-9)**
Systox (OSHA)	**(8065-48-3)**
Iso-Systox Sulfone	**(2496-91-5)**
Systox Sulfone	**(4891-54-7)**
Sytam	**(152-16-9)**
Sytasol	**(973-21-7)**
Sytobex	**(68-19-9)**
Sytobex-H	**(13422-51-0)**
Syton Blue B	**(1325-87-7)**
Syton Fast Orange G	**(3520-72-7)**
Syton Fast Red 2G	**(3468-63-1)**
Syton Fast Red R	**(2814-77-9)**
Syton Fast Scarlet RB	**(2425-85-6)**

Name	CAS No.
TCMTB	(2593-15-9)
TCNA	(2438-88-2)
TCNB	(117-18-0)
TCP	(1330-78-5)
TCP	(27355-22-2)
TCP	(58-90-2)
TCP	(6515-38-4)
TCPA	(85-34-7)
TCPE	(2122-77-2)
m-TCPN	(1897-45-6)
TCPO	(3083-23-6)
TCPP	(13674-87-8)
2,4,5-TCPPA	(93-72-1)
TCT	(9007-12-9)
TCTH	(13121-70-5)
TCTNB	(2631-68-7)
TCTP	(6012-97-1)
T-Cellit	(9004-35-7)
(T-4)Cerium hydroxide	(12014-56-1)
TC hydrochloride	(64-75-5)
TD 1771	(23564-05-8)
TD 1874	(35493-90-4)
TD-183	(6012-97-1)
TDA	(95-80-7)
TDBP (Czech)	(126-72-7)
TDBPP	(126-72-7)
TDCPP	(13674-87-8)
TDCPP	(78-43-3)
TDE	(1954-28-5)
o,p'-TDE	(53-19-0)
o,p-TDE	(53-19-0)
p,p'-TDE	(72-54-8)
TDE (DOT)	(72-54-8)
T-DET	(27193-86-8)
2,4-TDI	(584-84-9)
2,6-TDI	(91-08-7)
TDI	(18869-73-3)
TDI	(26471-62-5)
TDI 80-20	(26471-62-5)
TDI-80	(26471-62-5)
TDI-80	(584-84-9)
TDI (OSHA)	(584-84-9)
TDPA	(111-17-1)
2,5-TDS	(6369-59-1)
(T-4)-Disodium tetrakis(cyano-C)zincate(2-)	(15333-24-1)
TE	(315-37-7)
TE 194	(78744-33-9)
TE 33	(26266-58-0)
TEA	(102-71-6)
TEA	(71-91-0)
TEA	(97-93-8)
TEA chloride	(56-34-8)
TEA-dodecylbenzenesulfonate	(27323-41-7)
TEA lauryl sulfate	(139-96-8)
TEA-myristate	(41669-40-3)
TEA-oleate	(2717-15-9)
TEA-stearate	(4568-28-9)
TEA-sulfate	(7376-31-0)
TEAB	(71-91-0)
TEAC	(56-34-8)
TEC	(77-93-0)
TEDMA	(109-16-0)
TEDP (OSHA)	(3689-24-5)
TEDTP	(3689-24-5)
TEF	(545-55-1)
TEG	(112-27-6)
TEGDH	(70729-68-9)
TEGDN	(111-22-8)
TEGON	(111-22-8)
TEL	(78-00-2)
TEM	(51-18-3)
TEMED	(110-18-9)
TEM-Simes	(51-18-3)
TEMUR	(632-22-4)
TEM (Cytostatic)	(51-18-3)
TEN	(121-44-8)
TEOS	(78-10-4)
TEP	(107-49-3)
TEP	(78-40-0)
TEPA	(545-55-1)
TEPIC-G	(28825-96-9)
TEPP (ACGIH,OSHA)	(107-49-3)
2,4,5-TES	(3570-61-4)
TET	(51-18-3)
TET	(597-64-8)
TET	(997-50-2)
TETD	(97-77-8)
TF	(354-58-5)
TF 1169	(69806-50-4)
TFC	(369-77-7)
TFDU	(70-00-8)
TFE	(75-89-8)
T-Fluoride	(7681-49-4)
TG	(154-42-7)
TGA 2	(4074-88-8)
TGD 5161	(9003-53-6)
TGDR	(789-61-7)
β-TGDR	(789-61-7)
TGM 2	(2358-84-1)
TGM 3	(109-16-0)
TGM 4	(109-17-1)
TGM 3PC	(109-16-0)
TGM 3S	(109-16-0)
T-Gas	(75-21-8)
T-Gelb BZW. Grun 1	(117-39-5)
1314 TH	(536-33-4)
TH 052	(1982-55-4)
TH 1314	(536-33-4)
TH 346-1	(2597-03-7)
TH 367-1	(4301-50-2)
TH 6040	(35367-38-5)
TH/100	(317-34-0)
THA	(321-64-2)
THAM	(77-86-1)
THAM-E	(77-86-1)
THAM SET	(77-86-1)
THBP	(131-55-5)
THBP	(1421-63-2)
THC	(1972-08-3)
δ^1-THC	(1972-08-3)
δ^3-THC	(32904-22-6)
δ^9-THC	(1972-08-3)
THD	(67-03-8)
THEIC	(839-90-7)
THF	(109-99-9)
THFA	(97-99-4)
TH 052-H	(1982-55-4)
TH 073-H	(1953-89-5)
THP	(142-68-7)
THPA	(85-43-8)
THPC	(124-64-1)
THPOH	(512-82-3)
THPS	(55566-30-8)
THQ	(95-71-6)

Name	CAS
THS-839	(3734-33-6)
THU	(18771-50-1)
THU	(62-56-6)
THX	(51-48-9)
T4 (Hormone)	(51-48-9)
TI-1258	(15263-52-2)
TIB	(88-82-4)
2,3,5-TIBA	(88-82-4)
TIBA	(88-82-4)
TIC	(329-01-1)
TIC Mustard	(5034-77-5)
TJB	(86-30-6)
TK 10352	(2425-79-8)
TKB	(156-10-5)
TKED	(140-07-8)
TKPP	(7320-34-5)
TL 69	(696-28-6)
TL 70	(5714-22-7)
TL 78	(822-06-0)
TL 83	(2315-36-8)
TL 85	(100-22-1)
TL 86	(3563-36-8)
TL 145	(555-77-1)
TL 146	(51-75-2)
TL 156	(7784-35-2)
TL 189	(627-63-4)
TL 190	(681-84-5)
TL 207	(627-11-2)
TL 214	(598-14-1)
TL 262	(3982-91-0)
TL 264	(627-53-2)
TL 294	(593-89-5)
TL 314	(107-13-1)
TL 329	(538-07-8)
TL 337	(151-56-4)
TL 350	(109-75-1)
TL 389	(79-44-7)
TL 391	(16212-28-5)
TL 423	(541-41-3)
TL 438	(79-22-1)
TL 457	(51-83-2)
TL 465	(867-13-0)
TL 466	(55-91-4)
TL 476	(553-27-5)
TL 670	(359-06-8)
TL 741	(371-62-0)
TL 751	(462-27-1)
TL 792	(115-26-4)
TL 797	(77-77-0)
TL 819	(816-40-0)
TL 822	(506-68-3)
TL 855	(459-99-4)
TL 869	(62-74-8)
TL 898	(7487-94-7)
TL 906	(40709-82-8)
TL 907	(1115-15-7)
TL 944	(538-28-3)
TL 1026	(14486-19-2)
TL-1049	(2438-53-1)
TL 1149	(538-07-8)
TL 1163	(75-77-4)
TL 1450	(624-83-9)
TL 1473	(86-98-6)
TL 1578	(77-81-6)
TL 1618	(107-44-8)
TLA	(540-88-5)
TL-Azole	(127-69-5)
TLD 100	(7789-24-4)
TL4N	(111-46-6)
TM	(75-59-2)
TM 12008	(78-52-4)
TM 30	(1333-86-4)
TM 5	(2058-46-0)
TM-4049	(121-75-5)
TMA	(528-44-9)
TMA	(552-30-7)
TMA	(75-50-3)
TMAB	(75-22-9)
TMAN	(552-30-7)
TMB	(108-67-8)
TMB	(54827-17-7)
TMBAC	(56-93-9)
TM (9CI)	(95032-59-0)
TMDE 6500	(9003-53-6)
TME	(77-85-0)
TMEDA	(110-18-9)
3-TMI	(2094-99-7)
TML	(75-74-1)
5'-TMP	(365-07-1)
TMP	(365-07-1)
TMP	(512-56-1)
TMP	(77-99-6)
TMP	(9000-71-9)
TMPD	(100-22-1)
TMPD	(144-19-4)
TMPEA	(54-04-6)
TMP (Nucleotide)	(365-07-1)
TMP-P	(1005-93-2)
TMPTA	(15625-89-5)
TMP (Alcohol)	(77-99-6)
TMS 480	(8064-90-2)
TMSN	(3333-52-6)
TMTD	(137-26-8)
TMTDS	(137-26-8)
TMTM	(97-74-5)
TMTMS	(97-74-5)
TMTU	(2782-91-4)
TMU	(632-22-4)
T-2 Mycotoxin	(21259-20-1)
TN 80	(2939-80-2)
TNA	(3698-54-2)
TNB	(99-35-4)
TNCB	(88-88-0)
TNCS 53	(1332-03-2)
TNCS 53	(7758-98-7)
TNG	(55-63-0)
TN 3J	(78-04-6)
TNK 2G5	(25038-54-4)
TNM	(509-14-8)
α-TNT	(118-96-7)
TNT (OSHA)	(118-96-7)
TNT-tolite (French)	(118-96-7)
TO 10	(9005-65-6)
TOA BLA-S	(2079-00-7)
TOCE	(702-54-5)
TOCEN	(702-54-5)
TOCP	(78-30-8)
TOF	(78-42-2)
TOFK	(78-30-8)
TOK	(1836-75-5)
TOK-2	(1836-75-5)
TOK E	(1836-75-5)
TOK E 40	(1836-75-5)
TOK E-25	(1836-75-5)

TOK WP-50	(1836-75-5)	TU	(141-90-2)
TO NTU	(79-08-3)	TUS-1	(28981-97-7)
TOPE	(2303-25-5)	TUADS, ethyl	(97-77-8)
TOTM	(89-04-3)	TUR	(999-81-5)
TOTP	(78-30-8)	825TV	(9003-53-6)
2,4,5-TP	(93-72-1)	TVM 370	(9003-11-6)
TP	(57-85-2)	TVM 474	(151-21-3)
TP-21	(50-52-2)	TVS-MA 300	(78-04-6)
TP-95	(141-17-3)	TVS-N 2000E	(78-04-6)
TPA	(16561-29-8)	825TV-PS	(9003-53-6)
TPA (VAN)	(16561-29-8)	TVS 8105	(16091-18-2)
TP 90B	(143-29-3)	T-WD602	(354-58-5)
TPB	(143-66-8)	TX 100	(9002-93-1)
TPC	(78-32-0)	TXIB	(6846-50-0)
TPCP	(78-32-0)	TY 80	(14807-96-6)
2,4,5-TP (DOT)	(29990-39-4)	Tabalgin	(103-90-2)
2,4,5-T PGBE ester	(62922-39-8)	Tabatrex	(141-03-7)
TPIA	(3771-19-5)	Tabernanthe iboga	(83-74-9)
TPN-12	(14759-06-9)	Tabilin	(113-98-4)
TPN (Pesticide)	(1897-45-6)	Table Salt	(7647-14-5)
TPP	(115-86-6)	Tabloid	(154-42-7)
TPS23	(5588-33-0)	Taboon A	(77-81-6)
TPTA	(900-95-8)	Tabular Spar	(13983-17-0)
TPTC	(639-58-7)	Tabun	(77-81-6)
TPTH	(76-87-9)	Tabutrex	(141-03-7)
TPTOH	(76-87-9)	Tacaryl	(1229-35-2)
TPZA	(900-95-8)	Tacaryl	(1982-37-2)
TR	(108-78-1)	Tacazyl	(1982-37-2)
TR 201	(9003-55-8)	Tachigaren	(10004-44-1)
TR-495	(72-44-6)	Tachionin	(133-67-5)
TRI-6	(58-89-9)	Tachysterol, dihydro-(8CI)	(67-96-9)
TRIB	(723-46-6)	Tachysterol2, dihydro-	(67-96-9)
TRP	(73-22-3)	Tachystin	(67-96-9)
TRP-P-2	(62450-07-1)	Tackle	(62476-59-9)
TRP-P-1 (Acetate)	(68808-54-8)	Tackle 2AS	(62476-59-9)
TRP-P-2 (Acetate)	(72254-58-1)	Tackle 2S	(62476-59-9)
TS 160	(555-77-1)	Tacol	(62-44-2)
TS 219	(311-45-5)	Tacosal	(57-41-0)
TS-160	(817-09-4)	Tacosal	(630-93-3)
TS2	(9003-20-7)	Tacrine	(321-64-2)
TSA-HP	(104-15-4)	Tacryl	(1982-37-2)
TSA-MH	(104-15-4)	Tactaran	(113-59-7)
TSC	(79-19-6)	Tacumil	(8064-90-2)
TSF 431	(63148-62-9)	Tad	(317-34-0)
TSF 451	(63148-62-9)	Tadenan	(661-19-8)
TSF 451-1000	(63148-62-9)	Taeniatol	(97-23-4)
TSIZP 34	(62-56-6)	Tafasan	(123-88-6)
TSP	(7601-54-9)	Tafasan 6W	(123-88-6)
TSPA	(52-24-4)	Tafazine	(122-34-9)
TSPP	(7722-88-5)	Tafazine 50-W	(122-34-9)
TST	(64-02-8)	Tafclean	(71-55-6)
T-Serp	(50-55-5)	Tagamet	(51481-61-9)
T-State Hemoglobin	(9008-02-0)	Tagat	(9005-38-3)
T-Stuff	(7722-84-1)	Tagathen	(148-65-2)
TT87	(54-95-5)	Tahmabon	(10265-92-6)
TTA	(326-91-0)	Taifen	(12071-83-9)
TTB	(326-91-0)	Taifun	(66441-11-0)
TTD	(137-26-8)	Taigu Wood	(84-79-7)
TTD	(97-77-8)	Taiguic acid	(84-79-7)
L-TTP	(73-22-3)	Takamina	(51-43-4)
TTS	(97-77-8)	Takamine	(51-43-4)
TTS 5	(77-08-7)	Takanarumin	(315-30-0)
TTT	(368-66-1)	Takaoka Amaranth	(915-67-3)
TTX	(4368-28-9)	Takaoka Brilliant Scarlet 3R	(2611-82-7)
T-2 Toxin	(21259-20-1)	Takaoka Metanil Yellow	(587-98-4)
2,4,5-T-Tris(2-hydroxyethyl)ammonium	(3813-14-7)	Takaoka Rhodamine B	(81-88-9)
2-TU	(141-90-2)	Takathene	(9002-88-4)

Takathene P 12 (9002-88-4)
Takathene P 3 (9002-88-4)
Takimetol (443-48-1)
Takineocol (67-63-0)
Taktic (33089-61-1)
Taladren (58-54-8)
Talamo (57-43-2)
Talampicilina (Spanish) (47747-56-8)
Talampicillin (47747-56-8)
Talampicilline (French) (47747-56-8)
Talampicillinum (Latin) (47747-56-8)
Talan (973-21-7)
Talargan (50-35-1)
Talatrol (77-86-1)
Talbot (3687-31-8)
Talbot (7784-40-9)
Talbutal (115-44-6)
Talc (ACGIH,OSHA) (14807-96-6)
Talc, Containing asbestos fibers (14807-96-6)
Talc (Powder), Containing no asbestos fibers (14807-96-6)
Talcan PK-P (14807-96-6)
Talcord (52645-53-1)
Talcron CP 44-31 (14807-96-6)
Talculin Z (557-05-1)
Talcum (14807-96-6)
Talidomida (Spanish) (50-35-1)
Talidomide (50-35-1)
Talimol (50-35-1)
Talin (53850-34-3)
Talismol (50-35-1)
Tall Oil (8002-26-4)
Tall Oil Rosin (8002-26-4)
Tall Oil Rosin and Fatty Acids (8002-26-4)
Talleol (8002-26-4)
Tall oil acid (61790-12-3)
Tall oil acids (61790-12-3)
Tall oil acids, ethoxylated (61791-00-2)
Tall oil acids, phthalic anhydride, benzoic acid, pentaerythritol resin (66070-84-6)
2-(Tall oil alkyl)-1-benzyl-1-(2-hydroxyethyl)-2-imidazolinium chloride (68309-34-2)
2-(Tall oil alkyl)-1-(2-hydroxyethyl)-2-imidazoline (61791-39-7)
Tall oil, copper salt (61789-22-8)
Tall oil, diethylenetriamine imidazoline (68442-97-7)
Tall oil fatty acid, benzoic acid, phthalic anhydride, pentaerythritol polymer (66070-84-6)
Tall oil fatty acid, esters, ethoxylated (61791-00-2)
Tall oil fatty acid ethoxylate (61791-00-2)
Tall oil fatty acid, ethoxylated (61791-00-2)
Tall oil fatty acid, glycerine, phthalic anhydride resin (66070-71-1)
Tall oil fatty acid, glycerin, phthalic anhydride polymer (66070-71-1)
Tall oil fatty acid, glycerol, phthalic anhydride alkyd resin (66070-71-1)
Tall oil fatty acid monoester of polyethylene glycol (61791-00-2)
Tall oil fatty acid, pentaerythritol, benzoic acid, phthalic anhydride resin (66070-84-6)
Tall oil fatty acid pentaerythritol, phthalic anhydride alkyd resin benzoic acid modified (66070-84-6)
Tall oil fatty acid, pentaerythritol, phthalic anhydride, benzoic acid polymer (66070-84-6)
Tall oil fatty acid, phthalic anhydride, glycerin polymer (66070-71-1)
Tall oil fatty acid, phthalic anhydride, glycerol polymer (66070-71-1)
Tall oil fatty acid, polyethylene glycol ester (61791-00-2)
Tall oil fatty acids, 1,2-benzenedicarboxylic anhydride, glycerin polymer (66070-71-1)
Tall oil fatty acids, glycerine, phthalic anhydride polymer (66070-71-1)
Tall oil fatty acids, glycerin, phthalic anhydride alkyd resin (66070-71-1)
Tall oil fatty acids, glycerin, phthalic anhydride polymer (66070-71-1)
Tall oil fatty acids, pentaerythritol, benzoic acid, phthalic anhydride polymer (66070-84-6)
Tall oil fatty acids, pentaerythritol, phthalic anhydride, benzoic acid alkyd resin (66070-84-6)
Tall oil fatty acids, pentaerythritol, phthalic anhydride, benzoic acid polymer (66070-84-6)
Tall oil fatty acids, phthalic acid, glycerol polymer (66070-71-1)
Tall oil fatty acids, phthalic anhydride, benzoic acid, pentaerythritol polymer (66070-84-6)
Tall oil fatty acids, phthalic anhydride, glycerin polymer (66070-71-1)
Tall oil fatty acids, phthalic anhydride, glycerol polymer (66070-71-1)
Tall oil fatty acids, phthalic anhydride pentaerythritol, benzoic acid resin (66070-84-6)
Tall oil fatty acids, phthalic anhydride, pentaerythritol, benzoic acid polymer (66070-84-6)
Tall oil fatty acids, phthalic anhydride, pentaerythritol, benzoic acid resin (66070-84-6)
Tall oil fatty acids, polyethylene glycol ester (61791-00-2)
Tall oil hydroxyethyl imidazoline (61791-39-7)
Tall oil imidazoline (61791-36-4)
Tall oil, imidazoline deriv. (61791-36-4)
Tall oil monoester of polyethylene glycol (61791-00-2)
Tallol (8002-26-4)
Tallow (61789-97-7)
Tallow, Beef (61789-97-7)
Tallow, Mutton (61789-97-7)
Tallow acid (61790-37-2)
N-((Tallow alkyl)amidoethyl)-N-((tallow alkyl)ethyl)methyl ethylammonium, methosulfate (68153-35-5)
1-((Tallow alkyl)amino)-3-aminopropane diacetate (68911-78-4)
1-(2-Tallow amidoethyl)-1-methyl-2-nor(tallow alkyl)-2-imidazolinium methyl sulfate (68122-86-1)
1-(2-Tallowamidoethyl)-1-methyl-2-tallowalkylimidazolinium methylsulfate (68122-86-1)
Tallow benzyl dimethyl ammonium chloride (122-19-0)
Tallow fatty acid (61790-37-2)
Tallow nitrile (61790-28-1)
(Tallow)trimethylammonium chloride (8030-78-2)
N-Tallow-trimethylammonium surfactant (8030-78-2)
Talmon (118-71-8)
Talodex (55-38-9)
Talofloc (107-64-2)
β.-D-Talofuranose, 1,2:5,6-bis-O-(1-methylethylidene)- (9CI) (23262-79-5)
Talofuranose, 1,2:5,6-di-O-isopropylidene-, .β.-D- (8CI) (23262-79-5)
Talofuranose, 2,3:5,6-di-O-isopropylidene-, .α.-D- (8CI) (23262-78-4)
Talon (56073-10-0)
Talon Rodenticide (56073-10-0)
Talpheno (50-06-6)
Talwan (359-83-1)
Talwin (359-83-1)
Tamariz (28249-77-6)
Tamaron (10265-92-6)
Tamate (57-53-4)
Tametin (51481-61-9)
Tamex (33629-47-9)
2,4,5-T amines (6369-96-6)
Tampovagan stilboestrol (56-53-1)
Tampules (127-65-1)
Tamraghol (1332-40-7)
2,4,5-T amyl ester (120-39-8)
Tanafol (80-77-3)
Tanager Red X-761 (2814-77-9)
Tanakan (54-05-7)
Tanaphen P 500 (1401-55-4)
Tandacote (129-20-4)
Tandalgesic (129-20-4)
Tandearil (129-20-4)

Tanderal (129-20-4)
Tanderil (129-20-4)
Tandex (4849-32-5)
Tangantangan Oil (8001-79-4)
Tangarine Lake X-917 (633-96-5)
Tangerine-Oil (8016-85-1)
Tanidil (60-87-7)
Tanipent (78-11-5)
Tannex (53-86-1)
Tannic-acid (1401-55-4)
Tannic acids (1401-55-4)
Tannin (1401-55-4)
Tannin from Chestnut (1401-55-4)
Tannin from Mimosa (1401-55-4)
Tannin from Quebracho (1401-55-4)
Tannin from Sweet GUM (1401-55-4)
Tannins (1401-55-4)
Tanone (2597-03-7)
Tanrutin (153-18-4)
Tanston (61-68-7)
Tantalic acid anhydride (1314-61-0)
Tantalum-181 (7440-25-7)
Tantalum (ACGIH) (7440-25-7)
Tantalum, Metal (OSHA) (7440-25-7)
Tantalum-chloride (7721-01-9)
Tantalum-fluoride (7783-71-3)
Tantalum oxide (ACGIH) (8CI,9CI) (1314-61-0)
Tantalum oxide dust (1314-61-0)
Tantalum pentachloride (7721-01-9)
Tantalum penta oxide (1314-61-0)
Tantalum pentoxide (1314-61-0)
Tantalum potassium fluoride (16924-00-8)
Tanzene (4849-32-5)
Taomycin (79-57-2)
Taomyxin (79-57-2)
Tapar (103-90-2)
Tapazole (60-56-0)
Taphazine (122-34-9)
Tapioca Starch (9005-25-8)
Tapon (9005-25-8)
Tapuzole (60-56-0)
Tar (8007-45-2)
Tar Acids, Coal, Crude (65996-85-2)
Tar Acids, Cresylic, C8-rich, phosphates (68952-33-0)
Tar Acids, Cresylic, Ph phosphates (68952-35-2)
Tar Acids, Cresylic Residues (68555-24-8)
Tar Acids, Cresylic, Sodium Salts, Caustic Solns. (68815-21-4)
Tar Camphor (91-20-3)
Tar, Coal (8007-45-2)
Tar, Liquid [UN 1999] (8007-45-2)
Tar Oil (8001-58-9)
Tar Oils (8002-29-7)
Tar Oils, Coal (65996-82-9)
Tar, Pine (8011-48-1)
Tara Gum (39300-88-4)
Taractan (113-59-7)
Tarapacaite (7789-00-6)
Tarapon K 12 (151-21-3)
Tarasan (113-59-7)
Tar bases, Coal, Crude (65996-84-1)
Tarcortin (50-23-7)
Tardex 100 (1163-19-5)
Tardigal (71-63-6)
Tardocillin (1538-09-6)
Targa (76578-14-8)
Target MSMA (2163-80-6)
Tarichatoxin (4368-28-9)

Tarimyl (80-08-0)
Tarlon X-A (105-60-2)
Tarlon X-A (25038-54-4)
Tarlon XB (105-60-2)
Tarlon XB (25038-54-4)
Tarnamid T (105-60-2)
Tarnamid T (25038-54-4)
Tarnamid T 2 (105-60-2)
Tarnamid T 2 (25038-54-4)
Tarnamid T 27 (105-60-2)
Taroctyl (69-09-0)
Tarpamid T 27 (25038-54-4)
Tarquinor (8067-69-4)
Tarragon (140-67-0)
Tartar (868-14-4)
Tartar Cream (868-14-4)
Tartar Emetic (28300-74-5)
Tartar Yellow FS (1934-21-0)
Tartar Yellow N (1934-21-0)
Tartar Yellow PF (1934-21-0)
Tartar Yellow S (1934-21-0)
(+-)-Tartaric acid (133-37-9)
D-Tartaric acid (147-71-7)
DL-Tartaric acid (133-37-9)
Tartaric-acid (87-69-4)
l-(+)-Tartaric acid (87-69-4)
Tartaric acid, (+-)- (8CI) (133-37-9)
Tartaric acid D,L (133-37-9)
L-Tartaric acid, ammonium salt (3164-29-2)
DL-Tartaric acid, antimony potassium salt (64070-12-8)
L-Tartaric acid, antimony potassium salt (11071-15-1)
Tartaric acid, antimony potassium salt (28300-74-5)
Tartaric acid, copper(2+) salt (1:1), (+)- (8CI) (815-82-7)
Tartaric acid, diammonium salt (3164-29-2)
Tartaric acid, dibutyl ester (8CI) (87-92-3)
Tartaric acid, dimethyl ester, meso- (8CI) (5057-96-5)
Tartaric acid, disodium salt (868-18-8)
Tartaric acid, lead(2+) salt (1:1) (8CI) (815-84-9)
Tartaric acid, monopotassium monosodium salt (304-59-6)
Tartaric acid, monopotassium salt (868-14-4)
Tartaric acid, monosodium salt (526-94-3)
Tartaric acid, strontium salt (1:1) (8CI) (868-19-9)
Tartarized Antimony (28300-74-5)
Tartox (28300-74-5)
Tartran Yellow (1934-21-0)
Tartraphenine (1934-21-0)
Tartrate antimonio-potassique (French) (28300-74-5)
Tartrated antimony (28300-74-5)
Tartrate de nicotine (French) (65-31-6)
Tartrazine (1934-21-0)
Tartrazine A Expo T (1934-21-0)
Tartrazine B (1934-21-0)
Tartrazine B.P.C. (1934-21-0)
Tartrazine Extra Pure A (1934-21-0)
Tartrazine FD & C Yellow #5 (1934-21-0)
Tartrazine FQ (1934-21-0)
Tartrazine G (1934-21-0)
Tartrazine Lake (1934-21-0)
Tartrazine Lake Yellow N (1934-21-0)
Tartrazine M (1934-21-0)
Tartrazine MCGL (1934-21-0)
Tartrazine N (1934-21-0)
Tartrazine NS (1934-21-0)
Tartrazine O (1934-21-0)
Tartrazine T (1934-21-0)
Tartrazine XX (1934-21-0)
Tartrazine XXX (1934-21-0)

Name	CAS
Tartrazine Yellow	(1934-21-0)
Tartrazol BPC	(1934-21-0)
Tartrazol Yellow	(1934-21-0)
Tartrine Yellow O	(1934-21-0)
Tarzol	(14255-88-0)
Task	(62-73-7)
Task Tabs	(62-73-7)
Tat Chlor 4	(57-74-9)
Taterpex	(101-21-3)
Tathion	(70-18-8)
Tathione	(70-18-8)
Tatren 141	(9003-07-0)
Tatren EB 111	(9003-07-0)
Tattoo	(22781-23-3)
Taturil	(396-01-0)
Taurine	(107-35-7)
Taurine, N,N-bis(2-hydroxyethyl)-	(10191-18-1)
Taurine, N-choloyl- (8CI)	(81-24-3)
Taurine, N-choloyl-, sodium salt	(145-42-6)
Taurine, N-lauroyl-N-methyl-, sodium salt (8CI)	(4337-75-1)
Taurine, N-methyl- (8CI)	(107-68-6)
Taurine, N-methyl-N-oleoyl-	(97-80-3)
Taurine, N-methyl-N-oleoyl-, sodium salt	(137-20-2)
Taurocholate	(81-24-3)
Taurocholate sodium	(145-42-6)
Taurocholate sodium salt	(145-42-6)
Taurocholic acid	(81-24-3)
Taurocholic acid sodium salt	(145-42-6)
Tauroglycocholic acid	(11006-55-6)
Tavor	(846-49-1)
Tax-11-en-9-one, 5-β,20-epoxy-1,2-α,4,7-β,10-β,13-α-hexahydroxy-, 4,10- diacetate 2-benzoate 13-ester with (2R,3S)-N-benzoyl-3-phenylisoserine	(33069-62-4)
Taxicatigenin	(500-99-2)
Taxifolin	(480-18-2)
Taxifoliol	(480-18-2)
Taxin (German)	(12607-93-1)
Taxine	(12607-93-1)
Taxol	(33069-62-4)
Tayssato	(123-88-6)
Tazepam	(604-75-1)
Tazone	(50-33-9)
2,4,5-T butoxyethanol ester	(2545-59-7)
2,4,5-T butoxyethyl ester	(2545-59-7)
2,4,5-T butyl ester	(93-79-8)
2,4,5-T n-butyl ester	(93-79-8)
2,4,5-T n-butyl ester Mixed with 2,4-d n-butyl ester	(39277-47-9)
T-250 capsules	(64-75-5)
Tea LS	(8001-21-6)
Teaberry Oil	(119-36-8)
Tebacon	(466-90-0)
Tebacona (Spanish)	(466-90-0)
Tebacone	(466-90-0)
Tebecid	(54-85-3)
Tebenic	(54-85-3)
Teberus	(536-33-4)
Tebexin	(54-85-3)
Tebilon	(54-85-3)
Tebos	(54-85-3)
Tebrazid	(98-96-4)
Tebulan	(34014-18-1)
Tebuthiuron	(34014-18-1)
Tech DDT	(50-29-3)
Tech Pet F	(8012-95-1)
Techlofthalam	(76280-91-6)
Technetium	(7440-26-8)
Technetium, Isotope of mass 99 (8CI,9CI)	(14133-76-7)
Technetium Tc 99M Sulfur Colloid	(7704-34-9)
90 Technical glycerine	(56-81-5)
Techtoquinone	(84-54-8)
Tecnazen (German)	(117-18-0)
Tecnazene	(117-18-0)
Tecomin	(84-79-7)
Tecpol	(9003-01-4)
Tecquinol	(123-31-9)
Tecsol	(64-17-5)
Tectilon Orange 3GT	(6373-74-6)
Tectilon Red 2B	(61931-22-4)
Tecto	(148-79-8)
Tecto 60	(148-79-8)
Tecto RPH	(148-79-8)
Tectochinon	(84-54-8)
Tectoquinone	(84-54-8)
Tecza	(112-24-3)
Tedion	(116-29-0)
Tedion V-18	(116-29-0)
Teebaconin	(54-85-3)
Teepol CH 53	(50642-03-0)
Teepol CH 31 (9CI)	(50642-02-9)
Teepol CH 610 (9CI)	(50958-32-2)
Tefamin	(317-34-0)
Tefamin	(58-55-9)
Tefaserpina	(50-55-5)
Tefilan	(479-18-5)
Tefilin	(64-75-5)
Tefluthrin	(79538-32-2)
Tefluthrine	(79538-32-2)
Tegester	(110-27-0)
Tegester 504-D	(6938-94-9)
Tegester Isopalm	(142-91-6)
Tegester butyl stearate	(123-95-5)
Tegiloxan 50	(63148-62-9)
Tegin	(123-94-4)
Tegin	(31566-31-1)
Tegin 503	(31566-31-1)
Tegin 515	(123-94-4)
Tegin 515	(31566-31-1)
Tegin 55G	(123-94-4)
Tegin P	(1323-39-3)
Teginex FP 90	(63148-62-9)
Tegison	(54350-48-0)
Tego-Oleic 130	(112-80-1)
Tego-Stearate	(111-60-4)
Tegolan	(57-88-5)
Tegopen	(1397-89-3)
Tegosept B	(94-26-8)
Tegosept E	(120-47-8)
Tegosept M	(99-76-3)
Tegosept P	(94-13-3)
Tegosept butyl	(94-26-8)
Tegostearic 254	(57-11-4)
Tegostearic 255	(57-11-4)
Tegostearic 272	(57-11-4)
Tegretal	(298-46-4)
Tegretol	(298-46-4)
Tegretol	(51-52-5)
Teib	(68-76-8)
Tekazin	(54-85-3)
Tekkam	(86-87-3)
Teknazen (Czech)	(117-18-0)
Tekresol	(1319-77-3)
Tekwaisa	(298-00-0)
Telagan	(50-35-1)
Telar	(64902-72-3)

Name	CAS No.
Telargan	**(50-35-1)**
Telargean	**(50-35-1)**
Telcotene	**(9002-88-4)**
Teldrin	**(113-92-8)**
Telecothene	**(9002-88-4)**
Telefos	**(2275-18-5)**
Telepaque	**(96-83-3)**
Teleprin	**(8064-90-2)**
Telesmin	**(298-46-4)**
Teletrast	**(96-83-3)**
Telfairic acid	**(60-33-3)**
Telidal	**(129-20-4)**
Teline	**(64-75-5)**
Telipex	**(57-85-2)**
Telloy	**(13494-80-9)**
Tellur (Polish)	**(13494-80-9)**
Tellurac	**(20941-65-5)**
Telluric acid, disodium salt (8CI,9CI)	**(10102-20-2)**
Telluric acid, lead(2+) salt (1:1) (8CI,9CI)	**(13845-35-7)**
Telluric chloride	**(10026-07-0)**
Tellurium (ACGIH,OSHA)	**(13494-80-9)**
Tellurium(IV) diethyl dithiocarbamate	**(20941-65-5)**
Tellurium, Isotope of mass 131 (8CI,9CI)	**(14683-12-6)**
Tellurium bis(diethyldithiocarbamate)	**(20941-65-5)**
Tellurium-chloride	**(10026-07-0)**
Tellurium chloride, (T-4)-	**(10026-07-0)**
Tellurium diethyldithiocarbamate	**(20941-65-5)**
Tellurium hexafluoride (ACGIH,OSHA) [UN 2195]	**(7783-80-4)**
Tellurium tetrachloride	**(10026-07-0)**
Tellurium, tetrakis(diethylcarbamodithioato-S,S')-	**(20941-65-5)**
Tellurium, tetrakis(diethylcarbamodithioato-S,S')- (9CI)	**(20941-65-5)**
Tellurium, tetrakis(diethylcarbamodithioato-S,S')-, (DD-8-111''1''1'1'1'''1''')-	**(20941-65-5)**
Tellurium, tetrakis(diethyldithiocarbamato) (8CI)	**(20941-65-5)**
Tellurium, tetrakis(diethyldithiocarbamato)- (8CI)	**(20941-65-5)**
Tellurous acid, disodium salt	**(10102-20-2)**
Telmicid	**(514-73-8)**
Telmid	**(514-73-8)**
Telmide	**(514-73-8)**
Telmin	**(31431-39-7)**
Telodrin	**(297-78-9)**
Telon Blue RRL	**(4368-56-3)**
Telon Fast Black E	**(1937-37-7)**
Telon Fast Red GG	**(6459-94-5)**
Telon Fast Scarlet N	**(3567-65-5)**
Telone	**(542-75-6)**
Telone	**(8003-19-8)**
Telone II	**(542-75-6)**
Telone II Soil Fumigant	**(542-75-6)**
Telotrex	**(64-75-5)**
Teltozan	**(62-54-4)**
Telvar	**(150-68-5)**
Telvar	**(330-54-1)**
Telvar Diuron Weed Killer	**(330-54-1)**
Telvar Monuron Weedkiller	**(150-68-5)**
Telvar W. Monuron Weedkiller	**(150-68-5)**
Tem-Histine	**(135-23-9)**
Temasept	**(87-10-5)**
Temasept	**(87-12-7)**
Temasept II	**(87-10-5)**
Temasept IV	**(87-10-5)**
Temazepam	**(846-50-4)**
Temefos	**(3383-96-8)**
Tementil	**(58-38-8)**
Temephos (ACGIH,OSHA)	**(3383-96-8)**
Temesta	**(846-49-1)**
Temgesic	**(52485-79-7)**
Temic	**(116-06-3)**
Temik	**(116-06-3)**
Temik 10 G	**(116-06-3)**
Temik G10	**(116-06-3)**
Temik Oxime	**(1646-75-9)**
Temik Sulfoxide	**(1646-87-3)**
Temlo	**(103-90-2)**
Temophos	**(3383-96-8)**
Tempo-Reserpina	**(50-55-5)**
Tempanal	**(103-90-2)**
Temparin	**(66-76-2)**
Tempasept II	**(87-10-5)**
Tempo	**(68359-37-5)**
Tempodex	**(51-63-8)**
Temponitrin	**(55-63-0)**
Temposerpine	**(50-55-5)**
Tempra	**(103-90-2)**
Temus	**(28772-56-7)**
Temus W	**(81-81-2)**
Ten	**(78-11-5)**
Tenac	**(62-73-7)**
Tenalin	**(91-80-5)**
Tenamene	**(3081-14-9)**
Tenamene 2	**(101-96-2)**
Tenamene 3	**(128-37-0)**
Tenamene 31	**(139-60-6)**
Tenamene 60	**(94-91-7)**
Tenamine	**(24815-24-5)**
Tenaplas	**(9002-88-4)**
Tendearil	**(129-20-4)**
Tendimethalin	**(40487-42-1)**
Tendoscen-compr.	**(50-55-5)**
Tendust	**(54-11-5)**
Tenemene 30	**(103-96-8)**
Tenfidil	**(91-79-2)**
Teniathane	**(97-23-4)**
Teniatol	**(97-23-4)**
Tenite 423	**(9003-07-0)**
Tenite 423DF	**(9003-07-0)**
Tenite 800	**(9002-88-4)**
Tenite 1811	**(9002-88-4)**
Tenite 2910	**(9002-88-4)**
Tenite 2918	**(9002-88-4)**
Tenite 3300	**(9002-88-4)**
Tenite 3340	**(9002-88-4)**
Tenite 4231	**(9003-07-0)**
Tenite Butyrate Formula 264 H4	**(9004-36-8)**
Tenite I	**(9004-35-7)**
Tenite P 7673-079F	**(9003-07-0)**
Tenn-Plas	**(65-85-0)**
Tennessee Brand Tri-Basic Copper Sulfate	**(1332-03-2)**
Tennus 0565	**(9003-22-9)**
Tenociclidina (Spanish)	**(21500-98-1)**
Tenocyclidine	**(21500-98-1)**
Tenocyclidinum (Latin)	**(21500-98-1)**
Tenoran	**(1982-47-4)**
Tenormin	**(29122-68-7)**
Tenosin-wirkstoff	**(51-67-2)**
Tenox BHA	**(25013-16-5)**
Tenox BHT	**(128-37-0)**
Tenox HQ	**(123-31-9)**
Tenox PG	**(121-79-9)**
Tenox P Grain Preservative	**(79-09-4)**
Tenox TBHQ	**(1948-33-0)**
Tensanyl	**(50-55-5)**
Tenserlix	**(50-55-5)**
Tenserpine (Assia)	**(50-55-5)**

Tenserpinie (50-55-5)
Tensional (50-55-5)
Tensionorme (50-55-5)
Tensival (50-35-1)
Tensol (57-53-4)
Tensol 7 (9011-14-7)
Tensonal (57-53-4)
Tensopam (439-14-5)
1,2,3,6,7-Tentachlorodibenzofuran (57117-42-7)
2,3,4,6,7-Tentachlorodibenzofuran (57117-43-8)
Tentoxin (28540-82-1)
Tentrate-20 (78-11-5)
Tenurid (97-77-8)
Tenutex (97-77-8)
Teobromin (83-67-0)
Teodramin (523-87-5)
Teofilina (Polish) (58-55-9)
Teofyllamin (58-55-9)
Teoharn (108-78-1)
Teolaxin (50-06-6)
Teolit (968-93-4)
Teoloxin (50-06-6)
Teonanacatl (520-52-5)
Tepanil (90-84-6)
Tepidone (136-30-1)
Tepidone Rubber Accelerator (136-30-1)
Tepilta (126-27-2)
Tepserpine (50-55-5)
Tequinol (123-31-9)
Terabol (74-83-9)
Teralen (84-96-8)
Teralin (135-23-9)
Teralutil (630-56-8)
Teramethyl thiuram disulfide (137-26-8)
Teranol (6673-35-4)
Teraprint (17418-58-5)
Terasil Black S-RL (9CI) (59948-01-5)
Terasil Blue R (12222-78-5)
Terasil Brilliant Blue 3RL (12217-79-7)
Terasil Brilliant Pink 4BN (2872-48-2)
Terasil Brilliant Violet 3B (82-33-7)
Terasil Yellow GBA Extra (2832-40-8)
Terasil Yellow 2GC (2832-40-8)
Terbacil (5902-51-2)
Terbenol (128-62-1)
Terbenzene (26140-60-3)
Terbium (9CI) (7440-27-9)
Terbolan (50-55-5)
Terbucarb (1918-11-2)
Terbuconazole (107534-96-3)
Terbufos (13071-79-9)
Terbufos sulfone (56070-16-7)
Terbufos sulfoxide (10548-10-4)
Terbumeton (33693-04-8)
Terbutalin (23031-25-6)
Terbutaline (23031-25-6)
Terbuthylazine (5915-41-3)
Terbutol (1918-11-2)
Terbutrex (886-50-0)
Terbutryn (886-50-0)
Terbutryne (886-50-0)
Terbutylethylazine (5915-41-3)
Tercinin (149-29-1)
Tercyl (63-25-2)
Terebenthine (French) (8006-64-2)
Terephtahlic acid-ethylene glycol polyester (25038-59-9)
Terephtaldehyde (623-27-8)
Terephtaldehydes (French) (623-27-8)
Terephthalaldehyde (623-27-8)
Terephthalaldehydic-acid (619-66-9)
Terephthalaldehydic acid, methyl ester (1571-08-0)
Terephthalamic acid, N-octadecyl-, monosodium salt (5994-45-6)
Terephthalic acid-1,4-butanediol polymer (26062-94-2)
Terephthalic-acid (100-21-0)
Terephthalic acid, bis(2-ethylhexyl)ester (6422-86-2)
Terephthalic acid, bis(2-hydroxyethyl) ester (959-26-2)
Terephthalic acid chloride (100-20-9)
Terephthalic acid, copper salt (8CI) (34262-89-0)
Terephthalic acid, dibutyl ester (1962-75-0)
Terephthalic acid dichloride (100-20-9)
Terephthalic acid, diethyl ester (636-09-9)
Terephthalic acid, dimethyl ester (120-61-6)
Terephthalic acid, diphenyl ester (8CI) (1539-04-4)
Terephthalic acid, 2,5-di-p-toluidino- (8CI) (10291-28-8)
Terephthalic acid methyl ester (120-61-6)
Terephthalic acid, monomethyl ester (8CI) (1679-64-7)
Terephthalic acid, polyester with 1,4-butanediol (26062-94-2)
Terephthalic acid, potassium salt (8CI) (13427-80-0)
Terephthalic acid, tetrachloro- (8CI) (2136-79-0)
Terephthalic acid, tetrachloro-, dimethyl ester (1861-32-1)
Terephthalic acid, tetrachloro-, methyl ester (887-54-7)
Terephthalic acid, tetrachloro-, monomethyl ester (8CI) (887-54-7)
Terephthalic acid, tetrachlorothio-, O,S-dimethyl ester (8CI) (3765-57-9)
Terephthalic acid-tetramethylene glycol copolymer (26062-94-2)
Terephthalic acid-tetramethylene glycol polymer (26062-94-2)
Terephthalic aldehyde (623-27-8)
Terephthalic dichloride (100-20-9)
Terephthaloyl-chloride (100-20-9)
Terephthaloyl chloride, tetrachloro- (719-32-4)
Terephthaloyldiacetic acid, diethyl ester (93-94-7)
Terephthaloyl dichloride (100-20-9)
Tereton (79-20-9)
Terfan (25038-59-9)
Terfluzine (117-89-5)
Tergal (25038-59-9)
Tergemist (126-92-1)
Tergentol (3088-31-1)
Tergimist (126-92-1)
Tergitol (139-88-8)
Tergitol 08 (126-92-1)
Tergitol 4 (139-88-8)
Tergitol 7 (3282-85-7)
Tergitol Anionic 08 (126-92-1)
Tergitol Anionic 4 (139-88-8)
Tergitol Anionic P-28 (141-65-1)
Tergitol NP-14 (9016-45-9)
Tergitol NP-27 (9016-45-9)
Tergitol NP-35 (9016-45-9)
Tergitol NP-40 (9016-45-9)
Tergitol NP-33 (Nonionic) (9016-45-9)
Tergitol NPX (9016-45-9)
Tergitol No 7 (3282-85-7)
Tergitol Nonionic XD (9038-95-3)
Tergitol 12-P-9 (9014-92-0)
Tergitol Penetrant 4 (139-88-8)
Tergitol 15-S-3 (68131-40-8)
Tergitol 15-S-5 (68131-40-8)
Tergitol 15-S-9 (68131-40-8)
Tergitol 15-S-15 (68131-40-8)
Tergitol 15-S-20 (68131-40-8)
Tergitol 15S (68131-40-8)
Tergitol 15-S-7 (Nonionic) (68131-40-8)
Tergitol 15-S-9 (Nonionic) (68131-40-8)
Tergitol 15-S-12 (Nonionic) (68131-40-8)

Name	CAS
Tergitol TP-9	**(9016-45-9)**
Tergitol XD (Nonionic)	**(9038-95-3)**
Tergitol anionic 7	**(1120-01-0)**
Teriam	**(396-01-0)**
Teric 12A	**(9002-92-0)**
Teric GX 13	**(9036-19-5)**
Teric PE	**(9003-11-6)**
Teric PE 40	**(9003-11-6)**
Teric PE 60	**(9003-11-6)**
Teric PE 61	**(9003-11-6)**
Teric PE 62	**(9003-11-6)**
Teric PE 68	**(9003-11-6)**
Teric PE 70	**(9003-11-6)**
Teric SF	**(9004-99-3)**
Terics	**(27986-36-3)**
Teridin	**(396-01-0)**
Teridox	**(50563-36-5)**
Terinin	**(149-29-1)**
Terlon (Fiber)	**(63428-84-2)**
Term-I-Trol	**(87-86-5)**
Termi-Ded	**(57-74-9)**
Termil	**(1897-45-6)**
Termitkil	**(95-50-1)**
Termofleks A (Czech)	**(101-70-2)**
Termosolido Green FG Supra	**(1328-53-6)**
Termosolido Red LCG	**(5160-02-1)**
Terolut	**(152-62-5)**
Terom	**(25038-59-9)**
Teronac	**(22232-71-9)**
$\delta^{6,8}$-(9)-Terpadienone-2	**(99-49-0)**
Terpal	**(24307-26-4)**
Terpene-polychlorinates	**(8001-50-1)**
Terpene resin (9CI)	**(9003-74-1)**
Terpenes and Terpenoids, C10-30, distn. residues	**(70084-98-9)**
Terpenol	**(98-55-5)**
Terpentin Oel (German)	**(8006-64-2)**
Terpentinoel (German)	**(8002-09-3)**
Terphan	**(25038-59-9)**
1,3-Terphenyl	**(92-06-8)**
Terphenyl	**(26140-60-3)**
m-Terphenyl	**(92-06-8)**
o-Terphenyl	**(84-15-1)**
p-Terphenyl	**(92-94-4)**
1,1':3',1''-Terphenyl- (9CI)	**(612-71-5)**
Terphenyl, chlorinated	**(61788-33-8)**
m-Terphenyl, 5'-phenyl	**(612-71-5)**
Terphenyls (ACGIH,OSHA)	**(26140-60-3)**
Terphenyls, hydrogenated	**(61788-32-7)**
1,8-Terpin	**(80-53-5)**
Terpin	**(80-53-5)**
Terpin (VAN)	**(80-53-5)**
α-Terpinene	**(99-86-5)**
γ-Terpinene	**(99-85-4)**
Terpinene (9CI)	**(8013-00-1)**
4-Terpinenol	**(562-74-3)**
Terpinen-4-ol	**(562-74-3)**
Terpinenol-4	**(562-74-3)**
Terpinenolu-4 (Czech)	**(562-74-3)**
(-)-α-Terpineol	**(10482-56-1)**
(L)-α-Terpineol	**(10482-56-1)**
1-Terpineol	**(586-82-3)**
4-Terpineol	**(562-74-3)**
L-α-Terpineol	**(10482-56-1)**
Terpineol	**(8006-39-1)**
α-Terpineol	**(10482-56-1)**
α-Terpineol	**(8000-41-7)**
α-Terpineol	**(98-55-5)**
β-Terpineol	**(138-87-4)**
Terpineol (9CI)	**(8000-41-7)**
α-Terpineol, acetate	**(80-26-2)**
Terpineol, acetate (9CI)	**(8007-35-0)**
Terpineols	**(8006-39-1)**
Terpineol schlechthin	**(98-55-5)**
Terpin hydrate	**(2451-01-6)**
Terpinol	**(2451-01-6)**
Terpinolene [UN 2541]	**(586-62-9)**
Terpinyl acetate	**(80-26-2)**
Terpinyl acetate	**(8007-35-0)**
Terpinyl thiocyanoacetate	**(115-31-1)**
δ-1,8-Terpodiene	**(138-86-3)**
3,2':4',3''-Terpyridine (8CI,9CI)	**(494-04-2)**
Terr-O-Gas 100	**(74-83-9)**
Terr-O-Gas 67	**(74-83-9)**
Terr-O-Gel	**(8004-09-9)**
Terra-Systam	**(115-26-4)**
Terra-Sytam	**(115-26-4)**
Terrachlor-Super X	**(2593-15-9)**
Terrachlor	**(82-68-8)**
Terraclor	**(82-68-8)**
Terracoat	**(2593-15-9)**
Terracoat l21	**(2593-15-9)**
Terracur	**(3655-88-7)**
Terracur p	**(115-90-2)**
Terracydin	**(62-44-2)**
Terraflo	**(2593-15-9)**
Terrafun	**(82-68-8)**
Terrafungine	**(79-57-2)**
Terraklene	**(1910-42-5)**
Terramitsin	**(79-57-2)**
Terramycin	**(79-57-2)**
Terramycin hydrochloride	**(2058-46-0)**
Terrasan-Schneckentod Gekoernt	**(9002-91-9)**
Terrasytum	**(115-26-4)**
Terrazole	**(2593-15-9)**
Tersan	**(137-26-8)**
Tersan 75	**(137-26-8)**
Tersan 1991	**(17804-35-2)**
Tersan-LSR	**(12427-38-2)**
Tersan-SP	**(2675-77-6)**
Tersaseptic	**(70-30-4)**
Tersetile Blue 2BL	**(12222-78-5)**
Tersetile Blue RBL	**(12217-79-7)**
Tersetile Red 5GL	**(12223-35-7)**
Tersetile Rubine FL	**(17418-58-5)**
Tersetile Yellow 5R	**(6300-37-4)**
Tersetile Yellow 5RL	**(6300-37-4)**
Tertiary calcium phosphate	**(7758-87-4)**
Tertiary magnesium phosphate	**(7757-87-1)**
Tertracid Brilliant Light Blue R	**(4368-56-3)**
Tertracid Brilliant Orange P4G	**(1934-20-9)**
Tertracid Fast Blue SR	**(3861-73-2)**
Tertracid Light Orange G	**(1936-15-8)**
Tertracid Light Yellow 2R	**(6373-74-6)**
Tertracid Milling Red AGE	**(10169-02-5)**
Tertracid Milling Red G	**(3567-65-5)**
Tertracid Milling Yellow R	**(6375-55-9)**
Tertracid Orange I	**(523-44-4)**
Tertracid Orange II	**(633-96-5)**
Tertracid Orange IV	**(554-73-4)**
Tertracid Ponceau 2R	**(3761-53-3)**
Tertracid Red A	**(915-67-3)**
Tertracid Red CA	**(3567-69-9)**
Tertracid Red E	**(2302-96-7)**
Tertracid Red RO	**(1658-56-6)**

Name	CAS No.
Tertracid Yellow M	**(587-98-4)**
Tertracid Yellow TRO	**(547-57-9)**
Tertral D	**(106-50-3)**
Tertral ERN	**(90-15-3)**
Tertral G	**(95-80-7)**
Tertral P Base	**(123-30-8)**
Tertral eg	**(591-27-5)**
Tertranese Yellow N-2GL	**(2832-40-8)**
Tertrochrome Blue FB	**(3567-69-9)**
Tertrochrome Violet N	**(2092-55-9)**
Tertrodirect Black BH	**(2429-73-4)**
Tertrodirect Black BHS	**(2429-73-4)**
Tertrodirect Black E	**(1937-37-7)**
Tertrodirect Black RW	**(2429-83-6)**
Tertrodirect Blue 2B	**(2602-46-2)**
Tertrodirect Blue F	**(2429-74-5)**
Tertrodirect Blue FF	**(2610-05-1)**
Tertrodirect Brown TB	**(2429-81-4)**
Tertrodirect Fast Brown BR	**(16071-86-6)**
Tertrodirect Fast Red 5B	**(2610-11-9)**
Tertrodirect Fast Yellow 8G	**(10190-68-8)**
Tertrodirect Green BG	**(3626-28-6)**
Tertrodirect Red 4B	**(992-59-6)**
Tertrodirect Red C	**(573-58-0)**
Tertrodirect Violet N	**(2586-60-9)**
Tertrogras Orange SG	**(2051-85-6)**
Tertrogras Orange SV	**(842-07-9)**
Tertrogras Red N	**(85-83-6)**
Tertrophene Brilliant Green G	**(633-03-4)**
Tertrophene Brown CG	**(532-82-1)**
Tertrophene Green M	**(569-64-2)**
Tertropigment Orange LRN	**(3468-63-1)**
Tertropigment Orange PG	**(3520-72-7)**
Tertropigment Red HAB	**(2425-85-6)**
Tertropigment Red LR	**(1248-18-6)**
Tertropigment Red P2G	**(3468-63-1)**
Tertropigment Scarlet LRN	**(2425-85-6)**
Tertrosulphur Black PB	**(51-28-5)**
Tertrosulphur PBR	**(51-28-5)**
Tertroxane Red G	**(6408-31-7)**
Terulan KP 2540	**(9003-54-7)**
Terulan KR 2540	**(9003-54-7)**
Tervalon	**(132-60-5)**
Tescol	**(107-21-1)**
Teserene	**(84-17-3)**
Teslac	**(968-93-4)**
Teslak	**(968-93-4)**
Teslen	**(58-22-0)**
Tespa	**(52-24-4)**
Tespamin	**(52-24-4)**
Tespamine	**(52-24-4)**
Test Lead Dross	**(69029-52-3)**
Testaform	**(57-85-2)**
Testandrone	**(58-22-0)**
Testascorbic	**(50-81-7)**
Testate	**(315-37-7)**
Testavol	**(68-26-8)**
Testex	**(57-85-2)**
Testhormone	**(58-18-4)**
Testiculosterone	**(58-22-0)**
Testobase	**(58-22-0)**
Testodet	**(57-85-2)**
Testodrin	**(57-85-2)**
Testogen	**(57-85-2)**
Testolacton	**(968-93-4)**
Testolactona (Spanish)	**(968-93-4)**
Testolactone	**(968-93-4)**
δ(1)-Testolactone	**(968-93-4)**
Testolactonum (Latin)	**(968-93-4)**
Testolattone	**(968-93-4)**
δ(1)-Testololactone	**(968-93-4)**
δ-1-Testololactone	**(968-93-4)**
Testololactone, 1-dehydro-	**(968-93-4)**
Testololactone, 1,2-didehydro-	**(968-93-4)**
Testonique	**(57-85-2)**
Testopropon	**(58-22-0)**
Testora	**(58-18-4)**
Testoral	**(76-43-7)**
Testormol	**(57-85-2)**
Testosteroid	**(58-22-0)**
Testosteron	**(58-22-0)**
Testosterone	**(58-22-0)**
trans-Testosterone	**(58-22-0)**
Testosterone enantate	**(315-37-7)**
Testosterone enanthate	**(315-37-7)**
Testosterone ethanate	**(315-37-7)**
Testosterone, heptanoate	**(315-37-7)**
Testosterone heptoate	**(315-37-7)**
Testosterone heptylate	**(315-37-7)**
Testosterone hydrate	**(58-22-0)**
Testosterone oenanthate	**(315-37-7)**
Testosterone, propionate	**(57-85-2)**
Testosterone-17-β-propionate	**(57-85-2)**
Testosterone-17-propionate	**(57-85-2)**
Testosteron propionate	**(57-85-2)**
Testostosterone	**(58-22-0)**
Testostroval	**(315-37-7)**
Testoviron	**(57-85-2)**
Testoviron	**(58-18-4)**
Testoviron T	**(58-22-0)**
Testoviron schering	**(58-22-0)**
Testoxyl	**(57-85-2)**
Testred	**(58-18-4)**
Testrex	**(57-85-2)**
Testrone	**(58-22-0)**
Testryl	**(58-22-0)**
Tesuloid	**(7704-34-9)**
Tet-Cy	**(64-75-5)**
Teta	**(112-24-3)**
Tetacin	**(62-33-9)**
Tetacin-calcium	**(62-33-9)**
Tetamon iodide	**(68-05-3)**
Tetazine	**(62-33-9)**
2,4,5-T 2-ethylhexyl ester	**(1928-47-8)**
2,4,5-T ethylhexyl ester	**(1928-47-8)**
Tetidis	**(97-77-8)**
Tetlen	**(127-18-4)**
Tetnor	**(50-33-9)**
Tetra-Base	**(101-61-1)**
Tetra-D	**(64-75-5)**
Tetra Olive N2G	**(120-12-7)**
Tetra Sytam	**(115-26-4)**
O,O,O',O'-Tetraaethyl-bis(dithiophosphat) (German)	**(563-12-2)**
O,O,O,O-Tetraaethyl-dithionopyrophosphat (German)	**(3689-24-5)**
Tetraaethylpyrophosphorsaeureester (German)	**(107-49-3)**
1,4,5,8-Tetraaminoanthraquinone	**(2475-45-8)**
3,3',4,4'-Tetraaminobiphenyl	**(91-95-2)**
3,3',4,4'-Tetraaminobiphenyl tetrahydrochloride	**(7411-49-6)**
(T-4)-Tetraamminezinc(2+) carbonate (1:1)	**(38714-47-5)**
Tetraammonium hexacyanoferrate	**(14481-29-9)**
Tetraammonium hexacyanoferrate(4-)	**(14481-29-9)**
1,3,5,7-Tetraazaadamantane	**(100-97-0)**
1,3,5,7-Tetraazabicyclo(3.3.1)nonane, 3,7-dinitroso	**(101-25-7)**
1,4,7,10-Tetraazadecane	**(112-24-3)**

1,5,8,12-Tetraazadodecane (10563-26-5)
3,6,9,12-Tetraazatetradecane-1,14-diamine (4067-16-7)
2,4,11,13-Tetraazatetradecanediimidamide, N,N'-bis(4-chlorophenyl)-3,12-diimino-, diacetate (56-95-1)
2,4,11,13-Tetraazatetradecanediimidamide, N,N'-bis(4-chlorophenyl)-3,12-diimino-, dihydrochloride (3697-42-5)
2,4,11,13-Tetraazatetradecanediimidamide, N,N''-bis(4-chlorophenyl)-3,12-diimino-, diacetate (9CI) (56-95-1)
2,4,11,13-Tetraazatetradecanediimidamide, N,N''-bis(4-chlorophenyl)-3,12-diimino-, dihydrochloride (3697-42-5)
1,3,5,7-Tetraazatricyclo(3.3.1.1^{37}decane (100-97-0)
Tetraazido benzene quinone (22826-61-5)
Tetraazido-p-benzoquinone (22826-61-5)
Tetrabakat (64-75-5)
Tetrabenzo(a,c,m,o)naphtho(1,2,3,4-rst)pentaphene (9CI) (72382-91-3)
Tetrabenzo(b,g,l,q)porphyrazine (574-93-6)
Tetrabenzo(de,hi,mn,qr)naphthacene (7CI,8CI,9CI) (385-13-7)
Tetrabenzo(de,hi,op,st)pentacene (7CI,8CI,9CI) (191-79-7)
Tetrabid (64-75-5)
Tetrablet (64-75-5)
Tetrabon (60-54-8)
1,1,2,2-Tetrabromaethan (German) (79-27-6)
Tetrabromoacetylene (79-27-6)
1,2,4,5-Tetrabromobenzene (636-28-2)
Tetrabromobiphenyltetrabromo (40088-45-7)
2,2',6,6'-Tetrabromobisphenol A (79-94-7)
3,5,3',5'-Tetrabromobisphenol A (79-94-7)
Tetrabromobisphenol A (79-94-7)
Tetrabromobisphenol A, bis(allyl ether) (25327-89-3)
Tetrabromobisphenol A bis(2,3-dibromopropyl) ether (21850-44-2)
Tetrabromobisphenol A, bisphenol A, epichlorohydrin polymer (26265-08-7)
3,5,3',5'-Tetrabromobisphenol A, epichlorohydrin polymer (40039-93-8)
1,2,3,4-Tetrabromobutane (1529-68-6)
Tetrabromobutane (1529-68-6)
3,4,5,6-Tetrabromocatechol (488-47-1)
Tetrabromocatechol (488-47-1)
3,4,5,6-Tetrabromo-o-cresol (576-55-6)
Tetrabromo-o-cresol (576-55-6)
Tetrabromodian (79-94-7)
1,3,6,8-Tetrabromodibenzo(b,e)(1,4)dioxin (76584-71-9)
Tetrabromodibenzo(b,e)(1,4)dioxin (103456-39-9)
1,3,6,8-Tetrabromodibenzo-p-dioxin (76584-71-9)
2,3,7,8-Tetrabromodibenzo-p-dioxin, (50585-41-6)
Tetrabromodibenzo-p-dioxin (103456-39-9)
1,2,7,8-Tetrabromodibenzofuran (84761-80-8)
2,3,7,8-Tetrabromodibenzofuran (67733-57-7)
Tetrabromodichlorocyclohexane (30554-72-4)
2,4,5,7-Tetrabromo-12,15-dichlorofluorescein, dipotassium salt (6441-77-6)
Tetrabromodiphenyl ether (40088-47-9)
Tetrabromodiphenyl oxide (40088-47-9)
1,1,2,2-Tetrabromoetano (Italian) (79-27-6)
1,1,2,2-Tetrabromoethane (79-27-6)
s-Tetrabromoethane (79-27-6)
Tetrabromoethane [UN 2504] (79-27-6)
Tetrabromoethene (79-28-7)
Tetrabromoethylene (79-28-7)
2,4,5,7-Tetrabromo-3,6-fluorandiol (15086-94-9)
2,4,5,7-Tetrabromo-3,6-fluorandiol (17372-87-1)
2',4',5',7'-Tetrabromofluorescein (15086-94-9)
Tetrabromofluorescein (15086-94-9)
Tetrabromofluorescein (17372-87-1)
Tetrabromofluorescein D (17372-87-1)
Tetrabromofluorescein S (17372-87-1)
Tetrabromofluorescein Soluble (17372-87-1)
2',4',5',7'-Tetrabromofluorescein disodium salt (17372-87-1)
Tetrabromoguaiacol (35488-17-6)
4,5,6,7-Tetrabromo-1,3-isobenzofurandione (632-79-1)
Tetrabromomethane (558-13-4)
2,3,4,5-Tetrabromo-6-methylphenol (576-55-6)
3,4,5,6-Tetrabromo-2-methylphenol (576-55-6)
3',3'',5',5''-Tetrabromophenolsulfophthalein (115-39-9)
Tetrabromophenolsulfophthalein (115-39-9)
Tetrabromophenolsulfophthalein (71-67-0)
Tetrabromophenoxybenzene (40088-47-9)
3,4,5,6-Tetrabromophthalic anhydride (632-79-1)
Tetrabromophthalic anhydride (632-79-1)
1,2,2,3-Tetrabromopropane (54268-02-9)
1,3,6,8-Tetrabromopyrene (128-63-2)
Tetrabromopyrocatechol (488-47-1)
Tetrabromosulfophthalein (71-67-0)
Tetrabromo-p-xylen (Czech) (23488-38-2)
Tetrabromo-o-xylene (2810-69-7)
Tetrabromo-p-xylene (23488-38-2)
Tetrabromsulfthalein (71-67-0)
1,1,2,2-Tetrabroomethaan (Dutch) (79-27-6)
Tetrabutylammonium p-nitrophenoxide (3002-48-0)
Tetra-n-butylcin (Czech) (1461-25-2)
N,N,N',N'-Tetrabutyl-1,6-hexanediamine (27090-63-7)
Tetra-n-butylphosphonium bromide (3115-68-2)
Tetra-n-butylphosphonium chloride (2304-30-5)
Tetrabutylplumbane (1920-90-7)
Tetrabutylstannane (1461-25-2)
Tetrabutylthiuram disulphide (1634-02-2)
Tetrabutyltin (1461-25-2)
Tetrabutyltitanate (Czech) (5593-70-4)
Tetracaine hydrochloride (136-47-0)
Tetracap (127-18-4)
Tetracaps (64-75-5)
Tetracarbonylhydridocobalt (16842-03-8)
Tetracarbonylhydrocobalt (16842-03-8)
1,2,4,5-Tetracarboxybenzene (89-05-4)
Tetracemate disodium (139-33-3)
Tetracemate tetrasodium (64-02-8)
Tetracemin (64-02-8)
Tetracene (109-27-3)
Tetracene (92-24-0)
Tetracene Explosive (109-27-3)
Tetracene (hydrocarbon) (92-24-0)
Tetrachel (64-75-5)
3,4,6,4'-Tetrachlor-diphenylsulfid (German) (2227-13-6)
2,4,4',5-Tetrachlor-diphenyl-sulfon (German) (116-29-0)
3,4,6,4'-Tetrachlor-diphenylsulfon (German) (116-29-0)
2,3,5,6-Tetrachlor-3-nitrobenzol (German) (117-18-0)
2,4,4',5-Tetrachloor-difenyl-sulfon (Dutch) (116-29-0)
1,1,2,2-Tetrachloorethaan (Dutch) (79-34-5)
Tetrachlooretheen (Dutch) (127-18-4)
Tetrachloorkoolstof (Dutch) (56-23-5)
Tetrachloormetaan (56-23-5)
1,1,2,2-Tetrachloraethan (German) (79-34-5)
Tetrachloraethen (German) (127-18-4)
N-(1,1,2,2-Tetrachloraethylthio)-cyclohex-4-en-1,4-diacarboximid (German) (2425-06-1)
N-(1,1,2,2-Tetrachloraethylthio)-tetrahydrophthalamid (German) (2425-06-1)
Tetrachlordian (Czech) (79-95-8)
Tetrachlorethane (79-34-5)
1,1,2,2-Tetrachlorethane (French) (79-34-5)
Tetrachlorethylene (127-18-4)
Tetrachlorisoftalonitril (Czech) (1897-45-6)
Tetrachlorkohlenstoff, tetra (German) (56-23-5)
Tetrachlormethan (German) (56-23-5)
1,1,1,3-Tetrachloroacetone (16995-35-0)
1,1,3,3-Tetrachloroacetone (632-21-3)
3,3',4,4'-Tetrachloroazobenzene (14047-09-7)
3,4,3',4'-Tetrachloroazobenzene (14047-09-7)

3,3',4,4'-Tetrachloroazoxybenzene	**(21232-47-3)**
3,4,3',4'-Tetrachloroazoxybenzene	**(21232-47-3)**
1,2,3,4-Tetrachlorobenzene	**(634-66-2)**
1,2,3,5-Tetrachlorobenzene	**(634-90-2)**
1,2,4,5-Tetrachlorobenzene	**(95-94-3)**
Tetrachloro-1,2-benzenediol	**(1198-55-6)**
3,4,5,6-Tetrachloro-1,2-benzenediol (9CI)	**(1198-55-6)**
2,2',5,5'-Tetrachlorobenzidine	**(15721-02-5)**
3,3',6,6'-Tetrachlorobenzidine	**(15721-02-5)**
Tetrachlorobenzidine	**(15721-02-5)**
2,3,4,5-Tetrachlorobenzoic acid	**(50-74-8)**
2,3,5,6-Tetrachloro-1,4-benzoquinone	**(118-75-2)**
2,3,5,6-Tetrachloro-p-benzoquinone	**(118-75-2)**
Tetrachloro-1,4-benzoquinone	**(118-75-2)**
Tetrachloro-p-benzoquinone	**(118-75-2)**
Tetrachlorobenzoquinone	**(118-75-2)**
2,2',3,3'-Tetrachlorobiphenyl	**(38444-93-8)**
2,2',3,4-Tetrachlorobiphenyl	**(52663-59-9)**
2,2',3,5'-Tetrachlorobiphenyl	**(41464-39-5)**
2,2',3,5-Tetrachlorobiphenyl	**(70362-46-8)**
2,2',3,6'-Tetrachlorobiphenyl	**(41464-47-5)**
2,2',4,5'-Tetrachlorobiphenyl	**(41464-40-8)**
2,2',4,5-Tetrachlorobiphenyl	**(70362-47-9)**
2,2',4,6-Tetrachlorobiphenyl	**(62796-65-0)**
2,2',5,5'-Tetrachlorobiphenyl	**(35693-99-3)**
2,2',5,6'-Tetrachlorobiphenyl	**(41464-41-9)**
2,2',6,6'-Tetrachloro-1,1'-biphenyl	**(15968-05-5)**
2,3',4',5'-Tetrachlorobiphenyl	**(70362-48-0)**
2,3',4',5-Tetrachlorobiphenyl	**(32598-11-1)**
2,3',4',6-Tetrachlorobiphenyl	**(41464-46-4)**
2,3',4,4'-Tetrachloro-1,1'-biphenyl	**(32598-10-0)**
2,3',4,4'-Tetrachlorobiphenyl	**(32598-10-0)**
2,3',4,5'-Tetrachlorobiphenyl	**(73575-52-7)**
2,3',5,5'-Tetrachlorobiphenyl	**(41464-42-0)**
2,3,3',4'-Tetrachlorobiphenyl	**(41464-43-9)**
2,3,3',6-Tetrachlorobiphenyl	**(74472-33-6)**
2,3,4',5-Tetrachlorobiphenyl	**(74472-34-7)**
2,3,4',6-Tetrachlorobiphenyl	**(52663-58-8)**
2,3,4,4'-Tetrachlorobiphenyl	**(33025-41-1)**
2,3,4,5-Tetrachlorobiphenyl	**(33284-53-6)**
2,3,5,6-Tetrachlorobiphenyl	**(33284-54-7)**
2,4,2',4'-Tetrachlorobiphenyl	**(2437-79-8)**
2,4,3',4'-Tetrachlorobiphenyl	**(32598-10-0)**
2,4,4',5-Tetrachlorobiphenyl	**(32690-93-0)**
2,5,2',5'-Tetrachlorobiphenyl	**(35693-99-3)**
3,3',4,4'-Tetrachlorobiphenyl	**(32598-13-3)**
3,3',4,5-Tetrachlorobiphenyl	**(70362-49-1)**
3,4,2',4'-Tetrachlorobiphenyl	**(32598-10-0)**
3,4,3',4'-Tetrachlorobiphenyl	**(32598-13-3)**
3,4,4',5-Tetrachlorobiphenyl	**(70362-50-4)**
3,5,3',5'-Tetrachlorobiphenyl	**(33284-52-5)**
2,2',6,6'-Tetrachloro-(1,1'-biphenyl)-4,4'-diol	**(41363-16-0)**
3,3',5,5'-Tetrachloro-2,2'-biphenyldiol	**(14477-61-3)**
2,2',6,6'-Tetrachlorobisphenol A	**(79-95-8)**
1,1,2,3-Tetrachloro-1,3-butadiene	**(921-09-5)**
Tetrachlorobutadiene	**(58334-79-5)**
1,2,3,3-Tetrachlorobutane	**(13138-51-7)**
1,2,3,4-Tetrachlorobutane	**(3405-32-1)**
2,2,3,3-Tetrachlorobutane	**(14499-87-7)**
2,4,4,4-Tetrachloro-1-butanol	**(3290-70-8)**
2,4,4,4-Tetrachlorobutanol	**(3290-70-8)**
Tetrachlorocarbon	**(56-23-5)**
Tetrachlorocatechol	**(1198-55-6)**
2,4,5,6-Tetrachloro-3-cyanobenzonitrile	**(1897-45-6)**
Tetrachloro-2-cyanopyridine	**(17824-83-8)**
2,3,5,6-Tetrachloro-2,5-cyclohexadiene-1,4-dione	**(118-75-2)**
3,4,5,6-Tetrachloro-3,5-cyclohexadiene-1,2-dione	**(2435-53-2)**
1,2,3,4-Tetrachlorocyclopentadiene	**(695-77-2)**
Tetrachlorocyclopentadiene	**(695-77-2)**
Tetrachlorocyclopentane	**(59808-78-5)**
2,2',5,5'-Tetrachloro-4,4'-diaminodiphenyl	**(15721-02-5)**
1,2,3,4-Tetrachlorodibenzo-para-dioxin	**(30746-58-8)**
1,2,3,8-Tetrachlorodibenzo-para-dioxin	**(53555-02-5)**
1,3,6,8-Tetrachlorodibenzo-para-dioxin	**(33423-92-6)**
1,3,7,8-Tetrachlorodibenzo-para-dioxin	**(50585-46-1)**
2,3,7,8-Tetrachlorodibenzo-1,4-dioxin	**(1746-01-6)**
2,3,7,8-Tetrachlorodibenzo(b,e)(1,4)dioxan	**(1746-01-6)**
1,2,3,6-Tetrachlorodibenzo(b,e)(1,4)dioxin	**(71669-25-5)**
1,2,3,7-Tetrachlorodibenzo(b,e)(1,4)dioxin	**(67028-18-6)**
1,2,3,9-Tetrachlorodibenzo(b,e)(1,4)dioxin	**(71669-26-6)**
1,2,4,6-Tetrachlorodibenzo(b,e)(1,4)dioxin	**(71669-27-7)**
1,2,4,7-Tetrachlorodibenzo(b,e)(1,4)dioxin	**(71669-28-8)**
1,2,4,8-Tetrachlorodibenzo(b,e)(1,4)dioxin	**(71669-29-9)**
1,2,4,9-Tetrachlorodibenzo(b,e)(1,4)dioxin	**(71665-99-1)**
1,2,6,7-Tetrachlorodibenzo(b,e)(1,4)dioxin	**(40581-90-6)**
1,2,6,8-Tetrachlorodibenzo(b,e)(1,4)dioxin	**(67323-56-2)**
1,2,6,9-Tetrachlorodibenzo(b,e)(1,4)dioxin	**(40581-91-7)**
1,2,7,9-Tetrachlorodibenzo(b,e)(1,4)dioxin	**(71669-23-3)**
1,2,8,9-Tetrachlorodibenzo(b,e)(1,4)dioxin	**(62470-54-6)**
1,3,6,9-Tetrachlorodibenzo(b,e)(1,4)dioxin	**(71669-24-4)**
1,4,6,9-Tetrachlorodibenzo(b,e)(1,4)dioxin	**(40581-93-9)**
1,4,7,8-Tetrachlorodibenzo(b,e)(1,4)dioxin	**(40581-94-0)**
1,2,3,4-Tetrachlorodibenzodioxin	**(30746-58-8)**
1,2,3,6-Tetrachlorodibenzo-p-dioxin	**(71669-25-5)**
1,2,3,7-Tetrachlorodibenzo-p-dioxin	**(67028-18-6)**
1,2,3,8-Tetrachlorodibenzodioxin	**(53555-02-5)**
1,2,3,9-Tetrachlorodibenzo-p-dioxin	**(71669-26-6)**
1,2,4,6-Tetrachlorodibenzo-p-dioxin	**(71669-27-7)**
1,2,4,7-Tetrachlorodibenzo-p-dioxin	**(71669-28-8)**
1,2,4,8-Tetrachlorodibenzo-p-dioxin	**(71669-29-9)**
1,2,4,9-Tetrachlorodibenzo-p-dioxin	**(71665-99-1)**
1,2,6,7-Tetrachlorodibenzo-p-dioxin	**(40581-90-6)**
1,2,6,8-Tetrachlorodibenzo-p-dioxin	**(67323-56-2)**
1,2,6,9-Tetrachlorodibenzo-p-dioxin	**(40581-91-7)**
1,2,7,8-Tetrachlorodibenzo-p-dioxin	**(34816-53-0)**
1,2,7,9-Tetrachlorodibenzo-p-dioxin	**(71669-23-3)**
1,2,8,9-Tetrachlorodibenzo-p-dioxin	**(62470-54-6)**
1,3,6,8-Tetrachlorodibenzodioxin	**(33423-92-6)**
1,3,6,9-Tetrachlorodibenzo-p-dioxin	**(71669-24-4)**
1,3,7,8-Tetrachlorodibenzodioxin	**(50585-46-1)**
1,3,7,9-Tetrachlorodibenzo-p-dioxin	**(62470-53-5)**
1,4,6,9-Tetrachlorodibenzo-p-dioxin	**(40581-93-9)**
1,4,7,8-Tetrachlorodibenzo-p-dioxin	**(40581-94-0)**
2,3,6,7-Tetrachlorodibenzo-p-dioxin	**(1746-01-6)**
2,3,6,7-Tetrachlorodibenzodioxin	**(34816-53-0)**
2,3,7,8-Tetrachlorodibenzo-p-dioxin	**(1746-01-6)**
2,3,7,8-Tetrachlorodibenzo-p-dioxin	**(41903-57-5)**
Tetrachlorodibenzo-p-dioxin	**(41903-57-5)**
1,2,3,6-Tetrachlorodibenzofuran	**(83704-21-6)**
1,2,3,7-Tetrachlorodibenzofuran	**(83704-22-7)**
1,2,3,8-Tetrachlorodibenzofuran	**(62615-08-1)**
1,2,3,9-Tetrachlorodibenzofuran	**(83704-23-8)**
1,2,4,6-Tetrachlorodibenzofuran	**(71998-73-7)**
1,2,4,7-Tetrachlorodibenzofuran	**(83719-40-8)**
1,2,4,9-Tetrachlorodibenzofuran	**(83704-24-9)**
1,2,6,7-Tetrachlorodibenzofuran	**(83704-25-0)**
1,2,6,8-Tetrachlorodibenzofuran	**(83710-07-0)**
1,2,6,9-Tetrachlorodibenzofuran	**(70648-18-9)**
1,2,7,8-Tetrachlorodibenzofuran	**(58802-20-3)**
1,2,7,9-Tetrachlorodibenzofuran	**(83704-26-1)**
1,2,8,9-Tetrachlorodibenzofuran	**(70648-22-5)**
1,3,4,6-Tetrachlorodibenzofuran	**(83704-27-2)**
1,3,4,7-Tetrachlorodibenzofuran	**(70648-16-7)**
1,3,4,8-Tetrachlorodibenzofuran	**(92341-04-3)**

1,3,4,9-Tetrachlorodibenzofuran (83704-28-3)
1,3,6,7-Tetrachlorodibenzofuran (57117-36-9)
1,3,6,8-Tetrachlorodibenzofuran (71998-72-6)
1,3,6,9-Tetrachlorodibenzofuran (83690-98-6)
1,3,7,8-Tetrachlorodibenzofuran (57117-35-8)
1,3,7,9-Tetrachlorodibenzofuran (64560-17-4)
1,4,6,7-Tetrachlorodibenzofuran (66794-59-0)
1,4,6,8-Tetrachlorodibenzofuran (82911-58-8)
1,4,6,9-Tetrachlorodibenzofuran (70648-19-0)
1,4,7,8-Tetrachlorodibenzofuran (83704-29-4)
1,6,7,8-Tetrachlorodibenzofuran (83704-33-0)
2,3,4,6-Tetrachlorodibenzofuran (83704-30-7)
2,3,4,7-Tetrachlorodibenzofuran (83704-31-8)
2,3,4,8-Tetrachlorodibenzofuran (83704-32-9)
2,3,6,7-Tetrachlorodibenzofuran (57117-39-2)
2,3,6,8-Tetrachlorodibenzofuran (57117-37-0)
2,3,7,8-Tetrachlorodibenzofuran (51207-31-9)
2,4,6,7-Tetrachlorodibenzofuran (57117-38-1)
2,4,6,8-Tetrachlorodibenzofuran (58802-19-0)
3,4,6,7-Tetrachlorodibenzofuran (57117-40-5)
Tetrachlorodifluoroethane (28605-74-5)
1,1,1,2-Tetrachloro-2,2-difluoroethane (ACGIH,OSHA) (76-11-9)
1,1,2,2-Tetrachloro-1,2-difluoroethane (ACGIH,OSHA) (76-12-0)
1,1,2,2-Tetrachloro-1,2-dimethyldisilane (4518-98-3)
4,6,6,6-Tetrachloro-3,3-dimethylhexanoic acid, methyl ester (64667-33-0)
3,3',5,5'-Tetrachlorodiphenyl (33284-52-5)
1,1,2,2-Tetrachloro-1,2-diphenylethane (13700-81-7)
Tetrachloro-1,2-diphenylethane (13700-81-7)
Tetrachlorodiphenylethane (72-54-8)
2,2',4,4'-Tetrachlorodiphenyl ether (28076-73-5)
2,4,4',5-Tetrachlorodiphenyl ether (61328-45-8)
Tetrachloro diphenyl ether (31242-94-1)
Tetrachloro diphenyl oxide (31242-94-1)
2,4,4',5-Tetrachlorodiphenyl sulfide (2227-13-6)
2,4,5,4'-Tetrachlorodiphenyl sulfide (2227-13-6)
2,4,4',5-Tetrachlorodiphenyl sulfone (116-29-0)
2,4,5,4'-Tetrachlorodiphenyl sulfone (116-29-0)
2,4,5,4'-Tetrachlorodiphenylsulphone (116-29-0)
Tetrachloroepoxyethane (16650-10-5)
1,1,1,2-Tetrachloroethane (630-20-6)
s-Tetrachloroethane (79-34-5)
sym-Tetrachloroethane (79-34-5)
1,1,2,2-Tetrachloroethane (ACGIH,DOT,OSHA) (79-34-5)
Tetrachloroethane [UN 1702] (25322-20-7)
Tetrachloroethane [UN 1702] (79-34-5)
1,1,2,2-Tetrachloroethanesulfenyl chloride (1185-09-7)
Tetrachloroethene (127-18-4)
1,1,2,2,-Tetrachloroethylene (127-18-4)
Tetrachloroethylene (OSHA) [UN 1897] (127-18-4)
Tetrachloroethylene oxide (16650-10-5)
1,1'-(Tetrachloroethylidene)bis(chlorobenzene) (68631-02-7)
N-1,1,2,2-Tetrachloroethylmercapto-4-cyclohexene-1,2-carboximide (2425-06-1)
1,1,2,2-Tetrachloroethylsulfenyl chloride (1185-09-7)
N-((1,1,2,2-Tetrachloroethyl)sulfenyl)-cis-4-cyclohexene-1,2-dicarboximide (2425-06-1)
N-(1,1,2,2-Tetrachloroethylthio)-4-cyclohexene-1,2-dicarboximide (2425-06-1)
cis-N-((1,1,2,2-Tetrachloroethyl)thio)-4-cyclohexene-1,2-dicarboxymide (2939-80-2)
1,3,4,6-Tetrachloroglycoluril (776-19-2)
Tetrachloroguaiacol (2539-17-5)
3,3',4,4'-Tetrachlorohydrazobenzene (71753-42-9)
3,4,3',4'-Tetrachlorohydrazobenzene (71753-42-9)
Tetrachlorohydroquinone (87-87-6)
Tetrachloroisophthalonitrile (1897-45-6)
Tetrachloromethane (56-23-5)
Tetrachloromethane (OSHA) (56-23-5)
2,3,5,6-Tetrachloro-4-(methylsulfonyl)pyridine (13108-52-6)
1,2,3,4-Tetrachloronaphthalene (20020-02-4)
Tetrachloronaphthalene (ACGIH,OSHA) (1335-88-2)
2,3,5,6-Tetrachloro-4-nitroanisole (2438-88-2)
Tetrachloronitroanisole (2438-88-2)
1,2,3,4-Tetrachloro-5-nitrobenzene (879-39-0)
1,2,4,5-Tetrachloro-3-nitrobenzene (117-18-0)
2,3,4,5-Tetrachloronitrobenzene (879-39-0)
2,3,5,6-Tetrachloronitrobenzene (117-18-0)
Tetrachloronitrobenzene (28804-67-3)
Tetrachloronitrobenzene (VAN) (28804-67-3)
2,3,4,5-Tetrachloro-6-(pentachlorophenoxy)phenol (35245-80-8)
2,3,4,6-Tetrachloro-5-(pentachlorophenoxy)phenol (42255-14-1)
2,3,4,5-Tetrachlorophenol (4901-51-3)
2,3,4,6-Tetrachlorophenol (58-90-2)
2,3,5,6-Tetrachlorophenol (935-95-5)
2,4,5,6-Tetrachlorophenol (58-90-2)
Tetrachlorophenol (25167-83-3)
1,1,1',1'-Tetra(p-chlorophenyl)dimethyl ether (74562-99-5)
2,2',4,4'-Tetrachlorophenyl oxide (28076-73-5)
(Tetrachlorophenyl)trichlorosilane (33434-63-8)
Tetrachlorophthalic acid (632-58-6)
Tetrachlorophthalic anhydride (117-08-8)
4,5,6,7-Tetrachlorophthalide (27355-22-2)
meta-Tetrachlorophthalodinitrile (1897-45-6)
m-Tetrachlorophthalonitrile (1897-45-6)
Tetrachloropicolinic acid (10469-09-7)
1,1,1,2-Tetrachloropropane (812-03-3)
1,1,1,3-Tetrachloropropane (1070-78-6)
1,1,2,3-Tetrachloropropane (18495-30-2)
1,2,2,3-Tetrachloropropane (13116-53-5)
1,1,1,3-Tetrachloro-2-propanone (16995-35-0)
1,1,3,3-Tetrachloro-2-propanone (632-21-3)
1,1,3,3-Tetrachloropropanone (632-21-3)
1,1,2,3-Tetrachloropropene (10436-39-2)
1,2,3,3-Tetrachloro-1-propene (20589-85-9)
2,3,4,5-Tetrachloropyridine (2808-86-8)
2,3,5,6-Tetrachloropyridine (2402-79-1)
3,4,5,6-Tetrachloro-2-pyridinecarbonitrile (17824-83-8)
3,4,5,6-Tetrachloro-2-pyridinecarboxylic acid (10469-09-7)
Tetrachloropyrocatechol (1198-55-6)
Tetrachloro-p-quinone (118-75-2)
Tetrachloroquinone (118-75-2)
Tetrachlorosilane (10026-04-7)
Tetrachlorotellurium (10026-07-0)
Tetrachloroterephthalic acid (2136-79-0)
Tetrachloroterephthalic acid dimethyl ester (1861-32-1)
2,3,5,6-Tetrachloroterephthalic acid, dimethyl ether (1861-32-1)
Tetrachloroterephthalic dichloride (719-32-4)
2,3,5,6-Tetrachloroterephthaloyl chloride (719-32-4)
Tetrachloroterephthaloyl chloride (719-32-4)
2,3,5,6-Tetrachloroterephthaloyl dichloride (719-32-4)
Tetrachloroterephthaloyl dichloride (719-32-4)
Tetrachlorothiofene (6012-97-1)
Tetrachlorothiophene (6012-97-1)
Tetrachlorothorium (10026-08-1)
2,3,4,5-Tetrachlorotoluene (1006-32-2)
2,3,5,6-Tetrachlorotoluene (1006-31-1)
α,α,α,4-Tetrachlorotoluene (5216-25-1)
α,α2,6-Tetrachlorotoluene (81-19-6)
p,α,α,α-Tetrachlorotoluene (5216-25-1)
5-endo,6-exo-2,2,5,6-Tetrachloro-1,7,7-tris(chloromethyl)-bicyclo-(2.2.1)heptane (51775-36-1)
(T-4)-Tetrachlorozincate(2-) diammonium (14639-97-5)
2,3,5,6-Tetrachlorphthalsaure-dimethylester (German) (1861-32-1)
Tetrachlorpyrokatechin (Czech) (1198-55-6)
Tetrachlorpyrokatechol (Czech) (1198-55-6)

Tetrachlorthiofen (Czech) (6012-97-1)
Tetrachlorure d'acetylene (French) (79-34-5)
Tetrachlorure de carbone (French) (56-23-5)
Tetrachlorure de silicium (French) (10026-04-7)
Tetrachlorure de titane (French) (7550-45-0)
Tetrachlorvinphos (22248-79-9)
Tetraciclina cloridrato (Italian) (64-75-5)
Tetracid (144-82-1)
Tetracid Brilliant Violet 6B (1694-09-3)
Tetracid Carmine Blue V (129-17-9)
Tetracid Milling Red G (6459-94-5)
Tetracina (Spanish) (70816-59-0)
2,4,4',5-Tetracloro-difenil-solfone (Italian) (116-29-0)
1,1,2,2-Tetracloroetano (Italian) (79-34-5)
Tetracloroetene (Italian) (127-18-4)
Tetraclorometano (Italian) (56-23-5)
Tetracloruro di carbonio (Italian) (56-23-5)
Tetracompren (64-75-5)
Tetracontane (4181-95-7)
Tetracontane, 3,5,24-trimethyl- (9CI) (55162-61-3)
Tetracor (54-95-5)
Tetracosactid (16960-16-0)
Tetracosactide (16960-16-0)
β^{1-24}-Tetracosactide (16960-16-0)
Tetracosactrin (16960-16-0)
2,6,10,14,18,22-Tetracosahexaene, 2,6,10,15,19,23-hexamethyl-, (all-E) (111-02-4)
Tetracosane (9CI) (646-31-1)
Tetracosane-d50 (8CI,9CI) (16416-32-3)
Tetracosane, 2,6,10,15,19,23-hexamethyl (111-01-3)
Tetracosanoic acid (9CI) (557-59-5)
n-Tetracosanol (506-51-4)
1-Tetracosanol (9CI) (506-51-4)
1-Tetracosanol, aluminum salt (9CI) (67905-29-7)
Tetracosapeptide (16960-16-0)
Tetracosyl alcohol (506-51-4)
Tetracresyl titanate (28503-70-0)
Tetracyanoethylene (670-54-2)
Tetracycline (60-54-8)
Tetracycline I (60-54-8)
Tetracycline II (60-54-8)
Tetracycline chloride (64-75-5)
Tetracycline, 7-chloro-6-demethyl- (127-33-3)
Tetracycline hydrochloride (64-75-5)
Tetracycline, 5-hydroxy- (79-57-2)
Tetracyclotetramethylenecyclotetrasiloxane (177-49-1)
Tetracydin (62-44-2)
Tetracyn (60-54-8)
Tetracyn (64-75-5)
1,13-Tetradecadiene (21964-49-8)
Tetradecafluorohexane (355-42-0)
Tetradecahydrophenanthrene (5743-97-5)
Tetradecamethylhexasiloxane (107-52-8)
1-Tetradecanal (124-25-4)
Tetradecanal (124-25-4)
Tetradecanamide, N-(2-hydroxyethyl)- (9CI) (142-58-5)
1-Tetradecanamine (9CI) (2016-42-4)
1-Tetradecanamine, N,N-dimethyl- (9CI) (112-75-4)
Tetradecanamine, N,N-dimethyl-, N-oxide (3332-27-2)
1-Tetradecanamine, N,N-dimethyl-, N-oxide (9CI) (3332-27-2)
1-Tetradecanamine, (2,4,5-trichlorophenoxy)acetate (9CI) (53535-37-8)
1-Tetradecanaminium, N,N-bis(2-hydroxyethyl)-1-oxo-N-(1-oxododecyl)-, hydroxide, inner salt (9CI) (52900-12-6)
1-Tetradecanaminium, N,N-dimethyl-N-tetradecyl-, chloride (9CI) (10108-91-5)
1-Tetradecanaminium, N,N,N-trimethyl-, bromide (9CI) (1119-97-7)
1-Tetradecanaminium, N,N,N-trimethyl-, chloride (9CI) (4574-04-3)
Tetradecane (629-59-4)
Tetradecane, 1-bromo- (9CI) (112-71-0)
Tetradecane, 1-chloro- (9CI) (2425-54-9)
1,14-Tetradecanedicarboxylic acid (505-54-4)
Tetradecanedioic acid (9CI) (821-38-5)
Tetradecane, 4-ethyl- (9CI) (55045-14-2)
Tetradecane, 1-iodo- (9CI) (19218-94-1)
Tetradecane, 2-methyl- (8CI,9CI) (1560-95-8)
Tetradecanenitrile (9CI) (629-63-0)
n-Tetradecane-ω,ω'-dicarboxylic acid (505-54-4)
1,2-Tetradecaneoxide (3234-28-4)
Tetradecane, 1-phenyl- (1459-10-5)
Tetradecane, 2-phenyl- (8CI) (4534-59-2)
Tetradecane, 3-phenyl- (8CI) (4534-58-1)
Tetradecane, 4-phenyl- (8CI) (4534-57-0)
Tetradecane, 5-phenyl- (8CI) (4534-56-9)
Tetradecane, 6-phenyl- (8CI) (4534-55-8)
Tetradecane, 7-phenyl- (8CI) (4534-54-7)
tert-Tetradecanethiol (9CI) (28983-37-1)
Tetradecanoato chromic chloride hydroxide (1:2:4:1) (15659-56-0)
Tetradecanoic acid (544-63-8)
n-Tetradecanoic acid (544-63-8)
Tetradecanoic acid, Compd. with 2,2',2''-nitrilotris(ethanol) (41669-40-3)
Tetradecanoic acid, Compd. with 2,2',2''-nitrilotris(ethanol) (1:1) (9CI) (41669-40-3)
Tetradecanoic acid, 9a-(acetyloxy)-1a,1b,4,4a,5,7a,7b,8,9,9a-decahydro-4a,7b-dihydroxy-3-(hydroxymethyl)-1,1,6,8-tetramethyl-5-oxo-1H-cyclopropa(3,4)benz(1,2-e)azulen-9-yl ester, (1aR-(1aα,1bβ,4aβ,7aα,7bα,8α,9β,9aα))- (9CI) (16561-29-8)
Tetradecanoic acid, barium salt (9CI) (10196-66-4)
Tetradecanoic acid, butyl ester (9CI) (110-36-1)
Tetradecanoic acid, cadmium salt (9CI) (10196-67-5)
Tetradecanoic acid, calcium salt (9CI) (15284-51-2)
Tetradecanoic acid, 2-chloroethyl ester (9CI) (51479-36-8)
Tetradecanoic acid, ethyl ester (9CI) (124-06-1)
Tetradecanoic acid, hexadecyl ester (9CI) (2599-01-1)
Tetradecanoic acid, 3-hydroxy (1961-72-4)
Tetradecanoic acid, 2-hydroxy- (9CI) (2507-55-3)
Tetradecanoic acid, 10-hydroxy- (8CI,9CI) (16899-08-4)
Tetradecanoic acid, isopropyl (110-27-0)
Tetradecanoic acid, isopropyl ester (110-27-0)
Tetradecanoic acid, lead salt (9CI) (20403-41-2)
Tetradecanoic acid, 2-mercaptoethyl ester (9CI) (29946-28-9)
Tetradecanoic acid, 12-methyl- (8CI,9CI) (5502-94-3)
Tetradecanoic acid, methyl ester (9CI) (124-10-7)
Tetradecanoic acid, 1-methylethyl ester (110-27-0)
Tetradecanoic acid, 7-methyl-, ethyl ester (8CI) (17670-75-6)
Tetradecanoic acid, 12-methyl-, methyl ester (8CI,9CI) (5129-66-8)
Tetradecanoic acid, 2-methyl-, methyl ester (7CI,9CI) (55554-09-1)
Tetradecanoic acid, potassium salt (9CI) (13429-27-1)
Tetradecanoic acid, 1,2,3-propanetriyl ester (9CI) (555-45-3)
Tetradecanoic acid, 2-sulfo-, 1-methyl ester (9CI) (29454-23-7)
Tetradecanoic acid, 2-sulfo-, 1-methyl ester, sodium salt (9CI) (4016-22-2)
Tetradecanoic acid, tetradecyl ester (9CI) (3234-85-3)
1-Tetradecanol (112-72-1)
n-Tetradecanol-1 (112-72-1)
2-Tetradecanol (9CI) (4706-81-4)
Tetradecanol, Mixed isomers (27196-00-5)
1-Tetradecanol acetate (638-59-5)
1-Tetradecanol, acetate (638-59-5)
1-Tetradecanol, aluminum salt (9CI) (67905-32-2)
1-Tetradecanol, 2-decyl- (9CI) (58670-89-6)
1-Tetradecanol, dihydrogen phosphate (9CI) (10054-29-2)
1-Tetradecanol, hydrogen sulfate (9CI) (4754-44-3)
1-Tetradecanol, hydrogen sulfate, Compd. with 2,2',2''-nitrilo tris(ethanol) (1:1) (9CI) (4492-78-8)
1-Tetradecanol, hydrogen sulfate, sodium salt (1191-50-0)

1-Tetradecanol, propanoate (9CI) (6221-95-0)
12-O-Tetradecanoyl phorbol acetate (16561-29-8)
12-O-Tetradecanoylphorbol-13-acetate (16561-29-8)
12-Tetradecanoylphorbol 13-acetate (16561-29-8)
Tetradecanoylphorbol acetate (16561-29-8)
(Z)-9-Tetradecen-1-al (53939-27-8)
(Z)-9-Tetradecenal (53939-27-8)
9-Tetradecenal, (Z)- (53939-27-8)
Tetradecene (26952-13-6)
α-Tetradecene (1120-36-1)
n-Tetradec-1-ene (1120-36-1)
n-Tetradecene (26952-13-6)
1-Tetradecene (9CI) (1120-36-1)
7-Tetradecene (8CI,9CI) (10374-74-0)
1-Tetradecenedisulfonic acid, disodium salt (9CI) (68003-17-8)
Tetradecenesulfonic acid, sodium salt (9CI) (11066-21-0)
(Z)-11-Tetradecen-1-ol (34010-15-6)
(Z)-9-Tetradecen-1-ol (35153-15-2)
11-Tetradecen-1-ol, (Z)- (34010-15-6)
9-Tetradecen-1-ol, (Z)- (35153-15-2)
(Z)-9-Tetradecen-1-ol acetate (16725-53-4)
9-Tetradecen-1-ol, acetate, (Z)- (16725-53-4)
(Z)-9-Tetradecenyl acetate (16725-53-4)
Tetradecenylsuccinic anhydride (33806-58-5)
Tetradecin (60-54-8)
n-Tetradecoic acid (544-63-8)
Tetradecyl acetate (638-59-5)
Tetradecyl alcohol (112-72-1)
Tetradecyl alcohol (27196-00-5)
n-Tetradecyl alcohol (112-72-1)
1-Tetradecyl aldehyde (124-25-4)
Tetradecylamine (8CI) (2016-42-4)
Tetradecylamine, N,N-dimethyl- (8CI) (112-75-4)
Tetradecylbenzene (1459-10-5)
4-Tetradecylbenzenesulfonic acid (47377-16-2)
Tetradecylbenzenesulfonic acid (30776-59-1)
4-Tetradecylbenzenesulfonic acid, sodium salt (1797-33-7)
1-Tetradecyl bromide (112-71-0)
Tetradecyl bromide (112-71-0)
n-Tetradecyl bromide (112-71-0)
Tetradecylcyclohexane (1795-18-2)
Tetradecyldimethylamine (112-75-4)
Tetradecyl-dimethyl-benzylammonium chloride (139-08-2)
Tetradecyldimethyl(ethylbenzyl)ammonium chloride (27479-29-4)
1-Tetradecylene (1120-36-1)
Tetradecylene (26952-13-6)
n-Tetradecylformate (5451-63-8)
Tetradecyl iodide (19218-94-1)
tert-Tetradecyl mercaptan (28983-37-1)
Tetradecyl methacrylate (2549-53-3)
Tetradecyl 2-methyl-2-propenoate (2549-53-3)
Tetradecyl octadecanoate (17661-50-6)
2-Tetradecyloxyethyl sodium sulfate (3694-74-4)
((Tetradecyloxy)methyl)oxirane (38954-75-5)
1-Tetradecylpyridinium chloride (2785-54-8)
Tetradecyl sodium sulfate (1191-50-0)
Tetradecyl stearate (17661-50-6)
Tetradecyl sulfate, sodium salt (1191-50-0)
Tetradecyl sulfate (1:1), triethanolamine salt (4492-78-8)
Tetradecyl sulfuric acid (4754-44-3)
Tetradecyl tetradecanoate (3234-85-3)
Tetradecyltrimethylammonium bromide (1119-97-7)
7-Tetradecyne-6,9-diol, 5,10-diethyl- (9CI) (25430-52-8)
12-O-Tetradekanoylphorbol-13-acetat (German) (16561-29-8)
12-o-Tetradekanoylphorbol-13-acetat (German) (16561-29-8)
Tetradichlone (116-29-0)
Tetradifon (116-29-0)
Tetradin (97-77-8)
Tetradine (97-77-8)
Tetradioxin (1746-01-6)
Tetradiphon (116-29-0)
Tetradonium bromide (1119-97-7)
Tetraetatone (15438-85-4)
Tetra(ethanolamine) ethylenediaminetetraacetate (53404-52-7)
Tetraethoxysilane (78-10-4)
Tetraethyl ammonium bromide (71-91-0)
Tetraethylammonium chloride (56-34-8)
Tetraethylammonium iodide (68-05-3)
Tetraethylammonium perchlorate (2567-83-1)
Tetraethylammonium perchlorate (Dry) (2567-83-1)
Tetraethylammonium perchlorate, Dry (2567-83-1)
Tetraethylbenzene (33637-20-6)
Tetra(2-ethylbutoxy) silane (78-13-7)
Tetra (2-ethylbutyl) orthosilicate (78-13-7)
4,4'-(Tetraethyldiamino)benzophenone (90-93-7)
p,p'-(Tetraethyldiamino)benzophenone (90-93-7)
Tetraethyldiamino-o-carboxy-phenyl-xanthenyl chloride (81-88-9)
O,O,O,O-Tetraethyl-difosfaat (Dutch) (107-49-3)
Tetraethyldifosfat (Czech) (107-49-3)
Tetraethyl diphosphate (107-49-3)
O,O,O,O-Tetraethyl-dithio-difosfaat (Dutch) (3689-24-5)
Tetraethyldithiodifosfat (Czech) (3689-24-5)
Tetraethyl dithionopyrophosphate (3689-24-5)
O,O,O,O-Tetraethyl dithiopyrophosphate (3689-24-5)
Tetraethyl dithiopyrophosphate, Liquid (DOT) (3689-24-5)
Tetraethyl dithiopyrophosphate Mixture, Dry (DOT) (3689-24-5)
Tetraethyl dithiopyrophosphate Mixture, Liquid (DOT) (3689-24-5)
Tetraethyl dithiopyrophosphate [UN 1704] (3689-24-5)
Tetraethylene-glycol (112-60-7)
Tetraethylene glycol diacrylate (17831-71-9)
Tetraethylene glycol, dibutyl ether (112-98-1)
Tetraethylene glycol diethyl ether (4353-28-0)
Tetraethylene glycol di(2-ethylhexoate) (18268-70-7)
Tetraethylene glycol dimethacrylate (109-17-1)
Tetraethylene glycol dimethyl ether (143-24-8)
Tetraethylene glycol-di-n-heptanoate (70729-68-9)
Tetraethylene glycol, monobutyl ether (1559-34-8)
Tetraethylene glycol, monoethyl ether (5650-20-4)
Tetraethylene glycol monomethyl ether (23783-42-8)
Tetraethylenepentamine [UN 2320] (112-57-2)
Tetraethylenepentamine polyisobutylene succinimide, molybdenum complex, sulfurized (72269-41-1)
Tetraethylenetetramine pelargonamide (26392-63-2)
Tetraethyl lead (ACGIH,DOT,OSHA) (78-00-2)
Tetraethyl lead, Liquid (including flashpoint for export shipment by water) [NA 1649] (78-00-2)
Tetraethyl lead, Motor fuel anti-knock mixtures [UN 1649] (78-00-2)
O,O,O',O'-Tetraethyl S,S'-methylenebisphosphordithioate (563-12-2)
O,O,O',O'-Tetraethyl-S,S'-methylenebisphosphorodithioate (563-12-2)
Tetraethyl S,S'-methylene bis(phosphorothiolothionate) (563-12-2)
O,O,O',O'-Tetraethyl S,S'-methylene di(phosphorodithioate) (563-12-2)
Tetraethylolovo (Czech) (78-00-2)
Tetraethylplumbane (78-00-2)
Tetraethylplumbium (78-00-2)
Tetraethyl pyrofosfaat (Dutch) (107-49-3)
Tetraethylpyrofosfat (Czech) (107-49-3)
E Tetraethylpyronin (2150-48-3)
Tetraethylpyrophosphate (107-49-3)
Tetraethyl pyrophosphate, Liquid (DOT) (107-49-3)
Tetraethyl pyrophosphate Mixture, Dry (DOT) (107-49-3)
Tetraethyl pyrophosphate Mixture, Liquid (DOT) (107-49-3)
Tetraethylrhodamine (81-88-9)
Tetraethylsilane (631-36-7)
Tetraethyl silicate (78-10-4)

Tetraethyl orthosilicate [UN 1292] **(78-10-4)**
Tetraethyl orthosilicate polymer **(11099-06-2)**
Tetraethylsilikat (Czech) **(78-10-4)**
Tetraethylstannane **(597-64-8)**
Tetraethylthioperoxydicarbonic diamide **(97-77-8)**
Tetraethylthiram disulphide **(97-77-8)**
Tetraethylthiuram **(97-77-8)**
Tetraethylthiuram disulfide **(97-77-8)**
N,N,N',N'-Tetraethylthiuram disulphide **(97-77-8)**
Tetraethylthiuram disulphide **(97-77-8)**
Tetraethyl tin **(597-64-8)**
Tetraethyl titanate **(3087-36-3)**
Tetraetil **(97-77-8)**
O,O,O,O-Tetraetil-ditio-pirofosfato (Italian) **(3689-24-5)**
O,O,O,O-Tetraetil-pirofosfato (Italian) **(107-49-3)**
Tetrafidon **(116-29-0)**
Tetrafinol **(56-23-5)**
Tetrafluorethylene **(116-14-3)**
Tetrafluoroboric acid **(16872-11-0)**
1,1,1,2-Tetrafluoro-2,2-dichloroethane **(374-07-2)**
1,1,2,2-Tetrafluoro-1,2-dichloroethane **(76-14-2)**
Tetrafluorodichloroethane **(1320-37-2)**
1,1,1,2-Tetrafluoroethane **(811-97-2)**
Tetrafluoroethene (9CI) **(116-14-3)**
Tetrafluoroethylene **(116-14-3)**
Tetrafluoroethylene, Inhibited [UN 1081] **(116-14-3)**
Tetrafluoromethane [UN 1982] **(75-73-0)**
2,2,3,3-Tetrafluoro-3-methoxypropanoic acid, methyl ester **(755-73-7)**
2,2,3,3-Tetrafluoro-1-propanol **(76-37-9)**
2,2,3,3-Tetrafluoropropanol **(76-37-9)**
Tetrafluoropropyl alcohol **(76-37-9)**
Tetrafluorosulfurane **(7783-60-0)**
o,α,α,α-Tetrafluorotoluene **(392-85-8)**
Tetraform **(56-23-5)**
Tetrafosfor (Dutch) **(7723-14-0)**
Tetraglycidyl-4,4'-methylene dianiline **(28768-32-3)**
Tetraglycine hydroperiodide **(7097-60-1)**
Tetraglyme **(143-24-8)**
Tetrahelicene **(195-19-7)**
1,2,3,4-Tetrahydro-9-acridinamine **(321-64-2)**
Tetrahydroaminacrine **(321-64-2)**
1,2,3,4-Tetrahydro-5-aminoacridine **(321-64-2)**
Tetrahydroaminoacridine **(321-64-2)**
Tetrahydroaminocrin **(321-64-2)**
Tetrahydroaminocrine **(321-64-2)**
Tetrahydroanacardol **(501-24-6)**
6,7,8,9-Tetrahydro-5-azepotetrazole **(54-95-5)**
Tetrahydrobenzaldehyde **(1321-16-0)**
1,2,5,6-Tetrahydrobenzaldehyde [UN 2498] **(100-50-5)**
1,2,3,4-Tetrahydrobenzene **(110-83-8)**
Tetrahydrobenzene **(110-83-8)**
exo-Tetrahydrobicyclopentadiene **(2825-82-3)**
Tetrahydroborate(1-) lithium **(16949-15-8)**
(-)-δ^{1}-3,4-trans-Tetrahydrocannabinol **(1972-08-3)**
(-)-δ^{9}-trans-Tetrahydrocannabinol **(1972-08-3)**
(l)-δ^{1}-Tetrahydrocannabinol **(1972-08-3)**
1-trans-δ^{9}-Tetrahydrocannabinol **(1972-08-3)**
δ^{1}-Tetrahydrocannabinol **(1972-08-3)**
δ^{9}-Tetrahydrocannabinol **(1972-08-3)**
trans-δ^{9}-Tetrahydrocannabinol **(1972-08-3)**
exo-Tetrahydrodi(cyclopentadiene) **(2825-82-3)**
1-(Tetrahydrodicyclopentadienyl)-3,3-dimethylurea **(18530-56-8)**
1-(Tetrahydrodicyclopentadienyl)-3,3-dimethylurea **(2163-79-3)**
3a,4,7,7a-Tetrahydrodimethyl-4,7-methanoindene **(26472-00-4)**
Tetrahydro-2H-3,5-dimethyl-1,3,5-thiadiazine-2-thione **(533-74-4)**
Tetrahydro-3,5-dimethyl-2H-1,3,5-thiadiazine-2-thione **(533-74-4)**
Tetrahydro-1,4-dioxin **(123-91-1)**
Tetrahydro-p-dioxin **(123-91-1)**
Tetrahydro-2,5-dioxofuran **(108-30-5)**
1,2,3,6-Tetrahydro-3,6-dioxopyridazine **(123-33-1)**
3a,4,7,7a-Tetrahydro-1,2-epoxy-4,5,6,7,8,8-hexachloro-4,7-methanoindan **(27304-13-8)**
5,6,7,8-Tetrahydro-2-ethylanthrahydroquinone **(68279-54-9)**
Tetrahydroftalanhydrid (Czech) **(85-43-8)**
Tetrahydrofuraan (Dutch) **(109-99-9)**
Tetrahydrofuran (ACGIH,OSHA) [UN 2056] **(109-99-9)**
Tetrahydro-2-furancarbinol **(97-99-4)**
Tetrahydro-2-furanmethanol **(97-99-4)**
Tetrahydro-2-furanmethanol acetate **(637-64-9)**
Tetrahydrofuranne (French) **(109-99-9)**
Tetrahydro-2-furanone **(96-48-0)**
Tetrahydrofurfuryl acetate **(637-64-9)**
Tetrahydrofurfuryl acrylate **(2399-48-6)**
Tetrahydrofurfuryl alcohol **(97-99-4)**
Tetrahydrofurfurylalkohol (Czech) **(97-99-4)**
Tetrahydrofurfuryl methacrylate **(2455-24-5)**
Tetrahydrofurylalkohol (Czech) **(97-99-4)**
Tetrahydro-2-furylmethanol **(97-99-4)**
Tetrahydrogeraniol **(106-21-8)**
3a-α,4,7,7a-α-Tetrahydro-2-α,4-β,5,6,7-β,8,8-heptachloro-4,7-methanoindan **(14168-01-5)**
1-(5,6,7,8-Tetrahydro-3,5,5,6,8,8-hexamethyl-2-naphthalenyl)-ethanone **(1506-02-1)**
2,3-cis-1,2,3,4-Tetrahydro-5-((2-hydroxy-3-tert-butylamino)-propoxy)-2,3-naphthalenediol **(42200-33-9)**
Tetrahydro-5-(2-hydroxyethyl)-1,3-bis(hydroxymethyl)-1,3,5-triazin-2(1H)-one **(1852-21-7)**
Tetrahydro-7-α-(1-hydroxy-1-methylbutyl)-6,14-endo-etheno-oripavine **(14521-96-1)**
Tetrahydro-7-α-(2-hydroxy-2-pentyl)-6,14-endo-ethenooripavine **(14521-96-1)**
Tetrahydroimidaz(d)imidazole-2,5(1H,3H)-dione **(496-46-8)**
Tetrahydroimidazo(4,5-d)imidazole-2,5(1H,3H)-dione **(496-46-8)**
3a,4,7,7a-Tetrahydro-1H-indene **(3048-65-5)**
3a,4,7,7a-Tetrahydroindene **(3048-65-5)**
4,7,8,9-Tetrahydroindene **(3048-65-5)**
Tetrahydroindene (Russian) **(3048-65-5)**
Tetrahydro-1,4-isoxazine **(110-91-8)**
1,2,3,4-Tetrahydrolepidine **(19343-78-3)**
Tetrahydrolinalool **(78-69-3)**
3a,4,7,7a-Tetrahydro-4,7-methanoindene **(77-73-6)**
Tetrahydromethyl-1,3-isobenzofurandione **(11070-44-3)**
1,2,3,4-Tetrahydro-5-methylnaphthalene **(2809-64-5)**
1,2,3,4-Tetrahydro-6-methylnaphthalene **(1680-51-9)**
1,2,3,4-Tetrahydromethylnaphthalene **(31291-71-1)**
1,2,3,4-Tetrahydro-4-methylquinoline **(19343-78-3)**
1,2,3,4-Tetrahydro-6-methylquinoline **(91-61-2)**
E-1,4,5,6-Tetrahydro-1-methyl-2-(2-(2-thienyl)vinyl)pyrimidine **(15686-83-6)**
Tetrahydro-3-methylthiophene 1,1-dioxide **(872-93-5)**
Tetrahydromyrcenol **(18479-57-7)**
1,2,3,4-Tetrahydronaphthalene **(119-64-2)**
Tetrahydronaphthalene **(119-64-2)**
1,2,3,4-Tetrahydro-2-naphthol **(530-91-6)**
Tetrahydronaphthyl hydroperoxide **(771-29-9)**
3-(1,2,3,4-Tetrahydro-1-naphthyl)-4-hydroxycumarin (German) **(5836-29-3)**
3-(1,2,3,4-Tetrahydro-1-naphtyl)-4-hydroxycoumarine (French) **(5836-29-3)**
Tetrahydronicotyrine, dl- **(54-11-5)**
1,2,3,6-Tetrahydro-1-nitroso-2,3'-bipyridine **(71267-22-6)**
1,2,5,6-Tetrahydro-1-nitroso-3-pyridinecarboxylic acid **(55557-01-2)**
Tetrahydro-1,4-oxazine **(110-91-8)**
Tetrahydro-2H-1,4-oxazine **(110-91-8)**
Tetrahydro-1,4-oxazinylmethylcodeine **(509-67-1)**
N-(Tetrahydro-2-oxo-3-thienyl)acetamide **(1195-16-0)**
Tetrahydro-3-pentyl-2H-pyran-4-ol acetate **(18871-14-2)**
1,1,2,2-Tetrahydroperfluoro-1-decanol **(678-39-7)**

1,1,2,2-Tetrahydroperfluoro dodecanol (865-86-1)
1,1,2,2-Tetrahydroperfluoro-1-hexanol (2043-47-2)
1,1,2,2-Tetrahydroperfluorohexyl iodide (2043-55-2)
1,1,2,2-Tetrahydroperfluorohexyliodide (2043-55-2)
1,1,2,2-Tetrahydroperfluorooctadecyl iodide (65150-94-9)
1,1,2,2-Tetrahydroperfluoro-1-octanol (647-42-7)
1,1,2,2-Tetrahydroperfluorooctyl iodide (2043-57-4)
1,1,2,2-Tetrahydroperfluorooctyliodide (2043-57-4)
Tetrahydrophenobarbital (52-31-3)
1,2,3,4-Tetrahydro-1-phenylnaphthalene (3018-20-0)
1,2,3,6-Tetrahydrophthalic acid (88-98-2)
Tetrahydrophthalic acid (88-98-2)
δ4-Tetrahydrophthalic acid (88-98-2)
Tetrahydrophthalic acid anhydride (85-43-8)
Tetrahydrophthalic acid imide (85-40-5)
1,2,3,6-Tetrahydrophthalic anhydride (85-43-8)
Tetrahydrophthalic anhydride (85-43-8)
δ^4-Tetrahydrophthalic anhydride (85-43-8)
Tetrahydrophthalic anhydride (DOT) (85-43-8)
Tetrahydrophthalic anhydrides with more than 0.05 percent of maleic anhydride [UN 2698] (85-43-8)
1,2,3,6-Tetrahydrophthalimide (85-40-5)
Tetrahydrophthalimide (85-40-5)
δ^4-Tetrahydrophthalimide (85-40-5)
3,4,5,6-Tetrahydrophthalimidomethyl cis and trans dl chrysanthemummonocarboxylic acid (7696-12-0)
2,3,4,5-Tetrahydrophthalimidomethylchrysanthemate (7696-12-0)
N-(3,4,5,6-Tetrahydrophthalimido)-methyl dl-cis-trans-chrysanthemate (7696-12-0)
3,4,5,6-Tetrahydro-phthalimidomethylester der dl-cis-trans-chrysanthemumsaeure (German) (7696-12-0)
Tetrahydro-6-propyl-2H-pyran-2-one (104-50-7)
Tetrahydro-6-propyl-2H-pyran-2-one (698-76-0)
Tetrahydro-2H-pyran (142-68-7)
Tetrahydropyran (142-68-7)
Tetrahydropyran-2-carbinol (100-72-1)
Tetrahydropyrane (142-68-7)
2-Tetrahydropyranilcarbinol (100-72-1)
Tetrahydropyran-2-methanol (100-72-1)
Tetrahydro-2-pyranone (542-28-9)
Tetrahydropyranyl-2-methanol (100-72-1)
1,2,5,6-Tetrahydropyridine (694-05-3)
1,2,3,6-Tetrahydropyridine [UN 2410] (694-05-3)
Tetrahydropyrrole (123-75-1)
5,6,12,13-Tetrahydro-quino(2,3-b)acridine-7,14-dione (5862-38-4)
1,2,3,4-Tetrahydroquinoline (635-46-1)
5,6,7,8-Tetrahydroquinoxaline (34413-35-9)
5,6,7,7a-Tetrahydrospiro(benzothiazole-2(4H),1'-cyclohexane) (6675-28-1)
1,2,3,4-Tetrahydrostyrene (100-40-3)
cis-3a,4,7,7a-Tetrahydro-2-((1,1,2,2-tetrachloroethyl)thio)-1H-isoindole-1,3(2H)-dione (2939-80-2)
6,7,8,9-Tetrahydro-5H-tetrazoloazepine (54-95-5)
Tetrahydrothiofen-1,1-dioxid (Czech) (126-33-0)
Tetrahydrothiofen (Czech) (110-01-0)
Tetrahydrothiophene [UN 2412] (110-01-0)
2,3,4,5-Tetrahydrothiophene-1,1-dioxide (126-33-0)
Tetrahydrothiophene 1,1-dioxide (126-33-0)
Tetrahydrothiophene dioxide (126-33-0)
1,2,3,7-Tetrahydro-2-thioxo-6H-purin-6-one (2487-40-3)
3a,4,7,7a-Tetrahydro-N-(trichloromethanesulphenyl)phthalimide (133-06-2)
1,2,3,6-Tetrahydro-N-(trichloromethylthio)phthalimide (133-06-2)
3,4,5,6-Tetrahydrouridine (18771-50-1)
Tetrahydrouridine (18771-50-1)
Tetrahydroxyadipic acid (526-99-8)
1,2,5,8-Tetrahydroxyanthraquinone (81-61-8)
1,4,5,6-Tetrahydroxyanthraquinone (81-61-8)
2,2',4,4'-Tetrahydroxy-benzophenone (131-55-5)
2,4,2',4'-Tetrahydroxybenzophenone (131-55-5)
2,3,7,8-Tetrahydroxy(1)benzopyrano(5,4,3-cde)(1)benzopyran-5,10-dione (476-66-4)
Tetrahydroxybutane (149-32-6)
2',3,4,4'-Tetrahydroxychalcone (487-52-5)
Tetra(hydroxyethyl)ethylenediamine (140-07-8)
Tetrahydroxyethylethylenediamine (140-07-8)
3',4',5,7-Tetrahydroxyflavan-3-ol (117-39-5)
3,4',5,7-Tetrahydroxyflavone (520-18-3)
11-β,16-α,17-α,21-Tetrahydroxy-9-α-fluoro-1,4-pregnadiene-3,20-dione (124-94-7)
Tetrahydroxymethylmetane (115-77-5)
3-α,4-β,7-α,15-Tetrahydroxyscirp-9-en-8-one (23282-20-4)
L-2,3,5,6-Tetrahyro-6-phenylimidazo(2,1-b)thiazole (14769-73-4)
Tetraidrofurano (Italian) (109-99-9)
1,2,4,5-Tetraiodobenzene (636-31-7)
2',4',5',7'-Tetraiodofluorescein (15905-32-5)
Tetraiodofluorescein (15905-32-5)
Tetraiodofluorescein sodium salt (16423-68-0)
Tetraiodomethane (507-25-5)
Tetraiodothyronine (51-48-9)
Tetrairon tris(hexacyanoferrate) (14038-43-8)
Tetrairon(3+) tris(hexacyanoferrate(4-)) (14038-43-8)
Tetraisopropoxide titanium (546-68-9)
Tetraisopropoxy titanate, Reaction products with triethanolamine (74665-17-1)
Tetraisopropoxytitanium (546-68-9)
2,2',6,6'-Tetraisopropyldiphenylcarbodiimide (2162-74-5)
Tetraisopropyl orthotitanate (546-68-9)
Tetraisopropyl titanate (546-68-9)
Tetrakap (64-75-5)
N,N,N',N'-Tetrakis(4-aminophenyl)-1,4-benzenediamine (3283-07-6)
Tetrakis(carbonato(2-))dihydroxypentamagnesium (7760-50-1)
(T-4)-Tetrakis(cyano-C)zincate(2-) disodium (15333-24-1)
Tetrakis(diethylcarbamodithioato-S,S')selenium (5456-28-0)
Tetrakis(diethylcarbamodithioato-S,S')tellurium (20941-65-5)
Tetrakis(diethyldithiocarbamato)selenium (5456-28-0)
Tetrakis(diethyldithiocarbamato)tellurium (20941-65-5)
Tetrakisdimethylaminophosphonous anhydride (152-16-9)
Tetrakis(dimethylcarbamodithioato-S,S')selenium (144-34-3)
Tetrakis(2-ethylbutoxy)silane (78-13-7)
N,N,N',N'-Tetrakis(2-hydroxyethyl)-1,2-diaminoethane (140-07-8)
N,N,N',N'-Tetrakis(2-hydroxyethyl)ethylenediamine (140-07-8)
N,N,N',N'-Tetrakis(hydroxyethyl)ethylenediamine (140-07-8)
Tetrakis(hydroxyethyl)ethylenediamine (140-07-8)
Tetrakis-(hydroxymethyl)fosfoniumchlorid (Czech) (124-64-1)
Tetrakis(hydroxymethyl)methane (115-77-5)
Tetrakis(hydroxymethyl)phosphonium acetate (7580-37-2)
Tetrakis(hydroxymethyl)phosphonium chloride (124-64-1)
Tetrakis(hydroxymethyl) phosphonium hydroxide (512-82-3)
Tetrakis(hydroxymethyl)phosphonium sulfate (55566-30-8)
Tetrakis(hydroxymethyl)phosphonium sulfate-urea condensation product (63502-25-0)
Tetrakis(hydroxymethyl)phosphonium, sulfate (2:1), urea polymer (63502-25-0)
N,N,N',N'-Tetrakis(2-hydroxy-propyl)ethylenediamine (102-60-3)
Tetrakis(isopropoxy)titanium (546-68-9)
Tetrakis(methylene(3,5-di-tert-butyl-4-hydroxyhydrocinnamate)methane (6683-19-8)
Tetrakis(trimethylsiloxy)silane (3555-47-3)
Tetrakis(tris(methylphenyl) phosphite-P)nickel (35884-66-3)
Tetrakyanethylen (Czech) (670-54-2)
Tetraleno (127-18-4)
Tetralex (127-18-4)
Tetralin (119-64-2)
Tetralina (Polish) (119-64-2)
Tetraline (119-64-2)
Tetralinehydroperoxyde (Dutch) (771-29-9)

Name	CAS No.
Tetralinhydroperoxid (German)	(771-29-9)
Tetralin hydroperoxide	(771-29-9)
Tetralin hydroperoxide, Technically pure	(771-29-9)
Tetralin hydroperoxide, Technically pure	(771-29-9)
Tetralit	(479-45-8)
Tetralite	(479-45-8)
Tetrallobarbital	(77-26-9)
β-Tetralol	(530-91-6)
1-Tetralone	(529-34-0)
α-Tetralone	(529-34-0)
3-(α-Tetral)-4-oxycoumarin	(5836-29-3)
Tetralution	(64-75-5)
3-(α-Tetralyl)-4-hydroxycoumarin	(5836-29-3)
3-(d-Tetralyl)-4-hydroxycoumarin	(5836-29-3)
Tetram	(78-53-5)
Tetram 75	(3734-97-2)
Tetram, acid oxalate	(3734-97-2)
Tetramavan	(64-75-5)
Tetrameen	(110-18-9)
6,7,3',4'-Tetramethoxy-1-benzylisoquinoline hydrochloride	(61-25-6)
6',7',10,11-Tetramethoxyemetan	(483-18-1)
1,1,3,3-Tetramethoxypropane	(102-52-3)
Tetramethoxypropane	(102-52-3)
Tetramethoxysilane	(681-84-5)
Tetramethrin	(7696-12-0)
Tetramethrine (French)	(7696-12-0)
N,N,N',N'-Tetramethyl-3,6-acridinediamine	(494-38-2)
Tetramethylammonium bromide	(64-20-0)
Tetramethylammonium chloride	(75-57-0)
Tetramethyl ammonium hydroxide, Liquid (DOT)	(75-59-2)
Tetramethylammonium hydroxide [UN 1835]	(75-59-2)
1,3,4,10-Tetramethyl-7,8-benzacridine (French)	(51787-44-1)
7,8,9,11-Tetramethylbenz(c)acridine	(51787-44-1)
1,2,3,4-Tetramethylbenzene	(488-23-3)
1,2,3,5-Tetramethylbenzene	(527-53-7)
1,2,4,5-Tetramethylbenzene	(527-53-7)
1,2,4,5-Tetramethylbenzene	(95-93-2)
α,α,α',α'-Tetramethyl-1,4-benzenedimethanol	(2948-46-1)
α,α,α',α'-Tetramethyl-p-benzenedimethanol	(2948-46-1)
3,3',5,5'-Tetramethylbenzidine	(54827-17-7)
3,5,3',5'-Tetramethylbenzidine	(54827-17-7)
N,N,N',N'-Tetramethyl-p,p'-benzidine	(366-29-0)
N,N,N',N'-Tetramethylbenzidine	(366-29-0)
3,3',4,4'-Tetramethylbibenzyl	(34101-86-5)
3,3',4,4'-Tetramethylbiphenyl	(4920-95-0)
N,N,N',N'-Tetramethyl-1,3-butanediamine	(97-84-7)
Tetramethyl butanediamine	(97-84-7)
1,1,4,4-Tetramethyl-1,4-butanediol	(110-03-2)
1,1,3,3-Tetramethylbutylamine	(107-45-9)
1,1,3,3-Tetramethylbutyl isocyanide	(14542-93-9)
1,1,3,3-Tetramethylbutylisonitrile	(14542-93-9)
(1,1,3,3-Tetramethylbutyl)phenol	(27193-28-8)
2-(1,1,3,3-Tetramethylbutyl)phenol	(3884-95-5)
4-(1,1,3,3-Tetramethylbutyl)phenol	(140-66-9)
4-(1,1,3,3-Tetramethylbutyl)phenyl salicylate	(2553-08-4)
Tetramethylbutynediol	(142-30-3)
Tetramethylcin (Czech)	(594-27-4)
Tetramethylcyclobuta-1,3-dione	(933-52-8)
Tetramethyl-1,3-cyclobutanedione	(933-52-8)
1,5,5,6-Tetramethyl-1,3-cyclohexadiene	(514-94-3)
4-(2,5,6,6-Tetramethyl-2-cyclo-hexen-1-yl)-3-buten-2-one	(79-69-6)
2,4,7,9-Tetramethyl-5-decyne-4,7-diol	(126-86-3)
Tetramethyl decynediol	(126-86-3)
N,N,N',N'-Tetramethyl-diamido-fluor-phosphin-oxid (German)	(115-26-4)
N,N,N',N'-Tetramethyl-diamido-fosforzuur-fluoride (Dutch)	(115-26-4)
Tetramethyldiamidophosphoric fluoride	(115-26-4)
N,N,N',N'-Tetramethyl-diamido-phosphorsaeure-fluorid (German)	(115-26-4)
Tetramethyldiaminobenzhydrol	(119-58-4)
Tetramethyldiaminobenzophenone	(90-94-8)
Tetramethyldiaminodiphenylacetimine	(492-80-8)
N,N,N'N'-Tetramethyl-4,4'-diaminodiphenylmethane	(101-61-1)
N,N,N'N'-Tetramethyl-p,p'-diaminodiphenylmethane	(101-61-1)
Tetramethyldiaminodiphenylmethane	(101-61-1)
p,p-Tetramethyldiaminodiphenylmethane	(101-61-1)
N,N'-(N,N'-Tetramethyl)-1-diaminodiphenylnaphthylamino-methane hydrochloride	(2185-86-6)
N,N,N',N'-Tetramethyl-1,2-diaminoethane	(110-18-9)
N,N,N',N'-Tetramethyldiaminomethan (German)	(51-80-9)
N,N,N',N'-Tetramethyldiaminomethane	(51-80-9)
Tetramethyl diapara-amido-triphenyl carbinol	(569-64-2)
1,1,2,2-Tetramethyldichlorodisilane	(4342-61-4)
Tetramethyldichlorodisilane	(4342-61-4)
3,7,9,13-Tetramethyl-5,11-dioxa-2,8,14-trithia-4,7,9,12-tetra-azapentadeca-3,12-diene-6,10-dione	(59669-26-0)
3,3',4,4'-Tetramethyldiphenyl-1,2-ethane	(34101-86-5)
3,3',4,4'-Tetramethyldiphenylethane	(34101-86-5)
N,N,N',N'-Tetramethyldipropylenetriamine	(6711-48-4)
1,1,3,3-Tetramethyldisiloxane	(3277-26-7)
sym-Tetramethyldisiloxane	(3277-26-7)
Tetramethyldiurane sulphite	(137-26-8)
1,1,3,3-Tetramethyl-1,3-divinyldisiloxane	(2627-95-4)
Tetramethylene Blue	(61-73-4)
Tetramethylene bis(methanesulfonate)	(55-98-1)
Tetramethylene chlorohydrin	(928-51-8)
Tetramethylene cyanide	(111-69-3)
1,4-Tetramethylenediamine	(110-60-1)
Tetramethylenediamine	(110-60-1)
Tetramethylene dimethacrylate	(2082-81-7)
Tetramethylene dimethane sulfonate	(55-98-1)
Tetramethylene 1,4-diol	(110-63-4)
Tetramethylenediol-terephthalic acid polymer	(26062-94-2)
1,4-Tetramethylene glycol	(110-63-4)
Tetramethylene glycol dimethacrylate	(2082-81-7)
Tetramethylene glycol-terephthalic acid polymer	(26062-94-2)
Tetramethylene methacrylate	(2082-81-7)
Tetramethylene oxide	(109-99-9)
Tetramethyleneoxirane	(286-20-4)
Tetramethylenester kyseliny methansulfonove (Czech)	(55-98-1)
Tetramethylenesulfide	(110-01-0)
Tetramethylene sulfone	(126-33-0)
Tetramethylenetetranitramine	(2691-41-0)
Tetramethylenethiuram disulphide	(137-26-8)
Tetramethylenimine	(123-75-1)
N,N,N',N'-Tetramethylethanediamine	(110-18-9)
1,1,2,2-Tetramethylethylene	(563-79-1)
N,N,N',N'-Tetramethylethylenediamine	(110-18-9)
Tetramethyl ethylene diamine	(110-18-9)
Tetramethylethylene glycol	(76-09-5)
N,N,N',N'-Tetramethyl-p-fenylendiamin (Czech)	(100-22-1)
1,1,3,3-Tetramethylguanidine	(80-70-6)
N,N,N',N'-Tetramethylguanidine	(80-70-6)
2,6,10,14-Tetramethylhexadecane	(638-36-8)
2,6,10,14-Tetramethylhexadec-15-en-14-ol	(505-32-8)
3,7,11,15-Tetramethyl-1-hexadecen-3-ol	(505-32-8)
3,7,11,15-Tetramethyl-1-hexadecen-3-ol	(60046-87-9)
3,7,11,15-Tetramethyl-2-hexadecen-1-ol	(150-86-7)
N,N,N',N'-Tetramethylhexamethylene diamine	(111-18-2)
N,N,N',N'-Tetramethyl-1,6-hexanediamine	(111-18-2)
Tetramethylhydrazine hydrochloride	(61556-82-9)
1,1,3,3-Tetramethyl-2-indanone	(5689-12-3)
Tetramethyl lead (ACGIH,OSHA)	(75-74-1)
N,N,N',N'-Tetramethylmethanediamine	(51-80-9)
N,N,N',N'-Tetramethylmethylenediamine	(51-80-9)
Tetramethyl methylene diamine	(51-80-9)

Tetramethylmethylenediamine (51-80-9)
Tetramethyl-p-nitroaniline (13171-61-4)
4,4,5,5-Tetramethyl-2,7-octanedione (17663-27-3)
Tetramethylolmethane (115-77-5)
Tetramethylolovo (Czech) (75-74-1)
Tetramethyloxirane (5076-20-0)
2,6,10,14-Tetramethylpentadecane (1921-70-6)
2,2,3,3-Tetramethylpentane (7154-79-2)
2,2,3,4-Tetramethylpentane (1186-53-4)
2,3,3,4-Tetramethylpentane (16747-38-9)
2,2,4,4-Tetramethylpentanoic acid (3302-12-3)
2,2,4,4-Tetramethyl-3-pentanone (815-24-7)
1:2:3:4-Tetramethylphenanthrene (4466-77-7)
2,3,5,6-Tetramethylphenol (527-35-5)
N,N,N,α-Tetramethyl-10H-phenothiazine-10-ethanaminium methyl sulfate (58-34-4)
2-(2,3,5,6-Tetramethylphenoxy)propionic acid (111-17-1)
Tetramethyl-p-phenyldiamine (100-22-1)
N,N,N',N'-Tetramethyl-p-phenylenediamine (100-22-1)
Tetramethyl-p-phenylenediamine (100-22-1)
N,N,N,N-Tetramethylphosphorodiamidic fluoride (115-26-4)
Tetramethylphosphorodiamidic fluoride (115-26-4)
2,2,6,6-Tetramethyl-4-piperidinol (2403-88-5)
2,2,6,6-Tetramethylpiperidinone (826-36-8)
2,2,6,6-Tetramethyl-4-piperidone (826-36-8)
2,2,6,6-Tetramethylpiperidone (826-36-8)
Tetramethylplumbane (75-74-1)
2,3,5,6-Tetramethyl pyrazine (1124-11-4)
Tetramethylpyrazine (1124-11-4)
Tetramethylsilane [UN 2749] (75-76-3)
Tetramethyl silicate (681-84-5)
Tetramethylsilikat (Czech) (681-84-5)
Tetramethylstannane (594-27-4)
Tetramethylsuccinic acid dinitrile (3333-52-6)
Tetramethylsuccinodinitrile (3333-52-6)
Tetramethylsuccinonitrile (3333-52-6)
Tetramethyl succinonitrile (ACGIH,OSHA) (3333-52-6)
Tetramethylsukcinonitril (Czech) (3333-52-6)
2,2,5,7-Tetramethyltetralin (23342-25-8)
2,4,6,8-Tetramethyl-1,3,5,7-tetraoxacyclooctane or acetaldehyde homopolymer (9002-91-9)
1,3,3,5-Tetramethyl-1,1,5,5-tetraphenyltrisiloxane (3982-82-9)
1,3,5,7-Tetramethyl-1,3,5,7-tetravinylcyclotetrasiloxane (2554-06-5)
2,4,6,8-Tetramethyl-1,3,5,7-tetroxocane (108-62-3)
r-2,c-4,c-6,c-8-Tetramethyl-1,3,5,7-tetroxocane (108-62-3)
2,2,4,4-Tetramethyl-3-thiapentane (107-47-1)
Tetramethylthiocarbamoyldisulphide (137-26-8)
O,O,O',O'-Tetramethyl O,O'-thiodi-p-phenylene phosphorothioate (3383-96-8)
Tetramethyl-O,O'-thiodi-p-phenylene phosphorothioate (3383-96-8)
Tetramethylthionine chloride (61-73-4)
Tetramethylthioramdisulfide (Dutch) (137-26-8)
1,1,3,3-Tetramethylthiourea (2782-91-4)
Tetramethylthiourea (2782-91-4)
Tetramethyl-thiram disulfid (German) (137-26-8)
Tetramethylthiuram (137-26-8)
Tetramethylthiuram bisulfide (137-26-8)
Tetramethylthiuram bisulphide (137-26-8)
N,N,N',N'-Tetramethylthiuram disulfide (137-26-8)
Tetramethylthiuram disulfide (137-26-8)
N,N-Tetramethylthiuram disulphide (137-26-8)
Tetramethylthiuram disulphide (137-26-8)
Tetramethylthiurammonium sulfide (97-74-5)
Tetramethylthiuram monosulfide (97-74-5)
Tetramethylthiuramonosulfide (97-74-5)
Tetramethylthiuram sulfide (97-74-5)
Tetramethylthiuram tetrasulfide (97-91-6)
Tetramethylthiuran disulphide (137-26-8)
Tetramethyl thiurane disulfide (137-26-8)
Tetramethyl thiurane disulphide (137-26-8)
Tetramethylthiurum disulfide (137-26-8)
Tetramethylthiurum disulphide (137-26-8)
Tetramethyl tin (594-27-4)
1,1,3,3-Tetramethylurea (632-22-4)
Tetramethylurea (632-22-4)
Tetramethyluree (French) (632-22-4)
2,2,4,4-Tetramethylvaleric acid (3302-12-3)
α,α,α',α'-Tetramethyl-p-xylylenediol (2948-46-1)
N,N,N',N'-Tetrametil-fosforodiammido-fluoruro (Italian) (115-26-4)
Tetramine (2058-46-0)
Tetramine Fast Brown BRDN Extra (16071-86-6)
Tetramine Fast Brown BRP (16071-86-6)
Tetramine Fast Brown BRS (16071-86-6)
Tetramine Fast Red 8B (2610-11-9)
Tetramine Fast Yellow Extra-Greenish (10190-68-8)
1,4,5,8-Tetraminoanthraquinone (2475-45-8)
(-)-Tetramisole (14769-73-4)
l-Tetramisole (14769-73-4)
Tetramminezinc(2+), carbonate (38714-47-5)
Tetram monooxalate (3734-97-2)
Tetramon J (68-05-3)
Tetramycin (64-75-5)
Tetran (79-57-2)
Tetranactin (33956-61-5)
Tetranap (119-64-2)
Tetranaphtho(3,2,1-de:1',2',3'-jk:3'',2'',1''-op:1''',2''',3'''-uv)p entacene (9CI) (72382-92-4)
Tetranatrium ethylendiamintetraacetat (Czech) (64-02-8)
Tetranatriumpyrophosphat (German) (7722-88-5)
Tetran hydrochloride (2058-46-0)
Tetranitraniline (French) (3698-54-2)
2,3,4,6-Tetranitroaniline (3698-54-2)
N,2,4,6-Tetranitroaniline (4591-46-2)
Tetranitroaniline (3698-54-2)
Tetranitroaniline [UN 0207] (53014-37-2)
2,4,5,7-Tetranitrochrysazin (517-92-0)
Tetranitro diglycerin (20600-96-8)
2,3,4,6-Tetranitrofenol (Spanish) (641-16-7)
2,4,5,7-Tetranitro-9H-fluoren-9-one (746-53-2)
2,4,5,7-Tetranitrofluoren-9-one (746-53-2)
Tetranitromethane (ACGIH,OSHA) [UN 1510] (509-14-8)
1,3,6,8-Tetranitronaphthalene (28995-89-3)
Tetranitropentaerythrite (78-11-5)
2,3,4,6-Tetranitrophenol (641-16-7)
Tetranitro-2,3,4,6 phenol (French) (641-16-7)
1,3,6,8-Tetranitropyrene (28767-61-5)
Tetra-n-octylstannane (3590-84-9)
Tetraoctylstannane (3590-84-9)
Tetra-n-octyltin (3590-84-9)
Tetraoctyltin (3590-84-9)
1,4,7,10-Tetraoxacyclododecane (294-93-9)
cis-1,4,5,8-Tetraoxadecalin (13405-83-9)
2,4,8,10-Tetraoxa-3,9-diphosphaspiro(5.5)undecane, 3,9-bis-(2,4-bis(1,1-dimethylethyl)phenoxy)- (26741-53-7)
2,4,8,10-Tetraoxa-3,9-diphosphaspiro(5.5)undecane, 3,9-bis-(isodecyloxy)- (9CI) (26544-27-4)
2,4,8,10-Tetraoxa-3,9-diphosphaspiro(5.5)undecane, 3,9-di-phenoxy- (9CI) (144-35-4)
2,5,8,11-Tetraoxadodecane (112-49-2)
2,2'-(2,5,8,11-Tetraoxa-1,12-dodecane diyl)bisoxirane (1954-28-5)
2,2'-(2,5,8,11-Tetraoxa-1,2-dodecanediyl)bisoxirane (1954-28-5)
4,7,10,13-Tetraoxahexadecane, 1,2:15,16-diepoxy- (8CI) (1954-28-5)
3,6,9,12-Tetraoxahexadecan-1-ol (9CI) (1559-34-8)
2,4,8,10-Tetraoxaspiro(5,5)undecane, 3,9-di(3-cyclohexenyl) (6600-31-3)
3,6,9,12-Tetraoxatetracosan-1-ol (9CI) (5274-68-0)

Name	CAS No.
3,6,9,12-Tetraoxatetradecan-1-ol (9CI)	**(5650-20-4)**
2,5,8,11-Tetraoxatetradecan-13-ol, 4,7,10-trimethyl- (9CI)	**(20324-34-9)**
3,6,9,12-Tetraoxatetradocane-1,14-diol	**(4792-15-8)**
7,9,11,13-Tetraoxa-6,8,10,12-tetrasilaspiro(5.7)tridecane, 8,8,10,10,12,12-hexamethyl- (9CI)	**(35331-58-9)**
6,12,18,24-Tetraoxa-5,7,13,19-tetrasilatetraspiro(4.1.4.1.4.1.4.1)-tetracosane (7CI,8CI,9CI)	**(177-49-1)**
3,6,9,12-Tetraoxatridecan-1-ol	**(23783-42-8)**
2,5,8,11-Tetraoxatridecan-13-ol (9CI)	**(23783-42-8)**
2,5,8,10-Tetraoxatridec-12-enoic acid, 9-oxo-, 2-propenyl ester (9CI)	**(142-22-3)**
(+-)-(3,5,3',5'-Tetraoxo)-1,2-dipiperazinopropane	**(21416-87-5)**
2,4,5,6-Tetraoxohexahydropyrimidine	**(50-71-5)**
2,4,5,6-Tetraoxohexahydropyrimidine hydrate	**(2244-11-3)**
Tetraphene	**(56-55-3)**
p-Tetraphenyl	**(135-70-6)**
Tetraphenylborate(1-) sodium	**(143-66-8)**
Tetraphenyl boron sodium salt	**(143-66-8)**
1,1,3,3-Tetraphenyl-1,3-dimethyldisiloxane	**(807-28-3)**
1,1,1',1'-Tetraphenyldimethyl ether	**(574-42-5)**
Tetraphenylolethane, Epichlorohydrin epoxy resin	**(7328-97-4)**
(Tetraphenyl-29H,31H-phthalocyaninato(2-)-N29,N30, N31,N32)copper	**(1330-40-1)**
Tetraphenyl sodium borate	**(143-66-8)**
Tetraphenylstannane	**(595-90-4)**
1,1,5,5-Tetraphenyl-1,3,3,5-tetramethyltrisiloxane	**(3982-82-9)**
Tetraphenyl tin	**(595-90-4)**
Tetraphosphate hexaethylique (French)	**(757-58-4)**
Tetraphosphor (German)	**(7723-14-0)**
Tetraphosphoric acid	**(8017-16-1)**
Tetraphosphoric acid, hexaethyl ester	**(757-58-4)**
Tetraphosphoric acid, hexaethyl ester, Mixed with compressed gas	**(757-58-4)**
Tetraphosphoric acid, hexaethyl ester, 0% to 25%	**(757-58-4)**
Tetraphosphorus trisulfide	**(1314-85-8)**
Tetrapom	**(137-26-8)**
Tetrapotassium diphosphorate	**(7320-34-5)**
Tetrapotassium ethylenediaminetetraacetate	**(5964-35-2)**
Tetrapotassium pyrophosphate	**(7320-34-5)**
(Tetrapropenyl)butanedioic acid	**(27859-58-1)**
Tetrapropenylsuccinic anhydride	**(26544-38-7)**
Tetrapropoxysilane	**(682-01-9)**
Tetra-n-propylammonium iodide	**(631-40-3)**
Tetrapropylammonium iodide	**(631-40-3)**
Tetrapropyldithiodifosfat (Czech)	**(3244-90-4)**
Tetra-n-propyl dithionopyrophosphate	**(3244-90-4)**
O,O,O,O-Tetrapropyl dithiopyrophosphate	**(3244-90-4)**
Tetra-n-propyl dithiopyrophosphate	**(3244-90-4)**
Tetrapropylene	**(25378-22-7)**
Tetrapropylene	**(6842-15-5)**
Tetrapropylenebenenesulphonate, sodium salt	**(11067-82-6)**
Tetrapropylene benzenesulfonate	**(11067-81-5)**
Tetrapropylene glycol, monomethyl ether (8CI) (VAN)	**(20324-34-9)**
Tetrapropyl lead	**(3440-75-3)**
Tetrapropylplumbane	**(3440-75-3)**
Tetrapropyl silicate	**(682-01-9)**
Tetrapropyl zirconate	**(23519-77-9)**
Tetrasan	**(102-30-7)**
Tetra(silacyclopentyl)cyclotetrasiloxane	**(177-49-1)**
Tetrasiloxane, decamethyl- (9CI)	**(141-62-8)**
Tetrasiloxane, 1,1,1,3,3,5,7,7,7-nonamethyl-5-phenyl- (7CI,8CI, 9CI)	**(17906-09-1)**
Tetrasiloxane, 1,1,1,3,5,7,7,7-octamethyl-3,5-bis(3-(oxiranyl-methoxy)propyl)- (9CI)	**(69155-42-6)**
Tetrasiloxane, 1,1,1,3,5,7,7,7-octamethyl-3,5-bis(2-(trimethyl-silyl)ethyl)- (9CI)	**(91455-17-3)**
Tetrasiloxane, 1,1,1,3,5,7,7,7-octamethyl-3,5-diphenyl- (6CI,7CI,8CI,9CI)	**(13270-97-8)**
Tetrasiloxane, 1,3,5,7-tetramethyl-1,1,3,5,7,7-hexaphenyl- (9CI)	**(38421-40-8)**
Tetrasipton	**(137-26-8)**
Tetrasodium N-(1,2-dicarboxyethyl)-N-octadecyl sulfosuccinate	**(3401-73-8)**
Tetrasodium dicarboxyethyl stearyl sulfosuccinamate	**(3401-73-8)**
Tetrasodium diphosphate	**(7722-88-5)**
Tetrasodium edetate	**(64-02-8)**
Tetrasodium EDTA	**(64-02-8)**
Tetrasodium ethylenediaminetetraacetate	**(64-02-8)**
Tetrasodium ethylenediaminetetracetate	**(64-02-8)**
Tetrasodium (ethylenedinitrilo)tetraacetate	**(64-02-8)**
Tetrasodium etidronate	**(3794-83-0)**
Tetrasodium ferrocyanide	**(13601-19-9)**
Tetrasodium hexacyanoferrate	**(13601-19-9)**
Tetrasodium hexacyanoferrate(4-)	**(13601-19-9)**
Tetrasodium pyrophosphate (ACGIH,OSHA)	**(7722-88-5)**
Tetrasodium pyrophosphate, Anhydrous	**(7722-88-5)**
Tetrasodium salt EDTA	**(64-02-8)**
Tetrasodium salt of EDTA	**(64-02-8)**
Tetrasodium salt of ethylenediaminetetracetic acid	**(64-02-8)**
Tetrasol	**(54-95-5)**
Tetrasol	**(56-23-5)**
Tetrastigmine	**(107-49-3)**
Tetrasul	**(2227-13-6)**
Tetrasule	**(78-11-5)**
Tetrasulfide, bis((dimethylamino)thioxomethyl) (9CI)	**(97-91-6)**
Tetrasulfide, bis(dimethylthiocarbamoyl) (8CI)	**(97-91-6)**
Tetrasulfide, bis(pentamethylenethiuram)-	**(120-54-7)**
Tetrasulfide, bis(piperidinothiocarbonyl)	**(120-54-7)**
n-Tetratetracontane	**(7098-22-8)**
Tetratetracontane (9CI)	**(7098-22-8)**
Tetrathiin	**(55290-64-7)**
Tetrathiin (Desiccant)	**(55290-64-7)**
Tetrathionic acid, dipotassium salt (8CI,9CI)	**(13932-13-3)**
Tetrathiuram disulfide	**(137-26-8)**
Tetrathiuram disulphide	**(137-26-8)**
n-Tetratriacontane	**(14167-59-0)**
Tetratriacontane (9CI)	**(14167-59-0)**
Tetravec	**(127-18-4)**
Tetraverine	**(60-54-8)**
Tetravinyltetramethylcyclotetrasiloxane	**(2554-06-5)**
Tetravos	**(62-73-7)**
Tetra-wedel	**(64-75-5)**
7,8,9,10-Tetrazabicyclo(5.3.0)-8,10-decadiene	**(54-95-5)**
1,2,3,3a-Tetrazacyclohepta-8a,2-cyclopentadiene	**(54-95-5)**
1,3,5,7,2,4,6,8-Tetrazatetraphosphocine, 2,2,4,4,6,6,8,8-octa-chloro-2,2,4,4,6,6,8,8-octahydro- (9CI)	**(2950-45-0)**
Tetrazene (DOT)	**(109-27-3)**
1-Tetrazene, 4-amidino-1-(nitrsoaminoamidino)- (8CI)	**(109-27-3)**
1-Tetrazene-1-carboximidic acid, 4-(aminoiminomethyl)-, 2-nitrosohydrazide (9CI)	**(109-27-3)**
1-Tetrazene, 1-guanyl-4-nitrosaminoguanyl	**(109-27-3)**
Tetrazepam	**(10379-14-3)**
Tetrazepamum (Latin)	**(10379-14-3)**
Tetrazido-1,4-benzoquinone	**(22826-61-5)**
Tetrazine, Dry	**(70816-59-0)**
Tetrazine (French)	**(70816-59-0)**
Tetrazo Deep Black G	**(1937-37-7)**
Tetrazo Deep Black GC Extra	**(6428-31-5)**
Tetrazo Deep Black R	**(2429-83-6)**
Tetrazobenzene-β-naphthol	**(85-86-9)**
1,3,5,7-Tetrazocine, 1-acetyloctahydro-3,5,7-trinitro- (8CI,9CI)	**(13980-00-2)**
1,3,5,7-Tetrazocine, octahydro-1,3,5,7-tetranitro	**(2691-41-0)**
Tetrazol	**(54-95-5)**
1H-Tetrazol-5-amine	**(4418-61-5)**
1H-Tetrazole, 5-amino	**(4418-61-5)**
Tetrazole, pentamethylene-	**(54-95-5)**
2H-Tetrazole, 5-phenyl-	**(18039-42-4)**

1H-Tetrazole, 5-phenyl- (8CI,9CI)	(18039-42-4)
1H-Tetrazole-5-thiol, 1-methyl	(13183-79-4)
1H-Tetrazole-5-thiol, 1-phenyl	(86-93-1)
5H-Tetrazolo(1,5-a)azepine, 6,7,8,9-tetrahydro- (8CI,9CI)	(54-95-5)
5H-Tetrazoloazepine, 6,7,8,9-tetrahydro	(54-95-5)
Tetril	(479-45-8)
Tetrine	(64-02-8)
Tetrine acid	(60-00-4)
Tetron-100	(107-49-3)
Tetrodirect Black EFD	(1937-37-7)
Tetrodirect Black V	(6428-31-5)
Tetrodirect Yellow A	(1325-37-7)
Tetrodontoxin	(4368-28-9)
Tetrodotoxin	(4368-28-9)
Tetrodoxin	(4368-28-9)
Tetroguer	(127-18-4)
Tetrole	(110-00-9)
Tetron	(107-49-3)
Tetrone A	(120-54-7)
Tetronic 701	(11111-34-5)
Tetropil	(127-18-4)
Tetrosan	(8023-53-8)
Tetrosin LY	(92-52-4)
Tetrosin OE	(90-43-7)
Tetrosol	(64-75-5)
1,3,5,7-Tetroxocane, 2,4,6,8-tetramethyl-	(108-62-3)
1,2,4,5-Tetroxonane, 3,3,6,6,9,9-hexamethyl- (8CI,9CI)	(22397-33-7)
1,4,5,8-Tetroxyantraquinone	(117-10-2)
2,4,6-Tetryl	(479-45-8)
Tetryl (ACGIH,OSHA) [UN 0208]	(479-45-8)
Tetrylammonium bromide	(71-91-0)
Tetryl formate	(542-55-2)
Teturam	(97-77-8)
Teturamin	(97-77-8)
Tevabolin	(10418-03-8)
Tevcocin	(56-75-7)
Tevcodyne	(50-33-9)
Tevcosin	(56-75-7)
Tex-Wet 1001	(577-11-7)
Texa Pon A 400	(2235-54-3)
1,2,3,6,7,8-Texachlorodibenzofuran	(57117-44-9)
Texaco Lead Appreciator	(540-88-5)
Texacort Lotion 25	(50-23-7)
Texan Red Toner D	(5160-02-1)
Texanol	(25265-77-4)
Texapon DL Conc.	(151-21-3)
Texapon DLS	(143-00-0)
Texapon K-1296	(151-21-3)
Texapon K12	(151-21-3)
Texapon L 100	(151-21-3)
Texapon N 25	(9004-82-4)
Texapon N 40	(9004-82-4)
Texapon NSO	(9004-82-4)
Texapon Special	(2235-54-3)
Texapon T-35	(139-96-8)
Texapon T-42	(139-96-8)
Texapon TH	(139-96-8)
Texapon V HC	(151-21-3)
Texapon V HC Powder	(151-21-3)
Texapon ZHC	(151-21-3)
Texapon Z High Conc. Needles	(151-21-3)
Texas	(1333-86-4)
Texcryl	(9003-01-4)
Texofor B 30	(9002-92-0)
Texofor B 9	(9002-92-0)
Textile	(7758-19-2)
Textile Red WD-263	(6471-49-4)
Textone	(7758-19-2)
ThG	(154-42-7)
Thacapzol	(60-56-0)
Thalamonal	(548-73-2)
Thalidomide	(50-35-1)
Thalidomidum (Latin)	(50-35-1)
Thalin	(50-35-1)
Thalinette	(50-35-1)
Thallic-oxide	(1314-32-5)
Thallium (ACGIH,OSHA)	(7440-28-0)
Thallium(I) acetate	(563-68-8)
Thallium(I) carbonate (2:1)	(6533-73-9)
Thallium(I) chloride	(7791-12-0)
Thallium(I) chromate	(13473-75-1)
Thallium(I) malonate	(2757-18-8)
Thallium(I) nitrate (1:1)	(10102-45-1)
Thallium(I) sulfate (2:1)	(7446-18-6)
Thallium acetate	(563-68-8)
Thallium(1+) acetate	(563-68-8)
Thallium chloride	(7791-12-0)
Thallium(1+) chloride	(7791-12-0)
Thallium malonate	(2757-18-8)
Thallium monoacetate	(563-68-8)
Thallium monochloride	(7791-12-0)
Thallium mononitrate	(10102-45-1)
Thallium monoselenide	(12039-52-0)
Thallium nitrate [UN 2727]	(10102-45-1)
Thallium(111) oxide	(1314-32-5)
Thallium(3+) oxide	(1314-32-5)
Thallium oxide (8CI,9CI)	(1314-32-5)
Thallium peroxide	(1314-32-5)
Thallium-selenide	(12039-52-0)
Thallium sesquioxide	(1314-32-5)
Thallium sulfate	(7446-18-6)
Thallium(1) sulfate	(7446-18-6)
Thallium(II) sulfate (1:1)	(63906-56-9)
Thallium-sulfate	(10031-59-1)
Thallium sulfate, Solid [NA]	(10031-59-1)
Thallous acetate	(563-68-8)
Thallous carbonate	(6533-73-9)
Thallous chloride	(7791-12-0)
Thallous malonate	(2757-18-8)
Thallous nitrate	(10102-45-1)
Thallous sulfate	(7446-18-6)
Thalo Green No.1	(1328-53-6)
Thanisol	(115-31-1)
Thanite	(115-31-1)
Thanol SF 1500	(25791-96-2)
Thapsic acid	(505-54-4)
Thaumatin	(53850-34-3)
Theo-Dur	(58-55-9)
Theal	(479-18-5)
Theal Ampules	(479-18-5)
Theal Tabl.	(58-55-9)
Thebacon	(466-90-0)
Thebacone (French)	(466-90-0)
Thebaconum (Latin)	(466-90-0)
Thebaine	(115-37-7)
Thebaine, demethyldihydro-, acetate	(466-90-0)
Thebaine, picrate	(5967-77-1)
Thebenidine	(194-03-6)
Theed	(140-07-8)
Theelin	(53-16-7)
Theelin, dihydro-	(50-28-2)
Theelol	(50-27-1)
Thefanil	(91-79-2)
Thefylan	(479-18-5)

Thein (58-08-2)
Theine (58-08-2)
Theithylperazine (1420-55-9)
Thelestrin (53-16-7)
Thelykinin (53-16-7)
Themalon (86-14-6)
Thenalton (81-13-0)
Thenardite (7757-82-6)
Thenardol (124-43-6)
Thenfadil (91-79-2)
Thenfadil hydrochloride (958-93-0)
Thenobarbital (50-06-6)
3-Thenoic acid (88-13-1)
1-Thenoyl-3,3,3-trifluoroacetone (326-91-0)
2-Thenoyltrifluoroacetone (326-91-0)
Thenoyltrifluoroacetone (326-91-0)
α-Thenoyltrifluoroacetone (326-91-0)
Thenyl D.P.E. Hydrochloride (135-23-9)
2-Thenylamine (8CI) (27757-85-3)
2-Thenylamine, 5-chloro-N-(2-(dimethylamino)ethyl)-N-2-pyridyl- (148-65-2)
Thenyldiamine (91-79-2)
Thenyldiamine chloride (958-93-0)
Thenyldiamine hydrochloride (958-93-0)
Thenylene (135-23-9)
Thenylene (91-80-5)
Thenylene hydrochloride (135-23-9)
Thenylpyramine (91-80-5)
Thenylpyramine hydrochloride (135-23-9)
Theobroma Oil (8002-31-1)
Theobromine (83-67-0)
Theobromine, 1-methyl- (58-08-2)
Theocin (58-55-9)
Theodrox (317-34-0)
Theofol (58-55-9)
Theograd (58-55-9)
Theoharn (108-78-1)
Theolair (58-55-9)
Theolamine (317-34-0)
Theolix (58-55-9)
Theolone (317-34-0)
Theoloxin (50-06-6)
Theomin (317-34-0)
Theominal (50-06-6)
Theophilcholine (50-35-1)
Theophyl-225 (58-55-9)
Theophyldine (317-34-0)
Theophyline (58-55-9)
Theophyllamine (317-34-0)
Theophyllaminium (317-34-0)
Theophyllin (58-55-9)
Theophyllin aethylendiamin (German) (317-34-0)
Theophylline (58-55-9)
Theophylline, Anhydrous (58-55-9)
Theophylline, Compd. with ethylenediamine (2:1) (317-34-0)
Theophylline, 8-bromo- (10381-75-6)
Theophylline, 8-chloro-, Compd. with 2-(diphenylmethoxy)-N,N-dimethylethylamine (1:1) (523-87-5)
Theophylline, 7-(2,3-dihydroxypropyl) (479-18-5)
Theophylline ethylenediamine (317-34-0)
Theophylline, 7-methyl (58-08-2)
Theophyllin ethylenediamine (317-34-0)
Theosalvose (83-67-0)
Theostene (83-67-0)
Thephorin A-C (62-44-2)
Thephyldine (317-34-0)
Theptine (60-13-9)
Thera-Flur-N (7681-49-4)
Thera-Flur (7681-49-4)
Theraderm (94-36-0)
Theradiazine (68-35-9)
Theralax (603-50-9)
Theramine (51-45-6)
Therapav (61-25-6)
Therapol (63-74-1)
Therazone (50-33-9)
Therma-Atomic Black (1333-86-4)
Thermacure (1338-23-4)
Thermal Acetylene Black (1333-86-4)
Thermal Cracked Residuum (Petroleum) (64741-80-6)
Thermally Cracked Residue (64741-80-6)
Thermalox (1304-56-9)
Thermatomic (1333-86-4)
Thermax (1333-86-4)
Thermblack (1333-86-4)
Therm chek 820 (77-58-7)
Thermia-C (8012-95-1)
Therminol 66 (9CI) (54578-28-8)
Therminol FR-1 (1336-36-3)
Thermoase PC-10 (9014-01-1)
Thermocomp BF (9003-54-7)
Thermoguard B (1309-64-4)
Thermoguard S (1309-64-4)
Thermolite 813 (16091-18-2)
Thermolite 831 (26401-97-8)
Thermoplastic 125 (9003-55-8)
Thermphos (7758-29-4)
Thermphos L 50 (7758-29-4)
Thermphos N (7758-29-4)
Thermphos SPR (7758-29-4)
Thesal (83-67-0)
Thesat (9002-92-0)
Thesit (9002-92-0)
Thesodate (83-67-0)
Thetamid (77-67-8)
Thiaanthrene (92-85-3)
4-Thia-1-azabicyclo(3.2.0)heptane-2-carboxilic acid, 3,3-dimethyl-7-oxo-6-(2-phenyl acetamido)-, Compd. with 2-(diethylamino)ethyl p-aminobenzoate (1:1), monohydrate (6130-64-9)
4-Thia-1-azabicyclo(3.2.0)heptane-2-carboxylic acid, 6-((amino(4-hydroxyphenyl)acetyl) amino)-3,3-dimethyl-7-oxo (26787-78-0)
4-Thia-1-azabicyclo(3.2.0)heptane-2-carboxylic acid, 6-((aminophenylacetyl)amino)-3,3-dimethyl-7-oxo-, 1,3-dihydro-3-oxo-1-isobenzofuranyl ester, (2S-(2α,5α,6β(S*)))- (9CI) (47747-56-8)
4-Thia-1-azabicyclo(3.2.0)heptane-2-carboxylic acid, 6-(2-amino-2-phenylacetamido)- 3,3-dimethyl-7-oxo-, trihydrate, D- (-) (7177-48-2)
4-Thia-1-azabicyclo(3.2.0)heptane-2-carboxylic acid, 6-(2-amino-2-phenylacetamido)-3,3- dimethyl-7-oxo-, D-(-) (69-53-4)
4-Thia-1-azabicyclo(3.2.0)heptane-2-carboxylic acid, 6-(((3-(2-chloro-6-fluorophenyl)-5-methyl-4-isoxazolyl)-carbonyl)amino)-3,3-dimethyl-7-oxo-, (2S(2α,5α,6β)) (5250-39-5)
4-Thia-1-azabicyclo(3.2.0)heptane-2-carboxylic acid, 6-(3-(o-chlorophenyl)-s-methyl-4-isoxazolecarboxamido)-3,3-dimethyl-7-oxo (61-72-3)
4-Thia-1-azabicyclo(3.2.0)heptane-2-carboxylic acid, 6-(((3-(2,6-dichlorophenyl)-5-methyl-4-isoxazolyl)carbonyl)-amino)-3,3-dimethyl-7-oxo-, (2S-(2α,5α,6β))- (3116-76-5)
4-Thia-1-azabicyclo(3.2.0)heptane-2-carboxylic acid, 6-(2,6-dimethoxybenzamido)-3,3,- dimethyl-7-oxo (61-32-5)
4-Thia-1-azabicyclo(3.2.0)heptane-2-carboxylic acid, 3,3-dimethyl-6-(((((3-(methylsulfonyl)- 2-oxo-1-imidazolidinyl)carbonyl)amino)phenylacetyl)amino)-7-oxo-, sodium salt, (2s-(2-α,5-α,6-β)(S*)) (51481-65-3)

4-Thia-1-azabicyclo(3.2.0)heptane-2-carboxylic acid, 3,3-dimethyl-7-oxo-6- (2-phenylacetamido) **(61-33-6)**
4-Thia-1-azabicyclo(3.2.0)heptane-2-carboxylic acid, 3,3-dimethyl-7-oxo-6- (2-phenylacetamido)-, Compd. with N,N'-dibenzylethylenediamine (2:1) **(1538-09-6)**
4-Thia-1-azabicyclo(3.2.0)heptane-2-carboxylic acid, 3,3-dimethyl-7-oxo-6- (2-phenylacetamido)-, mono-potassium salt **(113-98-4)**
4-Thia-1-azabicyclo(3.2.0)heptane-2-carboxylic acid, 3,3-dimethyl-7-oxo-6-(2- phenoxypropionamido)-, mono-potassium salt **(132-93-4)**
4-Thia-1-azabicyclo(3.2.0)heptane-2-carboxylic acid, 3,3-dimethyl-7-oxo-6-(2- phenylacetamido)-, Compd. with 2-(diethylamino)ethyl p-aminobenzoate (1:1) **(54-35-3)**
4-Thia-1-azabicyclo(3.2.0)heptane-2-carboxylic acid, 3,3-dimethyl-7-oxo-6-(2-phenoxy- acetamido)-, monopotassium salt **(132-98-9)**
4-Thia-1-azabicyclo(3.2.0)heptane-2-carboxylic acid, 3,3-dimethyl-7-oxo-6-(2-phenyl acetamido)-, monosodium salt **(69-57-8)**
4-Thia-1-azabicyclo(3.2.0)heptane-2-carboxylic acid, 6-(2-ethoxy-1-naphthamido)-3,3- dimethyl-7-oxo **(147-52-4)**
5-Thia-1-azabicyclo(4.2.0)oct-2-ene-2-carboxylic acid, 3-((acetyl-oxy)methyl)-7-(((2-amino-4-thiazolyl)(methoxyimino)acetyl)-amino)-8-oxo-, (6R-trans)- (9CI) **(60846-21-1)**
5-Thia-1-azabicyclo(4.2.0)oct-2-ene-2-carboxylic acid, 3-(((amino-carbonyl)oxy)methyl)-7-methoxy-8-oxo-7-((2-thienylacetyl)-amino)-, (6R-cis) **(35607-66-0)**
5-Thia-1-azabicyclo(4.2.0)oct-2-ene-2-carboxylic acid, 3-(((amino-carbonyl)oxy)methyl)-7-((2-furanyl(methyoxyimino)acetyl)-amino)-8-oxo-, (6R-(6-α,7-β(Z))) **(55268-75-2)**
5-Thia-1-azabicyclo(4.2.0)oct-2-ene-2-carboxylic acid, 7-((amino-1,4-cyclohexadien-1-ylacetyl)amino)-3-methyl-8-oxo-, (6r-(6-α,7-β(R*))) **(38821-53-3)**
5-Thia-1-azabicyclo(4.2.0)oct-2-ene-2-carboxylic acid, 7-((amino-1,4-cyclohexadien-1-ylacetyl)amino)-3-methoxy-8-oxo-, (6R-(6-α,7-β(R*))) **(51762-05-1)**
5-Thia-1-azabicyclo(4.2.0)oct-2-ene-2-carboxylic acid, 7-(2-amino-2-phenylacetamido)- 3-(hydroxymethyl)-8-oxo-, acetate (ester), D **(3577-01-3)**
5-Thia-1-azabicyclo(4.2.0)oct-2-ene-2-carboxylic acid, 7-(2-amino-2-phenylacetamido)-3- methyl-8-oxo-, D **(15686-71-2)**
5-Thia-1-azabicyclo(4.2.0)oct-2-ene-2-carboxylic acid, 7-((aminophenylacetyl)amino)-3-chloro- 8-oxo-, (6R-(6-α,7-β(R*))) **(53994-73-3)**
5-Thia-1-azabicyclo(4.2.0)oct-2-ene-2-carboxylic acid, 7-(2-cyanoacetamido)-3-(hydroxy methyl)-8-oxo-, acetate (ester), monosodium salt **(23239-41-0)**
5-Thia-1-azabicyclo(4.2.0)oct-2-ene-2-carboxylic acid, 3-(hy-droxymethyl)-8-oxo-7-(2-(4-pyridylthio)acetamido)-, acetate (ester) **(21593-23-7)**
5-Thia-1-azabicyclo(4.2.0)oct-2-ene-2-carboxylic acid, 7-((hy-droxyphenylacetyl)amino)-3-(((1-methyl-1H-tetrazol-5-yl)thio)-methyl)-8-oxo-, (6R-(6-α,7-β(R*))) **(34444-01-4)**
5-Thia-1-azabicyclo(4.2.0)oct-2-ene-2-carboxylic acid, 7-((amino (4-hydroxyphenyl)acetyl) amino)-3-methyl-8-oxo-, (6R-(6-α,7-β(R*))) **(50370-12-2)**
5-Thia-1-azabicyclo(4.2.0)oct-2-ene-2-carboxylic acid, 3-(hydroxy-methyl)-8-oxo-7-(2-(2- thienyl)acetamido)-, acetate (ester) **(153-61-7)**
5-Thia-1-azabicyclo(4.2.0)oct-2-ene-2-carboxylic acid, 3-(((5-methyl-1,3,4-thiadiazol-2-yl)thio)methyl)-8-oxo-7-(2-(1H-tetrazol-1-yl)acetamido) **(25953-19-9)**
1-Thia-2-azaindene **(272-16-2)**
1-Thia-3-azaindene **(95-16-9)**
Thiaben **(148-79-8)**
Thiabendazol **(148-79-8)**
Thiabendazole **(148-79-8)**
Thiabendazole **(28558-32-9)**
Thiabendazole hypophosphite **(28558-32-9)**
Thiabenzazole **(148-79-8)**
Thiabenzole **(148-79-8)**
7-Thiabicyclo(4.1.0)heptane (9CI) **(286-28-2)**
3-Thiabutan-2-one, O-(methylcarbamoyl)oxime **(16752-77-5)**
Thiacetamide **(62-55-5)**
Thiacetic acid **(507-09-5)**
Thiacoccine **(72-14-0)**
Thiacyclobutane **(287-27-4)**
Thiacyclohexane **(1613-51-0)**
Thiacyclopentadiene **(110-02-1)**
Thiacyclopentane **(110-01-0)**
Thiacyclopentane dioxide **(126-33-0)**
Thiacyclopropane **(420-12-2)**
4H,6H-1,3,5-Thiadiazineacetic acid, 5-methyl-6-thioxo **(3655-88-7)**
2H-1,3,5-Thiadiazine-2-thione, tetrahydro-3,5-dimethyl **(533-74-4)**
4H-1,3,5-Thiadiazin-4-one, 2-((1,1-dimethylethyl)imino)tetra-hydro-5-(4-hydroxyphenyl)-3-(1-methylethyl)- (9CI) **(69329-95-9)**
4H-1,3,5-Thiadiazin-4-one, 2-((1,1-dimethylethyl)imino)tetra-hydro-3-(1-methylethyl)-5-phenyl **(69327-76-0)**
Thiadiazinthion **(3655-88-7)**
1,3,4-Thiadiazol-2-amine, 5-(ethylthio)- (9CI) **(25660-70-2)**
1,3,4-Thiadiazol-2-amine, 5-(5-nitro-2-furanyl)- (9CI) **(712-68-5)**
1,2,4-Thiadiazol-5-amine, 3-phenyl **(17467-15-1)**
1,3,4-Thiadiazol-dithiol-(2,5) (German) **(1072-71-5)**
1,3,4-Thiadiazole, 2-amino-5-(ethylthio)- (8CI) **(25660-70-2)**
1,3,4-Thiadiazole, 2-amino-5-(5-nitro-2-furyl) **(712-68-5)**
1,2,4-Thiadiazole, 5-chloro-3-phenyl- (8CI,9CI) **(24255-23-0)**
1,2,4-Thiadiazole, 5-chloro-3-(trichloromethyl)- (9CI) **(5848-93-1)**
1,3,4-Thiadiazole-2,5-dithiol **(1072-71-5)**
1,2,4-Thiadiazole, 5-ethoxy-3-(trichloromethyl) **(2593-15-9)**
1,2,3-Thiadiazole, 5-methyl- (7CI,9CI) **(50406-54-7)**
1,2,4-Thiadiazole, 3-phenyl- (6CI,9CI) **(50483-82-4)**
1,3,4-Thiadiazole-2-sulfonamide, 5-acetamido- **(59-66-5)**
1,3,4-Thiadiazole-2-thiol, 5-amino **(2349-67-9)**
1,3,4-Thiadiazole-2(3H)-thione, 5-amino- (9CI) **(2349-67-9)**
(N-1,2,3-Thiadiazolyl-5)-N'-phenylurea **(51707-55-2)**
Thiadoxine **(67-03-8)**
Thiaflavan **(5961-99-9)**
9-Thiafluorene **(132-65-0)**
α-Thiafluorene **(132-65-0)**
Thiafur **(712-68-5)**
2-Thiaheptane **(1741-83-9)**
4-Thiaheptane **(111-47-7)**
4-Thiaheptanedioic acid **(111-17-1)**
1-Thiaindene **(95-15-8)**
Thialdine **(638-17-5)**
Thiamazole **(60-56-0)**
Thiambutene **(86-14-6)**
Thiameton **(640-15-3)**
Thiamin **(59-43-8)**
Thiaminal **(67-03-8)**
Thiamin chloride **(67-03-8)**
Thiamin dichloride **(67-03-8)**
Thiamine **(59-43-8)**
Thiamine chloride **(67-03-8)**
Thiamine, chloride **(59-43-8)**
Thiamine, chloride, hydrochloride **(67-03-8)**
Thiamine dichloride **(67-03-8)**
Thiamine hydrochloride **(67-03-8)**
Thiamine monochloride **(59-43-8)**
Thiamine thiazole **(137-00-8)**
Thiamin hydrochloride **(67-03-8)**
Thiaminium chloride hydrochloride **(67-03-8)**
Thiamol **(67-03-8)**
Thianaphtene **(95-15-8)**
Thianaphthen **(95-15-8)**
Thianaphthene **(95-15-8)**

Name	CAS No.
Thiane	(1613-51-0)
Thianid	(536-33-4)
Thianide	(536-33-4)
5-Thianonane	(544-40-1)
Thianonane-5	(544-40-1)
Thianthren	(92-85-3)
Thianthrene (9CI)	(92-85-3)
4-Thiapentanal	(3268-49-3)
Thia-4-pentanal [UN 2785]	(3268-49-3)
3-Thiapentane	(352-93-2)
Thiapentane (9CI)	(82987-03-9)
Thiaphene	(110-02-1)
2-Thiapropane	(75-18-3)
Thiaquinox	(93-75-4)
Thiaretic	(58-93-5)
Thiasin	(127-69-5)
Thiate E	(2489-77-2)
Thiate H	(105-55-5)
Thiate U	(109-46-6)
Thiavit	(67-03-8)
Thiaxanthenone	(492-22-8)
Thiaxanthon	(492-22-8)
Thiaxanthone	(492-22-8)
Thiazamide	(72-14-0)
Thiazide	(58-94-6)
Thiazidico	(73-48-3)
Thiazinamium methyl sulfate	(58-34-4)
2H-1,3-Thiazine, tetrahydro-2-(nitromethylene)-	(58842-20-9)
2H-1,3-Thiazine, tetrahydro-2-(nitromethylene)-, (E)- (9CI)	(97190-65-3)
4H-1,3-Thiazin-4-one, 2-(p-chlorophenyl)tetrahydro-3-methyl-, 1,1,-dioxide	(80-77-3)
4H-1,3-Thiazin-4-one, tetrahydro-2-(p-chlorophenyl)-3-methyl-, 1,1-dioxide	(80-77-3)
2-Thiazolamine	(96-50-4)
2-Thiazolamine, 4,5-dihydro- (9CI)	(1779-81-3)
2-Thiazolamine, 4,5-dihydro-, monohydrochloride (9CI)	(3882-98-2)
2-Thiazolamine, 5-nitro- (9CI)	(121-66-4)
2-Thiazolamine, 4-phenyl-, monohydrobromide, monohydrate (9CI)	(52253-69-7)
Thiazole	(288-47-1)
Thiazole, 2-acetamido-4-(5-nitro-2-furyl)-	(531-82-8)
Thiazole, 2-acetamino-4-(5-nitro-2-furyl)-	(531-82-8)
Thiazole, 2-acetylamino-4-(5-nitro-2-furyl)-	(531-82-8)
Thiazole, 2-amino	(96-50-4)
Thiazole, 2-amino-5-nitro	(121-66-4)
Thiazole, 2-butyl-4,5-dimethyl- (9CI)	(76572-48-0)
4-Thiazolecarbonitrile (9CI)	(1452-15-9)
4-Thiazolecarboxamidine, N-phenyl-, hydrochloride (8CI)	(13631-64-6)
4-Thiazolecarboxylic acid, 2-amino-4,5-dihydro-	(2150-55-2)
5-Thiazolecarboxylic acid, 2-amino-4-(trifluoromethyl)-, ethyl ester	(344-72-9)
Thiazole, 5-(2-chloroethyl)-4-methyl	(533-45-9)
Thiazole, 2,4-diamino-5-phenyl	(490-55-1)
Thiazole, 2,4-dimethyl	(541-58-2)
Thiazole, 4,5-dimethyl-2-(1-methylethyl)- (9CI)	(53498-30-9)
5-Thiazoleethanol, 4-methyl- (9CI)	(137-00-8)
Thiazole, 4-ethyl-2,5-dimethyl- (8CI,9CI)	(32272-57-4)
Thiazole, 2-ethyl-4-methyl- (9CI)	(15679-12-6)
Thiazole, 4-ethyl-5-methyl- (9CI)	(52414-91-2)
Thiazole, 4-methyl- (9CI)	(693-95-8)
Thiazole, 2-methyl-4-propyl- (9CI)	(41981-63-9)
Thiazole, 4-methyl-5-vinyl	(1759-28-0)
Thiazole, 4-propyl- (9CI)	(41981-60-6)
Thiazole, 5-propyl- (9CI)	(52414-82-1)
2,4-Thiazolidinedione, 2-thio-	(141-84-4)
Thiazolidine, 2-imino-, hydrochloride (8CI)	(3882-98-2)
Thiazolidine-2-methanol, 3-nitroso	(92134-93-5)
2-Thiazolidinethione, 3-methyl- (9CI)	(1908-87-8)
2-Thiazolidinimine	(1779-81-3)
4-Thiazolidinone, 2-thioxo-	(141-84-4)
2-Thiazoline, 2-amino	(1779-81-3)
2-Thiazoline-4-carboxylic acid, 2-amino	(2150-55-2)
Thiazolium, 2,2'-((2-carboxy-1,4-phenylene)bis(imino-2,1-ethenediyl))bis(3-ethyl-4,5-dihydro-, diiodide (9CI)	(20328-87-4)
Thiazolium, 3-methyl-2-((1-methyl-2-phenyl-1H-indol-3-yl)azo)-, chloride	(42373-04-6)
4-Thiazolol, 2-(4-chlorophenyl)-5-(4-methylphenyl)-, acetate (ester) (9CI)	(65752-48-9)
2-Thiazolylamine	(96-50-4)
2-(4'-Thiazolyl)benzimidazole	(148-79-8)
2-(4-Thiazolyl)-1H-benzimidazole	(148-79-8)
2-(4-Thiazolyl)benzimidazole	(148-79-8)
2-(Thiazol-4-yl)benzimidazole	(148-79-8)
2-(4-Thiazolyl)benzimidazole, hypophosphite salt	(28558-32-9)
N^1-2-Thiazolylsulfanilamide	(72-14-0)
N^1-2-Thiazolylsulfanilamide sodium salt	(144-74-1)
Thiazon	(533-74-4)
Thiazone	(533-74-4)
2-Thiazylamine	(96-50-4)
Thibenzol	(148-79-8)
Thibenzole	(148-79-8)
Thibenzole ATT	(148-79-8)
Thidiazuron	(51707-55-2)
Thidicur	(144-82-1)
Thieno(2,3-c)pyridine (8CI,9CI)	(272-12-8)
1H-Thieno(3,4-d)imidazole-4-pentanoic acid, hexahydro-2-oxo-, (3as-(3a-α,4-β, 6a-α))	(58-85-5)
2H-Thieno(2,3e)(1,4)-diazepin-2-one, 5-(o-chlorophenyl)-7-ethyl-1,3-dihydro-1-methyl- (8CI)	(33671-46-4)
2H-Thieno(2,3-e)(1,4)diazepin-2-one, 1,3-dihydro-5-(o-chlorophenyl)-7-ethyl-1-methyl-	(33671-46-4)
N-(5-(2-Thienoyl)-2-benzimidazolyl)carbamic acid methyl ester	(31430-18-9)
7-(2-Thienylacetamido)cephalosporanic acid	(153-61-7)
7-((2-Thienyl)acetamido)-3-(1-pyridylmethyl)cephalosporanic acid	(50-59-9)
7-(α-(2-Thienyl)acetamido)-3-(1-pyridylmethyl)-3-cephem-4-carboxylic acid betaine	(50-59-9)
2-Thienylacetic acid	(1918-77-0)
3-Thienylacetic acid	(6964-21-2)
2-Thienylaldehyde	(98-03-3)
2-Thienylcarboxaldehyde	(98-03-3)
((Thienyl-2)-1 cyclohexyle)-N piperidine (French)	(21500-98-1)
1-(1-(2-Thienyl)-cyclohexyl)-piperidine	(21500-98-1)
1-(1-(2-Thienyl)cyclohexyl)piperidine	(21500-98-1)
Thienylphencyclidine	(21500-98-1)
Thiergan	(60-87-7)
Thietane (9CI)	(287-27-4)
Thiethylperazine	(1420-55-9)
Thifor	(115-29-7)
Thiirane	(420-12-2)
Thiirene, 2,3-dihydro-	(420-12-2)
Thilane	(110-01-0)
Thillate	(137-26-8)
Thilopemal	(77-67-8)
Thilophenyl	(57-41-0)
Thilophenyt	(630-93-3)
Thimazole	(60-56-0)
Thimecil	(56-04-2)
Thimenox	(298-02-2)
Thimer	(137-26-8)
Thimerfonate sodium	(5964-24-9)
Thimerosal	(54-64-8)
Thimerosalate	(54-64-8)
Thimerosol	(54-64-8)
Thimersalate	(54-64-8)
Thimet	(298-02-2)

Thimet sulfone	**(2588-04-7)**
Thimet sulfoxide	**(2588-03-6)**
Thimul	**(115-29-7)**
Thio-Cara	**(333-20-0)**
Thio-TEP	**(52-24-4)**
Thio-TEPA	**(52-24-4)**
Thio-TEPA S	**(52-24-4)**
Thioacetaldehyde	**(2765-04-0)**
Thioacetamide	**(62-55-5)**
Thioacetic acid [UN 2436]	**(507-09-5)**
Thioacetimidic acid	**(62-55-5)**
Thioalkofen BM 4	**(96-69-5)**
Thioalkofen BMCh	**(96-69-5)**
Thioalkofen BP	**(90-66-4)**
Thioalkofen MBCh	**(96-69-5)**
Thioalkophene BM-4	**(96-69-5)**
Thioalkophene BP	**(90-66-4)**
Thioallate	**(95-06-7)**
Thioallyl ether	**(592-88-1)**
Thioamide	**(536-33-4)**
4,4'-Thioaniline	**(139-65-1)**
Thioaniline	**(139-65-1)**
Thioanisole	**(100-68-5)**
Thioarsenious acid (H3AsS3), tripropyl ester (8CI)	**(5582-57-0)**
2-Thiobarbituric acid	**(504-17-6)**
Thiobarbituric acid	**(504-17-6)**
Thiobel	**(15263-52-2)**
Thiobencarb	**(28249-77-6)**
Thiobencarbe	**(28249-77-6)**
Thiobenzamide	**(2227-79-4)**
Thiobenzyl alcohol	**(100-53-8)**
4,4'-Thiobis(aniline)	**(139-65-1)**
4,4'-Thiobisbenzenamine	**(139-65-1)**
1,1'-Thiobis(benzene)	**(139-66-2)**
4,4'-Thiobis(6-tert-butyl-m-cresol) (ACGIH,OSHA)	**(96-69-5)**
4,4'-Thiobis(2-tert-butyl-5-methylphenol)	**(96-69-5)**
4,4'-Thiobis(6-tert-butyl-3-methylphenol)	**(96-69-5)**
6,6'-Thiobis(4-chloro-o-cresol)	**(4418-66-0)**
1,1'-Thiobis(2-chloroethane)	**(505-60-2)**
2,2'-Thiobis(4-chloro-6-methyl-phenol)	**(4418-66-0)**
2,2'-Thiobis(4-chloro-6-methylphenol)	**(4418-66-0)**
2,2'-Thiobis(4-chlorophenol)	**(97-24-5)**
2,2'-Thiobis(4,6-dichlorophenol)	**(97-18-7)**
Thiobis(dodecylphenol) calcium salt (1:1)	**(26998-97-0)**
Thiobis(dodecyl propionate)	**(123-28-4)**
2,2'-Thiobisethanethiol	**(3570-55-6)**
1,1'-Thiobis(3-methylbutane)	**(544-02-5)**
2,2'-Thiobis(4-methyl-6-tert-butylphenol)	**(90-66-4)**
4,4'-Thiobis(3-methyl-6-tert-butylphenol)	**(96-69-5)**
1,1'-(Thiobis(methylene))bisbenzene	**(538-74-9)**
1,1'-Thiobis(2-methyl-4-hydroxy-5-tert-butylbenzene)	**(96-69-5)**
N,N'-(Thiobis((methylimino)carbonyloxy))bisethanimidothioic acid dimethyl ester	**(59669-26-0)**
2,2'-Thiobis(2-methylpropane)	**(107-47-1)**
1,1'-Thiobispropane	**(111-47-7)**
2,2'-Thiobispropane	**(625-80-9)**
3,3'-Thiobispropanoic acid, dioctadecyl ester	**(693-36-7)**
3,3'-Thiobispropanoic acid, ditridecyl ester	**(10595-72-9)**
3,3-Thiobis(1-propene)	**(592-88-1)**
Thiobis((tetrapropenyl)phenol) magnesium salt	**(68974-78-7)**
Thiobis((tetrapropylene)phenol)	**(68815-67-8)**
Thiobis(tetrapropylenephenol)	**(68815-67-8)**
Thiobis(tetrapropylenephenol), magnesium salt	**(68974-78-7)**
Thiocarb	**(148-18-5)**
Thiocarbamic acid, N-diisopropyl-, S-2,3,3-trichloroallyl ester	**(2303-17-5)**
Thiocarbamic acid-S,S-(2-(dimethylamino)trimethylene) ester hydrochloride	**(15263-52-2)**
Thiocarbamide	**(62-56-6)**
Thiocarbamylhydrazine	**(79-19-6)**
Thiocarbanil	**(103-72-0)**
Thiocarbanilide	**(102-08-9)**
Thiocarbazide	**(2231-57-4)**
Thiocarbohydrazide	**(2231-57-4)**
Thiocarbonic acid diamide	**(62-56-6)**
Thiocarbonic dichloride	**(463-71-8)**
Thiocarbonic dihydrazide	**(2231-57-4)**
Thiocarbonohydrazide	**(2231-57-4)**
Thiocarbonyl chloride	**(463-71-8)**
Thiocarbonylchloride (DOT)	**(463-71-8)**
Thiocarbonyl dichloride	**(463-71-8)**
Thiochlorid fosforecny (Czech)	**(3982-91-0)**
Thiochlormethyl	**(33439-45-1)**
Thiochroman-4-one, oxime	**(6533-73-9)**
3-Thiocresol	**(108-40-7)**
4-Thiocresol	**(106-45-6)**
m-Thiocresol	**(108-40-7)**
o-Thiocresol	**(137-06-4)**
p-Thiocresol	**(106-45-6)**
Thiocron	**(919-76-6)**
Thiocron 30	**(919-76-6)**
Thioctacid	**(62-46-4)**
6,8-Thioctic acid	**(62-46-4)**
6-Thioctic acid	**(62-46-4)**
Thioctic acid	**(62-46-4)**
Thioctidase	**(62-46-4)**
Thioctsan	**(62-46-4)**
Thiocyanate	**(302-04-5)**
Thiocyanate ion	**(302-04-5)**
Thiocyanate ion (1-)	**(302-04-5)**
Thiocyanate sodium	**(540-72-7)**
Thiocyanatoacetic acid isobornyl ester	**(115-31-1)**
α-Thiocyanatotoluene	**(3012-37-1)**
Thiocyanic acid (9CI)	**(463-56-9)**
Thiocyanic acid, 2-amino-6-benzothiazolyl ester (9CI)	**(7170-77-6)**
Thiocyanic acid, ammonium salt	**(1762-95-4)**
Thiocyanic acid, 2-(benzothiazolylthio)methyl ester, 60%	**(21564-17-0)**
Thiocyanic acid, 2-(benzothiazolylthio)methyl ester, 80%	**(21564-17-0)**
Thiocyanic acid, benzyl ester	**(3012-37-1)**
Thiocyanic acid, 4,6-bis(allylamino)-s-triazin-2-yl ester (8CI)	**(30360-05-5)**
Thiocyanic acid, 4,6-bis(ethylamino)-s-triazin-2-yl ester (7CI,8CI)	**(30360-95-3)**
Thiocyanic acid, 4,6-bis(isopropylamino)-s-triazin-2-yl ester (7CI,8CI)	**(30360-92-0)**
Thiocyanic acid, 2-(2-butoxyethoxy)ethyl ester	**(112-56-1)**
Thiocyanic acid, 6-(o-chloroanilino)-s-triazine-2,4-diyl ester (7CI,8CI)	**(30362-24-4)**
Thiocyanic acid, cobalt(2+) salt (9CI)	**(3017-60-5)**
Thiocyanic acid, copper(1+) salt (8CI,9CI)	**(1111-67-7)**
Thiocyanic acid, 4,6-diamino-s-triazin-2-yl ester (7CI,8CI)	**(30360-93-1)**
Thiocyanic acid, 4,6-dianilino-s-triazin-2-yl ester (8CI)	**(30360-06-6)**
Thiocyanic acid, 2,4-dinitrophenyl ester	**(1594-56-5)**
Thiocyanic acid, ethyl ester (8CI,9CI)	**(542-90-5)**
Thiocyanic acid, 2-hydroxyethyl ester, laurate	**(301-11-1)**
Thiocyanic acid, 4-hydroxyphenyl ester (9CI)	**(3774-52-5)**
Thiocyanic acid, p-hydroxyphenyl ester (8CI)	**(3774-52-5)**
Thiocyanic acid, lead(2+) salt (9CI)	**(592-87-0)**
Thiocyanic acid, mercury(2+) salt	**(592-85-8)**
Thiocyanic acid, mercury(II) salt	**(592-85-8)**
Thiocyanic acid, 4-methoxy-2-nitrophenyl ester	**(59607-71-5)**
Thiocyanic acid, methylene ester	**(6317-18-6)**
Thiocyanic acid, methyl ester	**(556-64-9)**
Thiocyanic acid, phenyl ester	**(5285-87-0)**
Thiocyanic acid, phenylmethyl ester (9CI)	**(3012-37-1)**
Thiocyanic acid, potassium salt	**(333-20-0)**

Thiocyanic acid, sodium salt (540-72-7)
Thiocyanic acid, s-triazine-2,4,6-triyl ester (8CI) (30863-24-2)
2-Thiocyanoethyl coconate (301-11-1)
2-Thiocyanoethyl dodecanoate (301-11-1)
2-Thiocyanoethyl laurate (301-11-1)
β-Thiocyanoethyl laurate (301-11-1)
2-(Thiocyanomethylthio)benzothiazole (2593-15-9)
2-(Thiocyanomethylthio)benzothiazole, 60% (21564-17-0)
2-Thiocyanomethylthiobenzothiazole, 80% (21564-17-0)
Thiocyclam (ethanedioate 1:1) (31895-22-4)
Thiocyclam hydrogen oxalate (31895-22-4)
Thiocyclopentane-1,1-dioxide (126-33-0)
Thiocynamine (109-57-9)
Thiodan (115-29-7)
Thiodan 35 (115-29-7)
α-Thiodan (959-98-8)
β-Thiodan (33213-65-9)
Thiodemeton (298-03-3)
Thiodemeton (298-04-4)
Thiodemetron (298-04-4)
4,4'-Thiodianiline (139-65-1)
p,p-Thiodianiline (139-65-1)
Thiodicarb (59669-26-0)
Thiodicarbonic acid, diethyl ester (9CI) (3278-35-1)
2,2'-Thiodiethanol (111-48-8)
Thiodiethylene glycol (111-48-8)
2,2'-Thiodiethylkyanid (Czech) (111-97-7)
Thiodifenylamine (Dutch) (92-84-2)
Thiodiglycol (111-48-8)
β-Thiodiglycol (111-48-8)
2,2'-Thiodiglycolic acid (123-93-3)
Thiodiglycolic acid (123-93-3)
β,β'-Thiodiglycolic acid (123-93-3)
Thiodiglycollic acid (123-93-3)
Thiodihydracrylic acid (111-17-1)
2-Thio-3,5-dimethyltetrahydro-1,3,5-thiadiazine (533-74-4)
4,4'-Thiodiphenol (2664-63-3)
Thiodiphenylamin (German) (92-84-2)
Thiodiphenylamine (92-84-2)
O,O'-(Thiodi-4,1-phenylene)bis(O,O-dimethyl phosphorothioate) (3383-96-8)
Thiodi-p-phenylenediamine (139-65-1)
O,O'-(Thiodi-p-phenylene) O,O,O',O'-tetramethyl bis-(phosphorothioate) (3383-96-8)
Thiodiphosphonic acid (((HS)HP(S))2S), bis(2,6,6-trimethyl-bicyclo(3.1.1)hept-2-enyl)-, bis(2,6,6-trimethylbicyclo-(3.1.1)hept-2-enyl) ester (9CI) (68400-79-3)
Thiodiphosphoric acid tetraethyl ester (3689-24-5)
3,3'-Thiodipropionic acid (111-17-1)
Thiodipropionic acid (111-17-1)
β,β'-Thiodipropionic acid (111-17-1)
Thiodipropionic acid, distearyl ester (693-36-7)
β,β'-Thiodipropionitrile (111-97-7)
2-Thio-1,3-dithio(4,5-b)quinoxaline (93-75-4)
Thiodow (12122-67-7)
Thiodril (638-23-3)
2-Thioethanol (60-24-2)
Thioethanol (75-08-1)
Thioethanolamine (60-23-1)
Thioethyl alcohol (75-08-1)
Thioethyl ether (352-93-2)
Thiofaco M-50 (141-43-5)
Thiofaco T-35 (102-71-6)
Thiofan (Czech) (110-01-0)
Thiofanate (23564-06-9)
Thiofanocarb (39196-18-4)
Thiofanox (39196-18-4)
Thiofen (Czech) (110-02-1)
Thiofenol (Czech) (108-98-5)
Thiofentanyl (467-85-6)
Thiofide (120-78-5)
Thioflavan (8CI) (5961-99-9)
Thioflavin S (2390-54-7)
Thioflavin T (2390-54-7)
Thiofor (115-29-7)
Thioform (Czech) (291-21-4)
Thiofos (56-38-2)
Thiofosgen (Czech) (463-71-8)
Thiofozil (52-24-4)
Thiofuram (110-02-1)
Thiofuran (110-02-1)
Thiofurfuran (110-02-1)
(1-Thio-D-glucopyranosato)gold (12192-57-3)
1-Thio-D-glucopyranose, gold complex (12192-57-3)
1-Thio-D-glucopyranose, monogold(1+) salt (12192-57-3)
1-Thio-glucopyranose, monogold(1+) salt (12192-57-3)
Thioglucose D'or (French) (12192-57-3)
Thioglycerin (96-27-5)
1-Thioglycerol (96-27-5)
Thioglycerol (96-27-5)
α-Thioglycerol (96-27-5)
Thioglycol [UN 2966] (60-24-2)
Thioglycolatesodium (367-51-1)
2-Thioglycolic acid (68-11-1)
Thioglycolic acid (ACGIH,OSHA) [UN 1940] (68-11-1)
Thioglycolic acid ethyl ester (623-51-8)
Thioglycolic acid 2-ethylhexyl ester (7659-86-1)
Thioglycolic acid methyl ester (2365-48-2)
Thioglycollic acid (68-11-1)
Thioglycollic acid, ammonium salt (5421-46-5)
Thioglycollic acid, sodium salt (367-51-1)
Thioglykolsaeure-aethylester (German) (623-51-8)
Thioglykolsaeure-2-aethylhexyl ester (German) (7659-86-1)
Thioglykolsaeure-methylester (German) (2365-48-2)
6-Thioguanine (154-42-7)
Thioguanine (154-42-7)
Thioguanine deoxyriboside (789-61-7)
β-Thioguanine deoxyriboside (789-61-7)
Thioindigo Brilliant Pink ZH (2379-74-0)
Thioindigo Brilliant Pink ZHP (2379-74-0)
Thioindigo Orange KKh (3263-31-8)
Thiokarbanilid (Czech) (102-08-9)
Thiokarbonylchlorid (Czech) (463-71-8)
2-Thio-4-ketothiazolidine (141-84-4)
Thioktsaeure (German) (62-46-4)
2-Thiokyanatoethylester kyseliny laurove (Czech) (301-11-1)
Thiol Systox (126-75-0)
Thiol Systox Sulfone (2496-91-5)
Thiolacetic acid (507-09-5)
2-Thiol-5-amino-1,3,4-thiadiazole (2349-67-9)
Thiolane (110-01-0)
Thiolane-1,1-dioxide (126-33-0)
Thioldemeton (126-75-0)
2-Thiol-dihydroglyoxaline (96-45-7)
Thiole (110-02-1)
Thiolite (7778-18-9)
Thiolmecaptophos (298-03-3)
Thiols (143-10-2)
Thiols (94805-33-1)
Thiolux (7704-34-9)
Thiomebumal (76-75-5)
Thiomecil (56-04-2)
Thiomerin (20223-84-1)
Thiomerosal (54-64-8)
Thiomersal (54-64-8)

Thiomersalate **(54-64-8)**
Thiomethanol **(74-93-1)**
2-Thio-6-methyl-1,3-pyrimidin-4-one **(56-04-2)**
6-Thio-4-methyluracil **(56-04-2)**
Thiometon **(640-15-3)**
Thiomex **(56-38-2)**
Thiomid **(536-33-4)**
Thiomidil **(56-04-2)**
Thiomocovina (Czech) **(62-56-6)**
Thiomonoglycol **(60-24-2)**
Thiomorpholidophosphoric diethylenimide **(2168-68-5)**
Thiomorpholine, 4-(4-((2,6-dichloro-4-nitrophenyl)azo)phenyl)-, 1,1-dioxide (9CI) **(17741-62-7)**
Thiomorpholine, 4-phenyl-, 1,1-dioxide (9CI) **(17688-68-5)**
Thiomul **(115-29-7)**
Thionaphthene **(95-15-8)**
2-Thionaphthol **(91-60-1)**
Thio-β-naphthol **(91-60-1)**
Thionaphthol **(91-60-1)**
β-Thionaphthol **(91-60-1)**
Thionate **(115-29-7)**
Thionazin **(297-97-2)**
Thionazine **(297-97-2)**
Thione (Reagent) **(3811-73-2)**
Thionex **(115-29-7)**
Thionex **(97-74-5)**
α-Thionex **(33213-65-9)**
β-Thionex **(959-98-8)**
Thionex Rubber Accelerator **(97-74-5)**
Thioniden **(536-33-4)**
Thionoacetic acid **(507-09-5)**
Thionodemeton Sulfone **(4891-54-7)**
Thionylan **(91-80-5)**
Thionyl chloride (ACGIH,OSHA) [UN 1836] **(7719-09-7)**
Thionyl dichloride **(7719-09-7)**
Thionyl difluoride **(7783-42-8)**
Thionyl-fluoride **(7783-42-8)**
Thiooctanoic acid **(62-46-4)**
2-Thio-4-oxo-6-methyl-1,3-pyrimidine **(56-04-2)**
2-Thio-4-oxo-6-propyl-1,3-pyrimidine **(51-52-5)**
2-Thio-6-oxypurine **(2487-40-3)**
2-Thio-6-oxypyrimidine **(141-90-2)**
Thiopental **(76-75-5)**
Thiopentobarbital **(76-75-5)**
Thiopentone **(76-75-5)**
Thioperoxydicarbonic acid diethyl ester **(502-55-6)**
Thioperoxydicarbonic diamide (((H2N)C(S))2S2), N,N'-bis-(1-methylethyl)-N,N'-dioctadecyl- (9CI) **(35318-10-6)**
Thioperoxydiphosphoric acid (((HO)2P(S))2S2), tetramethyl ester (9CI) **(5930-71-2)**
Thioperoxydiphosphoric acid (((HO)2PS)2S2), tetramethyl ester (8CI) **(5930-71-2)**
Thioperoxydiphosphoric acid, O,O,O,O-tetramethyl- **(5930-71-2)**
Thiophal **(133-07-3)**
Thiophanat (German) **(23564-06-9)**
Thiophanate **(23564-06-9)**
Thiophanate M **(23564-05-8)**
Thiophanate ethyl **(23564-06-9)**
Thiophanate methyl **(23564-05-8)**
Thiophane **(110-01-0)**
Thiophane dioxide **(126-33-0)**
Thiophan sulfone **(126-33-0)**
Thiophen **(110-02-1)**
Thiophene [UN 2414] **(110-02-1)**
7-(Thiophene-2-acetamido)cephalosporanic acid **(153-61-7)**
7-(Thiophene-2-acetamido)-3-(1-pyridylmethyl)-3-cephem-4-carboxylic acid betaine **(50-59-9)**
3-Thiopheneacetic acid **(6964-21-2)**
2-Thiopheneacetic acid (9CI) **(1918-77-0)**
2-Thiopheneacetonitrile **(20893-30-5)**
2-Thiophenealdehyde **(98-03-3)**
Thiophene analog of phencyclidine **(21500-98-1)**
Thiophene, bis(1,1-dimethylethyl)- (9CI) **(30140-46-6)**
Thiophene, 2-bromo- (9CI) **(1003-09-4)**
Thiophene, 3-bromo- (9CI) **(872-31-1)**
Thiophene, 2-tert-butoxy- (6CI,7CI,8CI) **(23290-55-3)**
α-Thiophenecarboxaldehyde **(98-03-3)**
2-Thiophenecarboxaldehyde (9CI) **(98-03-3)**
3-Thiophenecarboxylic acid **(88-13-1)**
β-Thiophenecarboxylic acid **(88-13-1)**
2-Thiophenecarboxylic acid, chloro- (9CI) **(99165-92-1)**
2-Thiophenecarboxylic acid, 5-methyl- (8CI,9CI) **(1918-79-2)**
Thiophene, 2-chloro- (9CI) **(96-43-5)**
Thiophene, 3-(decyloxy)tetrahydro-, 1,1-dioxide (8CI) **(18760-44-6)**
Thiophene, 2,5-di(benzoxazol-2-yl)- **(2866-43-5)**
Thiophene, 2,5-dibromo- (9CI) **(3141-27-3)**
Thiophene, 3,4-dibromo- (9CI) **(3141-26-2)**
Thiophene, di-tert-butyl- **(30140-46-6)**
2,5-Thiophenedicarboxylic acid (9CI) **(4282-31-9)**
Thiophenedicarboxylic acid, chloro- (9CI) **(99165-93-2)**
Thiophene, 2,5-dichloro- (9CI) **(3172-52-9)**
Thiophene, 2,5-dihydro-, 1,1-dioxide **(77-79-2)**
Thiophene, dimethyl- (9CI) **(28632-15-7)**
Thiophene, 2,5-dimethyl- (8CI,9CI) **(638-02-8)**
Thiophene, 2-(1,1-dimethylethoxy)- (9CI) **(23290-55-3)**
Thiophene, 2-(1,1-dimethylethyl)- **(1689-78-7)**
2,2'-(2,5-Thiophenediyl)bisbenzoxazole **(2866-43-5)**
Thiophene, ethyl- (9CI) **(52006-63-0)**
Thiophene, 2-ethyl- (8CI,9CI) **(872-55-9)**
Thiophene, 2-iodo- (9CI) **(3437-95-4)**
2-Thiophenemethanamine (9CI) **(27757-85-3)**
Thiophene, 2-methyl **(554-14-3)**
Thiophene, 3-methyl **(616-44-4)**
Thiophene, 2-methyl-5-propyl- (8CI,9CI) **(33933-73-2)**
Thiophene, 2-nitro **(609-40-5)**
Thiophene, 2-pentyl- (9CI) **(4861-58-9)**
Thiophene, phenyl- (9CI) **(56842-14-9)**
Thiophene, 2-phenyl- (8CI,9CI) **(825-55-8)**
Thiophene, 2,3,4,5-tetrachloro **(6012-97-1)**
Thiophene, tetrachloro- **(6012-97-1)**
Thiophene, tetrahydro **(110-01-0)**
Thiophene, tetrahydro-2,5-dimethyl- (8CI,9CI) **(1551-31-1)**
Thiophene, tetrahydrodimethyl- (7CI,9CI) **(55569-78-3)**
Thiophene, tetrahydro-, 1,1-dioxide **(126-33-0)**
Thiophene, tetrahydro-2-methyl- (8CI,9CI) **(1795-09-1)**
Thiophene, tetrahydro-3-methyl-, 1,1-dioxide (9CI) **(872-93-5)**
Thiophene, 2,3,4-trimethyl- (8CI,9CI) **(1795-04-6)**
β-Thiophenic acid **(88-13-1)**
Thiophenit **(298-00-0)**
Thiophenite **(23564-06-9)**
Thiophenol (DOT) **(108-98-5)**
Thiophenyl sodium salt **(930-69-8)**
Thiophos **(56-38-2)**
Thiophos 3422 **(56-38-2)**
Thiophosgene **(463-71-8)**
Thiophosgene [UN 2474] **(463-71-8)**
Thiophosphamide **(52-24-4)**
Thiophosphate de O-2,4-dichlorophenyle et de O,O-diethyle (French) **(97-17-6)**
Thiophosphate de O,O-diethyle et de O-(3-chloro-4-methyl-7-coumarinyle) (French) **(56-72-4)**
Thiophosphate de O,O-diethyle et de O-(2,5-dichloro-4-bromo) phenyle (French) **(4824-78-6)**
Thiophosphate de O,O-diethyle et de S-(2-ethylthio-ethyle)

(French) **(126-75-0)**
Thiophosphate de O,O-diethyle et de O-2-isopropyl-4-methyl-6-pyrimidyle (French) **(333-41-5)**
Thiophosphate de O,O-diethyle et de O-(4-methyl-7-coumarinyle) (French) **(299-45-6)**
Thiophosphate de O,O-diethyle et de O-(4-nitrophenyle) (French) **(56-38-2)**
Thiophosphate de O,O-dimethyle et de O-4-bromo-2,5-dichlorophenyle (French) **(2104-96-3)**
Thiophosphate de O,O-dimethyle et de O-3-chloro-4-nitrophenyle (French) **(500-28-7)**
Thiophosphate de O,O-dimethyle et de O-4-chloro-3-nitrophenyle (French) **(2463-84-5)**
Thiophosphate de O,O-dimethyle et de S-2-ethylsulfinylethyle (French) **(301-12-2)**
Thiophosphate de O,O-dimethyle et de S-2-ethylthio-ethyle (French) **(919-86-8)**
Thiophosphate de O,O-dimethyle et de O-2-ethylthio-ethyle (French) **(867-27-6)**
Thiophosphate de O,O-dimethyle et de S-((5-methoxy-4-pyronyl)-methyle) (French) **(2778-04-3)**
Thiophosphate de O,O-dimethyle et de S-(N-methylcarbamoyl)-methyle (French) **(1113-02-6)**
Thiophosphate de O,O-dimethyle et de O-(3-methyl-4-methyl-thiophenyle) (French) **(55-38-9)**
Thiophosphate de O,O-dimethyle et de O-(3-methyl-4-nitrophenyle) (French) **(122-14-5)**
Thiophosphate de O,O-dimethyle et de O-(4-nitrophenyle) (French) **(298-00-0)**
Thiophosphate de O,O-dimethyle et de O-(2,4,5-trichlorophenyle) (French) **(299-84-3)**
Thiophosphate de O,S-dimethyl et de O-(3-methyl-4-nitrophenyle) (French) **(3344-14-7)**
Thiophosphoramide, N,N',N''-tri-1,2-ethanediyl- **(52-24-4)**
Thiophosphoramide, N,N',N''-triethylene- **(52-24-4)**
Thiophosphoric acid **(13598-51-1)**
Thiophosphoric anhydride **(1314-80-3)**
Thiophosphoric trichloride **(3982-91-0)**
Thiophosphorsaeure-O,S-dimethylesteramid (German) **(10265-92-6)**
Thiophosphoryl chloride [UN 1837] **(3982-91-0)**
Thiophosphoryl trichloride **(3982-91-0)**
Thiophthalimide **(18138-18-6)**
2-Thiopropane **(75-18-3)**
Thiopropazate **(84-06-0)**
2-Thio-6-propyl-1,3-pyrimidin-4-one **(51-52-5)**
6-Thio-4-propyluracil **(51-52-5)**
2-Thiopseudourea **(62-56-6)**
β-Thiopseudourea **(62-56-6)**
6-Thiopurine **(50-44-2)**
Thiopyran, 2-ethyltetrahydro- **(1613-52-1)**
2H-Thiopyran, 2-ethyltetrahydro- (8CI,9CI) **(1613-52-1)**
Thiopyran, tetrahydro- **(1613-51-0)**
2H-Thiopyran, tetrahydro- (8CI,9CI) **(1613-51-0)**
2H-Thiopyran, tetrahydro-4-methyl- (9CI) **(5161-17-1)**
2H-Thiopyran, tetrahydromethyl- (9CI) **(73180-15-1)**
Thiopyran, tetrahydro-4-methyl- (8CI) **(5161-17-1)**
2-Thio-1,3-pyrimidin-4-one **(141-90-2)**
Thiopyrophosphoric acid, tetraethyl ester **(3689-24-5)**
Thiopyrophosphoric acid, tetrapropyl ester **(3244-90-4)**
Thioquinox **(93-75-4)**
Thioridazien thiomethyl sulfoxide **(5588-33-0)**
Thioridazin **(50-52-2)**
Thioridazine **(50-52-2)**
Thioridazine 5-oxide **(7776-05-8)**
Thioridazine 5-sulfoxide **(7776-05-8)**
Thioridazine sulfoxide **(7776-05-8)**
Thioryl **(56-04-2)**
Thiosalicylic acid **(147-93-3)**
Thiosan **(137-26-8)**
Thiosan **(97-77-8)**
Thioscabin **(97-77-8)**
Thiosemicarbazide **(79-19-6)**
Thiosemicarbazone acetone **(1752-30-3)**
Thioseptal **(144-83-2)**
Thioserine **(52-90-4)**
Thiosinamin **(109-57-9)**
Thiosinamine **(109-57-9)**
Thiostop N **(128-04-1)**
Thiosulfan **(115-29-7)**
Thiosulfan tionel **(115-29-7)**
Thiosulfate (8CI,9CI) **(14383-50-7)**
Thiosulfil **(144-82-1)**
Thiosulfil-A Forte **(136-40-3)**
Thiosulfuric acid (H2S2O3), monoammonium salt (8CI,9CI) **(10103-43-2)**
Thiosulfuric acid, calcium salt (1:1) **(10124-41-1)**
Thiosulfuric acid, diammonium salt **(7783-18-8)**
Thiosulfuric acid, disodium salt **(7772-98-7)**
Thiosulfuric acid, disodium salt, monohydrate (9CI) **(55755-19-6)**
Thiosulfuric acid, disodium salt, pentahydrate **(10102-17-7)**
Thiosulfuric acid, ion(2-) **(14383-50-7)**
Thiosulfuric acid, lead salt **(26265-65-6)**
Thiosulfuric acid, lead(2+) salt (1:1) (8CI,9CI) **(13478-50-7)**
Thiosulfurous dichloride **(10025-67-9)**
Thiotef **(52-24-4)**
Thiotepp **(3689-24-5)**
Thioterephthalic acid, tetrachloro-, O,S-dimethyl ester **(3765-57-9)**
Thiotetrole **(110-02-1)**
Thiotex **(137-26-8)**
Thiothal **(76-75-5)**
Thiothixene **(5591-45-7)**
Thiothixine **(5591-45-7)**
Thiothymin **(56-04-2)**
2-Thiothymine **(636-26-0)**
Thiothymine **(636-26-0)**
Thiothyron **(56-04-2)**
6,8-Thiotic acid **(62-46-4)**
6-Thiotic acid **(62-46-4)**
3-Thiotolene **(616-44-4)**
Thiotox **(137-26-8)**
Thiotox **(640-15-3)**
Thiotriethylenephosphoramide **(52-24-4)**
2-Thiouracil **(141-90-2)**
6-Thiouracil **(141-90-2)**
Thiouracil **(141-90-2)**
2-Thiourea **(62-56-6)**
Thiourea (DOT) **(62-56-6)**
Thiourea, N,N'-dibutyl- **(109-46-6)**
Thiourea, N,N'-diethyl- **(105-55-5)**
Thiourea, N,N'-dimethyl- (9CI) **(534-13-4)**
Thiourea, sym-diphenyl- **(102-08-9)**
Thiourea, N,N'-diphenyl- (9CI) **(102-08-9)**
Thiourea, (2-methylphenyl)- (9CI) **(614-78-8)**
Thiourea, N,N''-(methylphosphinylidene)bis- (9CI) **(79064-73-6)**
Thiourea, 1-naphthalenyl- **(86-88-4)**
6-Thiouric acid **(2002-60-0)**
D-β-Thiovaline **(52-67-5)**
Thiovanic acid **(68-11-1)**
Thiovanol **(96-27-5)**
Thiovit **(7704-34-9)**
Thioxamyl **(23135-22-0)**
p-Thioxane **(15980-15-1)**
Thioxanthene-9-one **(492-22-8)**
9H-Thioxanthene, 9-oxo- **(492-22-8)**
Thioxanthene, 9-oxo- **(492-22-8)**
Thioxanthene-δ^9, γ-propylamine, 2-chloro-N,N-dimethyl **(113-59-7)**
9H-Thioxanthene-2-sulfonamide, N,N-dimethyl-9-(3-(4-methyl-

1-piperanizyl)propylidene)-, (Z)- (5591-45-7)
Thioxanthene-2-sulfonamide, N,N-dimethyl-9-(3-(4-methyl-1-piperazinyl)propylidene) (5591-45-7)
Thioxanthenone (492-22-8)
9H-Thioxanthen-9-one (9CI) (492-22-8)
Thioxanthen-9-one (8CI) (492-22-8)
9H-Thioxanthen-9-one, 1-((2-(diethylamino)ethyl)amino)-4-(hydroxymethyl) (3105-97-3)
9H-Thioxanthen-9-one, 1-((2-(diethylamino)ethyl)amino)-4-(hydroxymethyl)-, monomethanesulfonate (Salt) (23255-93-8)
9H-Thioxanthen-9-one, 2-dodecyl- (9CI) (82799-46-0)
Thioxanthen-9-one, 2-methyl- (15774-82-0)
9H-Thioxanthen-9-one, 2-methyl- (9CI) (15774-82-0)
9H-Thioxanthen-9-one, 2-(1-methylethyl)- (9CI) (5495-84-1)
2-Thioxanthine (2487-40-3)
6-Thioxanthine (2002-59-7)
9-Thioxanthone (492-22-8)
Thioxanthone (492-22-8)
Thioxidrene (1195-16-0)
6-Thioxopurine (50-44-2)
2-Thioxo-4-thiazolidinone (141-84-4)
Thiozamide (72-14-0)
Thiram 75 (137-26-8)
Thiram (ACGIH,DOT,OSHA) (137-26-8)
Thiram B (137-26-8)
Thiramad (137-26-8)
Thirame (French) (137-26-8)
Thirasan (137-26-8)
Thireranide (97-77-8)
Thistrol (6062-26-6)
Thistrol (94-81-5)
Thiulix (137-26-8)
Thiurad (137-26-8)
Thiuragyl (51-52-5)
Thiuram (137-26-8)
Thiuram D (137-26-8)
Thiuram E (97-77-8)
Thiuram M (137-26-8)
Thiuram M Rubber Accelerator (137-26-8)
Thiuram disulfide (504-90-5)
Thiuram disulfide tetrabutyl (1634-02-2)
Thiuram disulfide, tetraethyl- (97-77-8)
Thiuram disulfide, tetramethyl- (137-26-8)
Thiuramin (137-26-8)
Thiuram monosulfide, tetramethyl- (97-74-5)
Thiuram, tetrabutyl-, disulfide (1634-02-2)
Thiuram tetrasulfide, bis(piperidinothiocarbonyl) (120-54-7)
Thiuramyl (137-26-8)
Thiuranide (97-77-8)
Thiuretic (58-93-5)
Thiuryl (56-04-2)
Thlaretic (58-93-5)
Thomapyrin (8003-03-0)
Thompson-Hayward TH6040 (35367-38-5)
Thompson's Wood Fix (87-86-5)
Thonzylamine (91-85-0)
Thonzylamine hydrochloride (63-56-9)
Thonzylaminium chloride (63-56-9)
Thorazine (50-53-3)
Thorazine (69-09-0)
Thorazine hydrochloride (69-09-0)
Thoria (1314-20-1)
Thorium (7440-29-1)
Thorium-232 (7440-29-1)
Thorium-270 (55739-99-6)
Thorium(IV), nitrate (13823-29-5)
Thorium-chloride (10026-08-1)
Thorium dioxide (1314-20-1)
Thorium, Isotope of mass 270 (9CI) (55739-99-6)
Thorium metal, Pyrophoric [UN 2975] (7440-29-1)
Thorium(4+) nitrate (13823-29-5)
Thorium nitrate, Solid [UN 2976] (13823-29-5)
Thorium-oxide (1314-20-1)
Thorium tetrachloride (10026-08-1)
Thorium tetranitrate (13823-29-5)
Thorotrast (1314-20-1)
Thortrast (1314-20-1)
Threamine (78-96-6)
Threaric acid (87-69-4)
Three Elephant (10043-35-3)
L-Threitol-1,4-bismethanesulfonate (299-75-2)
D-Threitol, 1,2:3,4-dianhydro (30419-67-1)
Threitol, 1,2:3,4-dianhydro- (1464-53-5)
Threitol, 1,4-dimethanesulfonate (299-75-2)
Threne Brilliant Orange GR (4424-06-0)
L-Threo-hex-2-enonic acid, γ-lactone (50-81-7)
Threonin (72-19-5)
L-Threonine (72-19-5)
Threonine (72-19-5)
Threonine, L (72-19-5)
Threthylen (79-01-6)
Threthylene (79-01-6)
Thrombocytin (50-67-9)
Thromboliquine (9005-49-6)
Thrombolysin (9001-90-5)
Thrombotonin (50-67-9)
3-Thujanol (513-23-5)
3-Thujanone, (-)- (546-80-5)
3-Thujanone, (1S,4R,5R)-(-) (546-80-5)
γ-Thujaplicin (672-76-4)
γ-Thujaplicine (672-76-4)
α-Thujene (2867-05-2)
4(10)-Thujene (8CI) (3387-41-5)
Thujon (546-80-5)
(-)-Thujone (546-80-5)
Thujone (546-80-5)
α-Thujone (546-80-5)
l-Thujone (546-80-5)
Thulium (9CI) (7440-30-4)
Thulol (50-27-1)
Thuricide (68038-71-1)
Thycapsol (60-56-0)
Thycapzol (60-56-0)
Thylakentrin (9002-68-0)
Thylate (137-26-8)
Thylogen (91-84-9)
Thylogen maleate (59-33-6)
Thyloquinone (58-27-5)
Thylose (9000-11-7)
Thylpar M-50 (298-00-0)
Thyme camphor (89-83-8)
Thymic acid (89-83-8)
Thymidin (50-89-5)
Thymidine (50-89-5)
Thymidine, 3'-azido-3'-deoxy (30516-87-1)
Thymidine mononucleotide (365-07-1)
Thymidine 5'-monophosphate (365-07-1)
Thymidine monophosphate (365-07-1)
Thymidine-5'-monophosphoric acid (365-07-1)
Thymidine 5'-phosphate (365-07-1)
Thymidine phosphate (365-07-1)
Thymidine 5'-phosphoric acid (365-07-1)
Thymidine, α,α,α-trifluoro (70-00-8)
5'-Thymidylic acid (365-07-1)

Thymidylic acid (365-07-1)
Thymine (65-71-4)
Thymine-2-deoxyriboside (50-89-5)
Thyminedeoxyriboside (50-89-5)
Thymine, 2-thio (636-26-0)
Thymin (purine base) (65-71-4)
Thymiode (552-22-7)
Thymiodol (552-22-7)
Thymodin (552-22-7)
Thymol (89-83-8)
m-Thymol (89-83-8)
o-Thymol (499-75-2)
Thymol, 6,6'-(3H-2,1-benzoxathiol-3-ylidene)bis(2-bromo-, S,S-dioxide (8CI) (76-59-5)
Thymol, 2,6-dinitro (303-21-9)
Thymol iodide (552-22-7)
Thymotol (552-22-7)
Thymyl bromide (4478-10-8)
Thynestron (53-16-7)
Thynon (3347-22-6)
Thyractin (9010-34-8)
Thyradin (8028-36-2)
Thyreoideum (51-48-9)
Thyreonorm (56-04-2)
Thyreostat (56-04-2)
Thyreostat I (56-04-2)
Thyreostat II (51-52-5)
Thyril (56-04-2)
Thyrocalcitonin (9007-12-9)
Thyrocrine (8028-36-2)
Thyroglobulin (9010-34-8)
Thyroglobuline (French) (9010-34-8)
Thyroglobulins (9010-34-8)
Thyroglobulinum (Latin) (9010-34-8)
Thyroid (8028-36-2)
Thyroid Extract (8028-36-2)
Thyroid Gland (8028-36-2)
Thyroid Tablets (8028-36-2)
Thyroid globulin (9010-34-8)
Thyroprotein (9010-34-8)
L-Thyroxin (51-48-9)
Thyroxin (51-48-9)
(-)-Thyroxine (51-48-9)
L-Thyroxine (51-48-9)
Thyroxine (51-48-9)
Thyroxine, L- (51-48-9)
Tin(II) chloride (1:2) (7772-99-8)
Tin(II) chromate (38455-77-5)
Tin(II) 2-ethylhexanoate (301-10-0)
Ti-Pure (13463-67-7)
Tiafur (712-68-5)
Tiamidon (67-03-8)
Tiaminal (67-03-8)
Tiazon (533-74-4)
Tibamato (4268-36-4)
Tibazide (54-85-3)
Tibemid (54-85-3)
Tiberal (16773-42-5)
Tibinide (54-85-3)
Tibison (54-85-3)
Tibivis (54-85-3)
Tibizide (54-85-3)
Tibusan (54-85-3)
Tibutol (74-55-5)
Tin(3+), butyl- (9CI) (78763-54-9)
Ticarcilina (Spanish) (34787-01-4)
Ticarcillin (34787-01-4)
Ticarcilline (French) (34787-01-4)
Ticarcillinum (Latin) (34787-01-4)
Ticinil (50-33-9)
Ticlobran (637-07-0)
Tiempe (738-70-5)
Tin(2+) 2-ethylhexanoate (301-10-0)
Tiezene (12122-67-7)
Tiferron (149-45-1)
Tifomycin (56-75-7)
Tifomycine (56-75-7)
Tifosyl (52-24-4)
Tigason (54350-48-0)
Tiger Orange (2814-77-9)
Tiglaldehyde (497-03-0)
trans-Tiglaldehyde (497-03-0)
Tiglic acid (80-59-1)
Tiglic acid aldehyde (497-03-0)
Tiglic acid, methyl ester (6CI,7CI) (6622-76-0)
Tiglic aldehyde (497-03-0)
Tiglinic acid (80-59-1)
7-Tiglylretronecine viridiflorate (22571-95-5)
7-Tiglyl-9-viridiflorylretronecine (22571-95-5)
Tigrex (330-54-1)
Tiguvon (55-38-9)
Tikofuran (67-45-8)
Tilcarex (82-68-8)
Tildin (77-65-6)
Tilidina (20380-58-9)
Tilidine (20380-58-9)
Tilidino (Spanish) (20380-58-9)
Tilidinum (Latin) (20380-58-9)
Tillam-6-E (1114-71-2)
Tillam (Russian) (1114-71-2)
Tillantox (495-73-8)
Tillram (97-77-8)
Tilt (60207-90-1)
Timazin (51-21-8)
Timet (298-02-2)
Timolet (50-49-7)
Timolol (26839-75-8)
Timololum (Latin) (26839-75-8)
Timonil (298-46-4)
Timonox (1309-64-4)
Tin (α) (7440-31-5)
Tin (ACGIH,OSHA) (7440-31-5)
Tin Flake (7440-31-5)
Tin (IV) phosphide (25324-56-5)
Tin P (2440-22-4)
Tin Powder (7440-31-5)
Tin-San (56573-85-4)
Tina Brilliant Pink R (2379-74-0)
Tina Orange R (3263-31-8)
Tin,' acetoxytriphenyl- (900-95-8)
Tinactin (2398-96-1)
Tinaderm (2398-96-1)
Tin bifluoride (7783-47-3)
Tin(II) bis(2-ethylhexanoate) (301-10-0)
Tin, bis(mercaptoacetate)dioctyl-, bis(2-ethylhexyl) ester (15571-58-1)
Tin, bis(mercaptoacetate)dioctyl-, bis(isooctyl) ester (26401-97-8)
Tin, bis(tributyl)-, oxide (56-35-9)
Tin, n-butyl-, trichloride (1118-46-3)
Tin, butyl-, tris(isooctylthioglycollate) (25852-70-4)
Tin chloride (7647-18-9)
Tin(IV) chloride (1:4) (7646-78-8)
Tin chloride (9CI) (26903-07-1)
Tin chloride (8CI,9CI) (13931-79-8)
Tin chloride (DOT) (7646-78-8)

Name	CAS No.
Tin chloride (VAN)(9CI)	(1344-13-4)
Tin(2+), chloromethyl-, ion (8CI)	(23066-18-4)
Tin(IV) chromate	(10101-75-4)
Tinctura Opii Benzoica	(8029-99-0)
Tinctura Thebaica Benzoica	(8029-99-0)
Tin(2+), dibutyl- (9CI)	(14488-53-0)
Tin, dibutyl-, bis(isooctylthioglycollate)	(25168-24-5)
Tin, dibutylbis(lauroyloxy)-	(77-58-7)
Tin, dibutyl-, diacetate	(1067-33-0)
Tin, dibutyl-, dichloride	(683-18-1)
Tin, di-n-butyl-, di(dodecanoate)	(77-58-7)
Tin, dibutyl-, difluoride	(563-25-7)
Tin dibutyl dilaurate	(77-58-7)
Tin(2+), dibutyl-, ion (8CI)	(14488-53-0)
Tin dibutyl mercaptide	(4253-22-9)
Tin, dibutyl(3-mercaptopropionato(2-))-	(78-06-8)
Tin, dibutyl-, oxide	(818-08-6)
Tin, dibutyloxo-	(818-08-6)
Tin dichloride	(7772-99-8)
Tin difluoride	(7783-47-3)
Tin, dimethyl- (6CI)	(23120-99-2)
Tin, dimethyl-, bis(isooctylthioglycollate)	(26636-01-1)
Tin, dimethyl-, dichloride	(753-73-1)
Tin, dimethyl-, fluoride	(3582-17-0)
Tin(2+), dimethyl-, ion (8CI)	(16408-14-3)
Tin dioctoate	(301-10-0)
Tin, dioctyl-, bis(2-ethylhexylthioglycolate)	(15571-58-1)
Tin, dioctyl-, bis(isooctylthioglycollate)	(26401-97-8)
Tin, dioctyl-, dichloride	(3542-36-7)
Tin, dioctyl-, dilaurate	(3648-18-8)
Tin, dioctyl-, β-mercaptopropionate	(3033-29-2)
Tin, dioctyl-, oxide	(870-08-6)
Tin, diphenyl-, dichloride	(1135-99-5)
Tindurin	(58-14-0)
Tine	(63428-84-2)
Tinestan	(900-95-8)
Tinestan 60 WP	(900-95-8)
Tin 2-ethylhexanoate	(301-10-0)
Tin ethylhexanoate	(301-10-0)
Tin(II) 2-ethylhexylate	(301-10-0)
Tin-fluoride	(7783-47-3)
Tin, fluorotriphenyl-	(379-52-2)
Tin, hydroxytriphenyl-	(76-87-9)
Tinic	(59-67-6)
Tin(IV) iodide (1:4)	(7790-47-8)
Tinmate	(639-58-7)
Tin(3+), methyl- (9CI)	(16408-15-4)
Tin(3+), methyl-, ion (8CI)	(16408-15-4)
Tin, methyl-, trichloride	(993-16-8)
Tin, methyl-, tris(isooctyl thioglycollate)	(57583-34-3)
Tinning Flux (DOT)	(7646-85-7)
Tin octadecanoate	(7637-13-0)
Tin octoate	(301-10-0)
Tin, octyl-, trichloride	(3091-25-6)
Tinolite	(1333-86-4)
Tinon Blue GF	(130-20-1)
Tinon Blue RS	(81-77-6)
Tinon Blue RSN	(81-77-6)
Tinon Brilliant Green 2BF	(128-58-5)
Tinon Brilliant Green BF	(128-58-5)
Tinon Brilliant Green B2F-F	(128-58-5)
Tinon Brilliant Green BFP	(128-58-5)
Tinon Brilliant Orange GR	(4424-06-0)
Tinon Brown BR	(2475-33-4)
Tinon Golden Orange G	(128-70-1)
Tinon Golden Orange GN	(128-70-1)
Tinon Golden Yellow	(128-66-5)
Tinon Navy Blue RA	(1324-54-5)
Tinon Olive 2R	(2379-81-9)
Tinon Red 6B	(4203-77-4)
Tinon Violet B 4RP	(1324-55-6)
Tinon Violet 4R	(1324-55-6)
Tinon Violet 2RB	(1324-55-6)
Tinopal AMS	(16090-02-1)
Tinopal BHS	(81-11-8)
Tinopal 5BM	(13863-31-5)
Tinopal CBS	(27344-41-8)
Tinopal CBS-X	(27344-41-8)
Tinopal EMS	(16090-02-1)
Tinopal RBS	(6416-68-8)
Tinopal RBS 200	(6416-68-8)
Tinostat	(77-58-7)
Tinox	(2587-90-8)
Tinox	(8065-62-1)
Tin oxalate	(814-94-8)
Tin(2+) oxalate	(814-94-8)
Tin(II) oxalate	(814-94-8)
Tin-oxide	(18282-10-5)
Tin, oxybis(tributyl-	(56-35-9)
Tin perchloride (DOT)	(7646-78-8)
Tin phosphide	(25324-56-5)
Tin-potassium-tartrate	(73926-79-1)
Tin protochloride	(7772-99-8)
Tin-sodium-tartrate	(72378-89-3)
Tin, tetrabutyl-	(1461-25-2)
Tin tetrachloride	(7646-78-8)
Tin tetrachloride, Anhydrous (DOT)	(7646-78-8)
Tintetrachloride (Dutch)	(7646-78-8)
Tin, tetraethyl-	(597-64-8)
Tin tetraiodide	(7790-47-8)
Tin, tetramethyl-	(594-27-4)
Tin, tetraoctyl-	(3590-84-9)
Tintofen Violet R	(1325-82-2)
Tintofen Violet R Supra	(1325-82-2)
Tintorane	(129-06-6)
Tin, tributyl-, acetate	(56-36-0)
Tin, tributyl-, benzoate	(4342-36-3)
Tin, tri-n-butyl-, chloride	(1461-22-9)
Tin, tributyl-, fluoride	(1983-10-4)
Tin, tri-n-butyl-, hydride	(688-73-3)
Tin, tri-n-butyl-, iodide	(7342-47-4)
Tin(1+), tributyl-, ion (8CI)	(36643-28-4)
Tin, tributyl-, linoleate	(24124-25-2)
Tin tributylmethacrylate	(2155-70-6)
Tin, tributyl-, neodecanoate	(28801-69-6)
Tin, tricyclohexylhydroxy-	(13121-70-5)
Tin, triethyl-, chloride	(994-31-0)
Tin, trimethyl-, hydroxide	(56-24-6)
Tin(1+), trimethyl-, ion (8CI)	(5089-96-3)
Tin, (2,4,6-trioxo-s-triazine-1,3,5(2H,4H,6H)-triyl)tris(triphenyl- (7CI,8CI)	(752-74-9)
Tin triphenyl acetate	(900-95-8)
Tin, tripropyl-, chloride	(2279-76-7)
Tinuvin	(3896-11-5)
Tinuvin 144	(63843-89-0)
Tinuvin P	(2440-22-4)
Tio-Mid	(536-33-4)
Tio TEF	(52-24-4)
Tiobenzamide (Italian)	(2227-79-4)
Tiocarbazil	(36756-79-3)
2-Tiocianometiltiobenzotiazolo, 80% (Italian)	(21564-17-0)
Tioctacid	(62-46-4)
Tioctidasi	(62-46-4)
Tioctidasi acetate replacing factor	(62-46-4)

Tiodifenilamina (Italian) **(92-84-2)**
Tiofanate etile (Italian) **(23564-06-9)**
Tiofanate metile (Italian) **(23564-05-8)**
Tiofine **(13463-67-7)**
Tiofos **(56-38-2)**
Tiofosfamid **(52-24-4)**
Tiofosyl **(52-24-4)**
Tiofozil **(52-24-4)**
Tioguanin **(154-42-7)**
Tioguanine **(154-42-7)**
Tiomeracil **(56-04-2)**
Tiona T.D. **(13463-67-7)**
Tional **(76-20-0)**
Tiopentale (Italian) **(76-75-5)**
Tiorale M **(56-04-2)**
Tiotiron **(56-04-2)**
Tiotixene **(5591-45-7)**
Tiouracyl (Polish) **(141-90-2)**
Tiovel **(115-29-7)**
Tioxide **(13463-67-7)**
Tip-Nip **(112-42-5)**
Tip-Off **(86-87-3)**
Tipaque **(13463-67-7)**
Tipol **(11043-90-6)**
Tippon **(93-76-5)**
Tirampa **(137-26-8)**
Tiroglobulina (Spanish) **(9010-34-8)**
Tiroidina **(8028-36-2)**
Tiron **(149-45-1)**
Tirpate **(26419-73-8)**
Tisin **(54-85-3)**
Tisiodrazida **(54-85-3)**
Tiskan (Czech) **(127-68-4)**
2,4,5-T, isobutyl ester **(4938-72-1)**
Tisomycin **(68-41-7)**
2,4,5-T isooctyl ester **(25168-15-4)**
2,4,5-T, isopropyl ester **(93-78-7)**
Tisperse MB-58 **(155-04-4)**
Tisperse MB-2X **(93-46-9)**
Titaantetrachloride (Dutch) **(7550-45-0)**
Titafrance **(13463-67-7)**
Titan White **(13463-67-7)**
Titanate **(7440-32-6)**
Titanate(2-), bis(ethanedioato(2-)-O,O')oxo-, dipotassium, (SP-5-21)- (9CI) **(14481-26-6)**
Titanate, dipotassium (9CI) **(12030-97-6)**
Titanate, strontium (1:1) (9CI) **(12060-59-2)**
Titandioxid (Swedish) **(13463-67-7)**
Titane (tetrachlorure de) (French) **(7550-45-0)**
Titanic acid, lead salt **(12060-00-3)**
Titanic acid, tetrabutyl ester **(5593-70-4)**
Titaniium disulfate **(13693-11-3)**
Titanio (tetracloruro di) (Italian) **(7550-45-0)**
Titanium **(7440-32-6)**
Titanium 50A **(7440-32-6)**
Titanium (DOT) **(7440-32-6)**
Titanium(IV) sulfate **(13693-11-3)**
Titanium, Powder wetted with 20% or more water **(7440-32-6)**
Titanium, iso-Pr alc. triethanolamine complexes **(74665-17-1)**
Titanium, Wet, With less than 20% water (DOT) **(7440-32-6)**
Titanium acetonyl acetonate **(14024-64-7)**
Titanium alloy **(7440-32-6)**
Titanium, bis(ethyl 3-oxobutanoato-O1',O3)bis(2-propanolato)- (9CI) **(27858-32-8)**
Titanium, bis((2,2',2''-nitrilotris(ethanolato))(1-)-N,O)bis-(2-propanolato)- (9CI) **(36673-16-2)**
Titanium, bis(2,4-pentanedionato-O,O')bis(2-propanolato)- (9CI) **(17927-72-9)**
Titanium chloride **(7550-45-0)**
Titanium(III) chloride **(7705-07-9)**
Titanium-chloride **(7705-07-9)**
Titanium chloride (TiCl4) **(7550-45-0)**
Titanium, dichlorobis(eta^5-2,4-cyclopentadien-1-yl- (9CI) **(1271-19-8)**
Titanium, dichloro-di-pi-cyclopentadienyl **(1271-19-8)**
Titanium dioxide (ACGIH,OSHA) **(13463-67-7)**
Titanium disulfate **(13693-11-3)**
Titanium ferrocene **(1271-19-8)**
Titanium(4+) isopropoxide **(546-68-9)**
Titanium isopropylate **(546-68-9)**
Titanium metal, Powder, dry [UN 2546] **(7440-32-6)**
Titanium oxide **(1317-80-2)**
Titanium-oxide **(13463-67-7)**
Titanium oxide bis(acetylacetonate) **(14024-64-7)**
Titanium, oxobis(2,4-pentanedionato) **(14024-64-7)**
Titanium, oxobis(2,4-pentanedionato-O,O') **(14024-64-7)**
Titanium peroxide **(13463-67-7)**
Titanium phosphate **(15578-51-5)**
Titanium potassium oxalate **(14481-26-6)**
Titanium, powder wetted with not less than 25 per cent water (a visible excess of water must be present) (a) mechanically produced, particle size less than 53 microns; (b) chemically produced, particle size less than 840 microns [UN 1352] **(7440-32-6)**
Titanium sulfate **(13693-11-3)**
Titanium sulfate (1:2) **(13693-11-3)**
Titanium(4+) sulfate **(13693-11-3)**
Titanium sulfate, basic **(13693-11-3)**
Titanium sulfate solution **(13693-11-3)**
Titanium sulfate solution (Containing not more than 45% sulfuric acid) **(13693-11-3)**
Titanium tetrachloride [UN 1838] **(7550-45-0)**
Titanium tetraisopropoxide **(546-68-9)**
Titanium tetraisopropylate **(546-68-9)**
Titanium tetra-n-propoxide **(546-68-9)**
Titanium trichloride **(7705-07-9)**
Titanium trichloride, Pyrophoric [UN 2441] **(7705-07-9)**
Titanocene **(1271-19-8)**
Titanocene, dichloride **(1271-19-8)**
Titanous chloride **(7705-07-9)**
Titanox **(13463-67-7)**
Titanox 2010 **(13463-67-7)**
Titantetrachlorid (German) **(7550-45-0)**
Titanyl bis(acetylacetonate) **(14024-64-7)**
Titriplex **(60-00-4)**
Titriplex I **(139-13-9)**
Titriplex III **(139-33-3)**
Tiuram **(97-77-8)**
Tiuram (Polish) **(137-26-8)**
Tiuramyl **(137-26-8)**
Tiurolan **(34014-18-1)**
Tixogel VP **(68953-58-2)**
Tixoton **(1302-78-9)**
Tizide **(54-85-3)**
Tl-1380 **(57-64-7)**
Tnk 2G5 **(105-60-2)**
Toabond 2 **(9003-20-7)**
Toabond 6 **(9003-20-7)**
Toabond 40H **(9003-20-7)**
Tobacco Dust **(84961-66-0)**
Tobacco Leaf, Aqueous Extract **(84961-66-0)**
Tobacco Wood **(68916-39-2)**
Tobaz **(148-79-8)**
Tobias acid **(81-16-3)**
Tobias acid, 5-sulfo- **(117-62-4)**
Tobradistin **(32986-56-4)**
Tobramycin **(32986-56-4)**

Name	CAS
Tobrex	(32986-56-4)
Tocainide	(41708-72-9)
Tochlorine	(127-65-1)
(2R,4'R,8'R)-α-Tocopherol	(59-02-9)
(R,R,R)-α-Tocopherol	(59-02-9)
α-Tocopherol	(59-02-9)
d-α-Tocopherol	(59-02-9)
Tocopherols	(1406-66-2)
D-α Tocopheryl succinate	(4345-03-3)
DL-α Tocopheryl succinate	(4345-03-3)
Tocopheryl succinate	(4345-03-3)
Tocosine	(51-67-2)
Todalgil	(50-33-9)
Tofranil	(50-49-7)
Tofuron	(3688-53-7)
N-Toin	(67-20-9)
Toin	(57-41-0)
Toin unicelles	(57-41-0)
Tok-Ultra	(8075-80-7)
Tok Ultra B	(8075-80-7)
Tokamina	(51-43-4)
Tokiocillin	(69-53-4)
Tokkorn	(1836-75-5)
Tokokin	(53-16-7)
Tokopharm	(59-02-9)
Tokuthion	(34643-46-4)
Tokyo Aniline Brilliant Green	(633-03-4)
Tokyo Aniline Malachite Green	(569-64-2)
2,4-Tolamine	(95-80-7)
Tolamine	(127-65-1)
Tolan	(501-65-5)
Tolane	(501-65-5)
Tolane tetrachloride	(13700-81-7)
Tolazamide	(1156-19-0)
Tolban	(26399-36-0)
Tolbusal	(64-77-7)
Tolbutamid	(64-77-7)
Tolbutamide	(64-77-7)
Tolcasone	(133-67-5)
o-Tolidin	(119-93-7)
2-Tolidin (German)	(119-93-7)
2-Tolidina (Italian)	(119-93-7)
2-Tolidine	(119-93-7)
3,3'-Tolidine	(119-93-7)
Tolidine	(119-93-7)
o,o'-Tolidine	(119-93-7)
o-Tolidine (ACGIH)	(119-93-7)
Tolinase	(1156-19-0)
Tolit	(118-96-7)
Tolite	(118-96-7)
Tolivalin	(8048-52-0)
Tolkan	(34123-59-6)
Tolken	(34123-59-6)
Toll	(298-00-0)
Tolmetin	(26171-23-3)
Tolmetine	(26171-23-3)
Tolnaftate	(2398-96-1)
Tolnaphthate	(2398-96-1)
Tolomol	(69-53-4)
Toloxatone	(29218-27-7)
Tolsanil	(2398-96-1)
Tolseron	(93-14-1)
Tolu-Sol	(108-88-3)
2-Tolualdehyde	(529-20-4)
4-Tolualdehyde	(104-87-0)
Tolualdehyde	(1334-78-7)
α-Tolualdehyde	(122-78-1)
para-Tolualdehyde	(104-87-0)
m-Tolualdehyde (8CI)	(620-23-5)
o-Tolualdehyde (8CI)	(529-20-4)
p-Tolualdehyde (8CI)	(104-87-0)
α-Tolualdehyde, α-methyl-	(93-53-8)
p-Tolualdehyde, oxime, (E)- (8CI)	(3717-15-5)
α-Toluamide	(103-81-1)
o-Toluamide (8CI)	(527-85-5)
m-Toluamide, N,N-diethyl	(134-62-3)
o-Toluamide, 3,5-dinitro	(148-01-6)
p-Toluamide, N-isopropyl-α-(2-methylhydrazino)	(671-16-9)
p-Toluamide, N-isopropyl-α-(2-methylhydrazino)-, hydrochloride	(366-70-1)
p-Toluamide, N-isopropyl-α-(2-methylhydrazino)-, mono-hydrochloride	(366-70-1)
o-Toluanilide	(7055-03-0)
Toluazotoluidine	(97-56-3)
1,4-Toluchinon (Czech)	(553-97-9)
o-Tolueno-azo-β-naphthol	(2646-17-5)
Tolueen (Dutch)	(108-88-3)
Tolueen-diisocyanaat (Dutch)	(584-84-9)
Toluen (Czech)	(108-88-3)
Toluen-disocianato (Italian)	(584-84-9)
Toluene (ACGIH,OSHA) [UN 1294]	(108-88-3)
Toluene, o-allyl- (8CI)	(1587-04-8)
Toluene, 3-amino-α,α,α-trifluoro-	(98-16-8)
Toluene-2-azonaphthol-2	(2646-17-5)
o-Toluene-1-azo-2-naphthylamine	(131-79-3)
o-Tolueneazo-o-tolueneazo-β-naphthol	(85-83-6)
o-Tolueneazo-o-toluene-β-naphthol	(85-83-6)
o-Tolueneazo-o-toluidine	(97-56-3)
Toluene, α-bromo	(100-39-0)
Toluene, m-bromo	(591-17-3)
Toluene, o-bromo	(95-46-5)
Toluene, p-bromo	(106-38-7)
Toluene, α-bromo-o-chloro-	(611-17-6)
Toluene, m-bromo-α,α,α-trifluoro	(401-78-5)
Toluene, α-(2-(2-butoxyethoxy)ethoxy)-4,5-(methylene-dioxy)-2-propyl	(51-03-6)
Toluene, p-tert-butyl	(98-51-1)
Toluene, tert-butyl- (8CI)	(27138-21-2)
Toluene, 3-tert-butyl-5-ethyl-	(6630-01-9)
Toluenecarboxaldehyde	(1334-78-7)
Toluene, α-chloro	(100-44-7)
Toluene, m-chloro	(108-41-8)
Toluene, o-chloro	(95-49-8)
Toluene, p-chloro	(106-43-4)
Toluene, ar-chloro (8CI)	(25168-05-2)
Toluene, 4-chloro-3,5-dinitro-α,α,α-trifluoro	(393-75-9)
Toluene, α-chloro-o-fluoro-	(345-35-7)
Toluene, α-chloro-m-fluoro- (8CI)	(456-42-8)
Toluene, α-chloro-p-fluoro- (8CI)	(352-11-4)
Toluene, 2-chloro-4-nitro	(121-86-8)
Toluene, 2-chloro-6-nitro	(83-42-1)
Toluene, α-chloro-m-nitro	(619-23-8)
Toluene, α-chloro-o-nitro	(612-23-7)
Toluene, α-chloro-p-nitro	(100-14-1)
Toluene, 4-chloro-2-nitro- (8CI)	(89-59-8)
Toluene, 4-chloro-3-nitro- (8CI)	(89-60-1)
Toluene, 4-chloro-3-nitro-α,α,α-trifluoro	(121-17-5)
Toluene, m-chloro-α,α,α-trifluoro	(98-15-7)
Toluene, o-chloro-α,α,α-trifluoro	(88-16-4)
Toluene, p-chloro-α,α,α-trifluoro	(98-56-6)
Toluene, 2-chloro-α,α,α-trifluoro-5-nitro- (8CI)	(777-37-7)
Toluene, α-cyano-	(140-29-4)
Toluene-α,α,α-d3 (8CI)	(1124-18-1)
Toluene-d8 (8CI)	(2037-26-5)
2,4-Toluenediamine	(95-80-7)

Name	CAS
3,5-Toluenediamine	**(108-71-4)**
Toluene-2,3-diamine	**(2687-25-4)**
Toluene-2,4-diamine	**(95-80-7)**
Toluene-2,5-diamine	**(95-70-5)**
Toluene-2,6-diamine	**(823-40-5)**
Toluene-3,4-diamine	**(496-72-0)**
Toluene-3,5-diamine	**(108-71-4)**
m-Toluenediamine	**(95-80-7)**
p-Toluenediamine	**(95-70-5)**
Toluenediamine (DOT)	**(25376-45-8)**
Toluene-2,4-diamine, N^4-acetyl-N^2-trifluoromethylsulfonyl	**(47000-92-0)**
Toluene-2,4-diamine, N^4,N^4-diethyl-3,5-dinitro-α,α,α-trifluoro	**(29091-05-2)**
Toluene-2,4-diamine, N^4,N^4-diethyl-α,α,α-trifluoro-3,5-dinitro-	**(29091-05-2)**
2,4-Toluenediamine dihydrochloride	**(636-23-7)**
Toluene-2,4-diamine, dihydrochloride	**(636-23-7)**
Toluene-2,5-diamine, dihydrochloride	**(615-45-2)**
Toluene-2,6-diamine, dihydrochloride	**(15481-70-6)**
p-Toluenediamine dihydrochloride	**(615-45-2)**
Toluene-2,4-diamine, monohydrochloride (8CI)	**(5459-85-8)**
2,5-Toluenediamine sulfate	**(6369-59-1)**
Toluene-2,5-diamine, sulfate	**(6369-59-1)**
Toluene-2,5-diamine, sulfate (1:1)	**(615-50-9)**
p-Toluenediamine sulfate	**(6369-59-1)**
Toluene-2,5-diamine sulphate	**(615-50-9)**
p-Toluenediamine sulphate	**(615-50-9)**
Toluene, 3,5-diamino-	**(108-71-4)**
Toluene, 3,5-dibromo- (6CI,7CI,8CI)	**(1611-92-3)**
Toluene, 2,4-dichloro	**(95-73-8)**
Toluene, α,α-dichloro-	**(98-87-3)**
Toluene, p,α-dichloro	**(104-83-6)**
Toluene, 2,5-dichloro- (8CI)	**(19398-61-9)**
Toluene, 2,6-dichloro- (8CI)	**(118-69-4)**
Toluene, 3,4-dichloro- (8CI)	**(95-75-0)**
Toluene, 3,5-dichloro- (8CI)	**(25186-47-4)**
Toluene, 3,4-dichloro-α,α,α-trifluoro-	**(328-84-7)**
Toluene, 3,5-diethyl	**(2050-24-0)**
Toluene, diethyl-	**(25550-13-4)**
Toluene, ar,ar-diethyl- (8CI)	**(25550-13-4)**
2,4-Toluenediisocyanate	**(584-84-9)**
2,6-Toluene diisocyanate	**(91-08-7)**
Toluene 2,6-diisocyanate	**(91-08-7)**
Toluene diisocyanate	**(584-84-9)**
Toluene-2,4-diisocyanate (ACGIH,OSHA)	**(584-84-9)**
Toluene diisocyanate [UN 2078]	**(26471-62-5)**
Toluene diisocyanate dimer	**(26747-90-0)**
Toluene, 2,6-dimethoxy- (8CI)	**(5673-07-4)**
Toluene, 3,4-dimethoxy- (8CI)	**(494-99-5)**
Toluene, 2,3-dinitro	**(602-01-7)**
Toluene, 2,4-dinitro	**(121-14-2)**
Toluene, 2,5-dinitro	**(619-15-8)**
Toluene, 2,6-dinitro	**(606-20-2)**
Toluene, 3,4-dinitro	**(610-39-9)**
Toluene, 3,5-dinitro	**(618-85-9)**
Toluene, ar,ar-dinitro-	**(25321-14-6)**
Toluene, dinitro	**(25321-14-6)**
2,3-Toluenediol	**(488-17-5)**
2,5-Toluenediol	**(95-71-6)**
Toluene-3,4-diol	**(452-86-8)**
Toluene, o-ethyl	**(611-14-3)**
Toluene, p-ethyl	**(622-96-8)**
Toluene, m-ethyl- (8CI)	**(620-14-4)**
Toluene, p-(1-ethylpropyl)- (8CI)	**(22975-58-2)**
p-Tolueneethylsulfonamide	**(80-39-7)**
Toluene, m-fluoro	**(352-70-5)**
Toluene, o-fluoro	**(95-52-3)**
Toluene, p-fluoro	**(352-32-9)**
Toluene, ar-fluoro- (8CI)	**(25496-08-6)**
Toluene, m-fluoro- (8CI)	**(352-70-5)**
Toluene, p-fluoro- (8CI)	**(352-32-9)**
Toluene hexahydride	**(108-87-2)**
Toluene, hexahydro-	**(108-87-2)**
Toluene, α-iodo	**(620-05-3)**
Toluene, m-iodo- (8CI)	**(625-95-6)**
Toluene, o-iodo- (8CI)	**(615-37-2)**
Toluene, p-iodo- (8CI)	**(624-31-7)**
Toluene, p-isobutyl- (8CI)	**(5161-04-6)**
Toluene, m,α-dichloro- (8CI)	**(620-20-2)**
p-Toluenenitrile	**(104-85-8)**
Toluene, m-nitro	**(99-08-1)**
Toluene, o-nitro	**(88-72-2)**
Toluene, p-nitro	**(99-99-0)**
Toluene, 2-nitro-α,α,α-trifluoro	**(384-22-5)**
Toluene, 3-nitro-α,α,α-trifluoro	**(98-46-4)**
Toluene, 4-nitro-α,α,α-trifluoro	**(402-54-0)**
Toluene, octafluoro	**(434-64-0)**
Toluene, o,α-dichloro- (8CI)	**(611-19-8)**
Toluene, o,α,α,α-tetrachloro- (8CI)	**(2136-89-2)**
Toluene, 2,3,4,5,6-pentabromo-	**(87-83-2)**
Toluene, 2,3,4,5,6-pentachloro- (8CI)	**(877-11-2)**
Toluene, α,α,α,2,4-pentachloro- (8CI)	**(13014-18-1)**
Toluene, α,α,α,3,4-pentachloro- (8CI)	**(13014-24-9)**
Toluene, pentyl- (6CI, 8CI)	**(1320-01-0)**
Toluene, α-phenyl-	**(101-81-5)**
Toluene, m-propyl- (8CI)	**(1074-43-7)**
Toluene, o-propyl- (8CI)	**(1074-17-5)**
Toluene, p-propyl- (8CI)	**(1074-55-1)**
4-Toluenesulfanamide	**(70-55-3)**
p-Toluenesulfanamide	**(70-55-3)**
N-(2-Toluenesulfenyl)propoxur	**(50539-85-0)**
p-Toluenesulfinic acid, sodium salt	**(824-79-3)**
ortho-Toluenesulfochloride	**(133-59-5)**
p-Toluenesulfochloride	**(98-59-9)**
para-Toluenesulfochloride	**(98-59-9)**
Toluene-2-sulfonamide	**(88-19-7)**
Toluene-4-sulfonamide	**(70-55-3)**
Toluenesulfonamide	**(1333-07-9)**
ar-Toluenesulfonamide	**(1333-07-9)**
o-Toluenesulfonamide	**(88-19-7)**
p-Toluenesulfonamide	**(70-55-3)**
m-Toluenesulfonamide (8CI)	**(1899-94-1)**
p-Toluenesulfonamide, N-chloro-, sodium salt	**(127-65-1)**
p-Toluenesulfonamide, N-cyclohexyl- (8CI)	**(80-30-8)**
p-Toluenesulfonamide, N,N-dimethyl	**(599-69-9)**
o-Toluenesulfonamide, N-ethyl-	**(1077-56-1)**
p-Toluenesulfonamide, N-ethyl- (8CI)	**(80-39-7)**
p-Toluenesulfonamide, N-(ethylmercuri)-N-phenyl-+	**(517-16-8)**
p-Toluenesulfonamide, N-methyl- (8CI)	**(640-61-9)**
p-Toluenesulfonamide, N-methyl-N-nitroso	**(80-11-5)**
p-Toluenesulfonamide, N-(phenylsulfonyl)-	**(14706-41-3)**
4-(p-Toluenesulfonamido)diphenylamine	**(100-93-6)**
p-(p-Toluenesulfonamido)diphenylamine	**(100-93-6)**
p-Toluenesulfonanilide, 4'-anilino- (8CI)	**(100-93-6)**
p-Toluenesulfonanilide, N-(ethylmercuri)-	**(517-16-8)**
ortho-Toluenesulfonchloride	**(133-59-5)**
para-Toluenesulfonchloride	**(98-59-9)**
4-Toluenesulfonic acid	**(104-15-4)**
Toluenesulfonic acid	**(104-15-4)**
α-Toluenesulfonic acid	**(100-87-8)**
p-Toluenesulfonic acid	**(104-15-4)**
m-Toluenesulfonic acid, 6-amino	**(88-44-8)**
m-Toluenesulfonic acid, 4-amino- (8CI)	**(98-33-9)**
m-Toluenesulfonic acid, 6-amino-4-chloro	**(88-51-7)**
p-Toluenesulfonic acid, 2-amino-5-chloro-	**(88-53-9)**
4-Toluenesulfonic acid, chloride	**(98-59-9)**

p-Toluenesulfonic acid chloride **(98-59-9)**
p-Toluenesulfonic acid, cyclohexyl ester (8CI) **(953-91-3)**
p-Toluenesulfonic acid, 2-(4,6-dichloro-s-triazin-2-yl)hydrazide (6CI,8CI) **(30357-79-0)**
p-Toluenesulfonic acid, ethyl ester **(80-40-0)**
p-Toluenesulfonic acid, hydrazide **(1576-35-8)**
p-Toluenesulfonic acid, methyl ester **(80-48-8)**
p-Toluenesulfonic acid, phenethyl ester (7CI,8CI) **(4455-09-8)**
p-Toluenesulfonic acid, pyridine adduct **(24057-28-1)**
4-Toluenesulfonic acid sodium salt **(657-84-1)**
p-Toluenesulfonic acid, sodium salt **(657-84-1)**
p-Toluenesulfonylamide **(70-55-3)**
1-p-Toluenesulfonyl-3-butylurea **(64-77-7)**
4-Toluenesulfonyl chloride **(98-59-9)**
ortho-Toluenesulfonyl chloride **(133-59-5)**
p-Toluenesulfonyl chloride **(98-59-9)**
para-Toluenesulfonyl chloride **(98-59-9)**
o-Toluenesulfonyl chloride (8CI) **(133-59-5)**
Toluenesulfonyl chloride (VAN) **(98-59-9)**
o-Toluenesulfonyl chloride, 5-nitro **(121-02-8)**
p-Toluenesulfonyl-N-ethylamide **(80-39-7)**
N-(p-Toluenesulfonyl)-N'-hexamethyleniminourea **(1156-19-0)**
Toluene-p-sulfonylmethylnitrosamide **(80-11-5)**
4-Toluenesulfonyl semicarbazide **(10396-10-8)**
p-Toluenesulfonyl semicarbazide **(10396-10-8)**
2'-O-(p-Toluenesulfonyl)uridine **(6206-10-6)**
Toluene-p-sulphonamide **(70-55-3)**
p-Toluenesulphonic acid **(104-15-4)**
p-Toluenesulphonyl chloride **(98-59-9)**
Toluene, α,α,α,p-tetrachloro **(5216-25-1)**
Toluene, α,ar,ar,ar-tetrachloro **(1344-32-7)**
Toluene, α,α2,6-tetrachloro- (8CI) **(81-19-6)**
Toluene, 2,3,4,5-tetrachloro- (6CI,7CI,8CI) **(1006-32-2)**
Toluene, 2,3,5,6-tetrachloro- (6CI,7CI,8CI) **(1006-31-1)**
Toluene-2,3,5,6-tetrol (8CI) **(700-19-6)**
2-Toluenethiol **(137-06-4)**
4-Toluenethiol **(106-45-6)**
α-Toluenethiol **(100-53-8)**
m-Toluenethiol **(108-40-7)**
o-Toluenethiol **(137-06-4)**
p-Toluenethiol **(106-45-6)**
p-Toluenethiol, S-ester with O-ethyl ethylphosphonodithioate **(333-43-7)**
Toluene, tribromo- (8CI) **(27476-22-8)**
Toluene trichloride **(98-07-7)**
Toluene, 2,3,6-trichloro **(2077-46-5)**
Toluene, α, ar, ar-trichloro- **(38721-71-0)**
Toluene, α,α,α-trichloro **(98-07-7)**
Toluene, 2,4,5-trichloro- (8CI) **(6639-30-1)**
Toluene, 2,4,6-trichloro- (8CI) **(23749-65-7)**
Toluene, α,3,4-trichloro- (8CI) **(102-47-6)**
Toluene, α2,4-trichloro- (8CI) **(94-99-5)**
Toluene, 2,3,4-trichloro- (7CI,8CI) **(7359-72-0)**
Toluene, α,α,α-trifluoro **(98-08-8)**
Toluene, trifluoro- (6CI,7CI,8CI) **(27359-10-0)**
Toluene, α,α,α-trifluoro-3-bromo- **(401-78-5)**
Toluene, α,α,α-trifluoro-m-nitro- (8CI) **(98-46-4)**
Toluene, 2,3,4-trinitro **(602-29-9)**
Toluene, 2,3,6-trinitro **(18292-97-2)**
Toluene, 2,4,5-trinitro **(610-25-3)**
Toluene, 2,4,6-trinitro- (Wet) **(118-96-7)**
Toluene, vinyl- (Mixed isomers) **(25013-15-4)**
3-Toluenkarbonitril (Czech) **(620-22-4)**
4-Toluenkarbonitril (Czech) **(104-85-8)**
2-Toluenkarbonitril (Czech) **(529-19-1)**
α-Toluenol **(100-51-6)**
ar-Toluenol **(1319-77-3)**
p-Toluensulfinan sodny (Czech) **(824-79-3)**
p-Toluhydroquinol **(95-71-6)**
Toluhydroquinone **(95-71-6)**
p-Toluhydroquinone **(95-71-6)**
α-Toluic acid **(103-82-2)**
m-Toluic acid **(99-04-7)**
meta-Toluic acid **(99-04-7)**
o-Toluic acid **(118-90-1)**
p-Toluic acid **(99-94-5)**
m-Toluic acid, 4-amino- (8CI) **(2486-70-6)**
p-Toluic acid chloride **(874-60-2)**
p-Toluic acid, α-chloro-, methyl ester (8CI) **(34040-64-7)**
p-Toluic acid, 2,5-dichloro- (6CI,8CI) **(21460-88-8)**
m-Toluic acid diethylamide **(134-62-3)**
o-Toluic acid, 4,6-dimethoxy-, ester with 6-methoxy-4,2-eresotic acid bimol. ester, methyl ester **(19314-74-0)**
o-Toluic acid, 3,5-dinitro- **(28169-46-2)**
α-Toluic acid, ethyl ester **(101-97-3)**
α-Toluic acid, α-hydroxy- **(90-64-2)**
o-Toluic acid, α-hydroxy- (8CI) **(612-20-4)**
α-Toluic acid, α-(hydroxymethyl)- **(529-64-6)**
meta-Toluic acid, methyl ester **(99-36-5)**
p-Toluic acid, methyl ester **(99-75-2)**
m-Toluic acid, methyl ester (8CI) **(99-36-5)**
o-Toluic acid, 5-nitro- (8CI) **(1975-52-6)**
o-Toluic acid, 6-nitro- (8CI) **(13506-76-8)**
p-Toluic acid, 3-nitro- (8CI) **(96-98-0)**
o-Toluic acid, α-phenyl- (8CI) **(612-35-1)**
m-Toluic acid, α,α,α-trifluoro- (8CI) **(454-92-2)**
α-Toluic aldehyde **(122-78-1)**
o-Toluic aldehyde **(529-20-4)**
o-Toluic amide **(527-85-5)**
o-Toluic nitrile **(529-19-1)**
p-Toluic nitrile **(104-85-8)**
m-Toluidin (Czech) **(108-44-1)**
o-Toluidin (Czech) **(95-53-4)**
p-Toluidin (Czech) **(106-49-0)**
Toluidina (Spanish) **(26915-12-8)**
2-Toluidine **(95-53-4)**
3-Toluidine **(108-44-1)**
4-Toluidine **(106-49-0)**
Toluidine **(26915-12-8)**
p-Toluidine **(106-49-0)**
o-Toluidine (ACGIH,DOT,OSHA) **(95-53-4)**
m-Toluidine (ACGIH,OSHA) [UN 1708] **(108-44-1)**
p-Toluidine, Liquid (ACGIH,OSHA) [UN 1708] **(106-49-0)**
Toluidine Red **(2425-85-6)**
Toluidine Red 10451 **(2425-85-6)**
Toluidine Red 3B **(2425-85-6)**
Toluidine Red BFB **(2425-85-6)**
Toluidine Red BFGG **(2425-85-6)**
Toluidine Red D 28-3930 **(2425-85-6)**
Toluidine Red Light **(2425-85-6)**
Toluidine Red M 20-3785 **(2425-85-6)**
Toluidine Red 4R **(2425-85-6)**
Toluidine Red R **(2425-85-6)**
Toluidine Red RT-115 **(2425-85-6)**
Toluidine Red Toner **(2425-85-6)**
Toluidine Red XL 20-3050 **(2425-85-6)**
o-Toluidine, Solid [UN 1708] **(95-53-4)**
Toluidine Toner **(2425-85-6)**
Toluidine Toner Dark 5040 **(2425-85-6)**
Toluidine Toner HR X-2741 **(2425-85-6)**
Toluidine Toner Keep HR X-2742 **(2425-85-6)**
Toluidine Toner L 20-3300 **(2425-85-6)**
Toluidine Toner RT-252 **(2425-85-6)**
Toluidine Toner 4R X-2700 **(2425-85-6)**
o-Toluidine, 4-((p-aminophenyl)(4-imino-2,5-cyclohexadien-

1-ylidene)methyl)- (3248-93-9)
m-Toluidine, N,N-bis(2-hydroxyethyl)- (91-99-6)
o-Toluidine, N,N-bis(2-hydroxyethyl) (28005-74-5)
m-Toluidine, 4-bromo- (8CI) (6933-10-4)
p-Toluidine, N-butyl-N-ethyl-α,α,α-trifluoro-2,6-dinitro (1861-40-1)
o-Toluidine, 3-chloro (87-60-5)
o-Toluidine, 4-chloro (95-69-2)
o-Toluidine, 5-chloro (95-79-4)
o-Toluidine, 6-chloro (87-63-8)
p-Toluidine, 2-chloro (615-65-6)
p-Toluidine, 3-chloro (95-74-9)
m-Toluidine, 4-chloro- (8CI) (7149-75-9)
p-Toluidine, N-(2-chloroethyl)-2,6-dinitro-N-propyl-α,α,α-trifluoro (33245-39-5)
o-Toluidine, 4-chloro-, hydrochloride (3165-93-3)
p-Toluidine, 3-chloro-, hydrochloride (7745-89-3)
m-Toluidine, N-cyanoethyl-N-ethyl- (148-69-6)
p-Toluidine, N-(cyclopropylmethyl)-2,6-dinitro-N-propyl-α,α,α-trifluoro (26399-36-0)
m-Toluidine, N,N-diethyl- (8CI) (91-67-8)
o-Toluidine, N,N-dimethyl (609-72-3)
p-Toluidine, N,N-dimethyl (99-97-8)
o-Toluidine, N,N-dimethyl-4-((m-nitrophenyl)azo)- (8CI) (4313-14-8)
o-Toluidine, N,N-dimethyl-4-((p-nitrophenyl)azo)- (8CI) (4313-13-7)
m-Toluidine, 2,4-dinitro (19406-51-0)
m-Toluidine, 2,6-dinitro (70343-06-5)
m-Toluidine, 4,6-dinitro (5267-27-6)
o-Toluidine, 3,5-dinitro (35572-78-2)
o-Toluidine, 3,6-dinitro (56207-39-7)
p-Toluidine, 2,6-dinitro (6393-42-6)
p-Toluidine, 2,6-dinitro-N,N-dipropyl (1918-08-7)
p-Toluidine, 2,6-dinitro-N-ethyl-N-(2-methyl-2-propenyl)-α,α,α-trifluoro (55283-68-6)
o-Toluidine, 4,6-dinitro-N-methyl-N-(2,4,6-tribromophenyl)-α,α,α-trifluoro- (63333-35-7)
m-Toluidine, N-(2,4-dinitrophenyl)- (8CI) (964-79-4)
p-Toluidine, N-(2,4-dinitrophenyl)-.α.,.α.,.α.-trifluoro- (8CI) (13744-79-1)
m-Toluidine, N-(2,4-dinitrophenyl)-.α.,.α.,.α.-trifluoro- (7CI,8CI) (1869-67-6)
m-Toluidine, N-ethyl (102-27-2)
o-Toluidine, 6-ethyl- (24549-06-2)
o-Toluidine, N-ethyl (94-68-8)
2-Toluidine hydrochloride (636-21-5)
m-Toluidine, hydrochloride (638-03-9)
o-Toluidine, hydrochloride (636-21-5)
p-Toluidine, hydrochloride (540-23-8)
o-Toluidine, 4,4'-methylenedi- (838-88-0)
o-Toluidine, N-methyl-4-nitro- (8CI) (10439-77-7)
m-Toluidine, 6-nitro (578-46-1)
o-Toluidine, 3-nitro (603-83-8)
o-Toluidine, 4-nitro (99-52-5)
o-Toluidine, 5-nitro (99-55-8)
p-Toluidine, 3-nitro (119-32-4)
m-Toluidine, 4-nitro- (8CI) (611-05-2)
p-Toluidine, 2-nitro- (8CI) (89-62-3)
m-Toluidine, N-(p-nitrophenyl)- (8CI) (15979-82-5)
Toluidine, N-phenyl-α,α,α-trifluoro (101-23-5)
o-Toluidine-m-sulfonic acid (98-33-9)
p-Toluidine-m-sulfonic acid (88-44-8)
o-Toluidine, 4-(o-tolylazo) (97-56-3)
m-Toluidine, α,α,α-trifluoro (98-16-8)
o-Toluidine, α,α,α-trifluoro (88-17-5)
p-Toluidine, α,α,α-trifluoro-2,6-dinitro- (8CI) (445-66-9)
p-Toluidine, α,α,α-trifluoro-2,6-dinitro-N,N-dipropyl (1582-09-8)
p-Toluidine, α,α,α-trifluoro-2,6-dinitro-N-propyl- (8CI) (2077-99-8)
m-Toluidine, .α.,.α.,.α.-trifluoro-N-(p-nitrophenyl)- (8CI) (369-90-4)
o-Toluidin hydrochloride (636-21-5)
p-Toluidinium chloride (540-23-8)
2-o-Toluidinoethanol (136-80-1)
o-Toluidinomethanesulfonic acid (94-57-5)
o-Toluidyna (Polish) (95-53-4)
Toluilenodwuizocyjanian (Polish) (584-84-9)
α-Toluimidic acid (103-81-1)
Toluina (64-77-7)
Tolumid (64-77-7)
p-Tolunitril (Czech) (104-85-8)
4-Tolunitrile (104-85-8)
α-Tolunitrile (140-29-4)
m-Tolunitrile (620-22-4)
p-Tolunitrile (104-85-8)
o-Tolunitrile, 6-chloro- (8CI) (6575-09-3)
m-Toluol (108-39-4)
o-Toluol (95-48-7)
p-Toluol (106-44-5)
Toluol (DOT) (108-88-3)
o-Toluol-azo-o-toluidin (German) (97-56-3)
Toluolo (Italian) (108-88-3)
ortho-Toluol-sulfonamid (German) (88-19-7)
p-Toluolsulfonsaeure aethyl ester (German) (80-40-0)
p-Toluolsulfonsaeure methyl ester (German) (80-48-8)
N-Toluolsulphonyl hydrazine (1576-35-8)
α-Toluolthiol (100-53-8)
o-Toluonitrile (529-19-1)
p-Toluonitrile (104-85-8)
4-Toluoyl chloride (874-60-2)
p-Toluoyl chloride (874-60-2)
m-Toluquinaldine (93-37-8)
o-Toluquinaldine (1463-17-8)
Toluquinol (95-71-6)
p-Toluquinol (95-71-6)
m-Toluquinoline (612-60-2)
p-Toluquinoline (91-62-3)
1,4-Toluquinone (553-97-9)
Toluquinone (553-97-9)
p-Toluquinone (553-97-9)
Tolusafranine (477-73-6)
Toluvan (64-77-7)
o-Toluylaldehyde (529-20-4)
p-Toluyl aldehyde (104-87-0)
p-Toluylaldehyde (104-87-0)
para-Toluyl aldehyde (104-87-0)
p-Toluyl chloride (874-60-2)
m-Toluylendiamin (Czech) (95-80-7)
p-Toluylendiamine (95-70-5)
2,3-Toluylenediamine (2687-25-4)
2,6-Toluylenediamine (823-40-5)
3,4-Toluylenediamine (496-72-0)
Toluylene-2,5-diamine (95-70-5)
m-Toluylenediamine (95-80-7)
meta Toluylene diamine (95-80-7)
2,4-Toluylenediamine [UN 1709] (95-80-7)
Toluylene-2,5-diamine sulphate (615-50-9)
p-Toluylenediamine sulphate (615-50-9)
Toluylene-2,4-diisocyanate (584-84-9)
m-Toluylic acid (99-04-7)
o-Toluylic acid (118-90-1)
p-Toluylic acid (99-94-5)
p-Tolycarbamide (622-51-5)
m-Tolyenediamine (95-80-7)
Tolyene 2,4-diisocyanate (584-84-9)
Tolyfluanide (731-27-1)
N-m-Tolylacetamide (537-92-8)
m-Tolylacetamide (537-92-8)
o-Tolyl acetate (533-18-6)
p-Tolyl acetate (140-39-6)

(p-Tolyl)acetic acid	(622-47-9)
o-Tolyl alcohol	(89-95-2)
p-Tolyl alcohol	(106-44-5)
Tolyl aldehyde	(1334-78-7)
o-Tolylaldehyde	(529-20-4)
p-Tolylaldehyde	(104-87-0)
α-Tolyl aldehyde dimethyl acetal	(101-48-4)
o-Tolylamide	(527-85-5)
Tolylamine	(106-49-0)
m-Tolylamine	(108-44-1)
o-Tolylamine	(95-53-4)
p-Tolylamine	(106-49-0)
o-Tolylamine hydrochloride	(636-21-5)
2-o-Tolylaminoethanol	(136-80-1)
5-(o-Tolylazo)-2-aminotoluene	(97-56-3)
4-o-Tolylazo-o-diacetotoluide	(83-63-6)
4'-(o-Tolylazo)-o-diacetotoluidide	(83-63-6)
1-(o-Tolylazo)-β-naphthol	(2646-17-5)
1-o-Tolylazo-2-naphthol	(2646-17-5)
1-(o-Tolylazo)-2-naphthylamine	(131-79-3)
4-(o-Tolylazo)-o-toluidine	(97-56-3)
o-Tolylazo-o-tolylazo-β-naphthol	(85-83-6)
o-Tolylazo-o-tolylazo-2-naphthol	(85-83-6)
1-(4-o-Tolylazo-o-tolylazo)-2-naphthol	(85-83-6)
2-Tolyl bromide	(95-46-5)
m-Tolyl bromide	(591-17-3)
o-Tolylbromide	(95-46-5)
m-Tolylcarbamide	(63-99-0)
4-Tolylcarbinol	(589-18-4)
p-Tolylcarbinol	(589-18-4)
Tolyl chloride	(100-44-7)
m-Tolyl chloride	(108-41-8)
o-Tolyl chloride	(95-49-8)
p-Tolyl chloride	(106-43-4)
m-Tolyldiethanolamine	(91-99-6)
o-Tolyldiethanolamine	(28005-74-5)
Tolyl diphenyl phosphate	(26444-49-5)
2,3-Tolylenediamine	(2687-25-4)
2,4-Tolylenediamine	(95-80-7)
2,6-Tolylenediamine	(823-40-5)
3,4-Tolylenediamine	(496-72-0)
4-m-Tolylenediamine	(95-80-7)
Tolylene-2,4-diamine	(95-80-7)
Tolylenediamine	(25376-45-8)
m-Tolylenediamine	(95-80-7)
p,m-Tolylenediamine	(95-70-5)
p-Tolylenediamine sulphate	(615-50-9)
2,4-Tolylenediisocyanate	(584-84-9)
Tolylene 2,6-diisocyanate	(91-08-7)
Tolylene diisocyanate	(26471-62-5)
Tolylene-2,4-diisocyanate	(584-84-9)
meta-Tolylene diisocyanate	(584-84-9)
meta-Tolylene diisocyanate	(91-08-7)
Tolylene isocyanate	(26471-62-5)
m-Tolylester kyseliny methylkarbaminove (Czech)	(1129-41-5)
p-Tolyl ethanoate	(140-39-6)
1-(p-Tolyl)ethanol	(536-50-5)
o-Tolyl ethanolamine	(136-80-1)
1-p-Tolylethene	(622-97-9)
Tolylfluanid	(731-27-1)
9-m-Tolylfluorene	(18153-42-9)
o-Tolylformamide	(94-69-9)
Tolyl glycidyl ether	(26447-14-3)
Tolylhydroquinone	(95-71-6)
N-(p-Tolyl)-4-hydroxy-1-anthraquinonylamine	(81-48-1)
2-(N'-p-Tolyl-N'-m-hydroxyphenylaminomethyl)-2-imidazoline	(50-60-2)
2,2'-(m-Tolylimino)diethanol	(91-99-6)

1,1-(p-Tolylimino)dipropan-2-ol	(38668-48-3)
p-Tolyl isocyanide (6CI,7CI,8CI)	(7175-47-5)
p-Tolyl isonitrile	(7175-47-5)
α-Tolyl mercaptan	(100-53-8)
m-Tolylmercaptan	(108-40-7)
o-Tolyl mercaptan	(137-06-4)
p-Tolyl mercaptan	(106-45-6)
Tolylmercuric acetate	(1300-78-3)
3-Tolyl-N-methylcarbamate	(1129-41-5)
m-Tolyl N-methylcarbamate	(1129-41-5)
p-Tolylmethylcarbinol (German)	(536-50-5)
p-Tolyl methyl ether	(104-93-8)
1-(p-Tolyl)-3-methylpyrazolone-5	(86-92-0)
o-Tolylnitrile	(529-19-1)
p-Tolylnitrile	(104-85-8)
o-Tolyl phosphate	(78-30-8)
N-m-Tolylphthalamic acid	(85-72-3)
N-meta-Tolylphthalamic acid	(85-72-3)
Tolylsulfonamide	(70-55-3)
p-Tolylsulfonamide	(70-55-3)
p-Tolylsulfonic acid	(104-15-4)
p-(p-Tolylsulfonylamino)diphenylamine	(100-93-6)
N-(p-Tolylsulfonyl)-N'-butylcarbamide	(64-77-7)
3-(p-Tolyl-4-sulfonyl)-1-butylurea	(64-77-7)
Tolylsulfonylbutylurea	(64-77-7)
o-Tolylsulfonyl chloride	(133-59-5)
p-Tolylsulfonyl chloride	(98-59-9)
4-(p-Tolylsulfonyl)-1,1-hexamethylenesemicarbazide	(1156-19-0)
p-Tolylsulfonylhydrazine	(1576-35-8)
p-Tolylsulfonyl-methyl-nitrosamid (German)	(80-11-5)
p-Tolylsulfonylmethylnitrosamide	(80-11-5)
p-Tolylsulfonylmethylnitrosamine	(80-11-5)
p-Tolylthiol	(106-45-6)
1-o-Tolyl-2-thiourea	(614-78-8)
o-Tolyl thiourea	(614-78-8)
Tolyltriazole, sodium salt	(64665-57-2)
3-Tolylurea	(63-99-0)
m-Tolylurea	(63-99-0)
p-Tolylurea	(622-51-5)
Tolyn	(93-14-1)
m-Tolynitrile	(620-22-4)
p-Tolyurea	(622-51-5)
Tomarin	(117-52-2)
Tomaset	(85-72-3)
Tomathrel	(16672-87-0)
A''-Tomatidine	(17406-45-0)
Tomatidine, glycoside	(17406-45-0)
Tomatin	(17406-45-0)
Tomatine	(17406-45-0)
α-Tomatine	(17406-45-0)
Tomato Fix	(122-88-3)
Tomato Fix Concentrate	(122-88-3)
Tomato Hold	(122-88-3)
Tomatotone	(122-88-3)
Tomobil	(536-50-5)
Tomofan	(9004-34-6)
Tomorin	(81-82-3)
Tonalid	(21145-77-7)
Tonaril	(91-81-6)
Tonarsen	(144-21-8)
Tonarsin	(144-21-8)
Toncarine	(92-48-8)
Tonco-70	(112-96-9)
Tonedrin	(300-42-5)
Tonedron	(300-42-5)
Tonephin	(11000-17-2)
Toner Lake Red C	(5160-02-1)

Tonexol	(136-47-0)
Toney Red	(85-86-9)
Toning Blue MV	(1325-82-2)
Tonite	(78-95-5)
Tonix	(8065-62-1)
Tonka Bean Camphor	(91-64-5)
Tonkalide	(695-06-7)
Tonocard	(41708-72-9)
Tonocard	(59-26-7)
Tonocor	(59-26-7)
Tonoftal	(2398-96-1)
Tonogen	(51-43-4)
Tonolyt isopropyl meprobamate	(78-44-4)
Tonox	(101-77-9)
Tonsillosan	(598-82-3)
Tony Red	(85-86-9)
Tonzilamine	(91-85-0)
Top Cop Tri Basic	(1332-03-2)
Top Flake	(7647-14-5)
Top Form Wormer	(148-79-8)
Topane	(132-27-4)
Topanel	(123-73-9)
Topanol	(128-37-0)
Topanol 354	(489-01-0)
Topanol O	(128-37-0)
Topanol OC	(128-37-0)
Topazone	(67-45-8)
Topcaine	(94-09-7)
Topcide	(5251-93-4)
Top distillation cut by-product acids, monobasic (C1-C5)	(68937-68-8)
Topex	(94-36-0)
Tophol	(132-60-5)
Tophosan	(132-60-5)
Topicain	(126-27-2)
Topichlor 20	(57-74-9)
Topiclor	(57-74-9)
Topiclor 20	(57-74-9)
Topicort	(50-23-7)
Topicycline	(64-75-5)
Topitox	(3691-35-8)
Topitracin	(1405-87-4)
Topitrasin	(1405-87-4)
Topogard	(8066-11-3)
Topogard 3623	(8066-11-3)
Topokain	(51-05-8)
Toporex 500	(9003-53-6)
Toporex 550-02	(9003-53-6)
Toporex 830	(9003-53-6)
Toporex 850-51	(9003-53-6)
Toporex 855-51	(9003-53-6)
Toppel	(52315-07-8)
Topsin	(23564-06-9)
Topsin E	(23564-06-9)
Topsin M	(23564-05-8)
Topsin NF 35	(23564-06-9)
Topsin NF-44	(23564-05-8)
Topsin Turf and Ornamentals	(23564-05-8)
Topsin WP Methyl	(23564-05-8)
Topsym	(67-68-5)
Topusyn	(1014-69-3)
Torak	(10311-84-9)
Torayca N 6	(105-60-2)
Torayca N 6	(25038-54-4)
Torazina	(50-53-3)
Torazina	(69-09-0)
Torbin	(759-94-4)
Torch	(1689-84-5)
Torch brand	(1333-86-4)
Tordon	(1918-02-1)
Tordon K	(2545-60-0)
Tordon 10K	(2545-60-0)
Tordon 10K	(1918-02-1)
Tordon 22K	(2545-60-0)
Tordon 22K	(1918-02-1)
Tordon 50D	(8067-55-8)
Tordon 101 Mixture	(8067-55-8)
Tordon 101 Mixture	(1918-02-1)
Tordon 101	(8067-55-8)
Tordon 3220	(26952-20-5)
Torecan	(1420-55-9)
Torelle	(5598-52-7)
Torex G	(9011-14-7)
Torinal	(72-44-6)
Tormona	(93-76-5)
Tormona	(93-79-8)
Tormona 80	(120-39-8)
Tornade	(52645-53-1)
Torpedo	(26389-78-6)
Torque	(13356-08-6)
Torsite	(90-43-7)
Tosic Acid	(104-15-4)
Tostrin	(57-85-2)
Tosyl	(302-17-0)
Tosylamide	(70-55-3)
p-Tosylamide	(70-55-3)
Tosylchloramide sodium	(127-65-1)
Tosyl chloride	(98-59-9)
o-Tosyl chloride	(133-59-5)
p-Tosyl chloride	(98-59-9)
N-Tosylethylamine	(80-39-7)
3.β.-(Tosyloxy)-5.α.-cholestane	(3381-52-0)
2-Tosyloxyethylbenzene	(4455-09-8)
Totacillin	(69-53-4)
Totacol	(1910-42-5)
Total	(77182-82-2)
Totalciclina	(69-53-4)
Total clarified oil solvent extract (Petroleum)	(68782-98-9)
Totapen	(69-53-4)
Totazina	(1066-17-7)
Totomycin	(64-75-5)
Totril	(1689-83-4)
Totril	(3861-47-0)
Tough	(55512-33-9)
Toukalide	(695-06-7)
Tox 47	(56-38-2)
Tox-Hid	(81-81-2)
Toxadust	(8001-35-2)
Toxafeen (Dutch)	(8001-35-2)
Toxakil	(8001-35-2)
Toxan	(63-25-2)
Toxaphen (German)	(8001-35-2)
Toxaphene (DOT,OSHA)	(8001-35-2)
Toxaphene Toxicant B	(51775-36-1)
Toxaphene Toxicant C	(52819-39-3)
Toxer Total	(1910-42-5)
Toxichlor	(57-74-9)
Toxilic Anhydride	(108-31-6)
Toxilic acid	(110-16-7)
Toxin F2	(17924-92-4)
Toxin T2	(21259-20-1)
Toxolan P	(128-37-0)
Toxon 63	(8001-35-2)
Toxyphen	(8001-35-2)
Toyo Eosine G	(17372-87-1)

Toyo Fast Red GL Base (89-62-3)
Toyo Oil Orange (842-07-9)
Toyo Oil Red BB (85-83-6)
Toyo Oil Yellow G (60-11-7)
Toyo Oriental Oil Blue G (128-80-3)
Toyo Roccelline (1658-56-6)
Toyo acid phloxine (6441-77-6)
Toyo amaranth (915-67-3)
Toyocamycin (606-58-6)
Toyocamycin Nucleoside (606-58-6)
Toyodan (34643-46-4)
Toyofine A (1344-95-2)
Toyothion (34643-46-4)
2,4,5-T propylene glycol butyl ester (62922-39-8)
Trachosept (8048-52-0)
Tradon (2152-34-3)
Tradone (2152-34-3)
Trafarbiot (7177-48-2)
Tragacanth-Gum (9000-65-1)
Tragaya (9005-38-3)
Trakephon (51249-05-9)
Tralgon (103-90-2)
Tralomethrin (66841-25-6)
Tralomethrine (66841-25-6)
Tramacin (76-25-5)
Tramat (26225-79-6)
Trametan (137-26-8)
Tran-Q (68-88-2)
Trancalgyl (938-73-8)
Trancopal (80-77-3)
Tranid (15271-41-7)
Tranilast (53902-12-8)
Tranimul (439-14-5)
Tranite D-Lay (78-11-5)
Trank (956-90-1)
Trankvilan (57-53-4)
Tranlisant (57-53-4)
Tranmep (57-53-4)
Tranqdyn (439-14-5)
Tranquil (57-53-4)
Tranquilan (113-59-7)
Tranquilan (57-53-4)
Tranquilate (57-53-4)
Tranquilax (2898-12-6)
Tranquilax (57-53-4)
Tranquiline (57-53-4)
Tranquilsan (57-53-4)
Tranquinol (57-53-4)
Tranquirit (439-14-5)
Tranquisan (57-53-4)
Tranquo-buscopan-wirkstoff (604-75-1)
Trans-Aid (1762-95-4)
Trans-Vert (2163-80-6)
Transallyl CR 39 (142-22-3)
Transamine (155-09-9)
Transamine (93-76-5)
Transamine (94-75-7)
Transcutol (111-90-0)
Transentine (64-95-9)
Transerpin (50-55-5)
Transetile Blue P-FER (2475-46-9)
Transetile Red P 5GL (12223-35-7)
Transetile Rubine P-FL (17418-58-5)
Transetile Violet P 3R (128-95-0)
Transit (54-31-9)
Transparent Bronze Scarlet (5160-02-1)
Transplantone (86-87-3)
Tranylcypromine (155-09-9)
Tranzetil (64-95-9)
Tranzine (69-09-0)
Trapex (137-42-8)
Trapex (556-61-6)
Trapexide (556-61-6)
Traquilan (113-59-7)
Traquizine (68-88-2)
Trasan (94-74-6)
Trasentin (64-95-9)
Trasentine (64-95-9)
Tratul (51481-61-9)
Traubensaure (133-37-9)
Traubofan (132-60-5)
Traumaide (50-23-7)
Travad (7727-43-7)
Traveler's Joy LS (8001-21-6)
Travelin (523-87-5)
Travelmin (523-87-5)
Travelon (569-65-3)
Travert (8013-17-0)
Travex (7775-09-9)
Trawotox (302-17-0)
Trazentyna (Polish) (64-95-9)
Tre-Hold (86-87-3)
Trecalmo (33671-46-4)
Trecator (536-33-4)
Trefanocide (1582-09-8)
Treficon (1582-09-8)
Treflam (1582-09-8)
Treflan (1582-09-8)
Treflanocide elancolan (1582-09-8)
D-(+)-Trehalose (99-20-7)
α,α-Trehalose (99-20-7)
α-D-Trehalose (99-20-7)
α-Trehalose (99-20-7)
Trehalose (8CI) (99-20-7)
Trehalose, dihydrate (99-20-7)
Treitol, 1,4-dimethanesulfonate, (2S,3S)- (8CI) (299-75-2)
Trelmar (57-53-4)
Trementina (Spanish) (9005-90-7)
Tremin (13983-17-0)
Tremolite (14567-73-8)
Tremolite (OSHA) (77536-68-6)
Tremolite asbestos (77536-68-6)
Trenamine D-200 (143-18-0)
Trenamine D-201 (143-18-0)
Trenimon (68-76-8)
Treninon (68-76-8)
Treomicetina (56-75-7)
Treosulfan (299-75-2)
Trepenol WA (151-21-3)
Trepibutone (41826-92-0)
Trescatyl (536-33-4)
Trescazide (536-33-4)
Tresitope (6893-02-3)
Tresochin (54-05-7)
Tresortil (532-03-6)
Trespaphan (9003-07-0)
Trespaphan CEA (9003-07-0)
Trespaphan N 12 (9003-07-0)
Tresulfan (299-75-2)
Tret-O-Lite XC 511 (63449-41-2)
Tretamin (51-18-3)
Tretamine (51-18-3)
Trethylene (79-01-6)
Tretinoin (302-79-4)

Treupel (62-44-2)
Trevin (23564-05-8)
Tri (79-01-6)
Tri-BAN (83-26-1)
Tri-Clene (79-01-6)
Tri-Clor (76-06-2)
Tri-Cornox (3813-05-6)
Tri-Cornox Special (3813-05-6)
Tri-Endothal (129-67-9)
Tri-Endothal (145-73-3)
Tri-Fen (85-34-7)
Tri Fene (85-34-7)
TriML (1520-78-1)
Tri-Me (953-17-3)
Tri-PCNB (82-68-8)
Tri-Plus (79-01-6)
Tri-Plus M (79-01-6)
Tri-Rodazene (1594-56-5)
Tri-Sil (1343-88-0)
Tri-Span (396-01-0)
Tri-Tumine (154-69-8)
Tri-VC 13 (97-17-6)
Triabarb (50-06-6)
Triacetaldehyde (French) (123-63-7)
Triacetin (102-76-1)
Triacetine (102-76-1)
Triacetonamin (826-36-8)
Triacetonamine (826-36-8)
Triacetone amine (826-36-8)
Triacetoxy-methylsilane (4253-34-3)
Triacetoxyvinylsilane (4130-08-9)
Triacetyldiphenolisatin (18869-73-3)
Triacetylene (3161-99-7)
Triacetyl glycerine (102-76-1)
Triacetylpyrogallol (525-52-0)
Triacontane (9CI) (638-68-6)
Triacontane, 2-methyl- (8CI,9CI) (1560-72-1)
1-Triacontanoic acid (506-50-3)
n-Triacontanoic acid (506-50-3)
Triacontanoic acid (9CI) (506-50-3)
Triacontanoic acid, methyl ester (8CI,9CI) (629-83-4)
1-Triacontanol (9CI) (593-50-0)
17-Triacontanone (22986-69-2)
Triacrylformal (959-52-4)
1,3,5-Triacryloylhexahydrotriazine (959-52-4)
Tri(N-acryloyl)hexahydrotriazine (959-52-4)
Triacryloylhexahydro-s-triazine (959-52-4)
Triacryloylhexahydrotriazine (959-52-4)
Triacryloylperhydrotriazine (959-52-4)
Triad (79-01-6)
Triadenyl (56-65-5)
Triadimefon (43121-43-3)
Triadimefone (43121-43-3)
Triadimenol (123-88-6)
Triadimenol (55219-65-3)
Triaethanolamin-NG (102-71-6)
Triaethylamin (German) (121-44-8)
Triaethylenmelamin (German) (51-18-3)
Triaethylenphosphorsaeureamid (German) (545-55-1)
Triaethylzinnchlorid (German) (994-31-0)
Triafur (712-68-5)
Trial (79-01-6)
Triallat (German) (2303-17-5)
Triallate (2303-17-5)
Triallylamine [UN 2610] (102-70-5)
Triallyl cyanaurate (101-37-1)
Triallyl cyanurate (101-37-1)
Triallylfosfat (Czech) (1623-19-4)
Triallyl isocyanurate (1025-15-6)
N,N',N''-Triallylmelamine (30360-21-5)
Triallyl phosphate (1623-19-4)
Triallyl trimellitate (2694-54-4)
Triamcincolone acetonide (76-25-5)
Triamcinolone (124-94-7)
Triamcinolone 16,17-acetonide (76-25-5)
Triamcinolone acetonide (76-25-5)
Triamelin (51-18-3)
Triamifos (German, Dutch, Italian) (1031-47-6)
2,4,7-Triamino-6-fenilpteridina (Italian) (396-01-0)
2,4,7-Triamino-6-phenylpteridine (396-01-0)
2,4,6-Triaminopyrimidine (1004-38-2)
2,4,6-Triamino-1,3,5-triazine (108-78-1)
2,4,6-Triamino-s-triazine (108-78-1)
1,3,5-Triamino-2,4,6-trinitrobenzene (3058-38-6)
s-Triaminotrinitrobenzene (3058-38-6)
4,4',4''-Triaminotriphenylcarbinol (467-62-9)
4,4'4''-Triaminotriphenylmethan-hydrochlorid (German) (569-61-9)
Triamiphos (1031-47-6)
Triammonium hexakis-(cyano-C)ferrate(4-) (14481-29-9)
Triammonium tris-(ethanedioato(2-)-O,O')ferrate(3-1) (14221-47-7)
Triamphos (1031-47-6)
Triampur (396-01-0)
Triamsinolone acetonide (76-25-5)
Triamteren (396-01-0)
Triamterene (396-01-0)
Triamteril (396-01-0)
Triamteril Complex (396-01-0)
Triamylamine (621-77-2)
Triangle (1332-03-2)
Triangle (1333-86-4)
Triangle (7758-99-8)
Tri-p-anisylchloroethylene (569-57-3)
Trianol Direct Blue 3B (72-57-1)
Trianon (144-83-2)
Triantine Brown BRS (16071-86-6)
Triantine Fast Brown OG (16071-86-6)
Triantine Fast Brown OR (16071-86-6)
Triantine Fast Red 4BN (2610-11-9)
Triantine Light Brown BRS (16071-86-6)
Triantine Light Brown OG (16071-86-6)
Triantine Light Brown 3RN (2650-18-2)
Triantine Light Red 4BN (2610-11-9)
Triantoin (50-12-4)
Triarimol (26766-27-8)
Triascl (79-01-6)
Triasyn (101-05-3)
Triatix (33089-61-1)
Triatomic oxygen (10028-15-6)
Triatox (33089-61-1)
3,5,7-Triaza-1-azoniaadamantane, 1-(3-chloroallyl)-, chloride (4080-31-3)
1,5,10-Triazadecane (124-20-9)
1,2,3-Triaza-1H-indene (95-14-7)
1,2,3-Triazaindene (95-14-7)
3,5,7-Triazaindole (120-73-0)
1,3,5,2,4,6-Triazatriphosphorine, 2-chloro-2,2,4,4,6,6-hexahydro-2,4,4,6,6-pentakis(2,2,3,3-tetrafluor opropoxy)- (9CI) (59700-60-6)
1,3,5,2,4,6-Triazatriphosphorine, 2,2-dichloro-2,2,4,4, 6,6-hexahydro-4,4,6,6-tetrakis(2,2,3,3-tetrafluoropropoxy)- (9CI) (65601-42-5)
1,3,5,2,4,6-Triazatriphosphorine, 2,2,4,4,6,6-hexachloro-2,2,4,4,6,6-hexahydro- (8CI,9CI) (940-71-6)
1,3,5,2,4,6-Triazatriphosphorine, 2,2,4,4,6,6-hexahydro-2,2,4,4,6,6-hexakis(2,2,3,3-tetrafluoropropoxy)- (9CI) (58943-98-9)
1,3,5,2,4,6-Triazatriphosphorine, 2,2,4,4,6,6-hexakis(1-aziridinyl)-2,2,4,4,6,6-hexahydro (52-46-0)

1,3,5,2,4,6-Triazatriphosphorine, 2,2,4,4,6-pentachloro-2,2,4,4,6,6-hexahydro-6-(2,2,3,3-tetrafluoropr opoxy)- (9CI) **(59700-57-1)**
1,3,5,2,4,6-Triazatriphosphorine, 2,2,4,4-tetrachloro-2,2,4,4,6,6-hexahydro-6,6-bis(2,2,3,3-tetrafluor opropoxy)- (9CI) **(65601-40-3)**
1,3,5,2,4,6-Triazatriphosphorine, 2,2,4-trichloro-2,2,4,4,6,6-hexahydro-4,6,6-tris(2,2,3,3-tetrafluoro propoxy)- (9CI) **(65601-41-4)**
3,6,9-Triazaundecanedioic acid, 3,6,9-tris(carboxymethyl)- **(67-43-6)**
Triazbutil **(16227-10-4)**
Triazene, 1-(p-chlorophenyl)-3,3-dimethyl- **(7203-90-9)**
Triazene, 3,3-dimethyl-1-(p-chlorophenyl) **(7203-90-9)**
Triazene, 3,3-dimethyl-1-phenyl **(7227-91-0)**
1-Triazene, 3,3-dimethyl-1-phenyl- (9CI) **(7227-91-0)**
Triazene, 1,3-diphenyl **(136-35-6)**
Triazichon (German) **(68-76-8)**
Triazin **(101-05-3)**
1,3,5-Triazin-2-amine (9CI) **(4122-04-7)**
1,3,5-Triazin-2-amine, 4-azido-N-ethyl-6-(ethylthio)- (9CI) **(30362-29-9)**
1,3,5-Triazin-2-amine, 4-azido-N-(1-methylethyl)-6-(methylthio)- **(4658-28-0)**
1,3,5-Triazin-2-amine, 4,6-bis(butylthio)- (9CI) **(30362-15-3)**
1,3,5-Triazin-2-amine, 4,6-bis(4-chlorophenoxy)- (9CI) **(30358-16-8)**
1,3,5-Triazin-2-amine, 4,6-bis(ethylthio)- (9CI) **(30362-14-2)**
1,3,5-Triazin-2-amine, 4,6-bis(methylthio)- (9CI) **(30358-19-1)**
1,3,5-Triazin-2-amine, 4,6-bis(2-propenyloxy)- (9CI) **(30358-11-3)**
1,3,5-Triazin-2-amine, 4,6-bis(trichloromethyl)- (9CI) **(20376-31-2)**
1,3,5-Triazin-2-amine, N-(4-bromophenyl)-4,6-dichloro- (9CI) **(20376-36-7)**
1,3,5-Triazin-2-amine, N-(4-bromophenyl)-4,6-dimethoxy- (9CI) **(30358-04-4)**
1,3,5-Triazin-2-amine, 4-butoxy-N-ethyl-6-methyl- (9CI) **(5248-42-0)**
1,3,5-Triazin-2-amine, N-butyl-4-(methylthio)-6-(trichloromethyl)- (9CI) **(30369-45-0)**
1,3,5-Triazin-2-amine, 4-(butylthio)-6-(trichloromethyl)- (9CI) **(30369-52-9)**
1,3,5-Triazin-2-amine, 4-chloro- (9CI) **(7709-13-9)**
1,3,5-Triazin-2-amine, 4-chloro-6-(4-chlorophenyl)- (9CI) **(30369-33-6)**
1,3,5-Triazin-2-amine, 4-chloro-N,6-diethyl- (9CI) **(30369-29-0)**
1,3,5-Triazin-2-amine, 4-chloro-N,6-diphenyl- (9CI) **(3842-52-2)**
1,3,5-Triazin-2-amine, 4-chloro-6-ethoxy-N-ethyl- (9CI) **(4743-13-9)**
1,3,5-Triazin-2-amine, 4-chloro-N-ethyl-6-methoxy- (9CI) **(4653-94-5)**
1,3,5-Triazin-2-amine, 4-chloro-N-ethyl-6-methyl- (9CI) **(30369-24-5)**
1,3,5-Triazin-2-amine, 4-chloro-N-ethyl-6-(methylthio)- (9CI) **(30357-95-0)**
1,3,5-Triazin-2-amine, 4-chloro-N-ethyl-6-propoxy- (9CI) **(4417-64-5)**
1,3,5-Triazin-2-amine, 4-chloro-6-methoxy- (9CI) **(7254-11-7)**
1,3,5-Triazin-2-amine, 4-chloro-6-methoxy-N,N-dimethyl- (9CI) **(13882-55-8)**
1,3,5-Triazin-2-amine, 4-chloro-6-methoxy-N-(1-methylethyl)- (9CI) **(4446-75-7)**
1,3,5-Triazin-2-amine, 4-chloro-6-methoxy-N-phenyl- (9CI) **(30357-85-8)**
1,3,5-Triazin-2-amine, 4-chloro-6-methoxy-N-2-propenyl- (9CI) **(4446-76-8)**
1,3,5-Triazin-2-amine, 4-chloro-6-methoxy-N-propyl- (9CI) **(4407-72-1)**
1,3,5-Triazin-2-amine, 4-chloro-6-methyl- (9CI) **(21320-62-7)**
1,3,5-Triazin-2-amine, 4-chloro-N-(2-methylpropyl)-6-(methylthio)- (9CI) **(4658-30-4)**
1,3,5-Triazin-2-amine, 4-chloro-6-(methylthio)- (9CI) **(30357-94-9)**
1,3,5-Triazin-2-amine, 4-chloro-6-(methylthio)-N-phenyl- (9CI) **(30357-96-1)**
1,3,5-Triazin-2-amine, 4-(chloromethyl)-6-(trichloromethyl)- (9CI) **(30357-76-7)**
1,3,5-Triazin-2-amine, 4-chloro-6-phenoxy- (9CI) **(30357-91-6)**
1,3,5-Triazin-2-amine, 4-chloro-6-phenoxy-N-phenyl- (9CI) **(13838-29-4)**
1,3,5-Triazin-2-amine, 4-chloro-6-phenyl- (9CI) **(3842-53-3)**
1,3,5-Triazin-2-amine, N-(4-chlorophenyl)-4,6-bis(trichloromethyl)- (9CI) **(3599-75-5)**
1,3,5-Triazin-2-amine, N-(4-chlorophenyl)-4,6-difluoro- (9CI) **(30369-78-9)**
1,3,5-Triazin-2-amine, N-(2-chlorophenyl)-4,6-dimethoxy- (9CI) **(27315-26-0)**
1,3,5-Triazin-2-amine, N-(4-chlorophenyl)-4,6-dimethoxy- (9CI) **(27315-27-1)**
1,3,5-Triazin-2-amine, 4-((4-chlorophenyl)thio)-N-ethyl-6-(trichloromethyl)- (9CI) **(30369-59-6)**
1,3,5-Triazin-2-amine, 4-((4-chlorophenyl)thio)-N-(1-methylethyl)-6-(trichloromethyl)- (9CI) **(30369-60-9)**
1,3,5-Triazin-2-amine, 4-((4-chlorophenyl)thio)-N-methyl-6-(trichloromethyl)- (9CI) **(30369-58-5)**
1,3,5-Triazin-2-amine, 4-((4-chlorophenyl)thio)-6-(trichloromethyl)- (9CI) **(30369-57-4)**
1,3,5-Triazin-2-amine, 4-chloro-6-propyl- (9CI) **(30369-30-3)**
1,3,5-Triazin-2-amine, 4,6-dibromo-N-(4-chlorophenyl)- (9CI) **(30357-83-6)**
1,3,5-Triazin-2-amine, 4,6-dichloro- **(933-20-0)**
1,3,5-Triazin-2-amine, 4,6-dichloro-N-(3-chlorophenyl)- (9CI) **(20376-34-5)**
1,3,5-Triazin-2-amine, 4,6-dichloro-N-(4-chlorophenyl)- (9CI) **(2272-29-9)**
1,3,5-Triazin-2-amine, 4,6-dichloro-N-(2,5-dichlorophenyl)- (9CI) **(2272-33-5)**
1,3,5-Triazin-2-amine, 4,6-dichloro-N-(3,4-dichlorophenyl)- (9CI) **(18996-86-6)**
1,3,5-Triazin-2-amine, 4,6-dichloro-N,N-diphenyl- (9CI) **(16033-74-2)**
1,3,5-Triazin-2-amine, 4,6-dichloro-N-dodecyl- (9CI) **(26113-25-7)**
1,3,5-Triazin-2-amine, 4,6-dichloro-N-ethyl- (9CI) **(3440-19-5)**
1,3,5-Triazin-2-amine, 4,6-dichloro-N-(2-methoxyphenyl)- (9CI) **(30369-87-0)**
1,3,5-Triazin-2-amine, 4,6-dichloro-N-(4-methoxyphenyl)- (9CI) **(30377-27-6)**
1,3,5-Triazin-2-amine, 4,6-dichloro-N-(1-methylethyl)- (9CI) **(3703-10-4)**
1,3,5-Triazin-2-amine, 4-(dichloromethyl)-6-methyl- (9CI) **(30576-27-3)**
1,3,5-Triazin-2-amine, 4,6-dichloro-N-(2-methylphenyl)- (9CI) **(2272-23-3)**
1,3,5-Triazin-2-amine, 4,6-dichloro-N-(4-methylphenyl)- (9CI) **(2272-24-4)**
1,3,5-Triazin-2-amine, 4,6-dichloro-N-methyl-N-phenyl- (9CI) **(3995-42-4)**
1,3,5-Triazin-2-amine, 4,6-dichloro-N-(4-nitrophenyl)- (9CI) **(2352-36-5)**
1,3,5-Triazin-2-amine, 4,6-dichloro-N-phenyl- (9CI) **(2272-40-4)**
1,3,5-Triazin-2-amine, N-(3,4-dichlorophenyl)-4,6-bis(trichloromethyl)- (9CI) **(30356-55-9)**
1,3,5-Triazin-2-amine, 4,6-dichloro-N-(phenylmethyl)- (9CI) **(30369-82-5)**
1,3,5-Triazin-2-amine, 4-((2,4-dichlorophenyl)thio)-N-methyl-6-(trichloromethyl)- (9CI) **(30369-61-0)**
1,3,5-Triazin-2-amine, 4-((2,4-dichlorophenyl)thio)-6-(trichloromethyl)- (9CI) **(30576-31-9)**
1,3,5-Triazin-2-amine, 4,6-dichloro-N-2-propenyl- (9CI) **(30369-80-3)**
1,3,5-Triazin-2-amine, 4,6-dichloro-N-propyl- (9CI) **(25354-39-6)**
1,3,5-Triazin-2-amine, 4,6-diethoxy- (9CI) **(30576-32-0)**
1,3,5-Triazin-2-amine, N,N-diethyl-4-(methylthio)-6-(trichloromethyl)- (9CI) **(30369-48-3)**
1,3,5-Triazin-2-amine, 4,6-difluoro- (9CI) **(1652-36-4)**
1,3,5-Triazin-2-amine, 4,6-difluoro-N-phenyl- (9CI) **(717-90-8)**
1,3,5-Triazin-2-amine, 4,6-dimethoxy- (9CI) **(16370-63-1)**
1,3,5-Triazin-2-amine, 4,6-dimethoxy-N-methyl- (9CI) **(30357-98-3)**
1,3,5-Triazin-2-amine, 4,6-dimethoxy-N-methyl-N-phenyl- (9CI) **(30358-06-6)**
1,3,5-Triazin-2-amine, 4,6-dimethoxy-N-phenyl- (9CI) **(30358-01-1)**
1,3,5-Triazin-2-amine, 4,6-dimethyl- (9CI) **(1853-90-3)**
1,3,5-Triazin-2-amine, N,N-dimethyl- (9CI) **(4040-00-0)**
1,3,5-Triazin-2-amine, N,N-dimethyl-4,6-diphenyl- (9CI) **(18808-10-1)**
1,3,5-Triazin-2-amine, N,4-dimethyl-6-(trichloromethyl)- (9CI) **(24830-33-9)**
1,3,5-Triazin-2-amine, N,N-dimethyl-4,6-bis(trichloromethyl)- (9CI) **(24803-05-2)**
1,3,5-Triazin-2-amine, 4,6-diphenoxy- (9CI) **(1230-80-4)**
1,3,5-Triazin-2-amine, 4,6-diphenoxy-N-phenyl- (9CI) **(1973-08-6)**
1,3,5-Triazin-2-amine, 4,6-diphenyl- (9CI) **(5418-07-5)**
1,3,5-Triazin-2-amine, N-dodecyl-4-(methylthio)-6-(trichloromethyl)- (9CI) **(30369-47-2)**
1,3,5-Triazin-2-amine, 4-ethoxy-N,6-diethyl- (9CI) **(6864-23-9)**
1,3,5-Triazin-2-amine, 4-ethoxy-N-ethyl-6-methyl- (9CI) **(5248-41-9)**
1,3,5-Triazin-2-amine, N-ethyl- (9CI) **(4122-05-8)**
1,3,5-Triazin-2-amine, N-ethyl-4,6-bis(methylthio)- (9CI) **(30362-12-0)**
1,3,5-Triazin-2-amine, N-ethyl-4,6-bis(2-propenyloxy)- (9CI) **(30358-12-4)**
1,3,5-Triazin-2-amine, N-ethyl-4,6-bis(trichloromethyl)- (9CI) **(24803-29-0)**
1,3,5-Triazin-2-amine, N-ethyl-4,6-difluoro- (9CI) **(30369-76-7)**
1,3,5-Triazin-2-amine, N-ethyl-4-methoxy-6-methyl- (9CI) **(16399-10-3)**
1,3,5-Triazin-2-amine, N-ethyl-4-methoxy-6-phenyl- (9CI) **(30369-39-2)**
1,3,5-Triazin-2-amine, N-ethyl-4-(methylthio)-6-(trichloromethyl)- (9CI) **(30369-41-6)**
1,3,5-Triazin-2-amine, 4-(ethylthio)-N-methyl-6-(trichloromethyl)- (9CI) **(30369-50-7)**
1,3,5-Triazin-2-amine, 4-(ethylthio)-6-(trichloromethyl)- (9CI) **(30369-49-4)**
1,3,5-Triazin-2-amine, 4-heptadecyl-6-(trichloromethyl)- (9CI) **(30369-04-1)**
1,3,5-Triazin-2-amine, N-hexyl-4-(methylthio)-6-(trichloro-

methyl)- (9CI) **(30576-30-8)**
1,3,5-Triazin-2-amine, 4-(hexylthio)-N-methyl-6-(trichloromethyl)- (9CI) **(30369-54-1)**
1,3,5-Triazin-2-amine, 4-methoxy- (9CI) **(1122-73-2)**
1,3,5-Triazin-2-amine, 4-methoxy-6-methyl- (9CI) **(1668-54-8)**
1,3,5-Triazin-2-amine, 4-methoxy-6-(methylthio)- (9CI) **(30358-18-0)**
1,3,5-Triazin-2-amine, 4-methoxy-6-phenyl- (9CI) **(30369-38-1)**
1,3,5-Triazin-2-amine, 4-methyl- (9CI) **(27622-91-9)**
1,3,5-Triazin-2-amine, N-methyl- (9CI) **(4039-99-0)**
1,3,5-Triazin-2-amine, N-methyl-4,6-bis(methylthio)- (9CI) **(30362-11-9)**
1,3,5-Triazin-2-amine, N-(1-methylethyl)- (9CI) **(30360-48-6)**
1,3,5-Triazin-2-amine, N-(1-methylethyl)-4-(methylthio)-6-(trichloromethyl)- (9CI) **(30369-44-9)**
1,3,5-Triazin-2-amine, N-methyl-4-((1-methylethyl)thio)-6-(trichloromethyl)- (9CI) **(30369-51-8)**
1,3,5-Triazin-2-amine, 4-methyl-N-(2-methylpropyl)-6-(trichloromethyl)- (9CI) **(24803-53-0)**
1,3,5-Triazin-2-amine, N-methyl-4-(methylthio)-6-(trichloromethyl)- (9CI) **(30377-26-5)**
1,3,5-Triazin-2-amine, 4-methyl-6-phenyl- (9CI) **(1853-91-4)**
1,3,5-Triazin-2-amine, N-methyl-4-(phenylthio)-6-(trichloromethyl)- (9CI) **(30369-56-3)**
1,3,5-Triazin-2-amine, N-(2-methylpropyl)-4-(methylthio)-6-(trichloromethyl)- (9CI) **(30369-46-1)**
1,3,5-Triazin-2-amine, 4-((2-methylpropyl)thio)-6-(trichloromethyl)- (9CI) **(30369-53-0)**
1,3,5-Triazin-2-amine, 4-methyl-N-propyl-6-(trichloromethyl)- (9CI) **(24803-52-9)**
1,3,5-Triazin-2-amine, 4-(methylthio)-N-2-propenyl-6-(trichloromethyl)- (9CI) **(30369-43-8)**
1,3,5-Triazin-2-amine, 4-(methylthio)-N-propyl-6-(trichloromethyl)- (9CI) **(30369-42-7)**
1,3,5-Triazin-2-amine, 4-(methylthio)-6-(trichloromethyl)- (9CI) **(14946-02-2)**
1,3,5-Triazin-2-amine, 4-methyl-6-(trichloromethyl)- (9CI) **(21227-47-4)**
1,3,5-Triazin-2-amine, N-methyl-4-(trichloromethyl)-6-((2,4,5-trichlorophenyl)thio)- (9CI) **(30369-63-2)**
1,3,5-Triazin-2-amine, 4-phenyl- (9CI) **(1853-95-8)**
1,3,5-Triazin-2-amine, N-phenyl- (9CI) **(4040-07-7)**
1,3,5-Triazin-2-amine, N-(phenylmethyl)-4,6-bis(trichloromethyl)- (9CI) **(24803-16-5)**
1,3,5-Triazin-2-amine, 4-(phenylthio)-6-(trichloromethyl)- (9CI) **(30369-55-2)**
1,3,5-Triazin-2-amine, N-2-propenyl-4,6-bis(trichloromethyl)- (9CI) **(24803-17-6)**
1,3,5-Triazin-2-amine, 4-(trichloromethyl)- (9CI) **(30360-33-9)**
1,3,5-Triazin-2-amine, 4-(trichloromethyl)-6-((2,4,5-trichlorophenyl)thio)- (9CI) **(30369-62-1)**
1,3,5-Triazin-2-amine, N,N,4-trimethyl-6-(trichloromethyl)- (9CI) **(24803-57-4)**
1,3,5-Triazin-2-amine, N,4,6-triphenyl- (9CI) **(30369-20-1)**
Triazine **(101-05-3)**
s-Triazine **(290-87-9)**
sym-Triazine **(290-87-9)**
Triazine A 1294 **(1912-24-9)**
Triazine A 384 **(122-34-9)**
1,3,5-Triazine (9CI) **(290-87-9)**
1,3,5-Triazine-2-acetonitrile, 4,6-diamino- (9CI) **(13301-35-4)**
s-Triazine-2-acetonitrile, 4,6-diamino- (8CI) **(13301-35-4)**
1,3,5-Triazine, 1-acetylhexahydro-3,5-dinitro- (9CI) **(14168-42-4)**
s-Triazine, 1-acetylhexahydro-3,5-dinitro- (8CI) **(14168-42-4)**
s-Triazine, 2-(allylamino)-4-(allyloxy)-6-azido- (8CI) **(4480-45-9)**
s-Triazine, 2-(allylamino)-4-(allyloxy)-6-chloro- (7CI,8CI) **(4417-72-5)**
s-Triazine, 2-(allylamino)-4-amino-6-chloro- (6CI,8CI) **(30355-01-2)**
s-Triazine, 2-(allylamino)-4,6-bis(trichloromethyl)- (8CI) **(24803-17-6)**
s-Triazine, 2-(allylamino)-4-bromo-6-(ethylamino)- (7CI,8CI) **(30360-52-2)**
s-Triazine, 2-(allylamino)-4-chloro-6-methoxy- (7CI,8CI) **(4446-76-8)**
s-Triazine, 2-(allylamino)-4,6-dichloro- (8CI) **(30369-80-3)**
s-Triazine, 2-(allylamino)-4-(diethylamino)-6-methyl- (8CI) **(27431-12-5)**
s-Triazine, 2-(allylamino)-4-(ethylamino)-6-methyl- (8CI) **(27431-01-2)**
s-Triazine, 2-(allylamino)-4-(isopropylamino)-6-methyl- (8CI) **(27431-06-7)**
s-Triazine, 2-(allylamino)-4-(methylthio)-6-(trichloromethyl)- (8CI) **(30369-43-8)**
s-Triazine, 2-(allylamino)-4-methyl-6-(trichloromethyl)- (8CI) **(24803-63-2)**
s-Triazine, 2-(allyloxy)-4,6-bis(ethylamino)- (7CI,8CI) **(30360-68-0)**
s-Triazine, 2-(allyloxy)-4,6-bis(trichloromethyl)- (8CI) **(30863-54-8)**
s-Triazine, 2-(allyloxy)-4-chloro-6-(ethylamino)- (7CI,8CI) **(4417-70-3)**
s-Triazine, 2-(allyloxy)-4,6-diamino- (8CI) **(6291-87-8)**
s-Triazine, 2-(allyloxy)-4,6-dichloro- (8CI) **(26650-76-0)**
s-Triazine, 2-(allyloxy)-4-methyl-6-(trichloromethyl)- (7CI,8CI) **(949-43-9)**
s-Triazine, 2-(allylthio)-4,6-diamino- (8CI) **(30360-84-0)**
s-Triazine, 2-(allylthio)-4,6-dichloro- (6CI,7CI,8CI) **(25713-56-8)**
s-Triazine, 2-amino- (8CI) **(4122-04-7)**
s-Triazine, 2-amino-4-anilino- (8CI) **(537-17-7)**
s-Triazine, 2-amino-4-anilino-6-(bromomethyl)- (8CI) **(30359-64-9)**
Triazine, 2-amino-4-anilino-6-chloro- (6CI) **(16007-72-0)**
s-Triazine, 2-amino-4-anilino-6-chloro- (7CI,8CI) **(16007-72-0)**
s-Triazine, 2-amino-4-anilino-6-(chloromethyl)- (6CI,8CI) **(30355-60-3)**
s-Triazine, 2-amino-4-anilino-6-(dichloromethyl)- (6CI,8CI) **(30355-65-8)**
s-Triazine, 2-amino-4-anilino-6-ethyl- (6CI,8CI) **(30359-79-6)**
s-Triazine, 2-amino-4-anilino-6-methyl **(7426-35-9)**
s-Triazine, 2-amino-4-anilino-6-propyl- (8CI) **(30359-95-6)**
s-Triazine, 2-amino-4-anilino-6-(tribromomethyl)- (8CI) **(30359-71-8)**
s-Triazine, 2-amino-4-anilino-6-(trichloromethyl)- (8CI) **(30355-69-2)**
s-Triazine, 2-amino-4-anilino-6-(trifluoromethyl)- (6CI,7CI,8CI) **(368-61-6)**
s-Triazine, 2-amino-4-azido-6-chloro- (7CI,8CI) **(5248-70-4)**
s-Triazine, 2-amino-4-azido-6-methoxy- (7CI,8CI) **(5248-73-7)**
s-Triazine, 2-amino-4,6-bis(butylthio)- (8CI) **(30362-15-3)**
s-Triazine, 2-amino-4,6-bis(chloromethyl)- (6CI,8CI) **(30360-36-2)**
s-Triazine, 2-amino-4,6-bis(o-chlorophenoxy)- (8CI) **(30358-14-6)**
s-Triazine, 2-amino-4,6-bis(p-chlorophenoxy)- (8CI) **(30358-16-8)**
s-Triazine, 2-amino-4,6-bis(p-chlorophenyl)- (8CI) **(30369-21-2)**
s-Triazine, 2-amino-4,6-bis(1,1-dichloroethyl)- (6CI,8CI) **(30339-57-2)**
s-Triazine, 2-amino-4,6-bis(dichloromethyl)- (6CI,8CI) **(30360-41-9)**
s-Triazine, 2-amino-4,6-bis(ethylthio)- (8CI) **(30362-14-2)**
s-Triazine, 2-amino-4,6-bis(methylthio)- (8CI) **(30358-19-1)**
s-Triazine, 2-amino-4,6-bis(p-tolyloxy)- (7CI,8CI) **(4150-79-2)**
s-Triazine, 2-amino-4,6-bis(tribromomethyl)- (7CI,8CI) **(30339-36-7)**
s-Triazine, 2-amino-4,6-bis(1,1,2-trichloroethyl)- (8CI) **(30339-60-7)**
s-Triazine, 2-amino-4,6-bis(trichloromethyl)- (8CI) **(20376-31-2)**
s-Triazine, 2-amino-4,6-bis(trifluoromethyl)- (8CI) **(29181-67-7)**
s-Triazine, 2-amino-4-(2-bromoethyl)-6-(trichloromethyl)- (8CI) **(30339-52-7)**
s-Triazine, 2-amino-4-(bromomethyl)-6-(ethylamino)- (8CI) **(30377-16-3)**
s-Triazine, 2-amino-4-bromo-6-phenoxy- (8CI) **(30357-93-8)**
s-Triazine, 2-amino-4-(butylthio)-6-(trichloromethyl)- (8CI) **(30369-52-9)**
s-Triazine, 2-amino-4-butyl-6-(trichloromethyl)- (8CI) **(30339-70-9)**
s-Triazine, 2-amino-4-sec-butyl-6-(trichloromethyl)- (8CI) **(30339-73-2)**
s-Triazine, 2-amino-4-chloro- (8CI) **(7709-13-9)**
s-Triazine, 2-amino-4-(p-chloroanilino) **(500-42-5)**
s-Triazine, 2-amino-4-(p-chloroanilino)-6-(chloromethyl)- (6CI,8CI) **(30355-61-4)**
s-Triazine, 2-amino-4-(p-chloroanilino)-6-methyl- (8CI) **(645-18-1)**
s-Triazine, 2-amino-4-(p-chloroanilino)-6-(trichloromethyl)- (8CI) **(30355-71-6)**
s-Triazine, 2-amino-4-chloro-6-(p-chlorophenyl)- (8CI) **(30369-33-6)**
s-Triazine, 2-amino-4-chloro-6-(diazomethyl)- (6CI,7CI,8CI) **(30369-27-8)**
s-Triazine, 2-amino-4-chloro-6-ethyl- (8CI) **(30369-28-9)**
s-Triazine, 2-amino-4-chloro-6-(ethylamino)- (8CI) **(1007-28-9)**
s-Triazine, 2-amino-4-(2-chloroethyl)-6-(trichloromethyl)- (8CI) **(30339-47-0)**
s-Triazine, 2-amino-4-chloro-6-(isopropylamino)- **(6190-65-4)**
s-Triazine, 2-amino-4-chloro-6-methoxy- (8CI) **(7254-11-7)**
s-Triazine, 2-amino-4-chloro-6-methyl- (8CI) **(21320-62-7)**
s-Triazine, 2-amino-4-(chloromethyl)-6-(ethylamino)- (8CI) **(30355-59-0)**
s-Triazine, 2-amino-4-(chloromethyl)-6-phenyl- (7CI,8CI) **(1853-96-9)**
s-Triazine, 2-amino-4-chloro-6-(methylthio)- (8CI) **(30357-94-9)**
s-Triazine, 2-amino-4-(chloromethyl)-6-(trichloromethyl)- (8CI) **(30357-76-7)**

s-Triazine, 2-amino-4-chloro-6-phenoxy- (7CI,8CI) (30357-91-6)
s-Triazine, 2-amino-4-chloro-6-phenyl- (6CI,7CI,8CI) (3842-53-3)
s-Triazine, 2-amino-4-chloro-6-(phenylthio)- (8CI) (30357-97-2)
s-Triazine, 2-amino-4-((p-chlorophenyl)thio)-6-(trichloromethyl)- (8CI) (30369-57-4)
s-Triazine, 2-amino-4-(p-chlorophenyl)-6-(trichloromethyl)- (8CI) (30377-24-3)
s-Triazine, 2-amino-4-chloro-6-propyl- (8CI) (30369-30-3)
s-Triazine, 2-amino-4,6-dibromo- (7CI,8CI) (4514-62-9)
s-Triazine, 2-amino-4-(dibromomethyl)-6-(ethylamino)- (8CI) (30359-67-2)
s-Triazine, 2-amino-4,6-dibutoxy- (8CI) (30384-47-5)
s-Triazine, 2-amino-4,6-dichloro (933-20-0)
s-Triazine, 2-amino-4-(dichloromethyl)-6-(ethylamino)- (8CI) (30355-64-7)
s-Triazine, 2-amino-4-(dichloromethyl)-6-methyl- (8CI) (30576-27-3)
s-Triazine, 2-amino-4-(dichloromethyl)-6-(trichloromethyl)- (8CI) (30357-60-9)
s-Triazine, 2-amino-4-((2,4-dichlorophenyl)thio)-6-(trichloromethyl)- (8CI) (30576-31-9)
s-Triazine, 2-amino-4-(2,4-dichlorophenyl)-6-(trichloromethyl)- (8CI) (30369-15-4)
s-Triazine, 2-amino-4,6-diethoxy- (8CI) (30576-32-0)
s-Triazine, 2-amino-4,6-diethyl- (7CI,8CI) (5599-20-2)
s-Triazine, 2-amino-4,6-difluoro- (7CI,8CI) (1652-36-4)
s-Triazine, 2-amino-4,6-diisopropoxy- (8CI) (30358-13-5)
s-Triazine, 2-amino-4,6-dimethoxy- (8CI) (16370-63-1)
s-Triazine, 2-amino-4,6-dimethyl- (8CI) (1853-90-3)
s-Triazine, 2-amino-4,6-dipentyl- (8CI) (30369-01-8)
s-Triazine, 2-amino-4,6-diphenoxy- (6CI,7CI,8CI) (1230-80-4)
s-Triazine, 2-amino-4,6-diphenyl- (6CI,7CI,8CI) (5418-07-5)
s-Triazine, 2-amino-4,6-dipropoxy- (8CI) (30358-10-2)
s-Triazine, 2-amino-4,6-di-p-tolyl- (7CI,8CI) (5599-21-3)
s-Triazine, 2-amino-4-ethyl- (8CI) (30360-34-0)
s-Triazine, 2-amino-4-(ethylamino)- (8CI) (30368-49-1)
s-Triazine, 2-amino-4-(ethylamino)-6-methoxy- (8CI) (30360-56-6)
s-Triazine, 2-amino-4-(ethylamino)-6-methyl- (8CI) (30368-51-5)
s-Triazine, 2-amino-4-(ethylamino)-6-(tribromomethyl)- (8CI) (30359-70-7)
s-Triazine, 2-amino-4-(ethylamino)-6-(trichloromethyl)- (8CI) (27470-98-0)
s-Triazine, 2-amino-4-(ethylamino)-6-(trifluoromethyl)- (8CI) (30355-58-9)
s-Triazine, 2-amino-4-(ethylthio)-6-(trichloromethyl)- (8CI) (30369-49-4)
s-Triazine, 2-amino-4-ethyl-6-(trichloromethyl)- (8CI) (30339-40-3)
s-Triazine, 2-amino-4-fluoro-6-phenoxy- (8CI) (30357-90-5)
s-Triazine, 2-amino-4-heptadecyl-6-(trichloromethyl)- (8CI) (30369-04-1)
s-Triazine, 2-amino-4-(isobutylthio)-6-(trichloromethyl)- (8CI) (30369-53-0)
s-Triazine, 2-amino-4-isobutyl-6-(trichloromethyl)- (8CI) (30339-79-8)
s-Triazine, 2-amino-4-isopropyl-6-(trichloromethyl)- (8CI) (30339-67-4)
s-Triazine, 2-amino-4-methoxy- (8CI) (1122-73-2)
s-Triazine, 2-amino-4-methoxy-6-methyl (1668-54-8)
s-Triazine, 2-amino-4-methoxy-6-(methylthio)- (7CI,8CI) (30358-18-0)
s-Triazine, 2-amino-4-methoxy-6-phenyl- (8CI) (30369-38-1)
s-Triazine, 2-amino-4-(p-methoxyphenyl)-6-(trichloromethyl)- (8CI) (30369-17-6)
s-Triazine, 2-amino-4-methyl- (8CI) (27622-91-9)
s-Triazine, 2-amino-4-(methylamino)-6-(trichloromethyl)- (8CI) (27431-19-2)
s-Triazine, 2-amino-4-methyl-6-(methylamino)- (8CI) (21320-64-9)
s-Triazine, 2-amino-4-methyl-6-phenyl- (8CI) (1853-91-4)
s-Triazine, 2-amino-4-(methylthio)-6-(trichloromethyl)- (8CI) (14946-02-2)
s-Triazine, 2-amino-4-methyl-6-(tribromomethyl)- (7CI,8CI) (30339-34-5)
s-Triazine, 2-amino-4-methyl-6-(trichloromethyl)- (8CI) (21227-47-4)
s-Triazine, 2-amino-4-methyl-6-(trichloromethyl)-, 3-oxide (7CI,8CI) (30805-12-0)
s-Triazine, 2-amino-4-(m-nitrophenyl)-6-(trichloromethyl)- (8CI) (30369-18-7)
s-Triazine, 2-amino-4-nonyl-6-(trichloromethyl)- (8CI) (30369-02-9)
s-Triazine, 2-amino-4-pentyl-6-(trichloromethyl)- (8CI) (30339-74-3)
s-Triazine, 2-amino-4-phenyl- (7CI,8CI) (1853-95-8)
s-Triazine, 2-(m-aminophenyl)-4,6-diphenyl- (8CI) (30363-00-9)
s-Triazine, 2-amino-4-(phenylthio)-6-(trichloromethyl)- (8CI) (30369-55-2)
s-Triazine, 2-amino-4-phenyl-6-(trichloromethyl)- (6CI,8CI) (30369-07-4)
s-Triazine, 2-amino-4-propyl-6-(trichloromethyl)- (8CI) (30339-80-1)
s-Triazine, 2-amino-4-p-tolyl-6-(trichloromethyl)- (8CI) (30369-16-5)
s-Triazine, 2-amino-4-(1,1,2-trichloroethyl)-6-(trichloromethyl)- (8CI) (30339-50-5)
s-Triazine, 2-amino-4-(trichloromethyl)- (8CI) (30360-33-9)
s-Triazine, 2-amino-4-(trichloromethyl)-6-((2,4,5-trichlorophenyl)thio)- (8CI) (30369-62-1)
s-Triazine, 2-amino-4-(trichloromethyl)-6-vinyl- (8CI) (30339-54-9)
s-Triazine, 2-anilino-4,6-dibromo- (8CI) (30357-82-5)
s-Triazine, 2-anilino-4,6-dichloro- (8CI) (2272-40-4)
s-Triazine, 2-anilino-4,6-difluoro- (6CI,8CI) (717-90-8)
s-Triazine, 2-anilino-4,6-dimethoxy- (6CI,8CI) (30358-01-1)
s-Triazine, 2-anilino-4,6-dimethyl- (8CI) (30360-35-1)
s-Triazine, 2-anilino-4,6-diphenoxy- (6CI,7CI,8CI) (1973-08-6)
s-Triazine, 2-anilino-4,6-diphenyl- (6CI,8CI) (30369-20-1)
s-Triazine, 2-anilino-4-methoxy-6-methyl- (8CI) (30369-37-0)
s-Triazine, 2-anilino- (6CI,7CI,8CI) (4040-07-7)
s-Triazine, 2-anilino-4-azido-6-methoxy- (8CI) (30362-30-2)
s-Triazine, 2-anilino-4,6-bis(chloromethyl)- (8CI) (30360-39-5)
s-Triazine, 2-anilino-4,6-bis(o-chlorophenoxy)- (6CI,8CI) (30358-15-7)
s-Triazine, 2-anilino-4,6-bis(p-chlorophenoxy)- (6CI,8CI) (30358-17-9)
s-Triazine, 2-anilino-4,6-bis(dichloromethyl)- (8CI) (30360-44-2)
s-Triazine, 2-anilino-4,6-bis(methylthio)- (8CI) (30362-13-1)
s-Triazine, 2-anilino-4,6-bis(tribromomethyl)- (8CI) (30339-39-0)
s-Triazine, 2-anilino-4,6-bis(trichloromethyl)- (8CI) (30356-50-4)
s-Triazine, 2-anilino-4-chloro-6-(p-chlorophenyl)- (7CI,8CI) (7296-12-0)
s-Triazine, 2-anilino-4-chloro-6-methoxy- (6CI,8CI) (30357-85-8)
s-Triazine, 2-anilino-4-chloro-6-methyl- (7CI,8CI) (6023-57-0)
s-Triazine, 2-anilino-4-chloro-6-(methylthio)- (8CI) (30357-96-1)
s-Triazine, 2-anilino-4-chloro-6-phenoxy- (8CI) (13838-29-4)
s-Triazine, 2-anilino-4-chloro-6-phenyl- (6CI,7CI,8CI) (3842-52-2)
s-Triazine, 2-anilino-4,6-diethoxy- (8CI) (30358-08-8)
s-Triazine, 2-o-anisidino-4,6-dichloro- (8CI) (30369-87-0)
s-Triazine, 2-p-anisidino-4,6-dichloro- (6CI,8CI) (30377-27-6)
s-Triazine, 2-azido-4,6-bis(ethylamino)- (8CI) (2854-95-7)
s-Triazine, 2-azido-4,6-bis(methylthio)- (7CI,8CI) (4407-41-4)
s-Triazine, 2-azido-4,6-bis(trichloromethyl)- (8CI) (30805-09-5)
s-Triazine, 2-azido-4-bromo-6-(ethylamino)- (8CI) (30362-26-6)
s-Triazine, 2-azido-4-(tert-butylamino)-6-(ethylamino) (2854-70-8)
s-Triazine, 2-azido-4-chloro-6-(dimethylamino)- (7CI,8CI) (5248-69-1)
s-Triazine, 2-azido-4-chloro-6-(methylamino)- (7CI,8CI) (5248-65-7)
s-Triazine, 2-azido-4,6-dichloro (30805-06-2)
s-Triazine, 2-azido-4-(diethylamino)-6-(ethylamino)- (8CI) (2854-96-8)
1,3,5-Triazine, 2-azido-4,6-dimethoxy- (9CI) (30805-07-3)
s-Triazine, 2-azido-4,6-dimethoxy- (8CI) (30805-07-3)
s-Triazine, 2-azido-4,6-dimethyl- (8CI) (30805-08-4)
s-Triazine, 2-azido-4-(ethylamino)-6-(ethylthio)- (8CI) (30362-29-9)
s-Triazine, 2-azido-4-(ethylamino)-6-methoxy- (7CI,8CI) (4587-03-5)
s-Triazine, 2-azido-4-(ethylamino)-6-(methylthio)- (7CI,8CI) (4658-25-7)
s-Triazine, 2-azido-4-(ethylamino)-6-phenoxy- (8CI) (30362-27-7)
s-Triazine, 2-azido-4-(isopropylamino)-6-(methylthio) (4658-28-0)
s-Triazine, 2-(1-aziridinyl)-4,6-bis(trichloromethyl)- (6CI,8CI) (21384-33-8)
s-Triazine, 2-benzyl- (6CI,8CI) (30361-89-8)
s-Triazine, 2-(benzylamino)- (6CI,7CI,8CI) (4040-01-1)
s-Triazine, 2-(benzylamino)-4,6-bis(trichloromethyl)- (8CI) (24803-16-5)
s-Triazine, 2-(benzylamino)-4-chloro-6-methyl- (7CI,8CI) (5122-28-1)
s-Triazine, 2-(benzylamino)-4,6-dichloro- (6CI,7CI,8CI) (30369-82-5)
s-Triazine, 2-(benzylamino)-4,6-dimethoxy- (8CI) (30358-00-0)
s-Triazine, 2-(benzylamino)-4-methyl- (8CI) (30377-22-1)
s-Triazine, 2-benzyl-4,6-bis(trichloromethyl)- (7CI,8CI) (3584-24-5)
s-Triazine, 2-(benzyloxy)-4,6-dichloro- (8CI) (30886-24-9)
s-Triazine, 2-(benzylthio)-4,6-bis(trichloromethyl)- (7CI,8CI) (5516-51-8)
s-Triazine, 2-(benzylthio)-4,6-dichloro- (8CI) (25713-57-9)
s-Triazine, 2-(4-biphenylylamino)-4,6-dichloro- (6CI,8CI) (30369-86-9)
s-Triazine, 2,4-bis(allylamino)-6-chloro- (8CI) (15468-86-7)
s-Triazine, 2,4-bis(allylamino)-6-methyl- (8CI) (26234-39-9)
s-Triazine, 2,4-bis(allylamino)-6-(methylthio)- (7CI,8CI) (1020-53-7)
s-Triazine, 2,4-bis(allylamino)-6-(trichloromethyl)- (7CI,8CI) (26235-02-9)

Name	CAS
s-Triazine, 2,4-bis(allyloxy)-6-amino- (8CI)	(30358-11-3)
s-Triazine, 2,4-bis(allyloxy)-6-anilino- (8CI)	(30384-46-4)
s-Triazine, 2,4-bis(allyloxy)-6-(ethylamino)- (8CI)	(30358-12-4)
s-Triazine, 2,4-bis(benzylamino)-6-chloro- (6CI,8CI)	(30355-04-5)
s-Triazine, 2,4-bis(butylamino)-6-isobutyl- (8CI)	(26235-32-5)
s-Triazine, 2,4-bis(butylamino)-6-methyl- (8CI)	(26234-34-4)
s-Triazine, 2,4-bis(sec-butylamino)-6-methyl- (8CI)	(26234-36-6)
s-Triazine, 2,4-bis(butylamino)-6-pentyl- (8CI)	(26235-39-2)
s-Triazine, 2,4-bis(butylamino)-6-propyl- (8CI)	(26235-21-2)
s-Triazine, 2,4-bis(butylamino)-6-(trichloromethyl)- (8CI)	(26234-99-1)
s-Triazine, 2,4-bis(butylthio)-6-chloro- (6CI,7CI,8CI)	(17494-99-4)
s-Triazine, 2,4-bis(p-chloroanilino)-6-methyl- (8CI)	(30368-93-5)
s-Triazine, 2,4-bis(p-chloroanilino)-6-(tribromomethyl)- (8CI)	(30377-17-4)
s-Triazine, 2,4-bis(p-chloroanilino)-6-(trichloromethyl)- (8CI)	(30355-92-1)
s-Triazine, 2,4-bis(chloromethyl)-6-(dichloromethyl)- (8CI)	(30361-94-5)
s-Triazine, 2,4-bis(chloromethyl)-6-(dimethylamino)- (6CI,8CI)	(30360-38-4)
s-Triazine, 2,4-bis(chloromethyl)-6-(ethylamino)- (8CI)	(30360-37-3)
1,3,5-Triazine, 2,4-bis(chloromethyl)-6-(methylthio)- (9CI)	(33032-17-6)
s-Triazine, 2,4-bis(chloromethyl)-6-(methylthio)- (8CI)	(33032-17-6)
s-Triazine, 2,4-bis(p-chlorophenyl)-6-(trichloromethyl)- (8CI)	(30362-65-3)
s-Triazine, 2,4-bis(dibromomethyl)-6-(ethylamino)- (8CI)	(30339-33-4)
1,3,5-Triazine, 2,4-bis(dibromomethyl)-6-methyl- (9CI)	(30362-00-6)
s-Triazine, 2,4-bis(dibromomethyl)-6-methyl- (7CI,8CI)	(30362-00-6)
s-Triazine, 2,4-bis(1,1-dichloroethyl)-6-(ethylamino)- (8CI)	(30339-59-4)
s-Triazine, 2,4-bis(1,1-dichloroethyl)-6-(methylamino)- (8CI)	(30339-58-3)
1,3,5-Triazine, 2,4-bis(1,1-dichloroethyl)-6-(methylthio)- (9CI)	(30863-00-4)
s-Triazine, 2,4-bis(1,1-dichloroethyl)-6-(methylthio)- (8CI)	(30863-00-4)
s-Triazine, 2,4-bis(dichloromethyl)-6-(dimethylamino)- (6CI,8CI)	(30360-43-1)
s-Triazine, 2,4-bis(dichloromethyl)-6-(ethylamino)- (8CI)	(30360-42-0)
s-Triazine, 2,4-bis(dichloromethyl)-6-methyl- (6CI,7CI,8CI)	(5311-23-9)
s-Triazine, 2,4-bis(dichloromethyl)-6-(methylthio)- (7CI,8CI)	(5516-50-7)
s-Triazine, 2,4-bis(diethylamino)-6-ethoxy- (6CI,7CI,8CI)	(30360-60-2)
s-Triazine, 2,4-bis(diethylamino)-6-ethyl- (7CI,8CI)	(26235-16-5)
s-Triazine, 2,4-bis(diethylamino)-6-fluoro	(1598-99-8)
s-Triazine, 2,4-bis(diethylamino)-6-isopropoxy- (7CI,8CI)	(30360-65-7)
s-Triazine, 2,4-bis(diethylamino)-6-methoxy- (8CI)	(15438-85-4)
s-Triazine, 2,4-bis(diethylamino)-6-methyl- (7CI,8CI)	(26234-40-2)
s-Triazine, 2,4-bis(diethylamino)-6-(methylthio)- (8CI)	(30360-80-6)
s-Triazine, 2,4-bis(diethylamino)-6-(trichloromethyl)- (8CI)	(26235-03-0)
s-Triazine, 2,4-bis(dimethylamino)-6-hydrazino- (8CI)	(10409-78-6)
s-Triazine, 2,4-bis(dodecylamino)-6-methyl- (8CI)	(26234-38-8)
s-Triazine, 2,4-bis(dodecylamino)-6-(trichloromethyl)- (8CI)	(26235-01-8)
s-Triazine, 2,4-bis(N-ethylacetamido)-6-(trichloromethyl)- (7CI)	(30359-61-6)
s-Triazine, 2,4-bis(ethylamino)- (8CI)	(4150-59-8)
s-Triazine, 2,4-bis(ethylamino)-6-(ethylthio)- (8CI)	(30360-82-8)
s-Triazine, 2,4-bis(ethylamino)-6-fluoro- (8CI)	(658-65-1)
s-Triazine, 2,4-bis(ethylamino)-6-heptadecyl- (8CI)	(26235-90-5)
s-Triazine, 2,4-bis(ethylamino)-6-hydrazino- (7CI,8CI)	(10421-98-4)
s-Triazine, 2,4-bis(ethylamino)-6-(iodomethyl)- (7CI,8CI)	(30359-75-2)
s-Triazine, 2,4-bis(ethylamino)-6-isobutyl- (8CI)	(26235-30-3)
s-Triazine, 2,4-bis(ethylamino)-6-isopropyl- (7CI,8CI)	(26235-24-5)
s-Triazine, 2,4-bis(ethylamino)-6-methoxy	(673-04-1)
s-Triazine, 2,4-bis(ethylamino)-6-(2-methoxyethoxy)- (8CI)	(31858-13-6)
s-Triazine, 2,4-bis(ethylamino)-6-methyl- (8CI)	(1973-07-5)
s-Triazine, 2,4-bis(ethylamino)-6-(methylthio)	(1014-70-6)
s-Triazine, 2,4-bis(ethylamino)-6-(p-nitrophenoxy)- (7CI,8CI)	(6494-90-2)
s-Triazine, 2,4-bis(ethylamino)-6-nonyl- (8CI)	(26235-88-1)
s-Triazine, 2,4-bis(ethylamino)-6-pentyl- (8CI)	(26235-37-0)
s-Triazine, 2,4-bis(ethylamino)-6-phenoxy- (8CI)	(6494-91-3)
s-Triazine, 2,4-bis(ethylamino)-6-phenyl- (7CI,8CI)	(1972-98-1)
s-Triazine, 2,4-bis(ethylamino)-6-(phenylthio)- (8CI)	(6495-07-4)
s-Triazine, 2,4-bis(ethylamino)-6-propoxy- (7CI,8CI)	(30360-62-4)
s-Triazine, 2,4-bis(ethylamino)-6-propyl- (7CI,8CI)	(26235-18-7)
s-Triazine, 2,4-bis(ethylamino)-6-(p-tolylthio)- (8CI)	(6494-92-4)
s-Triazine, 3,4-bis(ethylamino)-6-(p-tolylthio)- (7CI)	(6494-92-4)
s-Triazine, 2,4-bis(ethylamino)-6-(trichloromethyl)- (7CI,8CI)	(26234-97-9)
s-Triazine, 2,4-bis(ethylamino)-6-(trifluoromethyl)- (8CI)	(721-61-9)
s-Triazine, 2,4-bis(hexylamino)-6-methyl- (8CI)	(26234-37-7)
s-Triazine, 2,4-bis(hexylamino)-6-(trichloromethyl)- (8CI)	(26235-00-7)
s-Triazine, 2,4-bis(isobutylamino)-6-isopropyl- (8CI)	(26235-26-7)
s-Triazine, 2,4-bis(isobutylamino)-6-methyl- (8CI)	(26234-35-5)
s-Triazine, 2,4-bis(isobutylamino)-6-pentyl- (8CI)	(26235-87-0)
s-Triazine, 2,4-bis(isobutylamino)-6-propyl- (8CI)	(26235-22-3)
s-Triazine, 2,4-bis(isobutylamino)-6-(trichloromethyl)- (8CI)	(26322-45-2)
s-Triazine, 2,4-bis(isopropylamino)-6-ethylthio	(4147-51-7)
s-Triazine, 2,4-bis(isopropylamino)-6-methoxy	(1610-18-0)
s-Triazine, 2,4-bis(isopropylamino)-6-methyl- (8CI)	(26263-49-0)
s-Triazine, 4,6-bis(isopropylamino)-2-methylmercapto-	(7287-19-6)
s-Triazine, 2,4-bis(isopropylamino)-6-(methylthio)	(7287-19-6)
s-Triazine, 2,4-bis(isopropylamino)-6-nonyl- (8CI)	(30354-83-7)
s-Triazine, 2,4-bis(isopropylamino)-6-propyl- (7CI,8CI)	(26235-20-1)
s-Triazine, 2,4-bis(isopropylamino)-6-(trichloromethyl)- (7CI,8CI)	(26263-50-3)
s-Triazine, 2,4-bis((2-methoxyethyl)amino)-6-methyl- (8CI)	(30368-96-8)
s-Triazine, 2,4-bis((2-methoxyethyl)amino)-6-(trichloromethyl)- (8CI)	(30388-85-3)
s-Triazine, 2,4-bis((3-methoxypropyl)amino)-6-(methylthio)	(845-52-3)
s-Triazine, 2,4-bis(methylamino)- (8CI)	(30368-50-4)
s-Triazine, 4,6-bis(methylamino)-2-chloro	(2911-36-6)
s-Triazine, 2,4-bis(methylamino)-6-pentyl- (8CI)	(26235-36-9)
s-Triazine, 2,4-bis(methylamino)-6-propyl- (8CI)	(26235-17-6)
s-Triazine, 2,4-bis(methylamino)-6-(tribromomethyl)- (8CI)	(26235-91-6)
s-Triazine, 2,4-bis(methylamino)-6-(trichloromethyl)- (7CI,8CI)	(26234-96-8)
s-Triazine, 2,4-bis(methylthio)-6-chloro	(4407-40-3)
s-Triazine, 2-(bis(methylthio)methyl)-4,6-bis(dichloromethyl)- (8CI)	(24478-18-0)
s-Triazine, 2-(bis(methylthio)methyl)-4-(dichloromethyl)-6-methyl- (8CI)	(30361-95-6)
1,3,5-Triazine, 2,4-bis(methylthio)-6-phenyl- (9CI)	(15067-64-8)
s-Triazine, 2,4-bis(methylthio)-6-phenyl- (8CI)	(15067-64-8)
1,3,5-Triazine, 2,4-bis(methylthio)-6-(trichloromethyl)- (9CI)	(14946-18-0)
s-Triazine, 2,4-bis(methylthio)-6-(trichloromethyl)- (8CI)	(14946-18-0)
1,3,5-Triazine, 2,4-bis(4-nitrophenyl)-6-phenyl- (9CI)	(30363-02-1)
s-Triazine, 2,4-bis(m-nitrophenyl)-6-phenyl- (8CI)	(30363-01-0)
s-Triazine, 2,4-bis(p-nitrophenyl)-6-phenyl- (8CI)	(30363-02-1)
s-Triazine, 4,6-bis(pentachlorophenoxy)-2-chloro	(26396-34-9)
s-Triazine, 2,4-bis(propylamino)-6-(trichloromethyl)- (8CI)	(26234-98-0)
1,3,5-Triazine, 2,4-bis(trichloromethyl)- (9CI)	(3599-74-4)
s-Triazine, 2,4-bis(trichloromethyl)- (7CI,8CI)	(3599-74-4)
s-Triazine, 2,4-bis(trichloromethyl)-6-(2,4,5-trichlorophenyl)- (8CI)	(24481-50-3)
1,3,5-Triazine, 2,4-bis(trichloromethyl)-6-((2,4,5-trichloro-phenyl)thio)- (9CI)	(24478-11-3)
s-Triazine, 2,4-bis(trichloromethyl)-6-((2,4,5-trichlorophenyl)-thio)- (8CI)	(24478-11-3)
s-Triazine, 2,4-bis(trichloromethyl)-6-vinyl- (7CI,8CI)	(4147-62-0)
s-Triazine, 2-(m-bromoanilino)-4,6-bis(trichloromethyl)- (8CI)	(30356-58-2)
s-Triazine, 2-(p-bromoanilino)-4,6-bis(trichloromethyl)- (8CI)	(30356-59-3)
s-Triazine, 2-(m-bromoanilino)-4,6-dichloro- (6CI,8CI)	(30369-85-8)
s-Triazine, 2-(o-bromoanilino)-4,6-dichloro- (6CI,8CI)	(30369-84-7)
s-Triazine, 2-(p-bromoanilino)-4,6-dichloro- (6CI,8CI)	(20376-36-7)
s-Triazine, 2-(p-bromoanilino)-4,6-dimethoxy- (8CI)	(30358-04-4)
s-Triazine, 2-bromo-4,6-bis(ethylamino)- (8CI)	(3084-94-4)
s-Triazine, 2-bromo-4,6-bis(isopropylamino)- (6CI,8CI)	(30360-54-4)
s-Triazine, 2-bromo-4,6-bis(1,1,2-trichloroethyl)- (8CI)	(30894-99-6)
s-Triazine, 2-bromo-4-(diethylamino)-6-(ethylamino)- (8CI)	(30360-55-5)
s-Triazine, 2-bromo-4-(ethylamino)-6-(isopropylamino)- (8CI)	(30360-53-3)
s-Triazine, 2-bromo-4-(ethylamino)-6-(propylamino)- (7CI,8CI)	(30360-51-1)
s-Triazine, 2-(2-bromoethyl)-4,6-bis(trichloromethyl)- (7CI,8CI)	(956-38-7)
s-Triazine, 2-(1-bromoethyl)-4,6-diethyl- (7CI,8CI)	(30362-67-5)
s-Triazine, 2-(1-bromoethyl)-4,6-diphenyl- (8CI)	(30362-79-9)
s-Triazine, 2-(2-bromoethyl)-4-(ethylamino)-6-(trichloromethyl)- (8CI)	(24802-96-8)
s-Triazine, 2-(bromomethyl)- (8CI)	(30361-84-3)

s-Triazine, 2-(bromomethyl)-4,6-bis(ethylamino)- (7CI,8CI) (30359-65-0)
s-Triazine, 2-(bromomethyl)-4,6-dichloro- (8CI) (30894-67-8)
s-Triazine, 2-(bromomethyl)-4,6-dimethoxy- (8CI) (30863-43-5)
s-Triazine, 2-(bromomethyl)-4,6-dimethyl- (7CI,8CI) (30361-98-9)
s-Triazine, 2-(p-bromophenyl)-4,6-bis(trichloromethyl)- (8CI) (24481-49-0)
1,3,5-Triazine, 2-((3-bromophenyl)thio)-4,6-bis(trichloromethyl)- (9CI) (24478-12-4)
s-Triazine, 2-((m-bromophenyl)thio)-4,6-bis(trichloromethyl)- (8CI) (24478-12-4)
s-Triazine, 2-butoxy-4,6-bis(ethylamino)- (7CI,8CI) (30360-67-9)
s-Triazine, 2-butoxy-4,6-bis(trichloromethyl)- (6CI,7CI,8CI) (30863-56-0)
s-Triazine, 2-butoxy-4-(butylthio)-6-chloro- (6CI,7CI,8CI) (30894-82-7)
s-Triazine, 2-butoxy-4-chloro-6-(ethylamino)- (7CI,8CI) (4402-46-4)
s-Triazine, 2-butoxy-4,6-diamino (25254-67-5)
1,3,5-Triazine, 2-butoxy-4,6-dichloro- (9CI) (13838-32-9)
s-Triazine, 2-butoxy-4,6-dichloro- (8CI) (13838-32-9)
s-Triazine, 2-butoxy-4-(ethylamino)-6-methyl- (7CI,8CI) (5248-42-0)
s-Triazine, 2-butoxy-4-methyl-6-(trichloromethyl)- (6CI,8CI) (30863-49-1)
s-Triazine, 2-(butylamino)-4,6-bis(trichloromethyl)- (8CI) (24802-82-2)
s-Triazine, 2-(sec-butylamino)-4,6-bis(trichloromethyl)- (8CI) (24802-83-3)
s-Triazine, 2-(tert-butylamino)-4,6-bis(trichloromethyl)- (8CI) (24802-85-5)
s-Triazine, 2-(sec-butylamino)-4-chloro-6-(ethylamino) (7286-69-3)
s-Triazine, 2-(tert-butylamino)-4-chloro-6-(ethylamino) (5915-41-3)
s-Triazine, 2-(sec-butylamino)-4-chloro-6-(methylamino)- (8CI) (33692-99-8)
s-Triazine, 2-(butylamino)-4-(diethylamino)-6-methyl- (8CI) (27431-11-4)
s-Triazine, 2-(butylamino)-4,6-dimethoxy- (8CI) (30357-99-4)
s-Triazine, 2-(butylamino)-4-(dimethylamino)-6-methyl- (8CI) (27431-09-0)
s-Triazine, 2-(sec-butylamino)-4-(ethylamino)-6-methoxy (26259-45-0)
s-Triazine, 2-tert-butylamino-4-ethylamino-6-methoxy (33693-04-8)
s-Triazine, 2-(butylamino)-4-(ethylamino)-6-methyl- (8CI) (27430-94-0)
s-Triazine, 2-(sec-butylamino)-4-(ethylamino)-6-methyl- (8CI) (27430-95-1)
s-Triazine, 2-(tert-butylamino)-4-(ethylamino)-6-methyl- (8CI) (27430-96-2)
s-Triazine, 2-(tert-butylamino)-4-(ethylamino)-6-(methylthio) (886-50-0)
s-Triazine, 2-(sec-butylamino)-4-(ethylamino)-6-(methylthio)- (8CI) (836-24-8)
s-Triazine, 2-(butylamino)-4-ethyl-6-(trichloromethyl)- (8CI) (24803-22-3)
s-Triazine, 2-(butylamino)-4-(isopropylamino)-6-methyl- (8CI) (27431-03-4)
s-Triazine, 2-(sec-butylamino)-4-(isopropylamino)-6-methyl- (8CI) (27470-65-1)
s-Triazine, 2-(butylamino)-4-methyl-6-(methylamino)- (8CI) (27430-91-7)
s-Triazine, 2-(butylamino)-4-(methylthio)-6-(trichloromethyl)- (8CI) (30369-45-0)
s-Triazine, 2-(butylamino)-4-methyl-6-(trichloromethyl)- (8CI) (24830-35-1)
s-Triazine, 2-(butylamino)-4-phenyl-6-(trichloromethyl)- (8CI) (24848-37-1)
s-Triazine, 2-sec-butyl-4,6-bis(ethylamino)- (8CI) (26235-34-7)
s-Triazine, 2-butyl-4,6-bis(ethylamino)- (7CI,8CI) (26235-27-8)
s-Triazine, 2-butyl-4,6-bis(isopropylamino)- (8CI) (26235-28-9)
s-Triazine, 2-sec-butyl-4,6-bis(propylamino)- (8CI) (26235-35-8)
1,3,5-Triazine, 2-butyl-4,6-bis(trichloromethyl)- (9CI) (24481-41-2)
s-Triazine, 2-butyl-4,6-bis(trichloromethyl)- (8CI) (24481-41-2)
s-Triazine, 2-butyl-4,6-dimethyl- (7CI,8CI) (30362-05-1)
s-Triazine, 2-butyl-4-(ethylamino)-6-(trichloromethyl)- (8CI) (24803-28-9)
s-Triazine, 2-sec-butyl-4-(ethylamino)-6-(trichloromethyl)- (8CI) (24802-89-9)
1,3,5-Triazine, 2-(butylthio)-4,6-bis(trichloromethyl)- (9CI) (24504-18-5)
s-Triazine, 2-(butylthio)-4,6-bis(trichloromethyl)- (8CI) (24504-18-5)
s-Triazine, 2-butylthio-4,6-dichloro (30894-61-2)
s-Triazine-2-carbamic acid, 4,6-dichlorothio-, O-isopropyl ester (7CI,8CI) (30863-26-4)
s-Triazine-2-carbamic acid, 4,6-dichlorothio-, O-methyl ester (7CI,8CI) (5097-58-5)
s-Triazine-2-carbonitrile, 4,6-bis(ethylamino)- (7CI,8CI) (20776-81-2)
1,3,5-Triazine-2-carbonitrile, 4,6-bis(phenylamino)- (9CI) (30360-09-9)
s-Triazine-2-carbonitrile, 4,6-dianilino- (8CI) (30360-09-9)
s-Triazine-2-carbonitrile, 4-(ethylamino)-6-(isopropylamino)- (7CI,8CI) (20776-86-7)
s-Triazine-2-carboxaldehyde, 4,6-bis(dichloromethyl)-, dimethyl mercaptal (8CI) (24478-18-0)
s-Triazine-2-carboxaldehyde, 4-(dichloromethyl)-6-methyl-, dimethyl mercaptal (6CI,8CI) (30361-95-6)
s-Triazine-2-carboxamide, 4-amino-6-hydroxy- (6CI,8CI) (30362-21-1)
s-Triazine-2-carboxamide, 4-amino-6-(trichloromethyl)- (6CI,8CI) (30362-20-0)
s-Triazine-2-carboxylic acid, 4-amino-6-(trichloromethyl)-, ethyl ester (6CI,8CI) (30362-19-7)
s-Triazine-2-carboxylic acid, 4,6-bis(trichloromethyl)-, ethyl ester (8CI) (30863-22-0)
s-Triazine, 2-(p-chloroanilino)- (8CI) (30360-30-6)
s-Triazine, 2-(p-chloroanilino)-4,6-bis(chloromethyl)- (8CI) (30360-40-8)
s-Triazine, 2-(p-chloroanilino)-4,6-bis(tribromomethyl)- (8CI) (30339-78-7)
s-Triazine, 2-(m-chloroanilino)-4,6-bis(trichloromethyl)- (8CI) (30356-52-6)
s-Triazine, 2-(o-chloroanilino)-4,6-bis(trichloromethyl)- (8CI) (30356-51-5)
s-Triazine, 2-(p-chloroanilino)-4,6-bis(trichloromethyl)- (7CI,8CI) (3599-75-5)
s-Triazine, 2-(o-chloroanilino)-4,6-diethoxy- (8CI) (30358-09-9)
s-Triazine, 2-(p-chloroanilino)-4,6-difluoro- (8CI) (30369-78-9)
s-Triazine, 2-(o-chloroanilino)-4,6-difluoro- (6CI,8CI) (2925-56-6)
s-Triazine, 2-(o-chloroanilino)-4,6-dimethoxy- (8CI) (27315-26-0)
s-Triazine, 2-(p-chloroanilino)-4,6-dimethoxy- (6CI,8CI) (27315-27-1)
s-Triazine, 2-(p-chloroanilino)-4,6-dimethyl- (8CI) (30360-47-5)
s-Triazine, 2-((p-chlorobenzyl)thio)-4,6-bis(trichloromethyl)- (8CI) (24481-73-0)
s-Triazine, 2-chloro-4,6-bis(p-chloroanilino)- (6CI,7CI,8CI) (2572-44-3)
s-Triazine, 2-chloro-4,6-bis(chloromethyl)- (6CI,8CI) (30894-87-2)
1,3,5-Triazine, 2-chloro-4,6-bis(2-chlorophenoxy)- (9CI) (30894-77-0)
s-Triazine, 2-chloro-4,6-bis(o-chlorophenoxy)- (6CI,8CI) (30894-77-0)
s-Triazine, 2-chloro-4,6-bis(p-chlorophenoxy)- (6CI,8CI) (30894-78-1)
1,3,5-Triazine, 2-chloro-4,6-bis(4-chlorophenyl)- (9CI) (30894-94-1)
s-Triazine, 2-chloro-4,6-bis(p-chlorophenyl)- (8CI) (30894-94-1)
s-Triazine, 2-chloro-4,6-bis(3,4-dichloroanilino)- (7CI,8CI) (30355-06-7)
s-Triazine, 2-chloro-4,6-bis(1,1-dichloroethyl)- (6CI,8CI) (30894-90-7)
s-Triazine, 2-chloro-4,6-bis(dichloromethyl)- (6CI,8CI) (30894-88-3)
s-Triazine, 2-chloro-4,6-bis(diethylamino) (580-48-3)
s-Triazine, 2-chloro-4,6-bis(dodecylamino)- (8CI) (30355-03-4)
s-Triazine, 2-chloro-4,6-bis(N-ethylacetamido)- (7CI) (4065-24-1)
s-Triazine, 2-chloro-4,6-bis(ethylamino) (122-34-9)
s-Triazine, 2-chloro-4,6-bis(ethylthio)- (6CI,8CI) (30894-83-8)
s-Triazine, 2-chloro-4,6-bis(hexylamino)- (8CI) (30355-02-3)
s-Triazine, 2-chloro-4,6-bis(isopropylamino) (139-40-2)
s-Triazine, 2-chloro-4,6-bis(methoxyamino)- (8CI) (5217-85-6)
s-Triazine, 2-chloro-4,6-bis((3-methoxypropyl)amino)- (6CI,7CI,8CI) (3426-63-9)
s-Triazine, 2-chloro-4,6-bis(N-methylanilino)- (6CI,7CI,8CI) (3995-43-5)
1,3,5-Triazine, 2-chloro-4,6-bis(4-methylphenyl)- (9CI) (21902-34-1)
s-Triazine, 2-chloro-4,6-bis(1-naphthylamino)- (6CI,8CI) (30355-07-8)
s-Triazine, 2-chloro-4,6-bis(p-nitroanilino)- (6CI,7CI,8CI) (2352-37-6)
s-Triazine, 2-chloro-4,6-bis(p-nitrophenoxy)- (8CI) (30894-80-5)
1,3,5-Triazine, 2-chloro-4,6-bis(3-nitrophenyl)- (9CI) (30894-95-2)
s-Triazine, 2-chloro-4,6-bis(m-nitrophenyl)- (8CI) (30894-95-2)
s-Triazine, 2-chloro-4,6-bis(propylamino)- (8CI) (3071-66-7)
s-Triazine, 2-chloro-4,6-bis(p-tolyloxy)- (6CI,8CI) (30894-79-2)
s-Triazine, 2-chloro-4,6-bis(1,1,2-trichloroethyl)- (7CI,8CI) (30894-91-8)
1,3,5-Triazine, 2-chloro-4,6-bis(trichloromethyl)- (9CI) (30894-89-4)
s-Triazine, 2-chloro-4,6-bis(trichloromethyl)- (8CI) (30894-89-4)
s-Triazine, 2-chloro-4-(p-chloroanilino)-6-(isopropylamino)- (8CI) (30355-05-6)
s-Triazine, 2-chloro-4-(p-chlorophenyl)-6-phenyl- (7CI,8CI) (30894-93-0)
s-Triazine, 2-chloro-4-cyclopropylamino-6-isopropylamino (22936-86-3)
s-Triazine, 2-chloro-4-(diazomethyl)-6-methyl- (6CI,7CI,8CI) (30894-86-1)
s-Triazine, 2-chloro-4-(diazomethyl)-6-phenyl- (6CI,7CI,8CI) (30894-92-9)
1,3,5-Triazine, 2-chloro-4,6-diethoxy- (9CI) (30894-75-8)
s-Triazine, 2-chloro-4,6-diethoxy- (6CI,8CI) (30894-75-8)
1,3,5-Triazine, 2-chloro-4,6-diethyl- (9CI) (701-77-9)
s-Triazine, 2-chloro-4,6-diethyl- (7CI,8CI) (701-77-9)
s-Triazine, 2-chloro-4-(diethylamino)-6-(ethylamino) (1912-26-1)
s-Triazine, 2-chloro-4-(diethylamino)-6-(isopropylamino) (1912-25-0)

s-Triazine, 2-chloro-4,6-diisopropoxy- (8CI) **(30894-76-9)**
1,3,5-Triazine, 2-chloro-4,6-dimethoxy- (9CI) **(3140-73-6)**
s-Triazine, 2-chloro-4,6-dimethoxy- (8CI) **(3140-73-6)**
1,3,5-Triazine, 2-chloro-4,6-dimethyl- (9CI) **(30894-84-9)**
s-Triazine, 2-chloro-4,6-dimethyl- (8CI) **(30894-84-9)**
s-Triazine, 2-chloro-4-(dimethylamino)-6-methoxy- (7CI,8CI) **(13882-55-8)**
1,3,5-Triazine, 2-chloro-4,6-diphenoxy- (9CI) **(2972-65-8)**
s-Triazine, 2-chloro-4,6-diphenoxy- (6CI,7CI,8CI) **(2972-65-8)**
1,3,5-Triazine, 2-chloro-4,6-diphenyl- (9CI) **(3842-55-5)**
s-Triazine, 2-chloro-4,6-diphenyl- (6CI,7CI,8CI) **(3842-55-5)**
s-Triazine, 2-chloro-4,6-di-p-tolyl- (8CI) **(21902-34-1)**
s-Triazine, 2-chloro-4-ethoxy-6-(ethylamino)- (6CI,7CI,8CI) **(4743-13-9)**
s-Triazine, 2-chloro-4-ethoxy-6-(ethylthio)- (6CI,8CI) **(30894-81-6)**
s-Triazine, 2-(1-chloroethyl)- (7CI,8CI) **(30361-85-4)**
s-Triazine, 2-chloro-4-ethylamino-6-(1-cyano-1-methyl)ethyl-amino- **(21725-46-2)**
s-Triazine, 2-chloro-4-(ethylamino)-6-(ethylthio)- (6CI,8CI) **(30388-95-5)**
s-Triazine, 2-chloro-4-(ethylamino)-6-isobutoxy- (8CI) **(30357-89-2)**
s-Triazine, 2-chloro-4-ethylamino-6-isopropylamino **(1912-24-9)**
s-Triazine, 2-chloro-4-(ethylamino)-6-methoxy- (7CI,8CI) **(4653-94-5)**
s-Triazine, 2-chloro-4-(ethylamino)-6-methyl- (8CI) **(30369-24-5)**
s-Triazine, 2-chloro-4-(ethylamino)-6-(methylamino)- (8CI) **(3084-92-2)**
s-Triazine, 2-chloro-4-(ethylamino)-6-(methylthio)- (8CI) **(30357-95-0)**
s-Triazine, 2-chloro-4-(ethylamino)-6-phenoxy- (8CI) **(30357-92-7)**
s-Triazine, 2-chloro-4-(ethylamino)-6-propoxy- (7CI,8CI) **(4417-64-5)**
s-Triazine, 2-(2-chloroethyl)-4,6-bis(ethylamino)- (8CI) **(26235-92-7)**
1,3,5-Triazine, 2-(2-chloroethyl)-4,6-bis(trichloromethyl)- (9CI) **(1145-44-4)**
s-Triazine, 2-(2-chloroethyl)-4,6-bis(trichloromethyl)- (8CI) **(1145-44-4)**
s-Triazine, 2-(1-chloroethyl)-4,6-diethyl- (7CI,8CI) **(31858-10-3)**
s-Triazine, 2-(1-chloroethyl)-4,6-diphenyl- (7CI,8CI) **(30362-73-3)**
s-Triazine, 2-chloro-4-ethyl-6-(ethylamino)- (8CI) **(30369-29-0)**
s-Triazine, 2-(2-chloroethyl)-4-(ethylamino)-6-(trichloro-methyl)- (8CI) **(24802-94-6)**
1,3,5-Triazine, 2-chloro-4-ethyl-6-phenyl- (9CI) **(13704-97-7)**
s-Triazine, 2-chloro-4-ethyl-6-phenyl- (7CI,8CI) **(13704-97-7)**
s-Triazine, 2-(2-chloroethyl)-4-(propylamino)-6-(trichloro-methyl)- (8CI) **(24802-95-7)**
s-Triazine, 2-chloro-4-(hexylamino)-6-methoxy- (8CI) **(30357-84-7)**
s-Triazine, 2-chloro-4-(isobutylamino)-6-(methylthio)- (7CI,8CI) **(4658-30-4)**
s-Triazine, 2-chloro-4-isopropoxy-6-(isopropylamino)- (7CI,8CI) **(4417-68-9)**
s-Triazine, 2-chloro-4-(isopropylamino)-6-methoxy- (8CI) **(4446-75-7)**
s-Triazine, 2-chloro-4-(isopropylamino)-6-((3-methoxy-propyl)amino)- (8CI) **(1824-09-5)**
s-Triazine, 2-chloro-4-(isopropylamino)-6-(methylamino) **(3004-71-5)**
s-Triazine, 2-chloro-4-methoxy-6-(methylthio)- (7CI,8CI) **(4407-44-7)**
s-Triazine, 2-chloro-4-methoxy-6-(propylamino)- (7CI,8CI) **(4407-72-1)**
s-Triazine, 2-(chloromethyl)- (7CI,8CI) **(30361-82-1)**
s-Triazine, 2-chloro-4-(methylamino)-6-phenyl- (7CI,8CI) **(1973-03-1)**
s-Triazine, 2-(chloromethyl)-4,6-bis(ethylamino)- (7CI,8CI) **(30355-62-5)**
1,3,5-Triazine, 2-(chloromethyl)-4,6-bis(trichloromethyl)- (9CI) **(30361-97-8)**
s-Triazine, 2-(chloromethyl)-4,6-bis(trichloromethyl)- (8CI) **(30361-97-8)**
s-Triazine, 2-(chloromethyl)-4,6-dimethoxy- (6CI,8CI) **(30863-39-9)**
1,3,5-Triazine, 2-(chloromethyl)-4,6-dimethyl- (9CI) **(15640-03-6)**
s-Triazine, 2-(chloromethyl)-4,6-dimethyl- (7CI,8CI) **(15640-03-6)**
s-Triazine, 2-(1-chloro-1-methylethyl)- (7CI,8CI) **(30361-88-7)**
1,3,5-Triazine, 2-chloro-4-methyl-6-phenyl- (9CI) **(30937-70-3)**
s-Triazine, 2-chloro-4-methyl-6-phenyl- (7CI,8CI) **(30937-70-3)**
1,3,5-Triazine, 2-((4-chloro-3-methylphenyl)thio)-4,6-bis-(trichloromethyl)- (9CI) **(24478-13-5)**
1,3,5-Triazine, 2-chloro-4-(methylthio)-6-phenyl- (9CI) **(13705-07-2)**
s-Triazine, 2-chloro-4-(methylthio)-6-phenyl- (7CI,8CI) **(13705-07-2)**
1,3,5-Triazine, 2-chloro-4-methyl-6-(trichloromethyl)- (9CI) **(30894-85-0)**
s-Triazine, 2-chloro-4-methyl-6-(trichloromethyl)- (8CI) **(30894-85-0)**
1,3,5-Triazine, 2-((4-chloro-3-nitrophenyl)thio)-4,6-bis-(trichloromethyl)- (9CI) **(24478-15-7)**
s-Triazine, 2-((4-chloro-3-nitrophenyl)thio)-4,6-bis(trichloro-methyl)- (8CI) **(24478-15-7)**
s-Triazine, 2-(o-chlorophenoxy)-4,6-bis(ethylamino)- (7CI,8CI) **(6495-03-0)**
s-Triazine, 2-(p-chlorophenoxy)-4,6-bis(ethylamino)- (7CI,8CI) **(6585-96-2)**
s-Triazine, 2-(o-chlorophenoxy)-4,6-difluoro- (6CI,8CI) **(886-45-3)**
1,3,5-Triazine, 2-(4-chlorophenyl)-4,6-bis(trichloromethyl)- (9CI) **(3712-60-5)**
s-Triazine, 2-(m-chlorophenyl)-4,6-bis(trichloromethyl)- (8CI) **(24481-54-7)**
s-Triazine, 2-(o-chlorophenyl)-4,6-bis(trichloromethyl)- (8CI) **(24481-55-8)**
s-Triazine, 2-(p-chlorophenyl)-4,6-bis(trichloromethyl)- (7CI,8CI) **(3712-60-5)**
1,3,5-Triazine, 2-(4-chlorophenyl)-4,6-dimethyl- (9CI) **(30362-08-4)**
s-Triazine, 2-(p-chlorophenyl)-4,6-dimethyl- (7CI,8CI) **(30362-08-4)**
1,3,5-Triazine, 2-(4-chlorophenyl)-4,6-diphenyl- (9CI) **(3114-52-1)**
s-Triazine, 2-(p-chlorophenyl)-4,6-diphenyl- (7CI,8CI) **(3114-52-1)**
s-Triazine, 2-(p-chlorophenyl)-4-(ethylamino)-6-(trichloro-methyl)- (8CI) **(24848-39-3)**
s-Triazine, 2-(p-chlorophenyl)-4-methoxy-6-(trichloromethyl)- (8CI) **(30863-62-8)**
1,3,5-Triazine, 2-(((4-chlorophenyl)methyl)thio)-4,6-bis-(trichloromethyl)- (9CI) **(24481-73-0)**
s-Triazine, 2-(p-chlorophenyl)-4-phenyl-6-(trichloromethyl)- (7CI,8CI) **(3599-80-2)**
s-Triazine, 2-((p-chlorophenyl)thio)-4,6-bis(ethylamino)- (7CI,8CI) **(6560-09-4)**
1,3,5-Triazine, 2-((2-chlorophenyl)thio)-4,6-bis(trichloromethyl)- (9CI) **(24478-06-6)**
1,3,5-Triazine, 2-((3-chlorophenyl)thio)-4,6-bis(trichloromethyl)- (9CI) **(24478-07-7)**
1,3,5-Triazine, 2-((4-chlorophenyl)thio)-4,6-bis(trichloromethyl)- (9CI) **(24478-08-8)**
s-Triazine, 2-((m-chlorophenyl)thio)-4,6-bis(trichloromethyl)- (8CI) **(24478-07-7)**
s-Triazine, 2-((o-chlorophenyl)thio)-4,6-bis(trichloromethyl)- (8CI) **(24478-06-6)**
s-Triazine, 2-((p-chlorophenyl)thio)-4,6-bis(trichloromethyl)- (8CI) **(24478-08-8)**
s-Triazine, 2-((p-chlorophenyl)thio)-4-(ethylamino)-6-(tri-chloromethyl)- (8CI) **(30369-59-6)**
s-Triazine, 2-((p-chlorophenyl)thio)-4-(isopropylamino)-6-(trichloro-methyl)- (8CI) **(30369-60-9)**
s-Triazine, 2-((p-chlorophenyl)thio)-4-(methylamino)-6-(trichloro-methyl)- (8CI) **(30369-58-5)**
1,3,5-Triazine, 2-chloro-4-phenyl-6-(trichloromethyl)- (9CI) **(13704-90-0)**
s-Triazine, 2-chloro-4-phenyl-6-(trichloromethyl)- (7CI,8CI) **(13704-90-0)**
s-Triazine, 2-chloro-4-propoxy-6-(propylamino)- (7CI,8CI) **(4417-65-6)**
s-Triazine, 2-(3-chloropropyl)-4,6-dimethyl- (8CI) **(30362-04-0)**
s-Triazine, 2-((4-chloro-m-tolyl)thio)-4,6-bis(trichloromethyl)- (8CI) **(24478-13-5)**
s-Triazine, 2-(cyclohexylamino)-4,6-bis(trichloromethyl)- (8CI) **(24803-15-4)**
s-Triazine, 2-cyclohexyl-4,6-bis(trichloromethyl)- (8CI) **(30362-48-2)**
s-Triazine, 2,4-diacetamido-6-methyl- (6CI) **(30355-54-5)**
1,3,5-Triazine-2,4-diamine (9CI) **(504-08-5)**
1,3,5-Triazine-2,4-diamine, 6-azido- (9CI) **(16376-89-9)**
1,3,5-Triazine-2,4-diamine, 6-azido-N,N'-diethyl- (9CI) **(2854-95-7)**
1,3,5-Triazine-2,4-diamine, 6-azido-N,N,N'-triethyl- (9CI) **(2854-96-8)**
1,3,5-Triazine-2,4-diamine, N,N'-bis(4-chlorophenyl)-6-methyl- (9CI) **(30368-93-5)**
1,3,5-Triazine-2,4-diamine, N,N'-bis(1-methylethyl)-6-(methyl-thio)-, Mixt. with 6-Chloro- N-ethyl-N'-(1-methylethyl)-1,3,5-triazine-2,4-diamine **(8073-77-6)**
1,3,5-Triazine-2,4-diamine, 6-bromo- (9CI) **(4649-67-6)**
1,3,5-Triazine-2,4-diamine, 6-bromo-N,N'-diethyl- (9CI) **(3084-94-4)**
1,3,5-Triazine-2,4-diamine, 6-bromo-N-ethyl-N'-(1-methylethyl)- (9CI) **(30360-53-3)**
1,3,5-Triazine-2,4-diamine, 6-(bromomethyl)- (9CI) **(4576-40-3)**
1,3,5-Triazine-2,4-diamine, 6-bromo-N,N,N'-triethyl- (9CI) **(30360-55-5)**
1,3,5-Triazine-2,4-diamine, 6-butyl- **(3232-26-6)**
1,3,5-Triazine-2,4-diamine, 6-chloro- (9CI) **(3397-62-4)**
1,3,5-Triazine-2,4-diamine, 6-chloro-N,N'-bis(4-chlorophenyl)- (9CI) **(2572-44-3)**

1,3,5-Triazine-2,4-diamine, 6-chloro-N,N'-bis(4-methoxyphenyl)- (9CI) **(6737-62-8)**
1,3,5-Triazine-2,4-diamine, 6-chloro-N,N'-bis(1-methylethyl)- **(139-40-2)**
1,3,5-Triazine-2,4-diamine, 6-chloro-N,N'-bis(4-nitrophenyl)- (9CI) **(2352-37-6)**
1,3,5-Triazine-2,4-diamine, 6-chloro-N,N'-bis(phenylmethyl)- (9CI) **(30355-04-5)**
1,3,5-Triazine-2,4-diamine, 6-chloro-N-cyclopropyl-N'-(1-methylethyl)- **(22936-86-3)**
1,3,5-Triazine-2,4-diamine, 6-chloro-N,N'-didodecyl- (9CI) **(30355-03-4)**
1,3,5-Triazine-2,4-diamine, 6-chloro-N,N'-dimethoxy- (9CI) **(5217-85-6)**
1,3,5-Triazine-2,4-diamine, 6-chloro-N,N'-dimethyl-N,N'-diphenyl- (9CI) **(3995-43-5)**
1,3,5-Triazine-2,4-diamine, 6-chloro-N-(1,1-dimethylethyl)-N'-ethyl-, Mixt. with N-(1,1-dimethylethyl)-N'-ethyl-6-(methylthio)-1,3,5-triazine-2,4-diamine (9CI) **(8066-11-3)**
1,3,5-Triazine-2,4-diamine, 6-chloro-N-(1,1-dimethylethyl)-N'-ethyl-, Mixt. with N-(1,1-dimethylethyl)-N'-ethyl-6-methoxy-1,3,5-triazine-2,4-diamine **(8072-81-9)**
1,3,5-Triazine-2,4-diamine, 6-chloro-N-(1,1-dimethylethyl)-N'-methyl- (9CI) **(34333-27-2)**
1,3,5-Triazine-2,4-diamine, 6-chloro-N,N'-di-1-naphthalenyl- (9CI) **(30355-07-8)**
1,3,5-Triazine-2,4-diamine, 6-chloro-N,N'-diphenyl- (9CI) **(1973-09-7)**
1,3,5-Triazine-2,4-diamine, 6-chloro-N,N'-di-2-propenyl- (9CI) **(15468-86-7)**
1,3,5-Triazine-2,4-diamine, 6-chloro-N,N'-dipropyl- (9CI) **(3071-66-7)**
1,3,5-Triazine-2,4-diamine, 6-chloro-N-ethyl- (9CI) **(1007-28-9)**
1,3,5-Triazine-2,4-diamine, 6-chloro-N-ethyl-N'-methyl- (9CI) **(3084-92-2)**
1,3,5-Triazine-2,4-diamine, 6-chloro-N-ethyl-N'-(1-methylethyl)- (9CI) **(1912-24-9)**
1,3,5-Triazine-2,4-diamine, 6-chloro-N-ethyl-N'-(1-methylpropyl)- (9CI) **(7286-69-3)**
1,3,5-Triazine-2,4-diamine, 6-(5-chloro-2-methoxyphenyl)-N,N-diethyl- (9CI) **(68215-98-5)**
1,3,5-Triazine-2,4-diamine, 6-chloro-N-(3-methoxypropyl)-N'-(1-methylethyl)- (9CI) **(1824-09-5)**
1,3,5-Triazine-2,4-diamine, 6-(chloromethyl)- (9CI) **(10581-62-1)**
1,3,5-Triazine-2,4-diamine, 6-chloro-n-(1-methylethyl)- **(6190-65-4)**
1,3,5-Triazine-2,4-diamine, 6-chloro-N-methyl-N'-(1-methylpropyl)- (9CI) **(33692-99-8)**
1,3,5-Triazine-2,4-diamine, 6-(chloromethyl)-N-phenyl- (9CI) **(30355-60-3)**
1,3,5-Triazine-2,4-diamine, 6-chloro-n,n,N'-triethyl- **(1912-26-1)**
1,3,5-Triazine-2,4-diamine, 6-(4-chlorophenyl)- (9CI) **(4514-53-8)**
1,3,5-Triazine-2,4-diamine, 6-chloro-N-phenyl- (9CI) **(16007-72-0)**
1,3,5-Triazine-2,4-diamine, 5-(4-chlorophenyl)-5,6-dihydro-6-methyl- (9CI) **(30805-24-4)**
1,3,5-Triazine-2,4-diamine, N-(4-chlorophenyl)-6-methyl- (9CI) **(645-18-1)**
1,3,5-Triazine-2,4-diamine, 6-chloro-N,N,N',N'-tetraethyl- **(565-33-3)**
1,3,5-Triazine-2,4-diamine, 6-(3,4-dichlorophenyl)- (9CI) **(30354-89-3)**
1,3,5-Triazine-2,4-diamine, 1-(3,4-dichlorophenyl)-1,6-dihydro-6,6-dimethyl- (9CI) **(13344-99-5)**
1,3,5-Triazine-2,4-diamine, N,N'-diethyl- (9CI) **(4150-59-8)**
1,3,5-Triazine-2,4-diamine, N,N'-diethyl-6-(ethylthio)- (9CI) **(30360-82-8)**
1,3,5-Triazine-2,4-diamine, N,N'-diethyl-6-fluoro- (9CI) **(658-65-1)**
1,3,5-Triazine-2,4-diamine, N,N'-diethyl-6-(2-methoxyethoxy)- (9CI) **(31858-13-6)**
1,3,5-Triazine-2,4-diamine, N,N'-diethyl-6-methyl- (9CI) **(1973-07-5)**
1,3,5-Triazine-2,4-diamine, N,N'-diethyl-6-phenoxy- (9CI) **(6494-91-3)**
1,3,5-Triazine-2,4-diamine, N,N'-diethyl-6-phenyl- (9CI) **(1972-98-1)**
1,3,5-Triazine-2,4-diamine, N,N'-diethyl-6-(phenylthio)- (9CI) **(6495-07-4)**
1,3,5-Triazine-2,4-diamine, N,N'-diethyl-6-(trifluoromethyl)- (9CI) **(721-61-9)**
1,3,5-Triazine-2,4-diamine, 1,6-dihydro-6,6-dimethyl-1-(4-nitrophenyl)- (9CI) **(30805-27-7)**
1,3,5-Triazine-2,4-diamine, 1,6-dihydro-6,6-dimethyl-1-phenyl- (9CI) **(4022-58-6)**
1,3,5-Triazine-2,4-diamine, N,N'-dimethyl- (9CI) **(30368-50-4)**
1,3,5-Triazine-2,4-diamine, N-(1,1-dimethylethyl)-N'-ethyl-6-methyl- (9CI) **(27430-96-2)**
1,3,5-Triazine-2,4-diamine, N,N'-dimethyl-6-(trichloromethyl)- (9CI) **(26234-96-8)**
1,3,5-Triazine-2,4-diamine, N,N'-diphenyl- (9CI) **(13107-54-5)**
1,3,5-Triazine-2,4-diamine, N,N'-diphenyl-6-(phenylthio)- (9CI) **(30360-87-3)**
1,3,5-Triazine-2,4-diamine, 6-ethoxy-N,N'-diethyl- (9CI) **(30360-59-9)**
1,3,5-Triazine-2,4-diamine, 6-ethoxy-N,N'-bis(1-methylethyl)- (9CI) **(30377-19-6)**
1,3,5-Triazine-2,4-diamine, 6-(2-ethoxyethoxy)-N,N'-diethyl- (9CI) **(30360-73-7)**
1,3,5-Triazine-2,4-diamine, 6-ethyl- **(934-75-8)**
1,3,5-Triazine-2,4-diamine, N-ethyl- (9CI) **(30368-49-1)**
1,3,5-Triazine-2,4-diamine, N-ethyl-6-methoxy- (9CI) **(30360-56-6)**
1,3,5-Triazine-2,4-diamine, N-ethyl-N'-(1-methylethyl)-6-phenoxy- (9CI) **(6494-99-1)**
1,3,5-Triazine-2,4-diamine, N-ethyl-6-methyl-N'-(1-methylethyl)- (9CI) **(27430-93-9)**
1,3,5-Triazine-2,4-diamine, N-ethyl-N'-(1-methylpropyl)-6-(methylthio)- (9CI) **(836-24-8)**
1,3,5-Triazine-2,4-diamine, 6-ethyl-N-phenyl- (9CI) **(30359-79-6)**
1,3,5-Triazine-2,4-diamine, 6-(ethylthio)- (9CI) **(30360-81-7)**
1,3,5-Triazine-2,4-diamine, 6-heptadecyl- (9CI) **(2533-20-2)**
1,3,5-Triazine-2,4-diamine, 6-methoxy-N,N'-dimethyl- (9CI) **(30360-57-7)**
1,3,5-Triazine-2,4-diamine, 6-methoxy-N,N'-diphenyl- (9CI) **(30360-58-8)**
1,3,5-Triazine-2,4-diamine, 6-methoxy-N,N'-bis(3-methoxypropyl)- (9CI) **(1771-07-9)**
1,3,5-Triazine-2,4-diamine, 6-methoxy-N-methyl-N'-(1-methylethyl)- (9CI) **(3035-45-8)**
1,3,5-Triazine-2,4-diamine, 6-(4-methoxyphenyl)- (9CI) **(30354-91-7)**
1,3,5-Triazine-2,4-diamine, 6-methyl- (9CI) **(542-02-9)**
1,3,5-Triazine-2,4-diamine, 6-methyl-N,N'-bis(1-methylethyl)- (9CI) **(26263-49-0)**
1,3,5-Triazine-2,4-diamine, 6-methyl-N,N'-diphenyl- (9CI) **(2039-34-1)**
1,3,5-Triazine-2,4-diamine, 6-(1-methylethoxy)-N,N'-bis-(1-methylethyl)- (9CI) **(30360-64-6)**
1,3,5-Triazine-2,4-diamine, 6-(4-methylphenyl)- (9CI) **(19338-12-6)**
1,3,5-Triazine-2,4-diamine, 6-(1-methylpropyl)- **(30354-74-6)**
1,3,5-Triazine-2,4-diamine, 6-(2-methylpropyl)- (9CI) **(30354-68-8)**
1,3,5-Triazine-2,4-diamine, 6-(methylthio)- **(5397-01-3)**
1,3,5-Triazine-2,4-diamine, 6-(3-nitrophenyl)- (9CI) **(29366-72-1)**
1,3,5-Triazine-2,4-diamine, 6-nonyl- (9CI) **(5921-65-3)**
1,3,5-Triazine-2,4-diamine, 6-pentyl- (9CI) **(3283-17-8)**
1,3,5-Triazine-2,4-diamine, 6-phenoxy- (9CI) **(1467-72-7)**
1,3,5-Triazine-2,4-diamine, 6-phenoxy-N,N'-diphenyl- (9CI) **(30360-76-0)**
1,3,5-Triazine-2,4-diamine, N-phenyl- (9CI) **(537-17-7)**
1,3,5-Triazine-2,4-diamine, 6-(phenylmethyl)- (9CI) **(1853-88-9)**
1,3,5-Triazine-2,4-diamine, 6-((phenylmethyl)thio)- (9CI) **(30360-85-1)**
1,3,5-Triazine-2,4-diamine, N-phenyl-6-propyl- (9CI) **(30359-95-6)**
1,3,5-Triazine-2,4-diamine, N-phenyl-6-(trifluoromethyl)- (9CI) **(368-61-6)**
1,3,5-Triazine-2,4-diamine, 6-(2-propenyloxy)- (9CI) **(6291-87-8)**
1,3,5-Triazine-2,4-diamine, 6-propoxy- (9CI) **(6295-15-4)**
1,3,5-Triazine-2,4-diamine, 6-propyl- **(5962-23-2)**
1,3,5-Triazine-2,4-diamine, N,N,N',N'-tetraethyl-6-methoxy- (9CI) **(15438-85-4)**
1,3,5-Triazine-2,4-diamine, N,N,N',N'-tetraethyl-6-(methylthio)- (9CI) **(30360-80-6)**
1,3,5-Triazine-2,4-diamine, N,N,N',N'-tetraethyl-6-(trichloromethyl)- (9CI) **(26235-03-0)**
1,3,5-Triazine-2,4-diamine, 6-(trichloromethyl)- (9CI) **(16088-73-6)**
1,3,5-Triazine-2,4-diamine, N,N,N'-triethyl-6-methoxy- (9CI) **(13532-26-8)**
1,3,5-Triazine-2,4-diamine, 6-(trifluoromethyl)- (9CI) **(705-25-9)**
1,3,5-Triazine-2,4-diamine, N,N',6-trimethyl- (9CI) **(1973-06-4)**
s-Triazine, 2,4-diamino- (8CI) **(504-08-5)**
s-Triazine, 2,4-diamino-6-azido- (8CI) **(16376-89-9)**
s-Triazine, 2,4-diamino-6-benzyl- (8CI) **(1853-88-9)**
s-Triazine, 2,4-diamino-6-(benzyloxy)- (8CI) **(30360-74-8)**
s-Triazine, 2,4-diamino-6-(benzylthio)- (6CI,8CI) **(30360-85-1)**

Name	CAS
s-Triazine, 2,4-diamino-6-bromo- (7CI,8CI)	**(4649-67-6)**
s-Triazine, 2,4-diamino-6-(bromomethyl)- (7CI,8CI)	**(4576-40-3)**
s-Triazine, 2,4-diamino-6-(butylthio)- (8CI)	**(30360-83-9)**
s-Triazine, 2,4-diamino-6-chloro- (8CI)	**(3397-62-4)**
s-Triazine, 2,4-diamino-6-(1-chloroethyl)- (8CI)	**(30359-86-5)**
s-Triazine, 2,4-diamino-6-(2-chloroethyl)- (8CI)	**(30359-87-6)**
s-Triazine, 2,4-diamino-6-(chloromethyl)- (6CI,7CI,8CI)	**(10581-62-1)**
s-Triazine, 2,4-diamino-6-(1-chloro-1-methylethyl)- (8CI)	**(30354-65-5)**
s-Triazine, 2,4-diamino-6-(o-chlorophenoxy)- (8CI)	**(30360-77-1)**
s-Triazine, 2,4-diamino-6-(p-chlorophenoxy)- (8CI)	**(30360-78-2)**
s-Triazine, 2,4-diamino-6-(p-chlorophenyl)- (8CI)	**(4514-53-8)**
s-Triazine, 4,6-diamino-1-(p-chlorophenyl)-1,2-dihydro-2-isopropyl- (8CI)	**(30805-25-5)**
s-Triazine, 4,6-diamino-1-(p-chlorophenyl)-1,2-dihydro-2-methyl- (8CI)	**(30805-24-4)**
s-Triazine, 2,4-diamino-6-(diazomethyl)- (6CI,7CI,8CI)	**(30355-56-7)**
s-Triazine, 2,4-diamino-6-(1,2-dibromoethyl)- (8CI)	**(30359-94-5)**
s-Triazine, 2,4-diamino-6-(dibromomethyl)- (7CI,8CI)	**(30359-66-1)**
s-Triazine, 2,4-diamino-6-(1,1-dichloroethyl)- (8CI)	**(30359-89-8)**
s-Triazine, 2,4-diamino-6-(1,2-dichloroethyl)- (8CI)	**(30359-90-1)**
s-Triazine, 2,4-diamino-6-(1,1-dichloroheptadecyl)- (8CI)	**(30354-86-0)**
s-Triazine, 2,4-diamino-6-(dichloromethyl)- (6CI,8CI)	**(30355-63-6)**
s-Triazine, 2,4-diamino-6-(3,4-dichlorophenyl)- (8CI)	**(30354-89-3)**
s-Triazine, 4,6-diamino-1-(3,4-dichlorophenyl)-1,2-dihydro-2,2-dimethyl- (8CI)	**(13344-99-5)**
s-Triazine, 4,6-diamino-1-(3,4-dichlorophenyl)-1,2-dihydro-2-isopropyl- (8CI)	**(30805-26-6)**
s-Triazine, 4,6-diamino-1,2-dihydro-2,2-dimethyl-1-(p-nitrophenyl)- (8CI)	**(30805-27-7)**
s-Triazine, 4,6-diamino-1,2-dihydro-2,2-dimethyl-1-phenyl- (8CI)	**(4022-58-6)**
s-Triazine, 2,4-diamino-6-(diiodomethyl)- (8CI)	**(30359-76-3)**
s-Triazine, 2,4-diamino-6-(dodecyloxy)- (6CI,8CI)	**(24126-22-5)**
s-Triazine, 4,6-diamino-2-ethoxy	**(2827-44-3)**
s-Triazine, 2,4-diamino-6-(2-ethoxyethyl)- (8CI)	**(30354-97-3)**
s-Triazine, 2,4-diamino-6-(ethylthio)- (8CI)	**(30360-81-7)**
s-Triazine, 2,4-diamino-6-fluoro	**(823-95-0)**
s-Triazine, 2,4-diamino-6-(2-furyl)	**(4685-18-1)**
s-Triazine, 2,4-diamino-6-heptadecyl- (8CI)	**(2533-20-2)**
s-Triazine, 4,6-diamino-2-hexoxy	**(24126-20-3)**
s-Triazine, 2,4-diamino-6-(iodomethyl)- (7CI,8CI)	**(30359-74-1)**
s-Triazine, 4,6-diamino-2-isobutoxy	**(27963-33-3)**
s-Triazine, 2,4-diamino-6-isobutyl- (8CI)	**(30354-68-8)**
s-Triazine, 4,6-diamino-2-isopropoxy	**(24860-40-0)**
s-Triazine, 2,4-diamino-6-isopropyl	**(5397-04-6)**
s-Triazine, 4,6-diamino-2-methoxy	**(2827-45-4)**
s-Triazine, 2,4-diamino-6-(p-methoxyphenyl)- (8CI)	**(30354-91-7)**
s-Triazine, 2,4-diamino-6-methyl- (8CI)	**(542-02-9)**
s-Triazine, 2,4-diamino-6-methyl-, 3-oxide (7CI,8CI)	**(4030-02-8)**
s-Triazine, 2,4-diamino-6-(1-naphthyl)- (8CI)	**(30354-93-9)**
s-Triazine, 2,4-diamino-6-(p-nitrophenoxy)- (8CI)	**(30360-79-3)**
s-Triazine, 2,4-diamino-6-(m-nitrophenyl)- (8CI)	**(29366-72-1)**
s-Triazine, 2,4-diamino-6-nonyl- (8CI)	**(5921-65-3)**
s-Triazine, 2,4-diamino-6-(pentachloroethyl)- (8CI)	**(30359-93-4)**
s-Triazine, 4,6-diamino-2-pentoxy	**(30360-70-4)**
s-Triazine, 2,4-diamino-6-pentyl- (8CI)	**(3283-17-8)**
s-Triazine, 2,4-diamino-6-phenoxy- (6CI,7CI,8CI)	**(1467-72-7)**
s-Triazine, 2,4-diamino-6-phenyl	**(91-76-9)**
s-Triazine, 2,4-diamino-6-(phenylthio)- (8CI)	**(30360-86-2)**
s-Triazine, 2,4-diamino-6-(2-propyl)	**(5397-04-6)**
s-Triazine, 4,6-diamino-2-propyl (8CI)	**(6295-15-4)**
s-Triazine, 2,4-diamino-6-(1,1,2,2-tetrachloroethyl)- (8CI)	**(30359-92-3)**
s-Triazine, 2,4-diamino-6-p-tolyl- (8CI)	**(19338-12-6)**
s-Triazine, 2,4-diamino-6-(p-tolyloxy)- (7CI,8CI)	**(1639-39-0)**
s-Triazine, 2,4-diamino-6-(tribromomethyl)- (8CI)	**(16088-72-5)**
s-Triazine, 2,4-diamino-6-(1,1,2-trichloroethyl)- (8CI)	**(30359-91-2)**
s-Triazine, 2,4-diamino-6-(trichloromethyl)- (8CI)	**(16088-73-6)**
s-Triazine, 2,4-diamino-6-(trichloromethyl)-, 3-oxide (7CI,8CI)	**(30805-11-9)**
s-Triazine, 2,4-diamino-6-(1,1,2-trichloropropyl)- (8CI)	**(30360-01-1)**
s-Triazine, 2,4-diamino-6-(trifluoromethyl)- (6CI,7CI,8CI)	**(705-25-9)**
s-Triazine, 2,4-dianilino- (7CI,8CI)	**(13107-54-5)**
s-Triazine, 2,4-dianilino-6-azido- (8CI)	**(30360-10-2)**
s-Triazine, 2,4-dianilino-6-chloro- (8CI)	**(1973-09-7)**
s-Triazine, 2,4-dianilino-6-fluoro- (6CI,8CI)	**(1650-76-6)**
s-Triazine, 2,4-dianilino-6-methoxy- (6CI,8CI)	**(30360-58-8)**
s-Triazine, 2,4-dianilino-6-methyl- (7CI,8CI)	**(2039-34-1)**
s-Triazine, 2,4-dianilino-6-phenoxy- (8CI)	**(30360-76-0)**
s-Triazine, 2,4-dianilino-6-(phenylthio)- (8CI)	**(30360-87-3)**
s-Triazine, 2,4-dianilino-6-(tribromomethyl)- (8CI)	**(30359-73-0)**
s-Triazine, 2,4-dianilino-6-(trichloromethyl)- (8CI)	**(30355-91-0)**
s-Triazine, 2,4-di-p-anisidino-6-chloro- (6CI,7CI,8CI)	**(6737-62-8)**
s-Triazine, 2-(diazomethyl)-4,6-bis(ethylamino)- (8CI)	**(30355-57-8)**
s-Triazine, 2-(diazomethyl)-4,6-bis(methylthio)- (6CI,7CI,8CI)	**(30863-65-1)**
1,3,5-Triazine, 2-(diazomethyl)-4,6-dimethoxy- (9CI)	**(4803-06-9)**
s-Triazine, 2-(diazomethyl)-4,6-dimethoxy- (6CI,7CI,8CI)	**(4803-06-9)**
s-Triazine, 2,4-dibenzyl- (6CI,7CI,8CI)	**(30361-92-3)**
s-Triazine, 2,4-dibromo-6-(o-chloroanilino)- (8CI)	**(30388-94-4)**
s-Triazine, 2,4-dibromo-6-(p-chloroanilino)- (8CI)	**(30357-83-6)**
s-Triazine, 2,4-dibromo-6-(ethylamino)- (8CI)	**(30357-80-3)**
s-Triazine, 2,4-dibromo-6-(isobutylamino)- (7CI,8CI)	**(4658-20-2)**
s-Triazine, 2-(dibromomethyl)-4,6-bis(ethylamino)- (7CI,8CI)	**(30359-68-3)**
s-Triazine, 2-(dibromomethyl)-4,6-dimethoxy- (6CI,7CI,8CI)	**(30863-44-6)**
s-Triazine, 2,4-dibromo-6-phenoxy- (8CI)	**(30894-96-3)**
s-Triazine, 2,4-dibromo-6-(phenylthio)- (8CI)	**(30894-97-4)**
s-Triazine, 2,4-dibromo-6-(1,1,2-trichloroethyl)- (8CI)	**(30894-98-5)**
s-Triazine, 2,4-dibutoxy-6-(trichloromethyl)- (6CI,7CI,8CI)	**(30863-42-4)**
s-Triazine, 2-(dibutylamino)-4,6-bis(trichloromethyl)- (8CI)	**(30356-34-4)**
s-Triazine-2,4-dicarboxylic acid, 6-amino-, diethyl ester (6CI,8CI)	**(30362-18-6)**
s-Triazine-2,4-dicarboxylic acid, 6-(trichloromethyl)-, diethyl ester (8CI)	**(30863-21-9)**
1,3,5-Triazine, 2,4-dichloro- (9CI)	**(2831-66-5)**
s-Triazine, 2,4-dichloro- (8CI)	**(2831-66-5)**
s-Triazine, 2-(2,5-dichloroanilino)-4,6-bis(trichloromethyl)- (8CI)	**(30356-54-8)**
s-Triazine, 2-(3,4-dichloroanilino)-4,6-bis(trichloromethyl)- (8CI)	**(30356-55-9)**
s-Triazine, 2-(3,4-dichloroanilino)-4,6-dimethoxy- (8CI)	**(13960-26-4)**
s-Triazine, 2,4-dichloro-6-(o-chloroanilino)	**(101-05-3)**
s-Triazine, 2,4-dichloro-6-(m-chloroanilino)- (8CI)	**(20376-34-5)**
s-Triazine, 2,4-dichloro-6-(p-chloroanilino)- (8CI)	**(2272-29-9)**
1,3,5-Triazine, 2,4-dichloro-6-(chloromethyl)- (9CI)	**(30894-64-5)**
s-Triazine, 2,4-dichloro-6-(chloromethyl)- (6CI,8CI)	**(30894-64-5)**
1,3,5-Triazine, 2,4-dichloro-6-(2-chlorophenoxy)- (9CI)	**(30886-25-0)**
1,3,5-Triazine, 2,4-dichloro-6-(4-chlorophenoxy)- (9CI)	**(30886-26-1)**
s-Triazine, 2,4-dichloro-6-(o-chlorophenoxy)- (8CI)	**(30886-25-0)**
s-Triazine, 2,4-dichloro-6-(p-chlorophenoxy)- (8CI)	**(30886-26-1)**
1,3,5-Triazine, 2,4-dichloro-6-(4-chlorophenyl)- (9CI)	**(10202-46-7)**
s-Triazine, 2,4-dichloro-6-(p-chlorophenyl)- (8CI)	**(10202-46-7)**
1,3,5-Triazine, 2,4-dichloro-6-((4-chlorophenyl)thio)- (9CI)	**(30894-63-4)**
s-Triazine, 2,4-dichloro-6-((p-chlorophenyl)thio)- (8CI)	**(30894-63-4)**
s-Triazine, 2,4-dichloro-6-(diazomethyl)- (6CI,7CI,8CI)	**(4803-05-8)**
s-Triazine, 2,4-dichloro-6-(dibromomethyl)- (8CI)	**(30894-68-9)**
s-Triazine, 2,4-dichloro-6-(3,4-dichloroanilino)- (8CI)	**(18996-86-6)**
s-Triazine, 2,4-dichloro-6-(2,5-dichloroanilino)- (6CI,7CI,8CI)	**(2272-33-5)**
s-Triazine, 2,4-dichloro-6-(1,1-dichloroethyl)- (8CI)	**(30894-71-4)**
s-Triazine, 2,4-dichloro-6-(dichloromethyl)- (6CI,7CI,8CI)	**(30894-65-6)**
1,3,5-Triazine, 2,4-dichloro-6-(2,4-dichlorophenoxy)- (9CI)	**(18247-77-3)**
s-Triazine, 2,4-dichloro-6-(2,4-dichlorophenoxy)- (8CI)	**(18247-77-3)**
s-Triazine, 2,4-dichloro-6-(diiodomethyl)- (6CI,7CI,8CI)	**(30894-70-3)**
s-Triazine, 2,4-dichloro-6-dimethylamino	**(2401-64-1)**
s-Triazine, 2,4-dichloro-6-(diphenylamino)- (6CI,7CI,8CI)	**(16033-74-2)**
s-Triazine, 2,4-dichloro-6-(dodecylamino)- (6CI,7CI,8CI)	**(26113-25-7)**
s-Triazine, 2,4-dichloro-6-(dodecyloxy)- (6CI,7CI,8CI)	**(4628-08-4)**
1,3,5-Triazine, 2,4-dichloro-6-ethoxy- (9CI)	**(18343-30-1)**
s-Triazine, 2,4-dichloro-6-ethoxy- (8CI)	**(18343-30-1)**
1,3,5-Triazine, 2,4-dichloro-6-ethyl- (9CI)	**(698-72-6)**

s-Triazine, 2,4-dichloro-6-ethyl- (7CI,8CI) **(698-72-6)**
s-Triazine, 2-(1,1-dichloroethyl)- (7CI,8CI) **(30361-86-5)**
s-Triazine, 2,4-dichloro-6-(ethylamino)- (8CI) **(3440-19-5)**
1,3,5-Triazine, 2,4-dichloro-6-(ethylthio)- (9CI) **(13733-90-9)**
s-Triazine, 2,4-dichloro-6-(ethylthio)- (6CI,8CI) **(13733-90-9)**
s-Triazine, 2,4-dichloro-6-(iodomethyl)- (8CI) **(30894-69-0)**
s-Triazine, 2,4-dichloro-6-isobutoxy- (6CI,8CI) **(30886-23-8)**
s-Triazine, 2,4-dichloro-6-isopropoxy- (8CI) **(6684-27-1)**
s-Triazine, 2,4-dichloro-6-isopropyl- (7CI,8CI) **(30894-74-7)**
s-Triazine, 2,4-dichloro-6-(isopropylamino)- (8CI) **(3703-10-4)**
s-Triazine, 2,4-dichloro-6-(isopropylthio)- (6CI,8CI) **(30894-60-1)**
1,3,5-Triazine, 2,4-dichloro-6-isothiocyanato- (9CI) **(4267-15-6)**
1,3,5-Triazine, 2,4-dichloro-6-methoxy- (9CI) **(3638-04-8)**
s-Triazine, 2,4-dichloro-6-methoxy- (8CI) **(3638-04-8)**
1,3,5-Triazine, 2,4-dichloro-6-methyl- (9CI) **(1973-04-2)**
s-Triazine, 2,4-dichloro-6-methyl- (8CI) **(1973-04-2)**
s-Triazine, 2,4-dichloro-6-(N-methylanilino)- (6CI,8CI) **(3995-42-4)**
s-Triazine, 2-(dichloromethyl)-4,6-bis(ethylamino)- (7CI,8CI) **(30355-66-9)**
1,3,5-Triazine, 2-(dichloromethyl)-4,6-bis(trichloromethyl)- (9CI) **(30362-31-3)**
s-Triazine, 2-(dichloromethyl)-4,6-bis(trichloromethyl)- (8CI) **(30362-31-3)**
s-Triazine, 2-(dichloromethyl)-4,6-dimethoxy- (6CI,7CI,8CI) **(5311-24-0)**
1,3,5-Triazine, 2,4-dichloro-6-(1-methylethoxy)- (9CI) **(6684-27-1)**
s-Triazine, 2-(dichloromethyl)-4-methyl-6-(trichloromethyl)- (8CI) **(15640-10-5)**
1,3,5-Triazine, 2,4-dichloro-6-(4-methylphenoxy)- (9CI) **(13838-34-1)**
1,3,5-Triazine, 2,4-dichloro-6-((4-methylphenyl)thio)- (9CI) **(13733-93-2)**
1,3,5-Triazine, 2,4-dichloro-6-(2-methylpropoxy)- (9CI) **(30886-23-8)**
s-Triazine, 2,4-dichloro-6-(methylthio) **(13705-05-0)**
1,3,5-Triazine, 2,4-dichloro-6-(1-naphthalenyloxy)- (9CI) **(30886-30-7)**
s-Triazine, 2,4-dichloro-6-(1-naphthylamino) **(30369-88-1)**
s-Triazine, 2,4-dichloro-6-(1-naphthyloxy)- (8CI) **(30886-30-7)**
s-Triazine, 2,4-dichloro-6-(p-nitroanilino)- (8CI) **(2352-36-5)**
1,3,5-Triazine, 2,4-dichloro-6-(4-nitrophenoxy)- (9CI) **(30886-29-4)**
s-Triazine, 2,4-dichloro-6-(p-nitrophenoxy)- (8CI) **(30886-29-4)**
1,3,5-Triazine, 2,4-dichloro-6-(pentachlorophenoxy)- (9CI) **(5599-24-6)**
s-Triazine, 2,4-dichloro-6-(pentachlorophenoxy)- (8CI) **(5599-24-6)**
1,3,5-Triazine, 2,4-dichloro-6-phenoxy- (9CI) **(4682-78-4)**
s-Triazine, 2,4-dichloro-6-phenoxy- (8CI) **(4682-78-4)**
1,3,5-Triazine, 2,4-dichloro-6-phenyl- (9CI) **(1700-02-3)**
s-Triazine, 2,4-dichloro-6-phenyl- (8CI) **(1700-02-3)**
1,3,5-Triazine, 2-(2,4-dichlorophenyl)-4,6-bis(trichloromethyl)- (9CI) **(24481-52-5)**
s-Triazine, 2-(2,4-dichlorophenyl)-4,6-bis(trichloromethyl)- (8CI) **(24481-52-5)**
s-Triazine, 2-(3,4-dichlorophenyl)-4,6-bis(trichloromethyl)- (8CI) **(24481-51-4)**
1,3,5-Triazine, 2,4-dichloro-6-(phenylmethoxy)- (9CI) **(30886-24-9)**
1,3,5-Triazine, 2,4-dichloro-6-(phenylthio)- (9CI) **(3019-16-7)**
s-Triazine, 2,4-dichloro-6-(phenylthio)- (8CI) **(3019-16-7)**
1,3,5-Triazine, 2-((2,4-dichlorophenyl)thio)-4,6-bis(trichloromethyl)- (9CI) **(24478-09-9)**
1,3,5-Triazine, 2-((3,4-dichlorophenyl)thio)-4,6-bis(trichloromethyl)- (9CI) **(24478-10-2)**
s-Triazine, 2-((2,4-dichlorophenyl)thio)-4,6-bis(trichloromethyl)- (8CI) **(24478-09-9)**
s-Triazine, 2-((3,4-dichlorophenyl)thio)-4,6-bis(trichloromethyl)- (8CI) **(24478-10-2)**
s-Triazine, 2-((2,4-dichlorophenyl)thio)-4-(methylamino)-6-(trichloromethyl)- (8CI) **(30369-61-0)**
1,3,5-Triazine, 2,4-dichloro-6-(2-propenyloxy)- (9CI) **(26650-76-0)**
1,3,5-Triazine, 2,4-dichloro-6-propoxy- (9CI) **(26650-75-9)**
s-Triazine, 2,4-dichloro-6-propoxy- (8CI) **(26650-75-9)**
1,3,5-Triazine, 2,4-dichloro-6-propyl- (9CI) **(30894-73-6)**
s-Triazine, 2,4-dichloro-6-propyl- (8CI) **(30894-73-6)**
s-Triazine, 2,4-dichloro-6-(propylamino)- (8CI) **(25354-39-6)**
1,3,5-Triazine, 2,4-dichloro-6-(propylthio)- (9CI) **(30894-58-7)**
s-Triazine, 2,4-dichloro-6-(propylthio)- (6CI,8CI) **(30894-58-7)**
s-Triazine, 2,4-dichloro-6-o-toluidino- (8CI) **(2272-23-3)**
s-Triazine, 2,4-dichloro-6-p-toluidino- (8CI) **(2272-24-4)**
s-Triazine, 2,4-dichloro-6-(p-tolyloxy)- (8CI) **(13838-34-1)**
s-Triazine, 2,4-dichloro-6-(p-tolylthio)- (8CI) **(13733-93-2)**
s-Triazine, 2,4-dichloro-6-(2,4,5-trichloroanilino)- (7CI,8CI) **(2272-39-1)**
s-Triazine, 2,4-dichloro-6-(1,1,2-trichloroethyl)- (7CI,8CI) **(30894-72-5)**
1,3,5-Triazine, 2,4-dichloro-6-(trichloromethyl)- (9CI) **(2113-00-0)**
s-Triazine, 2,4-dichloro-6-(trichloromethyl)- (6CI,7CI,8CI) **(2113-00-0)**
1,3,5-Triazine, 2,4-dichloro-6-(2,4,5-trichlorophenoxy)- (9CI) **(30886-27-2)**
s-Triazine, 2,4-dichloro-6-(2,4,5-trichlorophenoxy)- (6CI,7CI,8CI) **(30886-27-2)**
s-Triazine, 2,4-diethoxy-6-(ethylamino)- (6CI,8CI) **(30358-07-7)**
s-Triazine, 2,4-diethoxy-6-(trichloromethyl)- (6CI,7CI,8CI) **(30863-40-2)**
s-Triazine, 2,4-diethyl- (6CI,7CI,8CI) **(3599-60-8)**
s-Triazine, 2-(diethylamino)-4,6-bis(trichloromethyl)- (8CI) **(24803-06-3)**
s-Triazine, 2-(diethylamino)-4-(dimethylamino)-6-methyl- (8CI) **(30368-85-5)**
s-Triazine, 2-(diethylamino)-4-(ethylamino)-6-fluoro- (7CI,8CI) **(3797-36-2)**
s-Triazine, 2-(diethylamino)-4-(ethylamino)-6-methoxy- (8CI) **(13532-26-8)**
s-Triazine, 2-(diethylamino)-4-(ethylamino)-6-methyl- (8CI) **(27431-10-3)**
s-Triazine, 2-(diethylamino)-4-(ethylamino)-6-(trichloromethyl)- (8CI) **(27470-97-9)**
s-Triazine, 2-(diethylamino)-4-(isopropylamino)-6-methoxy **(3004-70-4)**
s-Triazine, 2-(diethylamino)-4-(methylamino)-6-(trichloromethyl)- (7CI,8CI) **(27470-96-8)**
s-Triazine, 2-(diethylamino)-4-methyl-6-(propylamino)- (8CI) **(27470-67-3)**
s-Triazine, 2-(diethylamino)-4-(methylthio)-6-(trichloromethyl)- (8CI) **(30369-48-3)**
s-Triazine, 2-(diethylamino)-4-methyl-6-(trichloromethyl)- (8CI) **(24803-58-5)**
1,3,5-Triazine, 2,4-diethyl-6-methyl- (9CI) **(30362-60-8)**
s-Triazine, 2,4-diethyl-6-methyl- (7CI,8CI) **(30362-60-8)**
s-Triazine, 2,4-diethyl-6-(methylamino)- (8CI) **(30339-56-1)**
s-Triazine, 2,4-diethyl-6-(methylthio)- (7CI,8CI) **(3846-74-0)**
1,3,5-Triazine, 2,4-diethyl-6-phenyl- (9CI) **(30362-69-7)**
s-Triazine, 2,4-diethyl-6-phenyl- (7CI,8CI) **(30362-69-7)**
s-Triazine, 2,4-diethyl-6-(trichloromethyl)- (7CI,8CI) **(30362-62-0)**
s-Triazine, 2,4-diethyl-6-vinyl- (7CI,8CI) **(30362-68-6)**
s-Triazine, 2,4-difluoro-6-(p-nitrophenoxy)- (6CI,8CI) **(2343-60-4)**
1,3,5-Triazine, 2,4-difluoro-6-phenoxy- (9CI) **(30886-18-1)**
s-Triazine, 2,4-difluoro-6-phenoxy- (8CI) **(30886-18-1)**
1,3,5-Triazine, 2,4-difluoro-6-(phenylthio)- (9CI) **(717-91-9)**
s-Triazine, 2,4-difluoro-6-(phenylthio)- (6CI,8CI) **(717-91-9)**
1,3,5-Triazine, 2,4-difluoro-6-(trifluoromethyl)- (9CI) **(368-55-8)**
s-Triazine, 2,4-difluoro-6-(trifluoromethyl)- (6CI,8CI) **(368-55-8)**
s-Triazine, 1,2-dihydro-1-(p-chlorophenyl)-4,6-diamino-2,2-dimethyl **(516-21-2)**
1,3,5-Triazine, 1,2-dihydro-2,2,4,6-tetraphenyl- (9CI) **(30805-23-3)**
s-Triazine, 1,2-dihydro-2,2,4,6-tetraphenyl- (6CI,8CI) **(30805-23-3)**
1,2,4-Triazine, 3-di(hydroxymethyl)amino-6-(5-nitro-2-furyl-ethenyl)- **(794-93-4)**
s-Triazine, 2-(diiodomethyl)-4,6-bis(ethylamino)- (8CI) **(30359-77-4)**
s-Triazine, 2-(diisobutylamino)-4,6-bis(trichloromethyl)- (8CI) **(30356-35-5)**
s-Triazine, 2,4-diisopropoxy-6-(trichloromethyl)- (8CI) **(30863-41-3)**
1,3,5-Triazine, 2,4-dimethoxy-6-methyl- (9CI) **(4000-78-6)**
s-Triazine, 2,4-dimethoxy-6-methyl- (6CI,7CI,8CI) **(4000-78-6)**
s-Triazine, 2,4-dimethoxy-6-(methylamino)- (8CI) **(30357-98-3)**
s-Triazine, 2,4-dimethoxy-6-(N-methylanilino)- (8CI) **(30358-06-6)**
s-Triazine, 2,4-dimethoxy-6-(p-nitroanilino)- (8CI) **(30358-05-5)**
1,3,5-Triazine, 2,4-dimethoxy-6-phenyl- (9CI) **(18213-73-5)**
s-Triazine, 2,4-dimethoxy-6-phenyl- (7CI,8CI) **(18213-73-5)**
s-Triazine, 2,4-dimethoxy-6-(2,4,5-trichloroanilino)- (8CI) **(13960-29-7)**
s-Triazine, 2,4-dimethoxy-6-(1,1,2-trichloroethyl)- (8CI) **(31120-23-7)**
s-Triazine, 2,4-dimethoxy-6-(trichloromethyl)- (6CI,7CI,8CI) **(5311-25-1)**
1,3,5-Triazine, 2,4-dimethyl- (9CI) **(1722-15-2)**
s-Triazine, 2,4-dimethyl- (6CI,7CI,8CI) **(1722-15-2)**
s-Triazine, 2-(dimethylamino)- (6CI,7CI,8CI) **(4040-00-0)**
s-Triazine, 2-(dimethylamino)-4,6-bis(trichloromethyl)- (6CI,8CI) **(24803-05-2)**
s-Triazine, 2-(dimethylamino)-4,6-diphenyl- (8CI) **(18808-10-1)**
s-Triazine, 2-((2-(dimethylamino)ethyl)amino)-4,6-bis(trichloromethyl)- (6CI,8CI) **(24803-14-3)**
s-Triazine, 2-(dimethylamino)-4-(ethylamino)-6-methyl- (8CI) **(27470-66-2)**

s-Triazine, 2-(dimethylamino)-4-(ethylamino)-6-(trichloromethyl)- (7CI,8CI) **(27431-17-0)**
s-Triazine, 2-(dimethylamino)-4-hydrazino-6-(trichloromethyl)- (8CI) **(30355-70-5)**
s-Triazine, 2-(dimethylamino)-4-(methylamino)-6-(trichloromethyl)- (7CI,8CI) **(27470-95-7)**
s-Triazine, 2-(dimethylamino)-4-methyl-6-(methylamino)- (8CI) **(27431-07-8)**
s-Triazine, 2-(dimethylamino)-4-methyl-6-(propylamino)- (8CI) **(27431-08-9)**
s-Triazine, 2-(dimethylamino)-4-methyl-6-(trichloromethyl)- (8CI) **(24803-57-4)**
s-Triazine, 2-(dimethylamino)-4-phenyl-6-(trichloromethyl)- (8CI) **(24848-38-2)**
1,3,5-Triazine, 2,4-dimethyl-6-(methylthio)- (9CI) **(3745-18-4)**
s-Triazine, 2,4-dimethyl-6-(methylthio)- (7CI,8CI) **(3745-18-4)**
s-Triazine, 2,4-dimethyl-6-(m-nitrophenyl)- (8CI) **(30362-09-5)**
s-Triazine, 2,4-dimethyl-6-phenethyl- (7CI,8CI) **(30362-07-3)**
1,3,5-Triazine, 2,4-dimethyl-6-phenyl- (9CI) **(3599-61-9)**
s-Triazine, 2,4-dimethyl-6-phenyl- (7CI,8CI) **(3599-61-9)**
s-Triazine, 2,4-dimethyl-6-propyl- (7CI,8CI) **(7559-34-4)**
s-Triazine, 2-((1,2-dimethylpropyl)amino)-4-ethylamino-6-methylthio **(22936-75-0)**
1,3,5-Triazine, 2,4-dimethyl-6-(trichloromethyl)- (9CI) **(15640-05-8)**
s-Triazine, 2,4-dimethyl-6-(trichloromethyl)- (7CI,8CI) **(15640-05-8)**
Triazine, 2,4-dimorpholino-6-methyl **(26234-42-4)**
s-Triazine-2,4-diol, 6-anilino- (6CI,7CI,8CI) **(20964-55-0)**
s-Triazine-2,4-diol, 6-(p-chlorophenyl)- (8CI) **(30885-98-4)**
s-Triazine-2,4-diol, 6-ethyl- (8CI) **(7501-27-1)**
s-Triazine-2,4-diol, 6-(ethylamino)- (8CI) **(2630-10-6)**
s-Triazine-2,4-diol, 6-isopropyl- (8CI) **(30885-97-3)**
s-Triazine-2,4-diol, 6-methoxy- (8CI) **(30885-95-1)**
s-Triazine-2,4-diol, 6-methyl- (6CI,7CI) **(933-19-7)**
s-Triazine-2,4-diol, 6-phenyl- **(7459-63-4)**
s-Triazine-2,4-diol, 6-styryl- (7CI) **(1917-34-6)**
1,3,5-Triazine-2,4(1H,3H)-dione (9CI) **(71-33-0)**
s-Triazine-2,4(1H,3H)-dione (8CI) **(71-33-0)**
1,3,5-Triazine-2,4(1H,3H)-dione, 6-amino- **(645-93-2)**
1,2,4-Triazine-3,5(2H,4H)-dione, 4-amino-6-(1,1-dimethylethyl)- (9CI) **(56507-37-0)**
s-Triazine-2,4(1H,3H)-dione, 3-cyclohexyl-6-(dimethylamino)-1-methyl **(51235-04-2)**
s-Triazine-2,4(1H,3H)-dione, dihydro-6-isopropyl- (8CI) **(25113-45-5)**
1,3,5-Triazine-2,4(1H,3H)-dione, dihydro-6-phenyl- (9CI) **(30805-30-2)**
s-Triazine-2,4(1H,3H)-dione, dihydro-6-phenyl- (6CI,8CI) **(30805-30-2)**
s-Triazine-2,4(1H,3H)-dione, 6-ethyl- (7CI) **(7501-27-1)**
1,3,5-Triazine-2,4(1H,3H)-dione, 6-(ethylamino)- (9CI) **(2630-10-6)**
1,3,5-Triazine-2,4(1H,3H)-dione, 6-methoxy- (9CI) **(30885-95-1)**
1,3,5-Triazine-2,4(1H,3H)-dione, 6-methyl- (9CI) **(933-19-7)**
s-Triazine-2,4(1H,3H)-dione, 6-methyl- (8CI) **(933-19-7)**
1,3,5-Triazine-2,4(1H,3H)-dione, 6-phenyl- (9CI) **(7459-63-4)**
s-Triazine-2,4(1H,3H)-dione, 6-phenyl- (8CI) **(7459-63-4)**
1,3,5-Triazine-2,4(1H,3H)-dione, 6-(phenylamino)- (9CI) **(20964-55-0)**
as-Triazine-3,5(2H,4H)-dione, 2-β-D-ribofuranosyl **(54-25-1)**
s-Triazine-2,4(1H,3H)-dione, 6-styryl- (8CI) **(1917-34-6)**
1,3,5-Triazine, 2,4-diphenyl- (9CI) **(1898-74-4)**
s-Triazine, 2,4-diphenyl- (6CI,7CI,8CI) **(1898-74-4)**
s-Triazine, 2,4-diphenyl-6-(phenylthio)- (8CI) **(30863-04-8)**
1,3,5-Triazine, 2,4-diphenyl-6-propyl- (9CI) **(30362-95-9)**
s-Triazine, 2,4-diphenyl-6-propyl- (8CI) **(30362-95-9)**
s-Triazine, 2,4-diphenyl-6-p-tolyl- (6CI,8CI) **(16107-88-3)**
1,3,5-Triazine, 2,4-diphenyl-6-(trichloromethyl)- (9CI) **(30362-63-1)**
s-Triazine, 2,4-diphenyl-6-(trichloromethyl)- (7CI,8CI) **(30362-63-1)**
Triazine, 2,4-dipiperidino-6-methyl **(26234-41-3)**
s-Triazine, 2,4-dipiperidino-6-(trichloromethyl)- (8CI) **(26235-04-1)**
s-Triazine, 2,4-dipropoxy-6-(trichloromethyl)- (8CI) **(30863-83-3)**
s-Triazine, 2-(dipropylamino)-4,6-bis(trichloromethyl)- (8CI) **(24803-07-4)**
s-Triazine-2,4-dithiol, 6-amino- (8CI) **(2770-75-4)**
s-Triazine-2,4-dithiol, 6-anilino- (8CI) **(13733-91-0)**
s-Triazine-2,4-dithiol, 6-(o-chloroanilino)- (8CI) **(30362-17-5)**
s-Triazine-2,4-dithiol, 6-chloro-, S,S'-diester with O,O-diethyl phosphorodithioate **(18895-89-1)**
s-Triazine-2,4-dithiol, 5,6-dihydro-6-phenyl- (6CI) **(30805-32-4)**
s-Triazine-2,4-dithiol, 6-methyl- (8CI) **(6237-86-1)**
s-Triazine-2,4-dithiol, 6-phenyl- (6CI,8CI) **(30886-13-6)**
1,3,5-Triazine-2,4(1H,3H)-dithione, 6-amino- (9CI) **(2770-75-4)**
1,3,5-Triazine-2,4(1H,3H)-dithione, dihydro-6-phenyl- (9CI) **(30805-32-4)**
s-Triazine-2,4(1H,3H)-dithione, dihydro-6-phenyl- (8CI) **(30805-32-4)**
s-Triazine-2,4(1H,3H)-dithione, 6-methyl- **(6237-86-1)**
1,3,5-Triazine-2,4(1H,3H)-dithione, 6-methyl- (9CI) **(6237-86-1)**
1,3,5-Triazine-2,4(1H,3H)-dithione, 6-phenyl- (9CI) **(30886-13-6)**
s-Triazine-2,4(1H,3H)-dithione, 6-phenyl- (7CI) **(30886-13-6)**
1,3,5-Triazine-2,4(1H,3H)-dithione, 6-(phenylamino)- (9CI) **(13733-91-0)**
s-Triazine, 2,4-di-p-tolyl-6-(trichloromethyl)- (7CI,8CI) **(30362-66-4)**
s-Triazine, 2-(dodecylamino)-4,6-bis(trichloromethyl)- (8CI) **(24803-04-1)**
s-Triazine, 2-(dodecylamino)-4-(ethylamino)-6-methyl- (8CI) **(27431-00-1)**
s-Triazine, 2-(dodecylamino)-4-(isopropylamino)-6-methyl- (8CI) **(27431-05-6)**
s-Triazine, 2-(dodecylamino)-4-(methylthio)-6-(trichloromethyl)- (8CI) **(30369-47-2)**
s-Triazine, 2-(dodecylamino)-4-methyl-6-(trichloromethyl)- (8CI) **(24803-56-3)**
s-Triazine, 2-(dodecyloxy)-4,6-bis(trichloromethyl)- (8CI) **(30863-59-3)**
1,3,5-Triazine, 2-(dodecylthio)-4,6-bis(trichloromethyl)- (9CI) **(24481-72-9)**
s-Triazine, 2-(dodecylthio)-4,6-bis(trichloromethyl)- (8CI) **(24481-72-9)**
s-Triazine, 2-ethoxy-4,6-bis(ethylamino)- (8CI) **(30360-59-9)**
s-Triazine, 2-ethoxy-4,6-bis(isopropylamino)- (7CI,8CI) **(30377-19-6)**
s-Triazine, 2-ethoxy-4,6-bis(trichloromethyl)- (6CI,7CI,8CI) **(30863-52-6)**
s-Triazine, 2-(2-ethoxyethoxy)-4,6-bis(ethylamino)- (7CI,8CI) **(30360-73-7)**
s-Triazine, 2-ethoxy-4-(ethylamino)-6-methyl- (7CI,8CI) **(5248-41-9)**
s-Triazine, 2-ethoxy-4-ethyl-6-(ethylamino)- (7CI,8CI) **(6864-23-9)**
s-Triazine, 2-ethoxy-4-methyl-6-(trichloromethyl)- (6CI,8CI) **(30863-46-8)**
1,3,5-Triazine, 2-((2-ethoxyphenyl)thio)-4,6-bis(trichloromethyl)- (9CI) **(24478-04-4)**
1,3,5-Triazine, 2-((4-ethoxyphenyl)thio)-4,6-bis(trichloromethyl)- (9CI) **(24478-05-5)**
s-Triazine, 2-((o-ethoxyphenyl)thio)-4,6-bis(trichloromethyl)- (8CI) **(24478-04-4)**
s-Triazine, 2-((p-ethoxyphenyl)thio)-4,6-bis(trichloromethyl)- (8CI) **(24478-05-5)**
s-Triazine, 2-ethoxy-4-phenyl-6-(trichloromethyl)- (6CI,8CI) **(30863-61-7)**
1,3,5-Triazine, 2-ethyl- (9CI) **(3599-59-5)**
s-Triazine, 2-ethyl- (6CI,7CI,8CI) **(3599-59-5)**
s-Triazine, 2-(ethylamino)- (8CI) **(4122-05-8)**
s-Triazine, 2-(ethylamino)-4,6-bis(methylthio)- (8CI) **(30362-12-0)**
s-Triazine, 2-(ethylamino)-4,6-bis(tribromomethyl)- (8CI) **(30339-38-9)**
s-Triazine, 2-(ethylamino)-4,6-bis(trichloromethyl)- (8CI) **(24803-29-0)**
s-Triazine, 2-(ethylamino)-4,6-difluoro- (8CI) **(30369-76-7)**
s-Triazine, 2-(ethylamino)-4,6-dimethyl- (8CI) **(30377-23-2)**
s-Triazine, 2-(ethylamino)-4,6-diphenoxy- (8CI) **(30384-48-6)**
s-Triazine, 2-(ethylamino)-4-((1-ethylhexyl)amino)-6-methyl- (8CI) **(27430-99-5)**
s-Triazine, 2-(ethylamino)-4-fluoro-6-(isopropylamino)- (7CI,8CI) **(777-56-0)**
s-Triazine, 2-(ethylamino)-4-fluoro-6-((3-methoxypropyl)amino)- (8CI) **(30388-86-4)**
s-Triazine, 2-(ethylamino)-4-heptadecyl-6-(trichloromethyl)- (8CI) **(24802-93-5)**
s-Triazine, 2-(ethylamino)-4-(hexylamino)-6-methyl- (8CI) **(27430-98-4)**
s-Triazine, 2-(ethylamino)-4-(isobutylamino)-6-methyl- (8CI) **(27470-63-9)**
s-Triazine, 2-(ethylamino)-4-(isobutylamino)-6-(trichloromethyl)- (8CI) **(27529-92-6)**
s-Triazine, 2-(ethylamino)-4-isobutyl-6-(trichloromethyl)- (8CI) **(24802-90-2)**
s-Triazine, 2-ethylamino-4-isopropylamino-6-methoxy **(1610-17-9)**
s-Triazine, 4-ethylamino-6-isopropylamino-2-methoxy- **(1610-17-9)**
s-Triazine, 2-(ethylamino)-4-(isopropylamino)-6-methyl- (8CI) **(27430-93-9)**
s-Triazine, 2-ethylamino-4-isopropylamino-6-methylthio **(834-12-8)**
s-Triazine, 2-(ethylamino)-4-(isopropylamino)-6-phenoxy- (8CI) **(6494-99-1)**
s-Triazine, 2-(ethylamino)-4-(isopropylamino)-6-(trichloromethyl)- (7CI,8CI) **(27470-69-5)**

s-Triazine, 2-(ethylamino)-4-isopropyl-6-(trichloromethyl)- (8CI) **(24803-27-8)**
s-Triazine, 2-(ethylamino)-4-methoxy-6-methyl- (8CI) **(16399-10-3)**
s-Triazine, 2-(ethylamino)-4-methoxy-6-phenyl- (8CI) **(30369-39-2)**
s-Triazine, 2-(ethylamino)-4-methyl- (8CI) **(30360-32-8)**
s-Triazine, 2-(ethylamino)-4-(methylamino)-6-(trichloromethyl)- (7CI,8CI) **(27431-14-7)**
s-Triazine, 2-(ethylamino)-4-methyl-6-(methylamino)- (8CI) **(27430-88-2)**
s-Triazine, 2-(ethylamino)-4-methyl-6-(pentylamino)- (8CI) **(27430-97-3)**
s-Triazine, 2-(ethylamino)-4-methyl-6-phenoxy- (8CI) **(30369-40-5)**
s-Triazine, 2-(ethylamino)-4-methyl-6-(propylamino)- (8CI) **(27430-92-8)**
s-Triazine, 2-(ethylamino)-4-(methylthio)-6-(trichloromethyl)- (8CI) **(30369-41-6)**
s-Triazine, 2-(ethylamino)-4-methyl-6-(tribromomethyl)- (8CI) **(30339-35-6)**
s-Triazine, 2-(ethylamino)-4-methyl-6-(trichloromethyl)- (8CI) **(24803-51-8)**
s-Triazine, 2-(ethylamino)-4-nonyl-6-(trichloromethyl)- (8CI) **(24802-92-4)**
s-Triazine, 2-(ethylamino)-4-phenyl-6-(trichloromethyl)- (8CI) **(24803-00-7)**
s-Triazine, 2-(ethylamino)-4-(propylamino)-6-(trichloromethyl)- (8CI) **(27431-16-9)**
s-Triazine, 2-(ethylamino)-4-propyl-6-(trichloromethyl)- (8CI) **(24803-25-6)**
s-Triazine, 2-(ethylamino)-4-(1,1,2-trichloroethyl)-6-(trichloromethyl)- (8CI) **(24802-97-9)**
s-Triazine, 2-(ethylamino)-4-(trichloromethyl)-6-vinyl- (8CI) **(24802-98-0)**
s-Triazine, 2-ethyl-4,6-bis(ethylamino)- (7CI,8CI) **(26235-13-2)**
s-Triazine, 2-ethyl-4,6-bis(isobutylamino)- (8CI) **(26235-15-4)**
s-Triazine, 2-ethyl-4,6-bis(isopropylamino)- (7CI,8CI) **(26235-14-3)**
s-Triazine, 2-ethyl-4,6-bis(methylamino)- (8CI) **(26235-12-1)**
1,3,5-Triazine, 2-ethyl-4,6-bis(trichloromethyl)- (9CI) **(3599-71-1)**
s-Triazine, 2-ethyl-4,6-bis(trichloromethyl)- (8CI) **(3599-71-1)**
s-Triazine, 2-ethyl-4,6-bis(trifluoromethyl)- (8CI) **(30362-10-8)**
1,3,5-Triazine, 2-ethyl-4,6-dimethoxy- (9CI) **(705-78-2)**
s-Triazine, 2-ethyl-4,6-dimethoxy- (7CI,8CI) **(705-78-2)**
1,3,5-Triazine, 2-ethyl-4,6-dimethyl- (9CI) **(30362-03-9)**
s-Triazine, 2-ethyl-4,6-dimethyl- (7CI,8CI) **(30362-03-9)**
s-Triazine, 2-ethyl-4,6-diphenoxy- (7CI,8CI) **(740-77-2)**
1,3,5-Triazine, 2-ethyl-4,6-diphenyl- (9CI) **(30362-71-1)**
s-Triazine, 2-ethyl-4,6-diphenyl- (6CI,7CI,8CI) **(30362-71-1)**
s-Triazine, 2-ethyl-4-(ethylamino)-6-(isopropylamino)- (8CI) **(27431-18-1)**
s-Triazine, 2-ethyl-4-ethylamino-6-methoxy **(5248-48-6)**
s-Triazine, 2-ethyl-4-(ethylamino)-6-(trichloromethyl)- (8CI) **(24803-19-8)**
s-Triazine, 2-((1-ethylhexyl)amino)-4,6-bis(trichloromethyl)- (8CI) **(24802-88-8)**
s-Triazine, 2-ethyl-4-(isobutylamino)-6-(trichloromethyl)- (8CI) **(24803-23-4)**
s-Triazine, 2-ethyl-4-(isopropylamino)-6-(trichloromethyl)- (8CI) **(24803-21-2)**
s-Triazine, 2-ethyl-4-(methylamino)-6-(trichloromethyl)- (8CI) **(24803-18-7)**
s-Triazine, 2-ethyl-4-(propylamino)-6-(trichloromethyl)- (8CI) **(24803-20-1)**
1,3,5-Triazine, 2-(ethylthio)-4,6-bis(trichloromethyl)- (9CI) **(5516-46-1)**
s-Triazine, 2-(ethylthio)-4,6-bis(trichloromethyl)- (7CI,8CI) **(5516-46-1)**
s-Triazine, 2-(ethylthio)-4-(methylamino)-6-(trichloromethyl)- (8CI) **(30369-50-7)**
1,3,5-Triazine, 2-fluoro-4,6-bis(trifluoromethyl)- (9CI) **(717-62-4)**
s-Triazine, 2-fluoro-4,6-bis(trifluoromethyl)- (6CI,8CI) **(717-62-4)**
s-Triazine, 2-fluoro-4,6-diisopropyl- (8CI) **(30355-00-1)**
s-Triazine, 2-(2-furyl)-4,6-diphenyl- (7CI,8CI) **(7753-01-7)**
s-Triazine, 2-heptadecyl-4,6-bis(isopropylamino)- (8CI) **(26322-48-5)**
1,3,5-Triazine, 2-heptadecyl-4,6-bis(trichloromethyl)- (9CI) **(24481-35-4)**
s-Triazine, 2-heptadecyl-4,6-bis(trichloromethyl)- (8CI) **(24481-35-4)**
s-Triazine, hexahydro-2-imino-4,6-dimethyl- (7CI,8CI) **(30805-34-6)**
s-Triazine, hexahydro-1,3,5-triacryloyl **(959-52-4)**
s-Triazine, hexahydro-1,3,5-triethyl **(7779-27-3)**
s-Triazine, hexahydro-1,3,5-triisopropyl- (8CI) **(10556-98-6)**
1,3,5-Triazine, hexahydro-1,3,5-trimethyl- **(108-74-7)**
s-Triazine, hexahydro-1,3,5-trimethyl **(108-74-7)**
s-Triazine, hexahydro-1,3,5-trinitro **(121-82-4)**
1,3,5-Triazine, hexahydro-1,3,5-trinitro- (9CI) **(121-82-4)**
1,3,5-Triazine, hexahydro-1,3,5-triphenyl- (9CI) **(91-78-1)**
s-Triazine, hexahydro-1,3,5-triphenyl- (8CI) **(91-78-1)**
s-Triazine, hexahydro-1,3,5-tripropionyl- (6CI,8CI) **(30805-19-7)**
s-Triazine, hexahydro-1,3,5-tris(cyanomethyl)- **(4560-87-6)**
s-Triazine, hexahydro-1,3,5-tris(dimethylaminopropyl) **(15875-13-5)**
1,3,5-Triazine, hexahydro-1,3,5-tris(1-methylethyl)- (9CI) **(10556-98-6)**
1,3,5-Triazine, hexahydro-1,3,5-tris(1-oxo-2-propenyl)- (9CI) **(959-52-4)**
1,3,5-Triazine, hexahydro-1,3,5-tris(1-oxopropyl)- (9CI) **(30805-19-7)**
s-Triazine, hexahydro-1,3,5-tris(trichloroacetyl)- (8CI) **(30805-18-6)**
s-Triazine, hexahydro-1,3,5-tris(trichloromethyl)- (8CI) **(30805-22-2)**
s-Triazine, hexahydro-1,3,5-tris(2,2,3-trichloropropionyl)- (8CI) **(30805-20-0)**
s-Triazine, 2-(hexylamino)-4,6-bis(trichloromethyl)- (8CI) **(24802-87-7)**
s-Triazine, 2-(hexylamino)-4-(isopropylamino)-6-methyl- (8CI) **(27431-04-5)**
s-Triazine, 2-(hexylamino)-4-(methylthio)-6-(trichloromethyl)- (8CI) **(30576-30-8)**
s-Triazine, 2-(hexylamino)-4-methyl-6-(trichloromethyl)- (8CI) **(24803-55-2)**
s-Triazine, 2-(hexyloxy)-4,6-bis(trichloromethyl)- (8CI) **(30863-58-2)**
s-Triazine, 2-(hexyloxy)-4-methyl-6-(trichloromethyl)- (8CI) **(30863-50-4)**
1,3,5-Triazine, 2-(hexylthio)-4,6-bis(trichloromethyl)- (9CI) **(24481-71-8)**
s-Triazine, 2-(hexylthio)-4,6-bis(trichloromethyl)- (8CI) **(24481-71-8)**
s-Triazine, 2-(hexylthio)-4-(methylamino)-6-(trichloromethyl)- (8CI) **(30369-54-1)**
s-Triazine, 2-hydrazino-4,6-dimethoxy- (6CI,7CI,8CI) **(13882-61-6)**
s-Triazine, 2-hydrazino-4-methyl-6-(methylamino)- (8CI) **(30368-52-6)**
s-Triazine, 2-(hydroxyamino)-4,6-bis(trichloromethyl)- (8CI) **(30357-61-0)**
s-Triazine, 2-(isobutylamino)-4,6-bis(trichloromethyl)- (8CI) **(24802-84-4)**
s-Triazine, 2-(isobutylamino)-4-(isopropylamino)-6-methyl- (8CI) **(27470-64-0)**
s-Triazine, 2-(isobutylamino)-4-(methylthio)-6-(trichloromethyl)- (8CI) **(30369-46-1)**
s-Triazine, 2-(isobutylamino)-4-methyl-6-(trichloromethyl)- (8CI) **(24803-53-0)**
s-Triazine, 2-isobutyl-4,6-bis(isobutylamino)- (8CI) **(26235-33-6)**
s-Triazine, 2-isobutyl-4,6-bis(methylamino)- (8CI) **(26235-29-0)**
s-Triazine, 2-isobutyl-4,6-bis(propylamino)- (8CI) **(26235-31-4)**
s-Triazine, 2-isobutyl-4,6-bis(trichloromethyl)- (8CI) **(24481-40-1)**
s-Triazine, 2-isobutyl-4,6-dimethyl- (8CI) **(30362-06-2)**
s-Triazine, 2-(isobutylthio)-4,6-bis(trichloromethyl)- (8CI) **(24481-69-4)**
s-Triazine, 2-isopropoxy-4,6-bis(isopropylamino)- (7CI,8CI) **(30360-64-6)**
s-Triazine, 2-isopropoxy-4,6-bis(trichloromethyl)- (6CI,7CI,8CI) **(30863-55-9)**
s-Triazine, 2-isopropoxy-4-methyl-6-(trichloromethyl)- (8CI) **(30863-48-0)**
s-Triazine, 2-isopropyl- (7CI,8CI) **(30361-87-6)**
s-Triazine, 2-(isopropylamino)- (8CI) **(30360-48-6)**
s-Triazine, 2-(isopropylamino)-4,6-bis(trichloromethyl)- (8CI) **(24803-31-4)**
s-Triazine, 2-isopropylamino-4-methoxy-6-methylamino **(3035-45-8)**
s-Triazine, 2-(isopropylamino)-4-((3-methoxypropyl)amino)-6-(methylthio) **(841-06-5)**
s-Triazine, 2-(isopropylamino)-4-(methylamino)-6-(methylthio) **(1014-69-3)**
s-Triazine, 2-(isopropylamino)-4-(methylamino)-6-(trichloromethyl)- (7CI,8CI) **(27431-15-8)**
s-Triazine, 2-(isopropylamino)-4-methyl-6-(methylamino)- (8CI) **(27430-90-6)**
s-Triazine, 2-(isopropylamino)-4-methyl-6-(propylamino)- (8CI) **(27431-02-3)**
s-Triazine, 2-(isopropylamino)-4-(methylthio)-6-(trichloromethyl)- (8CI) **(30369-44-9)**
s-Triazine, 2-(isopropylamino)-4-methyl-6-(trichloromethyl)- (8CI) **(24830-34-0)**
s-Triazine, 2-(isopropylamino)-4-phenyl-6-(trichloromethyl)- (8CI) **(24803-02-9)**
s-Triazine, 2-isopropyl-4,6-bis(isopropylamino)- (7CI,8CI) **(26235-25-6)**
s-Triazine, 2-isopropyl-4,6-bis(methylamino)- (8CI) **(26235-23-4)**
s-Triazine, 2-isopropyl-4,6-bis(trichloromethyl)- (8CI) **(24481-42-3)**
s-Triazine, 2-isopropyl-4-(methylamino)-6-(trichloromethyl)- (8CI) **(24803-26-7)**
s-Triazine, 2-(isopropylthio)-4,6-bis(trichloromethyl)- (8CI) **(24481-68-3)**
s-Triazine, 2-(isopropylthio)-4-(methylamino)-6-(trichloromethyl)- (8CI) **(30369-51-8)**
s-Triazine-2-methanethiol, 4,6-diamino-, S-ester with O,O-dimethylphosphorodithioate **(78-57-9)**
s-Triazine-2-methanol, 4,6-bis(trichloromethyl)-, acetate (7CI) **(30362-44-8)**
s-Triazine-2-methanol, 4,6-bis(trichloromethyl)-, acetate (ester) (8CI) **(30362-44-8)**
s-Triazine-2-methanol, 4,6-diethyl-.α.-methyl-, acetate (7CI) **(30362-70-0)**

s-Triazine-2-methanol, 4,6-diethyl-.α.-methyl-, acetate (ester) (8CI) (30362-70-0)
s-Triazine-2-methanol, .α.,.α.-dimethyl-4,6-bis(trichloromethyl)-, acetate (7CI) (30362-47-1)
s-Triazine-2-methanol, .α.,.α.-dimethyl-4,6-bis(trichloromethyl)-, acetate (ester) (8CI) (30362-47-1)
s-Triazine-2-methanol, .α.-methyl-4,6-bis(trichloromethyl)-, acetate (7CI) (30362-45-9)
s-Triazine-2-methanol, .α.-methyl-4,6-bis(trichloromethyl)-, acetate (ester) (8CI) (30362-45-9)
s-Triazine-2-methanol, .α.,4,6-tris(trichloromethyl)-, acetate (7CI) (30362-46-0)
s-Triazine-2-methanol, .α.,4,6-tris(trichloromethyl)-, acetate (ester) (8CI) (30362-46-0)
1,3,5-Triazine, 2-methoxy- (9CI) (17635-40-4)
s-Triazine, 2-methoxy- (6CI,8CI) (17635-40-4)
s-Triazine, 2-methoxy-4,6-bis((3-methoxypropyl)amino)- (8CI) (1771-07-9)
s-Triazine, 2-methoxy-4,6-bis(methylamino)- (8CI) (30360-57-7)
1,3,5-Triazine, 2-methoxy-4,6-bis(tribromomethyl)- (9CI) (30863-63-9)
s-Triazine, 2-methoxy-4,6-bis(tribromomethyl)- (7CI,8CI) (30863-63-9)
s-Triazine, 2-methoxy-4,6-bis(trichloromethyl)- (6CI,7CI,8CI) (30863-51-5)
1,3,5-Triazine, 2-methoxy-4,6-dimethyl- (9CI) (3599-66-4)
s-Triazine, 2-methoxy-4,6-dimethyl- (7CI,8CI) (3599-66-4)
s-Triazine, 2-((2-methoxyethyl)amino)-4,6-bis(trichloromethyl)- (8CI) (24803-13-2)
s-Triazine, 2-methoxy-4-methyl-6-(trichloromethyl)- (8CI) (30863-45-7)
s-Triazine, 2-((2-methoxy-4-nitrophenyl)thio)-4,6-bis-(trichloromethyl)- (8CI) (24478-16-8)
1,3,5-Triazine, 2-((2-methoxy-4-nitrophenyl)thio)-4,6-bis-(trichloromethyl)- (9CI) (24478-16-8)
1,3,5-Triazine, 2-(4-methoxyphenyl)-4,6-bis(trichloromethyl)- (9CI) (3584-23-4)
s-Triazine, 2-(p-methoxyphenyl)-4,6-bis(trichloromethyl)- (8CI) (3584-23-4)
s-Triazine, 2-(p-methoxyphenyl)-4-phenyl-6-(trichloromethyl)- (7CI,8CI) (3599-82-4)
1,3,5-Triazine, 2-((2-methoxyphenyl)thio)-4,6-bis(trichloro-methyl)- (9CI) (24478-02-2)
1,3,5-Triazine, 2-((4-methoxyphenyl)thio)-4,6-bis(trichloro-methyl)- (9CI) (24478-03-3)
s-Triazine, 2-((o-methoxyphenyl)thio)-4,6-bis(trichloro-methyl)- (8CI) (24478-02-2)
s-Triazine, 2-((p-methoxyphenyl)thio)-4,6-bis(trichloro-methyl)- (8CI) (24478-03-3)
s-Triazine, 2-methoxy-4-phenyl-6-(trichloromethyl)- (8CI) (30863-60-6)
1,3,5-Triazine, 2-methyl- (9CI) (3599-87-9)
s-Triazine, 2-methyl- (6CI,7CI,8CI) (3599-87-9)
s-Triazine, 2-(methylamino)- (8CI) (4039-99-0)
s-Triazine, 2-(methylamino)-4,6-bis(methylthio)- (8CI) (30362-11-9)
s-Triazine, 2-(methylamino)-4,6-bis(tribromomethyl)- (8CI) (30339-37-8)
s-Triazine, 2-(methylamino)-4,6-bis(trichloromethyl)- (7CI,8CI) (24803-64-3)
s-Triazine, 2-(methylamino)-4,6-bis(trifluoromethyl)- (8CI) (29181-68-8)
s-Triazine, 2-(methylamino)-4,6-diphenyl- (7CI,8CI) (2039-37-4)
s-Triazine, 2-(methylamino)-4-(methylthio)-6-(trichloro-methyl)- (8CI) (30377-26-5)
s-Triazine, 2-(methylamino)-4-(phenylthio)-6-(trichloro-methyl)- (8CI) (30369-56-3)
s-Triazine, 2-(methylamino)-4-phenyl-6-(trichloromethyl)- (8CI) (24802-99-1)
s-Triazine, 2-(methylamino)-4-(propylamino)-6-(trichloro-methyl)- (8CI) (27470-68-4)
s-Triazine, 2-(methylamino)-4-propyl-6-(trichloromethyl)- (8CI) (24803-24-5)
s-Triazine, 2-(methylamino)-4-(trichloromethyl)-6-((2,4,5-tri-chlorophenyl)thio)- (8CI) (30369-63-2)
s-Triazine, 2-methyl-4,6-bis(methylamino)- (6CI,7CI,8CI) (1973-06-4)
1,3,5-Triazine, 2-methyl-4,6-bis(methylthio)- (9CI) (5336-94-7)
s-Triazine, 2-methyl-4,6-bis(methylthio)- (6CI,8CI) (5336-94-7)
s-Triazine, 2-methyl-4,6-bis(m-nitrophenyl)- (8CI) (30362-61-9)
s-Triazine, 2-methyl-4,6-bis(propylamino)- (7CI,8CI) (26234-33-3)
1,3,5-Triazine, 2-methyl-4,6-bis(tribromomethyl)- (9CI) (30362-02-8)
s-Triazine, 2-methyl-4,6-bis(tribromomethyl)- (7CI,8CI) (30362-02-8)
1,3,5-Triazine, 2-methyl-4,6-bis(trichloromethyl)- (9CI) (949-42-8)
s-Triazine, 2-methyl-4,6-bis(trichloromethyl)- (8CI) (949-42-8)
s-Triazine, 2-methyl-4,6-bis(trifluoromethyl)- (8CI) (30361-93-4)
1,3,5-Triazine, 2-methyl-4,6-diphenoxy- (9CI) (738-71-6)
s-Triazine, 2-methyl-4,6-diphenoxy- (7CI,8CI) (738-71-6)
1,3,5-Triazine, 2-methyl-4,6-diphenyl- (9CI) (3599-62-0)
s-Triazine, 2-methyl-4,6-diphenyl- (6CI,7CI,8CI) (3599-62-0)
s-Triazine, 2-methyl-4,6-di-p-tolyl- (8CI) (30377-13-0)
1,3,5-Triazine, 2-(1-methylethyl)- (9CI) (30361-87-6)
1,3,5-Triazine, 2-(1-methylethyl)-4,6-bis(trichloromethyl)- (9CI) (24481-42-3)
1,3,5-Triazine, 2-((1-methylethyl)thio)-4,6-bis(trichloro-methyl)- (9CI) (24481-68-3)
s-Triazine, 2-methyl-4-(methylamino)-6-(propylamino)- (8CI) (27430-89-3)
s-Triazine, 2-methyl-4-(methylamino)-6-(trichloromethyl)- (8CI) (24830-33-9)
1,3,5-Triazine, 2-methyl-4,6-bis(4-methylphenyl)- (9CI) (30377-13-0)
1,3,5-Triazine, 2-methyl-4-(4-morpholinyl)-6-(trichloro-methyl)- (9CI) (24803-60-9)
s-Triazine, 2-methyl-4-morpholino-6-(trichloromethyl)- (8CI) (24803-60-9)
s-Triazine, 2-methyl-4-(pentylamino)-6-(trichloromethyl)- (8CI) (24803-54-1)
1,3,5-Triazine, 2-(4-methylphenyl)-4,6-bis(trichloromethyl)- (9CI) (3584-22-3)
1,3,5-Triazine, 2-(4-methylphenyl)-4,6-diphenyl- (9CI) (16107-88-3)
1,3,5-Triazine, 2-(4-methylphenyl)-4-phenyl-6-(trichloro-methyl)- (9CI) (3599-79-9)
1,3,5-Triazine, 2-((2-methylphenyl)thio)-4,6-bis(trichloromethyl)- (9CI) (24477-99-4)
1,3,5-Triazine, 2-((3-methylphenyl)thio)-4,6-bis(trichloromethyl)- (9CI) (24478-00-0)
1,3,5-Triazine, 2-((4-methylphenyl)thio)-4,6-bis(trichloromethyl)- (9CI) (24478-01-1)
s-Triazine, 2-methyl-4-phenyl-6-(trichloromethyl)- (7CI,8CI) (30362-33-5)
s-Triazine, 2-methyl-4-piperidino-6-(trichloromethyl)- (8CI) (24803-59-6)
s-Triazine, 2-methyl-4-(propylamino)-6-(trichloromethyl)- (8CI) (24803-52-9)
1,3,5-Triazine, 2-((2-methylpropyl)thio)-4,6-bis(trichloromethyl)- (9CI) (24481-69-4)
s-Triazine, 2-(methylthio)- (6CI,8CI) (26292-91-1)
1,3,5-Triazine, 2-(methylthio)-4,6-bis(pentachloroethyl)- (9CI) (30863-02-6)
s-Triazine, 2-(methylthio)-4,6-bis(pentachloroethyl)- (8CI) (30863-02-6)
1,3,5-Triazine, 2-(methylthio)-4,6-bis(1,1,2-trichloroethyl)- (9CI) (5516-48-3)
s-Triazine, 2-(methylthio)-4,6-bis(1,1,2-trichloroethyl)- (7CI,8CI) (5516-48-3)
1,3,5-Triazine, 2-(methylthio)-4,6-bis(trichloromethyl)- (9CI) (3599-76-6)
s-Triazine, 2-(methylthio)-4,6-bis(trichloromethyl)- (7CI,8CI) (3599-76-6)
1,3,5-Triazine, 2-(methylthio)-4,6-diphenyl- (9CI) (735-69-3)
s-Triazine, 2-(methylthio)-4,6-diphenyl- (7CI,8CI) (735-69-3)
s-Triazine, 2-(methylthio)-4-(propylamino)-6-(trichloromethyl)- (8CI) (30369-42-7)
s-Triazine, 2-methyl-4-(trichloromethyl)- (8CI) (30361-91-2)
s-Triazine, 2-morpholino-4,6-bis(trichloromethyl)- (8CI) (24803-10-9)
1,3,5-Triazine, 2-(4-morpholinyl)-4,6-bis(trichloromethyl)- (9CI) (24803-10-9)
1,3,5-Triazine, 2-(1-naphthalenyl)-4,6-bis(trichloromethyl)- (9CI) (24481-46-7)
1,3,5-Triazine, 2-(2-naphthalenyl)-4,6-bis(trichloromethyl)- (9CI) (24481-45-6)
s-Triazine, 2-(1-naphthyl)-4,6-bis(trichloromethyl)- (8CI) (24481-46-7)
s-Triazine, 2-(2-naphthyl)-4,6-bis(trichloromethyl)- (8CI) (24481-45-6)
s-Triazine, 2-((p-nitrobenzyl)thio)-4,6-bis(trichloromethyl)- (8CI) (24481-74-1)
1,3,5-Triazine, 2-(3-nitrophenyl)- (9CI) (30361-90-1)
s-Triazine, 2-(m-nitrophenyl)- (6CI,8CI) (30361-90-1)
s-Triazine, 2-(m-nitrophenyl)-4,6-bis(trichloromethyl)- (7CI,8CI) (3599-72-2)
1,3,5-Triazine, 2-(4-nitrophenyl)-4,6-diphenyl- (9CI) (13960-33-3)
s-Triazine, 2-(p-nitrophenyl)-4,6-diphenyl- (7CI,8CI) (13960-33-3)
s-Triazine, 2-(m-nitrophenyl)-4,6-diphenyl- (6CI,7CI,8CI) (7753-05-1)
1,3,5-Triazine, 2-(((4-nitrophenyl)methyl)thio)-4,6-bis(tri-chloromethyl)- (9CI) (24481-74-1)
s-Triazine, 2-(m-nitrophenyl)-4-phenyl-6-(trichloromethyl)- (7CI,8CI) (3599-81-3)
1,3,5-Triazine, 2-((4-nitrophenyl)thio)-4,6-bis(trichloromethyl)- (9CI) (24478-14-6)

s-Triazine, 2-((p-nitrophenyl)thio)-4,6-bis(trichloromethyl)- (8CI) **(24478-14-6)**
1,3,5-Triazine, 2-nonyl-4,6-bis(trichloromethyl)- (9CI) **(24481-36-5)**
s-Triazine, 2-nonyl-4,6-bis(trichloromethyl)- (8CI) **(24481-36-5)**
s-Triazine, 2-(2,3,4,5,6-pentachloroanilino)-4,6-bis(trichloromethyl)- (8CI) **(30356-57-1)**
s-Triazine, 2-(pentachloroethyl)-4,6-bis(trichloromethyl)- (8CI) **(30362-35-7)**
s-Triazine, 2-(pentachlorophenyl)-4,6-bis(trichloromethyl)- (8CI) **(30362-55-1)**
1,3,5-Triazine, 2-((pentachlorophenyl)thio)-4,6-bis(trichloromethyl)- (9CI) **(31007-95-1)**
s-Triazine, 2-((pentachlorophenyl)thio)-4,6-bis(trichloromethyl)- (8CI) **(31007-95-1)**
s-Triazine, 2-(pentylamino)-4,6-bis(trichloromethyl)- (8CI) **(24802-86-6)**
s-Triazine, 2-pentyl-4,6-bis(propylamino)- (8CI) **(26235-38-1)**
1,3,5-Triazine, 2-pentyl-4,6-bis(trichloromethyl)- (9CI) **(24481-37-6)**
s-Triazine, 2-pentyl-4,6-bis(trichloromethyl)- (8CI) **(24481-37-6)**
s-Triazine, 2-(pentyloxy)-4,6-bis(trichloromethyl)- (6CI,8CI) **(30863-57-1)**
s-Triazine, 2-pentyl-4-(propylamino)-6-(trichloromethyl)- (8CI) **(24802-91-3)**
1,3,5-Triazine, 2-(pentylthio)-4,6-bis(trichloromethyl)- (9CI) **(24481-70-7)**
s-Triazine, 2-(pentylthio)-4,6-bis(trichloromethyl)- (8CI) **(24481-70-7)**
Triazine (pesticide) **(101-05-3)**
1,3,5-Triazine, 2-phenyl- (9CI) **(1722-18-5)**
s-Triazine, 2-phenyl- (6CI,7CI,8CI) **(1722-18-5)**
1,3,5-Triazine, 2-phenyl-4,6-bis(trichloromethyl)- (9CI) **(24504-22-1)**
s-Triazine, 2-phenyl-4,6-bis(trichloromethyl)- (6CI,7CI,8CI) **(24504-22-1)**
s-Triazine, 2-phenyl-4,6-bis(trifluoromethyl)- (8CI) **(30362-32-4)**
s-Triazine, 2-phenyl-4,6-distyryl- (8CI) **(21620-54-2)**
1,3,5-Triazine, 2-((phenylmethyl)thio)-4,6-bis(trichloromethyl)- (9CI) **(5516-51-8)**
s-Triazine, 2-phenyl-4-(propylamino)-6-(trichloromethyl)- (8CI) **(24803-01-8)**
1,3,5-Triazine, 2-(phenylthio)-4,6-bis(1,1,2-trichloroethyl)- (9CI) **(30863-01-5)**
s-Triazine, 2-(phenylthio)-4,6-bis(1,1,2-trichloroethyl)- (8CI) **(30863-01-5)**
1,3,5-Triazine, 2-(phenylthio)-4,6-bis(trichloromethyl)- (9CI) **(5516-47-2)**
s-Triazine, 2-(phenylthio)-4,6-bis(trichloromethyl)- (7CI,8CI) **(5516-47-2)**
s-Triazine, 2-phenyl-4-p-tolyl-6-(trichloromethyl)- (7CI,8CI) **(3599-79-9)**
s-Triazine, 2-piperidino-4,6-bis(trichloromethyl)- (8CI) **(24803-09-6)**
1,3,5-Triazine-2-propanenitrile, 4,6-diamino- (9CI) **(4784-14-9)**
s-Triazine-2-propionitrile, 4,6-diamino- (7CI,8CI) **(4784-14-9)**
s-Triazine, 2-propoxy-4,6-bis(trichloromethyl)- (8CI) **(30863-53-7)**
s-Triazine, 2-(propylamino)-4,6-bis(trichloromethyl)- (8CI) **(24803-30-3)**
s-Triazine, 2-propyl-4,6-bis(propylamino)- (8CI) **(26235-19-8)**
1,3,5-Triazine, 2-propyl-4,6-bis(trichloromethyl)- (9CI) **(24481-43-4)**
s-Triazine, 2-propyl-4,6-bis(trichloromethyl)- (8CI) **(24481-43-4)**
1,3,5-Triazine, 2-(propylthio)-4,6-bis(trichloromethyl)- (9CI) **(24504-17-4)**
s-Triazine, 2-(propylthio)-4,6-bis(trichloromethyl)- (8CI) **(24504-17-4)**
s-Triazine, 2-(3-pyridyl)-4,6-bis(trichloromethyl)- (8CI) **(30362-59-5)**
s-Triazine, 2-(4-pyridyl)-4,6-bis(trichloromethyl)- (7CI,8CI) **(3599-73-3)**
s-Triazine, 2-((1,1,3,3-tetramethylbutyl)amino)-4,6-bis(trichloromethyl)- (8CI) **(30357-71-2)**
s-Triazine-2-thiol, 4-amino-6-benzyl- (8CI) **(30369-73-4)**
s-Triazine-2-thiol, 4-amino-6-(p-chlorophenyl)- (8CI) **(30369-75-6)**
s-Triazine-2-thiol, 4-amino-6-ethyl- (8CI) **(30369-71-2)**
s-Triazine-2-thiol, 4-amino-6-isobutyl- (8CI) **(30369-72-3)**
s-Triazine-2-thiol, 4-amino-6-methyl- (8CI) **(30369-70-1)**
s-Triazine-2-thiol, 4-amino-6-phenyl- (8CI) **(30369-74-5)**
s-Triazine-2-thiol, 4,6-bis(allylamino)- (7CI,8CI) **(5210-79-7)**
s-Triazine-2-thiol, 4,6-bis(diethylamino)- (6CI,7CI,8CI) **(4022-55-3)**
s-Triazine-2-thiol, 4,6-bis(ethylamino)- (8CI) **(1011-91-2)**
s-Triazine-2-thiol, 4,6-bis(isopropylamino)- (8CI) **(5133-47-1)**
s-Triazine-2-thiol, 4-chloro-6-methoxy-, diethyldithiocarbamate (ester) **(13733-96-5)**
s-Triazine-2-thiol, 4-chloro-6-phenyl-, diethyldithiocarbamate (ester) **(13733-97-6)**
s-Triazine-2-thiol, 4,6-diamino- **(767-17-9)**
s-Triazine-2-thiol, 4,6-dianilino- (8CI) **(15989-50-1)**
s-Triazine-2-thiol, 4,6-dichloro-, diethyldithiocarbamate (ester) **(13733-95-4)**
s-Triazine-2-thiol, 4,6-dichloro-, S-ester with O,O-diethyl phosphorodithioate **(14991-93-6)**
s-Triazine-2-thiol, 4,6-dimethoxy- (8CI) **(30886-14-7)**
s-Triazine-2-thiol, 4,6-dimethyl- (8CI) **(30886-15-8)**
s-Triazine-2-thiol, 4,6-diphenyl- (8CI) **(30886-16-9)**
s-Triazine-2-thiol, 4,6-di-p-tolyl- (7CI,8CI) **(30886-17-0)**
s-Triazine-2-thiol, 1,4,5,6-tetrahydro-5-methyl- (8CI) **(6746-27-6)**
1,3,5-Triazine-2(1H)-thione, 4-amino-6-methyl- (9CI) **(30369-70-1)**
1,3,5-Triazine-2(1H)-thione, 4-amino-6-phenyl- (9CI) **(30369-74-5)**
1,3,5-Triazine-2(1H)-thione, 4,6-bis(diethylamino)- (9CI) **(4022-55-3)**
1,3,5-Triazine-2(1H)-thione, 4,6-bis(ethylamino)- (9CI) **(1011-91-2)**
1,3,5-Triazine-2(1H)-thione, 4,6-bis((1-methylethyl)amino)- (9CI) **(5133-47-1)**
1,3,5-Triazine-2(1H)-thione, 4,6-bis(phenylamino)- (9CI) **(15989-50-1)**
1,3,5-Triazine-2(1H)-thione, 4,6-bis(2-propenylamino)- (9CI) **(5210-79-7)**
s-Triazine-2-thione, 4,6-diamino **(767-17-9)**
1,3,5-Triazine-2(1H)-thione, 4,6-diamino- (9CI) **(767-17-9)**
1,3,5-Triazine-2(1H)-thione, 4,6-dimethoxy- (9CI) **(30886-14-7)**
1,3,5-Triazine-2(1H)-thione, 4,6-diphenyl- (9CI) **(30886-16-9)**
s-Triazine-2(1H)-thione, 4,6-diphenyl- (7CI) **(30886-16-9)**
s-Triazine-2(1H)-thione, tetrahydro-4,6-bis(trichloromethyl)- (8CI) **(30805-37-9)**
s-Triazine-2(1H)-thione, tetrahydro-4,6-diimino- **(767-17-9)**
1,3,5-Triazine-2(1H)-thione, tetrahydro-4,6-dimethyl- (9CI) **(21306-32-1)**
s-Triazine-2(1H)-thione, tetrahydro-4,6-dimethyl- (8CI) **(21306-32-1)**
s-Triazine-2(1H)-thione, tetrahydro-5-methyl- **(6746-27-6)**
1,3,5-Triazine-2(1H)-thione, tetrahydro-5-methyl- (9CI) **(6746-27-6)**
s-Triazine, 2-p-tolyl-4,6-bis(trichloromethyl)- (7CI,8CI) **(3584-22-3)**
s-Triazine, 2-(m-tolylthio)-4,6-bis(trichloromethyl)- (8CI) **(24478-00-0)**
s-Triazine, 2-(o-tolylthio)-4,6-bis(trichloromethyl)- (8CI) **(24477-99-4)**
s-Triazine, 2-(p-tolylthio)-4,6-bis(trichloromethyl)- (8CI) **(24478-01-1)**
s-Triazine, 2,4,6-triacetamido- **(5637-82-1)**
s-Triazine-1,3,5(2H,4H,6H)-triacetamide, 2,4,6-trioxo- (7CI,8CI) **(1843-48-7)**
s-Triazine-1,3,5(2H,4H,6H)triacetonitrile **(4560-87-6)**
1,3,5-Triazine, 1,3,5-triacetylhexahydro- (9CI) **(26028-46-6)**
s-Triazine, 1,3,5-triacetylhexahydro- (8CI) **(26028-46-6)**
1,3,5-Triazine-2,4,6-triamine **(108-78-1)**
1,3,5-Triazine-2,4,6-triamine, Polymer with formaldehyde (9CI) **(9003-08-1)**
1,3,5-Triazine-2,4,6-triamine, N,N'-bis(4-chlorophenyl)- (9CI) **(30360-18-0)**
1,3,5-Triazine-2,4,6-triamine, N-(4-chlorophenyl)-N'-(1-methylethyl)- (9CI) **(30360-17-9)**
1,3,5-Triazine-2,4,6-triamine, N,N'-diethyl- **(5606-16-6)**
1,3,5-Triazine-2,4,6-triamine, N,N'-diethyl-N''-(1-methylethyl)- (9CI) **(30360-19-1)**
1,3,5-Triazine-2,4,6-triamine, N,N'-dimethyl-N,N'-diphenyl- (9CI) **(30377-20-9)**
1,3,5-Triazine-2,4,6-triamine, N,N'-diphenyl- (9CI) **(5606-18-8)**
1,3,5-Triazine-2,4,6-triamine, N,N'-di-2-propenyl- (9CI) **(30360-15-7)**
1,3,5-Triazine-2,4,6-triamine, N-ethyl- (9CI) **(5606-23-5)**
1,3,5-Triazine-2,4,6-triamine, N,N,N',N',N'',N''-hexachloro- **(2428-04-8)**
1,3,5-Triazine-2,4,6-triamine, N,N,N',N',N'',N''-hexaethyl- (9CI) **(2827-49-8)**
1,3,5-Triazine-2,4,6-triamine, N,N,N',N',N'',N''-hexakis-(methoxymethyl) **(3089-11-0)**
1,3,5-Triazine-2,4,6-triamine, N,N,N',N',N'',N''-hexamethyl- (9CI) **(645-05-6)**
1,3,5-Triazine-2,4,6-triamine, N-(1-methylethyl)- (9CI) **(16274-81-0)**
1,3,5-Triazine-2,4,6-triamine, N,N,N',N''-pentamethyl- (9CI) **(16268-62-5)**
1,3,5-Triazine-2,4,6-triamine, N-phenyl- (9CI) **(5606-27-9)**
1,3,5-Triazine-2,4,6-triamine, N,N,N',N'-tetraethyl- (9CI) **(5606-20-2)**
1,3,5-Triazine-2,4,6-triamine, N,N',N''-triacetyl- **(5637-82-1)**
1,3,5-Triazine-2,4,6-triamine, N,N',N''-tribenzoyl- **(5637-84-3)**
1,3,5-Triazine-2,4,6-triamine, N,N',N''-triethyl- (9CI) **(16268-92-1)**
1,3,5-Triazine-2,4,6-triamine, N,N',N''-trimethyl- (9CI) **(2827-46-5)**
1,3,5-Triazine-2,4,6-triamine, N,N',N''-triphenyl- (9CI) **(1973-05-3)**
1,3,5-Triazine-2,4,6-triamine, N,N',N''-tri-2-propenyl- (9CI) **(30360-21-5)**
s-Triazine, 2,4,6-triamino- **(108-78-1)**
1,3,5-Triazine, 2,4,6-triazido- **(5637-83-2)**
s-Triazine, 2,4,6-triazido **(5637-83-2)**
s-Triazine, 2,4,6-tribenzamido- (6CI) **(5637-84-3)**
1,3,5-Triazine, 1,3,5-tribenzoylhexahydro- (9CI) **(5434-82-2)**

Name	CAS RN
s-Triazine, 1,3,5-tribenzoylhexahydro- (8CI)	**(5434-82-2)**
s-Triazine, 2,4,6-tribenzyl- (6CI,7CI,8CI)	**(13960-31-1)**
1,3,5-Triazine, 2,4,6-tribromo- (9CI)	**(14921-00-7)**
s-Triazine, 2,4,6-tribromo- (8CI)	**(14921-00-7)**
1,3,5-Triazine, 2,4,6-tributoxy- (9CI)	**(19837-00-4)**
s-Triazine, 2,4,6-tributoxy- (8CI)	**(19837-00-4)**
s-Triazine, 2,4,6-tributyl- (7CI,8CI)	**(4803-17-2)**
s-Triazinetricarbonitrile	**(7615-57-8)**
1,3,5-Triazine-2,4,6-tricarbonitrile (9CI)	**(7615-57-8)**
s-Triazine-2,4,6-tricarbonitrile (7CI,8CI)	**(7615-57-8)**
s-Triazine-2,4,6-tricarboxamide (8CI)	**(23297-24-7)**
s-Triazine-2,4,6-tricarboxamide, N,N',N''-trimethyl- (8CI)	**(30863-20-8)**
1,3,5-Triazine-2,4,6-tricarboxylic acid, triethyl ester (9CI)	**(898-22-6)**
s-Triazine-2,4,6-tricarboxylic acid, triethyl ester (6CI,7CI,8CI)	**(898-22-6)**
s-Triazine-2,4,6-tricarboxylic acid, triisobutyl ester (8CI)	**(30863-19-5)**
s-Triazine-2,4,6-tricarboxylic acid, trimethyl ester (6CI,8CI)	**(30863-34-4)**
s-Triazine trichloride	**(108-77-0)**
s-Triazine, 2,4,6-trichloro	**(108-77-0)**
s-Triazine, 2-(2,4,5-trichloroanilino)-4,6-bis(trichloromethyl)- (8CI)	**(30356-56-0)**
1,3,5-Triazine, 2-(1,1,2-trichloroethyl)-4,6-bis(trichloromethyl)- (9CI)	**(24481-33-2)**
s-Triazine, 2-(1,1,2-trichloroethyl)-4,6-bis(trichloromethyl)- (8CI)	**(24481-33-2)**
1,3,5-Triazine, 2-(trichloromethyl)- (9CI)	**(30361-83-2)**
s-Triazine, 2-(trichloromethyl)- (6CI,7CI,8CI)	**(30361-83-2)**
1,3,5-Triazine, N,N',N''-trichloro-2,4,6-triamino	**(7673-09-8)**
s-Triazine-1,3,5(2H,4H,6H)-triethanol	**(4719-04-4)**
1,3,5-Triazine-1,3,5(2H,4H,6H)-triethanol (9CI)	**(4719-04-4)**
s-Triazine-2,4,6-triethanol, .α.,.α.',.α.''-tris(o-chlorophenyl)- (8CI)	**(30363-04-3)**
s-Triazine-2,4,6-triethanol, .α.,.α.',.α.''-tris(2,4-dichlorophenyl)- (8CI)	**(30363-05-4)**
1,3,5-Triazine, 2,4,6-triethoxy- (9CI)	**(884-43-5)**
s-Triazine, 2,4,6-triethoxy- (8CI)	**(884-43-5)**
1,3,5-Triazine, 2,4,6-triethyl- (9CI)	**(1009-74-1)**
s-Triazine, 2,4,6-triethyl- (6CI,7CI,8CI)	**(1009-74-1)**
s-Triazine, 2,4,6-trifluoro	**(675-14-9)**
s-Triazine, 2,4,6-triisobutoxy- (8CI)	**(30895-05-7)**
s-Triazine, 2,4,6-triisopropoxy- (8CI)	**(29263-11-4)**
s-Triazine, 2,4,6-triisopropyl- (6CI,8CI)	**(25176-37-8)**
1,3,5-Triazine-2,4,6-trimercaptan	**(638-16-4)**
1,3,5-Triazine, 2,4,6-trimethoxy- (9CI)	**(877-89-4)**
s-Triazine, 2,4,6-trimethoxy- (8CI)	**(877-89-4)**
1,3,5-Triazine, 2,4,6-trimethyl- (9CI)	**(823-94-9)**
s-Triazine, 2,4,6-trimethyl- (8CI)	**(823-94-9)**
s-Triazine, 2,4,6-trinonyl- (8CI)	**(30362-98-2)**
s-2,4,6-Triazinetriol	**(108-80-5)**
s-Triazine-2,4,6-triol	**(108-80-5)**
sym-Triazinetriol	**(108-80-5)**
s-Triazine-2,4,6(1H,3H,5H)-trione	**(108-80-5)**
s-Triazine-2,4,6(1H,3H,5H)-trione, 1,3-dichloro	**(2782-57-2)**
s-Triazine-2,4,6(1H,3H,5H)-trione, dichloro-, potassium deriv	**(2244-21-5)**
1,3,5-Triazine-2,4,6(1H,3H,5H)-trione, 1,3-dichloro-, potassium salt	**(2244-21-5)**
s-Triazine-2,4,6(1H,3H,5H)-trione, 1,3-dichloro-, potassium salt	**(2244-21-5)**
1,3,5-Triazine-2,4,6(1H,3H,5H)-trione, 1,3-dichloro-, sodium salt	**(2893-78-9)**
s-Triazine-2,4,6(1H,3H,5H)-trione, dichloro-, sodium salt	**(2893-78-9)**
1,3,5-Triazine-2,4,6(1H,3H,5H)-trione, disodium salt (9CI)	**(36452-21-8)**
1,3,5-Triazine-2,4,6(1H,3H,5H)-trione, 1-ethyl- (9CI)	**(30805-28-8)**
s-Triazine-2,4,6(1H,3H,5H)-trione, 1-ethyl- (8CI)	**(30805-28-8)**
s-Triazine-2,4,6(1H,3H,5H)-trione, monosodium salt	**(2624-17-1)**
1,3,5-Triazine-2,4,6(1H,3H,5H)-trione, monosodium salt (9CI)	**(2624-17-1)**
s-'Triazine-2,4,6(1H,3H,5H)-trione, tribenzyl-	**(606-03-1)**
s-Triazine-2,4,6(1H,3H,5H)-trione, 1,3,5-tribenzyl- (8CI)	**(606-03-1)**
1,3,5-Triazine-2,4,6(1H,3H,5H)-trione, 1,3,5-trichloro-	**(87-90-1)**
s-Triazine-2,4,6(1H,3H,5H)-trione, 1,3,5-trichloro	**(87-90-1)**
1,3,5-Triazine-2,4,6(1H,3H,5H)-trione, 1,3,5-trimethyl- (9CI)	**(827-16-7)**
s-Triazine-2,4,6(1H,3H,5H)-trione, 1,3,5-trimethyl- (8CI)	**(827-16-7)**
s-Triazine-2,4,6(1H,3H,5H)-trione, 1,3,5-triphenyl	**(1785-02-0)**
1,3,5-Triazine-2,4,6(1H,3H,5H)-trione, 1,3,5-tri-2-propenyl- (9CI)	**(1025-15-6)**
1,3,5-Triazine-2,4,6(1H,3H,5H)-trione, 1,3,5-tris(3-chloro-2-hydroxypropyl)- (9CI)	**(7423-53-2)**
s-Triazine-2,4,6(1H,3H,5H)-trione, tris(3-chloro-2-hydroxypropyl)- (7CI,8CI)	**(7423-53-2)**
s-Triazine-2,4,6(1H,3H,5H)-trione, 1,3,5-tris(2,3-epoxypropyl)- (8CI)	**(2451-62-9)**
s-Triazine-2,4,6(1H,3H,5H)-trione, 1,3,5-tris(2,3-epoxypropyl)-, Polymers (8CI)	**(28825-96-9)**
s-Triazine-2,4,6(1H,3H,5H)-trione, 1,3,5-tris(2-hydroxyethyl)-	**(839-90-7)**
1,3,5-Triazine-2,4,6(1H,3H,5H)-trione, 1,3,5-tris(2-hydroxyethyl)- (9CI)	**(839-90-7)**
1,3,5-Triazine-2,4,6(1H,3H,5H)-trione, 1,3,5-tris(6-isocyanatohexyl)- (9CI)	**(3779-63-3)**
1,3,5-Triazine-2,4,6(1H,3H,5H)-trione, 1,3,5-tris(3-isocyanatomethylphenyl)- (9CI)	**(26603-40-7)**
1,3,5-Triazine-2,4,6(1H,3H,5H)-trione, trisodium salt (9CI)	**(3047-33-4)**
1,3,5-Triazine-2,4,6(1H,3H,5H)-trione, 1,3,5-tris(oxiranylmethyl)- (9CI)	**(2451-62-9)**
1,3,5-Triazine-2,4,6(1H,3H,5H)-trione, 1,3,5-tris(oxiranylmethyl)-, Homopolymer (9CI)	**(28825-96-9)**
1,3,5-Triazine-2,4,6(1H,3H,5H)-trione, 1,3,5-tris(2-oxopropyl)- (9CI)	**(61050-97-3)**
1,3,5-Triazine-2,4,6(1H,3H,5H)-trione, 1,3,5-tris(phenylmethyl)- (9CI)	**(606-03-1)**
1,3,5-Triazine-2,4,6(1H,3H,5H)-trione, 1,3,5-tris(3-(trimethoxysilyl)propyl)- (9CI)	**(26115-70-8)**
s-Triazine-2,4,6(1H,3H,5H)-trione, tris(triphenylstannyl)-	**(752-74-9)**
1,3,5-Triazine, 2,4,6-triphenoxy- (9CI)	**(1919-48-8)**
1,3,5-Triazine, 2,4,6-triphenyl- (9CI)	**(493-77-6)**
s-Triazine, 2,4,6-triphenyl- (8CI)	**(493-77-6)**
1,3,5-Triazine, 2,4,6-tripropoxy- (9CI)	**(29263-10-3)**
s-Triazine, 2,4,6-tripropoxy- (8CI)	**(29263-10-3)**
1,3,5-Triazine, 2,4,6-tripropyl- (9CI)	**(1017-55-6)**
s-Triazine, 2,4,6-tripropyl- (7CI,8CI)	**(1017-55-6)**
s-Triazine, 2,4,6-tris(allyloxy)	**(101-37-1)**
s-Triazine, 2,4,6-tris(allylthio)- (8CI)	**(30863-37-7)**
1,3,5-Triazine, 2,4,6-tris(1-aziridinyl)-	**(51-18-3)**
s-Triazine, 2,4,6-tris(1-aziridinyl)	**(51-18-3)**
s-Triazine, 2,4,6-tris(1-bromoethyl)- (7CI,8CI)	**(30362-78-8)**
s-Triazine, 2,4,6-tris(bromomethyl)- (8CI)	**(30361-99-0)**
1,3,5-Triazine, 2,4,6-tris(4-bromophenyl)- (9CI)	**(30363-03-2)**
s-Triazine, 2,4,6-tris(p-bromophenyl)- (8CI)	**(30363-03-2)**
s-Triazine, 2,4,6-tris(2-chloroacetamido)- (6CI)	**(30360-27-1)**
s-Triazine, 2,4,6-tris(2-chloroethoxy)- (7CI,8CI)	**(959-60-4)**
s-Triazine, 2,4,6-tris(1-chloroethyl)- (6CI,8CI)	**(30362-72-2)**
1,3,5-Triazine, 1,3,5-tris(3-chloro-1-oxopropyl)hexahydro- (9CI)	**(20120-32-5)**
1,3,5-Triazine, 2,4,6-tris(4-chlorophenoxy)- (9CI)	**(1919-45-5)**
s-Triazine, 2,4,6-tris(p-chlorophenoxy)- (7CI,8CI)	**(1919-45-5)**
1,3,5-Triazine, 2,4,6-tris(4-chlorophenyl)- (9CI)	**(3114-54-3)**
s-Triazine, 2,4,6-tris(p-chlorophenyl)- (6CI,7CI,8CI)	**(3114-54-3)**
1,3,5-Triazine, 1,3,5-tris(4-chlorophenyl)hexahydro- (9CI)	**(30805-14-2)**
s-Triazine, 1,3,5-tris(p-chlorophenyl)hexahydro- (8CI)	**(30805-14-2)**
s-Triazine, 2,4,6-tris((p-chlorophenyl)thio)- (8CI)	**(13270-03-6)**
s-Triazine, 1,3,5-tris(3-chloropropionyl)hexahydro- (8CI)	**(20120-32-5)**
s-Triazine, 2,4,6-tris(p-chlorostyryl)- (8CI)	**(30362-99-3)**
s-Triazine, 2,4,6-tris(1,1-dibromoethyl)- (7CI,8CI)	**(30362-94-8)**
1,3,5-Triazine, 2,4,6-tris(dibromomethyl)- (9CI)	**(30362-01-7)**
s-Triazine, 2,4,6-tris(dibromomethyl)- (8CI)	**(30362-01-7)**
s-Triazine, 2,4,6-tris(2,2-dichloroacetamido)- (6CI)	**(30360-28-2)**
s-Triazine, 2,4,6-tris(1,1-dichloroethyl)- (7CI,8CI)	**(30362-74-4)**
1,3,5-Triazine, 2,4,6-tris(dichloromethyl)- (9CI)	**(5311-21-7)**
s-Triazine, 2,4,6-tris(dichloromethyl)- (8CI)	**(5311-21-7)**
s-Triazine, 2,4,6-tris(dimethylamino)-	**(645-05-6)**
1,3,5-Triazine, 2,4,6-tris(ethylthio)- (9CI)	**(30863-36-6)**

s-Triazine, 2,4,6-tris(ethylthio)- (8CI) **(30863-36-6)**
1,3,5-Triazine, 2,4,6-tris(4-methoxyphenyl)- (9CI) **(7753-12-0)**
s-Triazine, 2,4,6-tris(p-methoxyphenyl)- (6CI,7CI,8CI) **(7753-12-0)**
s-Triazine, 2,4,6-tris(methylamino)- **(2827-46-5)**
1,3,5-Triazine, 2,4,6-tris(1-methylethoxy)- (9CI) **(29263-11-4)**
1,3,5-Triazine, 2,4,6-tris(1-methylethyl)- (9CI) **(25176-37-8)**
1,3,5-Triazine, 2,4,6-tris(3-methylphenoxy)- (9CI) **(1919-46-6)**
1,3,5-Triazine, 2,4,6-tris(4-methylphenyl)- (9CI) **(6726-45-0)**
1,3,5-Triazine, 2,4,6-tris(2-methylpropoxy)- (9CI) **(30895-05-7)**
s-Triazine, 2,4,6-tris(methylthio) **(5759-58-0)**
s-Triazine, 2,4,6-tris(1-naphthyloxy)- (7CI,8CI) **(3949-34-6)**
s-Triazine, 2,4,6-tris(m-nitrophenyl)- (6CI,8CI) **(14043-38-0)**
s-Triazine, 2,4,6-tris(pentachloroethyl)- (8CI) **(30362-77-7)**
s-Triazine, 2,4,6-tris(2,2,3,3,3-pentachloropropoxy)- (8CI) **(30895-02-4)**
1,3,5-Triazine, 2,4,6-tris(2-phenylethenyl)- (9CI) **(21577-41-3)**
1,3,5-Triazine, 2,4,6-tris(phenylmethyl)- (9CI) **(13960-31-1)**
1,3,5-Triazine, 2,4,6-tris(phenylthio)- (9CI) **(30863-82-2)**
s-Triazine, 2,4,6-tris(phenylthio)- (7CI,8CI) **(30863-82-2)**
s-Triazine, 2,4,6-tris(1,1,2,2-tetrachloroethyl)- (8CI) **(30362-76-6)**
s-Triazine, 2,4,6-tris(m-tolyloxy)- (7CI,8CI) **(1919-46-6)**
s-Triazine, 2,4,6-tris(p-tolylthio)- (8CI) **(13270-05-8)**
1,3,5-Triazine, 2,4,6-tris(tribromomethyl)- (9CI) **(24687-55-6)**
s-Triazine, 2,4,6-tris(tribromomethyl)- (8CI) **(24687-55-6)**
s-Triazine, 2,4,6-tris(2,2,2-trichloroacetamido)- (6CI) **(30360-29-3)**
s-Triazine, 2,4,6-tris(2,2,2-trichloroethoxy)- (8CI) **(29808-66-0)**
s-Triazine, 2,4,6-tris(1,1,2-trichloroethyl)- (7CI,8CI) **(30362-75-5)**
1,3,5-Triazine, 2,4,6-tris(trichloromethyl)- (9CI) **(6542-67-2)**
s-Triazine, 2,4,6-tris(trichloromethyl)- (8CI) **(6542-67-2)**
s-Triazine, 2,4,6-tris(trifluoromethyl) **(368-66-1)**
s-Triazine, 2,4,6-tristyryl- (8CI) **(21577-41-3)**
2,4,6-Triazinetrithiol **(638-16-4)**
s-Triazine-2,4,6-trithiol **(638-16-4)**
1,3,5-Triazine-2,4,6(1H,3H,5H)-trithione **(638-16-4)**
s-Triazine, 2,4,6-tri-p-tolyl- (6CI,7CI,8CI) **(6726-45-0)**
1,1',1''-s-Triazine-2,4,6-triyltrisaziridine **(51-18-3)**
s-Triazin-2-ol, 4-amino- **(931-86-2)**
s-Triazin-2-ol, 4-amino-6-benzyl- (8CI) **(30369-68-7)**
s-Triazin-2-ol, 4-amino-6-(p-chlorophenyl)- (8CI) **(17584-14-4)**
s-Triazin-2-ol, 4-amino-6-ethyl- (8CI) **(16352-07-1)**
s-Triazin-2-ol, 4-amino-6-(ethylamino)- (8CI) **(7313-54-4)**
s-Triazin-2-ol, 4-amino-6-isobutyl- (8CI) **(30369-67-6)**
s-Triazin-2-ol, 4-amino-6-methyl- (8CI) **(16352-06-0)**
s-Triazin-2-ol, 4-amino-6-(p-nitrophenyl)- (8CI) **(30369-69-8)**
s-Triazin-2-ol, 4-amino-6-phenyl- (6CI,7CI) **(33957-63-0)**
s-Triazin-2-ol, 4-amino-6-(tribromomethyl)- (8CI) **(31949-57-2)**
s-Triazin-2-ol, 4-amino-6-(trichloromethyl)- (8CI) **(30369-65-4)**
s-Triazin-2-ol, 4,6-bis(p-chlorophenyl)- (8CI) **(30886-09-0)**
s-Triazin-2-ol, 4,6-bis(diethylamino)- (7CI,8CI) **(22305-35-7)**
s-Triazin-2-ol, 4,6-bis(ethylamino)- (8CI) **(2599-11-3)**
s-Triazin-2-ol, 4,6-bis(isopropylamino)- (8CI) **(7374-53-0)**
s-Triazin-2-ol, 4,6-bis(m-nitrophenyl)- (8CI) **(30886-11-4)**
s-Triazin-2-ol, 4,6-bis(trichloromethyl)-, Compd. with triethylamine (1:1) (6CI,7CI,8CI) **(30886-03-4)**
s-Triazin-2-ol, 4-chloro-6-(ethylamino)- **(49624-61-5)**
s-Triazin-2-ol, 4-(p-chlorophenyl)-6-phenyl- (7CI,8CI) **(1917-38-0)**
s-Triazin-2-ol, 4,6-diamino- (8CI) **(645-92-1)**
s-Triazin-2-ol, 4,6-dianilino- (6CI,7CI,8CI) **(30303-58-3)**
s-Triazin-2-ol, 4,6-dimethoxy- (8CI) **(1075-59-8)**
s-Triazin-2-ol, 4,6-dimethyl- (6CI,7CI,8CI) **(30885-99-5)**
s-Triazin-2-ol, 4-(ethylamino)-6-methyl- (8CI) **(30369-64-3)**
s-Triazin-2-ol, 4-ethyl-6-phenyl- (7CI) **(1917-40-4)**
s-Triazin-2-ol, 4-methyl-6-phenyl- (7CI,8CI) **(2519-50-8)**
s-Triazin-2-ol, 4-(methylthio)-6-phenyl- (7CI,8CI) **(1917-43-7)**
s-Triazin-2-ol, 4-phenyl-6-(trichloromethyl)- (7CI,8CI) **(1917-41-5)**
1,3,5-Triazin-2(1H)-one, 4-amino- (9CI) **(931-86-2)**
s-Triazin-2(1H)-one, 4-amino- (8CI) **(931-86-2)**
1,2,4-Triazin-5-one, 4-amino-6-tert-butyl-3-(methylthio)- **(21087-64-9)**
as-Triazin-5(4H)-one, 4-amino-6-tert-butyl-3-(methylthio) **(21087-64-9)**
1,2,4-Triazin-5(4H)-one, 4-amino-6-cyclohexyl-3-(methylthio)- (9CI) **(21085-19-8)**
as-Triazin-5(4H)-one, 4-amino-6-cyclohexyl-3-(methylthio)- (8CI) **(21085-19-8)**
s-Triazin-2(1H)-one, 4-amino-1-(2-deoxy-β-d-erythro-pentofuranosyl) **(2353-33-5)**
1,3,5-Triazin-2(1H)-one, 4-amino-1-(2-deoxy-β-d-erythro-pentofuranosyl)- (9CI) **(2353-33-5)**
1,3,5-Triazin-2(1H)-one, 6-amino-3,4-dihydro-4-phenyl- (9CI) **(38261-35-7)**
1,2,4-Triazin-5(2H)-one, 4-amino-6-(1,1-dimethylethyl)-3,4-dihydro-3-thioxo- (9CI) **(33509-43-2)**
1,2,4-Triazin-5(4H)-one, 4-amino-6-(1,1-dimethylethyl)-3-(methylthio)- **(21087-64-9)**
s-Triazin-2(1H)-one, 4-amino-6-ethyl- **(16352-07-1)**
1,3,5-Triazin-2(1H)-one, 4-amino-6-ethyl- (9CI) **(16352-07-1)**
1,3,5-Triazin-2(1H)-one, 4-amino-6-(ethylamino)- (9CI) **(7313-54-4)**
1,3,5-Triazin-2(1H)-one, 4-amino-6-methyl- (9CI) **(16352-06-0)**
1,2,4-Triazin-5(4H)-one, 4-amino-6-(1-methylethyl)-3-(methylthio)- **(21087-61-6)**
as-Triazin-5(4H)-one, 4-amino-3-methyl-6-phenyl **(41394-05-2)**
1,3,5-Triazin-2(1H)-one, 4-amino-6-phenyl- (9CI) **(33957-63-0)**
s-Triazin-2(1H)-one, 4-amino-6-phenyl- (8CI) **(33957-63-0)**
1,3,5-Triazin-2(1H)-one, 4-amino-1-β-D-ribofuranosyl- **(320-67-2)**
as-Triazin-3(2H)-one, 5-amino-2-β-D-ribofuranosyl **(3131-60-0)**
s-Triazin-2(1H)-one, 4-amino-1-β-D-ribofuranosyl **(320-67-2)**
1,3,5-Triazin-2(1H)-one, 4,6-bis(4-chlorophenyl)- (9CI) **(30886-09-0)**
1,3,5-Triazin-2(1H)-one, 4,6-bis(diethylamino)- (9CI) **(22305-35-7)**
1,3,5-Triazin-2(1H)-one, 4,6-bis(dimethylamino)-, hydrazone (9CI) **(10409-78-6)**
1,3,5-Triazin-2(1H)-one, 4,6-bis(ethylamino)- (9CI) **(2599-11-3)**
1,3,5-Triazin-2(1H)-one, 4,6-bis(ethylamino)-, hydrazone (9CI) **(10421-98-4)**
1,3,5-Triazin-2(1H)-one, 4,6-bis((1-methylethyl)amino)- (9CI) **(7374-53-0)**
1,3,5-Triazin-2(1H)-one, 4,6-bis(4-methylphenyl)- (9CI) **(30886-10-3)**
1,3,5-Triazin-2(1H)-one, 4,6-bis(phenylamino)- (9CI) **(30303-58-3)**
1,3,5-Triazin-2(1H)-one, 4-chloro-6-(ethylamino)- (9CI) **(49624-61-5)**
1,3,5-Triazin-2(1H)-one, 4-(4-chlorophenyl)-6-phenyl- (9CI) **(1917-38-0)**
1,3,5-Triazin-2(1H)-one, 4,6-diamino- (9CI) **(645-92-1)**
s-Triazin-2(1H)-one, 4,6-dianilino- **(30303-58-3)**
s-Triazin-2(1H)-one, 5,6-dihydro-4-amino-1-β-D-ribofuranosyl **(62488-57-7)**
1,3,5-Triazin-2(1H)-one, 4,6-dimethoxy- (9CI) **(1075-59-8)**
1,3,5-Triazin-2(1H)-one, 4,6-dimethoxy-, hydrazone (9CI) **(13882-61-6)**
1,3,5-Triazin-2(1H)-one, 4,6-dimethyl- (9CI) **(30885-99-5)**
1,2,4-Triazin-5(2H)-one, 6-(1,1-dimethylethyl)-3-(methylthio) **(35045-02-4)**
1,3,5-Triazin-2(1H)-one, 4,6-diphenyl- (9CI) **(1917-44-8)**
s-Triazin-2(1H)-one, 4,6-diphenyl- (8CI) **(1917-44-8)**
s-Triazin-2(1H)-one, 4,6-di-p-tolyl- (7CI,8CI) **(30886-10-3)**
1,3,5-Triazin-2(1H)-one, 4-(ethylamino)-6-methyl- (9CI) **(30369-64-3)**
1,3,5-Triazin-2(1H)-one, 4-ethyl-6-phenyl- (9CI) **(1917-40-4)**
s-Triazin-2(1H)-one, 6-ethyl-4-phenyl- (8CI) **(1917-40-4)**
1,3,5-Triazin-2(1H)-one, (hydroxyethyl)bis(hydroxymethyl)- (9CI) **(35503-54-9)**
1,3,5-Triazin-2(1H)-one, 4-(methylthio)-6-phenyl- (9CI) **(1917-43-7)**
s-Triazin-2(1H)-one, tetrahydro-4,6-bis(trichloromethyl)- (8CI) **(30805-36-8)**
s-Triazin-2(1H)-one, tetrahydro-4,6-diisopropyl- (8CI) **(30805-38-0)**
1,3,5-Triazin-2(1H)-one, tetrahydro-4,6-dimethyl- (9CI) **(30805-33-5)**
s-Triazin-2(1H)-one, tetrahydro-4,6-dimethyl- (7CI,8CI) **(30805-33-5)**
1,3,5-Triazin-2(1H)-one, tetrahydro-5-(2-hydroxyethyl)-1,3-bis-(hydroxymethyl)- (9CI) **(1852-21-7)**
s-Triazin-2(1H)-one, tetrahydro-4-imino-6-phenyl- (8CI) **(38261-35-7)**
s-Triazin-2(1H)-one, tetrahydro-5-isopropyl- (8CI) **(30913-44-1)**
1,3,5-Triazin-2(1H)-one, tetrahydro-5-methyl- (9CI) **(1910-89-0)**
s-Triazin-2(1H)-one, tetrahydro-5-methyl- (6CI,7CI,8CI) **(1910-89-0)**
N-(s-Triazin-2-yl)aniline **(4040-07-7)**
Triaziquinone **(68-76-8)**
Triaziquinonum **(68-76-8)**
Triaziquon **(68-76-8)**

Name	CAS No.
Triaziquon (German)	**(68-76-8)**
Triaziquone	**(68-76-8)**
Triaziridinophosphine oxide	**(545-55-1)**
2,3,5-Tri-(1-aziridinyl)-p-benzoquinone	**(68-76-8)**
Tris-(1-aziridinyl)fosfinoxid (Czech)	**(545-55-1)**
Tri(1-aziridinyl)phosphine oxide	**(545-55-1)**
Tri(aziridinyl)phosphine oxide	**(545-55-1)**
Tri-1-aziridinylphosphine oxide	**(545-55-1)**
Triaziridinylphosphine oxide	**(545-55-1)**
Tris-(1-aziridinyl)phosphine oxide (DOT)	**(545-55-1)**
Tris-(1-aziridinyl)phosphine oxide, Solution [UN 2501]	**(545-55-1)**
Triaziridinylphosphine sulfide	**(52-24-4)**
Triaziridinyl triazine	**(51-18-3)**
Triazofosz (Hungarian)	**(24017-47-8)**
Triazoic acid	**(7782-79-8)**
Triazol Brown B	**(2429-81-4)**
Triazol Fast Red 8B	**(6548-29-4)**
Triazol Fast Scarlet 3B	**(6358-29-8)**
Triazol Violet B	**(6426-67-1)**
Triazolam	**(28911-01-5)**
1H-1,2,4-Triazol-3-amine	**(61-82-5)**
Triazolamine	**(61-82-5)**
Triazolamum (Latin)	**(28911-01-5)**
Triazolblau 3BX	**(72-57-1)**
s-Triazole	**(288-88-0)**
1H-1,2,4-Triazole (9CI)	**(288-88-0)**
1H-1,2,4-Triazole, 1-acetyl- (6CI,7CI,8CI,9CI)	**(15625-88-4)**
s-Triazole, 3-amino	**(61-82-5)**
1H-1,2,4-Triazole, 1-benzoyl-3-phenyl- (9CI)	**(79746-00-2)**
4H-1,2,4-Triazole, 4-butyl	**(16227-10-4)**
s-Triazole, 4-butyl-	**(16227-10-4)**
1H-1,2,4-Triazole, 1-butyl- (8CI,9CI)	**(6086-22-2)**
1H-1,2,4-Triazole, 1-((tert-butylcarbonyl-4-chlorophenoxy)-methyl)-	**(43121-43-3)**
1,2,4-Triazole-3-carboxamide, 1-β-D-ribofuranosyl-	**(36791-04-5)**
1H-1,2,4-Triazole-3-carboxamide, 1-β-D-ribofuranosyl- (9CI)	**(36791-04-5)**
s-Triazole, 3,5-diamino	**(1455-77-2)**
1H-1,2,4-Triazole, 1-((2-(2,4-dichlorophenyl)-1,3-dioxolan-2-yl)methyl)	**(60207-31-0)**
1H-1,2,4-Triazole, 1-((2-(2,4-dichlorophenyl)-4-propyl-1,3-dioxolan-2-yl)methyl)	**(60207-90-1)**
1H-1,2,4-Triazole, 1-(2,2-dimethyl-1-oxopropyl)- (9CI)	**(60718-52-7)**
1H-1,2,4-Triazole-1-ethanol, β-((1,1'-biphenyl)-4-yloxy)-α-(1,1-dimethylethyl)	**(55179-31-2)**
1H-1,2,4-Triazole-1-ethanol, α-(2-(4-chlorophenyl)ethyl)-α-(1,1-dimethylethyl)-, (+-)	**(107534-96-3)**
1H-1,2,4-Triazole-1-ethanol, β-(cyclohexylmethylene)-α-(1,1-dimethylethyl)-, (E)- (9CI)	**(76608-88-3)**
1H-1,2,4-Triazole-1-ethanol, β-((2,4-dichlorophenyl)methyl)-α-(1,1-dimethylethyl)-, (R*,R*)-(+-)	**(75736-33-3)**
1H-1,2,4-Triazole-1-ethanol, α-(2-fluorophenyl)-α-(4-fluorophenyl)-, (+-)- (9CI)	**(87676-93-5)**
1H-1,2,4-Triazole, 3-ethyl- (9CI)	**(7411-16-7)**
s-Triazole, 3-ethyl- (8CI)	**(7411-16-7)**
1H-1,2,4-Triazole, 3-mercapto-	**(3179-31-5)**
1H-1,2,4-Triazole, 1-methyl- (8CI,9CI)	**(6086-21-1)**
1H-1,2,4-Triazole-3-thiol	**(3179-31-5)**
1,2,4-Triazolidine-3,5-dione (9CI)	**(3232-84-6)**
1H-1,2,4-Triazolium, 5-((4-(dimethylamino)phenyl)azo)-1,4-dimethyl-	**(12221-52-2)**
4H-(1,2,4)Triazolo(4,3-a)(1,4)benzodiazepine, 8-chloro-6-(2-chlorophenyl)-1-methyl-	**(28911-01-5)**
4H-s-Triazolo(4,3-a)(1,4)benzodiazepine, 8-chloro-6-(o-chlorophenyl)-1-methyl-	**(28911-01-5)**
4H-(1,2,4)Triazolo(4,3-a)(1,4)benzodiazepine, 8-chloro-1-methyl-6-phenyl-	**(28981-97-7)**
4H-s-Triazolo(4,3-a)(1,4)benzodiazepine, 8-chloro-1-methyl-6-phenyl-	**(28981-97-7)**
4H-(1,2,4)Triazolo(4,3-a)(1,4)-benzodiazepine, 8-chloro-6-phenyl-	**(29975-16-4)**
4H-(1,2,4)Triazolo(4,3-a)(1,4)benzodiazepine, 8-chloro-6-phenyl- (9CI)	**(29975-16-4)**
4H-s-Triazolo(4,3-a)(1,4)benzodiazepine, 8-chloro-6-phenyl- (8CI)	**(29975-16-4)**
(1,2,4)Triazolo(1,5-a)pyrimidin-7-ol, 5-methyl- (9CI)	**(2503-56-2)**
(1,2,4)Triazolo(1,5-a)pyrimidin-7-ol, 5-methyl-, sodium salt (9CI)	**(38299-08-0)**
s-Triazolo(3,4-b)benzothiazole, 5-methyl	**(41814-78-2)**
1H-v-Triazolo(4,5-d)pyrimidine-5,7(4H,6H)-dione	**(1468-26-4)**
v-Triazolo(4,5-d)pyrimidine-5,7-diol	**(1468-26-4)**
v-Triazolo(4,5-d)pyrimidin-7-ol, 5-amino-	**(134-58-7)**
7H-v-Triazolo(4,5-d)pyrimidin-7-one, 5-amino-1,6-dihydro	**(134-58-7)**
7H-1,2,3-Triazolo(4,5-d)pyrimidin-7-one, 5-amino-1,4-dihydro- (9CI)	**(134-58-7)**
Triazologuanine	**(134-58-7)**
1,2,4-Triazol-thiol-(3) (German)	**(3179-31-5)**
Triazophos	**(24017-47-8)**
Triazotion (Russian)	**(2642-71-9)**
Triazulenone	**(61197-73-7)**
Triazurol	**(500-42-5)**
Tribac	**(50-31-7)**
Tri-basic Copper Fungicide	**(1332-03-2)**
Tribasic aluminum stearate	**(637-12-7)**
Tribasic calcium citrate	**(813-94-5)**
Tribasic copper sulfate	**(1332-03-2)**
Tribasic copper sulfate monohydrate	**(1332-03-2)**
Tribasic magnesium phosphate	**(7757-87-1)**
Tribasic sodium phosphate	**(7601-54-9)**
Tribavirin	**(36791-04-5)**
2,4,6-Tribenzamido-s-triazine	**(5637-84-3)**
Tribenzo(a,e,i)pyrene	**(192-47-2)**
Tribenzo(a,i,l)pyrene	**(191-20-8)**
Tribenzofluoranthene (9CI)	**(65256-17-9)**
Tribenzoperylene (9CI)	**(65256-18-0)**
(1,2,4,5,7,8)-Tribenzopyrene	**(192-47-2)**
(1,2,4,5,8,9)-Tribenzopyrene	**(192-47-2)**
1,2:4,5:8,9-Tribenzopyrene	**(192-47-2)**
N2,N4,N6-Tribenzoylmelamine	**(5637-84-3)**
Tribenzylamine, hydrochloride	**(7673-07-6)**
Tribenzyl-s-triazine	**(13960-31-1)**
Tribomoneopentyl alcohol	**(1522-92-5)**
Tribrommethaan (Dutch)	**(75-25-2)**
Tribrommethan (German)	**(75-25-2)**
2,2,2-Tribromoacetaldehyde	**(115-17-3)**
Tribromoacetaldehyde	**(115-17-3)**
Tribromoacetic acid	**(75-96-7)**
Tribromoaluminum	**(7727-15-3)**
2,4,6-Tribromoaniline	**(147-82-0)**
sym-Tribromoaniline	**(147-82-0)**
Tribromoarsine	**(7784-33-0)**
2,3,7-Tribromobenzo-4-dioxin	**(51974-40-4)**
1,2,4-Tribromobenzene	**(615-54-3)**
1,3,5-Tribromobenzene	**(626-39-1)**
1,3,5-Tribromo-2-(2-bromoethoxy)benzene	**(68413-71-8)**
Tribromo-tert-butyl alcohol	**(76-08-4)**
3,4,5-Tribromocatechol	**(2747-17-3)**
1,2,8-Tribromodibenzofuran	**(84761-81-9)**
2,3,8-Tribromodibenzofuran	**(84761-82-0)**
Tribromodiphenyl ether	**(49690-94-0)**
Tribromodiphenyl oxide	**(49690-94-0)**
Tribromoethene	**(598-16-3)**
1,1,2-Tribromoethylene	**(598-16-3)**
Tribromoethylene	**(598-16-3)**
3,4,5-Tribromoguaiacol	**(38926-85-1)**

2,4,5-Tribromoimidazole	(2034-22-2)
Tribromometan (Italian)	(75-25-2)
Tribromomethane	(75-25-2)
1,3,5-Tribromo-2-methoxybenzene	(607-99-8)
1,1,1-Tribromo-2-methyl-2-propanol	(76-08-4)
2-Tribromomethyl-2-propanol	(76-08-4)
Tribromoneopentyl alcohol	(36483-57-5)
2,4,6-Tribromophenol	(118-79-6)
Tribromophenol	(118-79-6)
Tribromophenol	(25376-38-9)
2,4,6-Tribromophenol carbonate (2:1)	(67990-32-3)
Tribromophenoxybenzene	(49690-94-0)
1-(2,4,6-Tribromophenoxy)-2-bromoethane	(68413-71-8)
2-(2,4,6-Tribromophenoxy)ethanol	(23976-66-1)
Tri(p-bromophenyl)phosphate	(40946-60-9)
Tribromophosphine	(7789-60-8)
1,2,3-Tribromopropane	(96-11-7)
sym-Tribromopropane	(96-11-7)
1,1,3-Tribromo-1-propene	(36417-14-8)
3,4,5-Tribromopyrocatechol	(2747-17-3)
3,4',5-Tribromosalicylanide and 4,5-dibromosalicylanide mixtures	(87-10-5)
3,4',5-Tribromosalicylanilide	(87-10-5)
Tribromosalicylanilide	(1322-38-9)
Tribromosalicylanilide	(87-10-5)
Tribromostibine	(7789-61-9)
Tribromotoluene	(27476-22-8)
2,4,6-Tribromo-s-triazine	(14921-00-7)
2,4,6-Tribromo-sym-triazine	(14921-00-7)
1,2,4-Tribromo-3,5,6-trichlorobenzene	(13075-01-9)
Tribromotrichlorocyclohexane	(30554-73-5)
Tribromotrimethyldialuminum	(12263-85-3)
3,4,5-Tribromo-N,N,α-trimethyl-1H-pyrazole-1-acetamide	(34157-48-7)
Tribromsalan	(87-10-5)
Tribromsalanum (Latin)	(87-10-5)
Tribromsalen	(87-10-5)
Tribufon	(126-22-7)
Tribunil	(18691-97-9)
Tribunil-Combi	(39283-72-2)
Tributilamina (Romanian)	(102-82-9)
Tributilfosfato (Italian)	(126-73-8)
Tributon	(93-76-5)
Tributon	(94-75-7)
Tri-n-butoxyborane	(688-74-4)
Tributoxyborane	(688-74-4)
Tri(2-butoxyethanol)phosphate	(78-51-3)
Tris-(2-butoxyethyl)fosfat (Czech)	(78-51-3)
Tri(2-butoxyethyl) phosphate	(78-51-3)
Tributoxyethyl phosphate	(78-51-3)
2,4,6-Tributoxytriazine	(19837-00-4)
Tributyl 2-acetoxy-1,2,3-propanetricarboxylate	(77-90-7)
Tributyl O-acetylcitrate	(77-90-7)
Tributyl acetylcitrate	(77-90-7)
Tributyl 2-(acetyloxy)-1,2,3-propanetricarboxylic acid	(77-90-7)
Tri-n-butylaluminum	(1116-70-7)
Tributylaluminum	(1116-70-7)
Tri-n-butylamine	(102-82-9)
Tris-n-butylamine	(102-82-9)
Tributylamine [UN 2542]	(102-82-9)
Tributylamine, heptacosafluoro	(311-89-7)
1,2,4-Tri-tert-butylbenzene	(1459-11-6)
1,2,4-Tributylbenzene	(14800-16-9)
1,3,4-Tributylbenzene	(14800-16-9)
Tri-n-butyl borate	(688-74-4)
Tributyl borate	(688-74-4)
N,N,N-Tributyl-1-butanaminium bromide	(1643-19-2)
N,N,N-Tributyl-1-butanaminium sulfate (1:1)	(32503-27-8)
Tributyl cellosolve phosphate	(78-51-3)
Tributylchlorotin	(1461-22-9)
Tri-n-butyl citrate	(77-94-1)
Tributyl citrate	(77-94-1)
Tributyl citrate acetate	(77-90-7)
Tri(butylcresyl)butane	(1843-03-4)
Tributylcresylbutane	(1843-03-4)
Tributyl(2,4-dichlorobenzyl)phosphonium chloride	(115-78-6)
Tributyle (phosphate de) (French)	(126-73-8)
Tributylester kyseliny borite (Czech)	(688-74-4)
Tributylfosfaat (Dutch)	(126-73-8)
Tributylfosfat (Czech)	(126-73-8)
Tributylfosfin (Czech)	(998-40-3)
Tributylfosfinoxid (Czech)	(814-29-9)
Tributylfosfit (Czech)	(102-85-2)
Tributyl lead chloride	(13302-14-2)
Tributyl(methacryloxy)stannane	(2155-70-6)
Tributyl(methacryloyloxy)stannane	(2155-70-6)
Tributyl((2-methyl-1-oxo-2-propenyl)oxy)stannane	(2155-70-6)
N,N',N''-Tributyl-1-methylsilanetriamine	(16411-33-9)
2,4,6-Tributyloxy-s-triazine	(19837-00-4)
(Tributyl)peroxide	(110-05-4)
2,4,6-Tri-tert-butylphenol	(732-26-3)
Tri(tert-butylphenyl) phosphate	(28777-70-0)
Tributylphosphat (German)	(126-73-8)
Tri-n-butyl phosphate	(126-73-8)
Tributyl phosphate (ACGIH,OSHA)	(126-73-8)
Tri-n-butylphosphine	(998-40-3)
Tributylphosphine	(998-40-3)
Tributylphosphine oxide	(814-29-9)
Tributyl phosphite	(102-85-2)
S,S,S-Tributyl phosphorotrithioate	(78-48-8)
S,S,S-Tributyl phosphorotrithioite	(150-50-5)
Tributyl phosphorotrithioite	(150-50-5)
Tributylplumbium chloride	(13302-14-2)
Tributylstannane	(688-73-3)
Tributylstannane fluoride	(1983-10-4)
Tributylstannic hydride	(688-73-3)
Tributylstannium acetate	(56-36-0)
Tributylstannium chloride	(1461-22-9)
Tributylstannyl chloride	(1461-22-9)
Tri-n-butylstannylmethacrylate	(2155-70-6)
Tributylstannyl methacrylate	(2155-70-6)
Tributylthiofosfin (Czech)	(150-50-5)
Tributyltin	(688-73-3)
Tributyltin acetate	(56-36-0)
Tributyltin acrylate	(13331-52-7)
Tributyltin benzoate	(4342-36-3)
Tributyltin chloride	(1461-22-9)
Tributyltin chloride complex of ethylene oxide condensate of abietylamine	(56573-85-4)
Tributyltin chloride complex of ethylene oxide condensate of abietylamine (Give equivalent tin)	(56573-85-4)
Tributyltin fluoride	(1983-10-4)
Tri-n-butyltin hydride	(688-73-3)
Tributyltin hydride	(688-73-3)
Tri-n-butyl tin iodide	(7342-47-4)
Tributyltin isopropyl succinate	(53404-82-3)
Tributyltin isopropylsuccinate	(53404-82-3)
Tributyltin linoleate	(24124-25-2)
Tributyltin maleate	(14275-57-1)
Tributyltin methacrylate (8CI)	(2155-70-6)
Tributyltin monopropyleneglycol maleate	(53466-85-6)
Tributyltin neodecanoate	(28801-69-6)
S,S,S-Tributyltrithiofosfat (Czech)	(78-48-8)
S,S,S-Tributyl trithiophosphate	(78-48-8)
S,S,S-Tributyl trithiophosphite	(150-50-5)
Tri-n-butyl-zinn-acetat (German)	(56-36-0)

Tri-n-butyl-zinn benzoate (German) (4342-36-3)
Tri-n-butylzinn-chlorid (German) (1461-22-9)
Tributyrin (60-01-5)
Tributyroin (60-01-5)
Tricaine (582-33-2)
Tricalciumarsenat (German) (7778-44-1)
Tricalcium arsenate (7778-44-1)
Tricalcium citrate (813-94-5)
Tricalcium diphosphate (7758-87-4)
Tricalcium orthophosphate (7758-87-4)
Tricalcium phosphate (7758-87-4)
Tricalcium silicate (12168-85-3)
Tricamba (2307-49-5)
Tricapronin (621-70-5)
Tricaprylic glyceride (538-23-8)
Tricaprylin (538-23-8)
Tricaprylmethylammonium chloride (5137-55-3)
Tricaprylylamine (1116-76-3)
Tricaprylylmethylammonium chloride (5137-55-3)
Tricarballylic acid, β-acetoxytributyl ester (77-89-4)
Tricarbamix Z (137-30-4)
2,4,6-Tricarbamoyl-s-triazine (23297-24-7)
1,2,3-Tricarboxybenzene (569-51-7)
1,2,4-Tricarboxybenzene (528-44-9)
1,3,5-Tricarboxybenzene (554-95-0)
Tricarnam (63-25-2)
Tricesium tribromide (7787-69-1)
Tricesium trichloride (7647-17-8)
Tricesium triiodide (7789-17-5)
Trichazol (443-48-1)
Trichex (443-48-1)
O-(2,4,5-Trichloor-fenyl)-O,O-dimethyl-monothiofosfaat (Dutch) (299-84-3)
Trichloorazijnzuur (Dutch) (76-03-9)
1,1,1-Trichloor-2,2-bis(4-chloor fenyl)-ethaan (Dutch) (50-29-3)
2,2,2-Trichloor-1,1-bis(4-chloor fenyl)-ethanol (Dutch) (115-32-2)
1,1,1-Trichloorethaan (Dutch) (71-55-6)
Trichlooretheen (Dutch) (79-01-6)
Trichloorethyleen, tri (Dutch) (79-01-6)
(2,4,5-Trichloor-fenoxy)-azijnzuur (Dutch) (93-76-5)
2-(2,4,5-Trichloor-fenoxy)-propionzuur (Dutch) (93-72-1)
Trichloorfon (Dutch) (52-68-6)
Trichloormethaan (Dutch) (67-66-3)
Trichloormethylbenzeen (Dutch) (98-07-7)
Trichloornitromethaan (Dutch) (76-06-2)
Trichloorsilaan (Dutch) (10025-78-2)
(2,4,5-Trichlor-phenoxy)-essigsaeure (German) (93-76-5)
2-(2,4,5-Trichlor-phenoxy)-propionsaeure (German) (93-72-1)
Trichloracetaldehyd-hydrat (German) (302-17-0)
Trichloracetic acid (76-03-9)
Trichlor-acetonitril (German) (545-06-2)
1,1,1-Trichloraethan (German) (71-55-6)
Trichloraethen (German) (79-01-6)
Trichloraethylen, tri (German) (79-01-6)
2,3,3-Trichlorallyl-N,N-(diisopropyl)-thiocarbamat (German) (2303-17-5)
Trichloramine (10025-85-1)
Trichloran (79-01-6)
2,3,6-Trichlorbenzoesaeure (German) (50-31-7)
1,1,1-Trichlor-2,2-bis(4-chlor-phenyl)-aethan (German) (50-29-3)
1,1,1-Trichlor-2,2-bis(4-chlorphenyl)-aethanol (German) (115-32-2)
2,2,2-Trichlor-1,1-bis(4-chlor-phenyl)-aethanol (German) (115-32-2)
1,1,1-Trichlor-2,2-bis(4-methoxy-phenyl)-aethan (German) (72-43-5)
Trichlorbutylene (51023-22-4)
Trichlordene (21641-70-3)
Trichloren (79-01-6)
Trichloressigsaeure (German) (76-03-9)
Trichloressigsaures natrium (German) (650-51-1)
1,1,2-Trichlorethane (79-00-5)
Trichlorethanol (115-20-8)
Trichlorethene (French) (79-01-6)
Trichlorethylene (79-01-6)
Trichlorethylene, tri (French) (79-01-6)
Tris-(2-chlorethyl)fosfat (Czech) (115-96-8)
Trichlorethyl phosphate (115-96-8)
2,4,6-Trichlorfenol (Czech) (88-06-2)
2-(2,4,5-Trichlorfenoxy)ethylester kyseliny 2,2-dichlor-propionove (Czech) (136-25-4)
Trichlorfenson (Obs.) (80-33-1)
Trichlor-2-fenylethylsilan (Czech) (940-41-0)
Trichlorfon (German) (52-68-6)
Trichlorfon [UN 2783] (52-68-6)
Trichlorinated isocyanuric acid (87-90-1)
Trichlorine nitride (10025-85-1)
Trichlor-2-kyanethylsilan (Czech) (1071-22-3)
Trichlormetafos-3 (2633-54-7)
Trichlormetazid (133-67-5)
Trichlormethan (Czech) (67-66-3)
Trichlormethiazide (133-67-5)
Trichlormethine (555-77-1)
Trichlormethine (817-09-4)
Trichlormethinium chloride (817-09-4)
Trichlormethylbenzol (German) (98-07-7)
Trichlormethylester kyseliny chlormravenci (Czech) (503-38-8)
Trichlormethylkyanid (Czech) (545-06-2)
Trichlor-methylsilan (Czech) (75-79-6)
Trichlormethylthioamid kyseliny 1,2,3,6-tetrahydroftalove (Czech) (133-06-2)
Trichlormethylthioimid kyseliny ftalove (Czech) (133-07-3)
N-(Trichlor-methylthio)-phthalamid (German) (133-07-3)
N-(Trichlor-methylthio)-phthalimid (German) (133-06-2)
N-(Trichlormethylthio)phthalimide (133-07-3)
Trichlornitromethan (German) (76-06-2)
2,2,2-Trichloroacetaldehyde (75-87-6)
Trichloroacetaldehyde (75-87-6)
Trichloroacetaldehyde hydrate (302-17-0)
Trichloroacetaldehyde monohydrate (302-17-0)
2,2,2-Trichloroacetamide (594-65-0)
Trichloroacetamide (594-65-0)
α,α,α-Trichloroacetamide (594-65-0)
Trichloroacetic acid (ACGIH,OSHA) (76-03-9)
Trichloroacetic acid, Solid [UN 1839] (76-03-9)
Trichloroacetic acid, Solution [UN 2564] (76-03-9)
Trichloroacetic acid anhydride (4124-31-6)
Trichloroacetic acid chloride (76-02-8)
Trichloroacetic acid sodium salt (650-51-1)
Trichloroacetic anhydride (4124-31-6)
Trichloroacetochloride (76-02-8)
1,1,1-Trichloroacetone (918-00-3)
1,1,3-Trichloroacetone (921-03-9)
α,α',α'-Trichloroacetone (921-03-9)
α,α,α-Trichloroacetone (918-00-3)
Trichloroacetonitrile (545-06-2)
2,2',4'-Trichloroacetophenone (4252-78-2)
2,3,4-Trichloroacetophenone (13608-87-2)
Trichloroacetyl chloride [UN 2442] (76-02-8)
Trichloroacrylonitrile (16212-28-5)
α,β,β-Trichloroacrylonitrile (16212-28-5)
2,3,3-Trichloroallyl diisopropylthiocarbamate (2303-17-5)
S-2,3,3-Trichloroallyl N,N-diisopropylthiocarbamate (2303-17-5)
Trichloroaluminum (7446-70-0)
3,5,6-Trichloro-4-aminopicolinic acid (1918-02-1)
2,4,6-Trichloroaniline (634-93-5)
sym-Trichloroaniline (634-93-5)
3,5,6-Trichloro-o-anisic acid (2307-49-5)
2,4,6-Trichloroanisole (87-40-1)
Trichloroarsine (7784-34-1)

2,3,6-Trichlorobenzaldehyde **(4659-47-6)**
2,4,5-Trichlorobenzenamine **(636-30-6)**
2,4,6-Trichlorobenzenamine **(634-93-5)**
1,2,3-Trichlorobenzene **(87-61-6)**
1,2,5-Trichlorobenzene **(120-82-1)**
1,2,6-Trichlorobenzene **(87-61-6)**
1,3,4-Trichlorobenzene **(120-82-1)**
1,3,5-Trichlorobenzene **(108-70-3)**
Trichlorobenzene **(12002-48-1)**
s-Trichlorobenzene **(108-70-3)**
sym-Trichlorobenzene **(108-70-3)**
unsym-Trichlorobenzene **(120-82-1)**
vic-Trichlorobenzene **(87-61-6)**
1,2,4-Trichlorobenzene (ACGIH,OSHA) **(120-82-1)**
1,2,3-Trichlorobenzene, Liquid [UN 2321] **(87-61-6)**
1,2,4-Trichlorobenzene, Liquid [UN 2321] **(120-82-1)**
1,3,5-Trichlorobenzene, Liquid [UN 2321] **(108-70-3)**
Trichlorobenzene, Liquid [UN 2321] **(12002-48-1)**
2,3,6-Trichlorobenzeneacetic acid **(85-34-7)**
2,4,5-Trichlorobenzenesulfonyl chloride **(15945-07-0)**
2,4,5-Trichlorobenzenethiol **(3773-14-6)**
2,3,5-Trichlorobenzoic acid **(50-73-7)**
2,3,6-Trichlorobenzoic acid **(50-31-7)**
2,4,5-Trichlorobenzoic acid **(50-82-8)**
2,4,6-Trichlorobenzoic acid **(50-43-1)**
Trichlorobenzoic acid **(50-31-7)**
2,3,6-Trichlorobenzoic acid, dimethylamine salt **(3426-62-8)**
2,3,6-Trichlorobenzoic acid, sodium salt **(2078-42-4)**
1,2,4-Trichlorobenzol **(120-82-1)**
4,5,7-Trichloro-2,1,3-benzothiadiazole **(1982-55-4)**
4,5,7-Trichlorobenzthiadiazole-2,1,3 **(1982-55-4)**
Trichlorobenzyl chloride **(1344-32-7)**
1-(2,3,6-Trichlorobenzyloxy)propan-2-ol **(1861-44-5)**
2',3,4-Trichlorobiphenyl **(38444-86-9)**
2,2',3-Trichlorobiphenyl **(38444-78-9)**
2,2',4-Trichlorobiphenyl **(37680-66-3)**
2,2',5-Trichlorobiphenyl **(37680-65-2)**
2,2',6-Trichlorobiphenyl **(38444-73-4)**
2,3',4-Trichlorobiphenyl **(55712-37-3)**
2,3',5'-Trichlorobiphenyl **(37680-68-5)**
2,3',5-Trichlorobiphenyl **(38444-81-4)**
2,3',6-Trichlorobiphenyl **(38444-76-7)**
2,3,3'-Trichlorobiphenyl **(38444-84-7)**
2,3,4'-Trichlorobiphenyl **(38444-85-8)**
2,3,4-Trichlorobiphenyl **(55702-46-0)**
2,3,5-Trichlorobiphenyl **(55720-44-0)**
2,3,6-Trichlorobiphenyl **(55702-45-9)**
2,4',5-Trichloro-1,1'-biphenyl **(16606-02-3)**
2,4',5-Trichlorobiphenyl **(16606-02-3)**
2,4',6-Trichlorobiphenyl **(38444-77-8)**
2,4,4'-Trichlorobiphenyl **(7012-37-5)**
2,4,5-Trichlorobiphenyl **(15862-07-4)**
2,4,6-Trichlorobiphenyl **(35693-92-6)**
3,3',4-Trichlorobiphenyl **(37680-69-6)**
3,3',5-Trichlorobiphenyl **(38444-87-0)**
3,4',5-Trichlorobiphenyl **(38444-88-1)**
3,4,4'-Trichlorobiphenyl **(38444-90-5)**
4,2',5'-Trichlorobiphenyl **(16606-02-3)**
Trichlorobiphenyl **(25323-68-6)**
1,1,1-Trichloro-2,2-bis(p-anisyl)ethane **(72-43-5)**
1,1,1-Trichloro-2,2-bis(chlorophenyl)ethane **(33086-18-9)**
1,1,1-Trichloro-2,2-bis(p-chlorophenyl)ethane **(50-29-3)**
Trichlorobis(4-chlorophenyl)ethane **(50-29-3)**
2,2,2-Trichloro-1,1-bis(4-chlorophenyl)-ethanol (French) **(115-32-2)**
2,2,2-Trichloro-1,1-bis(4-cloro-fenil)-etanolo (Italian) **(115-32-2)**
1,1,1-Trichloro-2,2-bis(p-fluorophenyl) ethane **(475-26-3)**
1,1,1-Trichloro-2,2-bis(p-methoxyphenol)ethanol **(72-43-5)**
1,1,1-Trichloro-2,2-bis(4-methoxyphenyl)ethane **(72-43-5)**
1,1,1-Trichloro-2,2-bis(p-methoxyphenyl)ethane **(72-43-5)**
2,2,2-Trichloro-1,1-bis(4-methoxyphenyl)ethane **(72-43-5)**
Trichlorobromomethane **(75-62-7)**
1,1,2-Trichlorobutadiene **(2852-07-5)**
4,4,4-Trichloro-1,2-butadiene **(34819-62-0)**
1,2,4-Trichlorobutane **(1790-22-3)**
1,3,3-Trichlorobutane **(15187-71-0)**
Trichloro-tert-butanol **(57-15-8)**
1,1,4-Trichloro-2-butene **(41601-59-6)**
1,2,4-Trichlorobutene-2 **(2431-54-1)**
2,3,4-Trichlorobutene-1 **(2431-50-7)**
trans-1,1,4-Trichlorobut-2-ene **(57808-36-3)**
Trichlorobutene [UN 2322] **(51023-22-4)**
Trichloro-t-butyl alcohol **(57-15-8)**
β,β,β-Trichloro-tert-butyl alcohol **(57-15-8)**
t-Trichlorobutyl alcohol **(57-15-8)**
Trichlorobutylene oxide **(3083-25-8)**
Trichlorobutylsilane **(7521-80-4)**
3,4,4'-Trichlorocarbanilide **(101-20-2)**
3,4,6-Trichlorocatechol **(32139-72-3)**
Trichlorocatechol **(25167-84-4)**
2,4,5-Trichloro-α-(chloromethylene)benzyl phosphate **(961-11-5)**
2,4,5-Trichloro-α-(chloromethylene)benzyl phosphate ester **(22248-79-9)**
Trichloro(chloromethyl)-silane **(1558-25-4)**
Trichloro(chloromethyl)silane **(1558-25-4)**
1,1,1-Trichloro-2-(o-chlorophenyl)-2-(p-chlorophenyl)ethane **(789-02-6)**
Trichloro(3-chloropropyl)silane **(2550-06-3)**
Trichlorochromium **(10025-73-7)**
2,4,6-Trichloro-m-cresol **(551-76-8)**
Trichloroctan sodny (Czech) **(650-51-1)**
Trichlorocumene **(61465-79-0)**
Trichlorocyanidine **(108-77-0)**
Trichlorocyanuric acid **(87-90-1)**
Trichloro-3-cyclohexen-1-ylsilane **(10137-69-6)**
Trichloro-3-cyclohexenylsilane **(10137-69-6)**
Trichlorocyclohexylsilane **(98-12-4)**
1,2,4-Trichlorodibenzo-para-dioxin **(39227-58-2)**
1,3,7-Trichlorodibenzo(b,e)(1,4)dioxin **(67028-17-5)**
Trichlorodibenzo(b,e)(1,4)dioxin **(69760-96-9)**
1,2,4-Trichlorodibenzodioxin **(39227-58-2)**
1,3,7-Trichlorodibenzo-p-dioxin **(67028-17-5)**
2,3,7-Trichlorodibenzo-p-dioxin **(33857-28-2)**
Trichlorodibenzo-p-dioxin **(69760-96-9)**
1,2,3-Trichlorodibenzofuran **(83636-47-9)**
1,2,6-Trichlorodibenzofuran **(64560-15-2)**
1,3,4-Trichlorodibenzofuran **(82911-61-3)**
1,3,6-Trichlorodibenzofuran **(83704-39-6)**
1,3,7-Trichlorodibenzofuran **(64560-16-3)**
1,3,8-Trichlorodibenzofuran **(76621-12-0)**
1,4,6-Trichlorodibenzofuran **(82911-60-2)**
1,4,8-Trichlorodibenzofuran **(64560-14-1)**
1,4,9-Trichlorodibenzofuran **(70648-13-4)**
2,3,4-Trichlorodibenzofuran **(57117-34-7)**
2,3,7-Trichlorodibenzofuran **(58802-17-8)**
2,3,8-Trichlorodibenzofuran **(57117-32-5)**
2,4,6-Trichlorodibenzofuran **(58802-14-5)**
2,4,7-Trichlorodibenzofuran **(83704-42-1)**
2,4,8-Trichlorodibenzofuran **(54589-71-8)**
2,4,9-Trichlorodibenzofuran **(82911-59-9)**
2,6,7-Trichlorodibenzofuran **(83704-45-4)**
N-(2,2,2-Trichloro-1-(3,4-dichloroanilino)ethyl)formamide **(20856-57-9)**
1,1,1-Trichloro-2,2-di(4-chlorophenyl)-ethane **(50-29-3)**
2,2,2-Trichloro-1,1-di-(4-chlorophenyl)ethanol **(115-32-2)**
Trichloro(dichlorophenyl)silane **(27137-85-5)**
3,4,5-Trichloro-2,6-dimethoxyphenol **(2539-26-6)**
1,1,1-Trichloro-2,2-di(4-methoxyphenyl)ethane **(72-43-5)**

Name	CAS
1,2,3-Trichloro-4,6-dinitrobenzene	**(6379-46-0)**
Trichlorodiphenyl	**(25323-68-6)**
1,1,1-Trichloro-2,2-diphenylethane	**(2971-22-4)**
Trichloro diphenyl ether	**(31242-93-0)**
Trichloro diphenyl oxide	**(31242-93-0)**
3,4,4'-Trichlorodiphenylurea	**(101-20-2)**
Trichlorododecylsilane	**(4484-72-4)**
4,4,4-Trichloro-1,2-epoxybutane	**(3083-25-8)**
1,1,2-Trichloroepoxyethane	**(16967-79-6)**
Trichloroethanal	**(75-87-6)**
1,1,1-Trichloroethane	**(71-55-6)**
1,2,2-Trichloroethane	**(79-00-5)**
Trichloroethane	**(25323-89-1)**
Trichloroethane	**(71-55-6)**
α-Trichloroethane	**(71-55-6)**
β-Trichloroethane	**(79-00-5)**
1,1,2-Trichloroethane (ACGIH,OSHA)	**(79-00-5)**
Trichloro-1,1,1-ethane (French)	**(71-55-6)**
1,1,1-Trichloroethane (OSHA) [UN 2831]	**(71-55-6)**
1,1,1-Trichloroethane, 2,2-bis(p-fluorophenyl-	**(475-26-3)**
1,1,1-Trichloro-2,2-ethanediol	**(302-17-0)**
2,2,2-Trichloro-1,1-ethanediol	**(302-17-0)**
Trichloroethanoic acid	**(76-03-9)**
2,2,2-Trichloroethanol	**(115-20-8)**
Trichloroethanol	**(115-20-8)**
Trichloroethene	**(79-01-6)**
Trichloroethyl acetate	**(515-84-4)**
2,2,2-Trichloroethyl alcohol	**(115-20-8)**
Trichloroethyl alcohol	**(115-20-8)**
Tri-(2-chloroethyl)amine	**(555-77-1)**
Tri(β-chloroethyl)amine hydrochloride	**(817-09-4)**
Tri-(2-chloroethyl)amine hydrochloride	**(817-09-4)**
2-(1,1,2-Trichloroethyl)-4,6-bis(trichloromethyl)-1,3,5-triazine	**(24481-33-2)**
2,2,2-Trichloroethyl dimethyl phosphate	**(52-68-6)**
1,1,2-Trichloroethylene	**(79-01-6)**
1,2,2-Trichloroethylene	**(79-01-6)**
Trichloroethylene (ACGIH,OSHA) [UN 1710]	**(79-01-6)**
Trichloroethylene epoxide	**(16967-79-6)**
Trichloroethylene oxide	**(16967-79-6)**
1,1'-(2,2,2-Trichloroethylidene)bis(4-methoxybenzene)	**(72-43-5)**
4,4-(2,2,2-Trichloroethylidene)dianisole	**(72-43-5)**
(R)-1,2-O-(2,2,2-Trichloroethylidene)-α-D-glucofuranose	**(15879-93-3)**
1,2-O-(2,2,2-Trichloroethylidene)-α-D-glucofuranose	**(15879-93-3)**
2,2,2-Trichloroethyl N-phenylcarbamate	**(42864-21-1)**
2,2,2-Trichloroethyl phosphate	**(306-52-5)**
Tri(2-chloroethyl)phosphate	**(115-96-8)**
Tri-β-chloroethyl phosphate	**(115-96-8)**
Trichloroethyl phosphate	**(306-52-5)**
Trichloroethylsilane	**(115-21-9)**
Trichloroethylsilicane	**(115-21-9)**
1,1,2-Trichloro-2-fluoroethane	**(359-28-4)**
Trichlorofluoromethane (ACGIH,OSHA)	**(75-69-4)**
Trichlorofon	**(52-68-6)**
Trichloroform	**(67-66-3)**
Trichlorogallium	**(13450-90-3)**
4,5,6-Trichloroguaiacol	**(2668-24-8)**
Trichloroguaiacol	**(57057-83-7)**
Trichloroguaiacol	**(61966-36-7)**
Trichlorohydrin	**(96-18-4)**
2,4,4'-Trichloro-2'-hydroxydiphenyl ether	**(3380-34-5)**
((2,2,2-Trichloro-1-hydroxyethyl) dimethylphosphonate)	**(52-68-6)**
2,2,2-Trichloro-1-hydroxyethyl-phosphonate, dimethyl ester	**(52-68-6)**
(2,2,2-Trichloro-1-hydroxyethyl)phosphonic acid dimethyl ester	**(52-68-6)**
2,3,5-Trichloro-4-hydroxypyridine	**(1970-40-7)**
2,4,5-Trichloroimidazole	**(7682-38-4)**
Trichloroisocyanic acid	**(87-90-1)**
Trichloroisocyanurate	**(87-90-1)**
1,3,5-Trichloroisocyanuric acid	**(87-90-1)**
N,N',N''-Trichloroisocyanuric acid	**(87-90-1)**
Trichloroisocyanuric acid	**(87-90-1)**
Trichloroisocyanuric acid, Dry [UN 2468]	**(87-90-1)**
Trichlorol	**(2633-54-7)**
Trichloromelamine	**(7673-09-8)**
Trichlorometafos	**(299-84-3)**
Trichlorometaphos-3	**(2633-54-7)**
Trichloromethane	**(67-66-3)**
Trichloromethane (OSHA)	**(67-66-3)**
Trichloromethane sulfenyl chloride	**(594-42-3)**
Trichloromethanesulphenyl chloride	**(594-42-3)**
Trichloromethanethiol	**(75-70-7)**
Trichloro-3-methaphos	**(2633-54-7)**
Trichloromethiadiazide	**(133-67-5)**
Trichloromethiazide	**(133-67-5)**
1,2,4-Trichloro-5-methoxybenzene	**(6130-75-2)**
1,3,5-Trichloro-2-methoxybenzene	**(87-40-1)**
3,5,6-Trichloro-2-methoxybenzoic acid	**(2307-49-5)**
Trichloro-2-methoxyphenol	**(61966-36-7)**
1,2,4-Trichloro-5-methylbenzene	**(6639-30-1)**
1-(Trichloromethyl)benzene	**(98-07-7)**
Trichloromethylbenzene	**(30583-33-6)**
Trichloromethylbenzene	**(98-07-7)**
α-(Trichloromethyl)benzenemethanol	**(2000-43-3)**
α-(Trichloromethyl)benzyl alcohol	**(2000-43-3)**
Trichloromethyl bromide	**(75-62-7)**
p-Trichloromethylchlorobenzene	**(5216-25-1)**
Trichloromethyl chloroformate	**(503-38-8)**
Trichloromethyl cyanide	**(545-06-2)**
3-(Trichloromethyl)-5-ethoxy-1,2,4-thiadiazole	**(2593-15-9)**
Trichloromethyl mercaptan	**(75-70-7)**
N-Trichloromethylmercapto-4-cyclohexene-1,2-dicarboximide	**(133-06-2)**
N-(Trichloromethylmercapto)phthalimide	**(133-07-3)**
N-(Trichloromethylmercapto)-δ^{4}-tetrahydrophthalimide	**(133-06-2)**
Trichloromethylmethane	**(71-55-6)**
N-(α-Trichloromethyl-4-methoxybenzyl)-4-methoxyaniline	**(38766-64-2)**
N-(α-Trichloromethyl-p-methoxybenzyl)-p-methoxyaniline	**(38766-64-2)**
Trichloromethylnitrile	**(545-06-2)**
Trichloromethyl perchlorate	**(67632-66-0)**
Trichloromethylphenyl carbinol	**(2000-43-3)**
Trichloromethylphenylcarbinyl acetate	**(90-17-5)**
1,1,1-Trichloro-2-methyl-2-propanol	**(57-15-8)**
Trichloro(2-methylpropyl)silane	**(18169-57-8)**
Trichloromethylstannane	**(993-16-8)**
Trichloromethylsulfenyl chloride	**(594-42-3)**
Trichloromethylsulphenyl chloride	**(594-42-3)**
N-((Trichloromethyl)thio)-4-cyclohexene-1,2-dicarboximide	**(133-06-2)**
N-(Trichloromethylthio)cyclohex-4-ene-1,2-dicarboximide	**(133-06-2)**
N-Trichloromethylthio-cis-δ^{4}-cyclohexene-1,2-dicarboximide	**(133-06-2)**
N-Trichloromethylthiocyclohex-4-ene-1,2-dicarboximide	**(133-06-2)**
2-((Trichloromethyl)thio)-1H-isoindole-1,3(2H)-dione	**(133-07-3)**
N-(Trichloromethylthio)phthalimide	**(133-07-3)**
Trichloromethylthio-1,2,5,6-tetrahydrophthalamide	**(133-06-2)**
N-((Trichloromethyl)thio)tetrahydrophthalimide	**(133-06-2)**
N-Trichloromethylthio-3a,4,7,7a-tetrahydrophthalimide	**(133-06-2)**
Trichloromethyltin	**(993-16-8)**
(Trichloromethyl)triazine	**(30361-83-2)**
Trichloromonofluoromethane	**(75-69-4)**
Trichloromonosilane	**(10025-78-2)**
Trichloronaphthalene (ACGIH,OSHA)	**(1321-65-9)**
Trichloronat	**(327-98-0)**
Trichloronate	**(327-98-0)**
1,2,3-Trichloro-4-nitrobenzene	**(17700-09-3)**
1,2,4-Trichloro-5-nitrobenzene	**(89-69-0)**
1,3,5-Trichloro-2-nitrobenzene	**(18708-70-8)**
2,4,5-Trichloronitrobenzene	**(89-69-0)**

2,4,6-Trichloro-1-nitrobenzene **(18708-70-8)**
2,4,6-Trichloronitrobenzene **(18708-70-8)**
2',4',6'-Trichloro-4-nitrobiphenyl ether **(1836-77-7)**
2,4,6-Trichloro-4'-nitrodiphenyl ether **(1836-77-7)**
Trichloronitromethane **(76-06-2)**
3,4,6-Trichloro-2-nitrophenol **(82-62-2)**
Trichloro-oxirane **(16967-79-6)**
2,4,5-Trichlorophenacylidene chloride **(1203-86-7)**
2,4,5-Trichlorophenacylidene dichloride **(1203-86-7)**
Trichlorophene **(70-30-4)**
Trichlorophenethylsilane **(940-41-0)**
2,3,4-Trichlorophenol **(15950-66-0)**
2,3,5-Trichlorophenol **(933-78-8)**
2,3,6-Trichlorophenol **(933-75-5)**
2,4,5-Trichlorophenol **(95-95-4)**
2,4,6-Trichlorophenol **(88-06-2)**
3,4,5-Trichlorophenol **(609-19-8)**
Trichlorophenol (DOT) **(25167-82-2)**
2,4,5-Trichlorophenol, sodium salt **(136-32-3)**
(2,4,6-Trichlorophenoxy)acetic acid **(575-89-3)**
2,4,5-Trichlorophenoxyacetic acid (DOT) **(93-76-5)**
(2,4,5-Trichlorophenoxy)acetic acid 2-butoxyethyl ester **(2545-59-7)**
2,4,5-Trichlorophenoxyacetic acid 3-butoxypropyl ester **(1928-48-9)**
2,4,5-Trichlorophenoxyacetic acid, butyl ester **(93-79-8)**
2,4,5-Trichlorophenoxyacetic acid, N,N-diethylethanolamine salt **(53404-86-7)**
2,4,5-Trichlorophenoxyacetic acid dimethylamine salt **(6369-97-7)**
(2,4,5-Trichlorophenoxy)acetic acid 2-ethylhexyl ester **(1928-47-8)**
2,4,5-Trichlorophenoxyacetic acid isobutyl ester **(4938-72-1)**
2,4,5-Trichlorophenoxyacetic acid, isopropyl ester **(93-78-7)**
(2,4,5-Trichlorophenoxy)acetic acid phentyl ester **(120-39-8)**
2,4,5-Trichlorophenoxyacetic acid propylene glycol butyl ester **(62922-39-8)**
(2,4,5-Trichlorophenoxy)acetic acid sodium salt **(13560-99-1)**
2,4,5-Trichlorophenoxyacetic acid triethanolamine salt **(3813-14-7)**
2,4,5-Trichlorophenoxyacetic acid triethylamine salt **(2008-46-0)**
4-(2,4,5-Trichlorophenoxy)butyric acid **(93-80-1)**
2-(2,4,5-Trichlorophenoxy)ethanol **(2122-77-2)**
2,4,5-Trichlorophenoxyethyl-α,α-dichloropropionate **(136-25-4)**
2-(2,4,5-Trichlorophenoxy)ethyl 2,2-dichloropropionate **(136-25-4)**
2-(2,4,5-Trichlorophenoxy)ethyl sulfate, sodium salt **(3570-61-4)**
(2,4,5-Trichlorophenoxy)propionic acid **(29990-39-4)**
2,4,5-Trichlorophenoxy-α-propionic acid **(93-72-1)**
2-(2,4,5-Trichlorophenoxy)propionic acid **(93-72-1)**
α-(2,4,5-Trichlorophenoxy)propionic acid **(93-72-1)**
2,4,5-Trichlorophenoxypropionic acid (DOT) **(29990-39-4)**
2-(2,4,5-Trichlorophenoxy)propionic acid, diethanolamine salt **(51170-59-3)**
2-(2,4,5-Trichlorophenoxy)propionic acid isooctyl ester **(32534-95-5)**
2-(2,4,5-Trichlorophenoxy)propionic acid potassium salt **(2818-16-8)**
2-(2,4,5-Trichlorophenoxy)propionic acid, sodium salt **(37913-89-6)**
2,3,6-Trichlorophenylacetic acid **(85-34-7)**
2,3,6-Trichlorophenylacetic acid sodium salt **(2439-00-1)**
1-(2,4,6-Trichlorophenyl)-3-anilinopyrazolone **(27241-31-2)**
Trichloro-2-phenylethylsilane **(940-41-0)**
(2,4,6-Trichlorophenyl)hydrazine **(5329-12-4)**
2,4,5-Trichlorophenyl iodopropargyl ether **(777-11-7)**
2,4,5-Trichlorophenyl-γ-iodopropargyl ether **(777-11-7)**
Trichlorophenylmethane **(98-07-7)**
2,4,6-Trichlorophenyl 4-nitrophenyl ether **(1836-77-7)**
Trichlorophenylsilane **(98-13-5)**
Trichlorophon **(52-68-6)**
Trichlorophosphine sulfide **(3982-91-0)**
2,2,3-Trichloropropanal **(7789-90-4)**
1,1,1-Trichloropropane **(7789-89-1)**
1,1,2-Trichloropropane **(598-77-6)**
1,2,2-Trichloropropane **(3175-23-3)**
Trichloropropane **(25735-29-9)**
1,2,3-Trichloropropane (ACGIH,OSHA) **(96-18-4)**
1,1,1-Trichloropropane-2,3-oxide **(3083-23-6)**
Trichloropropane oxide **(3083-23-6)**
1,1,1-Trichloropropanone **(918-00-3)**
1,1,3-Trichloro-2-propanone **(921-03-9)**
1,2,3-Trichloropropene **(96-19-5)**
3,3,3-Trichloro-1-propene **(2233-00-3)**
3,3,3-Trichloropropene **(2233-00-3)**
1,1,1-Trichloropropene oxide **(3083-23-6)**
1,1,1-Trichloropropene-2,3-oxide **(3083-23-6)**
3,3,3-Trichloropropene oxide **(3083-23-6)**
Trichloropropene oxide **(3083-23-6)**
2,2,3-Trichloropropionaldehyde **(7789-90-4)**
2,2,3-Trichloropropionic acid **(3278-46-4)**
1,1,1-Trichloropropylene oxide **(3083-23-6)**
3,3,3-Trichloropropylene oxide **(3083-23-6)**
2,3,5-Trichloro-4-(n-propylsulfonyl)-pyridine **(38827-35-9)**
2,3,5-Trichloro-4-(propylsulfonyl)pyridine **(38827-35-9)**
2,6,8-Trichloropurine **(2562-52-9)**
2,3,5-Trichloropyridine **(16063-70-0)**
2,3,6-Trichloropyridine **(6515-09-9)**
2,4,6-Trichloropyridine **(16063-69-7)**
2,3,5-Trichloro-4-pyridinol **(1970-40-7)**
3,5,6-Trichloro-2-pyridinol **(6515-38-4)**
3,5,6-Trichloro-2-pyridinol, sodium salt **(37439-34-2)**
3,5,6-Trichloro-2(1H)-pyridinone **(6515-38-4)**
3,5,6-Trichloro-2(1H)-pyridinone sodium salt **(37439-34-2)**
((3,5,6-Trichloro-2-pyridinyl)oxy)acetic acid 2-butoxyethyl ester **(64700-56-7)**
((3,5,6-Trichloro-2-pyridinyl)oxy)acetic acid, Compd. with N,N-diethylethanamine (1:1) **(57213-69-1)**
3,5,6-Trichloro-2-pyridyloxyacetic acid **(55335-06-3)**
((3,5,6-Trichloro-2-pyridyl)oxy)acetic acid, methyl ester **(60825-26-5)**
Trichloropyrocatechol **(25167-84-4)**
Trichloropyrogallol **(56961-21-8)**
Trichlorosilane [UN 1295] **(10025-78-2)**
Trichlorostibine **(10025-91-9)**
3,4,5-Trichlorosyringol **(2539-26-6)**
Trichloro(tetrachlorophenyl)silane **(33434-63-8)**
2,4,5-Trichlorothiophenol **(3773-14-6)**
Trichlorotitanium **(7705-07-9)**
2,3,4-Trichlorotoluene **(7359-72-0)**
2,3,6-Trichlorotoluene **(2077-46-5)**
Trichlorotoluene **(30583-33-6)**
α,2,4-Trichlorotoluene **(94-99-5)**
α,3,4-Trichlorotoluene **(102-47-6)**
α,α,α-Trichlorotoluene **(98-07-7)**
ω,ω,ω-Trichlorotoluene **(98-07-7)**
N,N',N''-Trichloro-2,4,6-triamine-1,3,5-triazine **(7673-09-8)**
1,3,5-Trichlorotriazine **(108-77-0)**
2,4,6-Trichloro-1,3,5-triazine **(108-77-0)**
2,4,6-Trichloro-s-triazine **(108-77-0)**
2,4,6-Trichlorotriazine **(108-77-0)**
Trichloro-s-triazine **(108-77-0)**
sym-Trichlorotriazine **(108-77-0)**
1,3,5-Trichloro-s-triazine-2,4,6(1H,3H,5H)-trione **(87-90-1)**
Trichloro-s-triazine-2,4,6(1H,3H,5H)-trione **(87-90-1)**
Trichloro-s-triazinetrione **(87-90-1)**
Trichloro-s-triazinetrione, Dry, containing over 39% available chlorine [NA 2468] **(87-90-1)**
3,5-Trichloro-2-(trichloromethyl)pyridine **(1128-16-1)**
2,2',2''-Trichlorotriethylamine **(555-77-1)**
2,2',2''-Trichlorotriethylamine hydrochloride **(817-09-4)**
Trichlorotriethyldialuminium **(12075-68-2)**
Trichlorotriethyldialuminum **(12075-68-2)**
1,1,2-Trichlorotrifluoroethane **(76-13-1)**
1,2,2-Trichlorotrifluoroethane **(76-13-1)**
Trichlorotrifluoroethane **(354-58-5)**
Trichlorotrifluoroethane **(76-13-1)**
1,1,2-Trichloro-1,2,2-trifluoroethane (ACGIH,OSHA) **(76-13-1)**

Trichloro-1,2,3-trihydroxybenzene **(56961-21-8)**
Trichlorotrimethyldialuminum **(12542-85-7)**
1,1,2-Trichloro-1,2,2-trimethyldisilane **(13528-88-6)**
1,3,5-Trichloro-2,4,6-trinitrobenzene **(2631-68-7)**
Trichloro-1,3,5-trinitrobenzene **(2631-68-7)**
sym-Trichlorotrinitrobenzene **(2631-68-7)**
1,3,5-Trichloro-2,4,6-trioxohexahydro-s-triazine **(87-90-1)**
Trichloro(vinyl)silane **(75-94-5)**
Trichlorovinyl silicane **(75-94-5)**
Trichlorphene **(52-68-6)**
O-(2,4,5-Trichlor-phenyl)-O,O-dimethyl-monothiophosphat (German) **(299-84-3)**
2,3,6-Trichlorphenylessigsaeure (German) **(85-34-7)**
Trichlorphon **(52-68-6)**
Trichlorphon FN **(52-68-6)**
Trichlorsilan (German) **(10025-78-2)**
Trichlor-triaethylamin-hydrochlorid (German) **(817-09-4)**
Trichlorure d'antimoine (French) **(10025-91-9)**
Trichlorure d'arsenic (French) **(7784-34-1)**
Trichlorure d'azote (French) **(10025-85-1)**
syn-Trichlotriazin (Czech) **(108-77-0)**
Trichlouracetonitril (Dutch) **(545-06-2)**
Trichochromogenic Factor **(150-13-0)**
Trichocide **(443-48-1)**
Trichofuron **(67-45-8)**
Trichomol **(443-48-1)**
Trichomonacid (Pharmachim) **(443-48-1)**
Trichopal **(443-48-1)**
Trichopol **(443-48-1)**
T_2-Trichothecene **(21259-20-1)**
Trichothec-9-ene-3-α,4-β,8-α,15-tetrol, 12,13-epoxy-, 4,15-diacetate 8-isovalerate **(21259-20-1)**
Trichothec-9-en-8-one, 4-(acetyloxy)-12,13-epoxy-3,7,15-trihydroxy-, (3-α,4-β,7-β) **(23255-69-8)**
Trichothec-9-en-8-one, 12,13-epoxy-3,4,7,15-tetrahydroxy-, (3-α,4-β,7-α) **(23282-20-4)**
Trichothec-9-en-8-one, 12,13-epoxy-3,7,15-trihydroxy-, (3-α,7-α) **(51481-10-8)**
Triclocarban **(101-20-2)**
Triclofos **(306-52-5)**
Triclopyr **(55335-06-3)**
Triclopyr triethylamine **(57213-69-1)**
Triclopyr triethylamine salt **(57213-69-1)**
Triclordiuride **(133-67-5)**
Tricloretene (Italian) **(79-01-6)**
Triclormetiazide (Italian) **(133-67-5)**
1,1,1-Tricloro-2,2-bis(4-cloro-fenil)-etano (Italian) **(50-29-3)**
1,1,1-Tricloroetano (Italian) **(71-55-6)**
Tricloroetilene (Italian) **(79-01-6)**
O-(2,4,5-Tricloro-fenil)-O,O-dimetil-monotiofosfato (Italian) **(299-84-3)**
Triclorometano (Italian) **(67-66-3)**
Triclorometilbenzene (Italian) **(98-07-7)**
Tricloro-nitro-metano (Italian) **(76-06-2)**
Triclorosilano (Italian) **(10025-78-2)**
Triclorotoluene (Italian) **(98-07-7)**
Tricloruro de nitrogeno (Spanish) **(10025-85-1)**
Triclos **(306-52-5)**
Triclosan **(3380-34-5)**
Tricocet **(443-48-1)**
Tricofuron **(67-45-8)**
Tricom **(443-48-1)**
Tricon BW **(60-00-4)**
Tricop 50 **(1332-40-7)**
Tricosane (9CI) **(638-67-5)**
Tricosanoic acid (8CI,9CI) **(2433-96-7)**
Tricosanoic acid, ethyl ester (6CI,8CI,9CI) **(18281-07-7)**
1-Tricosanol (8CI,9CI) **(3133-01-5)**
12-Tricosanone **(540-09-0)**
(Z)-9-Tricosene **(27519-02-4)**
11-Tricosene (9CI) **(52078-56-5)**
9-Tricosene, (Z)- **(27519-02-4)**
Tricowas B **(443-48-1)**
Tricresilfosfati (Italian) **(1330-78-5)**
Tricresol **(1319-77-3)**
Tricresylfosfaten (Dutch) **(1330-78-5)**
Tri-m-cresyl phosphate **(563-04-2)**
Tri-o-cresyl phosphate **(78-30-8)**
Tri-p-cresyl phosphate **(78-32-0)**
Tricresyl phosphate **(1330-78-5)**
Tricresyl phosphate **(78-30-8)**
Tris-m-cresyl phosphate **(563-04-2)**
Triorthocresyl phosphate (ACGIH,OSHA) **(78-30-8)**
Tricresylphosphate, With more than 3% ortho isomer [UN 2574] **(1330-78-5)**
Tri-m-cresyl phosphite **(563-04-2)**
Tri-p-cresyl phosphite **(620-42-8)**
Trictal **(113-59-7)**
Tricuran **(65-29-2)**
Tricyanic acid **(108-80-5)**
Tricyanogen chloride **(108-77-0)**
1,3,6-Tricyanohexane **(1772-25-4)**
Tricyclazole **(41814-78-2)**
Tricyclazone **(41814-78-2)**
Tricyclene **(508-32-7)**
Tricyclo(3.3.1.1^{3,7})decan-1-amine **(768-94-5)**
Tricyclo(3.3.1.13,7)decan-1-amine, sulfate (9CI) **(32793-63-8)**
Tricyclo(5.2.1.02,6)decane **(6004-38-2)**
exo-Tricyclo(5.2.1.0^{2,6})decane **(2825-82-3)**
Tricyclo(3.3.1.13,7)decane (9CI) **(281-23-2)**
Tricyclo(3.3.1.1(3,7))decane, bromo- (9CI) **(102938-79-4)**
Tricyclodecanedimethanol (9CI) **(26896-48-0)**
Tricyclo(3.3.1.1(3,7))decane, 1,3-dimethyl- (9CI) **(702-79-4)**
Tricyclo(3.3.1.13,7)decane-2,6-dione, 4-amino- (9CI) **(56728-08-6)**
Tricyclo(4.4.0.0(2,7))dec-3-ene, 1,3-dimethyl-8-(1-methylethyl)-, stereoisomer (9CI) (VAN) **(3856-25-5)**
Tricyclo(4.4.0.0(2,7))dec-3-ene, 8-isopropyl-1,3-dimethyl- **(3856-25-5)**
Tricyclo(4.4.0.0(2,7))dec-3-ene, 8-isopropyl-1,3-dimethyl-, (1R,2S,6S,7S,8S)-(-)- (8CI) **(3856-25-5)**
Tricyclo(5.2.1.02,6)dec-3-en-9-yl acetate **(2500-83-6)**
Tricyclodecen-4-yl 8-acetate **(5413-60-5)**
Tricyclodecenyl propionate **(17511-60-3)**
Tricyclo(3.1.1.06,7)heptane (6CI) **(287-13-8)**
Tricyclo(4.1.0.02,4)heptane (7CI,8CI,9CI) **(187-26-8)**
Tricyclo(4.1.0.02,7)heptane (7CI,8CI,9CI) **(287-13-8)**
Tricyclo(2.2.1.0(2.6))heptane, 1,7,7-trimethyl- **(508-32-7)**
Tricyclo(2.2.1.02,6)heptane, 1,7,7-trimethyl- (9CI) **(508-32-7)**
Tricyclohexylhydroxystannane **(13121-70-5)**
Tricyclohexylhydroxytin **(13121-70-5)**
Tricyclohexylstannanol **(13121-70-5)**
Tricyclohexylstannium hydroxide **(13121-70-5)**
Tricyclohexyltin hydroxide **(13121-70-5)**
Tricyclohexylzinnhydroxid (German) **(13121-70-5)**
Tricyclo(5.2.0.02,5)nona-3,8-dien-6-ol, 6-phenyl-, (1.α.,2.β.,5.β.,6.β.,7.α.)- (9CI) **(56771-50-7)**
Tricyclo(4.3.1.13,8)undecane, 1-chloro- (8CI) **(27011-46-7)**
Tricyclo(4.3.1.13,8)undecan-1-ol (8CI,9CI) **(31061-64-0)**
1,12-Tridecadiene (8CI,9CI) **(21964-48-7)**
1,1,2,2,3,3,4,4,5,5,6,6,6-Tridecafluoro-1-hexanesulfonyl fluoride **(423-50-7)**
Tridecanal (9CI) **(10486-19-8)**
1-Tridecanamine (9CI) **(2869-34-3)**
1-Tridecanamine, (2,4,5-trichlorophenoxy)acetate (9CI) **(53404-85-6)**
Tridecane **(629-50-5)**
n-Tridecane **(629-50-5)**
Tridecane, 1-bromo **(765-09-3)**
1-Tridecanecarboxylic acid **(544-63-8)**

1-Tridecanecarboxylic acid, isopropyl ester	**(110-27-0)**
Tridecane, 1-chloro- (9CI)	**(822-13-9)**
Tridecane, chloro- (6CI,8CI,9CI)	**(34214-84-1)**
Tridecane, 2,5-dimethyl- (9CI)	**(56292-66-1)**
Tridecane, 4,8-dimethyl- (9CI)	**(55030-62-1)**
1,13-Tridecanedioic acid	**(505-52-2)**
Tridecanedioic acid (9CI)	**(505-52-2)**
Tridecanedioic acid, dimethyl ester (9CI)	**(1472-87-3)**
Tridecane, 2-methyl- (8CI,9CI)	**(1560-96-9)**
Tridecane, 7-methyl- (6CI,8CI,9CI)	**(26730-14-3)**
Tridecane, 6-methyl- (6CI,7CI,8CI,9CI)	**(13287-21-3)**
Tridecanenitrile (9CI)	**(629-60-7)**
Tridecane, 1-phenyl- (8CI)	**(123-02-4)**
Tridecane, 2-phenyl- (8CI)	**(4534-53-6)**
Tridecane, 3-phenyl- (8CI)	**(4534-52-5)**
Tridecane, 4-phenyl- (8CI)	**(4534-51-4)**
Tridecane, 5-phenyl- (8CI)	**(4534-50-3)**
Tridecane, 6-phenyl- (8CI)	**(4534-49-0)**
Tridecane, 7-phenyl- (8CI)	**(2400-01-3)**
Tridecane, 5-propyl- (9CI)	**(55045-11-9)**
1-Tridecanethiol (9CI)	**(19484-26-5)**
Tridecane, 2,6,10-trimethyl- (8CI,9CI)	**(3891-99-4)**
Tridecanoic-acid	**(638-53-9)**
Tridecanoic acid, 2,3-epoxypropyl ester	**(63978-73-4)**
Tridecanoic acid, methyl ester (9CI)	**(1731-88-0)**
Tridecanoic acid, 2-methyl-, methyl ester (9CI)	**(55955-78-7)**
Tridecanoic acid, 12-methyl-, methyl ester (8CI,9CI)	**(5129-58-8)**
1-Tridecanol	**(112-70-9)**
Tridecanol	**(112-70-9)**
Tridecanol	**(26248-42-0)**
n-Tridecanol	**(112-70-9)**
2-Tridecanol (8CI,9CI)	**(1653-31-2)**
Tridecanol Condensed with 6 moles ethylene oxide	**(24938-91-8)**
1-Tridecanol, acetate	**(1072-33-9)**
6-Tridecanol, 3,9-diethyl-, hydrogen sulfate, sodium salt	**(3282-85-7)**
1-Tridecanol, dihydrogen phosphate (9CI)	**(5116-94-9)**
1-Tridecanol, hydrogen phosphate (9CI)	**(5116-95-0)**
1-Tridecanol, hydrogen sulfate, sodium salt (9CI)	**(3026-63-9)**
1-Tridecanol, phthalate	**(119-06-2)**
Tridecanol stearate	**(31556-45-3)**
7-Tridecanone	**(462-18-0)**
2-Tridecanone (9CI)	**(593-08-8)**
Tridecanone (9CI)	**(57702-05-3)**
1-Tridecene	**(2437-56-1)**
Tridecene	**(25377-82-6)**
6-Tridecene, 7-methyl- (6CI,8CI,9CI)	**(24949-42-6)**
(E)-4-Tridecen-l-yl acetate	**(72269-48-8)**
4-Tridecen-1-ol, acetate, (E)-	**(72269-48-8)**
(E)-4-Tridecenyl acetate	**(72269-48-8)**
(Z)-4-Tridecen-1-yl acetate	**(65954-19-0)**
E-4-Tridecenyl acetate	**(72269-48-8)**
Trideceth-6 phosphate	**(9046-01-9)**
n-Tridecoic acid	**(638-53-9)**
Tridecyl acid phosphate	**(5116-94-9)**
Tridecyl acrylate	**(3076-04-8)**
Tridecyl alcohol	**(112-70-9)**
n-Tridecyl alcohol	**(112-70-9)**
Tridecyl alcohol, ethoxylated and phosphated, potassium salt	**(68186-36-7)**
Tridecyl alcohol, ethoxylated, phosphated, potassium salt	**(68186-36-7)**
Tridecylalcohol, ethoxylated, phosphated, potassium salt	**(68186-36-7)**
Tridecyl alcohol, ethoxylated, phosphate, potassium salt	**(68186-36-7)**
Tridecyl alcohol, ethoxylate, phosphate, potassium salt	**(68186-36-7)**
Tri-1-decylaluminum	**(1726-66-5)**
Tri-n-decylaluminum	**(1726-66-5)**
Tridecylaluminum	**(1726-66-5)**
Tridecylbenzene	**(123-02-4)**
Tridecylbenzenesulfonic acid	**(25496-01-9)**
Tridecylbenzenesulfonic acid	**(25496-01-9)**
Tridecylbenzene sulfonic acid sodium salt	**(26248-24-8)**
Tridecylbenzenesulfonic acid, sodium salt	**(26248-24-8)**
Tridecylchloride	**(822-13-9)**
Tridecylcyclohexane	**(6006-33-3)**
N-Tridecyl-2,6-dimethylmorpholin (German)	**(24602-86-6)**
4-Tridecyl-2,6-dimethylmorpholine	**(24602-86-6)**
N-Tridecyl-2,6-dimethylmorpholine	**(24602-86-6)**
Tridecylic acid	**(638-53-9)**
Tridecyl mercaptan	**(19484-26-5)**
Tridecyl methacrylate	**(2495-25-2)**
Tridecyl 2-methyl-2-propenoate	**(2495-25-2)**
Tridecyl octadecanoate	**(31556-45-3)**
Tridecyloxirane	**(18633-25-5)**
Tridecyl recycle alcohol	**(68515-39-9)**
Tridecyl sodium sulfate	**(3026-63-9)**
Tridecyl stearate	**(31556-45-3)**
Tridemorph	**(24602-86-6)**
Tridestrin	**(50-27-1)**
Tridezibarbitur	**(50-06-6)**
Tris-2,3-dibrompropyl ester kyseliny fosforecne (Czech)	**(126-72-7)**
Tris-(2,3-dibrompropyl)fosfat (Czech)	**(126-72-7)**
Tri(β,β'-dichloroisopropyl)phosphate	**(13674-87-8)**
Tri(2,4-dichlorophenoxyethyl) phosphite	**(94-84-8)**
Tris-(1,3-dichloro-2-propyl)-phosphate	**(13674-87-8)**
Tris-dichloropropylphosphate	**(78-43-3)**
1,2,3-Tri(β-diethylaminoethoxy)benzene triethiodide	**(65-29-2)**
Tri(β-diethylaminoethoxy)-1,2,3-benzenetri-iodoethylate	**(65-29-2)**
Tri-digitoxoside (German)	**(71-63-6)**
Tris-(dimethylamid) kyseliny fosforecne (Czech)	**(680-31-9)**
2,4,6-Tris-N,N-dimethylaminomethylfenol (Czech)	**(90-72-2)**
2,4,6-Tri(dimethylaminomethyl)phenol	**(90-72-2)**
Tri(dimethylamino)phosphineoxide	**(680-31-9)**
Tri(2,4-dimethylphenyl) phosphate	**(3862-12-2)**
Tri(2,5-dimethylphenyl) phosphate	**(19074-59-0)**
Tri(2,6-dimethylphenyl) phosphate	**(121-06-2)**
Tridimite (French)	**(15468-32-3)**
Tridipam	**(137-26-8)**
Tri(docosyl) aluminum	**(6651-25-8)**
Tridocosylaluminum	**(6651-25-8)**
Tri(dodecyl) aluminum	**(1529-59-5)**
Tridodecylaluminum	**(1529-59-5)**
Tri-n-dodecylamine	**(102-87-4)**
Tridodecylamine (8CI)	**(102-87-4)**
Tridymite	**(15468-32-3)**
Tri(eicosyl) aluminum	**(1529-57-3)**
Trieicosylaluminum	**(1529-57-3)**
Trielene	**(79-01-6)**
Trielin	**(79-01-6)**
Trielina (Italian)	**(79-01-6)**
Trieline	**(79-01-6)**
Trien	**(112-24-3)**
Trientine	**(112-24-3)**
Tri(epoxypropyl)isocyanurate	**(2451-62-9)**
Trieste Flowers	**(8003-34-7)**
Triestearato de sorbitano (Spanish)	**(26658-19-5)**
Trietatone	**(13532-26-8)**
Trietazine	**(1912-26-1)**
Tri-ethane	**(71-55-6)**
N,N',N''-Tri-1,2-ethanediylphosphoric triamide	**(545-55-1)**
N,N',N''-Tri-1,2-ethanediylphosphorothioic triamide	**(52-24-4)**
N,N',N''-Tri-1,2-ethanediylthiophosphoramide	**(52-24-4)**
Triethanolamin	**(102-71-6)**
Triethanolamine	**(102-71-6)**
Triethanolamine acetate	**(14806-72-5)**
Triethanolamine p-tert-butylbenzoate	**(59993-86-1)**
Triethanolamine 2,4-dichlorophenoxyacetate	**(2569-01-9)**

Name	CAS
Triethanolamine dodecylbenzene sulfonate	**(27323-41-7)**
Triethanolamine dodecylbenzenesulfonate	**(27323-41-7)**
Triethanolamine dodecyl sulfate	**(139-96-8)**
Triethanolamine ethosulfate	**(31774-90-0)**
Triethanolamine ethoxylated	**(36936-60-4)**
Triethanolamine laurate	**(2224-49-9)**
Triethanolamine lauryl sulfate	**(139-96-8)**
Triethanolamine myristate	**(41669-40-3)**
Triethanolamine myristyl sulfate	**(4492-78-8)**
Triethanolamine oleate	**(2717-15-9)**
Triethanolamine, oleic acid monoester	**(10277-04-0)**
Triethanolamine oleic acid salt	**(2717-15-9)**
Triethanolamine salt of oleic acid	**(2717-15-9)**
Triethanolamine stearate	**(4568-28-9)**
Triethanolamine, stearic acid salt	**(4568-28-9)**
Triethanolamine sulfate	**(7376-31-0)**
Triethanolamine tetradecyl sulfate	**(4492-78-8)**
Triethanolamine triacetate	**(3002-18-4)**
Triethanolamine tristearate	**(3002-22-0)**
Triethanolammonium oleate	**(2717-15-9)**
Triethanomelamine	**(51-18-3)**
Triethazine	**(1912-26-1)**
Triethoxyaluminum	**(555-75-9)**
Triethoxy(3-aminopropyl)silane	**(919-30-2)**
3-(2,4,5-Triethoxybenzoyl)propionic acid	**(41826-92-0)**
1,1,1-Triethoxyethane	**(78-39-7)**
Triethoxy-ethylsilane	**(78-07-9)**
Triethoxyfenylsilan (Czech)	**(780-69-8)**
Triethoxy-2-kyanethylsilan (Czech)	**(919-31-3)**
Triethoxymethane	**(122-51-0)**
Triethoxy-methylsilane	**(2031-67-6)**
Triethoxyoctylsilane	**(2943-75-1)**
2,4,5-Triethoxy-γ-oxobenzenebutanoic acid	**(41826-92-0)**
Triethoxyphenylsilane	**(780-69-8)**
1,1,1-Triethoxypropane	**(115-80-0)**
Triethoxysilane	**(998-30-1)**
3-(Triethoxysilyl)propylamine	**(919-30-2)**
(3-Triethoxysilyl)propylcarbamic acid ethyl ester	**(17945-05-0)**
(3-(Triethoxysilyl)propyl)urea	**(23779-32-0)**
N-(Triethoxysilylpropyl)urea	**(23779-32-0)**
Triethoxyvinylsilane	**(78-08-0)**
Triethoxyvinylsilicane	**(78-08-0)**
Triethyl orthoacetate	**(78-39-7)**
Triethyl acetylcitrate	**(77-89-4)**
Triethylaluminium (DOT)	**(97-93-8)**
Triethylaluminum	**(97-93-8)**
Triethylaluminum sesquichloride	**(12075-68-2)**
Triethylamine (ACGIH,OSHA) [UN 1296]	**(121-44-8)**
Triethylamine, 2-(p-(2-chloro-1,2-diphenylvinyl)phenoxy)	**(911-45-5)**
Triethylamine, 2-(p-(2-chloro-1,2-diphenylvinyl)phenoxy)-, citrate (1:1)	**(50-41-9)**
Triethylamine, 2-chloro-, hydrochloride	**(869-24-9)**
Triethylamine, 2,2'-dichloro	**(538-07-8)**
Triethylamine 2,4-dichlorophenoxyacetate	**(2646-78-8)**
Triethylamine, 1,1'-dimethyl- (8CI)	**(7087-68-5)**
Triethylamine, 2,2'''-dithiobis(1',1''-dimethyl-	**(65332-44-7)**
Triethylamine, hydrochloride	**(554-68-7)**
Triethylamine picloram	**(35832-11-2)**
Triethylamine, 2,2',2''-trichloro	**(555-77-1)**
Triethylamine, 2,2',2''-trichloro-, hydrochloride	**(817-09-4)**
Triethylamine, 2,2',2''-trihydroxy-	**(102-71-6)**
1,2,3-Triethylbenzene	**(42205-08-3)**
1,3,5-Triethylbenzene	**(102-25-0)**
Triethylbenzene	**(25340-18-5)**
Triethylborane	**(97-94-9)**
Triethyl borate	**(150-46-9)**
Triethylborine	**(97-94-9)**
Triethylcarbinol	**(597-49-9)**
Triethylchlorosilane	**(994-30-9)**
Triethylchlorostannane	**(994-31-0)**
Triethylchlorotin	**(994-31-0)**
Triethyl citrate	**(77-93-0)**
Triethyl citrate, acetate	**(77-89-4)**
Triethyldialuminum trichloride	**(12075-68-2)**
Triethylene-glycol	**(112-27-6)**
Triethylene glycol, bis(2-ethylbutyrate)	**(95-08-9)**
Triethylene glycol, diacetate	**(111-21-7)**
Triethylene glycol diacrylate	**(1680-21-3)**
Triethylene glycol, dibenzoate (8CI)	**(120-56-9)**
Triethylene glycol, bis(ethylhexanoate)	**(94-28-0)**
Triethylene glycol n-butyl ether	**(143-22-6)**
Triethylene glycol dicaprylate	**(106-10-5)**
Triethylene glycol dichloride	**(112-26-5)**
Triethylene glycol di(2-ethyl butyrate)	**(95-08-9)**
Triethylene glycol di(2-ethylhexoate)	**(94-28-0)**
Triethylene glycol diglycidyl ether	**(1954-28-5)**
Triethylene glycol dimethacrylate	**(109-16-0)**
Triethylene glycol dimethyl ether	**(112-49-2)**
Triethylene glycol, dinitrate	**(111-22-8)**
Triethylene glycol dioctanoate	**(106-10-5)**
Triethylene glycol dodecyl ether	**(3055-94-5)**
Triethylene glycol ethyl ether	**(112-50-5)**
Triethylene glycol monobutyl ether	**(143-22-6)**
Triethylene glycol monoethyl ether	**(112-50-5)**
Triethylene glycolmonomethyl ether	**(112-35-6)**
Triethylenediamine	**(280-57-9)**
Triethyleneglycol diethyl butyrate	**(95-08-9)**
Triethyleneglycol dimercaptan	**(14970-87-7)**
Triethyleneglycol distearate	**(9005-08-7)**
2,3,5-Triethyleneimino-1,4-benzoquinone	**(68-76-8)**
2,3,5-Triethyleneimino-p-benzoquinone	**(68-76-8)**
Triethyleneiminobenzoquinone	**(68-76-8)**
Tri(ethyleneimino)thiophosphoramide	**(52-24-4)**
2,4,6-Tri(ethyleneimino)-1,3,5-triazine	**(51-18-3)**
2,4,6-Triethyleneimino-s-triazine	**(51-18-3)**
Triethylenemelamine	**(51-18-3)**
N,N',N''-Triethylenephosphoramide	**(545-55-1)**
Triethylenephosphoramide	**(545-55-1)**
N,N',N''-Triethylenephosphoric triamide	**(545-55-1)**
Triethylenephosphoric triamide	**(545-55-1)**
N,N',N''-Triethylenephosphorothioic triamide	**(52-24-4)**
Triethylenephosphorotriamide	**(545-55-1)**
Triethylenetetramine [UN 2259]	**(112-24-3)**
N,N',N''-Triethylenethiophosphamide	**(52-24-4)**
N,N',N''-Triethylenethiophosphoramide	**(52-24-4)**
Triethylene thiophosphoramide	**(52-24-4)**
Triethylenethiophosphoramide	**(52-24-4)**
Triethylenethiophosphorotriamide	**(52-24-4)**
Triethylenfosforamid (Czech)	**(545-55-1)**
Triethylenglykol (Czech)	**(112-27-6)**
Triethyleniminobenzoquinone	**(68-76-8)**
2,4,6-Triethylenimino-1,3,5-triazine	**(51-18-3)**
2,4,6-Triethylenimino-s-triazine	**(51-18-3)**
Triethylester kyseliny acetylcitronove (Czech)	**(77-89-4)**
Triethylester kyseliny borite (Czech)	**(150-46-9)**
Triethylester kyseliny citronove (Czech)	**(77-93-0)**
Triethylester kyseliny orthomravenci (Czech)	**(122-51-0)**
O,O,O-Triethylester kyseliny thiofosforecne (Czech)	**(126-68-1)**
N,N,N-Triethylethanaminium perchlorate	**(2567-83-1)**
Triethyl orthoformate	**(122-51-0)**
Triethylfosfat (Czech)	**(78-40-0)**
Triethylhexadecylammonium bromide	**(13316-70-6)**
Tris-(2-ethylhexyl)fosfat (Czech)	**(78-42-2)**
Tri(2-ethylhexyl)phosphate	**(78-42-2)**

Name	CAS No.
Triethylhexyl phosphate	**(78-42-2)**
Tri-(2-ethylhexyl)trimellitate	**(3319-31-1)**
Tri-2-ethylhexyl trimellitate	**(3319-31-1)**
Triethyl lead	**(14570-15-1)**
Triethyl lead	**(5224-23-7)**
Triethyllead	**(5224-23-7)**
Triethyl lead chloride	**(1067-14-7)**
Triethyllead hydride (6CI)	**(5224-23-7)**
Triethylmethanol	**(597-49-9)**
Triethylmethyllead	**(1762-28-3)**
Triethylmethylplumbane	**(1762-28-3)**
1,1,3-Triethyl-3-nitrosourea	**(50285-70-6)**
Triethylolamine	**(102-71-6)**
Triethyl phosphate	**(78-40-0)**
Triethylphosphine	**(554-70-1)**
Triethyl phosphite [UN 2323]	**(122-52-1)**
Triethyl phosphonoacetate	**(867-13-0)**
O,O,O-Triethyl phosphorothioate	**(126-68-1)**
O,O,S-Triethyl phosphorothioate	**(1186-09-0)**
Triethyl phosphorothioate	**(126-68-1)**
Triethylplumbane	**(5224-23-7)**
Triethylplumbium chloride	**(1067-14-7)**
Triethyl o-propionate	**(115-80-0)**
Triethyl orthopropionate	**(115-80-0)**
Triethylsilane	**(617-86-7)**
Triethylstannane	**(997-50-2)**
Triethylstannium chloride	**(994-31-0)**
Triethylstannyl chloride	**(994-31-0)**
O,O,O-Triethylthiofosfat (Czech)	**(126-68-1)**
Triethylthiofosfat (Czech)	**(126-68-1)**
O,O,S-Triethyl thiophosphate	**(1186-09-0)**
Triethyltin	**(997-50-2)**
Triethyltin chloride	**(994-31-0)**
Triethyltin hydride	**(997-50-2)**
2,4,6-Triethyl-1,3,5-triazine	**(1009-74-1)**
Triethyl-s-triazine	**(1009-74-1)**
Triethyl 1,3,5-triazine-2,4,6-tricarboxylate	**(898-22-6)**
Triethyltrichlorodialuminum	**(12075-68-2)**
Trietilamina (Italian)	**(121-44-8)**
Trietol	**(2717-15-9)**
Trifaron	**(58-39-9)**
Trifenil stagno acetato (Italian)	**(900-95-8)**
Trifenmorph	**(1420-06-0)**
Trifenox	**(120-39-8)**
Trifenox 80	**(120-39-8)**
Trifenoxyfosfin (Czech)	**(101-02-0)**
Trifenson	**(80-38-6)**
Trifenylfosfat (Czech)	**(115-86-6)**
Trifenylfosfin (Czech)	**(603-35-0)**
Trifenylfosfit (Czech)	**(101-02-0)**
Trifenylstanniumhydroxid (Czech)	**(76-87-9)**
Trifenylstibin (Czech)	**(603-36-1)**
Trifenyl-tinacetaat (Dutch)	**(900-95-8)**
Trifenyl-tinhydroxyde (Dutch)	**(76-87-9)**
Triflumen	**(133-67-5)**
Triflumethazine	**(69-23-8)**
Trifluomethylthiazide	**(148-56-1)**
Trifluoperazina (Italian)	**(117-89-5)**
Trifluoperazine	**(117-89-5)**
Trifluoracetic acid	**(76-05-1)**
Trifluoralin	**(1582-09-8)**
Trifluorchlorethylen (Czech)	**(79-38-9)**
3-(5-Trifluormethylphenyl)-1,1-dimethylharnstoff (German)	**(2164-17-2)**
Trifluoroacetic acid [UN 2699]	**(76-05-1)**
Trifluoroacetic acid anhydride	**(407-25-0)**
Trifluoroacetic acid, ethyl ester	**(383-63-1)**
Trifluoroacetic anhydride	**(407-25-0)**
1,1,1-Trifluoroacetone	**(421-50-1)**
3,3,3-Trifluoroacetone	**(421-50-1)**
1,1,1-Trifluoroacetophenone	**(434-45-7)**
2,2,2-Trifluoroacetophenone	**(434-45-7)**
Trifluoroacetophenone	**(434-45-7)**
α,α,α-Trifluoroacetophenone	**(434-45-7)**
Trifluoroacetyl anhydride	**(407-25-0)**
Trifluoroacetyl chloride	**(354-32-5)**
Trifluoroantimony	**(7783-56-4)**
Trifluoroarsine	**(7784-35-2)**
1,2,3-Trifluorobenzene	**(1489-53-8)**
Trifluoroborane diethyl etherate	**(109-63-7)**
Trifluoroboron etherate	**(109-63-7)**
1,1,1-Trifluoro-2-bromo-2-chloroethane	**(151-67-7)**
1,1,2-Trifluoro-1-bromo-2-chloroethane	**(354-06-3)**
Trifluorobromoethylene	**(598-73-2)**
Trifluorobromomethane (ACGIH,DOT,OSHA)	**(75-63-8)**
1,1,1-Trifluoro-2-chloro-2-bromoethane	**(151-67-7)**
2,2,2-Trifluoro-1-chloro-1-bromoethane	**(151-67-7)**
1,1,1-Trifluoro-2-chloroethane	**(75-88-7)**
2,2,2-Trifluorochloroethane	**(75-88-7)**
1,1,2-Trifluoro-2-chloroethylene	**(79-38-9)**
Trifluorochloroethylene (DOT)	**(79-38-9)**
Trifluorochloroethylene, Inhibited [UN 1082]	**(79-38-9)**
Trifluorochloromethane (DOT)	**(75-72-9)**
1,1,1-Trifluoro-3-chloropropane	**(460-35-5)**
α,α,α-Trifluoro-4-chlorotoluene	**(98-56-6)**
5-Trifluoro-2'-deoxythymidine	**(70-00-8)**
1,1,1-Trifluoro-2,2-dichloroethane	**(306-83-2)**
1,1,2-Trifluoro-1,2-dichloroethane	**(354-23-4)**
α,α,α-Trifluoro-2,6-dinitro-N,N-dipropyl-p-toluidine	**(1582-09-8)**
α,α,α-Trifluoro-2,6-dinitro-N,N-ethylbutyl-p-toluidine	**(1861-40-1)**
1,1,1-Trifluoroethane	**(420-46-2)**
Trifluoroethanoic acid	**(76-05-1)**
2,2,2-Trifluoroethanol	**(75-89-8)**
Trifluoroethene	**(359-11-5)**
2,2,2-Trifluoroethyl carbanilate	**(370-32-1)**
1,1,1-Trifluoroethyl chloride	**(75-88-7)**
Trifluoroethylene	**(359-11-5)**
2,2,2-Trifluoroethyl methacrylate	**(352-87-4)**
Trifluoroethyl methacrylate	**(352-87-4)**
2,2,2-Trifluoroethyl N-phenylcarbamate	**(370-32-1)**
2,2,2-Trifluoroethyl vinyl ether	**(406-90-6)**
1,1,1-Trifluoroform	**(420-46-2)**
Trifluoroiodomethane	**(2314-97-8)**
Trifluoromethane [UN 1984]	**(75-46-7)**
Trifluoromethanesulfonic acid	**(1493-13-6)**
Trifluoromethanesulfonyl fluoride	**(335-05-7)**
3-Trifluoromethylacetanilide	**(351-36-0)**
m-Trifluoromethyl acetanilide	**(351-36-0)**
3-(Trifluoromethyl)aniline	**(98-16-8)**
m-(Trifluoromethyl)aniline	**(98-16-8)**
2-Trifluoromethyl aniline [UN 2942]	**(88-17-5)**
3-Trifluoromethyl aniline [UN 2948]	**(98-16-8)**
4-(Trifluoromethyl)benzaldehyde	**(455-19-6)**
p-(Trifluoromethyl)benzaldehyde	**(455-19-6)**
3-(Trifluoromethyl)benzenamine	**(98-16-8)**
(Trifluoromethyl)benzene	**(98-08-8)**
3-Trifluoromethylbenzoate	**(454-92-2)**
3-(Trifluoromethyl)benzoic acid	**(454-92-2)**
4-(Trifluoromethyl)benzoic acid	**(455-24-3)**
m-(Trifluoromethyl)benzoic acid	**(454-92-2)**
6-(Trifluoromethyl)-1,4,2-benzothiadiazine-7-sulfonamido 1,1-dioxide	**(148-56-1)**
6-(Trifluoromethyl)-1,2,4-benzo-thiadiazine-7-sulfonamide 1,1-dioxide	**(148-56-1)**
6-Trifluoromethyl-3-benzyl-7-sulfamyl-3,4-dihydro-1,2,4-benzo-	

thiadiazine, 1,1-dioxide (73-48-3)
3-(Trifluoromethyl)bromobenzene (401-78-5)
m-(Trifluoromethyl)bromobenzene (401-78-5)
Trifluoromethyl chloride (75-72-9)
(Trifluoromethyl)chlorobenzene (52181-51-8)
o-(Trifluoromethyl)chlorobenzene (88-16-4)
p-(Trifluoromethyl)chlorobenzene (98-56-6)
5-(Trifluoromethyl)-2'-deoxyuridine (70-00-8)
5-(Trifluoromethyl)deoxyuridine (70-00-8)
5-Trifluoromethyl-2-deoxyuridine (70-00-8)
Trifluoromethyldeoxyuridine (70-00-8)
8-Trifluoromethyl-7-deschloroglafenine (23779-99-9)
3-Trifluoromethyl-4,4'-dichloro-N,N'-diphenylurea (369-77-7)
3-(Trifluoromethyl)-4,4'-dichlorocarbanilide (369-77-7)
Trifluoromethyldichlorocarbanilide (369-77-7)
6-Trifluoromethyl-3,4-dihydro-7-sulfamoyl-2H-1,2,4-benzothiadiazine 1,1-dioxide (135-09-1)
α,α,α-Trifluoro-N-methyl-4,6-dinitro-N-(2,4,6-tribromophenyl)-o-toluidine (63333-35-7)
3-Trifluoromethyl diphenylamine (101-23-5)
Trifluoromethylethylene (677-21-4)
3-(Trifluoromethyl)-N-ethyl-α-methylphenethylamine (458-24-2)
Trifluoromethylhydrothiazide (135-09-1)
Trifluoromethyl iodide (2314-97-8)
Trifluoromethyl methyl ketone (421-50-1)
Trifluoromethyl-10-(3'-(1-methyl-4-piperazinyl)propyl)phenothiazine (117-89-5)
3-Trifluoromethylnitrobenzene (98-46-4)
4-(Trifluoromethyl)nitrobenzene (402-54-0)
m-(Trifluoromethyl)nitrobenzene (98-46-4)
o-(Trifluoromethyl)nitrobenzene (384-22-5)
3-Trifluoromethyl-4-nitrophenol (88-30-2)
Trifluoromethylperazine (117-89-5)
3-(Trifluoromethyl)phenol (98-17-9)
m-(Trifluoromethyl)phenol (98-17-9)
4-(3-(2-Trifluoromethyl-10-phenothiazyl)-propyl)-1-piperazineethanol (69-23-8)
3-(Trifluoromethyl)phenyl bromide (401-78-5)
m-(Trifluoromethyl)phenyl bromide (401-78-5)
m-Trifluoromethylphenyl chloride (98-15-7)
o-Trifluoromethylphenyl chloride (88-16-4)
p-Trifluoromethylphenyl chloride (98-56-6)
1-(m-Trifluoromethylphenyl)-3-chloro-4-chloromethyl-2-pyrrolidone (61213-25-0)
3-(m-Trifluoromethylphenyl)-1,1-dimethylurea (2164-17-2)
N-(3-Trifluoromethylphenyl)-N'-N'-dimethylurea (2164-17-2)
N-(m-Trifluoromethylphenyl)-N',N'-dimethylurea (2164-17-2)
1-(meta-Trifluoromethyl-phenyl)-2 ethylaminopropane (458-24-2)
Trifluoromethyl phenyl ketone (434-45-7)
1,1,1-Trifluoro-N-(2-methyl-4-(phenylsulfonyl)phenyl)-methanesulfonamide (37924-13-3)
(3-(Trifluoromethyl)phenyl)urea (13114-87-9)
3-Trifluoromethylphenylurea (13114-87-9)
N-(3-Trifluoromethylphenyl)urea (13114-87-9)
2-(8'-Trifluoromethyl-4'-quinolylamino)benzoic acid 2,3-dihydroxy propyl ester (23779-99-9)
6-Trifluoromethyl-7-sulfamoyl-4H-1,4,2-benzothiadiazine 1,1-dioxide (148-56-1)
6-Trifluoromethyl-7-sulfamyl-1,2,4-benzothiadiazine-1,1-dioxide (148-56-1)
6-Trifluoromethyl-7-sulfamyl-3,4-dihydro-1,2,4-benzothiadiazine-1,1-dioxide (135-09-1)
3-Trifluoromethylsulfonamido-p-acetotoluidide (47000-92-0)
Trifluoromethylthiazide (148-56-1)
(Trifluoromethyl)trifluorooxirane (428-59-1)
Trifluoromethyl trifluorovinyl ether (1187-93-5)
Trifluoromonobromomethane (75-63-8)
Trifluoromonochlorocarbon (75-72-9)
Trifluoromonochloroethylene (79-38-9)
2,2,2-Trifluoro-1-phenylethanone (434-45-7)
1,1,1-Trifluoro-2-propanol (374-01-6)
1,1,1-Trifluoro-2-propanone (421-50-1)
1,1,1-Trifluoropropene (677-21-4)
3,3,3-Trifluoropropene (677-21-4)
3,3,3-Trifluoropropionaldehyde (460-40-2)
Trifluoropropionaldehyde (460-40-2)
3,3,3-Trifluoropropylene (677-21-4)
1,1,1-Trifluoro-3-(2-thenoyl)acetone (326-91-0)
Trifluorothymidine (70-00-8)
α,α,α-Trifluorothymidine (70-00-8)
Trifluorotoluene (27359-10-0)
α,α,α-Trifluorotoluene (98-08-8)
ω-Trifluorotoluene (98-08-8)
(α,α,α-Trifluoro-m-tolyl) isocyanate (329-01-1)
3-(α,α,α-Trifluoro-m-tolyl)urea (13114-87-9)
2,4,6-Trifluoro-s-triazine (675-14-9)
1,1,2-Trifluoro-1,2,2-trichloroethane (76-13-1)
1,1,2-Trifluorotrichloroethane (76-13-1)
4,4'-(2,2,2-Trifluoro-1-(trifluoromethyl)ethylidene)bisphenol (1478-61-1)
Trifluoro(trifluoromethyl)oxirane (428-59-1)
3,3,3-Trifluoro-2-(trifluoromethyl)propene (382-10-5)
Trifluorovinyl bromide (598-73-2)
Trifluorovinyl chloride (79-38-9)
Trifluorure de bore et d'acide acetique, Complexe de (French) (7578-36-1)
Trifluorure de chlore (French) (7790-91-2)
Trifluoruro de boro y acido acetico, Complejo de (Spanish) (7578-36-1)
Trifluperazine (117-89-5)
Triflupromazine (146-54-3)
Trifluralin (1582-09-8)
Trifluraline (1582-09-8)
Trifluridine (70-00-8)
Trifocide (534-52-1)
Trifolex (94-81-5)
Trifolitin (520-18-3)
Triforine (26644-46-2)
Triformol (110-88-3)
Triformol (30525-89-4)
Trifrina (534-52-1)
Trifumuron (64628-44-0)
Trifungol (14484-64-1)
Trifungol (15339-38-5)
Trifurex (1582-09-8)
Trifurox (67-45-8)
Trigen (112-27-6)
Trigenolline (535-83-1)
Triglycerol (9CI) (56090-54-1)
Triglycidylglycerol (13236-02-7)
Triglycidyl isocyanurate (2451-62-9)
Triglycidylisocyanurate (2451-62-9)
1,3,5-Triglycidyl-s-triazinetrione (2451-62-9)
Triglycine (139-13-9)
Triglycol (112-27-6)
Triglycol, diacetate (111-21-7)
Triglycol dicaproate (95-08-9)
Triglycol dichloride (112-26-5)
Triglycol dihexoate (95-08-9)
Triglycol dimercaptan (14970-87-7)
Triglycol distearate (9005-08-7)
Triglycollamic acid (139-13-9)
Triglycol monobutyl ether (143-22-6)
Triglycol monoethyl ether (112-50-5)
Triglycol monomethyl ether (112-35-6)
Triglyme (112-49-2)
Trigonellin (535-83-1)
Trigonelline (535-83-1)

Trigonox 101-101/45 **(78-63-7)**
Trigonox 17/40 **(995-33-5)**
Trigonox 25/75 **(927-07-1)**
Trigonox 29 **(6731-36-8)**
Trigonox 29/40 **(6731-36-8)**
Trigonox 40 **(37187-22-7)**
Trigonox A-75 (Czech) **(75-91-2)**
Trigonox A-W70 **(75-91-2)**
Trigonox 29B50 **(6731-36-8)**
Trigonox 29B75 **(6731-36-8)**
Trigonox B **(110-05-4)**
Trigonox 23-C75 **(26748-41-4)**
Trigonox C **(614-45-9)**
Trigonox F-C50 **(107-71-1)**
Trigonox HM 80 **(37206-20-5)**
Trigonox T **(3457-61-2)**
Trigonyl **(8064-90-2)**
Trigosan **(62-38-4)**
Triherbicide CIPC **(101-21-3)**
Triherbide **(122-42-9)**
Triherbide-IPC **(122-42-9)**
Tri(hexacosyl) aluminum **(10449-71-5)**
Trihexacosylaluminum **(10449-71-5)**
Tri(hexadecyl) aluminum **(1726-65-4)**
Trihexadecylaluminum **(1726-65-4)**
Trihexylalane **(1116-73-0)**
Tri-n-hexylaluminum **(1116-73-0)**
Trihexylaluminum **(1116-73-0)**
Tri-n-hexylamine **(102-86-3)**
Trihexylamine (8CI) **(102-86-3)**
Trihexyphenidyl **(144-11-6)**
Trihydrated alumina **(21645-51-2)**
(Trihydrogen-2-hydroxy-1-((((6-amino-1-hydroxy-3-sulfo-2-naphthyl)azo)-3,3'-dihydroxy-4,4'-biphenylene)azo)-3,6-naphthalenedisulfonato-)dicopper trisodium salt **(66418-17-5)**
2,3,4-Trihydroxyacetophenone **(528-21-2)**
3,7,15-Trihydroxy-4-acetoxy-8-oxo-12,13-epoxy-δ^9-trichothecene **(23255-69-8)**
Trihydroxyaluminum **(21645-51-2)**
1,8,9-Trihydroxyanthracene **(480-22-8)**
1,2,4-Trihydroxyanthrachinon (Czech) **(81-54-9)**
1,2,4-Trihydroxyanthraquinone **(81-54-9)**
1,2,3-Trihydroxybenzen (Czech) **(87-66-1)**
1,2,3-Trihydroxybenzene **(87-66-1)**
1,2,4-Trihydroxybenzene **(533-73-3)**
1,3,5-Trihydroxybenzene **(108-73-6)**
s-Trihydroxybenzene **(108-73-6)**
sym-Trihydroxybenzene **(108-73-6)**
3,4,5-Trihydroxybenzene-1-propylcarboxylate **(121-79-9)**
3,4,5-Trihydroxybenzoic acid **(149-91-7)**
3,4,5-Trihydroxybenzoic acid n-propyl ester **(121-79-9)**
2,3',6-Trihydroxybiphenyl **(27949-30-0)**
2,4,3'-Trihydroxybiphenyl **(4190-05-0)**
1,2,4-Trihydroxybutane **(3068-00-6)**
2',4',5'-Trihydroxybutyrophenone **(1421-63-2)**
3α,7α,12α-Trihydroxy-5β-cholanic acid 24-taurine **(81-24-3)**
3,7,12-Trihydroxy-cholan-24-oic acid (3-α,5-β,7-α,12-α) **(81-25-4)**
3-α,7-α,12-α-Trihydroxy-5-β-cholan-24-oic acid **(81-25-4)**
3-α,7-α,12-α-Trihydroxycholansaeure (German) **(81-25-4)**
Trihydroxycyanidine **(108-80-5)**
1,3,5-Trihydroxycyclohexatriene **(108-73-6)**
3,4,5-Trihydroxy-1-cyclohexene-1-carboxylic acid **(138-59-0)**
2,3',4-Trihydroxydiphenyl **(4190-05-0)**
2,3',6-Trihydroxydiphenyl **(27949-30-0)**
3,16-α,17-β-Trihydroxy-δ-1,3,5-estratriene **(50-27-1)**
3,16-α,17-β-Trihydroxyestra-1,3,5(10)-triene **(50-27-1)**
Trihydroxyestrin **(50-27-1)**
Tri(hydroxyethyl)amine **(102-71-6)**
5,7,4'-Trihydroxyflavonol **(520-18-3)**
2',4',6'-Trihydroxy-3-(p-hydroxyphenyl)propiophenone **(60-82-2)**
Tris-hydroxymethyl-aminomethan (German) **(77-86-1)**
Tris-hydroxymethylaminomethane **(77-86-1)**
1,8,9-Trihydroxy-3-methylanthracene **(491-59-8)**
N,N',N''-Trihydroxymethylmelamine **(1017-56-7)**
2,4a,7-Trihydroxy-1-methyl-8-methylenegibb-3-ene-1,10-dicarboxylic acid 1,4a-lactone **(77-06-5)**
Trihydroxymethylnitromethane **(126-11-4)**
11-β,17,21-Trihydroxy-6-α-methylpregna-1,4-diene-3,20-dione **(83-43-2)**
11-β,17-α,21-Trihydroxy-6-α-methyl-1,4-pregnadiene-3,20-dione **(83-43-2)**
1,1,1-Tri(hydroxymethyl)propane **(77-99-6)**
3,16-α,17-β-Trihydroxy-δ-1,3,5-oestratriene **(50-27-1)**
3,16-α,17-β-Trihydroxyoestra-1,3,5(10)-triene **(50-27-1)**
Trihydroxyoestrin **(50-27-1)**
2-((3-α,7-α,12-α-Trihydroxy-24-oxo-5-beta-cholan-24-yl)amino)-ethanesulfonic acid **(81-24-3)**
Trihydroxyphosphine **(13598-36-2)**
11-β,17,21-Trihydroxypregna-1,4-diene-3,20-dione **(50-24-8)**
11-β,17-α,21-Trihydroxy-1,4-pregnadiene-3,20-dione **(50-24-8)**
11-β,17-α,21-Trihydroxypregna-1,4-diene-3,20-dione **(50-24-8)**
11-β,17,21-Trihydroxypregn-4-ene-3,20-dione **(50-23-7)**
11-β,17-α,21-Trihydroxy-4-pregnene-3,20-dione **(50-23-7)**
11-β,17-α,21-Trihydroxypregn-4-ene-3,20-dione **(50-23-7)**
1,2,3-Trihydroxypropane **(56-81-5)**
Trihydroxypropane **(56-81-5)**
(5Z,9-α,11-α,13E,15S)-9,11,15-Trihydroxyprosta-5,13-dien-1-oic acid **(551-11-1)**
9,11,15-Trihydroxyprosta-5,13-dien-1-oic acid **(551-11-1)**
2,6,8-Trihydroxypurine **(69-93-2)**
2,4,6-Trihydroxypyrimidine **(67-52-7)**
3,7,15-Trihydroxyscirp-4-acetoxy-9-en-8-one **(23255-69-8)**
2,4,6-Trihydroxy-1,3,5-triazine **(108-80-5)**
Trihydroxytriethylamine **(102-71-6)**
Triiodoarsine **(7784-45-4)**
1,2,3-Triiodobenzene **(608-29-7)**
1,3,5-Triiodobenzene **(626-44-8)**
2,3,5-Triiodobenzoic acid **(88-82-4)**
Triiodobenzoic acid **(88-82-4)**
Triiodoethylate de gallamine (French) **(65-29-2)**
Tri(iodoethylate) de tri (β diethylaminoethoxy)-1,2,3 benzene (French) **(65-29-2)**
Triiodoethylate of tri(diethylaminoethyloxy)-1,2,3-benzene **(65-29-2)**
Triiodo-l-thyronine **(6893-02-3)**
Triiodomethane **(75-47-8)**
2,4,6-Triiodophenol **(609-23-4)**
Triiodostibine **(7790-44-5)**
3,3',5-Triiodothyronine **(6893-02-3)**
3,5,3'-Triiodothyronine **(6893-02-3)**
L-Triiodothyronine **(6893-02-3)**
Triiodothyronine **(6893-02-3)**
Triiodure d'azote (French) **(13444-85-4)**
Triiodure de tri(β-triethylammoniumethoxy)-1,2,3 benzene (French) **(65-29-2)**
Triisobutoxytriazine **(30895-05-7)**
Triisobutylalane **(100-99-2)**
Triisobutylaluminium (DOT) **(100-99-2)**
Triisobutylaluminum **(100-99-2)**
Triisobutyl borate **(13195-76-1)**
Triisobutylene [UN 2324] **(7756-94-7)**
Triisobutylene oxide (Obs.) **(68955-06-6)**
Triisobutyl phosphate **(126-71-6)**
N,N,N-Tri(isodecyl)amine **(35723-89-8)**
Tri(isodecyl)amine **(35723-89-8)**
Triisooctylamine **(25549-16-0)**
Triisooctyl phosphite **(25103-12-2)**
Triisooctyl trimellitate **(27251-75-8)**

Triisopropanolamine	(122-20-3)
Triisopropanolamine 2,4-dichlorophenoxyacetate	(32341-80-3)
Triisopropanolamine picloram	(6753-47-5)
Triisopropoxyaluminum	(555-31-7)
1,3,5-Triisopropylbenzene	(717-74-8)
Triisopropyl borate [UN 2616]	(5419-55-6)
2,4,6-Triisopropyloxy-s-triazine	(29263-11-4)
Triisopropyl phosphite	(116-17-6)
2,4,6-Triisopropyl-1,3,5-triazine	(25176-37-8)
2,4,6-Trijodfenol (Czech)	(609-23-4)
Trijodmethane (Czech)	(75-47-8)
Trikacide	(443-48-1)
Trikamon	(443-48-1)
Trikepin	(1582-09-8)
o-Trikesylphosphate (German)	(78-30-8)
Triketohydrindene hydrate	(485-47-2)
Trikhlorfenolyat medi (Russian)	(25267-55-4)
Triklone	(79-01-6)
Trikojol	(443-48-1)
Trikozol	(443-48-1)
Trikresylfosfat (Czech)	(1330-78-5)
Trikresylphosphate (German)	(1330-78-5)
Trilafon	(58-39-9)
Trilan	(15676-16-1)
Trilaurin	(538-24-9)
Trilaurylamine	(102-87-4)
Trilax	(77-09-8)
Trilead dicitrate	(512-26-5)
Trilead phosphate	(7446-27-7)
Trilead tetroxide	(1314-41-6)
Trilen	(79-01-6)
Trilene	(79-01-6)
Triline	(79-01-6)
Trillekamin	(817-09-4)
Trilon 83	(77-81-6)
Trilon A	(139-13-9)
Trilon AO	(150-38-9)
Trilon B	(139-33-3)
Trilon B	(60-00-4)
Trilon B	(64-02-8)
Trilon BD	(139-33-3)
Trilon BW	(60-00-4)
Trilone 46	(107-44-8)
Trilostane	(13647-35-3)
Trilostano (Spanish)	(13647-35-3)
Trilostanum (Latin)	(13647-35-3)
Trim	(1582-09-8)
Trimagnesium diorthophosphate	(7757-87-1)
Trimagnesium diphosphate	(7757-87-1)
Trimagnesium phosphate	(7757-87-1)
Trimanganese tetraoxide	(1317-35-7)
Trimanganese tetroxide	(1317-35-7)
Trimangol	(12427-38-2)
Trimangol 80	(12427-38-2)
Trimanyl	(738-70-5)
Trimar	(79-01-6)
Trimaton	(137-42-8)
Trimaton	(6734-80-1)
Trimax	(1343-88-0)
Trimeks	(443-48-1)
Trimellic acid 1,2-anhydride	(552-30-7)
Trimellic acid anhydride	(552-30-7)
Trimellitic acid	(528-44-9)
Trimellitic acid cyclic 1,2-anhydride	(552-30-7)
Trimellitic acid trimethyl ester	(2459-10-1)
Trimellitic anhydride (ACGIH,OSHA)	(552-30-7)
Trimeperidina (Spanish)	(64-39-1)
Trimeperidine	(64-39-1)
Trimeperidinum (Latin)	(64-39-1)
Trimeprazine	(84-96-8)
Trimeprimina (Italian)	(739-71-9)
Trimeprimine	(739-71-9)
Trimeproprimine	(739-71-9)
1,3,5-Trimercaptotriazine	(638-16-4)
2,4,6-Trimercapto-s-triazine	(638-16-4)
Trimesic acid	(554-95-0)
Trimesinic acid	(554-95-0)
Trimesitinic acid	(554-95-0)
Trimesoyl chloride	(4422-95-1)
Trimesulf	(8064-90-2)
Trimet	(77-85-0)
Trimetaphosphate	(7785-84-4)
Trimethacarb	(12407-86-2)
1,1,1-Trimethanolethane	(77-85-0)
Trimethoate	(2275-18-5)
Trimethobenzamide	(138-56-7)
Trimethobenzamidum (Latin)	(138-56-7)
Trimethoprim	(738-70-5)
Trimethoprim and sulphamethoxazole	(8064-90-2)
Trimethoprimsulfa	(8064-90-2)
Trimethopriom	(738-70-5)
Trimethoprim-sulfamethoxazole	(8064-90-2)
3,4,5-Trimethoxyamphetamine	(1082-88-8)
3,4,5-Trimethoxyaniline	(24313-88-0)
2,4,5-Trimethoxybenzaldehyde	(4460-86-0)
3,4,5-Trimethoxybenzaldehyde	(86-81-7)
3,4,6-Trimethoxybenzaldehyde	(4460-86-0)
3,4,5-Trimethoxybenzenamine	(24313-88-0)
1,2,3-Trimethoxybenzene	(634-36-6)
1,3,5-Trimethoxybenzene	(621-23-8)
3,4,5-Trimethoxybenzeneethanamine	(54-04-6)
2,4,5-Trimethoxybenzoic acid	(490-64-2)
3,4,5-Trimethoxybenzoic acid	(118-41-2)
3,4,5-Trimethoxybenzoyl methyl reserpate	(50-55-5)
5-(3,4,5-Trimethoxybenzyl)-2,4-diaminopyrimidine	(738-70-5)
Trimethoxyborane	(121-43-7)
Trimethoxyborine	(121-43-7)
Trimethoxyboron	(121-43-7)
3,4,5-Trimethoxycinnamic acid	(90-50-6)
Trimethoxycinnamoyl methyl reserpate	(24815-24-5)
1,1,1-Trimethoxyethane	(1445-45-0)
Trimethoxyfosfin (Czech)	(121-45-9)
Trimethoxymethane	(149-73-5)
Trimethoxy(2-methylpropyl)silane	(18395-30-7)
Trimethoxymethylsilane	(1185-55-3)
3,4,5-Trimethoxyphenethylamine	(54-04-6)
3,4,5-Trimethoxyphenylacrylic acid	(90-50-6)
3,4,5-Trimethoxyphenyl-β-aminopropane	(1082-88-8)
Trimethoxyphenyl-β-aminopropane	(1082-88-8)
Trimethoxyphosphine	(121-45-9)
Trimethoxy silane	(2487-90-3)
3-(Trimethoxysilyl)-1-propanamine	(13822-56-5)
3-(Trimethoxysilyl)-1-propanol methacrylate	(2530-85-0)
3-(Trimethoxysilyl)propylamine	(13822-56-5)
N-(Trimethoxysilylpropyl)amine	(13822-56-5)
Trimethoxysilylpropyldidecylmethylammonium chloride	(68959-20-6)
(3-Trimethoxysilylpropyl)diethylenetriamine	(35141-30-1)
3-(Trimethoxysilyl)propyldimethyloctadecylammonium	(27668-52-6)
3-(Trimethoxysilyl)propyl dimethyl octadecyl ammonium chloride	(27668-52-6)
Trimethoxysilyl-3-propylester kyseliny methakrylove (Czech)	(2530-85-0)
N-(3-Trimethoxysilylpropyl)-ethylenediamine	(1760-24-3)
Trimethoxyvinylsilane	(2768-02-7)
Trimethylacetic acid	(75-98-9)

Trimethyl-β-acetoxypropylammonium chloride (62-51-1)
Trimethylacetyl chloride [UN 2438] (3282-30-2)
Trimethylalane (75-24-1)
Trimethylaluminium (DOT) (75-24-1)
Trimethylaluminum (75-24-1)
Trimethylamine (ACGIH,OSHA) (75-50-3)
Trimethylamine, Anhydrous [UN 1083] (75-50-3)
Trimethylamine, Aqueous solution (DOT) (75-50-3)
Trimethylamine, Aqueous solutions containing not more than 30% of trimethylamine (DOT) (75-50-3)
Trimethylamine, Aqueous solutions not more than 50 per cent trimethylamine by mass [UN 1297] (75-50-3)
Trimethylamine, Compd. with borane (1:1) (75-22-9)
Trimethylamine borane (75-22-9)
Trimethylamine, hydrochloride (593-81-7)
Trimethylamine, nitrate (8CI) (25238-43-1)
Trimethylamine, N-oxide (8CI) (1184-78-7)
1,2,4-Trimethyl-5-aminobenzene (137-17-7)
1,2,4-Trimethyl-5-aminobenzene hydrochloride (21436-97-5)
Trimethylaminomethane (75-64-9)
2,4,6-Tris-methylamino-1,3,5-triazine (2827-46-5)
S-(2-(N,N,N-Trimethylammonio)ethyl) O,O-diethyl-phosphorothiolate iodide (513-10-0)
Trimethylammonium chloride (593-81-7)
γ-Trimethyl-ammonium-β-hydroxybutirate (541-15-1)
4,4',6-Trimethylangelicin (90370-29-9)
2,4,5-Trimethylanilin (Czech) (137-17-7)
2,4,5-Trimethylaniline (137-17-7)
2,4,6-Trimethylaniline (88-05-1)
N,N,2-Trimethylaniline (609-72-3)
N,N,4-Trimethylaniline (99-97-8)
p,N,N-Trimethylaniline (99-97-8)
2,4,5-Trimethylaniline hydrochloride (21436-97-5)
2,4,6-Trimethylaniline hydrochloride (6334-11-8)
2,9,10-Trimethylanthracene (63018-94-0)
4,7,12-Trimethylbenz(a)anthracene (35187-24-7)
6,7,12-Trimethylbenz(a)anthracene (20627-33-2)
6,7,8-Trimethylbenz(a)anthracene (20627-32-1)
6,8,12-Trimethylbenz(a)anthracene (20627-34-3)
7,10,12-Trimethylbenz(a)anthracene (35187-27-0)
1,3,10-Trimethyl-5,6-benzacridine (French) (51787-43-0)
1,3,10-Trimethyl-7,8-benzacridine (French) (51787-42-9)
2,4,6-Trimethylbenzaldehyde (487-68-3)
4,9,10-Trimethyl-1,2-benzanthracene (20627-33-2)
7,9,11-Trimethylbenz(c)acridine (51787-42-9)
2,4,5-Trimethylbenzenamine (137-17-7)
1,2,3-Trimethylbenzene (526-73-8)
1,2,4-Trimethylbenzene (95-63-6)
as-Trimethylbenzene (95-63-6)
sym-Trimethylbenzene (108-67-8)
Trimethyl benzene (ACGIH) (108-67-8)
Trimethyl benzene (ACGIH) (95-63-6)
Trimethyl benzene (ACGIH,OSHA) (25551-13-7)
1,3,5-Trimethylbenzene [UN 2325] (108-67-8)
N,α,α-Trimethylbenzeneethanamine (100-92-5)
α,α,4-Trimethylbenzenemethanol (1197-01-9)
2,4,6-Trimethyl-benzenemethanol (9CI) (4170-90-5)
2,4,6-Trimethylbenzenesulfonyl chloride (773-64-8)
Trimethyl 1,2,4-benzenetricarboxylate (2459-10-1)
1,3,6-Trimethylbenzo(a)pyrene (16757-92-9)
2,4,5-Trimethylbenzoic acid (528-90-5)
2,4,6-Trimethylbenzoic acid (480-63-7)
Trimethylbenzol (108-67-8)
Trimethylbenzoyloxypiperidine (500-34-5)
2,4,6-Trimethylbenzyl alcohol (4170-90-5)
Trimethylbenzylammonium bromide (5350-41-4)
Trimethylbenzylammonium chloride (56-93-9)
Trimethylbenzylammonium hydroxide (100-85-6)
1,7,7-Trimethylbicyclo(2.2.1)hept-2-ene (464-17-5)
2,6,6-Trimethylbicyclo(3.1.1)-2-hept-2-ene (80-56-8)
(1R-cis)-4,6,6-Trimethylbicyclo(3.1.1)hept-3-en-2-one (18309-32-5)
2,6,6-Trimethylbicyclo(3.1.1)heptane (473-55-2)
1,3,3-Trimethylbicyclo(2.2.1)heptan-2-ol (1632-73-1)
1,7,7-Trimethyl-bicyclo(2.2.1)heptan-2-ol, endo- (507-70-0)
2,6,6-Trimethylbicyclo(3.1.1)heptan-2-ol (473-54-1)
2,6,6-Trimethylbicyclo(3.1.1)heptan-2-ol, (1α,2α,5α)- (4948-28-1)
1,7,7-Trimethylbicyclo(2.2.1)heptan-2-ol acetate (76-49-3)
(1 α,2 α,5 α)-2,6,6-Trimethylbicyclo(3.1.1)heptan-3-one (547-60-4)
1,3,3-Trimethylbicyclo(2.2.1)heptan-2-one (1195-79-5)
1,3,3-Trimethylbicyclo(2.2.1)heptan-2-one (4695-62-9)
1,7,7-Trimethylbicyclo(2.2.1)-2-heptanone (76-22-2)
1,7,7-Trimethylbicyclo(2.2.1)heptan-2-one-3-benzylidene (15087-24-8)
3,7,7-Trimethylbicyclo(4.1.0)-3-heptene (13466-78-9)
4,6,6-Trimethylbicyklo(3,1,1)hept-3-en (Czech) (80-56-8)
2,4,6-Trimethylbiphenyl (3976-35-0)
Trimethyl borate [UN 2416] (121-43-7)
Trimethylbromomethane (507-19-7)
2,2,3-Trimethylbutane (464-06-2)
2,3,3-Trimethyl-1-butene (594-56-9)
2,4,4-Trimethyl-2-butenolide (50598-50-0)
Trimethylcarbinol (75-65-0)
Trimethylcetylammonium bromide (57-09-0)
Trimethylcetylammonium chloride (112-02-7)
Trimethyl-β-chlorethylammoniumchlorid (Czech) (999-81-5)
Trimethylchloromethane (507-20-0)
Trimethyl chlorosilane (75-77-4)
Trimethylchlorosilane [UN 1298] (75-77-4)
Trimethylchlorostannane (1066-45-1)
Trimethylchlorotin (1066-45-1)
3,4,5-Trimethylcinnamic acid, ester with methyl reserpate (24815-24-5)
3,4,5-Trimethylcinnamoyl methyl reserpate (24815-24-5)
2,2,3-Trimethylcyclobutanone (1449-49-6)
2,3,5-Trimethyl-2,5-cyclohexadiene-1,4-dione (935-92-2)
1,1,3-Trimethylcyclohexane (3073-66-3)
2,2,6-Trimethyl-1,4-cyclohexanedione (20547-99-3)
2,6,6-Trimethyl-1,4-cyclohexanedione (20547-99-3)
3,5,5-Trimethyl-1,4-cyclohexanedione (20547-99-3)
3,3,5-Trimethyl-1-cyclohexanol (116-02-9)
3,3,5-Trimethylcyclohexanol (116-02-9)
3,5,5-Trimethylcyclohexanol (116-02-9)
Trimethylcyclohexanol (1321-60-4)
3,3,5-Trimethylcyclohexanol α-phenyl-α-hydroxyacetate (456-59-7)
2,2,6-Trimethylcyclohexanone (2408-37-9)
3,3,5-Trimethylcyclohexanone (873-94-9)
2,4,6-Trimethyl-3-cyclohexene-1-carboxaldehyde (1423-46-7)
2,4,6-Trimethyl-4-cyclohexene-1-carboxaldehyde (1423-46-7)
1,1,3-Trimethyl-3-cyclohexene-5-one (78-59-1)
3,5,5-Trimethyl-2-cyclohexene-1-one (78-59-1)
3,5,5-Trimethyl-2-cyclohexen-1-on (German, Dutch) (78-59-1)
2,5,6-Trimethyl-2-cyclohexen-1-one (20030-30-2)
3,3,5-Trimethyl-5-cyclohexen-1-one (78-59-1)
3,5,5-Trimethyl-2-cyclohexen-1-one (78-59-1)
3,5,5-Trimethyl-2-cyclohexenone (78-59-1)
4-(2,6,6-Trimethyl-1-cyclohexen-1-yl)-3-buten-2-one (14901-07-6)
4-(2,6,6-Trimethyl-1-cyclohexen-1-yl)-3-buten-2-one (79-77-6)
4-(2,6,6-Trimethyl-2-cyclohexen-1-yl)-3-buten-2-one (127-41-3)
1-(2,6,6-Trimethyl-2-cyclohexen-1-yl)-1,6-heptadien-3-one (79-78-7)
4-(2,6,6-Trimethyl-2-cyclohexen-1-yl)-3-methyl-3-buten-2-one (127-51-5)
5-(2,6,6-Trimethyl-1-cyclohexenyl)-4-penten-3-one (127-43-5)
5-(2,6,6-Trimethyl-2-cyclohexenyl)-4-penten-3-one (127-42-4)
3,5,5-Trimethylcyclohexyl amygdalate (456-59-7)
3,3,5-Trimethylcyclohexyl 2-hydroxybenzoate (118-56-9)
3,3,5-Trimethylcyclohexyl mandelate (456-59-7)
3,3,5-Trimethylcyclohexyl salicylate (118-56-9)

2,2,3-Trimethyl-3-cyclopentacetaldehyde **(4501-58-0)**
1,2,2-Trimethyl-1,3-cyclopentanedicarboxylic acid, (1R-cis)- **(124-83-4)**
2,2,3-Trimethyl-3-cyclopenten-1-acetaldehyde **(4501-58-0)**
3,5,5-Trimethyl-2-cyclopenten-1-one **(24156-95-4)**
3,5,5-Trimethyl-2-cyclopentenone **(24156-95-4)**
3,3,4-Trimethyldecane **(49622-18-6)**
2,2,4-Trimethyl-1,2-dihydrochinolin (Czech) **(147-47-7)**
1,4,7-Trimethyl-(2,3-dihydroindene) **(54340-87-3)**
2,2,4-Trimethyl-1,2-dihydroquinoline **(147-47-7)**
Trimethyl-1,2-dihydroquinoline **(147-47-7)**
2,2,4-Trimethyl-1,2-dihydroquinoline polymer **(26780-96-1)**
Trimethyldihydroquinoline polymer **(26780-96-1)**
2,2,4-Trimethyl-1,6-diisocyanatohexane **(16938-22-0)**
2,4,4-Trimethyl-1,6-diisocyanatohexane **(15646-96-5)**
N,N,N'-Trimethyl-N'-(3-(dimethylamino)propyl)-1,3-propanediamine **(3855-32-1)**
1,3,7-Trimethyl-2,6-dioxopurine **(58-08-2)**
N,N,1-Trimethyl-3,3-di-2-thienylallylamine **(524-84-5)**
N,N,1-Trimethyl-3,3-di(2-thienyl)-2-propenylamine **(524-84-5)**
(E,E)-3,7,11-Trimethyl-2,4-dodecadienoic acid, ethyl ester **(41096-46-2)**
N,N,N-Trimethyl-1-dodecanaminium **(10182-91-9)**
N,N,N-Trimethyl-1-dodecanaminium chloride **(112-00-5)**
3,7,11-Trimethyl-3-dodecanol **(7278-65-1)**
3,7,11-Trimethyldodecan-3-ol **(7278-65-1)**
2,6,10-Trimethyl-2,6,9,11-dodecatetraenal **(4955-32-2)**
3,7,11-Trimethyl-1,3,6,10-dodecatetraene **(502-61-4)**
3,7,11-Trimethyl-1,6,10-dodecatrien-3-ol **(7212-44-4)**
3,7,11-Trimethyl-2,6,10-dodecatrien-1-ol **(4602-84-0)**
3,7,11-Trimethyldodeca-1,6,10-trien-3-ol **(7212-44-4)**
Trimethyldodecylammonium chloride **(112-00-5)**
Trimethyldodecylammonium methosulfate **(13623-06-8)**
Trimethyl dodecylbenzyl ammonium chloride **(1330-85-4)**
Trimethyleentrinitramine (Dutch) **(121-82-4)**
Trimethylene **(75-19-4)**
Trimethylene bromide **(109-64-8)**
Trimethylene bromide chloride **(109-70-6)**
Trimethylene chlorobromide **(109-70-6)**
Trimethylene chlorohydrin **(627-30-5)**
Trimethylenediamine **(109-76-2)**
Trimethylene dibromide **(109-64-8)**
Trimethylene dichloride **(142-28-9)**
Trimethylene dimercaptan **(109-80-8)**
1,3-Trimethylenedinitrile **(544-13-8)**
4,4'-Trimethylenedipiperidine **(16898-52-5)**
Trimethylenedithioglycol **(109-80-8)**
Trimethylenedithiol **(109-80-8)**
Trimethylene glycol **(504-63-2)**
exo-5,6-Trimethylenenorbornane **(2825-82-3)**
exo-Trimethylenenorbornane **(2825-82-3)**
Trimethylene oxide **(503-30-0)**
Trimethylene sulfide (8CI) **(287-27-4)**
Trimethylenetrinitramine **(121-82-4)**
sym-Trimethylenetrinitramine **(121-82-4)**
Trimethylene trisulfide **(291-21-4)**
Trimethylenoxid (German) **(503-30-0)**
Trimethylentrisulfid (Czech) **(291-21-4)**
3,5,5-Trimethyl-2,3-epoxycyclohexanone **(10276-21-8)**
Trimethylester kyseliny borite (Czech) **(121-43-7)**
Trimethylester kyseliny orthomravenci (Czech) **(149-73-5)**
O,O,O-Trimethylester kyseliny thiofosforecne (Czech) **(152-18-1)**
N,N,N-Trimethylethanaminium chloride **(27697-51-4)**
N,N,N-Trimethylethenaminium hydroxide **(463-88-7)**
2,2,4-Trimethyl-6-ethoxy-1,2-dihydroquinoline **(91-53-2)**
Trimethylethylene **(513-35-9)**
Trimethylethylene oxide **(5076-19-7)**
Trimethylethyllead **(1762-26-1)**
3,4,5-Trimethylfenylester kyseliny methylkarbaminove (Czech) **(2686-99-9)**
Trimethylfluorosilane **(420-56-4)**
Trimethyl orthoformate **(149-73-5)**
1,3,3-Trimethyl-2-(formylmethylene)indoline **(84-83-3)**
Trimethylfosfat (Czech) **(512-56-1)**
Trimethylfosfit (Czech) **(121-45-9)**
2,3,4-Trimethylfuran **(10599-57-2)**
2,5,9-Trimethyl-7H-furo(3,2-g)(1)benzopyran-7-one **(3902-71-4)**
4,6,9-Trimethyl-2H-furo(2,3-h)(1)benzopyran-2-one **(90370-29-9)**
Tri-O-methylgallic acid **(118-41-2)**
Trimethylgallium **(1445-79-0)**
Trimethylglycine **(107-43-7)**
Trimethylglycocoll **(107-43-7)**
Trimethyl glycol **(57-55-6)**
2,4,6-Trimethylhendecanoic acid **(28484-22-2)**
N,N,N-Trimethyl-1-hexadecanaminium bromide **(57-09-0)**
N,N,N-Trimethyl-1-hexadecanaminium chloride **(112-02-7)**
Trimethylhexadecylammonium **(6899-10-1)**
Trimethylhexadecylammonium bromide **(57-09-0)**
Trimethylhexadecylammonium chloride **(112-02-7)**
1,3,5-Trimethylhexahydro-1,3,5-triazine **(108-74-7)**
1,3,5-Trimethylhexahydro-sym-triazine **(108-74-7)**
Trimethylhexamethylenediamine **(25620-58-0)**
Trimethylhexamethylene diamine [UN 2327] **(25620-58-0)**
2,2,5-Trimethylhexane **(3522-94-9)**
3,3,4-Trimethylhexane **(16747-31-2)**
Trimethyl-1,6-hexanediamine **(25620-58-0)**
3,5,5-Trimethyl-1-hexanol **(3452-97-9)**
3,5,5-Trimethylhexanol **(3452-97-9)**
3,5,5-Trimethylhexanoyl chloride **(36727-29-4)**
Trimethylhydrazine hydrochloride **(60597-20-8)**
2,3,5-Trimethylhydroquinone **(700-13-0)**
Trimethylhydroquinone **(700-13-0)**
γ-Trimethyl-β-hydroxybutyrobetaine **(541-15-1)**
N,N,N-Trimethyl(2-hydroxyethyl)ammonium bisulfite **(28427-24-9)**
N,N,N'-Trimethyl-N'-(2-hydroxyethyl)-1,2-ethanediamine **(2212-32-0)**
1,4,7-Trimethylindan **(54340-87-3)**
2,3,3-Trimethyl-3H-indole **(1640-39-7)**
2,3,3-Trimethylindolenine **(1640-39-7)**
1,3,3-Trimethyl-δ2,α-indolineacetaldehyde **(84-83-3)**
(1,3,3-Trimethylindolin-2-ylidene)acetaldehyde **(84-83-3)**
Trimethyl orthoisobutyrate **(52698-46-1)**
Trimethyllaurylammonium chloride **(112-00-5)**
Trimethyl lead **(14570-16-2)**
Trimethyl lead **(7442-13-9)**
Trimethyl lead chloride **(1520-78-1)**
N^2,N^4,N^6-Trimethylmelamine **(2827-46-5)**
Trimethylmethoxysilane **(1825-61-2)**
1,3,3-Trimethyl-2-methyleneindoline **(118-12-7)**
4,11,11-Trimethyl-8-methylene-5-oxatricyclo(8.2.0.0(4,6))dodecane **(1139-30-6)**
Trimethyl (1-methyl-2-phenothiazin-10-ylethyl)ammonium methyl sulfate **(58-34-4)**
Trimethyl(1-methyl-2-(10-phenothiazinyl)ethyl)ammonium methyl sulfate **(58-34-4)**
Trimethylmyristylammonium bromide **(1119-97-7)**
Trimethylnitrosoharnstoff (German) **(3475-63-6)**
Trimethylnitrosomocovina (Czech) **(3475-63-6)**
1,1,3-Trimethyl-3-nitrosourea **(3475-63-6)**
N-Trimethyl-N-nitrosourea **(3475-63-6)**
2,4,8-Trimethyl-6-nonanol **(123-17-1)**
2,6,8-Trimethyl-4-nonanol **(123-17-1)**
2,6,8-Trimethylnonanol-4 **(123-17-1)**
2,6,8-Trimethyl-4-nonanone **(123-18-2)**
Trimethyl nonanone **(1331-50-6)**
4,6,8-Trimethyl-1-nonene **(54410-98-9)**
1,3,3-Trimethyl-2-norbornanone **(1195-79-5)**
d-1,3,3-Trimethyl-2-norbornanone **(4695-62-9)**

1,7,7-Trimethyl-2-norbornene (464-17-5)
1,7,7-Trimethylnorbornene (464-17-5)
1,3,3-Trimethyl-2-norcamphanone (1195-79-5)
d-1,3,3-Trimethyl-2-norcamphanone (4695-62-9)
4,7,7-Trimethyl-3-norcarene (13466-78-9)
2,6,6-Trimethyl norpinanyl hydroperoxide, Technically pure (DOT) (28324-52-9)
Trimethyloctadecylammonium chloride (112-03-8)
Trimethylolaminomethane (77-86-1)
1,1,1-Trimethylolethane (77-85-0)
Trimethylolethane (77-85-0)
Trimethylolethane tribenzoate (4196-87-6)
1,1,1-Trimethylolethane trinitrate (3032-55-1)
Trimethylolethane trinonanoate (10535-50-9)
Trimethylolethane tripelargonate (10535-50-9)
Trimethylolmelamine (1017-56-7)
Trimethylolnitromethane (126-11-4)
Trimethylol nitromethane trinitrate (20820-44-4)
1,1,1-Trimethylolpropane (77-99-6)
Trimethylolpropane (77-99-6)
Trimethylolpropane, (chloromethyl)oxirane polymer (30499-70-8)
Trimethylolpropane diallyl ether (682-09-7)
Trimethylolpropane ethoxytriacrylate (28961-43-5)
Trimethylolpropane phosphite (824-11-3)
Trimethylolpropane propoxylate triacrylate (53879-54-2)
Trimethylolpropane triacrylate (15625-89-5)
Trimethylolpropane trimethacrylate (3290-92-4)
Trimethylolpropane trimethancrylate (3290-92-4)
Trimethylolpropane tripelargonate (126-57-8)
Trimethylolpropane tris(mercaptopropionate) (33007-83-9)
Trimethylopropane phosphate (1005-93-2)
3,4,4-Trimethyl-1-oxa-3-azacyclopentane (75673-43-7)
Trimethyloxacyclopropane (5076-19-7)
Trimethyloxamine (1184-78-7)
2,7,7-Trimethyl-3-oxatricyclo(4.1.1.02,4)octane (1686-14-2)
3,4,4-Trimethyloxazolidine (75673-43-7)
2,4,5-Trimethyl-3-oxazoline (22694-96-8)
2,2,3-Trimethyloxirane (5076-19-7)
Trimethyloxirane (5076-19-7)
N,N,N-Trimethyl-2-(1-oxo-2-methyl-2-propenyloxy)ethanaminium methyl sulfate (6891-44-7)
Trimethyloxonium hexachloroantimonate (54075-76-2)
6,10,14-Trimethyl-2-pentadecanone (502-69-2)
6,10,14-Trimethylpentadecan-2-one (502-69-2)
2,2,3-Trimethylpentane (564-02-3)
2,2,4-Trimethylpentane (540-84-1)
2,3,3-Trimethylpentane (560-21-4)
2,3,4-Trimethylpentane (565-75-3)
2,2,4-Trimethyl-1,3-pentanediol (144-19-4)
2,2,4-Trimethyl-1,3-pentanediol, dibenzoate (68052-23-3)
2,2,4-Trimethyl-1,3-pentanediol diisobutyrate (6846-50-0)
2,2,4-Trimethylpentanediol-1,3-diisobutyrate (6846-50-0)
2,2,4-Trimethyl-1,3-pentanediol, dioleate (68201-79-6)
2,2,4-Trimethyl-1,3-pentanediol monoisobutyrate (25265-77-4)
2,4,4-Trimethyl-2-pentanethiol (141-59-3)
2,2,4-Trimethylpentanol (123-44-4)
2,3,4-Trimethyl-3-pentanol (3054-92-0)
1,3,5-Trimethyl-1,1,3,5,5-pentaphenyltrisiloxane (3390-61-2)
2,4,4-Trimethyl-2-pentenal (53907-61-2)
2,2,4-Trimethyl-3-pentene (107-40-4)
2,3,3-Trimethyl-1-pentene (560-23-6)
2,3,4-Trimethyl-1-pentene (565-76-4)
2,4,4-Trimethyl pentene (25167-70-8)
2,4,4-Trimethyl-1-pentene (107-39-1)
2,4,4-Trimethyl-2-pentene (107-40-4)
2,4,4-Trimethylpentene-2 (107-40-4)
3,4,4-Trimethyl-2-pentene (598-96-9)
3,4,4-Trimethylpentene-2 (598-96-9)
6,6,9-Trimethyl-3-pentyl-7,8,9,10-tetrahydro-6H-dibenzo-(b,d)pyran-1-ol (1972-08-3)
N,α,α-Trimethyl-β-phenethylamine (100-92-5)
N,α,α-Trimethylphenethylamine (100-92-5)
2,3,5-Trimethylphenol (697-82-5)
2,3,6-Trimethylphenol (2416-94-6)
2,4,5-Trimethylphenol (496-78-6)
2,4,6-Trimethylphenol (527-60-6)
3,4,5-Trimethylphenol (527-54-8)
Trimethylphenol (26998-80-1)
2,3,5-Trimethyl-phenol methylcarbamate (2655-15-4)
2,4,6-Trimethylphenol phosphate (3:1) (56444-79-2)
2,4,6-Trimethylphenyl 2,6-dimethylphenyl phenyl phosphate (73179-41-6)
1-(2,4,6-Trimethylphenyl)ethanone (1667-01-2)
1,3,3-Trimethyl-1-phenylindane (3910-35-8)
Trimethylphenylmethane (98-06-6)
2,3,5-Trimethylphenyl methylcarbamate (2655-15-4)
3,4,5-Trimethylphenyl methylcarbamate (2686-99-9)
Trimethylphenyl methylcarbamate (12407-86-2)
1,7,7-Trimethyl-3-(phenylmethylene)bicyclo(2.2.1)heptan-2-one (15087-24-8)
Tri 2-methylphenyl phosphate (78-30-8)
1,2,5-Trimethyl-4-phenyl-4-piperidinol, propionate (ester) (64-39-1)
1,2,5-Trimethyl-4-phenyl-4-propionoxypiperidine (64-39-1)
O,O,O-Trimethyl phosphate (512-56-1)
Trimethyl phosphate (512-56-1)
Trimethyl phosphite (ACGIH,OSHA) [UN 2329] (121-45-9)
O,O,S-Trimethylphosphorodithioate (2953-29-9)
O,S,S-Trimethyl phosphorodithioate (22608-53-3)
O,O,O-Trimethyl phosphorothioate (152-18-1)
O,O,S-Trimethyl phosphorothioate (152-20-5)
Trimethylplumbane (7442-13-9)
Trimethylplumbane chloride (1520-78-1)
1,2,2-Trimethylpropylamine (3850-30-4)
1,2,2-Trimethylpropylester kyseliny methylfluorfosfonove (Czech) (96-64-0)
1,2,2-Trimethylpropyl methylphosphonofluoridate (96-64-0)
2,4,6-Tri(2-methylpropyloxy)-s-triazine (30895-05-7)
2',4,8-Trimethylpsoralen (3902-71-4)
4,5',8-Trimethylpsoralen (3902-71-4)
2,3,5-Trimethylpyrazine (14667-55-1)
Trimethylpyrazine (14667-55-1)
2,3,6-Trimethylpyridine (1462-84-6)
2,4,6-Trimethylpyridine (108-75-8)
Trimethylpyridine (29611-84-5)
2,3,5-Trimethylpyrrole (2199-41-9)
2,4,5-Trimethylpyrrole (2199-41-9)
Trimethylpyruvic acid (815-17-8)
2,3,4-Trimethylquinoline (2437-72-1)
3-(Trimethylsiloxy)-1-butyne (17869-76-0)
1-(Trimethylsilyl)-1-butyne (62108-37-6)
α-(Trimethylsilyl)-ω-methylpoly(oxy(dimethylsilylene)) (9006-65-9)
4-(Trimethylsilyl)phenol (13132-25-7)
p-Trimethylsilylphenol (13132-25-7)
α-(Trimethylsilyl)poly(oxy(dimethylsilylene))-ω-methyl (63148-62-9)
α-(Trimethylsilyl)-ω-((trimethylsilyl)oxy) (63148-62-9)
Trimethylstannane (1631-73-8)
Trimethylstannanol (56-24-6)
Trimethylstannyl chloride (1066-45-1)
Trimethylstearylammonium chloride (112-03-8)
3,5,24-Trimethyltetracontane (55162-61-3)
N,N,N-Trimethyl-1-tetradecanaminium bromide (1119-97-7)
N,N,N-Trimethyl-1-tetradecanaminium chloride (4574-04-3)
Trimethyltetradecylammonium bromide (1119-97-7)
1,4,6-Trimethyl-1,2,3,4-tetrahydronaphthalene (22824-32-4)
1,2,4-Trimethyl-1,4,5,6-tetrahydropyrimidine (53517-92-3)
O,O,O-Trimethylthiofosfat (Czech) (152-18-1)
Trimethylthiofosfat (Czech) (152-18-1)

O,O,O-Trimethyl thiophosphate **(152-18-1)**
Trimethylthiophosphate **(152-18-1)**
1,1,3-Trimethyl-2-thiourea **(2489-77-2)**
N,N,N'-Trimethylthiourea **(2489-77-2)**
Trimethylthiourea **(2489-77-2)**
Trimethyltin **(1631-73-8)**
Trimethyltin chloride **(1066-45-1)**
Trimethyl tin hydroxide **(56-24-6)**
5,7,8-Trimethyltocol **(59-02-9)**
2,5,8-Trimethyl-2,5,8-triazanonane **(3030-47-5)**
1,1,2-Trimethyltrichlorodisilane **(13528-88-6)**
1,7,7-Trimethyltricyclo(2.2.1.$0^{2,6}$)heptane **(508-32-7)**
4,7,7-Trimethyltricyclo(2.2.1.0(2,6))heptan-3-one **(875-99-0)**
2-(4,8,12-Trimethyltridecyl)buta-1,3-diene **(504-96-1)**
Trimethyl trimellitate **(2459-10-1)**
α,α,α'-Trimethyltrimethylene glycol **(107-41-5)**
2,4,6-Trimethyl-1,3,5-trioxaan (Dutch) **(123-63-7)**
2,4,6-Trimethyl-1,3,5-trioxane **(123-63-7)**
2,4,6-Trimethyl-s-trioxane **(123-63-7)**
s-Trimethyltrioxymethylene **(123-63-7)**
cis-2,4,6-Trimethyl-2,4,6-triphenylcyclotrisiloxane **(3424-57-5)**
1,1,3-Trimethyl-1,3,3-triphenyldisiloxane **(14920-93-5)**
1,3,3-Trimethyl-1,1,3-triphenyldisiloxane **(14920-93-5)**
1,3,5-Trimethyl-2,4,6-tris(3,5-di-tert-butyl-4-hydroxybenzyl)-benzene **(1709-70-2)**
2,4,6-Trimethyl-2,4,6-tris(3,3,3-trifluoropropyl)cyclotrisiloxane **(2374-14-3)**
2,4,6-Trimethylundecanoic acid **(28484-22-2)**
1,3,6-Trimethyluracil **(13509-52-9)**
Trimethyl vinyl ammonium hydroxide **(463-88-7)**
Trimethyl(vinylbenzyl)ammonium chloride **(26616-35-3)**
Trimethylvinylmethane **(558-37-2)**
2,6,10-Trimethyl-14-vinylpentadecan-14-ol **(505-32-8)**
1,3,7-Trimethylxanthine **(58-08-2)**
Trimethyolpropane phosphate **(1005-93-2)**
2-(Trimetil-acetil)-indan-1,3-dione (Italian) **(83-26-1)**
3,5,5-Trimetil-2-cicloesen-1-one (Italian) **(78-59-1)**
2,4,6-Trimetil-1,3,5-triossano (Italian) **(123-63-7)**
Trimetion **(60-51-5)**
Trimetobenzamida (Spanish) **(138-56-7)**
Trimeton **(86-21-5)**
Trimeton maleate **(132-20-7)**
Trimetoprim-Sulfa **(723-46-6)**
Trimetose **(132-20-7)**
2,4,6-Trimetylofenol (polish) **(527-60-6)**
Trimex N 08 **(89-04-3)**
Trimex T 08 **(3319-31-1)**
Trimforte **(8064-90-2)**
Trimidal **(26766-27-8)**
Trimipramine **(739-71-9)**
Trimitan **(817-09-4)**
Tri(mixed mono- and dinonylphenyl)phosphite **(58968-53-9)**
Trimopan **(738-70-5)**
Trimorfamid **(60029-23-4)**
Trimorfamide **(60029-23-4)**
Trimorphamid **(60029-23-4)**
Trimorphamide **(60029-23-4)**
Trimosulfa **(8064-90-2)**
Trimpex **(738-70-5)**
Trimustine **(817-09-4)**
Trimustine hydrochloride **(817-09-4)**
Trimyristin **(555-45-3)**
Trimysten **(23593-75-1)**
Trinagle **(7758-98-7)**
Trinalgon **(55-63-0)**
Trinatox D **(834-12-8)**
Trinatrium ethylendiamintetraacetat (Czech) **(150-38-9)**
Trinatriumphosphat (German) **(7601-54-9)**
Trinesium **(1343-88-0)**
Trinex **(52-68-6)**
Trinickel disulphide **(12035-72-2)**
Trinitrate de nitroisobutane triol (French) **(20820-44-4)**
Trinitrato de nitro isobutanotriol (Spanish) **(20820-44-4)**
Trinitrin **(55-63-0)**
Trinitroacetonitrile **(630-72-8)**
Trinitroacetonitrile (French) **(630-72-8)**
Trinitroacetonitrilo (Spanish) **(630-72-8)**
2,4,6-Trinitroaniline **(489-98-5)**
Trinitroaniline [UN 0153] **(26952-42-1)**
2,4,6-Trinitroanisole **(606-35-9)**
Trinitrobenzeen (Dutch) **(99-35-4)**
1,3,5-Trinitrobenzene **(99-35-4)**
Trinitrobenzene **(99-35-4)**
2,4,6-Trinitro-1,3-benzenediamine **(1630-08-6)**
2,4,6-Trinitro-1,3-benzenediol **(82-71-3)**
2,4,6-Trinitro-1,3-benzenediol, magnesium salt (1:1) **(13255-27-1)**
2,4,6-Trinitro-1,3,5-benzenetriamine **(3058-38-6)**
Trinitrobenzoic acid, Dry [UN 0215] **(129-66-8)**
Trinitrobenzoic acid, Wet, at least 10% water, over 25 lbs. in one outside packaging (DOT) **(129-66-8)**
Trinitrobenzoic acid, Wet, containing less than 30% water [UN 0215] **(129-66-8)**
Trinitrobenzoic acid, Wet, containing not less than 30% water [UN 1355] **(129-66-8)**
Trinitrobenzol (German) **(99-35-4)**
2,4,6-Trinitrochlorobenzene **(88-88-0)**
Trinitrochlorobenzene [UN 0155] **(28260-61-9)**
Trinitro-m-cresol **(602-99-3)**
Trinitro-meta-cresol [UN 0216] **(602-99-3)**
Trinitro-m-cresolic acid **(602-99-3)**
Trinitrocyclotrimethylene triamine **(121-82-4)**
2,4,6-Trinitro-1,3-dimethyl-5-tert-butylbenzene **(81-15-2)**
2,4,6-Trinitro-3,5-dimethyl-tert-butylbenzene **(81-15-2)**
2,3',4-Trinitrodiphenylamine **(970-91-2)**
2,4,3'-Trinitrodiphenylamine **(970-91-2)**
Trinitroetanol (Spanish) **(918-54-7)**
2,2,2-Trinitroethanol **(918-54-7)**
Trinitroethanol **(918-54-7)**
Trinitroethanol (French) **(918-54-7)**
Trinitroethylnitrate **(66849-71-6)**
2,4,6-Trinitrofenol (Dutch) **(88-89-1)**
2,4,6-Trinitrofenolo (Italian) **(88-89-1)**
2,4,7-Trinitro-9-fluorenone **(129-79-3)**
2,4,7-Trinitrofluoren-9-one **(129-79-3)**
Trinitroglycerin **(55-63-0)**
Trinitroglycerol **(55-63-0)**
1,3,5-Trinitrohexahydro-s-triazine **(121-82-4)**
Trinitrol **(55-63-0)**
Trinitromethane (DOT) **(517-25-9)**
2,4,6-Trinitrophenoate lead **(25721-38-4)**
1,3,5-Trinitrophenol **(88-89-1)**
2,4,6-Trinitrophenol **(88-89-1)**
Trinitrophenol, Dry [UN 0154] **(88-89-1)**
Trinitrophenol, Wetted with at least 10% water [NA 1344] **(88-89-1)**
Trinitrophenol, Wetted with at least 30% water [UN 1344] **(88-89-1)**
Trinitrophenol, Wetted with less than 30% water [UN 0154] **(88-89-1)**
2,4,6-Trinitrophenol ammonium salt **(131-74-8)**
2,4,6-Trinitrophenyl (OSHA) **(88-89-1)**
2,4,6-Trinitrophenyl-N-methylnitramine **(479-45-8)**
2,4,6-Trinitrophenylmethylnitramine (OSHA) **(479-45-8)**
Trinitrophenylmethylnitramine [UN 0208] **(479-45-8)**
2,4,6-Trinitrophenyl nitramine **(4591-46-2)**
2,4,6-Trinitrophenylnitramine **(4591-46-2)**
1,3,6-Trinitropyrene **(75321-19-6)**
2,4,6-Trinitroresorcinol **(82-71-3)**

Name	CAS No.
Trinitroresorcinol [UN 0219]	**(82-71-3)**
Trinitroresorcinol, Dry [UN 0219]	**(82-71-3)**
Trinitroresorcinol, Wetted with less than 20% water [UN 0219]	**(82-71-3)**
2,4,6-Trinitrotolueen (Dutch)	**(118-96-7)**
2,3,4-Trinitrotoluene	**(602-29-9)**
2,3,6-Trinitrotoluene	**(18292-97-2)**
2,4,5-Trinitrotoluene	**(610-25-3)**
Trinitrotoluene	**(118-96-7)**
s-Trinitrotoluene	**(118-96-7)**
sym-Trinitrotoluene	**(118-96-7)**
2,4,6-Trinitrotoluene (ACGIH,OSHA)	**(118-96-7)**
Trinitrotoluene, Dry or containing, by weight, less than 30% water [UN 0209]	**(118-96-7)**
Trinitrotoluene, Wet containing at least 10% water	**(118-96-7)**
Trinitrotoluene, Wet containing at least 10% water, over 16 ozs. in one outside packaging	**(118-96-7)**
Trinitrotoluene, Wetted with not less than 30% water [UN 1356]	**(118-96-7)**
s-Trinitrotoluol	**(118-96-7)**
sym-Trinitrotoluol	**(118-96-7)**
2,4,6-Trinitrotoluol (German)	**(118-96-7)**
1,3,5-Trinitro-1,3,5-triazacyclohexane	**(121-82-4)**
Tri-n-octylaluminum	**(1070-00-4)**
Tri-n-octylamine	**(1116-76-3)**
Tri-n-octyl trimellitate	**(89-04-3)**
28,29,30-Trinorlanostane	**(481-21-0)**
20,29,30-Trinorlupane, (17α)- (9CI)	**(55199-72-9)**
Trinoxol	**(2545-59-7)**
Trinoxol	**(93-76-5)**
Trinoxol	**(94-75-7)**
Tri-n-propylamine	**(102-69-2)**
Tri-n-propyltin chloride	**(2279-76-7)**
Triocil	**(141-94-6)**
Tri(octacosyl) aluminum	**(6651-27-0)**
Trioctacosylaluminum	**(6651-27-0)**
Trioctadecanoin	**(555-43-1)**
Tri(octadecyl) aluminum	**(3041-23-4)**
Trioctadecylaluminum	**(3041-23-4)**
Trioctadecyl citrate	**(7775-50-0)**
Trioctanoin	**(538-23-8)**
Trioctanoylglycerol	**(538-23-8)**
Trioctylaluminum	**(1070-00-4)**
Trioctylamine	**(1116-76-3)**
Trioctylmethylammonium chloride	**(5137-55-3)**
Trioctyl phosphate	**(78-42-2)**
Trioctylphosphine	**(4731-53-7)**
Trioctyl phosphine oxide	**(78-50-2)**
Trioctylphosphine oxide	**(78-50-2)**
Trioctyl trimellitate	**(89-04-3)**
Triodurin	**(50-27-1)**
Triol	**(79-01-6)**
Triol 124	**(3068-00-6)**
Triolein	**(122-32-7)**
Trioleoylglyceride	**(122-32-7)**
Trioleoylglycerol	**(122-32-7)**
Trional	**(76-20-0)**
Triothyrone	**(6893-02-3)**
Triovex	**(50-27-1)**
2,8,9-Trioxaadamantane	**(281-32-3)**
2,4,10-Trioxaadamantane (8CI)	**(281-32-3)**
2,8,9-Trioxa-5-aza-1-silabicyclo(3.3.3)undecane, 1-phenyl	**(2097-19-0)**
2,4,10-Trioxa-7-azaundecan-11-oic acid, 7-(3-methylphenyl)-3-oxo-, methyl ester (9CI)	**(25790-28-7)**
1,3,5-Trioxacycloheptane	**(5981-06-6)**
3,6,9-Trioxa-1-decanol	**(112-35-6)**
2,4,6-Trioxaheptane	**(628-90-0)**
Trioxalen	**(3902-71-4)**
Trioxan	**(110-88-3)**
1,3,5-Trioxane	**(110-88-3)**
Trioxane	**(110-88-3)**
s-Trioxane	**(110-88-3)**
sym-Trioxane	**(110-88-3)**
s-Trioxane, 2,4,6-trimethyl	**(123-63-7)**
Trioxanona	**(63-98-9)**
5,8,11-Trioxapentadecane	**(112-73-2)**
2,6,7-Trioxa-1-phosphabicyclo(2.2.2)octane, 4-ethyl- (9CI)	**(824-11-3)**
2,6,7-Trioxa-1-phosphabicyclo(2.2.2)octane, 4-ethyl-, 1-oxide (9CI)	**(1005-93-2)**
5,7,12-Trioxa-6-stannatetracosa-2,9-dienoic acid, 6,6-dibutyl-4,8,11-trioxo-, dodecyl ester, (Z,Z)- (9CI)	**(33466-31-8)**
5,7,12-Trioxa-6-stannatrideca-2,9-dienoic acid, 6,6-dibutyl-4,8,11-trioxo-, methyl ester	**(15546-11-9)**
2,4,10-Trioxatricyclo(3.3.1.13,7)decane (9CI)	**(281-32-3)**
3,6,9-Trioxa-1-tridecanol	**(143-22-6)**
3,6,9-Trioxaundecane	**(112-36-7)**
3,6,9-Trioxaundecane, 2-(3,4-(methylenedioxy)phenoxy)-	**(51-14-9)**
3,6,9-Trioxaundecanol	**(112-50-5)**
1,3,5-Trioxepane (8CI,9CI)	**(5981-06-6)**
Trioxide(S)	**(13463-67-7)**
1,3,6-Trioxocane (8CI,9CI)	**(1779-19-7)**
1,3,6-Trioxocin, tetrahydro-	**(1779-19-7)**
1,3,5-Trioxo-2,4-dioxa-1,5-dibismapentane	**(5892-10-4)**
2,4,6-Trioxohexahydropyrimidine	**(67-52-7)**
Trioxon	**(93-76-5)**
Trioxone	**(93-76-5)**
Trioxone	**(93-79-8)**
2,6,8-Trioxopurine	**(69-93-2)**
Trioxsalen	**(3902-71-4)**
Trioxymethylene	**(110-88-3)**
Trioxymethylene	**(30525-89-4)**
2,6,8-Trioxypurine	**(69-93-2)**
Trioxysalen	**(3902-71-4)**
Tripan Blue	**(72-57-1)**
Tripelenamine	**(91-81-6)**
Tripelenamine hydrochloride	**(154-69-8)**
Tripelennamina (Italian)	**(91-81-6)**
Tripelennamine	**(91-81-6)**
Tripelennamine hydrochloride	**(154-69-8)**
Tripelennamine monohydrochloride	**(154-69-8)**
Tripentaerythritol (8CI)	**(78-24-0)**
Tripentaerytritol	**(78-24-0)**
Tripentylamine (8CI)	**(621-77-2)**
N,N,N-Tripentyl-1-pentanaminium iodide	**(2498-20-6)**
Tri(perfluorobutyl)amine	**(311-89-7)**
Triphacyclin	**(64-75-5)**
Triphenatol	**(50-06-6)**
Triphenidyl	**(144-11-6)**
Triphenmorphe	**(1420-06-0)**
2,9-Triphenodioxazinedisulfonic acid, 6,13-dichloro-3,10-bis-(phenylamino)-, disodium salt (9CI)	**(6527-70-4)**
2,4,6-Triphenoxy-1,3,5-triazine	**(1919-48-8)**
2,4,6-Triphenoxy-s-triazine	**(1919-48-8)**
Triphenyl	**(26140-60-3)**
m-Triphenyl	**(92-06-8)**
p-Triphenyl	**(92-94-4)**
Triphenylaceto stannane	**(900-95-8)**
Triphenylamine	**(603-34-9)**
Triphenyl amine (ACGIH,OSHA)	**(603-34-9)**
Triphenylantimony	**(603-36-1)**
Triphenylarsene	**(603-32-7)**
Triphenylarsine	**(603-32-7)**
Triphenylarsine oxide	**(1153-05-5)**
1,3,5-Triphenylbenzene	**(612-71-5)**
Triphenylbenzene	**(612-71-5)**
symmetrical Triphenylbenzene	**(612-71-5)**

Triphenylborane (960-71-4)
Triphenylcarbinol (76-84-6)
Triphenylchlorostannane (639-58-7)
Triphenylchlorotin (639-58-7)
Triphenylene (217-59-4)
Triphenyleno(1,12-bcd)thiophene (9CI) (68558-73-6)
1,1,2-Triphenylethylene (58-72-0)
Triphenylethylene (58-72-0)
1,3,5-Triphenylhexahydro-1,3,5-triazine (91-78-1)
1,3,5-Triphenylhexahydro-s-triazine (91-78-1)
Triphenyl isocyanurate (1785-02-0)
Triphenyllead acetate (1162-06-7)
Triphenylmelamine (1973-05-3)
Triphenylmethane (519-73-3)
Triphenylmethanol (76-84-6)
Triphenylmethyl alcohol (76-84-6)
4-(Triphenylmethyl)morpholine (1420-06-0)
Triphenyl(phenylmethyl)phosphonium chloride (1100-88-5)
Triphenyl phosphate (ACGIH,OSHA) (115-86-6)
Triphenylphosphine (603-35-0)
Triphenylphosphine oxide (791-28-6)
Triphenyl phosphite (101-02-0)
Triphenylstannane (892-20-6)
Triphenylstannium acetate (900-95-8)
Triphenylstannium hydroxide (76-87-9)
Triphenylstannyl hydride (892-20-6)
Triphenyl stibine (603-36-1)
Triphenyltin (892-20-6)
Triphenyltin acetate (900-95-8)
Triphenyltin chloride (639-58-7)
Triphenyltin fluoride (379-52-2)
Triphenyltin hydride (892-20-6)
Triphenyltin hydroxide (76-87-9)
Triphenyltin oxide (76-87-9)
2,4,6-Triphenyl-1,3,5-triazine (493-77-6)
2,4,6-Triphenyl-s-triazine (493-77-6)
2,4,6-Triphenyltriazine (493-77-6)
Triphenyl-s-triazine (493-77-6)
s-Triphenyltriazine (493-77-6)
1,3,5-Triphenyl-s-triazine-2,4,6(1H,3H,5H)-trione (1785-02-0)
Triphenyl trithiocyanurate (30863-82-2)
Triphenyl-zinnacetat (German) (900-95-8)
Triphenyl-zinnhydroxid (German) (76-87-9)
Triphosaden (56-65-5)
Triphosphaden (56-65-5)
Triphosphonitrilic chloride (940-71-6)
Triphosphoric acid adenosine ester (56-65-5)
Triphosphoric acid, pentasodium salt (7758-29-4)
Triphosphoric acid, sodium salt (8CI,9CI) (13573-18-7)
Triphthazine (117-89-5)
Tripion CB (25319-90-8)
Tripla-Etilo (8048-52-0)
Triplex III (139-33-3)
Tripoli (1317-95-9)
Tripoly (7758-29-4)
Tri(polynonylphenyl)phosphite (58968-53-9)
Tripomol (137-26-8)
Tripotassium EDTA (17572-97-3)
Tripotassium citrate (866-84-2)
Tripotassium citrate monohydrate (866-84-2)
Tripotassium ethylenediaminetetraacetate (17572-97-3)
Tripotassium nitrilotriacetate (2399-85-1)
Tripotassium phosphate (7778-53-2)
Tripotassium tribromide (7758-02-3)
Tripotassium trichloride (7447-40-7)
Tripoton (86-21-5)
Triprolidin (486-12-4)
Triprolidine (486-12-4)
Triprolidine hydrochloride (6138-79-0)
Tripropargyl cyanurate (101-37-1)
Tri-2-propenyl 1,2,4-benzenetricarboxylate (2694-54-4)
Tripropionin (139-45-7)
Tripropionine (139-45-7)
1,3,5-Tripropionylhexahydro-s-triazine (30805-19-7)
Tripropyl aluminum (102-67-0)
Tripropylaluminum (102-67-0)
Tripropylamine (8CI) [UN 2260] (102-69-2)
Tripropylene (27215-95-8)
Tripropylene [UN 2057] (13987-01-4)
Tripropylene-glycol (24800-44-0)
Tripropyleneglycol diacrylate (42978-66-5)
Tripropylene glycol dimethacrylate (51247-87-1)
Tripropylene glycol methyl ether (10213-77-1)
Tripropylene glycol methyl ether (20324-33-8)
Tripropylene glycol methyl ether (25498-49-1)
Tripropylenglykolmonomethylether (Czech) (20324-33-8)
Tripropyl lead (6618-03-7)
Tripropylstannium acetate (3267-78-5)
Tripropylstannium chloride (2279-76-7)
Tripropyltin acetate (3267-78-5)
2,4,6-Tripropyl-1,3,5-triazine (1017-55-6)
Tripropyl-s-triazine (1017-55-6)
2,4,6-Triprop-2-ynyloxy-s-triazine (101-37-1)
Triptan (464-06-2)
Triptane (464-06-2)
Triptide (70-18-8)
Triptil (438-60-8)
Triptone (114-49-8)
Tripyrolidine (486-12-4)
Trirodazeen (1594-56-5)
Tris (126-72-7)
Tris (77-86-1)
Tris-Amino (77-86-1)
Tris-BP (126-72-7)
Tris Buffer (77-86-1)
Tris (Flame retardant) (126-72-7)
Tris Nitro (126-11-4)
Tris-N-Lost (German) (817-09-4)
Tris-Steril (77-86-1)
Tris(acetylacetonato)aluminum (13963-57-0)
Tris(acetylacetone)aluminum (13963-57-0)
Tris(acetylacetonyl)aluminum (13963-57-0)
Tris(N-acryloyl)hexahydrotriazine (959-52-4)
Tris(acryloyl)hexahydro-s-triazine (959-52-4)
Trisaethyleniminobenzochinon (German) (68-76-8)
2,4,6-Tris(allylamino)-s-triazine (30360-21-5)
2,4,6-Tris(allyloxy)triazine (101-37-1)
Trisamine (77-86-1)
N,N,N'-Tris(2-aminoethyl)-1,2-ethanediamine (31295-46-2)
Trisaminol (77-86-1)
Tris (4-aminophenyl) carbinol (467-62-9)
Tris(p-aminophenyl)methanol (467-62-9)
Trisanil (87-10-5)
Trisanyl (87-10-5)
Trisatin (18869-73-3)
Tris(1-aziridine)phosphine oxide (545-55-1)
2,3,5-Tris(1-aziridino)-p-benzoquinone (68-76-8)
2,3,5-Tris(aziridino)-1,4-benzoquinone (68-76-8)
2,3,5-Tris(1-aziridinyl)-p-benzoquinone (68-76-8)
2,3,5-Tris(aziridinyl)-1,4-benzoquinone (68-76-8)
Tris(1-aziridinyl)-p-benzoquinone (68-76-8)
Tris(aziridinyl)-p-benzoquinone (68-76-8)
2,3,5-Tris(1-aziridinyl)-2,5-cylohexadiene-1,4-dione (68-76-8)
Tris(1-aziridinyl)phosphine oxide (545-55-1)

Tris(aziridinyl)phosphine oxide	**(545-55-1)**
Tris(1-aziridinyl)phosphine sulfide	**(52-24-4)**
Tris(1-aziridinyl)phosphine sulphide	**(52-24-4)**
2,4,6-Tris(1'-aziridinyl)-1,3,5-triazine	**(51-18-3)**
2,4,6-Tris(1-aziridinyl)-s-triazine	**(51-18-3)**
Trisaziridinyltriazine	**(51-18-3)**
Trisben (Czech)	**(50-31-7)**
Tris bis-bifluoroamino diethoxy propane	**(39409-64-8)**
Tris, bis-bifluoroamino diethoxy propane	**(39409-64-8)**
Tris, bis-bifluoroamino dietoxi propano (Spanish)	**(39409-64-8)**
Tris, bis-fluoroamino diethoxy propane (French)	**(39409-64-8)**
Tris(1-bromomethyl-2-bromoethyl)phosphate	**(18713-51-4)**
2,2,2-Tris(bromomethyl)ethanol	**(1522-92-5)**
Tris(4-bromophenyl) phosphate	**(40946-60-9)**
Tris(p-bromophenyl) phosphate	**(40946-60-9)**
Tris(sec-butanolamine)	**(2421-02-5)**
Tris(2-butoxyethyl) phosphate	**(78-51-3)**
Tris(4-tert-butylphenyl) phosphate	**(78-33-1)**
Tris(p-t-butylphenyl) phosphate	**(78-33-1)**
Tris(p-tert-butylphenyl) phosphate	**(78-33-1)**
Trisbutylphosphine oxide	**(814-29-9)**
Triscabol	**(137-30-4)**
1,3,5-Tris(carbonyl-2-ethyl-1-azidine)benzene	**(7722-73-8)**
2,4,6-Tris(chloroamine)triazine	**(7673-09-8)**
Tris(2-chloroethyl)amine	**(555-77-1)**
Tris(β-chloroethyl)amine	**(555-77-1)**
Tris(2-chloroethyl)amine hydrochloride	**(817-09-4)**
Tris(β-chloroethyl)amine hydrochloride	**(817-09-4)**
Tris(2-chloroethyl)amine monohydrochloride	**(817-09-4)**
Tris(2-chloroethyl)ammonium chloride	**(817-09-4)**
Tris(2-chloroethyl)ester of phosphorus acid	**(140-08-9)**
Tris(2-chloroethyl) phosphate	**(115-96-8)**
Tris(β-chloroethyl) phosphate	**(115-96-8)**
Tris(2-chloroethyl)phosphite	**(140-08-9)**
Tris(3-chloro-2-hydroxypropyl) isocyanurate	**(7423-53-2)**
Tris(2-chloroisopropyl)phosphate	**(13674-84-5)**
Tris-1,2,3-(chloromethoxy)propane	**(38571-73-2)**
Tris(1-chloromethyl-2-chloroethyl)phosphate	**(13674-87-8)**
2,4,6-Tris(4-chlorophenoxy)-1,3,5-triazine	**(1919-45-5)**
2,4,6-Tris(p-chlorophenoxy)-s-triazine	**(1919-45-5)**
Tris(chloropropyl)phosphate	**(26248-87-3)**
Triscol	**(141-94-6)**
Tris(o-cresyl)-phosphate	**(78-30-8)**
1,3,5-Tris(cyanomethyl)hexahydro-s-triazine	**(4560-87-6)**
Tris(decyl)aluminum	**(1726-66-5)**
Tris(n-decyl)aluminum	**(1726-66-5)**
Tris(decyl) phosphate	**(4200-55-9)**
Tris(2,3-dibromopropyl) phosphate	**(126-72-7)**
Tris(dibromopropyl)phosphate	**(126-72-7)**
Tris(2,3-dibromopropyl) phosphoric acid ester	**(126-72-7)**
Tris(2,4-di-tert-butylphenyl) phosphite	**(31570-04-4)**
Tris(1,3-dichloroisopropyl)phosphate	**(13674-87-8)**
1,3,5-Tris(dichloromethyl)benzene	**(56682-87-2)**
Tris(2,4-dichlorophenoxyethyl) phosphite	**(94-84-8)**
Tris(2,4-dichlorophenoxyethyl)phosphite mixed with bis(2,4-dichlorophenoxyethyl)phosphite	**(8005-49-0)**
Tris(2,3-dichloropropyl)phosphate	**(78-43-3)**
1,2,3-Tris(2-diethylaminoethoxy)benzene triethiodide	**(65-29-2)**
1,2,3-Tris(2-diethylaminoethoxy)benzene tris(ethyliodide)	**(65-29-2)**
2,4,6-Tris(dimethylaminomethyl)phenol	**(90-72-2)**
Tris(dimethylamino)phosphine oxide	**(680-31-9)**
Tris(dimethylamino)phosphorus oxide	**(680-31-9)**
N,N',N''-Tris(dimethylaminopropyl)-s-hexahydrotriazine	**(15875-13-5)**
Tris(dimethylamino)silane	**(15112-89-7)**
2,4,6-Tris(dimethylamino)-1,3,5-triazine	**(645-05-6)**
2,4,6-Tris(dimethylamino)-s-triazine	**(645-05-6)**
Tris(dimethylcarbamodithioato-S,S')iron	**(14484-64-1)**
(OC-6-11)-Tris(dimethylcarbamodithioato-S,S')iron	**(14484-64-1)**
Tris(dimethylcarbamodthioato-S,S')iron	**(14484-64-1)**
Tris(dimethyldithiocarbamato)bismuth	**(21260-46-8)**
Tris(N,N-dimethyldithiocarbamato) iron(111)	**(14484-64-1)**
Tris(N,N-dimethyldithiocarbamato)iron(III)	**(14484-64-1)**
Tris(dimethyldithiocarbamato)iron	**(14484-64-1)**
2,4,6-Tris(1,1-dimethylethyl)phenol	**(732-26-3)**
Tris(dipropylene glycol)phosphine	**(36788-39-3)**
Tris(dipropylene glycol)phosphonate	**(36788-39-3)**
Tris(dodecyl)aluminum	**(1529-59-5)**
Triserpin	**(50-55-5)**
Tris(ethanedioato(2-))dilanthanum	**(537-03-1)**
Tris(ethanedioato(2-))dipraseodymium	**(3269-10-1)**
2,4,6-Tris(ethoxycarbonyl)-1,3,5-triazine	**(898-22-6)**
2,3,5-Tris(ethyleneimino)-1,4-benzoquinone	**(68-76-8)**
2,3,5-Tris(ethyleneimino)benzoquinone	**(68-76-8)**
2,3,5-Trisethyleneiminobenzoquinone	**(68-76-8)**
Tris(ethyleneimino)benzoquinone	**(68-76-8)**
Trisethyleneiminoquinone	**(68-76-8)**
2,4,6-Tris(ethyleneimino)-s-triazine	**(51-18-3)**
Tris(ethyleneimino)triazine	**(51-18-3)**
Trisethyleneimino-1,3,5-triazine	**(51-18-3)**
Tris(N-ethylene)phosphorotriamidate	**(545-55-1)**
2,3,5-Tris(ethylenimino)-p-benzoquinone	**(68-76-8)**
Tris(ethylenimino)thiophosphate	**(52-24-4)**
2,4,6-Tris(ethylenimino)-s-triazine	**(51-18-3)**
Tris(2-ethylhexyl)phosphate	**(78-42-2)**
Tris(2-ethylhexyl)phosphite	**(301-13-3)**
Tris(2-ethylhexyl) trimellitate	**(3319-31-1)**
Tris(2-hydroxyethyl)amine	**(102-71-6)**
1,3,5-Tris(2-hydroxyethyl) isocyanurate	**(839-90-7)**
N,N',N''-Tris(2-hydroxyethyl) isocyanurate	**(839-90-7)**
Tris(2-hydroxyethyl) isocyanurate	**(839-90-7)**
Tris(β-hydroxyethyl) isocyanurate	**(839-90-7)**
Tris(hydroxyethyl) isocyanurate	**(839-90-7)**
1,3,5-Tris(2-hydroxyethyl) isocyanuric acid	**(839-90-7)**
Tris(2-hydroxyethyl)methylammonium silicate	**(12687-85-3)**
Tris(2-hydroxyethyl)phenylmercuriammonium lactate	**(23319-66-6)**
Tris(2-hydroxyethyl)-s-triazine-2,4,6-trione	**(839-90-7)**
Tris(hydroxymethyl)(bromomethyl)methane	**(19184-65-7)**
1,1,1-Tris(hydroxymethyl)ethane	**(77-85-0)**
Tris(hydroxymethyl)ethane	**(77-85-0)**
Tris(hydroxymethyl)methanamine	**(77-86-1)**
Tris(hydroxymethyl)methylamine	**(77-86-1)**
Tris(hydroxymethyl)nitromethane	**(126-11-4)**
1,1,1-Tris(hydroxymethyl)propane	**(77-99-6)**
Tris(hydroxymethyl)propane	**(77-99-6)**
1,1,1-Trishydroxymethylpropane bicyclic phosphite	**(824-11-3)**
N,N',N''-Tris(hydroxymethyl)-1,3,5-triazine-2,4,6-triamine	**(1017-56-7)**
Tris(4-hydroxyphenyl)propanetriglycidyl ether	**(68517-02-2)**
Tris(2-hydroxy-1-propyl)amine	**(122-20-3)**
Tris(2-hydroxypropyl)amine	**(122-20-3)**
Trisilicalm	**(14987-04-3)**
Trisiloxane, 1,1,1,3,5,5,5-heptamethyl- (9CI)	**(1873-88-7)**
Trisiloxane, 1,1,1,3,5,5,5-heptamethyl-3-(3-(oxiranylmethoxy)propyl)- (9CI)	**(7422-52-8)**
Trisiloxane, 1,1,1,3,3,5,5-heptamethyl-5-phenyl- (8CI,9CI)	**(18407-16-4)**
Trisiloxane, 1,1,1,3,5,5,5-heptamethyl-3-(2-(trimethylsilyl)ethyl)- (6CI,8CI,9CI)	**(18077-53-7)**
Trisiloxane, 1,1,1,3,5,5,5-heptamethyl-3-((trimethylsilyl)oxy)- (9CI)	**(17928-28-8)**
Trisiloxane, 1,1,1,5,5,5-hexamethyl-3,3-bis(trimethylsiloxy)-	**(3555-47-3)**
Trisiloxane, 1,1,1,5,5,5-hexamethyl-3,3-bis((trimethylsilyl)oxy)- (9CI)	**(3555-47-3)**
Trisiloxane, 1,1,1,5,5,5-hexamethyl-3-phenyl-3-((trimethylsilyl)oxy)- (9CI)	**(2116-84-9)**
Trisiloxane, octamethyl- (9CI)	**(107-51-7)**

Trisiloxane, 1,3,3,5-tetramethyl-1,1,5,5-tetraphenyl- (9CI) **(3982-82-9)**
Trisiloxane, 1,3,5-trimethyl-1,1,3,5,5-pentaphenyl- (9CI) **(3390-61-2)**
Tris(isocyanatohexyl)biuret **(4035-89-6)**
Tris(4-isocyanatophenyl)thiophosphate **(4151-51-3)**
(Tris(β-methoxyethoxy))vinylsilane **(1067-53-4)**
Tris(2-methoxyethoxy)vinylsilane **(1067-53-4)**
Tris(methoxyethoxy)vinylsilane **(1067-53-4)**
Tris(p-methoxyphenyl)chloroethylene **(569-57-3)**
2,4,6-Tris(methylamino)-s-triazine **(2827-46-5)**
Tris(2-methyl-1-aziridinyl)phosphine oxide **(57-39-6)**
Tris(2-methylaziridin-1-yl)phosphine oxide **(57-39-6)**
1,3,5-Tris(1-methylethyl)benzene **(717-74-8)**
N,N',N''-Tris(1-methylethylene)phosphoramide **(57-39-6)**
Tris(1-methylethylene)phosphoric triamide **(57-39-6)**
1,1,3-Tris(2-methyl-4-hydroxy-5-tert-butylphenyl)butane **(1843-03-4)**
1,1,1-Tris(methylol)ethane **(77-85-0)**
Tris(3-methylphenyl) phosphate **(563-04-2)**
Tris(4-methylphenyl) phosphate **(78-32-0)**
Tris(methylphenyl) phosphate **(1330-78-5)**
Tris(o-methylphenyl)phosphate **(78-30-8)**
Tris(2-methylpropyl) phosphate **(126-71-6)**
2,4,6-Tris(methylthio)-s-triazine **(5759-58-0)**
Tris(nonafluorobutyl)amine **(311-89-7)**
Tris(nonylphenyl)phosphite **(26523-78-4)**
Trisodium EDTA **(150-38-9)**
Trisodium HEDTA **(139-89-9)**
Trisodium 4'-anilino-8-hydroxy-1,1'-azonaphthalene-3,6,5'-trisulfonate **(3861-73-2)**
Trisodium arsenate **(13464-38-5)**
Trisodium arsenite **(13464-37-4)**
Trisodium calcium diethylenetriaminepentaacetate **(12111-24-9)**
Trisodium 3-carboxy-5-hydroxy-1-p-sulfophenyl-4-p-sulfophenylazopyrazole **(1934-21-0)**
Trisodium carboxymethyloxysuccinate **(34128-01-3)**
Trisodium citrate **(68-04-2)**
Trisodium cyanurate **(3047-33-4)**
Trisodium edetate **(150-38-9)**
Trisodium ethylenediaminetetraacetate **(150-38-9)**
Trisodium etidronate **(2666-14-0)**
Trisodium hydrogen ethylenediaminetetraacetate **(150-38-9)**
Trisodium hydrogen (ethylenedinitrilo)tetraacetate **(150-38-9)**
Trisodium (2-hydroxyethyl)ethylenediaminetriacetate **(139-89-9)**
Trisodium N-(2-hydroxyethyl)-N,N',N'-ethylenediaminetriacetate **(139-89-9)**
Trisodium N-(2-hydroxyethyl)ethylenediamine-N,N',N'-triacetate **(139-89-9)**
Trisodium N-(2-hydroxyethyl)ethylenediaminetriacetate **(139-89-9)**
Trisodium N-(hydroxyethyl)ethylenediaminetriacetate **(139-89-9)**
Trisodium N-hydroxyethylethylenediamine-N,N',N'-triacetate **(139-89-9)**
Trisodium N-hydroxyethylethylenediaminetriacetate **(139-89-9)**
Trisodium hydroxyethyl ethylenediaminetriacetate **(139-89-9)**
Trisodium(2-hydroxyethyl)ethylenediaminetriacetate **(139-89-9)**
Trisodium 1-hydroxy-3,6,8-pyrenetrisulfonate **(6358-69-6)**
Trisodium nitrilotriacetate **(5064-31-3)**
Trisodium nitrilotriacetate monohydrate **(18662-53-8)**
Trisodium nitrilotriacetic acid **(5064-31-3)**
Trisodium orthophosphate **(7601-54-9)**
Trisodium phosphate **(7601-54-9)**
Trisodium salt N-hydroxyethyl ethylenediaminetriacetic acid **(139-89-9)**
Trisodium salt N-hydroxyethylethylenediaminetriacetic acid **(139-89-9)**
Trisodium salt of 3-carboxy-5-hydroxy-1-sulfophenylazopyrazole **(1934-21-0)**
Trisodium salt of (hydroxyethyl)ethylenediamine **(139-89-9)**
Trisodium salt of 1-(4-sulfo-1-naphthylazo)-2-naphthol-3,6-disulfonic acid **(915-67-3)**
Trisodium salt of 1-(4-sulpho-1-naphthylazo)-2-naphthol-3,6-disulphonic acid **(915-67-3)**
Trisodium sulfobutanedioate **(13419-59-5)**
Trisodium 1,3,5-triazine-2,4,6(1H,3H,5H)-trione **(3047-33-4)**
Trisodium trifluoride **(7681-49-4)**
Trisodium versenate **(150-38-9)**
Trisomin **(1343-88-0)**
Trisomnin **(76-73-3)**
Trisoralen **(3902-71-4)**
Tris(pentaerythritol) **(78-24-0)**
Tris(2,4-pentanedionato)aluminum **(13963-57-0)**
Tris(2,4-pentanedione)aluminum **(13963-57-0)**
Tris(phosphonomethyl)amine **(6419-19-8)**
Tris(propylthio)arsine **(5582-57-0)**
Trispuffer **(77-86-1)**
Tris(β-stearatoethyl)amine **(3002-22-0)**
Tristearate de sorbitan (French) **(26658-19-5)**
Tristearin **(555-43-1)**
Tristearyl citrate **(7775-50-0)**
Tris(tetradecyl)aluminum **(1529-58-4)**
Tris(tolyloxy)phosphine oxide **(1330-78-5)**
Tris(m-tolyl) phosphate **(563-04-2)**
Tris(o-tolyl)-phosphate **(78-30-8)**
1,2,3-Tris(2-triethylammonium ethoxy)benzene triiodide **(65-29-2)**
2,4,6-Tris(trifluoromethyl)-s-triazine **(368-66-1)**
1,3,5-Tris(trifluoropropyl)trimethylcyclotrisiloxane **(2374-14-3)**
1,3,5-Tris(γ-trimethoxysilylpropyl)isocyanurate **(26115-70-8)**
Trisulfide, dimethyl (9CI) **(3658-80-8)**
Trisulfon Congo Red **(573-58-0)**
Trisulfon Fast Blue B **(7082-31-7)**
Trisulfon Violet B **(6426-67-1)**
Trisulfon Violet N **(2586-60-9)**
Trisulfon Yellow G **(1325-37-7)**
Trisulfon Yellow GF **(1325-37-7)**
Trisulfurated phosphorus **(1314-85-8)**
Trisulphone Brown B **(2429-81-4)**
Tris(undecafluoropentyl)amine **(338-84-1)**
Tris(2,4-xylenyl)phosphate **(3862-12-2)**
Tris(2,5-xylenyl)phosphate **(19074-59-0)**
Tris(2,6-xylenyl)phosphate **(121-06-2)**
Tris(3,4-xylenyl)phosphate **(3862-11-1)**
Tris(3,5-xylenyl)phosphate **(25653-16-1)**
Tritac **(1861-44-5)**
Tritane **(519-73-3)**
Tritanol **(76-84-6)**
Triteren **(396-01-0)**
Tri(tetracosyl) aluminum **(6651-26-9)**
Tritetracosylaluminum **(6651-26-9)**
Tri(tetradecyl) aluminum **(1529-58-4)**
Tritetradecylaluminum **(1529-58-4)**
Tritex-Extra **(74051-80-2)**
Trithene **(79-38-9)**
1,3,5-Trithiacyclohexane **(291-21-4)**
sym-Trithian (Czech) **(291-21-4)**
1,2,3-Trithian-5-amine, N,N-dimethyl-, ethanedioate (1:1) **(31895-22-4)**
1,3,5-Trithiane **(291-21-4)**
s-Trithiane **(291-21-4)**
v-Trithiane, 5-(dimethylamino)-, oxalate **(31895-22-4)**
1,3,5-Trithiane, 2,4,6-trimethyl- (9CI) **(2765-04-0)**
s-Trithiane, 2,4,6-trimethyl- (8CI) **(2765-04-0)**
Trithioacetaldehyde **(2765-04-0)**
Trithiocarbonic acid, cyclic ester with 2,3-quinoxalinedithiol **(93-75-4)**
Trithiocyanuric acid **(638-16-4)**
Trithioformaldehyde **(291-21-4)**
1,2,4-Trithiolane (8CI,9CI) **(289-16-7)**
1,2,4-Trithiolane, 3,5-dimethyl- (9CI) **(23654-92-4)**
Trithion **(786-19-6)**
Trithion Miticide **(786-19-6)**
Tritisan **(82-68-8)**
Tritium (9CI) **(10028-17-8)**
Tritoftorol **(12122-67-7)**
Tritol **(118-96-7)**

Name	CAS No.
Tritolylfosfat (Czech)	**(1330-78-5)**
Tri-2-tolyl phosphate	**(78-30-8)**
Tri-m-tolyl phosphate	**(563-04-2)**
Tri-o-tolyl phosphate	**(78-30-8)**
Tri-p-tolyl phosphate	**(78-32-0)**
Tritolyl phosphate	**(1330-78-5)**
Tritolyl phosphite	**(25586-42-9)**
Triton	**(118-96-7)**
Triton AG 98	**(9036-19-5)**
Triton B	**(100-85-6)**
Triton GR 7	**(577-11-7)**
Triton GR-5	**(577-11-7)**
Triton K-60	**(8001-54-5)**
Triton N-100	**(9016-45-9)**
Triton X	**(9002-93-1)**
Triton X 15	**(9036-19-5)**
Triton X 35	**(9002-93-1)**
Triton X-40	**(122-19-0)**
Triton X 45	**(9002-93-1)**
Triton X 100	**(9002-93-1)**
Triton X100	**(9036-19-5)**
Triton X 102	**(9002-93-1)**
Triton X 114	**(9036-19-5)**
Triton X 165	**(9002-93-1)**
Triton X 207	**(9036-19-5)**
Triton X 305	**(9002-93-1)**
Triton X-400	**(122-19-0)**
Triton X 405	**(9002-93-1)**
Triton X 705	**(9002-93-1)**
Tritonyl 45	**(9004-78-8)**
Tritox	**(545-06-2)**
Tritriacontane (8CI,9CI)	**(630-05-7)**
17-Tritriacontanone (8CI,9CI)	**(22986-69-2)**
Tri(2,4,6-trimethylphenyl) phosphate	**(56444-79-2)**
Tritrol	**(94-81-5)**
Trityl alcohol	**(76-84-6)**
Trityl chloride	**(76-83-5)**
N-Tritylmorpholine	**(1420-06-0)**
N-Trityl-p-nitroaniline	**(20222-29-1)**
Triulose	**(96-26-4)**
Trivazol	**(443-48-1)**
Trivitan	**(67-97-0)**
Tri(2,4-xylenyl)phosphate	**(3862-12-2)**
Tri(2,5-xylenyl)phosphate	**(19074-59-0)**
Tri(2,6-xylenyl)phosphate	**(121-06-2)**
Tri(3,5-xylenyl)phosphate	**(25653-16-1)**
Tri-3,4-xylenyl phosphate	**(3862-11-1)**
Trixylenyl phosphate	**(25155-23-1)**
Tri-2,6-xylyl phosphate	**(121-06-2)**
Trixylyl phosphate	**(25155-23-1)**
Triyoduro de nitrogeno (Spanish)	**(13444-85-4)**
Trizilin	**(1836-75-5)**
Trizilin 25	**(1836-75-5)**
Triziman	**(8018-01-7)**
Triziman D	**(8018-01-7)**
Trizinc diphosphate	**(7779-90-0)**
Trizone	**(8000-21-3)**
Trochin	**(54-05-7)**
Troclosene	**(2782-57-2)**
Troclosene potassium	**(2244-21-5)**
Trofan	**(6912-86-3)**
Trofurit	**(54-31-9)**
Trogamid T	**(63428-83-1)**
Trogum	**(9005-25-8)**
Trojacetonoaminy (Polish)	**(826-36-8)**
Trojchlorek fosforu (Polish)	**(7719-12-2)**
Trojchlorobenzen (Polish)	**(120-82-1)**
Trojchloroetan(1,1,2) (Polish)	**(79-00-5)**
Trojkrezylu fosforan (Polish)	**(78-30-8)**
Trojnitrotoluen (Polish)	**(118-96-7)**
Trolamine	**(102-71-6)**
Trolen	**(299-84-3)**
Trolene	**(299-84-3)**
Trolitul	**(9003-53-6)**
Trolitul AN	**(9003-54-7)**
Trolitul EN	**(9003-54-7)**
Trolovol	**(52-67-5)**
Trombavar	**(1191-50-0)**
Trombosan	**(66-76-2)**
Trombovar	**(1191-50-0)**
Trombovar	**(139-88-8)**
Trometamol	**(77-86-1)**
Trometamole	**(77-86-1)**
Tromete	**(7601-54-9)**
Tromethamine	**(77-86-1)**
Tromethane	**(77-86-1)**
Tromethanmin	**(77-86-1)**
Trona	**(10294-33-4)**
Trona	**(497-19-8)**
Trona	**(533-96-0)**
Trona	**(7757-82-6)**
Tronabor	**(1303-96-4)**
Tronamang	**(7439-96-5)**
Tronox	**(13463-67-7)**
Tropacaine	**(537-26-8)**
Tropacocain	**(537-26-8)**
Tropacocaine	**(537-26-8)**
Tropaelin-O	**(547-57-9)**
Tropaeolene (Biological stain)	**(547-57-9)**
Tropaeolin	**(547-58-0)**
Tropaeolin 1	**(523-44-4)**
Tropaeolin D	**(140-56-7)**
Tropaeolin G	**(523-44-4)**
Tropaeolin G	**(587-98-4)**
Tropaeolin O	**(547-57-9)**
Tropaeolin OO	**(554-73-4)**
Tropaeolin OOO	**(633-96-5)**
Tropaeolin OOO 2	**(633-96-5)**
Tropaeolin OOO No. 1	**(523-44-4)**
Tropaeolin R	**(547-57-9)**
Tropaeoline	**(547-57-9)**
Tropaeoline O	**(547-57-9)**
1αH,5αH-Tropane-2β-carboxylic acid, 3β-hydroxy-, methyl ester, benzoate (ester) (8CI)	**(50-36-2)**
1-α-H,5-α-H-Tropane-2-β-carboxylic acid, 3-β-hydroxy-, methyl ester, benzoate	**(50-36-2)**
1-α-H,5-α-H-Tropane-2-β-carboxylic acid, 3-β-hydroxy-, methyl ester, benzoate (ester), hydrochloride	**(53-21-4)**
2-β-Tropanecarboxylic acid, 3-β-hydroxy-, methyl ester, benzoate (ester)	**(50-36-2)**
1-α-H,5-α-H-Tropane, 3-α-(diphenylmethoxy)	**(86-13-5)**
1-α-H,5-α-H-Tropanium, 6-β,7-β-epoxy-3-α-hydroxy-8-methyl-, bromide, (-)-tropate (ester)	**(155-41-9)**
1-α-H,5-α-H-Tropanium, 3-α-hydroxy-8-methyl-, bromide, 2-propylvalerate	**(80-50-2)**
1-α-H,5-α-H-Tropanium, 3-α-hydroxy-8-methyl-, bromide, (+-)-tropate (Ester)	**(2870-71-5)**
3α-Tropanol	**(120-29-6)**
1αH,5αH-Tropan-3α-ol (8CI)	**(120-29-6)**
1-α-H,5-α-H-Tropan-3-β-ol, benzoate (ester)	**(537-26-8)**
1-α-H,5-α-H-Tropan-3-α-ol, 6-β,7-β-epoxy-, (-)-tropate (ester), hydrobromide	**(114-49-8)**
1αH,5αH-Tropan-3α-ol, 6β,7β-epoxy-, (-)-tropate (ester), 8-oxide, hydrobromide (8CI)	**(6106-81-6)**

1-α-H,5-α-H-Tropan-3-α-ol, mandelate (ester) **(87-00-3)**
1-α-H,5-α-H-Tropan-3-α-ol, mandelate (ester), hydrobromide **(51-56-9)**
1-α-H,5-α-H-Tropan-3-α-ol (+-)-tropate (ester) (8CI) **(51-55-8)**
1-α-H,5-α-H-Tropan-3-α-ol (+-)-tropate (ester), sulfate (2:1) salt **(55-48-1)**
3-Tropanylbenzoate-2-carboxylic acid methyl ester **(50-36-2)**
DL-Tropanyl 2-hydroxy-1-phenylpropionate **(51-55-8)**
DL-Tropanyl 2-hydroxy-1-phenylpropionate sulfate **(55-48-1)**
Tropeolin **(3012-37-1)**
Tropeolin (Czech) **(523-44-4)**
Tropeolin O **(547-57-9)**
Tropeolin OO **(554-73-4)**
Trophicardyl **(58-63-9)**
Trophite **(67-03-8)**
Tropic acid **(529-64-6)**
Tropic acid, ester with scopine **(51-34-3)**
Tropic acid, (-)-, ester with tropine **(101-31-5)**
Tropic acid, ester with tropine **(51-55-8)**
Tropic acid, 9-methyl-3-oxa-9-azatricyclo(3.3.1.$O^{2,4}$)non-7-yl ester **(51-34-3)**
Tropic acid, 3-α-tropanyl ester **(51-55-8)**
Tropical Resins LS **(8001-21-6)**
Tropical Royal Blue Toner **(1325-87-7)**
Tropical Violet Toner **(1325-82-2)**
Tropilidene **(544-25-2)**
Tropilidin **(544-25-2)**
Tropin **(120-29-6)**
Tropin **(2870-71-5)**
Tropine **(120-29-6)**
psi-Tropine benzoate **(537-26-8)**
Tropine, mandelate (ester) **(87-00-3)**
Tropine tropate **(51-55-8)**
Tropine, (-)-tropate (ester) **(101-31-5)**
Tropine, tropate (ester) **(51-55-8)**
Tropintran **(55-48-1)**
Tropital **(5281-13-0)**
Tropolone **(533-75-5)**
Tropone **(539-80-0)**
Tropotox **(6062-26-6)**
Tropotox **(94-81-5)**
(+,-)-Tropyl tropate **(51-55-8)**
dl-Tropyltropate **(51-55-8)**
Trotox **(94-81-5)**
Trotyl **(118-96-7)**
Trotyl Oil **(118-96-7)**
Trovidur **(75-01-4)**
Trovidur PE **(9002-88-4)**
Troxilan **(357-56-2)**
Troykyd Anti-Skin B **(96-29-7)**
Troykyd Anti-Skin BTO **(110-69-0)**
Troysan 142 **(533-74-4)**
Troysan 174 **(34375-28-5)**
Troysan Anti-Mildew O **(133-07-3)**
Troysan Copper 8% **(1338-02-9)**
Troysan KK-108A **(55406-53-6)**
Troysan Polyphase Anti-Mildew **(55406-53-6)**
Trp-P-1 **(62450-06-0)**
Truban **(2593-15-9)**
Trucidor **(2275-23-2)**
True Ammonium Sulfide **(12124-99-1)**
True Ammonium Sulfide **(12135-76-1)**
Truflex 146 **(110-29-2)**
Truflex DOA **(103-23-1)**
Truflex DOP **(117-81-7)**
Truflex DTDP **(119-06-2)**
Truflex DOX **(103-24-2)**
Truflex OBP **(84-78-6)**
Trusono **(77-75-8)**
Truxal **(113-59-7)**
Truxaletten **(113-59-7)**
Truxil **(113-59-7)**
Tryben **(50-31-7)**
Trycite 1000 **(9003-53-6)**
Trycol LAL 12 **(9002-92-0)**
Trycol LAL Series **(9002-92-0)**
Trycol NP-1 **(9016-45-9)**
Trycol OAL 23 **(9004-98-2)**
Trydet OS Series **(9004-96-0)**
Trydet SA 40 **(9004-99-3)**
Trydet SA Series **(9004-99-3)**
Tryosine, 3,5-diiodo- **(300-39-0)**
Trypaflavin **(65431-33-6)**
Trypaflavin **(8048-52-0)**
Trypaflavin **(86-40-8)**
Trypaflavine **(65431-33-6)**
Trypaflavine **(8048-52-0)**
Trypaflavine Neutral **(8048-52-0)**
Trypaflavinum **(8048-52-0)**
Trypan Blue **(72-57-1)**
Trypan Blue BPC **(72-57-1)**
Trypan Blue sodium salt **(72-57-1)**
Trypanblau (German) **(72-57-1)**
Trypane Blue **(72-57-1)**
Trypoxyl **(127-85-5)**
Tryptacin **(6912-86-3)**
Tryptamine **(61-54-1)**
Tryptamine, 5-hydroxy- **(50-67-9)**
Tryptizol **(50-48-6)**
L-Tryptofan **(73-22-3)**
Tryptones **(73049-73-7)**
(-)-Tryptophan **(73-22-3)**
D-Tryptophan **(153-94-6)**
DL-Tryptophan **(54-12-6)**
L-Tryptophan **(6912-86-3)**
L-Tryptophan **(73-22-3)**
Tryptophan **(6912-86-3)**
Tryptophan **(73-22-3)**
levoTryptophan **(6912-86-3)**
Tryptophan, D **(153-94-6)**
Tryptophan, DL **(54-12-6)**
Tryptophan, L **(73-22-3)**
Tryptophan P1 **(62450-06-0)**
Tryptophan P2 **(62450-07-1)**
L-Tryptophane **(73-22-3)**
Tryptophane **(73-22-3)**
Tryptophan, 5-hydroxy **(56-69-9)**
l-Tryptophan, 5-hydroxy-, (9CI) **(4350-09-8)**
Tryptophan, 5-hydroxy-, L **(4350-09-8)**
Trysben **(50-31-7)**
Trysben 200 **(3426-62-8)**
Trysben 200 **(50-31-7)**
D-Trytophane **(153-94-6)**
TsL 431 **(9003-11-6)**
TsPB **(140-72-7)**
Tsapolak 964 **(9004-70-0)**
Tserenox **(495-73-8)**
Tsiazid **(140-87-4)**
Tsidial **(2597-03-7)**
Tsiklamid **(968-81-0)**
Tsiklomitsin **(60-54-8)**
Tsimat **(137-30-4)**
Tsineb (Russian) **(12122-67-7)**
Tsipromat (Russian) **(12071-83-9)**
Tsiram (Russian) **(137-30-4)**
Tsitrex **(2439-10-3)**
2,4,5-T sodium **(13560-99-1)**

T$_3$	(6893-02-3)
Tsudohmin	(15307-79-6)
Tsumacide	(1129-41-5)
Tu Cillin	(113-98-4)
Tuads	(137-26-8)
Tuamine	(123-82-0)
Tuaminoheptane	(123-82-0)
Tuareg	(24815-24-5)
Tuasal 100	(87-10-5)
Tuasol	(87-10-5)
Tuasol 100	(87-10-5)
Tuazole	(72-44-6)
Tuazolone	(72-44-6)
Tubatoxin	(83-79-4)
Tubazid	(54-85-3)
Tubazide	(54-85-3)
Tubeco	(54-85-3)
Tubercid	(54-85-3)
Tubercuprose	(544-19-4)
Tuberian	(54-85-3)
Tuberit	(122-42-9)
Tuberite	(122-42-9)
Tubermin	(536-33-4)
Tuberoid	(536-33-4)
Tuberoson	(536-33-4)
Tubicon	(54-85-3)
Tubocin	(13292-46-1)
Tubocurarin	(57-95-4)
(+)-Tubocurarine	(57-95-4)
Tubocurarine	(57-95-4)
Tubocurarine, (+)	(57-95-4)
d-Tubocurarine	(57-95-4)
Tubomel	(54-85-3)
Tubophan	(57-66-9)
Tubothane	(12427-38-2)
Tubotin	(76-87-9)
Tubotin	(900-95-8)
Tubotoxin	(83-79-4)
Tuclase	(77-23-6)
Tuex	(137-26-8)
Tuff-Lite	(9003-07-0)
Tugon	(52-68-6)
Tugon Fly Bait	(52-68-6)
Tugon Stable Spray	(52-68-6)
Tugon fliegenkugel	(114-26-1)
Tulabase Fast Blue BB	(120-00-3)
Tulabase Fast Garnet GB	(97-56-3)
Tulabase Fast Garnet GBC	(97-56-3)
Tulabase Fast Red GL	(89-62-3)
Tulabase Fast Red RL	(99-52-5)
Tulabase Fast Red TR	(95-69-2)
Tuladisperse Fast Yellow 2G	(2832-40-8)
Tulasteron Fast Yellow 5R-B	(6300-37-4)
Tulathol AS-BO	(132-68-3)
Tulathol AS-BS	(135-65-9)
Tulathol AS-D	(135-61-5)
Tulathol AS-G	(91-96-3)
Tulathol AS-OL	(135-62-6)
Tulathol AS	(92-77-3)
Tulathol AS-phenyl	(92-74-0)
Tulisan	(137-26-8)
Tuluylendiisocyanat (German)	(584-84-9)
Tulyl	(93-14-1)
Tumbleaf	(7775-09-9)
Tumescal OPE	(90-43-7)
Tumex	(148-24-3)
Tung Oil	(8001-20-5)
Tungsten (ACGIH)	(7440-33-7)
Tungsten Blue	(1314-35-8)
Tungsten carbide	(11130-73-7)
Tungsten-carbide	(12070-12-1)
(OC-6-11)Tungsten fluoride	(7783-82-6)
Tungsten fluoride, (OC-6-11)- (9CI)	(7783-82-6)
Tungsten hexafluoride	(7783-82-6)
Tungsten monocarbide	(12070-12-1)
Tungsten-oxide	(1314-35-8)
Tungsten trioxide	(1314-35-8)
Tungstic acid, disodium salt	(13472-45-2)
Tungstic acid, lead(2+) salt (1:1)	(7759-01-5)
Tungstic anhydride	(1314-35-8)
Tungstic oxide	(1314-35-8)
Tunic	(20354-26-1)
Tunicin	(9004-34-6)
Tupersan	(1982-49-6)
Tuphetamine	(51-63-8)
Turbacil	(5902-51-2)
Turbulethylazin (German)	(5915-41-3)
Turcam	(22781-23-3)
Turf-Cal	(7778-44-1)
3336 Turf Fungicide	(23564-06-9)
Turgex	(70-30-4)
Turkey Red	(72-48-0)
Turkey Red Oil	(8002-33-3)
Turkey-Red Oil	(8002-33-3)
Turkey Red Oil, Sodium salt	(68187-76-8)
Turkey Rhubarb	(481-74-3)
Turmeric Oil	(8024-37-1)
Turmeric Oleoresin	(8024-37-1)
Turpentine (ACGIH,OSHA) [UN 1299]	(8006-64-2)
Turpentine Oil	(8006-64-2)
Turpentine Oil, Rectifier	(8006-64-2)
Turpentine Steam Distilled	(8006-64-2)
Turpentine Substitute	(9005-90-7)
Turpentine [UN 1299]	(9005-90-7)
Turpeth mineral	(1312-03-4)
Turpinal SL	(2809-21-4)
Tussapap	(103-90-2)
Tusscapine	(128-62-1)
Tutane	(13952-84-6)
Tutofusin tris	(77-86-1)
Tuzet	(2445-07-0)
Tween 20	(9005-64-5)
Tween 40	(9005-66-7)
Tween 60	(9005-67-8)
Tween 80	(9005-65-6)
Tween 81	(9005-65-6)
Tween 80 A	(9005-65-6)
Twin Light Rat Away	(81-81-2)
Tybamate	(4268-36-4)
Tybatran	(4268-36-4)
Tyclarosol	(64-02-8)
Tydex	(51-63-8)
Tydex 12	(9002-98-6)
Tygon	(9003-22-9)
Tylan	(1401-69-0)
Tylandril	(50-55-5)
Tylenol	(103-90-2)
Tylinal	(90-84-6)
Tylorol LT 50	(139-96-8)
Tylose 444	(9004-67-5)
Tylose 666	(9004-32-4)
Tylose A4S	(9004-67-5)
Tylose C	(9004-32-4)
Tylose C 30	(9004-32-4)

Name	CAS No.
Tylose C 300	(9004-32-4)
Tylose C 600	(9004-32-4)
Tylose C 1000P	(9004-32-4)
Tylose CB 200	(9004-32-4)
Tylose CBR 400	(9004-32-4)
Tylose CBR Series	(9004-32-4)
Tylose CBS 30	(9004-32-4)
Tylose CBS 70	(9004-32-4)
Tylose CB Series	(9004-32-4)
Tylose CH 50	(9004-30-2)
Tylose CR	(9004-32-4)
Tylose CR 50	(9004-32-4)
Tylose DKL	(9004-32-4)
Tylose H 20	(9004-62-0)
Tylose H 300	(9004-62-0)
Tylose H Series	(9004-62-0)
Tylose MB	(9004-62-0)
Tylose MF	(9004-67-5)
Tylose MH	(9004-62-0)
Tylose MH	(9004-67-5)
Tylose MH20	(9004-67-5)
Tylose MH50	(9004-67-5)
Tylose MH300	(9004-67-5)
Tylose MH1000	(9004-67-5)
Tylose MH2000	(9004-67-5)
Tylose MH4000	(9004-67-5)
Tylose MHB	(9004-62-0)
Tylose MHB-Y	(9004-62-0)
Tylose MHB-YP	(9004-62-0)
Tylose MH300P	(9004-67-5)
Tylose MH-K	(9004-62-0)
Tylose MH-XP	(9004-62-0)
Tylose P	(9004-62-0)
Tylose P-X	(9004-62-0)
Tylose PS-X	(9004-62-0)
Tylose P-Z Series	(9004-62-0)
Tylose SAP	(9004-67-5)
Tylose SL	(9004-67-5)
Tylose SL 100	(9004-67-5)
Tylose SL 400	(9004-67-5)
Tylose SL 600	(9004-67-5)
Tylose TWA	(9004-67-5)
Tylosin	(1401-69-0)
Tylosterone	(56-53-1)
Tyox A	(111-17-1)
Tyox B	(123-28-4)
Typogen Brown N	(6416-57-5)
Typogen Carmine	(6368-72-5)
3-Tyramine	(588-05-6)
Tyramine	(51-67-2)
m-Tyramine	(588-05-6)
meta-Tyramine	(588-05-6)
p-Tyramine	(51-67-2)
Tyranton	(123-42-2)
Tyrazol	(22232-54-8)
Tyrene	(87-40-1)
Tyrian Blue I-RSN	(81-77-6)
Tyrian Brilliant Blue I-R	(81-77-6)
Tyrian Brilliant Green I-B	(128-58-5)
Tyrian Brilliant Pink I-R	(2379-74-0)
Tyrian Brilliant hreen I-FFB	(128-58-5)
Tyrian Brown I-BR	(2475-33-4)
Tyrian Golden Orange I-G	(128-70-1)
Tyrian Olive I-R	(2379-81-9)
Tyrian Orange A-RF	(3263-31-8)
Tyril	(9003-54-7)
Tyril 760	(9003-54-7)
Tyril 767	(9003-54-7)
Tyril 780	(9003-54-7)
Tyril 783	(9003-54-7)
Tyril 790	(9003-54-7)
Tyril 860	(9003-54-7)
Tyril 867	(9003-54-7)
Tyrilfoam	(9003-54-7)
Tyrin	(9002-88-4)
Tyrion Yellow	(128-66-5)
Tyrosamine	(51-67-2)
L-p-Tyrosine	(60-18-4)
Tyrosine	(60-18-4)
m-Tyrosine	(587-33-7)
p-Tyrosine	(60-18-4)
L-Tyrosine (9CI)	(60-18-4)
Tyrosine, L	(60-18-4)
m-Tyrosine, L- (8CI)	(587-33-7)
Tyrosine, 3,5-dibromo- (8CI,9CI)	(537-24-6)
L-Tyrosine, 3,5-diiodo- (9CI)	(300-39-0)
Tyrosine, 3,5-diiodo- (9CI)	(66-02-4)
Tyrosine, 3,5-diiodo-, L- (8CI)	(300-39-0)
l-Tyrosine, 3-hydroxy- (9CI)	(59-92-7)
L-Tyrosine, o-(4-hydroxy-3,5-diiodophenyl)-3,5-diiodo	(51-48-9)
L-Tyrosine, o-(4-hydroxy-3-iodophenyl)-3,5-diiodo- (9CI)	(6893-02-3)
L-Tyrosine, 3-hydroxy-α-methyl- (9CI)	(555-30-6)
L-Tyrosine, 3-hydroxy-α-methyl-, sesquihydrate	(41372-08-1)
Tyrylen	(653-03-2)
Tyvek	(9002-88-4)
Tyvid	(54-85-3)
Tyzor TBT	(5593-70-4)
Tyzor TPt	(546-68-9)
475U	(9003-53-6)
666U	(9003-53-6)
U 02	(7440-44-0)
U 46 KV-Fluid	(93-65-2)
U 46 M-Fluid	(94-74-6)
U 46 KV-Ester	(93-65-2)
U 46DP	(94-75-7)
U 46	(94-75-7)
U 46	(94-74-6)
U 46	(93-76-5)
U 46	(93-65-2)
U 46T	(25168-15-4)
U 963	(9011-05-6)
U 1363	(82-66-6)
U 4513	(957-51-7)
U 5963	(127-31-1)
U 6020	(53-03-2)
U 6040	(76-43-7)
U 6324	(103-85-5)
U 6987	(339-43-5)
U 7800	(53-34-9)
U 8210	(53-36-1)
U 8840	(520-85-4)
U 12927	(671-04-5)
U 15030	(105-55-5)
U 17835	(1156-19-0)
U 18496	(320-67-2)
U 19571	(109-57-9)
U 19920	(69-74-9)
U 23284	(18771-50-1)
U 25,354	(96-23-1)
U 27,462	(538-43-2)
U 28774	(27223-35-4)
U 31889	(28981-97-7)
U 33737	(29975-16-4)
U-32.104	(10605-21-7)

U-660 **(9038-95-3)**
U-1149 **(110-17-8)**
U-1434 **(67-21-0)**
U-2000 **(9038-95-3)**
U-2069 **(99-30-9)**
U-3818 **(94-20-2)**
U-3886 **(26628-22-8)**
U-4224 **(68-12-2)**
U-4527 **(66-81-9)**
U-4748 **(50-44-2)**
U-5043 **(94-75-7)**
U-5100 **(9038-95-3)**
U-5227 **(86-88-4)**
U-5897 **(96-24-2)**
U-5954 **(127-19-5)**
U-5965 **(64-75-5)**
U-6062 **(56-75-7)**
U-6233 **(95-14-7)**
U-6421 **(59-87-0)**
U-8344 **(66-75-1)**
U-8953 **(51-21-8)**
U-9889 **(18883-66-4)**
U-10149 **(154-21-2)**
U-14583 **(551-11-1)**
U-17004 **(62046-37-1)**
U-17004 **(671-04-5)**
U-19,646 **(886-74-8)**
U-19,920 **(147-94-4)**
U-19920a **(69-74-9)**
U-21,251 **(18323-44-9)**
U-22,550 **(17021-26-0)**
U-27,415 **(25939-05-3)**
U-33,030 **(28911-01-5)**
U-36059 **(33089-61-1)**
U46 **(120-36-5)**
U625 **(9003-53-6)**
U-19920 A **(147-94-4)**
UBE 1022B **(105-60-2)**
UBE 1022B **(25038-54-4)**
UBI-N 252 **(55290-64-7)**
UC 7744 **(63-25-2)**
UC 8305 **(2921-31-5)**
UC 9880 **(2631-37-0)**
UC 10854 **(64-00-6)**
UC 19786 **(973-21-7)**
UC 20,047A **(15271-41-7)**
UC 20047 **(15271-41-7)**
UC 22,463 **(62046-37-1)**
UC 26089 **(15271-41-7)**
UC 27867 **(12407-86-2)**
UC 51762 **(59669-26-0)**
UC 51769 **(59669-26-0)**
UC 80502 **(59669-26-0)**
UC-21149 **(116-06-3)**
UC-21865 **(1646-88-4)**
UC-25074 **(17702-57-7)**
UC-34096 **(17702-57-7)**
UC22463 **(1966-58-1)**
UCAR 15 **(9003-20-7)**
UCAR 17 **(107-21-1)**
UCAR 38 **(7782-42-5)**
UCAR 130 **(9003-20-7)**
UCAR Butylene Oxide 12 (Obs.) **(26249-20-7)**
UCAR Solvent 2LM **(34590-94-8)**
UCAR Solvent LM (Obs.) **(107-98-2)**
UCAR Solvent 2lM **(34590-94-8)**
UCAR amyl phenol 4T **(80-46-6)**
UCAR butylphenol 4-T **(98-54-4)**
UCAR butylphenol 4-T Flake **(98-54-4)**
U.C.B. 2543 **(77-23-6)**
U.CB 4492 **(68-88-2)**
UC 70480 (9CI) **(103737-38-8)**
UC 70667 (9CI) **(103737-39-9)**
UCB 170 **(569-65-3)**
UCB 492 **(68-88-2)**
UCC 974 **(533-74-4)**
UCB 5052 **(569-65-3)**
UCB 5062 **(569-65-3)**
UCC 6863 **(9003-53-6)**
UCET **(1333-86-4)**
UC Liquid G **(63148-62-9)**
UCON 112 **(76-12-0)**
U-Compound **(51-79-6)**
UDMH (DOT) **(57-14-7)**
U46 DP-Fluid **(120-36-5)**
UDVF **(62-73-7)**
UF 1 **(131-56-6)**
UF 2 **(131-53-3)**
UF 3 **(131-57-7)**
UF 33 **(9011-05-6)**
UF 240 **(9011-05-6)**
UF 10000A **(8061-51-6)**
UK 131 **(9003-20-7)**
UKJ 1506 **(83623-05-6)**
UKS 72 **(9011-05-6)**
UKS 73 **(9011-05-6)**
U46KW **(93-79-8)**
UL 52R **(9011-05-6)**
ULUP **(51-21-8)**
ULV **(8065-48-3)**
U46 MCPB **(94-81-5)**
UM-G **(9011-05-6)**
UML 491 **(361-37-5)**
UN 0004 [Ammonium picrate, Dry or wetted with less than 10 per cent water, by mass] **(131-74-8)**
UN 0072 [Cyclotrimethylenetrinitramine (Cyclonite; Hexogen; RDX) wetted with not less than 15 per cent water, by mass] **(121-82-4)**
UN 0075 [Diethyleneglycol dinitrate, desensitized with not less than 25 percent non-volatile water- insoluble phlegmatizer, by mass] **(693-21-0)**
UN 0076 [Dinitrophenol, Dry or wetted with less than 15 per cent water, by mass] **(25550-58-7)**
UN 0114 [Guanyl nitrosaminoguanyltetrazene (Tetrazene), Wetted with not less than 30 per cent water or mixture of alcohol and water, by mass] **(109-27-3)**
UN 0118 [Hexolite, Dry or wetted with less than 15 per cent water, by mass] **(121-82-4)**
UN 0129 [Lead azide, Wetted with not less than 20 per cent water or mixture of alcohol and water, by mass] **(13424-46-9)**
UN 0133 [Mannitol hexanitrate (Nitromannite), Wetted with not less than 40 per cent water, or mixture of alcohol and water, by mass] **(15825-70-4)**
UN 0135 [Mercury fulminate, Wetted with not less than 20 per cent water, or mixture of alcohol and water, by mass] **(628-86-4)**
UN 0143 [Nitroglycerin, Desensitized with not less than 40 per cent non-volatile water insoluble phlegmatizer, by mass] **(55-63-0)**
UN 0144 [Nitroglycerin, Solution in alcohol, with more than 5% but not more than 10% nitroglycerin] **(55-63-0)**
UN 0146 [Nitrostarch, Dry or wetted with less than 20 per cent water, by mass] **(9056-38-6)**
UN 0150 (DOT) **(78-11-5)**
UN 0153 [Trinitroaniline or Picramide] **(26952-42-1)**
UN 0154 [Trinitrophenol (Picric acid), Dry or wetted with less than 30 per cent water, by mass] **(88-89-1)**

UN 0155 [Trinitrochlorobenzene; (Picryl chloride)] **(28260-61-9)**
UN 0207 [Tetranitroaniline] **(53014-37-2)**
UN 0208 [Trinitrophenylmethylnitramine (Tetryl)] **(479-45-8)**
UN 0209 [Trinitrotoluene (TNT), Dry or wetted with less than 30 per cent water, by mass] **(118-96-7)**
UN 0215 [Trinitrobenzoic acid, Dry or wetted with less than 30 per cent water, by mass] **(129-66-8)**
UN 0216 [Trinitro-meta-cresol] **(602-99-3)**
UN 0219 [Trinitroresorcinol (Styphnic acid), Dry or wetted with less than 20 per cent water, or mixture of alcohol and water, by mass] **(82-71-3)**
UN 0220 [Urea nitrate, Dry or wetted with less than 20 per cent water, by mass] **(124-47-0)**
UN 0224 [Barium azide, Dry or wetted with less than 50 per cent water, by mass] **(18810-58-7)**
UN 0226 [Cyclotetramethylenetetranitramine (HMX; Octogen), Wetted with not less than 15 per cent water, by mass] **(2691-41-0)**
UN 0235 [Sodium picramate, Dry or wetted with less than 20 per cent water, by mass] **(831-52-7)**
UN 0236 [Zirconium picramate, Dry or wetted with less than 20 per cent water, by mass] **(63868-82-6)**
UN 0282 [Nitroguanidine; (Picrite), Dry or wetted with less than 20 per cent water, by mass] **(556-88-7)**
UN 0340 [Nitrocellulose, Dry or wetted with less than 25 per cent water (or alcohol), by mass] **(9004-70-0)**
UN 0342 [Nitrocellulose, Wetted with not less than 25 per cent alcohol, by mass] **(9004-70-0)**
UN 0402 [Ammonium perchlorate, explosive] **(7790-98-9)**
UN 0411 [Pentaerythrite tetranitrate; Pentaerythritol tetranitrate (PETN) with not less than 7 per cent wax by mass] **(78-11-5)**
UN 1001 [Acetylene, dissolved, liquefied] **(74-86-2)**
UN 1005 [Ammonia, anhydrous, Liquefied or ammonia solutions with more than 50 per cent ammonia] **(7664-41-7)**
UN 1006 [Argon, compressed] **(7440-37-1)**
UN 1008 [Boron trifluoride] **(7637-07-2)**
UN 1009 [Bromotrifluoromethane] **(75-63-8)**
UN 1010 [Butadienes, Inhibited] **(25339-57-5)**
UN 1011 [Butane or Butane mixtures see also Petroleum gases, liquified] **(106-97-8)**
UN 1012 [Butylene see also Petroleum gases, Liquefied] **(25167-67-3)**
UN 1013 [Carbon dioxide] **(124-38-9)**
UN 1014 [Carbon dioxide and oxygen mixtures] **(8063-77-2)**
UN 1015 [Carbon dioxide and nitrous oxide mixtures] **(53569-62-3)**
UN 1016 [Carbon monoxide] **(630-08-0)**
UN 1017 [Chlorine] **(7782-50-5)**
UN 1018 [Chlorodifluoromethane] **(75-45-6)**
UN 1020 [Chloropentafluoroethane] **(76-15-3)**
UN 1021 [Chlorotetrafluoroethane] **(63938-10-3)**
UN 1022 [Chlorotrifluoromethane] **(75-72-9)**
UN 1026 [Cyanogen, Liquefied] **(460-19-5)**
UN 1027 [Cyclopropane, Liquified] **(75-19-4)**
UN 1028 [Dichlorodifluoromethane] **(75-71-8)**
UN 1029 [Dichloromonofluoromethane] **(75-43-4)**
UN 1030 [Difluoroethane] **(25497-28-3)**
UN 1030 (DOT) **(75-37-6)**
UN 1032 [Dimethylamine, anhydrous] **(124-40-3)**
UN 1033 [Dimethyl ether] **(115-10-6)**
UN 1035 [Ethane, Compressed] **(74-84-0)**
UN 1036 [Ethylamine] **(75-04-7)**
UN 1037 [Ethyl chloride] **(75-00-3)**
UN 1038 [Ethylene, Refrigerated liquid (cryogenic liquid)] **(74-85-1)**
UN 1039 [Ethyl methyl ether] **(540-67-0)**
UN 1040 [Ethylene oxide, Pure or with nitrogen at 50 degrees C] **(75-21-8)**
UN 1045 [Fluorine, compressed] **(7782-41-4)**
UN 1046 [Helium, Compressed] **(7440-59-7)**
UN 1048 [Hydrogen bromide, Anhydrous] **(10035-10-6)**
UN 1049 [Hydrogen, Compressed] **(1333-74-0)**
UN 1050 [Hydrogen chloride, Anhydrous] **(7647-01-0)**
UN 1051 [Hydrogen cyanide, Anhydrous, Stabilized] **(74-90-8)**
UN 1052 [Hydrogen fluoride, Anhydrous] **(7664-39-3)**
UN 1055 [Isobutylene] **(115-11-7)**
UN 1056 [Krypton, Compressed] **(7439-90-9)**
UN 1060 [Methylacetylene and propadiene mixtures, Stabilized] **(59355-75-8)**
UN 1061 [Methylamine, anhydrous] **(74-89-5)**
UN 1062 [Methyl bromide] **(74-83-9)**
UN 1063 [Methyl chloride] **(74-87-3)**
UN 1064 [Methyl mercaptan] **(74-93-1)**
UN 1065 [Neon, Compressed] **(7440-01-9)**
UN 1066 [Nitrogen, Compressed] **(7727-37-9)**
UN 1067 [Dinitrogen tetraoxide; (Nitrogen dioxide), Liquefied] **(10544-72-6)**
UN 1067 [Nitrogen dioxide, Liquefied] **(10102-44-0)**
UN 1069 [Nitrosyl chloride] **(2696-92-6)**
UN 1070 [Nitrous oxide, Compressed] **(10024-97-2)**
UN 1072 [Oxygen, Compressed] **(7782-44-7)**
UN 1073 [Oxygen, Refrigerated liquid (cryogenic liquid)] **(7782-44-7)**
UN 1075 [Liquefied petroleum gas] **(106-97-8)**
UN 1075 [Liquefied petroleum gas] **(75-28-5)**
UN 1075 [Petroleum gases, Liquefied see also Liquefied petroleum gas] **(115-11-7)**
UN 1075 [Petroleum gases, Liquefied see also Liquefied petroleum gas] **(68476-85-7)**
UN 1075 [Petroleum gases, Liquefied] **(74-98-6)**
UN 1076 [Phosgene] **(503-38-8)**
UN 1076 [Phosgene] **(75-44-5)**
UN 1077 [Propylene see also Petroleum gases, Liquefied] **(115-07-1)**
UN 1079 [Sulfur dioxide, Liquefied] **(7446-09-5)**
UN 1080 [Sulfur hexafluoride] **(2551-62-4)**
UN 1081 [Tetrafluoroethylene, Inhibited] **(116-14-3)**
UN 1082 [Trifluorochloroethylene, Inhibited] **(79-38-9)**
UN 1083 [Trimethylamine, Anhydrous] **(75-50-3)**
UN 1085 [Vinyl bromide, Inhibited] **(593-60-2)**
UN 1086 [Vinyl chloride, Inhibited] **(75-01-4)**
UN 1087 [Vinyl methyl ether, Inhibited] **(107-25-5)**
UN 1088 [Acetal] **(105-57-7)**
UN 1089 [Acetaldehyde] **(75-07-0)**
UN 1090 [Acetone] **(67-64-1)**
UN 1092 [Acrolein, Inhibited] **(107-02-8)**
UN 1093 [Acrylonitrile, Inhibited] **(107-13-1)**
UN 1098 [Allyl alcohol] **(107-18-6)**
UN 1099 [Allyl bromide] **(106-95-6)**
UN 1100 [Allyl chloride] **(107-05-1)**
UN 1101 (DOT) **(96-10-6)**
UN 1102 (DOT) **(97-93-8)**
UN 1103 (DOT) **(75-24-1)**
UN 1104 [Amyl acetates] **(626-38-0)**
UN 1104 [Amyl acetates] **(628-63-7)**
UN 1105 [Amyl alcohols] **(6032-29-7)**
UN 1105 [Amyl alcohols] **(123-51-3)**
UN 1105 [Amyl alcohols] **(71-41-0)**
UN 1105 [Amyl alcohols] **(75-85-4)**
UN 1107 [Amyl chlorides] **(543-59-9)**
UN 1108 [n-Amylene] **(25377-72-4)**
UN 1109 [Amyl formates] **(638-49-3)**
UN 1109 [Amyl formates] **(110-45-2)**
UN 1110 [Amyl methyl ketone] **(110-43-0)**
UN 1111 [Amyl mercaptans] **(110-66-7)**
UN 1112 [Amyl nitrate] **(1002-16-0)**
UN 1113 [Amyl nitrites] **(463-04-7)**
UN 1114 [Benzene] **(71-43-2)**
UN 1120 [Butanols] **(78-92-2)**
UN 1120 [Butanols] **(71-36-3)**
UN 1120 [Butanols] **(75-65-0)**
UN 1123 [Butyl acetates] **(105-46-4)**
UN 1123 [Butyl acetates] **(540-88-5)**

UN 1123 [Butyl acetates] **(123-86-4)**
UN 1125 [n-Butylamine] **(109-73-9)**
UN 1126 [n-Butyl bromide] **(109-65-9)**
UN 1127 [Chlorobutanes] **(109-69-3)**
UN 1127 [Chlorobutanes] **(78-86-4)**
UN 1128 [n-Butyl formate] **(592-84-7)**
UN 1129 [Butyraldehyde] **(123-72-8)**
UN 1130 [Camphor oil] **(8008-51-3)**
UN 1130 [Camphor oil] **(76-22-2)**
UN 1131 [Carbon disulfide] **(75-15-0)**
UN 1134 [Chlorobenzene] **(108-90-7)**
UN 1135 [Ethylene chlorohydrin] **(107-07-3)**
UN 1136 [Coal tar distillates, Flammable] **(8001-58-9)**
UN 1136 [Coal tar distillates, Flammable] **(65996-92-1)**
UN 1137 **(65996-92-1)**
UN 1143 [Crotonaldehyde, Stabilized] **(4170-30-3)**
UN 1144 [Crotonylene] **(503-17-3)**
UN 1145 [Cyclohexane] **(110-82-7)**
UN 1146 [Cyclopentane] **(287-92-3)**
UN 1147 [Decahydronaphthalene] **(91-17-8)**
UN 1148 [Diacetone alcohol] **(123-42-2)**
UN 1149 [Dibutyl ethers] **(6863-58-7)**
UN 1149 [Dibutyl ethers] **(6163-66-2)**
UN 1149 [Dibutyl ethers] **(142-96-1)**
UN 1150 [Dichloroethylene] **(25323-30-2)**
UN 1152 [Dichloropentanes] **(30586-10-8)**
UN 1153 [Ethylene glycol diethyl ether] **(629-14-1)**
UN 1154 [Diethylamine] **(109-89-7)**
UN 1155 [Diethyl ether; (ethyl ether)] **(60-29-7)**
UN 1156 [Diethyl ketone] **(96-22-0)**
UN 1157 [Diisobutyl ketone] **(108-83-8)**
UN 1158 [Diisopropylamine] **(108-18-9)**
UN 1159 [Diisopropyl ether] **(108-20-3)**
UN 1160 [Dimethylamine solution] **(124-40-3)**
UN 1161 [Dimethyl carbonate] **(616-38-6)**
UN 1162 [Dimethyldichlorosilane] **(75-78-5)**
UN 1163 [Dimethylhydrazine, unsymmetrical] **(57-14-7)**
UN 1164 [Dimethyl sulfide] **(75-18-3)**
UN 1165 [Dioxane] **(123-91-1)**
UN 1166 [Dioxolane] **(100-79-8)**
UN 1167 [Divinyl ether, inhibited] **(109-93-3)**
UN 1170 [Ethanol; (ethyl alcohol) or Ethanol Solutions; (ethyl alcohol solutions)] **(64-17-5)**
UN 1171 [Ethylene glycol monoethyl ether] **(110-80-5)**
UN 1172 [Ethylene glycol monoethyl ether acetate] **(111-15-9)**
UN 1173 [Ethyl acetate] **(141-78-6)**
UN 1175 [Ethylbenzene] **(100-41-4)**
UN 1176 **(51845-86-4)**
UN 1176 [Ethyl borate] **(34099-73-5)**
UN 1177 [2-Ethylbutyl acetate] **(123-66-0)**
UN 1177 **(40780-64-1)**
UN 1178 [2-Ethylbutyraldehyde] **(97-96-1)**
UN 1179 [Ethyl butyl ether] **(628-81-9)**
UN 1180 [Ethyl butyrate] **(105-54-4)**
UN 1181 [Ethyl chloroacetate] **(105-39-5)**
UN 1182 [Ethyl chloroformate] **(541-41-3)**
UN 1183 [Ethyldichlorosilane] **(1789-58-8)**
UN 1184 [Ethylene dichloride] **(107-06-2)**
UN 1185 [Ethyleneimine, Inhibited] **(151-56-4)**
UN 1188 [Ethylene glycol monomethyl ether] **(109-86-4)**
UN 1189 [Ethylene glycol monomethyl ether acetate] **(110-49-6)**
UN 1190 [Ethyl formate] **(109-94-4)**
UN 1191 [Octyl aldehydes, Flammable] **(123-05-7)**
UN 1192 [Ethyl lactate] **(97-64-3)**
UN 1193 [Ethyl methyl ketone or methyl ethyl ketone] **(78-93-3)**
UN 1194 [Ethyl nitrite solutions] **(109-95-5)**
UN 1195 [Ethyl propionate] **(105-37-3)**
UN 1196 [Ethyltrichlorosilane] **(115-21-9)**
UN 1198 [Formaldehyde, Solutions, Flammable] **(50-00-0)**
UN 1199 [Furfural] **(98-01-1)**
UN 1201 [Fusel oil] **(8013-75-0)**
UN 1203 [Gasoline] **(8006-61-9)**
UN 1204 [Nitroglycerin solution in alcohol with not more than 1 per cent nitroglycerin] **(55-63-0)**
UN 1206 [Heptanes] **(31394-54-4)**
UN 1206 [Heptanes] **(142-82-5)**
UN 1207 [Hexaldehyde] **(66-25-1)**
UN 1208 [Hexanes] **(110-54-3)**
UN 1208 [Hexanes] **(107-83-5)**
UN 1208 [Hexanes] **(75-83-2)**
UN 1208 **(96-14-0)**
UN 1212 [Isobutanol or isobutyl alcohol] **(78-83-1)**
UN 1213 [Isobutyl acetate] **(110-19-0)**
UN 1214 [Isobutylamine] **(78-81-9)**
UN 1216 [Isooctenes] **(11071-47-9)**
UN 1218 [Isoprene, Inhibited] **(78-79-5)**
UN 1219 [Isopropanol or isopropyl alcohol] **(67-63-0)**
UN 1220 [Isopropyl acetate] **(108-21-4)**
UN 1221 [Isopropylamine] **(75-31-0)**
UN 1222 [Isopropyl nitrate] **(1712-64-7)**
UN 1223 [Kerosene] **(8008-20-6)**
UN 1229 [Mesityl oxide] **(141-79-7)**
UN 1230 [Methanol, or methyl alcohol] **(67-56-1)**
UN 1231 [Methyl acetate] **(79-20-9)**
UN 1232 **(8013-65-8)**
UN 1232 (DOT) **(78-93-3)**
UN 1233 [Methylamyl acetate] **(108-84-9)**
UN 1234 [Methylal] **(109-87-5)**
UN 1235 [Methylamine, aqueous solution] **(74-89-5)**
UN 1237 [Methyl butyrate] **(623-42-7)**
UN 1238 [Methyl chloroformate] **(79-22-1)**
UN 1239 [Methylchloromethyl ether] **(107-30-2)**
UN 1242 [Methyldichlorosilane] **(75-54-7)**
UN 1243 [Methyl formate] **(107-31-3)**
UN 1244 [Methylhydrazine] **(60-34-4)**
UN 1245 [Methyl isobutyl ketone] **(108-10-1)**
UN 1246 [Methyl isopropenyl ketone, Inhibited] **(814-78-8)**
UN 1247 [Methyl methacrylate monomer, Inhibited] **(80-62-6)**
UN 1248 [Methyl propionate] **(554-12-1)**
UN 1249 [Methyl propyl ketone] **(107-87-9)**
UN 1250 [Methyltrichlorosilane] **(75-79-6)**
UN 1251 [Methyl vinyl ketone] **(78-94-4)**
UN 1255 [Naphtha, Petroleum] **(8030-30-6)**
UN 1256 [Naphtha, Solvent] **(8030-30-6)**
UN 1257 [Natural gasoline] **(8006-61-9)**
UN 1259 **(12612-55-4)**
UN 1259 [Nickel carbonyl] **(13463-39-3)**
UN 1261 [Nitromethane] **(75-52-5)**
UN 1262 [Octanes] **(540-84-1)**
UN 1262 [Octanes] **(26635-64-3)**
UN 1262 [Octanes] **(111-65-9)**
UN 1264 [Paraldehyde] **(123-63-7)**
UN 1265 [n-Pentanes or isopentane] **(78-78-4)**
UN 1265 [n-Pentanes or isopentane] **(109-66-0)**
UN 1267 [Petroleum crude oil] **(8002-05-9)**
UN 1268 [Petroleum distillates, N.O.S.] **(8002-05-9)**
UN 1270 [Petroleum oil] **(8002-05-9)**
UN 1271 [Petroleum spirit] **(8032-32-4)**
UN 1272 [Pine oil] **(8002-09-3)**
UN 1274 [n-Propanol or propyl alcohol normal] **(71-23-8)**
UN 1275 [Propionaldehyde] **(123-38-6)**
UN 1276 [n-Propyl acetate] **(109-60-4)**
UN 1277 [Propylamine] **(107-10-8)**
UN 1278 [Propyl chloride] **(540-54-5)**

UN 1279 [Propylene dichloride] **(26638-19-7)**
UN 1280 [Propylene oxide] **(75-56-9)**
UN 1281 [Propyl formates] **(625-55-8)**
UN 1281 [Propyl formates] **(110-74-7)**
UN 1282 [Pyridine] **(110-86-1)**
UN 1286 [Rosin Oil] **(8002-16-2)**
UN 1288 [Shale oil] **(68308-34-9)**
UN 1292 [Tetraethyl silicate] **(78-10-4)**
UN 1294 [Toluene] **(108-88-3)**
UN 1295 [Trichlorosilane] **(10025-78-2)**
UN 1296 [Triethylamine] **(121-44-8)**
UN 1297 [Trimethylamine, Aqueous solutions not more than 50 per cent trimethylamine by mass] **(75-50-3)**
UN 1298 [Trimethylchlorosilane] **(75-77-4)**
UN 1299 [Turpentine] **(8006-64-2)**
UN 1299 [Turpentine] **(9005-90-7)**
UN 1301 [Vinyl acetate, Inhibited] **(108-05-4)**
UN 1302 [Vinyl ethyl ether, Inhibited] **(109-92-2)**
UN 1303 [Vinylidene chloride, Inhibited] **(75-35-4)**
UN 1304 [Vinyl isobutyl ether, Inhibited] **(109-53-5)**
UN 1305 [Vinyltrichlorosilane] **(75-94-5)**
UN 1307 [Xylenes] **(108-38-3)**
UN 1307 [Xylenes] **(106-42-3)**
UN 1307 [Xylenes] **(1330-20-7)**
UN 1307 [Xylenes] **(95-47-6)**
UN 1308 [Zirconium suspended in a liquid] **(7440-67-7)**
UN 1309 [Aluminum powder, coated] **(7429-90-5)**
UN 1310 [Ammonium picrate, Wetted with not less than 10 per cent water, by mass] **(131-74-8)**
UN 1312 [Borneol] **(507-70-0)**
UN 1313 [Calcium resinate] **(9007-13-0)**
UN 1314 [Calcium resinate, fused] **(9007-13-0)**
UN 1318 [Cobalt resinate, Precipitated] **(68956-82-1)**
UN 1320 [Dinitrophenol, Wetted with not less than 15 per cent water, by mass] **(25550-58-7)**
UN 1324 [Films, nitrocellulose base, gelatine coated (except scrap)] **(9004-70-0)**
UN 1325 [Flammable solids, N.O.S.] **(2567-83-1)**
UN 1326 [Hafnium powder, Wetted with not less than 25 per cent water (a visible excess of water must be present) (a) mechanically produced, particle size less than 53 microns; (b) chemically produced, particle size less than 840 microns] **(7440-58-6)**
UN 1328 [Hexamine] **(100-97-0)**
UN 1332 [Metaldehyde] **(108-62-3)**
UN 1333 [Cerium, slabs, ingots, or rods] **(7440-45-1)**
UN 1334 [Naphthalene, Crude or refined] **(91-20-3)**
UN 1336 [Nitroguanidine; (Picrite) Wetted with not less than 20 per cent water, by mass] **(556-88-7)**
UN 1337 [Nitrostarch, Wetted with not less than 20 per cent water, by mass] **(9056-38-6)**
UN 1339 [Phosphorus heptasulfide, Free from yellow or white phosphorus] **(12037-82-0)**
UN 1340 [Phosphorus pentasulfide, Free from yellow or white phosphorus] **(1314-80-3)**
UN 1341 [Phosphorus sesquisulfide, Free from yellow or white phosphorous] **(1314-85-8)**
UN 1343 [Phosphorus trisulfide, Free from yellow or white phosphorus] **(12165-69-4)**
UN 1344 [Trinitrophenol, Wetted with not less than 30 per cent water, by mass] **(88-89-1)**
UN 1346 [Silicon powder, amorphous] **(7440-21-3)**
UN 1347 [Silver picrate, Wetted with not less than 30 per cent water, by mass] **(146-84-9)**
UN 1349 [Sodium picramate, Wetted with not less than 20 per cent water, by mass] **(831-52-7)**
UN 1350 [Sulfur] **(7704-34-9)**
UN 1355 [Trinitrobenzoic acid, Wetted with not less than 30 per cent water, by mass] **(129-66-8)**
UN 1356 [Trinitrotoluene, Wetted with not less than 30% water] **(118-96-7)**
UN 1357 [Urea nitrate, Wetted with not less than 20 per cent water, by mass] **(124-47-0)**
UN 1358 [Zirconium powder, Wetted with not less than 25 per cent water (a visible excess of water must be present) (a) mechanically produced, particle size less than 53 microns; (b) chemically produced, particle size less than 840 microns] **(7440-67-7)**
UN 1360 [Calcium phosphide] **(1305-99-3)**
UN 1362 [Carbon, activated] **(7440-44-0)**
UN 1363 [Copra] **(8001-31-8)**
UN 1366 [Diethylzinc] **(557-20-0)**
UN 1368 (DOT) **(2999-74-8)**
UN 1369 [p-Nitrosodimethylaniline] **(138-89-6)**
UN 1370 [Dimethylzinc] **(544-97-8)**
UN 1376 [Iron oxide, spent, or Iron sponge, spent obtained from coal gas purification] **(1332-37-2)**
UN 1380 [Pentaborane] **(19624-22-7)**
UN 1382 [Potassium sulfide, Anhydrous or Potassium sulfide with less than 30 per cent water of crystallization] **(1312-73-8)**
UN 1383 [Pyrophoric metals, N.O.S., or Pyrophoric alloys, N.O.S.] **(7429-90-5)**
UN 1383 [Pyrophoric metals, N.O.S.] **(7440-66-6)**
UN 1384 [Sodium dithionite or sodium hydrosulfite] **(7775-14-6)**
UN 1385 [Sodium sulfide, Anhydrous or Sodium sulfide with less than 30 per cent water of crystallization] **(1313-82-2)**
UN 1394 [Aluminum carbide] **(12656-43-8)**
UN 1396 [Aluminum powder, uncoated] **(7429-90-5)**
UN 1397 [Aluminum phosphide] **(20859-73-8)**
UN 1399 (DOT) **(7440-39-3)**
UN 1400 [Barium] **(7440-39-3)**
UN 1401 [Calcium] **(7440-70-2)**
UN 1402 [Calcium carbide] **(75-20-7)**
UN 1403 [Calcium cyanamide with more than 0.1 per cent of calcium carbide] **(156-62-7)**
UN 1405 [Calcium silicide] **(12737-18-7)**
UN 1406 **(12737-18-7)**
UN 1407 [Cesium or Caesium] **(7440-46-2)**
UN 1408 [Ferrosilicon, with 30 percent or more but less than 90 percent silicon] **(8049-17-0)**
UN 1410 [Lithium aluminum hydride] **(16853-85-3)**
UN 1411 [Lithium aluminum hydride, ethereal] **(16853-85-3)**
UN 1412 (DOT) **(7782-89-0)**
UN 1413 [Lithium borohydride] **(16949-15-8)**
UN 1414 [Lithium hydride] **(7580-67-8)**
UN 1415 [Lithium] **(7439-93-2)**
UN 1417 **(68848-64-6)**
UN 1418 [Magnesium, powder or Magnesium alloys, powder] **(7439-95-4)**
UN 1420 [Potassium, metal alloys] **(7440-09-7)**
UN 1422 [Potassium sodium alloys] **(11135-81-2)**
UN 1423 [Rubidium] **(7440-17-7)**
UN 1425 (DOT) **(7782-92-5)**
UN 1426 [Sodium borohydride] **(16940-66-2)**
UN 1427 [Sodium hydride] **(7646-69-7)**
UN 1428 [Sodium] **(7440-23-5)**
UN 1429 (DOT) **(7440-23-5)**
UN 1431 [Sodium methylate] **(124-41-4)**
UN 1432 **(87835-45-8)**
UN 1433 [Stannic phosphide] **(25324-56-5)**
UN 1436 [Zinc powder or Zinc dust] **(7440-66-6)**
UN 1437 **(7704-99-6)**
UN 1438 [Aluminum nitrate] **(13473-90-0)**
UN 1439 [Ammonium dichromate] **(7789-09-5)**
UN 1442 [Ammonium perchlorate, oxidizer] **(7790-98-9)**
UN 1444 [Ammonium persulfate] **(7727-54-0)**
UN 1445 [Barium chlorate] **(13477-00-4)**
UN 1446 [Barium nitrate] **(10022-31-8)**

UN 1447 [Barium perchlorate] **(13465-95-7)**
UN 1448 [Barium permanganate] **(7787-36-2)**
UN 1449 [Barium peroxide] **(1304-29-6)**
UN 1452 [Calcium chlorate] **(10137-74-3)**
UN 1453 [Calcium chlorite] **(14674-72-7)**
UN 1454 [Calcium nitrate] **(10124-37-5)**
UN 1456 [Calcium permanganate] **(10118-76-0)**
UN 1457 [Calcium peroxide] **(1305-79-9)**
UN 1463 [Chromium trioxide, anhydrous] **(1333-82-0)**
UN 1466 [Ferric nitrate] **(10421-48-4)**
UN 1467 [Guanidine nitrate] **(506-93-4)**
UN 1469 [Lead nitrate] **(10099-74-8)**
UN 1470 [Lead perchlorate, Solid or solution] **(13637-76-8)**
UN 1471 [Lithium hypochlorite, Dry or Lithium hypochlorite mixtures, Dry] **(13840-33-0)**
UN 1472 [Lithium peroxide] **(12031-80-0)**
UN 1474 [Magnesium nitrate] **(10377-60-3)**
UN 1475 [Magnesium perchlorate] **(10034-81-8)**
UN 1476 [Magnesium peroxide] **(14452-57-4)**
UN 1484 [Potassium bromate] **(7758-01-2)**
UN 1485 [Potassium chlorate] **(3811-04-9)**
UN 1486 [Potassium nitrate] **(7757-79-1)**
UN 1488 [Potassium nitrite] **(7758-09-0)**
UN 1489 [Potassium perchlorate, Solid or solution] **(7778-74-7)**
UN 1490 [Potassium permanganate] **(7722-64-7)**
UN 1491 [Potassium peroxide] **(17014-71-0)**
UN 1492 [Potassium persulfate] **(7727-21-1)**
UN 1493 [Silver nitrate] **(7761-88-8)**
UN 1494 [Sodium bromate] **(7789-38-0)**
UN 1495 [Sodium chlorate] **(7775-09-9)**
UN 1496 [Sodium chlorite] **(7758-19-2)**
UN 1498 [Sodium nitrate] **(7631-99-4)**
UN 1500 [Sodium nitrite] **(7632-00-0)**
UN 1502 [Sodium perchlorate] **(7601-89-0)**
UN 1503 [Sodium permanganate] **(10101-50-5)**
UN 1504 [Sodium peroxide] **(1313-60-6)**
UN 1505 [Sodium persulfate] **(7775-27-1)**
UN 1506 [Strontium chlorate, Solid or solution] **(7791-10-8)**
UN 1507 [Strontium nitrate] **(10042-76-9)**
UN 1508 [Strontium perchlorate] **(13450-97-0)**
UN 1509 [Strontium peroxide] **(1314-18-7)**
UN 1510 [Tetranitromethane] **(509-14-8)**
UN 1510 [Tetranitromethane] **(20600-96-8)**
UN 1511 [Urea hydrogen peroxide] **(124-43-6)**
UN 1512 **(63885-01-8)**
UN 1513 [Zinc chlorate] **(10361-95-2)**
UN 1514 [Zinc nitrate] **(7779-88-6)**
UN 1515 [Zinc permanganate] **(23414-72-4)**
UN 1516 [Zinc peroxide] **(1314-22-3)**
UN 1517 [Zirconium picramate, Wetted with not less than 20 per cent water, by mass] **(63868-82-6)**
UN 1541 [Acetone cyanohydrin, stabilized] **(75-86-5)**
UN 1545 [Allyl isothiocyanate, stabilized] **(57-06-7)**
UN 1546 [Ammonium arsenate] **(7784-44-3)**
UN 1547 [Aniline] **(62-53-3)**
UN 1548 [Aniline hydrochloride] **(142-04-1)**
UN 1550 [Antimony lactate] **(58164-88-8)**
UN 1551 [Antimony potassium tartrate] **(28300-74-5)**
UN 1553 [Arsenic acid, Liquid] **(7778-39-4)**
UN 1553 (DOT) **(7778-39-4)**
UN 1554 (DOT) **(7778-39-4)**
UN 1554 [Arsenic acid, Solid] **(7778-39-4)**
UN 1555 [Arsenic bromide] **(7784-33-0)**
UN 1557 [Arsenic compounds, Solid, N.O.S. including arsenates, N.O.S.; arsenites, n.o.s.; arsenic sulfides, n.o.s.; and organic compounds of arsenic, n.o.s.] **(1303-33-9)**
UN 1558 [Arsenic] **(7440-38-2)**
UN 1559 [Arsenic pentoxide] **(1303-28-2)**
UN 1560 [Arsenic trichloride] **(7784-34-1)**
UN 1561 [Arsenic trioxide] **(1327-53-3)**
UN 1562 [Arsenical dust] **(8028-73-7)**
UN 1565 [Barium cyanide] **(542-62-1)**
UN 1567 [Beryllium, powder] **(7440-41-7)**
UN 1569 [Bromoacetone] **(598-31-2)**
UN 1570 [Brucine] **(357-57-3)**
UN 1571 [Barium azide, Wetted with not less than 50 per cent water, by mass] **(18810-58-7)**
UN 1572 [Cacodylic acid] **(75-60-5)**
UN 1573 [Calcium arsenate] **(7778-44-1)**
UN 1575 [Calcium cyanide] **(592-01-8)**
UN 1577 [Chlorodinitrobenzenes] **(25567-67-3)**
UN 1578 [Chloronitrobenzenes meta or para, Solid] **(100-00-5)**
UN 1578 [Chloronitrobenzenes meta or para, Solid] **(121-73-3)**
UN 1578 [Chloronitrobenzene, ortho, Liquid] **(88-73-3)**
UN 1579 [4-Chloro-o-toluidine hydrochloride] **(3165-93-3)**
UN 1580 [Chloropicrin] **(76-06-2)**
UN 1581 [Chloropicrin and methyl bromide mixtures] **(8004-09-9)**
UN 1582 [Chloropicrin mixture, Flammable] **(76-06-2)**
UN 1583 [Chloropicrin mixtures, N.O.S.] **(76-06-2)**
UN 1584 [Cocculus] **(124-87-8)**
UN 1585 [Copper acetoarsenite] **(12002-03-8)**
UN 1586 [Copper arsenite] **(10290-12-7)**
UN 1587 [Copper cyanide] **(544-92-3)**
UN 1587 [Copper cyanide] **(14763-77-0)**
UN 1589 [Cyanogen chloride, Inhibited] **(506-77-4)**
UN 1590 [Dichloroanilines, Solid or liquid] **(27134-27-6)**
UN 1591 [o-Dichlorobenzene] **(95-50-1)**
UN 1592 [p-Dichlorobenzene] **(106-46-7)**
UN 1593 [Dichloromethane] **(75-09-2)**
UN 1594 [Diethyl sulfate] **(64-67-5)**
UN 1595 [Dimethyl sulfate] **(77-78-1)**
UN 1596 [Dinitroanilines] **(26471-56-7)**
UN 1597 [Dinitrobenzenes, Solid] **(528-29-0)**
UN 1597 [Dinitrobenzenes, Liquid] **(99-65-0)**
UN 1597 [Dinitrobenzenes, Solid] **(100-25-4)**
UN 1597 [Dinitrobenzenes, Solid] **(25154-54-5)**
UN 1598 [Dinitro-o-cresol, Solid] **(1335-85-9)**
UN 1599 [Dinitrophenol solutions] **(25550-58-7)**
UN 1600 [Dinitrotoluenes, Molten] **(25321-14-6)**
UN 1603 [Ethyl bromoacetate] **(105-36-2)**
UN 1604 [Ethylenediamine] **(107-15-3)**
UN 1605 [Ethylene dibromide] **(106-93-4)**
UN 1606 [Ferric arsenate] **(10102-49-5)**
UN 1607 [Ferric arsenite] **(63989-69-5)**
UN 1608 [Ferrous arsenate] **(10102-50-8)**
UN 1611 [Hexaethyl tetraphosphate liquid] **(757-58-4)**
UN 1612 [Hexaethyl tetraphosphate and compressed gas mixtures] **(757-58-4)**
UN 1613 [Hydrocyanic acid, Aqueous solutions not more than 20% hydrocyanic acid] **(74-90-8)**
UN 1614 [Hydrogen cyanide, Anhydrous, Stabilized, absorbed in a porous inert material] **(74-90-8)**
UN 1616 [Lead acetate] **(301-04-2)**
UN 1617 [Lead arsenates] **(7645-25-2)**
UN 1617 [Lead arsenates] **(7784-40-9)**
UN 1617 [Lead arsenates] **(3687-31-8)**
UN 1618 **(70910-35-9)**
UN 1620 [Lead cyanide] **(592-05-2)**
UN 1621 [London Purple] **(8012-74-6)**
UN 1622 [Magnesium arsenate] **(10103-50-1)**
UN 1624 [Mercuric chloride] **(7487-94-7)**
UN 1625 [Mercuric nitrate] **(10045-94-0)**
UN 1626 [Mercuric potassium cyanide] **(591-89-9)**
UN 1627 [Mercurous nitrate] **(10415-75-5)**
UN 1628 (DOT) **(7783-36-0)**

UN 1629 [Mercury acetate] **(631-60-7)**
UN 1629 [Mercury acetate] **(1600-27-7)**
UN 1630 [Mercury ammonium chloride] **(10124-48-8)**
UN 1631 [Mercury benzoate] **(583-15-3)**
UN 1633 (DOT) **(7783-35-9)**
UN 1634 [Mercury bromides] **(7789-47-1)**
UN 1634 [Mercury bromides] **(10031-18-2)**
UN 1636 **(37020-93-2)**
UN 1636 [Mercury cyanide] **(592-04-1)**
UN 1637 [Mercury gluconate] **(63937-14-4)**
UN 1638 [Mercury iodide] **(15385-57-6)**
UN 1638 [Mercury iodide] **(7774-29-0)**
UN 1639 [Mercury nucleate] **(12002-19-6)**
UN 1640 [Mercury oleate] **(1191-80-6)**
UN 1641 [Mercury oxide] **(15829-53-5)**
UN 1641 [Mercury oxide] **(21908-53-2)**
UN 1642 [Mercury oxycyanide, Desensitized] **(1335-31-5)**
UN 1643 [Mercury potassium iodide] **(7783-33-7)**
UN 1644 [Mercury salicylate] **(5970-32-1)**
UN 1645 [Mercury sulfates] **(7783-35-9)**
UN 1646 [Mercury thiocyanate] **(592-85-8)**
UN 1648 [Methyl cyanide] **(75-05-8)**
UN 1649 [Motor fuel anti-knock mixtures] **(78-00-2)**
UN 1650 [β-Naphthylamine] **(91-59-8)**
UN 1651 [Naphthylthiourea] **(86-88-4)**
UN 1653 [Nickel cyanide] **(557-19-7)**
UN 1654 [Nicotine] **(54-11-5)**
UN 1655 [Nicotine compounds, Solid, N.O.S.] **(54-11-5)**
UN 1656 [Nicotine hydrochloride or Nicotine hydrochloride solution] **(2820-51-1)**
UN 1657 [Nicotine salicylate] **(29790-52-1)**
UN 1658 [Nicotine sulfate, Solid or solution] **(65-30-5)**
UN 1659 [Nicotine tartrate] **(65-31-6)**
UN 1660 [Nitric oxide] **(10102-43-9)**
UN 1661 **(29757-24-2)**
UN 1661 [Nitroanilines (o-; m-; p-;)] **(100-01-6)**
UN 1661 [Nitroanilines (o-; m-; p-;)] **(88-74-4)**
UN 1661 [Nitroanilines (o-; m-; p-;)] **(99-09-2)**
UN 1662 [Nitrobenzene] **(98-95-3)**
UN 1663 [Nitrophenols (o-; m-; p-;)] **(88-75-5)**
UN 1663 [Nitrophenols (o-; m-; p-;)] **(100-02-7)**
UN 1663 [Nitrophenols (o-; m-; p-;)] **(554-84-7)**
UN 1664 [Nitrotoluenes, Solid m-, or p-] **(99-08-1)**
UN 1664 [Nitrotoluenes, Solid m-, or p-] **(99-99-0)**
UN 1664 [Nitrotoluenes, Liquid o-; m-; p-;] **(88-72-2)**
UN 1669 [Pentachloroethane] **(76-01-7)**
UN 1670 [Perchloromethylmercaptan] **(594-42-3)**
UN 1671 [Phenol, Solid] **(108-95-2)**
UN 1672 [Phenylcarbylamine chloride] **(622-44-6)**
UN 1673 [Phenylenediamines (o-; m-; p-;)] **(108-45-2)**
UN 1673 [Phenylenediamines (o-; m-; p-;)] **(106-50-3)**
UN 1673 [Phenylenediamines (o-; m-; p-;)] **(95-54-5)**
UN 1674 [Phenylmercuric acetate] **(62-38-4)**
UN 1677 [Potassium arsenate] **(7784-41-0)**
UN 1678 [Potassium arsenite] **(10124-50-2)**
UN 1680 [Potassium cyanide] **(151-50-8)**
UN 1684 [Silver cyanide] **(506-64-9)**
UN 1685 **(13464-42-1)**
UN 1685 [Sodium arsenate] **(7631-89-2)**
UN 1686 [Sodium arsenite, Aqueous solutions] **(7784-46-5)**
UN 1686 [Sodium arsenite, Aqueous solutions] **(13464-37-4)**
UN 1687 (DOT) **(26628-22-8)**
UN 1687 [Sodium azide] **(26628-22-8)**
UN 1688 [Sodium cacodylate] **(124-65-2)**
UN 1689 **(70152-47-5)**
UN 1689 [Sodium cyanide] **(143-33-9)**
UN 1690 [Sodium fluoride] **(7681-49-4)**
UN 1691 [Strontium arsenite] **(91724-16-2)**
UN 1692 [Strychnine or Strychnine salts] **(57-24-9)**
UN 1694 [Bromobenzyl cyanides, Liquid] **(16532-79-9)**
UN 1694 [Bromobenzyl cyanides, Liquid] **(5798-79-8)**
UN 1695 [Chloroacetone, Stabilized] **(78-95-5)**
UN 1697 [Chloroacetophenone (CN), Liquid or solid] **(532-27-4)**
UN 1698 [Diphenylamine chloroarsine] **(578-94-9)**
UN 1699 [Diphenylchloroarsine, Solid or liquid] **(712-48-1)**
UN 1701 [Xylyl bromide] **(35884-77-6)**
UN 1702 [Tetrachloroethane] **(79-34-5)**
UN 1702 [Tetrachloroethane] **(25322-20-7)**
UN 1704 [Tetraethyl dithiopyrophosphate] **(3689-24-5)**
UN 1708 [Toluidines liquid or solid] **(108-44-1)**
UN 1708 [Toluidines solid] **(95-53-4)**
UN 1708 [Toluidines liquid] **(106-49-0)**
UN 1709 [2,4-Toluylenediamine] **(95-80-7)**
UN 1710 [Trichloroethylene] **(79-01-6)**
UN 1711 [Xylidines, Solid or solution] **(87-59-2)**
UN 1711 [Xylidines, Solid or solution] **(1300-73-8)**
UN 1711 [Xylidines, Solid or solution] **(95-78-3)**
UN 1711 [Xylidines, Solid or solution] **(95-68-1)**
UN 1712 [Zinc arsenate or Zinc arsenite or Zinc arsenate and Zinc arsenite mixtures.] **(1303-39-5)**
UN 1713 [Zinc cyanide] **(557-21-1)**
UN 1714 [Zinc phosphide] **(1314-84-7)**
UN 1715 [Acetic anhydride] **(108-24-7)**
UN 1716 [Acetyl bromide] **(506-96-7)**
UN 1717 [Acetyl chloride] **(75-36-5)**
UN 1718 [Butyl acid phosphate] **(12788-93-1)**
UN 1718 **(52933-01-4)**
UN 1722 [Allyl chloroformate] **(2937-50-0)**
UN 1723 [Allyl iodide] **(556-56-9)**
UN 1724 [Allyltrichlorosilane, stabilized] **(107-37-9)**
UN 1725 [Aluminum bromide, anhydrous] **(7727-15-3)**
UN 1726 [Aluminum chloride, anhydrous] **(7446-70-0)**
UN 1727 [Ammonium hydrogen fluoride, Solid] **(1341-49-7)**
UN 1728 [Amyltrichlorosilane] **(107-72-2)**
UN 1729 [Anisoyl chloride] **(100-07-2)**
UN 1730 [Antimony pentachloride, Liquid] **(7647-18-9)**
UN 1731 [Antimony pentachloride, Solutions] **(7647-18-9)**
UN 1732 [Antimony pentafluoride] **(7783-70-2)**
UN 1733 [Antimony trichloride, Liquid] **(10025-91-9)**
UN 1736 [Benzoyl chloride] **(98-88-4)**
UN 1737 [Benzyl bromide] **(100-39-0)**
UN 1738 [Benzyl chloride unstabilized] **(100-44-7)**
UN 1739 [Benzyl chloroformate] **(501-53-1)**
UN 1741 [Boron trichloride] **(10294-34-5)**
UN 1742 [Boron trifluoride acetic acid complex] **(7578-36-1)**
UN 1744 [Bromine] **(7726-95-6)**
UN 1745 [Bromine pentafluoride] **(7789-30-2)**
UN 1746 [Bromine trifluoride] **(7787-71-5)**
UN 1747 [Butyltrichlorosilane] **(7521-80-4)**
UN 1748 [Calcium hypochlorite, Dry or calcium hypochlorite mixtures dry with more than 39 per cent available chlorine (8.8 per cent available oxygen)] **(7778-54-3)**
UN 1749 [Chlorine trifluoride] **(7790-91-2)**
UN 1750 [Chloroacetic acid, Liquid] **(79-11-8)**
UN 1751 [Chloroacetic acid, Solid] **(79-11-8)**
UN 1752 [Chloroacetyl chloride] **(79-04-9)**
UN 1753 [Chlorophenyltrichlorosilane] **(26571-79-9)**
UN 1754 [Chlorosulfonic acid (with or without sulfur trioxide)] **(7790-94-5)**
UN 1755 [Chromic acid solution] **(1308-14-1)**
UN 1755 [Chromic acid solution] **(13530-68-2)**
UN 1755 [Chromic acid solution] **(1333-82-0)**
UN 1756 [Chromic fluoride, Solid] **(7788-97-8)**
UN 1757 [Chromic fluoride, Solution] **(7788-97-8)**
UN 1758 [Chromium oxychloride] **(14977-61-8)**

UN 1761 [Cupriethylenediamine solution] **(13426-91-0)**
UN 1762 [Cyclohexenyltrichlorosilane] **(10137-69-6)**
UN 1763 [Cyclohexyltrichlorosilane] **(98-12-4)**
UN 1764 [Dichloroacetic acid] **(79-43-6)**
UN 1765 [Dichloroacetyl chloride] **(79-36-7)**
UN 1766 [Dichlorophenyltrichlorosilane] **(27137-85-5)**
UN 1767 [Diethyldichlorosilane] **(1719-53-5)**
UN 1768 [Difluorophosphoric acid, Anhydrous] **(13779-41-4)**
UN 1769 [Diphenyldichlorosilane] **(80-10-4)**
UN 1770 [Diphenylmethyl bromide] **(776-74-9)**
UN 1771 [Dodecyltrichlorosilane] **(4484-72-4)**
UN 1773 [Ferric chloride] **(7705-08-0)**
UN 1775 [Fluoboric acid] **(16872-11-0)**
UN 1776 [Fluorophosphoric acid anhydrous] **(13537-32-1)**
UN 1777 [Fluorosulfonic acid] **(7789-21-1)**
UN 1778 [Fluorosilicic acid] **(16961-83-4)**
UN 1779 [Formic acid] **(64-18-6)**
UN 1780 [Fumaryl chloride] **(627-63-4)**
UN 1781 [Hexadecyltrichlorosilane] **(5894-60-0)**
UN 1782 [Hexafluorophosphoric acid] **(16940-81-1)**
UN 1783 [Hexamethylenediamine solution] **(124-09-4)**
UN 1784 [Hexyltrichlorosilane] **(928-65-4)**
UN 1787 [Hydriodic acid, solution] **(10034-85-2)**
UN 1788 [Hydrobromic acid solution] **(10035-10-6)**
UN 1789 [Hydrochloric acid, Solution] **(7647-01-0)**
UN 1790 [Hydrofluoric acid, Solution, more than 60 per cent strength] **(7664-39-3)**
UN 1792 [Iodine monochloride] **(7790-99-0)**
UN 1793 [Isopropyl acid phosphate] **(1623-24-1)**
UN 1794 [Lead sulfate with more than 3 per cent free acid] **(7446-14-2)**
UN 1794 [Lead sulfate with more than 3 per cent free acid] **(12036-76-9)**
UN 1794 [Lead sulfate with more than 3 per cent free acid] **(69029-52-3)**
UN 1798 [Nitrohydrochloric acid] **(8007-56-5)**
UN 1799 [Nonyltrichlorosilane] **(5283-67-0)**
UN 1800 [Octadecyltrichlorosilane] **(112-04-9)**
UN 1801 [Octyltrichlorosilane] **(5283-66-9)**
UN 1802 [Perchloric acid not more than 50 per cent acid by mass] **(7601-90-3)**
UN 1803 [Phenolsulfonic acid, Liquid] **(1333-39-7)**
UN 1804 [Phenyltrichlorosilane] **(98-13-5)**
UN 1805 [Phosphoric acid] **(7664-38-2)**
UN 1806 [Phosphorus pentachloride] **(10026-13-8)**
UN 1807 [Phosphorus pentoxide] **(1314-56-3)**
UN 1808 [Phosphorus tribromide] **(7789-60-8)**
UN 1809 [Phosphorus trichloride] **(7719-12-2)**
UN 1810 [Phosphorus oxychloride] **(10025-87-3)**
UN 1811 [Potassium bifluoride, Solution] **(7789-29-9)**
UN 1812 [Potassium fluoride] **(7789-23-3)**
UN 1813 [Potassium hydroxide, Solid] **(1310-58-3)**
UN 1814 [Potassium hydroxide, Solution] **(1310-58-3)**
UN 1815 [Propionyl chloride] **(79-03-8)**
UN 1816 [Propyltrichlorosilane] **(141-57-1)**
UN 1817 [Pyrosulfuryl chloride] **(7791-27-7)**
UN 1818 [Silicon tetrachloride] **(10026-04-7)**
UN 1819 [Sodium aluminate, Solution] **(12042-47-6)**
UN 1821 [Sodium hydrogen sulfate, Solid] **(7681-38-1)**
UN 1823 [Sodium hydroxide, Solid] **(1310-73-2)**
UN 1824 [Sodium hydroxide solution] **(1310-73-2)**
UN 1825 [Sodium monoxide] **(12401-86-4)**
UN 1827 [Stannic chloride, Anhydrous] **(7646-78-8)**
UN 1828 **(14989-32-3)**
UN 1828 [Sulfur chlorides] **(10025-67-9)**
UN 1828 [Sulfur chlorides] **(10545-99-0)**
UN 1829 [Sulfur trioxide, Inhibited] **(7446-11-9)**
UN 1830 [Sulfuric acid] **(7664-93-9)**
UN 1831 [Sulfuric acid, fuming] **(8014-95-7)**
UN 1832 [Sulfuric acid, spent] **(7664-93-9)**
UN 1833 [Sulfurous acid] **(7782-99-2)**
UN 1834 [Sulfuryl chloride] **(7791-25-5)**
UN 1835 [Tetramethylammonium hydroxide] **(75-59-2)**
UN 1836 [Thionyl chloride] **(7719-09-7)**
UN 1837 [Thiophosphoryl chloride] **(3982-91-0)**
UN 1838 [Titanium tetrachloride] **(7550-45-0)**
UN 1839 [Trichloroacetic acid] **(76-03-9)**
UN 1840 [Zinc chloride, Solution] **(7646-85-7)**
UN 1841 [Acetaldehyde ammonia] **(75-39-8)**
UN 1845 [Carbon dioxide, Solid; (dry ice)] **(124-38-9)**
UN 1846 [Carbon tetrachloride] **(56-23-5)**
UN 1847 [Potassium sulfide, Hydrated with not less than 30 per cent water of crystallization] **(1312-73-8)**
UN 1848 [Propionic acid] **(79-09-4)**
UN 1854 [Barium alloys, pyrophoric] **(7440-39-3)**
UN 1855 [Calcium, pyrophoric or Calcium alloys, pyrophoric] **(7440-70-2)**
UN 1858 [Hexafluoropropylene] **(116-15-4)**
UN 1859 [Silicon tetrafluoride] **(7783-61-1)**
UN 1860 [Vinyl fluoride, Inhibited] **(75-02-5)**
UN 1862 [Ethyl crotonate] **(623-70-1)**
UN 1865 [n-Propyl nitrate] **(627-13-4)**
UN 1868 [Decaborane] **(17702-41-9)**
UN 1869 [Magnesium or Magnesium alloys with more than 50 per cent magnesium in pellets, turnings or ribbons] **(7439-95-4)**
UN 1870 [Potassium borohydride] **(13762-51-1)**
UN 1872 [Lead dioxide] **(1309-60-0)**
UN 1873 [Perchloric acid more than 50 per cent but not more than 72 per cent acid, by mass] **(7601-90-3)**
UN 1884 [Barium oxide] **(1304-28-5)**
UN 1885 [Benzidine] **(92-87-5)**
UN 1886 [Benzylidene chloride] **(98-87-3)**
UN 1887 [Bromochloromethane] **(74-97-5)**
UN 1888 [Chloroform] **(67-66-3)**
UN 1889 [Cyanogen bromide] **(506-68-3)**
UN 1891 [Ethyl bromide] **(74-96-4)**
UN 1892 [Ethyldichloroarsine] **(598-14-1)**
UN 1894 [Phenylmercuric hydroxide] **(100-57-2)**
UN 1895 [Phenylmercuric nitrate] **(55-68-5)**
UN 1897 [Tetrachloroethylene] **(127-18-4)**
UN 1898 [Acetyl iodide] **(507-02-8)**
UN 1902 [Diisooctyl acid phosphate] **(27215-10-7)**
UN 1905 [Selenic acid] **(7783-08-6)**
UN 1907 [Soda lime with more than 4 per cent sodium hydroxide] **(8006-28-8)**
UN 1908 [Sodium chlorite solution with more than 5 per cent available chlorine] **(7758-19-2)**
UN 1910 [Calcium oxide] **(1305-78-8)**
UN 1911 [Diborane] **(19287-45-7)**
UN 1911 (DOT) **(19287-45-7)**
UN 1913 [Neon, Refrigerated liquid (cryogenic liquid)] **(7440-01-9)**
UN 1914 [Butylpropionate] **(590-01-2)**
UN 1915 [Cyclohexanone] **(108-94-1)**
UN 1916 [2,2-Dichlorodiethyl ether] **(111-44-4)**
UN 1917 [Ethyl acrylate, Inhibited] **(140-88-5)**
UN 1918 [Isopropylbenzene] **(98-82-8)**
UN 1919 [Methyl acrylate, Inhibited] **(96-33-3)**
UN 1920 [Nonanes] **(111-84-2)**
UN 1920 [Nonanes] **(34464-40-9)**
UN 1921 [Propyleneimine, Inhibited] **(75-55-8)**
UN 1922 [Pyrrolidine] **(123-75-1)**
UN 1923 [Calcium dithionite (calcium hydrosulfite)] **(13780-03-5)**
UN 1924 (DOT) **(563-43-9)**
UN 1925 (DOT) **(12075-68-2)**
UN 1926 (DOT) **(12263-85-3)**
UN 1927 **(12542-85-7)**
UN 1928 [Methyl magnesium bromide, in ethyl ether] **(75-16-1)**
UN 1930 (DOT) **(100-99-2)**

UN 1931 [Zinc dithionite or Zinc hydrosulfite] **(7779-86-4)**
UN 1932 [Zirconium scrap] **(7440-67-7)**
UN 1938 [Bromoacetic acid, Solution] **(79-08-3)**
UN 1939 [Phosphorus oxybromide] **(7789-59-5)**
UN 1940 [Thioglycolic acid] **(68-11-1)**
UN 1941 [Dibromodifluoromethane] **(75-61-6)**
UN 1951 [Argon, refrigerated liquid (cryogenic liquid)] **(7440-37-1)**
UN 1957 [Deuterium] **(7782-39-0)**
UN 1958 [Dichlorotetrafluoroethane] **(1320-37-2)**
UN 1959 [1,1-Difluoroethylene] **(75-38-7)**
UN 1961 [Ethane, Refrigerated liquid (cryogenic liquid)] **(74-84-0)**
UN 1962 [Ethylene, compressed] **(74-85-1)**
UN 1963 [Helium, Refrigerated liquid (cryogenic liquid)] **(7440-59-7)**
UN 1966 [Hydrogen, Refrigerated liquid (cryogenic liquid)] **(1333-74-0)**
UN 1969 [Isobutane or isobutane mixtures] **(75-28-5)**
UN 1970 [Krypton, Refrigerated liquid (cryogenic liquid)] **(7439-90-9)**
UN 1971 [Methane, Compressed or Natural gas, compressed (with high methane content)] **(74-82-8)**
UN 1972 [Methane, Refrigerated liquid (cryogenic liquid)or Natural gas, refrigerated liquid (cryogenic liquid)(with high methane content)] **(74-82-8)**
UN 1974 [Chlorodifluorobromomethane] **(353-59-3)**
UN 1975 [Nitric oxide and dinitrogen tetroxide mixtures (Nitric oxide and nitrogen dioxide mixtures)] **(63907-41-5)**
UN 1976 [Octafluorocyclobutane] **(115-25-3)**
UN 1977 [Nitrogen, Refrigerated liquid (cryogenic liquid)] **(7727-37-9)**
UN 1978 [Propane] **(74-98-6)**
UN 1982 [Tetrafluoromethane] **(75-73-0)**
UN 1984 [Trifluoromethane] **(75-46-7)**
UN 1991 [Chloropropene, Inhibited] **(126-99-8)**
UN 1994 [Iron pentacarbonyl] **(13463-40-6)**
UN 1999 [Tars, Liquid including road asphalt and oils, bitumen and cut backs] **(8007-45-2)**
UN 1999 [Tars, Liquid including road asphalt and oils, bitumen and cut backs] **(8052-42-4)**
UN 2001 [Cobalt naphthenates, Powder] **(61789-51-3)**
UN 2008 [Zirconium powder, dry] **(7440-67-7)**
UN 2009 [Zirconium, dry, finished sheets, strip or coiled wire] **(7440-67-7)**
UN 2011 [Magnesium phosphide] **(12057-74-8)**
UN 2014 [Hydrogen peroxide, Aqueous solutions with more than 40 per cent but not more than 60 per cent hydrogen peroxide (stabilized as necessary)] **(7722-84-1)**
UN 2015 [Hydrogen peroxide, Stabilized or Hydrogen peroxide aqueous solutions, stabilized with more than 60 per cent hydrogen peroxide] **(7722-84-1)**
UN 2018 [Chloroanilines, Solid] **(95-51-2)**
UN 2018 [Chloroanilines, Solid] **(106-47-8)**
UN 2018 [Chloroanilines, Solid] **(108-42-9)**
UN 2019 [Chloroanilines, Liquid] **(108-42-9)**
UN 2019 [Chloroanilines, Liquid] **(95-51-2)**
UN 2019 [Chloroanilines, Liquid] **(106-47-8)**
UN 2020 [Chlorophenols, Solid] **(95-57-8)**
UN 2020 [Chlorophenols, Solid] **(106-48-9)**
UN 2020 [Chlorophenols, Solid] **(108-43-0)**
UN 2021 [Chlorophenols, Liquid] **(95-57-8)**
UN 2021 [Chlorophenols, Liquid] **(108-43-0)**
UN 2021 [Chlorophenols, Liquid] **(106-48-9)**
UN 2022 [Cresylic acid] **(1319-77-3)**
UN 2023 [Epichlorohydrin] **(106-89-8)**
UN 2027 [Sodium arsenite, Solid] **(7784-46-5)**
UN 2029 [Hydrazine, Anhydrous or Hydrazine aqueous solutions with more than 64 per cent hydrazine, by mass] **(302-01-2)**
UN 2030 [Hydrazine hydrate or Hydrazine aqueous solutions, with not more than 64 per cent hydrazine, by mass] **(302-01-2)**
UN 2031 (DOT) **(7697-37-2)**
UN 2032 [Nitric acid, red fuming] **(7697-37-2)**
UN 2036 [Xenon] **(7440-63-3)**
UN 2038 [Dinitrotoluenes, Liquid] **(25321-14-6)**
UN 2044 [2,2-Dimethylpropane other than pentane and isopentane] **(463-82-1)**
UN 2045 (DOT) **(78-84-2)**
UN 2046 [Cymenes] **(535-77-3)**
UN 2046 [Cymenes] **(99-87-6)**
UN 2046 [Cymenes] **(527-84-4)**
UN 2047 [Dichloropropene] **(26952-23-8)**
UN 2048 [Dicyclopentadiene] **(77-73-6)**
UN 2049 [Diethylbenzene] **(25340-17-4)**
UN 2050 [Diisobutylene, Isomeric compounds] **(25167-70-8)**
UN 2051 [Dimethylethanolamine] **(108-01-0)**
UN 2052 [Dipentene] **(138-86-3)**
UN 2053 [Methyl isobutyl carbinol] **(108-11-2)**
UN 2054 [Morpholine] **(110-91-8)**
UN 2055 [Styrene monomer, Inhibited] **(100-42-5)**
UN 2056 [Tetrahydrofuran] **(109-99-9)**
UN 2057 [Tripropylene] **(13987-01-4)**
UN 2058 [Valeraldehyde] **(110-62-3)**
UN 2059 [Nitrocellulose, Solution, flammable with not more than 12.6 per cent nitrogen, by mass, and not more than 55 per cent nitrocellulose] **(9004-70-0)**
UN 2060 (DOT) **(9004-70-0)**
UN 2073 [Ammonia solutions, with more than 35 per cent but not more than 50 per cent ammonia] **(7664-41-7)**
UN 2074 [Acrylamide] **(79-06-1)**
UN 2075 [Chloral, Anhydrous, Inhibited] **(75-87-6)**
UN 2076 [Cresols (o-; m-; p-)] **(106-44-5)**
UN 2076 [Cresols (o-; m-; p-)] **(108-39-4)**
UN 2076 [Cresols (o-; m-; p-)] **(1319-77-3)**
UN 2076 [Cresols (o-; m-; p-)] **(95-48-7)**
UN 2077 [α-Naphthylamine] **(134-32-7)**
UN 2078 [Toluene diisocyanate] **(26471-62-5)**
UN 2079 [Diethylenetriamine] **(111-40-0)**
UN 2080 (DOT) **(37187-22-7)**
UN 2081 (DOT) **(644-31-5)**
UN 2082 **(3179-56-4)**
UN 2083 **(3179-56-4)**
UN 2084 (DOT) **(110-22-5)**
UN 2085 (DOT) **(94-36-0)**
UN 2086 **(94-36-0)**
UN 2087 [Benzoyl peroxide, Not more than 72% as a paste] **(94-36-0)**
UN 2088 **(94-36-0)**
UN 2089 [Benzoyl peroxide, Not less than 30% but not more than 52% with inert solid] **(94-36-0)**
UN 2090 [Benzoyl peroxide, Not more than 77% with water] **(94-36-0)**
UN 2091 (DOT) **(3457-61-2)**
UN 2093 (DOT) **(75-91-2)**
UN 2094 (DOT) **(75-91-2)**
UN 2095 (DOT) **(107-71-1)**
UN 2096 (DOT) **(107-71-1)**
UN 2097 (DOT) **(614-45-9)**
UN 2098 (DOT) **(614-45-9)**
UN 2099 (DOT) **(1931-62-0)**
UN 2100 (DOT) **(1931-62-0)**
UN 2101 (DOT) **(1931-62-0)**
UN 2102 (DOT) **(110-05-4)**
UN 2103 (DOT) **(2372-21-6)**
UN 2106 **(2155-71-7)**
UN 2107 **(2155-71-7)**
UN 2108 **(2155-71-7)**
UN 2110 (DOT) **(927-07-1)**
UN 2111 **(2167-23-9)**
UN 2113 (DOT) **(94-17-7)**
UN 2114 (DOT) **(94-17-7)**
UN 2115 (DOT) **(94-17-7)**
UN 2116 (DOT) **(80-15-9)**

UN 2118 **(4904-55-6)**
UN 2119 **(4904-55-6)**
UN 2120 (DOT) **(762-12-9)**
UN 2121 (DOT) **(80-43-3)**
UN 2122 (DOT) **(16111-62-9)**
UN 2123 (DOT) **(16111-62-9)**
UN 2124 (DOT) **(105-74-8)**
UN 2126 **(37206-20-5)**
UN 2127 (DOT) **(1338-23-4)**
UN 2129 (DOT) **(762-16-3)**
UN 2130 **(762-13-0)**
UN 2131 (DOT) **(79-21-0)**
UN 2132 **(3248-28-0)**
UN 2133 (DOT) **(105-64-6)**
UN 2134 (DOT) **(105-64-6)**
UN 2135 (DOT) **(123-23-9)**
UN 2136 **(771-29-9)**
UN 2137 (DOT) **(133-14-2)**
UN 2138 (DOT) **(133-14-2)**
UN 2139 (DOT) **(133-14-2)**
UN 2140 (DOT) **(995-33-5)**
UN 2141 (DOT) **(995-33-5)**
UN 2142 (DOT) **(109-13-7)**
UN 2143 **(62695-55-0)**
UN 2145 (DOT) **(6731-36-8)**
UN 2146 (DOT) **(6731-36-8)**
UN 2147 (DOT) **(6731-36-8)**
UN 2148 **(2407-94-5)**
UN 2149 **(2144-45-8)**
UN 2150 (DOT) **(19910-65-7)**
UN 2151 (DOT) **(19910-65-7)**
UN 2152 **(1561-49-5)**
UN 2153 **(1561-49-5)**
UN 2154 **(15520-11-3)**
UN 2155 (DOT) **(78-63-7)**
UN 2156 (DOT) **(78-63-7)**
UN 2157 (DOT) **(13052-09-0)**
UN 2158 (DOT) **(1068-27-5)**
UN 2159 (DOT) **(1068-27-5)**
UN 2162 (DOT) **(28324-52-9)**
UN 2163 **(54693-46-8)**
UN 2164 **(26322-14-5)**
UN 2165 **(22397-33-7)**
UN 2166 **(22397-33-7)**
UN 2167 **(22397-33-7)**
UN 2169 **(16215-49-9)**
UN 2170 **(16215-49-9)**
UN 2171 (DOT) **(26762-93-6)**
UN 2172 (DOT) **(2618-77-1)**
UN 2173 (DOT) **(2618-77-1)**
UN 2174 (DOT) **(3025-88-5)**
UN 2175 **(14666-78-5)**
UN 2176 (DOT) **(16066-38-9)**
UN 2177 (DOT) **(26748-41-4)**
UN 2178 **(2614-76-8)**
UN 2179 **(3006-86-8)**
UN 2180 **(3006-86-8)**
UN 2182 **(3437-84-1)**
UN 2183 **(23474-91-1)**
UN 2186 [Hydrogen chloride, Refrigerated liquid (cryogenic liquid)] **(7647-01-0)**
UN 2187 [Carbon dioxide, Refrigerated liquid (cryogenic liquid)] **(124-38-9)**
UN 2188 [Arsine] **(7784-42-1)**
UN 2189 [Dichlorosilane] **(4109-96-0)**
UN 2190 [Oxygen difluoride] **(7783-41-7)**
UN 2191 [Sulfuryl fluoride] **(2699-79-8)**
UN 2192 [Germane] **(7782-65-2)**
UN 2193 (DOT) **(76-16-4)**
UN 2194 [Selenium hexafluoride] **(7783-79-1)**
UN 2195 [Tellurium hexafluoride] **(7783-80-4)**
UN 2196 [Tungsten hexafluoride] **(7783-82-6)**
UN 2197 [Hydrogen iodide, Anhydrous] **(10034-85-2)**
UN 2198 [Phosphorus pentafluoride] **(7647-19-0)**
UN 2199 [Phosphine] **(7803-51-2)**
UN 2200 [Propadiene, Inhibited.] **(463-49-0)**
UN 2201 [Nitrous oxide, Refrigerated liquid (cryogenic liquid)] **(10024-97-2)**
UN 2202 [Hydrogen selenide, Anhydrous] **(7783-07-5)**
UN 2203 [Silane] **(7803-62-5)**
UN 2204 [Carbonyl sulfide] **(463-58-1)**
UN 2205 (DOT) **(111-69-3)**
UN 2208 [Calcium hypochlorite mixtures, Dry, with more than 10 per cent but not more than 39 per cent available chlorine] **(7778-54-3)**
UN 2209 [Formaldehyde solution] **(50-00-0)**
UN 2210 [Maneb or Maneb preparations with not less than 60 per cent Maneb] **(12427-38-2)**
UN 2211 [Polystyrene beads, expandable, evolving flammable vapor.] **(9003-53-6)**
UN 2212 [Blue Asbestos (Crocidolite) or Brown asbestos (amosite, mysorite)] **(12001-28-4)**
UN 2213 [Paraformaldehyde] **(30525-89-4)**
UN 2214 (DOT) **(85-44-9)**
UN 2215 (DOT) **(108-31-6)**
UN 2218 (DOT) **(79-10-7)**
UN 2219 (DOT) **(106-92-3)**
UN 2222 [Anisole] **(100-66-3)**
UN 2224 [Benzonitrile] **(100-47-0)**
UN 2225 [Benzene sulfonyl chloride] **(98-09-9)**
UN 2226 [Benzotrichloride] **(98-07-7)**
UN 2227 [n-Butyl methacrylate] **(97-88-1)**
UN 2228 [Butylphenols, Liquid] **(3180-09-4)**
UN 2228 [Butylphenols, Liquid] **(1638-22-8)**
UN 2229 [Butylphenols, Solid] **(1638-22-8)**
UN 2229 [Butylphenols, Solid] **(3180-09-4)**
UN 2232 [Chloroacetaldehyde] **(107-20-0)**
UN 2234 [Chlorobenzotrifluorides] **(98-15-7)**
UN 2234 [Chlorobenzotrifluorides] **(88-16-4)**
UN 2234 [Chlorobenzotrifluorides] **(98-56-6)**
UN 2235 [Chlorobenzylchlorides] **(104-83-6)**
UN 2238 [Chlorotoluenes] **(106-43-4)**
UN 2238 [Chlorotoluenes] **(108-41-8)**
UN 2238 [Chlorotoluenes] **(95-49-8)**
UN 2241 [Cycloheptane] **(291-64-5)**
UN 2242 [Cycloheptene] **(628-92-2)**
UN 2243 [Cyclohexyl acetate] **(622-45-7)**
UN 2244 [Cyclopentanol] **(96-41-3)**
UN 2245 [Cyclopentanone] **(120-92-3)**
UN 2246 [Cyclopentene] **(142-29-0)**
UN 2247 [n-Decane] **(124-18-5)**
UN 2248 [Di-n-butylamine] **(111-92-2)**
UN 2249 [Dichlorodimethyl ether, symmetrical] **(542-88-1)**
UN 2252 [1,2-Dimethoxyethane] **(110-71-4)**
UN 2253 [N,N-Dimethylaniline] **(121-69-7)**
UN 2256 [Cyclohexene] **(110-83-8)**
UN 2257 [Potassium] **(7440-09-7)**
UN 2258 [1,2-Propylenediamine] **(78-90-0)**
UN 2259 [Triethylenetetramine] **(112-24-3)**
UN 2260 [Tripropylamine] **(102-69-2)**
UN 2261 [Xylenols] **(95-87-4)**
UN 2261 [Xylenols] **(526-75-0)**
UN 2261 [Xylenols] **(1300-71-6)**
UN 2261 [Xylenols] **(105-67-9)**
UN 2262 [Dimethylcarbamoyl chloride] **(79-44-7)**
UN 2263 [Dimethylcyclohexanes] **(583-57-3)**

UN 2263 [Dimethylcyclohexanes] **(591-21-9)**
UN 2263 [Dimethylcyclohexanes] **(589-90-2)**
UN 2264 [Dimethylcyclohexylamine] **(98-94-2)**
UN 2265 [N,N-Dimethylformamide] **(68-12-2)**
UN 2267 [Dimethyl thiophosphoryl chloride] **(993-12-4)**
UN 2269 [3,3'-Iminodipropylamine] **(56-18-8)**
UN 2270 [Ethylamine, Aqueous solution with not less than 50 per cent but not more than 70 per cent ethylamine] **(75-04-7)**
UN 2271 [Ethyl amyl ketone] **(106-68-3)**
UN 2272 [N-Ethylaniline] **(103-69-5)**
UN 2273 [2-Ethylaniline] **(578-54-1)**
UN 2274 [N-Ethyl-N-benzylaniline] **(92-59-1)**
UN 2275 [2-Ethylbutanol] **(97-95-0)**
UN 2276 [2-Ethylhexylamine] **(104-75-6)**
UN 2277 [Ethyl methacrylate] **(97-63-2)**
UN 2279 [Hexachlorobutadiene] **(87-68-3)**
UN 2280 [Hexamethylenediamine, Solid] **(124-09-4)**
UN 2281 [Hexamethylenediisocyanate] **(822-06-0)**
UN 2282 [Hexanols] **(111-27-3)**
UN 2282 [Hexanols] **(97-95-0)**
UN 2283 [Isobutyl methacrylate] **(97-86-9)**
UN 2284 [Isobutyronitrile] **(78-82-0)**
UN 2288 [Isohexenes] **(27236-46-0)**
UN 2289 [Isophoronediamine] **(2855-13-2)**
UN 2290 [Isophoronediisocyanate] **(4098-71-9)**
UN 2293 [4-Methoxy-4-methylpentan-2-one] **(107-70-0)**
UN 2294 [N-Methylaniline] **(100-61-8)**
UN 2295 [Methyl chloroacetate] **(96-34-4)**
UN 2296 [Methyl cyclohexane] **(108-87-2)**
UN 2297 [Methyl cyclohexanone] **(1331-22-2)**
UN 2298 [Methyl cyclopentane] **(96-37-7)**
UN 2299 [Methyl dichloroacetate] **(116-54-1)**
UN 2300 [2-Methyl-5-ethylpyridine] **(104-90-5)**
UN 2301 [2-Methylfuran] **(534-22-5)**
UN 2302 [5-Methylhexan-2-one] **(110-12-3)**
UN 2303 [Isopropenylbenzene] **(98-83-9)**
UN 2304 [Naphthalene, Molten] **(91-20-3)**
UN 2305 [Nitrobenzenesulfonic acid] **(31212-28-9)**
UN 2306 [Nitrobenzotrifluorides] **(98-46-4)**
UN 2306 [Nitrobenzotrifluorides] **(402-54-0)**
UN 2306 [Nitrobenzotrifluorides] **(384-22-5)**
UN 2307 [3-Nitro-4-chlorobenzotrifluoride] **(121-17-5)**
UN 2310 [Pentan-2,4-dione] **(123-54-6)**
UN 2311 [Phenetidines] **(156-43-4)**
UN 2312 [Phenol, molten] **(108-95-2)**
UN 2313 [Picolines] **(108-99-6)**
UN 2313 [Picolines] **(108-89-4)**
UN 2313 [Picolines] **(109-06-8)**
UN 2313 [Picolines] **(1333-41-1)**
UN 2315 [Polychlorinated biphenyls] **(1336-36-3)**
UN 2316 [Sodium cuprocyanide, Solid] **(14264-31-4)**
UN 2317 [Sodium cuprocyanide, Solution] **(14264-31-4)**
UN 2318 [Sodium hydrosulfide, with less than 25 per cent water of crystallization] **(16721-80-5)**
UN 2320 [Tetraethylenepentamine] **(112-57-2)**
UN 2321 [Trichlorobenzenes, Liquid] **(87-61-6)**
UN 2321 [Trichlorobenzenes, Liquid] **(120-82-1)**
UN 2321 [Trichlorobenzenes, Liquid] **(108-70-3)**
UN 2321 [Trichlorobenzenes, Liquid] **(12002-48-1)**
UN 2322 [Trichlorobutene] **(51023-22-4)**
UN 2323 [Triethyl phosphite] **(122-52-1)**
UN 2324 [Triisobutylene] **(7756-94-7)**
UN 2325 [1,3,5-Trimethylbenzene] **(108-67-8)**
UN 2327 [Trimethylhexamethylenediamines] **(25620-58-0)**
UN 2329 [Trimethyl phosphite] **(121-45-9)**
UN 2330 [Undecane] **(1120-21-4)**
UN 2331 [Zinc chloride, Anhydrous] **(7646-85-7)**
UN 2332 [Acetaldehyde oxime] **(107-29-9)**
UN 2333 [Allyl acetate] **(591-87-7)**
UN 2334 [Allylamine] **(107-11-9)**
UN 2335 [Allyl ethyl ether] **(557-31-3)**
UN 2337 [Phenyl mercaptan] **(108-98-5)**
UN 2338 [Benzotrifluoride] **(98-08-8)**
UN 2339 [2-Bromobutane] **(78-76-2)**
UN 2341 [1-Bromo-3-methylbutane] **(107-82-4)**
UN 2342 [Bromomethylpropanes] **(507-19-7)**
UN 2342 [Bromomethylpropanes] **(78-77-3)**
UN 2343 [2-Bromopentane] **(107-81-3)**
UN 2344 [2-Bromopropane] **(75-26-3)**
UN 2344 **(106-94-5)**
UN 2345 [3-Bromopropyne] **(106-96-7)**
UN 2346 [Butanedione] **(431-03-8)**
UN 2347 [Butyl mercaptans] **(109-79-5)**
UN 2348 [Butylacrylate] **(141-32-2)**
UN 2350 [Butyl methyl ether] **(628-28-4)**
UN 2351 [Butyl nitrites] **(544-16-1)**
UN 2352 [Butyl vinyl ether, inhibited] **(111-34-2)**
UN 2353 [Butyryl chloride] **(141-75-3)**
UN 2354 [Chloromethyl ethyl ether] **(3188-13-4)**
UN 2356 [2-Chloropropane] **(75-29-6)**
UN 2357 [Cyclohexylamine] **(108-91-8)**
UN 2358 [Cyclooctatetraene] **(629-20-9)**
UN 2359 [Diallylamine] **(124-02-7)**
UN 2360 [Diallylether] **(557-40-4)**
UN 2361 [Diisobutylamine] **(110-96-3)**
UN 2362 [1,1-Dichloroethane] **(75-34-3)**
UN 2363 [Ethyl mercaptan] **(75-08-1)**
UN 2364 [n-Propyl benzene] **(103-65-1)**
UN 2366 [Diethyl carbonate] **(105-58-8)**
UN 2367 [α-Methylvaleraldehyde] **(123-15-9)**
UN 2368 [α-Pinene] **(80-56-8)**
UN 2368 [α-Pinene] **(1330-16-1)**
UN 2369 [Ethylene glycol monobutyl ether] **(111-76-2)**
UN 2370 [1-Hexene] **(592-41-6)**
UN 2371 [Isopentenes] **(563-46-2)**
UN 2371 [Isopentenes] **(513-35-9)**
UN 2371 [Isopentenes] **(563-45-1)**
UN 2371 [Isopentenes] **(26760-64-5)**
UN 2372 [1,2-Di-(dimethylamino) ethane] **(110-18-9)**
UN 2373 [Diethoxymethane] **(462-95-3)**
UN 2375 [Diethyl sulfide] **(352-93-2)**
UN 2376 [2,3-Dihydropyran] **(110-87-2)**
UN 2377 [1,1-Dimethoxyethane] **(534-15-6)**
UN 2378 [2-Dimethylaminoacetonitrile] **(926-64-7)**
UN 2379 [1,3-Dimethylbutylamine] **(108-09-8)**
UN 2381 [Dimethyl disulfide] **(624-92-0)**
UN 2382 [Dimethylhydrazine, symmetrical] **(540-73-8)**
UN 2383 [Dipropylamine] **(142-84-7)**
UN 2384 [Dipropyl ether] **(111-43-3)**
UN 2385 [Ethyl isobutyrate] **(97-62-1)**
UN 2386 [1-Ethylpiperidine] **(766-09-6)**
UN 2387 [Fluorobenzene] **(462-06-6)**
UN 2388 [Fluorotoluenes] **(352-70-5)**
UN 2388 [Fluorotoluenes] **(352-32-9)**
UN 2388 [Fluorotoluenes] **(95-52-3)**
UN 2389 [Furan] **(110-00-9)**
UN 2390 [2-Iodobutane] **(513-48-4)**
UN 2392 [Iodopropanes] **(26914-02-3)**
UN 2393 [Isobutyl formate] **(542-55-2)**
UN 2394 [Isobutyl propionate] **(540-42-1)**
UN 2396 [Methacrylaldehyde] **(78-85-3)**
UN 2397 [3-Methylbutan-2-one] **(563-80-4)**
UN 2398 [Methyl-tert-butylether] **(1634-04-4)**
UN 2399 [1-Methylpiperidine] **(626-67-5)**

UN 2400 [Methyl isovalerate] (556-24-1)
UN 2401 [Piperidine] (110-89-4)
UN 2402 [Propanethiols] (75-33-2)
UN 2402 [Propanethiols] (107-03-9)
UN 2403 [Isopropenyl acetate] (108-22-5)
UN 2404 [Propionitrile] (107-12-0)
UN 2406 [Isopropyl isobutyrate] (617-50-5)
UN 2407 [Isopropyl chloroformate] (108-23-6)
UN 2409 [Isopropyl propionate] (637-78-5)
UN 2410 [1,2,3,6-Tetrahydropyridine] (694-05-3)
UN 2411 [Butyronitrile] (109-74-0)
UN 2412 [Tetrahydrothiophene] (110-01-0)
UN 2414 [Thiophene] (110-02-1)
UN 2416 [Trimethyl borate] (121-43-7)
UN 2417 [Carbonyl fluoride] (353-50-4)
UN 2418 [Sulfur tetrafluoride] (7783-60-0)
UN 2419 [Bromotrifluoroethylene] (598-73-2)
UN 2420 [Hexafluoroacetone] (684-16-2)
UN 2421 [Nitrogen trioxide] (10544-73-7)
UN 2422 [Octafluorobut-2-ene] (360-89-4)
UN 2424 [Octafluoropropane] (76-19-7)
UN 2426 [Ammonium nitrate, Liquid (hot concentrated solution)] (6484-52-2)
UN 2427 [Potassium chlorate, Solution] (3811-04-9)
UN 2428 [Sodium chlorate, Solution] (7775-09-9)
UN 2429 [Calcium chlorate solution] (10137-74-3)
UN 2431 [Anisidines] (90-04-0)
UN 2432 [N,N-Diethyl aniline] (19-66-7)
UN 2434 [Dibenzyldichlorosilane] (18414-36-3)
UN 2435 [Ethylphenyldichlorosilane] (1125-27-5)
UN 2436 [Thioacetic acid] (507-09-5)
UN 2437 [Methylphenyldichlorosilane] (149-74-6)
UN 2438 [Trimethylacetyl chloride] (3282-30-2)
UN 2439 [Sodium hydrogen fluoride] (1333-83-1)
UN 2441 [Titanium trichloride, pyrophoric or Titanium trichloride mixtures, pyrophoric] (7705-07-9)
UN 2442 [Trichloroacetyl chloride] (76-02-8)
UN 2443 [Vanadium oxytrichloride] (7727-18-6)
UN 2444 [Vanadium tetrachloride] (7632-51-1)
UN 2446 [Nitrocresols] (12167-20-3)
UN 2448 [Sulfur, molten] (7704-34-9)
UN 2451 [Nitrogen trifluoride] (7783-54-2)
UN 2452 [Ethylacetylene, Inhibited] (107-00-6)
UN 2453 [Ethyl fluoride] (353-36-6)
UN 2454 [Methyl fluoride] (593-53-3)
UN 2456 [2-Chloropropene] (557-98-2)
UN 2457 [2,3-Dimethylbutane] (79-29-8)
UN 2458 [Hexadienes] (42296-74-2)
UN 2459 [2-Methyl-1-butene] (563-46-2)
UN 2460 [2-Methyl-2-butene] (26760-64-5)
UN 2460 [2-Methyl-2-butene] (513-35-9)
UN 2462 (DOT) (96-14-0)
UN 2462 (DOT) (107-83-5)
UN 2463 [Aluminum hydride] (7784-21-6)
UN 2464 [Beryllium nitrate] (13597-99-4)
UN 2465 [Dichloroisocyanuric acid, Dry or dichloroisocyanuric acid salts] (2244-21-5)
UN 2465 [Dichloroisocyanuric acid, Dry or dichloroisocyanuric acid salts] (2893-78-9)
UN 2465 (54579-28-1)
UN 2465 [Dichloroisocyanuric acid, Dry or dichloroisocyanuric acid salts] (2782-57-2)
UN 2466 [Potassium superoxide] (12030-88-5)
UN 2468 [Trichloroisocyanuric acid, dry] (87-90-1)
UN 2470 [Phenylacetonitrile, Liquid] (140-29-4)
UN 2471 [Osmium tetroxide] (20816-12-0)
UN 2472 (DOT) (83-26-1)
UN 2473 [Sodium arsanilate] (127-85-5)
UN 2474 [Thiophosgene] (463-71-8)
UN 2475 [Vanadium trichloride] (7718-98-1)
UN 2477 [Methyl isothiocyanate] (556-61-6)
UN 2480 [Methyl isocyanate] (624-83-9)
UN 2481 [Ethyl isocyanate] (109-90-0)
UN 2482 [n-Propyl isocyanate] (110-78-1)
UN 2485 [n-Butyl isocyanate] (111-36-4)
UN 2487 [Phenyl isocyanate] (103-71-9)
UN 2488 [Cyclohexyl isocyanate] (3173-53-3)
UN 2489 [Diphenylmethane-4,4'diisocyanate] (101-68-8)
UN 2490 [Dichloroisopropyl ether] (108-60-1)
UN 2491 [Ethanolamine or Ethanolamine solutions] (141-43-5)
UN 2493 [Hexamethyleneimine] (111-49-9)
UN 2495 [Iodine pentafluoride] (7783-66-6)
UN 2496 [Propionic anhydride] (123-62-6)
UN 2497 [Sodium phenolate, Solid] (139-02-6)
UN 2498 [1,2,3,6-Tetrahydrobenzaldehyde] (100-50-5)
UN 2501 [Tris-(1-aziridinyl)phosphine oxide, solution] (545-55-1)
UN 2502 [Valeryl chloride] (638-29-9)
UN 2503 [Zirconium tetrachloride] (10026-11-6)
UN 2504 [Tetrabromoethane] (79-27-6)
UN 2505 [Ammonium fluoride] (12125-01-8)
UN 2506 [Ammonium hydrogen sulfate] (7803-63-6)
UN 2507 [Chloroplatinic acid, Solid] (16941-12-1)
UN 2508 [Molybdenum pentachloride] (10241-05-1)
UN 2509 [Potassium hydrogen sulfate] (7646-93-7)
UN 2511 [α-Chloropropionic acid] (28554-00-9)
UN 2512 [Aminophenols (o-; m-; p-)] (591-27-5)
UN 2512 [Aminophenols (o-; m-; p-)] (95-55-6)
UN 2512 [Aminophenols (o-; m-; p-)] (123-30-8)
UN 2514 [Bromobenzene] (108-86-1)
UN 2515 [Bromoform] (75-25-2)
UN 2516 [Carbon tetrabromide] (558-13-4)
UN 2517 (DOT) (75-68-3)
UN 2517 [Chlorodifluoroethanes or Difluorochloroethanes] (25497-29-4)
UN 2518 [1,5,9-Cyclododecatriene] (4904-61-4)
UN 2521 [Diketene, Inhibited] (674-82-8)
UN 2522 [Dimethylaminoethyl methacrylate] (2867-47-2)
UN 2524 [Ethyl orthoformate] (122-51-0)
UN 2525 [Ethyl oxalate] (95-92-1)
UN 2526 [Furfurylamine] (617-89-0)
UN 2527 [Isobutyl acrylate] (106-63-8)
UN 2528 [Isobutyl isobutyrate] (97-85-8)
UN 2529 [Isobutyric acid] (79-31-2)
UN 2530 [Isobutyric anhydride] (97-72-3)
UN 2531 [Methacrylic acid, inhibited] (79-41-4)
UN 2533 [Methyl trichloroacetate] (598-99-2)
UN 2535 [Methylmorpholine] (109-02-4)
UN 2538 [Nitronaphthalene] (27254-36-0)
UN 2541 [Terpinolene] (586-62-9)
UN 2542 [Tributylamine] (102-82-9)
UN 2545 [Hafnium powder, Dry] (7440-58-6)
UN 2546 [Titanium powder, Dry] (7440-32-6)
UN 2547 [Sodium superoxide] (12034-12-7)
UN 2550 (DOT) (1338-23-4)
UN 2553 [Naphtha] (8030-30-6)
UN 2554 [Methyl allyl chloride] (563-47-3)
UN 2555 [Nitrocellulose with water not less than 25 per cent water, by mass] (9004-70-0)
UN 2556 [Nitrocellulose with alcohol not less than 25 per cent alcohol by mass, and not more than 12.6 per cent nitrogen, by dry mass] (9004-70-0)
UN 2558 [Epibromohydrin] (3132-64-7)
UN 2560 [2-Methylpentan-2-ol] (590-36-3)
UN 2561 [3-Methyl-1-butene] (563-45-1)
UN 2562 (DOT) (109-13-7)
UN 2564 [Trichloroacetic acid, solution] (76-03-9)

UN 2565 [Dicyclohexylamine] (101-83-7)
UN 2567 [Sodium pentachlorophenate] (131-52-2)
UN 2572 [Phenylhydrazine] (100-63-0)
UN 2574 [Tricresylphosphate, With more than 3% ortho isomer] (1330-78-5)
UN 2576 [Phosphorus oxybromide, molten] (7789-59-5)
UN 2577 [Phenylacetyl chloride] (103-80-0)
UN 2578 [Phosphorus trioxide] (1314-24-5)
UN 2579 [Piperazine] (110-85-0)
UN 2580 [Aluminum bromide, solution] (7727-15-3)
UN 2581 [Aluminum chloride, solution] (7446-70-0)
UN 2582 [Ferric chloride, Solution] (7705-08-0)
UN 2587 [Benzoquinone] (583-63-1)
UN 2587 [Benzoquinone] (106-51-4)
UN 2589 [Vinyl chloroacetate] (2549-51-1)
UN 2590 [White asbestos, (chrysotile, actinolite, anthophyllite, tremolite)] (12001-29-5)
UN 2591 [Xenon, Refrigerated liquid (cryogenic liquids)] (7440-63-3)
UN 2592 (52326-66-6)
UN 2593 (895-85-2)
UN 2594 (DOT) (26748-41-4)
UN 2595 (53220-22-7)
UN 2601 [Cyclobutane] (287-23-0)
UN 2603 [Cycloheptatriene] (544-25-2)
UN 2604 [Boron trifluoride diethyl etherate] (109-63-7)
UN 2606 [Methyl orthosilicate] (681-84-5)
UN 2607 [Acrolein dimer, Stabilized] (100-73-2)
UN 2608 [Nitropropanes] (25322-01-4)
UN 2610 [Triallylamine] (102-70-5)
UN 2611 [Propylene chlorohydrin] (78-89-7)
UN 2612 [Methyl propyl ether] (557-17-5)
UN 2614 [Methallyl alcohol] (513-42-8)
UN 2615 [Ethyl propyl ether] (628-32-0)
UN 2616 [Triisopropyl borate] (5419-55-6)
UN 2617 [Methyl cyclohexanols, flash point not more than 60.5 degrees C] (25639-42-3)
UN 2618 [Vinyl toluene, Inhibited mixed isomers] (25013-15-4)
UN 2619 [Benzyldimethylamine] (103-83-3)
UN 2620 [Amyl butyrates] (540-18-1)
UN 2620 [Amyl butyrates] (106-27-4)
UN 2621 [Acetyl methyl carbinol] (513-86-0)
UN 2622 [Glycidaldehyde] (765-34-4)
UN 2626 [Chloric acid solution, with not more than 10 per cent chloric acid] (7790-93-4)
UN 2629 [Sodium fluoroacetate] (62-74-8)
UN 2630 [Selenates or Selenites] (10102-18-8)
UN 2642 [Fluoroacetic acid] (144-49-0)
UN 2643 [Methyl bromoacetate] (96-32-2)
UN 2644 [Methyl iodide] (74-88-4)
UN 2645 [Phenacyl bromide] (70-11-1)
UN 2646 [Hexachlorocyclopentadiene] (77-47-4)
UN 2647 [Malononitrile] (109-77-3)
UN 2649 [1,3-Dichloroacetone] (534-07-6)
UN 2650 [1,1-Dichloro-1-nitroethane] (594-72-9)
UN 2651 [4,4'-Diaminodiphenyl methane] (101-77-9)
UN 2653 [Benzyl iodide] (620-05-3)
UN 2656 [Quinoline] (91-22-5)
UN 2657 [Selenium disulfide] (7488-56-4)
UN 2658 [Selenium powder] (7782-49-2)
UN 2659 [Sodium chloroacetate] (3926-62-3)
UN 2661 [Hexachloroacetone] (116-16-5)
UN 2662 [Hydroquinone, Solid or liquid] (123-31-9)
UN 2664 [Dibromomethane] (74-95-3)
UN 2666 [Ethyl cyanoacetate] (105-56-6)
UN 2668 [Chloroacetonitrile] (107-14-2)
UN 2669 [Chlorocresols, Solid] (54548-50-4)
UN 2670 [Cyanuric chloride] (108-77-0)
UN 2671 [Aminopyridines (o-; m-; p-)] (504-24-5)
UN 2671 [Aminopyridines (o-; m-; p-)] (462-08-8)
UN 2671 [Aminopyridines (o-; m-; p-)] (504-29-0)
UN 2672 [Ammonia solutions, With more than 10 per cent but not more than 35 percent ammonia by mass] (7664-41-7)
UN 2674 [Sodium fluorosilicate] (16893-85-9)
UN 2676 [Stibine] (7803-52-3)
UN 2679 [Lithium hydroxide, Solution] (1310-65-2)
UN 2681 [Caesium hydroxide solution] (21351-79-1)
UN 2682 [Caesium hydroxide, Solid] (21351-79-1)
UN 2683 [Ammonium sulfide solution] (12135-76-1)
UN 2684 [Diethylaminopropylamine] (104-78-9)
UN 2685 [N,N-Diethylethylenediamine] (100-36-7)
UN 2686 [Diethylaminoethanol] (100-37-8)
UN 2688 [1-Chloro-3-bromopropane] (109-70-6)
UN 2689 [Glycerol α-monochlorohydrin] (96-24-2)
UN 2690 [N-n-Butyl imidazole] (4316-42-1)
UN 2692 [Boron tribromide] (10294-33-4)
UN 2693 [Bisulfites, inorganic, aqueous solutions, N.O.S.] (7631-90-5)
UN 2698 [Tetrahydrophthalic anhydrides with more than 0.05 percent of maleic anhydride] (85-43-8)
UN 2699 [Trifluoroacetic acid] (76-05-1)
UN 2706 (DOT) (584-02-1)
UN 2707 [Dimethyldioxanes] (25136-55-4)
UN 2708 [Butoxyl] (4435-53-4)
UN 2709 [Butyl benzenes] (135-98-8)
UN 2709 [Butyl benzenes] (98-06-6)
UN 2709 [Butyl benzenes] (104-51-8)
UN 2710 [Dipropylketone] (123-19-3)
UN 2711 [Dibromobenzene] (26249-12-7)
UN 2713 [Acridine] (260-94-6)
UN 2714 [Zinc resinate] (9010-69-9)
UN 2716 [1,4-Butynediol] (110-65-6)
UN 2717 [Camphor, synthetic] (76-22-2)
UN 2718 (102-67-0)
UN 2722 [Lithium nitrate] (7790-69-4)
UN 2723 [Magnesium chlorate] (10326-21-3)
UN 2725 [Nickel nitrate] (13138-45-9)
UN 2727 [Thallium nitrate] (10102-45-1)
UN 2728 [Zirconium nitrate] (13746-89-9)
UN 2729 [Hexachlorobenzene] (118-74-1)
UN 2730 [Nitroanisole] (91-23-6)
UN 2730 [Nitroanisole] (555-03-3)
UN 2730 [Nitroanisole] (100-17-4)
UN 2732 [Nitrobromobenzenes liquid] (586-78-7)
UN 2732 [Nitrobromobenzenes solid] (585-79-5)
UN 2732 [Nitrobromobenzenes liquid] (577-19-5)
UN 2738 [N-Butylaniline] (1126-78-9)
UN 2739 [Butyric anhydride] (106-31-0)
UN 2740 [n-Propyl chloroformate] (109-61-5)
UN 2746 [Phenylchloroformate] (1885-14-9)
UN 2748 [2-Ethylhexylchloroformate] (24468-13-1)
UN 2749 [Tetramethylsilane] (75-76-3)
UN 2750 [1,3-Dichloropropanol-2] (96-23-1)
UN 2751 [Diethylthiophosphoryl chloride] (2524-04-1)
UN 2752 [1,2-Epoxy-3-ethoxypropane] (4016-11-9)
UN 2754 [N-Ethyltoluidines] (94-68-8)
UN 2754 [N-Ethyltoluidines] (102-27-2)
UN 2755 (937-14-4)
UN 2761 [Organochlorine pesticides, Solid toxic N.O.S.] (50-29-3)
UN 2783 (DOT) (757-58-4)
UN 2783 [Organophosphorus pesticides, Solid, toxic, N.O.S.] (52-68-6)
UN 2783 [Organophosphorus pesticides, Solid, toxic, N.O.S.] (56-72-4)
UN 2783 [Organophosphorus pesticides, Solid, toxic, N.O.S.] (371-86-8)
UN 2785 [Thia-4-pentanal] (3268-49-3)
UN 2789 [Acetic acid, glacial or Acetic acid solution, more than 80 per cent acid, by mass] (64-19-7)
UN 2790 [Acetic acid solution, more than 10 per cent but not

more than 80 per cent acid, by mass] **(64-19-7)**
UN 2798 [Phenyl phosphorus dichloride] **(644-97-3)**
UN 2798 [Phenyl phosphorus dichloride] **(824-72-6)**
UN 2799 [Phenyl phosphorus thiodichloride] **(3497-00-5)**
UN 2802 [Copper chloride] **(1344-67-8)**
UN 2803 [Gallium] **(7440-55-3)**
UN 2805 [Lithium hydride, Fused solid] **(7580-67-8)**
UN 2806 [Lithium nitride] **(26134-62-3)**
UN 2809 (DOT) **(7439-97-6)**
UN 2812 [Sodium aluminate, Solid] **(12042-47-6)**
UN 2815 [N-Aminoethylpiperazine] **(140-31-8)**
UN 2817 [Ammonium hydrogen fluoride, Solution] **(1341-49-7)**
UN 2818 [Ammonium polysulfide, Solution] **(9080-17-5)**
UN 2819 [Amyl acid phosphate] **(12789-46-7)**
UN 2820 [Butyric acid] **(107-92-6)**
UN 2821 [Phenol solutions] **(108-95-2)**
UN 2822 [2-Chloropyridine] **(109-09-1)**
UN 2823 [Crotonic acid, Solid] **(3724-65-0)**
UN 2825 (DOT) **(96-80-0)**
UN 2826 [Ethyl chlorothioformate] **(2812-73-9)**
UN 2826 [Ethyl chlorothioformate] **(2941-64-2)**
UN 2830 [Lithium ferrosilicon] **(64082-35-5)**
UN 2831 [1,1,1-Trichloroethane] **(71-55-6)**
UN 2834 [Phosphorous acid, ortho] **(13598-36-2)**
UN 2835 [Sodium aluminum hydride] **(13770-96-2)**
UN 2837 [Sodium hydrogen sulfate, Solution] **(7681-38-1)**
UN 2838 [Vinyl butyrate, Inhibited] **(123-20-6)**
UN 2839 [Aldol] **(107-89-1)**
UN 2840 [Butyraldoxime] **(110-69-0)**
UN 2841 [Di-n-amylamine] **(2050-92-2)**
UN 2842 [Nitroethane] **(79-24-3)**
UN 2849 [3-Chloropropanol-1] **(627-30-5)**
UN 2850 [Propylene tetramer] **(6842-15-5)**
UN 2854 [Ammonium fluorosilicate] **(16919-19-0)**
UN 2855 [Zinc fluorosilicate] **(16871-71-9)**
UN 2858 [Zirconium, Dry, coiled wire, finished metal sheets, strip (thinner than 254 microns but not thinner than 18 microns)] **(7440-67-7)**
UN 2859 [Ammonium metavanadate] **(7803-55-6)**
UN 2860 [Vanadium trioxide, nonfused form] **(1314-34-7)**
UN 2862 [Vanadium pentoxide, nonfused form] **(1314-62-1)**
UN 2865 [Hydroxylamine sulfate] **(10039-54-0)**
UN 2871 [Antimony powder] **(7440-36-0)**
UN 2872 [Dibromochloropropane] **(96-12-8)**
UN 2873 [Dibutylaminoethanol] **(102-81-8)**
UN 2874 [Furfuryl alcohol] **(98-00-0)**
UN 2875 [Hexachlorophene] **(70-30-4)**
UN 2876 [Resorcinol] **(108-46-3)**
UN 2877 (DOT) **(62-56-6)**
UN 2878 [Titanium sponge granules or Titanium sponge powders] **(7440-32-6)**
UN 2879 [Selenium oxychloride] **(7791-23-3)**
UN 2885 **(3006-86-8)**
UN 2886 **(62695-55-0)**
UN 2887 **(62695-55-0)**
UN 2888 **(62695-55-0)**
UN 2889 **(82065-80-3)**
UN 2890 (DOT) **(614-45-9)**
UN 2891 **(68299-16-1)**
UN 2892 **(53220-22-7)**
UN 2893 (DOT) **(105-74-8)**
UN 2894 **(15520-11-3)**
UN 2895 **(26322-14-5)**
UN 2896 **(4904-55-6)**
UN 2897 **(3006-86-8)**
UN 2898 (DOT) **(686-31-7)**
UN 2901 [Bromide chloride] **(13863-41-7)**
UN 2931 [Vanadyl sulfate] **(27774-13-6)**
UN 2933 [Methyl-2-chloropropionate] **(17639-93-9)**
UN 2935 [Ethyl-2-chloropropionate] **(535-13-7)**
UN 2937 [α-Methylbenzyl alcohol] **(98-85-1)**
UN 2938 [Methyl benzoate] **(93-58-3)**
UN 2941 [Fluoroanilines] **(348-54-9)**
UN 2942 [2-Trifluoromethylaniline] **(88-17-5)**
UN 2944 (DOT) **(371-40-4)**
UN 2945 [N-Methylbutylamine] **(110-68-9)**
UN 2947 [Isopropyl chloroacetate] **(105-48-6)**
UN 2948 [3-Trifluoromethylaniline] **(98-16-8)**
UN 2949 [Sodium hydrosulfide with not less than 25 per cent water of crystallization] **(16721-80-5)**
UN 2950 [Magnesium granules, Coated particle size not less than 149 microns] **(7439-95-4)**
UN 2952 [Azodiisobutyronitrile] **(78-67-1)**
UN 2953 [2,2'-Azodi-(2,4 dimethylvaleronitrile)] **(4419-11-8)**
UN 2954 [1,1'-Azodi-(hexahydrobenzonitrile)] **(2094-98-6)**
UN 2959 (DOT) **(2618-77-1)**
UN 2960 (DOT) **(16111-62-9)**
UN 2962 (DOT) **(123-23-9)**
UN 2965 [Boron trifluoride dimethyl etherate] **(353-42-4)**
UN 2966 [Thioglycol] **(60-24-2)**
UN 2967 [Sulfamic acid] **(5329-14-6)**
UN 2968 [Maneb stabilized or Maneb preparations, stabilized against self-heating] **(12427-38-2)**
UN 2975 [Thorium metal, pyrophoric] **(7440-29-1)**
UN 2976 [Thorium nitrate, Solid] **(13823-29-5)**
UN 2977 [Uranium hexafluoride, fissile (containing more than 1% U-235)] **(7783-81-5)**
UN 2978 [Uranium hexafluoride, fissile excepted or non-fissile] **(7783-81-5)**
UN 2979 [Uranium metal, pyrophoric] **(7440-61-1)**
UN 2980 [Uranyl nitrate hexahydrate solution] **(13520-83-7)**
UN 2981 [Uranyl nitrate, Solid] **(10102-06-4)**
UN 3064 [Nitroglycerin, Solution in alcohol, with more than 1% but not more than 5% nitroglycerin] **(55-63-0)**
UNADS **(97-74-5)**
UNT 51239 **(16268-62-5)**
UOP 26 **(4175-38-6)**
UOP 88 **(139-60-6)**
UOP 288 **(103-96-8)**
UOP 688 **(15233-47-3)**
UOP Copper Deactivator **(94-91-7)**
UP 1 **(9003-53-6)**
UP 1E **(9003-55-8)**
UP 2 **(9003-53-6)**
UP 27 **(9003-53-6)**
UP 925 **(9011-06-7)**
UPM **(9003-53-6)**
UPM703 **(9003-53-6)**
UPM508L **(9003-53-6)**
200U/P-RVM **(12174-11-7)**
URAB **(4482-55-7)**
U.S.-67 **(127-69-5)**
USACERT Blue No. 2 **(860-22-0)**
USACERT Red No. 2 **(915-67-3)**
USAF S-1 **(635-39-2)**
USAF A-233 **(1137-41-3)**
USAF A-4600 **(109-77-3)**
USAF A-6598 **(97-39-2)**
USAF A-8354 **(1072-71-5)**
USAF A-8564 **(111-94-4)**
USAF A-8565 **(107-16-4)**
USAF A-8798 **(542-76-7)**
USAF A-9442 **(110-61-2)**
USAF AM-1 **(100-36-7)**
USAF AM-3 **(96-29-7)**

USAF AM-5 (107-29-9)
USAF AM-6 (110-69-0)
USAF AM-8 (151-00-8)
USAF AN-11 (5902-95-4)
USAF AN-7 (150-76-5)
USAF AN-8 (75-57-0)
USAF AN-9 (150-78-7)
USAF B-12 (454-29-5)
USAF B-15 (96-69-5)
USAF B-17 (103-34-4)
USAF B-19 (102-06-7)
USAF B-21 (79-74-3)
USAF B-22 (97-18-7)
USAF B-24 (91-53-2)
USAF B-30 (137-26-8)
USAF B-31 (120-54-7)
USAF B-32 (97-74-5)
USAF B-33 (120-78-5)
USAF B-33 (97-77-8)
USAF B-35 (136-30-1)
USAF B-40 (96-27-5)
USAF B-44 (541-53-7)
USAF B-45 (767-17-9)
USAF B-51 (133-49-3)
USAF B-55 (142-46-1)
USAF CB-2 (54-85-3)
USAF CB-7 (1405-87-4)
USAF CB-10 (58-61-7)
USAF CB-11 (118-00-3)
USAF CB-13 (59-30-3)
USAF CB-17 (69-89-6)
USAF CB-18 (73-24-5)
USAF CB-19 (1405-10-3)
USAF CB-20 (67-03-8)
USAF CB-21 (62-55-5)
USAF CB-22 (91-59-8)
USAF CB-24 (7728-98-5)
USAF CB-26 (1468-26-4)
USAF CB-27 (50-55-5)
USAF CB-35 (68-11-1)
USAF CB-36 (123-93-3)
USAF CB-37 (96-27-5)
USAF CB-96 (56-69-9)
USAF CF-2 (522-12-3)
USAF CF-5 (153-18-4)
USAF CS-1 (7005-47-2)
USAF CY-2 (156-62-7)
USAF CY-4 (91-60-1)
USAF CY-5 (120-78-5)
USAF CY-6 (88-24-4)
USAF CY-7 (102-77-2)
USAF CY-9 (131-57-7)
USAF CY-10 (130-15-4)
USAF CY-14 (767-17-9)
USAF CZ-1 (123-76-2)
USAF D-3 (59-85-8)
USAF D-9 (122-42-9)
USAF DO-1 (99-91-2)
USAF DO-4 (90-98-2)
USAF DO-10 (623-33-6)
USAF DO-12 (119-84-6)
USAF DO-17 (262-20-4)
USAF DO-19 (103-00-4)
USAF DO-20 (98-52-2)
USAF DO-21 (1632-16-2)
USAF DO-22 (103-76-4)
USAF DO-28 (131-56-6)
USAF DO-29 (79-07-2)
USAF DO-30 (92-88-6)
USAF DO-32 (71-91-0)
USAF DO-36 (140-79-4)
USAF DO-37 (93-62-9)
USAF DO-40 (103-88-8)
USAF DO-41 (126-72-7)
USAF DO-42 (96-13-9)
USAF DO-44 (96-29-7)
USAF DO-43 (147-82-0)
USAF DO-44 (126-84-1)
USAF DO-45 (105-57-7)
USAF DO-46 (140-31-8)
USAF DO-49 (100-49-2)
USAF DO-50 (109-83-1)
USAF DO-52 (105-59-9)
USAF DO-54 (134-84-9)
USAF DO-55 (142-73-4)
USAF DO-60 (946-31-6)
USAF DO-61 (2050-47-7)
USAF DO-62 (87-87-6)
USAF DO-68 (320-72-9)
USAF E-1 (882-33-7)
USAF E-2 (123-93-3)
USAF EA-1 (67-45-8)
USAF EA-2 (67-20-9)
USAF EA-4 (59-87-0)
USAF EK (1421-63-2)
USAF EK-3 (103-84-4)
USAF EK-206 (541-69-5)
USAF EK-218 (91-22-5)
USAF EK-245 (102-08-9)
USAF EK-338 (60-11-7)
USAF EK-356 (123-31-9)
USAF EK-394 (106-50-3)
USAF EK-442 (142-04-1)
USAF EK-488 (75-05-8)
USAF EK-496 (98-86-2)
USAF EK-497 (62-56-6)
USAF EK-510 (106-45-6)
USAF EK-534 (102-07-8)
USAF EK-572 (88-67-5)
USAF EK-600 (86-74-8)
USAF EK-631 (99-92-3)
USAF EK-660 (504-17-6)
USAF EK-678 (615-28-1)
USAF EK-695 (141-82-2)
USAF EK-704 (103-33-3)
USAF EK-743 (587-84-8)
USAF EK-749 (50-01-1)
USAF EK-794 (148-24-3)
USAF EK-906 (96-29-7)
USAF EK-1047 (85-98-3)
USAF EK-1235 (93-55-0)
USAF EK-1239 (102-01-2)
USAF EK-1270 (102-06-7)
USAF EK-1275 (79-19-6)
USAF EK-1375 (60-09-3)
USAF EK-1509 (100-53-8)
USAF EK-1569 (103-85-5)
USAF EK-1597 (141-43-5)
USAF EK-1719 (62-55-5)
USAF EK-1803 (105-55-5)
USAF EK-1853 (120-75-2)
USAF EK-1860 (110-02-1)
USAF EK-1995 (420-04-2)
USAF EK-2070 (623-51-8)

USAF EK-2089 **(137-26-8)**
USAF EK-2122 **(1639-09-4)**
USAF EK-2124 **(538-28-3)**
USAF EK-2138 **(109-46-6)**
USAF EK-2219 **(90-43-7)**
USAF EK-2596 **(148-18-5)**
USAF EK-2676 **(137-06-4)**
USAF EK-2680 **(108-40-7)**
USAF EK-3302 **(70-70-2)**
USAF EK-3941 **(136-95-8)**
USAF EK-3967 **(615-21-4)**
USAF EK-4037 **(934-32-7)**
USAF EK-4196 **(60-24-2)**
USAF EK-4376 **(137-07-5)**
USAF EK-4394 **(79-40-3)**
USAF EK-4628 **(111-31-9)**
USAF EK-4733 **(288-32-4)**
USAF EK-4812 **(95-16-9)**
USAF EK-4890 **(591-08-2)**
USAF EK-5017 **(273-53-0)**
USAF EK-5185 **(58-27-5)**
USAF EK-5199 **(367-51-1)**
USAF EK-5426 **(5351-69-9)**
USAF EK-5429 **(59-49-4)**
USAF EK-5432 **(120-78-5)**
USAF EK-6454 **(56-04-2)**
USAF EK-6540 **(583-39-1)**
USAF EK-6561 **(121-66-4)**
USAF EK-6754 **(835-64-3)**
USAF EK-7094 **(91-19-0)**
USAF EK-7119 **(2365-48-2)**
USAF EK-7372 **(2231-57-4)**
USAF EK-P-433 **(1762-95-4)**
USAF EK-P-583 **(110-17-8)**
USAF EK-P-737 **(507-09-5)**
USAF EK-P-5501 **(96-50-4)**
USAF EK-P-5976 **(86-88-4)**
USAF EK-P-6255 **(97-74-5)**
USAF EK-P-6281 **(541-53-7)**
USAF EK-T-434 **(540-72-7)**
USAF EK-T-2805 **(147-93-3)**
USAF EK-T-6645 **(103-34-4)**
USAF EL-30 **(60-56-0)**
USAF EL-42 **(126-52-3)**
USAF EL-45 **(5351-69-9)**
USAF EL-62 **(96-45-7)**
USAF EL-78 **(1875-92-9)**
USAF EL-82 **(3287-99-8)**
USAF EL-101 **(592-35-8)**
USAF FA-5 **(441-38-3)**
USAF FO-1 **(592-35-8)**
USAF GE-12 **(58-32-2)**
USAF GE-13 **(57-96-5)**
USAF GE-14 **(129-20-4)**
USAF GE-15 **(50-33-9)**
USAF GY-2 **(74-31-7)**
USAF GY-3 **(149-30-4)**
USAF GY-5 **(136-23-2)**
USAF GY-7 **(155-04-4)**
USAF HA-2 **(141-84-4)**
USAF HA-5 **(111-97-7)**
USAF HC-1 **(111-20-6)**
USAF K-2801 **(2040-04-2)**
USAF KE-5 **(142-91-6)**
USAF KE-7 **(31566-31-1)**
USAF KE-8 **(106-11-6)**
USAF KE-9 **(9004-99-3)**
USAF KE-11 **(111-60-4)**
USAF KE-12 **(9004-99-3)**
USAF KE-13 **(1323-39-3)**
USAF KE-14 **(9004-99-3)**
USAF KE-20 **(118-93-4)**
USAF KF-2 **(147-93-3)**
USAF KF-5 **(107-14-2)**
USAF KF-11 **(2698-41-1)**
USAF KF-13 **(86-29-3)**
USAF KF-14 **(107-91-5)**
USAF KF-15 **(99-93-4)**
USAF KF-17 **(372-09-8)**
USAF KF-18 **(140-87-4)**
USAF KF-19 **(109-77-3)**
USAF KF-21 **(140-29-4)**
USAF KF-22 **(105-34-0)**
USAF KF-25 **(105-56-6)**
USAF M-4 **(103-72-0)**
USAF M-5 **(84-86-6)**
USAF M-6 **(123-08-0)**
USAF M-7 **(89-98-5)**
USAF MA-1 **(121-89-1)**
USAF MA-2 **(90-16-4)**
USAF MA-3 **(98-64-6)**
USAF MA-4 **(98-16-8)**
USAF MA-5 **(98-46-4)**
USAF MA-6 **(88-30-2)**
USAF MA-8 **(99-47-8)**
USAF MA-14 **(351-36-0)**
USAF MA-16 **(98-08-8)**
USAF MA-17 **(3120-74-9)**
USAF ME-1 **(59-52-9)**
USAF MK-6 **(79-40-3)**
USAF MO-2 **(5421-46-5)**
USAF NB-1 **(2457-76-3)**
USAF ND-09 **(91-15-6)**
USAF ND-54 **(131-56-6)**
USAF ND-59 **(117-10-2)**
USAF P-2 **(137-30-4)**
USAF P-5 **(137-26-8)**
USAF P-7 **(330-54-1)**
USAF P-8 **(150-68-5)**
USAF P-220 **(106-51-4)**
USAF PD-25 **(2349-67-9)**
USAF PD-57 **(1779-81-3)**
USAF PD-101 **(4946-22-9)**
USAF Q-1 **(617-89-0)**
USAF RH-1 **(79-39-0)**
USAF RH-3 **(2867-47-2)**
USAF RH-4 **(80-52-4)**
USAF RH-5 **(91-76-9)**
USAF RH-6 **(27193-28-8)**
USAF RH-7 **(109-78-4)**
USAF RH-8 **(75-86-5)**
USAF SN-9 **(72-14-0)**
USAF SC-2 **(126-07-8)**
USAF SO-1 **(84-51-5)**
USAF SO-2 **(15547-17-8)**
USAF ST-40 **(126-98-7)**
USAF TH-3 **(638-16-4)**
USAF UCTL-1791 **(150-78-7)**
USAF UCTL-1856 **(70-69-9)**
USAF VI-6 **(834-28-6)**
USAF XF-21 **(583-39-1)**
USAF XR-10 **(140-64-70)**
USAF XR-19 **(108-98-5)**
USAF XR-20 **(89-05-4)**

USAF XR-22 (61-82-5)
USAF XR-27 (136-95-8)
USAF XR-29 (149-30-4)
USAF XR-35 (147-93-3)
USAF XR-41 (150-68-5)
USAF XR-42 (330-54-1)
USAcert Red No. 1 (3564-09-8)
USB (9003-27-4)
USB-3584 (29091-05-2)
USI 11-4-0047 (9003-07-0)
USP 711 (2618-77-1)
U.S.P. Menthol (2216-51-5)
USP Methylcellulose (9004-67-5)
USP Sodium Chloride (7647-14-5)
USP XIII Stearyl alcohol (112-92-5)
USR 604 (117-80-6)
U.S. Rubber 604 (117-80-6)
U.S. Rubber D-014 (2312-35-8)
UST (9011-05-6)
UV 24 (131-53-3)
UV Absorber-1 (2440-22-4)
UV Absorber-3 (6197-30-4)
UV Absorber-5 (3147-75-9)
UV Absorber-6 (3896-11-5)
UV Absorber NL/5 (85-19-8)
UV Chek AM 104 (13927-77-0)
UVasorb DMO (21245-02-3)
Uantox ASCA (50-81-7)
Uantox sebate (6381-77-7)
Ubatol U 2001 (9003-53-6)
UcB 2073 (469-82-9)
Ucarsil DJ (63148-62-9)
Ucet Textile Finish 11-74 (Obs.) (106-87-6)
Ucon 12 (75-71-8)
Ucon 22 (75-45-6)
Ucon 113 (76-13-1)
Ucon 114 (76-14-2)
Ucon Fluid 50-HB-260 (9038-95-3)
Ucon Fluorocarbon 113 (76-13-1)
Ucon Flurocarbon 11 (75-69-4)
Ucon 25H (9003-11-6)
Ucon 75H (9003-11-6)
Ucon 75H1400 (9003-11-6)
Ucon 75H90000 (9003-11-6)
Ucon 50-HB-55 (9038-95-3)
Ucon 50-HB-100 (9038-95-3)
Ucon 50-HB-260 (9038-95-3)
Ucon 50-HB-280-X (9038-95-3)
Ucon 50-HB-400 (9038-95-3)
Ucon 50-HB-660 (9038-95-3)
Ucon 50-HB-2000 (9038-95-3)
Ucon 50-HB-3520 (9038-95-3)
Ucon 50-HB-5100 (9038-95-3)
Ucon 113/Halocarbon 113 (76-13-1)
Ucon LB 165 (9038-95-3)
Ucon LB-250 (9003-13-8)
Ucon LB-285 (Fluid) (9038-95-3)
Ucon LB 1145 (9003-13-8)
Ucon LB-1715 (37286-64-9)
Ucon LB 1800X (9003-13-8)
Ucon Refrigerant 11 (75-69-4)
Ucon 12/halocarbon 12 (75-71-8)
Ucon 22/halocarbon 22 (75-45-6)
Udicil (147-94-4)
Udolac (80-08-0)
Uformite 700 (9011-05-6)
Uformite F 240N (9011-05-6)

Uintahite (12002-43-6)
Uintaite (12002-43-6)
Ukarb (1333-86-4)
Ukopen (7177-48-2)
Ulacort (50-24-8)
Ulcedine (51481-61-9)
Ulcerban (54182-58-0)
Ulcerfen (50-35-1)
Ulcerlmin (54182-58-0)
Ulcimet (51481-61-9)
Ulcine (53-46-3)
Ulcogant (54182-58-0)
Ulcolind (52-86-8)
Ulcomet (51481-61-9)
Ulcudexter (53-46-3)
Ulexine (485-35-8)
Uliolind (52-86-8)
Uloid 100 (9011-05-6)
Uloid 22 (9011-05-6)
Uloid 301 (9011-05-6)
Ulstron (9003-07-0)
Ultandren (76-43-7)
Ultandrene (76-43-7)
Ultima White (8009-03-8)
Ultra Blue B (1325-87-7)
Ultra Brilliant Blue P (138-89-6)
Ultra Sulfate SL-1 (151-21-3)
Ultrabion (69-53-4)
Ultrabron (69-53-4)
Ultracid 40 (950-37-8)
Ultracide (950-37-8)
Ultracorten (53-03-2)
Ultracortene (53-03-2)
Ultracortene-H (50-24-8)
Ultradine (25655-41-8)
Ultrafur (139-91-3)
Ultralon PA 6 (63428-84-2)
Ultramarine Green (1308-38-9)
Ultramarine Yellow (10294-40-3)
Ultramid B 3 (105-60-2)
Ultramid B 3 (25038-54-4)
Ultramid B 4 (105-60-2)
Ultramid B 4 (25038-54-4)
Ultramid B 5 (105-60-2)
Ultramid B 5 (25038-54-4)
Ultramid BMK (105-60-2)
Ultramid BMK (25038-54-4)
Ultranox 624 (26741-53-7)
Ultranox 626 (26741-53-7)
Ultrapole H (142-78-9)
Ultrasul (144-82-1)
Ultrasulfate SE 5 (9004-82-4)
Ultrawet DS (1322-98-1)
Ultrawet 60K (25155-30-0)
Ultrawet K (25155-30-0)
Ultrawet KX (25155-30-0)
Ultrawet SK (25155-30-0)
Ultrawet 40SX (1300-72-7
Ultrazol D (135-61-5
Ultrazol G (91-96-3
Ultrazol I-AS (92-77-3
Ultrazol IV-BS (135-65-9
Ultrazol VII-BO (132-68-3
Ulvair (6923-22-4
Umalur (9011-05-6
Umbelliferon (93-35-6
Umbelliferone (93-35-6

Umbethion	**(56-72-4)**
Umbrathor	**(1314-20-1)**
Umbrium	**(439-14-5)**
Uni-Guar	**(9000-30-0)**
Unakalm	**(27223-35-4)**
Unal	**(123-30-8)**
Unamide J-56	**(120-40-1)**
Unamycin-B	**(606-58-6)**
1,2,3,4,5,5,6,7,9,10,10-Undecachloropentacyclo(5.3.0.0^{2,6}.0^{3,9}.0^{4,8})decane	**(39801-14-4)**
1,4-Undecadiene, (E)- (9CI)	**(55976-13-1)**
3,4-Undecadiene-2,10-dione, 6,6-dimethyl- (9CI)	**(52588-78-0)**
5,9-Undecadien-2-one, 6,10-dimethyl-, (E)- (9CI)	**(3796-70-1)**
1,1,2,2,3,3,4,4,5,5,5-Undecafluoro-1-pentanesulfonyl fluoride	**(375-81-5)**
γ-Undecalactone	**(104-67-6)**
1-Undecanal	**(112-44-7)**
Undecanal	**(112-44-7)**
n-Undecanal	**(112-44-7)**
Undecanaldehyde	**(112-44-7)**
Undecanal, 2-methyl- (9CI)	**(110-41-8)**
n-Undecane	**(1120-21-4)**
Undecane [UN 2330]	**(1120-21-4)**
Undecane, 1-bromo- (9CI)	**(693-67-4)**
1-Undecanecarboxylic acid	**(143-07-7)**
Undecane, 1-chloro- (9CI)	**(2473-03-2)**
Undecane, chloro- (6CI,8CI,9CI)	**(34214-82-9)**
Undecane, 1-cyclohexyl-	**(54105-66-7)**
1,1-Undecanedicarboxylic acid	**(4372-29-6)**
1,11-Undecanedicarboxylic acid	**(505-52-2)**
Undecanedioic acid (9CI)	**(1852-04-6)**
Undecane, 1-fluoro	**(506-05-8)**
Undecane, 1-iodo-	**(4282-44-4)**
Undecane, 2-methyl- (8CI,9CI)	**(7045-71-8)**
Undecane, 3-methyl- (8CI,9CI)	**(1002-43-3)**
Undecanenitrile (9CI)	**(2244-07-7)**
Undecane, 1-phenyl- (8CI)	**(6742-54-7)**
Undecane, 2-phenyl- (8CI)	**(4536-88-3)**
Undecane, 3-phenyl- (8CI)	**(4536-87-2)**
Undecane, 4-phenyl- (8CI)	**(4536-86-1)**
Undecane, 5-phenyl- (8CI)	**(4537-15-9)**
Undecane, 6-phenyl- (8CI)	**(4537-14-8)**
1-Undecanethiol (8CI,9CI)	**(5332-52-5)**
Undecane, 2,6,10-trimethyl- (8CI,9CI)	**(6864-53-5)**
2,5,10-Undecanetrione, 6,6-dimethyl- (9CI)	**(50464-96-5)**
Undecanoic-acid	**(112-37-8)**
n-Undecanoic acid	**(112-37-8)**
Undecanoic acid, 11-amino	**(2432-99-7)**
Undecanoic acid, butyl ester (8CI,9CI)	**(10580-24-2)**
Undecanoic acid, ethyl ester	**(627-90-7)**
Undecanoic acid, 4-hydroxy- (6CI,8CI,9CI)	**(7779-60-4)**
Undecanoic acid, 4-hydroxy-, γ-lactone	**(104-67-6)**
Undecanoic acid, 10-(p-iodophenyl)-, ethyl ester	**(99-79-6)**
Undecanoic acid, methyl ester (9CI)	**(1731-86-8)**
Undecanoic acid, 2,4,6-trimethyl- (8CI,9CI)	**(28484-22-2)**
2-Undecanol	**(1653-30-1)**
Undecanol	**(112-42-5)**
n-Undecanol	**(112-42-5)**
Undecanol (9CI)	**(30207-98-8)**
6-Undecanol (8CI,9CI)	**(23708-56-7)**
1-Undecanol, acetate (9CI)	**(1731-81-3)**
γ-Undecanolactone	**(104-67-6)**
4-Undecanol, 7-ethyl-2-methyl-, hydrogen sulfate, sodium salt	**(139-88-8)**
1,4-Undecanolide	**(104-67-6)**
4-Undecanolide	**(104-67-6)**
γ-Undecanolide	**(104-67-6)**
2-Undecanone	**(112-12-9)**
6-Undecanone	**(927-49-1)**
Undecan-2-one	**(112-12-9)**
Undecan-6-one	**(927-49-1)**
2-Undecanone, 6,10-dimethyl- (9CI)	**(1604-34-8)**
5-Undecanone, 2-methyl- (9CI)	**(50639-02-6)**
3,5,9-Undecatrien-2-one, 6,10-dimethyl-	**(141-10-6)**
4,5,9-Undecatrien-2-one, 6,10-dimethyl- (8CI,9CI)	**(16647-05-5)**
1-Undecen-10-al	**(112-45-8)**
10-Undecen-1-al	**(112-45-8)**
10-Undecenal (8CI,9CI)	**(112-45-8)**
α-Undecene	**(821-95-4)**
n-1-Undecene	**(821-95-4)**
1-Undecene (9CI)	**(821-95-4)**
Undecene (9CI)	**(28761-27-5)**
10-Undecenoic acid	**(112-38-9)**
10-Undecenoic acid, calcium salt (9CI)	**(1322-14-1)**
10-Undecenoic acid, ethyl ester	**(692-86-4)**
Undecenoic acid, methyl ester	**(111-81-9)**
10-Undecenoic acid, methyl ester (9CI)	**(111-81-9)**
10-Undecenoic acid, zinc (2+) salt	**(557-08-4)**
10-Undecenoic acid, zinc salt (9CI)	**(557-08-4)**
5-Undecen-4-one (6CI,9CI)	**(56312-55-1)**
5-Undecen-3-yne, (E)- (9CI)	**(74744-31-3)**
n-Undecoic acid	**(112-37-8)**
Undecyl-alcohol	**(112-42-5)**
n-Undecyl alcohol	**(112-42-5)**
Undecyl alcohol, acetate (8CI)	**(1731-81-3)**
Undecyl aldehyde	**(112-44-7)**
n-Undecyl aldehyde	**(112-44-7)**
Undecylbenzene	**(6742-54-7)**
n-Undecylbenzene	**(6742-54-7)**
Undecylbenzenesulfonic acid	**(50854-94-9)**
p-Undecylbenzenesulfonic acid, sodium salt	**(2211-98-5)**
Undecylenaldehyde	**(112-45-8)**
α-Undecylene	**(821-95-4)**
10-Undecylenealdehyde	**(112-45-8)**
10-Undecylenic acid	**(112-38-9)**
9-Undecylenic acid	**(112-38-9)**
Undecyl-10-enic acid	**(112-38-9)**
Undecylenic acid	**(112-38-9)**
Undecylenic acid, methyl ester	**(111-81-9)**
Undecylenic acid, zinc salt	**(557-08-4)**
Undecylenic aldehyde	**(112-45-8)**
Undecylic acid	**(112-37-8)**
n-Undecylic acid	**(112-37-8)**
Undecylic aldehyde	**(112-44-7)**
2-Undecyl-2-imidazoline-1-ethanol	**(136-99-2)**
γ-Undekalakton (Czech)	**(104-67-6)**
Undekan (Fiber)	**(63428-84-2)**
Unden	**(114-26-1)**
Unden	**(53-16-7)**
Undeveldoped Lithol Toner	**(1248-18-6)**
Unem 5005	**(27554-26-3)**
Unette-O	**(67762-27-0)**
Unette-W	**(67762-27-0)**
Unexan-Koeder	**(57-74-9)**
Unfinished Lubricating Oil	**(68308-34-9)**
Unibaryt	**(7727-43-7)**
Unicel-ND	**(101-25-7)**
Unicel NDX	**(101-25-7)**
Unicelles	**(125-40-6)**
Unichem AMAL	**(7784-25-0)**
Unichem ZO	**(1314-13-2)**
Unichlor	**(63449-39-8)**
Unichlor 50	**(63449-39-8)**
Unicin	**(64-75-5)**
Unicocyde	**(54-85-3)**
Unicol 123	**(67762-27-0)**

Unicol CA-2 **(9004-95-9)**
Unicol CA-4 **(9004-95-9)**
Unicol CA-10 **(9004-95-9)**
Unicol CA-20 **(9004-95-9)**
Unicol CPS **(67762-27-0)**
Unicol LA-4 **(5274-68-0)**
Unicol NP-4 **(7311-27-5)**
Unicol OA-2 **(9004-98-2)**
Unicol OA-4 **(9004-98-2)**
Unicol OA-10 **(9004-98-2)**
Unicol OA-20 **(9004-98-2)**
Unicol SA-2 **(9005-00-9)**
Unicol SA-10 **(9005-00-9)**
Unicol SA-13 **(9005-00-9)**
Unicol SA-15 **(9005-00-9)**
Unicol SA-20 **(9005-00-9)**
Unicol SA-40 **(9005-00-9)**
Unicozyde **(54-85-3)**
Unicrop CIPC **(101-21-3)**
Unicrop DNBP **(88-85-7)**
Unicrop Maneb **(12427-38-2)**
Uniderm A **(97-59-6)**
Uniderm Homsal **(118-56-9)**
Uniderm SSME **(8008-74-0)**
Unidigin **(71-63-6)**
Unidone **(117-37-3)**
Unidron **(330-54-1)**
Uniflex BYO **(142-77-8)**
Uniflex BYS **(123-95-5)**
Uniflex DOA **(103-23-1)**
Uniflex DOS **(122-62-3)**
Unifoam AZ **(123-77-3)**
Uniform AZ **(123-77-3)**
Unifos **(62-73-7)**
Unifos Dyob S **(9002-88-4)**
Unifos 50 EC **(62-73-7)**
Unifos EFD 0118 **(9002-88-4)**
Unifos (Pesticide) **(62-73-7)**
Unifume **(106-93-4)**
Unifur **(139-91-3)**
Unihydag WAX-O **(67762-27-0)**
Unihydag WAX-SX **(67762-27-0)**
Unihydol LS-2 **(3055-93-4)**
Unihydol LS-4 **(5274-68-0)**
Unibetaine BA-35 **(61789-40-0)**
Unibetaine BC-35 **(61789-40-0)**
Unibetaine K **(61789-40-0)**
Unibetaine LB **(683-10-3)**
Unibetaine OLB-30 **(871-37-4)**
Unibetaine OLB-50 **(871-37-4)**
Unilax **(18869-73-3)**
Unilex **(8002-43-5)**
Unilex DS **(8002-43-5)**
Unilex S **(8002-43-5)**
Unilex SH **(8002-43-5)**
Unilord **(50-55-5)**
Unilube 70DP950B **(9003-11-6)**
Unilube 50MB168X **(9003-11-6)**
Unilube 50MB26X **(9003-11-6)**
Unimate GMS **(31566-31-1)**
Unimate IPM **(110-27-0)**
Unimate IPP **(142-91-6)**
Unimoll 66 **(84-61-7)**
Unimoll BB **(85-68-7)**
Unimox OL **(14351-50-9)**
Unimul-05 **(9004-98-2)**
Unimul-10 **(9004-98-2)**
Unimul-CTV **(3687-46-5)**
Unimul-1002 Conc. **(67762-27-0)**
Unimul-G-16 **(36311-34-9)**
Unimulgade-F **(67762-27-0)**
Unimulgade-F Special **(67762-27-0)**
Unimulgade-1000NI **(67762-27-0)**
Unimycetin **(56-75-7)**
Unimycin **(64-75-5)**
Union Black EM **(1937-37-7)**
08-Union Carbide **(126-92-1)**
Union Carbide 10854 **(64-00-6)**
Union Carbide 19786 **(973-21-7)**
Union Carbide 21149 **(116-06-3)**
Union Carbide 7,744 **(63-25-2)**
Union Carbide A-15 **(78-07-9)**
Union Carbide A-150 **(75-94-5)**
Union Carbide A-151 **(78-08-0)**
Union Carbide A-162 **(2031-67-6)**
Union Carbide A-163 **(1185-55-3)**
Union Carbide A-174 **(2530-85-0)**
Union Carbide A-186 **(3388-04-3)**
Union Carbide A-187 **(2530-83-8)**
Union Carbide A-189 **(4420-74-0)**
Union Carbide Liquid G **(63148-62-9)**
Union Carbide UC-8305 **(2921-31-5)**
Union Carbide UC-9880 **(2631-37-0)**
Union Carbide UC 20047 **(15271-41-7)**
Union Carbide UC-10,854 **(64-00-6)**
Union Carbide UC-21149 **(116-06-3)**
Union Carbide UC-25074 **(17702-57-7)**
Union Dark Green B **(3626-28-6)**
Union Fast Navy Blue DS **(2429-73-4)**
Union Fast Scarlet 3B **(6358-29-8)**
Unipeg-S-40 **(9004-99-3)**
Unipeg-DGL **(141-20-8)**
Unipeg-400 DO **(9005-07-6)**
Unipeg-600 DO **(9005-07-6)**
Unipeg-400 DS **(9005-08-7)**
Unipeg-4000 DS **(9005-08-7)**
Unipeg-600 DS **(9005-08-7)**
Unipeg-6000 DS **(9005-08-7)**
Unipeg-EGDS **(627-83-8)**
Unipeg-ETG-12 **(31694-55-0)**
Unipeg-ETG-26 **(31694-55-0)**
Unipeg-400 MOT **(61791-00-2)**
Unipeg-1000 MS **(9004-99-3)**
Unipeg-1540 MS **(9004-99-3)**
Unipeg-400 MS **(9004-99-3)**
Unipeg-4000 MS **(9004-99-3)**
Unipeg-600 MS **(9004-99-3)**
Unipeg-6000 MS **(9004-99-3)**
Uniphat A20 **(111-11-5)**
Uniphat A30 **(110-42-9)**
Uniphat A40 **(111-82-0)**
Uniphat A50 **(124-10-7)**
Uniphat A60 **(112-39-0)**
Unipine **(8002-09-3)**
Unipol Conc. 7021 **(143-00-0)**
Unipol DEA **(143-00-0)**
Unipol 125-E **(9004-82-4)**
Unipol 130-E **(9004-82-4)**
Unipol ES-40 **(25446-80-4)**
Unipol EA-40 **(27731-61-9)**
Unipol SCO **(8002-33-3)**
Unipon **(127-20-8)**
Unipon **(75-99-0)**
Unipro CAL-CASE **(9000-71-9)**

Unipro CO-CASE **(9000-71-9)**
Uniprofen **(51234-28-7)**
Uniroyal **(117-80-6)**
Uniroyal D014 **(2312-35-8)**
Unisedil **(439-14-5)**
Unisept B **(94-26-8)**
Unisept BZ **(94-18-8)**
Unisil SF-V **(69430-24-6)**
Unisol S-22 **(15087-24-8)**
Unisol 4-O **(9004-96-0)**
Unisol RH **(9004-32-4)**
Unisom **(562-10-7)**
Unisomnia **(146-22-5)**
Unisperse Blue G-E **(147-14-8)**
Unistradiol **(50-50-0)**
Unisulf **(127-69-5)**
Unisweet 70 **(50-70-4)**
Unisweet CONC **(50-70-4)**
Unitane **(13463-67-7)**
Unitane O-110 **(13463-67-7)**
Unitane O-220 **(13463-67-7)**
Unitane OR-150 **(13463-67-7)**
Unitane OR-340 **(13463-67-7)**
Unitane OR-342 **(13463-67-7)**
Unitane OR-350 **(13463-67-7)**
Unitane OR-540 **(13463-67-7)**
Unitane OR-640 **(13463-67-7)**
United **(1333-86-4)**
United Chemical Defoliant No. 1 **(7775-09-9)**
Unitene **(138-86-3)**
Unitensen **(50-55-5)**
Unitensen **(69-09-0)**
Uniterge NP-4 **(7311-27-5)**
Unitertracid Light Blue AB **(2666-17-3)**
Unitertracid Light Orange G **(1936-15-8)**
Unitertracid Red 2G **(3734-67-6)**
Unitertracid Yellow TE **(1934-21-0)**
Uniteston **(57-85-2)**
Unitex 610-L **(14960-06-6)**
Unitex 710-L **(1462-54-0)**
Unitina AGS **(627-83-8)**
Unitina GMO **(111-03-5)**
Unitina HR **(8001-78-3)**
Unitina MD **(123-94-4)**
Unitina MD-A **(123-94-4)**
Unitolate **(3687-45-4)**
Unitolate 165-C **(123-94-4)**
Unitolate GS **(123-94-4)**
Unitolate V **(3687-46-5)**
Unitox **(470-90-6)**
Unitox **(62-73-7)**
Univerm **(56-23-5)**
Univol U 316S **(544-63-8)**
Unizeb **(12122-67-7)**
Unizeen C-2 **(61791-14-8)**
Unizeen C-5 **(61791-14-8)**
Unizeen C-10 **(61791-14-8)**
Unizeen SA **(7651-02-7)**
Unjecol 50 **(143-28-2)**
Unjecol 70 **(143-28-2)**
Unjecol 90 **(143-28-2)**
Unjecol 110 **(143-28-2)**
Unox 201 **(141-37-7)**
Unox 207 **(81-21-0)**
Unox Epoxide 201 **(141-37-7)**
Unox Epoxide 206 **(106-87-6)**
Unox Epoxide 207 **(81-21-0)**
Unox 207X **(81-21-0)**
Unoxat Epoxide 101 **(106-86-5)**
Unoxat Epoxide 269 **(96-08-2)**
Unsymmetrical dimethylhydrazine **(57-14-7)**
UpJohn U-32714 **(13067-93-1)**
Upamid Resin UPC-1283 **(32131-17-2)**
Upamide CD **(136-26-5)**
Upamide O-20 **(93-83-4)**
Upamide OD **(93-83-4)**
Upamide SME-M **(111-57-9)**
Upiwax 163 **(67762-27-0)**
Upiwax 163 R **(67762-27-0)**
Upjohn U-12,927 **(671-04-5)**
Upjohn U-36059 **(33089-61-1)**
Upper Coal Tar Distillate **(65996-91-0)**
Uro-Alvar **(14698-29-4)**
Uro-Septra **(8064-90-2)**
Uracil **(66-22-8)**
Uracil, 6-amino **(873-83-6)**
Uracil, 5-(bis(2-chloroethyl)amino) **(66-75-1)**
Uracil, 5-bromo **(51-20-7)**
Uracil, 5-bromo-3-sec-butyl-6-methyl **(314-40-9)**
Uracil, 5-bromo-3-tert-butyl-6-methyl **(7286-76-2)**
Uracil, 5-bromo-3-isopropyl-6-methyl **(314-42-1)**
Uracil, 3-tert-butyl-5-chloro-6-methyl **(5902-51-2)**
6-Uracilcarboxylic acid **(65-86-1)**
Uracil, 5-chloro **(1820-81-1)**
Uracil, 5-fluoro **(51-21-8)**
Uracil, 5-iodo **(696-07-1)**
Uracil lost (German) **(66-75-1)**
Uracil, 6-methyl **(626-48-2)**
Uracil, 6-methyl-2-thio **(56-04-2)**
Uracilmostaza **(66-75-1)**
Uracil mustard **(66-75-1)**
Uracil nitrogen mustard **(66-75-1)**
Uracil, 6-propyl-2-thio **(51-52-5)**
Uracil, 1-β-D-ribofuranosyl **(58-96-8)**
Uracil riboside **(58-96-8)**
Uracil, 2-thio **(141-90-2)**
Uractone **(52-01-7)**
Uradal **(77-65-6)**
Uragan **(314-40-9)**
Uragon **(314-40-9)**
Uralgin **(389-08-2)**
Uralite **(9011-05-6)**
Uralite (Polymer) **(9011-05-6)**
Uramine T 80 **(924-42-5)**
Uramine T101 **(9011-05-6)**
Uramine T105 **(9011-05-6)**
Uramine TSL 58 **(9011-05-6)**
Uramite **(9011-05-6)**
Uramustin **(66-75-1)**
Uramustine **(66-75-1)**
Uramycin B **(606-58-6)**
Uranin **(518-47-8)**
Uranine **(518-47-8)**
Uranine A Extra **(518-47-8)**
Uranine O **(518-47-8)**
Uranine SS **(518-47-8)**
Uranine USP XII **(518-47-8)**
Uranine WSS **(518-47-8)**
Uranine Yellow **(518-47-8)**
Uranium (ACGIH,OSHA) **(7440-61-1)**
Uranium(IV) chloride **(10026-10-5)**
Uranium acetate **(541-09-3)**
Uranium, bis(aceto)dioxo **(541-09-3)**
Uranium, bis(nitrato-O,O')dioxo-, (OC-6-11) **(36478-76-9)**

Uranium, bis(nitrato-O,O')dioxo- (Solid) **(10102-06-4)**
Uranium, bis(nitrato)dioxo-, hexahydrate **(13520-83-7)**
Uranium, difluorodioxo **(13536-84-0)**
Uranium, dinitratodioxo-, hexahydrate **(13520-83-7)**
Uranium, dioxosulfato- **(1314-64-3)**
Uranium, dioxo(sulfato(2-)-O)- (9CI) **(1314-64-3)**
Uranium fluoride (Fissle) **(7783-81-5)**
Uranium fluoride oxide **(13536-84-0)**
Uranium hexafluoride, Fissile (Containing more than 1% U-235) [UN 2977] **(7783-81-5)**
Uranium hexafluoride, Fissile excepted or non-fissile [UN 2978] **(7783-81-5)**
Uranium hexafluoride, Low specific activity (DOT) **(7783-81-5)**
Uranium metal, Pyrophoric [UN 2979] **(7440-61-1)**
Uranium oxide **(12036-71-4)**
Uranium oxide peroxide **(12036-71-4)**
Uranium oxide sulfate **(1314-64-3)**
Uranium oxyacetate **(541-09-3)**
Uranium oxyfluoride **(13536-84-0)**
Uranium oxysulfate **(1314-64-3)**
Uranium peroxide **(12036-71-4)**
Uranium tetrachloride **(10026-10-5)**
Uranyl acetate (DOT) **(541-09-3)**
Uranyl fluoride **(13536-84-0)**
Uranyl nitrate **(36478-76-9)**
Uranyl nitrate, Solid [UN 2981] **(10102-06-4)**
Uranylnitrate hexahydrate **(13520-83-7)**
Uranyl nitrate hexahydrate, Solution **(13520-83-7)**
Uranyl nitrate hexahydrate solution **(13520-83-7)**
Uranyl sulfate **(1314-64-3)**
Urao **(533-96-0)**
Urazium **(136-40-3)**
Urbacid **(2445-07-0)**
Urbacide **(2445-07-0)**
Urbanyl **(22316-47-8)**
Urbason **(83-43-2)**
Urbason crystal suspension **(53-36-1)**
Urbasone **(83-43-2)**
Urbasulf **(2533-82-6)**
Urbazid **(2445-07-0)**
Urbil **(57-53-4)**
Urbilat **(57-53-4)**
Urbol **(315-30-0)**
Urea **(57-13-6)**
Urea, 1-Butyl-3-(3,4-dichlorophenyl)-1-methyl **(555-37-3)**
Urea, Compd. with hydrogen peroxide (1:1) (8CI) **(124-43-6)**
Urea, Compd. with hydrogen peroxide (H_2O_2) (1:1) (9CI) **(124-43-6)**
Urea J Acid **(134-47-4)**
Urea, Polymer with formaldehyde **(9011-05-6)**
Urea, Reaction products with formaldehyde **(68611-64-3)**
Urea, 1-((p-acetylphenyl)sulfonyl)-3-cyclohexyl **(968-81-0)**
Urea, 1-acetyl-2-thio **(591-08-2)**
Urea-agua oxigenada (Spanish) **(124-43-6)**
Urea, allyl **(557-11-9)**
Urea, 1-allyl-2-thio **(109-57-9)**
Urea, amino- **(57-56-7)**
Urea, (aminocarbonyl)- **(108-19-0)**
Urea, (aminoiminomethyl)- (9CI) **(141-83-3)**
Urea, (aminoiminomethyl)-, phosphate (9CI) **(17675-60-4)**
Urea, (aminoiminomethyl)-, sulfate (2:1) (9CI) **(591-01-5)**
Urea, (2-benzimidazolyl) **(24370-25-0)**
Urea, 1H-benzimidazol-2-yl- (9CI) **(24370-25-0)**
Urea, 1-(2-benzothiazolyl)-1,3-dimethyl **(18691-97-9)**
Urea, 1-(2-benzothiazolyl)-3-methyl **(1929-88-0)**
Urea, 3-(2-benzothiazolyl)-1-methyl- **(1929-88-0)**
Urea, 1-benzoyl-1-(3,4-dichlorophenyl)-3,3-dimethyl- (8CI) **(3134-12-1)**
Urea, 1,1'-(4,4'-biphenylyene)bis(2-thio **(1614-30-8)**
Urea, N,N'-bis(2-benzothiazolylmercaptomethyl)- (9CI) **(64216-20-2)**
Urea, 1,3-bis(2-benzothiazolylthiomethyl) **(95-35-2)**
Urea, N,N'-bis(butoxymethyl)- (9CI) **(4981-47-9)**
Urea, 1,3-bis(2-chloroethyl)-1-nitroso **(154-93-8)**
Urea, N,N'-bis(2-chloroethyl)-N-nitroso- (9CI) **(154-93-8)**
Urea, 1,3-bis(p-chlorophenyl)-2-thio **(1220-00-4)**
Urea, N,N'-bis(3-(dimethylamino)propyl)- (9CI) **(52338-87-1)**
Urea, N,N'-bis(1,1-dimethylethyl)- (9CI) **(5336-24-3)**
Urea, 1,3-bis(hydroxymethyl) **(140-95-4)**
Urea, N,N'-bis(hydroxymethyl)- (9CI) **(140-95-4)**
Urea, bis(hydroxymethyl)- (9CI) **(25155-29-7)**
Urea, 1,3-bis(methoxymethyl)- (8CI) **(141-07-1)**
Urea, N,N'-bis(methoxymethyl)- (9CI) **(141-07-1)**
Urea, N,N'-bis(1-methylethyl)- (9CI) **(4128-37-4)**
Urea, N,N'-bis(1-methylpropyl)- (9CI) **(869-79-4)**
Urea, 1,3-bis(2,2,2-trichloro-1-hydroxyethyl) **(116-52-9)**
Urea, 3-(4,6-bis(trichloromethyl)-s-triazin-2-yl)-1,1-dimethyl- (8CI) **(30804-99-0)**
Urea, 1-(4,6-bis(trichloromethyl)-s-triazin-2-yl)-3-phenyl- (8CI) **(30805-00-6)**
Urea, 3-(4-bromo-3-chlorophenyl)-1-methoxy-1-methyl **(13360-45-7)**
Urea, N'-(4-bromo-3-chlorophenyl)-N-methoxy-N-methyl- **(13360-45-7)**
Urea, N'-(4-bromo-3,5-dimethylphenyl)-N,N-dimethyl- (9CI) **(78508-43-7)**
Urea, (2-bromo-ethylbutyryl) **(77-65-6)**
Urea, (2-bromo-2-ethylbutyryl)- (8CI) **(77-65-6)**
Urea, (4-bromo-3-methylphenyl)- (9CI) **(78508-46-0)**
Urea, (3-bromophenyl)- **(2989-98-2)**
Urea, (4-bromophenyl)- **(1967-25-5)**
Urea, 3-(p-bromophenyl)-1,1-dimethyl- (7CI, 8CI) **(3408-97-7)**
Urea, N'-(4-bromophenyl)-N,N-dimethyl- (9CI) **(3408-97-7)**
Urea, 3-(p-bromophenyl)-1-methoxy-1-methyl **(3060-89-7)**
Urea, butyl **(592-31-4)**
Urea, sec-butyl **(689-11-2)**
Urea, 1-butyl-1-nitroso **(869-01-2)**
Urea, n-butyl-n-nitroso- **(869-01-2)**
Urea, 1-butyl-3-sulfanilyl **(339-43-5)**
Urea, 1-(5-(t-butyl)-1,3,4-thiadiazol-2-yl)-1,3-dimethyl **(34014-18-1)**
Urea, 1-butyl-3-(p-tolylsulfonyl) **(64-77-7)**
Urea, (carboxymethyl)- **(462-60-2)**
Urea, 1-(3-chloro-4-(chlorodifluoromethylthio)phenyl)-3,3-dimethyl **(33439-45-1)**
Urea, 1-(2-chloroethyl)-3-cyclohexyl-1-nitroso **(13010-47-4)**
Urea, N-(2-chloroethyl)-N'-cyclohexyl-N-nitroso- (9CI) **(13010-47-4)**
Urea, 1-(2-chloroethyl)-3-(D-glucopyranos-2-yl)-1-nitroso- **(54749-90-5)**
Urea, 1-(2-chloroethyl)-3-(4-methylcyclohexyl)-1-nitroso-, (E) **(13909-09-6)**
Urea, 1-(2-chloroethyl)-3-(4-methylcyclohexyl)-1-nitroso-, trans- **(13909-09-6)**
Urea, N-(2-chloroethyl)-N'-(4-methylcyclohexyl)-N-nitroso- **(13909-09-6)**
Urea, (3-(chloromercuri)-2-methoxypropyl)- **(62-37-3)**
Urea, (3-(chloromercurio)-2-methoxypropyl)- **(62-37-3)**
Urea, (3-chloro-4-methoxyphenyl)- (8CI,9CI) **(25277-05-8)**
Urea, 3-(3-chloro-4-methoxyphenyl)-1,1-dimethyl- **(19937-59-8)**
Urea, N'-(3-chloro-4-methoxyphenyl)-N,N-dimethyl **(19937-59-8)**
Urea, 1-(3-chloro-4-methoxyphenyl)-3-methyl- (8CI) **(20782-57-4)**
Urea, N-(3-chloro-4-methoxyphenyl)-N'-methyl- (9CI) **(20782-57-4)**
Urea, (3-chloro-4-methylphenyl)- (9CI) **(13142-64-8)**
Urea, N'-(3-chloro-4-methylphenyl)-N,N-dimethyl-, Mixt. with N-(1,1-dimethylethyl)-N'-ethyl-6-(methylthio)-1,3,5-triazine-2,4-diamine (9CI) **(37341-11-0)**
Urea, N-(3-chloro-4-methylphenyl)-N'-methyl- (9CI) **(22175-22-0)**
Urea, 3-(chloro-2-norbornyl)-1,1-dimethyl **(1319-96-6)**
Urea, 3-(p-(p-chlorophenoxy)phenyl)-1,1-dimethyl **(1982-47-4)**
Urea, 1-(p-chlorophenyl) **(140-38-5)**
Urea, (2-chlorophenyl)- (9CI) **(114-38-5)**
Urea, (o-chlorophenyl)- (8CI) **(114-38-5)**
Urea, N'-(4-chlorophenyl)-N-((β-D-glucopyranosyloxy)methyl)-N-methoxy- (9CI) **(81393-48-8)**
Urea, N-(4-chlorophenyl)-N'-(4-chloro-3-(trifluoromethyl)-phenyl)- (9CI) **(369-77-7)**
Urea, N-(4-chlorophenyl)-N'-(3,4-dichlorophenyl)- (9CI) **(101-20-2)**

Urea, 1-(p-chlorophenyl)-3-(2,6-difluorobenzoyl) **(35367-38-5)**
Urea, 1-(m-chlorophenyl)-3,3-dimethyl **(587-34-8)**
Urea, 3-(p-chlorophenyl)-1,1-dimethyl **(150-68-5)**
Urea, 3-(p-chlorophenyl)-1,1-dimethyl-, Cmpd. with trichloro-acetic acid (1:1) **(140-41-0)**
Urea, 3-(p-chlorophenyl)-1-methoxy-1-methyl **(1746-81-2)**
Urea, N'-(4-chlorophenyl)-N-methoxy-N-methyl- **(1746-81-2)**
Urea, 1-(m-chlorophenyl)-3-methyl- (8CI) **(20940-42-5)**
Urea, N-(3-chlorophenyl)-N'-methyl- (9CI) **(20940-42-5)**
Urea, 3-(p-chlorophenyl)-1-methyl-1-(1-methyl-2-propynyl) **(3766-60-7)**
Urea, 1-(m-chlorophenyl)-3-(4-methyl-6-(trichloromethyl)-s-triazin-2-yl)- (8CI) **(30804-94-5)**
Urea, 1-(o-chlorophenyl)-3-(4-methyl-6-(trichloromethyl)-s-triazin-2-yl)- (8CI) **(30804-93-4)**
Urea, 1-(p-chlorophenyl)-3-(4-methyl-6-(trichloromethyl)-s-triazin-2-yl)- (8CI) **(30804-95-6)**
Urea, 1-((o-chlorophenyl)sulfonyl)-3-(4-methoxy-6-methyl-s-triazin-2-yl) **(64902-72-3)**
Urea, 1-((p-chlorophenyl)sulfonyl)-3-propyl **(94-20-2)**
Urea, 1-(o-chlorophenyl)-2-thio **(5344-82-1)**
Urea, (3-chloro-p-tolyl)- (8CI) **(13142-64-8)**
Urea, 3-(3-chloro-p-tolyl)-1,1-dimethyl **(15545-48-9)**
Urea, 1-(3-chloro-p-tolyl)-3-methyl- (8CI) **(22175-22-0)**
Urea, 1-cycloheptyl-3-phenyl- (8CI) **(19095-79-5)**
Urea, N-cycloheptyl-N'-phenyl- (9CI) **(19095-79-5)**
Urea, 1-cyclohexyl-3-(4-methylmetanilyl) **(565-33-3)**
Urea, 1-cyclohexyl-3-phenyl- (8CI) **(886-59-9)**
Urea, N-cyclohexyl-N'-phenyl- (9CI) **(886-59-9)**
Urea, 3-cyclooctyl-1,1-dimethyl **(2163-69-1)**
Urea, 3-cyclooctyl-1,1-dimethyl- mixed with butynyl-3N-3-chloro-phenylcarbamate **(8015-55-2)**
Urea, 1-cyclopentyl-3-phenyl- (8CI) **(13140-89-1)**
Urea, N-cyclopentyl-N'-phenyl- (9CI) **(13140-89-1)**
Urea, 1-cyclopropyl-3-phenyl- (8CI) **(13140-86-8)**
Urea, N-cyclopropyl-N'-phenyl- (9CI) **(13140-86-8)**
Urea, 1,3-diamino- **(497-18-7)**
Urea, 1,3-di-tert-butyl- (8CI) **(5336-24-3)**
Urea, 1,3-di-sec-butyl- (6CI,7CI,8CI) **(869-79-4)**
Urea, 1,3-dibutyl-3-nitroso **(56654-52-5)**
Urea,1,3-dibutyl-2-thio- (8CI) **(109-46-6)**
Urea, (3,4-dichlorophenyl)- **(2327-02-8)**
Urea, 3-(3,4-dichlorophenyl)-1,1-dimethyl **(330-54-1)**
Urea, 3-(3,4-dichlorophenyl)-1-methoxy-1-methyl **(330-55-2)**
Urea, N'-(3,4-dichlorophenyl)-N-methoxy-N-methyl-, Mixt. with 2,4-dichloro-1-(4-nitrophenoxy)benzene (9CI) **(8075-80-7)**
Urea, 1-(3,4-dichlorophenyl)-3-methyl- (8CI) **(3567-62-2)**
Urea, N-(3,4-dichlorophenyl)-N'-methyl- (9CI) **(3567-62-2)**
Urea, 1-(2,4-dichlorophenyl)-3-(4-methyl-6-(trichloromethyl)-s-triazin-2-y l)- (8CI) **(30804-96-7)**
Urea, 3-(4,6-dichloro-s-triazin-2-yl)-1,1-dimethyl- (7CI,8CI) **(5097-52-9)**
Urea, 1-(4,6-dichloro-s-triazin-2-yl)-3-isopropyl- (7CI,8CI) **(5097-54-1)**
Urea, 1-(4,6-dichloro-s-triazin-2-yl)-3-isopropyl-2-thio- (7CI,8CI) **(5097-56-3)**
Urea, 1-(4,6-dichloro-s-triazin-2-yl)-3-phenyl- (7CI,8CI) **(5097-51-8)**
Urea, 1-(4,6-dichloro-s-triazin-2-yl)-3-phenyl-2-thio- (7CI,8CI) **(5577-35-5)**
Urea, 1,3-dicyclohexyl-2-thio **(1212-29-9)**
Urea, 1,1-diethyl **(634-95-7)**
Urea, N,N-diethyl- (9CI) **(634-95-7)**
Urea, 1,3-diethyl-1,3-diphenyl- **(85-98-3)**
Urea, N,N'-diethyl-N,N'-diphenyl- (9CI) **(85-98-3)**
Urea, 1,1-diethyl-3-methyl-3-nitroso **(50285-72-8)**
Urea, 1,3-diethyl-2-thio **(105-55-5)**
Urea, 1,3-dihydroxy **(686-68-0)**
Urea, 1,3-diisopropyl- (8CI) **(4128-37-4)**
Urea, N-(2,2-dimethoxyethyl)-N'-(5-(1,1-dimethylethyl)-1,3,4-thiadiazol-2-yl)-N-methyl- (9CI) **(51461-71-3)**
Urea, 1,1-dimethyl **(598-94-7)**
Urea, 1,3-dimethyl **(96-31-1)**
Urea, 1-(α,α-dimethylbenyzl)-3-methyl-3-phenyl **(42609-52-9)**
Urea, N,N-dimethyl-N'-(5-(1,1-dimethylethyl)-3-isoxazolyl) **(55861-78-4)**
Urea, N,N'-dimethyl-N,N'-diphenyl- (9CI) **(611-92-7)**
Urea, 1,1-dimethyl-3-ethyl-3-nitroso **(50285-71-7)**
Urea, 1,1-dimethyl-3-(p-fluorophenyl) **(332-33-2)**
Urea, 1,1-dimethyl-3-(p-isopropylphenyl) **(34123-59-6)**
Urea, 1,1-dimethyl-3-(p-(p-methoxyphenoxy)phenyl) **(14214-32-5)**
Urea, N,N-dimethyl-N'-(4-methylphenyl)- **(7160-01-2)**
Urea, 1,1-dimethyl-3-(4-methyl-6-(trichloromethyl)-s-triazin-2-yl)-(8CI) **(30804-91-2)**
Urea, N,N-dimethyl-N'-(octahydro-4,7-methano-1H-inden-5-yl)- **(2163-79-3)**
Urea, 1,1-dimethyl-3-phenyl **(101-42-8)**
Urea, N'-(3,5-dimethylphenyl)-N,N-dimethyl- (9CI) **(36627-56-2)**
Urea, 1,1-dimethyl-3-phenyl-, trichloroacetate **(4482-55-7)**
Urea, 1,3-dimethyl-2-thio **(534-13-4)**
Urea, N,N-dimethyl-N'-(3-(trifluoromethyl)phenyl)- **(2164-17-2)**
Urea, 1,1-dimethyl-3-(α,α,α-trifluoro-m-tolyl) **(2164-17-2)**
Urea, 1,3-di-n-butyl-2-thio **(109-46-6)**
Urea dioxide **(124-43-6)**
Urea, (2,5-dioxo-4-imidazolidinyl)- (9CI) **(97-59-6)**
Urea, 1,3-diphenyl- **(102-07-8)**
Urea, N,N'-diphenyl- **(102-07-8)**
Urea, N,N-diphenyl- (9CI) **(603-54-3)**
Urea, diphenyl- (8CI,9CI) **(26763-63-3)**
Urea, 1,3-diphenyl-2-thio- **(102-08-9)**
Urea, dodecyl- (9CI) **(2158-09-0)**
Urea, N,N''-1,2-ethanediylbis- (9CI) **(1852-14-8)**
Urea, (p-ethoxyphenyl) **(150-69-6)**
Urea, 1-(p-ethoxyphenyl)-3-(6-ethoxy-2-pyridyl)-2-thio- (7CI,8CI) **(900-38-9)**
Urea, 1-ethyl- **(625-52-5)**
Urea, ethyl **(625-52-5)**
Urea, 3-ethyl-1,1-diphenyl- (8CI) **(18168-01-9)**
Urea, N'-ethyl-N,N-diphenyl- (9CI) **(18168-01-9)**
Urea, 1,3-ethylene- **(120-93-4)**
Urea, 1,1'-ethylenedi **(1852-14-8)**
Urea, 1,3-ethylene-2-thio- **(96-45-7)**
Urea, 1-ethyl-1-nitroso **(759-73-9)**
Urea, 1-ethyl-3-(5-nitro-2-thiazolyl) **(139-94-6)**
Urea, N-ethyl-N'-(5-nitro-2-thiazolyl)- **(139-94-6)**
Urea, 1-ethyl-2-thio **(625-53-6)**
Urea, 1,1'-(6-ethyl-s-triazine-2,4-diyl)bis(3-phenyl- (8CI) **(30805-03-9)**
Urea, (4-fluoro-3-methylphenyl)- (9CI) **(78508-45-9)**
Urea, 1-(p-fluorophenyl) **(659-30-3)**
Urea, (2-fluorophenyl)- (9CI) **(656-31-5)**
Urea, (3-fluorophenyl)- (9CI) **(770-19-4)**
Urea, (m-fluorophenyl)- (8CI) **(770-19-4)**
Urea, (o-fluorophenyl)- (8CI) **(656-31-5)**
Urea, 3-(m-fluorophenyl)-1,1-dimethyl- (8CI) **(330-39-2)**
Urea, N'-(3-fluorophenyl)-N,N-dimethyl- (9CI) **(330-39-2)**
Urea-formaldehyde adduct **(9011-05-6)**
Urea-formaldehyde condensate **(9011-05-6)**
Urea-formaldehyde copolymer **(9011-05-6)**
Urea-formaldehyde oligomer **(9011-05-6)**
Urea-formaldehyde polymer **(9011-05-6)**
Urea-formaldehyde precondensate **(9011-05-6)**
Urea-formaldehyde prepolymer **(9011-05-6)**
Urea-formaldehyde resin **(9011-05-6)**
Urea, 1-(hexahydro-1H-azepin-1-yl)-3-(p-tolylsulfonyl) **(1156-19-0)**
Urea, 3-(hexahydro-4,7-methanoindan-5-yl)-1,1-dimethyl- (8CI) **(2163-79-3)**
Urea, 3-(hexahydro-4,7-methanoindan-5-yl)-1,1-dimethyl-, endo,exo-5 **(18530-56-8)**
Urea hydrogen peroxide **(124-43-6)**
Urea hydrogen peroxide salt **(124-43-6)**
Urea hydroperoxide **(124-43-6)**
Urea, hydroxy **(127-07-1)**
Urea, (hydroxymethyl)- (9CI) **(1000-82-4)**
Urea, 1-hydroxy-3-methyl-1-phenyl **(6263-38-3)**

Urea, (3-hydroxyphenyl)- (9CI) **(701-82-6)**
Urea, (m-hydroxyphenyl)- (8CI) **(701-82-6)**
Urea, (imino(methylamino)methyl)- (9CI) **(44648-02-4)**
Urea, 1,1'-isobutylidenedi **(6104-30-9)**
Urea, isopropyl **(691-60-1)**
Urea, 1-isopropyl-3-(4-methyl-6-(trichloromethyl)-s-triazin-2-yl)- (8CI) **(30863-32-2)**
Urea, 1-isopropyl-3-phenyl- (8CI) **(19895-44-4)**
Urea, mesoxalyl- **(50-71-5)**
Urea, N'-(3-methoxyphenyl)-N,N-dimethyl- (9CI) **(28170-54-9)**
Urea, N'-(4-methoxyphenyl)-N,N-dimethyl- (9CI) **(7160-02-3)**
Urea, 3-(m-methoxyphenyl)-1,1-dimethyl- (8CI) **(28170-54-9)**
Urea, 3-(p-methoxyphenyl)-1,1-dimethyl- (8CI) **(7160-02-3)**
Urea, (2-methoxypropyl)-, mercury complex **(62-37-3)**
Urea, methyl **(598-50-5)**
Urea, 1-(2-methylcyclohexyl)-3-phenyl **(1982-49-6)**
Urea, N'-methyl-N,N-diphenyl- (9CI) **(13114-72-2)**
Urea, (1-methylethyl)- (9CI) **(691-60-1)**
Urea, N-(1-methylethyl)-N'-phenyl- (9CI) **(19895-44-4)**
Urea, 1-methyl-1-nitroso- **(684-93-5)**
Urea, N-methyl-N-nitroso **(684-93-5)**
Urea, 1-methyl-3-phenyl **(1007-36-9)**
Urea, 1,1'-(methyl-m-phenylene)bis(3,3-dimethyl- **(26604-41-1)**
Urea, N,N''-(4-methyl-1,3-phenylene)bis(N',N'-dimethyl- (9CI) **(17526-94-2)**
Urea, N,N''-(methyl-1,3-phenylene)bis(N',N'-dimethyl- (9CI) **(26604-41-1)**
Urea, (1-methyl-1-phenylethyl)- (9CI) **(58609-76-0)**
Urea, 1-methyl-3-phenyl-2-thio **(2724-69-8)**
Urea, N,N''-(2-methylpropylidene)bis-, (9CI) **(6104-30-9)**
Urea, 1-methyl-2-thio **(598-52-7)**
Urea, 1,1'-(6-methyl-s-triazine-2,4-diyl)bis(3-phenyl- (8CI) **(30805-01-7)**
Urea, 1-(4-methyl-6-(trichloromethyl)-s-triazin-2-yl)-3-(m-nitrophenyl)- (8CI) **(30804-98-9)**
Urea, 1-(4-methyl-6-(trichloromethyl)-s-triazin-2-yl)-3-phenyl-(8CI) **(30804-92-3)**
Urea, 1-(4-methyl-6-(trichloromethyl)-s-triazin-2-yl)-3-phenyl-2-thio- (8CI) **(30863-28-6)**
Urea, 1-(4-methyl-6-(trichloromethyl)-s-triazin-2-yl)-3-p-tolyl-(8CI) **(30804-97-8)**
Urea, mononitrate (8CI,9CI) **(124-47-0)**
Urea, 1-(1-naphthyl)-2-thio **(86-88-4)**
Urea nitrate **(124-47-0)**
Urea nitrate, Dry or containing less than 20% water [UN 0220] **(124-47-0)**
Urea nitrate (Wet) **(124-47-0)**
Urea nitrate, Wetted with not less than 20% water [UN 1357] **(124-47-0)**
Urea nitrate, Wet with 10% or more water **(124-47-0)**
Urea nitrate, Wet with 10% or more water, over 25 lbs. in one outside packaging **(124-47-0)**
Urea, 1-nitrophenyl-3-(3-pyridylmethyl) **(53558-25-1)**
Urea, 3-nitroso-1,1,3-triethyl **(50285-70-6)**
Ureapap W **(9011-05-6)**
Urea peroxide **(124-43-6)**
Urea, (4-phenoxyphenyl)- (9CI) **(78508-44-8)**
Urea, phenyl **(64-10-8)**
Urea, (phenylacetyl) (8CI) **(63-98-9)**
Urea (phenylmercuri)- **(2279-64-3)**
Urea, 1-phenyl-3-(1,2,3-thiadiazol-5-yl) **(51707-55-2)**
Urea, N-phenyl-N'-1,2,5-thiadiazol-3-yl- **(71769-74-9)**
Urea, 1-phenyl-2-thio **(103-85-5)**
Urea, 1,1'-(6-phenyl-s-triazine-2,4-diyl)bis(3-phenyl- (8CI) **(30805-04-0)**
Urea, N,N''-(6-phenyl-1,3,5-triazine-2,4-diyl)bis(N'-phenyl-(9CI) **(30805-04-0)**
Ureaphil **(57-13-6)**
Urea, phosphate (1:1) **(4861-19-2)**
Urea, 2-propenyl- (9CI) **(557-11-9)**
Urease **(9002-13-5)**
Urea, seleno **(630-10-4)**
Urea, 1,1,3,3-tetramethyl **(632-22-4)**
Urea, 1,1,3,3-tetramethyl-2-thio **(2782-91-4)**
Urea, 2-thio **(62-56-6)**
Urea, thio- (8CI) **(62-56-6)**
Urea, 2-thio-1-(thiocarbamoyl)- **(541-53-7)**
Urea, 2-thio-1-o-tolyl- (8CI) **(614-78-8)**
Urea, m-tolyl **(63-99-0)**
Urea, p-tolyl **(622-51-5)**
Urea, 1,1',1''-(s-triazine-2,4,6-triyl)tris(3-phenyl- (8CI) **(30805-05-1)**
Urea, 1,1'-(6-(trichloromethyl)-s-triazine-2,4-diyl)bis(3-phenyl-(8CI) **(30805-02-8)**
Urea, (3-(triethoxysilyl)propyl)- (9CI) **(23779-32-0)**
Urea, (3-(trifluoromethyl)phenyl)- (9CI) **(13114-87-9)**
Urea, (α,α,α-trifluoro-m-tolyl)- **(13114-87-9)**
Urea, 1,1,3-trimethyl-3-nitroso **(3475-63-6)**
Urea, 1,1,3-trimethyl-2-thio **(2489-77-2)**
Urea, ureido- **(110-21-4)**
Urecoli S **(9011-05-6)**
Urecoll K **(9011-05-6)**
Urecoll KL **(9011-05-6)**
Uree-eau oxygenee (French) **(124-43-6)**
Uregit **(58-54-8)**
p-Ureidobenzenearsonic acid **(121-59-5)**
2-Ureidobenzimidazole **(24370-25-0)**
Ureidoformamide **(108-19-0)**
5-Ureidohydantoin **(97-59-6)**
4-Ureido-1-phenylarsonic acid **(121-59-5)**
Ureidourea **(110-21-4)**
Urelit C **(9011-05-6)**
Urelit HM **(9011-05-6)**
Urelit R **(9011-05-6)**
Ureol P **(140-95-4)**
Ureophil **(57-13-6)**
Urepret **(9011-05-6)**
Urese **(91-33-8)**
Uretan **(51-79-6)**
Uretan etylowy (Polish) **(51-79-6)**
Urethan **(51-79-6)**
Urethane **(51-79-6)**
Urethane polymers **(9009-54-5)**
Urethan, phenyl- **(101-99-5)**
Urethylane **(598-55-0)**
Urevert **(57-13-6)**
Urex **(54-31-9)**
7,7'-Ureylenebis(4-hydroxy-3-((6-sulfo-2-naphthyl)azo)-2-naphthalenesulfonic acid), tetrasodium salt **(28706-25-4)**
3,3'-(Ureylenebis((3-methoxy-p-phenylene)azo))dibenzenesulfonic acid, disodium salt **(10114-86-0)**
6,6'-Ureylenebis(1-naphthol-3-sulfonic acid) **(134-47-4)**
Uri **(61-76-7)**
Uriben **(389-08-2)**
Uribest **(52570-16-8)**
Uric-acid **(69-93-2)**
Uric acid, 3-methyl- (8CI) **(605-99-2)**
Uric acid, 6-thio- (VAN) (8CI) **(2002-60-0)**
Uricemil **(315-30-0)**
Uriclar **(389-08-2)**
Uricosid **(57-66-9)**
Uridin **(58-96-8)**
Uridinal **(136-40-3)**
Uridinal **(94-78-0)**
Uridine **(58-96-8)**
β-Uridine **(58-96-8)**
Uridine, 5-chloro-2'-deoxy **(50-90-8)**
Uridine, 2'-deoxy-5-fluoro (8CI,9CI) **(50-91-9)**
Uridine, 2'-deoxy-5-(trifluoromethyl)- **(70-00-8)**
Uridine, 5-fluoro **(316-46-1)**
Uridine, 2'-(4-methylbenzenesulfonate) (9CI) **(6206-10-6)**

Uridine, 2'-p-toluenesulfonate (7CI,8CI) **(6206-10-6)**
Urinary Hebin **(9002-68-0)**
Urinex **(58-94-6)**
Uriplex **(136-40-3)**
Urisal **(389-08-2)**
Urisoxin **(127-69-5)**
Uritas **(315-30-0)**
Uritone **(100-97-0)**
Uritrate **(14698-29-4)**
Uritrisin **(127-69-5)**
Urizept **(67-20-9)**
Urlea **(73-48-3)**
Urner's liquid **(79-43-6)**
Urobenyl **(315-30-0)**
Urobiotic-250 **(136-40-3)**
Urodiaton **(144-82-1)**
Urodiazin **(58-93-5)**
Urodine **(136-40-3)**
Urodine **(94-78-0)**
Urodixin **(389-08-2)**
Urofeen **(136-40-3)**
Urofix **(9011-05-6)**
neo-Urofort **(500-42-5)**
Urogan **(127-69-5)**
Urografin acid **(117-96-4)**
Urogranoic acid **(117-96-4)**
Urolucosil **(144-82-1)**
Uromaline **(90-64-2)**
Uroman **(389-08-2)**
Uromide **(136-40-3)**
Uromiro **(440-58-4)**
Uromiron **(440-58-4)**
Uromycine **(1403-66-3)**
Uronal **(57-44-3)**
Uroneg **(389-08-2)**
Uropan **(389-08-2)**
Urophenyl **(136-40-3)**
Uropyridin **(136-40-3)**
Uropyrine **(136-40-3)**
Urosemide **(54-31-9)**
Urosin **(315-30-0)**
Urosulfon **(144-80-9)**
Urosulfone **(144-80-9)**
Urotrast **(117-96-4)**
Urotrate **(14698-29-4)**
Urotropin **(100-97-0)**
Urotropine **(100-97-0)**
Urox **(140-41-0)**
Urox 379 **(140-41-0)**
Urox B **(314-40-9)**
Urox B Water soluble concentrate weed killer **(314-40-9)**
Urox D **(330-54-1)**
Urox-HX **(314-40-9)**
Urox HX Granular Weed Killer **(314-40-9)**
Uroxin **(76-24-4)**
Uroxine **(76-24-4)**
Uroxol **(14698-29-4)**
Ursoferran **(9004-66-4)**
Ursol Brown O **(615-66-7)**
Ursol Brown RR **(5307-14-2)**
Ursol D **(106-50-3)**
Ursol EG **(591-27-5)**
Ursol ERN **(90-15-3)**
Ursol Olive 6G **(95-83-0)**
Ursol P **(123-30-8)**
Ursol P Base **(123-30-8)**
Ursol SLA **(39156-41-7)**
Ursol Yellow Brown A **(121-88-0)**
Urtosal **(65-45-2)**
Urushiol (9CI) **(53237-59-5)**
Urzan S **(8061-51-6)**
Usacert Blue No. 1 **(3844-45-9)**
Usacert FD & C Red No. 4 **(4548-53-2)**
Usacert FD & C Yellow No. 6 **(2783-94-0)**
Usacert Red No. 3 **(16423-68-0)**
Usacert Red No. 4 **(4548-53-2)**
Usacert Yellow No. 5 **(1934-21-0)**
Usacert Yellow No. 6 **(2783-94-0)**
Usalake FD & C Yellow No. 6 Lake **(2783-94-0)**
Usempax AP **(439-14-5)**
Usnein **(125-46-2)**
Usniacin **(125-46-2)**
Usnic acid **(125-46-2)**
Usninic acid **(125-46-2)**
Usninsaeure (German) **(125-46-2)**
Ustinex **(94-74-6)**
Utal **(38641-94-0)**
Utedrin **(50-56-6)**
Uteracon **(50-56-6)**
Uteramine **(51-67-2)**
Utibid **(14698-29-4)**
Uticillin VK **(132-98-9)**
Util **(77-75-8)**
Utostan **(136-40-3)**
Utrasul **(144-82-1)**
Uulcol Scarlet 2G **(1248-18-6)**
Uvic acid **(133-37-9)**
Uvinol D-50 **(131-55-5)**
Uvinul 400 **(131-56-6)**
Uvinul D 49 **(131-54-4)**
Uvinul D-50 **(131-55-5)**
Uvinul M 40 **(131-57-7)**
Uzone **(50-33-9)**
V 50 **(2997-92-4)**
V 101 **(2227-13-6)**
V 501 **(9003-20-7)**
V 17004 **(2425-10-7)**
V-18 **(116-29-0)**
V-101 **(2227-13-6)**
V-103 **(8057-49-6)**
VA **(121-34-6)**
VA 0112 **(9003-20-7)**
VA 3 (Copolymer) **(9003-22-9)**
VAGD **(9003-22-9)**
V-Brite **(7775-14-6)**
V-C 1-13 **(97-17-6)**
V-C 9-104 **(13194-48-4)**
V-C-13 **(97-17-6)**
VC **(75-01-4)**
V-C Chemical V-C 9-104 **(13194-48-4)**
VCM **(75-01-4)**
VCN **(107-13-1)**
VC13 Nemacide **(97-17-6)**
VCR **(57-22-7)**
VCR sulfate **(2068-78-2)**
V C.S. **(21609-90-5)**
VCS 438 **(20354-26-1)**
VCS-506 **(21609-90-5)**
V-Cil **(87-08-1)**
V-Cil-K **(132-98-9)**
V-Cillin **(87-08-1)**
V-Cillin K **(132-98-9)**
VDC **(75-35-4)**
VDF **(75-38-7)**

VDM (137-42-8)
VEL 3510 (51461-71-3)
VEL-4207 (56141-00-5)
VEL-5026 (55511-98-3)
VEL 4283 (Obs.) (31218-83-4)
VFR 3801 (25038-59-9)
VH (64-10-8)
VH 10/60 (9003-22-9)
VHR X (1334-78-7)
VI-Cad (10108-64-2)
VK 55 (94-19-9)
VKhVD 40 (9011-06-7)
VLB (865-21-4)
VLB monosulfate (143-67-9)
VLVF (9003-22-9)
VM & P Naphtha (ACGIH,OSHA) (8032-32-4)
VMCC (9003-22-9)
VMI 10-3 (504-24-5)
VM and P Naphtha (8032-32-4)
VP 114 (555-45-3)
VP 1940 (900-95-8)
VPI 300 (20227-92-3)
VPM (137-42-8)
VPM (6734-80-1)
V-Pyrol (88-12-0)
VR-8 (865-21-4)
V-Serp (50-55-5)
V-Sul (127-69-5)
VT 1 (7440-32-6)
VTI 1 (122-37-2)
VUAgT 866 (60029-23-4)
VUAgT 866/72 (60029-23-4)
VVP 66-95 (7782-42-5)
VX (50782-69-9)
VYAC (108-05-4)
VYGEN 220 (9003-22-9)
VYHH (9003-22-9)
VYNS (9003-22-9)
VYNW (9003-22-9)
Vaben (604-75-1)
Vabrocid (59-87-0)
Vac (108-05-4)
Vac-10 (50-33-9)
Vacate (94-74-6)
Vacor (53558-25-1)
Vacuum Gas Oil (Petroleum) (64741-57-7)
Vacuum residue (64741-56-6)
Vacuum residuum (64741-49-7)
Vadebex (128-62-1)
Vadrocid (59-87-0)
Vaflol (68-26-8)
Vafol (68-26-8)
Vagamin (53-46-3)
Vagantin (53-46-3)
Vagestrol (56-53-1)
Vagilen (443-48-1)
Vagilia (127-69-5)
Vagimid (443-48-1)
Vagostigmin (51-44-5)
Vagostigmine (59-99-4)
Vagostigmine bromide (114-80-7)
Val-Drop (7775-09-9)
Valadol (103-90-2)
Valamin (126-52-3)
Valamina (126-52-3)
Valamine (78-81-9)
Valaminettae (126-52-3)
Valaminetten (126-52-3)
Valan (58-34-4)
Valdrene (147-24-0)
Valecor (390-64-7)
Valentinite (1309-64-4)
Valentinite (1317-98-2)
Valeo (439-14-5)
Valeral (110-62-3)
n-Valeraldehyde (ACGIH,OSHA) (110-62-3)
Valeraldehyde [UN 2058] (110-62-3)
Valeraldehyde, 2,3-dimethyl- (8CI) (32749-94-3)
Valeraldehyde, 2,4-dimethyl- (8CI) (27944-79-2)
Valeraldehyde, 2-methyl (123-15-9)
Valeraldehyde, 4-methyl (1119-16-0)
Valeraldehyde, 3-methyl- (8CI) (15877-57-3)
Valeraldehyde, 3-oxo- (8CI) (623-38-1)
Valeramide (8CI) (626-97-1)
Valeramide, N-acetyl- (7CI,8CI) (10601-69-1)
Valeramide, N-methyl- (8CI) (6225-10-1)
Valeranilide, 4'-chloro-2,2-dimethyl (7287-36-7)
Valeranilide, 3',4'-dichloro-2-methyl (2533-89-3)
Valerian Root (8057-49-6)
Valeriana Officinalis l., Root extract (8057-49-6)
Valerianic acid (109-52-4)
Valerianic aldehyde (110-62-3)
Valeric-acid (109-52-4)
n-Valeric acid (109-52-4)
Valeric acid [NA 1760] (109-52-4)
Valeric acid aldehyde (110-62-3)
Valeric acid, 2-amino-3-methyl- (73-32-5)
Valeric acid, 2-amino-4-methyl- (61-90-5)
Valeric acid, 4,4'-azobis(4-cyano (2638-94-0)
Valeric acid, 4,4-bis(tert-butylperoxy)-, butyl ester (995-33-5)
Valeric acid, 4,4-bis(p-hydroxyphenyl)- (8CI) (126-00-1)
Valeric acid, 2-bromo (584-93-0)
Valeric acid, α-bromo- (584-93-0)
Valeric acid, 2-chloro- (8CI) (6155-96-0)
Valeric acid, 5-chloro- (8CI) (1119-46-6)
Valeric acid, 2,2-dimethyl (1185-39-3)
Valeric acid, 2,3-epoxy-3,4-dimethyl-, tert-butyl ester, cis- (8CI) (24222-06-8)
Valeric acid, 2,3-epoxy-3,4-dimethyl-, ethyl ester, cis- (8CI) (24222-05-7)
Valeric acid, 2-ethyl- (8CI) (20225-24-5)
Valeric acid, ethyl ester (8CI) (539-82-2)
Valeric acid, 5-hydroxy- (8CI) (13392-69-3)
Valeric acid, δ-hydroxy-, δ-lactone (542-28-9)
Valeric acid, 4-hydroxy-2-methyl-, γ-lactone (5145-01-7)
Valeric acid, 4-hydroxy-2-methyl-, lactone (5145-01-7)
Valeric acid, 4-hydroxy-4-methyl-, γ-lactone (3123-97-5)
Valeric acid, 4-hydroxy-4-methyl-, lactone (3123-97-5)
Valeric acid, 2-methyl (97-61-0)
Valeric acid, methyl ester (624-24-8)
Valeric acid, 4-methyl-4-nitro-, ethyl ester (8CI) (23102-02-5)
Valeric acid, 3-methyl-2-oxo- (8CI) (1460-34-0)
Valeric acid, 4-methyl-, pentyl ester (8CI) (25415-71-8)
Valeric acid, 4-oxo- (123-76-2)
Valeric acid, pentyl ester (8CI) (2173-56-0)
Valeric acid, 4-phenyl (16433-43-5)
Valeric acid, 5-phenyl (2270-20-4)
Valeric acid, 2-propyl (99-66-1)
Valeric acid, 2-propyl-, sodium salt (1069-66-5)
Valeric acid, sodium salt (8CI) (6106-41-8)
Valeric acid, 2,2,4,4-tetramethyl- (6CI,7CI,8CI) (3302-12-3)
Valeric aldehyde (110-62-3)
n-Valeric aldehyde (110-62-3)
1,5-Valerodinitrile, 2-methyl (4553-62-2)
4-Valerolactone (108-29-2)
5-Valerolactone (542-28-9)

δ-Valerolactone **(542-28-9)**
γ-Valerolactone **(108-29-2)**
γ-Valerolactone **(542-28-9)**
γ-Valerolakton (Czech) **(108-29-2)**
Valeron **(9002-88-4)**
Valerone **(108-83-8)**
Valeronitrile **(110-59-8)**
n-Valeronitrile **(110-59-8)**
Valeronitrile, 2,2'-azobis(2,4-dimethyl **(4419-11-8)**
Valeronitrile, 5-((3,4-dimethoxyphenethyl)methylamino)-2-(3,4-dimethoxyphenyl)-2-isopropyl **(52-53-9)**
Valeronitrile, 4-methyl **(542-54-1)**
δ-Valerosultone **(1633-83-6)**
p-Valerotoluidide. 3'-chloro-2-methyl **(2307-68-8)**
Valerylaldehyde **(110-62-3)**
Valeryl chloride **(638-29-9)**
δ-Valeryllactone **(542-28-9)**
Valetan **(15307-79-6)**
Valexon **(14816-18-3)**
Valexone **(14816-18-3)**
Valfor **(1302-76-7)**
Valgesic **(103-90-2)**
Valgis **(50-35-1)**
Valgraine **(50-35-1)**
D-Valine **(640-68-6)**
L-Valine **(72-18-4)**
DL-Valine (9CI) **(516-06-3)**
Valine, DL- (8CI) **(516-06-3)**
Valine, D **(640-68-6)**
Valine, L **(72-18-4)**
Valine aldehyde **(78-84-2)**
DL-Valine, N-(2-chloro-4-(trifluoromethyl)phenyl)-, cyano-(3-phenoxyphenyl)methyl ester **(69409-94-5)**
Valine, N-(N(2)-(N-(N-(N-(N-(N-glycyl-L-isoleucyl)glycyl)-L-alanyl)-L-valyl)-L-leucyl)-L-lysyl)-, monoacetate, L- (8CI) **(31062-69-8)**
Valine, N-lauroyl-, L- (8CI) **(14379-28-3)**
D-Valine, 3-mercapto- **(52-67-5)**
Valine, 3-mercapto-, D **(52-67-5)**
L-Valine, N-(1-oxododecyl)- (9CI) **(14379-28-3)**
Valinomicin **(2001-95-8)**
Valinomycin **(2001-95-8)**
Valioil **(129-20-4)**
Valitran **(439-14-5)**
Valium **(439-14-5)**
Vallene **(64-55-1)**
Vallergine **(60-87-7)**
Vallestril **(517-18-0)**
Valmid **(126-52-3)**
Valmidate **(126-52-3)**
Valmiran **(2259-96-3)**
Valoid **(82-92-8)**
Valone **(83-28-3)**
Valpin **(80-50-2)**
Valproate **(99-66-1)**
Valproate sodium **(1069-66-5)**
Valproic acid **(99-66-1)**
Valproic acid sodium salt **(1069-66-5)**
Valspex 155-53 **(9002-88-4)**
Valsyn **(139-91-3)**
Value 2205 **(9002-92-0)**
Value 3608 **(9036-19-5)**
Valylene **(78-80-8)**
Valzin **(150-69-6)**
Vamidoate **(2275-23-2)**
Vamidothion **(2275-23-2)**
Vampirinip II **(81-81-2)**
Vampirinip III **(81-81-2)**
Van Dyk 264 **(113-48-4)**
Vanadate, hexasodium (9CI) **(12200-88-3)**
Vanadic acid, ammonium salt **(11115-67-6)**
Vanadic acid, ammonium salt **(7803-55-6)**
Vanadic acid, monosodium salt **(13718-26-8)**
Vanadic anhydride **(1314-62-1)**
Vanadic oxide **(1314-34-7)**
Vanadio, pentossido di (Italian) **(1314-62-1)**
Vanadium **(7440-62-2)**
Vanadium Dust and Fume (ACGIH) **(1314-62-1)**
Vanadium chloride **(7632-51-1)**
Vanadium(III) chloride **(7718-98-1)**
Vanadium-dichloride **(10580-52-6)**
Vanadium, dichlorodi-pi-cyclopentadienyl **(12083-48-6)**
Vanadium oxide **(1314-34-7)**
Vanadium oxide (V2O5) **(1314-62-1)**
Vanadium, oxysulfato **(27774-13-6)**
Vanadium oxytrichloride [UN 2443] **(7727-18-6)**
Vanadium pentaoxide **(1314-62-1)**
Vanadiumpentoxid (German) **(1314-62-1)**
Vanadium pentoxide (Dust) **(1314-62-1)**
Vanadium pentoxide (Fume) **(1314-62-1)**
Vanadium pentoxide, Non-fused form [UN 2862] **(1314-62-1)**
Vanadiumpentoxyde (Dutch) **(1314-62-1)**
Vanadium, pentoxyde de (French) **(1314-62-1)**
Vanadium sesquioxide **(1314-34-7)**
Vanadium tetrachloride [UN 2444] **(7632-51-1)**
Vanadium trichloride [UN 2475] **(7718-98-1)**
Vanadium, trichlorooxo **(7727-18-6)**
Vanadium-trioxide **(1314-34-7)**
Vanadium trioxide, Nonfused form [UN 2860] **(1314-34-7)**
Vanadyl sulfate [UN 2931] **(27774-13-6)**
Vanadyl trichloride **(7727-18-6)**
Vanalate **(1929-77-7)**
Vananote **(100-06-1)**
Vanax NS **(95-31-8)**
Vanay **(102-76-1)**
Vancida TM-95 **(137-26-8)**
Vancide 40 **(2224-44-4)**
Vancide 89 **(133-06-2)**
Vancide BL **(97-18-7)**
Vancide F 5386 **(2224-44-4)**
Vancide FE95 **(14484-64-1)**
Vancide Fe95 **(14484-64-1)**
Vancide KS **(76-87-9)**
Vancide MZ-96 **(137-30-4)**
Vancide Maneb 80 **(12427-38-2)**
Vancide P **(13463-41-7)**
Vancide P-75 **(133-06-2)**
Vancide PA **(1113-14-0)**
Vancide PA Dispersion **(1113-14-0)**
Vancide PB **(6379-46-0)**
Vancide 89RE **(133-06-2)**
Vancide TBS **(87-10-5)**
Vancide TH **(7779-27-3)**
Vancide TM **(137-26-8)**
Vancocin **(1404-90-6)**
Vancomycin **(1404-90-6)**
Vandem VAC **(1314-13-2)**
Vandem VOC **(1314-13-2)**
Vandex **(7782-49-2)**
Vandyke Brown **(1317-34-6)**
Vangard K **(133-06-2)**
Vanguard K **(133-06-2)**
Vanguard N **(13927-77-0)**
Vanicell **(8061-51-6)**
Vanicide **(133-06-2)**

Name	CAS
Vanilla	(121-33-5)
Vanilla Plant	(8024-14-4)
Vanillacetic acid	(306-08-1)
Vanillal	(121-32-4)
Vanillaldehyde	(121-33-5)
Vanillic-acid	(121-34-6)
o-Vanillic acid	(877-22-5)
p-Vanillic acid	(121-34-6)
Vanillic acid, 5-chloro-	(62936-23-6)
Vanillic aldehyde	(121-33-5)
Vanillin	(121-33-5)
o-Vanillin	(148-53-8)
p-Vanillin	(121-33-5)
Vanillin Black Liquor	(68514-06-7)
Vanillin Still Bottoms	(73246-95-4)
Vanillin, 6-bromo- (6CI)	(60632-40-8)
Vanillin, 5-chloro- (8CI)	(19463-48-0)
Vanillin, 6-chloro- (8CI)	(18268-76-3)
Vanillin, 5,6-dichloro- (8CI)	(18268-69-4)
Vanillin, ethyl-	(121-32-4)
Vanillin methyl ether	(120-14-9)
Vanillyl acetone	(122-48-5)
Vanillyl alcohol, α-methyl- (8CI)	(2480-86-6)
Vanilmandelic acid	(55-10-7)
Vanilpyruvic acid	(1081-71-6)
Vanirex HW	(8061-51-6)
Vanirex N	(8061-51-6)
Vanirom	(121-32-4)
Vanisperse	(8061-51-6)
Vanlube 81	(101-67-7)
Vanlube PC	(128-37-0)
Vanlube PCX	(128-37-0)
Vanoxide	(94-36-0)
Vansil	(13983-17-0)
Vansil W 10	(13983-17-0)
Vansil W 20	(13983-17-0)
Vansil W 30	(13983-17-0)
Vantyl	(132-60-5)
Vanzoate	(120-51-4)
Vapam	(137-42-8)
Vapam	(6734-80-1)
Vapedrine	(60-13-9)
Vapin	(80-50-2)
Vapo-Iso	(51-30-9)
Vapona	(62-73-7)
Vapona insecticide	(62-73-7)
Vaponefrin	(51-43-4)
Vaponite	(62-73-7)
Vapophos	(56-38-2)
Vapora II	(62-73-7)
Vaporole	(110-46-3)
Vaporooter	(6734-80-1)
Vapotone	(107-49-3)
Varamid ML 1	(120-40-1)
Varcum 5169 (9CI)	(59800-48-5)
Vardhak	(86-87-3)
Varfine	(129-06-6)
Variamine Blue Salt RT	(101-54-2)
Varicol	(139-88-8)
Varioform I	(6484-52-2)
Varioform II	(57-13-6)
Variquat E 228	(112-02-7)
Varisoft 100	(107-64-2)
Varisoft 222 (9CI)	(59113-22-3)
Varisoft SDC	(122-19-0)
Varitox	(650-51-1)
Varitox	(76-03-9)
Varnish Marker's Naphtha	(8032-32-4)
Varonic K 215	(61791-14-8)
Varonic K 205LC	(61791-14-8)
Varox	(78-63-7)
Varox DCP-R	(80-43-3)
Varox DCP-T	(80-43-3)
Vasal	(61-25-6)
Vasazol	(54-95-5)
Vasazol	(59-26-7)
Vascardin	(87-33-2)
Vascuals	(59-02-9)
Vaseline	(8009-03-8)
Vasitol	(78-11-5)
Vaso-80 Unicelies	(78-11-5)
Vasocon	(550-99-2)
Vasoconstrictine	(51-43-4)
Vasoconstrictor	(51-43-4)
Vasodiatol	(78-11-5)
Vasodilan	(395-28-8)
Vasodilian	(395-28-8)
Vasodrine	(51-43-4)
Vasofilina	(317-34-0)
Vasoglyn	(55-63-0)
Vasolan	(52-53-9)
Vasoliment	(8009-03-8)
Vasoperif	(51-83-2)
Vasophysin	(11000-17-2)
Vasopressin	(11000-17-2)
Vasopressina	(11000-17-2)
Vasopressine	(11000-17-2)
Vasopressinum	(11000-17-2)
Vasorbate	(87-33-2)
Vasorex	(54-95-5)
Vasorome	(53-39-4)
Vasoton	(51-43-4)
Vasotonin	(51-43-4)
Vasoxine	(61-16-5)
Vasoxine hydrochloride	(61-16-5)
Vasoxyl hydrochloride	(61-16-5)
Vat Blue 1	(482-89-3)
Vat Blue 4	(81-77-6)
Vat Blue 6	(130-20-1)
Vat Blue 18	(1324-54-5)
Vat Blue O	(81-77-6)
Vat Blue OD	(81-77-6)
Vat Bright Violet K	(1324-55-6)
Vat Brilliant Green C	(128-58-5)
Vat Brilliant Green S	(128-58-5)
Vat Brilliant Orange	(4424-06-0)
Vat Brilliant Violet K	(1324-55-6)
Vat Brilliant Violet KD	(1324-55-6)
Vat Brilliant Violet KP	(1324-55-6)
Vat Fast Blue R	(81-77-6)
Vat Golden Yellow	(128-66-5)
Vat Green 3	(3271-76-9)
Vat Green B	(130-20-1)
Vat Orange 9	(128-70-1)
Vat Orange R	(3263-31-8)
Vat Orange RF	(3263-31-8)
Vat Pink FF	(2379-74-0)
Vat Printing Orange R	(3263-31-8)
Vat Printing Pink FF	(2379-74-0)
Vat Red 1	(2379-74-0)
Vat Red 13	(4203-77-4)
Vat Scarlet 2Zh	(4424-06-0)
Vat Sky Blue K	(130-20-1)
Vat Yellow 2	(129-09-9)

Vatran (439-14-5)
Vatrolite (7775-14-6)
Vatsol OT (577-11-7)
Vazadrine (54-85-3)
Vazo 64 (78-67-1)
Vebecillin (87-08-1)
Vectal (1912-24-9)
Vectal SC (1912-24-9)
Vederon (54-85-3)
Vedita 250 (290-87-9)
Vedril (9011-14-7)
Vedril 5 (9011-14-7)
Vedril 8 (9011-14-7)
Vee Gee Gelatin (9000-70-8)
Veetids (132-98-9)
Vegaben (133-90-4)
Vegadex (95-06-7)
Vegadex Super (95-06-7)
Veganine (62-44-2)
Vegantine (64-95-9)
Vegedex (95-06-7)
Vegentine Blue CSW (2610-05-1)
Vegentine Fast Brown B (2429-81-4)
Vegetable Oil (68956-68-3)
Vegetable Oil Mist (68956-68-3)
Vegetable Oil Mist, Respirable fraction (OSHA) (68956-68-3)
Vegetable Oil Mist, Total dust (OSHA) (68956-68-3)
Vegetox (15263-52-2)
Vegfru (298-02-2)
Vegfru Malatox (121-75-5)
Vegfru fosmite (563-12-2)
Vegiben (133-90-4)
Vel 88 (116-01-8)
Vel 3973 (53780-34-0)
Vel 4284 (2540-82-1)
Vel 59CS52 (3004-74-8)
Velban (143-67-9)
Velbe (143-67-9)
Veldopa (59-92-7)
Velium (439-14-5)
Velmol (577-11-7)
Velon (9011-06-7)
Velosef (38821-53-3)
Velpar (51235-04-2)
Velpar Weed Killer (51235-04-2)
Velsicol (2307-49-5)
Velsicol 104 (76-44-8)
Velsicol 506 (21609-90-5)
Velsicol 1068 (57-74-9)
Velsicol C (2307-49-5)
Velsicol 53-CS-17 (1024-57-3)
Velsicol 58-CS-11 (1918-00-9)
Velsicol 58-CS-25 (2307-49-5)
Velsicol Compound C (2307-49-5)
Velsicol Compound "R" (1918-00-9)
Velsicol Heptachlor (76-44-8)
Velsicol VCS 506 (21609-90-5)
Velustral KPA (9002-88-4)
Velvetex (1333-86-4)
Velvetol OE 2NT1 (9003-11-6)
Vena (147-24-0)
Vena (58-73-1)
Vencipon (299-42-3)
Vendacid Light Orange 2G (1936-15-8)
Vendex (13356-08-6)
Venetian Red (1309-37-1)
Vengicide (606-58-6)
Ventox (107-13-1)
Ventramine (59-26-7)
Ventrazol (54-95-5)
Venturol (2439-10-3)
Venzar (2164-08-1)
Venzonate (120-51-4)
Veon (93-76-5)
Veon 245 (93-76-5)
Vepen (132-98-9)
Verantin (81-54-9)
Verapamil (52-53-9)
Verapret AN (136-84-5)
Verapret DH (1854-26-8)
Verapret DKH (1854-26-8)
Veratraldehyde (120-14-9)
Veratramine (60-70-8)
Veratramine 3-glycoside (475-00-3)
Veratric acid (93-07-2)
Veratric acid, 6-chloro-, methyl ester (8CI) (30714-88-6)
o-Veratric acid, 6-((6-(2-(dimethylamino)ethyl)-2-methoxy-3,4-(methylenedioxy)phenyl)acetyl) (131-28-2)
Veratric acid, 5-methoxy- (118-41-2)
Veratric acid, methyl ester (8CI) (2150-38-1)
Veratric aldehyde (120-14-9)
p-Veratric aldehyde (120-14-9)
Veratridine (71-62-5)
Veratridine (8051-02-3)
Veratrin (German) (8051-02-3)
Veratrine (62-59-9)
Veratrine (8051-02-3)
Veratrine (Amorphous) (71-62-5)
Veratrine (Crystallized) (62-59-9)
Veratrol (91-16-7)
Veratrole (91-16-7)
Veratrole methyl ether (93-15-2)
Veratrosine (475-00-3)
3-Veratroylveracevine (71-62-5)
Veratrum Viride (65072-04-0)
Veratrum Viride Alkaloids Extract (65072-04-0)
Veratryl alcohol (93-03-8)
Veratryl alcohol, .α.-(2-hydroxy-1-(p-methoxyphenoxy)ethyl)-(8CI) (22676-00-2)
Veratryl aldehyde (120-14-9)
Veratrylglycerol .β.-phenyl ether (75217-43-5)
Verazinc (7733-02-0)
d-Verbenone (18309-32-5)
Vercidon (514-73-8)
Vercyte (54-91-1)
Verdican (62-73-7)
Verdict (69806-40-2)
Verdipor (62-73-7)
Verdone (94-74-6)
Verdyl acetate (5413-60-5)
Veresene disodium salt (139-33-3)
Vergemaster (94-75-7)
Vergfru foratox (298-02-2)
Vergonil (135-09-1)
Veriloid (50-55-5)
Veriloid (65072-04-0)
Verindal Ultra (58-89-9)
Vermadax (23564-06-9)
Vermicid (548-62-9)
Vermicide Bayer 2349 (52-68-6)
Vermiculite (1318-00-9)
Verminum (14255-87-9)
Vermirax (31431-39-7)
Vermitin (50-65-7)

Name	CAS No.
Vermitin	(92-84-2)
Vermoestricid	(56-23-5)
Vermox	(31431-39-7)
Vernam	(1929-77-7)
Vernam 7E	(1929-77-7)
Vernatzer 30/10	(104-38-1)
Vernine	(118-00-3)
Vernolate	(1929-77-7)
Verol	(55-55-0)
Veroletten	(57-44-3)
Verolettin	(57-44-3)
Verona Yellow X-1791	(6358-85-6)
Veronal	(57-44-3)
Verophen	(58-40-2)
Verosin	(1397-94-0)
Verospiron	(52-01-7)
Verospirone	(52-01-7)
Verrol	(59-02-9)
Versa TL 71	(9080-79-9)
Versa TL 400	(9080-79-9)
Versa TL 500	(9080-79-9)
Versal Blue B	(147-14-8)
Versal Blue BG	(147-14-8)
Versal Blue GGSL	(81-77-6)
Versal Fast Red R	(2814-77-9)
Versal Green G	(1328-53-6)
Versal Orange RNL	(3468-63-1)
Versal Scarlet PRNL	(2425-85-6)
Versal Scarlet RNL	(2425-85-6)
Versar DSMA LQ	(144-21-8)
Versatic 9-11	(71700-95-3)
Versatic 9-11 acid	(71700-95-3)
Versatic acid 911	(71700-95-3)
Versen-Ol	(139-89-9)
Versene	(60-00-4)
Versene	(64-02-8)
Versene 9	(150-38-9)
Versene 67	(64-02-8)
Versene 100	(64-02-8)
Versene 220	(64-02-8)
Versene Acid	(60-00-4)
Versene Beads	(64-02-8)
Versene FE 3	(64-02-8)
Versene Flake	(64-02-8)
Versene NTA Acid	(139-13-9)
Versene Powder	(64-02-8)
Versene Powder tetra sodium	(64-02-8)
Versene Sodium 2	(139-33-3)
Versenex 80	(140-01-2)
Versenol	(139-89-9)
Versenol	(150-39-0)
Versenol 120	(139-89-9)
Versenol 120	(150-39-0)
Versicol E 7	(9003-01-4)
Versicol E9	(9003-01-4)
Versicol E15	(9003-01-4)
Versicol S 25	(9003-01-4)
Versneller NL 49	(136-52-7)
Versneller NL 63/10	(121-69-7)
Versomnal	(50-06-6)
Verstran	(2955-38-6)
Vertac	(25013-16-5)
Vertac	(709-98-8)
Vertac 90%	(8001-35-2)
Vertac Dinitro Weed Killer	(88-85-7)
Vertac General Weed Killer	(88-85-7)
Vertac Methyl Parathion Technisch 80%	(298-00-0)
Vertac Selective Weed Killer	(88-85-7)
Vertac Toxaphene 90	(8001-35-2)
Vertafume	(8004-09-9)
Vertavis	(65072-04-0)
Vertenex	(32210-23-4)
Verthion	(122-14-5)
Vertisal	(443-48-1)
Vertolan	(57-68-1)
Verton	(94-75-7)
Verton 2D	(94-75-7)
Verton D	(94-75-7)
Verton 2T	(93-76-5)
Vertron 2D	(94-75-7)
Vesakontuho MCPA	(94-74-6)
Vesalium	(52-86-8)
Vesparaz-wirkstoff	(68-88-2)
Vesperal	(57-44-3)
Vesprin	(146-54-3)
Vestin	(136-40-3)
Vestinol AH	(117-81-7)
Vestinol DZ	(26761-40-0)
Vestinol NN	(28553-12-0)
Vestinol OA	(103-23-1)
Vestolen	(9002-88-4)
Vestolen 5200	(9003-07-0)
Vestolen A 6016	(9002-88-4)
Vestolen A 616	(9002-88-4)
Vestolen P 5500	(9003-07-0)
Vestolen P 5232g	(9003-53-6)
Vestrol	(79-01-6)
Vestyron	(9003-53-6)
Vestyron 114-12	(9003-53-6)
Vestyron 512	(9003-53-6)
Vestyron HL	(9003-55-8)
Vestyron MB	(9003-53-6)
Vestyron N	(9003-53-6)
Vetacalm	(113-59-7)
Vetaflavin	(8048-52-0)
Vetalar	(1867-66-9)
Vetalin S	(67-03-8)
Vetalog	(76-25-5)
Vetamox	(59-66-5)
Vetarcillin	(1538-09-6)
Vetbutal	(57-33-0)
Veterinary Nitrofurazone	(59-87-0)
Veticillin	(69-57-8)
Veticol	(56-75-7)
Vetidrex	(58-93-5)
Vetiol	(121-75-5)
Vetiver acetate	(117-98-6)
Vetiverol, acetate	(117-98-6)
Vetivert acetate	(117-98-6)
Vetiveryl acetate	(117-98-6)
Vetkalm	(548-73-2)
Vetol	(118-71-8)
Vetquamycin-324	(64-75-5)
Vetren	(9005-49-6)
Vetsin	(142-47-2)
Vetspen	(54-35-3)
Vetstrep	(3810-74-0)
Vi-A	(68-26-8)
Vi-Nicotyl	(98-92-0)
Vi-Noctyl	(98-92-0)
Vi-Par	(93-65-2)
Vi-Pex	(1929-86-8)
Vi-Pex	(93-65-2)
Vi-Twel	(68-19-9)

Vialidon (61-68-7)
Vialin (100-92-5)
Vianin (548-62-9)
Vianol (128-37-0)
Vibalt (68-19-9)
Vibatex S (9002-89-5)
Vibazine (569-65-3)
Vibeden (13422-51-0)
Vibisone (68-19-9)
Vibramycin (564-25-0)
Vicalin (1304-85-4)
Viccillin (69-53-4)
Viccillin S (69-53-4)
Vicelat (50-81-7)
Vicillin (69-53-4)
Vicin (1538-09-6)
Vicin (50-81-7)
Vicknite (7757-79-1)
Vicomin C (50-81-7)
Victor TSPP (7722-88-5)
Victoria Blue B (2580-56-5)
Victoria Blue R (2185-86-6)
Victoria Blue RS (2185-86-6)
Victoria Green (569-64-2)
Victoria Green B (569-64-2)
Victoria Green S (569-64-2)
Victoria Green WB (569-64-2)
Victoria Green WPB (569-64-2)
Victoria Lake Blue R (2185-86-6)
Victoria Orange (609-93-8)
Victoria Pure Blue B (1325-87-7)
Victoria Rubine O (915-67-3)
Victoria Rubine O for Food (915-67-3)
Victoria Scarlet 3R (2611-82-7)
Victoria Scarlet Red (2611-82-7)
Victoria Scharlach 4 R Extra (German) (2611-82-7)
Victoria Yellow (609-93-8)
Vidden D (542-75-6)
Vidden D (8003-19-8)
Viden (62-44-2)
Vidine (123-41-1)
Vidlon (105-60-2)
Vidlon (25038-54-4)
Vidon 638 (94-75-7)
Vidopen (7177-48-2)
Vienna Green (12002-03-8)
Viforcit (50-81-7)
D3-Vigantol (67-97-0)
Vigantol (50-14-6)
Vigil (75736-33-3)
Vigorsan (67-97-0)
Vikane (2699-79-8)
Vikane Fumigant (2699-79-8)
Vikasolum (130-37-0)
Vikh 65 (9011-06-7)
Vikor (814-71-1)
Vikrol RQ (8001-54-5)
Vilit 40 (9003-22-9)
Villiaumite (15096-52-3)
Villiaumite (7681-49-4)
Vilnite (13983-17-0)
Viloxazin (46817-91-8)
Viloxazine (46817-91-8)
Vinac (9003-20-7)
Vinac ASB 10 (9003-20-7)
Vinac B 7 (9003-20-7)
Vinac RP251 (9003-20-7)
Vinacet D (9003-20-7)
Vinacol MH (9002-89-5)
Vinadine (58-36-6)
Vinalak (9002-89-5)
Vinalite D 50N (9003-20-7)
Vinalite DS 41/11 (9003-20-7)
Vinamar (109-92-2)
Vinamul 9300 (9003-20-7)
Vinamul N 710 (9003-53-6)
Vinamul N 7700 (9003-53-6)
Vinapol A 16 (9003-20-7)
Vinarol (9002-89-5)
Vinarol DT (9002-89-5)
Vinarol ST (9002-89-5)
Vinarole (9002-89-5)
Vinavilol 2-98 (9002-89-5)
Vinbarbital (125-42-8)
Vinbarbital (VAN) (125-42-8)
Vinbarbitalum (Latin) (125-42-8)
Vinbarbitone (125-42-8)
Vinblastin (865-21-4)
Vinblastine (865-21-4)
Vinblastine sulfate (143-67-9)
Vincaleucoblastin (865-21-4)
Vincaleucoblastine (865-21-4)
Vincaleukoblastine (865-21-4)
Vincaleukoblastine, 4'-deoxy-3',4'-epoxy-, (3'-α,4'-α)- (9CI) (23360-92-1)
Vincaleukoblastine, 22-oxo- (57-22-7)
Vincaleukoblastine, 22-oxo-, sulfate (1:1) (Salt) (2068-78-2)
Vincaleukoblastine, sulfate (143-67-9)
Vincaleukoblastine, sulfate (1:1) (Salt) (143-67-9)
Vincamin compositum (637-07-0)
Vinces (80-35-3)
Vinclozolin (German) (50471-44-8)
Vinclozoline (50471-44-8)
Vincoblastine (865-21-4)
Vincristine (57-22-7)
Vincristine sulfate (2068-78-2)
Vincristinsulfat (German) (2068-78-2)
Vincrisul (2068-78-2)
Vincrystine (57-22-7)
Vincubina (826-36-8)
Vincubine (826-36-8)
Vinegar Naphtha (141-78-6)
Vinegar acid (64-19-7)
Vinegar salts (62-54-4)
Vinesthene (109-93-3)
Vinesthesin (109-93-3)
Vinethen (109-93-3)
Vinethene (109-93-3)
Vinether (109-93-3)
Vinfos (22248-79-9)
Vinicizer 80 (117-81-7)
Vinicizer 85 (117-84-0)
Viniden 60 (9011-06-7)
Vinidyl (109-93-3)
Vinile (acetato di) (Italian) (108-05-4)
Vinile (bromuro di) (Italian) (593-60-2)
Vinile (cloruro di) (Italian) (75-01-4)
Vinipaint 555 (9003-20-7)
Vinisil (9003-39-8)
Vinkeil 100 (60-00-4)
Vinkristin (57-22-7)
Vinleurosine (23360-92-1)
4-Vinlycyclohexene dioxide (106-87-6)
Vinnapas B (9003-20-7)
Vinnapas B 17 (9003-20-7)

Vinnapas B 100 (9003-20-7)
Vinnapas UW 50 (9003-20-7)
Vinnarol (9002-89-5)
Vinnol H 10/60 (9003-22-9)
Vinnol H 15/45 (9003-22-9)
Vinnol H 40/60 (9003-22-9)
Vinoflex MO 400* (109-53-5)
Vinol (9002-89-5)
Vinol 125 (9002-89-5)
Vinol 205 (9002-89-5)
Vinol 351 (9002-89-5)
Vinol 523 (9002-89-5)
Vinol Unisize (9002-89-5)
Vinothiam (59-43-8)
Vinothiam (67-03-8)
Vinstop (128-04-1)
Vinydan (109-93-3)
Vinyl A Monomer (108-05-4)
Vinyl Alcohol Polymer (9002-89-5)
Vinyl C monomer (75-01-4)
Vinyl Products R 10688 (9003-20-7)
Vinyl Products R 3612 (9003-53-6)
Vinylacetaat (Dutch) (108-05-4)
Vinylacetat (German) (108-05-4)
Vinyl acetate (ACGIH,DOT,OSHA) (108-05-4)
Vinyl acetate H.Q. (108-05-4)
Vinyl acetate, Inhibited [UN 1301] (108-05-4)
Vinyl acetate, butyl acrylate, N-methylolacrylamide polymer (26428-41-1)
Vinyl acetate, ethene polymer (24937-78-8)
Vinyl acetate, ethylene polymer (24937-78-8)
Vinyl acetate homopolymer (9003-20-7)
Vinyl acetate polymer (9003-20-7)
Vinyl acetate resin (9003-20-7)
Vinyl acetate-vinyl chloride copolymer (9003-22-9)
Vinyl acetate-vinyl chloride polymer (9003-22-9)
Vinylacetonitrile (109-75-1)
Vinylacetylene (689-97-4)
Vinyl alcohol, 2,2-dichloro-, dimethyl phosphate (62-73-7)
Vinyl allyl ether (3917-15-5)
Vinyl amide (79-06-1)
Vinylamine, N-ethyl-N-nitroso (13256-13-8)
Vinylamine, N-methyl-N-nitroso (4549-40-0)
Vinylbenzen (Czech) (100-42-5)
Vinylbenzene (100-42-5)
Vinyl benzene (OSHA) (100-42-5)
Vinylbenzene polymer (9003-53-6)
Vinylbenzol (100-42-5)
Vinylbenzyl chloride (30030-25-2)
Vinylbromid (German) (593-60-2)
Vinyl bromide (ACGIH,OSHA) (593-60-2)
Vinyl bromide, Inhibited [UN 1085] (593-60-2)
Vinyl 2-butenoate (14861-06-4)
Vinyl 2-(butoxyethyl) ether (4223-11-4)
Vinyl butyl ether (111-34-2)
Vinyl n-butyl ether (111-34-2)
Vinyl butyrate (123-20-6)
Vinyl butyrate, Inhibited [UN 2838] (123-20-6)
Vinylbutyrolactam (88-12-0)
n-Vinylbutyrolactam polymer (9003-39-8)
9-Vinylcarbazole (1484-13-5)
N-Vinylcarbazole (1484-13-5)
Vinylcarbazole (1484-13-5)
Vinylcarbinol (107-18-6)
Vinylchlorid (German) (75-01-4)
Vinyl chloride (DOT,OSHA) (75-01-4)
Vinyl chloride, Inhibited [UN 1086] (75-01-4)
Vinyl chloride acetate copolymer (34149-92-3)
Vinylchloride-acetate copolymer (34149-92-3)
Vinyl chloride copolymer with vinylidene chloride (9011-06-7)
Vinyl chloride-1,1-dichloroethylene copolymer (9011-06-7)
Vinyl chloride monomer (75-01-4)
Vinyl chloride vinyl acetate copolymer (34149-92-3)
Vinyl chloride-vinyl acetate copolymers (34149-92-3)
Vinyl chloride-vinyl acetate polymer (9003-22-9)
Vinyl chloride-vinylidene chloride copolymer (9011-06-7)
Vinyl chloride-vinylidene chloride polymer (9011-06-7)
Vinyl chloroacetate [UN 2589] (2549-51-1)
Vinylchlorodimethylsilane (1719-58-0)
Vinyl 2-chloroethyl ether (110-75-8)
Vinyl β-chloroethyl ether (110-75-8)
Vinyl crotonate (14861-06-4)
Vinyl cyanide (107-13-1)
Vinylcyanide (OSHA) (107-13-1)
4-Vinylcyclohexane, 1,2-epoxide (106-86-5)
Vinylcyclohexane monoxide (106-86-5)
1-Vinylcyclohex-3-ene (100-40-3)
1-Vinylcyclohexene (2622-21-1)
1-Vinylcyclohexene-3 (100-40-3)
4-Vinyl-1-cyclohexene (100-40-3)
4-Vinylcyclohexene (100-40-3)
4-Vinylcyclohexene-1 (100-40-3)
4-Vinyl-1,2-cyclohexene diepoxide (106-87-6)
4-Vinyl-1-cyclohexene diepoxide (106-87-6)
4-Vinylcyclohexene diepoxide (106-87-6)
Vinyl cyclohexene diepoxide (106-87-6)
1-Vinyl-3-cyclohexene dioxide (106-87-6)
4-Vinyl-1-cyclohexene dioxide (106-87-6)
Vinyl cyclohexene dioxide (ACGIH,OSHA) (106-87-6)
4-Vinylcyclohexene-1,2-epoxide (106-86-5)
4-Vinylcyclohexene monoxide (106-86-5)
Vinylcyclohexene monoxide (106-86-5)
Vinyl(diethyl)methylsilane (18292-29-0)
Vinyldimethylchlorosilane (1719-58-0)
Vinyle (acetate de) (French) (108-05-4)
Vinyle (bromure de) (French) (593-60-2)
Vinyle(chlorure de) (French) (75-01-4)
Vinylene carbonate (872-36-6)
2,2'-(Vinylenedi-4-phenylene)bis(benzoxazole) (1533-45-5)
1-Vinyl-3,4-epoxycyclohexane (106-86-5)
Vinylester kyseliny 2-ethylkapronove (Czech) (94-04-2)
Vinylester kyseliny krotonove (Czech) (14861-06-4)
Vinylester kyseliny maselne (Czech) (123-20-6)
Vinylester kyseliny mravenci (Czech) (692-45-5)
Vinylester kyseliny octove (Czech) (108-05-4)
Vinylester kyseliny propionove (Czech) (105-38-4)
Vinyl ethanoate (108-05-4)
Vinyl-ether (109-93-3)
Vinylethylene (106-99-0)
Vinyl ethyl ether (109-92-2)
Vinyl ethyl ether, Inhibited [UN 1302] (109-92-2)
Vinyl 2-ethylhexanoate (94-04-2)
Vinyl 2-ethylhexoate (94-04-2)
Vinyl-2-ethylhexoate (94-04-2)
Vinyl 2-ethylhexyl ether (103-44-6)
Vinylethylnitrosamin (German) (13256-13-8)
Vinylethylnitrosamine (13256-13-8)
2-Vinyl-5-ethylpyridine (5408-74-2)
Vinyl fluoride, Inhibited [UN 1860] (75-02-5)
Vinyl formate (692-45-5)
Vinylformic acid (79-10-7)
Vinylidene chloride (II) (75-35-4)
Vinylidene chloride (ACGIH,OSHA) (75-35-4)
Vinylidene chloride, Inhibited [UN 1303] (75-35-4)
Vinylidene chloride-vinyl chloride polymer (9011-06-7)

Vinylidene dichloride (75-35-4)
Vinylidene difluoride (75-38-7)
Vinylidene fluoride (75-38-7)
Vinylidine chloride (75-35-4)
Vinyl isobutyl ether (DOT) (109-53-5)
Vinyl isobutyl ether, Inhibited [UN 1304] (109-53-5)
Vinyl isopropyl ether (926-65-8)
Vinylite AYAF (9003-20-7)
Vinylite AYAT (9003-20-7)
Vinylite VGHH (9003-22-9)
Vinylite VYDR (9003-22-9)
Vinylite VYDR 21 (9003-22-9)
Vinylite VYFS (9003-22-9)
Vinylite VYHD (9003-22-9)
Vinylite VYHH (9003-22-9)
Vinylite VYHH-1 (9003-22-9)
Vinylite VYNS (9003-22-9)
Vinylite VYNW (9003-22-9)
N-Vinylkarbazol (Czech) (1484-13-5)
Vinylkyanid (Czech) (107-13-1)
Vinyl 2-methoxyethyl ether (1663-35-0)
Vinyl methyl ether (DOT) (107-25-5)
Vinyl methyl ether, Inhibited [UN 1087] (107-25-5)
Vinyl methyl ketone (78-94-4)
Vinyl neodecanate (45115-34-2)
2-Vinylnorbornene (3048-64-4)
5-Vinyl-2-norbornene (3048-64-4)
5-Vinylnorbornene (3048-64-4)
Vinylnorbornene (3048-64-4)
Vinylofos (62-73-7)
Vinylon Film 2000 (9002-89-5)
Vinylophos (62-73-7)
3-Vinyl-7-oxabicyclo(4.1.0)heptane (106-86-5)
Vinyloxirane (930-22-3)
Vinylphate (470-90-6)
4-Vinylphenol (2628-17-3)
p-Vinylphenol (2628-17-3)
4-Vinylphenol polymer (24979-70-2)
p-Vinylphenol polymer (24979-70-2)
Vinylphosphate (22248-79-9)
Vinyl propionate (105-38-4)
2-Vinylpyridine (100-69-6)
4-Vinylpyridine (100-43-6)
1-Vinyl-2-pyrrolidinone (88-12-0)
N-Vinyl-2-pyrrolidinone (88-12-0)
N-Vinylpyrrolidinone (88-12-0)
1-Vinyl-2-pyrrolidinone Polymer, Compd. with iodine (25655-41-8)
Vinylpyrrolidinone polymer (9003-39-8)
n-Vinylpyrrolidinone polymer (9003-39-8)
1-Vinyl-2-pyrrolidone (88-12-0)
N-Vinyl-2-pyrrolidone (88-12-0)
N-Vinylpyrrolidone (88-12-0)
Vinylpyrrolidone (88-12-0)
Vinylpyrrolidone polymer (9003-39-8)
n-Vinylpyrrolidone polymer (9003-39-8)
Vinylsilicon trichloride (75-94-5)
Vinyl stearate (111-63-7)
Vinylstyrene (1321-74-0)
m-Vinylstyrene (108-57-6)
Vinyl sulfide (8CI) (627-51-0)
Vinyl sulfone (77-77-0)
Vinyl-sulfoxide (1115-15-7)
2-Vinyltoluene (611-15-4)
3-Vinyltoluene (100-80-1)
4-Vinyltoluene (622-97-9)
Vinyltoluene (25013-15-4)
m-Vinyltoluene (100-80-1)
o-Vinyltoluene (611-15-4)
p-Vinyltoluene (622-97-9)
Vinyl toluene (ACGIH,OSHA) (25013-15-4)
Vinyl toluenes (Mixed isomers), Inhibited [UN 2618] (25013-15-4)
Vinyltriacetoxysilane (4130-08-9)
Vinyl trichloride (79-00-5)
Vinyltrichlorosilane (75-94-5)
Vinyl trichlorosilane, Inhibited (DOT) (75-94-5)
Vinyl trichlorosilane [UN 1305] (75-94-5)
Vinyltriethoxysilane (78-08-0)
Vinyl trimethoxy silane (2768-02-7)
Vinyltrimethylammonium hydroxide (463-88-7)
Vinyltris(2-methoxyethoxy)silane (1067-53-4)
Vinyltris(β-methoxyethoxy)silane (1067-53-4)
Vinyltris(methoxyethoxy)silane (1067-53-4)
Vinyon (9003-22-9)
Vinyzene BP 5 (58-36-6)
Vinyzene BP 5-2 (58-36-6)
Vinyzene (Pesticide) (58-36-6)
Vinyzene SB 1 (58-36-6)
Vio-Bamate (57-53-4)
Vio-Serpine (50-55-5)
Viocid (548-62-9)
Vioform (130-26-7)
Vioform N.N.R. (130-26-7)
Viofuragyn (67-45-8)
Violanthrone (116-71-2)
Violanthrone, amino- (26763-69-9)
Violanthrone, 16,17-dimethoxy- (128-58-5)
Violaquercitrin (153-18-4)
11092 Violet (81-48-1)
11386 Violet (1694-09-3)
12416 Violet (548-62-9)
Violet 2 (1694-09-3)
Violet 3 (1330-20-7)
Violet 5B (1694-09-3)
Violet 6B (1694-09-3)
Violet 5BN (1694-09-3)
Violet 6BN (548-62-9)
Violet BN Acid Anthraquinone (81-78-7)
Violet BNP (4129-84-4)
Violet 5BO (548-62-9)
Violet CP (548-62-9)
Violet Disperzni 4 (Czech) (1220-94-6)
Violet Gencianova (Czech) (548-62-9)
Violet Krystalova (Czech) (548-62-9)
Violet Kypova 1 (Czech) (1324-55-6)
Violet Kysela 49 (Czech) (1694-09-3)
Violet Lake (1325-82-2)
Violet Methylova (Czech) (8004-87-3)
Violet No. 1 (1694-09-3)
Violet Pigment 31 (1324-55-6)
Violet Potravinarska 2 (Czech) (1694-09-3)
Violet Powder H 2503 (8004-87-3)
Violet Prima 1 (Czech) (2586-60-9)
Violet Rozpoustedlova 12 (Czech) (1220-94-6)
Violet Rozpoustedlova 13 (Czech) (81-48-1)
Violet Rozpoustedlova 26 (Czech) (2872-48-2)
Violet 2S (82-33-7)
Violet Toner PTMA 55-2925 (1325-82-2)
Violet XXIII (548-62-9)
Violet Zasadita 1 (Czech) (8004-87-3)
Violet Zasadita 3 (Czech) (548-62-9)
Violet Zasadita 10 (Czech) (81-88-9)
Violet Zasadita 14 (Czech) (632-99-5)
Viologen, methyl- (1910-42-5)
Viophan (132-60-5)

Name	CAS No.
Viopsicol	(58-25-3)
Viosterol	(50-14-6)
Vioxan	(63-25-2)
Viozene	(299-84-3)
Viramid	(36791-04-5)
Virazole	(36791-04-5)
Virchem	(7775-14-6)
Virginia-Carolina VC 9-104	(13194-48-4)
Viricuivre	(1332-40-7)
Viridine	(101-48-4)
Viridiol	(23820-80-6)
Viroptic	(70-00-8)
Virormone	(58-22-0)
Virosterone	(58-22-0)
Virset 656-4	(108-78-1)
Virtex CC	(7775-14-6)
Virtex D	(7775-14-6)
Virtex L	(7775-14-6)
Virtex RD	(7775-14-6)
Virubra	(68-19-9)
Visadron	(59-42-7)
Visadron	(61-76-7)
Viscalex HV 30	(9003-01-4)
Viscarin	(9000-07-1)
Viscarin 402 & TP-4	(9000-07-1)
Viscasil	(63148-62-9)
Viscasil 5000	(63148-62-9)
Viscasil 10000	(63148-62-9)
Visco Black N	(6428-31-5)
Viscoemulsan	(80450-55-1)
Viscoform Navy Blue BG	(7082-31-7)
Viscoform Navy Blue 2GB	(7082-31-7)
Viscol	(9004-67-5)
Viscol 350P	(9003-07-0)
Viscol 550P	(9003-07-0)
Viscol 660 P	(9003-07-0)
Viscoleo-Oil	(68956-68-3)
Viscoleo Oil	(68956-68-3)
Viscon 103	(9003-01-4)
Viscontran L52	(9004-67-5)
Viscope	(6106-46-3)
Viscorin	(50-81-7)
Viscose Black G	(6428-31-5)
Viscose Black GNA	(6428-31-5)
Viscose Black J	(6428-31-5)
Viscose Black N	(6428-31-5)
Viscose Black NG	(6428-31-5)
Viscosol	(9004-67-5)
Visking cellophane	(9005-81-6)
Visko-Rhap	(120-36-5)
Visko-Rhap	(94-75-7)
Visko-Rhap Low Drift Herbicides	(94-75-7)
Visko Rhap Low Volatile Ester	(93-76-5)
Visko-Rhap Low Volatile 4L	(94-75-7)
Vistabamate	(57-53-4)
Vistalan	(142-78-9)
Vistar	(53780-34-0)
Vistar Herbicide	(53780-34-0)
Visubutina	(129-20-4)
Visumetazone	(50-02-2)
Vita-Rubra	(68-19-9)
Vitablend	(54-35-3)
Vitace	(50-81-7)
Vitacee	(50-81-7)
Vitacimin	(50-81-7)
Vitacin	(50-81-7)
Vitadurin	(13422-51-0)
Vitaflo 250	(5234-68-4)
Vitahexin P	(54-47-7)
Vitallium	(11114-92-4)
Vitaloid	(463-88-7)
Vitamin A	(11103-57-4)
Vitamin A	(68-26-8)
Vitamin A1	(11103-57-4)
Vitamin A1	(68-26-8)
13-cis-Vitamin A Acid	(4759-48-2)
Vitamin A Alcohol	(68-26-8)
Vitamin A1 Alcohol	(68-26-8)
all-trans-Vitamin A Alcohol	(68-26-8)
Vitamin AB	(9005-49-6)
Vitamin A (VAN)(9CI)	(11103-57-4)
Vitamin A acetate	(127-47-9)
trans-Vitamin A acetate	(127-47-9)
Vitamin A acid	(302-79-4)
Vitamin A alcohol acetate	(127-47-9)
Vitamin B-5	(137-08-6)
Vitamin B1	(59-43-8)
Vitamin B2	(83-88-5)
Vitamin B3	(79-83-4)
Vitamin B3	(98-92-0)
Vitamin B4	(73-24-5)
Vitamin B5	(79-83-4)
Vitamin B6	(65-23-6)
Vitamin B7	(58-85-5)
Vitamin B11	(59-30-3)
Vitamin B12	(68-19-9)
Vitamin B12A	(13422-51-0)
Vitamin B12B	(68-19-9)
Vitamin BC	(59-30-3)
Vitamin BT	(541-15-1)
Vitamin BX	(150-13-0)
Vitamin B12 complex	(68-19-9)
Vitamin B hydrochloride	(67-03-8)
Vitamin B1 hydrochloride	(67-03-8)
Vitamin B6 hydrochloride	(65-22-5)
Vitamin B6-hydrochloride	(58-56-0)
Vitamin B_1 hydrochloride	(67-03-8)
Vitamin B_{12} preparation	(68-19-9)
Vitamin C	(134-03-2)
Vitamin C	(50-81-7)
Vitamin C sodium	(134-03-2)
Vitamin D1	(67-96-9)
Vitamin D2	(50-14-6)
Vitamin D3	(67-97-0)
Vitamin-D	(1406-16-2)
Vitamin E	(59-02-9)
Vitamin E Hemisuccinate	(4345-03-3)
Vitamin E Succinate	(4345-03-3)
Vitamin G	(83-88-5)
Vitamin H	(150-13-0)
Vitamin H	(58-85-5)
Vitamin K1	(84-80-0)
Vitamin K2(0)	(58-27-5)
Vitamin K2(35)	(2124-57-4)
Vitamin K3	(58-27-5)
Vitamin L	(118-92-3)
Vitamin M	(59-30-3)
Vitamin MK 7	(2124-57-4)
Vitamin P	(153-18-4)
Vitamin PP	(98-92-0)
Vitamine E Succinate	(4345-03-3)
Vitamisin	(50-81-7)
Vitaneuron	(59-43-8)
Vitaneuron	(67-03-8)

Vitaplex E	(59-02-9)
Vitaplex N	(59-67-6)
Vitarubin	(68-19-9)
Vitascorbol	(50-81-7)
Vitavax	(5234-68-4)
Vitavax 100	(5234-68-4)
Vitavax sulfone	(5259-88-1)
Vitavel-A	(68-26-8)
Vitavel-D	(50-14-6)
Vitayonon	(59-02-9)
Vitazechs	(54-47-7)
Vitellin	(8002-43-5)
Viteolin	(59-02-9)
Vitestrol	(84-16-2)
Viticolor	(96-26-4)
Vitigran	(1332-40-7)
Vitigran Blue	(1332-40-7)
Vitinc Dan-Dee-3	(67-97-0)
Viton	(58-89-9)
Vitpex	(68-26-8)
Vitral	(68-19-9)
Vitran	(79-01-6)
Vitreous Quartz	(60676-86-0)
Vitrex	(56-38-2)
Vitriol Brown Oil	(7664-93-9)
Vitriol, Oil of (DOT)	(7664-93-9)
Vitriol Red	(1309-37-1)
Vitrolan Red GRE	(6408-31-7)
Vitrum AB	(9005-49-6)
Vituf	(25038-59-9)
Vivactil	(438-60-8)
Vival	(439-14-5)
Vivol	(439-14-5)
Vladicort	(53-34-9)
Vladipor	(9004-35-7)
Vniivlon M	(63428-84-2)
Vofatox	(298-00-0)
Vogan	(68-26-8)
Vogan-Neu	(68-26-8)
Vogel's Iron Red	(1309-37-1)
Volamin	(126-52-3)
Volatile Oil of Mustard	(57-06-7)
Volaton	(14816-18-3)
Volcaly Bentonite BC	(1302-78-9)
Volclay	(1302-78-9)
Volfartol	(52-68-6)
Volfazol	(7700-17-6)
Volid	(56073-10-0)
Volidan	(595-33-5)
Volital	(2152-34-3)
Volitol	(2152-34-3)
Volpo 20	(9004-98-2)
Voltaren	(15307-79-6)
Voltarol	(15307-79-6)
Vomex A	(523-87-5)
Vomissels	(569-65-3)
Vomitoxin	(51481-10-8)
Vonamycin Powder V	(1404-04-2)
Vondacel Black N	(1937-37-7)
Vondacel Black RW	(2429-83-6)
Vondacel Black VG	(6428-31-5)
Vondacel Black VN	(6428-31-5)
Vondacel Blue 2B	(2602-46-2)
Vondacel Blue FF	(2610-05-1)
Vondacel Blue HH	(2429-74-5)
Vondacel Brown S	(2429-81-4)
Vondacel Brown SP	(2429-81-4)
Vondacel Dark Blue BH	(2429-73-4)
Vondacel Green DB	(3626-28-6)
Vondacel Red CL	(573-58-0)
Vondacel Yellow RN	(1325-37-7)
Vondacid Fast Yellow AE	(6373-74-6)
Vondacid Green L	(4680-78-8)
Vondacid Light Red NG	(3734-67-6)
Vondacid Metanil Yellow G	(587-98-4)
Vondacid Orange II	(633-96-5)
Vondacid Red GN	(1658-56-6)
Vondacid Tartrazine	(1934-21-0)
Vondalhyde	(123-33-1)
Vondamol Brilliant Red G	(10169-02-5)
Vondamol Fast Blue R	(3861-73-2)
Vondamol Fast Red G	(3567-65-5)
Vondamol Fast Red RS	(6459-94-5)
Vondcaptan	(133-06-2)
Vondodine	(2439-10-3)
Vondozeb	(8018-01-7)
Vondrax	(123-33-1)
Vonduron	(330-54-1)
Vonedrine	(300-42-5)
Vonteryl Violet 2B	(82-33-7)
Vonteryl Yellow G	(2832-40-8)
Vonteryl Yellow 3R	(6300-37-4)
Vonteryl Yellow R	(2832-40-8)
Vontrol	(972-02-1)
Vopcolene 27	(112-80-1)
Voranol CP 260	(25791-96-2)
Voranol CP 301	(25791-96-2)
Voranol CP 450	(25791-96-2)
Voranol CP 700	(25791-96-2)
Voranol CP 1500	(25791-96-2)
Voranol CP 3000	(25791-96-2)
Voranol CP 3001	(25791-96-2)
Voranol P 2001	(9003-11-6)
Vorlex	(556-61-6)
Vorlex	(8066-01-1)
Voronit	(3878-19-1)
Voronite	(3878-19-1)
Vorox	(61-82-5)
Vorox AA	(61-82-5)
Vorox AS	(61-82-5)
Vortex	(556-61-6)
Votexit	(52-68-6)
Vuagt 179	(8066-11-3)
Vuagt 210	(65280-19-5)
Vuagt-I-4	(137-26-8)
Vul-Cup	(25155-25-3)
Vul-Cup 40KE	(25155-25-3)
Vul-Cup R	(25155-25-3)
Vulcacel B-40	(101-25-7)
Vulcacel BN	(101-25-7)
Vulcacid D	(102-06-7)
Vulcacur	(136-30-1)
Vulcacure	(136-23-2)
Vulcacure	(136-30-1)
Vulcacure	(137-30-4)
Vulcacure	(14324-55-1)
Vulcacure ZM	(137-30-4)
Vulcafix Blue R	(482-89-3)
Vulcafix Fast Blue SD	(81-77-6)
Vulcafix Orange J	(3520-72-7)
Vulcafix Orange JV	(3520-72-7)
Vulcafix Red J	(2814-77-9)
Vulcafix Red R	(1248-18-6)
Vulcafix Scarlet R	(5160-02-1)

Name	CAS
Vulcafix Scarlet R-D Masse	(5160-02-1)
Vulcafor BSM	(102-77-2)
Vulcafor Blue A	(482-89-3)
Vulcafor Fast Blue 3R	(81-77-6)
Vulcafor Fast Orange G	(3520-72-7)
Vulcafor Fast Orange GA	(3520-72-7)
Vulcafor Fast Yellow GT	(6358-85-6)
Vulcafor Fast Yellow GTA	(6358-85-6)
Vulcafor Orange R	(2814-77-9)
Vulcafor Red 2R	(5160-02-1)
Vulcafor Scarlet A	(2425-85-6)
Vulcafor TMTD	(137-26-8)
Vulcal Fast Green F2G	(1328-53-6)
Vulcalent A	(86-30-6)
Vulcamel TBN	(91-60-1)
Vulcan	(1333-86-4)
Vulcan Fast Orange G	(3520-72-7)
Vulcan Fast Orange GA	(3520-72-7)
Vulcan Fast Orange GN	(3520-72-7)
Vulcan Red LC	(5160-02-1)
Vulcan Red R	(2814-77-9)
Vulcan Violet BN	(81-78-7)
Vulcanosine Dark Blue L	(482-89-3)
Vulcanosine Fast Blue GG	(81-77-6)
Vulcanosine Fast Green G	(1328-53-6)
Vulcanosine Red RBKX	(1103-38-4)
Vulcanosine Red RCKX	(1103-39-5)
Vulcatard	(86-30-6)
Vulcatard A	(86-30-6)
Vulcol Fast Blue S	(81-77-6)
Vulcol Fast Green F2G	(1328-53-6)
Vulcol Fast Orange G	(3520-72-7)
Vulcol Fast Red L	(5160-02-1)
Vulcol Fast Yellow GR	(6358-85-6)
Vulcuren 2	(95-32-9)
Vulcuren-2	(95-32-9)
Vulkacit CA	(102-08-9)
Vulkacit 4010 (Czech)	(101-87-1)
Vulkacit D	(102-06-7)
Vulkacit D/C	(102-06-7)
Vulkacit DM	(120-78-5)
Vulkacit DM/MGC	(120-78-5)
Vulkacit DOTG/C	(97-39-2)
Vulkacit L	(137-30-4)
Vulkacit LDA	(14324-55-1)
Vulkacit LDB/C	(136-23-2)
Vulkacit MOZ	(102-77-2)
Vulkacit MS	(97-74-5)
Vulkacit MTIC	(137-26-8)
Vulkacit Mercapto	(149-30-4)
Vulkacit NPV/C2	(96-45-7)
Vulkacit NZ	(95-31-8)
Vulkacit ZM	(155-04-4)
Vulkacit ZP	(13878-54-1)
Vulkacite L	(137-30-4)
Vulkacit thiuram	(137-26-8)
Vulkacit thiuram/C	(137-26-8)
Vulkacit thiuram MS/C	(97-74-5)
Vulkalent A (Czech)	(86-30-6)
Vulkanox 4020	(106-50-3)
Vulkanox 4020	(793-24-8)
Vulkanox BKF	(119-47-1)
Vulkanox HS/LG	(147-47-7)
Vulkanox·HS/Powder	(147-47-7)
Vulkanox KB	(128-37-0)
Vulkanox Pan	(90-30-2)
Vulkasil	(7631-86-9)
Vulkasil A 1	(1344-00-9)
Vulkazit	(102-06-7)
Vulklor	(118-75-2)
Vulnoc AB	(1863-63-4)
Vulnopol NM	(128-04-1)
Vultrol	(86-30-6)
Vulvan	(57-85-2)
Vydate	(23135-22-0)
Vydate L Insecticide/Nematicide	(23135-22-0)
Vydate L Oxamyl Insecticide/Nematocide	(23135-22-0)
Vydyne	(63428-83-1)
Vylor	(63428-84-2)
Vynamon Blue A	(482-89-3)
Vynamon Blue B	(147-14-8)
Vynamon Blue 3R	(81-77-6)
Vynamon Claret Y	(5858-81-1)
Vynamon Green BE	(1328-53-6)
Vynamon Green BES	(1328-53-6)
Vynamon Green GNA	(1328-53-6)
Vynamon Green 6Y	(14302-13-7)
Vynamon Orange CR	(18454-12-1)
Vynamon Orange G	(3520-72-7)
2317-W	(1594-56-5)
80W	(957-51-7)
W 32	(114-86-3)
W 70	(9011-05-6)
W 101	(9003-07-0)
W 491	(535-89-7)
W 524	(26644-46-2)
W 583	(64-55-1)
W 713	(4268-36-4)
W 1655	(94-78-0)
W 1655	(136-40-3)
W 4020	(2955-38-6)
W 4565	(14698-29-4)
W 4744	(340-57-8)
W 6658	(122-34-9)
W 7618	(54-05-7)
W 1544-A	(51-71-8)
W 1164-3	(9002-69-1)
W 36095	(41708-72-9)
W-583	(64-55-1)
W1544	(51-71-8)
W5769	(14437-17-3)
W.A.R.F. 42	(81-81-2)
WBA 8119	(56073-10-0)
WEC 50	(52-68-6)
WEX 1242	(9003-07-0)
W-Gum	(9005-25-8)
WHR 169	(23787-97-5)
WHR 539	(36616-52-1)
WHR-169	(23787-97-5)
W-53 Hydrochloride	(135-23-9)
WIN 244	(54-05-7)
WIN 4692	(80-77-3)
WIN 5095	(115-44-6)
WIN 5512	(104-14-3)
WIN 16568	(3734-33-6)
WIN 17757	(17230-88-5)
WIN 18,320	(389-08-2)
WIN 20228	(359-83-1)
WIN 24450	(13647-35-3)
WIN 24933	(3105-97-3)
WIN-2848	(91-79-2)
WIN-2848	(958-93-0)
WJG 11	(9002-88-4)
WL 7	(156-51-4)

Name	CAS No.
WL 1650	**(297-78-9)**
WL-5792	**(1918-13-4)**
WL 8008	**(1420-06-0)**
WL 8517	**(52315-07-8)**
WL 9385	**(2854-70-8)**
WL 17731	**(33878-50-1)**
WL 18236	**(16752-77-5)**
WL 19805	**(21725-46-2)**
WL 25735	**(56549-12-3)**
WL 26738	**(38338-57-7)**
WL 29761	**(52756-25-9)**
WL 41706	**(39515-41-8)**
WL 43467	**(52315-07-8)**
WL 43479	**(52645-53-1)**
WL 43775	**(51630-58-1)**
WL 85871	**(67375-30-8)**
WN 12	**(556-61-6)**
WNF 15	**(9002-88-4)**
WNM	**(1854-26-8)**
53WP	**(1332-03-2)**
WP 155	**(1031-47-6)**
WP 40	**(565-33-3)**
WPSB Pheromone	**(16974-11-1)**
WP 40 (Polysaccharide)	**(37353-59-6)**
WR 448	**(80-08-0)**
WR 2978	**(58-14-0)**
WR 3396	**(1461-22-9)**
WR 5473	**(516-21-2)**
WR 14,997	**(52-52-8)**
WS 24	**(9003-01-4)**
WS 661	**(9003-11-6)**
WS 801	**(9003-01-4)**
WS 2434	**(83-98-7)**
WSCP	**(31512-74-0)**
WSR-301	**(25322-68-3)**
W-13 Stabilizer	**(9005-25-8)**
WT-5100	**(11138-66-2)**
WT-6500	**(11138-66-2)**
W-T SASP Oral	**(599-79-1)**
WVG 23	**(9002-88-4)**
WV 562 (German)	**(5350-41-4)**
W-VII/117	**(3878-19-1)**
WY-401	**(77-15-6)**
WY-509	**(60-87-7)**
WY-554	**(50-13-5)**
WY-585	**(100-92-5)**
WY-806	**(126-27-2)**
WY-1094	**(58-40-2)**
WY-1485	**(533-45-9)**
WY-2917	**(846-50-4)**
WY-3707	**(6533-00-2)**
WY-3917	**(846-50-4)**
WY-4036	**(846-49-1)**
WY-3263	**(5560-72-5)**
WY-3467	**(439-14-5)**
WY-3498	**(604-75-1)**
WY-4082	**(848-75-9)**
WY-5103	**(69-53-4)**
WY-23,409	**(37751-39-6)**
WYPAX	**(846-49-1)**
Watercarb	**(7440-44-0)**
Wacholderbeer Oel	**(8002-68-4)**
Wacholderbeer Oel (German)	**(8002-68-4)**
Wacker S 14/10	**(115-26-4)**
Wait's Green Mountain Antihistamine	**(91-84-9)**
Wakil	**(77732-09-3)**
Walnut Extract	**(481-39-0)**
Walnut Hull Extract	**(8024-09-7)**
Walnut Oil	**(8024-09-7)**
Walnut Stain	**(1317-34-6)**
Walsroder MC 20000S	**(9004-67-5)**
Wampocap	**(59-67-6)**
Wanadu pieciotlenek (Polish)	**(1314-62-1)**
Wandamin HM	**(1761-71-3)**
Wanin S	**(8061-51-6)**
Wapniowy tlenek (Polish)	**(1305-78-8)**
Waqe	**(151-21-3)**
Waran	**(129-06-6)**
Waran	**(81-81-2)**
Warbex	**(52-85-7)**
Warbexol	**(52-85-7)**
Warcoumin	**(129-06-6)**
Wardamate	**(57-53-4)**
Warduzide	**(58-94-6)**
Warecure C	**(96-45-7)**
Wareflex	**(141-17-3)**
Warf Compound 42	**(81-81-2)**
Warfarat	**(81-81-2)**
Warfarin (ACGIH,OSHA)	**(81-81-2)**
Warfarin Plus	**(81-81-2)**
Warfarin Q	**(81-81-2)**
Warfarine (French)	**(81-81-2)**
Warfarin sodium	**(129-06-6)**
Warfarin, sodium deriv.	**(129-06-6)**
Warfarin, sodium salt	**(129-06-6)**
Warficide	**(81-81-2)**
Warfilone	**(129-06-6)**
Warkeelate PS-42	**(64-02-8)**
Warkeelate PS-43	**(64-02-8)**
Warkeelate PS-47	**(64-02-8)**
Warkeelate S-42	**(64-02-8)**
Warkeelate acid	**(60-00-4)**
Wash Oil	**(8001-58-9)**
Wasserina	**(68-41-7)**
Wasserstoffperoxid (German)	**(7722-84-1)**
Waste treatment plant sludge	**(68188-15-8)**
Water	**(7732-18-5)**
Water Black 100	**(6428-31-5)**
Water-D2 (9CI)	**(7789-20-0)**
Water Glass	**(1344-09-8)**
Water Glass	**(6834-92-0)**
Water^{2H2}	**(7789-20-0)**
Water, Heavy (D_2O)	**(7789-20-0)**
Waterstofperoxyde (Dutch)	**(7722-84-1)**
Wattle Gum	**(9000-01-5)**
Wax C	**(110-30-5)**
Wax LE	**(9002-88-4)**
Wax, Yellow	**(8012-89-3)**
Waxakol Orange GL	**(842-07-9)**
Waxakol Red BL	**(85-83-6)**
Waxakol Vermilion L	**(3118-97-6)**
Waxakol Yellow NL	**(97-56-3)**
Waxes, Candelilla	**(8006-44-8)**
Waxes, Carnauba	**(8015-86-9)**
Waxoline Blue AP	**(14233-37-5)**
Waxoline Blue GA	**(128-85-8)**
Waxoline Green	**(128-80-3)**
Waxoline Green G	**(128-80-3)**
Waxoline Orange A	**(494-38-2)**
Waxoline Orange Y	**(495-54-5)**
Waxoline Purple A	**(81-48-1)**
Waxoline Red A	**(3248-93-9)**
Waxoline Red MAA	**(82-38-2)**
Waxoline Red MP	**(82-38-2)**

Waxoline Red O	(85-83-6)
Waxoline Red OM	(85-83-6)
Waxoline Red OS	(85-83-6)
Waxoline Rhodamine B	(509-34-2)
Waxoline Rhodamine BS	(509-34-2)
Waxoline Yellow AD	(60-11-7)
Waxoline Yellow ADS	(60-11-7)
Waxoline Yellow ED	(2481-94-9)
Waxoline Yellow I	(842-07-9)
Waxoline Yellow IM	(842-07-9)
Waxoline Yellow IP	(842-07-9)
Waxoline Yellow IS	(842-07-9)
Waxoline Yellow O	(492-80-8)
Waxoline Yellow T	(8003-22-3)
Waxsol	(577-11-7)
Wayne Red X-2486	(5160-02-1)
Weatherproofer	(13110-37-7)
Weckamine	(300-62-9)
Weckamine	(60-13-9)
Wecoline 1295	(143-07-7)
Wecoline OO	(112-80-1)
Wedding Bells	(50-37-3)
Weecon	(917-61-3)
Weed 108	(2163-80-6)
Weed-AG-Bar	(94-75-7)
Weed-B-Gon	(93-72-1)
Weed-B-Gon	(94-75-7)
Weed Broom	(144-21-8)
Weed Drench	(107-18-6)
Weed-E-Rad	(144-21-8)
Weed-E-Rad	(2163-80-6)
Weed-E-Rad 360	(144-21-8)
Weed-E-Rad DMA Powder	(144-21-8)
Weed-HOE	(144-21-8)
Weed-HOE	(2163-80-6)
Weed-Rhap	(94-74-6)
Weed-Rhap	(94-75-7)
Weed Tox	(94-75-7)
Weedanol Cyanol	(590-28-3)
Weedar	(93-76-5)
Weedar	(94-74-6)
Weedar	(94-75-7)
Weedar-64	(94-75-7)
Weedar ADS	(61-82-5)
Weedar AT	(61-82-5)
Weedar MCPA Concentrate	(94-74-6)
Weedatul	(94-75-7)
Weedazin	(61-82-5)
Weedazin arginit	(61-82-5)
Weedazol	(61-82-5)
Weedazol GP2	(61-82-5)
Weedazol T	(61-82-5)
Weedazol TL	(1762-95-4)
Weedazol TL	(61-82-5)
Weedazol super	(61-82-5)
Weedbeads	(131-52-2)
Weedex	(122-34-9)
Weedex A	(1912-24-9)
Weedex Granulat	(61-82-5)
Weedez Wonder Bar	(94-75-7)
Weedmaster Grass Killer	(650-51-1)
Weedoclor	(61-82-5)
Weedol	(1910-42-5)
Weedol	(85-00-7)
Weedone	(87-86-5)
Weedone	(93-76-5)
Weedone	(94-74-6)
Weedone	(94-75-7)
Weedone 40	(533-23-3)
Weedone 128	(94-11-1)
Weedone 170	(120-36-5)
Weedone Concentrate 48	(533-23-3)
Weedone Crab Grass Killer	(590-28-3)
Weedone DP	(120-36-5)
Weedone LV 4	(1929-73-3)
Weedone LV4	(94-75-7)
Weedone MCPA Ester	(94-74-6)
Weedone 2,4,5-T	(93-76-5)
Weedtrine-D	(85-00-7)
Weedtrine-II	(25168-26-7)
Weedtrol	(94-75-7)
Weeviltox	(75-15-0)
Wegantyna (Polish)	(64-95-9)
Wegla dwusiarczek (Polish)	(75-15-0)
Wegla tlenek (Polish)	(630-08-0)
Wehydryl	(147-24-0)
Weifacodine	(509-67-1)
Weiss phosphor (German)	(7723-14-0)
Weisspiessglanz (German)	(1309-64-4)
Welfurin	(67-20-9)
Wellcome U3B	(154-42-7)
Wellcome preparation 47-83	(82-92-8)
Wellcoprim	(738-70-5)
Wemcol	(25640-78-2)
Wepsin	(1031-47-6)
Wepsyn	(1031-47-6)
Wepsyn 155	(1031-47-6)
Wescozone	(50-33-9)
Wespuril	(97-23-4)
West Indian Lemongrass Oil	(8007-02-1)
West Indian Sandalwood Oil	(8015-65-4)
Westocaine	(51-05-8)
Weston 626	(26741-53-7)
Weston MDW 626	(26741-53-7)
Westron	(79-34-5)
Westrosol	(79-01-6)
Wetaid SR	(577-11-7)
Whale Oil	(977040-42-8)
Whatman CC-31	(9004-34-6)
Wheat LS	(8001-21-6)
Whey factor	(65-86-1)
Whip	(66441-23-4)
1700 White	(13463-67-7)
White Arsenic	(1327-53-3)
White Asbestos [UN 2590]	(12001-29-5)
White Camphor Oil	(8008-51-3)
White Caustic	(1310-73-2)
White Cocoa Essence	(8002-31-1)
White Copperas	(7733-02-0)
White Crystal	(7647-14-5)
White Lead	(1319-46-6)
White Lead	(598-63-0)
White Lead, Hydrocerussite	(1319-46-6)
White Mercury Precipitated	(10124-48-8)
White Mineral Oil	(8042-47-5)
White Oil of Camphor	(8008-51-3)
White Petrolatum USP	(8009-03-8)
White Petroleum Jelly	(8009-03-8)
White Precipitate	(10124-48-8)
White Protopet	(8009-03-8)
White Seal-7	(1314-13-2)
White Streptocide	(63-74-1)
White Stuff	(561-27-3)
White Tar	(91-20-3)

White Vaseline (8009-03-8)
White Vitriol (7446-20-0)
White phosphorus (DOT) (7723-14-0)
White vitriol (7733-02-0)
Whitex SKC (24019-80-5)
Whitmoyer Ethylene Diamine Dihydriodide (5700-49-2)
Whortleberry Red (915-67-3)
Wickenol 101 (110-27-0)
Wickenol 111 (142-91-6)
Wickenol 116 (6938-94-9)
Wickenol 122 (123-95-5)
Wickenol 127 (112-10-7)
Wickenol 155 (29806-73-3)
Wickenol 156 (22047-49-0)
Wickenol 158 (103-23-1)
Wickenol 303 (12042-91-0)
Wickenol 321 (12042-91-0)
Wickenol 323 (12042-91-0)
Wickenol 324 (12042-91-0)
Widlon (105-60-2)
Widlon (25038-54-4)
Wie Oben (13593-03-8)
Wigraine (62-44-2)
Wijs' Chloride (7790-99-0)
Wikol (9003-20-7)
Wilkinite (1302-78-9)
Willbutamide (64-77-7)
Willestrol (130-80-3)
Willnestrol (84-17-3)
Willosetten (128-44-9)
Wilpo (122-09-8)
Wiltz-65 (1338-02-9)
Wiltz-65 (26896-20-8)
Win 771 (468-56-4)
Win 1539 (469-79-4)
Win 1783 (466-40-0)
Win 40350 (1306-06-5)
Win 2848 Hydrochloride Salt (135-23-9)
Win 14833 (VAN) (10418-03-8)
Winacet D (9003-20-7)
Wine Ether (123-29-5)
Wine Oil (106-30-9)
Wing Stop B (128-04-1)
Wingstay 300 (793-24-8)
Winiden 60 (9011-06-7)
Winstroid (10418-03-8)
Winstrol (VAN) (10418-03-8)
Winstrol V (VAN) (10418-03-8)
Winter Bloom (68916-39-2)
Wintergreen Oil (119-36-8)
Wintergreen Oil, Synthetic (119-36-8)
Wintermin (50-53-3)
Wintersteiner's Compound F (53-06-5)
Winterwash (534-52-1)
Winthrop Red X 1666 (1103-39-5)
Wintomylon (389-08-2)
Winylophos (62-73-7)
Winylu chlorek (Polish) (75-01-4)
Winzer Solution (122-18-9)
Wipeout (67485-29-4)
Wirkstoff 37289 (327-98-0)
Witamina PP (98-92-0)
Witamol 150 (28553-12-0)
Witamol 320 (103-23-1)
Witcamide 61 (111-05-7)
Witcamide 70 (111-57-9)
Witcamide 4120 (93-83-4)
Witcamide 5085 (93-83-4)
Witcamide 511C (93-83-4)
Witcamide MM (142-58-5)
Witcarb 940 (7440-44-0)
Witch Hazel (68916-39-2)
Witcizer 200 (123-95-5)
Witcizer 201 (123-95-5)
Witcizer 300 (84-74-2)
Witcizer 312 (117-81-7)
Witco (1333-86-4)
Witco 31 (9004-96-0)
Witco G 339S (1592-23-0)
Witco 1298 Sulfonic Acid (27176-87-0)
Witco Zinc Soap #26 (2452-01-9)
Witcoblak No. 100 (1333-86-4)
Witcolate A (151-21-3)
Witcolate A Powder (151-21-3)
Witcolate C (151-21-3)
Witcolate DLS (143-00-0)
Witconate 5725 (27323-41-7)
Witconate 60L (27323-41-7)
Witconate NXS (26447-10-9)
Witconate 79S (27323-41-7)
Witconate S-1280 (27323-41-7)
Witconate STS (12068-03-0)
Witconate 60T (27323-41-7)
Witconate TAB (27323-41-7)
Witconate TDB (26248-24-8)
Witconol APB (9003-13-8)
Witconol F26-46 (31394-71-5)
Witconol H35A (9004-99-3)
Witconol L32-45 (9005-08-7)
Witconol MS (123-94-4)
Witconol MS (31566-31-1)
Witconol MST (123-94-4)
Witconol MST (31566-31-1)
Witophen P (87-86-5)
Witron 131 (9003-29-6)
Wittox C (1338-02-9)
Wochem No. 320 (112-80-1)
Wofatos (298-00-0)
Wofatox (298-00-0)
Wofotox (298-00-0)
Wogenal Yellow CG (1325-37-7)
Wojtab (53-03-2)
Wolfen (137-29-1)
Wolfram (7440-33-7)
Wolframite (1314-35-8)
Wollastokup (13983-17-0)
Wollastonite (13983-17-0)
Wonuk (1912-24-9)
Wood Alcohol (DOT) (67-56-1)
Wood Naphtha (67-56-1)
Wood Oil (8030-55-5)
Wood Rosin, Methyl Ester (127-25-3)
Wood Spirit (67-56-1)
Wood ether (115-10-6)
Woodlife (55406-53-6)
Wool Blue RL (3861-73-2)
Wool Bordeaux 6RK (915-67-3)
Wool Brilliant Green SF (5141-20-8)
Wool Fast Blue R (3861-73-2)
Wool Fat (8006-54-0)
Wool Grease (8006-54-0)
Wool Orange A (633-96-5)
Wool Orange 2G (1936-15-8)
Wool Orange G (1936-15-8)

Name	CAS No.
Wool Red	(915-67-3)
Wool Red 40F	(915-67-3)
Wool Violet	(1694-09-3)
Wool Violet 4BN	(1694-09-3)
Wool Violet 5BN	(1694-09-3)
Wool Wax	(8006-54-0)
Wool Wax, Refined	(8006-54-0)
Wool Yellow	(1934-21-0)
Worm-Agen	(136-77-6)
Worm Guard	(14255-87-9)
Wormwood	(110-15-6)
Wormwood Acid	(110-15-6)
Wotexit	(52-68-6)
Wuerzberg Green	(12002-03-8)
Wurm-Thional	(92-84-2)
Wurster's Blue	(100-22-1)
Wurster's Reagent	(100-22-1)
Wy-1359	(362-29-8)
Wy-2039	(469-82-9)
Wy-8678	(5051-62-7)
Wyacort	(83-43-2)
Wyamine	(100-92-5)
Wyandotte 7135	(9003-11-6)
Wycillin	(6130-64-9)
Wycillina	(1538-09-6)
Wyex	(1333-86-4)
Wyfentermina	(100-92-5)
Wynestron	(53-16-7)
Wyovin	(77-19-0)
Wyseals	(57-53-4)
3960-X14	(8001-50-1)
X 13	(13464-35-2)
X 27	(154-42-7)
X 41	(51-79-6)
X 97	(50-76-0)
X 119	(7227-91-0)
X 149	(55-98-1)
X 188	(123-39-7)
X 201	(52-52-8)
X 250	(12174-11-7)
X 423	(9003-11-6)
X 427	(9003-11-6)
X 600	(9003-53-6)
X 970	(2082-81-7)
X 20-201	(63148-62-9)
X 2-1163	(63148-62-9)
X-52	(32861-85-1)
X52	(39362-66-8)
X9-5700	(27668-52-6)
XA 2	(302-70-5)
XAD-4 Resin	(37380-42-0)
X-All Liquid	(61-82-5)
XB 2615	(28825-96-9)
XD 7818	(58145-38-3)
XD 8379	(9003-11-6)
XE 340	(7440-44-0)
XE-938	(39515-41-8)
XF 13-563	(63148-62-9)
XF 4175L	(7440-44-0)
XK-62-3	(69375-05-9)
XL 7	(97-18-7)
XL-50	(92-84-2)
XL-90	(93-14-1)
XL 335-1	(9002-88-4)
XL 1246	(9002-88-4)
XLV	(53973-98-1)
XL all insecticide	(54-11-5)
3,5-XMC	(2655-14-3)
XMC	(2655-14-3)
XNM 68	(9002-88-4)
XO 440	(9002-88-4)
XPA	(9003-01-4)
XPDR-A 288	(9003-17-2)
X-539-R	(9004-96-0)
XRDR-A-288	(9003-17-2)
XRM 3972	(1702-17-6)
XSA	(25321-41-9)
XX 78	(1314-13-2)
XU 238	(68012-07-7)
XX 203	(1314-13-2)
XX 601	(1314-13-2)
Xamamina	(523-87-5)
Xan	(69-89-6)
Xanax	(28981-97-7)
Xanteline	(53-46-3)
Xanthacridine	(86-40-8)
Xanthacridinum	(8048-52-0)
Xanthan	(11138-66-2)
Xanthan Gum (9CI)	(11138-66-2)
Xanthate	(151-01-9)
Xanthaurine	(117-39-5)
3H-Xanthen-6-amine, N-methyl-3-(methylimino)-, hydrochloride	(2465-29-4)
Xanthene	(92-83-1)
9H-Xanthene (9CI)	(92-83-1)
Xanthene-9-carboxylic acid, ester with diethyl(2-hydroxy-ethyl)methylammonium bromide	(53-46-3)
Xanthene-9-carboxylic acid, ester with (2-hydroxyethyl)-diisopropylmethylammonium bromide	(50-34-0)
9H-Xanthene, 1,2,4,5,7,8-hexachloro-	(38178-99-3)
9H-Xanthenol, methyl- (9CI)	(116211-89-3)
9-Xanthenone	(90-47-1)
Xanthen-9-one	(90-47-1)
Xanthic acid, ethyl-, potassium salt	(140-89-6)
Xanthic acid, isopropyl-, potassium salt	(140-92-1)
Xanthic acid, isopropyl-, sodium salt	(140-93-2)
Xanthic acid, pentyl-, potassium salt	(2720-73-2)
Xanthic oxide	(69-89-6)
Xanthin	(69-89-6)
Xanthine	(69-89-6)
Xanthine, 8-chloro- (8CI)	(13548-68-0)
Xanthine, 1,3-dimethyl-	(58-55-9)
Xanthine, 3,7-dimethyl-	(83-67-0)
Xanthine, 1-methyl	(6136-37-4)
Xanthine, 3-methyl	(1076-22-8)
Xanthine, 7-methyl	(552-62-5)
Xanthine, 3-methyl-2-thio- (6CI,7CI,8CI)	(28139-02-8)
Xanthine, 2-thio- (8CI)	(2487-40-3)
Xanthine, 1,3,7-trimethyl	(58-08-2)
Xanthogen, bis(ethyl-	(502-55-6)
Xanthotoxin	(298-81-7)
Xanthoxin	(298-81-7)
Xanthylium, 3,6-bis(ethylamino)-9-(2-(methoxycarbonyl)-phenyl)-2,7-dimethyl-, chloride (9CI)	(3068-39-1)
Xanthylium, 9-(2-carboxyphenyl)-3,6-bis(diethylamino)-, hydroxide, inner salt (9CI)	(3375-25-5)
Xanthylium, 9-(2-carboxyphenyl)-3,6-bis(diethylamino)-, molybdatetungstatephosphate (9CI)	(1326-03-0)
Xanthylium, 9-(2-(ethoxycarbonyl)phenyl)-3,6-bis(ethylamino)-2,7-dimethyl-, molybdatetungstatephosphate (9CI)	(12224-98-5)
Xanturat	(315-30-0)
Xaril	(62-44-2)
Xaxa	(50-78-2)
Xenene	(92-52-4)
o-Xenol	(90-43-7)

Name	CAS No.
Xenon, Refrigerated liquid [UN 2591]	**(7440-63-3)**
Xenon [UN]	**(7440-63-3)**
Xenylamin (Czech)	**(92-67-1)**
Xenylamine	**(992-67-1)**
Xerac	**(94-36-0)**
Xilenoli (Italian)	**(1300-71-6)**
Xilidine (Italian)	**(1300-73-8)**
Xilocaina (Italian)	**(137-58-6)**
Xiloli (Italian)	**(1330-20-7)**
Xiron Golden Yellow 2R-HD	**(12225-84-2)**
Xitix	**(50-81-7)**
Xoru-Ox	**(140-41-0)**
Xycaine	**(137-58-6)**
Xyduril	**(637-07-0)**
Xylecarb	**(2425-10-7)**
1,2-Xylene	**(95-47-6)**
1,3-Xylene	**(108-38-3)**
1,4-Xylene	**(106-42-3)**
o-Xylene (ACGIH,DOT,OSHA) [UN 1307]	**(95-47-6)**
Xylene (ACGIH,OSHA) [UN 1307]	**(1330-20-7)**
m-Xylene (ACGIH,OSHA) [UN 1307]	**(108-38-3)**
p-Xylene (ACGIH,OSHA) [UN 1307]	**(106-42-3)**
Xylene Black F	**(2519-30-4)**
Xylene Blue VS	**(129-17-9)**
Xylene Blue VSG	**(2650-18-2)**
Xylene Brilliant Cyanine G	**(6104-58-1)**
Xylene Fast Orange G	**(1936-15-8)**
Xylene Fast Yellow ES	**(6373-74-6)**
Xylene Fast Yellow GT	**(1934-21-0)**
Xylene Milling Red G	**(10169-02-5)**
Xylene Milling Yellow SH	**(6375-55-9)**
Xylene Musk	**(81-15-2)**
m-Xylene, benzylated	**(68908-87-2)**
Xylene, α-bromo- (8CI)	**(28777-60-8)**
m-Xylene, α-bromo- (8CI)	**(620-13-3)**
o-Xylene, α-bromo- (8CI)	**(89-92-9)**
p-Xylene, 2-bromo- (8CI)	**(553-94-6)**
p-Xylene, α-bromo- (8CI)	**(104-81-4)**
m-Xylene, 4-sec-butyl-	**(1483-60-9)**
m-Xylene, 5-tert-butyl	**(98-19-1)**
m-Xylene, 5-tert-butyl-2,4,6-trinitro-	**(81-15-2)**
o-Xylene, α-chloro-	**(552-45-4)**
Xylene, ar-chloro- (8CI)	**(25323-41-5)**
m-Xylene, α-chloro- (8CI)	**(620-19-9)**
p-Xylene, α-chloro- (8CI)	**(104-82-5)**
Xylene, (1-chloroethyl)- (7CI)	**(54411-21-1)**
m-Xylene α,α'-diamine (ACGIH,OSHA)	**(1477-55-0)**
Xylene, dibromo- (8CI)	**(28805-90-5)**
Xylene, α,α'-dichloro	**(28347-13-9)**
m-Xylene, α,α'-dichloro	**(626-16-4)**
o-Xylene, α,α'-dichloro	**(612-12-4)**
p-Xylene, 2,5-dichloro- (8CI)	**(1124-05-6)**
p-Xylene, 2,5-diisopropyl- (8CI)	**(10375-96-9)**
m-Xylene, 4,6-diisopropyl- (6CI,7CI,8CI)	**(5186-68-5)**
m-Xylene, 4,6-dinitro	**(616-72-8)**
p-Xylene-α,α'-diol (8CI)	**(589-29-7)**
p-Xylene-α,α'-diol, α,α,α',α'-tetramethyl-	**(2948-46-1)**
Xylene, ethyl- (8CI)	**(29224-55-3)**
m-Xylene, 4-ethyl- (8CI)	**(874-41-9)**
m-Xylene, 5-ethyl- (8CI)	**(934-74-7)**
o-Xylene, 4-ethyl- (8CI)	**(934-80-5)**
p-Xylene, 2-ethyl- (8CI)	**(1758-88-9)**
p-Xylene, α,α,α,α',α',α'-hexachloro	**(68-36-0)**
m-Xylene, α,α,α,α',α',α'-hexafluoro- (8CI)	**(402-31-3)**
p-Xylene, 2-isobutyl- (6CI)	**(55669-88-0)**
Xylenen (Dutch)	**(1330-20-7)**
Xylene, nitro	**(25168-04-1)**
m-Xylene, 2-nitro	**(81-20-9)**
o-Xylene, 3-nitro-	**(83-41-0)**
p-Xylene, 2-nitro	**(89-58-7)**
m-Xylene, 4-nitro- (8CI)	**(89-87-2)**
m-Xylene, 5-nitro- (8CI)	**(99-12-7)**
o-Xylene, 4-nitro- (8CI)	**(99-51-4)**
2,3-Xylenesulfonamide, 5-amino-N-(2-hydroxyethyl)- (8CI)	**(25797-78-8)**
2,3-Xylenesulfonamide, N-(2-hydroxyethyl)- (8CI)	**(25959-70-0)**
2,4-Xylenesulfonic acid	**(88-61-9)**
Xylenesulfonic-acid	**(25321-41-9)**
m-Xylene-4-sulfonic acid	**(88-61-9)**
m-Xylenesulfonic acid	**(88-61-9)**
Xylenesulfonic acid, sodium salt	**(1300-72-7)**
p-Xylene, α,α,α',α'-tetrabromo	**(23488-38-2)**
o-Xylene, 3,4,5,6-tetrabromo- (6CI,7CI)	**(2810-69-7)**
o-Xylene, α,α,α',α'-tetrachloro-	**(25641-99-0)**
p-Xylene, 2,3,5,6-tetrachloro- (8CI)	**(877-10-1)**
1,2,5-Xylenol	**(95-87-4)**
1,3,4-Xylenol	**(95-65-8)**
1,3,5-Xylenol	**(108-68-9)**
2,3-Xylenol	**(526-75-0)**
2,4-Xylenol	**(105-67-9)**
2,5-Xylenol	**(95-87-4)**
2,6-Xylenol	**(576-26-1)**
3,4-Xylenol	**(95-65-8)**
3,5-Xylenol	**(108-68-9)**
p-Xylenol	**(95-87-4)**
Xylenol [UN 2261]	**(1300-71-6)**
m-Xylenol [UN 2261]	**(105-67-9)**
o-Xylenol [UN 2261]	**(526-75-0)**
p-Xylenol [UN 2261]	**(95-87-4)**
2,4-Xylenol, 6-tert-butyl	**(1879-09-0)**
2,5-Xylenol, tert-butyl-	**(31391-49-8)**
Xylenol,tert-butyl	**(36812-13-2)**
3,5-Xylenol, 4-chloro	**(88-04-0)**
3,5-Xylenol, 2,4-dichloro- (8CI)	**(133-53-9)**
3,5-Xylenol, 4-(dimethylamino)-, methylcarbamate	**(315-18-4)**
Xylenolen (Dutch)	**(1300-71-6)**
3,5-Xylenol, methylcarbamate	**(2655-14-3)**
3,5-Xylenol, 4-(methylthio)-, methylcarbamate	**(2032-65-7)**
2,6-Xylenol, 4-nitro	**(2423-71-4)**
2,4-Xylenol, phosphate (3:1)	**(3862-12-2)**
3,4-Xylenol, phosphate (3:1)	**(3862-11-1)**
3,5-Xylenol, phosphate (3:1)	**(25653-16-1)**
Xylenol, phosphate (3:1)	**(25155-23-1)**
2,5-Xylenol, phosphate (3:1) (8CI)	**(19074-59-0)**
2,6-Xylenol, phosphate (3:1) (8CI)	**(121-06-2)**
3,5-Xylenyl N-methylcarbamate	**(2655-14-3)**
Xylestesin	**(137-58-6)**
L-Xylo-2-hexulose	**(87-79-6)**
2,3-Xylidine	**(87-59-2)**
2,4-Xylidine	**(95-68-1)**
2,5-Xylidine	**(95-78-3)**
2,6-Xylidine	**(87-62-7)**
3,4-Xylidine	**(95-64-7)**
3,5-Xylidine	**(108-69-0)**
Xylidine	**(1300-73-8)**
m-4-Xylidine	**(95-68-1)**
m-Xylidine	**(95-68-1)**
o-Xylidine	**(87-59-2)**
o-Xylidine	**(87-62-7)**
p-Xylidine	**(95-78-3)**
Xylidine (ACGIH,OSHA) [UN 1711]	**(1300-73-8)**
Xylidine Ponceau	**(3761-53-3)**
Xylidine Ponceau 3RS	**(3761-53-3)**
Xylidine Red	**(3761-53-3)**
m-Xylidine [UN 1711]	**(95-68-1)**

o-Xylidine [UN 1711] (87-59-2)
p-Xylidine [UN 1711] (95-78-3)
2,4-Xylidine, α^4-(p-aminophenyl)-α^4-(4-imino-2,5-cyclohexadien-1-ylidene) (3248-93-9)
3,4-Xylidine, 2,6-dinitro-N-(1-ethylpropyl)- (40487-42-1)
3,5-Xylidine, α,α,α,α',α',α'-hexafluoro (328-74-5)
2,4-Xylidine, hydrochloride (21436-96-4)
2,5-Xylidine, hydrochloride (51786-53-9)
m-4-Xylidine hydrochloride (21436-96-4)
m-Xylidine hydrochloride (21436-96-4)
para-Xylidine hydrochloride (51786-53-9)
2,4-Xylidine, N,N'-(methyliminodimethylidyne)bis (33089-61-1)
Xylidinen (Dutch) (1300-73-8)
2,6-Xylidine, 4-nitro- (8CI) (16947-63-0)
Xyligen B (60568-05-0)
Xylite (87-99-0)
Xylite (sugar) (87-99-0)
Xylitol (87-99-0)
L-Xyloascorbic acid (50-81-7)
Xyloascorbic acid, L- (50-81-7)
Xylocain (137-58-6)
Xylocaine (137-58-6)
Xylocitin (137-58-6)
Xylofop-Ethyl (76578-14-8)
α-D-Xylofuranose, 1,2:3,5-bis-O-(1-methylethylidene)- (9CI) (20881-04-3)
Xylofuranose, 1,2:3,5-di-O-isopropylidene- (20881-04-3)
Xylofuranose, 1,2:3,5-di-O-isopropylidene-, α-D- (8CI) (20881-04-3)
D-Xylo-hexonic acid, 3-deoxy- (8CI,9CI) (18521-63-6)
α-L-Xylo-2-hexulofuranosonic acid, 2,3:4,6-bis-O-(1-methylethylidene)-, sodium salt (52508-35-7)
Xyloidin (9004-70-0)
p-Xylol (106-42-3)
Xylol (DOT) (1330-20-7)
m-Xylol (DOT) (108-38-3)
p-Xylol (DOT) (106-42-3)
o-Xylol [UN 1307] (95-47-6)
Xylole (German) (1330-20-7)
5-(3,5-Xyloloxymethyl)oxazolidin-2-one (1665-48-1)
Xylon (63428-84-2)
o-Xylorcinol (527-55-9)
(D)-Xylose (58-86-6)
Xylose, D (58-86-6)
Xylotox (137-58-6)
2,3-Xylylamine (87-59-2)
2,6-Xylylamine (87-62-7)
3,4-Xylylamine (95-64-7)
3,5-Xylylamine (108-69-0)
N-(2,3-Xylyl)-2-aminobenzoic acid (61-68-7)
N-(2,3-Xylyl)anthranilic acid (61-68-7)
1-(2,4-Xylylazo)-2-naphthol (3118-97-6)
1-(o-Xylylazo)-2-naphthol (3118-97-6)
1-Xylylazo-2-naphthol (3118-97-6)
1-Xylylazo-2-naphthol-3,6-disulfonic acid, disodium salt (3761-53-3)
1-(2,4-Xylylazo)-2-naphthol-3,6-disulphonic acid, disodium salt (3761-53-3)
1-Xylylazo-2-naphthol-3,6-disulphonic acid, disodium salt (3761-53-3)
2,5-Xylyl bromide (553-94-6)
2-Xylyl bromide (89-92-9)
m-Xylyl bromide (620-13-3)
o-Xylyl bromide (89-92-9)
p-Xylyl bromide (104-81-4)
p-Xylyl-α-bromide (104-81-4)
Xylyl bromide [UN 1701] (35884-77-6)
Xylylcarb (2425-10-7)
((2,6-Xylylcarbamoyl)methyl diethyl benzyl ammonium benzoate (3734-33-6)
o-Xylyl chloride (552-45-4)
o-Xylyl-α-chloride (552-45-4)
p-Xylyl chloride (104-82-5)
p-Xylyl-α-chloride (104-82-5)
m-Xylylendiamin (Czech) (1477-55-0)
Xylylene chloride (28347-13-9)
m-Xylylenediamine (1477-55-0)
Xylylene dichloride (28347-13-9)
m-Xylylene dichloride (626-16-4)
o-Xylylene dichloride (612-12-4)
3,4-Xylylester kyseliny methylkarbaminove (Czech) (2425-10-7)
3,5-Xylylester kyseliny methylkarbaminove (Czech) (2655-14-3)
3,4-Xylyl methylcarbamate (2425-10-7)
3,5-Xylyl-N-methylcarbamate (2655-14-3)
5-((3,5-Xylyloxy)methyl)-2-oxazolidinone (1665-48-1)
Xylyl phosphate (25155-23-1)
2,6-Xylyl phosphate, (C8H9O)3PO (121-06-2)
3Y9 (94-84-8)
Y 2 (122-42-9)
Y 3 (101-21-3)
Y 6047 (33671-46-4)
Y 9208 (63148-62-9)
Y-4 (1934-21-0)
YF 3842 (63148-62-9)
YH 1 (2152-34-3)
YPH 103 (17341-40-1)
YX 500 (9002-92-0)
Yacca Gum (8000-48-4)
Yadalan (58-94-6)
Yalan (2212-67-1)
Yaltox (1563-66-2)
Yamaclean M (1713-12-8)
Yamada Fast Red RL Base (99-52-5)
Yamamoto Methylene Blue B (61-73-4)
Yamamoto Methylene Blue ZF (61-73-4)
Yanock (640-19-7)
Yanomite (351-05-3)
Yara Yara (93-04-9)
Yara-Yara (93-04-9)
Yarmor (8002-09-3)
Yarmor Pine Oil (8002-09-3)
Yasoknock (62-74-8)
Yatrocin (59-87-0)
Yeh-Yan-Ku (43222-48-6)
1351 Yellow (2783-94-0)
1409 Yellow (1934-21-0)
1504 Yellow (2051-85-6)
1899 Yellow (2783-94-0)
11363 Yellow (587-98-4)
11641 Yellow (83-08-9)
11712 Yellow (2321-07-5)
11824 Yellow (518-47-8)
12417 Yellow (518-47-8)
21310 Yellow (1934-21-0)
Yellow AB (85-84-7)
Yellow Beeswax (8012-89-3)
Yellow Cross Liquid (505-60-2)
Yellow Cuprocide (1317-39-1)
Yellow EMBL (1325-37-7)
Yellow Fast Dye 4K (6300-37-4)
Yellow Ferric oxide (1309-37-1)
Yellow G Soluble in grease (60-11-7)
Yellow Gentian (72968-42-4)
Yellow Ginger (458-37-7)
Yellow Lake 69 (1934-21-0)
Yellow Lead Ocher (1317-36-8)
Yellow M Soluble in Grease (2051-85-6)
Yellow Mercuric Oxide (21908-53-2)
Yellow Mercury Iodide (15385-57-6)
Yellow No. 2 (85-84-7)

Yellow No. 5 **(1934-21-0)**
Yellow No. 6 **(2783-94-0)**
Yellow No. 5 FDC **(1934-21-0)**
Yellow No. 201 (Japan) **(2321-07-5)**
Yellow OB **(131-79-3)**
Yellow Orange S **(2783-94-0)**
Yellow Orange S Specially Pure **(2783-94-0)**
Yellow Orange Specially Pure 85 **(2783-94-0)**
Yellow Oxide of Iron **(1309-37-1)**
Yellow Oxide of Mercury **(21908-53-2)**
Yellow Petrolatum **(8009-03-8)**
1903 Yellow Pink **(17372-87-1)**
Yellow Precipitate **(21908-53-2)**
Yellow Prussiate of Soda **(13601-19-9)**
Yellow Puccoon **(458-37-7)**
Yellow Pyoctanine **(492-80-8)**
Yellow Reliton G **(2832-40-8)**
Yellow Root **(458-37-7)**
Yellow SF for Food **(2783-94-0)**
Yellow SY for Food **(2783-94-0)**
Yellow Stable 4K **(6300-37-4)**
Yellow Sun **(2783-94-0)**
Yellow T **(547-57-9)**
Yellow Ultramarine **(13765-19-0)**
Yellow Z **(2832-40-8)**
Yellow phosphorus (DOT) **(7723-14-0)**
Yespazine **(69-23-8)**
Yetrazol **(54-95-5)**
Yetrium **(357-56-2)**
Ylestol **(57-63-6)**
Yo-Kin **(458-37-7)**
Yoclo **(637-07-0)**
Yodochrome Metanil Yellow **(587-98-4)**
Yodochrome Violet B **(2092-55-9)**
Yodomin **(50-35-1)**
Yodoxin **(83-73-8)**
Yoduro de Mercurio (Spanish) **(15385-57-6)**
Yohimban-16-carboxylic acid derivative of benz(g)indolo-(2,3-a)quinolizine **(50-55-5)**
3-β,20-α-Yohimban-16-β-carboxylic acid, 18-β-hydroxy-11,17-α-dimethoxy-, methyl ester, 3,4,5-trimethoxybenzoate (ester) **(50-55-5)**
3-β,20-α-Yohimban-16-β-carboxylic acid, 18-β-hydroxy-11,17-α-dimethoxy-, methyl ester, 3,4,5-trimethoxycinnamate (ester) **(24815-24-5)**
Yohimban-16-α-carboxylic acid, 17-α-hydroxy-, methyl ester **(146-48-5)**
Yohimban-16-α-carboxylic acid, 17-α-hydroxy-, methyl ester, hydrochloride **(65-19-0)**
Yohimbic acid methyl ester **(146-48-5)**
Yohimbin **(146-48-5)**
Yohimbine **(146-48-5)**
Yohimbine hydrochloride **(65-19-0)**
Yohimbine monohydrochloride **(65-19-0)**
Yohimbin hydrochloride **(65-19-0)**
Yomesan **(50-65-7)**
Yoracryl Blue G **(12217-41-3)**
Yoshi 864 **(3458-22-8)**
Yoshinox S **(96-69-5)**
Yoshinox SR **(96-69-5)**
Yperite **(505-60-2)**
Yperite Sulfone **(471-03-4)**
Yrodazin **(122-34-9)**
Ytterbium **(7440-64-4)**
Yttria **(1314-36-9)**
Yttrium-89 **(7440-65-5)**
Yttrium (ACGIH,OSHA) **(7440-65-5)**
Yttrium-oxide **(1314-36-9)**
Yuban 10HV **(9011-05-6)**
Yuban 10S **(9011-05-6)**
Yukalon EH 30 **(9002-88-4)**
Yukalon HE 60 **(9002-88-4)**
Yukalon K 3212 **(9002-88-4)**
Yukalon LK 30 **(9002-88-4)**
Yukalon MS 30 **(9002-88-4)**
Yukalon PS 30 **(9002-88-4)**
Yukalon YK 30 **(9002-88-4)**
Yulan **(2212-67-1)**
Yulon **(63428-84-2)**
Yunihomu AZ **(123-77-3)**
Yura Yara **(93-04-9)**
Z 11 **(140-93-2)**
Z 75 **(137-30-4)**
Z-78 **(12122-67-7)**
Z 200 **(141-98-0)**
Z-200 **(141-98-0)**
Z-905 **(52463-83-9)**
Z 2008 **(20091-61-6)**
Z 4942 **(3778-73-2)**
Z3 **(140-89-6)**
ZR-856 **(54460-46-7)**
ZB 112 **(1332-07-6)**
ZB 237 **(1332-07-6)**
ZC **(137-30-4)**
Z-C Spray **(137-30-4)**
ZDTP **(25103-54-2)**
ZF 36 **(9002-88-4)**
ZIP **(16509-79-8)**
ZMBT **(155-04-4)**
ZN 100 **(1332-07-6)**
ZP **(1314-84-7)**
ZPT **(13463-41-7)**
ZR 512 **(41096-46-2)**
ZR 515 **(40596-69-8)**
ZR77 **(42588-37-4)**
Z10-TR **(604-75-1)**
ZV8-253 **(8030-89-5)**
Zaclondiscoids **(74-90-8)**
Zactane **(77-15-6)**
Zactirin Compound **(62-44-2)**
Zactran **(315-18-4)**
Zadoletten **(50-06-6)**
Zadonal **(50-06-6)**
Zaffre **(1307-96-6)**
Zagreb **(55-86-7)**
Zaharina **(81-07-2)**
Zahlreiche bezeichnungen (German) **(534-52-1)**
Zambesi Dark Blue BH **(2429-73-4)**
Zambesil **(77-36-1)**
Zami 905 **(52463-83-9)**
Zamine **(51-63-8)**
Zanchol **(519-95-9)**
Zanosar **(18883-66-4)**
Zantac **(66357-35-5)**
Zaprawa Nasienna Plynna **(502-39-6)**
Zaprawa Nasienna Universal **(517-16-8)**
Zaprawa Nasienna Universal R **(517-16-8)**
Zaprawa Nasienna Uniwersalna **(517-16-8)**
Zaraondan **(77-67-8)**
Zardex **(54460-46-7)**
Zarlate **(137-30-4)**
Zarodan **(77-67-8)**
Zarondan-Saft **(77-67-8)**
Zarontin **(77-67-8)**
Zaroxolyn **(17560-51-9)**

Zartalin (77-67-8)
Zassol (917-61-3)
Zeaphos (1912-24-9)
Zeaprim (8073-77-6)
Zeapur (122-34-9)
Zearalanol (26538-44-3)
(-)-Zearalenone (17924-92-4)
(10S)-Zearalenone (17924-92-4)
(S)-Zearalenone (17924-92-4)
Zearalenone (17924-92-4)
trans-Zearalenone (17924-92-4)
Zearanol (26538-44-3)
Zeazin (1912-24-9)
Zeazine (1912-24-9)
Zebedassite (1319-41-1)
Zebenide (12122-67-7)
Zebtox (12122-67-7)
Zectane (315-18-4)
Zectran (315-18-4)
Zedrine (60-13-9)
Zeidane (50-29-3)
Zelan (94-74-6)
Zelen Alizarinova Brilantni G-Extra (Czech) (4403-90-1)
Zelen Brilantni (Czech) (633-03-4)
Zelen Kypova 1 (Czech) (128-58-5)
Zelen Kypova 3 (Czech) (3271-76-9)
Zelen Kysela 3 (Czech) (4680-78-8)
Zelen Kysela 5 (Czech) (5141-20-8)
Zelen Kysela 25 (Czech) (4403-90-1)
Zelen Kysela F (Czech) (5141-20-8)
Zelen Malachitova (Czech) (569-64-2)
Zelen Malachitova G (Czech) (633-03-4)
Zelen Moridlova 4 (Czech) (131-91-9)
Zelen Olivova Ostanthrenova B (Czech) (3271-76-9)
Zelen Ostanthrenova Brilantni FFB (Czech) (128-58-5)
Zelen Potravinarska 1 (Czech) (4680-78-8)
Zelen Potravinarska 2 (Czech) (5141-20-8)
Zelen Potravinarska 3 (Czech) (2353-45-9)
Zelen Rozpoustedlova 3 (Czech) (128-80-3)
Zelen Smaragdova (Czech) (633-03-4)
Zelen Stala FCF (Czech) (2353-45-9)
Zelen Sudan 4B (Czech) (128-80-3)
Zelen Svetla SF (Czech) (5141-20-8)
Zelen Zasadita 1 (Czech) (633-03-4)
Zelen Zasadita 4 (Czech) (569-64-2)
Zelio (10031-59-1)
Zellek (69806-40-2)
Zellek (87237-48-7)
Zenadrid (53-03-2)
Zenadrid (Veterinary) (53-03-2)
Zenalosyn (434-07-1)
Zenite (155-04-4)
Zenite Special (155-04-4)
Zenkor (21087-64-9)
Zentinic (81-13-0)
Zentronal (57-41-0)
Zentropil (57-41-0)
Zentropil (630-93-3)
Zeogel (12174-11-7)
Zeolex (1344-00-9)
Zeolex 100 (1344-00-9)
Zeolex 23A (1344-00-9)
Zeolex 25 (1344-00-9)
Zeolex 35 (1344-00-9)
Zeolex 23P (1344-00-9)
Zeolite (1318-02-1)
Zeolite A (1318-02-1)
Zeolite NaA (1318-02-1)
Zeolites (1318-02-1)
Zeolithe A (1318-02-1)
Zeorinane (1176-44-9)
Zepel (9CI) (62712-23-6)
Zephiral (8001-54-5)
Zephiran chloride (8001-54-5)
Zephrol (299-42-3)
Zeranol (26538-44-3)
Zerdane (50-29-3)
Zerlate (137-30-4)
Zero One (25319-90-8)
Zertell (5598-13-0)
Zeset S (136-84-5)
Zeset T (108-05-4)
Zest (142-47-2)
Zetar (8007-45-2)
Zetax (155-04-4)
Zetesol LES 2 (9004-82-4)
Zetesol NL (9004-82-4)
Zetifex ZN (126-72-7)
Zextran (315-18-4)
3ZhP (7439-89-6)
Ziarnik (62-38-4)
Zidan (12122-67-7)
Zide (58-93-5)
Zidovudine (30516-87-1)
Zimalloy (11114-92-4)
Ziman-Dithane (8018-01-7)
Zimanat (8018-01-7)
Zimaneb (8018-01-7)
Zimate (12122-67-7)
Zimate (137-30-4)
Zimate, butyl (136-23-2)
Zimate, ethyl (14324-55-1)
Zimate, methyl (137-30-4)
Zimco (121-33-5)
(Z)-Zimelidine (56775-88-3)
Zimelidine (56775-88-3)
cis-Zimelidine (56775-88-3)
Zimtaldehyde (104-55-2)
Zimtsaeure (German) (621-82-9)
Zinacef (55268-75-2)
Zinadon (54-85-3)
Zinc (7440-66-6)
Zinc, Ashes (DOT) (7440-66-6)
Zinc Butter (7646-85-7)
Zinc Chromate AM (13530-65-9)
Zinc Chromate C (13530-65-9)
Zinc Chromate O (13530-65-9)
Zinc Chromate T (13530-65-9)
Zinc Chromate Z (13530-65-9)
Zinc Chrome Yellow (13530-65-9)
Zinc Dust (7440-66-6)
Zinc, Isotope of mass 65 (8CI,9CI) (13982-39-3)
Zinc (2+) NTA (53113-57-8)
Zinc PT (13463-41-7)
Zinc Powder (7440-66-6)
Zinc, Powder or dust, Non-pyrophoric (DOT) (7440-66-6)
Zinc, Powder or dust, Pyrophoric [UN 1383] (7440-66-6)
Zinc Tetraoxychromate 76A (13530-65-9)
Zinc Tetraoxychromate 780B (13530-65-9)
Zinc-Tox (1314-84-7)
Zinc Uversol (12001-85-3)
Zinc Yellow (11103-86-9)
Zinc Yellow (13530-65-9)
Zinc Yellow (15930-94-6)

Zinc Yellow **(37300-23-5)**
Zinc Yellow 1 **(13530-65-9)**
Zinc Yellow 1425 **(13530-65-9)**
Zinc Yellow 40-9015 **(13530-65-9)**
Zinc Yellow AZ-16 **(13530-65-9)**
Zinc Yellow AZ-18 **(13530-65-9)**
Zinc Yellow KSH **(13530-65-9)**
Zinc Yellow 386N **(13530-65-9)**
Zinc Yellows **(13530-65-9)**
Zinca 20 **(1314-13-2)**
Zinc acetate **(557-34-6)**
Zinc acetate **(5970-45-6)**
Zinc acetate, dihydrate **(5970-45-6)**
Zinc acetoacetonate **(14024-63-6)**
Zinc ammonium chloride **(14639-97-5)**
Zinc ammonium chloride **(14639-98-6)**
Zinc ammonium chloride **(52628-25-8)**
Zinc ammonium nitrite **(63885-01-8)**
Zinc ammonium sulfate **(7783-24-6)**
Zinc arsenate **(1303-39-5)**
Zinc arsenate, Basic **(1303-39-5)**
Zinc arsenate, Solid **(1303-39-5)**
Zincate(4-), bis(carbonato(2-))pentaoxopenta-, tetrahydrogen (9CI) **(5970-61-6)**
Zincate(1-), (N,N-bis(carboxymethyl)glycinato(3-)-N,O,O',O'')-, hydrogen, (T-4)- **(53113-57-8)**
Zincate(2-), (7,12-diethenyl-3,8,13,17-tetramethyl-21H,23H-porphine-2,18-dipropanoato(4-)-N21,N22,N23,N24)-, dihydrogen, (SP-4-2)- (9CI) **(15442-64-5)**
Zincate(2-), ((N,N'-1,2-ethanediylbis(N-(carboxymethyl)-glycinato))(4-)-N,N',O,O',ON,ON')-, diammonium, (OC-6-21-)-(9CI) **(67859-51-2)**
Zincate(2-), ((N,N'-1,2-ethanediylbis(N-(carboxymethyl)-glycinato))(4-)-N,N',O,O',ON,ON')-, disodium, (OC-6-21)-(9CI) **(14025-21-9)**
Zincate(2-), ((N,N'-1,2-ethanediylbis(N-(carboxymethyl)-glycinato))(4-))-, disodium **(14025-21-9)**
Zincate(3-), pentachloro-, triammonium (9CI) **(14639-98-6)**
Zincate(2-), tetrachloro-, diammonium, (T-4)- (9CI) **(14639-97-5)**
Zincate(2-), tetrakis(cyano-C)-, disodium, (T-4)- (9CI) **(15333-24-1)**
Zinc benzoate **(553-72-0)**
Zinc 2-benzothiazolethiolate **(155-04-4)**
Zinc benzothiazolyl mercaptide **(155-04-4)**
Zinc benzothiazol-2-ylthiolate **(155-04-4)**
Zinc benzothiazyl-2-mercaptide **(155-04-4)**
Zinc beryllium silicate **(39413-47-3)**
Zinc bibutyldithiocarbamate **(136-23-2)**
Zinc bichromate **(14018-95-2)**
Zinc, bis(2-benzothiazolethiolato)- **(155-04-4)**
Zinc, bis(O,O-bis(1,3-dimethylbutyl) phosphorodithioato-S,S')-, (T-4)- (9CI) **(2215-35-2)**
Zinc, bis(O,O-bis(2-ethylhexyl) phosphorodithioato-S,S')-, (T-4)- (9CI) **(4259-15-8)**
Zinc, bis(O,O-bis(4-methylpentyl) phosphorodithioato-S,S')-, (T-4)- (9CI) **(15874-15-4)**
Zinc, bis(bis(phenylmethyl)carbamodithioato-S,S')-, (T-4)- (9CI) **(14726-36-4)**
Zinc, bis(dibutyldithiocarbamato) **(136-23-2)**
Zinc, bis(O,O-dibutyl phosphorodithioato-S,S')-, (T-4)- (9CI) **(6990-43-8)**
Zinc, bis(diethyldithiocarbamato) **(14324-55-1)**
Zinc, bis(O,O-dihexyl phosphorodithioato-S,S')-, (T-4)- (9CI) **(7282-28-2)**
Zinc bis(diisooctyl dithiophosphate) **(28629-66-5)**
Zinc bis(dimethyldithiocarbamate) **(137-30-4)**
Zinc bis(dimethyldithiocarbamate)cyclohexylamine complex **(16509-79-8)**
Zinc, bis(dimethyldithiocarbamato) **(137-30-4)**
Zinc bis(dimethyldithiocarbamoyl)disulphide **(137-30-4)**
Zinc bis(dimethylthiocarbamoyl)disulfide **(137-30-4)**
Zinc, bis(O,O-dioctyl phosphorodithioato-S,S')-, (T-4)- (9CI) **(7059-16-7)**
Zinc, bis(dipentylcarbamodithioato-S,S')-, (T-4)- (9CI) **(15337-18-5)**
Zinc, bis(O-(2-ethylhexyl) O-(2-methylpropyl) phosphorodithioato-S,S')-, (T-4)- (9CI) **(26566-95-0)**
Zinc, bis(D-gluconato-O^1,O^2)- (9CI) **(4468-02-4)**
Zinc, bis(hydroxymethanesulfinato-O,O')- **(24887-06-7)**
Zinc, bis(hydroxymethanesulfinato-OS,O1)-, (T-4)- (9CI) **(24887-06-7)**
Zinc, bis(1-hydroxy-2(1H)-pyridinethionato)- (8CI) **(13463-41-7)**
Zinc, bis(1-hydroxy-2(1H)-pyridinethionato-O,S)-, (T-4)- (9CI) **(13463-41-7)**
Zinc O,O-bisisodecyl dithiophosphate **(25103-54-2)**
Zinc, bis(O-(1-methylethyl) carbonodithioato-S,S')-, (T-4)- (9CI) **(1000-90-4)**
Zinc, bis(2,4-pentanedionato-O,O') **(14024-63-6)**
Zinc, bis(1-piperidinecarbodithioato) **(13878-54-1)**
Zinc, bis(1-piperidinecarbodithioato-S,S')-, (T-4)- (9CI) **(13878-54-1)**
Zinc, bis(2-pyridylthio)-, 1,1'-dioxide **(13463-41-7)**
Zinc, bis(2-pyridylthio)-, N,N'-dioxide **(13463-41-7)**
Zinc, bis(2,4,5-trichlorophenoxy)- **(136-24-3)**
Zinc blende **(1314-98-3)**
Zinc borate **(1332-07-6)**
Zinc-bromide **(7699-45-8)**
Zinc carbonate (1:1) **(3486-35-9)**
Zinc chlorate [UN 5113] **(10361-95-2)**
Zinc-chloride **(7646-85-7)**
Zinc chloride (ACGIH,OSHA) **(7646-85-7)**
Zinc chloride, Anhydrous [UN 2331] **(7646-85-7)**
Zinc chloride, Solution [UN 1840] **(7646-85-7)**
Zinc chloride (ZnCl) (6CI,7CI,8CI,9CI) **(18623-80-8)**
Zinc chloride ($ZnCl_2$), compd. with 3,7-bis(diethylamino)-2-methylphenazoxonium chloride (9CI) **(7199-02-2)**
Zinc chloride, Compd. with 3,7-bis(diethylamino)-2-methyl-phenazoxonium chloride **(7199-02-2)**
Zinc (chlorure de) (French) **(7646-85-7)**
Zinc chromate (ACGIH,OSHA) **(12018-19-8)**
Zinc chromate (ACGIH,OSHA) **(13530-65-9)**
Zinc chromate (ACGIH,OSHA) **(14018-95-2)**
Zinc chromate (VI) hydroxide **(15930-94-6)**
Zinc chromate(VI) hydroxide **(13530-65-9)**
Zinc chromate hydroxide **(15930-94-6)**
Zinc chrome **(11103-86-9)**
Zinc chrome **(13530-65-9)**
Zinc chrome (Anti-corrosion) **(13530-65-9)**
Zinc chromite **(12018-19-8)**
Zinc chromium oxide **(12018-19-8)**
Zinc chromium oxide **(13530-65-9)**
Zinc chromium oxide **(14018-95-2)**
Zinc citrate **(546-46-3)**
Zinc cyanide [UN 1713] **(557-21-1)**
Zinc, (cyclohexylamine)bis(dimethyldithiocarbamato)- **(16509-79-8)**
Zinc diacetate **(557-34-6)**
Zinc diacetate, dihydrate **(5970-45-6)**
Zinc diacrylate **(14643-87-9)**
Zinc dialkyl dithiophosphorate **(68457-79-4)**
Zinc diammonium disulfate hexahydrate **(7783-24-6)**
Zinc dibenzoate **(553-72-0)**
Zinc dibenzyldithiocarbamate **(14726-36-4)**
Zinc dibromide **(7699-45-8)**
Zinc N,N-dibutyldithiocarbamate **(136-23-2)**
Zinc dibutyldithiocarbamate **(136-23-2)**
Zinc dichloride **(7646-85-7)**
Zinc, dichloro(4,4-dimethyl-5-((((methylamino)carbonyl)oxy)-imino)pentanenitrile)-, (T-4)- (9CI) **(58270-08-9)**
Zinc dichromate **(14018-95-2)**
Zinc dichromate (VI) **(14018-95-2)**
Zinc dicyanide **(557-21-1)**
Zinc, diethyl **(557-20-0)**
Zinc N,N-diethyldithiocarbamate **(14324-55-1)**
Zinc diethyldithiocarbamate **(14324-55-1)**
Zinc difluoride tetrahydrate **(13986-18-0)**

Zinc diformate (557-41-5)
Zinc diformate dihydrate (5970-62-7)
Zinc O,O-dihexyl dithiophosphate (7282-28-2)
Zinc dihexyl dithiophosphate (7282-28-2)
Zinc dihydrogen phosphate (13598-37-3)
Zinc diiodide (10139-47-6)
Zinc diisoctyl dithiophosphate (28629-66-5)
Zinc O,O-diisodecyl dithiophosphate (25103-54-2)
Zinc O,O-diisooctyl dithiophosphate (28629-66-5)
Zinc diisooctyl dithiophosphate (28629-66-5)
Zinc, dimethyl (544-97-8)
Zinc N,N-dimethyldithiocarbamate (137-30-4)
Zinc dimethyldithiocarbamate (137-30-4)
Zinc, dimethyldithiocarbamate cyclohexylamine complex (16509-79-8)
Zincdimethyl dithiocarbamate cyclohexylamine complex (16509-79-8)
Zinc dinonylnaphthalenesulfonate (28016-00-4)
Zinc dipropenoate (14643-87-9)
Zinc dipropionate (557-28-8)
Zinc distearate (557-05-1)
Zinc dithionite (7779-86-4)
Zinc dodecanoate (2452-01-9)
Zinc, ((1,2-ethanediylbis(carbamodithioato))(2-))- (9CI) (12122-67-7)
Zinc ethide (557-20-0)
Zinc ethyl (DOT) (557-20-0)
Zinc ethylene bisdithiocarbamate (12122-67-7)
Zinc ethylene-1,2-bisdithiocarbamate (12122-67-7)
Zinc ethylenebis(dithiocarbamate) (12122-67-7)
Zinc ethylenebisdithiocarbamate (12122-67-7)
Zinc ethylenebis(dithiocarbamate) (Polymeric) (12122-67-7)
Zinc, (ethylenebis(dithiocarbamato)) (12122-67-7)
Zinc, (ethylenebis(dithiocarbamato))- (12122-67-7)
Zinc 2-ethylhexanoate (136-53-8)
Zinc flue dust (Secondary nonferrous plant) (69012-58-4)
Zinc-fluoride (7783-49-5)
Zinc fluoride, tetrahydrate (8CI,9CI) (13986-18-0)
Zinc fluorosilicate [UN 2855] (16871-71-9)
Zinc fluorure (French) (7783-49-5)
Zinc formaldehyde sulfoxylate (24887-06-7)
Zinc formate (557-41-5)
Zinc formate (Zn(HCO2)2) dihydrate (5970-62-7)
Zinc formate, dihydrate (7CI) (5970-62-7)
Zinc fume (69012-65-3)
Zinc-gluconate (4468-02-4)
Zinc hexadecanoate (4991-47-3)
Zinc hexafluorosilicate (16871-71-9)
Zinc hydrosulfite [UN 1931] (7779-86-4)
Zinc p-hydroxybenzenesulfonate (127-82-2)
Zinc hydroxychromate (13530-65-9)
Zinc hydroxychromate (15930-94-6)
Zinc iodide (9CI) (10139-47-6)
Zinc, ion (Zn(2+)) (8CI,9CI) (23713-49-7)
Zincite (1314-13-2)
Zinc laurate (2452-01-9)
Zincmate (137-30-4)
Zinc mercaptobenzothiazolate (155-04-4)
Zinc 2-mercaptobenzothiazole (155-04-4)
Zinc mercaptobenzothiazole salt (155-04-4)
Zinc 4-methylbenzenesulfinate (24345-02-6)
Zinc metiram (9006-42-2)
Zinc monosulfide (1314-98-3)
Zinc muriate, Solution (DOT) (7646-85-7)
Zinc naphthenate (12001-85-3)
Zinc nitrate [UN 1514] (7779-88-6)
Zinc(II) nitrate, hexahydrate (1:2:6) (10196-18-6)
Zinco (cloruro di) (Italian) (7646-85-7)
Zinc octadecanoate (557-05-1)
Zinco(fosfuro di) (Italian) (1314-84-7)
Zincoid (1314-13-2)
Zinc oleate (557-07-3)
Zinc oleate (1:2) (557-07-3)
Zinc omadine (13463-41-7)
Zinc orthophosphate (7779-90-0)
Zinc-oxide (1314-13-2)
Zinc oxide (ACGIH,OSHA) (1314-13-2)
Zinc oxide Fume (1314-13-2)
Zinc pentamethylenedithiocarbamate (13878-54-1)
Zinc permangante [UN 1515] (23414-72-4)
Zinc peroxide [UN 1516] (1314-22-3)
Zinc p-phenol sulfonate (127-82-2)
Zinc phenolsulfonate (127-82-2)
Zinc phosphate (7779-90-0)
Zinc orthophosphate, dihydrogen (13598-37-3)
Zinc phosphide [UN 1714] (1314-84-7)
Zinc(phosphure de) (French) (1314-84-7)
Zincpolyanemine (13463-41-7)
Zinc polyphosphate (68607-18-1)
Zinc propanoate (557-28-8)
Zinc 2-propenoate (14643-87-9)
Zinc propionate (557-28-8)
Zinc 1,2-propylene bisdithiocarbamate (12071-83-9)
Zinc, (N,N'-propylene-1,2-bis(dithiocarbamate)) (12071-83-9)
Zinc protoporphyrin (15442-64-5)
Zinc 2-pyridinethiol-1-oxide (13463-41-7)
Zinc pyridine-2-thiol-1-oxide (13463-41-7)
Zinc pyridinethione (13463-41-7)
Zinc pyrion (13463-41-7)
Zinc pyrithione (13463-41-7)
Zinc resinate [UN 2714] (9010-69-9)
Zinc silicofluoride (DOT) (16871-71-9)
Zinc-stearate (557-05-1)
Zinc stearate (ACGIH,OSHA) (557-05-1)
Zinc sulfate (7446-19-7)
Zinc sulfate (7446-20-0)
Zinc sulfate (7733-02-0)
Zinc sulfate (1:1) (7733-02-0)
Zinc sulfate, Basic (68813-94-5)
Zinc sulfate (1:1) heptahydrate (7446-20-0)
Zinc sulfate heptahydrate (1:1:7) (7446-20-0)
Zinc sulfide (9CI) (1314-98-3)
Zinc sulfocarbolate (127-82-2)
Zinc sulfophenate (127-82-2)
Zinc sulphate (7446-20-0)
Zinc sulphate (7733-02-0)
Zinc superoxide (1314-22-3)
Zinc(2+), tetraammine-, (T-4)-, carbonate (1:1) (9CI) (38714-47-5)
Zinc tetraoxychromate (13530-65-9)
Zinc tetroxychromate (13530-65-9)
Zinc 10-undecenoate (557-08-4)
Zinc undecenoate (557-08-4)
Zinc undecylenate (557-08-4)
Zinc vitriol (7446-20-0)
Zinc vitriol (7733-02-0)
Zinc white (1314-13-2)
Zineb (12122-67-7)
Zineb 75 (12122-67-7)
Zineb 80 (12122-67-7)
Zineb 75 WP (12122-67-7)
Zinebe (French) (12122-67-7)
Zineb-ethylene thiuram disulfide adduct (9006-42-2)
Zingerone (122-48-5)
Zingiberone (122-48-5)
Zink-(N,N'-aethylen-bis(dithiocarbamat)) (German) (12122-67-7)
Zink-bis(N,N-dimethyl-dithiocarbamaat) (Dutch) (137-30-4)
Zink-bis(N,N-dimethyl-dithiocarbamat) (German) (137-30-4)

Zinkcarbamate (137-30-4)
Zinkchlorid (German) (7646-85-7)
Zinkchloride (Dutch) (7646-85-7)
Zink-(N,N-dimethyl-dithiocarbamat) (German) (137-30-4)
Zinkfosfide (Dutch) (1314-84-7)
Zinkosite (7733-02-0)
Zinkphosphid (German) (1314-84-7)
Zink-(N,N'-propylen-1,2-bis(dithiocarbamat)) (German) (12071-83-9)
Zinn (German) (7440-31-5)
Zinniol (17811-28-8)
Zinntetrachlorid (German) (7646-78-8)
Zinochlor (101-05-3)
Zinophos (297-97-2)
Zinosan (12122-67-7)
Zinpol (9002-88-4)
Zinpol (9003-01-4)
Zipan (439-14-5)
Zipromat (12071-83-9)
Ziram (137-30-4)
Ziram Technical (137-30-4)
Ziram-cyclohexylamine complex (16509-79-8)
Ziram cyclohexylamine complex (16509-79-8)
Ziram, cyclohexylamine complex (16509-79-8)
Zirame (137-30-4)
Ziramvis (137-30-4)
Zirasan (137-30-4)
Zirasan 90 (137-30-4)
Zirberk (137-30-4)
Zircat (7440-67-7)
Zirconate(2-), hexafluoro-, diammonium (16919-31-6)
Zirconate(2-), hexafluoro-, diammonium, (OC-6-11)- (9CI) (16919-31-6)
Zirconate(2-), hexafluoro-, dipotassium (16923-95-8)
Zirconium (ACGIH,OSHA) (7440-67-7)
Zirconium, Borings (DOT) (7440-67-7)
Zirconium, Clippings (DOT) (7440-67-7)
Zirconium(IV) chloride (1:4) (10026-11-6)
Zirconium(IV) sulfate (1:2) (14644-61-2)
Zirconium, Isotope of mass 95 (8CI,9CI) (13967-71-0)
Zirconium, Scrap [UN 1932] (7440-67-7)
Zirconium, Suspended in fammable liquid (DOT) (7440-67-7)
Zirconium, Turnings (DOT) (7440-67-7)
Zirconium acetate (7585-20-8)
Zirconium aluminide (12004-83-0)
Zirconium aluminum (12004-83-0)
Zirconium ammonium fluoride (16919-31-6)
Zirconium chloride (10026-11-6)
Zirconium, chlorohydroxyoxo- (18428-88-1)
Zirconium, dichloro-di-pi-cyclopentadienyl (1291-32-3)
Zirconium, dichlorooxo (7699-43-6)
Zirconium, dichlorooxo-, hexahydrate (8CI,9CI) (25399-81-9)
Zirconium 2-ethylhexanoate (22464-99-9)
Zirconium hydride (8CI,9CI) (7704-99-6)
Zirconium metal (DOT) (7440-67-7)
Zirconium nitrate [UN 2728] (13746-89-9)
Zirconium oxide (9CI) (1314-23-4)
Zirconium oxychloride (7699-43-6)
Zirconium picramate, Dry or containing less than 20% water [UN 0236] (63868-82-6)
Zirconium picramate, Wet (with at least 20% water) [UN 1517] (63868-82-6)
Zirconium sodium lactate (10377-98-7)
Zirconium sulfate [NA 9163] (14644-61-2)
Zirconium sulphate (14644-61-2)
Zirconium tetrachloride, Solid (DOT) (10026-11-6)
Zirconium tetrachloride [UN 2503] (10026-11-6)
Zirconocene, dichloride (1291-32-3)
Zirconyl chloride (7699-43-6)
Zirconyl sulfate (14644-61-2)
Zirex 90 (137-30-4)
Ziride (137-30-4)
Zironate(2-), hexafluoro-, diammonium, (OC-6-11)- (16919-31-6)
Zirpon (57-53-4)
Zirthane (137-30-4)
Zisnet AF (13733-91-0)
Zisnet AS (13733-91-0)
Zithiol (121-75-5)
Zitox (137-30-4)
Zitronell Oel (German) (8000-29-1)
Zitronen Oel (German) (8008-56-8)
Zlut Anilinova (Czech) (60-09-3)
Zlut Chinolonova (Czech) (8004-92-0)
Zlut Disperzni 3 (Czech) (2832-40-8)
Zlut Kypova 2 (Czech) (129-09-9)
Zlut Kysela 3 (Czech) (8004-92-0)
Zlut Kysela 23 (Czech) (1934-21-0)
Zlut Kysela 36 (Czech) (587-98-4)
Zlut Kysela 73 (Czech) (2321-07-5)
Zlut Marciova (Czech) (605-69-6)
Zlut Maselna AB (Czech) (85-84-7)
Zlut Maselna (Czech) (60-11-7)
Zlut Maselna OB (Czech) (131-79-3)
Zlut Metanilova (Czech) (587-98-4)
Zlut Naftolova (Czech) (605-69-6)
Zlut Ostanthrenova GC (Czech) (129-09-9)
Zlut Pigment 12 (Czech) (6358-85-6)
Zlut Pigment 100 (Czech) (1934-21-0)
Zlut Potravinarska 3 (Czech) (2783-94-0)
Zlut Potravinarska 4 (Czech) (1934-21-0)
Zlut Potravinarska 13 (Czech) (8004-92-0)
Zlut Prirodni 3 (Czech) (458-37-7)
Zlut Prirodni 16 (Czech) (84-79-7)
Zlut Prirodni 26 (Czech) (7235-40-7)
Zlut Rozpoustedlova 1 (Czech) (60-09-3)
Zlut Rozpoustedlova 2 (Czech) (60-11-7)
Zlut Rozpoustedlova 3 (Czech) (97-56-3)
Zlut Rozpoustedlova 5 (Czech) (85-84-7)
Zlut Rozpoustedlova 6 (Czech) (131-79-3)
Zlut Rozpoustedlova 7 (Czech) (1689-82-3)
Zlut Rozpoustedlova 14 (Czech) (842-07-9)
Zlut Rozpoustedlova 77 (Czech) (2832-40-8)
Zlut Zasadita 2 (2465-27-2)
Zn-0401 E 3/16'' (1314-13-2)
ZnMB (155-04-4)
ZnPT (13463-41-7)
Zn 0701T (1314-13-2)
Zn-0312 T 1/4'' (12018-19-8)
Zoalene (148-01-6)
Zoamix (148-01-6)
Zoaquin (83-73-8)
Zoba Black D (106-50-3)
Zoba Brown P Base (123-30-8)
Zoba Brown RR (5307-14-2)
Zoba EG (591-27-5)
Zoba ERN (90-15-3)
Zoba 3GA (95-55-6)
Zoba GKE (95-80-7)
Zoba 4R (96-91-3)
Zoba SLE (39156-41-7)
Zobar (1338-32-5)
Zobar (50-31-7)
Zogen Developer H (95-80-7)
Zolaphen (50-33-9)
Zolidinum (50-33-9)
Zolon (2310-17-0)
Zolone (2310-17-0)

Zolone PM (2310-17-0)
Zomax (64092-48-4)
Zomepirac (64092-48-4)
Zomepirac sodium (64092-48-4)
Zomepirac sodium salt (64092-48-4)
Zomine Yellow R (1325-37-7)
Zonarez 7085 (9003-73-0)
Zonarez 7115 (9003-73-0)
Zonazide (54-85-3)
Zoocoumarin (Russian) (81-81-2)
Zoofurin (67-20-9)
Zoolobelin (63990-84-1)
Zoolon (2310-17-0)
Zopaque (13463-67-7)
Zorane (1665-48-1)
Zorane (50-33-9)
Zorial (27314-13-2)
Zoriflavin (8048-52-0)
Zotil (152-72-7)
Zotox (1303-28-2)
Zotox (7778-39-4)
Zotox Crab Grass Killer (7778-39-4)
Zutracin (1405-87-4)
Zwavelwaterstof (Dutch) (7783-06-4)
Zwavelzuuroplossingen (Dutch) (7664-93-9)
Zwickau Green (12002-03-8)
Zwitsalax (117-10-2)
Zwittergent 3-12 (14933-08-5)
Zyban (23564-05-8)
Zygomycin A1 (7542-37-2)
Zygosporin A (22144-77-0)
Zyklophosphamid (German) (50-18-0)
Zyloprim (315-30-0)
Zyloric (315-30-0)
Zymostanol (360-68-9)
Zymostanol (80-97-7)
Zytel 211 (105-60-2)
Zytel 211 (25038-54-4)
Zytox (74-83-9)
Zytron (299-85-4)

Ac
(7440-34-8)
Ag
(7440-22-4)
(14391-76-5)
$Ag.ClHO_2$
(7783-91-7)
$Ag.ClHO_4$
(7783-93-9)
$Ag.1/2CrH_2O_4$
(7784-01-2)
AgBr
(7785-23-1)
AgCl
(7783-90-6)
AgF
(7775-41-9)
AgF_2
(7783-95-1)
AgI
(7783-96-2)
AgN_3
(13863-88-2)
Ag_2S
(21548-73-2)
Al
(7429-90-5)
$Al.3ClHO_3$
(15477-33-5)
$Al.2H_2O_4S.K$
(10043-67-1)
$Al.2H_2O_4S.Na$
(10102-71-3)
$Al.H_3N.2H_2O_4S$
(7784-25-0)
Al.Zr
(12004-83-0)
AlB_3H_{12}
(13771-22-7)
$AlBr_3$
(7727-15-3)
$AlCl_3$
(7446-70-0)
$AlCl_3.6H_2O$
(7784-13-6)
$AlF_2O_{10}Si_3.K.3Mg$
(12003-38-2)
AlF_3
(7784-18-1)
$AlF_6.3Na$
(15096-52-3)
AlH_3
(7784-21-6)
AlH_3O_3
(21645-51-2)
$AlH_4.1/2Ca$
(16941-10-9)
$AlH_4.Li$
(16853-85-3)
$AlH_4.Na$
(13770-96-2)
$AlH_{24}NaO_{20}S_2$
(7784-28-3)
$AlNH_4(O_4S)_2.12H_2O$
(7784-26-1)
$AlO_2.Na$
(1302-42-7)
AlP
(20859-73-8)
$Al_2ClH_5O_5$
(12042-91-0)
Al_2Cl_6
(13845-12-0)
Al_2O_3
(1302-74-5)
(1344-28-1)
$Al_2O_3.Al.Ni.Si$
(12743-20-3)
$Al_2O_3.xH_2O$
(1318-16-7)
$Al_2O_{18}S_3.6Ca$
(12004-14-7)
$Al_2O_{18}S_3.6Ca.xH_2O$
(11070-82-9)
$Al_2O_{18}Si_6.3Be$
(1302-52-9)
$Al_2O_{18}Si_7.1/2Ca.7H_2O.-$
$1/2Na$
(66733-21-9)
$Al_6O_{13}Si_2$
(12068-56-3)
$Al_{11}O_{17}.Na$
(12005-48-0)
Am
(7440-35-9)
(14596-10-2)
Ar
(7440-37-1)
As
(7440-38-2)
$AsBr_3$
(7784-33-0)
$AsCl_3$
(7784-34-1)
AsF_3
(7784-35-2)
AsGa
(1303-00-0)
$AsHO_2.2H_2O.1/2Sr$
(10378-48-0)
$AsHO_2.K$
(13464-35-2)
$AsHO_3.1/2Ca$
(15194-99-7)
$AsHO_3.Cu$
(10290-12-7)
$AsHO_4.2H_4N$
(7784-44-3)
$AsHO_4.2Na$
(7778-43-0)
$AsHO_4.2Na.7H_2O$
(10048-95-0)
$AsHO_4.Pb$
(7784-40-9)
$AsH_2O_4.K$
(7784-41-0)
AsH_3
(7784-42-1)
$AsH_3O_3.Ca$
(52740-16-6)
$AsH_3O_3.3/2Ca$
(27152-57-4)
$AsH_3O_3.xK$
(10124-50-2)
$AsH_3O_3.3Na$
(13464-37-4)
AsH_3O_4
(7778-39-4)
$AsH_3O_4.3/2Co$
(24719-19-5)
$AsH_3O_4.xMg$
(10103-50-1)
$AsH_3O_4.3Na$
(13464-38-5)
$AsH_3O_4.xNa$
(7631-89-2)
$AsH_3O_4.Pb$
(14758-11-3)
$AsH_3O_4.xPb$
(7645-25-2)
AsI_3
(7784-45-4)
$AsO_2.Na$
(7784-46-5)
AsO_4
(15584-04-0)
$AsO_4.Fe$
(10102-49-5)
$As_2Fe_2O_6.Fe_2O_3.5H_2O$
(63989-69-5)
As_2O_3
(1327-53-3)
As_2O_4Sr
(91724-16-2)
As_2O_5
(1303-28-2)
$As_2O_8.3Ca$
(7778-44-1)
$As_2O_8.3Fe$
(10102-50-8)
$As_2O_8.3Pb$
(3687-31-8)
As_2S_3
(1303-33-9)
As_2Se_3
(1303-36-2)
$As_4O_{15}.5Zn$
(1303-39-5)
Au
(7440-57-5)
$AuCl_3$
(13453-07-1)
B
(7440-42-8)
BBr_3
(10294-33-4)
BCl_3
(10294-34-5)
BFH_4N
(13826-83-0)
BF_3
(7637-07-2)
$BF_4.H$
(16872-11-0)
$BF_4.3H_2O.1/2Ni$
(15684-36-3)
$BHO_2.1/2Ba$
(13701-59-2)
$BHO_2.Li$
(13453-69-5)
$BHO_2.Na$
(7775-19-1)
$BHO_2.1/2Pb$
(14720-53-7)
$BHO_3.Na$
(7632-04-4)
$BH_3.K$
(13762-51-1)
BH_3O_3
(10043-35-3)
$BH_4.Li$
(16949-15-8)
$BH_4.Na$
(16940-66-2)
BO_3
(14213-97-9)
B_2CdF_8
(14486-19-2)
$B_2F_8.Pb$
(13814-96-5)
B_2H_6
(19287-45-7)
B_2O_3
(1303-86-2)
$B_4Li_2O_7$
(12007-60-2)
$B_4Na_2O_7$
(1330-43-4)
$B_4O_7.2Na.5H_2O$
(11130-12-4)
$B_4O_7.2Na.10H_2O$
(1303-96-4)
$B_5HO_8.H_3N$
(12007-89-5)
B_5H_9
(19624-22-7)
$B_{10}H_{14}$
(17702-41-9)
Ba
(7440-39-3)
$Ba.2ClHO_4$
(13465-95-7)
$Ba.CrH_2O_4$
(10294-40-3)
$BaCl_2$
(10361-37-2)
$BaCl_2.2H_2O$
(10326-27-9)
BaF_2
(7787-32-8)
BaH_2O_2
(17194-00-2)
BaN_6
(18810-58-7)
BaO
(1304-28-5)
BaO_2
(1304-29-6)
Be
(7440-41-7)
Be.Cu
(64535-95-1)
$Be.1/2H_4O_4Si$
(13598-00-0)
$BeCl_2$
(7787-47-5)
BeF_2
(7787-49-7)
$BeHO_4P$
(13598-15-7)
$BeNO_3$
(7787-55-5)
BeN_2O_6
(13597-99-4)
BeO
(1304-56-9)
Bi
(7440-69-9)
(14331-79-4)
Bi_2Te_3
(1304-82-1)
$Bi_5H_9N_4O_{22}$
(1304-85-4)
Br
(15541-45-4)
BrCl
(13863-41-7)
BrClPb
(13778-36-4)
BrCs
(7787-69-1)
BrCu
(7787-70-4)
BrF_3
(7787-71-5)
BrF_5
(7789-30-2)
BrFe
(12514-32-8)
BrH
(10035-10-6)
$BrHO_3.H_3N$
(13843-59-9)
BrH_3Si
(13465-73-1)
BrH_4N
(12124-97-9)
BrHg
(10031-18-2)
BrK
(7758-02-3)
BrLi
(7550-35-8)
BrNa
(7647-15-6)
$BrO_3.K$
(7758-01-2)
$BrO_3.Na$
(7789-38-0)
BrPb

(15576-47-3)

Br_2
(7726-95-6)

Br_2Ca
(7789-41-5)

Br_2Cd
(7789-42-6)

Br_2Co
(7789-43-7)

Br_2Cu
(7789-45-9)

Br_2H
(11071-85-5)

Br_2Hg
(7789-47-1)

Br_2Ni
(13462-88-9)

$Br_2O_5Sb_4$
(12323-32-9)

Br_2Pb
(10031-22-8)

Br_2Zn
(7699-45-8)

Br_3In
(13465-09-3)

Br_3OP
(7789-59-5)

Br_3P
(7789-60-8)

Br_3Sb
(7789-61-9)

Br_4Ge
(13450-92-5)

C
(7440-44-0)
(7782-40-3)
(7782-42-5)
(14762-75-5)

C.Cu.Fe.Mn.Si
(39362-66-8)

CAgN
(506-64-9)

CB_4
(12069-32-8)

CBi_2O_5
(5892-10-4)

$CBrClF_2$
(353-59-3)

$CBrCl_2F$
(353-58-2)

$CBrCl_3$
(75-62-7)

$CBrF_3$
(75-63-8)

CBrN
(506-68-3)

CBr_2Cl_2
(594-18-3)

CBr_2F_2
(75-61-6)

CBr_3Cl
(594-15-0)

CBr_3NO_2
(464-10-8)

CBr_4
(558-13-4)

$CClF_3$
(75-72-9)

CClN
(506-77-4)

$CClN_3O_6$
(1943-16-4)

CCl_2F_2
(75-71-8)

CCl_2O
(75-44-5)

CCl_2S
(463-71-8)

CCl_3F
(75-69-4)

CCl_3NO_2
(76-06-2)

CCl_4
(56-23-5)

CCl_4O_4
(67632-66-0)

CCl_4S
(594-42-3)

CCuN
(544-92-3)

CF_2O
(353-50-4)

CF_3I
(2314-97-8)

CF_3NO_2
(335-02-4)

CF_4
(75-73-0)

CF_4O_2S
(335-05-7)

CHBrClI
(34970-00-8)

$CHBrCl_2$
(75-27-4)

$CHBr_2Cl$
(124-48-1)

$CHBr_2I$
(593-94-2)

$CHBr_3$
(75-25-2)

$CHClF_2$
(75-45-6)

$CHCl_2F$
(75-43-4)

$CHCl_2I$
(594-04-7)

$CHCl_3$
(67-66-3)

$CHCl_3S$
(75-70-7)

CHFO
(1493-02-3)

CHF_3
(75-46-7)

CHF_3O_3S
(1493-13-6)

CHI_3
(75-47-8)

CHN
(74-90-8)

CHNO
(75-13-8)
(420-05-3)

CHNS
(463-56-9)
(3129-90-6)

CHNS.1/2Co
(3017-60-5)

CHNS.Cu
(1111-67-7)

CHNS.1/2Pb
(592-87-0)

CHN_3O_6
(517-25-9)

CHN_3S_2
(4472-06-4)

CHO_2
(71-47-6)

$CHO_2.K$
(590-29-4)

$CHO_2.Na$
(141-53-7)

$CHO_3.Na$
(144-55-8)

CHO_3Pb
(13427-42-4)

CH_2BrCl
(74-97-5)

CH_2Br_2
(74-95-3)

CH_2Cl
(6806-86-6)

CH_2ClF
(593-70-4)

CH_2ClI
(593-71-5)

CH_2Cl_2
(75-09-2)

CH_2Cl_4Si
(1558-25-4)

CH_2F_2
(75-10-5)

CH_2I_2
(75-11-6)

$CH_2NO_2.H_4N$
(1111-78-0)

$CH_2NS_2.Na$
(4384-81-0)

CH_2N_2
(334-88-3)
(420-04-2)

$CH_2N_2.Pb$
(20837-86-9)

CH_2N_2Na
(19981-17-0)

$CH_2N_2O_4$
(625-76-3)

$CH_2N_2O_6$
(38483-28-2)

CH_2O
(50-00-0)

CH_2O_2
(64-18-6)

$CH_2O_2.1/3Al$
(7360-53-4)

$CH_2O_2.1/2Co$
(544-18-3)

$CH_2O_2.1/2Cu$
(544-19-4)

$CH_2O_2.H_2O.1/2Ni$
(15694-70-9)

$CH_2O_2.H_2O.1/2Zn$
(5970-62-7)

$CH_2O_2.H_3N$
(540-69-2)

$CH_2O_2.Li$
(556-63-8)

$CH_2O_2.1/2Pb$
(811-54-1)

$CH_2O_2.1/2Zn$
(557-41-5)

$CH_2O_3.Ca$
(13397-26-7)

$CH_2O_3.1/2Ca.1/2Mg$
(16389-88-1)

$CH_2O_3.2/3Ce$
(537-01-9)

$CH_2O_3.2Cu$
(3444-14-2)

$CH_2O_3.xCu.xH_3N$
(33113-08-5)

$CH_2O_3.2H_2O.Mg$
(68973-26-2)

$CH_2O_3.2H_3N$
(506-87-6)

$CH_2O_3.Fe$
(563-71-3)

$CH_2O_3.K$
(298-14-6)

$CH_2O_3.2/3La$
(587-26-8)

$CH_2O_3.1/2Mg$
(2090-64-4)

$CH_2O_3.Mn$
(598-62-9)

$CH_2O_3.3/2Na$
(533-96-0)

$CH_2O_3.2/3Nd$
(5895-46-5)

$CH_2O_3.2/3Pr$
(5895-45-4)

$CH_2O_3.Sr$
(1633-05-2)

CH_3AsCl_2
(593-89-5)

$CH_3AsO_3.2Na$
(144-21-8)

CH_3AsS
(2533-82-6)

CH_3Br
(74-83-9)

$CH_3Br.CCl_3NO_2$
(8004-09-9)

CH_3BrMg
(75-16-1)

CH_3Cl
(74-87-3)

CH_3ClHg
(115-09-3)

CH_3ClMg
(676-58-4)

CH_3ClO_2S
(124-63-0)

CH_3ClSn
(23066-18-4)

CH_3Cl_2OP
(676-97-1)

CH_3Cl_2PS
(676-98-2)

CH_3Cl_3Si
(75-79-6)

CH_3Cl_3Sn
(993-16-8)

CH_3CuO_3
(7492-68-4)

CH_3D
(676-49-3)

CH_3F
(593-53-3)

CH_3FO_2S
(558-25-8)

CH_3FO_3S
(421-20-5)

CH_3Hg
(16056-34-1)
(22967-92-6)

CH_3I
(74-88-4)

$CH_3IO_3S.Na$
(126-31-8)

CH_3NO
(75-12-7)

CH_3NO_2
(75-52-5)
(463-77-4)
(624-91-9)

CH_3NO_3
(598-58-3)

$CH_3NS_2.H_3N$
(513-74-6)

CH_3N_5
(4418-61-5)

CH_3NiO_2
(15843-02-4)

$CH_3O.Na$
(124-41-4)

$CH_3O_4S.Na$
(870-72-4)

CH_3Sn
(16408-15-4)

CH_4
(74-82-8)

$CH_4AsO_3.Na$
(2163-80-6)

CH_4Cl_2Si
(75-54-7)

CH_4CrO_3
(29689-14-3)

CH_4HgO
(1184-57-2)

CH_4N_2O
(57-13-6)
(624-84-0)
$CH_4N_2O.CH_2O$
(9011-05-6)
$CH_4N_2O.H_2O_2$
(124-43-6)
$CH_4N_2O.H_3O_4P$
(4861-19-2)
$CH_4N_2O_2$
(127-07-1)
$CH_4N_2O_2S$
(1758-73-2)
$CH_4N_2O_3$
(686-68-0)
CH_4N_2S
(62-56-6)
$CH_4N_2S_2.N_2H_4$
(20469-71-0)
CH_4N_2Se
(630-10-4)
CH_4N_4O
(674-81-7)
$CH_4N_4O_2$
(556-88-7)
$CH_4Ni_3O_7$
(12607-70-4)
$CH_4Ni_3O_7.4H_2O$
(39430-27-8)
CH_4O
(67-56-1)
CH_4O_2
(3031-73-0)
$CH_4O_2S_2.Na$
(1950-85-2)
CH_4O_3S
(75-75-2)
$CH_4O_3S.Na$
(149-44-0)
CH_4O_4S
(75-92-3)
(75-93-4)
$CH_4O_4S.Na$
(512-42-5)
$CH_4O_6S_2$
(503-40-2)
CH_4S
(74-93-1)
$CH_4S.Na$
(5188-07-8)
$CH_4S.1/2Pb$
(35029-96-0)
CH_5AsO_3
(124-58-3)
CH_5N
(74-89-5)
$CH_5N.CHN_3O_6$
(14147-71-8)
$CH_5N.ClH$
(593-51-1)
$CH_5N.HNO_3$
(22113-87-7)
CH_5NO
(67-62-9)
(593-77-1)
$CH_5NO.ClH$
(593-56-6)
$CH_5NO_7S_2.Na$
(67874-55-9)
$CH_5NO_7S_2.2Na$
(63450-73-7)
CH_5N_3
(113-00-8)
$CH_5N_3.ClH$
(50-01-1)
$CH_5N_3.H$
(25215-10-5)
$CH_5N_3.HNO_3$
(506-93-4)
$CH_5N_3.1/2CH_2O_3$
(593-85-1)
CH_5N_3O
(57-56-7)
$CH_5N_3O.ClH$
(563-41-7)
$CH_5N_3O_4$
(124-47-0)
CH_5N_3S
(79-19-6)
$CH_5O_2PS_2$
(102255-23-2)
CH_5O_3P
(993-13-5)
(13590-71-1)
CH_5O_4P
(812-00-0)
(2617-47-2)
CH_6NO_3P
(1066-51-9)
CH_6N_2
(60-34-4)
$CH_6N_2.ClH$
(7339-53-9)
CH_6N_4
(79-17-4)
$CH_6N_4.ClH$
(1937-19-5)
CH_6N_4O
(497-18-7)
$CH_6N_4O_3$
(52470-25-4)
CH_6N_4S
(2231-57-4)
$CH_6O_3Si.xNa$
(16589-43-8)
$CH_{13}AsFeN_3O_2$
(6585-53-1)
$CH_{16}Al_2Mg_6O_{19}$
(11097-59-9)
ClN
(506-78-5)
CI_4
(507-25-5)
CKO
(12397-35-2)
$CMnO_3.H$
(68013-64-9)
CN
(57-12-5)
$CN.K$
(151-50-8)
$CNNa$
(143-33-9)
CNO
(506-85-4)
(71000-82-3)
$CNO.Ag$
(5610-59-3)
$CNO.K$
(590-28-3)
$CNO.Na$
(917-61-3)
CNS
(302-04-5)
$CNS.H_4N$
(1762-95-4)
$CNS.K$
(333-20-0)
$CNS.Na$
(540-72-7)
$CN_2.Ca$
(156-62-7)
CN_4O_8
(509-14-8)
$CNiO_3$
(3333-67-3)
CO
(630-08-0)
$COH_{28}N_6$
(4067-16-7)
COS
(463-58-1)
CO_2
(124-38-9)
$CO_2.N_2O$
(53569-62-3)
$CO_2.O_2$
(8063-77-2)
CO_3
(3812-32-6)
$CO_3.Ba$
(513-77-9)
$CO_3.Be$
(13106-47-3)
$CO_3.Ca$
(471-34-1)
(1317-65-3)
$CO_3.Cd$
(513-78-0)
$CO_3.Co$
(513-79-1)
$CO_3.Cu$
(1184-64-1)
$CO_3.H_2O_2.2Cu$
(12069-69-1)
$CO_3.2H_4N$
(1066-33-7)
$CO_3.H_{12}N_4Zn$
(38714-47-5)
$CO_3.2K$
(584-08-7)
$CO_3.2Li$
(554-13-2)
$CO_3.Mg$
(546-93-0)
$CO_3.2Na$
(497-19-8)
$CO_3.Pb$
(598-63-0)
$CO_3.2Tl$
(6533-73-9)
$CO_3.Zn$
(3486-35-9)
CS_2
(75-15-0)
CSi
(409-21-2)
CW
(12070-12-1)
$C_2AgN_2.K$
(506-61-6)
C_2BaN_2
(542-62-1)
C_2BrCl_5
(79504-02-2)
C_2BrF_3
(598-73-2)
C_2Br_2
(624-61-3)
$C_2Br_2ClF_3$
(29256-79-9)
$C_2Br_2Cl_4$
(630-24-0)
$C_2Br_2F_4$
(124-73-2)
C_2Br_4
(79-28-7)
C_2Ca
(75-20-7)
C_2CaN_2
(592-01-8)
C_2ClF_3
(79-38-9)
C_2ClF_3O
(354-32-5)
C_2ClF_5
(76-15-3)
C_2Cl_2
(7572-29-4)
$C_2Cl_2F_2$
(79-35-6)
(598-88-9)
(27156-03-2)
$C_2Cl_2F_4$
(76-14-2)
(374-07-2)
(1320-37-2)
$C_2Cl_3F_3$
(76-13-1)
(354-58-5)
(26523-64-8)
C_2Cl_3N
(545-06-2)
$C_2Cl_3O_2.Na$
(650-51-1)
C_2Cl_4
(127-18-4)
$C_2Cl_4F_2$
(76-11-9)
(76-12-0)
(28605-74-5)
C_2Cl_4O
(76-02-8)
(16650-10-5)
$C_2Cl_4O_2$
(503-38-8)
C_2Cl_6
(67-72-1)
$C_2Cl_6O_2S$
(3064-70-8)
$C_2Co_2O_6.Co_3H_6O_6$
(12602-23-2)
C_2Cu
(12540-13-5)
C_2CuN_2
(14763-77-0)
C_2D_6O
(1516-08-1)
C_2D_6OS
(2206-27-1)
C_2F_4
(116-14-3)
$C_2F_4O_3S$
(677-67-8)
(697-18-7)
C_2F_5I
(354-64-3)
C_2F_6
(76-16-4)
C_2HAg
(13092-75-6)
$C_2HBrClF_3$
(151-67-7)
(354-06-3)
$C_2HBrClN$
(83463-62-1)
C_2HBr_2N
(3252-43-5)
C_2HBr_3
(598-16-3)
C_2HBr_3O
(115-17-3)
$C_2HBr_3O_2$
(75-96-7)
C_2HClF_2
(359-10-4)
C_2HClF_4
(2837-89-0)
(63938-10-3)
$C_2HCl_2F_3$
(306-83-2)
(354-23-4)
C_2HCl_2N
(3018-12-0)
C_2HCl_3
(79-01-6)
C_2HCl_3O
(75-87-6)
(79-36-7)
(16967-79-6)

$C_2HCl_3O.H_2O$
(302-17-0)
$C_2HCl_3O_2$
(76-03-9)
$C_2HCl_3O_2.C_9H_{11}ClN_2O$
(140-41-0)
$C_2HCl_3O_2.H_3N$
(7646-88-0)
C_2HCl_5
(76-01-7)
C_2HCl_5S
(1185-09-7)
C_2HF_3
(359-11-5)
$C_2HF_3O_2$
(76-05-1)
$C_2HF_3O_2.Ag$
(2966-50-9)
C_2HF_5
(354-33-6)
C_2HHg
(68833-55-6)
C_2HN_3
(504-66-5)
$C_2HO_4.K$
(127-95-7)
C_2H_2
(74-86-2)
$C_2H_2AsCl_3$
(541-25-3)
$C_2H_2Be_3O_8$
(66104-24-3)
C_2H_2BrN
(590-17-0)
$C_2H_2Br_2$
(540-49-8)
(593-92-0)
(25429-23-6)
$C_2H_2Br_2Cl_2$
(75-81-0)
(683-68-1)
$C_2H_2Br_2O$
(3039-13-2)
$C_2H_2Br_2O_2$
(631-64-1)
$C_2H_2Br_4$
(79-27-6)
C_2H_2ClFO
(359-06-8)
$C_2H_2ClF_3$
(75-88-7)
C_2H_2ClN
(107-14-2)
$C_2H_2ClO_2.Na$
(3926-62-3)
$C_2H_2Cl_2$
(75-35-4)
(156-59-2)
(156-60-5)
(540-59-0)
(25323-30-2)
$C_2H_2Cl_2F_2$
(1649-08-7)
(25915-78-0)
$C_2H_2Cl_2O$
(79-02-7)
(79-04-9)
$C_2H_2Cl_2O_2$
(79-43-6)
$C_2H_2Cl_3F$
(359-28-4)
(811-95-0)
$C_2H_2Cl_3NO$
(594-65-0)
$C_2H_2Cl_4$
(79-34-5)
(630-20-6)
(25322-20-7)
$C_2H_2FO_2.Na$
(62-74-8)
$C_2H_2F_2$
(75-38-7)
(1320-41-8)
(1630-77-9)
(1630-78-0)
(1691-13-0)
$C_2H_2F_2O_2$
(381-73-7)
$C_2H_2F_2O_2.Na$
(2218-52-2)
$C_2H_2F_4$
(359-35-3)
(811-97-2)
$C_2H_2I_2$
(590-26-1)
(590-27-2)
(20244-70-6)
$C_2H_2NO_4$
(61201-44-3)
$C_2H_2N_2S_3$
(1072-71-5)
$C_2H_2N_4$
(70816-59-0)
$C_2H_2N_8O_2$
(67880-17-5)
C_2H_2O
(463-51-4)
$C_2H_2O_2$
(107-22-2)
$C_2H_2O_3$
(298-12-4)
$C_2H_2O_4$
(144-62-7)
$C_2H_2O_4.2Ag$
(533-51-7)
$C_2H_2O_4.Ca$
(544-17-2)
(563-72-4)
$C_2H_2O_4.Cr$
(814-90-4)
$C_2H_2O_4.Cu$
(814-91-5)
$C_2H_2O_4.Fe$
(516-03-0)
$C_2H_2O_4.1/3Fe.H_3N$
(2944-67-4)
$C_2H_2O_4.1/3Fe.Na$
(555-34-0)
$C_2H_2O_4.2H_3N$
(1113-38-8)
$C_2H_2O_4.2H_3N.H_2O$
(6009-70-7)
$C_2H_2O_4.Pb$
(814-93-7)
$C_2H_2O_4.Sn$
(814-94-8)
$C_2H_2O_4.Sr$
(814-95-9)
$C_2H_2O_8Pb_3$
(1319-46-6)
$C_2H_3BF_3O_2.H$
(7578-36-1)
C_2H_3Br
(593-60-2)
$C_2H_3BrCl_2$
(683-53-4)
C_2H_3BrO
(506-96-7)
$C_2H_3BrO_2$
(79-08-3)
$C_2H_3Br_2Cl$
(598-20-9)
(27949-36-6)
C_2H_3Cl
(75-01-4)
$C_2H_3ClF_2$
(75-68-3)
(25497-29-4)
C_2H_3ClO
(75-36-5)
(107-20-0)
(7763-77-1)
C_2H_3ClOS
(2812-72-8)
$C_2H_3ClO_2$
(79-11-8)
(79-22-1)
$C_2H_3Cl_2NO_2$
(594-72-9)
$C_2H_3Cl_3$
(71-55-6)
(79-00-5)
(25323-89-1)
$C_2H_3Cl_3O$
(115-20-8)
$C_2H_3Cl_3Si$
(75-94-5)
C_2H_3F
(75-02-5)
$C_2H_3FO_2$
(144-49-0)
$C_2H_3FO_2.H_3N$
(60916-92-9)
$C_2H_3F_3$
(420-46-2)
(430-66-0)
$C_2H_3F_3O$
(75-89-8)
$C_2H_3F_3O_3S$
(333-27-7)
C_2H_3HgN
(2597-97-9)
C_2H_3IO
(507-02-8)
$C_2H_3IO_2$
(64-69-7)
C_2H_3N
(75-05-8)
(593-75-9)
C_2H_3NO
(107-16-4)
(624-83-9)
$C_2H_3NO_3$
(471-47-6)
$C_2H_3NO_5$
(2278-22-0)
C_2H_3NS
(556-61-6)
(556-64-9)
$C_2H_3N_3$
(288-88-0)
$C_2H_3N_3O_2$
(3232-84-6)
$C_2H_3N_3O_7$
(918-54-7)
$C_2H_3N_3S$
(3179-31-5)
$C_2H_3N_3S_2$
(2349-67-9)
$C_2H_3O_2$
(71-50-1)
$C_2H_3O_2.H_4N$
(631-61-8)
$C_2H_3O_2.Hg$
(631-60-7)
$C_2H_3O_2.K$
(127-08-2)
$C_2H_3O_2.Na$
(127-09-3)
$C_2H_3O_2.Tl$
(563-68-8)
$C_2H_3O_2S.H_3N$
(5421-46-5)
$C_2H_3O_2S.Na$
(367-51-1)
$C_2H_3O_3.Na$
(2836-32-0)
C_2H_4
(74-85-1)
C_2H_4BrCl
(107-04-0)
(593-96-4)
C_2H_4BrNO
(79-15-2)
$C_2H_4Br_2$
(106-93-4)
(557-91-5)
(25620-62-6)
$C_2H_4ClNO_2$
(598-92-5)
C_2H_4ClNO
(79-07-2)
$C_2H_4Cl_2$
(75-34-3)
(107-06-2)
(1300-21-6)
$C_2H_4Cl_2N_6$
(502-98-7)
$C_2H_4Cl_2O$
(542-88-1)
(598-38-9)
(4885-02-3)
$C_2H_4Cl_2O_2S$
(1622-32-8)
$C_2H_4Cl_3O_4P$
(306-52-5)
C_2H_4FNO
(640-19-7)
$C_2H_4F_2$
(75-37-6)
(624-72-6)
(25497-28-3)
$C_2H_4I_2$
(624-73-7)
$C_2H_4NO_3$
(53422-49-4)
$C_2H_4NS_2.Na$
(137-42-8)
$C_2H_4NS_2.Na.2H_2O$
(6734-80-1)
$C_2H_4N_2O_2$
(471-46-5)
(628-36-4)
$C_2H_4N_2O_4$
(600-40-8)
(7570-26-5)
$C_2H_4N_2O_5$
(4528-34-1)
$C_2H_4N_2O_6$
(628-96-6)
$C_2H_4N_2S_2$
(79-40-3)
$C_2H_4N_2S_4$
(504-90-5)
$C_2H_4N_4$
(61-82-5)
(461-58-5)
$C_2H_4N_4O_2$
(123-77-3)
$C_2H_4N_4S$
(13183-79-4)
$C_2H_4N_6$
(629-13-0)
C_2H_4O
(75-07-0)
(75-21-8)
$C_2H_4O.H_3N$
(75-39-8)
C_2H_4OS
(507-09-5)
$C_2H_4O_2$
(64-19-7)
(107-31-3)
(141-46-8)
$C_2H_4O_2.1/3Al$
(139-12-8)
$C_2H_4O_2.1/3Bi$
(22306-37-2)
$C_2H_4O_2.xCu.xH_3N$
(23087-46-9)

$C_2H_4O_2.H_4N_2$
(7335-65-1)
$C_2H_4O_2.Li$
(546-89-4)
$C_2H_4O_2.xPb$
(15347-57-6)
$C_2H_4O_2.1/2Na$
(126-96-5)
$C_2H_4O_2S$
(68-11-1)
$C_2H_4O_2S.1/2Ca$
(814-71-1)
$C_2H_4O_3$
(79-14-1)
(79-21-0)
$C_2H_4O_3.K$
(1932-50-9)
$C_2H_4O_3.xK$
(25904-89-6)
$C_2H_4O_3S.Na$
(3039-83-6)
$C_2H_4O_5S$
(123-43-3)
$C_2H_4O_7P_2.4Na$
(3794-83-0)
C_2H_4S
(420-12-2)
$C_2H_4S_3$
(289-16-7)
$C_2H_5AlCl_2$
(563-43-9)
$C_2H_5AsCl_2$
(598-14-1)
C_2H_5Br
(74-96-4)
C_2H_5BrO
(540-51-2)
$C_2H_5BrO_3S$
(26978-65-4)
C_2H_5Cl
(75-00-3)
C_2H_5ClHg
(107-27-7)
C_2H_5ClMg
(2386-64-3)
C_2H_5ClO
(107-07-3)
(107-30-2)
$C_2H_5ClO_3S$
(18024-00-5)
$C_2H_5ClO_4$
(22750-93-2)
$C_2H_5Cl_2O_2P$
(1498-51-7)
$C_2H_5Cl_2PS$
(993-43-1)
$C_2H_5Cl_3Si$
(115-21-9)
(1558-33-4)
C_2H_5F
(353-36-6)
C_2H_5FO
(371-62-0)
C_2H_5I
(75-03-6)
C_2H_5IO
(624-76-0)
C_2H_5N
(151-56-4)
C_2H_5NO
(60-35-5)
(107-29-9)
(123-39-7)
$C_2H_5NO_2$
(56-40-6)
(79-24-3)
(109-95-5)
(546-88-3)
(598-55-0)
(13052-19-2)
$C_2H_5NO_2.ClH$
(6000-43-7)
$C_2H_5NO_3$
(625-48-9)
(625-58-1)
$C_2H_5NO_4$
(14258-49-2)
C_2H_5NS
(62-55-5)
$C_2H_5N_3O_2$
(108-19-0)
(684-93-5)
$C_2H_5N_3S_2$
(541-53-7)
$C_2H_5N_5$
(1455-77-2)
$C_2H_5N_5O_3$
(70-25-7)
$C_2H_5O_3P$
(1746-03-8)
$C_2H_5O_5P$
(4408-78-0)
$C_2H_5O_7P_2.3Na$
(2666-14-0)
C_2H_6
(74-84-0)
$C_2H_6AlNO_4$
(13682-92-3)
$C_2H_6AsO_2.Na$
(124-65-2)
C_2H_6Cd
(506-82-1)
$C_2H_6ClO_2PS$
(2524-03-0)
$C_2H_6ClO_3P$
(813-77-4)
(16672-87-0)
C_2H_6ClPS
(993-12-4)
C_2H_6ClSn
(41079-92-9)
$C_2H_6Cl_2Pb$
(1520-77-0)
$C_2H_6Cl_2Si$
(75-78-5)
(1789-58-8)
$C_2H_6Cl_2Sn$
(753-73-1)
$C_2H_6Cl_4Si_2$
(4518-98-3)
$C_2H_6F_2Si$
(353-66-2)
$C_2H_6F_2Sn$
(3582-17-0)
C_2H_6GeS
(16090-49-6)
C_2H_6Hg
(593-74-8)
C_2H_6Mg
(2999-74-8)
$C_2H_6N_2$
(503-28-6)
$C_2H_6N_2O$
(62-75-9)
(598-50-5)
(758-17-8)
(25843-45-2)
$C_2H_6N_2O_2$
(590-96-5)
(1000-82-4)
(4164-28-7)
(6294-89-9)
$C_2H_6N_2S$
(598-52-7)
$C_2H_6N_4O$
(141-83-3)
$C_2H_6N_4O.1/2H_2O_4S$
(591-01-5)
$C_2H_6N_4O_2$
(110-21-4)
(996-98-5)
$C_2H_6N_4S_2$
(142-46-1)
C_2H_6O
(64-17-5)
(115-10-6)
$C_2H_6O.1/3Al$
(555-75-9)
$C_2H_6O.BF_3$
(353-42-4)
$C_2H_6O.1/2Mg$
(2414-98-4)
$C_2H_6O.Na$
(141-52-6)
$C_2H_6O.1/4Ti$
(3087-36-3)
C_2H_6OS
(60-24-2)
(67-68-5)
$C_2H_6OS.Na$
(37482-11-4)
$C_2H_6O_2$
(107-21-1)
(3031-74-1)
(4461-52-3)
$C_2H_6O_2.Li$
(23248-23-9)
$C_2H_6O_2PS_2.Na$
(26377-29-7)
$C_2H_6O_2S$
(67-71-0)
$C_2H_6O_3S$
(66-27-3)
(594-45-6)
$C_2H_6O_4S$
(77-78-1)
(87954-49-2)
$C_2H_6O_4S.Na$
(1562-00-1)
$C_2H_6O_6S_2Zn$
(24887-06-7)
$C_2H_6O_7P_2.2Na$
(7414-83-7)
C_2H_6Pb
(21774-13-0)
C_2H_6S
(75-08-1)
(75-18-3)
$C_2H_6S.BH_3$
(13292-87-0)
C_2H_6SSn
(13269-74-4)
$C_2H_6S_2$
(540-63-6)
(624-92-0)
(26914-40-9)
(29414-47-9)
(74004-30-1)
$C_2H_6S_3$
(3658-80-8)
$C_2H_6S_3Sn_2$
(33397-79-4)
C_2H_6Se
(593-79-3)
$C_2H_6Se_2$
(7101-31-7)
C_2H_6Sn
(16408-14-3)
(23120-99-2)
C_2H_6Zn
(544-97-8)
$C_2H_7AsO_2$
(75-60-5)
$C_2H_7AsO_2.1/3Fe$
(5968-84-3)
$C_2H_7AsO_4$
(34442-00-7)
$C_2H_7BF_3N$
(75-23-0)
$C_2H_7BO_3$
(34099-73-5)
C_2H_7ClSi
(1066-35-9)
$C_2H_7FeNO_4$
(55488-87-4)
$C_2H_7HgO_4P$
(2235-25-8)
C_2H_7N
(75-04-7)
(124-40-3)
$C_2H_7N.ClH$
(506-59-2)
(557-66-4)
$C_2H_7N.H_2O_4S$
(23307-05-3)
C_2H_7NO
(141-43-5)
(1117-97-1)
$C_2H_7NO.ClH$
(2002-24-6)
(16645-06-0)
$C_2H_7NO.HNO_3$
(20748-72-5)
$C_2H_7NO.1/2H_2O_3S$
(15535-29-2)
$C_2H_7NO.xH_3O_4P$
(29868-05-1)
$C_2H_7NO_3S$
(107-35-7)
$C_2H_7NO_4S$
(926-39-6)
$C_2H_7NO_4S.Na$
(5390-94-3)
$C_2H_7NO_5$
(5972-73-6)
C_2H_7NS
(60-23-1)
$C_2H_7N_3$
(471-29-4)
$C_2H_7N_3S$
(6610-29-3)
$C_2H_7N_5$
(56-03-1)
$C_2H_7N_5O_2$
(4381-07-1)
$C_2H_7O_2PS_2$
(756-80-9)
(32534-66-0)
$C_2H_7O_2Zr$
(7585-20-8)
$C_2H_7O_3P$
(868-85-9)
(6779-09-5)
(15845-66-6)
$C_2H_7O_3PS$
(1112-38-5)
(59401-04-6)
$C_2H_7O_3PS.K$
(28523-79-7)
$C_2H_7O_4P$
(813-78-5)
(1623-14-9)
$C_2H_8As_2O_6.Ca$
(5902-95-4)
$C_2H_8NO_2PS$
(10265-92-6)
$C_2H_8NO_4P$
(1071-23-4)
$C_2H_8N_2$
(57-14-7)
(107-15-3)
(540-73-8)
$C_2H_8N_2.ClH$
(593-82-8)
(56400-60-3)
$C_2H_8N_2.2ClH$
(306-37-6)
(333-18-6)
$C_2H_8N_2.2HI$
(5700-49-2

$C_2H_8N_2.2HNO_3$
(20829-66-7)
$C_2H_8N_2O$
(109-84-2)
$C_2H_8N_{10}O$
(109-27-3)
$C_2H_8O_2Si$
(1066-42-8)
$C_2H_8O_7P_2$
(2809-21-4)
$C_2H_8O_7P_2.2K$
(21089-06-5)
$C_2H_8O_7P_2.xK$
(67953-76-8)
$C_2H_9ClN_2$
(15467-15-9)
$C_2H_9N_4O_5P$
(17675-60-4)
$C_2H_9NaO_7P_2$
(29329-71-3)
$C_2H_{10}N_2.xCu$
(13426-91-0)
$C_2H_4O_2.1/2Cd$
(543-90-8)
C_2HgN_2
(592-04-1)
$C_2HgN_2.2CKN$
(591-89-9)
$C_2HgN_2O_2$
(628-86-4)
$C_2Hg_2N_2O$
(1335-31-5)
C_2I_2
(624-74-8)
C_2N_2
(460-19-5)
C_2N_2Ni
(557-19-7)
C_2N_2Pb
(592-05-2)
$C_2N_2S_2.Hg$
(592-85-8)
C_2N_2Zn
(557-21-1)
$C_2N_4O_6$
(630-72-8)
$C_2N_6O_{12}$
(918-37-6)
$C_2O_4.2Na$
(62-76-0)
$C_2O_{11}Zn_5.4H$
(5970-61-6)
$C_3Br_3N_3$
(14921-00-7)
C_3ClF_5O
(79-53-8)
$C_3Cl_2N_6$
(30805-06-2)
C_3Cl_3N
(16212-28-5)
$C_3Cl_3N_3$
(108-77-0)
$C_3Cl_3N_3O_3$
(87-90-1)
C_3Cl_4
(18608-30-5)
$C_3Cl_4N_2S$
(5848-93-1)
C_3Cl_6
(1888-71-7)
$C_3Cl_6N_6$
(2428-04-8)
C_3Cl_6O
(116-16-5)
C_3Cl_8
(594-90-1)
$C_3CuN_3.2Na$
(14264-31-4)
C_3D_6O
(666-52-4)
$C_3F_3N_3$
(675-14-9)
C_3F_6
(116-15-4)
C_3F_6O
(428-59-1)
(684-16-2)
(1187-93-5)
$C_3F_6O.3H_2O$
(34202-69-2)
$C_3F_6O.3/2H_2O$
(13098-39-0)
C_3F_7I
(27636-85-7)
C_3F_8
(76-19-7)
$C_3HBr_3N_2$
(2034-22-2)
C_3HClF_6
(422-55-9)
$C_3HCl_2N_3$
(2831-66-5)
$C_3HCl_2N_3O_3.K$
(2244-21-5)
$C_3HCl_2N_3O_3.Na$
(2893-78-9)
$C_3HCl_3N_2$
(7682-38-4)
$C_3HCl_3O_2$
(2257-35-4)
C_3HCl_5
(1600-37-9)
(6262-51-7)
C_3HCl_5O
(1768-31-6)
C_3HCl_7
(594-89-8)
$C_3H_2BrN_3O_2$
(6963-65-1)
(65902-59-2)
$C_3H_2Br_2N_2O$
(10222-01-2)
$C_3H_2Br_2N_4$
(4514-62-9)
$C_3H_2ClF_5O$
(13838-16-9)
(26675-46-7)
$C_3H_2ClF_5$
(422-02-6)
$C_3H_2ClN_3O_2$
(57531-37-0)
(57531-38-1)
$C_3H_2ClN_7$
(5248-70-4)
$C_3H_2Cl_2$
(25523-14-2)
(83682-32-0)
$C_3H_2Cl_2N_2O_3$
(2782-57-2)
$C_3H_2Cl_2N_4$
(933-20-0)
$C_3H_2Cl_2O_2$
(1561-20-2)
(99165-89-6)
$C_3H_2Cl_4$
(10436-39-2)
(20589-85-9)
$C_3H_2Cl_4O$
(632-21-3)
(16995-35-0)
$C_3H_2Cl_6$
(3607-78-1)
$C_3H_2F_2N_4$
(1652-36-4)
$C_3H_2F_6O$
(920-66-1)
(53520-89-1)
$C_3H_2N_2$
(109-77-3)
$C_3H_2N_2S_2$
(6317-18-6)
$C_3H_2N_4O_2$
(5213-49-0)
$C_3H_2N_4O_4$
(19183-14-3)
C_3H_2O
(624-67-9)
$C_3H_2O_2$
(471-25-0)
$C_3H_2O_3$
(872-36-6)
$C_3H_2O_4.2Tl$
(2757-18-8)
$C_3H_3B_2D_9$
(23797-84-4)
C_3H_3Br
(106-96-7)
$C_3H_3Br.CH_3Br.CCl_3NO_2$
(8000-21-3)
$C_3H_3BrF_4$
(70192-84-6)
$C_3H_3Br_3$
(36417-14-8)
C_3H_3Cl
(624-65-7)
(3223-70-9)
(7747-84-4)
$C_3H_3ClF_4$
(421-75-0)
$C_3H_3ClN_4$
(7709-13-9)
C_3H_3ClO
(683-51-2)
(814-68-6)
$C_3H_3ClO_2$
(598-79-8)
(1609-93-4)
(2345-61-1)
(26952-44-3)
$C_3H_3ClO_4$
(600-33-9)
$C_3H_3Cl_2O_2.Na$
(127-20-8)
$C_3H_3Cl_3$
(96-19-5)
(2233-00-3)
(2567-14-8)
(21400-25-9)
(37077-84-2)
$C_3H_3Cl_3N_6$
(7673-09-8)
$C_3H_3Cl_3O$
(918-00-3)
(921-03-9)
(3083-23-6)
(7789-90-4)
$C_3H_3Cl_3O_2$
(598-99-2)
(3278-46-4)
$C_3H_3Cl_4NO$
(3659-66-3)
$C_3H_3Cl_5$
(15104-61-7)
(16714-68-4)
$C_3H_3Cl_5F_4N_3OP_3$
(59700-57-1)
$C_3H_3F_3$
(677-21-4)
$C_3H_3F_3O$
(421-50-1)
(460-40-2)
$C_3H_3F_3O_2$
(431-47-0)
$C_3H_3F_5$
(1814-88-6)
$C_3H_3F_5O$
(422-05-9)
C_3H_3N
(107-13-1)
C_3H_3NO
(288-14-2)
(288-42-6)
$C_3H_3NOS_2$
(141-84-4)
$C_3H_3NO_2$
(372-09-8)
C_3H_3NS
(288-47-1)
$C_3H_3N_3$
(290-87-9)
$C_3H_3N_3O_2S$
(121-66-4)
$C_3H_3N_3O_2$
(71-33-0)
(527-73-1)
(3034-38-6)
(36877-68-6)
$C_3H_3N_3O_3$
(108-80-5)
$C_3H_3N_3O_3.Na$
(2624-17-1)
$C_3H_3N_3O_3.2Na$
(36452-21-8)
$C_3H_3N_3O_3.3Na$
(3047-33-4)
$C_3H_3N_3O_4$
(2825-15-2)
$C_3H_3N_3S_3$
(638-16-4)
$C_3H_3O_2$
(10344-93-1)
$C_3H_3O_2.Na$
(24382-04-5)
$C_3H_3O_4.Na$
(141-95-7)
C_3H_4
(74-99-7)
(463-49-0)
C_3H_4BrCl
(3737-00-6)
$C_3H_4BrN_5$
(4649-67-6)
$C_3H_4Br_2$
(513-31-5)
$C_3H_4Br_2O$
(563-76-8)
$C_3H_4Br_2O_2$
(600-05-5)
$C_3H_4Br_4$
(54268-02-9)
$C_3H_4ClFO_2$
(462-27-1)
$C_3H_4ClF_3$
(460-35-5)
C_3H_4ClN
(542-76-7)
$C_3H_4ClN_5$
(3397-62-4)
$C_3H_4Cl_2$
(78-88-6)
(542-75-6)
(563-54-2)
(563-57-5)
(563-58-6)
(10061-01-5)
(10061-02-6)
(26952-23-8)
$C_3H_4Cl_2F_2O$
(76-38-0)
$C_3H_4Cl_2O$
(513-88-2)
(534-07-6)
(7623-09-8)
$C_3H_4Cl_2O_2$
(75-99-0)
(116-54-1)
(565-64-0)
(627-11-2)
$C_3H_4Cl_2O_2.1/2Mg$
(29110-22-3)

$C_3H_4Cl_3NSi$
(1071-22-3)
$C_3H_4Cl_4$
(812-03-3)
(1070-78-6)
(13116-53-5)
(13116-60-4)
(18495-30-2)
$C_3H_4FN_5$
(823-95-0)
$C_3H_4F_4$
(40723-63-5)
$C_3H_4F_4O$
(76-37-9)
$C_3H_4N_2$
(288-13-1)
(288-32-4)
$C_3H_4N_2O$
(107-91-5)
(5918-93-4)
$C_3H_4N_2O_2$
(461-72-3)
(21729-98-6)
$C_3H_4N_2S$
(96-50-4)
(50406-54-7)
$C_3H_4N_2Se$
(17505-11-2)
$C_3H_4N_4$
(4122-04-7)
(5239-06-5)
$C_3H_4N_4O$
(931-86-2)
$C_3H_4N_4O_2$
(645-93-2)
$C_3H_4N_4S_2$
(2770-75-4)
$C_3H_4N_8$
(16376-89-9)
C_3H_4O
(107-02-8)
(107-19-7)
$C_3H_4O_2$
(57-57-8)
(78-98-8)
(79-10-7)
(542-78-9)
(692-45-5)
(765-34-4)
$C_3H_4O_2.H_3N$
(10604-69-0)
$C_3H_4O_2.K$
(10192-85-5)
$C_3H_4O_2.1/2Mg$
(5698-98-6)
$C_3H_4O_2.Na$
(7446-81-3)
$C_3H_4O_2.1/2Zn$
(14643-87-9)
$C_3H_4O_3$
(96-49-1)
(127-17-3)
(922-68-9)
$C_3H_4O_3.Na$
(113-24-6)
$C_3H_4O_4$
(141-82-2)
$C_3H_5AlO_5$
(34202-30-7)
C_3H_5Br
(106-95-6)
(557-93-7)
$C_3H_5BrCl_2$
(17759-88-5)
(33037-07-9)
(36668-45-8)
C_3H_5BrO
(598-31-2)
(3132-64-7)
$C_3H_5BrO_2$
(96-32-2)
(590-92-1)
(598-72-1)
$C_3H_5Br_2Cl$
(96-12-8)
(55162-35-1)
(67708-83-2)
$C_3H_5Br_3$
(96-11-7)
C_3H_5Cl
(107-05-1)
(557-98-2)
(590-21-6)
C_3H_5ClO
(78-95-5)
(79-03-8)
(106-89-8)
(5976-47-6)
(13403-37-7)
(19434-65-2)
(29560-84-7)
C_3H_5ClOS
(2812-73-9)
(2941-64-2)
$C_3H_5ClO_2$
(96-34-4)
(107-94-8)
(541-41-3)
(598-78-7)
(28554-00-9)
$C_3H_5Cl_2NO_2$
(13698-16-3)
$C_3H_5Cl_3$
(96-18-4)
(598-77-6)
(3175-23-3)
(7789-89-1)
(25735-29-9)
$C_3H_5Cl_3Si$
(107-37-9)
C_3H_5F
(818-92-8)
(1184-60-7)
C_3H_5FO
(430-51-3)
(503-09-3)
$C_3H_5FO_2$
(461-56-3)
$C_3H_5F_3O$
(374-01-6)
C_3H_5I
(556-56-9)
$C_3H_5IO_2$
(141-76-4)
C_3H_5N
(107-12-0)
(624-79-3)
(2450-71-7)
(38239-27-9)
C_3H_5NO
(78-97-7)
(79-06-1)
(109-78-4)
(109-90-0)
$C_3H_5NO_2$
(306-44-5)
$C_3H_5NO_4$
(504-88-1)
$C_3H_5NO_5$
(5796-89-4)
C_3H_5NS
(542-85-8)
(542-90-5)
$C_3H_5N_3$
(6086-21-1)
$C_3H_5N_3O$
(140-87-4)
$C_3H_5N_3O_2$
(27032-78-6)
$C_3H_5N_3O_9$
(55-63-0)
$C_3H_5N_5$
(504-08-5)
$C_3H_5N_5O$
(645-92-1)
$C_3H_5N_5S$
(767-17-9)
$C_3H_5OS_2.K$
(140-89-6)
$C_3H_5O_2.Na$
(137-40-6)
$C_3H_5O_3.Na$
(72-17-3)
C_3H_6
(75-19-4)
(115-07-1)
C_3H_6BrCl
(109-70-6)
(3017-95-6)
(3017-96-7)
(34652-54-5)
C_3H_6BrClO
(4540-44-7)
(73727-39-6)
$C_3H_6BrNO_4$
(52-51-7)
$C_3H_6Br_2$
(78-75-1)
(109-64-8)
(594-16-1)
$C_3H_6Br_2O$
(96-13-9)
(96-21-9)
C_3H_6ClN
(34508-68-4)
C_3H_6ClNS
(16420-13-6)
$C_3H_6ClNO_2$
(594-71-8)
(600-25-9)
(2425-66-3)
(2832-19-1)
C_3H_6ClNO
(79-44-7)
$C_3H_6Cl_2$
(78-87-5)
(78-99-9)
(142-28-9)
(594-20-7)
(26638-19-7)
$C_3H_6Cl_2.C_3H_4Cl_2$
(8003-19-8)
$C_3H_6Cl_2.C_3H_4Cl_2.C_2H_3NS$
(8066-01-1)
$C_3H_6Cl_2O$
(96-23-1)
(616-23-9)
(26545-73-3)
(63151-11-1)
(83682-72-8)
$C_3H_6Cl_2Si$
(124-70-9)
$C_3H_6Cl_4Si$
(2550-06-3)
$C_3H_6HgN_4$
(502-39-6)
$C_3H_6NOS_2.K$
(51026-28-9)
$C_3H_6NS_2.K.H_2O$
(128-03-0)
$C_3H_6NS_2.Na$
(128-04-1)
$C_3H_6N_2$
(151-18-8)
(1467-79-4)
$C_3H_6N_2O$
(120-93-4)
(4549-40-0)
$C_3H_6N_2OS$
(591-08-2)
$C_3H_6N_2O_2$
(68-41-7)
(7417-67-6)
$C_3H_6N_2O_3$
(462-60-2)
(3720-97-6)
(13256-22-9)
$C_3H_6N_2O_6$
(6423-43-4)
$C_3H_6N_2O_7$
(623-87-0)
$C_3H_6N_2S$
(96-45-7)
(1779-81-3)
$C_3H_6N_2S.ClH$
(3882-98-2)
$C_3H_6N_4$
(1609-07-0)
$C_3H_6N_4O$
(13101-26-3)
$C_3H_6N_6$
(108-78-1)
$C_3H_6N_6O_6$
(121-82-4)
C_3H_6O
(67-64-1)
(75-56-9)
(107-18-6)
(107-25-5)
(123-38-6)
(503-30-0)
C_3H_6OS
(1534-08-3)
$C_3H_6OS_2$
(151-01-9)
$C_3H_6OS_2.Na$
(140-90-9)
$C_3H_6O_2$
(79-09-4)
(79-20-9)
(109-94-4)
(116-09-6)
(556-52-5)
(646-06-0)
(1759-53-1)
(68937-68-8)
$C_3H_6O_2.1/2Zn$
(557-28-8)
$C_3H_6O_2S$
(107-96-0)
(2365-48-2)
$C_3H_6O_3$
(50-21-5)
(56-82-6)
(96-26-4)
(110-88-3)
(598-82-3)
(616-38-6)
(625-45-6)
$C_3H_6O_3.1/2Cu$
(814-81-3)
$C_3H_6O_3.1/3Sb$
(58164-88-8)
$C_3H_6O_3S$
(1120-71-4)
$C_3H_6O_3S.Na$
(2495-39-8)
$C_3H_6O_4$
(473-81-4)
C_3H_6S
(287-27-4)
(870-23-5)
$C_3H_6S_3$
(291-21-4)
C_3H_7Br
(75-26-3)
(106-94-5)
(26446-77-5)
C_3H_7BrHgO
(18832-83-2)

C_3H_7BrO
(598-18-5)
(627-18-9)
(19686-73-8)
$C_3H_7BrO_2$
(4704-77-2)
$C_3H_7Br_2O_4P$
(5324-12-9)
C_3H_7Cl
(75-29-6)
(540-54-5)
C_3H_7ClHg
(2440-40-6)
C_3H_7ClHgO
(123-88-6)
C_3H_7ClO
(78-89-7)
(127-00-4)
(627-30-5)
(627-42-9)
(3188-13-4)
$C_3H_7ClO_2$
(96-24-2)
(497-04-1)
$C_3H_7ClO_4S.Na$
(126-83-0)
C_3H_7ClS
(542-81-4)
$C_3H_7Cl_2O_2P$
(14212-91-0)
$C_3H_7Cl_2PS_2$
(5390-61-4)
$C_3H_7Cl_3Si$
(141-57-1)
C_3H_7F
(420-26-8)
(460-13-9)
C_3H_7I
(75-30-9)
(107-08-4)
(26914-02-3)
C_3H_7N
(75-55-8)
(107-11-9)
(503-29-7)
(765-30-0)
(1072-44-2)
C_3H_7NO
(68-12-2)
(79-05-0)
(79-16-3)
(127-06-0)
(627-45-2)
C_3H_7NOS
(10533-67-2)
(13749-94-5)
$C_3H_7NO_2$
(51-79-6)
(56-41-7)
(79-46-9)
(107-95-9)
(107-97-1)
(108-03-2)
(541-42-4)
(543-67-9)
(25322-01-4)
(71261-64-8)
$C_3H_7NO_2S$
(52-90-4)
$C_3H_7NO_2S.ClH$
(52-89-1)
$C_3H_7NO_2Se$
(18312-66-8)
$C_3H_7NO_3$
(56-45-1)
(56-45-1)
(565-71-9)
(627-13-4)
(1712-64-7)
(5395-01-7)
(25182-84-7)
$C_3H_7NO_5$
(624-43-1)
$C_3H_7NS_2$
(79-45-8)
$C_3H_7N_3O_2$
(759-73-9)
$C_3H_7O_2.F_6P$
(50318-32-6)
C_3H_8
(74-98-6)
C_3H_8ClN
(26245-56-7)
$C_3H_8ClNO_2S$
(26118-67-2)
$C_3H_8NO_4P.H_3N$
(69975-80-0)
$C_3H_8NO_5P$
(1071-83-6)
$C_3H_8NO_5P.Na$
(34494-03-6)
$C_3H_8N_2O$
(96-31-1)
(598-94-7)
(625-52-5)
(10595-95-6)
$C_3H_8N_2O_2$
(109-58-0)
$C_3H_8N_2O_3$
(140-95-4)
(25155-29-7)
$C_3H_8N_2S$
(534-13-4)
(625-53-6)
$C_3H_8N_4O$
(44648-02-4)
C_3H_8O
(67-63-0)
(71-23-8)
(540-67-0)
$C_3H_8O.1/4Ti$
(546-68-9)
$C_3H_8O.1/4Zr$
(23519-77-9)
$C_3H_8OS_2$
(59-52-9)
$C_3H_8O_2$
(57-55-6)
(109-86-4)
(109-87-5)
(504-63-2)
(3031-75-2)
$C_3H_8O_2S$
(96-27-5)
$C_3H_8O_3$
(56-81-5)
$C_3H_8O_3S$
(62-50-0)
$C_3H_8O_3S_3$
(74-61-3)
$C_3H_8O_4$
(2614-76-8)
C_3H_8S
(75-33-2)
(107-03-9)
(624-89-5)
$C_3H_8S_2$
(109-80-8)
(1618-26-4)
$C_3H_8S_3$
(42474-44-2)
C_3H_9Al
(75-24-1)
$C_3H_9Al_2Br_3$
(12263-85-3)
$C_3H_9Al_2Cl_3$
(12542-85-7)
$C_3H_9BO_3$
(121-43-7)
C_3H_9ClPb
(1520-78-1)
C_3H_9ClSi
(75-77-4)
C_3H_9ClSn
(1066-45-1)
$C_3H_9Cl_3Si_2$
(13528-88-6)
C_3H_9FSi
(420-56-4)
C_3H_9Ga
(1445-79-0)
C_3H_9N
(75-31-0)
(75-50-3)
(107-10-8)
(624-78-2)
$C_3H_9N.BH_3$
(75-22-9)
$C_3H_9N.C_3H_8NO_5P$
(38641-94-0)
$C_3H_9N.ClH$
(556-53-6)
(593-81-7)
(624-60-2)
$C_3H_9N.HNO_3$
(25238-43-1)
(87478-71-5)
C_3H_9NO
(78-96-6)
(109-83-1)
(109-85-3)
(156-87-6)
(1184-78-7)
(14002-21-2)
$C_3H_9NO_2$
(616-30-8)
(34375-28-5)
$C_3H_9NO_3S$
(107-68-6)
$C_3H_9NO_3S.Na$
(4316-74-9)
C_3H_9NS
(598-36-7)
$C_3H_9N_3$
(3324-71-8)
$C_3H_9N_3.ClH$
(22583-29-5)
$C_3H_9N_4OPS_2$
(79064-73-6)
$C_3H_9O.Cl_6Sb$
(54075-76-2)
$C_3H_9O_2P$
(16391-07-4)
$C_3H_9O_2PS$
(18005-40-8)
$C_3H_9O_2PS_2$
(2953-29-9)
(22608-53-3)
$C_3H_9O_3P$
(121-45-9)
(756-79-6)
(1832-53-7)
(4672-38-2)
$C_3H_9O_3PS$
(152-18-1)
(152-20-5)
$C_3H_9O_4P$
(512-56-1)
(1623-06-9)
(1623-24-1)
$C_3H_9O_6P$
(57-03-4)
(17181-54-3)
$C_3H_9O_6P.xK$
(1335-34-8)
$C_3H_9O_6P.2Na$
(1334-74-3)
C_3H_9Pb
(14570-16-2)
C_3H_9Sn
(5089-96-3)
$C_3H_{10}N_2$
(78-90-0)
(109-76-2)
$C_3H_{10}N_2.ClH$
(60597-20-8)
$C_3H_{10}N_2O$
(616-29-5)
$C_3H_{10}OSn$
(56-24-6)
$C_3H_{10}O_3Si$
(2487-90-3)
$C_3H_{10}Pb$
(7442-13-9)
$C_3H_{10}Sn$
(1631-73-8)
$C_3H_{12}NO_9P_3$
(6419-19-8)
$C_3H_{12}NO_9P_3.5Na$
(2235-43-0)
$C_3H_{12}NO_9P_3.xK$
(27794-93-0)
$C_3H_{13}NNaO_9P_3$
(20592-85-2)
C_3N_{12}
(5637-83-2)
$C_4CdN_4.2K$
(14402-75-6)
C_4ClD_9
(918-20-7)
$C_4Cl_2N_4S$
(4267-15-6)
$C_4Cl_3F_7$
(28984-80-7)
$C_4Cl_4H_2$
(921-09-5)
$C_4Cl_4N_4$
(877-83-8)
C_4Cl_4S
(6012-97-1)
$C_4Cl_5N_3$
(2113-00-0)
C_4Cl_6
(87-68-3)
(6130-82-1)
(56827-79-3)
(83682-47-7)
$C_4Cl_6O_3$
(4124-31-6)
C_4Cl_8
(2482-68-0)
(3050-42-8)
C_4Cl_{10}
(6820-74-2)
$C_4CuN_4.3K$
(14263-73-1)
$C_4D_{10}O$
(53001-22-2)
$C_4F_5N_3$
(368-55-8)
$C_4F_6O_3$
(407-25-0)
C_4F_8
(115-25-3)
(360-89-4)
(382-21-8)
C_4F_8O
(335-42-2)
$C_4F_8O_2$
(2927-83-5)
C_4F_9NO
(360-46-3)
C_4F_{10}
(355-25-9)
$C_4F_{10}O_2S$
(375-72-4)
$C_4HBr_2Cl_2N_3$
(30894-68-9)
C_4HClF_8
(71342-62-6)

$C_4HCl_2I_2N_3$
(30894-70-3)
$C_4HCl_2N_5$
(4803-05-8)
C_4HCl_3O
(78099-58-8)
$C_4HCl_4N_3$
(30894-65-6)
C_4HCl_5
(55880-77-8)
C_4HCl_7
(83682-39-7)
C_4HCl_9
(21483-62-5)
C_4HCoO_4
(16842-03-8)
$C_4HF_7O_2$
(375-22-4)
$C_4HF_9O_3S.K$
(29420-49-3)
$C_4HO_3.K.H_2O$
(52591-22-7)
C_4H_2
(460-12-8)
$C_4H_2BrCl_2N_3$
(30894-67-8)
$C_4H_2Br_2S$
(3141-26-2)
(3141-27-3)
$C_4H_2Br_6$
(36678-45-2)
$C_4H_2Cl_2IN_3$
(30894-69-0)
$C_4H_2Cl_2O$
(103339-60-2)
$C_4H_2Cl_2O_2$
(627-63-4)
$C_4H_2Cl_2O_4$
(43180-81-0)
$C_4H_2Cl_2S$
(3172-52-9)
$C_4H_2Cl_3N_3$
(30361-83-2)
(30894-64-5)
$C_4H_2Cl_4$
(1637-31-6)
(36038-53-6)
(58334-79-5)
$C_4H_2Cl_4N_2O_2$
(776-19-2)
$C_4H_2Cl_6$
(83682-38-6)
(83682-71-7)
$C_4H_2Cl_8$
(18791-19-0)
(20338-26-5)
(32694-76-1)
$C_4H_2F_6$
(382-10-5)
$C_4H_2Mg_5O_{14}$
(7760-50-1)
$C_4H_2N_2$
(764-42-1)
$C_4H_2N_2O_4$
(50-71-5)
$C_4H_2N_2O_4.H_2O$
(2244-11-3)
$C_4H_2N_2S$
(1452-15-9)
$C_4H_2O_3$
(108-31-6)
$C_4H_2O_4$
(142-45-0)
$C_4H_3BrN_2$
(4595-59-9)
$C_4H_3BrN_2O_2$
(51-20-7)
C_4H_3BrS
(872-31-1)
(1003-09-4)
$C_4H_3Br_3N_4O$
(31949-57-2)
$C_4H_3ClN_2O_2$
(1820-81-1)
$C_4H_3ClN_6$
(30369-27-8)
$C_4H_3ClO_4$
(617-42-5)
(617-43-6)
(19071-21-7)
C_4H_3ClS
(96-43-5)
$C_4H_3Cl_2N_3$
(56-05-3)
(1973-04-2)
$C_4H_3Cl_2N_3O$
(3638-04-8)
$C_4H_3Cl_2N_3S$
(13705-05-0)
$C_4H_3Cl_3$
(1573-58-6)
(2852-07-5)
(34819-62-0)
(39083-26-6)
(53317-48-9)
(53978-04-4)
(58679-08-6)
(83682-43-3)
(83682-46-6)
$C_4H_3Cl_3N_2$
(873-25-6)
$C_4H_3Cl_3N_4$
(30360-33-9)
$C_4H_3Cl_3N_4O$
(30369-65-4)
$C_4H_3Cl_5$
(14129-82-9)
$C_4H_3Cl_7$
(34973-41-6)
(83682-70-6)
$C_4H_3FN_2O_2$
(51-21-8)
$C_4H_3F_7O$
(375-01-9)
(56860-81-2)
$C_4H_3IN_2O_2$
(696-07-1)
C_4H_3IS
(3437-95-4)
$C_4H_3NO_2S$
(609-40-5)
$C_4H_3NO_3$
(609-39-2)
$C_4H_3N_5O_2$
(1468-26-4)
C_4H_4
(689-97-4)
$C_4H_4BrN_3$
(30361-84-3)
$C_4H_4BrN_3O_2$
(933-87-9)
(16681-63-3)
(18874-52-7)
(59177-47-8)
$C_4H_4Br_2F_4$
(18599-20-7)
$C_4H_4Br_2N_2O$
(63619-09-0)
$C_4H_4Br_3N_5$
(16088-72-5)
$C_4H_4CaCl_3NOS$
(57373-19-0)
$C_4H_4ClNO_2$
(128-09-6)
C_4H_4ClNOS
(26172-55-4)
$C_4H_4ClN_3$
(30361-82-1)
$C_4H_4ClN_3O_2$
(4897-31-8)
(63634-21-9)
(63662-67-9)
(86072-07-3)
$C_4H_4ClN_7$
(5248-65-7)
$C_4H_4Cl_2$
(821-10-3)
(1653-19-6)
(2984-42-1)
(6061-06-9)
(28577-62-0)
(83682-41-1)
(83682-42-2)
(83682-44-4)
(83682-45-5)
$C_4H_4Cl_2O_2$
(543-20-4)
(16502-88-8)
$C_4H_4Cl_2O_3$
(541-88-8)
$C_4H_4Cl_2O_4$
(15519-38-7)
$C_4H_4Cl_3N_5$
(16088-73-6)
$C_4H_4Cl_3N_5O$
(30805-11-9)
$C_4H_4Cl_6$
(1573-57-5)
(2431-55-2)
(79458-54-1)
(83682-69-3)
$C_4H_4FN_3O$
(2022-85-7)
$C_4H_4F_3N_5$
(705-25-9)
$C_4H_4N_2$
(110-61-2)
(289-80-5)
(289-95-2)
(290-37-9)
$C_4H_4N_2O$
(4341-85-9)
(4562-27-0)
(10111-08-7)
$C_4H_4N_2OS$
(141-90-2)
$C_4H_4N_2O_2$
(66-22-8)
(123-33-1)
$C_4H_4N_2O_2.K$
(28382-15-2)
$C_4H_4N_2O_2S$
(504-17-6)
$C_4H_4N_2O_3$
(67-52-7)
$C_4H_4N_2O_5$
(3237-50-1)
$C_4H_4N_2S_3$
(33813-20-6)
$C_4H_4N_4$
(1187-42-4)
$C_4H_4N_4O_4$
(5213-50-3)
(19183-15-4)
(19183-16-5)
$C_4H_4N_6O$
(134-58-7)
$C_4H_4N_6O_6$
(55510-04-8)
$C_4H_4N_6S$
(30360-93-1)
C_4H_4O
(110-00-9)
$C_4H_4O_2$
(497-23-4)
(674-82-8)
(922-67-8)
(3675-13-6)
(15506-53-3)
$C_4H_4O_3$
(108-30-5)
$C_4H_4O_4$
(110-16-7)
(110-17-8)
(6915-18-0)
$C_4H_4O_4.Cd$
(141-00-4)
$C_4H_4O_5$
(328-42-7)
$C_4H_4O_6.2H_4N$
(3164-29-2)
$C_4H_4O_6.Na$
(526-94-3)
$C_4H_4O_6.2Na$
(868-18-8)
$C_4H_4O_6.Pb$
(815-84-9)
$C_4H_4O_7Sb.K$
(28300-74-5)
(64070-12-8)
C_4H_4S
(110-02-1)
$C_4H_5Br_2N_5$
(30359-66-1)
$C_4H_5CaCl_2NOS$
(57373-20-3)
$C_4H_5CaCl_2NOS$-
$C_4H_4CaCl_3NOS$
(50815-77-5)
C_4H_5Cl
(126-99-8)
(627-22-5)
(627-23-6)
(21020-24-6)
(25790-55-0)
(34581-41-4)
(51908-64-6)
(62981-74-2)
$C_4H_5ClN_4$
(21320-62-7)
$C_4H_5ClN_4O$
(7254-11-7)
$C_4H_5ClN_4S$
(30357-94-9)
C_4H_5ClO
(683-70-5)
(920-46-7)
$C_4H_5ClO_2$
(80-63-7)
(2549-51-1)
(2937-50-0)
(16197-90-3)
$C_4H_5ClO_3$
(4755-77-5)
$C_4H_5ClO_4$
(16045-92-4)
$C_4H_5Cl_2$
(926-57-8)
$C_4H_5Cl_2N_5$
(30355-63-6)
$C_4H_5Cl_3$
(2431-50-7)
(2431-54-1)
(4749-27-3)
(13245-65-3)
(13279-86-2)
(31702-33-7)
(41601-59-6)
(51023-22-4)
(57808-36-3)
(60845-51-4)
$C_4H_5Cl_3N_2O$
(16977-58-5)
$C_4H_5Cl_3O$
(3083-25-8)
$C_4H_5Cl_3O_2$
(515-84-4)
$C_4H_5Cl_5$
(77753-24-3)
(83293-82-7)

$C_4H_5F_3O$
(406-90-6)
$C_4H_5F_3O_2$
(383-63-1)
$C_4H_5I_2N_5$
(30359-76-3)
C_4H_5N
(109-75-1)
(109-97-7)
(126-98-7)
(4786-20-3)
(5500-21-0)
C_4H_5NOS
(2682-20-4)
$C_4H_5NOS.C_4H_4ClNOS$
(55965-84-9)
$C_4H_5NO_2$
(105-34-0)
(123-56-8)
(1001-55-4)
(10004-44-1)
$C_4H_5NO_3$
(557-24-4)
(2987-87-3)
(116211-83-7)
$C_4H_5NO_4.2Na$
(928-72-3)
$C_4H_5NO_4S$
(33665-90-6)
C_4H_5NS
(57-06-7)
(693-92-5)
(693-95-8)
$C_4H_5N_3$
(109-12-6)
(3599-87-9)
$C_4H_5N_3O$
(71-30-7)
(15625-88-4)
(17635-40-4)
$C_4H_5N_3O_2$
(56-09-7)
(696-23-1)
(873-83-6)
(933-19-7)
(1671-82-5)
(3034-41-1)
(3034-42-2)
(14003-66-8)
$C_4H_5N_3O_3$
(30885-95-1)
(68019-78-3)
$C_4H_5N_3S$
(26292-91-1)
$C_4H_5N_3S_2$
(6237-86-1)
$C_4H_5N_5O_2$
(30362-21-1)
$C_4H_5N_7$
(30355-56-7)
$C_4H_5N_7O$
(5248-73-7)
$C_4H_5O_2$
(18358-13-9)
$C_4H_5O_6.K$
(868-14-4)
$C_4H_5O_7Sb.K$
(11071-15-1)
C_4H_6
(106-99-0)
(107-00-6)
(503-17-3)
(590-19-2)
(25339-57-5)
$C_4H_6As_6Cu_4O_{16}$
(12002-03-8)
C_4H_6BrCl
(54410-84-3)
$C_4H_6BrN_5$
(4576-40-3)
$C_4H_6Br_2O_2$
(3234-02-4)
(24442-57-7)
$C_4H_6Br_4$
(1529-68-6)
$C_4H_6ClN_5$
(10581-62-1)
$C_4H_6Cl_2$
(110-57-6)
(760-23-6)
(764-41-0)
(1476-11-5)
(1871-57-4)
(3375-22-2)
(7415-31-8)
(11069-19-5)
(31423-92-4)
(64037-54-3)
$C_4H_6Cl_2O$
(3511-19-1)
(3583-47-9)
(4635-59-0)
(10141-22-7)
$C_4H_6Cl_2O_2$
(95-59-0)
(535-15-9)
(2612-35-3)
(3883-43-0)
(10140-87-1)
(10411-52-6)
(13023-00-2)
$C_4H_6Cl_2O_2.Na$
(2517-16-0)
$C_4H_6Cl_4$
(3405-32-1)
(13138-51-7)
(14499-87-7)
(33455-24-2)
$C_4H_6Cl_4Cr_2O_3$
(15096-41-0)
$C_4H_6Cl_4O$
(3290-70-8)
(75536-53-7)
$C_4H_6F_2O_2$
(454-31-9)
(459-99-4)
$C_4H_6IN_5$
(30359-74-1)
$C_4H_6MnN_2S_4.C_4H_6N_2S_4Zn$
(8018-01-7)
$C_4H_6N_2$
(616-47-7)
(693-98-1)
$C_4H_6N_2O$
(108-26-9)
$C_4H_6N_2O_2$
(106-57-0)
(2763-96-4)
$C_4H_6N_2O_2S$
(2150-55-2)
$C_4H_6N_2O_3S$
(94751-62-9)
$C_4H_6N_2S$
(60-56-0)
(24340-76-9)
$C_4H_6N_2S_4.Mn$
(12427-38-2)
$C_4H_6N_2S_4.2Na$
(142-59-6)
$C_4H_6N_2S_4.Zn$
(12122-67-7)
$C_4H_6N_4$
(4039-99-0)
(27622-91-9)
$C_4H_6N_4O$
(1122-73-2)
(16352-06-0)
$C_4H_6N_4O_2$
(496-46-8)
(4531-54-8)
$C_4H_6N_4O_3$
(97-59-6)
$C_4H_6N_4O_3S_2$
(59-66-5)
$C_4H_6N_4O_{11}$
(20820-44-4)
$C_4H_6N_4S$
(30369-70-1)
$C_4H_6N_6O$
(13236-84-5)
C_4H_6O
(78-85-3)
(78-94-4)
(109-93-3)
(123-73-9)
(764-01-2)
(927-80-0)
(930-22-3)
(1191-95-3)
(1708-29-8)
(4170-30-3)
C_4H_6OS
(1115-15-7)
$C_4H_6O_2$
(79-41-4)
(96-33-3)
(96-48-0)
(107-93-7)
(108-05-4)
(110-65-6)
(298-18-0)
(431-03-8)
(503-64-0)
(564-00-1)
(625-38-7)
(1464-53-5)
(3068-88-0)
(3724-65-0)
(4401-11-0)
(30031-64-2)
(30419-67-1)
(36536-46-6)
$C_4H_6O_2.K$
(6900-35-2)
$C_4H_6O_2.Na$
(5536-61-8)
$C_4H_6O_2.1/2Pb$
(1068-61-7)
$C_4H_6O_2S$
(77-77-0)
(77-79-2)
$C_4H_6O_3$
(108-24-7)
(108-32-7)
(541-50-4)
(600-22-6)
(692-29-5)
$C_4H_6O_4$
(110-15-6)
(110-22-5)
(516-05-2)
(553-90-2)
$C_4H_6O_4.Ba$
(543-80-6)
$C_4H_6O_4.Be$
(543-81-7)
$C_4H_6O_4.Ca$
(62-54-4)
$C_4H_6O_4.Co$
(71-48-7)
(3267-76-3)
$C_4H_6O_4.Co.4H_2O$
(6147-53-1)
$C_4H_6O_4.Cu$
(142-71-2)
$C_4H_6O_4.Hg$
(1600-27-7)
$C_4H_6O_4.Mg$
(142-72-3)
$C_4H_6O_4.Mn$
(638-38-0)
$C_4H_6O_4.Mn.4H_2O$
(6156-78-1)
$C_4H_6O_4.xNa$
(14047-56-4)
$C_4H_6O_4.Ni$
(373-02-4)
$C_4H_6O_4.Ni.4H_2O$
(6018-89-9)
$C_4H_6O_4.Pb$
(301-04-2)
$C_4H_6O_4.Pb.3H_2O$
(6080-56-4)
$C_4H_6O_4.Sr$
(543-94-2)
$C_4H_6O_4.Zn$
(557-34-6)
$C_4H_6O_4.Zn.2H_2O$
(5970-45-6)
$C_4H_6O_4S$
(123-93-3)
$C_4H_6O_5$
(110-99-6)
(617-48-1)
(6915-15-7)
$C_4H_6O_6$
(87-69-4)
(133-37-9)
(147-71-7)
$C_4H_6O_6.Cu$
(815-82-7)
$C_4H_6O_6.K.Na$
(304-59-6)
$C_4H_6O_6.Sr$
(868-19-9)
$C_4H_6O_6U$
(541-09-3)
$C_4H_6O_7S$
(5138-18-1)
$C_4H_6O_7S.3Na$
(13419-59-5)
$C_4H_6O_8$
(76-30-2)
C_4H_6S
(627-51-0)
$C_4H_7AlO_5$
(142-03-0)
C_4H_7Br
(4784-77-4)
(5162-44-7)
(13294-71-8)
(31844-98-1)
C_4H_7BrO
(814-75-5)
(816-40-0)
(64341-49-7)
$C_4H_7BrO_2$
(80-58-0)
(105-36-2)
(927-68-4)
(2623-86-1)
$C_4H_7Br_2Cl$
(10474-14-3)
(69036-12-0)
$C_4H_7Br_2Cl_2O_4P$
(300-76-5)
$C_4H_7Br_3O$
(76-08-4)
C_4H_7Cl
(513-37-1)
(563-47-3)
(563-52-0)
(927-73-1)
(4461-41-0)
(4461-42-1)
C_4H_7ClO
(110-75-8)
(141-75-3)
(671-56-7)
(917-93-1)

(4091-39-8)
(75455-41-3)
C_4H_7ClOS
(13889-92-4)
(40709-82-8)
$C_4H_7ClO_2$
(105-39-5)
(108-23-6)
(109-61-5)
(542-58-5)
(627-00-9)
(1951-12-8)
(2568-30-1)
(4170-24-5)
(17639-93-9)
$C_4H_7Cl_2F_3Si$
(675-62-7)
$C_4H_7Cl_2O_4P$
(62-73-7)
$C_4H_7Cl_3$
(1790-22-3)
(15187-71-0)
$C_4H_7Cl_3O$
(57-15-8)
$C_4H_7Cl_3O_2$
(76-40-4)
$C_4H_7FO_2$
(459-72-3)
$C_4H_7IO_2$
(623-48-3)
C_4H_7N
(78-82-0)
(109-74-0)
C_4H_7NO
(75-86-5)
(79-39-0)
(110-67-8)
(110-78-1)
(616-45-5)
(1187-59-3)
(2567-01-3)
(6228-73-5)
$C_4H_7NO_2$
(57-71-6)
(625-77-4)
(924-42-5)
(2114-11-6)
(2133-34-8)
$C_4H_7NO_3$
(543-24-8)
(638-32-4)
(94818-85-6)
$C_4H_7NO_4$
(56-84-8)
(142-73-4)
$C_4H_7NS_2$
(1908-87-8)
$C_4H_7N_3$
(7411-16-7)
$C_4H_7N_3O$
(60-27-5)
(60153-49-3)
$C_4H_7N_3O_9$
(6659-60-5)
(84002-64-2)
$C_4H_7N_3S_2$
(25660-70-2)
$C_4H_7N_5$
(542-02-9)
(1004-38-2)
$C_4H_7N_5O$
(2827-45-4)
(4030-02-8)
$C_4H_7N_5S$
(5397-01-3)
$C_4H_7OS_2.K$
(140-92-1)
$C_4H_7OS_2.Na$
(140-93-2)
C_4H_8
(106-98-9)
(107-01-7)
(115-11-7)
(287-23-0)
(590-18-1)
(594-11-6)
(624-64-6)
(25167-67-3)
$C_4H_8Br_2$
(107-80-2)
(110-52-1)
(533-98-2)
(594-34-3)
(5408-86-6)
$C_4H_8Br_2O$
(5414-19-7)
C_4H_8Cl
(591-97-9)
C_4H_8ClF
(462-73-7)
C_4H_8ClNO
(23328-69-0)
$C_4H_8Cl_2$
(110-56-5)
(541-33-3)
(594-37-6)
(616-21-7)
(1190-22-3)
(4279-22-5)
(7581-97-7)
$C_4H_8Cl_2O$
(111-44-4)
(2419-74-1)
(6986-48-7)
$C_4H_8Cl_2OS$
(5819-08-9)
$C_4H_8Cl_2O_2$
(13483-18-6)
$C_4H_8Cl_2O_2S$
(471-03-4)
$C_4H_8Cl_2S$
(505-60-2)
$C_4H_8Cl_2Si$
(10138-21-3)
$C_4H_8Cl_3O_4P$
(52-68-6)
$C_4H_8HgO_2$
(109-62-6)
$C_4H_8N_2$
(534-26-9)
(926-64-7)
(19355-69-2)
$C_4H_8N_2O$
(557-11-9)
(930-55-2)
(4549-43-3)
(13256-13-8)
(16568-02-8)
$C_4H_8N_2O_2$
(59-89-2)
(95-45-4)
(110-14-5)
$C_4H_8N_2O_2S$
(92134-93-5)
$C_4H_8N_2O_3$
(70-47-3)
(556-50-3)
(592-62-1)
(615-53-2)
(626-36-8)
(4164-32-3)
(4202-74-8)
(56856-83-8)
$C_4H_8N_2O_4$
(60913-86-2)
$C_4H_8N_2O_5$
(24884-69-3)
$C_4H_8N_2O_7$
(693-21-0)
$C_4H_8N_2S$
(109-57-9)
(2122-19-2)
$C_4H_8N_2S_4$
(111-54-6)
(12656-69-8)
$C_4H_8N_2S_4.2H_3N$
(3566-10-7)
$C_4H_8N_4O_2$
(140-79-4)
$C_4H_8N_4O_3$
(32976-88-8)
$C_4H_8N_6O_4$
(3844-60-8)
$C_4H_8N_8O_8$
(2691-41-0)
C_4H_8O
(78-84-2)
(78-93-3)
(106-88-7)
(109-92-2)
(109-99-9)
(116-11-0)
(123-72-8)
(513-42-8)
(558-30-5)
(598-32-3)
(1758-33-4)
(2167-39-7)
(3266-23-7)
(4088-60-2)
(6117-91-5)
(21490-63-1)
C_4H_8OS
(625-60-5)
(3268-49-3)
(15980-15-1)
$C_4H_8OS_2$
(108-25-8)
$C_4H_8O_2$
(79-31-2)
(107-89-1)
(107-92-6)
(110-64-5)
(110-74-7)
(123-91-1)
(141-78-6)
(497-06-3)
(497-26-7)
(505-22-6)
(513-86-0)
(554-12-1)
(590-90-9)
(625-55-8)
(821-11-4)
(930-37-0)
(5878-19-3)
(6117-80-2)
(26063-00-3)
(50317-11-8)
$C_4H_8O_2.H_3N$
(14287-04-8)
(22228-82-6)
$C_4H_8O_2.Na$
(156-54-7)
$C_4H_8O_2S$
(126-33-0)
(623-51-8)
(2935-90-2)
$C_4H_8O_3$
(300-85-6)
(542-59-6)
(547-64-8)
(565-70-8)
(594-61-6)
(627-03-2)
(628-82-0)
(5981-06-6)
(19693-75-5)
$C_4H_8O_3S$
(1633-83-6)
$C_4H_8O_4$
(13382-47-3)
C_4H_8S
(110-01-0)
(5954-72-3)
(10152-76-8)
(10152-77-9)
$C_4H_8S_2$
(505-29-3)
(24298-49-5)
$C_4H_8S_3$
(23654-92-4)
$C_4H_9Al_2ClN_4O_7$
(1317-25-5)
$C_4H_9BO_2$
(10173-38-3)
C_4H_9Br
(78-76-2)
(78-77-3)
(109-65-9)
(507-19-7)
$C_4H_9BrClO_2P$
(24327-56-8)
C_4H_9Cl
(78-86-4)
(109-69-3)
(507-20-0)
(513-36-0)
$C_4H_9ClIO_2P$
(17052-15-2)
C_4H_9ClMg
(15366-08-2)
C_4H_9ClO
(563-84-8)
(628-34-2)
(928-51-8)
(1320-66-7)
(7081-78-9)
C_4H_9ClOS
(693-30-1)
$C_4H_9ClO_2$
(97-97-2)
(628-89-7)
$C_4H_9ClO_2S$
(2386-60-9)
C_4H_9ClS
(693-07-2)
C_4H_9ClSi
(1719-58-0)
$C_4H_9Cl_2O_2P$
(13274-84-5)
$C_4H_9Cl_3Si$
(7521-80-4)
(18169-57-8)
$C_4H_9Cl_3Sn$
(1118-46-3)
C_4H_9DO
(4712-38-3)
(4712-39-4)
C_4H_9F
(353-61-7)
(2366-52-1)
C_4H_9I
(513-38-2)
(513-48-4)
(542-69-8)
(25267-27-0)
$C_4H_9I_2O_2P$
(17052-17-4)
C_4H_9Li
(109-72-8)
(594-19-4)
(598-30-1)
C_4H_9N
(123-75-1)
(2549-67-9)
(2658-24-4)
C_4H_9NO
(96-29-7)
(110-69-0)

(110-91-8)
(127-19-5)
(151-00-8)
(541-35-5)
(563-83-7)
(625-50-3)
(1072-52-2)
(1187-58-2)
$C_4H_9NO.ClH$
(10024-89-2)
C_4H_9NOSe
(90030-80-1)
$C_4H_9NO_2$
(56-12-2)
(105-40-8)
(142-26-7)
(541-48-0)
(542-56-3)
(544-16-1)
(594-70-7)
(600-24-8)
(627-05-4)
(627-12-3)
(924-43-6)
(927-60-6)
(1746-77-6)
(2835-81-6)
$C_4H_9NO_2.ClH$
(623-33-6)
$C_4H_9NO_2S$
(454-29-5)
(7728-98-5)
$C_4H_9NO_3$
(72-19-5)
(76-39-1)
(543-29-3)
(609-31-4)
(672-15-1)
(928-45-0)
(1616-88-2)
$C_4H_9NO_4$
(77-49-6)
$C_4H_9NO_5$
(126-11-4)
$C_4H_9NO_6$
(14307-43-8)
C_4H_9NSe
(6474-16-4)
$C_4H_9N_3O$
(110-20-3)
(1910-89-0)
$C_4H_9N_3O_2$
(57-00-1)
(3475-63-6)
$C_4H_9N_3S$
(1752-30-3)
(6746-27-6)
$C_4H_9NaO_3S$
(68608-15-1)
$C_4H_9O_4P$
(1073-75-2)
$C_4H_9O_4PS_2$
(1113-01-5)
C_4H_9Sn
(78763-54-9)
C_4H_{10}
(75-28-5)
(106-97-8)
(68475-59-2)
$C_4H_{10}AlCl$
(96-10-6)
$C_4H_{10}AlI$
(2040-00-8)
$C_4H_{10}AuBr$
(26645-10-3)
$C_4H_{10}BF_3O$
(109-63-7)
$C_4H_{10}ClN.ClH$
(4584-46-7)
$C_4H_{10}ClOPS$
(1497-68-3)
$C_4H_{10}ClO_2PS$
(2524-04-1)
$C_4H_{10}ClO_3P$
(814-49-3)
$C_4H_{10}Cl_2Pb$
(13231-90-8)
$C_4H_{10}Cl_2Si$
(1719-53-5)
$C_4H_{10}FO_2P$
(107-44-8)
(650-20-4)
$C_4H_{10}Hg$
(627-44-1)
$C_4H_{10}HgO_2$
(26983-51-7)
$C_4H_{10}NO_3PS$
(30560-19-1)
(42072-27-5)
$C_4H_{10}N_2$
(110-85-0)
$C_4H_{10}N_2.2ClH$
(142-64-3)
$C_4H_{10}N_2O$
(55-18-5)
(691-60-1)
(16301-26-1)
(56375-33-8)
$C_4H_{10}N_2O_2$
(13147-25-6)
(17463-44-4)
$C_4H_{10}N_2O_3$
(1116-54-7)
$C_4H_{10}N_2S$
(2489-77-2)
$C_4H_{10}N_4O_2$
(1852-14-8)
$C_4H_{10}O$
(60-29-7)
(71-36-3)
(75-65-0)
(78-83-1)
(78-92-2)
(557-17-5)
(598-53-8)
(35296-72-1)
$C_4H_{10}O.1/3Al$
(2269-22-9)
(3085-30-1)
$C_4H_{10}O.K$
(865-47-4)
(3999-70-0)
$C_4H_{10}O.Na$
(2372-45-4)
$C_4H_{10}OS$
(110-77-0)
(505-10-2)
$C_4H_{10}OSn$
(51590-67-1)
$C_4H_{10}O_2$
(75-91-2)
(107-88-0)
(107-98-2)
(110-63-4)
(110-71-4)
(110-80-5)
(513-85-9)
(534-15-6)
(584-03-2)
(1320-67-8)
(1589-47-5)
(5341-95-7)
(6982-25-8)
(19132-06-0)
(24347-58-8)
(25265-75-2)
(28677-93-2)
$C_4H_{10}O_2.1/2Mg$
(14064-03-0)
$C_4H_{10}O_2PS_2.Na$
(3338-24-7)
$C_4H_{10}O_2S$
(111-48-8)
$C_4H_{10}O_2Sn$
(2273-43-0)
$C_4H_{10}O_3$
(111-46-6)
(149-73-5)
(628-90-0)
(3068-00-6)
(4435-50-1)
$C_4H_{10}O_3S$
(623-81-4)
(926-06-7)
(1912-30-7)
$C_4H_{10}O_4$
(149-32-6)
$C_4H_{10}O_4S$
(64-67-5)
(2580-77-0)
$C_4H_{10}O_8Pb_3$
(1335-32-6)
$C_4H_{10}S$
(75-66-1)
(109-79-5)
(352-93-2)
(513-44-0)
(513-53-1)
(1551-21-9)
(3877-15-4)
(82987-03-9)
$_4H_{10}S_2$
(110-81-6)
(2179-60-4)
(6628-18-8)
(26750-44-7)
$C_4H_{10}S_3$
(3570-55-6)
$C_4H_{10}Se$
(627-53-2)
$C_4H_{10}Se_2$
(628-39-7)
$C_4H_{10}Zn$
(557-20-0)
$C_4H_{11}Al$
(871-27-2)
$C_4H_{11}As$
(692-42-2)
$C_4H_{11}BO_2$
(4426-47-5)
$C_4H_{11}ClO_2Si$
(6485-91-2)
$C_4H_{11}ClO_2Sn$
(13355-96-9)
$C_4H_{11}ClSi$
(1609-19-4)
$C_4H_{11}N$
(75-64-9)
(78-81-9)
(109-73-9)
(109-89-7)
(513-49-5)
(598-56-1)
(627-35-0)
(13952-84-6)
$C_4H_{11}N.BH_3$
(7337-45-3)
$C_4H_{11}N.BrH$
(6274-12-0)
$C_4H_{11}N.ClH$
(660-68-4)
(3858-78-4)
(5041-09-8)
$C_4H_{11}N.HI$
(19833-78-4)
$C_4H_{11}NO$
(96-20-8)
(108-01-0)
(110-73-6)
(124-68-5)
(3710-84-7)
(5332-73-0)
(5856-62-2)
(13552-21-1)
$C_4H_{11}NO.ClH$
(3207-12-3)
$C_4H_{11}NO_2$
(111-42-2)
(115-69-5)
(929-06-6)
(22483-09-6)
$C_4H_{11}NO_2.C_2H_4O_2$
(23251-72-1)
$C_4H_{11}NO_2.C_4H_4N_2O_2$
(5716-15-4)
$C_4H_{11}NO_2.ClH$
(14426-21-2)
$C_4H_{11}NO_3$
(77-86-1)
$C_4H_{11}NO_8P_2$
(2439-99-8)
$C_4H_{11}O_2P$
(54423-73-3)
$C_4H_{11}O_2PS_2$
(298-06-6)
$C_4H_{11}O_2PS_2.K$
(3454-66-8)
$C_4H_{11}O_3P$
(762-04-9)
(1832-54-8)
(3321-64-0)
(7305-61-5)
$C_4H_{11}O_3PS$
(2465-65-8)
$C_4H_{11}O_3PS.K$
(5871-17-0)
$C_4H_{11}O_4P$
(598-02-7)
(1623-15-0)
$C_4H_{11}O_4P.K$
(25238-99-7)
$C_4H_{11}O_4P.2K$
(26290-70-0)
$C_4H_{11}O_4P.Hg$
(21504-45-0)
$C_4H_{11}Pb$
(103730-90-1)
$C_4H_{12}Cl_2Si_2$
(4342-61-4)
$C_4H_{12}FN_2OP$
(115-26-4)
$C_4H_{12}N.Br$
(64-20-0)
$C_4H_{12}N.Cl$
(75-57-0)
$C_4H_{12}N.HO$
(75-59-2)
$C_4H_{12}N_2$
(110-60-1)
(590-88-5)
(1615-80-1)
(6291-84-5)
$C_4H_{12}N_2.ClH$
(7400-27-3)
(56795-65-4)
(61556-82-9)
$C_4H_{12}N_2.2ClH$
(7699-31-2)
$C_4H_{12}N_2O$
(111-41-1)
$C_4H_{12}N_4S_2.H_2O_4S$
(867-44-7)
$C_4H_{12}OSi$
(1825-61-2)
$C_4H_{12}O_3Si$
(1185-55-3)
$C_4H_{12}O_4P.C_2H_3O_2$
(7580-37-2)
$C_4H_{12}O_4P.1/2C_2O_4$
(52221-67-7)

$C_4H_{12}O_4P.C_4H_{12}O_4P.C_2$-$H_3O_2.1/3O_4P$
(55818-96-7)
$C_4H_{12}O_4P.Cl$
(124-64-1)
$C_4H_{12}O_4P.HO$
(512-82-3)
$C_4H_{12}O_4P.1/3O_4P$
(22031-17-0)
$C_4H_{12}O_4P_2S_4$
(5930-71-2)
$C_4H_{12}O_4Si$
(681-84-5)
$C_4H_{12}Pb$
(75-74-1)
(24952-65-6)
$C_4H_{12}Si$
(75-76-3)
$C_4H_{12}Sn$
(594-27-4)
(2406-65-7)
$C_4H_{13}NO_7P_2$
(5995-42-6)
$C_4H_{13}N_3$
(111-40-0)
$C_4H_{14}OSi_2$
(3277-26-7)
$C_4N_4Zn.2Na$
(15333-24-1)
C_4NiO_4
(13463-39-3)
$C_4O_9Ti.2K$
(14481-26-6)
C_5Cl_5N
(2176-62-7)
$C_5Cl_5N_5$
(30863-30-0)
C_5Cl_6
(77-47-4)
$C_5Cl_6N_4$
(10243-82-0)
$C_5Cl_6N_6$
(30805-09-5)
$C_5Cl_7N_3$
(30894-89-4)
C_5Cl_8
(706-78-5)
$C_5F_7N_3$
(717-62-4)
$C_5F_{10}O$
(1623-05-8)
C_5F_{12}
(678-26-2)
$C_5F_{12}O_2S$
(375-81-5)
C_5FeO_5
(13463-40-6)
$C_5HCl_3N_4$
(2562-52-9)
$C_5HCl_3O_2$
(103354-08-1)
C_5HCl_4N
(2402-79-1)
(2808-86-8)
(33752-16-8)
C_5HCl_5
(25329-35-5)
$C_5HCl_5O_2$
(5659-41-6)
$C_5HCl_5O_3$
(99165-95-4)
$C_5HCl_6N_3$
(3599-74-4)
$C_5HCl_6N_3O.C_2H_3Cl_3N_2$
(30886-02-3)
$C_5H_2Br_2Cl_3N_3$
(30894-98-5)
$C_5H_2Br_6N_4$
(30339-36-7)
$C_5H_2Cl_2O_2$
(94650-97-2)
$C_5H_2Cl_3N$
(6515-09-9)
(16063-69-7)
(16063-70-0)
$C_5H_2Cl_3NO$
(1970-40-7)
(6515-38-4)
$C_5H_2Cl_3NO.Na$
(37439-34-2)
$C_5H_2Cl_4$
(695-77-2)
$C_5H_2Cl_4O_3$
(99165-91-0)
(99165-94-3)
(108082-06-0)
$C_5H_2Cl_5N_3$
(30894-72-5)
(30894-88-3)
$C_5H_2Cl_6$
(72030-26-3)
$C_5H_2Cl_6N_4$
(20376-31-2)
$C_5H_2Cl_6N_4O$
(30357-61-0)
$C_5H_2Cl_8$
(83682-40-0)
$C_5H_2F_6N_4$
(29181-67-7)
$C_5H_3Br_2N$
(625-92-3)
$C_5H_3ClN_4O_2$
(13548-68-0)
$C_5H_3ClO_2$
(103339-62-4)
$C_5H_3ClO_2S$
(99165-92-1)
$C_5H_3Cl_2N$
(2402-77-9)
(2402-78-0)
(2457-47-8)
(16110-09-1)
(26452-80-2)
$C_5H_3Cl_3NO_3PS$
(66651-97-6)
$C_5H_3Cl_3O_3$
(77439-76-0)
(99165-90-9)
(99165-97-6)
(115340-67-5)
$C_5H_3Cl_4N_3$
(30894-71-4)
(30894-85-0)
$C_5H_3Cl_4N_3O.C_2H_4Cl_2N_2$
(30886-01-2)
$C_5H_3Cl_5N_4$
(30357-60-9)
$C_5H_3NO_4$
(698-63-5)
$C_5H_3NO_5$
(645-12-5)
C_5H_4BrN
(109-04-6)
(626-55-1)
(1120-87-2)
$C_5H_4BrN.ClH$
(19524-06-2)
C_5H_4ClN
(109-09-1)
(626-60-8)
(626-61-9)
(29154-12-9)
$C_5H_4ClN.ClH$
(7379-35-3)
$C_5H_4ClN_3O_2$
(4897-25-0)
$C_5H_4ClN_5$
(10310-21-1)
(30894-86-1)
$C_5H_4Cl_2N_4OS$
(5097-58-5)
$C_5H_4Cl_3N_3$
(30361-91-2)
(30894-87-2)
$C_5H_4Cl_3N_5O$
(30362-20-0)
$C_5H_4Cl_4N_4$
(30357-76-7)
(30360-41-9)
$C_5H_4Cl_5N_5$
(30359-93-4)
$C_5H_4Cl_6$
(68258-91-3)
$C_5H_4N_2O_2$
(1122-61-8)
(15009-91-3)
$C_5H_4N_2O_3$
(1124-33-0)
$C_5H_4N_2O_4$
(65-86-1)
$C_5H_4N_2O_4.1/2Ca$
(22454-86-0)
$C_5H_4N_4$
(120-73-0)
$C_5H_4N_4O$
(68-94-0)
(315-30-0)
$C_5H_4N_4OS$
(2002-59-7)
(2487-40-3)
$C_5H_4N_4O_2$
(69-89-6)
(2465-59-0)
$C_5H_4N_4O_2S$
(2002-60-0)
$C_5H_4N_4O_3$
(69-93-2)
$C_5H_4N_4S$
(50-44-2)
$C_5H_4N_4S.H_2O$
(6112-76-1)
C_5H_4OS
(98-03-3)
$C_5H_4O_2$
(98-01-1)
(108-97-4)
(498-60-2)
(504-31-4)
$C_5H_4O_2S$
(88-13-1)
(527-72-0)
$C_5H_4O_3$
(88-14-2)
(488-93-7)
(616-02-4)
(15849-14-6)
(26447-28-9)
$C_5H_5BrN_2$
(13534-98-0)
$C_5H_5Br_3N_4$
(30339-34-5)
C_5H_5Cl
(41851-50-7)
$C_5H_5ClN_2$
(1072-98-6)
C_5H_5ClO
(53102-14-0)
$C_5H_5Cl_2N_3$
(698-72-6)
(30361-86-5)
$C_5H_5Cl_2N_3O$
(18343-30-1)
$C_5H_5Cl_2N_3O.C_2H_5ClN_2$
(30886-00-1)
$C_5H_5Cl_2N_3S$
(13733-90-9)
$C_5H_5Cl_3N_2OS$
(2593-15-9)
$C_5H_5Cl_3N_4$
(21227-47-4)
$C_5H_5Cl_3N_4O$
(30805-12-0)
$C_5H_5Cl_3N_4S$
(14946-02-2)
$C_5H_5Cl_4N_5$
(30359-92-3)
$C_5H_5Cl_6N_3O$
(30805-36-8)
$C_5H_5Cl_6N_3S$
(30805-37-9)
C_5H_5N
(110-86-1)
$C_5H_5N.ClH$
(628-13-7)
$C_5H_5N.ClHO_4$
(15598-34-2)
C_5H_5NO
(109-00-2)
(142-08-5)
(626-64-2)
(694-59-7)
(1003-29-8)
C_5H_5NOS
(1121-31-9)
(23003-22-7)
$C_5H_5NOS.Na$
(3811-73-2)
(15922-78-8)
$C_5H_5NO_2$
(137-05-3)
(634-97-9)
(16867-04-2)
$C_5H_5N_3O$
(98-96-4)
$C_5H_5N_4.xH_3O_4P$
(52175-10-7)
$C_5H_5N_5$
(73-24-5)
(452-06-2)
$C_5H_5N_5O$
(73-40-5)
$C_5H_5N_5O.ClH$
(635-39-2)
$C_5H_5N_5S$
(154-42-7)
C_5H_6
(78-80-8)
(542-92-7)
$C_5H_6BrClN_2O_2$
(126-06-7)
$C_5H_6BrN_3O_2$
(21117-52-2)
(21431-58-3)
$C_5H_6BrN_7$
(30362-26-6)
$C_5H_6Br_2N_2O_2$
(77-48-5)
$C_5H_6Br_2N_4$
(30357-80-3)
$C_5H_6ClN_3$
(5600-21-5)
(30361-85-4)
(30894-84-9)
$C_5H_6ClN_3OS$
(4407-44-7)
$C_5H_6ClN_3O_2$
(3140-73-6)
(91027-93-9)
(91027-94-0)
$C_5H_6ClN_3S_2$
(4407-40-3)
$C_5H_6ClN_7$
(5248-69-1)
$C_5H_6Cl_2$
(61626-71-9)
$C_5H_6Cl_2N_2O_2$
(118-52-5)
$C_5H_6Cl_2N_4$
(2401-64-1)
(3440-19-5)

(30360-36-2)
(30576-27-3)
$C_5H_6Cl_2O$
(103339-61-3)
$C_5H_6Cl_2O_2$
(2873-74-7)
$C_5H_6Cl_2O_4$
(50901-13-8)
$C_5H_6Cl_3N_5$
(27431-19-2)
(30359-91-2)
$C_5H_6Cl_4$
(59808-78-5)
$C_5H_6Cl_6N_2O_3$
(116-52-9)
$C_5H_6F_2N_4$
(30369-76-7)
$C_5H_6F_4O_3$
(755-73-7)
$C_5H_6N_2$
(109-08-0)
(462-08-8)
(504-24-5)
(504-29-0)
(544-13-8)
(1072-63-5)
(3438-46-8)
(7321-55-3)
$C_5H_6N_2O$
(3149-28-8)
(16867-03-1)
$C_5H_6N_2OS$
(56-04-2)
(636-26-0)
$C_5H_6N_2O_2$
(65-71-4)
(626-48-2)
$C_5H_6N_4O_4$
(19183-17-6)
$C_5H_6N_6$
(13301-35-4)
(30805-08-4)
$C_5H_6N_6O_2$
(30805-07-3)
$C_5H_6N_6S_2$
(4407-41-4)
C_5H_6O
(534-22-5)
(930-27-8)
(930-30-3)
$C_5H_6O_2$
(98-00-0)
(623-47-2)
(1333-38-6)
(4412-91-3)
(5729-47-5)
(22122-36-7)
$C_5H_6O_3$
(108-55-4)
(4100-80-5)
$C_5H_6O_4$
(97-65-4)
(498-23-7)
(5699-58-1)
$C_5H_6O_5$
(328-50-7)
$C_5H_6O_7$
(55203-12-8)
$C_5H_6O_7.3Na$
(41999-58-0)
C_5H_6S
(554-14-3)
(616-44-4)
$C_5H_7BrO_2$
(4823-47-6)
$C_5H_7Br_2N_5$
(30359-94-5)
C_5H_7Cl
(96-40-2)
$C_5H_7ClN_4$
(30369-28-9)
$C_5H_7ClN_4O$
(49624-61-5)
$C_5H_7ClN_4O_2S$
(78744-33-9)
$C_5H_7ClN_4S$
(68925-41-7)
(70958-50-8)
(83623-05-6)
$C_5H_7ClO_2$
(2206-89-5)
$C_5H_7ClO_3$
(4755-81-1)
$C_5H_7Cl_2N_5$
(30359-89-8)
(30359-90-1)
$C_5H_7Cl_3O$
(84987-77-9)
$C_5H_7D_3$
(69432-94-6)
C_5H_7N
(96-54-8)
(592-51-8)
(616-43-3)
(636-41-9)
(4403-61-6)
(4635-87-4)
(4786-19-0)
(13284-42-9)
(16529-56-9)
(20068-02-4)
(25899-50-7)
(26294-98-4)
(30574-97-1)
C_5H_7NO
(617-89-0)
(20662-83-3)
$C_5H_7NO_2$
(105-56-6)
(39201-33-7)
$C_5H_7NO_3$
(98-79-3)
(116211-84-8)
C_5H_7NS
(541-58-2)
(27757-85-3)
$C_5H_7N_3$
(54-96-6)
(108-52-1)
(141-86-6)
(452-58-4)
(1722-15-2)
(3599-59-5)
(4318-76-7)
$C_5H_7N_3O$
(30885-99-5)
$C_5H_7N_3O_2$
(551-92-8)
(7464-68-8)
(7501-27-1)
(13230-04-1)
$C_5H_7N_3O_2.ClH$
(25332-20-1)
$C_5H_7N_3O_2S$
(30886-14-7)
$C_5H_7N_3O_3$
(1075-59-8)
(30805-28-8)
(35687-41-3)
(35687-42-4)
$C_5H_7N_3O_4$
(115-02-6)
$C_5H_7N_3S$
(30886-15-8)
C_5H_8
(78-79-5)
(142-29-0)
(157-40-4)
(185-94-4)
(504-60-9)
(591-93-5)
(591-95-7)
(598-25-4)
(627-19-0)
(627-21-4)
(1120-56-5)
(1574-41-0)
(2004-70-8)
C_5H_8BrCl
(14376-82-0)
$C_5H_8ClNO_2$
(4116-10-3)
(58629-01-9)
$C_5H_8ClN_5$
(1007-28-9)
(2911-36-6)
(30359-86-5)
(30359-87-6)
$C_5H_8ClN_5O_2$
(5217-85-6)
$C_5H_8Cl_2O$
(78-71-7)
(4300-97-4)
$C_5H_8Cl_2O_2$
(64855-18-1)
(67329-11-7)
$C_5H_8NO_4$
(11070-68-1)
$C_5H_8NO_4.H_4N$
(7558-63-6)
$C_5H_8NO_4.Na$
(142-47-2)
$C_5H_8N_2$
(1739-84-0)
$C_5H_8N_2O$
(2749-59-9)
$C_5H_8N_2O_2$
(77-71-4)
(60754-24-7)
$C_5H_8N_2O_2S$
(58842-20-9)
(97190-65-3)
$C_5H_8N_2O_3$
(7519-36-0)
$C_5H_8N_2O_4$
(30310-80-6)
$C_5H_8N_4$
(1853-90-3)
(4040-00-0)
(4122-05-8)
(30360-34-0)
$C_5H_8N_4O$
(1668-54-8)
(16352-07-1)
$C_5H_8N_4OS$
(30358-18-0)
$C_5H_8N_4O_2$
(2630-10-6)
(16370-63-1)
(21677-57-6)
$C_5H_8N_4O_3S_2$
(554-57-4)
$C_5H_8N_4O_{12}$
(78-11-5)
$C_5H_8N_4S$
(30369-71-2)
$C_5H_8N_4S_2$
(30358-19-1)
C_5H_8O
(110-87-2)
(115-19-5)
(120-92-3)
(285-67-6)
(497-03-0)
(625-33-2)
(764-39-6)
(765-43-5)
(814-78-8)
(1115-11-3)
(1487-15-6)
(1517-15-3)
(1629-58-9)
(2100-17-6)
(3102-33-8)
(3917-15-5)
(34314-83-5)
$C_5H_8O_2$
(80-59-1)
(80-62-6)
(105-38-4)
(108-22-5)
(108-29-2)
(111-30-8)
(123-54-6)
(140-88-5)
(541-47-9)
(542-28-9)
(565-63-9)
(591-80-0)
(591-87-7)
(600-14-6)
(623-38-1)
(626-98-2)
(1617-32-9)
(1955-45-9)
(3188-00-9)
(3586-58-1)
(3721-95-7)
(3973-18-0)
(5204-64-8)
(13991-37-2)
(16666-42-5)
(18707-60-3)
(33698-87-2)
$C_5H_8O_3$
(105-45-3)
(123-76-2)
(592-20-1)
(617-35-6)
(818-61-1)
(4660-80-4)
(6387-89-9)
$C_5H_8O_3S$
(583-92-6)
$C_5H_8O_4$
(108-59-8)
(110-94-1)
(498-21-5)
(595-46-0)
(601-75-2)
(3878-55-5)
$C_5H_8O_4.xNa$
(32224-61-6)
$C_5H_8O_4S_2$
(3278-22-6)
C_5H_9Al
(24683-32-7)
C_5H_9Br
(137-43-9)
$C_5H_9BrO_2$
(565-74-2)
(584-93-0)
(2067-33-6)
$C_5H_9Br_3O$
(1522-92-5)
(36483-57-5)
C_5H_9Cl
(503-60-6)
(5166-35-8)
(13417-43-1)
(17773-65-8)
(21450-13-5)
C_5H_9ClO
(638-29-9)
(3003-84-7)
(3282-30-2)
$C_5H_9ClO_2$
(105-48-6)
(535-13-7)
(543-27-1)

(1119-46-6)
(6155-96-0)
(13511-38-1)
(17462-58-7)
$C_5H_9Cl_2NSi$
(1190-16-5)
$C_5H_9Cl_2N_3O_2$
(154-93-8)
$C_5H_9Cl_2O_2P$
(14212-97-6)
C_5H_9N
(110-59-8)
(630-18-2)
(694-05-3)
(2978-58-7)
(18936-17-9)
C_5H_9NO
(111-36-4)
(675-20-7)
(872-50-4)
(1192-28-5)
(1768-24-7)
(2141-62-0)
(2680-03-7)
(10431-98-8)
(27154-43-4)
(77311-02-5)
C_5H_9NOS
(17773-41-0)
$C_5H_9NO_2$
(147-85-3)
(609-36-9)
(923-02-4)
(1113-68-4)
(4394-85-8)
(6065-18-5)
(6065-19-6)
(15438-71-8)
(20306-75-6)
$C_5H_9NO_3$
(51-35-4)
(97-69-8)
(106-60-5)
(24935-97-5)
$C_5H_9NO_3S$
(616-91-1)
$C_5H_9NO_4$
(56-86-0)
(617-65-2)
(4408-64-4)
$C_5H_9NO_4.ClH$
(138-15-8)
$C_5H_9NO_4S$
(638-23-3)
C_5H_9NS
(590-42-1)
$C_5H_9N_3$
(51-45-6)
$C_5H_9N_3.2ClH$
(56-92-8)
$C_5H_9N_3O_2$
(1070-19-5)
$C_5H_9N_3O_9$
(3032-55-1)
$C_5H_9N_5$
(934-75-8)
(21320-64-9)
(30368-49-1)
(30368-50-4)
$C_5H_9N_5O$
(2827-44-3)
(7313-54-4)
$C_5H_9N_5O_2$
(13882-61-6)
$C_5H_9N_5O_5$
(14168-42-4)
$C_5H_9N_5S$
(30360-81-7)
C_5H_{10}
(109-67-1)
(109-68-2)
(287-92-3)
(513-35-9)
(563-45-1)
(563-46-2)
(598-61-8)
(627-20-3)
(646-04-8)
(25377-72-4)
(26760-64-5)
$C_5H_{10}BrClN_2O_2$
(16079-88-2)
$C_5H_{10}Br_2$
(111-24-0)
(5434-27-5)
(19398-53-9)
$C_5H_{10}Br_2O_2$
(3296-90-0)
$C_5H_{10}CaO_7$
(68568-63-8)
$C_5H_{10}ClNO$
(88-10-8)
$C_5H_{10}Cl_2$
(507-45-9)
(625-67-2)
(628-76-2)
(1674-33-5)
(17773-66-9)
(23010-07-3)
(30586-10-8)
$C_5H_{10}Cl_2O_2$
(111-91-1)
$C_5H_{10}HgO_3$
(151-38-2)
$C_5H_{10}NO_2S_2.K$
(23746-34-1)
$C_5H_{10}NO_7P$
(5994-61-6)
$C_5H_{10}NS_2.Na$
(148-18-5)
$C_5H_{10}NS_2.Na.3H_2O$
(20624-25-3)
$C_5H_{10}N_2$
(1738-25-6)
$C_5H_{10}N_2O$
(100-75-4)
$C_5H_{10}N_2O_2$
(3699-54-5)
(7119-94-0)
$C_5H_{10}N_2O_2S$
(16752-77-5)
(30558-43-1)
$C_5H_{10}N_2O_3$
(56-85-9)
(136-84-5)
(614-95-9)
$C_5H_{10}N_2O_4$
(7327-69-7)
$C_5H_{10}N_2O_5$
(1854-26-8)
$C_5H_{10}N_2S_2$
(533-74-4)
$C_5H_{10}N_6$
(1985-46-2)
(5606-23-5)
(30368-52-6)
$C_5H_{10}N_6O_2$
(101-25-7)
$C_5H_{10}O$
(96-17-3)
(96-22-0)
(96-41-3)
(96-47-9)
(107-87-9)
(110-62-3)
(115-18-4)
(142-68-7)
(556-82-1)
(557-31-3)
(563-80-4)
(590-86-3)
(616-25-1)
(625-31-0)
(630-19-3)
(763-32-6)
(821-09-0)
(926-65-8)
(928-55-2)
(1003-14-1)
(1569-50-2)
(1576-95-0)
(4675-87-0)
(5076-19-7)
(5614-38-0)
(6921-35-3)
(20273-24-9)
$C_5H_{10}OS_2$
(110-50-9)
(623-80-3)
(35200-02-3)
$C_5H_{10}OS_2.Na$
(25306-75-6)
$C_5H_{10}O_2$
(75-98-9)
(97-99-4)
(105-37-3)
(108-21-4)
(109-52-4)
(109-60-4)
(115-22-0)
(116-53-0)
(503-74-2)
(542-55-2)
(547-63-7)
(592-84-7)
(597-31-9)
(600-07-7)
(623-42-7)
(1120-97-4)
(1331-09-5)
(1663-35-0)
(2568-96-9)
(2806-85-1)
(3393-64-4)
(4016-11-9)
(4161-60-8)
(4221-03-8)
(7326-46-7)
(35915-22-1)
(36960-22-2)
(67755-97-9)
$C_5H_{10}O_2.Na$
(539-66-2)
(6106-41-8)
$C_5H_{10}O_2S$
(872-93-5)
(5466-06-8)
(7314-30-9)
$C_5H_{10}O_3$
(97-64-3)
(105-58-8)
(110-49-6)
(1331-11-9)
(1779-19-7)
(3852-09-3)
(4835-90-9)
(6342-56-9)
(13392-69-3)
(61996-25-6)
$C_5H_{10}O_3.Na$
(56974-57-3)
$C_5H_{10}O_3S$
(583-91-5)
$C_5H_{10}O_4$
(533-67-5)
(4767-03-7)
(16024-56-9)
(26446-35-5)
$C_5H_{10}O_5$
(50-69-1)
(58-86-6)
(87-72-9)
(147-81-9)
(21569-63-1)
$C_5H_{10}S$
(1613-51-0)
(1679-07-8)
(1795-09-1)
$C_5H_{11}AsO_2$
(64436-13-1)
$C_5H_{11}Br$
(107-81-3)
(107-82-4)
(110-53-2)
(630-17-1)
(1809-10-5)
(10422-35-2)
$C_5H_{11}BrO_3$
(19184-65-7)
$C_5H_{11}Cl$
(107-84-6)
(543-59-9)
(594-36-5)
(616-13-7)
(616-20-6)
(625-29-6)
(753-89-9)
$C_5H_{11}ClHgN_2O_2$
(62-37-3)
$C_5H_{11}Cl_2N$
(51-75-2)
$C_5H_{11}Cl_2N.ClH$
(55-86-7)
$C_5H_{11}Cl_2NO$
(126-85-2)
$C_5H_{11}Cl_2NO.ClH$
(302-70-5)
$C_5H_{11}Cl_3Si$
(107-72-2)
$C_5H_{11}F$
(592-50-7)
$C_5H_{11}I$
(594-38-7)
(628-17-1)
(637-97-8)
$C_5H_{11}N$
(110-89-4)
(120-94-5)
(765-38-8)
(1003-03-8)
$C_5H_{11}N.ClH$
(6091-44-7)
$C_5H_{11}NO$
(109-02-4)
(541-46-8)
(617-84-5)
(626-97-1)
(754-10-9)
(2425-74-3)
(5331-48-6)
(6859-99-0)
(51200-87-4)
$C_5H_{11}NOS$
(1646-75-9)
$C_5H_{11}NO_2$
(72-18-4)
(107-43-7)
(110-46-3)
(463-04-7)
(516-06-3)
(543-28-2)
(592-35-8)
(623-78-9)
(640-68-6)
(687-48-9)
(760-78-1)
(5417-42-5)
(18266-55-2)
(20073-50-1)
$C_5H_{11}NO_2S$

(52-67-5)
(59-51-8)
(63-68-3)
(348-67-4)
$C_5H_{11}NO_2Se$
(1464-42-2)
$C_5H_{11}NO_3$
(926-42-1)
(1002-16-0)
$C_5H_{11}NO_4$
(597-09-1)
$C_5H_{11}NO_4P.H_4N$
(77182-82-2)
$C_5H_{11}NS_2$
(147-84-2)
$C_5H_{11}NS_3.C_2H_2O_4$
(31895-22-4)
$C_5H_{11}N_2O_2P$
(77-81-6)
$C_5H_{11}N_3O$
(6281-42-1)
(30805-33-5)
$C_5H_{11}N_3O_2$
(869-01-2)
(35404-55-8)
(50285-71-7)
$C_5H_{11}N_3S$
(21306-32-1)
C_5H_{12}
(78-78-4)
(109-66-0)
(463-82-1)
(64771-72-8)
$C_5H_{12}ClN.ClH$
(4584-49-0)
$C_5H_{12}ClO_2P$
(24327-58-0)
$C_5H_{12}ClO_2PS_2$
(24934-91-6)
$C_5H_{12}N.HO$
(463-88-7)
$C_5H_{12}NO_2.Cl$
(590-46-5)
$C_5H_{12}NO_3PS_2$
(60-51-5)
$C_5H_{12}NO_4P$
(35597-44-5)
(51276-47-2)
$C_5H_{12}NO_4PS$
(1113-02-6)
$C_5H_{12}N_2$
(109-01-3)
(7422-92-6)
$C_5H_{12}N_2O$
(592-31-4)
(632-22-4)
(634-95-7)
(689-11-2)
(7068-83-9)
(16339-04-1)
(50816-31-4)
$C_5H_{12}N_2O_2$
(70-26-8)
(27076-30-8)
$C_5H_{12}N_2O_3$
(141-07-1)
$C_5H_{12}N_2O_3S$
(1982-67-8)
$C_5H_{12}N_2S$
(105-55-5)
(2782-91-4)
$C_5H_{12}N_4$
(30805-34-6)
$C_5H_{12}N_4O_3$
(543-38-4)
$C_5H_{12}N_4O_3.H_2O_4S$
(2219-31-0)
$C_5H_{12}N_4S$
(18801-52-0)
$C_5H_{12}O$
(71-41-0)
(75-84-3)
(75-85-4)
(123-51-3)
(137-32-6)
(584-02-1)
(598-75-4)
(625-44-5)
(625-54-7)
(628-28-4)
(628-32-0)
(1634-04-4)
(6032-29-7)
(6795-87-5)
$C_5H_{12}O_2$
(77-76-9)
(109-59-1)
(111-29-5)
(111-32-0)
(111-35-3)
(126-30-7)
(462-95-3)
(625-69-4)
(1569-02-4)
(2517-43-3)
(2807-30-9)
(3085-35-6)
(3425-61-4)
(4744-10-9)
(7778-85-0)
(10471-14-4)
(19089-47-5)
(53778-73-7)
$C_5H_{12}O_3$
(77-85-0)
(111-77-3)
(623-69-8)
(1445-45-0)
$C_5H_{12}O_3Si$
(2768-02-7)
$C_5H_{12}O_4$
(115-77-5)
$C_5H_{12}O_5$
(87-99-0)
(488-81-3)
$C_5H_{12}S$
(110-66-7)
(541-31-1)
(628-29-5)
(1679-09-0)
(1878-18-8)
(2084-18-6)
(4110-50-3)
(5145-99-3)
$C_5H_{12}S_2$
(6156-18-9)
(30453-31-7)
$C_5H_{13}ClN.Cl$
(999-81-5)
$C_5H_{13}N$
(96-15-1)
(107-85-7)
(110-58-7)
(110-68-9)
(616-24-0)
(616-39-7)
(625-30-9)
(625-43-4)
(63493-28-7)
$C_5H_{13}N.ClH$
(541-23-1)
$C_5H_{13}NO$
(109-56-8)
(16369-21-4)
$C_5H_{13}NO_2$
(105-59-9)
(115-70-8)
(122-07-6)
(52299-20-4)
$C_5H_{13}NO_3S.Na$
(25857-20-9)
$C_5H_{13}N_3$
(80-70-6)
$C_5H_{13}N_3S.2ClH$
(16111-27-6)
$C_5H_{13}O_2PS$
(2511-10-6)
(6996-81-2)
$C_5H_{13}O_2PS_2$
(3288-58-2)
$C_5H_{13}O_3P$
(683-08-9)
(4672-26-8)
(21921-96-0)
$C_5H_{13}O_3PS$
(2404-05-9)
$C_5H_{13}O_3PS_2$
(2587-90-8)
$C_5H_{13}O_3PS_2.C_5H_{13}O_3PS_2$
(8065-62-1)
$C_5H_{13}O_4P$
(2382-76-5)
(12789-46-7)
$C_5H_{13}O_{14}P_3$
(97-55-2)
C_5H_{14}
(68475-58-1)
$C_5H_{14}N.Cl$
(27697-51-4)
$C_5H_{14}NO$
(62-49-7)
$C_5H_{14}NO.CHO_3$
(78-73-9)
$C_5H_{14}NO.C_4H_5O_6$
(87-67-2)
$C_5H_{14}NO.Cl$
(67-48-1)
$C_5H_{14}NO.HO$
(123-41-1)
$C_5H_{14}NO.HO_3S$
(28427-24-9)
$C_5H_{14}N_2$
(51-80-9)
(109-55-7)
(462-94-2)
(7328-91-8)
$C_5H_{14}N_2O$
(108-16-7)
(10138-74-6)
$C_5H_{14}OSi$
(1825-62-3)
$C_5H_{14}Pb$
(1762-26-1)
$C_5H_{15}N_3$
(13531-52-7)
C_6BrF_5
(344-04-7)
$C_6Br_3Cl_3$
(13075-01-9)
C_6Br_6
(87-82-1)
$C_6Br_9N_3$
(24687-55-6)
C_6ClF_5
(344-07-0)
$C_6Cl_3N_3O_6$
(2631-68-7)
$C_6Cl_4N_2$
(17824-83-8)
$C_6Cl_4O_2$
(118-75-2)
(2435-53-2)
C_6Cl_5F
(319-87-9)
$C_6Cl_5NO_2$
(82-68-8)
$C_6Cl_5O.Na$
(131-52-2)
C_6Cl_6
(118-74-1)
$C_6Cl_6.C_6Cl_5NO_2$
(62180-90-9)
$C_6Cl_6N_4S$
(30863-25-3)
C_6Cl_7N
(1134-04-9)
C_6Cl_8O
(4024-81-1)
$C_6Cl_9N_3$
(6542-67-2)
C_6Cl_{14}
(83682-34-2)
C_6CrO_6
(13007-92-6)
C_6F_6
(392-56-3)
$C_6F_9N_3$
(368-66-1)
C_6F_{12}
(355-68-0)
(13429-24-8)
$C_6F_{12}O_2$
(2062-98-8)
C_6F_{14}
(355-42-0)
$C_6F_{14}O_2S$
(423-50-7)
$C_6FeN_6.Fe.H_4N$
(25869-00-5)
$C_6FeN_6.4/3Fe$
(14038-43-8)
$C_6FeN_6.10H_2O.4Na$
(14434-22-1)
$C_6FeN_6.4H_4N$
(14481-29-9)
$C_6FeN_6.4Na$
(13601-19-9)
$C_6FeO_{12}.3H_4N$
(14221-47-7)
C_6HBrCl_4
(90077-78-4)
C_6HBr_5
(608-90-2)
C_6HBr_5O
(608-71-9)
$C_6HCl_3N_2O_4$
(2678-21-9)
(6379-46-0)
(8003-46-1)
$C_6HCl_3N_2S$
(1982-55-4)
$C_6HCl_4NO_2$
(117-18-0)
(879-39-0)
(3714-62-3)
(10469-09-7)
(28804-67-3)
$C_6HCl_4NO_3$
(4824-72-0)
C_6HCl_5
(608-93-5)
C_6HCl_5O
(87-86-5)
$C_6HCl_5O.K$
(7778-73-6)
C_6HCl_5S
(133-49-3)
C_6HCl_6N
(1201-30-5)
C_6HCl_7Si
(33434-63-8)
$C_6HCl_8N_3$
(30362-31-3)
C_6HF_5
(363-72-4)
C_6HF_5O
(771-61-9)
C_6HF_5S
(771-62-0)
C_6HI_5

(608-96-8)
C_6H_2
(3161-99-7)
$C_6H_2Br_2ClNO$
(537-45-1)
$C_6H_2Br_4$
(634-89-9)
(636-28-2)
(22311-25-7)
$C_6H_2Br_4O$
(14400-94-3)
$C_6H_2Br_4O_2$
(488-47-1)
$C_6H_2ClF_3$
(75181-94-1)
$C_6H_2ClN_3O_6$
(88-88-0)
(28260-61-9)
$C_6H_2Cl_2NO_3.Na$
(64047-88-7)
$C_6H_2Cl_2O_4$
(87-88-7)
$C_6H_2Cl_3NO_2$
(89-69-0)
(17700-09-3)
(18708-70-8)
$C_6H_2Cl_3NO_3$
(82-62-2)
(20404-02-8)
$C_6H_2Cl_3N_2O_2.K$
(2545-60-0)
$C_6H_2Cl_3N_3$
(14143-60-3)
$C_6H_2Cl_3O.Na$
(136-32-3)
$C_6H_2Cl_4$
(95-94-3)
(634-66-2)
(634-90-2)
(12408-10-5)
$C_6H_2Cl_4O$
(58-90-2)
(58-90-2)
(935-95-5)
(4901-51-3)
(25167-83-3)
$C_6H_2Cl_4O.K$
(53535-27-6)
$C_6H_2Cl_4O.Na$
(25567-55-9)
$C_6H_2Cl_4O_2$
(87-87-6)
(1198-55-6)
(28520-00-5)
$C_6H_2Cl_4O_2S$
(15945-07-0)
(34732-09-7)
$C_6H_2Cl_5N$
(527-20-8)
(1128-16-1)
(1817-13-6)
(69045-83-6)
$C_6H_2Cl_6$
(55044-46-7)
$C_6H_2Cl_7N_3$
(30361-97-8)
$C_6H_2F_4$
(327-54-8)
(551-62-2)
(2367-82-0)
(28016-01-5)
$C_6H_2F_5N$
(771-60-8)
$C_6H_2FeN_6$
(13408-63-4)
$C_6H_2I_4$
(634-68-4)
(634-92-4)
(636-31-7)
$C_6H_2N_4O_5$
(4682-03-5)
$C_6H_2N_4O_6.K$
(29267-75-2)
$C_6H_2N_4O_9$
(641-16-7)
$C_6H_2N_6O_{10}$
(21985-87-5)
$C_6H_3BrClNO_3$
(58349-01-2)
$C_6H_3BrCl_2$
(1435-50-3)
(18282-59-2)
(19393-92-1)
(19752-55-7)
(56961-77-4)
$C_6H_3BrCl_2O$
(1940-42-7)
(3217-15-0)
(4524-77-0)
$C_6H_3BrN_2O_4$
(610-38-8)
(63460-06-0)
$C_6H_3Br_2ClO$
(4526-56-1)
(5324-13-0)
$C_6H_3Br_2F$
(1435-53-6)
$C_6H_3Br_2NO_3$
(99-28-5)
$C_6H_3Br_3$
(608-21-9)
(615-54-3)
(626-39-1)
(28779-08-0)
$C_6H_3Br_3O$
(118-79-6)
(25376-38-9)
$C_6H_3Br_3O_2$
(2747-17-3)
$C_6H_3Br_6N_3$
(30362-01-7)
(30362-02-8)
$C_6H_3Br_6N_3O$
(30863-63-9)
$C_6H_3ClFNO_2$
(350-30-1)
$C_6H_3ClN_2O_4$
(97-00-7)
(606-21-3)
(25567-67-3)
$C_6H_3ClN_2O_5$
(88-87-9)
(946-31-6)
$C_6H_3ClN_2O_7S.K$
(38185-06-7)
$C_6H_3ClN_3O_2$
(27165-22-6)
$C_6H_3ClN_3O_2.1/2Cl_4Zn$
(14263-89-9)
$C_6H_3ClO_4S$
(99165-93-2)
$C_6H_3Cl_2NO_2$
(89-61-2)
(99-54-7)
(601-88-7)
(611-06-3)
(618-62-2)
(1702-17-6)
(3209-22-1)
(27900-75-0)
$C_6H_3Cl_2NO_3$
(609-89-2)
(618-80-4)
(5847-57-4)
(39224-65-2)
$C_6H_3Cl_2NO_4S$
(97-08-5)
(4533-95-3)
(4533-96-4)
$C_6H_3Cl_3$
(87-61-6)
(108-70-3)
(120-82-1)
(12002-48-1)
$C_6H_3Cl_3N_2O_2$
(1918-02-1)
$C_6H_3Cl_3O$
(88-06-2)
(95-95-4)
(609-19-8)
(933-75-5)
(933-78-8)
(15950-66-0)
(25167-82-2)
$C_6H_3Cl_3O.1/2Zn$
(136-24-3)
$C_6H_3Cl_3O_2$
(608-94-6)
(32139-72-3)
(56961-20-7)
$C_6H_3Cl_3O_2S$
(98-31-7)
$C_6H_3Cl_3O_3$
(56961-21-8)
(94650-91-6)
$C_6H_3Cl_3S$
(3773-14-6)
$C_6H_3Cl_4N$
(634-83-3)
(1929-82-4)
(3481-20-7)
(69045-78-9)
$C_6H_3Cl_5Si$
(27137-85-5)
$C_6H_3Cl_6N_3O$
(30863-51-5)
$C_6H_3Cl_6N_3S$
(3599-76-6)
$C_6H_3Cl_6N_3$
(949-42-8)
(5311-21-7)
$C_6H_3FN_2O_4$
(70-34-8)
$C_6H_3F_3$
(372-38-3)
(1489-53-8)
$C_6H_3F_4N$
(700-17-4)
(5580-80-3)
$C_6H_3F_6N_3$
(30361-93-4)
$C_6H_3F_9$
(19430-93-4)
$C_6H_3I_2NO_3$
(305-85-1)
$C_6H_3I_3$
(608-29-7)
(626-44-8)
(61878-55-5)
$C_6H_3I_3O$
(609-23-4)
$C_6H_3N_3O_6$
(99-35-4)
$C_6H_3N_3O_7$
(88-89-1)
$C_6H_3N_3O_7.Ag$
(146-84-9)
$C_6H_3N_3O_7.xPb$
(25721-38-4)
$C_6H_3N_3O_8$
(82-71-3)
$C_6H_3N_3O_8.Mg$
(13255-27-1)
$C_6H_3N_3O_8.Pb$
(15245-44-0)
$C_6H_3N_3O_{10}Pb_2$
(12403-82-6)
$C_6H_3N_5O_8$
(3698-54-2)
(4591-46-2)
(53014-37-2)
C_6H_4Br
(2973-44-6)
C_6H_4BrCl
(106-39-8)
(108-37-2)
(694-80-4)
(28906-38-9)
C_6H_4BrClO
(2040-88-2)
(3964-56-5)
$C_6H_4BrClO_2S$
(98-58-8)
(2905-24-0)
$C_6H_4BrClN_2O_2$
(99-29-6)
$C_6H_4BrCl_2N$
(697-86-9)
(1940-29-0)
C_6H_4BrF
(460-00-4)
(1073-06-9)
C_6H_4BrI
(583-55-1)
(589-87-7)
(591-18-4)
$C_6H_4BrNO_3$
(5470-65-5)
(5847-59-6)
(7693-52-9)
$C_6H_4BrNO_2$
(577-19-5)
(585-79-5)
(586-78-7)
(61878-56-6)
$C_6H_4BrN_3O_4$
(1817-73-8)
(63460-09-3)
$C_6H_4Br_2$
(106-37-6)
(108-36-1)
(583-53-9)
(26249-12-7)
$C_6H_4Br_2N_2O_2$
(827-94-1)
$C_6H_4Br_2O$
(608-33-3)
(615-56-5)
(615-58-7)
(626-41-5)
(28165-52-8)
(28514-45-6)
(57383-80-9)
$C_6H_4Br_3N$
(147-82-0)
(52628-37-2)
$C_6H_4Br_6N_4$
(30339-37-8)
C_6H_4ClF
(348-51-6)
(352-33-0)
(625-98-9)
(55256-17-2)
$C_6H_4ClFO_2S$
(349-88-2)
C_6H_4ClI
(615-41-8)
(625-99-0)
(637-87-6)
$C_6H_4ClIO_2S$
(98-61-3)
(50702-38-0)
C_6H_4ClNO
(637-61-6)
(932-98-9)
$C_6H_4ClNO_2$
(88-73-3)
(100-00-5)
(121-73-3)
(4684-94-0)

(25167-93-5)

$C_6H_4ClNO_2S$
(7669-54-7)

$C_6H_4ClNO_3$
(89-64-5)
(491-11-2)
(603-86-1)
(611-07-4)
(619-08-9)

$C_6H_4ClNO_4S$
(98-74-8)
(121-51-7)
(1694-92-4)

$C_6H_4ClNO_5S$
(96-73-1)

$C_6H_4ClN_3$
(94-97-3)

$C_6H_4ClN_3O_4$
(3531-19-9)
(5388-62-5)
(10250-71-2)

$C_6H_4ClO_3S.Na$
(5138-90-9)

$C_6H_4Cl_2$
(95-50-1)
(106-46-7)
(541-73-1)
(25321-22-6)

$C_6H_4Cl_2N_2O_2$
(99-30-9)
(2683-43-4)
(6627-34-5)
(6641-64-1)
(66280-95-3)

$C_6H_4Cl_2N_2O_2.C_6Cl_5NO_2$
(37203-85-3)

$C_6H_4Cl_2N_2O_3$
(38116-59-5)

$C_6H_4Cl_2O$
(87-65-0)
(95-77-2)
(120-83-2)
(576-24-9)
(583-78-8)
(591-35-5)
(25167-81-1)

$C_6H_4Cl_2O.K$
(68938-81-8)

$C_6H_4Cl_2O.Na$
(3757-76-4)
(52166-72-0)

$C_6H_4Cl_2O_2$
(608-44-6)
(824-69-1)
(3428-24-8)
(3938-16-7)
(3978-67-4)
(13673-92-2)
(25167-85-5)

$C_6H_4Cl_2O_2S$
(98-60-2)
(2888-06-4)

$C_6H_4Cl_2O_3$
(94650-90-5)

$C_6H_4Cl_2S$
(24966-39-0)

$C_6H_4Cl_3N$
(634-67-3)
(634-91-3)
(634-93-5)
(636-30-6)
(18487-39-3)

$C_6H_4Cl_3NO$
(6358-15-2)

$C_6H_4Cl_4Si$
(26571-79-9)

$C_6H_4Cl_5N_3$
(15640-10-5)

$C_6H_4Cl_6N_4$
(24803-64-3)
(30339-50-5)

$C_6H_4Cl_6N_4S$
(30357-74-5)

$C_6H_4Cl_8$
(83682-30-8)

C_6H_4FI
(352-34-1)

$C_6H_4FNO_2$
(350-46-9)
(402-67-5)

$C_6H_4FNO_3$
(403-19-0)
(446-36-6)

$C_6H_4F_2$
(367-11-3)
(372-18-9)
(540-36-3)

$C_6H_4F_6N_4$
(29181-68-8)

$C_6H_4F_9I$
(2043-55-2)

$C_6H_4INO_2$
(636-98-6)
(645-00-1)

$C_6H_4I_2$
(615-42-9)
(624-38-4)
(626-00-6)

$C_6H_4NO_5S.Na$
(127-68-4)

$C_6H_4N_2$
(100-48-1)
(100-54-9)
(100-70-9)

$C_6H_4N_2O_2$
(105-12-4)
(480-96-6)
(7617-57-4)

$C_6H_4N_2O_4$
(99-65-0)
(100-25-4)
(528-29-0)
(25154-54-5)

$C_6H_4N_2O_5$
(51-28-5)
(66-56-8)
(329-71-5)
(573-56-8)
(577-71-9)
(586-11-8)
(25550-58-7)

$C_6H_4N_2O_6$
(519-44-8)

$C_6H_4N_3O_5.Na$
(831-52-7)

$C_6H_4N_4O_3S$
(712-68-5)

$C_6H_4N_4O_6$
(489-98-5)
(26952-42-1)

$C_6H_4N_6$
(2294-47-5)

$C_6H_4O_2$
(106-51-4)
(583-63-1)

$C_6H_4O_2.Na$
(54-86-4)

$C_6H_4O_4S$
(4282-31-9)

$C_6H_4O_7.2Cu$
(866-82-0)

$C_6H_4O_8S_2.2Na$
(149-45-1)

$C_6H_5AsCl_2$
(696-28-6)

$C_6H_5BHgO_3.2H$
(102-98-7)

C_6H_5Br
(108-86-1)

$C_6H_5BrN_2O_2$
(875-51-4)
(63460-07-1)

C_6H_5BrO
(95-56-7)
(106-41-2)
(591-20-8)
(32762-51-9)

$C_6H_5BrO_2$
(14381-51-2)

$C_6H_5Br_2N$
(615-57-6)
(63505-64-6)

$C_6H_5Br_4N_3$
(30362-00-6)

C_6H_5Cl
(108-90-7)

C_6H_5ClFN
(367-21-5)

C_6H_5ClHg
(100-56-1)

$C_6H_5ClHgO_3S$
(554-77-8)

C_6H_5ClMg
(100-59-4)

$C_6H_5ClN_2.H_2O_4S$
(6219-71-2)

$C_6H_5ClN_2O_2$
(89-63-4)
(121-87-9)
(635-22-3)
(769-11-9)
(825-41-2)
(1635-61-6)
(5344-44-5)
(6283-25-6)

$C_6H_5ClN_2O_3$
(6358-07-2)
(6358-08-3)
(6358-09-4)

$C_6H_5ClN_2O_3.ClH$
(62625-14-3)

$C_6H_5ClN_2O_4S$
(97-09-6)

C_6H_5ClO
(95-57-8)
(106-48-9)
(108-43-0)
(25167-80-0)

$C_6H_5ClO_2$
(95-88-5)
(615-67-8)
(2138-22-9)
(4018-65-9)

$C_6H_5ClO_2S$
(98-09-9)

$C_6H_5ClO_3S$
(98-66-8)
(56157-93-8)

C_6H_5ClS
(106-54-7)
(2037-31-2)

C_6H_5ClSe
(5707-04-0)

$C_6H_5Cl_2N$
(95-76-1)
(95-82-9)
(554-00-7)
(608-27-5)
(608-31-1)
(626-43-7)
(27134-27-6)

$C_6H_5Cl_2NO$
(527-62-8)

$C_6H_5Cl_2NO_3S$
(88-50-6)

$C_6H_5Cl_2N_3O$
(26650-76-0)

$C_6H_5Cl_2N_3S$
(25713-56-8)

$C_6H_5Cl_2OP$
(824-72-6)

$C_6H_5Cl_2O_2P$
(770-12-7)

$C_6H_5Cl_2P$
(644-97-3)

$C_6H_5Cl_2PS$
(3497-00-5)

$C_6H_5Cl_3N_2$
(5329-12-4)

$C_6H_5Cl_3N_4$
(30339-54-9)

$C_6H_5Cl_3Si$
(98-13-5)

$C_6H_5Cl_3Sn$
(1124-19-2)

$C_6H_5Cl_4N_3$
(5311-23-9)
(30361-94-5)

$C_6H_5Cl_4N_3S$
(5516-50-7)

$C_6H_5Cl_5$
(319-94-8)

C_6H_5F
(462-06-6)

$C_6H_5FN_2O_2$
(364-76-1)

C_6H_5FO
(367-12-4)
(371-41-5)
(372-20-3)

$C_6H_5F_2N$
(367-25-9)

$C_6H_5F_9O$
(2043-47-2)

$C_6H_5HgNO_3$
(55-68-5)

C_6H_5I
(591-50-4)

C_6H_5IO
(533-58-4)
(540-38-5)
(626-02-8)

C_6H_5NO
(500-22-1)
(586-96-9)
(872-85-5)
(1121-60-4)

$C_6H_5NO_2$
(55-22-1)
(59-67-6)
(98-95-3)
(98-98-6)
(104-91-6)
(637-62-7)
(32075-31-3)

$C_6H_5NO_2.Na$
(823-87-0)

$C_6H_5NO_2S$
(1849-36-1)
(4875-10-9)

$C_6H_5NO_3$
(88-75-5)
(100-02-7)
(554-84-7)
(874-24-8)
(2398-81-4)
(13602-12-5)
(25154-55-6)

$C_6H_5NO_3.Na$
(824-78-2)

$C_6H_5NO_4$
(99-11-6)
(601-89-8)
(3316-09-4)

$C_6H_5NO_4.xPb$
(51317-24-9)

$C_6H_5NO_5S$
(80-82-0)
(98-47-5)
(138-42-1)

(31212-28-9)
C_6H_5NS
(20893-30-5)
$C_6H_5N_2.Cl$
(100-34-5)
$C_6H_5N_2.F_6P$
(369-58-4)
$C_6H_5N_3$
(95-14-7)
(272-97-9)
$C_6H_5N_3.Ag$
(22257-44-9)
$C_6H_5N_3O_4$
(97-02-9)
(606-22-4)
(618-87-1)
(26471-56-7)
$C_6H_5N_3O_5$
(96-91-3)
$C_6H_5N_5O_2$
(529-69-1)
$C_6H_5N_5O_6$
(1630-08-6)
$C_6H_5O.Na$
(139-02-6)
$C_6H_5O_3S.Na$
(515-42-4)
$C_6H_5O_7$
(126-44-3)
$C_6H_5O_7.3K$
(866-84-2)
$C_6H_5O_7.3Na$
(68-04-2)
$C_6H_5O_7.3/2Zn$
(546-46-3)
C_6H_6
(71-43-2)
$C_6H_6AsNO_6$
(121-19-7)
$C_6H_6BrCl_3N_4$
(30339-52-7)
C_6H_6BrN
(106-40-1)
(591-19-5)
(615-36-1)
(55777-84-9)
$C_6H_6BrN.ClH$
(624-19-1)
C_6H_6BrNO
(84455-06-1)
$C_6H_6Br_2N_2$
(35691-65-7)
$C_6H_6Br_3Cl_3$
(30554-73-5)
$C_6H_6Br_3N_3$
(30361-99-0)
$C_6H_6Br_4Cl_2$
(30554-72-4)
$C_6H_6Br_5Cl$
(87-84-3)
$C_6H_6Br_6$
(1837-91-8)
$C_6H_6CdNO_6.H$
(49784-44-3)
C_6H_6ClN
(95-51-2)
(106-47-8)
(108-42-9)
(27134-26-5)
$C_6H_6ClN.ClH$
(137-04-2)
(141-85-5)
(6959-47-3)
(6959-48-4)
(20265-96-7)
C_6H_6ClNO
(84455-05-0)
$C_6H_6ClNO_2S$
(98-64-6)
$C_6H_6ClNO_2S.Na$
(127-52-6)
$C_6H_6ClNO_3S$
(88-43-7)
(98-36-2)
(133-74-4)
$C_6H_6ClNO_4S$
(88-23-3)
$C_6H_6ClN_3O_2$
(42389-30-0)
$C_6H_6Cl_2$
(69645-07-4)
(83682-37-5)
$C_6H_6Cl_2N_2$
(609-20-1)
(20103-09-7)
$C_6H_6Cl_2N_2O_2S$
(17901-16-5)
$C_6H_6Cl_2N_2O_4S_2$
(120-97-8)
$C_6H_6Cl_2N_4$
(30369-80-3)
$C_6H_6Cl_3N_3$
(15640-05-8)
$C_6H_6Cl_3N_3O$
(30863-45-7)
$C_6H_6Cl_3N_3O_2$
(5311-25-1)
$C_6H_6Cl_3N_3S_2$
(14946-18-0)
$C_6H_6Cl_4$
(1782-00-9)
$C_6H_6Cl_4F_8N_3O_2P_3$
(65601-40-3)
$C_6H_6Cl_4N_4$
(30339-47-0)
$C_6H_6Cl_6$
(58-89-9)
(319-84-6)
(319-85-7)
(319-86-8)
(608-73-1)
(1725-74-2)
(6108-10-7)
(6108-11-8)
(6108-12-9)
(6108-13-0)
(27154-44-5)
(83682-31-9)
$C_6H_6Cl_8O$
(127-90-2)
$C_6H_6Cl_9N_3$
(30805-22-2)
$C_6H_6CoNO_6.H$
(53108-50-2)
$C_6H_6CuNO_6.H$
(34831-02-2)
$C_6H_6CuNO_6.H_4N$
(71484-80-5)
C_6H_6FN
(348-54-9)
(371-40-4)
(372-19-0)
$C_6H_6FeNO_6$
(16448-54-7)
$C_6H_6HgNO_6.H$
(53113-61-4)
C_6H_6HgO
(100-57-2)
C_6H_6IN
(540-37-4)
(615-43-0)
(626-01-7)
$C_6H_6INO_2S$
(825-86-5)
$C_6H_6NNiO_6.H$
(34831-03-3)
$C_6H_6NO_6.3K$
(2399-85-1)
$C_6H_6NO_6.3Na$
(5064-31-3)
$C_6H_6NO_6.3Na.H_2O$
(18662-53-8)
$C_6H_6NO_6P$
(330-13-2)
$C_6H_6NO_6Pb.H$
(53113-59-0)
$C_6H_6NO_6Zn.H$
(53113-57-8)
$C_6H_6N_2$
(1119-85-3)
(13042-02-9)
$C_6H_6N_2O$
(98-92-0)
(696-54-8)
(873-69-8)
(1452-77-3)
(1453-82-3)
(22047-25-2)
(51892-16-1)
$C_6H_6N_2O_2$
(88-74-4)
(99-09-2)
(100-01-6)
(105-11-3)
(645-55-6)
(3167-49-5)
(116211-92-8)
$C_6H_6N_2O_2.ClH$
(15873-51-5)
$C_6H_6N_2O_2.H_4N$
(135-20-6)
$C_6H_6N_2O_2S$
(1615-06-1)
$C_6H_6N_2O_3$
(99-57-0)
(119-34-6)
(121-88-0)
(51037-30-0)
$C_6H_6N_2O_4S$
(6325-93-5)
$C_6H_6N_2O_5S$
(96-75-3)
$C_6H_6N_2O_5S.H_3N$
(4346-51-4)
$C_6H_6N_2O_6S$
(96-67-3)
(96-93-5)
$C_6H_6N_4$
(2004-03-7)
(7327-60-8)
$C_6H_6N_4O$
(2503-56-2)
$C_6H_6N_4O.Na$
(38299-08-0)
$C_6H_6N_4OS$
(28139-02-8)
$C_6H_6N_4O_2$
(552-62-5)
(1076-22-8)
(6136-37-4)
$C_6H_6N_4O_3$
(605-99-2)
$C_6H_6N_4O_3S$
(61-57-4)
$C_6H_6N_4O_4$
(59-87-0)
$C_6H_6N_4O_7$
(131-74-8)
$C_6H_6N_4S_2$
(58139-59-6)
$C_6H_6N_6O_3$
(23297-24-7)
$C_6H_6N_6O_6$
(3058-38-6)
C_6H_6O
(108-95-2)
(1487-18-9)
$C_6H_6O.1/3Al$
(15086-27-8)
$C_6H_6O.K$
(100-67-4)
C_6H_6OS
(88-15-3)
(1468-83-3)
$C_6H_6O_2$
(108-46-3)
(120-80-9)
(123-31-9)
(620-02-0)
(1192-62-7)
(4323-21-1)
$C_6H_6O_2.C_6H_6O_2$
(106-34-3)
$C_6H_6O_2S$
(618-41-7)
(1918-77-0)
(1918-79-2)
(5556-16-1)
(6964-21-2)
$C_6H_6O_2S.Na$
(873-55-2)
$C_6H_6O_2Se$
(6996-92-5)
$C_6H_6O_3$
(67-47-0)
(87-66-1)
(108-73-6)
(118-71-8)
(533-73-3)
(766-39-2)
(1334-76-5)
(4505-54-8)
$C_6H_6O_3S$
(98-11-3)
$C_6H_6O_4$
(505-70-4)
(762-42-5)
(1119-72-8)
(1124-48-7)
(3270-98-2)
$C_6H_6O_4S$
(98-67-9)
(609-46-1)
(1333-39-7)
$C_6H_6O_4S.Na$
(1300-51-2)
$C_6H_6O_5$
(24740-88-3)
$C_6H_6O_5S.K$
(21799-87-1)
$C_6H_6O_5S.Na$
(10021-55-3)
$C_6H_6O_6$
(499-12-7)
(585-84-2)
$C_6H_6O_6S_2$
(98-48-6)
$C_6H_6O_7.2Na$
(144-33-2)
$C_6H_6O_8S_2.2K$
(15763-57-2)
C_6H_6S
(108-98-5)
$C_6H_6S.Na$
(930-69-8)
$C_6H_7AsNO_3.Na$
(127-85-5)
$C_6H_7AsO_3$
(98-05-5)
$C_6H_7Br_2N_3O_2$
(30863-44-6)
$C_6H_7Br_3N_2$
(31250-78-9)
C_6H_7Cl
(83682-35-3)
$C_6H_7ClN_2$
(95-83-0)
(615-66-7)
(5131-60-2)
$C_6H_7ClN_2.2ClH$

(615-46-3)
$C_6H_7ClN_2.H_2O_4S$
(61702-44-1)
(68239-80-5)
(68459-98-3)
$C_6H_7ClN_6O$
(2609-46-3)
$C_6H_7Cl_2N_3$
(30894-73-6)
(30894-74-7)
$C_6H_7Cl_2N_3O$
(6684-27-1)
(26650-75-9)
$C_6H_7Cl_2N_3O_2$
(5311-24-0)
$C_6H_7Cl_2N_3S$
(30894-58-7)
(30894-60-1)
(33032-17-6)
$C_6H_7Cl_2N_5O$
(5097-52-9)
$C_6H_7Cl_3N_4$
(24830-33-9)
(30339-40-3)
$C_6H_7Cl_3N_4S$
(30369-49-4)
(30377-26-5)
$C_6H_7F_3O_2$
(352-87-4)
$C_6H_7F_3O_3$
(372-31-6)
C_6H_7N
(62-53-3)
(108-89-4)
(108-99-6)
(109-06-8)
(1333-41-1)
$C_6H_7N.BrH$
(542-11-0)
$C_6H_7N.ClH$
(142-04-1)
$C_6H_7N.1/2H_2O_4S$
(542-16-5)
C_6H_7NO
(95-55-6)
(100-55-0)
(100-65-2)
(123-30-8)
(586-95-8)
(586-98-1)
(591-27-5)
(931-19-1)
(1003-67-4)
(1003-73-2)
(1072-83-9)
(1121-78-4)
(1192-58-1)
(1192-79-6)
(7295-76-3)
$C_6H_7NO.ClH$
(51-19-4)
(51-78-5)
$C_6H_7NO.1/2H_2O_4S$
(63084-98-0)
(67845-79-8)
$C_6H_7NOS_2$
(7725-93-1)
$C_6H_7NO_2$
(106-71-8)
(7085-85-0)
(17825-86-4)
$C_6H_7NO_2S$
(98-10-2)
$C_6H_7NO_3S$
(88-21-1)
(121-47-1)
(121-57-3)
$C_6H_7NO_4S$
(98-37-3)
$C_6H_7NO_6.2Na$
(15467-20-6)
C_6H_7NS
(137-07-5)
(1193-02-8)
(1759-28-0)
(22948-02-3)
$C_6H_7N_2$
(4553-62-2)
$C_6H_7N_3O$
(54-85-3)
$C_6H_7N_3O_2$
(99-56-9)
(5042-55-7)
(5131-58-8)
(5307-14-2)
$C_6H_7N_3O_2.2ClH$
(6219-77-8)
(18266-52-9)
$C_6H_7N_3O_2.H_2O_4S$
(68239-82-7)
(68239-83-8)
$C_6H_7N_5O$
(10030-78-1)
$C_6H_7N_5O_2$
(4803-06-9)
(30345-27-8)
(30345-28-9)
$C_6H_7N_5S_2$
(30863-65-1)
$C_6H_7O_2.K$
(590-00-1)
(24634-61-5)
$C_6H_7O_2.Na$
(7757-81-5)
$C_6H_7O_2P$
(1779-48-2)
$C_6H_7O_2P.Na$
(4297-95-4)
$C_6H_7O_3P$
(1571-33-1)
$C_6H_7O_4P$
(701-64-4)
$C_6H_7O_7.Na$
(18996-35-5)
C_6H_7P
(638-21-1)
C_6H_8
(592-57-4)
(628-41-1)
(2235-12-3)
(26519-91-5)
(29797-09-9)
$C_6H_8AsNO_3$
(98-50-0)
$C_6H_8BrN_3$
(30361-98-9)
$C_6H_8BrN_3O_2$
(30863-43-5)
$C_6H_8Br_2O_2$
(19660-16-3)
$C_6H_8Br_3N_5$
(26235-91-6)
(30359-70-7)
C_6H_8ClNS
(533-45-9)
$C_6H_8ClN_3$
(15640-03-6)
(30361-88-7)
$C_6H_8ClN_3O_2$
(30863-39-9)
$C_6H_8ClN_5$
(30355-01-2)
$C_6H_8ClN_7O.ClH$
(2016-88-8)
$C_6H_8Cl_2$
(55667-43-1)
(67546-51-4)
(83682-33-1)
$C_6H_8Cl_2N_4$
(3703-10-4)
(25354-39-6)
$C_6H_8Cl_2O_2$
(111-50-2)
$C_6H_8Cl_2O_5$
(106-75-2)
$C_6H_8Cl_3N_5$
(26234-96-8)
(27470-98-0)
(30360-01-1)
$C_6H_8Cl_6$
(18585-38-1)
(83682-28-4)
(83682-29-5)
$C_6H_8F_3N_5$
(30355-58-9)
$C_6H_8N.Cl$
(7680-73-1)
$C_6H_8N_2$
(95-54-5)
(100-63-0)
(106-50-3)
(108-45-2)
(108-50-9)
(111-69-3)
(123-32-0)
(1558-17-4)
(1603-41-4)
(1824-81-3)
(5910-89-4)
(13925-00-3)
(17611-82-4)
(25265-76-3)
(31410-01-2)
$C_6H_8N_2.C_2H_2O_4$
(62654-17-5)
$C_6H_8N_2.ClH$
(59-88-1)
(540-24-9)
$C_6H_8N_2.2ClH$
(541-69-5)
(615-28-1)
(624-18-0)
$C_6H_8N_2.H_2O_4S$
(541-70-8)
(16245-77-5)
$C_6H_8N_2O$
(95-86-3)
(1656-48-0)
(2882-21-5)
$C_6H_8N_2O.2ClH$
(137-09-7)
$C_6H_8N_2O_2S$
(63-74-1)
(80-17-1)
(98-18-0)
$C_6H_8N_2O_3$
(2518-72-1)
(55557-01-2)
$C_6H_8N_2O_3S$
(88-63-1)
(98-32-8)
$C_6H_8N_2O_6$
(17977-09-2)
$C_6H_8N_2O_6S_2$
(137-50-8)
$C_6H_8N_2O_8$
(87-33-2)
$C_6H_8N_2S$
(111-97-7)
$C_6H_8N_4O_3S$
(139-94-6)
$C_6H_8N_6$
(4784-14-9)
$C_6H_8N_6O_{18}$
(15825-70-4)
C_6H_8O
(142-83-6)
(625-86-5)
(930-68-7)
(1120-73-6)
(2758-18-1)
(3208-16-0)
(3710-43-8)
(20521-42-0)
(28802-49-5)
C_6H_8OS
(1438-91-1)
$C_6H_8O_2$
(80-71-7)
(100-73-2)
(110-44-1)
(504-02-9)
(637-88-7)
(765-69-5)
(765-87-7)
(820-69-9)
(999-55-3)
(3234-54-6)
(4436-75-3)
(10048-32-5)
(14861-06-4)
(17559-81-8)
(20019-64-1)
$C_6H_8O_2.1/2Ca$
(7492-55-9)
$C_6H_8O_3$
(106-90-1)
(517-23-7)
(3658-77-3)
$C_6H_8O_4$
(488-21-1)
(624-48-6)
(3990-03-2)
(24615-84-7)
$C_6H_8O_4.2Na$
(7486-38-6)
$C_6H_8O_5$
(689-31-6)
(2044-00-0)
$C_6H_8O_6$
(50-81-7)
(89-65-6)
$C_6H_8O_6.1/2Ca$
(5743-27-1)
$C_6H_8O_6.Na$
(134-03-2)
(6381-77-7)
$C_6H_8O_7$
(77-92-9)
(38945-27-6)
$C_6H_8O_7.3/2Ca$
(813-94-5)
$C_6H_8O_7.3/2Co$
(866-81-9)
$C_6H_8O_7.xFe$
(23383-11-1)
$C_6H_8O_7.xFe.xH_4N$
(1185-57-5)
$C_6H_8O_7.H_2O.3Na$
(19287-96-8)
$C_6H_8O_7.2H_3N$
(3012-65-5)
$C_6H_8O_7.xH_3N$
(7632-50-0)
$C_6H_8O_7.Mn$
(5968-88-7)
$C_6H_8O_7.3Na$
(34128-01-3)
$C_6H_8O_7.xNa$
(994-36-5)
$C_6H_8O_7.3/2Pb$
(512-26-5)
C_6H_8S
(638-02-8)
(872-55-9)
(28632-15-7)
(52006-63-0)
$C_6H_9Br_2N_5$
(30359-67-2)
C_6H_9Cl

(930-65-4)
(930-66-5)
(2441-97-6)
$C_6H_9ClN_2O$
(29849-01-2)
$C_6H_9ClN_4$
(30369-24-5)
(30369-30-3)
$C_6H_9ClN_4O$
(4653-94-5)
(13882-55-8)
$C_6H_9ClN_4S$
(30357-95-0)
$C_6H_9ClO_3$
(609-15-4)
$C_6H_9Cl_2N_5$
(30355-64-7)
$C_6H_9Cl_3$
(99308-22-2)
$C_6H_9Cl_3N_6$
(30355-70-5)
$C_6H_9Cl_3Si$
(10137-69-6)
$C_6H_9D_3$
(69432-95-7)
$C_6H_9FeNO_6$
(68391-67-3)
C_6H_9N
(625-84-3)
(31551-28-7)
(49813-61-8)
C_6H_9NO
(88-12-0)
(10471-78-0)
(20662-84-4)
C_6H_9NOS
(137-00-8)
$C_6H_9NO_2$
(2314-78-5)
$C_6H_9NO_2S$
(1195-16-0)
$C_6H_9NO_6$
(139-13-9)
$C_6H_9NO_6$.Ca.Na
(60034-45-9)
$C_6H_9NO_6.H_2O$.2Na
(23255-03-0)
$C_6H_9NO_6$.xNa
(10042-84-9)
C_6H_9NS
(15679-12-6)
(41981-60-6)
(52414-82-1)
(52414-91-2)
$C_6H_9N_2O_2P$
(7450-69-3)
$C_6H_9N_3$
(111-94-4)
(767-15-7)
(823-94-9)
(30361-87-6)
$C_6H_9N_3O$
(3599-66-4)
$C_6H_9N_3O_2$
(71-00-1)
(3289-50-7)
(4000-78-6)
(4998-57-6)
(30885-97-3)
$C_6H_9N_3O_3$
(443-48-1)
(827-16-7)
(877-89-4)
(35687-44-6)
$C_6H_9N_3O_4$
(13551-92-3)
$C_6H_9N_3S$
(3745-18-4)
$C_6H_9N_3S_2$
(5336-94-7)
$C_6H_9N_3S_3$
(5759-58-0)
$C_6H_9N_5O$
(6291-87-8)
(30354-98-4)
(30355-53-4)
$C_6H_9N_5S$
(30360-84-0)
$C_6H_9N_7O$
(4587-03-5)
$C_6H_9N_7S$
(4658-25-7)
$C_6H_9NaO_6$
(7378-23-6)
$C_6H_9O_6$.Ce
(537-00-8)
$C_6H_9O_6$.Cr
(1066-30-4)
C_6H_{10}
(110-83-8)
(513-81-5)
(592-42-7)
(592-45-0)
(592-46-1)
(592-48-3)
(693-02-7)
(693-89-0)
(763-30-4)
(917-92-0)
(926-56-7)
(1115-08-8)
(1118-58-7)
(1120-62-3)
(1759-81-5)
(4549-74-0)
(5194-50-3)
(5194-51-4)
(20237-34-7)
(26856-30-4)
(42296-74-2)
$C_6H_{10}BrCl$
(51422-75-4)
$C_6H_{10}BrMgNO$
(17091-31-5)
$C_6H_{10}BrN_5$
(30377-16-3)
$C_6H_{10}Br_2$
(35076-92-7)
$C_6H_{10}Br_2O_2$
(13137-43-4)
(35243-89-1)
$C_6H_{10}ClNO_2$
(5810-11-7)
$C_6H_{10}ClN_5$
(3084-92-2)
(6190-65-4)
(30354-65-5)
(30355-59-0)
$C_6H_{10}Cl_2$
(822-86-6)
(1121-21-7)
(2108-92-1)
(24955-63-3)
$C_6H_{10}Cl_2O$
(2648-59-1)
(2648-60-4)
(22591-21-5)
$C_6H_{10}Cl_2Si$
(3651-23-8)
$C_6H_{10}Cl_4O$
(63283-80-7)
(99342-08-2)
$C_6H_{10}D_2O_6$
(18991-62-3)
$C_6H_{10}N_2$
(931-36-2)
(1072-91-9)
$C_6H_{10}N_2O$
(1632-26-4)
(5807-02-3)
(16338-97-9)
(37788-55-9)
$C_6H_{10}N_2O_2$
(5394-36-5)
$C_6H_{10}N_2O_2S$
(13253-44-6)
$C_6H_{10}N_2O_2S_2$
(3655-88-7)
$C_6H_{10}N_2O_3$
(116-25-6)
$C_6H_{10}N_4$
(54-95-5)
(30360-32-8)
(30360-48-6)
$C_6H_{10}N_4O$
(30369-64-3)
$C_6H_{10}N_4O_2$
(30357-98-3)
$C_6H_{10}N_4O_{13}$
(20600-96-8)
$C_6H_{10}N_4S_2$
(30362-11-9)
$C_6H_{10}N_6O$
(4342-03-4)
$C_6H_{10}O$
(77-75-8)
(105-31-7)
(108-94-1)
(109-49-9)
(109-50-2)
(111-28-4)
(141-79-7)
(279-49-2)
(286-20-4)
(505-57-7)
(557-40-4)
(623-36-9)
(758-87-2)
(763-93-9)
(822-67-3)
(1120-72-5)
(1335-39-3)
(1629-60-3)
(1757-42-2)
(1777-33-9)
(2497-21-4)
(3744-02-3)
(5362-50-5)
(5362-56-1)
(6728-26-3)
(6789-80-6)
(16015-11-5)
(16635-54-4)
(25044-01-3)
(25659-22-7)
(69112-21-6)
$C_6H_{10}O_2$
(97-63-2)
(106-92-3)
(110-13-4)
(123-20-6)
(502-44-3)
(533-60-8)
(623-70-1)
(689-12-3)
(695-06-7)
(823-22-3)
(924-50-5)
(925-60-0)
(1191-04-4)
(1577-22-6)
(1888-89-7)
(3031-66-1)
(3123-97-5)
(3142-72-1)
(3400-45-1)
(3588-31-6)
(4454-05-1)
(4478-63-1)
(4528-26-1)
(5145-01-7)
(6203-88-9)
(6622-76-0)
(7493-58-5)
(10544-63-5)
(13861-97-7)
(25465-18-3)
(41653-93-4)
$C_6H_{10}O_2S_4$
(502-55-6)
$C_6H_{10}O_3$
(123-62-6)
(141-97-9)
(332-77-4)
(815-17-8)
(816-66-0)
(868-77-9)
(999-61-1)
(1460-34-0)
(2238-07-5)
(3121-61-7)
(3128-06-1)
(4433-85-6)
(25584-83-2)
(58973-18-5)
$C_6H_{10}O_3S_2$
(3278-35-1)
$C_6H_{10}O_4$
(95-92-1)
(106-65-0)
(111-55-7)
(124-04-9)
(542-10-9)
(597-43-3)
(616-62-6)
(617-62-9)
(626-51-7)
(652-67-5)
(1501-27-5)
(3248-28-0)
(6705-89-1)
(13405-83-9)
$C_6H_{10}O_4$.Ca
(4075-81-4)
$C_6H_{10}O_4$.xNa
(23311-84-4)
$C_6H_{10}O_4S$
(111-17-1)
$C_6H_{10}O_4S_2$
(123-81-9)
(55290-64-7)
$C_6H_{10}O_5S$
(10595-80-9)
$C_6H_{10}O_5S$.Na
(1804-87-1)
$C_6H_{10}O_5S_2$
(26750-50-5)
$C_6H_{10}O_6$
(90-80-2)
(608-68-4)
(5057-96-5)
(14666-78-5)
$C_6H_{10}O_6$.Ca
(814-80-2)
$C_6H_{10}O_6$.Fe
(5905-52-2)
$C_6H_{10}O_6$.Sr
(29870-99-3)
$C_6H_{10}O_7$
(669-90-9)
$C_6H_{10}O_8$
(526-99-8)
$C_6H_{10}S$
(286-28-2)
(592-88-1)
$C_6H_{10}S_2$
(2179-57-9)
$C_6H_{11}Br$
(108-85-0)
$C_6H_{11}BrO$

(10226-29-6)
(32388-22-0)
$C_6H_{11}BrO_2$
(600-00-0)
(616-05-7)
(4224-70-8)
$C_6H_{11}CaNO_6$
(2399-94-2)
$C_6H_{11}Cl$
(542-18-7)
$C_6H_{11}ClMg$
(931-51-1)
$C_6H_{11}ClNO_2P$
(21310-38-3)
$C_6H_{11}ClO$
(1561-86-0)
(13547-70-1)
$C_6H_{11}ClO_2$
(590-02-3)
(3153-36-4)
$C_6H_{11}ClS$
(17797-03-4)
$C_6H_{11}Cl_2O_4P$
(72-00-4)
$C_6H_{11}Cl_3O_3$
(38571-73-2)
$C_6H_{11}Cl_3Si$
(98-12-4)
$C_6H_{11}I$
(626-62-0)
$C_6H_{11}IO_3$
(5634-39-9)
$C_6H_{11}N$
(124-02-7)
(542-54-1)
(628-73-9)
$C_6H_{11}NO$
(100-64-1)
(105-60-2)
(931-20-4)
(2210-25-5)
(2591-86-8)
(2687-91-4)
(22694-96-8)
$C_6H_{11}NO_2$
(52-52-8)
(535-75-1)
(1122-60-7)
(1696-20-4)
(2044-64-6)
(3445-11-2)
(6065-17-4)
(22534-71-0)
(35633-50-2)
$C_6H_{11}NO_3$
(6542-37-6)
$C_6H_{11}NO_4$
(505-47-5)
$C_6H_{11}NO_5$
(74754-55-5)
$C_6H_{11}NO_5.2Na$
(135-37-5)
$C_6H_{11}NS_2.C_5H_{11}N$
(98-77-1)
$C_6H_{11}N_2O_3PS_4$
(13432-51-4)
$C_6H_{11}N_2O_4PS_3$
(950-37-8)
$C_6H_{11}N_3$
(6086-22-2)
(16227-10-4)
$C_6H_{11}N_3O_2$
(25113-45-5)
$C_6H_{11}N_3O_4$
(556-33-2)
$C_6H_{11}N_5$
(1973-06-4)
(5397-04-6)
(5962-23-2)
(30368-51-5)
$C_6H_{11}N_5O$
(6295-15-4)
(24860-40-0)
(30360-56-6)
(30360-57-7)
$C_6H_{11}N_7O_7$
(13980-00-2)
$C_6H_{11}NaO_7$
(14906-97-9)
$C_6H_{11}OS_2.K$
(2720-73-2)
$C_6H_{11}O_2PS_2$
(5851-14-9)
$C_6H_{11}O_3P$
(824-11-3)
$C_6H_{11}O_4P$
(1005-93-2)
$C_6H_{11}O_5S.Au$
(12192-57-3)
$C_6H_{11}O_6PS_2$
(1190-28-9)
$C_6H_{11}O_7.Hg$
(63937-14-4)
$C_6H_{11}O_7.K$
(299-27-4)
$C_6H_{11}O_7.Na$
(527-07-1)
C_6H_{12}
(96-37-7)
(110-82-7)
(558-37-2)
(563-78-0)
(563-79-1)
(592-41-6)
(592-43-8)
(592-47-2)
(616-12-6)
(625-27-4)
(674-76-0)
(691-37-2)
(691-38-3)
(760-20-3)
(760-21-4)
(763-29-1)
(922-61-2)
(922-62-3)
(2415-72-7)
(4050-45-7)
(4461-48-7)
(7642-09-3)
(7688-21-3)
(13269-52-8)
(25264-93-1)
(27236-46-0)
(37275-41-5)
$C_6H_{12}Br_2O_4$
(488-41-5)
(10318-26-0)
$C_6H_{12}ClNO$
(2315-36-8)
$C_6H_{12}ClO_4P$
(311-47-7)
$C_6H_{12}Cl_2$
(2162-92-7)
(2163-00-0)
(13275-18-8)
$C_6H_{12}Cl_2O$
(108-60-1)
(629-36-7)
(39638-32-9)
$C_6H_{12}Cl_2O_2$
(112-26-5)
(619-33-0)
$C_6H_{12}Cl_2O_4S_2$
(3944-87-4)
$C_6H_{12}Cl_2S_2$
(3563-36-8)
$C_6H_{12}Cl_3N$
(555-77-1)
$C_6H_{12}Cl_3N.ClH$
(817-09-4)
$C_6H_{12}Cl_3O_3P$
(140-08-9)
(6294-34-4)
$C_6H_{12}Cl_3O_4P$
(115-96-8)
$C_6H_{12}FeN_2S_4$
(15339-38-5)
$C_6H_{12}NO_3PS_2$
(21548-32-3)
$C_6H_{12}NO_3S.Na$
(139-05-9)
$C_6H_{12}NO_4PS_2$
(2540-82-1)
$C_6H_{12}NO_5$
(93-62-9)
$C_6H_{12}N_2$
(280-57-9)
(931-35-1)
(2432-74-8)
(13893-53-3)
$C_6H_{12}N_2O$
(932-83-2)
(13603-07-1)
$C_6H_{12}N_2O_2$
(628-94-4)
(1456-28-6)
(10520-38-4)
(13534-15-1)
$C_6H_{12}N_2O_3$
(1596-84-5)
(66017-91-2)
$C_6H_{12}N_2O_4S_2$
(56-89-3)
$C_6H_{12}N_2O_5$
(5952-26-1)
$C_6H_{12}N_2O_6$
(283-66-9)
$C_6H_{12}N_2S_3$
(97-74-5)
$C_6H_{12}N_2S_4$
(137-26-8)
$C_6H_{12}N_2S_4.Cu$
(137-29-1)
$C_6H_{12}N_2S_4.Ni$
(15521-65-0)
$C_6H_{12}N_2S_4.Pb$
(19010-66-3)
$C_6H_{12}N_2S_4.Zn$
(137-30-4)
$C_6H_{12}N_2S_6$
(97-91-6)
$C_6H_{12}N_3OP$
(545-55-1)
$C_6H_{12}N_3PS$
(52-24-4)
$C_6H_{12}N_4$
(100-97-0)
$C_6H_{12}N_5O_2PS_2$
(78-57-9)
$C_6H_{12}N_6$
(2827-46-5)
(16274-81-0)
$C_6H_{12}N_6O_3$
(1017-56-7)
$C_6H_{12}O$
(66-25-1)
(75-97-8)
(97-96-1)
(108-10-1)
(108-93-0)
(109-53-5)
(111-34-2)
(123-15-9)
(544-12-7)
(565-61-7)
(565-69-5)
(589-38-8)
(591-78-6)
(592-90-5)
(928-95-0)
(928-96-1)
(928-97-2)
(1003-38-9)
(1119-16-0)
(1192-22-9)
(1471-03-0)
(2088-07-5)
(2144-41-4)
(2305-21-7)
(4798-44-1)
(4798-45-2)
(4798-58-7)
(5076-20-0)
(6124-90-9)
(15877-57-3)
(24070-77-7)
(25144-04-1)
(25144-05-2)
(35301-43-0)
(53778-61-3)
(53778-62-4)
(54774-27-5)
(69432-97-9)
(69432-98-0)
$C_6H_{12}O_2$
(88-09-5)
(97-61-0)
(97-62-1)
(100-72-1)
(105-43-1)
(105-46-4)
(105-54-4)
(106-36-5)
(110-19-0)
(110-45-2)
(123-42-2)
(123-86-4)
(126-39-6)
(142-62-1)
(504-01-8)
(540-88-5)
(556-24-1)
(556-48-9)
(595-37-9)
(598-98-1)
(624-24-8)
(637-78-5)
(638-49-3)
(646-07-1)
(766-15-4)
(822-83-3)
(868-57-5)
(931-17-9)
(1070-83-3)
(1331-17-5)
(3126-95-2)
(4016-14-2)
(7319-23-5)
(14287-61-7)
(25136-55-4)
(27583-37-5)
(36687-98-6)
(54774-28-6)
(61142-77-6)
(68603-84-9)
$C_6H_{12}O_2.Na$
(10051-44-2)
$C_6H_{12}O_2Sn$
(22771-18-2)
$C_6H_{12}O_3$
(100-79-8)
(107-71-1)
(108-65-6)
(111-15-9)
(123-34-2)
(123-63-7)
(617-51-6)
(1191-25-9)
(4026-20-4)

(7397-62-8)
(25136-53-2)
(70657-70-4)
(84540-57-8)
$C_6H_{12}O_4$
(624-47-5)
(17640-28-7)
(32555-29-6)
$C_6H_{12}O_4.N_2O_4$
(111-22-8)
$C_6H_{12}O_5$
(154-17-6)
(2438-80-4)
(3615-41-6)
(12441-09-7)
$C_6H_{12}O_6$
(50-99-7)
(57-48-7)
(59-23-4)
(87-79-6)
(87-89-8)
(498-43-1)
(1518-54-3)
(1518-56-5)
(1518-59-8)
(3458-28-4)
(7660-25-5)
(18521-63-6)
$C_6H_{12}O_6Sb_2$
(29736-75-2)
$C_6H_{12}O_7$
(526-95-4)
$C_6H_{12}O_7.xK$
(35087-77-5)
$C_6H_{12}O_7.1/2Mn$
(6485-39-8)
$C_6H_{12}O_7.1/2Sr$
(10101-21-0)
$C_6H_{12}S$
(1551-31-1)
(1569-69-3)
(5161-17-1)
(55569-78-3)
(73180-15-1)
$C_6H_{12}S_2$
(2179-59-1)
$C_6H_{12}S_3$
(2765-04-0)
$C_6H_{13}Br$
(111-25-1)
(3377-86-4)
(3377-87-5)
(51116-73-5)
$C_6H_{13}Cl$
(544-10-5)
(638-28-8)
(25495-90-3)
(68920-70-7)
$C_6H_{13}ClO_2$
(621-62-5)
$C_6H_{13}Cl_2N$
(538-07-8)
$C_6H_{13}Cl_2O_2P$
(14590-60-4)
(24767-66-6)
$C_6H_{13}Cl_3Si$
(928-65-4)
$C_6H_{13}F$
(373-14-8)
$C_6H_{13}I$
(638-45-9)
(25495-92-5)
$C_6H_{13}I_2O_2P$
(17052-18-5)
$C_6H_{13}N$
(108-91-8)
(109-05-7)
(111-49-9)
(626-67-5)
(1120-85-0)
(18328-90-0)
$C_6H_{13}NO$
(100-74-3)
(141-91-3)
(628-02-4)
(685-91-6)
(762-84-5)
(1119-49-9)
(6225-10-1)
(6850-38-0)
(75673-43-7)
$C_6H_{13}NOS$
(141-98-0)
$C_6H_{13}NO_2$
(60-32-2)
(61-90-5)
(73-32-5)
(319-78-8)
(327-57-1)
(328-38-1)
(328-39-2)
(543-86-2)
(616-06-8)
(622-40-2)
(646-14-0)
(39076-02-3)
(58695-42-4)
$C_6H_{13}NO_2S$
(67-21-0)
(535-32-0)
(13073-35-3)
$C_6H_{13}NO_3$
(20633-11-8)
(67953-32-6)
$C_6H_{13}NO_3S$
(100-88-9)
$C_6H_{13}NO_3S.K$
(7758-04-5)
$C_6H_{13}NO_3S_3.Na$
(18880-36-9)
$C_6H_{13}NO_4$
(4987-75-1)
$C_6H_{13}NO_4.Na$
(139-41-3)
$C_6H_{13}NO_5$
(3416-24-8)
(7535-00-4)
(10143-22-3)
$C_6H_{13}NS_2$
(638-17-5)
$C_6H_{13}N_3O$
(30913-44-1)
$C_6H_{13}N_3O_2$
(50285-72-8)
$C_6H_{13}O_5P$
(3095-95-2)
$C_6H_{13}O_5P.K$
(34170-84-8)
$C_6H_{13}O_5PS$
(2088-72-4)
$C_6H_{13}O_9P$
(56-73-5)
(59-56-3)
(573-35-3)
$C_6H_{13}P$
(822-68-4)
C_6H_{14}
(75-83-2)
(79-29-8)
(96-14-0)
(107-83-5)
(110-54-3)
$C_6H_{14}ClN$
(24948-81-0)
$C_6H_{14}ClN.ClH$
(869-24-9)
$C_6H_{14}ClO_2PS$
(2524-05-2)
(2524-06-3)
$C_6H_{14}FO_3P$
(55-91-4)
$C_6H_{14}N.Cl$
(1516-27-4)
$C_6H_{14}NO.Cl$
(3033-77-0)
$C_6H_{14}NO_2Se$
(7728-97-4)
$C_6H_{14}NO_3PS_2$
(116-01-8)
(13265-60-6)
$C_6H_{14}NO_4P$
(597-25-1)
$C_6H_{14}NO_5P$
(20120-33-6)
$C_6H_{14}N_2$
(106-55-8)
(106-58-1)
(694-83-7)
(3114-70-3)
(7144-05-0)
(21544-02-5)
$C_6H_{14}N_2.C_4H_6O_6$
(133-35-7)
$C_6H_{14}N_2O$
(103-76-4)
(601-77-4)
(621-64-7)
(2038-03-1)
(3398-69-4)
(4549-44-4)
(13256-07-0)
(25154-38-5)
$C_6H_{14}N_2O_2$
(56-87-1)
(70-54-2)
(3312-58-1)
$C_6H_{14}N_2O_2.ClH$
(657-27-2)
(7274-88-6)
$C_6H_{14}N_2O_2.2ClH$
(617-68-5)
$C_6H_{14}N_2O_3$
(53609-64-6)
$C_6H_{14}N_2S_2$
(51-82-1)
$C_6H_{14}N_4O_2$
(74-79-3)
(6104-30-9)
$C_6H_{14}N_4O_2.C_5H_9NO_4$
(4320-30-3)
$C_6H_{14}N_4O_2.ClH$
(1119-34-2)
$C_6H_{14}O$
(77-74-7)
(97-95-0)
(105-30-6)
(108-11-2)
(108-20-3)
(111-27-3)
(111-43-3)
(464-07-3)
(565-60-6)
(565-67-3)
(589-35-5)
(590-36-3)
(594-60-5)
(623-37-0)
(624-95-3)
(626-89-1)
(626-93-7)
(627-08-7)
(628-81-9)
(637-92-3)
(994-05-8)
(1185-33-7)
(1320-98-5)
(2679-87-0)
(19550-30-2)
(25917-35-5)
(36839-67-5)
(54206-54-1)
(54972-97-3)
(79956-98-2)
$C_6H_{14}O.1/3Al$
(23275-26-5)
$C_6H_{14}O_2$
(76-09-5)
(105-57-7)
(107-41-5)
(111-76-2)
(629-11-8)
(629-14-1)
(1569-01-3)
(2935-44-6)
(4439-24-1)
(4457-71-0)
(4799-62-6)
(7580-85-0)
(7795-91-7)
(10215-30-2)
(30136-13-1)
(41632-89-7)
(56539-66-3)
$C_6H_{14}O_2.Na$
(52663-57-7)
$C_6H_{14}O_2S_2$
(5244-34-8)
(14970-87-7)
$C_6H_{14}O_3$
(77-99-6)
(106-62-7)
(106-69-4)
(110-98-5)
(111-90-0)
(111-96-6)
(621-63-6)
(7564-64-9)
(25265-71-8)
$C_6H_{14}O_3S_2$
(36727-72-7)
$C_6H_{14}O_4$
(112-27-6)
$C_6H_{14}O_4S$
(2973-10-6)
$C_6H_{14}O_5$
(51266-87-6)
(59113-36-9)
$C_6H_{14}O_6$
(50-70-4)
(69-65-8)
(87-78-5)
(608-66-2)
$C_6H_{14}O_6.C_6H_8O_7.xFe$
(1338-16-5)
$C_6H_{14}O_6S_2$
(55-98-1)
(299-75-2)
$C_6H_{14}O_6S_4.2Na$
(27206-35-5)
$C_6H_{14}O_7S_2$
(36724-43-3)
$C_6H_{14}S$
(111-31-9)
(111-47-7)
(625-80-9)
(638-46-0)
(1741-83-9)
$C_6H_{14}S_2$
(629-19-6)
(76229-76-0)
$C_6H_{15}Al$
(97-93-8)
$C_6H_{15}AlO$
(1586-92-1)
$C_6H_{15}Al_2Cl_3$
(12075-68-2)
$C_6H_{15}B$
(97-94-9)
$C_6H_{15}BO_3$
(150-46-9)

(24072-75-1)
$C_7H_4Cl_2O$
(83-38-5)
(122-01-0)
(618-46-2)
(874-42-0)
(6287-38-3)
(31155-09-6)
$C_7H_4Cl_2O_2$
(50-30-6)
(50-45-3)
(50-79-3)
(50-84-0)
(51-36-5)
(51-44-5)
(2314-36-5)
(35915-19-6)
(75248-87-2)
(91930-03-9)
$C_7H_4Cl_2O_3$
(320-72-9)
(3336-41-2)
(3401-80-7)
$C_7H_4Cl_2O_3.K.Na$
(68938-79-4)
$C_7H_4Cl_2O_4$
(69845-51-8)
$C_7H_4Cl_3NO_3$
(55335-06-3)
$C_7H_4Cl_3NO_3.C_6H_{15}N$
(57213-69-1)
$C_7H_4Cl_4$
(81-19-6)
(134-25-8)
(1006-31-1)
(1006-32-2)
(1344-32-7)
(2136-89-2)
(5216-25-1)
(6936-40-9)
(8005-43-4)
$C_7H_4Cl_4O$
(938-22-7)
(938-86-3)
(53452-81-6)
$C_7H_4Cl_4OS$
(107409-52-9)
$C_7H_4Cl_4O_2$
(2539-17-5)
$C_7H_4Cl_4O_2S$
(98555-82-9)
$C_7H_4Cl_4S$
(53014-41-8)
$C_7H_4Cl_6$
(22039-38-9)
$C_7H_4Cl_6N_4$
(21384-33-8)
$C_7H_4Cl_6N_4O$
(30388-91-1)
$C_7H_4Cl_7N_3$
(1145-44-4)
(30894-91-8)
$C_7H_4FNO_4$
(403-24-7)
$C_7H_4F_2O_2$
(385-00-2)
$C_7H_4F_3NO_2$
(98-46-4)
(384-22-5)
(402-54-0)
$C_7H_4F_3NO_3$
(88-30-2)
(400-99-7)
$C_7H_4F_3N_3O_4$
(445-66-9)
$C_7H_4F_4$
(392-85-8)
$C_7H_4HgO_3$
(5970-32-1)
$C_7H_4I_2O_3$
(133-91-5)
(1321-04-6)
$C_7H_4NO_3S.Na$
(128-44-9)
$C_7H_4NS_2.Na$
(2492-26-4)
$C_7H_4N_2O_2$
(619-24-9)
(619-72-7)
$C_7H_4N_2O_6$
(99-34-3)
(528-45-0)
(610-28-6)
(610-30-0)
$C_7H_4N_2O_6.C_6H_{13}N$
(5473-16-5)
$C_7H_4N_2O_7$
(609-99-4)
$C_7H_4O_4S$
(81-08-3)
$C_7H_4O_6$
(99-32-1)
$C_7H_4O_7$
(497-59-6)
$C_7H_5BiO_4$
(14882-18-9)
$C_7H_5BrCl_2O$
(86006-43-1)
C_7H_5BrO
(618-32-6)
(3132-99-8)
$C_7H_5BrO_2$
(88-65-3)
(585-76-2)
(586-76-5)
$C_7H_5Br_2ClO$
(86006-44-2)
$C_7H_5Br_3$
(27476-22-8)
$C_7H_5Br_3O$
(607-99-8)
(65436-87-5)
$C_7H_5Br_3O_2$
(38926-85-1)
$C_7H_5ClF_3N$
(445-03-4)
$C_7H_5ClHgO_2$
(59-85-8)
$C_7H_5ClN_2S$
(19952-47-7)
C_7H_5ClO
(89-98-5)
(98-88-4)
(104-88-1)
(587-04-2)
$C_7H_5ClO_2$
(74-11-3)
(118-91-2)
(535-80-8)
(1885-14-9)
(2420-16-8)
(26264-09-5)
$C_7H_5ClO_2.1/2Hg$
(15516-76-4)
$C_7H_5ClO_3$
(321-14-2)
(937-14-4)
(3964-58-7)
(103339-64-6)
$C_7H_5ClO_3S$
(21639-41-8)
(85822-16-8)
$C_7H_5ClO_4$
(79188-95-7)
(103339-65-7)
$C_7H_5Cl_2N$
(622-44-6)
$C_7H_5Cl_2NO$
(2008-58-4)
(5980-23-4)
(5980-26-7)
$C_7H_5Cl_2NO_2$
(133-90-4)
(56961-25-2)
$C_7H_5Cl_2NO_4S$
(80-13-7)
$C_7H_5Cl_2NS$
(1918-13-4)
$C_7H_5Cl_3$
(94-99-5)
(98-07-7)
(102-47-6)
(2077-46-5)
(6639-30-1)
(7359-72-0)
(23749-65-7)
(30583-33-6)
(38721-71-0)
$C_7H_5Cl_3O$
(87-40-1)
(551-76-8)
(6130-75-2)
(50375-10-5)
(53452-80-5)
(54135-80-7)
(54135-81-8)
(54135-82-9)
$C_7H_5Cl_3O_2$
(938-23-8)
(2668-24-8)
(57057-83-7)
(60712-44-9)
(61966-36-7)
(108544-90-7)
$C_7H_5Cl_3O_2S$
(4163-80-8)
$C_7H_5Cl_4NO$
(7159-34-4)
(70439-96-2)
$C_7H_5Cl_5$
(5825-64-9)
(18317-90-3)
$C_7H_5Cl_6N_3$
(3599-71-1)
$C_7H_5Cl_6N_3O$
(30863-52-6)
$C_7H_5Cl_6N_3O.C_3H_5Cl_3N_2$
(30886-05-6)
$C_7H_5Cl_6N_3S$
(5516-46-1)
$C_7H_5D_3$
(1124-18-1)
$C_7H_5FO_2$
(445-29-4)
(455-38-9)
(456-22-4)
$C_7H_5F_3$
(98-08-8)
(27359-10-0)
$C_7H_5F_3O$
(98-17-9)
$C_7H_5F_6N_3$
(30362-10-8)
$C_7H_5IO_2$
(88-67-5)
(618-51-9)
(619-58-9)
C_7H_5N
(100-47-0)
(931-54-4)
C_7H_5NO
(103-71-9)
(273-53-0)
(611-20-1)
(767-00-0)
(873-62-1)
C_7H_5NOS
(934-34-9)
(2634-33-5)
(3774-52-5)
$C_7H_5NO_2$
(59-49-4)
$C_7H_5NO_3$
(99-61-6)
(552-89-6)
(555-16-8)
$C_7H_5NO_3S$
(81-07-2)
$C_7H_5NO_3S.H_3N$
(6381-61-9)
$C_7H_5NO_3S_2.Na$
(21465-51-0)
$C_7H_5NO_4$
(62-23-7)
(89-00-9)
(100-26-5)
(121-92-6)
(490-11-9)
(499-80-9)
(499-81-0)
(499-83-2)
(552-16-9)
(3011-34-5)
$C_7H_5NO_4.C_6H_{13}N$
(34139-62-3)
$C_7H_5NO_4.Na$
(827-95-2)
$C_7H_5NO_5$
(85-38-1)
(96-97-9)
(138-60-3)
(616-82-0)
(619-14-7)
(32368-69-7)
C_7H_5NS
(95-16-9)
(103-72-0)
(272-12-8)
(272-16-2)
(5285-87-0)
$C_7H_5NS_2$
(149-30-4)
$C_7H_5NS_2.C_4H_{11}N$
(63302-50-1)
$C_7H_5NS_2.C_6H_{13}N$
(37437-20-0)
$C_7H_5NS_2.xCu$
(32510-27-3)
$C_7H_5NS_2.K$
(7778-70-3)
$C_7H_5N_2O_2$
(17420-30-3)
$C_7H_5N_2O_5.Na$
(2312-76-7)
$C_7H_5N_3O$
(90-16-4)
(582-61-6)
(24410-19-3)
$C_7H_5N_3O_2$
(94-52-0)
(5401-94-5)
$C_7H_5N_3O_2S$
(6285-57-0)
$C_7H_5N_3O_6$
(118-96-7)
(602-29-9)
(610-25-3)
(18292-97-2)
$C_7H_5N_3O_7$
(602-99-3)
(606-35-9)
$C_7H_5N_5O_8$
(479-45-8)
$C_7H_5O_2$
(766-76-7)
$C_7H_5O_2.H_4N$
(1863-63-4)
$C_7H_5O_2.Na$
(532-32-1)
$C_7H_5O_3$

(63-36-5)
(456-23-5)
$C_7H_5O_3.Na$
(54-21-7)
$C_7H_5O_3.1/2Sr$
(526-26-1)
C_7H_6BrCl
(611-17-6)
$C_7H_6BrClO_2S$
(54091-06-4)
C_7H_6BrNO
(40086-66-6)
$C_7H_6BrNO_3$
(5798-94-7)
(20039-91-2)
$C_7H_6Br_2$
(615-59-8)
(1611-92-3)
(31543-75-6)
$C_7H_6Br_2O$
(609-22-3)
(86006-42-0)
$C_7H_6Br_6N_4$
(30339-38-9)
C_7H_6ClF
(345-35-7)
(352-11-4)
(443-83-4)
(456-42-8)
C_7H_6ClNO
(609-66-5)
(619-56-7)
(2617-79-0)
$C_7H_6ClNO_2$
(83-42-1)
(89-59-8)
(89-60-1)
(100-14-1)
(121-86-8)
(612-23-7)
(619-23-8)
(635-21-2)
(2457-76-3)
(2486-71-7)
$C_7H_6ClNO_3$
(7147-89-9)
$C_7H_6ClNO_4S$
(121-02-8)
$C_7H_6ClNO_5S$
(6973-13-3)
$C_7H_6ClN_3O_4S_2$
(58-94-6)
$C_7H_6Cl_2$
(95-73-8)
(95-75-0)
(98-87-3)
(104-83-6)
(118-69-4)
(611-19-8)
(620-20-2)
(19398-61-9)
(25186-47-4)
(29797-40-8)
(32768-54-0)
$C_7H_6Cl_2N_2O$
(2280-93-5)
(2327-02-8)
(23505-21-7)
$C_7H_6Cl_2O$
(553-82-2)
(1570-65-6)
(1777-82-8)
(1805-32-9)
(1984-58-3)
(1984-59-4)
(1984-65-2)
(2432-12-4)
(15258-73-8)
(33719-74-3)
(36404-30-5)
(54518-15-9)
(60211-57-6)
$C_7H_6Cl_2O_2$
(2423-72-5)
(2460-49-3)
(16766-31-7)
(18113-14-9)
(39542-65-9)
(56680-68-3)
(56680-89-8)
(65724-16-5)
(77102-93-3)
(77102-94-4)
$C_7H_6Cl_2O_2S$
(56157-92-7)
$C_7H_6Cl_2O_3$
(75562-93-5)
$C_7H_6Cl_5N_3$
(30894-90-7)
$C_7H_6Cl_6N_4$
(24803-05-2)
(24803-29-0)
(30339-60-7)
$C_7H_6Cl_6N_4O$
(24803-11-0)
(30863-15-1)
$C_7H_6F_3N$
(88-17-5)
(98-16-8)
$C_7H_6NO_2.Na$
(555-06-6)
$C_7H_6N_2$
(51-17-2)
(271-29-4)
(271-34-1)
(271-44-3)
(272-49-1)
(873-74-5)
(1885-29-6)
(2237-30-1)
$C_7H_6N_2O$
(524-40-3)
(767-98-6)
$C_7H_6N_2O_3$
(610-15-1)
(619-80-7)
(645-09-0)
$C_7H_6N_2O_3S_2$
(21951-32-6)
$C_7H_6N_2O_4$
(121-14-2)
(602-01-7)
(606-20-2)
(610-39-9)
(616-79-5)
(618-85-9)
(619-15-8)
(619-17-0)
(25321-14-6)
$C_7H_6N_2O_5$
(119-27-7)
(497-56-3)
(534-52-1)
(609-93-8)
(616-73-9)
(1335-85-9)
(1817-66-9)
(63989-82-2)
(68191-07-1)
$C_7H_6N_2S$
(136-95-8)
(583-39-1)
$C_7H_6N_3O_3$
(27165-17-9)
$C_7H_6N_4$
(18039-42-4)
$C_7H_6N_4S$
(86-93-1)
C_7H_6O
(100-52-7)
(539-80-0)
$C_7H_6O_2$
(65-85-0)
(90-02-8)
(100-83-4)
(123-08-0)
(274-09-9)
(533-75-5)
(553-97-9)
(623-30-3)
(3880-99-7)
$C_7H_6O_2.1/3Al$
(555-32-8)
$C_7H_6O_2.1/2Ba$
(533-00-6)
$C_7H_6O_2.C_6H_{13}N$
(3129-92-8)
$C_7H_6O_2.1/2Ca$
(2090-05-3)
$C_7H_6O_2.1/2Cd$
(3026-22-0)
$C_7H_6O_2.K$
(582-25-2)
$C_7H_6O_2.1/2Mg$
(553-70-8)
$C_7H_6O_2.1/2Zn$
(553-72-0)
$C_7H_6O_2S$
(147-93-3)
$C_7H_6O_3$
(69-72-7)
(95-01-2)
(99-06-9)
(99-96-7)
(139-85-5)
$C_7H_6O_4$
(89-86-1)
(99-10-5)
(99-50-3)
(149-29-1)
(303-07-1)
(303-38-8)
(490-79-9)
$C_7H_6O_4.1/2Pb$
(41453-50-3)
$C_7H_6O_4Pb$
(87903-39-7)
$C_7H_6O_4S$
(91-25-8)
$C_7H_6O_4S.Na$
(1008-72-6)
$C_7H_6O_5$
(149-91-7)
(610-02-6)
$C_7H_6O_5S$
(121-53-9)
(632-25-7)
(636-78-2)
$C_7H_6O_6$
(2547-45-7)
(16426-62-3)
$C_7H_6O_6S$
(97-05-2)
C_7H_7Br
(95-46-5)
(100-39-0)
(106-38-7)
(591-17-3)
(28807-97-8)
$C_7H_7BrN_2O$
(1967-25-5)
(2989-98-2)
$C_7H_7BrN_4O_2$
(10381-75-6)
C_7H_7BrO
(104-92-7)
(578-57-4)
(2398-37-0)
(55909-73-4)
C_7H_7Cl
(95-49-8)
(100-44-7)
(106-43-4)
(108-41-8)
(25168-05-2)
$C_7H_7ClN_2O$
(114-38-5)
(140-38-5)
(536-40-3)
(1967-27-7)
$C_7H_7ClN_2S$
(5344-82-1)
C_7H_7ClO
(59-50-7)
(87-64-9)
(615-74-7)
(623-12-1)
(766-51-8)
(873-76-7)
(1321-10-4)
(1570-64-5)
(2845-89-8)
(6640-27-3)
(17849-38-6)
(30915-79-8)
(54548-50-4)
(68137-05-3)
$C_7H_7ClO.Na$
(52106-86-2)
C_7H_7ClOS
(934-73-6)
$C_7H_7ClO_2$
(16766-30-6)
(18113-03-6)
$C_7H_7ClO_2S$
(98-57-7)
(98-59-9)
(133-59-5)
$C_7H_7ClO_3$
(103339-63-5)
$C_7H_7ClO_3S$
(98-68-0)
C_7H_7ClS
(123-09-1)
$C_7H_7Cl_2NO$
(2971-90-6)
$C_7H_7Cl_3NO_3PS$
(5598-13-0)
$C_7H_7Cl_3NO_4P$
(5598-52-7)
$C_7H_7Cl_3N_2O_2$
(30362-19-7)
$C_7H_7Cl_4N_3O.C_3H_6Cl_2N_2$
(30886-04-5)
$C_7H_7Cl_5O$
(68334-67-8)
C_7H_7F
(95-52-3)
(350-50-5)
(352-32-9)
(352-70-5)
(25496-08-6)
$C_7H_7FN_2O$
(656-31-5)
(659-30-3)
(770-19-4)
C_7H_7FO
(456-49-5)
(459-60-9)
$C_7H_7F_3N_2O_2S$
(344-72-9)
C_7H_7I
(615-37-2)
(620-05-3)
(624-31-7)
(625-95-6)
(61878-58-8)
C_7H_7IO
(57455-06-8)
C_7H_7N

(638-03-9)
(3287-99-8)
C_7H_9NO
(90-04-0)
(95-84-1)
(103-74-2)
(104-94-9)
(150-75-4)
(536-90-3)
(1073-23-0)
(2835-95-2)
(2836-00-2)
(3718-65-8)
(5344-27-4)
(29191-52-4)
$C_7H_9NO.ClH$
(134-29-2)
(20265-97-8)
$C_7H_9NO.O_4S$
(55-55-0)
$C_7H_9NO_2$
(20189-42-8)
(37491-68-2)
$C_7H_9NO_2S$
(70-55-3)
(88-19-7)
(1333-07-9)
(1899-94-1)
$C_7H_9NO_3$
(30674-80-7)
$C_7H_9NO_3S$
(88-44-8)
(98-30-6)
(98-33-9)
(103-06-0)
$C_7H_9NO_3S.Na$
(26021-90-9)
$C_7H_9NO_4S$
(13244-33-2)
$C_7H_9NO_4S.Na$
(19433-86-4)
$C_7H_9N_3O$
(103-03-7)
(537-47-3)
$C_7H_9N_3S$
(5351-69-9)
$C_7H_9O_2PS.K$
(67446-04-2)
$C_7H_9O_3P$
(10088-45-6)
C_7H_{10}
(187-26-8)
(287-13-8)
(498-66-8)
(4054-38-0)
(16554-83-9)
$C_7H_{10}Br_2N_4$
(4658-20-2)
$C_7H_{10}Br_2O_2$
(3066-70-4)
$C_7H_{10}ClN_3$
(535-89-7)
(701-77-9)
$C_7H_{10}ClN_3O_2$
(30894-75-8)
$C_7H_{10}ClN_3O_3$
(16773-42-5)
$C_7H_{10}ClN_3OS$
(30894-81-6)
$C_7H_{10}ClN_3S_2$
(30894-83-8)
$C_7H_{10}Cl_2N_3O_2PS_2$
(14991-93-6)
$C_7H_{10}Cl_2N_4$
(30360-37-3)
(30360-38-4)
$C_7H_{10}Cl_3N_5$
(27431-14-7)
(27470-95-7)
$C_7H_{10}N.Br$
(1906-79-2)
$C_7H_{10}N_2$
(95-70-5)
(95-80-7)
(108-71-4)
(496-72-0)
(538-08-9)
(646-20-8)
(823-40-5)
(1122-58-3)
(2687-25-4)
(5407-87-4)
(13360-64-0)
(13925-03-6)
(14667-55-1)
(15707-23-0)
(25376-45-8)
(29460-90-0)
$C_7H_{10}N_2.ClH$
(5459-85-8)
$C_7H_{10}N_2.2ClH$
(615-45-2)
(636-23-7)
(15481-70-6)
$C_7H_{10}N_2.H_2O_4S$
(615-50-9)
$C_7H_{10}N_2.xH_2O_4S$
(6369-59-1)
$C_7H_{10}N_2O$
(615-05-4)
(5307-02-8)
(15872-73-8)
(25680-58-4)
$C_7H_{10}N_2O.2ClH$
(614-94-8)
$C_7H_{10}N_2O.H_2O_4S$
(39156-41-7)
$C_7H_{10}N_2OS$
(51-52-5)
$C_7H_{10}N_2O_2$
(110-26-9)
(13509-52-9)
$C_7H_{10}N_2O_2S$
(1576-35-8)
(22232-54-8)
$C_7H_{10}N_2O_3$
(55557-02-3)
$C_7H_{10}N_2O_3S$
(80-23-9)
$C_7H_{10}N_4O_2S$
(57-67-0)
$C_7H_{10}N_4O_3$
(57966-95-7)
$C_7H_{10}N_4O_{14}$
(13225-10-0)
$C_7H_{10}N_6$
(50473-86-4)
$C_7H_{10}O$
(100-50-5)
(497-38-1)
(1121-05-7)
(1193-18-6)
(1321-16-0)
(3146-39-2)
(4041-11-6)
(4313-03-5)
(10599-57-2)
(53783-87-2)
(53783-88-3)
$C_7H_{10}O_2$
(96-05-9)
(689-89-4)
(3883-58-7)
(4771-80-6)
(50598-50-0)
$C_7H_{10}O_2Si$
(3959-13-5)
$C_7H_{10}O_3$
(106-91-2)
(281-32-3)
(15022-08-9)
(19900-46-0)
$C_7H_{10}O_4$
(869-29-4)
(924-83-4)
(28098-80-8)
(41654-09-5)
$C_7H_{10}O_5$
(138-59-0)
(13192-04-6)
$C_7H_{10}S$
(1795-04-6)
$C_7H_{11}Br$
(34825-93-9)
$C_7H_{11}ClN_2O$
(65879-44-9)
$C_7H_{11}ClN_4$
(30369-29-0)
$C_7H_{11}ClN_4O$
(4407-72-1)
(4446-75-7)
(4743-13-9)
$C_7H_{11}ClN_4S$
(30388-95-5)
$C_7H_{11}Cl_3N_2O_2$
(60029-23-4)
$C_7H_{11}Cl_3O_2Si$
(7351-61-3)
$C_7H_{11}D_3$
(69432-96-8)
$C_7H_{11}N$
(2199-41-9)
(53783-86-1)
$C_7H_{11}NO$
(931-97-5)
(3173-53-3)
(3731-38-2)
(19615-27-1)
$C_7H_{11}NO_2$
(77-67-8)
(156-56-9)
(3470-97-1)
$C_7H_{11}NO_3$
(115-67-3)
$C_7H_{11}NS$
(1122-82-3)
(32272-57-4)
(41981-63-9)
$C_7H_{11}N_3$
(3599-60-8)
(30362-03-9)
$C_7H_{11}N_3O$
(60718-52-7)
$C_7H_{11}N_3O_2$
(705-78-2)
$C_7H_{11}N_3O_4$
(13551-87-6)
(35503-54-9)
$C_7H_{11}N_5O$
(30354-99-5)
$C_7H_{11}N_7S$
(4658-28-0)
(30362-29-9)
$C_7H_{11}O$
(7379-12-6)
$C_7H_{11}O_9P$
(37971-36-1)
C_7H_{12}
(286-08-8)
(591-47-9)
(591-48-0)
(591-49-1)
(628-71-7)
(628-92-2)
(1192-37-6)
(1335-86-0)
(3070-53-9)
(4049-81-4)
(20479-71-4)
(20479-72-5)
$C_7H_{12}BrN_5$
(3084-94-4)
$C_7H_{12}ClN_5$
(122-34-9)
(3004-71-5)
$C_7H_{12}FN_5$
(658-65-1)
$C_7H_{12}N_2$
(4316-42-1)
$C_7H_{12}N_2O$
(4542-47-6)
$C_7H_{12}N_2O_2$
(5455-34-5)
$C_7H_{12}N_2O_3$
(29071-93-0)
$C_7H_{12}N_2O_4$
(6440-58-0)
$C_7H_{12}N_4$
(5599-20-2)
(30377-23-2)
$C_7H_{12}N_4O$
(16399-10-3)
(30369-67-6)
$C_7H_{12}N_4OS$
(21087-61-6)
(33509-43-2)
$C_7H_{12}N_4O_2$
(30576-32-0)
(56507-37-0)
$C_7H_{12}N_4S$
(30369-72-3)
$C_7H_{12}N_4S_2$
(30362-12-0)
(30362-14-2)
$C_7H_{12}N_8$
(2854-95-7)
$C_7H_{12}O$
(502-42-1)
(583-60-8)
(589-92-4)
(591-24-2)
(684-94-6)
(1119-44-4)
(1331-22-2)
(1449-49-6)
(1679-51-2)
(2043-61-0)
(2463-63-0)
(4643-25-8)
(5166-53-0)
(5204-80-8)
(6004-60-0)
(6714-00-7)
(7740-69-4)
(18829-55-5)
(23758-27-2)
(33061-16-4)
(34314-84-6)
(35200-79-4)
(35200-80-7)
(89896-73-1)
$C_7H_{12}O_2$
(98-89-5)
(103-75-3)
(105-21-5)
(106-63-8)
(141-32-2)
(1663-39-4)
(1703-51-1)
(2210-28-8)
(3720-22-7)
(4351-54-6)
(4655-34-9)
(5331-08-8)
(6945-35-3)
(7307-02-0)
(15120-99-7)
(29901-85-7)
(51945-98-3)
(55670-09-2)

(62243-57-6)
$C_7H_{12}O_2.xPb$
(61790-14-5)
$C_7H_{12}O_2Pb$
(50825-29-1)
$C_7H_{12}O_3$
(106-74-1)
(539-88-8)
(637-64-9)
(923-26-2)
(5185-97-7)
(5739-83-3)
(19424-29-4)
(27813-02-1)
(41654-04-0)
$C_7H_{12}O_4$
(105-53-3)
(111-16-0)
(534-59-8)
(623-84-7)
(626-70-0)
(627-91-8)
(681-57-2)
(1119-40-0)
(1604-11-1)
(4839-46-7)
(61898-58-6)
$C_7H_{12}O_5$
(25395-31-7)
$C_7H_{12}O_6$
(77-95-2)
(562-73-2)
(36413-60-2)
$C_7H_{12}O_6Si$
(4253-34-3)
$C_7H_{13}Br$
(2404-35-5)
(6294-40-2)
(89942-12-1)
$C_7H_{13}BrN_2O_2$
(77-65-6)
$C_7H_{13}BrO_2$
(2624-01-3)
$C_7H_{13}ClO$
(2528-61-2)
$C_7H_{13}ClO_2$
(6092-54-2)
$C_7H_{13}Cl_2N_2O_3P$
(27046-19-1)
$C_7H_{13}N$
(100-76-5)
(629-08-3)
(643-20-9)
(19424-34-1)
$C_7H_{13}NO$
(107-58-4)
(618-42-8)
(1121-92-2)
(3623-05-0)
$C_7H_{13}NO_2$
(1563-84-4)
(2439-35-2)
(6065-13-0)
(6065-14-1)
(6187-24-2)
(7747-35-5)
(10601-69-1)
$C_7H_{13}NO_3$
(3375-84-6)
(27317-59-5)
$C_7H_{13}NO_3.ClH$
(2318-25-4)
$C_7H_{13}NO_4$
(1119-33-1)
$C_7H_{13}NO_4S$
(15214-89-8)
(21861-11-0)
$C_7H_{13}NO_4S.Na$
(5165-97-9)
$C_7H_{13}N_2O_4PS_3$
(2669-32-1)
$C_7H_{13}N_3$
(21150-01-6)
$C_7H_{13}N_3O_3S$
(23135-22-0)
$C_7H_{13}N_5$
(3232-26-6)
(4150-59-8)
(26235-12-1)
(27430-88-2)
(27431-07-8)
(30354-68-8)
(30354-74-6)
$C_7H_{13}N_5O$
(2599-11-3)
(25254-67-5)
(27963-33-3)
(30354-97-3)
$C_7H_{13}N_5S$
(1011-91-2)
(30360-83-9)
$C_7H_{13}O_6P$
(298-01-1)
(338-45-4)
(7786-34-7)
C_7H_{14}
(108-87-2)
(291-64-5)
(592-76-7)
(592-77-8)
(592-78-9)
(594-56-9)
(625-65-0)
(690-08-4)
(762-62-9)
(762-63-0)
(816-79-5)
(822-50-4)
(1192-18-3)
(1638-26-2)
(1640-89-7)
(1759-58-6)
(2213-32-3)
(2452-99-5)
(2453-00-1)
(2532-58-3)
(2738-19-4)
(3404-61-3)
(3404-62-4)
(3404-65-7)
(3404-72-6)
(3404-73-7)
(3769-23-1)
(4038-04-4)
(6094-02-6)
(6443-92-1)
(7385-78-6)
(7642-10-6)
(10574-36-4)
(10574-37-5)
(14686-13-6)
(14686-14-7)
(17618-77-8)
(24910-63-2)
(25339-56-4)
(28729-52-4)
(42154-69-8)
(50819-06-2)
$C_7H_{14}Br_2$
(59104-79-9)
$C_7H_{14}Cl_2$
(821-25-0)
$C_7H_{14}Cl_3NO_4$
(2218-68-0)
$C_7H_{14}F_2$
(407-96-5)
$C_7H_{14}NO_3PS_2$
(947-02-4)
$C_7H_{14}NO_4PS_2$
(29173-31-7)
$C_7H_{14}NO_5P$
(919-44-8)
(6923-22-4)
$C_7H_{14}N_2$
(693-13-0)
(53517-92-3)
$C_7H_{14}N_2O_2$
(6062-02-8)
$C_7H_{14}N_2O_2S$
(116-06-3)
(34681-10-2)
$C_7H_{14}N_2O_3$
(13464-10-3)
(56986-36-8)
$C_7H_{14}N_2O_3S$
(1646-87-3)
$C_7H_{14}N_2O_4$
(7388-44-5)
(37658-95-0)
$C_7H_{14}N_2O_4S$
(1646-88-4)
(34681-23-7)
$C_7H_{14}N_2O_5$
(3001-61-4)
$C_7H_{14}N_5O_2PS_2$
(30863-35-5)
$C_7H_{14}N_6$
(5606-16-6)
$C_7H_{14}O$
(100-49-2)
(105-42-0)
(106-35-4)
(110-12-3)
(110-43-0)
(111-71-7)
(123-19-3)
(502-41-0)
(565-80-0)
(583-59-5)
(589-91-3)
(590-50-1)
(590-67-0)
(591-23-1)
(623-56-3)
(931-56-6)
(1860-39-5)
(3521-91-3)
(4938-52-7)
(5363-63-3)
(5454-79-5)
(7443-52-9)
(7443-55-2)
(7443-70-1)
(7731-28-4)
(7731-29-5)
(14925-96-3)
(17612-35-0)
(25639-42-3)
(27944-79-2)
(30801-96-8)
(32749-94-3)
(33467-76-4)
(41065-97-8)
(42328-43-8)
(53229-39-3)
(53897-31-7)
(53897-32-8)
(56052-83-6)
(56052-95-0)
$C_7H_{14}OSi$
(17869-76-0)
$C_7H_{14}O_2$
(105-66-8)
(106-70-7)
(107-70-0)
(108-64-5)
(111-14-8)
(123-92-2)
(539-82-2)
(540-42-1)
(590-01-2)
(617-50-5)
(624-41-9)
(625-16-1)
(626-38-0)
(628-46-6)
(628-63-7)
(644-49-5)
(926-41-0)
(1118-47-4)
(1185-39-3)
(1330-19-4)
(1561-11-1)
(2412-80-8)
(2426-08-6)
(3938-95-2)
(4352-98-1)
(4405-16-7)
(4536-23-6)
(7452-79-1)
(7665-72-7)
(19889-37-3)
(20225-24-5)
(29006-00-6)
(36687-99-7)
(53907-91-8)
$C_7H_{14}O_2.1/2Cu$
(5128-10-9)
$C_7H_{14}O_2.K$
(16761-12-9)
$C_7H_{14}O_3$
(138-22-7)
(763-69-9)
(4435-53-4)
(54063-14-8)
(54063-18-2)
(66822-98-8)
$C_7H_{14}O_4$
(557-25-5)
(629-38-9)
$C_7H_{14}O_6$
(97-30-3)
$C_7H_{14}O_8$
(23351-51-1)
$C_7H_{14}O_8.Na$
(13007-85-7)
(30080-50-3)
(31138-65-5)
$C_7H_{14}S$
(1613-52-1)
$C_7H_{14}Si$
(62108-37-6)
$C_7H_{15}AsN_2S_4$
(2445-07-0)
$C_7H_{15}Br$
(629-04-9)
(1974-04-5)
(1974-05-6)
$C_7H_{15}Cl$
(629-06-1)
(29756-37-4)
$C_7H_{15}ClO_2$
(16224-33-2)
$C_7H_{15}Cl_2N$
(621-68-1)
$C_7H_{15}Cl_2N_2O_2P$
(50-18-0)
(3778-73-2)
$C_7H_{15}Cl_2N_2O_2P.H_2O$
(6055-19-2)
$C_7H_{15}Cl_2N_2O_4P$
(22788-18-7)
$C_7H_{15}F$
(661-11-0)
$C_7H_{15}I$
(4282-40-0)
$C_7H_{15}N$
(100-60-7)
(671-36-3)
(766-09-6)

(5452-35-7)
(7003-32-9)
(26389-60-6)
$C_7H_{15}NO$
(1484-84-0)
$C_7H_{15}NOS$
(2941-55-1)
$C_7H_{15}NO_2$
(10220-23-2)
$C_7H_{15}NO_3$
(541-15-1)
(20227-92-3)
(20633-12-9)
$C_7H_{15}NO_4$
(52304-17-3)
$C_7H_{15}N_3O_2$
(50285-70-6)
$C_7H_{15}N_3O_2.ClH$
(65206-90-8)
$C_7H_{15}N_3O_2S_2.ClH$
(15263-52-2)
$C_7H_{15}N_3O_4$
(1852-21-7)
$C_7H_{15}N_7$
(10409-78-6)
(10421-98-4)
$C_7H_{15}O_4P$
(1067-71-6)
C_7H_{16}
(108-08-7)
(142-82-5)
(464-06-2)
(562-49-2)
(565-59-3)
(589-34-4)
(590-35-2)
(591-76-4)
(617-78-7)
(31394-54-4)
$C_7H_{16}FO_2P$
(96-64-0)
$C_7H_{16}N.Cl$
(24307-26-4)
$C_7H_{16}NO_4PS_2$
(919-76-6)
$C_7H_{16}N_2$
(15932-66-8)
(23159-07-1)
(27578-60-5)
$C_7H_{16}N_2O$
(123-00-2)
(4128-37-4)
$C_7H_{16}N_2O_2$
(143-06-6)
(90853-19-3)
$C_7H_{16}O$
(111-70-6)
(543-49-7)
(589-55-9)
(589-82-2)
(594-83-2)
(595-41-5)
(597-49-9)
(597-96-6)
(600-36-2)
(615-29-2)
(617-29-8)
(623-55-2)
(625-06-9)
(625-23-0)
(627-59-8)
(627-98-5)
(919-94-8)
(2313-65-7)
(2370-12-9)
(3121-79-7)
(3970-62-5)
(4911-70-0)
(6305-71-1)
(10143-23-4)
(13231-81-7)
(18641-81-1)
(36749-13-0)
(53535-33-4)
(61949-26-6)
$C_7H_{16}O_2$
(111-89-7)
(115-76-4)
(126-84-1)
(629-30-1)
(762-46-9)
(4744-08-5)
(5131-66-8)
(20680-10-8)
(23436-19-3)
(29387-86-8)
(40091-57-4)
(53907-95-2)
(55724-04-4)
$C_7H_{16}O_3$
(122-51-0)
(1002-67-1)
(6881-94-3)
(13429-07-7)
(13588-28-8)
(20324-32-7)
(34590-94-8)
(52698-46-1)
(55956-21-3)
(55956-22-4)
$C_7H_{16}O_4$
(102-52-3)
(112-35-6)
$C_7H_{16}O_4S_2$
(115-24-2)
$C_7H_{16}S$
(1639-09-4)
$C_7H_{16}S_3$
(69078-83-7)
$C_7H_{16}Si$
(18292-29-0)
$C_7H_{17}Cl_2N_2O_2P.C_6H_{13}N$
(4465-94-5)
$C_7H_{17}N$
(111-68-2)
(123-82-0)
$C_7H_{17}NO_2$
(121-93-7)
(55759-85-8)
$C_7H_{17}NO_5$
(6284-40-8)
$C_7H_{17}O_2PS$
(25371-75-9)
$C_7H_{17}O_2PS_3$
(298-02-2)
(36614-38-7)
$C_7H_{17}O_3P$
(616-52-4)
(1445-75-6)
$C_7H_{17}O_3PS$
(5301-73-5)
(22907-64-8)
$C_7H_{17}O_3PS_3$
(2588-03-6)
$C_7H_{17}O_4PS_2$
(2588-05-8)
$C_7H_{17}O_4PS_3$
(2588-04-7)
$C_7H_{18}N_2$
(104-78-9)
(52198-64-8)
$C_7H_{18}N_2O$
(2212-32-0)
$C_7H_{18}N_2O_2$
(4985-85-7)
$C_7H_{18}O_3Si$
(2031-67-6)
(18395-30-7)
$C_7H_{18}Pb$
(1762-28-3)
(54964-75-9)
$C_7H_{19}N_3$
(105-83-9)
(124-20-9)
$C_7H_{22}O_2Si_3$
(1873-88-7)
$C_8Br_4O_3$
(632-79-1)
$C_8Cl_4N_2$
(1897-45-6)
(90077-80-8)
$C_8Cl_4O_3$
(117-08-8)
$C_8Cl_6O_2$
(719-32-4)
C_8Cl_8
(29082-74-4)
$C_8Co_2O_8$
(10210-68-1)
$C_8F_{15}O_2.H_4N$
(3825-26-1)
$C_8F_{16}O$
(335-66-0)
$C_8F_{16}O_5S$
(4089-58-1)
C_8F_{18}
(307-34-6)
$C_8F_{18}O_2S$
(307-35-7)
$C_8HCl_3N_2O$
(28343-61-5)
C_8HCl_7
(29086-38-2)
(29086-39-3)
(61255-81-0)
(61593-44-0)
$C_8HF_{15}O_2$
(335-67-1)
$C_8HF_{17}O_3S.K$
(2795-39-3)
C_8H_2
(6165-96-4)
$C_8H_2Br_4O_4.1/2C_2H_8N_2$
(66046-78-4)
$C_8H_2Cl_4O_2$
(27355-22-2)
$C_8H_2Cl_4O_4$
(632-58-6)
(2136-79-0)
$C_8H_2Cl_6$
(61128-00-5)
$C_8H_3Cl_2F_3N_2$
(2338-25-2)
(3615-21-2)
$C_8H_3Cl_3N_2O_2$
(115044-73-0)
$C_8H_3Cl_5$
(83484-75-7)
$C_8H_3Cl_5O$
(1203-86-7)
$C_8H_3Cl_5O_2$
(1441-02-7)
$C_8H_3Cl_5O_3$
(2877-14-7)
$C_8H_3Cl_{10}N_3S$
(30863-02-6)
$C_8H_3NO_5$
(5466-84-2)
$C_8H_3N_2O_5$
(42540-91-0)
$C_8H_4ClNO_2$
(3481-09-2)
$C_8H_4Cl_2N_2$
(2213-63-0)
$C_8H_4Cl_2O_2$
(88-95-9)
(99-63-8)
(100-20-9)
$C_8H_4Cl_2O_4$
(41906-38-1)
$C_8H_4Cl_3O_3.Na$
(13560-99-1)
$C_8H_4Cl_4$
(71489-58-2)
$C_8H_4Cl_4N_2O_2$
(1786-86-3)
$C_8H_4Cl_6$
(68-36-0)
$C_8H_4Cl_8$
(83682-58-0)
$C_8H_4F_3NO$
(329-01-1)
$C_8H_4F_6$
(402-31-3)
$C_8H_4F_{13}I$
(2043-57-4)
$C_8H_4N_2$
(91-15-6)
(626-17-5)
$C_8H_4N_2O_4$
(89-40-7)
(603-62-3)
$C_8H_4N_2S_2$
(4044-65-9)
$C_8H_4O_3$
(85-44-9)
$C_8H_4O_4$
(116211-86-0)
$C_8H_5BrN_2O_2$
(13183-09-0)
$C_8H_5Br_5$
(85-22-3)
$C_8H_5ClN_2S$
(24255-23-0)
$C_8H_5ClO_4$
(89-20-3)
$C_8H_5Cl_2NO_4$
(34408-25-8)
(40188-83-8)
$C_8H_5Cl_2O_3.Li$
(3766-27-6)
$C_8H_5Cl_2O_3.Na$
(2702-72-9)
$C_8H_5Cl_3$
(54935-00-1)
$C_8H_5Cl_3O$
(4252-78-2)
(13608-87-2)
$C_8H_5Cl_3O_2$
(85-34-7)
$C_8H_5Cl_3O_2.Na$
(2439-00-1)
$C_8H_5Cl_3O_3$
(93-76-5)
(575-89-3)
(2307-49-5)
(33433-95-3)
$C_8H_5Cl_3O_3.C_2H_7N$
(6369-97-7)
$C_8H_5Cl_3O_3.C_3H_9N$
(6369-96-6)
$C_8H_5Cl_3O_3.C_3H_9NO$
(1319-72-8)
$C_8H_5Cl_3O_3.C_6H_{15}N$
(2008-46-0)
$C_8H_5Cl_3O_3.C_6H_{15}NO$
(53404-86-7)
$C_8H_5Cl_3O_3.C_6H_{15}NO_3$
(3813-14-7)
$C_8H_5Cl_5$
(606-07-5)
$C_8H_5Cl_6N_3O$
(30863-54-8)
$C_8H_5Cl_6N_3O_2$
(30362-44-8)
(30863-22-0)
$C_8H_5Cl_6N_5$
(24848-40-6)
$C_8H_5FO_4$
(1583-67-1)

$C_8H_5F_3O$
(434-45-7)
(455-19-6)
$C_8H_5F_3O_2$
(454-92-2)
(455-24-3)
$C_8H_5F_3O_2S$
(326-91-0)
$C_8H_5F_6N$
(328-74-5)
$C_8H_5F_{13}O$
(647-42-7)
$C_8H_5KO_4$
(13427-80-0)
$C_8H_5MnO_3$
(12079-65-1)
C_8H_5NO
(613-90-1)
C_8H_5NOS
(18138-18-6)
$C_8H_5NO_2$
(85-41-6)
(91-56-5)
(619-65-8)
(1877-72-1)
$C_8H_5NO_2.K$
(1074-82-4)
$C_8H_5NO_3$
(118-48-9)
(524-38-9)
$C_8H_5NO_3S$
(116211-93-9)
$C_8H_5NO_3S.Na$
(80789-74-8)
$C_8H_5NO_6$
(618-88-2)
$C_8H_5N_3O_2$
(116211-87-1)
$C_8H_5N_3O_4$
(3682-19-7)
$C_8H_5N_3O_4S$
(24554-26-5)
$C_8H_5N_3S_2$
(7170-77-6)
$C_8H_5O_7S.Na$
(6362-79-4)
C_8H_6
(536-74-3)
C_8H_6BrClO
(41011-01-2)
C_8H_6BrN
(5798-79-8)
(10075-50-0)
(16532-79-9)
$C_8H_6BrNO_2$
(7166-19-0)
$C_8H_6Br_2O$
(99-73-0)
$C_8H_6Br_2O_3$
(7507-35-9)
$C_8H_6Br_4$
(2810-69-7)
(23488-38-2)
(36059-21-9)
$C_8H_6Br_4O$
(68413-71-8)
C_8H_6ClN
(140-53-4)
(6575-09-3)
C_8H_6ClNO
(19219-99-9)
$C_8H_6ClNO_3$
(99-47-8)
$C_8H_6ClNO_4$
(42087-80-9)
(51282-49-6)
$C_8H_6ClNS_2$
(28908-00-1)
$C_8H_6ClN_3O$
(24310-41-6)
$C_8H_6Cl_2$
(1123-84-8)
(2123-27-5)
(2123-28-6)
(6607-45-0)
(28469-92-3)
$C_8H_6Cl_2O$
(2234-16-4)
$C_8H_6Cl_2O_2$
(2905-69-3)
(21460-88-8)
(35112-28-8)
(61031-72-9)
$C_8H_6Cl_2O_2S$
(6274-27-7)
$C_8H_6Cl_2O_3$
(94-75-7)
(575-90-6)
(587-64-4)
(588-22-7)
(1918-00-9)
(2976-74-1)
(18268-69-4)
(22775-37-7)
(108545-01-3)
$C_8H_6Cl_2O_3.C_2H_7N$
(2300-66-5)
$C_8H_6Cl_2O_3.C_4H_{11}N$
(20940-37-8)
$C_8H_6Cl_2O_3.C_4H_{11}NO_2$
(25059-78-3)
$C_8H_6Cl_2O_3.C_8H_5Cl_3O_3$
(8015-35-8)
$C_8H_6Cl_2O_3.C_8H_6Cl_2O_3$
(8068-77-7)
$C_8H_6Cl_2O_3.H_3N$
(2307-55-3)
$C_8H_6Cl_2O_3.Na$
(1982-69-0)
$C_8H_6Cl_2O_4$
(108544-97-4)
$C_8H_6Cl_3NO_3$
(60825-26-5)
$C_8H_6Cl_3O_5S.Na$
(3570-61-4)
$C_8H_6Cl_4$
(877-10-1)
(25641-99-0)
$C_8H_6Cl_4O_2$
(944-61-6)
(944-77-4)
(944-78-5)
$C_8H_6Cl_6$
(83682-57-9)
$C_8H_6Cl_6N_4$
(24803-17-6)
$C_8H_6Cl_9N_3O_2$
(30863-17-3)
$C_8H_6F_3N_3O_4S_2$
(148-56-1)
$C_8H_6N_2$
(91-19-0)
(253-52-1)
(253-66-7)
(253-69-0)
(253-82-7)
$C_8H_6N_2O$
(119-39-1)
(825-52-5)
$C_8H_6N_2O_2$
(610-66-2)
(60544-75-4)
$C_8H_6N_2O_3S$
(59607-71-5)
$C_8H_6N_2O_6$
(643-43-6)
(4232-27-3)
(28169-46-2)
$C_8H_6N_2O_7$
(22633-33-6)
$C_8H_6N_2S$
(50483-82-4)
$C_8H_6N_4O_4S$
(3570-75-0)
$C_8H_6N_4O_5$
(67-20-9)
$C_8H_6N_4O_8$
(76-24-4)
C_8H_6O
(271-89-6)
C_8H_6OS
(3610-02-4)
$C_8H_6O_2$
(87-41-2)
(623-27-8)
(643-79-8)
(7169-34-8)
(30025-33-3)
$C_8H_6O_3$
(120-57-0)
(611-73-4)
(619-21-6)
(619-66-9)
$C_8H_6O_4$
(88-99-3)
(94-53-1)
(100-21-0)
(121-91-5)
(29010-86-4)
$C_8H_6O_4.xCo$
(34262-88-9)
$C_8H_6O_4.K$
(877-24-7)
$C_8H_6O_4.Na$
(54537-30-3)
$C_8H_6O_4.Pb$
(6838-85-3)
$C_8H_6O_5$
(610-35-5)
(636-46-4)
$C_8H_6O_6$
(63958-66-7)
$C_8H_6O_7S$
(89-08-7)
$C_8H_6O_7S.Na$
(33562-89-9)
C_8H_6S
(95-15-8)
(11095-43-5)
C_8H_7Br
(103-64-0)
(2039-82-9)
(2039-88-5)
C_8H_7BrFNO
(351-05-3)
C_8H_7BrO
(70-11-1)
(99-90-1)
(2142-63-4)
$C_8H_7BrO_2$
(610-94-6)
(618-89-3)
(619-42-1)
(2491-38-5)
(6232-88-8)
$C_8H_7BrO_3$
(2973-76-4)
(60632-40-8)
$C_8H_7Br_3O$
(58170-32-4)
$C_8H_7Br_3O_2$
(23976-66-1)
C_8H_7Cl
(622-25-3)
(1073-67-2)
(1331-28-8)
(2039-85-2)
(2039-87-4)
C_8H_7ClO
(99-91-2)
(103-80-0)
(532-27-4)
(874-60-2)
(1341-24-8)
$C_8H_7ClO_2$
(100-07-2)
(501-53-1)
(610-96-8)
(620-73-5)
(876-27-7)
(1126-46-1)
(1878-65-5)
(1878-66-6)
(2444-36-2)
(2905-65-9)
(5335-05-7)
(34025-32-6)
$C_8H_7ClO_2S$
(18619-18-6)
$C_8H_7ClO_3$
(122-88-3)
(492-86-4)
(588-32-9)
(614-61-9)
(18268-76-3)
(19463-48-0)
(61670-76-6)
(82668-20-0)
$C_8H_7ClO_4$
(62936-23-6)
(62936-24-7)
(69845-52-9)
$C_8H_7ClO_4S$
(26638-43-7)
$C_8H_7Cl_2NO$
(2150-93-8)
$C_8H_7Cl_2NOS$
(1953-89-5)
$C_8H_7Cl_2NO_2$
(1918-18-9)
(7286-84-2)
$C_8H_7Cl_2O_5S.Na$
(136-78-7)
$C_8H_7Cl_3O$
(2000-43-3)
(27683-60-9)
$C_8H_7Cl_3O_2$
(2122-77-2)
(16766-29-3)
(61465-81-4)
(85298-07-3)
$C_8H_7Cl_3O_3$
(2539-26-6)
$C_8H_7Cl_6N_3$
(24481-42-3)
(24481-43-4)
$C_8H_7Cl_6N_3O$
(30863-53-7)
(30863-55-9)
$C_8H_7Cl_6N_3S$
(5516-48-3)
(24481-68-3)
(24504-17-4)
$C_8H_7Cl_6N_5O$
(30804-99-0)
C_8H_7FO
(403-42-9)
(455-36-7)
$C_8H_7FO_2$
(405-50-5)
$C_8H_7F_3N_2O$
(13114-87-9)
C_8H_7N
(104-85-8)
(120-72-9)
(140-29-4)
(274-40-8)
(529-19-1)
(620-22-4)
(7175-47-5)

C_8H_7NO
(480-91-1)
(532-28-5)
(614-68-6)
(2380-94-1)
(14191-95-8)
(69594-78-1)
C_8H_7NOS
(5325-20-2)
(40991-38-6)
$C_8H_7NO_2$
(81-20-9)
(102-96-5)
(50700-49-7)
$C_8H_7NO_3$
(88-97-1)
(100-19-6)
(121-89-1)
(577-59-3)
$C_8H_7NO_3S$
(4083-64-1)
(15448-99-4)
$C_8H_7NO_4$
(96-98-0)
(610-69-5)
(619-50-1)
(830-03-5)
(1523-06-4)
(1975-50-4)
(1975-52-6)
(3113-71-1)
(3113-72-2)
(5437-38-7)
(13506-76-8)
$C_8H_7NO_5$
(99-42-3)
(1878-87-1)
C_8H_7NS
(120-75-2)
(622-78-6)
(3012-37-1)
(3048-48-4)
$C_8H_7NS_2$
(615-22-5)
$C_8H_7N_3$
(3468-11-9)
(19064-69-8)
$C_8H_7N_3O_2$
(521-31-3)
$C_8H_7N_3O_5$
(67-45-8)
(148-01-6)
$C_8H_7N_3S$
(17467-15-1)
$C_8H_7O_2.Na$
(114-70-5)
$C_8H_7O_3.Na$
(3598-16-1)
$C_8H_7O_4$
(6746-48-1)
$C_8H_7O_4.Na$
(4418-26-2)
C_8H_8
(100-42-5)
(629-20-9)
(694-87-1)
C_8H_8BrCl
(6529-53-9)
$C_8H_8BrCl_2O_3PS$
(2104-96-3)
C_8H_8BrNO
(103-88-8)
(621-38-5)
$C_8H_8Br_2$
(93-52-7)
(28805-90-5)
(30812-87-4)
C_8H_8ClNO
(533-17-5)
(539-03-7)
(587-65-5)
(588-07-8)
$C_8H_8ClNO_2$
(20668-13-7)
(2150-88-1)
(2620-53-3)
(3942-54-9)
$C_8H_8ClNO_3S$
(121-60-8)
$C_8H_8Cl_2$
(612-12-4)
(626-16-4)
(1074-11-9)
(1124-05-6)
(28347-13-9)
$C_8H_8Cl_2IO_3PS$
(18181-70-9)
$C_8H_8Cl_2N_2O$
(3567-62-2)
$C_8H_8Cl_2N_4$
(5051-62-7)
$C_8H_8Cl_2O$
(133-53-9)
$C_8H_8Cl_2O_2$
(120-67-2)
(2675-77-6)
(2772-46-5)
(90283-00-4)
(90283-01-5)
(90283-02-6)
$C_8H_8Cl_2O_3$
(35869-50-2)
(75248-88-3)
(78782-46-4)
(108548-71-6)
$C_8H_8Cl_3NO_2S$
(38827-35-9)
$C_8H_8Cl_3N_3O$
(949-43-9)
$C_8H_8Cl_3N_3O_4S_2$
(133-67-5)
$C_8H_8Cl_3N_5$
(24848-44-0)
$C_8H_8Cl_3N_5O_2$
(30359-60-5)
$C_8H_8Cl_3O_3PS$
(299-84-3)
$C_8H_8Cl_4$
(83682-55-7)
$C_8H_8Cl_6$
(83682-53-5)
$C_8H_8Cl_6N_4$
(24802-97-9)
(24803-30-3)
(24803-31-4)
$C_8H_8Cl_6N_4O$
(24803-12-1)
(24803-13-2)
$C_8H_8Cu_2O_4$
(34262-89-0)
C_8H_8FNO
(351-28-0)
(351-83-7)
$C_8H_8F_3N_3O_4S_2$
(135-09-1)
$C_8H_8HgO_2$
(62-38-4)
$C_8H_8I_2O_2S$
(20018-09-1)
$C_8H_8NO_3PS$
(31328-15-1)
$C_8H_8NO_4P$
(31328-16-2)
$C_8H_8N_2$
(615-15-6)
$C_8H_8N_2OS$
(1747-60-0)
(5464-79-9)
$C_8H_8N_2O_2$
(88-96-0)
$C_8H_8N_2O_2S_2$
(17557-67-4)
$C_8H_8N_2O_3$
(104-04-1)
(122-28-1)
(552-32-9)
(583-08-4)
(19013-11-7)
(63892-06-8)
$C_8H_8N_2O_3S$
(2225-40-3)
$C_8H_8N_2O_4$
(616-72-8)
$C_8H_8N_2O_5$
(610-54-8)
$C_8H_8N_2O_6$
(2831-60-9)
$C_8H_8N_2O_6S$
(3337-70-0)
$C_8H_8N_2S$
(1477-42-5)
(16954-69-1)
$C_8H_8N_4$
(86-54-4)
$C_8H_8N_4.ClH$
(304-20-1)
$C_8H_8N_4O$
(24370-25-0)
$C_8H_8N_4O_4$
(555-84-0)
C_8H_8O
(96-09-3)
(98-86-2)
(104-87-0)
(122-78-1)
(496-14-0)
(496-16-2)
(529-20-4)
(620-23-5)
(1334-78-7)
(2628-17-3)
$C_8H_8O_2$
(93-58-3)
(99-04-7)
(99-93-4)
(99-94-5)
(103-82-2)
(104-57-4)
(118-90-1)
(118-93-4)
(121-71-1)
(122-79-2)
(123-11-5)
(135-02-4)
(493-09-4)
(591-31-1)
(623-15-4)
(7339-87-9)
(25567-10-6)
$C_8H_8O_2.K$
(13005-36-2)
$C_8H_8O_3$
(50-85-1)
(83-40-9)
(85-43-8)
(89-56-5)
(89-84-9)
(90-64-2)
(99-76-3)
(100-09-4)
(102-29-4)
(119-36-8)
(121-33-5)
(122-59-8)
(148-53-8)
(156-38-7)
(490-78-8)
(495-76-1)
(567-61-3)
(579-75-9)
(586-38-9)
(611-72-3)
(612-20-4)
(614-75-5)
(621-37-4)
(621-59-0)
(1197-09-7)
(1335-08-6)
$C_8H_8O_3S.Na$
(2695-37-6)
(27457-28-9)
$C_8H_8O_4$
(102-32-9)
(121-34-6)
(451-13-8)
(520-45-6)
(528-21-2)
(645-08-9)
(877-22-5)
(1878-84-8)
(2150-47-2)
(2237-36-7)
(3929-89-3)
(5981-39-5)
(6324-11-4)
(16715-77-8)
(108548-68-1)
(108548-69-2)
$C_8H_8O_5$
(99-24-1)
(3934-84-7)
$C_8H_8O_5.2K$
(59985-42-1)
$C_8H_8O_5.2Na$
(129-67-9)
C_8H_9Br
(89-92-9)
(103-63-9)
(104-81-4)
(553-94-6)
(585-71-7)
(620-13-3)
(1585-07-5)
(1973-22-4)
(28777-60-8)
(35884-77-6)
$C_8H_9BrN_2O$
(78508-46-0)
C_8H_9BrO
(52289-93-7)
(58170-30-2)
(73908-23-3)
$C_8H_9BrO_3S$
(54322-31-5)
$C_8H_9BrO_3S.Na$
(65036-65-9)
C_8H_9Cl
(89-96-3)
(104-82-5)
(552-45-4)
(620-19-9)
(622-24-2)
(672-65-1)
(25323-41-5)
$C_8H_9ClNO_5PS$
(500-28-7)
(2463-84-5)
$C_8H_9ClN_2O$
(13142-64-8)
(20940-42-5)
$C_8H_9ClN_2O_2$
(25277-05-8)
$C_8H_9ClN_2O_4S$
(137-47-3)
C_8H_9ClO
(88-04-0)
(622-86-6)
(3391-10-4)
(13524-04-4)
(18979-90-3)

(18980-00-2)
$C_8H_9ClO_2$
(2100-42-7)
(16766-27-1)
(90282-99-8)
$C_8H_9ClO_2S$
(609-60-9)
(938-09-0)
$C_8H_9ClO_3$
(108545-00-2)
C_8H_9ClS
(5535-49-9)
$C_8H_9Cl_2N$
(56961-05-8)
$C_8H_9Cl_2NO$
(67883-07-2)
$C_8H_9Cl_2O_2P$
(14212-98-7)
$C_8H_9Cl_3$
(83682-63-7)
$C_8H_9Cl_3N_4$
(24802-98-0)
(24803-63-2)
$C_8H_9Cl_3N_4S$
(30369-43-8)
$C_8H_9Cl_3O$
(52314-67-7)
$C_8H_9Cl_3Si$
(940-41-0)
$C_8H_9Cl_4N_3S$
(30863-00-4)
$C_8H_9Cl_4N_3S_2$
(24478-18-0)
$C_8H_9Cl_5O_2$
(75147-20-5)
$C_8H_9Cl_6N_5O_2$
(30863-16-2)
$C_8H_9FN_2O$
(78508-45-9)
$C_8H_9FO_2S$
(455-20-9)
(34586-49-7)
$C_8H_9HgO_3S_2.Na$
(5964-24-9)
C_8H_9N
(140-76-1)
(496-15-1)
C_8H_9NO
(94-69-9)
(99-03-6)
(99-92-3)
(103-81-1)
(103-84-4)
(527-85-5)
(551-93-9)
(613-93-4)
(619-55-6)
(3717-15-5)
$C_8H_9NO_2$
(83-41-0)
(85-40-5)
(89-58-7)
(89-87-2)
(99-12-7)
(99-51-4)
(100-12-9)
(103-90-2)
(134-20-3)
(614-80-2)
(619-45-4)
(621-42-1)
(875-74-1)
(1197-55-3)
(1570-45-2)
(1943-79-9)
(2486-70-6)
(2524-52-9)
(2603-10-3)
(2835-06-5)
(4746-61-6)
(7214-61-1)
(25168-04-1)
$C_8H_9NO_2.K$
(19525-59-8)
$C_8H_9NO_2.Na$
(10265-69-7)
$C_8H_9NO_2S$
(10302-15-5)
$C_8H_9NO_3$
(66-72-8)
(100-27-6)
(100-29-8)
(119-10-8)
(2423-71-4)
(15121-84-3)
(22818-40-2)
$C_8H_9NO_3.ClH$
(65-22-5)
$C_8H_9NO_4S$
(121-62-0)
$C_8H_9NO_4$
(16365-27-8)
$C_8H_9N_3O_2$
(17433-31-7)
$C_8H_9O_3PS$
(3811-49-2)
C_8H_{10}
(95-47-6)
(100-41-4)
(106-42-3)
(108-38-3)
(1330-20-7)
(29759-77-1)
$C_8H_{10}BrCl_3N_4$
(24802-96-8)
$C_8H_{10}Br_2O_4$
(20679-58-7)
$C_8H_{10}Br_3N_3O$
(34157-48-7)
$C_8H_{10}ClN$
(698-01-1)
(6848-13-1)
$C_8H_{10}ClNO_2$
(6358-64-1)
$C_8H_{10}ClNO_2S$
(7463-22-1)
$C_8H_{10}ClNO_3S$
(88-56-2)
$C_8H_{10}ClN_3$
(7203-90-9)
$C_8H_{10}ClOPS$
(5075-13-8)
$C_8H_{10}Cl_2$
(13547-06-3)
(13547-07-4)
(14112-00-6)
(65122-21-6)
(83682-54-6)
(83682-56-8)
(83682-65-9)
$C_8H_{10}Cl_2N_4S_2$
(13733-95-4)
$C_8H_{10}Cl_2Si$
(1125-27-5)
$C_8H_{10}Cl_3N_3$
(30362-62-0)
$C_8H_{10}Cl_3N_3O$
(30863-48-0)
$C_8H_{10}Cl_3N_3O_2$
(30863-40-2)
$C_8H_{10}Cl_3N_5O$
(30804-91-2)
$C_8H_{10}Cl_4$
(83682-50-2)
(83682-52-4)
$C_8H_{10}Cl_4N_4$
(24802-94-6)
(30339-58-3)
$C_8H_{10}Cl_6$
(83682-64-8)
$C_8H_{10}F_9NO_3S$
(34449-89-3)
$C_8H_{10}NO_5PS$
(298-00-0)
(597-89-7)
$C_8H_{10}NO_6P$
(54-47-7)
(950-35-6)
$C_8H_{10}NO_6P.H_2O$
(41468-25-1)
$C_8H_{10}N_2$
(23747-48-0)
(32736-90-6)
(34413-35-9)
$C_8H_{10}N_2O$
(63-99-0)
(102-28-3)
(114-83-0)
(122-80-5)
(138-89-6)
(555-48-6)
(612-64-6)
(622-51-5)
(937-40-6)
(1007-36-9)
(19406-86-1)
$C_8H_{10}N_2O.ClH$
(621-35-2)
(4801-39-2)
$C_8H_{10}N_2O_2$
(100-23-2)
(2556-36-7)
(3665-80-3)
(6263-38-3)
(10439-77-7)
(16947-63-0)
(34761-82-5)
$C_8H_{10}N_2O_3S$
(80-11-5)
(144-80-9)
$C_8H_{10}N_2O_4S$
(3337-71-1)
$C_8H_{10}N_2O_4$
(500-44-7)
$C_8H_{10}N_2O_4S.Na$
(2302-17-2)
$C_8H_{10}N_2S$
(536-33-4)
(614-78-8)
(2724-69-8)
$C_8H_{10}N_2S.ClH$
(538-28-3)
$C_8H_{10}N_3O_3S.Na$
(140-56-7)
$C_8H_{10}N_4O_2$
(58-08-2)
(2760-98-7)
$C_8H_{10}O$
(60-12-8)
(89-95-2)
(90-00-6)
(95-65-8)
(95-87-4)
(98-85-1)
(100-84-5)
(103-73-1)
(104-93-8)
(105-67-9)
(108-68-9)
(123-07-9)
(526-75-0)
(538-86-3)
(576-26-1)
(587-03-1)
(589-18-4)
(620-17-7)
(1300-71-6)
(25429-37-2)
(26897-24-5)
$C_8H_{10}O.Na$
(19277-91-9)
$C_8H_{10}OS$
(1879-16-9)
(3120-74-9)
$C_8H_{10}O_2$
(91-16-7)
(93-51-6)
(93-56-1)
(94-71-3)
(105-13-5)
(122-99-6)
(150-78-7)
(151-10-0)
(501-94-0)
(527-55-9)
(589-29-7)
(622-62-8)
(933-99-3)
(1331-81-3)
(2785-75-3)
(2785-76-4)
(2896-60-8)
(3071-32-7)
(14548-60-8)
(27598-81-8)
$C_8H_{10}O_2S$
(3185-99-7)
(72428-03-6)
$C_8H_{10}O_3$
(85-42-7)
(91-10-1)
(500-99-2)
(623-68-7)
(760-93-0)
(2033-89-8)
$C_8H_{10}O_3S$
(80-48-8)
(88-61-9)
(98-69-1)
(515-46-8)
(609-54-1)
(25321-41-9)
(57352-34-8)
$C_8H_{10}O_3S.H_3N$
(26447-10-9)
$C_8H_{10}O_3S.K$
(30346-73-7)
$C_8H_{10}O_3S.Na$
(827-19-0)
(1300-72-7)
(14995-38-1)
(30995-65-4)
$C_8H_{10}O_4$
(88-98-2)
(90-65-3)
(635-08-5)
(2305-26-2)
$C_8H_{10}O_5$
(145-73-3)
$C_8H_{10}O_5.2K$
(2164-07-0)
$C_8H_{10}O_8$
(123-23-9)
(1703-58-8)
$C_8H_{10}O_9$
(7408-18-6)
$C_8H_{10}S$
(622-38-8)
(766-92-7)
(25550-52-1)
$C_8H_{11}AsO_2$
(24582-52-3)
$C_8H_{11}BrN_2O_2$
(314-42-1)
$C_8H_{11}ClN_4O$
(4417-70-3)
$C_8H_{11}ClSi$
(768-33-2)
$C_8H_{11}Cl_2NO$
(37764-25-3)

$C_8H_{11}Cl_2N_3O_2$
(66-75-1)
$C_8H_{11}Cl_2N_3S_2$
(30361-95-6)
$C_8H_{11}Cl_3N_4$
(24803-19-8)
(24803-24-5)
(24803-26-7)
(24803-52-9)
(24830-34-0)
(30339-70-9)
(30339-73-2)
(30339-79-8)
$C_8H_{11}Cl_3N_4O$
(24803-62-1)
$C_8H_{11}Cl_3N_4S$
(30369-42-7)
(30369-44-9)
(30369-51-8)
(30369-52-9)
(30369-53-0)
$C_8H_{11}Cl_3O_6$
(15879-93-3)
$C_8H_{11}Cl_4N_5S$
(3028-00-0)
$C_8H_{11}FSi$
(454-57-9)
$C_8H_{11}N$
(64-04-0)
(87-59-2)
(87-62-7)
(95-64-7)
(95-68-1)
(95-78-3)
(98-84-0)
(103-67-3)
(103-69-5)
(104-90-5)
(108-69-0)
(108-75-8)
(121-69-7)
(578-54-1)
(587-02-0)
(589-16-2)
(618-36-0)
(623-08-5)
(695-98-7)
(1122-39-0)
(1122-69-6)
(1122-81-2)
(1300-73-8)
(1462-84-6)
(2233-29-6)
(2627-86-3)
(3886-69-9)
(3999-78-8)
(6975-71-9)
(27987-10-6)
(29611-84-5)
$C_8H_{11}N.ClH$
(21436-96-4)
(51786-53-9)
$C_8H_{11}N.1/2H_2O_4S$
(5471-08-9)
$C_8H_{11}NO$
(51-67-2)
(94-70-2)
(99-07-0)
(102-50-1)
(104-10-9)
(120-71-8)
(122-98-5)
(156-43-4)
(588-05-6)
(621-31-8)
(621-33-0)
(2454-37-7)
(2859-67-8)
(5339-85-5)
(5961-59-1)
(6265-13-0)
(7568-93-6)
(16452-01-0)
(63460-04-8)
$C_8H_{11}NO_2$
(51-61-6)
(102-56-7)
(104-14-3)
(2735-04-8)
(26400-24-8)
(27396-39-0)
$C_8H_{11}NO_2.ClH$
(54150-69-5)
$C_8H_{11}NO_2S$
(640-61-9)
$C_8H_{11}NO_3$
(51-41-2)
(65-23-6)
$C_8H_{11}NO_3.ClH$
(55-27-6)
(58-56-0)
$C_8H_{11}NO_3S$
(94-57-5)
(5246-57-1)
(15471-17-7)
$C_8H_{11}NO_4S$
(93-13-0)
(6471-78-9)
(17601-96-6)
$C_8H_{11}NO_6S_2$
(2494-89-5)
$C_8H_{11}NO_7S_2$
(4726-22-1)
$C_8H_{11}NS$
(4946-22-9)
$C_8H_{11}N_3$
(7227-91-0)
$C_8H_{11}N_3O_3$
(2871-01-4)
$C_8H_{11}N_3O_3S$
(10396-10-8)
$C_8H_{11}N_3O_6$
(54-25-1)
$C_8H_{11}N_5O_2$
(30355-54-5)
$C_8H_{11}N_7S$
(539-21-9)
$C_8H_{11}OS.Cl$
(1005-35-2)
C_8H_{12}
(100-40-3)
(111-78-4)
(659-84-7)
(931-64-6)
(1552-12-1)
(1700-10-3)
(2422-85-7)
(2622-21-1)
(4916-63-6)
(4982-20-1)
(22038-69-3)
(25168-07-4)
(62338-00-5)
$C_8H_{12}BrN_5$
(30360-52-2)
$C_8H_{12}Br_4$
(3322-93-8)
$C_8H_{12}ClNO$
(93-71-0)
$C_8H_{12}ClN_3$
(30362-04-0)
$C_8H_{12}Cl_2$
(45803-84-7)
(83682-48-8)
(83682-49-9)
(83682-51-3)
(83682-61-5)
$C_8H_{12}Cl_2N_6O$
(5034-77-5)
$C_8H_{12}Cl_2O$
(83682-66-0)
$C_8H_{12}Cl_3N_5$
(26234-97-9)
(27431-15-8)
(27431-17-0)
(27470-68-4)
$C_8H_{12}Cl_3N_5O_2$
(26235-06-3)
$C_8H_{12}Cl_4$
(51962-63-1)
(83682-62-6)
$C_8H_{12}F_3N_5$
(721-61-9)
$C_8H_{12}INO_2$
(55406-53-6)
$C_8H_{12}NO_3PS$
(13306-70-2)
$C_8H_{12}NO_5PS_2$
(115-93-5)
$C_8H_{12}NO_6P$
(447-05-2)
$C_8H_{12}N_2$
(51-71-8)
(99-98-9)
(105-10-2)
(629-40-3)
(1124-11-4)
(1477-55-0)
(3333-52-6)
(10309-79-2)
(13238-84-1)
(13360-65-1)
(13925-07-0)
(15707-24-1)
(15707-34-3)
$C_8H_{12}N_2.H_2O_4S$
(156-51-4)
$C_8H_{12}N_2O$
(1197-37-1)
(2814-20-2)
(25773-40-4)
$C_8H_{12}N_2O.2ClH$
(67801-06-3)
$C_8H_{12}N_2O.H_2O_4S$
(68015-98-5)
$C_8H_{12}N_2O_2$
(85-87-0)
(822-06-0)
(20461-60-3)
(28175-98-6)
$C_8H_{12}N_2O_2.2ClH$
(524-36-7)
(66422-95-5)
$C_8H_{12}N_2O_3$
(57-44-3)
(7391-61-9)
$C_8H_{12}N_2O_4S_2$
(73928-09-3)
$C_8H_{12}N_4$
(78-67-1)
$C_8H_{12}N_4O_4$
(2353-33-5)
$C_8H_{12}N_4O_5$
(320-67-2)
(3131-60-0)
(36791-04-5)
$C_8H_{12}N_6$
(20776-81-2)
$C_8H_{12}N_6S$
(30360-07-7)
(30360-95-3)
$C_8H_{12}O$
(95-12-5)
(106-86-5)
(932-66-1)
(1123-09-7)
(4466-24-4)
(16647-04-4)
(24156-95-4)
(24480-99-7)
(25172-06-9)
(28790-86-5)
(30086-02-3)
(54562-24-2)
(55449-70-2)
(55683-21-1)
$C_8H_{12}OS$
(23290-55-3)
$C_8H_{12}O_2$
(106-87-6)
(126-81-8)
(874-23-7)
(933-52-8)
(1920-21-4)
(2396-84-1)
(3508-78-9)
(4342-60-3)
(16523-06-1)
$C_8H_{12}O_3$
(2399-48-6)
$C_8H_{12}O_4$
(141-05-9)
(623-91-6)
(925-21-3)
(1687-30-5)
(10476-95-6)
$C_8H_{12}O_5$
(1931-62-0)
(10420-33-4)
$C_8H_{12}O_6Si$
(4130-08-9)
$C_8H_{12}S$
(1689-78-7)
(33933-73-2)
$C_8H_{12}Si$
(766-77-8)
$C_8H_{13}Br_2N_5$
(30359-68-3)
$C_8H_{13}ClN_4O$
(4417-64-5)
$C_8H_{13}ClN_4S$
(4658-30-4)
$C_8H_{13}Cl_2N_5$
(30355-66-9)
$C_8H_{13}I_2N_5$
(30359-77-4)
$C_8H_{13}N$
(16967-61-6)
$C_8H_{13}NO$
(19519-45-0)
$C_8H_{13}NO_2$
(480-85-3)
(520-63-8)
(13861-99-9)
$C_8H_{13}NO_3$
(702-54-5)
$C_8H_{13}NS$
(53498-30-9)
$C_8H_{13}N_2O_2P$
(1754-58-1)
$C_8H_{13}N_2O_3PS$
(297-97-2)
$C_8H_{13}N_2O_5P$
(529-96-4)
$C_8H_{13}N_2O_5P.2H_2O$
(84878-64-8)
$C_8H_{13}N_3$
(7559-34-4)
(30362-60-8)
$C_8H_{13}N_3OS$
(35045-02-4)
$C_8H_{13}N_3O_6$
(13433-11-9)
$C_8H_{13}N_3S$
(3846-74-0)
$C_8H_{13}N_7$
(30355-57-8)
C_8H_{14}
(280-33-1)
(627-58-7)

(629-05-0)
(764-13-6)
(931-88-4)
(1603-01-6)
(1674-10-8)
(1942-45-6)
(2080-89-9)
(2809-67-8)
(3710-30-3)
(3710-41-6)
(5623-78-9)
(14072-86-7)
(18669-52-8)
(34067-75-9)
$C_8H_{14}BrN_5$
(30359-65-0)
(30360-51-1)
(30360-53-3)
$C_8H_{14}ClNS_2$
(95-06-7)
$C_8H_{14}ClN_5$
(1912-24-9)
(30355-62-5)
(33692-99-8)
(34333-27-2)
$C_8H_{14}ClO_2PS$
(2921-31-5)
$C_8H_{14}Cl_3O_5P$
(126-22-7)
$C_8H_{14}FN_5$
(777-56-0)
$C_8H_{14}IN_5$
(30359-75-2)
$C_8H_{14}N_2O_2S_2$
(26419-73-8)
$C_8H_{14}N_4$
(3217-00-3)
(30339-56-1)
$C_8H_{14}N_4O$
(5248-41-9)
(5248-48-6)
$C_8H_{14}N_4OS$
(21087-64-9)
$C_8H_{14}N_4O_4$
(62488-57-7)
$C_8H_{14}N_8$
(2854-94-6)
$C_8H_{14}O$
(107-54-0)
(110-93-0)
(409-02-9)
(502-49-8)
(645-62-5)
(823-76-7)
(1004-24-6)
(1187-87-7)
(1669-44-9)
(1965-38-4)
(2363-89-5)
(2816-57-1)
(3664-60-6)
(4423-94-3)
(4579-31-1)
(19600-63-6)
(20685-45-4)
(22287-11-2)
(22319-24-0)
(22319-25-1)
(22319-31-9)
(25447-69-2)
(26266-68-2)
(53907-61-2)
$C_8H_{14}O_2$
(97-86-9)
(97-88-1)
(104-50-7)
(140-77-2)
(142-30-3)
(177-10-6)
(585-07-9)
(585-25-1)
(622-45-7)
(698-76-0)
(1460-16-8)
(1540-38-1)
(1577-19-1)
(1708-82-3)
(2426-07-5)
(2497-18-9)
(2835-39-4)
(3681-71-8)
(4331-54-8)
(5169-51-7)
(5292-21-7)
(5309-52-4)
(7299-91-4)
(18294-89-8)
(20628-36-8)
(34003-77-5)
(40991-93-3)
(56922-75-9)
(63721-05-1)
$C_8H_{14}O_2S_2$
(62-46-4)
$C_8H_{14}O_2S_4Zn$
(1000-90-4)
(1000-90-4)
$C_8H_{14}O_3$
(97-72-3)
(106-31-0)
(591-60-6)
(1694-31-1)
(2370-63-0)
(4305-26-4)
(7487-28-7)
(10138-34-8)
(14112-98-2)
(23474-91-1)
(56009-36-0)
(83682-67-1)
(83682-68-2)
$C_8H_{14}O_4$
(123-25-1)
(505-48-6)
(609-08-5)
(616-88-6)
(627-93-0)
(828-00-2)
(1606-85-5)
(2224-15-9)
(3437-84-1)
$C_8H_{14}O_4S$
(4131-74-2)
$C_8H_{14}O_4S_2$
(15441-06-2)
$C_8H_{14}O_5$
(7554-12-3)
$C_8H_{14}O_5S$
(3179-56-4)
$C_8H_{14}O_6$
(105-64-6)
(16066-38-9)
$C_8H_{14}O_6Si$
(17689-77-9)
$C_8H_{15}Br$
(1647-26-3)
$C_8H_{15}BrO_2$
(2623-82-7)
(17696-11-6)
$C_8H_{15}ClO$
(111-64-8)
(760-67-8)
$C_8H_{15}ClO_3$
(5330-17-6)
$C_8H_{15}Cl_3$
(85269-46-1)
$C_8H_{15}N$
(124-12-9)
(3399-73-3)
(13618-93-4)
$C_8H_{15}NO$
(120-29-6)
(1124-53-4)
(5809-41-6)
$C_8H_{15}NOS$
(6961-73-5)
$C_8H_{15}NO_2$
(1852-16-0)
(2235-46-3)
(2867-47-2)
(6065-09-4)
(6065-10-7)
(6065-11-8)
(6949-77-5)
(10601-70-4)
(15862-72-3)
(16669-59-3)
(58695-41-3)
(65882-19-1)
(65882-21-5)
$C_8H_{15}NO_2.C_2H_4O_2$
(10018-87-8)
$C_8H_{15}NO_3$
(1522-00-5)
(16067-01-9)
$C_8H_{15}NO_4$
(23102-02-5)
$C_8H_{15}NS_2.Na$
(77100-49-3)
$C_8H_{15}N_2O_3PS$
(108-35-0)
$C_8H_{15}N_2O_4P$
(108-34-9)
$C_8H_{15}N_3O_2$
(30979-48-7)
(67292-92-6)
$C_8H_{15}N_3O_7$
(18883-66-4)
$C_8H_{15}N_5$
(1973-07-5)
(3283-17-8)
(26235-17-6)
(26235-23-4)
(27430-89-3)
(27430-90-6)
(27470-66-2)
$C_8H_{15}N_5O$
(673-04-1)
(2163-68-0)
(3035-45-8)
(30360-70-4)
$C_8H_{15}N_5O_2$
(5944-07-0)
$C_8H_{15}N_5S$
(1014-69-3)
(1014-70-6)
$C_8H_{15}O_2.Na$
(1069-66-5)
$C_8H_{15}O_6PS_2$
(35884-76-5)
$C_8H_{15}P$
(13396-80-0)
(13887-02-0)
C_8H_{16}
(107-39-1)
(107-40-4)
(111-66-0)
(111-67-1)
(292-64-8)
(560-23-6)
(565-77-5)
(583-57-3)
(589-90-2)
(590-66-9)
(591-21-9)
(592-99-4)
(598-96-9)
(624-29-3)
(627-97-4)
(638-04-0)
(930-90-5)
(1632-16-2)
(1678-91-7)
(2040-96-2)
(2207-01-4)
(2207-03-6)
(2207-04-7)
(2815-58-9)
(3404-67-9)
(3404-78-2)
(3726-47-4)
(3875-51-2)
(4126-78-7)
(4516-69-2)
(4810-09-7)
(5026-76-6)
(6876-23-9)
(7116-86-1)
(7145-20-2)
(7642-04-8)
(11071-47-9)
(13389-42-9)
(14850-23-8)
(15870-10-7)
(15890-40-1)
(15910-22-2)
(16106-59-5)
(16747-50-5)
(18923-87-0)
(25167-70-8)
(25377-83-7)
(26266-69-3)
(30498-64-7)
(30951-95-2)
(61141-83-1)
(61593-45-1)
(61665-19-8)
$C_8H_{16}Br_2$
(62168-26-7)
$C_8H_{16}Cl_2$
(20395-24-8)
$C_8H_{16}Cl_2N_2O_2$
(6865-97-0)
$C_8H_{16}Cl_2O$
(6334-96-9)
$C_8H_{16}Cl_2OS_2$
(63918-89-8)
$C_8H_{16}Cl_2O_3$
(638-56-2)
$C_8H_{16}F_2$
(61350-03-6)
$C_8H_{16}N.Cl$
(7398-69-8)
$C_8H_{16}NO_2.Cl$
(44992-01-0)
$C_8H_{16}NO_3PS_2$
(950-10-7)
$C_8H_{16}NO_4PS_2$
(144-41-2)
$C_8H_{16}NO_5P$
(141-66-2)
$C_8H_{16}N_2O_2S_2$
(103-34-4)
$C_8H_{16}N_2O_3$
(2224-44-4)
$C_8H_{16}N_2O_7$
(14901-08-7)
$C_8H_{16}N_3OPS$
(2168-68-5)
$C_8H_{16}N_6$
(16268-62-5)
$C_8H_{16}O$
(96-06-0)
(106-68-3)
(111-13-7)
(123-05-7)
(124-13-0)
(541-85-5)
(589-63-9)
(624-42-0)

(696-71-9)
(928-68-7)
(1004-29-1)
(1193-81-3)
(1335-09-7)
(1888-57-9)
(1940-18-7)
(2105-40-0)
(2984-50-1)
(3391-86-4)
(3760-20-1)
(4442-79-9)
(4630-06-2)
(5337-72-4)
(5363-64-4)
(5441-52-1)
(5932-91-2)
(6137-11-7)
(13019-20-0)
(15045-43-9)
(15726-15-5)
(18185-81-4)
(18217-12-4)
(18409-17-1)
(26118-38-7)
(27457-18-7)
(50639-00-4)
(53907-72-5)
(55320-58-6)
(57648-55-2)
(64275-73-6)
$C_8H_{16}O_2$
(78-72-8)
(97-85-8)
(97-87-0)
(99-66-1)
(105-08-8)
(105-68-0)
(106-73-0)
(108-84-9)
(109-21-7)
(123-66-0)
(124-07-2)
(142-92-7)
(149-57-5)
(496-03-7)
(539-90-2)
(557-00-6)
(624-54-4)
(819-97-6)
(933-40-4)
(1188-02-9)
(4223-11-4)
(6135-54-2)
(6299-66-7)
(7250-85-3)
(25086-25-3)
(25103-52-0)
(41065-91-2)
(42329-90-8)
$C_8H_{16}O_2$.1/3Al
(6028-57-5)
$C_8H_{16}O_2$.1/2Ba
(2457-01-4)
$C_8H_{16}O_2$.1/2Ca
(136-51-6)
$C_8H_{16}O_2$.1/2Cd
(2420-98-6)
$C_8H_{16}O_2$.1/2Co
(136-52-7)
$C_8H_{16}O_2$.xCu
(22221-10-9)
$C_8H_{16}O_2$.xFe
(19583-54-1)
$C_8H_{16}O_2$.K
(764-71-6)
(3164-85-0)
$C_8H_{16}O_2$.xMn
(15956-58-8)
$C_8H_{16}O_2$.Na
(1984-06-1)
(19766-89-3)
$C_8H_{16}O_2$.1/2Ni
(4454-16-4)
$C_8H_{16}O_2$.1/2Pb
(301-08-6)
$C_8H_{16}O_2$.xPb
(15696-43-2)
(16996-40-0)
$C_8H_{16}O_2$.1/2Sn
(301-10-0)
$C_8H_{16}O_2$.1/2Zn
(136-53-8)
$C_8H_{16}O_2$.xZr
(22464-99-9)
$C_8H_{16}O_3$
(109-13-7)
(112-07-2)
(14292-27-4)
(55724-73-7)
$C_8H_{16}O_4$
(108-62-3)
(112-15-2)
(294-93-9)
(2372-21-6)
$C_8H_{16}O_4S_2$
(1113-14-0)
$C_8H_{16}O_5$
(42064-17-5)
$C_8H_{16}O_8$.Pb
(546-67-8)
$C_8H_{17}Br$
(111-83-1)
$C_8H_{17}Cl$
(111-85-3)
(123-04-6)
(628-61-5)
(26655-49-2)
$C_8H_{17}ClO_2$
(1120-23-6)
$C_8H_{17}ClO_2S$
(7795-95-1)
$C_8H_{17}Cl_3Si$
(5283-66-9)
$C_8H_{17}Cl_3Sn$
(3091-25-6)
$C_8H_{17}CoO_2$
(13586-82-8)
$C_8H_{17}F$
(463-11-6)
$C_8H_{17}FO_2S$
(40630-63-5)
$C_8H_{17}I$
(557-36-8)
(629-27-6)
$C_8H_{17}N$
(98-94-2)
(104-89-2)
(458-88-8)
(5459-93-8)
(5470-02-0)
$C_8H_{17}NO$
(533-15-3)
(629-01-6)
(1005-67-0)
(2842-38-8)
(7501-79-3)
(24448-89-3)
(64890-90-0)
$C_8H_{17}NO_2$
(28770-01-6)
$C_8H_{17}NO_3$
(629-39-0)
(27247-96-7)
$C_8H_{17}NO_3S$.Na
(3076-05-9)
$C_8H_{17}NO_5$
(68413-83-2)
$C_8H_{17}N_5$
(79787-65-8)
$C_8H_{17}O_4P$
(1067-90-9)
$C_8H_{17}O_4P$.2K
(19045-79-5)
$C_8H_{17}O_4PS_2$
(919-54-0)
$C_8H_{17}O_4S$.Na
(142-31-4)
$C_8H_{17}O_5P$
(867-13-0)
$C_8H_{17}O_5PS$
(2425-25-4)
C_8H_{18}
(111-65-9)
(540-84-1)
(560-21-4)
(563-16-6)
(564-02-3)
(565-75-3)
(583-48-2)
(584-94-1)
(589-43-5)
(589-53-7)
(589-81-1)
(590-73-8)
(592-13-2)
(592-27-8)
(594-82-1)
(609-26-7)
(619-99-8)
(1067-08-9)
(26635-64-3)
(28777-67-5)
(29222-48-8)
$C_8H_{18}AlCl$
(1779-25-5)
$C_8H_{18}ClN$
(999-33-7)
$C_8H_{18}ClO_2S_2$.Cl
(64036-91-5)
$C_8H_{18}Cl_2Sn$
(683-18-1)
$C_8H_{18}CrO_4$
(1189-85-1)
$C_8H_{18}F_2Sn$
(563-25-7)
$C_8H_{18}NO_2$.Cl
(62-51-1)
$C_8H_{18}NO_4PS_2$
(2275-23-2)
$C_8H_{18}N_2$
(3529-08-6)
(5700-53-8)
(51388-00-2)
$C_8H_{18}N_2O$
(924-16-3)
(4385-05-1)
(57910-79-9)
$C_8H_{18}N_2OS$.ClH
(19622-19-6)
$C_8H_{18}N_2O_2$
(122-96-3)
(3817-11-6)
(6841-96-9)
(90853-18-2)
(90853-22-8)
$C_8H_{18}N_4S$
(59653-29-1)
$C_8H_{18}N_6$.2ClH
(2997-92-4)
$C_8H_{18}NiO_2$
(7580-31-6)
$C_8H_{18}O$
(104-76-7)
(106-67-2)
(111-87-5)
(123-44-4)
(123-96-6)
(142-96-1)
(589-62-8)
(589-98-0)
(597-05-7)
(597-76-2)
(598-01-6)
(598-06-1)
(625-25-2)
(628-55-7)
(1653-40-3)
(3054-92-0)
(3730-60-7)
(4128-31-8)
(4209-91-0)
(4730-22-7)
(5162-48-1)
(5582-82-1)
(6163-66-2)
(6863-58-7)
(7294-05-5)
(13432-25-2)
(14979-39-6)
(19550-03-9)
(26952-21-6)
(29063-28-3)
(31367-46-1)
(32970-45-9)
(54630-50-1)
$C_8H_{18}O$.1/3Al
(14624-13-6)
$C_8H_{18}O$.1/4Ti
(1070-10-6)
$C_8H_{18}OSi_2$
(2627-95-4)
$C_8H_{18}OSn$
(818-08-6)
$C_8H_{18}O_2$
(94-96-2)
(110-03-2)
(110-05-4)
(112-25-4)
(144-19-4)
(1117-86-8)
(1338-23-4)
(1741-41-9)
(51422-54-9)
$C_8H_{18}O_2S$
(1886-75-5)
$C_8H_{18}O_3$
(78-39-7)
(112-34-5)
(112-36-7)
(54446-78-5)
(97419-16-4)
$C_8H_{18}O_3S$.Na
(5324-84-5)
$C_8H_{18}O_3Si$
(78-08-0)
$C_8H_{18}O_4$
(112-49-2)
(112-50-5)
$C_8H_{18}O_4S$.Na
(126-92-1)
$C_8H_{18}O_4S_2$
(76-20-0)
(3563-34-6)
$C_8H_{18}O_5$
(112-60-7)
$C_8H_{18}S$
(107-47-1)
(111-88-6)
(141-59-3)
(544-40-1)
(626-26-6)
(3001-66-9)
$C_8H_{18}SSn$
(4253-22-9)
$C_8H_{18}S_2$
(110-06-5)
(629-45-8)
$C_8H_{18}S_3Sn_2$
(15666-29-2)

$C_8H_{18}Sn$
(14488-53-0)
$C_8H_{19}Al$
(1191-15-7)
$C_8H_{19}N$
(104-75-6)
(107-45-9)
(110-96-3)
(111-86-4)
(111-92-2)
(626-23-3)
(693-16-3)
(4444-68-2)
(7087-68-5)
(20810-06-4)
(42966-64-3)
$C_8H_{19}N.CH_5AsO_3$
(6379-37-9)
$C_8H_{19}NO$
(96-80-0)
$C_8H_{19}NO_2$
(102-79-4)
(140-82-9)
(2160-93-2)
(21838-75-5)
$C_8H_{19}NO_6S_2.ClH$
(3458-22-8)
$C_8H_{19}N_3$
(90853-14-8)
$C_8H_{19}OP$
(684-19-5)
$C_8H_{19}O_2PS_2$
(107-55-1)
(2253-52-3)
(13194-48-4)
$C_8H_{19}O_2PS_2.Na$
(33619-92-0)
$C_8H_{19}O_2PS_3$
(78-52-4)
(298-04-4)
$C_8H_{19}O_3P$
(1809-19-4)
$C_8H_{19}O_3PS_2$
(126-75-0)
(298-03-3)
$C_8H_{19}O_3PS_2.C_8H_{19}O_3PS_2$
(8065-48-3)
$C_8H_{19}O_3PS_3$
(2497-07-6)
$C_8H_{19}O_4P$
(107-66-4)
(1070-03-7)
(3991-73-9)
(12645-53-3)
(26403-12-3)
(39407-03-9)
$C_8H_{19}O_4P.K$
(25238-98-6)
$C_8H_{19}O_4P.2Na$
(15505-13-2)
$C_8H_{19}O_4PS_3$
(2497-06-5)
$C_8H_{19}O_5PS_2$
(2496-91-5)
(4891-54-7)
$C_8H_{20}N.Br$
(71-91-0)
$C_8H_{20}N.Cl$
(56-34-8)
$C_8H_{20}N.ClO_4$
(2567-83-1)
$C_8H_{20}N.I$
(68-05-3)
$C_8H_{20}NO_3.C_2H_5O_4S$
(31774-90-0)
$C_8H_{20}N_2$
(97-84-7)
$C_8H_{20}N_2.2ClH$
(7613-16-3)
$C_8H_{20}N_2O$
(3033-62-3)
(90853-17-1)
$C_8H_{20}N_2O.C_5H_{10}N_2$
(62765-93-9)
$C_8H_{20}N_4$
(6531-38-0)
$C_8H_{20}NaO_4P$
(31044-12-9)
$C_8H_{20}O_3Si$
(78-07-9)
$C_8H_{20}O_4Si$
(78-10-4)
$C_8H_{20}O_5P_2S_2$
(3689-24-5)
$C_8H_{20}O_7P_2$
(107-49-3)
$C_8H_{20}Pb$
(78-00-2)
(65122-14-7)
$C_8H_{20}Si$
(631-36-7)
$C_8H_{20}Sn$
(597-64-8)
(1002-53-5)
$C_8H_{21}NO_2Si$
(3179-76-8)
$C_8H_{22}N_2O_3Si$
(1760-24-3)
$C_8H_{22}N_4$
(10563-26-5)
$C_8H_{23}N_5$
(112-57-2)
(31295-46-2)
$C_8H_{24}N_4O_3P_2$
(152-16-9)
$C_8H_{24}O_2Si_3$
(107-51-7)
$C_8H_{24}O_4Si_4$
(556-67-2)
$C_8H_{24}O_8P_2.O_4S$
(55566-30-8)
$C_9Cl_7N_3O$
(5599-24-6)
$C_9Cl_{15}N_3$
(30362-77-7)
$C_9H_2Cl_5N_3O$
(30886-27-2)
$C_9H_2Cl_6O_3$
(115-27-5)
$C_9H_3Cl_3O_3$
(4422-95-1)
$C_9H_3Cl_4N_3O$
(18247-77-3)
$C_9H_3Cl_5N_4$
(2272-39-1)
$C_9H_3Cl_9N_6O_3$
(30360-29-3)
$C_9H_3Cl_{12}N_3$
(30362-76-6)
$C_9H_4ClF_2N_3O$
(886-45-3)
$C_9H_4Cl_2N_4O_3$
(30886-29-4)
$C_9H_4Cl_3IO$
(777-11-7)
$C_9H_4Cl_3NO_2S$
(133-07-3)
$C_9H_4Cl_3N_3$
(10202-46-7)
$C_9H_4Cl_3N_3O$
(30886-25-0)
(30886-26-1)
$C_9H_4Cl_3N_3S$
(30894-63-4)
$C_9H_4Cl_4N_4$
(2272-33-5)
(18996-86-6)
$C_9H_4Cl_4O_4$
(887-54-7)
$C_9H_4Cl_6O_4$
(115-28-6)
$C_9H_4Cl_8O$
(297-78-9)
$C_9H_4Cl_9N_3O_2$
(30362-46-0)
$C_9H_4F_2N_4O_3$
(2343-60-4)
$C_9H_4F_{17}NO_2S$
(31506-32-8)
$C_9H_4N_2S_3$
(93-75-4)
$C_9H_4O_5$
(552-30-7)
$C_9H_5BrCl_2N_4$
(20376-36-7)
(30369-84-7)
(30369-85-8)
$C_9H_5Br_2ClN_4$
(30357-83-6)
(30388-94-4)
$C_9H_5Br_2N_3O$
(30894-96-3)
$C_9H_5Br_2N_3S$
(30894-97-4)
$C_9H_5ClF_2N_4$
(2925-56-6)
(30369-78-9)
C_9H_5ClINO
(130-26-7)
$C_9H_5ClO_2$
(3240-10-6)
(7396-28-3)
(24654-08-8)
$C_9H_5Cl_2N$
(86-98-6)
$C_9H_5Cl_2N_3$
(1700-02-3)
$C_9H_5Cl_2N_3O$
(4682-78-4)
$C_9H_5Cl_2N_3S$
(3019-16-7)
$C_9H_5Cl_2N_5O_2$
(2352-36-5)
$C_9H_5Cl_3N_4$
(101-05-3)
(2272-29-9)
(20376-34-5)
$C_9H_5Cl_3N_4O_2S$
(17752-71-5)
$C_9H_5F_2N_3O$
(30886-18-1)
$C_9H_5F_2N_3S$
(717-91-9)
$C_9H_5I_2NO$
(83-73-8)
C_9H_6BrN
(5332-24-1)
$C_9H_6Br_2N_4$
(30357-82-5)
C_9H_6ClNO
(130-16-5)
$C_9H_6ClNO_2$
(17564-64-6)
$C_9H_6ClNO_3S$
(3813-05-6)
$C_9H_6ClN_3O_2$
(30885-98-4)
$C_9H_6Cl_2N_2O_3$
(20354-26-1)
$C_9H_6Cl_2N_4$
(2272-40-4)
(30369-33-6)
$C_9H_6Cl_2N_4O_2S$
(30369-89-2)
$C_9H_6Cl_2N_4O_3S$
(16110-89-7)
$C_9H_6Cl_2O_2$
(1201-99-6)
$C_9H_6Cl_3N_3O$
(27241-31-2)
$C_9H_6Cl_3N_5O_2S$
(30357-78-9)
$C_9H_6Cl_6$
(56682-87-2)
$C_9H_6Cl_6N_6O_3$
(30360-28-2)
$C_9H_6Cl_6O_3S$
(115-29-7)
(959-98-8)
(33213-65-9)
$C_9H_6Cl_9N_3$
(30362-75-5)
$C_9H_6Cl_9N_3O_3$
(29808-66-0)
(30805-18-6)
$C_9H_6F_2N_4$
(717-90-8)
$C_9H_6N_2O_2$
(91-08-7)
(584-84-9)
(607-34-1)
(607-35-2)
(613-50-3)
(1321-38-6)
(12408-11-6)
(26471-62-5)
$C_9H_6N_2O_3$
(56-57-5)
$C_9H_6N_2O_4$
(41663-84-7)
$C_9H_6N_2S_3$
(21564-17-0)
$C_9H_6N_4O_2$
(30361-90-1)
$C_9H_6O_2$
(91-64-5)
(606-23-5)
$C_9H_6O_3$
(93-35-6)
(703-59-3)
(19438-61-0)
(30140-42-2)
$C_9H_6O_4$
(305-01-1)
(485-47-2)
(116211-88-2)
$C_9H_6O_5$
(528-46-1)
$C_9H_6O_6$
(528-44-9)
(554-95-0)
(569-51-7)
$C_9H_7BrN_4O$
(30357-93-8)
$C_9H_7BrO_2$
(32862-97-8)
$C_9H_7ClN_4$
(3842-53-3)
(30360-30-6)
$C_9H_7ClN_4O$
(17584-14-4)
(30357-91-6)
$C_9H_7ClN_4S$
(30357-97-2)
(30369-75-6)
$C_9H_7ClN_4S_2$
(30362-17-5)
C_9H_7ClO
(102-92-1)
(42180-82-5)
(73908-22-2)
$C_9H_7ClO_2$
(1615-02-7)
$C_9H_7Cl_2N$
(17849-64-8)
$C_9H_7Cl_2N_5$
(30354-89-3)
$C_9H_7Cl_2N_5O_2S$
(18237-29-1)
$C_9H_7Cl_3N_6$

(30388-90-0)
$C_9H_7Cl_3O_2$
(29990-39-4)
$C_9H_7Cl_3O_3$
(93-72-1)
(1928-37-6)
$C_9H_7Cl_3O_3.C_2H_7N$
(55617-85-1)
$C_9H_7Cl_3O_3.C_2H_7NO$
(7374-47-2)
$C_9H_7Cl_3O_3.C_3H_9NO$
(53404-13-0)
$C_9H_7Cl_3O_3.C_4H_{11}NO_2$
(51170-59-3)
$C_9H_7Cl_3O_3.C_6H_{15}N$
(53404-74-3)
$C_9H_7Cl_3O_3.C_6H_{15}NO_2$
(53404-09-4)
$C_9H_7Cl_3O_3.K$
(2818-16-8)
$C_9H_7Cl_3O_3.Na$
(37913-89-6)
$C_9H_7Cl_5N_2O$
(20856-57-9)
$C_9H_7Cl_6N_3O_2$
(30362-45-9)
$C_9H_7Cl_6N_5$
(24848-41-7)
$C_9H_7FN_4O$
(30357-90-5)
$C_9H_7MnO_3$
(12108-13-3)
C_9H_7N
(91-22-5)
(119-65-3)
(495-10-3)
$C_9H_7N.H_2O_4S$
(530-66-5)
C_9H_7NO
(59-31-4)
(148-24-3)
(491-30-5)
(1321-40-0)
$C_9H_7NO.C_6H_8O_7$
(134-30-5)
C_9H_7NOS
(3819-18-9)
$C_9H_7NO_2$
(550-44-7)
$C_9H_7NO_3$
(118-29-6)
$C_9H_7NO_3S$
(85-48-3)
$C_9H_7NO_4S$
(2149-36-2)
$C_9H_7N_3$
(1722-18-5)
$C_9H_7N_3O_2$
(7459-63-4)
$C_9H_7N_3O_4S$
(531-82-8)
$C_9H_7N_3S$
(41814-78-2)
$C_9H_7N_3S_2$
(30886-13-6)
$C_9H_7N_5O_3$
(30369-69-8)
$C_9H_7N_7O_2S$
(446-86-6)
C_9H_8
(95-13-6)
$C_9H_8Br_2O_2$
(20217-01-0)
$C_9H_8ClN_5$
(500-42-5)
(4514-53-8)
(16007-72-0)
$C_9H_8ClN_5O$
(30360-77-1)
(30360-78-2)
$C_9H_8ClO_3.H_4N$
(19480-39-8)
$C_9H_8ClO_3.Na$
(3653-48-3)
$C_9H_8Cl_2O_2$
(14437-20-8)
(56882-52-1)
$C_9H_8Cl_2O_3$
(120-36-5)
(1928-38-7)
(3004-74-8)
(3307-41-3)
(24295-27-0)
$C_9H_8Cl_2O_4$
(76330-06-8)
$C_9H_8Cl_2O_5$
(20624-96-8)
$C_9H_8Cl_3NO_2$
(42864-21-1)
$C_9H_8Cl_3NO_2S$
(133-06-2)
$C_9H_8Cl_6N_4O$
(24803-10-9)
$C_9H_8F_3NO$
(351-36-0)
$C_9H_8F_3NO_2$
(370-32-1)
$C_9H_8N_2$
(578-66-5)
(580-15-4)
(580-17-6)
(580-22-3)
(611-34-7)
(7251-61-8)
(12794-10-4)
$C_9H_8N_2O$
(5004-48-8)
$C_9H_8N_2O_2$
(2152-34-3)
(4637-56-3)
$C_9H_8N_2O_5$
(2645-07-0)
$C_9H_8N_2S.BrH.H_2O$
(52253-69-7)
$C_9H_8N_4$
(1853-95-8)
(4040-07-7)
$C_9H_8N_4O$
(33957-63-0)
$C_9H_8N_4OS$
(51707-55-2)
(71769-74-9)
$C_9H_8N_4O_2$
(20964-55-0)
$C_9H_8N_4S$
(30369-74-5)
$C_9H_8N_4S_2$
(13733-91-0)
$C_9H_8N_6O_2$
(29366-72-1)
$C_9H_8N_6O_3$
(30360-79-3)
C_9H_8O
(83-33-0)
(104-55-2)
(615-13-4)
(4265-25-2)
(14371-10-9)
(25586-38-3)
(30286-23-8)
(43145-54-6)
C_9H_8OS
(3528-17-4)
$C_9H_8O_2$
(119-84-6)
(140-10-3)
(579-07-7)
(621-82-9)
(669-04-5)
(1075-49-6)
(3453-64-3)
(20895-41-4)
(54120-64-8)
$C_9H_8O_3$
(156-06-9)
(501-98-4)
(586-89-0)
(588-30-7)
(614-60-8)
(1571-08-0)
(7400-08-0)
(25429-38-3)
$C_9H_8O_4$
(50-78-2)
(156-39-8)
(331-39-5)
(644-31-5)
(1679-64-7)
(2345-34-8)
(4316-23-8)
(4376-18-5)
(5780-07-4)
(30497-87-1)
(37102-74-2)
$C_9H_8O_6$
(2134-91-0)
(63035-28-9)
C_9H_8S
(31393-23-4)
C_9H_9Br
(1124-14-7)
(36617-02-4)
$C_9H_9BrO_2$
(5798-75-4)
$C_9H_9Br_2NO_3$
(537-24-6)
$C_9H_9Br_6N_3$
(30362-94-8)
C_9H_9Cl
(30030-25-2)
$C_9H_9ClN_6$
(30360-11-3)
(30360-12-4)
C_9H_9ClO
(145-94-8)
(53299-53-9)
$C_9H_9ClO_2$
(140-18-1)
(34040-64-7)
$C_9H_9ClO_2S$
(94-76-8)
$C_9H_9ClO_3$
(94-74-6)
(101-10-0)
(3307-39-9)
$C_9H_9ClO_3.C_4H_{11}NO_2$
(20405-19-0)
$C_9H_9ClO_3.C_9H_6ClNO_3S.Na$
(65280-19-5)
$C_9H_9ClO_4$
(76341-69-0)
$C_9H_9Cl_2NO$
(709-98-8)
$C_9H_9Cl_2NO_2$
(587-56-4)
(1966-58-1)
(35661-56-4)
(62046-37-1)
$C_9H_9Cl_2N_3$
(4205-90-7)
$C_9H_9Cl_3$
(61465-79-0)
$C_9H_9Cl_3F_{12}N_3O_3P_3$
(65601-41-4)
$C_9H_9Cl_3N_6O_3$
(30360-27-1)
$C_9H_9Cl_3O_3$
(2539-13-1)
$C_9H_9Cl_6N_3$
(24481-40-1)
(24481-41-2)
(30362-74-4)
$C_9H_9Cl_6N_3O$
(30863-56-0)
$C_9H_9Cl_6N_3S$
(24481-69-4)
(24504-18-5)
$C_9H_9HgO_2S.Na$
(54-64-8)
$C_9H_9I_2NO_3$
(66-02-4)
(300-39-0)
C_9H_9N
(83-34-1)
(95-20-5)
(603-76-9)
(614-96-0)
(1823-91-2)
(27323-28-0)
C_9H_9NO
(104-47-2)
(1006-94-6)
(1438-94-4)
(54698-11-2)
(116211-85-9)
$C_9H_9NO_3$
(89-52-1)
(495-69-2)
(556-08-1)
(21715-90-2)
$C_9H_9NO_4$
(99-77-4)
(5251-93-4)
(5255-75-4)
$C_9H_9NO_5$
(53606-41-0)
(67851-29-0)
$C_9H_9NO_5S$
(5031-74-3)
$C_9H_9N_3OS$
(1929-88-0)
$C_9H_9N_3O_2$
(10605-21-7)
(30805-30-2)
$C_9H_9N_3O_2.H_3O_4P$
(52316-55-9)
$C_9H_9N_3O_2S_2$
(72-14-0)
$C_9H_9N_3O_2S_2.Na$
(144-74-1)
$C_9H_9N_3O_6$
(30863-34-4)
$C_9H_9N_3S$
(490-55-1)
$C_9H_9N_3S_2$
(30805-32-4)
$C_9H_9N_5$
(91-76-9)
(537-17-7)
$C_9H_9N_5O$
(1467-72-7)
$C_9H_9N_5S$
(30360-86-2)
$C_9H_9N_7O_2$
(30360-13-5)
C_9H_{10}
(98-83-9)
(100-80-1)
(300-57-2)
(496-11-7)
(611-15-4)
(622-97-9)
(637-50-3)
(1319-73-9)
(25013-15-4)
$C_9H_{10}BrClN_2O_2$
(13360-45-7)
$C_9H_{10}ClNO$
(2759-54-8)
(57058-33-0)

$C_9H_{10}ClNO_2$
(3747-48-6)
(5825-87-6)
(7073-42-9)
(34197-98-3)
$C_9H_{10}Cl_2$
(54789-29-6)
$C_9H_{10}Cl_2N_2O$
(330-54-1)
$C_9H_{10}Cl_2N_2O_2$
(330-55-2)
$C_9H_{10}Cl_2O_3$
(99849-00-0)
(108544-93-0)
$C_9H_{10}Cl_3N_5$
(24848-45-1)
$C_9H_{10}Cl_3O_3PS$
(2633-54-7)
$C_9H_{10}Cl_6N_4$
(24802-82-2)
(24802-83-3)
(24802-84-4)
(24802-85-5)
(24803-06-3)
$C_9H_{10}FNO_2$
(51-65-0)
(60-17-3)
$C_9H_{10}F_2O_3S$
(882-71-3)
$C_9H_{10}HgO_3$
(122-64-5)
$C_9H_{10}NO_3PS$
(2636-26-2)
$C_9H_{10}NO_4P$
(61090-94-6)
$C_9H_{10}N_2$
(532-12-7)
(582-60-5)
(936-49-2)
$C_9H_{10}N_2O$
(92-43-3)
$C_9H_{10}N_2OS$
(94-45-1)
$C_9H_{10}N_2O_2$
(63-98-9)
(28141-13-1)
(37704-51-1)
$C_9H_{10}N_2O_3S_2$
(452-35-7)
$C_9H_{10}N_2O_3$
(61-78-9)
(56222-10-7)
$C_9H_{10}N_2O_4$
(608-50-4)
(7304-99-6)
$C_9H_{10}N_2O_5$
(4097-47-6)
$C_9H_{10}N_2S$
(28291-69-2)
(29927-08-0)
$C_9H_{10}N_4O$
(38261-35-7)
$C_9H_{10}N_4O_2S_2$
(144-82-1)
$C_9H_{10}N_6$
(5606-27-9)
$C_9H_{10}O$
(93-53-8)
(93-55-0)
(103-79-7)
(104-09-6)
(104-54-1)
(122-00-9)
(577-16-2)
(585-74-0)
(637-69-4)
(1470-94-6)
(1745-81-9)
(2085-88-3)
(4748-78-1)
(5779-94-2)
(5973-71-7)
(6351-10-6)
(26444-19-9)
(28351-09-9)
(53951-50-1)
(103453-97-0)
$C_9H_{10}O_2$
(70-70-2)
(93-89-0)
(99-36-5)
(99-75-2)
(100-06-1)
(101-41-7)
(122-60-1)
(140-11-4)
(140-39-6)
(492-37-5)
(499-06-9)
(501-52-0)
(533-18-6)
(586-37-8)
(603-79-2)
(610-72-0)
(611-01-8)
(613-69-4)
(619-04-5)
(619-64-7)
(621-36-3)
(622-47-9)
(632-46-2)
(637-27-4)
(935-92-2)
(1450-72-2)
(4647-42-1)
(4647-43-2)
(10031-82-0)
(25567-11-7)
(28134-31-8)
(30587-19-0)
$C_9H_{10}O_2.Ag$
(75112-79-7)
$C_9H_{10}O_3$
(93-02-7)
(93-25-4)
(104-01-8)
(118-61-6)
(120-14-9)
(120-47-8)
(121-32-4)
(121-98-2)
(498-02-2)
(501-97-3)
(515-30-0)
(529-64-6)
(619-86-3)
(621-51-2)
(2065-23-8)
(3480-87-3)
(5333-84-6)
(7781-98-8)
(11070-44-3)
(14199-15-6)
(23287-26-5)
(42058-59-3)
$C_9H_{10}O_4$
(91-52-1)
(93-07-2)
(134-96-3)
(306-08-1)
(708-53-2)
(1078-61-1)
(1132-21-4)
(1466-76-8)
(1521-38-6)
(1878-85-9)
(3247-75-4)
(7507-89-3)
(57244-88-9)
$C_9H_{10}O_5$
(55-10-7)
(530-57-4)
(831-61-8)
$C_9H_{10}O_8$
(3786-91-2)
$C_9H_{11}Br$
(586-61-8)
(637-59-2)
(7073-94-1)
$C_9H_{11}BrN_2O$
(3408-97-7)
$C_9H_{11}BrN_2O_2$
(3060-89-7)
$C_9H_{11}Cl$
(104-52-9)
(1667-04-5)
(2077-13-6)
(2621-46-7)
(26968-58-1)
$C_9H_{11}ClN_2O$
(150-68-5)
(587-34-8)
(22175-22-0)
$C_9H_{11}ClN_2O_2$
(1746-81-2)
(14273-76-8)
(20782-57-4)
$C_9H_{11}ClN_2O_5$
(50-90-8)
$C_9H_{11}ClN_4O$
(4417-72-5)
$C_9H_{11}ClO_2$
(36220-29-8)
$C_9H_{11}ClO_2S$
(773-64-8)
$C_9H_{11}ClO_3$
(2675-80-1)
$C_9H_{11}Cl_2FN_2O_2S_2$
(1085-98-9)
$C_9H_{11}Cl_2N_3O_4S_2$
(135-07-9)
$C_9H_{11}Cl_2O_2PS_3$
(3735-23-7)
$C_9H_{11}Cl_3NO_3PS$
(2921-88-2)
$C_9H_{11}Cl_3NO_3PS$-
$C_4H_7Cl_2O_4P$
(70840-42-5)
$C_9H_{11}Cl_3N_4O$
(24803-60-9)
$C_9H_{11}Cl_6N_5$
(24803-14-3)
$C_9H_{11}Cl_6N_5S$
(30377-15-2)
$C_9H_{11}FN_2O$
(330-39-2)
(332-33-2)
$C_9H_{11}FN_2O_5$
(50-91-9)
$C_9H_{11}FN_2O_6$
(316-46-1)
$C_9H_{11}N$
(155-09-9)
(635-46-1)
(5408-74-2)
(6872-06-6)
$C_9H_{11}NO$
(70-69-9)
(100-10-7)
(103-89-9)
(120-66-1)
(537-92-8)
(588-46-5)
(611-74-5)
(620-71-3)
(6830-82-6)
(28602-27-9)
$C_9H_{11}NO_2$
(51-66-1)
(63-91-2)
(85-91-6)
(87-25-2)
(94-09-7)
(99-64-9)
(101-99-5)
(150-30-1)
(582-33-2)
(619-84-1)
(673-06-3)
(938-73-8)
(1129-41-5)
(1817-47-6)
(5279-14-1)
(6375-17-3)
(22367-76-6)
(28537-55-5)
$C_9H_{11}NO_3$
(60-18-4)
(587-33-7)
(709-93-3)
$C_9H_{11}NO_4$
(59-92-7)
$C_9H_{11}N_3$
(1772-25-4)
$C_9H_{11}N_3O$
(16543-55-8)
$C_9H_{11}N_7O$
(4480-45-9)
C_9H_{12}
(95-63-6)
(98-82-8)
(103-65-1)
(108-67-8)
(526-73-8)
(611-14-3)
(620-14-4)
(622-96-8)
(2396-65-8)
(2806-45-3)
(3048-64-4)
(3048-65-5)
(16219-75-3)
(25550-14-5)
(25551-13-7)
(55956-43-9)
$C_9H_{12}Br_3N_3$
(30362-78-8)
$C_9H_{12}ClN_5$
(15468-86-7)
$C_9H_{12}ClN_5S_4$
(30863-05-9)
$C_9H_{12}ClO_2PS_3$
(953-17-3)
$C_9H_{12}ClO_4P$
(34783-40-9)
$C_9H_{12}Cl_2O_2$
(61898-95-1)
$C_9H_{12}Cl_2Si$
(772-65-6)
$C_9H_{12}Cl_3N_3$
(30362-72-2)
$C_9H_{12}Cl_3N_3O$
(30863-49-1)
$C_9H_{12}Cl_3N_3O_3$
(959-60-4)
$C_9H_{12}Cl_3N_5O$
(30863-32-2)
$C_9H_{12}Cl_4N_4$
(24802-95-7)
(30339-59-4)
$C_9H_{12}FN_3O_4$
(10356-76-0)
$C_9H_{12}NO_3S$
(62571-86-2)
$C_9H_{12}NO_5PS$
(122-14-5)
(3344-14-7)
$C_9H_{12}NO_6P$
(2255-17-6)
$C_9H_{12}N_2O$

(101-42-8)
(69481-32-9)
$C_9H_{12}N_2O.C_2HCl_3O_2$
(4482-55-7)
$C_9H_{12}N_2O_2$
(150-69-6)
(6265-11-8)
(6375-47-9)
(19962-04-0)
(25186-43-0)
$C_9H_{12}N_2O_6$
(58-96-8)
$C_9H_{12}N_4O_2$
(30358-11-3)
$C_9H_{12}N_4O_3S$
(19077-97-5)
$C_9H_{12}N_4O_4$
(30362-18-6)
$C_9H_{12}N_6$
(51-18-3)
(4560-87-6)
$C_9H_{12}N_6O_3$
(5637-82-1)
(30863-20-8)
$C_9H_{12}N_6O_6$
(1843-48-7)
$C_9H_{12}O$
(88-69-7)
(93-54-9)
(99-89-8)
(122-97-4)
(496-78-6)
(527-54-8)
(527-60-6)
(536-50-5)
(539-30-0)
(617-94-7)
(618-45-1)
(621-32-9)
(622-85-5)
(644-35-9)
(645-56-7)
(697-82-5)
(698-71-5)
(768-59-2)
(1123-85-9)
(2416-94-6)
(4013-34-7)
(4844-11-5)
(6966-10-5)
(16308-92-2)
(25168-06-3)
(26998-80-1)
(27129-87-9)
(30230-52-5)
(43142-43-4)
(43212-86-8)
$C_9H_{12}OS$
(7379-51-3)
$C_9H_{12}O_2$
(80-15-9)
(494-99-5)
(551-45-1)
(622-08-2)
(700-13-0)
(770-35-4)
(1855-09-0)
(2138-43-4)
(2138-48-9)
(2785-89-9)
(4169-04-4)
(4812-20-8)
(5673-07-4)
(18979-50-5)
$C_9H_{12}O_3$
(93-03-8)
(538-43-2)
(621-23-8)
(634-36-6)
(2480-86-6)
(16354-95-3)
(25550-51-0)
(32954-58-8)
(35692-98-9)
(46005-09-8)
$C_9H_{12}O_3S$
(80-40-0)
(80-42-2)
(15592-74-2)
(16066-35-6)
$C_9H_{12}O_3S.H_3N$
(37475-88-0)
$C_9H_{12}O_3S.Na$
(28348-53-0)
$C_9H_{12}O_4$
(530-56-3)
(35340-00-2)
$C_9H_{12}O_5$
(25876-47-5)
$C_9H_{13}AsO$
(24582-56-7)
$C_9H_{13}BrN_2O_2$
(314-40-9)
(7286-76-2)
$C_9H_{13}BrN_2O_2.Li$
(53404-19-6)
$C_9H_{13}BrN_2O_2.Na$
(69484-12-4)
$C_9H_{13}Cl$
(55723-99-4)
$C_9H_{13}ClN_2O_2$
(5902-51-2)
$C_9H_{13}ClN_4OS_2$
(13733-96-5)
$C_9H_{13}ClN_6$
(21725-46-2)
$C_9H_{13}ClO$
(72175-27-0)
$C_9H_{13}Cl_3N_4$
(24803-20-1)
(24803-21-2)
(24803-25-6)
(24803-27-8)
(24803-53-0)
(24803-58-5)
(24830-35-1)
(30339-74-3)
$C_9H_{13}Cl_3N_4S$
(30369-45-0)
(30369-46-1)
(30369-48-3)
$C_9H_{13}FNO_3PS$
(39624-86-7)
$C_9H_{13}HgNS_2$
(32407-99-1)
$C_9H_{13}N$
(51-64-9)
(60-15-1)
(88-05-1)
(94-68-8)
(99-88-7)
(99-97-8)
(102-27-2)
(103-83-3)
(137-17-7)
(300-62-9)
(539-32-2)
(609-72-3)
(613-97-8)
(622-80-0)
(643-28-7)
(768-52-5)
(2696-84-6)
(3748-84-3)
(3978-81-2)
(24549-06-2)
(36917-36-9)
$C_9H_{13}N.ClH$
(1875-92-9)
(6334-11-8)
(21436-97-5)
$C_9H_{13}NO$
(93-90-3)
(104-63-2)
(119-31-3)
(120-37-6)
(136-80-1)
(492-39-7)
(14838-15-4)
(25338-55-0)
$C_9H_{13}NO.ClH$
(154-41-6)
$C_9H_{13}NO_2$
(54-49-9)
(59-42-7)
(126-52-3)
$C_9H_{13}NO_2.ClH$
(61-76-7)
$C_9H_{13}NO_2S$
(80-39-7)
(599-69-9)
(1077-56-1)
(8047-99-2)
$C_9H_{13}NO_3$
(51-43-4)
(329-65-7)
(24313-88-0)
$C_9H_{13}NO_3.C_4H_6O_6$
(51-42-3)
$C_9H_{13}NO_3.ClH$
(55-31-2)
(329-63-5)
$C_9H_{13}N_2O_2.Br$
(101-26-8)
$C_9H_{13}N_3$
(30362-68-6)
$C_9H_{13}N_3O_3$
(7481-89-2)
$C_9H_{13}N_3O_5$
(65-46-3)
(147-94-4)
$C_9H_{13}N_3O_5.ClH$
(69-74-9)
$C_9H_{13}N_5S$
(5210-79-7)
$C_9H_{13}N_7$
(57657-42-8)
$C_9H_{13}O_4P$
(46355-07-1)
$C_9H_{13}O_4PS$
(3254-63-5)
$C_9H_{13}O_6PS$
(2778-04-3)
C_9H_{14}
(529-16-8)
(5664-17-5)
$C_9H_{14}BrN_3$
(30362-67-5)
$C_9H_{14}ClN_3$
(31858-10-3)
$C_9H_{14}ClN_3O_2$
(30894-76-9)
$C_9H_{14}ClN_5$
(22936-86-3)
$C_9H_{14}Cl_3N_5$
(27431-16-9)
(27470-69-5)
(27470-96-8)
$C_9H_{14}Cl_4O$
(99308-23-3)
$C_9H_{14}Cl_4O_2$
(64667-33-0)
$C_9H_{14}Cl_6O$
(99308-25-5)
$C_9H_{14}FN_3$
(30355-00-1)
$C_9H_{14}NO_3PS$
(13306-69-9)
$C_9H_{14}N_2$
(13925-06-9)
(18138-04-0)
(18138-05-1)
(18433-97-1)
(31626-02-5)
(32736-91-7)
(68966-84-7)
$C_9H_{14}N_2O$
(24168-70-5)
(24683-00-9)
$C_9H_{14}N_2O_2.Na$
(65086-97-7)
(65208-42-6)
$C_9H_{14}N_2O_3$
(50-11-3)
$C_9H_{14}N_6$
(20776-86-7)
(30360-15-7)
$C_9H_{14}N_6S$
(30360-89-5)
$C_9H_{14}O$
(78-59-1)
(471-01-2)
(504-20-1)
(768-50-3)
(874-68-0)
(3777-69-3)
(6750-03-4)
(13747-73-4)
(14377-11-8)
(17429-29-7)
(20030-30-2)
(24903-95-5)
(27939-60-2)
(54458-61-6)
(55282-90-1)
(64079-01-2)
(92366-34-2)
$C_9H_{14}OSi$
(13132-25-7)
$C_9H_{14}O_2$
(4840-76-0)
(10276-21-8)
(20547-99-3)
(30964-01-3)
(38653-34-8)
(45955-66-6)
(54244-72-3)
$C_9H_{14}O_3$
(2455-24-5)
(6994-95-2)
(24588-61-2)
(55402-04-5)
$C_9H_{14}O_3Si$
(2996-92-1)
$C_9H_{14}O_6$
(102-76-1)
$C_9H_{14}S$
(4861-58-9)
$C_9H_{15}AlO_9$
(18917-91-4)
$C_9H_{15}Br$
(697-40-5)
$C_9H_{15}Br_6O_4P$
(126-72-7)
(18713-51-4)
$C_9H_{15}ClN_4O$
(4402-46-4)
(4417-65-6)
(4417-68-9)
(30357-89-2)
$C_9H_{15}Cl_2N_3O_2Zn$
(58270-08-9)
$C_9H_{15}Cl_5O$
(99308-24-4)
$C_9H_{15}Cl_6O_4P$
(78-43-3)
(13674-87-8)
(26604-51-3)
(40120-74-9)
(68460-03-7)

$C_9H_{15}N$
(102-70-5)
(520-69-4)
$C_9H_{15}NO_2$
(2873-97-4)
$C_9H_{15}NO_5$
(1068-90-2)
$C_9H_{15}NS$
(76572-48-0)
$C_9H_{15}N_3$
(1009-74-1)
(30362-05-1)
(30362-06-2)
$C_9H_{15}N_3O_3$
(884-43-5)
(26028-46-6)
$C_9H_{15}N_3O_6$
(839-90-7)
$C_9H_{15}N_3S_3$
(30863-36-6)
$C_9H_{15}N_5$
(27431-01-2)
$C_9H_{15}O_4P$
(1623-19-4)
$C_9H_{15}O_8P$
(122-10-1)
C_9H_{16}
(496-10-6)
(3452-09-3)
(6191-90-8)
(23097-98-5)
(23799-25-9)
(32273-77-1)
(56700-77-7)
(61142-36-7)
$C_9H_{16}BrN_5$
(30360-54-4)
(30360-55-5)
$C_9H_{16}ClNOS$
(107348-44-7)
$C_9H_{16}ClN_3O_2$
(13010-47-4)
$C_9H_{16}ClN_3O_7$
(54749-90-5)
$C_9H_{16}ClN_4.Cl$
(4080-31-3)
$C_9H_{16}ClN_5$
(139-40-2)
(1912-26-1)
(3071-66-7)
(5915-41-3)
(7286-69-3)
(26235-92-7)
$C_9H_{16}ClN_5O_2$
(30084-25-4)
$C_9H_{16}FN_5$
(3797-36-2)
$C_9H_{16}FN_5O$
(30388-86-4)
$C_9H_{16}N_2$
(6674-22-2)
$C_9H_{16}N_2O_2S_2$
(13752-51-7)
$C_9H_{16}N_2O_4$
(26850-24-8)
$C_9H_{16}N_2O_6$
(18771-50-1)
$C_9H_{16}N_3O_2PS_2$
(30863-14-0)
$C_9H_{16}N_3O_4PS_2$
(30863-13-9)
$C_9H_{16}N_4O$
(6864-23-9)
$C_9H_{16}N_4OS$
(34014-18-1)
$C_9H_{16}N_4O_2$
(30357-99-4)
(30358-07-7)
(30358-10-2)
(30358-13-5)
$C_9H_{16}N_8$
(2854-70-8)
(2854-96-8)
$C_9H_{16}O$
(94-65-5)
(873-94-9)
(2277-16-9)
(2408-37-9)
(2463-53-8)
(5009-32-5)
(14309-57-0)
(16525-05-6)
(19860-71-0)
(33046-81-0)
(54549-81-4)
(56805-23-3)
$C_9H_{16}OS_2$
(2956-12-9)
(15288-12-7)
$C_9H_{16}O_2$
(104-61-0)
(123-68-2)
(2499-95-8)
(3301-94-8)
(3953-10-4)
(14090-88-1)
(24985-48-6)
(31502-23-5)
(31642-67-8)
(55013-32-6)
(69668-85-5)
$C_9H_{16}O_3$
(2052-15-5)
(6064-52-4)
(17526-74-8)
(24222-05-7)
(29304-40-3)
$C_9H_{16}O_4$
(123-99-9)
(133-13-1)
(818-38-2)
(1619-62-1)
(7328-17-8)
$C_9H_{16}O_5$
(3568-29-4)
(27043-36-3)
$C_9H_{17}ClN_3O_3PS$
(42509-80-8)
$C_9H_{17}ClO$
(764-85-2)
(36727-29-4)
$C_9H_{17}ClO_2$
(24468-13-1)
$C_9H_{17}ClO_3$
(40137-60-8)
$C_9H_{17}N$
(2243-27-8)
(14542-93-9)
$C_9H_{17}NO$
(826-36-8)
(4419-57-2)
$C_9H_{17}NOS$
(2212-67-1)
(52372-17-5)
(67587-12-6)
(107348-45-8)
$C_9H_{17}NO_2S$
(112-56-1)
$C_9H_{17}NO_2$
(2426-54-2)
(6065-01-6)
(6065-04-9)
(26872-84-4)
$C_9H_{17}NO_5$
(79-83-4)
$C_9H_{17}NO_5.Na$
(867-81-2)
$C_9H_{17}N_5$
(26235-13-2)
(26235-29-0)
(27430-91-7)
(27430-92-8)
(27430-93-9)
(27431-08-9)
$C_9H_{17}N_5O$
(1610-17-9)
(7374-53-0)
(24126-20-3)
(30360-59-9)
$C_9H_{17}N_5S$
(834-12-8)
(5133-47-1)
(30360-82-8)
C_9H_{18}
(124-11-8)
(696-29-7)
(1502-38-1)
(1678-92-8)
(1678-97-3)
(1795-27-3)
(2040-95-1)
(3073-66-3)
(3728-56-1)
(7667-55-2)
(13427-43-5)
(13987-01-4)
(19489-10-2)
(27215-95-8)
(30498-63-6)
(30677-34-0)
(33933-74-3)
(54549-80-3)
(56728-10-0)
(95461-54-4)
$C_9H_{18}Br_2$
(4549-33-1)
(62168-27-8)
$C_9H_{18}Cl_2$
(821-88-5)
(821-99-8)
$C_9H_{18}Cl_3O_4P$
(1067-98-7)
(6145-73-9)
(13674-84-5)
(26248-87-3)
$C_9H_{18}F_2$
(62127-42-8)
$C_9H_{18}NO_2.CH_3O_4S$
(6891-44-7)
$C_9H_{18}NO_2.Cl$
(5039-78-1)
$C_9H_{18}NS_2.Na$
(136-30-1)
$C_9H_{18}N_2O$
(5205-93-6)
$C_9H_{18}N_2O_2$
(17341-40-1)
$C_9H_{18}N_2O_2S$
(39196-18-4)
$C_9H_{18}N_2O_4$
(57-53-4)
$C_9H_{18}N_2O_5$
(4356-60-9)
$C_9H_{18}N_3OP$
(57-39-6)
$C_9H_{18}N_3S_6.Bi$
(21260-46-8)
$C_9H_{18}N_3S_6.Fe$
(14484-64-1)
$C_9H_{18}N_6$
(645-05-6)
(16268-92-1)
(16274-44-5)
$C_9H_{18}O$
(108-83-8)
(116-02-9)
(124-19-6)
(502-56-7)
(815-24-7)
(821-55-6)
(925-78-0)
(1321-60-4)
(4621-04-9)
(6137-08-2)
(18521-07-8)
(18641-71-9)
(19549-80-5)
(20809-46-5)
(36712-20-6)
(55702-60-8)
(55956-37-1)
(58654-67-4)
$C_9H_{18}O_2$
(106-27-4)
(106-30-9)
(109-19-3)
(111-11-5)
(112-05-0)
(112-06-1)
(504-99-4)
(540-18-1)
(589-59-3)
(626-77-7)
(935-45-5)
(2311-46-8)
(2445-76-3)
(3274-29-1)
(3302-12-3)
(5926-90-9)
(26896-18-4)
(51149-70-3)
(60148-94-9)
(68134-06-5)
$C_9H_{18}O_2.1/2Ca$
(29813-38-5)
$C_9H_{18}O_2.xPb$
(27253-41-4)
$C_9H_{18}O_3$
(927-07-1)
(10213-74-8)
(13483-47-1)
(29006-06-2)
(33796-87-1)
(55956-25-7)
$C_9H_{19}Br$
(693-58-3)
$C_9H_{19}Cl$
(2473-01-0)
(34314-35-7)
$C_9H_{19}Cl_3Si$
(5283-67-0)
$C_9H_{19}F$
(463-18-3)
$C_9H_{19}I$
(4282-42-2)
$C_9H_{19}N$
(1195-42-2)
(3592-81-2)
(4396-27-4)
$C_9H_{19}NO$
(103-00-4)
(761-65-9)
(1115-96-4)
(2403-88-5)
(31121-12-7)
$C_9H_{19}NOS$
(759-94-4)
$C_9H_{19}NOS.C_8H_{11}Cl_2NO$
(51990-04-6)
$C_9H_{19}NO_3Si$
(919-31-3)
$C_9H_{19}NO_4$
(81-13-0)
(16485-10-2)
$C_9H_{19}N_3O$
(30805-38-0)
$C_9H_{19}N_3O_2$
(56654-52-5)
C_9H_{20}
(111-84-2)

(921-47-1)
(922-28-1)
(926-82-9)
(1067-20-5)
(1068-87-7)
(1069-53-0)
(1070-87-7)
(1071-26-7)
(1072-05-5)
(1186-53-4)
(2213-23-2)
(2216-30-0)
(2216-32-2)
(2216-33-3)
(2216-34-4)
(3074-75-7)
(3074-76-8)
(3074-77-9)
(3221-61-2)
(3522-94-9)
(7154-79-2)
(16747-25-4)
(16747-26-5)
(16747-28-7)
(16747-30-1)
(16747-31-2)
(16747-32-3)
(16747-33-4)
(16747-38-9)
(30498-66-9)
(34464-40-9)
(73507-01-4)

$C_9H_{20}NO_2.Cl$
(2303-35-7)

$C_9H_{20}NO_3PS_2$
(2275-18-5)

$C_9H_{20}N_2$
(3312-60-5)
(3437-33-0)
(25560-00-3)
(90853-11-5)

$C_9H_{20}N_2O$
(869-79-4)
(5336-24-3)

$C_9H_{20}N_2O_2$
(24579-73-5)

$C_9H_{20}N_2S$
(109-46-6)

$C_9H_{20}N_4S$
(41361-12-0)

$C_9H_{20}O$
(108-82-7)
(143-08-8)
(623-93-8)
(624-51-1)
(628-99-9)
(1573-28-0)
(2430-22-0)
(3452-97-9)
(5340-36-3)
(5932-79-6)
(13254-34-7)
(14202-62-1)
(18636-66-3)
(19549-73-6)
(19549-77-0)
(19549-79-2)
(19780-40-6)
(27458-94-2)
(28473-21-4)
(62238-02-2)

$C_9H_{20}O_2$
(115-84-4)
(2568-90-3)
(2568-92-5)
(3842-03-3)
(3937-56-2)
(10138-47-3)
(54340-89-5)

$C_9H_{20}O_3$
(115-80-0)
(124-16-3)
(54518-04-6)

$C_9H_{20}O_4$
(24800-44-0)
(62005-54-3)

$C_9H_{20}O_5$
(23783-42-8)

$C_9H_{20}O_5Si$
(2530-83-8)

$C_9H_{20}O_7$
(56090-54-1)

$C_9H_{20}S$
(1455-21-6)
(25360-10-5)

$C_9H_{21}Al$
(102-67-0)

$C_9H_{21}AlO_3$
(555-31-7)

$C_9H_{21}AsS_3$
(5582-57-0)

$C_9H_{21}BO_3$
(5419-55-6)

$C_9H_{21}ClSn$
(2279-76-7)

$C_9H_{21}N$
(102-69-2)
(112-20-9)
(20193-23-1)
(66793-76-8)
(67953-04-2)

$C_9H_{21}NO_3$
(122-20-3)

$C_9H_{21}NO_3.C_9H_7Cl_3O_3$
(53404-75-4)

$C_9H_{21}NO_3.C_9H_{21}NO_3.C_8H_6Cl_2O_3.C_6H_3Cl_3N_2O_2$
(8067-55-8)

$C_9H_{21}N_3$
(7779-27-3)

$C_9H_{21}N_3O_3$
(4719-04-4)

$C_9H_{21}O_2PS_2$
(102255-22-1)

$C_9H_{21}O_2PS_3$
(13071-79-9)

$C_9H_{21}O_3P$
(116-17-6)
(2404-73-1)

$C_9H_{21}O_3PS$
(20822-30-4)

$C_9H_{21}O_3PS_3$
(5827-05-4)
(10548-10-4)

$C_9H_{21}O_4P$
(513-08-6)
(20195-08-8)

$C_9H_{21}O_4PS_3$
(56070-16-7)

$C_9H_{21}Pb$
(44910-38-5)

$C_9H_{22}NO_2PS$
(21770-86-5)
(73207-98-4)

$C_9H_{22}NO_5P$
(2781-11-5)

$C_9H_{22}N_2$
(25620-58-0)

$C_9H_{22}N_2O$
(90853-15-9)
(90853-16-0)

$C_9H_{22}O_4P_2S_4$
(563-12-2)

$C_9H_{22}Pb$
(6618-03-7)
(65122-13-6)

$C_9H_{23}NO_2Si$
(3037-72-7)

$C_9H_{23}NO_3PS.I$
(513-10-0)

$C_9H_{23}NO_3Si$
(919-30-2)

$C_9H_{23}N_3$
(3030-47-5)

$C_9H_{28}N_3O_{15}P_5$
(15827-60-8)

$C_9H_{29}N_3NaO_{15}P_5$
(22042-96-2)

$C_{10}Cl_8$
(2234-13-1)

$C_{10}Cl_{10}$
(2227-17-0)

$C_{10}Cl_{10}O$
(143-50-0)

$C_{10}Cl_{12}$
(2385-85-5)
(14979-34-1)

$C_{10}D_{12}$
(75840-23-2)

$C_{10}F_{21}I$
(2050-77-3)

$C_{10}F_{22}$
(3021-63-4)

$C_{10}HBr_7$
(55688-01-2)

$C_{10}HCl_7$
(32241-08-0)

$C_{10}HCl_{11}$
(39801-14-4)

$C_{10}H_2Br_6$
(56480-06-9)
(75625-24-0)

$C_{10}H_2Cl_6$
(1335-87-1)

$C_{10}H_2Cl_{10}O$
(1034-41-9)

$C_{10}H_2O_6$
(89-32-7)

$C_{10}H_3Cl_5$
(1321-64-8)

$C_{10}H_3Cl_9$
(21641-70-3)

$C_{10}H_4Br_3ClN_2O$
(15287-32-8)

$C_{10}H_4Cl_2O_2$
(117-80-6)

$C_{10}H_4Cl_4$
(1335-88-2)
(20020-02-4)

$C_{10}H_4Cl_6N_4$
(3599-73-3)
(30362-59-5)

$C_{10}H_4Cl_6N_4S$
(30369-62-1)

$C_{10}H_4Cl_8O$
(27304-13-8)

$C_{10}H_4N_4O_8$
(28995-89-3)

$C_{10}H_4O_4$
(11063-25-5)

$C_{10}H_5ClN_2$
(2698-41-1)
(35254-70-7)

$C_{10}H_5ClN_2O_4$
(2401-85-6)

$C_{10}H_5ClN_2O_3S$
(3770-97-6)

$C_{10}H_5Cl_3$
(1321-65-9)

$C_{10}H_5Cl_4N_3$
(13704-90-0)

$C_{10}H_5Cl_5N_4$
(30369-15-4)

$C_{10}H_5Cl_5N_4S$
(30576-31-9)

$C_{10}H_5Cl_7$
(76-44-8)
(21161-58-0)
(33442-83-0)

$C_{10}H_5Cl_7O$
(1024-57-3)

$C_{10}H_5Cl_9$
(3734-49-4)
(5103-73-1)
(39765-80-5)
(98318-97-9)

$C_{10}H_5F_7N_2O_4$
(2712-83-6)

$C_{10}H_5F_{17}O$
(678-39-7)

$C_{10}H_5NO_4$
(80267-67-0)

$C_{10}H_5N_3O_6S$
(84-91-3)
(130-59-6)

$C_{10}H_6ClNO_2$
(2797-51-5)

$C_{10}H_6ClN_5$
(30894-92-9)

$C_{10}H_6Cl_2$
(1825-30-5)
(1825-31-6)
(2050-69-3)
(2050-72-8)
(2050-73-9)
(2050-74-0)
(2050-75-1)
(2065-70-5)
(2198-75-6)
(2198-77-8)
(28699-88-9)

$C_{10}H_6Cl_2N_2O_2$
(1018-71-9)

$C_{10}H_6Cl_2O$
(2050-76-2)

$C_{10}H_6Cl_2O_4S_2$
(1928-01-4)

$C_{10}H_6Cl_3N_3O$
(1917-41-5)

$C_{10}H_6Cl_3N_3O_2$
(30369-18-7)

$C_{10}H_6Cl_4N_4$
(30377-24-3)

$C_{10}H_6Cl_4N_4S$
(30369-57-4)

$C_{10}H_6Cl_4O_3S$
(3765-57-9)

$C_{10}H_6Cl_4O_4$
(1861-32-1)
(69158-26-5)

$C_{10}H_6Cl_6$
(3734-48-3)

$C_{10}H_6Cl_6O$
(12408-14-9)
(24009-05-0)

$C_{10}H_6Cl_6O_2$
(39660-14-5)

$C_{10}H_6Cl_8$
(57-74-9)
(5103-71-9)
(5103-74-2)
(5566-34-7)

$C_{10}H_6N_2OS_2$
(2439-01-2)

$C_{10}H_6N_2O_4$
(525-47-3)
(602-38-0)
(605-71-0)
(606-37-1)
(27478-34-8)

$C_{10}H_6N_2O_4S$
(84-23-1)

$C_{10}H_6N_2O_4S.Na$
(2657-00-3)

$C_{10}H_6N_2O_5$
(605-69-6)

$C_{10}H_6O_2$
(130-15-4)
(524-42-5)
(12679-43-5)

$C_{10}H_6O_3$
(83-72-7)
(481-39-0)
$C_{10}H_6O_3S$
(83-31-8)
$C_{10}H_6O_5S.Na$
(7045-83-2)
$C_{10}H_6O_7S_2.2Na$
(135-51-3)
$C_{10}H_6O_8$
(89-05-4)
(476-73-3)
(479-47-0)
$C_{10}H_7Br$
(90-11-9)
(580-13-2)
$C_{10}H_7Cl$
(90-13-1)
(91-58-7)
(25586-43-0)
$C_{10}H_7ClO$
(604-44-4)
(606-40-6)
$C_{10}H_7ClO_2S$
(93-11-8)
$C_{10}H_7Cl_2N_3O$
(13838-34-1)
(30886-24-9)
$C_{10}H_7Cl_2N_3S$
(13733-93-2)
(25713-57-9)
$C_{10}H_7Cl_2N_5O$
(5097-51-8)
$C_{10}H_7Cl_2N_5S$
(5577-35-5)
$C_{10}H_7Cl_3N_4$
(30369-07-4)
$C_{10}H_7Cl_3N_4S$
(30369-55-2)
$C_{10}H_7Cl_4N_5$
(30355-71-6)
$C_{10}H_7Cl_5O_3$
(55773-90-5)
$C_{10}H_7Cl_7$
(14168-01-5)
$C_{10}H_7Cl_9$
(115384-94-6)
$C_{10}H_7F$
(321-38-0)
$C_{10}H_7F_7N_2O_2$
(847-51-8)
$C_{10}H_7I$
(90-14-2)
$C_{10}H_7NO_2$
(86-57-7)
(131-91-9)
(132-53-6)
(581-89-5)
(941-69-5)
(27254-36-0)
$C_{10}H_7NO_3$
(492-27-3)
(82322-43-8)
$C_{10}H_7N_3S$
(148-79-8)
$C_{10}H_7N_3S.H_3O_2P$
(28558-32-9)
$C_{10}H_8$
(91-20-3)
(275-51-4)
$C_{10}H_8BrNO_2$
(574-98-1)
$C_{10}H_8Br_3N_5$
(30359-71-8)
$C_{10}H_8ClN$
(4377-41-7)
$C_{10}H_8ClNOS_2$
(6012-92-6)
$C_{10}H_8ClN_3$
(30937-70-3)
$C_{10}H_8ClN_3O$
(1698-60-8)
$C_{10}H_8ClN_3O_2$
(5707-69-7)
(5707-73-3)
$C_{10}H_8ClN_3S$
(13705-07-2)
$C_{10}H_8Cl_2N_2O_4S$
(84-57-1)
$C_{10}H_8Cl_2N_4$
(2272-23-3)
(2272-24-4)
(3995-42-4)
(30369-82-5)
$C_{10}H_8Cl_2N_4O$
(30369-87-0)
(30377-27-6)
$C_{10}H_8Cl_2O_6$
(75315-50-3)
$C_{10}H_8Cl_3N_5$
(30355-69-2)
$C_{10}H_8Cl_4O_3$
(88927-42-8)
$C_{10}H_8Cl_8$
(1319-80-8)
$C_{10}H_8F_3N_5$
(368-61-6)
$C_{10}H_8N_2$
(366-18-7)
(553-26-4)
(581-50-0)
(37275-48-2)
$C_{10}H_8N_2O_2$
(25854-16-4)
$C_{10}H_8N_2O_2S_2$
(3696-28-4)
$C_{10}H_8N_2O_2S_2.O_4S.Mg$
(43143-11-9)
$C_{10}H_8N_2O_2S_2.Zn$
(13463-41-7)
$C_{10}H_8N_2O_3$
(14073-00-8)
$C_{10}H_8N_2O_4$
(116211-91-7)
$C_{10}H_8N_2O_6S$
(21951-33-7)
$C_{10}H_8N_4$
(67730-10-3)
$C_{10}H_8O$
(90-15-3)
(135-19-3)
(1321-67-1)
$C_{10}H_8O_2$
(83-56-7)
(92-44-4)
(92-48-8)
(132-86-5)
(571-60-8)
(574-00-5)
(575-44-0)
(582-17-2)
$C_{10}H_8O_3$
(90-33-5)
(5463-50-3)
$C_{10}H_8O_3S$
(85-47-2)
(120-18-3)
(25155-18-4)
$C_{10}H_8O_3S.Na$
(532-02-5)
(1321-69-3)
$C_{10}H_8O_4$
(529-84-0)
(552-86-3)
(2107-76-8)
(2107-77-9)
(2373-80-0)
$C_{10}H_8O_4S$
(84-87-7)
(92-40-0)
(93-01-6)
(567-18-0)
$C_{10}H_8O_4S.K$
(833-66-9)
$C_{10}H_8O_4S.Na$
(135-76-2)
$C_{10}H_8O_5S$
(92-27-3)
$C_{10}H_8O_6S_2$
(81-04-9)
(525-37-1)
(581-75-9)
$C_{10}H_8O_6S_2.2Na$
(1655-35-2)
$C_{10}H_8O_7S_2$
(118-32-1)
(148-75-4)
$C_{10}H_8O_7S_2.2K$
(842-18-2)
$C_{10}H_8O_7S_2.Na$
(61931-87-1)
$C_{10}H_8O_8S_2.2Na$
(129-96-4)
$C_{10}H_8O_9S_3.xNa$
(19437-42-4)
$C_{10}H_8S$
(91-60-1)
(825-55-8)
(56842-14-9)
$C_{10}H_8S_2$
(5325-88-2)
$C_{10}H_9ClN_2$
(22752-98-3)
$C_{10}H_9ClN_4$
(1853-96-9)
(1973-03-1)
(6023-57-0)
$C_{10}H_9ClN_4O$
(30357-85-8)
$C_{10}H_9ClN_4S$
(30357-96-1)
$C_{10}H_9ClO_4$
(5438-40-4)
$C_{10}H_9Cl_2N$
(21342-85-8)
$C_{10}H_9Cl_2NO$
(2164-09-2)
(2759-71-9)
$C_{10}H_9Cl_2N_5$
(30355-61-4)
(30355-65-8)
$C_{10}H_9Cl_2N_5O_2S$
(30357-79-0)
$C_{10}H_9Cl_2O_3.Na$
(10433-59-7)
$C_{10}H_9Cl_3O_2$
(90-17-5)
$C_{10}H_9Cl_3O_3$
(93-80-1)
(1928-39-8)
$C_{10}H_9Cl_3O_4$
(108548-70-5)
$C_{10}H_9Cl_4NO_2S$
(2425-06-1)
(2939-80-2)
$C_{10}H_9Cl_4O_4P$
(961-11-5)
(22248-79-9)
$C_{10}H_9Cl_6N_3O_2$
(30362-47-1)
$C_{10}H_9Cl_6N_5$
(24848-42-8)
$C_{10}H_9Cl_9$
(52819-39-3)
$C_{10}H_9F_9O_2$
(1799-84-4)
$C_{10}H_9HgNO$
(86-85-1)
$C_{10}H_9N$
(91-59-8)
(91-62-3)
(91-63-4)
(134-32-7)
(491-35-0)
(611-32-5)
(612-58-8)
(612-60-2)
(635-90-5)
(1125-80-0)
(1721-93-3)
(25168-10-9)
(27601-00-9)
$C_{10}H_9N.ClH$
(552-46-5)
(612-52-2)
(3007-43-0)
(62763-89-7)
$C_{10}H_9NO$
(606-43-9)
(607-30-7)
(826-81-3)
(2834-90-4)
(2834-92-6)
(5263-87-6)
(42884-33-3)
(95609-86-2)
$C_{10}H_9NO.ClH$
(1198-27-2)
(41772-23-0)
$C_{10}H_9NO_2$
(87-51-4)
$C_{10}H_9NO_2S$
(1079-33-0)
$C_{10}H_9NO_3S$
(81-06-1)
(81-16-3)
(82-75-7)
(84-86-6)
(84-89-9)
(93-00-5)
(119-28-8)
(119-79-9)
(6528-46-7)
$C_{10}H_9NO_3S.H_3N$
(68540-41-0)
$C_{10}H_9NO_3S.Na$
(130-13-2)
$C_{10}H_9NO_4S$
(87-02-5)
(90-51-7)
(116-63-2)
$C_{10}H_9NO_6$
(5292-45-5)
(13290-96-5)
$C_{10}H_9NO_6S_2$
(86-65-7)
(117-62-4)
(131-27-1)
$C_{10}H_9NO_6S_2.K$
(842-15-9)
$C_{10}H_9NO_6S_2.Na$
(118-33-2)
(19532-03-7)
$C_{10}H_9NO_7S_2$
(82-47-3)
(90-20-0)
$C_{10}H_9NO_7S_2.Na$
(5460-09-3)
$C_{10}H_9N_3$
(1202-34-2)
(30361-89-8)
$C_{10}H_9N_3O$
(2519-50-8)
$C_{10}H_9N_3OS$
(1917-43-7)
$C_{10}H_9N_5O$
(525-79-1)
$C_{10}H_9N_7O$
(30362-30-2)
$C_{10}H_9O_7S.Na$

(3965-55-7)
$C_{10}H_{10}$
(105-06-6)
(108-57-6)
(447-53-0)
(767-59-9)
(767-60-2)
(1321-74-0)
(2177-47-1)
(29036-25-7)
(29828-28-2)
$C_{10}H_{10}.Ni$
(1271-28-9)
$C_{10}H_{10}BrCl_2O_4P$
(13104-21-7)
$C_{10}H_{10}BrN_5$
(30359-64-9)
$C_{10}H_{10}Br_2O_2$
(22421-59-6)
(75150-13-9)
$C_{10}H_{10}ClNO_2$
(93-70-9)
(101-92-8)
(140-49-8)
$C_{10}H_{10}ClN_3S$
(13631-64-6)
$C_{10}H_{10}ClN_5$
(645-18-1)
(30355-60-3)
$C_{10}H_{10}ClO_3.K$
(1929-86-8)
$C_{10}H_{10}Cl_2F_2N_2OS$
(33439-45-1)
$C_{10}H_{10}Cl_2O_2$
(14437-17-3)
$C_{10}H_{10}Cl_2O_3$
(94-82-6)
(533-23-3)
(67883-08-3)
$C_{10}H_{10}Cl_2O_4$
(108544-99-6)
(108548-73-8)
(108673-04-7)
$C_{10}H_{10}Cl_2O_5$
(108572-08-3)
$C_{10}H_{10}Cl_2O_6$
(75315-51-4)
$C_{10}H_{10}Cl_2Ti$
(1271-19-8)
$C_{10}H_{10}Cl_2V$
(12083-48-6)
$C_{10}H_{10}Cl_2Zr$
(1291-32-3)
$C_{10}H_{10}Cl_3NO_2$
(52819-37-1)
$C_{10}H_{10}Cl_3N_3O_4$
(30863-21-9)
$C_{10}H_{10}Cl_3N_7$
(26235-11-0)
$C_{10}H_{10}Cl_3O_2PS$
(56549-12-3)
$C_{10}H_{10}Cl_3O_4P$
(2274-67-1)
$C_{10}H_{10}Cl_6N_4$
(24803-09-6)
$C_{10}H_{10}Cl_8$
(8001-35-2)
$C_{10}H_{10}Co$
(1277-43-6)
$C_{10}H_{10}Cr$
(1271-24-5)
$C_{10}H_{10}F_3N_3O_4$
(2077-99-8)
$C_{10}H_{10}F_{13}NO_3S$
(34455-03-3)
$C_{10}H_{10}F_{15}O_5P$
(71181-76-5)
$C_{10}H_{10}Fe$
(102-54-5)
$C_{10}H_{10}N_2$
(479-27-6)
(487-19-4)
(2243-62-1)
$C_{10}H_{10}N_2O$
(89-25-8)
$C_{10}H_{10}N_2O_4$
(17311-31-8)
$C_{10}H_{10}N_2O_4S$
(89-36-1)
$C_{10}H_{10}N_4$
(1853-91-4)
(4040-01-1)
$C_{10}H_{10}N_4O$
(30369-38-1)
(30369-68-7)
(41394-05-2)
$C_{10}H_{10}N_4OS$
(21087-63-8)
$C_{10}H_{10}N_4O_2S$
(68-35-9)
(116-44-9)
$C_{10}H_{10}N_4S$
(30369-73-4)
$C_{10}H_{10}N_6O_5$
(1037-57-6)
$C_{10}H_{10}O$
(101-39-3)
(122-57-6)
(529-34-0)
(3782-00-1)
(4411-89-6)
(25586-39-4)
(28715-26-6)
(37442-55-0)
(87259-53-8)
$C_{10}H_{10}O_2$
(94-59-7)
(103-26-4)
(120-58-1)
(583-04-0)
(1009-61-6)
(1199-77-5)
(1504-74-1)
(1866-39-3)
(2243-52-9)
(6781-42-6)
(13524-76-0)
(20895-45-8)
(30773-71-8)
$C_{10}H_{10}O_2S_2$
(4437-20-1)
$C_{10}H_{10}O_3$
(458-36-6)
(614-27-7)
(830-09-1)
(943-89-5)
(1603-79-8)
(2051-95-8)
(2243-35-8)
(4219-55-0)
(5208-87-7)
(5669-15-8)
(5894-79-1)
(15482-54-9)
(16824-02-5)
(25134-21-8)
(25832-09-1)
$C_{10}H_{10}O_4$
(120-61-6)
(131-11-3)
(326-61-4)
(537-98-4)
(1014-83-1)
(1135-24-6)
(1459-93-4)
(7500-53-0)
(26638-01-7)
(59901-91-6)
$C_{10}H_{10}O_5$
(1081-71-6)
$C_{10}H_{10}O_7S$
(138-25-0)
$C_{10}H_{10}S$
(30027-44-2)
(54385-63-6)
$C_{10}H_{11}BrO$
(73908-28-8)
$C_{10}H_{11}Cl$
(16608-68-7)
$C_{10}H_{11}ClO$
(7473-99-6)
(73908-29-9)
$C_{10}H_{11}ClO_3$
(93-65-2)
(882-09-7)
(3547-07-7)
(7085-19-0)
(14426-42-7)
(17413-73-9)
$C_{10}H_{11}ClO_3.C_4H_{11}NO_2$
(1432-14-0)
$C_{10}H_{11}ClO_3$-
$C_9H_6ClNO_3S.C_8H_6Cl_2O_3$
(93746-34-0)
$C_{10}H_{11}ClO_4$
(30714-88-6)
(108544-98-5)
$C_{10}H_{11}Cl_2NO_2$
(2150-25-6)
(2150-28-9)
(2150-32-5)
$C_{10}H_{11}Cl_2NO_3$
(2008-39-1)
$C_{10}H_{11}Cl_3O_2$
(1861-44-5)
$C_{10}H_{11}Cl_3O_3$
(108544-94-1)
$C_{10}H_{11}Cl_6N_3$
(24481-37-6)
$C_{10}H_{11}Cl_6N_3O$
(30863-57-1)
$C_{10}H_{11}Cl_6N_3S$
(24481-70-7)
$C_{10}H_{11}Cl_7$
(51775-36-1)
$C_{10}H_{11}F_3N_2O$
(2164-17-2)
$C_{10}H_{11}F_3N_2O_3S$
(47000-92-0)
$C_{10}H_{11}F_3N_2O_5$
(70-00-8)
$C_{10}H_{11}F_7O_2$
(17587-22-3)
$C_{10}H_{11}N$
(875-79-6)
$C_{10}H_{11}NO$
(6328-48-9)
$C_{10}H_{11}NO_2$
(102-01-2)
(7493-63-2)
$C_{10}H_{11}NO_3$
(1205-08-9)
(2623-33-8)
(2719-08-6)
(13414-54-5)
(13414-55-6)
(13414-58-9)
(64061-59-2)
$C_{10}H_{11}NO_4$
(94-22-4)
(104-30-3)
(833-43-2)
(73622-98-7)
$C_{10}H_{11}N_2O.HO_4S$
(38420-60-9)
$C_{10}H_{11}N_2O.1/2O_4S$
(68298-47-5)
$C_{10}H_{11}N_3$
(1131-18-6)
$C_{10}H_{11}N_3O$
(71267-22-6)
$C_{10}H_{11}N_3OS$
(18691-97-9)
$C_{10}H_{11}N_3OS.C_9H_8Cl_2O_3$
(39283-72-2)
$C_{10}H_{11}N_3O_3S$
(89-29-2)
(723-46-6)
$C_{10}H_{11}N_5$
(1853-88-9)
(7426-35-9)
(19338-12-6)
$C_{10}H_{11}N_5O$
(1639-39-0)
(30354-91-7)
(30360-74-8)
$C_{10}H_{11}N_5S$
(30360-85-1)
$C_{10}H_{11}N_7O_3$
(30355-55-6)
$C_{10}H_{11}O_2$
(774-40-3)
$C_{10}H_{12}$
(77-73-6)
(119-64-2)
(767-58-8)
(768-49-0)
(768-56-9)
(824-22-6)
(824-63-5)
(874-35-1)
(1005-64-7)
(1560-06-1)
(1587-04-8)
(2039-89-6)
(2039-90-9)
(2234-20-0)
(3290-53-7)
(3454-07-7)
(5379-20-4)
(7525-62-4)
(7564-63-8)
(27133-93-3)
(27576-03-0)
(27831-13-6)
(28106-30-1)
(33877-87-1)
$C_{10}H_{12}BrCl_2O_3PS$
(4824-78-6)
$C_{10}H_{12}BrNO_3$
(17199-23-4)
$C_{10}H_{12}CaN_2O_8.2Na$
(62-33-9)
$C_{10}H_{12}ClNO_2$
(101-21-3)
(671-04-5)
(2239-92-1)
(35600-63-6)
$C_{10}H_{12}ClNO_3$
(14593-28-3)
$C_{10}H_{12}ClNO_4$
(91-43-0)
(886-74-8)
$C_{10}H_{12}ClNO_5S$
(13279-58-8)
$C_{10}H_{12}ClN_3O_2$
(15271-41-7)
$C_{10}H_{12}ClN_3O_3S$
(73-49-4)
$C_{10}H_{12}ClN_5$
(30805-24-4)
$C_{10}H_{12}Cl_2$
(65724-12-1)
$C_{10}H_{12}Cl_2O$
(34593-75-4)
(92366-35-3)
$C_{10}H_{12}Cl_2O_2$
(56894-91-8)
$C_{10}H_{12}Cl_2O_3$
(108545-02-4)

$C_{10}H_{12}Cl_2O_4$
(108544-91-8)
$C_{10}H_{12}Cl_3N_5$
(24848-46-2)
(26235-02-9)
$C_{10}H_{12}Cl_3O_2PS$
(327-98-0)
$C_{10}H_{12}Cl_6N_4$
(24802-86-6)
$C_{10}H_{12}CuN_2O_8.2H$
(54453-03-1)
$C_{10}H_{12}CuN_2O_8.2Na$
(14025-15-1)
$C_{10}H_{12}FeN_2O_8.H_4N$
(21265-50-9)
$C_{10}H_{12}MgN_2O_8.2Na$
(14402-88-1)
$C_{10}H_{12}MnN_2O_8.2Na$
(15375-84-5)
$C_{10}H_{12}N_2$
(61-54-1)
(94-34-8)
(581-49-7)
(5851-43-4)
$C_{10}H_{12}N_2O$
(50-67-9)
(486-56-6)
(2654-57-1)
(13140-86-8)
$C_{10}H_{12}N_2O_2$
(2828-42-4)
$C_{10}H_{12}N_2O_3$
(52-43-7)
(343-65-7)
$C_{10}H_{12}N_2O_3S$
(25057-89-0)
$C_{10}H_{12}N_2O_4$
(1777-84-0)
$C_{10}H_{12}N_2O_5$
(88-85-7)
(303-21-9)
(1420-07-1)
(4097-49-8)
$C_{10}H_{12}N_2O_5.H_3N$
(6365-83-9)
$C_{10}H_{12}N_2O_6S$
(1773-37-1)
$C_{10}H_{12}N_2O_8.Na.Fe$
(15708-41-5)
$C_{10}H_{12}N_2O_8.4Na$
(64-02-8)
$C_{10}H_{12}N_2O_8Zn.2Na$
(14025-21-9)
$C_{10}H_{12}N_2O_8Zn.2H_4N$
(67859-51-2)
$C_{10}H_{12}N_3O_3PS_2$
(86-50-0)
$C_{10}H_{12}N_3O_4PS$
(961-22-8)
$C_{10}H_{12}N_4O_2S_2$
(94-19-9)
$C_{10}H_{12}N_4O_5$
(58-63-9)
$C_{10}H_{12}N_5O_6P$
(60-92-4)
$C_{10}H_{12}N_6$
(30360-14-6)
$C_{10}H_{12}N_6S$
(30360-05-5)
$C_{10}H_{12}O$
(89-74-7)
(104-46-1)
(122-03-2)
(140-67-0)
(487-68-3)
(495-40-9)
(530-91-6)
(611-70-1)
(937-30-4)
(1007-32-5)
(1335-42-8)
(2142-64-5)
(2142-71-4)
(2142-76-9)
(2550-26-7)
(4180-23-8)
(5337-93-9)
(5779-72-6)
(6169-78-4)
(13037-71-3)
(16277-67-1)
(18328-11-5)
(20944-88-1)
(22699-70-3)
(103528-31-0)
$C_{10}H_{12}O_2$
(81-21-0)
(90-27-7)
(93-91-4)
(93-92-5)
(94-58-6)
(97-53-0)
(97-54-1)
(101-97-3)
(103-25-3)
(103-45-7)
(122-63-4)
(480-63-7)
(528-90-5)
(536-66-3)
(672-76-4)
(771-29-9)
(826-55-1)
(939-48-0)
(1076-47-7)
(1563-38-8)
(1821-12-1)
(2186-24-5)
(2210-79-9)
(2315-68-6)
(2438-04-2)
(2529-36-4)
(4346-18-3)
(4593-90-2)
(5912-86-7)
(5932-68-3)
(7364-20-7)
(7473-98-5)
(7495-84-3)
(15012-36-9)
(22084-89-5)
(23617-71-2)
(25081-39-4)
(26447-14-3)
(28655-62-1)
(51410-44-7)
(56588-40-0)
(59204-74-9)
$C_{10}H_{12}O_3$
(94-13-3)
(104-21-2)
(458-35-5)
(607-85-2)
(1131-62-0)
(1835-14-9)
(2040-04-2)
(2211-94-1)
(2503-46-0)
(2555-49-9)
(4223-84-1)
(6303-58-8)
(6828-41-7)
(17138-28-2)
(53744-50-6)
(54966-51-7)
$C_{10}H_{12}O_4$
(56-25-7)
(86-81-7)
(93-40-3)
(999-21-3)
(1135-23-5)
(1421-63-2)
(2150-38-1)
(2478-38-8)
(4460-86-0)
(4707-47-5)
(4707-50-0)
(19420-61-2)
(24607-12-3)
$C_{10}H_{12}O_4S$
(50397-64-3)
$C_{10}H_{12}O_4S.Na$
(1208-67-9)
$C_{10}H_{12}O_5$
(118-41-2)
(121-79-9)
(490-64-2)
(573-11-5)
$C_{10}H_{12}O_5S$
(15267-77-3)
$C_{10}H_{12}O_6$
(6289-46-9)
$C_{10}H_{13}Br$
(4478-10-8)
(65724-11-0)
$C_{10}H_{13}Cl$
(515-40-2)
(4395-79-3)
(54411-19-7)
(54411-21-1)
$C_{10}H_{13}ClN_2$
(6164-98-3)
$C_{10}H_{13}ClN_2.ClH$
(19750-95-9)
$C_{10}H_{13}ClN_2O$
(15545-48-9)
$C_{10}H_{13}ClN_2O_2$
(19937-59-8)
$C_{10}H_{13}ClN_2O_3S$
(94-20-2)
$C_{10}H_{13}ClN_6$
(32889-48-8)
$C_{10}H_{13}ClO$
(98-28-2)
(18980-02-4)
(19010-45-8)
$C_{10}H_{13}ClO_2$
(28693-00-7)
(52196-74-4)
$C_{10}H_{13}Cl_2FN_2O_2S_2$
(731-27-1)
$C_{10}H_{13}Cl_2N$
(553-27-5)
$C_{10}H_{13}Cl_2NO_4$
(3599-58-4)
$C_{10}H_{13}Cl_2O_3PS$
(97-17-6)
$C_{10}H_{13}Cl_3N_4$
(24803-59-6)
$C_{10}H_{13}N$
(91-61-2)
(19343-78-3)
(35203-06-6)
$C_{10}H_{13}NO$
(92-53-5)
(1129-50-6)
(2050-43-3)
(25079-96-3)
(68298-46-4)
$C_{10}H_{13}NO_2$
(62-44-2)
(94-12-2)
(122-42-9)
(536-69-6)
(591-33-3)
(2425-10-7)
(2655-14-3)
(4764-17-4)
(5532-90-1)
(6962-44-3)
(15118-60-2)
(20642-93-7)
(35103-34-5)
(56728-08-6)
$C_{10}H_{13}NO_2.C_9H_8O_4-$
$.C_8H_{10}N_4O_2$
(8003-03-0)
$C_{10}H_{13}NO_2S$
(3566-00-5)
(17688-68-5)
$C_{10}H_{13}NO_3$
(3279-07-0)
(3467-59-2)
(3555-18-8)
$C_{10}H_{13}NO_4$
(136-44-7)
(555-30-6)
$C_{10}H_{13}NO_4.3/2H_2O$
(41372-08-1)
$C_{10}H_{13}NO_4S$
(27375-52-6)
$C_{10}H_{13}N_2O_8.3Na$
(150-38-9)
$C_{10}H_{13}N_3O$
(1133-64-8)
(84237-39-8)
$C_{10}H_{13}N_3O_2$
(64091-90-3)
(64091-91-4)
$C_{10}H_{13}N_3O_3$
(24458-48-8)
$C_{10}H_{13}N_5O_3S$
(789-61-7)
$C_{10}H_{13}N_5O_4$
(58-61-7)
(961-07-9)
(30516-87-1)
$C_{10}H_{13}N_5O_5$
(118-00-3)
$C_{10}H_{13}N_7$
(30368-97-9)
$C_{10}H_{14}$
(95-93-2)
(98-06-6)
(99-87-6)
(104-51-8)
(105-05-5)
(135-01-3)
(135-98-8)
(141-93-5)
(488-23-3)
(527-53-7)
(527-84-4)
(535-77-3)
(538-93-2)
(874-41-9)
(933-98-2)
(934-74-7)
(934-80-5)
(1074-17-5)
(1074-43-7)
(1074-55-1)
(1758-88-9)
(2870-04-4)
(4116-93-2)
(25155-15-1)
(25340-17-4)
(25619-60-7)
(29224-55-3)
(49826-53-1)
(49826-54-2)
$C_{10}H_{14}BrN$
(2052-06-4)
(63460-05-9)
$C_{10}H_{14}ClN$
(92-49-9)
(461-78-9)
(10389-73-8)
$C_{10}H_{14}ClNO_2$
(92-00-2)

$C_{10}H_{14}ClOP$
(4923-85-7)
$C_{10}H_{14}ClOPS_2$
(2984-64-7)
$C_{10}H_{14}Cl_2NO_2PS$
(299-85-4)
$C_{10}H_{14}Cl_3N_3O_2$
(30863-41-3)
(30863-83-3)
$C_{10}H_{14}Cl_6N_4O_2$
(26644-46-2)
$C_{10}H_{14}CoO_4$
(14024-48-7)
$C_{10}H_{14}FOP$
(55236-56-1)
$C_{10}H_{14}NO_3S.Na$
(5123-63-7)
$C_{10}H_{14}NO_5PS$
(56-38-2)
(597-88-6)
(3270-86-8)
$C_{10}H_{14}NO_6P$
(311-45-5)
$C_{10}H_{14}N_2$
(54-11-5)
(92-54-6)
(148-87-8)
(494-52-0)
(10278-71-4)
(22083-74-5)
$C_{10}H_{14}N_2.2C_4H_6O_6$
(65-31-6)
$C_{10}H_{14}N_2.C_7H_6O_3$
(29790-52-1)
$C_{10}H_{14}N_2.ClH$
(21361-93-3)
$C_{10}H_{14}N_2.xClH$
(2820-51-1)
$C_{10}H_{14}N_2O$
(59-26-7)
(7160-01-2)
(19895-44-4)
(58609-76-0)
$C_{10}H_{14}N_2O_2$
(4138-38-9)
(6358-20-9)
(7160-02-3)
(13171-61-4)
(17026-81-2)
(28170-54-9)
$C_{10}H_{14}N_2O_2.ClH$
(25953-06-4)
$C_{10}H_{14}N_2O_3$
(77-02-1)
$C_{10}H_{14}N_2O_5$
(50-89-5)
(52551-67-4)
$C_{10}H_{14}N_2O_5S$
(25959-70-0)
$C_{10}H_{14}N_2O_8.2Na$
(139-33-3)
$C_{10}H_{14}N_3.1/2Cl_4Zn$
(5149-85-9)
$C_{10}H_{14}N_4O_4$
(479-18-5)
$C_{10}H_{14}N_5O_6P$
(653-63-4)
$C_{10}H_{14}N_5O_7P$
(61-19-8)
$C_{10}H_{14}N_5O_8P$
(85-32-5)
$C_{10}H_{14}O$
(88-18-6)
(89-72-5)
(89-83-8)
(98-54-4)
(99-49-0)
(99-71-8)
(100-86-7)
(104-45-0)
(499-75-2)
(527-35-5)
(536-60-7)
(585-34-2)
(875-99-0)
(1126-79-0)
(1197-01-9)
(1638-22-8)
(2111-75-3)
(2244-16-8)
(3180-09-4)
(3228-04-4)
(3299-05-6)
(4170-90-5)
(5379-19-1)
(6485-40-1)
(18309-32-5)
(26967-65-7)
(27178-34-3)
(27577-96-4)
(33967-19-0)
(54264-96-9)
$C_{10}H_{14}O.Na$
(5787-50-8)
$C_{10}H_{14}O_2$
(98-29-3)
(101-48-4)
(122-94-1)
(1948-33-0)
(2785-87-7)
(4230-32-4)
(5426-78-8)
(7452-01-9)
(10373-78-1)
(22607-13-2)
(27213-78-1)
(27593-23-3)
(34883-05-1)
$C_{10}H_{14}O_3$
(104-68-7)
(2186-92-7)
(2305-13-7)
(2386-90-5)
(41654-27-7)
(54346-06-4)
$C_{10}H_{14}O_3S$
(80-44-4)
(18521-59-0)
$C_{10}H_{14}O_3S.Na$
(26746-29-2)
$C_{10}H_{14}O_4$
(93-14-1)
(97-90-5)
(104-38-1)
(19485-03-1)
$C_{10}H_{14}O_4.Cu$
(13395-16-9)
$C_{10}H_{14}O_4Zn$
(14024-63-6)
$C_{10}H_{14}O_5$
(1208-42-0)
(4074-88-8)
(21282-97-3)
$C_{10}H_{14}O_5Ti$
(14024-64-7)
$C_{10}H_{14}O_6$
(57230-48-5)
$C_{10}H_{15}AsO_2$
(3141-11-5)
$C_{10}H_{15}AsS_2$
(5582-58-1)
$C_{10}H_{15}Br$
(102938-79-4)
$C_{10}H_{15}Cl_3N_4$
(24802-89-9)
(24802-90-2)
(24803-22-3)
(24803-23-4)
(24803-28-9)
(24803-54-1)
$C_{10}H_{15}FeN_2O_7$
(17084-02-5)
$C_{10}H_{15}FeN_2O_7.Na$
(16485-47-5)
$C_{10}H_{15}HgNO_3$
(5822-97-9)
$C_{10}H_{15}N$
(91-66-7)
(104-13-2)
(122-09-8)
(537-46-2)
(579-66-8)
(1126-78-9)
(1701-68-4)
(2294-76-0)
(2449-49-2)
(5146-66-7)
(5585-39-7)
(14719-47-2)
(31983-27-4)
(54675-14-8)
(68400-78-2)
$C_{10}H_{15}N.ClH$
(300-42-5)
$C_{10}H_{15}NO$
(90-82-4)
(91-68-9)
(92-50-2)
(103-62-8)
(299-42-3)
(23239-32-9)
(31198-76-2)
$C_{10}H_{15}NO_2$
(120-07-0)
(6967-70-0)
$C_{10}H_{15}NO_7S_2$
(21635-69-8)
$C_{10}H_{15}NO_8S_2$
(26672-24-2)
$C_{10}H_{15}N_3$
(55-73-2)
$C_{10}H_{15}N_5$
(114-86-3)
(26234-39-9)
$C_{10}H_{15}N_5.ClH$
(834-28-6)
$C_{10}H_{15}N_5S$
(1020-53-7)
$C_{10}H_{15}OPS_2$
(944-22-9)
$C_{10}H_{15}OPSe$
(51584-27-1)
(51584-28-2)
$C_{10}H_{15}O_2PS_2$
(2703-13-1)
$C_{10}H_{15}O_3P$
(33684-08-1)
$C_{10}H_{15}O_3PS$
(32345-29-2)
$C_{10}H_{15}O_3PS_2$
(55-38-9)
$C_{10}H_{15}O_4P$
(2510-86-3)
$C_{10}H_{15}O_4PS_2$
(3761-41-9)
$C_{10}H_{15}O_5PS_2$
(3761-42-0)
$C_{10}H_{16}$
(79-92-5)
(80-56-8)
(99-83-2)
(99-85-4)
(99-86-5)
(123-35-3)
(127-91-3)
(138-86-3)
(281-23-2)
(464-17-5)
(499-03-6)
(508-32-7)
(514-94-3)
(554-60-9)
(555-10-2)
(586-62-9)
(673-84-7)
(1330-16-1)
(2825-82-3)
(2867-05-2)
(3338-55-4)
(3387-41-5)
(3779-61-1)
(4221-98-1)
(5989-27-5)
(5989-54-8)
(6004-38-2)
(7705-14-8)
(7785-26-4)
(8013-00-1)
(13466-78-9)
(13837-95-1)
(13877-91-3)
(15402-94-5)
(18172-67-3)
(29714-87-2)
(53282-47-6)
(86853-03-4)
$C_{10}H_{16}Br_2N_2O_2$
(54-91-1)
$C_{10}H_{16}ClN_2O_2P$
(56185-01-4)
$C_{10}H_{16}Cl_2O_2$
(111-19-3)
$C_{10}H_{16}Cl_3NOS$
(2303-17-5)
$C_{10}H_{16}Cl_3N_5$
(26234-98-0)
(26263-50-3)
(27470-97-9)
(27529-92-6)
$C_{10}H_{16}Cl_3N_5O_2$
(26235-07-4)
(26235-08-5)
(26235-09-6)
(30388-85-3)
$C_{10}H_{16}N.Br$
(5350-41-4)
$C_{10}H_{16}N.Cl$
(56-93-9)
$C_{10}H_{16}N.HO$
(100-85-6)
$C_{10}H_{16}N.I$
(4525-46-6)
$C_{10}H_{16}NO_3PS$
(3735-01-1)
$C_{10}H_{16}NO_5PS_2$
(52-85-7)
$C_{10}H_{16}N_2$
(93-05-0)
(100-22-1)
(1871-96-1)
(3663-23-8)
$C_{10}H_{16}N_2.ClH$
(2198-58-5)
$C_{10}H_{16}N_2NaO_8$
(7379-28-4)
$C_{10}H_{16}N_2O_2.H_2O_4S$
(54381-16-7)
$C_{10}H_{16}N_2O_3$
(77-28-1)
(125-40-6)
$C_{10}H_{16}N_2O_3.Na$
(143-81-7)
$C_{10}H_{16}N_2O_3S$
(58-85-5)
(25797-78-8)
$C_{10}H_{16}N_2O_8$
(60-00-4)
$C_{10}H_{16}N_2O_8.2H_3N$
(20824-56-0)
$C_{10}H_{16}N_2O_8.4H_3N$

(22473-78-5)
$C_{10}H_{16}N_2O_8.xH_3N$
(7379-26-2)
$C_{10}H_{16}N_2O_8.3K$
(17572-97-3)
$C_{10}H_{16}N_2O_8.4K$
(5964-35-2)
$C_{10}H_{16}N_2O_8.xK$
(7379-27-3)
$C_{10}H_{16}N_2O_8.Na$
(17421-79-3)
$C_{10}H_{16}N_4$
(13472-08-7)
$C_{10}H_{16}N_4OS$
(21085-19-8)
$C_{10}H_{16}N_4O_2S$
(55511-98-3)
$C_{10}H_{16}N_4O_3$
(644-64-4)
$C_{10}H_{16}N_5O_{13}P_3$
(56-65-5)
$C_{10}H_{16}N_6S$
(21689-84-9)
(30360-08-8)
(30360-90-8)
(30360-92-0)
(51481-61-9)
$C_{10}H_{16}O$
(76-22-2)
(89-81-6)
(99-48-9)
(106-26-3)
(141-27-5)
(432-25-7)
(464-48-2)
(464-49-3)
(499-74-1)
(515-00-4)
(536-59-4)
(546-80-5)
(547-60-4)
(1195-79-5)
(1423-46-7)
(1686-14-2)
(2363-88-4)
(4501-58-0)
(4573-50-6)
(4695-62-9)
(4884-24-6)
(5392-40-5)
(5524-05-0)
(7764-50-3)
(10292-98-5)
(20053-88-7)
(21399-51-9)
(25152-84-5)
(26532-24-1)
(29171-20-8)
(40702-26-9)
(69103-20-4)
$C_{10}H_{16}O_2$
(96-08-2)
(101-43-9)
(459-80-3)
(512-85-6)
(704-76-7)
(2704-78-1)
(3491-27-8)
(7003-48-7)
(10453-89-1)
(12001-40-0)
(21766-50-7)
(28645-03-6)
(32511-06-1)
(53690-92-9)
(54789-11-6)
(100428-67-9)
$C_{10}H_{16}O_3$
(473-72-3)
(54774-94-6)
$C_{10}H_{16}O_4$
(94-60-0)
(124-83-4)
(777-95-7)
(5394-83-2)
$C_{10}H_{16}O_4S$
(3144-16-9)
$C_{10}H_{16}O_5$
(1115-30-6)
$C_{10}H_{16}O_6$
(10043-09-1)
$C_{10}H_{17}ClN_2O$
(1319-96-6)
$C_{10}H_{17}ClN_4O$
(30357-84-7)
$C_{10}H_{17}Cl_2NOS$
(2303-16-4)
(17708-57-5)
$C_{10}H_{17}N$
(768-94-5)
$C_{10}H_{17}N.xH_2O_4S$
(32793-63-8)
$C_{10}H_{17}NO$
(13372-77-5)
$C_{10}H_{17}NOS$
(16516-78-2)
(107348-43-6)
$C_{10}H_{17}NO_2$
(125-64-4)
$C_{10}H_{17}NO_6$
(554-35-8)
$C_{10}H_{17}N_2O_4PS$
(38260-54-7)
$C_{10}H_{17}N_2O_8P$
(365-07-1)
$C_{10}H_{17}N_3O_2$
(119-38-0)
(55861-78-4)
$C_{10}H_{17}N_3O_6S$
(70-18-8)
$C_{10}H_{17}N_5$
(27431-06-7)
$C_{10}H_{17}N_5O$
(30360-68-0)
$C_{10}H_{17}N_7O_4.2ClH$
(35554-08-6)
$C_{10}H_{18}$
(91-17-8)
(473-55-2)
(493-01-6)
(493-02-7)
(764-93-2)
(935-31-9)
(1124-27-2)
(1636-39-1)
(1647-16-1)
(1942-46-7)
(2198-20-1)
(2228-98-0)
(2384-86-3)
(2436-90-0)
(4104-56-7)
(5502-88-5)
(6876-13-7)
(13837-71-3)
(24524-52-5)
$C_{10}H_{18}ClNOS$
(79720-82-4)
$C_{10}H_{18}ClN_3O_2$
(13909-09-6)
$C_{10}H_{18}ClN_5$
(1912-25-0)
$C_{10}H_{18}ClN_5O$
(1824-09-5)
$C_{10}H_{18}N_2O_3$
(64346-47-0)
$C_{10}H_{18}N_2O_7$
(150-39-0)
$C_{10}H_{18}N_2O_7.3Na$
(139-89-9)
$C_{10}H_{18}N_4O$
(5248-42-0)
$C_{10}H_{18}N_4O_4S_3$
(59669-26-0)
$C_{10}H_{18}O$
(78-70-6)
(89-80-5)
(98-53-3)
(98-55-5)
(106-23-0)
(106-24-1)
(106-25-2)
(124-76-5)
(138-87-4)
(470-67-7)
(470-82-6)
(473-54-1)
(491-04-3)
(491-07-6)
(499-70-7)
(507-70-0)
(512-13-0)
(513-23-5)
(543-39-5)
(562-74-3)
(586-82-3)
(624-15-7)
(825-51-4)
(1632-73-1)
(2903-23-3)
(3913-71-1)
(3913-81-3)
(4099-07-4)
(4819-67-4)
(4948-28-1)
(5877-42-9)
(5949-05-3)
(6555-95-9)
(7785-53-7)
(7786-67-6)
(8000-41-7)
(8006-39-1)
(10137-73-2)
(10458-14-7)
(10482-56-1)
(15404-57-6)
(16409-43-1)
(17983-22-1)
(18479-54-4)
(32064-70-3)
(38049-26-2)
(49833-96-7)
(54244-89-2)
$C_{10}H_{18}O_2$
(78-66-0)
(94-04-2)
(142-09-6)
(705-86-2)
(706-14-9)
(1321-87-5)
(1365-19-1)
(5989-33-3)
(7367-82-0)
(14009-71-3)
(14049-11-7)
(28324-52-9)
(39028-58-5)
(54382-58-0)
(69668-83-3)
$C_{10}H_{18}O_3$
(1422-26-0)
(1538-75-6)
$C_{10}H_{18}O_4$
(111-20-6)
(141-28-6)
(759-36-4)
(1732-09-8)
(2049-70-9)
(2050-60-4)
(2163-48-6)
(2425-79-8)
(3775-85-7)
(3971-33-3)
(17219-21-5)
$C_{10}H_{18}O_4.C_6H_{16}N_2$
(6422-99-7)
$C_{10}H_{18}O_4.Pb$
(29473-77-6)
$C_{10}H_{18}O_4S_2$
(10194-00-0)
$C_{10}H_{18}O_5$
(4206-61-5)
(6335-02-0)
$C_{10}H_{18}O_6$
(111-21-7)
(16215-49-9)
(19910-65-7)
$C_{10}H_{19}ClNO_5P$
(297-99-4)
(13171-21-6)
(23783-98-4)
$C_{10}H_{19}ClO$
(112-13-0)
(40292-82-8)
$C_{10}H_{19}ClO_3$
(67969-81-7)
$C_{10}H_{19}N$
(1975-78-6)
$C_{10}H_{19}NOS$
(3134-66-5)
(107348-46-9)
$C_{10}H_{19}NO_2$
(105-16-8)
(3775-90-4)
(25457-47-0)
(65882-20-4)
$C_{10}H_{19}N_5$
(26234-33-3)
(26235-18-7)
(26235-24-5)
(26235-36-9)
(26263-49-0)
(27430-94-0)
(27430-95-1)
(27430-96-2)
(27431-02-3)
(27431-09-0)
(27431-10-3)
(27431-18-1)
(27470-63-9)
(30368-85-5)
$C_{10}H_{19}N_5O$
(1610-18-0)
(13532-26-8)
(26259-45-0)
(30360-62-4)
(33693-04-8)
$C_{10}H_{19}N_5O_2$
(5943-83-9)
(26234-95-7)
(26322-44-1)
(30368-96-8)
(31858-13-6)
$C_{10}H_{19}N_5S$
(836-24-8)
(886-50-0)
(7287-19-6)
$C_{10}H_{19}N_5S.C_8H_{14}ClN_5$
(8073-77-6)
$C_{10}H_{19}N_5S.C_9H_{16}ClN_5$
(8066-11-3)
$C_{10}H_{19}N_5S.C_{10}H_{13}ClN_2O$
(37341-11-0)
$C_{10}H_{19}O_2.Na$
(1002-62-6)
$C_{10}H_{19}O_6PS_2$
(121-75-5)
(3344-12-5)
$C_{10}H_{19}O_7PS$
(1634-78-2)

(24478-11-3)
$C_{11}H_3Cl_7N_4O_2S$
(24478-15-7)
$C_{11}H_3Cl_8N_3$
(24481-51-4)
(24481-52-5)
$C_{11}H_3Cl_8N_3S$
(24478-09-9)
(24478-10-2)
$C_{11}H_3Cl_9N_4$
(30356-56-0)
$C_{11}H_4BrCl_6N_3$
(24481-49-0)
$C_{11}H_4BrCl_6N_3S$
(24478-12-4)
$C_{11}H_4Cl_6N_4O_2$
(3599-72-2)
$C_{11}H_4Cl_6N_4O_2S$
(24478-14-6)
$C_{11}H_4Cl_7N_3$
(3712-60-5)
(24481-54-7)
(24481-55-8)
$C_{11}H_4Cl_7N_3S$
(24478-06-6)
(24478-07-7)
(24478-08-8)
$C_{11}H_4Cl_8N_4$
(30356-54-8)
(30356-55-9)
$C_{11}H_5BrCl_6N_4$
(30356-58-2)
(30356-59-3)
$C_{11}H_5Br_6ClN_4$
(30339-78-7)
$C_{11}H_5ClN_6S_2$
(30362-22-2)
(30362-23-3)
(30362-24-4)
$C_{11}H_5Cl_6N_3$
(24504-22-1)
$C_{11}H_5Cl_6N_3S$
(5516-47-2)
$C_{11}H_5Cl_7N_4$
(3599-75-5)
(30356-51-5)
(30356-52-6)
$C_{11}H_5F_6N_3$
(30362-32-4)
$C_{11}H_6Br_6N_4$
(30339-39-0)
$C_{11}H_6Cl_6N_4$
(30356-50-4)
$C_{11}H_6Cl_6N_4S$
(30369-63-2)
$C_{11}H_6Cl_6O_2$
(61217-08-1)
$C_{11}H_6O_3$
(66-97-7)
(523-50-2)
$C_{11}H_6O_4$
(116211-95-1)
$C_{11}H_6O_{10}$
(1585-40-6)
$C_{11}H_7Cl_4N_3O$
(30863-62-8)
$C_{11}H_7Cl_5N_4S$
(30369-61-0)
$C_{11}H_7Cl_6HgNO_2$
(2597-93-5)
$C_{11}H_7N$
(86-53-3)
(613-46-7)
(1984-04-9)
$C_{11}H_7NS$
(551-06-4)
$C_{11}H_8BrO_4.Na$
(21739-91-3)
$C_{11}H_8Cl_3N_3$
(30362-33-5)
$C_{11}H_8Cl_3N_3O$
(30863-60-6)
$C_{11}H_8Cl_4N_4$
(30360-44-2)
$C_{11}H_8Cl_4N_4S$
(30369-58-5)
$C_{11}H_8Cl_6O$
(21858-40-2)
(61167-23-5)
$C_{11}H_8F_{17}NO_3S$
(24448-09-7)
$C_{11}H_8N_2$
(244-63-3)
$C_{11}H_8N_2O$
(3878-19-1)
$C_{11}H_8N_2O_5$
(3688-53-7)
$C_{11}H_8O$
(66-77-3)
(66-99-9)
(30678-61-6)
$C_{11}H_8O_2$
(58-27-5)
(86-55-5)
(93-09-4)
$C_{11}H_8O_3$
(86-48-6)
(92-70-6)
(483-55-6)
(2283-08-1)
(16712-64-4)
$C_{11}H_8O_3.Na$
(14206-62-3)
$C_{11}H_9Cl$
(86-52-2)
$C_{11}H_9ClO_2S_2$
(53135-95-8)
$C_{11}H_9Cl_2NO_2$
(101-27-9)
$C_{11}H_9Cl_2NO_3$
(24201-58-9)
$C_{11}H_9Cl_3N_4$
(24802-99-1)
(30360-40-8)
(30369-16-5)
$C_{11}H_9Cl_3N_4O$
(30369-17-6)
$C_{11}H_9Cl_3N_4O_2$
(13960-29-7)
$C_{11}H_9Cl_3N_4S$
(30369-56-3)
$C_{11}H_9Cl_3O_3$
(136-25-4)
$C_{11}H_9I_3N_2O_4$
(117-96-4)
$C_{11}H_9N$
(939-23-1)
(1008-88-4)
(1008-89-5)
(52642-16-7)
$C_{11}H_9NO_2$
(880-93-3)
(881-03-8)
(1204-72-4)
(5959-52-4)
(13615-38-8)
(41037-13-2)
(54357-08-3)
(54755-20-3)
(54755-21-4)
(63017-87-8)
(91137-27-8)
(91137-28-9)
(116530-07-5)
$C_{11}H_9NO_3$
(117-57-7)
(2608-48-2)
(5455-98-1)
$C_{11}H_9NO_4$
(22509-74-6)
$C_{11}H_9N_3$
(26148-68-5)
$C_{11}H_9N_3O_2$
(1917-34-6)
$C_{11}H_{10}$
(90-12-0)
(91-57-6)
(1321-94-4)
(2471-83-2)
$C_{11}H_{10}BrNO_2$
(5460-29-7)
$C_{11}H_{10}ClNO_2$
(1967-16-4)
$C_{11}H_{10}ClN_3$
(13704-97-7)
(30362-08-4)
$C_{11}H_{10}Cl_2N_4$
(30360-39-5)
$C_{11}H_{10}Cl_2N_4O_2$
(13960-26-4)
$C_{11}H_{10}Cl_3N_7O_3$
(30576-26-2)
$C_{11}H_{10}N_2$
(64859-47-8)
$C_{11}H_{10}N_2O$
(14491-59-9)
$C_{11}H_{10}N_2S$
(86-88-4)
$C_{11}H_{10}N_4$
(67730-11-4)
(76180-96-6)
$C_{11}H_{10}N_4O_2$
(30362-09-5)
$C_{11}H_{10}N_4O_4$
(6804-07-5)
$C_{11}H_{10}O$
(93-04-9)
(2216-69-5)
(4780-79-4)
(7469-77-4)
$C_{11}H_{10}O_2$
(17369-59-4)
$C_{11}H_{10}O_3$
(31005-02-4)
$C_{11}H_{10}O_4$
(120-08-1)
(480-34-2)
$C_{11}H_{10}O_5S.Na$
(130-37-0)
$C_{11}H_{11}BrN_2O$
(603-65-6)
$C_{11}H_{11}BrN_4O_2$
(30358-04-4)
$C_{11}H_{11}ClN_4$
(5122-28-1)
(30360-47-5)
$C_{11}H_{11}ClN_4O$
(30357-92-7)
$C_{11}H_{11}ClN_4O_2$
(27315-26-0)
(27315-27-1)
$C_{11}H_{11}Cl_2NO$
(2860-64-2)
$C_{11}H_{11}Cl_3O_3$
(93-78-7)
$C_{11}H_{11}Cl_6N_3$
(30362-48-2)
$C_{11}H_{11}Cl_6N_5$
(24848-43-9)
$C_{11}H_{11}N$
(93-37-8)
(118-31-0)
(877-43-0)
(1198-37-4)
(1463-17-8)
(1721-89-7)
(2436-92-2)
(2623-50-9)
(7661-47-4)
(28351-04-4)
(53123-73-2)
$C_{11}H_{11}NO_2$
(86-34-0)
(304-17-6)
(830-96-6)
$C_{11}H_{11}NO_3$
(3692-90-8)
$C_{11}H_{11}N_3$
(3599-61-9)
$C_{11}H_{11}N_3O$
(715-99-1)
(1917-40-4)
$C_{11}H_{11}N_3O_2$
(18213-73-5)
$C_{11}H_{11}N_3O_2S$
(144-83-2)
$C_{11}H_{11}N_3S_2$
(15067-64-8)
$C_{11}H_{11}N_5$
(94-78-0)
(77500-04-0)
$C_{11}H_{11}N_5.ClH$
(136-40-3)
$C_{11}H_{11}N_5.xClH$
(10393-51-8)
$C_{11}H_{11}N_5O_4$
(25962-77-0)
(30358-05-5)
(55738-54-0)
$C_{11}H_{11}N_5O_5$
(794-93-4)
$C_{11}H_{11}N_7O$
(30362-27-7)
$C_{11}H_{12}$
(18636-55-0)
(39292-53-0)
(53204-57-2)
$C_{11}H_{12}ClN$
(2012-81-9)
$C_{11}H_{12}ClNO_2$
(20139-55-3)
$C_{11}H_{12}ClNO_3S$
(80-77-3)
$C_{11}H_{12}ClO_3.Na$
(6062-26-6)
$C_{11}H_{12}Cl_2N_2O_5$
(56-75-7)
$C_{11}H_{12}Cl_2O_3$
(94-11-1)
(1928-61-6)
$C_{11}H_{12}Cl_2O_4$
(75315-44-5)
(75315-46-7)
$C_{11}H_{12}Cl_2O_5$
(75315-45-6)
$C_{11}H_{12}Cl_3O_2PS$
(41491-52-5)
$C_{11}H_{12}Cl_6N_4$
(24803-15-4)
$C_{11}H_{12}I_3NO_2$
(96-83-3)
$C_{11}H_{12}NO_4PS_2$
(732-11-6)
$C_{11}H_{12}N_2O$
(60-80-0)
(86-92-0)
$C_{11}H_{12}N_2OS_2$
(102-77-2)
$C_{11}H_{12}N_2OS_3$
(95-32-9)
$C_{11}H_{12}N_2O_2$
(54-12-6)
(73-22-3)
(153-94-6)
(6912-86-3)
$C_{11}H_{12}N_2O_3$
(56-69-9)
(4350-09-8)
$C_{11}H_{12}N_2S$
(14769-73-4)

$C_{11}H_{12}N_4$
(30360-35-1)
(30377-22-1)
$C_{11}H_{12}N_4O$
(30369-37-0)
$C_{11}H_{12}N_4O_2$
(30358-01-1)
$C_{11}H_{12}N_4O_2S.Na$
(127-58-2)
$C_{11}H_{12}N_4O_3S$
(80-35-3)
$C_{11}H_{12}N_4S_2$
(30362-13-1)
$C_{11}H_{12}O$
(16440-97-4)
(56667-10-8)
(103433-72-3)
$C_{11}H_{12}O_2$
(103-36-6)
(1078-19-9)
(2495-37-6)
(3318-61-4)
(5660-60-6)
$C_{11}H_{12}O_3$
(94-02-0)
(121-39-1)
(607-91-0)
(3160-37-0)
(10031-96-6)
(54815-21-3)
$C_{11}H_{12}O_4$
(2316-26-9)
(10487-92-0)
(14186-60-8)
(20116-65-8)
(34006-77-4)
$C_{11}H_{12}O_5$
(530-59-6)
$C_{11}H_{13}ClF_3N_3O_4S_3$
(346-18-9)
$C_{11}H_{13}ClO$
(13347-42-7)
$C_{11}H_{13}ClO_2$
(2012-74-0)
$C_{11}H_{13}ClO_2S$
(25319-90-8)
$C_{11}H_{13}ClO_3$
(94-81-5)
(55162-41-9)
$C_{11}H_{13}Cl_2N_3O_4$
(26389-78-6)
$C_{11}H_{13}Cl_2N_5$
(13344-99-5)
$C_{11}H_{13}Cl_2O_2PS$
(38338-57-7)
$C_{11}H_{13}Cl_2O_2PS_3$
(2275-14-1)
$C_{11}H_{13}Cl_6N_3O$
(30863-58-2)
$C_{11}H_{13}Cl_6N_3S$
(24481-71-8)
$C_{11}H_{13}F_3N_2O_3S$
(53780-34-0)
$C_{11}H_{13}F_3N_4O_4$
(29091-05-2)
$C_{11}H_{13}N$
(1640-39-7)
(30642-36-5)
$C_{11}H_{13}NO$
(20200-86-6)
$C_{11}H_{13}NO_2$
(93-68-5)
(2415-85-2)
$C_{11}H_{13}NO_3$
(92-15-9)
(7244-67-9)
(29218-27-7)
$C_{11}H_{13}NO_4$
(6988-21-2)
(22781-23-3)
$C_{11}H_{13}NS$
(100182-85-2)
$C_{11}H_{13}N_3O_3S$
(127-69-5)
$C_{11}H_{13}N_5$
(30359-79-6)
$C_{11}H_{14}$
(769-25-5)
(1559-81-5)
(1680-51-9)
(1685-82-1)
(2809-64-5)
(3877-19-8)
(4165-78-0)
(4165-86-0)
(4175-53-5)
(4489-84-3)
(4830-99-3)
(6682-71-9)
(17057-82-8)
(17059-48-2)
(19947-22-9)
(31291-71-1)
(53563-67-0)
(56147-63-8)
$C_{11}H_{14}ClNO$
(1918-16-7)
$C_{11}H_{14}ClNO_2$
(2164-13-8)
$C_{11}H_{14}ClNO_3$
(94483-57-5)
$C_{11}H_{14}ClNO_4$
(51487-69-5)
$C_{11}H_{14}ClN_5$
(516-21-2)
$C_{11}H_{14}Cl_2O_4$
(75315-54-7)
(108545-03-5)
$C_{11}H_{14}Cl_3N_5$
(24848-51-9)
(24863-53-4)
$C_{11}H_{14}Cl_6N_4$
(24802-87-7)
(24803-07-4)
$C_{11}H_{14}N_2O$
(92-64-8)
(485-35-8)
(2654-58-2)
(28291-83-0)
$C_{11}H_{14}N_2O_2$
(7206-76-0)
(13047-13-7)
$C_{11}H_{14}N_2O_5$
(3996-59-6)
(4097-36-3)
$C_{11}H_{14}N_2S$
(95-31-8)
(15686-83-6)
$C_{11}H_{14}N_4O_4S$
(342-69-8)
$C_{11}H_{14}N_6O_2$
(30805-27-7)
$C_{11}H_{14}O$
(645-13-6)
(939-97-9)
(1200-14-2)
(1518-83-8)
(1667-01-2)
(2142-65-6)
(2235-83-8)
(38393-92-9)
$C_{11}H_{14}O_2$
(93-15-2)
(93-16-3)
(95-39-6)
(98-73-7)
(103-37-7)
(122-70-3)
(136-60-7)
(1010-48-6)
(1320-16-7)
(2021-28-5)
(2270-20-4)
(4920-92-7)
(5669-17-0)
(7315-68-6)
(16433-43-5)
(20115-23-5)
(20651-71-2)
(54549-72-3)
$C_{11}H_{14}O_2.C_6H_{15}NO_3$
(59993-86-1)
$C_{11}H_{14}O_3$
(87-19-4)
(94-26-8)
(122-48-5)
(614-45-9)
(1835-04-7)
$C_{11}H_{14}O_4$
(1136-86-3)
(39589-98-5)
$C_{11}H_{15}BrClO_3PS$
(41198-08-7)
$C_{11}H_{15}BrN_2O$
(78508-43-7)
$C_{11}H_{15}ClN_2S.ClH$
(34863-74-6)
$C_{11}H_{15}Cl_2NO_3$
(5742-17-6)
$C_{11}H_{15}Cl_2O_2PS_2$
(34643-46-4)
$C_{11}H_{15}Cl_2O_3PS_2$
(21923-23-9)
(60238-56-4)
$C_{11}H_{15}N$
(4096-20-2)
(35203-08-8)
$C_{11}H_{15}NO$
(120-21-8)
(134-49-6)
(1696-17-9)
(89151-70-2)
$C_{11}H_{15}NO_2$
(64-00-6)
(94-25-7)
(1013-75-8)
(1538-74-5)
(2021-20-7)
(2631-40-5)
(2655-15-4)
(2686-99-9)
(7756-96-9)
(7779-77-3)
(10287-53-3)
(12407-86-2)
(17754-90-4)
$C_{11}H_{15}NO_2S$
(2032-65-7)
(29973-13-5)
$C_{11}H_{15}NO_3$
(114-26-1)
(13674-05-0)
$C_{11}H_{15}NO_5$
(532-03-6)
$C_{11}H_{15}N_3O_2$
(22259-30-9)
$C_{11}H_{15}N_3O_2.ClH$
(23422-53-9)
$C_{11}H_{15}N_3O_4$
(1882-26-4)
$C_{11}H_{15}N_5$
(4022-58-6)
$C_{11}H_{15}N_5O_4$
(15763-06-1)
$C_{11}H_{16}$
(98-51-1)
(538-68-1)
(700-12-9)
(1007-26-7)
(1196-58-3)
(2049-94-7)
(2049-95-8)
(2050-24-0)
(2719-52-0)
(4218-48-8)
(4706-89-2)
(4706-90-5)
(4920-99-4)
(5161-04-6)
(17851-27-3)
(25550-13-4)
(27138-21-2)
(29316-05-0)
(33156-92-2)
(54120-62-6)
$C_{11}H_{16}BrNO_2.BrH$
(53581-53-6)
$C_{11}H_{16}ClN_5O_2$
(4065-24-1)
$C_{11}H_{16}ClN_5S_4$
(30863-07-1)
$C_{11}H_{16}ClO_2PS_3$
(786-19-6)
$C_{11}H_{16}Cl_3N_3O$
(30863-50-4)
$C_{11}H_{16}N.C_6H_2N_3O_9S$
(3806-34-6)
$C_{11}H_{16}N_2$
(19730-04-2)
$C_{11}H_{16}N_2O$
(36627-56-2)
(41708-72-9)
$C_{11}H_{16}N_2O_2$
(92-13-7)
(2032-59-9)
(6265-09-4)
$C_{11}H_{16}N_2O_2.ClH$
(54-71-7)
$C_{11}H_{16}N_2O_3$
(77-26-9)
(115-44-6)
(125-42-8)
(17872-56-9)
$C_{11}H_{16}N_4O_2$
(30358-12-4)
$C_{11}H_{16}N_4O_4$
(21416-87-5)
$C_{11}H_{16}O$
(80-46-6)
(88-60-8)
(94-06-4)
(98-27-1)
(103-05-9)
(136-81-2)
(488-10-8)
(877-65-6)
(1333-13-7)
(2219-72-9)
(2219-82-1)
(2409-55-4)
(2819-86-5)
(3279-27-4)
(3280-08-8)
(14938-35-3)
(25567-40-2)
(29225-54-5)
(36731-23-4)
$C_{11}H_{16}O.K$
(53404-18-5)
$C_{11}H_{16}O.Na$
(31366-95-7)
$C_{11}H_{16}O_2$
(88-32-4)
(121-00-6)
(500-66-3)
(15356-74-8)
(15359-99-6)
(17092-92-1)
(18979-53-8)
(25013-16-5)

(53894-31-8)
$C_{11}H_{16}O_3$
(1133-03-5)
$C_{11}H_{16}O_4$
(2223-82-7)
(4316-33-0)
$C_{11}H_{16}O_5$
(25876-07-7)
$C_{11}H_{16}O_6$
(52942-64-0)
(53417-29-1)
$C_{11}H_{16}S$
(54576-36-2)
(54576-37-3)
(54576-41-9)
$C_{11}H_{17}AsS$
(23486-02-4)
$C_{11}H_{17}Cl$
(27011-46-7)
$C_{11}H_{17}Cl_3N_4$
(24803-55-2)
$C_{11}H_{17}Cl_3N_4S$
(30369-54-1)
(30576-30-8)
$C_{11}H_{17}N$
(91-67-8)
(100-92-5)
(457-87-4)
(2049-92-5)
(6311-92-8)
$C_{11}H_{17}NO$
(91-88-3)
$C_{11}H_{17}NO_2$
(91-99-6)
(101-32-6)
(28005-74-5)
$C_{11}H_{17}NO_2.ClH$
(24973-25-9)
$C_{11}H_{17}NO_3$
(54-04-6)
(122-15-6)
$C_{11}H_{17}NO_3.ClH$
(51-30-9)
(61-16-5)
(2801-68-5)
$C_{11}H_{17}N_2O_2$
(17752-10-2)
$C_{11}H_{17}N_3O_2$
(30362-70-0)
$C_{11}H_{17}N_3O_3S$
(339-43-5)
$C_{11}H_{17}N_3O_4$
(2784-94-3)
$C_{11}H_{17}N_3O_8$
(4368-28-9)
$C_{11}H_{17}OPS_2$
(333-43-7)
$C_{11}H_{17}O_3PS$
(13286-32-3)
$C_{11}H_{17}O_3PS_2$
(3070-15-3)
$C_{11}H_{17}O_4PS_2$
(115-90-2)
$C_{11}H_{17}O_5PS_2$
(14255-72-2)
$C_{11}H_{18}$
(22822-99-7)
(56030-49-0)
(56324-70-0)
(74744-31-3)
$C_{11}H_{18}Br_2Cl_3O_4P$
(66108-37-0)
$C_{11}H_{18}ClN_3S_2$
(17494-99-4)
$C_{11}H_{18}ClN_3OS$
(30894-82-7)
$C_{11}H_{18}N.Cl$
(5197-80-8)
$C_{11}H_{18}N_2$
(148-71-0)
(19248-13-6)
(34176-71-1)
(68479-98-1)
$C_{11}H_{18}N_2.ClH$
(2051-79-8)
$C_{11}H_{18}N_2O.H_2O_4S$
(25646-77-9)
$C_{11}H_{18}N_2O_2$
(15646-96-5)
(16938-22-0)
$C_{11}H_{18}N_2O_2S$
(76-75-5)
$C_{11}H_{18}N_2O_3$
(57-43-2)
(76-74-4)
(115-58-2)
$C_{11}H_{18}N_2O_3.Na$
(57-33-0)
$C_{11}H_{18}N_2O_9$
(3148-72-9)
$C_{11}H_{18}N_4O_2$
(23103-98-2)
$C_{11}H_{18}N_6S_4$
(30863-10-6)
$C_{11}H_{18}O$
(3777-71-7)
(31061-64-0)
(55702-54-0)
$C_{11}H_{18}O_2$
(105-86-2)
(3021-73-6)
$C_{11}H_{18}O_5$
(20881-04-3)
$C_{11}H_{19}NOS$
(26530-20-1)
$C_{11}H_{19}NO_2$
(13361-34-7)
$C_{11}H_{19}N_3O$
(23947-60-6)
$C_{11}H_{19}N_5$
(27431-12-5)
$C_{11}H_{19}N_7$
(67704-68-1)
$C_{11}H_{20}$
(55402-13-6)
(55976-13-1)
$C_{11}H_{20}ClN_3O_4P_2S_4$
(18895-89-1)
$C_{11}H_{20}ClN_5$
(580-48-3)
$C_{11}H_{20}ClN_5O_2$
(3426-63-9)
$C_{11}H_{20}FN_5$
(1598-99-8)
$C_{11}H_{20}N_3O_3PS$
(5221-49-8)
(29232-93-7)
$C_{11}H_{20}N_4O_2$
(30384-47-5)
$C_{11}H_{20}N_4S_2$
(30362-15-3)
$C_{11}H_{20}O$
(112-45-8)
(878-13-7)
(2371-42-8)
(3664-64-0)
(16587-71-6)
(40648-26-8)
(54411-03-9)
(56312-55-1)
$C_{11}H_{20}OSi_2$
(14920-92-4)
$C_{11}H_{20}O_2$
(103-11-7)
(104-67-6)
(112-38-9)
(2499-59-4)
(7367-83-1)
(25279-09-8)
(29590-42-9)
(35194-39-9)
$C_{11}H_{20}O_2.1/2Ca$
(1322-14-1)
$C_{11}H_{20}O_2.1/2Zn$
(557-08-4)
$C_{11}H_{20}O_3$
(2575-07-7)
(3433-16-7)
(24222-06-8)
$C_{11}H_{20}O_4$
(760-55-4)
(1732-10-1)
(1852-04-6)
(2050-20-6)
(10203-58-4)
(17557-23-2)
(55898-43-6)
(58447-69-1)
$C_{11}H_{20}O_5$
(20721-76-0)
$C_{11}H_{21}N$
(2244-07-7)
$C_{11}H_{21}NO$
(4223-03-4)
(10213-75-9)
(16728-49-7)
$C_{11}H_{21}NOS$
(1134-23-2)
(3134-71-2)
(55573-38-1)
$C_{11}H_{21}N_5$
(26235-14-3)
(26235-27-8)
(26235-30-3)
(26235-34-7)
(27430-97-3)
(27431-03-4)
(27470-64-0)
(27470-65-1)
(27470-67-3)
$C_{11}H_{21}N_5O$
(3004-70-4)
(22305-35-7)
(30360-67-9)
(30377-19-6)
$C_{11}H_{21}N_5OS$
(841-06-5)
$C_{11}H_{21}N_5O_2$
(30360-73-7)
$C_{11}H_{21}N_5S$
(4022-55-3)
(4147-51-7)
(22936-75-0)
$C_{11}H_{22}$
(821-95-4)
(4292-92-6)
(28761-27-5)
(41977-34-8)
$C_{11}H_{22}FN_2O_3PS_2$
(78051-43-1)
$C_{11}H_{22}N_2O$
(2163-69-1)
$C_{11}H_{22}N_2O.C_{11}H_{10}ClNO_2$
(8015-55-2)
$C_{11}H_{22}N_3O_6P$
(35597-43-4)
$C_{11}H_{22}N_6$
(5606-20-2)
$C_{11}H_{22}O$
(112-12-9)
(112-44-7)
(927-49-1)
(16624-06-9)
(16782-30-2)
$C_{11}H_{22}O_2$
(110-42-9)
(112-37-8)
(123-29-5)
(143-13-5)
(540-07-8)
(2198-61-0)
(2461-15-6)
(2553-96-0)
(3385-66-8)
(3613-30-7)
(5454-28-4)
(5458-59-3)
(5601-60-5)
(10032-15-2)
(25415-71-8)
(68134-07-6)
$C_{11}H_{22}O_2SSn$
(78-06-8)
$C_{11}H_{22}O_3$
(1322-17-4)
(7779-60-4)
$C_{11}H_{22}O_4$
(22397-33-7)
$C_{11}H_{22}O_4Si$
(3388-04-3)
$C_{11}H_{22}O_6$
(3149-65-3)
$C_{11}H_{23}Br$
(693-67-4)
(55162-38-4)
$C_{11}H_{23}Cl$
(2473-03-2)
(34214-82-9)
$C_{11}H_{23}ClOS$
(3569-57-1)
$C_{11}H_{23}ClO_4Si$
(71808-64-5)
$C_{11}H_{23}F$
(506-05-8)
$C_{11}H_{23}I$
(4282-44-4)
$C_{11}H_{23}NOS$
(2008-41-5)
$C_{11}H_{23}NO_2$
(2432-99-7)
$C_{11}H_{24}$
(1120-21-4)
(2847-72-5)
(6975-98-0)
(34464-43-2)
(62016-34-6)
$C_{11}H_{24}N_2$
(71326-18-6)
$C_{11}H_{24}N_2O$
(64819-51-8)
$C_{11}H_{24}N_2O_3$
(4981-47-9)
$C_{11}H_{24}O$
(103-08-2)
(112-42-5)
(1653-30-1)
(23708-56-7)
(30207-98-8)
(32357-83-8)
$C_{11}H_{24}O_2Sn$
(3267-78-5)
$C_{11}H_{24}O_4$
(20178-34-1)
$C_{11}H_{24}O_6S_2$
(4248-77-5)
$C_{11}H_{24}O_4Si$
(2897-60-1)
$C_{11}H_{24}O_6Si$
(1067-53-4)
$C_{11}H_{24}S$
(5332-52-5)
$C_{11}H_{25}NO$
(2109-64-0)
(5397-31-9)
(15930-66-2)
$C_{11}H_{25}O_3P$
(1000-36-8)
$C_{11}H_{26}N.Cl$
(10108-86-8)
$C_{11}H_{26}NO_2S$

(50782-69-9)
$C_{11}H_{26}N_4O$
(52338-87-1)
$C_{11}H_{27}N_3$
(3855-32-1)
$C_{11}H_{28}O_4Si_4$
(35331-58-9)
$C_{12}Br_6Cl_2O_2$
(2170-44-7)
$C_{12}Br_8O$
(103582-29-2)
$C_{12}Br_8O_2$
(2170-45-8)
$C_{12}Br_{10}$
(13654-09-6)
$C_{12}Br_{10}O$
(1163-19-5)
$C_{12}Cl_2F_{24}$
(103188-54-1)
$C_{12}Cl_8O$
(39001-02-0)
$C_{12}Cl_8O_2$
(3268-87-9)
$C_{12}Cl_{10}$
(2051-24-3)
$C_{12}Cl_{10}O$
(31710-30-2)
$C_{12}F_{26}$
(103188-55-2)
$C_{12}F_{27}N$
(311-89-7)
$C_{12}HBrCl_6O$
(107207-47-6)
$C_{12}HBr_7O$
(62994-32-5)
$C_{12}HBr_7O_2$
(103456-43-5)
$C_{12}HBr_9$
(27753-52-2)
$C_{12}HBr_9O$
(63936-56-1)
$C_{12}HCl_7O$
(38998-75-3)
(55673-89-7)
(67562-39-4)
(69698-58-4)
(70648-25-8)
$C_{12}HCl_7O_2$
(35822-46-9)
(37871-00-4)
(58200-70-7)
$C_{12}HCl_9$
(40186-72-9)
(52663-77-1)
(53742-07-7)
$C_{12}HCl_9O_2$
(35245-80-8)
(42255-14-1)
$C_{12}H_2Br_6O$
(103456-33-3)
$C_{12}H_2Br_6O_2$
(103456-42-4)
(116490-11-0)
$C_{12}H_2Br_8$
(27858-07-7)
$C_{12}H_2Br_8O$
(32536-52-0)
$C_{12}H_2Cl_6O$
(55684-94-1)
(57117-44-9)
(60851-34-5)
(67562-40-7)
(69698-59-5)
(69698-60-8)
(70648-26-9)
(71998-75-9)
(72918-21-9)
(75198-38-8)
(75627-02-0)
(79060-60-9)
(91538-83-9)
(91538-84-0)
(92341-05-4)
(92341-06-5)
(92341-07-6)
$C_{12}H_2Cl_6O_2$
(19408-74-3)
(34465-46-8)
(39227-28-6)
(39227-62-8)
(57653-85-7)
(58200-66-1)
(58200-67-2)
(58200-68-3)
(58200-69-4)
(58802-09-8)
(64461-98-9)
$C_{12}H_2Cl_8$
(2136-99-4)
(31472-83-0)
(33091-17-7)
(35694-08-7)
(40186-71-8)
(42740-50-1)
(52663-73-7)
(52663-75-9)
(52663-76-0)
(52663-78-2)
(55722-26-4)
(68194-17-2)
$C_{12}H_3Br_5O$
(68795-14-2)
$C_{12}H_3Br_5O_2$
(103456-36-6)
$C_{12}H_3Br_7$
(67733-52-2)
$C_{12}H_3Br_7O$
(68928-80-3)
$C_{12}H_3Cl_5O$
(30402-15-4)
(57117-31-4)
(57117-41-6)
(57117-42-7)
(57117-43-8)
(58802-15-6)
(58802-16-7)
(67481-22-5)
(67517-48-0)
(69433-00-7)
(69698-57-3)
(70648-15-6)
(70648-20-3)
(70648-21-4)
(70648-23-6)
(70648-24-7)
(70872-82-1)
(71998-74-8)
(83704-35-2)
(83704-36-3)
(83704-47-6)
(83704-48-7)
(83704-49-8)
(83704-50-1)
(83704-51-2)
(83704-52-3)
(83704-53-4)
(83704-54-5)
(83704-55-6)
$C_{12}H_3Cl_5O_2$
(36088-22-9)
(39227-61-7)
(40321-76-4)
(58802-08-7)
(67028-19-7)
(71925-15-0)
(71925-16-1)
(71925-17-2)
(71925-18-3)
(71998-76-0)
(82291-34-7)
(82291-35-8)
(82291-36-9)
(82291-37-0)
(82291-38-1)
$C_{12}H_3Cl_7$
(28655-71-2)
(35065-29-3)
(35065-30-6)
(38411-25-5)
(39635-31-9)
(40186-70-7)
(41411-64-7)
(52663-64-6)
(52663-67-9)
(52663-68-0)
(52663-69-1)
(52663-70-4)
(52663-71-5)
(52663-74-8)
(52712-05-7)
(60145-23-5)
(68194-16-1)
(69782-91-8)
(74472-47-2)
(74472-48-3)
(74472-50-7)
$C_{12}H_3Cl_7O$
(55684-92-9)
$C_{12}H_3Cl_7O_2$
(94888-12-7)
(94888-13-8)
$C_{12}H_4BrCl_3O$
(107227-56-5)
$C_{12}H_4BrCl_3O_2$
(107227-75-8)
$C_{12}H_4Br_2Cl_2O_2$
(50585-40-5)
$C_{12}H_4Br_2F_2O_2$
(50585-43-8)
$C_{12}H_4Br_4O$
(67733-57-7)
(84761-80-8)
$C_{12}H_4Br_4O_2$
(50585-41-6)
(76584-71-9)
(103456-39-9)
$C_{12}H_4Br_6$
(36355-01-8)
(59080-40-9)
$C_{12}H_4Br_6O$
(36483-60-0)
$C_{12}H_4Cl_2F_2O_2$
(50585-42-7)
$C_{12}H_4Cl_4$
(90077-79-5)
$C_{12}H_4Cl_4O$
(30402-14-3)
(51207-31-9)
(57117-35-8)
(57117-36-9)
(57117-37-0)
(57117-38-1)
(57117-39-2)
(57117-40-5)
(58802-19-0)
(58802-20-3)
(62615-08-1)
(64126-87-0)
(64560-17-4)
(66794-59-0)
(70648-16-7)
(70648-18-9)
(70648-19-0)
(70648-22-5)
(71998-72-6)
(71998-73-7)
(82911-58-8)
(83690-98-6)
(83704-21-6)
(83704-22-7)
(83704-23-8)
(83704-24-9)
(83704-25-0)
(83704-26-1)
(83704-27-2)
(83704-28-3)
(83704-29-4)
(83704-30-7)
(83704-31-8)
(83704-32-9)
(83704-33-0)
(83710-07-0)
(83719-40-8)
(92341-04-3)
$C_{12}H_4Cl_4O_2$
(1746-01-6)
(30746-58-8)
(33423-92-6)
(34816-53-0)
(40581-90-6)
(40581-91-7)
(40581-93-9)
(40581-94-0)
(41903-57-5)
(50585-46-1)
(53555-02-5)
(62470-53-5)
(62470-54-6)
(67028-18-6)
(67323-56-2)
(71665-99-1)
(71669-23-3)
(71669-24-4)
(71669-25-5)
(71669-26-6)
(71669-27-7)
(71669-28-8)
(71669-29-9)
$C_{12}H_4Cl_6$
(26601-64-9)
(32774-16-6)
(33979-03-2)
(35065-27-1)
(35065-28-2)
(35694-04-3)
(35694-06-5)
(38380-04-0)
(38380-05-1)
(38380-07-3)
(38380-08-4)
(38411-22-2)
(41411-63-6)
(51908-16-8)
(52663-63-5)
(52663-66-8)
(52663-72-6)
(52704-70-8)
(52712-04-6)
(52744-13-5)
(55215-18-4)
(59291-64-4)
(68194-08-1)
(68194-13-8)
(68194-14-9)
(69782-90-7)
(74472-42-7)
(74472-46-1)
$C_{12}H_4Cl_6O$
(31242-93-0)
(55720-99-5)
(71859-30-8)
$C_{12}H_4Cl_6O_2$
(90986-11-1)
(94888-10-5)
(94888-11-6)
(94897-81-1)
$C_{12}H_4N_8O_{12}$
(19159-68-3)
$C_{12}H_5Br_3O$
(84761-81-9)

(84761-82-0)
$C_{12}H_5Br_3O_2$
(51974-40-4)
(80246-33-9)
$C_{12}H_5Br_5$
(67888-96-4)
$C_{12}H_5Br_5O$
(32534-81-9)
$C_{12}H_5Cl_3O$
(24478-73-7)
(43048-00-6)
(54589-71-8)
(57117-32-5)
(57117-33-6)
(57117-34-7)
(58802-14-5)
(58802-17-8)
(58802-18-9)
(64560-14-1)
(64560-15-2)
(64560-16-3)
(70648-13-4)
(76621-12-0)
(82911-59-9)
(82911-60-2)
(82911-61-3)
(83636-47-9)
(83704-34-1)
(83704-37-4)
(83704-38-5)
(83704-39-6)
(83704-40-9)
(83704-41-0)
(83704-42-1)
(83704-43-2)
(83704-44-3)
(83704-45-4)
(83704-46-5)
$C_{12}H_5Cl_3O_2$
(33857-28-2)
(39227-58-2)
(67028-17-5)
(69760-96-9)
$C_{12}H_5Cl_5$
(18259-05-7)
(25429-29-2)
(31508-00-6)
(32598-14-4)
(37680-73-2)
(38379-99-6)
(38380-01-7)
(38380-02-8)
(38380-03-9)
(39485-83-1)
(41464-51-1)
(52663-60-2)
(52663-61-3)
(52663-62-4)
(55215-17-3)
(55312-69-1)
(57465-28-8)
(60233-25-2)
(65510-44-3)
(65510-45-4)
(68194-05-8)
(68194-07-0)
(68194-12-7)
(70424-68-9)
(70424-69-0)
(73575-57-2)
(74472-36-9)
(74472-37-0)
$C_{12}H_5Cl_5O$
(42279-29-8)
(60123-64-0)
(60123-65-1)
$C_{12}H_5Cl_5O_2$
(61639-90-5)
(90986-10-0)
(94888-09-2)
$C_{12}H_5Cl_6NO$
(64630-65-5)
$C_{12}H_5F_{21}O$
(865-86-1)
$C_{12}H_5N_7O_{12}$
(131-73-7)
$C_{12}H_6Br_2O$
(65489-80-7)
$C_{12}H_6Br_2O_2$
(39073-07-9)
(50585-37-0)
(91371-14-1)
$C_{12}H_6Br_4$
(40088-45-7)
$C_{12}H_6Br_4O$
(40088-47-9)
$C_{12}H_6Br_4O_2$
(36511-35-0)
$C_{12}H_6Cl_2FNO_3$
(13738-63-1)
$C_{12}H_6Cl_2N_2O_5$
(13867-27-1)
$C_{12}H_6Cl_2O$
(5409-83-6)
(24478-74-8)
(43047-99-0)
(58802-21-4)
(60390-27-4)
(64126-85-8)
(64126-86-9)
(64560-13-0)
(70648-14-5)
(74918-40-4)
(74992-97-5)
(74992-98-6)
(81638-37-1)
(94538-00-8)
(94538-01-9)
(94538-02-0)
(94570-83-9)
$C_{12}H_6Cl_2O_2$
(29446-15-9)
(33857-26-0)
(38178-38-0)
(38964-22-6)
(50585-39-2)
(54536-18-4)
(54536-19-5)
(64501-00-4)
$C_{12}H_6Cl_3NO_3$
(1836-77-7)
$C_{12}H_6Cl_4$
(1336-36-3)
(2437-79-8)
(15968-05-5)
(26914-33-0)
(32598-10-0)
(32598-11-1)
(32598-13-3)
(32690-93-0)
(33025-41-1)
(33284-52-5)
(33284-53-6)
(33284-54-7)
(35693-99-3)
(38444-93-8)
(41464-39-5)
(41464-40-8)
(41464-41-9)
(41464-42-0)
(41464-43-1)
(41464-46-4)
(41464-47-5)
(52663-58-8)
(52663-59-9)
(62796-65-0)
(70362-46-8)
(70362-47-9)
(70362-48-0)
(70362-49-1)
(70362-50-4)
(73575-52-7)
(73575-53-8)
(74472-33-6)
(74472-34-7)
$C_{12}H_6Cl_4N_2$
(14047-09-7)
$C_{12}H_6Cl_4N_2O$
(21232-47-3)
$C_{12}H_6Cl_4O$
(28076-73-5)
(31242-94-1)
(56348-72-2)
(61328-45-8)
$C_{12}H_6Cl_4O_2$
(14477-61-3)
(41363-16-0)
$C_{12}H_6Cl_4O_2S$
(97-18-7)
(116-29-0)
$C_{12}H_6Cl_4S$
(2227-13-6)
$C_{12}H_6Cl_5NO$
(64630-64-4)
$C_{12}H_6Cl_6N_4O_2S$
(24481-74-1)
$C_{12}H_6Cl_6N_4O_3S$
(24478-16-8)
$C_{12}H_6Cl_6O_2.Cu$
(25267-55-4)
$C_{12}H_6Cl_6O_4$
(25167-84-4)
$C_{12}H_6Cl_7N_3S$
(24478-13-5)
(24481-73-0)
$C_{12}H_6Cl_{15}N_3O_3$
(30895-02-4)
$C_{12}H_6F_2O_2$
(50585-38-1)
$C_{12}H_6N_2O_2$
(3173-72-6)
$C_{12}H_6O_2$
(82-86-0)
$C_{12}H_6O_3$
(81-84-5)
(5343-99-7)
(34314-32-4)
$C_{12}H_6O_{12}$
(517-60-2)
$C_{12}H_7BrO$
(86-76-0)
$C_{12}H_7Br_3O$
(49690-94-0)
$C_{12}H_7ClN_2O_5$
(20115-34-8)
$C_{12}H_7ClO$
(25074-67-3)
(51230-49-0)
(74992-96-4)
(84761-86-4)
$C_{12}H_7ClO_2$
(35656-51-0)
(39227-53-7)
(39227-54-8)
$C_{12}H_7Cl_2N$
(28804-85-5)
$C_{12}H_7Cl_2NO_3$
(135-12-6)
(1836-75-5)
$C_{12}H_7Cl_2NO_3.C_9H_{10}Cl_2N_2O_2$
(8075-80-7)
$C_{12}H_7Cl_3$
(7012-37-5)
(15862-07-4)
(16606-02-3)
(25323-68-6)
(35693-92-6)
(37680-65-2)
(37680-66-3)
(37680-68-5)
(37680-69-6)
(38444-73-4)
(38444-76-7)
(38444-77-8)
(38444-78-9)
(38444-81-4)
(38444-84-7)
(38444-85-8)
(38444-86-9)
(38444-87-0)
(38444-88-1)
(38444-90-5)
(55702-45-9)
(55702-46-0)
(55712-37-3)
(55720-44-0)
$C_{12}H_7Cl_3O$
(52322-80-2)
(57321-63-8)
(61328-44-7)
$C_{12}H_7Cl_3O_2$
(3380-34-5)
$C_{12}H_7Cl_6N_3$
(3584-22-3)
(3584-24-5)
$C_{12}H_7Cl_6N_3O$
(3584-23-4)
$C_{12}H_7Cl_6N_3OS$
(24478-02-2)
(24478-03-3)
$C_{12}H_7Cl_6N_3S$
(5516-51-8)
(24477-99-4)
(24478-00-0)
(24478-01-1)
$C_{12}H_7Cl_6N_5O$
(30805-00-6)
$C_{12}H_7F_{17}NO_4S.K$
(2991-51-7)
$C_{12}H_7IO_2$
(101714-96-9)
$C_{12}H_7NOS$
(581-30-6)
$C_{12}H_7NO_2$
(81-83-4)
$C_{12}H_7NO_4$
(550-82-3)
$C_{12}H_8$
(208-96-8)
(259-79-0)
(28109-99-1)
$C_{12}H_8Br_2$
(92-86-4)
$C_{12}H_8Br_2O$
(2050-47-7)
$C_{12}H_8ClN_3O_4$
(16220-58-9)
$C_{12}H_8ClN_3O_5$
(6358-18-5)
$C_{12}H_8Cl_2$
(2050-67-1)
(2050-68-2)
(2974-90-5)
(2974-92-7)
(13029-08-8)
(16605-91-7)
(25512-42-9)
(25569-80-6)
(33039-81-5)
(33146-45-1)
(33284-50-3)
(34883-39-1)
(34883-41-5)
(34883-43-7)
$C_{12}H_8Cl_2N_2$
(1602-00-2)
(7334-33-0)
(15426-14-9)
$C_{12}H_8Cl_2N_2O$
(614-26-6)

$C_{12}H_8Cl_2N_2O_3S.Na$
(62959-41-5)
$C_{12}H_8Cl_2O$
(2444-89-5)
(5335-24-0)
(6903-65-7)
(28675-08-3)
$C_{12}H_8Cl_2O_2S$
(80-07-9)
(97-24-5)
$C_{12}H_8Cl_2O_3S$
(80-33-1)
(97-16-5)
$C_{12}H_8Cl_2O_5S_2$
(121-63-1)
$C_{12}H_8Cl_2S_2$
(1142-19-4)
$C_{12}H_8Cl_4N_2$
(15721-02-5)
(71753-42-9)
$C_{12}H_8Cl_5DO$
(60468-28-2)
$C_{12}H_8Cl_5N_5O$
(30804-96-7)
$C_{12}H_8Cl_6$
(124-96-9)
(309-00-2)
(465-73-6)
(13350-71-5)
$C_{12}H_8Cl_6N_4$
(24803-16-5)
$C_{12}H_8Cl_6O$
(60-57-1)
(72-20-8)
(128-10-9)
(7421-93-4)
(13366-73-9)
(33058-12-7)
(53494-70-5)
$C_{12}H_8F_2$
(398-23-2)
$C_{12}H_8F_{17}NO_4S.Na$
(3871-50-9)
$C_{12}H_8I_2$
(3001-15-8)
$C_{12}H_8N_2$
(66-71-7)
(92-82-0)
(230-17-1)
$C_{12}H_8N_2O_2$
(95273-11-3)
$C_{12}H_8N_2O_4$
(1528-74-1)
$C_{12}H_8N_2O_4S_2$
(1155-00-6)
$C_{12}H_8N_2O_5$
(101-63-3)
(731-92-0)
$C_{12}H_8N_4O_6$
(970-76-3)
(970-91-2)
$C_{12}H_8O$
(132-64-9)
$C_{12}H_8OS$
(262-20-4)
(1013-23-6)
$C_{12}H_8O_2$
(262-12-4)
(70848-82-7)
$C_{12}H_8O_2S$
(1016-05-3)
$C_{12}H_8O_3$
(5656-90-6)
(73459-03-7)
(78982-40-8)
$C_{12}H_8O_4$
(298-81-7)
(484-20-8)
(1141-38-4)
$C_{12}H_8S$
(132-65-0)
$C_{12}H_8S_2$
(92-85-3)
$C_{12}H_9AsClN$
(578-94-9)
$C_{12}H_9Br$
(92-66-0)
(2052-07-5)
(2113-57-7)
$C_{12}H_9BrO$
(101-55-3)
(36563-47-0)
$C_{12}H_9Cl$
(2051-60-7)
(2051-61-8)
(2051-62-9)
(27323-18-8)
$C_{12}H_9ClF_3N_3O$
(27314-13-2)
$C_{12}H_9ClN_2O_2$
(15979-85-8)
$C_{12}H_9ClN_2$
(4340-77-6)
$C_{12}H_9ClN_2O$
(6657-05-2)
$C_{12}H_9ClN_2O_3S.Na$
(2777-05-1)
$C_{12}H_9ClO$
(85-97-2)
(92-04-6)
(607-12-5)
(7005-72-3)
$C_{12}H_9ClO.Na$
(5578-88-1)
(10605-10-4)
$C_{12}H_9ClO_2S$
(80-00-2)
$C_{12}H_9ClO_3S$
(80-38-6)
$C_{12}H_9Cl_2N$
(15979-79-0)
$C_{12}H_9Cl_2NO_3$
(50471-44-8)
$C_{12}H_9Cl_3N_6O_3$
(30804-98-9)
$C_{12}H_9Cl_4N_5O$
(30804-93-4)
(30804-94-5)
(30804-95-6)
$C_{12}H_9Cl_9$
(83682-60-4)
$C_{12}H_9F$
(321-60-8)
(324-74-3)
$C_{12}H_9F_{13}O_2$
(2144-53-8)
$C_{12}H_9I$
(1591-31-7)
$C_{12}H_9N$
(86-74-8)
(132-75-2)
(244-99-5)
(4523-49-3)
(77417-07-3)
(97485-90-0)
$C_{12}H_9NO$
(14548-46-0)
$C_{12}H_9NO_2$
(86-00-0)
(92-93-3)
(602-87-9)
(2113-58-8)
(28984-85-2)
$C_{12}H_9NO_3$
(620-88-2)
(885-82-5)
$C_{12}H_9NO_4$
(116211-97-3)
$C_{12}H_9NS$
(92-84-2)
$C_{12}H_9N_2O_8P$
(645-15-8)
$C_{12}H_9N_3O_2$
(2491-52-3)
$C_{12}H_9N_3O_3$
(1435-60-5)
(55936-40-8)
$C_{12}H_9N_3O_4$
(74-39-5)
(961-68-2)
(15979-87-0)
$C_{12}H_9N_3O_5$
(119-15-3)
(6358-23-2)
$C_{12}H_9N_3O_5S.Na$
(2491-72-7)
$C_{12}H_9N_9O_5S_2$
(64690-01-3)
$C_{12}H_9O.Na$
(132-27-4)
$C_{12}H_9O_2P$
(35948-25-5)
$C_{12}H_{10}$
(83-32-9)
(92-52-4)
(827-54-3)
$C_{12}H_{10}.C_{12}H_{10}O$
(8004-13-5)
$C_{12}H_{10}AsCl$
(712-48-1)
$C_{12}H_{10}Ca_3O_{14}$
(7693-13-2)
$C_{12}H_{10}ClN$
(101-17-7)
(1205-71-6)
$C_{12}H_{10}ClNO_2S$
(4750-28-1)
$C_{12}H_{10}ClP$
(1079-66-9)
$C_{12}H_{10}Cl_2F_3NO$
(61213-25-0)
$C_{12}H_{10}Cl_2N_2$
(84-68-4)
(91-94-1)
(782-74-1)
$C_{12}H_{10}Cl_2N_2.2ClH$
(612-83-9)
$C_{12}H_{10}Cl_2N_2O$
(28434-86-8)
$C_{12}H_{10}Cl_2Si$
(80-10-4)
$C_{12}H_{10}Cl_2Sn$
(1135-99-5)
$C_{12}H_{10}Cl_3N_3O$
(30863-61-7)
$C_{12}H_{10}Cl_3N_5O$
(30804-92-3)
$C_{12}H_{10}Cl_3N_5S$
(30863-28-6)
$C_{12}H_{10}Cl_4N_4$
(24848-39-3)
$C_{12}H_{10}Cl_4N_4S$
(30369-59-6)
$C_{12}H_{10}F_{15}NO_3$
(42268-97-3)
$C_{12}H_{10}F_{17}NO_3S$
(1691-99-2)
$C_{12}H_{10}Hg$
(587-85-9)
$C_{12}H_{10}I.AsF_6$
(62613-15-4)
$C_{12}H_{10}I.F_6P$
(58109-40-3)
$C_{12}H_{10}N_2$
(103-33-3)
(486-84-0)
$C_{12}H_{10}N_2O$
(86-30-6)
(156-10-5)
(495-48-7)
(1689-82-3)
(2362-57-4)
$C_{12}H_{10}N_2O.Na$
(63451-40-1)
$C_{12}H_{10}N_2O_2$
(119-75-5)
(836-30-6)
(2051-85-6)
(4531-79-7)
$C_{12}H_{10}N_2O_3S$
(2484-88-0)
$C_{12}H_{10}N_2O_3S.Na$
(42975-18-8)
$C_{12}H_{10}N_2O_4S.Na$
(2623-36-1)
$C_{12}H_{10}N_2O_5$
(28657-80-9)
$C_{12}H_{10}N_2O_5S.Na$
(547-57-9)
$C_{12}H_{10}N_3.HO_4S$
(4477-28-5)
$C_{12}H_{10}N_4$
(655-86-7)
$C_{12}H_{10}N_4O_2$
(730-40-5)
$C_{12}H_{10}N_4O_4$
(6373-73-5)
$C_{12}H_{10}N_4O_6S$
(19757-13-2)
$C_{12}H_{10}O$
(90-43-7)
(92-69-3)
(93-08-3)
(101-84-8)
(580-51-8)
(941-98-0)
(1322-20-9)
(1333-52-4)
(6306-07-6)
(77468-37-2)
$C_{12}H_{10}O.K$
(13707-65-8)
$C_{12}H_{10}O_2$
(86-87-3)
(92-05-7)
(92-88-6)
(581-96-4)
(713-68-8)
(830-81-9)
(831-82-3)
(1079-21-6)
(1133-63-7)
(1806-29-7)
(18855-13-5)
(26445-01-2)
(26983-52-8)
(31835-45-7)
(68334-50-9)
$C_{12}H_{10}O_2.Na$
(61-31-4)
$C_{12}H_{10}O_2S$
(127-63-9)
(2664-63-3)
$C_{12}H_{10}O_3$
(883-99-8)
(948-03-8)
(4190-05-0)
(27949-30-0)
(29222-39-7)
$C_{12}H_{10}O_4$
(4371-31-7)
(10441-36-8)
(63037-96-7)
$C_{12}H_{10}O_4S$
(80-09-1)
$C_{12}H_{10}S$
(139-66-2)
$C_{12}H_{10}S_2$
(882-33-7)
$C_{12}H_{11}BO$

(2622-89-1)
$C_{12}H_{11}ClN_2O_5S$
(54-31-9)
$C_{12}H_{11}ClN_2OS$
(97073-94-4)
$C_{12}H_{11}Cl_2NO$
(23950-58-5)
$C_{12}H_{11}Cl_2N_3O_2$
(60207-31-0)
$C_{12}H_{11}Cl_3N_4$
(24803-00-7)
(24848-38-2)
$C_{12}H_{11}Cl_3O_3$
(2971-38-2)
$C_{12}H_{11}I_3N_2O_4$
(440-58-4)
$C_{12}H_{11}N$
(90-41-5)
(92-67-1)
(101-82-6)
(122-39-4)
(620-95-1)
(2116-65-6)
(4657-93-6)
(41674-04-8)
(64828-54-2)
$C_{12}H_{11}N.ClH$
(2185-92-4)
$C_{12}H_{11}N.H_2O_4S$
(587-84-8)
$C_{12}H_{11}NO$
(86-86-2)
(101-18-8)
(122-37-2)
(139-59-3)
(1134-36-7)
(3400-33-7)
(19434-42-5)
$C_{12}H_{11}NO_2$
(63-25-2)
$C_{12}H_{11}NO_5S$
(6334-97-0)
$C_{12}H_{11}N_3$
(60-09-3)
(136-35-6)
(62450-07-1)
(68006-83-7)
$C_{12}H_{11}N_3.C_2H_4O_2$
(72254-58-1)
$C_{12}H_{11}N_3O$
(103-18-4)
$C_{12}H_{11}N_3O_2$
(6149-34-4)
$C_{12}H_{11}N_3O_3S$
(104-23-4)
(97073-93-3)
$C_{12}H_{11}N_3O_3S.Na$
(2491-71-6)
$C_{12}H_{11}N_3O_5S$
(91-29-2)
$C_{12}H_{11}N_3O_6S_2$
(101-50-8)
$C_{12}H_{11}N_4O_3S.Na$
(10190-66-6)

$C_{12}H_{11}N_5$
(1214-39-7)
$C_{12}H_{11}N_7$
(396-01-0)
$C_{12}H_{11}O_3P$
(4712-55-4)
$C_{12}H_{11}O_4P$
(838-85-7)
$C_{12}H_{12}$
(569-41-5)
(571-58-4)
(571-61-9)
(573-98-8)
(575-37-1)
(575-41-7)
(575-43-9)
(581-40-8)
(581-42-0)
(582-16-1)
(939-27-5)
(1127-76-0)
(4794-05-2)
(13703-52-1)
(27138-19-8)
(28804-88-8)
$C_{12}H_{12}BrNO$
(80045-50-7)
$C_{12}H_{12}Br_2N_2$
(66630-68-0)
$C_{12}H_{12}Cd_3N_2O_{12}$
(50648-02-7)
$C_{12}H_{12}ClNO$
(21267-72-1)
(24911-15-7)
$C_{12}H_{12}ClN_5O_4S$
(64902-72-3)
$C_{12}H_{12}Cl_2F_{16}N_3O_4P_3$
(65601-42-5)
$C_{12}H_{12}Cl_2O_6$
(108544-95-2)
$C_{12}H_{12}Cl_6$
(83682-59-1)
$C_{12}H_{12}Cl_9N_3O_3$
(30805-20-0)
$C_{12}H_{12}FNO$
(80045-51-8)
$C_{12}H_{12}FeO$
(1271-55-2)
$C_{12}H_{12}N.Cl$
(2876-13-3)
$C_{12}H_{12}N_2$
(92-87-5)
(101-54-2)
(122-66-7)
(492-17-1)
(530-50-7)
(2764-72-9)
(38622-18-3)
$C_{12}H_{12}N_2.2Cl$
(4032-26-2)
$C_{12}H_{12}N_2.ClH$
(530-47-2)
$C_{12}H_{12}N_2.2ClH$
(531-85-1)

$C_{12}H_{12}N_2.2Br$
(85-00-7)
$C_{12}H_{12}N_2.H_2O_4S$
(531-86-2)
$C_{12}H_{12}N_2O$
(101-80-4)
$C_{12}H_{12}N_2OS$
(97073-95-5)
$C_{12}H_{12}N_2O_2$
(2373-98-0)
$C_{12}H_{12}N_2O_2S$
(80-08-0)
$C_{12}H_{12}N_2O_3$
(50-06-6)
(89-33-8)
(389-08-2)
(39098-01-6)
$C_{12}H_{12}N_2O_3.Na$
(57-30-7)
$C_{12}H_{12}N_2O_4$
(379-34-0)
$C_{12}H_{12}N_2O_6S_2$
(117-61-3)
$C_{12}H_{12}N_2S$
(139-65-1)
$C_{12}H_{12}N_4$
(495-54-5)
(538-41-0)
(77094-11-2)
$C_{12}H_{12}N_4.ClH$
(532-82-1)
$C_{12}H_{12}N_6O_{18}$
(105554-30-1)
$C_{12}H_{12}O$
(93-18-5)
(773-99-9)
(5328-01-8)
$C_{12}H_{12}O_2S_2$
(54616-10-3)
$C_{12}H_{12}O_4$
(4620-47-7)
$C_{12}H_{12}O_4S_2$
(53135-94-7)
$C_{12}H_{12}O_5$
(10088-95-6)
(13988-26-6)
(59901-90-5)
$C_{12}H_{12}O_6$
(525-52-0)
(2459-10-1)
$C_{12}H_{12}O_8S_2.Zn$
(127-82-2)
$C_{12}H_{12}S_2$
(10075-74-8)
$C_{12}H_{13}ClF_3N_3O_4$
(33245-39-5)
$C_{12}H_{13}ClN_2O$
(3766-60-7)
$C_{12}H_{13}ClN_4$
(58-14-0)
$C_{12}H_{13}ClO_3$
(68683-30-7)
$C_{12}H_{13}Cl_2N_5$
(30355-05-6)

$C_{12}H_{13}Cl_3O_3$
(93-79-8)
(4938-72-1)
(30431-53-9)
(61792-07-2)
$C_{12}H_{13}Cl_6N_5$
(24863-51-2)
(24863-52-3)
$C_{12}H_{13}I_3N_2O_2$
(5587-89-3)
$C_{12}H_{13}I_3N_2O_3$
(16034-77-8)
$C_{12}H_{13}N$
(86-56-6)
(118-44-5)
(2437-72-1)
(51366-52-0)
(76602-24-9)
$C_{12}H_{13}NO$
(33244-86-9)
$C_{12}H_{13}NO_2$
(77-41-8)
(87-01-4)
(133-32-4)
(1515-72-6)
$C_{12}H_{13}NO_2S$
(5234-68-4)
$C_{12}H_{13}NO_3S$
(17757-70-9)
$C_{12}H_{13}NO_4$
(16709-30-1)
$C_{12}H_{13}NO_4S$
(5259-88-1)
$C_{12}H_{13}NO_6$
(1152-61-0)
$C_{12}H_{13}N_3$
(537-65-5)
(1555-66-4)
$C_{12}H_{13}N_3O_2$
(68-76-8)
$C_{12}H_{13}N_3O_3S$
(119-70-0)
$C_{12}H_{13}N_5$
(6364-34-7)
$C_{12}H_{13}N_5O_4$
(606-58-6)
$C_{12}H_{14}$
(4994-16-5)
(31017-40-0)
$C_{12}H_{14}ClNO$
(2884-69-7)
$C_{12}H_{14}ClNO_2$
(81777-89-1)
$C_{12}H_{14}ClNO_4$
(4433-79-8)
$C_{12}H_{14}ClN_2O_3PS$
(14816-20-7)
$C_{12}H_{14}Cl_2O_3$
(94-80-4)
(1713-15-1)
$C_{12}H_{14}Cl_2O_3.C_{12}H_{13}Cl_3O_3$
(39277-47-9)
$C_{12}H_{14}Cl_2O_4$
(75315-56-9)

$C_{12}H_{14}Cl_2O_5$
(75315-55-8)
$C_{12}H_{14}Cl_2O_6$
(75315-57-0)
$C_{12}H_{14}Cl_3O_3PS$
(1224-63-1)
(1757-18-2)
$C_{12}H_{14}Cl_3O_4P$
(470-90-6)
$C_{12}H_{14}NO_4PS$
(5131-24-8)
$C_{12}H_{14}N_2$
(4685-14-7)
$C_{12}H_{14}N_2.2CH_3O_4S$
(2074-50-2)
$C_{12}H_{14}N_2.2Cl$
(1910-42-5)
$C_{12}H_{14}N_2.2ClH$
(1465-25-4)
$C_{12}H_{14}N_2O_2$
(50-12-4)
(125-33-7)
$C_{12}H_{14}N_2O_5$
(131-89-5)
(4097-58-9)
$C_{12}H_{14}N_2O_5.Na$
(130-60-9)
$C_{12}H_{14}N_2O_6$
(2813-95-8)
$C_{12}H_{14}N_4$
(91-95-2)
$C_{12}H_{14}N_4.4ClH$
(7411-49-6)
$C_{12}H_{14}N_4O$
(30369-39-2)
(30369-40-5)
$C_{12}H_{14}N_4O_2$
(30358-00-0)
(30358-06-6)
$C_{12}H_{14}N_4O_2S$
(57-68-1)
$C_{12}H_{14}N_4O_3S$
(17784-12-2)
$C_{12}H_{14}N_4O_4S_2$
(23564-05-8)
$C_{12}H_{14}N_4O_5S_2$
(80-51-3)
$C_{12}H_{14}O$
(76379-66-3)
$C_{12}H_{14}O_2$
(7778-83-8)
$C_{12}H_{14}O_3$
(77-83-8)
(93-28-7)
(19464-92-7)
(40552-84-9)
(63919-02-8)
(101038-68-0)
$C_{12}H_{14}O_4$
(84-66-2)
(101-90-6)
(131-70-4)
(636-09-9)
(636-53-3)

(2425-01-6)
(24968-12-5)
$C_{12}H_{14}O_5$
(90-50-6)
$C_{12}H_{14}O_6$
(959-26-2)
$C_{12}H_{14}O_9S.Na$
(24019-46-3)
$C_{12}H_{15}ClNO_3PS$
(84704-01-8)
$C_{12}H_{15}ClNO_4PS_2$
(2310-17-0)
$C_{12}H_{15}ClN_6$
(30360-17-9)
$C_{12}H_{15}ClO$
(55012-69-6)
$C_{12}H_{15}ClO_3$
(637-07-0)
(52716-17-3)
$C_{12}H_{15}Cl_2NO$
(2533-89-3)
$C_{12}H_{15}Cl_2N_5$
(30805-26-6)
$C_{12}H_{15}Cl_3$
(18226-46-5)
$C_{12}H_{15}N$
(118-12-7)
(147-47-7)
$C_{12}H_{15}NO_2$
(97-36-9)
(2095-06-9)
(27179-64-2)
$C_{12}H_{15}NO_3$
(122-82-7)
(1563-66-2)
(1665-48-1)
$C_{12}H_{15}NO_4$
(6375-27-5)
(14309-42-3)
(16655-82-6)
$C_{12}H_{15}N_2O_3PS$
(13593-03-8)
(14816-18-3)
$C_{12}H_{15}N_3O_2$
(26408-28-6)
$C_{12}H_{15}N_3O_3$
(101-37-1)
(959-52-4)
(1025-15-6)
$C_{12}H_{15}N_3O_6$
(81-15-2)
(898-22-6)
(2451-62-9)
(61050-97-3)
$C_{12}H_{15}N_3S_3$
(30863-37-7)
$C_{12}H_{15}N_5$
(30359-95-6)
$C_{12}H_{16}$
(827-52-1)
(1076-61-5)
(1746-23-2)
(1985-59-7)
(2388-14-9)
(2613-76-5)
(6682-06-0)
(13065-07-1)
(21564-92-1)
(22531-20-0)
(25419-33-4)
(26472-00-4)
(32367-54-7)
(36541-18-1)
(54340-87-3)
(54340-88-4)
(56298-75-0)
(81598-29-0)
$C_{12}H_{16}ClNOS$
(28249-77-6)
(34622-58-7)
$C_{12}H_{16}ClNO_2$
(50563-41-2)
$C_{12}H_{16}ClN_5$
(30805-25-5)
$C_{12}H_{16}Cl_2N_2O$
(555-37-3)
$C_{12}H_{16}Cl_2O_4$
(108544-96-3)
$C_{12}H_{16}Cl_3N_5$
(24848-47-3)
(24848-48-4)
(24848-52-0)
$C_{12}H_{16}Cl_3N_5O_2$
(30359-61-6)
$C_{12}H_{16}F_3N$
(458-24-2)
$C_{12}H_{16}NO_3PS$
(100253-12-1)
$C_{12}H_{16}N_2$
(61-50-7)
(148-69-6)
(15686-61-0)
$C_{12}H_{16}N_2O$
(487-93-4)
(13140-89-1)
$C_{12}H_{16}N_2O_3$
(52-31-3)
(56-29-1)
(16118-45-9)
(16118-49-3)
$C_{12}H_{16}N_2O_5$
(83-66-9)
(4099-65-4)
$C_{12}H_{16}N_3O_3PS$
(24017-47-8)
$C_{12}H_{16}N_3O_3PS_2$
(2642-71-9)
$C_{12}H_{16}N_4O_4$
(2638-94-0)
$C_{12}H_{16}N_8O_3$
(10422-01-2)
$C_{12}H_{16}O$
(119-42-6)
(943-27-1)
(1131-60-8)
(1444-64-0)
(1943-95-9)
(6853-57-2)
(14171-89-2)
(25870-62-6)
(26570-85-4)
$C_{12}H_{16}O_2$
(102-13-6)
(151-05-3)
(1453-06-1)
(2049-96-9)
(2500-83-6)
(5413-60-5)
(24306-23-8)
(59094-71-2)
$C_{12}H_{16}O_3$
(87-20-7)
(103-60-6)
(1798-04-5)
(2050-08-0)
(2883-98-9)
(4511-39-1)
(6175-45-7)
(22313-62-8)
$C_{12}H_{16}O_4$
(9003-01-4)
$C_{12}H_{17}N$
(2201-24-3)
$C_{12}H_{17}NO$
(91-49-6)
(134-62-3)
(634-03-7)
(42498-33-9)
$C_{12}H_{17}NO_2$
(2631-37-0)
(3766-81-2)
(13110-37-7)
(56961-11-6)
(63075-06-9)
$C_{12}H_{17}NO_3$
(32953-89-2)
$C_{12}H_{17}NO_3Si$
(2097-19-0)
$C_{12}H_{17}N_2O_4P$
(520-52-5)
$C_{12}H_{17}N_3O_2$
(17702-57-7)
$C_{12}H_{17}N_3O_3$
(4178-93-2)
$C_{12}H_{17}N_3O_4$
(2757-90-6)
$C_{12}H_{17}N_4OS.Cl$
(59-43-8)
$C_{12}H_{17}N_4OS.ClH.Cl$
(67-03-8)
$C_{12}H_{17}O_4PS_2$
(2597-03-7)
$C_{12}H_{17}O_7.Na$
(52508-35-7)
$C_{12}H_{18}$
(87-85-4)
(98-19-1)
(99-62-7)
(100-18-5)
(102-25-0)
(577-55-9)
(676-22-2)
(877-44-1)
(1077-16-3)
(1320-01-0)
(1483-60-9)
(1985-57-5)
(4468-42-2)
(4815-57-0)
(4904-61-4)
(6031-02-3)
(19219-85-3)
(19262-20-5)
(22975-58-2)
(25321-09-9)
(25340-18-5)
(27070-59-3)
(28080-86-6)
(42205-08-3)
(54410-74-1)
(55669-88-0)
(55682-73-0)
$C_{12}H_{18}Be_4O_{13}$
(19049-40-2)
$C_{12}H_{18}BrNO$
(63455-63-0)
$C_{12}H_{18}Br_6$
(3194-55-6)
(25637-99-4)
$C_{12}H_{18}ClN$
(17243-57-1)
$C_{12}H_{18}Cl_3N_3O_2$
(30863-42-4)
$C_{12}H_{18}Cl_3N_3O_3$
(20120-32-5)
$C_{12}H_{18}Cl_3N_3O_6$
(7423-53-2)
$C_{12}H_{18}N.Cl$
(26616-35-3)
$C_{12}H_{18}N_2O$
(6375-46-8)
(34123-59-6)
$C_{12}H_{18}N_2O_2$
(315-18-4)
(4098-71-9)
$C_{12}H_{18}N_2O_3$
(76-73-3)
$C_{12}H_{18}N_2O_3S$
(64-77-7)
$C_{12}H_{18}N_2O_4$
(15336-82-0)
$C_{12}H_{18}N_4O_6S$
(19044-88-3)
$C_{12}H_{18}N_6$
(30360-21-5)
$C_{12}H_{18}O$
(96-70-8)
(122-73-6)
(879-97-0)
(1011-12-7)
(1879-09-0)
(2078-54-8)
(2446-69-7)
(2934-05-6)
(4237-25-6)
(13485-66-0)
(17696-37-6)
(24142-77-6)
(26886-05-5)
(31391-49-8)
(35946-91-9)
(36812-13-2)
(54518-11-5)
(58443-82-6)
(63452-61-9)
$C_{12}H_{18}O_2$
(97-42-7)
(136-77-6)
(713-46-2)
(2948-46-1)
(18979-55-0)
$C_{12}H_{18}O_3$
(7757-96-2)
(26680-54-6)
$C_{12}H_{18}O_3S$
(58425-67-5)
$C_{12}H_{18}O_4$
(532-34-3)
(1189-08-8)
(2082-81-7)
(2998-04-1)
(7204-16-2)
(13048-33-4)
$C_{12}H_{18}O_4S_2$
(50512-35-1)
$C_{12}H_{18}O_5$
(2358-84-1)
(57472-68-1)
$C_{12}H_{18}O_6$
(1680-21-3)
(2754-17-8)
$C_{12}H_{18}O_7$
(142-22-3)
$C_{12}H_{19}ClNO_3P$
(299-86-5)
$C_{12}H_{19}Cl_3N_4$
(24802-91-3)
$C_{12}H_{19}Cl_2O_2$
(99308-28-8)
$C_{12}H_{19}N$
(24544-04-5)
(33228-45-4)
$C_{12}H_{19}NO$
(1864-92-2)
(3046-94-4)
$C_{12}H_{19}NO_2$
(92-11-5)
$C_{12}H_{19}NO_3$
(1082-88-8)
(23031-25-6)
$C_{12}H_{19}NS$
(6675-28-1)
$C_{12}H_{19}N_2O_2$
(59-99-4)
$C_{12}H_{19}N_2O_2.Br$
(114-80-7)
$C_{12}H_{19}N_2O_4PS$
(14151-45-2)
(33857-23-7)
$C_{12}H_{19}N_3O$

(671-16-9)
$C_{12}H_{19}N_3O.ClH$
(366-70-1)
$C_{12}H_{19}N_3O_3S$
(56046-62-9)
(63494-59-7)
$C_{12}H_{19}N_3O_5$
(33229-34-4)
$C_{12}H_{19}N_5O_2$
(26234-42-4)
$C_{12}H_{19}N_6OP$
(1031-47-6)
$C_{12}H_{19}O_2$
(26762-93-6)
$C_{12}H_{19}O_2PS_3$
(35400-43-2)
$C_{12}H_{19}O_4P$
(51496-03-8)
$C_{12}H_{20}$
(702-79-4)
(16336-83-7)
(30496-78-7)
(40482-18-6)
(55976-10-8)
$C_{12}H_{20}Cl_3N_5$
(26234-99-1)
(26235-03-0)
(26322-45-2)
$C_{12}H_{20}Cl_6O$
(99308-26-6)
$C_{12}H_{20}Cl_6O_2$
(99308-27-7)
$C_{12}H_{20}HgNO_3.C_3H_5O_3$
(23319-66-6)
$C_{12}H_{20}N_2$
(3867-15-0)
(4251-01-8)
$C_{12}H_{20}N_2O_2S.Na$
(27159-90-6)
$C_{12}H_{20}N_2O_3S$
(3930-20-9)
$C_{12}H_{20}N_2S_4Zn$
(13878-54-1)
$C_{12}H_{20}N_2S_6$
(120-54-7)
$C_{12}H_{20}N_2S_8$
(971-15-3)
$C_{12}H_{20}N_4O_2$
(51235-04-2)
$C_{12}H_{20}O$
(90-42-6)
(19377-97-0)
(21662-15-7)
(21662-16-8)
(55103-65-6)
(56025-96-8)
(62338-24-3)
$C_{12}H_{20}O_2$
(76-49-3)
(80-26-2)
(97-41-6)
(105-87-3)
(115-95-7)
(125-12-2)
(141-12-8)
(1118-39-4)
(2705-87-5)
(4821-04-9)
(8007-35-0)
(26896-48-0)
$C_{12}H_{20}O_3Si$
(780-69-8)
$C_{12}H_{20}O_4$
(105-75-9)
(105-76-0)
(7423-42-9)
(14234-82-3)
$C_{12}H_{20}O_4Sn$
(78-04-6)
$C_{12}H_{20}O_6$
(139-45-7)
(4064-06-6)
(13236-02-7)
(20880-93-7)
(23262-78-4)
(23262-79-5)
$C_{12}H_{20}O_7$
(77-93-0)
$C_{12}H_{20}O_{12}.4Na.Zr$
(10377-98-7)
$C_{12}H_{20}S$
(30140-46-6)
$C_{12}H_{21}Br$
(112-29-8)
$C_{12}H_{21}F_9O_3Si_3$
(2374-14-3)
$C_{12}H_{21}NO_6$
(3002-18-4)
$C_{12}H_{21}N_2O_3PS$
(333-41-5)
(5826-91-5)
$C_{12}H_{21}N_2O_4P$
(962-58-3)
$C_{12}H_{21}N_3$
(1017-55-6)
(25176-37-8)
$C_{12}H_{21}N_3O_2S.3/2H_2O_4S$
(25646-71-3)
$C_{12}H_{21}N_3O_3$
(29263-10-3)
(29263-11-4)
(30805-19-7)
$C_{12}H_{22}$
(92-51-3)
(181-15-7)
(1008-80-6)
(1501-82-2)
(1618-22-0)
(4431-89-4)
(5876-87-9)
(21293-02-7)
(24524-53-6)
(25551-49-9)
(41977-45-1)
(55638-50-1)
$C_{12}H_{22}CuO_{14}$
(527-09-3)
$C_{12}H_{22}N_2O$
(947-92-2)
$C_{12}H_{22}N_4O_3S$
(51461-71-3)
$C_{12}H_{22}O$
(137-03-1)
(286-99-7)
(830-13-7)
(4645-15-2)
(4826-62-4)
(6531-86-8)
(13019-16-4)
(19700-21-1)
(51953-10-7)
$C_{12}H_{22}OSn$
(22771-17-1)
$C_{12}H_{22}O_2$
(88-41-5)
(111-81-9)
(150-84-5)
(688-84-6)
(713-95-1)
(2157-01-9)
(2305-05-7)
(2382-40-3)
(7070-15-7)
(7367-84-2)
(7549-37-3)
(13757-90-9)
(16409-45-3)
(17663-27-3)
(32210-23-4)
(38421-90-8)
(51000-52-3)
(67446-07-5)
(67452-27-1)
(68760-70-3)
(81634-99-3)
$C_{12}H_{22}O_3$
(682-09-7)
(18871-14-2)
$C_{12}H_{22}O_4$
(106-19-4)
(106-79-6)
(141-03-7)
(693-23-2)
(2050-23-9)
(2407-94-5)
(6938-94-9)
(53268-44-3)
(55114-29-9)
(58447-70-4)
$C_{12}H_{22}O_4.C_6H_{16}N_2$
(13188-60-8)
$C_{12}H_{22}O_4Si_2$
(55557-21-6)
$C_{12}H_{22}O_4Zn$
(4468-02-4)
$C_{12}H_{22}O_5$
(78-18-2)
(13519-20-5)
(20591-89-3)
(41638-13-5)
$C_{12}H_{22}O_6$
(87-92-3)
(1954-28-5)
$C_{12}H_{22}O_7S.Na$
(127-39-9)
$C_{12}H_{22}O_{11}$
(57-50-1)
(63-42-3)
(69-79-4)
(99-20-7)
(528-50-7)
$C_{12}H_{22}O_{14}.Ca$
(299-28-5)
$C_{12}H_{22}O_{14}.Fe$
(299-29-6)
$C_{12}H_{22}S$
(7133-46-2)
$C_{12}H_{22}S_2$
(2550-40-5)
$C_{12}H_{23}AlO_5$
(14782-75-3)
$C_{12}H_{23}ClO$
(112-16-3)
$C_{12}H_{23}ClO_2$
(15175-04-9)
$C_{12}H_{23}N$
(101-83-7)
(2437-25-4)
(17943-83-8)
$C_{12}H_{23}N.HNO_2$
(3129-91-7)
$C_{12}H_{23}NO$
(947-04-6)
$C_{12}H_{23}NOS$
(55852-95-4)
(75013-55-7)
(107348-42-5)
$C_{12}H_{23}NO_2$
(16715-83-6)
$C_{12}H_{23}N_5$
(5921-65-3)
(26234-34-4)
(26234-35-5)
(26234-36-6)
(26234-40-2)
(26235-19-8)
(26235-20-1)
(26235-25-6)
(26235-37-0)
(27430-98-4)
(27431-11-4)
$C_{12}H_{23}N_5O$
(15438-85-4)
(30360-64-6)
$C_{12}H_{23}N_5O_2S$
(845-52-3)
$C_{12}H_{23}N_5O_3$
(1771-07-9)
$C_{12}H_{23}N_5S$
(30360-80-6)
$C_{12}H_{23}O_2.K$
(10124-65-9)
$C_{12}H_{23}O_2.Na$
(629-25-4)
$C_{12}H_{23}P$
(829-84-5)
$C_{12}H_{24}$
(112-41-4)
(123-48-8)
(294-62-2)
(4292-75-5)
(6842-15-5)
(7756-94-7)
(14031-86-8)
(25378-22-7)
(54410-98-9)
(62199-50-2)
$C_{12}H_{24}B_2O_5$
(14697-50-8)
$C_{12}H_{24}B_2O_6$
(2665-13-6)
$C_{12}H_{24}N_2O$
(36060-61-4)
$C_{12}H_{24}N_2O_2$
(90853-20-6)
(90853-23-9)
$C_{12}H_{24}N_2O_3$
(6425-39-4)
$C_{12}H_{24}N_2O_4$
(78-44-4)
$C_{12}H_{24}N_2O_6S_2.Ca$
(139-06-0)
$C_{12}H_{24}N_4S_8.Se$
(144-34-3)
$C_{12}H_{24}N_6$
(5465-03-2)
$C_{12}H_{24}N_9P_3$
(52-46-0)
$C_{12}H_{24}O$
(110-41-8)
(112-54-9)
(123-18-2)
(765-05-9)
(1331-50-6)
(1534-27-6)
(1724-39-6)
(2855-19-8)
(4354-58-9)
(6175-49-1)
(19780-10-0)
(20056-92-2)
(50639-02-6)
(68955-06-6)
$C_{12}H_{24}O_2$
(110-38-3)
(112-17-4)
(143-07-7)
(589-75-3)
(1731-86-8)
(6378-65-0)
$C_{12}H_{24}O_2.1/2Ba$
(4696-57-5)
$C_{12}H_{24}O_2.C_4H_{11}NO_2$
(7487-79-8)
$C_{12}H_{24}O_2.C_6H_{15}NO_3$
(2224-49-9)
$C_{12}H_{24}O_2.H_3N$
(2437-23-2)
$C_{12}H_{24}O_2.1/2Zn$
(2452-01-9)

$C_{12}H_{24}O_3$
(77-68-9)
(2388-12-7)
(3006-82-4)
(7779-95-5)
(25265-77-4)
(35875-13-9)
$C_{12}H_{24}O_3S.Na$
(99744-82-8)
$C_{12}H_{24}O_4$
(62695-55-0)
$C_{12}H_{24}O_4Si_4$
(2554-06-5)
$C_{12}H_{24}O_4Sn$
(1067-33-0)
$C_{12}H_{24}O_5S$
(3054-88-4)
$C_{12}H_{24}O_5S.C_4H_{11}NO_2$
(65520-66-3)
$C_{12}H_{24}O_6$
(17455-13-9)
$C_{12}H_{25}Br$
(143-15-7)
$C_{12}H_{25}Cl$
(112-52-7)
(2350-11-0)
(28519-07-5)
$C_{12}H_{25}ClO_2SSn$
(57807-89-3)
$C_{12}H_{25}Cl_3Si$
(4484-72-4)
$C_{12}H_{25}F$
(334-68-9)
$C_{12}H_{25}I$
(4292-19-7)
$C_{12}H_{25}NO$
(1120-16-7)
(10264-29-6)
(14433-76-2)
$C_{12}H_{25}NO_3$
(3077-30-3)
$C_{12}H_{25}N_3S_4Zn$
(16509-79-8)
$C_{12}H_{25}O_4S.H_4N$
(2235-54-3)
$C_{12}H_{25}O_4S.Na$
(151-21-3)
$C_{12}H_{26}$
(112-40-3)
(1002-43-3)
(7045-71-8)
(13475-82-6)
(31807-55-3)
(55499-04-2)
(64771-71-7)
$C_{12}H_{26}N_2O_4$
(15511-81-6)
$C_{12}H_{26}O$
(112-53-8)
(112-58-3)
(123-17-1)
(3913-02-8)
(6836-38-0)
(10203-28-8)
(10203-33-5)
(27342-88-7)
$C_{12}H_{26}O.1/3Al$
(14624-15-8)
$C_{12}H_{26}O_2$
(5921-80-2)
(70709-97-6)
$C_{12}H_{26}O_3$
(112-73-2)
$C_{12}H_{26}O_3S$
(1510-16-3)
$C_{12}H_{26}O_3S.Na$
(2386-53-0)
$C_{12}H_{26}O_4$
(2167-23-9)
$C_{12}H_{26}O_4S$
(151-41-7)
$C_{12}H_{26}O_4S.C_4H_{11}NO_2$
(143-00-0)
$C_{12}H_{26}O_4S.C_6H_{15}NO_3$
(139-96-8)
$C_{12}H_{26}O_4S.1/2Mg$
(3097-08-3)
$C_{12}H_{26}O_5$
(1559-34-8)
(4353-28-0)
$C_{12}H_{26}O_6P_2S_4$
(78-34-2)
$C_{12}H_{26}O_7$
(2615-15-8)
$C_{12}H_{26}S$
(112-55-0)
(25103-58-6)
$C_{12}H_{27}Al$
(100-99-2)
(1116-70-7)
$C_{12}H_{27}BO_3$
(688-74-4)
(13195-76-1)
$C_{12}H_{27}ClPb$
(13302-14-2)
$C_{12}H_{27}ClSn$
(1461-22-9)
$C_{12}H_{27}FSn$
(1983-10-4)
$C_{12}H_{27}ISn$
(7342-47-4)
$C_{12}H_{27}N$
(102-82-9)
(124-22-1)
(143-16-8)
(1116-40-1)
(1120-24-7)
$C_{12}H_{27}N.C_2H_4O_2$
(2016-56-0)
$C_{12}H_{27}N.C_6H_5Cl_3O_3$
(53404-84-5)
$C_{12}H_{27}N.ClH$
(929-73-7)
$C_{12}H_{27}NO_3$
(2421-02-5)
$C_{12}H_{27}NO_5Si$
(17945-05-0)
$C_{12}H_{27}N_3$
(10556-98-6)
$C_{12}H_{27}OP$
(814-29-9)
$C_{12}H_{27}OPS_3$
(78-48-8)
$C_{12}H_{27}O_2PS_2$
(6028-47-3)
$C_{12}H_{27}O_2PS_2.K$
(3287-87-4)
$C_{12}H_{27}O_2PS_2.Na$
(26537-89-3)
$C_{12}H_{27}O_3P$
(78-46-6)
(102-85-2)
(5137-70-2)
$C_{12}H_{27}O_4P$
(126-71-6)
(126-73-8)
(2627-35-2)
$C_{12}H_{27}P$
(998-40-3)
$C_{12}H_{27}PS_3$
(150-50-5)
$C_{12}H_{27}Sn$
(20763-88-6)
(36643-28-4)
$C_{12}H_{28}N.I$
(631-40-3)
$C_{12}H_{28}N_2$
(2783-17-7)
$C_{12}H_{28}N_2O_5Si$
(1067-66-9)
$C_{12}H_{28}O_2S_2Sn$
(3026-81-1)
$C_{12}H_{28}O_4Si$
(682-01-9)
$C_{12}H_{28}O_5P_2S_2$
(3244-90-4)
$C_{12}H_{28}Pb$
(3440-75-3)
$C_{12}H_{28}Sn$
(688-73-3)
$C_{12}H_{29}N_3$
(143-23-7)
$C_{12}H_{30}OSn_2$
(1112-63-6)
$C_{12}H_{30}O_7Si_2$
(2157-42-8)
$C_{12}H_{30}O_{13}P_4$
(757-58-4)
$C_{12}H_{34}O_2Si_4$
(18077-53-7)
$C_{12}H_{36}O_4Si_5$
(141-63-9)
(3555-47-3)
$C_{12}H_{36}O_6Si_6$
(540-97-6)
$C_{12}H_{54}Al_{16}O_{75}S_8$
(54182-58-0)
$C_{13}H_3Cl_5O$
(90077-77-3)
$C_{13}H_4Br_6O_3$
(67990-32-3)
$C_{13}H_4Cl_4O$
(90077-76-2)
$C_{13}H_4Cl_6O$
(38178-99-3)
$C_{13}H_4N_4O_9$
(746-53-2)
$C_{13}H_5Cl_3O$
(90077-75-1)
$C_{13}H_5Cl_3O_2.Na$
(5736-15-2)
$C_{13}H_5Cl_5O_2$
(97534-06-0)
(97534-07-1)
$C_{13}H_5N_3O_7$
(129-79-3)
$C_{13}H_6Cl_2O$
(90077-74-0)
$C_{13}H_6Cl_6N_2O$
(20632-35-3)
$C_{13}H_6Cl_6O_2$
(70-30-4)
(90985-96-9)
(97534-03-7)
(97534-04-8)
(97534-05-9)
$C_{13}H_6N_2O_5$
(31551-45-8)
$C_{13}H_7Br_2N_3O_6$
(13181-17-4)
$C_{13}H_7ClO$
(3096-47-7)
(85897-29-6)
$C_{13}H_7Cl_2N_3O$
(30886-30-7)
$C_{13}H_7Cl_3O_2$
(63709-64-8)
(90985-94-7)
(97534-02-6)
$C_{13}H_7F_3N_2O_5$
(15457-05-3)
$C_{13}H_7NO_3$
(42135-22-8)
$C_{13}H_8Br_3NO_2$
(87-10-5)
(1322-38-9)
$C_{13}H_8ClFO$
(1806-23-1)
$C_{13}H_8Cl_2N_2O_4$
(50-65-7)
$C_{13}H_8Cl_2N_2O_4.C_2H_7NO$
(1420-04-8)
$C_{13}H_8Cl_2N_4$
(30369-88-1)
$C_{13}H_8Cl_2O$
(85-29-0)
(90-98-2)
$C_{13}H_8Cl_4N_2$
(25939-05-3)
$C_{13}H_8Cl_8N_6$
(30359-63-8)
$C_{13}H_8F_3N_3O_4$
(1869-67-6)
(13744-79-1)
$C_{13}H_8N_2O_4$
(5405-53-8)
(15110-74-4)
$C_{13}H_8O$
(486-25-9)
(548-39-0)
$C_{13}H_8OS$
(492-22-8)
$C_{13}H_8O_2$
(90-47-1)
(119620-42-7)
$C_{13}H_8O_3$
(79075-22-2)
$C_{13}H_9Br$
(1133-80-8)
$C_{13}H_9BrO_4S$
(106939-92-8)
(106939-97-3)
$C_{13}H_9Br_2NO_2$
(87-12-7)
(2577-72-2)
$C_{13}H_9Cl$
(2523-44-6)
$C_{13}H_9ClFNO_2$
(94370-36-2)
$C_{13}H_9ClN_2O_4$
(5099-06-9)
(36016-30-5)
$C_{13}H_9ClO$
(134-85-0)
$C_{13}H_9ClO_2$
(85-19-8)
(55299-12-2)
$C_{13}H_9Cl_2NO$
(10286-75-6)
$C_{13}H_9Cl_2NO_2$
(36016-27-0)
$C_{13}H_9Cl_2NO_4$
(32861-85-1)
$C_{13}H_9Cl_3N_2O$
(101-20-2)
$C_{13}H_9Cl_3O$
(21571-58-4)
$C_{13}H_9Cl_6N_3OS$
(24478-04-4)
(24478-05-5)
$C_{13}H_9Cl_6N_3S$
(30863-01-5)
$C_{13}H_9F_3N_2O_2$
(369-90-4)
$C_{13}H_9IO_2$
(1488-42-2)
$C_{13}H_9N$
(85-02-9)
(229-87-8)
(230-27-3)
(260-94-6)
(2920-38-9)
(4440-33-9)
(28804-96-8)
(39327-16-7)
(83536-56-5)
$C_{13}H_9NO$
(578-95-0)
(2508-20-5)
(14548-01-7)

(91311-51-2)
$C_{13}H_9NOS$
(3411-95-8)
$C_{13}H_9NO_2$
(607-57-8)
(835-64-3)
(55345-04-5)
$C_{13}H_9NO_3$
(1144-74-7)
$C_{13}H_9NO_6S$
(106939-93-9)
(106939-98-4)
$C_{13}H_9N_3O_5$
(6333-15-9)
$C_{13}H_{10}$
(86-73-7)
(203-80-5)
(244-36-0)
(58548-38-2)
$C_{13}H_{10}BrCl_2O_2PS$
(21609-90-5)
$C_{13}H_{10}BrNO$
(10282-57-2)
$C_{13}H_{10}ClFS$
(405-30-1)
$C_{13}H_{10}ClNO$
(719-59-5)
(1020-39-9)
(6833-13-2)
$C_{13}H_{10}ClNO_2$
(1528-82-1)
(16400-09-2)
$C_{13}H_{10}Cl_2$
(101-76-8)
(134-83-8)
(2051-90-3)
(25249-39-2)
(56961-47-8)
$C_{13}H_{10}Cl_2N_2S$
(1220-00-4)
$C_{13}H_{10}Cl_2O$
(90-97-1)
$C_{13}H_{10}Cl_2O_2$
(97-23-4)
(555-89-5)
$C_{13}H_{10}Cl_2O_2.2Na$
(22232-25-3)
$C_{13}H_{10}Cl_2S$
(103-17-3)
$C_{13}H_{10}Cl_3NO_2S_2$
(2280-49-1)
$C_{13}H_{10}FNO$
(1747-80-4)
$C_{13}H_{10}F_3N$
(101-23-5)
$C_{13}H_{10}INO$
(15310-01-7)
$C_{13}H_{10}NO_2PS$
(61073-10-7)
$C_{13}H_{10}N_2$
(90-45-9)
(581-28-2)
(581-29-3)
(883-40-9)
$C_{13}H_{10}N_2.ClH$
(134-50-9)
$C_{13}H_{10}N_2.ClH.H_2O$
(52417-22-8)
$C_{13}H_{10}N_2O$
(2963-66-8)
$C_{13}H_{10}N_2O_2$
(1562-93-2)
(3682-56-2)
(116232-63-4)
$C_{13}H_{10}N_2O_3S$
(27503-81-7)
$C_{13}H_{10}N_2O_4$
(50-35-1)
(1817-74-9)
(35289-89-5)
$C_{13}H_{10}O$
(92-83-1)
(119-61-9)
(60826-62-2)
(84412-11-3)
$C_{13}H_{10}OS$
(3988-77-0)
$C_{13}H_{10}O_2$
(117-99-7)
(1137-42-4)
(39515-51-0)
$C_{13}H_{10}O_3$
(102-09-0)
(118-55-8)
(131-56-6)
(136-36-7)
(886-77-1)
(3739-38-6)
(4063-41-6)
(22975-76-4)
$C_{13}H_{10}O_4$
(1143-72-2)
$C_{13}H_{10}O_4S$
(106939-91-7)
(106939-96-2)
$C_{13}H_{10}O_5$
(131-55-5)
(478-42-2)
$C_{13}H_{10}S$
(7372-88-5)
(16587-52-3)
(20928-02-3)
(30995-64-3)
(31317-07-4)
$C_{13}H_{11}Br$
(776-74-9)
(56961-07-0)
$C_{13}H_{11}BrO$
(51632-16-7)
$C_{13}H_{11}Cl$
(90-99-3)
(831-81-2)
$C_{13}H_{11}ClF_3N_3O$
(23576-23-0)
$C_{13}H_{11}ClO$
(119-56-2)
(120-32-1)
(949-38-2)
(19962-25-5)
$C_{13}H_{11}ClO.K$
(35471-49-9)
$C_{13}H_{11}ClO.Na$
(3184-65-4)
$C_{13}H_{11}Cl_2NO_2$
(32809-16-8)
$C_{13}H_{11}Cl_2NO_5$
(72391-46-9)
$C_{13}H_{11}Cl_2O_3P$
(40911-36-2)
$C_{13}H_{11}N$
(92-81-9)
(153-78-6)
(538-51-2)
(1484-12-4)
(3652-91-3)
(3770-48-7)
(4630-20-0)
(6510-65-2)
(27323-29-1)
$C_{13}H_{11}NO$
(53-94-1)
(93-98-1)
(1137-41-3)
$C_{13}H_{11}NO_2$
(87-17-2)
(91-40-7)
(15990-43-9)
$C_{13}H_{11}NO_3$
(2303-25-5)
(2444-29-3)
$C_{13}H_{11}NO_5$
(14698-29-4)
$C_{13}H_{11}N_2O_7PS$
(39004-94-9)
$C_{13}H_{11}N_2O_8P$
(799-87-1)
$C_{13}H_{11}N_3$
(92-62-6)
$C_{13}H_{11}N_3.ClH$
(952-23-8)
$C_{13}H_{11}N_3.2ClH$
(531-73-7)
$C_{13}H_{11}N_3.H_2O_4S$
(553-30-0)
$C_{13}H_{11}N_3.1/2H_2O_4S$
(1811-28-5)
$C_{13}H_{11}N_3O$
(2440-22-4)
$C_{13}H_{11}N_3O_2$
(495-73-8)
(7030-18-4)
$C_{13}H_{11}N_3O_2.CH_3AsS$
(8066-69-1)
$C_{13}H_{11}N_3O_3$
(1435-71-8)
(29418-59-5)
$C_{13}H_{11}N_3O_3.Na$
(6470-98-0)
$C_{13}H_{11}N_3O_4$
(964-79-4)
$C_{13}H_{11}N_5$
(30354-93-9)
$C_{13}H_{12}$
(101-81-5)
(643-58-3)
(643-93-6)
(644-08-6)
(2489-86-3)
(28652-72-4)
(36541-21-6)
$C_{13}H_{12}Br_2Cl_6$
(51936-55-1)
$C_{13}H_{12}Cl_2N_2$
(101-14-4)
$C_{13}H_{12}Cl_2O_4$
(58-54-8)
$C_{13}H_{12}Cl_3N_5O$
(30804-97-8)
$C_{13}H_{12}Cl_4N_4S$
(30369-60-9)
$C_{13}H_{12}NO_2.Cl$
(16214-98-5)
$C_{13}H_{12}NO_4PS$
(2665-30-7)
$C_{13}H_{12}N_2$
(949-87-1)
(6676-90-0)
$C_{13}H_{12}N_2O$
(101-15-5)
(102-07-8)
(603-54-3)
(952-47-6)
(2396-60-3)
(6319-21-7)
(26763-63-3)
(63460-08-2)
(98404-93-4)
$C_{13}H_{12}N_2O_2$
(15979-82-5)
(78508-44-8)
$C_{13}H_{12}N_2O_3S.Na$
(62959-39-1)
$C_{13}H_{12}N_2S$
(102-08-9)
$C_{13}H_{12}N_3O_6S.Na$
(23239-41-0)
$C_{13}H_{12}N_4$
(7621-86-5)
$C_{13}H_{12}N_4O_2$
(62308-10-5)
$C_{13}H_{12}N_4O_3$
(101-52-0)
(53558-25-1)
$C_{13}H_{12}O$
(86-26-0)
(91-01-0)
(101-53-1)
(613-37-6)
(3586-14-9)
(3597-91-9)
(28994-41-4)
$C_{13}H_{12}O_2$
(103-16-2)
(620-92-8)
(3121-71-9)
(13826-35-2)
$C_{13}H_{12}O_3$
(93-43-6)
(2224-00-2)
$C_{13}H_{12}S$
(831-91-4)
$C_{13}H_{13}AsO$
(24582-54-5)
$C_{13}H_{13}ClSi$
(144-79-6)
$C_{13}H_{13}Cl_2N_3O_3$
(36734-19-7)
$C_{13}H_{13}Cl_3N_4$
(24803-01-8)
(24803-02-9)
$C_{13}H_{13}Cl_3N_6$
(30359-62-7)
$C_{13}H_{13}F_{17}N_2O_2S$
(13417-01-1)
$C_{13}H_{13}N$
(552-82-9)
(620-84-8)
(1205-64-7)
$C_{13}H_{13}NO$
(101-16-6)
(1208-86-2)
$C_{13}H_{13}NO_2$
(2619-00-3)
$C_{13}H_{13}NO_2S$
(15979-81-4)
$C_{13}H_{13}NO_4S_2$
(14706-41-3)
$C_{13}H_{13}N_3$
(102-06-7)
(62450-06-0)
$C_{13}H_{13}N_3.C_2H_4O_2$
(68808-54-8)
$C_{13}H_{13}N_3O$
(785-30-8)
$C_{13}H_{13}N_3O_4S.Na$
(6300-07-8)
$C_{13}H_{13}O_4P$
(115-89-9)
$C_{13}H_{14}$
(829-26-5)
(2027-17-0)
(2131-38-6)
(2131-41-1)
(2131-42-2)
(2245-38-7)
(3031-08-1)
(3876-97-9)
(6158-45-8)
(17057-91-9)
(28652-77-9)
$C_{13}H_{14}ClNO_2$
(31793-07-4)
$C_{13}H_{14}F_3N_3O_4$
(55283-68-6)
$C_{13}H_{14}N.Cl$
(26747-91-1)
$C_{13}H_{14}N_2$
(101-77-9)
(321-64-2)
(1208-52-2)

(6582-52-1)
$C_{13}H_{14}N_2.2ClH$
(13552-44-8)
$C_{13}H_{14}N_2O$
(6219-89-2)
$C_{13}H_{14}N_2O_3$
(115-38-8)
$C_{13}H_{14}N_2O_4S_2$
(67-99-2)
$C_{13}H_{14}N_4$
(5042-54-6)
$C_{13}H_{14}N_4O$
(140-22-7)
$C_{13}H_{14}O$
(33583-02-7)
$C_{13}H_{14}OSi$
(778-25-6)
$C_{13}H_{14}O_2$
(50976-02-8)
$C_{13}H_{14}O_3$
(6075-11-2)
(36112-95-5)
$C_{13}H_{14}O_5$
(518-75-2)
$C_{13}H_{14}O_6$
(85-71-2)
$C_{13}H_{15}ClN_4O_2$
(30358-09-9)
$C_{13}H_{15}Cl_3O_3$
(120-39-8)
$C_{13}H_{15}N$
(68228-10-4)
$C_{13}H_{15}NO$
(84-83-3)
(2094-99-7)
(2889-58-9)
(39108-91-3)
$C_{13}H_{15}NO_2$
(77-21-4)
(24691-76-7)
(39108-81-1)
$C_{13}H_{15}NO_4$
(13551-17-2)
$C_{13}H_{15}N_3$
(30362-07-3)
(30362-69-7)
$C_{13}H_{15}N_3O_2$
(87-47-8)
$C_{13}H_{15}N_3O_3$
(81334-34-1)
$C_{13}H_{15}N_5O_2S$
(54289-46-2)
$C_{13}H_{16}$
(4506-36-9)
(30316-19-9)
(30316-23-5)
(55682-80-9)
$C_{13}H_{16}ClNO.ClH$
(1867-66-9)
$C_{13}H_{16}ClN_3O$
(6495-03-0)
(6585-96-2)
$C_{13}H_{16}ClN_5S$
(6560-09-4)
$C_{13}H_{16}Cl_3NO_4$
(64700-56-7)
$C_{13}H_{16}Cl_{12}O_8$
(78-12-6)
$C_{13}H_{16}F_3N_3O_4$
(1582-09-8)
(1861-40-1)
$C_{13}H_{16}NO_4PS$
(18854-01-8)
$C_{13}H_{16}N_2O$
(119-97-1)
$C_{13}H_{16}N_2O_2$
(73-31-4)
(22031-33-0)
$C_{13}H_{16}N_2S_2$
(95-33-0)
$C_{13}H_{16}N_4O_2$
(30358-08-8)
$C_{13}H_{16}N_4O_6$
(139-91-3)
(3795-88-8)
$C_{13}H_{16}N_4O_6.ClH$
(3031-51-4)
$C_{13}H_{16}N_6O_3$
(6494-90-2)
$C_{13}H_{16}O$
(5689-12-3)
(54789-45-6)
$C_{13}H_{16}O_2$
(126-15-8)
(538-65-8)
(947-19-3)
(33791-58-1)
$C_{13}H_{16}O_4$
(34006-76-3)
$C_{13}H_{17}ClN_4O_6$
(13146-28-6)
$C_{13}H_{17}ClN_6$
(30360-20-4)
$C_{13}H_{17}ClO_3$
(1713-11-7)
(1713-12-8)
(10443-70-6)
$C_{13}H_{17}Cl_2NO$
(40164-69-0)
$C_{13}H_{17}NO$
(28178-42-9)
$C_{13}H_{17}NO_2$
(7779-16-0)
$C_{13}H_{17}NO_4$
(136-45-8)
$C_{13}H_{17}N_3O$
(58-15-1)
$C_{13}H_{17}N_3O_2$
(14255-87-9)
$C_{13}H_{17}N_3S$
(68406-57-5)
$C_{13}H_{17}N_5$
(1972-98-1)
$C_{13}H_{17}N_5O$
(6494-91-3)
$C_{13}H_{17}N_5S$
(6495-07-4)
$C_{13}H_{18}$
(829-99-2)
(21693-51-6)
(22824-32-4)
(26447-63-2)
$C_{13}H_{18}ClNO$
(2307-68-8)
(3785-20-4)
(7287-36-7)
$C_{13}H_{18}ClNO_2$
(50563-36-5)
$C_{13}H_{18}Cl_2N_2O_2$
(148-82-3)
(531-76-0)
(13045-94-8)
$C_{13}H_{18}Cl_3N_5$
(24848-53-1)
(24848-59-7)
$C_{13}H_{18}Cl_6N_4$
(24802-88-8)
(30356-34-4)
(30356-35-5)
(30357-71-2)
$C_{13}H_{18}N_2O$
(886-59-9)
(34484-77-0)
$C_{13}H_{18}N_2OS_2$
(95-29-4)
$C_{13}H_{18}N_2O_2$
(2164-08-1)
$C_{13}H_{18}N_2O_3$
(509-86-4)
$C_{13}H_{18}N_2O_4$
(145-39-1)
$C_{13}H_{18}N_4O_5$
(68897-50-7)
$C_{13}H_{18}N_4O_6$
(73215-09-5)
$C_{13}H_{18}O$
(103-95-7)
(6502-20-1)
(14128-61-1)
(32281-79-1)
$C_{13}H_{18}O_2$
(3101-60-8)
(6789-88-4)
(6938-45-0)
(15687-27-1)
(17511-60-3)
(26447-45-0)
(27185-77-9)
(30275-76-4)
(30316-14-4)
(55683-10-8)
$C_{13}H_{18}O_3$
(5451-76-3)
(6259-76-3)
(54889-98-4)
$C_{13}H_{18}O_3S$
(953-91-3)
$C_{13}H_{18}O_5S$
(26225-79-6)
$C_{13}H_{18}O_7$
(138-52-3)
$C_{13}H_{18}O_7.C_{15}H_{16}O_6$
(124-87-8)
$C_{13}H_{19}ClNO_3PS_2$
(83733-82-8)
$C_{13}H_{19}ClN_2O_2$
(133-16-4)
$C_{13}H_{19}ClO$
(18979-96-9)
(18980-06-8)
(23500-79-0)
$C_{13}H_{19}NO$
(90-84-6)
$C_{13}H_{19}NO_2$
(672-04-8)
(2282-34-0)
(3329-91-7)
$C_{13}H_{19}NO_2.C_{13}H_{19}NO_2$
(8065-36-9)
$C_{13}H_{19}NO_2S$
(80-30-8)
(115-31-1)
(62850-32-2)
$C_{13}H_{19}NO_3$
(13898-68-5)
(46817-91-8)
$C_{13}H_{19}NO_3S$
(103614-75-1)
$C_{13}H_{19}NO_4S$
(57-66-9)
$C_{13}H_{19}NO_4S.C_6H_{15}NO_3$
(26919-50-6)
$C_{13}H_{19}N_3O_4$
(1918-08-7)
(40487-42-1)
$C_{13}H_{19}N_3O_6S$
(4726-14-1)
$C_{13}H_{20}$
(98-23-7)
(1078-71-3)
(2132-86-7)
(6630-01-9)
(21777-84-4)
$C_{13}H_{20}ClN_5S_4$
(30863-06-0)
$C_{13}H_{20}F_6O_2$
(38274-67-8)
$C_{13}H_{20}N_2O$
(721-50-6)
$C_{13}H_{20}N_2O_2$
(59-46-1)
(19433-93-3)
$C_{13}H_{20}N_2O_2.ClH$
(51-05-8)
$C_{13}H_{20}N_4O_2$
(17526-94-2)
(26604-41-1)
$C_{13}H_{20}O$
(79-77-6)
(127-41-3)
(141-10-6)
(1320-21-4)
(1987-50-4)
(7597-97-9)
(14901-07-6)
(16647-05-5)
(26997-02-4)
(37677-14-8)
$C_{13}H_{20}O_2$
(128-51-8)
(3457-61-2)
(5888-33-5)
(6819-19-8)
(7493-57-4)
(13037-86-0)
(30026-92-7)
(52588-78-0)
(57289-16-4)
$C_{13}H_{20}O_3S.Na$
(33660-91-2)
$C_{13}H_{20}O_4$
(1985-51-9)
$C_{13}H_{21}Cl_3N_4$
(30369-02-9)
$C_{13}H_{21}N$
(56038-89-2)
$C_{13}H_{21}NO$
(51219-00-2)
$C_{13}H_{21}NO_2$
(15588-95-1)
(38668-48-3)
$C_{13}H_{21}N_2O.I$
(66290-87-7)
$C_{13}H_{21}N_2O_4PS$
(36335-67-8)
$C_{13}H_{21}N_3O$
(51-06-9)
$C_{13}H_{21}O_3PS$
(26087-47-8)
$C_{13}H_{21}O_4PS$
(7292-16-2)
$C_{13}H_{22}$
(5744-03-6)
(50746-55-9)
$C_{13}H_{22}NO_3PS$
(22224-92-6)
$C_{13}H_{22}NO_4PS$
(31972-43-7)
$C_{13}H_{22}NO_5PS$
(31972-44-8)
$C_{13}H_{22}N_2$
(538-75-0)
$C_{13}H_{22}N_2O$
(2163-79-3)
(18530-56-8)
$C_{13}H_{22}O$
(3796-70-1)
(55103-68-9)
$C_{13}H_{22}O_2$
(2756-56-1)
(31906-04-4)
$C_{13}H_{22}O_3$
(24851-98-7)
(28928-97-4)
(50464-96-5)
$C_{13}H_{24}$
(21964-48-7)
$C_{13}H_{24}Cl_6O_8P_2$
(38051-10-4)
$C_{13}H_{24}N_2S$

(1212-29-9)
$C_{13}H_{24}N_3O_3PS$
(23505-41-1)
$C_{13}H_{24}N_4$
(30369-01-8)
$C_{13}H_{24}N_4O_3S$
(26839-75-8)
$C_{13}H_{24}N_4O_3S.ClH$
(66357-35-5)
$C_{13}H_{24}O_2$
(141-14-0)
(692-86-4)
(1330-61-6)
(1725-04-8)
(2156-96-9)
(61444-39-1)
$C_{13}H_{24}O_3$
(26761-45-5)
(71206-09-2)
$C_{13}H_{24}O_4$
(505-52-2)
(624-17-9)
(4372-29-6)
$C_{13}H_{24}O_5$
(20591-90-6)
(20721-77-1)
$C_{13}H_{25}CuO_2$
(1338-02-9)
$C_{13}H_{25}F_6O_3P$
(102040-47-1)
$C_{13}H_{25}N$
(629-60-7)
(7560-83-0)
$C_{13}H_{25}NO$
(16728-51-1)
$C_{13}H_{25}N_2S.Cl$
(68279-00-5)
$C_{13}H_{25}N_3O_4$
(1854-23-5)
$C_{13}H_{25}N_3O_4.C_8H_{16}N_2O_3$
(37304-88-4)
$C_{13}H_{25}N_5$
(26235-15-4)
(26235-16-5)
(26235-28-9)
(26235-31-4)
(26235-35-8)
(27431-04-5)
$C_{13}H_{25}N_5O$
(30360-60-2)
$C_{13}H_{26}$
(2437-56-1)
(5617-41-4)
(25377-82-6)
$C_{13}H_{26}N_2$
(1761-71-3)
(6693-29-4)
(6693-30-7)
(6693-31-8)
(16898-52-5)
(24650-10-0)
$C_{13}H_{26}N_2O_2$
(90853-21-7)
(90853-24-0)
$C_{13}H_{26}N_2O_3$
(499-48-9)
(23315-05-1)
$C_{13}H_{26}N_2O_4$
(4268-36-4)
$C_{13}H_{26}O$
(462-18-0)
(593-08-8)
(1604-34-8)
(10486-19-8)
(54852-75-4)
(57702-05-3)
$C_{13}H_{26}O_2$
(111-82-0)
(627-90-7)
(638-53-9)
(1731-81-3)
(2035-99-6)
(2311-59-3)
(25448-24-2)
(26403-14-5)
(50623-57-9)
(68002-90-4)
$C_{13}H_{26}O_2Si_3$
(18407-16-4)
$C_{13}H_{26}O_3$
(686-31-7)
$C_{13}H_{26}O_4Si_4$
(10448-09-6)
$C_{13}H_{26}O_5S.Na$
(4016-21-1)
$C_{13}H_{27}Br$
(765-09-3)
$C_{13}H_{27}Cl$
(822-13-9)
(34214-84-1)
$C_{13}H_{27}NO_3$
(3077-37-0)
$C_{13}H_{27}N_3O_3Si$
(22984-54-9)
$C_{13}H_{28}$
(629-50-5)
(1560-97-0)
(49622-18-6)
(62108-21-8)
(62108-25-2)
(62108-31-0)
(62185-54-0)
(62237-99-4)
(62338-16-3)
$C_{13}H_{28}N_2O$
(2158-09-0)
(55090-44-3)
$C_{13}H_{28}O$
(112-70-9)
(1653-31-2)
(26248-42-0)
(27458-92-0)
(57289-26-6)
(68551-07-5)
$C_{13}H_{28}O_2$
(70709-95-4)
$C_{13}H_{28}O_4S.Na$
(3026-63-9)
$C_{13}H_{28}O_5$
(20324-34-9)
$C_{13}H_{28}S$
(19484-26-5)
$C_{13}H_{29}N$
(2869-34-3)
$C_{13}H_{29}N.C_8H_5Cl_3O_3$
(53404-85-6)
$C_{13}H_{29}NO$
(7617-78-9)
(30113-45-2)
$C_{13}H_{29}NO.C_2H_4O_2$
(28701-67-9)
$C_{13}H_{29}N_3$
(112-65-2)
$C_{13}H_{29}N_3.C_2H_4O_2$
(2439-10-3)
$C_{13}H_{29}N_3.ClH$
(13590-97-1)
$C_{13}H_{29}O_3P$
(7040-58-6)
$C_{13}H_{29}O_4P$
(5116-94-9)
$C_{13}H_{30}N.Br$
(2082-84-0)
$C_{13}H_{30}N.Cl$
(10108-87-9)
$C_{13}H_{30}N_2O_4$
(139-90-2)
$C_{13}H_{30}Sn$
(1528-01-4)
$C_{13}H_{32}O_4Si_3$
(7422-52-8)
$C_{13}H_{33}N_3Si$
(16411-33-9)
$C_{14}H_3Cl_5F_6$
(95998-68-8)
$C_{14}H_4Cl_4F_6$
(95998-67-7)
$C_{14}H_4Cl_{12}O$
(31107-44-5)
$C_{14}H_4N_2O_2S_2$
(3347-22-6)
$C_{14}H_4N_4O_{12}$
(517-92-0)
$C_{14}H_4O_6$
(81-30-1)
$C_{14}H_5Cl_3F_6$
(95998-66-6)
$C_{14}H_5Cl_6NO_3$
(76280-91-6)
$C_{14}H_5Cl_{10}O_3PS$
(95114-66-2)
$C_{14}H_6ClF_3NO_5.Na$
(62476-59-9)
$C_{14}H_6ClNO_4$
(129-40-8)
$C_{14}H_6Cl_2F_6$
(99686-52-9)
$C_{14}H_6Cl_2N_2O_4$
(66121-41-3)
$C_{14}H_6Cl_2O_2$
(82-43-9)
(82-46-2)
$C_{14}H_6Cl_4O_4$
(133-14-2)
$C_{14}H_6N_2O_6$
(57875-61-3)
$C_{14}H_6N_2O_8$
(81-55-0)
(128-91-6)
$C_{14}H_6N_8O_{14}$
(29135-62-4)
$C_{14}H_6O_8$
(476-66-4)
$C_{14}H_6O_8S_2.2Na$
(853-35-0)
$C_{14}H_7BrO_2$
(632-83-7)
$C_{14}H_7Br_2NO_2$
(81-49-2)
$C_{14}H_7Br_3F_3N;N_4$
(63333-35-7)
$C_{14}H_7ClF_6$
(95998-64-4)
$C_{14}H_7ClO_2$
(82-44-0)
(131-09-9)
$C_{14}H_7Cl_2F_3O$
(95998-69-9)
$C_{14}H_7Cl_2F_5$
(95998-70-2)
$C_{14}H_7NO_3$
(85878-62-2)
$C_{14}H_7NO_4$
(82-34-8)
$C_{14}H_7O_5S.Na$
(128-56-3)
(131-08-8)
$C_{14}H_7O_7S.Na$
(130-22-3)
$C_{14}H_8BrNO_3$
(116-82-5)
(63460-11-7)
$C_{14}H_8BrNO_5S$
(116-81-4)
$C_{14}H_8Br_2F_3NO_2$
(4776-06-1)
$C_{14}H_8Br_2O_4$
(523-88-6)
$C_{14}H_8Br_6O_2$
(37853-59-1)
$C_{14}H_8ClF_3O_3$
(63734-62-3)
$C_{14}H_8ClF_3O_3.K$
(72252-48-3)
$C_{14}H_8ClNO_2$
(117-11-3)
$C_{14}H_8ClNO_3$
(2478-67-3)
$C_{14}H_8Cl_2$
(605-48-1)
(17219-94-2)
(59116-88-0)
$C_{14}H_8Cl_2FNO_5$
(51937-92-9)
$C_{14}H_8Cl_2N_2O_2$
(81-42-5)
$C_{14}H_8Cl_2O_2$
(3457-46-3)
$C_{14}H_8Cl_2O_4$
(94-17-7)
$C_{14}H_8Cl_4$
(72-55-9)
(3424-82-6)
(68679-99-2)
$C_{14}H_8Cl_6$
(68631-02-7)
$C_{14}H_8N_2O_4$
(3006-93-7)
$C_{14}H_8N_2O_5$
(116212-00-1)
$C_{14}H_8N_2O_6$
(57474-29-0)
$C_{14}H_8N_2S_4$
(120-78-5)
$C_{14}H_8N_2S_4.Zn$
(155-04-4)
$C_{14}H_8O_2$
(84-11-7)
(84-65-1)
$C_{14}H_8O_3$
(129-43-1)
(605-32-3)
(27938-76-7)
$C_{14}H_8O_4$
(72-48-0)
(81-64-1)
(84-60-6)
(117-10-2)
(117-12-4)
(1322-60-7)
$C_{14}H_8O_5$
(81-54-9)
$C_{14}H_8O_5S$
(82-49-5)
$C_{14}H_8O_6$
(81-61-8)
$C_{14}H_8O_8S_2$
(82-48-4)
(117-14-6)
(53123-81-2)
$C_{14}H_8O_8S_2.2Na$
(853-67-8)
$C_{14}H_8S$
(30796-92-0)
$C_{14}H_9BrN_2O_4$
(27312-17-0)
(27733-08-0)
$C_{14}H_9Cl$
(947-72-8)
(10271-57-5)
(24423-11-8)
(50602-11-4)
(90077-73-9)
$C_{14}H_9ClFNO_5$
(51282-69-0)
$C_{14}H_9ClF_2N_2O_2$
(35367-38-5)
$C_{14}H_9ClN_2O_4$
(12217-79-7)
$C_{14}H_9ClO_3$

(2464-37-1)
$C_{14}H_9Cl_2F_3N_2O$
(369-77-7)
$C_{14}H_9Cl_2NO_5$
(42576-02-3)
$C_{14}H_9Cl_3$
(1022-22-6)
$C_{14}H_9Cl_3F_2$
(475-26-3)
$C_{14}H_9Cl_3O_2$
(108548-72-7)
$C_{14}H_9Cl_5$
(50-29-3)
(789-02-6)
(8017-34-3)
(33086-18-9)
$C_{14}H_9Cl_5O$
(115-32-2)
(10606-46-9)
$C_{14}H_9Cl_6O_3PS$
(95150-15-5)
$C_{14}H_9F_{17}O_2$
(1996-88-9)
$C_{14}H_9N$
(203-65-6)
(2523-48-0)
$C_{14}H_9NO_2$
(82-45-1)
(117-79-3)
(602-60-8)
(17024-18-9)
(25620-59-1)
(54738-93-1)
(68455-92-5)
$C_{14}H_9NO_3$
(116-85-8)
(52738-29-1)
(116211-98-4)
$C_{14}H_9N_3O_4$
(82-33-7)
$C_{14}H_{10}$
(85-01-8)
(120-12-7)
(501-65-5)
$C_{14}H_{10}BrN_3O$
(1812-30-2)
$C_{14}H_{10}Br_4O_4$
(49693-09-6)
$C_{14}H_{10}Cl_2$
(951-86-0)
$C_{14}H_{10}Cl_2NO_2.Na$
(15307-79-6)
$C_{14}H_{10}Cl_2O_2$
(83-05-6)
$C_{14}H_{10}Cl_2O_3$
(23851-46-9)
$C_{14}H_{10}Cl_2O_4$
(51338-10-4)
$C_{14}H_{10}Cl_4$
(53-19-0)
(72-54-8)
(13700-81-7)
(76253-60-6)
$C_{14}H_{10}F_{17}NO_4S$
(25268-77-3)
$C_{14}H_{10}MgO_6$
(18917-89-0)
$C_{14}H_{10}N_2O_2$
(128-95-0)
(129-42-0)
(129-44-2)
(37070-85-2)
(62245-47-0)
$C_{14}H_{10}N_2O_4$
(128-94-9)
(145-49-3)
(586-91-4)
(6275-02-1)
$C_{14}H_{10}N_2O_6S_2.Ca$
(6485-34-3)
$C_{14}H_{10}N_2O_{10}S_2$
(128-42-7)
(128-86-9)
$C_{14}H_{10}N_2O_{10}S_2.2Na$
(3709-43-1)
$C_{14}H_{10}N_4O_5$
(7261-97-4)
$C_{14}H_{10}N_4O_6S_2.2Cl$
(13954-62-6)
$C_{14}H_{10}O$
(90-44-8)
(484-17-3)
(2840-51-9)
(30774-95-9)
(77468-39-4)
(79147-47-0)
$C_{14}H_{10}OS$
(15774-82-0)
$C_{14}H_{10}O_2$
(134-81-6)
(1210-05-5)
(4981-66-2)
$C_{14}H_{10}O_3$
(85-52-9)
(93-97-0)
(480-22-8)
(1143-38-0)
(27458-06-6)
$C_{14}H_{10}O_4$
(94-36-0)
(482-05-3)
(17648-03-2)
$C_{14}H_{10}O_4.Hg$
(583-15-3)
$C_{14}H_{10}O_5$
(552-94-3)
(20073-24-9)
$C_{14}H_{10}O_6S_2$
(61736-91-2)
(61736-92-3)
$C_{14}H_{11}Cl$
(1460-06-6)
$C_{14}H_{11}ClN_2O_4S$
(77-36-1)
$C_{14}H_{11}ClO$
(33184-55-3)
$C_{14}H_{11}ClO_2$
(108548-66-9)
$C_{14}H_{11}Cl_3$
(2642-80-0)
(2971-22-4)
$C_{14}H_{11}Cl_4O_3PS$
(53066-66-3)
$C_{14}H_{11}N$
(85-06-3)
(86-29-3)
(256-96-2)
(613-13-8)
(948-65-2)
(1484-13-5)
(31603-77-7)
(54116-90-4)
(88813-63-2)
$C_{14}H_{11}NO_2$
(4003-94-5)
$C_{14}H_{11}NO_3$
(135-69-3)
(3271-05-4)
(37070-87-4)
(37076-88-3)
$C_{14}H_{11}NO_4$
(23787-97-5)
$C_{14}H_{11}NO_5S.Na$
(10359-69-0)
$C_{14}H_{11}N_2NaO_{10}S_2$
(15883-59-7)
$C_{14}H_{11}N_3O_3S$
(31430-18-9)
$C_{14}H_{12}$
(103-30-0)
(530-48-3)
(588-59-0)
(613-31-0)
(645-49-8)
(776-35-2)
(1430-97-3)
(1556-99-6)
(1730-37-6)
(2350-89-2)
(2523-37-7)
(26914-17-0)
$C_{14}H_{12}ClF_{12}O_4P$
(102040-62-0)
$C_{14}H_{12}ClNO_2$
(36016-24-7)
$C_{14}H_{12}ClNO_3$
(77915-81-2)
$C_{14}H_{12}Cl_2$
(5963-49-5)
$C_{14}H_{12}Cl_2O$
(80-06-8)
(2642-82-2)
(56428-00-3)
(56960-97-5)
$C_{14}H_{12}Cl_2O_2S$
(4418-66-0)
$C_{14}H_{12}F_3NO_4S_2$
(37924-13-3)
$C_{14}H_{12}F_{17}NO_4S$
(1869-77-8)
$C_{14}H_{12}N_2O_2$
(81-63-0)
(86-20-4)
(4629-58-7)
$C_{14}H_{12}N_2O_3S_2$
(130-17-6)
$C_{14}H_{12}N_2O_4$
(97-32-5)
(2130-56-5)
(49839-35-2)
$C_{14}H_{12}N_2O_7S_2$
(7277-87-4)
$C_{14}H_{12}N_2O_8S_2$
(119-72-2)
$C_{14}H_{12}N_2O_8S_2.2Na$
(6634-82-8)
$C_{14}H_{12}N_2S$
(92-36-4)
$C_{14}H_{12}N_4O_2$
(2475-45-8)
(20282-70-6)
$C_{14}H_{12}N_4O_2S$
(59-40-5)
$C_{14}H_{12}N_6O_6$
(804-36-4)
$C_{14}H_{12}O$
(92-91-1)
(134-84-9)
(451-40-1)
(947-91-1)
(2523-46-8)
(17619-97-5)
$C_{14}H_{12}O_2$
(117-34-0)
(119-53-9)
(120-51-4)
(612-35-1)
(116211-89-3)
$C_{14}H_{12}O_2.2Na$
(73347-80-5)
$C_{14}H_{12}O_3$
(76-93-7)
(94-18-8)
(118-58-1)
(131-57-7)
(3902-71-4)
(90370-29-9)
$C_{14}H_{12}O_4$
(131-53-3)
(840-65-3)
(5697-00-7)
$C_{14}H_{12}O_4S$
(106939-90-6)
(106939-95-1)
$C_{14}H_{12}O_5S$
(106939-89-3)
(106939-94-0)
$C_{14}H_{12}O_6S$
(4065-45-6)
$C_{14}H_{12}O_8$
(479-66-3)
$C_{14}H_{12}S$
(70021-47-5)
$C_{14}H_{13}Cl$
(4714-14-1)
$C_{14}H_{13}Cl_2N_2O_2PS$
(4104-14-7)
$C_{14}H_{13}Cl_2O_3PS$
(53066-65-2)
(53066-68-5)
$C_{14}H_{13}F_{12}O_4P$
(63325-06-4)
$C_{14}H_{13}N$
(86-28-2)
(494-19-9)
$C_{14}H_{13}NO$
(4075-79-0)
(7055-03-0)
$C_{14}H_{13}NO_2$
(101-71-3)
(441-38-3)
(13599-69-4)
(16323-13-0)
$C_{14}H_{13}NO_3$
(1630-17-7)
(19219-48-8)
$C_{14}H_{13}NO_3S$
(6265-01-6)
$C_{14}H_{13}N_2O_7PS$
(7508-73-8)
(16604-76-5)
$C_{14}H_{13}N_2O_8P$
(905-14-6)
$C_{14}H_{13}N_3O$
(4128-71-6)
$C_{14}H_{13}N_3O_2$
(5302-39-6)
$C_{14}H_{13}O_3.1/2Ca$
(23710-76-1)
$C_{14}H_{14}$
(103-29-7)
(605-39-0)
(612-00-0)
(612-75-9)
(613-33-2)
(1812-51-7)
(4433-11-8)
(7383-90-6)
(28013-11-8)
(30551-09-8)
(38888-98-1)
(40529-66-6)
$C_{14}H_{14}Cl_2N_2O$
(35554-44-0)
$C_{14}H_{14}Cl_2Si$
(18414-36-3)
$C_{14}H_{14}F_{17}NO_3S$
(2263-09-4)
$C_{14}H_{14}Hg$
(537-64-4)
$C_{14}H_{14}NO_4PS$
(2104-64-5)
$C_{14}H_{14}N_2$
(132-32-1)
(501-60-0)
(584-90-7)
(588-04-5)
(621-09-0)
(835-31-4)
(17590-87-3)

(26444-20-2)
(29418-21-1)
(29418-22-2)
$C_{14}H_{14}N_2.ClH$
(550-99-2)
(6109-97-3)
$C_{14}H_{14}N_2O$
(54-36-4)
(3366-61-8)
(13114-72-2)
(29268-77-7)
(29268-78-8)
$C_{14}H_{14}N_2O_2$
(120-35-4)
(501-58-6)
(613-55-8)
(6319-23-9)
(92245-57-3)
$C_{14}H_{14}N_2O_3$
(1562-94-3)
$C_{14}H_{14}N_2O_3S.Na$
(62959-40-4)
$C_{14}H_{14}N_2O_6$
(10103-89-6)
$C_{14}H_{14}N_2O_6S_2$
(81-11-8)
$C_{14}H_{14}N_2O_6S_2.2Na$
(7336-20-1)
$C_{14}H_{14}N_2O_6S_2.xNa$
(25394-13-2)
$C_{14}H_{14}N_3.C_{13}H_{11}N_3.2ClH.Cl$
(8018-07-3)
$C_{14}H_{14}N_3.Cl$
(86-40-8)
$C_{14}H_{14}N_3.Cl.C_{13}H_{11}N_3$
(8048-52-0)
$C_{14}H_{14}N_3O_3S.Na$
(547-58-0)
$C_{14}H_{14}N_4O_2$
(2491-74-9)
(3010-38-6)
$C_{14}H_{14}N_4O_5S$
(74223-56-6)
$C_{14}H_{14}N_4S_2$
(1614-30-8)
$C_{14}H_{14}N_8O_4S_3$
(25953-19-9)
$C_{14}H_{14}O$
(103-50-4)
(834-14-0)
(28299-41-4)
$C_{14}H_{14}O_2$
(104-66-5)
(781-74-8)
(2122-70-5)
(2132-80-1)
$C_{14}H_{14}O_3$
(83-26-1)
(83-28-3)
(22204-53-1)
(36429-48-8)
$C_{14}H_{14}O_4$
(131-17-9)
(1087-21-4)
$C_{14}H_{14}O_6$
(7195-45-1)
$C_{14}H_{14}S$
(538-74-9)
$C_{14}H_{15}AsO$
(24582-55-6)
$C_{14}H_{15}AsS$
(24582-60-3)
$C_{14}H_{15}ClN_4S_2$
(13733-97-6)
$C_{14}H_{15}Cl_2N$
(494-03-1)
$C_{14}H_{15}Cl_3N_4$
(24848-37-1)
$C_{14}H_{15}N$
(103-49-1)
(2057-49-0)
$C_{14}H_{15}N.ClH$
(20455-68-9)
$C_{14}H_{15}NO_2$
(101-70-2)
(63455-65-2)
$C_{14}H_{15}NO_2S$
(17796-82-6)
$C_{14}H_{15}NS$
(64142-01-4)
$C_{14}H_{15}N_3$
(60-11-7)
(97-56-3)
$C_{14}H_{15}N_3O_5S$
(22025-44-1)
$C_{14}H_{15}N_5O$
(18371-12-5)
$C_{14}H_{15}N_5O_6S$
(74223-64-6)
$C_{14}H_{15}O_2PS_2$
(17109-49-8)
(27157-94-4)
$C_{14}H_{15}O_3PS$
(16611-66-8)
$C_{14}H_{16}$
(490-65-3)
(1134-62-9)
(1634-09-9)
(2876-35-9)
(16727-91-6)
(17085-91-5)
(54774-89-9)
$C_{14}H_{16}ClN_3O$
(67129-08-2)
$C_{14}H_{16}ClN_3O_2$
(43121-43-3)
$C_{14}H_{16}ClN_3O_4S_2$
(2259-96-3)
$C_{14}H_{16}ClN_5O_5S$
(82097-50-5)
$C_{14}H_{16}ClO_5PS$
(56-72-4)
$C_{14}H_{16}Cl_2O_2$
(36616-52-1)
$C_{14}H_{16}F_3N_3O_4$
(26399-36-0)
$C_{14}H_{16}F_3N_3O_5$
(34129-07-2)
$C_{14}H_{16}F_{17}N_2O_2S.Cl$
(38006-74-5)
$C_{14}H_{16}F_{17}N_2O_2S.I$
(1652-63-7)
$C_{14}H_{16}N.Br$
(10551-21-0)
$C_{14}H_{16}N_2$
(119-93-7)
$C_{14}H_{16}N_2.2ClH$
(612-82-8)
$C_{14}H_{16}N_2O_2$
(119-90-4)
(2778-41-8)
(2778-42-9)
(7652-64-4)
$C_{14}H_{16}N_2O_2.2ClH$
(20325-40-0)
$C_{14}H_{16}N_2O_2S$
(81-10-7)
$C_{14}H_{16}N_4$
(539-17-3)
$C_{14}H_{16}O_2Si$
(3440-02-6)
(6843-66-9)
$C_{14}H_{16}O_3S$
(26761-78-4)
$C_{14}H_{16}O_3S.H_3N$
(27478-24-6)
$C_{14}H_{16}O_3S.Na$
(25638-17-9)
$C_{14}H_{16}O_6$
(84-72-0)
$C_{14}H_{17}ClNO_4PS_2$
(10311-84-9)
$C_{14}H_{17}Cl_3O_4$
(2545-59-7)
$C_{14}H_{17}N$
(3290-24-2)
$C_{14}H_{17}NO_2$
(91-44-1)
$C_{14}H_{17}NO_2S$
(88683-38-9)
$C_{14}H_{17}NO_4.K$
(961-69-3)
$C_{14}H_{17}NO_5$
(15772-26-6)
$C_{14}H_{17}NS_2$
(524-84-5)
$C_{14}H_{17}N_2O_4PS$
(119-12-0)
$C_{14}H_{17}O_5PS$
(299-45-6)
$C_{14}H_{18}$
(5325-97-3)
(66289-74-5)
$C_{14}H_{18}.C_{10}H_{16}.C_7H_{14}$
(82863-50-1)
$C_{14}H_{18}ClN_3O_2$
(55219-65-3)
$C_{14}H_{18}ClN_3S$
(148-65-2)
$C_{14}H_{18}ClN_5O$
(68215-98-5)
$C_{14}H_{18}Cl_2O_3$
(1917-95-9)
(6753-24-8)
$C_{14}H_{18}Cl_2O_4$
(1929-73-3)
$C_{14}H_{18}FeN_3O_{10}.H.Na$
(12389-75-2)
$C_{14}H_{18}N_2O_3$
(151-83-7)
$C_{14}H_{18}N_2O_3.Na$
(309-36-4)
$C_{14}H_{18}N_2O_4$
(116-66-5)
(77732-09-3)
$C_{14}H_{18}N_2O_5$
(81-14-1)
(22839-47-0)
$C_{14}H_{18}N_2O_7$
(973-21-7)
$C_{14}H_{18}N_3O_{10}.Ca.3Na$
(12111-24-9)
$C_{14}H_{18}N_4O_3$
(738-70-5)
(17804-35-2)
$C_{14}H_{18}N_4O_3.C_{10}H_{11}N_3O_3S$
(8064-90-2)
$C_{14}H_{18}N_4O_4S_2$
(23564-06-9)
$C_{14}H_{18}O$
(122-40-7)
(13586-68-0)
$C_{14}H_{18}O_2$
(55591-12-3)
$C_{14}H_{18}O_4$
(104-28-9)
(131-16-8)
(605-45-8)
(53188-07-1)
$C_{14}H_{18}O_6$
(117-82-8)
$C_{14}H_{18}O_7$
(3524-68-3)
$C_{14}H_{19}ClO_3$
(1713-14-0)
$C_{14}H_{19}Cl_2NO_2$
(305-03-3)
$C_{14}H_{19}Cl_3N_2O_2$
(26952-20-5)
$C_{14}H_{19}Cl_6N_3$
(24481-36-5)
$C_{14}H_{19}F_6O_4P$
(102040-45-9)
$C_{14}H_{19}N$
(35331-89-6)
$C_{14}H_{19}NO$
(91-53-2)
$C_{14}H_{19}NOS$
(6996-88-9)
$C_{14}H_{19}NO_2$
(77-17-8)
(113-45-1)
$C_{14}H_{19}NO_2.ClH$
(298-59-9)
$C_{14}H_{19}NO_4$
(19249-34-4)
(22862-76-6)
$C_{14}H_{19}N_3O.C_4H_4O_4$
(5429-41-4)
$C_{14}H_{19}N_3O_2$
(19433-94-4)
(63505-65-7)
$C_{14}H_{19}N_3O_3$
(22588-78-9)
$C_{14}H_{19}N_3S$
(91-79-2)
(91-80-5)
$C_{14}H_{19}N_3S.ClH$
(135-23-9)
(958-93-0)
$C_{14}H_{19}N_5O$
(6494-99-1)
$C_{14}H_{19}N_5S$
(6494-92-4)
$C_{14}H_{19}N_7O_4S$
(103947-07-5)
$C_{14}H_{19}O_6P$
(7700-17-6)
$C_{14}H_{20}$
(81-03-8)
(1203-17-4)
(23342-25-8)
$C_{14}H_{20}BrNO_3$
(17199-22-3)
$C_{14}H_{20}Br_2N_2$
(3572-43-8)
$C_{14}H_{20}Br_6$
(18122-77-5)
$C_{14}H_{20}ClNO_2$
(15972-60-8)
$C_{14}H_{20}ClNO_2.C_9H_{19}NOS$
(8074-35-9)
$C_{14}H_{20}ClNO_3$
(17199-21-2)
$C_{14}H_{20}Cl_2O$
(73986-52-4)
$C_{14}H_{20}Cl_3N_5$
(24848-54-2)
(24863-54-5)
(24863-55-6)
(26235-04-1)
$C_{14}H_{20}NO_5PS$
(28789-80-2)
$C_{14}H_{20}N_2$
(61-51-8)
$C_{14}H_{20}N_2O$
(1982-49-6)
(19095-79-5)
$C_{14}H_{20}N_2O_5$
(3687-22-7)
(4097-33-0)
(37224-61-6)
$C_{14}H_{20}N_3O_3.1/2Cl_4Zn$
(6023-29-6)
$C_{14}H_{20}N_3O_5PS$
(13457-18-6)
$C_{14}H_{20}N_4$
(2094-98-6)
$C_{14}H_{20}O$
(80-54-6)

$C_{14}H_{20}O_2$
(719-22-2)
(2460-77-7)
(2611-00-9)
(54644-40-5)
$C_{14}H_{20}O_3$
(4693-19-0)
$C_{14}H_{20}O_4$
(2386-87-0)
$C_{14}H_{20}O_6$
(5493-45-8)
$C_{14}H_{21}ClO$
(17199-24-5)
$C_{14}H_{21}Cl_2NO_6$
(2569-01-9)
$C_{14}H_{21}N$
(2201-15-2)
$C_{14}H_{21}NO_2$
(644-26-8)
(14309-40-1)
(14779-78-3)
$C_{14}H_{21}NO_3$
(60568-05-0)
$C_{14}H_{21}NO_4$
(87130-20-9)
$C_{14}H_{21}N_2O_5PS$
(22941-83-9)
$C_{14}H_{21}N_3O_3$
(4849-32-5)
$C_{14}H_{21}N_3O_3S$
(565-33-3)
(1156-19-0)
$C_{14}H_{21}N_3O_4$
(33629-47-9)
$C_{14}H_{22}$
(1012-72-2)
(1014-41-1)
(1079-96-5)
(2189-60-8)
(5186-68-5)
(10375-96-9)
(33637-20-6)
(33962-13-9)
(54789-15-0)
$C_{14}H_{22}ClN_3O_2$
(364-62-5)
$C_{14}H_{22}ClOP$
(25097-44-3)
$C_{14}H_{22}N_2O$
(137-58-6)
$C_{14}H_{22}N_2O_2$
(5180-59-6)
$C_{14}H_{22}N_2O_3$
(6673-35-4)
(29122-68-7)
$C_{14}H_{22}N_2O_5$
(32568-89-1)
$C_{14}H_{22}N_2O_8$
(482-54-2)
$C_{14}H_{22}O$
(79-69-6)
(96-76-4)
(127-42-4)
(127-43-5)
(127-51-5)
(128-39-2)
(140-66-9)
(949-13-3)
(1138-52-9)
(1335-46-2)
(1806-26-4)
(1849-18-9)
(3884-95-5)
(5875-45-6)
(7388-22-9)
(26651-96-7)
(27193-28-8)
(27515-66-8)
(27985-70-2)
(31291-60-8)
(33704-61-9)
(54932-78-4)
$C_{14}H_{22}O_2$
(88-58-4)
(7534-94-3)
(16219-25-3)
$C_{14}H_{22}O_3S$
(17012-98-5)
(25321-43-1)
(99422-01-2)
$C_{14}H_{22}O_3S.Na$
(6149-03-7)
$C_{14}H_{22}O_4$
(6606-59-3)
$C_{14}H_{22}O_6$
(109-16-0)
(1561-49-5)
$C_{14}H_{22}O_7$
(17831-71-9)
$C_{14}H_{22}O_8$
(77-89-4)
$C_{14}H_{23}N$
(613-29-6)
(10137-80-1)
(16245-79-7)
$C_{14}H_{23}NO_2$
(63455-64-1)
$C_{14}H_{23}N_3O_4PS_3$
(72542-56-4)
$C_{14}H_{23}N_3O_{10}$
(67-43-6)
$C_{14}H_{23}N_3O_{10}.5Na$
(140-01-2)
$C_{14}H_{23}N_5$
(26234-41-3)
$C_{14}H_{23}O_4P$
(2528-36-1)
$C_{14}H_{24}$
(5743-97-5)
(6596-35-6)
(33704-59-5)
$C_{14}H_{24}NO_4PS_3$
(741-58-2)
$C_{14}H_{24}NS.Cl$
(68278-98-8)
$C_{14}H_{24}N_2$
(101-96-2)
$C_{14}H_{24}N_4$
(4419-11-8)
$C_{14}H_{24}O$
(28219-61-6)
(54594-42-2)
$C_{14}H_{24}O_2$
(64726-91-6)
$C_{14}H_{24}O_4$
(1471-17-6)
(10099-71-5)
(14228-73-0)
$C_{14}H_{26}$
(21964-49-8)
(54823-94-8)
(54823-95-9)
(54823-98-2)
(56292-64-9)
$C_{14}H_{26}O$
(35194-22-0)
(53939-27-8)
$C_{14}H_{26}O_2$
(126-86-3)
(1333-17-1)
(3179-47-3)
(7492-66-2)
(14959-86-5)
(16974-11-1)
(29964-84-9)
(35148-19-7)
(56438-08-5)
(61444-41-5)
$C_{14}H_{26}O_4$
(105-99-7)
(110-40-7)
(141-04-8)
(821-38-5)
(4337-65-9)
(56051-60-6)
$C_{14}H_{26}O_5$
(20591-91-7)
$C_{14}H_{26}O_7$
(17626-93-6)
$C_{14}H_{26}O_7S.Na$
(922-80-5)
$C_{14}H_{26}O_7S.2Na$
(37294-49-8)
$C_{14}H_{27}AlO_5$
(24772-51-8)
$C_{14}H_{27}ClO_2$
(64919-15-9)
$C_{14}H_{27}N$
(629-63-0)
$C_{14}H_{27}N_5$
(26235-21-2)
(26235-22-3)
(26235-26-7)
(26235-38-1)
(27430-99-5)
$C_{14}H_{27}N_5O$
(30360-65-7)
$C_{14}H_{27}O_5S.Na$
(1847-58-1)
$C_{14}H_{28}$
(1120-36-1)
(1795-15-9)
(2883-05-8)
(10374-74-0)
(24949-42-6)
(26952-13-6)
$C_{14}H_{28}Cl_4Cr_2O_3$
(15659-56-0)
$C_{14}H_{28}NO_3PS_2$
(24151-93-7)
$C_{14}H_{28}N_2$
(25707-70-4)
$C_{14}H_{28}O$
(124-25-4)
(3234-28-4)
(34010-15-6)
(35153-15-2)
$C_{14}H_{28}O_2$
(106-33-2)
(112-66-3)
(544-63-8)
(1731-88-0)
(4887-30-3)
(28484-22-2)
(30673-36-0)
(32399-56-7)
(70709-99-8)
$C_{14}H_{28}O_2.1/2Ba$
(10196-66-4)
$C_{14}H_{28}O_2.C_6H_{15}NO_3$
(41669-40-3)
$C_{14}H_{28}O_2.1/2Ca$
(15284-51-2)
$C_{14}H_{28}O_2.1/2Cd$
(10196-67-5)
$C_{14}H_{28}O_2.K$
(13429-27-1)
$C_{14}H_{28}O_2.xPb$
(20403-41-2)
$C_{14}H_{28}O_2S$
(3746-39-2)
$C_{14}H_{28}O_3$
(1961-72-4)
(2507-55-3)
(16899-08-4)
(26748-41-4)
$C_{14}H_{28}O_3S$
(18760-44-6)
$C_{14}H_{28}O_3S.Na$
(11066-21-0)
$C_{14}H_{28}O_4$
(3006-86-8)
$C_{14}H_{28}O_6S_2.2Na$
(68003-17-8)
$C_{14}H_{29}Br$
(112-71-0)
$C_{14}H_{29}Cl$
(2425-54-9)
$C_{14}H_{29}I$
(19218-94-1)
$C_{14}H_{29}N$
(5432-61-1)
$C_{14}H_{29}NO_2$
(142-78-9)
$C_{14}H_{29}NO_3$
(136-26-5)
$C_{14}H_{29}NaO_3S$
(68037-49-0)
$C_{14}H_{29}O_4S.Na$
(139-88-8)
(1191-50-0)
$C_{14}H_{30}$
(629-59-4)
(1560-96-9)
(6864-53-5)
(7225-67-4)
(13287-21-3)
(26730-14-3)
(56292-65-0)
$C_{14}H_{30}N_2O_4$
(306-40-1)
$C_{14}H_{30}O$
(112-72-1)
(4706-81-4)
(27196-00-5)
$C_{14}H_{30}O.1/3Al$
(67905-32-2)
$C_{14}H_{30}O_2$
(4536-30-5)
(5405-58-3)
(29718-44-3)
$C_{14}H_{30}O_2Sn$
(56-36-0)
$C_{14}H_{30}O_4S$
(4754-44-3)
$C_{14}H_{30}O_4S.C_6H_{15}NO_3$
(4492-78-8)
$C_{14}H_{30}O_5S.Na$
(15826-16-1)
$C_{14}H_{30}O_6$
(23601-39-0)
$C_{14}H_{30}S$
(28983-37-1)
$C_{14}H_{31}N$
(112-18-5)
(2016-42-4)
$C_{14}H_{31}N.C_8H_5Cl_3O_3$
(53535-37-8)
$C_{14}H_{31}N.C_8H_6Cl_2O_3$
(28685-18-9)
$C_{14}H_{31}N.ClH$
(2016-48-0)
$C_{14}H_{31}NO$
(1643-20-5)
$C_{14}H_{31}NO_2$
(28482-15-7)
$C_{14}H_{31}O_2PS_2$
(68187-41-7)
$C_{14}H_{31}O_4P$
(10054-29-2)
$C_{14}H_{32}N_2$
(4835-11-4)
$C_{14}H_{32}N_2O_4$
(102-60-3)
$C_{14}H_{32}O_3Si$
(2943-75-1)
$C_{14}H_{32}Pb$
(65151-10-2)
$C_{14}H_{42}O_5Si_6$
(107-52-8)

(59700-60-6)
$C_{15}H_{15}ClN_2O_2$
(1982-47-4)
$C_{15}H_{15}Cl_2N_2O_8.Na$
(982-57-0)
$C_{15}H_{15}F_{12}O_4P$
(102040-58-4)
(102040-61-9)
$C_{15}H_{15}F_{12}O_5P$
(102040-59-5)
$C_{15}H_{15}N$
(6196-54-9)
(6267-02-3)
$C_{15}H_{15}NO$
(530-44-9)
$C_{15}H_{15}NO_2$
(61-68-7)
(133-18-6)
$C_{15}H_{15}NO_3$
(26171-23-3)
$C_{15}H_{15}N_3O_2$
(493-52-7)
(2832-40-8)
(6268-49-1)
(20691-84-3)
$C_{15}H_{16}$
(778-22-3)
(1081-75-0)
(3976-35-0)
(5814-85-7)
(13540-56-2)
(25640-78-2)
(42504-54-1)
$C_{15}H_{16}N_2$
(6319-26-2)
(29418-23-3)
(29418-24-4)
(29418-26-6)
$C_{15}H_{16}N_2O$
(611-92-7)
(18168-01-9)
$C_{15}H_{16}N_2O_2$
(12771-68-5)
(35438-85-8)
$C_{15}H_{16}N_2O_4S$
(20241-68-3)
$C_{15}H_{16}N_2O_6S_2$
(34787-01-4)
$C_{15}H_{16}N_4$
(63690-09-5)
$C_{15}H_{16}N_4O_2$
(4313-13-7)
(4313-14-8)
(30384-46-4)
$C_{15}H_{16}N_4O_5S$
(74222-97-2)
$C_{15}H_{16}O$
(599-64-4)
(20017-67-8)
(22239-54-9)
(27576-86-9)
$C_{15}H_{16}O_2$
(80-05-7)
(5574-34-5)
(67701-06-8)
$C_{15}H_{16}O_2.2Na$
(2444-90-8)
$C_{15}H_{16}O_3S$
(4455-09-8)
$C_{15}H_{16}O_4$
(74798-20-2)
$C_{15}H_{17}Br_2NO_2$
(1689-99-2)
$C_{15}H_{17}Cl_2N_3O_2$
(60207-90-1)
$C_{15}H_{17}HgNO_2S$
(517-16-8)
$C_{15}H_{17}I_2NO_2$
(3861-47-0)
$C_{15}H_{17}N$
(92-59-1)
(5650-10-2)
(55030-65-4)
$C_{15}H_{17}NO$
(101-73-5)
$C_{15}H_{17}NO_3S$
(101-11-1)
$C_{15}H_{17}N_3$
(55-80-1)
(97-39-2)
(102-63-6)
(3731-39-3)
$C_{15}H_{17}N_3.C_{12}H_8BO_4.H$
(16971-82-7)
$C_{15}H_{17}N_3O_2$
(75104-43-7)
$C_{15}H_{17}O_2PS$
(21722-85-0)
$C_{15}H_{17}O_4P$
(60763-39-5)
$C_{15}H_{17}P$
(7650-84-2)
$C_{15}H_{18}$
(483-78-3)
$C_{15}H_{18}ClNO_4$
(67932-85-8)
$C_{15}H_{18}Cl_2N_2O_3$
(19666-30-9)
$C_{15}H_{18}F_3N_3O_5$
(34128-99-9)
$C_{15}H_{18}N_2$
(101-72-4)
(104-69-8)
(838-88-0)
(1807-55-2)
(3085-82-3)
$C_{15}H_{18}N_2O_2$
(38353-82-1)
$C_{15}H_{18}N_2O_6$
(485-31-4)
$C_{15}H_{18}N_4O_5$
(50-07-7)
$C_{15}H_{18}O_2$
(40693-04-7)
$C_{15}H_{18}O_3$
(481-06-1)
$C_{15}H_{18}O_4$
(6754-13-8)
(13561-08-5)
$C_{15}H_{19}Cl_2N_3O$
(75736-33-3)
$C_{15}H_{19}Cl_3O_4$
(1928-48-9)
(53466-86-7)
(62922-39-8)
$C_{15}H_{19}F_{13}N_2O_4S$
(34455-29-3)
$C_{15}H_{19}NO$
(80045-52-9)
$C_{15}H_{19}NO.ClH$
(51-02-5)
$C_{15}H_{19}NO_2$
(537-26-8)
$C_{15}H_{19}NO_4$
(5026-74-4)
$C_{15}H_{19}N_3O_3$
(81335-77-5)
$C_{15}H_{20}Cl_2O_4$
(1320-18-9)
(1928-45-6)
(3966-11-8)
(53404-31-2)
$C_{15}H_{20}N_2O_4S$
(968-81-0)
$C_{15}H_{20}O$
(101-86-0)
$C_{15}H_{20}O_2$
(97165-23-6)
$C_{15}H_{20}O_3$
(65983-31-5)
$C_{15}H_{20}O_4$
(55785-58-5)
$C_{15}H_{20}O_6$
(15625-89-5)
(51481-10-8)
$C_{15}H_{20}O_7$
(23282-20-4)
$C_{15}H_{21}AlO_6$
(13963-57-0)
$C_{15}H_{21}ClN_2O_8$
(81393-48-8)
$C_{15}H_{21}ClO_4$
(19480-43-4)
$C_{15}H_{21}F_6O_4P$
(102040-46-0)
$C_{15}H_{21}MnO_6$
(14284-89-0)
$C_{15}H_{21}N$
(1209-98-9)
$C_{15}H_{21}NO$
(3734-52-9)
$C_{15}H_{21}NO_2$
(57-42-1)
(469-79-4)
(500-34-5)
$C_{15}H_{21}NO_2.ClH$
(50-13-5)
$C_{15}H_{21}NO_2S$
(71868-10-5)
$C_{15}H_{21}NO_3$
(468-56-4)
$C_{15}H_{21}NO_4$
(21615-36-1)
(57837-19-1)
$C_{15}H_{21}NO_4.C_4H_6MnN_2S_4$-
$.C_4H_6N_2S_4Zn$
(75701-74-5)
$C_{15}H_{21}NO_6$
(25790-28-7)
$C_{15}H_{21}N_3O_2$
(57-47-6)
(63460-10-6)
$C_{15}H_{21}N_3O_2.C_7H_6O_3$
(57-64-7)
$C_{15}H_{22}$
(483-77-2)
(6617-49-8)
(22339-23-7)
(38393-97-4)
$C_{15}H_{22}BrNO$
(74712-19-9)
$C_{15}H_{22}BrNO_2$
(2163-81-7)
$C_{15}H_{22}ClNO_2$
(51218-45-2)
$C_{15}H_{22}Cl_2N_2O_3S$
(63905-03-3)
$C_{15}H_{22}Cl_3N_5$
(24848-60-0)
$C_{15}H_{22}N_2O_2$
(5124-30-1)
(28605-81-4)
$C_{15}H_{22}O$
(77-61-2)
(4674-50-4)
(4955-32-2)
(6008-36-2)
(17909-77-2)
(60066-88-8)
$C_{15}H_{22}O_2$
(5444-75-7)
(16202-79-2)
(22117-06-2)
$C_{15}H_{22}O_3$
(1421-49-4)
(2493-84-7)
$C_{15}H_{22}O_4$
(3388-03-2)
(17811-28-8)
$C_{15}H_{22}O_5$
(57377-32-9)
$C_{15}H_{22}O_6$
(51-14-9)
$C_{15}H_{23}BaO$
(28987-17-9)
$C_{15}H_{23}ClO$
(60044-33-9)
$C_{15}H_{23}ClO_4S$
(140-57-8)
$C_{15}H_{23}NO$
(25033-65-2)
$C_{15}H_{23}NO_2$
(13655-52-2)
(14309-41-2)
$C_{15}H_{23}NO_3$
(6452-71-7)
$C_{15}H_{23}NO_4$
(66-81-9)
(58882-17-0)
$C_{15}H_{23}NS$
(21500-98-1)
$C_{15}H_{23}N_3O_4$
(33820-53-0)
$C_{15}H_{23}N_3O_4S$
(15676-16-1)
$C_{15}H_{24}$
(88-84-6)
(469-61-4)
(469-92-1)
(473-13-2)
(475-20-7)
(483-76-1)
(489-39-4)
(502-61-4)
(511-59-1)
(523-47-7)
(717-74-8)
(1081-77-2)
(1460-97-5)
(1461-03-6)
(3856-25-5)
(6753-98-6)
(13877-93-5)
(17066-67-0)
(18431-82-8)
(33880-83-0)
(79554-39-5)
(86825-83-4)
$C_{15}H_{24}NO_4PS$
(25311-71-1)
$C_{15}H_{24}NO_5P$
(31120-85-1)
$C_{15}H_{24}N_2O_2.ClH$
(136-47-0)
$C_{15}H_{24}O$
(104-40-5)
(115-71-9)
(128-37-0)
(136-83-4)
(497-39-2)
(1139-30-6)
(14840-89-2)
(25154-52-3)
(52427-13-1)
(57716-72-0)
$C_{15}H_{24}O.1/2Ca$
(30977-64-1)
$C_{15}H_{24}O_2$
(88-26-6)
(489-01-0)
$C_{15}H_{24}O_3$
(17844-07-4)
$C_{15}H_{24}O_3S$
(79419-43-5)
$C_{15}H_{24}O_6$
(42978-66-5)
$C_{15}H_{25}Cl_2N_3O$
(4628-08-4)
$C_{15}H_{25}Cl_3N_4$
(24802-92-4)

$C_{15}H_{25}NO_3$
(37350-58-6)
$C_{15}H_{25}NO_5$
(480-82-0)
(6029-84-1)
$C_{15}H_{25}NO_6$
(41708-76-3)
$C_{15}H_{25}N_3O$
(76608-88-3)
$C_{15}H_{26}$
(87-44-5)
(19078-35-4)
(29350-73-0)
$C_{15}H_{26}Cl_2N_4$
(26113-25-7)
$C_{15}H_{26}N.Cl$
(1609-21-8)
$C_{15}H_{26}NO_3P$
(52670-79-8)
$C_{15}H_{26}NS.Cl$
(70700-59-3)
$C_{15}H_{26}N_2$
(90-39-1)
$C_{15}H_{26}N_2.H_2O_4S$
(299-39-8)
$C_{15}H_{26}O$
(77-53-2)
(142-50-7)
(3790-71-4)
(4602-84-0)
(7212-44-4)
(15352-77-9)
(55332-02-0)
(55332-03-1)
$C_{15}H_{26}O_3$
(38963-91-6)
$C_{15}H_{26}O_6$
(60-01-5)
(68959-23-9)
$C_{15}H_{26}O_6S_3$
(33007-83-9)
$C_{15}H_{26}O_7S_3$
(88083-39-0)
$C_{15}H_{26}O_{11}$
(12002-22-1)
$C_{15}H_{27}NO_2S$
(301-11-1)
$C_{15}H_{27}N_3$
(4803-17-2)
$C_{15}H_{27}N_3O$
(90-72-2)
$C_{15}H_{27}N_3O_3$
(19837-00-4)
(30895-05-7)
$C_{15}H_{28}$
(15404-63-4)
(30824-81-8)
(31230-13-4)
(31624-59-6)
(54832-83-6)
(54934-92-8)
$C_{15}H_{28}ClN_5$
(30355-02-3)
$C_{15}H_{28}O$
(502-72-7)
(1335-48-4)
(1604-35-9)
$C_{15}H_{28}O_2$
(80-04-6)
(106-02-5)
(2156-97-0)
(72269-48-8)
$C_{15}H_{28}O_3$
(1984-77-6)
$C_{15}H_{28}O_4$
(1472-87-3)
$C_{15}H_{29}NO_3.Na$
(137-16-6)
$C_{15}H_{29}N_5$
(26235-32-5)
(26235-33-6)
$C_{15}H_{29}N_5O$
(24126-22-5)
$C_{15}H_{30}$
(2883-02-5)
(13360-61-7)
$C_{15}H_{30}N_6$
(2827-49-8)
$C_{15}H_{30}N_6O_6$
(3089-11-0)
$C_{15}H_{30}O$
(818-23-5)
(2765-11-9)
(16424-67-2)
(18633-25-5)
$C_{15}H_{30}O_2$
(124-10-7)
(1002-84-2)
(1072-33-9)
(2461-18-9)
(5129-58-8)
(5502-94-3)
(10233-13-3)
(10580-24-2)
(55955-78-7)
$C_{15}H_{30}O_3$
(68299-16-1)
$C_{15}H_{30}O_5S$
(29454-23-7)
$C_{15}H_{30}O_5S.Na$
(4016-22-2)
$C_{15}H_{31}Br$
(629-72-1)
$C_{15}H_{31}Cl$
(4862-03-7)
(34214-86-3)
$C_{15}H_{31}NO_2$
(142-54-1)
(1462-54-0)
$C_{15}H_{31}NO_4S.Na$
(4337-75-1)
$C_{15}H_{31}NO_5S$
(59997-81-8)
$C_{15}H_{32}$
(629-62-9)
(1560-95-8)
(3891-98-3)
(31295-56-4)
(55030-62-1)
(56292-66-1)
$C_{15}H_{32}O$
(629-76-5)
(7278-65-1)
(31389-11-4)
$C_{15}H_{32}OS$
(67124-09-8)
$C_{15}H_{32}O_2$
(924-06-1)
(38471-49-7)
(54661-98-2)
$C_{15}H_{32}O_3Si_4$
(2116-84-9)
(17906-09-1)
$C_{15}H_{32}O_{10}$
(78-24-0)
$C_{15}H_{32}O_{11}$
(51555-31-8)
$C_{15}H_{33}N$
(621-77-2)
(2570-26-5)
$C_{15}H_{33}NO.C_8H_{10}O_5$
(35493-90-4)
$C_{15}H_{33}NO_2$
(821-91-0)
$C_{15}H_{34}N$
(10182-91-9)
$C_{15}H_{34}N.CH_3O_4S$
(13623-06-8)
$C_{15}H_{34}N.Cl$
(112-00-5)
$C_{15}H_{34}OSi$
(6221-88-1)
$C_{15}H_{36}N_4$
(33329-35-0)
$C_{16}H_4Cl_4O_2S_2$
(14295-43-3)
$C_{16}H_6Br_4$
(128-63-2)
$C_{16}H_6Br_4N_2O_2$
(2475-31-2)
$C_{16}H_6N_4O_8$
(28767-61-5)
$C_{16}H_6O_8S$
(2540-99-0)
$C_{16}H_7N_3O_6$
(75321-19-6)
$C_{16}H_7O_{10}S_3.3Na$
(6358-69-6)
$C_{16}H_8Cl_2$
(68402-20-0)
(86329-60-4)
$C_{16}H_8Cl_5N_3$
(30362-65-3)
$C_{16}H_8N_2O_4$
(22506-53-2)
(42397-64-8)
(42397-65-9)
(75321-20-9)
(78432-19-6)
(105735-71-5)
$C_{16}H_8N_2O_8S_2.2Na$
(860-22-0)
$C_{16}H_8O_2$
(1785-51-9)
(2304-85-0)
$C_{16}H_9BrN_2O_2$
(6492-73-5)
$C_{16}H_9Cl$
(25911-51-7)
(34244-14-9)
$C_{16}H_9Cl_3N_4O_2$
(3599-81-3)
$C_{16}H_9Cl_4N_3$
(3599-80-2)
$C_{16}H_9F_{21}O_2$
(2144-54-9)
$C_{16}H_9NO_2$
(789-07-1)
(892-21-7)
(5522-43-0)
(13177-28-1)
(13177-29-2)
(57835-92-4)
(63021-86-3)
(77468-36-1)
(114790-09-9)
$C_{16}H_9N_3O_4$
(128-81-4)
$C_{16}H_9N_4O_9S_2.3Na$
(1934-21-0)
$C_{16}H_9N_5O_6$
(36888-99-0)
$C_{16}H_{10}$
(129-00-0)
(201-06-9)
(206-44-0)
$C_{16}H_{10}Br_3Cl_2N_5$
(30377-17-4)
$C_{16}H_{10}ClCrN_4O_9S_2.2Na$
(6408-31-7)
$C_{16}H_{10}ClN_3O_3$
(2814-77-9)
$C_{16}H_{10}Cl_2N_2O_4S.Na$
(5850-81-7)
$C_{16}H_{10}Cl_3N_3$
(30362-63-1)
$C_{16}H_{10}Cl_5N_5$
(30355-92-1)
$C_{16}H_{10}CrN_4O_7S.Na$
(10241-21-1)
$C_{16}H_{10}N_2Na_2O_7S_2$
(2347-72-0)
$C_{16}H_{10}N_2O$
(482-89-3)
$C_{16}H_{10}N_2O_7S_2.2Na$
(1936-15-8)
(2783-94-0)
$C_{16}H_{10}N_2O_8S_2$
(483-20-5)
$C_{16}H_{10}N_4O_5$
(3468-63-1)
$C_{16}H_{10}N_4O_6$
(4998-82-7)
$C_{16}H_{10}N_4O_8S_2$
(130-34-7)
$C_{16}H_{10}O$
(205-39-0)
(5315-79-7)
$C_{16}H_{10}O_{12}S_4$
(6528-53-6)
$C_{16}H_{10}S$
(205-43-6)
(239-35-0)
(243-46-9)
(61523-34-0)
$C_{16}H_{11}ClN_2O_3$
(23887-31-2)
$C_{16}H_{11}ClN_2O_4S$
(4531-71-9)
$C_{16}H_{11}ClN_2O_8S_2.2Na$
(3624-68-8)
$C_{16}H_{11}ClN_2O_9S_2.2Na$
(1058-92-0)
$C_{16}H_{11}ClN_4$
(29975-16-4)
$C_{16}H_{11}N$
(239-01-0)
(243-28-7)
(1606-67-3)
(34777-33-8)
(67526-84-5)
$C_{16}H_{11}NO_2$
(132-60-5)
$C_{16}H_{11}N_2O_4S.Na$
(523-44-4)
$C_{16}H_{11}N_3O_3$
(6410-10-2)
$C_{16}H_{11}N_3O_6S$
(5859-04-1)
$C_{16}H_{11}N_5O_4$
(30362-61-9)
$C_{16}H_{12}$
(605-02-7)
(612-94-2)
(2444-68-0)
(28779-32-0)
(35465-71-5)
(41593-24-2)
$C_{16}H_{12}Br_3N_5$
(30359-73-0)
$C_{16}H_{12}ClFN_2O$
(3900-31-0)
$C_{16}H_{12}ClNO_3$
(51234-28-7)
$C_{16}H_{12}Cl_2N_2O_2$
(848-75-9)
$C_{16}H_{12}Cl_2N_4O_4$
(6486-23-3)
$C_{16}H_{12}Cl_2N_4O_4S.Na$
(6359-97-3)
$C_{16}H_{12}Cl_2N_4O_7S_2.2Na$
(6359-98-4)
$C_{16}H_{12}Cl_3N_5$
(30355-91-0)
$C_{16}H_{12}CoF_2N_2O_2$
(62207-76-5)
$C_{16}H_{12}FN_3O_3$
(1622-62-4)
$C_{16}H_{12}N_2O$
(842-07-9)

$C_{16}H_{12}N_2O_2$
(91-97-4)
(6537-68-4)
$C_{16}H_{12}N_2O_4$
(91-93-0)
$C_{16}H_{12}N_2O_4S$
(573-89-7)
(574-69-6)
(6616-62-2)
(22080-08-6)
(23481-33-6)
$C_{16}H_{12}N_2O_4S.Na$
(633-96-5)
(1934-20-9)
$C_{16}H_{12}N_2O_5S.Na$
(2092-55-9)
$C_{16}H_{12}N_2O_6S.Na$
(2052-25-7)
$C_{16}H_{12}N_2O_7S_2$
(5859-11-0)
$C_{16}H_{12}N_2O_7S_2.2Na$
(5859-00-7)
$C_{16}H_{12}N_2O_8S_2.2Na$
(4197-07-3)
$C_{16}H_{12}N_2O_{10}S_3$
(2657-89-8)
$C_{16}H_{12}N_2O_{13}S_4$
(78335-10-1)
$C_{16}H_{12}N_4O_2$
(3025-77-2)
$C_{16}H_{12}N_4O_{10}S_2$
(13301-33-2)
$C_{16}H_{12}N_6$
(30360-09-9)
$C_{16}H_{12}N_6S$
(30360-06-6)
$C_{16}H_{12}O$
(79075-27-7)
$C_{16}H_{12}O_2$
(84-51-5)
$C_{16}H_{12}O_3$
(117-37-3)
$C_{16}H_{12}O_4$
(6448-90-4)
$C_{16}H_{12}O_6$
(530-75-6)
$C_{16}H_{13}BrO_4$
(92387-49-0)
$C_{16}H_{13}ClFNO_3$
(58667-63-3)
$C_{16}H_{13}ClF_3NO_4$
(69806-40-2)
$C_{16}H_{13}ClN_2O$
(439-14-5)
(22232-71-9)
$C_{16}H_{13}ClN_2O_2$
(846-50-4)
(22316-47-8)
$C_{16}H_{13}ClN_4O_4S.Na$
(6359-90-6)
$C_{16}H_{13}ClO_4$
(92387-48-9)
$C_{16}H_{13}Cl_2N_5$
(30368-93-5)
$C_{16}H_{13}Cl_2N_5O_3S$
(12239-15-5)
$C_{16}H_{13}F_2N_3O$
(87676-93-5)
$C_{16}H_{13}N$
(90-30-2)
(135-88-6)
(4789-76-8)
(6907-59-1)
(27356-46-3)
(103837-23-6)
$C_{16}H_{13}NO$
(93-45-8)
$C_{16}H_{13}NO_2$
(80182-27-0)
(80191-45-3)
$C_{16}H_{13}NO_3$
(4465-58-1)
(56594-27-5)
$C_{16}H_{13}NO_3S$
(82-76-8)
$C_{16}H_{13}NO_3S.H_3N$
(28836-03-5)
$C_{16}H_{13}NO_3S.Na$
(1445-19-8)
$C_{16}H_{13}NO_4S$
(119-40-4)
$C_{16}H_{13}N_3$
(85-84-7)
(131-22-6)
(3599-62-0)
$C_{16}H_{13}N_3O$
(79746-01-3)
$C_{16}H_{13}N_3O_2$
(738-71-6)
$C_{16}H_{13}N_3O_3$
(2011-67-8)
(31431-39-7)
$C_{16}H_{13}N_3O_4S$
(16432-45-4)
$C_{16}H_{13}N_3O_7S_2.2Na$
(3567-66-6)
$C_{16}H_{13}N_3S$
(735-69-3)
$C_{16}H_{14}$
(483-87-4)
(613-26-3)
(781-43-1)
(886-65-7)
(1576-67-6)
(1576-69-8)
(2497-54-3)
(3674-65-5)
(3674-74-6)
(3674-75-7)
(7372-87-4)
(7469-40-1)
(16664-45-2)
(17980-09-5)
(17980-16-4)
(20291-74-1)
(22349-59-3)
(29062-98-4)
(29063-00-1)
(30997-38-7)
(52251-71-5)
(66291-32-5)
(66291-33-6)
$C_{16}H_{14}ClN_3O$
(58-25-3)
$C_{16}H_{14}ClN_3O_7$
(2542-29-2)
$C_{16}H_{14}Cl_2N_2O_2$
(3134-12-1)
$C_{16}H_{14}Cl_2N_4O_4S$
(17741-62-7)
$C_{16}H_{14}Cl_2O_3$
(510-15-6)
$C_{16}H_{14}Cl_2O_4$
(51338-27-3)
$C_{16}H_{14}Cl_3O_5P$
(74548-80-4)
$C_{16}H_{14}CoN_2O_2$
(14167-18-1)
$C_{16}H_{14}F_{17}NO_4S$
(376-14-7)
$C_{16}H_{14}N.Cl$
(15619-48-4)
(35674-56-7)
$C_{16}H_{14}N_2O$
(72-44-6)
$C_{16}H_{14}N_2O_2$
(2475-44-7)
$C_{16}H_{14}N_2O_4$
(2449-05-0)
(3860-63-7)
$C_{16}H_{14}N_2O_{16}S_4.4Na$
(6370-93-0)
$C_{16}H_{14}N_4$
(2039-37-4)
(6054-48-4)
(6416-57-5)
$C_{16}H_{14}N_4O_2$
(6629-10-3)
$C_{16}H_{14}OS$
(5495-84-1)
$C_{16}H_{14}O_2$
(103-41-3)
$C_{16}H_{14}O_3$
(22071-15-4)
$C_{16}H_{14}O_4$
(482-44-0)
(792-74-5)
(33533-53-8)
$C_{16}H_{14}O_6$
(517-28-2)
(2144-45-8)
$C_{16}H_{14}O_8$
(2134-90-9)
(2555-99-9)
$C_{16}H_{15}BrCl_2N_4O_4$
(17464-91-4)
$C_{16}H_{15}ClN_2$
(2898-12-6)
$C_{16}H_{15}ClN_2OS$
(33671-46-4)
$C_{16}H_{15}ClO_4$
(31934-88-0)
$C_{16}H_{15}Cl_2NO_2$
(117-26-0)
$C_{16}H_{15}Cl_2NO_2.C_{15}H_{13}Cl_2NO_2$
(8027-00-7)
$C_{16}H_{15}Cl_3$
(4413-31-4)
$C_{16}H_{15}Cl_3N_4O_4$
(23355-64-8)
$C_{16}H_{15}Cl_3O_2$
(72-43-5)
(30667-99-3)
$C_{16}H_{15}Cl_3OS$
(34197-16-5)
$C_{16}H_{15}Cl_3S_2$
(19679-38-0)
$C_{16}H_{15}FO_2$
(4301-50-2)
$C_{16}H_{15}NO_2$
(87-29-6)
$C_{16}H_{15}NO_3$
(21528-31-4)
$C_{16}H_{15}NO_4$
(37070-83-0)
$C_{16}H_{15}N_5$
(2039-34-1)
$C_{16}H_{15}N_5O$
(30360-58-8)
$C_{16}H_{15}N_5O_2$
(31464-38-7)
$C_{16}H_{15}N_5O_4S$
(5264-47-1)
$C_{16}H_{16}$
(3018-20-0)
(3299-99-8)
(6416-39-3)
(7614-93-9)
$C_{16}H_{16}ClN_3O_3S$
(17560-51-9)
$C_{16}H_{16}Cl_2O_2$
(6012-83-5)
$C_{16}H_{16}Cl_3NO_2$
(38766-64-2)
$C_{16}H_{16}N_2$
(2064-28-0)
$C_{16}H_{16}N_2O_2$
(613-35-4)
$C_{16}H_{16}N_2O_4$
(13684-56-5)
(13684-63-4)
$C_{16}H_{16}N_2O_4.C_{13}H_{18}N_2O_2$
(53028-35-6)
$C_{16}H_{16}N_2O_6S_2$
(153-61-7)
$C_{16}H_{16}N_4O_2$
(15446-39-6)
$C_{16}H_{16}N_4O_5S$
(21811-92-7)
$C_{16}H_{16}N_4O_8S$
(55268-75-2)
$C_{16}H_{16}O_2$
(102-20-5)
(4705-34-4)
(15547-17-8)
(41295-28-7)
(67923-88-0)
$C_{16}H_{16}O_3$
(24650-42-8)
$C_{16}H_{16}O_8$
(22268-16-2)
$C_{16}H_{17}BrN_2$
(56775-88-3)
$C_{16}H_{17}ClN_2O$
(10379-14-3)
$C_{16}H_{17}ClN_4O_3$
(3180-81-2)
$C_{16}H_{17}ClN_4O_4$
(4540-00-5)
$C_{16}H_{17}Cl_2O_3PS$
(95114-71-9)
$C_{16}H_{17}NO$
(957-51-7)
$C_{16}H_{17}NO_3$
(466-97-7)
$C_{16}H_{17}N_2.CH_3O_4S$
(43222-48-6)
$C_{16}H_{17}N_2O_4S.K$
(113-98-4)
$C_{16}H_{17}N_2O_4S.Na$
(69-57-8)
$C_{16}H_{17}N_2O_5S.K$
(132-98-9)
$C_{16}H_{17}N_3O$
(478-94-4)
$C_{16}H_{17}N_3O_4S$
(15686-71-2)
$C_{16}H_{17}N_3O_5S$
(50370-12-2)
$C_{16}H_{17}N_3O_7S_2$
(35607-66-0)
$C_{16}H_{17}N_5O_5$
(62570-20-1)
$C_{16}H_{17}N_5O_6$
(60129-67-1)
$C_{16}H_{17}N_5O_7S_2$
(60846-21-1)
$C_{16}H_{18}$
(1520-44-1)
(4613-11-0)
(4920-95-0)
(6196-95-8)
(26130-84-7)
(26137-53-1)
(38171-97-0)
(41638-55-5)
(61141-66-0)
(64800-83-5)
$C_{16}H_{18}ClN$
(51-50-3)
$C_{16}H_{18}F_{12}NO_4P$
(102040-60-8)
$C_{16}H_{18}N_2$
(6311-44-0)
(24526-64-5)
(29418-25-5)
(61653-33-6)
$C_{16}H_{18}N_2O_3$
(14214-32-5)
$C_{16}H_{18}N_2O_4S$

(61-33-6)
$C_{16}H_{18}N_2O_4S.C_{13}H_{20}N_2O_2$
(54-35-3)
$C_{16}H_{18}N_2O_4S$-$C_{13}H_{20}N_2O_2.H_2O$
(6130-64-9)
$C_{16}H_{18}N_2O_5S$
(87-08-1)
$C_{16}H_{18}N_2O_8S$
(6206-10-6)
$C_{16}H_{18}N_3S.Cl$
(61-73-4)
$C_{16}H_{18}N_4O_2$
(51-12-7)
(3025-52-3)
$C_{16}H_{18}N_4O_3$
(2872-52-8)
$C_{16}H_{18}N_4O_4$
(2734-52-3)
$C_{16}H_{18}N_4O_7S$
(83055-99-6)
$C_{16}H_{18}O$
(93-96-9)
$C_{16}H_{18}O_2$
(68279-54-9)
$C_{16}H_{18}O_4$
(17088-28-7)
$C_{16}H_{18}O_6$
(93-94-7)
$C_{16}H_{18}O_9$
(327-97-9)
$C_{16}H_{19}AsS$
(24582-61-4)
$C_{16}H_{19}BrN_2$
(86-22-6)
$C_{16}H_{19}ClN_2$
(132-22-9)
(25523-97-1)
$C_{16}H_{19}ClN_2.C_4H_4O_4$
(113-92-8)
$C_{16}H_{19}N$
(10024-74-5)
$C_{16}H_{19}N.ClH$
(14148-99-3)
$C_{16}H_{19}NO_4$
(519-09-5)
(18717-72-1)
$C_{16}H_{19}N_3$
(2481-94-9)
$C_{16}H_{19}N_3O_2$
(2452-84-8)
$C_{16}H_{19}N_3O_2S$
(900-38-9)
$C_{16}H_{19}N_3O_4S$
(69-53-4)
(38821-53-3)
$C_{16}H_{19}N_3O_4S.3H_2O$
(7177-48-2)
$C_{16}H_{19}N_3O_5S$
(26787-78-0)
(51762-05-1)
$C_{16}H_{19}N_3O_6$
(4055-39-4)
(4055-40-7)

$C_{16}H_{19}N_9$
(30355-52-3)
$C_{16}H_{19}O_4P$
(2752-95-6)
(21993-11-3)
$C_{16}H_{19}O_3PS$
(95114-67-3)
(95114-68-4)
(95114-69-5)
$C_{16}H_{20}$
(24157-81-1)
(38640-62-9)
$C_{16}H_{20}N_2$
(86-21-5)
(366-29-0)
(788-17-0)
(10029-31-9)
(54827-17-7)
$C_{16}H_{20}N_2.2C_2H_4O_2$
(122-75-8)
$C_{16}H_{20}N_2.C_4H_4O_4$
(132-20-7)
$C_{16}H_{20}N_4O_2$
(20721-50-0)
$C_{16}H_{20}O_2$
(31991-61-4)
$C_{16}H_{20}O_3S$
(28757-00-8)
$C_{16}H_{20}O_3S.C_6H_{13}N$
(68425-61-6)
$C_{16}H_{20}O_3S.Na$
(1322-93-6)
$C_{16}H_{20}O_5$
(35818-31-6)
$C_{16}H_{20}O_6$
(30270-60-1)
$C_{16}H_{20}O_6P_2S_3$
(3383-96-8)
$C_{16}H_{21}Cl_2NO_2$
(3478-94-2)
$C_{16}H_{21}Cl_3O_3$
(1928-47-8)
(25168-15-4)
(69462-12-0)
$C_{16}H_{21}Cl_3O_4$
(25537-26-2)
(28903-26-6)
(53466-84-5)
$C_{16}H_{21}NO$
(1531-12-0)
$C_{16}H_{21}NO_2$
(525-66-6)
$C_{16}H_{21}NO_3$
(87-00-3)
$C_{16}H_{21}NO_3.BrH$
(51-56-9)
$C_{16}H_{21}NS_2$
(86-14-6)
$C_{16}H_{21}N_3$
(91-81-6)
(637-31-0)
$C_{16}H_{21}N_3.ClH$
(154-69-8)
$C_{16}H_{21}N_3O_3$

(27550-64-7)
$C_{16}H_{22}ClNO_3$
(38727-55-8)
$C_{16}H_{22}Cl_2O_3$
(1928-43-4)
(1928-44-5)
(25168-26-7)
$C_{16}H_{22}Cl_2O_4$
(32357-46-3)
$C_{16}H_{22}N_2O_4S$
(55285-05-7)
$C_{16}H_{22}N_2O_5$
(27059-08-1)
$C_{16}H_{22}N_2O_6$
(800-24-8)
$C_{16}H_{22}N_4O$
(91-85-0)
$C_{16}H_{22}N_4O.ClH$
(63-56-9)
$C_{16}H_{22}OSi_2$
(56-33-7)
$C_{16}H_{22}O_3$
(118-56-9)
(68586-19-6)
$C_{16}H_{22}O_4$
(84-69-5)
(84-74-2)
(1962-75-0)
(4376-20-9)
(17851-53-5)
(30849-48-0)
$C_{16}H_{22}O_6$
(2155-71-7)
(41826-92-0)
$C_{16}H_{23}ClN_2O$
(107534-96-3)
$C_{16}H_{23}ClO_3S$
(3021-31-6)
$C_{16}H_{23}N$
(2201-39-0)
$C_{16}H_{23}NO_2$
(77-15-6)
(77-20-3)
(468-59-7)
(561-76-2)
$C_{16}H_{23}NO_6$
(315-22-0)
$C_{16}H_{23}N_3OS$
(69327-76-0)
$C_{16}H_{23}N_3O_2S$
(69329-95-9)
$C_{16}H_{24}Cl_3N_5$
(24848-61-1)
(24848-62-2)
$C_{16}H_{24}N_2O$
(1491-59-4)
$C_{16}H_{24}N_2O_2$
(7416-34-4)
$C_{16}H_{24}Ni$
(1295-35-8)
$C_{16}H_{24}O$
(79-78-7)
(6048-82-4)
$C_{16}H_{24}O_2$

(1942-71-8)
(18017-73-7)
(81325-80-6)
$C_{16}H_{24}O_3$
(15234-85-2)
$C_{16}H_{24}O_4$
(141-37-7)
(20350-15-6)
$C_{16}H_{24}O_6$
(30145-51-8)
$C_{16}H_{25}HgNO_6S.2H$
(20223-84-1)
$C_{16}H_{25}NOS$
(36756-79-3)
$C_{16}H_{25}NO_2$
(2655-19-8)
$C_{16}H_{25}N_3O$
(15686-91-6)
$C_{16}H_{25}O_3S.Na$
(1322-98-1)
$C_{16}H_{26}$
(104-72-3)
(605-01-6)
(4537-11-5)
(4537-12-6)
(4537-13-7)
(4621-36-7)
(68648-86-2)
$C_{16}H_{26}N_2O_4$
(68444-05-3)
$C_{16}H_{26}O$
(120-95-6)
(138-00-1)
(4130-42-1)
(21078-95-5)
(27157-66-0)
(28652-04-2)
(28983-26-8)
(62607-69-6)
$C_{16}H_{26}O_2$
(79-74-3)
(1322-97-0)
(2349-85-1)
$C_{16}H_{26}O_3$
(19780-11-1)
(25377-73-5)
(26544-38-7)
$C_{16}H_{26}O_3S$
(140-60-3)
(67716-07-8)
(68584-22-5)
$C_{16}H_{26}O_3S.Na$
(2627-06-7)
(73602-65-0)
(73602-67-2)
$C_{16}H_{26}O_7$
(109-17-1)
$C_{16}H_{26}O_7S.Na$
(23386-52-9)
$C_{16}H_{27}N$
(37529-30-9)
$C_{16}H_{27}NO_5$
(303-33-3)
$C_{16}H_{27}N_3O_8$

(6420-47-9)
$C_{16}H_{28}Cl_3N_5$
(26235-00-7)
$C_{16}H_{28}NS.Cl$
(70700-62-8)
$C_{16}H_{28}O$
(19377-95-8)
$C_{16}H_{28}O_4$
(105-52-2)
(27859-58-1)
(29658-97-7)
$C_{16}H_{28}O_6Ti$
(17927-72-9)
$C_{16}H_{30}$
(629-74-3)
(6165-44-2)
(54965-61-6)
$C_{16}H_{30}Cl_3O_5P$
(4414-15-7)
$C_{16}H_{30}O$
(541-91-3)
(53939-28-9)
$C_{16}H_{30}O_2$
(142-90-5)
(373-49-9)
(2091-29-4)
(3076-04-8)
(10030-73-6)
(16725-53-4)
$C_{16}H_{30}O_3$
(63978-73-4)
$C_{16}H_{30}O_4$
(505-54-4)
(762-16-3)
(1068-27-5)
(6846-50-0)
(14027-78-2)
(56438-07-4)
$C_{16}H_{30}O_4.Cd$
(2191-10-8)
$C_{16}H_{30}O_7S.Na$
(2373-38-8)
(3006-15-3)
(6001-97-4)
$C_{16}H_{30}O_7S.2Na$
(26838-05-1)
(36409-57-1)
$C_{16}H_{31}BrO_2$
(18263-25-7)
$C_{16}H_{31}ClO$
(112-67-4)
$C_{16}H_{31}ClO_2$
(51479-36-8)
$C_{16}H_{31}N$
(629-79-8)
$C_{16}H_{31}N_5$
(26234-37-7)
(26235-39-2)
(26235-87-0)
(26235-88-1)
$C_{16}H_{32}$
(629-73-2)
(1795-16-0)
(15220-85-6)

(41977-40-6)
$C_{16}H_{32}Cl_2S_2$
(70776-26-0)
$C_{16}H_{32}N_2O$
(136-99-2)
$C_{16}H_{32}O$
(629-80-1)
(7320-37-8)
(10378-01-5)
(56683-54-6)
$C_{16}H_{32}O_2$
(57-10-3)
(124-06-1)
(638-59-5)
(2306-88-9)
(5129-66-8)
(7132-64-1)
(7425-14-1)
(32844-67-0)
(55554-09-1)
$C_{16}H_{32}O_2.1/3Al$
(555-35-1)
$C_{16}H_{32}O_2.1/2Ca$
(542-42-7)
$C_{16}H_{32}O_2.H_3N$
(593-26-0)
$C_{16}H_{32}O_2.Na$
(408-35-5)
$C_{16}H_{32}O_2.1/2Zn$
(4991-47-3)
$C_{16}H_{32}O_2S$
(29946-28-9)
$C_{16}H_{32}O_2Sn$
(2155-70-6)
$C_{16}H_{32}O_3$
(106-13-8)
(506-13-8)
$C_{16}H_{32}O_4$
(141-20-8)
$C_{16}H_{32}O_4Si_4$
(177-49-1)
$C_{16}H_{33}Br$
(112-82-3)
$C_{16}H_{33}Cl$
(4860-03-1)
(34214-79-4)
$C_{16}H_{33}ClO_2S$
(38775-38-1)
$C_{16}H_{33}Cl_3Si$
(5894-60-0)
$C_{16}H_{33}F_9O_5Si_5$
(22474-57-3)
$C_{16}H_{33}I$
(544-77-4)
$C_{16}H_{33}NO$
(1541-81-7)
(3352-87-2)
$C_{16}H_{33}NO_2$
(142-58-5)
(683-10-3)
$C_{16}H_{33}NO_3$
(120-40-1)
$C_{16}H_{33}NO_5S$
(59997-83-0)
$C_{16}H_{33}NO_6S$
(59997-79-4)
(65520-65-2)
$C_{16}H_{33}N_3$
(10595-60-5)
$C_{16}H_{33}O_4S.Na$
(1120-01-0)
$C_{16}H_{34}$
(544-76-3)
(1560-93-6)
(2801-87-8)
(2882-96-4)
(3891-99-4)
(4390-04-9)
(6165-40-8)
(10105-38-1)
(22306-28-1)
(25117-33-3)
(55045-11-9)
(55045-14-2)
$C_{16}H_{34}Cl_2Sn$
(3542-36-7)
$C_{16}H_{34}O$
(629-82-3)
(10143-60-9)
(29354-98-1)
(36311-34-9)
(36653-82-4)
$C_{16}H_{34}O.1/3Al$
(19141-82-3)
$C_{16}H_{34}OSn$
(870-08-6)
$C_{16}H_{34}O_2$
(2136-70-1)
(19102-74-0)
$C_{16}H_{34}O_3$
(3055-93-4)
$C_{16}H_{34}O_4$
(78-63-7)
$C_{16}H_{34}O_4S$
(143-02-2)
$C_{16}H_{34}O_4S.H_3N$
(4696-47-3)
$C_{16}H_{34}O_5$
(112-98-1)
$C_{16}H_{34}O_5S.Na$
(3694-74-4)
$C_{16}H_{34}O_5Si_2$
(126-80-7)
$C_{16}H_{34}O_6S.Na$
(3088-31-1)
$C_{16}H_{34}S$
(2917-26-2)
$C_{16}H_{35}N$
(106-20-7)
(112-75-4)
(143-27-1)
(1120-48-5)
$C_{16}H_{35}N.C_2H_4O_2$
(2016-52-6)
$C_{16}H_{35}NO$
(3332-27-2)
$C_{16}H_{35}NO_2$
(1541-67-9)
$C_{16}H_{35}O_2PS_2$
(2253-57-8)
(5810-88-8)
(26999-29-1)
$C_{16}H_{35}O_2PS_2.1/3Sb$
(15874-52-9)
$C_{16}H_{35}O_2PS_2.1/2Zn$
(28629-66-5)
$C_{16}H_{35}O_3P$
(3658-48-8)
$C_{16}H_{35}O_4P$
(298-07-7)
(3115-39-7)
(3539-43-3)
(27215-10-7)
$C_{16}H_{35}O_4P.Na$
(141-65-1)
$C_{16}H_{35}O_9P_3$
(53529-45-6)
$C_{16}H_{36}N.Br$
(1643-19-2)
$C_{16}H_{36}N.C_4H_6NO_2$
(33684-09-2)
$C_{16}H_{36}N.C_6H_4NO_3$
(3002-48-0)
$C_{16}H_{36}N.C_6H_4NO_4$
(11072-43-8)
$C_{16}H_{36}N.C_7H_6NO_2$
(33684-10-5)
$C_{16}H_{36}N.C_{10}H_{11}O_2$
(33684-11-6)
$C_{16}H_{36}N.HO_4S$
(32503-27-8)
$C_{16}H_{36}N_2S_2$
(65332-44-7)
$C_{16}H_{36}O_4.Ti$
(5593-70-4)
$C_{16}H_{36}O_4P_2S_4Zn$
(6990-43-8)
$C_{16}H_{36}O_4Si$
(5089-76-9)
$C_{16}H_{36}O_7P_2$
(26658-09-3)
(26836-28-2)
$C_{16}H_{36}P.Br$
(3115-68-2)
$C_{16}H_{36}P.Cl$
(2304-30-5)
$C_{16}H_{36}Pb$
(1920-90-7)
$C_{16}H_{36}Sn$
(1461-25-2)
$C_{16}H_{38}N_2$
(156-74-1)
$C_{16}H_{48}O_6Si_7$
(541-01-5)
$C_{16}H_{48}O_8Si_8$
(556-68-3)
$C_{17}H_6O_7$
(2421-28-5)
$C_{17}H_9N$
(83536-57-6)
$C_{17}H_{10}Cl_3N_5S_4$
(30863-09-3)
$C_{17}H_{10}O$
(82-05-3)
(479-79-8)
(3074-03-1)
(6051-98-5)
(68967-09-9)
(86853-88-5)
(100647-29-8)
(116232-62-3)
$C_{17}H_{10}O_2$
(93-99-2)
$C_{17}H_{11}ClF_3N_3O_4S.Na$
(57741-47-6)
$C_{17}H_{11}ClF_4N_2S$
(36735-22-5)
$C_{17}H_{11}ClN_2O_6S.1/2Ca.Na$
(5850-80-6)
$C_{17}H_{11}F_3N_2O_2$
(36783-34-3)
$C_{17}H_{11}N$
(84-56-0)
(218-02-0)
(218-08-6)
(225-11-6)
(225-51-4)
$C_{17}H_{11}N_3O_4S$
(78335-09-8)
$C_{17}H_{12}$
(205-12-9)
(238-84-6)
(243-17-4)
(1706-01-0)
(2381-21-7)
(3353-12-6)
(3442-78-2)
(19561-31-0)
(27577-90-8)
(30777-18-5)
(30777-19-6)
(30997-39-8)
(33543-31-6)
$C_{17}H_{12}ClF_3N_2O$
(23092-17-3)
$C_{17}H_{12}ClN_2O_4S.1/2Ba$
(5160-02-1)
$C_{17}H_{12}ClN_5S_4$
(30863-08-2)
$C_{17}H_{12}Cl_2N_2O$
(60168-88-9)
$C_{17}H_{12}Cl_2N_4$
(28911-01-5)
$C_{17}H_{12}Cl_3N_3$
(3599-79-9)
$C_{17}H_{12}Cl_3N_3O$
(3599-82-4)
$C_{17}H_{12}Cl_{10}O_4$
(4234-79-1)
$C_{17}H_{12}I_2O_3$
(68-90-6)
$C_{17}H_{12}N_2$
(18936-75-9)
$C_{17}H_{12}N_2O_3$
(21184-58-7)
(29128-56-1)
(32624-40-1)
(32624-41-2)
(69644-64-0)
$C_{17}H_{12}N_2O_4$
(135-65-9)
$C_{17}H_{12}N_2O_6S$
(69644-65-1)
(69644-66-2)
$C_{17}H_{12}N_2O_6S.2Na$
(10114-96-2)
(10114-97-3)
$C_{17}H_{12}N_2O_9S_2.2Ba$
(1325-16-2)
$C_{17}H_{12}N_2O_9S_2.3/2Ba$
(15782-06-6)
$C_{17}H_{12}O_2S$
(14814-89-2)
$C_{17}H_{12}O_6$
(1162-65-8)
$C_{17}H_{12}O_7$
(1165-39-5)
(6795-23-9)
$C_{17}H_{13}ClN_2O_4S.Na$
(5850-90-8)
$C_{17}H_{13}ClN_4$
(28981-97-7)
$C_{17}H_{13}ClN_6O_5$
(12236-62-3)
$C_{17}H_{13}N$
(56842-43-4)
$C_{17}H_{13}NO_2$
(92-77-3)
$C_{17}H_{13}NO_5S$
(132-87-6)
$C_{17}H_{13}N_3O_3$
(2425-85-6)
$C_{17}H_{13}O_5.Na$
(34490-93-2)
$C_{17}H_{14}$
(613-59-2)
(38620-92-7)
(97232-29-6)
$C_{17}H_{14}BrFN_2O_2$
(59128-97-1)
$C_{17}H_{14}BrN_3$
(30362-79-9)
$C_{17}H_{14}ClF_7O_2$
(79538-32-2)
$C_{17}H_{14}ClNO_2$
(6260-97-5)
$C_{17}H_{14}ClN_3$
(21902-34-1)
(30362-73-3)
$C_{17}H_{14}ClN_3O_2$
(30894-79-2)
$C_{17}H_{14}Cl_2N_2O_2$
(24166-13-0)
$C_{17}H_{14}Cl_2N_4O_4$
(32432-45-4)
$C_{17}H_{14}N_2O$
(2646-17-5)
$C_{17}H_{14}N_2O_2$
(1229-55-6)
$C_{17}H_{14}N_2O_3$

(20007-87-8)
$C_{17}H_{14}N_2O_4S$
(5859-07-4)
$C_{17}H_{14}N_2O_4S.Na$
(5850-86-2)
$C_{17}H_{14}N_2O_5S$
(27959-50-8)
$C_{17}H_{14}N_2O_5S.Na$
(5858-39-9)
$C_{17}H_{14}N_4OS_4$
(95-35-2)
(64216-20-2)
$C_{17}H_{14}O_5$
(117-52-2)
$C_{17}H_{14}O_6$
(7220-81-7)
$C_{17}H_{14}O_7$
(6885-57-0)
(7241-98-7)
(64330-03-6)
$C_{17}H_{15}ClFNO_3$
(52756-25-9)
$C_{17}H_{15}ClN_2O$
(37751-39-6)
$C_{17}H_{15}ClN_4O_5$
(13515-40-7)
$C_{17}H_{15}Cl_2N_5O_2$
(13301-61-6)
$C_{17}H_{15}NO_3$
(56594-28-6)
$C_{17}H_{15}N_3$
(131-79-3)
(30361-92-3)
(30362-71-1)
$C_{17}H_{15}N_3O$
(30886-10-3)
$C_{17}H_{15}N_3O_2$
(740-77-2)
$C_{17}H_{15}N_3O_6S_2$
(2494-93-1)
$C_{17}H_{15}N_3S$
(30886-17-0)
$C_{17}H_{15}N_5O_3$
(51083-28-4)
$C_{17}H_{16}$
(3674-73-5)
(23189-64-2)
(27358-28-7)
(30232-26-9)
(63018-94-0)
$C_{17}H_{16}Br_2O_3$
(18181-80-1)
$C_{17}H_{16}Br_4O_2$
(37853-61-5)
$C_{17}H_{16}ClN_3O_2$
(61443-78-5)
$C_{17}H_{16}ClN_5$
(3995-43-5)
(30355-04-5)
$C_{17}H_{16}ClN_5O_2$
(6737-62-8)
$C_{17}H_{16}ClN_5O_3$
(6657-33-6)
$C_{17}H_{16}Cl_2N_6$
(30360-22-6)
$C_{17}H_{16}Cl_2O_3$
(5836-10-2)
$C_{17}H_{16}F_3NO_2$
(66332-96-5)
$C_{17}H_{16}F_{17}NO_4S$
(383-07-3)
$C_{17}H_{16}N_2$
(5062-67-9)
$C_{17}H_{16}N_2.ClH$
(1497-49-0)
$C_{17}H_{16}N_2O_3$
(2475-46-9)
(86722-66-9)
$C_{17}H_{16}N_2S$
(1628-58-6)
$C_{17}H_{16}N_4$
(5599-21-3)
(18808-10-1)
$C_{17}H_{16}N_4O_2$
(4150-79-2)
(30384-48-6)
$C_{17}H_{16}N_4O_4$
(2512-29-0)
$C_{17}H_{16}N_4O_7S.1/2Ca$
(12286-66-7)
$C_{17}H_{16}O_4$
(19224-26-1)
(23005-56-3)
$C_{17}H_{16}O_5$
(33533-57-2)
$C_{17}H_{17}ClO_6$
(126-07-8)
$C_{17}H_{17}Cl_2N_5O_4$
(71617-28-2)
$C_{17}H_{17}Cl_3O$
(34197-05-2)
$C_{17}H_{17}NO_2$
(58-00-4)
$C_{17}H_{17}NO_2.ClH$
(314-19-2)
$C_{17}H_{17}N_2$
(49866-87-7)
$C_{17}H_{17}N_3O_3$
(81335-37-7)
$C_{17}H_{17}N_3O_6S_2$
(21593-23-7)
$C_{17}H_{17}N_5O_2$
(31482-56-1)
$C_{17}H_{18}$
(17114-78-2)
$C_{17}H_{18}ClN_3$
(36945-03-6)
$C_{17}H_{18}ClN_3O$
(3896-11-5)
$C_{17}H_{18}ClN_5O_6$
(65125-87-3)
$C_{17}H_{18}Cl_2N_4O_4$
(58528-60-2)
$C_{17}H_{18}Cl_2O_5$
(75217-44-6)
$C_{17}H_{18}N_2O_2$
(94-91-7)
$C_{17}H_{18}N_2O_6$
(21829-25-4)
$C_{17}H_{18}N_2O_6S$
(4697-36-3)
$C_{17}H_{18}N_6$
(13486-13-0)
(30377-20-9)
$C_{17}H_{18}O$
(5396-91-8)
$C_{17}H_{18}O_3$
(87-18-3)
$C_{17}H_{18}O_4$
(126-00-1)
$C_{17}H_{18}O_6P_2$
(144-35-4)
$C_{17}H_{19}ClN_2S$
(50-53-3)
$C_{17}H_{19}ClN_2S.ClH$
(69-09-0)
$C_{17}H_{19}ClN_4O_4$
(3769-57-1)
$C_{17}H_{19}ClO_2$
(59900-47-9)
$C_{17}H_{19}NO$
(13669-70-0)
$C_{17}H_{19}NO_2$
(55814-41-0)
$C_{17}H_{19}NO_3$
(57-27-2)
(94-62-2)
(466-99-9)
(467-15-2)
$C_{17}H_{19}NO_3.C_2H_4O_2$
(596-15-6)
$C_{17}H_{19}NO_3.1/2C_4H_6O_6$
(302-31-8)
$C_{17}H_{19}NO_4$
(76-41-5)
(639-46-3)
(34197-26-7)
(72490-01-8)
$C_{17}H_{19}NO_7$
(81826-15-5)
$C_{17}H_{19}N_2O_5S.K$
(132-93-4)
$C_{17}H_{19}N_2S.Cl$
(2390-54-7)
$C_{17}H_{19}N_3$
(494-38-2)
$C_{17}H_{19}N_3.ClH$
(65-61-2)
$C_{17}H_{19}N_3O$
(50-60-2)
$C_{17}H_{19}N_5O$
(2312-73-4)
$C_{17}H_{19}N_5O_6$
(41541-13-3)
$C_{17}H_{20}ClO_3PS_2$
(59010-86-5)
$C_{17}H_{20}N_2O$
(85-98-3)
(90-94-8)
(29103-58-0)
(42609-52-9)
$C_{17}H_{20}N_2O_3$
(120-00-3)
$C_{17}H_{20}N_2O_6S$
(61-32-5)
$C_{17}H_{20}N_2S$
(58-40-2)
(60-87-7)
$C_{17}H_{20}N_2S.ClH$
(53-60-1)
(58-33-3)
$C_{17}H_{20}N_4O_4$
(3179-89-3)
$C_{17}H_{20}N_4O_5$
(41541-14-4)
$C_{17}H_{20}N_4O_6$
(83-88-5)
$C_{17}H_{20}O$
(15087-24-8)
$C_{17}H_{20}O_5$
(75217-43-5)
$C_{17}H_{20}O_6$
(7382-59-4)
$C_{17}H_{20}O_8$
(4986-89-4)
$C_{17}H_{21}N$
(156-08-1)
$C_{17}H_{21}N.ClH$
(5411-22-3)
$C_{17}H_{21}NO$
(58-73-1)
(15299-99-7)
$C_{17}H_{21}NO.C_7H_7ClN_4O_2$
(523-87-5)
$C_{17}H_{21}NO.ClH$
(147-24-0)
$C_{17}H_{21}NO_2$
(427-00-9)
$C_{17}H_{21}NO_3$
(509-60-4)
$C_{17}H_{21}NO_4$
(50-36-2)
(51-34-3)
(2183-56-4)
$C_{17}H_{21}NO_4.BrH$
(114-49-8)
$C_{17}H_{21}NO_4.CH_3NO_3$
(6106-46-3)
$C_{17}H_{21}NO_4.ClH$
(53-21-4)
$C_{17}H_{21}NO_5.BrH$
(6106-81-6)
$C_{17}H_{21}N_3$
(492-80-8)
$C_{17}H_{21}N_3.ClH.H_2O$
(2465-27-2)
$C_{17}H_{21}N_3O_2$
(3771-38-8)
$C_{17}H_{22}$
(38622-51-4)
$C_{17}H_{22}NO$
(7181-73-9)
$C_{17}H_{22}N_2$
(101-61-1)
(961-71-7)
$C_{17}H_{22}N_2.ClH$
(2045-52-5)
$C_{17}H_{22}N_2O$
(119-58-4)
(469-21-6)
$C_{17}H_{22}N_2O.C_4H_6O_4$
(562-10-7)
$C_{17}H_{22}O$
(31215-04-0)
$C_{17}H_{22}O_2$
(505-75-9)
(52275-04-4)
(57345-30-9)
$C_{17}H_{22}O_8$
(23255-69-8)
$C_{17}H_{23}Cl_3O_3$
(32534-95-5)
(53404-10-7)
(53404-14-1)
(53404-76-5)
$C_{17}H_{23}Cl_3O_5$
(1928-58-1)
$C_{17}H_{23}FO_3$
(58327-09-6)
$C_{17}H_{23}NO$
(77-07-6)
(297-90-5)
$C_{17}H_{23}NO_2$
(7149-26-0)
$C_{17}H_{23}NO_3$
(51-55-8)
(101-31-5)
(13269-35-7)
$C_{17}H_{23}N_3O$
(91-84-9)
(2531-04-6)
$C_{17}H_{23}N_3O.C_4H_4O_4$
(59-33-6)
$C_{17}H_{23}N_7O_5$
(2005-98-3)
$C_{17}H_{24}NO_5.Na$
(55635-13-7)
$C_{17}H_{24}N_2O_6$
(23128-51-0)
$C_{17}H_{24}O$
(13171-00-1)
(15340-76-8)
$C_{17}H_{24}O_3$
(456-59-7)
(1334-99-2)
$C_{17}H_{25}ClN_2O_4$
(75096-86-5)
$C_{17}H_{25}ClO_3$
(26544-20-7)
$C_{17}H_{25}Cl_6N_3O$
(30863-59-3)
$C_{17}H_{25}Cl_6N_3S$
(24481-72-9)
$C_{17}H_{25}N$
(77-10-1)
$C_{17}H_{25}N.ClH$
(956-90-1)
$C_{17}H_{25}NO_2$
(64-39-1)
(113-48-4)

$C_{17}H_{26}ClNO_2$
(23184-66-9)
$C_{17}H_{26}Cl_6N_4$
(24803-04-1)
$C_{17}H_{26}N_2O$
(129-83-9)
$C_{17}H_{26}N_8O_5$
(2079-00-7)
$C_{17}H_{26}O$
(1217-08-9)
(32388-55-9)
$C_{17}H_{26}O_2$
(81325-79-3)
$C_{17}H_{26}O_3$
(3115-49-9)
$C_{17}H_{26}O_4Si_4$
(13093-12-4)
$C_{17}H_{27}NO_2$
(1918-11-2)
(21245-02-3)
$C_{17}H_{27}NO_4$
(42200-33-9)
$C_{17}H_{27}O_2$
(117-98-6)
$C_{17}H_{28}$
(4536-86-1)
(4536-87-2)
(4536-88-3)
(4537-14-8)
(4537-15-9)
(6742-54-7)
$C_{17}H_{28}N_2O$
(36637-18-0)
$C_{17}H_{28}N_2O_2$
(32687-77-7)
$C_{17}H_{28}O$
(28679-13-2)
$C_{17}H_{28}O_2$
(77-54-3)
(104-35-8)
(27986-36-3)
$C_{17}H_{28}O_3S$
(39156-49-5)
(50854-94-9)
$C_{17}H_{28}O_3S.Na$
(27636-75-5)
$C_{17}H_{28}O_4$
(1471-18-7)
$C_{17}H_{28}O_6$
(51247-87-1)
$C_{17}H_{28}O_8S_4$
(7575-23-7)
$C_{17}H_{29}Cl_3N_4$
(24803-56-3)
$C_{17}H_{29}Cl_3N_4S$
(30369-47-2)
$C_{17}H_{29}NO_3S$
(74051-80-2)
$C_{17}H_{29}O_4P.1/2Ni$
(30947-30-9)
$C_{17}H_{30}N.Cl$
(104-74-5)
(959-55-7)
$C_{17}H_{30}N.I$
(3026-66-2)
$C_{17}H_{30}N_2O_3Si.ClH$
(33401-49-9)
$C_{17}H_{30}N_6S_4$
(30863-11-7)
$C_{17}H_{30}O$
(22117-09-5)
$C_{17}H_{30}O_2$
(41096-46-2)
$C_{17}H_{32}NO_2.Br$
(80-50-2)
$C_{17}H_{32}O_2$
(1120-25-8)
(2495-25-2)
$C_{17}H_{32}O_4$
(105-80-6)
(2917-73-9)
$C_{17}H_{32}O_6$
(87257-05-4)
$C_{17}H_{33}N$
(5399-02-0)
$C_{17}H_{33}NO_3$
(14379-28-3)
$C_{17}H_{33}N_2S.Cl$
(68279-02-7)
$C_{17}H_{34}$
(6765-39-5)
(26266-05-7)
(41977-41-7)
(54105-66-7)
$C_{17}H_{34}N_2O$
(16058-17-6)
$C_{17}H_{34}N_2O.ClH$
(71242-00-7)
$C_{17}H_{34}O$
(2922-51-2)
(22092-38-2)
$C_{17}H_{34}O_2$
(110-27-0)
(112-39-0)
(506-12-7)
(5129-60-2)
(5918-29-6)
(6221-95-0)
(17670-75-6)
(38954-75-5)
(41114-00-5)
$C_{17}H_{34}O_4$
(6731-36-8)
$C_{17}H_{34}O_5S$
(58849-75-5)
$C_{17}H_{34}O_5S.Na$
(4016-24-4)
$C_{17}H_{34}O_6$
(995-33-5)
$C_{17}H_{35}Br$
(3508-00-7)
$C_{17}H_{35}Cl$
(62016-75-5)
$C_{17}H_{35}NO_5S$
(65520-64-1)
$C_{17}H_{35}O_4S.Na$
(3282-85-7)
$C_{17}H_{36}$
(629-78-7)
(1560-92-5)
$C_{17}H_{36}N_2O$
(3179-80-4)
$C_{17}H_{36}N_2O.ClH$
(71732-95-1)
$C_{17}H_{36}O$
(1454-85-9)
(2490-48-4)
(57289-07-3)
$C_{17}H_{36}O_2$
(70709-94-3)
(70709-96-5)
$C_{17}H_{36}O_6$
(143-29-3)
$C_{17}H_{37}N$
(4455-26-9)
$C_{17}H_{37}NO_3S$
(14933-08-5)
$C_{17}H_{37}O_3P$
(1832-68-4)
$C_{17}H_{38}N.Br$
(1119-97-7)
$C_{17}H_{38}N.Cl$
(4574-04-3)
$C_{17}H_{38}O_4Si_5$
(60587-10-2)
$C_{17}H_{39}N_5O$
(26392-63-2)
$C_{17}H_{44}N_3O_{15}P_5$
(34690-00-1)
$C_{17}H_{44}N_3O_{15}P_5.xNa$
(35657-77-3)
$C_{18}F_{39}N$
(432-08-6)
$C_{18}H_4Br_8N_2O_4$
(32588-76-4)
$C_{18}H_4F_{33}I$
(65150-94-9)
$C_{18}H_9F_{25}O_2$
(6014-75-1)
$C_{18}H_9NO_8S_2.2Na$
(8004-92-0)
$C_{18}H_9N_3O_2$
(116212-02-3)
$C_{18}H_{10}$
(203-12-3)
(27208-37-3)
$C_{18}H_{10}BrNO_3$
(10319-14-9)
$C_{18}H_{10}Cl_2O_2S_2$
(2379-74-0)
(5462-29-3)
$C_{18}H_{10}Cl_5N_3$
(27125-68-4)
$C_{18}H_{10}N_2O$
(6925-69-5)
$C_{18}H_{10}N_2O_2S$
(2866-43-5)
$C_{18}H_{10}O$
(790-60-3)
(86853-91-0)
$C_{18}H_{10}O_2$
(1090-13-7)
(2498-66-0)
(18508-00-4)
$C_{18}H_{10}S$
(68558-73-6)
$C_{18}H_{11}ClMnN_2O_6S$
(12688-94-7)
$C_{18}H_{11}NO_2$
(83-08-9)
(6365-50-0)
(7496-02-8)
(20268-51-3)
(63021-85-2)
$C_{18}H_{11}NO_3$
(7576-65-0)
(116212-01-2)
$C_{18}H_{12}$
(56-55-3)
(92-24-0)
(195-19-7)
(217-59-4)
(218-01-9)
(65777-08-4)
$C_{18}H_{12}Br_3O_4P$
(40946-60-9)
$C_{18}H_{12}Cl_{12}$
(13560-89-9)
$C_{18}H_{12}CuN_2O_2$
(10380-28-6)
$C_{18}H_{12}NO_3.Na$
(132-67-2)
$C_{18}H_{12}N_2$
(119-91-5)
(51913-96-3)
$C_{18}H_{12}N_2O_5$
(69579-72-2)
$C_{18}H_{12}N_3O_{10}P$
(60337-47-5)
$C_{18}H_{12}N_4O_4$
(26747-90-0)
$C_{18}H_{12}N_5O_6$
(1898-66-4)
$C_{18}H_{12}O$
(82-03-1)
(962-32-3)
$C_{18}H_{13}AlCl_6O_7$
(69622-82-8)
$C_{18}H_{13}ClFN_3$
(59467-70-8)
$C_{18}H_{13}ClN_2O$
(52463-83-9)
$C_{18}H_{13}ClN_2O_6S.Ba$
(7585-41-3)
$C_{18}H_{13}ClN_2O_6S.Ca$
(7023-61-2)
(17852-99-2)
$C_{18}H_{13}ClN_2O_6S.Sr$
(15782-05-5)
$C_{18}H_{13}Cl_2NO$
(17781-31-6)
$C_{18}H_{13}Cl_2N_2O$
(26766-27-8)
$C_{18}H_{13}CrN_4O_{10}S_2.2Na$
(10127-05-6)
$C_{18}H_{13}N$
(2381-18-2)
(2642-98-0)
(3519-87-7)
(56961-60-5)
$C_{18}H_{13}NO_3$
(132-66-1)
$C_{18}H_{13}N_4O_7S.Na$
(6373-74-6)
$C_{18}H_{13}N_3O_8S.2Na$
(3734-67-6)
$C_{18}H_{13}N_5O_6S_2.Na$
(35294-62-3)
$C_{18}H_{14}$
(84-15-1)
(92-06-8)
(92-94-4)
(15679-24-0)
(26140-60-3)
(41593-31-1)
$C_{18}H_{14}ClFN_2O_3$
(29177-84-2)
$C_{18}H_{14}ClNO_2$
(135-63-7)
$C_{18}H_{14}ClNO_2S$
(65752-48-9)
$C_{18}H_{14}ClNO_3$
(137-52-0)
$C_{18}H_{14}Cl_3N_3$
(30362-66-4)
$C_{18}H_{14}Cl_3N_7O_2$
(30805-02-8)
$C_{18}H_{14}NO_6P$
(10359-36-1)
$C_{18}H_{14}N_2$
(7466-42-4)
$C_{18}H_{14}N_2O$
(86-72-6)
(6201-64-5)
$C_{18}H_{14}N_2O_2$
(1041-00-5)
(17233-65-7)
$C_{18}H_{14}N_2O_2.H_2O_4S$
(134-31-6)
$C_{18}H_{14}N_2O_3S_2$
(3181-86-0)
$C_{18}H_{14}N_2O_6S$
(16014-23-6)
$C_{18}H_{14}N_2O_6S.Ba$
(17852-98-1)
$C_{18}H_{14}N_2O_6S.Ca$
(5281-04-9)
$C_{18}H_{14}N_2O_6S.2Na$
(5858-81-1)
$C_{18}H_{14}N_2O_7S_2.2Na$
(3257-28-1)
(3761-53-3)
(4548-53-2)
$C_{18}H_{14}N_3O_3S.Na$
(587-98-4)
$C_{18}H_{14}N_4O$
(6250-23-3)
$C_{18}H_{14}N_4O_2$
(2581-69-3)
$C_{18}H_{14}N_4O_5S$

(599-79-1)
$C_{18}H_{14}N_4S$
(34367-95-8)
$C_{18}H_{14}O$
(28984-89-6)
$C_{18}H_{14}O.C_{12}H_{10}O$
(62587-63-7)
$C_{18}H_{15}AlO_9$
(23413-80-1)
$C_{18}H_{15}As$
(603-32-7)
$C_{18}H_{15}AsO$
(1153-05-5)
$C_{18}H_{15}AsS_2$
(24582-59-0)
$C_{18}H_{15}B$
(960-71-4)
$C_{18}H_{15}ClN_6O_2$
(4058-30-4)
$C_{18}H_{15}ClSn$
(639-58-7)
$C_{18}H_{15}Cl_3N_2O.HNO_3$
(24169-02-6)
$C_{18}H_{15}FSn$
(379-52-2)
$C_{18}H_{15}N$
(603-34-9)
$C_{18}H_{15}NO_2$
(135-61-5)
$C_{18}H_{15}NO_3$
(135-62-6)
$C_{18}H_{15}N_3O_3S$
(4005-68-9)
$C_{18}H_{15}N_3O_3S.Na$
(554-73-4)
$C_{18}H_{15}N_3O_4S$
(5124-25-4)
$C_{18}H_{15}N_5O_4S_2.Na$
(21493-04-9)
$C_{18}H_{15}N_5O_5$
(31837-42-0)
$C_{18}H_{15}OP$
(791-28-6)
$C_{18}H_{15}O_3P$
(101-02-0)
$C_{18}H_{15}O_4P$
(115-86-6)
$C_{18}H_{15}O_{21}.In$
(4194-69-8)
$C_{18}H_{15}P$
(603-35-0)
$C_{18}H_{15}Sb$
(603-36-1)
$C_{18}H_{16}BO.Na$
(12113-07-4)
$C_{18}H_{16}ClNO_4S$
(66441-11-0)
(93921-16-5)
$C_{18}H_{16}ClNO_5$
(66441-23-4)
$C_{18}H_{16}F_9O_3P$
(76943-21-0)
$C_{18}H_{16}N_2$
(74-31-7)
$C_{18}H_{16}N_2O$
(101-74-6)
(3118-97-6)
$C_{18}H_{16}N_2O_2$
(6410-20-4)
$C_{18}H_{16}N_2O_3$
(6358-53-8)
$C_{18}H_{16}N_2O_8S_2.2Na$
(25956-17-6)
$C_{18}H_{16}N_6O_2$
(16889-10-4)
$C_{18}H_{16}N_6O_2S$
(25510-81-0)
$C_{18}H_{16}OSi$
(791-31-1)
$C_{18}H_{16}OSn$
(76-87-9)
$C_{18}H_{16}O_3$
(435-97-2)
$C_{18}H_{16}O_5$
(16432-81-8)
$C_{18}H_{16}O_6$
(2316-10-1)
(10048-13-2)
$C_{18}H_{16}O_7$
(125-46-2)
$C_{18}H_{16}Sn$
(892-20-6)
$C_{18}H_{17}ClN_2O_2$
(24143-17-7)
$C_{18}H_{17}Cl_2NO_3$
(22212-55-1)
(33878-50-1)
$C_{18}H_{17}NO_3$
(56594-29-7)
(63610-08-2)
$C_{18}H_{17}NO_5$
(3224-15-5)
(53902-12-8)
$C_{18}H_{17}N_3$
(30362-95-9)
(30377-13-0)
$C_{18}H_{17}N_3O_7S_2$
(78335-11-2)
$C_{18}H_{17}N_7O_2$
(30805-01-7)
$C_{18}H_{18}$
(483-65-8)
(4466-77-7)
(7343-06-8)
(7396-38-5)
(71607-70-0)
$C_{18}H_{18}ClNS$
(113-59-7)
$C_{18}H_{18}ClN_3O$
(1977-10-2)
$C_{18}H_{18}ClN_3O_2$
(61443-77-4)
$C_{18}H_{18}ClN_5O_2$
(16586-43-9)
$C_{18}H_{18}Cl_2N_2O_5S_2.Na$
(35441-13-5)
$C_{18}H_{18}F_{24}N_3O_6P_3$
(58943-98-9)
$C_{18}H_{18}N_2O_2$
(6994-46-3)
$C_{18}H_{18}N_2O_4$
(4471-41-4)
$C_{18}H_{18}N_2O_6$
(3179-90-6)
$C_{18}H_{18}N_4O_6$
(6358-31-2)
(6528-34-3)
$C_{18}H_{18}N_6O_5S_2$
(34444-01-4)
$C_{18}H_{18}N_8$
(1052-38-6)
$C_{18}H_{18}O_2$
(84-17-3)
(122-68-9)
$C_{18}H_{18}O_3.C_{17}H_{25}ClO_3$
(53568-85-7)
$C_{18}H_{18}O_4$
(103-43-5)
$C_{18}H_{18}O_5$
(120-55-8)
$C_{18}H_{18}O_6$
(2694-54-4)
$C_{18}H_{19}Cl_3O_2$
(4329-03-7)
$C_{18}H_{19}F_3N_2S$
(146-54-3)
$C_{18}H_{19}NO$
(96334-91-7)
$C_{18}H_{19}NO.ClH$
(2887-91-4)
$C_{18}H_{19}NO_4$
(5809-23-4)
$C_{18}H_{19}N_3O_2$
(83-63-6)
$C_{18}H_{19}N_3O_6S$
(3577-01-3)
$C_{18}H_{19}N_5O_3$
(6054-58-6)
$C_{18}H_{19}NaO_5S$
(16680-47-0)
$C_{18}H_{19}O_2$
(6639-99-2)
$C_{18}H_{20}$
(3910-35-8)
(6144-04-3)
(21113-55-3)
(23586-64-3)
(33508-02-0)
(33603-39-3)
(33611-16-4)
(91741-91-2)
$C_{18}H_{20}Cl_2$
(72-56-0)
$C_{18}H_{20}Cl_2O_2$
(7388-32-1)
$C_{18}H_{20}Cl_3NO_2$
(36236-41-6)
$C_{18}H_{20}N_2S$
(1982-37-2)
$C_{18}H_{20}N_2S.ClH$
(1229-35-2)
$C_{18}H_{20}N_4O_2$
(6237-24-7)
$C_{18}H_{20}N_4O_2S$
(13486-43-6)
$C_{18}H_{20}O_2$
(56-53-1)
(474-86-2)
(22499-12-3)
(61578-04-9)
$C_{18}H_{21}ClN_2$
(82-93-9)
$C_{18}H_{21}ClN_2.ClH$
(14362-31-3)
$C_{18}H_{21}ClO$
(56265-27-1)
$C_{18}H_{21}ClO_4$
(70-43-9)
$C_{18}H_{21}NO$
(467-60-7)
(26227-73-6)
$C_{18}H_{21}NO_3$
(76-57-3)
(125-29-1)
(143-52-2)
(54010-81-0)
$C_{18}H_{21}NO_3.ClH$
(1422-07-7)
$C_{18}H_{21}NO_3.H_3O_4P$
(52-28-8)
$C_{18}H_{21}NO_3S$
(50539-85-0)
$C_{18}H_{21}NO_4$
(76-42-6)
(3688-65-1)
$C_{18}H_{22}$
(10224-91-6)
(34101-86-5)
(36876-13-8)
(56667-01-7)
(57364-79-1)
(68398-19-6)
(69009-90-1)
$C_{18}H_{22}ClNO$
(59-96-1)
$C_{18}H_{22}ClNO.ClH$
(63-92-3)
$C_{18}H_{22}NO_3.Br$
(125-23-5)
$C_{18}H_{22}N_2$
(50-47-5)
(82-92-8)
(101-87-1)
(5692-66-0)
$C_{18}H_{22}N_2O_3$
(10155-47-2)
$C_{18}H_{22}N_2S$
(84-96-8)
$C_{18}H_{22}N_4.2ClH$
(19395-62-1)
$C_{18}H_{22}N_4O_5$
(41541-11-1)
$C_{18}H_{22}O_2$
(53-16-7)
(80-43-3)
(84-16-2)
(651-55-8)
(3563-27-7)
(4483-62-9)
(76379-67-4)
$C_{18}H_{22}O_3$
(517-18-0)
(23031-38-1)
$C_{18}H_{22}O_4$
(500-38-9)
$C_{18}H_{22}O_5$
(17924-92-4)
$C_{18}H_{22}O_5S$
(481-97-0)
(75383-83-4)
$C_{18}H_{22}O_5S.C_4H_{10}N_2$
(7280-37-7)
$C_{18}H_{22}O_5S.Na$
(438-67-5)
$C_{18}H_{22}O_6$
(22676-00-2)
$C_{18}H_{22}O_8P_2$
(522-40-7)
$C_{18}H_{23}Cl_5O_2$
(3772-94-9)
$C_{18}H_{23}NO$
(83-98-7)
$C_{18}H_{23}NO_3$
(125-28-0)
(395-28-8)
(795-38-0)
(7732-92-5)
$C_{18}H_{23}NO_5$
(480-81-9)
$C_{18}H_{23}NO_6$
(23246-96-0)
$C_{18}H_{23}N_2S.CH_3O_4S$
(58-34-4)
$C_{18}H_{23}N_3S$
(67465-67-2)
$C_{18}H_{23}O_3PS$
(95114-72-0)
(95114-73-1)
(95114-74-2)
(95114-75-3)
(95114-76-4)
(95114-77-5)
$C_{18}H_{24}$
(3905-64-4)
(10275-58-8)
(41638-56-6)
$C_{18}H_{24}NO_4.Br$
(155-41-9)
$C_{18}H_{24}N_2$
(793-24-8)
(61931-82-6)
$C_{18}H_{24}N_2O_6$
(131-72-6)
(39300-45-3)
(49794-90-3)
(49794-91-4)
$C_{18}H_{24}O$
(74851-17-5)
$C_{18}H_{24}O_2$
(50-28-2)

(57-91-0)
(3650-04-2)
$C_{18}H_{24}O_3$
(50-27-1)
$C_{18}H_{24}O_3S.Na$
(25417-20-3)
(27213-90-7)
$C_{18}H_{24}O_4$
(84-64-0)
(21213-69-4)
(54699-44-4)
$C_{18}H_{24}O_6$
(85-70-1)
$C_{18}H_{25}Cl_3O_5$
(53535-31-2)
$C_{18}H_{25}N$
(34562-31-7)
$C_{18}H_{25}NO$
(125-70-2)
(125-71-3)
(510-53-2)
$C_{18}H_{25}NO_4$
(62251-96-1)
$C_{18}H_{25}NO_5$
(130-01-8)
$C_{18}H_{25}NO_6$
(480-54-6)
(6870-67-3)
$C_{18}H_{25}NO_7$
(15503-86-3)
$C_{18}H_{25}O_2$
(25384-17-2)
$C_{18}H_{26}$
(17465-58-6)
(17465-59-7)
(20273-27-2)
$C_{18}H_{26}ClN_3$
(54-05-7)
$C_{18}H_{26}Cl_2O_3$
(1320-15-6)
$C_{18}H_{26}NO_3.Br$
(2870-71-5)
$C_{18}H_{26}N_2.H_2O_4S$
(51-63-8)
(60-13-9)
$C_{18}H_{26}N_2O_3S$
(65907-30-4)
$C_{18}H_{26}O$
(1222-05-5)
(1506-02-1)
(21145-77-7)
$C_{18}H_{26}O_2$
(434-22-0)
(87818-31-3)
$C_{18}H_{26}O_3$
(5466-77-3)
$C_{18}H_{26}O_4$
(131-18-0)
(31047-64-0)
$C_{18}H_{26}O_5$
(26538-44-3)
(33374-28-6)
$C_{18}H_{26}O_5Si_4$
(40169-27-5)
$C_{18}H_{26}O_6$
(3290-92-4)
(63915-78-6)
$C_{18}H_{26}O_{12}$
(7208-47-1)
$C_{18}H_{27}N$
(10594-03-3)
$C_{18}H_{27}NO_3$
(404-86-4)
$C_{18}H_{27}NO_4$
(469-82-9)
$C_{18}H_{27}N_3O_6$
(30863-19-5)
$C_{18}H_{27}N_3O_{18}.C_6H_8BiNO_7.7Na$
(5798-43-6)
$C_{18}H_{28}$
(42981-76-0)
$C_{18}H_{28}N_2$
(4175-38-6)
$C_{18}H_{28}O_2$
(6178-32-1)
$C_{18}H_{28}O_3$
(6386-38-5)
(13402-96-5)
$C_{18}H_{28}O_3S$
(120-62-7)
$C_{18}H_{28}O_4Si_4$
(4657-20-9)
(18604-02-9)
(30026-85-8)
$C_{18}H_{28}O_8Sn$
(15546-11-9)
$C_{18}H_{29}ClO_5S$
(3761-60-2)
$C_{18}H_{29}NO_2$
(58353-63-2)
$C_{18}H_{29}O_3S.Na$
(11067-82-6)
(25155-30-0)
$C_{18}H_{30}$
(123-01-3)
(604-88-6)
(1459-11-6)
(1460-02-2)
(2400-00-2)
(2719-61-1)
(2719-62-2)
(2719-63-3)
(2719-64-4)
(14800-16-9)
(68648-87-3)
$C_{18}H_{30}N_6S_6$
(30863-12-8)
$C_{18}H_{30}O$
(732-26-3)
(17540-75-9)
(26537-71-3)
(27193-86-8)
$C_{18}H_{30}O_2$
(463-40-1)
(4906-91-6)
$C_{18}H_{30}O_3$
(33806-58-5)
$C_{18}H_{30}O_3S$
(121-65-3)
(11067-81-5)
(16577-13-2)
(18777-54-3)
(24271-16-7)
(24271-17-8)
(24271-18-9)
(24271-19-0)
(27176-87-0)
(47221-31-8)
$C_{18}H_{30}O_3S.1/2Ca$
(26264-06-2)
$C_{18}H_{30}O_3S.C_2H_7NO$
(26836-07-7)
$C_{18}H_{30}O_3S.C_3H_9N$
(26264-05-1)
$C_{18}H_{30}O_3S.C_3H_9NO$
(42504-46-1)
$C_{18}H_{30}O_3S.C_4H_{11}NO_2$
(26545-53-9)
$C_{18}H_{30}O_3S.C_6H_{15}NO_3$
(27323-41-7)
$C_{18}H_{30}O_3S.H_3N$
(1331-61-9)
$C_{18}H_{30}O_3S.K$
(27177-77-1)
$C_{18}H_{30}O_3S.Na$
(2211-98-5)
(2211-99-6)
(2212-52-4)
(68628-60-4)
$C_{18}H_{30}O_4$
(849-99-0)
$C_{18}H_{31}N$
(104-42-7)
$C_{18}H_{32}NS.Cl$
(68315-17-3)
$C_{18}H_{32}O$
(1604-32-6)
(18202-28-3)
(56554-35-9)
$C_{18}H_{32}O_2$
(60-33-3)
(503-05-9)
(506-21-8)
(2197-52-6)
(50933-33-0)
(51607-94-4)
(52207-99-5)
$C_{18}H_{32}O_2.1/2Ca$
(19704-83-7)
$C_{18}H_{32}O_2.xCu$
(7721-15-5)
$C_{18}H_{32}O_2.xPb$
(16996-51-3)
$C_{18}H_{32}O_3S$
(56554-67-7)
$C_{18}H_{32}O_5S$
(68201-84-3)
$C_{18}H_{32}O_7$
(77-94-1)
$C_{18}H_{32}O_8Ti$
(27858-32-8)
$C_{18}H_{32}O_{14}$
(37206-01-2)
$C_{18}H_{32}O_{16}$
(512-69-6)
$C_{18}H_{33}ClN_2O_5S$
(18323-44-9)
$C_{18}H_{33}ClO$
(112-77-6)
$C_{18}H_{33}N$
(112-91-4)
$C_{18}H_{33}O_2.Na$
(143-19-1)
$C_{18}H_{33}O_3.Na$
(5323-95-5)
$C_{18}H_{34}Cl_2O_2$
(31135-63-4)
$C_{18}H_{34}N_2O_6S$
(154-21-2)
$C_{18}H_{34}O$
(2423-10-1)
(5090-41-5)
(56554-86-0)
(56554-96-2)
(56554-98-4)
(56554-99-5)
$C_{18}H_{34}OSn$
(13121-70-5)
$C_{18}H_{34}O_2$
(112-79-8)
(112-80-1)
(593-39-5)
(693-72-1)
(2027-47-6)
(2549-53-3)
(25430-52-8)
(26764-26-1)
$C_{18}H_{34}O_2.1/3Al$
(688-37-9)
$C_{18}H_{34}O_2.CH_5N_3$
(53048-47-8)
$C_{18}H_{34}O_2.C_4H_9NO$
(1095-66-5)
$C_{18}H_{34}O_2.C_4H_{11}N$
(26094-13-3)
$C_{18}H_{34}O_2.C_4H_{11}NO_2$
(13961-86-9)
$C_{18}H_{34}O_2.C_6H_{15}NO_3$
(2717-15-9)
$C_{18}H_{34}O_2.C_{17}H_{19}NO_3$
(6033-05-2)
$C_{18}H_{34}O_2.1/2Ca$
(142-17-6)
$C_{18}H_{34}O_2.1/2Cd$
(10468-30-1)
$C_{18}H_{34}O_2.H_3N$
(544-60-5)
$C_{18}H_{34}O_2.K$
(143-18-0)
$C_{18}H_{34}O_2.1/2Mg$
(1555-53-9)
$C_{18}H_{34}O_2.1/2Pb$
(1120-46-3)
$C_{18}H_{34}O_3$
(141-22-0)
(2443-39-2)
(24560-98-3)
$C_{18}H_{34}O_3.K$
(7492-30-0)
$C_{18}H_{34}O_4$
(109-43-3)
(110-33-8)
(762-13-0)
(10022-60-3)
$C_{18}H_{34}O_6$
(95-08-9)
(141-18-4)
(1338-39-2)
(5959-89-7)
(16111-62-9)
$C_{18}H_{35}AlO_3$
(13419-15-3)
$C_{18}H_{35}BrO_2$
(142-94-9)
$C_{18}H_{35}ClO$
(112-76-5)
$C_{18}H_{35}ClO_2$
(929-16-8)
$C_{18}H_{35}CoO_2$
(14666-94-5)
$C_{18}H_{35}CrO_2$
(13308-40-2)
$C_{18}H_{35}CuO_2$
(10402-16-1)
$C_{18}H_{35}N$
(638-65-3)
$C_{18}H_{35}NO$
(301-02-0)
$C_{18}H_{35}NO_2$
(65882-22-6)
$C_{18}H_{35}NO_4.Na$
(14960-06-6)
$C_{18}H_{35}NO_4.2Na$
(3655-00-3)
$C_{18}H_{35}N_2O_3.HO.Na$
(68647-44-9)
$C_{18}H_{35}N_5$
(27431-00-1)
(30354-83-7)
$C_{18}H_{35}O_2.H_4N$
(1002-89-7)
$C_{18}H_{35}O_2.K$
(593-29-3)
$C_{18}H_{35}O_2.Li$
(4485-12-5)
$C_{18}H_{35}O_2.Na$
(822-16-2)
$C_{18}H_{36}$
(112-88-9)
(1795-17-1)
(27070-58-2)
$C_{18}H_{36}Cl_4Cr_2O_3$
(15242-96-3)
$C_{18}H_{36}NO.C_2H_4O_2$
(31717-87-0)
$C_{18}H_{36}N_2S_4$
(1634-02-2)
$C_{18}H_{36}N_2S_4.Ni$
(13927-77-0)

$C_{18}H_{36}N_4O_{11}$
(59-01-8)
$C_{18}H_{36}O$
(143-28-2)
(502-69-2)
(638-66-4)
(7390-81-0)
(16825-16-4)
$C_{18}H_{36}O_2$
(57-11-4)
(110-36-1)
(628-97-7)
(629-70-9)
(1731-92-6)
(2306-89-0)
(2306-92-5)
(2490-49-5)
(5638-12-0)
(30399-84-9)
(70710-00-8)
$C_{18}H_{36}O_2.Ag$
(3507-99-1)
$C_{18}H_{36}O_2.C_6H_{15}NO_3$
(4568-28-9)
$C_{18}H_{36}O_2.xCe$
(10119-53-6)
$C_{18}H_{36}O_2.xCo$
(13586-84-0)
$C_{18}H_{36}O_2.1/2Cu$
(660-60-6)
$C_{18}H_{36}O_2.1/3Fe$
(555-36-2)
$C_{18}H_{36}O_2.1/2Pb$
(1072-35-1)
$C_{18}H_{36}O_2.1/2Sr$
(10196-69-7)
$C_{18}H_{36}O_2.xSn$
(7637-13-0)
$C_{18}H_{36}O_3$
(106-14-9)
(638-26-6)
(1330-70-7)
(3384-24-5)
$C_{18}H_{36}O_3.1/2Ca$
(3159-62-4)
$C_{18}H_{36}O_3.Li$
(7620-77-1)
$C_{18}H_{36}O_3S.Na$
(15075-85-1)
$C_{18}H_{36}O_4S.Na$
(1847-55-8)
$C_{18}H_{37}AlO_4$
(7047-84-9)
$C_{18}H_{37}Br$
(112-89-0)
$C_{18}H_{37}Cl$
(3386-33-2)
$C_{18}H_{37}Cl_3Si$
(112-04-9)
$C_{18}H_{37}I$
(629-93-6)
$C_{18}H_{37}N$
(112-90-3)
(1838-19-3)
$C_{18}H_{37}N.C_2H_4O_2$
(3811-68-5)
(10460-00-1)
$C_{18}H_{37}N.C_9H_7Cl_3O_3$
(53404-73-2)
$C_{18}H_{37}NO$
(124-26-5)
$C_{18}H_{37}NO_3P$
(51249-05-9)
$C_{18}H_{37}NO_5S$
(59997-74-9)
$C_{18}H_{37}N_5O_9$
(32986-56-4)
$C_{18}H_{37}O_4P$
(7722-71-6)
(24613-61-4)
$C_{18}H_{37}O_4S.Na$
(1120-04-3)
$C_{18}H_{37}O_7S.Na$
(13150-00-0)
$C_{18}H_{38}$
(593-45-3)
(1560-89-0)
(3892-00-0)
(20959-33-5)
$C_{18}H_{38}N_2S_4Zn$
(136-23-2)
$C_{18}H_{38}O$
(112-92-5)
(2456-27-1)
(26762-44-7)
(55723-93-8)
$C_{18}H_{38}O.1/3Al$
(3985-81-7)
$C_{18}H_{38}O_2$
(2136-71-2)
(70709-98-7)
$C_{18}H_{38}O_4$
(3055-94-5)
$C_{18}H_{38}O_4S$
(143-03-3)
$C_{18}H_{38}O_4S.H_3N$
(4696-46-2)
$C_{18}H_{38}O_5S.Na$
(14858-54-9)
$C_{18}H_{38}O_{10}$
(3386-18-3)
$C_{18}H_{38}O_{13}$
(36675-34-0)
$C_{18}H_{38}S$
(2885-00-9)
$C_{18}H_{39}Al$
(1116-73-0)
$C_{18}H_{39}N$
(102-86-3)
(112-69-6)
(124-30-1)
$C_{18}H_{39}N.C_2H_4O_2$
(2190-04-7)
$C_{18}H_{39}N.ClH$
(1838-08-0)
$C_{18}H_{39}NO$
(7128-91-8)
$C_{18}H_{39}NO_2$
(1541-66-8)
$C_{18}H_{39}N_2O.CH_3O_4S$
(10595-49-0)
$C_{18}H_{39}N_3O_2$
(6843-97-6)
$C_{18}H_{39}O_4P$
(2958-09-0)
$C_{18}H_{39}O_7P$
(78-51-3)
$C_{18}H_{39}O_9P$
(36788-39-3)
$C_{18}H_{40}N.Cl$
(5538-94-3)
$C_{18}H_{41}N_7$
(13516-27-3)
$C_{18}H_{41}N_7.3C_2H_4O_2$
(57520-17-9)
$C_{18}H_{42}N_2O_8Ti$
(36673-16-2)
$C_{18}H_{42}N_6$
(15875-13-5)
$C_{18}H_{50}O_3Si_6$
(91455-17-3)
$C_{18}H_{54}O_7Si_8$
(556-69-4)
$C_{18}H_{54}O_9Si_9$
(556-71-8)
$C_{19}H_{10}Br_4O_5S$
(115-39-9)
$C_{19}H_{10}Cl_2N_6O_7S_2.2Na$
(17804-49-8)
$C_{19}H_{10}O$
(3074-00-8)
(80398-28-3)
(86853-89-6)
(86853-90-9)
$C_{19}H_{11}Cl_4N_2O_5S.Na$
(3567-25-7)
$C_{19}H_{11}N$
(189-90-2)
(189-92-4)
$C_{19}H_{12}Br_8O_4$
(34372-18-4)
$C_{19}H_{12}Cl_2N_6O_7S_2.2Na$
(18886-16-3)
$C_{19}H_{12}Cl_2N_6O_{10}S_3.3Na$
(17752-85-1)
$C_{19}H_{12}Cl_2O_5S$
(4430-20-0)
$C_{19}H_{12}O_2$
(6051-87-2)
$C_{19}H_{12}O_6$
(66-76-2)
$C_{19}H_{13}Cl$
(21846-07-1)
(32377-11-0)
$C_{19}H_{13}F$
(2498-63-7)
(2541-68-6)
$C_{19}H_{13}NO_2$
(80182-33-8)
$C_{19}H_{13}N_3O$
(7753-01-7)
$C_{19}H_{14}$
(316-49-4)
(789-24-2)
(1705-85-7)
(2381-15-9)
(2381-16-0)
(2381-31-9)
(2422-79-9)
(2498-75-1)
(2498-76-2)
(2498-77-3)
(2541-69-7)
(3351-28-8)
(3351-30-2)
(3351-31-3)
(3351-32-4)
(3697-24-3)
(6111-78-0)
(41637-90-5)
(43178-22-9)
(61089-87-0)
(78328-47-9)
$C_{19}H_{14}F_3NO$
(59756-60-4)
$C_{19}H_{14}O_3$
(603-45-2)
$C_{19}H_{14}O_5S$
(143-74-8)
$C_{19}H_{15}Cl$
(76-83-5)
$C_{19}H_{15}ClO_4$
(81-82-3)
$C_{19}H_{15}N$
(53-69-0)
(963-89-3)
(2381-40-0)
(3518-05-6)
(17401-48-8)
(32740-01-5)
$C_{19}H_{15}NO_6$
(152-72-7)
$C_{19}H_{15}N_5O_4$
(19800-42-1)
$C_{19}H_{15}O_4.Na$
(129-06-6)
$C_{19}H_{16}$
(519-73-3)
$C_{19}H_{16}ClNO_4$
(53-86-1)
$C_{19}H_{16}Cl_2N_2O_4S$
(58011-68-0)
$C_{19}H_{16}N_2O_7S_2.2Na$
(3564-09-8)
$C_{19}H_{16}N_4O$
(6300-37-4)
$C_{19}H_{16}O$
(76-84-6)
$C_{19}H_{16}O_3$
(5836-29-3)
$C_{19}H_{16}O_4$
(81-81-2)
$C_{19}H_{17}ClFN_3O_5S$
(5250-39-5)
$C_{19}H_{17}ClN_2O$
(2955-38-6)
$C_{19}H_{17}ClN_2O_4$
(76578-14-8)
$C_{19}H_{17}Cl_2N_3O_5S$
(3116-76-5)
$C_{19}H_{17}Cl_2N_5O_4$
(5261-31-4)
$C_{19}H_{17}NOS$
(2398-96-1)
$C_{19}H_{17}NO_2$
(52570-16-8)
$C_{19}H_{17}NO_3$
(92-74-0)
$C_{19}H_{17}N_3$
(479-73-2)
$C_{19}H_{17}N_3.ClH$
(569-61-9)
$C_{19}H_{17}N_3.HNO_3$
(61467-64-9)
$C_{19}H_{17}N_3O_4S_2$
(50-59-9)
$C_{19}H_{17}N_3O_8S_2.2Na$
(6625-46-3)
$C_{19}H_{17}N_4S.Cl$
(42373-04-6)
$C_{19}H_{17}O_4P$
(78-31-9)
(26444-49-5)
$C_{19}H_{18}ClN_3O_3$
(36104-80-0)
$C_{19}H_{18}ClN_3O_5S$
(61-72-3)
$C_{19}H_{18}N_2$
(897-55-2)
$C_{19}H_{18}N_2O_2$
(101657-77-6)
$C_{19}H_{18}N_2O_2S$
(100-93-6)
$C_{19}H_{18}N_6O_2S$
(16586-42-8)
$C_{19}H_{18}N_8O_7$
(29291-35-8)
$C_{19}H_{18}O_2$
(13936-21-5)
$C_{19}H_{19}ClFNO_3$
(52756-22-6)
$C_{19}H_{19}ClF_3NO_5$
(87237-48-7)
$C_{19}H_{19}N_3O$
(467-62-9)
$C_{19}H_{19}N_7O_2$
(30805-03-9)
$C_{19}H_{19}N_7O_5$
(24170-60-3)
$C_{19}H_{19}N_7O_6$
(59-30-3)
$C_{19}H_{20}Br_4O_4$
(4162-45-2)
$C_{19}H_{20}F_3NO_4$
(69806-50-4)
(79241-46-6)
$C_{19}H_{20}N_2O_2$
(50-33-9)
$C_{19}H_{20}N_2O_3$
(129-20-4)

$C_{19}H_{20}N_8O$
(6358-83-4)
$C_{19}H_{20}N_8O_5$
(54-62-6)
$C_{19}H_{20}O_4$
(85-68-7)
(2095-03-6)
(4196-89-8)
(39817-09-9)
(54208-63-8)
$C_{19}H_{21}N$
(72-69-5)
(438-60-8)
(41122-70-7)
$C_{19}H_{21}NO$
(1668-19-5)
$C_{19}H_{21}NO_3$
(62-67-9)
(115-37-7)
$C_{19}H_{21}NO_3.C_6H_3N_3O_7$
(5967-77-1)
$C_{19}H_{21}NO_4$
(465-65-6)
$C_{19}H_{21}NS$
(113-53-1)
$C_{19}H_{21}N_5O_2$
(69472-19-1)
$C_{19}H_{21}N_5O_4$
(19216-56-9)
$C_{19}H_{22}Cl_2O_2$
(56265-23-7)
$C_{19}H_{22}N_2$
(486-12-4)
$C_{19}H_{22}N_2.ClH.H_2O$
(6138-79-0)
$C_{19}H_{22}N_2O$
(118-10-5)
(441-91-8)
(485-71-2)
$C_{19}H_{22}N_6O_6$
(29757-24-2)
$C_{19}H_{22}O_6$
(77-06-5)
$C_{19}H_{22}O_6.K$
(125-67-7)
$C_{19}H_{22}O_7$
(11024-67-2)
$C_{19}H_{23}ClN_2$
(303-49-1)
$C_{19}H_{23}ClN_2O_2S$
(55512-33-9)
$C_{19}H_{23}ClO_2$
(56265-22-6)
$C_{19}H_{23}ClO_7$
(41787-75-1)
$C_{19}H_{23}NO_3$
(76-58-4)
$C_{19}H_{23}NO_3.ClH$
(125-30-4)
$C_{19}H_{23}NO_3.ClH.2H_2O$
(6746-59-4)
$C_{19}H_{23}NO_4$
(26258-70-8)
$C_{19}H_{23}N_3$
(33089-61-1)
$C_{19}H_{23}N_3O_2$
(60-79-7)
$C_{19}H_{23}N_4O_2S.CH_3O_4S$
(12270-13-2)
$C_{19}H_{24}NO_3.Br$
(125-27-9)
$C_{19}H_{24}N_2$
(50-49-7)
$C_{19}H_{24}N_2O_2$
(55268-74-1)
$C_{19}H_{24}N_2O_3$
(36894-69-6)
$C_{19}H_{24}N_2O_3Si$
(16230-35-6)
$C_{19}H_{24}N_3O.xCl_2Zn.Cl$
(7199-02-2)
$C_{19}H_{24}N_4O_2.C_4H_{12}O_8S_2$
(140-64-7)
$C_{19}H_{24}N_6O_4S_3$
(68971-14-2)
$C_{19}H_{24}N_{10}O.2CH_4O_3S$
(15427-93-7)
$C_{19}H_{24}O$
(56265-26-0)
$C_{19}H_{24}O_2$
(4741-74-6)
$C_{19}H_{24}O_3$
(119-89-1)
(968-93-4)
(27223-49-0)
$C_{19}H_{24}O_4$
(901-44-0)
$C_{19}H_{25}ClN_5O_2$
(14097-03-1)
$C_{19}H_{25}NO$
(152-02-3)
$C_{19}H_{25}NO_2$
(447-41-6)
$C_{19}H_{25}NO_4$
(7696-12-0)
$C_{19}H_{26}N_2$
(3081-01-4)
$C_{19}H_{26}N_2O_2$
(2519-77-9)
$C_{19}H_{26}N_2S_2$
(4979-32-2)
$C_{19}H_{26}N_8O_{14}P_2$
(15648-73-4)
$C_{19}H_{26}O_2$
(70-38-2)
$C_{19}H_{26}O_3$
(584-79-2)
(28434-00-6)
$C_{19}H_{26}O_4S$
(2312-35-8)
$C_{19}H_{26}O_8S$
(13964-21-1)
$C_{19}H_{26}O_{12}$
(490-67-5)
$C_{19}H_{27}Cl_3O_5$
(53535-26-5)
$C_{19}H_{27}NO$
(359-83-1)
$C_{19}H_{27}NO_4$
(3176-03-2)
$C_{19}H_{27}NO_7$
(26782-43-4)
(60102-37-6)
$C_{19}H_{28}NO_6$
(2318-18-5)
$C_{19}H_{28}N_2$
(5560-72-5)
$C_{19}H_{28}O_2$
(58-22-0)
$C_{19}H_{28}O_4$
(6600-31-3)
(22527-63-5)
$C_{19}H_{29}IO_2$
(99-79-6)
$C_{19}H_{30}O_2$
(18202-24-9)
(57156-91-9)
$C_{19}H_{30}O_3$
(53-39-4)
(26748-47-0)
$C_{19}H_{30}O_4$
(106807-78-7)
$C_{19}H_{30}O_5$
(51-03-6)
$C_{19}H_{31}N_2O_2.Cl$
(2438-53-1)
$C_{19}H_{32}$
(104-41-6)
(123-02-4)
(2400-01-3)
(4534-49-0)
(4534-50-3)
(4534-51-4)
(4534-52-5)
(4534-53-6)
$C_{19}H_{32}Cl_2P.Cl$
(115-78-6)
$C_{19}H_{32}O_2$
(301-00-8)
$C_{19}H_{32}O_2Sn$
(4342-36-3)
$C_{19}H_{32}O_3$
(17904-23-3)
(20427-84-3)
(27176-93-8)
$C_{19}H_{32}O_3S$
(25496-01-9)
$C_{19}H_{32}O_3S.Na$
(26248-24-8)
$C_{19}H_{33}Cl_5O_2$
(26638-28-8)
$C_{19}H_{34}N.Cl$
(965-32-2)
(2785-54-8)
$C_{19}H_{34}NS.Cl$
(70700-60-6)
$C_{19}H_{34}N_2O_{10}.Ca$
(137-08-6)
$C_{19}H_{34}O_2$
(112-63-0)
(738-87-4)
(13038-47-6)
(26543-36-2)
$C_{19}H_{34}O_3$
(40596-69-8)
$C_{19}H_{35}ClN_{10}S$
(8072-81-9)
$C_{19}H_{35}NO_2$
(77-19-0)
$C_{19}H_{35}NO_2.C_{17}H_{22}N_2O.C_8H_{11}NO_3.C_4H_6O_4.2ClH$
(8064-77-5)
$C_{19}H_{36}$
(2090-15-5)
$C_{19}H_{36}O_2$
(112-62-9)
(1937-62-8)
(6140-74-5)
(13038-45-4)
(13402-02-3)
(31627-33-5)
(52355-31-4)
(52380-33-3)
$C_{19}H_{36}O_3$
(141-24-2)
$C_{19}H_{37}N$
(28623-46-3)
$C_{19}H_{37}NO$
(112-96-9)
$C_{19}H_{37}N_5$
(27431-05-6)
$C_{19}H_{38}$
(6006-33-3)
(18435-45-5)
$C_{19}H_{38}N_2O_3$
(4292-10-8)
$C_{19}H_{38}N_2S_4$
(10254-57-6)
$C_{19}H_{38}O$
(504-57-4)
(29804-22-6)
(67860-04-2)
$C_{19}H_{38}O_2$
(109-32-0)
(112-61-8)
(142-91-6)
(646-30-0)
(2239-78-3)
(14010-23-2)
(15965-99-8)
$C_{19}H_{38}O_2SSn$
(3033-29-2)
$C_{19}H_{38}O_3$
(111-07-9)
(141-23-1)
$C_{19}H_{38}O_3.Li$
(53422-16-5)
$C_{19}H_{38}O_4Sn$
(53404-82-3)
$C_{19}H_{38}O_5S$
(3076-26-4)
$C_{19}H_{38}O_5S.Na$
(4062-78-6)
$C_{19}H_{38}O_6S.Na$
(139-99-1)
$C_{19}H_{39}Br$
(4434-66-6)
$C_{19}H_{39}NO$
(24602-86-6)
$C_{19}H_{39}NO_2$
(3370-35-2)
(16545-85-0)
$C_{19}H_{39}N_3Si$
(15901-40-3)
$C_{19}H_{39}N_5O_7$
(1403-66-3)
$C_{19}H_{40}$
(629-92-5)
(1560-88-9)
(1921-70-6)
(55000-52-7)
$C_{19}H_{40}BrNO$
(20091-61-6)
$C_{19}H_{40}N_2O_2$
(38848-76-9)
$C_{19}H_{40}O$
(52783-43-4)
$C_{19}H_{40}O_7S.Na$
(25446-78-0)
$C_{19}H_{41}N$
(22020-14-0)
$C_{19}H_{41}N_5O_{11}S$
(1405-41-0)
$C_{19}H_{42}N$
(6899-10-1)
$C_{19}H_{42}N.BH_4$
(19710-01-1)
$C_{19}H_{42}N.Br$
(57-09-0)
$C_{19}H_{42}N.Cl$
(112-02-7)
$C_{19}H_{42}NO_2S.Cl$
(78865-85-7)
$C_{19}H_{42}OSi$
(6221-90-5)
$C_{19}H_{43}N_3O_5$
(17121-34-5)
$C_{20}D_{16}$
(32976-87-7)
$C_{20}H_6Br_4Cl_2O_5.2K$
(6441-77-6)
$C_{20}H_6I_4O_5.2Na$
(16423-68-0)
$C_{20}H_8Br_4O_5$
(15086-94-9)
$C_{20}H_8Br_4O_5.2Na$
(548-26-5)
(17372-87-1)
$C_{20}H_8Br_4O_{10}S_2.2Na$
(71-67-0)
$C_{20}H_8Cl_4O_5$
(6262-21-1)
$C_{20}H_8I_4O_5$
(15905-32-5)
$C_{20}H_8I_4O_5.xAl$
(12227-78-0)
$C_{20}H_{10}Br_6N_2O_2$
(49693-20-1)
$C_{20}H_{10}Cl_2N_2O_2$
(3089-16-5)

(3089-17-6)
$C_{20}H_{10}CrN_2O_8S_2.Na$
(70942-15-3)
$C_{20}H_{10}N_2O_4$
(1503-48-6)
$C_{20}H_{10}O_2$
(3067-12-7)
(3067-13-8)
(3067-14-9)
(42286-46-4)
$C_{20}H_{10}O_5.2Na$
(518-47-8)
$C_{20}H_{11}Br$
(21248-00-0)
$C_{20}H_{11}CrN_2O_9S_2.2Na$
(6370-08-7)
$C_{20}H_{11}NO_2$
(20589-63-3)
(63041-90-7)
(70021-42-0)
(70021-98-6)
(70021-99-7)
$C_{20}H_{11}NO_3$
(82039-09-6)
(82039-10-9)
$C_{20}H_{11}N_2O_{10}S_3.3Na$
(915-67-3)
$C_{20}H_{12}$
(50-32-8)
(192-97-2)
(198-55-0)
(203-33-8)
(205-82-3)
(205-99-2)
(207-08-9)
(73467-76-2)
$C_{20}H_{12}Cl_2N_2O_2$
(15715-19-2)
$C_{20}H_{12}CrN_4O_8S_2.Na$
(52677-44-8)
$C_{20}H_{12}N_2O_2$
(1047-16-1)
$C_{20}H_{12}N_2O_6$
(20241-76-3)
$C_{20}H_{12}N_2O_7S_2.2Na$
(3567-69-9)
$C_{20}H_{12}O$
(37574-47-3)
$C_{20}H_{12}O_5$
(518-45-6)
(2321-07-5)
$C_{20}H_{13}F_3$
(32377-12-1)
$C_{20}H_{13}N$
(194-59-2)
(239-64-5)
(32377-09-6)
(71012-25-4)
$C_{20}H_{13}NO_4$
(17418-58-5)
$C_{20}H_{13}N_2O_4S.Na$
(1248-18-6)
$C_{20}H_{13}N_2O_5S.Na$
(2538-85-4)
$C_{20}H_{13}N_3O_7S.Na$
(1787-61-7)
(3618-58-4)
$C_{20}H_{14}$
(479-23-2)
(602-55-1)
(604-53-5)
(4325-76-2)
(5779-79-3)
(11068-27-2)
$C_{20}H_{14}NO_4$
(2447-54-3)
$C_{20}H_{14}N_2O_2$
(4395-65-7)
(5862-38-4)
$C_{20}H_{14}N_2O_4$
(20241-77-4)
$C_{20}H_{14}N_2O_4S.1/2Ba$
(1103-38-4)
$C_{20}H_{14}N_2O_4S.1/2Ca$
(1103-39-5)
$C_{20}H_{14}N_2O_4S.Na$
(1658-56-6)
$C_{20}H_{14}N_2O_5$
(7098-08-0)
(31529-83-6)
$C_{20}H_{14}N_2O_5S$
(2786-71-2)
$C_{20}H_{14}N_2O_5S.Na$
(6408-78-2)
$C_{20}H_{14}N_2O_7S_2.2Na$
(2302-96-7)
(5858-93-5)
$C_{20}H_{14}N_2O_{10}S_3.3Na$
(2611-82-7)
$C_{20}H_{14}N_2O_{13}S_4.4Na$
(5850-44-2)
$C_{20}H_{14}N_2O_{14}S_4.4Na$
(29637-28-3)
$C_{20}H_{14}O$
(5471-63-6)
$C_{20}H_{14}O_2S_2$
(6088-51-3)
$C_{20}H_{14}O_3$
(519-95-9)
$C_{20}H_{14}O_4$
(77-09-8)
(84-62-8)
(744-45-6)
(1539-04-4)
$C_{20}H_{14}O_6$
(88899-62-1)
$C_{20}H_{14}O_{10}$
(6219-66-5)
$C_{20}H_{15}Br$
(16238-56-5)
$C_{20}H_{15}N_3O_8S_2$
(128-99-4)
$C_{20}H_{15}N_3O_{10}S_3.3Na$
(5045-23-8)
$C_{20}H_{16}$
(57-97-6)
(58-72-0)
(568-81-0)
(572-89-4)
(572-89-4)
(1572-46-9)
(3697-27-6)
(3697-30-1)
(15914-23-5)
(18153-42-9)
(18153-43-0)
(18429-70-4)
(18868-66-1)
(20627-31-0)
(35187-28-1)
(41637-92-7)
(56961-62-7)
(58429-99-5)
(63018-79-1)
$C_{20}H_{16}N_2O_2$
(2478-20-8)
$C_{20}H_{16}N_2O_4$
(10109-95-2)
$C_{20}H_{16}N_4O$
(4751-43-3)
$C_{20}H_{16}O$
(21846-08-2)
(32377-13-2)
$C_{20}H_{16}O_4S_2$
(3263-31-8)
$C_{20}H_{17}ClN_2O_3$
(27223-35-4)
$C_{20}H_{17}ClO_5$
(31265-39-1)
$C_{20}H_{17}FO_3S$
(38194-50-2)
$C_{20}H_{17}F_3N_2O_4$
(23779-99-9)
$C_{20}H_{17}N$
(51787-42-9)
(51787-43-0)
$C_{20}H_{17}N_3O_4$
(3176-88-3)
$C_{20}H_{17}N_3O_5$
(12217-80-0)
$C_{20}H_{17}N_3O_9S_3.2Na$
(3244-88-0)
$C_{20}H_{18}ClNO_6$
(303-47-9)
$C_{20}H_{18}N_4O_4$
(13418-49-0)
$C_{20}H_{18}N_4O_5S.Na$
(1320-07-6)
$C_{20}H_{18}N_4O_6S.Na$
(6222-63-5)
$C_{20}H_{18}N_6O_2S$
(19745-44-9)
$C_{20}H_{18}O_2Pb$
(1162-06-7)
$C_{20}H_{18}O_2Sn$
(900-95-8)
$C_{20}H_{18}O_6$
(23820-80-6)
$C_{20}H_{18}O_6S_2$
(32527-15-4)
$C_{20}H_{19}NO_2$
(1096-48-6)
$C_{20}H_{19}NO_2S$
(3622-84-2)
$C_{20}H_{19}NO_3$
(7008-42-6)
$C_{20}H_{19}NO_5$
(130-86-9)
$C_{20}H_{19}NO_5.ClH$
(6164-47-2)
$C_{20}H_{19}NO_6$
(4825-86-9)
$C_{20}H_{19}N_2O_5S.Na$
(4368-56-3)
$C_{20}H_{19}N_3$
(3248-93-9)
$C_{20}H_{19}N_3.ClH$
(632-99-5)
$C_{20}H_{19}N_3O_{11}S_3.2Na$
(20262-58-2)
$C_{20}H_{19}N_4.Cl$
(477-73-6)
$C_{20}H_{19}N_7O_3$
(41642-51-7)
$C_{20}H_{19}O_4P$
(23666-94-6)
(26446-73-1)
(52784-49-3)
(73179-40-5)
$C_{20}H_{20}BrClN_4O_6$
(59709-38-5)
$C_{20}H_{20}BrN_7O_6$
(22578-86-5)
$C_{20}H_{20}Br_4N_2O_4$
(52907-07-0)
$C_{20}H_{20}Cl_2N_8O_5$
(528-74-5)
$C_{20}H_{20}N_2O_4$
(25502-52-7)
$C_{20}H_{20}N_2O_5S$
(5617-28-7)
$C_{20}H_{20}P.C_2H_3O_2$
(35835-94-0)
$C_{20}H_{20}P.I$
(4736-60-1)
$C_{20}H_{21}ClOSi_2$
(53634-34-7)
$C_{20}H_{21}NO_3$
(1165-48-6)
$C_{20}H_{21}NO_4$
(58-74-2)
$C_{20}H_{21}NO_4.ClH$
(61-25-6)
$C_{20}H_{21}NO_5$
(34231-26-0)
$C_{20}H_{21}N_3$
(25834-80-4)
$C_{20}H_{21}N_5O_2S_2$
(16588-67-3)
$C_{20}H_{21}N_8O_5.Na$
(15475-56-6)
$C_{20}H_{22}N_2O_2$
(84-31-1)
(509-15-9)
(14233-37-5)
$C_{20}H_{22}N_2O_2.ClH$
(35306-33-3)
$C_{20}H_{22}N_4O_{12}S_3.2K$
(20317-19-5)
$C_{20}H_{22}N_6NiO_6$
(29204-84-0)
$C_{20}H_{22}N_8O_5$
(59-05-2)
$C_{20}H_{22}O_3$
(3771-19-5)
(70356-09-1)
$C_{20}H_{22}O_4$
(33533-56-1)
$C_{20}H_{22}O_5$
(27138-31-4)
$C_{20}H_{22}O_6$
(120-56-9)
$C_{20}H_{23}N$
(50-48-6)
(10262-69-8)
$C_{20}H_{23}NO_3$
(3321-80-0)
(71626-11-4)
$C_{20}H_{23}NO_4$
(466-90-0)
$C_{20}H_{23}N_3O_2$
(55179-31-2)
$C_{20}H_{23}N_3O_3$
(84268-33-7)
$C_{20}H_{24}CaO_{10}S_2$
(8061-52-7)
$C_{20}H_{24}ClN_3O$
(3864-99-1)
$C_{20}H_{24}ClN_3S$
(58-38-8)
$C_{20}H_{24}I_2O_2$
(552-22-7)
$C_{20}H_{24}N_2OS$
(362-29-8)
$C_{20}H_{24}N_2O_2$
(56-54-2)
(130-95-0)
(19125-99-6)
$C_{20}H_{24}N_2O_2.BrH$
(549-49-5)
$C_{20}H_{24}N_2O_2.C_3H_6O_3$
(749-49-5)
$C_{20}H_{24}N_2O_2.ClH$
(130-89-2)
(1668-99-1)
$C_{20}H_{24}N_2O_2.H_2O_4S$
(549-56-4)
(747-45-5)
$C_{20}H_{24}N_2O_2.H_3O_2P$
(6119-53-5)
$C_{20}H_{24}N_2O_2.O_4S$
(804-63-7)
$C_{20}H_{24}N_2O_2S$
(3105-97-3)
$C_{20}H_{24}N_2O_2S.CH_4O_3S$
(23255-93-8)
$C_{20}H_{24}N_2O_3$
(70116-00-6)
$C_{20}H_{24}N_2O_3.C_4H_4O_4$
(70096-14-9)

$C_{20}H_{24}N_3O.CH_3O_4S$
(54060-92-3)
$C_{20}H_{24}O_2$
(57-63-6)
$C_{20}H_{25}ClO_2$
(56265-24-8)
$C_{20}H_{25}NO$
(467-85-6)
(511-45-5)
$C_{20}H_{25}NO_2$
(64-95-9)
(42576-07-8)
$C_{20}H_{25}NO_3$
(302-40-9)
(509-78-4)
$C_{20}H_{25}NO_4$
(3861-72-1)
(26258-71-9)
$C_{20}H_{25}N_3O$
(50-37-3)
(3147-75-9)
$C_{20}H_{25}N_3O_2$
(113-42-8)
$C_{20}H_{26}$
(79606-18-1)
$C_{20}H_{26}Cl_2O_2$
(57055-39-7)
(65281-77-8)
$C_{20}H_{26}N_2$
(739-71-9)
$C_{20}H_{26}N_2O$
(83-74-9)
$C_{20}H_{26}N_2O_2$
(522-66-7)
$C_{20}H_{26}N_4.O_4S$
(65-30-5)
$C_{20}H_{26}O$
(2668-47-5)
$C_{20}H_{26}O_2$
(68-22-4)
(68-23-5)
(33145-10-7)
(56265-21-5)
$C_{20}H_{26}O_3$
(18684-55-4)
$C_{20}H_{26}O_4$
(84-61-7)
$C_{20}H_{26}O_5S$
(78749-45-8)
$C_{20}H_{26}O_6$
(83-59-0)
$C_{20}H_{27}ClO_2$
(57055-38-6)
$C_{20}H_{27}N$
(150-59-4)
(4175-37-5)
$C_{20}H_{27}NO_3$
(13647-35-3)
$C_{20}H_{27}NO_{11}$
(29883-15-6)
$C_{20}H_{27}O_3P$
(15647-08-2)
(26401-27-4)
$C_{20}H_{27}O_3PS$
(95114-78-6)
$C_{20}H_{27}O_4P$
(115-88-8)
(1241-94-7)
$C_{20}H_{27}O_6$
(17673-25-5)
$C_{20}H_{28}$
(55000-53-8)
$C_{20}H_{28}NO.CH_3O_4S$
(52793-97-2)
$C_{20}H_{28}N_2$
(15233-47-3)
$C_{20}H_{28}N_6P_2.2Br$
(10310-38-0)
$C_{20}H_{28}O$
(52-76-6)
(24035-50-5)
$C_{20}H_{28}O_2$
(72-63-9)
(302-79-4)
(4759-48-2)
(5155-70-4)
$C_{20}H_{29}FO_3$
(76-43-7)
$C_{20}H_{29}N_3O_2$
(85-79-0)
$C_{20}H_{30}N_2O_2.H_2O_4S$
(134-72-5)
$C_{20}H_{30}N_2O_3$
(469-81-8)
$C_{20}H_{30}O$
(68-26-8)
$C_{20}H_{30}O_2$
(58-18-4)
(79-54-9)
(127-27-5)
(471-74-9)
(471-77-2)
(514-10-3)
(1945-53-5)
(5835-26-7)
$C_{20}H_{30}O_2.Na$
(14351-66-7)
$C_{20}H_{30}O_4$
(84-75-3)
(84-78-6)
(85-69-8)
(146-50-9)
(21577-80-0)
(68515-50-4)
$C_{20}H_{30}O_6$
(117-83-9)
(3130-19-6)
$C_{20}H_{30}O_8$
(87625-62-5)
$C_{20}H_{31}N$
(1446-61-3)
$C_{20}H_{31}N.C_2H_4O_2$
(2026-24-6)
$C_{20}H_{31}NO$
(144-11-6)
$C_{20}H_{31}NO_3$
(77-23-6)
$C_{20}H_{31}NO_6$
(22571-95-5)
$C_{20}H_{31}NO_7.ClH$
(520-68-3)
$C_{20}H_{31}N_2O_3S$
(55285-14-8)
$C_{20}H_{32}$
(55255-58-8)
$C_{20}H_{32}O$
(965-90-2)
$C_{20}H_{32}O_2$
(58-19-5)
(506-32-1)
(57342-02-6)
$C_{20}H_{33}ClO_3$
(66028-01-1)
$C_{20}H_{34}$
(1459-10-5)
(4534-54-7)
(4534-55-8)
(4534-56-9)
(4534-57-0)
(4534-58-1)
(4534-59-2)
$C_{20}H_{34}O$
(5749-44-0)
$C_{20}H_{34}O_2$
(1783-84-2)
$C_{20}H_{34}O_3$
(32072-96-1)
$C_{20}H_{34}O_3S$
(30776-59-1)
(47377-16-2)
$C_{20}H_{34}O_3S.Na$
(1797-33-7)
(13419-31-3)
(28348-61-0)
$C_{20}H_{34}O_3Si_4$
(13270-97-8)
$C_{20}H_{34}O_4$
(2212-81-9)
(2781-00-2)
(25155-25-3)
$C_{20}H_{34}O_5$
(551-11-1)
$C_{20}H_{34}O_6S.Na$
(2917-94-4)
$C_{20}H_{34}O_8$
(77-90-7)
$C_{20}H_{35}ClO_2$
(25525-76-2)
$C_{20}H_{36}NS.Cl$
(70700-63-9)
$C_{20}H_{36}N_2$
(3081-14-9)
$C_{20}H_{36}O$
(13393-93-6)
$C_{20}H_{36}O_2$
(53120-26-6)
(53120-27-7)
(56666-38-7)
$C_{20}H_{36}O_4$
(141-02-6)
(142-16-5)
(1330-76-3)
(2915-53-9)
$C_{20}H_{36}O_4Sn$
(16091-18-2)
$C_{20}H_{37}Cl_2N_5$
(30354-86-0)
$C_{20}H_{37}O_7S.Na$
(1639-66-3)
$C_{20}H_{38}$
(504-96-1)
(765-27-5)
(2437-92-5)
(54964-83-9)
$C_{20}H_{38}O$
(29171-23-1)
$C_{20}H_{38}O_2$
(111-62-6)
(111-63-7)
(2495-27-4)
(54460-46-7)
$C_{20}H_{38}O_2S$
(59118-78-4)
$C_{20}H_{38}O_4$
(762-12-9)
(14697-48-4)
$C_{20}H_{38}O_5S.Na$
(142-15-4)
$C_{20}H_{38}O_7S$
(2373-23-1)
$C_{20}H_{38}O_7S.Na$
(577-11-7)
(20727-33-7)
$C_{20}H_{39}N.C_8H_5Cl_3O_3$
(53404-88-9)
$C_{20}H_{39}NO$
(2664-42-8)
$C_{20}H_{39}NO_2$
(65882-23-7)
$C_{20}H_{39}N_5$
(2533-20-2)
$C_{20}H_{40}$
(1786-12-5)
(1795-18-2)
(3452-07-1)
(56009-20-2)
$C_{20}H_{40}N_2.C_2H_4O_2$
(556-22-9)
$C_{20}H_{40}N_4S_8.Se$
(5456-28-0)
$C_{20}H_{40}N_4S_8.Te$
(20941-65-5)
$C_{20}H_{40}O$
(150-86-7)
(505-32-8)
(930-02-9)
(2955-56-8)
(60046-87-9)
$C_{20}H_{40}O_2$
(110-34-9)
(111-06-8)
(111-61-5)
(506-30-9)
(1654-86-0)
(1731-94-8)
(5353-25-3)
(57274-46-1)
$C_{20}H_{40}O_2.Ag$
(24687-57-8)
$C_{20}H_{40}O_3$
(111-60-4)
$C_{20}H_{41}Br$
(4276-49-7)
$C_{20}H_{41}Cl$
(42217-02-7)
$C_{20}H_{41}N$
(28061-69-0)
$C_{20}H_{41}N.C_8H_5Cl_3O_3$
(53404-89-0)
$C_{20}H_{41}N.C_{20}H_{39}N$-
$.C_8H_5Cl_3O_3.C_8H_5Cl_3O_3$
(55256-33-2)
$C_{20}H_{41}NO$
(14351-50-9)
$C_{20}H_{41}NO_2$
(111-57-9)
(693-33-4)
$C_{20}H_{41}NO_6S$
(59997-80-7)
$C_{20}H_{41}N_5O_7$
(69375-05-9)
$C_{20}H_{42}$
(112-95-8)
(638-36-8)
$C_{20}H_{42}N_2O_5S$
(19223-55-3)
$C_{20}H_{42}O$
(629-96-9)
(2456-28-2)
(5333-42-6)
$C_{20}H_{42}O.1/3Al$
(67905-31-1)
$C_{20}H_{42}O_5$
(5274-68-0)
$C_{20}H_{42}O_7S.Na$
(25446-80-4)
$C_{20}H_{43}N$
(124-28-7)
(1120-49-6)
$C_{20}H_{43}N.C_2H_4O_2$
(19855-61-9)
$C_{20}H_{43}NO$
(2571-88-2)
$C_{20}H_{43}NO_4$
(28865-36-3)
$C_{20}H_{43}O_2PS_2$
(28631-44-9)
$C_{20}H_{43}O_2PS_2.1/2Zn$
(25103-54-2)
$C_{20}H_{43}O_3P$
(25371-54-4)
$C_{20}H_{44}N.Br$
(124-03-8)
$C_{20}H_{44}N.Cl$
(32426-11-2)
$C_{20}H_{44}N.I$
(2498-20-6)
$C_{20}H_{46}O_7Si_4$
(69155-42-6)
$C_{20}H_{60}O_8Si_9$

(2652-13-3)
$C_{21}H_{10}O$
(83622-91-7)
(86853-92-1)
(86862-68-2)
$C_{21}H_{12}Br_3N_3$
(30363-03-2)
$C_{21}H_{12}Cl_3N_3O_3$
(1919-45-5)
$C_{21}H_{12}Cl_3N_3S_3$
(13270-03-6)
$C_{21}H_{12}Cl_3N_3$
(3114-54-3)
$C_{21}H_{12}N_6O_6$
(14043-38-0)
$C_{21}H_{12}N_3O_6PS$
(4151-51-3)
$C_{21}H_{12}O$
(4599-92-2)
(4599-94-4)
(5623-32-5)
(28609-66-7)
(53223-75-9)
(60848-01-3)
(62716-20-5)
(63041-47-4)
(80440-44-4)
(83589-46-2)
(86853-93-2)
(86853-94-3)
(86853-95-4)
(86853-96-5)
(86853-97-6)
(86853-98-7)
(86853-99-8)
(86854-00-4)
(86854-01-5)
(86854-02-6)
(86854-03-7)
(86854-04-8)
(86854-05-9)
(86854-06-0)
$C_{21}H_{13}N$
(215-62-3)
(224-42-0)
(224-53-3)
(226-36-8)
$C_{21}H_{13}N_5O_4$
(30363-01-0)
(30363-02-1)
$C_{21}H_{14}$
(207-83-0)
(2381-39-7)
(4514-19-6)
(10350-33-1)
(16757-80-5)
(16757-81-6)
(16757-82-7)
(16757-83-8)
(23143-01-3)
(24471-47-4)
(64031-91-0)
$C_{21}H_{14}ClN_3$
(3114-52-1)
$C_{21}H_{14}Cl_2N_4O_2$
(30358-15-7)
(30358-17-9)
$C_{21}H_{14}N_2O_4$
(13676-54-5)
$C_{21}H_{14}N_2O_6S.Ca$
(6417-83-0)
$C_{21}H_{14}N_4O_2$
(7753-05-1)
(13960-33-3)
$C_{21}H_{15}ClN_2$
(1707-67-1)
$C_{21}H_{15}Cl_3N_6$
(2272-28-8)
(2748-40-5)
$C_{21}H_{15}NO_2$
(132-68-3)
$C_{21}H_{15}NO_3$
(81-48-1)
$C_{21}H_{15}NO_4$
(23552-76-3)
$C_{21}H_{15}NO_6S$
(1594-08-7)
$C_{21}H_{15}N_3$
(493-77-6)
$C_{21}H_{15}N_3O$
(3202-86-6)
$C_{21}H_{15}N_3O_3$
(1785-02-0)
(1919-48-8)
$C_{21}H_{15}N_3S$
(30863-04-8)
$C_{21}H_{15}N_3S_3$
(30863-82-2)
$C_{21}H_{16}$
(56-49-5)
$C_{21}H_{16}N_2O_3$
(27341-33-9)
$C_{21}H_{16}N_2O_5$
(4702-64-1)
(4702-65-2)
(31288-44-5)
$C_{21}H_{16}N_2O_5S$
(34114-36-8)
$C_{21}H_{16}N_2O_9S_2$
(134-47-4)
$C_{21}H_{16}N_2O_9S_2.2Na$
(20324-87-2)
$C_{21}H_{16}N_4$
(30363-00-9)
(30369-20-1)
$C_{21}H_{16}N_4O_2$
(1973-08-6)
$C_{21}H_{16}O$
(3342-98-1)
$C_{21}H_{16}O_2$
(1096-84-0)
$C_{21}H_{16}O_6S_2.2Na$
(26545-58-4)
$C_{21}H_{17}N_5O$
(30360-76-0)
$C_{21}H_{17}N_5S$
(30360-87-3)
$C_{21}H_{18}$
(20627-32-1)
(20627-33-2)
(20627-34-3)
(35187-24-7)
(35187-27-0)
$C_{21}H_{18}Cl_3N_3$
(30805-14-2)
$C_{21}H_{18}N_6$
(1973-05-3)
$C_{21}H_{19}N$
(32377-15-4)
(51787-44-1)
$C_{21}H_{19}N_3O_3S$
(51264-14-3)
$C_{21}H_{19}N_5O_3$
(6465-02-7)
$C_{21}H_{20}$
(26898-17-9)
$C_{21}H_{20}BrN_7O_6$
(68877-63-4)
$C_{21}H_{20}Br_4O_2$
(25327-89-3)
$C_{21}H_{20}Br_8O_2$
(21850-44-2)
$C_{21}H_{20}Cl_2O_3$
(52341-32-9)
(52341-33-0)
(52645-53-1)
(61949-76-6)
(61949-77-7)
$C_{21}H_{20}NP$
(52112-04-6)
$C_{21}H_{20}N_2O_3$
(5254-41-1)
$C_{21}H_{20}N_3.Br$
(1239-45-8)
$C_{21}H_{20}N_4O_5S.Na$
(68555-86-2)
$C_{21}H_{20}O_6$
(458-37-7)
$C_{21}H_{20}O_{11}$
(522-12-3)
$C_{21}H_{21}ClN_2O_8$
(127-33-3)
$C_{21}H_{21}ClN_5O_2.Cl$
(36986-04-6)
$C_{21}H_{21}N$
(129-03-3)
$C_{21}H_{21}N.ClH$
(7673-07-6)
$C_{21}H_{21}NO_6$
(118-08-1)
$C_{21}H_{21}N_3$
(91-78-1)
$C_{21}H_{21}N_5O_6$
(30124-94-8)
$C_{21}H_{21}N_9$
(3786-23-0)
$C_{21}H_{21}O_3P$
(25586-42-9)
$C_{21}H_{21}O_4P$
(78-30-8)
(78-32-0)
(563-04-2)
(1330-78-5)
(28108-99-8)
(55864-04-5)
(64532-94-1)
(73179-43-8)
$C_{21}H_{21}O_4P.C_{18}H_{15}O_4P$
(96300-97-9)
$C_{21}H_{21}O_6P$
(620-42-8)
$C_{21}H_{22}$
(65755-17-1)
(74229-83-7)
$C_{21}H_{22}N_2O_2$
(57-24-9)
$C_{21}H_{22}N_2O_2.AsH_3O_4$
(10476-82-1)
$C_{21}H_{22}N_2O_2.HNO_3$
(66-32-0)
$C_{21}H_{22}N_2O_2.1/2H_2O_4S$
(60-41-3)
$C_{21}H_{22}N_2O_2.HI$
(52748-69-3)
$C_{21}H_{22}N_2O_5S$
(147-52-4)
$C_{21}H_{22}O_9$
(1415-73-2)
$C_{21}H_{23}ClFN_3O$
(17617-23-1)
$C_{21}H_{23}ClFNO_2$
(52-86-8)
$C_{21}H_{23}ClFN_3O.2ClH$
(1172-18-5)
$C_{21}H_{23}NO_5$
(561-27-3)
$C_{21}H_{23}NO_5.ClH$
(1502-95-0)
$C_{21}H_{23}N_2O_3.Cl$
(6272-74-8)
$C_{21}H_{24}F_3N_3S$
(117-89-5)
$C_{21}H_{24}N_5O_8S_2.Na$
(51481-65-3)
$C_{21}H_{24}N_8.2ClH$
(5421-66-9)
$C_{21}H_{24}OSi_2$
(14920-93-5)
$C_{21}H_{24}O_2$
(3739-67-1)
$C_{21}H_{24}O_3Si_3$
(3424-57-5)
(6138-53-0)
$C_{21}H_{24}O_4$
(77-08-7)
(1675-54-3)
$C_{21}H_{24}O_{10}$
(60-81-1)
$C_{21}H_{25}ClO_5$
(5251-34-3)
$C_{21}H_{25}Cl_2N_3O.2ClH$
(146-59-8)
$C_{21}H_{25}NO$
(86-13-5)
$C_{21}H_{25}NO_2$
(64-52-8)
$C_{21}H_{25}NO_3$
(3567-12-2)
$C_{21}H_{25}NO_5$
(477-30-5)
$C_{21}H_{25}N_2O_2.Cl$
(4208-80-4)
$C_{21}H_{25}N_5O_4$
(51249-07-1)
$C_{21}H_{26}ClN_3OS$
(58-39-9)
$C_{21}H_{26}NO_3.Br$
(53-46-3)
$C_{21}H_{26}N_2OS_2$
(5588-33-0)
(7776-05-8)
$C_{21}H_{26}N_2O_2S_2$
(14759-06-9)
$C_{21}H_{26}N_2O_3$
(146-48-5)
$C_{21}H_{26}N_2O_3.ClH$
(65-19-0)
$C_{21}H_{26}N_2S_2$
(50-52-2)
$C_{21}H_{26}O_2$
(72-33-3)
(521-35-7)
$C_{21}H_{26}O_3$
(1843-05-6)
(2553-08-4)
$C_{21}H_{26}O_4$
(17080-02-3)
$C_{21}H_{26}O_5$
(53-03-2)
$C_{21}H_{27}ClN_2O_2$
(68-88-2)
$C_{21}H_{27}FO_5$
(53-34-9)
$C_{21}H_{27}FO_6$
(124-94-7)
$C_{21}H_{27}NO$
(76-99-3)
(466-40-0)
(972-02-1)
$C_{21}H_{27}N_2O.Cl$
(2150-48-3)
$C_{21}H_{27}N_3O_2$
(361-37-5)
$C_{21}H_{27}N_3O_3$
(7722-73-8)
$C_{21}H_{28}N_2O$
(90-93-7)
(552-25-0)
$C_{21}H_{28}N_2O_5$
(138-56-7)
$C_{21}H_{28}N_2S_2$
(51308-54-4)
$C_{21}H_{28}N_4O_2S_2.2I$
(20328-87-4)
$C_{21}H_{28}O_2$
(152-62-5)
(797-63-7)
(6533-00-2)
(27955-87-9)
$C_{21}H_{28}O_3$

(128-62-1)
(6035-40-1)
$C_{22}H_{23}N_3.ClH$
(3248-91-7)
$C_{22}H_{23}O_4P$
(981-40-8)
(23666-93-5)
(28109-02-6)
(56803-37-3)
(72121-83-6)
(73179-45-0)
(75675-48-8)
(83242-23-3)
$C_{22}H_{23}O_4P.C_{18}H_{15}O_4P$
(96300-96-8)
$C_{22}H_{24}ClN_5O_7$
(1533-78-4)
$C_{22}H_{24}N_2O_4$
(91-96-3)
$C_{22}H_{24}N_2O_8$
(60-54-8)
(564-25-0)
$C_{22}H_{24}N_2O_8.ClH$
(64-75-5)
$C_{22}H_{24}N_2O_9$
(79-57-2)
$C_{22}H_{24}N_2O_9.ClH$
(2058-46-0)
$C_{22}H_{24}N_3O_2.CH_3O_4S$
(35869-60-4)
$C_{22}H_{24}N_3O_2.Cl$
(29556-33-0)
$C_{22}H_{25}NO_3$
(24140-30-5)
$C_{22}H_{25}NO_6$
(64-86-8)
$C_{22}H_{26}$
(76501-51-4)
$C_{22}H_{26}F_3N_3OS$
(69-23-8)
$C_{22}H_{26}N_4O_8S$
(29426-52-6)
$C_{22}H_{26}O_3$
(10453-86-8)
(28434-01-7)
$C_{22}H_{26}O_4$
(68052-23-3)
$C_{22}H_{26}O_5$
(3188-83-8)
$C_{22}H_{26}O_6$
(2618-77-1)
$C_{22}H_{27}NO$
(127-35-5)
$C_{22}H_{27}NO_2$
(90-69-7)
(17230-88-5)
$C_{22}H_{27}NO_2.ClH$
(63990-84-1)
$C_{22}H_{28}N_2O$
(437-38-7)
$C_{22}H_{28}N_2O_2$
(144-14-9)
$C_{22}H_{28}N_4O_2.2ClH$
(80547-56-4)
$C_{22}H_{28}N_4O_3$
(911-65-9)
$C_{22}H_{28}O_3$
(51-98-9)
$C_{22}H_{28}O_5$
(121-29-9)
$C_{22}H_{29}FO_5$
(50-02-2)
(53-33-8)
(378-44-9)
$C_{22}H_{29}NO_2$
(469-62-5)
(2338-37-6)
(37339-32-5)
$C_{22}H_{29}N_3O$
(25973-55-1)
$C_{22}H_{29}N_3O_3$
(52184-19-7)
$C_{22}H_{29}N_3S_2$
(1420-55-9)
$C_{22}H_{30}Cl_2N_{10}$
(55-56-1)
$C_{22}H_{30}Cl_2N_{10}.2C_2H_4O_2$
(56-95-1)
$C_{22}H_{30}Cl_2N_{10}.2C_6H_{12}O_7$
(18472-51-0)
$C_{22}H_{30}Cl_2N_{10}.2ClH$
(3697-42-5)
$C_{22}H_{30}N_2O_2S$
(56030-54-7)
$C_{22}H_{30}N_4O_4$
(28540-82-1)
$C_{22}H_{30}O_2S$
(90-66-4)
(96-69-5)
(3818-54-0)
$C_{22}H_{30}O_3$
(3562-63-8)
$C_{22}H_{30}O_5$
(83-43-2)
$C_{22}H_{30}O_7S_2.2Na$
(36445-71-3)
$C_{22}H_{31}NO_3$
(5633-20-5)
$C_{22}H_{31}O_3P$
(3287-06-7)
(26544-23-0)
$C_{22}H_{31}O_3PS$
(95114-70-8)
(95150-16-6)
$C_{22}H_{31}O_4P$
(29761-21-5)
$C_{22}H_{32}$
(38641-16-6)
$C_{22}H_{32}N_2O_5$
(63-12-7)
$C_{22}H_{32}N_4O_8$
(38304-52-8)
$C_{22}H_{32}O_2$
(117-51-1)
(127-47-9)
$C_{22}H_{32}O_3$
(57-85-2)
(520-85-4)
$C_{22}H_{33}Cl_3O_6$
(53535-30-1)
$C_{22}H_{34}O_4$
(89-19-0)
(3648-21-3)
(42343-36-2)
(68515-44-6)
$C_{22}H_{35}Cl_6N_3$
(24481-35-4)
$C_{22}H_{36}O_2$
(1808-26-0)
$C_{22}H_{36}O_7$
(4720-09-6)
$C_{22}H_{38}$
(1459-09-2)
(54934-71-3)
$C_{22}H_{38}O$
(2589-78-8)
(25401-86-9)
(29988-16-7)
$C_{22}H_{38}O_3$
(28777-98-2)
$C_{22}H_{38}O_3S$
(16722-32-0)
$C_{22}H_{38}O_4$
(25724-58-7)
$C_{22}H_{38}O_6$
(15520-11-3)
$C_{22}H_{38}O_7$
(137-66-6)
$C_{22}H_{39}O_3P$
(1754-47-8)
(3164-60-1)
$C_{22}H_{39}O_4P$
(16368-97-1)
$C_{22}H_{40}N.Cl$
(1330-85-4)
$C_{22}H_{40}N_2$
(103-96-8)
(139-60-6)
(28633-36-5)
$C_{22}H_{40}N_2O_4S_2$
(24928-72-1)
$C_{22}H_{40}O_4Si_5$
(20252-66-8)
$C_{22}H_{41}NO_3$
(27883-12-1)
(28984-69-2)
$C_{22}H_{42}N_2O$
(95-38-5)
(27136-73-8)
$C_{22}H_{42}N_2O.C_2H_4O_2$
(3388-72-5)
$C_{22}H_{42}O_2$
(112-86-7)
(142-77-8)
(10024-47-2)
(25378-26-1)
(32360-05-7)
$C_{22}H_{42}O_3$
(106-83-2)
(151-13-3)
$C_{22}H_{42}O_4$
(103-23-1)
(106-12-7)
(108-63-4)
(123-79-5)
(1330-86-5)
(22707-35-3)
$C_{22}H_{42}O_6$
(94-28-0)
(106-10-5)
(110-32-7)
(141-19-5)
(7790-07-0)
(26266-57-9)
$C_{22}H_{42}O_6S$
(38621-44-2)
$C_{22}H_{42}O_7$
(70729-68-9)
$C_{22}H_{42}O_8$
(141-17-3)
$C_{22}H_{43}IO_2.1/2Ca$
(1319-91-1)
$C_{22}H_{43}NO$
(112-84-5)
$C_{22}H_{43}NO_2$
(871-37-4)
$C_{22}H_{43}NO_3$
(93-83-4)
$C_{22}H_{43}NO_6S.2Na$
(14481-60-8)
$C_{22}H_{43}N_3$
(3528-63-0)
(27476-93-3)
$C_{22}H_{43}N_5O_{13}$
(37517-28-5)
$C_{22}H_{44}$
(1599-67-3)
(6812-38-0)
$C_{22}H_{44}N_2O$
(10212-58-5)
$C_{22}H_{44}N_2O.C_2H_4O_2$
(28832-11-3)
$C_{22}H_{44}N_2O_2S_2$
(33312-01-5)
$C_{22}H_{44}N_2PbS_4$
(36501-84-5)
$C_{22}H_{44}N_2S_4Zn$
(15337-18-5)
$C_{22}H_{44}O_2$
(112-85-6)
(123-95-5)
(646-13-9)
(6064-90-0)
(18281-05-5)
(55193-79-8)
$C_{22}H_{44}O_2.Ag$
(2489-05-6)
$C_{22}H_{44}O_2.Li$
(4499-91-6)
$C_{22}H_{44}O_4$
(106-11-6)
$C_{22}H_{44}O_4S_2Sn$
(26636-01-1)
(57583-35-4)
$C_{22}H_{44}O_4Sn$
(68928-76-7)
$C_{22}H_{44}O_6S.Na$
(67859-39-6)
$C_{22}H_{45}NO$
(3061-75-4)
$C_{22}H_{45}NO_2$
(25307-17-9)
$C_{22}H_{45}NO_3$
(93-82-3)
$C_{22}H_{45}N_3O$
(15566-80-0)
$C_{22}H_{46}$
(629-97-0)
$C_{22}H_{46}NO.C_2H_5O_4S$
(78-21-7)
$C_{22}H_{46}N_2O$
(95-19-2)
$C_{22}H_{46}N_2O_2$
(120-41-2)
(141-21-9)
$C_{22}H_{46}O$
(661-19-8)
$C_{22}H_{46}O.1/3Al$
(67905-30-0)
$C_{22}H_{46}O_2Sn$
(28801-69-6)
$C_{22}H_{46}O_6$
(3055-95-6)
$C_{22}H_{47}N$
(14130-06-4)
$C_{22}H_{47}NO_2$
(2269-21-8)
(10213-78-2)
$C_{22}H_{48}N.Br$
(2390-68-3)
(13316-70-6)
$C_{22}H_{48}N.Cl$
(7173-51-5)
$C_{22}H_{48}N_2$
(27090-63-7)
$C_{22}H_{48}O_2$
(68516-18-7)
$C_{23}H_8Cl_8N_4O_2$
(5045-40-9)
$C_{23}H_{10}O$
(86854-07-1)
$C_{23}H_{12}Cl_2N_6O_8S_2.2Na$
(4499-01-8)
$C_{23}H_{12}O$
(7267-90-5)
(83484-79-1)
(86854-08-2)
(86854-09-3)
(86854-10-6)
(86854-11-7)
(86854-12-8)
(86854-13-9)
(86854-14-0)
(86854-15-1)
(86854-16-2)
(86854-17-3)
(86854-18-4)
(86854-19-5)
(86854-20-8)
(86854-21-9)

(86854-22-0)
(86854-23-1)
(86854-24-2)
$C_{23}H_{14}$
(41699-09-6)
$C_{23}H_{14}O_{11}.2Na$
(15826-37-6)
$C_{23}H_{15}ClO_3$
(3691-35-8)
$C_{23}H_{15}Cl_2N_3O_2$
(6041-94-7)
$C_{23}H_{15}O_3.Na$
(42721-99-3)
$C_{23}H_{16}$
(30283-95-5)
(32377-10-9)
$C_{23}H_{16}ClN_5$
(30355-07-8)
$C_{23}H_{16}O_3$
(82-66-6)
$C_{23}H_{16}O_{11}$
(16110-51-3)
$C_{23}H_{17}N$
(63021-33-0)
(63021-35-2)
$C_{23}H_{18}$
(16757-92-9)
$C_{23}H_{18}N_3O_6S.Na$
(2666-17-3)
$C_{23}H_{19}ClF_3NO_3$
(68085-85-8)
$C_{23}H_{19}N_7O_2$
(30805-04-0)
$C_{23}H_{20}N_2O_3S$
(57-96-5)
$C_{23}H_{21}ClN_6O_3$
(61197-73-7)
$C_{23}H_{21}ClO_3$
(569-57-3)
$C_{23}H_{21}F_6O_2P$
(102040-53-9)
$C_{23}H_{21}F_6O_3P$
(102040-44-8)
(102040-51-7)
$C_{23}H_{21}N_5O_6S_2.Na$
(6359-85-9)
$C_{23}H_{22}O_6$
(83-79-4)
$C_{23}H_{23}NO$
(1420-06-0)
$C_{23}H_{23}N_2S_2.I$
(514-73-8)
$C_{23}H_{24}ClN_2.Cl$
(3521-06-0)
$C_{23}H_{24}N_6O_4$
(72828-64-9)
$C_{23}H_{24}O$
(1817-68-1)
$C_{23}H_{24}O_4S$
(58769-20-3)
$C_{23}H_{24}O_6$
(6659-45-6)
$C_{23}H_{25}BrN_6O_{10}$
(3618-72-2)
$C_{23}H_{25}ClN_6O_{10}$
(3618-73-3)
$C_{23}H_{25}F_3N_2OS$
(53772-82-0)
$C_{23}H_{25}N$
(13042-18-7)
$C_{23}H_{25}N_2.Cl$
(569-64-2)
$C_{23}H_{25}N_3$
(68189-23-1)
$C_{23}H_{25}O_4P$
(73179-41-6)
(73179-42-7)
$C_{23}H_{26}N_2O_4$
(357-57-3)
$C_{23}H_{26}N_4O_6S_2.Na$
(32846-21-2)
$C_{23}H_{26}O_3$
(26002-80-2)
$C_{23}H_{27}NO_3$
(56625-58-2)
$C_{23}H_{27}NO_8$
(131-28-2)
$C_{23}H_{27}N_2O.Cl$
(27326-17-6)
$C_{23}H_{27}N_3O_7$
(10118-90-8)
$C_{23}H_{27}N_5O_7$
(1533-76-2)
$C_{23}H_{27}N_5O_8$
(1533-77-3)
$C_{23}H_{28}ClN_3O_2S$
(84-06-0)
$C_{23}H_{28}Cl_3N_3O.2ClH$
(4213-45-0)
$C_{23}H_{29}ClO_4$
(302-22-7)
$C_{23}H_{29}NO$
(561-48-8)
$C_{23}H_{29}NO_2$
(467-84-5)
$C_{23}H_{29}NO_3$
(562-26-5)
(3691-78-9)
$C_{23}H_{29}NO_3.Br$
(50-34-0)
$C_{23}H_{29}N_2.Cl$
(6359-45-1)
$C_{23}H_{29}N_3O_2S_2$
(5591-45-7)
$C_{23}H_{30}ClN_3O$
(83-89-6)
$C_{23}H_{30}ClN_3O.2ClH$
(69-05-6)
$C_{23}H_{30}NO_3$
(298-50-0)
$C_{23}H_{30}N_2O$
(42045-86-3)
$C_{23}H_{30}N_2O_2$
(78995-14-9)
$C_{23}H_{30}N_2O_4$
(509-67-1)
$C_{23}H_{30}O_3$
(54350-48-0)
$C_{23}H_{30}O_4Si_4$
(10448-10-9)
$C_{23}H_{30}O_6$
(50-04-4)
$C_{23}H_{31}NO_2$
(509-74-0)
$C_{23}H_{32}Cl_2NO_6P$
(4891-15-0)
$C_{23}H_{32}O_2$
(79-64-1)
(79-96-9)
(119-47-1)
$C_{23}H_{33}N_5O_6SSi$
(115094-43-4)
$C_{23}H_{34}O_5$
(73573-88-3)
$C_{23}H_{38}N_6O_5$
(4035-89-6)
$C_{23}H_{38}O_3.1/2Ca$
(68540-40-9)
$C_{23}H_{40}$
(14752-75-1)
$C_{23}H_{40}O$
(4306-88-1)
(86812-27-3)
$C_{23}H_{40}O_3S$
(39735-13-2)
$C_{23}H_{40}O_5$
(7311-27-5)
$C_{23}H_{41}Cl_3N_4$
(24802-93-5)
$C_{23}H_{42}N.Cl$
(139-08-2)
(27479-28-3)
$C_{23}H_{42}O_5$
(140-05-6)
$C_{23}H_{43}N_3O_3$
(51365-70-9)
$C_{23}H_{44}O_4$
(15834-05-6)
$C_{23}H_{44}O_5$
(10332-32-8)
$C_{23}H_{45}N_5O_{14}$
(7542-37-2)
$C_{23}H_{46}$
(19781-73-8)
(27519-02-4)
(52078-56-5)
$C_{23}H_{46}N_6O_{13}$
(1404-04-2)
$C_{23}H_{46}O$
(540-09-0)
$C_{23}H_{46}O_2$
(929-77-1)
(2433-96-7)
$C_{23}H_{46}O_5$
(78-23-9)
$C_{23}H_{48}$
(638-67-5)
(55124-80-6)
$C_{23}H_{48}NO_2.Cl$
(18448-65-2)
$C_{23}H_{48}N_2O$
(7651-02-7)
$C_{23}H_{48}N_2O_2$
(25066-20-0)
$C_{23}H_{48}O$
(3133-01-5)
$C_{23}H_{50}NO_2S.Cl$
(78865-87-9)
$C_{24}D_{50}$
(16416-32-3)
$C_{24}H_8Cl_{10}$
(89590-79-4)
$C_{24}H_8O_6$
(128-69-8)
$C_{24}H_{10}Cl_8$
(89590-80-7)
$C_{24}H_{10}N_2O_4$
(81-33-4)
$C_{24}H_{12}$
(191-07-1)
$C_{24}H_{12}Cl_4CuN_4$
(69742-55-8)
$C_{24}H_{12}Cl_6$
(89590-81-8)
$C_{24}H_{12}O$
(84665-41-8)
$C_{24}H_{12}O_2$
(128-66-5)
$C_{24}H_{14}$
(189-55-9)
(189-64-0)
(189-96-8)
(191-30-0)
(192-51-8)
(192-65-4)
(193-09-9)
(196-42-9)
(197-70-6)
(203-20-3)
(205-97-0)
(207-18-1)
(238-04-0)
(2997-45-7)
(5385-75-1)
$C_{24}H_{16}As_2O_3$
(58-36-6)
$C_{24}H_{16}Cl_2N_6O_{10}S_3.3Na$
(70616-90-9)
$C_{24}H_{16}Cl_3N_3O_2$
(6535-46-2)
$C_{24}H_{16}Hg_2O_9$
(3570-80-7)
$C_{24}H_{16}N_2O_2$
(1806-34-4)
$C_{24}H_{16}N_2O_4$
(4170-07-4)
$C_{24}H_{16}N_3O_3S.Na$
(6416-68-8)
$C_{24}H_{17}ClN_4O_4$
(6410-30-6)
$C_{24}H_{17}Cl_2N_3O_3$
(6410-38-4)
$C_{24}H_{17}N_3O_3S.Na$
(56776-27-3)
$C_{24}H_{17}N_5O_7$
(6471-49-4)
$C_{24}H_{18}$
(135-70-6)
(612-71-5)
(641-96-3)
(1166-18-3)
(1679-02-3)
(29036-02-0)
$C_{24}H_{18}N_2O_4$
(6448-95-9)
$C_{24}H_{18}N_6O_3$
(5637-84-3)
$C_{24}H_{18}O$
(58841-70-6)
$C_{24}H_{19}F_{34}N_2O_8PS_2.H_3N$
(30381-98-7)
$C_{24}H_{19}N_5O_7S.2Na$
(10482-43-6)
$C_{24}H_{19}N_{12}O_{20}.Zr$
(63868-82-6)
$C_{24}H_{19}O_4P$
(132-29-6)
$C_{24}H_{20}B.Na$
(143-66-8)
$C_{24}H_{20}N_4O$
(85-83-6)
$C_{24}H_{20}O_6$
(614-33-5)
$C_{24}H_{20}Sn$
(595-90-4)
$C_{24}H_{21}Cl_2N_5O_6S_2.Na$
(12217-38-8)
$C_{24}H_{21}Cl_6O_6P$
(94-84-8)
$C_{24}H_{21}Cl_6O_6P.C_{10}H_{12}N_2O_5$
(8005-49-0)
$C_{24}H_{21}N_3$
(6726-45-0)
(13960-31-1)
$C_{24}H_{21}N_3O_3$
(606-03-1)
(1919-46-6)
(5434-82-2)
(7753-12-0)
$C_{24}H_{21}N_3S_3$
(13270-05-8)
$C_{24}H_{21}N_5$
(6368-72-5)
$C_{24}H_{21}N_7O_{11}S_3$
(12239-00-8)
$C_{24}H_{21}N_9O_3$
(30805-05-1)
$C_{24}H_{22}$
(74229-81-5)
$C_{24}H_{22}N_2O_3$
(24460-06-8)
$C_{24}H_{23}F_6O_4P$
(102040-55-1)
$C_{24}H_{23}N_3O_6S$
(47747-56-8)
$C_{24}H_{23}N_3O_8S_2$
(36897-88-8)
$C_{24}H_{23}N_3O_8S_2.Na$
(25797-81-3)
$C_{24}H_{24}$

(28213-80-1)
$C_{24}H_{24}F_6NO_2P$
(102040-48-2)
$C_{24}H_{24}N_2O_4$
(3688-66-2)
$C_{24}H_{25}NO_3$
(14297-87-1)
$C_{24}H_{26}O$
(2772-45-4)
$C_{24}H_{26}O_6$
(57230-49-6)
$C_{24}H_{27}BrN_6O_{10}$
(12239-34-8)
$C_{24}H_{27}N$
(390-64-7)
$C_{24}H_{27}NO_2$
(6197-30-4)
$C_{24}H_{27}O_4P$
(121-06-2)
(3862-11-1)
(3862-12-2)
(5770-08-1)
(19074-59-0)
(25155-23-1)
(25653-16-1)
(28109-00-4)
(58570-87-9)
(64532-96-3)
(69500-29-4)
(69682-29-7)
(73179-44-9)
(73179-48-3)
(73179-49-4)
$C_{24}H_{28}O_3$
(22431-62-5)
$C_{24}H_{28}O_4$
(130-80-3)
$C_{24}H_{29}N$
(51772-35-1)
$C_{24}H_{29}NO$
(468-07-5)
$C_{24}H_{30}Cl_2N_5O_7$
(12663-46-6)
$C_{24}H_{30}F_2O_6$
(67-73-2)
$C_{24}H_{30}N_2O_2$
(309-29-5)
$C_{24}H_{30}N_2O_3$
(59708-52-0)
$C_{24}H_{30}O_4$
(140-24-9)
$C_{24}H_{31}FO_6$
(76-25-5)
$C_{24}H_{31}N_3OS$
(653-03-2)
$C_{24}H_{32}O_3$
(979-32-8)
$C_{24}H_{32}O_4$
(113-38-2)
(297-76-7)
(595-33-5)
$C_{24}H_{32}O_4S$
(52-01-7)
$C_{24}H_{32}O_6$
(53-36-1)
$C_{24}H_{33}FO_6$
(1524-88-5)
$C_{24}H_{33}O_4$
(1172-82-3)
$C_{24}H_{34}O$
(25619-63-0)
$C_{24}H_{34}O_4$
(71-58-9)
$C_{24}H_{34}O_7S_2$
(30260-72-1)
$C_{24}H_{34}O_7S_2.2Na$
(28519-02-0)
$C_{24}H_{34}O_9$
(21259-20-1)
$C_{24}H_{36}N_6O_6$
(3779-63-3)
$C_{24}H_{36}O_3$
(51022-71-0)
$C_{24}H_{36}O_5$
(75330-75-5)
$C_{24}H_{38}HgO_2$
(104-60-9)
$C_{24}H_{38}O_4$
(117-81-7)
(117-84-0)
(131-15-7)
(137-89-3)
(6422-86-2)
(27554-26-3)
(68515-41-3)
(68515-42-4)
$C_{24}H_{39}N$
(89-28-1)
$C_{24}H_{39}NO_4$
(468-76-8)
$C_{24}H_{39}NO_5$
(36150-73-9)
$C_{24}H_{40}N_8O_4$
(58-32-2)
$C_{24}H_{40}O_3$
(434-13-9)
$C_{24}H_{40}O_4$
(83-44-3)
(474-25-9)
$C_{24}H_{40}O_5$
(81-25-4)
$C_{24}H_{40}O_8$
(5281-13-0)
$C_{24}H_{40}O_8Sn$
(15546-16-4)
$C_{24}H_{41}NO_2$
(103-99-1)
$C_{24}H_{42}$
(4445-07-2)
$C_{24}H_{42}O$
(137-99-5)
(1323-65-5)
$C_{24}H_{44}O_4$
(84-71-9)
(140-04-5)
$C_{24}H_{44}O_6$
(1338-43-8)
$C_{24}H_{46}O_2$
(45294-18-6)
$C_{24}H_{46}O_3$
(109-39-7)
$C_{24}H_{46}O_4$
(105-74-8)
(110-29-2)
(151-32-6)
(31474-57-4)
(33703-08-1)
(54050-62-3)
(68052-04-0)
$C_{24}H_{46}O_5Si_6$
(60573-45-7)
$C_{24}H_{46}O_6$
(106-06-9)
(1338-41-6)
(13052-09-0)
(68937-69-9)
$C_{24}H_{46}O_7$
(18268-70-7)
$C_{24}H_{46}O_7S.Na$
(29857-13-4)
$C_{24}H_{46}O_{13}$
(25339-99-5)
$C_{24}H_{47}NO_4$
(10277-04-0)
$C_{24}H_{47}NO_4S.Na$
(132-43-4)
$C_{24}H_{47}N_2O.C_2H_5O_4S$
(26266-76-2)
$C_{24}H_{47}N_5$
(26235-90-5)
$C_{24}H_{48}$
(4445-06-1)
(27195-67-1)
$C_{24}H_{48}N_2O_4$
(139-92-4)
$C_{24}H_{48}O_2$
(557-59-5)
(5908-87-2)
(29806-73-3)
$C_{24}H_{48}O_3$
(109-38-6)
$C_{24}H_{48}O_4.Cd$
(2605-44-9)
$C_{24}H_{48}O_4Sn$
(2781-10-4)
$C_{24}H_{49}NO_4$
(10248-74-5)
$C_{24}H_{50}$
(646-31-1)
$C_{24}H_{50}N_2O$
(16889-14-8)
$C_{24}H_{50}O$
(506-51-4)
(4542-57-8)
(58670-89-6)
$C_{24}H_{50}O.1/3Al$
(67905-29-7)
$C_{24}H_{50}O_7$
(3055-96-7)
(14149-99-6)
$C_{24}H_{50}S_2$
(27458-90-8)
$C_{24}H_{50}Se$
(5819-01-2)
$C_{24}H_{51}Al$
(1070-00-4)
$C_{24}H_{51}N$
(1116-76-3)
(3007-31-6)
(25549-16-0)
$C_{24}H_{51}NO_2$
(28137-64-6)
$C_{24}H_{51}OP$
(78-50-2)
$C_{24}H_{51}O_3P$
(301-13-3)
(21302-09-0)
$C_{24}H_{51}O_4P$
(78-42-2)
(1806-54-8)
(7057-92-3)
$C_{24}H_{51}O_{10}P$
(7332-46-9)
$C_{24}H_{51}P$
(4731-53-7)
$C_{24}H_{52}O_4P_2S_4Zn$
(2215-35-2)
(7282-28-2)
(15874-15-4)
(26566-95-0)
$C_{24}H_{52}O_4Si$
(78-13-7)
$C_{24}H_{54}OSn_2$
(56-35-9)
$C_{24}H_{54}O_3P$
(25103-12-2)
$C_{25}H_{12}O$
(86854-25-3)
$C_{25}H_{14}O$
(86854-26-4)
$C_{25}H_{15}Cl_2N_5O_{11}S_3.3Na$
(2407-13-8)
$C_{25}H_{15}N_3O_2$
(3333-62-8)
$C_{25}H_{18}$
(17165-86-5)
$C_{25}H_{18}O$
(72776-75-1)
$C_{25}H_{19}Cl_2N_3O_2$
(6471-51-8)
$C_{25}H_{19}N_3$
(21620-54-2)
$C_{25}H_{20}N_2O_2$
(20222-29-1)
$C_{25}H_{20}N_2O_5$
(25177-16-6)
$C_{25}H_{20}N_4O_3$
(16403-84-2)
$C_{25}H_{20}N_4O_4$
(6655-84-1)
(36968-27-1)
$C_{25}H_{22}ClNO_3$
(51630-58-1)
(66230-04-4)
(66267-77-4)
(67614-32-8)
(67614-33-9)
$C_{25}H_{22}N_4O_9S_3.2Na$
(10190-68-8)
$C_{25}H_{22}P.Cl$
(1100-88-5)
$C_{25}H_{23}NO_7S$
(28517-81-9)
$C_{25}H_{24}$
(1242-76-8)
$C_{25}H_{24}F_6N_4$
(67485-29-4)
$C_{25}H_{24}N_6O.3ClH$
(23491-45-4)
$C_{25}H_{25}F_6O_2P$
(102040-50-6)
(102040-52-8)
$C_{25}H_{25}F_6O_5P$
(102040-49-3)
$C_{25}H_{26}O_3$
(2393-53-5)
$C_{25}H_{27}ClN_2$
(569-65-3)
$C_{25}H_{28}O_3$
(50-50-0)
$C_{25}H_{29}O_4P$
(73179-37-0)
(73179-46-1)
(73179-47-2)
$C_{25}H_{30}N_2O_4$
(28768-32-3)
$C_{25}H_{30}N_3.Cl$
(548-62-9)
$C_{25}H_{31}N_3$
(603-48-5)
$C_{25}H_{32}N_2O_2$
(357-56-2)
(545-59-5)
(5666-11-5)
$C_{25}H_{32}OS$
(82799-46-0)
$C_{25}H_{33}NO_4$
(14521-96-1)
$C_{25}H_{33}NO_4.ClH$
(13764-49-3)
$C_{25}H_{34}N_2$
(2162-74-5)
$C_{25}H_{34}O_3$
(2985-59-3)
$C_{25}H_{35}NO_9$
(15662-33-6)
$C_{25}H_{36}Cl_6O_4$
(4827-55-8)
$C_{25}H_{36}O_2$
(88-24-4)
$C_{25}H_{38}O_2$
(32904-22-6)
$C_{25}H_{39}NO_6.ClH$
(23451-24-3)
$C_{25}H_{40}N.Cl$
(1733-96-6)
$C_{25}H_{40}O_6$
(55320-02-0)
$C_{25}H_{42}$
(119973-28-3)

$C_{25}H_{42}O_3$
(14230-52-5)

$C_{25}H_{42}O_3S$
(80387-97-9)

$C_{25}H_{44}$
(29136-19-4)

$C_{25}H_{44}N_2O.2ClH$
(1249-84-9)

$C_{25}H_{45}NO_8S.2Na$
(67815-88-7)

$C_{25}H_{46}N.Cl$
(122-18-9)
(27479-29-4)

$C_{25}H_{48}N_6O_8$
(70-51-9)

$C_{25}H_{48}O_4$
(103-24-2)
(26544-17-2)
(29733-18-4)
(99562-17-1)

$C_{25}H_{50}O_2$
(506-38-7)
(18281-07-7)

$C_{25}H_{50}O_6P_2$
(26544-27-4)

$C_{25}H_{52}$
(629-99-2)

$C_{25}H_{53}N_2O_2.NO_3$
(2764-13-8)

$C_{25}H_{54}N.Cl$
(5137-55-3)

$C_{25}H_{54}NO_2S.Cl$
(78865-89-1)

$C_{25}H_{54}N_2$
(15268-40-3)

$C_{26}H_6Cl_8N_2O_4$
(30125-47-4)

$C_{26}H_{10}N_6O_{16}$
(116-78-9)

$C_{26}H_{12}N_4O_2$
(4216-02-8)
(4424-06-0)
(8005-56-9)

$C_{26}H_{14}N_2O_4$
(5521-31-3)

$C_{26}H_{16}$
(216-00-2)
(217-37-8)
(217-54-9)
(226-86-8)
(227-04-3)

$C_{26}H_{16}N_2O_5S_2.Na$
(4121-67-9)

$C_{26}H_{16}O_4$
(82-17-7)
(82-21-3)

$C_{26}H_{18}$
(1499-10-1)

$C_{26}H_{18}Cl_4O$
(74562-99-5)

$C_{26}H_{18}N_4O_8S_2.2Na$
(5850-16-8)

$C_{26}H_{18}O_6.Cu$
(5328-04-1)

$C_{26}H_{19}N_3O_{10}S_3.3Na$
(3861-73-2)

$C_{26}H_{19}N_5O_5$
(6985-92-8)

$C_{26}H_{19}N_5O_8S_2.2Na$
(6227-02-7)

$C_{26}H_{20}$
(632-51-9)

$C_{26}H_{20}ClN_5O_4$
(12236-64-5)

$C_{26}H_{20}N_2$
(93-46-9)

$C_{26}H_{20}N_2O_2S_2$
(135-57-9)

$C_{26}H_{20}N_4O_8S_2$
(91-34-9)

$C_{26}H_{20}N_4O_8S_2.2Na$
(3051-11-4)

$C_{26}H_{20}N_6O_{10}$
(29398-96-7)

$C_{26}H_{21}NO_6$
(18869-73-3)

$C_{26}H_{22}ClF_3N_2O_3$
(69409-94-5)

$C_{26}H_{22}N_4O_4$
(2786-76-7)

$C_{26}H_{22}N_4O_8S_2.Na$
(12220-06-3)

$C_{26}H_{22}O$
(574-42-5)

$C_{26}H_{23}Cl_4NO_6$
(56141-00-5)

$C_{26}H_{23}F_2NO_4$
(70124-77-5)

$C_{26}H_{24}Cl_6O_8$
(30431-54-0)

$C_{26}H_{24}O_6$
(2652-25-7)
(4196-87-6)

$C_{26}H_{25}N_5O_{19}S_6.4Na$
(17095-24-8)

$C_{26}H_{26}N_2O_2S$
(7128-64-5)

$C_{26}H_{26}OSi_2$
(807-28-3)

$C_{26}H_{27}ClN_4O_8S$
(12225-18-2)

$C_{26}H_{28}$
(29253-36-9)

$C_{26}H_{28}ClNO$
(911-45-5)

$C_{26}H_{28}ClNO.C_6H_8O_7$
(50-41-9)

$C_{26}H_{28}N_3$
(7187-62-4)

$C_{26}H_{29}N_3O_2$
(1552-42-7)

$C_{26}H_{30}N_2O_2$
(15958-68-6)

$C_{26}H_{30}O_{11}$
(21794-01-4)

$C_{26}H_{31}O_4P$
(115-87-7)
(65652-41-7)
(73179-38-1)
(73195-13-8)

$C_{26}H_{32}N_2O$
(34913-07-0)

$C_{26}H_{32}O_{11}$
(22467-31-8)

$C_{26}H_{33}NO_2$
(65646-68-6)

$C_{26}H_{33}N_3.2Cl$
(82-94-0)

$C_{26}H_{35}NO_4$
(14357-78-9)

$C_{26}H_{38}O_2$
(85-60-9)

$C_{26}H_{40}$
(56388-47-7)
(119973-31-8)

$C_{26}H_{40}MnN_6O_6S_8Zn$
(8064-42-4)

$C_{26}H_{40}O_3$
(315-37-7)

$C_{26}H_{42}$
(119973-29-4)

$C_{26}H_{42}O_4$
(84-76-4)
(119-07-3)
(1330-96-7)
(28553-12-0)
(42343-35-1)
(68515-45-7)
(68515-48-0)
(68515-51-5)
(68648-91-9)
(71549-78-5)

$C_{26}H_{43}NO_3.Na$
(5994-45-6)

$C_{26}H_{44}$
(119973-30-7)

$C_{26}H_{44}NO_7S.Na$
(145-42-6)

$C_{26}H_{45}NO_7S$
(81-24-3)

$C_{26}H_{46}$
(2398-64-3)
(2398-65-4)
(2398-66-5)
(2398-68-7)
(2400-02-4)
(2400-03-5)
(2400-04-6)

$C_{26}H_{46}O_7$
(2497-58-7)

$C_{26}H_{47}NO_{10}S.4Na$
(3401-73-8)

$C_{26}H_{47}O_3P$
(1254-78-0)
(25550-98-5)

$C_{26}H_{47}O_4P$
(51363-64-5)

$C_{26}H_{48}O_2$
(67874-38-8)

$C_{26}H_{48}O_4$
(63907-12-0)

$C_{26}H_{50}O_2$
(26761-50-2)

$C_{26}H_{50}O_3$
(106-84-3)
(141-38-8)

$C_{26}H_{50}O_4$
(105-97-5)
(122-62-3)
(27178-16-1)

$C_{26}H_{51}N_5$
(26322-48-5)

$C_{26}H_{52}$
(4443-55-4)
(4443-61-2)
(6703-81-7)
(6703-82-8)
(18835-33-1)

$C_{26}H_{52}O_2$
(506-46-7)
(22047-49-0)
(55373-89-2)

$C_{26}H_{54}$
(630-01-3)

$C_{26}H_{54}O$
(506-52-5)

$C_{26}H_{54}O.1/3Al$
(67905-28-6)

$C_{26}H_{54}O_8$
(3055-97-8)

$C_{26}H_{55}N$
(5910-75-8)

$C_{26}H_{55}O_4P$
(5116-95-0)
(27073-01-4)

$C_{26}H_{56}N.Br$
(3282-73-3)

$C_{26}H_{56}N.Cl$
(3401-74-9)

$C_{26}H_{56}S_2Sn$
(51287-84-4)

$C_{26}H_{58}NO_3Si.Cl$
(27668-52-6)

$C_{27}H_{18}Cl_3N_3$
(30362-99-3)

$C_{27}H_{18}N_6O_6$
(26603-40-7)

$C_{27}H_{21}Cl_6N_3O_3$
(30363-05-4)

$C_{27}H_{21}N_3$
(21577-41-3)
(30805-23-3)

$C_{27}H_{22}ClN_7O_{10}S_3.3Na$
(70210-46-7)

$C_{27}H_{22}N_2O_4$
(101-65-5)

$C_{27}H_{22}N_6O_8S.2Na$
(7248-45-5)

$C_{27}H_{24}Cl_3N_3O_3$
(30363-04-3)

$C_{27}H_{24}N_2O_5$
(19014-53-0)

$C_{27}H_{24}N_4O_7S_3.2Na$
(2498-95-5)

$C_{27}H_{24}N_6O_7S$
(12225-08-0)

$C_{27}H_{24}N_6O_9S_2$
(25712-08-7)

$C_{27}H_{24}N_6O_9S_2.2Na$
(10114-86-0)

$C_{27}H_{25}O_4P$
(34364-42-6)
(50851-28-0)

$C_{27}H_{27}N_5O_7$
(29765-00-2)

$C_{27}H_{28}Br_2O_5S$
(76-59-5)

$C_{27}H_{28}N_2O_4$
(52235-18-4)

$C_{27}H_{29}NO_{10}$
(20830-81-3)

$C_{27}H_{29}NO_{11}$
(23214-92-8)

$C_{27}H_{29}NO_{11}.ClH$
(25316-40-9)

$C_{27}H_{29}N_2O_3.Cl$
(3068-39-1)

$C_{27}H_{30}Hg_3O_6$
(1300-78-3)

$C_{27}H_{30}O_{16}$
(153-18-4)

$C_{27}H_{31}N_2O_6S_2.Na$
(129-17-9)

$C_{27}H_{32}O_6$
(24448-20-2)

$C_{27}H_{32}O_{14}$
(10236-47-2)

$C_{27}H_{33}N_2.C_2H_2O_4.1/2C_2O_4$
(36351-18-5)

$C_{27}H_{33}N_2.HO_4S$
(633-03-4)

$C_{27}H_{33}O_4P$
(2502-15-0)
(26967-76-0)
(38638-05-0)
(56444-79-2)
(64532-97-4)
(72668-27-0)

$C_{27}H_{34}N_4O$
(302-41-0)

$C_{27}H_{34}O_6$
(16883-83-3)

$C_{27}H_{35}NO_5$
(25333-77-1)

$C_{27}H_{38}N_2O_4$
(52-53-9)

$C_{27}H_{39}NO_2$
(60-70-8)

$C_{27}H_{39}NO_3$
(469-59-0)

$C_{27}H_{40}O_4$
(630-56-8)

$C_{27}H_{41}NO_2$
(4449-51-8)

$C_{27}H_{42}NO_2.Cl$
(121-54-0)

$C_{27}H_{42}O$
(567-72-6)

$C_{27}H_{42}O_6$
(1528-49-0)

(34870-88-7)
(67989-23-5)
$C_{27}H_{43}NO_2$
(126-17-0)
$C_{27}H_{44}$
(747-90-0)
(69760-73-2)
(81546-39-6)
$C_{27}H_{44}O$
(67-97-0)
(434-16-2)
(601-54-7)
(15459-85-5)
$C_{27}H_{44}O_3$
(32222-06-3)
$C_{27}H_{45}NO_3$
(7333-86-0)
$C_{27}H_{46}$
(570-74-1)
(28338-69-4)
(53584-59-1)
(55199-72-9)
$C_{27}H_{46}O$
(57-88-5)
$C_{27}H_{46}O_2$
(1250-95-9)
$C_{27}H_{48}$
(481-21-0)
(56975-84-9)
(57030-15-6)
$C_{27}H_{48}N.Cl$
(37139-99-4)
$C_{27}H_{48}O$
(80-97-7)
(360-68-9)
$C_{27}H_{48}O_2$
(3347-60-2)
$C_{27}H_{50}N.Cl$
(122-19-0)
(29656-52-8)
$C_{27}H_{50}O_6$
(78-16-0)
(538-23-8)
$C_{27}H_{52}ClN_5$
(30355-03-4)
$C_{27}H_{54}$
(55282-34-3)
$C_{27}H_{54}O$
(542-50-7)
$C_{27}H_{54}O_2$
(5802-82-4)
(7138-40-1)
$C_{27}H_{55}N_2.Cl$
(71729-96-9)
$C_{27}H_{56}$
(593-49-7)
$C_{27}H_{57}O_4P$
(13018-37-6)
$C_{27}H_{58}NO_2S.Cl$
(78865-90-4)
$C_{27}H_{60}NO_3Si.Cl$
(68959-20-6)
$C_{28}H_{12}Cl_2N_2O_4$
(130-20-1)
$C_{28}H_{13}ClN_2O_4$
(1324-27-2)
$C_{28}H_{14}$
(190-70-5)
$C_{28}H_{14}N_2O_2S_2$
(129-09-9)
$C_{28}H_{14}N_2O_4$
(81-77-6)
(4424-87-7)
$C_{28}H_{16}$
(191-20-8)
(192-47-2)
(192-59-6)
(193-11-3)
(196-45-2)
(385-14-8)
$C_{28}H_{16}N_2O_4$
(4051-63-2)
$C_{28}H_{17}N_9O_{16}S_2.2Na$
(6637-87-2)
$C_{28}H_{18}$
(58615-36-4)
(60382-88-9)
$C_{28}H_{18}N_2O_2$
(1533-45-5)
$C_{28}H_{18}N_2O_6$
(81-78-7)
$C_{28}H_{18}N_3NaO_7S$
(1328-24-1)
$C_{28}H_{20}N_2O_8S_2.2Na$
(4403-90-1)
$C_{28}H_{20}O_6S_2.2Na$
(27344-41-8)
$C_{28}H_{21}N_5O_{14}S_4.4Na$
(2519-30-4)
$C_{28}H_{22}N_2O_2$
(128-80-3)
(8005-40-1)
$C_{28}H_{23}N_9O_7S_2.2Na$
(68877-33-8)
$C_{28}H_{26}N_6O_6S.2Na$
(6637-88-3)
$C_{28}H_{27}N_2.Cl$
(4657-00-5)
$C_{28}H_{28}N_2O_2$
(28782-42-5)
$C_{28}H_{28}O_6$
(67786-03-2)
$C_{28}H_{30}N_2O_3$
(509-34-2)
(3375-25-5)
$C_{28}H_{31}N_2O_3.Cl$
(81-88-9)
$C_{28}H_{30}N_2O_3.ClH$
(989-38-8)
$C_{28}H_{32}NO_7$
(36011-19-5)
$C_{28}H_{32}O_2Si_3$
(3982-82-9)
$C_{28}H_{32}O_4Si_4$
(77-63-4)
$C_{28}H_{35}NO_4.ClH$
(67652-39-5)
$C_{28}H_{36}Hg_2O_4$
(27236-65-3)
$C_{28}H_{36}O_{15}$
(20702-77-6)
$C_{28}H_{38}N_2O_4.2ClH$
(5853-29-2)
$C_{28}H_{38}O_{19}$
(126-14-7)
$C_{28}H_{40}N_2O_9$
(1397-94-0)
$C_{28}H_{40}NiO_2S$
(27574-34-1)
$C_{28}H_{41}N_3O_3$
(126-27-2)
$C_{28}H_{42}O_2S$
(3294-03-9)
$C_{28}H_{42}O_7$
(61235-00-5)
$C_{28}H_{43}N$
(101-67-7)
$C_{28}H_{44}$
(63512-64-1)
$C_{28}H_{44}O$
(50-14-6)
(57-87-4)
$C_{28}H_{44}O_3S$
(25322-17-2)
$C_{28}H_{44}O_3S.1/2Ca$
(57855-77-3)
$C_{28}H_{44}O_3S.1/2Ba$
(25619-56-1)
$C_{28}H_{44}O_3S.Li$
(28214-91-7)
$C_{28}H_{44}O_3S.Na$
(26834-28-6)
$C_{28}H_{44}O_3S.1/2Zn$
(28016-00-4)
$C_{28}H_{46}$
(77327-07-2)
$C_{28}H_{46}O$
(67-96-9)
(474-67-9)
$C_{28}H_{46}O_2$
(19356-17-3)
$C_{28}H_{46}O_4$
(84-77-5)
(26761-40-0)
(53306-54-0)
$C_{28}H_{48}$
(65636-26-2)
$C_{28}H_{48}O$
(474-62-4)
(17105-75-8)
(62014-96-4)
$C_{28}H_{48}O_2$
(1406-66-2)
$C_{28}H_{50}O$
(43217-65-8)
$C_{28}H_{52}Cl_3N_5$
(26235-01-8)
$C_{28}H_{52}O_4$
(2915-52-8)
$C_{28}H_{52}O_{12}$
(26446-38-8)
$C_{28}H_{54}O_2$
(3687-46-5)
$C_{28}H_{54}O_4$
(4121-16-8)
$C_{28}H_{54}O_6$
(82065-80-3)
$C_{28}H_{54}O_6Sn$
(23850-94-4)
$C_{28}H_{54}O_7S.Na$
(4229-35-0)
$C_{28}H_{55}N_5$
(26234-38-8)
$C_{28}H_{55}O_6PS_2$
(68413-48-9)
$C_{28}H_{55}P$
(13886-99-2)
(13887-00-8)
$C_{28}H_{56}O_2$
(506-48-9)
(3234-85-3)
(31565-38-5)
(55682-91-2)
$C_{28}H_{56}O_4S_2Sn$
(25168-24-5)
$C_{28}H_{58}$
(630-02-4)
(14167-66-9)
$C_{28}H_{58}O$
(557-61-9)
$C_{28}H_{58}O.1/3Al$
(67905-27-5)
$C_{29}H_{10}Br_8N_2O_4$
(32588-74-2)
$C_{29}H_{12}Cl_8N_6O_2$
(40716-47-0)
$C_{29}H_{14}N_2O_5$
(2379-79-5)
$C_{29}H_{19}N_5O_8S_2.2Na$
(2610-11-9)
$C_{29}H_{21}N_5O_6S.2Na$
(2429-79-0)
$C_{29}H_{21}N_5O_7S.2Na$
(2429-82-5)
(2429-84-7)
$C_{29}H_{22}N_6O_7S.2Na$
(13164-93-7)
$C_{29}H_{24}N_6$
(4197-25-5)
$C_{29}H_{25}N_3O_5$
(639-48-5)
$C_{29}H_{25}N_5O_5$
(31778-10-6)
$C_{29}H_{26}O_{10}$
(35082-49-6)
$C_{29}H_{29}N_2O_5.Cl.2Na$
(37299-86-8)
$C_{29}H_{30}O_{10}$
(19314-74-0)
$C_{29}H_{32}ClN_3$
(2185-86-6)
$C_{29}H_{33}ClN_2O_2$
(53179-11-6)
$C_{29}H_{36}O_8$
(1565-94-2)
$C_{29}H_{37}NO_5$
(14930-96-2)
$C_{29}H_{38}N_2O_4$
(4914-30-1)
$C_{29}H_{40}N_2O_4$
(483-18-1)
$C_{29}H_{40}N_2O_4.2ClH$
(316-42-7)
$C_{29}H_{40}O_2$
(77-62-3)
$C_{29}H_{40}O_9$
(1986-70-5)
(20304-47-6)
$C_{29}H_{40}O_{10}$
(20304-49-8)
$C_{29}H_{41}NO_4$
(52485-79-7)
$C_{29}H_{42}O_3$
(4221-80-1)
$C_{29}H_{42}O_{10}$
(508-75-8)
$C_{29}H_{42}O_{11}$
(639-13-4)
$C_{29}H_{44}O_2$
(118-82-1)
(27725-17-3)
$C_{29}H_{44}O_4$
(38134-94-0)
$C_{29}H_{44}O_{12}$
(630-60-4)
$C_{29}H_{46}$
(86709-50-4)
$C_{29}H_{46}CaO_2$
(68527-62-8)
$C_{29}H_{48}$
(79897-80-6)
$C_{29}H_{48}NO_3$
(36069-45-1)
$C_{29}H_{48}O$
(83-48-7)
(18472-36-1)
$C_{29}H_{48}O_2$
(604-35-3)
$C_{29}H_{48}O_3$
(1256-83-3)
$C_{29}H_{50}$
(3258-87-5)
(36728-72-0)
(53584-60-4)
$C_{29}H_{50}O$
(83-46-5)
(83-47-6)
$C_{29}H_{50}O_2$
(59-02-9)
$C_{29}H_{50}O_6$
(99554-33-3)
$C_{29}H_{52}$
(67597-34-6)
(67597-35-7)
$C_{29}H_{56}O_4$
(28472-97-1)
$C_{29}H_{58}O_2$
(4250-38-8)
(55682-92-3)
$C_{29}H_{60}$

(630-03-5)
$C_{30}H_{12}Br_2O_2$
(1324-35-2)
$C_{30}H_{14}O_2$
(128-70-1)
$C_{30}H_{16}$
(385-13-7)
$C_{30}H_{18}Cl_2N_4O_8S_2.2Na$
(6527-70-4)
$C_{30}H_{22}N_6O_6S_2$
(37069-54-8)
$C_{30}H_{22}O_{10}$
(23537-16-8)
$C_{30}H_{22}O_{12}$
(21884-44-6)
$C_{30}H_{23}BrO_4$
(28772-56-7)
$C_{30}H_{24}N_4O_8S_2.2Na$
(3530-19-6)
$C_{30}H_{25}N_5O_8S_2.2Na$
(6227-14-1)
$C_{30}H_{28}N_2S_4Zn$
(14726-36-4)
$C_{30}H_{28}N_4O_8S_2.2Na$
(2870-32-8)
$C_{30}H_{28}N_6$
(3283-07-6)
$C_{30}H_{29}N_3O$
(70321-86-7)
$C_{30}H_{29}N_3O_3$
(70693-50-4)
$C_{30}H_{31}N$
(10081-67-1)
$C_{30}H_{32}N_2O_2$
(915-30-0)
$C_{30}H_{32}O_6$
(6130-72-9)
(68517-02-2)
$C_{30}H_{37}NO_6$
(22144-77-0)
$C_{30}H_{39}O_4P$
(78-33-1)
(28777-70-0)
$C_{30}H_{42}N_6O_4.O_4S$
(64-47-1)
$C_{30}H_{46}ClO_2P$
(63302-49-8)
$C_{30}H_{46}O_2S$
(28503-85-7)
$C_{30}H_{47}N$
(24925-59-5)
(36878-20-3)
$C_{30}H_{50}$
(111-02-4)
(546-99-6)
(1615-91-4)
(7683-64-9)
$C_{30}H_{50}N_4O_{19}$
(1398-61-4)
$C_{30}H_{50}O$
(120056-15-7)
$C_{30}H_{50}O_2$
(633-31-8)
(3648-20-2)
$C_{30}H_{50}O_4$
(68515-49-1)
$C_{30}H_{52}$
(471-62-5)
(1176-44-9)
(13849-96-2)
$C_{30}H_{52}O$
(473-03-0)
$C_{30}H_{54}O$
(25482-47-7)
$C_{30}H_{56}O_{12}$
(25168-73-4)
$C_{30}H_{57}N_3$
(30362-98-2)
$C_{30}H_{58}O_2Sn$
(24124-25-2)
$C_{30}H_{58}O_4$
(2432-89-5)
$C_{30}H_{58}O_4S$
(123-28-4)
$C_{30}H_{58}O_6$
(53220-22-7)
$C_{30}H_{58}O_7S.Na$
(2673-22-5)
$C_{30}H_{59}NO_4$
(52900-12-6)
$C_{30}H_{60}N_3O_3.3I$
(65-29-2)
$C_{30}H_{60}O_2$
(506-50-3)
(2599-01-1)
(4082-55-7)
$C_{30}H_{62}$
(111-01-3)
(638-68-6)
(14167-67-0)
$C_{30}H_{62}O$
(593-50-0)
$C_{30}H_{62}O_{21}$
(9041-07-0)
$C_{30}H_{63}Al$
(1726-66-5)
$C_{30}H_{63}N$
(35723-89-8)
$C_{30}H_{63}O_3P$
(2929-86-4)
(25448-25-3)
$C_{30}H_{63}O_4P$
(4200-55-9)
$C_{30}H_{64}N.Cl$
(10108-91-5)
$C_{31}H_{15}NO_3$
(3271-76-9)
$C_{31}H_{20}N_6O_9S.Cu.2Na$
(16071-86-6)
$C_{31}H_{20}O_8$
(38103-06-9)
$C_{31}H_{22}N_6O_8S.2Na$
(2893-80-3)
$C_{31}H_{23}BrO_3$
(56073-10-0)
$C_{31}H_{23}N_5O_6$
(6448-96-0)
$C_{31}H_{26}N_6O_{11}S_3.3Na$
(28706-19-6)
$C_{31}H_{28}N_2O_3$
(29512-49-0)
$C_{31}H_{32}N_4O_2$
(15301-48-1)
$C_{31}H_{40}O$
(15860-96-5)
$C_{31}H_{41}NO_3$
(32180-75-9)
$C_{31}H_{42}N_2O_6$
(23509-16-2)
$C_{31}H_{42}N_3.Cl$
(2390-59-2)
$C_{31}H_{46}O_2$
(84-80-0)
$C_{31}H_{52}O_3$
(58-95-7)
$C_{31}H_{54}$
(53584-62-6)
(60305-22-8)
(60305-23-9)
$C_{31}H_{60}O_6S_3Sn$
(54849-38-6)
(57583-34-3)
$C_{31}H_{62}O_2$
(629-83-4)
(31556-45-3)
(31565-37-4)
(38232-01-8)
$C_{31}H_{64}$
(630-04-6)
(1560-72-1)
$C_{32}Br_6Cl_{10}CuN_8$
(14302-13-7)
$C_{32}Cl_{16}CuN_8$
(14832-14-5)
$C_{32}H_8Cl_8CuN_8$
(1330-37-6)
$C_{32}H_{12}Cl_2CuN_8O_{10}S_4.2H$
(31361-57-6)
$C_{32}H_{12}CoN_8O_{12}S_4.4H$
(14285-59-7)
$C_{32}H_{13}Cl_3CuN_8O_6S_3$
(27121-30-8)
$C_{32}H_{13}CuN_8O_9S_3.3H$
(30638-09-6)
$C_{32}H_{13}CuN_8O_9S_3.3H_4N$
(25512-11-2)
$C_{32}H_{13}CuN_8O_9S_3.3Na$
(1330-39-8)
$C_{32}H_{14}CoN_8O_6S_2.2H$
(29383-29-7)
$C_{32}H_{14}CuN_8O_6S_2.2Na$
(1330-38-7)
$C_{32}H_{14}Cu_2N_4O_{16}S_4.4Na$
(12222-00-3)
$C_{32}H_{15}ClCuN_8$
(12239-87-1)
$C_{32}H_{15}CoN_8O_3S.H$
(30638-08-5)
$C_{32}H_{16}CoN_8$
(3317-67-7)
$C_{32}H_{16}CuN_8$
(147-14-8)
$C_{32}H_{16}Cu_2N_5O_{13}S_3.3Na$
(66418-17-5)
$C_{32}H_{16}Cu_2N_6O_{16}S_4.4Na$
(16143-79-6)
(28407-37-6)
$C_{32}H_{16}N_8Ni$
(14055-02-8)
$C_{32}H_{18}$
(189-43-5)
(189-45-7)
(192-54-1)
(192-60-9)
(196-46-3)
(313-71-3)
(31541-02-3)
(31541-03-4)
(31541-07-8)
$C_{32}H_{18}Cl_2CrN_4O_4.H$
(31714-55-3)
$C_{32}H_{18}CrN_6O_8.H$
(32517-36-5)
$C_{32}H_{18}CuN_8O_3S$
(28901-96-4)
$C_{32}H_{18}CuN_8O_6S_2$
(29188-28-1)
$C_{32}H_{18}N_8$
(574-93-6)
$C_{32}H_{20}$
(65256-40-8)
$C_{32}H_{20}N_4O_8S_2.2Na$
(10169-02-5)
$C_{32}H_{20}N_6O_{14}S_4.4Na$
(2602-46-2)
$C_{32}H_{22}Cl_2N_6O_6S_2.2Na$
(6470-31-1)
$C_{32}H_{22}Cl_2N_6O_{12}S_4.4Na$
(6548-29-4)
$C_{32}H_{22}N_4O_2$
(4203-77-4)
$C_{32}H_{22}N_4O_{11}S_3.3Na$
(6426-67-1)
$C_{32}H_{22}N_6O_8S_2.2Na$
(2586-60-9)
$C_{32}H_{23}N_5O_6S_2.2Na$
(3351-05-1)
$C_{32}H_{23}N_5O_7S_2.2Na$
(6406-45-7)
$C_{32}H_{24}Cl_2N_8O_2$
(3520-72-7)
$C_{32}H_{24}Cl_2N_{10}O_6S_2.2Na$
(37138-23-1)
$C_{32}H_{24}N_6O_5$
(12225-06-8)
$C_{32}H_{24}N_6O_6S_2.2Na$
(573-58-0)
$C_{32}H_{24}N_6O_{11}S_3.3Na$
(2429-73-4)
$C_{32}H_{24}N_8O_6S.2Na$
(2586-58-5)
$C_{32}H_{24}N_8O_8S_2.2Na$
(6375-55-9)
$C_{32}H_{26}Cl_2N_6O_4$
(6358-85-6)
$C_{32}H_{26}CoN_{10}O_8S_2.Na$
(34664-47-6)
$C_{32}H_{28}CoN_8O_{10}S_2.2H$
(12715-61-6)
$C_{32}H_{28}CoN_8O_{10}S_2.Na$
(72496-88-9)
$C_{32}H_{28}N_4O_8S_2.2Na$
(6358-29-8)
$C_{32}H_{30}N_2O_2$
(116-75-6)
$C_{32}H_{30}N_2O_8S_2.2Na$
(4474-24-2)
$C_{32}H_{36}N_2O_{10}$
(80090-30-8)
$C_{32}H_{36}N_4O_8S_2.C_{16}H_{20}N_2$
(1538-09-6)
$C_{32}H_{41}F_7N_2O_4$
(2923-93-5)
$C_{32}H_{43}N$
(68586-20-9)
$C_{32}H_{44}O_6Si_6$
(60573-48-0)
$C_{32}H_{44}O_{12}$
(507-60-8)
$C_{32}H_{46}O_8$
(6199-67-3)
$C_{32}H_{48}O_8$
(5988-76-1)
$C_{32}H_{48}O_9$
(465-16-7)
$C_{32}H_{49}NO_9$
(62-59-9)
(8051-02-3)
$C_{32}H_{51}NNiO_2S$
(14516-71-3)
$C_{32}H_{52}N_4O_4.2Br$
(56-94-0)
$C_{32}H_{54}O_4$
(2432-90-8)
$C_{32}H_{56}$
(67069-15-2)
(67069-25-4)
$C_{32}H_{56}O_8Sn$
(25168-21-2)
$C_{32}H_{60}O_6$
(10535-50-9)
$C_{32}H_{62}O_4$
(16958-92-2)
$C_{32}H_{62}O_4S$
(10595-72-9)
$C_{32}H_{64}O_2$
(540-10-3)
(3625-52-3)
(17661-50-6)
(77630-51-4)
$C_{32}H_{64}O_4Sn$
(77-58-7)
$C_{32}H_{66}$
(544-85-4)
(55401-55-3)
$C_{32}H_{67}N.C_2H_4O_2$
(71764-17-5)
$C_{32}H_{67}O_4P$
(2197-63-9)
$C_{32}H_{68}O_4P_2S_4Zn$

(4259-15-8)
(7059-16-7)
$C_{32}H_{68}S_2Sn$
(1185-81-5)
$C_{32}H_{68}Sn$
(3590-84-9)
$C_{33}H_{16}Cu_2N_6O_{17}S_4.4Na$
(15418-16-3)
$C_{33}H_{21}N_3O_3$
(3949-34-6)
$C_{33}H_{24}N_6$
(30360-24-8)
$C_{33}H_{25}N_3O_3$
(991-42-4)
$C_{33}H_{26}N_2O_6$
(54395-52-7)
$C_{33}H_{27}ClN_4O_6$
(5280-68-2)
$C_{33}H_{28}N_8O_6S.2Na$
(6360-54-9)
$C_{33}H_{28}O_8$
(4196-86-5)
$C_{33}H_{32}N_3.Cl$
(2580-56-5)
$C_{33}H_{33}N_3O$
(6786-83-0)
$C_{33}H_{34}O_2Si_3$
(3390-61-2)
$C_{33}H_{35}N_5O_5$
(113-15-5)
$C_{33}H_{36}N_4O_6$
(635-65-4)
$C_{33}H_{37}CrN_2O_{11}$
(31303-42-1)
$C_{33}H_{38}ClNO_5$
(98611-44-0)
$C_{33}H_{40}N_2O_9$
(50-55-5)
$C_{33}H_{40}N_3.Cl$
(2390-60-5)
$C_{33}H_{40}O_{19}$
(301-19-9)
$C_{33}H_{47}ClN_2O_4$
(26110-32-7)
$C_{33}H_{48}O_2$
(58003-48-8)
$C_{33}H_{49}NO_7$
(475-00-3)
$C_{33}H_{50}O_6P_2$
(26741-53-7)
$C_{33}H_{51}NO_7$
(23185-94-6)
$C_{33}H_{52}O_2$
(7786-17-6)
$C_{33}H_{54}O_5$
(4345-03-3)
$C_{33}H_{54}O_6$
(89-04-3)
(3319-31-1)
(27251-75-8)
$C_{33}H_{58}$
(67069-16-3)
(67069-26-5)
$C_{33}H_{60}O_7Si_8$
(60617-40-5)
$C_{33}H_{62}O_6$
(126-57-8)
$C_{33}H_{66}N_3S_6Sb$
(15890-25-2)
$C_{33}H_{66}O$
(22986-69-2)
$C_{33}H_{68}$
(630-05-7)
(1720-11-2)
(20129-49-1)
$C_{33}H_{69}N$
(67700-99-6)
$C_{34}H_{12}Cl_4O_2$
(6373-20-2)
$C_{34}H_{14}Cl_2O_2$
(1324-55-6)
$C_{34}H_{15}BrO_2$
(1324-17-0)
$C_{34}H_{15}NO_4$
(128-60-9)
$C_{34}H_{16}O_2$
(116-71-2)
$C_{34}H_{17}NO_2$
(26763-69-9)
$C_{34}H_{18}$
(191-79-7)
$C_{34}H_{18}CrN_8O_{26}S_6.H.6Na$
(6408-22-6)
$C_{34}H_{22}Cl_2N_4O_2$
(6358-30-1)
$C_{34}H_{22}Cl_2N_4O_{11}S_3.3Na$
(1324-58-9)
$C_{34}H_{24}$
(751-38-2)
$C_{34}H_{24}N_6O_{11}S_2.3Na$
(6420-40-2)
$C_{34}H_{24}N_6O_{14}S_4.4Na$
(314-13-6)
$C_{34}H_{24}N_8O_{10}S_2.2Na$
(4335-09-5)
$C_{34}H_{25}N_7O_8S_2.2Na$
(3626-28-6)
$C_{34}H_{25}N_9O_7S_2.2Na$
(1937-37-7)
$C_{34}H_{26}N_4O_{10}S_2.2Na$
(2429-71-2)
$C_{34}H_{26}N_4O_{16}S_4.4Na$
(2150-54-1)
$C_{34}H_{26}N_4O_{18}S_4.4Na$
(4198-19-0)
$C_{34}H_{26}N_6O_6S_2.2Na$
(992-59-6)
$C_{34}H_{27}N_5O_9S_2.2Na$
(6428-94-0)
$C_{34}H_{27}N_5O_{10}S_2.2Na$
(2586-57-4)
(6449-35-0)
$C_{34}H_{28}Cl_2N_8O_2$
(15793-73-4)
$C_{34}H_{28}Cl_4N_6O_4$
(5979-28-2)
$C_{34}H_{28}N_6O_8S_2.2Na$
(2868-75-9)
$C_{34}H_{28}N_6O_{14}S_4.4Na$
(72-57-1)
$C_{34}H_{28}N_6O_{16}S_4.4Na$
(2429-74-5)
(2610-05-1)
$C_{34}H_{29}N_6NaO_{16}S_4$
(68966-50-7)
$C_{34}H_{29}N_{13}O_7S_2.2Na$
(6428-31-5)
$C_{34}H_{30}Cl_2N_6O_4$
(5468-75-7)
(6358-37-8)
$C_{34}H_{30}Cl_2N_6O_6$
(4531-49-1)
$C_{34}H_{30}N_4O_4Zn.2H$
(15442-64-5)
$C_{34}H_{32}N_6O_6$
(6505-28-8)
$C_{34}H_{32}N_{12}O_6S_2$
(35632-99-6)
$C_{34}H_{34}N_2O_2$
(20241-74-1)
$C_{34}H_{34}N_4O_4$
(553-12-8)
$C_{34}H_{37}Cl_3N_4O_4$
(31188-91-7)
$C_{34}H_{38}N_2O_6.H_2O_4S$
(64-31-3)
$C_{34}H_{46}N_2O_6.H_2O_4S$
(55-48-1)
$C_{34}H_{47}NO_{11}$
(302-27-2)
$C_{34}H_{48}O_2$
(25485-34-1)
$C_{34}H_{49}N_3$
(4482-70-6)
$C_{34}H_{50}O_5Si_6$
(13271-58-4)
$C_{34}H_{50}O_7$
(5697-56-3)
$C_{34}H_{50}O_8$
(36443-68-2)
$C_{34}H_{51}NO$
(55493-86-2)
$C_{34}H_{52}N_2O_4$
(32687-78-8)
$C_{34}H_{52}O_3S$
(1182-65-6)
$C_{34}H_{54}O_3S$
(3381-52-0)
$C_{34}H_{58}O_4$
(119-06-2)
(68515-47-9)
$C_{34}H_{62}O_{11}$
(2315-66-4)
$C_{34}H_{64}O_8S_2Sn$
(63397-60-4)
$C_{34}H_{66}O_4S$
(16545-54-3)
$C_{34}H_{66}O_6$
(26322-14-5)
$C_{34}H_{66}O_6S_3Sn$
(25852-70-4)
$C_{34}H_{67}NO$
(16260-09-6)
$C_{34}H_{68}O_2$
(25339-09-7)
$C_{34}H_{70}$
(14167-59-0)
$C_{34}H_{70}O_9$
(13149-87-6)
$C_{34}H_{72}N.Cl$
(1812-53-9)
$C_{35}H_{24}N_8O_{12}S_2.3Na$
(5422-17-3)
$C_{35}H_{25}N_5O_7S.2Na$
(3476-90-2)
$C_{35}H_{26}N_4O_{10}S_3.2Na$
(3567-65-5)
$C_{35}H_{27}N_7O_{10}S_2.2Na$
(3441-14-3)
$C_{35}H_{28}N_6O_{13}S_3.3Na$
(6420-44-6)
$C_{35}H_{28}N_6O_{13}S_4.4Na$
(3214-47-9)
(28706-21-0)
$C_{35}H_{28}N_6O_{15}S_4.4Na$
(28706-22-1)
$C_{35}H_{29}N_9O_7S_2.2Na$
(2429-83-6)
$C_{35}H_{42}N_2O_9$
(24815-24-5)
$C_{35}H_{48}N_8O_{10}S$
(28227-92-1)
$C_{35}H_{48}N_8O_{11}S$
(17466-45-4)
$C_{35}H_{52}O_5$
(467-81-2)
(467-82-3)
$C_{35}H_{52}O_8$
(71033-08-4)
$C_{35}H_{60}N_2O_4.2Br$
(15500-66-0)
$C_{35}H_{62}O_3$
(2082-79-3)
$C_{35}H_{64}O_{11}$
(65455-72-3)
$C_{35}H_{65}N_9O_9.C_2H_4O_2$
(31062-69-8)
$C_{35}H_{68}O_4$
(26719-40-4)
$C_{35}H_{70}O$
(504-53-0)
$C_{35}H_{72}$
(630-07-9)
$C_{36}H_{17}AlO_5$
(300-92-5)
$C_{36}H_{18}O_4$
(6424-76-6)
$C_{36}H_{20}$
(192-46-1)
(31541-10-3)
(72382-90-2)
$C_{36}H_{20}O_4$
(128-58-5)
$C_{36}H_{25}N_5O_6S_2.2Na$
(3071-73-6)
$C_{36}H_{26}N_8O_{11}S_3.3Na$
(8003-69-8)
$C_{36}H_{28}Cl_2N_8O_6$
(6358-87-8)
$C_{36}H_{28}CrN_8O_{10}S_2.Na$
(24305-97-3)
$C_{36}H_{28}N_2O_6$
(91-92-9)
$C_{36}H_{28}N_4O_{10}S_3.2Na$
(71701-30-9)
$C_{36}H_{29}N_7O_{12}S_3.3Na$
(68318-35-4)
$C_{36}H_{30}CrO_4Si_2$
(1624-02-8)
$C_{36}H_{32}Cl_4N_6O_4$
(22094-93-5)
$C_{36}H_{32}Cl_4N_6O_8$
(5567-15-7)
$C_{36}H_{34}Cl_2N_6O_4$
(5102-83-0)
$C_{36}H_{34}Cl_2N_6O_6$
(20139-66-6)
$C_{36}H_{35}O_4P$
(63302-95-4)
$C_{36}H_{36}N_6O_6$
(7147-42-4)
$C_{36}H_{40}N_6O_4$
(25857-05-0)
$C_{36}H_{42}N_2$
(4844-10-4)
$C_{36}H_{42}N_2O_6.O_4S$
(1420-53-7)
$C_{36}H_{43}O_4P$
(63340-28-3)
$C_{36}H_{51}NO_{11}$
(71-62-5)
$C_{36}H_{51}O_4P$
(63302-94-3)
$C_{36}H_{52}ClN_3O_3$
(33956-01-3)
$C_{36}H_{56}O_8$
(16561-29-8)
$C_{36}H_{58}O$
(69834-19-1)
$C_{36}H_{58}O_2S$
(68815-67-8)
$C_{36}H_{58}O_2S.Ca$
(26998-97-0)
$C_{36}H_{58}O_7S_2$
(30260-73-2)
$C_{36}H_{58}O_7S_2.2Na$
(25167-32-2)
$C_{36}H_{58}S_2$
(28986-55-2)
$C_{36}H_{59}O_2PS_2$
(30304-41-7)
$C_{36}H_{60}MgO_2S$
(68974-78-7)
$C_{36}H_{61}O_{11}.Na$
(22373-78-0)
$C_{36}H_{62}O_{11}$
(17090-79-8)
$C_{36}H_{66}O_4.Hg$
(1191-80-6)
$C_{36}H_{68}O_2$

(3687-45-4)
$C_{36}H_{68}O_4.Zn$
(557-07-3)
$C_{36}H_{70}O_4.Ba$
(6865-35-6)
$C_{36}H_{70}O_4.Ca$
(1592-23-0)
$C_{36}H_{70}O_4.Mg$
(557-04-0)
$C_{36}H_{70}O_4.Pb$
(7428-48-0)
$C_{36}H_{70}O_4.Zn$
(557-05-1)
$C_{36}H_{70}O_4S$
(13103-52-1)
$C_{36}H_{70}O_6Pb_2$
(56189-09-4)
$C_{36}H_{71}O_3P$
(25088-57-7)
$C_{36}H_{71}O_4P$
(14450-07-8)
$C_{36}H_{72}O_2$
(2778-96-3)
$C_{36}H_{72}O_4.Cd$
(2223-93-0)
$C_{36}H_{72}O_4S_2Sn$
(15571-58-1)
(26401-97-8)
(28570-24-3)
$C_{36}H_{74}$
(630-06-8)
$C_{36}H_{75}Al$
(1529-59-5)
$C_{36}H_{75}N$
(102-87-4)
$C_{36}H_{75}O_3P$
(3076-63-9)
$C_{36}H_{75}O_4P$
(3037-89-6)
$C_{36}H_{75}PS_3$
(1656-63-9)
$C_{36}H_{76}N.Cl$
(1118-41-8)
$C_{37}H_{25}N_7O_{10}S_2.3Na$
(3841-15-4)
$C_{37}H_{29}N_3.ClH$
(2152-64-9)
$C_{37}H_{29}N_3O_3S$
(1324-76-1)
$C_{37}H_{30}N_4O_{10}S_3.2Na$
(6358-57-2)
(6459-94-5)
$C_{37}H_{31}N_3O$
(23681-60-9)
$C_{37}H_{32}N_6O_{15}S_4.4Na$
(6420-33-3)
$C_{37}H_{36}N_2O_6S_2.Na$
(4680-78-8)
$C_{37}H_{36}N_2O_9S_3.2/3Al$
(15792-67-3)
$C_{37}H_{36}N_2O_9S_3$-
$.C_{16}H_{12}N_4O_9S_2.2H_3N.3Na$
(92170-50-8)
$C_{37}H_{36}N_2O_9S_3.2H_3N$
(2650-18-2)
$C_{37}H_{36}N_2O_9S_3.2Na$
(3844-45-9)
(5141-20-8)
$C_{37}H_{36}N_2O_{10}S_3.2Na$
(2353-45-9)
$C_{37}H_{49}N_7O_9S$
(5534-95-2)
$C_{37}H_{51}NO_{10}$
(12607-93-1)
$C_{37}H_{52}O_3$
(1843-03-4)
$C_{37}H_{67}NO_{13}$
(114-07-8)
$C_{37}H_{67}NO_{13}.C_{18}H_{36}O_2$
(643-22-1)
$C_{37}H_{68}O_8$
(3008-50-2)
$C_{37}H_{76}$
(7194-84-5)
$C_{37}H_{77}N$
(4088-22-6)
$C_{38}H_{16}Cl_2N_2O_{14}S_4.4Na$
(33700-25-3)
$C_{38}H_{20}$
(435-02-9)
$C_{38}H_{20}Cu_2N_5O_{13}S_3.3Na$
(6656-03-7)
$C_{38}H_{22}N_2O_6$
(6424-77-7)
$C_{38}H_{28}N_6O_{12}S_3.3Na$
(6460-01-1)
$C_{38}H_{28}N_6O_{13}S_3.3Na$
(10114-24-6)
$C_{38}H_{28}N_8O_{13}S.3Na$
(6739-62-4)
$C_{38}H_{28}N_8O_{14}S_4.4Na$
(32651-66-4)
$C_{38}H_{30}N_{10}O_9S.2Na$
(70210-28-5)
$C_{38}H_{32}CrN_8O_{10}S_2.H$
(12218-95-0)
$C_{38}H_{34}N_2O_3$
(34372-72-0)
$C_{38}H_{36}N_{12}O_8S_2.2Na$
(13863-31-5)
$C_{38}H_{38}O_8$
(7328-97-4)
(27043-37-4)
$C_{38}H_{38}O_{10}$
(34807-41-5)
$C_{38}H_{40}N_{12}O_8S_2.2Na$
(32694-95-4)
(56776-29-5)
$C_{38}H_{44}N_2O_6$
(57-95-4)
$C_{38}H_{60}O_9$
(53202-98-5)
$C_{38}H_{72}N_2O_2$
(110-31-6)
$C_{38}H_{74}O_4$
(627-83-8)
(26720-21-8)
$C_{38}H_{74}O_6$
(52326-66-6)
$C_{38}H_{74}O_6S_3Sn$
(26401-86-5)
$C_{38}H_{76}N_2O_2$
(110-30-5)
$C_{38}H_{76}O_2$
(22413-02-1)
$C_{38}H_{78}$
(7194-85-6)
$C_{38}H_{80}N.Cl$
(107-64-2)
$C_{38}H_{80}O_2Si$
(29043-70-7)
$C_{39}H_{41}N_3O_6S_2.Na$
(1694-09-3)
$C_{39}H_{53}N_9O_{14}S$
(21150-21-0)
$C_{39}H_{53}N_9O_{15}S$
(21150-22-1)
$C_{39}H_{54}H_{10}O_{12}S$
(13567-11-8)
$C_{39}H_{54}N_{10}O_{14}S$
(23109-05-9)
$C_{39}H_{59}Cl_2NO_2$
(3546-10-9)
$C_{39}H_{66}O_6$
(36631-30-8)
$C_{39}H_{72}O_4$
(105-62-4)
$C_{39}H_{72}O_5$
(25637-84-7)
$C_{39}H_{74}O_6$
(538-24-9)
$C_{40}H_{20}CrN_6O_{14}S_2.H.2Na$
(5610-64-0)
$C_{40}H_{22}CrN_4O_{10}S_2.H.2Na$
(12392-64-2)
$C_{40}H_{23}Cl_3N_8O_8$
(35869-64-8)
$C_{40}H_{23}Cl_5N_6O_4$
(5280-78-4)
$C_{40}H_{24}Cl_4N_6O_4$
(3905-19-9)
$C_{40}H_{26}N_2O_4$
(4948-15-6)
$C_{40}H_{26}N_2O_6$
(24108-89-2)
$C_{40}H_{27}N_7O_{13}S_4.4Na$
(4399-55-7)
$C_{40}H_{35}N_3O_3S$
(6417-46-5)
$C_{40}H_{36}N_2O_8.O_4S$
(316-41-6)
$C_{40}H_{38}N_{12}O_8S_2.2Na$
(16090-02-1)
$C_{40}H_{40}N_{12}O_8S_2.2Na$
(56776-30-8)
$C_{40}H_{40}N_{12}O_{10}S_2.2Na$
(4193-55-9)
$C_{40}H_{42}O_3Si_4$
(38421-40-8)
$C_{40}H_{44}N_{12}O_{10}S_2$
(4404-43-7)
$C_{40}H_{44}N_{12}O_{10}S_2.2Na$
(59453-69-9)
$C_{40}H_{44}N_{12}O_{16}S_4.4Na$
(16470-24-9)
$C_{40}H_{46}O_8P_2$
(96300-95-7)
$C_{40}H_{48}N_4O_4.H_2O_4S$
(50-54-4)
$C_{40}H_{56}$
(7235-40-7)
$C_{40}H_{60}P_2S_5$
(68400-79-3)
$C_{40}H_{62}O_{19}$
(126-13-6)
$C_{40}H_{64}N_2O_4$
(23128-74-7)
$C_{40}H_{71}NO_{14}$
(134-36-1)
$C_{40}H_{72}O_8Sn$
(33466-31-8)
$C_{40}H_{78}O_5$
(109-30-8)
$C_{40}H_{79}NO$
(10094-45-8)
$C_{40}H_{80}O_4Sn$
(3648-18-8)
$C_{40}H_{82}$
(4181-95-7)
$C_{40}H_{82}O_2$
(56554-64-4)
$C_{41}H_{28}N_6O_{15}S_4.4C_6H_{15}NO_3$
(36596-36-8)
$C_{41}H_{28}N_6O_{15}S_4.4Na$
(28706-25-4)
$C_{41}H_{30}N_8O_{14}S_2.3Na$
(71566-41-1)
$C_{41}H_{32}N_4O_4$
(6483-64-3)
$C_{41}H_{43}Cl_3N_6O_5$
(28279-36-9)
$C_{41}H_{45}N_3O_6S_2.Na$
(4129-84-4)
$C_{41}H_{64}O_{13}$
(71-63-6)
$C_{41}H_{64}O_{14}$
(20830-75-5)
$C_{41}H_{66}O_{13}$
(3786-76-3)
$C_{41}H_{76}O_8$
(14450-05-6)
$C_{41}H_{80}O_6$
(13081-97-5)
$C_{41}H_{84}O$
(40710-42-7)
$C_{42}H_{18}N_2O_6$
(2172-33-0)
(2475-33-4)
$C_{42}H_{22}$
(190-09-0)
$C_{42}H_{23}N_3O_6$
(131-92-0)
(2379-81-9)
$C_{42}H_{25}N_3O_6$
(128-89-2)
$C_{42}H_{28}Cl_2N_4$
(1707-68-2)
$C_{42}H_{29}N_7O_{13}S_4.4Na$
(2503-73-3)
(6428-60-0)
$C_{42}H_{42}N_{12}O_8S_2.2Na$
(24019-80-5)
$C_{42}H_{46}N_{14}O_{10}S_2.2Na$
(27344-06-5)
$C_{42}H_{46}N_{14}O_{16}S_4.4Na$
(29637-52-3)
$C_{42}H_{47}N_3O_6S_2.Na$
(5863-46-7)
$C_{42}H_{50}Cl_4N_2O_4$
(22966-79-6)
$C_{42}H_{62}O_{16}$
(1405-86-3)
$C_{42}H_{63}O_3P$
(31570-04-4)
$C_{42}H_{72}N_2O_5$
(63843-89-0)
$C_{42}H_{74}O_6$
(4252-85-1)
$C_{42}H_{74}O_9Si_{10}$
(60573-46-8)
$C_{42}H_{76}O_4S_2Sn$
(67859-64-7)
$C_{42}H_{78}N_{14}O_{24}.H_6O_{12}S_3$
(3810-74-0)
$C_{42}H_{78}O_5$
(25111-05-1)
$C_{42}H_{80}O_4S_2Sn$
(67859-63-6)
$C_{42}H_{80}O_7$
(36521-89-8)
$C_{42}H_{82}N_{14}O_{24}.3H_2O_4S$
(5490-27-7)
$C_{42}H_{82}O_4S$
(693-36-7)
$C_{42}H_{85}N_3O_2$
(13998-73-7)
$C_{42}H_{86}$
(55470-97-8)
$C_{42}H_{87}Al$
(1529-58-4)
$C_{43}H_{22}N_6O_{13}S_3$
(5610-94-6)
$C_{43}H_{27}CrN_6O_8S.2Na$
(27425-58-7)
$C_{43}H_{35}Cl_5N_8O_6$
(5580-57-4)
$C_{43}H_{48}N_2O_6S_2.Na$
(3599-32-4)
$C_{43}H_{49}N_3O_6S_2.Na$
(6505-30-2)
$C_{43}H_{58}N_4O_{12}$
(13292-46-1)
$C_{43}H_{66}N_{12}O_{12}S_2$
(50-56-6)
$C_{43}H_{75}NO_{16}$
(1264-62-6)
$C_{43}H_{88}$
(55162-61-3)
$C_{44}H_{24}$
(72382-91-3)

$C_{44}H_{32}N_6O_{14}S_3.3Na$ (7082-31-7)
$C_{44}H_{35}N_{13}O_{11}S_3.3Na$ (6473-13-8)
$C_{44}H_{38}Cl_4N_8O_6$ (5280-80-8)
$C_{44}H_{46}CuN_{11}.3Cl$ (26854-10-4)
$C_{44}H_{72}O_{12}$ (33956-61-5)
$C_{44}H_{78}O_4$ (14117-96-5)
$C_{44}H_{82}O_4$ (68201-79-6)
$C_{44}H_{88}N_2S_4$ (35318-10-6)
$C_{44}H_{88}O_4Sn$ (5847-55-2)
$C_{44}H_{90}$ (7098-22-8)
$C_{45}H_{22}N_2O_5$ (4395-53-3)
$C_{45}H_{32}N_{10}O_{21}S_6.6Na$ (2610-10-8)
$C_{45}H_{45}O_4P$ (63302-98-7)
$C_{45}H_{51}N_3O_6S$ (37921-74-7)
$C_{45}H_{69}O_3P$ (26523-78-4)
$C_{45}H_{69}O_4P$ (26569-53-9)
$C_{45}H_{73}NO_{15}$ (20562-02-1)
$C_{45}H_{74}O$ (13190-97-1)
$C_{45}H_{77}NO_{17}$ (1401-69-0)
$C_{45}H_{85}N_{13}O_{10}$ (1066-17-7)
$C_{45}H_{86}O_6$ (555-45-3)
$C_{45}H_{91}N_2O_5.CH_3O_4S$ (13441-22-0)
$C_{46}H_{28}Cl_4N_6O_4$ (5280-74-0)
$C_{46}H_{30}Cl_2N_6O_4$ (41709-76-6)
$C_{46}H_{30}N_{10}O_{13}S_3.4Na$ (2429-81-4)
$C_{46}H_{50}N_4.C_2H_2O_4.2C_2HO_4$ (2437-29-8)
$C_{46}H_{56}N_4O_9$ (23360-92-1)
$C_{46}H_{56}N_4O_{10}$ (57-22-7)
$C_{46}H_{56}N_4O_{10}.H_2O_4S$ (2068-78-2)
$C_{46}H_{58}N_4O_9$ (865-21-4)
$C_{46}H_{58}N_4O_9.H_2O_4S$ (143-67-9)
$C_{46}H_{60}N_6$ (3956-73-8)
$C_{46}H_{64}O_2$ (2124-57-4)
$C_{46}H_{83}NO_{18}$ (1400-61-9)
$C_{46}H_{90}N_4O_3$ (72050-94-3)
$C_{47}H_{49}N_3O_7S_2.Na$ (6104-58-1)
$C_{47}H_{51}NO_{14}$ (33069-62-4)
$C_{47}H_{73}NO_{17}$ (1397-89-3)
$C_{48}H_{26}N_6O_4$ (3049-71-6)
$C_{48}H_{32}N_8O_{18}S_6.6Na$ (12222-60-5)
$C_{48}H_{40}O_4Si_4$ (546-56-5)
$C_{48}H_{72}O_{14}$ (65195-55-3)
$C_{48}H_{74}O_4$ (36388-36-0)
$C_{48}H_{92}O_4S_2Sn$ (67361-77-7)
$C_{48}H_{99}Al$ (1726-65-4)
$C_{49}H_{79}N_3O_3.xCa$ (68568-82-1)
$C_{50}H_{26}$ (72382-92-4)
$C_{50}H_{38}N_{12}O_{18}S_4.5Na$ (6388-26-7)
$C_{50}H_{83}NO_{21}$ (17406-45-0)
$C_{50}H_{94}Cl_3N_3$ (1399-80-0)
$C_{51}H_{46}O_{20}$ (1402-68-2)
$C_{52}H_{61}ClN_2O_8S$ (30744-85-5)
$C_{52}H_{76}O_{24}$ (18378-89-7)
$C_{53}H_{44}N_{14}O_{13}S_4.4Na$ (6854-81-5)
$C_{53}H_{58}O_9$ (38122-80-4)
$C_{53}H_{77}O_4P$ (20227-53-6)
$C_{54}H_{78}O_3$ (1709-70-2)
$C_{54}H_{87}O_3P$ (54771-30-1)
$C_{54}H_{90}N_6O_{18}$ (2001-95-8)
$C_{54}H_{105}O_6.Al$ (637-12-7)
$C_{54}H_{111}Al$ (3041-23-4)
$C_{55}H_{72}MgN_4O_5$ (479-61-8)
$C_{56}H_{32}CuN_8$ (1330-40-1)
$C_{56}H_{92}O_{29}$ (11024-24-1)
$C_{57}H_{45}N_3O_3Sn_3$ (752-74-9)
$C_{57}H_{64}O_8$ (72319-24-5)
$C_{57}H_{104}O_6$ (122-32-7)
$C_{57}H_{104}O_{12}$ (106-81-0)
$C_{57}H_{110}O_6$ (555-43-1)
$C_{58}H_{114}O_{26}$ (9005-64-5)
$C_{60}H_{78}OSn_2$ (13356-08-6)
$C_{60}H_{87}N_3O_{12}$ (34137-09-2)
$C_{60}H_{108}O_8$ (26266-58-0)
$C_{60}H_{114}O_8$ (26658-19-5)
$C_{60}H_{116}O_7$ (7775-50-0)
$C_{60}H_{117}NO_6$ (3002-22-0)
$C_{60}H_{122}$ (7667-80-3)
$C_{60}H_{123}Al$ (1529-57-3)
$C_{61}H_{114}O_6S_3Sn$ (59118-79-5)
$C_{62}H_{86}N_{12}O_{16}$ (50-76-0)
$C_{62}H_{89}CoN_{13}O_{15}P$ (13422-51-0)
$C_{62}H_{111}N_{11}O_{12}$ (59865-13-3)
$C_{62}H_{122}O_{26}$ (9005-66-7)
$C_{63}H_{88}CoN_{14}O_{14}P$ (68-19-9)
$C_{63}H_{98}O_6$ (125-93-9)
$C_{64}H_{120}O_6S_3Sn$ (67361-76-6)
$C_{64}H_{126}O_{26}$ (9005-67-8)
$C_{65}H_{124}O_8S_4$ (29598-76-3)
$C_{66}H_{70}N_{10}O_{10}.C_4H_6O_6$ (379-79-3)
$C_{66}H_{75}Cl_2N_9O_{24}$ (1404-90-6)
$C_{66}H_{135}Al$ (6651-25-8)
$C_{68}H_{92}O_4P_2$ (38613-77-3)
$C_{69}H_{138}O_2$ (40710-32-5)
$C_{72}H_{116}O_4P_2S_4Zn$ (54261-67-5)
$C_{72}H_{123}O_3P$ (1333-21-7)
$C_{72}H_{140}O_8.Ba.Cd$ (1191-79-3)
$C_{72}H_{147}Al$ (6651-26-9)
$C_{73}H_{108}O_{12}$ (6683-19-8)
$C_{76}H_{52}O_{46}$ (1401-55-4)
$C_{77}H_{129}N_{27}O_{36}S_7$ (98526-74-0)
$C_{77}H_{148}O_8$ (115-83-3)
$C_{78}H_{159}Al$ (10449-71-5)
$C_{80}H_{52}N_{16}O_{20}S_4.6Na$ (8014-91-3)
$C_{84}H_{84}NiO_{12}P_4$ (35884-66-3)
$C_{84}H_{171}Al$ (6651-27-0)
$C_{136}H_{210}N_{40}O_{31}S$ (16960-16-0)
$C_{153}H_{225}N_{43}O_{49}S$ (16941-32-5)
$C_{207}H_{309}N_{57}O_{57}S$ (39319-42-1)
Ca (7440-70-2)
$Ca.Cr_2O_7$ (14307-33-6)
$Ca.2HIO_3$ (7789-80-2)
$Ca.2HNO_2$ (13780-06-8)
$Ca.2H_2O_3S$ (13780-03-5)
$Ca.H_2O_3Si$ (10101-39-0)
$Ca.2H_3O_2P$ (7789-79-9)
$Ca.H_3O_4P$ (7757-93-9)
$Ca.2/3H_3O_4P$ (7758-87-4)
$Ca.2H_3O_4P.H_2O$ (7758-23-8)
$Ca.1/2H_4O_7P_2$ (7790-76-3)
$CaCl_2$ (10043-52-4)
$CaCl_2O_4$ (14674-72-7)
CaF_2 (7789-75-5) (14542-23-5)
CaH_2 (7789-78-8)
CaH_2O_2 (1305-62-0)
CaH_2O_3Si (13983-17-0)
CaO (1305-78-8)
CaO_2 (1305-79-9)
CaS (20548-54-3)
$CaS.x$ (1344-81-6)
Ca_3P_2 (1305-99-3)
$Ca_5HO_{13}P_3$ (1306-06-5)
Ca_3O_5Si (12168-85-3)
Cd (7440-43-9) (14336-68-6) (22537-48-0)
$CdCl$ (13966-86-4)
$CdCl_2$ (10108-64-2)
CdF_2 (7790-79-6)
$CdMoO_4$ (13972-68-4)
CdN_2O_6 (10325-94-7)
CdO (1306-19-0)
CdS (1306-23-6)
$CdSe$ (1306-24-7)
Ce (7440-45-1) (13967-74-3) (14762-78-8)
CeF_3 (7758-88-5)
CeH_4O_4 (12014-56-1)
CeO_2 (1306-38-3)
Ce_2O_3 (1345-13-7)
Cl (16887-00-6)
$ClCs$ (7647-17-8)
$ClCu$ (7758-89-6)
$ClCu_2H_3O_3$ (1332-65-6)
$ClFO_3$ (7616-94-6)
ClF_3 (7790-91-2)
ClH (7647-01-0)
$ClH.HNO_3$ (8007-56-5)
$ClHO.Li$ (13840-33-0)
$ClHO_2Zr$ (18428-88-1)
$ClHO_3$ (7790-93-4)
$ClHO_3S$

(7790-94-5)
$ClHO_4$
(7601-90-3)
$ClHO_4.xH_4N_2$
(27978-54-7)
$ClHO_4.1/2Ni$
(13637-71-3)
ClH_2HgN
(10124-48-8)
ClH_2N
(10599-90-3)
ClI
(7790-99-0)
ClK
(7447-40-7)
$ClLi$
(7447-41-8)
$ClMn$
(50646-06-5)
$ClNO$
(2696-92-6)
ClN_3
(13973-88-1)
$ClNa$
(7647-14-5)
$ClNaO_2$
(7758-19-2)
$ClNa_{13}O_{17}P_4$
(11084-85-8)
$ClO.Na$
(7681-52-9)
$ClO.Na.5H_2O$
(10022-70-5)
ClO_2
(10049-04-4)
$ClO_3.H_3N$
(10192-29-7)
$ClO_3.K$
(3811-04-9)
$ClO_3.Na$
(7775-09-9)
$ClO_4.H_4N$
(7790-98-9)
$ClO_4.K$
(7778-74-7)
$ClO_4.Na$
(7601-89-0)
$ClSn$
(13931-79-8)
$ClTl$
(7791-12-0)
$ClZn$
(18623-80-8)
Cl_2
(7782-50-5)
Cl_2Co
(7646-79-9)
Cl_2Cr
(10049-05-5)
Cl_2CrO_2
(14977-61-8)
Cl_2Cu
(7447-39-4)
$Cl_2Cu_4H_6O_6$
(1332-40-7)
Cl_2Fe
(7758-94-3)
$Cl_2Fe.4H_2O$
(13478-10-9)
Cl_2HN
(3400-09-7)
$Cl_2H_6N_2Pt$
(14913-33-8)
(15663-27-1)
Cl_2Hg
(7487-94-7)
Cl_2Hg_2
(10112-91-1)
$Cl_2Hg_2.Cl_2Hg$
(8065-83-6)
Cl_2Mg
(7786-30-3)
$Cl_2Mg.6H_2O$
(7791-18-6)
Cl_2Mn
(7773-01-5)
$Cl_2Mn.4H_2O$
(13446-34-9)
Cl_2Ni
(7718-54-9)
$Cl_2Ni.6H_2O$
(7791-20-0)
Cl_2O
(7791-21-1)
Cl_2OS
(7719-09-7)
Cl_2OSe
(7791-23-3)
Cl_2OZr
(7699-43-6)
$Cl_2OZr.6H_2O$
(25399-81-9)
$Cl_2O_2.Ca$
(7778-54-3)
Cl_2O_2S
(7791-25-5)
$Cl_2O_5S_2$
(7791-27-7)
$Cl_2O_6.Ba$
(13477-00-4)
$Cl_2O_6.Ca$
(10137-74-3)
$Cl_2O_6.Mg$
(10326-21-3)
$Cl_2O_6.Zn$
(10361-95-2)
Cl_2O_6Sr
(7791-10-8)
$Cl_2O_8.Mg$
(10034-81-8)
$Cl_2O_8.Ni.6H_2O$
(13520-61-1)
$Cl_2O_8.Pb$
(13637-76-8)
$Cl_2O_8.Sr$
(13450-97-0)
Cl_2Pb
(7758-95-4)
Cl_2Pd
(7647-10-1)
Cl_2Pt
(10025-65-7)
Cl_2S
(10545-99-0)
Cl_2S_2
(10025-67-9)
Cl_2Sn
(7772-99-8)
Cl_2Sr
(10476-85-4)
Cl_2V
(10580-52-6)
Cl_2Zn
(7646-85-7)
Cl_3Cr
(10025-73-7)
Cl_3Fe
(7705-08-0)
$Cl_3Fe.6H_2O$
(10025-77-1)
Cl_3Ga
(13450-90-3)
Cl_3HSi
(10025-78-2)
Cl_3In
(10025-82-8)
Cl_3La
(10099-58-8)
Cl_3Lu
(10099-66-8)
Cl_3N
(10025-85-1)
Cl_3OP
(10025-87-3)
Cl_3OV
(7727-18-6)
Cl_3P
(7719-12-2)
Cl_3PS
(3982-91-0)
Cl_3Rh
(10049-07-7)
Cl_3Sb
(10025-91-9)
Cl_3Sn
(26903-07-1)
Cl_3Ti
(7705-07-9)
Cl_3V
(7718-98-1)
$Cl_4H_{14}O_6$
(7195-43-9)
Cl_4Ir
(10025-97-5)
Cl_4Pt
(13454-96-1)
Cl_4Si
(10026-04-7)
Cl_4Sn
(7646-78-8)
Cl_4Te
(10026-07-0)
Cl_4Th
(10026-08-1)
Cl_4Ti
(7550-45-0)
Cl_4U
(10026-10-5)
Cl_4V
(7632-51-1)
$Cl_4Zn.2H_4N$
(14639-97-5)
Cl_4Zr
(10026-11-6)
Cl_5Mo
(10241-05-1)
Cl_5Nb
(10026-12-7)
Cl_5P
(10026-13-8)
Cl_5Sb
(7647-18-9)
Cl_5Ta
(7721-01-9)
$Cl_5Zn.3H_4N$
(14639-98-6)
$Cl_6N_3P_3$
(940-71-6)
$Cl_6Pt.2H$
(16941-12-1)
$Cl_6Pt.2H_4N$
(16919-58-7)
$Cl_8N_4P_4$
(2950-45-0)
$Cl_{10}N_5P_5$
(13596-41-3)
$Cl_{11}H_{15}N_3O_2$
(26396-34-9)
$Cl_{12}N_6P_6$
(2851-52-7)
Co
(7440-48-4)
(10198-40-0)
$Co.2H_3N.2H_2O_4S$
(13596-46-8)
$Co.2H_3NO_3S$
(14017-41-5)
$Co.CrH_2O_4$
(13455-25-9)
CoF_2
(10026-17-2)
CoO
(1307-96-6)
$CoO_4S.7H_2O$
(10026-24-1)
CoS
(1317-42-6)
Co_2O_4Ti
(12017-38-8)
Co_3O_4
(1308-06-1)
Cr
(7440-47-3)
(14392-02-0)
(16065-83-1)
(18540-29-9)
$Cr.2H_2O_4S.K$
(10141-00-1)
$Cr.H_3O_4P$
(7789-04-0)
CrF_3
(7788-97-8)
$CrHO_5S$
(12336-95-7)
CrH_2O_4
(7738-94-5)
$CrH_2O_4.Cu$
(13548-42-0)
$CrH_2O_4.2Cs$
(13454-78-9)
$CrH_2O_4.xCr$
(41261-95-4)
$CrH_2O_4.2Li$
(14307-35-8)
$CrH_2O_4.2Rb$
(13446-72-5)
$CrH_2O_4.Mg$
(13423-61-5)
$CrH_2O_4.Pb$
(14654-05-8)
$CrH_2O_4.Zn$
(13530-65-9)
CrH_3O_3
(1308-14-1)
CrO_2
(12018-01-8)
CrO_3
(1333-82-0)
CrO_4
(13907-45-4)
$CrO_4.Ca$
(13765-19-0)
$CrO_4.H_2O_2.Zn_2.H_2O$
(15930-94-6)
$CrO_4.2H_4N$
(7788-98-9)
$CrO_4.2K$
(7789-00-6)
$CrO_4.2Na$
(7775-11-3)
$CrO_4.2Na.4H_2O$
(10034-82-9)
$CrO_4.2Na.10H_2O$
(13517-17-4)
$CrO_4.Pb$
(7758-97-6)
$CrO_4.Sr$
(7789-06-2)
$CrO_4.Zn.H_4O_2Zn.CrO_3$
(37300-23-5)
$CrO_4Pb.OPb$
(18454-12-1)
$CrO_4Zn.H_2O_2Zn$
(50922-29-7)
Cr_2FeO_4
(1308-31-2)
$Cr_2HO_9Zn_2.K$
(11103-86-9)
$Cr_2H_2O_7$
(13530-68-2)

$Cr_2H_2O_7.2K$
(27020-65-1)
Cr_2O_3
(1308-38-9)
Cr_2O_4Zn
(12018-19-8)
Cr_2O_7
(13907-47-6)
$Cr_2O_7.2H_4N$
(7789-09-5)
$Cr_2O_7.2K$
(7778-50-9)
$Cr_2O_7.2Na$
(10588-01-9)
$Cr_2O_7.Zn$
(14018-95-2)
$Cr_2O_7Rb_2$
(13446-73-6)
$Cr_3O_{12}.2Cr$
(24613-89-6)
$Cr_4O_{17}Zn_4.2K$
(12433-50-0)
Cs
(7440-46-2)
(10045-97-3)
(13967-70-9)
(14234-29-8)
CsHO
(21351-79-1)
CsI
(7789-17-5)
Cu
(7440-50-8)
(15158-11-9)
$Cu.2/3H_3O_4P$
(7798-23-4)
Cu.Ni
(11102-90-2)
$CuCl_2.2H_2O$
(10125-13-0)
CuI
(7681-65-4)
CuO
(1317-38-0)
CuO_4S
(12400-75-8)
CuS
(1317-40-4)
Cu_2O
(1317-39-1)
Cu_2S
(22205-45-4)
$Cu_4H_6O_{10}S.H_2O$
(1332-03-2)
D_2
(7782-39-0)
D_2O
(7789-20-0)
Dy
(7429-91-6)
Er
(7440-52-0)
Eu
(7440-53-1)
F
(16984-48-8)
FH
(7664-39-3)
FH.FNa
(51273-71-3)
FHO_3S
(7789-21-1)
FH_2O_3P
(13537-32-1)
FK
(7789-23-3)
FK.FH
(7789-29-9)
FLi
(7789-24-4)
FNa
(7681-49-4)
$FO_3P.2Na$
(10163-15-2)
F_2
(7782-41-4)
F_2HN
(10405-27-3)
F_2HNa
(1333-83-1)
F_2HOP
(14939-34-5)
F_2HO_2P
(13779-41-4)
F_2H_5N
(1341-49-7)
F_2Mn
(7782-64-1)
F_2Ni
(10028-18-9)
F_2O
(7783-41-7)
F_2OS
(7783-42-8)
F_2O_2S
(2699-79-8)
F_2O_2U
(13536-84-0)
F_2Pb
(7783-46-2)
F_2Sn
(7783-47-3)
F_2Sr
(7783-48-4)
F_2Zn
(7783-49-5)
$F_2Zn.4H_2O$
(13986-18-0)
F_3Fe
(7783-50-8)
F_3La
(13709-38-1)
F_3N
(7783-54-2)
F_3Sb
(7783-56-4)
F_4S
(7783-60-0)
F_4Si
(7783-61-1)
F_5I
(7783-66-6)
F_5P
(7647-19-0)
F_5Sb
(7783-70-2)
F_5Ta
(7783-71-3)
F_6HP
(16940-81-1)
F_6S
(2551-62-4)
F_6Se
(7783-79-1)
$F_6Si.Ba$
(17125-80-3)
$F_6Si.2H$
(16961-83-4)
$F_6Si.2H_4N$
(16919-19-0)
$F_6Si.Mg$
(16949-65-8)
$F_6Si.2Na$
(16893-85-9)
$F_6Si.Pb$
(25808-74-6)
$F_6Si.Zn$
(16871-71-9)
F_6Te
(7783-80-4)
F_6U
(7783-81-5)
F_6W
(7783-82-6)
$F_6Zr.2H_4N$
(16919-31-6)
$F_6Zr.2K$
(16923-95-8)
$F_7Ta.2K$
(16924-00-8)
$F_{10}S_2$
(5714-22-7)
Fe
(7439-89-6)
(14596-12-4)
(14681-59-5)
(20074-52-6)
$Fe.4H_2O_3Si.3/2Mg.Na$
(61105-31-5)
$Fe.H_3O_4P$
(10045-86-0)
$Fe.2H_3N.2H_2O_4S$
(10045-89-3)
$Fe.H_4O_7P_2.Na$
(10045-87-1)
$Fe.3/4H_4O_7P_2$
(10058-44-3)
Fe.Mo.Ni
(37304-37-3)
Fe.Ni
(11148-32-6)
FeH_2O_2
(18624-44-7)
FeLiSi
(64082-35-5)
FeN_3O_9
(10421-48-4)
FeO
(1345-25-1)
FeS
(1317-37-9)
FeS_2
(1309-36-0)
FeSe
(1310-32-3)
Fe_2O_3
(1309-37-1)
(1317-60-8)
$Fe_2O_{12}S_3$
(10028-22-5)
Fe_2P
(1310-43-6)
Fe_3O_4
(1317-61-9)
$Fe_9Ni_9S_{16}$
(53809-87-3)
$Fe_{12}O_{19}.Sr$
(12023-91-5)
Ga
(7440-55-3)
Ga_2O_3
(12024-21-4)
Gd
(7440-54-2)
Gd_2O_3
(12064-62-9)
Ge
(7440-56-4)
GeH_4
(7782-65-2)
GeO_2
(1310-53-8)
HBr
(24959-67-9)
HI
(10034-85-2)
$HIO_3.1/2Pb$
(25659-31-8)
HKO
(1310-58-3)
HLi
(7580-67-8)
HLiO
(1310-65-2)
HNO_2
(7782-77-6)
$HNO_2.1/2Pb$
(13826-65-8)
HNO_3
(7697-37-2)
$HNO_3.H_3N$
(6484-52-2)
$HNO_3.Li$
(7790-69-4)
HN_3
(7782-79-8)
HNa
(7646-69-7)
HNaO
(1310-73-2)
HNaS
(16721-80-5)
HO
(3352-57-6)
HO_2
(3170-83-0)
$HO_3P.Ca.Na$
(23209-59-8)
$HO_3S.H_4N$
(10192-30-0)
$HO_3S.Na$
(7631-90-5)
$HO_3Se.Na$
(7782-82-3)
$HO_4P.2Na$
(7558-79-4)
$HO_4P.2Na.12H_2O$
(10039-32-4)
HO_4PbS
(15739-80-7)
$HO_4S.H_4N$
(7803-63-6)
$HO_4S.K$
(7646-93-7)
$HO_4S.Na$
(7681-38-1)
$HPO_4.2H_4N$
(7783-28-0)
H_2
(1333-74-0)
H_2AsO_4Pb
(10102-48-4)
H_2Cl_2Si
(4109-96-0)
H_2LiN
(7782-89-0)
H_2MgO_2
(1309-42-8)
H_2NNa
(7782-92-5)
H_2NNiO_3
(14216-75-2)
$H_2NO_3S.H_4N$
(7773-06-0)
H_2N_2O
(35576-91-1)
$H_2N_2O_2$
(7782-94-7)
(14448-38-5)
H_2NiO_2
(11113-74-9)
(12054-48-7)
H_2O
(7732-18-5)
H_2O_2
(7722-84-1)
$H_2O_2.Be$
(13327-32-7)
$H_2O_2.Cu$
(20427-59-2)

$H_2O_2P.Na$
(7681-53-0)

H_2O_2Pb
(19783-14-3)

H_2O_2Sr
(18480-07-4)

H_2O_3S
(7782-99-2)

$H_2O_3S.2K$
(10117-38-1)

$H_2O_3S.K$
(7773-03-7)

$H_2O_3S.Zn$
(13597-44-9)

$H_2O_3S_2.H_2O.2Na$
(55755-19-6)

$H_2O_3S_2.Pb$
(13478-50-7)

$H_2O_3S_2.1/2Pb$
(26265-65-6)

H_2O_3Se
(7783-00-8)

$H_2O_3Se.Pb$
(7488-51-9)

H_2O_3Si
(12627-13-3)

$H_2O_3Si.2Li$
(10102-24-6)

$H_2O_3Si.5/6H_2O.3/4Mg$
(18307-23-8)

$H_2O_3Te.2Na$
(10102-20-2)

$H_2O_4P.Na$
(7558-80-7)

H_2O_4S
(7664-93-9)

$H_2O_4S.O_3S$
(8014-95-7)

$H_2O_4S.Sr$
(7759-02-6)

$H_2O_4S.1/2Ti$
(13693-11-3)

H_2O_4Se
(7783-08-6)

$H_2O_4Se.Pb$
(7446-15-3)

$H_2O_4Te.Pb$
(13845-35-7)

$H_2O_4W.Pb$
(7759-01-5)

$H_2O_5S.K$
(10058-23-8)

$H_2O_6S_2$
(14970-71-9)

$H_2O_6S_4.2K$
(13932-13-3)

$H_2O_7P_2.2Na$
(7758-16-9)

$H_2O_8S_2.2K$
(7727-21-1)

H_2S
(7783-06-4)

H_2Se
(7783-07-5)

H_2Si
(13825-90-6)

H_2Zr
(7704-99-6)

H_3HgNO_4
(7782-86-7)

H_3N
(7664-41-7)

$H_3N.1/2H_2O_3S$
(10196-04-0)

$H_3N.H_2O_3S_2$
(10103-43-2)

$H_3N.H_2O_4S.3H_2O.1/2Zn$
(7783-24-6)

$H_3N.H_3O_2P$
(7803-65-8)

H_3NO
(7803-49-8)

$H_3NO.ClH$
(5470-11-1)

$H_3NO.HI$
(59917-23-6)

$H_3NO.HNO_3$
(13465-08-2)

$H_3NO.H_2O_4S$
(10046-00-1)

H_3NO_3S
(5329-14-6)

H_3N_5
(14546-44-2)

$H_3Na_2O_5P$
(10140-65-5)

$H_3Na_2O_{10}P_3$
(13573-18-7)

H_3NiO_3
(12125-56-3)

H_3O_2P
(6303-21-5)

$H_3O_2P.K$
(7782-87-8)

H_3O_3P
(10294-56-1)
(13598-36-2)

H_3O_3PS
(13598-51-1)

H_3O_4P
(7664-38-2)

$H_3O_4P.H_3N$
(7722-76-1)

$H_3O_4P.K$
(7778-77-0)

$H_3O_4P.2K$
(7758-11-4)

$H_3O_4P.3K$
(7778-53-2)

$H_3O_4P.3/2Mg$
(7757-87-1)

$H_3O_4P.xMn$
(10124-54-6)

$H_3O_4P.xNa$
(7632-05-5)

$H_3O_4P.Pb$
(15845-52-0)

$H_3O_4P.3/4Ti$
(15578-51-5)

$H_3O_4P.1/2Zn$
(13598-37-3)

$H_3O_9P_3.3Na$
(7785-84-4)

H_3P
(7803-51-2)

H_3Sb
(7803-52-3)

H_4CuO_4P
(10103-48-7)

H_4IN
(12027-06-4)

H_4N
(14798-03-9)

$H_4N.C_3H_7NO_4P$
(25954-13-6)

$H_4N.Cl$
(12125-02-9)

$H_4N.F$
(12125-01-8)

$H_4N.HO$
(1336-21-6)

$H_4N.HS$
(12124-99-1)

$H_4N.1/2Mo_2O_7$
(27546-07-2)

H_4N_2
(302-01-2)

$H_4N_2.ClH$
(2644-70-4)

$H_4N_2.H_2O$
(7803-57-8)

$H_4N_2.H_2O_4S$
(10034-93-2)

$H_4N_2.1/2H_2O_4S$
(13464-80-7)

$H_4N_2.xH_2O_4Se$
(73506-32-8)

$H_4N_2NiO_6S_2$
(13770-89-3)

H_4N_4
(12164-94-2)

$H_4N_4O_{12}.Th$
(13823-29-5)

$H_4O_4Si.4Na$
(13472-30-5)

H_4O_5SZn
(7446-19-7)

$H_4O_7P_2.4K$
(7320-34-5)

$H_4O_6P_2.2Na$
(7782-95-8)

$H_4O_7P_2.2Pb$
(13453-66-2)

$H_4O_8Si_3.2Mg$
(14987-04-3)

$H_4Se_2Si_2$
(287-68-3)

H_4Si
(7803-62-5)

$H_5CuO_7P_2$
(10102-90-6)

$H_6CaN_2O_3$
(15245-12-2)

H_6Cu_6
(8012-69-9)

$H_6N_2O_2.H_2O_4S$
(10039-54-0)

$H_6O_6Sb.K$
(12208-13-8)

H_7AlNaO_4P
(7785-88-8)

H_8N_2S
(12135-76-1)

$H_{10}O_9Si_2.H_2O.Be_4$
(12161-82-9)

$H_{11}N_2O_8PS$
(12593-60-1)

$H_{12}N_4O_7S$
(12436-94-1)

$H_{14}CuN_4O_4S$
(10380-29-7)

$H_{20}Na_3O_{14}P$
(10361-89-4)

$H_{203}Si.3/4Mg$
(14807-96-6)

He
(7440-59-7)
(14762-55-1)

HeO_2
(58933-55-4)

Hf
(7440-58-6)

Hg
(7439-97-6)
(14302-87-5)

HgI_2
(7774-29-0)

$HgI_4.2K$
(7783-33-7)

HgO
(21908-53-2)

Hg_2I_2
(15385-57-6)

Hg_2N_6
(38232-63-2)

Hg_2O
(15829-53-5)

Hg_3N_2
(12136-15-1)

Hg_3O_6S
(1312-03-4)

Ho
(7440-60-0)

I
(10043-66-0)
(14683-16-0)
(14834-67-4)
(15046-84-1)
(20461-54-5)

IK
(7681-11-0)

ILi
(10377-51-2)

IN_3
(14696-82-3)

INa
(7681-82-5)

$IO_3.K$
(7758-05-6)

$IO_3.Na$
(7681-55-2)

I_2
(7553-56-2)

I_2Ni
(13462-90-3)

$I_2O_5Sb_4$
(12196-43-9)

I_2Pb
(10101-63-0)

I_2Zn
(10139-47-6)

I_3N
(13444-85-4)

I_3Sb
(7790-44-5)

I_4Sn
(7790-47-8)

In
(7440-74-6)

InN_3O_9
(13770-61-1)

InP
(22398-80-7)

K
(7440-09-7)

$K.1/2O_3Ti$
(12030-97-6)

KNO_3
(7757-79-1)

KN_3
(20762-60-1)

KO_2
(12030-88-5)

KO_8Ti_4
(12056-53-0)

K_2O_2
(17014-71-0)

K_2S
(1312-73-8)

Kr
(7439-90-9)

La
(7439-91-0)
(13981-28-7)

La_2O_3
(1312-81-8)

Li
(7439-93-2)

Li_2O
(12057-24-8)

Li_2O_2
(12031-80-0)

Li_3N
(26134-62-3)

Lu
(7439-94-3)

Mg
(7439-95-4)

MgO
(1309-48-4)

MgO_2
(14452-57-4)
Mg_3P_2
(12057-74-8)
Mn
(7439-96-5)
(16397-91-4)
MnO
(1344-43-0)
MnO_2
(1313-13-9)
$MnO_4.H_4N$
(13446-10-1)
$MnO_4.K$
(7722-64-7)
$MnO_4.Na$
(10101-50-5)
$MnO_4S.H_2O$
(10034-96-5)
Mn_2O_3
(1317-34-6)
$Mn_2O_8.Ba$
(7787-36-2)
$Mn_2O_8.Ca$
(10118-76-0)
$Mn_2O_8.Zn$
(23414-72-4)
Mn_3O_4
(1317-35-7)
Mo
(7439-98-7)
(14119-15-4)
MoO_3
(1313-27-5)
$MoO_4.2H_4N$
(13106-76-8)
$MoO_4.2Na$
(7631-95-0)
MoO_4Pb
(10190-55-3)
MoS_2
(1317-33-5)
$Mo_7O_{24}.6H_4N$
(12027-67-7)
NO
(10102-43-9)
NO_2
(10102-44-0)
(14797-65-0)
$NO_2.H_4N$
(13446-48-5)
$NO_2.K$
(7758-09-0)
$NO_2.Na$
(7632-00-0)
NO_3
(14797-55-8)
$NO_3.Ag$
(7761-88-8)
$NO_3.Hg$
(10415-75-5)
$NO_3.Na$
(7631-99-4)
$NO_3.Tl$
(10102-45-1)
NSe
(12033-59-9)
N_2
(7727-37-9)
N_2O
(10024-97-2)
N_2O_3
(10544-73-7)
N_2O_4
(10544-72-6)
$N_2O_6.Ba$
(10022-31-8)
$N_2O_6.Ca$
(10124-37-5)
$N_2O_6.Co$
(10141-05-6)
$N_2O_6.Cu$
(3251-23-8)
$N_2O_6.Hg$
(10045-94-0)
$N_2O_6.Mg$
(10377-60-3)
$N_2O_6.Ni$
(13138-45-9)
$N_2O_6.Ni.6H_2O$
(13478-00-7)
$N_2O_6.Pb$
(10099-74-8)
$N_2O_6.Sr$
(10042-76-9)
$N_2O_6.Zn$
(7779-88-6)
$N_2O_6.Zn.6H_2O$
(10196-18-6)
N_2O_8U
(10102-06-4)
(36478-76-9)
$N_2O_8U.6H_2O$
(13520-83-7)
N_3Na
(26628-22-8)
$N_3O_9.Al$
(13473-90-0)
$N_3O_9.Ce$
(10108-73-3)
$N_3O_9.Cr.9H_2O$
(7789-02-8)
$N_3O_9.La$
(10099-59-9)
$N_3O_9.La.6H_2O$
(10277-43-7)
$N_4O_{12}.Zr$
(13746-89-9)
N_6Pb
(13424-46-9)
Na
(7440-23-5)
$Na.1/6O_{28}V_{10}$
(12200-88-3)
$Na.Tl$
(12136-93-5)
NaO_2
(12034-12-7)
Na_2O
(12401-86-4)
Na_2O_2
(1313-60-6)
Na_2S
(1313-82-2)
Nb
(7440-03-1)
Nb_2O_6Pb
(12034-88-7)
Nd_2O_3
(1313-97-9)
Ne
(7440-01-9)
Ni
(7440-02-0)
NiO
(1313-99-1)
NiO_2
(12035-36-8)
$NiO_4S.6H_2O$
(10101-97-0)
NiS
(11113-75-0)
(16812-54-7)
Ni_2O_3
(1314-06-3)
Ni_3S_2
(12035-72-2)
No
(10028-14-5)
$ONa_2Fe_2O_{33}FeO_8SiO_2H_2O$
(12001-28-4)
OPb
(1317-36-8)
OZn
(1314-13-2)
O_2
(7782-44-7)
(11062-77-4)
O_2Pb
(1309-60-0)
O_2S
(7446-09-5)
O_2Se
(7446-08-4)
O_2Si
(7631-86-9)
(14464-46-1)
(14808-60-7)
(15468-32-3)
(60676-86-0)
O_2Sn
(18282-10-5)
O_2Sr
(1314-18-7)
O_2Th
(1314-20-1)
O_2Ti
(1317-70-0)
(1317-80-2)
(13463-67-7)
O_2Zn
(1314-22-3)
O_2Zr
(1314-23-4)
O_3
(10028-15-6)
$O_3P.Na$
(10361-03-2)
O_3P_2
(1314-24-5)
O_3PbZr
(12060-01-4)
O_3Pb_2
(1314-27-8)
O_3S
(7446-11-9)
(14265-45-3)
$O_3S.2Na$
(7757-83-7)
O_3S_2
(14383-50-7)
$O_3S_2.Ca$
(10124-41-1)
$O_3S_2.2H_4N$
(7783-18-8)
$O_3S_2.2Na$
(7772-98-7)
$O_3S_2.2Na.5H_2O$
(10102-17-7)
O_3Sb_2
(1309-64-4)
O_3Se
(14124-67-5)
$O_3Se.2Na$
(10102-18-8)
$O_3Si.2Na$
(6834-92-0)
O_3Sm_2
(12060-58-1)
$O_3Ti.Pb$
(12060-00-3)
$O_3Ti.Sr$
(12060-59-2)
O_3Tl_2
(1314-32-5)
$O_3V.H_4N$
(7803-55-6)
$O_3V.Na$
(13718-26-8)
O_3V_2
(1314-34-7)
O_3W
(1314-35-8)
O_3Y_2
(1314-36-9)
O_4Os
(20816-12-0)
O_4P
(14265-44-2)
$O_4P.Al$
(7784-30-7)
$O_4P.3Na$
(7601-54-9)
$O_4P.3Na.12H_2O$
(10101-89-0)
O_4Pb_3
(1314-41-6)
O_4S
(14808-79-8)
$O_4S.Ba$
(7727-43-7)
$O_4S.Be$
(13510-49-1)
$O_4S.Be.4H_2O$
(7787-56-6)
$O_4S.Ca$
(7778-18-9)
$O_4S.Ca.2H_2O$
(13397-24-5)
$O_4S.Cd$
(10124-36-4)
$O_4S.Co$
(10124-43-3)
$O_4S.Cu$
(7758-98-7)
$O_4S.Cu.5H_2O$
(7758-99-8)
$O_4S.Fe$
(7720-78-7)
$O_4S.Fe.7H_2O$
(7782-63-0)
$O_4S.2H_4N$
(7783-20-2)
$O_4S.Hg$
(7783-35-9)
$O_4S.2Hg$
(7783-36-0)
$O_4S.2K$
(7778-80-5)
$O_4S.2Li$
(10377-48-7)
$O_4S.Mg$
(7487-88-9)
$O_4S.Mn$
(7785-87-7)
$O_4S.2Na$
(7757-82-6)
$O_4S.Ni$
(7786-81-4)
$O_4S.Pb$
(7446-14-2)
$O_4S.Tl$
(63906-56-9)
$O_4S.2Tl$
(7446-18-6)
$O_4S.xTl$
(10031-59-1)
$O_4S.Zn$
(7733-02-0)
$O_4SZn.7H_2O$
(7446-20-0)
$O_4S_2.2Na$
(7775-14-6)
$O_4S_2.Zn$
(7779-86-4)
O_4Sb_2
(1332-81-6)
O_4Se
(14124-68-6)
$O_4Se.2Na$